W9-AEG-026

Science in Medicine

Science in Medicine
The *JCI* Textbook of Molecular Medicine

Editors

Andrew R. Marks and Ushma S. Neill

JONES AND BARTLETT PUBLISHERS

Sudbury, Massachusetts

BOSTON   TORONTO   LONDON   SINGAPORE

World Headquarters
Jones and Bartlett Publishers
40 Tall Pine Drive
Sudbury, MA 01776
978-443-5000
info@jbpub.com
www.jbpub.com

Jones and Bartlett Publishers Canada
6339 Ormindale Way
Mississauga, Ontario L5V 1J2
Canada

Jones and Bartlett Publishers
International
Barb House, Barb Mews
London W6 7PA
United Kingdom

Jones and Bartlett's books and products are available through most bookstores and online booksellers. To contact Jones and Bartlett Publishers directly, call 800-832-0034, fax 978-443-8000, or visit our website www.jbpub.com.

Substantial discounts on bulk quantities of Jones and Bartlett's publications are available to corporations, professional associations, and other qualified organizations. For details and specific discount information, contact the special sales department at Jones and Bartlett via the above contact information or send an email to specialsales@jbpub.com.

The authors, editor, and publisher have made every effort to provide accurate information. However, they are not responsible for errors, omissions, or for any outcomes related to the use of the contents of this book and take no responsibility for the use of the products and procedures described. Treatments and side effects described in this book may not be applicable to all people; likewise, some people may require a dose or experience a side effect that is not described herein. Drugs and medical devices are discussed that may have limited availability controlled by the Food and Drug Administration (FDA) for use only in a research study or clinical trial. Research, clinical practice, and government regulations often change the accepted standard in this field. When consideration is being given to use of any drug in the clinical setting, the health care provider or reader is responsible for determining FDA status of the drug; reading the package insert; and reviewing prescribing information for the most up-to-date recommendations on dose, precautions, and contraindications; and determining the appropriate usage for the product. This is especially important in the case of drugs that are new or seldom used.

Journal of Clinical Investigation
Executive Editor: Ushma S. Neill
Science Editor: Brooke Grindlinger
News and Reviews Editor: Karen Honey
Executive Director: John B. Hawley
Managing Director: Karen G. Kosht

Production Editors: Lara L. McCarron, Martha L. Sidhu
Scientific Illustrator: Bruce Worden
Senior Copy Editor: Maya Hoptman
Copy Editors: Clare Cross, Carolyn Hayes, Autumn Kelly
Associate Copy Editor: Rachel Nelson
Assistant Copy Editor: Catherine Ahmann

Production Credits
Chief Executive Officer: Clayton Jones
Chief Operating Officer: Don W. Jones, Jr.
President, Higher Education and Professional Publishing: Robert W. Holland, Jr.
V.P., Sales and Marketing: William J. Kane
V.P., Design and Production: Anne Spencer
V.P., Manufacturing and Inventory Control: Therese Connell
Executive Editor: Cathleen Sether

Acquisition Editor: Shoshanna Grossman
Associate Editor: Molly Steinbach
Senior Production Editor: Louis C. Bruno, Jr.
Production Assistant: Leah E. Corrigan
Senior Marketing Manager: Andrea DeFronzo
Composition: Philip Regan
Cover Design: Bruce Worden
Printing and Binding: Replika Press
Cover Printing: Replika Press

Library of Congress Cataloging-in-Publication Data
Science in medicine / American Society for Clinical Investigation. — 1st ed.
    p. ; cm.
  Includes bibliographical references.
  ISBN 978-0-7637-5083-1 (alk. paper)
  1. Medicine—Miscellanea. 2. Science—Miscellanea. I. American
Society for Clinical Investigation.
  [DNLM: 1. Medicine—Review. WB 100 S412 2007]
  R708.S36 2007
  610—dc22
                        2007004765
6048

Printed in India
11 10 09 08 07 10 9 8 7 6 5 4 3 2 1

Contents

PART 4 – INFECTIOUS DISEASES

Clinical syndromes

Gram-positive bacteria in health and disease

Gram-negative bacteria in health and disease

PART 14 – GENETICS AND DISEASE

Preface

When I was a medical student doing my internal medicine rotation, I would spend hours every night after finishing my patient work-up in the hospital library browsing through the stacks looking for articles that would help me understand the diseases afflicting my patients. This was a certainly an enriching, but highly inefficient learning experience. Today, medical students, house staff, and faculty merely have to find a computer terminal, and the world's literature is instantly available in searchable databases. This plethora of easily accessed information has lead to a new problem — there is too much information available online! This textbook addresses both problems. It brings together under a single cover state-of-the art brief reviews on diseases in all the major areas of medicine written by the world's leading authorities.

Given the pressures of clinical services these days, academic medicine is in retreat, with less time available for scholarly discussions that can help with patient care. Each of the chapters in this book addresses current practice with an eye toward understanding how modern biology has changed the way we diagnose and treat diseases. In addition to the revolution in information technology, the past 30 years have brought a revolution in medicine: diseases are now understood at a mechanistic level that could only be dreamt of when I was in medical school. These new understandings form the basis for improved patient care but at the same time require that modern physicians have a much broader comprehension of science and technology than was necessary for providing the best care decades ago.

Unlike other textbooks of medicine that focus on disease processes, diagnosis, and management, the *JCI Textbook of Molecular Medicine* focuses on the science of medicine and the societal issues that affect how we apply that science to care for our patients. The chapters have evolved from two series started in the journal five years ago: "Science in Medicine," which challenged leading experts in each field of medicine to clearly explain how molecular biology had advanced diagnosis and treatment; and "Science and Society," which has explored key issues that influence medical care. Topics covered in the "Science and Society" articles include payment of clinical research subjects, genetic counseling, stem cells, and umbilical cord blood banking.

This is a book that has something in it for everyone. All the major divisions of medicine are covered. Busy medical students or house officers will be able to find a scholarly chapter relevant to almost any patient they have on their service. In addition to being written by a renowned expert in the field, each chapter has been peer reviewed and contains beautiful illustrations that can be used for presentations. The practicing physician can use this book to rapidly get up-to-date information that will enhance understanding of even the most complex disorders and provide important background for patient care.

Scientists involved in biomedical research can use this book to gain insight into the pathogenesis of most of the common diseases. The information conveyed is directly relevant to those involved in research because each chapter is written by practicing scientists who understand the key mechanisms underpinning the disease at a molecular level and also point out the challenges for the future. Thus, scientists who need to rapidly get up to speed on a disease, those who are writing grants or papers or giving lectures can use this text knowing that the information will be reliable, precise, and useful.

For readers who want to catch up on advances in medical science over the past five years, this book is an ideal resource. It does not replace the classic textbooks; rather, the book takes our understanding to the next level. Nothing is glossed over, because unlike in most textbooks of medicine, no chapter is written by a generalist who has had to read others' work and distill it to its essence. Rather, each chapter is written by those who have done the research on which the advances have been based. This removes a level of uncertainty and should make readers confident that they can rely on the information to be state-of-the-art.

Thus, the book has value and a unique purpose for medical students, graduate students, house staff, attending and practicing physicians, and biomedical researchers. Moreover, each chapter is carefully referenced and indexed, allowing the reader to pursue topics in even greater depth easily by doing online literature searches based on the reference lists.

The book has the added advantage that each chapter can be found online at the *JCI* Web site, so figures and charts can be downloaded for use in presentations.

I would like to thank Ushma Neill, executive editor, and Brooke Grindlinger, science editor at the *JCI* for bringing together the material in this text and for making an idea become a reality. I would also like to thank all of the authors who contributed chapters. Their high level of expertise and careful attention to detail have resulted in chapters of the highest quality and greatest utility in their respective fields.

Information technology is rapidly evolving, yet there is still no substitute for a well-prepared scholarly text that covers the broad scope of the science of medicine. This book fills such a niche and should be a welcome addition to any library of medical texts.

Andrew R. Marks
Editor-in-chief, *The Journal of Clinical Investigation*, and
Columbia University College of Physicians and Surgeons,
New York, New York, USA.

Pharmacological manipulation of cell death: clinical applications in sight?

Douglas R. Green[1] and Guido Kroemer[2]

[1]Department of Immunology, St. Jude Children's Research Hospital, Memphis, Tennessee, USA.
[2]CNRS, UMR8125, Institut Gustave Roussy, Villejuif, France.

This series of Reviews on cell death explores the creation of new therapies for correcting excessive or deficient cell death in human disease. Signal transduction pathways controlling cell death and the molecular core machinery responsible for cellular self-destruction have been elucidated with unprecedented celerity during the last decade, leading to the design of novel strategies for blocking pathological cell loss or for killing unwanted cells. Thus, an increasing number of compounds targeting a diverse range of apoptosis-related molecules are being explored at the preclinical and clinical levels. Beyond the agents that are already FDA approved, a range of molecules targeting apoptosis-regulatory transcription factors, regulators of mitochondrial membrane permeabilization, and inhibitors or activators of cell death–related proteases are under close scrutiny for drug development.

Introduction

Apoptosis was originally defined based on morphological features seen as the cell dies: nuclear condensation, nuclear fragmentation, membrane blebbing, cellular fragmentation into membrane-bound bodies, phagocytosis of the dying cell, and lack of an ensuing inflammatory response (1). The final outcome — apoptosis — is generally the result of the activation of a subset of caspase proteases, in particular caspase-3, -6, and -7 (2). These are the "executioner" caspases, and they mediate their effects by cleavage of specific substrates in the cell. The executioner caspases, and, indeed, all of the core components of the apoptosis machinery, often preexist in healthy cells in inactive forms. Activation of the executioner caspase-3 and -7 by initiator caspase-8, -9, and -10 defines the best-understood apoptotic pathways, and we focus on these: the death receptor (extrinsic) pathway, and the mitochondrial (intrinsic) pathway.

In the extrinsic pathway, ligation of death receptors (a subset of the TNF receptor [TNFR] family, including TNFR1, CD95, TNF-related apoptosis-inducing ligad receptor-1 and -2 [TRAIL-R1 and -R2], and probably DR3/TRAMP) causes recruitment and oligomerization of the adapter molecule FADD within the death-inducing signaling complex (DISC). The oligomerized FADD binds initiator caspase-8 and -10, causing their dimerization and activation (3).

Most cell death in vertebrates proceeds via the mitochondrial pathway of apoptosis (Figure 1) (4). Here, the executioner caspases are cleaved and activated by the initiator caspase-9. Like other initiator caspases, caspase-9 can only be activated by dimerization on the adapter molecule Apaf-1. Apaf-1 preexists in the cytosol as a monomer, and its activation depends on the presence of holocytochrome c. Upon binding of cytochrome c to Apaf-1, dATP gains access to a nucleotide-binding site in Apaf-1, inducing a conformational change in the adapter molecule (5). Apaf-1 then oligomerizes into an "apoptosome" that recruits and activates caspase-9 (6). The release of holo-

cytochrome, which is normally present only in the mitochondrial intermembrane space, is rate-limiting for the generation of the apoptosome. Hence, mitochondrial outer membrane permeabilization (MOMP) is the critical event responsible for caspase activation in the intrinsic pathway. However, MOMP, which represents the "point of no return" of cell death, can even commit a cell to die when caspases are not activated. This "caspase-independent death" (7, 8) can occur due to irreversible loss of mitochondrial function and mitochondrial release of caspase-independent death effectors such as apoptosis-inducing factor (AIF) (9), endonuclease G (10), and others (7, 8).

In ischemic injury and other clinical conditions, MOMP can occur as a consequence of the mitochondrial permeability transition (MPT). Channels in the inner mitochondrial membrane open to allow movement of solutes and ions, resulting in a loss of inner membrane function and swelling of the matrix. This can lead to matrix remodeling, perhaps freeing proteins of the intermembrane space for release, and can lead to outer membrane rupture. The opening of the MPT involves the adenosine nuclear transporter (ANT) in the inner mitochondrial membrane, although this is not absolutely required for MPT (11), and the voltage-dependent anion channels (VDACs) in the outer membrane. Cyclophilin D, located in the matrix, facilitates MPT, probably through enhancing conformational change of inner membrane proteins. Recent studies in cyclophilin D–deficient animals show that this molecule is important for MPT (12–14) and that these mice are resistant to cell death caused by ischemic injury (12, 13). However, MOMP and apoptosis induced by several proapoptotic agents remained intact, indicating that a cyclophilin D–independent mechanism (presumably MPT-independent) can also operate in apoptotic death.

This second major mechanism of MOMP involves the Bcl-2 family of proteins. These proteins share 1 or more Bcl-2 homology (BH) domains and control MOMP, mostly at the mitochondrial outer membrane. The proapoptotic multidomain proteins Bax and Bak (containing BH-1, -2, and -3) probably mediate MOMP, as these can permeabilize vesicles, are composed of mitochondrial lipids (15, 16), and are required for MOMP to occur during apoptosis (17). Bax and/or Bak can be activated by other Bcl-2 family proteins, the BH3-only proteins (which share BH3). Some of the proteins, notably Bid and Bim, can activate Bax (and probably Bak) directly, causing MOMP (16). Antiapop-

Nonstandard abbreviations used: AIF, apoptosis-inducing factor; BH, Bcl-2 homology (domain); DISC, death-inducing signaling complex; MLK, mixed-lineage kinase; MOMP, mitochondrial outer membrane permeabilization; MPT, mitochondrial permeability transition; PARP, poly(ADP-ribose) polymerase; TNFR, TNF receptor; TRAIL, TNF-related apoptosis-inducing ligand; TRAIL-R, TRAIL receptor.

Conflict of interest: The authors have declared that no conflict of interest exists.

Citation for this article: J. Clin. Invest. 115:2610–2617 (2005). doi:10.1172/JCI26321.

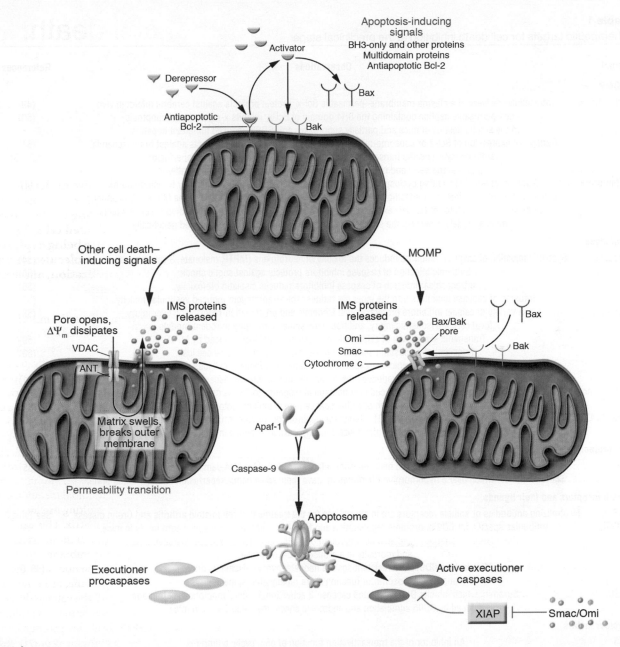

Figure 1
Checkpoints for apoptosis in the mitochondrial pathway. Most mammalian cell death proceeds via the mitochondrial pathway, as illustrated. Stimuli for the induction of apoptosis predominantly act by engaging proapoptotic members of the Bcl-2 family, which work to cause MOMP, and this is countered by the antiapoptotic Bcl-2 family members. Other cell death stimuli can cause MOMP by the induction of a mitochondrial permeability transition. In either case, release of proteins from the intermembrane space triggers the activation of caspases via the formation of an Apaf-1 apoptosome, which recruits and activates caspase-9. This, in turn, cleaves and activates the executioner caspases. The activation of caspase-3, -7, and -9 is antagonized by XIAP, which in turn can be inhibited by Smac, Omi, and other proteins released upon MOMP. Not shown here are other pathways of caspase activation and apoptosis, including the death receptor pathway, and those resulting in activation of caspase-1 and -2 (see text). ANT, adenosine nuclear transporter; VDAC, voltage-dependent anion channel; IMS, intermembrane space; $\Delta\Psi_m$, mitochondrial transmembrane potential.

totic Bcl-2 family proteins (including Bcl-2, Bcl-x_L, Mcl-1, and others) sequester the BH3-only proteins and probably the activated multidomain proteins, preventing MOMP. Other BH3-only proteins can antagonize the antiapoptotic proteins, thus sensitizing cells for death. These sensitizing or "derepressor" BH3-only proteins include Puma, Noxa, and Bad, among others (16, 18, 19). Peptides and drugs that mimic the BH3-only

proteins therefore have promise as agents that can cause MOMP and apoptosis, or sensitize cells for death (20).

The caveats: apoptosis inhibition without cell death inhibition

One major problem concerning the development of cell death inhibitors is the definition of cell death and hence the experimental

Table 1

Therapeutic targets for cell death inhibition at the preclinical stage

Target	Observations	References
MOMP		
Bcl-2, Bcl-x$_L$	Intracellular delivery of a plasma membrane–permeable Bcl-x$_L$ protein protects against cerebral infarct in vivo.	(49)
	A cell-permeable peptide containing the BH4 domain of Bcl-x$_L$ inhibits x-ray–induced apoptosis in the small intestine of mice and partially suppresses anti-CD95–induced fulminant hepatitis.	(50)
	Transgenic expression of Bcl-2 or liposome-mediated delivery of the Bcl-2 gene protects against brain ischemia.	(51)
	Cardiomyocyte-specific transgenic expression of Bcl-2 reduces ischemia/reperfusion injury in the heart and mitigates a mouse model of chronic cardiomyopathy.	(52, 53)
Other proteins	A variety of agents including cyclosporin A (the prototypic inhibitor of the permeability transition pore) and inhibitors of mitochondrion-specific ion channels (such as diazoxide, an inhibitor of the MitoKATP channel, and diltiazem, an inhibitor of Na$^+$Ca^{2+} exchange) inhibit MOMP and exert neuro- and cardioprotective effects in mouse models. However, the specificity of these agents has not been validated genetically.	(4)
Proteases		
Caspases	Systemic injection of caspase inhibitors reduces the toxicity of neurotoxins (MPTP, malonate, 3-nitrophenol) in vivo.	(54)
	Systemic injection of caspase inhibitors protects against septic shock.	(55)
	Intracochlear infusion of caspase inhibitors reduces cisplatin ototoxicity.	(56)
	Infusion of caspase inhibitors into the inner ear reduces the streptomycin-induced vestibular toxicity.	(57)
Calpain	A variety of calpain inhibitors protect against ischemic and preservation-reperfusion liver injury, focal cerebral ischemic injury, and traumatic spinal cord injury in rodent models.	(58)
	Transgenic expression of calpastatin reduces neuronal loss induced by MPTP.	(59)
	Knockout of calpastatin sensitizes to kainate-induced neuronal cell loss, while transgenic overexpression of calpastatin reduces cell loss.	(60)
	Transgenic expression of calpastatin reduces muscle atrophy due to muscle disuse or dystrophin deficiency.	(61)
Cathepsins	Cathepsin B knockout mice exhibit reduced liver damage in response to TNF-α or cholestasis, and this effect can be mimicked by injection of a chemical cathepsin inhibitor into wild-type mice.	(62)
Omi/HtrA2	The Omi/HtrA2–specific inhibitor ucf-101 inhibits the nephrotoxic, proapoptotic effect of cisplatin in mice.	(63)
	Ucf-101 reduces infarct size after myocardial ischemia in mice.	(64)
Nonproteolytic death effectors		
AIF	Neutralization of AIF can rescue neurons from cell death induced by DNA damage or excitotoxins. Harlequin mice, which bear a hypomorphic AIF mutation, have reduced ischemia/reperfusion damage of the brain.	(65)
Death receptors and their ligands		
TNF-α	Neutralizing antibodies or soluble receptors are in clinical use for the treatment of rheumatoid arthritis and Crohn disease.	See Table 2
CD95L	Antibodies specific for CD95L promote regeneration and functional recovery after spinal cord injury in mice.	(66)
	Simultaneous, antibody-mediated neutralization of CD95L and TNF-α reduces infarct volume and mortality in a mouse model of stroke.	(67)
	Mice lacking CD95L are resistant against neonatal hypoxic-ischemic brain injury and recover locomotor function more rapidly after spinal injury.	(68, 69)
DISC	Suramin, which inhibits DISC-mediated caspase-8 activation, inhibits apoptotic liver damage induced by CD95 stimulation and endotoxic shock–mediated TNF-α in mice.	(70)
Nuclear factors		
p53	An inhibitor of the transactivation function of p53, cyclic pifithrin-α is radioprotective in mice, mimicking the effects of the p53 knockout.	(71)
PARP	PARP inhibitors and PARP-1 knockout reduce brain damage induced by stroke or MPTP in mice.	(72)
	PARP inhibitors and PARP-1 knockout protect against cardiac failure induced by transverse aortic constriction in mice.	(73)
	Pharmacological inhibition of PARP and PARP-1 knockout protect against ischemic renal injury in mice.	(74)
Stress kinases		
JNK3	JNK inhibitors reduce infarct size induced by focal ischemia of the rodent brain.	(75)
	JNK3 knockout protects the adult mouse brain from glutamate-induced excitotoxicity and cerebral ischemia-hypoxia.	(76)
MLK	Inhibitors of MLK inhibit MLK3/MKK7/JNK3 activation in vivo in a variety of models of acute neurodegeneration.	(77)

verification of cell death inhibition. It has long been assumed that two decisive events in cellular destruction are apoptotic chromatin condensation and the fragmentation of nuclear DNA detectable with the TUNEL technique. In many instances, the suppression of these nuclear signs of apoptosis has been interpreted as inhibition of cell death. However, inhibition of some signal-transducing events

required for apoptotic DNA degradation frequently does not prevent cell death. Thus many studies involving, for instance, caspase inhibitors are flawed when they claim that caspase inhibition prevents cell death without clear demonstration of cellular survival. For example, one of the major DNA-degrading enzymes, caspase-activated DNase (CAD), strictly relies on caspase activation yet is not required for cell death, as demonstrated by knockout studies (21). So when CAD is inhibited (which often results in abolition of TUNEL staining), this may indicate that a given compound has achieved caspase inactivation, yet it does not demonstrate that the cell will actually survive. If cell death is evaluated by other techniques — by measurement of MOMP, by assessment of the long-term survival, or by functional assays — then caspase inhibition is often without any effect.

In the context of caspase inhibitors, cells often do not die by apoptosis but succumb to a delayed cell death (7, 8) that can have an apoptosis-like morphology (with peripheral chromatin condensation yet without karyorrhexis) (22) or manifest as autophagic cell death (with accumulation of autophagic vacuoles) (23) or necrotic death (with swelling of cytoplasmic organelles) (24). Such "caspase-independent cell death" is nevertheless under the same regulatory mechanisms upstream of MOMP that control apoptosis (7).

Inhibition of MOMP may have a wider range of cytoprotective actions than inhibition of caspases (4). Accordingly, the antiapoptotic proteins of the Bcl-2 family (which prevent MOMP) have a strong cytoprotective action, even in systems in which caspases play no role in cell death induction (25). However, inhibition of MOMP is not a panacea against cell death. Thus, in conditions in which caspase activation is enforced through the ligation of death receptors or, for instance, by inhibitors of XIAP, cell death can occur without signs of MOMP, and genetic interventions on MOMP-regulatory proteins (such as overexpression of Bcl-2 or knockout of Bax and Bak) may not affect apoptotic cell death. Reportedly, successful MOMP inhibition can also cause a shift from apoptotic to autophagic cell death (26).

Inhibition of cell death at the preclinical and clinical levels

The death of postmitotic cells in the CNS and myocardium, be it the result of an acute or a chronic degenerative process, is a pathogenic event. Similarly, septic shock, ischemia, and intoxication can cause massive apoptosis in multiple organs. Infections can cause an elevated apoptotic turnover of specific cell types, for instance, lymphocytes in AIDS or gastric mucosa cells in *Helicobacter pylori* infection. These diseases are therefore candidates for therapeutic cell death suppression.

A cornucopia of putative drugs and targets for cytoprotection are being explored at the preclinical and clinical levels (Tables 1 and 2). MOMP inhibitors have attracted interest, with encouraging results in animal models. Similarly, a variety of protease inhibitors are being designed to prevent cell death. Among this class of agents, caspase inhibitors have been extensively explored. This exploration has been driven by the often-erroneous belief that caspases would generally dictate the "point of no return" of cell death. Although some preclinical results are promising, many of the specific caspase inhibitors (such as zVAD-fmk and zDEVD-fmk) can actually inhibit calpains and cathepsins (27, 28). Calpains and cathepsin can also be involved in pathogenic cell death, as indicated by pharmacological studies in rodents and genetic validation in mice (Table 2). Genetic methods to inhibit caspases, for instance by overexpressing the baculovirus protein p35, have not validated the role of caspases in cell death, for instance in kainate-induced neurotoxicity (29). Moreover, a proof-of-principle experiment illustrating that the conditional or tissue-specific knockout of 1 or several caspase genes can confer true cytoprotection is still elusive. In this sense, it is not surprising that clinical studies of caspase inhibitors assess the treatment of inflammatory diseases such as

hepatitis C, in which the inhibition of inflammatory caspases (caspase-1 and perhaps caspase-4 and -5) may be the therapeutic goal.

Other strategies of cytoprotection target death-inducing molecules such as TNF-α and CD95L, the proapoptotic transcription factor p53 (which is activated by genotoxic stress as well as by some kinds of metabolic stress), poly(ADP-ribose) polymerase (PARP) (which is activated by DNA damage and hypoxia), or a variety of stress kinases (such as JNK3 or the mixed-lineage kinases [MLKs]) (Tables 1 and 2). Antioxidants may also blunt acute and unwarranted cell death. Some antiapoptotic compounds have been FDA approved (Table 2). This applies in particular to agents targeting TNF-α, used to treat rheumatoid arthritis and Crohn disease. Whether such agents act as true cell death blockers or through their antiinflammatory effects, however, is an open question.

Cell damage sensors as targets for cell death induction

Damage to a variety of cellular organelles can lead to apoptosis. It is conceivable that each organelle possesses sensors that detect specific alterations, locally activate signal transduction pathways, emit signals that ensure interorganellar cross-talk, and ultimately stimulate the common apoptotic pathway, presumably by activating the central executioner (30).

Proteins that belong to the family of PI3K-like kinases sense the DNA damage response. This family includes ATM (ataxia telangiectasia mutated), ATR (ATM- and Rad-3–related), and DNA-dependent protein kinase (DNA-PK). In human cells, this directly or indirectly results in the phosphorylation of p53, modulating the DNA-binding activity of p53 and enhancing its stability. The tumor suppressor protein p53 mediates part of the response of mammalian cells to DNA damage, either by stimulating DNA repair or, beyond a certain threshold of DNA damage, by initiating apoptosis. Transcription factor p53 transactivates a number of proapoptotic proteins from the Bcl-2 family (in particular Bax, Bid, Puma, and Noxa), which induce MOMP and release apoptogenic factors from the mitochondrial intermembrane space. p53 may also induce apoptosis in a transcription-independent manner, by direct physical interactions with members of the Bcl-2 family (31–33). Reactivation of mutated p53 hence is one of the strategies used to kill tumor cells, in which p53 is frequently mutated or inactivated. Thus a reactivation may be mediated by p53-expressing adenovirus or by small molecules that reestablish the normal conformation of mutated p53.

While DNA damage often induces p53-dependent apoptosis (which implies that p53-mutated tumor cells do not respond to genotoxic agents), there are a variety of techniques to induce apoptosis by damaging organelles other than the nucleus, a priori in a p53-independent fashion. One possibility is to use agents that cause lysosomal membrane permeabilization, leading to the cytosolic release of cathepsins B and D, which are normally secluded in the lumen of lysosomes (34, 35). Cathepsins then trigger an apoptotic response that involves MOMP, followed by caspase activation (36). Alternatively, one can attempt to trigger MOMP by agents directly acting on mitochondria, again bypassing p53 dependency (4). Finally, it is possible to trigger apoptosis via the ER, by inducing the unfolded protein response or by stimulating lethal Ca^{2+} fluxes (30).

Although there are many ways to stimulate apoptotic responses by compounds inducing organelle-specific damage, thus far no agent specifically acting on cytoplasmic organelles has made it into clinical trials for the treatment of neoplasia. However, ligands of so-called death receptors (such as CD95/Fas, TNFR, and the 2 receptors of

Table 2
Pharmacological inhibitors of apoptosis that are FDA approved or in clinical development

Drug	Company or institution	Type of compound	Target	Status	Indication
MOMP					
Minocycline	Danbury Pharmacal	Small compound	Mitochondria	Phase I	Huntington disease
Rasagiline (Agilect)	Teva	Small compound	Peripheral benzodiazepine receptor?	FDA approved	Parkinson disease
p53					
Amifostine (Ethyol)		Small molecule	p53	FDA approved	Reduction of renal cisplatin toxicity in ovarian or non–small cell lung carcinoma; reduction of radiation effects on the parotid gland
Caspases and their endogenous inhibitors					
IDN-6556 liver transplantation	Pfizer	Small molecule	Caspases	Phase II	Hepatitis C, acute alcoholic hepatitis,
IDN-6734	Pfizer	Small molecule	Caspases	Phase I	Acute myocardial infarction
VX-740	Vertex/Aventis	Small molecule	Caspase-1	Phase II	Rheumatoid arthritis
Death receptors and their ligands					
Adalimumab (HUMIRA)	Abbott	Neutralizing mAb	TNF-α	FDA approved	Rheumatoid arthritis, psoriasis, ankylosing spondylitis, Crohn disease
Infliximab (Remicade)	Centocor/ Schering-Plough	Neutralizing mAb	TNF-α	FDA approved	Rheumatoid arthritis, Crohn disease
Etanercept (Enbrel)	Amgen/Wyeth	TNFR2/IgG fusion protein	TNF-α	FDA approved	Rheumatoid arthritis, Crohn disease
ISIS 104828	Isis	Antisense oligonucleotide	TNF-α	Phase II	Rheumatoid arthritis, Crohn disease, psoriasis
PARP					
INO-1001	Inotek	Small molecule	PARP	Phase I	Ischemia/reperfusion damage
Nicotine amide	Johns Hopkins University	Small molecule	PARP	Phase I	Ataxia telangiectasia
Antioxidants					
Edaravone	Mitsubishi-Tokyo	Small molecule	ROS	Phase III; approved in Japan for treatment of stroke	Reperfusion injury after acute myocardial infarction
Idebenone		Small molecule	ROS	Phase I	Friedreich ataxia
Kinase inhibitors					
CEP-1347	Cephalon/H. Lundbeck	Small molecule	MLK inhibitor	Phase II/III	Parkinson disease

TRAIL) are being evaluated as potential anticancer agents that could bypass the resistance against conventional chemotherapy (Table 3).

Cell death induction: the problem of specificity

It is easy to trigger apoptosis in most vertebrate cells; the problem is to do this in such a way that only those cells we wish to target die. This is the basis for most cancer therapies but also applies to any situation in which a particular cell population is responsible for disease. For example, selective elimination of a subpopulation of immune cells can form the basis of therapy for inflammatory or autoimmune disease, while specific death of hyperplastic fibroblasts can impact connective tissue diseases. Similarly, we can protect cells from death, but again, specificity is required to ensure that such protection does not promote undesirable cell survival, leading to hyperaccumulation or cancer.

One key to specificity may come with a deeper understanding of why a disease-causing cell does not undergo cell death through normal homeostatic mechanisms. For example, it is well estab-lished that signals to enter the cell cycle simultaneously sensitize cells for apoptosis, such that tissue expansion depends on availability of exogenous survival factors (37). A cell that bypasses the homeostatic apoptotic mechanisms (e.g., through mutation or overexpression of cell death regulators) will, in theory, only have interrupted one physiologically relevant pathway of apoptosis, while leaving other pathways intact and with heightened sensitivity. This is likely the reason why agents that broadly target cells for DNA damage, microtubule breakdown, protease dysfunction, or defects in metabolism can have relatively specific effects on cancer cells, forming the basis of many established approaches to cancer therapy. By understanding and targeting such pathways specifically, we can improve therapies with fewer side effects.

Therapeutic cell death induction at the preclinical and clinical levels

Most cytotoxic agents used in anticancer chemotherapy induce apoptosis, although very few actually directly target the apoptotic

Table 3
Pharmacological inducers of apoptosis that are FDA approved or in clinical development

Drug	Company or institution	Type of compound	Target	Status	Indication
MOMP					
Oblimersen (Genasense)	Genta	Antisense oligonucleotide	Bcl-2	Phase III	Chronic lymphocytic leukemia, multiple myeloma, non–small cell lung carcinoma
				Phase II	Hormone-refractory prostate cancer
				Phase I	Advanced breast cancer
EGCG	Burnham Institute/Mayo Clinic	Small compound	Bcl-2	Phase I	Chronic lymphocytic leukemia
Gossypol	University of Michigan	Small compound	Bcl-2	Phase I/II	Breast cancer
LY2181308	Isis/Eli Lilly	Antisense oligonucleotide	Survivin	Phase I	Cancer
Arsenic trioxide		Small compound	Mitochondria?	FDA approved	Relapsed acute promyelocytic leukemia
				Phase II	Acute T cell leukemia/lymphoma
p53					
Advexin (INGN201)	Invitrogen	Adenovirus	p53	Phase III	Head and neck cancer
				Phase II	Breast, lung, colorectal, and ovarian cancer
SCH58500	Schering-Plough	Adenovirus	p53	Phase III	Ovarian and peritoneal cancer
				Phase II	Head/neck and liver cancer
ONYX-015	Onyx	Adenovirus (E1B mutant)	Mutated p53	Phase III	Pancreatic, colorectal, head and neck, and liver cancer
				Phase II/III	Non–small cell lung cancer
Caspases and their endogenous inhibitors					
AEG35156/ GEM640	Aegera/Hybridon	Antisense oligonucleotide	XIAP	Phase I	Cancer
Death receptors and their ligands					
TNF-α		Recombinant protein	TNF-αR	FDA approved	Isolated limb perfusion therapy of nonresectable melanoma (78)
HGS-ETR1	Human Genome Sciences	Agonistic mAb	TRAIL-R1	Phase II	Non-Hodgkin lymphoma, colorectal cancer
HGS-ETR2	Human Genome Sciences	Agonistic mAb	TRAIL-R2	Phase I	Cancer
HGS-TR2J	Human Genome Sciences/ Kirin Pharmaceuticals	Agonistic mAb	TRAIL-R2	Phase I	Solid tumors
PRO1764	Genentech/Amgen	Soluble TRAIL ligand	TRAIL-R	Phase I	Cancer
PARP					
AG014699	Cancer Research Technology	Small molecule	PARP	Phase I	Brain cancer
Proteasome					
Bortezomib (Velcade)	Millennium	Small molecule	26S proteasome	FDA approved	Multiple myeloma
				Phase II/III	Mantle cell lymphoma, lymphoma, lung, breast, prostate, and ovarian cancer
Kinase inhibitors					
Trastuzumab (Herceptin)	Roche	mAb	HER2	FDA approved	Breast cancer with her2neu overexpression
Cetuximab (Erbitux)	ImClone/ Bristol-Myers Squibb	mAb	HER1	FDA approved	Colon cancer
Gefitinib (Iressa)	AstraZeneca	Small molecule	HER1	FDA approved	Non–small cell lung cancer
				Phase III	Head and neck, breast, ovarian, prostate, pancreatic, and colorectal cancer; glioma
Erlotinib (Tarceva)	Genentech/OSI Pharmaceuticals/Roche	Small molecule	HER1	FDA approved	Non–small cell lung cancer
				Phase III	Pancreatic cancer
CCI-779	Novartis	Small molecule	mTOR	Phase II	Solid tumors
BAY 43-9006	Onyx/Bayer	Small molecule	Raf, VEGFR	Phase III	Renal cell cancer
Imatinib mesylate (Gleevec; STI-571)	Novartis	Small molecule	cKit, PDGFR, Bcr-Abl	FDA approved	Gastrointestinal stromal tumors, chronic myeloid leukemia

machinery (Table 3). Oblimersen (G3139, Genasense) is an anti-sense oligonucleotide that targets human *bcl-2* mRNA. This compound, which has been tested in clinical trials with encouraging results, constitutes the prototype of a new therapeutic concept targeting antiapoptotic molecules. Small molecules targeting Bcl-2 and its relatives are also being developed (20).

p53-expressing adenoviruses as well as an adenovirus designed to replicate in p53-deficient tumor cells (ONYX-015) are also being clinically tested for cancer treatment. TNF-α has been approved for isolated limb perfusion, in which the recombinant protein is injected into an artery of a limb affected by nonresectable melanoma. TRAIL or agonistic antibodies targeting TRAIL-R1 or TRAIL-R2 have entered the early stage of clinical evaluation. A specific PARP inhibitor has also been tested for chemosensitization of brain cancers. More importantly, an inhibitor of the proteasome bortezomib (Velcade) has been approved for chemotherapy of multiple myeloma and is under evaluation as a single agent or in combination chemotherapy for the treatment of hematopoietic and solid cancers (Table 3).

However, the true revolution in cancer therapy concerns the therapeutic use of kinase inhibitors targeting protein kinases whose activity is required for cancer cell proliferation and/or survival (Table 3). As an example, one of the prototypic kinase inhibitors, imatinib mesylate (Gleevec), can eradicate BCR/ABL–positive leukemia cells, presumably through the induction of apoptotic (38) or nonapoptotic, caspase-independent (39) cell death. The FDA has approved several other specific inhibitors of survival kinases recently, and dozens are currently under preclinical development.

Conclusions and perspectives: this series

The current Review Series contains 9 Reviews of areas in cell death research that hold promise for the development of therapeutics.

The Review by Amaravadi and Thompson (40) considers this problem from the perspective of the signaling pathways, in particular the kinases, that control cell survival. This includes not only regulation of apoptosis, per se, but also the signals that control metabolic demand and, in turn, control life and death. Other survival signals, notably those controlled by NF-κB, are reviewed by Luo et al. (41) and considered with respect to cancer therapy. In addition, cellular survival can be preserved by the functions of members of the heat shock protein family, an active area of research that is reviewed by Beere (42). In all of these cases, important targets have been identified, and agents to modulate them are in different stages of development and testing for use in human disease.

As outlined above, the mitochondria sit at the center of a major pathway for cell death controlled by multiple inputs, and this can result in apoptosis or other forms of cell death. Bouchier-Hayes

and colleagues (43) consider the mitochondria as a target for pharmacological intervention, especially with respect to control of the Bcl-2 family proteins. Strategies for elaborating specific apoptosis modulators that mimic Bcl-2 family proteins or act on such proteins are discussed by Letai (44). Another organelle that appears to control cell life and death is the ER, which coordinates complex signaling events around calcium storage and the unfolded protein response as well as other forms of ER stress. Xu et al. (45) consider the signal transduction pathways that center on the ER with respect to cell death and its control.

The regulation of caspase activation is, of course, highly relevant to the manipulation of apoptosis, and this subject is reviewed by Lavrik and colleagues (46). Similarly, the IAPs have for some time been thought of predominantly as caspase inhibitors. Recent discoveries have shown that these also play roles distinct from the control of caspases, acting as ubiquitin ligases for other proteins, passenger molecules, and signal transduction proteins. Wright and Duckett (47) consider IAPs as pharmacological targets and how these affect various outcomes in the cell and organism.

Finally, the process of autophagy, distinct from that of apoptosis, accompanies cell death in many cases, and it is possible that this "self-eating" may result in death, or, alternatively, may represent a mechanism for cell survival. Levine and Yuan (48) discuss this process and consider pharmacological approaches to its manipulation.

Many drug targets elucidated in studies of cell death have led to approved pharmacological agents that are currently in use, while many more are in clinical trial. These are promising, of course, but we are only beginning to understand these processes and their potential for clinical benefit. The application of our understanding is now in sight, but the potential extends beyond our vision.

Acknowledgments

G. Kroemer receives a special grant from the Ligue Nationale contre le cancer as well as grants from the European Community, Fondation de France, Agence Nationale pour la Recherche sur le SIDA, Sidaction, Cancéropôle Ile-de-France, and the French Ministry of Science. D.R. Green is supported by grants from the NIH.

Address correspondence to: Douglas R. Green, Department of Immunology, St. Jude Children's Research Hospital, 332 North Lauderdale Street, Memphis, Tennessee 38105, USA. Phone: (901) 495-3378; Fax: (901) 495-3107; E-mail: douglas.green@stjude.org. Or to: Guido Kroemer, Centre National de la Recherche Scientifique, UMR8125, Institut Gustave Roussy, 39 rue Camille-Desmoulins, F-94805 Villejuif, France. Phone: 331-4211-6046; Fax: 331-4211-6047; E-mail: kroemer@igr.fr.

1. Wyllie, A.H., Kerr, J.F., and Currie, A.R. 1980. Cell death: the significance of apoptosis. *Int. Rev. Cytol.* **68**:251–306.

2. Fuentes-Prior, P., and Salvesen, G.S. 2004. The protein structures that shape caspase activity, specificity, activation and inhibition. *Biochem. J.* **384**:201–232.

3. Debatin, K.M., and Krammer, P.H. 2004. Death receptors in chemotherapy and cancer. *Oncogene.* **23**:2950–2966.

4. Green, D.R., and Kroemer, G. 2004. The pathophysiology of mitochondrial cell death. *Science.* **305**:626–629.

5. Bao, Q., Riedl, S.J., and Shi, Y. 2005. Structure of Apaf-1 in the auto-inhibited form: a critical role for ADP. *Cell Cycle.* **4**:1001–1003.

6. Acehan, D., et al. 2002. Three-dimensional structure of the apoptosome: implications for assem-

bly, procaspase-9 binding, and activation. *Mol. Cell.* **9**:423–432.

7. Chipuk, J.E., and Green, D.R. 2005. Do inducers of apoptosis trigger caspase-independent cell death? *Nat. Rev. Mol. Cell Biol.* **6**:268–275.

8. Kroemer, G., and Martin, S.J. 2005. Caspase-independent cell death. *Nat. Med.* **11**:725–730.

9. Susin, S.A., et al. 1999. Molecular characterization of mitochondrial apoptosis-inducing factor. *Nature.* **397**:441–446.

10. Li, L.Y., Luo, X., and Wang, X. 2001. Endonuclease G is an apoptotic DNase when released from mitochondria. *Nature.* **412**:95–99.

11. Kokoszka, J.E., et al. 2004. The ADP/ATP translocator is not essential for the mitochondrial permeability transition pore. *Nature.* **427**:461–465.

12. Nakagawa, T., et al. 2005. Cyclophilin D-dependent

mitochondrial permeability transition regulates some necrotic but not apoptotic cell death. *Nature.* **434**:652–658.

13. Baines, C.P., et al. 2005. Loss of cyclophilin D reveals a critical role for mitochondrial permeability transition in cell death. *Nature.* **434**:658–662.

14. Basso, E., et al. 2005. Properties of the permeability transition pore in mitochondria devoid of Cyclophilin D. *J. Biol. Chem.* **280**:18558–18561.

15. Kuwana, T., et al. 2002. Bid, Bax, and lipids cooperate to form supramolecular openings in the outer mitochondrial membrane. *Cell.* **111**:331–342.

16. Kuwana, T., et al. 2005. BH3 domains of BH3-only proteins differentially regulate Bax-mediated mitochondrial membrane permeabilization both directly and indirectly. *Mol. Cell.* **17**:525–535.

17. Wei, M.C., et al. 2001. Proapoptotic BAX and BAK:

a requisite gateway to mitochondrial dysfunction and death. *Science.* **292**:727–730.

18. Letai, A., et al. 2002. Distinct BH3 domains either sensitize or activate mitochondrial apoptosis, serving as prototype cancer therapeutics. *Cancer Cell.* **2**:183–192.

19. Chen, L., et al. 2005. Differential targeting of prosurvival Bcl-2 proteins by their BH3-only ligands allows complementary apoptotic function. *Mol. Cell.* **17**:393–403.

20. Oltersdorf, T., et al. 2005. An inhibitor of Bcl-2 family proteins induces regression of solid tumours. *Nature.* **435**:677–681.

21. Zhang, J., Liu, X., Scherer, D.C., van Kaer, L., Wang, X., and Xu, M. 1998. Resistance to DNA fragmentation and chromatin condensation in mice lacking the DNA fragmentation factor 45. *Proc. Natl. Acad. Sci. U. S. A.* **95**:12480–12485.

22. Susin, S.A., et al. 2000. Two distinct pathways leading to nuclear apoptosis. *J. Exp. Med.* **192**:571–579.

23. Yu, L., et al. 2004. Regulation of an ATG7-beclin 1 program of autophagic cell death by caspase-8. *Science.* **304**:1500–1502.

24. Hirsch, T., et al. 1997. The apoptosis-necrosis paradox. Apoptogenic proteases activated after mitochondrial permeability transition determine the mode of cell death. *Oncogene.* **15**:1573–1582.

25. Xu, P., Rogers, S.J., and Roossinck, M.J. 2004. Expression of antiapoptotic genes bcl-xL and ced-9 in tomato enhances tolerance to viral-induced necrosis and abiotic stress. *Proc. Natl. Acad. Sci. U. S. A.* **101**:15805–15810.

26. Shimizu, S., et al. 2004. A role of Bcl-2 family of proteins in non-apoptotic programmed cell death dependent on autophagy genes. *Nat. Cell Biol.* **6**:1221–1228.

27. Schotte, P., Declercq, W., Van Huffel, S., Vandenabeele, P., and Beyaert, R. 1999. Non-specific effects of methyl ketone peptide inhibitors of caspases. *FEBS Lett.* **442**:117–121.

28. Knoblach, S.M., et al. 2004. Caspase inhibitor z-DEVD-fmk attenuates calpain and necrotic cell death in vitro and after traumatic brain injury. *J. Cereb. Blood Flow Metab.* **24**:1119–1132.

29. Tomioka, M., et al. 2002. In vivo role of caspases in excitotoxic neuronal death: generation and analysis of transgenic mice expressing baculoviral caspase inhibitor p35, in postnatal neurons. *Brain Res. Mol. Brain Res.* **108**:18–32.

30. Ferri, K.F., and Kroemer, G.K. 2001. Organelle-specific initiation of cell death pathways. *Nat. Cell Biol.* **3**:E255–E263.

31. Marchenko, N.D., Zaika, A., and Moll, U.M. 2000. Death signal-induced localization of p53 protein to mitochondria. A potential role in apoptotic signaling. *J. Biol. Chem.* **275**:16202–16212.

32. Chipuk, J.E., et al. 2004. Direct activation of Bax by p53 mediates mitochondrial membrane permeabilization and apoptosis. *Science.* **303**:1010–1014.

33. Leu, J.I., Dumont, P., Hafey, M., Murphy, M.E., and George, D.L. 2004. Mitochondrial p53 activates Bak and causes disruption of a Bak-Mcl1 complex. *Nat. Cell Biol.* **6**:443–450.

34. Boya, P., et al. 2003. Lysosomal membrane permeabilization induces cell death in a mitochondrion-dependent fashion. *J. Exp. Med.* **197**:1323–1334.

35. Erdal, H., et al. 2005. Induction of lysosomal membrane permeabilization by compounds that activate p53-independent apoptosis. *Proc. Natl. Acad. Sci. U. S. A.* **102**:192–197.

36. Boya, P., et al. 2003. Lysosomal membrane permeabilization induces cell death in a mitochondrion-dependent fashion. *J. Exp. Med.* **197**:1323–1334.

37. Green, D.R., and Evan, G.I. 2002. A matter of life and death. *Cancer Cell.* **1**:19–30.

38. le Coutre, P., et al. 1999. In vivo eradication of human BCR/ABL-positive leukemia cells with an ABL kinase inhibitor. *J. Natl. Cancer Inst.* **91**:163–168.

39. Okada, M., et al. 2004. A novel mechanism for imatinib mesylate-induced cell death of BCR-ABL-positive human leukemic cells: caspase-independent, necrosis-like programmed cell death mediated by serine protease activity. *Blood.* **103**:2299–2307.

40. Amaravadi, R., and Thompson, C.B. 2005. The survival kinases Akt and Pim as potential pharmacological targets. *J. Clin. Invest.* **115**:2618–2624. doi:10.1172/JCI26273.

41. Luo, J.-L., Kamata, H., and Karin, M. 2005. IKK/NF-κB signaling: balancing life and death a new approach to cancer therapy. *J. Clin. Invest.* **115**:2625–2632. doi:10.1172/JCI26322.

42. Beere, H.M. 2005. Death versus survival: functional interaction between the apoptotic and stress-inducible heat shock protein pathways. *J. Clin. Invest.* **115**:2633–2639. doi:10.1172/JCI26471.

43. Bouchier-Hayes, L., Lartigue, L., and Newmeyer, D.D. 2005. Mitochondria: pharmacological manipulation of cell death. *J. Clin. Invest.* **115**:2640–2647. doi:10.1172/JCI26274.

44. Letai, A. 2005. Pharmacological manipulation of Bcl-2 family members to control cell death. *J. Clin. Invest.* **115**:2648–2655. doi:10.1172/JCI26250.

45. Xu, C., Bailly-Maitre, B., and Reed, J.C. 2005. Endoplasmic reticulum stress: cell life and death decisions. *J. Clin. Invest.* **115**:2656–2664. doi:10.1172/JCI26373.

46. Lavrik, I.N., Golks, A., and Krammer, P.H. 2005. Caspases: pharmacological manipulation of cell death. *J. Clin. Invest.* **115**:2665–2672. doi:10.1172/JCI26252.

47. Wright, C.W., and Duckett, C.S. 2005. Reawakening the cellular death program in neoplasia through the therapeutic blockade of IAP function. *J. Clin. Invest.* **115**:2673–2678. doi:10.1172/JCI26251.

48. Levine, B., and Yuan, J. 2005. Autophagy in cell death: an innocent convict? *J. Clin. Invest.* **115**:2679–2688. doi:10.1172/JCI26390.

49. Cao, G., et al. 2002. In vivo delivery of a Bcl-xL fusion protein containing the TAT protein transduction domain protects against ischemic brain injury and neuronal apoptosis. *J. Neurosci.* **22**:5423–5431.

50. Sugioka, R., et al. 2003. BH4-domain peptide from Bcl-xL exerts anti-apoptotic activity in vivo. *Oncogene.* **22**:8432–8440.

51. Akhtar, R.S., Ness, J.M., and Roth, K.A. 2004. Bcl-2 family regulation of neuronal development and neurodegeneration. *Biochim. Biophys. Acta.* **1644**:189–203.

52. Tanaka, M., et al. 2004. Cardiomyocyte-specific Bcl-2 overexpression attenuates ischemia-reperfusion injury, immune response during acute ischemia, and graft coronary artery disease. *Blood.* **104**:3789–3796.

53. Weisleder, N., Taffet, G.E., and Capetanaki, Y. 2004. Bcl-2 overexpression corrects mitochondrial defects and ameliorates inherited desmin null cardiomyopathy. *Proc. Natl. Acad. Sci. U. S. A.* **101**:769–774.

54. Yang, L., et al. 2004. A novel systemically active caspase inhibitor attenuates the toxicities of MPTP, malonate, and 3NP in vivo. *Neurobiol. Dis.* **17**:250–259.

55. Methot, N., et al. 2004. Differential efficacy of caspase inhibitors on apoptosis markers during sepsis in rats and implication for fractional inhibition requirements for therapeutics. *J. Exp. Med.* **199**:199–207.

56. Wang, J., et al. 2004. Caspase inhibitors, but not c-Jun NH2-terminal kinase inhibitor treatment, prevent cisplatin-induced hearing loss. *Cancer Res.* **64**:9217–9224.

57. Matsui, J.I., et al. 2003. Caspase inhibitors promote vestibular hair cell survival and function after aminoglycoside treatment in vivo. *J. Neurosci.* **23**:6111–6122.

58. Liu, X., Van Vleet, T., and Schnellmann, R.G. 2004. The role of calpain in oncotic cell death. *Annu. Rev. Pharmacol. Toxicol.* **44**:349–370.

59. Crocker, S.J., et al. 2003. Inhibition of calpains prevents neuronal and behavioral deficits in an MPTP mouse model of Parkinson's disease. *J. Neurosci.* **23**:4081–4091.

60. Takano, J., et al. 2005. Calpain mediates excitotoxic DNA fragmentation via mitochondrial pathways in adult brains: evidence from calpastatin-mutant mice. *J. Biol. Chem.* **280**:16175–16184.

61. Spencer, M.J., and Mellgren, R.L. 2002. Overexpression of a calpastatin transgene in mdx muscle reduces dystrophic pathology. *Hum. Mol. Genet.* **11**:2645–2655.

62. Guicciardi, M.A., Leist, M., and Gores, G.J. 2004. Lysosomes in cell death. *Oncogene.* **23**:2881–2890.

63. Cilenti, L., et al. 2005. Omi/HtrA2 protease mediates cisplatin-induced cell death in renal cells. *Am. J. Physiol. Renal Physiol.* **288**:F371–F379.

64. Liu, H.R., et al. 2005. Role of Omi/HtrA2 in apoptotic cell death after myocardial ischemia and reperfusion. *Circulation.* **111**:90–96.

65. Cregan, S.P., et al. 2002. Apoptosis-inducing factor is involved in the regulation of caspase-independent neuronal cell death. *J. Cell Biol.* **158**:507–517.

66. Demjen, D., et al. 2004. Neutralization of CD95 ligand promotes regeneration and functional recovery after spinal cord injury. *Nat. Med.* **10**:389–395.

67. Martin-Villalba, A., et al. 2001. Therapeutic neutralization of CD95-ligand and TNF attenuates brain damage in stroke. *Cell Death Differ.* **8**:679–686.

68. Graham, E.M., et al. 2004. Neonatal mice lacking functional Fas death receptors are resistant to hypoxic-ischemic brain injury. *Neurobiol. Dis.* **17**:89–98.

69. Yoshino, O., et al. 2004. The role of Fas-mediated apoptosis after traumatic spinal cord injury. *Spine.* **29**:1394–1404.

70. Eichhorst, S.T., et al. 2004. Suramin inhibits death receptor-induced apoptosis in vitro and fulminant apoptotic liver damage in mice. *Nat. Med.* **10**:602–609.

71. Komarov, P.G., et al. 1999. A chemical inhibitor of p53 that protects mice from the side effects of cancer therapy. *Science.* **285**:1733–1737.

72. Hong, S.J., Dawson, T.M., and Dawson, V.L. 2004. Nuclear and mitochondrial conversations in cell death: PARP-1 and AIF signaling. *Trends Pharmacol. Sci.* **25**:259–264.

73. Xiao, C.V., et al. 2005. Poly(ADP-ribose) polymerase promotes cardiac remodeling, contractile failure and translocation of apoptosis-inducing factor in a murine experimental model of aortic banding and heart failure. *J. Pharmacol. Exp. Ther.* **312**:891–898.

74. Zheng, J., Devalaraja-Narashimha, K., Singaravelu, K., and Padanilam, B.J. 2005. Poly(ADP-ribose) polymerase-1 gene ablation protects mice from ischemic renal injury. *Am. J. Physiol. Renal Physiol.* **288**:F387–F398.

75. Hirt, L., et al. 2004. D-JNKI1, a cell-penetrating c-Jun-N-terminal kinase inhibitor, protects against cell death in severe cerebral ischemia. *Stroke.* **35**:1738–1743.

76. Kuan, C.V., et al. 2003. A critical role of neural-specific JNK3 for ischemic apoptosis. *Proc. Natl. Acad. Sci. U. S. A.* **100**:15184–15189.

77. Wang, L.H., Besirli, C.G., and Johnson, E.M., Jr. 2004. Mixed-lineage kinases: a target for the prevention of neurodegeneration. *Annu. Rev. Pharmacol. Toxicol.* **44**:451–474.

78. Grunhagen, D.J., et al. 2004. One hundred consecutive isolated limb perfusions with TNF-alpha and melphalan in melanoma patients with multiple in-transit metastases. *Ann. Surg.* **240**:939–947.

Pharmacological manipulation of Bcl-2 family members to control cell death

Anthony Letai

Department of Medical Oncology, Dana-Farber Cancer Institute, Harvard Medical School, Boston, Massachusetts, USA.

The commitment to programmed cell death involves complex interactions among pro- and antiapoptotic members of the Bcl-2 family of proteins. The physiological result of a decision by these proteins to undergo cell death is permeabilization of the mitochondrial outer membrane. Pharmacologic manipulation of proteins in this family appears both feasible and efficacious, whether the goal is decreased cell death, as in ischemia of the myocardium or brain, or increased cell death, as in cancer.

The Bcl-2 family of proteins

Bcl-2 was initially cloned from the breakpoint of the t(14;18) chromosomal translocation found in the vast majority of patients with follicular lymphoma, an indolent B cell non-Hodgkin lymphoma (1–3). Expression of Bcl-2 blocked cell death following numerous cell insults (4, 5). As a test of its oncogenic function, a minigene bearing the *bcl-2*–immunoglobulin gene fusion, after a period of follicular hyperplasia (4, 5), induced lymphoma in transgenic mice (6). While previously characterized oncogenes shared the ability to increase cellular proliferation, Bcl-2 established a new class of oncogenes: inhibitors of programmed cell death.

Since the cloning of Bcl-2, an entire family of proteins related by sequence homology and participation in the control of apoptosis has been identified. Certain proteins share Bcl-2's ability to oppose programmed cell death: Bcl-x_L (7), Bcl-w (8), Mcl-1 (9), and Bfl-1 (A1) (10). These proteins share sequence homology in 4 α-helical Bcl-2 homology (BH) regions, BH1–BH4. Bax (11) and Bak (12), which promote cell death, share only the BH1–BH3 domains. Later, a third class of protein was discovered (13). These include Bid, Bad, Bik, Puma, Noxa, Bmf, and Hrk, which are called "BH3-only" proteins and demonstrate homology only in the BH3 region (14, 15). Like Bax and Bak, the proapoptotic BH3-only proteins require an intact BH3 domain to promote apoptosis (14, 16).

Bcl-2 proteins control mitochondrial permeabilization. Complex interactions among Bcl-2 family members govern mitochondrial outer membrane permeabilization (MOMP), the final common endpoint for execution of death signals by the Bcl-2 family (17) (Figure 1). Data show that activation of either Bax or Bak is required for MOMP (18–20), suggesting that Bax and Bak are the effectors in the Bcl-2 family most proximal to MOMP. In healthy cells, inactive Bax monomers reside either in the cytosol or in loose association with the mitochondrial outer membrane (21), while monomeric Bak is inserted in the outer membrane. Activation of Bax and/or Bak is accompanied by an allosteric change detectable by conformation-specific antibodies (22–24). Following activation, Bax inserts into the membrane, Bax and Bak homo-oligomerize, and then MOMP occurs (19, 25–27). Permeabilization releases proapoptotic factors, including cytochrome *c* (28), omi/htra2 (29),

Smac/DIABLO (30, 31), endonuclease G (29), and AIF (32), to the cytosol. Released cytochrome *c* participates in a holoenzyme complex with Apaf-1 and caspase-9 that activates effector caspases by cleavage, resulting in widespread proteolysis. An extrinsic pathway of caspase activation initiated by cell surface death receptor signaling, which operates independently of the mitochondrion and Bcl-2 family members, exists but is beyond the scope of this discussion (33). While oligomers of Bax can form pores in artificial membranes that permit the passage of cytochrome *c* (34) or high-molecular weight dextran (35, 36), it is not clear whether activated Bax and/or Bak independently form pores in vivo, or whether Bax and/or Bak cooperate with other factors (17).

Activation of Bax and/or Bak. The mechanism of Bax and/or Bak activation has been controversial. Recent evidence supports activation of Bax and/or Bak via interaction with select BH3-only proteins. Bid protein and BH3 domains from Bid and Bim, but not other BH3-only proteins, induce MOMP in a Bax and/or Bak–dependent fashion and induce Bak and Bax oligomerization (19, 37, 38). Induction of these apoptotic changes requires an intact BH3 domain. The ability of Bid, Bim$_s$, or their BH3 domains to stimulate Bax oligomerization and pore formation required no other proteins in a defined synthetic liposomal system; this supports a direct interaction (35, 36). It has been hypothesized that Bid performs primarily as an inhibitor of Bcl-2. However, a Bid mutant that lacks the ability to interact with Bcl-2 but maintains interaction with Bax is still potently proapoptotic, which suggests that interaction with Bax and/or Bak is important in Bid's function (14).

Complexes of Bid or Bim with Bax or Bak have been difficult, but not impossible, to isolate (14, 39–41). Interactions between BH3 domains and Bax and/or Bak may be transient, with the BH3 domain leaving after allosteric activation of Bax and/or Bak in a "hit and run" model. While the BH3 domains of Bid and Bim are necessary for interaction with Bax and Bak, their most efficient presentation to Bax and Bak may require conformational changes, post-translational modifications, and/or certain isoforms of the entire protein (39, 40, 42–45). In addition, proteins outside of the Bcl-2 family bind and modulate function of Bax (46–48) and Bak (49).

Bcl-2 blocks Bax and/or Bak–dependent MOMP. Bcl-2 and the related antiapoptotic proteins Bcl-x_L, Mcl-1, Bcl-w, and Bfl-1 inhibit MOMP by binding pro-death family members. Like the proapoptotic multidomain proteins Bax and Bak, the antiapoptotic proteins possess a hydrophobic pocket made from the α-helices BH1, BH2, and BH3, where the hydrophobic face of amphipathic α-helical BH3 domains from proapoptotic members binds

Nonstandard abbreviations used: BH, Bcl-2 homology (domain); MOMP, mitochondrial outer membrane permeabilization.

Conflict of interest: The author has declared that no conflict of interest exists.

Citation for this article: *J. Clin. Invest.* **115**:2648–2655 (2005).
doi:10.1172/JCI26250.

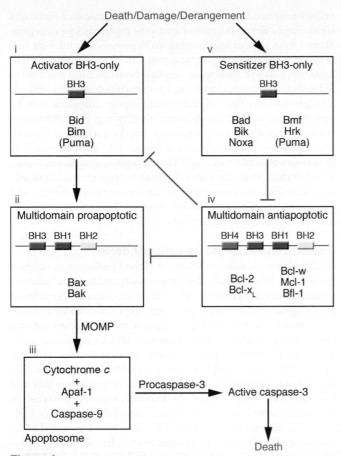

Figure 1
A model of Bcl-2 family member control over programmed cell death. In response to myriad death, damage, or derangement signals, BH3-only family members are activated (i). Activator BH3-only proteins interact with multidomain proapoptotic Bax and/or Bak (Bax/Bak), inducing their oligomerization (ii) and thus resulting in MOMP, release of cytochrome *c*, apoptosome formation, and caspase activation (iii) Bcl-2 and other multidomain antiapoptotic proteins interrupt the death signal by binding and sequestering activator BH3-only family members, and perhaps also Bax/Bak (iv). Bcl-2 antiapoptotic function may be antagonized by the competitive displacement of activator BH3-only molecules by sensitizer BH3-only proteins (v).

(50–53). While earlier work focused on the ability of antiapoptotic proteins to bind Bax or Bak, a Bcl-x$_L$ mutant lacking Bax or Bak binding, but still binding BH3-only proteins, retained the majority of its antiapoptotic function. This result suggests that binding and sequestration of BH3-only family members prior to their interaction with Bax and Bak is also an important function of the antiapoptotic proteins (26, 54). It is also consistent with the finding that Bcl-2 inhibits apoptosis upstream of Bax and/or Bak conformational change, membrane insertion (in the case of Bax), and oligomerization (26, 37, 55, 56). By binding and sequestering activator BH3 domains like Bid and Bim, the antiapoptotic proteins inhibit Bax and/or Bak activation and subsequent MOMP.

Antagonism of Bcl-2 by sensitizer BH3-only proteins. While all BH3-only proteins are proapoptotic, it is likely that only a subset interacts with Bax and Bak. Using a series of BH3 peptides, BH3 domains have been divided into 2 classes: the activators (including Bid and Bim), which can induce Bax and/or Bak oligomerization

and MOMP, and the sensitizers, which cannot (37). The ability of p53 to activate Bax suggests there may be other "cryptic" activator molecules outside the Bcl-2 family (57). The sensitizer BH3 domains interact with the antiapoptotic molecules and only indirectly induce Bax and/or Bak activation by competitive displacement of activator BH3 proteins from the Bcl-2–binding cleft (37). These so-called sensitizer BH3 domains are thus prototypes of selective inhibitors of the antiapoptotic proteins (see below) (37, 58). Binding of a given BH3 protein to antiapoptotic proteins is not necessarily promiscuous. For instance, while BH3 domains from Bid, Bim, and Puma interact with all of the antiapoptotic proteins tested, the remainder interact only with select antiapoptotic partners, suggesting that antiapoptotic proteins have biophysically distinct binding pockets (36, 37, 59, 60). In theory, therefore, individual antiapoptotic family members can be selectively targeted by small molecules that mimic sensitizer BH3 domain behavior.

BH3-only proteins' response to death stimuli. There remains the question of how BH3-only molecules are triggered to respond to death stimuli (61). In some cases, activation of BH3-only molecules is transcriptional. Noxa and Puma are p53-inducible genes that are transcriptionally induced in response to numerous DNA-damaging agents (62–66). While Bim can be transcriptionally regulated (67, 68), its regulation may also depend on cytoskeletal interaction (69) and phosphorylation, which may affect interaction with Bax (40) as well as ubiquitination and proteosomal degradation (70–72). In response to death signals that activate the extrinsic apoptotic pathway, Bid is first cleaved by caspase-8 (42, 43), and then the new amino terminus is myristoylated to facilitate targeting to the mitochondria (44). Bad is controlled, at least in part, by phosphorylation that mediates its sequestration by cytoplasmic 14-3-3 protein (73). Interaction of Bad with members of the glycolytic pathway suggests a role for Bad in glucose homeostasis (74). It appears that BH3-only family members serve individual but overlapping roles in sensing different types of cellular derangement and communicating these to the core death pathway.

To summarize, conduction of a death signal by the Bcl-2 family members begins with activation of the BH3-only proteins, which act as sentinels of myriad damage signals. Activator BH3 domains then trigger allosteric activation of Bax and/or Bak, which oligomerize at the mitochondria, inducing MOMP. Proapoptotic factors are released, including cytochrome *c*, which forms a holoenzyme complex with Apaf-1 and caspase-9, termed the apoptosome. This complex then catalytically cleaves effector caspases like caspase-3, resulting in widespread proteolysis and commitment to cell death. This pathway can be interrupted by Bcl-2 and related antiapoptotic proteins that bind and sequester the activator BH3 molecules. Sensitizer BH3-only molecules can further assist the progression of a death signal by competitive displacement of activator BH3 signals from the Bcl-2 pocket (Figure 1).

Alternative models of the control of apoptosis by Bcl-2 family members exist. In one, Bcl-2 regulates caspases, directly or via an as-yet undiscovered mammalian adaptor (13, 61, 75). In another, Bcl-2 tonically inhibits oligomerization of a Bax and/or Bak that is ready to induce MOMP without BH3 interaction (76). In a variation of this model, combinations of particular Bcl-2–like proteins must be neutralized by BH3 ligands to allow Bax and/or Bak activation (60). In another, Bax and/or Bak inhibit a dominantly acting Bcl-2 to induce death (13). However, recent results that emphasize the centrality of Bax and Bak (18, 20) and their activation by activator BH3 domains (19, 26, 35–37) to MOMP

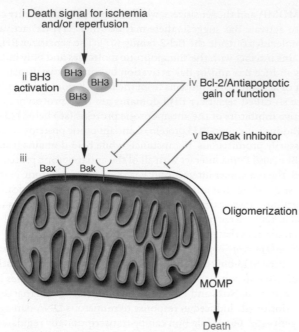

Figure 2
Model of interventions to reduce ischemic and ischemia/reperfusion injury. (i and ii) Following ischemia or ischemia/reperfusion, cell death signals are initiated (i) and conducted to the intrinsic pathway via activated BH3-only family members (ii). (iii) These BH3 family members interact with Bax/Bak, inducing oligomerization, MOMP, and commitment to cell death. (iv) Treatment with a viral vector expressing Bcl-2 or other antiapoptotic gain of function might prevent BH3-only activation of Bax/Bak and/or oligomerization of Bax/Bak. (v) Treatment with a Bax/Bak inhibitor might prevent Bax/Bak induction of MOMP.

may call into question the consistency of the above alternatives with mounting experimental data.

Diseases of excessive cell death

Bcl-2 proteins in acute ischemic diseases. Myocardial ischemia and cerebral ischemia are 2 of the 3 leading causes of death in the developed world. In acute ischemia, sudden loss of blood perfusion, and/or the acute restoration of perfusion, can result in extensive death in the cells of the supplied organ, as in stroke or acute myocardial infarction. In mouse models of cerebral ischemia as well as ischemia/reperfusion injury, loss of proapoptotic Bax or Bid and gain of antiapoptotic Bcl-2 function all reduced infarct size (77, 78). In murine models of cardiac ischemia/reperfusion, interference with Bax activation, genetic Bax loss-of-function, or overexpression of Bcl-2 attenuated apoptosis and reduced infarct size, while reduction of Bcl-2 levels via antisense oligonucleotides eliminated adaptive protection from injury (79–82). These studies paint a consistent picture of Bcl-2 family members controlling cell fate during ischemia and reperfusion in the myocardium and in the brain. Inhibition of the intrinsic apoptotic pathway by Bcl-2 overexpression also attenuates the phenotype in animal models of chronic neurodegenerative disease and cardiomyopathy (83–85).

Effective pharmacologic intervention would seem to require either antiapoptotic gain of function or proapoptotic loss of function (Figure 2). In an effort to use the former strategy, viral delivery of a Bcl-2 transgene to cerebral infarcts decreased infarct size (86). Pharmacologic inhibition of Bax has been reported using an assay of Bax-induced

MOMP triggered by recombinant Bid (87). Whether such molecules are safe and effective in vivo remains to be reported. A pentapeptide derived from the Bax-modulating Ku70 protein also inhibits Bax-dependent apoptosis (88). In an attempt to block signaling further upstream, a class of small molecules have been described that bind to Bid and inhibit its activation of Bax in mitochondrial assays (89).

In general, the field of therapeutic apoptosis inhibition by manipulation of Bcl-2 family members is young, but mechanistic studies suggest that such a strategy may be useful and feasible. There are many additional chronic degenerative diseases involving apoptosis whose patients might benefit from apoptotic inhibition. Whether therapeutic long-term inhibition of apoptosis might predispose to cancer, as suggested by numerous mouse models, and whether such deleterious side effects might be overcome by pulsatile dosing schedules, remain to be seen.

Diseases characterized by deficient death

Bcl-2 family proteins in cancer. The third of the 3 leading causes of mortality in the developed world is cancer, characterized by a failure of programmed cell death (90). While Bcl-2 was initially identified as an oncogene in follicular lymphoma, its expression has been identified in many cancers, including melanoma, myeloma, small-cell lung cancer, and prostate and acute leukemias (58). Expression of other antiapoptotic proteins has been detected in many cancers, including Bfl-1 in diffuse large-cell lymphoma (91), Mcl-1 in myeloma (92), and Bcl-x$_L$ in lung adenocarcinoma (93). The oncogenic EBV and human herpes virus-8 (HHV-8; also known as Kaposi sarcoma herpes virus) encode Bcl-2 homologs that oppose cell death from multiple stimuli, analogous to Bcl-2 (94, 95). EBV has been implicated in the causation of HIV-related lymphoma, Burkitt lymphoma, nasopharyngeal cancer, and post-transplantation lymphomas, and HHV-8 in the causation of Kaposi sarcoma, Castleman disease, and body cavity lymphomas. Blocking the intrinsic pathway to programmed cell death is apparently important enough in viral infection and perhaps oncogenesis for evolutionary selection for Bcl-2 homologs.

While it seems clear that Bcl-2 can be important in oncogenesis, the essential therapeutic question is whether Bcl-2 is necessary for tumor maintenance. To formally test this question requires the establishing of a tumor that expresses Bcl-2 followed by the induction of loss of Bcl-2 function. To this end, a mouse model of leukemia was generated that contained a Bcl-2 transgene that could be silenced by doxycycline administration (96). To accelerate oncogenesis, mice bearing the conditional Bcl-2 expression system were bred with those bearing the Eμ-myc transgene. Mice coexpressing both c-myc and Bcl-2 developed a B-lymphoblastic leukemia in the first days of life (97). After diagnosis of leukemia, doxycycline was administered to turn off Bcl-2, and a rapid drop in wbc count, from as high as 500,000–1,000,000 cells per microliter to the normal range of 5,000–10,000 cells per microliter, was seen in a matter of days. Bcl-2 downregulation also fostered significantly longer survival (96). This in vivo experiment established the proof of principle that under certain conditions, correction of an apoptotic defect by itself could be lethal to cancer cells, and thus that Bcl-2 was a valid target for cancer therapy. An implication is that Bcl-2 antagonists may have the potential to be efficacious in cancer therapy, even as single agents.

Peptide-based Bcl-2 antagonist compounds

If Bcl-2 is a valid therapeutic target, how might it be antagonized? Studies suggest that a molecule behaving like a sensitizer BH3

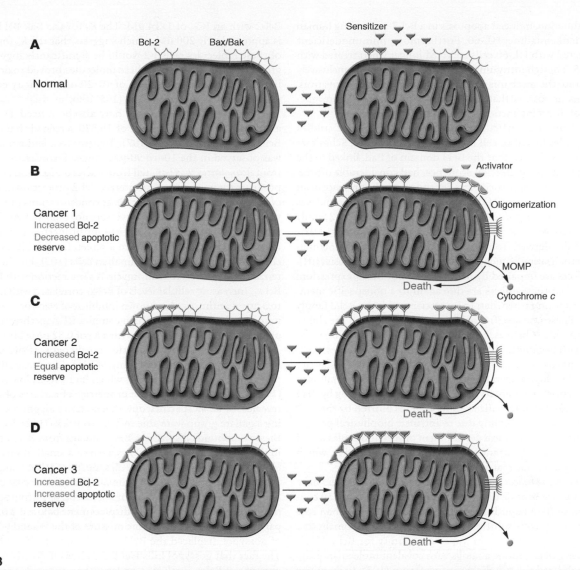

Figure 3

Model for targeting cancer cells with sensitizer BH3 mimetics. (**A**) Mitochondrion from a normal cell has some Bax/Bak and Bcl-2. Bcl-2 is unoccupied; normal cell behavior is provoking no death signals. (**B–D**) Mitochondria from cancer cells have equal Bax/Bak and overexpress Bcl-2 in this model. Antiapoptotic reserve is defined as the number of unoccupied antiapoptotic Bcl-2 family member binding pockets per cell. Compared with normal mitochondria, those that overexpress Bcl-2 may provide decreased (**B**), equal (**C**), or increased (**D**) antiapoptotic reserve. Because of genomic instability, oncogene activation, cell cycle checkpoint violation, or perhaps cancer-specific response to cytotoxic chemotherapy, activator BH3 domains have been triggered and are sequestered by Bcl-2. After exposure to a sensitizer BH3 mimetic (a protein, peptide, or small molecule), activator BH3 domains are displaced from cancer cells, but not normal cells, activating Bax/Bak and allowing selective cancer cell killing, perhaps even as a single agent. It can be seen why sensitizer mimetics might offer a greater therapeutic window than an activator, as an activator molecule would provide selective killing only at low doses and only for cancer cells in condition 1 (Cancer 1). At higher doses, or if the cancer cells were in condition 2 or 3, there would be killing of normal and cancer cells. It is unclear whether activator- or sensitizer-type BH3-only family members predominate in the response to conventional chemotherapy agents, and it is likely that a mixture is present. These models also speculate why certain cancers, such as follicular lymphoma and chronic lymphocytic leukemia, despite expressing higher levels of Bcl-2, are more prone to apoptosis than normal cells after DNA-damaging chemotherapy.

domain peptide, binding with high affinity to Bcl-2's pocket, would act as a competitive inhibitor of Bcl-2 function (37), which might induce selective killing of cancer cells (96) (Figure 3). Unmodified BH3 domain peptides are cell impermeant, so the problem of intracellular delivery must first be overcome. It has been shown that tagging peptides with a poly-D-arginine tag facilitates cell internalization in vitro and in vivo (98). We linked BH3 peptides from Bad, Bid, and a double point mutant of Bid to an N-terminal poly-D-arginine octomer. While r8BidBH3 killed a human leukemia cell

line that expresses Bcl-2, the r8BidBH3 double point mutant did not. Furthermore, 10 μM of the r8BadBH3 peptide, which caused no apoptosis on its own, increased the killing induced by the r8BidBH3 peptide. These results suggest that the moiety did indeed facilitate internalization, and that an intact BH3 domain was necessary for killing (37). In a separate study, a 27–amino acid peptide derived from the BH3 domain of Bad was linked to decanoic acid to facilitate intracellular entry (99). This compound (cpm-1285), but not a peptide bearing a point mutation at a residue necessary for BH3 function, induced apoptosis in a Bcl-2–expressing human myeloid leukemia line, HL-60. Furthermore, immunodeficient mice injected with HL-60 cells survived longer when treated with cpm-1285. Though intriguing, these 2 studies do not conclusively demonstrate the mechanism of action of the peptide derivatives, and it remains possible that some of the cytotoxic effects were independent of direct interaction with Bcl-2 family members.

Highlighting the difficulty in interpreting results of studies using peptides linked to cell internalization moieties, Schimmer and coworkers showed that the BH3 domain of Bad, linked to the Antennapedia internalization sequence, had considerable off-target toxicity (100). This toxicity was dependent on the presence of an α-helix, but independent of the Bcl-2 pathway. The compound was toxic to a wide variety of cells, including yeast, wherein Bcl-2 family members have yet to be identified. Others have demonstrated that BH3 peptides derived from Bax and Bcl-2 linked to an Antennapedia internalization sequence induce MOMP and apoptosis (101). These effects are impaired by neither Bcl-2 nor Bcl-x$_L$ expression. It may be that these effects are due again to a nonspecific membrane disruption rather than to interaction with the Bcl-2 family pathway. These nonspecific toxicities may be linked to the ability of amphipathic α-helices, especially when positively charged, to interact with negatively charged mitochondrial membranes. Such interactions may disrupt the lipid matrix and membrane barrier function of biological membranes independent of Bcl-2 family protein interaction. In summary, interpretation of cell killing by BH3 peptides linked to internalization moieties is blurred by the possibility of nonspecific killing due to intrinsic biophysical properties. Further pharmaceutical development of such molecules would require considerable attention to reducing this toxicity, which would likely affect normal as well as cancer cells.

Stabilizing the α-helical conformation of BH3 peptides. Others have tried to improve peptide function by stabilizing the α-helical conformation of BH3 peptides, which generally show less than 25% α-helicity in aqueous solution. Grafting a Bak BH3 domain to a helix-stabilizing miniprotein improved affinity for Bcl-2 (102). Synthesizing BH3 peptide analogs with covalent molecular bridges also stabilized the α-helical conformation (103). Perhaps the most dramatic example of the potential of α-helix stabilization was provided by a Bid BH3 peptide stabilized by an all-hydrocarbon "staple" (104). This modification enhanced α-helicity, affinity for Bcl-2, cell entry, protease resistance, and leukemia cell line toxicity in vitro and in vivo. Mice bearing leukemia cell line xenografts enjoyed statistically significant survival improvement of 6 days; treatment duration was limited to 7 days by sufficiency of material. Normal tissues seemed unaffected as measured by histological analysis. It is important to note that this molecule does not behave as a selective Bcl-2 antagonist. As expected, since it was modeled on a Bid peptide previously shown to be an activator (37), the compound was able to directly induce cytochrome *c* release in a Bak-dependent fashion in vitro. While this molecule behaved as an activator in vitro, it was still able to exploit an apparent therapeutic window between the tumor xenograft and the normal tissues (Figure 3). Subsequent attempts using sensitizer BH3-based compounds may provide an even greater therapeutic window.

Small-molecule Bcl-2 antagonists
Small molecules that bind to antiapoptotic Bcl-2 family members have been identified using structure-based computer screening. The molecules isolated displaced the Bak BH3 peptide from

Bcl-2 with an IC$_{50}$ of 1–14 μM. The K_d for the Bak BH3 peptide is approximately 200 nM, which suggests that the K_d for binding of these molecules to Bcl-2 would be significantly higher, likely greater than 1 μM. One of these molecules proved toxic to 4 cell lines tested at concentrations of 10–20 μM. Toxicity correlated with amount of Bcl-2 expressed (105, 106).

Screens of chemical libraries have also been used. Degterev et al. identified 2 molecules out of 16,320 screened that disrupt a Bcl-x$_L$/Bak BH3 complex (107). Toxicity to a leukemia cell line was observed in the 10- to 90-μM range. Tetrocarcin A, derived from *Actinomyces*, was isolated from a screen of a library of natural products for its ability to counteract Bcl-2 protection of anti-Fas/cycloheximide–treated HeLa cells at concentrations in the micromolar range (108). Binding assays were not reported. Antimycin A, an antibacterial agent with antitumor properties in experimental systems, was identified from a screen for inhibitors of mitochondrial respiration in mammalian cells (109). Further characterization demonstrated antimycin A's interaction with Bcl-2 and Bcl-x$_L$. Increasing cellular levels of Bcl-x$_L$ correlated with increasing toxicity of antimycin A, a known inhibitor of electron transport at mitochondrial respiratory chain complex III. A methoxy derivative of antimycin A binds to Bcl-2 with a K_d of 0.82 μM (110).

Kutzki et al. used a nonpeptide terphenyl scaffold to design molecules that mimic the binding of BH3 domains in the hydrophobic pocket of Bcl-x$_L$, one of which had a K_d as low as 114 nM (111). An investigation of the properties of extracts of green tea revealed, by NMR spectroscopy, that certain polyphenols containing a gallate group were able to bind to Bcl-x$_L$ (112). In addition, these compounds displaced a BH3 domain from Bcl-x$_L$ and Bcl-2 in the submicromolar range. A screen of a small library of natural products identified 2 molecules, purpurogallin and gossypol, that inhibit binding of a BH3 domain (resembling that of human Bad) to Bcl-x$_L$ (113). Chemical modification of purpurogallin did not lower the IC$_{50}$ of peptide displacement below 2.2 μM of the parental compound. A racemic mixture of the (+) and (–) isomers of gossypol displaced the BH3 peptide with an IC$_{50}$ of 0.5 μM. The fact that gossypol kills HeLa cells less efficiently when they overexpress Bcl-x$_L$ is taken as evidence that gossypol targets Bcl-x$_L$. Removal of 2 aldehyde groups from gossypol was performed after molecular modeling suggested that this might reduce steric hindrance in binding the hydrophobic pocket of Bcl-x$_L$ (114). The so-named apogossypol, however, demonstrated inferior binding to Bcl-2 family members.

Validation of antagonists of antiapoptotic family members
Many of the studies cited above provide solid evidence that a particular molecule binds with reasonable affinity (K_d roughly 1 μM or less) to Bcl-2 or Bcl-x$_L$. Mechanistic evidence demonstrating that killing of living cells depends on specific targeting of Bcl-2 or Bcl-x$_L$ is generally lacking, however. Studies lacking a defined Bcl-2–dependent system can be difficult to interpret. This is evident in that different studies have been interpreted to show, variously, that (a) the compound is working specifically via Bcl-2 or Bcl-x$_L$ inhibition because cells expressing Bcl-2 or Bcl-x$_L$ are *more* sensitive (106, 109) or (b) the compound is working specifically via Bcl-2 or Bcl-x$_L$ inhibition because cells expressing Bcl-2 or Bcl-x$_L$ are *less* sensitive to compound treatment (113). While, in fact, either statement is potentially true of selective Bcl-2 inhibition, detailed, controlled understanding of the cellular context is necessary for proper interpretation.

The essential problem of assaying Bcl-2 function is that the endpoint of cell death is shared by the targeting of countless cellular pathways, and Bcl-2 lacks quantifiable enzymatic activity. Some researchers find the use of defined Bcl-2–dependent systems indispensable for demonstration of specific Bcl-2 inhibition. One useful example is the murine FL5.12 cell line, which is dependent on IL-3 for survival. Programmed cell death following IL-3 removal is prevented by overexpression of a human Bcl-2 transgene (5). Therefore, the IL-3–starved Bcl-2–overexpressing FL5.12 cell is a model of dependence on Bcl-2 for survival. A compound working selectively via Bcl-2 inhibition should kill IL-3–starved, but not unstarved, Bcl-2–overexpressing cells; critically, the parental FL5.12 cells should also be spared death following treatment (115). Testing can be extended to isolated mitochondria, with cytochrome c release as the readout of MOMP (37). Peptides derived from the BH3 domains of Bad or Noxa, while cell impermeant, may then be used as validated positive and negative controls of Bcl-2 antagonism. Insensitivity of mitochondria and cells deficient in both Bax and Bak to compound treatment would provide further support for the test compound's operating through the mitochondrial pathway controlled by Bcl-2 family members (20, 37).

Bcl-2 antagonists in clinical development

Antisense strategy. Oblimersen (Genasense; Genta Inc.) is an 18-mer phosphorothioated oligonucleotide directed against the first 6 codons of the human Bcl-2 open reading frame (116). It has advanced through clinical trials with tolerable side effects, mainly thrombocytopenia, fatigue, back pain, weight loss, and dehydration (117). Demonstration of efficacy has been difficult. In a randomized phase III study of patients with metastatic melanoma, treatment with dacarbazine and oblimersen showed no significant benefit in overall survival compared with dacarbazine alone (274 vs. 238 days, $P = 0.18$). There was significant benefit in progression-free survival (74 vs. 49 days, $P = 0.0003$), but since overall survival was the primary endpoint, an FDA panel stated that the clinical benefit was inadequate to offset the increased toxicity.

In chronic lymphocytic leukemia, despite early problems with a cytokine release syndrome, a phase III trial showed improved major responses (complete response plus nodular partial responses) in patients treated with oblimersen, fludarabine, and cyclophosphamide compared with those treated with fludarabine and cyclophosphamide alone (16% vs. 7%, $P = 0.039$). A phase III trial in myeloma in which oblimersen plus high dexamethasone was compared with dexamethasone alone failed to meet its primary endpoint, which was time to disease progression. Furthermore, in another myeloma trial, response to oblimersen did not correlate with reduction in Bcl-2 protein levels.

In general, efficacy has not been overwhelming, and FDA approval of oblimersen as an anticancer agent remains in doubt. There are several reasons why this does not necessarily augur poorly for strategies targeting Bcl-2 in general. First, the cellular effects of the lowering of Bcl-2 levels by antisense oligonucelotides are likely very differ-

ent from the cellular effects of functional antagonism of the protein and include possible undesirable coregulation of other Bcl-2 family members (118). Furthermore, the reductions in Bcl-2 protein levels tend to be modest, in the 10–50% range. Finally, oblimersen likely has immune system effects via its 2 CpG dinucleotides. While potentially beneficial, these may well limit its maximum tolerated dose, because of side effects including cytokine release syndrome.

Competitive antagonists of Bcl-2. Other compounds in clinical development are small molecules that bind the hydrophobic pocket of Bcl-2 analogously to our sensitizer BH3 peptides. The biotechnology company Gemin X has isolated a compound (GX01) from a high-throughput screen of chemical libraries, reported (in abstract form) to possess the ability to bind Bcl-2 and Bcl-x_L and displace BH3 domains from their binding pockets (119). There is little public information regarding its mechanism of action in living cells. It is currently in phase I clinical trials in both chronic lymphocytic leukemia (at UCSD) and solid tumors (at Georgetown University). Ascenta Therapeutics has an orally administered gossypol derivative in a phase I cancer trial. There is little publicly available information on this compound.

Using a strategy of high-throughput screening combined with iterative modulation of chemical structure based on NMR, Abbott Laboratories has developed compounds reported to displace BH3 domains from Bcl-2, Bcl-x_L, and Bcl-w with an IC_{50} of no more than 1 nM. One lead molecule, ABT-737, is reported to have significant activity in mouse xenograft models of lung cancer and lymphoma (115). This series of compounds has not yet entered clinical trials.

Conclusions

Our current understanding of the mechanisms by which Bcl-2 family members control commitment to cell death gives good theoretical backing to strategies aimed at manipulating this system for clinical benefit. Studies presented above support both the feasibility and the utility of targeting pro- and antiapoptotic proteins. Clinical trials of Bcl-2 antagonists are under way, with more to be expected in the next few years. Given that different antiapoptotic proteins demonstrate distinct binding specificities, it seems likely that in the future the other antiapoptotic proteins might also be individually targeted pharmacologically. Twenty years after the cloning of Bcl-2, we are starting to reap the translational harvest of much fundamental research into the molecular mechanisms controlling apoptosis.

Acknowledgments

A. Letai is supported by NIH grant K08 CA10254, the Claudia Adams Barr Foundation, and the Richard and Susan Smith Family Foundation (Chestnut Hill, Massachusetts, USA).

Address correspondence to: Anthony Letai, Dana 530B, Department of Medical Oncology, Dana-Farber Cancer Institute, Harvard Medical School, 44 Binney Street, Boston, Massachusetts 02115, USA. Phone: (617) 632-2348; Fax: (617) 582-8160; E-mail: anthony_letai@dfci.harvard.edu.

1. Tsujimoto, Y., Cossman, J., Jaffe, E., and Croce, C.M. 1985. Involvement of the bcl-2 gene in human follicular lymphoma. *Science.* **228**:1440–1443.
2. Cleary, M.L., and Sklar, J. 1985. Nucleotide sequence of a t(14;18) chromosomal breakpoint in follicular lymphoma and demonstration of a breakpoint-cluster region near a transcriptionally active locus on chromosome 18. *Proc. Natl. Acad. Sci. U. S. A.* **82**:7439–7443.
3. Bakhshi, A., et al. 1985. Cloning the chromosomal breakpoint of t(14;18) human lymphomas: clustering around JH on chromosome 14 and near a transcriptional unit on 18. *Cell.* **41**:899–906.
4. McDonnell, T.J., et al. 1989. bcl-2-immunoglobulin transgenic mice demonstrate extended B cell survival and follicular lymphoproliferation. *Cell.* **57**:79–88.
5. Vaux, D.L., Cory, S., and Adams, J.M. 1988. Bcl-2 gene promotes haemopoietic cell survival and cooperates with c-myc to immortalize pre-B cells. *Nature.* **335**:440–442.
6. McDonnell, T.J., and Korsmeyer, S.J. 1991. Progression from lymphoid hyperplasia to high-grade malignant lymphoma in mice transgenic for the t(14; 18). *Nature.* **349**:254–256.
7. Boise, L.H., et al. 1993. bcl-x, a bcl-2-related gene that functions as a dominant regulator of apop-

totic cell death. *Cell.* **74**:597–608.

8. Gibson, L., et al. 1996. bcl-w, a novel member of the bcl-2 family, promotes cell survival. *Oncogene.* **13**:665–675.

9. Kozopas, K.M., Yang, T., Buchan, H.L., Zhou, P., and Craig, R.W. 1993. MCL1, a gene expressed in programmed myeloid cell differentiation, has sequence similarity to BCL2. *Proc. Natl. Acad. Sci. U. S. A.* **90**:3516–3520.

10. Choi, S.S., et al. 1995. A novel Bcl-2 related gene, Bfl-1, is overexpressed in stomach cancer and preferentially expressed in bone marrow. *Oncogene.* **11**:1693–1698.

11. Oltvai, Z.N., Milliman, C.L., and Korsmeyer, S.J. 1993. Bcl-2 heterodimerizes in vivo with a conserved homolog, Bax, that accelerates programmed cell death. *Cell.* **74**:609–619.

12. Chittenden, T., et al. 1995. Induction of apoptosis by the Bcl-2 homologue Bak. *Nature.* **374**:733–736.

13. Huang, D.C., and Strasser, A. 2000. BH3-only proteins-essential initiators of apoptotic cell death. *Cell.* **103**:839–842.

14. Wang, K., Yin, X.M., Chao, D.T., Milliman, C.L., and Korsmeyer, S.J. 1996. BID: a novel BH3 domain-only death agonist. *Genes Dev.* **10**:2859–2869.

15. Boyd, J.M., et al. 1995. Bik, a novel death-inducing protein shares a distinct sequence motif with Bcl-2 family proteins and interacts with viral and cellular survival-promoting proteins. *Oncogene.* **11**:1921–1928.

16. Chittenden, T., et al. 1995. A conserved domain in Bak, distinct from BH1 and BH2, mediates cell death and protein binding functions. *EMBO J.* **14**:5589–5596.

17. Green, D.R., and Kroemer, G. 2004. The pathophysiology of mitochondrial cell death. *Science.* **305**:626–629.

18. Lindsten, T., et al. 2000. The combined functions of proapoptotic Bcl-2 family members Bak and Bax are essential for normal development of multiple tissues. *Mol. Cell.* **6**:1389–1399.

19. Wei, M.C., et al. 2000. tBID, a membrane-targeted death ligand, oligomerizes BAK to release cytochrome c. *Genes Dev.* **14**:2060–2071.

20. Wei, M.C., et al. 2001. Proapoptotic BAX and BAK: a requisite gateway to mitochondrial dysfunction and death. *Science.* **292**:727–730.

21. Suzuki, M., Youle, R.J., and Tjandra, N. 2000. Structure of Bax: coregulation of dimer formation and intracellular localization. *Cell.* **103**:645–654.

22. Griffiths, G.J., et al. 1999. Cell damage-induced conformational changes of the pro-apoptotic protein Bak in vivo precede the onset of apoptosis. *J. Cell Biol.* **144**:903–914.

23. Desagher, S., et al. 1999. Bid-induced conformational change of Bax is responsible for mitochondrial cytochrome c release during apoptosis. *J. Cell Biol.* **144**:891–901.

24. Hsu, Y.T., and Youle, R.J. 1997. Nonionic detergents induce dimerization among members of the Bcl-2 family. *J. Biol. Chem.* **272**:13829–13834.

25. Wolter, K.G., et al. 1997. Movement of Bax from the cytosol to mitochondria during apoptosis. *J. Cell Biol.* **139**:1281–1292.

26. Cheng, E.H., et al. 2001. BCL-2, Bcl-X(L) sequester BH3 domain-only molecules preventing BAX- and BAK-mediated mitochondrial apoptosis. *Mol. Cell.* **8**:705–711.

27. Gross, A., Jockel, J., Wei, M.C., and Korsmeyer, S.J. 1998. Enforced dimerization of BAX results in its translocation, mitochondrial dysfunction and apoptosis. *EMBO J.* **17**:3878–3885.

28. Liu, X., Kim, C.N., Yang, J., Jemmerson, R., and Wang, X. 1996. Induction of apoptotic program in cell-free extracts: requirement for dATP and cytochrome c. *Cell.* **86**:147–157.

29. Li, L.Y., Luo, X., and Wang, X. 2001. Endonuclease G is an apoptotic DNase when released from mitochondria. *Nature.* **412**:95–99.

30. Du, C., Fang, M., Li, Y., Li, L., and Wang, X. 2000. Smac, a mitochondrial protein that promotes cytochrome c-dependent caspase activation by eliminating IAP inhibition. *Cell.* **102**:33–42.

31. Verhagen, A.M., et al. 2000. Identification of DIABLO, a mammalian protein that promotes apoptosis by binding to and antagonizing IAP proteins. *Cell.* **102**:43–53.

32. Susin, S.A., et al. 1999. Molecular characterization of mitochondrial apoptosis-inducing factor. *Nature.* **397**:441–446.

33. Danial, N.N., and Korsmeyer, S.J. 2004. Cell death: critical control points [review]. *Cell.* **116**:205–219.

34. Saito, M., Korsmeyer, S.J., and Schlesinger, P.H. 2000. BAX-dependent transport of cytochrome c reconstituted in pure liposomes. *Nat. Cell Biol.* **2**:553–555.

35. Kuwana, T., et al. 2002. Bid, Bax, and lipids cooperate to form supramolecular openings in the outer mitochondrial membrane. *Cell.* **111**:331–342.

36. Kuwana, T., et al. 2005. BH3 domains of BH3-only proteins differentially regulate Bax-mediated mitochondrial membrane permeabilization both directly and indirectly. *Mol. Cell.* **17**:525–535.

37. Letai, A., et al. 2002. Distinct BH3 domains either sensitize or activate mitochondrial apoptosis, serving as prototype cancer therapeutics. *Cancer Cell.* **2**:183–192.

38. Eskes, R., Desagher, S., Antonsson, B., and Martinou, J.C. 2000. Bid induces the oligomerization and insertion of Bax into the outer mitochondrial membrane. *Mol. Cell. Biol.* **20**:929–935.

39. Marani, M., Tenev, T., Hancock, D., Downward, J., and Lemoine, N.R. 2002. Identification of novel isoforms of the BH3 domain protein Bim which directly activate Bax to trigger apoptosis. *Mol. Cell. Biol.* **22**:3577–3589.

40. Harada, H., Quearry, B., Ruiz-Vela, A., and Korsmeyer, S.J. 2004. Survival factor-induced extracellular signal-regulated kinase phosphorylates BIM, inhibiting its association with BAX and proapoptotic activity. *Proc. Natl. Acad. Sci. U. S. A.* **101**:15313–15317.

41. Cartron, P.F., et al. 2004. The first alpha helix of Bax plays a necessary role in its ligand-induced activation by the BH3-only proteins Bid and PUMA. *Mol. Cell.* **16**:807–818.

42. Li, H., Zhu, H., Xu, C.J., and Yuan, J. 1998. Cleavage of BID by caspase 8 mediates the mitochondrial damage in the Fas pathway of apoptosis. *Cell.* **94**:491–501.

43. Luo, X., Budihardjo, I., Zou, H., Slaughter, C., and Wang, X. 1998. Bid, a Bcl2 interacting protein, mediates cytochrome c release from mitochondria in response to activation of cell surface death receptors. *Cell.* **94**:481–490.

44. Zha, J., Weiler, S., Oh, K.J., Wei, M.C., and Korsmeyer, S.J. 2000. Posttranslational N-myristoylation of BID as a molecular switch for targeting mitochondria and apoptosis. *Science.* **290**:1761–1765.

45. Terradillos, O., Montessuit, S., Huang, D.C., and Martinou, J.C. 2002. Direct addition of BimL to mitochondria does not lead to cytochrome c release. *FEBS Lett.* **522**:29–34.

46. Sawada, M., et al. 2003. Ku70 suppresses the apoptotic translocation of Bax to mitochondria. *Nat. Cell Biol.* **5**:320–329.

47. Guo, B., et al. 2003. Humanin peptide suppresses apoptosis by interfering with Bax activation. *Nature.* **423**:456–461.

48. Ohtsuka, T., et al. 2004. ASC is a Bax adaptor and regulates the p53-Bax mitochondrial apoptosis pathway. *Nat. Cell Biol.* **6**:121–128.

49. Cheng, E.H., Sheiko, T.V., Fisher, J.K., Craigen, W.J., and Korsmeyer, S.J. 2003. VDAC2 inhibits BAK activation and mitochondrial apoptosis. *Science.* **301**:513–517.

50. Muchmore, S.W., et al. 1996. X-ray and NMR structure of human Bcl-xL, an inhibitor of programmed cell death. *Nature.* **381**:335–341.

51. Sattler, M., et al. 1997. Structure of Bcl-xL-Bak peptide complex: recognition between regulators of apoptosis. *Science.* **275**:983–986.

52. Petros, A.M., et al. 2001. Solution structure of the antiapoptotic protein bcl-2. *Proc. Natl. Acad. Sci. U. S. A.* **98**:3012–3017.

53. Petros, A.M., et al. 2000. Rationale for Bcl-xL/Bad peptide complex formation from structure, mutagenesis, and biophysical studies. *Protein Sci.* **9**:2528–2534.

54. Cheng, E.H., Levine, B., Boise, L.H., Thompson, C.B., and Hardwick, J.M. 1996. Bax-independent inhibition of apoptosis by Bcl-XL. *Nature.* **379**:554–556.

55. Murphy, K.M., Streips, U.N., and Lock, R.B. 2000. Bcl-2 inhibits a Fas-induced conformational change in the Bax N terminus and Bax mitochondrial translocation. *J. Biol. Chem.* **275**:17225–17228.

56. Mikhailov, V., et al. 2001. Bcl-2 prevents Bax oligomerization in the mitochondrial outer membrane. *J. Biol. Chem.* **276**:18361–18374.

57. Chipuk, J.E., et al. 2004. Direct activation of Bax by p53 mediates mitochondrial membrane permeabilization and apoptosis. *Science.* **303**:1010–1014.

58. Letai, A. 2003. BH3 domains as BCL-2 inhibitors: prototype cancer therapeutics. *Expert Opin. Biol. Ther.* **3**:293–304.

59. Opferman, J.T., et al. 2003. Development and maintenance of B and T lymphocytes requires antiapoptotic MCL-1. *Nature.* **426**:671–676.

60. Chen, L., et al. 2005. Differential targeting of prosurvival Bcl-2 proteins by their BH3-only ligands allows complementary apoptotic function. *Mol. Cell.* **17**:393–403.

61. Puthalakath, H., and Strasser, A. 2002. Keeping killers on a tight leash: transcriptional and post-translational control of the pro-apoptotic activity of BH3-only proteins [review]. *Cell Death Differ.* **9**:505–512.

62. Oda, E., et al. 2000. Noxa, a BH3-only member of the Bcl-2 family and candidate mediator of p53-induced apoptosis. *Science.* **288**:1053–1058.

63. Nakano, K., and Vousden, K.H. 2001. PUMA, a novel proapoptotic gene, is induced by p53. *Mol. Cell.* **7**:683–694.

64. Yu, J., Zhang, L., Hwang, P.M., Kinzler, K.W., and Vogelstein, B. 2001. PUMA induces the rapid apoptosis of colorectal cancer cells. *Mol. Cell.* **7**:673–682.

65. Han, J., et al. 2001. Expression of bbc3, a proapoptotic BH3-only gene, is regulated by diverse cell death and survival signals. *Proc. Natl. Acad. Sci. U. S. A.* **98**:11318–11323.

66. Jeffers, J.R., et al. 2003. Puma is an essential mediator of p53-dependent and -independent apoptotic pathways. *Cancer Cell.* **4**:321–328.

67. Dijkers, P.F., Medema, R.H., Lammers, J.W., Koenderman, L., and Coffer, P.J. 2000. Expression of the pro-apoptotic Bcl-2 family member Bim is regulated by the forkhead transcription factor FKHR-L1. *Curr. Biol.* **10**:1201–1204.

68. Putcha, G.V., et al. 2001. Induction of BIM, a proapoptotic BH3-only BCL-2 family member, is critical for neuronal apoptosis. *Neuron.* **29**:615–628.

69. Puthalakath, H., Huang, D.C., O'Reilly, L.A., King, S.M., and Strasser, A. 1999. The proapoptotic activity of the Bcl-2 family member Bim is regulated by interaction with the dynein motor complex. *Mol. Cell.* **3**:287–296.

70. Tan, T.T., et al. 2005. Key roles of BIM-driven apoptosis in epithelial tumors and rational chemotherapy. *Cancer Cell.* **7**:227–238.

71. Ley, R., et al. 2004. Extracellular signal-regulated kinases 1/2 are serum-stimulated "Bim(EL) kinases" that bind to the BH3-only protein Bim(EL) causing its phosphorylation and turnover. *J. Biol. Chem.* **279**:8837–8847.

72. Luciano, F., et al. 2003. Phosphorylation of Bim-EL by Erk1/2 on serine 69 promotes its degradation via the proteasome pathway and regulates its pro-apoptotic function. *Oncogene.* **22**:6785–6793.

73. Zha, J., Harada, H., Yang, E., Jockel, J., and Korsmeyer, S.J. 1996. Serine phosphorylation of death agonist BAD in response to survival factor results in binding to 14-3-3 not Bcl-X(L). *Cell.* **87**:619–628.

74. Danial, N.N., et al. 2003. BAD and glucokinase reside in a mitochondrial complex that integrates glycolysis and apoptosis. *Nature.* **424**:952–956.

75. Cory, S., and Adams, J.M. 2002. The Bcl2 family: regulators of the cellular life-or-death switch [review]. *Nat. Rev. Cancer.* **2**.647–656.

76. Adams, J.M. 2003. Ways of dying: multiple pathways to apoptosis [review]. *Genes Dev.* **17**:2481–2495.

77. Gibson, M.E., et al. 2001. BAX contributes to apoptotic-like death following neonatal hypoxia-ischemia: evidence for distinct apoptosis pathways. *Mol. Med.* **7**:644–655.

78. Plesnila, N., et al. 2001. BID mediates neuronal cell death after oxygen/glucose deprivation and focal cerebral ischemia. *Proc. Natl. Acad. Sci. U. S. A.* **98**:15318–15323.

79. Gustafsson, A.B., Tsai, J.G., Logue, S.E., Crow, M.T., and Gottlieb, R.A. 2004. Apoptosis repressor with caspase recruitment domain protects against cell death by interfering with Bax activation. *J. Biol. Chem.* **279**:21233–21238.

80. Chen, Z., Chua, C.C., Ho, Y.S., Hamdy, R.C., and Chua, B.H. 2001. Overexpression of Bcl-2 attenuates apoptosis and protects against myocardial I/R injury in transgenic mice. *Am. J. Physiol. Heart Circ. Physiol.* **280**:H2313–H2320.

81. Hochhauser, E., et al. 2003. Bax ablation protects against myocardial ischemia-reperfusion injury in transgenic mice. *Am. J. Physiol. Heart Circ. Physiol.* **284**:H2351–H2359.

82. Hattori, R., et al. 2001. An essential role of the antioxidant gene Bcl 2 in myocardial adaptation to ischemia: an insight with antisense Bcl-2 therapy. *Antioxid. Redox Signal.* **3**:403–413.

83. Azzouz, M., et al. 2000. Increased motoneuron survival and improved neuromuscular function in transgenic ALS mice after intraspinal injection of an adeno-associated virus encoding Bcl-2. *Hum. Mol. Genet.* **9**:803–811.

84. Kostic, V., Jackson-Lewis, V., de Bilbao, F., Dubois-Dauphin, M., and Przedborski, S. 1997. Bcl-2: prolonging life in a transgenic mouse model of familial amyotrophic lateral sclerosis. *Science.* **277**:559–562.

85. Weisleder, N., Taffet, G.E., and Capetanaki, Y. 2004. Bcl-2 overexpression corrects mitochondrial defects and ameliorates inherited desmin null cardiomyopathy. *Proc. Natl. Acad. Sci. U. S. A.* **101**:769–774.

86. Lawrence, M.S., Ho, D.Y., Sun, G.H., Steinberg, G.K., and Sapolsky, R.M. 1996. Overexpression of Bcl-2 with herpes simplex virus vectors protects CNS neurons against neurological insults in vitro and in vivo. *J. Neurosci.* **16**:486–496.

87. Bombrun, A., et al. 2003. 3,6-Dibromocarbazole piperazine derivatives of 2-propanol as first inhibitors of cytochrome c release via Bax channel modulation. *J. Med. Chem.* **46**:4365–4368.

88. Sawada, M., Hayes, P., and Matsuyama, S. 2003. Cytoprotective membrane-permeable peptides designed from the Bax-binding domain of Ku70. *Nat. Cell Biol.* **5**:352–357.

89. Becattini, B., et al. 2004. Targeting apoptosis via chemical design: inhibition of bid-induced cell death by small organic molecules. *Chem. Biol.* **11**:1107–1117.

90. Green, D.R., and Evan, G.I. 2002. A matter of life and death. *Cancer Cell.* **1**:19–30.

91. Shipp, M.A., et al. 2002. Diffuse large B-cell lymphoma outcome prediction by gene-expression profiling and supervised machine learning. *Nat. Med.* **8**:68–74.

92. Derenne, S., et al. 2002. Antisense strategy shows that Mcl-1 rather than Bcl-2 or Bcl-x(L) is an essential survival protein of human myeloma cells. *Blood.* **100**:194–199.

93. Berrieman, H.K., et al. 2005. The expression of Bcl-2 family proteins differs between nonsmall cell lung carcinoma subtypes. *Cancer.* **103**:1415–1419.

94. Henderson, S., et al. 1993. Epstein-Barr virus-coded BHRF1 protein, a viral homologue of Bcl-2, protects human B cells from programmed cell death. *Proc. Natl. Acad. Sci. U. S. A.* **90**:8479–8483.

95. Cheng, E.H., et al. 1997. A Bcl-2 homolog encoded by Kaposi sarcoma-associated virus, human herpesvirus 8, inhibits apoptosis but does not heterodimerize with Bax or Bak. *Proc. Natl. Acad. Sci. U. S. A.* **94**:690–694.

96. Letai, A., Beard, C., Sorcinelli, M., and Korsmeyer, S.J. 2004. Anti-apoptotic BCL-2 is required for maintenance of a model leukemia. *Cancer Cell.* **6**:241–249.

97. Strasser, A., Harris, A.W., Bath, M.L., and Cory, S. 1990. Novel primitive lymphoid tumours induced in transgenic mice by cooperation between myc and bcl-2. *Nature.* **348**:331–333.

98. Rothbard, J.B., et al. 2000. Conjugation of arginine oligomers to cyclosporin A facilitates topical delivery and inhibition of inflammation. *Nat. Med.* **6**:1253–1257.

99. Wang, J.L., et al. 2000. Cell permeable Bcl-2 binding peptides: a chemical approach to apoptosis induction in tumor cells. *Cancer Res.* **60**:1498–1502.

100. Schimmer, A.D., et al. 2001. The BH3 domain of BAD fused to the Antennapedia peptide induces apoptosis via its alpha helical structure and independent of Bcl-2. *Cell Death Differ.* **8**:725–733.

101. Vieira, H.L., et al. 2000. Permeabilization of the mitochondrial inner membrane during apoptosis: impact of the adenine nucleotide translocator. *Cell Death Differ.* **7**:1146–1154.

102. Chin, J.W., and Schepartz, A. 2001. Design and evolution of a miniature Bcl-2 binding protein. *Angew. Chem. Int. Ed. Engl.* **40**:3806–3809.

103. Yang, B., Liu, D., and Huang, Z. 2004. Synthesis and helical structure of lactam bridged BH3 peptides derived from pro-apoptotic Bcl-2 family proteins. *Bioorg. Med. Chem. Lett.* **14**:1403–1406.

104. Walensky, L.D., et al. 2004. Activation of apoptosis in vivo by a hydrocarbon-stapled BH3 helix. *Science.* **305**:1466–1470.

105. Wang, J.L., et al. 2000. Structure-based discovery of an organic compound that binds Bcl-2 protein and induces apoptosis of tumor cells. *Proc. Natl. Acad. Sci. U. S. A.* **97**:7124–7129.

106. Enyedy, I.J., et al. 2001. Discovery of small-molecule inhibitors of Bcl-2 through structure-based computer screening. *J. Med. Chem.* **44**:4313–4324.

107. Degterev, A., et al. 2001. Identification of small-molecule inhibitors of interaction between the BH3 domain and Bcl-xL. *Nat. Cell Biol.* **3**:173–182.

108. Nakashima, T., Miura, M., and Hara, M. 2000. Tetrocarcin A inhibits mitochondrial functions of Bcl-2 and suppresses its anti-apoptotic activity. *Cancer Res.* **60**:1229–1235.

109. Tzung, S.P., et al. 2001. Antimycin A mimics a cell-death-inducing Bcl-2 homology domain 3. *Nat. Cell Biol.* **3**:183–191.

110. Kim, K.M., et al. 2001. Biophysical characterization of recombinant human Bcl-2 and its interactions with an inhibitory ligand, antimycin A. *Biochemistry.* **40**:4911–4922.

111. Kutzki, O., et al. 2002. Development of a potent Bcl-x(L) antagonist based on alpha-helix mimicry. *J. Am. Chem. Soc.* **124**:11838–11839.

112. Leone, M., et al. 2003. Cancer prevention by tea polyphenols is linked to their direct inhibition of antiapoptotic Bcl-2-family proteins. *Cancer Res.* **63**:8118–8121.

113. Kitada, S., et al. 2003. Discovery, characterization, and structure-activity relationships studies of pro-apoptotic polyphenols targeting B-cell lymphocyte/leukemia-2 proteins. *J. Med. Chem.* **46**:4259–4264.

114. Becattini, B., et al. 2004. Rational design and real time, in-cell detection of the proapoptotic activity of a novel compound targeting Bcl-X(L). *Chem. Biol.* **11**:389–395.

115. Oltersdorf, T., et al. 2005. An inhibitor of Bcl-2 family proteins induces regression of solid tumours. *Nature.* **435**:677–681.

116. Jansen, B., et al. 1998. bcl-2 antisense therapy chemosensitizes human melanoma in SCID mice. *Nat. Med.* **4**:232–234.

117. Waters, J.S., et al. 2000. Phase I clinical and pharmacokinetic study of bcl-2 antisense oligonucleotide therapy in patients with non-Hodgkin's lymphoma. *J. Clin. Oncol.* **18**:1812–1823.

118. Konopleva, M., et al. 2000. Liposomal Bcl-2 antisense oligonucleotides enhance proliferation, sensitize acute myeloid leukemia to cytosine-arabinoside, and induce apoptosis independent of other antiapoptotic proteins. *Blood.* **95**:3929–3938.

119. Murthy, M.S., et al. 2001. A small molecule inhibitor of BCL-2 protein-protein interactions specifically induces apoptosis in cancer cells [abstract]. *Clin. Cancer Res.* **7**:313.

Caspases: pharmacological manipulation of cell death

Inna N. Lavrik, Alexander Golks, and Peter H. Krammer

Division of Immunogenetics, Tumor Immunology Program, German Cancer Research Center, Heidelberg, Germany.

Caspases, a family of cysteine proteases, play a central role in apoptosis. During the last decade, major progress has been made to further understand caspase structure and function, providing a unique basis for drug design. This Review gives an overview of caspases and their classification, structure, and substrate specificity. We also describe the current knowledge of how interference with caspase signaling can be used to pharmacologically manipulate cell death.

Introduction

Apoptosis, or programmed cell death, is a common property of all multicellular organisms (1, 2). It can be triggered by a number of factors, including ultraviolet or γ-irradiation, growth factor withdrawal, chemotherapeutic drugs, or signaling by death receptors (DRs) (3, 4). The central role in the regulation and the execution of apoptotic cell death belongs to caspases (5–7). Caspases, a family of cysteinyl aspartate–specific proteases, are synthesized as zymogens with a prodomain of variable length followed by a large subunit (p20) and a small subunit (p10). The caspases are activated through proteolysis at specific asparagine residues that are located within the prodomain, the p20 and p10 subunits (8). This results in the generation of mature active caspases that consist of the heterotetramer $p20_2$–$p10_2$. Subsequently, active caspases specifically process various substrates that are implicated in apoptosis and inflammation. Their important function in these processes makes caspases potential targets for drug development. In this Review, we discuss the structures and functions of caspases as well as their role in novel approaches for treating cancer, autoimmune diseases, degenerative disorders, and stroke.

Structure of caspases

General overview. Caspases are zymogens (inactive enzyme precursors, which require a biochemical change to become an active enzyme) that consist of an N-terminal prodomain followed by a large subunit of about 20 kDa, p20, and a small subunit of about 10 kDa, p10 (Figure 1A) (5). In a number of procaspases, the p20 and p10 subunits are separated by a small linker sequence. Depending on the structure of the prodomain and their function, caspases are typically divided into 3 major groups (Figure 1A). The caspases with large prodomains are referred to as inflammatory caspases (group I) and initiator of apoptosis caspases (group II), while caspases with a short prodomain of 20–30 amino acids are named effector caspases (group III).

Caspase prodomains. The large prodomains of procaspases contain structural motifs that belong to the so-called death domain superfamily (9, 10). Death domains are 80- to 100-residue-long motifs

Nonstandard abbreviations used: BIR, baculoviral IAP repeat; CARD, caspase recruitment domain; c-FLIP, cellular FLICE-inhibitory protein; DD, death domain; DED, death effector domain; DISC, death-inducing signaling complex; DR, death receptor; FLIP, FLICE-inhibitory protein; fmk, fluoromethyl ketone; IAP, inhibitor of apoptosis.

Conflict of interest: The authors have declared that no conflict of interest exists.

Citation for this article: *J. Clin. Invest.* **115**:2665–2672 (2005). doi:10.1172/JCI26252.

involved in the transduction of the apoptotic signal. This superfamily consists of the death domain (DD), the death effector domain (DED), and the caspase recruitment domain (CARD) (11). Each of these motifs interacts with other proteins by homotypic interactions. All members of the death domain superfamily are characterized by similar structures that comprise 6 or 7 antiparallel amphipathic α-helices. Structural similarity suggests a common evolutionary origin for all recruitment domains (12). However, the nature of the homotypic interactions differs within the superfamily. DD and CARD contacts are based on electrostatic interactions, while DED contacts use hydrophobic interactions (13).

Procaspase-8 and -10 possess 2 tandem DEDs in their prodomain (14, 15). The CARD is found in procaspase-1, -2, -4, -5, -9, -11, and -12 (16, 17). DEDs and CARDs are responsible for the recruitment of initiator caspases into death- or inflammation-inducing signaling complexes, resulting in proteolytic autoactivation of caspases that subsequently initiates inflammation or apoptosis.

Structure of active caspase heterotetramers. Cleavage of a procaspase at the specific Asp-X bonds results in the formation of the mature caspase, which comprises the heterotetramer $p20_2$–$p10_2$ and causes release of the prodomain (Figure 1B). X-ray structures have been determined for mature caspase-1 (18, 19), caspase-2 (20), caspase-3 (21–23), caspase-7 (24–26), caspase-8 (27), and caspase-9 (28). The overall architecture of all caspases is similar. Each heterodimer (p10–p20) is formed by hydrophobic interactions resulting in the formation of several parallel β-sheets, composed of 6 antiparallel β-strands. Two heterodimers interact via a 12-stranded β-sheet that is surrounded by α-helices (Figure 1C). This so-called caspase fold is a unique quaternary structure among proteases and has been described only for caspases and for gingipain R, the cysteine protease from *Porphyromonas gingivalis* (29). In the caspase heterotetramer, the 2 heterodimers align in a head-to-tail configuration. Correspondingly, 2 active sites are positioned at opposite ends of the molecule. The architecture of the active center comprises amino acid residues from both subunits. The catalytic machinery involves a diad composed of a cysteine sulfohydryl group (Cys285) and a histidine imidazol ring (His237) (19). Both of them are located in the p20 subunit. The tetrahedral transition state of the cysteine protease is stabilized through hydrogen bonding with the backbone amide protons of Cys285 and Gly238. The asparagine of the substrate seems to be stabilized by 4 residues: Arg179 and Gln283 from the p20 subunit and Arg341 and Ser347 from the p10 subunit.

Substrate specificity and synthetic peptide inhibitors of caspases. Caspases are specific cysteine proteases, recognizing 4 amino acids, named S4-S3-S2-S1. The cleavage takes place typically after the

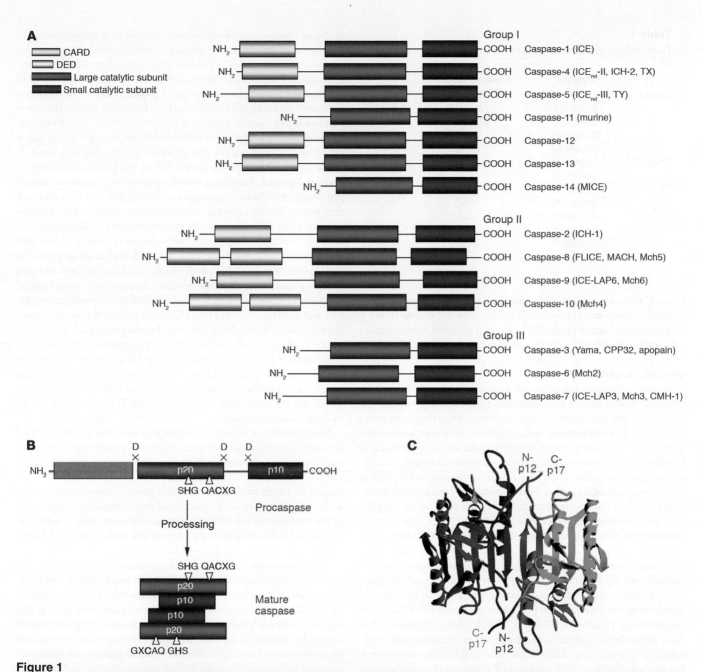

Figure 1

Caspase structure. (**A**) The caspase family. Three major groups of caspases are presented. Group I: inflammatory caspases; group II: apoptosis initiator caspases; group III: apoptosis effector caspases. The CARD, the DED, and the large (p20) and small (p10) catalytic subunits are indicated. (**B**) Scheme of procaspase activation. Cleavage of the procaspase at the specific Asp-X bonds leads to the formation of the mature caspase, which comprises the heterotetramer p20$_2$–p10$_2$, and the release of the prodomain. The residues involved in the formation of the active center are shown. (**C**) The 3D structure of caspase-3 heterotetramer. Each heterodimer is formed by hydrophobic interactions resulting in the formation of mostly parallel β-sheets, composed of 6 antiparallel β-strands. Two heterodimers fit together with formation of a 12-stranded β-sheet that is sandwiched by α-helices. N and C termini of the small and large protease subunits are indicated.

C-terminal residue (S1), which is usually an asparagine (30). A list of substrate specificities of caspases is presented in Table 1. Interestingly, the preferred S3 position is an invariant glutamine for all mammalian caspases. Thus, specificity of caspase cleavage can be described as X-Glu-X-Asp. Caspase-1, -4, and -5 (group I; Figure 1) prefer the tetrapeptide sequence WEHD. Caspase-2, -3, and -7 have a preference for the substrate DEXD, whereas caspase-6, -8, and -9 prefer the sequence (L/V)EXD. Interestingly, the cleavage

site between the large and small subunits for initiator caspases carries its own tetrapeptide recognition motif, which is remarkably consistent with the proposed mechanism of autoactivation of initiator caspases (31, 32).

Most of the synthetic peptide caspase inhibitors were developed based on the tetrapeptide caspase recognition motif. Therefore, the selectivity of inhibitors matches the caspase substrate specificities described above (Table 1). The introduction of an aldehyde group

Table 1
The substrate specificity of caspases

	Caspase	Substrate specificity
Group I	Caspase-1	WEHD
	Caspase-4	(W/L)EHD
	Caspase-5	(W/L)EHD
	Caspase-13	WEHD
	Caspase-14	WEHD
Group II	Caspase-2	DEHD
	Caspase-8	LETD
	Caspase-9	LEHD
	Caspase-10	LEXD
Group III	Caspase-3	DEVD
	Caspase-6	VEHD
	Caspase-7	DEVD

at the C terminus of the tetrapeptide results in the generation of reversible inhibitors (33), whereas a fluoromethyl ketone (fmk), a chloromethyl ketone (cmk) (34), or a diazomethyl ketone (dmk) (35) at this position irreversibly inactivates the enzyme.

Caspase signaling

Mechanisms of caspase activation. All caspases are produced in cells as catalytically inactive zymogens and must undergo proteolytic processing and activation during apoptosis. The effector caspases are activated by initiator or apical caspases. However, one central question of apoptosis is how initiator caspases are activated and how this activation is regulated to prevent spontaneous cell death. It is generally accepted that apical caspase activation takes place in large protein complexes that bring together several caspase zymogens (22–25, 36). All initiator caspases are characterized by the presence of a member of the DD superfamily (DED or CARD), which enables their recruitment into the initiation complex. Several activating complexes for initiator caspases have been reported so far.

The death-inducing signaling complex as an activating complex for procaspase-8 and -10. Procaspase-8 and -10 are apical caspases in apoptotic pathways triggered by engagement of death receptors. Several members of the TNF receptor (TNFR) superfamily (TNFR1, CD95 [Fas/APO-1], TRAIL-R1, TRAIL-R2, DR3, DR6) comprise DD in their intracellular domain and are, therefore, termed death receptors (3, 37). Triggering of CD95 and TRAIL-R1/R2 by corresponding ligands results in the formation of a death-inducing signaling complex (DISC) (38–42). CD95 and TRAIL-R1/R2 DISCs consist of oligomerized, probably trimerized, receptors, the DD-containing adaptor molecule FADD/MORT1 (Fas-associated death domain), 2 isoforms of procaspase-8 (procaspase-8/a [FLICE, MACHα1, Mch5] and procaspase-8/b [Machα2]), procaspase-10, and the cellular FLICE-inhibitory proteins (c-FLIP$_{L/S/R}$) (Figure 2A) (43, 44). The interactions between the molecules at the DISC are based on homotypic contacts. The DD of the receptor interacts with the DD of FADD, while the DED of FADD interacts with the N-terminal tandem DEDs of procaspase-8 and -10 and FLIP$_{L/S/R}$. Thus, DISC formation results in assembly of procaspase-8 and -10 molecules in close proximity to each other.

Activation of procaspase-8 is believed to follow an "induced proximity" model in which high local concentrations and favorable mutual orientation of procaspase-8 molecules at the DISC lead to their autoproteolytic activation (31, 45–47). There is strong

evidence from a number of in vitro studies that autoproteolytic activation of procaspase-8 occurs upon oligomerization at the receptor complex (45–47). Furthermore, it has been demonstrated that dimers formed by procaspase-8 molecules possess proteolytic activity, and proteolytic processing of procaspase-8 occurs between precursor dimers (45). Interestingly, it has been shown that procaspase-8 and mature caspase-8 possess different substrate specificities (45). It is likely that conformational changes in the active center of caspase-8 occur upon processing to mature caspase-8.

The processing of procaspase-8/a/b at the DISC is depicted in detail in Figure 2. According to the 2-step model, the processing of procaspase-8 includes 2 cleavage events (39, 45). The first cleavage step occurs between the protease domains, and the second cleavage step takes place between the prodomain and the large protease subunit. Correspondingly, after the first cleavage step, p43/p41 and p10 subunits are formed (Figure 2, A and B). Both cleavage products remain bound to the DISC, p43/p41 by DED interactions and p10 by interactions with the large protease domain (48). As a result of the second cleavage step, p43/41 is processed to the prodomain p26/p24 and p18 (Figure 2C). Thus, the active caspase-8 heterotetramer is formed at the DISC. Subsequently, the mature caspase-8 heterotetramer is released to the cytosol to trigger apoptotic processes.

Procaspase-10 is also activated at the DISC, forming an active heterotetramer (15, 49). However, whether caspase-10 can trigger cell death in the absence of caspase-8 in response to CD95 or TRAIL-R1/R2 stimulation is controversial. Thus, the exact role of caspase-10 remains elusive.

The apoptosome as activating complex for procaspase-9. A number of apoptotic stimuli, such as cytotoxic stress, heat shock, oxidative stress, and DNA damage, lead to the release of cytochrome c from mitochondria. The release of cytochrome c is followed by the formation of a high–molecular mass cytoplasmic complex referred to as the apoptosome (50). In mammals the central scaffold protein of the apoptosome is a 140-kDa protein known as Apaf-1 (apoptotic protease–activating factor-1), which is a homologue of CED-4, a key protein involved in programmed cell death in the nematode *Caenorhabditis elegans* (51, 52). In the presence of cytochrome c and dATP, Apaf-1 oligomerizes to form a very large (700–1,400 kDa) apoptosome complex. Procaspase-9 is recruited to the complex by CARD interactions, which results in its activation (53). It has been biochemically demonstrated that activation of procaspase-9 occurs by dimerization (31). Moreover, it has been shown that proteolytic activation of procaspase-9 takes place upon dimerization and subsequent cleavage within an interdomain linker, which itself is important for stabilization of caspase-9 dimers.

The inflammosome as activating complex for caspase-1 and -5. The activation of the initiator caspase-1 and -5 takes place in a complex that was named the inflammosome (54). The inflammosome comprises procaspase-1 and -5 as well as the CARD-containing protein NALP-1. The formation of this complex results in the processing and activation of the cytokines IL-1β and IL-18, which play a central role in the immune response to microbial pathogens.

Effector caspase cascade. The activation of the effector caspase cascade differs between extrinsic (death receptor–mediated) and intrinsic (mitochondria-mediated) pathways.

In death receptor–mediated apoptosis, 2 types of signaling pathways have been established (55). So-called type I cells are characterized by high levels of DISC formation and increased amounts of active caspase-8 (Figure 3). Activated caspase-8 directly leads to the activation of downstream, effector caspase-3 and -7. In type II cells,

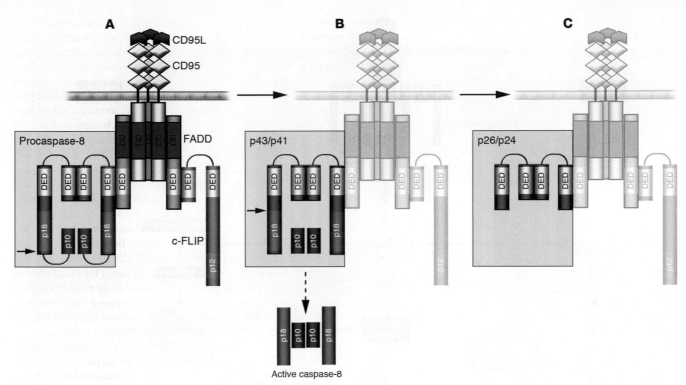

Figure 2

Scheme of procaspase-8 processing at the CD95 DISC. CD95 DISC formation is triggered by extracellular cross-linking with CD95L (depicted in red), which is followed by oligomerization of the receptor. FADD/MORT1 is recruited to the DISC by DD interactions (shown in red); procaspase-8 and -10 as well as c-FLIP proteins are recruited to the DISC by homophilic DED interactions (yellow). Upon recruitment to the DISC, procaspase-8 undergoes processing by forming dimers (depicted in green). (**A**) The first step of procaspase-8 cleavage occurs between 2 protease subunits. The site of cleavage is shown by a black arrow. As a result of the first cleavage step the p10 subunit is formed, which is not released into the cytosol but remains bound to the DISC as it is involved in the interactions with the large protease subunits. (**B**) The second cleavage step takes place between the prodomain and the large protease subunit at Asp216. As a result of this cleavage step the active caspase-8 heterotetramer is formed, which is then released into the cytosol. (**C**) Prodomain p26/p24 remains bound to the DISC.

there are lower levels of DISC formation and, thus, lower levels of active caspase-8. In this case, signaling requires an additional amplification loop that involves the cleavage of the Bcl-2–family protein Bid by caspase-8 to generate tBid and a subsequent tBid-mediated release of cytochrome c from mitochondria (56). The release of cytochrome c from mitochondria results in apoptosome formation, followed by the activation of procaspase-9, which in turn cleaves downstream, effector caspase-3 and -7. Type II signaling might be blocked by Bcl-2 family members such as Bcl-2 and Bcl-x_L.

In the intrinsic pathway, which is triggered by a number of factors, including UV or γ-irradiation, growth factor withdrawal, and chemotherapeutic drugs, the release of cytochrome c from mitochondria leads to apoptosome formation and activation of procaspase-9 (53). Subsequently, procaspase-9 cleaves effector caspase-3 and -7, which, correspondingly, initiate the death cascade. There are reports pointing toward a role of procaspase-2 in genotoxic stress acting upstream of mitochondria; however, this question requires further clarification (57, 58).

Cellular inhibitors of caspases. The action of caspases is regulated on several levels, including blockade of activation of caspases at the DISC as well as inhibition of enzymatic caspase activity (Figure 3). c-FLIP proteins are well-known inhibitors of death receptor–induced apoptosis (44, 59, 60). c-FLIPs possess 2 tandem DEDs at their N termini that facilitate their recruitment to the DISC. There are 3 c-FLIP isoforms described on the protein level, c-FLIP$_L$,

c-FLIP$_S$, and c-FLIP$_R$ (61). Under conditions of overexpression, all isoforms inhibit activation of procaspase-8 at the DISC by blocking its processing (62) (Figure 3). At the same time, there is increasing evidence that c-FLIP$_L$, when present at the DISC at low concentrations, facilitates the cleavage of procaspase-8 at the DISC by forming c-FLIP$_L$–procaspase-8 heterodimers (45, 63).

The IAP (inhibitor of apoptosis) family of proteins includes 8 mammalian family members, including XIAP, c-IAP1, c-IAP2, and ML-IAP/livin (64–66). They specifically inhibit the initiator caspase-9 and the effector caspase-3 and -7 (Figure 3). The functional unit in IAP is the baculoviral IAP repeat (BIR), which contains about 80 amino acids folded around a central zinc atom. XIAP, c-IAP-1, and c-IAP2 contain 3 BIR domains each. The third BIR domain (BIR3) is involved in interactions with caspase-9 resulting in the inhibition of its activity (67). The linker region between BIR1 and BIR2 selectively targets caspase-3 and -7. The activity of IAPs is regulated by Smac/DIABLO, a structural homologue of the *Drosophila* proteins Reaper, Hid, and Grim (68, 69) (Figure 3). Smac/DIABLO is released from mitochondria and inhibits IAPs, which facilitates caspase activation during apoptosis. Omi/HtrA2 has been recently identified as another modulator of IAP function (70). Omi/HtrA2 is a mitochondria-located serine protease, which is released into the cytosol and inhibits IAPs by a mechanism similar to that of Smac.

IAPs are not the only natural inhibitors of caspases. The baculoviral p35 protein is a pan-caspase inhibitor, and it targets most

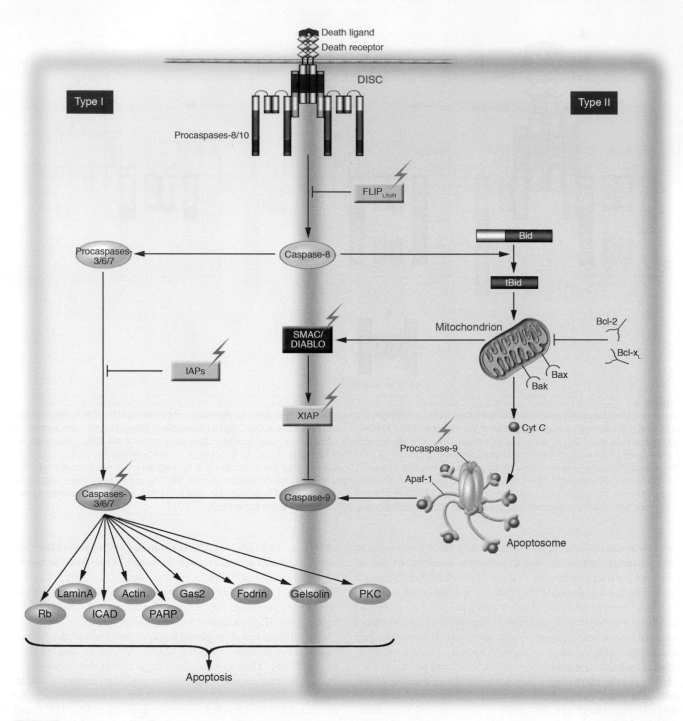

Figure 3

Caspase signaling and its modulation. In the extrinsic pathway, DISC formation leads to caspase-8 activation. Two signaling pathways downstream from the receptor were established. In type I cells (shown in light blue) caspase-8 directly cleaves caspase-3, which starts the death cascade. In type II cells (shown in light red) an additional amplification loop is required, which involves tBid-mediated cytochrome *c* release from mitochondria followed by apoptosome formation. Initiation of the intrinsic pathway results in mitochondria-mediated apoptosome formation, followed by caspase-9 and -3 activation, leading to destruction of the cell. Caspase action can be modulated on several levels. Activation of caspases at the DISC is inhibited by c-FLIP proteins; activation of effector caspases is inhibited by IAPs (see text). Effector caspases are shown in light green; cellular caspase inhibitors are presented in yellow. The targets for pharmacological modulation are shown with an orange arrow.

caspases, in contrast to IAPs, which affect only caspase-3, -7, and -9 (71). The mechanism of caspase inhibition by p35 involves the formation of an inhibitory complex that is characterized by a protected thioester link between the caspase and p35 (72). Structural analysis of the inhibitory complex between p35 and caspase-8 reveals a unique active-site conformation that protects the intermediary thioester link from hydrolysis. Another pan-caspase inhibitor, serpin CrmA, is derived from the cowpox virus (73). The mechanism of CrmA inhibition is likely to involve covalent modification of the caspase active center.

Caspases as targets in drug development

Caspases, being the key effector molecules in apoptosis, are potential targets for pharmacological modulation of cell death (Figure 3). First, increased levels of caspase activity are often observed at sites of cellular damage in a number of diseases, including myocardial infarction, stroke, sepsis, and Alzheimer, Parkinson, and Huntington diseases. Inhibition of caspase activity for these diseases is predicted to be therapeutically beneficial. Second, discovery of drugs that selectively inhibit inflammatory caspases (caspase-1, -4, and -5) may help to control autoimmune diseases like rheumatoid arthritis. Finally, selective activation of caspases would be an approach in the treatment of cancer and chronic viral infections.

The approach of direct caspase inhibition is currently in the center of investigations. Proof-of-concept data have been obtained in several animal models, using suboptimal peptidyl inhibitors of caspases, such as zVAD-fmk, which shows substantial protection in rodent models of stroke, myocardial infarction, hepatic injury, sepsis, amyotrophic lateral sclerosis, and several other diseases (74–76). Furthermore, so-called small-molecule inhibitors of caspases, which have a nonpeptidyl origin, have been developed. One of these caspase inhibitors, IDN6556 (Pfizer), is already in phase II of clinical studies. It is a broad-spectrum caspase inhibitor that forms irreversible adducts with cysteine residues from the active site of caspases (77). IDN6556 is considered to be a candidate for the treatment of acute–tissue injury diseases, which are characterized by excessive apoptosis. This inhibitor is currently being tested in treatment of liver diseases, including HCV and ischemia/reperfusion injury in liver transplantation.

Among the inhibitors of inflammatory caspases, VX-740 (Vertex Pharmaceuticals Inc.) is also in phase II of clinical studies (78). It is a selective and reversible inhibitor of caspase-1 that is developed for treatment of rheumatoid arthritis.

Another promising strategy involves selective activation of caspases in cancer cells, leading to induction of apoptosis. An important contribution to this strategy was achieved by the approach of "forward chemical genetics" (79). This involves screening of small molecules for their ability to perturb cellular pathways and subsequently identifying the specific targets of the active compounds. A number of potential drugs for selective induction of apoptosis were found by high-throughput screening of the compounds activating caspase-3 as a central suicide caspase. Among them are a small molecule, PETCM [α-(trichloromethyl)-4-pyridine ethanol] (80); carbamate and indolone classes of compounds (81); and MX2167, MX116407, MX128504, and MX90745 (82). These compounds engage different pathways; e.g., PETCM accelerates apoptosome formation by interacting with the inhibitor prothymosin-α (80); carbamate and indolone classes of compounds promote Apaf-1 oligomerization and, thereby, apoptosome formation with the subsequent activation of caspase-3 and -9 (81); the action of MX2167 is mediated via the transferrin receptor that is highly overexpressed in a number of tumors; MX116407 is a vascular targeting agent; and MX128504 targets a novel cytoplasmic protein affecting IGF growth signaling pathways. At the same time, all these drugs are characterized by selective induction of apoptosis, and some of them are in preclinical and phase I studies.

Strategies to selectively target cancer cells also involve the application of antibodies coupled to active caspases. There were reports on a chimeric protein, referred to as immunocasp-3, that comprises a single-chain anti–erbB2/HER2 antibody, a translocation domain of *Pseudomonas* exotoxin A, and constitutively active caspase-3 (83). Immunocasp-3 was shown to selectively bind to erbB2/HER2–overexpressing cancer cells, followed by its internalization and lysosomal cleavage. As a result a COOH-terminal caspase-3–containing fragment is released, which translocates to the cytosol and induces apoptosis. Following this study there was a report on erbB2/HER2–targeted immunocasp-6, which can directly cleave lamin A, leading to nuclear damage, and induce programmed cell death (84). These studies provide the platform for the development of novel therapeutic protocols against tumors that overexpress erbB2/HER2.

In addition to direct targeting of caspases, other strategies involve modulation of cellular caspase inhibitors, such as IAPs, c-FLIPs, and Smac (Figure 3). Some cancers are characterized by overexpression of IAPs that are associated with resistance to apoptosis. The prototypic example is melanoma IAP (MILAP, also known as livin/KIAP), which is found at high levels in melanomas (66). Therefore, strategies of downregulating IAPs play an important role, as this would result in caspase activation and subsequent apoptosis induction in cancer cells. XIAP antisense molecules can directly induce apoptosis as well as sensitize cells to chemotherapy and irradiation (85). Antisense XIAP oligonucleotides are currently in phase I clinical studies.

Another approach to target IAPs is the identification of small molecules that bind to IAPs and prevent their inhibition of caspases. There are 3 main screening strategies to search for IAP antagonists. First, peptides that structurally resemble the N terminus of Smac were shown to bind to the BIR3 pocket of XIAP, subsequently leading to inhibition of XIAP (86). The development of these peptides is promising; however, further work is required to determine whether these peptides can induce apoptosis or sensitize cells to chemotherapy. In a second strategy, phage display was used to identify XIAP-binding peptides. This screen identified peptides unrelated to Smac that bind to the BIR2 domain and can directly induce cell death in leukemia cells (87). In a third approach, biochemically based assays were used to identify small molecules and peptides that inhibit XIAP. The screening was based on the reversed XIAP-mediated repression of caspase-3. Through these screens, both peptidyl and nonpeptidyl XIAP inhibitors were identified (85, 88). The small molecules identified by Wu and colleagues appear to be capable of sensitizing TRAIL-resistant cells toward TRAIL-induced apoptosis.

In addition to targeting of IAPs in cancer, IAPs might be used in gene therapy strategies to reduce neuronal cell death that is followed by stroke or brain injury. Virus-mediated delivery of IAP into the brain preserved neurons in rodent models of stroke and other types of experimental injury (89). These data indicate that modulation of IAP action might be also used in treating neurological diseases.

Given the current status of research on caspases as targets for pharmacological manipulation, the value of certain strategies is variable. Notably, despite the attraction of using caspase inhibitors, this approach might be ineffective. Blocking caspases often results in triggering of caspase-independent cell death accompanied by the release of cytochrome *c*, AIF, and EndoG from mitochondria. Thus, how to prevent cross-talk between different pathways involved is not really clear. Strategies of selective activation of caspases in cancer cells by various drugs or site-directed delivery, however, may be more promising and might find their way into the clinic. Moreover, certain compounds might be applied in combination with standard chemotherapy and are likely to be more efficient. Finally, targeting of IAPs by, e.g., IAP antagonists appears to be a potential clinical application and may become a tool in cancer therapy.

Acknowledgments
We would like to thank Lars Weingarten for critical comments, Heidi Sauter for excellent secretarial work, and the Wilhelm Sander Stiftung, the Sonderforschungsbereich 405, and the Tumorzentrum Heidelberg/Mannheim for supporting our work.

Address correspondence to: Peter H. Krammer, Division of Immunogenetics, Tumor Immunology Program, German Cancer Research Center, Im Neuenheimer Feld 280, D-69120 Heidelberg, Germany. Phone: 49-6221-423717; Fax: 49-6221-411715; E-mail: p.krammer@dkfz-heidelberg.de.

1. Krammer, P.H. 2000. CD95's deadly mission in the immune system. *Nature.* **407**:789–795.
2. Danial, N.N., and Korsmeyer, S.J. 2004. Cell death: critical control points. *Cell.* **116**:205–219.
3. Ashkenazi, A., and Dixit, V.M. 1998. Death receptors: signaling and modulation. *Science.* **281**:1305–1308.
4. Krammer, P.H. 1998. The CD95(APO-1/Fas)/CD95L system [review]. *Toxicol. Lett.* **102–103**:131–137.
5. Thornberry, N.A., and Lazebnik, Y. 1998. Caspases: enemies within. *Science.* **281**:1312–1316.
6. Cohen, G.M. 1997. Caspases: the executioners of apoptosis. *Biochem. J.* **326**:1–16.
7. Nicholson, D.W. 1999. Caspase structure, proteolytic substrates, and function during apoptotic cell death. *Cell Death Differ.* **6**:1028–1042.
8. Alnemri, E.S., et al. 1996. Human ICE/CED-3 protease nomenclature [letter]. *Cell.* **87**:171.
9. Weber, C.H., and Vincenz, C. 2001. The death domain superfamily: a tale of two interfaces [review]? *Trends Biochem. Sci.* **26**:475–481.
10. Martinon, F., Hofmann, K., and Tschopp, J. 2001. The pyrin domain: a possible member of the death domain-fold family implicated in apoptosis and inflammation. *Curr. Biol.* **11**:R118–R120.
11. Fesik, S.W. 2000. Insights into programmed cell death through structural biology. *Cell.* **103**:273–282.
12. Hofmann, K. 1999. The modular nature of apoptotic signaling proteins. *Cell. Mol. Life Sci.* **55**:1113–1128.
13. Eberstadt, M., et al. 1998. NMR structure and mutagenesis of the FADD (Mort1) death-effector domain. *Nature.* **392**:941–945.
14. Muzio, M., et al. 1996. FLICE, a novel FADD-homologous ICE/CED-3-like protease, is recruited to the CD95 (Fas/APO-1) death–inducing signaling complex. *Cell.* **85**:817–827.
15. Sprick, M., et al. 2002. Caspase-10 is recruited to and activated at the native TRAIL and CD95 death-inducing signalling complexes in a FADD-dependent manner but can not functionally substitute caspase-8. *EMBO J.* **21**:4520–4530.
16. Fuentes-Prior, P., and Salvesen, G.S. 2004. The protein structures that shape caspase activity, specificity, activation and inhibition. *Biochem. J.* **384**:201–232.
17. Lamkanfi, M., et al. 2005. A novel caspase-2 complex containing TRAF2 and RIP1. *J. Biol. Chem.* **280**:6923–6932.
18. Walker, N.P., et al. 1994. Crystal structure of the cysteine protease interleukin-1 beta-converting enzyme: a (p20/p10)2 homodimer. *Cell.* **78**:343–352.
19. Wilson, K.P., et al. 1994. Structure and mechanism of interleukin-1 beta converting enzyme. *Nature.* **370**:270–275.
20. Schweizer, A., Briand, C., and Grutter, M.G. 2003. Crystal structure of caspase-2, apical initiator of the intrinsic apoptotic pathway. *J. Biol. Chem.* **278**:42441–42447.
21. Rotonda, J., et al. 1996. The three-dimensional structure of apopain/CPP32, a key mediator of apoptosis. *Nat. Struct. Biol.* **3**:619–625.
22. Mittl, P.R., et al. 1997. Structure of recombinant human CPP32 in complex with the tetrapeptide acetyl-Asp-Val-Ala-Asp fluoromethyl ketone. *J. Biol. Chem.* **272**:6539–6547.
23. Riedl, S.J., et al. 2001. Structural basis for the inhibition of caspase-3 by XIAP. *Cell.* **104**:791–800.
24. Wei, Y., et al. 2000. The structures of caspases-1, -3, -7 and -8 reveal the basis for substrate and inhibitor selectivity. *Chem. Biol.* **7**:423–432.
25. Chai, J., et al. 2001. Structural basis of caspase-7 inhibition by XIAP. *Cell.* **104**:769–780.
26. Huang, Y., et al. 2001. Structural basis of caspase inhibition by XIAP: differential roles of the linker versus the BIR domain. *Cell.* **104**:781–790.
27. Blanchard, H., et al. 1999. The three-dimensional structure of caspase-8: an initiator enzyme in apoptosis. *Structure Fold. Des.* **7**:1125–1133.
28. Renatus, M., Stennicke, H.R., Scott, F.L., Liddington, R.C., and Salvesen, G.S. 2001. Dimer formation drives the activation of the cell death protease caspase 9. *Proc. Natl. Acad. Sci. U. S. A.* **98**:14250–14255.
29. Eichinger, A., et al. 1999. Crystal structure of gingipain R: an Arg-specific bacterial cysteine proteinase with a caspase-like fold. *EMBO J.* **18**:5453–5462.
30. Thornberry, N.A., et al. 1997. A combinatorial approach defines specificities of members of the caspase family and granzyme B. Functional relationships established for key mediators of apoptosis. *J. Biol. Chem.* **272**:17907–17911.
31. Boatright, K.M., et al. 2003. A unified model for apical caspase activation. *Mol. Cell.* **11**:529–541.
32. Shi, Y. 2002. Mechanisms of caspase activation and inhibition during apoptosis. *Mol. Cell.* **9**:459–470.
33. Graybill, T.L., Dolle, R.E., Helaszek, C.T., Miller, R.E., and Ator, M.A. 1994. Preparation and evaluation of peptidic aspartyl hemiacetals as reversible inhibitors of interleukin-1 beta converting enzyme (ICE). *Int. J. Pept. Protein Res.* **44**:173–182.
34. Estrov, Z., et al. 1995. Effect of interleukin-1 beta converting enzyme inhibitor on acute myelogenous leukemia progenitor proliferation. *Blood.* **86**:4594–4602.
35. Thornberry, N.A., et al. 1992. A novel heterodimeric cysteine protease is required for interleukin-1 beta processing in monocytes. *Nature.* **356**:768–774.
36. Chang, D.W., et al. 2002. c-FLIP(L) is a dual function regulator for caspase-8 activation and CD95-mediated apoptosis. *EMBO J.* **21**:3704–3714.
37. Tartaglia, L.A., Ayres, T.M., Wong, G.H., and Goeddel, D.V. 1993. A novel domain within the 55 kd TNF receptor signals cell death. *Cell.* **74**:845–853.
38. Kischkel, F.C., et al. 1995. Cytotoxicity-dependent APO-1 (Fas/CD95)-associated proteins form a death-inducing signaling complex (DISC) with the receptor. *EMBO J.* **14**:5579–5588.
39. Scaffidi, C., Medema, J.P., Krammer, P.H., and Peter, M.E. 1997. FLICE is predominantly expressed as two functionally active isoforms, caspase-8/a and caspase-8/b. *J. Biol. Chem.* **272**:26953–26958.
40. Scaffidi, C., et al. 1999. Differential modulation of apoptosis sensitivity in CD95 type I and type II cells. *J. Biol. Chem.* **274**:22532–22538.
41. Scaffidi, C., Krammer, P.H., and Peter, M.E. 1999. Isolation and analysis of components of CD95 (APO-1/Fas) death-inducing signaling complex. *Methods.* **17**:287–291.
42. Scaffidi, C., Kischkel, F.C., Krammer, P.H., and Peter, M.E. 1999. Analysis of the CD95 (APO-1/Fas) death-inducing signaling complex (DISC) by high resolution two-dimensional gel electrophoresis. *Methods Enzymol.* **322**:363–373.
43. Peter, M.E., and Krammer, P.H. 2003. The CD95(APO-1/Fas) DISC and beyond. *Cell Death Differ.* **10**:26–35.
44. Golks, A., Brenner, D., Fritsch, C., Krammer, P.H., and Lavrik, I.N. 2005. C-FLIPR: a new regulator of death receptor-induced apoptosis. *J. Biol. Chem.* **280**:14507–14513.
45. Chang, D.W., Xing, Z., Capacio, V.L., Peter, M.E., and Yang, X. 2003. Interdimer processing mechanism of procaspase-8 activation. *EMBO J.* **22**:4132–4142.
46. Muzio, M., Stockwell, B.R., Stennicke, H.R., Salvesen, G.S., and Dixit, V.M. 1998. An induced proximity model for caspase-8 activation. *J. Biol. Chem.* **273**:2926–2930.
47. Martin, D.A., Siegel, R.M., Zheng, L., and Lenardo, M.J. 1998. Membrane oligomerization and cleavage activates the caspase-8 (FLICE/MACHalpha1) death signal. *J. Biol. Chem.* **273**:4345–4349.
48. Lavrik, I., et al. 2003. The active caspase-8 heterotetramer is formed at the CD95 DISC. *Cell Death Differ.* **10**:144–145.
49. Kischkel, F.C., et al. 2001. Death receptor recruitment of endogenous caspase-10 and apoptosis initiation in the absence of caspase-8. *J. Biol. Chem.* **276**:46639–46646.
50. Acehan, D., et al. 2002. Three-dimensional structure of the apoptosome: implications for assembly, procaspase-9 binding, and activation. *Mol. Cell.* **9**:423–432.
51. Hengartner, M.O. 1999. Programmed cell death in the nematode C. elegans [review]. *Recent Prog. Horm. Res.* **54**:213–222; discussion 222–214.
52. Chinnaiyan, A.M. 1999. The apoptosome: heart and soul of the cell death machine. *Neoplasia.* **1**:5–15.
53. Jiang, X., and Wang, X. 2000. Cytochrome c promotes caspase-9 activation by inducing nucleotide binding to Apaf-1. *J. Biol. Chem.* **275**:31199–31203.
54. Martinon, F., and Tschopp, J. 2004. Inflammatory caspases: linking an intracellular innate immune system to autoinflammatory diseases. *Cell.* **117**:561–574.
55. Scaffidi, C., et al. 1998. Two CD95 (APO-1/Fas) signaling pathways. *EMBO J.* **17**:1675–1687.
56. Korsmeyer, S.J., et al. 2000. Pro-apoptotic cascade activates BID, which oligomerizes BAK or BAX into pores that result in the release of cytochrome c. *Cell Death Differ.* **7**:1166–1173.
57. Robertson, J.D., Enoksson, M., Suomela, M., Zhivotovsky, B., and Orrenius, S. 2002. Caspase-2 acts upstream of mitochondria to promote cytochrome c release during etoposide-induced apoptosis. *J. Biol. Chem.* **277**:29803–29809.
58. Lassus, P., Opitz-Araya, X., and Lazebnik, Y. 2002. Requirement for caspase-2 in stress-induced apoptosis before mitochondrial permeabilization. *Science.* **297**:1352–1354.
59. Thome, M., et al. 1997. Viral FLICE-inhibitory proteins (FLIPs) prevent apoptosis induced by death receptors. *Nature.* **386**:517–521.
60. Scaffidi, C., Schmitz, I., Krammer, P.H., and Peter, M.E. 1999. The role of c-FLIP in modulation of CD95-induced apoptosis. *J. Biol. Chem.* **274**:1541–1548.
61. Golks, A., Brenner, D., Fritsch, C., Krammer, P.H., and Lavrik, I.N. 2005. C-FLIPR: a new regulator of death receptor-induced apoptosis. *J. Biol. Chem.* **280**:14507–14513.
62. Krueger, A., Schmitz, I., Baumann, S., Krammer, P.H., and Kirchhoff, S. 2001. Cellular flice-inhibitory protein splice variants inhibit different steps of caspase-8 activation at the cd95 death-inducing

signaling complex. *J. Biol. Chem.* **276**:20633–20640.

63. Micheau, O., et al. 2002. The long form of FLIP is an activator of caspase-8 at the Fas death-inducing signaling complex. *J. Biol. Chem.* **277**:45162–45171.

64. Deveraux, Q.L., and Reed, J.C. 1999. IAP family proteins: suppressors of apoptosis. *Genes Dev.* **13**:239–252.

65. Ashhab, Y., Alian, A., Polliack, A., Panet, A., and Ben Yehuda, D. 2001. Two splicing variants of a new inhibitor of apoptosis gene with different biological properties and tissue distribution pattern. *FEBS Lett.* **495**:56–60.

66. Vucic, D., Stennicke, H.R., Pisabarro, M.T., Salvesen, G.S., and Dixit, V.M. 2000. ML-IAP, a novel inhibitor of apoptosis that is preferentially expressed in human melanomas. *Curr. Biol.* **10**:1359–1366.

67. Srinivasula, S.M., et al. 2001. A conserved XIAP-interaction motif in caspase-9 and Smac/DIABLO regulates caspase activity and apoptosis. *Nature.* **410**:112–116.

68. Du, C., Fang, M., Li, Y., Li, L., and Wang, X. 2000. Smac, a mitochondrial protein that promotes cytochrome c-dependent caspase activation by eliminating IAP inhibition. *Cell.* **102**:33–42.

69. Ekert, P.G., Silke, J., Hawkins, C.J., Verhagen, A.M., and Vaux, D.L. 2001. DIABLO promotes apoptosis by removing MIHA/XIAP from processed caspase 9. *J. Cell Biol.* **152**:483–490.

70. Suzuki, Y., et al. 2001. A serine protease, HtrA2, is released from the mitochondria and interacts with XIAP, inducing cell death. *Mol. Cell.* **8**:613–621.

71. Miller, L.K. 1999. An exegesis of IAPs: salvation and surprises from BIR motifs. *Trends Cell Biol.* **9**:323–328.

72. Xu, G., et al. 2001. Covalent inhibition revealed by the crystal structure of the caspase-8/p35 complex. *Nature.* **410**:494–497.

73. Renatus, M., et al. 2000. Crystal structure of the apoptotic suppressor CrmA in its cleaved form. *Structure Fold. Des.* **8**:789–797.

74. Endres, M., et al. 1998. Attenuation of delayed neuronal death after mild focal ischemia in mice by inhibition of the caspase family. *J. Cereb. Blood Flow Metab.* **18**:238–247.

75. Wiessner, C., Sauer, D., Alaimo, D., and Allegrini, P.R. 2000. Protective effect of a caspase inhibitor in models for cerebral ischemia in vitro and in vivo. *Cell. Mol. Biol. (Noisy-le-grand).* **46**:53–62.

76. Rabuffetti, M., et al. 2000. Inhibition of caspase-1-like activity by Ac-Tyr-Val-Ala-Asp-chloromethyl ketone induces long-lasting neuroprotection in cerebral ischemia through apoptosis reduction and decrease of proinflammatory cytokines. *J. Neurosci.* **20**:4398–4404.

77. Hoglen, N.C., et al. 2004. Characterization of IDN-6556 (3-{2-(2-tert-butyl-phenylaminooxa-lyl)-amino]-propionylamino}-4-oxo-5-(2,3,5,6-tetrafluoro-phenoxy)-pentanoic acid): a liver-targeted caspase inhibitor. *J. Pharmacol. Exp. Ther.* **309**:634–640.

78. Leung-Toung, R., Li, W., Tam, T.F., and Karimian, K. 2002. Thiol-dependent enzymes and their inhibitors: a review. *Curr. Med. Chem.* **9**:979–1002.

79. Stockwell, B.R. 2000. Chemical genetics: ligand-based discovery of gene function. *Nat. Rev. Genet.* **1**:116–125.

80. Jiang, X., et al. 2003. Distinctive roles of PHAP proteins and prothymosin-alpha in a death regulatory pathway. *Science.* **299**:223–226.

81. Nguyen, J.T., and Wells, J.A. 2003. Direct activation of the apoptosis machinery as a mechanism to target cancer cells. *Proc. Natl. Acad. Sci. U. S. A.* **100**:7533–7538.

82. Maxim Pharmaceuticals. 2004. Annual Report 2004.

83. Jia, L.T., et al. 2003. Specific tumoricidal activity of a secreted proapoptotic protein consisting of HER2 antibody and constitutively active caspase-3. *Cancer Res.* **63**:3257–3262.

84. Xu, Y.M., et al. 2004. A caspase-6 and anti-human epidermal growth factor receptor-2 (HER2) antibody chimeric molecule suppresses the growth of HER2-overexpressing tumors. *J. Immunol.* **173**:61–67.

85. Schimmer, A.D., et al. 2004. Small-molecule antagonists of apoptosis suppressor XIAP exhibit broad antitumor activity. *Cancer Cell.* **5**:25–35.

86. Glover, C.J., et al. 2003. A high-throughput screen for identification of molecular mimics of Smac/DIABLO utilizing a fluorescence polarization assay. *Anal. Biochem.* **320**:157–169.

87. Tamm, I., et al. 2003. Peptides targeting caspase inhibitors. *J. Biol. Chem.* **278**:14401–14405.

88. Wu, T.Y., Wagner, K.W., Bursulaya, B., Schultz, P.G., and Deveraux, Q.L. 2003. Development and characterization of nonpeptidic small molecule inhibitors of the XIAP/caspase-3 interaction. *Chem. Biol.* **10**:759–767.

89. Reed, J. 2002. Apoptosis-based therapies. *Nat. Rev. Drug Discov.* **1**:111–121.

IKK/NF-κB signaling: balancing life and death — a new approach to cancer therapy

Jun-Li Luo, Hideaki Kamata, and Michael Karin

Laboratory of Gene Regulation and Signal Transduction, Department of Pharmacology and Cancer Center, School of Medicine, UCSD, La Jolla, California, USA.

IκB kinase/NF-κB (IKK/NF-κB) signaling pathways play critical roles in a variety of physiological and pathological processes. One function of NF-κB is promotion of cell survival through induction of target genes, whose products inhibit components of the apoptotic machinery in normal and cancerous cells. NF-κB can also prevent programmed necrosis by inducing genes encoding antioxidant proteins. Regardless of mechanism, many cancer cells, of either epithelial or hematopoietic origin, use NF-κB to achieve resistance to anticancer drugs, radiation, and death cytokines. Hence, inhibition of IKK-driven NF-κB activation offers a strategy for treatment of different malignancies and can convert inflammation-induced tumor growth to inflammation-induced tumor regression.

NF-κB proteins and IκB kinase signaling pathways

The mammalian NF-κB family contains 5 members: NF-κB1 (p105 and p50), NF-κB2 (p100 and p52), c-Rel, RelB, and RelA (p65). These proteins share a Rel homology domain (RHD), which mediates DNA binding, dimerization, and interactions with specific inhibitory factors, the IκBs, which retain NF-κB dimers in the cytoplasm. Many stimuli activate NF-κB, mostly through IκB kinase–dependent (IKK-dependent) phosphorylation and subsequent degradation of IκB proteins. The liberated NF-κB dimers enter the nucleus, where they regulate transcription of diverse genes encoding cytokines, growth factors, cell adhesion molecules, and pro- and antiapoptotic proteins (1, 2). The IKK complex consists of 2 highly homologous kinase subunits, IKKα and IKKβ, and a nonenzymatic regulatory component, IKKγ/NEMO (3).

Two NF-κB activation pathways exist (Figure 1). The first, the classical pathway, is normally triggered in response to microbial and viral infections or exposure to proinflammatory cytokines that activate the tripartite IKK complex, leading to phosphorylation-induced IκB degradation. This pathway, which mostly targets p50:RelA and p50:c-Rel dimers, depends mainly on IKKβ activity (4). The other pathway, the alternative pathway, leads to selective activation of p52:RelB dimers by inducing processing of the NF-κB2/p100 precursor protein, which mostly occurs as a heterodimer with RelB in the cytoplasm. This pathway is triggered by certain members of the TNF cytokine family, through selective activation of IKKα homodimers by the upstream kinase NIK (5). Both pathways regulate cell survival and death (6); the classical pathway is responsible for inhibition of programmed cell death (PCD) under most conditions (2, 3). The alternative pathway is important for survival of premature B cells and development of secondary lymphoid organs (7). The antiapoptotic activity of the IKKβ-driven classical pathway is important for various immu-noreceptors, including T and B cell receptors, TLR4, and type 1 TNF-α receptor (TNFR1), all of which generate pro-survival and pro-death signals upon ligation (8, 9). Under most circumstances, the survival signals dominate, but under conditions where IKKβ or NF-κB activities have been compromised, receptor activation results in cell death (10–12).

The survival function of NF-κB: mechanisms and mediators

Pathways of cell death. PCD can be either apoptotic or necrotic. Apoptosis is characterized by membrane blebbing, shrinking, and condensation of the cell and its organelles (13, 14). Two well-established pathways lead to apoptosis: the death receptor (DR) (extrinsic) pathway and the mitochondrial (intrinsic) pathway (15). Both pathways depend on cysteine proteases called caspases (15, 16). However, apoptosis-like PCD can sometimes proceed without caspase activation (17, 18). Furthermore, caspase activation does not always lead to cell death (19), and caspase-8 also has pro-survival functions (20, 21). Necrosis is characterized by swelling of the cell and its organelles, culminating in membrane disruption and cell lysis, often accompanied by inflammation. Failure of energy metabolism and massive generation of ROS are each thought to cause necrosis (22).

NF-κB suppresses both PCD types, although initially it was thought to antagonize only apoptosis. The first clear evidence for NF-κB as a PCD inhibitor was provided by RelA knockout mice that die mid-gestation by massive liver apoptosis (23). The role of NF-κB in embryonic liver survival, brought about by inhibition of TNFR1-mediated apoptosis (24), is underscored by the very similar phenotypes of mice lacking IKKβ (4, 25) or IKKγ (26). A protective role for NF-κB in adult liver was confirmed in mouse models of liver damage (10, 27, 28) and involves inhibition of both apoptosis and necrosis (9). We will discuss the various mechanisms by which NF-κB suppresses PCD (Figure 2).

NF-κB and caspases. There are 2 groups of DRs, based on their signaling complexes. The first group comprises Fas, DR4, and DR5, which directly recruit the death domain–containing (DD-containing) adaptor Fas-associated death domain (FADD), pro-caspase-8, procaspase-10, and the cellular FLICE-inhibitory protein (FLIP) to form death-inducing signaling complexes (DISCs) (29). The second group comprises TNFR1, DR3, DR6, and ecto-dysplasin A receptor (EDAR). TNFR1 forms a signaling complex

---

Nonstandard abbreviations used: ATO, arsenic trioxide; DD, death domain; DISC, death-inducing signaling complex; DR, death receptor; FADD, Fas-associated death domain; FHC, ferritin heavy chain; FLIP, FLICE-inhibitory protein; IKK, IκB kinase; MKP, MAPK phosphatase; MnSOD, manganese superoxide dismutase; PCD, programmed cell death; RHD, Rel homology domain; RIP, receptor-interacting protein; SOD, superoxide dismutase; TNFR1, type 1 TNF-α receptor; TRAF, TNFR-associated factor; XIAP, X chromosome–linked inhibitor of apoptosis.

Conflict of interest: The authors have declared that no conflict of interest exists.

Citation for this article: *J. Clin. Invest.* 115:2625–2632 (2005). doi:10.1172/JCI26322.

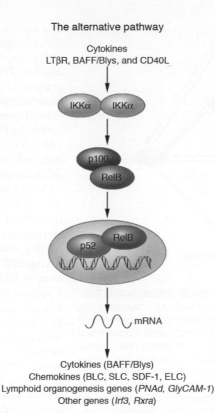

The classical pathway

Proinflammatory cytokines (TNF-α, IL-1β)
Viruses, TLRs, antigen receptors

Cytokines (TNF-α, IL-1β, IL-6, GM-CSF)
Chemokines (IL-8, RANTES, MIP-1α, MCP-1)
Adhesion molecules (VCAM-1, ICAM-1, E-selectin)
Enzymes (iNOS, COX-2, PLA2)

The alternative pathway

Cytokines
LTβR, BAFF/Blys, and CD40L

Cytokines (BAFF/Blys)
Chemokines (BLC, SLC, SDF-1, ELC)
Lymphoid organogenesis genes (*PNAd, GlyCAM-1*)
Other genes (*Irf3, Rxra*)

Figure 1

IKK/NF-κB signaling pathways. The classical pathway is activated by a variety of inflammatory signals, resulting in coordinate expression of multiple inflammatory and innate immune genes. The alternative pathway is strictly dependent on IKKα homodimers and is activated by lymphotoxin β receptor (LTβR), B cell–activating factor belonging to the TNF family (BAFF), and CD40 ligand (CD40L). The alternative pathway plays a central role in the expression of genes involved in development and maintenance of secondary lymphoid organs. BLC, B lymphocyte chemoattractant; ELC, Epstein-Barr virus–induced molecule 1 ligand CC chemokine; MCP-1, monocyte chemoattractant protein-1; MIP-1α, macrophage inflammatory protein-1α; PLA2, phospholipase A2; SDF-1, stromal cell–derived factor-1α; SLC, secondary lymphoid tissue chemokine.

(complex I) at the plasma membrane by recruiting the adaptor TNFR1-associated DD protein (TRADD) and the signaling proteins TNFR-associated factor 2 (TRAF2), TRAF5, and receptor-interacting protein 1 (RIP1). After assembly, complex I dissociates from TNFR1, which can then recruit FADD and caspase-8 and trigger an apoptotic response (30).

NF-κB as a transcription factor induces genes whose products prevent PCD. An elicitor of NF-κB activation is TNF-α, which is a rather poor inducer of PCD. TNF-α triggers PCD only when new protein or RNA synthesis is inhibited or in NF-κB–deficient cells. NF-κB exerts its pro-survival activity through several anti-apoptotic proteins, including FLIP, Bcl-X$_L$, A1/Bfl-1, cellular inhibitor of apoptosis (c-IAP), X chromosome–linked inhibitor of apoptosis (XIAP), TRAF1, and TRAF2 (2, 31). FLIP inhibits apoptosis by interfering with caspase-8 activation (30). c-IAP and XIAP directly bind and inhibit effector caspases, acting downstream of initiator caspases.

NF-κB and Bcl-2 family members. NF-κB induces expression of several Bcl-2 family members, most notably Bcl-X$_L$ and A1/Bfl-1, which prevent apoptosis by inhibiting permeability transition and depolarization of mitochondria, and cytochrome *c* release (2, 31). DRs can trigger apoptosis through different pathways (32). In certain cells, activated caspase-8 directly activates effector caspases, while in cells with poor DISC formation, death signaling requires an additional amplification loop, based on caspase-8–mediated Bid cleavage and generation of truncated tBid that triggers cytochrome *c* release (33, 34) and activation of caspase-9 and caspase-3 (15). This type of DR signaling can be blocked by antiapoptotic Bcl-2 family members, such as Bcl-2 and Bcl-X$_L$ (15).

ROS and the NF-κB–JNK cross-talk. The role of JNK in PCD has been controversial, because it has both survival and death-enhancing effects. The clearest evidence for JNK as regulator of PCD

comes from analysis of knockout mice: JNK1- or JNK2-deficient mice are relatively resistant to induction of fulminant hepatitis in response to concanavalin A, a pathology that depends on activation of TNFR1 and other DRs (10).

The ratio between JNK and NF-κB activities controls cell survival or death, not only in response to TNFR1 but also in response to other death stimuli (35–37). Whereas TNF-α leads to transient JNK activation in WT cells, it leads to prolonged JNK activation in cells that cannot activate NF-κB (9, 38–40). The pro-survival activity of NF-κB depends on this ability to prevent prolonged JNK activation (9, 38–40). Prolonged JNK activation following concanavalin A administration was also seen in mice lacking IKKβ in liver cells, resulting in massive TNFR1-dependent hepatocyte death (10). In the liver, however, TNFR1 and JNK signaling is also required for regeneration or compensatory hepatocyte proliferation following partial hepatectomy or chemically induced injury (41, 42). Thus, NF-κB may be a critical regulator of cell survival and death through its ability to control the duration of JNK activation (Figure 2).

Prolonged JNK activation in NF-κB–deficient cells implies that NF-κB induces expression of JNK inhibitors. Such a function was proposed for GADD45β (43) and XIAP (39). However, analysis of GADD45β- or XIAP-deficient fibroblasts failed to reveal changes in the kinetics of JNK activation (31, 44), suggesting that NF-κB regulates JNK activation through a different mediator.

ROS are likely the mediators (40). ROS, including H_2O_2, O_2^-, and HO$^•$ radicals, are generated through many enzymatic pathways, but their major source is leakage of electrons from the mitochondrial respiratory chain (45). ROS activate kinases through oxidation of kinase-interacting molecules. For instance, ROS activate tyrosine kinases by inactivating protein tyrosine phosphatases through oxidation of a highly reactive cysteine residue at their

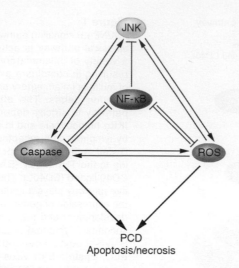

Figure 2
Control of cell survival and death through NF-κB–JNK cross-talk. Positive feedback loops exist between ROS and caspases, caspases and JNK, and JNK and ROS. Negative feedback loops exist between NF-κB and caspases, and NF-κB and ROS. NF-κB functions as a pro-survival transcription factor by inducing the expression of antiapoptotic genes, such as the Bcl-2 family members and caspase inhibitors, and antioxidant genes, such as MnSOD and FHC. Activation of NF-κB also results in inhibition of prolonged JNK activation, mostly through inhibition of ROS accumulation. Inhibition of NF-κB enhances PCD, which can be either apoptotic or necrotic, by removing the negative feedback loops.

catalytic site (46–49). Similarly, ROS mediate the NF-κB–JNK cross-talk through their ability to inactivate various MAPK phosphatases (MKPs) involved in JNK inactivation (45).

TNF-α induces ROS accumulation in many cell types, and these ROS are important mediators of PCD (22). TNF-α–induced ROS accumulation is seen in NF-κB–deficient cells, but not in NF-κB–competent cells (9, 40). Treatment of cells with the antioxidant butylated hydroxyanisole (BHA) has no effect on transient JNK activation triggered by TNF-α, but it suppresses prolonged JNK activation and PCD in TNF-α–treated NF-κB–deficient cells (9, 40). This protective effect is due to BHA's ability to prevent oxidation of MKPs, ensuring transient JNK activation (9). Expression of dominant-negative mutants of MKPs leads to prolonged JNK activation and allows killing by TNF-α of NF-κB–competent cells, which otherwise are TNF-α–resistant (9).

The loss of NF-κB activity results in ROS accumulation because NF-κB induces expression of several antioxidant genes such as manganese superoxide dismutase (MnSOD), ferritin heavy chain (FHC), glutathione-S-transferase, and metallothionein (50). Overexpression of mitochondrial MnSOD protects cells from TNF-α–induced cytotoxicity (9, 51). Overexpression of FHC also suppresses TNF-α–induced PCD along with attenuation of prolonged JNK activation (52). Another interesting observation is that TNF-α induces expression of a number of cytochrome p450 family members, such as CYP1B1, that enhance ROS production in NF-κB–deficient fibroblasts (37). Taken together, these findings show that NF-κB protects cells from oxidative stress by activating expression of various antioxidant systems, whose failure enhances TNF-α–induced PCD.

The mechanism of TNF-α–induced ROS production is unclear. One possible source of ROS is the cytosolic phospholipase A2 (53). However, several lines of evidence suggest that mitochon-

dria are the main source of ROS during TNF-α–induced PCD (22). Compared with our understanding of DR-induced caspase activation, the mechanism of DR-induced ROS production is obscure. TNF-α does not induce ROS accumulation and programmed necrosis in FADD- or RIP1-deficient cells, indicating essential roles for FADD and RIP1 (54). In contrast to the established function of RIP1 as an adaptor molecule in NF-κB activation, its kinase activity is necessary for Fas-induced necrosis, which mostly occurs in caspase-8–deficient cells (55). Interestingly, inhibition of caspases potentiates ROS accumulation and cell death (53, 56). Although DRs can induce ROS accumulation without caspase activation in certain cell types, caspase activation can also lead to mitochondrial damage and ROS accumulation (33, 34, 57–59). Thus, caspase-dependent and -independent mechanisms might be involved in ROS accumulation.

JNK activation may also enhance ROS accumulation, potentiating TNF-α–stimulated necrosis (60). Although the mechanism by which JNK potentiates ROS accumulation is unclear, a positive feedback loop between ROS accumulation and JNK activation may exist (Figure 2). Such a loop may also involve caspase activation. Although caspases are not involved in TNF-α–induced prolonged JNK activation in NF-κB–deficient cells (40), caspase-mediated cleavage of upstream MAP3Ks may cause constitutive JNK activation (61). JNK activation also contributes to caspase activation, an effect mediated through enhanced cytochrome c release, during UV-induced apoptosis (62). Alternatively, JNK causes caspase activation through jBid formation during TNFR1 signaling (63). Importantly, NF-κB suppresses all of these amplification loops by inducing expression of caspase inhibitors, Bcl-2 family members, and antioxidants (Figure 2). Interestingly, negative feedback loops exist between NF-κB and various death-promoting proteins. Caspase-mediated cleavage of RelA and IKKβ can prevent NF-κB activation (64, 65). Caspases can also cleave IκB to generate a degradation-resistant NF-κB inhibitor (66). Oxidation of a cysteine residue in the RHD of RelA prevents its binding to DNA (67), whereas oxidation of another cysteine within the activation loop of IKKβ interferes with its activation (68). It is unlikely, however, that all of these regulatory loops and modifications take place simultaneously, and a major challenge for the future is to sort out the events that do take place during different physiological and pathophysiological conditions. It is possible to use some of these regulatory loops in designing drugs and therapeutic strategies to kill cancer cells. Fas and TNF-α can induce both apoptosis and necrosis, and so do anticancer drugs. In L929 cells, for instance, TNF-α triggers mostly necrosis, whereas Fas can induce necrosis only when the apoptotic pathway is suppressed (69). FADD and RIP play central roles in controlling the choice between the 2 death pathways (70). NF-κB activation also inhibits programmed necrosis, in addition to its role in prevention of apoptosis.

Proapoptotic functions of NF-κB?

NF-κB may induce apoptosis in a cell type– and stimulus-dependent manner. Most commonly, NF-κB activation inhibits PCD, as evidenced by several knockout mouse models (4, 23, 26, 71). However, under certain circumstances activation of NF-κB may promote cell death. For instance, NF-κB may mediate doxorubicin-induced cell death in N-type neuroblastoma cells (72). NF-κB is also required for anti-CD3–induced apoptosis of double-positive thymocytes (73). Apoptosis in HL-60 cells induced by etoposide or

1-β-D-arabinofuranosylcytosine correlates with NF-κB activation (74). Human melanoma cells were protected from UV-induced apoptosis by NF-κB downregulation (75). More recently, it was reported that NF-κB induced by UV light or daunorubicin/doxorubicin is functionally distinct from the response elicited by TNF-α, and under such conditions NF-κB may become a repressor of antiapoptotic genes (76). Furthermore, UV light and daunorubicin inhibit TNF-α–induced NF-κB transcriptional activity, which is antiapoptotic, by enhancing association of RelA with histone deacetylases (76). These results suggest that NF-κB may mediate apoptosis under certain conditions. However, the pathophysiological relevance of these observations is not clear, and it remains to be demonstrated that NF-κB has proapoptotic functions in vivo. It appears that those agents or stimuli that were reported to induce apoptosis by activating NF-κB are neither strong nor typical NF-κB activators, as opposed to TNF-α, IL-1, or LPS, and that they also activate another signaling pathway(s), which may be more relevant to cell killing than NF-κB.

Tumor suppressors interact with NF-κB pathway. Suppression of cell proliferation, induction of premature senescence, and/or induction of apoptosis are some mechanisms through which tumor suppressors inhibit cancer development. In general, NF-κB acts antagonistically to tumor suppressors, based on its ability to promote cell survival, inhibit PCD, and enhance cell proliferation (77). However, in some cases NF-κB may collaborate with, rather than antagonize, certain tumor suppressors.

Although p53 stabilization decreases upon NF-κB activation (78), under special circumstances apoptosis induced by p53 may involve activation of NF-κB (79). Similar to situations in which NF-κB activation promotes apoptosis, NF-κB induction by p53 does not involve classical IKK activation and IκB degradation. Instead, p53 may stimulate the serine/threonine kinase ribosomal S6 kinase 1 (RSK1), which in turn phosphorylates RelA (80). The lower affinity of RSK1-phosphorylated RelA for IκBα decreases IκBα-mediated nuclear export, prolonging RelA nuclear residence (80). NF-κB also plays an essential role in activation of p53 to initiate proapoptotic signaling in response to ROS accumulation. Consequently, NF-κB–dependent p53 activity induces p53-regulated genes, such as Puma and $p21^{waf1}$ (81). However, a more common observation, seen in vivo, is that NF-κB activation counteracts p53-induced apoptosis by destabilizing p53, perhaps through enhanced Mdm2 expression (78, 82).

Another tumor suppressor, BRCA1, can bind RelA to serve as a coactivator (83). Treatment of 293T cells with TNF-α induces an interaction between endogenous RelA and BRCA1, mediated by the RHD of RelA and the N-terminal region of BRCA1. Forced BRCA1 expression significantly enhances the ability of TNF-α or IL-1β to induce NF-κB target genes, and inhibition of NF-κB by the chemical inhibitor SN-50 blocks this effect (83). Nonetheless, it remains to be seen whether any of these responses documented in vitro occurs in vivo.

NF-κB and proapoptotic genes. NF-κB has been implicated as a transcriptional activator of some proapoptotic genes, such as Fas/CD95 (84), FasL (85), DR4, and DR5 (86). FasL is expressed in activated T cells and represents a major cytotoxic effector through which T cells kill their targets. FasL expression is under the stringent control of various transcription factors, including NF-κB (85). Recently, it was reported that certain types of cancer cells also express FasL, which may contribute to their ability to escape immune surveillance and resist immunotherapy. Overexpression of the Myc family member Max in non–small cell lung cancer cell lines markedly increases basal FasL promoter activity and enhances NF-κB–mediated FasL induction. Thus, high levels of Max and stress-induced NF-κB activation may elevate FasL expression in human lung cancer cells (85). TNF-α combined with IFN-α accelerates NF-κB–mediated apoptosis by enhancing Fas expression in human colon adenocarcinoma RPMI4788 cells (84). However, there may be another explanation for these results, as type I IFNs and related cytokines, such as IL-10, may actually function as NF-κB inhibitors (87). Another TNF family member, TRAIL, triggers apoptosis through engagement of DR4 and DR5. The c-Rel subunit of NF-κB induces expression of both receptors, while a degradation-resistant mutant of IκBα (IκB super-repressor) or a transactivation-deficient mutant of c-Rel reduces DR expression (86). However, NF-κB was shown to be a major impediment to TRAIL-mediated tumor killing (88).

IKK/NF-κB and cancer

IKK/NF-κB links inflammation to cancer. Based on many functions of NF-κB target genes, a close relationship between NF-κB and cancer was proposed (89) and recently reviewed (89–94). The association of NF-κB activation with inflammation-associated tumor promotion, progression, and metastasis is well documented and was demonstrated in several mouse models (88, 95, 96). The IKKβ-dependent NF-κB activation pathway is a critical molecular link between inflammation and colon cancer in a mouse model (95). Activation of IKKβ in enterocytes, which give rise to the malignant component of this tumor, suppresses apoptosis of preneoplastic cells, whereas its activation in myeloid cells promotes production of various cytokines that serve as growth factors for the transformed enterocytes. Inhibition of 1 of these factors, IL-6, interferes with tumor growth but has no effect on tumor cell survival (97). Conversely, inactivation of IKKβ in enterocytes results in a dramatic decrease in tumor number due to increased apoptosis but has no effect on proliferation of transformed enterocytes or tumor growth (95).

The role of NF-κB in inflammation-associated cancer was also demonstrated in Mdr2-deficient mice, which develop cholestatic hepatitis followed by hepatocellular carcinoma (96). In this model, the inflammatory process triggered chronic activation of NF-κB in hepatocytes, most likely through enhanced production of TNF-α by adjacent endothelial and inflammatory cells. Switching NF-κB off in *Mdr2*$^{-/-}$ mice from birth to 7 months of age had no effect on the course of hepatitis or early phases of tumorigenesis (96). By contrast, suppressing chronic NF-κB activation at later stages resulted in the apoptotic death of transformed hepatocytes and failure to progress to hepatocellular carcinoma (96).

NF-κB activation also plays a critical role in inflammation-driven tumor progression as demonstrated in a syngeneic colon and mammary cancer xenograft mouse model (88). Cancer cells in this model were introduced into syngeneic immunocompetent mice to form metastatic growths in the lungs. Once the metastases were established, the mice were given a sublethal dose of LPS to elicit systemic inflammation, which stimulated tumor growth. Remarkably, inhibition of NF-κB in cancer cells converted LPS-induced tumor growth to LPS-induced tumor regression without affecting the ability of the cancer cells to migrate to the lung and establish metastatic growths (88). Further investigation revealed that inflammation-induced tumor growth in this model was mediated by TNF-α produced by host immune cells, whereas LPS-induced

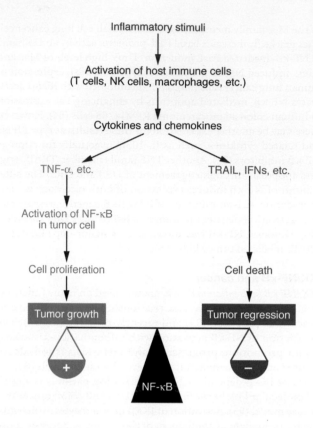

Figure 3
Inhibition of NF-κB in cancer cells converts inflammation-induced
tumor growth to tumor regression. Activation of the innate and adaptive immune system can have profound influence on tumor growth and
development. In addition to its role in activation of immune cells, NF-κB
within the malignant cell is a major modulator of the tumor response to
inflammation. Activation of NF-κB promotes tumor growth and confers
resistance to death cytokines, such as TRAIL. Conversely, inhibition of
NF-κB prevents inflammation-stimulated tumor growth and enhances
inflammation-induced tumor regression mediated by TRAIL.

in epidermal keratinocytes promotes keratinocyte growth arrest
and differentiation to maintain the barrier function of the epidermis, whose perturbation may result in severe inflammation (105).
Interestingly, formation of mouse squamous cell carcinomas in
response to a chemical carcinogen (106) and following inhibition
of NF-κB (107) is dependent on TNF-α. Similar observations were
recently made in a mouse model of chemically induced hepatocellular carcinoma, where deletion of IKKβ in hepatocytes promoted tumor development by enhancing compensatory proliferation, whereas an additional deletion of IKKβ in liver myeloid
cells prevented tumor development by depriving the transformed
hepatocytes of essential growth factors (108).

NF-κB inhibitors in cancer therapy. The pivotal role of the IKKβ/NF-κB
signaling pathway in inhibition of PCD, tumor promotion, and
tumor progression, together with the occurrence of constitutively
activated NF-κB in various solid and hematopoietic malignancies,
strongly suggests that IKKβ and/or NF-κB inhibitors would be
useful in cancer therapy. In fact, much effort is currently invested
in developing various IKKβ and/or NF-κB inhibitors and testing their efficacy in both animal models and human cancer (109,
110). Many inhibitors currently available are not specific for either
IKKβ or NF-κB. These include antiinflammatory agents such as
sulfasalazine and trans-resveratrol, NSAIDs such as aspirin and
sulindac sulfide, cyclopentenone prostaglandins, proteasome
inhibitors, and glucocorticoids (90, 109–112). However, specific
IKKβ inhibitors are being developed, and a few publications have
documented their efficacy in triggering apoptosis in cancer cell
lines in combination with either death-inducing cytokines or chemotherapeutic drugs (113–115).

Even nonspecific IKKβ/NF-κB inhibitors may be effective when
used as adjuvants with conventional anticancer treatments. As
many signaling pathways may be simultaneously activated and/or
inactivated in a given malignant cell, collectively contributing to
its neoplastic phenotype, nonspecific IKKβ/NF-κB inhibitors may
affect several signaling pathways at once and lead to much more
effective killing of such cells. The anticancer drug arsenic trioxide
(ATO), which is useful for treating promyelocytic leukemia (116)
and possibly multiple myeloma (117), is a noteworthy example.
ATO is not a specific inhibitor for IKKβ or NF-κB and may have
several molecular targets, since it was found that trivalent arsenicals, a chemical class to which ATO belongs, are potent JNK activators (118) as well as IKKβ inhibitors (68). JNK activation in this
case is mostly due to the ability of trivalent arsenicals to directly
interact with the catalytic cysteine of JNK phosphatases, whereas
in the case of IKKβ the target is the aforementioned reactive cysteine within the activation loop. An additional effect of ATO on JNK
activity may be due to NF-κB inhibition and accumulation of ROS
(35). Thus, by inhibiting IKK and activating JNK, ATO may trigger
apoptosis in many different types of cancers. Since NF-κB inhibi-

regression of NF-κB–deficient tumors was mediated by TRAIL,
whose synthesis is induced by IFNs in host inflammatory cells in
response to LPS-mediated activation of TLR4 (Figure 3). These
results indicate that NF-κB is a major mediator of inflammation-
induced tumor progression through overcoming the potential
tumor-killing by TRAIL induction (88). Given that NF-κB activation in cancer cells may be a major hindrance to TRAIL-induced
apoptosis, NF-κB or IKK inhibitors may potentiate the activity of
either recombinant TRAIL or TRAIL inducers, such as type I and
type II IFNs, to achieve enhanced tumor killing (Figure 3).

Whereas the role of NF-κB in inflammation-induced tumor promotion, growth, and progression is becoming clear (88, 95, 96), its
role in tumor initiation is still ambiguous (98–101). As NF-κB regulates a large group of genes that have different functions, some of
which display cell-type specificity, NF-κB may have distinct roles
in different cell types. For instance, in normal epidermal keratinocytes, NF-κB proteins are present in the cytoplasm of basal cells
but are nuclear in more differentiated suprabasal cells, suggesting
that NF-κB activation is linked to growth arrest (98, 102). Indeed,
inhibition of NF-κB signaling in the murine epidermis results in
an increased apoptosis, hyperproliferation of surviving cells, and
spontaneous development of squamous cell carcinomas (99, 103).
Correspondingly, application of a pharmacological NF-κB inhibitor to mouse skin induced epidermal hyperplasia (98). In contrast,
overexpression of active NF-κB subunits in transgenic epithelium
produced hypoplasia and growth inhibition (98). Contrary to the
requirement of an intact IKK/NF-κB pathway for H-ras–mediated fibroblast transformation (104), NF-κB inhibition synergized
with oncogenic H-ras to induce transformation of primary human
keratinocytes (100). Congruously, activation of NF-κB in normal
human epidermal keratinocytes triggered cell-cycle arrest (100).
These results suggest that the IKKβ-dependent NF-κB pathway

tion usually does not result in spontaneous apoptosis, it is unlikely that even specific IKKβ/NF-κB inhibitors would be functional as monotherapeutic agents in most cancers. Indeed, using Jurkat cells as a model, the IKKβ inhibitor AS602868 was not cytocidal on its own but strongly potentiated killing by TNF-α (113). Based on our analysis of knockout mice and tumor models, we predict that IKKβ/NF-κB inhibitors will be useful adjuvants for conventional chemotherapeutic drugs, ionizing radiation, or tumoricidal cytokines, such as IFNs or TRAIL (Figure 3).

NF-κB regulates PCD through a cross-talk with JNK, ROS, and caspases, and an important pro-survival factor regulated by NF-κB is the antioxidant enzyme MnSOD (9, 51). Thus, MnSOD2 inhibitors may target a particular NF-κB function, the suppression of ROS production and PCD, while leaving other functions, such as innate immunity, intact. In fact, inhibition of superoxide dismutase (SOD) in human leukemia cells caused accumulation of O_2^-, which was followed by ROS-mediated mitochondrial damage, cytochrome c release, and apoptosis (119). Given its regulation by NF-κB, whose activity is elevated in most types of cancer, it is likely that MnSOD expression is higher in malignant cells than in normal cells, and therefore the former may be more sensitive to SOD inhibitors. In fact, certain estrogen derivatives, acting as SOD inhibitors, selectively kill human leukemia cells but not normal lymphocytes (119). In case such compounds are not sufficiently potent on their own, they need to be tested as adjuvants for more conventional chemotherapeutic and radiotherapeutic approaches.

Concluding remarks

The inhibition of IKKβ/NF-κB appears to be a promising strategy for cancer therapy when combined with established cytocidal drugs, death cytokines, or therapeutic radiation. Certain anti-

cancer drugs may work much better with IKKβ/NF-κB inhibitors than others. For instance, the combined application of TRAIL or TRAIL inducers, such as IFNs (Figure 3), with antiinflammatory or anti–TNF-α therapy alongside IKKβ/NF-κB inhibitors may result in selective killing of malignant cells not achieved by either agent alone (88). An important advantage of IKKβ/NF-κB inhibitors over conventional therapeutics is their ability to block NF-κB activation also in infiltrating inflammatory cells, which are an important source of tumor growth and survival factors. It should be noted, however, that, given the critical role of NF-κB in innate and adaptive immune responses, there may be a certain amount of risk due to induced immunodeficiency caused by long-term use of IKKβ/NF-κB inhibitors. Hence, alternative approaches should be considered. For instance, an approach based on selective inhibition of antiapoptotic targets of NF-κB, without affecting target genes required for immune responses, would be particularly attractive.

Acknowledgments

Work in the authors' laboratory was supported by grants from the NIH (ES04151, AI43477, CA76188), the US Army Medical Research and Materiel Command (W8IXWH-04-1-0120), and the Prostate Cancer Foundation. M. Karin is an American Cancer Society Research Professor.

Address correspondence to: Michael Karin, Laboratory of Gene Regulation and Signal Transduction, Department of Pharmacology and Cancer Center, School of Medicine, University of California, San Diego, 9500 Gilman Drive, La Jolla, California 92093, USA. Phone: (858) 534-1361; Fax: (858) 534-8158; E-mail: karinoffice@ucsd.edu.

1. Ghosh, S., and Karin, M. 2002. Missing pieces in the NF-kappaB puzzle [review]. *Cell.* **109**(Suppl.): S81–S96.

2. Karin, M., and Lin, A. 2002. NF-kappaB at the crossroads of life and death. *Nat. Immunol.* **3**:221–227.

3. Karin, M., and Delhase, M. 2000. The I kappa B kinase (IKK) and NF-kappa B: key elements of proinflammatory signalling. *Semin. Immunol.* **12**:85–98.

4. Li, Z.W., et al. 1999. The IKKbeta subunit of Ikappa B kinase (IKK) is essential for nuclear factor kappaB activation and prevention of apoptosis. *J. Exp. Med.* **189**:1839–1845.

5. Senftleben, U., et al. 2001. Activation by IKKalpha of a second, evolutionary conserved, NF-kappa B signaling pathway. *Science.* **293**:1495–1499.

6. Senftleben, U., and Karin, M. 2002. The IKK/NF-kappa B pathway. *Crit. Care Med.* **30**(1 Suppl.):S18–S26.

7. Bonizzi, G., and Karin, M. 2004. The two NF-kappaB activation pathways and their role in innate and adaptive immunity. *Trends Immunol.* **25**:280–288.

8. Varfolomeev, E.E., and Ashkenazi, A. 2004. Tumor necrosis factor: an apoptosis JuNKie [review]? *Cell.* **116**:491–497.

9. Kamata, H., et al. 2005. Reactive oxygen species promote TNFalpha-induced death and sustained JNK activation by inhibiting MAP kinase phosphatases. *Cell.* **120**:649–661.

10. Maeda, S., et al. 2003. IKKbeta is required for prevention of apoptosis mediated by cell-bound but not by circulating TNFalpha. *Immunity.* **19**:725–737.

11. Weil, R., and Israel, A. 2004. T-cell-receptor- and B-cell-receptor-mediated activation of NF-kappaB in lymphocytes. *Curr. Opin. Immunol.* **16**:374–381.

12. Hsu, L.C., et al. 2004. The protein kinase PKR is required for macrophage apoptosis after activation of Toll-like receptor 4. *Nature.* **428**:341–345.

13. Kerr, J.F., Wyllie, A.H., and Currie, A.R. 1972. Apoptosis: a basic biological phenomenon with wide-ranging implications in tissue kinetics [review]. *Br. J. Cancer.* **26**:239–257.

14. Wyllie, A.H., Kerr, J.F., and Currie, A.R. 1980. Cell death: the significance of apoptosis [review]. *Int. Rev. Cytol.* **68**:251–306.

15. Danial, N.N., and Korsmeyer, S.J. 2004. Cell death: critical control points. *Cell.* **116**:205–219.

16. Hengartner, M.O. 2000. The biochemistry of apoptosis. *Nature.* **407**:770–776.

17. Leist, M., and Jaattela, M. 2001. Four deaths and a funeral: from caspases to alternative mechanisms [review]. *Nat. Rev. Mol. Cell Biol.* **2**:589–598.

18. Jaattela, M., and Tschopp, J. 2003. Caspase-independent cell death in T lymphocytes. *Nat. Immunol.* **4**:416–423.

19. Abraham, M.C., and Shaham, S. 2004. Death without caspases, caspases without death. *Trends Cell Biol.* **14**:184–193.

20. Yu, L., et al. 2004. Regulation of an ATG7-beclin 1 program of autophagic cell death by caspase-8. *Science.* **304**:1500–1502.

21. Kang, T.B., et al. 2004. Caspase-8 serves both apoptotic and nonapoptotic roles. *J. Immunol.* **173**:2976–2984.

22. Fiers, W., Beyaert, R., Declercq, W., and Vandenabeele, P. 1999. More than one way to die: apoptosis, necrosis and reactive oxygen damage [review]. *Oncogene.* **18**:7719–7730.

23. Beg, A.A., Sha, W.C., Bronson, R.T., Ghosh, S., and Baltimore, D. 1995. Embryonic lethality and liver degeneration in mice lacking the RelA component of NF-kappa B. *Nature.* **376**:167–170.

24. Doi, T.S., et al. 1999. Absence of tumor necrosis factor rescues RelA-deficient mice from embryonic lethality. *Proc. Natl. Acad. Sci. U. S. A.* **96**:2994–2999.

25. Li, Q., Van Antwerp, D., Mercurio, F., Lee, K.F., and Verma, I.M. 1999. Severe liver degeneration in mice lacking the IkappaB kinase 2 gene. *Science.* **284**:321–325.

26. Makris, C., et al. 2000. Female mice heterozygous for IKK gamma/NEMO deficiencies develop a dermatopathy similar to the human X-linked disorder incontinentia pigmenti. *Mol. Cell.* **5**:969–979.

27. Chaisson, M.L., Brooling, J.T., Ladiges, W., Tsai, S., and Fausto, N. 2002. Hepatocyte-specific inhibition of NF-kappaB leads to apoptosis after TNF treatment, but not after partial hepatectomy. *J. Clin. Invest.* **110**:193–202. doi:10.1172/JCI200215295.

28. Lavon, I., et al. 2000. High susceptibility to bacterial infection, but no liver dysfunction, in mice compromised for hepatocyte NF-kappaB activation. *Nat. Med.* **6**:573–577.

29. Peter, M.E., and Krammer, P.H. 2003. The CD95 (APO-1/Fas) DISC and beyond. *Cell Death Differ.* **10**:26–35.

30. Micheau, O., and Tschopp, J. 2003. Induction of TNF receptor I-mediated apoptosis via two sequential signaling complexes. *Cell.* **114**:181–190.

31. Kucharczak, J., Simmons, M.J., Fan, Y., and Gelinas, C. 2003. To be, or not to be: NF-kappaB is the answer. Role of Rel/NF-kappaB in the regulation of apoptosis [review]. *Oncogene.* **22**:8961–8982.

32. Scaffidi, C., et al. 1998. Two CD95 (APO-1/Fas) signaling pathways. *EMBO J.* **17**:1675–1687.

33. Luo, X., Budihardjo, I., Zou, H., Slaughter, C., and Wang, X. 1998. Bid, a Bcl2 interacting protein, mediates cytochrome c release from mitochondria in response to activation of cell surface death receptors. *Cell.* **94**:481–490.

34. Li, H., Zhu, H., Xu, C.J., and Yuan, J. 1998. Cleavage of BID by caspase 8 mediates the mitochondrial damage in the Fas pathway of apoptosis. *Cell.*

94:491–501.

35. Zhang, Y., and Chen, F. 2004. Reactive oxygen species (ROS), troublemakers between nuclear factor-kappaB (NF-kappaB) and c-Jun NH(2)-terminal kinase (JNK). *Cancer Res.* **64**:1902–1905.

36. Papa, S., Zazzeroni, F., Pham, C.G., Bubici, C., and Franzoso, G. 2004. Linking JNK signaling to NF-kappaB: a key to survival. *J. Cell Sci.* **117**:5197–5208.

37. Chen, F., Castranova, V., Li, Z., Karin, M., and Shi, X. 2003. Inhibitor of nuclear factor kappaB kinase deficiency enhances oxidative stress and prolongs c-Jun NH2-terminal kinase activation induced by arsenic. *Cancer Res.* **63**:7689–7693.

38. De Smaele, E., et al. 2001. Induction of gadd45beta by NF-kappaB downregulates pro-apoptotic JNK signalling. *Nature.* **414**:308–313.

39. Tang, G., et al. 2001. Inhibition of JNK activation through NF-kappaB target genes. *Nature.* **414**:313–317.

40. Sakon, S., et al. 2003. NF-kappaB inhibits TNF-induced accumulation of ROS that mediate prolonged MAPK activation and necrotic cell death. *EMBO J.* **22**:3898–3909.

41. Schwabe, R.F., et al. 2003. c-Jun-N-terminal kinase drives cyclin D1 expression and proliferation during liver regeneration. *Hepatology.* **37**:824–832.

42. Yamada, Y., Kirillova, I., Peschon, J.J., and Fausto, N. 1997. Initiation of liver growth by tumor necrosis factor: deficient liver regeneration in mice lacking type I tumor necrosis factor receptor. *Proc. Natl. Acad. Sci. U. S. A.* **94**:1441–1446.

43. Papa, S., et al. 2004. Gadd45 beta mediates the NF-kappa B suppression of JNK signalling by targeting MKK7/JNKK2. *Nat. Cell Biol.* **6**:146–153.

44. Amanullah, A., et al. 2003. Cell signalling: cell survival and a Gadd45-factor deficiency [comment]. *Nature.* **424**:741; discussion, 742.

45. Kamata, H., and Hirata, H. 1999. Redox regulation of cellular signalling. *Cell. Signal.* **11**:1–14.

46. Salmeen, A., et al. 2003. Redox regulation of protein tyrosine phosphatase 1B involves a sulphenyl-amide intermediate. *Nature.* **423**:769–773.

47. van Montfort, R.L., Congreve, M., Tisi, D., Carr, R., and Jhoti, H. 2003. Oxidation state of the active-site cysteine in protein tyrosine phosphatase 1B. *Nature.* **423**:773–777.

48. Meng, T.C., Fukada, T., and Tonks, N.K. 2002. Reversible oxidation and inactivation of protein tyrosine phosphatases in vivo. *Mol. Cell.* **9**:387–399.

49. Meng, T.C., Buckley, D.A., Galic, S., Tiganis, T., and Tonks, N.K. 2004. Regulation of insulin signaling through reversible oxidation of the protein-tyrosine phosphatases TC45 and PTP1B. *J. Biol. Chem.* **279**:37716–37725.

50. Sasazuki, T., et al. 2004. Genome wide analysis of TNF-inducible genes reveals that antioxidant enzymes are induced by TNF and responsible for elimination of ROS. *Mol. Immunol.* **41**:547–551.

51. Wong, G.H., Elwell, J.H., Oberley, L.W., and Goeddel, D.V. 1989. Manganous superoxide dismutase is essential for cellular resistance to cytotoxicity of tumor necrosis factor. *Cell.* **58**:923–931.

52. Pham, C.G., et al. 2004. Ferritin heavy chain upregulation by NF-kappaB inhibits TNFalpha-induced apoptosis by suppressing reactive oxygen species. *Cell.* **119**:529–542.

53. Cauwels, A., Janssen, B., Waeytens, A., Cuvelier, C., and Brouckaert, P. 2003. Caspase inhibition causes hyperacute tumor necrosis factor-induced shock via oxidative stress and phospholipase A2. *Nat. Immunol.* **4**:387–393.

54. Lin, Y., et al. 2004. Tumor necrosis factor-induced nonapoptotic cell death requires receptor-interacting protein-mediated cellular reactive oxygen species accumulation. *J. Biol. Chem.* **279**:10822–10828.

55. Holler, N., et al. 2000. Fas triggers an alternative, caspase-8-independent cell death pathway using the kinase RIP as effector molecule. *Nat. Immunol.*

1:489–495.

56. Liu, C.Y., et al. 2003. Broad-spectrum caspase inhibition paradoxically augments cell death in TNF-alpha-stimulated neutrophils. *Blood.* **101**:295–304.

57. Goldstein, J.C., Waterhouse, N.J., Juin, P., Evan, G.I., and Green, D.R. 2000. The coordinate release of cytochrome c during apoptosis is rapid, complete and kinetically invariant. *Nat. Cell Biol.* **2**:156–162.

58. Waterhouse, N.J., et al. 2001. Cytochrome c maintains mitochondrial transmembrane potential and ATP generation after outer mitochondrial membrane permeabilization during the apoptotic process. *J. Cell Biol.* **153**:319–328.

59. Ricci, J.E., et al. 2004. Disruption of mitochondrial function during apoptosis is mediated by caspase cleavage of the p75 subunit of complex I of the electron transport chain. *Cell.* **117**:773–786.

60. Ventura, J.J., Cogswell, P., Flavell, R.A., Baldwin, A.S., Jr., and Davis, R.J. 2004. JNK potentiates TNF-stimulated necrosis by increasing the production of cytotoxic reactive oxygen species. *Genes Dev.* **18**:2905–2915.

61. Cardone, M.H., Salvesen, G.S., Widmann, C., Johnson, G., and Frisch, S.M. 1997. The regulation of anoikis: MEKK-1 activation requires cleavage by caspases. *Cell.* **90**:315–323.

62. Tournier, C., et al. 2000. Requirement of JNK for stress-induced activation of the cytochrome c-mediated death pathway. *Science.* **288**:870–874.

63. Deng, Y., Ren, X., Yang, L., Lin, Y., and Wu, X. 2003. A JNK-dependent pathway is required for TNFalpha-induced apoptosis. *Cell.* **115**:61–70.

64. Levkau, B., Scatena, M., Giachelli, C.M., Ross, R., and Raines, E.W. 1999. Apoptosis overrides survival signals through a caspase-mediated dominant-negative NF-kappa B loop. *Nat. Cell Biol.* **1**:227–233.

65. Tang, G., Yang, J., Minemoto, Y., and Lin, A. 2001. Blocking caspase-3-mediated proteolysis of IKKbeta suppresses TNF-alpha-induced apoptosis. *Mol. Cell.* **8**:1005–1016.

66. Reuther, J.Y., and Baldwin, A.S., Jr. 1999. Apoptosis promotes a caspase-induced amino-terminal truncation of IkappaBalpha that functions as a stable inhibitor of NF-kappaB. *J. Biol. Chem.* **274**:20664–20670.

67. Toledano, M.B., and Leonard, W.J. 1991. Modulation of transcription factor NF-kappa B binding activity by oxidation-reduction in vitro. *Proc. Natl. Acad. Sci. U. S. A.* **88**:4328–4332.

68. Kapahi, P., et al. 2000. Inhibition of NF-kappa B activation by arsenite through reaction with a critical cysteine in the activation loop of Ikappa B kinase. *J. Biol. Chem.* **275**:36062–36066.

69. Denecker, G., et al. 2001. Death receptor-induced apoptotic and necrotic cell death: differential role of caspases and mitochondria. *Cell Death Differ.* **8**:829–840.

70. Vanden Berghe, T., et al. 2004. Differential signaling to apoptotic and necrotic cell death by Fas-associated death domain protein FADD. *J. Biol. Chem.* **279**:7925–7933.

71. Schmidt-Supprian, M., et al. 2000. NEMO/IKK gamma-deficient mice model incontinentia pigmenti. *Mol. Cell.* **5**:981–992.

72. Bian, X., et al. 2001. NF-kappa B activation mediates doxorubicin-induced cell death in N-type neuroblastoma cells. *J. Biol. Chem.* **276**:48921–48929.

73. Hettmann, T., DiDonato, J., Karin, M., and Leiden, J.M. 1999. An essential role for nuclear factor kappaB in promoting double positive thymocyte apoptosis. *J. Exp. Med.* **189**:145–158.

74. Bessho, R., et al. 1994. Pyrrolidine dithiocarbamate, a potent inhibitor of nuclear factor kappa B (NF-kappa B) activation, prevents apoptosis in human promyelocytic leukemia HL-60 cells and thymocytes. *Biochem. Pharmacol.* **48**:1883–1889.

75. Ivanov, V.N., and Ronai, Z. 2000. p38 protects human melanoma cells from UV-induced apoptosis through down-regulation of NF-kappaB activity

and Fas expression. *Oncogene.* **19**:3003–3012.

76. Campbell, K.J., Rocha, S., and Perkins, N.D. 2004. Active repression of antiapoptotic gene expression by RelA(p65) NF-kappa B. *Mol. Cell.* **13**:853–865.

77. Webster, G.A., and Perkins, N.D. 1999. Transcriptional cross talk between NF-kappaB and p53. *Mol. Cell. Biol.* **19**:3485–3495.

78. Tergaonkar, V., Pando, M., Vafa, O., Wahl, G., and Verma, I. 2002. p53 stabilization is decreased upon NFkappaB activation: a role for NFkappaB in acquisition of resistance to chemotherapy. *Cancer Cell.* **1**:493–503.

79. Ryan, K.M., Ernst, M.K., Rice, N.R., and Vousden, K.H. 2000. Role of NF-kappaB in p53-mediated programmed cell death. *Nature.* **404**:892–897.

80. Bohuslav, J., Chen, L.F., Kwon, H., Mu, Y., and Greene, W.C. 2004. p53 induces NF-kappaB activation by an IkappaB kinase-independent mechanism involving phosphorylation of p65 by ribosomal S6 kinase 1. *J. Biol. Chem.* **279**:26115–26125.

81. Fujioka, S., et al. 2004. Stabilization of p53 is a novel mechanism for proapoptotic function of NF-kappaB. *J. Biol. Chem.* **279**:27549–27559.

82. Egan, L.J., et al. 2004. IkappaB-kinasebeta-dependent NF-kappaB activation provides radioprotection to the intestinal epithelium. *Proc. Natl. Acad. Sci. U. S. A.* **101**:2452–2457.

83. Benezra, M., et al. 2003. BRCA1 augments transcription by the NF-kappaB transcription factor by binding to the Rel domain of the p65/RelA subunit. *J. Biol. Chem.* **278**:26333–26341.

84. Kimura, M., et al. 2003. TNF combined with IFN-alpha accelerates NF-kappaB-mediated apoptosis through enhancement of Fas expression in colon cancer cells. *Cell Death Differ.* **10**:718–728.

85. Wiener, Z., et al. 2004. Synergistic induction of the Fas (CD95) ligand promoter by Max and NFkappaB in human non-small lung cancer cells. *Exp. Cell Res.* **299**:227–235.

86. Ravi, R., et al. 2001. Regulation of death receptor expression and TRAIL/Apo2L-induced apoptosis by NF-kappaB. *Nat. Cell Biol.* **3**:409–416.

87. Driessler, F., Venstrom, K., Sabat, R., Asadullah, K., and Schottelius, A.J. 2004. Molecular mechanisms of interleukin-10-mediated inhibition of NF-kappaB activity: a role for p50. *Clin. Exp. Immunol.* **135**:64–73.

88. Luo, J.L., Maeda, S., Hsu, L.C., Yagita, H., and Karin, M. 2004. Inhibition of NF-kappaB in cancer cells converts inflammation-induced tumor growth mediated by TNFalpha to TRAIL-mediated tumor regression. *Cancer Cell.* **6**:297–305.

89. Karin, M., Cao, Y., Greten, F.R., and Li, Z.W. 2002. NF-kappaB in cancer: from innocent bystander to major culprit [review]. *Nat. Rev. Cancer.* **2**:301–310.

90. Lin, A., and Karin, M. 2003. NF-kappaB in cancer: a marked target. *Semin. Cancer Biol.* **13**:107–114.

91. Greten, F.R., and Karin, M. 2004. The IKK/NF-kappaB activation pathway: a target for prevention and treatment of cancer. *Cancer Lett.* **206**:193–199.

92. Gilmore, T.D. 2003. The Re1/NF-kappa B/I kappa B signal transduction pathway and cancer. *Cancer Treat. Res.* **115**:241–265.

93. Shishodia, S., and Aggarwal, B.B. 2004. Nuclear factor-kappaB: a friend or a foe in cancer [review]? *Biochem. Pharmacol.* **68**:1071–1080.

94. Perkins, N.D. 2004. NF-kappaB: tumor promoter or suppressor? *Trends Cell Biol.* **14**:64–69.

95. Greten, F.R., et al. 2004. IKKbeta links inflammation and tumorigenesis in a mouse model of colitis-associated cancer. *Cell.* **118**:285–296.

96. Pikarsky, E., et al. 2004. NF-kappaB functions as a tumour promoter in inflammation-associated cancer. *Nature.* **431**:461–466.

97. Becker, C., et al. 2004. TGF-beta suppresses tumor progression in colon cancer by inhibition of IL-6 trans-signaling. *Immunity.* **21**:491–501.

98. Seitz, C.S., Lin, Q., Deng, H., and Khavari, P.A.

1998. Alterations in NF-kappaB function in transgenic epithelial tissue demonstrate a growth inhibitory role for NF-kappaB. *Proc. Natl. Acad. Sci. U. S. A.* **95**:2307-2312.

99. van Hogerlinden, M., Rozell, B.L., Ahrlund-Richter, L., and Toftgard, R. 1999. Squamous cell carcinomas and increased apoptosis in skin with inhibited Rel/nuclear factor-kappaB signaling. *Cancer Res.* **59**:3299-3303.

100. Dajee, M., et al. 2003. NF-kappaB blockade and oncogenic Ras trigger invasive human epidermal neoplasia. *Nature.* **421**:639-643.

101. Zhang, J.Y., Green, C.L., Tao, S., and Khavari, P.A. 2004. NF-kappaB RelA opposes epidermal proliferation driven by TNFR1 and JNK. *Genes Dev.* **18**:17-22.

102. Seitz, C.S., Freiberg, R.A., Hinata, K., and Khavari, P.A. 2000. NF-κB determines localization and features of cell death in epidermis. *J. Clin. Invest.* **105**:253-260.

103. van Hogerlinden, M., Auer, G., and Toftgard, R. 2002. Inhibition of Rel/nuclear factor-kappaB signaling in skin results in defective DNA damage-induced cell cycle arrest and Ha-ras- and p53-independent tumor development. *Oncogene.* **21**:4969-4977.

104. Finco, T.S., et al. 1997. Oncogenic Ha-Ras-induced signaling activates NF-kappaB transcriptional activity, which is required for cellular transformation. *J. Biol. Chem.* **272**:24113-24116.

105. Pasparakis, M., et al. 2002. TNF-mediated inflammatory skin disease in mice with epidermis-specific deletion of IKK2. *Nature.* **417**:861-866.

106. Moore, R.J., et al. 1999. Mice deficient in tumor necrosis factor-alpha are resistant to skin carcinogenesis. *Nat. Med.* **5**:828-831.

107. Lind, M.H., et al. 2004. Tumor necrosis factor receptor 1-mediated signaling is required for skin cancer development induced by NF-kappaB inhibition. *Proc. Natl. Acad. Sci. U. S. A.* **101**:4972-4977.

108. Maeda, S., Kamata, H., Luo, J.L., Leffert, H., and Karin, M. 2005. IKKbeta couples hepatocyte death to cytokine-driven compensatory proliferation that promotes chemical hepatocarcinogenesis. *Cell.* **121**:977-990.

109. Monks, N.R., Biswas, D.K., and Pardee, A.B. 2004. Blocking anti-apoptosis as a strategy for cancer chemotherapy: NF-kappaB as a target [review]. *J. Cell. Biochem.* **92**:646-650.

110. Ravi, R., and Bedi, A. 2004. NF-kappaB in cancer: a friend turned foe. *Drug Resist. Updat.* **7**:53-67.

111. Karin, M., Yamamoto, Y., and Wang, Q.M. 2004. The IKK NF-kappa B system: a treasure trove for drug development [review]. *Nat. Rev. Drug Discov.* **3**:17-26.

112. Orlowski, R.Z., and Baldwin, A.S., Jr. 2002. NF-kappaB as a therapeutic target in cancer. *Trends Mol. Med.* **8**:385-389.

113. Frelin, C., et al. 2003. AS602868, a pharmacological inhibitor of IKK2, reveals the apoptotic potential of TNF-alpha in Jurkat leukemic cells. *Oncogene.* **22**:8187-8194.

114. Frelin, C., et al. 2005. Targeting NF-kappaB activation via pharmacologic inhibition of IKK2-induced apoptosis of human acute myeloid leukemia cells. *Blood.* **105**:804-811.

115. Ziegelbauer, K., et al. 2005. A selective novel low-molecular-weight inhibitor of IkappaB kinase-beta (IKK-beta) prevents pulmonary inflammation and shows broad anti-inflammatory activity. *Br. J. Pharmacol.* **145**:178-192.

116. Niu, C., et al. 1999. Studies on treatment of acute promyelocytic leukemia with arsenic trioxide: remission induction, follow-up, and molecular monitoring in 11 newly diagnosed and 47 relapsed acute promyelocytic leukemia patients. *Blood.* **94**:3315-3324.

117. Rousselot, P., et al. 2004. A clinical and pharmacological study of arsenic trioxide in advanced multiple myeloma patients. *Leukemia.* **18**:1518-1521.

118. Cavigelli, M., et al. 1996. The tumor promoter arsenite stimulates AP-1 activity by inhibiting a JNK phosphatase. *EMBO J.* **15**:6269-6279.

119. Huang, P., Feng, L., Oldham, E.A., Keating, M.J., and Plunkett, W. 2000. Superoxide dismutase as a target for the selective killing of cancer cells. *Nature.* **407**:390-395.

Reawakening the cellular death program in neoplasia through the therapeutic blockade of IAP function

Casey W. Wright[1] and Colin S. Duckett[1,2]

[1]Department of Pathology, and [2]Department of Internal Medicine, University of Michigan, Ann Arbor, Michigan, USA.

Recent studies have shown that members of the inhibitor of apoptosis (IAP) protein family are highly expressed in several classes of cancer. The primary implication of these findings is that the elevated expression of IAPs is not coincidental but actually participates in oncogenesis by helping to allow the malignant cell to avoid apoptotic cell death. This concept, together with the discovery of several IAP-regulatory proteins that use a conserved mode of action, has stimulated a major effort by many research groups to devise IAP-targeting strategies as a means of developing novel antineoplastic drugs. In this Review, we consider the evidence both for and against the IAPs being valid therapeutic targets, and we describe the types of strategies being used to neutralize their functions.

IAPs: structure and function

Inhibitor of apoptosis (*iap*) genes were first described in insect viruses through a genetic screen to identify compensatory replacements for the loss of the baculoviral antiapoptotic p35 protein, which functions to block baculovirus-induced apoptosis during infection (1–3). Since the initial identification in baculoviruses, *iap* homologs have been identified across phyla from *Caenorhabditis elegans* and yeast to insects and mammals. As many as 8 human *iap* gene products have been identified (reviewed in ref. 4), XIAP (hILP/MIHA/BIRC4), hILP-2 (TS-IAP), cellular IAP1 (c-IAP1/HIAP2/MIHB/BIRC2), cellular IAP2 (c-IAP2/HIAP1/MIHC/BIRC3), melanoma-associated IAP (ML-IAP/Livin/KIAP1/BIRC7), neuronal apoptosis–inhibitory protein (NAIP/BIRC1), survivin (TIAP1/BIRC5), and Apollon (BRUCE/BIRC6) (Figure 1A). These IAP proteins have been shown to play, for the most part, nonredundant cellular roles that range from apoptotic inhibition to the formation of the mitotic spindle during cytokinesis. For this reason, the term "IAP" is somewhat misleading, and these factors are often referred to as BIR-containing proteins, or BIRPs (5), a term derived from the presence of what has become the defining motif of this family, an approximately 65-residue domain rich in histidines and cysteines known as the baculovirus *iap* repeat (BIR). BIRPs contain 1–3 imperfectly repeated BIRs (Figure 1). Most of the IAP family members also harbor a RING domain at the carboxy terminus. The RING functions as an E3 ubiquitin ligase, which is preceded by ubiquitin-activating enzymes (E1) and ubiquitin-conjugating enzymes (E2) in the cascade of protein ubiquitination (6–9). E3 ubiquitin ligases provide specificity for the transfer of ubiquitin moieties onto the target protein. Therefore, IAP-mediated protein ubiquitination has a pivotal role in the regulation of apoptosis, allowing the IAP to control stability of itself and other proteins (10).

Survivin is a small (17 kDa) protein composed of a single BIR motif and is similar in structure to the IAPs/BIRPs of yeasts and nematodes (11). Targeted deletion of the murine *survivin* gene revealed a critical role for this protein in the cell cycle through regulation of the spindle formation during mitosis (12, 13), a role that is similar to that found in yeasts and the nematode *C. elegans* (14, 15). Additionally, a number of reports have also implicated survivin in apoptotic inhibition, although the details of this role are not entirely clear (16). Nevertheless, *survivin* is highly expressed in dividing cells and cancer-derived cell lines and has become a valid target for anticancer drugs, including those that use antisense approaches.

In addition to survivin, much attention has been focused on XIAP, in large part because its antiapoptotic properties have been best demonstrated (17). XIAP is a ubiquitously expressed 56-kDa protein that contains 3 BIRs, as well as a RING domain at the carboxy terminus that has been shown to exhibit E3 ubiquitin ligase activity (9, 18). A number of reviews have examined the clinical utility of survivin in more detail (19); therefore, this Review will focus on XIAP.

Ectopic expression of XIAP has been shown to confer protection from a wide range of apoptotic stimuli, and this protection is presumed to be largely mediated by the ability of XIAP to directly bind and enzymatically inhibit key components of the apoptotic machinery, the caspases (20, 21). This group of cysteine proteases with a specificity for aspartate residues (reviewed in this issue of the *JCI*, ref. 22) play essential roles in apoptotic cell death. They are initially produced as inactive zymogens, which are typically activated in a hierarchical sequence in which initiator caspases, such as caspase-8, are activated following, for example, engagement of a death receptor signal. Processed initiator caspases subsequently activate, through cleavage, downstream or effector caspases, such as caspase-3, which then disassemble the cell through the orchestrated proteolysis of essential cellular proteins. XIAP has been shown to directly bind and inhibit caspase-3, caspase-7, and caspase-9, which are important mediators of the apoptotic program (20, 23). c-IAP1 and c-IAP2 can also block caspase activity; however, they inhibit caspases 100- to 1,000-fold less efficiently than XIAP (24).

IAPs in cancer

Many studies have revealed a circumstantial association of IAPs and neoplasia (25). For example, XIAP levels are elevated in many

---

Nonstandard abbreviations used: ASO, antisense oligonucleotide; BIR, baculovirus *iap* repeat; BIRP, BIR-containing protein; c-IAP, cellular IAP; Hid, head involution defective; IAP, inhibitor of apoptosis; IBM, IAP-binding motif; ML-IAP, melanoma-associated IAP; Rpr, reaper; Smac/DIABLO, second mitochondria-derived activator of caspase/direct IAP-binding protein with low pI; TRAIL, TNF-related apoptosis-inducing ligand.

Conflict of interest: The authors have declared that no conflict of interest exists.

Citation for this article: *J. Clin. Invest.* **115**:2673–2678 (2005).
doi:10.1172/JCI26251.

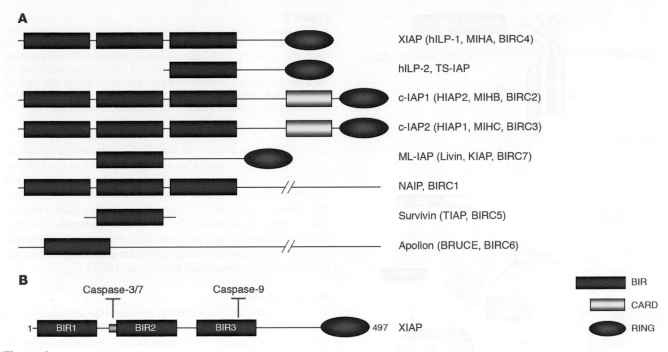

Figure 1
The IAP family members. (**A**) All IAP members contain 1 or more imperfect baculovirus *iap* repeats (BIRs), the defining motif of the IAP family. Many of the IAP proteins also posses an E3 ubiquitin ligase RING domain at the carboxy terminus. (**B**) XIAP can bind and enzymatically inhibit caspase-3, caspase-7, and caspase-9. XIAP binds caspase-3 and caspase-7 at a sequence directly upstream of BIR2, while it binds caspase-9 in a region of BIR3. CARD, caspase recruitment domain.

cancer cell lines, and several reports have shown that suppression of XIAP protein levels can sensitize cancer cells to chemotherapeutic drugs (26–29). IAP proteins appear to regulate the transcriptional activator NF-κB family, which itself has been associated with malignancy (30–32). NF-κB activation in turn upregulates expression of c-IAP2, providing a positive feedback loop for cell survival that may be important in the development of some cancers (32). Also, the presence of increased levels of c-IAP2 protein is correlated with carcinogenesis and chemotherapeutic resistance in malignant pleural mesothelioma, a tumor that attacks the pleura of the lung (33). Survivin expression is normally limited to cells of the developing fetus and is not expressed in differentiated adult tissue (11). However, aberrant expression of *survivin* has been detected in a number of different cancers and lymphomas (11), and expression of a dominant-negative form of *survivin* induced apoptosis in cancer but not normal cell lines (27). Furthermore, ML-IAP was identified as an IAP that is highly expressed in the majority of melanoma cell lines tested, but undetectable in primary melanocytes (34).

Although heightened expression of IAPs has been reported in and may contribute to the progression of many cancers, it is important to caution that many of these results are correlative in nature. For example, to our knowledge, no solid evidence has been reported to show that XIAP is itself an oncogene, or specifically that mutations, translocations, or amplifications at the XIAP locus have ever been found in natural tumors. This is true for all of the IAP/BIRP family, with the single exception of c-IAP2, which has been found to be translocated in mucosal-associated lymphoid tissue (MALT) lymphoma, where a RING-deleted form of c-IAP2 is fused to the MALT1 protein (35, 36). Furthermore, while elevated expression of IAPs has been reported in many cancers, it has also been shown

that IAP levels are not always correlated with disease progression or prognosis (37–39). However, this does not invalidate the targeting of, for example, XIAP, for therapeutic intervention, but it does emphasize that the relevance of XIAP as an anticancer target should be scrutinized more rigorously than that of a typical oncogene. It is quite possible that future studies will identify alterations in IAPs as a major causative genetic lesion in specific malignancies, and indeed a recent study reported the presence of the closely linked c-IAP1 and c-IAP2 genes in an amplicon associated with esophageal squamous cell carcinoma (40). Nevertheless, the current model of targeting IAPs, particularly XIAP, relies exclusively on expression differences between malignant and normal cells and presumes that tumor cells have selected for enhanced expression levels in the absence of apparent genetic alterations at the XIAP locus.

Two pathways of caspase activation regulated by the IAP proteins

Two major apoptotic signaling cascades have been described and are generally referred to as the extrinsic (or receptor-mediated) and intrinsic (or mitochondrial) pathways (41) (Figure 2). The extrinsic pathway transduces an intracellular signal into an apoptotic response and is exemplified by proapoptotic members of the TNF receptor superfamily, such as the TNF-related apoptosis-inducing ligand (TRAIL) receptor and the Fas receptor. Ligand-mediated activation of these receptors results in the binding of adaptor molecules that subsequently recruit and promote the activation of procaspase-8 (42–44). In this scheme, activated caspase-8 is thought to function as an initiator caspase, leading to the subsequent cleavage and activation of effector caspases, such as caspase-3, and thus leading to cell death (Figure 2).

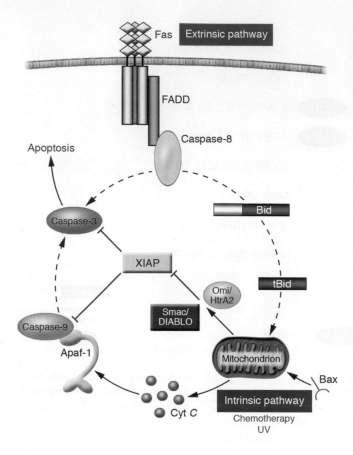

Figure 2
XIAP in apoptosis regulation. Following an apoptotic stimulus, enzymes known as caspases are activated and initiate a cascade leading to the destruction of the cell. The caspases are activated via 2 main avenues, by the stimulation of death receptors (the extrinsic pathway) and by the release of apoptogenic factors from the mitochondria (the intrinsic pathway). XIAP regulates both the extrinsic and the intrinsic apoptotic pathways through direct inhibition of caspase-3 and caspase-9.

distinct points (caspase-9 and caspase-3) in the intrinsic pathway. Indeed, biochemical studies have shown that XIAP can be found in the apoptosome (51), presumably in complex with caspase-9, and a number of elegant structural studies have revealed the presence of 2 distinct domains of XIAP that interact with caspase-3 and caspase-7, and with caspase-9 (52). The caspase-3/caspase-7–binding domain is located directly aminoterminal to the second BIR within XIAP, while the caspase-9–binding domain is contained within the third (most carboxyterminal) BIR (Figure 1B).

IAP antagonists

After the realization that XIAP can directly bind caspases, the crucial discovery that enforced the concept that XIAP is an attractive therapeutic target came from studies that identified endogenous regulators of its activity. The first, and best, characterized of these is second mitochondria-derived activator of caspase/direct IAP-binding protein with low pI (Smac/DIABLO), a nuclear-encoded protein that in healthy cells is localized to the mitochondria in a mature form lacking the aminoterminal 55 residues, which are removed during mitochondrial translocation (53, 54). Smac/DIABLO is a functional homolog of 3 proapoptotic factors in *Drosophila*, namely *reaper* (*rpr*), *head involution defective* (*hid*), and *grim* (55, 56). The main functional motif of these IAP antagonists is at the extreme amino terminus of Rpr, Hid, and Grim, and the mature form of Smac/DIABLO. Thus, this sequence has been termed the Rpr/Hid/Grim (RHG) motif, or IAP-binding motif (IBM) as a more general term (57). With kinetics that appear to be identical to those of cytochrome *c*, Smac/DIABLO is released from the mitochondria into the cytosol, where it can bind XIAP. Importantly, Smac/DIABLO has been shown to bind precisely into the same 2 grooves within XIAP that can be occupied by caspase-3 and -7 or caspase-9 (Figure 3), and this leads to an attractive model in which Smac/DIABLO can act as a proapoptotic molecule that functions through the displacement of XIAP from caspases (23).

Subsequent studies in mammalian cells have revealed the existence of additional molecules that target IAPs in a manner akin to that of Smac/DIABLO. Specifically, 2 factors have been described, designated Omi/HtrA2 (58–62) and GSPT1/eRF3 (63), that share with Smac/DIABLO the properties of mitochondrial localization and cytoplasmic release, as well as the presence of an IBM at the amino terminus of the mature protein. Beyond the tetrapeptide sequence, however, there is virtually no similarity between these proteins, and at least in the case of Omi/HtrA2, which appears to function as a chaperone and a regulator of oxidative stress (64), the primary physiological role appears to be very distinct from caspase/XIAP regulation.

The intrinsic pathway is activated by a range of apoptotic stimuli, including DNA damage, treatment of cells with chemotherapeutic agents, and growth factor withdrawal. The mitochondria serve as the control point for this pathway (45). A pivotal step in this process is the release of cytochrome *c* from the mitochondria into the cytosol (46). The Bcl-2 family of proteins very carefully regulates this release.

Once released into the cytosol, cytochrome *c* binds directly to a key cellular component of the apoptotic cascade, apoptotic protease–activating factor-1 (Apaf-1), which subsequently oligomerizes to form a high–molecular weight structure designated the apoptosome (47). This complex is able to recruit and activate caspase-9, which is the key initiator caspase in the intrinsic pathway. Apoptosome-activated caspase-9 is subsequently able to act on downstream effector caspases, such as caspase-3 and caspase-7, to generate the apoptotic signal (Figure 2).

Signaling crosstalk exists between the intrinsic and extrinsic cell death pathways. For example, the proapoptotic Bcl-2–related protein Bid can be cleaved by caspase-8 to produce an active form, known as tBid, which can subsequently translocate to mitochondria and lead to the induction of the intrinsic pathway (48, 49). Thus, in certain situations, the intrinsic pathway is required for the full induction of receptor-mediated apoptosis and is thought to exert these effects through an amplification loop.

Numerous studies of XIAP have revealed its ability to directly bind to caspase-3, caspase-7, and caspase-9 (50). Consequently, in the scheme of apoptosis shown in Figure 2, XIAP can function directly at the most downstream effector caspase (caspase-3 or caspase-7) in the extrinsic pathway and can block cell death at 2

Mimetics of IAP antagonists in cancer treatment

The development of small molecules that mimic Smac/DIABLO and therefore interfere with caspase-9 binding to XIAP is proving to be a promising avenue for the treatment of cancer (25, 65).

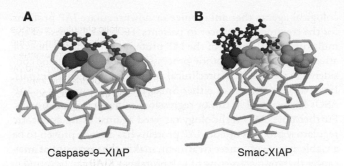

Caspase-9–XIAP   Smac-XIAP

Figure 3
Smac/DIABLO displaces caspase-9 from BIR3 of XIAP. (**A**) Crystal structure of processed caspase-9 bound to BIR3 of XIAP. The purple peptide represents the first 4 amino acids that contact XIAP, with the amino terminus near the orange residue of XIAP. Coordinates were obtained from Brookhaven Protein Databank file 1NW9 (82). (**B**) NMR structure of the Smac/DIABLO IBM (purple peptide with the amino terminus near the orange residue of XIAP) bound to the same groove on BIR3 of XIAP that binds caspase-9, thus abrogating the ability of XIAP to block caspase-9 activity. Coordinates were obtained from Brookhaven Protein Databank file 1G3F (83). The 4 residues displayed in van der Waals radii spacefill on XIAP BIR3 are Gly306 (red), Leu307 (green), Trp310 (yellow), and Glu314 (orange). These 4 residues are critical in forming the Smac/DIABLO–binding groove on XIAP BIR3 (84). Images were created using Protein Explorer (85, 86).

Furthermore, since the extrinsic and intrinsic pathways are both regulated by XIAP at the step of caspase-3 activation, small molecules that interfere with this interaction are also being tested. Smac/DIABLO IBM peptides, when used in conjunction with TRAIL, proved to promote apoptosis in cancer cells through caspase-9 and resulted in complete tumor regression in a murine intracranial malignant glioma xenograft model (27, 66, 67). However, the use of peptides in therapy is not feasible, because of the proteolytic instability and low cellular permeability of the peptide. Thus, variations on the first 4 residues of Smac/DIABLO have been synthesized to make it a more stable moiety (68–70). These alterations have provided reagents with higher affinity for XIAP than Smac/DIABLO, and lower concentrations of the compounds were required for activity. One of these molecules bound to and antagonized c-IAP1 and c-IAP2, in addition to XIAP, and sensitized HeLa cells to TNF-induced apoptosis (71). This finding suggests that c-IAP1 and c-IAP2 are downstream of NF-κB activity and function to inhibit caspase-8 activation, allowing prosurvival signals to proceed. Thus Smac/DIABLO mimetics could also provide therapy for inflammation disorders that arise from increased TNF stimulation.

Continual efforts to increase the efficacy of Smac/DIABLO peptide agonists have led other groups to devise small-molecule nonpeptidic compounds that target BIR2 of XIAP, liberating caspase-3 (72, 73). There may be an advantage to pharmacological agents that antagonize XIAP to release caspase-3, since this caspase is at the convergence of both the extrinsic and the intrinsic pathway. The use of these compounds resulted in apoptosis in cancer cells without the requirement for concomitant chemotherapeutic treatment. Furthermore, targeting the release of caspase-3 would eliminate the requirement for mitochondria-mediated activation, which is blocked in many cancers by the deregulation of the Bcl-2 pathway (72).

IAP antisense technology in cancer treatment
Another approach to blocking IAP activity in order to promote caspase activation in cancer cells is to downregulate protein levels by delivering IAP antisense into the cell. Downregulation of XIAP protein levels by adenovirus-mediated introduction of antisense induced apoptosis in chemoresistant ovarian cancer cells (26). Furthermore, a similar antisense approach against c-IAP2 showed c-IAP2 to be a major contributor of chemotherapeutic resistance in pleural mesothelioma (33).

More recently, antisense oligonucleotides (ASOs) have been used in studies to reduce IAP protein levels. ASO technology uses approximately 20-bp oligonucleotide sequences targeted to the mRNA of the protein of interest; this interferes with expression of the transcript. Reduction of XIAP protein levels in bladder cancer using XIAP ASOs was reported to induce apoptosis and contribute to doxorubicin cytotoxicity (28). Combining ASOs targeted against XIAP and survivin with radiotherapy delayed tumor growth in a mouse lung cancer model and decreased the survival of H460 lung cancer cells (74).

In contrast, a recent study in which ASO technology was used to suppress XIAP levels in cells over time determined that not all chemotherapeutic drugs are effective in combination with decreased XIAP protein levels; this knowledge will prove beneficial in the development of treatment schemes for patients in clinical trials (29). A survivin ASO (ISIS 23722; Isis Pharmaceuticals Inc.) is in preclinical development, and ASOs that target XIAP are in phase I clinical trials (29).

Other functions of IAPs
It is widely assumed that the primary physiological role of XIAP is as an apoptotic suppressor, and XIAP is often referred to as the only known endogenous caspase inhibitor in mammals. The biochemical evidence for this is compelling, and the caspase-suppressing effects are so striking (with inhibitory constants in the nanomolar range) that the argument has been made that the XIAP-caspase interaction cannot be accidental. It is, however, worth bearing in mind that the biological evidence for XIAP being an essential endogenous inhibitor lags behind the in vitro data. For example, Xiap-deficient mice exhibit no overt apoptotic abnormalities that might be expected (75), although a sensitivity of sympathetic neurons following cytochrome c injection has been reported (76). This is not to say that Xiap-null mice are indistinguishable from control animals: recent reports have revealed alterations in intracellular copper levels (77), and in mammary gland development (78), but these phenotypic differences are unlikely to be caused by alterations in apoptotic sensitivity. The prevailing view is that Xiap-deficient mice survive through a degenerate mechanism, such as compensatory expression of other Iap family members, particularly of c-IAP1, which has been reported in embryonic fibroblasts derived from these animals (75). However, the caspase-inhibitory properties of the c-IAP proteins are much weaker than those of XIAP; this indicates nonredundant roles for these proteins. Other studies from several laboratories have implicated XIAP in a number of cellular functions unrelated to caspases, such as the activation of signal transduction cascades including JNK, NF-κB, TGF-β, and Akt (30, 31, 79, 80). To date, however, these studies have all relied on ectopic expression of XIAP, and to our knowledge no studies have been described to suggest alterations in these signaling pathways in Xiap-null animals. Thus, the evidence supporting a physiological role for XIAP as a signal transduction intermediate is no stronger than that suggest-

ing a function in apoptotic regulation. Recently, our laboratory has linked XIAP to copper homeostasis (77), through an interaction with MURR1, a mammalian factor whose gene was identified by positional cloning in an inbred canine strain affected by a non-Wilsonian inherited copper toxicosis disorder (81). MURR1 was found to be a target of ubiquitination by the RING-mediated E3 ubiquitin ligase activity of XIAP, and, consistent with a model in which XIAP can function to regulate MURR1 and copper, tissues from *Xiap*-null mice were found to contain at the same time higher levels of Murr1 and lower levels of intracellular copper, compared with control animals (77). Thus, clearly much work remains to elucidate the true physiological functions of XIAP. The concept that XIAP participates in oncogenesis and is an anticancer target remains valid if XIAP's participation is due not to caspase inhibition, but to an alternative regulatory function in, for example, intracellular signaling or copper metabolism. The possibility that XIAP is exerting noncaspase-regulatory properties may, however, affect the experimental approaches that need be taken to develop therapeutically effective antagonists of XIAP function.

Conclusions

The multifunctional IAP proteins are involved in apoptosis regulation, cell signaling, and cell division. Substantial progress has been made in identifying the IAP proteins as factors in cancer development, progression, and desensitization to a wide array of chemotherapeutic drugs. This knowledge should lead to pharmacological agents that antagonize or downregulate IAP proteins for the treatment of cancer in patients. However, because of the multifunctional nature of the IAP proteins, the consequences of affecting other nonapoptotic processes in the cell need to be considered in the design of preclinical agents for use in clinical trials. Nevertheless, the use of either Smac/DIABLO mimetics or IAP ASOs has resulted in tumor regression in mouse cancer models. Furthermore, ASO technology targeted against other apoptotic regulatory proteins besides IAP proteins has already proven to be a viable avenue of cancer treatment, making the promise of anticancer therapy by targeting of IAP proteins a reality.

Acknowledgments

We would like to express our gratitude to the Duckett laboratory for helpful discussions, and in particular to John Wilkinson for critical reading of the manuscript. We apologize to those researchers whose work we could not cite because of space constraints. This work was supported in part by NIH grant T32 HL07517 to C.W. Wright and by the University of Michigan Biological Scholars Program, Department of Defense Idea Award PC040215, and NIH grant GM067827 to C.S. Duckett.

Address correspondence to: Colin S. Duckett, Medical Science I, Room 5315, 1301 Catherine Street, Ann Arbor, Michigan 48109-0602, USA. Phone: (734) 615-6414; Fax: (734) 615-7012; E-mail: colind@umich.edu.

1. Crook, N.E., Clem, R.J., and Miller, L.K. 1993. An apoptosis-inhibiting baculovirus gene with a zinc finger-like motif. *J. Virol.* **67**:2168–2174.

2. Birnbaum, M.J., Clem, R.J., and Miller, L.K. 1994. An apoptosis-inhibiting gene from a nuclear polyhedrosis virus encoding a polypeptide with Cys/His sequence motifs. *J. Virol.* **68**:2521–2528.

3. Clem, R.J., Fechheimer, M., and Miller, L.K. 1991. Prevention of apoptosis by a baculovirus gene during infection of insect cells. *Science.* **254**:1388–1390.

4. Salvesen, G.S., and Duckett, C.S. 2002. IAP proteins: blocking the road to death's door [review]. *Nat. Rev. Mol. Cell Biol.* **3**:401–410.

5. Uren, A.G., Coulson, E.J., and Vaux, D.L. 1998. Conservation of baculovirus inhibitor of apoptosis repeat proteins (BIRPs) in viruses, nematodes, vertebrates and yeasts. *Trends Biochem. Sci.* **23**:159–162.

6. Huang, H.-K., et al. 2000. The inhibitor of apoptosis, cIAP2, functions as a ubiquitin-protein ligase and promotes in vitro monoubiquitination of caspases 3 and 7. *J. Biol. Chem.* **275**:26661–26664.

7. Olson, M.R., et al. 2003. A GH3-like domain in reaper is required for mitochondrial localization and induction of IAP degradation. *J. Biol. Chem.* **278**:44758–44768.

8. Suzuki, Y., Nakabayashi, Y., and Takahashi, R. 2001. Ubiquitin-protein ligase activity of X-linked inhibitor of apoptosis protein promotes proteasomal degradation of caspase-3 and enhances its anti-apoptotic effect in Fas-induced cell death. *Proc. Natl. Acad. Sci. U. S. A.* **98**:8662–8667.

9. Yang, Y., Fang, S., Jensen, J.P., Weissman, A.M., and Ashwell, J.D. 2000. Ubiquitin protein ligase activity of IAPs and their degradation in proteasomes in response to apoptotic stimuli. *Science.* **288**:874–877.

10. Jesenberger, V., and Jentsch, S. 2002. Deadly encounter: ubiquitin meets apoptosis. *Nat. Rev. Mol. Cell Biol.* **3**:112–121.

11. Ambrosini, G., Adida, C., and Altieri, D.C. 1997. A novel anti-apoptosis gene, *survivin*, expressed in cancer and lymphoma. *Nat. Med.* **3**:917–921.

12. Uren, A.G., et al. 2000. Survivin and the inner centromere protein INCENP show similar cell-cycle localization and gene knockout phenotype. *Curr. Biol.* **10**:1319–1328.

13. Okada, H., et al. 2004. Survivin loss in thymocytes triggers p53-mediated growth arrest and p53-independent cell death. *J. Exp. Med.* **199**:399–410.

14. Uren, A.G., et al. 1999. Role for yeast inhibitor of apoptosis (IAP)-like proteins in cell division. *Proc. Natl. Acad. Sci. U. S. A.* **96**:10170–10175.

15. Fraser, A.G., James, C., Evan, G.I., and Hengartner, M.O. 1999. Caenorhabditis elegans inhibitor of apoptosis protein (IAP) homologue BIR-1 plays a conserved role in cytokinesis. *Curr. Biol.* **9**:292–301.

16. Altieri, D.C. 2003. Survivin, versatile modulation of cell division and apoptosis in cancer. *Oncogene.* **22**:8581–8589.

17. Holcik, M., Gibson, H., and Korneluk, R.G. 2001. XIAP: apoptotic brake and promising therapeutic target. *Apoptosis.* **6**:253–261.

18. Li, X., Yang, Y., and Ashwell, J.D. 2002. TNF-RII and c-IAP1 mediate ubiquitination and degradation of TRAF2. *Nature.* **416**:345–347.

19. Altieri, D.C. 2004. Molecular circuits of apoptosis regulation and cell division control: the survivin paradigm. *J. Cell. Biochem.* **92**:656–663.

20. Deveraux, Q.L., Takahashi, R., Salvesen, G.S., and Reed, J.C. 1997. X-linked IAP is a direct inhibitor of cell-death proteases. *Nature.* **388**:300–304.

21. Shi, Y. 2004. Caspase activation, inhibition, and reactivation: a mechanistic view. *Protein Sci.* **13**:1979–1987.

22. Lavrik, I.N., Golks, A., and Krammer, P.H. 2005. Caspases: pharmacological manipulation of cell death. *J. Clin. Invest.* **115**:2665–2672. doi:10.1172/JCI26252.

23. Shiozaki, E.N., and Shi, Y. 2004. Caspases, IAPs and Smac/DIABLO: mechanisms from structural biology. *Trends Biochem. Sci.* **29**:486–494.

24. Roy, N., Deveraux, Q.L., Takahashi, R., Salvesen, G.S., and Reed, J.C. 1997. The c-IAP-1 and c-IAP-2 proteins are direct inhibitors of specific caspases. *EMBO J.* **16**:6914–6925.

25. Schimmer, A.D. 2004. Inhibitor of apoptosis proteins: translating basic knowledge into clinical practice. *Cancer Res.* **64**:7183–7190.

26. Sasaki, H., Sheng, Y., Kotsuji, F., and Tsang, B.K. 2000. Down-regulation of X-linked inhibitor of apoptosis protein induces apoptosis in chemoresistant human ovarian cancer cells. *Cancer Res.* **60**:5659–5666.

27. Yang, L., Cao, Z., Yan, H., and Wood, W.C. 2003. Coexistence of high levels of apoptotic signaling and inhibitor of apoptosis proteins in human tumor cells: implication for cancer specific therapy. *Cancer Res.* **63**:6815–6824.

28. Bilim, V., Kasahara, T., Hara, N., Takahashi, K., and Tomita, Y. 2003. Role of XIAP in the malignant phenotype of transitional cell cancer (TCC) and therapeutic activity of XIAP antisense oligonucleotides against multidrug-resistant TCC in vitro. *Int. J. Cancer.* **103**:29–37.

29. McManus, D.C., et al. 2004. Loss of XIAP protein expression by RNAi and antisense approaches sensitizes cancer cells to functionally diverse chemotherapeutics. *Oncogene.* **23**:8105–8117.

30. Hofer-Warbinek, R., et al. 2000. Activation of NF-κB by XIAP, the X chromosome-linked inhibitor of apoptosis, in endothelial cells involves TAK1. *J. Biol. Chem.* **275**:22064–22068.

31. Birkey Reffey, S., Wurthner, J.U., Parks, W.T., Roberts, A.B., and Duckett, C.S. 2001. X-linked inhibitor of apoptosis protein functions as a cofactor in transforming growth factor-β signaling. *J. Biol. Chem.* **276**:26542–26549.

32. Chu, Z.-L., et al. 1997. Suppression of tumor necrosis factor-induced cell death by inhibitor of apoptosis c-IAP2 is under NF-κB control. *Proc. Natl. Acad. Sci. U. S. A.* **94**:10057–10062.

33. Gordon, G.J., et al. 2002. Inhibitor of apoptosis protein-1 promotes tumor cell survival in mesothelioma. *Carcinogenesis.* **23**:1017–1024.

34. Vucic, D., Stennicke, H.R., Pisabarro, M.T., Salvesen, G.S., and Dixit, V.M. 2000. ML-IAP, a novel inhibitor of apoptosis that is preferentially expressed in human melanomas. *Curr. Biol.* **10**:1359–1366.

35. Dierlamm, J., et al. 1999. The apoptosis inhibitor gene *API2* and a novel 18q gene, *MLT*, are recurrently rearranged in the t(11;18)(q21;q21) associated with mucosa-associated lymphoid tissue lymphomas. *Blood.* **93**:3601–3609.

36. Uren, A.G., et al. 2000. Identification of paracaspases and metacaspases: two ancient families of caspase-like proteins, one of which plays a key role in MALT lymphoma. *Mol. Cell.* **6**:961–967.

37. Tamm, I., et al. 2000. Expression and prognostic significance of IAP-family genes in human cancers and myeloid leukemias. *Clin. Cancer Res.* **6**:1796–1803.

38. Ferreira, C.G., et al. 2001. Expression of X-linked inhibitor of apoptosis as a novel prognostic marker in radically resected non-small cell lung cancer patients. *Clin. Cancer Res.* **7**:2468–2474.

39. Carter, B.Z., et al. 2003. Caspase-independent cell death in AML: caspase inhibition in vitro with pan-caspase inhibitors or in vivo by XIAP or Survivin does not affect cell survival or prognosis. *Blood.* **102**:4179–4186.

40. Imoto, I., et al. 2001. Identification of *cIAP1* as a candidate target gene within an amplicon at 11q22 in esophageal squamous cell carcinomas. *Cancer Res.* **61**:6629–6634.

41. Budihardjo, I., Oliver, H., Lutter, M., Luo, X., and Wang, X. 1999. Biochemical pathways of caspase activation during apoptosis. *Annu. Rev. Cell Dev. Biol.* **15**:269–290.

42. Hsu, H., Shu, H.-B., Pan, M.-G., and Goeddel, D.V. 1996. TRADD-TRAF2 and TRADD-FADD interactions define two distinct TNF receptor 1 signal transduction pathways. *Cell.* **84**:299–308.

43. Kischkel, F.C., et al. 1995. Cytotoxicity-dependent APO-1 (Fas/CD95)-associated proteins form a death-inducing signaling complex (DISC) with the receptor. *EMBO J.* **14**:5579–5588.

44. Boldin, M.P., Goncharov, T.M., Goltsev, Y.V., and Wallach, D. 1996. Involvement of MACH, a novel MORT1/FADD-interacting protease, in Fas/APO-1- and TNF receptor-induced death. *Cell.* **85**:803–815.

45. Desagher, S., and Martinou, J.C. 2000. Mitochondria as the central control point of apoptosis. *Trends Cell Biol.* **10**:369–377.

46. Li, P., et al. 1997. Cytochrome *c* and dATP-dependent formation of Apaf-1/caspase-9 complex initiates an apoptotic protease cascade. *Cell.* **91**:479–489.

47. Zou, H., Henzel, W.J., Liu, X., Lutschg, A., and Wang, X. 1997. Apaf-1, a human protein homologous to *C. elegans* CED-4, participates in cytochrome *c*-dependent activation of caspase-3. *Cell.* **90**:405–413.

48. Luo, X., Budihardjo, I., Zou, H., Slaughter, C., and Wang, X. 1998. Bid, a Bcl2 interacting protein, mediates cytochrome *c* release from mitochondria in response to activation of cell surface death receptors. *Cell.* **94**:481–490.

49. Li, H., Zhu, H., Xu, C.J., and Yuan, J. 1998. Cleavage of BID by caspase 8 mediates the mitochondrial damage in the Fas pathway of apoptosis. *Cell.* **94**:491–501.

50. Salvesen, G.S., and Abrams, J.M. 2004. Caspase activation: stepping on the gas or releasing the brakes? Lessons from humans and flies [review]. *Oncogene.* **23**:2774–2784.

51. Bratton, S.B., Lewis, J., Butterworth, M., Duckett, C.S., and Cohen, G.M. 2002. XIAP inhibition of caspase-3 preserves its association with the Apaf-1 apoptosome and prevents CD95- and Bax-induced apoptosis. *Cell Death Differ.* **9**:881–892.

52. Stennicke, H.R., Ryan, C.A., and Salvesen, G.S. 2002. Reprieval from execution: the molecular basis of caspase inhibition. *Trends Biochem. Sci.* **27**:94–101.

53. Verhagen, A.M., et al. 2000. Identification of DIABLO, a mammalian protein that promotes apoptosis by binding to and antagonizing IAP proteins. *Cell.* **102**:43–53.

54. Du, C., Fang, M., Li, Y., Li, L., and Wang, X. 2000. Smac, a mitochondrial protein that promotes cytochrome *c*-dependent caspase activation by eliminating IAP inhibition. *Cell.* **102**:33–42.

55. Wu, J.W., Cocina, A.E., Chai, J., Hay, B.A., and Shi, Y. 2001. Structural analysis of a functional DIAP1 fragment bound to grim and hid peptides. *Mol. Cell.* **8**:95–104.

56. Wright, C.W., and Clem, R.J. 2001. Sequence requirements for hid binding and apoptosis regulation in the anti-apoptotic baculovirus protein Op-IAP: hid binds Op-IAP in a manner similar to Smac binding of XIAP. *J. Biol. Chem.* **277**:2454–2462.

57. Shi, Y. 2002. A conserved tetrapeptide motif: potentiating apoptosis through IAP-binding. *Cell Death Differ.* **9**:93–95.

58. Hegde, R., et al. 2001. Identification of Omi/HtrA2 as a mitochondrial apoptotic serine protease that disrupts IAP-caspase interaction. *J. Biol. Chem.* **277**:432–438.

59. Martins, L.M., et al. 2001. The serine protease Omi/HtrA2 regulates apoptosis by binding XIAP through a Reaper-like motif. *J. Biol. Chem.* **277**:439–444.

60. Suzuki, Y., et al. 2001. A serine protease, HtrA2, is released from the mitochondria and interacts with XIAP, inducing cell death. *Mol. Cell.* **8**:613–621.

61. van Loo, G., et al. 2002. The serine protease Omi/HtrA2 is released from mitochondria during apoptosis. Omi interacts with caspase-inhibitor XIAP and induces enhanced caspase activity. *Cell Death Differ.* **9**:20–26.

62. Vaux, D.L., and Silke, J. 2003. Mammalian mitochondrial IAP binding proteins. *Biochem. Biophys. Res. Commun.* **304**:499–504.

63. Hegde, R., et al. 2003. The polypeptide chain-releasing factor GSPT1/eRF3 is proteolytically processed into an IAP-binding protein. *J. Biol. Chem.* **278**:38699–38706.

64. Martins, L.M., et al. 2004. Neuroprotective role of the reaper-related serine protease HtrA2/Omi revealed by targeted deletion in mice. *Mol. Cell. Biol.* **24**:9848–9862.

65. Huang, Y., Lu, M., and Wu, H. 2004. Antagonizing XIAP-mediated caspase-3 inhibition. Achilles' heel of cancers? [review]. *Cancer Cell.* **5**:1–2.

66. Fulda, S., Wick, W., Weller, M., and Debatin, K.M. 2002. Smac agonists sensitize for Apo2L/TRAIL- or anticancer drug-induced apoptosis and induce regression of malignant glioma in vivo. *Nat. Med.* **8**:808–815.

67. Arnt, C.R., Chiorean, M.V., Heldebrant, M.P., Gores, G.J., and Kaufmann, S.H. 2002. Synthetic Smac/DIABLO peptides enhance the effects of chemotherapeutic agents by binding XIAP and cIAP1 in situ. *J. Biol. Chem.* **277**:44236–44243.

68. Li, C.J., Friedman, D.J., Wang, C., Metelev, V., and Pardee, A.B. 1995. Induction of apoptosis in uninfected lymphocytes by HIV-1 Tat protein. *Science.* **268**:429–431.

69. Sun, H., et al. 2004. Structure-based design, synthesis, and evaluation of conformationally constrained mimetics of the second mitochondria-derived activator of caspase that target the X-linked inhibitor of apoptosis protein/caspase-9 interaction site. *J. Med. Chem.* **47**:4147–4150.

70. Oost, T.K., et al. 2004. Discovery of potent antagonists of the antiapoptotic protein XIAP for the treatment of cancer. *J. Med. Chem.* **47**:4417–4426.

71. Li, L., et al. 2004. A small molecule Smac mimic potentiates TRAIL- and TNFalpha-mediated cell death. *Science.* **305**:1471–1474.

72. Wu, T.Y., Wagner, K.W., Bursulaya, B., Schultz, P.G., and Deveraux, Q.L. 2003. Development and characterization of nonpeptidic small molecule inhibitors of the XIAP/caspase-3 interaction. *Chem. Biol.* **10**:759–767.

73. Schimmer, A.D., et al. 2004. Small-molecule antagonists of apoptosis suppressor XIAP exhibit broad antitumor activity. *Cancer Cell.* **5**:25–35.

74. Cao, C., Mu, Y., Hallahan, D.E., and Lu, B. 2004. XIAP and survivin as therapeutic targets for radiation sensitization in preclinical models of lung cancer. *Oncogene.* **23**:7047–7052.

75. Harlin, H., Reffey, S.B., Duckett, C.S., Lindsten, T., and Thompson, C.B. 2001. Characterization of XIAP-deficient mice. *Mol. Cell. Biol.* **21**:3604–3608.

76. Potts, P.R., Singh, S., Knezek, M., Thompson, C.B., and Deshmukh, M. 2003. Critical function of endogenous XIAP in regulating caspase activation during sympathetic neuronal apoptosis. *J. Cell Biol.* **163**:789–799.

77. Burstein, E., et al. 2004. A novel role for XIAP in copper homeostasis through regulation of MURR1. *EMBO J.* **23**:244–254.

78. Olayioye, M.A., et al. 2005. XIAP-deficiency leads to delayed lobuloalveolar development in the mammary gland. *Cell Death Differ.* **12**:87–90.

79. Sanna, M.G., Duckett, C.S., Richter, B.W.M., Thompson, C.B., and Ulevitch, R.J. 1998. Selective activation of JNK1 is necessary for the anti-apoptotic activity of hILP. *Proc. Natl. Acad. Sci. U. S. A.* **95**:6015–6020.

80. Asselin, E., Mills, G.B., and Tsang, B.K. 2001. XIAP regulates Akt activity and caspase-3-dependent cleavage during cisplatin-induced apoptosis in human ovarian epithelial cancer cells. *Cancer Res.* **61**:1862–1868.

81. van De Sluis, B., Rothuizen, J., Pearson, P.L., van Oost, B.A., and Wijmenga, C. 2002. Identification of a new copper metabolism gene by positional cloning in a purebred dog population. *Hum. Mol. Genet.* **11**:165–173.

82. Shiozaki, E.N., et al. 2003. Mechanism of XIAP-mediated inhibition of caspase-9. *Mol. Cell.* **11**:519–527.

83. Liu, Z., et al. 2000. Structural basis for binding of Smac/DIABLO to the XIAP BIR3 domain. *Nature.* **408**:1004–1008.

84. Wu, G., et al. 2000. Structural basis of IAP recognition by Smac/DIABLO. *Nature.* **408**:1008–1012.

85. Martz, E. 2002. Protein explorer: easy yet powerful macromolecular visualization. *Trends Biochem. Sci.* **27**:107–109.

86. Protein Explorer. http://proteinexplorer.org.

The survival kinases Akt and Pim as potential pharmacological targets

Ravi Amaravadi and Craig B. Thompson

Abramson Family Cancer Research Institute, Department of Cancer Biology and Medicine, University of Pennsylvania, Philadelphia, Pennsylvania, USA.

The Akt and Pim kinases are cytoplasmic serine/threonine kinases that control programmed cell death by phosphorylating substrates that regulate both apoptosis and cellular metabolism. The PI3K-dependent activation of the Akt kinases and the JAK/STAT–dependent induction of the Pim kinases are examples of partially overlapping survival kinase pathways. Pharmacological manipulation of such kinases could have a major impact on the treatment of a wide variety of human diseases including cancer, inflammatory disorders, and ischemic diseases.

Introduction

There is increasing evidence that serine/threonine kinases exist that directly regulate cell survival. Therapeutics that directly target these survival kinases have not yet been developed for clinical use. Activated survival kinases contribute to the pathogenesis of a wide variety of malignancies. In addition, reduced survival kinase signaling may contribute to organ damage following ischemic insults. Selective therapies such as imatinib (1) and gefitinib (2) elicit tumor cell death by indirect inactivation of survival kinases. Would direct inhibition of survival kinases result in better therapeutic efficacy? Alternatively, could therapies that activate survival kinases lead to better organ preservation in ischemic diseases? Many drug discovery programs have begun to develop lead compounds to address these questions. This Review will explore the potential risks and benefits of targeting survival kinases by outlining (a) Akt and Pim kinase action in malignancy, immunity, and vascular disease, (b) the common substrates that survival kinases share, (c) recent advances in the understanding of survival kinase regulation, and (d) investigational agents that target survival kinases.

Kinases that promote cell survival and control cell metabolism

For this Review survival kinases will be defined as cytoplasmic serine/threonine kinases that phosphorylate substrates that collectively contribute to the control of the programmed cell death machinery and cellular metabolism (Figure 1). This coordinated control ensures the maintenance of mitochondrial membrane potential and prevents the mitochondrial release of cytochrome c and other proapoptotic mediators. This coordinated control also maintains cellular ATP production, preventing cells from dying by necrosis (3) or autophagy (4). The best-characterized survival kinases were identified in screens to find suppressors of myc-induced apoptosis. *myc* is a protooncogene whose overexpression leads to increased proliferation as well as increased apoptosis in nonmalignant cells. Defects in pathways that control apoptosis prevent myc-induced apoptosis and

allow myc to act as an oncogene, leading to a malignant phenotype. While deficiency in the tumor suppressor gene *p53* and constitutive activation of the antiapoptosis gene *bcl-2* are well characterized events that block myc-induced apoptosis, screens using retroviral mutagenesis have uncovered several serine/threonine kinases, including the Akt (5) and Pim (6) kinases, as potent suppressors of myc-induced apoptosis. As described below, these kinases coordinately regulate both apoptosis and cellular metabolism. The ability to reproducibly suppress the strong apoptotic stimulus of myc expression might serve as a criterion to identify other survival kinases.

Another characteristic of survival kinases is that they are activated by extracellular survival signals through cell surface receptors. Most receptors that can promote cell survival engage multiple signal transduction pathways. Many signaling pathways associated with activated receptor tyrosine kinases — including Src, phospholipase Cγ (PLCγ), and Ras/Raf/MEK/MAPK signaling — appear to promote cell survival. However, the central role of PI3K and Akt in receptor-mediated regulation of cell survival has been demonstrated in a variety of cell types. For example, in VSMCs expressing a number of PDGFR genes that are mutant for 1 or multiple binding sites necessary to activate the Src, Ras, PLCγ, or PI3K signaling pathways, growth factor–induced activation of PI3K/Akt signaling is the only kinase pathway that can prevent cell death induced by diverse stimuli when other kinase pathways are inactivated (7). These findings suggest that many kinase signaling pathways impact cell survival by direct or indirect contributions to PI3K/Akt signaling.

Another family of kinases that satisfies the criteria for survival kinases, and whose function does not appear to be dependent on PI3K/Akt signaling, is the Pim kinase family. The Pim kinases were originally implicated in cell survival by their ability to suppress myc-induced apoptosis in a mouse model of lymphoma (6, 8). Unlike the other serine/threonine kinases mentioned thus far, these kinases are not regulated by membrane recruitment or phosphorylation. The Pim kinases are unusual in that they are regulated primarily by transcription. Activated cytokine receptors recruit JAKs to induce STAT-dependent transcription of the Pim genes. While the role of Akt in promoting the survival of both normal and malignant cells is well established, the role of Pim signaling for cell survival in nontransformed cells has only recently been identified (9).

Although there are numerous pharmacological agents in preclinical and clinical development that can induce cell death by targeting other serine/threonine kinases, discussion of these

Nonstandard abbreviations used: ILK, integrin-linked kinase; mTOR, mammalian target of rapamycin; PDK, phosphoinositide-dependent kinase; PIP$_3$, phosphatidylinositide 3,4,5-triphosphate; PLCγ, phospholipase Cγ; PTEN, phosphatase and tensin homolog; SGK1, serum- and glucocorticoid-inducible kinase-1.

Conflict of interest: The authors have declared that no conflict of interest exists.

Citation for this article: *J. Clin. Invest.* **115**:2618–2624 (2005). doi:10.1172/JCI26273.

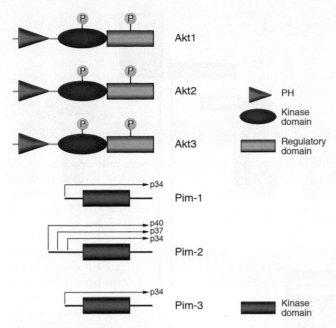

Figure 1
Domain structure of the Akt and Pim kinases. The structures of human Akt1, Akt2, and Akt3 consist of a pleckstrin homology domain (PH) that binds to PIP$_3$ at membrane surfaces, the kinase domain, and the regulatory domain. The 2 phosphorylation sites necessary for Akt activation are shown. The structures of human Pim-1, Pim-2, and Pim-3 demonstrate a conserved kinase domain and no regulatory domain. There are no required phosphorylation sites for Pim activation. Alternate start codons are depicted in Pim-2 leading to multiple Pim-2 isoforms that retain kinase activity.

agents and their targets is beyond the scope of this Review. We will focus on the structure, activation, and pharmacological manipulation of kinases that promote cell survival through the PI3K/Akt and JAK/STAT/Pim pathways.

Structure and regulation of survival kinase activation
There are 8 mammalian isoforms of PI3K, separated into class IA, class IB, class II, and class III. Class I PI3Ks are the only kinases that generate phosphatidylinositide 3,4,5-triphosphate (PIP$_3$). The most common form of class I PI3K associated with receptor tyrosine kinases is a heterodimer consisting of a p110 catalytic subunit and a p85 regulatory subunit. There are 3 p110 isoforms, p110α, p110β, and p110δ, the first 2 of which are widely expressed and the third of which is restricted to lymphocytes (10).

There are 3 isoforms of both Akt and Pim kinases, each encoded by distinct genes (Figure 1). The 3 Akt isoforms, Akt1, Akt2, and Akt3, share a highly conserved pleckstrin homology domain that binds these kinases to membrane surfaces through PIP$_3$. They have highly conserved catalytic and regulatory domains that must undergo phosphorylation for complete kinase activation. When overexpressed and constitutively membrane-localized through myristoylation, all 3 isoforms have the capacity to transform cells in vitro and in vivo (11). Most tissues express all 3 isoforms but at variable levels.

The regulation of Akt is primarily posttranslational. Once translated, the Akt kinases bind to heat shock protein 90 (Hsp90), protecting the inactive Akt proteins from proteosomal

degradation (12). Akt is recruited to the cell membrane through PIP$_3$ produced by the lipid kinase activity of PI3K (for review see ref. 13). PI3K is directly associated with many cell surface growth factor and cytokine receptors, and upon ligand binding, PI3K activation generates PIP$_3$. In addition to Akt, PIP$_3$ recruits phosphoinositide-dependent kinase-1 (PDK1) and integrin-linked kinase (ILK) to the cell membrane. Generation of PIP$_3$ is negatively regulated by the activity of phosphatase and tensin homolog (PTEN). *PTEN* deletion is the most common mechanism of inappropriate Akt activation in human malignancy (14). Akt activation requires 2 phosphorylation events: (a) PDK1 phosphorylation of Akt, and (b) phosphorylation of Akt by a kinase activity referred to as PDK2. Candidate kinases whose activities have been associated with PDK2 activity include ILK (15), DNA-dependent protein kinase (16), and PKCα (17, 18). Recently, the rictor–mammalian target of rapamycin (rictor-mTOR) complex has been suggested as the major contributor to PDK2 kinase activity (19). While ILK activity may contribute to Akt-dependent cell survival, ILK has Akt-independent survival functions as well. Activation of ILK by the cytoplasmic kinase domains of integrin and growth factor receptors (20) maintains cell structure through its cytoskeletal binding partners.

One PI3K-dependent survival kinase that does not phosphorylate Akt directly but augments the activity of Akt is the serum- and glucocorticoid-inducible kinase-1 (SGK1). SGK1 expression is regulated by the transcriptional activity of ligand-bound glucocorticoid receptor (21). In addition, diverse cellular insults such as osmotic stress, ultraviolet radiation, heat, and H$_2$O$_2$ result in induction of SGK1 expression (22). Although SGK1 does not require binding to PIP$_3$, it is PI3K-dependent, because, like Akt, it depends on phosphorylation by PDK1 and PDK2 kinase activities for activation.

In contrast to the Akt kinases, the Pim kinases do not have a regulatory domain and, based on recent crystallography findings, are likely constitutively active when expressed (23). Pim kinase regulation occurs at the level of transcription, translation, and proteosomal degradation. In lymphocytes, upon cytokine engagement of its receptor, JAK phosphorylates and activates STAT proteins. Once phosphorylated, STATs translocate to the nucleus and serve as transcription factors for the Pim genes. In addition to transcriptional control, regulation of *pim* mRNA stability is also a determinant of Pim activity (24). Adding to the complexity of Pim regulation and activity is the fact that the gene for *pim-2* encodes multiple proteins that have the same catalytic activity. Pim proteins are rapidly turned over by proteosomal degradation (25).

Survival kinases regulate common substrates
Once activated, the PI3K-dependent survival kinases and the Pim kinases phosphorylate common substrates that are involved in apoptosis and metabolism (Figure 2). Akt and Pim both directly phosphorylate and inactivate the proapoptotic Bcl-2 protein Bad (26–28). Both Akt and Pim kinases phosphorylate different components of proteins that are critical for maintaining a high rate of protein translation. For example, Akt phosphorylates TSC2, which controls mTOR activity, and mTOR and Pim kinases phosphorylate and inactivate the translational repressor 4EBP1. Akt and SGK1 act in concert to phosphorylate and inactivate FKHRL1, a transcription factor that upregulates proapoptotic Bcl-2 proteins such as Bim and death receptor compo-

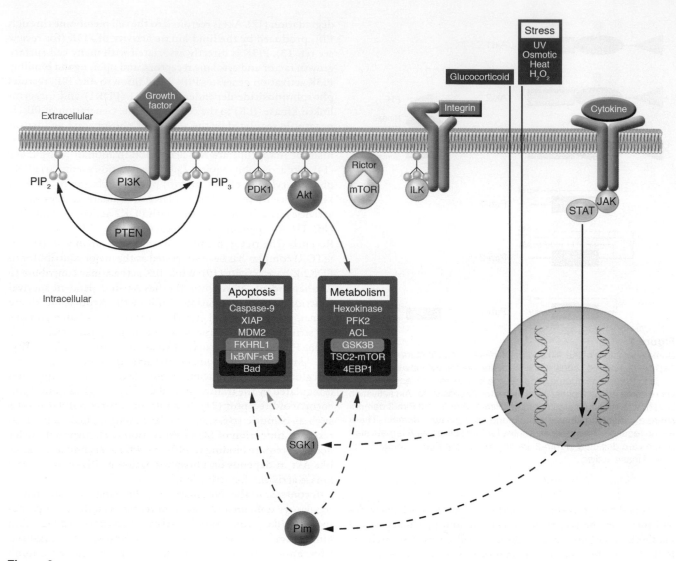

Figure 2
Survival kinases regulate cell death through the phosphorylation of common substrates of the apoptotic machinery and cellular metabolism. PI3K generates PIP$_3$, and PTEN converts PIP$_3$ back to PIP$_2$, negatively regulating PI3K signaling. PIP$_3$ recruits PDK1, ILK, and Akt to the cell membrane. PDK1, ILK, and the rictor-mTOR complex are important for the activation of Akt. Expression of both SGK1 and Pim kinases is inducible. Akt, SGK1, and Pim kinases share common substrates of the apoptosis machinery and cellular metabolism, depicted as color-coded overlapping boxes (see text for full description). PFK2, phosphofructokinase-2.

nents such as DR5 (29). Akt (30) and Pim kinases (31) regulate the IκB/NF-κB transcription factor complex by phosphorylating the serine/threonine kinase Cot (32), resulting in the proteasomal degradation of IκB, the activation of NF-κB, and the transcription of an array of antiapoptotic genes. Both Akt and Pim kinases also phosphorylate GSK3B, a regulator of cellular glucose metabolism. Both the Akt and the Pim kinase maintenance of cell survival is dependent on their ability to stimulate glucose uptake and metabolism (33, 34).

In addition to these common substrates, Akt has the potential to inactivate 3 pathways of apoptosis initiation: (a) p53-mediated apoptosis, by phosphorylation and activation of MDM2, a protein that binds p53 and facilitates its degradation (35); (b) mitochondrial-dependent apoptosis, by phosphorylation and inactivation of caspase-9 (36) and Bad, and phosphorylation and stabilization of the antiapoptotic protein XIAP (37); and (c)

death receptor–mediated apoptosis, by inhibition of the Forkhead family of transcription factors (38). Akt directly controls cellular metabolism by maintaining the association of hexokinase with mitochondria (39) and by phosphorylating and activating ATP-citrate lyase and phosphofructokinase-2 (40). Akt controls the translation of nutrient transporters through the activation of mTOR (41), and the localization of glucose transporters to the plasma membrane (42).

A number of drugs have emerged that target pathways downstream of both Akt and Pim signaling. One example is bortezomib, which inhibits proteasome function, leading to the inhibition of NF-κB, and is now used for multiple myeloma (43). Rapamycin derivatives, which inhibit mTOR, are now being developed for treatment of a broad range of solid tumor malignancies. Strategies such as RNA interference could be used to further validate other survival kinase substrates as pharmacological targets.

Table 1
Survival kinase inhibitors

Drug	Target(s)	Level of evidence	Comments	References
ATP-competitive inhibitors				
LY294002	PI3K, CK2, Pim-1	Mouse xenograft model	Broad specificity	43, 73
Wortmannin	PI3K	Mouse xenograft model	Short half-life	43
Methylxanthines	p110δ (PI3K)	Theophylline used in humans	Weak isoform-specific inhibitor	43
IC87114	p110δ (PI3K)	Prevents anaphylaxis in mice	Isoform-specific PI3K inhibitor	44
NSAIDs: sulindac, celecoxib	PDK1	Prevents colonic polyps in FAP	Trials terminated because of cardiovascular events	56–59
OSU-03012	PDK1	Human cancer cell lines	Coxib derivative with no anti-COX2 activity	60, 61
UCN-01	PDK1, CHK1, others	Human phase I trials	Staurosporine analog	55
KP-392	ILK	Mouse xenograft models	Antiangiogenic effect	66
AG490	JAK2	Mouse xenograft models	Synergy with imatinib	84
Non–ATP-competitive inhibitors				
API-2 (triciribine)	Akt1, Akt2, Akt3	Human cancer trials in 1980s	Caused hyperglycemia, hepatotoxicity; revisited at a lower dose	47
Pyridine derivatives	Akt1, Akt2, Akt1/2	Apoptosis in cancer cell lines	Akt1 and Akt2 both need to be inhibited	54
Perifosine	Akt	Phase II trials in breast cancer	Severe gastrointestinal side effects may be limiting	48–50
Rapamycin derivatives	mTOR	Phase II/III trials in multiple malignancies	Antitumor activity at low doses	67
Cucurbitacin I	JAK/STAT3	Mouse xenograft model	Natural product	78

A summary of some reported kinase inhibitors, their specificity, and their current role as research tools or therapeutics. CK2, casein kinase II; FAP, familial adenomatous polyposis.

Pharmacological inhibition of survival kinases

The problem of multiple isoforms: PI3K and Akt. Class I PI3K inhibitors such as wortmannin and LY294002 (Table 1) compete at the ATP binding site of the lipid kinase catalytic domain of all PI3Ks. Although these inhibitors have shown some efficacy in xenograft tumor models, they have not been developed for clinical use because of the broad specificity of kinase inhibition, poor pharmacokinetics, and relatively weak inhibition (44). Moreover, the recent finding that LY294002 binds and inhibits the activity of Pim-1 suggests that LY294002 can no longer be considered a selective tool to study PI3K-dependent biology. Currently a number of initiatives are under way that have identified isoform-selective ATP-competitive inhibitors of the PI3K. A potential validation of PI3K as a targetable survival kinase comes in the form of IC87114, a p110δ-specific PI3K inhibitor, which has been shown to prevent anaphylaxis in a mouse model (45).

Development of a specific inhibitor of Akt has proven difficult because of issues of toxicity. Concerns have been raised about Akt's role in insulin signaling, and the implications of chronic inhibition of this pathway. There is some evidence that the order of importance in insulin signaling for Akt isoforms is Akt2 > Akt1 > Akt3 (46, 47). Although mild hyperglycemia induced by an Akt inhibitor may be a tolerable side effect for cancer patients, drugs that induce overt diabetes in patients with refractory malignancy might not be seen as worthy of development. An example is the old drug API-2, also known as triciribine, which was tested in human cancer trials in the 1980s as a nucleoside analog. This drug was never fully developed, because it caused hepatotoxicity, hypertriglyceridemia, and hyperglycemia. More recently, this drug was shown to be a non–isoform-specific Akt inhibitor at much lower doses than previously tested (48). Another Akt inhibitor that has now entered phase II trials is the orally bioavailable alkylphospholipid perifosine. Perifosine likely interferes with proper Akt membrane local-

ization, leading to Akt dephosphorylation (49). In phase I trials of perifosine for patients with refractory solid tumors, a few patients benefited from partial responses and disease stabilization with no evidence of hyperglycemia, but gastrointestinal side effects were dose-limiting for the majority of patients (50, 51).

Which Akt isoform would serve as the most effective target in human cancers is also an unanswered question. Gene amplification of *AKT2* has been reported in pancreatic (52), breast, and ovarian cancers (53), while gene amplification of *AKT3* was found to contribute to the progression of sporadic melanomas (54). Numerous discovery programs are actively engaged in identifying isoform-selective Akt inhibitors. A series of pyridine derivatives were found to selectively inhibit Akt1, or Akt1 and Akt2, by binding to the pleckstrin homology domain (55). Currently, no Akt3-specific inhibitors have been reported.

Single-isoform survival kinases PDK1, ILK, and mTOR. PDK1 is a PI3K-dependent kinase that has been identified as the target of numerous inhibitors. Staurosporine analogs UCN-01 and CGP41251 inhibit PDK1, among other kinases, and have entered phase I cancer trials (56). Another example of an effective multitarget drug class that inhibits PDK1 is NSAIDs. NSAIDs such as sulindac (57) and celecoxib (58), a selective COX-2 inhibitor, were found to have the ability to induce apoptosis in colon cancer cells through inhibition of PDK1. A chemoprevention trial demonstrated that celecoxib reduced the incidence of polyp formation in patients with familial adenomatous polyposis (59). Unfortunately, further clinical evaluation of NSAIDs as chemopreventatives will require caution. An increased incidence of myocardial infarction and stroke in trial participants taking rofecoxib and high-dose celecoxib led to a moratorium on new chemopreventative trials with coxibs (COX-2 inhibitors) (60). Although the mechanism of increased thrombosis induced by COX-2 inhibitors remains unknown, many believe it is a COX-dependent phenomenon, opening the door for coxib-derived molecules that more selectively target PDK1 and not the COX enzymes. Recently, a coxib derivative, OSU-03012, which has no activity against COX enzymes, was found to inhibit PDK1/Akt activity in prostate cancer cells (61) and sensitize imatinib-resistant *BCR-ABL* clones to imatinib (62).

ILK, another PI3K-dependent kinase implicated in the activation of Akt, has been found to play an important role in malignant pathogenesis. Overexpression of ILK correlates with the clinical stage of many epithelial neoplasms (63–65). The importance of targeting ILK is further supported by the finding that ILK activity is essential for VEGF-dependent tumor angiogenesis (66). Two small-molecule inhibitors of ILK activity have been identified and are currently being developed for clinical trials (67).

While the rictor-mTOR complex responsible for Akt phosphorylation is rapamycin-insensitive, rapamycin-sensitive mTOR activity contributes to Akt-dependent cell survival. Rapamycin and the rapamycin derivatives CCI-779, RAD001, and AP23573 are currently in multiple phase II and phase III clinical trials for both solid tumors and hematological malignancies (68) and have been found to be most effective in tumors with *PTEN* deletion or Akt activation; this suggests that mTOR inhibitors act to suppress PI3K/Akt–induced cell survival.

Novel targets: Pim kinases. Pim overexpression has been reported in diffuse B cell lymphoma, chronic lymphocytic leukemia, and prostate cancer, and FLT3-mediated acute myelogenous leukemia (69–72). Besides cancer, Pim kinase activity has been shown to be important in the pathogenesis of vascular smooth muscle proliferation in vessel injury models (73). The only reported inhibitor of Pim function is LY294002, which was originally identified as a specific PI3K inhibitor (74). As knockout of all 3 *pim* genes leads to a mild phenotype, isoform-specific Pim inhibitors may not be necessary to avoid toxicity (75).

Targeting of the JAK/STAT pathway upstream of Pim expression is an active area of drug discovery. Recently, multiple groups have identified activating *JAK2* mutations in a large number of patients with myeloproliferative disorders (76–78). A natural product, cucurbitacin I, was found to specifically inhibit JAK/STAT3 signaling and lead to tumor cell death in a xenograft model (79). Further work is necessary to understand the relative contribution of Pim inhibition to the therapeutic efficacy of JAK/STAT inhibitors.

Activating survival kinases to preserve organ function

A few strategies involving survival kinase activation have emerged to address the problem of postischemic cardiac remodeling leading to heart failure. In rodent models of myocardial infarction, injection of bone marrow–derived mesenchymal stem cells expressing constitutively activated Akt (80) and injection of purified thymosin B4 (81), a naturally secreted peptide that activates ILK, were both found to enhance cardiomyocyte survival and organ function.

These data suggest that pharmacological activation of survival kinase activity for postinfarct remodeling could be a promising strategy. The empirical benefit of glucocorticoids, such as prednisone or dexamethasone, seen in a variety of diseases involving epithelial injury may in fact be an unappreciated example of this strategy. In addition to suppression of an exuberant immune response, glucocorticoid activation of SGK1 in epithelial cells may contribute to organ preservation in scenarios such as radiation injury.

In contrast to pharmacological activation of survival kinases, pharmacological inhibition of protein phosphatases could be another fruitful strategy for organ preservation. Inhibition of PTEN or PP2A could be a means of activating Akt pharmacologically. Phosphatase structure and regulation are complex, however. Recently, α4, a noncatalytic subunit of PP2A, was shown to be essential for cell survival, which suggests that certain conformations of phosphatase activity can mimic survival kinase function (82). Further consideration of the regulation of protein phosphatases is required before they can be pursued as therapeutic targets.

Finally, the PPARs are intriguing candidates for inducing Akt activation. PPARβ/δ was recently shown to coordinately upregulate PDK1 and ILK while downregulating expression of PTEN in keratinocytes, which implicates this nuclear receptor as an Akt regulator (83). Currently, selective PPARβ/δ agonists such as GW1505 are being developed for treatment of mucosal injury.

Conclusions

Recent advances in survival kinase structure and regulation have identified many potential targets for novel agents that can pharmacologically manipulate cell death. In turn, the basic understanding of the complexities of PI3K/Akt signaling with relation to cell survival will be enriched by the emergence of a new generation of more specific kinase inhibitors. The role of Pim kinases in cell survival will be better understood with the development of specific inhibitors. Evidence already exists to suggest that inhibitors of survival kinases can contribute to cancer therapy. Therapies designed to activate survival kinases may also be useful in preventing organ damage following injury. As our understanding of programmed cell death evolves, additional kinases that contribute to the regulation of cell survival will undoubtedly emerge.

Acknowledgments

We thank Casey Fox and Peter Hammerman for numerous helpful comments. R. Amaravadi is supported by NIH grant R25-CA87812.

Address correspondence to: Craig B. Thompson, University of Pennsylvania, Abramson Family Cancer Research Institute, 421 Curie Boulevard, Room 450 BRB II/III, Philadelphia, Pennsylvania 19104-6160, USA. Phone: (215) 746-5527; Fax: (215) 746-5511; E-mail: craig@mail.med.upenn.edu.

1. Kawauchi, K., Ogasawara, T., Yasuyama, M., and Ohkawa, S. 2003. Involvement of Akt kinase in the action of STI571 on chronic myelogenous leukemia cells. *Blood Cells Mol. Dis.* **31**:11–17.

2. Sordella, R., Bell, D.W., Haber, D.A., and Settleman, J. 2004. Gefitinib-sensitizing EGFR mutations in lung cancer activate anti-apoptotic pathways. *Science.* **305**:1163–1167.

3. Zong, W.X., Ditsworth, D., Bauer, D.E., Wang, Z.Q., and Thompson, C.B. 2004. Alkylating DNA damage stimulates a regulated form of necrotic cell death. *Genes Dev.* **18**:1272–1282.

4. Levine, B., and Yuan, J. 2005. Autophagy in cell death: an innocent convict? *J. Clin. Invest.* **115**:2679–2688. doi:10.1172/JCI26390.

5. Wendel, H.G., et al. 2004. Survival signalling by Akt and eIF4E in oncogenesis and cancer therapy. *Nature.* **428**:332–337.

6. van Lohuizen, M., et al. 1989. Predisposition to lymphomagenesis in pim-1 transgenic mice: cooperation with c-myc and N-myc in murine leukemia virus-induced tumors. *Cell.* **56**:673–682.

7. Vantler, M., Caglayan, E., Zimmermann, W.H., Baumer, A.T., and Rosenkranz, S. 2005. Systematic evaluation of anti-apoptotic growth factor signaling in vascular smooth muscle cells: only phospha-

tidylinositol 3′-kinase is important. *J. Biol. Chem.* **280**:14168–14176.

8. Hammerman, P.S., Fox, C.J., Birnbaum, M.J., and Thompson, C.B. 2005. The Pim and Akt oncogenes are independent regulators of hematopoietic cell growth and survival. *Blood.* **105**:4477–4483.

9. Fox, C.J., Hammerman, P.S., and Thompson, C.B. 2005. The Pim kinases control rapamycin-resistant T cell survival and activation. *J. Exp. Med.* **201**:259–266.

10. Fruman, D.A. 2004. Towards an understanding of isoform specificity in phosphoinositide 3-kinase signalling in lymphocytes. *Biochem. Soc. Trans.*

32:315–319.

11. Mende, I., Malstrom, S., Tsichlis, P.N., Vogt, P.K., and Aoki, M. 2001. Oncogenic transformation induced by membrane-targeted Akt2 and Akt3. *Oncogene.* **20**:4419–4423.

12. Munster, P.N., Marchion, D.C., Basso, A.D., and Rosen, N. 2002. Degradation of HER2 by ansamycins induces growth arrest and apoptosis in cells with HER2 overexpression via a HER3, phosphatidylinositol 3′-kinase-AKT-dependent pathway. *Cancer Res.* **62**:3132–3137.

13. Vivanco, I., and Sawyers, C.L. 2002. The phosphatidylinositol 3-kinase AKT pathway in human cancer. *Nat. Rev. Cancer.* **2**:489–501.

14. Ali, I.U., Schriml, L.M., and Dean, M. 1999. Mutational spectra of PTEN/MMAC1 gene: a tumor suppressor with lipid phosphatase activity. *J. Natl. Cancer Inst.* **91**:1922–1932.

15. Hannigan, G., Troussard, A.A., and Dedhar, S. 2005. Integrin-linked kinase: a cancer therapeutic target unique among its ILK. *Nat. Rev. Cancer.* **5**:51–63.

16. Feng, J., Park, J., Cron, P., Hess, D., and Hemmings, B.A. 2004. Identification of a PKB/Akt hydrophobic motif Ser-473 kinase as DNA-dependent protein kinase. *J. Biol. Chem.* **279**:41189–41196.

17. Partovian, C., and Simons, M. 2004. Regulation of protein kinase B/Akt activity and Ser473 phosphorylation by protein kinase Calpha in endothelial cells. *Cell. Signal.* **16**:951–957.

18. Fukuda, T., Guo, L., Shi, X., and Wu, C. 2003. CH-ILKBP regulates cell survival by facilitating the membrane translocation of protein kinase B/Akt. *J. Cell Biol.* **160**:1001–1008.

19. Sarbassov, D.D., Guertin, D.A., Ali, S.M., and Sabatini, D.M. 2005. Phosphorylation and regulation of Akt/PKB by the rictor-mTOR complex. *Science.* **307**:1098–1101.

20. Fukuda, T., Chen, K., Shi, X., and Wu, C. 2003. PINCH-1 is an obligate partner of integrin-linked kinase (ILK) functioning in cell shape modulation, motility, and survival. *J. Biol. Chem.* **278**:51324–51333.

21. Wu, W., et al. 2004. Microarray analysis reveals glucocorticoid regulated survival genes that are associated with inhibition of apoptosis in breast epithelial cells. *Cancer Res.* **64**:1757–1764.

22. Leong, M.L., Maiyar, A.C., Kim, B., O'Keeffe, B.A., and Firestone, G.L. 2003. Expression of the serum- and glucocorticoid-inducible protein kinase, Sgk, is a cell survival response to multiple types of environmental stress stimuli in mammary epithelial cells. *J. Biol. Chem.* **278**:5871–5882.

23. Qian, K.C., et al. 2004. Structural basis of constitutive activity and a unique nucleotide binding mode of human Pim-1 kinase. *J. Biol. Chem.* **280**:6130–6137.

24. Yip-Schneider, M.T., Horie, M., and Broxmeyer, H.E. 1995. Transcriptional induction of pim-1 protein kinase gene expression by interferon gamma and posttranscriptional effects on costimulation with steel factor. *Blood.* **85**:3494–3502.

25. Wang, Z., et al. 2001. Pim-1: a serine/threonine kinase with a role in cell survival, proliferation, differentiation and tumorigenesis. *J. Vet. Sci.* **2**:167–179.

26. Aho, T.L., et al. 2004. Pim-1 kinase promotes inactivation of the pro-apoptotic Bad protein by phosphorylating it on the Ser112 gatekeeper site. *FEBS Lett.* **571**:43–49.

27. Yan, B., et al. 2003. The PIM-2 kinase phosphorylates BAD on serine 112 and reverses BAD-induced cell death. *J. Biol. Chem.* **278**:45358–45367.

28. Datta, S.R., et al. 1997. Akt phosphorylation of BAD couples survival signals to the cell-intrinsic death machinery. *Cell.* **91**:231–241.

29. Brunet, A., et al. 2001. Protein kinase SGK mediates survival signals by phosphorylating the forkhead transcription factor FKHRL1 (FOXO3a). *Mol. Cell. Biol.* **21**:952–965.

30. Kane, L.P., Shapiro, V.S., Stokoe, D., and Weiss, A. 1999. Induction of NF-kappaB by the Akt/PKB kinase. *Curr. Biol.* **9**:601–604.

31. Hammerman, P.S., et al. 2004. Lymphocyte transformation by Pim-2 is dependent on nuclear factor-kappaB activation. *Cancer Res.* **64**:8341–8348.

32. Kane, L.P., Mollenauer, M.N., Xu, Z., Turck, C.W., and Weiss, A. 2002. Akt-dependent phosphorylation specifically regulates cot induction of NF-kappaB-dependent transcription. *Mol. Cell Biol.* **22**:5962–5974.

33. Plas, D.R., Talapatra, S., Edinger, A.L., Rathmell, J.C., and Thompson, C.B. 2001. Akt and Bcl-xL promote growth factor-independent survival through distinct effects on mitochondrial physiology. *J. Biol. Chem.* **276**:12041–12048.

34. Fox, C.J., et al. 2003. The serine/threonine kinase Pim-2 is a transcriptionally regulated apoptotic inhibitor. *Genes Dev.* **17**:1841–1854.

35. Mayo, L.D., and Donner, D.B. 2001. A phosphatidylinositol 3-kinase/Akt pathway promotes translocation of Mdm2 from the cytoplasm to the nucleus. *Proc. Natl. Acad. Sci. U. S. A.* **98**:11598–11603.

36. Cardone, M.H., et al. 1998. Regulation of cell death protease caspase-9 by phosphorylation. *Science.* **282**:1318–1321.

37. Kops, G.J., et al. 1999. Direct control of the Forkhead transcription factor AFX by protein kinase B. *Nature.* **398**:630–634.

38. Accili, D., and Arden, K.C. 2004. FoxOs at the crossroads of cellular metabolism, differentiation, and transformation. *Cell.* **117**:421–426.

39. Majewski, N., et al. 2004. Hexokinase-mitochondria interaction mediated by Akt is required to inhibit apoptosis in the presence or absence of Bax and Bak. *Mol. Cell.* **16**:819–830.

40. Deprez, J., Vertommen, D., Alessi, D.R., Hue, L., and Rider, M.H. 1997. Phosphorylation and activation of heart 6-phosphofructo-2-kinase by protein kinase B and other protein kinases of the insulin signaling cascades. *J. Biol. Chem.* **272**:17269–17275.

41. Edinger, A.L., and Thompson, C.B. 2002. Akt maintains cell size and survival by increasing mTOR-dependent nutrient uptake. *Mol. Biol. Cell.* **13**:2276–2288.

42. van Dam, E.M., Govers, R., and James, D.E. 2005. Akt activation is required at a late stage of insulin-induced GLUT4 translocation to the plasma membrane. *Mol. Endocrinol.* **19**:1067–1077.

43. Rajkumar, S.V., Richardson, P.G., Hideshima, T., and Anderson, K.C. 2005. Proteasome inhibition as a novel therapeutic target in human cancer. *J. Clin. Oncol.* **23**:630–639.

44. Workman, P. 2004. Inhibiting the phosphoinositide 3-kinase pathway for cancer treatment. *Biochem. Soc. Trans.* **32**:393–396.

45. Ali, K., et al. 2004. Essential role for the p110delta phosphoinositide 3-kinase in the allergic response. *Nature.* **431**:1007–1011.

46. Cho, H., et al. 2001. Insulin resistance and a diabetes mellitus-like syndrome in mice lacking the protein kinase Akt2 (PKB beta). *Science.* **292**:1728–1731.

47. Masure, S., et al. 1999. Molecular cloning, expression and characterization of the human serine/threonine kinase Akt-3. *Eur. J. Biochem.* **265**:353–360.

48. Yang, L., et al. 2004. Akt/protein kinase B signaling inhibitor-2, a selective small molecule inhibitor of Akt signaling with antitumor activity in cancer cells overexpressing Akt. *Cancer Res.* **64**:4394–4399.

49. Kondapaka, S.B., Singh, S.S., Dasmahapatra, G.P., Sausville, E.A., and Roy, K.K. 2003. Perifosine, a novel alkylphospholipid, inhibits protein kinase B activation. *Mol. Cancer Ther.* **2**:1093–1103.

50. Van Ummersen, L., et al. 2004. A phase I trial of perifosine (NSC 639966) on a loading dose/maintenance dose schedule in patients with advanced cancer. *Clin. Cancer Res.* **10**:7450–7456.

51. Crul, M., et al. 2002. Phase I and pharmacological study of daily oral administration of perifosine (D-21266) in patients with advanced solid tumours. *Eur. J. Cancer.* **38**:1615–1621.

52. Cheng, J.Q., et al. 1992. AKT2, a putative oncogene encoding a member of a subfamily of protein-serine/threonine kinases, is amplified in human ovarian carcinomas. *Proc. Natl. Acad. Sci. U. S. A.* **89**:9267–9271.

53. Bellacosa, A., et al. 1995. Molecular alterations of the AKT2 oncogene in ovarian and breast carcinomas. *Int. J. Cancer.* **64**:280–285.

54. Stahl, J.M., et al. 2004. Deregulated Akt3 activity promotes development of malignant melanoma. *Cancer Res.* **64**:7002–7010.

55. Barnett, S.F., et al. 2005. Identification and characterization of pleckstrin-homology-domain-dependent and isoenzyme-specific Akt inhibitors. *Biochem. J.* **385**:399–408.

56. Senderowicz, A.M. 2002. The cell cycle as a target for cancer therapy: basic and clinical findings with the small molecule inhibitors flavopiridol and UCN-01. *Oncologist.* **7**(Suppl. 3):12–19.

57. Piazza, G.A., et al. 1997. Apoptosis primarily accounts for the growth-inhibitory properties of sulindac metabolites and involves a mechanism that is independent of cyclooxygenase inhibition, cell cycle arrest, and p53 induction. *Cancer Res.* **57**:2452–2459.

58. Arico, S., et al. 2002. Celecoxib induces apoptosis by inhibiting 3-phosphoinositide-dependent protein kinase-1 activity in the human colon cancer HT-29 cell line. *J. Biol. Chem.* **277**:27613–27621.

59. Steinbach, G., et al. 2000. The effect of celecoxib, a cyclooxygenase-2 inhibitor, in familial adenomatous polyposis. *N. Engl. J. Med.* **342**:1946–1952.

60. Solomon, S.D., et al. 2005. Cardiovascular risk associated with celecoxib in a clinical trial for colorectal adenoma prevention. *N. Engl. J. Med.* **352**:1071–1080.

61. Zhu, J., et al. 2004. From the cyclooxygenase-2 inhibitor celecoxib to a novel class of 3-phosphoinositide-dependent protein kinase-1 inhibitors. *Cancer Res.* **64**:4309–4318.

62. Tseng, P.H., et al. 2005. Synergistic interactions between imatinib and the novel phosphoinositide-dependent kinase-1 inhibitor OSU-03012 in overcoming imatinib resistance. *Blood.* **105**:4021–4027.

63. Takanami, I. 2005. Increased expression of integrin-linked kinase is associated with shorter survival in non-small cell lung cancer. *BMC Cancer.* **5**:1.

64. Ito, R., et al. 2003. Expression of integrin-linked kinase is closely correlated with invasion and metastasis of gastric carcinoma. *Virchows Arch.* **442**:118–123.

65. Graff, J.R., et al. 2001. Integrin-linked kinase expression increases with prostate tumor grade. *Clin. Cancer Res.* **7**:1987–1991.

66. Tan, C., et al. 2004. Regulation of tumor angiogenesis by integrin-linked kinase (ILK). *Cancer Cell.* **5**:79–90.

67. Persad, S., et al. 2000. Inhibition of integrin-linked kinase (ILK) suppresses activation of protein kinase B/Akt and induces cell cycle arrest and apoptosis of PTEN-mutant prostate cancer cells. *Proc. Natl. Acad. Sci. U. S. A.* **97**:3207–3212.

68. Vignot, S., Faivre, S., Aguirre, D., and Raymond, E. 2005. mTOR-targeted therapy of cancer with rapamycin derivatives [review]. *Ann. Oncol.* **16**:525–537.

69. Alizadeh, A.A., et al. 2000. Distinct types of diffuse large B-cell lymphoma identified by gene expression profiling. *Nature.* **403**:503–511.

70. Valdman, A., Fang, X., Pang, S.T., Ekman, P., and Egevad, L. 2004. Pim-1 expression in prostatic intraepithelial neoplasia and human prostate cancer. *Prostate.* **60**:367–371.

71. Cohen, A.M., et al. 2004. Increased expression of the hPim-2 gene in human chronic lymphocytic

leukemia and non-Hodgkin lymphoma. *Leuk. Lymphoma.* **45**:951–955.

72. Mizuki, M., et al. 2003. Suppression of myeloid transcription factors and induction of STAT response genes by AML-specific Flt3 mutations. *Blood.* **101**:3164–3173.

73. Katakami, N., et al. 2004. Role of pim-1 in smooth muscle cell proliferation. *J. Biol. Chem.* **279**:54742–54749.

74. Jacobs, M.D., et al. 2005. PIM-1 ligand-bound structures reveal the mechanism of serine/threonine kinase inhibition by LY294002. *J. Biol. Chem.* **280**:13728–13734.

75. Mikkers, H., et al. 2004. Mice deficient for all PIM kinases display reduced body size and impaired responses to hematopoietic growth factors. *Mol. Cell. Biol.* **24**:6104–6115.

76. Jones, A.V., et al. 2005. Widespread occurrence of the JAK2 V617F mutation in chronic myeloproliferative disorders. *Blood.* doi:10.1182/blood-2005-03-1320.

77. Kralovics, R., et al. 2005. A gain-of-function mutation of JAK2 in myeloproliferative disorders. *N. Engl. J. Med.* **352**:1779–1790.

78. Levine, R.L., et al. 2005. Activating mutation in the tyrosine kinase JAK2 in polycythemia vera, essential thrombocythemia, and myeloid metaplasia with myelofibrosis. *Cancer Cell.* **7**:387–397.

79. Blaskovich, M.A., et al. 2003. Discovery of JSI-124 (cucurbitacin I), a selective Janus kinase/signal transducer and activator of transcription 3 signaling pathway inhibitor with potent antitumor activity against human and murine cancer cells in mice. *Cancer Res.* **63**:1270–1279.

80. Mangi, A.A., et al. 2003. Mesenchymal stem cells modified with Akt prevent remodeling and restore performance of infarcted hearts. *Nat. Med.* **9**:1195–1201.

81. Bock-Marquette, I., Saxena, A., White, M.D., Dimaio, J.M., and Srivastava, D. 2004. Thymosin beta4 activates integrin-linked kinase and promotes cardiac cell migration, survival and cardiac repair. *Nature.* **432**:466–472.

82. Kong, M., et al. 2004. The PP2A-associated protein alpha4 is an essential inhibitor of apoptosis. *Science.* **306**:695–698.

83. Di-Poi, N., Tan, N.S., Michalik, L., Wahli, W., and Desvergne, B. 2002. Antiapoptotic role of PPAR-beta in keratinocytes via transcriptional control of the Akt1 signaling pathway. *Mol. Cell.* **10**:721–733.

84. Sun, X., Layton, J.E., Elefanty, A., and Lieschke, G.J. 2001. Comparison of effects of the tyrosine kinase inhibitors AG957, AG490, and STI571 on BCR-ABL-expressing cells, demonstrating synergy between AG490 and STI571. *Blood.* **97**:2008–2015.

Mitochondria: pharmacological manipulation of cell death

Lisa Bouchier-Hayes, Lydia Lartigue, and Donald D. Newmeyer

La Jolla Institute for Allergy and Immunology, Department of Cellular Immunology, San Diego, California, USA.

Cell death by apoptosis or necrosis is often important in the etiology and treatment of disease. Since mitochondria play important roles in cell death pathways, these organelles are potentially prime targets for therapeutic intervention. Here we discuss the mechanisms through which mitochondria participate in the cell death process and also survey some of the pharmacological approaches that target mitochondria in various ways.

Introduction

Mitochondria are involved in many processes essential for cell survival, including energy production, redox control, calcium homeostasis, and certain metabolic and biosynthetic pathways. In addition, mitochondria often play an essential role in physiological cell death mechanisms. These genetically controlled mitochondrial pathways are often subject to dysfunction. Hence mitochondria can be central players in pathological conditions as diverse as cancer, diabetes, obesity, ischemia/reperfusion injury, and neurodegenerative disorders such as Parkinson and Alzheimer diseases. Research aimed at elucidating the role of mitochondria in cell death has become one of the fastest growing disciplines in biomedicine (1).

Mitochondria and cell death

The caspases are an apoptosis-related family of proteases that, upon activation, cleave specific substrates, leading to the demise of the cell (2). Although the signaling events upstream of caspases often remain obscure, in general, apoptotic stimuli engage caspases either through death receptor stimulation or through mitochondrial outer membrane permeabilization (MOMP) (Figure 1). With rare exceptions, MOMP leads to cell death even if caspases are inhibited (3). MOMP results in the release of multiple proteins from the mitochondrial intermembrane space (IMS) into the cytosol. This leads to caspase activation in the cytosol, loss of $\Delta\Psi_m$ (mitochondrial membrane potential), cellular ATP depletion, and free radical production (3). One of the released IMS proteins is cytochrome c, an important component of the mitochondrial respiratory chain. When translocated into the cytoplasm, cytochrome c stimulates the assembly of a multiprotein complex known as the Apaf-1 apoptosome. Caspase-9 is recruited to the apoptosome and activated, initiating a cascade of effector caspase activation (4). Other proteins released from the mitochondria during apoptosis include Smac/DIABLO, endonuclease G (Endo G), apoptosis-inducing factor (AIF), and HtrA2/Omi. Smac promotes caspase activation indirectly by neutralizing the effects of XIAP, a natural caspase inhibitor

(5, 6). The apoptotic roles of some IMS proteins remain controversial; a complicating issue is that these proteins may have essential functions in mitochondria whose dysregulation could affect cell survival through action upstream of MOMP (7, 8). Moreover, the release of Endo G and AIF from the mitochondria may require further caspase-independent events subsequent to MOMP (9, 10), such as cleavage of AIF by calpain (11).

Molecular mechanisms of MOMP

MOMP is thought to be a "point of no return" in the mitochondrial apoptotic pathway (12). As mitochondria are a potential therapeutic target for disorders connected with apoptosis dysregulation, it will be important to gain a detailed understanding of MOMP and its regulation. The mechanisms of MOMP have been controversial (12–19), and there are 2 principal hypotheses: in the first, MOMP is regulated by the Bcl-2 family of proteins, and in the second, by the permeability transition pore (PTP) (Figure 1). The first model considers MOMP as a process that is essentially intrinsic to the outer membrane and requires members of the Bcl-2 family of proteins to promote or prevent the formation of pores. Studies using vesicles formed from isolated mitochondrial outer membranes (MOMs) have shown that Bcl-2–family proteins can regulate the permeability of the MOM in the absence of interior structures of the mitochondria; moreover, many features of this process of membrane permeabilization can be reproduced using defined liposomes and recombinant Bcl-2–family proteins (20). However, these cell-free systems probably do not represent all the complexity of the permeabilization process as it occurs in the native MOM; other proteins of the MOM could modulate or potentiate the function of Bax and Bak. Moreover, the release of specific IMS proteins into the cytoplasm could be influenced by anchorage of these proteins to internal structures or entrapment in mitochondrial cristae (21). Proteins normally involved in mitochondrial fission and fusion may participate in MOMP (reviewed in refs. 3, 13, 22–24). However, this is a controversial topic, made especially challenging by the complex network of protein interactions involved in mitochondrial dynamics. The disruption of any element of this system could have pleiotropic and indirect effects that are difficult to interpret.

The second prominent model for MOMP is based on a phenomenon known as the mitochondrial permeability transition (PT). PT involves the rapid permeabilization of the inner mitochondrial membrane to solutes smaller than 1.5 kDa, through formation of the PT pore (PTP). The PTP complex is a "megapore" thought to span the contact sites between the inner and outer mitochondrial

Nonstandard abbreviations used: AIF, apoptosis-inducing factor; ANT, adenine nucleotide translocase; BH, Bcl-2 homology (domain); CsA, cyclosporin A; Cyp D, cyclophilin D; Endo G, endonuclease G; IMS, intermembrane space; MnSOD, manganese superoxide dismutase; MOM, mitochondrial outer membrane; MOMP, MOM permeabilization; PT, permeability transition; PTP, PT pore; VDAC, voltage-dependent anion channel.

Conflict of interest: The authors have declared that no conflict of interest exists.

Citation for this article: *J. Clin. Invest.* **115**:2640–2647 (2005).
doi:10.1172/JCI26274.

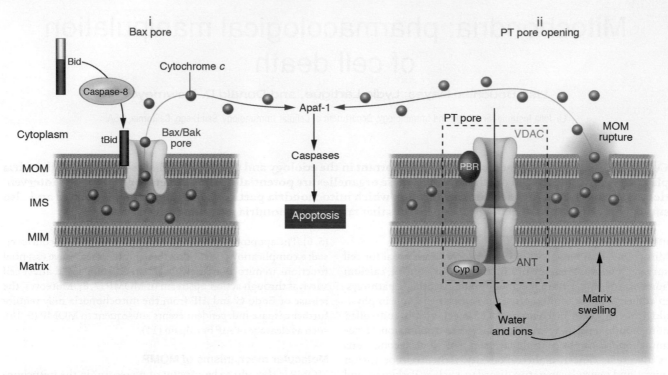

Figure 1
Molecular mechanisms of MOMP. The proposed models of MOMP leading to cytochrome *c* release are represented. (i) Bax pore. Bax or Bak forms a pore in the MOM after activation by a BH3-only protein such as Bid. (ii) PT pore opening. Opening of the PT pore allows an influx of water and ions into the matrix, causing matrix swelling; this leads to rupture of the MOM, releasing IMS proteins such as cytochrome *c*. MIM, mitochondrial inner membrane; PBR, peripheral benzodiazepine receptor.

membranes and is presumably composed of at least 3 proteins, namely the voltage-dependent anion channel (VDAC) in the outer membrane, the soluble matrix protein cyclophilin D (Cyp D), and the adenine nucleotide translocase (ANT) in the inner membrane (18), although the requirement for ANT is controversial (25, 26).

Under normal calcium homeostasis, the PTP exists in a state of low conductance; however, when excess Ca^{2+} is released from the endoplasmic reticulum and overloads the mitochondria, the pore transitions to a high-conductance state. This passage from low to high conductance is irreversible and strictly depends on the saturation of the calcium-binding sites of the PTP (27). The high-conductance conformation allows free diffusion of water and ions between the cytosol and the matrix, causing collapse of $\Delta\Psi_m$ (the inner membrane gradient), uncoupling of oxidative phosphorylation, and swelling of the mitochondrial matrix. This leads to rupture of the MOM and consequent release of IMS proteins.

A few proapoptotic stimuli, such as calcium overload, oxidative stress, or ischemia/reperfusion, seem to mediate cytochrome *c* release directly through the PT (28). However, the generality of PT as a primary mechanism for MOMP and apoptosis has been questioned, as mitochondrial matrix swelling is not always observed in apoptotic cells (29, 30); moreover, mitochondrial depolarization does not always precede the release of cytochrome *c* and is often blocked by caspase inhibition (31–33). Indeed, in such cases the loss of $\Delta\Psi_m$ does not result from PT, but rather from the caspase-dependent cleavage of at least 1 substrate in the respiratory chain: p75, a subunit of complex I (34). Cleavage of p75 disrupts oxygen consumption, increases ROS production, and is 1 of several events that may preclude recovery for the cell (35).

Fortunately, a clear resolution to this controversy over mechanisms of MOMP may be at hand. Two recent papers (36, 37) reported use of a genetic approach to determine how important PT is for cell death in vivo (for critical commentaries see refs. 38, 39). These groups produced mice defective in the gene encoding cyclophilin D (Cyp D) and observed that isolated mitochondria from these animals were defective in the PT response, except to extremely high concentrations of Ca^{2+}, consistent with observations that the drug cyclosporin A (CsA) also blocks PT only at lower Ca^{2+} concentrations. Strikingly, apoptotic cell death in these animals was entirely normal. On the other hand, some forms of cell death, such as those induced by oxygen radicals and Ca^{2+} overload, were defective. Thus, apoptotic death involving the actions of Bcl-2 family members at mitochondria does not involve Cyp D–dependent PT, whereas other forms of death expected to involve the PT response do depend on Cyp D. Despite the lack of evidence for PT as a mechanism of apoptotic death in general, it remains important to consider therapeutic strategies for inducing PT-mediated cell death selectively in certain cell types, as discussed below.

A unified model for the roles of Bcl-2–family proteins in regulating MOMP

The Bcl-2 family of proteins (Table 1) includes both proapoptotic and antiapoptotic members, each containing 1 or more Bcl-2 homology domains (BH1 through BH4) (40). Antiapoptotic members such as Bcl-2, Bcl-x_L, and Mcl-1 contain all 4 BH domains. The proapoptotic members are divided into 2 subgroups. One of these consists of Bax, Bak, and Bok, which possess domains BH1, BH2,

Table 1
Subdivisions of Bcl-2–family proteins, based on function and sequence homology

Category	Subcategory		Proteins	Domains
Antiapoptotic			Bcl-2, Bcl-x$_L$, Mcl-1, A1, Bcl-w	BH1, BH2, BH3, BH4
Proapoptotic	Multidomain		Bax, Bak, Bok	BH1, BH2, BH3
	BH3-only	Direct activator	Bid, Bim	BH3
		Derepressor	Bad, Bik, Bnip3, Puma, Noxa, Bmf, Hrk, others	BH3

and BH3; the other group consists of the more numerous BH3-only proteins, which include Bid, Bad, Puma, and several others (40).

While several Bcl-2–family proteins possess ion channel activity in lipid bilayers, only the multidomain proapoptotic proteins Bax and Bak can render membranes permeable to cytochrome c or larger macromolecules (15, 20). These 2 proteins exhibit a degree of functional redundancy in the mouse, as the ablation of both *bax* and *bak* genes results in a much more dramatic apoptotic phenotype than single *bax* or *bak* knockouts. Indeed, tissues and cells deficient in both genes are resistant to most stimuli that proceed through mitochondrial and endoplasmic reticulum–dependent pathways (41–44). The 2 proteins do not behave identically, however, since in nonapoptotic cells Bax is mostly free in the cytosol while Bak is constitutively localized in membranes of the mitochondria and endoplasmic reticulum (45).

Regardless of localization, Bax and Bak both seem to be in an inactive state in nonapoptotic cells. An activation process, which is not completely understood, is required for Bax to oligomerize and insert stably into the membrane (20, 46, 47). The tertiary structure of the antiapoptotic Bcl-2 family member Bcl-x$_L$ shows that its helical domains form a hydrophobic groove where the BH3-only proteins bind. Bax has a similar structural fold; however, in unactivated, soluble Bax, the C-terminal helix α9 occupies the BH3-binding groove (48), and helix α8 is also positioned to interfere with potential binding of BH3 domains (49). Thus, in order to interact with BH3 domains in other proteins, Bax must undergo a conformational alteration. Indeed, a Bax conformational change is commonly detected through the binding of conformation-sensitive antibodies (50), and the interaction of Bax with membranes is sufficient to enable a conformational change but not membrane insertion or pore formation (51). Bax oligomerization does not occur in solution (20), except in the presence of certain detergents (46), consistent with the idea that interaction with a hydrophobic environment is necessary to destabilize the Bax molecule and allow activation and oligomerization.

In what is commonly known as the "rheostat" model, cell survival is determined by the balance between antiapoptotic Bcl-2–family proteins such as Mcl-1, Bcl-x$_L$, or Bcl-2, and proapoptotic BH3-only proteins. Based on recent findings, we can now consider a more complex "switched rheostat" model (Figure 2) with several additional elements: First, Bax and Bak are the positive effector molecules, based on the biochemical and genetic studies mentioned above. Second, the antiapoptotic relatives act as inhibitors of this permeabilization process, not effectors of an independent survival mechanism, at least with regard to direct effects on MOMP. Third, MOMP mediated by Bax and Bak must be triggered by "direct activator" proteins, such as the BH3-only proteins Bim

or Bid; indeed, synthetic peptides corresponding to the BH3 domains of these proteins are able to activate Bax (20, 47). The p53 protein can also directly induce Bax-mediated membrane permeabilization in cells and liposomes, although it lacks a clearly identifiable BH3 domain (52, 53). Fourth, the antiapoptotic family members can oppose Bax and Bak as well as the BH3-only proteins (20). Fifth, BH3-only proteins fall into 2 functional classes: the direct activators Bim and Bid mentioned above, and the others, which act as "derepressors" by binding competitively to the antiapoptotic family members and thus freeing up the direct activator proteins to induce Bax/Bak–mediated MOMP (47, 54, 55). Finally, interactions between various BH3-only proteins and the antiapoptotic Bcl-2–family proteins exhibit a pattern of overlapping specificities that was previously unappreciated. Recent studies have shown that some of the BH3-only proteins bind to only a subset of the antiapoptotic proteins, and conversely the antiapoptotic proteins each bind to a subset of the BH3-only proteins (47, 56). For example, the BH3-only protein Noxa binds specifically to Mcl-1 and A1. In contrast, Puma, another BH3-only protein, binds to at least 5 different antiapoptotic relatives (56). Moreover, Bak is differentially inhibited by antiapoptotic Bcl-2–family proteins, further illustrating the intricacy of this protein network (57).

Adding to this complexity, a number of proteins unrelated to the Bcl-2 family interact with Bax and either inhibit (58–61) or enhance (58, 62) its activation. Bak has been reported to associate with VDAC2 at the mitochondrial membrane, and this inhibits the oligomerization and proapoptotic activity of Bak (63). Although their roles in cell death are not well understood, these

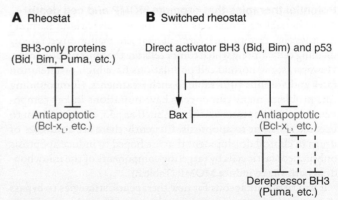

A Rheostat

BH3-only proteins (Bid, Bim, Puma, etc.)

Antiapoptotic (Bcl-x$_L$, etc.)

B Switched rheostat

Direct activator BH3 (Bid, Bim) and p53

Bax

Antiapoptotic (Bcl-x$_L$, etc.)

Derepressor BH3 (Puma, etc.)

Figure 2
Models of Bcl-2–family function at the mitochondrion during apoptosis. (**A**) The traditional simple rheostat model assumes that the antiapoptotic Bcl-2–family proteins antagonize the BH3-only proteins, in an equal and opposite manner. (**B**) Recent results suggest a more inclusive and detailed model, which we term the "switched rheostat." Bax and Bak are the effectors of MOMP. Certain BH3-only proteins ("direct activators") and p53 switch on Bax (and possibly Bak) directly and are antagonized by antiapoptotic Bcl-2–family proteins. Other BH3-only proteins ("derepressors") do not activate Bax directly but act by antagonizing the antiapoptotic family members, thereby freeing the direct activators to trigger Bax/Bak–induced MOMP. The dashed lines indicate that the derepressor BH3-only proteins have differing specificities for antiapoptotic family members.

proteins that modulate Bax or Bak activation could be important therapeutic targets (64).

Pathological processes involving MOMP

As MOMP is a crucial step for many pathways that induce apoptosis, disruption of this event would likely affect cell fate. For example, certain viruses kill the host cell by causing MOMP, while others prevent MOMP to allow propagation of the virus. The HIV protein Vpr (viral accessory protein R) induces MOMP; a mutation in this protein (R77Q) confers resistance to MOMP and is associated with a reduced risk of developing AIDS (65). In contrast, the cytomegalovirus vMIA protein blocks MOMP by forming inactive complexes with Bax in the MOM (66, 67). Ischemia/reperfusion–mediated injury results from excessive redox stress and mitochondrial uptake of Ca^{2+}, leading to the induction of PT. Some neurodegenerative diseases may involve typical apoptotic pathways and specifically MOMP (68). Manganese has been observed to cause CNS injury resulting in symptoms of Parkinson disease and hepatic encephalopathy leading to chronic liver failure. This involves manganese-induced oxidative stress, leading to MOMP and ultimately apoptosis or necrosis (69).

Mutations in mitochondrial DNA result in a number of metabolic disorders that are mostly due to impairment of oxidative phosphorylation, causing reductions in $\Delta\Psi_m$ and the rate of ATP synthesis (70). Tumorigenesis can be aided by defects in apoptotic pathways enhancing cell survival and transformation, giving cells time to accumulate genetic alterations that deregulate proliferation and differentiation and also favor angiogenesis and cell migration (71). Furthermore, neoplastic cells can evade immune surveillance because their destruction by CTLs and NK cells depends on the integrity of the apoptotic machinery (72). As MOMP plays such a prominent role in these diseases, the components of this pathway are the targets of currently used and potential therapies.

Potential therapies that promote MOMP and cell death

A goal in treating cancer is to target tumor cells without harming healthy cells. Traditional treatment regimens mainly target rapidly dividing cells, among which tumor cells are the most predominant. However, some normal cell populations have high proliferation rates and are thus vulnerable to such treatments. Compounding this problem, many tumor cells have mutations in key components of the apoptotic machinery, such as p53, allowing them to become resistant to apoptosis. Currently there are a number of drugs in clinical development that are hoped to induce apoptosis only in neoplastic cells by targeting components of the mitochondrial pathway to induce MOMP (Table 2).

It will be advantageous for new therapeutic strategies to bypass the survival adaptations, or take advantage of the distinguishing features, of tumor cells.

Tumor cells in general have higher plasma and mitochondrial membrane potentials compared with normal epithelial cells (73). As a consequence, certain compounds known as cationic lipophilic toxins can preferentially accumulate in the mitochondrial matrix of cancer cells. Malignant cells tend to possess higher numbers of mitochondria, perhaps further enhancing the specificity of these drugs for tumor cells (73, 74). MKT-077, a cationic rhodacyanine dye (75), selectively kills malignant cells, in part because of its ability to inhibit mitochondrial respiration, and is the basis of pharmacological studies and phase I clinical trials in chemoresistant solid tumors (76, 77).

Photodynamic therapy (PDT) (78) is a therapeutic approach that similarly makes use of the ability of lipophilic cations to target tumor cells preferentially. In PDT, a photosensitizer is irradiated in the presence of molecular oxygen. The resulting ROS production causes cell death. Cationic photosensitizers, such as rhodamine-123, krypto/phthalocyanines, and cationic porphyrins, preferentially accumulate in mitochondria in proportion to $\Delta\Psi_m$; many of these compounds show tumor-destructive activity in vivo. Anionic photosensitizers often have specific affinity for mitochondrial proteins or lipids; for example, verteporfin targets ANT (79). Clinical trials are in progress to evaluate similar drugs for the treatment of several types of tumors (80). Other "mitochondriotoxic" molecules, such as arsenic trioxide (ATO), used to treat acute promyelocytic leukemia (81), work through ROS production. The combination of ATO with ascorbic acid, a substrate of cytochrome c oxidase (82), was successful in treating myeloma cells and phorbol ester–differentiated leukemic cells resistant to classical treatment (83).

The antineoplastic drug lonidamine may induce PT through direct action on ANT (84). This compound was shown to be safe and effective in clinical trials for the treatment of solid tumors (85), as well as in phase II–III trials for the treatment of advanced breast, ovarian, and lung cancer. The combination of lonidamine with diazepam also increased the therapeutic index in the treatment of recurrent glioblastoma multiforme (86). Many other anticancer drugs also directly target components of the PTP. The thiol cross-linker 4-[N-(S-glutathionylacetyl)amino] phenylarsenoxide (GSAO) works similarly to lonidamine, binding to ANT to induce PT. Its killing activity seems to exclusively target proliferating, but not growth-quiescent, endothelial cells in vitro and prevents angiogenesis in solid tumors in mice (87). Betulinic acid (88), a natural pentacyclic triterpenoid, also acts via the PTP (89) and was shown to be effective against neuroectodermal (90) and malignant head and neck tumors (91) among others. Betulinic acid may also cooperate with TRAIL to induce apoptosis in different tumor cell lines and in primary tumor cells, without affecting normal human fibroblasts (92). PK11195 and RO5-4864, 2 ligands for the mitochondrial benzodiazepine receptor, a PTP regulator, are able to resensitize tumor cells to etoposide, doxorubicin, and γ-irradiation (93–95).

Therapeutic approaches that inhibit MOMP and cell death

In contrast to cancer, some disease states arise from an excess of apoptosis, and in these cases MOMP is also a potential therapeutic target. The use of antioxidants is an effective method of counteracting the effects of excess ROS production and shows promise for prevention of ischemia/reperfusion injury (96). Manganese superoxide dismutase (MnSOD), a mitochondrial antioxidant enzyme that can inhibit apoptosis, is a survival factor for cancer cells and, when present in small amounts, mediates resistance to inflammatory stimuli and anticancer drugs. MnSOD can be beneficial in reperfusion-induced injury, as it protects mitochondria and thus limits ROS-induced cell death (97).

The Cyp D–deficient mouse was shown to have significant resistance to ischemia/reperfusion–induced cardiac injury, which suggests that Cyp D is a useful drug target for treatment of stroke or myocardial infarction (36, 37). Indeed the undecapeptide CsA, which through binding to Cyp D can prevent PTP assembly, has protective properties with respect to ischemic injury. In a model

Table 2

Summary of the literature concerning various compounds that act on mitochondria and their experimental or clinical stages of development

Drug type	Examples	Target	Cytotoxic activity	Disease indicated	Experimental or clinical use
DNA damage inducers	Etoposide	Topoisomerase II	Cell cycle arrest, p53 activation (S1, S2)	Tumors	Phase I trials for solid tumors, phase II trials for ovarian carcinoma, gliomas
	Cisplatin	Induces DNA modifications	p53 activation (S3)	Germ-cell, breast cancer, head and neck, lung carcinoma	Widely prescribed for a variety of tumors
	Cytosine b-D-arabino-furanoside (AraC), 5-fluorouracil (5FU)	Selective inhibitors of DNA synthesis	p53 activation (S4, S5)	Acute myeloid leukemia, ovarian cancer (AraC) (S6); breast, gastrointestinal carcinomas (5FU)	Phase I, II, III clinical trials; prescribed for certain cancers
Microtubule inhibitors	Taxol (paclitaxel)	β-Tubulin (stabilizes microtubules)	Cell cycle arrest, Bim-dependent apoptosis	Breast, lung, and ovarian cancer	Prescribed
	Vincristine, vinblastine	Tubulin (destabilizes microtubules)	Cell cycle arrest, p53 activation, Bcl-2 phosphorylation (S7)	Leukemia, lymphoma, and breast and lung cancers	Prescribed
Avicins		Ubiquitination/ Proteasome pathway? PI3K?	Cancer cell cycle arrest, MOMP (S8)	Cancer cell lines, chemically induced mouse skin carcinogenesis (S9)	Preclinical
Bcl-2/Bcl-x inhibitors	HA14-1, antimycin A, BH3Is, YC-137	Bcl-2 or Bcl-x_L	Bcl-2 or Bcl-x_L inhibition, MOMP (S10–S13)	Antitumor against Bcl-2– overexpressing cells (S10, S11)	Cell lines overexpressing Bcl-2 (HA14-1) (S10) or Bcl-2/Bcl-x_L (antimycin A) (S11); in vitro and in vivo (BH3Is) (S12); in vitro and in cancer cells lines overexpressing Bcl-2 (YC-137) (S13)
BH3-only mimetics	Stapled Bid BH3 peptide, lactam-bridged BH3s (S14), Bax BH3 (S15)	Bax/Bak or Bax/Bcl-2	Bax or Bak activation, induction of MOMP (S16)	Antitumor in cell culture and in mouse leukemic model (SAHB BH3) (S16)	Mouse model (SAHB BH3) (S16)
Delocalized lipophilic cations (DLCs)	MKT-077, dequalinium (DEQ-B) (S17), F16	Mitochondrial matrix (ΔΨ_m)	Cell cycle arrest, inhibition of respiration, MOMP (S18–S20)	Breast, ovary, endometrial, colon, non–small cell lung carcinoma	Phase I trials for chemoresistant) solid tumors (MKT-077) (S21, S22); in vitro (F16) (S19)
Lipophilic agents	Hyperforin	Mitochondria	Loss of ΔΨ_m, MOMP (S23)	Inhibit tumor growth, cancer invasion, and metastasis (S24)	Prevent cancer spread and metastatic growth in mice and rats (S23, S24)
Cationic photosensitizers	Rhodamine 123, krypto/ phthalocyanines, cationic porphyrins	Mitochondrial matrix (ΔΨ_m)	Mitochondrial photodamage, respiratory chain enzyme inhibition, MOMP (S25)	Rat and mouse cancer cell lines (S25)	Preclinical
Hydroxy-chalcones	Phloretin, isoliquiritigen, butein	Mitochondria	Mitochondrial uncoupling, ΔΨ_m loss, increased oxygen uptake (S26)	K562 leukemia, breast cancer, and melanoma cells (S26)	Preclinical
Cytotoxic peptides	(KLAKLAK)_2	Mitochondrial matrix	Loss of membrane barrier function, mitochondrial swelling	Antitumor activity in mice, selective toxicity for angiogenic endothelial cells (S27)	In mice
	GSAO	ANT	Ca^{2+}-dependent PT, ATP depletion, MOMP (S28)	Inhibits tumor growth in mice (S28)	In mice
Respiratory chain inhibitors	Benzopyran-based inhibitors, arsenic trioxide (ATO; see below)	Complex I (S29)	Not described	Cancer	Potential chemotherapeutic/ chemopreventive candidates (S29)
	Polyketide inhibitors	F(0)-F(1)-ATPase (S30)	Not described	Cancer cell lines (S30)	Human cancer cell lines (S30)
	Tamoxifen	Complex III/IV (S31)	Mitochondrial uncoupling, MOMP	Breast cancer	Prescribed
Anionic photo-sensitizers	Verteporfin	ANT (S32)	ROS generation, PT, and MOMP (S32)	Age-related macular degeneration	Prescribed
ROS generators	CD437, short-chain fatty acids (propionate, acetate)	ANT (S33, S34)	ROS generation, PT, and MOMP (S34–S36)	Solid tumors, cutaneous carcinomas and leukemias (CD437) (S33, S35, S36), colorectal carcinoma cells (short-chain fatty acids) (S34)	In vitro and in malignant cell lines (CD437) (S33, S35, S36); both in vitro and in human colorectal carcinoma cell lines (S34)
	ATO	ANT (S33)	ROS generation, PT, and MOMP (S33, S37)	Acute promyelocytic leukemia (S38, S39)	Phase I/II trials of dual therapy ATO/ascorbic acid for chemoresistant myeloma cells, phorbol ester-differentiated leukemic cells (S40)
	Bortezomib (Velcade)	Proteasomes	ROS generation, PT, and MOMP (S41)	Multiple myeloma and chronic lymphocytic leukemia	Phase I/II trials for multiple myeloma and some solid tumors (S42)
Benzodiazepine ligands	PK11195, RO5-4864	Mitochondrial benzodiazepine receptor	PT and MOMP (S43)	Increases cytotoxicity of doxorubicin etoposide, ionizing radiation in human cancer cell lines (S43–S45)	Assays in human tumor cell lines and in mice (S43–S45)
Other drugs that target the PTP	MT-21	ANT (S46)	PT and MOMP (S46, S47)	Human leukemia HL-60 cells	In vitro and in cancer cell lines (S46, S47)
	Betulinic acid	PTP	PT and MOMP	Malignant brain tumor cells, neuroectodermal tumors and melanoma (S48–S52)	In tumor cell lines and in mice (S48–S52)
	Lonidamine	ANT (S33)	PT and MOMP (S33)	Tumors (S53)	Trials in combination with standard chemotherapy for solid tumors, phase II/III trials for advanced breast, ovarian, and lung cancers (S40, S54, S55)
	CsA	Cyp D	Prevention of PT and MOMP	Ischemia/reperfusion (S56, S57)	Used as an immunosuppressant for transplants (S58); neuroprotection in rats
Antioxidants	MnSOD	Mitochondria	Prevention of MOMP (S59)	Ischemia/reperfusion (S59)	In transgenic mice (S59)
	MCI-186	Mitochondria	Inhibition of PT and MOMP (S60)	Reperfusion injury in rats (S61)	In rats (S60, S61)
	MitoVitE, MitoQ (targeted to mitochondria with the TPP lipophilic cation)	Mitochondria (ΔΨ_m)	MOMP inhibition; prevention of lipid peroxidation, oxidative stress, and damage (S62–S64)	Prevents mitochondrial damage induced by H$_2$O$_2$ (S64)	In human cell lines (S64)

"In vitro" refers to biochemical (cell-free) studies. References S1–S64 are available online with this article; doi:10.1172/JCI26274DS1.

of transient focal cerebral ischemia, CsA is neuroprotective when administered before or immediately after ischemia, resulting in a smaller infarct (98, 99). CsA is widely used as an immunosuppressant to prevent organ rejection following transplantation. This immunosuppressant function of CsA is due not to its effects on Cyp D, but rather to its ability to inhibit the calcineurin-mediated upregulation of a number of inflammatory cytokines. NIM811, an analog of CsA, retains the neuroprotective but not the immunosuppressive function, suggesting that the ability of CsA to protect cells from ischemic injury results from the neutralization of Cyp D and hence PT (100). Although CsA has protective effects in ischemia/reperfusion models, this drug also caused neurotoxicity in up to 60% of transplants (101). Indeed, studies have shown that CsA can be both neurotoxic to normal cells and neuroprotective, in similar dose ranges, to cells that have received an ischemic injury (102). Thus, although CsA is a prototype drug for the prevention or treatment of ischemia/reperfusion injury, its functions and effects are still not fully understood.

The ability of viruses to induce apoptosis also provides potential targets for drugs that can be used in the treatment of certain viral infections. The HIV-encoded protein Vpr, for example, induces apoptosis by causing MOMP and is thought to do so through interaction with ANT, as Vpr fails to kill ANT-deficient cells (103). HIV protease inhibitors can inhibit Vpr-induced PT and MOMP, and inhibitors of Vpr or the Vpr/ANT interaction may have benefits in the clinic (104).

Bcl-2 inhibitors

Often, tumors overexpress Bcl-2 or related antiapoptotic proteins and as a result are resistant to apoptosis induced by many chemotherapeutic agents. Therefore, there is interest in developing small molecules and peptides that mimic proapoptotic BH3-only proteins to overcome Bcl-2–associated resistance. The nonpeptidic compound HA14-1 was identified using a computer screening approach exploiting the 3D structure of Bcl-2 (105). HA14-1 can bind to the BH3-binding site of Bcl-2, thus mimicking a BH3-only protein, and can induce apoptosis in tumor cells. Chemical screens for molecules that can inhibit the antiapoptotic activity of Bcl-2 and Bcl-x_L have identified antimycin A (coincidentally, already known as an inhibitor of respiratory complex III) and inhibitors known as BH3Is (106, 107). Antimycin A, like HA14-1 and the BH3Is, binds the BH3-binding pocket to induce apoptosis in cells that overexpress Bcl-2 or Bcl-x_L. While these compounds serve as a proof of principle for this approach, compounds with improved affinities for Bcl-2 and Bcl-x_L must be identified if they are to be considered for use in a clinical setting. Indeed, very recently, ABT-737, a novel inhibitor of Bcl-2, Bcl-x_L, and Bcl-w, was reported to bind to these molecules with an affinity orders of magnitude higher than that of previously described Bcl-2 antagonist compounds (108). ABT-737 can sensitize cells to killing by other apoptosis-inducing stimuli but also showed significant efficacy as a single agent against lymphoma and small-cell lung carcinoma cell lines and primary cells derived from patients, and in animal models. The inability of ABT-737 to antagonize Mcl-1 (another antiapoptotic relative of Bcl-2) will likely limit the effectiveness of this compound against tumors in which Mcl-1 is elevated or stabilized.

Synthetic BH3 peptides can also function as antagonists of Bcl-x_L, sensitizing cells to proapoptotic stimuli such as anti-Fas treatment (109). However, peptides tend to be degraded and lose their secondary structure easily, severely limiting their efficacy. Korsmeyer and colleagues recently described a method of overcoming these problems called hydrocarbon stapling (110). A stapled Bid BH3 peptide was found to be both cell permeable and more stable than the unmodified peptide and showed efficacy in a mouse leukemic model.

The Nur77 protein induces apoptosis by binding to Bcl-2. Nur77 binds not to the BH3-binding pocket but rather to the loop region between the BH4 and BH3 domains (111). Deletion of this loop blocks paclitaxel-induced death, and thus the loop may be required for certain anticancer drugs to be effective (112). Also, deamidation of residues in this loop in Bcl-x_L reduces its protective activity (113). Therefore, drugs that target this region may be effective inhibitors of Bcl-2. The use of antisense Bcl-2 oligonucleotides is another potential method of inhibiting Bcl-2 in treatment of cancer. However, studies have shown that when used alone to treat lymphoma, antisense Bcl-2 has dose-limiting toxicities (114). But when used in combination with existing chemotherapeutic drugs (e.g., mitoxantrone in the treatment of prostate cancer [ref. 115]), lower doses may be effective, thus reducing toxic side effects.

A question arises, however: How can compounds targeting Bcl-2 and related proteins selectively kill cancer cells? The unified model shown in Figure 2 reflects a greater flexibility for cell type–specific regulation of apoptosis by the Bcl-2 family than previously assumed. As a corollary, we can envision therapeutic applications in which a certain cell type is targeted selectively based on the subsets of BH3-only proteins and antiapoptotic Bcl-2 relatives it expresses. In particular, the model in Figure 2 predicts that cells lacking expression of direct activator BH3-only proteins should not undergo apoptosis when the antiapoptotic proteins are downregulated. In contrast, tumor cells, which may be under constant oncogenic stress, may constitutively express a direct Bax/Bak activator protein such as Bim (116), which then is presumably held in check by antiapoptotic proteins such as Bcl-2 (117). If this turns out to be true generally, the introduction of a cell-permeable derepressor-type peptide or a small-compound inhibitor of antiapoptotic Bcl-2–family proteins might induce apoptosis selectively in a variety of tumors (108).

Conclusion

Mitochondria hold great promise as targets for therapeutic intervention. However, only a fraction of this potential has so far been realized. Ironically, it may turn out that some therapies succeed by unintentionally affecting MOMP, despite being directed at other biochemical targets. Indeed, a recent study showed that a promising proteasome inhibitor, bortezomib, restored paclitaxel sensitivity to Ras-transformed cells by protecting the BH3-only protein, Bim, from proteasomal degradation (118). Although we are certainly glad for such serendipity, there is hope that, in time, a variety of mitochondria-directed therapies, such as more potent inhibitors of Bcl-2 and related proteins (108), will be put into common use.

Address correspondence to: Donald D. Newmeyer, La Jolla Institute for Allergy and Immunology, Department of Cellular Immunology, 10355 Science Center Drive, San Diego, California 92121, USA. Phone: (858) 558-3539; Fax: (858) 558-3526; E-mail: don@liai.org.

Lisa Bouchier-Hayes and Lydia Lartigue contributed equally to this work.

1. Weissig, V., Boddapati, S.V., D'Souza, G.G.M., and Cheng, S.M. 2004. Targeting of low-molecular weight drugs to mammalian mitochondria. *Drug Design Reviews – Online.* **1**:15–28.

2. Martin, S.J., and Green, D.R. 1995. Protease activation during apoptosis: death by a thousand cuts? *Cell.* **82**:349–352.

3. Newmeyer, D.D., and Ferguson-Miller, S. 2003. Mitochondria: releasing power for life and unleashing the machineries of death. *Cell.* **112**:481–490.

4. Adrain, C., and Martin, S.J. 2001. The mitochondrial apoptosome: a killer unleashed by the cytochrome seas. *Trends Biochem. Sci.* **26**:390–397.

5. Du, C., Fang, M., Li, Y., Li, L., and Wang, X. 2000. Smac, a mitochondrial protein that promotes cytochrome c-dependent caspase activation by eliminating IAP inhibition. *Cell.* **102**:33–42.

6. Suzuki, Y., et al. 2001. A serine protease, HtrA2, is released from the mitochondria and interacts with XIAP, inducing cell death. *Mol. Cell.* **8**:613–621.

7. Klein, J.A., et al. 2002. The harlequin mouse mutation downregulates apoptosis-inducing factor. *Nature.* **419**:367–374.

8. Jones, J.M., et al. 2003. Loss of Omi mitochondrial protease activity causes the neuromuscular disorder of mnd2 mutant mice. *Nature.* **425**:721–727.

9. Uren, R.T., et al. 2005. Mitochondrial release of pro-apoptotic proteins: electrostatic interactions can hold cytochrome c but not Smac/DIABLO to mitochondrial membranes. *J. Biol. Chem.* **280**:2266–2274.

10. Otera, H., Ohsakaya, S., Nagaura, Z.I., Ishihara, N., and Mihara, K. 2005. Export of mitochondrial AIF in response to proapoptotic stimuli depends on processing at the intermembrane space. *EMBO J.* **24**:1375–1386.

11. Polster, B.M., Basanez, G., Etxebarria, A., Hardwick, J.M., and Nicholls, D.G. 2005. Calpain I induces cleavage and release of apoptosis-inducing factor from isolated mitochondria. *J. Biol. Chem.* **280**:6447–6454.

12. Von Ahsen, O., Waterhouse, N.J., Kuwana, T., Newmeyer, D.D., and Green, D.R. 2000. The 'harmless' release of cytochrome c. *Cell Death Differ.* **7**:1192–1199.

13. Kuwana, T., and Newmeyer, D.D. 2003. Bcl-2-family proteins and the role of mitochondria in apoptosis. *Curr. Opin. Cell Biol.* **15**:691–699.

14. Tsujimoto, Y., and Shimizu, S. 2000. VDAC regulation by the Bcl-2 family of proteins. *Cell Death Differ.* **7**:1174–1181.

15. Korsmeyer, S.J., et al. 2000. Pro-apoptotic cascade activates BID, which oligomerizes BAK or BAX into pores that result in the release of cytochrome c. *Cell Death Differ.* **7**:1166–1173.

16. Harris, M.H., and Thompson, C.B. 2000. The role of the Bcl-2 family in the regulation of outer mitochondrial membrane permeability. *Cell Death Differ.* **7**:1182–1191.

17. Martinou, J.C., and Green, D.R. 2001. Breaking the mitochondrial barrier. *Nat. Rev. Mol. Cell Biol.* **2**:63–67.

18. Zamzami, N., and Kroemer, G. 2001. The mitochondrion in apoptosis: how Pandora's box opens [review]. *Nat. Rev. Mol. Cell Biol.* **2**:67–71.

19. Martinou, J.C., Desagher, S., and Antonsson, B. 2000. Cytochrome c release from mitochondria: all or nothing. *Nat. Cell Biol.* **2**:E41–E43.

20. Kuwana, T., et al. 2002. Bid, Bax, and lipids cooperate to form supramolecular openings in the outer mitochondrial membrane. *Cell.* **111**:331–342.

21. Scorrano, L., et al. 2002. A distinct pathway remodels mitochondrial cristae and mobilizes cytochrome c during apoptosis. *Dev. Cell.* **2**:55–67.

22. Bossy-Wetzel, E., Barsoum, M.J., Godzik, A., Schwarzenbacher, R., and Lipton, S.A. 2003. Mitochondrial fission in apoptosis, neurodegeneration and aging. *Curr. Opin. Cell Biol.* **15**:706–716.

23. Scorrano, L. 2003. Divide et impera: Ca2+ signals, mitochondrial fission and sensitization to apoptosis. *Cell Death Differ.* **10**:1287–1289.

24. Lee, Y., Jeong, S.-Y., Karbowski, M., Smith, C.L., and Youle, R.J. 2004. Roles of the mammalian mitochondrial fission and fusion mediators Fis1, Drp1, and Opa1 in apoptosis. *Mol. Biol. Cell.* **15**:5001–5011.

25. Kokoszka, J.E., et al. 2004. The ADP/ATP translocator is not essential for the mitochondrial permeability transition pore. *Nature.* **427**:461–465.

26. Halestrap, A.P. 2004. Mitochondrial permeability: dual role for the ADP/ATP translocator? *Nature.* **430**:1 page following 983.

27. Ichas, F., and Mazat, J.P. 1998. From calcium signaling to cell death: two conformations for the mitochondrial permeability transition pore. Switching from low- to high-conductance state. *Biochim. Biophys. Acta.* **1366**:33–50.

28. Crompton, M. 1999. The mitochondrial permeability transition pore and its role in cell death. *Biochem. J.* **341**:233–249.

29. Jurgensmeier, J.M., et al. 1998. Bax directly induces release of cytochrome c from isolated mitochondria. *Proc. Natl. Acad. Sci. U. S. A.* **95**:4997–5002.

30. De Giorgi, F., et al. 2002. The permeability transition pore signals apoptosis by directing Bax translocation and multimerization. *FASEB J.* **16**:607–609.

31. Bossy-Wetzel, E., Newmeyer, D.D., and Green, D.R. 1998. Mitochondrial cytochrome c release in apoptosis occurs upstream of DEVD-specific caspase activation and independently of mitochondrial transmembrane depolarization. *EMBO J.* **17**:37–49.

32. von Ahsen, O., et al. 2000. Preservation of mitochondrial structure and function after Bid- or Bax-mediated cytochrome c release. *J. Cell Biol.* **150**:1027–1036.

33. Waterhouse, N.J., et al. 2001. Cytochrome c maintains mitochondrial transmembrane potential and ATP generation after outer mitochondrial membrane permeabilization during the apoptotic process. *J. Cell Biol.* **153**:319–328.

34. Ricci, J.E., et al. 2004. Disruption of mitochondrial function during apoptosis is mediated by caspase cleavage of the p75 subunit of complex I of the electron transport chain. *Cell.* **117**:773–786.

35. Ricci, J.E., Gottlieb, R.A., and Green, D.R. 2003. Caspase-mediated loss of mitochondrial function and generation of reactive oxygen species during apoptosis. *J. Cell Biol.* **160**:65–75.

36. Baines, C.P., et al. 2005. Loss of cyclophilin D reveals a critical role for mitochondrial permeability transition in cell death. *Nature.* **434**:658–662.

37. Nakagawa, T., et al. 2005. Cyclophilin D-dependent mitochondrial permeability transition regulates some necrotic but not apoptotic cell death. *Nature.* **434**:652–658.

38. Green, D.R. 2005. Apoptotic pathways: ten minutes to dead. *Cell.* **121**:671–674.

39. Halestrap, A. 2005. Biochemistry: a pore way to die. *Nature.* **434**:578–579.

40. Adams, J.M., and Cory, S. 1998. The Bcl-2 protein family: arbiters of cell survival. *Science.* **281**:1322–1326.

41. Scorrano, L., et al. 2003. BAX and BAK regulation of endoplasmic reticulum Ca2+: a control point for apoptosis. *Science.* **300**:135–139.

42. Wei, M.C., et al. 2001. Proapoptotic BAX and BAK: a requisite gateway to mitochondrial dysfunction and death. *Science.* **292**:727–730.

43. Lindsten, T., et al. 2000. The combined functions of proapoptotic Bcl-2 family members bak and bax are essential for normal development of multiple tissues. *Mol. Cell.* **6**:1389–1399.

44. Knudson, C.M., Tung, K.S., Tourtellotte, W.G., Brown, G.A., and Korsmeyer, S.J. 1995. Bax-deficient mice with lymphoid hyperplasia and male germ cell death. *Science.* **270**:96–99.

45. Nechushtan, A., Smith, C.L., Lamensdorf, I., Yoon, S.H., and Youle, R.J. 2001. Bax and Bak coalesce into novel mitochondria-associated clusters during apoptosis. *J. Cell Biol.* **153**:1265–1276.

46. Eskes, R., Desagher, S., Antonsson, B., and Martinou, J.C. 2000. Bid induces the oligomerization and insertion of Bax into the outer mitochondrial membrane. *Mol. Cell. Biol.* **20**:929–935.

47. Kuwana, T., et al. 2005. BH3 domains of BH3-only proteins differentially regulate Bax-mediated mitochondrial membrane permeabilization both directly and indirectly. *Mol. Cell.* **17**:525–535.

48. Suzuki, M., Youle, R.J., and Tjandra, N. 2000. Structure of Bax: coregulation of dimer formation and intracellular localization. *Cell.* **103**:645–654.

49. Liu, X., Dai, S., Zhu, Y., Marrack, P., and Kappler, J.W. 2003. The structure of a Bcl-xL/Bim fragment complex: implications for Bim function. *Immunity.* **19**:341–352.

50. Desagher, S., et al. 1999. Bid-induced conformational change of Bax is responsible for mitochondrial cytochrome c release during apoptosis. *J. Cell Biol.* **144**:891–901.

51. Yethon, J.A., Epand, R.F., Leber, B., Epand, R.M., and Andrews, D.W. 2003. Interaction with a membrane surface triggers a reversible conformational change in Bax normally associated with induction of apoptosis. *J. Biol. Chem.* **278**:48935–48941.

52. Chipuk, J.E., Maurer, U., Green, D.R., and Schuler, M. 2003. Pharmacologic activation of p53 elicits Bax-dependent apoptosis in the absence of transcription. *Cancer Cell.* **4**:371–381.

53. Chipuk, J.E., et al. 2004. Direct activation of Bax by p53 mediates mitochondrial membrane permeabilization and apoptosis. *Science.* **303**:1010–1014.

54. Letai, A., et al. 2002. Distinct BH3 domains either sensitize or activate mitochondrial apoptosis, serving as prototype cancer therapeutics. *Cancer Cell.* **2**:183–192.

55. Cartron, P.F., et al. 2004. The first alpha helix of Bax plays a necessary role in its ligand-induced activation by the BH3 only proteins Bid and PUMA. *Mol. Cell.* **16**:807–818.

56. Chen, L., et al. 2005. Differential targeting of prosurvival Bcl-2 proteins by their BH3-only ligands allows complementary apoptotic function. *Mol. Cell.* **17**:393–403.

57. Willis, S.N., et al. 2005. Proapoptotic Bak is sequestered by Mcl-1 and Bcl-xL, but not Bcl-2, until displaced by BH3-only proteins. *Genes Dev.* **19**:1294–1305.

58. Tan, K.O., et al. 2001. MAP-1, a novel proapoptotic protein containing a BH3-like motif that associates with Bax through its Bcl-2 homology domains. *J. Biol. Chem.* **276**:2802–2807.

59. Sawada, M., et al. 2003. Ku70 suppresses the apoptotic translocation of Bax to mitochondria. *Nat. Cell Biol.* **5**:320–329.

60. Guo, B., et al. 2003. Humanin peptide suppresses apoptosis by interfering with Bax activation. *Nature.* **423**:456–461.

61. Cohen, H.Y., et al. 2004. Acetylation of the C terminus of Ku70 by CBP and PCAF controls Bax-mediated apoptosis. *Mol. Cell.* **13**:627–638.

62. Cuddeback, S.M., et al. 2001. Molecular cloning and characterization of Bif-1. A novel Src homology 3 domain-containing protein that associates with Bax. *J. Biol. Chem.* **276**:20559–20565.

63. Cheng, E.H., Sheiko, T.V., Fisher, J.K., Craigen, W.J., and Korsmeyer, S.J. 2003. VDAC2 inhibits BAK activation and mitochondrial apoptosis. *Science.* **301**:513–517.

64. Sawada, M., Hayes, P., and Matsuyama, S. 2003. Cytoprotective membrane-permeable peptides designed from the Bax-binding domain of Ku70. *Nat. Cell Biol.* **5**:352–357.

65. Lum, J.J., et al. 2003. Vpr R77Q is associated with long-term nonprogressive HIV infection and impaired induction of apoptosis. *J. Clin. Invest.* **111**:1547–1554. doi:10.1172/JCI200316233.

66. Poncet, D., et al. 2004. An anti-apoptotic viral protein that recruits Bax to mitochondria. *J. Biol. Chem.* **279**:22605–22614.

67. Arnoult, D., et al. 2004. Cytomegalovirus cell death suppressor vMIA blocks Bax- but not Bak-mediated apoptosis by binding and sequestering Bax at mitochondria. *Proc. Natl. Acad. Sci. U. S. A.* **101**:7988–7993.

68. Lindholm, D., Eriksson, O., and Korhonen, L. 2004. Mitochondrial proteins in neuronal degeneration. *Biochem. Biophys. Res. Commun.* **321**:753–758.

69. Rao, K.V., and Norenberg, M.D. 2004. Manganese induces the mitochondrial permeability transition in cultured astrocytes. *J. Biol. Chem.* **279**:32333–32338.

70. Lowell, B.B., and Shulman, G.I. 2005. Mitochondrial dysfunction and type 2 diabetes. *Science.* **307**:384–387.

71. Reed, J.C. 1999. Dysregulation of apoptosis in cancer. *J. Clin. Oncol.* **17**:2941–2953.

72. Tschopp, J., Martinon, F., and Hofmann, K. 1999. Apoptosis: silencing the death receptors [review]. *Curr. Biol.* **9**:R381–R384.

73. Davis, S., Weiss, M.J., Wong, J.R., Lampidis, T.J., and Chen, L.B. 1985. Mitochondrial and plasma membrane potentials cause unusual accumulation and retention of rhodamine 123 by human breast adenocarcinoma-derived MCF-7 cells. *J. Biol. Chem.* **260**:13844–13850.

74. Leprat, P., Ratinaud, M.H., and Julien, R. 1990. A new method for testing cell ageing using two mitochondria specific fluorescent probes. *Mech. Ageing Dev.* **52**:149–167.

75. Modica-Napolitano, J.S., et al. 1996. Selective damage to carcinoma mitochondria by the rhodacyanine MKT-077. *Cancer Res.* **56**:544–550.

76. Propper, D.J., et al. 1999. Phase I trial of the selective mitochondrial toxin MKT077 in chemo-resistant solid tumours. *Ann. Oncol.* **10**:923–927.

77. Britten, C.D., et al. 2000. A phase I and pharmacokinetic study of the mitochondrial-specific rhodacyanine dye analog MKT 077. *Clin. Cancer Res.* **6**:42–49.

78. Morgan, J., and Oseroff, A.R. 2001. Mitochondria-based photodynamic anti-cancer therapy. *Adv. Drug Deliv. Rev.* **49**:71–86.

79. Belzacq, A.S., et al. 2001. Apoptosis induction by the photosensitizer verteporfin: identification of mitochondrial adenine nucleotide translocator as a critical target. *Cancer Res.* **61**:1260–1264.

80. Brown, S.B., Brown, E.A., and Walker, I. 2004. The present and future role of photodynamic therapy in cancer treatment. *Lancet Oncol.* **5**:497–508.

81. Zhang, T.D., et al. 2001. Arsenic trioxide, a therapeutic agent for APL. *Oncogene.* **20**:7146–7153.

82. Grad, J.M., et al. 2001. Ascorbic acid enhances arsenic trioxide-induced cytotoxicity in multiple myeloma cells. *Blood.* **98**:805–813.

83. Sordet, O., et al. 2001. Mitochondria-targeting drugs arsenic trioxide and lonidamine bypass the resistance of TPA-differentiated leukemic cells to apoptosis. *Blood.* **97**:3931–3940.

84. Belzacq, A.S., et al. 2001. Adenine nucleotide translocator mediates the mitochondrial membrane permeabilization induced by lonidamine, arsenite

and CD437. *Oncogene.* **20**:7579–7587.

85. Di Cosimo, S., et al. 2003. Lonidamine: efficacy and safety in clinical trials for the treatment of solid tumors [review]. *Drugs Today (Barc.).* **39**:157–174.

86. Oudard, S., et al. 2003. Phase II study of lonidamine and diazepam in the treatment of recurrent glioblastoma multiforme. *J. Neurooncol.* **63**:81–86.

87. Don, A.S., et al. 2003. A peptide trivalent arsenical inhibits tumor angiogenesis by perturbing mitochondrial function in angiogenic endothelial cells. *Cancer Cell.* **3**:497–509.

88. Pisha, E., et al. 1995. Discovery of betulinic acid as a selective inhibitor of human melanoma that functions by induction of apoptosis. *Nat. Med.* **1**:1046–1051.

89. Fulda, S., Susin, S.A., Kroemer, G., and Debatin, K.M. 1998. Molecular ordering of apoptosis induced by anticancer drugs in neuroblastoma cells. *Cancer Res.* **58**:4453–4460.

90. Fulda, S., and Debatin, K.M. 2000. Betulinic acid induces apoptosis through a direct effect on mitochondria in neuroectodermal tumors. *Med. Pediatr. Oncol.* **35**:616–618.

91. Thurnher, D., et al. 2003. Betulinic acid: a new cytotoxic compound against malignant head and neck cancer cells. *Head Neck.* **25**:732–740.

92. Fulda, S., Jeremias, I., and Debatin, K.M. 2004. Cooperation of betulinic acid and TRAIL to induce apoptosis in tumor cells. *Oncogene.* **23**:7611–7620.

93. Hirsch, T., et al. 1998. PK11195, a ligand of the mitochondrial benzodiazepine receptor, facilitates the induction of apoptosis and reverses Bcl-2-mediated cytoprotection. *Exp. Cell Res.* **241**:426–434.

94. Banker, D.E., et al. 2002. PK11195, a peripheral benzodiazepine receptor ligand, chemosensitizes acute myeloid leukemia cells to relevant therapeutic agents by more than one mechanism. *Leuk. Res.* **26**:91–106.

95. Decaudin, D., et al. 2002. Peripheral benzodiazepine receptor ligands reverse apoptosis resistance of cancer cells in vitro and in vivo. *Cancer Res.* **62**:1388–1393.

96. Marczin, N., El-Habashi, N., Hoare, G.S., Bundy, R.E., and Yacoub, M. 2003. Antioxidants in myocardial ischemia-reperfusion injury: therapeutic potential and basic mechanisms. *Arch. Biochem. Biophys.* **420**:222–236.

97. Chen, Z., et al. 1998. Overexpression of MnSOD protects against myocardial ischemia/reperfusion injury in transgenic mice. *J. Mol. Cell. Cardiol.* **30**:2281–2289.

98. Yoshimoto, T., Uchino, H., He, Q.P., Li, P.A., and Siesjo, B.K. 2001. Cyclosporin A, but not FK506, prevents the downregulation of phosphorylated Akt after transient focal ischemia in the rat. *Brain Res.* **899**:148–158.

99. Matsumoto, S., Isshiki, A., Watanabe, Y., and Wieloch, T. 2002. Restricted clinical efficacy of cyclosporin A on rat transient middle cerebral artery occlusion. *Life Sci.* **72**:591–600.

100. Waldmeier, P.C., Feldtrauer, J.J., Qian, T., and Lemasters, J.J. 2002. Inhibition of the mitochondrial permeability transition by the nonimmunosuppressive cyclosporin derivative NIM811. *Mol. Pharmacol.* **62**:22–29.

101. Serkova, N.J., Christians, U., and Benet, L.Z. 2004. Biochemical mechanisms of cyclosporine neurotoxicity. *Mol. Interv.* **4**:97–107.

102. Niemann, C.U., et al. 2002. Close association between the reduction in myocardial energy metabolism and infarct size: dose-response assessment of cyclosporine. *J. Pharmacol. Exp. Ther.* **302**:1123–1128.

103. Jacotot, E., et al. 2001. Control of mitochondrial membrane permeabilization by adenine nucleotide translocator interacting with HIV-1 viral protein rR and Bcl-2. *J. Exp. Med.* **193**:509–519.

104. Phenix, B.N., Lum, J.J., Nie, Z., Sanchez-Dardon, J., and Badley, A.D. 2001. Antiapoptotic mechanism of HIV protease inhibitors: preventing mitochondrial transmembrane potential loss. *Blood.* **98**:1078–1085.

105. Wang, J.L., et al. 2000. Structure-based discovery of an organic compound that binds Bcl-2 protein and induces apoptosis of tumor cells. *Proc. Natl. Acad. Sci. U. S. A.* **97**:7124–7129.

106. Tzung, S.P., et al. 2001. Antimycin A mimics a cell-death-inducing Bcl-2 homology domain 3. *Nat. Cell Biol.* **3**:183–191.

107. Degterev, A., et al. 2001. Identification of small-molecule inhibitors of interaction between the BH3 domain and Bcl-xL. *Nat. Cell Biol.* **3**:173–182.

108. Oltersdorf, T., et al. 2005. An inhibitor of Bcl-2 family proteins induces regression of solid tumors. *Nature.* **435**:677–681.

109. Holinger, E.P., Chittenden, T., and Lutz, R.J. 1999. Bak BH3 peptides antagonize Bcl-xL function and induce apoptosis through cytochrome c-independent activation of caspases. *J. Biol. Chem.* **274**:13298–13304.

110. Walensky, L.D., et al. 2004. Activation of apoptosis in vivo by a hydrocarbon-stapled BH3 helix. *Science.* **305**:1466–1470.

111. Lin, B., et al. 2004. Conversion of Bcl-2 from protector to killer by interaction with nuclear orphan receptor Nur77/TR3. *Cell.* **116**:527–540.

112. Rodi, D.J., et al. 1999. Screening of a library of phage-displayed peptides identifies human bcl-2 as a taxol-binding protein. *J. Mol. Biol.* **285**:197–203.

113. Deverman, B.E., et al. 2002. Bcl-xL deamidation is a critical switch in the regulation of the response to DNA damage. *Cell.* **111**:51–62.

114. Waters, J.S., et al. 2000. Phase I clinical and pharmacokinetic study of bcl-2 antisense oligonucleotide therapy in patients with non-Hodgkin's lymphoma. *J. Clin. Oncol.* **18**:1812–1823.

115. Chi, K.N., et al. 2001. A phase I dose-finding study of combined treatment with an antisense Bcl-2 oligonucleotide (Genasense) and mitoxantrone in patients with metastatic hormone-refractory prostate cancer. *Clin. Cancer Res.* **7**:3920–3927.

116. Egle, A., Harris, A.W., Bouillet, P., and Cory, S. 2004. Bim is a suppressor of Myc-induced mouse B cell leukemia. *Proc. Natl. Acad. Sci. U. S. A.* **101**:6164–6169.

117. Letai, A., Sorcinelli, M.D., Beard, C., and Korsmeyer, S.J. 2004. Antiapoptotic BCL-2 is required for maintenance of a model leukemia. *Cancer Cell.* **6**:241–249.

118. Tan, T.-T., et al. 2005. Key roles of BIM-driven apoptosis in epithelial tumors and rational chemotherapy. *Cancer Cell.* **7**:227–238.

Death versus survival: functional interaction between the apoptotic and stress-inducible heat shock protein pathways

Helen M. Beere

St. Jude Children's Research Hospital, Department of Immunology, Memphis, Tennessee, USA.

Induction of heat shock proteins (Hsps) following cellular damage can prevent apoptosis induced by both the intrinsic and the extrinsic pathways. The intrinsic pathway is characterized by mitochondrial outer membrane permeabilization (MOMP), cytochrome *c* release, apoptosome assembly, and caspase activation. Hsps promote cell survival by preventing MOMP or apoptosome formation as well as via regulation of Akt and JNK activities. Engagement of the TNF death receptors induces the extrinsic pathway that is characterized by Fas-associated death domain–dependent (FADD-dependent) caspase-8 activation or induction of NF-κB to promote cellular survival. Hsps can directly suppress proapoptotic signaling events or stabilizing elements of the NF-κB pathway to promote cellular survival.

Introduction

Exposure of cells to potentially damaging stresses such as UV or nutrient withdrawal induces signals able to mediate cell death, or alternatively, survival pathways that allow cells to tolerate and/or to recover from the damage imposed. This paradoxical activation of both pro- and antiapoptotic events in response to the same stimulus ensures that neither aberrant cellular survival nor inappropriate cell death arises and, in doing so, averts the onset and persistence of the pathological state.

The network of heat shock or stress proteins represents an emerging paradigm for the coordinated, multistep regulation of apoptotic signaling events to provide protection from and to facilitate cellular recovery after exposure to damaging stimuli (1–3). Heat shock proteins (Hsps) constitute a highly conserved and functionally interactive network of chaperone proteins, some of which are constitutively expressed and others of which are rapidly induced in response to a variety of chemical, environmental, and physiological stresses. Their collective ability to disaggregate, refold, and renature misfolded proteins offsets the otherwise fatal consequences of damaging stimuli (4, 5). This protective function of Hsps has been suggested to reflect their ability to suppress several forms of cell death, including apoptosis.

The Hsp70 and Hsp90 subfamilies are composed of an N-terminal ATPase domain and a C-terminal peptide-binding region that is further characterized by an extreme C-terminal EEVD motif, required for interdomain communication and peptide-binding capacity (6). The C-terminus of Hsp70 is absolutely required for many of its antiapoptotic effects (7). However, the role of the ATPase domain is less clear. Some studies suggest that the ATPase domain, although required, is not sufficient to protect against apoptosis (7), while others indicate that the ATPase domain alone is independently able to inhibit aspects of apoptotic signaling (8).

Nonstandard abbreviations used: DR, death receptor; FADD, Fas-associated death domain; Hsp, heat shock protein; IAP, inhibitor of apoptosis; IKK, IκB kinase; MOMP, mitochondrial outer membrane permeabilization; TRADD, TNFR1-associated death domain.

Conflict of interest: The author has declared that no conflict of interest exists.

Citation for this article: *J. Clin. Invest.* **115**:2633–2639 (2005). doi:10.1172/JCI26471.

Biochemical characterization of apoptosis

The primary biochemical signature of apoptosis is the activation of caspases, a family of cysteine proteases with specificity for aspartate residues (9, 10). Caspases are produced as inactive zymogens and are broadly characterized according to their mode of activation — (a) "initiator" caspases, including caspase-2, -8, and -9, and (b) "executioner" caspases, exemplified by caspase-3, -6, and -7.

The initiator caspases form active catalytic dimers via association of their long pro-domains with 1 of several adapter molecules with selectivity for individual caspases (11). In contrast, the executioner caspases become catalytically active only after cleavage by the active initiator caspases (11). Once activated, executioner caspases can cleave the initiator caspases. This does not necessarily lead to their activation, but it may serve to stabilize active initiator caspase dimers (11) or render them subject to regulation via the inhibitors of apoptosis (IAPs) (12, 13). This hierarchy of caspase activation generates a cascade culminating in executioner-mediated cleavage of substrates including iCAD (14), PARP (15), fodrin (16), p75 (17), actin (18), and focal adhesion kinase (FAK) (19), to display an array of phenotypic characteristics including loss of mitochondrial membrane potential, cell blebbing, condensation and fragmentation of chromatin, and redistribution of lipids in the outer plasma membrane.

The mechanism by which active caspases are generated has led to the categorization of the intrinsic (Figure 1) and extrinsic (Figure 2) apoptotic pathways. The intrinsic pathway is associated with mitochondrial outer membrane permeabilization (MOMP), cytochrome *c* release, and the activation of procaspase-9. In contrast, the extrinsic pathway can (but does not always) proceed independently of any alteration in mitochondrial function but is instead characterized by the ligation of cell surface death receptors via specific death ligands to generate catalytically active caspase-8. Both pathways result in activation of the executioner caspases and cleavage of their target substrates to induce apoptosis.

Antiapoptotic role of Hsps

The Hsp27, Hsp70, and Hsp90 subfamilies have been implicated in the protection against apoptosis induced by a variety of stimuli, including chemotherapeutically induced DNA damage, UV radiation (20), polyglutamine repeat expansion (21), death recep-

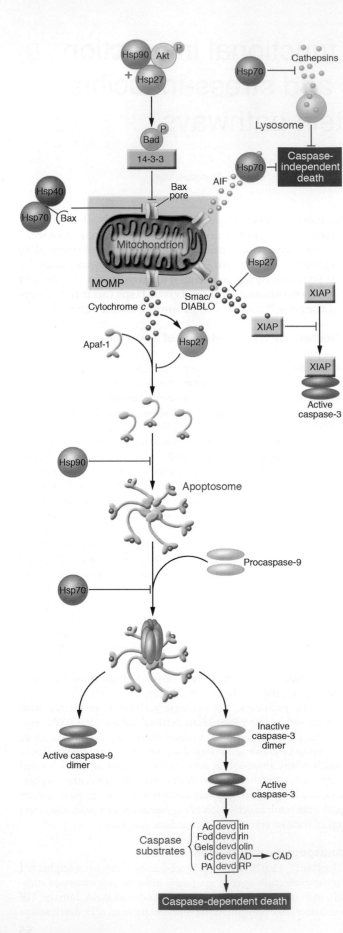

tor ligation (22, 23), heat shock (7, 24), nutrient withdrawal (25), ceramide (26), reactive oxygen species, ER stresses, proteasome inhibition, and cytoskeletal perturbation. In many cases this inhibition is observed as a suppression of caspase activation and the consequent inhibition of substrate cleavage (7, 27, 28). Because these events represent the functional end point of upstream apoptotic signals, it is difficult to decipher the likely mechanisms that mediate the survival-promoting effects of the Hsps.

Regulation of mitochondria-mediated apoptosis: the intrinsic pathway

One defining feature of the intrinsic apoptotic pathway is perturbation of mitochondrial function to induce MOMP and the release of proapoptotic factors that normally reside in the intermembrane space. These include cytochrome c (29, 30), the serine protease HtrA2/Omi (31), AIF (32), Smac/DIABLO (33, 34), and endonuclease G (35) (Figure 1). Although the Bcl-2 family of proteins play a pivotal role in regulating MOMP (36, 37), the physiological significance of each of the intermembrane proteins, once released, remains unclear. This may be attributed to the reported dependence on caspase activity to mediate the release of some of these factors (38), suggesting an ancillary rather than requisite role in cell death.

Regulation of MOMP

Permeability of the outer mitochondrial membrane is regulated via the opposing activities of the pro- and antiapoptotic members of the Bcl-2 family. The ability of the multidomain proapoptotic Bcl-2 proteins Bax and Bak to induce MOMP is regulated by a number of mechanisms, including changes in localization (Bax ordinarily resides in the cytosol but translocates to the mitochondria following apoptotic stimuli) and alterations in their conformation and oligomeric status. These changes in Bax and Bak, concomitant with their acquisition of permeabilization ability, are likely mediated via the BH3-only proteins, e.g., Bid, Bim, Bmf, and Bad, each of which is subject to differential regulation according to the type of apoptotic stimulus imposed on the cell (39, 40). For example, Bid is cleaved and activated by caspase-8 following death receptor ligation (41, 42); Bim (43, 44) and Bmf (45) are constitutively sequestered to cytoskeletal components and are released to exert their proapoptotic activities after treatment with drugs such as taxol, or cellular detachment (anoikis), respectively; and the ability of Bad to induce apoptosis is revealed by changes in its phosphorylation status following cytokine withdrawal (46, 47).

Hsps can inhibit the mitochondrial release of cytochrome c (7, 48, 49), consistent with their ability to inhibit caspase activation and substrate cleavage during heat shock–induced death (7). Whether the ability of Hsp70 to block cytochrome c release reflects a direct effect on mitochondrial integrity or the indirect consequences of its intervention at 1 or more points upstream

Figure 1

Regulation of the intrinsic pathway by Hsps. Hsps regulate several aspects of the intrinsic apoptotic pathway. These include both direct mediators — e.g., Bax — and indirect regulators — e.g., Akt — of mitochondrial membrane permeabilization to prevent MOMP as well as events downstream of mitochondrial disruption to regulate apoptosome assembly. Caspase-independent cell death may also be affected via Hsp-mediated suppression of AIF activity and inhibition of lysosome permeabilization and cathepsin release.

of MOMP is unclear (7, 48, 49). However, based on recent studies, the latter seems more likely.

Both Hsp27 and Hsp70 can modulate Bid-dependent apoptosis (49, 50). Caspase-8–dependent cleavage of Bid and subsequent Bax/Bak–dependent release of cytochrome c integrates the extrinsic and intrinsic pathways (Figure 2). Hsp70 inhibits TNF-induced release of cytochrome c by suppressing caspase-8–mediated cleavage and activation of Bid, independently of its chaperoning ability (50). Hsp27 can also prevent the translocation of Bid to the mitochondria, which may be related to the ability of Hsp27 to stabilize the cytoskeleton.

In addition, Hsp70 and its co-chaperone Hsp40 prevent the translocation of Bax to the mitochondria during NO-mediated apoptosis, an event that requires both the ATPase and peptide-binding activities of Hsp70 (51) (Figure 1). The co-chaperone proteins play an essential role in regulating the peptide-folding capacity of the Hsps by mediating the interconversion of the ADP-versus ATP-bound states of the Hsps and represent an aspect of Hsp function yet to be fully integrated into our understanding of the role of Hsps in negatively regulating cell death.

MOMP also releases AIF, a factor involved in the induction of caspase-independent cell death (32, 52) (Figure 1). Although Hsp70 has not been shown to regulate the release of AIF from mitochondria, it does interact with AIF to prevent the nuclear translocation and pro-death signaling of this molecule (53, 54). In a recent study, an AIF-derived protein that retained the Hsp70-interacting domain but not its death-inducing activity conferred chemosensitivity to several human cancer cell lines. This was attributed to its ability to sequester and neutralize the antiapoptotic activities of Hsp70 (55). In addition, antisense for Hsp70 can induce a caspase-independent tumor cell death (56) that has been attributed to the ability of Hsp70 to maintain the integrity of lysosomal membranes to prevent cathepsin release into the cytosol (57) (Figure 1).

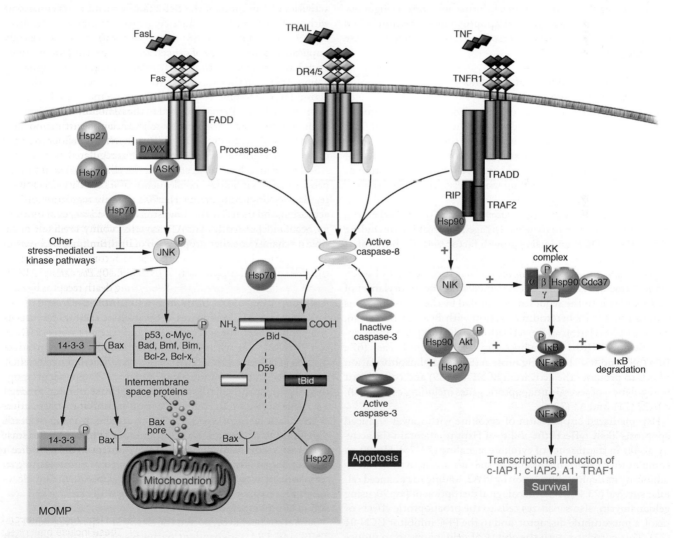

Figure 2
Regulation of the extrinsic pathway by Hsps. Hsps regulate at multiple points within the signaling pathways activated by ligation of a cell surface death receptor by the appropriate ligand. These include the maintenance of prosurvival signals generated via TNF-mediated activation of NF-κB and suppression of proapoptotic signaling events, e.g., JNK activity and Bid cleavage. Integration of the extrinsic and intrinsic pathways is mediated via the caspase-8–mediated cleavage and activation of Bid as well as activation of JNK, which can impact on numerous molecules that regulate mitochondrial integrity (shown in the shaded area).

Negative regulation of apoptosome assembly

Cytochrome c, integral to the maintenance of mitochondrial respiration, is rapidly released into the cytosol following permeabilization of the outer mitochondrial membrane by the insertion of the multidomain Bcl-2 proteins Bax and Bak (36, 37). Once released, cytochrome c induces assembly of the apoptosome, a complex composed of cytochrome c, oligomerized Apaf-1 (58, 59), and procaspase-9 (60) (Figure 1). Formation of this complex relies on the interaction between Apaf-1 and procaspase-9 via their respective caspase recruitment domains (CARDs) to generate active caspase-9 dimers, which then cleave and activate the executioner caspases (60). Apoptosome assembly represents a key regulatory point at which both Hsp70 and Hsp90 can halt the progression of apoptotic signaling, by directly interacting with Apaf-1 to prevent its oligomerization and/or association with procaspase-9 (61, 62). Although these observations were recently challenged (48), it is nevertheless intriguing to note that the abilities of Hsp70 and Hsp90 to negatively regulate apoptosome assembly parallel their role in the regulation of protein conformation and/or oligomerization. Hsp27 may also disrupt apoptosome formation via its interaction with and sequestration of cytochrome c after its release from the mitochondria (63).

Hsp-mediated regulation of pro- and antiapoptotic signaling

Hsps have a profound impact on the activities of several prosurvival signaling cascades, including those mediated by Akt, JNK, and NF-κB. By regulating the activities of these upstream signaling molecules, the Hsps can exert an effect on a diverse number of events both distal and proximal to mitochondrial disruption as well as elements of the signaling machinery within the extrinsic pathway (Figure 2).

PI3K and its kinase effector protein kinase B (PKB/Akt) exert a potent prosurvival activity when engaged by cytokines including IGF-1, IL-2, PDGF, and nerve growth factor (64). Withdrawal of these cytokines can elicit an apoptotic cell death that is blocked via Hsps (25, 65). Akt can promote cellular survival via 1 of several mechanisms. These include Akt-mediated phosphorylation of Bad that leads to the sequestration of Bad by the 14-3-3 proteins (66) to prevent its heterodimerization with Bcl-2 or Bcl-x_L (46, 47); phosphorylation and inactivation of the forkhead transcription factor FKHRL1 (67), whose target genes include FasL (67), Bim (68), and IGF-1–binding protein (67); and phosphorylation of IκB to promote the activation of NF-κB (69) and consequent upregulation of several antiapoptotic genes including c-IAP1 (70), c-IAP2 (71), and A1 (72).

Hsp-mediated suppression of cytokine withdrawal–induced apoptosis likely reflects the ability of Hsps to maintain the activity of Akt in the absence of cytokine signaling (73–75). Hsp90, in complex with Cdc37, maintains the stability and activity of Akt by inhibiting its dephosphorylation by PPA2, leading to enhanced cellular survival (73, 76). Pharmacological disruption of Hsp90, using geldanamycin, also sensitizes cells to the proapoptotic effects of taxol, a microtubule disruptor, and to the PKC inhibitor UCN-01 (77). This correlates with the ability of geldanamycin to induce the destabilization of Akt and an enhancement of Bax-dependent release of cytochrome c and Smac/DIABLO from mitochondria (78). Hsp27 may also maintain the stability of Akt following apoptotic stimuli that would ordinarily induce its destabilization (75, 79). Hsp27 is phosphorylated by MAPK-activated protein kinase 2

(MAPKAP-K2), an event essential for the ability of Hsp27 to promote survival. Hsp27, recently reported to form a signaling complex with Akt, p38 MAPK, and MAPKAP-K2, is also subject to phosphorylation by Akt. This results in the promotion of an interaction between Akt and Hsp27 and a consequent enhancement of Akt stabilization and cellular survival (75, 79).

Activation of the stress-inducible JNK signaling pathway mediates both pro- and antiapoptotic effects depending on the type and duration of the damaging signal as well as cell type (80). Direct targets of JNK activity include Bcl-2 and Bcl-x_L, both of which can be phosphorylated and inactivated by JNK (81–83). JNK may also engage the mitochondrial death pathway via direct phosphorylation and activation of Bim and Bmf (84). Recent data also indicate that JNK-mediated phosphorylation of 14-3-3 proteins can facilitate the translocation of Bax to the mitochondria (85). The proapoptotic BH3-only protein Bad is also directly phosphorylated via JNK, although the biological consequence of this event remains unclear (86). This ability of JNK signaling to directly regulate the activities of 1 or more of the Bcl-2 family members is consistent with the failure of UV to induce cytochrome c release in cells lacking JNK (87). Both p53 (88) and c-Myc (89), genes commonly implicated in the etiology of cancer, are regulated by JNK; likewise, both have been shown to directly modulate apoptotic signaling events. At least in the case of p53, this reflects its ability to directly activate Bax or to bind to and sequester antiapoptotic Bcl-x_L in a manner analogous to that ascribed to the BH3-only Bcl-2 proteins, leading to induced cytochrome c release, caspase activation, and apoptotic cell death (90). The ability of JNK to influence mitochondrial integrity is summarized in Figure 2.

Several studies have suggested that the ability of Hsp70 to suppress apoptosis is a direct consequence of its ability to negatively regulate JNK signaling events. Hsp70's negative regulation of JNK activation, and therefore the downstream consequences of its activity, occur independently of its ATPase activity (24, 91) via 1 of 2 possible mechanisms: either suppression of its direct phosphorylation and activation by the upstream kinase SEK (92), or inhibition of the stress-induced suppression of JNK dephosphorylation (93).

Antiapoptotic role of Hsps in the extrinsic pathway

The death receptors belong to a large family of TNF receptor–related members, including TNFR1, Fas, and death receptor 5 (DR5, or TRAILR2), that, when engaged by trimers of their appropriate ligands (TNF, FasL, and TRAIL, respectively), trimerize to elicit an effective and rapid induction of caspase-8–dependent apoptosis that frequently bypasses the requirement for mitochondrial involvement (Figure 2). Following ligand-induced oligomerization of the death receptor, specialized recruitment domains — death domains (DDs) — recruit DD-containing adapter molecules such as Fas-associated death domain (FADD). FADD, in turn, via death effector domains (DEDs), associates with procaspase-8, leading to formation of active caspase-8 dimers (94). Subsequent autocleavage of procaspase-8 is observed (11), along with cleavage and activation of procaspase-3 (Figure 2).

Several studies have reported the ability of Hsps to regulate Fas-, TNF-, and TRAIL-dependent pathways of apoptotic cell death (22, 23, 95–101). However, their underlying mechanisms of action are likely extremely complex, with studies concluding that Hsps can elicit either a positive or a negative effect to promote or to inhibit apoptosis via a number of suggested pathways including suppression of phospholipase A2 activation (99), inhibition of the generation of reactive oxygen species (102), and suppression of phosphatase activity (95).

The diversity of death receptors and ligands along with the complexity of their adapter proteins generates an equally complex series of signaling pathways, some of which can mediate cell death, and others of which paradoxically mediate enhanced cellular survival. For example, TNFR1, via a mechanism dependent on TNFR1-associated death domain (TRADD) and FADD, can mediate the activation of caspase-8 and consequent apoptotic cell death (103–105). Alternatively, the adapter molecule TRADD can mediate the FADD-independent recruitment of a TRAF/RIP kinase complex (104, 105), leading to the derepression and nuclear translocation of NF-κB. NF-κB is then able to induce the expression of several survival proteins, including c-IAP1 (70), c-IAP2 (71), TRAF1 and TRAF2 (106), and the antiapoptotic Bcl-2 homolog A1 (107).

NF-κB is ordinarily sequestered in an inactive cytosolic form via its interaction with IκB. Once phosphorylated by the multisubunit IκB kinase (IKK) complex, IκB is subject to ubiquitination and proteasome-dependent degradation, leading to the release of NF-κB. The ability of Hsp27 and Hsp90 to regulate TNF-mediated cellular survival has been linked to their ability to regulate the stability and activity of a number of components of the NF-κB activation pathway, primarily at the level of the IKK complex (108–110). For example, Hsp90, in complex with Cdc37, a co-chaperone that functions in cooperation with Hsp90 (111, 112), binds to the IKK complex via interaction with both the regulatory subunit IKKγ (NEMO) and the kinase domains of the catalytic subunits IKKα and IKKβ (108). This Hsp90-mediated IKK-containing complex appears essential for the efficient TNF-mediated activation of NF-κB and its survival-promoting effect (108). Accordingly, geldanamycin, a proposed anticancer agent with specificity for Hsp90, confers an enhanced sensitivity of cells to TNF-induced death, consistent with its ability to disrupt Hsp90-containing signaling complexes (113). In contrast, the ability of Hsp27 to interact with both catalytic subunits of the IKK complex, IKKα and IKKβ, the latter of which is enhanced via the TNF-induced activation of MAPK-dependent phosphorylation of Hsp27, leads to the enhanced inhibition of IKK activity, reduced IκB degradation, and consequent suppression of NF-κB activation (109). These effects of Hsp27 appear to contradict previous findings that it can suppress TNF-induced apoptosis (23, 114), but they may simply indicate how poorly we understand the underlying complexity of TNF-induced signaling events and how these might be differentially offset according to the relative expression levels of different Hsps and their predominant mode of antiapoptotic activity. Indeed, a more recent study attributes the survival-promoting activity of Hsp27 to the maintenance of NF-κB activity via its ability to promote the proteasome-dependent degradation of IκB (110).

The ability of the Hsps to regulate elements of the death receptor–mediated pathways also includes effects at the level of the individual receptor-bound complexes. For example, ligation of the Fas receptor, in addition to leading to the FADD-dependent recruitment and activation of procaspase-8, has also been reported to elicit a FADD-independent pathway via the recruitment of the adapter DAXX, which in turn leads to the activation of the MAPKKK apoptosis signal–regulating kinase 1 (ASK1) and downstream induction of JNK (115, 116). Although the role of DAXX in mediating Fas-induced death is controversial (117), the survival-promoting effects of both Hsp27 and Hsp90 have been linked to their ability to regulate the recruitment of DAXX (116) and ASK1 (118) to the Fas signaling complex. Hsp90 can also bind directly to, and stabilize, RIP1 (119, 120), thereby providing a more robust signal to induce NF-κB activation and enhanced cellular survival (113) (Figure 2).

Hsps as pharmacological targets

In normal cells under nonstressed conditions, the inducible members of the Hsp family are poorly expressed and are only induced after changes in cellular environment or in response to damaging stimuli. In contrast, the expression levels of several members of the Hsp family are significantly elevated in many cancers, and in the case of some tumor types this is linked with poor prognosis and a muted response to chemotherapy (121, 122). It is unclear precisely why Hsp levels are elevated in tumor cells. It may reflect an elevation in the expression of misfolded proteins and a consequent increase in the demand for Hsp-mediated refolding capacity; or, alternatively, the microenvironment within the tumor, often hypoxic and glucose restricted, may favor the hyperexpression of several Hsps. It is also unclear whether the overexpression of Hsps in tumors simply supports the malignant phenotype (56), or whether they play a more fundamental role in the development and manifestation of the transformed phenotype (123).

Whatever their precise role, the correlation between Hsp overexpression and chemoresistance in many tumor types is likely due, at least in part, to the ability of Hsps to inhibit apoptosis. For that reason, the pharmacological manipulation of Hsp levels likely represents an as-yet underexploited opportunity either to render tumor cells susceptible to the induction of apoptosis by chemotherapeutics and/or UV irradiation or, alternatively, to directly and selectively disrupt their survival.

Geldanamycin and an analog, 17AAG, represent the only Hsp-targeted pharmacological agents so far described. These agents bind to and block the nucleotide-binding pocket of Hsp90 and, by doing so, disrupt interactions between Hsp90 and its target proteins (124, 125). Targets of Hsp90 that are directly regulated by their ATP-dependent association with Hsp90 include Ras, p53, and Akt, all of which are frequently dysregulated in cancer and play an essential role in the maintenance and progression of the transformed state. Whether this mode of activity might be applied for the disruption of interactions between other Hsp family members and their target proteins has yet to be explored. It does, however, remain a tantalizing possibility that Hsp-targeted pharmacological strategies may herald the advent of a novel approach to rendering tumor cells sensitive to conventional chemotherapies.

Conclusion

The rationale for a functional interaction between an ancient and highly conserved cellular protective mechanism and a similarly evolutionarily maintained series of pathways to engage cell death seems somewhat intuitive. Just as all species harbor an inducible protective Hsp or stress response, elements of the core apoptotic machinery are likewise found in both uni- and multicellular organisms. It has yet to be determined whether aspects of the Hsp-mediated regulation of apoptotic events observed in mammalian systems are common to lower-order organisms. However, it remains a distinct possibility that to ensure homeostatic regulation of cell number during development and in response to damaging events, evolution has incorporated a vital and necessary susceptibility of death-inducing signals to the opposing regulatory effects of the Hsp stress pathway.

Acknowledgments
The author made every effort to include the key references in each of the areas discussed but apologizes to those whose citations were omitted due to space constraints. H.M. Beere is supported by NIH grants GM066914 and AI47891.

Address correspondence to: Helen M. Beere, St. Jude Children's Research Hospital, Department of Immunology, 332 North Lauderdale Street, Memphis, Tennessee 38105, USA. Phone: (901) 495-3470; Fax: (901) 495-3107; E-mail: Helen.Beere@stjude.org.

1. Beere, H.M. 2001. Stressed to death: regulation of apoptotic signaling pathways by the heat shock proteins. *Sci. STKE.* doi:10.1126/stke.2001.93.re1.
2. Beere, H.M., and Green, D.R. 2001. Stress management: heat shock protein-70 and the regulation of apoptosis. *Trends Cell Biol.* **11**:6–10.
3. Beere, H.M. 2004. "The stress of dying": the role of heat shock proteins in the regulation of apoptosis. *J. Cell Sci.* **117**:2641–2651.
4. Parsell, D.A., and Lindquist, S. 1993. The function of heat-shock proteins in stress tolerance: degradation and reactivation of damaged proteins. *Annu. Rev. Genet.* **27**:437–496.
5. Nollen, E.A.A., and Morimoto, R.I. 2002. Chaperoning signaling pathways: molecular chaperones as stress-sensing 'heat shock' proteins. *J. Cell Sci.* **115**:2809–2816.
6. Freeman, B.C., et al. 1995. Identification of a regulatory motif in Hsp70 that affects ATPase activity, substrate binding and interaction with HDJ-1. *EMBO J.* **14**:2281–2292.
7. Mosser, D.D., et al. 2000. The chaperone function of Hsp70 is required for protection against stress-induced apoptosis. *Mol. Cell. Biol.* **20**:7146–7159.
8. Gabai, V.L., et al. 2000. Hsp72-mediated suppression of c-Jun N-terminal kinase is implicated in development of tolerance to caspase-independent cell death. *Mol. Cell. Biol.* **20**:6826–6836.
9. Boatright, K.M., and Salvesen, G.S. 2003. Mechanisms of caspase activation. *Curr. Opin. Cell Biol.* **15**:725–731.
10. Thornberry, N.A. 1998. Caspases: key mediators of apoptosis. *Chem. Biol.* **5**:R97–R103.
11. Boatright, K.M., et al. 2003. A unified model for apical caspase activation. *Mol. Cell.* **11**:529–541.
12. Deveraux, Q.L., et al. 1997. X-linked IAP is a direct inhibitor of cell-death proteases. *Nature.* **388**:300–304.
13. Deveraux, Q.L., and Reed, J.C. 1999. IAP family proteins: suppressors of apoptosis. *Genes Dev.* **13**:239–252.
14. Enari, M., et al. 1998. A caspase-activated DNase that degrades DNA during apoptosis, and its inhibitor ICAD [erratum 1998, **393**:396]. *Nature.* **391**:43–50.
15. Lazebnik, Y.A., et al. 1994. Cleavage of poly(ADP-ribose) polymerase by a proteinase with properties like ICE. *Nature.* **371**:346–347.
16. Martin, S.J., et al. 1995. Proteolysis of fodrin (non-erythroid spectrin) during apoptosis. *J. Biol. Chem.* **270**:6425–6428.
17. Ricci, J.E., et al. 2004. Disruption of mitochondrial function during apoptosis is mediated by caspase cleavage of the p75 subunit of complex I of the electron transport chain. *Cell.* **117**:773–786.
18. Mashima, T., et al. 1997. Actin cleavage by CPP-32/apopain during the development of apoptosis. *Oncogene.* **14**:1007–1012.
19. Gervais, F.G., et al. 1998. Caspases cleave focal adhesion kinase during apoptosis to generate a FRNK-like polypeptide. *J. Biol. Chem.* **273**:17102–17108.
20. Simon, M.M., et al. 1995. Heat shock protein 70 overexpression affects the response to ultraviolet light in murine fibroblasts. Evidence for increased cell viability and suppression of cytokine release. *J. Clin. Invest.* **95**:926–933.
21. Warrick, J.M., et al. 1999. Suppression of polyglutamine-mediated neurodegeneration in Drosophila by the molecular chaperone HSP70. *Nat. Genet.* **23**:425–428.
22. Clemons, N.J., et al. 2005. Hsp72 inhibits Fas-mediated apoptosis upstream of the mitochondria in type II cells. *J. Biol. Chem.* **280**:9005–9012.
23. Mehlen, P., Schulze-Osthoff, K., and Arrigo, A.P. 1996. Small stress proteins as novel regulators of apoptosis. Heat shock protein 27 blocks Fas/APO-1- and staurosporine-induced cell death. *J. Biol. Chem.* **271**:16510–16514.
24. Mosser, D.D., et al. 1997. Role of the human heat shock protein hsp70 in protection against stress-induced apoptosis. *Mol. Cell. Biol.* **17**:5317–5327.
25. Mailhos, C., Howard, M.K., and Latchman, D.S. 1993. Heat shock protects neuronal cells from programmed cell death by apoptosis. *Neuroscience.* **55**:621–627.
26. Ahn, J.H., et al. 1999. Suppression of ceramide-mediated apoptosis by HSP70. *Mol. Cell.* **9**:200–206.
27. Mao, H., et al. 2003. Hsp72 inhibits focal adhesion kinase degradation in ATP-depleted renal epithelial cells. *J. Biol. Chem.* **278**:18214–18220.
28. Garrido, C., et al. 1999. HSP27 inhibits cytochrome c-dependent activation of procaspase-9. *FASEB J.* **13**:2061–2070.
29. Kluck, R.M., et al. 1997. The release of cytochrome c from mitochondria: a primary site for Bcl-2 regulation of apoptosis. *Science.* **275**:1132–1326.
30. Yang, J., et al. 1997. Prevention of apoptosis by Bcl-2: release of cytochrome c from mitochondria blocked. *Science.* **275**:1129–1132.
31. Suzuki, Y., et al. 2001. A serine protease, HtrA2, is released from the mitochondria and interacts with XIAP, inducing cell death. *Mol. Cell.* **8**:613–621.
32. Susin, S.A., et al. 1999. Molecular characterization of mitochondrial apoptosis-inducing factor. *Nature.* **397**:441–446.
33. Du, C., et al. 2000. Smac, a mitochondrial protein that promotes cytochrome c-dependent caspase activation by eliminating IAP inhibition. *Cell.* **102**:33–42.
34. Verhagen, A.M., et al. 2000. Identification of DIABLO, a mammalian protein that promotes apoptosis by binding to and antagonizing IAP proteins. *Cell.* **102**:43–53.
35. Li, L., Luo, X., and Wang, X. 2001. Endonuclease G is an apoptotic DNase when released from mitochondria. *Nature.* **412**:95–99.
36. Sharpe, J.C., Arnoult, D., and Youle, R.J. 2004. Control of mitochondrial permeability by Bcl-2 family members. *Biochim. Biophys. Acta.* **1644**:107–113.
37. Willis, S., et al. 2003. The Bcl-2-regulated apoptotic pathway. *J. Cell Sci.* **116**:4053–4056.
38. Arnoult, D., Karbowski, M., and Youle, R.J. 2003. Caspase inhibition prevents the mitochondrial release of apoptosis-inducing factor. *Cell Death Differ.* **10**:845–849.
39. Chao, D.T., and Korsmeyer, S.J. 1998. BCL-2 family: regulators of cell death. *Annu. Rev. Immunol.* **16**:395–419.
40. Marsden, V.S., and Strasser, A. 2003. Control of apoptosis in the immune system: Bcl-2, BH3-only proteins and more. *Annu. Rev. Immunol.* **21**:71–105.
41. Gross, A., et al. 1999. Caspase cleaved BID targets mitochondria and is required for cytochrome c release, while BCL-XL prevents this release but not tumor necrosis factor-R1/Fas death. *J. Biol. Chem.* **274**:1156–1163.
42. Luo, X., et al. 1998. Bid, a Bcl2 interacting protein, mediates cytochrome c release from mitochondria in response to activation of cell surface death receptors. *Cell.* **94**:481–490.
43. O'Connor, L., et al. 1998. Bim: a novel member of the Bcl-2 family that promotes apoptosis. *EMBO J.* **17**:384–395.
44. Puthalakath, H., et al. 1999. The proapoptotic activity of the Bcl-2 family member Bim is regulated by interaction with the dynein motor complex. *Mol. Cell.* **3**:287–296.
45. Puthalakath, H., et al. 2001. Bmf: a proapoptotic BH3-only protein regulated by interaction with the myosin V actin motor complex, activated by anoikis. *Science.* **293**:1829–1832.
46. Zha, J., et al. 1996. Serine phosphorylation of death agonist BAD in response to survival factor results in binding to 14-3-3 not BCL-X(L). *Cell.* **87**:619–628.
47. del Peso, L., et al. 1997. Interleukin-3-induced phosphorylation of BAD through the protein kinase Akt. *Science.* **278**:687–689.
48. Steel, R., et al. 2004. Hsp72 inhibits apoptosis upstream of the mitochondria and not through interactions with Apaf-1. *J. Biol. Chem.* **279**:51490–51499.
49. Paul, C., et al. 2002. Hsp27 as a negative regulator of cytochrome c release. *Mol. Cell. Biol.* **22**:816–834.
50. Gabai, V.L., et al. 2002. Hsp72 and stress kinase c-jun N-terminal kinase regulate the bid-dependent pathway in tumor necrosis factor-induced apoptosis. *Mol. Cell. Biol.* **22**:3415–3424.
51. Gotoh, T., et al. 2004. hsp70-DnaJ chaperone pair prevents nitric oxide- and CHOP-induced apoptosis by inhibiting translocation of Bax to mitochondria. *Cell Death Differ.* **11**:390–402.
52. Susin, S.A., et al. 1996. Bcl-2 inhibits the mitochondrial release of an apoptogenic protease. *J. Exp. Med.* **184**:1331–1341.
53. Ravagnan, L., et al. 2001. Heat-shock protein 70 antagonizes apoptosis-inducing factor. *Nat. Cell Biol.* **3**:839–843.
54. Gurbuxani, S., et al. 2003. Heat shock protein 70 binding inhibits nuclear import of apoptosis-inducing factor. *Oncogene.* **22**:6669–6678.
55. Schmitt, E., et al. 2003. Chemosensitization by a non-apoptogenic heat shock protein 70-binding apoptosis-inducing factor mutant. *Cancer Res.* **63**:8233–8240.
56. Nylandsted, J., et al. 2000. Selective depletion of heat shock protein 70 (Hsp70) activates a tumor-specific death program that is independent of caspases and bypasses Bcl-2. *Proc. Natl. Acad. Sci. U. S. A.* **97**:7871–7876.
57. Nylandsted, J., et al. 2004. Heat shock protein 70 promotes cell survival by inhibiting lysosomal membrane permeabilization. *J. Exp. Med.* **200**:425–435.
58. Zou, H., et al. 1997. Apaf-1, a human protein homologous to C. elegans CED-4, participates in cytochrome c-dependent activation of caspase-3. *Cell.* **90**:405–413.
59. Zou, H., et al. 1999. An APAF-1·cytochrome c multimeric complex is a functional apoptosome that activates procaspase-9. *J. Biol. Chem.* **274**:11549–11556.
60. Srinivasula, S.M., et al. 1998. Autoactivation of procaspase-9 by Apaf-1-mediated oligomerization. *Mol. Cell.* **1**:949–957.
61. Beere, H.M., et al. 2000. Heat-shock protein 70 inhibits apoptosis by preventing recruitment of

procaspase-9 to the apaf-1 apoptosome. *Nat. Cell Biol.* **2**:469–475.

62. Pandey, P., et al. 2000. Negative regulation of cytochrome c-mediated oligomerization of apaf-1 and activation of procaspase-9 by heat shock protein 90. *EMBO J.* **19**:4310–4322.

63. Bruey, J.-M., et al. 2000. Hsp27 negatively regulates cell death by interacting with cytochrome c. *Nat. Cell Biol.* **2**:645–652.

64. Datta, S.D., Brunet, A., and Greenberg, M. 1999. Cellular survival: a play in three Akts. *Genes Dev.* **13**:2905–2927.

65. Mearow, K.M., et al. 2002. Stress-mediated signaling in PC12 cells: the role of the small heat shock protein, Hsp27, and Akt in protecting cells from heat stress and nerve growth factor withdrawal. *J. Neurochem.* **83**:452–462.

66. Dougherty, M.K., and Morrison, D.K. 2004. Unlocking the code of 14-3-3. *J. Cell Sci.* **117**:1875–1884.

67. Brunet, A., et al. 1999. Akt promotes cell survival by phosphorylating and inhibiting a forkhead transcription factor. *Cell.* **96**:857–868.

68. Dijkers, P.F., et al. 2000. Expression of the proapoptotic family member Bim is regulated by the forkhead transcription factor, FKHR-L1. *Curr. Biol.* **10**:1201–1204.

69. Kane, L.P., et al. 1999. Induction of NF-κB by the Akt/PKB kinase. *Curr. Biol.* **9**:601–604.

70. You, M., et al. 1997. ch-IAP1, a member of the inhibitor-of-apoptosis protein family, is a mediator of the antiapoptotic activity of the v-Rel oncoprotein. *Mol. Cell. Biol.* **17**:7328–7341.

71. Chu, Z.L., et al. 1997. Suppression of tumor necrosis factor induced death by inhibitor of apoptosis c-IAP2 is under NF-κB control. *Proc. Natl. Acad. Sci. U. S. A.* **94**:10057–10062.

72. Zong, W.-X., et al. 1999. The pro-survival Bcl-2 homolog Bfl-1/A1 is a direct transcriptional target of NF-κB that blocks TNF-α-induced apoptosis. *Genes Dev.* **13**:382–387.

73. Sato, S., Fujita, N., and Tsuruo, T. 2000. Modulation of Akt kinase activity by binding to Hsp90. *Proc. Natl. Acad. Sci. U. S. A.* **97**:10832–10837.

74. Nakagomi, S., et al. 2003. Expression of the activating transcription factor 3 prevents c-Jun N-terminal kinase-induced neuronal death by promoting heat shock protein 27 expression and Akt activation. *J. Neurosci.* **23**:5187–5196.

75. Rane, M.J., et al. 2003. Heat shock protein 27 controls apoptosis by regulating Akt activation. *J. Biol. Chem.* **278**:27826–27835.

76. Basso, A.D., et al. 2002. Akt forms an intracellular complex with heat shock protein 90 (Hsp90) and Cdc37 and is destabilized by inhibitors of Hsp90 function. *J. Biol. Chem.* **277**:39858–39866.

77. Jia, W., et al. 2003. Synergistic antileukemic interactions between 17-AAG and UCN-01 interruption of RAF-MEK- and AKT-related pathways. *Blood.* **102**:1824–1832.

78. Nimmanapalli, R., et al. 2003. Regulation of 17-AAG-induced apoptosis: role of Bcl-2, Bcl-XL and Bax downstream of 17-AAG-mediated down regulation of Akt, Raf-1 and Src kinases. *Blood.* **102**:269–275.

79. Rane, M.J., et al. 2001. p38 kinase-dependent MAPKAPK-2 activation functions as 3-phosphoinositide-dependent kinase-2 for Akt in human neutrophils. *J. Biol. Chem.* **276**:3517–3523.

80. Davis, R.J. 2000. Signal transduction by the JNK group of MAP kinases. *Cell.* **103**:239–252.

81. Fan, M., et al. 2000. Vinblastine-induced phosphorylation of Bcl-2 and Bcl-XL is mediated by JNK and occurs in parallel with inactivation of the Raf-1/MEK/ERK cascade. *J. Biol. Chem.* **275**:29980–29985.

82. Yamamoto, K., Ichijo, H., and Korsmeyer, S.J. 1999. BCL-2 is phosphorylated and inactivated by an ASK1/Jun N-terminal protein kinase pathway normally activated at G(2)/M. *Mol. Biol. Cell.*

19:8469–8478.

83. Maundrell, K., et al. 1997. Bcl-2 undergoes phosphorylation by c-Jun N-terminal kinase/stress-activated protein kinases in the presence of the constitutively active GTP binding protein Rac-1. *J. Biol. Chem.* **272**:25238–25242.

84. Lei, K., and Davis, R.J. 2003. JNK phosphorylation of Bim-related members of the Bcl2 family induces Bax-dependent apoptosis. *Proc. Natl. Acad. Sci. U. S. A.* **100**:2432–2437.

85. Tsuruta, F., et al. 2004. JNK promotes Bax translocation to mitochondria through phosphorylation of 14-3-3 proteins. *EMBO J.* **23**:1889–1899.

86. Yu, C., et al. 2004. JNK suppresses apoptosis via phosphorylation of the proapoptotic Bcl-2 family protein BAD. *Mol. Cell.* **13**:329–340.

87. Tournier, C., et al. 2000. Requirement of JNK for stress-induced activation of the cytochrome c-mediated death pathway. *Science.* **288**:870–874.

88. Fuchs, S.Y., et al. 1998. JNK targets p53 ubiquitination and degradation in non-stressed cells. *Genes Dev.* **12**:2658–2663.

89. Noguchi, K., et al. 1999. Regulation of c-Myc through phosphorylation at Ser-62 and Ser-71 by c-Jun N-terminal kinase. *J. Biol. Chem.* **274**:32580–32587.

90. Chipuk, J.E., et al. 2004. Direct activation of Bax by p53 mediates mitochondrial membrane permeabilization and apoptosis. *Science.* **303**:1010–1014.

91. Volloch, V., et al. 1999. ATPase activity of the heat shock protein Hsp72 is dispensable for its effects on dephosphorylation of stress kinase JNK and on heat-induced apoptosis. *FEBS Lett.* **461**:73–76.

92. Park, H.-S., et al. 2001. Hsp72 functions as a natural inhibitory protein of c-Jun N-terminal kinase. *EMBO J.* **20**:446–456.

93. Meriin, A.B., et al. 1999. Protein-damaging stresses activate c-Jun N-terminal kinase via inhibition of its dephosphorylation: a novel pathway controlled by HSP72 [erratum 1999, **19**:5235]. *Mol. Cell. Biol.* **19**:2547–2555.

94. Chinnaiyan, A.M., et al. 1995. FADD, a novel death domain containing protein, interacts with the death domain of Fas and initiates apoptosis. *Cell.* **81**:505–512.

95. Liossis, S.N., et al. 1997. Overexpression of the heat shock protein 70 enhances the TCR/CD3- and Fas/Apo-1/CD95-mediated apoptotic cell death in Jurkat T cells. *J. Immunol.* **158**:5668–5675.

96. Creagh, E.M., and Cotter, T.G. 1999. Selective protection by Hsp70 against cytotoxic drug-, but not Fas-induced T-cell apoptosis. *Immunology.* **97**:36–44.

97. Kamradt, M.C., et al. 2005. The small heat shock protein alpha B-crystallin is a novel inhibitor of TRAIL-induced apoptosis that suppresses the activation of caspase-3. *J. Biol. Chem.* **280**:11059–11066.

98. Van Molle, W., et al. 2002. HSP70 protects against TNF-induced lethal inflammatory shock. *Immunity.* **16**:685–695.

99. Jaattela, M. 1993. Overexpression of major heat shock protein hsp70 inhibits tumor necrosis factor-induced activation of phospholipase A2. *J. Immunol.* **151**:4286–4294.

100. Jaattela, M., et al. 1992. Major heat shock protein hsp70 protects tumor cells from tumor necrosis factor cytotoxicity. *EMBO J.* **11**:3507–3512.

101. Galea-Lauri, J., et al. 1996. Increased heat shock protein 90 (hsp90) expression leads to increased apoptosis in the monoblastoid cell line U937 following induction with TNF-alpha and cycloheximide: a possible role in immunopathology. *J. Immunol.* **157**:4109–4118.

102. Mehlen, P., et al. 1996. Human Hsp27, Drosophila hsp27 and human αB-crystallin expression-mediated increase in gluathione is essential for the protective activity of these proteins against TNFα-induced cell death. *EMBO J.* **15**:2695–2706.

103. Chinnaiyan, A.M., et al. 1996. FADD/MORT is a common mediator of CD95 (Fas/APO1)- and

TNF-receptor-induced apoptosis. *J. Biol. Chem.* **271**:4961–4965.

104. Hsu, H., et al. 1996. TNF-dependent recruitment of the protein kinase RIP to the TNF receptor 1 signaling complex. *Immunity.* **4**:387–396.

105. Hsu, H., et al. 1996. TRADD-TRAF2 and TRADD-FADD interactions define two distinct TNF receptor signal transduction pathways. *Cell.* **84**:299–308.

106. Wang, C.-Y., et al. 1998. NF-κB antiapoptosis: induction of TRAF1 and TRAF2 and c-IAP1 and c-IPA2 to suppress caspase-8 activation. *Science.* **281**:1680–1683.

107. Wang, C.Y., et al. 1999. NF-kappaB induces expression of the Bcl-2 homologue A1/Bfl-1 to preferentially suppress chemotherapy-induced apoptosis. *Mol. Cell. Biol.* **19**:5923–5929.

108. Chen, G., Cao, P., and Goeddel, D.V. 2002. TNF-induced recruitment and activation of the IKK complex require Cdc37 and Hsp90. *Mol. Cell.* **9**:401–410.

109. Park, K.-J., Gaynor, R.B., and Kwak, Y.T. 2003. Heat shock protein 27 association with the IκB kinase complex regulates tumor necrosis factor α-induced NF-κB activation. *J. Biol. Chem.* **278**:35273–35278.

110. Parcellier, A., et al. 2003. HSP27 is a ubiquitin-binding protein involved in I-κBα proteasomal degradation. *Mol. Cell. Biol.* **23**:5790–5802.

111. Kimura, Y., et al. 1997. Cdc37 is a molecular chaperone with specific functions in signal transduction. *Genes Dev.* **11**:1775–1785.

112. Septanova, L., et al. 1996. Mammalian p50Cdc37 is a protein kinase-targeting subunit of Hsp90 that binds and stabilizes Cdk4. *Genes Dev.* **10**:1491–1502.

113. Lewis, J., et al. 2000. Disruption of Hsp90 function results in degradation of the death domain kinase, receptor-interacting protein (RIP), and blockage of tumor necrosis factor-induced nuclear factor-κB activation. *J. Biol. Chem.* **275**:10519–10526.

114. Mehlen, P., et al. 1995. Constitutive expression of human hsp27, drosophila hsp27, or human αB-crystallin confers resistance to TNF- and oxidative stress-induced cytotoxicity in stably transfected murine L929 fibroblasts. *J. Immunol.* **154**:363–374.

115. Chang, H.Y., et al. 1998. Activation of apoptosis signal-regulating kinase (ASK-1) by the adapter protein daxx. *Science.* **281**:1860–1863.

116. Yang, X., et al. 1997. Daxx, a novel fas binding protein that activates JNK and apoptosis. *Cell.* **89**:1066–1076.

117. Torii, S., et al. 1999. Human Daxx regulates Fas-induced apoptosis from nuclear PML oncogenic domains (PODs). *EMBO J.* **18**:6037–6049.

118. Park, H.-S., et al. 2002. Heat shock protein Hsp72 is a negative regulator of apoptosis signal-regulating kinase 1. *Mol. Cell. Biol.* **22**:7721–7730.

119. Devin, A., et al. 2000. The distinct roles of TRAF2 and RIP in IKK activation by TNF-R1: TRAF2 recruits IKK to TNF-R1 while RIP mediates IKK activation. *Immunity.* **12**:419–429.

120. Zhang, S.Q., et al. 2000. Recruitment of the IKK signalosome to the p55 TNF receptor: RIP and A20 bind to NEMO (IKKγ) upon receptor stimulation. *Immunity.* **12**:301–311.

121. Jaattela, M. 1999. Escaping cell death: survival proteins in cancer. *Exp. Cell Res.* **248**:30–43.

122. Ciocca, D.R., and Calderwood, S.K. 2005. Heat shock proteins in cancer: diagnostic, prognostic, predictive, and treatment implications. *Cell Stress Chaperones.* **10**:86–103.

123. Volloch, V., and Sherman, M.Y. 1999. Oncogenic potential of Hsp72. *Oncogene.* **18**:3648–3651.

124. Workman, P. 2002. Pharmacogenomics in cancer drug discovery and development: inhibitors of the Hsp90 molecular chaperone. *Cancer Detect. Prev.* **26**:405–410.

125. Neckers, L., and Ivy, S.P. 2003. Heat shock protein 90. *Curr. Opin. Oncol.* **15**:419–424.

Autophagy in cell death: an innocent convict?

Beth Levine[1] and Junying Yuan[2]

[1]Departments of Internal Medicine and Microbiology, University of Texas Southwestern Medical Center, Dallas, Texas, USA.
[2]Department of Cell Biology, Harvard Medical School, Boston, Massachusetts, USA.

The visualization of autophagosomes in dying cells has led to the belief that autophagy is a nonapoptotic form of programmed cell death. This concept has now been evaluated using cells and organisms deficient in autophagy genes. Most evidence indicates that, at least in cells with intact apoptotic machinery, autophagy is primarily a pro-survival rather than a pro-death mechanism. This review summarizes the evidence linking autophagy to cell survival and cell death, the complex interplay between autophagy and apoptosis pathways, and the role of autophagy-dependent survival and death pathways in clinical diseases.

Cell biologists have long recognized the possibility that eukaryotic cells may undergo nonapoptotic forms of programmed cell death. Autophagy, a lysosomal pathway involving the bulk degradation of cytoplasmic contents, has been identified as a prime suspect in such death, and recent studies have implicated the autophagy pathway as a cause of nonapoptotic cellular demise. However, most evidence linking autophagy to cell death is circumstantial. Now, with new tools to assess causality, it is an opportune time to revisit the case of autophagy in cell death. Is autophagy an innocent bystander, a direct death execution pathway, a defense mechanism that ultimately fails in its mission to preserve cell viability, and/or a garbage disposal mechanism that cleans up remnants of a cell already committed to die?

Autophagy is a regulated lysosomal degradation pathway

The term *autophagy* (Greek, "to eat oneself") does not refer to a death process; it denotes the process of self-cannibalization through a lysosomal degradation pathway. Autophagy is the cell's major regulated mechanism for degrading long-lived proteins and the only known pathway for degrading organelles (reviewed in refs. 1, 2). During autophagy, an isolation membrane forms, presumably arising from a vesicular compartment known as the preautophagosomal structure, invaginates, and sequesters cytoplasmic constituents including mitochondria, endoplasmic reticulum, and ribosomes (Figure 1). The edges of the membrane fuse to form a double or multimembranous structure, known as the autophagosome or autophagic vacuole. The outer membrane of the autophagosome fuses with the lysosome (in mammalian cells) or vacuole (in yeast and plants) to deliver the inner membranous vesicle to the lumen of the degradative compartment. Degradation of the sequestered material generates nucleotides, amino acids, and free fatty acids that are recycled for macromolecular synthesis and ATP generation.

Autophagy occurs at low basal levels in all cells to perform homeostatic functions (e.g., cytoplasmic and organelle turnover) but is rapidly upregulated when cells need to generate intracellular nutrients and energy (e.g., during starvation or trophic factor withdrawal), undergo architectural remodeling (e.g., during developmental transitions),

or rid themselves of damaging cytoplasmic components (e.g., during oxidative stress, infection, and accumulation of protein aggregates). Nutritional status, hormonal factors, and other cues like temperature, oxygen concentrations, and cell density are important in the control of autophagy. Two evolutionarily conserved nutrient sensors play roles in autophagy regulation: (a) the target of rapamycin (TOR) kinase is the major inhibitory signal that shuts off autophagy during nutrient abundance (reviewed in ref. 3), and (b) the eukaryotic initiation factor 2α (eIF2α) kinase Gcn2 and its downstream target Gcn4, a transcriptional transactivator of autophagy genes, turn on autophagy during nutrient depletion (4). The class I PI3K/Akt signaling molecules link receptor tyrosine kinases to TOR activation and thereby repress autophagy in response to insulin-like and other growth factor signals (reviewed in ref. 3).

Downstream of TOR kinase, there are approximately 17 gene products essential for autophagy and related pathways in yeast, referred to as the *ATG* genes (5), and most yeast *ATG* genes have orthologs in higher eukaryotes (reviewed in ref. 2). The *ATG* genes encode proteins needed for the induction of autophagy, and the generation, maturation, and recycling of autophagosomes. These proteins are composed of 4 functional groups, including a protein serine/threonine kinase complex that responds to upstream signals such as TOR kinase (Atg1, Atg13, Atg17), a lipid kinase signaling complex that mediates vesicle nucleation (Atg6, Atg14, Vps34, and Vps15), 2 novel ubiquitin-like conjugation pathways that mediate vesicle expansion (the Atg8 and Atg12 systems), and a recycling pathway that mediates the disassembly of Atg proteins from matured autophagosomes (Atg2, Atg9, Atg18) (Figure 2).

Autophagy as a cell death mechanism

The term "autophagic cell death" describes a form of programmed cell death morphologically distinct from apoptosis and presumed to result from excessive levels of cellular autophagy (6). In classical apoptosis, or type I programmed cell death, there is early collapse of cytoskeletal elements but preservation of organelles until late in the process. In contrast, in autophagic, or type II, programmed cell death, there is early degradation of organelles but preservation of cytoskeletal elements until late stages. Whereas apoptotic cell death is caspase-dependent and characterized by internucleosomal DNA cleavage, caspase activation and DNA fragmentation occur very late (if at all) in autophagic cell death (Figure 3). In contrast with necrosis, both apoptotic and autophagic cell death are characterized by the lack of a tissue inflammatory response.

Large numbers of autophagic vacuoles have been observed in dying cells of animals of diverse taxa (reviewed in refs. 6–9) (Table 1).

Nonstandard abbreviations used: Htt, Huntingtin; 3-MA, 3-methyladenine; MEF, murine embryonic fibroblast; MPT, mitochondrial permeability transition; polyQ, polyglutamine; RNAi, RNA interference; TOR, target of rapamycin; TRAIL, TNF-related apoptosis-inducing ligand.

Conflict of interest: The authors have declared that no conflict of interest exists.

Citation for this article: *J. Clin. Invest.* **115**:2679–2688 (2005). doi:10.1172/JCI26390.

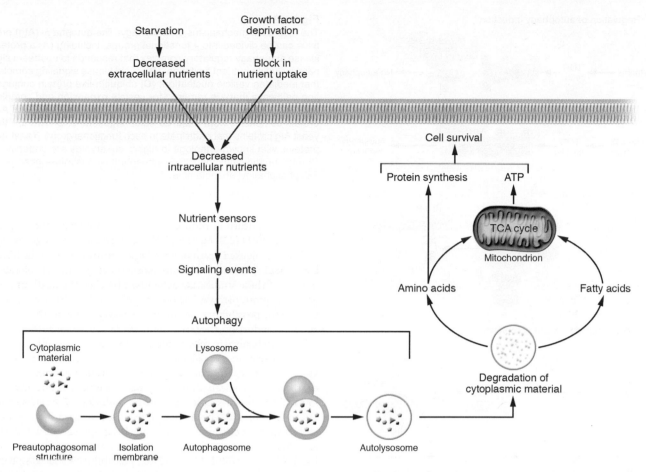

Figure 1
The autophagy pathway and its role in cellular adaptation to nutrient deprivation. Starvation or growth factor deprivation results in a decrease in intracellular nutrients and activation of nutrient-sensing signaling pathways (reviewed in ref. 97) that stimulate autophagy. Autophagy involves the sequestration of cytoplasmic material by an isolation membrane (derived from the preautophagosomal structure) to form a double-membrane vacuole, the autophagosome. The autophagosome undergoes fusion with a late endosome or lysosome, to form an autolysosome, in which the sequestered material is degraded. Degradation of membrane lipids and proteins by the autolysosome generates free fatty acids and amino acids that can be reused by the cell to maintain mitochondrial ATP energy production and protein synthesis and thereby promote cell survival. Disruption of this pathway by autophagy gene inactivation prevents cell survival in diverse organisms (Table 2). The same molecular machinery and overlapping dynamic membrane rearrangement events that occur during starvation may also be used in other settings to degrade unwanted cytoplasmic contents, including damaged mitochondria, protein aggregates, and intracellular pathogens. See text for discussion. TCA cycle, tricarboxylic acid cycle.

The consensus view has been that autophagic cell death occurs primarily when the developmental program (e.g., insect metamorphosis) or homeostatic processes in adulthood (e.g., mammary gland postlactational involution, prostate involution following castration) require massive cell elimination. Recently, studies have also described autophagic cell death in diseased mammalian tissues and in tumor cell lines treated with chemotherapeutic agents (Table 1). In many of these cases, morphologic features of autophagic and apoptotic cell death or of autophagic and necrotic cell death are observed in the same cell.

What is the evidence that autophagy is a death execution mechanism in autophagic cell death? If cell death is truly due to autophagy, then pharmacologic or genetic inhibition of autophagy should prevent the death. Yet, for most of the developmental, disease-associated, and toxic stimulus-induced deaths that are presumed to be autophagic (Table 1), the evidence for its role is only correlative. Moreover, in certain cases of autophagic cell death, the available evidence calls into question a causative role of autophagy. For exam-

ple, in *Drosophila*, autophagic cell death but not autophagy observed during salivary gland regression is prevented by mutations in the ecdysone-regulated transcription factors *BR-C* and *E74A* (10). In the slime mold *Dictyostelium*, a null mutation in the autophagy gene *atg1* blocks vacuolization but not cell death in an in vitro model of autophagic cell death (11). Thus, in these model systems, autophagy per se is neither sufficient nor required for autophagic cell death. Furthermore, the caspase inhibitor p35 blocks metamorphic cell death in *Drosophila* without complete inhibition of autophagy, suggesting that it is caspase-mediated apoptosis, rather than autophagy, that plays a key role in this death process (10).

There is, however, some evidence in certain in vitro settings that pharmacologic or genetic inhibition of autophagy can prevent cell death. The pharmacologic inhibitor of autophagy 3-methyladenine (3-MA), a nucleotide derivative that blocks class III PI3K activity (12–14), delays or partially inhibits death in starved hepatocytes from carcinogen-treated rats (15), in anti-estrogen–treated human mammary carcinoma cells (16), in chloroquine-treated cortical neu-

A Regulation of autophagy induction

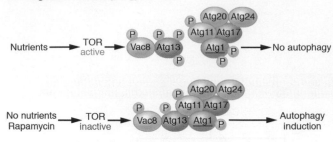

B Vesicle nucleation

C Vesicle expansion and completion

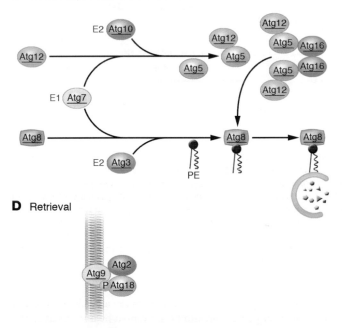

D Retrieval

Figure 2
The molecular mechanisms of autophagy. The autophagy (Atg) proteins can be divided into 4 functional groups, including (**A**) a protein kinase autophagy regulatory complex that responds to upstream signals, including nutrient limitation; (**B**) a lipid kinase signaling complex that mediates vesicle nucleation; (**C**) ubiquitin-like protein conjugation pathways that are required for vesicle expansion and completion; and (**D**) a retrieval pathway required for the disassembly of Atg protein complexes from matured autophagosomes. Shown are the yeast Atg proteins that participate in each functional group. Yeast Atg proteins with known orthologs in higher eukaryotes are underlined. PI, phosphatidylinositol; PI3-P, phosphatidylinositol 3-phosphate; PE, phosphatidylethanolamine.

blocked cell death in mouse L929 cells treated with the caspase inhibitor zVAD (23). Further, RNAi against autophagy genes *atg5* and *beclin 1* blocked death of *bax⁻/⁻*, *bak⁻/⁻* murine embryonic fibroblasts (MEFs) treated with staurosporine or etoposide (24). Notably, in both of these studies, *atg* gene RNAi blocked the death of cells whose apoptotic pathway had been crippled. Although these findings exclude the possibility that autophagy is triggering death through apoptosis induction, they raise the question of whether autophagy is a death mechanism in cells whose apoptotic machinery is intact.

Interestingly, in etoposide-treated wild-type MEFs (which die by apoptosis), only minimal autophagic activity and no inhibition of death by 3-MA is seen, indicating that autophagy is not involved in the death process unless apoptosis is blocked (24). These data are consistent with the theory previously proposed by Lockshin and Zakeri that cells preferentially die by apoptosis but will die by any alternative available route, including autophagy, if exposed to harsh enough stimuli (9). A related possibility is that apoptotic death is faster than autophagic death and, therefore, autophagy is only witnessed playing a role in cell death in apoptotic-deficient cells. This hypothesis is consistent with recent data indicating that growth factor–deprived wild-type cells undergo a rapid apoptotic death, whereas growth factor–deprived *bax⁻/⁻*, *bak⁻/⁻* cells undergo a slow demise characterized by progressive self-cannibalization (25).

Given the uncertain physiologic relevance of autophagy gene–dependent cell death in zVAD-treated cells or in *bax⁻/⁻*, *bak⁻/⁻* cells, it seems premature to conclude that autophagy is a physiologically important cause of cell death. To prove that autophagy is an important cell death pathway in normal cells, it will be necessary to demonstrate cell death resistance phenotypes in apoptotic-competent cells lacking autophagy genes.

Autophagy as a cell survival mechanism

A contradictory, but equally plausible, explanation for the presence of autophagy in dying cells is that activation of autophagy is a cellular survival strategy. This concept was first proposed in 1977 (26) and considered radical (8) but now is supported by studies demonstrating increased death in cells or organisms lacking gene products essential for autophagy (Table 2). The pro-survival function of autophagy is an evolutionarily ancient process, conserved from yeast to mammals, and best characterized in nutrient deficiency. During nutrient deficiency, degradation of membrane lipids and proteins by the autolysosome generates free fatty acids and amino acids that can be reused to fuel mitochondrial ATP energy production and maintain protein synthesis (Figure 1). Presumably, this recycling function of autophagy is linked mechanistically to its ability to sustain life during starvation.

rons (17), in nerve growth factor–deprived sympathetic neurons (18), in serum- and potassium-deprived cerebellar granule cells (19), in serum-deprived PC12 cells (20), and in TNF-treated human T lymphoblastic leukemia cells (21). However, in several of these studies, autophagy occurred in cells thought to die by apoptosis, and it was presumed that autophagy triggered apoptosis, rather than playing a direct role in the death process. Moreover, 3-MA can inhibit kinases other than class III PI3K (18), some of which may independently affect death signaling, as well as inhibit the permeability transition in mitochondria (22). Thus, it is not possible to directly implicate autophagy in death execution from these 3-MA inhibitor studies.

Two recent studies provide the first genetic evidence that the autophagy pathway is capable of killing cells (Table 2). RNA interference (RNAi) directed against 2 autophagy genes, *atg7* and *beclin 1*,

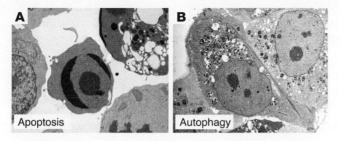

Figure 3
Ultrastructural examples of apoptotic and autophagic cell death. Electron micrographs of a FasL-treated Jurkat cell undergoing cell death with apoptotic features (**A**) and of a tamoxifen-treated MCF7 human breast carcinoma cell undergoing cell death with autophagic features (**B**). In **A**, note chromatin condensation (cell in center) and cytoplasmic vacuolization (cell in upper right). In **B**, note absence of chromatin condensation and presence of numerous autophagosomes. Images in **A** and **B** reproduced with permission from *Nature Chemical Biology* (98) and Landes Bioscience (90), respectively.

Unicellular organisms with null mutations in autophagy genes are viable in normal growth conditions; however, unlike their wild-type counterparts, they die rapidly during starvation (Table 2) (27–29). In plants, deletion of autophagy genes (e.g., *ATG6, ATG7,* and *ATG9*) results in loss of chlorophyll and accelerated senescence following nutrient deprivation (30–32). Mice lacking Atg5, an acceptor molecule for the ubiquitin-like molecule Atg12, die during the neonatal period, when the placental blood supply is interrupted and they undergo a form of starvation (33). *Atg5*−/− mice have decreased amino acid levels, decreased cardiac ATP production, and myocardial damage. Although the death of individual cells has not yet been assessed in *atg5*−/− mice, it is predicted that the recycling function of autophagy is critical to maintain cellular energy homeostasis and cellular survival during the neonatal period. This is particularly likely in tissues such as the heart and diaphragm that have sudden increases in energy needs and exhibit increased autophagy immediately following birth. Similarly, the inability of *Caenorhabditis elegans* with RNAi-silenced autophagy genes (e.g., *unc-51, bec-1, atg8,* and *atg18*) to survive during dauer diapause (34) likely reflects an inability to recycle nutrients at the organismal level.

Autophagy genes may also be critical for maintaining cellular bioenergetics and survival when cells are unable to take up external nutrients (i.e., during growth factor deprivation). In the absence of growth factors such as IL-3, there is decreased surface expression of nutrient transporters, decreased nutrient uptake, and an intracellular deficiency of nutrients (35). Growth factor withdrawal usually results in rapid apoptotic cell death, but recent studies in apoptotic-deficient *bax*−/−, *bak*−/− cells have unraveled an essential role for autophagy genes (e.g., *atg5, atg7*) in maintaining cellular survival following IL-3 deprivation (25). As in nutrient starvation in yeast, autophagy is a self-limited survival strategy during growth factor deprivation. IL-3–deprived *bax*−/−, *bak*−/− cells eventually die, presumably because of excessive self-consumption and bioenergetic failure. However, at any point before death, the addition of growth factor reverses the catabolic process and maintains cell viability. These observations are consistent with the concept that autophagy is a self-limited survival strategy, rather than a primary or irreversible death execution program.

Like the pro-death function of autophagy genes in etoposide-treated *bax*−/−, *bak*−/− knockout cells, it will be important to determine whether this pro-survival function of autophagy genes during growth factor deprivation in *bax*−/−, *bak*−/− cells is conserved in cells with intact apoptotic machinery. It will also be interesting to examine whether autophagy genes play a similar cytoprotective role during withdrawal of hormonal support or growth factors besides IL-3. Perhaps, rather than contributing to death execution, autophagy delays initiation of the apoptotic death pathway in cells deprived of trophic support. Although studies have not been performed in apoptosis-competent cells deprived of trophic factors, autophagy genes prevent the onset of apoptosis during nutrient deprivation. RNAi against *beclin 1, atg5, atg10,* and *atg12* enhances starvation-induced, but not staurosporine-induced, apoptotic cell death (36). Thus, the mechanism by which autophagy genes promote survival during nutrient deprivation may involve suppression of the canonical apoptotic death pathway.

The mechanisms by which autophagy promotes cell survival are not restricted to its role in maintaining cellular energy homeostasis during starvation. Autophagy is also involved in removing damaged mitochondria and other organelles, in degrading intracellular pathogens, and in degrading protein aggregates too large to be removed by the ubiquitin-proteasomal system. These functions of autophagy could promote cellular survival during aging, infectious diseases, and neurodegenerative processes. In addition to a cell-autonomous role for autophagy in promoting survival, autophagy may regulate programmed cell death during physiologic processes in vivo. For example, during the plant innate immune response, silencing of autophagy genes *beclin 1, vps34, atg3,* and *atg7* does not alter the death of infected cells or pathogen spread but results in uncontrolled spread of programmed cell death beyond sites of pathogen infection (32). This suggests that autophagy limits cell death to the site of infection, allowing plant innate immunity to contain pathogen spread without death of innocent bystander cells. It is not known whether autophagy alters the production of death-promoting signals, prevents the movement of death-promoting signals into uninfected tissues, or protects uninfected tissues against death induced by these signals, or whether a similar function of autophagy plays a role in the spatial restriction of development and stress-induced programmed cell death in other eukaryotic organisms.

Autophagy as a self-clearance mechanism

A third explanation for high levels of autophagy in dying cells is that it is a clean-up or self-clearance mechanism in cells committed to die by apoptosis or necrosis. This theory might explain why only selected populations of dying apoptotic cells have morphologic features of autophagy. The dogma is that most apoptotic cells are engulfed by phagocytes, with the lysosomes of the phagocyte responsible for the final degradation of dead cell bodies. However, in some forms of developmental programmed cell death (e.g., embryogenesis, insect metamorphosis), the availability of engulfment cells may be insufficient for clearance of dead cells. In such cases, dying cells may activate autophagy to target the cell's contents for degradation by its own lysosomes.

This need might contribute to the overlap between signaling pathways that activate apoptosis and autophagy. It has been shown that the proapoptotic signaling molecule TNF-related apoptosis-inducing ligand (TRAIL) regulates autophagy in an in vitro model of mammary gland formation (37). Here, TRAIL-dependent induction of autophagy occurs in parallel with apoptosis. Suppression

Table 1

Examples of cell death with morphologic features of autophagy[A]

Species	Tissue/cell type	Setting	References
Developmental programmed cell death			
Insects			
Chironomus tentans	Salivary gland	Metamorphosis	(8)
Manduca sexta	Intersegmental muscle, labial gland, prothoracic gland, larval fat body, motoneuron	Metamorphosis	(99, 100; reviewed in ref. 7)
Gryllus bimaculatus	Prothoracic gland	Metamorphosis	(7, 101)
Calliphora vomitoria	Salivary gland	Metamorphosis	(103)
Drosophila melanogaster	Salivary gland, midgut cells, larval fat body	Metamorphosis	(10, 103–105)
Orgyia leucostigma	Epithelial wing	Adult reduction in females	(reviewed in ref. 7)
Birds			
Chicken	Limb bud, neurons, mesonephros, müllerian duct epithelium, feather melanocytes, heart	Embryonic development	(reviewed in refs. 7, 8)
Quail	Ovarian follicle	Atresia	(7)
Amphibians			
Frog	Tail nerve cord	Metamorphosis	(8)
	Gills, neurons	Larval development	
Mammals			
Mouse	Palatal epithelium	Embryonic development	(8)
Human	Mammary epithelial cells	Lumen formation	(37, 106)
Disease-associated cell death			
Mammals			
Mouse	Brain (cerebellar Purkinje cells)	*Lurcher* mutation	(67)
	Striatal neuronal cell line	Mutant Huntingtin expression	(107)
	Macrophages	*Salmonella* infection	(83)
Rat	PC12 neuronal cell line	Mutant α-synuclein expression	(108)
Hamster	Brain	Experimentally induced transmissible spongiform	(65)
Gerbil	Hippocampus	Ischemic injury	(109)
Human	Heart	Dilated cardiomyopathy	(110)
		Aortic stenosis	(111)
	Dopaminergic neurons	Parkinson disease	(63)
Drug-, toxin-, or stress stimulus–associated cell death			
Protozoans			
Leishmania donovani		Antimicrobial peptide treatment	(112)
Tetrahymena thermophila		Staurosporine treatment	(113)
Dictyostelium discoideum		Dual exposure to starvation and differentiation-induced factor	(114)
Mammals			
Mouse	Heart	Diphtheria toxin treatment	(115)
	Cortical neurons	Chloroquine treatment	(17)
	Neural precursor cells	FGF withdrawal	(50)
Rat	Sympathetic neurons	Nerve growth factor withdrawal	(18)
	Cultured cerebellar granule cells	Serum and potassium deprivation	(19)
	Retinal explant	Anisomycin treatment	(116)
	Hippocampal neurons	*N*-methyl-D-aspartate treatment	(117)
	PC12 pheochromocytoma cells	Serum deprivation	(118)
	Oral keratinocytes	5-Fluorouracil treatment	(119)
Human	Ovarian carcinoma cell lines	Resveratrol treatment	(120)
	Mammary carcinoma cell line	Anti-estrogen treatment	(16, 121)
	Glioma cell lines	Arsenic trioxide treatment	(95, 122)
		Ceramide treatment	(43)
	Endothelial cells	Endostatin treatment	(123)

[A]Includes only those references in which (a) autophagy is observed at the ultrastructural level, and (b) autophagy is postulated by the authors to be a cause of death; excludes references in which (a) autophagy is assessed only at the light microscopic or biochemical level (e.g., by monodansylcadaverine staining, GFP-Atg8/LC3 staining, or LC3I-to-LC3II conversion), or (b) autophagy is not postulated to be mechanistically important in death.

Table 2
Examples of autophagy gene–dependent cell death, cell survival, and metazoan survival

Species	Tissue/cell type	*ATG* gene	Setting[A]	References
Cell death				
Mammals				
Mouse	bax−/−, bak−/− embryonic fibroblasts	atg5, beclin 1	Etoposide or staurosporine treatment	(24)
Human	L929 fibroblast cells, U937 monocytoid cells	atg7, beclin 1	Caspase inhibition	(23)
Cell survival				
Yeast				
Saccharomyces cerevisiae		All *ATG* genes	Starvation	(27)
Protozoans				
Dictyostelium discoideum		ATG1, ATG5, ATG6, ATG8, ATG12	Starvation	(28, 29)
Plants				
Arabidopsis thaliana	Leaves	ATG7, ATG9	Starvation	(30, 31)
Nicotiana benthamiana	Leaves	ATG3, ATG7, beclin 1, vps34	Pathogen infection/innate immune response	(32)
Mammals				
Mouse	bax−/−, bak−/− hematopoietic cells	atg5, atg7	IL-3 withdrawal	(25)
	Brain	beclin 1[B]	Alphavirus encephalitis	(79)
Human	HeLa cells	atg5, beclin 1, atg10, atg12	Starvation	(36)
Metazoan survival				
Caenorhabditis elegans		unc-51, bec-1, atg7, atg8, atg18	Dauer development	(34)
		bec-1, atg8, atg18	Larval development	(34)
Drosophila		ATG1, ATG3	Larval/pupal development	(85, 124)
Mouse		beclin 1	Embryonic development	(88, 89)
		atg5	Early postnatal starvation	(33)

[A]Setting indicates conditions in which mutation in the indicated autophagy gene blocks cell death, cell survival, or metazoan survival. [B]The role of this gene was assessed by overexpression; in all other studies referenced in the table, the role of each gene was assessed by RNAi silencing or loss-of-function mutation.

of either apoptosis alone or TRAIL signaling does not prevent lumen formation, but simultaneous inhibition of apoptosis and TRAIL signaling prevents cell clearance.

These findings suggest that both apoptosis and autophagy may be involved in cavitation during mammary gland morphogenesis. However, to confirm a role for autophagy, it will be important to observe whether luminal filling occurs if autophagy genes are inactivated in cells with intact TRAIL signaling. It is not clear whether autophagy is required for luminal cell death or for removal of cells committed to death by an apoptotic pathway. Similarly, it is not known whether autophagy is required for caspase-dependent death and/or for the clearance of dying cells during insect metamorphosis.

Cross-talk between apoptotic signaling, autophagy, and mitochondria

Several proapoptotic signals induce autophagy — e.g., components of the extrinsic apoptosis pathway, TRAIL, TNF, and FADD (21, 37–40); the calcium/calmodulin–regulated serine/threonine kinases DRP-1 and DAPk (41); and ceramide (42, 43). Conversely, antiapoptotic signaling pathways suppress autophagy — e.g., the class I PI3K/Akt/TOR signaling pathway (reviewed in refs. 3, 44). Coordinated regulation of apoptosis and autophagy is also reflected in the results of genome-wide analyses of transcriptional changes during developmental programmed cell death of the *Drosophila* salivary gland (45, 46).

The mitochondrion may integrate cell death signals and autophagy activation. Mitochondria generate apoptotic signals but are removed when damaged by autophagy; therefore, mitochondria represent a nexus at which autophagy and apoptosis pathways may interact. Accordingly, genes involved in mitochondrial physiology and/or mitochondrial regulation of apoptosis interact with the autophagy pathway. One example is the yeast gene, *UTH1*, that encodes an outer mitochondrial membrane protein involved in mitochondrial biogenesis and stress responses. *Uth1* mutants are defective in degrading mitochondria during autophagy (47) and survive and proliferate when expressing the mammalian proapoptotic cell death gene *bax* or when treated with the autophagy inducer rapamycin (48). These findings led Camougrand and colleagues to suggest that Uth1p mediates mitochondrial autophagy and autophagic death. However, it is not yet clear whether rapamycin induces cell death versus cell cycle arrest in wild-type yeast, and whether the phenotype of rapamycin-treated *uth1* mutant yeasts is due to direct effects of *UTH1* and the autophagy pathway in death regulation.

In mammalian cells, Bcl-2 family members in the outer mitochondrial membrane modulate autophagy. Bcl-2 downregulation increases autophagy in a caspase-independent manner in human leukemic cells (49), and Bcl-2 overexpression inhibits both autophagy and caspase-independent death in growth factor–deprived neural progenitor cells and in serum- and potassium-deprived cultured cerebellar granule cells (19, 50). Recent evidence suggests that Bcl-2 inhibits autophagy through a direct interaction with the Beclin 1 autophagy protein and that the interaction between Bcl-2 and Beclin 1 may function as a rheostat that maintains autophagy at levels that are compatible with cell survival rather than cell death (51). In contrast, Bcl-2 or Bcl-x$_L$ overexpression potentiates autophagy and autophagy gene–dependent death in MEFs treated with the proapoptotic stimulus etoposide (24). The basis for the opposite effects of Bcl-2 family members on autophagy in different settings is unclear. Furthermore, it is not yet clear that Bcl-2 proteins function at the mitochondrion to regulate autophagy, since autophagy is inhibited by Bcl-2 targeted to endoplasmic reticulum but not by Bcl-2 targeted to mitochondria (51).

The role of proapoptotic Bcl-2 family members in autophagy gene–dependent life-and-death decisions is also controversial. As discussed earlier, *bax*$^{-/-}$, *bak*$^{-/-}$ cells undergo autophagy gene–dependent death when treated with etoposide (24) but undergo autophagy gene–dependent survival when deprived of trophic factor support (25). It is possible that in the setting of *bax/bak* deficiency, the stimulus plays a critical role in determining cell fate, and that etoposide, but not growth factor deprivation, can target an intracellular pathway that turns autophagy into a deadly process. Some atypical Bcl-2 family members, including BNIP3 and Hspin, also activate autophagy and nonapoptotic cell death (52–54), but it is not yet known whether this caspase-independent cell death requires autophagy genes.

Another question is how the autophagy pathway recognizes damaged mitochondria. The mitochondrial permeability transition (MPT) may trigger the engulfment of depolarized mitochondria by autophagy (55). However, it is not known whether inhibition of autophagy increases the numbers of depolarized mitochondria in mammalian cells or how a depolarized mitochondrion might be targeted to autophagosomes. The MPT may represent a point of convergence of apoptotic and autophagy pathways, since Bcl-2 family members regulate the MPT. The proapoptotic family member Bax interacts with the voltage-dependent anion channel (56) and/or the adenine nucleotide translocator (57) to induce the MPT in cells and isolated mitochondria upon induction of apoptosis. It is currently not clear, however, whether the MPT regulated by Bax triggers mitochondrial turnover by autophagy.

In many canonical apoptosis pathways, the MPT is caspase-dependent (58). Thus, if the MPT is a critical signal for mitochondrial degradation through autophagy, inhibition of caspases should prevent the loss of mitochondria. However, Tolkovsky and coworkers reported that, although caspase inhibitors effectively inhibit neuronal cell death, they fail to prevent the formation of autophagosomes or the degradation of mitochondria. In fact, the long-term culturing of neurons in the presence of proapoptotic stimuli and caspase inhibitors leads to the loss of mitochondria (59, 60). Thus, the relationship among MPT, caspase-dependent cell death, and mitochondrial autophagy remains unclear.

Autophagy in neurodegenerative diseases

The accumulation of mutant or toxic proteins plays a major role in chronic neurodegenerative diseases (61). Morphologic evidence of autophagy has been reported in neurodegenerative diseases including Parkinson, Huntington, and Alzheimer diseases, and transmissible spongiform encephalopathies (62–65). It is possible that autophagy activation contributes to neurodegeneration (66, 67), but the evidence is correlative. A contrasting view is that autophagy may be a protective mechanism to degrade mutant or toxic proteins. According to this model, defects in autophagy-related pathways contribute to the accumulation of neurotoxic proteins and the ensuing neuronal cell death. Although the exact roles of autophagy in neurodegenerative diseases are not fully defined, recent studies have provided some insights.

The protein α-synuclein is a major component of neuronal cytoplasmic inclusions that characterize Parkinson and other neurodegenerative diseases (68). Although earlier studies suggested that α-synuclein is degraded through both the proteasome and classical autophagy pathways (69), a recent study demonstrated that the turnover of α-synuclein is regulated by chaperone-mediated autophagy, which involves the direct lysosomal targeting of proteins containing specific pentapeptide recognition motifs (70). Interestingly, pathogenic α-synuclein mutants associated with familial, autosomal-dominant forms of Parkinson disease (71, 72) are inefficiently degraded by chaperone-mediated autophagy. Since the accumulation of wild-type α-synuclein in neuronal inclusions is common in adult-onset neurodegenerative diseases, these experiments suggest that defects in autophagy-related pathways may contribute to multiple neurodegenerative diseases.

Consistent with this hypothesis, autophagy has also been implicated in regulating the turnover of Huntingtin (Htt), the protein involved in Huntington disease, an autosomal-dominant neurodegenerative disorder caused by the expansion of a polyglutamine (polyQ) tract in Htt. Although the mechanism of neurotoxicity mediated by expanded polyQ is still controversial, expanded polyQ provokes a dominant gain-of-function neurotoxicity, regardless of the specific protein context within which it resides. The accumulation of expanded polyQ-containing proteins in insoluble aggregates in affected neurons is a hallmark feature of Huntington and other polyQ expansion diseases (73).

Although neuronal Htt proteins in inclusions are highly ubiquitinated, polyQ is a poor substrate for proteasomes (74). Thus, the highly ubiquitinated state of Htt inclusions may indicate the inability of proteasomes in affected neurons to clear abnormal Htt proteins. In contrast, there is pharmacologic evidence to suggest a role for autophagy in the degradation of the N-terminus of Htt. For example, 3-MA increases the aggregation of Htt with expanded polyQ in clonal striatal cells (62). Rapamycin, an inducer of autophagy, reduces the aggregation of expanded polyQ in transfected cells (75), protects against neurodegeneration in a fly model of Huntington disease, and improves performance on behavioral tests and decreases aggregate formation in a mouse model of Huntington disease (76). These results suggest a possible role of autophagy in the turnover of expanded polyQ proteins and in protection of neurons against their toxicity.

Autophagy and infectious diseases

The autophagic machinery is used to degrade intracellular pathogens (reviewed in refs. 77, 78) including intracellular bacteria (e.g., *Shigella flexneri* and *Mycobacterium tuberculosis*), mammalian viruses that produce encephalitis (e.g., alphaviruses and herpes simplex virus), and plant viruses (32, 78–81). It also may be used to degrade invading extracellular pathogens such as group A *Streptococcus* (82). It is reasonable to propose that autophagy might promote cellular survival during pathogen invasion because of either enhanced degradation of intracellular pathogens and consequent decreases in microbial replication; enhanced degradation of specific cytotoxic microbial virulence products; or preservation of cellular nutrient status during a period of microbial parasitism (which mimics nutrient starvation). However, with the exception of the finding that forced expression of the *beclin 1* autophagy gene protects against Sindbis virus–induced apoptosis in mouse brains (in parallel with decreasing viral replication) (79), direct proof of a cell-autonomous, pro-survival role of autophagy in pathogen infection is lacking. Moreover, it has been proposed that a virulence protein, SipB, of the intracellular pathogen *Salmonella enterica* causes macrophage death by inducing autophagy, perhaps by triggering mitochondrial fusion with autophagosomes (83). Yet, in this study, there was no direct evidence that macrophage cell death was caused by, rather than simply associated with, autophagy. Further studies in autophagy-deficient host organisms are required to determine the role of autophagy in life-and-death decisions during pathogen infection.

Autophagy and cancer

Cancer results from the dysregulation of pathways that regulate cell differentiation, cell proliferation, and cell survival. Autophagy may protect against cancer by sequestering damaged organelles, permitting cellular differentiation, increasing protein catabolism, and/or promoting autophagic death. Alternatively, autophagy may contribute to cancer by promoting the survival of nutrient-starved cells. Recent data are most consistent with a model in which autophagy contributes to tumor suppression and defects in autophagy contribute to oncogenesis. Biochemical evidence in mammalian cells and genetic evidence in *C. elegans* and *Drosophila* indicate that autophagy is positively regulated by the PTEN tumor suppressor gene and negatively regulated by the oncogenic class I PI3K signaling pathway (14, 34, 84, 85). Furthermore, the mammalian autophagy gene *beclin 1* has tumor suppressor activity in breast carcinoma cells (86), is commonly deleted in human breast ovarian and prostate cancer (87), and is a haploinsufficient tumor suppressor gene in mice (88, 89).

Several theories regarding the role of autophagy-dependent death and autophagy-dependent survival in cancer biology have been proposed (3, 66, 90–94). One is that autophagy-dependent death is a mechanism of tumor suppression. However, there are no direct data to support this hypothesis. In contrast, studies in cells and animals with a deficiency in *beclin 1* suggest that death induction may not be involved in the tumor suppressor function of this autophagy gene. *Beclin 1⁻/⁻* ES cells are not resistant to death triggered by UV irradiation or serum withdrawal, and *beclin 1⁻/⁻* null animals die early during embryogenesis with massive cell death (89). Moreover, in *beclin 1* heterozygous-deficient mice (with reduced tissue levels of autophagy), there is hyperproliferation of mammary epithelial cells during glandular morphogenesis and increased antigen-driven proliferation of B cells without decreased cell death (88). Together, these observations suggest that the role of the *beclin 1* autophagy gene in tumor suppression is related not to cell death induction, but rather to inhibition of cellular proliferation.

It is possible that autophagy is involved in the spontaneous or chemotherapy-induced death of existing tumor cells. Although the role of autophagy in cell death in apoptosis-competent cells is unclear, autophagy gene–dependent death in cells crippled in apoptosis (e.g., zVAD-treated cells; *bax⁻/⁻*, *bak⁻/⁻* cells) may have relevance for cancer biology and therapy, since human tumor cells frequently contain mutations that render them resistant to apoptosis. One prediction is that such cells have an increased dependency on autophagy pathways for self-destruction, and that the impact of decreased autophagy-dependent cell death on tumor progression may be greater in tumor cells that are resistant to apoptosis.

Another prediction is that the enhanced autophagy-dependent death potential of apoptosis-resistant tumor cells might be exploited therapeutically by the administration of autophagy-inducing agents. Indeed, there are examples of putative autophagic cell death in cancer cell lines treated with chemotherapeutic agents (Table 1). During tamoxifen-induced death of MCF7 cells (a cell type that contains a mutation in caspase-3), there is a marked upregulation of Beclin 1 autophagy protein expression (42, 90), and, in some examples, chemotherapy-induced autophagic cell death is inhibited by 3-MA (16, 42, 95). However, evidence proving that autophagy is a bona fide death pathway in chemotherapy-treated cancer cells is lacking. In addition, rapamycin, an inhibitor of TOR kinase that has promising antitumor effects in human clinical trials (96), is one of the most potent known inducers of autophagy but is not known to induce autophagic cell death.

In contrast to potential pro-death effects, more clearly established pro-survival effects of autophagy during nutrient starvation might foster tumor initiation and/or progression (3, 91, 93). As tumor cells grow beyond their blood supply, they are exposed to nutrient-limiting conditions, and it is possible that transformed cells use autophagy as a survival strategy in this setting. It has been proposed that such a need for autophagy in tumor initiation might explain the retention of the wild-type allele in all tumors arising in *beclin 1⁺/⁻* mice (92). However, the role of autophagy in tumor cell survival in vivo has not been tested experimentally. Moreover, in considering the net effect of autophagy on tumorigenesis, it is important to recognize its other functions that could contribute to restricting tumorigenesis (e.g., the degradation of certain proteins or organelles required for cell growth and/or the degradation of damaged mitochondria and other organelles that generate genotoxic stress and increase the likelihood of oncogenic mutations).

Conclusion

Autophagy functions across a diverse range of species as a pro-survival pathway during nutrient deprivation and other forms of cellular stress. Paradoxically, in cells that cannot die by apoptosis and, more speculatively, in cells that cannot be removed by engulfment cells, the autophagic machinery may also be used for self-destruction. The challenge for scientists will be to understand the molecular basis of this paradox. The challenge for clinicians will be to selectively turn on or turn off autophagy gene–dependent survival and death pathways in the treatment of different clinical diseases.

Acknowledgments

The work from the authors' laboratories was supported by NIH grants RO1 CA084254, RO1 AI151367, and RO1 CA109618 (to B. Levine) and R37 NIA12859 (to J. Yuan); American Cancer Society grant RSG 98-339 (to B. Levine); and an Ellison Medical Foundation Senior Scholar Award in Infectious Diseases (to B. Levine). We thank Alexi Degetrev for providing electron micrographs and Sophie Pattingre and Renee Talley for help with manuscript preparation.

Address correspondence to: Beth Levine, Division of Infectious Diseases, University of Texas Southwestern Medical Center, 5323 Harry Hines Boulevard, Dallas, Texas 75390-9113, USA. Phone: (214) 648-0493; Fax: (214) 648-0284; E-mail: beth.levine@utsouthwestern.edu.

1. Klionsky, D.J., and Emr, S.D. 2000. Autophagy as a regulated pathway of cellular degradation. *Science.* **290**:1717–1721.

2. Levine, B., and Klionsky, D.J. 2004. Development by self-digestion: molecular mechanisms and biological functions of autophagy. *Dev. Cell.* **6**:463–477.

3. Lum, J.J., DeBarardinis, R.J., and Thompson, C.B. 2005. Autophagy in metazoans: cell survival in the land of plenty [review]. *Nat. Rev. Mol. Cell Biol.* **6**:439–448.

4. Talloczy, Z., et al. 2002. Regulation of starvation- and virus-induced autophagy by the eIF2α kinase signaling pathway. *Proc. Natl. Acad. Sci. U. S. A.* **99**:190–195.

5. Klionsky, D.J., et al. 2003. A unified nomenclature for yeast autophagy-related genes. *Dev. Cell.* **5**:539–545.

6. Schweichel, J.-U., and Merker, H.-J. 1973. The morphology of various types of cell death in prenatal tissues. *Teratology.* **7**:253–266.

7. Bursch, W. 2001. The autophagosomal-lysosomal compartment in programmed cell death. *Cell Death Differ.* **8**:569–581.

8. Clarke, P.G. 1990. Developmental cell death: morphological diversity and multiple mechanisms [review]. *Anat. Embryol. (Berl.)* **181**:195–213.

9. Lockshin, R.A., and Zakeri, Z. 2004. Apoptosis, autophagy, and more. *Int. J. Biochem. Cell Biol.* **36**:2405–2419.

10. Lee, C.Y., and Baehrecke, E.H. 2001. Steroid regulation of autophagic programmed cell death during development. *Development.* **128**:1443–1455.

11. Kosta, A., et al. 2004. Autophagy gene disruption reveals a non-vacuolar cell death pathway in dictyotelium. *J. Biol. Chem.* **279**:48404–48409.

12. Seglen, P.O., and Gordon, P.B. 1982. 3-Methyladenine: specific inhibitor of autophagic/lysosomal protein degradation in isolated rat hepatocytes. *Proc. Natl. Acad. Sci. U. S. A.* **79**:1889–1892.

13. Blommaart, E.F., Krause, U., Schellens, J.P., Vreeling-Sindelarova, H., and Meijer, A.J. 1997. The phosphatidylinositol 3-kinase inhibitors wortmannin and LY294002 inhibit autophagy in isolated rat hepatocytes. *Eur. J. Biochem.* **243**:240–246.

14. Petiot, A., Ogier-Denis, E., Blommaart, E.F., Meijer, A.J., and Codogno, P. 2000. Distinct classes of phosphatidylinositol 3'-kinases are involved in signaling pathways that control macroautophagy in HT-29 cells. *J. Biol. Chem.* **275**:992–998.

15. Schwarze, P.E., and Seglen, P.O. 1985. Reduced autophagic activity, improved protein balance and enhanced in vitro survival of hepatocytes isolated from carcinogen-treated rats. *Exp. Cell Res.* **157**:15–28.

16. Bursch, W., et al. 1996. Active cell death induced by the anti-estrogens tamoxifen and ICI 164 384 in human mammary carcinoma cells (MCF-7) in culture: the role of autophagy. *Carcinogenesis.* **17**:1595–1607.

17. Kaidi, A.U., et al. 2001. Chloroquine-induced neuronal cell death is p53 and Bcl-2 family-dependent but caspase-independent. *J. Neuropathol. Exp. Neurol.* **60**:937–945.

18. Xue, L., Fletcher, G.C., and Tolkovsky, A.M. 1999. Autophagy is activated by apoptotic signalling in sympathetic neurons: an alternative mechanism of death execution. *Mol. Cell. Neurosci.* **14**:180–198.

19. Canu, N., et al. 2005. Role of the autophagic-lysosomal system on low potassium-induced apoptosis in cultured cerebellar granule cells. *J. Neurochem.* **92**:1228–1242.

20. Uchiyama, Y. 2001. Autophagic cell death and its execution by lysosomal cathepsins. *Arch. Histol. Cytol.* **64**:233–246.

21. Jia, L., et al. 1997. Inhibition of autophagy abrogates tumour necrosis factor alpha induced apoptosis in human T-lymphoblastic leukaemic cells. *Br. J. Haematol.* **98**:673–685.

22. Xue, L., Borutaite, V., and Tolkovsky, A.M. 2002. Inhibition of mitochondrial permeability transition and release of cytochrome c by anti-apoptotic nucleoside analogues. *Biochem. Pharmacol.* **64**:441–449.

23. Yu, L., et al. 2004. Regulation of an *ATG7-beclin 1* program of autophagic cell death by caspase 8. *Science.* **304**:1500–1502.

24. Shimizu, S., et al. 2004. Role of Bcl-2 family proteins in a non-apoptotic programmed cell death dependent on autophagy genes. *Nat. Cell Biol.* **6**:1221–1228.

25. Lum, J.J., et al. 2005. Growth factor regulation of autophagy and cell survival in the absence of autophagy. *Cell.* **120**:237–248.

26. Hourdry, J. 1977. Cytological and cytochemical changes in the intestinal epithelium during anuran metamorphosis. *Int. Rev. Cytol. Suppl.* **5**:337–385.

27. Tsukada, M., and Ohsumi, Y. 1993. Isolation and characterization of autophagy-defective mutants of *Saccharomyces cerevisiae. FEBS Lett.* **333**:169–174.

28. Otto, G.P., Wu, M.Y., Kazgan, N., Anderson, O.R., and Kessin, R.H. 2003. Macroautophagy is required for multicellular development of the social amoeba *Dictyostelium discoideum. J. Biol. Chem.* **278**:17636–17645.

29. Otto, G.P., Wu, M.Y., Kazgan, N., Anderson, O.R., and Kessin, R.H. 2004. *Dictyostelium* macroautophagy mutants vary in the severity of their developmental defects. *J. Biol. Chem.* **279**:15621–15629.

30. Doelling, J.H., Walker, J.M., Friedman, E.M., Thompson, A.R., and Vierstra, R.D. 2002. The APG8/12-activating enzyme APG7 is required for proper nutrient recycling and senescence in *Arabidopsis thaliana. J. Biol. Chem.* **277**:33105–33114.

31. Hanaoka, H., et al. 2002. Leaf senescence and starvation-induced chlorosis are accelerated by the disruption of an *Arabidopsis* autophagy gene. *Plant Physiol.* **129**:1181–1193.

32. Liu, Y., et al. 2005. Autophagy genes are essential for limiting the spread of programmed cell death associated with plant innate immunity. *Cell.* **121**:567–577.

33. Kuma, A., et al. 2004. The role of autophagy during the early neonatal starvation period. *Nature.* **432**:1032–1036.

34. Melendez, A., et al. 2003. Autophagy genes are essential for dauer development and lifespan extension in *C. elegans. Science.* **301**:1387–1391.

35. Edinger, A.L., Cinnalli, R.M., and Thompson, C.B. 2003. Rab7 prevents growth factor-independent survival by inhibiting cell-autonomous nutrient transport expression. *Dev. Cell.* **5**:571–582.

36. Boya, P., et al. 2005. Inhibition of macroautophagy triggers apoptosis. *Mol. Cell. Biol.* **25**:1025–1040.

37. Mills, K.R., Reginato, M., Debnath, J., Queenan, B., and Brugge, J.S. 2004. Tumor necrosis factor-related apoptosis-inducing ligand (TRAIL) is required for induction of autophagy during lumen formation in vitro. *Proc. Natl. Acad. Sci. U. S. A.* **101**:3438–3443.

38. Thorburn, J., et al. 2005. Selective inactivation of FADD-dependent apoptosis and autophagy pathway in immortal epithelial cells. *Mol. Biol. Cell.* **16**:1189–1199.

39. Prins, J., et al. 1998. Tumour necrosis factor induced autophagy and mitochondrial morphological abnormalities are mediated by TNFR-1 and/or TNFR-II and do not invariably lead to cell death. *Biochem. Soc. Trans.* **26**:S314.

40. Pyo, J.-O., et al. 2005. Essential roles of Atg5 and FADD in autophagic cell death: dissection of autophagic cell death into vacuole formation and cell death. *J. Biol. Chem.* **280**:20722–20729.

41. Inbal, B., Bialik, S., Sabanay, I., Shani, G., and Kimchi, A. 2002. DAP kinase and DRP-1 mediate membrane blebbing and the formation of autophagic vesicles during programmed cell death. *J. Cell Biol.* **157**:455–468.

42. Scarlatti, F., et al. 2004. Ceramide-mediated macroautophagy involves inhibition of protein kinase B and upregulation of Beclin 1. *J. Biol. Chem.* **279**:18384–18391.

43. Daido, S., et al. 2004. Pivotal role of the cell death factor BNIP3 in ceramide-induced autophagic cell death in malignant glioma cells. *Cancer Res.* **64**:4286–4293.

44. Meijer, A.J., and Codogno, P. 2004. Regulation and role of autophagy in mammalian cells. *Int. J. Biochem. Cell Biol.* **36**:2445–2462.

45. Lee, C.-Y., et al. 2003. Genome-wide analyses of steroid- and radiation-triggered programmed cell death in *Drosophila. Curr. Biol.* **13**:350–357.

46. Gorski, S.M., et al. 2003. A SAGE approach to discovery of genes involved in autophagic cell death. *Curr. Biol.* **13**:358–363.

47. Kissova, I., Deffieu, M., Manon, S., and Camougrand, N. 2004. Uth1p is involved in the autophagic degradation of mitochondria. *J. Biol. Chem.* **279**:39068–39074.

48. Camougrand, N., et al. 2003. The product of the *UTH1* gene, required for Bax-induced cell death in yeast, is involved in the response to rapamycin. *Mol. Microbiol.* **47**:495–506.

49. Saeki, K., et al. 2000. Bcl-2 down-regulation causes autophagy in caspase-independent manner in human leukemic HL60 cells. *Cell Death Differ.* **7**:1263–1269.

50. Cardenas-Aguayo, M.D.C., Santa-Olalla, J., Baizabal, J.M., Salgado, L.M., and Covarrubias, L. 2003. Growth factor deprivation induces an alternative non-apoptotic death mechanism that is inhibited by Bcl2 in cells derived from neural precursor cells. *J. Hematother. Stem Cell Res.* **12**:735–748.

51. Pattingre, S., et al. 2005. Bcl-2 antiapoptotic proteins inhibit Beclin 1-dependent autophagy. *Cell.* In press.

52. Velde, C.V., et al. 2000. BNIP3 and genetic control of necrosis-like cell death through the mitochondrial permeability transition pore. *Mol. Cell.* **20**:5454–5468.

53. Yanagisawa, H., Miyashita, T., Nakano, Y., and Yamamoto, D. 2003. HSpin1, a transmembrane protein interacting with Bc-2/Bcl-xL, induces a caspase-independent autophagic cell death. *Cell Death Differ.* **10**:798–807.

54. Nakano, Y., et al. 2001. Mutations in the novel membrane protein spinster interfere with programmed cell death and cause neural degeneration in *Drosophila melanogaster. Mol. Cell. Biol.* **21**:3775–3788.

55. Elmore, S.P., Qian, T., Grissom, S.F., and Lemasters, J.J. 2001. The mitochondrial permeability transition initiates autophagy in rat hepatocytes. *FASEB J.* **15**:2286–2287.

56. Jacotot, E., et al. 1999. Mitochondrial membrane permeabilization during the apoptotic process. *Ann. N. Y. Acad. Sci.* **887**:18–30.

57. Marzo, I., et al. 1998. Bax and adenine nucleotide translocator cooperate in the mitochondrial control of apoptosis. *Science.* **281**:2027–2031.

58. Li, H., Zhu, H., Xu, C.J., and Yuan, J. 1998. Cleavage of BID by caspase 8 mediates the mitochondrial damage in the Fas pathway of apoptosis. *Cell.* **94**:491–501.

59. Tolkovsky, A.M., Xue, L., Fletcher, G.C., and Borutaite, V. 2002. Mitochondrial disappearance from cells: a clue to the role of autophagy in programmed cell death and disease [review]? *Biochimie.* **84**:233–240.

60. Xue, L., Fletcher, G.C., and Tolkovsky, A.M. 2001. Mitochondria are selectively eliminated from eukaryotic cells after blockade of caspases during apoptosis. *Curr. Biol.* **11**:361–365.

61. Grune, T., Jung, T., Merker, K., and Davies, K.J. 2004. Decreased proteolysis caused by protein aggregates, inclusion bodies, plaques, lipofuscin, ceroid, and "aggresomes" during oxidative stress, aging, and disease. *Int. J. Biochem. Cell Biol.* **36**:2519–2530.

62. Qin, Z.H., et al. 2003. Autophagy regulates the processing of amino terminal huntingtin fragments. *Hum. Mol. Genet.* **12**:3231–3244.

63. Anglade, P., et al. 1997. Apoptosis and autophagy in nigral neurons of patients with Parkinson's dis-

ease. *Histol. Histopathol.* **12**:25–31.

64. Yu, W.H., et al. 2004. Autophagic vacuoles are enriched in amyloid precursor protein-secretase activities: implications for beta-amyloid peptide over-production and localization in Alzheimer's disease. *Int. J. Biochem. Cell Biol.* **36**:2531–2540.

65. Liberski, P.P., Sikorska, B., Bratosiewicz-Wasik, J., Gajdusek, D.C., and Brown, P. 2004. Neuronal cell death in transmissible spongiform encephalopathies (prion diseases) revisited: from apoptosis to autophagy. *Int. J. Biochem. Cell Biol.* **36**:2473–2490.

66. Shintani, T., and Klionsky, D.J. 2004. Autophagy in health and disease: a double-edged sword. *Science.* **306**:990–995.

67. Yue, Z., et al. 2002. A novel protein complex linking the δ2 glutamate receptor and autophagy: implications for neurodegeneration in lurcher mice. *Neuron.* **35**:921–933.

68. Maries, E., Dass, B., Collier, T.J., Kordower, J.H., and Steece-Collier, K. 2003. The role of alpha-synuclein in Parkinson's diseases: insights from animal models. *Nat. Rev. Neurosci.* **4**:727–738.

69. Webb, J.L., Ravikumar, B., Atkins, J., Skepper, J.N., and Rubinsztein, D.C. 2003. Alpha-synuclein is degraded by both autophagy and the proteasome. *J. Biol. Chem.* **278**:25009–25013.

70. Cuervo, A.M., Stefanis, L., Fredenburg, R., Lansbury, P.T., and Sulzer, D. 2004. Impaired degradation of mutant α-synuclein by chaperone-mediated autophagy. *Science.* **305**:1292–1295.

71. Polymeropoulos, M.H., et al. 1997. Mutation in the α-synuclein gene identified in families with Parkinson's disease. *Science.* **276**:2045–2047.

72. Kruger, R., et al. 1998. Ala30Pro mutation in the gene encoding alpha-synuclein in Parkinson's disease. *Nat. Genet.* **18**:106–108.

73. DiFiglia, M., et al. 1997. Aggregation of huntingtin in neuronal intranuclear inclusions and dystrophic neurites in brain. *Science.* **277**:1990–1993.

74. Venkatraman, P., Wetzel, R., Tanaka, M., Nukina, N., and Goldberg, A.L. 2004. Eukaryotic proteasomes cannot digest polyglutamine sequences and release them during degradation of polyglutamine-containing proteins. *Mol. Cell.* **14**:95–104.

75. Ravikumar, B., Duden, R., and Rubinsztein, D.C. 2002. Aggregate-prone proteins with polyglutamine and polyalanine expansions are degraded by autophagy. *Hum. Mol. Genet.* **11**:1107–1117.

76. Ravikumar, B., et al. 2004. Inhibition of mTOR induces autophagy and reduces toxicity of polyglutamine expansions in fly and mouse models of Huntington disease. *Nat. Genet.* **36**:585–595.

77. Kirkegaard, K., Taylor, M.P., and Jackson, W.T. 2004. Cellular autophagy: surrender, avoidance and subversion by microorganisms [review]. *Nat. Rev. Microbiol.* **2**:301–314.

78. Seay, M., Dinesh-Kumar, S., and Levine, B. 2005. Digesting oneself and digesting microbes: autophagy as a host response to viral infection. In *Modulation of host gene expression and innate immunity by viruses.* P. Palese, editor. Springer. Dordrecht, The Netherlands. 245–279.

79. Liang, X.H., et al. 1998. Protection against fatal Sindbis virus encephalitis by Beclin, a novel Bcl-2-interacting protein. *J. Virol.* **72**:8586–8596.

80. Ogawa, M., et al. 2005. Escape of intracellular Shigella from autophagy. *Science.* **307**:727–731.

81. Gutierrez, M.G., et al. 2004. Autophagy is a defense mechanism inhibiting BCG and *Mycobacterium tuberculosis* survival in infected macrophages. *Cell.* **119**:753–766.

82. Nakagawa, I., et al. 2004. Autophagy defends cells against invading group A *Streptococcus*. *Science.* **306**:1037–1040.

83. Hernandez, L.D., Pypaert, M., Flavell, R.A., and Galan, J.E. 2003. A *Salmonella* protein causes macrophage cell death by inducing autophagy. *J. Cell*

84. Rusten, T.E., et al. 2004. Programmed autophagy in the *Drosophila* fat body is induced by ecdysone through regulation of the PI3K pathway. *Dev. Cell.* **7**:179–192.

85. Scott, R.C., Schuldiner, O., and Neufeld, T.P. 2004. Role and regulation of starvation-induced autophagy in the *Drosophila* fat body. *Dev. Cell.* **7**:167–178.

86. Liang, X.H., et al. 1999. Induction of autophagy and inhibition of tumorigenesis by *beclin 1*. *Nature.* **402**:672–676.

87. Aita, V.M., et al. 1999. Cloning and genomic organization of *beclin 1*, a candidate tumor suppressor gene on chromosome 17q21. *Genomics.* **59**:59–65.

88. Qu, X., et al. 2003. Promotion of tumorigenesis by heterozygous disruption of the *beclin 1* gene. *J. Clin. Invest.* **112**:1809–1820. doi:10.1172/JCI200320039.

89. Yue, Z., Jin, S., Yang, C., Levine, A.J., and Heintz, N. 2003. Beclin 1, an autophagy gene essential for early embryonic development, is a haploinsufficient tumor suppressor. *Proc. Natl. Acad. Sci. U. S. A.* **100**:15077–15082.

90. Furuya, N., Liang, X.H., and Levine, B. 2004. Autophagy and cancer. In *Autophagy*. D.J. Klionsky, editor. Landes Bioscience. Georgetown, Texas, USA. 244–253.

91. Ogier-Denis, E., and Codogno, P. 2003. Autophagy: a barrier or an adaptive response to cancer? *Biochim. Biophys. Acta.* **1603**:113–128.

92. Edinger, A.L., and Thompson, C.B. 2004. Death by design: apoptosis, necrosis and autophagy. *Curr. Opin. Cell Biol.* **16**:663–669.

93. Gozuacik, D., and Kimchi, A. 2004. Autophagy as a cell death and tumor suppressor mechanism. *Oncogene.* **23**:2891–2906.

94. Nelson, D.A., and White, E. 2004. Exploiting different ways to die. *Genes Dev.* **18**:1223–1226.

95. Kanzawa, T., Kondo, Y., Ito, H., Kondo, S., and Germano, I. 2003. Induction of autophagic cell death in malignant glioma cells by arsenic trioxide. *Cancer Res.* **63**:2103–2108.

96. Huang, S., and Houghton, P.J. 2003. Targeting mTOR signaling for cancer therapy. *Curr. Opin. Pharmacol.* **3**:371–377.

97. Codogno, P., and Meijer, A.J. 2004. Signaling pathways in mammalian autophagy. In *Autophagy*. D.J. Klionsky, editor. Landes Bioscience. Georgetown, Texas, USA. 26–47.

98. Degterev, A., et al. 2005. Chemical inhibitor of nonapoptotic cell death with therapeutic potential for ischemic brain injury *Nat. Chem. Biol.* **1**:112–119.

99. Kinch, G., Hoffman, K.L., Rodrigues, E.M., Zee, M.C., and Weeks, J.C. 2003. Steroid-triggered programmed cell death of a motoneuron is autophagic and involves structural changes in mitochondria. *J. Comp. Neurol.* **457**:384–403.

100. Muller, F., Adori, C., and Sass, M. 2004. Autophagic and apoptotic features during programmed cell death in the fat body of the tobacco hornworm (*Manduca sexta*). *Eur. J. Cell Biol.* **83**:67–78.

101. Romer, F., and Martau, T. 1998. Degeneration of moulting glands in male crickets. *J. Insect Physiol.* **44**:981–989.

102. Bowen, I.D., Mullarkey, K., and Morgan, S.M. 1996. Programmed cell death during metamorphosis in the blow-fly *Calliphora vomitoria*. *Microsc. Res. Tech.* **34**:202–217.

103. Jones, H.E., and Bown, I.D. 1993. Acid phosphatase activity in the larval salivary glands of developing *Drosophila melanogaster*. *Cell Biol. Int.* **17**:305–315.

104. Lee, C.-Y., Cooksey, B.A.K., and Baehrecke, E.H. 2002. Steroid regulation of midgut cell death during *Drosophila* development. *Dev. Biol.* **250**:101–111.

105. Butterworth, F.M., and LaTendresse, B.L. 1973. Quantitative studies of cytochemical and cytological changes during cell death in the larval fat body of *Drosophila melanogaster*. *J. Insect. Physiol.*

19:1487–1500.

106. Debnath, J., et al. 2002. The role of apoptosis in creating and maintaining luminal space within normal and oncogene-expressing mammary acini. *Cell.* **111**:29–40.

107. Kegel, K.B., et al. 2000. Huntingtin expression stimulates endolysosomal-activity, endosome tubulation, and autophagy. *J. Neurosci.* **20**:7268–7278.

108. Stefanis, L., Larsen, K.E., Rideout, H.J., Sulzer, D., and Greene, L.A. 2001. Expression of A53T mutant but not wild-type α-synuclein in PC12 cells induces alterations of the ubiquitin-dependent degradation, loss of dopamine release, and autophagic cell death. *J. Neurosci.* **21**:9549–9560.

109. Nitatori, T., et al. 1995. Delayed neuronal death in the CA1 pyramidal cell layer of the gerbil hippocampus following transient ischemia is apoptosis. *J. Neurosci.* **2**:1001–1011.

110. Kostin, S. 2004. Types and mechanisms of myocyte cell death in diseased human heart. *Cardiovasc. J. S. Afr.* **15**(4 Suppl. 1):S1.

111. Hein, S., et al. 2003. Progression from compensated hypertrophy to failure in the pressure-overloaded human heart: structural deterioration and compensatory mechanisms. *Circulation.* **107**:984–991.

112. Bera, A., Singh, S., Nagaraj, R., and Vaidya, T. 2003. Induction of autophagic cell death in *Leishmania donovani* by antimicrobial peptides. *Mol. Biochem. Parasitol.* **127**:23–35.

113. Christensen, S.T., et al. 1998. Staurosporine-induced cell death in *Tetrahymena thermophila* has mixed characteristics of both apoptotic and autophagic degeneration. *Cell Biol. Int.* **22**:591–598.

114. Cornillon, S., et al. 1994. Programmed cell death in *Dictyostelium*. *J. Cell Sci.* **107**:2691–2704.

115. Akazawa, H., et al. 2004. Diphtheria toxin-induced autophagic cardiomyocyte death plays a pathogenic role in mouse model of heart failure. *J. Biol. Chem.* **279**:41095–41103.

116. Guimaraes, C.A., Benchimo, M., Amarante-Mendes, G.P., and Linden, R. 2003. Alternative programs of cell death in developing retinal tissue. *J. Biol. Chem.* **278**:41938–41946.

117. Borsello, T., Croquelois, K., Hornung, J.P., and Clarke, P.G.H. 2003. N-methyl-D-aspartate triggered neuronal death in organotypic hippocampal cultures is endocytic, autophagic and mediated by the c-Jun N terminal kinase pathway. *Eur. J. Neurosci.* **18**:473–485.

118. Ohsawa, Y., et al. 1998. An ultrastructural and immunohistochemical study of PC12 cells during apoptosis induced by serum deprivation with special reference to autophagy and lysosomal cathepsins. *Arch. Histol. Cytol.* **61**:395–403.

119. von Bultzingslowen, I., Jontell, M., Hurst, P., Nannmark, U., and Kardos, T. 2001. 5-Fluorouracil induces autophagic degeneration in rat oral keratinocytes. *Oral Oncol.* **37**:537–544.

120. Opipari, A.W., Jr., et al. 2004. Resveratrol-induced autophagocytosis in ovarian cancer cells. *Cancer Res.* **64**:696–703.

121. Bursch, W., et al. 2000. Autophagic and apoptotic types of programmed cell death exhibit different fates of cytoskeletal filaments. *J. Cell Sci.* **113**:1189–1198.

122. Kanzawa, T., et al. 2005. Arsenic trioxide induces autophagic cell death in malignant glioma cells by upregulation of mitochondrial cell death protein BNIP3. *Oncogene.* **24**:980–991.

123. Chau, Y.-P., Lin, S.-Y., Chen, J.H.-C., and Tai, M.-H. 2003. Endostatin induced autophagic cell death in EAhy926 human endothelial cells. *Histol. Histopathol.* **18**:715–726.

124. Juhasz, G., Csikos, G., Sinka, R., Erdelyi, M., and Sass, M. 2003. The *Drosophila* homolog of Aut1 is essential for autophagy and development. *FEBS Lett.* **543**:154–158.

Endoplasmic reticulum stress: cell life and death decisions

Chunyan Xu, Beatrice Bailly-Maitre, and John C. Reed

The Burnham Institute for Medical Research, La Jolla, California, USA.

Disturbances in the normal functions of the ER lead to an evolutionarily conserved cell stress response, the unfolded protein response, which is aimed initially at compensating for damage but can eventually trigger cell death if ER dysfunction is severe or prolonged. The mechanisms by which ER stress leads to cell death remain enigmatic, with multiple potential participants described but little clarity about which specific death effectors dominate in particular cellular contexts. Important roles for ER-initiated cell death pathways have been recognized for several diseases, including hypoxia, ischemia/reperfusion injury, neurodegeneration, heart disease, and diabetes.

Introduction

The ER fulfills multiple cellular functions (reviewed in refs. 1–4). The lumen of the ER is a unique environment, containing the highest concentration of Ca^{2+} within the cell because of active transport of calcium ions by Ca^{2+} ATPases. The lumen is an oxidative environment, critical for formation of disulfide bonds and proper folding of proteins destined for secretion or display on the cell surface. Because of its role in protein folding and transport, the ER is also rich in Ca^{2+}-dependent molecular chaperones, such as Grp78, Grp94, and calreticulin, which stabilize protein folding intermediates (reviewed in refs. 1, 5–7).

Many disturbances, including those of cellular redox regulation, cause accumulation of unfolded proteins in the ER, triggering an evolutionarily conserved response, termed the unfolded protein response (UPR). Glucose deprivation also leads to ER stress, by interfering with N-linked protein glycosylation. Aberrant Ca^{2+} regulation in the ER causes protein unfolding, because of the Ca^{2+}-dependent nature of Grp78, Grp94, and calreticulin (6). Viral infection may also trigger the UPR, representing one of the ancient evolutionary pressures for linking ER stress to cell suicide in order to avoid spread of viruses. Further, because a certain amount of basal protein misfolding occurs in the ER, normally ameliorated by retrograde transport of misfolded proteins into the cytosol for proteasome-dependent degradation, situations that impair proteasome function can create a veritable protein traffic jam and can even cause inclusion body diseases associated with neurodegeneration.

The initial intent of the UPR is to adapt to the changing environment, and reestablish normal ER function. These adaptive mechanisms involve transcriptional programs that induce expression of genes that enhance the protein folding capacity of the ER, and promote ER-associated protein degradation to remove misfolded proteins. Translation of mRNAs is also initially inhibited, reducing the influx of new proteins into the ER for hours until mRNAs encoding UPR proteins are produced. When adaptation fails,

ER-initiated pathways signal alarm by activating NF-κB, a transcription factor that induces expression of genes encoding mediators of host defense. Excessive and prolonged ER stress triggers cell suicide, usually in the form of apoptosis, representing a last resort of multicellular organisms to dispense of dysfunctional cells. Progress in understanding the mechanisms underlying these 3 phases of adaptation, alarm, and apoptosis has improved our knowledge of ER stress, and its role in disease.

Adaptation to ER stress: mechanisms to restore homeostasis

When unfolded proteins accumulate in the ER, resident chaperones become occupied, releasing transmembrane ER proteins involved in inducing the UPR. These proteins straddle ER membranes, with their N-terminus in the lumen of the ER and their C-terminus in the cytosol, providing a bridge that connects these 2 compartments. Normally, the N-termini of these transmembrane ER proteins are held by ER chaperone Grp78 (BiP), preventing their aggregation. But when misfolded proteins accumulate, Grp78 releases, allowing aggregation of these transmembrane signaling proteins, and launching the UPR. Among the critical transmembrane ER signaling proteins are PERK, Ire1, and ATF6 (Figure 1) (reviewed in refs. 1, 2, 8).

PERK (PKR-like ER kinase) is a Ser/Thr protein kinase, the catalytic domain of which shares substantial homology to other kinases of the eukaryotic initiation factor 2α (eIF2α) family (9, 10). Upon removal of Grp78, PERK oligomerizes in ER membranes, inducing its autophosphorylation and activating the kinase domain. PERK phosphorylates and inactivates eIF2α, thereby globally shutting off mRNA translation and reducing the protein load on the ER. However, certain mRNAs gain a selective advantage for translation under these conditions, including the mRNA encoding transcription factor ATF4. The ATF4 protein is a member of the bZIP family of transcription factors, which regulates the promoters of several genes implicated in the UPR. The importance of PERK-initiated signals for protection against ER stress has been documented by studies of *perk*$^{-/-}$ cells and of knock-in cells that express non-phosphorylatable eIF2α(S51A), both of which display hypersensitivity to ER stress (11, 12).

Ire1 similarly oligomerizes in ER membranes when released by Grp78. The Ire1α protein is a type I transmembrane protein, which contains both a Ser/Thr kinase domain and an endoribonuclease domain; the latter processes an intron from X box protein-1 (XBP-1)

Nonstandard abbreviations used: AD, Alzheimer disease; Aβ P, amyloid β-peptide; DED-L, death effector domain–like; eIF2α, eukaryotic initiation factor 2α; Htt, Huntingtin; IP3, inositol triphosphate; IP3R, IP3 receptor; NOS, nitric oxide synthase; PD, Parkinson disease; PERK, PKR-like ER kinase; polyQ, polyglutamine; PS-1, presenilin-1; SERCA, sarcoplasmic/endoplasmic reticulum Ca^{2+} ATPase; UPR, unfolded protein response; XBP-1, X box protein-1.

Conflict of interest: The authors have declared that no conflict of interest exists.

Citation for this article: *J. Clin. Invest.* **115**:2656–2664 (2005). doi:10.1172/JCI26373.

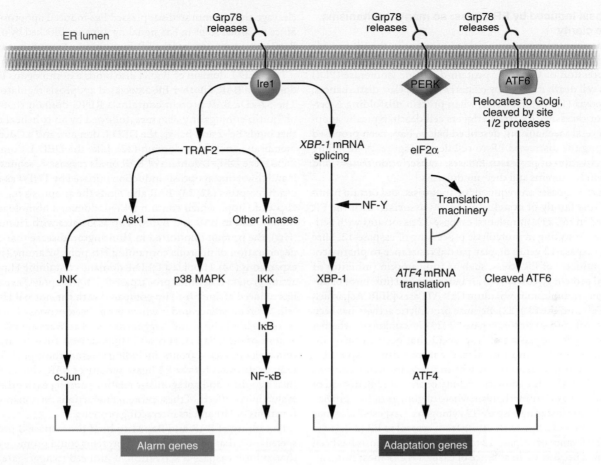

Figure 1
Signal transduction events associated with ER stress. Chaperone Grp78 binds the N-termini of Ire1, PERK, and ATF6, preventing their activation. Unfolded proteins in the ER cause Grp78 to release Ire1, PERK, and ATF6. Upon Grp78 release, Ire1 and PERK oligomerize in ER membranes. Oligomerized Ire1 binds TRAF2, signaling downstream kinases that activate NF-κB and c-Jun (AP-1), causing expression of genes associated with host defense (alarm). The intrinsic ribonuclease activity of Ire1 also results in production of XBP-1, a transcription factor that induces expression of genes involved in restoring protein folding or degrading unfolded proteins. Oligomerization of PERK activates its intrinsic kinase activity, resulting in phosphorylation of eIF2α and suppression of mRNA translation. Under these conditions, only selected mRNAs, including ATF4, are translated. ATF4 induces expression of genes involved in restoring ER homeostasis. Release of Grp78 from ATF6 allows this protein to translocate to the Golgi apparatus for proteolytic processing to release active ATF6, which controls expression of UPR genes.

mRNA, rendering it competent for translation to produce the 41-kDa XBP-1 protein, a bZIP-family transcription factor. XBP-1 binds to promoters of several genes involved in retrograde transport of misfolded proteins from ER to cytosol and in ER-induced protein degradation (reviewed in ref. 8). XBP-1 heterodimerizes with protein NF-Y and binds at least 2 types of *cis*-acting elements in gene promoters, including the ER stress enhancer (ERSE) and unfolded protein response element (UPRE) (13). Ablation of Ire1α in mice produces an embryonic lethal phenotype. Fibroblasts from *Ire1α*−/− embryos are defective in activation of UPRE-driven reporter genes, thus showing a cause-and-effect linkage of Ire1α to this *cis*-acting element (14).

Release of Grp78 from the N-terminus of ATF6 triggers a different mechanism of protein activation, compared with PERK and Ire1. Instead of oligomerizing, release of Grp78 frees ATF6 to translocate to the Golgi apparatus, where resident proteases (site 1 and site 2 protease) cleave ATF6 at a juxtamembrane site, releasing this transcription factor into the cytosol and allowing it to migrate into the nucleus to regulate gene expression (15). ATF6 collaborates

with Ire1, where ATF6 induces transcription to increase XBP-1 mRNA, and Ire1's endoribonuclease activity then processes that mRNA so that XBP-1 protein is produced.

Sounding the alarm in response to ER stress: NF-κB activation mechanisms

Given the massive glycoprotein production associated with many viral infections, it is not surprising that ER stress activates some of the same signal transduction pathways associated with innate immunity. In this regard, Ire1 shares in common with many members of the TNF receptor family the ability to bind adapter protein TRAF2. TRAF2 is an E3 ligase that binds Ubc13, resulting in non-canonical polyubiquitination of substrates involving lysine 63 rather than the canonical lysine 48 as a linking site (16). TRAF2 activates protein kinases implicated in immunity and inflammation, including Ask1, which activates JNK, and kinases linked to NF-κB activation. Also recruited to Ire1 is the c-Jun N-terminal inhibitory kinase (JIK), responsible for posttranslational modification of components of the Ire1α/TRAF2/Ask1 complex (17, 18).

Apoptosis induced by ER stress: so many mechanisms, so little clarity

The adaptive responses to misfolded proteins in the ER provide protection from cell death, inasmuch as gene transfer–mediated overexpression of Grp78 or protein-disulfide isomerase (PDI) reduces cell death induced by oxidative stress, Ca^{2+} disturbances, and hypoxia (19, 20). However, when protein misfolding is persistent or excessive, ER stress triggers cell death, typically apoptosis. Several mechanisms, described below, have been proposed for linking the distressed ER to cell death (Figure 2), including direct activation of proteases, kinases, transcription factors, and Bcl-2–family proteins and their modulators.

Proteases. Caspases are required for apoptosis, and certain members of this family of cysteine proteases associate with the ER (reviewed in ref. 21). In rodents, caspase-12 associates with activated Ire1, resulting in proteolytic processing of caspase-12. Mice lacking caspase-12 genes display partial resistance to pharmacological inducers of ER stress, such as tunicamycin (inhibitor of N-linked protein glycosylation) and thapsigargin (inhibitor of sarcoplasmic/endoplasmic reticulum Ca^{2+} ATPases [SERCAs], which pump Ca^{2+} into the ER) (22). Because proteolytic activity has been difficult to demonstrate for caspase-12 (23), it is unknown whether the proteolytic processing of caspase-12 that occurs during ER stress results in its activation. Also, the mechanisms responsible for proteolysis of caspase-12 may be indirect, involving calpains activated by Ca^{2+} released in the vicinity of the ER (24), instead of an induced proximity mechanism where oligomers of Ire1 provide a scaffold for clustering caspase-12 zymogens. Caspase-7 also may activate caspase-12 by translocating from cytosol to ER (25). However, the relevance of caspase-12 to ER-induced apoptosis has been questioned because of an absence of caspase-12 in most humans. In this regard, the ancestral human *CASPASE-12* gene is disrupted by a termination codon and thus is inactive (26). For persons with hereditary polymorphisms that leave the open reading frame intact (estimated at ~1% of African populations), caspase-12 operates as a *trans*-dominant inhibitor of proinflammatory caspases, lacking conserved residues required for catalytic activity (23).

Human caspase-4, one of the closest paralogs of rodent caspase-12, may associate with ER (27), raising the possibility that this protease can perform the functions normally ascribed to rodent caspase-12 in the context of ER stress. But caspase-4 belongs to the group of proinflammatory caspases responsible for proteolytic activation of cytokines, rather than the apoptotic caspases. Nevertheless, small interfering RNA–mediated knock down of caspase-4 in human neuroblastoma cells partially reduces cell death caused by the ER stress inducers thapsigargin and amyloid β-peptide (AβP), but not inducers of mitochondria-dependent cell death (e.g., UV irradiation, DNA-damaging drugs). However, caspase-4 knock down in HeLa cells had little effect on apoptosis induced by ER stress, implying that the relevance of this protease to ER stress is tissue-specific.

The ER resident protein Bap31 contains 3 predicted transmembrane domains, followed by a leucine zipper and a death effector domain–like (DED-L) region that associates with certain isoforms of procaspase-8 in the cytosol (28). Bap31 can display either pro- or antiapoptotic phenotypes, depending on whether its cytosolic tail is removed by cleavage by caspases. Overexpression of full-length Bap31 blocks apoptosis induced by anti-Fas antibody and cycloheximide, while expression of the truncated 20-kDa protein induces apoptosis (29, 30). A mutant of Bap31 in which the caspase-8 cleavage site was mutated suppressed Fas-induced apoptosis (29). Since proximal steps in Fas signaling were not blocked by mutant Bap31, this suggests that the ER participates as an intermediary in death receptor–induced apoptosis in some cells.

The DED-L domain of Bap31 also binds a homologous DED-L domain in BAR, another ER-associated apoptosis regulator (31). The 52-kDa BAR protein contains a RING domain that binds ubiquitin-conjugating enzymes, followed by an α-helical region that binds Bcl-2 and Bcl-x$_L$, the DED-L domain, and a C-terminal membrane-anchoring domain (32). Like the DED-L domain of Bap31, the DED-L domain of BAR binds caspase-8, sequestering it and thwarting apoptosis induction initiated by TNF/Fas-family death receptors (32, 33). BAR also binds the apoptosis regulators Hip and Hippi, which contain DED-L domains homologous to those found in BAR and Bap31. Hip associates with Huntingtin (Htt), the protein implicated in Huntington disease that causes degeneration of neurons containing Htt polyglutamine (polyQ) expansions (34). Hippi is a DED-L domain–containing Hip-interacting protein that binds procaspase-8. Htt with polyQ expansion has reduced affinity for Hip compared with the normal Htt protein, a circumstance under which it has been proposed that Hip is free to bind Hippi and trigger caspase-8 activation (35). Interactions of BAR, Bap31, Hip, and Hippi deserve further investigation on a number of fronts, including whether these proteins represent substrates for the E3 ligase activity of BAR, elucidation of their agonistic and antagonistic relations among each other, and evaluation of effects of these protein interactions on nonapoptotic functions of Htt and its interacting proteins.

The ability of BAR and Bap31 to bind procaspase-8 prompts speculation that perhaps these ER proteins could promote rather than inhibit caspase-8 activation, if induced to aggregate in ER membranes, thereby constituting a novel ER-associated "apoptosome." If so, then the parallel ability of BAR and Bap31 to bind Bcl-2 and Bcl-x$_L$ through domains separate from the DED-L domain might supply a mechanism for preventing caspase activation, providing a long-sought analogy to the paradigm for caspase regulation seen in *Caenorhabditis elegans*, where the Bcl-2 ortholog Ced9 binds caspase activator Ced4, preventing activation of Ced3 protease (36).

Kinases. The kinase Ask1 has been implicated in apoptosis induction in the context of signaling by TNF-family receptors (reviewed in ref. 37). During ER stress, Ask1 is recruited to oligomerized Ire1 complexes containing TRAF2, activating this kinase and causing downstream activation of JNK and p38 MAPK. Consistent with a key role for Ask1 in apoptosis induced by ER stress, studies of *ask1$^{-/-}$* neurons subjected to inducers of ER stress indicate a requirement for this kinase for JNK activation and cell death (38). The downstream death effectors of Ask1 are not clear. The kinase pathway initiated by Ask1 leads to JNK activation, and JNK-mediated phosphorylation activates the proapoptotic protein Bim (39–41), while inhibiting the antiapoptotic protein Bcl-2 (42).

Thus, Ire1 plays roles in all 3 of the ER responses to unfolded proteins (adaptation, alarm, and apoptosis), through its actions upon XBP-1 (adaptation), TRAF2 (alarm [NF-κB]), and apoptosis effectors caspase-12 and Ask1. How these 3 functions of Ire1 are integrated remains unclear.

The protein tyrosine kinase c-Abl can translocate from the ER surface to mitochondria in response to ER stress (43). Moreover, a functional role for c-Abl has been suggested by studies of *c-Abl$^{-/-}$* fibroblasts, which display resistance to cell death induced by Ca^{2+} ionophores, brefeldin A, and tunicamycin (43). How c-Abl promotes apoptosis is unknown at present.

Transcription factors. CHOP (GADD153) is a member of the C/EBP family of bZIP transcription factors, and its expression is induced to high levels by ER stress (reviewed in ref. 44). The *chop* gene promoter contains binding sites for all of the major inducers of the UPR, including ATF4, ATF6, and XBP-1, and these transcription factors play causative roles in inducing *chop* gene transcription. Cause-and-effect roles in *chop* gene induction have been demonstrated for signaling molecules involved in ER stress by genetic manipulation of mice, showing that *perk$^{-/-}$* and *atf4$^{-/-}$* cells and eIF2α(S51A) knock-in cells fail to induce *chop* during ER stress (11, 12, 45). Cross-talk between the PERK/eIF2α pathway and the Ire1/TRAF2/Ask1 pathway may also enhance CHOP activity at a posttranscriptional level, given that Ask1 activates both JNK and p38 MAPKs, and phosphorylation of the CHOP protein on serine 78 and serine 81 by p38 MAPKs increases its transcriptional and apoptotic activity (46). In addition to the aforementioned regulators, upstream activators of *chop* also include ATF2, which is induced by hypoxia and which is required for *chop* induction during amino acid starvation (47).

Overexpression of CHOP protein induces apoptosis, through a Bcl-2–inhibitable mechanism (48, 49). Moreover, *chop$^{-/-}$* mice are resistant to kidney damage induced by tunicamycin and to brain injury resulting from cerebral artery occlusion, demonstrating a role for CHOP in cell destruction when ER stress is involved (48, 50). How CHOP induces apoptosis is unclear. CHOP forms heterodimers with other C/EBP-family transcription factors via bZIP-domain interactions, which suppresses their binding to C/EBP sites in DNA, while promoting binding to alternative DNA sequences for target gene activation (51). Consequently, CHOP inhibits expression of genes responsive to C/EBP-family transcription factors, while enhancing expression of other genes containing the consensus motif 5'-(A/G)(A/G)(A/G)TGCAAT(A/C)CCC-3'. One relevant target may be *bcl-2*, whose expression is suppressed by CHOP, at least in some cellular contexts (49). CHOP may also have nontranscriptional actions, still poorly defined (44). While capable of inducing apoptosis and contributing to cell death in several scenarios involving ER stress, CHOP is not essential for cell death induced by ER stress, as demonstrated by the observation that *perk$^{-/-}$* and eIF2α(S51A) knock-in cells are hypersensitive to ER stress–induced apoptosis but fail to induce *chop* gene expression (12, 45).

Scotin is another ER-targeted apoptosis inducer (52). The gene encoding Scotin is a direct target of p53, suggesting a way to link DNA damage to ER-mediated cell death mechanisms.

Bcl-2–family proteins and their modulators. Association of certain Bcl-2/Bax–family proteins with ER membranes dates back to the initial discovery of Bcl-2 (53). Though better known for their actions upon mitochondria, Bcl-2/Bax–family proteins also integrate into ER membranes, where they modulate ER Ca^{2+} homeostasis and control cell death induced by ER stress agents, including tunicamycin, brefeldin A (an inhibitor of ER-Golgi transport), thapsigargin, and oxidants (reviewed in ref. 54). Experiments in which the normal C-terminal transmembrane domain of Bcl-2 was swapped with membrane-targeting domains from ER resident proteins suggested that Bcl-2 targeted exclusively to the ER (as opposed to both ER and mitochondria) is more restricted in its antiapoptotic actions, suppressing cell death induced by ER stress agents and by c-Myc. Recent findings that apoptosis induced by c-Myc may be attributable to its induction of Bim suggest that ER-targeted Bcl-2 may

sequester this BH3-only protein, preventing it from interacting with other members of the Bcl-2/Bax family (55).

Spike is a BH3-only protein anchored in the ER (56). The BH3-like domain of Spike is required for apoptosis induction, but dimerization partners among Bcl-2/Bax–family proteins have yet to be found. Several other Bcl-2/Bax–family proteins reside at least in part in association with or integrated into ER membranes, with some, such as the antiapoptotic protein Mcl-1 and proapoptotic Bik, found predominantly in the ER (57, 58). Given the preferences of certain BH3 domains for interactions with particular members of the Bcl-2/Bax family (59), it seems likely that a network of interactions among a subset of this family of apoptosis regulators takes place on ER membranes, the functional consequences of which are not yet fully understood. Recently, expression of at least 1 of the Bcl-2/Bax–family genes was linked to ER stress. The BH3-only protein Puma is induced by tunicamycin and thapsigargin in a p53-independent manner, with *Puma$^{-/-}$* cells showing resistance to apoptosis induced by ER stress (60).

The BI-1 protein contains 6 transmembrane domains, resides in the ER (61), interacts functionally or physically with Bcl-2–family members, and is induced by hypoxia (62). This protein blocks cell death induced by oxidative stress in yeast, plants, and animals (63). Mice lacking BI-1 display increased sensitivity to tunicamycin-induced kidney damage and to stroke injury, implying a role for BI-1 in protection from insults known to trigger ER stress. In cultured cells, overexpression of BI-1 selectively reduces, while BI-1 deficiency selectively increases, sensitivity to cell death induced by agents that trigger ER stress, while having far less effect on apoptosis induced by agents that trigger cell death pathways linked to mitochondria (intrinsic pathway) or TNF/Fas-family death receptors (extrinsic pathway) (64). BI-1 associates with the antiapoptotic proteins Bcl-2 and Bcl-x$_L$, but not proapoptotic Bax and Bak (61). Nevertheless, BI-1 inhibits cell death induced by Bax overexpression, in animal cells, plants, and yeast.

The ER protein Bap31 lacks homology with Bcl-2/Bax–family proteins and contains no recognizable BH3 dimerization domain, but it binds Bcl-2 and Bcl-x$_L$ and regulates apoptosis. BAR also binds Bcl-2 and Bcl-x$_L$, and the responsible domain is required for BAR-mediated suppression of cell death (32). Interestingly, BAR is capable of suppressing Bax-induced death of yeast, implying caspase-independent functions for this protein, given that yeast lack bona fide caspases. This suggests that, mechanistically, BAR may share something in common with Bcl-2 and BI-1, which also suppress Bax-induced killing of yeast.

Other apoptosis regulators. Given that mitochondria release apoptogenic proteins into the cytosol, the ER might use similar mechanisms for linking ER stress to cell death. In insect cells, at least one example has been uncovered of a protein, called Jafrac2, that is normally sequestered in the ER but is released into the cytosol during apoptosis induced by certain stimuli (65). Like most proteins imported into the ER, the N-terminal leader peptide of Jafrac2 is removed by proteolysis. This proteolytic processing exposes an IAP-binding motif in Jafrac2, poising it to attack antiapoptotic IAP-family proteins upon accessing the cytosol, thereby freeing caspases. It remains to be determined whether examples of apoptogenic protein release from the ER of mammalian cells will be discovered.

Ca^{2+} and apoptosis induced by ER stress

Release of Ca^{2+} from the ER plays critical roles in cellular signaling mediated by the second messengers inositol triphosphate

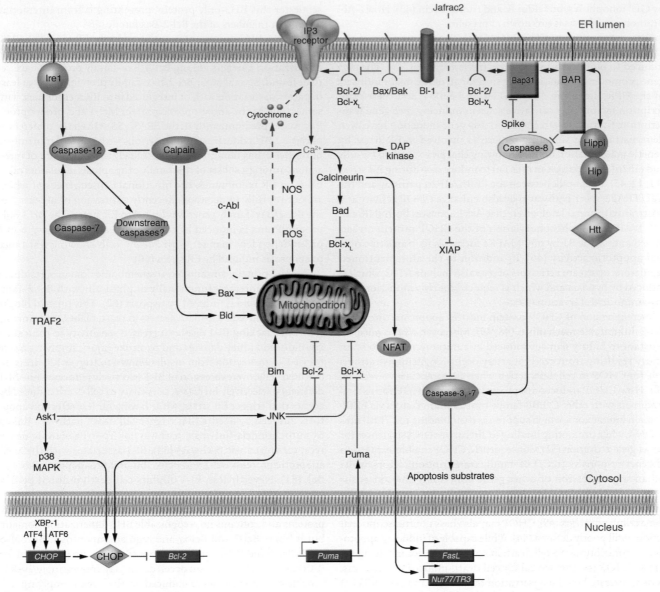

Figure 2
Cell death mechanisms induced by ER stress. Several of the proposed pathways linking ER stress to cell death are depicted. Dashed lines indicate protein translocation events (c-Abl, Jafrac2). The mitochondrial permeability transition pore complex, which is Ca²⁺-sensitive, is not shown in the diagram. See the text for additional details.

(IP3) and cytosolic ADP-ribose and other regulators via effects on IP3 receptors (IP3Rs) and ryanodine receptors (66, 67). Opposing these gated Ca²⁺ channels are the SERCA-family proteins, Ca²⁺ ATPases that pump Ca²⁺ into the ER, which are regulated by phosphorylation and interactions with other proteins (e.g., phospholamban and sarcolipin). Various stimuli that cause the ER to dump Ca²⁺ precipitate cell death, including hypoxia, oxidants, stimulators of IP3 production, and pharmacological antagonists of SERCA. The downstream effectors of Ca²⁺-induced cell death are potentially myriad and could minimally include (a) induction of mitochondrial permeability transition, induced upon entry of excessive amounts of Ca²⁺ into the matrix of mitochondria (68, 69); (b) local activation near the ER of calpains, a family of Ca²⁺-dependent cysteine proteases implicated in pathological cell death (70, 71),

whose substrates include Bax and Bid (which are activated) (72–74), Bcl-2 and Bcl-x_L (which are inhibited), and several caspases (reviewed in ref. 4); (c) alterations in Ca²⁺-dependent phospholipid scramblases, which alter membrane biology to promote apoptosis or necrosis, including transferring phosphatidylserine to the outer leaflet of the plasma membrane (a signal for clearance of cells by phagocytosis) and transferring cardiolipin from the inner to outer membrane of mitochondria (a signal for targeted insertion of proapoptotic Bcl-2–family proteins Bid and Bax into membranes) (75–77); (d) Ca²⁺/calmodulin–mediated activation of the protein phosphatase calcineurin, which dephosphorylates the proapoptotic protein Bad, allowing it to dimerize with and antagonize Bcl-x_L (78), and which dephosphorylates NFAT-family transcription factors, allowing entry into the nucleus and transactivation of proapoptotic genes encoding FasL and Nur77/TR3 (79); (e) stimulation of Ca^{2+}-sensitive isoforms of nitric oxide synthase (NOS), exacerbating oxidative stress (reviewed in ref. 5); (f) activation of death-associated protein kinase (DAP kinase) and its relative DRP-1, which contain calmodulin-binding domains (reviewed in ref. 80); (g) activation of Ca^{2+}-sensitive mitochondrial fission protein DRP-1 (81), which has been implicated in Bax-induced release of cytochrome c from mitochondria; and possibly (h) alterations in the Ca^{2+}-binding protein TCTP (fortilin), a putative modulator of antiapoptotic Bcl-2/Bax–family proteins such as Mcl-1 (82). In addition, ectopic expression of the proapoptotic mammalian protein Bak in yeast induces cell death through a calnexin-dependent pathway, correlating with Bak binding to this Ca^{2+}-dependent ER chaperone (83).

A role for Bcl-2 in modulating intracellular Ca^{2+} was first demonstrated over a decade ago (84), but only recently have clues about the mechanisms involved begun to emerge. Based on data from a variety of techniques, it appears that overexpression of antiapoptotic proteins Bcl-2 and Bcl-x_L lowers the basal amounts of Ca^{2+} in the ER, because of increased leakage of Ca^{2+} under resting conditions. The consequence of this is that upon exposure to stimuli that precipitously dump Ca^{2+} from internal stores, less Ca^{2+} enters the cytosol, resulting in lower peak concentrations of cytosolic Ca^{2+} and less overall cytosolic Ca^{2+} accumulation (54, 85–88). Downstream, less Ca^{2+} enters mitochondria, which possibly explains the inhibition of mitochondrial depolarization and the suppression of cytochrome c release. Like overexpression of Bcl-2 or Bcl-x_L, ablation of the expression of proapoptotic Bax and Bak also reduces basal Ca^{2+} in the ER, implying a role for these proapoptotic proteins in setting cellular ER Ca^{2+} concentrations (89). Interestingly, Bcl-2 remains competent in its ability to reduce ER Ca^{2+}, even in cells lacking Bax and Bak, implying that Bcl-2 operates downstream of Bax/Bak with respect to ER Ca^{2+} regulation, unlike the situation with mitochondria-dependent cell death, where genetic evidence indicates that Bcl-x_L and Bcl-2 function upstream of Bax/Bak (90).

Attempts to establish whether these changes in ER Ca^{2+} handling mediated by Bcl-2/Bax–family proteins are causally linked to cell death regulation have failed to provide firm answers, but supporting evidence has been obtained from a variety of experimental approaches, including genetic manipulations of Ca^{2+}-regulating proteins in the ER (88, 89, 91, 92).

Because several Bcl-2/Bax–family proteins share structural similarity with the pore-forming domains of bacterial toxins, they may function as ion channels, thus explaining the ability of Bcl-2/Bax–family proteins to modulate ER Ca^{2+} (reviewed in ref. 93). However, mutations designed to impair the putative pore-forming regions of Bcl-2 do not affect its ability to regulate ER Ca^{2+} (94); this suggests alternative mechanisms. In this regard, Bcl-2 was reported to bind IP3Rs several years ago (95), and recently Bcl-2 has been implicated in regulating IP3R activity (96). IP3R knockdown inhibits the ability of Bcl-2 to promote leakage of Ca^{2+} from the ER, suggesting that Bcl-2 relies on IP3Rs to reduce luminal ER Ca^{2+}. The mechanism by which Bcl-2 modulates IP3Rs has yet to be defined, particularly the issue of whether this is a direct effect of these proteins on IP3Rs or an indirect effect on unidentified IP3R-interacting proteins present in ER membranes. Interestingly, cytochrome c, an apoptosis-inducing protein released from mitochondria, binds IP3Rs and induces Ca^{2+} release from ER, thereby triggering ER stress (97) and providing another potential link between IP3Rs and cell death regulation. Also, reduction in or ablation of expression of certain IP3Rs (e.g., IP3R1 and IP3R3) decreases sensitivity of some types of cells (e.g., lymphocytes, neurons) to apoptosis (98–100), suggesting further links between Ca^{2+} dysregulation by IP3Rs and apoptosis induction.

Curiously, the antiapoptotic protein BI-1 also regulates ER Ca^{2+} homeostasis in a manner analogous to that of Bcl-2 and Bcl-x_L. Overexpression of BI-1 reduces basal ER Ca^{2+} concentrations, while ablation of the genes encoding BI-1 increases amounts of thapsigargin-releasable Ca^{2+} from internal stores (64). Since BI-1 associates with Bcl-2 and Bcl-x_L in ER membranes (61), it will be interesting to determine whether BI-1 also interacts with and regulates IP3Rs.

The truncated Bap31, resembling the caspase-cleavage product, induces Ca^{2+} efflux from the ER and induces apoptosis (30), providing further correlative connections between modulation of ER Ca^{2+} dynamics and cell death regulation. Interestingly, Bap31 was reported to bind an ER-associated putative ion channel called A4, but the relevance of this protein interaction to regulation of ER Ca^{2+} remains unclear (101).

ER stress and diseases

ER stress is associated with a range of diseases, including ischemia/reperfusion injury, neurodegeneration, and diabetes (reviewed in ref. 44), making ER stress a probable instigator of pathological cell death and dysfunction.

ER stress and neurodegeneration. AβP is a proteolytic product of amyloid β-precursor protein that is causally associated with Alzheimer disease (AD). Mice lacking caspase-12 are partially resistant to apoptosis induced by exposure to AβP (22), raising the possibility of a functional link between ER stress and AβP-induced toxicity. Mutant versions of the AβP-interacting protein presenilin-1 (PS-1), previously associated with AD, interfere with the UPR (102) and may render neurons more susceptible to cell death induced by ER stress (103). The brains of mice harboring AD mutants of PS-1 also have increased CHOP protein (104). Interestingly, PS-1 induces cleavage of Ire1α, releasing the cytosolic domain to translocate to the nucleus, suggesting further interactions between molecules involved in AD and ER stress responses (105).

Hereditary mutations in the ER-associated E3 ubiquitin ligase Parkin have also been associated with ER stress–induced cell death and are found in patients with familial Parkinson disease (PD) (106, 107). Overexpression of wild-type Parkin suppresses cell death induced by several ER stress–inducing agents, and by α-synuclein, the principal component of pathological Lewy bodies seen in PD (107). Parkin expression is induced by ER stress, suggesting a role for it in adaptation to ER stress, presumably functioning in the ER-associated protein degradation pathway to clear misfolded proteins.

Neurodegenerative diseases associated with inclusion body formation and protein aggregation have also been linked to ER stress, including amyotrophic lateral sclerosis, PD, Huntington disease, and others (reviewed in refs. 44, 108). Htt variants with polyQ expansions, for example, induce classical signal transduction events associated with the UPR and cause proteolytic processing of caspase-12 (109), as well as cause global reductions in proteasome activity (110). Thus, by exhausting the cytosolic protein degradation machinery, inclusion body diseases probably cause a back-up of misfolded proteins in the ER, triggering ER stress.

ER stress and ischemia/reperfusion injury. Reduced blood flow resulting from arterial occlusion or cardiac arrest results in tissue hypoxia and hypoglycemia, which cause protein misfolding and ER stress. Reperfusion of the affected tissues then triggers oxidative stress, with production of NO, and other reactive oxygen species that result in protein misfolding. NO and other reactive molecules also may modify oxidizable residues (cysteines) in ER-associated Ca^{2+} channels, including ryanodine receptors and SERCAs, causing ER Ca^{2+} depletion, yet another cause of protein misfolding.

Brain ischemia/reperfusion injury activates the PERK/eIF2α pathway and induces *chop* expression in rodents (111, 112). Moreover, *chop*$^{-/-}$ mice suffer less tissue loss after stroke, implying a causal role for this mediator of ER stress in neuronal cell death (113). NO, a known mediator of brain injury during stroke, induces *chop* expression in cultured neurons. Furthermore, a NOS inhibitor is protective in a rodent model of brain ischemia (114), and mice lacking the gene encoding iNOS display decreased sensitivity to brain ischemia (115), suggesting a causal role for this ER stress inducer in stroke damage.

A role for the antiapoptotic protein BI-1 in protection from cerebral ischemia has been demonstrated by studies of *bi-1*$^{-/-}$ mice, which suffer larger infarcts following cerebral artery occlusion (64). Given that hypoxia has been implicated in *bi-1* gene induction (62), these findings imply a role for endogenous *bi-1* in survival of cells traumatized by ER stress.

ER stress and heart disease. The role of ER stress in heart disease has not been extensively studied, but Ask1 kinase activity increases in mice following myocardial infarction or aortic constriction, and *ask1*$^{-/-}$ mice showed reduced cardiomyocyte apoptosis, in addition to better preservation of left ventricular function, compared with wild-type animals (116).

ER stress in diabetes. Pancreatic β cells have a well-developed ER, reflecting their role in secreting large amounts of insulin and various glycoproteins. This function of β cells may explain why mice lacking PERK are susceptible to diabetes, showing apoptosis of their β cells and progressive hyperglycemia with aging (117). Moreover, *PERK* gene mutations in association with infant-onset diabetes occur in humans with the autosomal recessive disorder Wolcott-Rallison syndrome (118). At autopsy, these patients show massive β cell loss, resembling the pathology of *perk*$^{-/-}$ mice. Similarly, eIF2α(S51A) knock-in mice suffer from β cell depletion, which begins in utero, suggesting a more rapid course than that in *perk*$^{-/-}$ mice (12). The failure of *perk*$^{-/-}$ to phenocopy eIF2α(S51A) raises the possibility that other kinases besides PERK inhibit eIF2α during ER stress. Pancreatic β cell apoptosis induced by NO, a mediator of inflammation relevant to autoimmune diabetes, is CHOP-dependent, further implicating ER stress as an instigator of β cell death (50). Also, in rodent models of diabetes caused by a nonsecreted insulin mutant, homozygous deletion of *chop* delays disease onset (119), implying a role for this gene in β cell depletion in vivo. Recently, *xbp-1*$^{+/-}$ heterozygous mice have been shown to be more sensitive to diabetes caused by obesity and high-fat diet (120). The underlying mechanism is related to the requirement of XBP-1 for dampening of JNK activation caused by ER stress, which correlates with phosphorylation of IRS-1 and reduced tyrosine phosphorylation of IRS-1 in insulin-stimulated cells. Interestingly, signs of ER stress were found in liver and adipose tissue of obese mice and mice fed high-fat diets, indicating that the metabolic abnormalities associated with obesity and unhealthy diets cause ER stress in vivo.

Other cells that secrete proteins in large quantities may also be at risk for ER stress–induced apoptosis. For example, studies of *perk*$^{-/-}$ mice indicate a requirement for differentiation (or survival) of plasma cells, known for their production of immunoglobulins (121).

Therapeutic targets. Several mediators of ER-initiated cell death are candidates for drug discovery efforts, though some are better validated than others. Gene ablation studies in mice argue that agents inhibiting Ask1 and CHOP are attractive, because mice lacking these genes are phenotypically normal but exhibit reduced sensitivity to cell death induced by ER stressors, such as stroke and polyQ-expanded proteins associated with neurodegeneration (reviewed in refs. 38, 44). Presumably Ask1 is also responsible for the hyperactivity of JNK associated with insulin resistance in the context of ER stress caused by high-fat diet. Ask1 theoretically could be attacked by small molecules targeting the ATP-binding site of the kinase domain, analogous to other kinase inhibitors recently approved for other indications. The CHOP protein may be difficult to attack with small-molecule drugs. However, since p38 MAPK augments CHOP activity, small-molecule antagonists of this kinase currently in development for inflammatory diseases might find utility as cytoprotective agents in clinical scenarios involving ER stress. Also, c-Abl inhibitors such as imatinib (Gleevec) could be examined for cytoprotective activity in ischemic and degenerative diseases, given recent evidence that c-Abl may relay death signals from ER to mitochondria (43).

Compounds that augment the PERK/eIF2α pathway may also protect against cell death by ER stress. Indeed, a recent screen for inhibitors of neuronal death induced by tunicamycin identified compounds that suppress protein phosphatases responsible for dephosphorylation of eIF2α on serine 51, thus increasing accumulation of phosphorylated eIF2α and providing protection from apoptosis induced by several inducers of ER stress (122). Interestingly, the prototype compound characterized (Salubrinal) apparently is not an active-site inhibitor of the phosphatase but rather specifically disrupts complexes containing GADD35 and protein phosphatase-1 (PP1), thereby preventing GADD34-mediated targeting of PP1 onto substrate phospho-eIF2α.

It remains to be determined whether broad-spectrum inhibitors of caspase-family cell death proteases would preserve cell survival in the face of ER stress, given that culture experiments have shown that nonapoptotic cell death still occurs in the presence of compounds such as benzoyl-valinyl-alaninyl-aspartyl-fluoromethylketone (zVAD-fmk), at least when strong pharmacological inducers of ER stress are used (64). However, mice lacking various individual caspases, including caspase-1, caspase-2, and caspase-11 (a probable caspase-4 or -5 ortholog), are resistant to stroke injury (123), a condition in which ER stress participates in the cell death mechanism.

Provided that side effects from vascular instability are not an issue, inhibitors of NOS are also attractive, since mice that lack iNOS show decreased sensitivity to brain ischemia and reduced CHOP expression (115). Finally, inhibitors of proapoptotic Bcl-2/Bax–family proteins that operate upon ER membranes could be useful for ameliorating tissue loss due to stimulators of ER stress.

Conclusions

ER stress has been implicated in several diseases, and pathways linking ER stress to cell death have been reported. The principal challenge with any strategy for blocking cell death caused by ER stress lies with the multitude of parallel pathways potentially leading to downstream cell death mechanisms. Thus, blocking only 1 cell death pathway emanating from the ER may be inadequate to preserve cell survival. Further studies of genes and gene products involved in ER-initiated cell death are needed to fully validate targets for drug discovery.

Acknowledgments
We thank J. Valois for manuscript preparation, M. Hanaii for artwork, and the California Breast Cancer Research Program, the Fondation pour la Recherche Médicale, and the NIH (grants NS047855 and AG15393) for their generous support.

Address correspondence to: John C. Reed, The Burnham Institute for Medical Research, 10901 North Torrey Pines Road, La Jolla, California 92037, USA. Phone: (858) 646-3140; Fax: (858) 646-3194; E-mail: reedoffice@burnham.org.

1. Schroder, M., and Kaufman, R.J. 2005. ER stress and the unfolded protein response. *Mutat. Res.* **569**:29–63.
2. Shen, X., Zhang, K., and Kaufman, R.J. 2004. The unfolded protein response: a stress signaling pathway of the endoplasmic reticulum [review]. *J. Chem. Neuroanat.* **28**:79–92.
3. Rao, R.V., Ellerby, H.M., and Bredesen, D.E. 2004. Coupling endoplasmic reticulum stress to the cell death program. *Cell Death Differ.* **11**:372–380.
4. Breckenridge, D.G., Germain, M., Mathai, J.P., Nguyen, M., and Shore, G.C. 2003. Regulation of apoptosis by endoplasmic reticulum pathways. *Oncogene.* **22**:8608–8618.
5. Orrenius, S., Zhivotovsky, B., and Nicotera, P. 2003. Regulation of cell death: the calcium-apoptosis link [review]. *Nat. Rev. Mol. Cell Biol.* **4**:552–565.
6. Ma, Y., and Hendershot, L.M. 2004. ER chaperone functions during normal and stress conditions. *J. Chem. Neuroanat.* **28**:51–65.
7. Rizzuto, R., Duchen, M.R., and Pozzan, T. 2004. Flirting in little space: the ER/mitochondria Ca2+ liaison [review]. *Sci. STKE.* **2004**:re1.
8. Rao, R.V., and Bredesen, D.E. 2004. Misfolded proteins, endoplasmic reticulum stress and neurodegeneration. *Curr. Opin. Cell Biol.* **16**:653–662.
9. Shi, Y., et al. 1998. Identification and characterization of pancreatic eukaryotic initiation factor 2 alpha-subunit kinase, PEK, involved in translational control. *Mol. Cell. Biol.* **18**:7499–7509.
10. Harding, H.P., Zhang, Y., and Ron, D. 1999. Protein translation and folding are coupled by an endoplasmic-reticulum-resident kinase. *Nature.* **397**:271–274.
11. Harding, H.P., Zhang, Y., Bertolotti, A., Zeng, H., and Ron, D. 2000. Perk is essential for translational regulation and cell survival during the unfolded protein response. *Mol. Cell.* **5**:897–904.
12. Scheuner, D., et al. 2001. Translational control is required for the unfolded protein response and in vivo glucose homeostasis. *Mol. Cell.* **7**:1165–1176.
13. Yoshida, H., Matsui, T., Yamamoto, A., Okada, T., and Mori, K. 2001. XBP1 mRNA is induced by ATF6 and spliced by IRE1 in response to ER stress to produce a highly active transcription factor. *Cell.* **107**:881–891.
14. Lee, K., et al. 2002. IRE1-mediated unconventional mRNA splicing and S2P-mediated ATF6 cleavage merge to regulate XBP1 in signaling the unfolded protein response. *Genes Dev.* **16**:452–466.
15. Ye, J., et al. 2000. ER stress induces cleavage of membrane-bound ATF6 by the same proteases that process SREBPs. *Mol. Cell.* **6**:1355–1364.
16. Habelhah, H., et al. 2004. Ubiquitination and translocation of TRAF2 is required for activation of JNK but not of p38 or NF-kappaB. *EMBO J.* **23**:322–332.
17. Urano, F., et al. 2000. Coupling of stress in the ER to activation of JNK protein kinases by transmembrane protein kinase IRE1. *Science.* **287**:664–666.
18. Yoneda, T., et al. 2001. Activation of caspase-12, an endoplasmic reticulum (ER) resident caspase, through tumor necrosis factor receptor-associated factor 2-dependent mechanism in response to the ER stress. *J. Biol. Chem.* **276**:13935–13940.
19. Liu, H., et al. 1997. Endoplasmic reticulum chaperones GRP78 and calreticulin prevent oxidative stress, Ca2+ disturbances, and cell death in renal epithelial cells. *J. Biol. Chem.* **272**:21751–21759.
20. Tanaka, S., Uehara, T., and Nomura, Y. 2000. Up-regulation of protein-disulfide isomerase in response to hypoxia/brain ischemia and its protective effect against apoptotic cell death. *J. Biol. Chem.* **275**:10388–10393.
21. Momoi, T. 2004. Caspases involved in ER stress-mediated cell death. *J. Chem. Neuroanat.* **28**:101–105.
22. Nakagawa, T., et al. 2000. Caspase-12 mediates endoplasmic-reticulum-specific apoptosis and cytotoxicity by amyloid-β. *Nature.* **403**:98–103.
23. Saleh, M., et al. 2004. Differential modulation of endotoxin responsiveness by human caspase-12 polymorphisms. *Nature.* **429**:75–79.
24. Nakagawa, T., and Yuan, J. 2000. Cross-talk between two cysteine protease families. Activation of caspase-12 by calpain in apoptosis. *J. Cell Biol.* **150**:887–894.
25. Rao, R.V., et al. 2001. Coupling endoplasmic reticulum stress to the cell death program: mechanism of caspase activation. *J. Biol. Chem.* **276**:33869–33874.
26. Fischer, H., Koenig, U., Eckhart, L., and Tschachler, E. 2002. Human caspase-12 has acquired deleterious mutations. *Biochem. Biophys. Res. Commun.* **293**:722–726.
27. Hitomi, J., et al. 2004. Involvement of caspase-4 in endoplasmic reticulum stress-induced apoptosis and Abeta-induced cell death. *J. Cell Biol.* **165**:347–356.
28. Ng, F.W.H., et al. 1997. p28 Bap31, a Bcl-2/Bcl-X$_L$- and procaspase-8-associated protein in the endoplasmic reticulum. *J. Cell Biol.* **139**:327–338.
29. Nguyen, M., Breckenridge, D.G., Ducret, A., and Shore, G.C. 2000. Caspase-resistant BAP31 inhibits fas-mediated apoptotic membrane fragmentation and release of cytochrome c from mitochondria. *Mol. Cell. Biol.* **20**:6731–6740.
30. Breckenridge, D.G., Stojanovic, M., Marcellus, R., and Shore, G.C. 2003. Caspase cleavage product of BAP31 induces mitochondrial fission through endoplasmic reticulum calcium signals, enhancing cytochrome c release to the cytosol. *J. Cell Biol.* **160**:1115–1127.
31. Roth, W., et al. 2003. Bifunctional apoptosis inhibitor (BAR) protects neurons from diverse cell death pathways. *Cell Death Differ.* **10**:1178–1187.
32. Zhang, H., et al. 2000. BAR: an apoptosis regulator at the intersection of caspase and bcl-2 family proteins. *Proc. Natl. Acad. Sci. U. S. A.* **97**:2597–2602.
33. Stegh, A.H., et al. 2002. Inactivation of caspase-8 on mitochondria of Bcl-x$_L$ expressing MCF7-Fas cells. *J. Biol. Chem.* **277**:4351–4360.
34. Kalchman, M.A., et al. 1997. HIP1, a human homologue of S. cerevisiae Sla2p, interacts with membrane-associated huntingtin in the brain. *Nat. Genet.* **16**:44–53.
35. Gervais, F.G., et al. 2002. Recruitment and activation of caspase-8 by the Huntingtin-interacting protein Hip-1 and a novel partner Hippi. *Nat. Cell Biol.* **4**:95–105.
36. Hengartner, M.O. 2000. The biochemistry of apoptosis. *Nature.* **407**:770–776.
37. Matsukawa, J., Matsuzawa, A., Takeda, K., and Ichijo, H. 2004. The ASK1-MAP kinase cascades in mammalian stress response. *J. Biochem. (Tokyo).* **136**:261–265.
38. Nishitoh, H., et al. 2002. ASK1 is essential for endoplasmic reticulum stress-induced neuronal cell death triggered by expanded polyglutamine repeats. *Genes Dev.* **16**:1345–1355.
39. Lei, K., and Davis, R.J. 2003. JNK phosphorylation of Bim-related members of the Bcl2 family induces Bax-dependent apoptosis. *Proc. Natl. Acad. Sci. U. S. A.* **100**:2432–2437.
40. Putcha, G.V., et al. 2003. JNK-mediated BIM phosphorylation potentiates BAX-dependent apoptosis. *Neuron.* **38**:899–914.
41. Luciano, F., et al. 2003. Phosphorylation of Bim-EL by Erk1/2 on serine 69 promotes its degradation via the proteasome pathway and regulates its proapoptotic function. *Oncogene.* **22**:6785–6793.
42. Yamamoto, K., Ichijo, H., and Korsmeyer, S.J. 1999. BCL-2 is phosphorylated and inactivated by an ASK1/Jun N-terminal protein kinase pathway normally activated at G(2)/M. *Mol. Cell. Biol.* **19**:8469–8478.
43. Ito, Y., et al. 2001. Targeting of the c-Abl tyrosine kinase to mitochondria in endoplasmic reticulum stress-induced apoptosis. *Mol. Cell. Biol.* **21**:6233–6242.
44. Oyadomari, S., and Mori, M. 2004. Roles of CHOP/GADD153 in endoplasmic reticulum stress. *Cell Death Differ.* **11**:381–389.
45. Harding, H.P., et al. 2003. An integrated stress response regulates amino acid metabolism and resistance to oxidative stress. *Mol. Cell.* **11**:619–633.
46. Wang, X.Z., and Ron, D. 1996. Stress-induced phosphorylation and activation of the transcription factor CHOP (GADD153) by p38 MAP Kinase. *Science.* **272**:1347–1349.
47. Bruhat, A., et al. 2000. Amino acids control mammalian gene transcription: activating transcription factor 2 is essential for the amino acid responsiveness of the CHOP promoter. *Mol. Cell. Biol.* **20**:7192–7204.
48. Zinszner, H., et al. 1998. CHOP is implicated in programmed cell death in response to impaired function of the endoplasmic reticulum. *Genes Dev.* **12**:982–995.
49. McCullough, K.D., Martindale, J.L., Klotz, L.O., Aw, T.Y., and Holbrook, N.J. 2001. Gadd153 sensitizes cells to endoplasmic reticulum stress by down-regulating Bcl2 and perturbing the cellular redox state. *Mol. Cell. Biol.* **21**:1249–1259.
50. Oyadomari, S., et al. 2001. Nitric oxide-induced apoptosis in pancreatic beta cells is mediated by the endoplasmic reticulum stress pathway. *Proc. Natl. Acad. Sci. U. S. A.* **98**:10845–10850.
51. Ubeda, M., et al. 1996. Stress-induced binding of the transcriptional factor CHOP to a novel DNA control element. *Mol. Cell. Biol.* **16**:1479–1489.
52. Bourdon, J.C., Renzing, J., Robertson, P.L., Fernandes, K.N., and Lane, D.P. 2002. Scotin, a novel p53-inducible proapoptotic protein located in the ER and the nuclear membrane. *J. Cell Biol.* **158**:235–246.
53. Tsujimoto, Y., and Croce, C. 1986. Analysis of the structure, transcripts, and protein products of Bcl-2, the gene involved in human follicular lymphoma. *Proc. Natl. Acad. Sci. U. S. A.* **83**:5214–5218.
54. Thomenius, M.J., and Distelhorst, C.W. 2003. Bcl-2 on the endoplasmic reticulum: protecting the mitochondria from a distance [review]. *J. Cell Sci.* **116**:4493–4499.
55. Egle, A., Harris, A.W., Bouillet, P., and Cory, S. 2004. Bim is a suppressor of Myc-induced mouse B cell leukemia. *Proc. Natl. Acad. Sci. U. S. A.* **101**:6164–6169.
56. Mund, T., Gewies, A., Schoenfeld, N., Bauer, M.K., and Grimm, S. 2003. Spike, a novel BH3-only protein, regulates apoptosis at the endoplasmic reticulum. *FASEB J.* **17**:696–698.
57. Yang, T., Kozopas, K.M., and Craig, R.W. 1995. The

intracellular distribution and pattern of expression of Mcl-1 overlap with, but are not identical to, those of Bcl-2. *J. Cell Biol.* **128**:1173–1184.

58. Mathai, J.P., Germain, M., Marcellus, R.C., and Shore, G.C. 2002. Induction and endoplasmic reticulum location of BIK/NBK in response to apoptotic signaling by E1A and p53. *Oncogene.* **21**:2534–2544.

59. Chen, L., et al. 2005. Differential targeting of prosurvival Bcl-2 proteins by their BH3-only ligands allows complementary apoptotic function. *Mol. Cell.* **17**:393–403.

60. Reimertz, C., Kogel, D., Rami, A., Chittenden, T., and Prehn, J.H. 2003. Gene expression during ER stress-induced apoptosis in neurons: induction of the BH3-only protein Bbc3/PUMA and activation of the mitochondrial apoptosis pathway. *J. Cell Biol.* **162**:587–597.

61. Xu, Q., and Reed, J.C. 1998. BAX inhibitor-1, a mammalian apoptosis suppressor identified by functional screening in yeast. *Mol. Cell.* **1**:337–346.

62. Blais, J.D., et al. 2004. Activating transcription factor 4 is translationally regulated by hypoxic stress. *Mol. Cell. Biol.* **24**:7469–7482.

63. Chae, H.-J., et al. 2003. Evolutionarily conserved cytoprotection provided by Bax Inhibitor-1 homologs from animals, plants, and yeast. *Gene.* **323**:101–113.

64. Chae, H.-J., et al. 2004. BI-1 regulates an apoptosis pathway linked to endoplasmic reticulum stress. *Mol. Cell.* **15**:355–366.

65. Tenev, T., Zachariou, A., Wilson, R., Paul, A., and Meier, P. 2002. Jafrac2 is an IAP antagonist that promotes cell death by liberating Dronc from DIAP1. *EMBO J.* **21**:5118–5129.

66. Berridge, M.J., Lipp, P., and Bootman, M.D. 2000. The versatility and universality of calcium signaling [review]. *Nat. Rev. Mol. Cell Biol.* **1**:11–21.

67. Benkusky, N.A., Farrell, E.F., and Valdivia, H.H. 2004. Ryanodine receptor channelopathies. *Biochem. Biophys. Res. Commun.* **322**:1280–1285.

68. Bernardi, P. 1999. Mitochondrial transport of cations: channels, exchangers, and permeability transition [review]. *Physiol. Rev.* **79**:1127–1155.

69. Kroemer, G., and Reed, J.C. 2000. Mitochondrial control of cell death [review]. *Nat. Med.* **6**:513–519.

70. Chan, S.L., and Mattson, M.P. 1999. Caspase and calpain substrates: roles in synaptic plasticity and cell death [review]. *J. Neurosci. Res.* **58**:167–190.

71. Huang, Y., and Wang, K.K. 2001. The calpain family and human disease. *Trends Mol. Med.* **7**:355–362.

72. Wood, D.E., et al. 1998. Bax cleavage is mediated by calpain during drug-induced apoptosis. *Oncogene.* **17**:1069–1078.

73. Wood, D.E., and Newcomb, E.W. 1999. Caspase-dependent activation of calpain during drug-induced apoptosis. *J. Biol. Chem.* **274**:8309–8315.

74. Chen, M., et al. 2001. Bid is cleaved by calpain to an active fragment in vitro and during myocardial ischemia/reperfusion. *J. Biol. Chem.* **276**:30724–30728.

75. McMillin, J.B., and Dowhan, W. 2002. Cardiolipin and apoptosis. *Biochem. Biophys. Acta.* **1585**:97–107.

76. Lutter, M., et al. 2000. Cardiolipin provides specificity for targeting of tBid to mitochondria. *Nat. Cell Biol.* **2**:754–756.

77. Kuwana, T., et al. 2002. Bid, bax, and lipids cooperate to form supramolecular openings in the outer mitochondrial membrane. *Cell.* **111**:331–342.

78. Wang, H.G., et al. 1999. Ca2+-induced apoptosis through calcineurin dephosphorylation of BAD. *Science.* **284**:339–343.

79. Youn, H., Sun, L., Prywes, R., and Liu, J. 1999. Apoptosis of T cells mediated by Ca2+-induced release of the transcription factor MEF2. *Science.* **286**:790–793.

80. Shohat, G., Shani, G., Eisenstein, M., and Kimchi, A. 2002. The DAP-kinase family of proteins: study of a novel group of calcium-regulated death-promoting kinases [review]. *Biochim. Biophys. Acta.* **1600**:45–50.

81. Szabadkai, G., et al. 2004. Drp-1-dependent division of the mitochondrial network blocks intra-

organellar Ca2+ waves and protects against Ca2+-mediated apoptosis. *Mol. Cell.* **16**:59–68.

82. Liu, H., Peng, H.W., Cheng, Y.S., Yuan, H.S., and Yang-Yen, H.F. 2005. Stabilization and enhancement of the antiapoptotic activity of mcl-1 by TCTP. *Mol. Cell. Biol.* **25**:3117–3126.

83. Torgler, C.N., et al. 1997. Expression of bak in *S. pombe* results in a lethality mediated through interaction with the calnexin homologue Cnx1. *Cell Death Differ.* **4**:263–271.

84. Baffy, G., Miyashita, T., Williamson, J.R., and Reed, J.C. 1993. Apoptosis induced by withdrawal of interleukin-3 (IL-3) from an IL-3-dependent hematopoietic cell line is associated with repartitioning of intracellular calcium and is blocked by enforced Bcl-2 oncoprotein production. *J. Biol. Chem.* **268**:6511–6519.

85. He, H., Lam, M., McCormick, T.S., and Distelhorst, C.W. 1997. Maintenance of calcium homeostasis in the reticulum by Bcl-2. *J. Cell Biol.* **138**:1219–1228.

86. Pinton, P., et al. 2000. Reduced loading of intracellular Ca2+ stores and downregulation of capacitative Ca2+ influx in Bcl-2-overexpressing cells. *J. Cell Biol.* **148**:857–862.

87. Pinton, P., et al. 2001. The Ca^{2+} concentration of the endoplasmic reticulum is a key determinant of ceramide-induced apoptosis: significance for the molecular mechanism of Bcl-2 action. *EMBO J.* **20**:2690–2701.

88. Demaurex, N., and Distelhorst, C. 2003. Cell biology: apoptosis. The calcium connection. *Science.* **300**:65–67.

89. Scorrano, L., et al. 2003. BAX and BAK regulation of endoplasmic reticulum Ca2+: a control point for apoptosis. *Science.* **300**:135–139.

90. Wei, M.C., et al. 2001. Proapoptotic BAX and BAK: a requisite gateway to mitochondrial dysfunction and death. *Science.* **292**:727–730.

91. Nakamura, K., et al. 2000. Changes in endoplasmic reticulum luminal environment affect cell sensitivity to apoptosis. *J. Cell Biol.* **150**:731–740.

92. Palmer, A.E., Jin, C., Reed, J.C., and Tsien, R.Y. 2004. Bcl-2-mediated alterations in endoplasmic reticulum Ca2+ analyzed with an improved genetically encoded fluorescent sensor. *Proc. Natl. Acad. Sci. U. S. A.* **101**:17404–17409.

93. Schendel, S., Montal, M., and Reed, J.C. 1998. Bcl-2 family proteins as ion-channels. *Cell Death Differ.* **5**:372–380.

94. Chami, M., et al. 2004. Bcl-2 and Bax exert opposing effects on Ca2+ signaling, which do not depend on their putative pore-forming region. *J. Biol. Chem.* **279**:54581–54589.

95. Kuo, T.H., et al. 1998. Modulation of endoplasmic reticulum calcium pump by Bcl-2. *Oncogene.* **17**:1903–1910.

96. Oakes, S.A., et al. 2005. Proapoptotic BAX and BAK regulate the type 1 inositol trisphosphate receptor and calcium leak from the endoplasmic reticulum. *Proc. Natl. Acad. Sci. U. S. A.* **102**:105–110.

97. Boehning, D., et al. 2003. Cytochrome c binds to inositol (1,4,5) trisphosphate receptors, amplifying calcium-dependent apoptosis. *Nat. Cell Biol.* **5**:1051–1061.

98. Jayaraman, T., and Marks, A.R. 1997. T cells deficient in inositol 1,4,5-trisphosphate receptor are resistant to apoptosis. *Mol. Cell. Biol.* **17**:3005–3012.

99. Khan, A.A., et al. 1996. Lymphocyte apoptosis: mediation by increased type 3 inositol 1,4,5-trisphosphate receptor. *Science.* **273**:503–507.

100. Blackshaw, S., et al. 2000. Type 3 inositol 1,4,5-trisphosphate receptor modulates cell death. *FASEB J.* **14**:1375–1379.

101. Wang, B., et al. 2003. Uncleaved BAP31 in association with A4 protein at the endoplasmic reticulum is an inhibitor of Fas-initiated release of cytochrome c from mitochondria. *J. Biol. Chem.* **278**:14461–14468.

102. Katayama, T., et al. 1999. Presenilin-1 mutations

downregulate the signalling pathway of the unfolded-protein response. *Nat. Cell Biol.* **1**:479–485.

103. Terro, F., et al. 2002. Neurons overexpressing mutant presenilin-1 are more sensitive to apoptosis induced by endoplasmic reticulum-Golgi stress. *J. Neurosci. Res.* **69**:530–539.

104. Milhavet, O., et al. 2002. Involvement of Gadd153 in the pathogenic action of presenilin-1 mutations. *J. Neurochem.* **83**:673–681.

105. Niwa, M., Sidrauski, C., Kaufman, R., and Walter, P. 1999. A role for presenilin-1 in nuclear accumulation of ire1 fragments and induction of the mammalian unfolded protein response. *Cell.* **99**:691–702.

106. Dawson, T.M., and Dawson, V.L. 2003. Rare genetic mutations shed light on the pathogenesis of Parkinson disease. *J. Clin. Invest.* **111**:145–151. doi:10.1172/JCI200317575.

107. Takahashi, R., Imai, Y., Hattori, N., and Mizuno, Y. 2003. Parkin and endoplasmic reticulum stress. *Ann. N. Y. Acad. Sci.* **991**:101–106.

108. Kakizuka, A. 1998. Protein precipitation: a common etiology in neurodegenerative disorders? *Trends Genet.* **14**:396–402.

109. Kouroku, Y., et al. 2002. Polyglutamine aggregates stimulate ER stress signals and caspase-12 activation. *Hum. Mol. Genet.* **11**:1505–1515.

110. Bence, N.F., Sampat, R.M., and Kopito, R.R. 2001. Impairment of the ubiquitin-proteasome system by protein aggregation. *Science.* **292**:1552–1555.

111. Kumar, R., et al. 2001. Brain ischemia and reperfusion activates the eukaryotic initiation factor 2alpha kinase, PERK. *J. Neurochem.* **77**:1418–1421.

112. Paschen, W., Gissel, C., Linden, T., Althausen, S., and Doutheil, J. 1998. Activation of gadd153 expression through transient cerebral ischemia: evidence that ischemia causes endoplasmic reticulum dysfunction. *Brain Res. Mol. Brain Res.* **60**:115–122.

113. Tajiri, S., et al. 2004. Ischemia-induced neuronal cell death is mediated by the endoplasmic reticulum stress pathway involving CHOP. *Cell Death Differ.* **11**:403–415.

114. Kohno, K., Higuchi, T., Ohta, S., Kumon, Y., and Sakaki, S. 1997. Neuroprotective nitric oxide synthase inhibitor reduces intracellular calcium accumulation following transient global ischemia in the gerbil. *Neurosci. Lett.* **224**:17–20.

115. Iadecola, C., Zhang, F., Casey, R., Nagayama, M., and Ross, M.E. 1997. Delayed reduction of ischemic brain injury and neurological deficits in mice lacking the inducible nitric oxide synthase gene. *J. Neurosci.* **17**:9157–9164.

116. Yamaguchi, O., et al. 2003. Targeted deletion of apoptosis signal-regulating kinase 1 attenuates left ventricular remodeling. *Proc. Natl. Acad. Sci. U. S. A.* **100**:15883–15888.

117. Harding, H.P., et al. 2001. Diabetes mellitus and exocrine pancreatic dysfunction in perk-/- mice reveals a role for translational control in secretory cell survival. *Mol. Cell.* **7**:1153–1163.

118. Araki, E., Oyadomari, S., and Mori, M. 2003. Endoplasmic reticulum stress and diabetes mellitus. *Intern. Med.* **42**:7–14.

119. Oyadomari, S., et al. 2002. Targeted disruption of the *Chop* gene delays endoplasmic reticulum stress-mediated diabetes. *J. Clin. Invest.* **109**:525–532. doi:10.1172/JCI200214550.

120. Ozcan, U., et al. 2004. Endoplasmic reticulum stress links obesity, insulin action, and type 2 diabetes. *Science.* **306**:457–461.

121. Zhang, K., et al. 2005. The unfolded protein response sensor IRE1α is required at 2 distinct steps in B cell lymphopoiesis. *J. Clin. Invest.* **115**:268–281. doi:10.1172/JCI200521848.

122. Boyce, M., et al. 2005. A selective inhibitor of eIF2alpha dephosphorylation protects cells from ER stress. *Science.* **307**:935–939.

123. Yuan, J., and Yankner, B.A. 2000. Apoptosis in the nervous system. *Nature.* **407**:802–809.

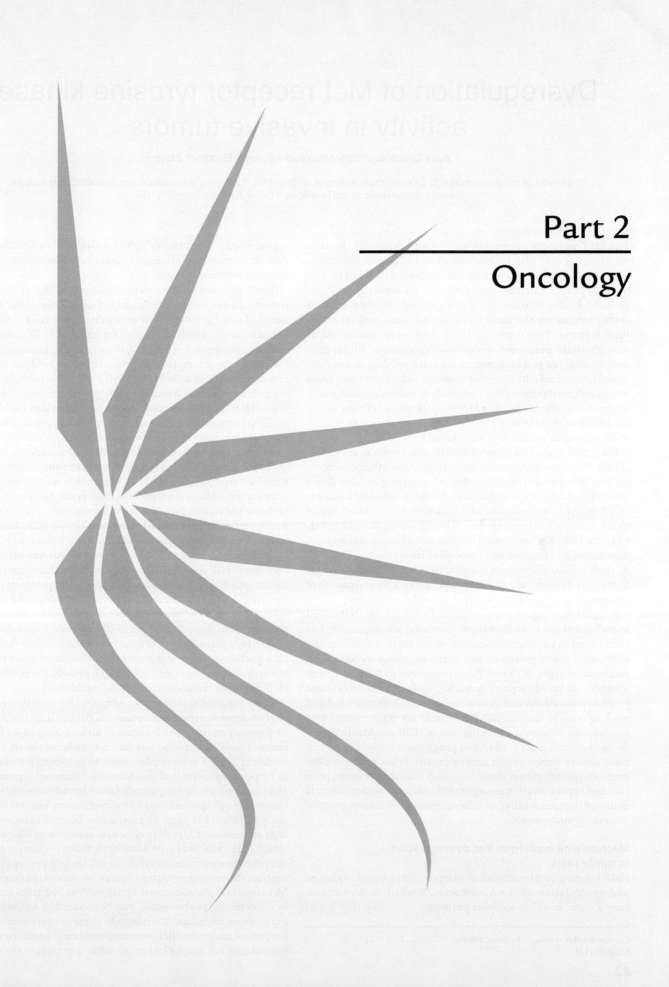

Part 2
Oncology

Dysregulation of Met receptor tyrosine kinase activity in invasive tumors

Alla Danilkovitch-Miagkova[1] and Berton Zbar[2]

[1]Department of Immunochemistry, PPD Development, Richmond, Virginia, USA. [2]Laboratory of Immunobiology, National Cancer Institute, Frederick Cancer Research and Development Center, Frederick, Maryland, USA.

The *MET* protooncogene was discovered because of the ability of oncogenic Met to mediate chemically induced transformation of a human osteogenic sarcoma cell line (1). The normal product of this gene, Met, is an unusual receptor tyrosine kinase that can be distinguished from most other such proteins on the basis of its biosynthesis and its structural features. This transmembrane protein is synthesized as a single-chain precursor, which undergoes intracellular proteolytic cleavage at a basic amino acid site, yielding a disulfide-linked heterodimer. Its C-terminal, intracellular region contains a multifunctional docking site that binds to various signaling molecules. These features define a Met receptor tyrosine kinase family consisting of three related proteins, Met, Ron, and c-Sea, the last of which may be the chicken ortholog of Ron.

The ligand of the Met receptor is HGF, also known as scatter factor (2). HGF was discovered simultaneously as a mitogenic factor for liver cells and as a fibroblast-derived scattering/motility factor for epithelial cells. It is a multifunctional factor affecting a number of cell targets including epithelium, endothelium, myoblasts, spinal motor neurons, and hematopoietic cells. Signaling pathways activated by the HGF-Met interaction mediate cell adhesion and motility. As is argued by Trusolino and Comoglio (this Perspective series, ref. 3), these cellular phenotypes, coupled to tightly regulated changes in cell growth, morphology, and survival, define a general pattern of invasive growth that occurs widely in normal development.

In addition to regulating normal cell functions, Met is involved in malignant cell transformation. Increased Met expression has been found in papillary carcinomas of the thyroid gland, in carcinomas of colon, pancreas, and ovary, in osteogenic sarcomas, and in other types of cancer. Point mutations in *MET* have been identified in hereditary and sporadic papillary renal carcinomas (4–6), hepatocellular and gastric carcinomas (7, 8), and head and neck squamous carcinomas (9). Numerous experimental and clinical data indicate a particular role of HGF and Met in tumor invasive growth, a stage of tumor progression leading to metastases. Dysregulation of Met activity in cells is thought to be a key event underlying tumor metastasis, and indeed, Met overexpression and hyperactivation are reported to correlate with metastatic ability of the tumor cells (see sidebar, *Invasive but indolent growth of Met-transformed tumors*).

Mechanisms leading to Met dysregulation in tumor cells

HGF binding to Met results in receptor autophosphorylation and upregulation of Met kinase activity, which in turn stimulates a number of intracellular pathways mediating HGF's bio-logical effects. In normal cells, Met activation is a ligand-dependent transient event, whereas in tumor cells Met activity is often constitutively upregulated.

Ligand-dependent mechanisms of Met activation. Met activation in tumor cells can occur through any of several molecular mechanisms (Table 1), the simplest of which involve HGF-dependent Met activation, much as occurs in normal cells. In some cases, tumor cells express both HGF and its receptor, setting the stage for an autocrine loop in which secreted HGF binds to Met and causes constitutive activation of Met and its downstream signaling pathways, thus enhancing tumor growth and invasive behavior. Such HGF-Met autocrine loops have been detected in gliomas, osteosarcomas, and mammary, prostate, breast, lung, and other carcinomas; they are often associated with malignant progression of tumors and correlate with poor prognosis. Interference with either HGF or Met expression can inhibit tumorigenic transformation, angiogenesis, tumor growth, and invasion (10).

Under physiological conditions HGF is not an autocrine, but rather a paracrine, factor: Mesenchymal cells produce HGF, which acts on epithelial and other cells that express Met. Similarly, Met-positive tumor cells that do not produce HGF may nevertheless respond to HGF produced by stromal cells. However, since HGF is secreted by cells as a single-chain inactive precursor (pro-HGF), which must be activated by proteolytic cleavage, HGF-Met autocrine and paracrine loops depend on a third component — an enzyme capable of processing pro-HGF to produce HGF. A number of serine-like proteases, including urokinase-type plasminogen activator and coagulation factor XII, have such an activity and have been detected in some tumors. Nevertheless, the mechanism by which pro-HGF is converted to HGF in tumor tissues remains to be established.

Ligand-independent mechanisms. Met can also be activated in an HGF-independent manner in tumors, particularly as a result of Met overexpression, which occurs in almost every case of differentiated papillary carcinomas. Increased Met expression can be mediated by *MET* gene amplification, by enhanced transcription, or by posttranscriptional mechanisms. Increased expression of Met on the cell surface apparently favors ligand-independent activation through spontaneous Met dimerization, but it is not generally sufficient to trigger Met activation. In some cases, even very high expression of Met does not cause constitutive receptor activation (11). Noncovalently associated, inactive clusters of these receptors have been identified on the cell surface, perhaps explaining the cells' resistance to transformation, even in the face of high Met levels (12). An additional signal, such as Met transactivation by other membrane receptors, may be required to activate signaling by these receptors. Alternatively, these clusters may contain suppressor molecules that prevent spontaneous Met activation in normal cells but may be lost or inactivated in tumor cells.

Citation for this article: *J. Clin. Invest.* **109**:863–867 (2002). DOI:10.1172/JCI200215418.

Table 1

Mechanisms of Met dysregulation in tumor cells

	With wild-type *MET*	With *MET* gene alterations
HGF-dependent		
	Autocrine loop	
	Paracrine loop	
HGF-independent		
	Overexpression	Overexpression (due to amplification)
	Abnormal Met processing	Gene rearrangement (*TPR-MET*)
	Defects of negative regulators	Mutations
	Truncation (cytoplasmic Met)	Truncation (cytoplasmic Met)

One well-known oncogenic form of Met, first identified in the chemically transformed human osteosarcoma cell line HOS (1), is the product of the *TPR-MET* fusion, which arises through a chromosomal rearrangement. The resulting chimeric gene contains the promoter and the N-terminal sequence of the *TPR* gene from chromosome 1, fused with the C-terminal sequence of *MET*, which maps to chromosome 7. The *TPR-MET* chimeric gene encodes a cytoplasmic protein with molecular weight 65 kDa comprising the TPR leucine zipper domain and the Met kinase domain. This protein is constitutively active as a result of TPR leucine zipper interactions, which allow for Met kinase dimerization, transphosphorylation, and activation (13), and it is potently oncogenic in vitro and in vivo.

Abnormal processing or the absence of normal negative regulators can also lead to constitutive Met activation and tumorigenesis. The mature Met consists of two subunits, α and β, arising from proteolytic cleavage of the single-chain precursor. As a result of defective posttranslational processing, the precursor

fails to be cleaved in the colon carcinoma cell line LoVo; consequently, Met is expressed on the cell surface as a single-chain molecule, which is constitutively tyrosine-phosphorylated (14). In metastatic B16 melanoma cells, on the other hand, cytosolic phosphatases that normally mediate Met dephosphorylation, internalization, and degradation are downregulated, leading to constitutive Met activation (15).

Finally, a large class of somatic or inherited mutations in the *MET* gene can lead to active, typically ligand-independent, Met signaling in tumor cells. For instance, a mutant in which the Met cytoplasmic domain is truncated immediately below the transmembrane domain encodes a constitutively active signaling domain that can transform rodent fibroblasts (16). A similar, naturally occurring truncation has been detected in malignant human musculoskeletal tumors. This short 85-kDa N-terminally truncated Met is tyrosine-phosphorylated and located in the cytoplasm (17). The mechanism by which this truncated Met is produced is not known, but its constitutive activation suggests a role in tumorigenesis.

Missense point mutations in *MET* have been identified in hereditary and sporadic papillary renal carcinomas (4-6), childhood hepatocellular carcinomas (7), gastric carcinomas (9), and head and neck squamous cell carcinomas (8, 9). At present, 21 such mutations have been described, as summarized in Table 2. All identified Met mutations in the kinase domain increase Met tyrosine kinase activity. Although mutations in the juxtamembrane domain do not trigger ligand-independent Met activation, receptors carrying the P1009I mutation show persistent Met activation in response to HGF (7). This mutated form of Met demonstrates transforming potential and invasive activity in vitro and in vivo.

Invasive but indolent growth of Met-transformed tumors

In clinical oncology, the term "invasive growth" refers to the ability of some cancer cells to leave their site of origin, to traverse anatomic barriers to cell movement — the basement membrane in the case of epithelial cancers — and to move to locations distant from the site of origin. Invasive growth is further classified as local or systemic, referring to the ability of some cancer cells to grow in locations adjacent to the site of origin or in locations distant from the site of origin. For most clinical classifications of stage of cancer, the presence of local or systemic tumor deposits (metastases) makes a major difference in the cancer stage.

Given the well-known and profound effects of HGF-Met signaling on cells' morphology and invasive character, it is interesting to consider the biological characteristics of cells in patients with inherited mutations in the Met tyrosine kinase domain. Such germline mutations are clearly compatible with normal growth and development. Cancer in patients who carry such kinase-activated *MET* alleles typically occurs late in life and is confined to a single organ, the kidney. In genetic terms, the penetrance (the proportion of individuals with germline mutations in the *MET* gene who exhibit detectable renal tumors) is 40% by age 50. These

renal cancers are indolent, slow-growing neoplasms. Even in the condition known as hereditary papillary renal carcinoma (one form of which is associated with *MET* mutations), the multiple, independently developing tumors display the clinical properties of invasive growth only late in the course of the disease. For this reason, urologic surgeons managing patients with this disorder generally wait until the largest renal tumor is 3 cm in diameter before operating.

These papillary renal cancers display a number of consistent somatic genetic changes, including duplication of the chromosome bearing the mutated *MET* gene and trisomy of chromosome 17. Hence, the development of papillary renal cancer likely requires a number of additional genetic changes along with the inherited *MET* mutation. Conversely, other human cancers, found in individuals with neither somatic nor germline mutations in *MET*, often show increased expression of Met protein. It is currently thought that this increased expression of Met contributes to invasive growth, but it must be acknowledged that these cells carry multiple genetic and epigenetic changes, making it difficult to isolate the contribution of Met to tumor progression and invasion.

Table 2

MET mutations identified in human tumors

Mutation	Type of tumor	Homologous mutations in other RTKs	Ligand-independent Met activation	References
In the juxtamembrane domain:				
P1009S (g)	Primary gastric carcinoma	Unknown	No, but persistent in response to HGF	8
T1010I (g,polymorphism)	breast cancer biopsy; HOP92 cell line; RCC	Unknown	No	5, 8
In the kinase domain:				
ATP-binding region				
V1110I (g)	RCC	c-erbB V157I	Yes (weak)	5, 6
H1112L (s)	RCC	Unknown	Yes (weak)	5
H1112Y (g, s)	RCC	Unknown	Yes	5
H1112R (g)	RCC	Unknown	Yes (weak)	5
H1124D (s?)	RCC	Unknown	Yes (weak)	5
Hinge region				
M1149T (g)	RCC	Unknown	Yes (weak)	4, 5, 8, 29, 30
T1191I (s)	Childhood HCC	Unknown	ND	7
V1206L (g)	RCC	Unknown	Yes (weak)	4, 29, 30
L1213V (s)	RCC	Unknown	Yes (weak)	4, 29, 30, 35
V1238I (g)	RCC	Unknown	Yes	4, 29, 30
Activation loop				
D1246H (s)	RCC	c-kit D816V	Yes	4, 29, 30
D1246N (g)	RCC	c-kit D816V	Yes	4, 30
Y1248H (s)	RCC	c-kit D820G	Yes	4, 5, 29, 30
Y1248D (g)	RCC	c-kit D820G	Yes	4, 5, 29, 30, 35
Y1248C (g)	RCC, HNSCC	c-kit D820G	Yes	4, 9, 29, 30
K1262R (s)	Childhood HCC	Unknown	ND	7
P+1 loop				
Y1253D (s)	HNSCC	Unknown	Yes	9
M1268T (s)	RCC	Ret M918T	Yes (strong)	4, 5, 29–31, 35
M1268I (s)	Childhood HCC	Ret M918T	ND	7

g, germline mutation; s, somatic mutation; ND, not determined; HCC, hepatocellular carcinoma; HNSCC, head and neck squamous cell carcinoma; RCC, renal cell carcinoma; RTK, receptor tyrosine kinase.

Met transactivation via other membrane receptors. Recent investigations have shown that Met kinase activity can be regulated through other receptors by HGF-independent mechanisms. Thus, Met can be activated by stimuli that do not directly interact with Met, including adhesive receptors, such as various integrins and CD44, and signal transducing receptors like Ron and the EGF receptor.

Integrins (discussed by Brakebusch et al., this Perspective series, ref. 18) are cell surface receptors that mediate cell adhesion to the ECM. Plating of Met-expressing cells on ECM, and the consequent ligation of cell surface integrins, can cause ligand-independent Met tyrosine phosphorylation (19). Interestingly, transgenic mice expressing Met in hepatocytes have activated Met and develop hepatocellular carcinoma (20), despite the absence of detectable HGF expression, perhaps as a result of cellular adhesion in this tissue (20).

CD44, a cell surface receptor for hyaluronic acid (a major glycosaminoglycan component of the ECM), regulates a number of normal cell functions and has been implicated in tumor progression and metastasis. This receptor can promote Met activation by two mechanisms. First, binding of CD44 to hyaluronic acid causes HGF-independent Met activation, leading to cell growth and migration (21). Second, a heparan sulfate proteoglycan iso-

form of CD44 binds HGF and presents it to Met in the form of a multivalent complex inducing a high level of Met activation in comparison with soluble nonbound HGF (22).

Because Ron belongs to the same family of receptor tyrosine kinases as Met and shares many common structural features, it is perhaps not surprising that activated Ron can transphosphorylate Met, and vice versa, as was recently shown (23). Pre-existing, ligand-independent heterodimers between Met and Ron have been detected on the cell surface (23), indicating that these receptors are colocalized and may be able to transphosphorylate and to activate one another. In addition, some human hepatoma cell lines — but not normal hepatocytes — are activated by a TGF-α–EGF receptor autocrine loop, which leads to constitutive, ligand-independent tyrosine phosphorylation of Met (24).

Consequences of Met dysregulation by various mechanisms

Although increased Met kinase activity represents a common feature of many tumors, the specific consequences of Met dysregulation are not uniform but reflect the molecular mechanism involved.

Cellular localization and the oncogenic potential of mutant Met. The effect of Met intracellular localization on its transforming ability

Table 3

Effects of oncogenic Met localized to the cytoplasm or the plasma membrane Met

	Cytoplasmic Met	Transmembrane Met
Met activation:		
by HGF	no	yes
through other receptors	no	yes
Activation of signaling molecules by Met:		
Membrane-associated molecules	no	yes
Cytoplasmic molecules	Yes, possibly by direct activation	Yes, but indirect
Transgenic expression in mice:		
in liver	Nontumorigenic	Tumorigenic, HCC
in epithelia	Tumorigenic, mammary tumors	ND

has been investigated using the chimeric oncoprotein TPR-Met, which is found in soluble form in the cytoplasm but can be targeted to the cellular membrane by the addition of an Src myristoylation signal. Membrane-localized TPR-Met, but not cytoplasmic TPR-Met, stimulates the phosphatidylinositol 3-kinase–dependent (PI 3-kinase–dependent) induction of hyaluronic acid and its receptor and enhances cellular transformation (25).

Compared with two other, membrane-associated forms of oncogenic Met, the soluble TPR-Met chimera is much more dependent on its ability to interact with the binding site of the adaptor protein Grb2 for its biological activity (26). Transgenic expression of TPR-Met in the mouse liver causes hepatocyte immortalization and protects cells from apoptosis, but it does not result in tumor formation (27), whereas overexpression of the full-sized transmembrane form of Met induces hepatocellular carcinoma in transgenic mouse liver (20). Conversely, cytoplasmic TPR-Met is indeed oncogenic in mammary epithelial cells, since its transgenic expression induces mammary tumors in mice (28). Distinctions between transmembrane and cytoplasmic Met are described in Table 3.

The catalytic activity and substrate specificity of the Met M1268T mutant. Point mutations in the kinase domain convert Met to an oncogenic receptor. Such mutants are highly active catalytically, which correlates with more efficient Met autophosphorylation, phosphorylation of substrates, and transforming ability (29, 30). For example, the constitutive binding of c-Src to the cytoplasmic domain of the Met M1268T mutant (identified in human renal papillary carcinomas) elevates c-Src phosphorylation and activity — an effect that is considerably more dramatic and longer-lasting than the transient activation of c-Src by HGF in normal cells. Dominant negative c-Src constructs effectively inhibit the oncogenic effect of Met M1268T, indicating that c-Src is required for transformation by this pathway (31). In NIH 3T3 cells, M1268 Met also activates the β-catenin pathway. Expression of Met M1268T mutant induces β-catenin tyrosine phosphorylation and accumulation; induces constitutive activation of the transcription factor Tcf, which acts in concert with β-catenin in the nucleus; and increases expression of the β-catenin/Tcf target genes *Myc* and *Cyclin D1* (see Conacci-Sorrell et al., this Perspective series, ref. 32). Activation of the β-catenin pathway correlates with Met M1268T mutant–mediated cell transformation (33).

Although the hyperactive catalytic function of mutated Met is clearly associated with cellular transformation, a switch in substrate specificity may play a primary role in oncogenesis mediated by mutant forms of Met. In particular, Met M1268T phosphorylates substrates of the cytosolic kinase c-Abl, whereas

wild-type Met does not (30), suggesting that mutated Met can activate signaling pathways distinct from those induced by wild-type Met. In addition to changes of substrate specificity, a set of phosphorylated tyrosine residues in wild-type and mutated Met molecules might be overlapping but distinct. These unique tyrosine phosphorylation sites may appear as a result of unusual conformations adopted by mutant Met kinase and may be phosphorylated by Met itself or by other kinases.

Catalytic and structural effects of Met missense mutations. Analyses of three-dimensional structures of Met wild-type and mutated Met catalytic core domains show that mutations can activate Met by multiple mechanisms (34). Mutations such as V1110I, Y1248H/D/C, and M1268T directly affect contacts between residues in the protein's activation loop; a region of the protein must undergo a regulated conformational change to permit the activation of the Met kinase. Mutations M1149T and L1213V may increase flexibility at the critical points of the Met tertiary structure leading to subdomain movements, whereas the D1246N mutation can stabilize the active form of the Met kinase (34). The M1268T mutation, which permits the efficient phosphorylation of c-Abl substrates (30), affects substrate binding sites within the Met kinase domain (30). These various structural changes are associated with functional diversity among oncogenic Met mutants. Thus, the D1246H/N and M1268T mutants have a high transforming ability, which correlates with activation of the Ras pathway, whereas the Y1248C and L1213V mutants are weakly transforming but promote cell migration, invasion, and resistance to apoptosis through activation of the PI 3-kinase/AKT pathway (35).

Some open questions

Data accumulating over the past few years represent significant progress in our understanding of the role of Met in oncogenesis, but a number of important questions remain. First, elucidating the mechanisms of Met dysregulation in various tumor types remains a high priority for both basic and clinical researchers. Met dysregulation may be a primary event in transformation, as a result of mutation, rearrangement, or amplification of the *MET* gene. In other cases, dysregulation of Met may be secondary to effects on other molecules. The study of signaling pathways by which Met expression and activity are regulated in normal cells and dysregulated in tumors suggests a number of promising therapeutic targets for anticancer drug development.

Another important direction for future investigations will involve solving the three-dimensional structure of Met, which will permit a clearer understanding of the structural and biological effects of the various Met kinase domain point mutations — particularly the pattern of tyrosine phosphorylation sites in Met and consequent switching of Met substrate specificity. Based on such analysis, it may be possible to develop agents that can selectively inhibit activity of the mutated Met kinase without interfering with the HGF-stimulated activity of normal Met.

Finally, it will be essential to deepen our understanding of receptor cross-talk and its contribution to Met activation and the propagation of Met-dependent oncogenesis. The dysregulated cell motility and the resulting tumor metastasis that follow from inappropriate Met activation involve collaborations with many other receptors and multiple signaling pathways. While the details remain obscure, it is clear that Met can function as part of other receptor complexes and can respond to stimuli that do not impinge on it directly. Conversely, dysregulated Met may promote activation of these other receptors. Both kinds of interactions undoubtedly have profound consequences for the invasive growth of tumors and, indeed, of healthy cells and tissues.

Acknowledgments

The authors would like to thank all researchers whose scientific contributions in the field helped to write this review. We apologize for failing to refer to many primary sources due to space limitations.

Address correspondence to: Alla Danilkovitch-Miagkova, PPD Development, 2244 Dabney Road, Richmond, Virginia 23230, USA. Phone: (804) 359-1900; Fax: (804) 253-1112; E-mail: Alla.Danilkovitch@richmond.ppdi.com.

1. Cooper, C.S., et al. 1984. Molecular cloning of a new transforming gene from a chemically transformed human cell line. *Nature.* **311**:29–33.
2. Bottaro, D.P., et al. 1991. Identification of the hepatocyte growth factor receptor as the c-met proto-oncogene product. *Science.* **251**:802–804.
3. Comoglio, P.M., and Trusolino, L., 2002. Invasive growth: from development to metastasis. *J. Clin. Invest.* **109**:857–862. DOI:10.1172/JCI200215392.
4. Schmidt, L., et al. 1997. Germline and somatic mutations in the tyrosine kinase domain of the *MET* proto-oncogene in papillary renal carcinomas. *Nat. Genet.* **16**:68–73.
5. Schmidt, L., et al. 1999. Novel mutations of the *MET* proto-oncogene in papillary renal carcinomas. *Oncogene.* **18**:2343–2350.
6. Olivero, M., et al. 1999. Novel mutation in the ATP-binding site of the *MET* oncogene tyrosine kinase in a HPRCC family. *Int. J. Cancer.* **82**:640–643.
7. Park, W.S., et al. 1999. Somatic mutations in the kinase domain of the Met/hepatocyte growth factor receptor gene in childhood hepatocellular carcinomas. *Cancer Res.* **59**:307–310.
8. Lee, J.H., et al. 2000. A novel germ line juxtamembrane Met mutation in human gastric cancer. *Oncogene.* **19**:4947–4953.
9. Di Renzo, M.F., et al. 2000. Somatic mutations of the MET oncogene are selected during metastatic spread of human HNSC carcinomas. *Oncogene.* **19**:1547–1555.
10. Abounader, R., et al. 1999. Reversion of human glioblastoma malignancy by U1 small nuclear RNA/ribozyme targeting of scatter factor/hepatocyte growth factor and c-met expression. *J. Natl. Cancer Inst.* **91**:1548–1556.
11. Ferracini, R., et al. 1995. The Met/HGF receptor is over-expressed in human osteosarcomas and is activated by either a paracrine or an autocrine circuit. *Oncogene.* **10**:739–749.
12. Faletto, D.L., et al. 1992. Evidence for non-covalent clusters of the c-met proto-oncogene product. *Oncogene.* **7**:1149–1157.
13. Rodrigues, G.A., and Park, M. 1993. Dimerization mediated through a leucine zipper activates the oncogenic potential of the met receptor tyrosine kinase. *Mol. Cell. Biol.* **13**:6711–6722.
14. Mondino, A., Giordano, S., and Comoglio, P.M. 1991. Defective posttranslational processing activates the tyrosine kinase encoded by the MET proto-oncogene (hepatocyte growth factor receptor). *Mol. Cell. Biol.* **11**:6084–6092.
15. Rusciano, D., Lorenzoni, P., and Burger, M.M. 1996. Constitutive activation of c-Met in liver metastatic B16 melanoma cells depends on both substrate adhesion and cell5 density and is regulated by a cytosolic tyrosine phosphatase activity. *J. Biol. Chem.* **271**:20763–20769.
16. Zhen, Z., et al. 1994. Structural and functional domains critical for constitutive activation of the HGF-receptor (Met). *Oncogene.* **9**:1691–1697.
17. Wallenius, V., et al. 2000. Overexpression of the hepatocyte growth factor (HGF) receptor (Met) and presence of a truncated and activated intracellular HGF receptor fragment in locally aggressive/malignant human musculoskeletal tumors. *Am. J. Pathol.* **156**:821–829.
18. Brakebusch, C., Bouvard, D., Stanchi, F., Sakai, T., and Fässler, R. 2002. Integrins in invasive growth. *J. Clin. Invest.* In press.
19. Wang, R., Kobayashi, R., and Bishop, J.M. 1996. Cellular adherence elicits ligand-independent activation of the Met cell-surface receptor. *Proc. Natl. Acad. Sci. USA.* **93**:8425–8430.
20. Wang, R., Ferrell, L.D., Faouzi, S., Maher, J.J., and Bishop, J.M. 2001. Activation of the Met receptor by cell attachment induces and sustains hepatocellular carcinomas in transgenic mice. *J. Cell Biol.* **153**:1023–1034.
21. Taher, T.E., et al. 1999. Cross-talk between CD44 and c-Met in B cells. *Curr. Top. Microbiol. Immunol.* **246**:31–37.
22. Van der Voort, R., et al. 1999. Heparan sulfate-modified CD44 promotes hepatocyte growth factor/scatter factor-induced signal transduction through the receptor tyrosine kinase c-Met. *J. Biol. Chem.* **274**:6499–6506.
23. Follenzi, A., et al. 2000. Cross-talk between the proto-oncogenes Met and Ron. *Oncogene.* **19**:3041–3049.
24. Jo, M., et al. 2000. Cross-talk between epidermal growth factor receptor and c-Met signal pathways in transformed cells. *J. Biol. Chem.* **275**:8806–8811.
25. Kamikura, D.M., Khoury, H., Maroun, C., Naujokas, M.A., and Park, M. 2000. Enhanced transformation by a plasma membrane-associated met oncoprotein: activation of a phosphoinositide 3′-kinase-dependent autocrine loop involving hyaluronic acid and CD44. *Mol. Cell. Biol.* **20**:3482–3496.
26. Jeffers, M., Koochekpour, S., Fiscella, M., Sathyanarayana, B.K., and Vande, W.G. 1998. Signaling requirements for oncogenic forms of the Met tyrosine kinase receptor. *Oncogene.* **17**:2691–2700.
27. Amicone, L., et al. 1997. Transgenic expression in the liver of truncated Met blocks apoptosis and permits immortalization of hepatocytes. *EMBO J.* **16**:495–503.
28. Liang, T.J., Reid, A.E., Xavier, R., Cardiff, R.D., and Wang, T.C. 1996. Transgenic expression of tpr-met oncogene leads to development of mammary hyperplasia and tumors. *J. Clin. Invest.* **97**:2872–2877.
29. Jeffers, M., et al. 1997. Activating mutations for the met tyrosine kinase receptor in human cancer. *Proc. Natl. Acad. Sci. USA.* **94**:11445–11450.
30. Bardelli, A., et al. 1998. Uncoupling signal transducers from oncogenic MET mutants abrogates cell transformation and inhibits invasive growth. *Proc. Natl. Acad. Sci. USA.* **95**:14379–14383.
31. Nakaigawa, N., Weirich, G., Schmidt, L., and Zbar, B. 2000. Tumorigenesis mediated by MET mutant M1268T is inhibited by dominant-negative Src. *Oncogene.* **19**:2996–3002.
32. Conacci-Sorrell, M., Zhurinsky, J., Ben-Ze'ev, A. 2002. The cadherin-catenin adhesion system in signaling and cancer. *J. Clin. Invest.* In press.
33. Danilkovitch-Miagkova, A., et al. 2001. Oncogenic mutants of RON and MET receptor tyrosine kinases cause activation of the β-catenin pathway. *Mol. Cell. Biol.* **21**:5857–5868.
34. Miller, M., et al. 2001. Structural basis of oncogenic activation caused by point mutations in the kinase domain of the MET proto-oncogene: modeling studies. *Proteins.* **44**:32–43.
35. Giordano, S., et al. 2000. Different point mutations in the met oncogene elicit distinct biological properties. *FASEB J.* **14**:401–408.

Integrins in invasive growth

Cord Brakebusch,[1,2] Daniel Bouvard,[1] Fabio Stanchi,[1] Takao Sakai,[1]
and Reinhard Fässler[1,2]

[1]Lund University Hospital, Department of Experimental Pathology, Lund, Sweden. [2]Max Planck Institute for Biochemistry, Martinsried, Germany.

Interactions between tumor cells and the ECM strongly influence tumor development, affecting cell proliferation and survival, as well as the ability to migrate beyond the original location into other tissues to form metastases. Many of these interactions are mediated by integrins, a ubiquitously expressed family of adhesion receptors. Integrins are essential for cell attachment and control cell migration, cell cycle progression, and programmed cell death, responses that they regulate in synergy with other signal transduction pathways.

This large group of transmembrane proteins is formed from 18 α and 8 β subunits, which dimerize to yield at least 24 different integrin heterodimers, each with distinct ligand binding and signaling properties. With their extracellular domain, integrins can bind to different ECM molecules, such as collagens and laminins, or to cellular receptors, such as VCAM-1. Their intracellular domains connect directly or indirectly to the actin cytoskeleton, thus linking the cytoskeleton to the ECM. Integrins also serve as bidirectional signaling receptors, inducing changes in protein activities or gene expression in response to ligand binding, while also modulating adhesive affinity on the cell surface in response to changes in cellular physiology. Here, we describe how integrins affect migration, proliferation, and survival of both transformed and normal cells, and we discuss how they modulate invasive growth in vivo. Throughout, we stress that many of these functions are restricted to particular cell types and may be altered upon transformation.

Integrin avidity and affinity changes in cell migration

Cell migration is essential not only for tissue infiltration and the formation of metastases, but also for nonpathological processes, such as angiogenesis and leukocyte extravasation. In order to migrate, a cell has to pass through a sequence of distinct processes. Migration is initiated by cell polarization and the formation of membrane protrusions at the leading edge. Integrins fix cellular protrusions to the ECM, interact with the actin cytoskeleton, and trigger the association of many different signaling molecules at the so-called focal contacts. Thereafter, integrin signals stimulate cell contraction, which allows the movement of the cell body on the adhesive contacts. Finally, the rear of the cell detaches from the substratum by inactivation of the integrins and disassembly of the adhesion complexes. Integrins further facilitate cell movement through the tissue by activating ECM-degrading enzymes. Although these basic mechanisms are considered to be similar among the various types of migratory cells, there are also clear distinctions, since, for example, fibroblasts are about 3–20 times more adhesive than nonactivated leukocytes, move 10–60 times more slowly, and have a different cytoskeletal organization. These differences in the migration mechanism might explain at least some of the known cell-specific effects of integrins on cell migration.

Integrin-mediated cell attachment depends not just on the expression of these receptors in a given cell type, but also on their affinity for various ligands and on their lateral mobility within the plasma membrane, which allows the formation of high-avidity clusters. Integrins adopt low- and high-affinity conformations, which can be distinguished by their binding to soluble ligands or by conformation-specific antibodies. High-affinity binding, which is important for the firm attachment of the leading edge to the ECM, can result from the binding of regulatory intracellular molecules to the cytoplasmic domains of the α and β subunits. Overexpression in osteosarcoma cells or normal human fibroblasts of chimeric proteins consisting of the β1, β3, or β5 intracellular domains, fused to the extracellular and transmembrane part of the IL-2 receptor, significantly reduces the affinity of the endogenous β1 integrins (1). This effect is most likely caused by sequestration of intracellular molecules that would otherwise bind the cytoplasmic domain of the integrin β subunits and alter the integrins' extracellular conformation. Such an effect, called transdominant inhibition, has also been reported for α2 integrin cytoplasmic domains (2). More than 20 proteins are currently known to bind intracellular domains of integrin subunits (3), but it is not known which of them are involved in the modulation of integrin affinity. Potential candidates for such regulation are ICAP-1 and TAP-20, which decrease, and β3-endonexin, which increases integrin-mediated adhesion in cells overexpressing these molecules. Intracellular signaling mechanisms, for example phosphorylation of ICAP-1 by CaMKII, are suggested to modulate the binding of these molecules to integrin and thereby the affinity state.

Cell attachment can be induced not only by affinity regulation, but also by clustering of integrins, which leads to increased adhesive avidity. Such clusters are present in focal adhesions, readily detectable cell-matrix contacts that have been extensively studied in cultured fibroblasts. Increased lateral mobility of integrins in the membrane might also be important for efficient ligand binding. In leukocytes, even prior to ligand binding, integrins seem to be associated with the actin network, which constrains integrin mobility and ligand binding. Releasing this contact by proteolytic digestion of the integrin connections with the cytoskeleton results in a rapid increase in lateral mobility of integrins, which increases the chance of ligand encounters and facilitates integrin aggregation into high-avidity clusters (4).

Signaling pathways regulating integrin affinity

Growth factor and chemokine signaling can also modulate the affinity and avidity of integrins. In particular, upregulation of integrin activity by growth factors often depends on phosphatidylinositol 3-kinase (PI3-K) activation. Thus, in mast cells, activation of PI3-K by FcεRI, c-kit, or PDGF-R increases the affinity of α5β1 integrin. In metastatic breast cancer cells, increased cell adhesion and migration upon stimulation of EGF-R or erbB3 are

Citation for this article: *J. Clin. Invest.* **109**:999–1006 (2002). DOI:10.1172/JCI200215468.

also dependent on PI3-K. Integrin avidity is upregulated in carcinoma cells by treatment with HGF (see Danilkovitch-Miagkova and Zbar, this Perspective series, ref. 5), but not with EGF. This increased integrin avidity is dependent on PI3-K and promotes invasive growth of these cells, suggesting an important role of integrin avidity regulation in metastasis (6).

Signaling initiated by chemokine receptors can induce a rapid and transient upregulation of integrin affinity, which is important for the tight adhesion of leukocytes to the endothelium during their extravasation into inflamed tissues. At least in lymphocytes, this activation is independent of PI3-K and involves activation of RhoA (4). Remarkably, chemokines can independently regulate the affinity and avidity of different integrins within the same cell. In eosinophils, the chemokines RANTES and monocyte chemoattractant protein–3 induce a transient upregulation of $\alpha4\beta1$ avidity and a long-lasting affinity increase of $\alpha M\beta2$ (Mac-1). Chemokines such as secondary lymphoid tissue cytokine, EBI1-ligand chemokine, and stroma cell-derived factor 1α stimulate integrin motility in lymphocytes through the coordinated action of cytosolic proteases and PI3-K. In contrast, PMA-stimulated integrin mobility is dependent on proteases, but independent of PI3-K (4).

The Ras and Rho GTPases can also regulate integrin affinity, although their effects are cell type– and integrin-specific. In Chinese hamster ovary (CHO) cells, for example, H-Ras inhibits activation of $\beta1$ and $\beta3$ integrins, while in a pro-B cell line it increases $\alpha L\beta2$ (LFA-1) activity in a PI3-K–dependent manner (7). R-Ras promotes integrin activation in myeloid cells but has no effect in a lymphoid cell line (7). The integrin-specificity of Ras effects is also nicely demonstrated in T47D cells, where R-Ras promotes migration mediated by $\alpha2\beta1$ but not by $\alpha5\beta1$ (2). Ras signaling can affect many signaling cascades, including the activation of Erk and PI3-K. The integrin activating functions of both H-Ras and R-Ras require PI3-K activation.

Recently, Rap1 was shown to activate integrins on lymphoid and endothelial cells by inducing a high-affinity conformation (7, 8). Overexpression of constitutively active Rac1 in lymphoid cells increases integrin affinity (7). In mammary epithelial cells, however, neither Rac1 nor Cdc42 induces obvious changes in integrin avidity or affinity, although they promote integrin-mediated motility and invasiveness through PI3-K.

Several reports have demonstrated cross-talk between integrins, in which signaling by one integrin influences the affinity or avidity of others. For example, stimulation of $\alpha IIb\beta3$ signaling has been shown to downregulate $\alpha2\beta1$- and $\alpha5\beta1$-mediated adhesion.

A few reports have also shown that direct interaction of secreted molecules with the extracellular domain of integrins can influence integrin affinity. Galectin-8, for example, a secreted galactoside-binding protein, binds $\beta1$ integrins and induces a conformational change that decreases their affinity for other ligands (9).

Detachment of the cell rear

Downregulation of integrin binding at the rear is required for directed movement. This process is poorly understood, but it is clear that cell migration can be inhibited by locking integrins into a high-affinity state by mutations or activating antibodies. Detachment of the cell rear also requires a contraction of the cell body. Activation of RhoA and its downstream targets ROCKI and II, which are essential for the detachment of migrating leukocytes at the cell rear, mediates both decreased adhesion by $\alpha4\beta1$ and $\beta2$ integrins and increased intracellular force generation, resulting from stimulation of myosin light chain phosphorylation (10). Migrating fibroblasts can leave integrins, but not integrin-associated proteins like vinculin and α-actinin, in fibrous structures behind the cell, suggesting a disruption of the integrin-cytoskeleton connection (11). This might be caused by contraction or by intracellular proteolytic cleavage of integrins and other focal adhesion proteins and a concomitant disintegration of focal adhesions. Activation of Src may play a pivotal role in the regulation of focal adhesion turnover during migration, since the constitutively active v-Src oncoprotein is located in focal adhesions, where it induces focal adhesion kinase (FAK) phosphorylation, calpain-mediated degradation of focal adhesion proteins, and, eventually, disassembly of focal adhesions during transformation (12). It is conceivable that growth factors like EGF promote cell migration by reducing cell adhesion and the number of focal adhesions through the same mechanism. Cleavage products of FAK or the focal adhesion–associated docking protein HEF-1 can induce cell rounding, which is important for cell division and might also be involved in migration (13).

All of the detachment mechanisms discussed above appear to be sensitive to Ca^{2+} transients. Migrating fibroblasts have increased Ca^{2+} levels in the rear, perhaps as a consequence of stretch-activated calcium channels opening in response to cell contraction at the trailing edge. In neutrophils, Ca^{2+}-induced activation of calcineurin disrupts interactions between $\alpha v\beta3$ and the cytoskeleton, allowing detachment of the rear. Ca^{2+} transients are also essential to release $\alpha5\beta1$-mediated attachment of neutrophils to fibronectin (14). This latter mechanism, however, is independent of calcineurin and may involve modulation of integrin affinity or of cellular contractility. Ca^{2+} influx may also lead to the activation of μ-calpain, which is proposed to sever the connection between the integrin and the cytoskeleton proteolytically.

During rear detachment, unbound integrins accumulate at the trailing edge. Endocytosis of these detached integrins and their transport to the leading edge are thought to be needed to prevent integrin depletion at the front (14). Such endocytotic transport of green-fluorescent protein–tagged $\alpha5$ integrin subunits has indeed been observed in migrating fibroblasts (11). Integrin molecules observed in endocytotic vesicles are not attached to α-actinin or vinculin, as they are in the focal contacts prior to

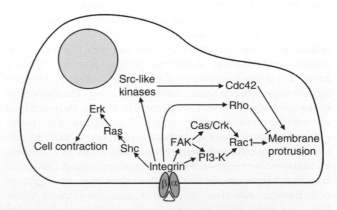

Figure 1
Major integrin-linked signaling pathways affecting migration. Integrin signaling stimulates Rho GTPases and thereby induces changes in the cytoskeletal organization that are required for cell migration. Ligand binding to integrins also promotes cell contractility through the activation of the tyrosine kinase Erk.

endocytosis. In monocyte chemoattractant protein–7 breast cancer cells, protein kinase Cα participates in the recycling of the activated pool of β1 integrins (15).

Integrin signaling in migration

Integrin signaling promotes cell migration by inducing changes in the cytoskeletal organization and by increasing cellular contractility (Figure 1). Activation of FAK plays a prominent role among the different integrin signaling pathways which affect migration. FAK is a non–receptor tyrosine kinase that is indirectly and perhaps also directly associated with integrins in focal contacts. Ligand binding to integrins leads to phosphorylation of FAK on at least seven different tyrosines in vivo, allowing the interaction of FAK with Src, Grb2, and PI3-K, and also promoting the phosphorylation of associated proteins, such as Cas and paxillin. FAK therefore functions as an important adaptor molecule that recruits various other signaling molecules to focal contacts. FAK is also a target for tyrosine phosphorylation induced by growth factor receptors. For this reason, FAK serves as an important integration point of growth factor and integrin signaling with respect to cell migration (16). Experiments in which FAK mutants are expressed in otherwise FAK-deficient fibroblasts have shown that FAK kinase activity and FAK autophosphorylation of tyrosine 397 are required for integrin-stimulated cell migration, whereas FAK's association with paxillin is dispensable.

FAK can promote cell movement by activating PI3-K and regulators of cytoskeletal dynamics like Rac1. Autophosphorylation of FAK on tyrosine 397 allows this molecule to bind and activate PI3-K, which can then influence integrin affinity and avidity, as discussed above. In addition, PI3-K contributes to the activation of Rac1, which mediates membrane ruffling. In adherent but not in suspended cells, activated Rac interacts with and activates p21-activated kinase (PAK), which can then stimulate migration by increasing the turnover of focal adhesions (17). Tyrosine-phosphorylated FAK also promotes Rac activation via an alternative pathway involving the adaptor proteins p130^Cas and Crk (18, 19).

Integrin signaling also stimulates Cdc42 and RhoA activity and facilitates the interaction of these Rho-family GTPases with their downstream targets. These GTPases cooperate and influence each other's activity in a complex, cell-specific manner. In astrocytes, integrin-dependent stimulation of Cdc42 via Src-like kinases is crucial for cell polarity, protrusion formation, and migration (20). In CHO cells, α5β1-mediated activation of Rac1 and Cdc42 is maximal already at intermediate fibronectin levels, whereas Rho activity continues to increase with increasing fibronectin levels (21). Since Rac1 and Cdc42 promote, but RhoA inhibits, membrane protrusion, increased integrin-mediated Rho GTPase activation halts migration in these cells preferentially at high fibronectin concentrations.

Recently, Chen et al. (22) found that the cytoplasmic tyrosine kinase Etk, which is highly expressed in migratory cells including endothelial cells and metastatic cell lines, is activated by integrin signaling. Integrin-mediated activation and autophosphorylation of FAK lead to interaction of FAK with the pleckstrin domain of Etk, and subsequent phosphorylation and activation of Etk. Etk may play an important role in migration, since changes in its expression directly correlate to changes in migration. Downstream targets of Etk mediating such a promigratory role are not yet known.

Integrin-triggered activation of Erk can contribute to migration via phosphorylation of MLCK, leading to increased phosphoryla-

tion of MLC and cell contraction (23). Erk activation furthermore leads to changes in gene expression, which have been suggested to promote migration. After integrin engagement or stimulation of v-Src, active Erk can also be targeted to newly forming focal adhesions, where it may phosphorylate focal adhesion proteins and influence cytoskeletal organization and migration (24). Erk can be activated via the interaction of FAK with Grb2. This pathway, however, apparently does not play a major role in integrin-mediated migration, since mutation of the Grb2-binding tyrosine 925 of FAK does not affect cell migration (18). Certain integrins can induce Erk activation independently of FAK via the Shc/Grb2/Sos/Ras cascade. α1β1, α5β1, and αvβ3 integrins activate this alternative pathway via an interaction of their α subunit with the membrane-associated protein caveolin, which helps define a distinctive membrane subdomain in which Shc becomes phosphorylated by the Src-like kinase Fyn (25).

Integrins and matrix metalloproteinases

Cell migration in vivo is often facilitated by a partial destruction of the surrounding ECM. Such degradation is catalyzed by matrix metalloproteinases (MMPs), a family of more than 20 substrate-specific zinc-dependent endoproteases, most of them soluble, but some of them transmembrane proteins. MMP activity, which is controlled by regulated expression, by proteolytic activation of inactive precursors (zymogens), and by the expression of a family of inhibitors, is crucial for tumor invasion, metastasis, and angiogenesis. Integrins can regulate the expression and activation of MMPs and can guide them to their targets by simultaneous binding of MMPs and ECM molecules.

Various integrin-induced signaling pathways are involved in the control of MMP expression. In collagen gels, for example, α1β1- and α2β1-dependent expression of MMP-13 requires p38 activity and is inhibited by Erk. In other settings, overexpression of the integrin-linked kinase (ILK) can result in AP-1–dependent expression of MMP-9 and an invasive phenotype. Finally, stimulation of the α3β1-tetraspanin complex on mammary epithelial cells upregulates the expression of MMP-2 and increases their invasive potential. Some of these transcriptional effects might be mediated by the transcription factor Cas-interacting zinc finger protein (CIZ), which can shuttle between the focal adhesions and the nucleus (26). CIZ activates the expression of several MMPs, and its effect is enhanced in the presence of Cas, which mediates some of the effects of FAK as mentioned above. However, integrins can also reduce the expression of certain MMPs. Deletion of the mouse gene for α1 integrin, for example, leads to a marked increase of MMP-7 and MMP-9 in the serum (27). MMP activation by integrins can also occur indirectly, through the increased expression of zymogen-activating proteases. Signaling by β1 integrin on ovarian carcinoma cells stimulates MMP-2 activity, perhaps by upregulating the activating protease MT1-MMP.

Yet another mechanism for modulating MMP function involves the simultaneous binding of integrins to ECM molecules and to MMPs, thus bringing the proteases close to their target. Complex formation of integrins and MMPs is crucial for wound healing and angiogenesis, as has been shown in basal keratinocytes of injured skin, where α2β1 integrin not only induces MMP-1 but also forms a complex with pro–MMP-1 (28). Similarly, in endothelial cells, αvβ3 induces the production of MMP-2 and subsequently interacts with the newly synthesized MMP-2 to promote vascular invasion. Without binding to αvβ3, MMP-2 is not functional (29). Inhibition of

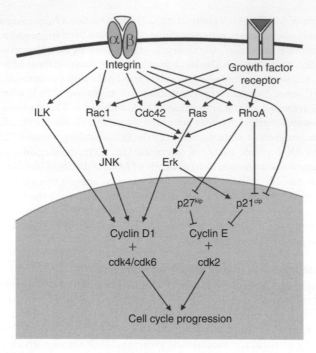

Figure 2
Integrin control of proliferation. Synergistic activation of Ras and Rho GTPases by integrins and growth factors is necessary for cell cycle progression. Activation of Rac1, Cdc42, and RhoA can facilitate activation of Erk via Ras. Erk, as well as JNK and ILK, can induce expression of cyclin D1, thus promoting cell cycle progression. In addition, integrins can promote proliferation by inhibiting cell cycle inhibitors like p21^{cip} and p27^{kip}.

the αvβ3–MMP-2 interaction by small organic compounds or viral vector–encoded peptides might therefore allow for reduced metastasis without blocking the binding of other molecules to αvβ3.

Integrin-mediated proliferation

The cell cycle is controlled by cyclin-dependent kinases (cdks). These proteins are constitutively expressed, but their activity is regulated by binding to cyclins and by the action of cdk inhibitory proteins (CKIs). Progression through the G1 phase is controlled by cdk4 and cdk6, which interact with cyclin D1, and by cdk2, which binds to cyclin E. The CKIs p21^{cip1} and p27^{kip1} inactivate cdk2. Cdks are responsible for the phosphorylation of Rb, which induces other cell cycle proteins including cyclin A. Association of cyclin A with cdk2 initiates then the G1/S transition.

Integrin signaling can regulate the G1/S transition in tight synergism with growth factors (Figure 2). Here again, however, different cell types require distinct integrins and growth factors for proliferation. In endothelial cells, for example, signaling of both bFGF and αvβ3 is required for cell cycle progression, whereas in fibroblasts bFGF can also synergize with β1 integrins (30). Integrin-mediated proliferation depends on the activation of the Erk pathway, which controls expression of cyclin D1 and p21^{cip}. Integrins can induce Erk via FAK or via caveolin, Fyn, and Shc signaling, both converging on the Ras/Erk cascade.

Since growth factor receptors also activate Erk via the Ras pathway, it was initially unclear why both integrin and growth factor receptor signaling are necessary for proliferation. Recent data, however, indicate that only combined signaling of both pathways

results in the appropriate kinetic and extent of Erk activation. Growth factors activate Erk strongly but transiently, which is not sufficient for the induction of cyclin D1. On the other hand, integrin-mediated activation of Erk is too weak to stimulate expression of cyclin D1. Only cells attached to the ECM via integrins show a sustained and robust activation of Erk in response to growth factors, which is necessary to induce cyclin D1. In fibroblasts, strong sustained (5–20 hours), but not transient (<1 hour) or weak, Erk activation results in the expression of a distinct repertoire of fos and jun proteins, which might be responsible for the differential effects on cyclin D1 expression (31). Indeed, the *Cyclin D* promoter contains an AP-1 site, which is recognized by several jun/fos heterodimers, and a CRE site, to which heterodimers of jun/fos and certain ATF/CREB family members might bind.

A second integrin-dependent proliferation checkpoint exists at the level of the cdk inhibitors. Thus, transient, high-level Erk activation increases p21^{cip} expression, which inhibits cyclin E/cdk2. Integrin-dependent signals, but not Erk, downregulate p21^{cip} in mid- to late G1, allowing cell cycle progression.

The molecular mechanisms by which integrin and growth factor receptor signaling pathways synergize to influence Erk activation are not yet clear, and several possible models should be considered. First, integrin signaling can influence the phosphorylation state of EGF and PDGF receptors as well as their association with signaling molecules such as SHP-2, Ras-GAP, IRS-1, and PI3-K. Interestingly, FAK activation is not required for the integrin-mediated increase of EGF-R phosphorylation. Alternatively, integrins may facilitate signaling of growth factor receptors via the Ras cascade. In some cells, for example, growth factor activation of Raf or MEK, two serine/threonine kinases upstream of Erk, is dependent on cell adhesion. This effect might be due to the integrin-mediated activation of PAK, which phosphorylates both Raf and MEK (32). Such crosstalk between integrin and growth factor signaling can be essential for the transformation of cells. The malignant behavior of a breast cancer cell line in a three-dimensional culture is crucially dependent on both integrin and EGF-R signaling, which together result in sustained Erk activation. Inhibition of integrins, EGF-R, or Erk leads to growth arrest and normal breast tissue morphogenesis. Interestingly, these effects are not seen in two-dimensional cell cultures, suggesting that the connections between signaling pathways depend on the spatial organization of the cytoskeleton.

Integrin-mediated cell cycle control has also been demonstrated in vivo. Thus, Faraldo et al. (33) have found that inhibition of integrin signaling in mice using a dominant negative integrin fusion protein reduces proliferation of mammary cells and attenuates Shc, Erk, and JNK activation without perturbing FAK activity.

Rho GTPases in integrin-mediated proliferation

In fibroblasts, but not in endothelial cells, the synergism between integrin and growth factor signaling is dependent on cell spreading, indicating an important role of the cytoskeleton. Rho GTPases might be key players in this regulation since they are activated by growth factors and, weakly, by integrins, and since they are pivotal for the organization of the cytoskeleton. Rho GTPases are crucial for cell cycle progression, as seen in the impairment of growth factor–induced production of cyclin D1 following treatment of adherent cells with the Rho inhibitor toxin A. RhoA downregulates the transcription of p21^{cip} and promotes degradation of the CDK inhibitor p27^{kip1} by stimulation of cyclin E/cdk2 activity. Binding to fibronectin can induce cyclin D1 and suppress

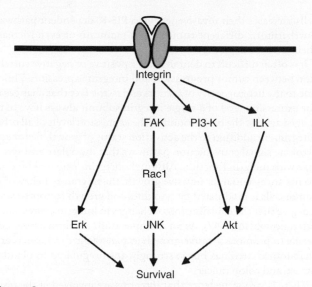

Figure 3
Integrin control of cell survival. Integrin activation of Akt via PI3-K or ILK plays an important role in the antiapoptotic effect of integrin-mediated adhesion. Activation of Erk and JNK likewise contributes to increased cell survival.

p21^{cip} in a Rho-dependent manner (34). Activation of RhoA and Cdc42 contributes to the integrin-mediated activation of Erk2. Activated Rac1 can induce cyclin D1 expression in serum-starved cells correlating with the activation of PAK-1, JNK, and NF-κB. Furthermore, in endothelial cells Rac1 can facilitate translation of cyclin D1 mRNA upon activation by α5β1 integrin (35). Here, the integrin-induced Rac activation depends on Shc, FAK, and PI3-K signals, which converge to regulate the guanine nucleotide exchange factor Sos.

Other integrin-dependent proliferation pathways have also been documented. Integrin and growth factor receptors both activate PI3-K, which downregulates p27^{kip} and contributes to the induction of cyclin D1. These effects are at least partly mediated by the PDK-1–dependent stimulation of the p70 S6 kinase, a key regulator of cell growth that controls protein translation and induces cyclin D3 expression (36). Many human tumors carry mutations in proteins of the PI3-K signaling pathway, underlining the importance of this mechanism in the prevention of malignant proliferation. In some cells, integrin-dependent activation of JNK also contributes to proliferation (37). JNK phosphorylates several members of the jun/fos family, causing increased transcriptional activity and reduced degradation of AP-1 proteins, which are important for the induction of cyclin D1. Finally, integrins might induce cell cycle progression by ILK, which is proposed to induce cyclin D1 expression by activating the transcription factor LEF-1.

Integrins and apoptosis

Apoptosis or programmed cell death regulates the lifespan of normal cells and eliminates cells after contact with certain toxic insults or following their detachment from the ECM. This last form of apoptosis is also called "anoikis." Reduced apoptosis can contribute to tumorigenesis, since it helps tumor cells to escape natural elimination.

Apoptosis is initiated by either the signaling of death receptors or the release of cytochrome *c* from the mitochondria, which then activates a cascade of intracellular proteases of the caspase family. The first pathway is controlled by the expression of death receptors and soluble or inactive "decoy" receptors. Induction of programmed cell death through the latter pathway is regulated especially by the balance of pro- and antiapoptotic members of the Bcl-2 family. Although anoikis is initiated by cell detachment, increased death receptor signaling also contributes to cell death. Integrin signaling can modulate apoptotic induction, leading in most cases to increased, but sometimes also to decreased, survival. Integrins can promote survival by activating PI3-K, ILK, Erk, and JNK (Figure 3). Integrin-mediated activation of PI3-K can trigger several antiapoptotic mechanisms. Activated PI3-K produces PI(3,4,5)P3 and PI(3,4)P2, which promote the relocation of Akt to the plasma membrane and stimulate its phosphorylation.

Akt blocks apoptosis by phosphorylating, and thus inactivating, a number of proapoptotic molecules, including the Bcl-2 family member Bad, caspase-9, and transcription factors of the forkhead family. Akt inhibits apoptosis via different pathways. It can also activate NF-κB, via phosphorylation of IκB, and thereby induce the expression of a set of survival factors, such as osteoprotegerin in endothelial cells. Finally, it prevents the release of cytochrome *c* from the mitochondria by an unknown mechanism. In CHO cells, integrin-mediated activation of PI3-K and Akt, and upregulation of antiapoptotic Bcl-2, are dependent on Shc and FAK-mediated Ras activation, but not on Erk (38).

Inhibition of β1 integrin function in the mammary gland in vivo using a dominant negative receptor leads to increased cell death and reduced phosphorylation of Akt and the Akt substrates Bad and the forkhead transcription factor FKHR. FAK phosphorylation is normal under these conditions (33), indicating that integrin-mediated activation of PI3-K in this tissue is independent of FAK, although it may be influenced by growth factor signaling, as has been described for mammary cells in culture. Some evidence suggests that the serine/threonine kinase ILK is activated after integrin binding and directly binds and phosphorylates Akt (39). In prostate carcinoma cells lacking the tumor suppressor PTEN, ILK is constitutively active and contributes to the survival of these cells.

Elicitation of Erk activity by integrins can prevent apoptosis, an effect that is limited to particular cell types and in some cases requires the additional activation of Rac1 by the FAK/Cas/Crk pathway (40). For example, in intestinal epithelial cells, α2β1-mediated Erk activation does not prevent apoptosis induced by serum deprivation (41). Integrin-mediated activation of JNK inhibits apoptosis of fibroblasts in the absence of growth factors (42). This effect is dependent on FAK, but independent of Akt and Erk. In the presence of growth factors, however, integrins prevent apoptosis of fibroblasts by FAK-mediated activation of PI3-K and Akt.

Loss of these integrin-dependent survival signals after cell detachment can result in apoptosis. Detachment also leads to direct changes in the cytoskeleton and release of the proapoptotic molecule Bmf, which is normally bound to the cytoskeleton (43), an event that is considered to help initiate apoptosis, since it occurs before caspase activation. Furthermore, unligated integrin can transmit apoptosis-stimulating signals. Overexpression of unligated, but not of ligand-bound, α5β1 integrin can induce apoptosis in a variety of detached human cancer cells.

As in migration and proliferation, the antiapoptotic functions of integrins depend on the context of other signaling pathways. Integrin signaling can enhance the survival effect of growth factors by facilitating downstream signaling events, as has been shown for α6β1 on oligodendrocytes and for α5β1 on intestinal epithelial cells and on mammary epithelium (41). On the other hand, signaling by the growth factor receptor c-erbB2 in a mammary epithelial cell line reduces the affinity of α2β1 integrin, decreases integrin-mediated survival signals, and results in apoptosis when the cells are growing on collagen (44). The tumor suppressor p53 modulates survival signals provided by α6β4 integrin in dramatic ways. α6β4 induces caspase-mediated inactivation of Akt and apoptosis in carcinoma cells expressing wild-type p53, while it promotes Akt-dependent survival in the absence of p53 (45).

In endothelial cells, shear stress induces the upregulation of α5β1 integrin expression and inhibits programmed death of these cells. Interestingly, abrogation of shear stress causes apoptosis mediated by an autocrine loop: It induces the secretion of thrombospondin, which binds to αvβ3, which acts as a death receptor. In neutrophils as well, integrins can have a proapoptotic role, since attachment via β2 integrins renders the cells susceptible to TNF-α–induced apoptosis.

Integrins and invasive growth in vivo

Many antibody and RGD peptide inhibition studies have suggested an important role for integrins in invasive growth during development. Analyses of mice lacking integrins, however, often fail to confirm such a role. For example, αvβ3 integrin has been shown to be essential for angiogenesis, although deletion of the mouse gene for the αv subunit does not impair angiogenesis. These discordant results might be explained by side effects of the antibody/peptide treatments or, alternatively, by compensatory mechanisms that are activated in the mutant mice, such as the upregulation of other, functionally similar molecules. Defects in cell migration have been described during the development of animals lacking integrins. Murine primordial germ cells lacking β1 integrin, for example, show impaired migration to the gonads. Interestingly, in *Drosophila*, the PS1 integrin is required on tracheal cells of the visceral branch for normal migration of tracheal cells on the visceral endoderm; the interacting PS2 integrin is required on the visceral endoderm in the same migratory process (46).

A plethora of experimental evidence documents the importance of integrins in tumor progression, invasive growth, and metastasis. Using β1-null tumor cells, for instance, it has been shown that β1 integrin promotes but is not essential for metastasis. Mutations of the intracellular domain of the β1 integrin chain differentially affect cell adhesion, invasion, and metastasis, indicating that integrin signaling is important for invasion. Different observations suggest an important role of α6β4 integrin in tumorigenesis. Expression of α6β4 integrin in β4-deficient tumor cells increases their invasiveness via a PI3-K–dependent pathway. Furthermore, different tumor types maintain or even increase their levels of α6β4 expression.

It is often difficult to demonstrate a positive or negative correlation between tumor progression and integrin expression. This is due to the heterogeneity of tumors and to the fact that changes in the expression level of a single integrin subunit always have to be judged against the background of the expression levels of all other integrins, in addition to the activation status of growth factor and cytokine signal transduction pathways that modulate and synergize with integrin function. Although integrin-mediated binding events are essential for invasive growth, the migratory behavior of tumor cells is modulated by cytokines and growth factors to such a degree that the contribution of changes in integrin expression is often not obvious (47). As an example, α2β1 integrin expression seems to promote invasive growth in pancreatic carcinomas and rhabdomyosarcomas but is strongly downregulated in bladder, breast, and colon cancer.

There is strong evidence that integrins are involved in the resistance of tumor cells to chemotherapy-induced apoptosis. Adhesion of small-cell lung cancer cells to ECM molecules protects them from chemotherapeutic, apoptosis-inducing agents (48). This effect is mediated by β1 integrins, which activate phosphotyrosine kinases in response to chemotherapy-induced DNA damage. In human myeloma cells, overexpression of α4β1 results in increased drug resistance. Finally, β1 integrin–mediated activation of PI3-K in breast cancer cells increases the resistance of tumor cells to apoptosis-inducing drugs (49).

Integrins thus clearly offer attractive drug targets to fight tumor growth and metastasis and drug resistance, but the variability of their functions in different cell types is daunting. Much more will need to be learned about integrin function, cross-talk with other signaling pathways, and tumor-specific roles to ensure that such drugs will be effective and safe.

Acknowledgments

We thank Michael Dictor, Martin Pfaff, and Kristiina Vuori for critically reading the manuscript. We apologize that, due to space limitations, we could not cite primary references for all the work mentioned and had to omit many interesting contributions to the field. D. Bouvard is supported by a Marie Curie fellowship.

Address correspondence to: Cord Brakebusch, Max Planck Institute for Biochemistry, Department of Molecular Medicine, Am Klopferspitz 18a, 82152 Martinsried, Germany. Phone: 49-89-8578-2466; Fax: 49-89-8578-3777; E-mail: brakebus@biochem.mpg.de.

1. Mastrangelo, A.M., Homan, S.M., Humphries, M.J., and LaFlamme, S.E. 1999. Amino acid motifs required for isolated β cytoplasmic domains to regulate 'in trans' β1 integrin conformation and function in cell attachment. *J. Cell Sci.* **112**:217–229.

2. Keely, P.J., Rusyn, E.V., Cox, A.D., and Parise, L.V. 1999. R-Ras signals through specific integrin α cytoplasmic domains to promote migration and invasion of breast epithelial cells. *J. Cell Biol.* **145**:1077–1088.

3. Liu, S., Calderwood, D.A., and Ginsberg, M.H. 2000. Integrin cytoplasmic domain-binding proteins. *J. Cell Sci.* **113**:3563–3571.

4. Constantin, G., et al. 2000. Chemokines trigger immediate β2 integrin affinity and mobility changes: differential regulation and roles in lymphocyte arrest under flow. *Immunity.* **13**:759–769.

5. Danilkovitch-Miagkova, A., and Zbar, B. 2002. Dysregulation of Met receptor tyrosine kinase activity in invasive tumors. *J. Clin. Invest.* **109**:863–867. DOI:10.1172/JCI200215418.

6. Trusolino, L., et al. 2000. HGF/scatter factor selectively promotes cell invasion by increasing integrin avidity. *FASEB J.* **14**:1629–1640.

7. Katagiri, K., et al. 2000. Rap1 is a potent activation signal for leukocyte function-associated antigen 1 distinct from protein kinase C and phosphatidylinositol-3-OH kinase. *Mol. Cell. Biol.* **20**:1956–1969.

8. Reedquist, K.A., et al. 2000. The small GTPase, Rap1, mediates CD31-induced integrin adhesion. *J. Cell Biol.* **148**:1151–1158.

9. Hadari, Y.R., et al. 2000. Galectin-8 binding to integrins inhibits cell adhesion and induces apoptosis.

J. Cell Sci. **113**:2385–2397.

10. Worthylake, R.A., Lemoine, S., Watson, J.M., and Burridge, K. 2001. RhoA is required for monocyte tail retraction during transendothelial migration. *J. Cell Biol.* **154**:147–160.

11. Laukaitis, C.M., Webb, D.J., Donais, K., and Horwitz, A.F. 2001. Differential dynamics of α5 integrin, paxillin, and α-actinin during formation and disassembly of adhesions in migrating cells. *J. Cell Biol.* **153**:1427–1440.

12. Carragher, N.O., Fincham, V.J., Riley, D., and Frame, M.C. 2001. Cleavage of focal adhesion kinase by different proteases during SRC-regulated transformation and apoptosis. Distinct roles for calpain and caspases. *J. Biol. Chem.* **276**:4270–4275.

13. O'Neill, G.M., and Golemis, E.A. 2001. Proteolysis of the docking protein HEF1 and implications for focal

adhesion dynamics. *Mol. Cell. Biol.* **21**:5094–5108.

14. Pierini, L.M., et al. 2000. Oriented endocytic recycling of α5β1 in motile neutrophils. *Blood.* **95**:2471–2480.

15. Ng, T., et al. 1999. PKCα regulates β1 integrin-dependent cell motility through association and control of integrin traffic. *EMBO J.* **18**:3909–3923.

16. Sieg, D.J., et al. 2000. FAK integrates growth-factor and integrin signals to promote cell migration. *Nat. Cell Biol.* **2**:249–256.

17. Kiosses, W.B., et al. 1999. A role for p21-activated kinase in endothelial cell migration. *J. Cell Biol.* **147**:831–844.

18. Cary, L.A., et al. 1998. Identification of p130Cas as a mediator of focal adhesion kinase-promoted cell migration. *J. Cell Biol.* **140**:211–221.

19. Klemke, R.L., et al. 1998. CAS/Crk coupling serves as a "molecular switch" for induction of cell migration. *J. Cell Biol.* **140**:961–972.

20. Etienne-Manneville, S., and Hall, A. 2001. Integrin-mediated activation of Cdc42 controls cell polarity in migrating astrocytes through PKCζ. *Cell.* **106**:489–498.

21. Cox, E.A., Sastry, S.K., and Huttenlocher, A. 2001. Integrin-mediated adhesion regulates cell polarity and membrane protrusion through the Rho family of GTPases. *Mol. Biol. Cell.* **12**:265–277.

22. Chen, R., et al. 2001. Regulation of the PH-domain-containing tyrosine kinase Etk by focal adhesion kinase through the FERM domain. *Nat. Cell Biol.* **3**:439–444.

23. Klemke, R.L., et al. 1997. Regulation of cell motility by mitogen-activated protein kinase. *J. Cell Biol.* **137**:481–492.

24. Fincham, V.J., James, M., Frame, M.C., and Winder, S.J. 2000. Active ERK/MAP kinase is targeted to newly forming cell-matrix adhesions by integrin engagement and v-Src. *EMBO J.* **19**:2911–2923.

25. Wary, K.K., Mariotti, A., Zurzolo, C., and Giancotti, F.G. 1998. A requirement for caveolin-1 and associated kinase Fyn in integrin signaling and anchorage-dependent cell growth. *Cell.* **94**:625–634.

26. Nakamoto, T., et al. 2000. CIZ, a zinc finger protein that interacts with p130(cas) and activates the expression of matrix metalloproteinases. *Mol. Cell. Biol.* **20**:1649–1658.

27. Pozzi, A., et al. 2000. Elevated matrix metalloprotease and angiostatin levels in integrin α1 knockout mice cause reduced tumor vascularization. *Proc. Natl. Acad. Sci. USA.* **97**:2202–2207.

28. Dumin, J.A., et al. 2001. Pro-collagenase-1 (matrix metalloproteinase-1) binds the α2β1 integrin upon release from keratinocytes migrating on type I collagen. *J. Biol. Chem.* **276**:29368–29374.

29. Silletti, S., et al. 2001. Disruption of matrix metalloproteinase 2 binding to integrin αvβ3 by an organic molecule inhibits angiogenesis and tumor growth in vivo. *Proc. Natl. Acad. Sci. USA.* **98**:119–124.

30. Roovers, K., and Assoian, R.K. 2000. Integrating the MAP kinase signal into the G1 phase cell cycle machinery. *Bioessays.* **22**:818–826.

31. Cook, S.J., Aziz, N., and McMahon, M. 1999. The repertoire of fos and jun proteins expressed during the G1 phase of the cell cycle is determined by the duration of mitogen-activated protein kinase activation. *Mol. Cell. Biol.* **19**:330–341.

32. Howe, A.K., and Juliano, R.L. 2000. Regulation of anchorage-dependent signal transduction by protein kinase A and p21-activated kinase. *Nat. Cell Biol.* **2**:593–600.

33. Faraldo, M.M., Deugnier, M.A., Thiery, J.P., and Glukhova, M.A. 2001. Growth defects induced by perturbation of β1-integrin function in the mammary gland epithelium result from a lack of MAPK activation via the Shc and Akt pathways. *EMBO Rep.* **2**:431–437.

34. Danen, E.H., Sonneveld, P., Sonnenberg, A., and Yamada, K.M. 2000. Dual stimulation of Ras/mitogen-activated protein kinase and RhoA by cell adhesion to fibronectin supports growth factor-stimulated cell cycle progression. *J. Cell Biol.* **151**:1413–1422.

35. Mettouchi, A., et al. 2001. Integrin-specific activation of Rac controls progression through the G(1) phase of the cell cycle. *Mol. Cell.* **8**:115–127.

36. Feng, L.X., Ravindranath, N., and Dym, M. 2000. Stem cell factor/c-kit up-regulates cyclin D3 and promotes cell cycle progression via the phosphoinositide 3-kinase/p70 S6 kinase pathway in spermatogonia. *J. Biol. Chem.* **275**:25572–25576.

37. Oktay, M., et al. 1998. Integrin-mediated activation of focal adhesion kinase is required for signaling to Jun NH2-terminal kinase and progression through the G1 phase of the cell cycle. *J. Cell Biol.* **145**:1461–1469.

38. Matter, M.L., and Ruoslahti, E. 2001. A signaling pathway from the alpha5beta1 and αvβ3 integrins that elevates bcl-2 transcription. *J. Biol. Chem.* **276**:27757–27763.

39. Persad, S., et al. 2001. Regulation of protein kinase B/Akt-serine 473 phosphorylation by integrin-linked kinase: critical roles for kinase activity and amino acids arginine 211 and serine 343. *J. Biol. Chem.* **276**:27462–27469.

40. Cho, S.Y., and Klemke, R.L. 2000. Extracellular-regulated kinase activation and CAS/Crk coupling regulate cell migration and suppress apoptosis during invasion of the extracellular matrix. *J. Cell Biol.* **149**:223–236.

41. Lee, J.W., and Juliano, R.L. 2000. α5β1 integrin protects intestinal epithelial cells from apoptosis through a phosphatidylinositol 3-kinase and protein kinase B-dependent pathway. *Mol. Biol. Cell.* **11**:1973–1987.

42. Almeida, E.A., et al. 2000. Matrix survival signaling: from fibronectin via focal adhesion kinase to c-Jun NH(2)-terminal kinase. *J. Cell Biol.* **149**:741–754.

43. Puthalakath, H., et al. 2001. Bmf: a proapoptotic BH3-only protein regulated by interaction with the myosin V actin motor complex, activated by anoikis. *Science.* **293**:1829–1832.

44. Baeckstrom, D., Lu, P.J., and Taylor-Papadimitriou, J. 2000. Activation of the α2β1 integrin prevents c-erbB2-induced scattering and apoptosis of human mammary epithelial cells in collagen. *Oncogene.* **19**:4592–4603.

45. Bachelder, R.E., et al. 1999. p53 inhibits α6β4 integrin survival signaling by promoting the caspase 3-dependent cleavage of AKT/PKB. *J. Cell Biol.* **147**:1063–1072.

46. Boube, M., Martin-Bermudo, M.D., Brown, N.H., and Casanova, J. 2001. Specific tracheal migration is mediated by complementary expression of cell surface proteins. *Genes Dev.* **15**:1554–1562.

47. Kassis, J., Lauffenburger, D.A., Turner, T., and Wells, A. 2001. Tumor invasion as dysregulated cell motility. *Semin. Cancer Biol.* **11**:105–117.

48. Sethi, T., et al. 1999. Extracellular matrix proteins protect small cell lung cancer cells against apoptosis: a mechanism for small cell lung cancer growth and drug resistance in vivo. *Nat. Med.* **5**:662–668.

49. Aoudjit, F., and Vuori, K. 2001. Integrin signaling inhibits paclitaxel-induced apoptosis in breast cancer cells. *Oncogene.* **20**:4995–5004.

Stat proteins and oncogenesis

Jacqueline Bromberg

Memorial Sloan-Kettering Cancer Center, New York, New York, USA.

The discovery of Stat proteins' key role in IFN signaling, initially described over ten years ago, provided the first molecular link of growth factor receptor stimulation to the direct activation of a transcription factor (1). Since that time a large number of growth factor receptors and some nonreceptor tyrosine kinases have been found to lead to the activation of these transcription factors (2).

The contributions of individual Stat proteins to normal cytokine signaling and development have been studied in various cell culture systems and in vivo in mice made deficient for one or more of these proteins (3). This approach has identified some related roles, as well as many unique, nonoverlapping physiological roles, for the various members of the Stat family. In summary, Stat1-deficient mice are unable to respond to IFNs and are subsequently susceptible to bacterial and viral pathogens. Likewise, disruption of Stat2 gives rise to animals unable to respond to type 1 IFNs, with increased susceptibility to viral infections (see Candotti et al., this Perspective series, ref. 4). Stat4- and Stat6-deficient animals reveal a requirement for IL-12– or IL-4–mediated proliferation of T cells, respectively (see Decker et al., this series, ref. 5). The phenotypes of Stat5A and Stat5B individual knockouts reveal the importance of Stat5A in breast development and lactation and the importance of Stat5B in the development of sexually dimorphic patterns of gene expression within the liver. In addition to these phenotypes, Stat5A/5B double knockouts are abnormal in their T cell and B cell development. Because Stat3-deficient animals die early in embryogenesis, the role of this protein in a number of biological functions had to be determined in conditional knockouts. As discussed by Levy and Lee in this series (6), Stat3 is implicated in keratinocyte migration, T cell apoptosis, IL-10–mediated signaling in macrophages, and the induction of apoptosis in the involuting breast.

Beyond these various roles in normal cellular and physiological processes, the Stat proteins are now known to participate in cellular transformation and oncogenesis. Here, I consider the evidence implicating these molecules, particularly Stats 1, 3, and 5, in tumor formation and progression.

Stats and tumorigenesis

During the multistep process of tumorigenesis, cells lose their normal ability to sense and repair DNA damage and to regulate cell cycle progression and apoptosis. In parallel, they acquire abnormal patterns of growth factor signaling, angiogenesis, and invasive growth. While the Stats are not known to contribute directly to cell cycle checkpoint regulation or DNA repair, they contribute to tumorigenesis through their intimate connection to growth factor signaling, apoptosis, and angiogenesis. In addition, because these molecules play key roles in immune responses, defective Stat signaling can favor tumor development by compromising immune surveillance.

STAT1 as a tumor suppressor

STAT1, the first STAT to be discovered, is required for signaling by the IFNs (7, 8), which, in addition to their requirement in innate immunity, serve as potent inhibitors of growth and promoters of apoptosis. Although Stat1-deficient mice develop no spontaneous tumors, they are highly susceptible to chemical carcinogen-induced tumorigenesis (9). Crossing the *Stat1* mutation into a p53-deficient background yields animals that develop tumors more rapidly, and with a broader spectrum of tumor types, than is seen with *p53* single mutants. Because tumors from carcinogen-treated wild-type animals also grow far more rapidly when transplanted into the Stat1-deficient animals than they do in a wild-type host, it appears that Stat1 is needed for the host's tumor surveillance capabilities (9).

The requirement of Stat1 for apoptosis and growth arrest (10–14) in some cell types may be explained by its ability to upregulate caspases and the cdk inhibitor p21. Interestingly, p21 upregulation by STAT1 in mammary cells appears to involve BRCA1, which is often lost in familial and other forms of breast cancer (15). Determining whether STAT1 itself is functionally impaired in primary cancers is not an easy task, particularly in solid tumors, where signaling function may be sensitive to the local environment, and ex vivo studies of STAT1 activity in isolated cells are of questionable significance. Nevertheless, it would not be surprising to find that *STAT1* is mutated in some human tumors. It will be interesting to determine whether a recently described (16) human *STAT1* mutation, found in families with an inherited susceptibility to mycobacterial infections, also leads to an increased risk of cancer.

Constitutive activation of Stats in cellular transformation

The first reports of persistently tyrosine-phosphorylated (that is, persistently activated) STAT proteins in primary cancers and tumor-derived cell lines came shortly after the discovery of the STATs (Table 1). Subsequent work showed that, in a number of tumor-derived cell lines, the STATs, particularly STAT3, are required to maintain a transformed phenotype. STAT5 is also commonly found to be constitutively activated in certain malignancies, especially leukemias and lymphomas (Table 1). The expression of fusion proteins that cause heightened or unrestrained JAK2, PDGF-R, or ABL signaling can lead to the constitutive activation of STAT5. Work in murine models shows that bcr-abl– and v-abl–induced leukemias do not require Stat5 (17) but that this protein is required for a myeloproliferative disorder that results from a TEL-JAK fusion (18).

Stat3-deficient murine T cells, mammary epithelial cells, macrophages, fibroblasts, and keratinocytes are viable, relatively normal cells with subtle defects typically involving the regulation of apoptosis (19). Thus, in these few examples, Stat3 is not essential for viability of normal cells. In contrast, many cancer-derived cell lines that contain consitutively activated STAT3 are dependent on this protein and undergo growth arrest or apoptosis when treated with antisense or dominant negative constructs directed at STAT3.

Citation for this article: *J. Clin. Invest.* **109**:1139–1142 (2002). DOI:10.1172/JCI200215617.

Table 1
Constitutive activation of STATs in primary cancers and tumor-derived cell lines

Solid tumors	Activated STAT	Liquid tumors	Activated STAT
Breast cancer	1,3	Chronic myelogenous leukemia	5
Head and neck cancer	1,3	Acute myeloid leukemia	1,3,5
Prostate cancer	3	Chronic lymphocytic leukemia	1,3
Melanoma	3	Mycosis fungoides	3
Ovarian cancer	3	Acute lymphoblastic leukemia	1,5
Lung cancer	1,3	Erythroleukemia	1,5
Brain tumors	1,3	Burkitt's lymphoma	3
Pancreatic cancer	3	Large granular lymphocyte leukemia	3
Renal carcinoma	3	Myeloma	3
		Hodgkins lymphoma	3
		Anaplastic large cell lymphoma	3

		STAT activation required for transformation	
Cell lines and/or primary tumor	Required STAT	Oncogenes	Required STAT
Breast cancer	3	src	3
Head and neck cancer	3	eyk	3
Prostate cancer	3	ret	3
Melanoma	3	lck	3,5
Thyroid cancer	3	$G_\alpha o$	3
Myeloma	3	Npm-alk	3
Hodgkins	3		
Large granular lymphocyte leukemia	3		
Hepatocellular carcinoma	3		

Black, Stats constitutively activated in a number of primary liquid and solid tumors; Blue, Stats required for the transformed phenotype of the tumor-derived cell lines, primary tumors, or oncogene transformed cell lines.

Direct evidence that STAT3 signaling is oncogenic comes from work with a spontaneously dimerizing-mutant form of STAT3, STAT3-C, which does not require tyrosine phosphorylation to be activated yet is capable of transforming fibroblasts (20). There are no known naturally occurring mutations of *STAT3* that lead to its constitutive activation and subsequent transformation of cells. In all naturally occurring tumors and in oncogene-transformed cells, STAT3 activation is typically dependent upon dysregulated growth factor receptor tyrosine kinases or their associated Jak kinases. Thus, in the case of thyroid cancers associated with an aberrantly regulated Ret receptor tyrosine kinase, Schuringa et al. (21) have found that transformation by Ret requires phosphorylation and activation of STAT3. Similarly, expression of a dominant negative STAT3 abrogates cellular transformation in the acute myelogenous leukemia (AML) and gastrointestinal stromal cell tumors (GISTs) associated with activating mutations in the receptor c-kit (22). STAT3 is also persistently activated in Hodgkin disease, where AG490, an inhibitor of JAK2 and STAT3 phosphorylation, can be used to inhibit tumor growth (23). In primary prostate cancer specimens and prostate cancer–derived cell lines, likewise, STAT3 is activated, and the introduction of antisense to STAT3 provokes tumor cell apoptosis (24, 25). Finally, persistently activated STAT3 and JAK2 are found in the rare malignancy large granular lymphocyte leukemia; treatments that block STAT3 expression or function cause cancer cell death by upregulating the proapoptotic protein Fas and downregulating the antiapoptotic Mcl-1 (26, 27).

Antisense reagents and dominant negative constructs directed at STAT3, as well as Jak inhibitors such as AG490, thus hold the promise of tumor-specific growth inhibition and appear to be useful against a range of cancer cells (Table 1). In addition, Jove and colleagues recently developed a novel approach to blocking STAT3 function, using a phosphopeptide tethered to a protein transduction domain (28). This peptide enters cells, recognizes the SH2 domain of STAT3, and prevents de novo STAT3 phosphorylation. Consistent with the requirement of activated STAT3 in cellular transformation, this peptide blocks the transformation of NIH-3T3 cells by v-src (28).

Routes to constitutive STAT activation

Dysregulated tyrosine kinase activity and increased levels of tyrosine phosphorylation are common in cancer cells and often result from the overexpression of growth factor receptors and their ligands. Salient examples are seen in multiple myeloma, which characteristically overexpresses IL-6 and its receptor, and in head and neck cancers, where EGF and its receptor are found at high levels (29). Increased expression of a growth factor receptor, such as the PDGF-R in mesotheliomas or gliomas and Erb2 (Her2neu) in breast cancer, is felt to be a critical step in the formation of these cancers. High levels of c-kit may be critical for the abnormal phenotype of GISTs. Excessive Jak kinase activity in such tumors is perhaps the most common mechanism for constitutive phosphorylation and activation of the STATs. The basis of receptor tyrosine kinase or growth factor receptor overexpression is not understood but may involve the loss of internalization and turnover that normally follows ligand binding. Proteins such as the JAKs and c-Src, which associate with these receptors, may also accumulate to abnormal levels and thus may promote STAT hyperactivation in cancer cells.

A number of human malignancies also express the TEL-JAK fusion protein described above, which induces the persistent dimerization and constitutive activation of JAK2 (18, 30–32). The

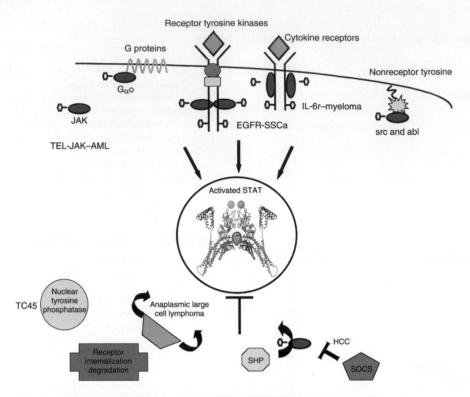

Figure 1

Routes to persistent activation of Stats in cancer cells. Stat proteins, in particular Stat3,are persistently activated in a wide variety of different cancers. Explanations for Stat activation are potentially quite varied and include over-expression or dysregulation of kinases or inhibition of the negative regulators. Some examples are listed in red. Jak kinases are indicated as blue ovals. The TEL-JAK fusion protein is constitutively active as a kinase and leads to AML through its effects on STATs 1, 3, or 5. The oncogenic G-protein subunit G$_\alpha$o, associates with Jak proteins, leading to Stat3 hyperactivation and the transformation of NIH3T3. Overexpression and dysregulation of the EGFR and its ligand TGF-α stimulates the receptor's intrinsic tyrosine kinase activity and those of the associate Jak and Src protein kinases, thereby inducing persistent Stat3 activity in squamous cell carcinomas. Overexpression and dysregulation of the IL-6 receptor and production of its ligand IL-6 stimulates Stat 3 activity in multiple myeloma. Abnormal regulation of the oncogenic proteins Src and Abl likewise causes excess. Stat3 phosphorylation in sarcomas and chronic myelogenous leukemia, among other malignancies. *SOCS1* gene methylation and silencing constitutively activates JAKs1 and 2 and leads to persistent STAT3 phosphorylation in hepatocellular carcinomas. Loss of PIAS3 expression allows for unregulated STAT3 activity in anaplastic large cell lymphoma.

JAK2 inhibitor AG490, first used against leukemia cells in which the kinase is constitutively active (33), has more recently been found to induce growth arrest or apoptosis in many tumor-derived cell lines, suggesting that this kinase is important in a variety of cancers (34). Endogenous inhibitors of the JAKs include SHP1 and SHP2, as well as the SOCS proteins, STAT-regulated factors that feed back to suppress the upstream Jaks (35). Interestingly, Yoshikawa et al. recently showed that methylation-induced silencing of the *SOCS1* gene leads to constitutive STAT3 activation in a hepatocellular carcinoma–derived cell line (36).

After the dimeric Stat has entered the nucleus and has bound its target DNA, it is eventually dephosphorylated by a still-unidentified tyrosine phosphatase and shuttles back to the cytoplasm. These interactions suggest still other mechanisms by which the normal regulation of Stat signaling might be lost in tumorigenesis, some of which have indeed been demonstrated in cancer cells or tumor-derived cell lines (Figure 1). In particular, a group of STAT inhibitors known as PIAS proteins have been identified that bind tyrosine-phosphorylated, dimeric STATs and prevent them from binding DNA, perhaps by potentiating tyrosine dephosphorylation (37). Downregulation of PIAS3 protein, as is seen in anaplastic large cell lymphoma, may be in part responsible for maintaining high levels of activated STAT3 (38).

Concluding thoughts

Cellular transformation by activated Stat3 and its relatives undoubtedly occurs through the transcriptional regulation of specific genes. Many Stat3 target genes are known, including those encoding the antiapoptotic proteins Bcl-xl, Mcl-1, and Bcl-2, the proliferation-associated proteins Cyclin D1 and Myc, and the proangiogenic factor VEGF (20, 27, 29, 39–41). In addition, STAT3 cooperates with c-Jun to repress expression of FAS, presumably interfering with cancer cell apoptosis (42). Still other genes must be regulated indirectly by STAT3, many of which may contribute to oncogenesis or tumor progression.

It is important to recognize that Stats, including Stat3, can be persistently activated under various circumstances in which cellular transformation is not the ultimate phenotype — in macrophages within an inflamed joint, for example, and in neuronal hypoxia (43–45). Because tumorigenesis is a multistep process, constitutive activation of Stat proteins alone need not lead to transformation. However, in the appropriate context it clearly is a critical molecule in tumorigenesis. The growing list of cases where suppression of STAT signaling leads to the demise of tumor cells establishes that these molecules contribute to the cancerous phenotype and provides hope that the near future will bring therapeutics targeting activated STAT molecules.

Address correspondence to: Jacqueline Bromberg, Memorial Sloan-Kettering Cancer Center, 1230 York Avenue, Box 397, New York, New York 10021, USA. Phone: (212) 639-2577; Fax: (646) 422-2045; E-mail: bromberj@mskcc.org.

1. Darnell, J.E., Jr. 1997. STATs and gene regulation. *Science.* **277**:1630–1635.

2. Bromberg, J., and Darnell, J.E., Jr. 2000. The role of STATs in transcriptional control and their impact on cellular function. *Oncogene.* **19**:2468–2473.

3. Levy, D.E. 1999. Physiological significance of STAT proteins: investigations through gene disruption in vivo. *Cell. Mol. Life Sci.* **55**:1559–1567.

4. Candotti, F., Notarangelo, L., Visconti, R., and O'Shea, J. 2002. Molecular aspects of primary immunodeficiencies. *J. Clin. Invest.* In press.

5. Decker, T., Stockinger, S., Karaghiosoff, M., Müller, M., and Kovarik, P. 2002. Interferons and Stats in innate immunity to microorganisms. *J. Clin. Invest.* In press.

6. Levy, D.E., and Lee, C. 2002. What does Stat3 do? *J. Clin. Invest.* **109**:1143–1148. DOI:10.1172/JCI200215650.

7. Meraz, M.A., et al. 1996. Targeted disruption of the Stat1 gene in mice reveals unexpected physiologic specificity in the JAK-STAT signaling pathway. *Cell.* **84**:431–442.

8. Durbin, J.E., Hackenmiller, R., Simon, M.C., and Levy, D.E. 1996. Targeted disruption of the mouse Stat1 gene results in compromised innate immunity to viral disease. *Cell.* **84**:443–450.

9. Kaplan, D.H., et al. 1998. Demonstration of an interferon gamma-dependent tumor surveillance system in immunocompetent mice. *Proc. Natl. Acad. Sci. USA.* **95**:7556–7561.

10. Bromberg, J.F., Horvath, C.M., Wen, Z., Schreiber, R.D., and Darnell, J.E., Jr. 1996. Transcriptionally active Stat1 is required for the antiproliferative effects of both IFN-α and IFN-γ. *Proc. Natl. Acad. Sci. USA.* **93**:7673–7678.

11. Kumar, A., Commane, M., Flickinger, T.W., Horvath, C.M., and Stark, G.R. 1997. Defective TNF-alpha-induced apoptosis in STAT1-null cells due to low constitutive levels of caspases. *Science.* **278**:1630–1632.

12. Lee, C.K., Smith, E., Gimeno, R., Gertner, R., and Levy, D.E. 2000. STAT1 affects lymphocyte survival and proliferation partially independent of its role downstream of IFN-gamma. *J. Immunol.* **164**:1286–1292.

13. Chin, Y.E., et al. 1996. Cell growth arrest and induction of cyclin-dependent kinase inhibitor p21 WAF1/CIP1 mediated by STAT1. *Science.* **272**:719–722.

14. Bromberg, J.F., Fan, Z., Brown, C., Mendelsohn, J., and Darnell, J.E., Jr. 1998. Epidermal growth factor-induced growth inhibition requires Stat1 activation. *Cell Growth Differ.* **9**:505–512.

15. Ouchi, T., Lee, S.W., Ouchi, M., Aaronson, S.A., and Horvath, C.M. 2000. Collaboration of signal transducer and activator of transcription 1 (STAT1) and BRCA1 in differential regulation of IFN-gamma target genes. *Proc. Natl. Acad. Sci. USA.* **97**:5208–5213.

16. Dupuis, S., et al. 2001. Impairment of mycobacterial but not viral immunity by a germline human STAT1 mutation. *Science.* **293**:300–303.

17. Sexl, V., et al. 2000. Stat5a/b contribute to interleukin 7-induced B-cell precursor expansion, but abl- and bcr/abl-induced transformation are independent of stat5. *Blood.* **96**:2277–2283.

18. Schwaller, J., et al. 2000. Stat5 is essential for the myelo- and lymphoproliferative disease induced by TEL/JAK2. *Mol. Cell.* **6**:693–704.

19. Akira, S. 2000. Roles of STAT3 defined by tissue-specific gene targeting. *Oncogene.* **19**:2607–2611.

20. Bromberg, J., et al. 1999. Stat3 as an oncogene. *Cell.* **98**:295–303.

21. Schuringa, J.J., et al. 2001. MEN2A-RET-induced cellular transformation by activation of STAT3. *Oncogene.* **20**:5350–5358.

22. Ning, Z.Q., Li, J., McGuinness, M., and Arceci, R.J. 2001. STAT3 activation is required for Asp(816) mutant c-Kit induced tumorigenicity. *Oncogene.* **20**:4528–4536.

23. Kube, D., et al. 2001. STAT3 is constitutively activated in Hodgkin cell lines. *Blood.* **98**:762–770.

24. Gao, B., et al. 2001. Constitutive activation of JAK-STAT3 signaling by BRCA1 in human prostate cancer cells. *FEBS Lett.* **488**:179–184.

25. Campbell, C.L., Jiang, Z., Savarese, D.M., and Savarese, T.M. 2001. Increased expression of the interleukin-11 receptor and evidence of STAT3 activation in prostate carcinoma. *Am. J. Pathol.* **158**:25–32.

26. Epling-Burnette, P.K., et al. 2001. Inhibition of STAT3 signaling leads to apoptosis of leukemic large granular lymphocytes and decreased Mcl-1 expression. *J. Clin. Invest.* **107**:351–362.

27. Epling-Burnette, P.K., et al. 2001. Cooperative regulation of Mcl-1 by Janus kinase/stat and phosphatidylinositol 3-kinase contribute to granulocyte-macrophage colony-stimulating factor-delayed apoptosis in human neutrophils. *J. Immunol.* **166**:7486–7495.

28. Turkson, J., et al. 2001. Phosphotyrosyl peptides block Stat3-mediated DNA binding activity, gene regulation, and cell transformation. *J. Biol. Chem.* **276**:45443–45455.

29. Bowman, T., Garcia, R., Turkson, J., and Jove, R. 2000. STATs in oncogenesis. *Oncogene.* **19**:2474–2488.

30. Lacronique, V., et al. 1997. A TEL-JAK2 fusion protein with constitutive kinase activity in human leukemia. *Science.* **278**:1309–1312.

31. Peeters, P., et al. 1997. Fusion of TEL, the ETS-variant gene 6 (ETV6), to the receptor-associated kinase JAK2 as a result of t(9;12) in a lymphoid and t(9;15;12) in a myeloid leukemia. *Blood.* **90**:2535–2540.

32. Lacronique, V., et al. 2000. Transforming properties of chimeric TEL-JAK proteins in Ba/F3 cells. *Blood.* **95**:2076–2083.

33. Meydan, N., et al. 1996. Inhibition of acute lymphoblastic leukaemia by a Jak-2 inhibitor. *Nature.* **379**:645–648.

34. Turkson, J., and Jove, R. 2000. STAT proteins: novel molecular targets for cancer drug discovery. *Oncogene.* **19**:6613–6626.

35. Hilton, D.J. 1999. Negative regulators of cytokine signal transduction. *Cell. Mol. Life Sci.* **55**:1568–1577.

36. Yoshikawa, H., et al. 2001. SOCS-1, a negative regulator of the JAK/STAT pathway, is silenced by methylation in human hepatocellular carcinoma and shows growth-suppression activity. *Nat. Genet.* **28**:29–35.

37. Shuai, K. 2000. Modulation of STAT signaling by STAT-interacting proteins. *Oncogene.* **19**:2638–2644.

38. Zhang, Q., et al. 2002. Multilevel dysregulation of STAT3 activation in anaplastic lymphoma kinase-positive T/null-cell lymphoma. *J. Immunol.* **168**:466–474.

39. Bowman, T., et al. 2001. Stat3-mediated Myc expression is required for Src transformation and PDGF-induced mitogenesis. *Proc. Natl. Acad. Sci. USA.* **98**:7319–7324.

40. Catlett-Falcone, R., et al. 1999. Constitutive activation of Stat3 signaling confers resistance to apoptosis in human U266 myeloma cells. *Immunity.* **10**:105–115.

41. Niu, G., et al. 2002. Constitutive Stat3 activity upregulates VEGF expression and tumor angiogenesis. *Proc. Natl. Acad. Sci. USA.* In press.

42. Ivanov, V.N., et al. 2001. Cooperation between STAT3 and c-jun suppresses Fas transcription. *Mol. Cell.* **7**:517–528.

43. Wen, T.C., Peng, H., Hata, R., Desaki, J., and Sakanaka, M. 2001. Induction of phosphorylated Stat3 following focal cerebral ischemia in mice. *Neurosci. Lett.* **303**:153–156.

44. Ivashkiv, L.B. 2000. Jak-STAT signaling pathways in cells of the immune system. *Rev. Immunogenet.* **2**:220–230.

45. Deon, D., et al. 2001. Cross-talk between IL-1 and IL-6 signaling pathways in rheumatoid arthritis synovial fibroblasts. *J. Immunol.* **167**:5395–5403.

The cadherin-catenin adhesion system in signaling and cancer

Maralice Conacci-Sorrell, Jacob Zhurinsky, and Avri Ben-Ze'ev

Department of Molecular Cell Biology, The Weizmann Institute of Science, Rehovot, Israel.

The adhesion of cells to their neighbors determines cellular and tissue morphogenesis and regulates major cellular processes including motility, growth, differentiation, and survival. Cell-cell adherens junctions (AJs), the most common (indeed, essentially ubiquitous) type of intercellular adhesions, are important for maintaining tissue architecture and cell polarity and can limit cell movement and proliferation. AJs assemble via homophilic interactions between the extracellular domains of calcium-dependent cadherin receptors on the surface of neighboring cells. The cytoplasmic domains of cadherins bind to the submembranal plaque proteins β-catenin or plakoglobin (γ-catenin), which are linked to the actin cytoskeleton via α-catenin (Figure 1; refs. 1, 2). The transmembrane assembly of cadherin receptors with the cytoskeleton is necessary for the stabilization of cell-cell adhesions and normal cell physiology.

Malignant transformation is often characterized by major changes in the organization of the cytoskeleton, decreased adhesion, and aberrant adhesion-mediated signaling. Disruption of normal cell-cell adhesion in transformed cells may contribute to tumor cells' enhanced migration and proliferation, leading to invasion and metastasis. This disruption can be achieved by downregulating the expression of cadherin or catenin family members or by activation of signaling pathways that prevent the assembly of AJs. The importance of the major epithelial cell cadherin, E-cadherin (E-cad, the product of the *CDH1* gene), in the maintenance of normal cell architecture and behavior is underscored by the observation that hereditary predisposition to gastric cancer results from germline mutations in *CDH1*. Loss of E-cad expression eliminates AJ formation and is associated with the transition from adenoma to carcinoma and acquisition of metastatic capacity (3). Re-establishment of AJs in cancer cells by restoration of cadherin expression (4) exerts tumor-suppressive effects, including decreased proliferation and motility. In this Perspective, we discuss the molecular mechanisms underlying the role of the cadherin-catenin system in the regulation of cell proliferation, invasion, and intracellular signaling during cancer progression.

Downregulation of AJ assembly by mutations, hypermethylation, and transcriptional repression of E-cad expression

Mutations in *CDH1* that compromise the adhesive function of E-cad have been observed in human gastric carcinoma cell lines, lobular breast cancer, and familial gastric cancer (5). Certain tumors, for example invasive lobular carcinoma of the breast, and tumor cell lines that display mutations in one allele of *CDH1* also acquire a deletion in the other allele, consistent with a two-hit mechanism for the loss of E-cad and suggesting that *CDH1* behaves as a classical tumor suppressor gene. While acquisition of loss-of-function mutations and the subsequent loss of heterozygosity are important mechanisms for silencing E-cad expression in tumor cells, progression to the metastatic phenotype can also involve a reversible downregulation of E-cad expression at the transcriptional level, sometimes achieved by methylation of the *CDH1* promoter.

DNA methylation often causes downregulation of tumor suppressor genes in cancer cells by changing chromatin structure, thereby making the DNA inaccessible for transcription factors and RNA polymerase II (6). Hypermethylation of the *CDH1* promoter has been observed in human breast, prostate, and hepatocellular tumors that carry a wild-type *CDH1* gene. This methylation is reversible and can vary according to changes in the tumor microenvironment. For example, Graff et al. (7) found that when primary cultures of human breast carcinoma cells displaying a methylated *CDH1* promoter were cultured as spheroids, which requires homotypic cell-cell adhesion, promoter methylation decreased, allowing expression of E-cad. In some cases, for example in patients with hereditary diffuse gastric cancers that carry germline mutations in one allele of the *CDH1* gene, the remaining allele is inactivated by DNA methylation (8). Finally, inhibition of DNA methylation can suppress the initiation of tumor development in a mouse model system for colorectal cancer, suggesting that methylation provides an attractive target for anticancer therapy.

Transcriptional silencing of *CDH1* may also result from aberrant expression of transcription factors that repress its promoter. These include snail, E12/E47, and SIP1, which bind to the *CDH1* promoter, inhibit E-cad transcription, and induce an epithelial-to-mesenchymal transition (EMT), leading to acquisition of invasiveness (9–12). The transcriptional repressor snail, a zinc finger protein originally identified as a regulator of mesoderm formation in developing *Drosophila*, was recently shown to be also critical for EMT in mouse development. In human carcinoma and melanoma, the expression of snail correlates with the absence of E-cad expression. Forced expression of snail in cultured epithelial cells represses the *CDH1* promoter and induces a program of invasive growth characterized by the loss of AJs and the induction of EMT and tumorigenesis (9, 10). Interestingly, the invasive areas of mouse skin tumors express the highest levels of snail, supporting the view that activation of snail confers invasive properties on cells (9). E12/E47 and SIP1, which also repress the *CDH1* promoter, can likewise induce EMT, increase cell motility and invasion, and confer tumorigenicity. Moreover, SIP1 expression is elevated in several E-cad–negative human carcinoma cells (11). Thus, cancer cells employ reversible or irreversible mechanisms to silence the expression of E-cad.

Citation for this article: *J. Clin. Invest.* **109**:987–991 (2002). DOI:10.1172/JCI200215429.

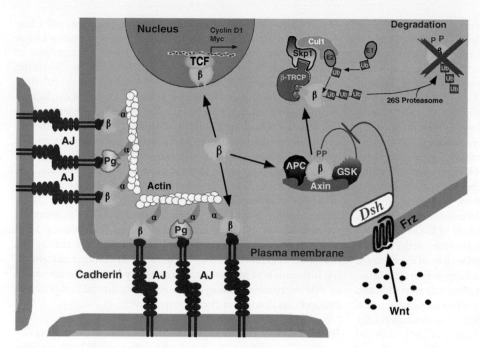

Figure 1
The dual role of β-catenin in cell adhesion and transcriptional activation. β-Catenin (β) and plako-globin (Pg) bind to cadherin adhesion receptors, and via α-catenin (α) they associate with the actin cytoskeleton to form AJs. When the Wnt signaling pathway is inactive, free β-catenin is degraded by a complex including glycogen synthase kinase (GSK), adenomatous polyposis coli (APC), and Axin, which phosphorylate β-catenin (PP). This protein complex recruits β-TrCP, which, togeth-er with Skp1, Cul1, and the E1 and E2 ubiquitination components, mediates the ubiquitination of β-catenin (Ub) and directs it to degradation by the 26S proteasome. The binding of Wnt to Frizzled (Frz) receptors activates Wnt signaling, and dishcveled (Dsh) inhibits β-catenin phosphorylation by GSK. This results in β-catenin accumulation in the nucleus, where it complexes with T cell factor (TCF) and transactivates target genes such as *Cyclin D1* and *Myc*. Modified from ref. 56.

their growth rate, and decreases their tumorigenic capacity (15). Deletions in the α-catenin binding site of β-catenin have been identified in a signet ring cell carcinoma of the stomach. In these cells, β-catenin–cadherin complexes are formed, but the lack of α-catenin in these complexes prevents their linkage to the actin cytoskeleton and the assem-bly of AJs. Moreover, in mice where *CTNNA1* is eliminated in the epider-mis, AJs are disrupted and the epithe-lial skin cells display hyperproliferation (16). This hyperproliferation has been attributed to aberrant adhesion-medi-ated signaling, indicating that disrup-tion of AJ assembly directly contributes to oncogenesis.

The reciprocal relationship between cadherin-mediated adhesion and β-catenin signaling
In addition to providing a physical link between cells, the mode of AJ assembly can influence various signal-ing pathways. A major route for signal transduction by AJs involves the regula-tion of β-catenin–T cell factor (TCF) sig-naling. β-Catenin plays a dual role in the cells: In addition to its structural role in AJs, β-catenin can act as a transcrip-tion factor in the nucleus by serving as a coactivator of the lymphoid enhancer factor (LEF)/TCF family of DNA-bind-

These mechanisms may contribute to enhanced cancer cell motility and invasiveness. The transcriptional mechanisms involved in down-regulation of E-cad are similar to those characteristic of typical stages during embryonic development, when changes in cell adhesion and motility take place. Since the loss of E-cad expression is a common feature in many types of carcinoma, and since E-cad reintroduction into cancer cells reduces invasion and metastasis, cancer therapies tar-geting the regulation of E-cad expression may offer powerful means to control the spread of cancer. A recent study demonstrated that E-cad mRNA expression and cell-cell adhesion are induced in SW480 human colon cancer cells by vitamin D3 (13). This makes vitamin D3 a potential candidate for treating colon cancer by inhibiting cell growth, similar to its effect on melanoma and soft tissue sarcoma.

Regulation of catenins and AJ assembly in cancer cells
The interaction between the cytoplasmic tail of cadherins with the catenins and the actin cytoskeleton (Figure 1; ref. 2) is critical for the establishment of stable and functional AJs. Catenin mutations that prevent this interaction have been reported in different types of cancer. A homozygous deletion of the α-catenin gene, *CTNNA1*, leading to loss of cell-cell adhesion, has been found in a human lung cancer cell line (14). Introduction of wild-type α-catenin into these cells restores normal adhesion. Likewise, AJ formation is blocked in an ovarian carcinoma cell line with a deletion in α-catenin that abolishes binding to β-catenin (15). Expression of full-length α-catenin in these cells restores their epithelial morphology, reduces

ing proteins (Figure 1; ref. 17). β-Catenin–mediated transcription is activated by the Wnt pathway, which is crucial during various stages of embryonal development. Activation of Wnt signaling involves the inhibition of β-catenin degradation by the proteasomes, resulting in its nuclear accumulation and transcriptional activation of LEF/TCF target genes (Figure 1). Mutations in components that regulate β-catenin turnover (e.g., adenomatous polyposis coli [APC] or Axin) (Figure 1), as well as N-terminal mutations in β-catenin itself that compromise the protein's degradation, have been found in a variety of human cancers (18). These mutations cause β-catenin to accu-mulate in the nucleus and activate target genes in an aberrant man-ner. Since the interaction of β-catenin with cadherins and LEF/TCF family members is mediated by the same domain (the so-called arm repeat) on the β-catenin molecule, these interactions are mutually exclusive. Thus, recruitment of β-catenin into AJs by elevating the expression of cadherin can decrease its nuclear pool and antagonize β-catenin–LEF/TCF transactivation (19–22).

β-Catenin can activate genes that stimulate cell proliferation, such as *Cyclin D1* (23, 24) and *Myc* (25), whose promoters contain LEF/TCF–binding sites. Downregulation of β-catenin–LEF/TCF signaling may therefore lead to decreased proliferation, as dem-onstrated in experiments where cadherin-dependent inhibition of growth was suggested to involve the sequestration of the signaling pool of β-catenin (26, 27). Another means to antagonize aberrant β-catenin signaling would be to enhance the antiproliferative p53-mediated response. Indeed, recent studies demonstrate that p53 expression is induced in cells displaying abnormal β-catenin levels (28). This induction is mediated by the elevation of p19^ARF and is analogous to the cellular response to other oncoproteins, such as Ras and Myc (29). In addition to growth arrest, cells with aberrant β-catenin signaling show enhanced turnover of β-catenin, again as a result of elevated p53 expression (30–32).

Disruption of AJs releases β-catenin from the AJ pool. When it is not rapidly degraded, β-catenin translocates into the nucleus, where it induces the transcription not only of cyclin D1 and c-myc, but also of a variety of other genes that contribute to cancer progression. These include the ECM protein fibronectin (33), which contributes to adhesion and motility; the metalloproteinase matrilysin (34), which stimulates extravasation and metastasis; and the multidrug resistance protein Mdr1 (35). In colorectal tumors, the central part of the tumor displays high levels of E-cad at the cell membrane that colocalizes with β-catenin. However, at the invasive front of these tumors, β-catenin is localized mainly in the nuclei of dissociated tumor cells that have lost E-cad expression (36). Interestingly, the loss of E-cad in such cells is apparently reversible and regulated by the tumor cell environment, since metastatic nodes from the same tumor display well-developed E-cad–containing AJs. Thus, the loss of E-cad that causes disruption of cell adhesion and polarity allows tumor cell metastasis, while the translocation of β-catenin into the nucleus might be required to induce the expression of genes that promote cell proliferation and invasion.

Activation of β-catenin signaling also occurs during EMT mediated either by c-Fos or by the IGF II signaling pathway. Tyrosine phosphorylation of β-catenin, which promotes disruption of AJs, also increases the affinity of β-catenin to the TATA-binding protein, a general positive regulator of transcription (37). Nevertheless, analysis of breast cancer cell lines lacking cadherin expression does not support a correlation between loss of E-cad expression and activation of β-catenin signaling (38). Additional events, including compromised proteasomal degradation of β-catenin, its tyrosine phosphorylation, and possibly a release from transcriptional inhibition, may be required to activate β-catenin signaling in such cancer cells.

Cross-talk between AJs and growth factor receptors

Growth factors such as EGF, HGF/scatter factor (SF), and FGF can induce the dismantling of AJs and can cause a dramatic change in cell morphology and gene expression in which cells shift from an epithelial to a fibroblastic phenotype and initiate a program of invasive growth (39, 40). Such growth factors operate via receptor tyrosine kinases, which often localize to sites of AJ assembly, as demonstrated for the EGF and HGF receptors. It is conceivable that these activated growth factor receptors tyrosine-phosphorylate various AJ components, including β-catenin, thereby contributing to the disruption of AJs observed in cancer cells. For example, the poor AJ organization of some Ras-transformed cells has been attributed to enhanced tyrosine phosphorylation of β-catenin (41), which, at least in vitro, results in its decreased binding to cadherin (42). Disruption of AJs by HGF stimulation requires, in addition, the activation of phosphatidylinositol (PI) 3-kinase and mitogen-activated protein kinase (MAPK) by HGF/SF, since inhibition of these kinases prevents the disassembly of AJs in these cells (43). The regulation of AJ assembly may also be achieved via phosphorylation by CKII on serine residues of the cadherin cytoplasmic tail, a modification that enhances its binding to β-catenin (44). Point mutations in the phosphorylation sites of the cadherin cytoplasmic tail have been shown to reduce its binding to β-catenin (44, 45).

Whereas tyrosine kinase receptor signaling regulates cell adhesion, it is also possible that AJs, in turn, influence tyrosine kinase signaling. The reduction in cell proliferation characteristic of dense cell cultures is mediated, in part, by cadherin-containing AJs that render the cells insensitive to growth factor stimulation. For example, in dense cultures that assemble extensive AJs, ligand-dependent dimerization and activation of the EGF receptor tyrosine kinase are inhibited (46). In addition, in cells maintained in a three-dimensional culture system, cadherin-dependent growth inhibition is mediated by the activation of tyrosine phosphatases, which suppress signaling pathways initiated by receptor tyrosine kinases. In such cells, the assembly of AJs prevents the activation of the EGF receptor pathway by TGF-α (47).

Interestingly, N-cadherin, in contrast to E-cadherin, enhances tyrosine kinase signaling and thus enhances cell motility. High levels of N-cadherin have been found in invasive tumor cell lines (48, 49). The ability of N-cadherin to enhance cell motility, invasion, and metastasis is apparently due to the activation of FGF receptor signaling and involves its interaction with the extracellular domain of N-cadherin (48, 49).

Rho GTPases and the regulation of cell-cell adhesion

Small GTPases of the Rho family (Rho, Rac, and Cdc42) are regulatory proteins that coordinate remodeling of the actin cytoskeleton in response to various stimuli (50). In the active, GTP-bound form, they interact with and activate target proteins that regulate actin polymerization, cell motility, and gene expression. The Rho GTPases are involved in cell transformation, and some of their activators are oncogenes, including Vav, Net, and Dbl (51). The assembly and maintenance of AJs require the activity of Rho and Rac (52, 53), and deregulation of these small GTPases in transformed cells has been shown to interfere with cadherin function. For example, in Ras-transformed cells that are inefficient in the assembly of AJs, expression of Tiam1, a Rac activator, can restore AJ assembly and epithelial morphology and can inhibit invasiveness (54). While the molecular mechanisms by which small GTPases affect cell-cell adhesion are still poorly understood, they likely involve regulating actin cytoskeletal organization and AJ assembly (55).

Conclusions and future perspectives

The assembly of cadherin-containing AJs is essential for the formation and maintenance of cellular and tissue integrity. In addition to their structural-mechanical function, these junctions play a pivotal role in regulating cellular responses to adhesion- and growth factor–mediated cues, and other environmental signals. The ability of excessive AJ assembly to decrease cell growth and motility is probably related to the action of AJs as tumor suppressors. During cancer progression, cells evade adhesion-mediated regulatory mechanisms by either acquiring mutations in AJ proteins, or inactivating their function by various signaling mechanisms. The loss of AJs in cooperation with deregulation of other pathways leads to enhanced cell proliferation, motility, and, eventually, metastasis.

The mechanisms regulating the assembly of AJs and those involved in cadherin-mediated signaling via β-catenin are poorly understood. Future efforts will probably be devoted to understanding the details of how AJs are disrupted in response to MAPK or PI3-kinase activation and to changes in the activity of the small GTPases of the Rho family. The signaling pathways that repress E-cad expression through the transcription factors snail, E12/E47, and others, as well as their regulation in cancer cells, require further study. These subjects are currently under intensive investigation, and we expect that such studies will provide new information that will allow the identification of novel targets for inhibiting tumor development.

Acknowledgments

The studies from the authors' laboratory are supported by a grant from the Israel Science Foundation.

Address correspondence to: Avri Ben-Ze'ev, Department of Molecular Cell Biology, The Weizmann Institute of Science, Rehovot 76100, Israel. Phone: 972-8-9342422; Fax: 972-8-9465261; E-mail: avri.ben-zeev@weizmann.ac.il.

Maralice Conacci-Sorrell and Jacob Zhurinsky contributed equally to this work.

1. Ben-Ze'ev, A., and Geiger, B. 1998. Differential molecular interactions of β-catenin and plakoglobin in adhesion, signaling and cancer. *Curr. Opin. Cell Biol.* **10**:629–639.

2. Nagafuchi, A. 2001. Molecular architecture of adherens junctions. *Curr. Opin. Cell Biol.* **13**:600–603.

3. Perl, A., Wilgenbus, P., Dahl, U., Semb, H., and Christofori, G. 1998. A causal role for E-cadherin in the transition from adenoma to carcinoma. *Nature.* **392**:190–193.

4. Vleminckx, K., Vakaet, L., Mareel, M., Fiers, W., and van Roy, F. 1991. Genetic manipulation of E-cadherin expression by epithelial tumor cells reveals an invasion suppressor role. *Cell.* **66**:107–119.

5. Hirohashi, S. 1998. Inactivation of the E-cadherin-mediated cell adhesion system in human cancers. *Am. J. Pathol.* **153**:333–339.

6. Jones, P., and Laird, P. 1999. Cancer-epigenetics comes of age. *Nat. Genet.* **21**:163–167.

7. Graff, J.R., Gabrielson, E., Fujii, H., Baylin, S.B., and Herman, J.G. 2000. Methylation patterns of the E-cadherin 5′ CpG island are unstable and reflect the dynamic, heterogeneous loss of E-cadherin expression during metastatic progression. *J. Biol. Chem.* **275**:2727–2732.

8. Grady, W., et al. 2000. Methylation of the CDH1 promoter as the second genetic hit in hereditary diffuse gastric cancer. *Nat. Genet.* **26**:16–17.

9. Cano, A., et al. 2000. The transcription factor snail controls epithelial-mesenchymal transitions by repressing E-cadherin expression. *Nat. Cell Biol.* **2**:76–83.

10. Batlle, E., et al. 2000. The transcription factor snail is a repressor of E-cadherin gene expression in epithelial tumour cells. *Nat. Cell Biol.* **2**:84–89.

11. Comijn, J., et al. 2001. The two-handed E box binding zinc finger protein SIP1 downregulates E-cadherin and induces invasion. *Mol. Cell.* **7**:1267–1278.

12. Perez-Moreno, M., et al. 2001. A new role for E12/E47 in the repression of E-cadherin expression and epithelial-mesenchymal transitions. *J. Biol. Chem.* **276**:27424–27431.

13. Palmer, H.G., et al. 2001. Vitamin D(3) promotes the differentiation of colon carcinoma cells by the induction of E-cadherin and the inhibition of β-catenin signaling. *J. Cell Biol.* **154**:369–387.

14. Hirano, S., Kimoto, N., Shimoyama, Y., Hirohashi, S., and Takeichi, M. 1992. Identification of a neural α-catenin as a key regulator of cadherin function and multicellular organization. *Cell.* **70**:293–301.

15. Bullions, L., Notterman, D., Chung, L., and Levine, A. 1997. Expression of wild-type α-catenin protein in cells with a mutant α-catenin gene restores both growth regulation and tumor suppressor activities. *Mol. Cell. Biol.* **17**:4501–4508.

16. Vasioukhin, V., Bauer, C., Degenstein, L., Wise, B., and Fuchs, E. 2001. Hyperproliferation and defects in epithelial polarity upon conditional ablation of α-catenin in skin. *Cell.* **104**:605–617.

17. Bienz, M., and Clevers, H. 2000. Linking colorectal cancer to Wnt signaling. *Cell.* **103**:311–320.

18. Polakis, P. 2000. Wnt signaling and cancer. *Genes Dev.* **14**:1837–1851.

19. Heasman, J., et al. 1994. Overexpression of cadherins and underexpression of β-catenin inhibit dorsal mesoderm induction in early Xenopus embryos. *Cell.* **79**:791–803.

20. Fagotto, F., Funayama, N., Gluck, U., and Gumbiner, B. 1996. Binding to cadherins antagonizes the signaling activity of β-catenin during axis formation in Xenopus. *J. Cell Biol.* **132**:1105–1114.

21. Sadot, E., Simcha, I., Shtutman, M., Ben-Ze'ev, A., and Geiger, B. 1998. Inhibition of β-catenin-mediated transactivation by cadherin derivatives. *Proc. Natl. Acad. Sci. USA.* **95**:15339–15344.

22. Simcha, I., et al. 1998. Differential nuclear translocation and transactivation potential of β-catenin and plakoglobin. *J. Cell Biol.* **141**:1433–1448.

23. Shtutman, M., et al. 1999. The cyclin D1 gene is a target of the β-catenin/LEF-1 pathway. *Proc. Natl. Acad. Sci. USA.* **96**:5522–5527.

24. Tetsu, O., and McCormick, F. 1999. β-catenin regulates expression of cyclin D1 in colon carcinoma cells. *Nature.* **398**:422–426.

25. He, T., et al. 1998. Identification of c-MYC as a target of the APC pathway. *Science.* **281**:1509–1512.

26. Gottardi, C., Wong, E., and Gumbiner, B. 2001. E-cadherin suppresses cellular transformation by inhibiting β-catenin signaling in an adhesion independent manner. *J. Cell Biol.* **153**:1049–1060.

27. Stockinger, A., Eger, A., Wolf, J., Beug, H., and Foisner, R. 2001. E-cadherin regulates cell growth by modulating proliferation-dependent β-catenin transcriptional activity. *J. Cell Biol.* **154**:1185–1196.

28. Damalas, A., et al. 1999. Excess β-catenin promotes accumulation of transcriptionally active p53. *EMBO J.* **18**:3054–3063.

29. Damalas, A., Kahan, S., Shtutman, M., Ben-Ze'ev, A., and Oren, M. 2001. Deregulated β-catenin induces a p53- and ARF-dependent growth arrest and cooperates with Ras in transformation. *EMBO J.* **20**:4912–4922.

30. Sadot, E., Geiger, B., Oren, M., and Ben-Ze'ev, A. 2001. Down-regulation of β-catenin by activated p53. *Mol. Cell. Biol.* **21**:6768–6781.

31. Liu, J., et al. 2001. Siah-1 mediates a novel β-catenin degradation pathway linking p53 to the adenomatous polyposis coli protein. *Mol. Cell.* **7**:927–936.

32. Matsuzawa, S.I., and Reed, J.C. 2001. Siah-1, SIP, and Ebi collaborate in a novel pathway for β-catenin degradation linked to p53 responses. *Mol. Cell.* **7**:915–926.

33. Gradl, D., Kuhl, M., and Wedlich, D. 1999. The Wnt/Wg signal transducer β-catenin controls fibronectin expression. *Mol. Cell. Biol.* **19**:5576–5587.

34. Crawford, H., et al. 1999. The metalloproteinase matrilysin is a target of β-catenin transactivation in intestinal tumors. *Oncogene.* **18**:2883–2891.

35. Yamada, T., et al. 2000. Transactivation of the multidrug resistance 1 gene by T-cell factor 4/β-catenin complex in early colorectal carcinogenesis. *Cancer Res.* **60**:4761–4766.

36. Brabletz, T., et al. 2001. Variable β-catenin expression in colorectal cancers indicates tumor progression driven by the tumor environment. *Proc. Natl. Acad. Sci. USA.* **98**:10356–10361.

37. Piedra, J., et al. 2001. Regulation of β-catenin structure and activity by tyrosine phosphorylation. *J. Biol. Chem.* **276**:20436–20443.

38. van de Wetering, M., et al. 2001. Mutant E-cadherin breast cancer cells do not display constitutive Wnt signaling. *Cancer Res.* **61**:278–284.

39. Savagner, P. 2001. Leaving the neighborhood: molecular mechanisms involved during epithelial-mesenchymal transition. *Bioessays.* **23**:912–923.

40. Hay, E. 1995. An overview of epithelio-mesenchymal transformation. *Acta Anat. (Basel).* **154**:8–20.

41. Kinch, M., Clark, G., Der, C., and Burridge, K. 1995. Tyrosine phosphorylation regulates the adhesions of ras-transformed breast epithelia. *J. Cell Biol.* **130**:461–471.

42. Roura, S., Miravet, S., Piedra, J., Garcia de Herreros, A., and Dunach, M. 1999. Regulation of E-cadherin/Catenin association by tyrosine phosphorylation. *J. Biol. Chem.* **274**:36734–36740.

43. Potempa, S., and Ridley, A. 1998. Activation of both MAP kinase and phosphatidylinositide 3-kinase by Ras is required for hepatocyte growth factor/scatter factor-induced adherens junction disassembly. *Mol. Biol. Cell.* **9**:2185–2200.

44. Lickert, H., Bauer, A., Kemler, R., and Stappert, J. 2000. Casein kinase II phosphorylation of E-cadherin increases E-cadherin/β-catenin interaction and strengthens cell-cell adhesion. *J. Biol. Chem.* **275**:5090–5095.

45. Simcha, I., et al. 2001. Cadherin sequences that inhibit β-catenin signaling: a study in yeast and mammalian cells. *Mol. Biol. Cell.* **12**:1177–1188.

46. Takahashi, K., and Suzuki, K. 1996. Density-dependent inhibition of growth involves prevention of EGF receptor activation by E-cadherin-mediated cell-cell adhesion. *Exp. Cell Res.* **226**:214–222.

47. St. Croix, B., et al. 1998. E-Cadherin-dependent growth suppression is mediated by the cyclin dependent kinase inhibitor p27(KIP1). *J. Cell Biol.* **142**:557–571.

48. Nieman, M., Prudoff, R., Johnson, K., and Wheelock, M. 1999. N-cadherin promotes motility in human breast cancer cells regardless of their E-cadherin expression. *J. Cell Biol.* **147**:631–644.

49. Hazan, R., Phillips, G., Qiao, R., Norton, L., and Aaronson, S. 2000. Exogenous expression of N-cadherin in breast cancer cells induces cell migration, invasion, and metastasis. *J. Cell Biol.* **148**:779–790.

50. Ridley, A. 2001. Rho family proteins: coordinating cell responses. *Trends Cell Biol.* **11**:471–477.

51. Van Aelst, L., and D'Souza-Schorey, C. 1997. Rho GTPases and signaling networks. *Genes Dev.* **11**:2295–2322.

52. Braga, V., Machesky, L., Hall, A., and Hotchin, N. 1997. The small GTPases Rho and Rac are required for the establishment of cadherin-dependent cell-cell contacts. *J. Cell Biol.* **137**:1421–1431.

53. Takaishi, K., Sasaki, T., Kotani, H., Nishioka, H., and Takai, Y. 1997. Regulation of cell-cell adhesion by rac and rho small G proteins in MDCK cells. *J. Cell Biol.* **139**:1047–1059.

54. Hordijk, P., et al. 1997. Inhibition of invasion of epithelial cells by Tiam1-Rac signaling. *Science.* **278**:1464–1466.

55. Fukata, M., and Kaibuchi, K. 2001. Rho-family GTPases in cadherin-mediated cell-cell adhesion. *Nat. Rev. Mol. Cell Biol.* **2**:887–897.

56. Zhurinsky, J., Shtutman, M., and Ben-Ze'ev, A. 2000. Plakoglobin and β-catenin: protein interactions, regulation and biological roles. *J. Cell Sci.* **113**:3127–3139.

Do tumor-suppressive mechanisms contribute to organism aging by inducing stem cell senescence?

Pier Giuseppe Pelicci

Department of Experimental Oncology, European Institute of Oncology and FIRC Institute of Molecular Oncology, Milan, Italy

Stem/progenitor cells ensure tissue and organism homeostasis and might represent a frequent target of transformation. Although these cells are potentially immortal, their life span is restrained by signaling pathways (p19-p53; p16-Rb) that are activated by DNA damage (telomere dysfunction, environmental stresses) and lead to senescence or apoptosis. Execution of these checkpoint programs might lead to stem cell depletion and organism aging, while their inactivation contributes to tumor formation.

Telomeres form high-order chromatin structures that cap the ends of eukaryotic chromosomes. They contain thousands of double-stranded repeats (TTAGGG) and terminate with a single-stranded 3′-extension, which, together with specific telomeric proteins, participates in the formation of the terminal loop structure (telomere cap). Telomeres are synthesized by telomerase, a cellular reverse transcriptase that adds TTAGGG repeats onto preexisting telomeres. In cells that do not express telomerase, TTAGGG repeats are lost at each cell division, and, when telomeres reach a critical length, a checkpoint is triggered that drives cells into a metabolically active state of irreversible growth arrest, termed replicative senescence. Therefore, telomere shortening has generally been regarded as a counting mechanism that limits the mitotic potential of any cell type. In this view, cellular senescence can be considered a potent tumor-protection mechanism. Indeed, telomerase is not expressed in most human somatic cells, while it is expressed in most cancer cell types (except in a minority of tumors in which telomeres are stabilized through alternative mechanisms) (1, 2). Constitutive expression of telomerase in primary, telomerase-negative cells induces immortalization or cooperates with specific oncogenes to induce transformation (3–5).

Cellular senescence, however, might also contribute to a decline in tissue homeostasis by exhausting the supply of progenitors or stem cells, which suggests that organism aging is the trade-off of the evolved adaptation to tumor suppression (antagonistic pleiotropy). As expected from their high replicative potential, stem cells express telomerase. They are not, however, immortal and undergo telomere erosion during aging. Notably, overexpression of telomerase in mouse hematopoietic stem cells prevents telomere erosion yet has no effect on the cells' lifespan suggesting that telomere-independent mechanisms regulate replicative senescence of stem cells in vivo (6, 7).

Telomerase-deficient mice show progressive telomere shortening. However, they have no obvious defects for the first few generations, most likely because of the unusually long telomeres in the germ line of laboratory mouse strains. After four to six generations, these mutant mice show reduced lifespan, though they lack the full spectrum of classical symptoms of aging. They demonstrate reduced tumor development but, surprisingly, enhanced tumor

initiation. This dual effect of telomere shortening has been interpreted to be the consequence of activation of an antiproliferative checkpoint response and induction of chromosomal instability, due to inappropriate fusions of the uncapped telomeres (8–10).

Recent findings shed new light on the molecular pathways associated with the execution of the cellular senescence program and suggest that the tumor-suppressive mechanisms involved may directly contribute to organismal aging, possibly acting at the level of stem/progenitor cells. These topics are addressed by the four review articles (11–14) of this *JCI* Perspective series.

Both telomere dysfunction and other forms of DNA damage activate p53 and induce senescence

The signal transduction pathways that activate cellular senescence are now better understood. Division of telomerase-negative cells causes the erosion of the telomeric single-stranded 3′-extension, preventing extension of the double-stranded region and, perhaps more importantly, causing functional uncapping of the telomeres. The chromosomal DNA ends are then exposed and recognized by the cell as a double-stranded break. As expected when a double-stranded break forms, DNA-repair and -damage checkpoint factors are recruited to the site of damaged DNA, and a p53-dependent checkpoint is initiated. So replicative senescence is a p53-dependent checkpoint response to DNA damage (11, 12).

p53 is a tumor suppressor that is activated by a variety of stressful cellular conditions, including DNA damage, oxidative stress, and oncogenic (hyperproliferative) signals. It is not surprising, then, that the senescence program can be activated by the same stresses, even in telomerase-positive cells — a phenomenon known as premature, or stress-induced, senescence. It is also well recognized that in vitro culturing can be, by itself, a powerful stress (culture shock) that induces premature senescence. For example, Schwann cells and oligodendrocyte precursor cells (which express telomerase) can either grow indefinitely or undergo senescence, depending on culture conditions (15, 16). Similarly, human epithelial cells expressing telomerase become immortal only if cultured on feeder layers (17). Therefore, replicative senescence and stress-induced senescence cannot be separated, as they "merely reflect the spectrum of different stimuli that feed into one response program" (11). Moreover, recent findings demonstrate that the rate of telomere shortening of growing cells is influenced by the culture conditions, suggesting that telomere shortening is not simply an intrinsic mitotic

Conflict of interest: The author has declared that no conflict of interest exists.

Citation for this article: *J. Clin. Invest.* **113**:4–7 (2004). doi:10.1172/JCI200420750.

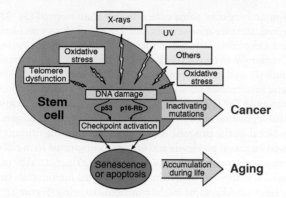

Figure 1
DNA damage accumulates as the consequence of endogenous (telomere dysfunction, oxidative stress) or exogenous (oxidative stress, γ-irradiation, UV light, and others) attacks. Damaged DNA activates checkpoint responses that are mediated by the p53 and p16-Rb pathways and that result in apoptosis or cellular senescence. If these events occur in stem/progenitor cells, tissue homeostasis is altered — a phenomenon that might contribute to aging. If, instead, DNA mutations that inactivate these checkpoint pathways accumulate, then cancer can arise.

counter but is itself regulated by stress signals. In general, it is as if every cell possessed a memory of its history, in terms of both cell divisions and the type of environment to which it was exposed. It is not surprising, then, that damaged DNA is the senescence-activating signal upon telomere shortening. Remarkably, it has been suggested recently that DNA damage is also the signal that limits proliferation of primary cells cultured under standard conditions (18). DNA, indeed, is the only cellular molecule where historical information can be permanently stored, in the form of mutations.

In the absence of p16-Rb or p53, telomere dysfunction induces genomic instability

Telomerase and the p53 pathways are not the sole genetic determinants of cellular senescence. The p16-Rb pathway is also involved. Expression of p16 increases during replicative or oncogene-induced senescence and, when increased experimentally, induces premature senescence. Heterochromatin foci are not formed in quiescent cells but are formed in senescent cells, where promoter-bound, hypophosphorylated Rb permanently represses transcription of a number of cell cycle genes (E2F-target genes) (13). Though this is not definitively ascertained, the p16-Rb pathway can also be activated by telomere dysfunction (11–14). In conclusion, telomere dysfunction, or other DNA damage, activates the senescence program through the p53 and p16-Rb checkpoint pathways.

Based on this simple model, it is not surprising that in vitro, in the absence of functional p53 or p16-Rb — which is characteristic of most if not all tumors — telomere dysfunction causes genomic instability and cells accumulate complex chromosome rearrangements. Something similar may also occur during colon and breast tumor development in vivo, where, early in cancer progression, low telomerase activity (and presumably telomere dysfunction) correlates with the presence of complex chromosomal rearrangements (19–21). Subsequently, high levels of telomerase expression are selected to ensure that the cancer cell maintains proper telomere function and growth potential, and the continuous telomerase expression that is needed for tumor growth.

Telomerase, however, may also contribute to tumorigenesis via mechanisms that are independent of its telomere-lengthening effect. Enhanced telomerase expression in telomerase-positive cells promotes growth and survival. When telomerase is overexpressed in mice with telomeres of normal length, it increases the incidence of spontaneous tumors (22). The frequency of mutagen-induced tumors is reduced in early-generation telomere-deficient mice, when telomeres are still long (23). Furthermore, telomerase is periodically expressed in normal human fibroblasts (previously considered telomere-negative), where it participates in the active maintenance of the length of the single-stranded 3'-extension and regulates each cell division (24).

Stem cells are frequent targets of transformation

How is telomerase expression reactivated in cancer cells, and how are high levels of telomerase expression selected during transformation? Despite intensive investigations, little is known about the physiological mechanisms that regulate telomerase expression. Notably, none of the genes that are most frequently altered in cancer are known to affect telomerase expression (with the possible exception of the myc oncogene). A simple answer to these questions is that the target cell for the initial transforming mutations is telomerase-positive. For most cancers, the target cell of transformation is unknown. Telomerase is expressed in a restricted subset of normal cells: germ cells, stem/progenitor cells, and proliferating lymphocytes. Stem/progenitor cells have a long lifespan, and the machinery for self-renewal is already activated. Therefore, as compared with differentiated cells, they have greater risks of accumulating mutations and may require fewer events to sustain uncontrolled growth. For leukemias, convincing genetic and biological data support the notion that stem/progenitor cells are the most frequent cellular target of transformation (25). Furthermore, it has been recently demonstrated that different types of malignant tumors (breast cancers, glioblastomas, neuroblastomas, and leukemias) contain rare cells with indefinite proliferation potential that support the formation and growth of the remaining tumor population (26). These cells possess the same functional and phenotypic features as the corresponding tissue stem cells and have therefore been named tumor stem cells. According to these findings, a tumor can be seen as an abnormal tissue that is initiated by a single, transformed stem cell that retains parts of its program of differentiation.

Inhibition of the senescence program in stem cells contributes to cancer formation

Regardless of how the problem of telomere lengthening is solved in cancer cells — transformation of telomerase-positive cells or reactivation of telomerase expression in telomere-negative cells — the senescence program has to be suppressed for a cell to evade tumor suppression. In fact, environmental and hyperproliferative stresses are also potent inducers of senescence (and apoptosis) in telomerase-positive cells. The two pathways that execute the program of stress-induced senescence are, as mentioned previously, p19-p53 and p16-Rb, which are invariably altered in cancer (Figure 1). However, since activation of the same two pathways can also trigger apoptosis, it has been difficult to assess the relative contribution of senescence as a tumor-suppressive barrier. Recent findings regarding the *Bmi-1* oncogene might help elucidate this issue.

Bmi-1 is a transcriptional repressor that functions as an inhibitor of the senescence program in cultured fibroblasts: its expression is downregulated during spontaneous senescence, and Bmi-1–null fibroblasts enter premature senescence, while enforced Bmi-1 expression extends the replicative lifespan. Bmi-1 regulates senescence through the p16-Rb pathway, by repressing p16 expression. More recently, it has been demonstrated that Bmi-1 functions in vivo to maintain the pool of neuronal and hematopoietic stem cells, by inhibiting their senescence program, again through p16, while it has no effect on growth and differentiation of precursors (27, 28). Bmi-1 might, therefore, contribute to tumorigenesis by inhibiting p16-mediated senescence, specifically in stem cells. Notably, Bmi-1 is frequently amplified in mantle cell lymphomas, and its expression is indispensable for the maintenance of leukemic stem cells (13).

Succesful activation of DNA-damage checkpoint pathways in stem cells (telomere dysfunction, p53) might lead to organism aging

If inactivation of the senescence program is indeed a relevant mechanism of tumor formation, then its reactivation might represent a successful antitumor strategy. Treatment with various chemotherapeutic agents, such as doxorubicin or other DNA-damage agents, was shown to induce cellular senescence of cancer cells in vitro and, more recently, in vivo (29, 30). This event seems to require functional p53 and p16. However, at higher doses, doxorubicin induces senescence of p53-null cancer cells. Other compounds are now being tested that induce senescence in p53- or p16-null cells, possibly through activation of downstream effectors of these pathways. Reactivation of the senescence program might turn out to be a relevant anticancer strategy, particularly in those tumors where the apoptotic pathways are severely compromised (14).

What happens if, instead, the telomere and the p19-p53 and p16-Rb checkpoints are properly activated in normal stem/progenitor cells? Cells will enter senescence (or apoptosis). If this occurs constantly over the lifespan of an organism, one would intuitively expect a reduction of the pool of stem/progenitor cells, altered tissue homeostasis, and compromised tissue repair, features which may well contribute to organism aging. Remarkably, recent findings suggest that some components of these checkpoint pathways are indeed genetic determinants of lifespan in mammals. I have already commented on the effects of telomerase mutations in mice; in humans, equivalent mutations cause a rare progeroid syndrome. Three different types of transgenic mice have been recently reported, in which p53 expres-

sion or activity is to some extent higher than normal (31–33). As expected, all three mouse lines had a markedly lower incidence of cancer. Despite the reduced incidence, these mice did not live longer, and, strikingly, in two of the three lines, lifespan was shorter and the animals showed signs of premature aging. In another model system, genetic intercrosses showed a marked progeric effect of the ATM mutation (ATM is another critical player in cell cycle checkpoints and the regulation of DNA damage) on the telomerase-null genotype (34). Notably, the progeric ATM-telomerase double-mutant mice showed increased p53 levels and apoptosis in various stem cell compartments. Consistently, loss of p53 or p16-p19 increases the pool of stem cells. It appears, in conclusion, that some mammalian tumor-suppressive mechanisms might contribute to aging (Figure 1).

If some tumor-suppressor genes show antagonistic pleiotropy and contribute to aging, then it would be impossible to improve tumor-suppressive mechanisms without accelerating aging, and to retard aging without accelerating tumor formation. The few available mutant mice with increased lifespan, however, have no increased risk of cancer. In the case of mice that lack the 66-kDa Shc isoform (*p66Shc*^{−/−} mice), available data suggest the existence of a defect in a selective branching of the p53 checkpoint pathway (35, 36). p66Shc is a downstream target of p53 and regulates the generation of oxygen radicals, the major source of intracellular oxidative stress. Genetic evidence indicates that the p53-p66Shc signaling pathway is specifically involved in the propagation of proapoptotic oxidative signals (p66Shc). Other functions of p53 are not influenced by p66Shc expression. It appears, therefore, that the tumor-suppressive and aging effects of p53 are mediated by different pathways. If aging is not the cost of tumor suppression, we are then left with one major question: Why do we age? It may be that we are yet too young to provide an answer.

Acknowledgments

I would like to thank G.F. Draetta, K. Helin, B. Amati, C. Basilico, and J.C. Marine for many helpful and exciting discussions.

Address correspondence to: Pier Giuseppe Pelicci, Department of Experimental Oncology, European Institute of Oncology, Via Ripamonti 435, 20141 Milan, Italy. Phone: 39-2-57489831; Fax: 39-2-57489851; E-mail: pgpelicci@ieo.it.

1. Kim, N.W., et al. 1994. Specific association of human telomerase activity with immortal cells and cancer. *Science.* **266**:2011-2015.

2. Bryan, T.M., Englezou, A., Dalla-Pozza, L., Dunham, M.A., and Reddel, R.R. 1997. Evidence for an alternative mechanism for maintaining telomere length in human tumors and tumor-derived cell lines. *Nat. Med.* **3**:1271-1274.

3. Bodnar, A.G., et al. 1998. Extension of life-span by introduction of telomerase into normal human cells. *Science.* **279**:349-352.

4. Vaziri, H., and Benchimol, S. 1998. Reconstitution of telomerase activity in normal human cells leads to elongation of telomeres and extended replicative life span. *Curr. Biol.* **8**:279-282.

5. Hahn, W.C., et al. 1999. Creation of human tumor cells with defined genetic elements. *Nature.* **400**:464-468.

6. Yui, J., Chiu, C.P., and Lansdorp, P.M. 1998. Telomerase activity in candidate stem cells from fetal liver and adult bone marrow. *Blood.* **91**:3255-3262.

7. Allsopp, R.C., and Weissman, I.L. 2002. Replicative senescence of hematopoietic stem cells during serial transplantation: does telomere shortening play a role? *Oncogene.* **21**:3270-3273.

8. Rudolph, K.L., et al. 1999. Longevity, stress response, and cancer in aging telomerase-deficient mice. *Cell.* **96**:701-712.

9. Hande, M.P., Samper, E., Lansdorp, P., and Blasco, M.A. 1999. Telomere length dynamics and chromosomal instability in cells derived from telomerase null mice. *J. Cell Biol.* **144**:589-601.

10. Niida, H., et al. 1998. Severe growth defect in mouse cells lacking the telomerase RNA component. *Nat. Genet.* **19**:203-206.

11. Ben-Porath, I., and Weinberg, R.A. 2004. When cells get stressed: an integrative view of cellular senescence. *J. Clin. Invest.* **113**:8-13. doi:10.1172/JCI200420663.

12. Sharpless, N.E., and DePinho, R.A. 2004. Telomeres, stem cells, senescence, and cancer. *J. Clin. Invest.* In press.

13. Park, I.-K., Morrison, S.J., and Clarke, M.F. 2004. *Bmi-1*, stem cells, and senescence regulation. *J. Clin. Invest.* In press.

14. Kahlem, P., Dörken, B., and Schmitt, C.A. 2004. Cellular senescence in cancer treatment: friend or foe? *J. Clin. Invest.* In press.

15. Mathon, N.F., Malcolm, D.S., Harrisingh, M.C., Cheng, L., and Lloyd, A.C. 2001. Lack of replica-

tive senescence in normal rodent glia. *Science.* **291**:872-875.

16. Tang, D.G., Tokumoto, Y.M., Apperly, J.A., Lloyd, A.C. and Raff, M.C. 2001. Lack of replicative senescence in cultured rat oligodendrocyte precursor cells. *Science.* **291**:868-871.

17. Ramirez, R.D., et al. 2001. Putative telomere-independent mechanisms of replicative aging reflect inadequate growth conditions. *Genes Dev.* **15**:398-403.

18. Parrinello, S., et al. 2003. Oxygen sensitivity severely limits the replicative lifespan of murine fibroblasts. *Nat. Cell Biol.* **5**:741-747.

19. Chin, L., et al. 1999. p53 deficiency rescues the adverse effects of telomere loss and cooperates with telomere dysfunction to accelerate carcinogenesis. *Cell.* **97**:527-538.

20. Artandi, S.E., et al. 2002. Telomere dysfunction promotes non-reciprocal translocations and epithelial cancers in mice. *Nature.* **406**:641-645.

21. Buerger, H., et al. 1999. Comparative genomic hybridization of ductal carcinoma in situ of the breast-evidence of multiple genetic pathways. *J. Pathol.* **187**:396-402.

22. Gonzalez-Suarez, E., et al. 2001. Increased epidermal

tumors and increased skin wound healing in transgenic mice overexpressing the catalytic subunit of telomerase, mTERT, in basal keratinocytes. *EMBO J.* **20**:2619–2630.

23. Gonzalez-Suarez, E., Samper, E., Flores, J.M., and Blasco, M.A. 2000. Telomerase-deficient mice with short telomeres are resistant to skin tumorigenesis. *Nat. Genet.* **26**:114–117.

24. Masutomi, K., et al. 2003. Telomerase maintains telomere structure in normal human cells. *Cell.* **114**:241–253.

25. Passegue, E., Jamieson, C.H., Ailles, L.E., and Weissman, I.L. 2003. Normal and leukemic hematopoiesis: are leukemias a stem cell disorder or a reacquisition of stem cell characteristics? *Proc. Natl. Acad. Sci. U. S. A.* **100**(Suppl 1):11842–11849.

26. Al-Hajj, M., Wicha, M.S., Benito-Hernandez, A.,

Morrison, S.J., and Clarke, M.F. 2003. Prospective identification of tumorigenic breast cancer cells. *Proc. Natl. Acad. Sci. U. S. A.* **100**:3983–3988.

27. Park, I.K., et al. 2003. Bmi-1 is required for maintenance of adult self-renewing haematopoietic stem cells. *Nature.* **423**:302–305.

28. Molofsky, A.V., et al. 2003. Bmi-1 dependence distinguishes neural stem cell self-renewal from progenitor proliferation. *Nature.* **425**:962–967.

29. Schmitt, C.A., et al. 2002. A senescence program controlled by p53 and p16INK4a contributes to the outcome of cancer therapy. *Cell.* **109**:335–346.

30. te Poele, R.H., Okorokov, A.L., Jardine, L., Cummings, J., and Joel, S.P. 2002. DNA damage is able to induce senescence in tumor cells in vitro and in vivo. *Cancer Res.* **62**:1876–1883.

31. Tyner, S.D., et al. 2002. p53 mutant mice that dis-

play early ageing-associated phenotypes. *Nature.* **415**:45–53.

32. Garcia-Cao, I., et al. 2002. "Super p53" mice exhibit enhanced DNA damage response, are tumor resistant and age normally. *EMBO J.* **21**:6225–6235.

33. Davenport, R.J. 2002. Tumor-free, but not in the clear. *Sci. Aging Knowl. Environ.* **40**:nw139.

34. Wong, K.K., et al. 2003. Telomere dysfunction and Atm deficiency compromises organ homeostasis and accelerates ageing. *Nature.* **421**:643–648.

35. Migliaccio, E., et al. 1999. The p66shc adaptor protein controls oxidative stress response and life span in mammals. *Nature.* **402**:309–313.

36. Trinei, M., et al. 2002. A p53-p66Shc signalling pathway controls intracellular redox status, levels of oxidation-damaged DNA and oxidative stress-induced apoptosis. *Oncogene.* **21**:3872–3878.

When cells get stressed: an integrative view of cellular senescence

Ittai Ben-Porath and Robert A. Weinberg

The Whitehead Institute for Biomedical Research, Cambridge, Massachusetts, USA

Cells entering a state of senescence undergo a permanent cell cycle arrest, accompanied by a set of functional and morphological changes. Senescence of cells occurs following an extended period of proliferation in culture or in response to various physiologic stresses, yet little is known about the role this phenomenon plays in vivo. The study of senescence has focused largely on its hypothesized role as a barrier to extended cell division, governed by a division-counting mechanism in the form of telomere length. Here, we discuss the biological functions of cellular senescence and suggest that it should be viewed in terms of its role as a general cellular stress response program, rather than strictly as a barrier to unlimited cycles of cell growth and division. We also discuss the relative roles played by telomere shortening and telomere uncapping in the induction of senescence.

Is senescence a biological program?

Cellular senescence, one of the most fundamental aspects of cell behavior, was first described and formalized in the work of Hayflick in 1961 (1). Most types of primary normal cells that are grown in culture do not proliferate indefinitely. Instead, after a period of rapid proliferation, their division rate slows, ultimately ceasing altogether. Such cells become unresponsive to mitogenic stimuli yet can remain viable for extended periods of time. Upon entering the state of senescence, cells undergo a dramatic change in morphology — their volume increases and they lose their original shape, acquiring a flattened cytoplasm. This shift is accompanied by changes in nuclear structure, gene expression, protein processing, and metabolism (2–4). This form of senescence, which follows an extended period of propagation of cells in culture, has been termed "replicative senescence."

The study of the molecular mechanisms underlying senescence has shed light on central aspects of tumor development and has contributed to the research on organismal aging. Nonetheless, the role that cellular senescence itself plays in the living organism is still poorly understood. This stands in contrast to another cellular response that serves to constrain cell proliferation — apoptosis — for which the molecular mechanisms and biological roles have been elucidated in minute detail over the past decade.

Can senescence be placed side-by-side with apoptosis as a fundamental biological program? Subsequent to Hayflick's discovery of replicative senescence, various studies have demonstrated that normal cells can undergo senescence rapidly in response to various physiologic stresses (5). This later work yielded a second category of senescence, often referred to as "stress-induced senescence." In fact, cells that are exposed to stress in culture will respond either by entry into senescence, by apoptosis, or by a transient growth arrest; the choice among these three responses depends on the cell type, the type of stress, and the level of stress. Hence, senescence seems to represent one of several programs that can be activated by the cell when physiologic stress is encountered.

The observation that a variety of stressors can bring about the senescence phenotype supports the notion that this phenotype represents a general cellular response mechanism rather than an idiosyncratic response to a specific type of physical or biochemical insult. Moreover, upon the activation of senescence, a plethora of changes in cellular morphology and function are induced in parallel, a task that is executed by specific molecular pathways.

These observations converge on the conclusion that senescence is a carefully orchestrated cellular program, indeed one that is likely to play an important role in the physiology of cells within living tissues. In truth, to date, only a handful of studies have reported the detection of senescence of cells in vivo (including recent contributions described below). Thus, it is not clear that the functional and morphological changes that cells undergo upon entering senescence in culture also occur in vivo.

Mouse mutants of the molecular activators of senescence — *p53*, *Rb*, and *p16/Ink4a* — are tumor prone (6–9), suggesting that senescence serves as a tumor-suppressing mechanism. However, it has been difficult to directly demonstrate that it is the inactivation of the senescence program, rather than of other functions performed by these proteins, that leads to tumor development in these mice. Due to these difficulties, it remains possible that senescence is largely a phenomenon of cells growing in the culture dish and does not serve a physiologic role in vivo (10, 11).

However, two recent studies provide convincing demonstrations of senescence occurring in cells in vivo. In one of these studies, mice carrying eroded telomeres due to a mutant telomerase enzyme were subjected to partial hepatectomy (12). The hepatocytes of these mice displayed limited ability to proliferate and regenerate the liver; instead, they underwent senescence. In the other study, mice carrying lymphomas were treated by a chemotherapeutic agent (13). These lymphoma cells underwent senescence in vivo, but only when the apoptosis program was blocked by overexpression of the *bcl2* gene. Thus, given the proper experimental setting, senescence can clearly be observed in vivo, both in response to an internal signal — telomere attrition — and in response to an externally inflicted stress in the form of an alkylating agent. In both studies senescence was characterized by an arrest of division and by the appearance of senescence-associated β-galactosidase activity, a commonly used marker for senescence in vitro.

Nonstandard abbreviations used: human telomerase reverse transcriptase (hTERT); mouse embryo fibroblast (MEF); short interfering RNA (siRNA).

Conflict of interest: The authors have declared that no conflict of interest exists.

Citation for this article: *J. Clin. Invest.* **113**:8–13 (2004). doi:10.1172/JCI200420663.

A limit for proliferation, a response to stress, or both

The initial discovery of the replicative senescence of human fibroblasts has led to the view that senescence serves as a mechanism whose purpose is to limit the proliferative capacity of normal cells (11, 14). According to this thinking, it is undesirable for cells to be capable of dividing beyond what is required for their participation in normal development and tissue maintenance. The capacity of cells to divide is therefore limited by an intrinsic mechanism that counts the number of divisions through which cell lineages have undergone, and triggers senescence when the predetermined limit for division is reached. While the senescence of cells might contribute to the aging of tissues, the breakdown of this division-restricting mechanism can lead to cancer.

The finding that telomeres, the nucleoprotein structures protecting chromosome ends, shorten with every cycle of cell growth and division suggested a molecular mechanism that could record the number of divisions that a lineage of cells has undergone. Erosion of telomeres to a critical length could serve to activate the senescence program (15). In accord with this mechanistic model, ectopic expression of the catalytic subunit of the telomerase enzyme, hTERT (human telomerase reverse transcriptase), halts the erosion of telomeres in human cells; in some cell types the expression of this gene prevents the entrance into replicative senescence, suggesting that indeed telomere shortening is the cause for senescence (16, 17).

As mentioned above, yet other work has demonstrated that normal cells that are exposed to various physiologic stresses rapidly enter into a state of senescence, doing so within a period as short as several days. Such stresses include DNA-damaging agents, oxidative stress, "oncogenic stress" (due to oncogene overexpression), and other metabolic perturbations (5, 18–22). Typically, these forms of senescence do not involve significant telomere shortening and cannot be prevented by ectopic hTERT expression (23, 24). Accordingly, the hypothesized telomere-based mechanism of division counting could not be invoked to explain these acute responses. Moreover, these situations of stress-induced senescence could not be accommodated by a model proposing that senescence functions exclusively as a barrier to extended growth-and-division cycles.

These observations raised the question of whether replicative senescence and stress-induced senescence serve the same biological function, and which of these mechanisms operates in vivo (10, 11, 25). In fact, the conditions that induce these two responses are not as distinct as the above description would suggest. The onset of replicative senescence exhibited by some types of normal cells is dependent on the conditions in which they are propagated. This suggests that certain culture conditions are physiologically stressful to the cells, and that cells can undergo replicative senescence due to the cumulative effect of this stress, rather than the progressive erosion of their telomeres. For example, populations of human mammary epithelial cells encounter their first growth barrier following 10–20 divisions in culture. This stage of senescence can be avoided if these cells are grown on fibroblast feeder layers or in a different type of medium (26, 27). Similarly, mouse embryo fibroblasts (MEFs) senesce after approximately ten divisions. It was recently demonstrated that when MEFs are propagated in 3% oxygen, rather than the commonly used 20% oxygen conditions, they can avoid senescence (28). This study also demonstrated that when grown in 20% oxygen, MEFs suffer from the rapid accumulation of DNA damage. Consequently, the cumulative oxidative damage induced by growth in conditions that are hyperoxic (by the standards of living tissues) leads to the onset of senescence in these cells. These findings provide a direct demonstration of how extrinsic physiologic stress experienced by cells can lead to replicative senescence.

In many cases, both types of mechanisms for the induction of replicative senescence — a telomere-based one and a stress-based one — seem to function together in the same cell population. Human fibroblasts are the best-studied example of a cell type in which the cause of senescence is attributed to critical telomere attrition, as their senescence can usually be prevented by ectopic expression of hTERT. However, these cells are hardly indifferent to their growth conditions in vitro: the timing of their entry into senescence is affected by various parameters of culturing, such as plating density, media composition, and others. Thus, when propagated in 1–3% oxygen instead of 20% oxygen, human fibroblasts are able to undergo more divisions prior to senescence; conversely, oxygen levels higher than 20% will shorten their lifespan (29–31). Clearly, a stress-based clock contributes to the effect of the telomere-based clock in these cells, and the combined effects of the two dictate the onset time of senescence.

Physiologic stress could hasten the onset of senescence in these cells in a manner independent of the process of telomere shortening. Alternatively, it could act by increasing the rates of telomere shortening. However, if the latter were true, then telomeric DNA could no longer be viewed as a counting device that advances autonomously at its own rate, but should rather be viewed as a cellular structure that responds to stress. In fact, it has been reported in the past that the telomeres of some cells shorten more quickly in high oxygen conditions (31). This observation received further substantiation in a recent study, demonstrating that in 2–5% oxygen conditions the rates of telomere shortening of some commonly studied human fibroblast lines — WI38 and IMR90 — are slower than the rates observed in 21% oxygen (32). Moreover, when the hTERT gene is ectopically introduced into various human fibroblast lines, resulting in the elongation of their telomeres, the rate of telomere elongation is much slower in 21% oxygen than in 2–5% oxygen growth conditions. This study clearly demonstrates that telomere shortening is not an intrinsic clock-like mechanism that operates independently of extrinsic physiologic stresses. This recent study also underscores a little-regarded observation, namely that many normal human fibroblast lines cannot be immortalized by ectopic expression of hTERT. Telomerase activity enables these cells to proliferate longer, but eventually they do undergo senescence, even though their telomeres have been elongated well beyond the lengths observed in early passage cells (32). Thus, the immortalizing capabilities of telomerase are only limited to a subset of cell types.

Different types of intrinsic and extrinsic stress signals are likely to converge on the activation of the p53 protein, the Rb protein, or both. In this manner, these two key tumor suppressor proteins might act as integrators of stress signals, and their combined level of activation would determine the onset of senescence (Figure 1). Recent studies have demonstrated that some cultured human fibroblast lines indeed express higher levels of the p16/INK4a gene than do others, presumably reflecting a higher degree of stress that these cells experience in vitro (33, 34). This stress-induced expression of p16, an activator of Rb, apparently acts to hasten the replicative senescence of cells, doing so in a manner independent of telomere length (34, 35). These and other findings suggest that telomere attrition leads mainly to the activation of the p53 protein, while culture stresses mainly activate Rb through p16.

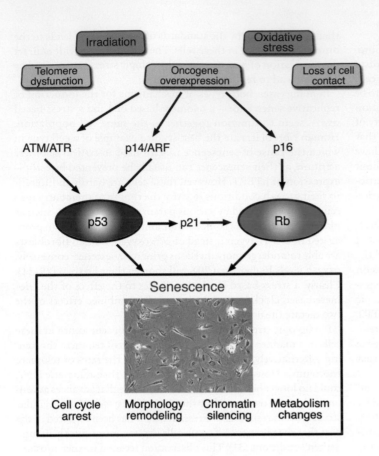

Figure 1
Senescence as a general stress-response program. A variety of physiologic stresses, intrinsic and extrinsic, lead to the onset of senescence. These stresses stimulate various cellular signaling pathways, which are funneled down to activate either the p53 protein, the Rb protein, or both. p53 can be activated by the DNA damage signaling pathways, the ATM and ATR proteins, or by the p14/ARF protein, which responds to oncogene overexpression and other stresses. p21, a target of p53, can cause the activation of Rb. Most cellular stresses will activate the p16/INK4a gene, also leading to Rb activation. Different stress signals will have different relative effects on the p53 and Rb arms, and their combined level of activation dictates the onset of senescence. Once this program is activated, a series of changes in cell function and morphology take place. ATM, ataxia telangiectasia mutated; ATR, ATM-related; p14/ARF, alternative reading frame product of *INK4a* gene locus.

The emerging picture is that various types of intrinsic and extrinsic stress stimuli can activate the senescence program, and whether this occurs rapidly, or gradually following a period of proliferation, is mainly determined by the combined levels of these stresses. Moreover, extrinsic signals may affect cell-intrinsic processes such as telomere shortening. It seems, then, that there is no necessity in a functional distinction between replicative senescence and stress-induced senescence, as these titles merely reflect the fact that a spectrum of different stimuli feed into one response program.

Telomeres — length or structure

The telomeres of human cells that are propagated until they reach senescence erode, on average, to about half of their original length — from 10–20 kb to 5–10 kb in most cell types (15, 36). A situation of almost complete erosion of telomeres can be reached if senescence is bypassed through the inactivation the p53 and Rb pathways (37). When this is done, cells continue to divide beyond their normal senescence point, and their telomeres continue to shorten until a subsequent growth barrier, termed "crisis," is reached. At crisis, massive cell death occurs due to multiple chromosomal fusions and consequent genomic catastrophe. Senescence is thus induced at a midway point of telomere shortening. This provokes questions regarding the nature of the molecular changes that occur at the telomeres at this point, and the manner by which these changes activate the senescence program.

Different models have been proposed to explain how telomere shortening triggers senescence. Some models suggest that the onset of senescence is dictated by the length of the shortest telomere within a cell, rather than by the average length of telomeres (38). These models argue that due to variability in telomere lengths, there exists at least one telomere per cell that has eroded to a very short length at the time of senescence. Loss of telomeric repeats may cause the exposure of the chromosomal DNA end, which is recognized as a double-stranded break (39). Indeed, several studies have indicated that telomere lengths within a cell are highly variable, and that cells accumulate short telomeres as they approach senescence, some being as short as 2 kb or less (40, 41).

Other studies have argued, however, that the relationship between telomere length and senescence is more complex. Telomere length does not always strictly correlate with the induction of senescence, and it is difficult to point to a consistent critical length that induces the senescence response (42, 43). Some cell lines are capable of dividing with very short telomeres, a situation that is apparently sustained by ongoing telomerase expression (44). It has been recently shown that human fibroblasts that overexpress the telomere-binding protein TRF2 display an accelerated rate of telomere shortening; surprisingly, these cells undergo senescence following the same number of divisions as control cells, even though their telomeres at senescence are shorter (45). This study clearly demonstrates that, at least in the case of these cells, it is not the length of the double-stranded region per se that dictates the timing of senescence.

Many groups have therefore arrived at the hypothesis that it is probably not the actual telomere length — the number of double-stranded hexameric repeat units at the telomeres — that is the molecular feature directly dictating the timing of senescence. Instead, the structure of telomeres — the nucleoprotein complex that serves as a protective "cap" for the chromosome end — may undergo changes during extended cell proliferation, possibly as a result of telomere shortening. "Uncapping" of telomeres, rather

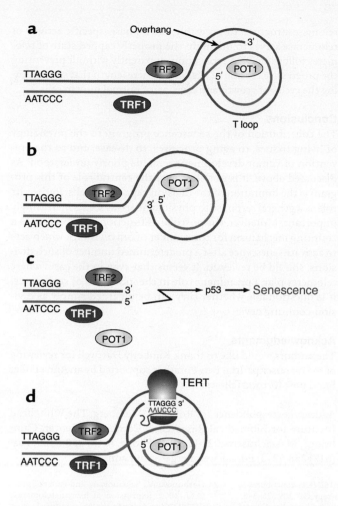

Figure 2
Telomere uncapping at senescence. The nucleoprotein structure at the end of telomeres presumably forms a protective cap. This structure may be composed of the T-loop (**a**), which is formed by the invasion of the single-stranded overhang into an upstream double-stranded region of the telomere, and of protective telomere-binding proteins such as TRF1, TRF2, and POT1. As cells approach replicative senescence, the double-stranded portion of the telomere shortens, and the single-stranded overhang is eroded (**b**). This may cause the collapse of the telomere cap and the exposure of the telomere end, which is detected by the DNA damage machinery and leads to the activation of the senescence program (**c**). Telomerase activity, apart from stabilizing overall telomere length, can prevent overhang erosion and protect the telomere cap, thereby circumventing senescence (**d**). TERT, telomerase enzyme reverse transcriptase.

to assess overhang lengths has been facilitated by a recently developed technique — the telomeric oligonucleotide ligation assay (T-OLA) (51). Use of this assay has revealed that in senescent cells the telomeric overhang is significantly eroded (Figure 2, b and c), when compared to overhang lengths in early-passage cells (52). The extent of overhang erosion suggests that a large proportion of the telomeres in a cell have lost much of their single-stranded overhang DNA, even though most of these telomeres still carry a significant number of repeats in their double-stranded portion.

It is reasonable to assume that telomeres that have lost their single-stranded overhangs are functionally uncapped. Such telomeres may have lost their ability to form T-loops, and in addition may have lost the protection afforded by certain telomeric proteins. This process may lead to exposure of the chromosome end, activation of the DNA damage machinery (39), and the triggering of senescence (Figure 2c). The fact that telomere uncapping can lead to senescence has been demonstrated experimentally. Expression of a dominant-negative form of the TRF2 protein causes loss of the telomeric overhang and the formation of telomeric fusions, without loss of overall telomere length in the double-stranded region (53, 54). When this protein is expressed in normal fibroblasts, the result is the induction of rapid senescence, demonstrating that the forced uncapping of telomeres activates the senescence program (55).

On the basis of these findings, it can be argued that as cells approach senescence, the physical structure of telomeres undergoes changes that result in the loss of protective capping. Perhaps the most central question regarding this process is whether telomere uncapping is a consequence of telomere shortening, or whether it is an event that occurs independently of overall telomere length, being induced by other cellular signals.

Telomerase as a cap protector

More light on the role played by the telomere cap in senescence has been shed by recent discoveries regarding the function of the telomerase enzyme. Introduction of the hTERT catalytic subunit into presenescent fibroblasts results in an elongation of the double-stranded region of the telomere (16, 17). The recent examination of the telomeric overhang in such cells has revealed that ectopic hTERT expression also protects the telomeric overhang from erosion, and can, in fact, elongate it (52) (Figure 2d). This finding is not surprising, since the direct biochemical activity of telomerase is to add repeat units to the 3' end of the telomere, i.e. to the overhang, whereas the extension of the double-stranded region occurs only upon complementation of the elongated over-

than shortening of telomeres, may, according to this thinking, lead to the exposure of the chromosome ends and to the activation of the senescence program (42, 46).

Telomere uncapping at senescence

Relatively little is known about the structure that is formed by the DNA and protein components of the telomeric cap. The double-stranded stretch of hexameric repeats that comprises several kilobases of telomeric DNA ends with a 3' overhang of 100–400 nucleotides (47–49) (Figure 2a). This overhang is considered to be a critical component of telomere structure, and appears to be actively generated on every telomere following DNA replication. It has been shown that the ends of telomeric DNA can exist in a structure termed the "T-loop," which is formed by the invasion of this single-stranded overhang into an upstream double-stranded region of telomeric DNA (50). Such a second-order DNA structure, bound and held together by specific telomeric proteins, notably TRF1, TRF2, and POT1, appears to function as the protective cap for the telomere.

Do telomeres undergo uncapping at senescence? The lack of direct molecular markers for the capped or uncapped state of a telomere has limited the ability to address this question. The detection of the T-loop structure itself is technically extremely challenging, and does not easily facilitate quantitative assessment of T-loop numbers in different cells. The telomeric overhang, however, being a putative key component of telomere structure, could well serve as an indicator of the telomere capping state. The ability

hang, possibly by conventional DNA polymerases. By protecting overhang DNA from erosion, telomerase activity can maintain proper telomere structure. This raises the possibility that it is this aspect of telomerase activity, and not the elongation of the double stranded telomere region, which is critical for the prevention of senescence (Figure 2d).

Yet another recent study indicates that maintenance of proper telomere structure is in fact performed by the endogenous telomerase enzyme, operating in presenescent normal cells (56). It has long been believed that most normal human cells do not express the *hTERT* gene and therefore do not possess telomerase activity. This recent work has shown, using sensitive detection methods, that *hTERT* is in fact expressed and active in normal human fibroblasts. This expression is transient, appearing only during S-phase (56). When this transient endogenous hTERT expression was eliminated in these cells through the use of a small interfering RNA (siRNA) vector, the cells underwent premature replicative senescence. Strikingly, the rate of overall telomere shortening did not change in these cells. Instead, their telomeric overhangs eroded more rapidly than did those of control cells, so that fibroblasts expressing the siRNA against hTERT and entering senescence carried significantly eroded overhangs, whereas control cells that underwent the same number of divisions carried overhangs of normal length (56).

These findings hold several important implications. First, they demonstrate that it is possible to accelerate the erosion of telomeric overhangs (and attendant telomere uncapping) without affecting overall telomere shortening rates. Second, they indicate that it is the capping state of telomeres that correlates with the induction of senescence, and not the overall length. Third, they point to a role for the telomerase enzyme in normal cells — maintenance of

telomere structure. Conceivably, the S-phase–specific activity of telomerase serves to maintain the properly capped state of telomeres following DNA replication, apparently without preventing the progressive erosion of overall telomere length that occurs during the cycles of growth and division of normal human cells.

Conclusions

The contribution of the senescence program to the physiology of living tissues, to aging processes, to disease, and to the prevention of tumor development remains poorly understood. As discussed above, it is not clear that the central role of this program is the limitation of the division capacity of cells. Rather, its role as a general response to physiologic stress seems increasingly important. Moreover, the postulated function of telomeres as a counting mechanism for the number of cell divisions, which acts to activate senescence after a predetermined number of such divisions, should be reviewed. It seems that even in the cases where telomeres play a prominent role in the induction of senescence, it is questionable whether they can be regarded simply as division-counting devices.

Acknowledgments

The authors would like to thank Kimberly Hartwell for reviewing of the manuscript. Ittai Ben-Porath is supported by an Anna Fuller Fund postdoctoral fellowship.

Address correspondence to: Robert A. Weinberg, The Whitehead Institute for Biomedical Research, 9 Cambridge Center, Cambridge, Massachusetts 02142, USA. Phone: (617) 258-5159; Fax: (617) 258-5213; E-mail: weinberg@wi.mit.edu.

1. Hayflick, L., and Moorhead, P.S. 1961. The serial cultivation of human diploid cell strains. *Exp. Cell Res.* **25**:585–621.
2. Campisi, J. 2000. Cancer, aging and cellular senescence. *In Vivo.* **14**:183–188.
3. Sitte, N., Merker, K., Von Zglinicki, T., Grune, T., and Davies, K.J. 2000. Protein oxidation and degradation during cellular senescence of human BJ fibroblasts. I. Effects of proliferative senescence. *FASEB J.* **14**:2495–2502.
4. Narita, M., et al. 2003. Rb-mediated heterochromatin formation and silencing of E2F target genes during cellular senescence. *Cell.* **113**:703–716.
5. Lloyd, A.C. 2002. Limits to lifespan. *Nat. Cell Biol.* **4**:E25–E27.
6. Donehower, L.A., et al. 1992. Mice deficient for p53 are developmentally normal but susceptible to spontaneous tumours. *Nature.* **356**:215–221.
7. Jacks, T., et al. 1992. Effects of an Rb mutation in the mouse. *Nature.* **359**:295–300.
8. Krimpenfort, P., Quon, K.C., Mooi, W.J., Loonstra, A., and Berns, A. 2001. Loss of p16Ink4a confers susceptibility to metastatic melanoma in mice. *Nature.* **413**:83–86.
9. Sharpless, N.E., et al. 2001. Loss of p16Ink4a with retention of p19Arf predisposes mice to tumorigenesis. *Nature.* **413**:86–91.
10. Sherr, C.J., and DePinho, R.A. 2000. Cellular senescence: mitotic clock or culture shock? *Cell.* **102**:407–410.
11. Wright, W.E., and Shay, J.W. 2002. Historical claims and current interpretations of replicative aging. *Nat. Biotechnol.* **20**:682–688.
12. Satyanarayana, A., et al. 2003. Telomere shortening impairs organ regeneration by inhibiting cell cycle re-entry of a subpopulation of cells. *EMBO J.* **22**:4003–4013.
13. Schmitt, C.A., et al. 2002. A senescence program

controlled by p53 and p16INK4a contributes to the outcome of cancer therapy. *Cell.* **109**:335–346.
14. Hayflick, L. 1998. How and why we age. *Exp. Gerontol.* **33**:639–653.
15. Harley, C.B., Futcher, A.B., and Greider, C.W. 1990. Telomeres shorten during ageing of human fibroblasts. *Nature.* **345**:458–460.
16. Bodnar, A.G., et al. 1998. Extension of life-span by introduction of telomerase into normal human cells. *Science.* **279**:349–352.
17. Vaziri, H., and Benchimol, S. 1998. Reconstitution of telomerase activity in normal human cells leads to elongation of telomeres and extended replicative life span. *Curr. Biol.* **8**:279–282.
18. Fairweather, D.S., Fox, M., and Margison, G.P. 1987. The in vitro lifespan of MRC-5 cells is shortened by 5-azacytidine-induced demethylation. *Exp. Cell Res.* **168**:153–159.
19. Chen, Q., and Ames, B.N. 1994. Senescence-like growth arrest induced by hydrogen peroxide in human diploid fibroblast F65 cells. *Proc. Natl. Acad. Sci. U. S. A.* **91**:4130–4134.
20. Venable, M.E., Lee, J.Y., Smyth, M.J., Bielawska, A., and Obeid, L.M. 1995. Role of ceramide in cellular senescence. *J. Biol. Chem.* **270**:30701–30708.
21. Serrano, M., Lin, A.W., McCurrach, M.E., Beach, D., and Lowe, S.W. 1997. Oncogenic ras provokes premature cell senescence associated with accumulation of p53 and p16INK4a. *Cell.* **88**:593–602.
22. Robles, S.J., and Adami, G.R. 1998. Agents that cause DNA double strand breaks lead to p16INK4a enrichment and the premature senescence of normal fibroblasts. *Oncogene.* **16**:1113–1123.
23. Wei, S., and Sedivy, J.M. 1999. Expression of catalytically active telomerase does not prevent premature cell senescence caused by overexpression of oncogenic Ha-Ras in normal human fibroblasts. *Cancer Res.* **59**:1539–1543.

24. Gorbunova, V., Seluanov, A., and Pereira-Smith, O.M. 2002. Expression of human telomerase (hTERT) does not prevent stress-induced senescence in normal human fibroblasts but protects the cells from stress-induced apoptosis and necrosis. *J. Biol. Chem.* **277**:38540–38549.
25. Wright, W.E., and Shay, J.W. 2000. Telomere dynamics in cancer progression and prevention: fundamental differences in human and mouse telomere biology. *Nat. Med.* **6**:849–851.
26. Ramirez, R.D., et al. 2001. Putative telomere-independent mechanisms of replicative aging reflect inadequate growth conditions. *Genes Dev.* **15**:398–403.
27. Herbert, B.S., Wright, W.E., and Shay, J.W. 2002. p16(INK4a) inactivation is not required to immortalize human mammary epithelial cells. *Oncogene.* **21**:7897–7900.
28. Parrinello, S., et al. 2003. Oxygen sensitivity severely limits the replicative lifespan of murine fibroblasts. *Nat. Cell Biol.* **5**:741–747.
29. Packer, L., and Fuehr, K. 1977. Low oxygen concentration extends the lifespan of cultured human diploid cells. *Nature.* **267**:423–425.
30. Chen, Q., Fischer, A., Reagan, J.D., Yan, L.J., and Ames, B.N. 1995. Oxidative DNA damage and senescence of human diploid fibroblast cells. *Proc. Natl. Acad. Sci. U. S. A.* **92**:4337–4341.
31. von Zglinicki, T., Saretzki, G., Docke, W., and Lotze, C. 1995. Mild hyperoxia shortens telomeres and inhibits proliferation of fibroblasts: a model for senescence? *Exp. Cell Res.* **220**:186–193.
32. Forsyth, N.R., Evans, A.P., Shay, J.W., and Wright, W.E. 2003. Developmental differences in the immortalization of lung fibroblasts by telomerase. *Aging Cell.* **2**:235–243.
33. Beausejour, C.M., et al. 2003. Reversal of human cellular senescence: roles of the p53 and p16 pathways. *EMBO J.* **22**:4212–4222.

34. Itahana, K., et al. 2003. Control of the replicative life span of human fibroblasts by p16 and the polycomb protein Bmi-1. *Mol. Cell. Biol.* **23**:389–401.

35. Wei, W., Herbig, U., Wei, S., Dutriaux, A., and Sedivy, J.M. 2003. Loss of retinoblastoma but not p16 function allows bypass of replicative senescence in human fibroblasts. *EMBO Rep.* **4**:1061–1065.

36. Allsopp, R.C., and Harley, C.B. 1995. Evidence for a critical telomere length in senescent human fibroblasts. *Exp. Cell Res.* **219**:130–136.

37. Shay, J.W., Pereira-Smith, O.M., and Wright, W.E. 1991. A role for both RB and p53 in the regulation of human cellular senescence. *Exp. Cell Res.* **196**:33–39.

38. Harley, C.B., Vaziri, H., Counter, C.M., and Allsopp, R.C. 1992. The telomere hypothesis of cellular aging. *Exp. Gerontol.* **27**:375–382.

39. d'Adda di Fagagna, F., et al. 2003. A DNA damage checkpoint response in telomere-initiated senescence. *Nature.* **426**:194–198.

40. Martens, U.M., Chavez, E.A., Poon, S.S., Schmoor, C., and Lansdorp, P.M. 2000. Accumulation of short telomeres in human fibroblasts prior to replicative senescence. *Exp. Cell Res.* **256**:291–299.

41. Baird, D.M., Rowson, J., Wynford-Thomas, D., and Kipling D. 2003. Extensive allelic variation and ultrashort telomeres in senescent human cells. *Nat. Genet.* **33**:203–207.

42. Blackburn, E.H. 2000. Telomere states and cell fates. *Nature.* **408**:53–56.

43. Rubin, H. 2002. The disparity between human cell senescence in vitro and lifelong replication in vivo. *Nat. Biotechnol.* **20**:675–681.

44. Zhu, J., Wang, H., Bishop, J.M., and Blackburn, E.H. 1999. Telomerase extends the lifespan of virus-transformed human cells without net telomere lengthening. *Proc. Natl. Acad. Sci. U. S. A.* **96**:3723–3728.

45. Karlseder, J., Smogorzewska, A., and de Lange, T. 2002. Senescence induced by altered telomere state, not telomere loss. *Science.* **295**:2446–2449.

46. Blackburn, E.H. 2001. Switching and signaling at the telomere. *Cell.* **106**:661–673.

47. McElligott, R., and Wellinger, R.J. 1997. The terminal DNA structure of mammalian chromosomes. *EMBO J.* **16**:3705–3714.

48. Wellinger, R.J., and Sen, D. 1997. The DNA structures at the ends of eukaryotic chromosomes. *Eur. J. Cancer.* **33**:735–749.

49. Wright, W.E., Tesmer, V.M., Huffman, K.E., Levene, S.D., and Shay, J.W. 1997. Normal human chromosomes have long G-rich telomeric overhangs at one end. *Genes Dev.* **11**:2801–2809.

50. Griffith, J.D., et al. 1999. Mammalian telomeres end in a large duplex loop. *Cell.* **97**:503–514.

51. Cimino-Reale, G., et al. 2001. The length of telomeric G-rich strand 3'-overhang measured by oligonucleotide ligation assay. *Nucleic Acids Res.* **29**:E35.

52. Stewart, S.A., et al. 2003. Erosion of the telomeric single-strand overhang at replicative senescence. *Nat. Genet.* **33**:492–496.

53. van Steensel, B., Smogorzewska, A., and de Lange, T. 1998. TRF2 protects human telomeres from end-to-end fusions. *Cell.* **92**:401–413.

54. Takai, H., Smogorzewska, A., and de Lange, T. 2003. DNA damage foci at dysfunctional telomeres. *Curr. Biol.* **13**:1549–1556.

55. Smogorzewska, A., and de Lange, T. 2002. Different telomere damage signaling pathways in human and mouse cells. *EMBO J.* **21**:4338–4348.

56. Masutomi, K., et al. 2003. Telomerase maintains telomere structure in normal human cells. *Cell.* **114**:241–253.

Telomeres, stem cells, senescence, and cancer

Norman E. Sharpless[1] and Ronald A. DePinho[2]

[1]Department of Medicine and Genetics, Lineberger Comprehensive Cancer Center, University of North Carolina School of Medicine, Chapel Hill, North Carolina, USA. [2]Department of Adult Oncology, Dana-Farber Cancer Institute, Department of Medicine and Genetics, Harvard Medical School, Boston, Massachusetts, USA.

Mammalian aging occurs in part because of a decline in the restorative capacity of tissue stem cells. These self-renewing cells are rendered malignant by a small number of oncogenic mutations, and overlapping tumor suppressor mechanisms (e.g., p16^{INK4a}-Rb, ARF-p53, and the telomere) have evolved to ward against this possibility. These beneficial antitumor pathways, however, appear also to limit the stem cell life span, thereby contributing to aging.

Every day, we sacrifice many varied cell types such as granulocytes, keratinocytes, hepatocytes, and erythrocytes at the altar of organismal homeostasis. For the individual to thrive, lost cells must be constantly replaced, and recent evidence has identified significant capacities for repair and regeneration even in organs once thought to be postmitotic such as the pancreatic islet and the brain. Given this continuous cellular attrition, normal tissue function requires that the rate of cell loss be matched by the rate of renewal. Aging is hastened by forces that either accelerate cellular loss or retard tissue repair. When loss exceeds repair, ensuing cellular attrition eventually leads to a decline in organ function and ultimately failure. When restricted to specific organs, this condition would be expected to result in one of the many chronic degenerative diseases such as liver cirrhosis. If, however, the process operates across multiple organ systems, then this progressive multisystem functional compromise may manifest clinically as frailty, accelerated aging, and death.

Elegant experiments from lower metazoans with postmitotic soma (e.g., *Drosophila* and *Caenorhabditis elegans*) have identified many of the pathways that influence the rate of cellular turnover and loss (reviewed in refs. 1, 2). These data have firmly established a link between the rate of cellular metabolism, the rate of production of unstable oxygen species, and longevity in these species. There appears little doubt that many of these conserved pathways (e.g., insulin-like growth factor-1 signaling) also influence the rate of mammalian metabolism and cellular decay, and therefore mammalian aging. As extensive tissue replacement does not occur in adults of these lower organisms, however, these model systems have been less helpful with regard to the genes that regulate tissue repair in the adult and thereby influence this aspect of aging. Therefore, one advantage of mammalian genetic systems is that they permit the investigation of the pathways responsible for the repair half of the aging equation.

To be sure, adult mammals require extensive proliferation and tissue replacement to survive. Even in the absence of pathology, the intestinal lining replaces itself entirely on a weekly basis, and the bone marrow produces trillions of new blood cells daily. An obvious cost of this massive and obligate proliferation, however, is that even under physiological conditions, it is presumed that stem cell genomes are showered with somatic mutations, some of which may target cancer-relevant genes. In accord with this view is the remarkable observation that roughly 1% of neonatal cord blood collections contain significant numbers of myeloid clones harboring oncogenic fusions such as the AML-ETO fusion associated with acute leukemia (3); similarly, as many as one in three adults possess detectable IgH-BCL2 translocations, which are commonly associated with follicular lymphoma (4). As the prevalence of these cancers is far lower in the general population, it would appear that potent tumor suppressor mechanisms function to monitor and constrain the growth and survival of these aspiring cancer cells. In humans, three principal and overlapping tumor suppressor barriers appear to be operative; they are represented by the p16^{INK4a}–retinoblastoma protein (p16^{INK4a}-Rb) pathway, the ARF–p53 pathway, and telomeres. The combined effect of these tumor suppressor mechanisms is to place a limit on the replicative life span of cells in the compartment capable of contributing to tissue regeneration (hence termed stem cells). In this Perspective, we will discuss these tumor suppressor mechanisms and the hypothesis that their anticancer roles come at the cost of a decline in stem cell number and their proliferative reserve, thereby compromising tissue repair and promoting the aging phenotype. We will detail the human and murine data in support of this hypothesis, and discuss the implications of the intimate link between cancer and aging.

Tumor suppressor pathways engender senescence and apoptosis

A common endpoint for these major tumor suppressor mechanisms is senescence. This specialized form of terminal differentiation is induced by a variety of stimuli including alterations of telomere length and structure, some forms of DNA damage (for example, oxidative stress), and activation of certain oncogenes (reviewed in refs. 5, 6). Senescence requires activation of the Rb and/or p53 protein, and expression of their regulators such as p16^{INK4a}, p21, and ARF (7–10) (Figure 1). An important form of senescence is induced by p53, which has several antiproliferative activities including stimulation of the expression of p21, a cyclin-dependent kinase inhibitor (CDKI). CDKIs inhibit progression through the cell cycle by inhibiting cyclin-dependent kinases that phosphorylate and thereby inactivate Rb and related proteins p107 and p130 (11). The activity of p53 is predominantly mediated by inhibiting its murine double minute 2 protein (MDM2)–mediated degradation, and p53 is stabilized by diverse stimuli including DNA damage signals (e.g., resulting from oxidative stress or telomeric shortening) and oncogene activation

Nonstandard abbreviations used: retinoblastoma protein (Rb); cyclin-dependent kinase inhibitor (CDKI); murine double minute 2 protein (MDM2); plasminogen activator inhibitor (PAI); senescence-associated heterochromatic foci (SAHF); telomerase RNA component (TERC); telomerase reverse transcriptase (TERT); double-strand break (DSB); telomeric repeat binding factor 2 (TRF2); alternative lengthening of telomeres (ALT); ataxia telangiectasia (AT); ataxia-telangiectasia mutated kinase (ATM); peripheral blood leukocyte (PBL); comparative genome hybridization (CGH); hematopoietic stem cell (HSC).

Conflict of interest: The authors have declared that no conflict of interest exists.

Citation for this article: *J. Clin. Invest.* **113**:160–168 (2004). doi:10.1172/JCI200420761.

Table 1
Senescence vs. quiescence

	Senescence	Quiescence
Stability	Permanent[A]	Reversible
Induced by:	Telomere shortening	Serum starvation
	Prolonged DNA damage	Growth factor deprivation
	Oxidative stress	Growth at high density
	Oncogene (e.g., RAS) activation	Transient DNA damage
SA β-gal expression	Present	Absent
SA heterochromatic foci	Present	Absent
Molecular markers	Uninducible c-fos expression to serum stimulation	Inducible c-fos expression to serum stimulation
	Increased PAI expression	Low PAI expression
Cell cycle inhibitors	p16^{INK4a}, ARF, p53, p21, Rb	p53, p21, p27, Rb

[A]In the setting of sustained p53 and/or Rb function (see text for details). SA, senescence-associated; SA β-gal, senescence-associated β-galactosidase.

(reviewed in refs. 12–14). The stabilization of p53 by oncogenes is in part mediated by ARF (also designated p19^{ARF} in the mouse), which is induced by inappropriate cell cycle entry (15–17), and binds to MDM2, thereby inhibiting the destruction of p53 (15, 18–20). Another CDKI, p16^{INK4a} increases markedly in senescent cells, and correlates with increasing Rb hypophosphorylation during this process (8, 9). The regulation of p16^{INK4a} in senescence is not as well understood as that of p53, although it appears to be induced by several stimuli including MAP kinase signaling, oncogene activation, and growth in culture (21–24). Either p53-p21 or p16^{INK4a} are able to produce Rb hypophosphorylation and initiate senescence. Some senescence-inducing stimuli (e.g., activation of the RAS oncogene) appear to induce both pathways, while others (e.g., DNA damage) appear to preferentially activate one or the other.

Senescence differs from other physiologic forms of cell cycle arrest such as quiescence in two important ways (Table 1). First, senescence in somatic cells is generally irreversible, barring the inactivation of p53 and/or Rb, which appear to be required for its maintenance in certain settings (10, 25–27). Second, it is associated with distinctive molecular and morphologic alterations such as cellular flattening and increased adherence, a loss of c-fos induction to serum stimulation, an increased expression of plasminogen activator inhibitor (PAI) and the expression of senescence-associated β-galactosidase activity (reviewed in refs. 6, 28). Recently, senescence has been shown to correlate with the establishment of an unusual form of heterochromatin present in discrete nuclear foci, known as senescence-associated heterochromatic foci (SAHF) (29). In aggregate, these data suggest that senescence results from the durable repression of promoters associated with growth control genes. This repression is enforced by the construction of stable, heterochromatin-like complexes, the formation of which is directed in part by hypophosphorylated Rb.

Recent evidence suggests senescence may differ qualitatively depending on the stimulus that leads to its establishment. Several groups have shown that either Rb or p53 inactivation could reverse the senescence in murine cells or certain types of human fibroblasts (10, 26, 27). Therefore, it seems clear that persistent p53 and Rb function are required to maintain certain forms of senescence. Campisi and colleagues, however, have reported that certain human cell types senesced in a p16^{INK4a}-dependent manner, while in other cells senescence was p53-mediated (25). Intriguingly, while the authors could revert p53-mediated senescence using lentiviral vectors expressing SV40 Large T Antigen

(which inactivates Rb and p53), p16^{INK4a}-mediated senescence was durable and not reversible by T Antigen. Of note, the formation of SAHF was more marked in p16^{INK4a}-mediated senescence, suggesting the possibility that Rb and p53 are no longer required after the establishment of these repressive chromatin structures (29). As both p16^{INK4a} and p21 induce Rb hypophosphorylation, however, the molecular basis for the difference between these forms of senescence is unclear. Therefore, robust and validated senescence markers are needed to dissect these issues.

In addition to senescence, it is worth noting that cancer-related stimuli such as oncogene activation, DNA damage, and telomere shortening can also induce an entirely distinct anticancer mechanism, namely apoptosis. Apoptotic loss of progenitor cells in response to such stimuli has been clearly demonstrated in animal models; for example, mice with shortened, dysfunctional telomeres demonstrate increased apoptosis in germ cells of the testes and crypt cells of the intestine (30–32). In these systems, an increase in apoptosis correlates with tissue atrophy and other phenotypes associated with premature aging. The role of p53 in mediating apoptosis is well-documented (33, 34), and this activity seems its major anticancer function in certain animal models (35, 36). Correspondingly, p53 loss greatly attenuates the apoptotic phenotype seen in proliferative organs in the setting of telomere dysfunction in animal models (37). While loss of p53 in these animals affords resistance to the effects of telomere dysfunction, these mice also demonstrate a marked increase in epithelial tumor formation (37, 38), reinforcing the view that aging and cancer are closely linked in this model system. Therefore, telomere shortening and p53 activation modulate two potent anticancer mechanisms: senescence and apoptosis. While the molecular biology of this fate-decision is incompletely understood, the specific response appears to depend on many variables including cell type, genetic context, and proliferation state. Although a role for p16^{INK4a} in inducing apoptosis has been suggested (39–42), the issue of whether p16^{INK4a} also has senescence-independent anticancer functions in vivo remains an area of active investigation.

Telomeres, telomerase, and checkpoints

Telomeres are nucleoprotein complexes at the chromosome ends which consist of many double-stranded TTAGGG repeats, a 3' single strand overhang, and associated telomere binding proteins (Figure 2) (reviewed in refs. 43, 44). It has been known since the early days of maize genetics that telomeres play a critical role in the maintenance of chromosomal integrity — a fact that has since

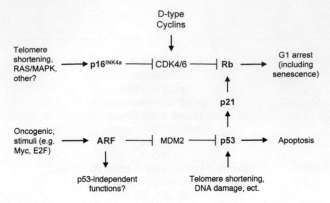

Figure 1

The p53 and Rb pathways. p53 activity is predominantly regulated at the protein level. In the unstressed state, p53 is rapidly degraded by MDM2; a process which is inhibited by ARF. Also, p53 can be stabilized by N-terminal serine phosphorylation in response to genotoxic stresses, and this phosphorylation inhibits its interaction with MDM2. p53 activation potently induces either growth arrest or apoptosis depending on cellular context. The antiproliferative activity of p53 in part results from p21 expression, which is a p53 transcriptional target. Rb is inactivated by phosphorylation as a result of the cyclin-dependent kinases CDK4 and CDK6. Hypophosphorylated Rb binds E2F and represses proliferation. CDK activity is inhibited by both p16^{INK4a} and p21.

been confirmed in diverse model systems from yeast to plants to humans. Interestingly, however, telomere lengths differ greatly, even among mammals. For example, human telomeres are 5 to 15 kb long while those of inbred mice may exceed 60 kb. Several lines of evidence have established that telomeres adopt a complex secondary and tertiary structure that relies on DNA-DNA, DNA-protein, and protein-protein interactions. With regard to senescence, telomere structure appears at least as important to telomere function as absolute telomere length (45–47).

DNA polymerase cannot fully replicate the ends of linear DNA duplexes, and, in the absence of telomerase, chromosomes shorten slightly with every cell division. Telomerase is a specialized RNA-protein complex that is responsible for the de novo synthesis and maintenance of telomere repeats. The telomerase holoenzyme consists of a telomerase RNA component (TERC) that serves as a template for the addition of repeats, and a protein component, telomerase reverse transcriptase (TERT). Telomerase expression is low or absent in most human somatic tissues, with its expression principally restricted in the adult to activated lymphocytes, germ cells, and tissue stem cells (48–51). The restricted pattern of telomerase activity relates primarily to the strict regulation of *TERT* gene transcription, whereas *TERC* gene expression is broader in distribution. The most widely accepted model holds that progressive telomere shortening with each cell division eventually triggers an alteration in telomere structure. From the perspective of senescence and apoptosis, the most important result of this telomere dysfunction is that the deprotected telomere end becomes, for all intents, a double-strand break (DSB) of DNA. Classical DSBs potently induce p53, and telomere dysfunction in cultured human and mouse cells has been shown to induce p53-mediated senescence or other checkpoint responses depending on the cell type.

The relationship between telomere dysfunction and activation of the p16^{INK4a}-Rb pathway is less clear. For example, TERT expression in cultured human fibroblasts abrogates progressive p16^{INK4a} accu-

mulation and the onset of senescence (52), a result that has generally been interpreted to suggest that dysfunctional telomeres produce a signal that activates p16^{INK4a}. Alternatively, it is possible that TERT per se possesses functions, independent of its telomere lengthening effects, capable of quelling p16^{INK4a} activation. The expression of a dominant negative form of telomeric repeat binding factor 2 (TRF2), a telomere binding protein, will induce an acute alteration in telomere structure accompanied by growth arrest in human and murine cells (45). In human cells, senescence can be induced by a dominant negative TRF2 (DN-TRF2) through either p16^{INK4a}- or p53-dependent mechanisms, while, in murine cells, p53 loss alone is sufficient to abrogate growth arrest by DN-TRF2 (53). Likewise, loss of p16^{INK4a} and ARF does not rescue the gonadal failure and other in vivo apoptotic responses seen in mice harboring telomere dysfunction (54). On the other hand, the placement of human epithelial cells and murine cells in culture has been shown to rapidly induce p16^{INK4a} in the absence of telomere shortening (16, 55–57). Therefore, like p53, p16^{INK4a} appears to be induced by telomere-dependent and -independent stimuli. The accumulation of p16^{INK4a} in many tissues is noted with aging in both mice and humans (24, 58), but whether this results from in vivo alterations of telomere structure remains to be established. Similarly, it is unclear if the major in vivo tumor suppressor function of p16^{INK4a} in humans results as a response to telomere dysfunction or other stimuli.

In the absence of the p53- and p16^{INK4a}-Rb-mediated checkpoints, cultured cells with dysfunctional telomeres continue to proliferate, entering a period of slow growth called "crisis" that is characterized by genomic instability. Clones emerging from crisis invariably either reactivate telomerase (59–61) or the alternative lengthening of telomeres (ALT) mechanism (62, 63). During crisis, the deprotected telomere ends in proliferating cells can be illegitimately fused through DNA repair mechanisms, ultimately leading to the generation of complex non-reciprocal translocations, a hallmark feature of adult solid tumors. In a tissue stem cell, telomere dysfunction would appear to have two outcomes depending on the integrity of checkpoint mechanisms: respond to checkpoint activation with senescence/apoptosis or proliferate and engender genomic instability. Evidence suggests that

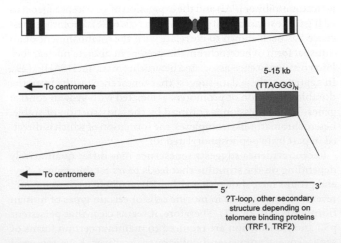

Figure 2

Telomere structure. Telomeres are present at chromosome ends. They consist of linear arrays of repeat sequences that are 5–15 kb in humans but considerably larger in mice. Telomeres also harbor a G-rich 3′ overhang that is important for the adoption of proper secondary structure.

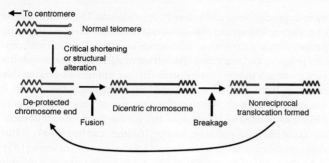

Figure 3

Bridge-fusion-breakage. Under normal circumstances, the chromosome ends are protected by telomeres. When normal cells develop telomere shortening, they undergo growth arrest or apoptosis in a p53- or p16INK4a-dependent manner. In cells with checkpoint inactivation, however, fusion between telomere-free ends leads to the formation of dicentric chromosomes, which are then broken during anaphase. These breakages produce non-reciprocal translocations and broken chromosome ends that are substrates for further fusion events.

most cells facing this decision respond to checkpoints, producing a progressive diminution of stem cell reserve throughout the human lifespan. This decline in tissue regeneration capacity eventually manifests as aging. The rare cells that continue to proliferate in the setting of telomere dysfunction and emerge from crisis having accumulated additional, transforming mutations are equally problematic from the organismal perspective. These cells necessarily have incurred proliferative genetic lesions and inactivated tumor suppressor checkpoints, and appear well on their way down the path to full-fledged cancer.

Telomere-mediated checkpoints prevent cancer but contribute to aging

The vast majority of cells exhibiting telomere dysfunction undergo senescence or apoptosis, and disparate lines of evidence have convincingly shown that these processes are significant barriers to cancer formation. Reactivation of telomerase is one of the most commonly observed features of cancer seen in greater than 80% of all human tumors (64–66). Telomerase expression greatly enhances the transformation of human cells in vitro (67), and tumorigenesis is inhibited by telomere shortening in animal models of cancer (54, 68, 69). In aggregate, these data suggest that telomere-induced checkpoint activation is a major in vivo tumor suppressor mechanism.

The tumor-promoting effects of telomerase activity appear to be more complex than initially believed. Certainly, tumor clones must resolve the telomere length problem to traverse crisis, and therefore telomerase reactivation or the development of ALT is a prerequisite of malignant growth. For this reason, inhibitors of telomerase have been considered promising therapeutic candidates for novel antineoplastic agents (70). Several lines of evidence, however, also suggest that telomerase activity may contribute to tumorigenesis in a manner independent of its telomere lengthening effects. For example, telomerase expression has been shown, independent of telomere length, to produce resistance to the antiproliferative signals of TGF-β in cultured mammary cells lacking p16INK4a (71). Additionally, animals overexpressing telomerase are more prone to neoplasia than littermate wild-type animals (72, 73), even though telomere length per se does not appear to limit

tumor growth in these models. Lastly, transformed ALT+ murine clones demonstrate further enhanced growth in a mouse model of metastasis after transduction with telomerase (74). These disparate lines of evidence suggest that the catalytic activity of telomerase contributes to malignant growth by influencing a cellular feature in addition to absolute telomere length, but the precise mechanism of these effects has remained elusive. One possibility has been suggested by the observation that telomerase activity appears to affect the length of the telomere 3' overhang (Figure 2), which plays a role in assumption of proper telomere structure, thereby modulating growth arrest or apoptosis independent of absolute telomere length (46, 47).

In addition to preventing cancer, however, telomere-mediated checkpoints appear to contribute to aging, and, as stated, mice with shortened, dysfunctional telomeres exhibit many characteristics of premature aging (31). A consideration of ataxia telangectasia (AT) is particularly intriguing in this regard. Humans with deficiency of the AT tumor suppressor (ATM) (ataxia-telangiectasia mutated kinase), which plays a role in DNA damage signaling, develop a progeroid syndrome in the setting of premature telomere shortening (75, 76). In comparison to humans, however, mice lacking ATM harbor a more modest phenotype in terms of aging. Such animals are markedly susceptible to thymic lymphoma (77–80), consistent with a role of ATM in sensing failed VDJ recombination intermediates (81). Strikingly, however, mice lacking ATM in the setting of telomere shortening developed accelerated aging and recapitulate more faithfully many features of the human ATM deficiency (30). Furthermore, Atm−/− mice with short telomeres developed less thymic lymphoma despite increased mortality to non-cancer causes. Although human AT patients are tumor prone, these mice are presumably resistant to cancer because they possess an otherwise intact p53 checkpoint, and are not able to reactivate telomerase because of its germline inactivation. Therefore, many of the age-associated phenotypes seen in AT patients appear to result from telomere dysfunction, which in turn reflects premature telomere shortening. The mechanism that produces premature telomere attrition in AT patients, however, remains unclear.

Further direct genetic evidence for a role of telomere dysfunction in human aging comes from the recent discovery that germline mutations of the telomerase complex cause the progeroid syndrome dyskeratosis congenita (82). Also, telomere shortening has been shown to precede the development of overt cirrhosis in patients with chronic hepatitis of various etiologies (83–86). In addition, several studies have demonstrated a relationship between telomere length in peripheral blood leukocytes (PBL) and the onset of certain diseases associated with aging. Such studies in non-neoplastic diseases have shown that PBL telomere lengths can provide predictive information on the risk of developing atherosclerosis (87, 88) and on overall mortality (89). In aggregate, these data indicate that although telomerase plays a clear role in malignant progression, telomere-induced checkpoints also contribute to certain aspects of human aging.

Telomere dysfunction contributes to cancer

As telomere-mediated checkpoints are no doubt major barriers to malignant progression in most would-be cancer cells, it therefore appears somewhat paradoxical that telomere dysfunction may also fuel tumorigenesis. Nonetheless, recent cytogenetic and molecular studies have provided strong evidence that telomere dynamics

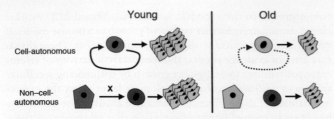

Figure 4
Cell-autonomous vs. non–cell-autonomous aging. Two models for impaired tissue repair are suggested: target cells of aging are shown in gray and senescent cell are shown in blue. In the cell-autonomous case, senescence (or apoptosis) of a progenitor with self-renewal capacity leads to impaired tissue regeneration in old animals. In the non–cell-autonomous case, however, a support cell supplies a factor (X), which is critical for the maintenance of tissue repair. X could be a hormone (e.g., estrogen) acting at a distance or cell-cell interactions (e.g., costimulatory signals from antigen-presenting cells) acting in a paracrine manner. In this model, aging results from the functional loss (e.g., by senescence) of the support cell, which may not be detectable in the tissue of interest.

can contribute to genomic instability, particularly early in tumorigenesis. In this model, a transient period of telomere dysfunction contributes to carcinogenesis by engendering large numbers of genome-wide changes (reviewed in refs. 43, 90) (Figure 3). This occurs through breakage-fusion-bridge cycles that result from the formation of dicentric chromosomes after inappropriate fusions of deprotected telomeres. This rapid reshuffling of the genetic deck produces rare cells with a threshold number of relevant changes to become full-fledged cancer.

Strong evidence that telomere-associated events are indeed relevant to carcinogenesis derives from the analysis of mice compound deficient for *TERC* and *p53* (37, 38). In this system, p53 deficient mice with telomere dysfunction demonstrate a marked shift in the tumor spectrum towards epithelial cancers. Additionally, these cancers exhibit the complex cytogenetic profiles found in human epithelial cancers as opposed to the more bland cytogenetic profiles of spontaneous mouse cancers (38, 91). While most cancers of adult humans are epithelial in origin, spontaneous carcinomas are unusual in mouse tumor models. Therefore, these data suggest that telomere dysfunction is an important event in carcinoma formation, and telomere dynamics play a major role in the species difference in carcinoma frequency between humans and mice.

This model of telomere dysfunction–driven carcinogenesis matches with the timing of telomerase activation and appearance of genomic changes during various stages of epithelial tumorigenesis. For example, comparative genome hybridization (CGH) has demonstrated that early breast (92–94) and esophageal (95, 96) lesions sustain widespread gains and losses of regions of chromosomes early in their development — such ploidy changes detected by CGH correlate tightly with the presence of complex chromosomal rearrangements. This genomic instability phase is evident in stages of high-grade dysplasia, and can be found before the acquisition of a frankly malignant phenotype.

Correspondingly, the measurement of telomerase activity in corresponding preneoplastic lesions has demonstrated a consistent pattern with the ploidy and cytogenetic changes (97–103). For example, studies of adenomatous polyps and colorectal cancers have established that telomerase activity is low or undetectable in small- and intermediate-sized polyps with marked increase in telomerase activ-

ity in large adenomas and colorectal carcinomas. These data suggest that there is widespread chromosomal instability early in neoplastic progression at a time when telomerase activity is low. As these cancers progress and reactivate telomerase or ALT, genomic instability continues at a moderate rate, with further mutations presumably resulting from non–telomere-based mechanisms. Additionally, it has recently been shown that the presence of short telomeres in PBLs is associated with increased risk for the development of carcinomas of the head and neck, kidney, bladder, and lung (104). While these correlative results can be interpreted in several ways (105), substantiation of the utility of PBL telomere lengths as a reliable surrogate marker for neoplasia would prove invaluable in assessing patient risk of later developing an epithelial malignancy.

It should be noted, however, that certain cancers, particularly those malignancies with simple cytogenetics such as pediatric acute leukemia, occur without going through an obligate period of genomic instability induced by telomere dysfunction. Such tumors likely accomplish this by solving the telomere-lengthening problem prior to the onset of telomere dysfunction. Perhaps these malignancies target a telomerase-expressing compartment like the hematopoietic stem cell (HSC), or involve potent activation of oncogenes like myc that induce telomerase expression (106, 107). It is likely, however, that these tumors represent a special, good-prognosis case of cancer as they may neither require nor possess the large number of genomic changes seen in adult solid tumors resulting from telomere dysfunction. It is tempting to speculate that the improved prognosis and response to therapy of hematopoietic compared with epithelial cancers could relate to the fact that solid tumors experience a DNA damage phase as part of their life history. This phase may result in the obligate deactivation of DNA damage checkpoints in order to permit cancer cell survival, yet at the same time position these cancers for resistance to a large class of DNA damage–inducing cancer therapies.

How are cancer and aging linked in vivo?

The aforementioned data suggest that senescence is induced in cultured cells by a variety of telomere-dependent and telomere-independent mechanisms. Likewise, the in vivo effects of inactivation (for p53 see ref. 108, for p16^{INK4a} see refs. 109, 110, and for telomerase see refs. 21, 32, 111) and hyperfunction (for p53 see refs. 112, 113, and for telomerase see refs. 72, 73) have been determined in mice, confirming the hypotheses that these mechanisms link cancer and aging. A problem, however, from the use of germline knockout animals and broadly expressing transgenics strains is that it is not necessarily straightforward to determine which effects are cell-autonomous and which are not (Figure 4). For example, lymphocytes from adult p16^{INK4a}-deficient mice demonstrate an enhanced mitogenic response to CD3 and CD28 (109), and loss of this response is a hallmark feature of immunosenescence in mice (114). Problematically, however, the expression of p16^{INK4a} in lymphocytes is low relative to other adult tissues (N.E. Sharpless, unpublished observation). Therefore, this resistance to aging seen in the immune system of *p16^{INK4a}* null mice could represent a direct antiproliferative effect of p16^{INK4a} on aged T-cells, or more likely could result from a role for p16^{INK4a} in other cell types, such as an antigen-presenting compartment. Therefore, a future challenge will be to analyze animals with conditional inactivation of these pathways in order to resolve the complex effects of their loss on organismal fitness and aging.

An additional conceptual problem stems from the observation that the expression of p53 and telomere shortening do not strictly correlate with the onset of tissue aging (115–117). While undoubtedly telomere shortening is seen in certain progeroid syndromes and in states characterized by chronic hyperproliferation such as cirrhosis and myelodysplasia, it also appears that telomere length is heterogeneous in the human population, and shorter lengths do not always correlate with aging. A caveat to this conclusion, however, is that these studies necessarily have analyzed mean telomere length, although it appears that checkpoint activation may occur in response to the shortest telomeres in the cell (118). Nonetheless, while telomere shortening can be seen immediately preceding organ failure in states of high cellular turnover, it may not be an obligate feature of all normal physiologic aging. This finding points to the existence of telomere-independent causes of aging, as is almost certainly the case in normal mice.

There are two important caveats to this sort of analysis. First, only surviving cells are considered in these molecular characterizations of aged tissues. Therefore, one would not expect to detect telomere shortening and p53 expression in cell types where these stimuli are proapoptotic. Nonetheless, apoptotic loss of stem cells may play an important role in organismal aging. Additionally, it is possible that alterations in telomere structure, which can potently induce senescence and apoptosis (45–47), occur in vivo. Therefore, alterations in telomere dynamics may contribute to a decline in tissue function even in the absence of an overall decrease in telomere length.

Alternatively, p53 or p16^{INK4a} activation may be contributing to aging in a very restricted compartment such as tissue stem cells. In this model, expression of p16^{INK4a} or p53 need not be detected throughout the tissue to exert its senescence-related effects. There are data from genetic systems in support of this view. For example, loss of bmi-1, a repressor of p16^{INK4a} and p19^{ARF}, leads to a decrease in HSC number (119), and HSCs from Ink4a/Arf-deficient animals demonstrate enhanced long-term function (120). These data collectively suggest that one or both products of the Ink4a/Arf locus, both potent inducers of senescence, play a role in determining the long-term replicative potential of murine HSCs. Similarly, the effects of telomere shortening in mice are most immediately obvious in the spermatagonia and colonic crypts (32), which harbor the stem cells that give rise to the large bowel mucosa. Mice with critical telomere shortening demonstrate marked anaphase bridging and apoptosis in these progenitor-enriched compartments. Therefore, the effects of these tumor suppressor pathways may be disproportionately felt in the stem cell compartment, which represents a rare minority of cells in any given tissue.

Summary

Given the aging populations in western nations, the presently enormous problems of aging and cancer are poised to become even greater public health and socioeconomic challenges (121, 122). A recent study of Americans receiving Medicare support estimated the entire future lifetime health-related costs of the average healthy 70-year-old to be approximately $150,000 per individual (122), and the aggregate cost of this health burden will only grow larger as increasing numbers of Americans age and retire. A ray of hope from this analysis, however, was suggested by the observation that the healthiest 70-year-olds with the longest life expectancy still had the lowest future health costs. Therefore, if a goal of modern medicine is to extend patient survival, then the future solvency of Medicare appears to hinge on our ability not merely to extend overall survival, but to extend disease-free survival (123).

Our new molecular understanding of cancer and aging is a major tool in this effort, for two reasons. First, we are beginning to develop molecular surrogate markers of aging and future cancer risk. Such markers will revolutionize medical advice regarding nutrition and wellness. Amazingly, physicians have very little hard evidence to support a given dose of vitamin supplements, exercise, or diet; and most wellness therapeutics of proven benefit are directed toward a measurable predisease state like hypercholesterolemia or hypertension. By examining markers such as telomere function or p53 / p16^{INK4a} expression in a given tissue, however, we may be able to better predict the future onset of cancer and/or aging, and likewise determine the beneficial or harmful effects of a therapeutic intervention with regard to these surrogate endpoints.

Secondly, we will be able to improve our understanding of environmental or lifestyle exposures that cause or contribute to aging. Likewise, we will be able to identify which tissue compartments are targeted by these exposures, allowing a delineation of the cell-autonomous and non–cell-autonomous effects. Whether such an understanding of the environmental triggers of aging will allow us to boost the maximum human lifespan is highly controversial, but certainly at the minimum such information would help extend the healthy life expectancy, free of cancer and other adverse consequences of aging.

Acknowledgments

We thank Steven Artandi for advice and ideas regarding the manuscript. N.E. Sharpless is a Sydney Kimmel Scholar in Cancer Research and a Paul Beeson Scholar in Aging Research, and is supported by the NIH. R.A. DePinho is an American Cancer Society Professor and an Ellison Medical Foundation Senior Scholar in Aging, and is supported by the NIH.

Address correspondence to: Norman E. Sharpless, Lineberger Comprehensive Cancer Center, University of North Carolina, Chapel Hill, North Carolina 27599-7295, USA. Phone: (919) 966-1185; Fax: (919) 966-8212; E-mail: nes@med.unc.edu.

1. Guarente, L., and Kenyon, C. 2000. Genetic pathways that regulate ageing in model organisms. *Nature.* **408**:255–262.
2. Gems, D., and Partridge, L. 2001. Insulin/IGF signalling and ageing: seeing the bigger picture. *Curr. Opin. Genet. Dev.* **11**:287–292.
3. Mori, H., et al. 2002. Chromosome translocations and covert leukemic clones are generated during normal fetal development. *Proc. Natl. Acad. Sci. U. S. A.* **99**:8242–8247.
4. Liu, Y., Hernandez, A.M., Shibata, D., and Cortopassi, G.A. 1994. BCL2 translocation frequency rises with age in humans. *Proc. Natl. Acad. Sci. U. S. A.* **91**:8910–8914.
5. Wright, W.E., and Shay, J.W. 2002. Historical claims and current interpretations of replicative aging. *Nat. Biotechnol.* **20**:682–688.
6. Campisi, J. 2001. Cellular senescence as a tumor-suppressor mechanism. *Trends Cell Biol.* **11**:S27–S31.
7. Kamijo, T., et al. 1997. Tumor suppression at the mouse INK4a locus mediated by the alternative reading frame product p19ARF. *Cell.* **91**:649–659.
8. Stein, G.H., Drullinger, L.F., Soulard, A., and Dulic, V. 1999. Differential roles for cyclin-dependent kinase inhibitors p21 and p16 in the mechanisms of senescence and differentiation in human fibroblasts. *Mol. Cell. Biol.* **19**:2109–2117.
9. Alcorta, D.A., et al. 1996. Involvement of the cyclin-dependent kinase inhibitor p16 (INK4a) in replicative senescence of normal human fibroblasts. *Proc. Natl. Acad. Sci. U. S. A.* **93**:13742–13747.
10. Sage, J., Miller, A.L., Perez-Mancera, P.A., Wysocki, J.M., and Jacks, T. 2003. Acute mutation of retinoblastoma gene function is sufficient for cell cycle re-entry. *Nature.* **424**:223–228.
11. Classon, M., and Harlow, E. 2002. The retinoblastoma tumour suppressor in development and cancer. *Nat. Rev. Cancer.* **2**:910–917.
12. Vousden, K.H. 2000. p53: death star. *Cell.* **103**:691–694.
13. Sharpless, N.E., and DePinho, R.A. 2002. p53: good cop/bad cop. *Cell.* **110**:9–12.
14. Levine, A.J. 1997. p53, the cellular gatekeeper for growth and division. *Cell.* **88**:323–331.
15. Pomerantz, J., et al. 1998. The Ink4a tumor suppressor gene product, p19Arf, interacts with MDM2 and neutralizes MDM2's inhibition of p53. *Cell.* **92**:713–723.

16. Zindy, F., et al. 1998. Myc signaling via the ARF tumor suppressor regulates p53-dependent apoptosis and immortalization. *Genes Dev.* **12**:2424–2433.

17. de Stanchina, E., et al. 1998. E1A signaling to p53 involves the p19(ARF) tumor suppressor. *Genes Dev.* **12**:2434–2442.

18. Zhang, Y., Xiong, Y., and Yarbrough, W.G. 1998. ARF promotes MDM2 degradation and stabilizes p53: ARF-INK4a locus deletion impairs both the Rb and p53 tumor suppression pathways. *Cell.* **92**:725–734.

19. Stott, F.J., et al. 1998. The alternative product from the human CDKN2A locus, p14(ARF), participates in a regulatory feedback loop with p53 and MDM2. *EMBO J.* **17**:5001–5014.

20. Kamijo, T., et al. 1998. Functional and physical interactions of the ARF tumor suppressor with p53 and Mdm2. *Proc. Natl. Acad. Sci. U. S. A.* **95**:8292–8297.

21. Zhu, J., Woods, D., McMahon, M., and Bishop, J.M. 1998. Senescence of human fibroblasts induced by oncogenic Raf. *Genes Dev.* **12**:2997–3007.

22. Serrano, M., Lin, A.W., McCurrach, M.E., Beach, D., and Lowe, S.W. 1997. Oncogenic ras provokes premature cell senescence associated with accumulation of p53 and p16INK4a. *Cell.* **88**:593–602.

23. Lin, A.W., et al. 1998. Premature senescence involving p53 and p16 is activated in response to constitutive MEK/MAPK mitogenic signaling. *Genes Dev.* **12**:3008–3019.

24. Zindy, F., Quelle, D.E., Roussel, M.F., and Sherr, C.J. 1997. Expression of the p16INK4a tumor suppressor versus other INK4 family members during mouse development and aging. *Oncogene.* **15**:203–211.

25. Beausejour, C.M., et al. 2003. Reversal of human cellular senescence: Roles of the p53 and p16 pathways. *EMBO J.* **22**:4212–4222.

26. Dirac, A.M., and Bernards, R. 2003. Reversal of senescence in mouse fibroblasts through lentiviral suppression of p53. *J. Biol. Chem.* **278**:11731–11734.

27. Gire, V., and Wynford-Thomas, D. 1998. Reinitiation of DNA synthesis and cell division in senescent human fibroblasts by microinjection of anti-p53 antibodies. *Mol. Cell. Biol.* **18**:1611–1621.

28. Sharpless, N.E. 2003. The persistence of senescence. *Sci. Aging Knowledge Environ.* 2003:PE24.

29. Narita, M., et al. 2003. Rb-mediated heterochromatin formation and silencing of E2F target genes during cellular senescence. *Cell.* **113**:703–716.

30. Wong, K.K., et al. 2003. Telomere dysfunction and Atm deficiency compromises organ homeostasis and accelerates ageing. *Nature.* **421**:643–648.

31. Rudolph, K.L., et al. 1999. Longevity, stress response, and cancer in aging telomerase-deficient mice. *Cell.* **96**:701–712.

32. Lee, H.W., et al. 1998. Essential role of mouse telomerase in highly proliferative organs. *Nature.* **392**:569–574.

33. Lowe, S.W., Ruley, H.E., Jacks, T., and Housman, D.E. 1993. p53 dependent apoptosis modulates the cytotoxicity of anticancer agents. *Cell.* **74**:957–967.

34. Yonish-Rouach, E., et al. 1991. Wild-type p53 induces apoptosis of myeloid leukaemic cells that is inhibited by interleukin-6. *Nature.* **352**:345–347.

35. Schmitt, C.A., et al. 2002. Dissecting p53 tumor suppressor functions in vivo. *Cancer Cell.* **1**:289–298.

36. Lu, X., et al. 2001. Selective inactivation of p53 facilitates mouse epithelial tumor progression without chromosomal instability. *Mol. Cell. Biol.* **21**:6017–6030.

37. Chin, L., et al. 1999. p53 deficiency rescues the adverse effects of telomere loss and cooperates with telomere dysfunction to accelerate carcinogenesis. *Cell.* **97**:527–538.

38. Artandi, S.E., et al. 2000. Telomere dysfunction promotes non-reciprocal translocations and epithelial cancers in mice. *Nature.* **406**:641–645.

39. Ausserlechner, M.J., et al. 2001. The cell cycle inhibitor p16(INK4A) sensitizes lymphoblastic leukemia cells to apoptosis by physiologic glucocorticoid levels. *J. Biol. Chem.* **276**:10984–10989.

40. Lagresle, C., et al. 2002. Transgenic expression of the p16(INK4a) cyclin-dependent kinase inhibitor leads to enhanced apoptosis and differentiation arrest of CD4-CD8-immature thymocytes. *J. Immunol.* **168**:2325–2331.

41. Plath, T., et al. 2000. A novel function for the tumor suppressor p16(INK4a): induction of anoikis via upregulation of the alpha(5)beta(1) fibronectin receptor. *J. Cell Biol.* **150**:1467–1478.

42. Tamm, I., et al. 2002. Adenovirus-mediated gene transfer of P16INK4/CDKN2 into bax-negative colon cancer cells induces apoptosis and tumor regression in vivo. *Cancer Gene Ther.* **9**:641–650.

43. Maser, R.S., and DePinho, R.A. 2002. Connecting chromosomes, crisis, and cancer. *Science.* **297**:565–569.

44. de Lange, T. 2002. Protection of mammalian telomeres. *Oncogene.* **21**:532–540.

45. van Steensel, B., Smogorzewska, A., and de Lange, T. 1998. TRF2 protects human telomeres from end-to-end fusions. *Cell.* **92**:401–413.

46. Stewart, S.A., et al. 2003. Erosion of the telomeric single-strand overhang at replicative senescence. *Nat. Genet.* **33**:492–496.

47. Masutomi, K., et al. 2003. Telomerase maintains telomere structure in normal human cells. *Cell.* **114**:241–253.

48. Broccoli, D., Young, J.W., and de Lange, T. 1995. Telomerase activity in normal and malignant hematopoietic cells. *Proc. Natl. Acad. Sci. U. S. A.* **92**:9082–9086.

49. Counter, C.M., Gupta, J., Harley, C.B., Leber, B., and Bacchetti, S. 1995. Telomerase activity in normal leukocytes and in hematologic malignancies. *Blood.* **85**:2315–2320.

50. Wright, W.E., Piatyszek, M.A., Rainey, W.E., Byrd, W., and Shay, J.W. 1996. Telomerase activity in human germline and embryonic tissues and cells. *Dev. Genet.* **18**:173–179.

51. Weng, N.P., Levine, B.L., June, C.H., and Hodes, R.J. 1996. Regulated expression of telomerase activity in human T lymphocyte development and activation. *J. Exp. Med.* **183**:2471–2479.

52. Bodnar, A.G., et al. 1998. Extension of life-span by introduction of telomerase into normal human cells. *Science.* **279**:349–352.

53. Smogorzewska, A., and de Lange, T. 2002. Different telomere damage signaling pathways in human and mouse cells. *EMBO J.* **21**:4338–4348.

54. Greenberg, R.A., et al. 1999. Short dysfunctional telomeres impair tumorigenesis in the INK4a(delta2/3) cancer-prone mouse. *Cell.* **97**:515–525.

55. Foster, S.A., Wong, D.J., Barrett, M.T., and Galloway, D.A. 1998. Inactivation of p16 in human mammary epithelial cells by CpG island methylation. *Mol. Cell. Biol.* **18**:1793–1801.

56. Huschtscha, L.I., et al. 1998. Loss of p16INK4 expression by methylation is associated with lifespan extension of human mammary epithelial cells. *Cancer Res.* **58**:3508–3512.

57. Kiyono, T., et al. 1998. Both Rb/p16INK4a inactivation and telomerase activity are required to immortalize human epithelial cells. *Nature.* **396**:84–88.

58. Nielsen, G.P., et al. 1999. Immunohistochemical survey of p16INK4A expression in normal human adult and infant tissues. *Lab. Invest.* **79**:1137–1143.

59. Counter, C.M., et al. 1992. Telomere shortening associated with chromosome instability is arrested in immortal cells which express telomerase activity. *EMBO J.* **11**:1921–1929.

60. Coursen, J.D., Bennett, W.P., Gollahon, L., Shay, J.W., and Harris, C.C. 1997. Genomic instability and telomerase activity in human bronchial epithelial cells during immortalization by human papillomavirus-16 E6 and E7 genes. *Exp. Cell Res.* **235**:245–253.

61. Counter, C.M., Botelho, F.M., Wang, P., Harley, C.B., and Bacchetti, S. 1994. Stabilization of short telomeres and telomerase activity accompany immortalization of Epstein-Barr virus-transformed human B lymphocytes. *J. Virol.* **68**:3410–3414.

62. Bryan, T.M., Englezou, A., Gupta, J., Bacchetti, S., and Reddel, R.R. 1995. Telomere elongation in immortal human cells without detectable telomerase activity. *EMBO J.* **14**:4240–4248.

63. Bryan, T.M., Marusic, L., Bacchetti, S., Namba, M., and Reddel, R.R. 1997. The telomere lengthening mechanism in telomerase-negative immortal human cells does not involve the telomerase RNA subunit. *Hum. Mol. Genet.* **6**:921–926.

64. Hiyama, K., et al. 1995. Telomerase activity in small-cell and non-small-cell lung cancers. *J. Natl. Cancer Inst.* **87**:895–902.

65. Kim, N.W., et al. 1994. Specific association of human telomerase activity with immortal cells and cancer. *Science.* **266**:2011–2015.

66. Hiyama, E., et al. 1995. Correlating telomerase activity levels with human neuroblastoma outcomes. *Nat. Med.* **1**:249–255.

67. Hahn, W.C., et al. 1999. Creation of human tumour cells with defined genetic elements. *Nature.* **400**:464–468.

68. Gonzalez-Suarez, E., Samper, E., Flores, J.M., and Blasco, M.A. 2000. Telomerase-deficient mice with short telomeres are resistant to skin tumorigenesis. *Nat. Genet.* **26**:114–117.

69. Rudolph, K.L., Millard, M., Bosenberg, M.W., and DePinho, R.A. 2001. Telomere dysfunction and evolution of intestinal carcinoma in mice and humans. *Nat. Genet.* **28**:155–159.

70. Shay, J.W., and Wright, W.E. 2002. Telomerase: a target for cancer therapeutics. *Cancer Cell.* **2**:257–265.

71. Stampfer, M.R., et al. 2001. Expression of the telomerase catalytic subunit, hTERT, induces resistance to transforming growth factor beta growth inhibition in p16INK4A(-) human mammary epithelial cells. *Proc. Natl. Acad. Sci. U. S. A.* **98**:4498–4503.

72. Artandi, S.E., et al. 2002. Constitutive telomerase expression promotes mammary carcinomas in aging mice. *Proc. Natl. Acad. Sci. U. S. A.* **99**:8191–8196.

73. Gonzalez-Suarez, E., et al. 2001. Increased epidermal tumors and increased skin wound healing in transgenic mice overexpressing the catalytic subunit of telomerase, mTERT, in basal keratinocytes. *EMBO J.* **20**:2619–2630.

74. Chang, S., Khoo, C.M., Naylor, M.L., Maser, R.S., and DePinho, R.A. 2003. Telomere-based crisis: functional differences between telomerase activation and ALT in tumor progression. *Genes Dev.* **17**:88–100.

75. Metcalfe, J.A., et al. 1996. Accelerated telomere shortening in ataxia telangiectasia. *Nat. Genet.* **13**:350–353.

76. Xia, S.J., Shammas, M.A., and Shmookler Reis, R.J. 1996. Reduced telomere length in ataxia-telangiectasia fibroblasts. *Mutat. Res.* **364**:1–11.

77. Borghesani, P.R., et al. 2000. Abnormal development of Purkinje cells and lymphocytes in Atm mutant mice. *Proc. Natl. Acad. Sci. U. S. A.* **97**:3336–3341.

78. Barlow, C., et al. 1996. Atm-deficient mice: a paradigm of ataxia telangiectasia. *Cell.* **86**:159–171.

79. Xu, Y., et al. 1996. Targeted disruption of ATM leads to growth retardation, chromosomal fragmentation during meiosis, immune defects, and thymic lymphoma. *Genes Dev.* **10**:2411–2422.

80. Elson, A., et al. 1996. Pleiotropic defects in ataxia-telangiectasia protein-deficient mice. *Proc. Natl. Acad. Sci. U. S. A.* **93**:13084–13089.

81. Liao, M.J., and Van Dyke, T. 1999. Critical role for Atm in suppressing V(D)J recombination-driven thymic lymphoma. *Genes Dev.* **13**:1246–1250.

82. Vulliamy, T., et al. 2001. The RNA component of telomerase is mutated in autosomal dominant dyskeratosis congenita. *Nature.* **413**:432–435.

83. Kitada, T., Seki, S., Kawakita, N., Kuroki, T., and Monna, T. 1995. Telomere shortening in chronic liver diseases. *Biochem. Biophys. Res. Commun.* **211**:33–39.

84. Miura, N., et al. 1997. Progressive telomere shortening and telomerase reactivation during hepatocellular carcinogenesis. *Cancer Genet. Cytogenet.* **93**:56–62.

85. Urabe, Y., et al. 1996. Telomere length in human liver diseases. *Liver.* **16**:293–297.

86. Wiemann, S.U., et al. 2002. Hepatocyte telomere shortening and senescence are general markers of human liver cirrhosis. *FASEB J.* **16**:935–942.

87. Samani, N.J., Boulrhy, R., Butler, R., Thompson, J.R., and Goodall, A.H. 2001. Telomere shortening in atherosclerosis. *Lancet.* **358**:472–473.

88. Obana, N., et al. 2003. Telomere shortening of peripheral blood mononuclear cells in coronary disease patients with metabolic disorders. *Intern. Med.* **42**:150–153.

89. Cawthon, R.M., Smith, K.R., O'Brien, E., Sivatchenko, A., and Kerber, R.A. 2003. Association between telomere length in blood and mortality in people aged 60 years or older. *Lancet.* **361**:393–395.

90. Artandi, S.E., and DePinho, R.A. 2000. A critical role for telomeres in suppressing and facilitating carcinogenesis. *Curr. Opin. Genet. Dev.* **10**:39–46.

91. O'Hagan, R.C., et al. 2002. Telomere dysfunction provokes regional amplification and deletion in cancer genomes. *Cancer Cell.* **2**:149–155.

92. Buerger, H., et al. 1999. Comparative genomic hybridization of ductal carcinoma in situ of the breast-evidence of multiple genetic pathways. *J. Pathol.* **187**:396–402.

93. Waldman, F.M., et al. 2000. Chromosomal alterations in ductal carcinomas in situ and their in situ recurrences. *J. Natl. Cancer Inst.* **92**:313–320.

94. Moore, E., Magee, H., Coyne, J., Gorey, T., and Dervan, P.A. 1999. Widespread chromosomal abnormalities in high-grade ductal carcinoma in situ of the breast. Comparative genomic hybridization study of pure high-grade DCIS. *J. Pathol.* **187**:403–409.

95. Walch, A.K., et al. 2000. Chromosomal imbalances in Barrett's adenocarcinoma and the metaplasia-dysplasia-carcinoma sequence. *Am. J. Pathol.* **156**:555–566.

96. van Dekken, H., Vissers, C.J., Tilanus, H.W., Tanke, H.J., and Rosenberg, C. 1999. Clonal analysis of a case of multifocal oesophageal (Barrett's) adenocarcinoma by comparative genomic hybridization. *J. Pathol.* **188**:263–266.

97. Lord, R.V., et al. 2000. Telomerase reverse transcriptase expression is increased early in the Barrett's metaplasia, dysplasia, adenocarcinoma sequence. *J. Gastrointest. Surg.* **4**:135–142.

98. Chadeneau, C., Hay, K., Hirte, H.W., Gallinger, S., and Bacchetti, S. 1995. Telomerase activity associated with acquisition of malignancy in human colorectal cancer. *Cancer Res.* **55**:2533–2536.

99. Engelhardt, M., Drullinsky, P., Guillem, J., and Moore, M.A. 1997. Telomerase and telomere length in the development and progression of premalignant lesions to colorectal cancer. *Clin. Cancer Res.* **3**:1931–1941.

100. Yan, P., Saraga, E.P., Bouzourene, H., Bosman, F.T., and Benhattar, J. 1999. Telomerase activation in colorectal carcinogenesis. *J. Pathol.* **189**:207–212.

101. Tang, R., Cheng, A.J., Wang, J.Y., and Wang, T.C. 1998. Close correlation between telomerase expression and adenomatous polyp progression in multistep colorectal carcinogenesis. *Cancer Res.* **58**:4052–4054.

102. Poremba, C., et al. 1998. Telomerase activity in human proliferative breast lesions. *Int. J. Oncol.* **12**:641–648.

103. Tsao, J., et al. 1997. Telomerase activity in normal and neoplastic breast. *Clin. Cancer Res.* **3**:627–631.

104. Wu, X., et al. 2003. Telomere dysfunction: a potential cancer predisposition factor. *J. Natl. Cancer Inst.* **95**:1211–1218.

105. Wong, K.K., and DePinho, R.A. 2003. Walking the telomere plank into cancer. *J. Natl. Cancer Inst.* **95**:1184–1186.

106. Greenberg, R.A., et al. 1999. Telomerase reverse transcriptase gene is a direct target of c-Myc but is not functionally equivalent in cellular transformation. *Oncogene.* **18**:1219–1226.

107. Wang, J., Xie, L.Y., Allan, S., Beach, D., and Hannon, G.J. 1998. Myc activates telomerase. *Genes Dev.* **12**:1769–1774.

108. Donehower, L.A., et al. 1992. Mice deficient for p53 are developmentally normal but susceptible to spontaneous tumours. *Nature.* **356**:215–221.

109. Sharpless, N.E., et al. 2001. Loss of p16Ink4a with retention of p19Arf predisposes mice to tumorigenesis. *Nature.* **413**:86–91.

110. Krimpenfort, P., Quon, K.C., Mooi, W.J., Loonstra, A., and Berns, A. 2001. Loss of p16Ink4a confers susceptibility to metastatic melanoma in mice. *Nature.* **413**:83–86.

111. Blasco, M.A., et al. 1997. Telomere shortening and tumor formation by mouse cells lacking telomerase RNA. *Cell.* **91**:25–34.

112. Tyner, S.D., et al. 2002. p53 mutant mice that display early ageing-associated phenotypes. *Nature.* **415**:45–53.

113. Garcia-Cao, I., et al. 2002. "Super p53" mice exhibit enhanced DNA damage response, are tumor resistant and age normally. *EMBO J.* **21**:6225–6235.

114. Engwerda, C.R., Handwerger, B.S., and Fox, B.S. 1994. Aged T cells are hyporesponsive to costimulation mediated by CD28. *J. Immunol.* **152**:3740–3747.

115. Hastie, N.D., et al. 1990. Telomere reduction in human colorectal carcinoma and with ageing. *Nature.* **346**:866–868.

116. Allsopp, R.C., et al. 1992. Telomere length predicts replicative capacity of human fibroblasts. *Proc. Natl. Acad. Sci. U. S. A.* **89**:10114–10118.

117. Frenck, R.W., Jr., Blackburn, E.H., and Shannon, K.M. 1998. The rate of telomere sequence loss in human leukocytes varies with age. *Proc. Natl. Acad. Sci. U. S. A.* **95**:5607–5610.

118. Hemann, M.T., Strong, M.A., Hao, L.Y., and Greider, C.W. 2001. The shortest telomere, not average telomere length, is critical for cell viability and chromosome stability. *Cell.* **107**:67–77.

119. Park, I.K., et al. 2003. Bmi-1 is required for maintenance of adult self-renewing haematopoietic stem cells. *Nature.* **423**:302–305.

120. Lewis, J.L., et al. 2001. The influence of INK4 proteins on growth and self-renewal kinetics of hematopoietic progenitor cells. *Blood.* **97**:2604–2610.

121. Spillman, B.C., and Lubitz, J. 2000. The effect of longevity on spending for acute and long-term care. *N. Engl. J. Med.* **342**:1409–1415.

122. Lubitz, J., Cai, L., Kramarow, E., and Lentzner, H. 2003. Health, life expectancy, and health care spending among the elderly. *N. Engl. J. Med.* **349**:1048–1055.

123. Cutler, D.M. 2003. Disability and the future of Medicare. *N. Engl. J. Med.* **349**:1084–1085.

Bmi1, stem cells, and senescence regulation

In-Kyung Park,[1] Sean J. Morrison,[1,2,3] and Michael F. Clarke[1,2]

[1]Department of Internal Medicine, [2]Department of Cellular and Developmental Biology, and
[3]Howard Hughes Medical Institute, University of Michigan, School of Medicine, Ann Arbor, Michigan, USA.

Stem cells generate the differentiated cell types within many organs throughout the lifespan of an organism and are thus ultimately responsible for the longevity of multicellular organisms. Therefore, senescence of stem cells must be prevented. *Bmi1* is required for the maintenance of adult stem cells in some tissues partly because it represses genes that induce cellular senescence and cell death.

Many tissues are maintained throughout the lifespan of an organism by a small number of adult stem cells. These cells are unique in that they have both the ability to give rise to new stem cells via a process called self-renewal and the ability to differentiate into the mature cells of a tissue. To maintain tissue homeostasis, stem cells have developed strict regulatory mechanisms to self-renew, differentiate, and prevent premature senescence and apoptosis (see review, ref. 1). The recent observation that *Bmi1*, a *Polycomb* group repressor, is essential for the self-renewal of adult murine hematopoietic stem cells (HSCs) and neuronal stem cells, in part via repression of genes involved in senescence, suggests that stem cells have evolved specific mechanisms to repress senescence and to prolong their capacity to proliferate. In this Perspective, we discuss the possible role of *Bmi1* in the prevention of senescence in stem cells.

What makes a cell a stem cell?

HSCs are among the best-characterized stem cells. The existence of these cells was proven using clonal assays and retroviral marking (2, 3). Flow cytometry was then used to isolate HSCs based on cell-surface marker expression (4, 5). Subsequently, other types of somatic stem cells such as neuronal stem cells from the peripheral and central nervous systems have been identified (6, 7).

Stem cells possess three fundamental properties (1). First, they must self-renew, allowing the maintenance of the original stem cell population. Self-renewal is a cell division in which one or both of the daughter cells are stem cells that retain the same developmental potential as the mother cell. In contrast, proliferation is a more general term that refers to all types of mitosis, whether they yield stem cells, restricted progenitors, or terminally differentiated cells. Second, stem cells must be able to differentiate into multiple types of mature cells in order to replace the mature cells that turn over in adult tissues. Third, the total number of stem cells is strictly regulated via both extrinsic and intrinsic mechanisms, resulting in the stability of a stable stem cell pool (8–11).

Stem cells and senescence

Senescence is a state in which a cell no longer has the ability to proliferate. Since stem cells maintain many tissues during the lifetime of an animal, it follows that stem cell senescence must be prevented to maintain an organ throughout life. Several studies suggest that cellular senescence is accompanied by changes in gene expression, which might be regulated by epigenetic mechanisms. In support of this hypothesis, histone deacetylase inhibitors, which decondense chromatin and activate the transcription of some genes, can induce a senescence-like state in human fibroblasts (12), suggesting that conversion of some heterochromatin to euchromatin may be a feature of replicative senescence (13, 14). Other studies suggest that chromatin condensation and subsequent downregulation of certain genes might regulate senescence. Senescence accompanies changes in nuclear morphology and formation of a distinct chromatin structure, called senescence-associated heterochromatic foci (SAHF) (15). SAHF do not contain active transcription sites, and they recruit heterochromatin proteins to the genes that are to be stably repressed during senescence. It was shown that SAHF contained the retinoblastoma protein (pRB) in the E2F-responsive promoters, such as cyclin A and proliferating cell nuclear antigen promoters, and silenced the expression of E2F-responsive genes during senescence but not during quiescence (15). Formation of SAHF and silencing require an intact pRB pathway, since inhibition of *p16^{Ink4a}* prevents SAHF formation and leads to DNA replication. These results provide a molecular mechanism for the maintenance of the senescent state and demonstrate the importance of pRB as a tumor suppressor.

HSCs have an impressive regenerative potential, as demonstrated by transplantation experiments using limited numbers of cells. In mice, serial transplantation is possible for four to six passages, suggesting that individual HSCs are capable of extensive self-renewal but may not be immortal. Even though HSCs express telomerase (16, 17), it is not sufficient to completely prevent telomere erosion during aging (18). Overexpression of the catalytic subunit of the telomerase enzyme in hematopoietic cells prevents telomeres from shortening during serial transplantation of bone marrow. However, even HSCs overexpressing telomerase could be serially transplanted no more than four times, as is the case with wild-type HSCs; this suggests that a telomere-independent mechanism regulates replicative senescence of mouse HSCs during serial transplantation (19). On the other hand, telomerase-deficient HSCs can be serially transplanted only twice, accompanied by an increased rate of telomere shortening, indicating that telomerase is nonetheless needed to prevent premature loss of telomere function during serial transplantation (20, 21).

Role of Bmi1 in stem cell self-renewal

Since epigenetic events such as histone modification have been implicated in senescence, it follows that genes involved in chromatin remodeling and gene expression, such as members of the *Polycomb* and *Trithorax* families, might be directly involved in deci-

Nonstandard abbreviations used: hematopoietic stem cell (HSC); senescence-associated heterochromatic foci (SAHF); retinoblastoma protein (pRB); acute myeloid leukemia (AML); mouse embryonic fibroblast (MEF); cyclin-dependent kinase (Cdk); mouse double minute 2 (MDM2); mammary epithelial cell (MEC).

Conflict of interest: The authors have declared that no conflict of interest exists.

Citation for this article: *J. Clin. Invest.* **113**:175–179 (2004). doi:10.1172/JCI200420800.

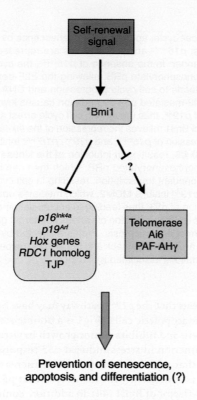

Prevention of senescence, apoptosis, and differentiation (?)

Figure 1
Postulated Bmi1 targets. Extrinsic signals for a stem cell to self-renew result in elevation of the Bmi1 level in stem cells. This allows repression of various genes including the *Ink4a* locus genes, *p16^Ink4a* and *p19^Arf*, and possibly activation, via indirect mechanisms, of some genes including telomerase, apoptosis inhibitor-6 (Ai6), and platelet-activating factor acetylhydrolase (PAF-AHγ). These genes are likely play a role in stem cell fate decisions including self-renewal and differentiation. *Sites of frequent mutations associated with cancer. TJP, tight junction protein. RDC1, chemokine orphan receptor 1.

Bmi1 may also play a key role in some types of cancer (33–35). In approximately 11% of cases of mantle cell lymphoma, the malignant cells have a three- to sevenfold amplification of *Bmi1* DNA and express high levels of the protein, implicating this gene in this invariably lethal form of lymphoma. In a mouse model of leukemia, *Bmi1* was essential for the maintenance of leukemic cells (36). Enforced expression of *Hoxa9/Meis-1* in both normal and *Bmi1*-deficient mouse fetal liver cells, followed by transplantation, initially resulted in infiltration of the bone marrow by cells that looked like acute myeloid leukemia (AML) blasts, and mice developed a bone marrow infiltrate that resembled AML. However, only *Bmi1* wild-type AML could be serially transplanted. Taken together with the detection of high levels of *Bmi1* in human AML stem cells (25), these results suggest that *Bmi1* is also required for the self-renewal of leukemic stem cells.

Bmi1 and senescence
In WI-38 human fetal lung fibroblasts, *Bmi1* is downregulated when the cells undergo replicative senescence, but not when they are quiescent. Additionally, *Bmi1* extends replicative lifespan but does not induce immortalization when overexpressed (37). In the absence of Bmi1, both the *p16^Ink4a* and the *p19^Arf* genes from the Ink4a locus are expressed (38). Lifespan extension by *Bmi1* is mediated in part by suppression of the *p16^Ink4a*-dependent senescence pathway and requires an intact pRB pathway, but not the p53 tumor-suppressor protein. The RING finger and helix-turn-helix domains of *Bmi1* were required for lifespan extension and *p16^Ink4a* suppression. Furthermore, a RING finger deletion mutant acted as a dominant negative, inducing *p16^Ink4a* and premature senescence (37).

Normal mouse embryonic fibroblasts (MEFs) reach replicative senescence after seven passages in culture, whereas MEFs from *Bmi1^−/−* mice show a premature-senescence phenotype at the third passage. This was correlated with increased expression of *p16^Ink4a*. Re-expression of *Bmi1* in *Bmi1^−/−* MEFs prevented premature senescence (28). Overexpression of *Bmi1* gave a proliferative advantage and extended MEF lifespan. Furthermore, unlike human fibroblasts, *Bmi1* could immortalize MEFs.

Downstream targets of Bmi1
Gene-profiling studies suggest that *Bmi1* modulates HSC self-renewal through the regulation of genes important for stem cell fate decisions, as well as survival genes, antiproliferative genes, and stem cell–associated genes (Figure 1) (25). The previously mentioned *Bmi1* target, the *Ink4a* locus (28), encodes *p16^Ink4a* and *p19^Arf* using different promoters (38). Enforced expression of *p16^Ink4a* and *p19^Arf* in HSCs led to senescence and apoptosis, respectively (25). In neural stem cells, *p16^Ink4a* deficiency partially restored the ability of *Bmi1*-deficient stem cells to self-renew (26). Figure 2 illustrates regulation of the cell cycle and senescence by

sions that affect stem cell fate, including self-renewal, senescence, and possibly aging. *Polycomb* and *Trithorax* proteins form large multimeric structures, which can lead to repression or activation of gene expression, respectively, via a concerted process of chromatin modifications (22, 23).

Both HSCs and neuronal stem cells express high levels of *Bmi1* (24–26), a member of the *Polycomb* group of transcription repressors that was initially identified as an oncogene cooperating with *c-myc* in a murine model of lymphoma (27, 28). Bmi1 has a *RING* finger at the amino-terminus and a central helix-turn-helix domain. The RING finger domain is required for the generation of lymphoma in Eμ-*Bmi1* transgenic mice (29, 30). Postnatal mice lacking *Bmi1* exhibit defects in hematopoiesis, skeletal patterning, neurological functions, and development of the cerebellum (31).

It has recently been shown that *Bmi1* is necessary for efficient self-renewing cell divisions of adult HSCs as well as adult peripheral and central nervous system neural stem cells, but that it is less critical for the generation of differentiated progeny (25, 26). Transplantation of *Bmi1^−/−* fetal liver cells resulted in only transient hematopoietic cell reconstitution, suggesting that the transplanted mutant fetal liver HSCs failed to generate more HSCs but gave rise to multipotent progenitors that could sustain hematopoiesis for up to 4–8 weeks. Similarly, *Bmi1* is needed for the maintenance of neural stem cells found in both the central and peripheral nervous systems. As with HSCs, the reduced self-renewal of *Bmi1*-deficient neural stem cells led to their postnatal depletion in vivo, but the proliferation and survival of committed progenitor cells were essentially normal (26). Given the broad ranges of phenotypic changes in *Bmi1*-deficient mice, including posterior transformation and neurological abnormalities (31), and its broad tissue distribution (32), it is likely that *Bmi1* regulates the self-renewal of other types of somatic stem cells.

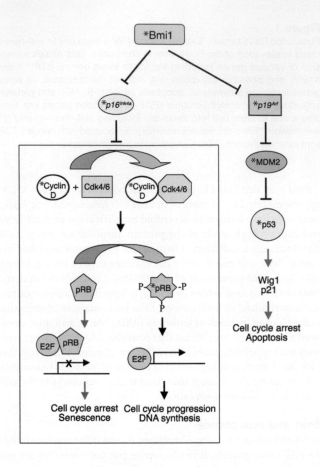

Figure 2
Regulation of cell cycle, apoptosis, and senescence by Bmi1. In normal stem cells, p16^{Ink4a} and p19^{Arf} genes are repressed in a Bmi1-dependent manner. In the absence of p16^{Ink4a}, the cyclin D/Cdk4/6 complex can phosphorylate pRB, allowing the E2F-dependent transcription that leads to cell cycle progression and DNA synthesis. In addition, MDM2-mediated p53 degradation causes low p53 levels in the absence of p19^{Arf}, thus preventing cell cycle arrest and apoptosis. The absence of Bmi1 relieves the repression of the Ink4a locus, resulting in the expression of p16^{Ink4a} and p19^{Arf}. p16^{Ink4a} inhibits binding of cyclin D to Cdk4/6, resulting in inhibition of the kinase activity. This leads to a hypophosphorylated pRB, which then can bind E2F and inhibit E2F-dependent transcription, resulting in cell cycle arrest and senescence. p19^{Arf} inhibits MDM2, which mediates ubiquitin-dependent degradation of p53, thus leading to accumulation of p53 protein in the cell. This leads to induction of various p53 target genes involved in cell cycle arrest and apoptosis. Proteins affected by high and low levels of Bmi1 are shown by black and red arrows, respectively. *Sites of frequent mutations associated with cancer.

p16^{Ink4a} and p19^{Arf}. During the cell cycle, pRB is hyperphosphorylated by the cyclin D/cyclin-dependent kinases 4 and 6 (cyclin D/Cdk4/6) complex (39). The hyperphosphorylated pRB is unable to bind and inhibit E2F transcription factor, allowing transcription of E2F target genes that are important for the G1/S transition, such as DNA polymerase II, cyclin E, p19, myb, and dihydrofolate reductase (40). This allows cell cycle progression. In the absence of Bmi1, p16^{Ink4a} is upregulated and prevents binding of Cdk4/6 to cyclin D, inhibiting the kinase activity. This results in hypophosphorylated pRB, which then binds E2F and inhibits E2F-mediated transcription, leading to cell cycle arrest and senescence (39). p19^{Arf} sequesters mouse double minute 2 (MDM2) and inhibits p53 degradation, resulting in p53-mediated cell cycle arrest and apoptosis (41, 42). Point mutations and deletion of p16^{Ink4a} and p19^{Arf} are frequently found in many types of human cancers, which implicates them as key regulators of immortalization and/or senescence checkpoints.

Mice lacking Bmi1 showed induction of both p16^{Ink4a} and p19^{Arf} in various hematopoietic and neuronal tissues (25). Overexpression of p16^{Ink4a} and p19^{Arf} in adult HSCs induced cell cycle arrest and apoptosis via the pRB and the p53-dependent pathway, respectively. Double deletion of the Bmi1 and p16^{Ink4a}/p19^{Arf} genes partially rescued the phenotypes observed in Bmi1-deficient mice (28), suggesting that p16^{Ink4a}, p19^{Arf}, and p53 are downstream effectors of Bmi1 that are involved in the control of the proliferation and survival of HSCs during self-renewing cell divisions (Figure 2). Therefore, Bmi1 maintains the HSC pool in part by repressing genes involved in cellular senescence. Increased expression of the p53 target gene Wig1 in Bmi1$^{-/-}$ bone

marrow suggests that the p19^{Arf} pathway may have been activated in Bmi1$^{-/-}$ hematopoietic cells. Wig1 is a double-stranded RNA-binding protein and inhibits tumor growth in vitro, suggesting that it may function in stress-induced p53 responses (43). The observation that p53-deficient mice have increased numbers of stem cells is consistent with the notion that p53 might be a downstream effector of Bmi1 (44). In addition, some of the Hox9 family of genes are also affected in Bmi1-deficient hematopoietic tissues and neurospheres (25, 26). Determination of the relative contribution of each of these pathways to the regulation of HSC self-renewal will require careful analysis of the HSCs from double- or triple-knockout mice.

There is evidence that Bmi1 might regulate telomerase expression in human mammary epithelial cells (MECs) and might play a role in the development of human breast cancer. Bmi1 is overexpressed in several breast cancer cell lines and postselection human MECs immortalized with human papilloma virus E6 oncogene, which abrogates the p53/p21^{waf} pathway (45), suggesting that Bmi1 might be involved in immortalization. Postselection MECs can be obtained by regular feeding of a heterogeneous population of MECs from primary mammary tissue. During this process, the p16^{Ink4a} gene is progressively silenced and not expressed in postselection MECs (46, 47). Overexpression of Bmi1 in postselection MECs bypasses senescence, extending replicative lifespan and immortalizing MECs. This is associated with human telomerase reverse transcriptase (hTERT) expression, which leads to induction of telomerase activity. Although hTERT is a direct target of c-Myc–induced transcription in MECs (48, 49), Bmi1 appeared to act independently of c-Myc. Since Bmi1 is a transcription repressor, induction of telomerase is probably mediated by an indirect mechanism. Deletion analysis of the Bmi1 protein suggested that the RING finger, as well as the conserved helix-turn-helix domain, was required for its ability to induce telomerase and immortalization. These data suggest that Bmi1 directly or indirectly regulates telomerase expression in MECs and might play a role in the development of human breast cancer. However, Bmi1 induction of telomerase is cell type specific; Bmi1 fails to induce telomerase in fibroblasts (45). This is consistent with the observation that Bmi1 overexpression did not immortalize human fibroblasts (37). It is not known whether Bmi1 is involved in telomere function in normal breast stem cells.

ated content.

Future directions

Bmi1 maintains the stem cell pool by preventing premature senescence, either through repression of genes involved in senescence or perhaps through induction of telomerase to prevent telomere shortening. It is very likely that *Bmi1* is important for maintenance of multiple types of somatic stem cells, since it is widely expressed and *Bmi1*-deficient mice have developmental defects in other organs. *Bmi1* is also important for maintenance of leukemic stem cells and perhaps other tumorigenic stem cells; therefore, *Bmi1* could be used as a molecular target to induce senescence in cancer stem cells (50).

Since *Bmi1* maintains the HSC pool size and regulates key genes implicated in senescence and aging, it is of interest to determine whether expression of *Bmi1* and its target genes changes during stem cell transplantation and/or aging. Whether stem cells undergo senescence during aging is controversial (51–53). In C57BL mice, in which most HSC studies have been performed, HSC numbers increase with age without losing overall function (54–56). However, HSC senescence might occur during aging in certain other strains of mice (57, 58). The number of times that HSCs can reconstitute the bone marrow of lethally irradiated mice is limited in serial-transplantation experiments. This observation might be either a result of an intrinsic stem cell aging program that occurs only when stem cell proliferation far exceeds that seen during normal aging, or a result of damage to the stem cells that is secondary to the stress of the transplant. In either model, it is possible that the loss of stem cell activity is mediated by *Bmi1* or its downstream targets.

Address correspondence to: Michael F. Clarke, University of Michigan, School of Medicine, 1500 E. Medical Center Drive, Ann Arbor, Michigan 48109, USA. Phone: (734) 764-8195; Fax: (734) 647-9654; E-mail: mclarke@med.umich.edu.

1. Reya, T., Morrison, S., Clarke, M., and Weissman, I. 2001. Stem cells, cancer, and cancer stem cells. *Nature.* **414**:105–111.
2. Till, J., and McCulloch, E. 1961. A direct measurement of the radiation sensitivity of normal mouse bone marrow cells. *Radiat. Res.* **14**:1419–1430.
3. Williams, D., Lemischka, I., Nathan, D., and Mulligan, R. 1984. Introduction of new genetic material into pluripotent haematopoietic stem cells of the mouse. *Nature.* **310**:476–480.
4. Spangrude, G.J., Heimfeld, S., and Weissman, I.L. 1988. Purification and characteristics of mouse hematopoietic stem cells. *Science.* **241**:58–62.
5. Morrison, S.J., and Weissman, I.L. 1994. The long-term repopulating subset of hematopoietic stem cells is deterministic and isolatable by phenotype. *Immunity.* **1**:661–673.
6. Stemple, D., and Anderson, D. 1992. Isolation of a stem cell for neurons and glia from the mammalian neural crest. *Cell.* **71**:973–985.
7. Davis, A., and Temple, S. 1994. A self-renewing multipotential stem cell in embryonic rat cerebral cortex. *Nature.* **372**:263–266.
8. Domen, J., Cheshier, S.H., and Weissman, I.L. 2000. The role of apoptosis in the regulation of hematopoietic stem cells: overexpression of BCL-2 increases both their number and repopulation potential. *J. Exp. Med.* **191**:253–264.
9. Lemischka, I.R., and Moore, K.A. 2003. Stem cells: interactive niches. *Nature.* **425**:778–779.
10. Calvi, L.M., et al. 2003. Osteoblastic cells regulate the haematopoietic stem cell niche. *Nature.* **425**:841–846.
11. Zhang, J., et al. 2003. Identification of the haematopoietic stem cell niche and control of the niche size. *Nature.* **425**:836–841.
12. Ogryzko, V., Hirai, T., Russanova, V., Barbie, D., and Howard, B. 1996. Human fibroblast commitment to a senescence-like state in response to histone deacetylase inhibitors is cell cycle dependent. *Mol. Cell. Biol.* **16**:5210–5218.
13. Howard, B. 1996. Replicative senescence: considerations relating to the stability of heterochromatin domains. *Exp. Gerontol.* **31**:281–293.
14. Villeponteau, B. 1997. The heterochromatin loss model of aging. *Exp. Gerontol.* **32**:383–394.
15. Narita, M., et al. 2003. Rb-mediated heterochromatin formation and silencing of E2F target genes during cellular senescence. *Cell.* **113**:703–716.
16. Pathak, S. 2002. Organ- and tissue-specific stem cells and carcinogenesis. *Anticancer Res.* **22**:1353–1356.
17. Morrison, S., Prowse, K., Ho, P., and Weissman, I. 1996. Telomerase activity in hematopoietic cells is associated with self-renewal potential. *Immunity.* **5**:207–216.
18. Yui, J., Chiu, C.-P., and Lansdorp, P.M. 1998. Telomerase activity in candidate stem cells from fetal liver and adult bone marrow. *Blood.* **91**:3255–3262.
19. Allsopp, R.C., and Weissman, I.L. 2002. Replicative senescence of hematopoietic stem cells during serial transplantation: does telomere shortening play a role? *Oncogene.* **21**:3270–3273.
20. Samper, E., et al. 2002. Long-term repopulating ability of telomerase-deficient murine hematopoietic stem cells. *Blood.* **99**:2767–2775.
21. Allsopp, R.C., Morin, G.B., DePinho, R., Harley, C.B., and Weissman, I.L. 2003. Telomerase is required to slow telomere shortening and extend replicative lifespan of HSCs during serial transplantation. *Blood.* **102**:517–520.
22. Simon, J., and Tamkun, J. 2002. Programming off and on states in chromatin: mechanisms of Polycomb and trithorax group complexes. *Curr. Opin. Genet. Dev.* **12**:210–218.
23. Orlando, V. 2003. Polycomb, epigenomes, and control of cell identity. *Cell.* **112**:599–606.
24. Lessard, J., Baban, S., and Sauvageau, G. 1999. Stage-specific expression of Polycomb group genes in human bone marrow cells. *Blood.* **91**:1216–1224.
25. Park, I.-K., et al. 2003. Bmi1 is required for maintenance of adult self-renewing haematopoietic stem cells. *Nature.* **423**:302–305.
26. Molofsky, A.V., et al. 2003. Bmi1 dependence distinguishes neural stem cell self-renewal from progenitor proliferation. *Nature.* **425**:962–967.
27. Haupt, Y., Bath, M., Harris, A., and Adams, J. 1993. Bmi1 transgene induces lymphomas and collaborates with myc in tumorigenesis. *Oncogene.* **8**:316–314.
28. Jacob, J., Kieboom, K., Marino, S., Depinho, R., and van Lohuizen, M. 1999. The oncogene and Polycomb-group gene Bmi1 regulates cell proliferation and senescence through the ink4a locus. *Nature.* **397**:164–168.
29. van Lohuizen, M., et al. 1991. Identification of cooperating oncogenes in E mu-myc transgenic mice by provirus tagging. *Cell.* **65**:737–752.
30. Alkema, M., Jacobs, H., van Lohuizen, M., and Berns, A. 1997. Perturbation of B and T cell development and predisposition to lymphomagenesis in Emu Bmi1 transgenic mice require the Bmi1 RING finger. *Oncogene.* **15**:899–910.
31. van der Lugt, N.M., et al. 1994. Posterior transformation, neurological abnormalities, and severe hematopoietic defects in mice with a targeted deletion of the Bmi1 proto-oncogene. *Genes Dev.* **8**:757–769.
32. Haupt, Y., Alexander, W., Barri, G., Klinken, S., and Adams, J. 1991. Novel zinc finger gene implicated as myc collaborator by retrovirally accelerated lymphomagenesis in E mu-myc transgenic mice. *Cell.* **65**:753–763.
33. Bea, S., et al. 2001. BMI1 gene amplification and overexpression in hematological malignancies occur mainly in mantle cell lymphomas. *Cancer Res.* **61**:2409–2412.
34. van Kemenade, F.J., et al. 2001. Coexpression of BMI1 and EZH2 polycomb-group proteins is associated with cycling cells and degree of malignancy in B-cell non-Hodgkin lymphoma. *Blood.* **97**:3896–3901.
35. Vonlanthen, S., et al. 2001. The Bmi1 oncoprotein is differentially expressed in non-small cell lung cancer and correlates with INK4A-ARF locus expression. *Br. J. Cancer.* **84**:1372–1376.
36. Lessard, J., and Sauvageau, G. 2003. Bmi1 determines the proliferative capacity of normal and leukaemic stem cells. *Nature.* **423**:255–260.
37. Itahana, K., et al. 2003. Control of the replicative life span of human fibroblasts by p16 and the polycomb protein Bmi1. *Mol. Cell. Biol.* **23**:389–401.
38. Quelle, D.E., Zindy, F., Ashmun, R.A., and Sherr, C.J. 1995. Alternative reading frames of the INK4a tumor suppressor gene encode two unrelated proteins capable of inducing cell cycle arrest. *Cell.* **84**:993–1000.
39. Sharpless, N., and DePinho, R. 1999. The INK4A/ARF locus and its two gene products. *Curr. Opin. Genet. Dev.* **9**:22–30.
40. Vernell, R., Helin, K., and Muller, H. 2003. Identification of target genes of the p16INK4A-pRB-E2F pathway. *J. Biol. Chem.* **278**:46124–46137.
41. Weber, J.D., Taylor, L.J., Roussel, M.F., Sherr, C.J., and Bar-Sagi, D. 1999. Nucleolar Arf sequesters Mdm2 and activates p53. *Nat. Cell Biol.* **1**:20–26.
42. Honda, R., and Yasuda, H. 1999. Association of p19ARF with Mdm2 inhibits ubiquitin ligase activity of Mdm2 for tumor suppressor p53. *EMBO J.* **18**:22–27.
43. Mendez-Vidal, C., Wilhelm, M.T., Hellborg, F., Qian, W., and Wiman, K.G. 2002. The p53-induced mouse zinc finger protein wig-1 binds double-stranded RNA with high affinity. *Nucleic Acids Res.* **30**:1991–1996.
44. TeKippe, M., Harrison, D.E., and Chen, J. 2003. Expansion of hematopoietic stem cell phenotype and activity in Trp53-null mice. *Exp. Hematol.* **31**:521–527.
45. Dimri, G.P., et al. 2002. The Bmi1 oncogene induces telomerase activity and immortalizes human mammary epithelial cells. *Cancer Res.* **62**:4736–4745.
46. Brenner, A., Stampfer, M., and Aldaz, C. 1998. Increased p16 expression with first senescence arrest in human mammary epithelial cells and extended growth capacity with p16 inactivation. *Oncogene.* **17**:199–205.
47. Wong, D., Foster, S., Galloway, D., and Reid, B. 1999. Progressive region-specific de novo methylation of the p16 CpG island in primary human mammary epithelial cell strains during escape from

M(0) growth arrest. *Mol. Biol. Cell.* **19**:5642–5651.

48. Wu, K., et al. 1999. Direct activation of TERT transcription by c-MYC. *Nat. Genet.* **21**:220–224.

49. Greenberg, R., et al. 1999. Telomerase reverse transcriptase gene is a direct target of c-Myc but is not functionally equivalent in cellular transformation. *Oncogene.* **18**:1219–1226.

50. Schmitt, C. 2003. Senescence, apoptosis and therapy: cutting the lifelines of cancer. *Nat. Rev. Cancer.* **3**:286–295.

51. Van Zant, G., and Liang, Y. 2003. The role of stem cells in aging. *Exp. Hematol.* **31**:659–672.

52. Liang, Y., and Van Zant, G. 2003. Genetic control of stem-cell properties and stem cells in aging. *Curr. Opin. Hematol.* **10**:195–202.

53. Geiger, H., True, J.M., de Haan, G., and Van Zant, G. 2001. Age- and stage-specific regulation patterns in the hematopoietic stem cell hierarchy. *Blood.* **98**:2966–2972.

54. Morrison, S., Prowse, K., Ho, P., and Weissman, I. 1996. The aging of hematopoietic stem cells. *Nat. Med.* **2**:1011–1016.

55. Sudo, K., Ema, H., Morita, Y., and Nakauchi, H. 2000. Age-associated characteristics of murine hematopoietic stem cells. *J. Exp. Med.* **192**:1273–1280.

56. Kim, M., Moon, H.-B., and Spangrude, G.J. 2003. Major age-related changes of mouse hematopoietic stem/progenitor cells. *Ann N. Y. Acad. Sci.* **996**:195–208.

57. Van Zant, G., Holland, B., Eldridge, P., and Chen, J. 1990. Genotype-restricted growth and aging patterns in hematopoietic stem cell populations of allophenic mice. *J. Exp. Med.* **171**:1547–1565.

58. Chen, J., Astle, C.M., and Harrison, D.E. 2000. Genetic regulation of primitive hematopoietic stem cell senescence. *Exp. Hematol.* **28**:442–450.

Cellular senescence in cancer treatment: friend or foe?

Pascal Kahlem,[1] Bernd Dörken,[1,2] and Clemens A. Schmitt[1,2]

[1]Humboldt University, Charité, Department of Hematology, Oncology, and Tumor Immunology, Berlin, Germany.
[2]Max Delbrück Center for Molecular Medicine, Berlin, Germany.

Damage to DNA, the prime target of anticancer therapy, triggers programmed cellular responses. In addition to apoptosis, therapy-mediated premature senescence has been identified as another drug-responsive program that impacts the outcome of cancer therapy. Here, we discuss whether induction of senescence is a beneficial or, rather, a detrimental consequence of the therapeutic intervention.

Achieving lasting remissions in patients suffering from non-localized malignancies remains the central problem of clinical oncology. Although the decades-old arsenal of classic anticancer treatment modalities such as surgery, chemotherapy, irradiation, and hormone ablation has been augmented by strategies including immunotherapy, gene therapy, inhibition of angiogenesis, hyperthermia, and a number of novel lesion-based approaches such as the administration of the Bcr-Abl kinase inhibitor STI-571, the goal to eradicate all cancer cells in a metastasized condition is rarely within reach. Anticancer treatment strategies may be insufficient for many reasons: potentially efficient therapies might not always find their way to virtually inaccessible tumor sites. Moreover, the multitude of tumor entities are known to differ remarkably in their susceptibilities to conventional DNA-damaging anticancer agents – particularly solid tumors, which often are largely refractory to chemotherapy or rapidly re-emerge from a remission. In addition, primarily susceptible tumors select for genetic defects during the course of therapy, which may render them resistant over time. While dose escalation can overcome the problem of insufficient chemosensitivity in some entities, its clinical applicability is limited by the severe toxicity codelivered to the normal cell compartment.

Traditional cytotoxic treatment strategies are driven by the assumption that quantitative execution of cell death is required to eliminate the malignant cell population. Interestingly, strategies to blunt properties of malignant growth by disabling proliferation without primarily targeting cancer viability have been less recognized, although forcing cells to exit the cell cycle by an irreversible arrest should terminate their contribution to disease progression just as effectively. In fact, recent evidence underscores the theory that premature senescence may act as an acute, drug-inducible arrest program that may contribute to the outcome of cancer therapy (1, 2). In this Perspective, we will review the role of drug-induced effector programs and discuss to what extent induction of cellular senescence may be a beneficial result of anticancer therapy.

Nonstandard abbreviations used: senescence-associated β-galactosidase (SA-β-Gal); ataxia telangiectasia mutated kinase (ATM); cyclin-dependent kinase (CDK); retinoblastoma protein (pRB); small interfering RNA (siRNA); mouse embryonic fibroblast (MEF).

Conflict of interest: The authors have declared that no conflict of interest exists.

Citation for this article: *J. Clin. Invest.* **113**:169–174 (2004). doi:10.1172/JCI200420784.

Anticancer therapy induces programmed cellular responses

Chemotherapy remains the mainstay in the treatment of systemic or metastasized malignancies. Although these highly toxic agents interfere with a plethora of cellular functions and may damage a variety of cellular structures, their pivotal cellular target is genomic DNA. It is now a well-accepted concept that drug-mediated DNA damage is not invariably lethal per se but provokes genetically encoded cellular responses. Hence, unrelated chemotherapeutic anticancer agents – in spite of their different pharmacological features and their individual target molecules participating in DNA replication and integrity – initiate common downstream mechanisms. Upon sensing DNA damage, cellular transducers activate pathways that either temporarily halt the cell cycle to allow the DNA repair machinery to fix the damage, or execute lethal programs such as apoptosis or mitotic catastrophe to restrain the damaged cell from further expansion (see ref. 3 for review). Ultimate, i.e., irreversible responses to DNA damage do not always determine the fate of cancer cells by programmed forms of cell death but may blunt their uncontrolled proliferative capacity by provoking a terminal cell-cycle arrest termed premature senescence (4–6). Moreover, recent reports demonstrated that tumor cell senescence is detectable following DNA-damaging treatment in vivo and significantly increases overall survival of the host (1, 2). In turn, the fact that different anticancer agents share genetic effector cascades renders the genetically encoded programs of apoptosis and senescence highly susceptible to inactivating mutations as a potential cause of chemoresistance. Hence, thorough dissection and mutational analysis of the pathways leading to cell death or cellular senescence are expected to identify specific genetic lesions that may be utilized by novel targeting therapies.

Drug-inducible senescence is a p53- and p16INK4a-controlled program

Premature senescence recapitulates cellular and molecular features of replicative senescence (7), which is a safeguard program that limits the growth potential, but not necessarily the viability, of a dividing cell as a consequence of the progressive shortening of its telomeres. Senescent cells, arrested in the G1 phase of the cell cycle, typically appear flattened and enlarged with increased cytoplasmic granularity. In addition to the characteristic morphology, senescent cells display enhanced activity of senescence-associated β-galactosidase (SA-β-Gal) when assessed at an acidic pH (8, 9). While refractory to mitogenic stimuli, senescent cells remain viable and metabolically active and possess a typical transcriptional profile that distinguishes them from quiescent cells (10). At the protein level, numerous regulators of cell-cycle progression, checkpoint control, and cellular integrity such as p53 or p16^{INK4a} have been found to be induced in response to various prosenescent stimuli (4, 11). Although the molecular mechanisms underlying the senescent phenotype remain largely unknown, there is increasing evidence that formation of heterochromatin in the vicinity of promoters that control gene expression related to cell-cycle progression might be implicated in the maintenance of an irreversible growth arrest (12).

Extrinsic factors such as anticancer agents, γ-irradiation, or UV light have been shown to induce premature senescence as a DNA damage–mediated cellular stress response (4–6). DNA lesions are sensed and transduced via protein complexes involved in DNA maintenance and repair, associated with members of the PI3K superfamily that includes the ataxia telangiectasia mutated kinase (ATM) (13), the ATM-related kinase (ATR), and DNA-protein kinase, among others (14–18). Besides a network of other downstream substrates, these kinases directly or indirectly phosphorylate the gatekeeper of cellular integrity (19), the p53 protein, at certain residues. Although a cascade of posttranslational modifications has been proposed to control p53 activity in response to DNA damage (20), and different DNA-damaging stimuli such as UV light or γ-irradiation can produce distinguishable signatures of posttranslational p53 modifications (21), the actual contribution of distinct phospho-residues to p53-mediated DNA-damage responses is still under debate (22, 23). p53 controls a plethora of effector functions (24), and the precise mechanisms by which specific downstream pathways are regulated in response to p53 activation have not been elucidated yet (25, 26). While it is now clear that p53 participates not only in apoptosis but also in the induction of DNA damage–mediated senescence (1, 5, 6), the signals that convert p53 from an apoptosis executor to a senescence inducer in response to anticancer therapy still need to be identified. Moreover, the posttranslational p53 modifications found in cells that entered replicative senescence shared only partial overlap with p53 modifications typically induced by DNA damage (27).

Comparable to p53, which functions as a fail-safe mediator, the cyclin-dependent kinase (CDK) inhibitor p16^{INK4a} has been implicated in both response to DNA damage and control of stress-induced senescence (1, 4, 28, 29). Although the molecular mechanism used by p16^{INK4a} to control not only a temporary but a permanent G1 arrest is largely unclear (30), p16^{INK4a} responds to DNA damage in a delayed manner and appears to be indispensable for the maintenance of cellular senescence (1, 2). For example, treatment of normal human foreskin and lung fibroblasts with DNA-damaging drugs such as bleomycin or actinomycin D induced an irreversible cell-cycle arrest with a senescence-like phenotype including a transient upregulation of p53 and p21 protein levels, followed by increased p16^{INK4a} protein expression and detectable SA-β-Gal activity (4). Similarly, exposure of adenocarcinoma cells to topoisomerase inhibitors produced premature senescence that was initially accompanied by overexpression of p53 and p21, whereby the subsequent overexpression of p16^{INK4a} persisted after drug withdrawal, highlighting the role of p16^{INK4a} in maintenance of the growth arrest (2). In an in vivo model of drug-senescent mouse lymphomas, repeated anticancer therapy eventually selected against senescence-controlling genes such as the p16^{INK4a}-encoding INK4a alleles or p53 genes, thereby producing relapse tumors that resumed growth in an aggressive manner (1). A broader drug screen based on several tumor cell lines demonstrated inducibility of a senescence-like arrest in numerous p53-proficient cancer cell lines in response to doxorubicin and other DNA-damaging agents, but it also suggested that p53-independent pathways leading to senescence might exist when some p53-mutated cell lines were exposed to escalated doses of doxorubicin (5). Interestingly, not only DNA-damaging agents, but also compounds that primarily target microtubules or the differentiating agent retinoic acid, were found to promote an SA-β-Gal–positive arrest phenotype. Moreover, a senescence-like growth arrest was also observed after γ-irradiation of normal human diploid cells (6, 31). p53 deficiency, previously reported to disable apoptosis in response to γ-irradiation (32, 33), also accounted for impaired γ-irradiation–induced senescence in a human carcinoma cell line wherein senescence could be restored upon introduction of wild-type p53 (6). Thus, in a variety of different test conditions, DNA damage–induced senescence was confirmed as a p53- and p16^{INK4a}-co-controlled safeguard program.

Drug-induced senescence: substitute player or powerful reliever?

The possibility of alternative outcomes in response to drug-induced DNA damage — apoptosis, mitotic catastrophe, cellular senescence, or simply necrosis — raises the question of whether additional stimuli or specific contextual scenarios determine the ultimate cell fate. Xenotransplant and transgenic mouse models have been used to visualize premature senescence as a quantitative response to anticancer agents in vivo (1, 5). Importantly, lymphomas generated in the Eμ-myc transgenic mouse model were prone to massive apoptosis as a default response following therapy with the alkylating agent cyclophosphamide, but they uniformly displayed premature senescence when apoptosis was blocked by overexpression of the strictly antiapoptotic mediator Bcl2 (1). Although mice harboring senescent lymphomas ultimately succumbed to their disease, they lived much longer than those bearing lymphomas with a defect in both apoptosis and senescence as a consequence of p53 loss. Given the rapid induction of apoptosis and the rather delayed detectability of an SA-β-Gal–positive long-term arrest in response to anticancer therapy in vivo, the data suggested that the senescence machinery may act as a back-up program to substitute for or to reinforce an insufficient apoptotic response (34). The actual impact of a fictitious population of cells that chose to primarily enter senescence is difficult to assess, since apoptotically competent tumors typically regress to a remission with no apparent residual tumor mass left. Senescent cells have also been detected in archival tumor samples from breast cancer patients who underwent neoadjuvant chemotherapy prior to surgical removal of their tumors. In contrast to untreated tumor samples, these samples revealed significant positive staining for SA-β-Gal activity associated with high expression of p16^{INK4a} (2). Hence, intrinsically rather chemosensitive entities such as hematological malignancies may use senescence as a secondary mechanism, while typically less susceptible solid tumors might rely on premature senescence as the chief drug-response program available.

According to many naturally occurring mutations in apoptosis-related genes, disruption of apoptosis is, at least in conjunction with certain oncogenic scenarios, a pivotal step in tumor development. Given the complex overlap between apoptosis and senescence as cellular fail-safe systems on one hand and tumor-suppressor mecha-

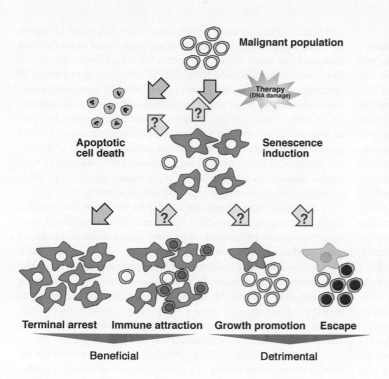

Figure 1
Drug-inducible senescence: friend or foe? In response to DNA-damaging agents, cancer cells can rapidly undergo apoptosis or may enter premature senescence as a potential back-up mechanism. Whether cells re-enter the cycle or execute apoptosis out of drug-mediated senescence remains unclear. A terminal arrest of the entire cancer cell population, possibly augmented through increased immunogenicity of senescent cells, is beneficial for the host. In contrast, feeder-like growth that reflects paracrine activity of senescent cells on their non-senescent neighbors, or escape from senescence based on acquired or preexisting mutations, is considered a detrimental outcome.

nisms and drug-effector programs on the other hand, the capability to execute senescence might be disabled in cancer cells for numerous reasons. In fact, mutations in genes that control cellular senescence may not only be selected for during therapy but might have been acquired already during tumor development (1). Oncogenes such as activated ras that are known to provoke premature senescence as the primary fail-safe mechanism may rely on defects in this program as a prerequisite to a fully transformed phenotype (28, 35), possibly inactivating senescence as a drug response as well. Nevertheless, tumors that preserve both an intact apoptotic and a functional senescence program may display a particularly robust drug response consisting of acutely inducible cell death in a first phase, corroborated by delayed apoptosis out of senescence at a later point (Figure 1). Although apoptosis is a fast-acting response mode, little is known about the possibility that apoptosis-competent cells might be sent into senescence following DNA-damaging therapy — and whether a senescent tumor could ever undergo apoptosis upon an additional proapoptotic signal (36). Finally, senescence could be recruited as an amplifier mechanism to lock temporarily arrested tumor cells — with their reduced susceptibility to checkpoint-licensed apoptosis — into irreversible cytostasis.

No effect without side effects
The fate of apoptotic cells is determined by their acute disruption of metabolic processes, rapid disintegration, and engulfment by attracted macrophages. In stark contrast, induction of cellular senescence as a formally irreversible growth arrest results in the preservation of a potentially malignant cell population locked into a nondividing state, yet possessing at least some metabolic activity (8). Although apoptotic cells provoke rather little inflammatory reaction, tumor-infiltrating immune cells reportedly can recognize altered autoantigens presented by apoptotic cancer cells (37). To what extent altered senescent tumor cells that previously managed to escape immunosurveillance can now challenge an antitumor immune response requires further investigation (Figure 1). Cor-

relative evidence points toward a link between dermal autoimmunity in the elder population and an age-dependent increase of SA-β-Gal–positive keratinocytes in human skin samples (8).

Although it is likely that senescent cells will ultimately be cleared by phagocytosis, no "eat-me" signals, as recently described for apoptotic cells, have been identified yet for the senescent state (38). While senescent neutrophils, like apoptotic granulocytes (39), might ultimately face phagocytosis through a yet unknown recognition mechanism (40), focal enrichment of lysosome-related β-galactosidase activity at autodigestive vacuoles indicated that aged human fibroblasts arrested in replicative senescence may eventually eliminate themselves by autophagy (41). As a side effect of anticancer therapy, DNA damage can also force susceptible normal cells to enter an acute SA-β-Gal–positive arrest in vivo (S. Lee and C.A. Schmitt, unpublished observations). Hence, it is conceivable that senescent cancer and normal bystander cells reside for some time next to their non-senescent malignant neighbors. In line with the role of irradiated fibroblasts as feeder cells, and with evidence reported by Waldman and colleagues, who described improved clonogenic growth in the presence of irradiation-arrested bystander cells (42), interspersed senescent — i.e., metabolically active — cells may support survival and growth of tumor cells in their vicinity. Campisi and coworkers demonstrated that senescent human fibroblasts stimulated proliferation of epithelial cell lines in vitro (43). Moreover, coimplantation of senescent fibroblasts together with preneoplastic epithelial cells in nude mice accelerated tumor formation in vivo, mainly via soluble factors secreted by the senescence-activated fibroblasts. Thus, the outcome of anticancer therapy is not only determined by a quantitative effect on cancer cells forced to irreversibly exit the cell cycle but may also depend on novel capabilities acquired by senescent cells that can impact their malignant neighbors in different ways (Figure 1). It is quite possible that, because of functionally compromised cell-cell interactions, senescent tumor cells can even exert a tumor-suppressive effect on their bystander cells.

Rest in peace?

Cellular senescence has become an attractive therapeutic concept because it qualitatively equals programmed cell death by excluding cells from active progression through the cell cycle. However, its therapeutic potential strongly relies on the irreversibility of this process. Unlike apoptosis, which acutely eliminates a potentially harmful cell, cellular senescence represents an operational change in a still-viable cell. Any cellular switch that could revert senescent cells into dividing cells implies the threat of a tumor relapse — at least as long as senescent cells have not been cleared by other processes such as phagocytosis. In fact, several experiments have provided evidence that senescence is a formally reversible process if proteins involved in its maintenance are lost. For example, reversal of replicative senescence in human lung fibroblasts was achieved via functional inactivation of both p53 and the retinoblastoma protein (pRB) by the expression of simian virus 40 large T antigen protein, or, as an alternative, by a combination of p53 inactivation and knockdown of p16^{INK4a} expression using small interfering RNA (siRNA) molecules (44). In contrast, the mere suppression of p53 function proved already sufficient to revert the senescent phenotype of human foreskin fibroblasts (44). Likewise, siRNA-driven inactivation of p53 in mouse embryonic fibroblasts (MEFs) enabled the cells to resume proliferation out of replicative senescence (45). These findings are in accordance with the putative cooperativity of p53 and p16^{INK4a} — and possibly pRB — in the induction and maintenance of premature senescence in vivo (1). Accordingly, the acute inactivation of pRB via recombinase-mediated gene deletion in MEFs that entered senescence upon replicative exhaustion or in response to oncogenic ras allowed the cells to re-emerge from the arrested phenotype (46). However, it needs to be shown that this cell-cycle re-entry truly reflects restored proliferative capacity and will not simply promote cell death by apoptosis or mitotic catastrophe within a few additional divisions.

Importantly, these experimental scenarios have not formally tested whether drug-inducible senescence is a reversible process as well. In the p53- and pRB-deficient cell line Saos-2, premature senescence — in addition to substantial cell death — was observed in response to doxorubicin treatment, suggesting a p53- and pRB-independent pathway to senescence (5). Furthermore, overexpression of a temperature-sensitive pRB transgene in Saos-2 cells produced a senescent phenotype. As a result of transgene inactivation, reinitiation of DNA synthesis was observed, and the cells underwent apoptotic cell death via the p53 homologue p73, indicating that active pRB is required to maintain the senescent state of these cells (47). This somewhat artificial setting underscores the complex wiring of signaling pathways into senescence, their relative responsiveness to different stimuli, their dependency on the cell type, and their sensitivity to experimental approaches based on nonphysiological overexpression of candidate regulators. Nevertheless, some of the experiments demonstrating successful reversal of cellular senescence that senescent cancer cells could actually encounter in vivo. While the acquisition of spontaneous mutations that disable p53 or pRB in a resting cell without DNA replication seems rather unlikely, epigenetic changes, for example, promoter methylation to silence p16^{INK4a} expression, might occur in senescent cells (Figure 1).

Exploiting cellular senescence for cancer therapy

The uncertainties regarding control and irreversibility of drug-induced senescence raise concerns as to what extent this effector mechanism reflects a desirable outcome of cancer therapy, particularly in light of therapy-inducible apoptosis as the alternate and possibly safer outcome in response to DNA damage. However, an intact apoptotic machinery is often unavailable in established malignancies. Since anticancer agents kill mainly by apoptotic cell death and, in turn, achieve little clinical efficacy in the presence of apoptotic defects (see, for example, refs. 48–52), promising treatment alternatives must use effector mechanisms that do not rely on an intact apoptotic machinery. Importantly, in vivo analyses of treatment responses in primary lymphomas harboring defined genetic lesions demonstrated that induction of senescence despite the presence of an apoptotic block improved the outcome of anticancer therapy (1) — regardless of a potential reversal or possible emergence of preexisting escape mutants at a later time.

Appropriate test systems are critical to elucidate the complex implications of drug-inducible senescence. Not surprisingly, drug sensitivity assays performed on primary tumor material in culture unveiled that adaptation to the nonphysiological culture conditions selected for apoptotic defects and chemoresistance (50). Paradoxically, many cell lines retain the ability to enter senescence following drug exposure in vitro (5), although inactivation of the terminal-arrest program appears to be a key prerequisite for any primary tumor during successful establishment as a continuous cell line. Irrespective of the technical inducibility of a senescence-like phenotype in culture-adapted cells, a Petri dish setting cannot mimic the complexity and interactivity of a natural tumor environment in vivo. Hence, many of the questions raised about the role of drug-induced senescence need to be addressed in vivo using physiological model systems. Tractable mouse models of cancer, such as the transgenic *Eμ-myc* lymphoma model (1), in alliance with sophisticated genetic tools will allow researchers to dissect the pathways and the impact of drug-inducible senescence in vivo. Moreover, large-scale analyses of the transcriptome and proteome of primary human tumor samples will expand our understanding of the molecular mechanisms that underlie drug responses in sensitive and resistant conditions.

Given the impact of apoptotic defects on tumor biology and treatment outcome, it is a research priority to invent small compound- or gene therapy–based approaches that may resensitize cancer cells to death signals. Likewise, one can envision lesion-based strategies to restore a defective senescence response. Ultimately, direct activation of pro-senescent pathways without induction of deleterious DNA damage seems to be a particularly appealing concept. For instance, cDNA microarray analysis of human diploid fibroblasts revealed that cGMP synthesis was inhibited during replicative senescence. Exposure of tumor cells to a guanylate cyclase inhibitor induced, by activation of the CDK inhibitor p21, a senescent phenotype that was independent of its upstream regulator p53, indicating that induction of p21 activity might be sufficient to halt cell proliferation, even in the absence of functional p53 and without an additional DNA-damage signal (53). Furthermore, a synthetic inhibitor of CDK4 — possibly mimicking the role of p16^{INK4a} to maintain a senescent phenotype — also produced a DNA damage–independent form of premature senescence in cells lacking proper p16^{INK4a} expression and inhibited the growth of xenotransplant tumors in mice (54). Inactivation of the papilloma virus oncoproteins E6 and E7, which deregulate the p53 and Rb proteins, respectively, by siRNA molecules restored cellular senescence in cervical cancer cells (55).

Cellular senescence and its potential use as a drug-effector program remains a complex biological phenomenon with unknown significance in cancer therapy. Whether cellular senescence is rather friend or foe most likely depends on accompanying lesions, first of all in apoptotic response programs, and on the cellular context.

In further preclinical investigations, it will be of particular interest to explore therapies that do not deliver devastating DNA damage to the cell, that do not rely on functional DNA-damage transducer systems, and that do not target pathways already mutated to cancel apoptosis, but that directly prompt a senescence response.

Acknowledgments

Research in the authors' laboratory is supported in part by grants from the Deutsche Forschungsgemeinschaft and Deutsche Krebshilfe. We apologize to our colleagues whose work has not been covered in this article because of space limitations. We thank Soyoung Lee for helpful discussions and editorial advice.

Address correspondence to: Clemens A. Schmitt, Max Delbrück Center for Molecular Medicine, and Humboldt University, Charité, Department of Hematology, Oncology, and Tumor Immunology, Augustenburger Platz 1, 13353 Berlin, Germany. Phone: 49-30-450-553-687; Fax: 49-30-450-553-986; E-mail: clemens.schmitt@charite.de.

1. Schmitt, C.A., et al. 2002. A senescence program controlled by p53 and p16INK4a contributes to the outcome of cancer therapy. *Cell.* **109**:335–346.
2. te Poele, R.H., Okorokov, A.L., Jardine, L., Cummings, J., and Joel, S.P. 2002. DNA damage is able to induce senescence in tumor cells in vitro and in vivo. *Cancer Res.* **62**:1876–1883.
3. Roninson, I.B., Broude, E.V., and Chang, B.D. 2001. If not apoptosis, then what? Treatment-induced senescence and mitotic catastrophe in tumor cells. *Drug Resist. Updat.* **4**:303–313.
4. Robles, S.J., and Adami, G.R. 1998. Agents that cause DNA double strand breaks lead to p16INK4a enrichment and the premature senescence of normal fibroblasts. *Oncogene.* **16**:1113–1123.
5. Chang, B.D., et al. 1999. A senescence-like phenotype distinguishes tumor cells that undergo terminal proliferation arrest after exposure to anticancer agents. *Cancer Res.* **59**:3761–3767.
6. Suzuki, K., et al. 2001. Radiation-induced senescence-like growth arrest requires TP53 function but not telomere shortening. *Radiat. Res.* **155**:248–253.
7. Hayflick, L., and Moorhead, P.S. 1961. The serial cultivation of human diploid cell strains. *Exp. Cell Res.* **25**:585–621.
8. Dimri, G.P., et al. 1995. A biomarker that identifies senescent human cells in culture and in aging skin in vivo. *Proc. Natl. Acad. Sci. U. S. A.* **92**:9363–9367.
9. Campisi, J., Dimri, G., and Hara, E. 1996. Control of replicative senescence. In *Handbook of the biology of aging.* S.A.J. Rowe, editor. Academic Press. New York, New York, USA. 121–149.
10. Shelton, D.N., Chang, E., Whittier, P.S., Choi, D., and Funk, W.D. 1999. Microarray analysis of replicative senescence. *Curr. Biol.* **9**:939–945.
11. Serrano, M. 1997. The tumor suppressor protein p16INK4a. *Exp. Cell Res.* **237**:7–13.
12. Narita, M., et al. 2003. Rb-mediated heterochromatin formation and silencing of E2F target genes during cellular senescence. *Cell.* **113**:703–716.
13. Canman, C.E., and Lim, D.S. 1998. The role of ATM in DNA damage responses and cancer. *Oncogene.* **17**:3301–3308.
14. Tee, A.R., and Proud, C.G. 2000. DNA-damaging agents cause inactivation of translational regulators linked to mTOR signalling. *Oncogene.* **19**:3021–3031.
15. Abraham, R.T. 2001. Cell cycle checkpoint signaling through the ATM and ATR kinases. *Genes Dev.* **15**:2177–2196.
16. Durocher, D., and Jackson, S.P. 2001. DNA-PK, ATM and ATR as sensors of DNA damage: variations on a theme? *Curr. Opin. Cell Biol.* **13**:225–231.
17. Rouse, J., and Jackson, S.P. 2002. Interfaces in the detection, signaling, and repair of DNA damage. *Science.* **297**:547–551.
18. Dent, P., Yacoub, A., Fisher, P.B., Hagan, M.P., and Grant, S. 2003. MAPK pathways in radiation responses. *Oncogene.* **22**:5885–5896.
19. Levine, A.J. 1997. p53, the cellular gatekeeper for growth and division. *Cell.* **88**:323–331.
20. Sakaguchi, K., et al. 1998. DNA damage activates p53 through a phosphorylation-acetylation cascade. *Genes Dev.* **12**:2831–2841.
21. Chehab, N.H., Malikzay, A., Stavridi, E.S., and Halazonetis, T.D. 1999. Phosphorylation of Ser-20 mediates stabilization of human p53 in response to DNA damage. *Proc. Natl. Acad. Sci. U. S. A.* **96**:13777–13782.
22. Shieh, S.Y., Taya, Y., and Prives, C. 1999. DNA damage-inducible phosphorylation of p53 at N-terminal sites including a novel site, Ser20, requires tetramerization. *EMBO J.* **18**:1815–1823.
23. Chao, C., Saito, S., Anderson, C.W., Appella, E., and Xu, Y. 2000. Phosphorylation of murine p53 at ser-18 regulates the p53 responses to DNA damage. *Proc. Natl. Acad. Sci. U. S. A.* **97**:11936–11941.
24. Vousden, K.H., and Lu, X. 2002. Live or let die: the cell's response to p53. *Nat. Rev. Cancer.* **2**:594–604.
25. Oda, K., et al. 2000. p53AIP1, a potential mediator of p53-dependent apoptosis, and its regulation by Ser-46-phosphorylated p53. *Cell.* **102**:849–862.
26. D'Orazi, G., et al. 2002. Homeodomain-interacting protein kinase-2 phosphorylates p53 at Ser 46 and mediates apoptosis. *Nat. Cell Biol.* **4**:11–19.
27. Webley, K., et al. 2000. Posttranslational modifications of p53 in replicative senescence overlapping but distinct from those induced by DNA damage. *Mol. Cell. Biol.* **20**:2803–2808.
28. Serrano, M., Lin, A.W., McCurrach, M.E., Beach, D., and Lowe, S.W. 1997. Oncogenic ras provokes premature cell senescence associated with accumulation of p53 and p16INK4a. *Cell.* **88**:593–602.
29. Shapiro, G.I., Edwards, C.D., Ewen, M.E., and Rollins, B.J. 1998. p16INK4A participates in a G1 arrest checkpoint in response to DNA damage. *Mol. Cell. Biol.* **18**:378–387.
30. Agami, R., and Bernards, R. 2000. Distinct initiation and maintenance mechanisms cooperate to induce G1 cell cycle arrest in response to DNA damage. *Cell.* **102**:55–66.
31. Seidita, G., Polizzi, D., Costanzo, G., Costa, S., and Di Leonardo, A. 2000. Differential gene expression in p53-mediated G(1) arrest of human fibroblasts after gamma-irradiation or N-phosphoacetyl-L-aspartate treatment. *Carcinogenesis.* **21**:2203–2210.
32. Lowe, S.W., Schmitt, E.M., Smith, S.W., Osborne, B.A., and Jacks, T. 1993. p53 is required for radiation-induced apoptosis in mouse thymocytes. *Nature.* **362**:847–849.
33. Schmitt, C.A., Wallace-Brodeur, R.R., Rosenthal, C.T., McCurrach, M.E., and Lowe, S.W. 2000. DNA damage responses and chemosensitivity in the E mu-myc mouse lymphoma model. *Cold Spring Harb. Symp. Quant. Biol.* **65**:499–510.
34. Lee, S., and Schmitt, C.A. 2003. Chemotherapy response and resistance. *Curr. Opin. Genet. Dev.* **13**:90–96.
35. Brookes, S., et al. 2002. INK4a-deficient human diploid fibroblasts are resistant to RAS-induced senescence. *EMBO J.* **21**:2936–2945.
36. Wang, E. 1995. Senescent human fibroblasts resist programmed cell death, and failure to suppress bcl2 is involved. *Cancer Res.* **55**:2284–2292.
37. Hansen, M.H., Nielsen, H., and Ditzel, H.J. 2001. The tumor-infiltrating B cell response in medullary breast cancer is oligoclonal and directed against the autoantigen actin exposed on the surface of apoptotic cancer cells. *Proc. Natl. Acad. Sci. U. S. A.* **98**:12659–12664.
38. Lauber, K., et al. 2003. Apoptotic cells induce migration of phagocytes via caspase-3-mediated release of a lipid attraction signal. *Cell.* **113**:717–730.
39. Savill, J.S., et al. 1989. Macrophage phagocytosis of aging neutrophils in inflammation. Programmed cell death in the neutrophil leads to its recognition by macrophages. *J. Clin. Invest.* **83**:865–875.
40. Murphy, J.F., McGregor, J.L., and Leung, L.L. 1998. Senescent human neutrophil binding to thrombospondin (TSP): evidence for a TSP-independent pathway of phagocytosis by macrophages. *Br. J. Haematol.* **102**:957–964.
41. Gerland, L.M., et al. 2003. Association of increased autophagic inclusions labeled for beta-galactosidase with fibroblastic aging. *Exp. Gerontol.* **38**:887–895.
42. Waldman, T., et al. 1997. Cell-cycle arrest versus cell death in cancer therapy. *Nat. Med.* **3**:1034–1036.
43. Krtolica, A., Parrinello, S., Lockett, S., Desprez, P.Y., and Campisi, J. 2001. Senescent fibroblasts promote epithelial cell growth and tumorigenesis: a link between cancer and aging. *Proc. Natl. Acad. Sci. U. S. A.* **98**:12072–12077.
44. Beausejour, C.M., et al. 2003. Reversal of human cellular senescence: roles of the p53 and p16 pathways. *EMBO J.* **22**:4212–4222.
45. Dirac, A.M., and Bernards, R. 2003. Reversal of senescence in mouse fibroblasts through lentiviral suppression of p53. *J. Biol. Chem.* **278**:11731–11734.
46. Sage, J., Miller, A.L., Perez-Mancera, P.A., Wysocki, J.M., and Jacks, T. 2003. Acute mutation of retinoblastoma gene function is sufficient for cell cycle re-entry. *Nature.* **424**:223–228.
47. Alexander, K., Yang, H.S., and Hinds, P.W. 2003. pRb inactivation in senescent cells leads to an E2F-dependent apoptosis requiring p73. *Mol. Cancer Res.* **1**:716–728.
48. Schmitt, C.A., McCurrach, M.E., de Stanchina, E., Wallace-Brodeur, R.R., and Lowe, S.W. 1999. INK4a/ARF mutations accelerate lymphomagenesis and promote chemoresistance by disabling p53. *Genes Dev.* **13**:2670–2677.
49. Dohner, H., et al. 2000. Genomic aberrations and survival in chronic lymphocytic leukemia. *N. Engl. J. Med.* **343**:1910–1916.
50. Schmitt, C.A., Rosenthal, C.T., and Lowe, S.W. 2000. Genetic analysis of chemoresistance in primary murine lymphomas. *Nat. Med.* **6**:1029–1035.
51. Chang, J.C., et al. 2003. Gene expression profiling for the prediction of therapeutic response to docetaxel in patients with breast cancer. *Lancet.* **362**:362–369.
52. Holleman, A., Den Boer, M.L., Kazemier, K.M., Janka-Schaub, G.E., and Pieters, R. 2003. Resistance to different classes of drugs is associated with impaired apoptosis in childhood acute lymphoblastic leukaemia. *Blood.* **102**:4541–4546.
53. Lodygin, D., Menssen, A., and Hermeking, H. 2002. Induction of the Cdk inhibitor p21 by LY83583 inhibits tumor cell proliferation in a p53-independent manner. *J. Clin. Invest.* **110**:1717–1727. doi:10.1172/JCI200216588.
54. Soni, R., et al. 2001. Selective in vivo and in vitro effects of a small molecule inhibitor of cyclin-dependent kinase 4. *J. Natl. Cancer Inst.* **93**:436–446.
55. Hall, A.H., and Alexander, K.A. 2003. RNA interference of human papillomavirus type 18 E6 and E7 induces senescence in HeLa cells. *J. Virol.* **77**:6066–6069.

Mechanisms for pituitary tumorigenesis: the plastic pituitary

Shlomo Melmed

Cedars-Sinai Medical Center, David Geffen School of Medicine at the University of California, Los Angeles, Los Angeles, California, USA.

The anterior pituitary gland integrates the repertoire of hormonal signals controlling thyroid, adrenal, reproductive, and growth functions. The gland responds to complex central and peripheral signals by trophic hormone secretion and by undergoing reversible plastic changes in cell growth leading to hyperplasia, involution, or benign adenomas arising from functional pituitary cells. Discussed herein are the mechanisms underlying hereditary pituitary hypoplasia, reversible pituitary hyperplasia, excess hormone production, and tumor initiation and promotion associated with normal and abnormal pituitary differentiation in health and disease.

Introduction

The pituitary gland responds to complex central and peripheral signals by two mechanisms. First, trophic hormone secretion is exquisitely controlled to regulate homeostasis. Second, developmental or acquired pituitary signals may elicit plastic pituitary growth responses, consisting of either hypoplasia, hyperplasia, or adenoma formation. These clinically apparent plastic changes of pituitary mass are indicative of physiologic or pathologic responses to extrapituitary or intrapituitary signals. Pituitary proliferative changes are usually accompanied by functional disorders of hormone secretion, leading to syndromes of hormone deficiency or excess. Mutations of early developmental genes (including *Rpx*, *Lhx3*, *Lhx4*, and *Pitx2*) pleiotropically affect adjacent midline structures, resulting in pituitary hypoplasia and pituitary hormone deficits, while mutations in genes determining specific pituitary lineages (including *Prop1*, *Pit1*, and *Tpit*) are involved in pituitary hormone deficiencies with hypoplasia. Excess pituitary hormone secretion is usually associated with invariably benign monoclonal adenomas arising from a specific cell type, and although pituitary chromosome instability is an early hallmark of pituitary adenoma development and growth, pituitary carcinoma is very rare, further supporting the concept that pituitary adenomas have the capacity for reversible plasticity.

The hypothalamic–anterior pituitary unit integrates stimulatory and inhibitory central and peripheral signals to synthesize and secrete hormones by five highly differentiated cell types: somatotrophs, gonadotrophs, lactotrophs, thyrotrophs, and corticotrophs (Figure 1). Each of these cell types expresses unique G protein–coupled receptors (GPCRs), which are specific for hypothalamic releasing and inhibiting hormones. These peptides traverse the hypophyseal portal system and impinge upon their cognate pituicytes to regulate the synthesis and secretion of anterior pituitary trophic

hormones that regulate growth (including growth hormone [GH]), sexual development and function (including luteinizing hormone [LH] and follicle-stimulating hormone [FSH]), lactation (including prolactin [PRL]), metabolism (including thyroid-stimulating hormone [TSH]), and stress responses (including adrenocorticotropic hormone [ACTH]). The gland itself responds to central and peripheral signals by undergoing reversible changes in cell growth leading to hyperplasia, involution or true adenoma formation. Using double-labeling with both BrdU and specific anterior pituitary hormone markers, it is apparent that even after their differentiation, pituitary cells continue mitosis, which may be augmented under certain conditions in the adult (e.g., pregnancy). About 30% of rat pituitary cells arise from "self-mitosis" of already differentiated cells, while others are produced by differentiation of hitherto undifferentiated cells, or possibly from pituitary stem cells. Thus, proliferative and apoptotic changes in the pituitary are observed during the first year of rodent life (1), and likely also occur in the human pituitary (2).

Tumors may arise from any of these cells, and their secretory products depend upon the cell of origin. Functional classification of pituitary tumors is facilitated by immunocytochemical or in situ mRNA detection of cell gene products, as well as by measurement of circulating trophic and target hormone concentrations (Figure 2). ACTH oversecretion results in Cushing disease, with features of hypercortisolism; GH hypersecretion leads to acral overgrowth and metabolic dysfunction associated with acromegaly; and PRL hypersecretion leads to gonadal failure, secondary infertility, and galactorrhea. More rarely, TSH hypersecretion leads to hyperthyroxinemia and goiter, and hypersecreted gonadotropins (or their respective subunits) lead to gonadal dysfunction. Mixed tumors cosecreting GH with PRL, TSH, or ACTH may also arise from single cells. In contrast, tumors arising from gonadotroph cells do not efficiently secrete their gene products, and they are usually clinically silent (3–5).

Hormone secretion from pituitary tumors, although excessive and associated with unique phenotypic features, often retains intact trophic control. For example, dopaminergic agents appropriately suppress PRL secretion by prolactinomas, and dexamethasone may suppress ACTH secretion in patients with pituitary Cushing disease. Excessive secretory patterns are also not uniform, and may in fact cycle between normal and excessive hormone release.

Several characteristic hallmarks of pituitary neoplasia point to a unique growth behavior distinct from that of other endocrine and nonendocrine malignancies. Pituitary tumors are invariably benign, and although aggressive local growth may occur, their general fail-

---

Nonstandard abbreviations used: G protein–coupled receptor (GPCR); growth hormone (GH); luteinizing hormone (LH); follicle-stimulating hormone (FSH); prolactin (PRL); thyroid-stimulating hormone (TSH); adrenocorticotropic hormone (ACTH); GH-releasing hormone (GHRH); nerve growth factor (NGF); proopiomelanocortin (POMC); cyclin-dependent kinase (CDK); estrogen receptor (ER); pituitary tumor transforming gene (PTTG); corticotropin-releasing hormone (CRH); dopamine D2 receptor (DRD2); somatostatin receptor subtype (SSTR); cAMP response element–binding protein (CREB); Carney complex (CNC); protein kinase A (PKA); PKA type I-α regulatory subunit (PRKAR1α); multiple endocrine neoplasia type I (MEN I); loss of heterozygosity (LOH).

Conflict of interest: The author has declared that no conflict of interest exists.

Citation for this article: *J. Clin. Invest.* **112**:1603–1618 (2003). doi:10.1172/JCI200320401.

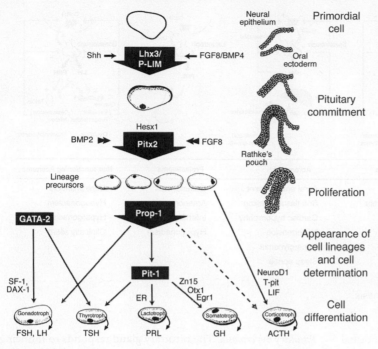

Figure 1

Model for development of human anterior pituitary cell lineage determination by a temporally controlled cascade of transcription factors. Trophic cells are depicted with transcription factors known to determine cell-specific human or murine gene expression. Adapted with permission from W.B. Saunders (139), Humana Press (140), and Karger Publishing (141).

ure to proceed to true malignancy with demonstrable extracranial metastases is intriguing. These adenomas grow slowly, and are discovered in up to 25% of unselected autopsy specimens. Although the natural history of pituitary microadenoma growth is difficult to ascertain because of the intrinsic inaccessibility of pituitary tissue for study, it is clear that microadenomas do not invariably progress to macroadenomas; furthermore, macroadenomas are stable or exhibit very slow growth, and may in fact resolve spontaneously (6). Oncogene mutations commonly encountered in nonendocrine neoplasms (e.g., ras and p53) are not generally present in pituitary adenomas, yet disturbed intrapituitary paracrine growth factor expression and action have been extensively documented (3–5). Study of human pituitary tissue is challenging due to several limitations, including the anatomic inaccessibility of the pituitary gland, the lack of functional human cell lines in culture, the paucity of faithful animal models, and unique differentiated tumor behaviors. Although mouse models may differ from human counterparts, their study provides important insights into human pituitary tumor pathogenesis. Animal studies of pituitary proliferative changes have largely followed two approaches. First, disruption of known tumor suppressor genes tested in transgenic animal models has revealed unexpected pituitary hyperplasia and tumor phenotypes (e.g., Rb and p27). Second, transgenic animals have been used to test genes known for their pituitary-regulatory functions (e.g., GH-releasing hormone [GHRH], nerve growth factor [NGF], and cytokines). These lines of investigation have also made it possible to understand the role of pituitary cell cycle proteins (Table 1).

In light of these observations, the approach to understanding pituitary adenoma pathogenesis requires insights into factors

regulating pituitary growth, from hypoplasia through hyperplasia, and ultimately true adenoma development (Table 2).

Pituitary hypoplasia: transcription factor regulation of pituitary development. Hormone-specific anterior pituicytes are embryologically derived from a pluripotent precursor, and arise as a consequence of concerted temporal and anatomic control of homeodomain repressor and activator transcription factor expression. Figure 1 depicts the developmental lineages of anterior pituitary subtypes (7). Functional disruption of this cellular cascade by transcription factor mutations may lead to hormone deficiencies due to disordered pituitary cell development and differentiation and resultant pituitary hypoplasia.

Early pituitary differentiation requires Rpx and Ptx expression. Ptx2 is mutated in patients with Rieger syndrome (8), which comprises maldevelopment of the anterior eye, teeth, and umbilicus, and GH deficiency. Lim homeobox (Lhx3) is defective in hypopituitary patients with features of a rigid cervical spine (9), and Lhx4 mutations are associated with combined pituitary hormone deficits in patients with Chiari-type cerebellar malformations (10). *Rpx (Hesx1)* is critical for the development of a committed Rathke's pouch, and rare Rpx mutations are found in subjects with septo-optic dysplasia (midline forebrain abnormalities, optic nerve hypoplasia, and pituitary dysplasia) (11). A T-box factor (12), Tbx19/Tpit, interacts cooperatively with PitX1 in corticotrophs and loss of T-pit function has been reported in patients with isolated ACTH deficiency (13). T-pit disruption in transgenic mice also leads to hypoplasia of the proopiomelanocortin (POMC)-expressing intermediate lobe (14).

Developmental PROP-1 defects produce variable phenotypes and include defects in all five anterior pituitary hormones (15). GH-, PRL-, and TSH-expressing cells share a common developmental pathway, and *Pit-1* mutations can affect all three cell types. Ames dwarf (*Prop1^{df/df}*) and *Pit-1^{dw/dw}* mice display growth deficiency, hypothyroidism, and infertility (16, 17). Inappropriate timing of pituitary transcription factor expression may also lead to developmental consequences. Persistent transgenic murine pituitary *Prop-1* expression results in delayed murine gonadotrope differentiation, persistent Rathke's cleft cysts, pituitary enlargement, and null cell nonsecreting pituitary adenomas (18).

Pituitary growth is altered dramatically by transcription factor mutations (19). In a study of 52 patients with PROP-1 mutations assessed by MRI, the pituitary was found to be hypoplastic in 34, hyperplastic in 14, and normal in 4 subjects (19). Pituitary height, as assessed by MRI, was diminished in over two thirds of 76 patients with both idiopathic and genetic GH deficiencies due to GH, GHRH receptor, or PROP-1 deficiency (20). Interestingly, subjects with idiopathic GH deficiency, likely due to perinatal damage (hypoxia), exhibited pituitary stalk thinning with an ectopic posterior lobe, presumably reflecting functional hypothalamic disruption. Paradoxically, some patients with inherited multiple pituitary hormone deficiencies may in fact develop nonhomogenous cystic pituitary hyperplasia, which must be distinguished from a pituitary adenoma (21). This discordant pituitary growth may reflect enhanced sensitivity to a pituitary growth signal unmasked by PROP-1 disruption.

Although *Pit-1* mRNA is increased up to fivefold in pituitary adenomas expressing GH and or PRL, the cell type distribution,

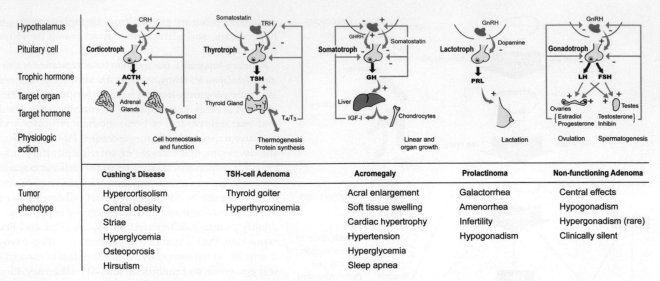

Figure 2

Hypothalamic-pituitary regulation and pituitary tumor pathogenesis.

	Cushing's Disease	TSH-cell Adenoma	Acromegaly	Prolactinoma	Non-functioning Adenoma
Tumor phenotype	Hypercortisolism	Thyroid goiter	Acral enlargement	Galactorrhea	Central effects
	Central obesity	Hyperthyroxinemia	Soft tissue swelling	Amenorrhea	Hypogonadism
	Striae		Cardiac hypertrophy	Infertility	Hypergonadism (rare)
	Hyperglycemia		Hypertension	Hypogonadism	Clinically silent
	Osteoporosis		Hyperglycemia		
	Hirsutism		Sleep apnea		

size, and sequence of *Pit-1* transcripts are unaltered from normal pituitary tissue (22). Ratios of adenoma Pit-1 and Pit-1α isoforms are also normal, further suggesting that pituitary tumorigenesis is not associated with altered *Pit-1* expression. Human *PROP-1* is persistently expressed beyond embryonic development, although prolonged expression is tumorigenic in transgenic mice (18). Persistent *PROP-1* expression in both normal and adenomatous human pituitary tissue and the absence of *PROP-1* coding mutations (23) from pituitary tumors suggest that this factor may be required to maintain mature pituitary cell types. Recently, a murine model of pituitary hypoplasia has been developed by the disruption of the cyclin-dependent kinase 4 (CDK4), leading to profound selective lactotroph hypoplasia (24).

Pituitary hyperplasia. The pituitary gland responds to inducing signals, as well as their withdrawal, by regulating both trophic hormone secretion and mitotic and apoptotic growth changes. Pituitary hyperplasia is characterized by increased proliferation of a single cell type, which may be focal, nodular, or diffuse. There is an absolute increase in numbers of specific cells, with pituitary enlargement visible on MRI. Pituitary hyperplasia may range from modest cell type increases to large glandular expansion with grossly altered tissue architecture and morphology (25). The pathological diagnosis of hyperplasia is difficult and is best made by demonstrating intact acinar structures utilizing a reticulin stain. Specifically, corticotroph hyperplasia may be associated with Crooke's hyaline changes, and thyrotroph hyperplasia, with

Table 1

Anterior pituitary hormone secretion and action

	Gonadotroph	Thyrotroph	Lactotroph	Somatotroph	Corticotroph
Fetal appearance	12 weeks	12 weeks	12 weeks	8 weeks	6 weeks
Hormone	FSH, LH	TSH	PRL	GH	POMC
Chromosomal gene locus	β-11p; β-19q	α-6q; β-1p	6	17q	2p
Protein	Glycoprotein α, β subunits	Glycoprotein α, β subunits	Polypeptide	Polypeptide	Polypeptide
AA	210; 204	211	199	191	266 (ACTH 1–39)
Stimulators	GnRH, estrogen	TRH	Estrogen, TRH	GHRH, GHS	CRH, AVP, gp-130 cytokines
Inhibitors	Sex steroids, Inhibin	T3, T4, dopamine, somatostatin, glucocorticoids	Dopamine	Somatostatin, IGF, activins	Glucocorticoids
Target gland	Ovary, testis	Thyroid	Breast, other tissues	Liver, bones, other tissues	Adrenal
Trophic effect	Sex steroid, follicle growth, germ cell maturation; M: 5–20 IU/l, F (basal): 5–20 IU/l	T4 synthesis and secretion	Milk production	IFG-I production, growth induction, insulin antagonism	Steroid production
Normal range	M: 5–20 IU/l, F (basal): 5–20 IU/l	0.1–5 mIU/l	M <15 µg/l, F <20 µg/l	<0.5 µg/l	ACTH: 4–22 pg/l

T3, triiodothyronine; T4, thyroxine; M, male; F, female, AVP, vasopressin. GHS, GH secretagogues.

Table 2
Factors regulating human pituitary growth and tumor formation

	Factor
Pituitary hypoplasia	
Transcription factor mutation	Pit-1, Prop-1, T-pit
Structural defect	Hesx1, Pitx2, Lhx3, Lhx4
Hypothalamic hormone disruption	GHRH receptor
Idiopathic GH deficiency	
Pituitary hyperplasia	
Lactotroph	
Pregnancy	Estrogen
Lactation	Estrogen
Excessive estrogen exposure	
Stalk-section	D2R disruption
Somatotroph	
Eutopic or ectopic GHRH production	GHRH
Hypothalamic tumor, chest and abdominal carcinoid, phaeochromocytoma	
McCune-Albright syndrome	*Gsp*
Mammosomatotroph hyperplasia (gigantism)	
Corticotroph	
Eutopic or ectopic CRH production	CRH
Hypothalamic tumor	
Thymic tumor	
Untreated adrenal failure	Adrenal steroid feedback
Cushing disease (~10%)	
Nelson syndrome	
Thyrotroph	
Untreated thyroid failure	Thyroid feedback
Gonadotroph	
Untreated gonadal failure	Gonadal steroid feedback
Klinefelter syndrome	
Pituitary adenoma	
Hereditary	
MEN-1	
CNC	
Signal transduction mutations	*gsp*, CREB
Loss of tumor suppressor gene function	Rb, p16, p27, GADD45γ
Disrupted paracrine growth factor or cytokine action	FGFs, EGF, NGF, cytokines, CAMS
Chromosomal instability	PTTG
Epigenetic events	Methylation, deacetylation

CAMS, cell adhesion molecules.

periodic acid Schiff–positive lysosomes. Rarely, pituitary hyperplasia may be of primary origin, and is usually secondary to extrinsic signals. Normal pituitary height as assessed by MRI is up to 9 mm in healthy subjects, while adolescent females tend to have larger pituitary glands. Clearly, hormonal and clinical evaluations are required for all enlarged pituitary images, as "primary" pituitary enlargement is rarely encountered, especially in male subjects. Invariably, an enlarged pituitary discovered incidentally on MRI can be ascribed to a pituitary adenoma (26).

Target hormones (sex and adrenal steroids, and thyroid hormones) exert powerful negative feedback inhibition of their respective trophic hormone gene transcription and hormone secretion, as well as suppression of pituicyte growth. Failure of target glands (thyroid, adrenal, and gonads) leads to loss of negative feedback inhibition and resultant compensatory hyperplasia of the respective pituitary trophic hormone cells (Figure 2). Thus, longstand-

ing primary hypothyroidism, hypogonadism, or hypoadrenalism may be associated with a clinically enlarged pituitary gland visible on MRI, with involution of the gland occurring after appropriate target hormone replacement and restoration of negative feedback (Figure 3).

The pituitary gland enlarges approximately twofold during pregnancy with most growth accounted for by diffuse lactotroph hyperplasia. Pre-existing lactotrophs proliferate, and somatotrophs are also recruited to switch from GH to PRL production (27). After birth or cessation of lactation, the pituitary size involutes and hyperactive "pregnancy cells" regress, and the number of lactotrophs reverts to almost normal. Nevertheless, both pituitary weight and cell numbers are higher in nonpregnant multiparous women with no demonstrable increased incidence of prolactinoma formation (28). Interestingly, in postmenopausal women, despite enhanced gonadotropin (FSH and LH) production due to loss of ovarian function, the pituitary gland is usually small.

Since the original observations of absent or minimal true mitotic activity in hyperplastic pituitary glands (29), the origin of hyperplastic cells has been debated. Although the origin of most such entities has been ascribed to expanded clones arising from a stem cell, several lines of evidence support the concept of "reversible transdifferentiation" whereby cells are recruited from heterologous cell types (30). Early in development, GH-secreting cells have the capacity to transdifferentiate to gonadotrophs (31). Reversible phenotypic switching of GH and PRL gene expression has long been reported in experimental rat pituitary tumor cells (32), and is reflective of a common acidophilic stem cell precursor for both PRL and GH cells. Several rodent models of pituitary hyperplasia exist, including pregnancy and administered antithyroid medication; these exhibit plastic interchange of PRL and GH, as well as TSH- and GH-secreting cell populations, respectively. In humans, during pregnancy, lactotroph cells are recruited from GH-secreting cells, and the hyperplastic cell population may be bihormonal, secreting both hormones. Similarly, hypothyroid patients exhibit thyrotroph hyperplasia with recruitment of GH-secreting cells, leading to bihormonal TSH- and GH-cell hyperplasia (33). It is unclear whether these bihormonal, hypertrophic cells arise as a consequence of transdifferentiation of already committed cells, or whether earlier more primitive stem cells undergo expansion.

Several lines of evidence corroborate the concept of a hypothalamic role for pituitary tumorigenesis. True pituitary adenomas often retain the capacity to respond to hypothalamic trophic stimuli. Furthermore, pituitary adenomas, especially prolactinomas, may resolve spontaneously, demonstrating the plasticity of adenomatous pituitary cell growth (6). Finally, the clear demonstration of pituitary tumor mass shrinkage in patients receiving somatostatin analogs for acromegaly or dopaminergic therapy for prolactinomas (34, 35) supports the notion that some of the transformed pituitary cells retain a measure of hypothalamic control, with the capacity to reverse adenoma growth.

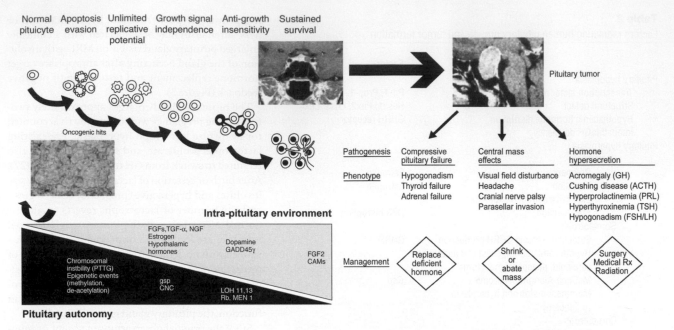

Figure 3
Pathogenesis of pituitary tumors. The spectrum of cellular changes leading from reversible hyperplasia to a committed pituitary microadenoma and ultimately to a macroadenoma. Pituitary cell types undergo proliferative and secretory changes as a consequence of a graded sequence of intrinsic and extrinsic signals. The cascade of growth-promoting signals and oncogenic "hits" may revert early, as observed with reversible pituitary hyperplasia (e.g., pregnancy or end-organ failure). Factors leading to enhanced pituicyte growth autonomy, with evasion of apoptosis and unrestrained replicative potential, are potentiated by intrapituitary changes in hormone, growth factor, or receptor functions. Figure based on a hypothesis proposed in (142). Rx, therapy.

Estrogen. Lactotroph hyperplasia is encountered in rodents and humans receiving high estrogen doses, and several models of estrogen action provide insight into the role of hyperplasia as a precursor for pituitary adenoma formation. Estrogen is mitogenic for lactotrophs and gonadotrophs, and is a ligand for the estrogen receptor (ER) encoded by two genes: ERα, expressed in 70–100% of prolactinomas, and ERβ, detectable in 60% of these tumors (36). Estradiol ligand binding leads to activation of estrogen-responsive genes, with a stronger estrogen response due to ERα than to ERβ. High doses of estrogen induce rat lactotroph hyperplasia and adenoma formation (37), and prenatal murine exposure to diesthylstilbestrol markedly enhances prolactinoma development in female offspring (38). The female preponderance of prolactinomas and their increased size during pregnancy may be ascribed to high estradiol levels, especially since prolactinomas express estrogen receptors most abundantly. In addition to cell trophic effects, estrogen induces the prolactin promoter, and activates the pituitary tumor transforming gene (*PTTG*), *FGF-β*, *FGF-β* receptors, and *TGF-β* and α expression, all of which are implicated in pituitary tumorigenesis. Alternatively spliced ERα mRNAs encode isoforms with altered responsiveness to estrogens and antiestrogens, suggesting a mechanism for tumorigenesis. ER transcripts with deletions of exons 2 and 5, which behave as stimulatory or ligand-independent isoforms, have been detected in nearly all prolactinomas. In contrast, deletions that encode dominant negative ERα isoforms lacking DNA binding or transactivation functions, respectively (exon 7), are less commonly encountered in prolactinomas (36, 39). PTTG is expressed in the female rat pituitary in an estrous cycle–dependent fashion, suggesting PTTG involvement in pituicyte proliferation after proestrous (40). Rat pituitary expression of PTTG is induced

early by estrogen and precedes estrogen-induced pituitary hyperplasia and adenoma formation (Figure 4). Antiestrogens attenuate pituitary expression of PTTG concomitantly with their blocking of the cellular effects of estrogen (41), and these agents also block pituitary tumor cell proliferation in vitro and in vivo (41). Although animal models depict estrogen-induced pituitary hyperplasia as preceding adenoma formation, only rare cases of prolactinoma formation have been reported in patients receiving high estrogen doses. A transsexual patient exposed to excess estrogen (42) and a young female with lactotroph hyperplasia (43) were reported to have developed coexistent prolactinomas.

Hypothalamic signaling. Generalized pituitary hyperplasia would be expected if the adenomatous growth were induced in response to hypothalamic hormone overstimulation. The hypothalamus secretes hormones through the portal vein; these factors impinge upon anterior pituitary cells to regulate anterior pituitary hormone synthesis and secretion (Figure 2). Disrupted stimulatory hypothalamic signals or defects in their cognate anterior pituitary receptors leads to pituitary hormone deficiencies. Mice with disrupted corticotropin-releasing hormone (CRH) (44) have impaired ACTH synthesis. In transgenic rodents, pituitary hyperplasia together with adenoma formation is observed in conjunction with overexpressed hypothalamic growth factors, including CRH or GHRH (45, 46), or those pituitaries deprived of dopamine inhibition, as observed in mice with a disrupted dopamine D2 receptor (DRD2) transgene (47). In contrast, human pituitary hyperplasia and hormone hypersecretion associated with rare hypothalamic or ectopic GHRH- or CRH-producing tumors (especially carcinoids) or Nelson syndrome are very rarely associated with adenoma development, suggesting that hyperplasia is not a prerequisite for ade-

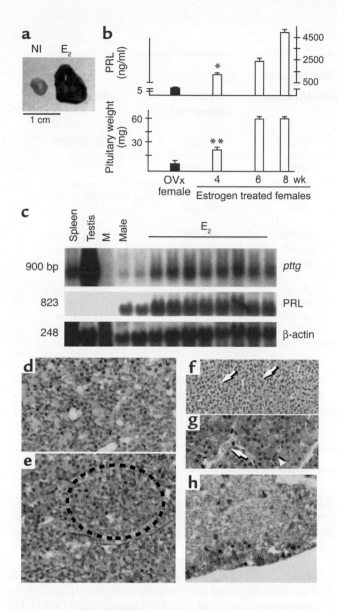

Figure 4
In vivo estrogen induction of PTTG and rat lactotroph tumors. (**a**) Representative normal rat pituitary (NI) and rat pituitary tumor (E₂). (**b** and **c**) Serum PRL and pituitary wet weight (**b**) and Northern blot analysis (**c**) of pituitary tissue extracts derived from estrogen-treated rats. β-Actin was utilized as the internal control. Ovx, ovariectomized controls. M, marker lane. *$P < 0.001$; **$P < 0.01$. (**d** and **e**) Reticulin fiber staining (broken circle) of rat anterior pituitary tissue at 24 hours (**d**) and 1 week (**e**) after commencement of estrogen infusion. (**f** and **g**) Reticulin stain (arrows) (**f**) and hematoxylin and eosin stain (**g**) of rat anterior pituitary tissue 4 weeks after estrogen infusion began. Widespread vacuolation, vascular lakes (**g**, arrow), nuclear pleomorphism and frequent mitosis (**g**, arrowhead) are visible. (**h**) pituitary bFGF immunoreactivity after 4 weeks of estrogen treatment. Original magnification, ×200. Reproduced with permission from *Nature Medicine* (37).

Somatostatin membrane receptors are encoded by five distinct somatostatin receptor (SSTR) subtypes (SSTR1–SSTR5), and SSTR2 and SSTR5 are mainly involved in mediating GH suppression. Patients exhibiting resistance to somatostatin analog therapy for acromegaly have demonstrated decreased tumor receptor expression, although inactivating mutations in the genes encoding SSTR2 and SSTR5 are uncommonly encountered (53, 54).

Hypothalamic TRH stimulates TSH thyrotroph release and lactotrophs. Although mutations in the TRH receptor gene were not detected in 50 pituitary adenomas, a mutated ligand-binding domain of thyroid hormone receptor β (TRβ) was reported in a TSH-secreting adenoma resistant to thyroid hormone (55); the gonadotrophin-releasing hormone receptor (*GnRH-R*) gene sequence was normal in 10 tumors. Despite extensive searches, activating mutations in hypothalamic hormone receptor genes (including *GHRH-R*, *GnRH-R*, *TRH-R*, *D2R*, *SSTR2*, and *SSTR5*) are rarely observed in human pituitary tumors (56–58).

GHRH signaling defects. GHRH is required both for somatotroph differentiation and proliferation and for regulation of GH expression. GHRH induces GH gene transcription and release, and also stimulates somatotroph cell DNA synthesis (59, 60). Inactivating defects in the pituitary GHRH receptor, identified in the *little* mouse strain (61) are similarly associated with pituitary hypoplasia, GH deficiency, and short stature in humans. Although activating GHRH mutations with constitutive activation of the cAMP signaling pathway have been excluded in somatotropinomas (58, 62), mutations of several key GHRH signaling pathways have been associated with pituitary hyperplasia and tumor pathogenesis.

Gsp mutations. The McCune-Albright syndrome comprises asymmetric defects in bony skeleton and skin (precocious puberty, thyrotoxicosis, acromegaly, gigantism, or Cushing syndrome). Pituitary lesions are usually hyperplastic and are less commonly adenomatous. The molecular defect in McCune-Albright syndrome is the activating *gsp* mutation in the *GNAS1* gene (guanine nucleotide–activating α subunit) identified on chromosome 20q13.2, which bypasses ligand-dependent signaling (63), resulting in constitutive hormone gene activation.

The *gsp* mutation identified in McCune-Albright syndrome also occurs in approximately 30% of sporadic GH-secreting tumors with constitutively activated G$_s$α protein–stimulating adenylate cyclase (64). Reduced GTPase activity leads to stabilization of the active form of the G$_s$α protein and increased adenylyl cyclase activity. Increased cAMP caused by *gsp* enhances GH secretion independently of hypothalamic GHRH ligand. The oncogenic *gsp* mutation correlates with constitutively increased cAMP response element–

noma development (47–51). In contrast, sporadic nonfunctioning pituitary tumors (α subunit positive) or prolactinomas are commonly well circumscribed, and the surrounding pituitary tissue is usually normal or even hypoplastic.

Hypothalamic dopamine exerts tonic inhibition of PRL synthesis and secretion as well as lactotroph proliferation, and DRD2 mediates these actions. Focal lactotroph hyperplasia is encountered when the pituitary is deprived of hypothalamic dopamine, as seen with compressive stalk disruption by a parasellar mass or pharmacologic dopamine antagonism. *Drd2*-disrupted mice are not responsive to dopamine, and develop lactotroph hyperplasia and ultimately prolactinomas (51). Prolonged lactotroph hyperplasia (for up to 18 months) precedes adenoma development in female *Drd2⁻/⁻* mice, likely by allowing an enlarged pool of lactotrophs to acquire initiating tumorigenic changes; alternatively, pituitary hyperplasia might not be a prerequisite for tumorigenesis. Clinically, dopamine agonist resistance in patients with prolactinomas reflects reduced adenoma dopamine receptor expression. However, in one study, the *DRD2* gene was not mutated in human pituitary tumors, in 46 prolactinomas, and in 19 mixed GH/PRL adenomas (52).

Table 3

Pathologic and clinical characteristics of pituitary adenomas

| Adenoma type | Pathological incidence (%) | Prevalence (total per10⁶) | Gene expression | | | Clinical syndrome |
			mRNA	Immunohistochemistry	EM secretory granules (nm)	
Lactotroph		60–100				
Sparsely granulated galactorrhea	28		PRL	PRL	150–500	Hypogonadism,
Densely granulated	1		PRL	PRL	400–1200	
Somatotroph		40–60				
Sparsely granulated	5		GH	GH	100–250	Acromegaly or gigantism
Densely granulated	5		GH	GH	300–700	
Combined GH/PRL cells						
Mixed GH/PRL	5		GH/PRL	GH/PRL	100–600	Hypogonadism
Mammosomatotroph	1		GH/PRL	GH/PRL	350–2000	Acromegaly
Acidophil stem cell	3		GH/PRL	GH/PRL	50–300	galactorrhea
Corticotroph		20–30				
Cushing	10		POMC	ACTH	250–700	Cushing disease
Silent corticotroph	3		POMC	ACTH	Variable	None
Nelson	2		POMC	ACTH	250–700	Local signs
Thyrotroph	1		TSH	TSH	50–250	Hyperthyroidism
Plurihormonal	10		GH/PRL	GH/PRL/Glycoprotein	Mixed	Mixed
Nonfunctioning/null cell/gonadotroph		70–90				
Non-oncocytic	14		FSH/LHαSU	Glycoprotein	<25% of cells 100–250	Silent or pituitary failure
Oncocytic	6		FSH/LHαSU	Glycoprotein many mitochondrial	<25% of cells 100–250	Pituitary failure
Gonadotroph	7–15		FSH/LH	FSH/LH	50–200	Silent or pituitary failure

Data are derived from studying a relatively stable 1 million catchment population surrounding Stoke-on-Trent, United Kingdom (144–148). Modified with permission from W.B. Saunders (139).

binding protein (CREB) phosphorylation and activity (65) leading to increased *Pit-1* transcription, and enhanced GH synthesis. Targeted CREB deletion in transgenic mice leads to somatotroph cell depletion and dwarfism (66). However, additional mechanisms inducing CREB phosphorylation appear to be required, as one study demonstrated that 6 of 15 GH-secreting tumors had altered $G_s\alpha$ gene sequences or expression levels, while all GH-secreting tumors studied had increased phosphorylated CREB, compared with nonfunctioning pituitary adenomas (65).

As phosphodiesterase activity is also elevated approximately sevenfold in *gsp*-positive pituitary adenomas, the effects of elevated intracellular cAMP levels appear to be counteracted by additional intracellular mechanisms (67). Intriguingly, there are few in vivo phenotypic features distinguishing GH-cell adenomas bearing the *gsp* mutation. *Gsp*-positive tumors show enhanced responsiveness to octreotide, a somatostatin analog, and can be biochemically controlled even when a tumor remnant is present after surgery.

Carney complex. The Carney complex (CNC) syndrome comprises cardiac myxomas, spotty skin pigmentation, and tumors of the adrenal gland and anterior pituitary. Pituitary lesions are hyperplastic, with multifocal microadenomas usually arising from GH-cells. CNC exhibits genetic heterogeneity, mapping to two chromosomal regions. Families with a putative 2p16 defect exhibit tumor chromosome instability. The 17q24 gene, *CNC1*, encodes protein kinase A (PKA) type I-α regulatory subunit (PRKAR1α) (68). Truncated PRKAR1α is detectable in families with CNC linked to 17q24. PRKAR1α defects likely act through haploinsufficiency, with an altered ratio of the PKA regulatory α and β

subunits resulting in lower basal PKA activity but increased PKA responses to cAMP stimulation in CNC tumors (69). As with multiple endocrine neoplasia type I (MEN I), some families with CNC exhibit only a subset of tumor types, with defects in PRKAR1α also identified in isolated familial cardiac myxomas (70).

Pituitary adenomas in both McCune-Albright syndrome and CNC tend to be multifocal and are often preceded by hyperplasia, similar to those observed in animal models expressing excess GHRH. The GHRH signaling pathway may therefore be implicated at several levels leading to pituitary tumorigenesis. GH-producing tumors arise in transgenic mice overexpressing GHRH, and GH-cell adenomas express activating G protein α-subunit mutations or inactivating mutations of PRKAR1α. Elevated cAMP levels lead to enhanced CREB phosphorylation, with subsequent activation of GH transcription and somatotroph cell transformation. Functional inactivation of CNC1-encoded PRKAR1α enhances cAMP sensitivity.

Growth factors. Several pituitary-driven growth factors have been shown to induce pituitary hyperplasia with or without ultimate adenoma development when expressed in transgenic mice.

bFGF. FGF-β is abundantly expressed in the pituitary and brain, induces angiogenesis, and is also a potent mitogen for neuroectoderm cells. FGF-β synthesis is induced in NIH3T3 cells overexpressing PTTG. During the hyperplastic phase, prior to the development of prolactinomas (37), pituitary expression of both PTTG and FGF-β is increased in a time- and dose-dependent manner in estrogen-treated rats.

Hst. The heparin-binding secretory transforming gene *hst*, encoding FGF-4, was identified from human sequences derived from

Table 4

Genetic syndromes involving pituitary tumors

Syndrome	Clinical Features	Chromosome location	Gene	Protein	Proposed function/defect
MEN I	Parathyroid, endocrine pancreas, anterior pituitary (mostly prolactinomas) tumors	11q13	*MEN1*	MENIN	Nuclear, tumor suppressor protein interacts with junD
Familial acromegaly	GH-cell adenomas, acromegaly/gigantism	11q13 and other loci	Not men1	---	---
McCune-Albright syndrome	Polyostotic fibrous dysplasia, pigmented skin patches; endocrine abnormalities: precocious puberty, GH-cell adenomas, acromegaly/gigantism Cushing syndrome	20q13.2 (mosaic)	*GNAS1* (gsp)	$G_s\alpha$	Signal transduction/inactive GTPase results in constitutive cAMP elevation independent of GHRH
CNC syndrome	Skin and cardiac myxomas, Cushing disease, acromegaly	2p16	---	---	PKA signaling defect for activating GH

Adapted with permission from (149).

prolactinoma DNA and was shown to be transforming in NIH3T3 cells (71). GH4 cells transfected with *hst* form tumors that are more aggressive when transplanted, and FGF-4 stimulates lactotroph proliferation and prolactin transcription and secretion (72). About 30% of human prolactinomas selectively express strong FGF-4 immunoreactivity (73), and FGF-4 abundance correlates with tumor invasiveness. A truncated FGF receptor isoform has also been shown to result in lactotroph tumors in transgenic mice when driven by the prolactin promoter (74).

TGFs. Pituitary TGF-α mRNA levels increase before initiation of lactotroph hyperplasia by estrogen administration. Transgenic female mice with pituitary TGF-α transgene expression driven by the PRL promoter develop lactotroph hyperplasia and, ultimately, prolactinomas by 12 months of age (75). TGF effects are likely potentiated in vivo by estrogen, and TGFs do not appear to induce other pituitary cell type tumors. TGF-β is a pituitary growth inhibitor, and the estrogen receptor has been shown to coimmunoprecipitate with Smad4 and Smad1, implicating a TGF-β interaction with estrogen in the pathogenesis of prolactinoma pathogenesis. (76).

NGF. Mammosomatotroph pituitary cells express NGF and its receptors. Mice, in which transgenic NGF is driven by the prolactin promoter, develop lactotroph hyperplasia without adenomas, despite having markedly enlarged pituitary glands (77).

Pituitary adenoma. Pituitary adenomas are common benign monoclonal neoplasms accounting for approximately 15% of intracranial tumors, while occult adenomas are discovered in as many as 25% of unselected autopsies. Pituitary tumors are usually benign, but cause significant morbidity due to their critical location, expanding size, and/or inappropriate pituitary hormone expression. Local compressive effects include headaches, visual disorders, cranial nerve dysfunction and/or altered hormone expression due to pituitary stalk disruption with compromised hypothalamic hormone access, and pituitary failure due to compression of normal pituitary tissue. Factors underlying pituitary tumorigenesis include both intrinsic pituicyte alterations and altered availability of regulatory factors including hypothalamic hormones, peripheral hormones, and paracrine growth factors.

Reflective of the specific cell type origin of the adenoma, unique clinical features are determined by the specific hormone hypersecreted (Table 3). Somatotropinomas overexpress GH, causing acro-

megaly in adults, with bony acral changes in soft tissues and bone, and increased risk of hypertension, cardiac disease, and diabetes. Prolactinomas are the most common of all functional pituitary adenomas, and patients harboring prolactinomas overexpressing PRL usually present with amenorrhea, infertility, and galactorrhea in females, and impotence or infertility in males. Tumors expressing both PRL and GH may originate from a common mammosomatotroph precursor cell. Corticotropinomas lead to ACTH hypersecretion (Cushing disease) and adrenal steroid overstimulation. Features of hypercortisolism include truncal obesity, striae, muscle wasting, hirsutism, cardiovascular complications, osteoporosis, and psychiatric disturbances. Pure gonadotropinomas secreting intact FSH or LH are rarely encountered and may cause sexual dysfunction and hypogonadism. Thyrotropinomas cause a mild increase in thyroxine levels with inappropriate TSH levels.

The etiology of these tumors is unresolved and likely involves multiple initiating and promoting factors. The literature is replete with heterogenous descriptions of overexpressed or underexpressed specific oncogenes or growth factors in these adenomas, and they may not necessarily be the direct cause of the tumor, especially as many of these factors are induced only after cells are transformed (78) and may in fact be epiphenomena rather than etiologic molecules.

Clonality of anterior pituitary tumors. Abundant evidence suggests that pituitary adenomas are derived from clonal expansion of mutated somatic cells (79, 80). Evidence supporting this "intrinsic defect" hypothesis includes the observed well-circumscribed discrete adenomas surrounded by normal nonhyperplastic tissue, as well as results of X-inactivation studies and loss-of-heterozygosity (LOH) analysis. Most pituitary tumors likely initiate with expansion of a single cell, as they exhibit nonrandom methylation patterns. Occasionally, multifocal polyclonal pituitary adenomas associated with hyperprolactinemia may arise either due to extrinsic changes in hypothalamic factors, or to pituitary stalk compression, blocking lactotroph inhibition by dopamine.

Familial pituitary tumor syndromes. Three specific genes have been identified that predispose to pituitary tumorigenesis, including MEN I, CNC, and McCune-Albright syndrome (Table 4). Four pedigrees in which MEN I was excluded have been reported with isolated prolactinomas (81). Linkage analysis and *MEN I* mutation screening show that familial hyperprolactinemia is genetically distinct from MEN I. Although isolated familial GH-secreting adenomas are linked to at least two chromosomal regions, 11q13 and a distinct locus unrelated to 11q13, mutations of *MEN I* have been excluded in these families (82, 83). Interestingly, early onset familial acromegaly has been linked to 11q13 and 2p16, chromosomal loci for both MEN I and CNC genes.

LOH and the role of tumor suppressor genes. Using genome-wide scanning and allelotyping with 122 microsatellite markers in 100 nonfunctioning and GH-secreting adenomas, multiple hot-spots have been detected, suggesting widespread LOH with enhanced susceptibility to epigenetic events (84). Deletions of chromosomal loci, including 11q13, 9p, 10q, and 13q14, occur in 12–30% of invasive pituitary tumors (85). Retention of heterozygosity has also been reported in subsequent tumor tissue whose earlier specimens displayed LOH. However, few of these tumors were examined for X-inactivation, to confirm the clonal populations, and in some, contamination by adjacent normal pituitary tissue was not rigorously excluded.

MEN I. The familial syndrome MEN I affects the parathyroid gland and endocrine pancreas, and, less frequently, the anterior pituitary, with mostly prolactinomas. MEN I behaves as an autosomal dominant trait with reduced penetrance. *MEN1* encodes a 610-amino-acid nuclear protein, MENIN, which interacts with JunD to repress transactivation. The chromosome 11q13 germline mutation is unmasked by a "second somatic hit" on the remaining allele and is visualized as LOH for polymorphic DNA markers spanning the *MEN1* gene. The "two-hit" requirement for phenotypic expression is also consistent with the presence of truncated *MEN1* mutations present in up to 85% of MEN I families. Transgenic mice bearing a *Men1* heterozygous disruption recapitulate the human MEN I syndrome (86). Heterozygotes develop tumors with LOH of the wild-type chromosome, including pancreatic (40% by 9 months), parathyroid (24% by 9 months), and pituitary (26% by 16 months) tumors.

The role of *MEN1* mutations in pituitary tumorigenesis in humans is not readily apparent. The preponderance of prolactinomas occurring in familial MEN I suggests that pituitary tumors might be caused by kindred-specific mutations. Furthermore, pituitary tumors develop with varying frequency in obligate carriers of *MEN1* mutations.

As LOH for 11q13 is observed in up to 30% of sporadic pituitary tumors (especially invasive), a role for *MEN1* in the progression but not the initiation of sporadic pituitary tumors has been suggested. However, sequence analysis shows that *MEN1* mutations occur in less than 2% of sporadic pituitary adenomas. Although reduced *MEN1* expression has been observed with decreased copy number, the overwhelming majority of sporadic pituitary tumors express both *MEN1* alleles equally, excluding genomic imprinting as a general mechanism to silence this chromosomal region (87). Taken together, evidence from studies of sporadic pituitary tumors predicts the putative presence of a distinct tumor suppressor gene for these adenomas in the 11q13 region.

Rb. The retinoblastoma protein Rb, a key cell cycle regulator, is differentially phosphorylated by cell cycle–dependent kinases and cyclins (88). Activated hypophosphorylated Rb binds E2F, blocking entry into the cell cycle and suppressing proliferation, while phosphorylated Rb releases E2F. Transgenic mice with disrupted *Rb* do not develop retinoblastoma, but surprisingly exhibit large pituitary tumors arising from pituitary ACTH-producing cells in the intermediate lobe (89). This observation led to a search for *Rb* as a candidate for involvement in human pituitary tumors. Although LOH for polymorphic markers on chromosome 13q14 is detected in aggressive pituitary tumors and their metastases, immunoreactive Rb protein is still detectable. Several lines of evidence indicate that Rb itself is not implicated in pituitary tumorigenesis. First, the *Rb* locus was demonstrated to be intact in more than 50 benign pituitary adenomas (90, 91). Wild-type hypophosphorylated (active) Rb has been detected in a large cohort of 24 pituitary tumors, and single-strand conformational polymorphism mobility shifts indicative of sequence changes were not detected for exons 20–24, which encode the pocket domain (92). These results are suggestive of a distinct tumor suppressor gene on 13q14 unmasked by LOH in aggressive sporadic pituitary tumors. Some reports indicate that Rb expression might be decreased by promoter methylation in a subset of invasive pituitary tumors.

p27. Mice disrupted for *p27* develop enhanced growth, with multiorgan hyperplasia and increased cell proliferation. Intriguingly, these mice frequently develop tumors of POMC-positive cells of the pituitary intermediate lobe (93). The *p27* gene behaves as a tumor suppressor gene both in animal models and also in sporadic human pituitary tumors. Immunodetectable p27 protein is underexpressed or absent in most human pituitary tumors, and is undetectable in rare pituitary carcinomas (94). Pituitary adenoma expression of p27 is likely regulated by post-transcriptional and post-translational mechanisms, including ubiquitin-dependent protein degradation (95), and p27 mutations have been excluded in these tumors. Doubly disrupted *p27* and *Rb* mice have enhanced pituitary tumor development, and p27 loss retards apoptosis in *Rb*$^{-/-}$ tumor cells (96). Galectin-3, which induces p27, is upregulated in neoplastic pituitary tissue derived from p27-null mice, and inhibition of galactin-3 expression decreases pituitary tumor cell proliferation (97).

p18. Disruption of the cyclin inhibitor gene *p18*^{Ink4c} results in widespread organomegaly and pituitary hyperplasia, with ACTH-cell intermediate lobe tumors (98). GH levels are normal and IGF-1 is only slightly elevated in these animals, indicating that tissue overgrowth is likely due to an intrinsic p18 defect rather than endocrine hypersecretion. The powerful suppressive effect of CDK inhibitors on pituitary cell growth and tumorigenesis is exemplified by disruption of both p18 and p27, leading to synergistic development of pituitary tumors, with greatly accelerated pituitary tumorigenesis (99).

p16. The protein p16^{Ink4a}, encoded by the *CDKN2A* gene on chromosome 9p21, maintains Rb in an unphosphorylated, active state by blocking CDK4. There was no detection of p16 by Western blot analysis of 25 pituitary tumors (95), and homozygous *p16* gene deletions have been described in pituitary adenomas. Moreover, p16 also appears to be inactivated in pituitary adenomas by methylation, and introduction of inducible *p16* into AtT20 murine ACTH-secreting cells caused reversible growth inhibition and G1 arrest (100).

p53. Although *p53* is commonly mutated in human tumors, it does not appear to undergo mutation in pituitary adenomas (101, 102).

Gadd45γ. The normal human pituitary expresses GADD45γ, a growth suppressor gene, while less than 10% of pituitary adenomas express the gene. Interestingly, transfected GADD45γ also reduces pituitary tumor cell proliferation in vitro (103).

Prolactin. PRL itself exhibits autocrine regulation of pituitary cell growth, and female mice in which the PRL gene is disrupted develop early pituitary hyperplasia, with definitive pituitary adenoma

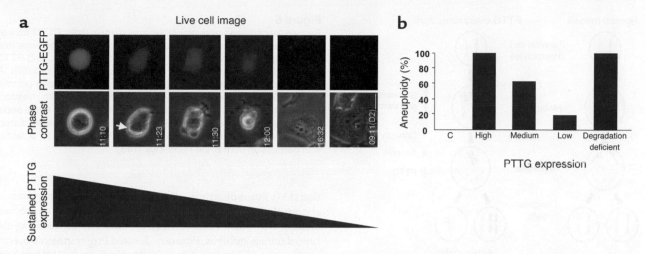

Figure 5

Chromosome nonsegregation and aneuploidy results from failure of PTTG-EGFP fusion protein degradation after transfection of a degradation-deficient PTTG mutant. (**a**) Single live H1299 cells with persistent PTTG-EGFP expression were continuously observed and show absence of chromosome segregation with completed cytokinesis. Phase contrast (bright field) and PTTG-EGFP images (green) are shown; D2, second day of observation. Arrow, cell entering mitosis. Scale bar, 10 μm. (**b**) Aneuploidy correlation with transfected PTTG expression levels or with failure to degrade PTTG in a degradation-deficient mutant. C, control. Panel **a** modified with permission from *Endocrinology* (129).

formation by 8 months (104). These tumors appear to arise from authentic lactotroph cells, albeit devoid of PRL, and may be secondary to deficient central dopaminergic tone in the absence of PRL.

Activating oncogenes and growth factors. Multiple growth factor and oncogene expression alterations have been described for pituitary tumors, and these changes may all contribute to the disordered intrapituitary growth milieu. Several reports have excluded involvement of oncogenes, including *ras, c-myc, c-myb,* and *c-fos,* in most pituitary tumors, although alterations are sometimes associated with local tumor invasiveness or aggression (78). Pituitary tumors do not commonly bear mutations of the *ras* gene family, and oncogenic mutations in codons 12, 13, and 61 were reported in highly invasive pituitary tumors, suggesting that *ras* activation is limited to pituitary tumor aggression or the rare metastatic occurrence (105). Although pituitary PPARγ expression is restricted to corticotroph cells, this receptor is abundantly expressed in most pituitary tumors (40). PPARγ ligands also block growth of experimental pituitary tumors (106). Galanin is induced by estrogen and stimulates lactotroph tumor proliferation, and mice with disrupted galanin expression exhibit lactotroph hypoplasia (107).

Chromosomal instability. About half of all pituitary tumors are grossly aneuploid (2), but the reported gain or loss of chromosomes is inconsistent (108). Chromosomal instability occurs as a result of disruption of cell cycle checkpoints, which control mitotic fidelity, resulting in chromosomal instability. These lesions may include mutation of genes responsible for (a) cell cycle regulation, (b) mitotic spindle assembly checkpoint signaling, and (c) attachment of kinetochores to microtubules (109). Although not uniformly accepted (110), aneuploidy and chromosomal instability have been implicated as prerequisites for, rather than resulting from, tumor development.

Pttg was isolated by mRNA differential display of rat pituitary tumor cells and normal pituitary tissue (111). PTTG is expressed in selected normal tissues undergoing active cell turnover, including testis and lymphopoietic tissues, and is abundantly expressed in several tumor types, including pituitary, thyroid, colon, and breast (37, 112–115). PTTG is overexpressed in pituitary tumors, correlates with pituitary tumor invasiveness, behaves as a transactivator and a potent transforming gene (116), stimulates FGF2 production (117), and induces angiogenesis (118).

Stimulation of FGF by PTTG theoretically would suffice to explain PTTG tumorigenesis. It is not clear, though, which is the initiator, but increased expression of either PTTG or FGF-β results in enhanced expression of the other. Although the positive autofeedback of *PTTG* and FGF-β suggests important roles in pituitary tumorigenesis, the role of vascularity (119) in pituitary tumorigenesis is controversial. Some investigators have observed increased pituitary adenoma vascularity, while others have reported fewer intratumoral vessels.

PTTG tumor overexpression is not associated with coding or promoter mutations, including homologous SP1, CCAAT, cell cycle–dependent element, and cell cycle homology region motifs. The human *PTTG* promoter also contains sites important for tissue-specific, cAMP-mediated, and estrogen-responsive regulation.

Point mutations of several key proline residues in the PTTG C terminus result in a dominant negative PTTG mutant defective in transforming activity, which suppresses PRL gene expression and inhibits growth of experimental rat pituitary adenomas (120).

Securin function of PTTG and aneuploidy. PTTG has been identified as a mammalian securin protein critical in mitosis (121). The process of cell cycle progression is controlled by cell cycle and checkpoint mediators ensuring genomic stability and faithful, diploid daughter cell production during mitosis. Cohesion of sister chromatids and their faithful segregation to ensure a normal daughter cell gene complement is a fundamental step whose disruption leads to chromosome instability and aneuploidy. A complex series of events ensures timely and equal separation of sister chromatids during mitosis, when sister chromatids are synchronously, equally, and irreversibly segregated to daughter cells. Cohesins bind sister chromatids, and are degraded by separins upon completion of metaphase (122). PTTG behaves as a vertebrate securin that binds to and inactivates separase during metaphase (123). Nonvertebrate

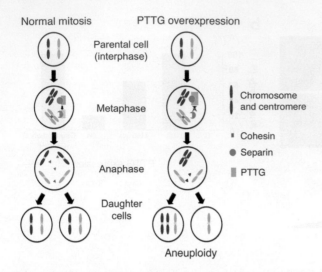

Normal mitosis   PTTG overexpression

Parental cell (interphase)

Metaphase

Chromosome and centromere

Cohesin

Separin

PTTG

Anaphase

Daughter cells

Aneuploidy

Figure 6

Securin function and aneuploidy. Normal mitosis (left): PTTG acts as a mammalian securin that maintains sister chromatid adherence during mitosis. Sister chromatids are bound with cohesions, and PTTG inactivates separin, an enzyme that regulates cohesin degradation. At the end of metaphase, securin degradation by an anaphase-promoting complex releases tonic separin inhibition, which in turn mediates cohesin degradation, thus releasing sister chromatids for equal separation into daughter cells. PTTG overexpression (right) may disrupt equal sister chromatid separation and result in aneuploidy. Adapted with permission from *Brain Pathology* (143).

securin proteins bind separin, and inhibit its function, thus preventing cohesin degradation. Securins share sequence homology for a destruction box, which targets the genes for cell cycle–dependent degradation. Separase activation occurs by phosphorylation or by PTTG degradation at the metaphase-anaphase transition (124).

During anaphase, sister chromatids are separated by separase, which cleaves chromosomal cohesin. Separase is inactivated by securin, which is degraded at the metaphase-anaphase transition by the anaphase-promoting complex (122). This complex behaves as an ubiquitin ligase, ensuring appropriate chromosome segregation. High CDK1 activity also prevents sister chromatid separation by inhibiting separase phosphorylation, even in the absence of binding of securin to separase (124). Dual inhibition of sister chromatid separation at metaphase may therefore control separase activation via two distinct mechanisms: First, cyclin B may be partially proteolysed to reduce CDK1 activity, and second, securin is destroyed. This may explain why securin-deficient human cells are indeed viable with relatively normal anaphase timing (123) and also why the *Pttg*/- mouse is viable (125). When separase cleavage sites in human cohesin are mutated (126), similar anaphase defects are observed, as when PTTG loss or overexpression attenuates separase activity (123). Several lines of evidence thus support the concept of the securin function of mammalian PTTG. Consistent with its role in regulating chromosome segregation, *PTTG* mRNA and protein levels vary with the cell cycle, disappearing at the end of the G2/M phase, and PTTG is also phosphorylated by CDC2 at the serine residue (127). *PTTG* overexpression in human JEG-3 and MG-63 cells causes a partial G2/M block, suggesting a pause in mitosis, and results in disturbed chromatid separation and aneuploidy (128, 129). Paradoxically, PTTG also causes p53-dependent and p53-independent apoptosis (130). p53 also interacts with securin, and a recent report indicates p53 inhibition by securin (131). *PTTG* is downregulated by degradation, and observations in single live cells support the concept that PTTG has to be degraded for a faithful cell cycle to occur (Figure 5). Similar delayed mitosis is reported in cells expressing nondegradable securin (132).

In adult mice, Pttg is required for tissue self-renewal, as *Pttg*-disrupted mice are viable and fertile, but display testicular, splenic (125), and pituitary hypoplasia (133). *Pttg*/- male animals more than 7 months of age develop hyperglycemia, accompanied by hypoinsulinemia, low pancreatic β cell mass, and decreased β cell replica-

tion (134). Pttg is directly implicated in chromosome segregation, as abnormal nuclei, increased aneuploidy, and premature centromere division are evident in cultured fibroblasts derived from *Pttg*-disrupted mouse embryos. Pituitary-directed Pttg transgenic overexpression also results in pituitary focal hyperplasia (133). Thus, both Pttg excess, as observed in tumors, and Pttg loss, as exemplified in the *Pttg*/- mouse, lead to cell cycle disruption and aneuploidy.

The mechanisms involved in the oncogenic function for PTTG are still obscure. Genetic instability could therefore underly the transforming and tumorigenic effects of PTTG overexpression as well as its association with tumor aggressiveness. Chromosomal instability of hyperplastic pituitary cells could provide a growth advantage for ultimate progression of tumor growth. In a recent report of genes associated with malignant cell behavior, PTTG was identified as one of nine genes comprising the "expression signature" for metastatic potential of solid tumors (135). Nevertheless, true pituitary malignancy is exceedingly rare.

PTTG thus functions as a securin that regulates chromosome separation (Figure 6), and aberrant PTTG expression leading to chromosome mis-segregation appears to be an early link in the multistep initiation and progression of pituitary tumors.

Clinical implications. Several clinical benefits are now apparent as a consequence of the unraveling of the nature of genetic changes associated with pituitary tumor formation. Ascertainment of carrier status for known pituitary-associated germline mutations (e.g., MEN1 and CNC) should precede the screening of families known to harbor such mutations. Screening protocols for the surveillance of patients identified as being at risk for tumor development should be initiated to more clearly delineate the onset of a pituitary tumor. These include serial pituitary MRI imaging and, if relevant, biochemical screening markers, such as measuring serum PRL (for prolactinoma) or IGF-I levels (for acromegaly). Ultimately, understanding the nature of the genetic lesion will contribute to therapeutic decision-making. For example, patients harboring MEN I–associated pituitary adenomas appear to have tumors that are more aggressive and are more resistant to treatment than sporadic tumors (136). Molecular profiling of pituitary tumor specimens will allow rational approaches to the timing of and requirement for postoperative therapies (medical or irradiation), especially if clinical outcomes will be correlated with these profiles. The identification of structurally and functionally intact somatostatin receptor subtypes and D2 receptors has resulted in development of selective peptide analogs as effective new therapies for pituitary tumors (137). Finally, unraveling the mechanisms underlying disordered pituitary growth will allow development of subcellular therapies, as has been demonstrated experimentally for direct Rb gene therapy (138) and for targeted PTTG inactivation for experimental pituitary tumors (120).

Conclusions. Pituitary cells are highly differentiated and are committed very early to synthesize unique hormone products. Key regulators of pituitary lineage development may sustain mutations resulting in pituitary hormone deficiency and pituitary hypoplasia. Defects in genes that function early in development have pleiotropic effects and result in multiple hormone deficiencies associated with aberrant structural development, whereas defects in genes that act as differentiating factors after lineage specification are associated with selective hormone-deficiency phenotypes. Early changes leading to pituitary tumorigenesis involve both intrinsic pituicyte alterations and altered availability of paracrine or endocrine regulatory factors regulating both hormone secretion and cell growth (Figure 6). Familial syndromes are associated with at least three genes that predispose to pituitary hyperplasia and tumorigenesis. Factors resulting in pituitary hyperplasia, including hypothalamic hormones, estrogens, and growth factors, likely facilitate a permissive intrapituitary milieu, potentiating genetic instability, cell mutation, and subsequent monoclonal growth expansion.

Acknowledgments

Work in the author's laboratory is supported by NIH grants CA75979 and DK64169.

Address correspondence to: Shlomo Melmed, Cedars-Sinai Medical Center, 8700 Beverly Boulevard, Room 2015, Los Angeles, California 90048, USA. Phone: (310) 423-4691; Fax: (310) 423-0119; E-mail: melmed@csmc.edu.

1. Taniguchi, Y., Yasutaka, S., Kominami, R., and Shinohara, H. 2002. Proliferation and differentiation of rat anterior pituitary cells. *Anat. Embryol. (Berl.).* **206**:1–11.

2. Levy, A., and Lightman, S. 2003. Molecular defects in the pathogenesis of pituitary tumours. *Front. Neuroendocrinol.* **24**:94–127.

3. Asa, S.L., and Ezzat, S. 1998. The cytogenesis and pathogenesis of pituitary adenomas. *Endocr. Rev.* **19**:798–827.

4. Faglia, G., and Spada, A. 2001. Genesis of pituitary adenomas: state of the art. *J. Neurooncol.* **54**:95–110.

5. Prezant, T.R., and Melmed, S. 2002. Molecular pathogenesis of pituitary disorders. *Current Opinion in Endocrinology & Diabetes.* **9**:61–78.

6. Jeffcoate, W.J., Pound, N., Sturrock, N.D., and Lambourne, J. 1996. Long-term follow-up of patients with hyperprolactinaemia. *Clin. Endocrinol. (Oxf.).* **45**:299–303.

7. Olson, L.E., and Rosenfeld, M.G. 2002. Perspective: genetic and genomic approaches in elucidating mechanisms of pituitary development. *Endocrinology.* **143**:2007–2011.

8. Semina, E.V., et al. 1996. Cloning and characterization of a novel bicoid-related homeobox transcription factor gene, RIEG, involved in Rieger syndrome. *Nat. Genet.* **14**:392–399.

9. Netchine, I., et al. 2000. Mutations in LHX3 result in a new syndrome revealed by combined pituitary hormone deficiency. *Nat. Genet.* **25**:182–186.

10. Machinis, K., et al. 2001. Syndromic short stature in patients with a germline mutation in the lim homeobox lhx4. *Am. J. Hum. Genet.* **69**:961–968.

11. Dattani, M.T., et al. 1998. Mutations in the homeobox gene HESX1/Hesx1 associated with septo-optic dysplasia in human and mouse. *Nat. Genet.* **19**:125–133.

12. Liu, J., et al. 2001. Tbx19, a tissue-selective regulator of POMC gene expression. *Proc. Natl. Acad. Sci. U. S. A.* **98**:8674–8679.

13. Lamolet, B., et al. 2001. A pituitary cell-restricted T box factor, Tpit, activates POMC transcription in cooperation with Pitx homeoproteins. *Cell.* **104**:849–859.

14. Pulichino, A.M., et al. 2003. Tpit determines alternate fates during pituitary cell differentiation. *Genes Dev.* **17**:738–747.

15. Cohen, L.E., and Radovick, S. 2002. Molecular basis of combined pituitary hormone deficiencies. *Endocr. Rev.* **23**:431–442.

16. Sornson, M.W., et al. 1996. Pituitary lineage determination by the Prophet of Pit-1 homeodomain factor defective in Ames dwarfism. *Nature.* **384**:327–333.

17. Li, S., et al. 1990. Dwarf locus mutants lacking three pituitary cell types result from mutations in the POU-domain gene pit-1. *Nature.* **347**:528–533.

18. Cushman, L.J., et al. 2001. Persistent Prop1 expression delays gonadotrope differentiation and enhances pituitary tumor susceptibility. *Hum. Mol. Genet.* **10**:1141–1153.

19. Riepe, F.G., et al. 2001. Longitudinal imaging reveals pituitary enlargement preceding hypoplasia in two brothers with combined pituitary hormone deficiency attributable to PROP1 mutation. *J. Clin. Endocrinol. Metab.* **86**:4353–4357.

20. Osorio, M.G., et al. 2002. Pituitary magnetic resonance imaging and function in patients with growth hormone deficiency with and without mutations in GHRH-R, GH-1, or PROP-1 genes. *J. Clin. Endocrinol. Metab.* **87**:5076–5084.

21. Teinturier, C., et al. 2002. Pseudotumor of the pituitary due to PROP-1 deletion. *J. Pediatr. Endocrinol. Metab.* **15**:95–101.

22. Pellegrini, I., et al. 1994. Pit-1 gene expression in the human pituitary and pituitary adenomas. *J. Clin. Endocrinol. Metab.* **79**:189–196.

23. Nakamura, S., et al. 1999. Prop-1 gene expression in human pituitary tumors. *J. Clin. Endocrinol. Metab.* **84**:2581–2584.

24. Moons, D.S., et al. 2002. Pituitary hypoplasia and lactotroph dysfunction in mice deficient for cyclin-dependent kinase-4. *Endocrinology.* **143**:3001–3008.

25. Horvath, E., Kovacs, K., and Scheithauer, B.W. 1999. Pituitary hyperplasia. *Pituitary.* **1**:169–179.

26. Chanson, P., et al. 2001. Normal pituitary hypertrophy as a frequent cause of pituitary incidentaloma: a follow-up study. *J. Clin. Endocrinol. Metab.* **86**:3009–3015.

27. Scheithauer, B.W., et al. 1990. The pituitary gland in pregnancy: a clinicopathologic and immunohistochemical study of 69 cases. *Mayo Clin. Proc.* **65**:461–474.

28. Coogan, P.F., Baron, J.A., and Lambe, M. 1995. Parity and pituitary adenoma risk. *J. Natl. Cancer Inst.* **87**:1410–1411.

29. Severinghaus, A.E. 1937. Cellular changes in the anterior hypophysis with special references to its secretory activities. *Physiol. Rev.* **17**:556–566.

30. Vidal, S., Horvath, E., Kovacs, K., Lloyd, R.V., and Smyth, H.S. 2001. Reversible transdifferentiation: interconversion of somatotrophs and lactotrophs in pituitary hyperplasia. *Mod. Pathol.* **14**:20–28.

31. Childs, G.V. 2002. Development of gonadotropes may involve cyclic transdifferentiation of growth hormone cells. *Arch. Physiol. Biochem.* **110**:42–49.

32. Ivarie, R., and Morris, J. 1983. Phenotypic switching in GH3 rat pituitary tumor cells: linked expression of growth hormone and another hormonally responsive protein. *DNA.* **2**:113–120.

33. Vidal, S., et al. 2000. Transdifferentiation of somatotrophs to thyrotrophs in the pituitary of patients with protracted primary hypothyroidism. *Virchows Arch.* **436**:43–51.

34. Bevan, J.S., et al. 2002. Primary medical therapy for acromegaly: an open, prospective, multicenter study of the effects of subcutaneous and intramuscular slow-release octreotide on growth hormone, insulin-like growth factor-I, and tumor size. *J. Clin. Endocrinol. Metab.* **87**:4554–4563.

35. Corsello, S.M., et al. 2003. Giant prolactinomas in men: efficacy of cabergoline treatment. *Clin. Endocrinol. (Oxf.).* **58**:662–670.

36. Shupnik, M.A., et al. 1998. Selective expression of estrogen receptor alpha and beta isoforms in human pituitary tumors. *J. Clin. Endocrinol. Metab.* **83**:3965–3972.

37. Heaney, A.P., Horwitz, G.A., Wang, Z., Singson, R., and Melmed, S. 1999. Early involvement of estrogen-induced pituitary tumor transforming gene and fibroblast growth factor expression in prolactinoma pathogenesis. *Nat. Med.* **5**:1317–1321.

38. Walker, B.E., and Kurth, L.A. 1993. Pituitary tumors in mice exposed prenatally to diethylstilbestrol. *Cancer Res.* **53**:1546–1549.

39. Chaidarun, S.S., Swearingen, B., and Alexander, J.M. 1998. Differential expression of estrogen receptor-β (ERβ) in human pituitary tumors: functional interactions with ERα and a tumor-specific splice variant. *J. Clin. Endocrinol. Metab.* **83**:3308–3315.

40. Heaney, A.P., Fernando, M., and Melmed, S. 2002. Functional role of estrogen in pituitary tumor pathogenesis. *J. Clin. Invest.* **109**:277–283. doi:10.1172/JCI200214264.

41. Heaney, A.P., Fernando, M., Yong, W.H., and Melmed, S. 2002. Functional PPAR-γ receptor is a novel therapeutic target for ACTH-secreting pituitary adenomas. *Nat. Med.* **8**:1281–1287.

42. Kovacs, K., Stefaneanu, L., Ezzat, S., and Smyth, H.S. 1994. Prolactin-producing pituitary adenoma in a male-to-female transsexual patient with protracted estrogen administration. A morphologic study. *Arch. Pathol. Lab. Med.* **118**:562–565.

43. Vidal, S., et al. 2002. Prolactin-producing pituitary adenoma associated with prolactin cell hyperplasia. *Endocr. Pathol.* **13**:157–165.

44. Muglia, L.J., Bethin, K.E., Jacobson, L., Vogt, S.K., and Majzoub, J.A. 2000. Pituitary-adrenal axis regulation in CRH-deficient mice. *Endocr. Res.* **26**:1057–1066.

45. Mayo, K.E., et al. 1988. Dramatic pituitary hyperplasia in transgenic mice expressing a human growth hormone-releasing factor gene. *Mol. Endocrinol.* **2**:606–612.

46. Asa, S.L., et al. 1992. Pituitary adenomas in mice transgenic for growth hormone-releasing hormone. *Endocrinology.* **131**:2083–2089.

47. Asa, S.L., Kelly, M.A., Grandy, D.K., and Low, M.J. 1999. Pituitary lactotroph adenomas develop after prolonged lactotroph hyperplasia in dopamine D2 receptor-deficient mice. *Endocrinology.* **140**:5348–5355.

48. Carey, R.M., et al. 1984. Ectopic secretion of corticotropin-releasing factor as a cause of Cushing's syndrome. A clinical, morphologic, and biochemical study. *N. Engl. J. Med.* **311**:13–20.

49. Sano, T., Asa, S.L., and Kovacs, K. 1988. Growth

hormone-releasing hormone-producing tumors: clinical, biochemical, and morphological manifestations. *Endocr. Rev.* **9**:357–373.

50. Thorner, M.O., et al. 1982. Somatotroph hyperplasia. Successful treatment of acromegaly by removal of a pancreatic islet tumor secreting a growth hormone-releasing factor. *J. Clin. Invest.* **70**:965–977.

51. Schuff, K.G., et al. 2002. Lack of prolactin receptor signaling in mice results in lactotroph proliferation and prolactinomas by dopamine-dependent and -independent mechanisms. *J. Clin. Invest.* **110**:973–981. doi:10.1172/JCI200215912.

52. Friedman, E., et al. 1994. Normal structural dopamine type 2 receptor gene in prolactin-secreting and other pituitary tumors. *J. Clin. Endocrinol. Metab.* **78**:568–574.

53. Petersenn, S., Heyens, M., Ludecke, D.K., Beil, F.U., and Schulte, H.M. 2000. Absence of somatostatin receptor type 2 A mutations and gip oncogene in pituitary somatotroph adenomas. *Clin. Endocrinol. (Oxf.)* **52**:35–42.

54. Ballare, E., et al. 2001. Mutation of somatostatin receptor type 5 in an acromegalic patient resistant to somatostatin analog treatment. *J. Clin. Endocrinol. Metab.* **86**:3809–3814.

55. Ando, S., Sarlis, N.J., Oldfield, E.H., and Yen, P.M. 2001. Somatic mutation of TRβ can cause a defect in negative regulation of TSH in a TSH-secreting pituitary tumor. *J. Clin. Endocrinol. Metab.* **86**:5572–5576.

56. Chanson, P., et al. 1998. Absence of activating mutations in the GnRH receptor gene in human pituitary gonadotroph adenomas. *Eur. J. Endocrinol.* **139**:157–160.

57. Faccenda, E., Melmed, S., Bevan, J.S., and Eidne, K.A. 1996. Structure of the thyrotrophin-releasing hormone receptor in human pituitary adenomas. *Clin. Endocrinol. (Oxf.).* **44**:341–347.

58. Lee, E.J., et al. 2001. Absence of constitutively activating mutations in the GHRH receptor in GH-producing pituitary tumors. *J. Clin. Endocrinol. Metab.* **86**:3989–3995.

59. Barinaga, M., Bilezikjian, L.M., Vale, W.W., Rosenfeld, M.G., and Evans, R.M. 1985. Independent effects of growth hormone releasing factor on growth hormone release and gene transcription. *Nature.* **314**:279–281.

60. Billestrup, N., Swanson, L.W., and Vale, W. 1986. Growth hormone-releasing factor stimulates proliferation of somatotrophs in vitro. *Proc. Natl. Acad. Sci. U. S. A.* **83**:6854–6857.

61. Lin, S.C., et al. 1993. Molecular basis of the little mouse phenotype and implications for cell type-specific growth. *Nature.* **364**:208–213.

62. Salvatori, R., et al. 2001. Absence of mutations in the growth hormone (GH)–releasing hormone receptor gene in GH-secreting pituitary adenomas. *Clin. Endocrinol. (Oxf.)* **54**:301–307.

63. Weinstein, L.S., et al. 1991. Activating mutations of the stimulatory G protein in the McCune-Albright syndrome. *N. Engl. J. Med.* **325**:1688–1695.

64. Vallar, L., Spada, A., and Giannattasio, G. 1987. Altered Gs and adenylate cyclase activity in human GH-secreting pituitary adenomas. *Nature.* **330**:566–568.

65. Bertherat, J., Chanson, P., and Montminy, M. 1995. The cyclic adenosine 3′,5′-monophosphate-responsive factor CREB is constitutively activated in human somatotroph adenomas. *Mol. Endocrinol.* **9**:777–783.

66. Struthers, R.S., Vale, W.W., Arias, C., Sawchenko, P.E., and Montminy, M.R. 1991. Somatotroph hypoplasia and dwarfism in transgenic mice expressing a non-phosphorylatable CREB mutant. *Nature.* **350**:622–624.

67. Lania, A., et al. 1998. Constitutively active Gsα is associated with an increased phosphodiesterase activity in human growth hormone-secreting adenomas. *J. Clin. Endocrinol. Metab.* **83**:1624–1628.

68. Kirschner, L.S., et al. 2000. Mutations of the gene encoding the protein kinase A type I-α regulatory subunit in patients with the Carney complex. *Nat. Genet.* **26**:89–92.

69. Stratakis, C.A., et al. 1996. Cytogenetic and microsatellite alterations in tumors from patients with the syndrome of myxomas, spotty skin pigmentation, and endocrine overactivity (Carney complex). *J. Clin. Endocrinol. Metab.* **81**:3607–3614.

70. Casey, M., et al. 2000. Mutations in the protein kinase A R1α regulatory subunit cause familial cardiac myxomas and Carney complex. *J. Clin. Invest.* **106**:R31–38.

71. Gonsky, R., Herman, V., Melmed, S., and Fagin, J. 1991. Transforming DNA sequences present in human prolactin-secreting pituitary tumors. *Mol. Endocrinol.* **5**:1687–1695.

72. Shimon, I., Huttner, A., Said, J., Spirina, O.M., and Melmed, S. 1996. Heparin-binding secretory transforming gene (hst) facilitates rat lactotrope cell tumorigenesis and induces prolactin gene transcription. *J. Clin. Invest.* **97**:187–195.

73. Shimon, I., Hinton, D.R., Weiss, M.H., and Melmed, S. 1998. Prolactinomas express human heparin-binding secretory transforming gene (hst) protein product: marker of tumour invasiveness. *Clin. Endocrinol. (Oxf.)* **48**:23–29.

74. Ezzat, S., Zheng, L., Zhu, X.-F., Wu, G.E., and Asa, S.L. 2002. Targeted expression of a human pituitary tumor-derived isoform of FGF receptor-4 recapitulates pituitary tumorigenesis. *J. Clin. Invest.* **109**:69–78. doi:10.1172/JCI200214306.

75. McAndrew, J., Paterson, A.J., Asa, S.L., McCarthy, K.J., and Kudlow, J.E. 1995. Targeting of transforming growth factor-α expression to pituitary lactotrophs in transgenic mice results in selective lactotroph proliferation and adenomas. *Endocrinology.* **136**:4479–4488.

76. Paez-Pereda, M., et al. 2003. Involvement of bone morphogenetic protein 4 (BMP-4) in pituitary prolactinoma pathogenesis through a Smad/estrogen receptor crosstalk. *Proc. Natl. Acad. Sci. U. S. A.* **100**:1034–1039.

77. Borrelli, E., Sawchenko, P.E., and Evans, R.M. 1992. Pituitary hyperplasia induced by ectopic expression of nerve growth factor. *Proc. Natl. Acad. Sci. U. S. A.* **89**:2764–2768.

78. Woloschak, M., Roberts, J.L., and Post, K. 1994. c-myc, c-fos, and c-myb gene expression in human pituitary adenomas. *J. Clin. Endocrinol. Metab.* **79**:253–257.

79. Herman, V., Fagin, J., Gonsky, R., Kovacs, K., and Melmed, S. 1990. Clonal origin of pituitary adenomas. *J. Clin. Endocrinol. Metab.* **71**:1427–1433.

80. Alexander, J.M., et al. 1990. Clinically nonfunctioning pituitary tumors are monoclonal in origin. *J. Clin. Invest.* **86**:336–340.

81. Berezin, M., and Karasik, A. 1995. Familial prolactinoma. *Clin. Endocrinol. (Oxf.)* **42**:483–486.

82. Tanaka, C., et al. 1998. Absence of germ-line mutations of the multiple endocrine neoplasia type 1 (MEN1) gene in familial pituitary adenoma in contrast to MEN1 in Japanese. *J. Clin. Endocrinol. Metab.* **83**:960–965.

83. Gadelha, M.R., et al. 2000. Isolated familial somatotropinomas: establishment of linkage to chromosome 11q13.1-11q13.3 and evidence for a potential second locus at chromosome 2p16-12. *J. Clin. Endocrinol. Metab.* **85**:707–714.

84. Simpson, D.J., et al. 2003. Genome-wide amplification and allelotyping of sporadic pituitary adenomas identify novel regions of genetic loss. *Genes Chromosomes Cancer.* **37**:225–236.

85. Bates, A.S., et al. 1997. Allelic deletion in pituitary adenomas reflects aggressive biological activity and has potential value as a prognostic marker. *J. Clin. Endocrinol. Metab.* **82**:818–824.

86. Crabtree, J.S., et al. 2001. A mouse model of multiple endocrine neoplasia, type 1, develops multiple endocrine tumors. *Proc. Natl. Acad. Sci. U. S. A.* **98**:1118–1123.

87. Zhuang, Z., et al. 1997. Mutations of the MEN1 tumor suppressor gene in pituitary tumors. *Cancer Res.* **57**:5446–5451.

88. Chau, B.N., and Wang, J.Y. 2003. Coordinated regulation of life and death by RB. *Nat. Rev. Cancer.* **3**:130–138.

89. Jacks, T., et al. 1992. Effects of an Rb mutation in the mouse. *Nature.* **359**:295–300.

90. Zhu, J., et al. 1994. Human pituitary adenomas show no loss of heterozygosity at the retinoblastoma gene locus. *J. Clin. Endocrinol. Metab.* **79**:922–927.

91. Woloschak, M., Roberts, J.L., and Post, K.D. 1994. Loss of heterozygosity at the retinoblastoma locus in human pituitary tumors. *Cancer.* **74**:693–696.

92. Woloschak, M., Yu, A., Xiao, J., and Post, K.D. 1996. Abundance and state of phosphorylation of the retinoblastoma gene product in human pituitary tumors. *Int. J. Cancer.* **67**:16–19.

93. Kiyokawa, H., et al. 1996. Enhanced growth of mice lacking the cyclin-dependent kinase inhibitor function of p27Kip1. *Cell.* **85**:721–732.

94. Lidhar, K., et al. 1999. Low expression of the cell cycle inhibitor p27Kip1 in normal corticotroph cells, corticotroph tumors, and malignant pituitary tumors. *J. Clin. Endocrinol. Metab.* **84**:3823–3830.

95. Shirane, M., et al. 1999. Down-regulation of p27Kip1 by two mechanisms, ubiquitin-mediated degradation and proteolytic processing. *J. Biol. Chem.* **274**:13886–13893.

96. Carneiro, C., et al. 2003. p27 deficiency desensitizes Rb−/− cells to signals that trigger apoptosis during pituitary tumor development. *Oncogene.* **22**:361–369.

97. Riss, D., et al. 2003. Differential expression of galectin-3 in pituitary tumors. *Cancer Res.* **63**:2251–2255.

98. Fero, M.L., et al. 1996. A syndrome of multiorgan hyperplasia with features of gigantism, tumorigenesis, and female sterility in p27Kip1-deficient mice. *Cell.* **85**:733–744.

99. Franklin, D.S., et al. 1998. CDK inhibitors p18INK4c and p27Kip1 mediate two separate pathways to collaboratively suppress pituitary tumorigenesis. *Genes Dev.* **12**:2899–2911.

100. Frost, S.J., Simpson, D.J., Clayton, R.N., and Farrell, W.E. 1999. Transfection of an inducible p16/CDKN2A construct mediates reversible growth inhibition and G1 arrest in the AtT20 pituitary tumor cell line. *Mol. Endocrinol.* **13**:1801–1810.

101. Thapar, K., Scheithauer, B.W., Kovacs, K., Pernicone, P.J., and Laws, E.R., Jr. 1996. p53 expression in pituitary adenomas and carcinomas: correlation with invasiveness and tumor growth fractions. *Neurosurgery.* **38**:763–770; discussion 770–771.

102. Levy, A., Hall, L., Yeudall, W.A., and Lightman, S.L. 1994. p53 gene mutations in pituitary adenomas: rare events. *Clin. Endocrinol. (Oxf.)* **41**:809–814.

103. Zhang, X., et al. 2002. Loss of expression of GADD45γ, a growth inhibitory gene, in human pituitary adenomas: implications for tumorigenesis. *J. Clin. Endocrinol. Metab.* **87**:1262–1267.

104. Cruz-Soto, M.E., Scheiber, M.D., Gregerson, K.A., Boivin, G.P., and Horseman, N.D. 2002. Pituitary tumorigenesis in prolactin gene-disrupted mice. *Endocrinology.* **143**:4429–4436.

105. Pei, L., Melmed, S., Scheithauer, B., Kovacs, K., and Prager, D. 1994. H-ras mutations in human pituitary carcinoma metastases. *J. Clin. Endocrinol. Metab.* **78**:842–846.

106. Heaney, A.P., Fernando, M., and Melmed, S. 2003. PPAR-γ receptor ligands: novel therapy for pituitary adenomas. *J. Clin. Invest.* **111**:1381–1388. doi:10.1172/JCI200316575.

107. Wynick, D., et al. 1998. Galanin regulates prolactin release and lactotroph proliferation. *Proc. Natl. Acad. Sci. U. S. A.* **95**:12671–12676.

108. Hui, A.B., Pang, J.C., Ko, C.W., and Ng, H.K. 1999. Detection of chromosomal imbalances in growth hormone-secreting pituitary tumors by comparative genomic hybridization. *Hum. Pathol.* **30**:1019–1023.

109. Lengauer, C., Kinzler, K.W., and Vogelstein, B. 1998. Genetic instabilities in human cancers. *Nature.* **396**:643–649.

110. Zimonjic, D., Brooks, M.W., Popescu, N., Weinberg, R.A., and Hahn, W.C. 2001. Derivation of human tumor cells in vitro without widespread genomic instability. *Cancer Res.* **61**:8838–8844.

111. Pei, L., and Melmed, S. 1997. Isolation and characterization of a pituitary tumor-transforming gene (PTTG). *Mol. Endocrinol.* **11**:433–441.

112. Heaney, A.P., et al. 2000. Expression of pituitary-tumour transforming gene in colorectal tumours. *Lancet.* **355**:716–719.

113. Saez, C., et al. 1999. hpttg is over-expressed in pituitary adenomas and other primary epithelial neoplasias. *Oncogene.* **18**:5473–5476.

114. Zhang, X., et al. 1999. Pituitary tumor transforming gene (PTTG) expression in pituitary adenomas. *J. Clin. Endocrinol. Metab.* **84**:761–767.

115. Zhang, X., et al. 1999. Structure, expression, and function of human pituitary tumor-transforming gene (PTTG). *Mol. Endocrinol.* **13**:156–166.

116. Pei, L. 2000. Activation of mitogen-activated protein kinase cascade regulates pituitary tumor-transforming gene transactivation function. *J. Biol. Chem.* **275**:31191–31198.

117. McCabe, C.J., et al. 2003. Expression of pituitary tumour transforming gene (PTTG) and fibroblast growth factor-2 (FGF-2) in human pituitary adenomas: relationships to clinical tumour behaviour. *Clin. Endocrinol. (Oxf.)* **58**:141–150.

118. McCabe, C.J., et al. 2002. Vascular endothelial growth factor, its receptor KDR/Flk-1, and pituitary tumor transforming gene in pituitary tumors. *J. Clin. Endocrinol. Metab.* **87**:4238–4244.

119. Turner, H.E., et al. 2000. Angiogenesis in pituitary adenomas - relationship to endocrine function, treatment and outcome. *J. Endocrinol.* **165**:475–481.

120. Horwitz, G.A., Miklovsky, I., Heaney, A.P., Ren, S.G., and Melmed, S. 2003. Human pituitary tumor-transforming gene (PTTG1) motif suppresses prolactin expression. *Mol. Endocrinol.* **17**:600–609.

121. Zou, H., McGarry, T.J., Bernal, T., and Kirschner, M.W. 1999. Identification of a vertebrate sister-chromatid separation inhibitor involved in transformation and tumorigenesis. *Science.* **285**:418–422.

122. Nasmyth, K. 2002. Segregating sister genomes: the molecular biology of chromosome separation. *Science.* **297**:559–565.

123. Jallepalli, P.V., et al. 2001. Securin is required for chromosomal stability in human cells. *Cell.* **105**:445–457.

124. Stemmann, O., Zou, H., Gerber, S.A., Gygi, S.P., and Kirschner, M.W. 2001. Dual inhibition of sister chromatid separation at metaphase. *Cell.* **107**:715–726.

125. Wang, Z., Yu, R., and Melmed, S. 2001. Mice lacking pituitary tumor transforming gene show testicular and splenic hypoplasia, thymic hyperplasia, thrombocytopenia, aberrant cell cycle progression, and premature centromere division. *Mol. Endocrinol.* **15**:1870–1879.

126. Hauf, S., Waizenegger, I.C., and Peters, J.M. 2001. Cohesin cleavage by separase required for anaphase and cytokinesis in human cells. *Science.* **293**:1320–1323.

127. Ramos-Morales, F., et al. 2000. Cell cycle regulated expression and phosphorylation of hpttg proto-oncogene product. *Oncogene.* **19**:403–409.

128. Yu, R., Ren, S.G., Horwitz, G.A., Wang, Z., and Melmed, S. 2000. Pituitary tumor transforming gene (PTTG) regulates placental JEG-3 cell division and survival: evidence from live cell imaging. *Mol. Endocrinol.* **14**:1137–1146.

129. Yu, R., Lu, W., and Melmed, S. 2003. Overexpressed pituitary tumor transforming gene (PTTG) induces aneuploidy in live human cells. *Endocrinology.* **144**:4991–4998.

130. Yu, R., Heaney, A.P., Lu, W., Chen, J., and Melmed, S. 2000. Pituitary tumor transforming gene causes aneuploidy and p53-dependent and p53-independent apoptosis. *J. Biol. Chem.* **275**:36502–36505.

131. Bernal, J.A., et al. 2002. Human securin interacts with p53 and modulates p53-mediated transcriptional activity and apoptosis. *Nat. Genet.* **32**:306–311.

132. Zur, A., and Brandeis, M. 2001. Securin degradation is mediated by fzy and fzr, and is required for complete chromatid separation but not for cytokinesis. *Embo J.* **20**:792–801.

133. Abbud, R., et al. 2003. Molecular pathogenesis. In *8th International Pituitary Congress.* Pituitary Society. New York, New York, USA. (Abstr. S1)

134. Wang, Z., Moro, E., Kovacs, K., Yu, R., and Melmed, S. 2003. Pituitary tumor transforming gene-null male mice exhibit impaired pancreatic beta cell proliferation and diabetes. *Proc. Natl. Acad. Sci. U. S. A.* **100**:3428–3432.

135. Ramaswamy, S., Ross, K.N., Lander, E.S., and Golub, T.R. 2003. A molecular signature of metastasis in primary solid tumors. *Nat. Genet.* **33**:49–54.

136. Verges, B., et al. 2002. Pituitary disease in MEN type 1 (MEN1): data from the France-Belgium MEN1 multicenter study. *J. Clin. Endocrinol. Metab.* **87**:457–465.

137. Shimon, I., et al. 1997. Somatostatin receptor (SSTR) subtype-selective analogues differentially suppress in vitro growth hormone and prolactin in human pituitary adenomas. Novel potential therapy for functional pituitary tumors. *J. Clin. Invest.* **100**:2386–2392.

138. Riley, D.J., Nikitin, A.Y., and Lee, W.H. 1996. Adenovirus-mediated retinoblastoma gene therapy suppresses spontaneous pituitary melanotroph tumors in Rb+/- mice. *Nat. Med.* **2**:1316–1321.

139. Melmed, S., and Kleinberg, D.L. 2003. Anterior pituitary. In *Williams textbook of endocrinology.* P. Larsen, H.M. Kronenberg, S. Melmed, and K.S. Polonsky, editors. W.B. Saunders. Philadelphia, Pennsylvania, USA. 177–279.

140. Shimon, I., and Melmed, S. 1997. Anterior pituitary hormones. In *Endocrinology: basic and clinical principles.* P.M. Conn, and S. Melmed, editors. Humana Press. Totowa, New Jersey, USA. 211–222.

141. Amselem, S. 2001. Perspectives on the molecular basis of developmental defects in the human pituitary region. In *Hypothalamic-pituitary development: genetic and clinical aspects.* R. Rappaport, and S. Amselem, editors. S. Karger Publishing. Basel, Switzerland. 30–47.

142. Schmitt, C.A. 2003. Senescence, apoptosis and therapy—cutting the lifelines of cancer. *Nat. Rev. Cancer.* **3**:286–295.

143. Yu, R., and Melmed, S. 2001. Oncogene activation in pituitary tumors. *Brain Pathology.* **11**:328–341.

144. Clayton, R.N. 1999. Sporadic pituitary tumors: from epidemiology to use of databases. *Baillieres Best Pract. Res. Clin. Endocrinol. Metab.* **13**:451–460.

145. Kovacs, K., and Horvath, E. 1986. Pathology of growth hormone–producing tumors of the human pituitary. *Semin. Diagn. Pathol.* **3**:18–33.

146. Scheithauer, B.W., Horvath, E., Lloyd, R.V., and Kovacs, K. 1994. Pathology of pituitary adenomas and pituitary hyperplasia. In *Diagnosis and management of pituitary tumors.* K. Thapar, K. Kovacs, B. Scheithauer, and R. Lloyd, editors. Humana Press Inc. Totowa, New Jersey, USA. 91–154.

147. Minderman, T., and Wilson, C.B. 1994. Age-related and gender-related occurrence of pituitary adenomas. *Clin. Endocrinol. (Oxf).* **41**:359–364.

148. Asa, S.L., Horvath, E., and Kovacs, K. 1993. Pituitary neoplasms: an overview of the clinical presentation, diagnosis, treatment and pathology. In *Endocrine tumours.* E.L. Mazzaferri and N.A. Samaan, editors. Blackwell Scientific Publications. Boston, Massachusetts, USA. 77–112.

149. Prezant, T., and Melmed, S. 1998. Pituitary oncogenes. In *Pituitary tumours: epidemiology, pathogenesis and management.* S.M. Webb, editor. Bioscientifica. Bristol, United Kingdom. 81–93.

Melanoma genetics and the development of rational therapeutics

Yakov Chudnovsky, Paul A. Khavari, and Amy E. Adams

Veterans Affairs Palo Alto Healthcare System, Palo Alto, California. Program in Epithelial Biology, Stanford University School of Medicine, Stanford, California, USA.

Melanoma is a cancer of the neural crest–derived cells that provide pigmentation to skin and other tissues. Over the past 4 decades, the incidence of melanoma has increased more rapidly than that of any other malignancy in the United States. No current treatments substantially enhance patient survival once metastasis has occurred. This review focuses on recent insights into melanoma genetics and new therapeutic approaches being developed based on these advances.

History and clinical features of melanoma

The first accredited report of melanoma is found in the writings of Hippocrates (born c. 460 BC), where he described "fatal black tumors with metastases." Paleopathologists discovered diffuse bony metastases and round melanotic masses in the skin of Peruvian mummies of the fourth century BC (1). However, it was not until 1806, when René Laennec described "la melanose" to the Faculté de Médecine in Paris, that the disease was characterized in detail and named (2). General practitioner William Norris suggested that melanoma may be hereditary in an 1820 manuscript describing a family with numerous moles and several family members with metastatic lesions (3). Molecular insights over the past 20 years have confirmed Norris's theory of a significant genetic contribution to the etiology of melanoma.

Melanoma develops from the malignant transformation of melanocytes, the pigment-producing cells that reside in the basal epidermal layer in human skin (Figure 1). Recognized as the most common fatal skin cancer, melanoma incidence has increased 15-fold in the past 40 years in the United States, a rate more rapid than that described for any other malignancy (4). Every hour, an American will die from melanoma (5), and it remains one of the most common types of cancer among young adults (6). Furthermore, according to US statistics for 1973–1997, the increase in the mortality rate for melanoma in individuals 65 years of age and older, especially men, was the second highest among all cancers (4).

As in many cancers, both genetic predisposition and exposure to environmental agents are risk factors for melanoma development. Case-control studies have identified several risk factors in populations susceptible to developing melanoma (7). Melanoma primarily affects fair-haired and fair-skinned individuals, and those who burn easily or have a history of severe sunburn are at higher risk than their darkly pigmented, age-matched controls. The UV component of sunlight causes skin damage and increas-

es the risk for skin cancers such as melanoma. It appears that melanoma risk is typically associated with intermittent, intense sun exposure rather than cumulative sun exposure (an exception is lentigo maligna melanoma). The exact mechanism and wavelengths of UV light that are the most critical remain controversial, but both UV-A (wavelength 320–400 nm) and UV-B (290–320 nm) have been implicated (4, 8). This is in contrast to the nonmelanoma skin cancers, basal cell carcinoma and squamous cell carcinoma, which arise from epidermal keratinocytes and are more strongly associated with cumulative sun exposure. Melanoma incidence in fair-skinned people is inversely related to latitude of residence, with the highest incidence found in Australia, which supports the role of UV-induced damage in melanoma pathogenesis (9). In the 1920s, women's fashions became more revealing, and French fashion designer Coco Chanel, who developed a suntan when cruising from Paris to Cannes, is credited with initiating the modern sunbathing trend (10). As our social dress has moved from petticoat and parasol or topcoat and hat to tank top and sunglasses, the incidence of skin cancers, including melanoma, has increased significantly.

Family history of melanoma, increased numbers of both common and dysplastic moles, and a tendency to freckle also increase risk (11). Ten percent of melanoma patients have an affected relative. In a small number of cases, melanomas occur in the setting of the familial atypical multiple mole and melanoma syndrome, also referred to as the dysplastic nevus syndrome (DNS) (12, 13). DNS-affected kindreds develop many atypical moles (dysplastic nevi) at a young age and acquire melanoma with a higher penetrance and earlier onset than are typical of sporadic melanoma. Some evidence suggests that dysplastic nevi may be melanoma precursors in a subset of cases; however, this correlation is controversial and difficult to clearly document (4, 12). More than 50% of melanomas likely arise de novo without a precursor lesion.

Cutaneous melanoma can be subdivided into several subtypes, primarily based on anatomic location and patterns of growth (see Table 1 for key clinical features of subtypes; reviewed in ref. 4). The majority of melanoma subtypes are observed to progress through distinct histologic phases (Figure 1). As melanomas progress from the radial growth phase (RGP) to the vertical growth phase (VGP), treatment options, cure rates, and survival rates decrease dramatically. Most melanoma subtypes demonstrate a slow RGP restricted to the epidermis, followed by a potentially more rapid VGP (14).

Nonstandard abbreviations used: ADI, arginine deiminase; CDK, cyclin-dependent kinase; *CDKN2A, cyclin-dependent kinase inhibitor 2A;* DNS, dysplastic nevus syndrome; MEK, MAPK kinase; PEG, polyethylene glycol; pRb, Rb protein; PTEN, phosphatase and tensin homolog; Rb, retinoblastoma; RCC, renal cell carcinoma; RGP, radial growth phase; siRNA, small interfering RNA; SLNB, sentinel lymph node biopsy; VGP, vertical growth phase.

Conflict of interest: The authors have declared that no conflict of interest exists.

Citation for this article: *J. Clin. Invest.* **115:**813–824 (2005).
doi:10.1172/JCI200524808.

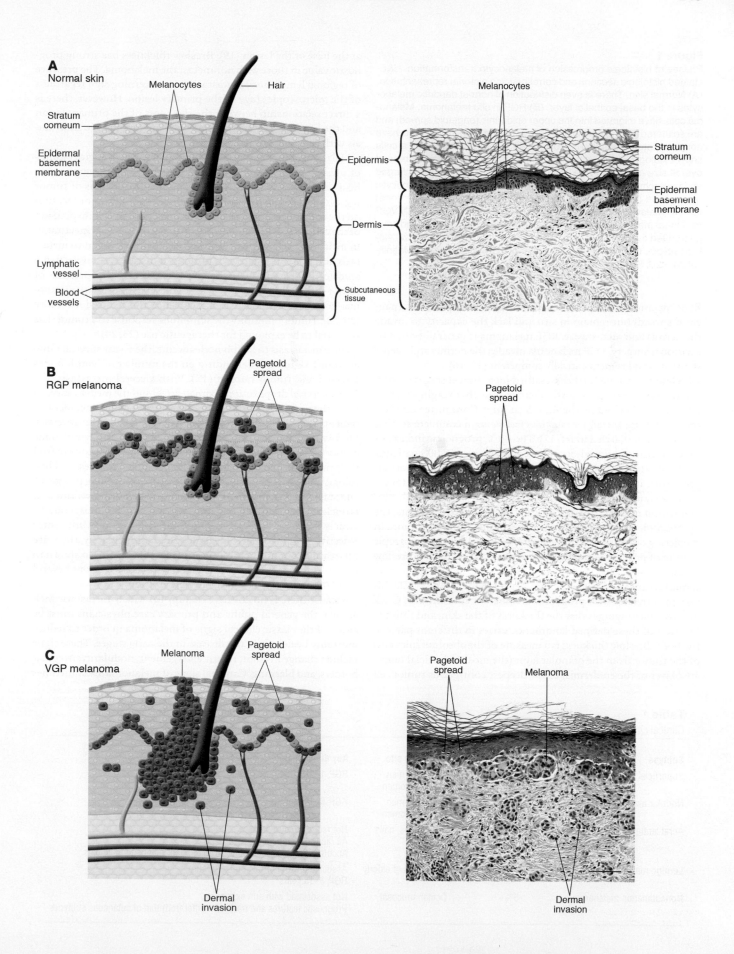

A Normal skin

Melanocytes — Hair

Stratum corneum

Epidermal basement membrane

Lymphatic vessel

Blood vessels

Melanocytes

Stratum corneum

Epidermal basement membrane

Epidermis

Dermis

Subcutaneous tissue

B RGP melanoma

Pagetoid spread

Pagetoid spread

C VGP melanoma

Melanoma

Pagetoid spread

Pagetoid spread

Melanoma

Dermal invasion

Dermal invasion

Figure 1

Phases of histologic progression of melanocyte transformation. H&E-stained histologic sections and corresponding pictorial representation. (**A**) Normal skin. There is even distribution of normal dendritic melanocytes in the basal epithelial layer. (**B**) RGP in situ melanoma. Melanoma cells have migrated into the upper epidermis (pagetoid spread) and are scattered among epithelial cells in a "buckshot" manner. Cells have not penetrated the epidermal basement membrane. Melanoma cells show cytologic atypia, with large abundant cytoplasm and increased overall size compared with normal melanocytes. Nuclei are enlarged and hyperchromatic. Commonly, there is more junctional melanocytic hyperplasia (nests of tumor cells at the basement membrane zone) in RGP melanoma than portrayed in the histologic example. (**C**) VGP malignant melanoma. Melanoma cells show pagetoid spread and have penetrated the dermal-epidermal junction. Melanoma cells show cytologic atypia. Cells in the dermis cluster or individually invade. Magnification, ×20. Scale bar: 20 μm.

RGP melanoma cells extend upward into the epidermis (pagetoid spread) but remain in situ and lack the capacity to invade the dermis and metastasize. RGP melanoma is generally cured by excisional surgery. VGP melanoma invades the dermis and deeper structures and is metastatically competent (15, 16).

Melanoma can be further classified into clinical stages according to significant prognostic factors, and this staging system was recently revised by the American Joint Committee on Cancer (AJCC) (see *Staging for cutaneous melanoma*; a complete staging system is summarized in ref. 17). The AJCC prognostic indicators were confirmed by analysis of outcomes in over 17,000 patients. In the absence of known distant metastasis, the most important prognostic indicator is regional lymph node involvement. However, the majority of melanoma patients present with clinically normal lymph nodes. Thus, in clinically node-negative patients, the microscopic degree of invasion of melanoma is of importance in predicting outcome. There are 2 systems described for microscopic staging of primary cutaneous melanoma, Clark level and Breslow thickness (4). Clark levels classify melanoma according to anatomic landmarks in the epidermis, dermis, and fat (18). While this system correlates with prognosis, an inherent concern with Clark microscopic staging is that the thickness of the skin, and thus the location of these defined landmarks, varies in different parts of the body. Breslow thickness is a measure of the absolute thickness of the tumor from the granular layer (the most superficial nucleated layer of the epidermis) to the deepest contiguous tumor cell

at the base of the lesion (19). Breslow thickness has strong prognostic value in those with nonmetastatic melanoma. The presence of regional lymph node metastasis is a concerning sign regardless of the microscopic stage of the primary lesion. However, there is a direct relationship between the thickness of the primary lesion and the likelihood of microscopic nodal involvement in individuals with clinically normal nodes.

Melanoma prognosis also worsens with the histologic findings of ulceration, high tumor cell mitotic rate, sparse lymphocytic host response, vascular invasion, and histologic signs of tumor regression (17, 20). Increasing age, male sex, and tumor location on the trunk, head, or neck also worsen prognosis. The expression of the cellular marker and melanocyte-specific protein, melastatin, in melanoma cells appears to be inversely proportional to metastasis and was correlated with prolonged disease-free survival in a study of 150 patients with localized disease (21). The presence of tumor-infiltrating T lymphocytes in the VGP of primary melanomas correlates with decreased recurrence and reduced mortality (20, 22). Tumor immunity (host immune response to a tumor) has potential to be exploited for therapeutic use (23, 24).

Once metastasis to lymph nodes occurs, the 5-year survival ranges from 13% to 69%, depending on the number of lymph nodes affected and tumor burden (17). With visceral metastasis, the 5-year survival drops to approximately 6%, and the median survival from time of diagnosis is 7.5 months (25). The management of patients with clinically normal lymph nodes remains controversial. Elective node dissection has been replaced by sentinel lymph node biopsy (SLNB). In this method, lymphatic mapping is done to find the primary (or sentinel) draining lymph node or nodes, and histologic analysis is performed to assist with determining prognosis and staging. Currently, SLNB remains a diagnostic procedure, as it is unclear whether SLNB improves survival. SLNB's impact on survival is the subject of a large, randomized study, the Multicenter Selective Lymphadenectomy Trial (26). Unfortunately, there are no treatment options currently available that have been shown to increase life expectancy once melanoma spreads to regions beyond which it can be cured by local surgical excision.

Because there is no effective therapy for widely metastatic melanoma, the general public and primary care physicians must be aware of the classic clinical signs of melanoma in order to reduce mortality by detecting the disease in the early stages. These signs include change in color, recent enlargement, nodularity, irregular borders, and bleeding. Cardinal signs of melanoma are sometimes

Table 1

Clinical classification of melanoma

Subtype	Frequency	Common site	Key distinguishing features
Superficial spreading melanoma	70%	Trunk of men Legs of women	RGP, 1–5 years
Nodular melanoma	10–25%	Trunk of men Legs of women	RGP, 6–18 months
Acral lentiginous melanoma	5%	Palms, soles, nails	Not related to sun damage All races affected Accounts for 30–70% of melanoma in dark-skinned individuals
Lentigo maligna melanoma	<1%	Head and neck of elderly	Associated with chronic sun exposure RGP, 3–15 years
Noncutaneous melanoma	5%	Ocular, mucosal	Not associated with sun exposure Prognostic features and treatment differ from that of cutaneous subtypes

Cancers such as melanoma arise due to accumulation of mutations in genes critical for cell proliferation, differentiation, and cell death (30). In addition, cancer cells acquire the ability to initiate and sustain angiogenesis, invade across tissue planes, and metastasize. The clinical and histologic progression observed in the growth phases of melanoma (Figure 1) is hypothesized to correspond to the accumulation of these genetic mutations (14). In order to develop treatments for advanced disease and to increase survival from

referred to by the mnemonic ABCDEs (asymmetry, border irregularity, color, diameter, elevation).

Histologic diagnosis of melanoma depends on a combination of certain characteristic architectural features and cellular atypia (reviewed in ref. 27). The discovery of histologic markers unique to melanocytes or melanoma, such as differentiation antigen melanoma antigen recognized by T cells 1 (MART-1), also known as Melan-A, and human melanoma, black-45 (HMB-45), has aided melanoma diagnosis, but there is no marker that is 100% specific or sensitive (28, 29). Thus, fulfillment of histopathologic criteria (reviewed in ref. 4), combined with multiple positive histologic markers, provides the most reliable method of diagnosis.

metastatic melanoma, it is critical to understand the genetic changes leading to each progressive step of the cancer, especially the transition from RGP to VGP. Furthermore, an understanding of the changes that permit invasion through the epidermal basement membrane and thus allow for subsequent metastasis will permit the rational design of treatments for early stages of melanoma and potentially the design of chemopreventive treatments for patients with premalignant lesions or who are at high risk for melanoma development. In addition, an understanding of tumor biology and immunology will aid in the rational design of biochemotherapy and immunotherapy agents for more advanced stages of disease. Mutations observed in human melanoma patients provide starting points for a genetic analysis of melanoma.

Table 2

Nomenclature of genes and proteins discussed in the text

Human gene	Mouse gene	Protein	Gene name	Function
CDK4	Cdk4	CDK4	Cyclin-dependent kinase 4	Promotes G_1-to-S phase transition by phosphorylating and thereby inactivating pRb
CDK6	Cdk6	CDK6	Cyclin-dependent kinase 6	Same as that of CDK4
CDKN2A	Cdkn2a	p16^{INK4a}, p14^{ARF} (p19^{ARF} in mouse)	Cyclin-dependent kinase inhibitor 2A	Encodes 2 tumor suppressors, INK4a and ARF (see below)
INK4a	Ink4a	p16^{INK4a}	Cyclin-dependent kinase inhibitor 2A (see above)	Tumor suppressor that blocks CDK4/6-mediated inactivation of pRb
ARF	Arf	p14^{ARF} (p19^{ARF} in mouse)	Alternative reading frame	Tumor suppressor that blocks HDM2-mediated degradation of p53
BRAF		B-Raf	v-raf murine sarcoma viral oncogene homolog B1	Protooncogene that activates MEK family of MAPK kinases; the gene most often mutated in melanoma and melanocytic nevi
CRAF		C-Raf	v-raf-1 murine leukemia viral oncogene homolog 1	Protooncogene that activates MEK family of MAPK kinases
HRAS	Hras	H-Ras	v-Ha-ras Harvey rat sarcoma viral oncogene homolog	Protooncogene mutated frequently in human tumors; in oncogenic state, activates many pathways important for cancer development, particularly Raf-MEK-ERK and PI3K-Akt
NRAS	Nras	N-Ras	Neuroblastoma RAS viral (v-ras) oncogene homolog	Same as that of HRAS; most frequently mutated RAS isoform in melanoma and nevi
PTEN	Pten	PTEN	Phosphatase and tensin homolog	Tumor suppressor that blocks PI3K-mediated activation of Akt
AKT3		Akt3	v-akt murine thymoma viral oncogene homolog 3	Protooncogene normally activated by PI3K; in melanoma, often constitutively activated via AKT3 gene amplification or loss/mutation of PTEN
TP53		p53	Tumor protein p53	Tumor suppressor activated by many stimuli, particularly genetic damage; promotes apoptosis or cell cycle arrest

HRAS and NRAS (along with KRAS) are members of the RAS gene family (protein family designated Ras). BRAF and CRAF (along with ARAF) are members of the RAF gene family (protein family designated Raf). Raf proteins are kinases that are activated by the Ras proteins and in turn activate MEK 1 and MEK2. AKT3 (along with AKT1 and AKT2) is a member of the AKT gene family (protein family designated Akt).

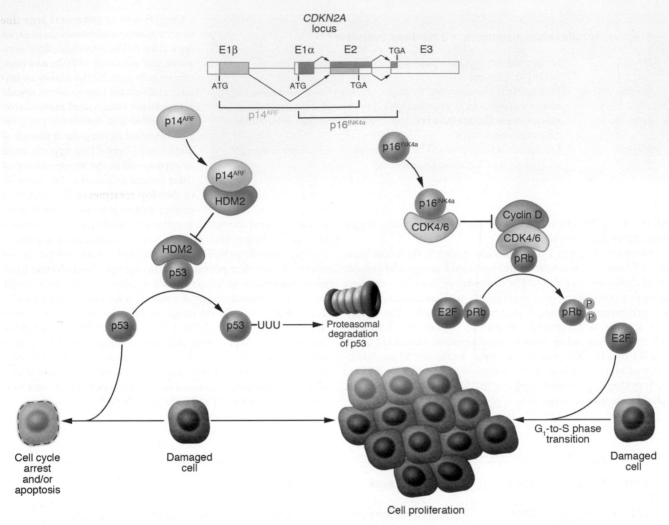

Figure 2

Genetic encoding and mechanism of action of tumor suppressors p16INK4a and p14ARF. The *CDKN2A* locus on chromosome 9p21 is composed of 4 exons (E) – 1α, 1β, 2, and 3 – and encodes 2 tumor suppressors, p16INK4a and p14ARF (termed p19ARF in the mouse), via alternative reading frames. p16INK4a is translated from the splice product of E1α, E2, and E3, while p14ARF is translated from the splice product of E1β, E2, and E3. Normally p16INK4a sequesters CDK4 and CDK6 thereby keeping pRb in an active hypophosphorylated state. In the absence of p16INK4a, CDK4 and CDK6 bind cyclin D and phosphorylate Rb. Phosphorylated pRb then releases the E2F transcription factor, which promotes the G1-to-S phase transition of the cell cycle. p14ARF, on the other hand, sequesters the p53-specific ubiquitin ligase HDM2. In the absence of p14ARF, HDM2 targets p53 for ubiquitination (UUU) and subsequent proteosomal degradation, and the loss of p53 impairs mechanisms that normally target genetically damaged cells for cell cycle arrest and/or apoptosis, which leads to proliferation of damaged cells. Loss of *CDKN2A* therefore contributes to tumorigenesis by disruption of both the pRb and p53 pathways. Figure modified with permission from *N. Engl. J. Med.* (S30).

Identification of melanoma-associated genes from patient samples

Much of our knowledge of melanoma susceptibility loci derives from genetic studies of melanoma-prone families. Although these families are rare, they have proven invaluable in the identification of genetic pathways that play roles in the development of familial as well as sporadic melanoma (31). See Table 2 for the symbols and definitions of melanoma-associated genes described herein.

Tumor suppressor pathways. Melanoma development has been shown to be strongly associated with inactivation of the p16INK4a/cyclin dependent kinases 4 and 6/retinoblastoma protein (p16INK4a/CDK4,6/pRb) and p14ARF/human double minute 2/p53 (p14ARF/HMD2/p53) tumor suppressor pathways. These pathways help control the G1 phase of the cell cycle and are most often inhibited

via deletions or mutations in the *cyclin-dependent kinase inhibitor 2A (CDKN2A)* locus on chromosome 9p21 (32, 33). This locus encodes 2 tumor suppressors, p16INK4a and p14ARF, via alternative reading frames (34) (Figure 2). p16INK4a sequesters CDK4 and CDK6 and thereby blocks phosphorylation of pRb by these kinases, keeping pRb in its active state inhibiting cell-cycle progression (35). p14ARF, on the other hand, stabilizes p53 by preventing HMD2-mediated degradation of the otherwise transiently expressed p53 protein (36–39) (Figure 2). Germline *CDKN2A* mutations were identified in 25–50% of familial melanoma kindreds (40–42). The importance of this locus in melanoma susceptibility was confirmed by studies showing that the penetrance of *CDKN2A* mutations significantly correlated with residence in a geographical location with a high population incidence rate of melanoma (43) and that *CDKN2A*

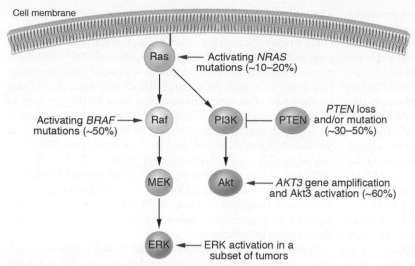

Cell membrane

Figure 3
Schematic of the canonical Ras effector pathways Raf-MEK-ERK and PI3K-Akt and the mutations that most often activate these pathways in melanoma patients. Oncogenic *NRAS* mutations, found in approximately 10–20% of melanomas, activate both effector pathways. The Raf-MEK-ERK pathway may also be activated via mutations in the *BRAF* gene, found in approximately 50% of melanomas. In a subset of melanomas, the ERK kinases have been shown to be constitutively active even in the absence of *NRAS* or *BRAF* mutations. The PI3K-Akt pathway may be activated through loss or mutation of the inhibitory tumor suppressor gene *PTEN*, occurring in 30–50% of melanomas, or through gene amplification of the *AKT3* isoform. While the exact frequency of *AKT3* gene amplification is unknown, the Akt3 kinase is constitutively activated in approximately 60% of melanomas. Activation of ERK and/or Akt3 promotes the development of melanoma by various mechanisms, including stimulation of cell proliferation and enhanced resistance to apoptosis.

cascades. The importance of both of these pathways in melanoma development has been further confirmed by the identification of genetic alterations that activate each one independently.

The first indication of involvement of the PI3K-Akt pathway in the progression of melanoma was the discovery of frequent loss of *phosphatase and tensin homolog (PTEN)* in melanocytic lesions. PTEN is a protein and lipid phosphatase that inhibits Akt activation by PI3K (reviewed in ref. 71) (Figure 3). Loss of heterozygosity on regions of chromosome 10q, which harbors the *PTEN* locus, was demonstrated in 30–50% of malignant melanomas (72–74), and approximately 10% of melanomas were shown to have *PTEN* mutations (75–77). Interestingly, patients with disorders such as Cowden disease, which arise from inherited *PTEN* mutations and are characterized by the presence of benign hamartomatous tumors, are not widely reported to display an increased susceptibility to melanoma. More recently, 2 studies directly demonstrated that Akt is constitutively activated in more than 60% of melanomas, with higher frequencies of activation at later stages of the disease (78, 79). The latter study, by Stahl et al. (79), identified amplification of the *AKT3* gene as an additional mechanism, distinct from *PTEN* loss, that may be responsible for these high frequencies of Akt activation.

The role of the Raf-MEK-ERK cascade in melanoma development has received a great deal of attention in the past 3 years, due to the identification of mutations in the *BRAF* gene in a high proportion of melanomas. The initial study by Davies et al. estimated that 67% of melanomas harbor *BRAF* mutations (66). By far the most frequent mutation is V599E; this mutation, found in the activation segment of the B-Raf kinase domain, renders the protein constitutively active and leads to elevated ERK1/2 activation. A great number of follow-up studies (e.g., refs. 67, 69, 80–83) have further examined the frequency of *BRAF* mutations in benign and malignant melanocytic lesions. The resulting estimates have varied from 30% to 70%, and taken together, the published data suggest that approximately 50% of melanomas carry the *BRAF*[V599E] mutation (84). B-Raf activation is thus one of the most frequent melanoma-associated genetic events, on par with activation of the PI3K-Akt pathway.

Several aspects of the association between *BRAF* mutation and melanoma bear further discussion. *BRAF* mutations are extremely rare in uveal (ocular) and mucosal melanomas (85, 86), which suggests that sun-related, UV-induced carcinogenesis may play a role in causing these mutations. Importantly, although approximately half of all melanomas carry the *BRAF*[V599E] mutation, this alteration is also found in 70–80% of common acquired melanocytic nevi (67, 83, 87). This finding strongly suggests that *BRAF* mutation is an early event in the development of melanocytic neoplasia and alone may not be sufficient to drive melanoma development. While all studies published to date have found a higher rate of *BRAF* mutations in nevi than in melanomas, contradictory findings have been reported concerning the frequencies of *BRAF* mutations at different stages of malignancy. Some groups have found *BRAF* mutations in a higher proportion of metastatic melanomas than

mutation carriers have increased total nevus number and total nevus density — known risk factors for melanoma — compared with noncarriers within the same family (44). Inheritance of variants of the *melanocortin-1 receptor* gene seen in red-haired patients, along with *CDKN2A* mutations, further increases the risk of melanoma in melanoma-prone families (31, 45). Subsequent studies showed a similar frequency of somatic *CDKN2A* alterations in sporadic melanomas (e.g., refs. 46–49). Notably, melanoma-associated *CDKN2A* mutations and deletions may ablate only the *INK4a* transcript (42, 50), the *ARF* transcript (51, 52), or both (47–49, 53, 54), which indicates that each of these tumor suppressors plays an important individual role as a guardian against melanoma development. Less frequently, pRb may also be inactivated via *CDK4* mutations — R24C or R24H — that occur in both sporadic and familial melanomas (55–58). These mutations prevent binding of p16[INK4a] to CDK4, rendering the kinase constitutively active. Although *TP53* mutations are not universal in melanoma (59–61), several groups have reported a mutation frequency in the range of 10–25% (62–65), with at least 1 recent study (65) suggesting a correlation between *TP53* mutation and poor prognosis.

Oncogenic Ras and its effectors. Melanoma-associated oncogene activation most often targets Ras and its canonical effector pathways, Raf-MAPK kinase-ERK (RAF-MEK-ERK) and PI3K-Akt (Figure 3). The mutation frequency of the *RAS* gene family, particularly *NRAS*, in human melanoma has been estimated in the majority of recent studies as 10–20% (65–69). A subset of melanomas has also been shown to harbor constitutive Ras activation without underlying mutations (70). Oncogenic Ras activates both the Raf and PI3K

early-stage primary melanomas (65, 81), but others have reported a decreasing mutation frequency in association with more advanced disease (67). Likewise, the presence of *BRAF* mutations in metastatic melanoma has been reported to correlate with a shorter survival in one study (88) and with a longer survival in another (89). Thus, the genetic evidence regarding the role of *BRAF* late in melanoma progression is currently inconclusive and merits further investigation. Finally, it is important to note that *BRAF* mutations may occur concurrently with loss or mutation of *PTEN*, but neither *BRAF* nor *PTEN* alterations are found together with *NRAS* mutations (65, 90, 91). Thus, melanomas that have arisen in the absence of *NRAS* mutations nearly always harbor activated *BRAF*, inactivating alterations of *PTEN*, or both. Conversely, tumors with oncogenic *NRAS* mutations typically retain wild-type *BRAF* and *PTEN*. These results reiterate the importance of both the PI3K-Akt and Raf-MEK-ERK cascades in melanoma development; they also suggest that each of these pathways plays a significant role in melanoma development and that, in a subset of melanomas, the 2 pathways may cooperate in promoting cancer progression.

Investigation of melanoma-associated mutations in experimental models

The functional importance of genes mutated in a high proportion of human melanomas has begun to be elucidated via studies in vitro and in mice. Initial experiments showed that mice carrying a deletion of the *Cdkn2a* locus have increased predisposition to tumor development (92). In subsequent studies designed to clarify the roles of the 2 genes encoded by this locus, mice with a selective knockout of the *Arf* gene (93, 94) or the *Ink4a* gene (95, 96) were generated and determined to be cancer prone. These findings directly showed that both *Ink4a* and *Arf* are important tumor suppressor genes, but in all of these studies, the knockout mice developed predominantly lymphomas and sarcomas, with very few spontaneous melanomas observed (92–96). However, when the *Cdkn2a*-null mice were engineered to express oncogenic *Hras^{G12V}* from a melanocyte-specific promoter (*Tyr-ras*), more than 50% of these mice developed melanomas by the age of 6 months (97). In fact, while almost no *Tyr-ras* mice on a wild-type background developed melanomas (97), approximately half of the mice expressing *Tyr-ras* on either a selective *Ink4a$^{-/-}$* or a selective *Arf$^{-/-}$* background succumbed to the cancer by 1 year of age (33, 98). The formation of Ras-driven melanomas in *Arf*-null mice was accelerated by UV treatment, and the resulting tumors harbored *Cdk6* amplification or loss of p16^{INK4a}, which suggested that UV exposure promoted melanoma development by inactivating the pRb pathway (98). This series of studies suggested that the p16^{INK4a}-pRb and p19^{ARF}-p53 tumor suppressor pathways indeed serve as guardians against melanoma formation, and it also emphasized the importance of oncogenic Ras signaling in promoting the development of this cancer. *Hras^{G12V}*, in fact, was shown to be essential not only for the initiation but also for the maintenance of the melanocytic tumors in this model (99), which confirmed its role as a driving oncogene of melanoma.

Functional demonstration of the role played by Akt activation, particularly occurring via loss of *PTEN*, in melanoma development at first proved elusive. *Pten*-knockout mice died at an early embryonic stage, and *Pten$^{+/-}$* mice were highly predisposed to tumor formation and developed prostate, gastrointestinal, thyroid, and various other types of cancer, but melanomas were not observed among the spectrum of tumors in these mice (100, 101). However,

when the *Pten$^{+/-}$* mice were crossed into the *Cdkn2a$^{-/-}$* background, approximately 10% of the resulting mice developed melanoma (102), which indicates that *Pten* loss, like Ras activation, can cooperate with inactivation of the p16^{INK4a}-pRb and p19^{ARF}-p53 pathways to induce melanocytic neoplasia. A separate line of investigation, in which human melanoma cell lines were used, showed that introduction of an intact chromosome 10 or a wild-type copy of *PTEN* into melanoma cells lacking *PTEN* expression inhibited in vitro growth of these cells (103) and in vivo tumor formation in immunodeficient mice (104) and that the introduced *PTEN* gene was targeted for loss of heterozygosity in these cells, both in vitro (103) and in vivo (104). Finally, specific inhibition of the *AKT3* isoform in these cells, much like introduction of wild-type *PTEN*, stimulated apoptosis and retarded tumor growth (79). These experiments suggest involvement of the PI3K-Akt pathway in the development of melanoma.

To date, there have been no published transgenic murine studies on the role of B-Raf in melanoma development, but a number of experiments have been carried out with mouse and human cell lines. *BRAF^{V599E}* has been shown to act as a transforming oncogene both in NIH3T3 cells (66) and in an immortalized murine melanocyte cell line (105). Furthermore, knockdown of B-Raf expression by small interfering RNA (siRNA) in human melanoma cell lines harboring the *BRAF^{V599E}* mutation decreased the levels of MEK and ERK activation, inhibited cell proliferation, increased the rates of apoptosis, and blocked colony formation in soft agar (105–107). There was a specific correlation between the presence of the *BRAF* mutation and effects of the B-Raf knockdown on cell growth and cell death. For example, depletion of B-Raf expression in human fibrosarcoma cells carrying wild-type *BRAF* (106) or in melanoma cells containing an oncogenic *HRAS* mutation rather than an activating *BRAF* mutation (105) failed to block MEK-ERK signaling, inhibit cell proliferation, or promote apoptosis. Additionally, siRNA-mediated knockdown of C-Raf did not affect ERK activation, cell growth, or apoptosis in the cell lines carrying the *BRAF^{V599E}* mutation (105–107). Finally, in the first preclinical study targeting the Raf-MEK-ERK pathway for small molecule–mediated inhibition in melanoma cells, the MEK inhibitor CI 1040 was shown to inhibit in vitro growth, soft agar colony formation, and Matrigel invasion of a human melanoma cell line bearing the *BRAF^{V599E}* mutation; more importantly, it blocked the formation of pulmonary metastases and caused regression of established metastases when administered orally to immunodeficient mice injected intravenously with this cell line (108). These studies established that mutant B-Raf does have oncogenic properties, at least in a subset of melanomas, that it can serve as a potential therapeutic target (reviewed in ref. 109), and that small-molecule inhibition of the MAPK pathway is a viable therapeutic strategy.

Melanoma therapeutics

Surgical treatment of cutaneous melanoma employs specific surgical margins depending on the depth of invasion of the tumor. Specific surgical treatment guidelines of primary, nodal, and metastatic melanoma are beyond the scope of this article and are reviewed elsewhere (refs. 4, 110). Numerous agents have been used in the treatment of late-stage melanoma, but to date no single agent has significantly changed survival rates. Development of adjuvant therapies that increase survival beyond that achieved following surgery alone has been a long-standing goal of melanoma researchers and clinicians.

Chemotherapy. Among traditional chemotherapeutic agents, only dacarbazine is FDA approved for the treatment of advanced melanoma (111). Combination chemotherapy regimens (such as cisplatin, vinblastine, and dacarbazine) are often employed, though no clinical trials have demonstrated a survival advantage for combination therapy over optimal single-agent therapy (111). Temozolomide is an oral alkylating agent approved for the treatment of malignant gliomas with activity comparable to that of dacarbazine in melanoma. While there is not yet supportive clinical trial data, temozolomide has been substituted in many combination regimens and is under investigation in combination with radiation therapy for patients with melanoma metastatic to the brain (112).

Immunotherapy. A large, randomized multicenter study in high-risk melanoma patients performed by the Eastern Cooperative Oncology Group (ECOG) showed significant improvements in relapse-free and overall survival with postoperative adjuvant IFN-α-2b therapy, compared with standard observation (ECOG 1684) (4). The outcome of this study led to FDA approval of IFN-α-2b for treatment of melanoma. This study was performed on patients with deep primary tumors without lymph node involvement and node-positive melanomas. In other studies, little antitumor activity has been demonstrated in IFN-α-2b–treated metastatic stage IV melanoma. Furthermore, while a large follow-up study (Intergroup E1690) (4) confirmed relapse-free survival with high-dose IFN-α-2b, it did not show an increase in overall survival. Toxicities of high-dose IFN include constitutional (flu-like) symptoms and neuropsychiatric (depression, suicidal intention), hematologic, and hepatic effects (4). Other studies of low-dose IFN versus either observation or high-dose IFN have not shown a statistically significant increase in overall survival (summarized in ref. 4). Based on patient immune responses to a vaccine containing GM₂, a ganglioside found predominantly in melanoma cells, a subsequent intergroup trial (ECOG 1694) was designed to compare high-dose IFN-α-2b with a GM₂-containing vaccine administered with an immune-stimulating adjuvant (113). This trial was closed early, as interim analysis indicated significantly superior disease-free survival with IFN treatment. IL-2 has activity against melanoma, and high-dose therapy is FDA approved (114). Because high-dose IL-2 side-effects are significant and may be life threatening, IL-2 therapy requires inpatient management.

Biochemotherapies, which combine traditional chemotherapy with immunotherapies, such as IL-2 and IFN-α-2b, seemed promising in phase II trials, but phase III studies have not yet confirmed statistically significant survival benefits in patients with stage IV metastasis, and studies in resected, advanced stage III disease (Southwest Oncology Group/Intergroup trial S0008) are ongoing (reviewed in ref. 115). Seven large studies failed to show statistically significant increased overall survival rates for various biochemotherapy regimens (115).

Because most current adjuvant treatments have significant side effects and many have limited benefit with respect to overall survival in late-stage melanoma, there exists a strong need for the rational design of mechanism-based treatments for this devastating metastatic cancer, whose median survival rate with visceral metastases has not surpassed the 1-year barrier. Current approaches include inhibition of tumor-activated signaling cascades, exploiting the host cancer immune response, targeting of melanocyte and tumor-specific antigens and activities (such as blockade of angiogenesis and of proteins that resist apoptosis), and blockade of tumor growth and tumor-stroma interactions.

Many treatment regimens are now in varying stages of development (from preclinical studies to phase II trials). This review will focus primarily on recent efforts to target the Ras cascade as an example of a rational approach to therapeutics.

Signal transduction inhibition by Raf inhibitor BAY 43-9006. Small-molecule kinase inhibitors have potential as novel therapies for cancer and inflammatory conditions, as exemplified by currently marketed compounds imatinib mesylate (Gleevec; Novartis Pharmaceuticals) and gefitinib (Iressa; AstraZeneca) (116). BAY 43-9006 is a bis-aryl urea that was discovered in a screen for C-Raf inhibitors (117). Related urea compounds have been described as inhibitors of several other kinases (118). In addition to blocking C-Raf kinase activity with an IC_{50} of 3–12 nM (107, 117, 119), this compound inhibits both wild-type and mutant B-Raf, albeit less potently, with an IC_{50} of 16–28 nM for wild-type B-Raf and an IC_{50} of 29–47 nM for B-Raf^{V599E} (105, 107, 119). It is also able to inhibit several receptor tyrosine kinases involved in cancer progression, although the inhibition of these kinases is even less robust than that of the Raf proteins (119). In vitro studies showed that BAY 43-9006 is able to block ERK activation and cell proliferation in immortalized murine melanocytes transformed with either oncogenic *RAS* or *BRAF*^{V599E} (105), as well as in several human melanoma cell lines carrying the *BRAF*^{V599E} mutation (107). More importantly, oral administration of the drug to mice carrying tumor xenografts of one of these cell lines significantly retarded tumor growth (107). A recent study extended these findings to a panel of human tumor cell lines derived from several different cancer types and determined that orally administered BAY43-9006 had broad antitumor activity against xenografts formed from breast, colon, and lung cancer cell lines, with no adverse toxicity (119). These encouraging findings have led to the use of BAY 43-9006 in a number of clinical trials in patients with various solid tumors.

Initial phase I trials showed that BAY 43-9006 was well tolerated when administered either as a monotherapy (120) or in combination with traditional chemotherapy (121), with only mild adverse side effects observed. Phase II trials are ongoing, and to date BAY 43-9006 monotherapy has shown promise against several types of cancer, particularly renal cell carcinoma (122, 123). Unfortunately, in patients with advanced melanoma, clear evidence of the drug's efficacy has yet to be demonstrated. After 12 weeks of treatment, fewer than 20% of patients had stable disease, and fewer than 5% had a partial response, with the vast majority of patients succumbing to progressive disease (123, 124). Phase II trials of BAY 43-9006 in combination with chemotherapy agents such as carboplatin and paclitaxel (125) are currently in progress. This combination regimen has shown some preliminary evidence of anticancer activity — of 35 patients in the initial trial, 11 had a partial response and 19 had stable disease (125).

There are several possible explanations for the rather modest therapeutic efficacy of BAY 43-9006 in melanoma patients. As a signal transduction inhibitor, this compound is predicted to act as a cytostatic agent (122), and it may need to be combined with one or more cytotoxic agents in order to exert significant antitumor activity. Alternatively, the inhibition of B-Raf^{V599E} activity by this drug may not be sufficiently potent. The progression of renal cell carcinoma (RCC) has been shown to involve *CRAF* alterations (S1) but not *BRAF* mutations (S2); a potentially important issue, given that BAY 43-9006 demonstrates 5- to 10-fold stronger inhibition of C-Raf than B-Raf^{V599E}. Taken together, these findings suggest that the greater therapeutic benefit enjoyed by RCC patients treated with BAY 43-9006, as compared with melanoma patients, may stem from more efficient inhibition of C-Raf or another target in RCC patients. Thus, when more specific and potent B-Raf inhibitors are developed, these compounds may show greater efficacy against melanoma. It is also possible that the cohort of melanoma patients treated with BAY 43-9006 did not have the proper mutation status to allow therapeutic benefit from the inhibitor. The drug has been shown to inhibit the growth of a human melanoma cell line carrying the $BRAF^{V599E}$ mutation (107) and several cell lines harboring oncogenic RAS (105, 119), although the latter result was demonstrated in transformed murine melanocytes (105) and in human cell lines derived from tumors other than melanoma (119) and has not been directly shown in human melanoma cells. It does not appear that all patients selected for the clinical trial of BAY 43-9006 carried the $BRAF^{V599E}$ mutation (123, 124). It is unclear whether patients whose melanomas harbor oncogenic RAS mutations rather than $BRAF$ mutations would benefit from this drug, and therapeutic benefit is even less likely in patients whose tumors arose through other mechanisms, such as $PTEN$ loss or $AKT3$ amplification. Efforts to classify the clinical trial participants by $BRAF$ mutation status and correlate the mutation status with treatment outcome are currently underway (124, 125). Finally, B-Raf may not represent the ideal therapeutic target in patients with advanced melanoma. As discussed above, the high frequency of $BRAF$ mutations in premalignant melanocytic lesions (67, 83, 87) suggests that the major effects of this oncogene may be limited to the initiation of melanoma development and that inhibition of B-Raf activity may have therapeutic benefit for patients with early-stage disease but not for those whose cancer has progressed to the metastatic stage. This theory is contradicted by data showing that melanoma cell lines carrying the $BRAF^{V599E}$ mutation depend on B-Raf expression for growth and survival (105–107), but the applicability of results from studies involving cell lines to actual patient tumors may be limited. More relevant experimental models of human melanoma may lead to more accurate preclinical assessments of potential therapeutic targets.

Angiogenesis inhibition and depletion of tumor nutrition. In addition to signal transduction inhibition, several other rationally designed melanoma treatments are currently in clinical trials. In lieu of direct cytotoxic approaches, these therapies are aimed at starving the tumor by inhibiting angiogenesis or depleting nutrients essential for cancer growth. Of the antiangiogenic compounds, VEGFR inhibitors SU5416 and AG-013736 demonstrated broad-spectrum antitumor activity in mice bearing xenografts of human cancer cell lines originating from various tissues, including melanoma (S3, S4). Both compounds showed acceptable safety profiles in phase I trials (S5, S6) and progressed to phase II. Currently, no phase II data are available on AG-013736, while the efficacy of SU5416 against metastatic melanoma has proven disappointing. Of 26 evaluable patients in the trial, only 1 had a partial response and 1 had stable disease, with all patients eventually succumbing to melanoma after a median survival time of 29 weeks (S7). In preclinical studies, angiogenesis inhibitors that function by antagonizing integrin signaling, such as EMD 121974, have also demonstrated activity against human melanoma xenografts (S8). EMD 121974 had tolerable adverse effects in phase I trials (S9), but there is not yet available published data for phase II trials in melanoma. Angiogenesis inhibition is theoretically an attractive strategy for inhibition of tumor growth, but the efficacy of antiangiogenic agents against melanoma remains to be determined.

Melanoma cells, unlike normal human cells, require arginine for survival (S10). The arginine-degrading enzyme arginine deiminase (ADI), when formulated with polyethylene glycol (PEG) to make ADI-PEG, was shown to inhibit the growth of human melanoma xenografts in mice (S11). Initial phase I and II trials of ADI-PEG in melanoma patients demonstrated that the enzyme was well tolerated and had modest anticancer activity: of 41 patients in the trial, 6 had a partial response and 2 had stable disease (S12). ADI-PEG is currently in phase II trials (see www.clinicaltrials.gov). The results for ADI-PEG and the angiogenesis inhibitors, much like those for BAY 43-9006, highlight the difficulties in translating data from preclinical models into clinical efficacy. However, it is possible that the phase II trials of these agents that are currently in progress will show more evidence of antitumor activity and that future studies combining signal transduction inhibitors, angiogenesis inhibitors, and nutrient depleting enzymes with each other and/or with traditional chemotherapeutic agents will bring out the effectiveness of these novel agents.

Future directions

Melanoma is strongly associated with inactivation of the retinoblastoma (Rb) and p53 tumor suppressor pathways and activation of oncogenes in the Ras cascade. Gene therapy to directly replace lost and/or mutated tumor suppressors at physiologic levels and in the appropriate tissue context remains a technical challenge, but it is a potential long-range scientific goal in the treatment of melanoma. Small-molecule human double minute 2 (HDM2) antagonists have been developed and have potential to treat tumors with wild-type $TP53$ (S13). B-Raf and other Ras pathway effectors are potential therapeutic targets in melanoma, and more potent B-Raf–specific inhibitors may prove more effective than the general Raf inhibitor BAY 43-9006. There are efforts currently underway to target the PI3K Ras-effector pathway with small-molecule inhibitors (S14). In addition, secondary analyses of trials designed to evaluate the effects of statins on coronary disease suggest that these agents may also have potential chemopreventive effects in melanoma (S15). One proposed mechanism of action of statins in melanoma prevention is blockade of prenylation, which is required for Ras localization to the cell membrane (S16). Case-control studies and metaanalyses are ongoing to evaluate this preliminary observation.

In addition to kinase pathways, other molecules unique to melanoma or melanocytes are viable drug targets. Although results of initial efforts using peptide and whole-cell extracts to stimulate an immune response to cell surface proteins specific to melanoma, including GM_2, were promising, large-scale studies have not revealed significant antitumor activity (S17). However, efforts to enhance host response to cellular vaccines by augmenting tumor immunity, with agents such as recombinant GM-CSF, may provide promising therapeutic agents (24). Since melanoma development is associated with telomerase activation (S18, S19), gene therapy approaches using anti-telomerase ribozymes (S20) and telomere homolog oligonucleotides (S21, S22) are currently in development for the treatment of melanoma and have shown antitumor activity in animal models. Targeting of pathways specific to melanocytes and critical to melanoma survival may also prove to be a viable therapeutic strategy, as suggested by recent preclinical data on the unique role of CDK2 in melanoma survival and CDK2 regulation by the microphthalmia-associated transcription factor in melanocytes (S23). Blockade of CDK2 suppressed growth and cell cycle progression in melanoma cell lines but not in other cancer cell lines.

Because early-stage melanoma is often curable by surgery, additional research efforts focus on blocking tumor-stroma and tumor-microvascular interactions that facilitate invasion and metastasis (S24, S25). Tumors such as melanoma often display resistance to apoptosis. Current efforts to inhibit the resistance of tumors to apoptosis target blockade of human B cell leukemia/lymphoma-2 (Bcl-2, an inhibitor of apoptosis) in melanoma (S26–S28). In addition to small-molecule inhibitors, vaccines, and gene therapies, humanized antibodies are a tantalizing therapeutic agent due to their low clinical safety profile. Also, because all melanomas may not have equal response to treatment modalities due to different mechanisms of tumorigenesis, the ability to perform complex genetic mutational analyses will allow for individually tailored patient prognoses and therapeutics (84). Finally, ongoing proteomics screens may identify other novel therapeutic targets in melanoma (S29).

Prevention should always be at the forefront of the fight to reduce melanoma. Protection from UV irradiation is the key preventive measure. In the final analysis, increased patient survival without a significant concurrent increase in morbidity remains a challenging treatment goal for metastatic melanoma. The design of rational therapeutics targeting key players in disease pathways tailored to tumor-specific mutations will certainly be the focus of translational cancer research in the coming years.

Acknowledgments

This work was supported by the US Veteran's Affairs Office of Research and Development. A.E. Adams is supported by a National Institute of Arthritis and Musculoskeletal and Skin Diseases (NIAMS) NIH training grant (Harvard Department of Dermatology). Y. Chudnovsky is a Howard Hughes Medical Institute predoctoral fellow. We thank Susan Swetter for manuscript review and helpful suggestions and Uma Sundram and the Stanford Dermatology Department for melanoma histology slides.

Note: References S1–S30 are available online with this article; doi:10.1172/JCI200524808DS1.

Address correspondence to: Amy E. Adams, Department of Dermatology, Stanford University, 269 Campus Drive, CCSR 2150, Stanford, California 94305, USA. Phone: (650) 728-6295; Fax: (650) 723-8762; E-mail: aadams91@stanford.edu.

1. Urteaga, O., and Pack, G.T. 1966. On the antiquity of melanoma. *Cancer.* **19**:607–610.
2. Laennec, R.T.H. 1806. Sur les melanoses. *Bulletin de Faculte de Medecine Paris.* **1**:24.
3. Norris, W. 1820. A case of fungoid disease. *Edinb. Med. Surg. J.* **16**:562–565.
4. Nestle, M., and Carol, H. 2003. Melanoma. In *Dermatology.* J. Bolognia, J. Jorizzo, and R. Rapini, editors. Mosby. New York, New York, USA. 1789–1815.
5. Greenlee, R.T., Hill-Harmon, M.B., Murray, T., and Thun, M. 2001. Cancer statistics, 2001. *CA Cancer J. Clin.* **51**:15–36.
6. Weinstock, M.A. 2001. Epidemiology, etiology, and control of melanoma. *Med. Health R. I.* **84**:234–236.
7. MacKie, R.M., Freudenberger, T., and Aitchison, T.C. 1989. Personal risk-factor chart for cutaneous melanoma. *Lancet.* **2**:487–490.
8. Jhappan, C., Noonan, F.P., and Merlino, G. 2003. Ultraviolet radiation and cutaneous malignant melanoma. *Oncogene.* **22**:3099–3112.
9. Marks, R. 2000. Epidemiology of melanoma. *Clin. Exp. Dermatol.* **25**:459–463.
10. Gilchrest, B. 1999. Anti-sunshine vitamin A. *Nat. Med.* **5**:376–377.
11. Tucker, M.A., et al. 1997. Clinically recognized dysplastic nevi. A central risk factor for cutaneous melanoma. *JAMA.* **277**:1439–1444.
12. Chin, L., Merlino, G., and DePinho, R.A. 1998. Malignant melanoma: modern black plague and genetic black box. *Genes Dev.* **12**:3467–3481.
13. Haluska, F.G., and Hodi, F.S. 1998. Molecular genetics of familial cutaneous melanoma. *J. Clin. Oncol.* **16**:670–682.
14. Clark, W.H., Jr., et al. 1984. A study of tumor progression: the precursor lesions of superficial spreading and nodular melanoma. *Hum. Pathol.* **15**:1147–1165.
15. Meier, F., et al. 1998. Molecular events in melanoma development and progression. *Front. Biosci.* **3**:D1005–D1010.
16. Rusciano, D. 2000. Differentiation and metastasis in melanoma. *Crit. Rev. Oncog.* **11**:147–163.
17. Balch, C.M., et al. 2001. Final version of the American Joint Committee on Cancer staging system for cutaneous melanoma. *J. Clin. Oncol.* **19**:3635–3648.
18. Clark, W.H., Jr., From, L., Bernardino, E.A., and Mihm, M.C. 1969. The histogenesis and biologic behavior of primary human malignant melanomas

of the skin. *Cancer Res.* **29**:705–727.
19. Breslow, A. 1979. Prognostic factors in the treatment of cutaneous melanoma. *J. Cutan. Pathol.* **6**:208–212.
20. Clark, W.H., Jr., et al. 1989. Model predicting survival in stage I melanoma based on tumor progression. *J. Natl. Cancer Inst.* **81**:1893–1904.
21. Duncan, L.M., et al. 2001. Melastatin expression and prognosis in cutaneous malignant melanoma. *J. Clin. Oncol.* **19**:568–576.
22. Clemente, C.G., et al. 1996. Prognostic value of tumor infiltrating lymphocytes in the vertical growth phase of primary cutaneous melanoma. *Cancer.* **77**:1303–1310.
23. Ramirez-Montagut, T., Turk, M.J., Wolchok, J.D., Guevara-Patino, J.A., and Houghton, A.N. 2003. Immunity to melanoma: unraveling the relation of tumor immunity and autoimmunity. *Oncogene.* **22**:3180–3187.
24. Dranoff, G. 2003. GM-CSF-secreting melanoma vaccines. *Oncogene.* **22**:3188–3192.
25. Barth, A., Wanek, L.A., and Morton, D.L. 1995. Prognostic factors in 1,521 melanoma patients with distant metastases. *J. Am. Coll. Surg.* **181**:193–201.
26. Morton, D.L., et al. 1999. Validation of the accuracy of intraoperative lymphatic mapping and sentinel lymphadenectomy for early-stage melanoma: a multicenter trial. Multicenter Selective Lymphadenectomy Trial Group. *Ann. Surg.* **230**:453–463; discussion 463–455.
27. Weedon, D. 2002. *Skin Pathology.* 2nd edition. Churchill Livingstone. London, United Kingdom. 1100 pp.
28. Busam, K.J., and Jungbluth, A.A. 1999. Melan-A, a new melanocytic differentiation marker. *Adv. Anat. Pathol.* **6**:12–18.
29. Gown, A.M., Vogel, A.M., Hoak, D., Gough, F., and McNutt, M.A. 1986. Monoclonal antibodies specific for melanocytic tumors distinguish subpopulations of melanocytes. *Am. J. Pathol.* **123**:195–203.
30. Hanahan, D., and Weinberg, R.A. 2000. The hallmarks of cancer. *Cell.* **100**:57–70.
31. Hayward, N.K. 2003. Genetics of melanoma predisposition. *Oncogene.* **22**:3053–3062.
32. Chin, L. 2003. The genetics of malignant melanoma: lessons from mouse and man. *Nat. Rev. Cancer.* **3**:559–570.

33. Sharpless, N.E., Kannan, K., Xu, J., Bosenberg, M.W., and Chin, L. 2003. Both products of the mouse Ink4a/Arf locus suppress melanoma formation in vivo. *Oncogene.* **22**:5055–5059.
34. Quelle, D.E., Zindy, F., Ashmun, R.A., and Sherr, C.J. 1995. Alternative reading frames of the INK4a tumor suppressor gene encode two unrelated proteins capable of inducing cell cycle arrest. *Cell.* **83**:993–1000.
35. Serrano, M., Hannon, G.J., and Beach, D. 1993. A new regulatory motif in cell-cycle control causing specific inhibition of cyclin D/CDK4. *Nature.* **366**:704–707.
36. Kamijo, T., et al. 1998. Functional and physical interactions of the ARF tumor suppressor with p53 and Mdm2. *Proc. Natl. Acad. Sci. U. S. A.* **95**:8292–8297.
37. Pomerantz, J., et al. 1998. The Ink4a tumor suppressor gene product, p19Arf, interacts with MDM2 and neutralizes MDM2's inhibition of p53. *Cell.* **92**:713–723.
38. Stott, F.J., et al. 1998. The alternative product from the human CDKN2A locus, p14(ARF), participates in a regulatory feedback loop with p53 and MDM2. *EMBO J.* **17**:5001–5014.
39. Zhang, Y., Xiong, Y., and Yarbrough, W.G. 1998. ARF promotes MDM2 degradation and stabilizes p53: ARF-INK4a locus deletion impairs both the Rb and p53 tumor suppression pathways. *Cell.* **92**:725–734.
40. Hussussian, C.J., et al. 1994. Germline p16 mutations in familial melanoma. *Nat. Genet.* **8**:15–21.
41. Kamb, A., et al. 1994. Analysis of the p16 gene (CDKN2) as a candidate for the chromosome 9p melanoma susceptibility locus. *Nat. Genet.* **8**:23–26.
42. Gruis, N.A., et al. 1995. Homozygotes for CDKN2 (p16) germline mutation in Dutch familial melanoma kindreds. *Nat. Genet.* **10**:351–353.
43. Bishop, D.T., et al. 2002. Geographical variation in the penetrance of CDKN2A mutations for melanoma. *J. Natl. Cancer Inst.* **94**:894–903.
44. Florell, S.R., et al. 2004. Longitudinal assessment of the nevus phenotype in a melanoma kindred. *J. Invest. Dermatol.* **123**:576–582.
45. Box, N.F., et al. 2001. MC1R genotype modifies risk of melanoma in families segregating CDKN2A mutations. *Am. J. Hum. Genet.* **69**:765–773.
46. Flores, J.F., et al. 1996. Loss of the p16INK4a and p15INK4b genes, as well as neighboring

9p21 markers, in sporadic melanoma. *Cancer Res.* **56**:5023–5032.

47. Kumar, R., Smeds, J., Lundh Rozell, B., and Hemminki, K. 1999. Loss of heterozygosity at chromosome 9p21 (INK4-p14ARF locus): homozygous deletions and mutations in the p16 and p14ARF genes in sporadic primary melanomas. *Melanoma Res.* **9**:138–147.

48. Fujimoto, A., Morita, R., Hatta, N., Takehara, K., and Takata, M. 1999. p16INK4a inactivation is not frequent in uncultured sporadic primary cutaneous melanoma. *Oncogene.* **18**:2527–2532.

49. Cachia, A.R., Indsto, J.O., McLaren, K.M., Mann, G.J., and Arends, M.J. 2000. CDKN2A mutation and deletion status in thin and thick primary melanoma. *Clin. Cancer Res.* **6**:3511–3515.

50. Brookes, S., et al. 2002. INK4a-deficient human diploid fibroblasts are resistant to RAS-induced senescence. *EMBO J.* **21**:2936–2945.

51. Randerson-Moor, J.A., et al. 2001. A germline deletion of p14(ARF) but not CDKN2A in a melanoma-neural system tumour syndrome family. *Hum. Mol. Genet.* **10**:55–62.

52. Rizos, H., et al. 2001. A melanoma-associated germline mutation in exon 1beta inactivates p14ARF. *Oncogene.* **20**:5543–5547.

53. Bahuau, M., et al. 1998. Germ-line deletion involving the INK4 locus in familial proneness to melanoma and nervous system tumors. *Cancer Res.* **58**:2298–2303.

54. Petronzelli, F., et al. 2001. CDKN2A germline splicing mutation affecting both p16(ink4) and p14(arf) RNA processing in a melanoma/neurofibroma kindred. *Genes Chromosomes Cancer.* **31**:398–401.

55. Wolfel, T., et al. 1995. A p16INK4a-insensitive CDK4 mutant targeted by cytolytic T lymphocytes in a human melanoma. *Science.* **269**:1281–1284.

56. Zuo, L., et al. 1996. Germline mutations in the p16INK4a binding domain of CDK4 in familial melanoma. *Nat. Genet.* **12**:97–99.

57. Soufir, N., et al. 1998. Prevalence of p16 and CDK4 germline mutations in 48 melanoma-prone families in France. The French Familial Melanoma Study Group. *Hum. Mol. Genet.* **7**:209–216.

58. Tsao, H., Benoit, E., Sober, A.J., Thiele, C., and Haluska, F.G. 1998. Novel mutations in the p16/CDKN2A binding region of the cyclin-dependent kinase-4 gene. *Cancer Res.* **58**:109–113.

59. Castresana, J.S., et al. 1993. Lack of allelic deletion and point mutation as mechanisms of p53 activation in human malignant melanoma. *Int. J. Cancer.* **55**:562–565.

60. Lubbe, J., Reichel, M., Burg, G., and Kleihues, P. 1994. Absence of p53 gene mutations in cutaneous melanoma. *J. Invest. Dermatol.* **102**:819–821.

61. Papp, T., Jafari, M., and Schiffmann, D. 1996. Lack of p53 mutations and loss of heterozygosity in non-cultured human melanocytic lesions. *J. Cancer Res. Clin. Oncol.* **122**:541–548.

62. Albino, A.P., et al. 1994. Mutation and expression of the p53 gene in human malignant melanoma. *Melanoma Res.* **4**:35–45.

63. Sparrow, L.E., Soong, R., Dawkins, H.J., Iacopetta, B.J., and Heenan, P.J. 1995. p53 gene mutation and expression in naevi and melanomas. *Melanoma Res.* **5**:93–100.

64. Zerp, S.F., van Elsas, A., Peltenburg, L.T., and Schrier, P.I. 1999. p53 mutations in human cutaneous melanoma correlate with sun exposure but are not always involved in melanomagenesis. *Br. J. Cancer.* **79**:921–926.

65. Daniotti, M., et al. 2004. BRAF alterations are associated with complex mutational profiles in malignant melanoma. *Oncogene.* **23**:5968–5977.

66. Davies, H., et al. 2002. Mutations of the BRAF gene in human cancer. *Nature.* **417**:949–954.

67. Pollock, P.M., et al. 2003. High frequency of BRAF mutations in nevi. *Nat. Genet.* **33**:19–20.

68. Demunter, A., Stas, M., Degreef, H., De Wolf-Peeters, C., and van den Oord, J.J. 2001. Analysis of N- and K-ras mutations in the distinctive tumor progression phases of melanoma. *J. Invest. Dermatol.* **117**:1483–1489.

69. Gorden, A., et al. 2003. Analysis of BRAF and N-RAS mutations in metastatic melanoma tissues. *Cancer Res.* **63**:3955–3957.

70. Satyamoorthy, K., et al. 2003. Constitutive mitogen-activated protein kinase activation in melanoma is mediated by both BRAF mutations and autocrine growth factor stimulation. *Cancer Res.* **63**:756–759.

71. Wu, H., Goel, V., and Haluska, F.G. 2003. PTEN signaling pathways in melanoma. *Oncogene.* **22**:3113–3122.

72. Isshiki, K., Elder, D.E., Guerry, D., and Linnenbach, A.J. 1993. Chromosome 10 allelic loss in malignant melanoma. *Genes Chromosomes Cancer.* **8**:178–184.

73. Herbst, R.A., Weiss, J., Ehnis, A., Cavenee, W.K., and Arden, K.C. 1994. Loss of heterozygosity for 10q22-10qter in malignant melanoma progression. *Cancer Res.* **54**:3111–3114.

74. Healy, E., Rehman, I., Angus, B., and Rees, J.L. 1995. Loss of heterozygosity in sporadic primary cutaneous melanoma. *Genes Chromosomes Cancer.* **12**:152–156.

75. Guldberg, P., et al. 1997. Disruption of the MMAC1/PTEN gene by deletion or mutation is a frequent event in malignant melanoma. *Cancer Res.* **57**:3660–3663.

76. Teng, D.H., et al. 1997. MMAC1/PTEN mutations in primary tumor specimens and tumor cell lines. *Cancer Res.* **57**:5221–5225.

77. Tsao, H., Zhang, X., Benoit, E., and Haluska, F.G. 1998. Identification of PTEN/MMAC1 alterations in uncultured melanomas and melanoma cell lines. *Oncogene.* **16**:3397–3402.

78. Dhawan, P., Singh, A.B., Ellis, D.L., and Richmond, A. 2002. Constitutive activation of Akt/protein kinase B in melanoma leads to up-regulation of nuclear factor-kappaB and tumor progression. *Cancer Res.* **62**:7335–7342.

79. Stahl, J.M., et al. 2004. Deregulated Akt3 activity promotes development of malignant melanoma. *Cancer Res.* **64**:7002–7010.

80. Brose, M.S., et al. 2002. BRAF and RAS mutations in human lung cancer and melanoma. *Cancer Res.* **62**:6997–7000.

81. Dong, J., et al. 2003. BRAF oncogenic mutations correlate with progression rather than initiation of human melanoma. *Cancer Res.* **63**:3883–3885.

82. Uribe, P., Wistuba, I.I., and Gonzalez, S. 2003. BRAF mutation: a frequent event in benign, atypical, and malignant melanocytic lesions of the skin. *Am. J. Dermatopathol.* **25**:365–370.

83. Yazdi, A.S., et al. 2003. Mutations of the BRAF gene in benign and malignant melanocytic lesions. *J. Invest. Dermatol.* **121**:1160–1162.

84. Rodolfo, M., Daniotti, M., and Vallacchi, V. 2004. Genetic progression of metastatic melanoma. *Cancer Lett.* **214**:133–147.

85. Cohen, Y., et al. 2003. Lack of BRAF mutation in primary uveal melanoma. *Invest. Ophthalmol. Vis. Sci.* **44**:2876–2878.

86. Edwards, R.H., et al. 2004. Absence of BRAF mutations in UV-protected mucosal melanomas. *J. Med. Genet.* **41**:270–272.

87. Kumar, R., Angelini, S., Snellman, E., and Hemminki, K. 2004. BRAF mutations are common somatic events in melanocytic nevi. *J. Invest. Dermatol.* **122**:342–348.

88. Houben, R., et al. 2004. Constitutive activation of the Ras-Raf signaling pathway in metastatic melanoma is associated with poor prognosis [abstract]. *J. Carcinog.* **3**:6.

89. Kumar, R., et al. 2003. BRAF mutations in metastatic melanoma: a possible association with clinical outcome. *Clin. Cancer Res.* **9**:3362–3368.

90. Tsao, H., Zhang, X., Fowlkes, K., and Haluska, F.G. 2000. Relative reciprocity of NRAS and PTEN/MMAC1 alterations in cutaneous melanoma cell lines. *Cancer Res.* **60**:1800–1804.

91. Tsao, H., Goel, V., Wu, H., Yang, G., and Haluska, F.G. 2004. Genetic interaction between NRAS and BRAF mutations and PTEN/MMAC1 inactivation in melanoma. *J. Invest. Dermatol.* **122**:337–341.

92. Serrano, M., et al. 1996. Role of the INK4a locus in tumor suppression and cell mortality. *Cell.* **85**:27–37.

93. Kamijo, T., et al. 1997. Tumor suppression at the mouse INK4a locus mediated by the alternative reading frame product p19ARF. *Cell.* **91**:649–659.

94. Kamijo, T., Bodner, S., van de Kamp, E., Randle, D.H., and Sherr, C.J. 1999. Tumor spectrum in ARF-deficient mice. *Cancer Res.* **59**:2217–2222.

95. Krimpenfort, P., Quon, K.C., Mooi, W.J., Loonstra, A., and Berns, A. 2001. Loss of p16Ink4a confers susceptibility to metastatic melanoma in mice. *Nature.* **413**:83–86.

96. Sharpless, N.E., et al. 2001. Loss of p16Ink4a with retention of p19Arf predisposes mice to tumorigenesis. *Nature.* **413**:86–91.

97. Chin, L., et al. 1997. Cooperative effects of INK4a and ras in melanoma susceptibility in vivo. *Genes Dev.* **11**:2822–2834.

98. Kannan, K., et al. 2003. Components of the Rb pathway are critical targets of UV mutagenesis in a murine melanoma model. *Proc. Natl. Acad. Sci. U. S. A.* **100**:1221–1225.

99. Chin, L., et al. 1999. Essential role for oncogenic Ras in tumour maintenance. *Nature.* **400**:468–472.

100. Di Cristofano, A., Pesce, B., Cordon-Cardo, C., and Pandolfi, P.P. 1998. Pten is essential for embryonic development and tumour suppression. *Nat. Genet.* **19**:348–355.

101. Podsypanina, K., et al. 1999. Mutation of Pten/Mmac1 in mice causes neoplasia in multiple organ systems. *Proc. Natl. Acad. Sci. U. S. A.* **96**:1563–1568.

102. You, M.J., et al. 2002. Genetic analysis of Pten and Ink4a/Arf interactions in the suppression of tumorigenesis in mice. *Proc. Natl. Acad. Sci. U. S. A.* **99**:1455–1460.

103. Robertson, G.P., et al. 1998. In vitro loss of heterozygosity targets the PTEN/MMAC1 gene in melanoma. *Proc. Natl. Acad. Sci. U. S. A.* **95**:9418–9423.

104. Stahl, J.M., et al. 2003. Loss of PTEN promotes tumor development in malignant melanoma. *Cancer Res.* **63**:2881–2890.

105. Wellbrock, C., et al. 2004. V599EB-RAF is an oncogene in melanocytes. *Cancer Res.* **64**:2338–2342.

106. Hingorani, S.R., Jacobetz, M.A., Robertson, G.P., Herlyn, M., and Tuveson, D.A. 2003. Suppression of BRAF(V599E) in human melanoma abrogates transformation. *Cancer Res.* **63**:5198–5202.

107. Karasarides, M., et al. 2004. B-RAF is a therapeutic target in melanoma. *Oncogene.* **23**:6292–6298.

108. Collisson, E.A., De, A., Suzuki, H., Gambhir, S.S., and Kolodney, M.S. 2003. Treatment of metastatic melanoma with an orally available inhibitor of the Ras-Raf-MAPK cascade. *Cancer Res.* **63**:5669–5673.

109. Tuveson, D.A., Weber, B.L., and Herlyn, M. 2003. BRAF as a potential therapeutic target in melanoma and other malignancies. *Cancer Cell.* **4**:95–98.

110. Balch, C.M., et al. 2000. Long-term results of a multi-institutional randomized trial comparing prognostic factors and surgical results for intermediate thickness melanomas (1.0 to 4.0 mm). Intergroup Melanoma Surgical Trial. *Ann. Surg. Oncol.* **7**:87–97.

111. Eggermont, A.M., and Kirkwood, J.M. 2004. Re-evaluating the role of dacarbazine in metastatic melanoma: what have we learned in 30 years? *Eur. J. Cancer.* **40**:1825–1836.

112. Douglas, J.G., and Margolin, K. 2002. The treatment of brain metastases from malignant melanoma. *Semin. Oncol.* **29**:518–524.

113. Kirkwood, J.M., et al. 2001. High-dose interferon alfa-2b significantly prolongs relapse-free and overall survival compared with the GM2-KLH/QS-21 vaccine in patients with resected stage IIB-III melanoma: results of intergroup trial E1694/S9512/C509801. *J. Clin. Oncol.* **19**:2370–2380.

114. Agarwala, S. 2003. Improving survival in patients with high-risk and metastatic melanoma: immunotherapy leads the way. *Am. J. Clin. Dermatol.* **4**:333–346.

115. Margolin, K.A. 2004. Biochemotherapy for melanoma: rational therapeutics in the search for weapons of melanoma destruction. *Cancer.* **101**:435–438.

116. Dumas, J., Smith, R.A., and Lowinger, T.B. 2004. Recent developments in the discovery of protein kinase inhibitors from the urea class. *Curr. Opin. Drug Discov. Devel.* **7**:600–616.

117. Lyons, J.F., Wilhelm, S., Hibner, B., and Bollag, G. 2001. Discovery of a novel Raf kinase inhibitor. *Endocr. Relat. Cancer.* **8**:219–225.

118. Dumas, J. 2002. Protein kinase inhibitors from the urea class. *Curr. Opin. Drug Discov. Devel.* **5**:718–727.

119. Wilhelm, S.M., et al. 2004. BAY 43-9006 exhibits broad spectrum oral antitumor activity and targets the RAF/MEK/ERK pathway and receptor tyrosine kinases involved in tumor progression and angiogenesis. *Cancer Res.* **64**:7099–7109.

120. Strumberg, D., et al. 2002. Results of phase I pharmacokinetic and pharmacodynamic studies of the Raf kinase inhibitor BAY 43-9006 in patients with solid tumors. *Int. J. Clin. Pharmacol. Ther.* **40**:580–581.

121. Richly, H., et al. 2003. A phase I clinical and pharmacokinetic study of the Raf kinase inhibitor (RKI) BAY 43-9006 administered in combination with doxorubicin in patients with solid tumors. *Int. J. Clin. Pharmacol. Ther.* **41**:620–621.

122. Ahmad, T., and Eisen, T. 2004. Kinase inhibition with BAY 43-9006 in renal cell carcinoma. *Clin. Cancer Res.* **10**:6388–6392.

123. Ratain, M.J., et al. 2004. Preliminary antitumor activity of BAY 43-9006 in metastatic renal cell carcinoma and other advanced refractory solid tumors in a phase II randomized discontinuation trial (RDT) [abstract]. *ASCO Annual Meeting Proceedings.* June 5–8, 2004. *J. Clin. Oncol.* **22**(Suppl.):4501.

124. Ahmad, T., et al. 2004. BAY 43-9006 in patients with advanced melanoma: The Royal Marsden Experience [abstract]. *2004 ASCO Annual Meeting Proceedings.* June 5–8, 2004. *J. Clin. Oncol.* **22**(Suppl.):7506.

125. Flaherty, K.T., et al. 2004. Phase I/II trial of BAY 43-9006, carboplatin (C) and paclitaxel (P) demonstrates preliminary antitumor activity in the expansion cohort of patients with metastatic melanoma [abstract]. *2004 ASCO Annual Meeting Proceedings.* June 5–8, 2004. *J. Clin. Oncol.* **22**(Suppl.):7507.

Immunopathogenesis and therapy of cutaneous T cell lymphoma

Ellen J. Kim, Stephen Hess, Stephen K. Richardson, Sara Newton,
Louise C. Showe, Bernice M. Benoit, Ravi Ubriani, Carmela C. Vittorio,
Jacqueline M. Junkins-Hopkins, Maria Wysocka, and Alain H. Rook

Department of Dermatology, University of Pennsylvania School of Medicine and the Wistar Institute, Philadelphia, Pennsylvania, USA.

Cutaneous T cell lymphomas (CTCLs) are a heterogenous group of lymphoproliferative disorders caused by clonally derived, skin-invasive T cells. Mycosis fungoides (MF) and Sézary syndrome (SS) are the most common types of CTCLs and are characterized by malignant CD4+/CLA+/CCR4+ T cells that also lack the usual T cell surface markers CD7 and/or CD26. As MF/SS advances, the clonal dominance of the malignant cells results in the expression of predominantly Th2 cytokines, progressive immune dysregulation in patients, and further tumor cell growth. This review summarizes recent insights into the pathogenesis and immunobiology of MF/SS and how these have shaped current therapeutic approaches, in particular the growing emphasis on enhancement of host antitumor immune responses as the key to successful therapy.

Historical perspective. Cutaneous T cell lymphomas (CTCLs) have a wide variety of clinical and histopathological manifestations but are all characterized as extranodal non-Hodgkin lymphomas of malignant, mature T lymphocytes that target and persist in the skin (1, 2). Recently the WHO-EORTC (WHO–European Organization for Research and Treatment of Cancer) jointly issued a new classification of the primary cutaneous lymphomas, and the variety of CTCLs are shown in *WHO-EORTC classification of cutaneous T cell and NK/T cell lymphomas* (3). Mycosis fungoides (MF) is the most common and best studied of the CTCLs. The term mycosis fungoides originally referred to the mushroom-like nodules seen in the tumor stage of disease (4). MF accounts for approximately 1% of all non-Hodgkin lymphomas, with a median age of presentation of 57 years and a male/female ratio of 2:1 (5). Its incidence is estimated at 0.36–0.90 per 100,000 person-years (6, 7). This rate appears to be stable, although cases of non-Hodgkin lymphoma overall appear to have doubled in the past 2 decades and likely MF/SS is underreported due to nomenclature variances (8).

As shown in Figure 1, patients with classical MF, as originally described by Alibert and Bazin 2 centuries ago, have pink or erythematous scaly patches and plaques that typically appear on sun-protected areas of the skin, such as the proximal extremities, trunk, and buttocks, and have variable degrees of scaling and pruritus (2). It has since been noted that MF actually has protean clinical and histopathological presentations and many variants (see *MF variants*).

MF can mimic benign skin conditions such as chronic eczema, allergic contact dermatitis, or psoriasis, and it is not unusual for the diagnosis to remain elusive for years. It is typically an indolent disorder, but the disease may progress toward or present de novo in more advanced forms, such as with tumors, erythroderma (> 80% body surface area involved with patches/plaques but without overt leukemic involvement), and blood or organ involvement, accompanied by an increase in morbidity and mortality (See Table 1 for MF lesion definitions). Sézary syndrome (SS) is the leukemic form of the disease in which erythroderma is accompanied by measurable blood involvement by malignant lymphocytes with hyperconvoluted, cerebriform nuclei known as Sézary cells.

The diagnosis of MF/SS is dependent on confirmatory tissue biopsy showing atypical, skin-homing (epidermotropic), malignant CD4+ lymphocytes, especially those clustering around Langerhans cells in the epidermis, forming the hallmark Pautrier microabcesses on routine histopathology (Figure 2). In MF plaques and tumors, the ratio of CD4+ to CD8+ T lymphocytes in the skin can also become elevated (>4:1). The malignant CD4+ T lymphocytes also frequently lack expression of certain T cell surface markers, such as CD7 and CD26 (9, 10). However, early disease in particular may not show these fully developed histopathological findings and, thus, accurate diagnosis frequently requires biopsies from multiple sites over time. Overt blood involvement has traditionally been measured by examination of a peripheral blood buffy coat smear for hyperconvoluted, cerebriform malignant lymphocytes. Known as the Sézary cell count, the quantification of malignant cells by this method is time consuming and has significant interobserver variability. Recently it has been supplanted by newer, more specific, and more sensitive flow cytometric and molecular techniques (see "The origins of the malignant T cell," below).

MF/SS has a tumor-node-metastasis-blood (TNMB) classification and staging system, originally developed during a 1978 National Cancer Institute Workshop (see *TNMB classification for MF and SS* and Table 2), with recently proposed modifications by the International Society for Cutaneous Lymphomas (ISCL) (11). Staging evaluation of MF/SS patients typically involves comprehensive skin and physical examination; complete blood count and quantification of atypical circulating T lymphocytes; comprehensive metabolic panel; lactate dehydrogenase level; and, in the case of significant clinical lymphadenopathy (> 1.5 cm) or

Nonstandard abbreviations used: ATLL, adult T cell leukemia/lymphoma; CCL17, CC-chemokine ligand 17; CCR4, CC-chemokine receptor 4; CLA, cutaneous lymphocyte antigen; CpG-ODN, CpG oligodeoxynucleotide; CTCL, cutaneous T cell lymphoma; CTLA-4, cytotoxic T lymphocyte antigen 4; CXCR3, CXC-chemokine receptor 3; ECP, extracorporeal photopheresis; HDAC, histone deacetylase; HTLV, human T cell lymphotrophic virus; ISCL, International Society for Cutaneous Lymphomas; MF, mycosis fungoides; PUVA, psoralen plus ultraviolet A phototherapy; RAR, retinoic acid receptor; SS, Sézary syndrome; TLR, Toll-like receptor; TNMB, tumor-node-metastasis-blood.

Conflict of interest: The authors have declared that no conflict of interest exists.

Citation for this article: *J. Clin. Invest.* **115**:798–812 (2005).
doi:10.1172/JCI200524826.

WHO-EORTC classification of cutaneous T cell and NK/T cell lymphomas

MF and variants/subtypes
SS
Primary cutaneous CD30+ lymphoproliferative disorders
Subcutaneous panniculitis-like T cell lymphoma
Extranodal NK/T cell lymphoma, nasal type
ATLL
Primary cutaneous peripheral T cell lymphoma, unspecified
     Aggressive epidermotropic CD8+ T cell lymphoma
     Cutaneous γδ T cell lymphoma
     Primary cutaneous CD4+ small/medium-sized pleomorphic T cell lymphoma
     Peripheral T cell lymphoma, other

Adapted with permission from *Blood* (3).

T3/T4 disease, chest, abdomen, and pelvic computed tomography (CT) scan with contrast and/or whole body positron emission tomography (PET) scanning.

Several studies have demonstated that early MF is typically indolent, and patients with limited patches or plaques who have received skin-directed treatment have median survival similar to that of matched control populations (Table 3). Survival decreases at more advanced stages, and 10% of all patients will progress to more advanced disease (12). SS patients traditionally have had a poor prognosis, with only approximately 30% of patients surviving beyond 5 years after diagnosis (13). Important clinical prognostic indicators in MF/SS are T classification, clinical stage, presence of extracutaneous disease, and patient age. An elevated Sézary count, elevated lactate dehydrogenase level, and peripheral eosinophilia also portend poorer prognosis. Histopathological features that are indicative of a poor prognosis include the presence of significant numbers of large, atypical CD30+ cells suggestive of large cell transformation of the tumor and certain variants (follicular MF, granulomatous) whereas the presence of CD8+ tumor-infiltrating lymphocytes indicates a more favorable prognosis.

In the past, there were no curative therapies for MF, and the goals of treatment were symptom relief and cosmetic improvement, as early aggressive systemic chemotherapy has not resulted in improved overall survival. The recent trend toward the use of systemic immune-modifying therapeutics appears to be beneficially altering the long-term prognosis of most patients (see "Current and emerg-

ing therapies for MF," below). With the exception of human T cell lymphotrophic virus (HTLV) — associated adult T cell leukemia/lymphoma (ATLL), the etiology of the vast majority of CTCLs remains unknown, although various viral and environmental agents have been studied (14).

Recent molecular biology and immunology studies have led to remarkable insights into the origin and function of the malignant T cell, the abnormalities in cytokine expression when tumor burden increases, the immune suppression that characterizes advanced MF/SS, and the realization that the host immune response is probably crucial for controlling disease progression. This has resulted in significant advances in diagnostic techniques (flow cytometric analysis, clonality studies) and disease staging as well as therapeutic approaches (biologic and targeted therapies), as will be discussed below. Although CTCLs are relatively uncommon lymphoproliferative disorders, these recent advances in the diagnosis and treatment of MF/SS vividly demonstrate the impact of the basic understanding of T cell function that has led us to strive for more rational, targeted, and, ultimately, effective therapies for these patients.

The origins of the malignant T cell: understanding the role of T cells in skin immune surveillance. What is the origin of the skin-homing behavior of the malignant T cell in MF/SS? To answer this question, one can look at seminal immunological work on the role of normal skin-homing lymphocytes in skin immune surveillance, recently reviewed in ref. 15. As mentioned above, MF/SS patient T cells are typically mature memory CD4+ helper T cells. In normal skin homeostasis, these T cells are recruited to the skin following injury. Environmental or infectious damage results in keratinocyte cytokine release and also triggers innate immune responses from skin-resident immune cells, such as DCs, mast cells, and macrophages. This response may be mediated by pathogen components (bacterial CpG-oligodeoxynucleotides [CpG-ODNs], cell wall components, and viral DNA and RNA) that are recognized by pattern-recognition receptors, most notably the Toll-like receptors (TLRs), on these cells.

The downstream results of the activation of these pathways via upregulation of NF-κB signaling are proinflammatory respons-

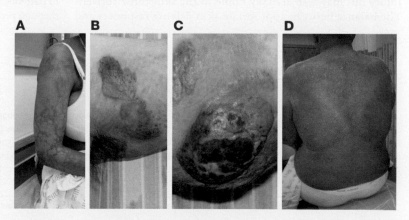

Figure 1
Cutaneous lesions of MF and SS. (**A**) Hypopigmented patches of MF on the proximal arm. Patches can be pink, red, hypopigmented, or hyperpigmented and are often scaly. (**B**) Two plaques near the axilla. MF lesions can have annular and gyrate configurations. (**C**) A tumor on the arm. Tumors frequently ulcerate. (**D**) Erythroderma in a patient with SS; more than 80% of his body surface area is affected with confluent erythematous scaly patches.

MF variants

Classic (Alibert-Bazin)
Erythrodermic/SS
Unilesional
Hypopigmented
Pagetoid reticulosis
Follicular (+/– follicular mucinosis)
Syringotropic
Granulomatous and granulomatous slack skin
Bullous/vesicular
Palmoplantar
Pigmented purpuric dermatosis–like
Interstitial MF
Icthyosiform
Hyperkeratotis/verrucous
Vegetating/papillomatous

es that have direct effects on pathogens and result in additional inflammatory cell recruitment and migration of activated APCs to local skin-draining lymph nodes. Once these APCs encounter the appropriate naive T cells in the lymph nodes, adaptive immune surveillance is called into action. The T cells become antigen-specific effector/memory cells and acquire the ability to home to the original site of inflammation, the skin. Antigen-specific central memory T cells are subsequently generated and circulate through lymph nodes to provide a long-term reservoir for specific immune surveillance.

These skin effector/memory T cells express cutaneous lymphocyte antigen (CLA) and normally make up 30% of all circulating memory T cells. CLA-expressing skin-homing T cells can interact with the receptor E-selectin on dermal postcapillary venules. These CLA+ cells also typically coexpress CC-chemokine receptor 4 (CCR4), which binds to the skin-manufactured chemokines CC-chemokine ligand 17 (CCL17) (also known as TARC) and CCL22 (also known as MDC), among others. A subset of skin-homing CLA+/CCR4+ lymphocytes also appear to coexpress CCR10 (16). All of these interactions are also crucial to the rolling/tethering interaction between the T cell and the endothelium that is necessary for the T cell to gain entry into the dermis and epidermis. It is believed that these memory skin-homing T cells can not only respond to skin inflammation and injury but may also actually home to the skin constitutively. The balance between appropriate skin defense responses and

inappropriate or dysregulated responses appears to be the key to understanding the pathogenesis and treatment of acquired inflammatory skin conditions, including MF/SS.

The malignant T cells in MF/SS patients have been shown to express the skin-homing receptors CLA and CCR4(Figure 3). Furthermore, MF lesions can express high levels of CCL17 and CCL22 (17, 18). Other chemokine receptors expressed by skin-infiltrating T cells found in MF lesions, such as CXC-chemokine receptor 3 (CXCR3) and CXCR4, and surface molecules, such as integrin $\alpha_E\beta_7$, have corresponding ligands expressed by epidermal keratinocytes and Langerhans cells. Their expression levels appear to correlate with earlier stage lesions and may reflect cells of the host immune response (19), as tumor stage lesions frequently lose expression of these markers. One can also observe less epidermotropism in the skin biopsies of advanced forms of MF, such as erythrodermic MF and SS. In a study comparing chemokine receptor expression in different stages of MF, tumor-stage tissue demonstrated decreased expression levels of CCR4 or CXCR3 but increased levels of CCR7, a receptor necessary for lymphatic entry (20).

Thus, the MF/SS malignant T cell clone probably arises out of the normal skin immune surveillance arsenal. Immunohistochemistry and ultrastructural studies have demonstrated that the hallmark Pautrier microabcess seen on histopathology consists of clusters of malignant T cells surrounding and in contact with Langerhans cells, the immature resident DCs in the skin. Because of these observations, the potential role of the epidermal Langerhans cells in the pathogenesis of MF/SS has been the subject of intense study (21). Epidermal Langerhans cells may be a source of persistent antigenic stimulation of T cells, resulting in activation and subsequent clonal expansion (22).

The state of activation of the malignant T cell in MF/SS is evidenced by the constitutive phosphorylation of Stat3 that is observed (23). In addition, these T cells express CD45RO and proliferating-cell nuclear antigen and may express CD25, the IL-2α receptor (1). Unlike malignant cells in ATLL, the majority of early MF lesions do not express CD25. However, in more advanced MF/SS, the soluble form of the IL-2 receptor (sIL-2R) can be detected in the blood and has been demonstrated as an important negative prognostic factor (24). Furthermore, Tregs also express CD25, and recent data is emerging to suggest that MF/SS may involve Tregs (see "Immune defects in MF/SS," below).

Although it is unknown what precisely triggers the constitutive activation of a population of skin-homing, mature CD4+ T cells in MF/SS, this may result in a skewed Th2-cytokine expression profile and subsequent inflammatory disease and immune dysregulation. In patients, this Th2 profile may contribute to disease progression through production of IL-4 and IL-10, which compromise the host

Table 1

Lesion terminology in MF[A]

Lesion	Definition
Patch	Any size skin lesion without significant elevation or induration; may show pigment changes, scaling, crusting, wrinkling
Plaque	Any size skin lesion that is elevated or indurated; may show pigment changes, scaling, crusting, or follicular prominence
Tumor	A solid/nodular lesion >1 cm or ulcerated plaque with evidence of depth and/or vertical growth
Erythroderma	Confluence of erythematous lesions covering ≥80% of the body surface area
Erythrodermic MF	Erythroderma without clinically significant blood involvement[B]
SS	Erythroderma with clinically significant blood involvement

[A]Unpublished 2005 provisional definitions, ISCL. [B]See text for the ISCL criteria of blood involvement in MF/SS.

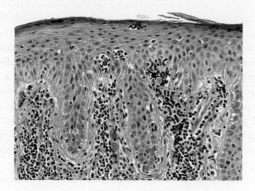

Figure 2
Histopathology of classic MF. This skin biopsy specimen demonstrates an atypical lymphocytic infilitrate going up into the epidermis (epidermotropism) in the absence of epidermal edema (spongiosis). The collection of atypical lymphocytes surrounding a Langerhans cell is a Pautrier microabcess, the hallmark of classic MF.

immune response (25) (see "Immune defects in MF/SS," below).

Malignant T cell clonality: a new diagnostic tool for MF/SS. As yet, MF/SS has not been clearly associated with a single or specific chromosomal mutation in either tumor suppressor genes or oncogenes. However, disease progression in MF/SS has been associated with several genetic abnormalities that enable a dominant clonal population to emerge. Chromosomal rearrangement hot spots have been detected in MF/SS and include deletions on 1p, 17p, 10q, and 19 and gains at 4q, 18, and 17q (26). Early patch and plaque MF lesions may demonstrate defects in Fas expression or constitutive STAT activation that result in impaired tumor cell apoptosis (27, 28). More advanced lesions, such as tumors, have been shown to have p53-expression defects (29, 30), microsatellite instability due to hypermethylation of mismatch repair genes, such as *hMLH1* (31), and p16/p15 alterations (32).

The clonal nature of MF/SS has been demonstrated by T cell receptor gene rearrangements (β or γ chain rearrangements) detectable by either Southern blotting or PCR analysis. In particular, PCR methods are extremely sensitive and can detect monoclonality even in very early disease (33). However, false-negative PCR testing can occur, especially when the lymphocyte infiltrate in the tissue is sparse. Furthermore, clonal TCR gene rearrangements can be demonstrated in a variety of benign inflammatory skin conditions (34) and occasionally in controls. In addition, the nonmalignant skin-infiltrating lymphocytes seen in MF lesions could also be an oligoclonal or monoclonal population and contribute to positive gene rearrangement results. Because of this, the presence of monoclonality by molecular testing should not be automatically equated with neoplasia.

Nonetheless, demonstration of clonality in skin lesions of MF/SS patients has become an established component of the diagnostic evaluation of patients and may be helpful, especially in early MF cases. In particular, for each patient, in the appropriate clinical and histopathological context, the ability to demonstrate the same clonal population in more than one skin lesion or in the skin, the lymph node, and the peripheral blood as well as longitudinally is viewed as significant. Microdissection of tumor cells can increase the sensitivity of detection (35, 36). In addition, PCR testing for clonality may be useful in detecting minimal residual disease after treatment (37).

In advancing disease, expansion of the malignant clone in the peripheral blood results in a concomitant decrease in the normal lymphocyte populations, in particular cytotoxic CD8+ T cells and CD56+ NK cells as well as both myeloid and plasmacytoid DCs. Quantification of the peripheral blood tumor burden was traditionally accomplished by manual examination of a peripheral blood buffy coat smear for the malignant Sézary cells. As mentioned earlier, this method is not highly sensitive or accurately reproducible among different laboratories (11). Activated normal lymphocytes can look phenotypically similar to the atypical Sézary cells. More specific and sensitive means of detecting the malignant T cells include using anti–TCR Vβ-specific monoclonal antibodies (38) or antibodies to other relevant T cell surface markers (e.g., CD3, CD4, CD8, CD7, and CD26) (9) and using fluorescence-activated cell-sorting analysis/flow cytometry. Flow cytometry has largely supplanted the Sézary count for the evaluation of MF/SS patient blood involvement in most medical centers.

The 1978 National Cancer Institute staging system for MF divides the B (Blood) classification into B0 (no circulating atypical cells) and B1 (circulating atypical cells) (see *TNMB classification for MF and SS*). To reflect the above advances in detecting peripheral blood involvement, updated criteria for significant blood involvement in erythrodermic patients was proposed by the ISCL in 2002 (11). The B classification was expanded to include B0 (no clinically significant blood involvement), B1 (clinically significant nonleukemic involvement, with Sézary cells < 1.0 K/μl), and B2 (leukemic involvement as indicated by [a] Sézary cells ≥ 1.0 K/μl; [b] CD4/CD8 ratio > 10 with CD4+CD7− population > 40% or CD4+CD26− population > 30%; [c] lymphocytosis with molecular genetic evidence of a T cell clone; or [d] evidence of a chromosomally abnormal T cell clone) (39).

More recent work has suggested that MF/SS patients not only have clonal expansion of the malignant population of T cells, but that their entire T cell repertoire may be profoundly

---

TNMB classification for MF and SS

T (skin)
T1 Limited patch/plaque (<10% body surface area)
T2 Generalized patch/plaque (≥ 10% body surface area)
T3 Tumors
T4 Generalized erythroderma
N (lymph node)
N0 Clinically uninvolved
N1 Clinically abnormal, histologically uninvolved
N2 Clinically uninvolved, histologically involved
N3 Clinically abnormal, histologically involved
M (viscera)
M0 No visceral involvement
M2 Visceral involvement
B (blood)
B0 No circulating atypical (Sézary) cells
B1 Circulating atypical (Sézary) cells

See ISCL 2002 consensus report on the B classification changing this to B0, B1, B2 as discussed in "The origins of the malignant T cell." Adapted with permission from *Hematology/Oncology Clinics of North America* (11).

Table 2

Clinical staging system for MF and SS, National Cancer Institute Workshop 1978

Clinical Stages

IA	T1	N0	M0
IB	T2	N0	M0
IIA	T1–2	N1	M0
IIB	T3	N0-1	M0
IIIA	T4	N0	M0
IIIB	T4	N1	M0
IVA	T1–4	N2-3	M0
IVB	T1–4	N0-3	M1

Adapted with permission from *Hematology/Oncology Clinics of North America* (11).

disrupted. β-variable–complementarity–determining region 3 spectratyping studies of the peripheral blood of MF/SS patients revealed decreased complexity of the T cell repertoire, reminiscent of spectratype patterns observed among patients with advanced HIV infection (40). This reduction of TCR complexity was observed among virtually all advanced-stage patients as well as among 50% of patients with even the earliest stages of MF. The true significance of this finding in early-stage patients remains unclear, but it certainly provides partial explanation of the immunosuppression and susceptibility to infectious agents seen in advanced MF and SS cases. Furthermore, the authors of this study suggest that a factor, either infectious or produced as a soluble agent by the malignant T cells, compromises the growth of normal T cell populations. Potential candidates that have been suggested include a member of the Th2 cytokine family, TGF-β, or a retrovirus (21, 40).

Immune defects in MF/SS: Th2 predominance and clues to understanding disease progression. The malignant CD4⁺ T cells observed in most cases of MF/SS appear to exhibit a Th2 phenotype (Figure 4). In vitro stimulation of peripheral blood cells derived from SS patients routinely results in increased levels of measurable IL-4 expression (25). Moreover, Vowels et al. demonstrated levels of IL-4 and IL-5 mRNA in clinically involved skin, even among patients with early patches or plaques, while uninvolved skin and the skin of normal volunteers did not have detectable levels of Th2 cytokine mRNA (41). Assadullah et al. were also able to demonstrate increasing levels of IL-10 mRNA in parallel with an increasing density of the malignant T cell infiltrate as lesions progressed from patch to plaque to tumor (42). In SS, Th2 cell–specific transcription factors, such as GATA-3 and Jun B are highly overexpressed, as detected by cDNA microarray analysis (43). Thus, despite evidence of a vigorous host response in skin lesions in early disease, characterized by the presence of IFN-γ–secreting CD8⁺/TiA-1⁺ T cells (44), the chronic production of Th2 cytokines, such as IL-4, IL-5, and IL-10 by the malignant T cell population likely represents one mechanism by which the tumor cells circumvent the antitumor immune response.

Even in advanced disease, the host response, although effete, may play a role, albeit small, in containment of disease progression. This has been evidenced by the rapid progression of SS associated with the use of immunosuppressive agents, such as cyclosporine (45). In addition to enhanced Th2 cytokine production during disease progression, patients with overt evidence of circulating malignant

T cells also manifest defects in Th1 cytokine production. Wysocka et al. observed a progressive decline in the production of IL-12 and IFN-α by peripheral blood cells that paralleled an increase in the peripheral blood burden of malignant T cells (46). The decline in production of these cytokines directly correlated with a decline in the numbers of peripheral blood myeloid and plasmacytoid DCs, respectively. Accompanying the decrease in DCs and a defect in IL-12 production is a presumed deficit in production of other products of myeloid DCs, including IL-15 and IL-18, all of which are important IFN-γ–stimulating agents and powerful boosters of Th1 responses. Indeed, peripheral blood cells of leukemic CTCL patients clearly manifest marked decreases in IFN-γ production in response to a multiplicity of stimuli (25).

Underlying the decline in Th1 cytokine production and numbers of circulating DCs is the enhanced Th2 cytokine production that can compromise both DC maturation and IL-12/IFN-γ production (47). Other mechanisms responsible for diminished IL-12 production have been highlighted by recent studies by French et al., which demonstrated a defect in expression of CD40 ligand on malignant T cells derived from patients with SS (48). CD40 ligand is not expressed on resting T cells but is normally upregulated on the cell surface upon engagement of the TCR. By contrast, malignant CD4⁺ T cells fail to express CD40 ligand upon engagement of the TCR by anti-CD3. Clearly, the absence of CD40 ligand interaction with CD40 on APCs during an immune response can lead to a profound reduction in DC activation and cytokine production. Through the in vitro addition of recombinant hexameric CD40 ligand, French and colleagues demonstrated reconstitution of IL-12 and TNF production by the cells of patients with SS (48). These findings provide obvious insights regarding potential strategies for circumventing the immune deficiency associated with advancing MF/SS.

Recently, Berger et al. demonstrated that under certain in vitro conditions, malignant CD4⁺ T cells derived from patients with CTCL can be induced to demonstrate a CD25⁺ Treg phenotype (49). Thus, when malignant T cells are cultured with DCs that have had the opportunity to process significant numbers of apoptotic T cells, they upregulate cytotoxic T lymphocyte antigen 4 (CTLA-4) expression, express Foxp3, and develop the enhanced capacity to produce IL-10 and TGF-β. Previously, only peripheral blood cells derived from ATLL patients who were infected with HTLV-1 were observed to have similar Treg characteristics (50). If indeed MF/SS is a tumor of Tregs, this may be an additional compelling explanation behind the immunosuppression seen in advanced disease.

The end results of alterations in cytokine expression pathways during progressive MF/SS are multiple abnormalities in cellular immunity (Figure 5). Our group routinely observes defects in cellu-

Table 3

Five-year survival of MF/SS patients as reported by a single center, retrospective cohort study

Stage	5-year survival (%)
IA	96
IB/IIA	73
IIB/III	44
IV	27

Adapted with permission from *Archives of Dermatology* (5).

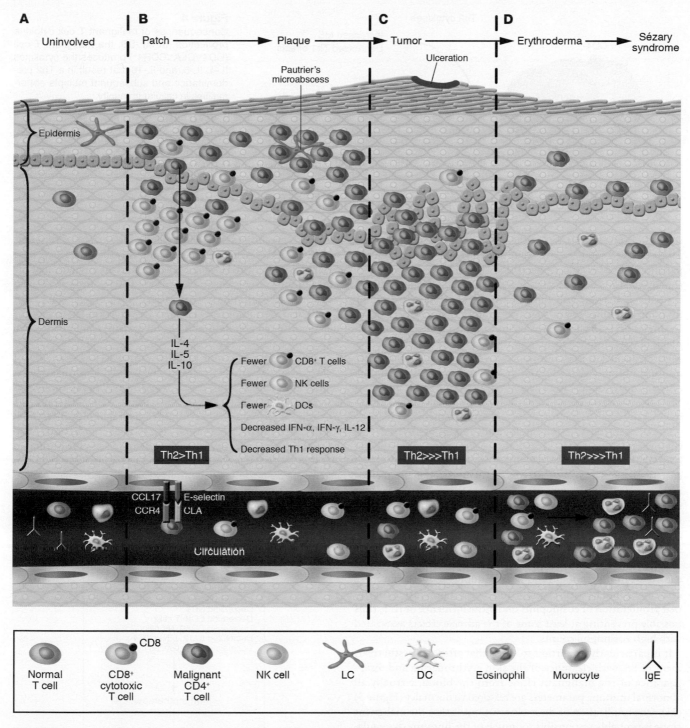

Figure 3

The skin microenvironment in MF progression. (**A**) Normal skin showing resident Langerhans cells in the epidermis and skin-homing T cells in the dermis and circulation. (**B**) Patch and plaque MF in which the CD4+ malignant T cells home to the epidermis and collect around Langerhans cells. Of note, in these stages, the epidermal and dermal infiltrate frequently have abundant CD8+ T cells as part of the host immune response. (**C**) Tumor MF in which the tumor occupies the dermis and subcutaneous tissue and is comprised of primarily malignant T cells and few CD8+ T cells. (**D**) Erythrodermic MF and SS with detectable circulating malignant T cells that elaborate Th2 cytokines that affect CD8+ T cell, NK cell, and DC numbers and function, and consequently, the host immune response.

lar cytotoxicity that correlate with the burden of circulating malignant T cells (51). Progression from early to more advanced MF/SS is typically associated with a marked decline in NK cell numbers

and activity. Similarly, a decline in the number of peripheral blood CD8+ T cells accompanies an increasing burden of circulating malignant T cells. Furthermore, the percentage of these cytotoxic

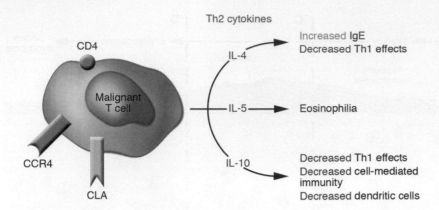

Figure 4
Consequences of malignant T cell cytokine production. In MF/SS, the malignant T cell (CD4+/CLA+/CCR4+) produces the cytokines IL-4, IL-5, and IL-10 that result in a Th2 predominance and subsequent multiple abnormalities in cellular immunity.

cells that express activation markers, such as CD69, is also significantly reduced compared to that of NK and CD8+ T cells from MF patients without overt peripheral blood involvement (52). A reduction in the number of functioning NK and CD8+ T cells is almost certainly associated with a deterioration of both host antitumor immunity and immune surveillance against microbial organisms. Examples of these phenomena include infections such as disseminated herpes simplex/zoster among SS patients who have never been iatrogenically immunosuppressed by chemotherapy or other immunosuppressive medications (53, 54). Secondary melanoma and nonmelanoma skin cancers may also be more common in MF/SS patients independent of the history of previous predisposing therapy (such as phototherapy or radiation therapy) (55, 56).

Other characteristic immunological findings associated with the progression of MF/SS include development of peripheral eosinophilia and elevated levels of serum IgE (25, 57). Peripheral eosinophilia has been determined to be an independent marker for poor prognosis and disease progression (58). In one study, peripheral blood cells from patients with SS and eosinophilia produced markedly higher levels of IL-5 upon stimulation than did the cells of patients or normal volunteers without eosinophilia (59). It is noteworthy that in this study, culture of the patients' cells with either recombinant IFN-α or IL-12 significantly inhibited the excess production of IL-5. These findings suggest that these cytokines could be useful therapeutic tools for preventing continued proliferation of eosinophils under the influence of IL-5, thus possibly preventing at least some of the adverse effects associated with high eosinophil counts.

It is particularly important to stress that after successful treatment of SS, with clearance of the skin erythroderma and disappearance of the malignant clone from the blood, virtually all abnormal immune parameters are restored to normalcy (Figure 5) (51). One implication of these observations is that the malignant clone is probably responsible for much of the immune dysregulation that occurs in SS. Moreover, these findings indicate that SS patients who experience remission will have their immune systems at least partially reconstituted. Thus, such patients should be less likely to experience severe consequences of microbial infection in comparison with those with advanced SS.

Current and emerging therapies for MF: a multimodality approach

Early stage MF. A variety of skin-directed and systemic therapies are available to treat MF/SS (Table 4), and the stage of disease guides the therapeutic choices (Table 5). In the past, traditional

systemic chemotherapy has not resulted in durable remissions in MF/SS. As a consequence, emerging therapeutic efforts have focused upon targeted biological agents and manipulation of the host immune response using a multimodality approach. Numerous arms of the immune system must cooperate to generate a sufficient host antitumor response such that the proliferation of the malignant T cell population in MF/SS patients can be controlled and, ideally, eradicated (Figure 6).

Patients with patches or plaques limited to less than 10% of their skin surface area (T1 disease) tend to exhibit normal cellular immune responses. Thus, use of skin-directed therapies, such as superpotent topical corticosteroids, topical chemotherapy (60, 61), topical retinoid application (62), psoralen plus ultraviolet A phototherapy (PUVA) (63), or electron beam radiation therapy (64), which target the vast majority of the tumor burden in the skin by directly inducing apoptosis of malignant T cells, is often sufficient to induce complete clearing of disease. Topical corticosteroids and PUVA also decrease the number of resident epidermal Langerhans cells and subsequently interrupt their

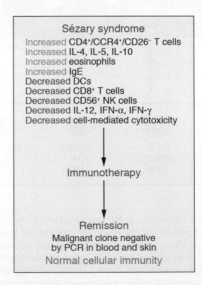

Figure 5
Elimination of the malignant T cell clone during immunotherapy leads to a restoration of a normal immune response. Studies of numerous patients with SS have demonstrated that induction of complete remission with clearing of the malignant T cell clone during multimodality immunotherapy leads to a restoration of normal host immune function.

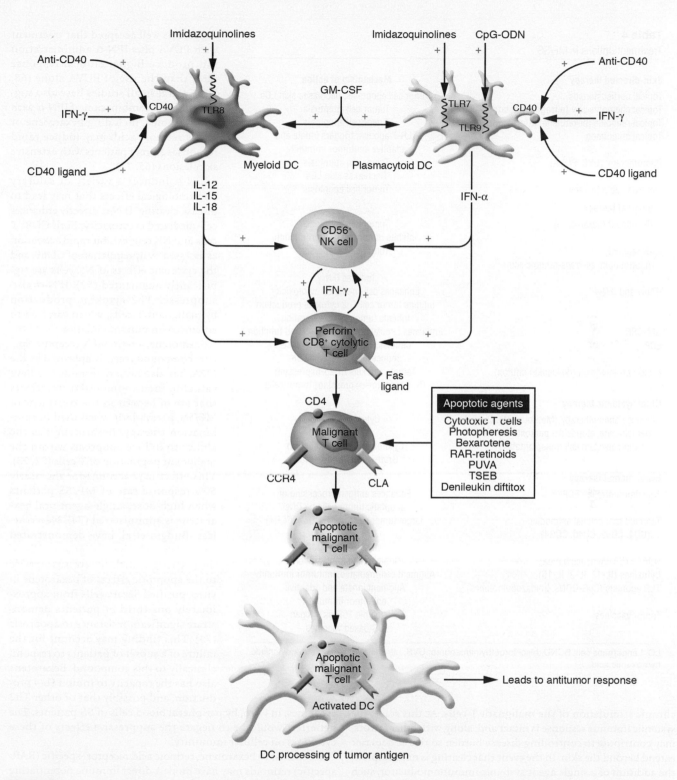

Figure 6

Multimodality strategy for enhancing the antitumor response using immunotherapeutics. A multimodality approach encompasses the activation of multiple arms of the immune response through the use of agents to activate DCs, CTLs (CD8+), and NK cells (CD56+). GM-CSF may enhance the numbers of DCs while agents that enhance CD40 expression (IFN-γ) and activation (CD40 ligand; activating anti-CD40 antibody) and TLR ligands (CpG-ODNs; imidazoquinolines) lead to DC cytokine production and to enhanced DC processing of apoptotic malignant T cells. Cytokines produced by DCs as well as exogenously administered cytokines augment CD8+ T cell and possibly NK cell cytolytic activity against the tumor cells. RAR-specific retinoids stimulate DC and CD8+ T cell cytokine production. Proapoptotic agents, including bexarotene, RAR-specific retinoids, PUVA, photopheresis, topical chemotherapy, total skin electron beam irradiation (TSEB), and denileukin diftitox, can assist in the development of an antitumor immune response by reducing the overall tumor burden and by providing a source of apoptotic malignant cells and tumor antigens for uptake by DCs.

Table 4
Treatment options in MF/SS

Skin-directed therapy	Mechanism of action
Topical corticosteroids	Tumor cell apoptosis; decreases skin LCs
Topical chemotherapy (nitrogen mustard, BCNU)	Tumor cell apoptosis
Topical retinoids (bexarotene, tazarotene)	Tumor cell apoptosis
Topical imiquimod	TLR-7 agonist; triggers innate and adaptive antitumor immunity
Phototherapy (UVB, PUVA)	Tumor cell apoptosis
	Decreases skin LCs
Electron beam therapy	Tumor cell apoptosis
Biological therapy	
RXR retinoid (bexarotene)	Tumor cell apoptosis
	Inhibits tumor cell IL-4 production
RAR retinoid	Tumor cell apoptosis
(isotretinoin, all-*trans*-retinoic acid)	
	Induces IFN-γ
IFN-α and IFN-γ	Enhances cell-mediated cytotoxicity
	Inhibits tumor cell Th2 cytokine production
	Inhibits tumor cell proliferation
GM-CSF	Enhances circulating DC numbers and function
ECP	Circulating tumor cell apoptosis
	Induces DC differentiation
Fusion protein/toxin (denileukin diftitox)	Targets and kills CD25-expressing (IL-2 receptor–expressing) tumor cells
Other systemic therapy	
Cytotoxic chemotherapy (Methotrexate, Doxil, gemcitabine, etoposide, pentostatin)	Cytotoxic agents
Bone marrow/stem cell transplantation	Cytotoxic agents (induction)
	Graft-versus-tumor effect
Experimental therapy	
Transimmunization ECP	Enhances antigen processing of apoptotic tumor cells by DCs
Targeted monoclonal antibodies (CD4, CD52, CD40, CCR4)	Target tumor cells (CD4, CD52, CCR4)
	Activate DCs (CD40)
HDACs (SAHA, depsipeptide)	Inhibit gene transcription
Cytokines (IL-12, IL-2, IL-15)	Augment cell-mediated antitumor immunity
TLR agonists (CpG-ODNs, imidazoquinolones)	Augment innate and adaptive antitumor immunity
Tumor vaccines	Clonotypic TCR as antigen
	DC-based vaccines

LC, Langerhans cell; BCNU, bischloroethylnitrosourea; UVB, ultraviolet B; SAHA, suberoylanilide hydroxamic acid.

chronic stimulation of the malignant T cells. At this stage, the systemic immune response is intact and, along with other factors, may contribute to controlling disease burden so that it does not extend beyond the skin. In the event that clearing is not complete, the addition of a single agent systemic immunomodulator, such as recombinant IFN-α or the retinoid bexarotene (Targretin), typically leads to a better clinical response.

For patients who do not yet manifest overt peripheral blood disease but who exhibit more extensively infiltrated cutaneous plaques on a greater skin surface area, combined therapeutic approaches appear to result in more rapid responses (65–68). IFN-α, produced by plasmacytoid DCs, is a product of the innate immune response, and appears to be one of the most highly active biologic agents used in the therapy of MF/SS (69). As suggested

above, it is well accepted that treatment with PUVA plus IFN-α administration can produce higher clinical response rates than the use of PUVA alone (68, 70). Several small studies have also suggested that a combination of IFN-α with oral retinoids, including bexarotene or 13-cis retinoic acid, may induce rapid responses among patients with extensive skin lesions (65, 71).

IFN-α induces a variety of salutary immunological effects that may lead to disease clearing. IFN-α directly enhances cell-mediated cytotoxicity; both CD8[+] T cells and NK cells exhibit rapid activation as assessed by upregulation of CD69, and the cytotoxic effects of NK cells are significantly augmented (52). IFN-α also suppresses Th2 cytokine production by malignant T cells, which can lead to enhanced immunomodulation.

Bexarotene, a retinoid X receptor–specific compound recently approved by the FDA, has also been determined to have valuable immunomodulatory effects that are of benefit in the treatment of MF/SS, particularly when used in combination therapy. Bexarotene has the ability to induce apoptosis within the malignant population of T cells (72, 73). This effect may account for the nearly 50% response rate of MF/SS patients when high-dose, single-agent oral bexarotene is administered (74). Nevertheless, Budgin et al. have demonstrated that, although the malignant cells of most patients with SS are susceptible to the apoptotic effects of bexarotene in vitro, purified Sézary cells from approximately one-third of patients demonstrate significant resistance to apoptosis (73). This finding may account for the failure of a subset of patients to respond clinically to this compound. Bexarotene also has the capacity to inhibit IL-4 production, and possibly that of other Th2 cytokines, in vitro, by peripheral blood cells of SS patients. The net effect would be to negate the suppressive effects of these cytokines on cellular immunity.

In contrast to bexarotene, retinoic acid receptor–specific (RAR-specific) retinoids may have modest direct immune potentiating properties. Using a number of different RAR-specific retinoids, including all-*trans*-retinoic acid and 13-*cis*-retinoic acid, Fox et al., demonstrated that these compounds exhibited the capacity to induce IL-12–dependent IFN-γ production (75). Moreover, synergistic production of IFN-γ occurred when low concentrations of IL-2 were added to the RAR-specific retinoids. In contrast, bexarotene does not induce IFN-γ production (73). These findings support the use of an RAR-specific retinoid as another component of the combined therapeutic approach.

Table 5
Treatment of MF/SS by clinical stage

Stage	Initial therapy	Subsequent therapy	Therapy for refractory disease
IA	Skin-directed therapies		

IB/IIA	Topical chemotherapy
	Phototherapy
	Electron beam therapy
	(+/– low-dose biological agents)
		IFNs
		Retinoids
		Multimodality therapy
		    Topical chemotherapy + biological agent
		    Phototherapy + biological agent
		    2 biological agents
		Denileukin diftitox
			Experimental therapies

IIB	Few tumors
	    Localized EBT
	    Intralesional IFN
	    Topical chemotherapy + biological agent
	Generalized tumors
	    Total skin EBT
	    Denileukin diftitox
	    Multimodality therapy
		Multimodality therapy
		Denileukin diftitox
		Single-agent chemotherapy
			Experimental therapies

IIIA, B	PUVA Retinoids IFNs Methotrexate ECP Multimodality therapy	Multimodality therapy Denileukin diftitox Single-agent chemotherapy	Experimental therapies
IVA, B	Single-agent systemic therapy Multimodality therapy (+ skin-directed therapy)	Adjuvant palliative local radiation for extracutaneous disease Bone marrow/stem cell transplant Chemotherapy	Experimental therapies

Advanced MF and SS

Current therapies. In marked contrast to patients with early MF, patients with advanced MF/SS manifest abnormalities of virtually every arm of the immune response that participates in antitumor immunity (CD8+ T cells, NK cells, and DCs). Thus, more aggressive therapy is required at these later stages of MF/SS to adequately reinvigorate the host response. Accordingly, evidence is now emerging indicating that multimodality immunotherapy using biologic agents can frequently induce complete clinical responses in advanced disease that are both durable and sufficient to eradicate the malignant clone (51, 76, 77).

As shown in Figure 6, central to the strategy for elimination of the malignant T cell population is the use of agents that can induce apoptosis of these cells while simultaneously enhancing the host's ability to process the apoptotic cells so that a robust cytotoxic T cell response can be generated. For patients with circulating malignant T cells, extracorporeal photopheresis (ECP) can result in massive apoptosis of cells within the peripheral blood (78, 79). ECP is a leukapheresis procedure approved by the FDA for the treatment of SS, in which approximately 10^{10} peripheral blood mononuclear cells are collected from the patient, treated with 8-methoxypsoralen, and exposed to 1–2 Joules of ultraviolet A light in the photopheresis machine, then reinfused back to the patient. In addition to inducing malignant T cell apoptosis, ECP also induces monocytes to differentiate into DCs capable of phagocytosing and processing the apoptotic tumor cell antigens. Repeated cycles of ECP for 2 consecutive days every 3 to 4 weeks with readministration of the treated cells is occasionally sufficient to induce a complete clinical response in SS. One potential modification to ECP, called transimmunization, is currently being studied; during ECP, the apoptotic malignant T cells and the newly formed DCs are coincubated prior to reinfusion to optimize the above antigen processing and stimulate a more efficient induction of tumor-targeted immunity (80).

Nevertheless, administration of large numbers of apoptotic cells as generated by ECP can compromise certain APC functions, including cytokine production, and therefore, has the potential to exacerbate the preexisting immune depressed state (81). Such observations support the rationale for the adjunctive use of multiple agents that can enhance both the afferent immune response and events related to processing of apoptotic malignant cells, as well as therapeutics that will boost the efferent response or the direct cytolytic attack on the tumor cells.

In support of this approach, Richardson and colleagues have recently demonstrated high response rates in SS patients when ECP was combined with the administration of multiple immune adjuvants (76). As part of this regimen, IFN-α and bexarotene were routinely used in combination with ECP. In some cases, GM-CSF was administered following each ECP treatment to enhance APC function. The monthly administration of 125 µg of GM-CSF on 2 consecutive days resulted in significant increases in circulating DC

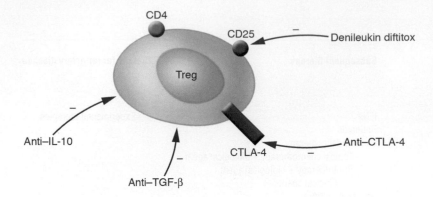

Figure 7
Inhibiting the effects of regulatory T cells. By inhibiting soluble factors produced by regulatory T cells, preventing CTLA-4 engagement of CD80 or CD86, and inducing apoptosis of regulatory T cells by agents such as denileukin diftitox, the overall effects of regulatory T cells on the immune response can be diminished.

numbers compared with DC numbers in ECP-treated SS patients who did not receive GM-CSF (46). In one patient, administration of GM-CSF 3 times per week for 6 months resulted in the persistent normalization of DC numbers, indicating that APC functions might be markedly augmented in this patient population with the long-term use of GM-CSF.

In addition to IFN-α, other approaches to enhancing the effector phase of antitumor immunity are presently being utilized (Figure 6). Although limited studies have been reported following the administration of recombinant IFN-γ, emerging evidence suggests that it has significant potential for the treatment of MF/SS, and it is clearly better tolerated than IFN-α, particularly by the elderly, who experience frequent cognitive dysfunction and fatigue with IFN-α treatment (82). In addition to enhancing the cytolytic lymphocyte function of patients with MF/SS, IFN-γ suppresses excess Th2 production, enhances CD40 expression, and primes abnormal DCs for IL-12 production, particularly in response to CD40 ligation (46). In the Cutaneous Lymphoma Program at the University of Pennsylvania, when possible, IFN-γ is routinely used for SS patients who appear to be refractory to IFN-α. In several cases, addition of IFN-γ to a multimodality regimen that included photopheresis and bexarotene appeared to be associated with the induction of a sustained complete clinical response (82).

Alternative mechanisms exist for the administration of IFN-γ. In a small pilot study, Dummer and colleagues demonstrated clinical efficacy of IFN-γ cDNA administered subcutaneously in an adenoviral vector (83). The local intralesional injection resulted in significant response rates of individual lesions in both MF and SS patients. Moreover, elevated serum levels of IFN-γ were observed and appeared to be associated with regression of lesions distant from the injection sites. This study further suggests that elevation of IFN-γ levels can beneficially alter disease progression.

Current therapy but a new potential mechanism: elimination of Treg activity. Although evidence for enhanced CD4⁺CD25⁺ Treg function in CTCLs is just beginning to emerge (49) and strategies for the elimination of suppressor activity in these diseases remains untested, several potential approaches are possible (Figure 7). Currently, denileukin diftitox (Ontak), a diptheria toxin–IL-2 protein conjugate, is available for targeting IL-2 receptor–bearing T cells (84). After binding to the IL-2 receptor, it undergoes endocytosis followed by release of the diptheria toxin, which results in arrest of protein synthesis and, ultimately, apoptosis of T cells. Intravenous administration of denileukin diftitox to patients with MF results in the regression of plaques and tumors (85). Its major mechanism of action is thought to be mediated by direct killing of malignant T cells. However, it is entirely possible that at

least a portion of its activity is mediated through the elimination of CD25-bearing Tregs.

Because Tregs express CTLA-4, another potential approach to their inhibition is through the use of anti–CTLA-4 antibody (86). This therapeutic approach for CTCL remains untested so far. Tregs may also suppress immunoreactivity by production of IL-10 or TGF-β. Antibodies with neutralizing activity for these factors could be utilized to reverse their inhibitory effects on the immune response.

Emerging immunomodulatory therapies. Numerous alternative strategies exist for the enhancement of APC function (Figure 6). These include the administration of a variety of TLR ligands, which are presently in clinical development (87, 88). Imiquimod, a member of the imidazoquinoline family, which has been approved by the FDA for the treatment of basal cell carcinoma, actinic keratoses, and condyloma, has recently been demonstrated to exhibit substantial clinical activity when applied topically to skin lesions of MF patients (89, 90). Imiquimod potently triggers TLR-7, which results in IFN-α and TNF-α production (91, 92). Imiquimod may also directly induce cells of some tumor types to undergo apoptosis (93).

Several newer members of the imidazoquinoline family have the capacity to trigger TLR-8 in addition to TLR-7 (94). This would be expected to result in a broader activation of both myeloid and plasmacytoid DCs with the release of a more extensive array of immune-activating cytokines. Thus, such compounds would likely broadly activate multiple compartments of the immune response and make newer members of the imidazoquinoline family highly desirable compounds to use as systemic therapeutic agents either alone or as part of a multimodality approach.

Clinical investigation has turned to the effects of an alternative class of DC-activating agents, CpG-ODNs. CpG-ODNs have been recognized as immune stimulatory agents by virtue of activation of DCs following binding to TLR-9 (95). The immunostimulatory potential of CpG-ODNs has been tested in murine tumor models and has been observed to lead to the generation of strong antitumor T cell responses resulting in complete remission of certain established solid tumors (96). Thus, there is substantial rationale for studying the activity of CpG-ODNs in human tumor systems. In this regard, in vitro data indicate that CpG-ODNs can potently activate CTCL patient DCs, leading to IFN-α production, increased expression of critical immune accessory molecules, and enhanced cell-mediated cytotoxicity (52). Moreover, Kim and colleagues have shown in a phase II clinical trial for refractory advanced CTCL that CpG-ODNs administered subcutaneously as a single agent demonstrate therapeutic efficacy, in some cases inducing complete clinical responses (97). Therefore, application of CpG-ODNs in a multimodality therapeutic approach that uses photopheresis might also yield significant therapeutic benefit. It is noteworthy that antiviral vaccination strategies that incorporate the use of CpG-ODNs along with viral antigen appear to be markedly superior to those that use antigen alone (98). Since ECP represents an immunization procedure using apoptotic tumor cells, such findings support the use of CpG-ODNs at the time of reinfusion of the treated tumor cells in an effort to directly target the tumor antigens to DCs for processing.

Other mechanisms for activating the DCs of patients with MF/SS that are under preclinical investigation include strategies for the engagement of CD40. As stressed above, there is substantial evidence that a defect in CD40-ligand expression by malignant T cells plays some role in the depressed production of DC-dependent cytokines (48). Furthermore, coculture of peripheral blood cells of SS patients with hexameric recombinant CD40 ligand resulted in substantial production of IL-12. Clinical trials using this approach have yet to be undertaken. Another strategy that also awaits clinical testing involves the use of an activating anti-CD40 antibody. Animal model studies that have used this approach indicate that enhanced generation of tumor-specific T cells can occur (99, 100).

IL-12 is a cytokine known to enhance cytolytic T cell and NK cell activities and to induce IFN-γ release and thus has the potential to enhance the antitumor immune response of MF/SS patients. In a small phase I study followed by a limited phase II study, the subcutaneous administration of recombinant IL-12 to a total of 32 patients with MF resulted in a response rate of approximately 50% (101, 102). Since malignant T cells lack the IL-12 β2 receptor (103) and are thus incapable of signaling in response to IL-12, it is presumed that the clinical response was not due to the direct effects of the cytokine on the malignant cells. Indeed, serial biopsies of cutaneous plaques during treatment revealed dense infiltrates of CD8⁺ T cells that appeared near the time of initial signs of lesional regression (101). Thus, it is believed that CD8⁺ T cells with augmented cytolytic activity are the predominant "workhorses" activated in response to IL-12. Whether IL-12 administration has advantages over IFN-γ use is presently unknown, but it is hoped that in the future, IL-12 will also find its place in a multimodality therapeutic approach.

Future therapeutic directions for enhancing the host immune response

Tumor vaccines. A number of different vaccination strategies are being applied to patients with MF/SS, most of which make use of the clonotypic TCR as a source of tumor-specific antigen. Immunogenic epitopes have been identified within both the variable (V) and the constant (C) regions of the clonotypic TCR-α and TCR-β receptors (104, 105). In some experiments, immunogenic peptides have been directly isolated from the MHC class I molecules on the surface of the malignant clone (104). This confirms that the antigen-processing pathway for endogenous proteins remains intact in the malignant clone and that the clonotypic TCR is subjected to antigen processing and presentation by MHC class I. Consequently, the TCR peptide–MHC complex on the surface of a malignant cell can serve as a target for recognition by CD8⁺ CTLs. This has been demonstrated in both healthy donors and patients with MF/SS, in whom immunogenic peptides derived from the clonotypic TCR-induced tumor-specific CD8⁺ T cells, which were capable of secreting TNF-α (104) as well as lysing autologous tumor cells in vitro (105).

DC-based vaccines are also being investigated for treatment of CTCL (106). Sources of antigen used to pulse DCs prior to vacci-

nation may include tumor-cell lysates, peptides, mimotopes, tumor-derived DNA or RNA, and even tumor cell–DC fusions. Maier et al. recently reported on 10 CTCL patients treated with weekly intranodal injections of autologous DCs pulsed with tumor lysate (107). In this study, 50% of patients had a clinical response to the vaccine, accompanied by an infiltration of CD8⁺ cytotoxic cells at the site of regressing lesions as well as molecular remission in some cases. It should be mentioned that clinical responses in this study were associated with a low tumor burden, which underscores the importance of instituting immunotherapy early in the course of the disease (i.e., prior to the development of significant immune dysregulation) (107).

As with any vaccination scheme, the addition of an immune adjuvant(s) may be used in an attempt to enhance the efficacy of the vaccine. Cytokines such as IL-12 (101, 102), IL-15 (52, 108), IL-18 (109), and IL-21 (110) may serve to augment the development, the effectiveness, and/or the maintenance of antitumor CTL responses. Moreover, these same cytokines have also been shown to enhance NK cell activity, which may play an important role in controlling tumor growth in vivo. GM-CSF is another cytokine that has been used as a cancer vaccine adjuvant to enhance both the number and function of DCs (111, 112). As discussed earlier, other immune-activating agents, including TLR agonists (e.g., imidizoquinolines, CPG-ODNs), anti-CD40, and anti–CTLA-4, could be used in conjunction with a tumor vaccine for patients with CTCL.

Other strategies. Monoclonal antibodies and fusion toxins directed against a variety of cell-surface markers have been tested in patients with SS/MF. In addition to denileukin diftitox, described earlier, examples of such antibodies include anti-CD4 and anti-CD52 (known as CamPath) (113, 114). These 2 agents have direct potent antitumor activity yet probably induce immunosuppression as well.

CCR4 may represent another possible target for antibody-based therapy for CTCL by virtue of the fact that its expression is relatively specific for skin-homing T cells, including the malignant cells in both CTCL and ATLL (18, 115). Preclinical studies have shown that a monoclonal antibody to CCR4 induces potent antibody-mediated cellular cytotoxicity against malignant T cell clones (116, 117). In addition to lysing CCR4⁺ T cells in vitro, this antibody reduced the expression of Foxp3 mRNA, suggesting a possible role in depleting Tregs (116).

Histone deacetylase (HDAC) inhibitors represent a novel class of compounds that modulate gene expression by shifting the balance toward acetylation of nucleosomal histones (118). Compounds such as suberoylanilide hydroxamic acid (118, 119) and depsipeptide (120, 121) have been shown to induce differentiation and/or apoptosis of malignant lymphocytes. In addition, recent data indicate that HDAC inhibitors may upregulate the expression of the IL-2 receptor on malignant T cells, resulting in enhanced susceptibility to killing by denileukin diftitox (121, 122).

Finally, bone marrow transplant offers hope to patients with advanced MF/SS that proves refractory to other therapeutic approaches. Allogeneic stem cells from an HLA-identical sibling (123, 124) or a matched unrelated donor (125) have induced complete and durable remissions in some patients, characterized by the disappearance of the malignant clone from the peripheral blood (as determined by gene rearrangement studies, flow cytometry, and the lack of circulating Sézary cells). Even in those patients who do not achieve complete remission, a reduction in disease severity may be observed (123), which is likely attributable to a graft-versus–MF antitumor effect. The obvious advantages to the use of allogeneic bone marrow transplant include its ability to aggressively deplete the malignant cells from the patient, to reconstitute the immune system, and to induce a graft-versus-tumor response.

Conclusions. A greater understanding of T cell function and the immunobiology of MF/SS has led to expansion of our therapeutic armamentarium against these diseases. Recent gene expression studies suggest that in the future we may be able to identify subsets of patients who would benefit most from initiation of immunomodulatory therapy earlier in the course of the disease (43). Future challenges include the development of well-designed clinical trials to elucidate the optimal combination and timing of these therapies in our patients. Looking back at the past 30 years, it is remarkable to survey the progress in the approach to treatment of MF/SS, from the past paradigm of cytotoxic chemotherapy to the current focus on targeted therapy that activates host immunity.

Acknowledgments

The authors thank John R. Stanley for reviewing the manuscript and Sam Dulay for his valuable assistance with the figures.

Address correspondence to: Ellen J. Kim, 2 Maloney Building, 2M31, Department of Dermatology, University of Pennsylvania Health System, 3600 Spruce Street, Philadelphia, Pennsylvania 19104, USA. Phone: (215) 349-5060; Fax: (215) 615-0047; E-mail: Ellen.kim@uphs.upenn.edu.

1. Girardi, M., Heald, P.W., and Wilson, L.D. 2004. The pathogenesis of mycosis fungoides. *N. Engl. J. Med.* **350**:1978–1988.

2. Kazakov, D.V., Burg, G., and Kempf, W. 2004. Clinicopathological spectrum of mycosis fungoides. *J. Eur. Acad. Dermatol. Venereol.* **18**:397–415.

3. Willemze, R., et al. 2005. WHO-EORTC classification for cutaneous lymphomas. *Blood.* doi:10.1182/blood-2004-09-3502.

4. Leboit, P.E., and McCalmont, T.H. 1997. Cutaneous lymphomas and leukemias. In *Lever's histopathology of the skin.* D. Elder, C. Jaworsky, and B. Johnson, editors. Lippincott-Raven. Philadelphia, Pennsylvania, USA. 820 pp.

5. Kim, Y.H., et al. 2003. Long-term outcome of 525 patients with mycosis fungoides and Sezary syndrome: clinical prognostic factors and risk for disease progression. *Arch. Dermatol.* **139**:857–866.

6. Weinstock, M.A., and Gardstein, B. 1999. Twenty-year trends in the reported incidence of mycosis fungoides and associated mortality. *Am. J. Public Health.* **89**:1240–1244.

7. Chuang, T.Y., Su, W.P., and Muller, S.A. 1990. Incidence of cutaneous T cell lymphoma and other rare skin cancers in a defined population. *J. Am. Acad. Dermatol.* **23**:254–256.

8. Fisher, S.G., and Fisher, R.I. 2004. The epidemiology of non-Hodgkin's lymphoma. *Oncogene.* **23**:6524–6534.

9. Bernengo, M.G., et al. 2001. The relevance of the CD4+ CD26- subset in the identification of circulating Sezary cells. *Br. J. Dermatol.* **144**:125–135.

10. Wood, G.S., et al. 1990. Leu-8/CD7 antigen expression by CD3+ T cells: comparative analysis of skin and blood in mycosis fungoides/Sezary syndrome relative to normal blood values. *J. Am. Acad. Dermatol.* **22**:602–607.

11. Vonderheid, E.C., and Bernengo, M.G. 2003. The Sezary syndrome: hematologic criteria. *Hematol. Oncol. Clin. North Am.* **17**:1367–1389.

12. Kim, Y.H., et al. 1996. Clinical stage IA (limited patch and plaque) mycosis fungoides. A long-term outcome analysis. *Arch. Dermatol.* **132**:1309–1313.

13. Diamandidou, E., Cohen, P.R., and Kurzrock, R. 1996. Mycosis fungoides and Sezary syndrome. *Blood.* **88**:2385–2409.

14. Lessin, S.R., Vowels, B.R., and Rook, A.H. 1994. Retroviruses and cutaneous T-cell lymphoma. *Dermatol. Clin.* **12**:243–253.

15. Kupper, T.S., and Fuhlbrigge, R.C. 2004. Immune surveillance in the skin: mechanisms and clinical consequences. *Nat. Rev. Immunol.* **4**:211–222.

16. Hudak, S., et al. 2002. Immune surveillance and effector functions of CCR10(+) skin homing T cells. *J. Immunol.* **169**:1189–1196.

17. Kakinuma, T., et al. 2003. Thymus and activation-regulated chemokine (TARC/CCL17) in mycosis fungoides: serum TARC levels reflect the disease activity of mycosis fungoides. *J. Am. Acad. Dermatol.* **48**:23–30.

18. Ferenczi, K., et al. 2002. Increased CCR4 expression in cutaneous T cell lymphoma. *J. Invest. Dermatol.* **119**:1405–1410.

19. Lu, D., et al. 2001. The T-cell chemokine receptor CXCR3 is expressed highly in low-grade mycosis fungoides. *Am. J. Clin. Pathol.* **115**:413–421.

20. Kallinich, T., et al. 2003. Chemokine receptor expression on neoplastic and reactive T cells in the skin at different stages of mycosis fungoides. *J. Invest. Dermatol.* **121**:1045–1052.

21. Edelson, R.L. 2001. Cutaneous T cell lymphoma: the helping hand of dendritic cells. *Ann. N. Y. Acad. Sci.* **941**:1–11.

22. Berger, C.L., et al. 2002. The growth of cutaneous T-cell lymphoma is stimulated by immature dendritic cells. *Blood.* **99**:2929–2939.

23. Zhang, Q., et al. 1996. Activation of Jak/STAT proteins involved in signal transduction pathway mediated by receptor for interleukin 2 in malignant T lymphocytes derived from cutaneous anaplastic large T-cell lymphoma and Sezary syndrome. *Proc. Natl. Acad. Sci. U. S. A.* **93**:9148–9153.

24. Wasik, M.A., et al. 1996. Increased serum concentration of the soluble interleukin-2 receptor in cutaneous T-cell lymphoma. Clinical and prognostic implications. *Arch. Dermatol.* **132**:42–47.

25. Vowels, B.R., et al. 1992. Aberrant cytokine production by Sezary syndrome patients: cytokine secretion pattern resembles murine Th2 cells. *J. Invest. Dermatol.* **99**:90–94.

26. Mao, X., et al. 2002. Molecular cytogenetic analysis of cutaneous T-cell lymphomas: identification of common genetic alterations in Sezary syndrome and mycosis fungoides. *Br. J. Dermatol.* **147**:464–475.

27. Sommer, V.H., et al. 2004. In vivo activation of STAT3 in cutaneous T-cell lymphoma. Evidence for an antiapoptotic function of STAT3. *Leukemia.* **18**:1288–1295.

28. Dereure, O., et al. 2002. Infrequent Fas mutations but no Bax or p53 mutations in early mycosis fungoides: a possible mechanism for the accumulation of malignant T lymphocytes in the skin. *J. Invest. Dermatol.* **118**:949–956.

29. Whittaker, S. 2001. Molecular genetics of cutaneous lymphomas. *Ann. N. Y. Acad. Sci.* **941**:39–45.

30. McGregor, J.M., et al. 1999. Spectrum of p53 gene mutations suggests a possible role for ultraviolet radiation in the pathogenesis of advanced cutaneous lymphomas. *J. Invest. Dermatol.* **112**:317–321.

31. Scarisbrick, J.J., et al. 2003. Microsatellite instability is associated with hypermethylation of the hMLH1 gene and reduced gene expression in mycosis fungoides. *J. Invest. Dermatol.* **121**:894–901.

32. Navas, I.C., et al. 2002. p16(INK4a) is selectively silenced in the tumoral progression of mycosis fungoides. *Lab. Invest.* **82**:123–132.

33. Wood, G.S., et al. 1994. Detection of clonal T-cell receptor gamma gene rearrangements in early mycosis fungoides/Sezary syndrome by poly-

merase chain reaction and denaturing gradient gel electrophoresis (PCR/DGGE). *J. Invest. Dermatol.* **103**:34–41.

34. Muche, J.M., et al. 2003. Peripheral blood T-cell clonality in mycosis fungoides and nonlymphoma controls. *Diagn. Mol. Pathol.* **12**:142–150.

35. Dereure, O., et al. 2003. Improved sensitivity of T-cell clonality detection in mycosis fungoides by hand microdissection and heteroduplex analysis. *Arch. Dermatol.* **139**:1571–1575.

36. Yazdi, A.S., et al. 2003. Improved detection of clonality in cutaneous T-cell lymphomas using laser capture microdissection. *J. Cutan. Pathol.* **30**:486–491.

37. Delfau-Larue, M.H., et al. 1998. Prognostic significance of a polymerase chain reaction-detectable dominant T-lymphocyte clone in cutaneous lesions of patients with mycosis fungoides. *Blood.* **92**:3376–3380.

38. Schwab, C., et al. 2002. The use of anti-T-cell receptor-Vbeta antibodies for the estimation of treatment success and phenotypic characterization of clonal T-cell populations in cutaneous T-cell lymphomas. *Br. J. Haematol.* **118**:1019–1026.

39. Vonderheid, E.C., et al. 2002. Update on erythrodermic cutaneous T-cell lymphoma: report of the International Society for Cutaneous Lymphomas. *J. Am. Acad. Dermatol.* **46**:95–106.

40. Yawalkar, N., et al. 2003. Profound loss of T-cell receptor repertoire complexity in cutaneous T-cell lymphoma. *Blood.* **102**:4059–4066.

41. Vowels, B.R., et al. 1994. Th2 cytokine mRNA expression in skin in cutaneous T-cell lymphoma. *J. Invest. Dermatol.* **103**:669–673.

42. Asadullah, K., et al. 1996. Progression of mycosis fungoides is associated with increasing cutaneous expression of interleukin-10 mRNA. *J. Invest. Dermatol.* **107**:833–837.

43. Kari, L., et al. 2003. Classification and prediction of survival in patients with the leukemic phase of cutaneous T cell lymphoma. *J. Exp. Med.* **197**:1477–1488.

44. Hoppe, R.T., et al. 1995. CD8-positive tumor-infiltrating lymphocytes influence the long-term survival of patients with mycosis fungoides. *J. Am. Acad. Dermatol.* **32**:448–453.

45. Zackheim, H.S., et al. 2002. Psoriasiform mycosis fungoides with fatal outcome after treatment with cyclosporine. *J. Am. Acad. Dermatol.* **47**:155–157.

46. Wysocka, M., et al. 2002. Sezary syndrome patients demonstrate a defect in dendritic cell populations: effects of CD40 ligand and treatment with GM-CSF on dendritic cell numbers and the production of cytokines. *Blood.* **100**:3287–3294.

47. De Smedt, T., et al. 1997. Effect of interleukin-10 on dendritic cell maturation and function. *Eur. J. Immunol.* **27**:1229–1235.

48. French, L.E., et al. 2005. Impaired CD40L signaling is a cause of defective IL-12 and TNF-{alpha} production in Sezary syndrome: circumvention by

hexameric soluble CD40L. *Blood.* **105**:219–225.

49. Berger, C.L., et al. 2004. Cutaneous T cell lymphoma, malignant proliferation of T-regulatory cells. *Blood.* **105**:1640–1647.
50. Karube, K., et al. 2004. Expression of FoxP3, a key molecule in CD4CD25 regulatory T cells, in adult T-cell leukaemia/lymphoma cells. *Br. J. Haematol.* **126**:81–84.
51. Yoo, E.K., et al. 2001. Complete molecular remission during biologic response modifier therapy for Sezary syndrome is associated with enhanced helper T type 1 cytokine production and natural killer cell activity. *J. Am. Acad. Dermatol.* **45**:208–216.
52. Wysocka, M., et al. 2004. Enhancement of the host immune responses in cutaneous T-cell lymphoma by CpG oligodeoxynucleotides and IL-15. *Blood.* **104**:4142–4149.
53. Axelrod, P.I., Lorber, B., and Vonderheid, E.C. 1992. Infections complicating mycosis fungoides and Sezary syndrome. *JAMA.* **267**:1354–1358.
54. Goldgeier, M.H., et al. 1981. An unusual and fatal case of disseminated cutaneous herpes simplex. Infection in a patient with cutaneous T cell lymphoma (mycosis fungoides). *J. Am. Acad. Dermatol.* **4**:176–180.
55. Evans, A.V., et al. 2004. Cutaneous malignant melanoma in association with mycosis fungoides. *J. Am. Acad. Dermatol.* **50**:701–705.
56. Pielop, J.A., Brownell, I., and Duvic, M. 2003. Mycosis fungoides associated with malignant melanoma and dysplastic nevus syndrome. *Int. J. Dermatol.* **42**:116–122.
57. Molin, L., Thomsen, K., and Volden, G. 1978. Serum IgE in mycosis fungoides. *Br. Med. J.* **1**:920–921.
58. Tancrede-Bohin, E., et al. 2004. Prognostic value of blood eosinophilia in primary cutaneous T-cell lymphomas. *Arch. Dermatol.* **140**:1057–1061.
59. Suchin, K.R., et al. 2001. Increased interleukin 5 production in eosinophilic Sezary syndrome: regulation by interferon alfa and interleukin 12. *J. Am. Acad. Dermatol.* **44**:28–32.
60. Kim, Y.H., et al. 2003. Topical nitrogen mustard in the management of mycosis fungoides: update of the Stanford experience. *Arch. Dermatol.* **139**:165–173.
61. Zackheim, H.S. 2003. Topical carmustine (BCNU) in the treatment of mycosis fungoides. *Dermatol. Ther.* **16**:299–302.
62. Zhang, C., and Duvic, M. 2003. Retinoids: therapeutic applications and mechanisms of action in cutaneous T-cell lymphoma. *Dermatol. Ther.* **16**:322–330.
63. Herrmann, J.J., et al. 1995. Treatment of mycosis fungoides with photochemotherapy (PUVA): long-term follow-up. *J. Am. Acad. Dermatol.* **33**:234–242.
64. Jones, G., Wilson, L.D., and Fox-Goguen, L. 2003. Total skin electron beam radiotherapy for patients who have mycosis fungoides. *Hematol. Oncol. Clin. North Am.* **17**:1421–1434.
65. McGinnis, K.S., et al. 2003. Psoralen plus long-wave UV-A (PUVA) and bexarotene therapy: an effective and synergistic combined adjunct to therapy for patients with advanced cutaneous T-cell lymphoma. *Arch. Dermatol.* **139**:771–775.
66. McGinnis, K.S., et al. 2004. Low-dose oral bexarotene in combination with low-dose interferon alfa in the treatment of cutaneous T-cell lymphoma: clinical synergism and possible immunologic mechanisms. *J. Am. Acad. Dermatol.* **50**:375–379.
67. Singh, F., and Lebwohl, M.G. 2004. Cutaneous T-cell lymphoma treatment using bexarotene and PUVA: a case series. *J. Am. Acad. Dermatol.* **51**:570–573.
68. Kuzel, T.M., et al. 1995. Effectiveness of interferon alfa-2a combined with phototherapy for mycosis fungoides and the Sezary syndrome. *J. Clin. Oncol.* **13**:257–263.
69. Olsen, E.A., and Bunn, P.A. 1995. Interferon in the treatment of cutaneous T-cell lymphoma. *Hematol. Oncol. Clin. North Am.* **9**:1089–1107.
70. Chiarion-Sileni, V., et al. 2002. Phase II trial of interferon-alpha-2a plus psolaren with ultraviolet light A in patients with cutaneous T-cell lymphoma. *Cancer.* **95**:569–575.
71. Knobler, R.M., et al. 1991. Treatment of cutaneous T cell lymphoma with a combination of low-dose interferon alfa-2b and retinoids. *J. Am. Acad. Dermatol.* **24**:247–252.
72. Zhang, C., et al. 2002. Induction of apoptosis by bexarotene in cutaneous T-cell lymphoma cells: relevance to mechanism of therapeutic action. *Clin. Cancer Res.* **8**:1234–1240.
73. Budgin, J.B., et al. 2005. Biological effects of bexarotene in cutaneous T-cell lymphoma. *Arch. Dermatol.* **141**:315–321.
74. Duvic, M., et al. 2001. Bexarotene is effective and safe for treatment of refractory advanced-stage cutaneous T-cell lymphoma: multinational phase II-III trial results. *J. Clin. Oncol.* **19**:2456–2471.
75. Fox, F.E., et al. 1999. Retinoids synergize with interleukin-2 to augment IFN-gamma and interleukin-12 production by human peripheral blood mononuclear cells. *J. Interferon Cytokine Res.* **19**:407–415.
76. Richardson, S.K., et al. 2003. Extracorporeal photopheresis and multimodality immunomodulatory therapy in the treatment of cutaneous T-cell lymphoma. *J. Cutan. Med. Surg.* **7**(Suppl. 2):8–12.
77. Suchin, K.R., et al. 2002. Treatment of cutaneous T-cell lymphoma with combined immunomodulatory therapy: a 14-year experience at a single institution. *Arch. Dermatol.* **138**:1054–1060.
78. Yoo, E.K., et al. 1996. Apoptosis induction of ultraviolet light A and photochemotherapy in cutaneous T-cell lymphoma: relevance to mechanism of therapeutic action. *J. Invest. Dermatol.* **107**:235–242.
79. Heald, P.W., and Edelson, R.L. 1988. Photopheresis for T cell mediated diseases. *Adv. Dermatol.* **3**:25–40.
80. Girardi, M., et al. 2002. Transimmunization and the evolution of extracorporeal photochemotherapy. *Transfus Apheresis Sci.* **26**:181–190.
81. Kim, S., Elkon, K.B., and Ma, X. 2004. Transcriptional suppression of interleukin-12 gene expression following phagocytosis of apoptotic cells. *Immunity.* **21**:643–653.
82. Shapiro, M., et al. 2002. Novel multimodality biologic response modifier therapy, including bexarotene and long-wave ultraviolet A for a patient with refractory stage IVa cutaneous T-cell lymphoma. *J. Am. Acad. Dermatol.* **47**:956–961.
83. Dummer, R., et al. 2004. Adenovirus-mediated intralesional interferon-gamma gene transfer induces tumor regressions in cutaneous lymphomas. *Blood.* **104**:1631–1638.
84. vanderSpek, J.C., et al. 1993. Structure/function analysis of the transmembrane domain of DAB389-interleukin-2, an interleukin-2 receptor-targeted fusion toxin. The amphipathic helical region of the transmembrane domain is essential for the efficient delivery of the catalytic domain to the cytosol of target cells. *J. Biol. Chem.* **268**:12077–12082.
85. Olsen, E., et al. 2001. Pivotal phase III trial of two dose levels of denileukin diftitox for the treatment of cutaneous T-cell lymphoma. *J. Clin. Oncol.* **19**:376–388.
86. Camacho, L.H., et al. 2004. Phase 1 clinical trial of anti-CTLA4 human monoclonal antibody CP-675,206 in patients with advanced solid malignancies [abstract]. *J. Clin. Oncol.* **22**(Suppl.):2505.
87. Wu, J.J., Huang, D.B., and Tyring, S.K. 2004. Resiquimod: a new immune response modifier with potential as a vaccine adjuvant for Th1 immune responses. *Antiviral Res.* **64**:79–83.
88. Dockrell, D.H., and Kinghorn, G.R. 2001. Imiquimod and resiquimod as novel immunomodulators. *J. Antimicrob. Chemother.* **48**:751–755.
89. Suchin, K.R., Junkins-Hopkins, J.M., and Rook, A.H. 2002. Treatment of stage IA cutaneous T-cell lymphoma with topical application of the immune response modifier imiquimod. *Arch. Dermatol.* **138**:1137–1139.
90. Dummer, R., et al. 2003. Imiquimod induces complete clearance of a PUVA-resistant plaque in mycosis fungoides. *Dermatology.* **207**:116–118.
91. Hurwitz, D.J., Pincus, L., and Kupper, T.S. 2003. Imiquimod: a topically applied link between innate and acquired immunity. *Arch. Dermatol.* **139**:1347–1350.
92. Kawai, T., et al. 2004. Interferon-alpha induction through Toll-like receptors involves a direct interaction of IRF7 with MyD88 and TRAF6. *Nat. Immunol.* **5**:1061–1068.
93. Schon, M.P., and Schon, M. 2004. Immune modulation and apoptosis induction: two sides of the antitumoral activity of imiquimod. *Apoptosis.* **9**:291–298.
94. Jones, T. 2003. Resiquimod 3M. *Curr. Opin. Investig. Drugs.* **4**:214–218.
95. Krieg, A.M. 2003. CpG motifs: the active ingredient in bacterial extracts? *Nat. Med.* **9**:831–835.
96. Lonsdorf, A.S., et al. 2003. Intratumor CpG-oligodeoxynucleotide injection induces protective antitumor T cell immunity. *J. Immunol.* **171**:3941–3946.
97. Kim, Y., et al. 2004. TLR9 agonist immunomodulator treatment of cutaneous T-cell lymphoma (CTCL) with CPG7909 [abstract]. American Society of Hematology Meeting. December 4–7, 2004. San Diego, California, USA. http://www.abstracts-2view.com/hem_sandiego2004/.
98. Tritel, M., et al. 2003. Prime-boost vaccination with HIV-1 Gag protein and cytosine phosphate guanosine oligodeoxynucleotide, followed by adenovirus, induces sustained and robust humoral and cellular immune responses. *J. Immunol.* **171**:2538–2547.
99. Bergstrom, R.T., et al. 2004. CD40 monoclonal antibody activation of antigen-presenting cells improves therapeutic efficacy of tumor-specific T cells. *Otolaryngol. Head Neck Surg.* **130**:94–103.
100. Watanabe, S., et al. 2003. The duration of signaling through CD40 directs biological ability of dendritic cells to induce antitumor immunity. *J. Immunol.* **171**:5828–5836.
101. Rook, A.H., et al. 1999. Interleukin-12 therapy of cutaneous T-cell lymphoma induces lesion regression and cytotoxic T cell responses. *Blood.* **94**:902–908.
102. Rook, A.H., et al. 1996. The potential therapeutic role of interleukin-12 in cutaneous T-cell lymphoma. *Ann. N. Y. Acad. Sci.* **795**:310–318.
103. Zaki, M.H., et al. 2001. Dysregulation of lymphocyte interleukin-12 receptor expression in Sezary syndrome. *J. Invest. Dermatol.* **117**:119–127.
104. Berger, C.L., et al. 1998. Tumor-specific peptides in cutaneous T-cell lymphoma: association with class I major histocompatibility complex and possible derivation from the clonotypic T-cell receptor. *Int. J. Cancer.* **76**:304–311.
105. Winter, D., et al. 2003. Definition of TCR epitopes for CTL-mediated attack of cutaneous T cell lymphoma. *J. Immunol.* **171**:2714–2724.
106. Muche, J.M., and Sterry, W. 2002. Vaccination therapy for cutaneous T-cell lymphoma. *Clin. Exp. Dermatol.* **27**:602–607.
107. Maier, T., et al. 2003. Vaccination of patients with cutaneous T-cell lymphoma using intranodal injection of autologous tumor-lysate-pulsed dendritic cells. *Blood.* **102**:2338–2344.
108. Berard, M., et al. 2003. IL-15 promotes the survival of naive and memory phenotype CD8+ T cells. *J. Immunol.* **170**:5018–5026.
109. Son, Y.I., et al. 2001. Interleukin-18 (IL-18) synergizes with IL-2 to enhance cytotoxicity, interferon-gamma production, and expansion of natural killer cells. *Cancer Res.* **61**:884–888.
110. Strengell, M., et al. 2003. IL-21 in synergy with IL-15

or IL-18 enhances IFN-gamma production in human NK and T cells. *J. Immunol.* **170**:5464–5469.

111. Miller, G., et al. 2002. Endogenous granulocyte-macrophage colony-stimulating factor overexpression in vivo results in the long-term recruitment of a distinct dendritic cell population with enhanced immunostimulatory function. *J. Immunol.* **169**:2875–2885.

112. Chang, D.Z., et al. 2004. Granulocyte-macrophage colony stimulating factor: an adjuvant for cancer vaccines. *Hematology.* **9**:207–215.

113. Lundin, J., et al. 1998. CAMPATH-1H monoclonal antibody in therapy for previously treated low-grade non-Hodgkin's lymphomas: a phase II multicenter study. European Study Group of CAMPATH-1H Treatment in Low-Grade Non-Hodgkin's Lymphoma. *J. Clin. Oncol.* **16**:3257–3263.

114. Lundin, J., et al. 2003. Phase 2 study of alemtuzumab (anti-CD52 monoclonal antibody) in patients with advanced mycosis fungoides/Sezary syndrome. *Blood.* **101**:4267–4272.

115. Ishida, T., et al. 2003. Clinical significance of CCR4 expression in adult T-cell leukemia/lymphoma: its close association with skin involvement and unfavorable outcome. *Clin. Cancer Res.* **9**:3625–3634.

116. Ishida, T., et al. 2004. The CC chemokine receptor 4 as a novel specific molecular target for immunotherapy in adult T-cell leukemia/lymphoma. *Clin. Cancer Res.* **10**:7529–7539.

117. Niwa, R., et al. 2004. Defucosylated chimeric anti-CC chemokine receptor 4 IgG1 with enhanced antibody-dependent cellular cytotoxicity shows potent therapeutic activity to T-cell leukemia and lymphoma. *Cancer Res.* **64**:2127–2133.

118. Mitsiades, N., et al. 2003. Molecular sequelae of histone deacetylase inhibition in human malignant B cells. *Blood.* **101**:4055–4062.

119. Kelly, W.K., et al. 2003. Phase I clinical trial of histone deacetylase inhibitor: suberoylanilide hydroxamic acid administered intravenously. *Clin. Cancer Res.* **9**:3578–3588.

120. Piekarz, R.L., et al. 2001. Inhibitor of histone deacetylation, depsipeptide (FR901228), in the treatment of peripheral and cutaneous T-cell lymphoma: a case report. *Blood.* **98**:2865–2868.

121. Piekarz, R.L., et al. 2004. T-cell lymphoma as a model for the use of histone deacetylase inhibitors in cancer therapy: impact of depsipeptide on molecular markers, therapeutic targets, and mechanisms of resistance. *Blood.* **103**:4636–4643.

122. Shao, R.H., et al. 2002. Arginine butyrate increases the cytotoxicity of DAB(389)IL-2 in leukemia and lymphoma cells by upregulation of IL-2Rbeta gene. *Leuk. Res.* **26**:1077–1083.

123. Burt, R.K., et al. 2000. Allogeneic hematopoietic stem cell transplantation for advanced mycosis fungoides: evidence of a graft-versus-tumor effect. *Bone Marrow Transplant.* **25**:111–113.

124. Soligo, D., et al. 2003. Treatment of advanced mycosis fungoides by allogeneic stem-cell transplantation with a nonmyeloablative regimen. *Bone Marrow Transplant.* **31**:663–666.

125. Molina, A., et al. 1999. Remission of refractory Sezary syndrome after bone marrow transplantation from a matched unrelated donor. *Biol. Blood Marrow Transplant.* **5**:400–404.

Cancer vaccines: progress reveals new complexities

Zhiya Yu and Nicholas P. Restifo

National Cancer Institute, National Institutes of Health, Bethesda, Maryland, USA

A decade ago, it seemed clear that our burgeoning knowledge of the molecular identities of tumor-associated antigens and a deeper understanding of basic immunology would point the way to an effective therapeutic cancer vaccine. Significant progress has been made and objective regressions after immune-based treatments are observed in some patients — even in those with bulky, metastatic disease. Notwithstanding this progress, we do not yet have a cancer vaccine in hand that can reliably increase patient survival or induce tumor destruction.

The creation of therapeutic cancer vaccines has proven to be an enormous challenge, and many of the strategies learned in the development of highly successful vaccines against infectious agents simply do not apply to cancer vaccines. Issues of antigenic change and immune escape are present in both antitumor and antiviral situations. However, one big difference between antiviral and antitumor vaccines is that the former are preventative whereas the latter are generally expected to be therapeutic.

Another problem with the targeting of tumor antigens relates to their poor immunogenicity. Tumor antigens appear to be relatively well tolerated in the host, perhaps because many of these antigens are also expressed in normal tissues. In this context, a successful cancer vaccine raises the specter of autoimmune attack if the vaccines are ultimately powerful enough to eliminate cancer cells. In this review, we highlight new challenges that have been revealed by recent progress in the field of tumor immunology. We then attempt to outline a future plan for cancer immunotherapy.

The existence of tumor-specific immune responses in cancer patients

The first reports that immune responses might result in tumor regression came over a century ago from William Coley, who treated cancer patients with live bacterial cultures, nonspecifically activating their immune systems. In the 1980s, Rosenberg et al. pioneered the use of high doses of the T cell growth factor IL-2 in individuals with metastatic kidney cancer or melanoma and achieved objective cancer regressions in 15–20% of treated patients (1). Because IL-2 is not known to have direct effects on the growth of solid tumors, its antitumor activities are most likely associated with its ability to expand lymphocytes, including the low-affinity T cells.

The most striking evidence for naturally occurring antitumor immune responses comes from rare anecdotal observations of spontaneous regressions of tumors in patients with cancer. Also rare, but somewhat more amenable to study, is the observation of paraneoplastic autoimmunity that can accompany often occult malignancies. For example, high titers of serum IgG specific for immunogenic proteins expressed in both carcinomas and normal tissues have been detected in cancer patients with paraneoplastic neurological diseases (2). Because these antigens are generally found in the nucleus and cytoplasm, it is unlikely that the autoantibodies have direct effects on either tumors or normal tissues. However, they are proposed to be surrogate markers for activated cellular immune responses. Indeed, cytotoxic T lymphocytes specific for cdr-2 have been detected and linked to cerebellar degeneration in those seropositive patients with breast and ovarian cancers (2).

Autoimmunity has been modeled to a limited extent in animals, where vitiligo, the patchy and permanent depigmentation that results from the destruction of dermal melanocytes, has been found to accompany the regression of the experimental B16 melanoma in C57BL/6 mice. Vitiligo is also positively correlated with a favorable response to IL-2 in patients with metastatic melanoma (3, 4).

Identification of tumor-associated antigens suitable for therapeutic targeting

A variety of techniques have been applied to identify tumor antigens recognizable by tumor-specific T cells. None has been more successful than an approach that uses transient transfection of pools of genes from a tumor-derived cDNA library to confer recognition upon a target cell, thus identifying the gene encoding the target epitope. While cloning efforts have been prodigiously successful, protein chemists have also made inroads into the identification of target antigens by pushing the limits of high-performance liquid chromatography and tandem mass spectrometry. Peptides can now be eluted from MHC complexes derived from tumor cell membranes and characterized directly (5). In addition, it is possible to test candidate tumor antigens by the so-called reverse immunology method, specifically by sensitizing immune cells with the candidate antigen, then testing the ability of sensitized cells to specifically kill tumor cells that are known to express the antigen.

To date, approximately 70 MHC class I– and class II–associated tumor antigens have been discovered, while more than 1,700 have been identified by antibodies in cancer patients. About ten antigens are currently known to be recognizable by both T cells and antibodies, although the actual number of antigens for which IgG production requires Th cells is probably much greater.

We do not know for certain how many of the candidate tumor antigens are suitable targets for tumor immunotherapy. A "valid" target antigen must be expressed specifically in the tumor, or at least be expressed at levels sufficiently higher there than in vital organs. For T cell–based therapy, it must be processed and presented in the context of MHC molecules. The need for positive and negative control tumor lines is often overlooked; rigorous (and numerous) controls are needed to convincingly demonstrate that a candidate antigen is a suitable target for use in an immunotherapy trial.

It has been estimated that 10^5 to 10^6 MHC/peptide complexes are present on the surface of a typical cell (6), although some anti-

Citation for this article: *J. Clin. Invest.* **110**:289–294 (2002). doi:10.1172/JCI200216216.

gen-presenting cells may express more. Considering the redundancy of individual complexes in a given cell and the poor MHC expression and antigen processing in the majority of tumor cells, a maximum of approximately 10,000 different MHC/peptide complexes is likely to be presented on a tumor cell. With the assumption that dozens of peptide epitopes could be derived from an expressed protein, and additional antigens could be generated from alternative open reading frames, a conservative estimate of the total number of peptides that are actually able to bind any given MHC may exceed 1 million. Therefore, the chances that any given peptide will be presented on the tumor cell surface are approximately 1:100, which leaves 99% of the potentially recognizable T cell epitopes simply absent from the surface of a given target tumor cell. Therefore, it is clearly not valid to assume that expression of a mutated candidate antigen will result in the MHC-restricted presentation of that antigen on the tumor cell surface.

Despite practical and theoretical concerns about some putative tumor-associated antigens, there is already a large and growing list of antigens that have been convincingly shown to be valid targets for immunotherapy. The success of tumor antigen identification approaches put to rest the notion that spontaneous human tumors — unlike their experimental mouse counterparts — simply lacked the expression of antigens recognizable by the immune system.

Enhancing tumor antigen immunogenicity by modifying epitope sequences

Tumor antigens in their original form generally bind poorly to their restricting MHC molecules. In addition, peripheral autoreactive T cell precursors recognize their cognate peptide/MHC complexes with low affinity. Thus, most tumor antigens identified so far are poorly immunogenic in vivo. Dramatic increases in the magnitude of T cell responses and sensitivity to antigen stimulation have been observed in both human and mouse models using agonistic altered peptide ligands. These altered peptides are capable of enhancing the stability of peptide/MHC complexes because of modifications in the MHC anchor residues (7) or as the result of favorable and generally conservative changes to the peptide at the interface with the T cell receptor (TCR) (8, 9). Most importantly, the enhanced T cell responses can retain their specificity to the native antigen, which allows them to kill target tumor cells ex vivo and, presumably, in vivo.

Both the association and the dissociation rates contribute to the steady-state stability of the interactions between peptide/MHC and ultimately TCR/peptide/MHC complexes. However, we have found that the enhancement of immunogenicity might be more affected by the off-rate than by the on-rate. For example, gp100$_{209(2M)}$, an anchor-fixing modification of a native antigenic peptide derived from human melanoma antigen, gp100$_{209-217}$, is 100-fold more potent in activation of naive T cells than is wild-type peptide. The steady-state binding affinity of the modified peptide is nine times higher than that of the wild-type peptide. In contrast, the dissociation rate of modified peptide from HLA-A2 molecule is more than 100-fold slower than that measured with the wild-type peptide (our unpublished data).

The stability of the target peptide/MHC complex is not only important in order to achieve the required antigenic density for naive T cell activation but may also alter the quality of the T cell response. A study by Mark Davis's group concluded that a TCR bound to antagonist ligands with lower affinities because of an increased off-rate (10). Conversely, extending the off-rate by amino

acid substitutions may augment T cell "affinity maturation." Immunizing mice with agonistic peptide ligands elicits high-avidity T cells that can recognize the relevant tumor cells, and even target cells pulsed at relatively low concentrations with weakly binding peptides (8). Studies from our laboratory have shown that although the human (hgp100$_{25-33}$) and mouse (mgp100$_{25-33}$) epitopes are homologous, differences in the three NH$_2$-terminal amino acids result in a 2-log increase in the ability of the human peptide to stabilize "empty" D^b MHC molecules and a 3-log increase in its ability to trigger IFN-γ release by T cells (11). In a clinical trial, using anchor-modified gp100 peptide immunization in melanoma patients resulted in dramatic increase of tumor-reactive T cell responses (12). Thus, modified antigenic peptide based on these ideas could be of significant value in vaccination against tolerant or weakly immunogenic tumor-associated cells.

The nature of antitumor effector cells

More than 40 years ago, Prehn and Main obtained evidence of specificity in the immune response to tumors. Mice immunized with irradiated methylcholanthrene-induced sarcoma cells, they showed, are fully protected against a subsequent challenge with that same tumor, but not with other tumors. Humoral responses against solid tumors may play some role in the effective killing of tumor targets either by augmenting antigen presentation or by the ligation of a growth factor receptor on cancer cells, such as HER-2/Neu. A large body of work using antibody depletion and gene knockout mice has revealed that both CD8$^+$ and CD4$^+$ T lymphocytes are crucial for therapeutic antitumor immune responses.

Compared with the comprehensive studies using CD8$^+$ T cells in tumor models, relatively little is known about how CD4$^+$ T cells influence antitumor immunity. Very early work demonstrated that disseminated murine leukemia could be eradicated by a combination of cyclophosphamide and adoptively transferred cells, now known to be CD4$^+$ T cells (13). The most dramatic examples of the power of CD4$^+$ T cells in the immune response to self-proteins can be found in murine models of autoimmune diseases, such as experimental allergic encephalomyelitis, systemic lupus erythematosus, and diabetes. In these models, disease can often be transferred to naive mice with purified, self-reactive CD4$^+$ splenocytes or specific CD4$^+$ T lymphocyte clones. Antigen-specific CD4$^+$ T lymphocyte clones can also treat tumor through CD8$^+$ T cells specific for the cognate antigen (14). Tumor antigen–specific CD4$^+$ T cells have been isolated from tumor-infiltrating lymphocytes from melanoma patients (15). Adoptive transfer into patients of unfractionated tumor-specific T cells has been shown to promote tumor regression (16). These studies and others suggest that the full activation of autoreactive CD4$^+$ T cells may be an important immune component that is currently missing from many current clinical cancer vaccine trials.

Natural killer cells kill many tumor cell lines and may also play a critical role in antitumor immunity. A recent study in transgenic mice lacking NK1.1$^+$CD3$^-$ cells linked an impaired acute tumor rejection to deficiency of NK activity (17). NKG2D receptors expressed by NK cells and activated CD8$^+$ T cells and macrophages can be stimulated by their ligands, which are often overexpressed on cancer cell lines (18). Most surprisingly, clinical results from hematopoietic cell transplantation revealed that alloreactive NK cells in the donor graft prevented leukemia relapse in leukemic recipients (19).

Although the innate immune system is often ignored as an important component of the antitumor immune response, lessons learned about immunity to infectious microorganisms indicate that early host defenses help determine the nature of downstream adaptive immune responses. One recent line of investigation pursued in our laboratory concerns the initiation of apoptotic death that results in the engagement of key innate immune pathways; the consequences of cell death may induce dendritic cell activation and benefit immune induction (ref. 20; see also Steinman and Pope, this Perspective series, ref. 21). A new generation of nucleic acid vaccines encoding an alphaviral replicase enzyme, together with tumor antigens, induces apoptotic death coupled with antigen production. This approach leads to quantitative and qualitative enhancement of the therapeutic antitumor immune response in animal models (22, 23).

Unexpected obstacles in early clinical trials

With antigens in hand, immunotherapists set off to create a new class of therapeutic vaccines based on defined antigens. Recombinant immunogens were created using the same viruses that have proven themselves to be so successful in the realm of infectious diseases — including vaccinia, polio, and influenza A — as well as some others including adenoviruses and bird poxviruses (canarypox and fowlpox) (24, 25). In animal models, these vaccines can prime T cell responses and elicit powerful immune responses that lead to tumor cell destruction. However, when these viruses were tried in the clinic, it became apparent that experiments in animal models had failed to predict key aspects of recombinant vaccine function in people.

One reason for these disappointing results was pre-existing neutralizing antibody. In one recent study in which patients were immunized with recombinant adenoviruses encoding the melanoma-associated antigens gp100 or MART-1, only 6 of 54 patients had neutralizing antibody titers of less than 100, with the majority having neutralizing titers of more than 400 (26). Similar problems might be observed using clinical-grade recombinant vaccinia viruses. Patients thus retain strong anti-vaccinia antibodies for many decades after immunization — a lasting legacy of the worldwide immunization program to eradicate smallpox. One would fully expect pre-existing immunity to be a problem in the use of several other vectors under consideration, including recombinant versions of polio and transfectant influenza A viruses of the more commonly observed strains. One way to circumvent this problem would be to use viruses whose natural hosts are non-mammalian, such as the avian poxviruses (Letvin, this Perspective series, ref. 27). These viruses are antigenically distinct from poxviruses and are incapable of replicating in mammalian cells. Strategies employing genetically engineered influenza viruses may prove useful in the development of live virus vaccines against cancer (24).

Another potential obstacle uncovered in early clinical trials is immunity to an antigenically complex vaccine that is immunodominant over a response to a transgene-encoded weak tumor antigen. This problem, which was not adequately studied in early experimental animal models and remains exceedingly difficult to model in ongoing preclinical work, may be summarized as follows: Vectors may interfere with the induction of reactivity to the encoded tumor antigen through the poorly understood mechanisms of immunodominance. T cell responses elicited by protein immunization tend to focus on one or a few sites in the antigen. Whether this phenomenon is driven by the predetermined TCR repertoire, the competition among T cells based on their affinity to antigenic determinants, or the characteristics of antigen processing is not known (28, 29). In viral vector–based vaccines, self-antigens are coexpressed with viral proteins. If the immunodominant sites reside in the viral components, the vaccine will fail to elicit the desired immune responses. The use of vaccines based on "naked" plasmid DNA vaccines (i.e., DNA without associated protein) may circumvent both pre-existing immunity and immunodominance. Despite these advantages, our own clinical work has shown no evidence of immunization or antitumor effect of naked DNA immunization against the gp100 tumor antigen (unpublished data), although naked DNA is effective in many animal models.

Consistent increases in tumor-specific T cells without consistent clinical responses

There is now incontrovertible evidence that precursor frequencies of tumor-specific T cells can be increased after immunization using several different tumor-associated antigens — including those antigens that are nonmutated "self" tissue differentiation antigens (12). The presence of increased antitumor T cell precursors after vaccination has been convincingly demonstrated in both mice and humans, using tetramer or ELISPOT analysis, real-time RT-PCR, and other techniques.

Thus far, the most effective immunization strategy in our patients with advanced melanoma has been vaccination with peptide emulsified in incomplete Freund's adjuvant. Immunization with a gp100-derived peptide modified to enhance its binding to HLA-A2 dramatically increased levels of peptide-specific CD8+ T cells in the peripheral blood. Importantly, these T cells recognized and killed a variety of melanoma cells that expressed the gp100 melanoma antigen and the restriction element HLA-A*0201 after culture ex vivo. Administration of IL-2 following peptide immunization resulted in significantly more objective tumor regressions than seen after IL-2 treatment alone (12). However, most of these responses turned out to be partial and transient, and most responding patients eventually succumbed to progressively growing tumor.

Proposed mechanisms of tumor escape

The current notion that tumor cells must "escape" immune recognition is based largely on the idea that neoantigens expressed by tumor cells as a consequence of their genetic instability will be immunogenic. There is little doubt that the tumor contains a large number of mutations (30) that can potentially generate new antigens recognizable by the immune system, but there is considerable doubt about what the immunological response to these potential immunogens will be. A number of groups have conducted experiments in which highly immunogenic foreign antigens, such as the hemagglutinin protein from influenza (31), the β-galactosidase enzyme from *Escherichia coli* (32), and the ovalbumin protein from the chicken, are expressed in tumor cells (33). The results are fairly uniform: tumors tend to grow progressively, retaining their lethality despite the expression of a foreign and highly immunogenic protein by the tumor cell.

Proposed mechanisms for tumor escape include those relating to the inherent genetic instability of tumor cells (34–38) and others that might be shared by many normal cells in the body. The latter include the lack of expression of costimulatory molecules (B7-1/CD80, B7-2/CD86, and CD40 ligand), the induction of suppressor cell activity, and the production of immunoinhibitory substances such as TGF-β or IL-10. Many of these theories are intuitively appealing but lack direct experimental evidence or consistent results. For example, Fas ligand (FasL) has been proposed as a mediator of the tumor "counterattack." However, controlled experiments show that FasL expressed in animal tumor models results not in escape, but in more rapid rejection (39–42).

Several groups have proposed the loss of β_2-microglobulin (β_2m) as a mechanism of immune escape. However, work by Karre, Snook, and colleagues in animal models demonstrated that the loss of β_2m, an essential and invariant subunit of class I MHC complexes, results in exquisite sensitivity to NK cell–mediated killing and leads to tumor elimination, not escape (43, 44). Although some human melanoma cells have also been shown to lose β_2m with clinical progression, human β_2m–deficient cells are also susceptible to NK cell–mediated killing (45). This evidence need not indicate that β_2m loss represents a mechanism of immune escape, since this molecule could be lost as a result of increasing derangement in the transformed genome and a mutation "hot spot" at the $\beta_2 m$ locus (46). Indeed, the mutability of this locus may have unexplored protective functions. Clearly, similar arguments can be advanced with regard to other events that decrease or eliminate MHC class I expression on the surface, such as loss of the MHC class I heavy chain or of transporters associated with antigen processing or low–molecular-weight proteins complex components.

To take another example, tumor cells (along with most normal cells) generally lack costimulatory molecules, such as B7-1 (CD80) and B7-2 (CD86), which are expressed on professional antigen-presenting cells and on a variety of other tissues after exposure to inflammatory cytokines (47). In the absence of costimulation, T cells tend to become anergic. In the non–tumor-bearing setting, the absence of B7 molecule expression has been hypothesized to protect normal cells against autoreactivity. Does this finding help explain tumor cell escape from immune recognition? Transfection of tumor cells with both isoforms has been used successfully to trigger their immune-mediated rejection of experimental mouse tumors, which have some inherent immunogenicity. However, rejection is not observed when B7 molecules are inserted into less immunogenic tumors (48). Nonimmunogenicity is a category into which most, if not all, human tumors would fall; thus a lack of expression of the CD80 and CD86 costimulatory molecules is unlikely to be a global explanation for immune escape. Nonetheless, a greater understanding of the interactions of costimulatory molecules with negative regulatory molecules, such as cytotoxic T lymphocyte–associated antigen 4 (CTLA-4), may enable more directed interventions (4).

Controlled unresponsiveness and negative regulation of antitumor T cells

Low-affinity autoreactive T cells can avoid negative selection in the thymus. Indeed a low level of autoreactivity is required for positive thymic selection. In normal circumstances, after maturation is complete, these autoreactive T cells are likely to be either ignorant (that is, they simply do not "see" their target epitope) or anergic (defined as a state of *induced* unresponsiveness). In the first case, they do not have any contact with the antigen that alters their phenotype or function. In the latter case, they are negatively regulated by host factors, such as cytokines, accessory molecules on antigen-presenting cells, and suppressor cells (49). Although there is undoubtedly some degree of ignorance to tumor antigens, there is clear evidence that tumor cells sensitize host T cells to tumor antigens (our unpublished data).

The spontaneous activation of host tumor-specific T cells is rarely sufficient to lead to tumor eradication. Interestingly, a very recent study has revealed a molecular mechanism mediated by a negative regulatory protein, Tob, a member of an antiproliferative gene family, in anergized T cells (50). These data may point the way to a new understanding of how cells maintain unresponsiveness to antigen.

Other candidate negative regulatory mechanisms that may keep an incipient antitumor response in check include active negative regulatory mechanisms mediated by CD4+CD25+ suppressor cells, IL-13–secreting NKT cells, and CD11b+Gr-1+ suppressor cells. The CD4+CD25+ T cell population was first found to inhibit proliferation of CD4+CD25− T cells ex vivo as a result of TCR ligation and IL-2 activation. Extensive studies of suppressor/regulatory T cells in mouse models have demonstrated their importance in inhibition of autoimmunity (51). CD4+CD25+ suppressor T cells also exist in humans and can inhibit proliferation and cytokine release in CD4+CD25− T cells by unknown mechanisms (52). IL-13 production following CD1 molecule ligation on NKT cells may also limit antitumor responses. In tumor-bearing mice, CD1 is upregulated on NKT cells. Knocking out CD1 promotes tumor resistance in mice. Similar results were also seen in anti–IL-13–treated mice (53). We do not yet know the relevance of this type of negative regulation to the immunotherapy of human cancer. Finally, like CD4+CD25+ T cells and NKT cells, CD11b+Gr-1+ myeloid cells may send negative regulatory signals to T cells, triggering apoptotic death in CD8+ T cells following vaccination with powerful immunogens (54). Surprisingly, many mouse and human tumors produce GM-CSF, which can stimulate highly inhibitory CD11b+Gr-1+ cells (55). The manipulation of each of these types of negative regulatory cells could be useful in the treatment of autoimmune disease and cancer.

Future directions: a focus on T cell activation and death

Significant evidence indicates that the central reasons for the failed antitumor immune response may be deficiencies in the maintenance of sustained tumor-specific T cell activation. It is now clear that there are many ways in which triggering a TCR can result in the ultimate inactivation or even demise of the T cell bearing it. The difference between antigen presentation in the tumor environment and that in a virally infected tissue is likely the activation of resident antigen-presenting cells, the scavengers and "danger" sensors of the immune system. The lack of proinflammatory mediators that induce maturation of dendritic cells, in conjunction with the abundant antigen presentation by noncostimulatory, tolerizing tumor cells, is the factor that may tip the T cell activation-inactivation balance in favor of tumor-specific T cell tolerance. On the other hand, overstimulation can terminate an otherwise effective T cell response through activation-induced cell death, fratricide, or exhaustion (56).

Although new antigen discovery and epitope mapping continue to be an important part of tumor immunology, few would dispute that several excellent targets expressed on a range of tumor histologies are now available. The next important breakthrough in cancer immunotherapy may come from an understanding of how to enhance T cell avidity, how to maintain T cell activation while preventing T cell apoptosis, and how to reduce or eliminate the effects of negative regulatory factors.

In animal models, a number of new transgenic mouse models are now available, allowing for a reductionistic study of tumor interactions with elements of the innate and adaptive immune system. One particularly fruitful area currently under development involves the use of TCR transgenic mice. It is now clear that very large numbers of tumor-specific transgenic CD8[+] and CD4[+] T cells have little effect on the growth rate or lethality of syngeneic tumor cells that express the antigens targeted by these transgenic T cells. These transgenic mouse systems model key aspects of increased tumor-specific T cells found in some patients with cancer after active immunization. Using these models, we and others are evaluating cellular and molecular mechanisms in T cell activation, death, and anergy as they relate to the development of more effective cancer vaccines.

Conclusions

Tumor immunologists have made great strides in understanding the components of the successful immunotherapy of cancer. We have cloned antigens that are expressed by tumors, processed and presented in the context of MHC class I and class II molecules, and recognized by cells from the patient's own T cell repertoire. We have also learned how to immunize and are now capable of significantly expanding precursor T cells with vaccination. Still, the proper and continued activation of antitumor T cells remains a crucial missing piece of the immunotherapy puzzle and a significant barrier to developing an effective therapeutic vaccine. Thus, the focus of tumor immunotherapy is shifting. The challenge now is to learn how to promote T cell activation and proliferation while abrogating T cell anergy and death in the context of a profoundly tolerogenic tumor environment.

Address correspondence to: Nicholas P. Restifo, National Cancer Institute, Building 10, Room 2B/42, National Institutes of Health, Bethesda, Maryland 20892-1502, USA. Phone: (301) 496-4904; Fax: (301) 402-0922; E-mail: restifo@nih.gov.

1. Rosenberg, S.A., et al. 1985. Observations on the systemic administration of autologous lymphokine-activated killer cells and recombinant interleukin-2 to patients with metastatic cancer. *N. Engl. J. Med.* **313**:1485–1492.
2. Albert, M.L., et al. 1998. Tumor-specific killer cells in paraneoplastic cerebellar degeneration. *Nat. Med.* **4**:1321–1324.
3. Overwijk, W.W., et al. 1999. Vaccination with a recombinant vaccinia virus encoding a "self" antigen induces autoimmune vitiligo and tumor cell destruction in mice: requirement for CD4(+) T lymphocytes. *Proc. Natl. Acad. Sci. USA.* **96**:2982–2987.
4. van Elsas, A., Hurwitz, A.A., and Allison, J.P. 1999. Combination immunotherapy of B16 melanoma using anticytotoxic T lymphocyte-associated antigen 4 (CTLA-4) and granulocyte/macrophage colony-stimulating factor (GM-CSF)-producing vaccines induces rejection of subcutaneous and metastatic tumors accompanied by autoimmune depigmentation. *J. Exp. Med.* **190**:355–366.
5. Cox, A.L., et al. 1994. Identification of a peptide recognized by five melanoma-specific human cytotoxic T cell lines. *Science.* **264**:716–719.
6. Meunier, L., et al. 1996. Quantification of CD1a, HLA-DR, and HLA class I expression on viable human Langerhans cells and keratinocytes. *Cytometry.* **26**:260–264.
7. Parkhurst, M.R., et al. 1996. Improved induction of melanoma-reactive CTL with peptides from the melanoma antigen gp100 modified at HLA-A*0201-binding residues. *J. Immunol.* **157**:2539–2548.
8. Tangri, S., et al. 2001. Structural features of peptide analogs of human histocompatibility leukocyte antigen class I epitopes that are more potent and immunogenic than wild-type peptide. *J. Exp. Med.* **194**:833–846.
9. Slansky, J.E., et al. 2000. Enhanced antigen-specific antitumor immunity with altered peptide ligands that stabilize the MHC-peptide-TCR complex. *Immunity.* **13**:529–538.
10. Lyons, D.S., et al. 1996. A TCR binds to antagonist ligands with lower affinities and faster dissociation rates than to agonists. *Immunity.* **5**:53–61.
11. Overwijk, W.W., et al. 1998. gp100/pmel 17 is a murine tumor rejection antigen: induction of "self"-reactive, tumoricidal T cells using high-affinity, altered peptide ligand. *J. Exp. Med.* **188**:277–286.
12. Rosenberg, S.A., et al. 1998. Immunologic and therapeutic evaluation of a synthetic peptide vaccine for the treatment of patients with metastatic melanoma. *Nat. Med.* **4**:321–327.
13. Greenberg, P.D., Kern, D.E., and Cheever, M.A. 1985. Therapy of disseminated murine leukemia with cyclophosphamide and immune Lyt-1+,2- T cells. Tumor eradication does not require participation of cytotoxic T cells. *J. Exp. Med.* **161**:1122–1134.
14. Surman, D.R., Dudley, M.E., Overwijk, W.W., and Restifo, N.P. 2000. Cutting edge: CD4+ T cell control of CD8+ T cell reactivity to a model tumor antigen. *J. Immunol.* **164**:562–565.
15. Topalian, S.L., et al. 1994. Human CD4+ T cells specifically recognize a shared melanoma-associated antigen encoded by the tyrosinase gene. *Proc. Natl. Acad. Sci. USA.* **91**:9461–9465.
16. Rosenberg, S.A., et al. 1990. Gene transfer into humans: immunotherapy of patients with advanced melanoma, using tumor-infiltrating lymphocytes modified by retroviral gene transduction. *N. Engl. J. Med.* **323**:570–578.
17. Kim, S., et al. 2000. In vivo natural killer cell activities revealed by natural killer cell-deficient mice. *Proc. Natl. Acad. Sci. USA.* **97**:2731–2736.
18. Diefenbach, A., Jensen, E.R., Jamieson, A.M., and Raulet, D.H. 2001. Rae1 and H60 ligands of the NKG2D receptor stimulate tumour immunity. *Nature.* **413**:165–171.
19. Ruggeri, L., et al. 2002. Effectiveness of donor natural killer cell alloreactivity in mismatched hematopoietic transplants. *Science.* **295**:2097–2100.
20. Restifo, N.P. 2001. Vaccines to die for. *Nat. Biotechnol.* **19**:527–528.
21. Steinman, R.M., and Pope, M. 2002. Exploiting dendritic cells to improve vaccine efficacy. *J. Clin. Invest.* **109**:1519–1526. doi:10.1172/JCI200215962.
22. Ying, H., et al. 1999. Cancer therapy using a self-replicating RNA vaccine. *Nat. Med.* **5**:823–827.
23. Leitner, W.W., et al. 2000. Enhancement of tumor-specific immune response with plasmid DNA replicon vectors. *Cancer Res.* **60**:51–55.
24. Palese, P., Zavala, F., Muster, T., Nussenzweig, R.S., and Garcia-Sastre, A. 1997. Development of novel influenza virus vaccines and vectors. *J. Infect. Dis.* **176**(Suppl. 1):S45–S49.
25. Restifo, N.P. 1996. The new vaccines: building viruses that elicit antitumor immunity. *Curr. Opin. Immunol.* **8**:658–663.
26. Rosenberg, S.A., et al. 1998. Immunizing patients with metastatic melanoma using recombinant adenoviruses encoding MART-1 or gp100 melanoma antigens. *J. Natl. Cancer Inst.* **90**:1894–1900.
27. Letvin, N.L. 2002. Strategies for an HIV vaccine. *J. Clin. Invest.* **110**:15–20. doi:10.1172/JCI200215985.
28. Kedl, R.M., et al. 2000. T cells compete for access to antigen-bearing antigen-presenting cells. *J. Exp. Med.* **192**:1105–1113.
29. Chen, W., Anton, L.C., Bennink, J.R., and Yewdell, J.W. 2000. Dissecting the multifactorial causes of immunodominance in class I-restricted T cell responses to viruses. *Immunity.* **12**:83–93.
30. Stoler, D.L., et al. 1999. The onset and extent of genomic instability in sporadic colorectal tumor progression. *Proc. Natl. Acad. Sci. USA.* **96**:15121–15126.
31. Staveley-O'Carroll, K., et al. 1998. Induction of antigen-specific T cell anergy: an early event in the course of tumor progression. *Proc. Natl. Acad. Sci. USA.* **95**:1178–1183.
32. Wang, M., et al. 1995. Active immunotherapy of cancer with a nonreplicating recombinant fowlpox virus encoding a model tumor-associated antigen. *J. Immunol.* **154**:4685–4692.
33. McCabe, B.J., et al. 1995. Minimal determinant expressed by a recombinant vaccinia virus elicits therapeutic antitumor cytolytic T lymphocyte responses. *Cancer Res.* **55**:1741–1747.
34. D'Urso, C.M., et al. 1991. Lack of HLA class I antigen expression by cultured melanoma cells FO-1 due to a defect in B2m gene expression. *J. Clin. Invest.* **87**:284–292.
35. Maio, M., Altomonte, M., Tatake, R., Zeff, R.A., and Ferrone, S. 1991. Reduction in susceptibility to natural killer cell-mediated lysis of human FO-1 melanoma cells after induction of HLA class I antigen expression by transfection with B2m gene. *J. Clin. Invest.* **88**:282–289.
36. Restifo, N.P., et al. 1993. Molecular mechanisms used by tumors to escape immune recognition: immunogenetherapy and the cell biology of major histocompatibility complex class I. *J. Immunother.* **14**:182–190.
37. Restifo, N.P., et al. 1993. Identification of human cancers deficient in antigen processing. *J. Exp. Med.* **177**:265–272.
38. Restifo, N.P., et al. 1996. Loss of functional beta 2-microglobulin in metastatic melanomas from five patients receiving immunotherapy. *J. Natl. Cancer Inst.* **88**:100–108.
39. Restifo, N.P. 2001. Countering the 'counterattack' hypothesis. *Nat. Med.* **7**:259.
40. Restifo, N.P. 2000. Not so Fas: re-evaluating the mechanisms of immune privilege and tumor escape. *Nat. Med.* **6**:493–495.
41. Restifo, N.P. 2000. Building better vaccines: how apoptotic cell death can induce inflammation and activate innate and adaptive immunity. *Curr. Opin. Immunol.* **12**:597–603.
42. Chappell, D.B., Zaks, T.Z., Rosenberg, S.A., and Restifo, N.P. 1999. Human melanoma cells do not express Fas (Apo-1/CD95) ligand. *Cancer Res.* **59**:59–62.
43. Kambayashi, T., et al. 2001. Purified MHC class I molecules inhibit activated NK cells in a cell-free system in vitro. *Eur. J. Immunol.* **31**:869–875.
44. Smyth, M.J., and Snook, M.B. 1999. Perforin-dependent cytolytic responses in beta2-microglobulin-deficient mice. *Cell. Immunol.* **196**:51–59.
45. Porgador, A., Mandelboim, O., Restifo, N.P., and Strominger, J.L. 1997. Natural killer cell lines kill autologous beta2-microglobulin-deficient melanoma cells: implications for cancer immunotherapy. *Proc. Natl. Acad. Sci. USA.* **94**:13140–13145.

46. Parnes, J.R., Sizer, K.C., Seidman, J.G., Stallings, V., and Hyman, R. 1986. A mutational hot-spot within an intron of the mouse beta 2-microglobulin gene. *EMBO J.* **5**:103–111.

47. Anderson, D.E., Sharpe, A.H., and Hafler, D.A. 1999. The B7-CD28/CTLA-4 costimulatory pathways in autoimmune disease of the central nervous system. *Curr. Opin. Immunol.* **11**:677–683.

48. Chen, L., et al. 1994. Tumor immunogenicity determines the effect of B7 costimulation on T cell-mediated tumor immunity. *J. Exp. Med.* **179**:523–532.

49. Antony, P.A., and Restifo, N.P. 2002. Do CD4+ CD25+ immunoregulatory T cells hinder tumor immunotherapy? *J. Immunother.* **25**:202–206.

50. Tzachanis, D., et al. 2001. Tob is a negative regulator of activation that is expressed in anergic and quiescent T cells. *Nat. Immunol.* **2**:1174–1182.

51. Shevach, E.M. 2000. Regulatory T cells in autoimmunity. *Annu. Rev. Immunol.* **18**:423–449.

52. Dieckmann, D., Plottner, H., Berchtold, S., Berger, T., and Schuler, G. 2001. Ex vivo isolation and characterization of CD4(+)CD25(+) T cells with regulatory properties from human blood. *J. Exp. Med.* **193**:1303–1310.

53. Terabe, M., et al. 2000. NKT cell-mediated repression of tumor immunosurveillance by IL-13 and the IL-4R-STAT6 pathway. *Nat. Immunol.* **1**:515–520.

54. Bronte, V., et al. 1998. Apoptotic death of CD8+ T lymphocytes after immunization: induction of a suppressive population of Mac-1+/Gr-1+ cells. *J. Immunol.* **161**:5313–5320.

55. Bronte, V., et al. 1999. Unopposed production of granulocyte-macrophage colony-stimulating factor by tumors inhibits CD8+ T cell responses by dysregulating antigen-presenting cell maturation. *J. Immunol.* **162**:5728–5737.

56. Overwijk, W.W., and Restifo, N.P. 2001. Creating therapeutic cancer vaccines: notes from the battlefield. *Trends Immunol.* **22**:5–7.

Part 3
Immune System and Immune-Mediated Injury

IFNs and STATs in innate immunity to microorganisms

Thomas Decker,[1] Silvia Stockinger,[1] Marina Karaghiosoff,[2] Mathias Müller,[2] and Pavel Kovarik[1]

[1]Vienna Biocenter, Institute of Microbiology and Genetics, Vienna, Austria. [2]Institute of Animal Breeding and Genetics, Veterinary University of Vienna, Vienna, Austria

Innate immune responses derive from the ability of cells to rapidly combat invading microorganisms without the requirement for an antigen-specific adaptation. These mechanisms have evolved to recognize common microbe-associated molecular patterns and to interfere with conserved replication and survival strategies that support the propagation of microbial invaders. Here, we consider the contributions of one cytokine family, the IFNs, to these innate defense mechanisms. IFNs were first recognized for their ability to impede viral replication, a function that is indeed critical for host survival in response to viral infection. In addition, IFN signaling is now known to play key roles in defending the host from bacteria and other pathogens and to help integrate early, innate responses with later events mediated by the adaptive immune system.

The two recognized types of IFN exhibit distinct immunological properties (1). In humans and mice, type I IFNs include a number of IFN-α subtypes and a single species of IFN-β. (The immunological impact of another type I IFN, IFN-ω, is poorly understood and will not be considered here). While type I IFNs can be produced by all cells under appropriate conditions, a subpopulation of immature dendritic cells (DCs) that will be described in more detail below stands out for the extent of its contribution to overall IFN production during infections. IFN-γ is a type II IFN and serves not only to induce antiviral function, but also to activate macrophages, which strengthens innate responses to unicellular microorganisms (2). Unlike the type I IFNs, IFN-γ is produced by a limited number of cell types: activated NK cells, activated Th1 cells, and, in the presence of IL-12 and IL-18, activated DCs and macrophages. Expression of IFN-γ by Th1 cells provides an important link by which the adaptive immune response reinforces macrophage-based innate immunity.

JAK-STAT signaling in IFN responses

A common property of both IFN types is to induce immediate transcriptional responses through a JAK-STAT signal transduction pathway (3). All type I IFNs bind to a class II cytokine receptor composed of IFN-α receptor 1 (IFNAR1) and IFNAR2 chains, which are associated with the Janus kinases (JAKs) TYK2 and JAK1, respectively. Ligand-bound, tyrosine-phosphorylated receptor complexes bind the SH2 domains of signal transducers and activators of transcription (STATs) 1 and 2, causing phosphorylation of the proteins on tyrosines 701 and 692, respectively. Interaction of STATs through reciprocal SH2 domain–phosphotyrosine binding results in formation of two distinct transcription factor complexes. ISGF3, a heteromeric complex consisting of STAT1 and STAT2 in association with a third protein, p48 or IRF9, associates specifically with and transactivates genes with interferon-stimulated response elements (ISRE) in their promoter or enhancer regions. A simpler complex, consisting solely of the STAT1 homodimer, is also active as a transcription factor and binds to different DNA sequences, termed IFN-γ–activated site (GAS) elements (Figure 1).

The receptor for IFN-γ is structurally related to that for type I IFN. It consists of IFN-γ receptor 1 (IFNGR1) and IFNGR2 chains in association with JAK1 and JAK2 kinases. Once tyrosine-phosphorylated in the presence of ligand, it binds STAT1 and causes phosphorylation of Y701. STAT1 homo-dimers are formed, move to the nucleus, and regulate transcription of promoters containing GAS sequences (Figure 1; ref. 4). Recent evidence indicates that nuclear responses can be stimulated by IFN-γ in the absence of STAT1 (5), but the transcription factors mediating the STAT1-independent response remain to be identified. In addition, some genes are induced by IFN-γ only in the absence of STAT1, suggesting that this protein can also mediate transcriptional repression.

Innate immunity to viruses

Cells in culture respond to viruses by deploying a complex network of signaling molecules that initiate and then amplify the production of type I IFNs. At an early step in this pathway, the transcription factor IFN regulatory factor 3 (IRF3) becomes phosphorylated on serine residues by a still-unidentified kinase. Activated IRF3 then stimulates transcription of IFN-α4 and IFN-β, and these secreted "early" IFNs activate ISGF3 and its target gene *Irf7*. IRF7 closes a positive feedback loop of IFN production by initiating a second, larger wave of IFN gene expression, consisting of non-α4 IFN-α subtypes. Studies in virus-infected mice confirm the importance of the IRF7-dependent positive feedback loop but show that its relative contribution to overall IFN production varies. Some viruses cause secretion of the non-α4 IFN-α subtypes already early during infection. Constitutive expression of IRF7, or the employment of another IRF family member, e.g., IRF5, in the producing cell have been proposed as potential molecular causes for immediate production of all IFN-α subtypes (reviewed in ref. 6). In support of the positive feedback model, deletion of genes encoding the ISGF3 subunits blocks IFN production and thereby reduces antiviral responses (7, 8). Loss of IFN-β, likewise, reduces IFN-α production by virus-infected fibroblasts and increases sensitivity of mice to Vaccinia virus (9).

Although a number of type I IFN–stimulated genes (ISGs) show a clear link to the antiviral state (Table 1), mechanisms of action for most of these are poorly understood. Many have been cloned by cDNA subtraction or related methods and have not been shown to contribute to the antiviral state. Studies in gene-targeted mice clearly support a decisive role of the ISGF3

Citation for this article: *J. Clin. Invest.* **109**:1271–1277 (2002). doi:10.1172/JCI200215770.

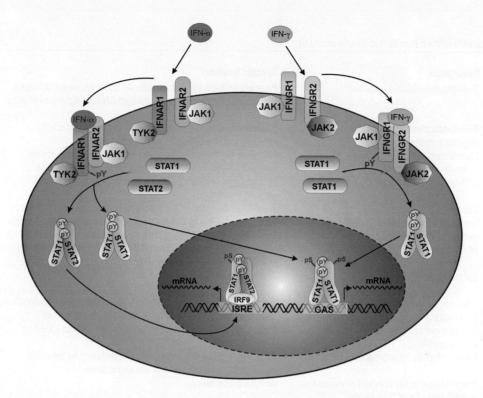

Figure 1

JAK-STAT signal transduction in response to type I and type II IFNs. Upon binding of ligand, IFN receptor–associated Janus kinases (JAKs) are activated and phosphorylate receptor chains on tyrosine. Cytoplasmic signal transducers and activators of transcription (STATs) bind to the phosphorylated receptors with their SH2 domains. JAKs associated with the type I IFN receptor (IFNAR) then phosphorylate STAT1 and STAT2 on tyrosine, causing the formation of predominantly STAT1-STAT2 heterodimers, and of STAT1 homodimers. IFN-γ receptor–associated (IFNGR-associated) JAKs phosphorylate STAT1, leading to the formation of STAT1 homodimers. STAT dimers translocate to the cell nucleus. Thereafter, STAT1-STAT2 heterodimers associate with a third protein, IRF9, and bind one class of type I IFN response elements, the ISRE, whereas STAT1 homodimers activate gene expression by binding to another class of IFN response elements, the GAS.

complex and its target genes in type I IFN–dependent innate immunity to virus. The individual knockouts of all three ISGF3 subunits cause severe loss of resistance to viruses, apparently of a similar magnitude to that resulting from disruption of *Ifnar1* (10–13). Since STAT1 is a component of ISGF3 and contributes to both IFN-α/β and IFN-γ responses, it has been difficult to assess the role of STAT1 dimer target genes specifically in type I IFN responses. Experiments in genetically modified human cell lines suggest it may be of minor importance, but IFN-γ–treated cells are thought to achieve their potent antiviral state primarily through the action of the STAT1 dimer, rather than through activation of ISGF3 target genes.

The importance of type I IFN in innate antiviral responses is clear from studies in mice with targeted disruption of either *Ifnar1* or *Stat2* (13, 14). Such mice fail to clear a variety of viruses, such as vesicular stomatitis virus (VSV), Semliki Forest virus (SFV), Vaccinia virus (VV), and lymphocytic choriomeningitis virus (LCMV) and are prone to die from the corresponding viral infections. The loss of virus-specific cytotoxic T lymphocyte (CTL) responses seen in *Ifnar1*−/− mice is thought to result from exhaustion of cytotoxic T cells due to overwhelming viral burden (14). Since IFN-nonresponsive mice do not display a general impediment in mounting adaptive immune responses, the main defect in antiviral immunity is likely to result from the innate component, the establishment of an antiviral state. Besides producing an antiviral state, the IFNs (particularly the type I IFNs) can affect the rate at which infected

cells undergo apoptosis (15). The contribution of this apoptotic response to viral clearance not known.

A fairly complex picture of IFN function has emerged from studies with influenza virus. Mouse lines with a wild-type allele of the type I IFN–inducible *Mx* gene show increased resistance compared with those with a mutated *Mx*, corroborating cell culture studies indicating that Mx can efficiently interfere with influenza virus replication (16). In Mx-deficient mice (including most inbred mouse lines), type I IFN and STAT1 determine the tissue tropism of influenza infection (17, 18). In infected lungs, where the antiviral state makes only a minor contribution to viral clearance, the major effect of IFN is to regulate the inflammatory environment. When a hemorrhagic form of the influenza virus progresses beyond the respiratory system in an Mx-deficient animal, the type I IFN– and STAT1-dependent antiviral state of infected cells appears to be a major determinant for its clearance.

The importance of IFN-γ in innate antiviral immunity depends greatly on the type of infecting virus. While *IFN-γ*–deficient mice and *IFNGR1*-deficient mice show an unimpaired ability to cope with VSV or SFV infections, their susceptibility to VV or to LCMV is strongly increased (14, 19, 20). Indeed, mice in which both *Ifnar1* and *Ifngr1* are disrupted display an additive susceptibility to VV and LCMV, compared with the individual knockouts (21). Taken together, the studies suggest that in animals, the two IFN types are complementary and to some extent nonredundant with respect to innate resistance to different viral pathogens. Whether

Science in Medicine

Table 1

IFN-induced genes with known function in innate immunity

Gene	Description	Specific function	Inducing IFN
2-5A synthetase	2′-5′-oligoadenylate synthetase	Products of 2–5A synthetase activate latent riboendonuclease RNaseL that degrades single-stranded RNA molecules of viral and cellular origin	Types I and II
RNaseL	Riboendonuclease activated by 2′,5′-linked oligonucleotides	Degrades both viral and cellular RNA upon viral infection, resulting in apoptosis and virus inhibition	Type I
PKR	Double-stranded RNA–activated protein serine/threonine kinase	Important for the induction of antiviral state; inhibits protein synthesis, plays a role in regulating (IFN) gene expression	Types I and II
Mx1 and 2	GTP-binding proteins with GTPase activity	Possess strong antiviral potential; mechanism of action not well understood	Type I
GBP1 and 2	Guanylate-binding proteins	Exhibit antiviral properties; mechanism of action not understood	Type II
PML, SP100	Localized to promyelocytic oncogenic domains (PODs) in the nucleus	Required for IFN-induced transcriptional repression of retroviral genes; inhibit replication of DNA tumor viruses	Type I
p56	56-kDa translation inhibitor	Virus, double-stranded RNA and IFN-induced inhibitor of the eukaryotic translation initiation factor 3	Type I
β_2m	β_2-microglobulin	Antigen presentation	Type I and II
TAP1			
TAP2	Transporters associated with antigen presentation	Essential for transport of peptides from cytosol to the endoplasmic reticulum and peptide loading of MHC class I molecules	Type II
LMP2	Proteasomal component, low–molecular weight protein	Important for production of peptides for MHC class I loading	Type II
GILT	IFN-γ–inducible lysosomal thiol reductase	Facilitates antigen processing and presentation by reducing disulfide bonds of antigens in late endosomes	Type II
CIITA	Transcription factor involved in expression of MHC class I and II molecules	Antigen presentation	Predominantly type II
gp91-phox	Heavy-chain subunit of the cytochrome b558 that is a component of NADPH oxidase	Involved in respiratory burst	Type II
iNOS (NOS2)	Inducible nitric oxide synthase	Protective role against protozoan, bacterial, and some viral infections. iNOS negatively influences the clearance of infection with *Mycobacterium avium*	Types I and II
NRAMP1	Natural resistance–associated macrophage protein	Confers resistance to infection with intracellular pathogens, such as *Mycobacterium*, *Salmonella*, and *Leishmania*; involved in the delivery of intracellular bactericidal agents to endosomes	Type II
Rab5a	GTPase	Facilitates translocation of Rac2 to bacteria-containing phagosomes (shown for *Listeria monocytogenes*), which governs the NADPH oxidase activity and, hence, oxidative burst	Type II
IGTP			
LRG-47			
GTPI			
IRG-47			
TGTP/Mg2			
IIGP	Group of 47- to 48-kDa GTPases likely to be involved in protein trafficking	Protection against protozoan and bacterial pathogens. The requirement for each member depends on the type of pathogen	Mostly type II
IDO	Indoleamine 2,3-dioxygenase	Catalyzes decyclization of L-tryptophan thereby limiting the availability of this amino acid to intracellular microorganisms	Type II
RANTES			
MIP-1α and MIP-1β			
MCP-1			
IP-10	Chemokines	8- to 10-kDa secreted proteins capable of regulating migration, activation, and maturation of leukocytes; important for orchestrating interactions between innate and adaptive immune systems	Mostly type II

this reflects different conditions of synthesis, a qualitative difference of the respective antiviral states, or distinct immunoregulatory properties still needs to be clarified.

Innate immunity to bacteria and protozoa

Many cell types, particularly macrophages and DCs, produce type I IFN following exposure to bacteria and protozoa (22). In most cases, the molecular mechanism by which type I IFN synthesis is stimulated is unknown. However, IRF3 has been shown to be activated by LPS (23, 24) or by infection of macrophages with the intracellular pathogen *Listeria monocytogenes* (S. Stockinger and T. Decker unpublished observations), suggesting that, as in viral infections, this transcription factor may be an important mediator of bacteria-induced IFN production.

The contribution of type I IFNs to innate immunity against bacteria or protozoa appears to vary with the particular microorganism. The mortality of IFNAR1- or STAT2-deficient mice after infection with *L. monocytogenes* is similar to that of wild-type mice (13, 21), but the bacterial loads found in livers and spleens are significantly lower in STAT2-deficient than in wild-type mice (C. Schindler, personal communication). Studies from our laboratories show that type I IFNs sensitize infected macrophages to *L. monocytogenes*–induced cell death in a STAT1-dependent manner (S. Stockinger et al., unpublished observations). Whether this is the underlying cause of the adverse effect of type I IFN in vivo remains to be shown. Likewise, there are few experimental data to indicate whether type I IFNs generally delay the clearance of intracellular bacteria. In support of this notion, a study of mice infected with *Mycobacterium tuberculosis* reported that bacterial virulence correlates directly with the ability to produce type I IFNs, and that the administration of recombinant IFN-α exacerbates lung disease (25). However, an independent examination of *M. tuberculosis* infection found a slightly enhanced replication of the bacteria in the lungs of IFNAR1-deficient mice compared with wild-type controls (26). Therefore the impact of type I IFN may vary with yet undefined parameters of the host-pathogen interplay.

Like *L. monocytogenes*, the protozoan parasite *Leishmania major* multiplies inside macrophages. However, in this case, mice benefit from the synthesis of and response to type I IFNs (27), which synergize with the parasite in stimulating the expression of inducible nitric oxide synthase (NOS2 or iNOS); NO production appears to limit the spread of the pathogen in the infected host. Exposure to LPS or viruses, as well as to *L. major*, induces expression of NOS2, particularly in the presence of type I IFN. The importance of this synergistic response is clear in the case of LPS, where the bacterial product stimulates macrophages to produce IFN and both together then stimulate NOS2 expression. This model of IFN-β's role as a secondary mediator of responses to bacterial stimuli of toll-like receptor 4 (TLR4) appears to apply more generally to genes linked to JAK-STAT signaling (28) and has been validated by work in *Tyk2* deficient mice (29). Due to their lower production of, and lower responsiveness to, type I IFNs, macrophages from these mice fail to produce NO following LPS treatment. Taken together, these findings suggest that type I IFNs and their stimulation of JAK-STAT signaling may be advantageous where NO is crucial in clearing or limiting an infection.

IFN-γ appears to act by at least two distinct mechanisms to augment innate cellular immunity. The main target of IFN-γ activity is the macrophage, and a plethora of investigations document the increased ability of IFN-γ–activated macrophage cultures to kill ingested bacterial or protozoan pathogens (2). The NADPH oxidase subunits, NOS2, lysosomal enzymes, and tryptophan-metabolizing enzymes — involved in the killing of ingested microorganisms by, respectively, reactive oxygen species, NO radicals, breakdown, and tryptophan depletion — are among the gene products induced by IFN-γ (Table 1). Some of these factors have been shown to be regulated by STAT1, explaining the increased susceptibility of STAT1-deficient mice to bacterial pathogens. In addition to arming macrophages for enhanced killing of microbes, a second important function of IFN-γ is to enhance the synthesis of cytokines that contribute to antimicrobial immunity in vivo. The IL-12 p40 subunit represents one well-documented example, but a similar situation may apply to TNF-α, IL-1, and possibly other cytokines.

Converging evidence from a number of different experimental approaches demonstrates the importance of IFN-γ in host defense, particularly against pathogens spending at least part of their life cycle inside cells, such as species of the genera *Listeria*, *Mycobacterium*, *Salmonella*, *Chlamydia*, and *Leishmania* (11, 19–21, 30–33). Treatment with recombinant IFN-γ or, conversely, abrogation of IFN-γ responsiveness (either genetically or by treating with neutralizing antibodies) consistently reveals the cytokine to be a major determinant of pathogen persistence or clearance and of host survival. Remarkably, the impact of IFN-γ can vary considerably among species of a given bacterial genus, as in the case of *Mycobacterium* (34), or even among different attenuated variants of *Salmonella typhimurium* carrying mutations in different virulence genes (35). Hence, there must be a close relationship between particular virulence strategies and the IFN-γ–dependent defense mechanisms.

Infection studies in mice make it hard to distinguish the effects of IFN-γ on innate immunity, like the activation of macrophages, from its effects on adaptive immunity, like the generation of Th1 cells and CTLs. However, the importance of the former responses, at least during the initial exposure to a pathogen, is seen in studies like that of Harty and Bevan (36), who found that naive animals resist virulent *L. monocytogenes* mainly through IFN-γ–dependent innate immunity, with NK cells being the main source of IFN-γ (37). By contrast, immune animals combat the bacterium in a CTL-dependent, but largely IFN-γ–independent, fashion. The crucial roles of IFN-γ and STAT1 in defense against intracellular pathogens have also been demonstrated in humans with mutations in STAT1 or the IFN-γ receptor chain genes (38, 39). In all cases, these mutations result in an increased susceptibility to infection with *Mycobacterium*. The STAT1 mutation (L706S) reported by Dupuis and colleagues (39) is particularly interesting because it behaves as a dominant with regard to the formation of a STAT1 dimer in response to IFN-γ, but as a recessive with regard to the formation of ISGF3 upon IFN-α treatment. As a result, heterozygous individuals display an increased risk of contracting mycobacterial and other bacterial infections, but an apparently normal innate response against viral pathogens. This finding is consistent with previous suggestions that ISGF3 target genes are sufficient to build up an antiviral state against most viruses, but that STAT1 dimer target genes are required for innate immunity against intracellular bacteria.

Bacteria and their products can directly influence STAT1's activity as a transcription factor. In addition to phosphorylation on Y701, STAT1 must be phosphorylated on S727, in its carboxy-terminal transactivation domain, for full activity (40). Whereas bacteria do not directly induce the former modification, they can rapidly stimulate STAT1 S727 phosphorylation via p38^{MAPK} (41), leaving STAT1 primed for increased transcription factor activity following IFN-stimulated tyrosine phosphorylation. Complicating this model of the effects of microbial infection is the finding that a prolonged encounter of macrophages with bacteria leads to the synthesis of suppressors of cytokine signaling (SOCS), proteins that can block the action of specific STATs (41, 42). Therefore, the duration of the bacterial stimulus on macrophages is of critical importance for its effect on transcriptional responses to IFN. The complexity of the situation makes it hard to assess the importance of the interplay of bacterial and IFN stimuli on JAK-STAT signaling in an infected host.

Regulation of dendritic cell and NK cell function

An important property of the innate immune system is that it translates its encounter with microorganisms into an appropriate stimulation of the adaptive immune system. For several reasons, DCs represent a major link between the innate and adaptive immune responses (43). DCs are the most potent antigen-presenting cells and the dominant cell type in the activation of naive T cells. Different subsets of DCs are thought to determine which type of immunity dominates the response to a given antigen (44). In humans, DC1 (myeloid) and DC2 (plasmacytoid) subsets have been defined based on their ability to produce, respectively, high or low amounts of IL-12 and thus to promote the generation of Th1 or Th2 cells. However, the simple beauty of this model has not held up on further study: DC1 and DC2 cells, maturing from their pre-DC1 and pre-DC2 progenitors under various conditions, show a high degree of plasticity with respect to their ability to stimulate Th1 or Th2 differentiation (45).

Among human leukocytes, pre-DC2 cells represent the major type I IFN–producing cells (46, 47). Production of IFN can be stimulated by microbes or their products — viral surface components; double-stranded RNA (48, 49), a ligand of TLR3; and nonmethylated CpG oligonucleotides (50), a ligand for TLR9. The vast quantities of type I IFNs produced following exposure to any of these inducers influence the biology of DCs in several respects. First, they promote pre-DC2 differentiation to DC2 by serving as a survival factor (51). In addition, they alter the character of the downstream DC and T cell response. Unlike DC2 cells maturing in the presence of IL-3, which cause Th2 differentiation, virus-induced DC2 cells induce the generation of a Th subset producing large amounts of IFN-γ as well as IL-10 (51). In this situation, DC2-derived IFN-α most likely stimulates the synthesis of IFN-γ by T cells, bypassing the usual requirement for IL-12. Because IFN-α suppresses the ability of CD14+ monocytes to mature to DCs (52) and promotes the apoptosis of monocyte-derived DCs in response to bacterial stimuli (53), it appears that the maturation-promoting effect of type I IFN is restricted to plasmacytoid pre-DCs. As in humans, an immature DC (CD11c+, Gr-1+, B220+) synthesizing large quantities of IFN has been identified in mice and shown to belong with a plasmacytoid, CD8– DC subpopulation (54, 55). CD8+ and CD8– DCs in mice are usually taken to correspond to the human DC1 and DC2 subpopulations, respectively. Finally, IFN-α has been implicated in DC cross-inhibition, whereby DC2s suppress IL-12 production by DC1s (56).

Work with IFNAR-deficient mice confirms the importance of DC-derived type I IFNs and DC responses to these cytokines for priming a clonal immune response. In wild-type mice, IFN-α efficiently substitutes for CFA in the generation of an efficient humoral immune response. Consistent with this, CFA itself fails to show normal adjuvant activity in the absence of IFNAR expression, and adoptively transferred wild-type DCs can rescue the adjuvant activity of type I IFN in *Ifnar*–/– mice (57). Together, the data suggest a model for adjuvant responses in which the bacterial components (in CFA, for example) stimulate TLRs, which promote type I IFN production. This in turn stimulates DCs to generate and activate Th cells. The

finding that human DC subsets differ in their expression of individual TLRs (50) suggests that type I IFN production and the subsequent generation of adaptive immune responses vary depending on which microbe stimulates one or another DC subset.

NK cells play an important role as producers of IFN-γ. NK cell–derived IFN-γ activates macrophages and serves later in the course of an immune response as a regulator of Th differentiation. Human and murine NK cells behave differently when stimulated by type I IFNs. As with Th cells, human but not murine NK cells produce IFN-γ in response to these cytokines (58). The behavior of STAT4 in human and murine cells explains this species difference: STAT4 is required in Th1 and NK cells to activate IFN-γ transcription in response to IL-12. Type I IFNs stimulate human NK cells to activate STAT4, because the protein interacts via the STAT2 C-terminus with the type I IFN receptor complex and becomes phosphorylated on tyrosine. By contrast, the C-terminus of the murine STAT2 carries a short insertion that disrupts this interaction, preventing the recruitment and activation of STAT4 (59). Curiously, IFN-α is not simply inert with respect to IFN-γ synthesis in the mouse but actually decreases the ability of murine NK cells to synthesize IFN-γ in a STAT1-dependent manner (60). A molecular mechanism for the antagonism between STAT1, activated by type I IFN, and STAT4, activated by IL-12, remains elusive.

Concluding remarks

In recent years the role of type I IFN as antiviral agents and the prominent role of IFN-γ as a macrophage-activating cytokine have been confirmed and extended in mice deficient in their response to one or both IFN types, but much still remains to be learned about IFN target genes that establish the antiviral state in response to different viruses, or that alter the ability of macrophages to control various pathogens. The importance of STAT-dependent (and perhaps also STAT-independent) immunoregulation of macrophage, DC, and NK cell responses by the IFNs is only beginning to be understood and will require further attention. Recent evidence for an IFNAR1-, STAT1-dependent signal occurring in the absence of ongoing immune responses to regulate MHC I expression suggests that a weak constitutive production of type I IFN keeps the host organism in a state of alertness to incoming virus (61). Despite the intensity of IFN research over several decades, the mechanisms of IFN signaling and their impact on immune responses are far from being understood.

Note. Due to space constraints, a number of important references could not be included in this article. Interested readers can find a supplementary reading list at www.jci.org/cgi/content/full/109/10/1271/DC1.

Address correspondence to: Thomas Decker, Vienna Biocenter, Institute of Microbiology and Genetics, Dr. Bohr-Gasse 9, A1030 Vienna, Austria. Phone: 43-1-4277-54605; Fax: 43-1-4277 9546; E-mail: decker@gem.univie.ac.at.

1. Pestka, S., Langer, J.A., Zoon, K.C., and Samuel, C.E. 1987. Interferons and their actions. *Annu. Rev. Biochem.* **56**:727–777.
2. Murray, H.W. 1992. The interferons, macrophage activation, and host defense against nonviral pathogens. *J. Interferon Res.* **12**:319–322.
3. Schindler, C., and Darnell, J.E., Jr. 1995. Transcriptional responses to polypeptide ligands: the JAK-STAT pathway. *Annu. Rev. Biochem.* **64**:621–651.
4. Decker, T., Kovarik, P., and Meinke, A. 1997. Gas

elements: a few nucleotides with a major impact on cytokine-induced gene expression. *J. Interferon Cytokine Res.* **17**:121–134.
5. Ramana, C.V. 2000. Regulation of c-myc expression by IFN-gamma through Stat1-dependent and -independent pathways. *EMBO J.* **19**:263–272.
6. Levy, D.E. 2002. Whence interferon? Variety in the production of interferon in response to viral infection. *J. Exp. Med.* **195**:F15–F18.
7. Sato, M., et al. 1998. Positive feedback regulation of

type I IFN genes by the IFN-inducible transcription factor IRF-7. *FEBS Lett.* **441**:106–110.
8. Marie, I., Durbin, J.E., and Levy, D.E. 1998. Differential viral induction of distinct interferon-alpha genes by positive feedback through interferon regulatory factor-7. *EMBO J.* **17**:6660–6669.
9. Deonarain, R. 2000. Impaired antiviral response and alpha/beta interferon induction in mice lacking beta interferon. *J. Virol.* **74**:3404–3409.
10. Durbin, J.E., Hackenmiller, R., Simon, M.C., and

Levy, D.E. 1996. Targeted disruption of the mouse Stat1 gene results in compromised innate immunity to viral disease. *Cell.* **84**:443–450.

11. Meraz, M.A., et al. 1996. Targeted disruption of the Stat1 gene in mice reveals unexpected physiologic specificity in the JAK-STAT signaling pathway. *Cell.* **84**:431–442.

12. Kimura, T., et al. 1996. Essential and non-redundant roles of p48 (ISGF3 gamma) and IRF-1 in both type I and type II interferon responses, as revealed by gene targeting studies. *Genes Cells.* **1**:115–124.

13. Park, C., Li, S., Cha, E., and Schindler, C. 2000. Immune response in Stat2 knockout mice. *Immunity.* **13**:795–804.

14. Muller, U., et al. 1994. Functional role of type I and type II interferons in antiviral defense. *Science.* **264**:1918–1921.

15. Tanaka, N., et al. 1998. Type I interferons are essential mediators of apoptotic death in virally infected cells. *Genes Cells.* **3**:29–37.

16. Pavlovic, J., and Staeheli, P. 1991. The antiviral potentials of Mx proteins. *J. Interferon Res.* **11**:215–219.

17. Garcia-Sastre, A., et al. 1998. The role of interferon in influenza virus tissue tropism. *J. Virol.* **72**:8550–8558.

18. Durbin, J.E., et al. 2000. Type I IFN modulates innate and specific antiviral immunity. *J. Immunol.* **164**:4220–4228.

19. Dalton, D.K., et al. 1993. Multiple defects of immune cell function in mice with disrupted interferon-gamma genes. *Science.* **259**:1739–1742.

20. Huang, S., et al. 1993. Immune response in mice that lack the interferon-gamma receptor. *Science.* **259**:1742–1745.

21. van den Broek, M.F., Muller, U., Huang, S., Zinkernagel, R.M., and Aguet, M. 1995. Immune defence in mice lacking type I and/or type II interferon receptors. *Immunol. Rev.* **148**:5–18.

22. Bogdan, C. 2000. The function of type I interferons in antimicrobial immunity. *Curr. Opin. Immunol.* **12**:419–424.

23. Navarro, L., and David, M. 1999. p38-dependent activation of interferon regulatory factor 3 by lipopolysaccharide. *J. Biol. Chem.* **274**:35535–35538.

24. Kawai, T., et al. 2001. Lipopolysaccharide stimulates the MyD88-independent pathway and results in activation of IFN-regulatory factor 3 and the expression of a subset of lipopolysaccharide-inducible genes. *J. Immunol.* **167**:5887–5894.

25. Manca, C., et al. 2001. Virulence of a *Mycobacterium tuberculosis* clinical isolate in mice is determined by failure to induce Th1 type immunity and is associated with induction of IFN-alpha/beta. *Proc. Natl. Acad. Sci. USA.* **98**:5752–5757.

26. Cooper, A.M., Pearl, J.E., Brooks, J.V., Ehlers, S., and Orme, I.M. 2000. Expression of the nitric oxide synthase 2 gene is not essential for early control of *Mycobacterium tuberculosis* in the murine lung. *Infect. Immun.* **68**:6879–6882.

27. Diefenbach, A., et al. 1998. Type 1 interferon (IFNalpha/beta) and type 2 nitric oxide synthase regulate the innate immune response to a protozoan parasite. *Immunity.* **8**:77–87.

28. Toshchakov, V., et al. 2002. TLR4, but not TLR2, mediates IFN-beta-induced STAT1alpha/beta-dependent gene expression in macrophages. *Nat. Immunol.* **3**:392–398.

29. Karaghiosoff, M., et al. 2000. Partial impairment of cytokine responses in Tyk2-deficient mice. *Immunity.* **13**:549–560.

30. Kiderlen, A.F., Kaufmann, S.H., and Lohmann-Matthes, M.L. 1984. Protection of mice against the intracellular bacterium *Listeria monocytogenes* by recombinant immune interferon. *Eur. J. Immunol.* **14**:964–967.

31. Nauciel, C., and Espinasse-Maes, F. 1992. Role of gamma interferon and tumor necrosis factor alpha in resistance to *Salmonella typhimurium* infection. *Infect. Immun.* **60**:450–454.

32. Cooper, A.M., et al. 1993. Disseminated tuberculosis in interferon gamma gene-disrupted mice. *J. Exp. Med.* **178**:2243–2247.

33. Rottenberg, M.E., et al. 2000. Regulation and role of IFN-gamma in the innate resistance to infection with *Chlamydia pneumoniae. J. Immunol.* **164**:4812–4818.

34. Doherty, T.M., and Sher, A. 1997. Defects in cell-mediated immunity affect chronic, but not innate, resistance of mice to *Mycobacterium avium* infection. *J. Immunol.* **158**:4822–4831.

35. VanCott, J.L., et al. 1998. Regulation of host immune responses by modification of *Salmonella* virulence genes. *Nat. Med.* **4**:1247–1252.

36. Harty, J.T., and Bevan, M.J. 1995. Specific immunity to *Listeria monocytogenes* in the absence of IFN gamma. *Immunity.* **3**:109–117.

37. Bancroft, G.J., Schreiber, R.D., and Unanue, E.R. 1991. Natural immunity: a T-cell-independent pathway of macrophage activation, defined in the scid mouse. *Immunol. Rev.* **124**:5–24.

38. Dupuis, S., et al. 2000. Human interferon-gamma-mediated immunity is a genetically controlled continuous trait that determines the outcome of mycobacterial invasion. *Immunol. Rev.* **178**:129–137.

39. Dupuis, S., et al. 2001. Impairment of mycobacterial but not viral immunity by a germline human STAT1 mutation. *Science.* **293**:300–303.

40. Decker, T., and Kovarik, P. 2000. Serine phosphorylation of STATs. *Oncogene.* **19**:2628–2637.

41. Stoiber, D., Stockinger, S., Steinlein, P., Kovarik, J., and Decker, T. 2001. *Listeria monocytogenes* modulates macrophage cytokine responses through STAT serine phosphorylation and the induction of suppressor of cytokine signaling 3. *J. Immunol.* **166**:466–472.

42. Stoiber, D., et al. 1999. Lipopolysaccharide induces in macrophages the synthesis of the suppressor of cytokine signaling 3 and suppresses signal transduction in response to the activating factor IFN-gamma. *J. Immunol.* **163**:2640–2647.

43. Lanzavecchia, A., and Sallusto, F. 2001. Regulation of T cell immunity by dendritic cells. *Cell.* **106**:263–266.

44. Lanzavecchia, A., and Sallusto, F. 2001. The instructive role of dendritic cells on T cell responses: lineages, plasticity and kinetics. *Curr. Opin. Immunol.* **13**:291–298.

45. Liu, Y.J., Kanzler, H., Soumelis, V., and Gilliet, M. 2001. Dendritic cell lineage, plasticity and cross-regulation. *Nat. Immunol.* **2**:585–589.

46. Cella, M., et al. 1999. Plasmacytoid monocytes migrate to inflamed lymph nodes and produce large amounts of type I interferon. *Nat. Med.* **5**:919–923.

47. Siegal, F.P., et al. 1999. The nature of the principal type 1 interferon-producing cells in human blood. *Science.* **284**:1835–1837.

48. Cella, M., et al. 1999. Maturation, activation, and protection of dendritic cells induced by double-stranded RNA. *J. Exp. Med.* **189**:821–829.

49. Kadowaki, N., Antonenko, S., and Liu, Y.J. 2001. Distinct CpG DNA and polyinosinic-polycytidylic acid double-stranded RNA, respectively, stimulate CD11c- type 2 dendritic cell precursors and CD11c+ dendritic cells to produce type I IFN. *J. Immunol.* **166**:2291–2295.

50. Kadowaki, N., et al. 2001. Subsets of human dendritic cell precursors express different toll-like receptors and respond to different microbial antigens. *J. Exp. Med.* **194**:863–869.

51. Kadowaki, N., Antonenko, S., Lau, J.Y., and Liu, Y.J. 2000. Natural interferon alpha/beta-producing cells link innate and adaptive immunity. *J. Exp. Med.* **192**:219–226.

52. McRae, B.L., Nagai, T., Semnani, R.T., van Seventer, J.M., and van Seventer, G.A. 2000. Interferon-alpha and -beta inhibit the in vitro differentiation of immunocompetent human dendritic cells from CD14(+) precursors. *Blood.* **96**:210–217.

53. Lehner, M., Felzmann, T., Clodi, K., and Holter, W. 2001. Type I interferons in combination with bacterial stimuli induce apoptosis of monocyte-derived dendritic cells. *Blood.* **98**:736–742.

54. Nakano, H., Yanagita, M., and Gunn, M.D. 2001. Cd11c(+)b220(+)gr-1(+) cells in mouse lymph nodes and spleen display characteristics of plasmacytoid dendritic cells. *J. Exp. Med.* **194**:1171–1178.

55. Asselin Paturel, C., et al. 2001. Mouse type I IFN-producing cells are immature APCs with plasmacytoid morphology. *Nat. Immunol.* **2**:1144–1150.

56. McRae, B.L., Semnani, R.T., Hayes, M.P., and van Seventer, G.A. 1998. Type I IFNs inhibit human dendritic cell IL-12 production and Th1 cell development. *J. Immunol.* **160**:4298–4304.

57. Le Bon, A., et al. 2001. Type i interferons potently enhance humoral immunity and can promote isotype switching by stimulating dendritic cells in vivo. *Immunity.* **14**:461–470.

58. Sinigaglia, F., D'Ambrosio, D., and Rogge, L. 1999. Type I interferons and the Th1/Th2 paradigm. *Dev. Comp. Immunol.* **23**:657–663.

59. Farrar, D.J., et al. 2000. Selective loss of type I interferon-induced STAT4 activation caused by a minisatellite insertion in mouse Stat2. *Nat. Immunol.* **1**:65–69.

60. Nguyen, K.B., et al. 2000. Interferon α/β-mediated inhibition and promotion of interferon γ: STAT1 resolves a paradox. *Nat. Immunol.* **1**:70–76.

61. Taniguchi, T., and Takaoka, A. 2001. A weak signal for strong responses: interferon-alpha/beta revisited. *Nat. Rev. Mol. Cell Biol.* **2**:378–386.

Molecular aspects of primary immunodeficiencies: lessons from cytokine and other signaling pathways

Fabio Candotti,[1] Luigi Notarangelo,[2] Roberta Visconti,[3] and John O'Shea[4]

[1]Genetics and Molecular Biology Branch, National Human Genome Research Institute, NIH, Bethesda, Maryland, USA.
[2]Department of Pediatrics, University of Brescia, Brescia, Italy. [3]Department of Pathology, University Frederico II, Naples, Italy.
[4]Molecular Immunology and Inflammation Branch, National Institute of Arthritis and Musculoskeletal and Skin Diseases, NIH, Bethesda, Maryland, USA.

Numerous examples indicate that mammalian development can be entirely normal in the absence of the immune system; indeed, prior to birth, individuals with even very severe immunodeficiencies are developmentally unaffected. Perhaps it is not surprising, therefore, that more than 95 different primary immunodeficiency syndromes have been identified, encompassing defects in lymphocytes, phagocytes, and complement proteins. Indeed, within the past several years, more than 70 separate genes have been identified whose mutations cause immunodeficiency. These discoveries have been made both by using candidate gene approaches and by positional cloning. In some cases, the generation of gene-targeted mice preceded the identification of human mutations, whereas in other cases, the reverse was true. Therapy for these disorders, ranging from replacement therapy to bone marrow transplantation and gene therapy, has also moved at a rapid pace. This field therefore provides outstanding examples of the power of molecular medicine, with tremendous opportunities for interplay between basic and clinical science.

Many of the processes that govern development of lymphoid and hematopoietic cells are now understood in some detail. We know that the growth and development of hematopoietic precursors, which develop in the fetal liver and bone marrow, are dependent upon a panoply of cytokines. In addition, lymphocytes require appropriate signals from antigen receptors to mature properly (Figure 1). Additionally, other receptors and counter-receptors on lymphoid and antigen-presenting cells are critical for initiating immune responses (Figure 2). We will provide examples in which mutations affect each of these steps and consequently result in immunodeficiency (Table 1). Several excellent reviews provide comprehensive discussion of the genetic basis of primary immunodeficiencies (1, 2). Our goal is not to summarize this field in its entirety; rather, a major focus of this review will be the role of cytokines and their receptors in the pathogenesis of primary immunodeficiencies. Additionally, the identification of new genes associated with immunodeficiency disorders, insights from patient-derived mutations, the heterogeneity of clinical presentations, the significance of revertants, and advances in gene therapy will be highlighted. To an extent, the areas emphasized are also a reflection of our interests, but in general the lessons are applicable to most of the diseases encompassed by primary immunodeficiencies.

Disorders of cytokines and cytokine receptors

Cytokines and receptors of different classes are important in lymphoid development and function. We will briefly consider

cytokines that bind the common γ chain, γ_c; cytokines involved in cell-mediated immunity; and members of the TNF superfamily.

γ_c cytokines, Jak3, and lymphocyte development

IL-2, IL-4, IL-7, IL-9, IL-15, and IL-21 bind γ_c in association with a unique ligand-specific receptor chain. The involvement in signaling via multiple cytokines and its mapping to the X-chromosome led to the identification of mutations of γ_c in boys with X-linked severe combined immunodeficiency (SCID). The disease is now termed SCIDX1 and is the most common type of SCID, accounting for about 50% of cases (2, 3). Through the generation of knockout mice, one γ_c cytokine, IL-7, was shown to be an essential cytokine for T cell development; SCID patients with IL-7 receptor (IL-7R) mutations were subsequently identified (4). However, absence of γ_c interferes with both IL-7 and IL-15 signaling; loss of IL-15 or its receptor abrogates NK cell development. Thus, IL-7R deficiency is associated with absence of T cells and residual NK cells, whereas γ_c deficiency typically results in absence of both lymphoid subsets.

Intracellularly, these receptor subunits associate with the Janus kinases Jak1 and Jak3. Jak1 is widely expressed and is used by many different cytokine receptors, and its knockout is perinatally lethal. In contrast, Jak3 is predominantly expressed in hematopoietic cells and uniquely binds γ_c. Consequently, JAK3 deficiency, the second most common cause of T$^-$B$^+$ SCID, has a phenotype identical to that of SCIDX1 (5, 6). More than 30 patients with more than 35 different mutations spanning the entire *JAK3* gene have been identified (7). These mutations are stochastic, with no evidence of founder effects or hot spots. About half of the patients are compound heterozygotes, whereas half are homozygous for their mutations, due to parental consanguinity. These mutations appear to be fully recessive in that the level of JAK3 protein is normal in heterozygotes; the factors that control JAK3 expression are presently being investigated and likely include both transcriptional and posttranscriptional mechanisms.

Most mutant *JAK3* alleles fail to encode any stable protein, but several missense or small in-frame deletion alleles have been identified that permit near-normal levels of expression of the mutant gene product. The analysis of the effect of these mutations on JAK3 function has greatly improved our understanding of how this tyrosine kinase works. Jaks comprise a C-terminal catalytic domain, preceded by a unique "pseudokinase" domain, which is a frequent site of mutations. Analysis of these mutant alleles has demonstrated the critical function of this domain (8). The pseudokinase domain is not catalytically active but instead has important regulatory functions, perhaps through binding to the kinase domain. The N-terminus of the Jaks has homology to a con-

Citation for this article: *J. Clin. Invest.* **109**:1261–1269 (2002). doi:10.1172/JCI200215769.

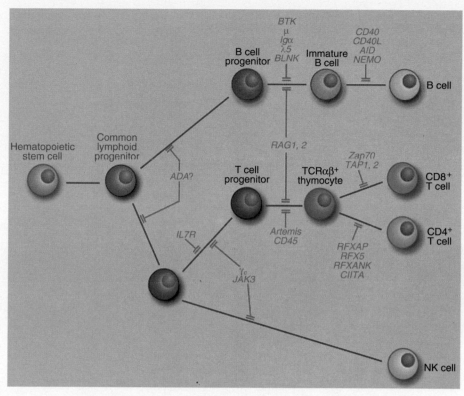

Figure 1
Schematic representation of lymphoid development and genetic lesions leading to immunodeficiency.

served domain in band four point one, ezrin, radixin, and moesin (FERM). Mutations in the JAK3 FERM domain have two important consequences: they disrupt binding to γ_c, and they interfere with catalytic activity (9).

The clinical presentations associated with *JAK3* mutations are surprisingly variable, with about one-third of patients developing T cells. Similarly, a missense mutation of γ_c has been associated with normal numbers of peripheral T and B cells, grossly normal T cell receptor repertoire, normal response to mitogenic stimuli, and the presence of a normal thymus (10). Taken together with the identification a kindred with a different missense mutation of γ_c that exhibited attenuated immunodeficiency (11), this indicates that clinical presentations associated with JAK3 and γ_c mutations can be remarkably broad.

Lack of γ_c or Jak3 interferes with signaling by not only IL-7 and IL-15 but also IL-2, a cytokine that acts in vivo to constrain lymphoid growth by promoting activation-induced cell death and maintaining peripheral tolerance. These effects are best illustrated in humans and mice with a lymphoproliferative disease resulting from mutations in the IL-2Rα and IL-2Rβ subunits (12). Recently, we described a SCID kindred with mutations in *JAK3*. One child in this family developed severe lymphoproliferative disease, but curiously, a sibling was clinically nearly normal but had lymphopenia with oligoclonal T cell expansion and increased numbers of activated and memory T cells. These cells also had poor expression of the proapoptotic molecule Fas ligand, which is typically upregulated by IL-2. Indeed, the T cells that develop in γ_c- and Jak3-deficient mice and humans often express activation markers (13).

Thus, a feature of JAK3 deficiency appears to be impaired lymphoid homeostasis, further complicating the clinical phenotype. This disorder can range from pure immunodeficiency to a mixed picture with varying degrees of autoimmunity and immunodeficiency and, in some cases, minimal clinical consequences. Indeed, it is becoming increasingly clear that T⁻B⁺ SCID represents just one extreme of a range of clinical presentations associated with Jak3 and γ_c mutations. There are other examples of syndromes with a mixed picture of autoimmune and immunodeficiency disease — immune dysregulation polyendocrinopathy X-link and syndrome (IPEX) and autoimmune polyendocrinopathy–candidiasis–ectodermal dystrophy (APECED) being examples. The former is due to mutations of the *FOXP3* gene, which encodes a putative winged-helix/forkhead transcription factor and is thought to serve as a transcriptional repressor of cytokine genes (14). APECED is associated with mutations of the *AIRE* (autoimmune regulator) gene, which is also thought to be a transcriptional regulator (15, 16).

IL-12 and IFN-γ: cytokines that regulate Th1 differentiation and cell-mediated immunity

After development in the thymus, naive CD4⁺ cells differentiate to Th1 cells that produce IFN-γ, a cytokine that promotes cell-mediated immunity; IFN-γ activates the transcription factor Stat1 (signal transducer and activator of transcription-1). One cytokine that regulates this process, IL-12, is composed of two subunits, IL-12p40 and IL-12p35. Its receptor is also a heterodimer, consisting of IL-12Rβ1 and IL-12Rβ2. Patients with mutations in the genes for either IFN-γ receptor subunit (IFNGR1 or IFNGR2), or for STAT1, IL-12p40, or IL-12Rβ1, present with atypical mycobac-

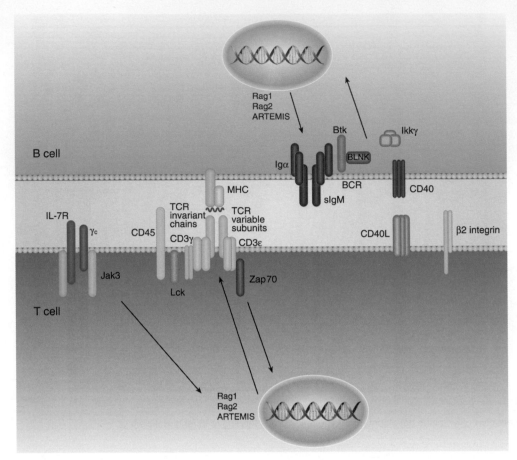

Figure 2
The development of lymphoid cells is dependent upon the expression and signaling by cytokine receptors, antigen receptors, and adhesion/accessory molecules. Mutations of many different genes can interfere with proper lymphoid development and function and, consequently, can lead to immunodeficiency.

terial and Salmonella infections (17, 18). Mice with IFN-γ, IFN-γR, and Stat1 deficiencies show increased incidence of tumors, but whether human mutations of these constituents are associated with increased risk of cancer remains to be determined.

The TNF and TNF receptor superfamilies

Another large family of cytokine receptors, with more than 20 members that bind both soluble and cell-bound ligands, is the TNF superfamily. Mutation of TNF receptor I (TNFRI or TNFSFRIA) is associated with a periodic fever and autoinflammatory disease termed TNFR-associated periodic syndrome (TRAPS) (19). TRAPS is one of a group of heritable periodic fever syndromes that also includes familial Mediterranean fever, Muckle-Wells syndrome, and hyper-IgD syndrome (20). Another receptor/ligand pair in the TNFR family is Fas/FasL, which regulates lymphocyte apoptosis through the activation of intracellular caspases. Deficiency of Fas and FasL was first identified in mice, where their essential pro-apoptotic functions are evident. Later, humans with mutations of *FAS* and *CASP10* (which encodes another proptotic molecule, caspase-10) were identified (21–23); these patients have autoimmune lymphoproliferative disorder and are at increased risk of developing lymphomas. In terms of immunodeficiency, the TNF-family receptor/ligand pair that is most relevant and best characterized is CD40/CD40L (CD154).

CD40/CD40L and hyper-IgM syndrome

CD40L (CD154 or TNFSF5) is expressed on activated CD4+ T cells, whereas its counter-receptor, CD40, is widely expressed on B cells, macrophages, dendritic cells (DCs), and other cells. Maturation of antibody responses occurs in the germinal centers (GCs) and involves the mechanisms of somatic hypermutation (SHM) and class switch recombination (CSR). Through SHM, B cells modify their rearranged Ig genes, generating diversity and high-affinity antibodies. With CSR, B lymphocytes modify their Ig constant region and produce antibodies with distinct biological properties. The cognate interaction between CD40 ligand (CD40L), expressed by activated CD4+ T cells, and CD40 on B lymphocytes provides a key signal. This is disrupted in patients with *CD40L* or *CD40* mutations, who suffer from X-linked immunodeficiency with hyper-IgM (HIGM1) and autosomal-recessive hyper-IgM, respectively (24, 25). HIGM1 is characterized by a lack of GCs and impaired CSR due to lack of T cell stimulation. The finding that SHM occurs normally in affected individuals suggests that SHM is not strictly dependent on CD40L-CD40 interaction and may occur outside of the GCs. Unlike HIGM1, CSR is precluded in CD40-deficient patients due to an intrinsic B cell defect. The concerted action of genes expressed by T and B lymphocytes and follicular DCs oversees the GC reaction, thus exposing terminal B cell differentiation to the consequences of inherited gene mutations.

Table 1
Defects leading to primary immune deficiency

Class of defect	Gene	Disease
Disorders of cytokines and cytokine signaling		
	γc, Jak3, IL-7R	T⁻B⁺ SCID
	IL-2Rα	Lymphoproliferative disease
	IFN-γR1, IFN-γR2, IL-12p40, IL-12Rβ1, Stat1	Atypical mycobacterial infection
	TNFR1	TRAPS
	Fas, caspase-10	ALPS
	CD40, CD40L, AID	HIGM
	IKKγ	Hypohidrotic ectodermal dysplasia
	FOXP3	IPEX
Antigen presentation, receptors, and signaling		
	Tap1, Tap2	Impaired MHC class I expression and CD8⁺ T cell development
	RFXAP, CIITA, RFX5, RFXANK	Impaired MHC class II expression and CD4⁺ T cell development
	Rag1, Rag2	SCID
	Artemis	SCID with radiosensitivity
	DNA ligase IV	SCID
	Nijmegen breakage syndrome	Immunodeficiency
	ATM	Ataxia-telangiectasia
	Igα, λ5, μ-chain	Agammaglobulinemia
	BLNK	Agammaglobulinemia
	Btk	X-linked agammaglobulinemia
	CD3γ, CD3ε,	T cell deficiency
	Lck	SCID
	Zap70	CD8⁺ T cell deficiency
	CD45	SCID
	WASP	WAS
Accessory and adhesion molecules	CD18 (β2 integrin)	LAD I
	GDP-fucose transporter	LAD II
	SH2D1A	XLP
Metabolic	ADA	SCID
	PNP	CID

Signaling through CD40 involves activation of TNF receptor associated factor 6 (TRAF6), activation of the inhibitor of κB kinase (IKK) complex, and induction of NF-κB. Mutations of the γ chain of the IKK complex (also known as NF-κB essential modulator, or NEMO) have been shown to impair NF-κB signaling (26, 27). Affected patients suffer from immune deficiency with hyper-IgM Ig profile due to impaired CSR and hypohidrotic ectodermal dysplasia.

Another form of autosomal-recessive hyper-IgM (HIGM2) has been recently described in patients with mutations of activation-induced cytidine deaminase (AID), a novel enzyme that likely plays a main role in the control of both CSR and SHM (28). AID is selectively expressed in the GC and can be induced in B cells following stimulation with LPS or CD40L and appropriate cytokines. AID-deficient patients show recurrent infections and enlarged GCs in lymph nodes and tonsils containing abundant proliferating IgM⁺ IgD⁺ CD38⁺ GC founder B cells.

Antigen receptors and receptor-mediated signaling

It is important to emphasize that in addition to cytokine signals, signals provided by antigen receptors represent a key event that allows lymphoid differentiation to proceed to the production of mature T and B lymphocytes. Accordingly, deficiencies of components of T and B cell antigen receptors can result in failure to develop these subsets. Multiple types of mutations can interfere with antigen receptor signaling, ranging from impairment antigen

presentation and defects in the assembly of the antigen receptor complex to deficiency of signaling molecules.

Defective MHC class I and class II presentation

Most T cells require antigen to be presented by MHC molecules, class I for CD8⁺ cells and class II for CD4⁺ cells. Mutation of TAP1 or TAP2 impairs expression of HLA class I molecules on the cell surface; as predicted, this interferes with the development of CD8⁺ but not CD4⁺ T cells. Patients with TAP deficiency suffer from recurrent infections and deep skin ulcers. A variety of transcription factors, including RFXAP, CIITA, RFX5, and RFXANK, control MHC class II expression, and mutation of any these leads to combined immunodeficiency. Recently a missense mutation of *CIITA* was identified in three sisters, one of whom was asymptomatic despite a profound reduction in class II expression (29), a finding that underscores once again the heterogeneity that can be seen in patients with primary immunodeficiencies.

Defects of V(D)J recombination and DNA repair

Lymphoid antigen receptors comprise invariant subunits associated with gene products generated by the rearrangement of single variable (V), diversity (D), and joining (J) elements with constant region genes. The invariant subunits are important for signal transduction, whereas the rearranged subunits are responsible for antigen recognition; both, however, are required for lymphoid development. A series of proteins encoded by lymphoid-restricted recombinase-activating genes 1 and 2 (RAG1 and RAG2) or ubiquitously expressed genes (DNA-PKcs, Ku70, Ku80, XRCC4, and DNA ligase IV) mediate the recombination process. Inactivation of any of these gene products results in a block in T and B cell development and profound immune deficiency. In particular, disruption of either *Rag1* or *Rag2* in mice and humans causes SCID with a virtual lack of circulating mature T and B lymphocytes (T⁻B⁻ SCID) (30). Interestingly, mutations that impair but do not abrogate RAG1 and RAG2 protein expression and function result in a distinct human phenotype (Omenn syndrome), characterized by a lack of circulating B cells, with presence of a substantial number of oligoclonal, activated, and poorly functioning T lymphocytes, mimicking graft-versus-host disease (31). In addition, background genetic or perhaps environmental factors also seem to play a role in determining the clinical phenotype of RAG mutations, as T⁻B⁻ SCID, Omenn syndrome, and atypical forms of SCID may coexist within the same family, or in distinct families with the same mutation.

Additionally, a newly identified player, ARTEMIS, which complexes with DNA-PKcs and is thought to provide hairpin-opening activity in V(D)J recombination, has been recently linked to the clinical subgroup of patients with T⁻B⁻ SCID associated with increased radiosensitivity (32, 33). Interestingly, *ARTEMIS* maps at 10p, in a region where the locus for a similar SCID condition found in Athabascan-speaking Native Americans had been mapped. Whether defects in the *ARTEMIS* gene account for all cases of radiation-sensitive T⁻B⁻ SCID in humans, and whether mutations that allow residual expression of the ARTEMIS protein result in a milder phenotype, remain to be established.

Although disruption of genes for DNA-PKcs, Ku70, Ku80, or XRCC4 has been shown to cause SCID in mice, mutations of the corresponding human genes have not been identified thus far. However, DNA ligase IV, which functions in DNA nonhomologous end-joining and V(D)J recombination, was recently demonstrated to be mutated in patients with features of immunodeficiency and developmental delay (34), a syndrome resembling that of the DNA damage response disorder Nijmegen breakage syndrome. Another related disorder, ataxia-telangiectasia (AT), is an autosomal-recessive condition characterized by cerebellar ataxia and progressive neuromotor disability associated with oculo-cutaneous telangiectasia, immunodeficiency of variable degree with absence or degeneration of the thymus, and high predisposition to lymphoid malignancies. AT is also associated with chromosomal instability and increased sensitivity to ionizing radiation. The gene responsible for AT is designated *ATM*; the predicted ATM protein has homology to mammalian phosphatidylinositol 3'-kinase and DNA-dependent protein kinase (35, 36).

Mutations of invariant chains and signaling molecules

Both T and B cell antigen receptors employ invariant chains to initiate signal transduction. For the T cell receptor (TCR), CD3 proteins and TCRζ serve this function; mutations of CD3γ and CD3ε have been identified. Similarly, the invariant chains for the B cell receptor (BCR) are Igα (CD79a) and Igβ (CD79b), and mutation of the former has been found in some patients with agammaglobulinemia. These invariant chains are phosphorylated in response to receptor cross-linking on tyrosine residue motifs termed ITAMs (immunoreceptor-based tyrosine activation motif). In T cells, this is mediated by the protein tyrosine kinase (PTK)

Lck, and phosphorylation of the receptor allows recruitment of a second kinase, Zap70; deficiency of both Lck and Zap70 is also associated with SCID (37, 38). Zap70 deficiency is characteristic in that CD4⁺ T cells are produced but CD8⁺ T cells are absent. The tyrosine phosphatase CD45 is also a key regulator of TCR signaling. Mutation of it was recently demonstrated, and this leads to T⁻B⁺ SCID in humans (39).

Bruton's tyrosine kinase (BTK), a key PTK expressed in B cells, is mutated in X-linked agammaglobulinemia (XLA), a severe immunodeficiency that affects approximately 1 in 50,000 male live births (40). Typically, XLA patients manifest a block in B cell development at the pre-B cell stage and do not produce Ig's. However, the clinical presentations associated with *BTK* mutations are quite heterogeneous. A considerable number of patients have a milder disorder in which B cells develop and variable amounts of Ig's are produced. Here again, there is no clear genotype-phenotype correlation, and a range of clinical phenotypes can be found within a given family. Roughly 10% of patients with agammaglobulinemia do not have mutations of *BTK*. Such patients may have mutations in the autosomal genes for Ig chains, such as the μ heavy chain or a surrogate light chain gene (λ5). Still others carry mutations in BLNK, an adapter molecule that contributes to BCR signaling and is required for the development of pro-B to pre-B cells (41).

Another X-linked disorder, Wiskott-Aldrich syndrome (WAS), is characterized by thrombocytopenia with small-sized platelets, eczema, and immune deficiency, leading to increased susceptibility to infection from all classes of pathogens and high susceptibility to lymphoid cancer. The disease gene was discovered by positional cloning (42) and encodes the WAS protein (WASP), which is expressed in all nonerythroid hematopoietic cells. Together with N-WASP and WAVE, WASP is a member of a family of human proteins responsible for the transduction of signals from the cell membrane to the actin cytoskeleton. WASP interactions with Rho family GTPases and the Arp2/3 complex are critical for this function and affected by WASP mutations, which provide an explanation for several of the biological defects characteristic of WAS hematopoietic cells, including abnormal signaling, polarization, migration, and phagocytosis (reviewed in ref. 43). Intramolecular interactions maintain WASP in an autoinhibited status that is relieved by binding of Cdc42 to the WASP GTPase-binding domain. This event releases the C-terminal region of the protein and enables its interaction with the Arp2/3 complex, thus resulting in localized polymerization of new actin filaments (44). Interestingly, recent findings have shown that mutations of WASP disrupting the autoinhibitory domain do not result in the WAS phenotype, but in an X-linked severe congenital neutropenia (45).

Accessory and adhesion molecules

In addition to antigen receptors, lymphocytes have an array of accessory and adhesion molecules that are important for providing costimulatory signals for T cell activation. Some adhesion molecules are also important for the function of phagocytic cells. Mutations of CD18, a β2 integrin, have long been known to result in the disorder leukocyte adhesion deficiency (LAD I). Additionally, a second form of leukocyte adhesion deficiency (LAD II) is caused by deficient expression of sialyl Lewis X determinants, the neutrophil ligands for E- and P-selectins. Patients affected by LAD II present with less severe infections than LAD I patients, along with developmental abnormalities, short stature, and mental retardation.

The transport of GDP-fucose into isolated Golgi vesicles of LAD II cells is reduced in patient cells, and a point mutation in a GDP-fucose transporter has been recently demonstrated to be responsible for the disease in one patient (46, 47).

The disorder X-linked lymphoproliferative disease is characterized by fatal Epstein-Barr virus (EBV) infection, lymphomas, immunodeficiency, aplastic anemia, and lymphohistiocytic disorders. It appears to result from the failure to control the proliferation of cytotoxic T cells following EBV infection; as such, it may be considered both an autoimmune and an immunodeficiency disease. The gene underlying X-linked lymphoproliferative syndrome, *SH2D1A* (SH2 domain–containing gene 1A), which is expressed in T and NK cells (48), encodes a small adapter protein with a single SH2 domain. SH2D1A binds SLAM (CD150), 2B4 (CD244), and perhaps other lymphoid receptors. The ligand for CD244 is CD48, which is upregulated with EBV infection. Whether SH2D1A is a positive or a negative regulator of signaling is still unclear, and the pathogenesis of this disorder is still incompletely understood. The disease has some features similar to familial hemophagocytic lymphohistiocytosis, which results from mutations of the gene encoding perforin (49), and Griscelli syndrome, which is due to *RAB27A* and *MYO5A* mutations that interfere with granule exocytosis (50, 51).

Metabolic defects

Genetic deficiency of adenosine deaminase (ADA) results in a clinical spectrum of immunodeficiency that ranges from infants with classical SCID to adult-onset presentation with milder combined immunodeficiency (CID) phenotype (reviewed in refs. 52, 53). The molecular basis of ADA deficiency has been known for three decades; however, important biological and clinical challenges still exist that make investigation of this disease current and attractive. The pathogenic mechanisms responsible for immunodeficiency in ADA deficiency remain uncertain. In the purine salvage pathway, ADA catalyzes the deamination of adenosine (Ado) and deoxyadenosine (dAdo) to inosine and deoxyinosine, respectively. In the absence of the enzyme, accumulation of dAdo results in a massive increase of dATP levels in the patients' red blood cells and lymphocytes. High concentrations of dAdo and dATP have profound metabolic consequences, including chromosome breakage, activation of apoptosis, inhibition of methylation reactions and DNA synthesis, and activation of ATP catabolism, all of which are thought to mediate lymphotoxicity and account for the severe depletion of both T and B lymphocytes in affected patients (T⁻B⁻ SCID). Recent studies in ADA knockout mice have shown abundant T cell apoptosis in the thymus (but not in the spleen and lymph nodes), accompanied by reduction of tyrosine phosphorylation of TCR-associated signaling molecules and block of TCR-mediated calcium flux. These findings suggest that, in addition to the direct apoptotic effects of dAdo and dATP, blocks in TCR-driven thymocyte maturation may play a role in the T cell depletion observed in ADA deficiency.

Spontaneous revertants and gene therapy

Primary immunodeficiencies fully fit Lord Garrod's 1924 definition of "Experiments of Nature" and have recently provided fascinating examples of Nature's attempts to correct its own missteps. Reversion of mutations to normal sequence has in fact been demonstrated in a series of immunodeficient patients with mild phenotype or progressive improvement of their clinical presentation. In vivo reversion to normal of an inherited mutation has been described in an ADA-deficient child with progressive clinical improvement and unusually mild biochemical and immunologic phenotype (54). A second case of in vivo reversion in ADA deficiency was recently identified in a subject with a relatively mild biochemical and immunological phenotype, who had acquired a deletion that abrogated the deleterious effects of the original mutation (55).

A reversion of a point mutation to normal sequence was also demonstrated in a boy with an attenuated form of SCIDX1 (56). In this case, the reversion appears to have occurred in a committed T cell progenitor, as the patient presented with low-to-normal numbers of circulating T cells expressing the normal γc molecule, in contrast to B lymphocytes, monocytes, and granulocytes, which all carried the original point mutation and had no detectable γc expression.

Somatic mosaicism sustained by true back mutations in T lymphocytes has also been described in two patients affected with WAS (57, 58). In both cases, WASP-positive and WASP-negative circulating T lymphocytes coexisted, but no revertants could be detected among other lymphoid or myeloid populations. In one case, the proband's condition ameliorated. This individual's peripheral T cells were largely of the revertant genotype, suggesting that the reversion event had resulted in clinical improvement, although this was ascertained retrospectively.

The observation of these immunodeficient patients with attenuated phenotype and carrying reversions of their original mutation adds a level of complexity to the clinical interpretation of primary immunodeficiencies as it broadens the spectrum of possible presentation. Whenever normal cells enjoy a selective advantage over cells with a mutant phenotype, reversions to the wild type can lead to substantial somatic mosaicism and relatively mild symptoms — a possibility that should be considered each time we are confronted with atypical presentations of known immunodeficiency syndromes.

Such in vivo reversions can be considered as spontaneous forms of gene therapy and can provide important information as to the legitimate chances of success of genetic correction strategies for immunodeficiency. In particular, the above-described case of somatic mosaicism due to reversion of γc mutation had suggested that gene therapy should be an efficient form of treatment for SCIDX1, a prediction that was recently confirmed by the results of a successful clinical trial (59). In this recent report, Cavazzana-Calvo and coworkers reported on the results from five patients treated with autologous CD34⁺ bone marrow cells in which the cDNA encoding for human γc had been transferred using a retroviral vector. This procedure resulted in the appearance of normal numbers of circulating T lymphocytes expressing γc in four out of five patients. These T lymphocytes were polyclonal and functionally competent, as demonstrated by normal responses to stimulation with mitogens and specific antigens. In addition, despite the fact that less than 1% of the patients' B cells became γc-positive, the first two treated patients developed specific antibody responses to tetanus and diphtheria toxins, as well as polioviruses. Not surprisingly, based on the predicted lack of selective advantage in myeloid lineages, only 0.01–1% of the patients' monocytes and granulocytes showed evidence of genetic correction. The observation of adequate humoral immune responses in the presence of very low numbers of gene-corrected B lymphocytes may indicate that the humoral immunodeficiency observed in untreated SCIDX1 patients is due mostly to lack of T cell helper function and less to intrinsic deficiencies of their B lymphocytes.

These results are regarded as sensational, especially when compared with previous outcomes of hematopoietic stem gene therapy for such other immunodeficiency diseases as ADA deficiency (60, 61), chronic granulomatous disease (62), and LAD I (63), which failed to demonstrate clear clinical benefit. There is now high hope that the application of the same technical progress — including the use of fibronectin cell culture support, novel cytokines, and modern vector design — will lead to similar successes in gene therapy for other immunodeficiencies. Indeed, novel gene therapy approaches for ADA deficiency have generated promising preliminary results (A.S. Aiuti, personal communication).

Conclusions

In this review, we have tried to emphasize several themes. First, since most genes that regulate immune responses are not essential for normal organogenesis and development, mutations of numerous components can be found in the population. Many more such mutations are likely to be found affecting innate and adaptive immune response. It would not be surprising at all if the present number of seventy-odd immunodeficiency-related genes doubled in the next few years. Mutations in other TNF receptors, Toll receptors, and the numerous chemokine receptors may well be found. It is intriguing to speculate how such mutations might present in humans.

Our second point is that the clinical phenotype associated with these mutations is amazingly unpredictable. While knockout mice have been useful in providing clues to the importance of a given gene, the clinical significance of the corresponding human mutations vary even within a single family. Thus, more than one gene can cause similar immunodeficiency, and a single gene can have rather variable clinical presentation. Given the surprisingly atypical presentations associated with these disorders, mutations may be far more common than we think. Improvements in sequencing technology will permit easier and more rapid analysis of larger numbers of patients. It will be of great interest to determine just how frequent mutations and polymorphisms are. It will be impor-

tant as well to try to determine what modifier genes influence the severity of primary immunodeficiencies.

Third, immunodeficiency and autoimmunity are not the opposites that one might think them to be a priori; indeed, mixed pictures of autoimmunity and immunodeficiency are quite common. The more we learn about the immune system, the more apparent are the complex functions of many of its components, which often play both positive and negative roles. IL-2 is an outstanding example, but TNF and other cytokines can also be proinflammatory and immunosuppressive in different contexts. Consequently, it is not surprising that primary immunodeficiency and autoimmunity can go hand in hand. The common autoimmune diseases are not simple mendelian disorders but rather are polygenic. It is tempting to speculate that we will ultimately recognize a continuum of diseases that encompasses primary immunodeficiencies, autoimmune disorders, and autoinflammatory disorders, which might share subsets of genetic lesions.

Primary immunodeficiency diseases have played a main role in the development of novel therapeutic strategies. In 1968, SCIDX1 was the first disease to be cured by bone marrow transplantation (64), and in 2000 the same disease was the first true success of gene therapy. It is foreseeable that gene therapy will soon achieve further therapeutic successes in those immunodeficiencies where corrected cells have a strong selective advantage over unmodified populations. On the other hand, gene therapy involving molecules not providing such selective advantage will certainly be more challenging and will require more efficient targeting systems.

Address correspondence to: John O'Shea, National Institute of Arthritis and Musculoskeletal and Skin Diseases, NIH, 10 Center Drive, Building 10, Room 9N262, Bethesda, Maryland 20892-1820, USA. Phone: (301) 496-6026; Fax: (301) 402-0012; E-mail: osheajo@mail.nih.gov.

1. Fischer, A. 2001. Primary immunodeficiency diseases: an experimental model for molecular medicine. *Lancet.* **357**:1863–1869.
2. Buckley, R.H. 2000. Primary immunodeficiency diseases due to defects in lymphocytes. *N. Engl. J. Med.* **343**:1313–1324.
3. Noguchi, M., et al. 1993. Interleukin-2 receptor gamma chain mutation results in X-linked severe combined immunodeficiency in humans. *Cell.* **73**:147–157.
4. Puel, A., Ziegler, S.F., Buckley, R.H., and Leonard, W.J. 1998. Defective IL7R expression in T(-)B(+)NK(+) severe combined immunodeficiency. *Nat. Genet.* **20**:394–397.
5. Macchi, P., et al. 1995. Mutations of Jak-3 gene in patients with autosomal severe combined immune deficiency (SCID). *Nature.* **377**:65–68.
6. Russell, S.M., et al. 1995. Mutation of Jak3 in a patient with SCID: essential role of Jak3 in lymphoid development. *Science.* **270**:797–800.
7. Notarangelo, L.D., et al. 2001. Mutations in severe combined immune deficiency (SCID) due to JAK3 deficiency. *Hum. Mutat.* **18**:255–263.
8. Chen, M., et al. 2000. Complex effects of naturally occurring mutations in the JAK3 pseudokinase domain: evidence for interactions between the kinase and pseudokinase domains. *Mol. Cell. Biol.* **20**:947–956.
9. Zhou, Y.J., et al. 2001. Unexpected effects of FERM domain mutations on catalytic activity of Jak3: structural implication for Janus kinases. *Mol. Cell.* **8**:959–969.
10. Sharfe, N., Shahar, M., and Roifman, C.M. 1997. An interleukin-2 receptor gamma chain mutation

with normal thymus morphology. *J. Clin. Invest.* **100**:3036–3043.
11. Schmalstieg, F.C., et al. 1995. Missense mutation in exon 7 of the common gamma chain gene causes a moderate form of X-linked combined immunodeficiency. *J. Clin. Invest.* **95**:1169–1173.
12. Sharfe, N., Dadi, H.K., Shahar, M., and Roifman, C.M. 1997. Human immune disorder arising from mutation of the alpha chain of the interleukin-2 receptor. *Proc. Natl. Acad. Sci. USA.* **94**:3168–3171.
13. Frucht, D.M., et al. 2001. Unexpected and variable phenotypes in a family with JAK3 deficiency. *Genes Immun.* **2**:422–432.
14. Wildin, R.S., et al. 2001. X-linked neonatal diabetes mellitus, enteropathy and endocrinopathy syndrome is the human equivalent of mouse scurfy. *Nat. Genet.* **27**:18–20.
15. 1997. An autoimmune disease, APECED, caused by mutations in a novel gene featuring two PHD-type zinc-finger domains. The Finnish-German APECED Consortium. Autoimmune Polyendocrinopathy-Candidiasis-Ectodermal Dystrophy. *Nat. Genet.* **17**:399–403.
16. Nagamine, K., et al. 1997. Positional cloning of the APECED gene. *Nat. Genet.* **17**:393–398.
17. Dupuis, S., et al. 2000. Human interferon-gamma-mediated immunity is a genetically controlled continuous trait that determines the outcome of mycobacterial invasion. *Immunol. Rev.* **178**:129–137.
18. Dupuis, S., et al. 2001. Impairment of mycobacterial but not viral immunity by a germline human STAT1 mutation. *Science.* **293**:300–303.
19. McDermott, M.F., et al. 1999. Germline mutations in the extracellular domains of the 55 kDa TNF recep-

tor, TNFR1, define a family of dominantly inherited autoinflammatory syndromes. *Cell.* **97**:133–144.
20. Kastner, D.L., and O'Shea, J.J. 2001. A fever gene comes in from the cold. *Nat. Genet.* **29**:241–242.
21. Rieux-Laucat, F., et al. 1995. Mutations in Fas associated with human lymphoproliferative syndrome and autoimmunity. *Science.* **268**:1347–1349.
22. Fisher, G.H., et al. 1995. Dominant interfering Fas gene mutations impair apoptosis in a human autoimmune lymphoproliferative syndrome. *Cell.* **81**:935–946.
23. Wang, J., et al. 1999. Inherited human Caspase 10 mutations underlie defective lymphocyte and dendritic cell apoptosis in autoimmune lymphoproliferative syndrome type II. *Cell.* **98**:47–58.
24. Notarangelo, L.D., and Hayward, A.R. 2000. X-linked immunodeficiency with hyper-IgM (XHIM). *Clin. Exp. Immunol.* **120**:399–405.
25. Ferrari, S., et al. 2001. Mutations of CD40 gene cause an autosomal recessive form of immunodeficiency with hyper IgM. *Proc. Natl. Acad. Sci. USA.* **98**:12614–12619.
26. Doffinger, R., et al. 2001. X-linked anhidrotic ectodermal dysplasia with immunodeficiency is caused by impaired NF-kappaB signaling. *Nat. Genet.* **27**:277–285.
27. Jain, A., et al. 2001. Specific missense mutations in NEMO result in hyper-IgM syndrome with hypohydrotic ectodermal dysplasia. *Nat. Immunol.* **2**:223–228.
28. Revy, P., et al. 2000. Activation-induced cytidine deaminase (AID) deficiency causes the autosomal recessive form of the Hyper-IgM syndrome (HIGM2). *Cell.* **102**:565–575.

29. Villard, J., et al. 2001. MHC class II deficiency: a disease of gene regulation. *Medicine (Baltimore).* **80**:405–418.

30. Schwarz, K., et al. 1996. RAG mutations in human B cell-negative SCID. *Science.* **274**:97–99.

31. Villa, A., et al. 1998. Partial V(D)J recombination activity leads to Omenn syndrome. *Cell.* **93**:885–896.

32. Moshous, D., et al. 2001. ARTEMIS, a novel DNA double-strand break repair/V(D)J recombination protein, is mutated in human severe combined immune deficiency. *Cell.* **105**:177–186.

33. Ma, Y., Pannicke, U., Schwarz, K., and Lieber, M.R. 2002. Hairpin opening and overhang processing by an ARTEMIS/DNA-dependent protein kinase complex in nonhomologous end joining and V(D)J recombination. *Cell.* **108**:781–794.

34. O'Driscoll, M., et al. 2001. DNA ligase IV mutations identified in patients exhibiting developmental delay and immunodeficiency. *Mol. Cell.* **8**:1175–1185.

35. Savitsky, K., et al. 1995. A single ataxia telangiectasia gene with a product similar to PI-3 kinase. *Science.* **268**:1749–1753.

36. Hartley, K.O., et al. 1995. DNA-dependent protein kinase catalytic subunit: a relative of phosphatidylinositol 3-kinase and the ataxia telangiectasia gene product. *Cell.* **82**:849–856.

37. Goldman, F.D., et al. 1998. Defective expression of p56lck in an infant with severe combined immunodeficiency. *J. Clin. Invest.* **102**:421–429.

38. Elder, M.E., et al. 1994. Human severe combined immunodeficiency due to a defect in ZAP-70, a T cell tyrosine kinase. *Science.* **264**:1596–1599.

39. Kung, C., et al. 2000. Mutations in the tyrosine phosphatase CD45 gene in a child with severe combined immunodeficiency disease. *Nat. Med.* **6**:343–345.

40. Satterthwaite, A.B., and Witte, O.N. 2000. The role of Bruton's tyrosine kinase in B-cell development and function: a genetic perspective. *Immunol. Rev.* **175**:120–127.

41. Minegishi, Y., et al. 1999. An essential role for BLNK in human B cell development. *Science.* **286**:1954–1957.

42. Derry, J.M., Ochs, H.D., and Francke, U. 1994. Isolation of a novel gene mutated in Wiskott-Aldrich syndrome. *Cell.* **78**:635–644.

43. Thrasher, A.J., Burns, S., Lorenzi, R., and Jones, G.E. 2000. The Wiskott-Aldrich syndrome: disordered actin dynamics in haematopoietic cells. *Immunol. Rev.* **178**:118–128.

44. Kim, A.S., Kakalis, L.T., Abdul-Manan, N., Liu, G.A., and Rosen, M.K. 2000. Autoinhibition and activation mechanisms of the Wiskott-Aldrich syndrome protein. *Nature.* **404**:151–158.

45. Devriendt, K., et al. 2001. Constitutively activating mutation in WASP causes X-linked severe congenital neutropenia. *Nat. Genet.* **27**:313–317.

46. Sturla, L., et al. 2001. Impairment of the Golgi GDP-L-fucose transport and unresponsiveness to fucose replacement therapy in LAD II patients. *Pediatr. Res.* **49**:537–542.

47. Luhn, K., et al. 2001. The gene defective in leukocyte adhesion deficiency II encodes a putative GDP-fucose transporter. *Nat. Genet.* **28**:69–72.

48. Sayos, J., et al. 1998. The X-linked lymphoproliferative-disease gene product SAP regulates signals induced through the co-receptor SLAM. *Nature.* **395**:462–469.

49. Stepp, S.E., et al. 1999. Perforin gene defects in familial hemophagocytic lymphohistiocytosis. *Science.* **286**:1957–1959.

50. Menasche, G., et al. 2000. Mutations in RAB27A cause Griscelli syndrome associated with haemophagocytic syndrome. *Nat. Genet.* **25**:173–176.

51. Pastural, E., et al. 2000. Two genes are responsible for Griscelli syndrome at the same 15q21 locus. *Genomics.* **63**:299–306.

52. Hirschhorn, R. 1999. Immunodeficiency disease due to deficiency of adenosine deaminase. In *Primary immunodeficiency diseases: a molecular and genetic approach.* H.D. Ochs, C.I.E. Smith, and J.M. Puck, editors. Oxford University Press. New York, New York, USA. 121–139.

53. Hershfield, M.S., and Mitchell, B.S. 2001. Immunodeficiency diseases caused by adenosine deaminase deficiency and purine nucleoside phosphorylase deficiency. In *The metabolic and molecular bases of inherited disease.* C.R. Scriver, A.L. Beaudet, W.S. Sly, and D. Valle, editors. McGraw-Hill. New York, New York, USA. 2585–2625.

54. Hirschhorn, R. 1996. Spontaneous in vivo reversion to normal of an inherited mutation in a patient with adenosine deaminase deficiency. *Nat. Genet.* **13**:290–295.

55. Arredondo-Vega, F.X., et al. 2001. Adenosine deaminase deficiency with mosaicism for a "second-site suppressor" of a splicing mutation: decline in revertant T lymphocytes during enzyme replacement therapy. *Blood.* **99**:1005–1013.

56. Stephan, V., et al. 1996. Atypical X-linked severe combined immunodeficiency due to possible spontaneous reversion of the genetic defect in T cells. *N. Engl. J. Med.* **335**:1563–1567.

57. Ariga, T., et al. 2001. Spontaneous in vivo reversion of an inherited mutation in the Wiskott-Aldrich syndrome. *J. Immunol.* **166**:5245–5249.

58. Wada, T., et al. 2001. Somatic mosaicism in Wiskott-Aldrich syndrome suggests in vivo reversion by a DNA slippage mechanism. *Proc. Natl. Acad. Sci. USA.* **98**:8697–8702.

59. Cavazzana-Calvo, M., et al. 2000. Gene therapy of human severe combined immunodeficiency (SCID)-X1 disease. *Science.* **288**:669–672.

60. Bordignon, C., et al. 1995. Gene therapy in peripheral blood lymphocytes and bone marrow for ADA-immunodeficient patients. *Science.* **270**:470–475.

61. Kohn, D.B., et al. 1998. T lymphocytes with a normal ADA gene accumulate after transplantation of transduced autologous umbilical cord blood CD34+ cells in ADA-deficient SCID neonates. *Nat. Med.* **4**:775–780.

62. Malech, H.L., et al. 1997. Prolonged production of NADPH oxidase-corrected granulocytes after gene therapy of chronic granulomatous disease. *Proc. Natl. Acad. Sci. USA.* **94**:12133–12138.

63. Bauer, T.R., and Hickstein, D.D. 2000. Gene therapy for leukocyte adhesion deficiency. *Curr. Opin. Mol. Ther.* **2**:383–388.

64. Gatti, R.A., Meuwissen, H.J., Allen, H.D., Hong, R., and Good, R.A. 1968. Immunological reconstitution of sex-linked lymphopenic immunological deficiency. *Lancet.* **2**:1366–1369.

An integrated view of suppressor T cell subsets in immunoregulation

Hong Jiang and Leonard Chess

Department of Medicine and Pathology, Columbia University College of Physicians and Surgeons, New York, New York, USA.

The immune system evolved to protect organisms from a virtually infinite variety of disease-causing agents but to avoid harmful responses to self. Because immune protective mechanisms include the elaboration of potent inflammatory molecules, antibodies, and killer cell activation — which together can not only destroy invading microorganisms, pathogenic autoreactive cells, and tumors, but also mortally injure normal cells — the immune system is inherently a "double-edged sword" and must be tightly regulated. Immune response regulation includes homeostatic mechanisms intrinsic to the activation and differentiation of antigen-triggered immunocompetent cells and extrinsic mechanisms mediated by suppressor cells. This review series will focus on recent advances indicating that distinct subsets of regulatory CD4+ and CD8+ T cells as well as NK T cells control the outgrowth of potentially pathogenic antigen-reactive T cells and will highlight the evidence that these suppressor T cells may play potentially important clinical roles in preventing and treating immune-mediated disease. Here we provide a historical overview of suppressor cells and the experimental basis for the existence of functionally and phenotypically distinct suppressor subsets. Finally, we will speculate on how the distinct suppressor cell subsets may function in concert to regulate immune responses.

The potential capacity for the immune response to induce or activate disease was clearly recognized at the turn of the 20th century by Paul Ehrlich, who emphasized that the immune system must carefully distinguish between self and non-self in order to avoid autoimmunity. Ehrlich envisioned that during the ontogeny and outgrowth of the immunocompetent clones responsive to foreign antigens, there had to be mechanisms to control the outgrowth of clones reactive with self (1). Moreover, the failure to control the outgrowth of autoreactive cells would lead to a state of "horror autotoxicus," or autoimmunity. Ehrlich's ideas were amplified and developed with the elaboration of the clonal selection hypothesis (2–4). This hypothesis was further refined with identification of the antigen receptors on T and B cells and discovery that the antigen specificity of these receptors is a consequence of random recombination of the many V, D, and J genes encoding the antigen-binding sites of these receptors, a process that could generate more than 10^9 distinct receptors (5).

According to a modern interpretation of the clonal selection hypothesis, multiple clones of immunocompetent cells displaying unique antigen-specific receptors exist prior to interaction with antigens and in the case of T cells get selected on the basis of interaction with self-peptides bound to MHC molecules in the thymus. The majority of thymocytes bearing high-affinity receptors for self-antigens are eliminated centrally during thymic differentiation by an apoptotic mechanism termed negative selection. However, many self-reactive T cells with low to intermediate affinity for self-antigen escape thymic negative selection and are released into the periphery, where they are capable of autoantigen-driven activation, proliferation, and differentiation into potentially pathogenic effector cells (6–9). Thus, mechanisms that normally regulate the outgrowth or function of these self-reactive T cells ultimately control the initiation and progression of autoimmune disease. A corollary of these notions is that autoimmune diseases arise from either the failure to eliminate or inactivate high-affinity immunocompetent cells during their ontogeny and/or the failure of the immune system to control the outgrowth or function of intermediate self-reactive clones that escape into the periphery.

In addition to direct autoimmune attack, the immune system can also induce disease, because the very protective immunologic mechanisms that are employed to limit the outgrowth of invading foreign pathogens or tumor cells can induce "collateral damage" on normal, uninfected cells in the vicinity of immune attack. This collateral damage is mediated in large part by inflammatory cytokines and is thought to account for the destruction of normal tissues clinically observed during the physiologic immune attack aimed at the elimination of foreign pathogens or tumors. To prevent immune disease either induced directly by autoimmune attack or to control collateral damage occurring during all immune responses, a complex network of interacting regulatory peripheral mechanisms has coevolved to prevent or dampen immune-mediated diseases. These regulatory systems include mechanisms intrinsic to the antigen activation and differentiation of T cells as well as those mediated by regulatory "suppressor" T cells. This *JCI* review series on regulatory T cells will focus on recent advances indicating that distinct subsets of regulatory CD4+ (10–12) and CD8+ T cells (refs. 13, 14 and the present article) as well as NK T cells (NKT cells) (15) function to suppress the outgrowth of potentially pathogenic antigen-reactive T cells. The series will highlight the resurrection of the idea that suppressor T cells play potentially important clinical roles in the prevention and treatment of immune-mediated disease. The articles will review evidence that suppressor cells are essential for the control of autoimmunity and are also involved in the control of the immune response to transplanted allografts (12), allergens (16), and infectious pathogens. The series will also

---

Nonstandard abbreviations used: AICD, activation-induced cell death; CD40L, CD40 ligand; EAE, experimental autoimmune encephalomyelitis; α-GalCer, α-galactosylceramide; ILT3, Ig-like transcript 3; IPEX syndrome, immunodysregulation, polyendocrinopathy, enteropathy, X-linked syndrome; MBP, myelin basic protein; NKT cells, NK T cells; PLP, proteolipid protein; T1D, type 1 diabetes.

Conflict of interest: The authors have declared that no conflict of interest exists.

Citation for this article: *J. Clin. Invest.* **114**:1198–1208 (2004). doi:10.1172/JCI200423411.

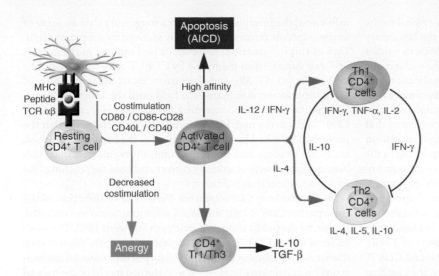

Figure 1
Homeostatic control of the outgrowth of antigen-activated CD4+ T cells. Control of the peripheral immunity is accomplished by mechanisms intrinsic to antigen activation of CD4+ T cells; including apoptosis, induction of anergy, and differentiation into Th subsets that are independent of other types of regulatory cells. In addition, superimposed on these intrinsic mechanisms are control mechanisms mediated by distinct subsets of NKT, CD4+, and CD8+ regulatory (suppressor) T cells.

include a review of T cell vaccination, which may be mediated by some or all of the Tregs (17).

In this series introductory article, we will emphasize that the regulation mediated by these suppressor T cell subsets is superimposed on intrinsic regulatory mechanisms induced by antigen activation alone. Thus, we will first describe the intrinsic homeostatic mechanisms that control immune responses independent of suppressor cells. We will then briefly describe the experimental historical basis for the existence of the CD4+, CD8+, and NKT suppressor subsets and attempt to give an overview of their unique functions. Finally, as a prelude to the rest of the articles in the series, we will speculate on how the distinct subsets of suppressor cells function in concert to regulate immune responses.

Homeostatic regulatory mechanisms intrinsic to antigen activation and differentiation that function independently of suppressor cells

TCR affinity, apoptosis, and antigen-induced cell death. Immune responses are initiated when the antigen-specific TCRs expressed by resting CD4+ T cells are triggered by MHC-peptide complexes in concert with costimulatory molecules expressed by APCs (18–20). This triggering induces CD4+ T cells to proliferate, secrete cytokines, and express cell-surface molecules including the IL-2 receptor (CD25), CTLA-4, and CD40 ligand (CD40L) critical for the subsequent growth and functional differentiation of T cells. To regulate the immune response, the immune system has also evolved several homeostatic feedback mechanisms to downregulate and control the outgrowth, differentiation, and function of peripheral antigen-activated CD4+ T cells (Figure 1). One level of control resides at the initial clonal activation of the T cell receptor itself by MHC-peptide complexes. The functional consequences of T cell signaling are dependent on the affinity and duration of binding of the TCRs with MHC-peptide complexes (21, 22). For example, there is a threshold of affinity necessary for optimal activation and differentiation of T cells, whereas triggering of very high-affinity TCRs can induce apoptotic pathways leading to activation-induced cell death (AICD) (23).

The homeostatic role of costimulatory molecules in immunoregulation. During the initial T cell-MHC-peptide interaction, other receptor ligand interactions also become pivotal in ultimately dictating the functional fate of T cells. For example, one of the earliest

antigen activation–induced cell-surface molecules expressed by T cells is CD40L (24). CD40L interacts with CD40 expressed on B cells to induce antibody formation and with CD40 expressed on antigen-presenting DCs to induce cellular immune responses (25, 26). Another critical consequence of the interaction of CD40L with CD40 expressed on APCs is the upregulation of other key costimulatory molecules, including CD80 (B71) and CD86 (B72) (27, 28). These molecules interact with CD28 expressed on naive T cells to provide the costimulatory signal required for optimal activation of T cells. However, following activation, a structurally related molecule termed CD152 or CTLA-4 is expressed that also binds CD80 and CD86 but delivers inhibitory signals to T cells. The consequence of CTLA-4 negative signaling is abrogation of functional activation. The resulting nonfunctional T cells are termed anergic or tolerant. Thus, blockade of either the CD40L/CD40 pathway or the CD28/B7 pathway can lead to inhibition of immune responses and tolerance induction (20, 28–30).

The differentiation of CD4+ T cells into subsets expressing different arrays of cytokines. A third general set of regulatory mechanisms, also a consequence of the initial triggering of CD4+ T cells by MHC-peptide complexes, is the further differentiation of the CD4+ T cells into the functionally distinct Th1 and Th2 subsets phenotypically distinguished, in part, by the elaboration of distinct sets of cytokines (31–33). In this regard, IFN-γ secreted by Th1 cells is known to downregulate the differentiation and function of TH2 cells and, conversely, IL-4. TGF-β and IL-10 inhibit Th1 cell differentiation (33–35). In addition, other cytokine-secreting Th subsets capable of secreting the immunosuppressive cytokines IL-10 and/or TGF-β but not IL-4 (termed Tr1 or Th3 cells) (36–38) have been observed and will be reviewed in detail in several articles in this review series. A widely prevalent view is that the balance between the emergence of Th1, Th2, as well as the other Th cytokine–secreting CD4+ Tregs following antigen activation plays a major role in the outgrowth and functions of self-reactive and foreign reactive clones (39–41).

Homeostatic regulatory mechanisms mediated by dominant suppression of the immune response

The origin of the idea of suppressor T cells and its initial demise. Superimposed on the intrinsic mechanisms of homeostatic regulation are extrinsic mechanisms mediated by suppressor T cells that control the induction and/or outgrowth of antigen-activated T cells. In this regard, the idea that suppressor T cells may be critical to the control of virtually all immune responses arose in the laboratory of Richard Gershon at Yale University in the late 1960s in studies of immune tolerance to foreign antigens. Gershon showed initially that adoptive transfer of T cells from animals made tolerant to foreign antigen "X" could specifically suppress the production of anti-X antibodies in recipient animals (42, 43). These studies were rapidly extended in numerous laboratories, which documented that suppressor cells were not only involved in the peripheral regulation of cells responding to foreign antigens but also participated in the regulation of self-reactive cells and in the control of autoimmunity (44, 45). The first models of the pathways by which suppressor T cells might function in the specific regulation of immunity arose from the seminal studies by Cantor and Boyse, who showed that genetically well-defined alloantisera could be used to identify phenotypically stable and functionally distinct subsets of T cells. These studies ultimately led to the discovery of CD4+ and CD8+ T cell subsets (46–50). Implicit in these discoveries was the idea that the CD4+ and CD8+ T cell subsets were genetically programmed during ontogeny, prior to interaction with antigen, to mediate distinct functional programs. The CD4+ T cells were functionally programmed to induce both antibody responses and cell-mediated immune reactions. In contrast, the CD8+ T cells were not genetically programmed as inducer cells but instead were programmed to differentiate into killer cells capable of destroying tumor cells or cells infected with intracellular pathogens. However, following interaction with antigen-activated CD4+ T cells, CD8+ T cells could be induced in vitro to differentiate into suppressor cells, which in turn downregulate the activity of the CD4+ T cell population (51, 52). Experiments employing allosera to the MHC molecule Qa-1 suggested that Qa-1 was expressed on the suppressor-inducer subset of CD4+ T cells (51, 52). This is of interest because eventually Qa-1 was cloned and found to be an MHC class Ib molecule capable of presenting endogenous as well as exogenous peptides (53) to CD8+ T cells. However, the potential significance of Qa-1 expression on suppressor-inducer cells in vivo was not delineated until the last several years, with experiments employing monoclonal antibodies to Qa-1 and Qa-1–knockout mice (54–56) (see below). Precise characterization of the specificity and functional phenotype of suppressor cells generated in these complex cellular experiments was impossible, largely because there were no phenotypic markers at the time that definitively distinguished the CD8 suppressor cells from the more conventional CD8 or CD4 T cells. Moreover, the biological significance of these in vitro findings was not placed on firm in vivo footing because of the lack of monoclonal antibodies to the murine CD4 and CD8 populations and the lack of molecular genetics–based approaches to study CD4- or CD8-deficient animals.

In this regard, it is important to emphasize that the original suppressor cell circuits were initially conceived and/or deduced at a time when molecular immunology was in its infancy. For example, the nature of the TCR was unknown, as was the precise structure and function of MHC molecules in restricting T cell activity. In addition, the great majority of the cytokines that are now known to regulate immune functions had not yet been identified. It was also unknown that antigen activation of the TCRs induced intrinsic homeostatic mechanisms, including the differentiation of CD4+ T cells into Th1 and Th2 subsets which elaborated distinct sets of regulatory cytokines. Clearly, understanding the precise role of these lymphokines in suppression would have significantly

influenced the interpretation of data suggesting that an array of antigen-specific factors was uniquely secreted by suppressor cells. Lack of this information, however, caused interest in the models of T cell suppression mediated by CD8+ T cells to wane by the mid-1980s (45, 57, 58), and skepticism about the importance of T cell suppression dominated the field until the 1990s. Although some of the skepticism concerning T cell suppression mediated by CD8+ T cells in the mid-1980s was justified, it is clear now that the proverbial baby had been thrown out with the bath water. Indeed, many of the ideas and models of immunoregulation that have been dismissed were essentially correct and have far reaching biological and clinical significance.

The resurrection of CD8+ suppressor T cells. The resurrection of the concept that CD8+ T cells mediate T cell suppression was initiated, in part, by the publication of 2 articles in *Science* in 1992. The studies showed that CD8+ T cells participate in vivo in the resistance to disease induced during the natural history of experimental autoimmune encephalomyelitis (EAE), a well-studied model of the human disease multiple sclerosis (59, 60). For example, EAE is induced by myelin proteins such myelin basic protein (MBP), which activates encephalitogenic CD4+ Th1 cells. Mice of the B10PL strain completely recover from the first episode of EAE and become highly resistant to the reinitiation of EAE by secondary immunization. If these protected mice are then depleted of CD8+ T cells through the use of monoclonal anti-CD8, the protection is reversed, and the mice develop clinical EAE upon reimmunization with MBP (59). Furthermore, mice depleted of CD8+ T cells during the initial induction of EAE and allowed to recover normal levels of CD8+ T cells are not resistant and develop EAE again upon rechallenge with MBP. Thus, CD8+ T cells require priming during the first episode of EAE to regulate CD4+ T cells triggered by secondary MBP stimulation in vivo. Moreover, when CD8-/- mice are bred with EAE-susceptible PL/J mice, the progeny of the CD8-/- and PL/J mating develop more chronic EAE than the wild-type PL/J mice, as reflected by a higher frequency of relapses (60). These experiments provide evidence that CD8+ T cells play a key role in both inducing resistance to autoimmune EAE and in abrogating or suppressing recurrent relapsing episodes of pathogenic autoimmunity in vivo.

These experiments set the stage for a series of studies designed to further delineate the cellular pathways involved in CD8+ T cell suppression. An unexpected consequence of these studies was the initial reemergence of the Qa-1 component of the suppressor story alluded to above. The key experiments involved the isolation of CD8+ T cells from EAE-recovered mice and showed that they functioned as suppressor cells and specifically downregulated or killed some but not all MBP-activated CD4+ T cell clones. Importantly, the CD8+ suppressor cells were shown to preferentially suppress the potentially pathogenic autoreactive clones, and deletion of the CD8+ suppressor T cell was associated with recurrence of disease. Conversely, adoptive transfer or induction of the suppressor cells prevents disease. Thus, during the natural history of EAE, the CD8+ suppressor T cells fine tune the MBP-reactive TCR repertoire even within the TCR Vβ families that are preferentially activated by MBP in vivo (9). Analogous experiments using the superantigen staphylococcal enterotoxin B (SEB) showed that CD8+ T cells were involved in the downregulation of the CD4+ T cell response to SEB in vivo and in vitro (54). The specific inhibition of CD4+ TCR Vβ8 target T cells in both the EAE and SEB experiments was blocked by monoclonal antibodies to the TCR αβ and CD8 but was not blocked by antibodies to MHC class 1a molecules. This was

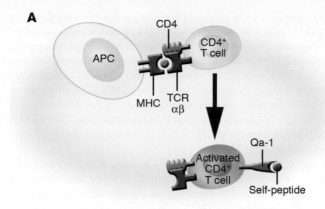

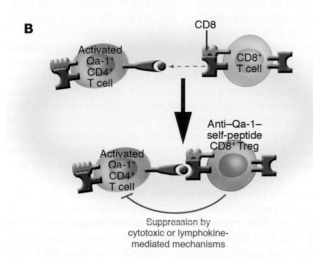

Suppression by cytotoxic or lymphokine-mediated mechanisms

Figure 2
Model of cognate interactions in the induction and function of Qa-1–restricted regulatory CD8+ T cells. (**A**) Initial activation of CD4+ T cell TCRs with peptide–MHC complexes induces the expression of Qa-1 bound with a variety of self-peptides on the surface of the CD4+ T cells. (**B**) Anti–Qa-1–self-peptide CD8+ precursor T cells are activated by Qa-1–expressing CD4+ T cells. The Qa-1–restricted CD8+ Tregs selectively downregulate certain but not all antigen-activated CD4+ T cells based on the specific recognition of Qa-1–self-peptide complexes expressed on certain CD4+ T cells by TCR αβ on the CD8+ T cells. In this regard, we have demonstrated in the EAE model that self-reactive CD4+ T cells, which are selectively downregulated by the CD8+ T cells, are enriched in potentially pathogenic self-reactive T cell clones (9).

CD4+ Th1 cells that induce ocular keratitis and blindness. These Th1 CD4+ T cells are specific for HSV peptides cross-reactive with self-corneal tissues and thus confirm the findings in another relevant autoimmune disease model. Moreover, these studies also show that the effect of Qa-1 on regulatory CD8+ T cells in the control of autoimmune disease in vivo is only observed during the secondary, but not the primary, immune response.

A number of biologic features of the Qa-1 molecule make it particularly interesting with respect to its role as a restricting element in immunoregulation. First, Qa-1 is preferentially expressed on activated, but not resting, T cells. Moreover, the fact that surface expression of Qa-1 on activated T cells is short lived may exclude resting T cells from downregulation by Qa-1–dependent CD8+ T cells. Second, Qa-1 is of limited polymorphism, with the potential to present self- and foreign peptides to CD8+ T cells. The predominant self-peptide presented by Qa-1 is Qdm, a hydrophobic peptide (AMAPRTLLL) derived from the leader sequence of certain MHC class Ia molecules (63). However, during T cell activation, Qa-1 can also bind other hydrophobic self-peptides that may serve as target antigens for the CD8+ T cells, as described above. Because the Qa-1–self-peptide complexes can interact with the CD94-NKG2 receptors expressed on NK or classical CD8+ T cells to either positively or negatively regulate function, Qa-1 may not only serve as the target of CD8 suppression, but also regulate the function of CD8+ suppressor T cells via the CD94-NKG2 receptors.

In summary, we envision that the CD8 T cell regulatory pathway comprises a series of sequential cellular events (Figure 2). It is initiated by the activation of naive CD4+ T cells during the primary immune response, in which the TCRs on CD4+ T cells interact with MHC class II–peptide complexes presented by conventional APCs. This initial interaction induces the surface expression of activation-dependent Qa-1–self-peptide complexes which is not observed in resting T cells. These activation-dependent Qa-1–self-peptide complex serve as both the inducer and target structures recognized by TCR αβ on regulatory CD8+ T cells. Thus, the Qa-1–self-peptide complex expressed by activated CD4+ T cells triggers TCR αβ on regulatory CD8+ T cells. These CD8+ T cells then differentiate into effector cells, which in turn suppress CD4+ T cells expressing the same target Qa-1–self-peptide complex. This suppression is linked to specific antigen activation of CD4+ T cells, and as a consequence the peptides bound by Qa-1 will likely be those induced by antigen activation. Whether they represent specific peptides derived from the CD4 T cell TCR (TCR Vβ or Vα peptides) or peptides derived from activation molecules induced by antigen triggering is not known. The mechanism of effecting suppression of the CD4+ T cells by the Qa-1–restricted CD8+ T cells has not been definitively

intriguing and suggested that perhaps the nonclassical MHC class 1b molecules may play a role in vivo in restricting the suppression mediated by CD8+ T cells. Indeed, subsequent experiments in vivo showed that the suppression mediated by CD8+ suppressor T cells is blocked by antibodies to the MHC class 1b Qa-1 molecule (9, 54, 61). Importantly, the Qa-1+ CD4+ T cells could be employed as vaccine T cells and used to induce regulatory CD8+ T cells in vivo. This T cell vaccination procedure protected animals from developing EAE, and this protection was abrogated by depletion of the Qa-1–restricted CD8+ T cells in vivo (61, 62).

In very recent experiments, the Cantor laboratory has created Qa-1–deficient mice and used them to directly demonstrate, in vivo, the important role that Qa-1 plays in the regulatory pathway mediated by CD8+ T cells in the control of autoimmunity (56). The authors convincingly demonstrated that the Qa-1–deficient mice develop severe EAE when exposed to the myelin-associated self–proteolipid protein (self-PLP) peptide and fail to develop the resistance to EAE that normally develops in wild-type mice after immunization with PLP peptide. Furthermore, the failure of resistance to EAE is associated with the escape of Qa-1–deficient CD4 cells from CD8+ T cell suppression, which could be restored by lentiviral-based expression of the syngeneic Qa-1 allele. These results form the heart of the functional data on Qa-1–deficient mice, demonstrating the in vivo relevance of Qa-1 in the control of autoimmunity. Furthermore, the authors also showed that Qa-1– deficient mice fail to control expansion of herpes simplex virus 1–induced (HSV-1–induced)

delineated; however, the consequences of TCR triggering of CD8+ T cell precursors may be conventional and involve differentiation into specific CTLs that lyse target cells and/or secrete lymphokines that downregulate the target cells.

Furthermore, it has also been shown that a distinct population of antigen-specific, non–Qa-1–restricted CD8+CD28- cells can suppress immune responses by directly interacting with antigen-presenting DCs and rendering these cells tolerogenic (64–66). The suppression involves the upregulation of inhibitory Ig-like transcript 3 (ILT3) and ILT4 receptors expressed on the DCs. APCs tolerized by CD8+ T cells show reduced expression of costimulatory molecules and induce antigen-specific unresponsiveness in CD4+ T helper cells. The precise function of these cells in vivo is not clear, and it is not known whether they interact with the Qa-1–restricted CD8+ suppressor cells or, alternatively, whether Qa-1–restricted T cells induce tolerogenic DCs.

Finally, the Qa-1–dependent regulatory CD8 pathway has begun to be translated from mice to humans with the in vitro findings that human CD8+ T cells can be induced to differentiate into regulatory cells whose function is dependent on HLA-E, the human homolog of Qa-1 (67, 68). These data support the idea that clinically relevant methods to induce and/or enhance this suppressive pathway in humans may prove useful in the prevention and treatment of human autoimmune disease. The details of several of the points in this section will be expanded upon in the series reviews from the Cantor and Kumar laboratories (13, 14).

The rapid rise of CD4+ suppressor T cells. Although suppressor cells were initially identified within the CD8+ T cell population (49, 50, 69), it was later found that suppressor cell function could be mediated by CD4+ T cells independent of CD8+ T cells (50, 70–72). For example, in the early 1980s, it was shown that coculture of graded numbers of polyclonally activated human CD4+ T cells to autologous resting CD4+ T cells inhibited the capacity of the resting CD4+ T cells to induce B cell differentiation and Ig synthesis in vitro (73, 74). The in vivo significance of these findings in humans were unknown. However, in vivo experiments in mice in the mid-1980s placed the idea of CD4+ suppressor cells on a firmer biological foundation. In particular, studies by Sakaguchi and colleagues on the organ-specific autoimmune disease induced in mice following neonatal thymectomy were quite revealing. These thymectomized mice were shown to have reduced numbers of CD4+ as well as CD8+ T cells. Furthermore, reconstitution of thymectomized mice by highly enriched populations of CD4+ but not CD8+ T cells from syngeneic normal mice completely inhibited disease development (70).

A decade later in 1995, Sakaguchi showed for the first time that the suppression mediated by CD4+ T cells is a function of the small subset of CD4+CD25+ cells (71). The transfer of T cells depleted in the CD4+CD25+ population into nude (athymic) mice led to a variety of autoimmune diseases, which could be prevented by injection of purified CD4+CD25+ T cells but not CD4+CD25- T cells. The field of CD4+CD25+ regulator cells, termed "CD4+CD25+ Tregs," was thus born and has experienced an explosive growth over the past few years (75, 76). Thus, the general experimental protocols employed to characterize these suppressor cells in vivo took advantage of the observation that mice deficient in T cells (neonatally thymectomized, *nu/nu* mice or *Rag-/-* mice) do not develop autoimmune disease following adoptive transfer of normal syngeneic spleen cells unless the spleen cells are depleted of CD4+CD25+ T cells. Thus, when

CD4+ splenic T cells prepared from normal mice were depleted of CD25+ cells and the remaining CD4+ T cells were transferred to syngeneic T cell–deficient mice, the recipients spontaneously developed various organ-specific autoimmune diseases (including type 1 diabetes [T1D], thyroiditis, and gastritis) and systemic wasting disease. Reconstitution of the CD4+CD25+ population inhibited the autoimmune development (72, 76). However, it is becoming increasingly clear that in many situations, CD4+CD25- T cells are as effective as CD4+CD25+ T cells in controlling T cell–mediated disease (77, 78).

In this regard, seminal studies of the potency of suppressor T cells by Lafaille and colleagues unequivocally showed that very small numbers of CD4+ suppressor cells can regulate immune responses in vivo independent of CD8+ T cells. This set of studies evaluated TCR-transgenic mice expressing the MBP specific for the TCR expressed on pathogenic clones and known to induce EAE (T/R+ mice). Lafaille found that these T/R+ mice rarely develop spontaneous EAE. However, when the T/R+ mice are crossed with *Rag1-/-* mice to obtain mice that have only T cells expressing the transgenic MBP-specific TCR (T/R- mice), almost all mice develop spontaneous EAE (79). Because both T/R+ and T/R- mice have large numbers of the potentially encephalitogenic CD4+ anti-MBP T cells, these results suggest that a very small number of nontransgenic lymphocytes, which are present in T/R+ but absent in T/R- mice, can potently suppress the in situ activation of CD4+ anti-MBP T cells mediating EAE. To identify the cellular regulatory mechanisms important in this suppression, T/R+ mice were crossed into mice deficient in either B cells, CD8+ T cells, NKT cells, TCR γδ cells, or TCR αβ cells (80, 81). Only mice that were deficient in CD4+ αβ T cells developed EAE. Moreover, T/R- mice were protected from EAE by the early (that is, prior to the onset of EAE) adoptive transfer of purified CD4+ T cells from normal donors. These results support the view that under certain experimental conditions CD4+ cells alone can suppress the initiation of autoimmunity. Moreover, it was subsequently shown that both CD4+CD25+ and CD4+CD25- cells could mediate this suppression (78, 82).

The precise relationship between the CD4+CD25+ and CD4+CD25- regulatory cells is not clear. As described above, the intrinsic homeostatic regulatory mechanisms pertaining to all CD4+ T cells — including AICD, costimulatory molecules (CD40L/CD40, CD28, or CTLA-4/B7), cytokine secretion and differentiation into Th1, Th2, and Tr1 subsets — are involved to some degree in the regulation of all immune responses independent of suppressor cells, so that the relationship between these known intrinsic mechanisms and suppressor cell function is of paramount importance. This is a particularly important issue with respect to regulation by CD4+CD25+ suppressor T cells, because at the present time, there are no known cell surface molecules that uniquely distinguish the CD4+ suppressor cells from conventional activated CD4+ cells. For example, the CD25 molecule, which is the α-chain of the IL-2 receptor, is expressed on all peripheral antigen-reactive CD4+ T cells from one to several days following antigen activation. Moreover, many of the other cell-surface molecules in addition to CD25 that seem to distinguish CD4+CD25+ from CD4+CD25- Tregs are upregulated on CD4+CD25- T cells following antigen activation. These molecules are therefore not unique differentiation antigens that define functional subsets but in fact are T cell activation molecules that are like CTLA-4, the glucocorticoid-induced TNF receptor family–related gene (GITR), and CD45RO (76, 78, 83, 84), which are expressed on the majority of CD4+ T cells follow-

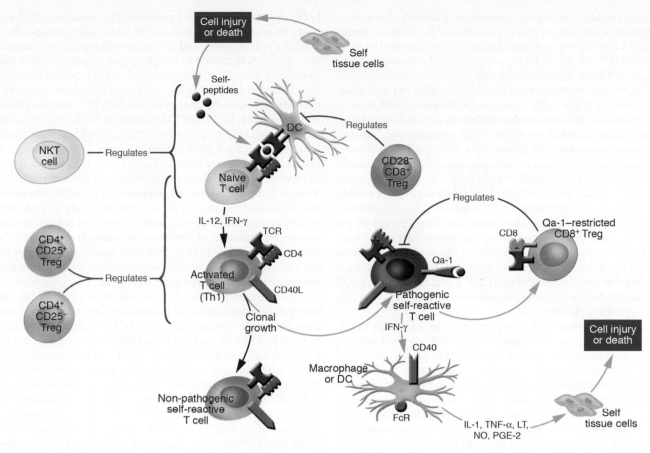

Figure 3
Tregs control the peripheral induction and clonal outgrowth of antigen-reactive T cells. This illustration shows various pathways of immunoregulation mediated by suppressor subsets of NKT, CD4+, and CD8+ T cell subsets. Each of the regulatory T cell subsets expresses distinct receptors, employs different effector mechanisms, and functions predominately at different stages during the course of the peripheral immune response. The NKT and CD4+CD25+ regulatory cells are "natural suppressor cells"; they are present prior to antigen activation and primarily function during the early "innate" and/or primary adaptive immune responses. In contrast, the CD8+ regulatory cells are induced to differentiate into suppressor effector cells during the primary immune response, and they function as effector-suppressor cells predominately during the secondary and memory phases of immunity.

ing antigen triggering of the TCR. The CTLA-4 may be of special interest with respect to the function of CD4+CD25+ regulatory cells because, like CD25, CTLA-4 is thought to be constitutively expressed on the CD4+CD25+ suppressor populations (85), and blockade of the CTLA-4 interactions with their CD80 or CD86 receptors in vivo can abrogate suppression. Moreover, T cells triggered via CTLA-4 predominantly secrete TGF-β, a cytokine with suppressive functions (86). On the other hand, in vitro experiments have not been able to document a blocking effect of anti–CTLA-4 on suppressor cell function (87). The search is still on to define specific markers that distinguish the subsets of CD4+ suppressor subsets (78, 83, 84). This general issue will be further discussed in several review articles in this series (particularly refs. 10–12).

With respect to the general role of cytokines in the suppression mediated by CD4+ Tregs, the cytokine profiles found in the various populations of CD4+ suppressor cells include a variety of combinations of the already-known immunoregulatory cytokines (i.e., IL-10, TGF-β, IL-4, IFN-γ). For example, there are CD4+CD45Rb^{low} activated suppressor cells that secrete large quantities of either IL-10 and IL-4 (termed Tr1 cells) and other CD4+CD45Rb^{low} suppressor T cells that secrete large quantities of

TGF-β (termed Th3 cells) (38). (The role of IL-10 cytokine expression will be discussed in detail in the series review by O'Garra et al., ref. 11) The functional significance of these cytokine-secreting CD4+ T cells is supported by the findings that TGF-β–deficient mice develop autoimmune disease (88) and that administration of neutralizing antibodies to IL-4 or TGF-β abrogates the in vivo prevention of autoimmunity or tolerance-inducing activity of CD4+ T cells in some models (89, 90). The relationship of these cells to the CD4+CD25+CTLA-4+ cells is unclear. For example, the in vitro capacity of CD4+CD25+ T cells to suppress immune responses is known to be contact dependent and not due to the IL-10, TGF-β, and IL-4 cytokines alone. It is possible the Th3 and Tr1 populations arise from "conventional" resting CD4+CD25– T cells, which, following antigen activation, express CD25. Taken together, these data may suggest that CD4+CD25– T cells, which function to suppress immune responses in vivo, may not represent a lineage-specific suppressor population. In contrast, the naturally occurring CD4+CD25+ T cells responsible for preventing autoimmunity in neonatal immune-deficient mice are thought to represent a lineage-specific suppressor population arising directly from the thymus (76).

In this regard, it is of great interest that a recently cloned transcription factor, termed Foxp3, a member of the forkhead family of DNA binding transcription factors, is not expressed in naive CD4+CD25- cells but is highly expressed in the naturally occurring CD4+CD25+ regulatory cells. Importantly, mutational defects in the *Foxp3* gene result in the fatal autoimmune and inflammatory disorder of the scurfy mouse and in the clinical and molecular features of the immunodysregulation, polyendocrinopathy, enteropathy, X-linked syndrome (IPEX syndrome) in humans. Both scurfy mice and IPEX patients have defects in T cell activation and reduced numbers and reduced suppressor functions mediated by the CD4+CD25+ T cells (91–93). In Foxp3–overexpressing mice, both CD4+CD25- and CD4-CD8+ T cells show suppressive activity, which suggests that expression of Foxp3 is linked to suppressor functions (94). Taken together, these data strongly support the idea that Foxp3 may uniquely define the subset of naturally occurring CD4+ suppressor T cells. However, the recent findings that Foxp3 can be expressed in CD4+CD25- cells following activation and are also expressed in activated CD8+ T cells suggest that Foxp3 is linked to functional suppression but not necessarily as a specific lineage marker (95–98).

Finally, the identity of the receptors that enable the CD4+CD25+ suppressor T cells to preferentially suppress self-reactive T cells and yet preserve normal immune functions remains a major unresolved issue. This apparent cognitive capacity to distinguish self from non-self is on the surface difficult to reconcile with a number of studies that have provided convincing evidence that the suppression mediated by CD4+CD25+ Tregs is not antigen specific (87). Thus, in elegant experiments, the Shevach laboratory showed that when T cells from TCR transgenic mice are activated with their peptide-MHC ligand and expanded in vitro in IL-2, the activated suppressors are subsequently capable of suppressing the responses of T cells from mice that express a different transgenic TCR (87, 99). Moreover, no MHC restriction is observed in the interaction of the activated suppressors and the responding targets (100). Thus, the TCRs employed by these regulatory CD4+ T cells are likely to be quite diverse, and it is unknown whether their capacity to distinguish self from non-self is dependent of the TCR αβ they express. Insight into the mechanism by which these antigen-nonspecific T cells can still distinguish self from non-self may come from recent studies showing that the CD4+ T regulatory cells when nonspecifically activated by LPS express toll-like receptors (101). Thus, these naturally occurring Tregs may represent the regulatory component of the innate immune system responsive to "danger-like signals." Perhaps self-reactive T cells preferentially express ligands that are recognized by toll-like receptors expressed by the regulatory cells. Clearly, the elucidation of the precise target structures recognized by CD4+CD25+ Tregs may help define the receptors employed by the regulatory cells to distinguish self from non-self. In contrast, as noted above, the Qa-1–restricted CD8+ suppressor T cells are not naturally occurring cells but instead are specifically induced during the primary adaptive immune response and are triggered to differentiate into effector-suppressor cells, which distinguish among the clones of autoreactive cells responding to a single peptide. Unlike the naturally occurring CD4+CD25+ suppressor T cells, Qa-1–restricted, CD8+ suppressor T cells employ their TCR αβ to recognize and distinguish targets of suppression.

NKT cells. NKT cells are a unique population of cells that express receptors of the NK lineage as well as a TCR αβ. Murine NKT cells express an invariant TCR α chain encoded by the Vα14-Jα281 gene segment, paired preferentially to various Vβ chains (102). Human NKT cells express the Va24–JaQ invariant chain. The NKT cells recognize and kill tumor cells expressing lipid antigens structurally related to the glycolipid α-galactosylceramide (α-GalCer). These lipids are presented to NKT cells by the MHC class Ib molecule CD1d (103). Analogous CD1d-restricted NKT cells, which express the invariant Vα24-JαQ TCR, are also present in humans (104, 105). The CD1d-restricted NKT cells are mainly of CD4+ or CD4-CD8- phenotype (106), and although originally functionally defined, both in vivo and vitro, by their capacity to lyse a variety of tumor cells (107), they were later found also to be involved in the regulation of autoimmune diseases (108, 109). These in vivo roles in immune responses are thought to be linked to the fact that following TCR antigen triggering in vivo, NKT cells were observed to develop augmented killer cell activity and secrete large amounts of cytokines, including IL-4 and IFN-γ, as well as TGF-β and IL-10 (103, 110–112), known to be involved in the activation of cell types important in mediating both innate immunity and Th2-type adaptive immunity.

As a consequence, NKT cells have been shown to influence the course of autoimmune disease in a variety of animal models. Prominent among the diseases affected by NKT cells are those primarily induced by Th1 cells, including the NOD diabetes model and the EAE models of multiple sclerosis (109, 113, 114). In these diseases, the evidence strongly suggests that the Th2-favoring cytokines, IL-4 and IL-10, secreted by NKT cells play an important role (111, 115, 116). Adoptive transfer of cell populations enriched for NKT cells prevents T1D in NOD recipients (115, 117, 118). Moreover, depletion of NKT cells early in the evolution of diabetes in the NOD mice accelerates the onset of diabetes (119). For example, it was found that lack of CD1-restricted NKT cells promotes the development of diabetes, whereas activation of Vα14+ NKT cells by α-GalCer suppresses disease in the NOD model. Similarly, models of colitis or multiple sclerosis (EAE) depletion of NKT cells accelerates the onset of disease, whereas in vivo activation of NKT cells by treatment with the glycolipid ligand induces significant improvement or prevents disease. These effects are abrogated in CD1d-deficient mice.

A reduction in number or altered function of NKT cells has also been correlated with autoimmune disease in humans. In patients with multiple sclerosis who are in relapse or remission, the frequency of Vα24-JαQ NKT cells is reduced in comparison with normal donors or patients with other autoimmune/inflammatory neurological diseases. In addition, diabetic individuals had lower frequencies of Vα24-JαQ NKT cells in comparison with their nondiabetic monozygotic twins (120). The few Vα24-JαQ NKT cell clones that could be isolated from these diabetic patients were deficient in IL-4 production. Taken together, these studies suggested the hypothesis that NKT cells play a role in natural protection against destructive Th1-mediated autoimmunity in T1D. However, a more recent study comparing diabetic patients and healthy controls, including discordant twin pairs, demonstrates that NKT cell frequency and IL-4 production are conserved during the course of T1D. These results do not necessarily refute the hypothesis that NKT cell defects underlie T1D but may indicate that immunoregulation of autoimmune disease is mediated by several subsets of immunoregulatory cells functioning in concert (121). Thus, in any particular patient, autoimmunity may not reflect not a single deficiency in one subset, but instead a defect in an integrated system of immunoregulation mediated at different levels by distinct T cell subsets.

Table 1
Properties of Treg subsets

Subsets of Tregs	Target cells of suppression	Molecular interaction between regulatory cells and inducer/target cells		Stage of immunity affected	Regulatory mechanisms	In vivo function
		Induction phase	Effector phase			
NKT cells	Tumor cells, pathogen-activated T cells, and/or APCs	TCR recognizing CD1d/glycolipid; restricted by CD1d	Same as induction phase	Natural; innate	IL-4, IL-10, TGF-β, IFN-γ; cytotoxicity	Destruction of tumors and pathogens; regulation of Th1-mediated autoimmune diseases
CD4+CD25− Tregs	T and B cells; ?APCs	Activated by MHC class II–peptide nonspecifically	May function by elaborating cytokines	Primary early	Predominately mediated by cytokines	Suppression of a variety of autoimmune diseases
CD4+CD25+ Tregs	T cells; ?APCs	Activated by MHC class II–peptides nonspecifically	Target and specificity is unknown; suppression is not MHC restricted	Primary early[A]	Requires cell-cell contact, cytokines	Prevention of a variety of autoimmune diseases, regulation of allograft rejection; immune response to pathogens
Qa-1–restricted CD8+ Tregs	Antigen-activated T cells differentially expressing Qa-1–self-peptide complexes	TCR recognizing Qa-1/hydrophobic self-peptides; restricted by Qa-1	Same as induction phase	Secondary late[A]	Cytotoxicity; requires cell-cell contact, ?cytokines	Fine tuning peripheral TCR repertoire; maintaining self-tolerance and controlling autoimmune disease
CD8+CD28− Tregs	DCs	Activated by classical MHC class Ia–peptide, nonspecifically?	Target of suppression is unknown	Primary early	Upregulation of ILT3 and ILT4 on DCs	Possibly regulation of autoimmunity

[A]CD4+CD25+ Tregs isolated from naive unprimed mice protect recipient animals from autoimmune diseases when adoptively transferred. In contrast, Qa-1–restricted CD8+ Tregs require priming during primary immune response in order to regulate the secondary immune response in vivo.

Thus, as a member of the family of immunocompetent cells participating in the innate immune response, NKT cells are positioned to influence and interact with other Tregs during the early phases of the autoimmune response (122). Their interaction with CD4+ T cells is implicit in their capacity to secrete cytokines such as IL-4 and IL-10 that may shift the Th1-Th2 balance. In addition, there is evidence that NKT cells also interact with CD8+ T cells in immunoregulation (123). The mechanisms through which NKT cells may regulate tolerance induction are complex and involve interactions with CD4+ and CD8+ regulatory cells. The complexity of how NKT cells are involved in regulating immune responses is the topic of the series review by Godfrey and Kronenberg (15).

An integrated overview of immunoregulation by NKT, CD4+CD25+, and Qa-1–restricted CD8+ T cell subsets

The resurgence of interest in immunosuppression mediated by T cells during the last decade has come from several distinct lines of investigation that have led to the concept that the immune system has evolved a variety of regulatory mechanisms mediated by distinct T cell subsets to suppress the outgrowth of potentially pathogenic self-reactive T cells. The regulation mediated by these suppressor subsets is superimposed on intrinsic regulatory mechanisms induced by the initial encounter of the TCR with antigen (Figure 3). These mechanisms include the induction of: (a) cell death if the encounter is of very high affinity; (b) nonresponsiveness if the encounter occurs in the absence of appropriate

costimulatory signals; and (c) the antigen-triggered differentiation into various Th subsets that secrete distinct arrays of Th1, Th2, Tr1, or Tr3 regulatory cytokines.

Given this intricate set of intrinsic regulatory mechanisms, one can legitimately ask why the immune system evolved an extrinsic set of regulatory mechanisms dependent on suppressor cells. The answer is simply that in the absence of T cell suppression, the intrinsic mechanisms are not sufficient to prevent autoimmunity or dampen immune responses to prevent collateral immune injury. To begin to understand the important interplay between the intrinsic and extrinsic mechanisms, it is important to emphasize that each of the extrinsic T cell suppressor subsets express distinct receptors, employ different effector mechanisms, and function at different stages during the evolution of immune responses (see Table 1). Thus, the NKT cells and CD4+CD25+ regulatory cells exist from the very early stages of life as "natural suppressor cells" prior to antigen activation and primarily function during the "innate" and/or primary immune responses. The NKT cells are endowed with pauciclonal TCR αβs consisting of an invariant Vα chain that specifically permits recognition of glycolipid molecules often expressed by various pathogens and presumably also expressed by tumor cells, activated blasts, and injured apoptotic cells that arise at the inception of immune responses. These NKT cells are thus poised to secrete IL-4 and IL-10, which are known to influence the balance of Th1 or Th2 cells that emerge during the primary immune response.

Cells of the naturally occurring CD4⁺CD25⁺ suppressor subset, which are capable of suppressing the outgrowth of autoreactive cells, also exist in the peripheral lymphoid system. These cells, like the NKT cells, can function during the primary immune response and do not require specific induction. In vitro, the suppressor function of these cells can be shown to dependent of cell-cell contact, but they can also express immunoregulatory cytokines including TGF-β, which may be involved in the their suppressor function in vivo. The precise specificity of these cells for their targets remains unknown, and it is not clear whether APCs and/or T cells are the targets of CD4⁺CD25⁺-mediated suppression. Moreover, although the CD4⁺CD25⁺ suppressor cells express conventional TCR αβ, the evidence suggests that these TCRs are not involved in the direct recognition of the targets of suppression.

In contrast to both the NKT cells and the naturally occurring CD4⁺CD25⁺ regulatory cells, the Qa-1-restricted CD8⁺ Tregs are not prevalent in naive animals prior to antigen encounter. As a consequence, adoptive transfer of CD8⁺ T cells from naive animals has no effect on the outcome of autoimmune responses, and depletion of CD8⁺ T cells prior the first induction of autoimmunity has no effect on the first episode of disease either. However, the CD8⁺ regulatory cells function like classical immunocompetent cells activated during adaptive immune responses. Thus, the they are induced by autologous CD4⁺ T cells activated during the primary immune response and differentiate into effector-suppressor cells, which function predominately during the secondary and memory phases of immunity. Thus, adoptive transfer of CD8⁺ T cells from self-antigen–primed mice will retard the outgrowth of the potentially pathogenic self-reactive T cells. In this regard, it is of interest that CD8⁺ Tregs are known to mediate resistance to autoimmunity following initial recovery from disease and to decrease the incidence and severity of relapse of the disease. In contrast to the CD4⁺CD25⁺ T cells, the CD8⁺ Tregs utilize their TCR αβs to directly recognize target cells in an MHC-restricted fashion. Thus, the CD8⁺ Tregs are Qa-1 restricted and selectively downregulate certain but not all activated T cells that preferentially express Qa-1–self-peptide(s) on their surface. They are poised to fine tune the immune response to suppress new or renewed outgrowth of autoreactive cells during episodes of relapse from autoimmune disease. These suppressor cells thus represent more the adaptive suppressive response to immunity, whereas the CD4⁺ and NKT suppressor subsets mediate natural immunity to the outgrowth of potentially pathogenic cells.

In summary, in this introductory *JCI* series article, we emphasized the idea that immunoregulation is carried out in an integrated manner by distinct suppressor T cells subsets that are superimposed on intrinsic homeostatic control mechanisms. Understanding the precise interplay between these intrinsic and suppressor mechanisms will undoubtedly be involved in the immunopathogenesis of a variety of diseases associated with abnormalities in immunoregulation. The evidence that distinct suppressor subsets are involved in many autoimmune states has been alluded to above and will be highlighted in greater detail in subsequent articles in this series (10, 13–15). In addition, the series will review the mounting evidence that suppressor cells are important in the immunopathogenesis of infectious and allergic diseases (11, 16) as well as allograft rejection (12). Finally, the idea that clinical induction of suppressor mechanisms may be important in the control of immunologic disease was conceived in the early days of the study of suppressor T cells. Induction of suppressor mechanisms by T cell vaccination to prevent autoimmunity, the study of which was pioneered by Irun Cohen in the early 1980s, was recently shown to involve suppressor T cells, as noted above. In this regard, the series will include a review of recent ideas concerning the immunobiology of T cell vaccination (17).

Acknowledgments

Research was supported by NIH grants AI39630 and AI39675 and National Multiple Sclerosis Society grant RG2938A (to H. Jiang), and by NIH grant U19 AI/46132 (to L. Chess).

Address correspondence to: Leonard Chess, Department of Medicine and Pathology, Rheumatology Division, Columbia University College of Physicians and Surgeons, 630 West 168th Street, PH8E, Suite 101, New York, New York 10032, USA. Phone: (212) 305-9986; Fax: (212) 305-4943; E-mail: LC19@Columbia.edu.

1. Ehrlich, P. 1900. The Croonian lecture: on immunity with special reference to cell life. *Proc. Royal Soc. London.* **66**:424.
2. Jerne, N.K. 1955. The natural selection theory of antibody formation. *Proc. Natl. Acad. Sci. U. S. A.* **41**:849.
3. Jerne, N.K. 1976. The immune system: a web of V domains. In *The Harvey Lectures 70.* Academic Press. New York, New York, USA. 93–110.
4. Burnet, F.M. 1957. A modification of Jerne's theory of antibody production using the concept of clonal selection. *Australian Journal of Science.* **20**:67–69.
5. Alt, F.W., Blackwell, T.K., DePinho, R.A., Reth, M.G., and Yancopoulos, G.D. 1986. Regulation of genome rearrangement events during lymphocyte differentiation. *Immunol. Rev.* **89**:5–30.
6. Goldrath, A.W., and Bevan, M.J. 1999. Selecting and maintaining a diverse T-cell repertoire. *Nature.* **402**:255–262.
7. Bouneaud, C., Kourilsky, P., and Bousso, P. 2000. Impact of negative selection on the T cell repertoire reactive to a self-peptide: a large fraction of T cell clones escapes clonal deletion. *Immunity.* **13**:829–840.
8. Kuchroo, V.K., et al. 2002. T cell response in experimental autoimmune encephalomyelitis (EAE): role of self and cross-reactive antigens in shaping, tuning, and regulating the autopathogenic T cell repertoire. *Annu. Rev. Immunol.* **20**:101–123.
9. Jiang, H., et al. 2003. Regulatory CD8+ T cells fine-tune the myelin basic protein-reactive T cell receptor V beta repertoire during experimental autoimmune encephalomyelitis. *Proc. Natl. Acad. Sci. U. S. A.* **100**:8378–8383.
10. Fehérvari, Z., and Sakaguchi, S. 2004. CD4⁺ Tregs and immune control. *J. Clin. Invest.* **114**:1209–1217. doi:10.1172/200423395.
11. O'Garra, A., Vieira, P.L., Vieira, P., and Goldfeld, A.E. 2004. IL-10 producing and naturally occurring Tregs: limiting collateral damage. *J. Clin. Invest.* In press. doi:10.1172/JCI200423215.
12. Walsh, P.T., Taylor, D.K., and Turka, L.A. 2004. Tregs and transplantation tolerance. *J. Clin. Invest.* In press. doi:10.1172/200423238.
13. Sarantopoulos, S., Lu, L., and Cantor, H. 2004. Qa-1 restriction of CD8⁺ suppressor T cells. *J. Clin. Invest.* **114**:1218–1221. doi:10.1172/JCI200423152.
14. Kumar, V. 2004. Homeostatic control of immunity by TCR peptide-specific Tregs. *J. Clin. Invest.* **114**:1222–1226. doi:10.1172/JCI200423166.
15. Godfrey, D.I., and Kronenberg, M. 2004. Going both ways: immune regulation via CD1d-dependent NKT cells. *J. Clin. Invest.* In press. doi:10.1172/JCI200423594.
16. Robinson, D.S., Larché, M., and Durham, S.R. 2004. Tregs and allergic disease. *J. Clin. Invest.* In press. doi:10.1172/JCI200423595.
17. Cohen, I.R., Quintana, F.J., and Mimran, A. 2004. Tregs in T cell vaccination: exploring the regulation of regulation. *J. Clin. Invest.* **114**:1227–1232. doi:10.1172/JCI200423396.
18. Mueller, D.L., Jenkins, M.K., and Schwartz, R.H. 1989. Clonal expansion versus functional clonal inactivation: a costimulatory signaling pathway determines the outcome of T cell antigen receptor occupancy [review]. *Annu. Rev. Immunol.* **7**:445–480.
19. Janeway, C., Jr., and Bottomly, K. 1994. Signals and signs for lymphocyte responses [review]. *Cell.* **76**:275–285.
20. Lenschow, D.J., Walunas, T.L., and Bluestone, J.A. 1996. CD28/B7 system of T cell costimulation. *Annu. Rev. Immunol.* **14**:233–258.
21. Savage, P.A., Boniface, J.J., and Davis, M.M. 1999. A kinetic basis for T cell receptor repertoire selection during an immune response. *Immunity.* **10**:485–492.
22. Davis, M.M., et al. 1998. Ligand recognition by alpha beta T cell receptors. *Annu. Rev. Immunol.* **16**:523–544.
23. Lenardo, M., et al. 1999. Mature T lymphocyte apoptosis--immune regulation in a dynamic and unpredictable antigenic environment. *Annu. Rev. Immunol.* **17**:221–253.
24. Lederman, S., et al. 1992. Identification of a novel surface protein on activated CD4+ T cells that

induces contact-dependent B cell differentiation (help). *J. Exp. Med.* **175**:1091–1101.

25. Foy, T.M., Aruffo, A., Bajorath, J., Buhlmann, J.E., and Noelle, R.J. 1996. Immune regulation by CD40 and its ligand GP39 [review]. *Annu. Rev. Immunol.* **14**:591–617.

26. Chess, L. 2000. Blockade of the CD40L/CD40 pathway. In *Therapeutic Immunology*. 2nd edition. K.F. Austen, S.J. Burakoff, F.S. Rosen, and T.B. Strom, editors. Blackwell Science, Inc. Cambridge, Massachusetts, USA. 441–456.

27. Caux, C., et al. 1994. Activation of human dendritic cells through CD40 cross-linking. *J. Exp. Med.* **180**:1263–1272.

28. Klaus, S.J., et al. 1994. Costimulation through CD28 enhances T cell-dependent B cell activation via CD40-CD40L interaction. *J. Immunol.* **152**:5643–5652.

29. Durie, F.H., Foy, T.M., Masters, S.R., Laman, J.D., and Noelle, R.J. 1994. The role of CD40 in the regulation of humoral and cell-mediated immunity [review]. *Immunol. Today.* **15**:406–411.

30. Koulova, L., Clark, E.A., Shu, G., and Dupont, B. 1991. The CD28 ligand B7/BB1 provides costimulatory signal for alloactivation of CD4+ T cells. *J. Exp. Med.* **173**:759–762.

31. Coffman, R.L., and Mosmann, T.R. 1991. CD4+ T-cell subsets: regulation of differentiation and function [review]. *Res. Immunol.* **142**:7–9.

32. Mosmann, T.R., Cherwinski, H., Bond, M.W., Giedlin, M.A., and Coffman, R.L. 1986. Two types of murine helper T cell clone. I. Definition according to profiles of lymphokine activities and secreted proteins. *J. Immunol.* **136**:2348–2357.

33. Mosmann, T.R., and Coffman, R.L. 1989. TH1 and TH2 cells: different patterns of lymphokine secretion lead to different functional properties. *Annu. Rev. Immunol.* **7**:145–173.

34. Fitch, F.W., McKisic, M.D., Lancki, D.W., and Gajewski, T.F. 1993. Differential regulation of murine T lymphocyte subsets. *Annu. Rev. Immunol.* **11**:29–48.

35. Seder, R.A., and Paul, W.E. 1994. Acquisition of lymphokine-producing phenotype by CD4+ T cells. *Annu. Rev. Immunol.* **12**:635–673.

36. Groux, H., et al. 1997. A CD4+ T-cell subset inhibits antigen-specific T-cell responses and prevents colitis. *Nature.* **389**:737–742.

37. Levings, M.K., and Roncarolo, M.G. 2000. T-regulatory 1 cells: a novel subset of CD4 T cells with immunoregulatory properties. *J. Allergy Clin. Immunol.* **106**:S109–S112.

38. Roncarolo, M.G., and Levings, M.K. 2000. The role of different subsets of T regulatory cells in controlling autoimmunity. *Curr. Opin. Immunol.* **12**:676–683.

39. Charlton, B., and Lafferty, K.J. 1995. The Th1/Th2 balance in autoimmunity. *Curr. Opin. Immunol.* **7**:793–798.

40. Del Prete, G. 1998. The concept of type-1 and type-2 helper T cells and their cytokines in humans. *Int. Rev. Immunol.* **16**:427–455.

41. O'Garra, A., and Vieira, P. 2004. Regulatory T cells and mechanisms of immune system control. *Nat. Med.* **10**:801–805.

42. Gershon, R.K., and Kondo, K. 1970. Cell interactions in the induction of tolerance: the role of thymic lymphocytes. *Immunology.* **18**:723–737.

43. Gershon, R.K., and Kondo, K. 1971. Infectious immunological tolerance. *Immunology.* **21**:903–914.

44. Green, D.R., Flood, P.M., and Gershon, R.K. 1983. Immunoregulatory T-cell pathways. *Annu. Rev. Immunol.* **1**:439–463.

45. Bloom, B.R., Salgame, P., and Diamond, B. 1992. Revisiting and revising suppressor T cells. *Immunol. Today.* **13**:131–136.

46. Cantor, H., and Boyse, E.A. 1975. Functional subclasses of T lymphocytes bearing different Ly antigens. II. Cooperation between subclasses of Ly+ cells in the generation of killer activity. *J. Exp. Med.* **141**:1390–1399.

47. Cantor, H., and Boyse, E.A. 1975. Functional subclasses of T-lymphocytes bearing different Ly antigens. I. The generation of functionally distinct T-cell subclasses is a differentiative process independent of antigen. *J. Exp. Med.* **141**:1376–1389.

48. Chess, L., and Schlossman, S.F. 1977. Human lymphocyte subpopulations. *Adv. Immunol.* **25**:213–241.

49. Reinherz, E.L., and Schlossman, S.F. 1980. The differentiation and function of human T lymphocytes. *Cell.* **19**:821–827.

50. Thomas, Y., et al. 1980. Functional analysis of human T cell subsets defined by monoclonal antibodies. I. Collaborative T-T interactions in the immunoregulation of B cell differentiation. *J. Immunol.* **125**:2402–2408.

51. Stanton, T.H., et al. 1978. The Qa-1 antigenic system. Relation of Qa-1 phenotypes to lymphocyte sets, mitogen responses, and immune functions. *J. Exp. Med.* **148**:963–973.

52. Eardley, D.D., et al. 1978. Immunoregulatory circuits among T-cell sets. I. T-helper cells induce other T-cell sets to exert feedback inhibition. *J. Exp. Med.* **147**:1106–1115.

53. Shawar, S.M., Vyas, J.M., Rodgers, J.R., and Rich, R.R. 1994. Antigen presentation by major histocompatibility complex class I-B molecules. *Annu. Rev. Immunol.* **12**:839–880.

54. Jiang, H., et al. 1995. Murine CD8+ T cells that specifically delete autologous CD4+ T cells expressing V beta 8 TCR: a role of the Qa-1 molecule. *Immunity.* **2**:185–194.

55. Jiang, H., et al. 1998. T cell vaccination induces T cell receptor Vbeta-specific Qa-1- restricted regulatory CD8(+) T cells. *Proc. Natl. Acad. Sci. U. S. A.* **95**:4533–4537.

56. Hu, D., et al. 2004. Analysis of regulatory CD8 T cells in mice deficient in the Qa-1 class Ib molecule, H2-T23. *Nat. Immunol.* **5**:516–523.

57. Moller, G. 1988. Do suppressor T cells exist? *Scand. J. Immunol.* **27**:247–250.

58. Janeway, C.A., Jr. 1988. Do suppressor T cells exist? A reply. *Scand. J. Immunol.* **27**:621–623.

59. Jiang, H., Zhang, S.I., and Pernis, B. 1992. Role of CD8+ T cells in murine experimental allergic encephalomyelitis. *Science.* **256**:1213–1215.

60. Koh, D.-R., et al. 1992. Less mortality but more relapses in experimental allergic encephalomyelitis in CD8-/- mice. *Science.* **256**:1210–1213.

61. Jiang, H., Braunstein, N.S., Yu, B., Winchester, R., and Chess, L. 2001. CD8+ T cells control the TH phenotype of MBP-reactive CD4+ T cells in EAE mice. *Proc. Natl. Acad. Sci. U. S. A.* **98**:6301–6306.

62. Panoutsakopoulou, V., et al. 2004. Suppression of autoimmune disease after vaccination with autoreactive T cells that express Qa-1 peptide complexes. *J. Clin. Invest.* **113**:1218–1224. doi:10.1172/JCI200420772.

63. Aldrich, C.J., et al. 1994. Identification of a Tap-dependent leader peptide recognized by alloreactive T cells specific for a class Ib antigen. *Cell.* **79**:649–658.

64. Jiang, S., et al. 1998. Induction of MHC-class I restricted human suppressor T cells by peptide priming in vitro. *Hum. Immunol.* **59**:690–699.

65. Chang, C.C., et al. 2002. Tolerization of dendritic cells by T(S) cells: the crucial role of inhibitory receptors ILT3 and ILT4. *Nat. Immunol.* **3**:237–243.

66. Najafian, N., et al. 2003. Regulatory functions of CD8+CD28- T cells in an autoimmune disease model. *J. Clin. Invest.* **112**:1037–1048. doi:10.1172/JCI200317935.

67. Jiang, H., and Chess, L. 2000. The specific regulation of immune responses by CD8+ T cells restricted by the MHC class IB molecule, QA-1. *Annu. Rev. Immunol.* **18**:185–216.

68. Li, J., Goldstein, I., Glickman-Nir, E., Jiang, H., and Chess, L. 2001. Induction of TCR Vbeta-specific CD8+ CTLs by TCR Vbeta-derived peptides bound to HLA-E. *J. Immunol.* **167**:3800–3808.

69. Cantor, H., Shen, F.W., and Boyse, E.A. 1976. Separation of helper T cells from suppressor T cells expressing different Ly components. II. Activation by antigen: after immunization, antigen-specific suppressor and helper activities are mediated by distinct T-cell subclasses. *J. Exp. Med.* **143**:1391–1340.

70. Sakaguchi, S., Fukuma, K., Kuribayashi, K., and Masuda, T. 1985. Organ-specific autoimmune diseases induced in mice by elimination of T cell subset. I. Evidence for the active participation of T cells in natural self tolerance; deficit of a T cell subset as a possible cause of autoimmune disease. *J. Exp. Med.* **161**:72–87.

71. Sakaguchi, S., Sakaguchi, N., Asano, M., Itoh, M., and Toda, M. 1995. Immunologic self-tolerance maintained by activated T cells expressing IL-2 receptor alpha-chains (CD25). Breakdown of a single mechanism of self-tolerance causes various autoimmune diseases. *J. Immunol.* **155**:1151–1164.

72. Shevach, E.M. 2000. Regulatory T cells in autoimmunity. *Annu. Rev. Immunol.* **18**:423–449.

73. Thomas, Y., et al. 1981. Functional analysis of human T cell subsets defined by monoclonal antibodies. IV. Induction of suppressor cells within the OKT4+ population. *J. Exp. Med.* **154**:459–467.

74. Thomas, Y., et al. 1982. Functional analysis of human T cell subsets defined by monoclonal antibodies. V. Suppressor cells within the activated OKT4+ population belong to a distinct subset. *J. Immunol.* **128**:1386–1390.

75. Shevach, E.M. 2001. Certified professionals: CD4(+)CD25(+) suppressor T cells. *J. Exp. Med.* **193**:F41–F46.

76. Sakaguchi, S. 2000. Regulatory T cells: key controllers of immunologic self-tolerance. *Cell.* **101**:455–458.

77. Apostolou, I., Sarukhan, A., Klein, L., and von Boehmer, H. 2002. Origin of regulatory T cells with known specificity for antigen. *Nat. Immunol.* **3**:756–763.

78. Curotto de Lafaille, M.A., and Lafaille, J.J. 2002. CD4(+) regulatory T cells in autoimmunity and allergy. *Curr. Opin. Immunol.* **14**:771–778.

79. Lafaille, J.J., Nagashima, K., Katsuki, M., and Tonegawa, S. 1994. High incidence of spontaneous autoimmune encephalomyelitis in immunodeficient anti-myelin basic protein T cell receptor transgenic mice. *Cell.* **78**:399–408.

80. Lafaille, J.J. 1998. The role of helper T cell subsets in autoimmune diseases. *Cytokine Growth Factor Rev.* **9**:139–151.

81. Van de Keere, F., and Tonegawa, S. 1998. CD4(+) T cells prevent spontaneous experimental autoimmune encephalomyelitis in anti-myelin basic protein T cell receptor transgenic mice. *J. Exp. Med.* **188**:1875–1882.

82. Furtado, G.C., et al. 2001. Regulatory T cells in spontaneous autoimmune encephalomyelitis. *Immunol. Rev.* **182**:122–134.

83. Shevach, E.M. 2002. CD4+ CD25+ suppressor T cells: more questions than answers. *Nat. Rev. Immunol.* **2**:389–400.

84. Piccirillo, C.A., and Thornton, A.M. 2004. Cornerstone of peripheral tolerance: naturally occurring CD4+CD25+ regulatory T cells. *Trends Immunol.* **25**:374–380.

85. Takahashi, T., et al. 2000. Immunologic self-tolerance maintained by CD25(+)CD4(+) regulatory T cells constitutively expressing cytotoxic T lymphocyte-associated antigen 4. *J. Exp. Med.* **192**:303–310.

86. Salomon, B., et al. 2000. B7/CD28 costimulation is essential for the homeostasis of the CD4+CD25+ immunoregulatory T cells that control autoimmune diabetes. *Immunity.* **12**:431–440.

87. Shevach, E.M., McHugh, R.S., Piccirillo, C.A., and Thornton, A.M. 2001. Control of T-cell activation

by CD4+ CD25+ suppressor T cells. *Immunol. Rev.* **182**:58–67.

88. Gorelik, L., and Flavell, R.A. 2000. Abrogation of TGFbeta signaling in T cells leads to spontaneous T cell differentiation and autoimmune disease. *Immunity.* **12**:171–181.

89. Seddon, B., and Mason, D. 1999. Regulatory T cells in the control of autoimmunity: the essential role of transforming growth factor beta and interleukin 4 in the prevention of autoimmune thyroiditis in rats by peripheral CD4(+)CD45RC- cells and CD4(+)CD8(-) thymocytes. *J. Exp. Med.* **189**:279–288.

90. Zhai, Y., and Kupiec-Weglinski, J.W. 1999. What is the role of regulatory T cells in transplantation tolerance? *Curr. Opin. Immunol.* **11**:497–503.

91. Bennett, C.L., et al. 2001. The immune dysregulation, polyendocrinopathy, enteropathy, X-linked syndrome (IPEX) is caused by mutations of FOXP3. *Nat. Genet.* **27**:20–21.

92. Brunkow, M.E., et al. 2001. Disruption of a new forkhead/winged-helix protein, scurfin, results in the fatal lymphoproliferative disorder of the scurfy mouse. *Nat. Genet.* **27**:68–73.

93. Wildin, R.S., et al. 2001. X-linked neonatal diabetes mellitus, enteropathy and endocrinopathy syndrome is the human equivalent of mouse scurfy. *Nat. Genet.* **27**:18–20.

94. Fontenot, J.D., Gavin, M.A., and Rudensky, A.Y. 2003. Foxp3 programs the development and function of CD4+CD25+ regulatory T cells. *Nat. Immunol.* **4**:330–336.

95. Walker, M.R., et al. 2003. Induction of FoxP3 and acquisition of T regulatory activity by stimulated human CD4+CD25- T cells. *J. Clin. Invest.* **112**:1437–1443. doi:10.1172/JCI200319441.

96. Chen, W., et al. 2003. Conversion of peripheral CD4+CD25- naive T cells to CD4+CD25+ regulatory T cells by TGF-beta induction of transcription factor Foxp3. *J. Exp. Med.* **198**:1875–1886.

97. Cosmi, L., et al. 2003. Human CD8+CD25+ thymocytes share phenotypic and functional features with CD4+CD25+ regulatory thymocytes. *Blood.* **102**:4107–4114.

98. Manavalan, J.S., et al. 2004. Alloantigen specific CD8+CD28- FOXP3+ T suppressor cells induce ILT3+ ILT4+ tolerogenic endothelial cells, inhibiting alloreactivity. *Int. Immunol.* **16**:1055–1068.

99. Thornton, A.M., and Shevach, E.M. 2000. Suppressor effector function of CD4+CD25+ immunoregulatory T cells is antigen nonspecific. *J. Immunol.* **164**:183–190.

100. Piccirillo, C.A., and Shevach, E.M. 2001. Cutting edge: control of CD8+ T cell activation by CD4+CD25+ immunoregulatory cells. *J. Immunol.* **167**:1137–1140.

101. Sakaguchi, S. 2003. Control of immune responses by naturally arising CD4+ regulatory T cells that express toll-like receptors. *J. Exp. Med.* **197**:397–401.

102. Bendelac, A. 1995. Mouse NK1+ T cells. *Curr. Opin. Immunol.* **7**:367–374.

103. Bendelac, A., Rivera, M.N., Park, S.H., and Roark, J.H. 1997. Mouse CD1-specific NK1 T cells: development, specificity, and function. *Annu. Rev. Immunol.* **15**:535–562.

104. Exley, M., Garcia, J., Balk, S.P., and Porcelli, S. 1997. Requirements for CD1d recognition by human invariant Valpha24+ CD4-CD8- T cells. *J. Exp. Med.* **186**:109–120.

105. Lee, P.T., Benlagha, K., Teyton, L., and Bendelac, A. 2002. Distinct functional lineages of human V(alpha)24 natural killer T cells. *J. Exp. Med.* **195**:637–641.

106. Eberl, G., et al. 1999. Tissue-specific segregation of CD1d-dependent and CD1d-independent NK T cells. *J. Immunol.* **162**:6410–6419.

107. Cui, J., et al. 1997. Requirement for Valpha14 NKT cells in IL-12-mediated rejection of tumors. *Science.* **278**:1623–1626.

108. Gombert, J.M., et al. 1996. Early quantitative and functional deficiency of NK1+-like thymocytes in the NOD mouse. *Eur. J. Immunol.* **26**:2989–2998.

109. Godfrey, D.I., Hammond, K.J., Poulton, L.D., Smyth, M.J., and Baxter, A.G. 2000. NKT cells: facts, functions and fallacies. *Immunol. Today.* **21**:573–583.

110. Bendelac, A., et al. 1995. CD1 recognition by mouse NK1+ T lymphocytes. *Science.* **268**:863–865.

111. Sharif, S., Arreaza, G.A., Zucker, P., and Delovitch, T.L. 2002. Regulatory natural killer T cells protect against spontaneous and recurrent type 1 diabetes. *Ann. N. Y. Acad. Sci.* **958**:77–88.

112. D'Orazio, T.J., and Niederkorn, J.Y. 1998. A novel role for TGF-beta and IL-10 in the induction of immune privilege. *J. Immunol.* **160**:2089–2098.

113. Furlan, R., et al. 2003. Activation of invariant NKT cells by alphaGalCer administration protects mice from MOG35-55-induced EAE: critical roles for administration route and IFN-gamma. *Eur. J. Immunol.* **33**:1830–1838.

114. Singh, A.K., et al. 2001. Natural killer T cell activation protects mice against experimental autoimmune encephalomyelitis. *J. Exp. Med.* **194**:1801–1811.

115. Baxter, A.G., Kinder, S.J., Hammond, K.J., Scollay, R., and Godfrey, D.I. 1997. Association between alphabetaTCR+CD4-CD8- T-cell deficiency and IDDM in NOD/Lt mice. *Diabetes.* **46**:572–582.

116. Hammond, K.J., et al. 1998. Alpha/beta-T cell receptor (TCR)+CD4-CD8- (NKT) thymocytes prevent insulin-dependent diabetes mellitus in nonobese diabetic (NOD)/Lt mice by the influence of interleukin (IL)-4 and/or IL-10. *J. Exp. Med.* **187**:1047–1056.

117. Sharif, S., et al. 2001. Activation of natural killer T cells by alpha-galactosylceramide treatment prevents the onset and recurrence of autoimmune type 1 diabetes. *Nat. Med.* **7**:1057–1062.

118. Falcone, M., Yeung, B., Tucker, L., Rodriguez, E., and Sarvetnick, N. 1999. A defect in interleukin 12-induced activation and interferon gamma secretion of peripheral natural killer T cells in nonobese diabetic mice suggests new pathogenic mechanisms for insulin-dependent diabetes mellitus. *J. Exp. Med.* **190**:963–972.

119. Frey, A.B., and Rao, T.D. 1999. NKT cell cytokine imbalance in murine diabetes mellitus. *Autoimmunity.* **29**:201–214.

120. Wilson, S.B., et al. 1998. Extreme Th1 bias of invariant Valpha24JalphaQ T cells in type 1 diabetes. *Nature.* **391**:177–181.

121. Lee, P.T., et al. 2002. Testing the NKT cell hypothesis of human IDDM pathogenesis. *J. Clin. Invest.* **110**:793–800. doi:10.1172/JCI200215832.

122. Bendelac, A., and Fearon, D.T. 1997. Innate pathways that control acquired immunity. *Curr. Opin. Immunol.* **9**:1–3.

123. Wilbanks, G.A., and Streilein, J.W. 1990. Distinctive humoral immune responses following anterior chamber and intravenous administration of soluble antigen. Evidence for active suppression of IgG2-secreting B lymphocytes. *Immunology.* **71**:566–572.

CD4+ Tregs and immune control

Zoltán Fehérvari[1] and Shimon Sakaguchi[1,2,3]

[1]Department of Experimental Pathology, Institute for Frontier Medical Sciences, Kyoto University, Kyoto, Japan. [2]Laboratory of Immunopathology, Research Center for Allergy and Immunology, The Institute for Physical and Chemical Research (RIKEN), Yokohama, Japan. [3]Core Research for Evolutional Science and Technology, Japan Science and Technology Agency, Kawaguchi, Japan.

Recent years have seen Tregs become a popular subject of immunological research. Abundant experimental data have now confirmed that naturally occurring CD25+CD4+ Tregs in particular play a key role in the maintenance of self tolerance, with their dysfunction leading to severe or even fatal immunopathology. The sphere of influence of Tregs is now known to extend well beyond just the maintenance of immunological tolerance and to impinge on a host of clinically important areas from cancer to infectious diseases. The identification of specific molecular markers in both human and murine immune systems has enabled the unprecedented investigation of these cells and should prove key to ultimately unlocking their clinical potential.

Introduction

Naturally occurring CD25+CD4+ suppressor or Tregs cells play an active part in establishing and maintaining immunological unresponsiveness to self constituents (i.e., immunological self tolerance) and negative control of various immune responses to non-self antigens (1). Although not a new idea for immunologists, the existence of Tregs as a definite cellular entity has been of great controversy until recently because of the paucity of reliable markers for defining the cell, the ambiguity in the molecular basis of suppressive phenomena, the lack of ample evidence for their roles in immunological disease, and even the elusive nature of some suppressive phenomena themselves (2). Recent years, however, have witnessed increasing interest in Tregs in many fields of basic and clinical immunology. Among the several types of Tregs so far reported, naturally occurring CD25+CD4+ Tregs are the main focus of current research, because accumulating evidence indicates that this population plays a crucial role in the maintenance of immunological self tolerance and negative control of pathological as well as physiological immune responses. A prominent feature of CD25+CD4+ Tregs is that the majority, if not all, of them are naturally produced by the normal thymus as a functionally distinct and mature subpopulation of T cells and persist in the periphery with stable function, and that their generation is, at least in part, developmentally controlled (1). Congenital deficiency of this population, therefore, results in serious impairment of self tolerance and immunoregulation, leading to severe autoimmunity, immunopathology, and allergy in humans (3). On the other hand, their natural presence in the immune system as a phenotypically distinct population makes them a good target for designing ways to treat or prevent immunological diseases and to control pathological as well as physiological immune responses. In addition to this naturally arising "professional" Treg population, there are several other types of Tregs that can be induced from naive T cells by antigenic stimulation under specialized conditions in the periphery (4, 5). Although physiological roles for

these inducible or "adaptive" Tregs need to be fully established, they can still be exploited as a therapeutic tool (6). In this article, we shall review recent progress in our understanding of the roles of natural and adaptive CD4+ Tregs in immune tolerance and negative control of immune responses. We shall also touch briefly on their possible clinical applications.

Naturally occurring CD25+CD4+ Tregs in self tolerance and their production by the normal thymus

Experimental evidence for the existence of Tregs with autoimmune-inhibitory activity has been suggested in various animal models of autoimmune disease for many years (7, 8). Neonatal thymectomy, for example, leads to spontaneous development of autoimmune diseases including gastritis, thyroiditis, and oophoritis in selected strains of mice (7–9). Adult thymectomy and subsequent sublethal X-irradiation produced thyroiditis and type 1 diabetes (T1D) in selected strains of rats (10, 11). In NOD mice or Bio-Breeding (BB) rats, which spontaneously develop T1D and autoimmune thyroiditis, inoculation with CD4+ T cells from histocompatible normal animals effectively prevented T1D (12, 13). On the other hand, characterization of effector T cells mediating these organ-specific autoimmune diseases firmly documented that CD4+ Th cells destroy the target organs/tissues by helping B cells to form specific autoantibodies and by inducing cell-mediated immune responses to the target self antigens. Collectively, these findings suggested that normal individuals harbor 2 functionally distinct populations of CD4+ T cells, one capable of mediating autoimmune disease and the other capable of dominantly inhibiting it in the normal physiological state (8). To test this hypothesis directly, attempts were made from the mid-1980s onward to dissect these 2 CD4+ T cell populations by expression levels of particular cell surface molecules and to examine their potential correlation with autoimmune induction or inhibition. When CD4+ splenic T cell suspensions prepared from normal mice or rats were depleted of CD25+, RT6.1+, CD5high, or CD45RB/RClow cells and the remaining CD4+ T cells transferred to syngeneic T cell–deficient mice or rats, the recipients spontaneously developed various organ-specific autoimmune diseases (including T1D, thyroiditis, and gastritis) and systemic wasting disease in a few months; reconstitution of the eliminated population inhibited the development of autoimmune disease (1, 8, 14). A similar transfer experiment also induced inflammatory bowel disease (IBD), which appeared to result from an excessive immune response of T cells to com-

Nonstandard abbreviations used: GITR, glucocorticoid-induced TNF receptor family–related gene; GITRL, GITR ligand; IBD, inflammatory bowel disease; IDO, indoleamine 2,3-dioxygenase; IPEX, immune dysregulation, polyendocrinopathy, enteropathy, X-linked syndrome; RAG, recombinase-activating gene; T1D, type 1 diabetes; Tr1, T regulatory cell type 1.

Conflict of interest: The authors have declared that no conflict of interest exists.

Citation for this article: *J. Clin. Invest.* **114**:1209–1217 (2004). doi:10.1172/JCI200423395.

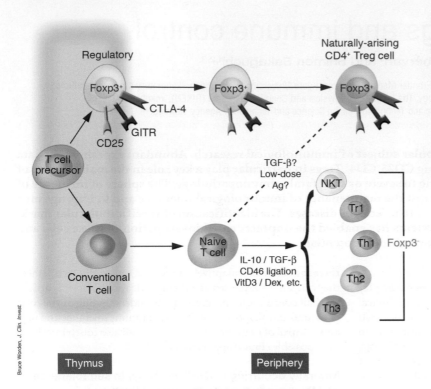

Bruce Worden, J. Clin. Invest.

Figure 1
Regulatory CD4+ cells can develop in a number of ways, although the mechanisms by which these occur and the relationship of the resulting cells to one another are contestable. Thymically generated Treg cells, otherwise known as natural T_R cells or CD25+CD4+ T_R cells, develop intrathymically according to a specialized combination of TCR and costimulatory signals. Extrathymically generated T_R cells, e.g., Tr1 cells or Th3 cells, can be generated under a whole host of conditions. Whether a conventional naive CD4+ T cell can be converted in the periphery to a de facto *Foxp3+* T_R cell remains controversial.

mensal bacteria in the intestine (15). Currently CD25 is the most specific cell surface marker for such autoimmune- and IBD-preventive CD4+ T cells, because CD25+CD4+ T cells, which constitute 5–10% of CD4+ T cells in normal naive mice, are included in the CD5^{high} or CD45RB^{low}CD4+ population, and furthermore their depletion alone is sufficient to cause autoimmune disease/IBD, while their reconstitution is effectively able to inhibit autoimmune disease/IBD, in various models (1, 16). Additionally, the various immunological properties of natural Tregs, including their in vitro suppressive activity, are assigned to CD25+CD4+ T cells naturally arising in the immune system (1, 8, 14). It should be noted, however, that CD25 is not an absolute marker for naturally occurring Tregs, since it is also expressed at high levels on activated but otherwise conventional nonregulatory T cells. We shall discuss this caveat later in the article.

The normal thymus produces the majority, if not all, of CD25+CD4+ Tregs as a functionally mature T cell subpopulation, which appears to constitute a distinct cellular lineage and to be contiguous with those found in the periphery (17) (Figure 1). As shown with the transfer of CD25−CD4+ spleen cells described above, transfer of mature thymocyte suspensions depleted of CD25+ thymocytes produced various autoimmune diseases in syngeneic T cell–deficient mice (17). This indicates that the normal thymus is continuously producing pathogenic self-reactive CD4+ T cells as well as functionally mature CD25+CD4+ Tregs capable of controlling them. This centralized production of Tregs has been referred to as "the third function of the thymus" (18).

Accumulating evidence also indicates that thymic development of CD25+CD4+ Tregs requires unique interactions of their TCR with self-peptide/MHC complexes expressed by thymic stromal cells (19). In TCR transgenic mice, for example, a large number of CD25+CD4+ T cells express endogenous TCR α chains paired with transgenic β chains; recombinase-activating gene-2 (RAG-2) defi-

ciency, which blocks the gene rearrangement of the endogenous TCR α chain locus, abrogates the development of CD25+CD4+ Tregs in such TCR transgenic mice (17, 20). Furthermore, compared with thymic selection of other T cells, the development of CD25+CD4+ Tregs requires higher-avidity interactions of their TCRs with self peptide/MHC or class II MHC itself expressed on the thymic stromal cells (especially cortical epithelial cells), yet the required avidity must not be so high as to lead to their deletion (19–23). Accessory molecules, such as CD28, B7, and CD40, expressed on developing thymocytes and thymic stromal cells also contribute to the thymic generation of CD25+CD4+ Tregs (24, 25).

The naturally occurring CD4+ Treg phenotype
Naturally occurring CD4+ Tregs constitutively express a variety of cell surface molecules more commonly associated with activated/memory cells, most significantly CD25, CD45RB^{low}, CD62L, CD103, cytotoxic T lymphocyte antigen-4 (CTLA-4, or CD152), and glucocorticoid-induced TNF receptor family–related gene (GITR) (15, 16, 26–30). Neuropilin-1, a molecule more usually associated with axon guidance, was very recently reported to be constitutively expressed by natural Tregs and, interestingly, is *downregulated* on conventional T cells following activation (31). Even though none of these markers is *uniquely* expressed by naturally occurring CD4+ Tregs, their level of expression and constitutive nature have still made them useful as functional descriptors by enabling the consistent isolation and investigation of CD4+ T cells with regulatory properties. The naturally occurring Treg surface phenotype indicates that they are in an antigen-primed state and are, at least superficially, similar to memory-type T cells. Judging from the finding that CD25+CD4+ Tregs require a high-avidity interaction with self peptide/MHC for their thymic development and become functional within the thymus, one could speculate that they are broad in antigen specificity yet more capable of recognizing self antigens than other T cells are.

Although CD25 has so far proven to be the best surface marker for thymically produced CD4+ Tregs, it can be expressed on any T cell following activation (1). In the human system, where there are relatively large numbers of activated T cells, this is especially problematic. Currently, therefore, the best way to select natural human CD4+ Tregs is to sort the population that is very high in CD25 (32). The high constitutive expression of CD25 by Tregs begs the question of whether it is simply a convenient marker or a molecule essential for their function. Several lines of evidence indicate that CD25 is indispensable for the maintenance of natural CD25+CD4+ Tregs in the immune system. For example, it has been shown that mice deficient in IL-2, IL-2Rα (CD25), or IL-2Rβ (CD122) develop lethal inflammatory disease, termed IL-2 deficiency syndrome, which can be prevented by inoculation of normal CD25+CD4+ T cells as long as a source of IL-2 is made available experimentally (33–35). Our own experiments indicated that neutralization of IL-2 selectively reduced numbers of CD25+CD4+ T cells in normal mice and consequently produced organ-specific autoimmune diseases similar to those produced by depletion of natural Tregs (R. Setoguchi et al., manuscript submitted for publication). Collectively, these results suggest that IL-2 is essential for the development, maintenance, and function of CD25+CD4+ Tregs.

GITR and its role in CD25+CD4+ Treg function is an interesting area. This molecule was identified as a constitutively expressed marker for naturally occurring Tregs, but, like most such candidate molecules involved in Treg identification, it is also upregulated on conventional activated CD4+ T cells (29, 30). An anti–mouse GITR mAb (DTA-1) is able to block CD25+CD4+ Treg suppression in vitro, and, furthermore, its injection leads to the induction of autoimmunity in vivo as well as enhancing the proliferation of CD25–CD4+ cells by transducing a costimulatory signal (29, 36). Since DTA-1 is nondepleting, it was originally presumed to primarily transmit a suppression-blocking signal to the CD25+CD4+ Treg. However, some very recent data instead suggest that ligation of GITR on *activated T cells*, not Tregs, renders them resistant to suppression (37). The natural ligand for GITR (GITRL) has now also been cloned and its distribution elucidated (38). GITRL is expressed on APCs (DCs, macrophages, and B cells) but is downregulated following maturation. Therefore the relative distribution patterns of GITR on activated T cells and Tregs, and of GITRL on APCs, suggest a complex dynamic of interaction, which is only just being elucidated.

Identification of an unambiguous surface marker for naturally occurring CD4+ Tregs remains something of a Holy Grail, especially where the isolation of human Tregs for clinical purposes is concerned. Efforts in this direction may well guide the progress of research in naturally occurring human Tregs.

FOXP3 as a master control gene for Treg development

A deeper understanding of the developmental processes of natural Tregs, as suggested by the neonatal thymectomy model of autoimmune disease, evolved out of studies on the Scurfy mouse and the human disease IPEX (immune dysregulation, polyendocrinopathy, enteropathy, X-linked syndrome). IPEX is an X-linked immunodeficiency syndrome associated with autoimmune disease in multiple endocrine organs (such as T1D and thyroiditis), IBD, atopic dermatitis, and fatal infections (3). The Scurfy mouse strain exhibits a fatal X-linked lymphoproliferation characterized by a multiorgan immunopathology very similar to the human disease IPEX (39–41). The causative gene, *Foxp3*

(*FOXP3* in humans), which underlies both syndromes, encodes a forkhead/winged-helix family transcriptional repressor called Scurfin (42–44). The striking similarities seen between mutations in *Foxp3/FOXP3* and depletion of CD25+CD4+ Tregs led several groups to investigate the relationship of this gene to Treg development and function. Experiments in mice indeed demonstrated *Foxp3* mRNA and Scurfin protein to be specifically expressed in CD25+CD4+ Tregs, and, in contrast to the cell surface markers used to date, they were never observed in non-Tregs following conventional activation or differentiation into Th1 and Th2, nor in natural killer T cells (45–47).Subsequent studies, also in mice, have further demonstrated the existence of a small population of CD25–CD4+ T cells that are nevertheless still *Foxp3*+ and have a regulatory function (ref. 1, and M. Ono et al., manuscript submitted for publication). Scurfy mutant mice, or those with a targeted deletion of *Foxp3*, were unable to support the development of natural CD25+CD4+ Tregs, although they contained large numbers of chronically activated CD25+ nonregulatory T cells (45, 47). By contrast, the number of CD25+CD4+ Tregs increased significantly in transgenic mice overexpressing *Foxp3* (45). A final critical observation showed that retroviral transduction of *Foxp3* into *Foxp3*– nonregulatory CD25–CD4+ T cells bestowed on them a fully functional Treg phenotype; e.g., cotransfer of *Foxp3*-transduced T cells with CD25–CD4+ T cells prevented autoimmune disease and IBD in SCID mice (see above) (46, 47).

Broadly speaking, an equivalent pattern of *FOXP3* expression has now also been reported in human cells, with Treg-like properties being similarly transferable by retroviral transduction (48–51). Already, however, some discrepancies are beginning to emerge between the behavior of human and that of mouse *FOXP3/Foxp3* expression. For instance, there is at least 1 example of *FOXP3* being apparently induced following standard antibody-mediated activation of normal CD25– human T cells; this has not been observed thus far in the murine model (50). Similarly, some instances of CD25– human T cell activation by DCs have also resulted in *FOXP3* upregulation (refs. 52, 53, and see below). The possibility remains, however, that the induction of *FOXP3* expression in human CD25– cells may simply be a result of the expansion of the human counterpart to the murine CD25–CD4+ *Foxp3*+ population described above, as these studies all isolated their Tregs solely on the basis of CD25 (51).

Thus, *Foxp3/FOXP3* appears to be a master control gene for the development and function of natural CD25+CD4+ Tregs. Given that humans bear natural CD25+CD4+ Tregs with a phenotype and function comparable to those found in rodents (32), it is most likely that in IPEX, disruption of the *FOXP3* gene abrogates the development of thymic Tregs, leading to hyperactivation of T cells reactive with self antigens, commensal bacteria in the intestine, or innocuous environmental substances, and thus causing autoimmune polyendocrinopathy, IBD, or allergy, respectively. This has several implications for self tolerance and autoimmune/inflammatory disease in humans. First, this is so far the clearest example that an abnormality in naturally arising Tregs is a primary cause of human autoimmune disease, IBD, and allergy. Second, the development of natural Tregs is, at least in part, genetically and developmentally programmed. Third, hemizygous defects of the *FOXP3* gene in females illustrate that the mechanism of dominant self tolerance is physiologically operating in humans. Owing to random inactivation of the X chromosome during lyonization of individual Tregs, hemizygous females have *FOXP3*-defective Tregs and *FOXP3*-normal ones as a genetic mosaic, yet they are nevertheless completely normal (54). This observation demonstrates that even reduced numbers of *Foxp3*+ Tregs are able to dominantly control pathogenic T cells, and, further, that even a partial restoration of Tregs could be sufficient to cure IPEX or, indeed, other autoimmune pathologies. Mechanistic data on Foxp3 are thus far lacking; it is therefore currently unclear how it exerts its effects at a molecular level. The molecular interactions of Foxp3, and indeed the signals triggering its expression, are now an intensely investigated area, and unraveling them may well prove critical to exploiting natural Tregs in a therapeutic setting.

Functional characteristics of natural CD25+CD4+ Tregs and their mechanisms of suppression

Without question the most remarkable feature of CD25+CD4+ Tregs is their ability to dampen immune responses. They appear capable of suppressing a wide variety of immune cells, encompassing those of both the innate (55–57) and the adaptive immune systems (58–60). This suppressive ability can be modeled in vitro by mixing of titrated numbers of highly purified CD25+CD4+ Tregs and responder cells, typically CD25−CD4+ T cells plus a T cell stimulus. Under such conditions, the CD25+ population suppresses both the proliferation and, more fundamentally, the IL-2 production of the CD25− cells in a dose-dependent manner (58, 59, 61). CD25+CD4+ Tregs themselves require TCR stimulation, and, it now seems, IL-2, to actually trigger their suppressive effects, but once this condition has been satisfied their ensuing suppression can act non–antigen-specifically (58, 59, 61). Therefore, suppression is an active process and can be directed against bystander cells. Curiously, CD25+CD4+ Tregs themselves are anergic in vitro, i.e., they do not proliferate or produce IL-2 in response to conventional T cell stimuli such as plate-or bead–bound anti-CD3, concanavlin A (ConA), or splenic APCs. This anergy can, however, be broken by a sufficiently potent stimulus, e.g., the addition of high-dose exogenous IL-2 or anti-CD28, or the use of mature DCs as APCs (27, 58, 59, 61–63). Some of these strong stimuli, particularly mature DCs, also perturb CD25+CD4+ Treg suppression both in vitro and in vivo (63, 64). At least in vitro, anergy seems to be the default state of naturally occurring Tregs, since they revert back to it once potent stimulation is withdrawn (58). In vivo, however, CD25+CD4+ Treg anergy is not readily observed; instead they seem to have a highly active rate of turnover (33, 65). It seems likely, then, that CD25+CD4+ Treg anergy is an in vitro phenomenon, merely reflecting an exacting set of activation requirements generally absent from cell culture.

Given that the ability to control immune responses is the cardinal feature of CD25+CD4+ Tregs, it is surprising that their mechanism(s) of suppression remains elusive. Essentially, Treg suppression can be divided into those mechanisms mediated by relatively far-reaching soluble factors and those requiring intimate cell contact. In vivo experiments based chiefly on the IBD model mentioned previously have demonstrated the importance of the immunomodulatory cytokines IL-10 and TGF-β (66). By blocking IL-10 signaling in vivo with an anti–IL-10 receptor mAb, it was possible to abrogate the normal colitis-preventative action of CD45RB^{low} cells (66). Similarly, CD45RB^{low} T cells from IL-10−/− mice lacked their otherwise intrinsic ability to protect from colitis and, moreover, were even colitogenic themselves when transferred alone (66). The importance of IL-10 is further underscored by the observation that IL-10−/− mice spontaneously develop colitis (67, 68). Examination of the in vivo role

of TGF-β has generally painted a similar picture to that of IL-10, with Treg function being blocked by the presence of neutralizing anti–TGF-β mAbs (69). Some data also suggest that TGF-β may not necessarily act as a soluble factor but can also be found on the surface of activated CD25+CD4+ Tregs and may therefore act in a membrane-proximal manner (70). Interestingly, virtually all TGF-β+ CD25+CD4+ Tregs also express thrombospondin, a factor capable of converting normally latent TGF-β into its active form (71). There should be a note of caution regarding these in vitro studies on TGF-β, since a comprehensive analysis by a second group failed to demonstrate any role for it in vitro (72).

The confusion over a definitive CD25+CD4+ Treg suppression mechanism is compounded when viewed in the context of the in vitro data, since here the overwhelming evidence highlights direct cell-cell interaction, and not cytokines, as being critical (58, 59, 73). Several lines of evidence lead to this conclusion: with the exception of the study on membrane-bound TGF-β alluded to above, both anti–IL-10 and anti–TGF-β fail to perturb CD25+CD4+ Treg suppression (58, 59, 70, 72). Similarly, supernatants from suppressed cultures or activated CD25+CD4+ Tregs show no inherent suppressive activity, nor can suppression be observed across a semipermeable membrane (58, 59). Collectively, these in vitro observations therefore appear to obviate a role not just for IL-10 and TGF-β but for soluble factors in general.

The actual membrane events occurring during suppression that depends on CD25+CD4+ Treg contact have yet to be clarified. The most simplistic models propose competition for APCs and specific MHC/peptide antigenic complexes. Additionally, the constitutive expression of the high-affinity IL-2 receptor could make naturally occurring Tregs into an effective IL-2 sink, depriving potential autoreactive T cells of this essential growth factor (74). However, given the relative physiological scarcity of naturally occurring Tregs, it is perhaps unlikely that a simple competitive-adsorptive model alone could account for their suppressive action in vivo. Other models of CD25+CD4+ Treg suppression propose a more proactive and antagonistic form of suppression that relies on the expression of specific "inhibitory" molecules. The identity and indeed even the very existence of such an inhibitory molecule are uncertain, but 1 potential molecule could be Treg-expressed CTLA-4. Aside from its well-established high affinity for the costimulatory molecules B7.1 and B7.2 (CD80 and CD86, respectively), CTLA-4 has also recently been shown to trigger the induction of the enzyme indoleamine 2,3-dioxygenase (IDO) when interacting with its ligands on DCs (75–78). IDO catalyzes the conversion of tryptophan to kynurenine and other metabolites, which have potent immunosuppressive effects in the local environment of the DC. In this way, CD25+CD4+ Tregs may exert their suppression by proxy through their action on APCs. Another APC-centric mode of suppression could be via the perturbation of antigen-presenting capacity. In support of this concept, one report has demonstrated that purified CD25+CD4+ Tregs are able to downregulate the expression of both CD80 and CD86 on DCs, converting them into inefficient APCs (57). At any rate, CD25+CD4+ Tregs need not act exclusively via the APC, since they are quite capable of suppressing in the context of "APC-free" systems such as plate- or bead–bound antibodies or MHC/peptide tetramers (58, 79). At least in vitro, *direct* suppression of the target cell is therefore still also possible.

A provocative investigation into the membrane events involved in CD25+CD4+ Treg suppression was recently reported (80). This

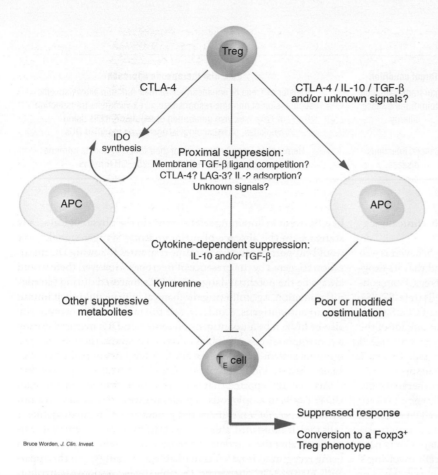

Bruce Worden, *J. Clin. Invest.*

Figure 2
Possible CD25+CD4+ Treg suppression mechanisms in vivo. CD25+CD4+ Tregs may suppress
their effector T cell targets (T_E) by a number of proposed mechanisms. In vivo CD25+CD4+ Tregs may
act in a cell contact–dependent manner by competing directly for stimulatory ligands on the APC, by
sinking essential growth factors such as IL-2, or
by directly transmitting an as-yet uncharacterized
negative signal. Alternatively, they may use longer-
range suppressive mechanisms by means of the
cytokines IL-10 and TGF-β. Finally, CD25+CD4+
Tregs may act through the APC either by triggering
IDO activity, resulting in the generation of immunosuppressive metabolites, or by perturbing the APC's
presenting capacity. Such mechanisms are not necessarily mutually exclusive, and more than 1 might
operate in tandem.

study suggested that engagement of CD80, and to a lesser extent
CD86, on the *responder* T cell and not the APC was responsible for
the transmission of a negative signal, and therefore these were the
molecular targets through which Tregs exert their function (80).
In support of this, the authors demonstrated that B7$^{-/-}$ responder
cells were resistant to suppression in vitro and induced a fatal
wasting disease refractory to cotransferred CD25+CD4+ Tregs
(80). Again the obvious candidate Treg molecule for this inhibitory interaction would be CTLA-4, although this fails to explain
the paradoxically intact suppression mediated by CTLA-4$^{-/-}$
CD25+CD4+ Tregs (27). The presence of B7 on conventional T cells
has been known for several years, and it would be interesting, then,
if this hitherto puzzling expression pattern were shown to play a
role in Treg–mediated suppression (81, 82). While the identification of a membrane-bound CD25+CD4+ Treg–specific inhibitory
molecule remains inconclusive, some very recent work has suggested that the CD4-related molecule LAG-3 may be important
(CD223), though this awaits independent confirmation (83, 84).
Proving a negative hypothesis is always a difficult task, but it may
yet be shown that there are no truly unique Treg–associated molecules responsible for inhibition. Rather, the specialized functions
of Tregs could simply be the product of known molecules acting
semi-redundantly, which together generate a suppressive phenotype. An integrated summary of CD25+CD4+ Treg suppressive
mechanisms is shown in Figure 2.

Given the relative physiological scarcity of CD25+CD4+ Tregs, it
seems likely that in vivo they would use mechanisms to amplify their
suppressive action, and ones that are not normally fully appreciated
under in vitro analysis. This could occur either by the modification

of APCs as outlined above or by the "infectious" spreading of tolerance to conventional T cells. In accordance with this, some recent
work has demonstrated that human CD25+CD4+ Tregs can confer
a suppressive phenotype to conventional CD4+ T cells in a contact-
dependent manner (52). These newly generated regulatory like cells
then suppress by means of IL-10 or TGF-β. This would constitute a
mechanism of not only spreading a suppressive phenotype but also
making it more efficient on a per-cell basis by engaging the action of
soluble mediators. Most satisfyingly, this scenario could also finally
reconcile some of the disparities observed between the in vitro and
the in vivo mechanisms of CD25+CD4+ Treg suppression.

Extrathymic generation of CD4+ Tregs
The possibility of extrathymic CD25+CD4+ Treg generation is
currently contentious. An interesting recent study in this area
readdressed so-called "low-zone" tolerance (85). This phenomenon was first observed decades ago and was described as the
antigen-specific tolerance resulting from subimmunogenic doses
of antigen given i.v. (86). Using an osmotic pump to deliver minute controlled quantities of antigenic peptide to TCR transgenic mice, the authors of this updated study demonstrated the
appearance of CD25+CD4+ Tregs measurable by their function,
surface phenotype, and *Foxp3* expression. Importantly, the study
showed that Treg development could still occur in thymectomized TCR transgenic mice on a RAG knockout background,
which are normally wholly lacking in CD25+CD4+ Tregs (17, 20,
46, 85). It therefore seems possible that CD25+CD4+ Treg development, as measured by *Foxp3* and suppressive function, can
occur in conventional T cells under specific in vivo activation

Table 1
Potential clinical applications of CD25+CD4+ Tregs

	Target condition	Potential therapeutic approach
Enhancement of CD25+CD4+ Treg function	Organ transplantation, autoimmune disease, allergy	Transfer of Tregs or enhancement of their function allows specific suppression of immune responses; e.g., ex vivo gene transduction of *Foxp3*; ex vivo generation of regulatory cells using cytokines, pharmacological agents, or modified DCs
Reduction of CD25+CD4+ Treg function	Cancer, infectious disease	Removal of Tregs or blocking of their function boosts immune responses; anti–CTLA-4, -CD25, -GITR mAbs

conditions, therefore demonstrating a potentially clinically significant developmental plasticity.

The involvement of TGF-β in many facets of Treg behavior is well known (87), but a recent report has also suggested that its exogenous addition results in development of *Foxp3*+ Tregs, from conventional and even from RAG knockout CD4+ T cells (refs. 87, 88). However, it remains to be seen whether generation of CD25+CD4+ Tregs by the mechanisms described above can occur outside of the *relatively* artificial confines of a TCR transgenic system, and indeed whether the initial cell populations contain only "truly" naive T cells and no potential *Foxp3*+CD25+CD4+ Treg precursors.

In the original demonstration of extrathymically generated regulatory cells, the cells were termed T regulatory cell type 1 (Tr1) or Th3 cells (89, 90). It is likely that these cells form a cell type distinct from their thymically generated CD25+CD4+ Treg counterparts that have been elaborated on above. Tr1 and related cells have been generated using a variety of approaches, typically involving T cell activation in the presence of immunomodulating cytokines or repetitive stimulation with nonprofessional APCs. Tr1 cells were initially generated by chronic stimulation of normal nonregulatory T cells in the presence of IL-10 (89). Such cells secrete a pattern of cytokines distinct from that of the more usual Th1 or Th2 profile and are characterized by high levels of IL-10 and generally low levels of TGF-β and IL-5 (89). Moreover, Tr1 cells are functionally suppressive in vivo and able to prevent the development of Th1 autoimmune diseases such as colitis (89, 91). Th3 cells, on the other hand, were cloned from the mesenteric lymph nodes of mice orally tolerized with myelin basic protein (90). The majority of such cells produce TGF-β and varying levels of Th2 cytokines and suppress the induction of experimental autoimmune encephalitis (90). In vitro treatment of human and mouse T cells with a combination of the immunosuppressants vitamin D$_3$ and dexamethasone has also resulted in the generation of *Foxp3*− regulatory cells, but with properties somewhat distinct from those reported for Tr1 or Th3 cells (92–94). Finally, there is also a study suggesting that signaling through the complement receptor (CD46) concomitant to more conventional TCR activation can trigger the peripheral induction of human CD4+ regulatory cells (95).

Much attention has also focused on the influence that DCs may have on the extrathymic development of regulatory cells. Stimulation with immature DCs (i.e., low levels of costimulatory molecules) and stimulation with DCs modified by pretreatment with IL-10 or TGF-β have both been shown to result in the induction of anergic cells with suppressive capabilities in vitro and in vivo (96–98). Current models of DC-based tolerance state that T cell antigen recognition on immature DCs results in tolerization whereas mature DCs elicit effector responses (99). A system structured in this way would be effective at maintaining self tolerance in the physiological steady state, i.e., in the absence of inflammatory "danger signals," yet would support productive immune responses following DC maturation triggered by the presence of microbes. However, there would always be the potential danger that DCs matured during "sterile" inflammation, e.g., following mechanical injury, could elicit immunity to autoantigens. Similarly, self tolerance could theoretically also be broken by autoantigens presented on DCs matured during a contemporaneous microbial infection. It seems, though, that the immune system may have yet another level of control to protect against just the kind of scenarios outlined above. In support of this, it was recently reported that the response of conventional human CD4+ T cells to autologous peptides presented by *mature*, but not by immature, DCs results in the generation of regulatory-like T cells (53). If confirmed, this ability of the immune system to so dramatically alter the outcome of a response depending on the antigen being recognized is rather remarkable, especially given the apparently matured DC phenotype. Possibly the net response is attributable to the nature or source of ancillary signals, e.g., which toll-like receptors (TLRs) are being engaged, alone or in combination.

Extrathymically generated regulatory cells represent a heterogeneous assemblage whose ontogenic relationship to naturally occurring Tregs is still being determined. The only really clear point of convergence between the 2 broad families of regulatory cells is that they share a suppressive capability. One interpretation would suggest that peripherally generated regulatory cells merely represent a specialized activation state of conventional CD4+ cells (i.e., "adaptive regulatory cells") whereas CD25+CD4+ Tregs are a de facto lineage by virtue of their distinctive *Foxp3* expression, although the most recent data stemming from the use of TGF-β or low-zone tolerance induction protocols are perhaps blurring even this distinction (85, 88). The use of *Foxp3/FOXP3* to disentangle this conundrum has been only partially successful. As far as mice are concerned, *Foxp3* seems, by and large, to be a stable marker expressed only in naturally occurring CD25+CD4+ Tregs and thus far is not in most models of extrathymically generated regulatory cells (but see refs. 85, 88). In contrast, human *FOXP3* expression appears far less stringent, with some reports already demonstrating upregulation in extrathymically generated regulatory cells following even basic activation (52, 53). Whether the apparent variability in human *FOXP3* and, to a much lesser extent, mouse *Foxp3* expression undermines its importance as an unambiguous marker for naturally occurring Tregs remains to be seen.

Conclusion and clinical perspective

Abundant evidence now strongly supports the once controversial existence of Tregs as key controllers of self tolerance. It also now seems that their roles can be expanded to many areas of immunology, in fact, potentially, to any scenario where the suppression and/or tuning of an immune response is required. A strategic manipulation of Tregs, either naturally occurring or extrathymically generated, to dampen or enhance their functions as appropriate may prove to have great clinical benefit (Table 1). Already manipulation of CD25+CD4+ Tregs in various animal models has provided encouraging results for both enhancement of tumor immunity and maintenance of allograft tolerance (100–104). In the case of organ transplantation in particular, CD25+CD4+ Tregs seem to offer a flexible and adaptive form of immunological control apparently not achievable with standard small-molecule immunosuppression (see for example refs. 102, 104, 105). These practical applications will be expanded upon in other Reviews in this series. Informed by the murine studies, recent experiments are also increasingly demonstrating the significant roles CD25+CD4+ Tregs can play in human pathologies as varied as rheumatoid arthritis, multiple sclerosis, HIV infection, and allergy (106–112). Recent advances in our understanding of CD25+CD4+ Treg development and important functional markers such as the association with Foxp3/FOXP3 have permitted the accurate isolation and manipulation of these cells in mice and, importantly, their human counterparts. Understanding the events both upstream and downstream of Foxp3/FOXP3 may enable us to "tailor-make" large numbers of CD25+CD4+ Tregs to specifically suppress immune responses in autoimmunity and allergy or to antagonize them where a boost of immunity is required, e.g., in microbial and antitumor responses. The potential clinical focus, though, need not be solely on thymically produced CD25+CD4+ Tregs, since peripherally generated regulatory cells such as Tr1, with their potent cytokine-mediated suppressive capacity, may also hold great therapeutic promise.

Acknowledgments

We apologize to those researchers whose work, because of space restrictions, has not been cited in this review. We thank our colleagues who have allowed the prepublication mention of their work and engaging discussion. Z. Fehérvari is supported by the Japan Society for the Promotion of Science and S. Sakaguchi by grants-in-aid from the Ministry of Education, Culture, Sports, Science and Technology and the Ministry of Human Welfare of Japan.

Address correspondence to: Zoltán Fehérvari, Department of Experimental Pathology, Institute for Frontier Medical Sciences, Kyoto University, Shogo-in 53, Sakyo-ku, Kyoto 606-8507, Japan. Phone: 81-75-751-3888; Fax: 81-75-751-3820; E-mail: zed72@frontier.kyoto-u.ac.jp.

1. Sakaguchi, S. 2004. Naturally arising CD4+ regulatory t cells for immunologic self-tolerance and negative control of immune responses. *Annu. Rev. Immunol.* **22**:531–562.

2. Bloom, B.R., Salgame, P., and Diamond, B. 1992. Revisiting and revising suppressor T cells. *Immunol. Today.* **13**:131–136.

3. Gambineri, E., Torgerson, T.R., and Ochs, H.D. 2003. Immune dysregulation, polyendocrinopathy, enteropathy, and X-linked inheritance (IPEX), a syndrome of systemic autoimmunity caused by mutations of FOXP3, a critical regulator of T-cell homeostasis. *Curr. Opin. Rheumatol.* **15**:430–435.

4. Roncarolo, M.G., Bacchetta, R., Bordignon, C., Narula, S., and Levings, M.K. 2001. Type 1 T regulatory cells. *Immunol. Rev.* **182**:68–79.

5. Weiner, H.L. 2001. Induction and mechanism of action of transforming growth factor-beta-secreting Th3 regulatory cells. *Immunol. Rev.* **182**:207–214.

6. Bluestone, J.A., and Abbas, A.K. 2003. Natural versus adaptive regulatory T cells. *Nat. Rev. Immunol.* **3**:253–257.

7. Sakaguchi, S., Takahashi, T., and Nishizuka, Y. 1982. Study on cellular events in post-thymectomy autoimmune oophoritis in mice. II. Requirement of Lyt-1 cells in normal female mice for the prevention of oophoritis. *J. Exp. Med.* **156**:1577–1586.

8. Sakaguchi, S., Fukuma, K., Kuribayashi, K., and Masuda, T. 1985. Organ-specific autoimmune diseases induced in mice by elimination of T cell subset. I. Evidence for the active participation of T cells in natural self-tolerance; deficit of a T cell subset as a possible cause of autoimmune disease. *J. Exp. Med.* **161**:72–87.

9. Asano, M., Toda, M., Sakaguchi, N., and Sakaguchi, S. 1996. Autoimmune disease as a consequence of developmental abnormality of a T cell subpopulation. *J. Exp. Med.* **184**:387–396.

10. Penhale, W.J., Farmer, A., McKenna, R.P., and Irvine, W.J. 1973. Spontaneous thyroiditis in thymectomized and irradiated Wistar rats. *Clin. Exp. Immunol.* **15**:225–236.

11. Fowell, D., and Mason, D. 1993. Evidence that the T cell repertoire of normal rats contains cells with the potential to cause diabetes. Characterization of the CD4+ T cell subset that inhibits this autoimmune potential. *J. Exp. Med.* **177**:627–636.

12. Boitard, C., Yasunami, R., Dardenne, M., and Bach, J.F. 1989. T cell-mediated inhibition of the transfer of autoimmune diabetes in NOD mice. *J. Exp. Med.* **169**:1669–1680.

13. Greiner, D.L., et al. 1987. Depletion of RT6.1+ T lymphocytes induces diabetes in resistant biobreeding/Worcester (BB/W) rats. *J. Exp. Med.* **166**:461–475.

14. Powrie, F., and Mason, D. 1990. OX-22high CD4+ T cells induce wasting disease with multiple organ pathology: prevention by the OX-22low subset. *J. Exp. Med.* **172**:1701–1708.

15. Powrie, F., Leach, M.W., Mauze, S., Caddle, L.B., and Coffman, R.L. 1993. Phenotypically distinct subsets of CD4+ T cells induce or protect from chronic intestinal inflammation in C. B-17 scid mice. *Int. Immunol.* **5**:1461–1471.

16. Sakaguchi, S., Sakaguchi, N., Asano, M., Itoh, M., and Toda, M. 1995. Immunologic self-tolerance maintained by activated T cells expressing IL-2 receptor alpha-chains (CD25). Breakdown of a single mechanism of self-tolerance causes various autoimmune diseases. *J. Immunol.* **155**:1151–1164.

17. Itoh, M., et al. 1999. Thymus and autoimmunity: production of CD25+CD4+ naturally anergic and suppressive T cells as a key function of the thymus in maintaining immunologic self-tolerance. *J. Immunol.* **162**:5317–5326.

18. Seddon, B., and Mason, D. 2000. The third function of the thymus. *Immunol. Today.* **21**:95–99.

19. Jordan, M.S., et al. 2001. Thymic selection of CD4+CD25+ regulatory T cells induced by an agonist self-peptide. *Nat. Immunol.* **2**:301–306.

20. Kawahata, K., et al. 2002. Generation of CD4(+)CD25(+) regulatory T cells from autoreactive T cells simultaneously with their negative selection in the thymus and from nonautoreactive T cells by endogenous TCR expression. *J. Immunol.* **168**:4399–4405.

21. Stephens, G.L., and Ignatowicz, L. 2003. Decreasing the threshold for thymocyte activation biases CD4+ T cells toward a regulatory (CD4+CD25+) lineage. *Eur. J. Immunol.* **33**:1282–1291.

22. Sakaguchi, S., et al. 2003. Thymic generation and selection of CD25+CD4+ regulatory T cells: implications of their broad repertoire and high self-reactivity for the maintenance of immunological self-tolerance. *Novartis Found. Symp.* **252**:6–16.

23. Hsieh, C.S., et al. 2004. Recognition of the peripheral self by naturally arising CD25(+) CD4(+) T cell receptors. *Immunity.* **21**:267–277.

24. Salomon, B., et al. 2000. B7/CD28 costimulation is essential for the homeostasis of the CD4+CD25+ immunoregulatory T cells that control autoimmune diabetes. *Immunity.* **12**:431–440.

25. Kumanogoh, A., et al. 2001. Increased T cell autoreactivity in the absence of CD40-CD40 ligand interactions: a role of CD40 in regulatory T cell development. *J. Immunol.* **166**:353–360.

26. Lehmann, J., et al. 2002. Expression of the integrin alpha Ebeta 7 identifies unique subsets of CD25+ as well as CD25- regulatory T cells. *Proc. Natl. Acad. Sci. U. S. A.* **99**:13031–13036.

27. Takahashi, T., et al. 2000. Immunologic self-tolerance maintained by CD25(+)CD4(+) regulatory T cells constitutively expressing cytotoxic T lymphocyte-associated antigen 4. *J. Exp. Med.* **192**:303–310.

28. Read, S., Malmstrom, V., and Powrie, F. 2000. Cytotoxic T lymphocyte-associated antigen 4 plays an essential role in the function of CD25(+)CD4(+) regulatory cells that control intestinal inflammation. *J. Exp. Med.* **192**:295–302.

29. Shimizu, J., Yamazaki, S., Takahashi, T., Ishida, Y., and Sakaguchi, S. 2002. Stimulation of CD25(+)CD4(+) regulatory T cells through GITreg breaks immunological self-tolerance. *Nat. Immunol.* **3**:135–142.

30. McHugh, R.S., et al. 2002. CD4(+)CD25(+) immunoregulatory T cells: gene expression analysis reveals a functional role for the glucocorticoid-induced TNF receptor. *Immunity.* **16**:311–323.

31. Bruder, D., et al. 2004. Neuropilin-1: a surface marker of regulatory T cells. *Eur. J. Immunol.* **34**:623–630.

32. Baecher-Allan, C., Brown, J.A., Freeman, G.J., and Hafler, D.A. 2001. CD4+CD25high regulatory cells in human peripheral blood. *J. Immunol.* **167**:1245–1253.

33. Almeida, A.R., Legrand, N., Papiernik, M., and Freitas, A.A. 2002. Homeostasis of peripheral CD4+ T cells: IL-2R alpha and IL-2 shape a population of regulatory cells that controls CD4+ T cell numbers. *J. Immunol.* **169**:4850–4860.

34. Malek, T.R., Yu, A., Vincek, V., Scibelli, P., and Kong,

L. 2002. CD4 regulatory T cells prevent lethal autoimmunity in IL-2Rbeta-deficient mice. Implications for the nonredundant function of IL-2. *Immunity.* **17**:167–178.

35. Papiernik, M., de Moraes, M.L., Pontoux, C., Vasseur, F., and Penit, C. 1998. Regulatory CD4 T cells: expression of IL-2R alpha chain, resistance to clonal deletion and IL-2 dependency. *Int. Immunol.* **10**:371–378.

36. Kanamaru, F., et al. 2004. Costimulation via glucocorticoid-induced TNF receptor in both conventional and CD25+ regulatory CD4+ T cells. *J. Immunol.* **172**:7306–7314.

37. Stephens, G.L. et al. 2004. Engagement of glucocorticoid-induced TNFR family-related receptor on effector T cells by its ligand mediates resistance to suppression by CD4+CD25+ T cells. *J. Immunol.* **173**:5008–5020.

38. Tone, M., et al. 2003. Mouse glucocorticoid-induced tumor necrosis factor receptor ligand is costimulatory for T cells. *Proc. Natl. Acad. Sci. U. S. A.* **100**:15059–15064.

39. Blair, P.J., et al. 1994. CD4+CD8- T cells are the effector cells in disease pathogenesis in the scurfy (sf) mouse. *J. Immunol.* **153**:3764–3774.

40. Godfrey, V.L., Wilkinson, J.E., Rinchik, E.M., and Russell, L.B. 1991. Fatal lymphoreticular disease in the scurfy (sf) mouse requires T cells that mature in a sf thymic environment: potential model for thymic education. *Proc. Natl. Acad. Sci. U. S. A.* **88**:5528–5532.

41. Lyon, M.F., Peters, J., Glenister, P.H., Ball, S., and Wright, E. 1990. The scurfy mouse mutant has previously unrecognized hematological abnormalities and resembles Wiskott-Aldrich syndrome. *Proc. Natl. Acad. Sci. U. S. A.* **87**:2433–2437.

42. Brunkow, M.E., et al. 2001. Disruption of a new forkhead/winged-helix protein, scurfin, results in the fatal lymphoproliferative disorder of the scurfy mouse. *Nat. Genet.* **27**:68–73.

43. Bennett, C.L., et al. 2001. The immune dysregulation, polyendocrinopathy, enteropathy, X-linked syndrome (IPEX) is caused by mutations of FOXP3. *Nat. Genet.* **27**:20–21.

44. Schubert, L.A., Jeffery, E., Zhang, Y., Ramsdell, F., and Ziegler, S.F. 2001. Scurfin (FOXP3) acts as a repressor of transcription and regulates T cell activation. *J. Biol. Chem.* **276**:37672–37679.

45. Khattri, R., Cox, T., Yasayko, S.A., and Ramsdell, F. 2003. An essential role for Scurfin in CD4+CD25+ T regulatory cells. *Nat. Immunol.* **4**:337–342.

46. Hori, S., Nomura, T., and Sakaguchi, S. 2003. Control of regulatory T cell development by the transcription factor Foxp3. *Science.* **299**:1057–1061.

47. Fontenot, J.D., Gavin, M.A., and Rudensky, A.Y. 2003. Foxp3 programs the development and function of CD4+CD25+ regulatory T cells. *Nat. Immunol.* **4**:330–336.

48. Oswald-Richter, K., et al. 2004. HIV infection of naturally occurring and genetically reprogrammed human regulatory T-cells. *PLoS Biol.* **2**:E198.

49. Weiss, L., et al. 2004. Human immunodeficiency virus-driven expansion of CD4+CD25+ regulatory T cells which suppress HIV-specific CD4 T-cell responses in HIV-infected patients. *Blood.* doi:10.1182/blood-2004-01-0365.

50. Walker, M.R., et al. 2003. Induction of FoxP3 and acquisition of T regulatory activity by stimulated human CD4+CD25- T cells. *J. Clin. Invest.* **112**:1437–1443. doi:10.1172/JCI200319441.

51. Yagi, H., et al. 2004. Crucial role of FOXP3 in the development and function of human CD25+CD4+ regulatory T cells. *Int. Immunol.* In press.

52. Stassen, M., et al. 2004. Human CD25+ regulatory T cells: two subsets defined by the integrins alpha 4 beta 7 or alpha 4 beta 1 confer distinct suppressive properties upon CD4+ T helper cells. *Eur. J. Immunol.* **34**:1303–1311.

53. Verhasselt, V., et al. 2004. Induction of FOXP3-expressing regulatory CD4pos T cells by human mature autologous dendritic cells. *Eur. J. Immunol.* **34**:762–772.

54. Tommasini, A., et al. 2002. X-chromosome inactivation analysis in a female carrier of FOXP3 mutation. *Clin. Exp. Immunol.* **130**:127–130.

55. Maloy, K.J., et al. 2003. CD4+CD25+ T(R) cells suppress innate immune pathology through cytokine-dependent mechanisms. *J. Exp. Med.* **197**:111–119.

56. Serra, P., et al. 2003. CD40 ligation releases immature dendritic cells from the control of regulatory CD4+CD25+ T cells. *Immunity.* **19**:877–889.

57. Cederbom, L., Hall, H., and Ivars, F. 2000. CD4+CD25+ regulatory T cells down-regulate costimulatory molecules on antigen-presenting cells. *Eur. J. Immunol.* **30**:1538–1543.

58. Takahashi, T., et al. 1998. Immunologic self-tolerance maintained by CD25+CD4+ naturally anergic and suppressive T cells: induction of autoimmune disease by breaking their anergic/suppressive state. *Int. Immunol.* **10**:1969–1980.

59. Thornton, A.M., and Shevach, E.M. 1998. CD4+CD25+ immunoregulatory T cells suppress polyclonal T cell activation in vitro by inhibiting interleukin 2 production. *J. Exp. Med.* **188**:287–296.

60. Janssens, W., Carlier, V., Wu, B., VanderElst, L., Jacquemin, M.G., and Saint-Remy, J.M. 2003. CD4+CD25+ T cells lyse antigen-presenting B cells by Fas-Fas ligand interaction in an epitope-specific manner. *J. Immunol.* **171**:4604–4612.

61. Thornton, A.M., Donovan, E.E., Piccirillo, C.A., and Shevach, E.M. 2004. Cutting edge: IL-2 is critically required for the in vitro activation of CD4+CD25+ T cell suppressor function. *J. Immunol.* **172**:6519–6523.

62. Thornton, A.M., Piccirillo, C.A., and Shevach, E.M. 2004. Activation requirements for the induction of CD4+CD25+ T cell suppressor function. *Eur. J. Immunol.* **34**:366–376.

63. Yamazaki, S., et al. 2003. Direct expansion of functional CD25+ CD4+ regulatory T cells by antigen-processing dendritic cells. *J. Exp. Med.* **198**:235–247.

64. Pasare, C., and Medzhitov, R. 2003. Toll pathway-dependent blockade of CD4+CD25+ T cell-mediated suppression by dendritic cells. *Science.* **299**:1033–1036.

65. Gavin, M.A., Clarke, S.R., Negrou, E., Gallegos, A., and Rudensky, A. 2002. Homeostasis and anergy of CD4(+)CD25(+) suppressor T cells in vivo. *Nat. Immunol.* **3**:33–41.

66. Asseman, C., Mauze, S., Leach, M.W., Coffman, R.L., and Powrie, F. 1999. An essential role for interleukin 10 in the function of regulatory T cells that inhibit intestinal inflammation. *J. Exp. Med.* **190**:995–1004.

67. Suri-Payer, E., and Cantor, H. 2001. Differential cytokine requirements for regulation of autoimmune gastritis and colitis by CD4(+)CD25(+) T cells. *J. Autoimmun.* **16**:115–123.

68. Berg, D.J., et al. 1996. Enterocolitis and colon cancer in interleukin-10-deficient mice are associated with aberrant cytokine production and CD4(+) TH1-like responses. *J. Clin. Invest.* **98**:1010–1020.

69. Powrie, F., Carlino, J., Leach, M.W., Mauze, S., and Coffman, R.L. 1996. A critical role for transforming growth factor-beta but not interleukin 4 in the suppression of T helper type 1-mediated colitis by CD45RB(low) CD4+ T cells. *J. Exp. Med.* **183**:2669–2674.

70. Nakamura, K., Kitani, A., and Strober, W. 2001. Cell contact-dependent immunosuppression by CD4(+)CD25(+) regulatory T cells is mediated by cell surface-bound transforming growth factor beta. *J. Exp. Med.* **194**:629–644.

71. Oida, T., et al. 2003. CD4+CD25- T cells that express latency-associated peptide on the surface

suppress CD4+CD45RBhigh-induced colitis by a TGF-beta-dependent mechanism. *J. Immunol.* **170**:2516–2522.

72. Piccirillo, C.A., et al. 2002. CD4(+)CD25(+) regulatory T cells can mediate suppressor function in the absence of transforming growth factor beta1 production and responsiveness. *J. Exp. Med.* **196**:237–246.

73. Zhang, X., Izikson, L., Liu, L., and Weiner, H.L. 2001. Activation of CD25(+)CD4(+) regulatory T cells by oral antigen administration. *J. Immunol.* **167**:4245–4253.

74. De La Rosa, M., Rutz, S., Dorninger, H., and Scheffold, A. 2004. Interleukin-2 is essential for CD4(+)CD25(+) regulatory T cell function. *Eur. J. Immunol.* **34**:2480–2488.

75. Grohmann, U., et al. 2002. CTLA-4-Ig regulates tryptophan catabolism in vivo. *Nat. Immunol.* **3**:1097–1101.

76. Fallarino, F., et al. 2003. Modulation of tryptophan catabolism by regulatory T cells. *Nat. Immunol.* **4**:1206–1212.

77. Munn, D.H., Sharma, M.D., and Mellor, A.L. 2004. Ligation of B7-1/B7-2 by human CD4(+) T cells triggers indoleamine 2,3-dioxygenase activity in dendritic cells. *J. Immunol.* **172**:4100–4110.

78. Collins, A.V., et al. 2002. The interaction properties of costimulatory molecules revisited. *Immunity.* **17**:201–210.

79. Piccirillo, C.A., and Shevach, E.M. 2001. Cutting edge: control of CD8+ T cell activation by CD4+CD25+ immunoregulatory cells. *J. Immunol.* **167**:1137–1140.

80. Paust, S., Lu, L., McCarty, N., and Cantor, H. 2004. Engagement of B7 on effector T cells by regulatory T cells prevents autoimmune disease. *Proc. Natl. Acad. Sci. U. S. A.* **101**:10398–10403.

81. Greenfield, E.A., et al. 1997. B7.2 expressed by T cells does not induce CD28-mediated costimulatory activity but retains CTLA4 binding: implications for induction of antitumor immunity to T cell tumors. *J. Immunol.* **158**:2025–2034.

82. Prabhu Das, M.R., et al. 1995. Reciprocal expression of co-stimulatory molecules, B7-1 and B7-2, on murine T cells following activation. *Eur. J. Immunol.* **25**:207–211.

83. Workman, C.J., and Vignali, D.A. 2003. The CD4-related molecule, LAG-3 (CD223), regulates the expansion of activated T cells. *Eur. J. Immunol.* **33**:970–979.

84. Huang, C.-T., et al. Role of LAG-3 in regulatory T cells. *Immunity.* In press.

85. Apostolou, I., and Von Boehmer, H. 2004. In vivo instruction of suppressor commitment in naive T cells. *J. Exp. Med.* **199**:1401–1408.

86. Mitchison, N.A. 1964. Induction of immunological paralysis in two zones of dosage. *Proc. R. Soc. Lond. B Biol. Sci.* **161**:275–292.

87. Gorelik, L., and Flavell, R.A. 2002. Transforming growth factor-beta in T-cell biology. *Nat. Rev. Immunol.* **2**:46–53.

88. Chen, W., et al. 2003. Conversion of peripheral CD4+CD25- naive T cells to CD4+CD25+ regulatory T cells by TGF-beta induction of transcription factor Foxp3. *J. Exp. Med.* **198**:1875–1886.

89. Groux, H., et al. 1997. A CD4+ T-cell subset inhibits antigen-specific T-cell responses and prevents colitis. *Nature.* **389**:737–742.

90. Chen, Y., Kuchroo, V.K., Inobe, J., Hafler, D.A., and Weiner, H.L. 1994. Regulatory T cell clones induced by oral tolerance: suppression of autoimmune encephalomyelitis. *Science.* **265**:1237–1240.

91. Groux, H. 2003. Type 1 T-regulatory cells: their role in the control of immune responses. *Transplantation.* **75**(Suppl. 9):8S–12S.

92. Barrat, F.J., et al. 2002. In vitro generation of interleukin 10-producing regulatory CD4(+) T cells is induced by immunosuppressive drugs and

inhibited by T helper type 1 (Th1)- and Th2-inducing cytokines. *J. Exp. Med.* **195**:603–616.

93. Vieira, P.L., et al. 2004. IL-10-secreting regulatory T cells do not express Foxp3 but have comparable regulatory function to naturally occurring CD4+CD25+ regulatory T cells. *J. Immunol.* **172**:5986–5993.

94. Gregori, S., et al. 2001. Regulatory T cells induced by 1 alpha,25-dihydroxyvitamin D3 and mycophenolate mofetil treatment mediate transplantation tolerance. *J. Immunol.* **167**:1945–1953.

95. Kemper, C., et al. 2003. Activation of human CD4+ cells with CD3 and CD46 induces a T-regulatory cell 1 phenotype. *Nature.* **421**:388–392.

96. Sato, K., Yamashita, N., Baba, M., and Matsuyama, T. 2003. Regulatory dendritic cells protect mice from murine acute graft-versus-host disease and leukemia relapse. *Immunity.* **18**:367–379.

97. Sato, K., Yamashita, N., Baba, M., and Matsuyama, T. 2003. Modified myeloid dendritic cells act as regulatory dendritic cells to induce anergic and regulatory T cells. *Blood.* **101**:3581–3589.

98. Jonuleit, H., Schmitt, E., Schuler, G., Knop, J., and Enk, A.H. 2000. Induction of interleukin 10-producing, nonproliferating CD4(+) T cells with regulatory properties by repetitive stimulation with allogeneic immature human dendritic cells. *J. Exp. Med.* **192**:1213–1222.

99. Steinman, R.M., Hawiger, D., and Nussenzweig, M.C. 2003. Tolerogenic dendritic cells. *Annu. Rev. Immunol.* **21**:685–711.

100. Shimizu, J., Yamazaki, S., and Sakaguchi, S. 1999. Induction of tumor immunity by removing CD25+CD4+ T cells: a common basis between tumor immunity and autoimmunity. *J. Immunol.* **163**:5211–5218.

101. Sutmuller, R.P., et al. 2001. Synergism of cytotoxic T lymphocyte-associated antigen 4 blockade and depletion of CD25(+) regulatory T cells in antitumor therapy reveals alternative pathways for suppression of autoreactive cytotoxic T lymphocyte responses. *J. Exp. Med.* **194**:823–832.

102. Edinger, M., et al. 2003. CD4+CD25+ regulatory T cells preserve graft-versus-tumor activity while inhibiting graft-versus-host disease after bone marrow transplantation. *Nat. Med.* **9**:1144–1150.

103. Trenado, A., et al. 2003. Recipient-type specific CD4+CD25+ regulatory T cells favor immune reconstitution and control graft-versus-host disease while maintaining graft-versus-leukemia. *J. Clin. Invest.* **112**:1688–1696. doi:10.1172/JCI200317702.

104. Nishimura, E., Sakihama, T., Setoguchi, R., Tanaka, K., and Sakaguchi, S. 2004. Induction of antigen-specific immunologic tolerance by in vivo and in vitro antigen-specific expansion of naturally arising Foxp3+CD25+CD4+ regulatory T cells. *Int. Immunol.* **16**:1189–1201.

105. Hoffmann, P., Ermann, J., Edinger, M., Fathman, C.G., and Strober, S. 2002. Donor-type CD4(+)CD25(+) regulatory T cells suppress lethal acute graft-versus-host disease after allogeneic bone marrow transplantation. *J. Exp. Med.* **196**:389–399.

106. Baecher-Allan, C., and Hafler, D.A. 2004. Suppressor T cells in human diseases. *J. Exp. Med.* **200**:273–276.

107. Viglietta, V., Baecher-Allan, C., Weiner, H.L., and Hafler, D.A. 2004. Loss of functional suppression by CD4+CD25+ regulatory T cells in patients with multiple sclerosis. *J. Exp. Med.* **199**:971–979.

108. Karlsson, M.R., Rugtveit, J., and Brandtzaeg, P. 2004. Allergen-responsive CD4+CD25+ regulatory T cells in children who have outgrown cow's milk allergy. *J. Exp. Med.* **199**:1679–1688.

109. Ehrenstein, M.R., et al. 2004. Compromised function of regulatory T cells in rheumatoid arthritis and reversal by anti-TNFα therapy. *J. Exp. Med.* **200**:277–285.

110. Kinter, A.L., et al. 2004. CD25+CD4+ regulatory T cells from the peripheral blood of asymptomatic HIV-infected individuals regulate CD4+ and CD8+ HIV-specific T cell immune responses in vitro and are associated with favorable clinical markers of disease status. *J. Exp. Med.* **200**:331–343.

111. Cao, D., et al. 2003. Isolation and functional characterization of regulatory CD25brightCD4+ T cells from the target organ of patients with rheumatoid arthritis. *Eur. J. Immunol.* **33**:215–223.

112. Cao, D., van Vollenhoven, R., Klareskog, L., Trollmo, C., and Malmstrom, V. 2004. CD25brightCD4+ regulatory T cells are enriched in inflamed joints of patients with chronic rheumatic disease. *Arthritis Res. Ther.* **6**:R335–R346.

Tregs and allergic disease

Douglas S. Robinson,[1,2] Mark Larché,[1] and Stephen R. Durham[1]

[1]Department of Allergy and Clinical Immunology, National Heart and Lung Institute, and [2]Section of Leukocyte Biology, Biomedical Sciences Division, Faculty of Medicine, Imperial College London, London, United Kingdom.

Allergic diseases such as asthma, rhinitis, and eczema are increasing in prevalence and affect up to 15% of populations in Westernized countries. The description of Tregs as T cells that prevent development of autoimmune disease led to considerable interest in whether these Tregs were also normally involved in prevention of sensitization to allergens and whether it might be possible to manipulate Tregs for the therapy of allergic disease. Current data suggest that Th2 responses to allergens are normally suppressed by both CD4+CD25+ Tregs and IL-10 Tregs. Furthermore, suppression by these subsets is decreased in allergic individuals. In animal models, Tregs could be induced by high- or low-dose inhaled antigen, and prior induction of such Tregs prevented subsequent development of allergen sensitization and airway inflammation in inhaled challenge models. For many years, allergen-injection immunotherapy has been used for the therapy of allergic disease, and this treatment may induce IL-10 Tregs, leading to both suppression of Th2 responses and a switch from IgE to IgG4 antibody production. Improvements in allergen immunotherapy, such as peptide therapy, and greater understanding of the biology of Tregs hold great promise for the treatment and prevention of allergic disease.

Allergic disease

Atopic allergic sensitization is defined by production of IgE against environmental antigens such as house dust mites, grass pollen, and animal proteins and can lead to diseases that include asthma, rhinitis and atopic dermatitis (1). These disorders affect 10–15% of Western populations, and their prevalence has doubled in the last 10–15 years. Triggering an immune response through allergen-specific IgE on the surface of mast cells and basophils can lead to immediate symptoms through release of histamine and other mediators while eosinophilic airway inflammation contributes to airway hyperresponsiveness (AHR) in asthma (1). Switching of B cells to IgE production and accumulation of eosinophils are under the control of Th2 lymphocytes through production of IL-4 and IL-5. These cells also produce IL-9 and IL-13, which contribute to AHR in asthma (2, 3).

Regulatory or suppressive T cells have increasingly been defined as important in the prevention of autoimmune disease and other immunopathologies (4); the biology and mechanism of suppression by these cells is reviewed elsewhere in this series. Whether Tregs normally also prevent atopic sensitization and how this regulatory process becomes defective or is bypassed in those individuals who develop allergic disease are areas of much current research. Additionally, the potential for manipulation of Tregs for therapy is clearly attractive for many disease types. Allergen-injection immunotherapy has been used for control of allergic disease for many years and appears to act through modulation of the Th2 response to the allergen, either by immune deviation of allergen-specific Th2 responses in favor of Th1 responses and/or through the induction of Tregs. Understanding the mechanism of therapeutic benefit in this treatment may hold important lessons for immunoregulation in other diseases.

Interest in active T cell suppression as a mechanism for immunological tolerance was reawakened by the demonstration by Sakaguchi and others that neonatal thymectomy of mice leads to organ-specific autoimmune pathology that can be prevented by transfer of CD4+CD25+ T cells from nonirradiated animals (5, 6). Shevach and coworkers developed an in vitro system, which has been used widely, showing that mouse CD4+CD25+ T cells do not proliferate in response to either anti-CD3 or antigenic stimulation and, furthermore, can inhibit the proliferative responses of CD4+CD25- T cells (7). Use of this in vitro assay for the study of T cell regulation allowed demonstration of suppression by human peripheral blood and thymic CD4+CD25+ T cells (8–13). Thymus-derived CD4+CD25+ regulatory cells have been termed naturally occurring Tregs. A number of other regulatory subsets have been described (see Table 1). These are largely T cell populations induced by in vitro or in vivo manipulation and have been termed adaptive Tregs (14), but it is of note that several reports also suggest the existence of naturally occurring Tregs within the CD4+CD25- T cell population (15). Currently there is intense interest in the interrelationship of naturally occurring and induced/adaptive CD4+CD25+ T cells, IL-10–producing Tregs, and other regulatory subsets, such as Th3 cells described as occurring after induction of oral tolerance (16). It will be important to determine whether these cell types are distinct and how data from mouse models can be related to the development or treatment of human disease.

The mechanism of suppression by different Treg subsets remains controversial; different in vivo and in vitro studies raise possible roles for the immunosuppressive cytokines IL-10 and TGF-β (including a membrane-bound form), the negative costimulatory molecules CTLA4 and PD-1, and the glucocorticoid-induced TNF receptor (6–12, 17–25). However, in vitro suppression by both mouse and human CD4+CD25+ T cells can be detected in the absence or blockade of these pathways. An important advance in the understanding of the biology of regulation by CD4+CD25+ T cells occurred in the demonstration of the dependence of their suppressive phenotype on expression of the transcription factor Foxp3. Gene deletion of *Foxp3* abrogated suppression by CD4+CD25+ T cells whereas ectopic gene expression in CD25- T cells rendered these cells suppressive (26–28). In addition, acquisition of regulatory phenotype by CD4+CD25- T cells, either

Nonstandard abbreviations used: AHR, airway hyperresponsiveness; LAR, late asthmatic reaction; PLA₂, phospholipase A2; TLR, toll-like receptor.

Conflict of interest: D.S. Robinson is a consultant to Lorantis Ltd. S.R. Durham is a consultant to ALK Abello and has received honoraria and research funding from ALK Abello, a manufacturer of allergy vaccines.

Citation for this article: *J. Clin. Invest.* 114:1389–1397 (2004).
doi:10.1172/JCI200423595.

Table 1
Different regulatory cell types implicated in the prevention of atopic sensitization in mouse models and humans

Type of Treg	Mechanism of suppression	Allergy active in (refs.)
CD4+CD25+ (naturally occurring, i.e., thymically derived)	Foxp3 contact dependent, IL-10, TGF-β, CTLA4, PD-1[A]	Mouse (41) Human (52–55)
CD4+CD25+ (induced)	IL-10, TGF-β	Mouse (29, 44) Human (52–54)
Tr1	IL-10, TGF-β	Mouse (45)
IL-10 Treg	Contact dependent, IL-10 (in vivo)	Mouse (46–48) Human (55)
Th1 cells	IFN-γ, IL-10	Mouse (33)

It remains unclear how these subtypes are interrelated and whether human CD4+CD25+ T cells represent naturally occurring or adaptive Tregs (see ref. 14). Which subtypes are induced (if any) by whole allergen immunotherapy or peptide immunotherapy also remains to be determined. [A]IL-10, TGF-β, CTLA4, and PD-1 are implicated in suppression by naturally occurring CD4+CD25+ T cells in some, but not all, reports.

through the action of TGF-β or coculture with CD4+CD25+ Tregs, was associated with acquisition of Foxp3 expression (29). Mutation of *Foxp3* was described in patients with immune dysregulation, polyendocrinopathy, enteropathy, and X-linked syndrome (IPEX syndrome), a syndrome which includes the development of atopic dermatitis; thus Foxp3 Tregs may be involved in the prevention of allergic sensitization (30). It has been suggested that Foxp3 acts as a master switch for the regulatory phenotype in much the same way as do GATA3 and T-bet in Th2 and Th1 T cell phenotypes respectively (31). However, it is of note that IL-10–producing Tregs do not show overexpression of *Foxp3*, which suggests that regulation can occur through different pathways (32).

In addition, Th1 cells have inhibitory effects on Th2 function (partially through production of the immunomodulatory cytokine IL-10) and have been shown to prevent or diminish airway inflammation in at least some mouse allergen challenge models (33). This was also shown by in vitro overexpression of the Th1-associated transcription factor T-bet, which was reported to reduce expression of GATA3 and Th2 cytokines in Th2 cells and is underexpressed in the airway in asthmatics compared to control subjects (34, 35). Thus in the context of allergic disease, Th1 cells may be suppressive, and immune deviation from a Th2 to a Th1 response may be a legitimate regulatory strategy for treatment of allergic disease. In addition, NK T cells and TCR γδ–bearing cells have also been suggested as having regulatory roles (36).

Tregs in mouse models of allergic disease
A number of reports of experiments using mouse models have suggested a potential role for Tregs in the prevention and control of Th2-mediated pathology.

Mouse CD4+CD25+ T cells were shown to suppress the in vitro differentiation of Th2 cells from naive CD4+ T cells but required preactivation to inhibit cytokine production from differentiated Th2 cells (37). In transgenic mice with monoclonal populations of T and B cells, a single immunization of antigen resulted in very high serum IgE levels, which could be prevented by both CD25+ and CD4+CD25- T cells. This suggested that Tregs normally prevent IgE responses (38). However, depletion of CD4+CD25+ T cells actually reduced airway inflammation in a murine model of allergic sensitization (39), and transfer of CD4+CD25+ T cells had no effect on inflammation or AHR

in another report (40). In a double-transgenic model, CD4+CD25+ OVA-specific TCR transgenic T cells infiltrated the lungs of animals expressing the OVA antigen under a lung-specific promoter. Although these cells suppressed airway inflammation, they did not prevent AHR in response to inhaled OVA (41). These data suggest that, at least in animal models, naturally occurring CD4+CD25+ T cells which develop during normal thymic maturation may not be sufficient to prevent allergic airway disease. In other models, CD4+CD25+ Tregs have been induced from CD4+CD25- T cells. In one report, these adaptive CD4+CD25+ T cells were derived after antigen exposure in the presence of TGF-β and expressed the transcription factor Foxp3. These cells could prevent house dust mite–induced airway inflammation in an inhaled allergen challenge model, although no data was presented on AHR (29). Repeated low-dose inhaled exposure to antigen induced tolerance to subsequent Th2 sensitization and airway pathology in mouse models (42, 43). Recently this was shown to depend on the induction of Foxp3+ CD4+ T cells that expressed surface TGF-β and suppressed development of airway inflammation in vivo, and antigen-induced T cell proliferation in vivo, through a contact-dependent mechanism (44). In human peripheral blood one might expect that the CD4+CD25+ T cell population will contain both naturally occurring and adaptive Tregs, and it will be of interest to determine how such subsets develop upon natural allergen exposure (see below).

In addition to the roles for CD4+CD25+ T cells, the use of mouse models suggests potential roles for IL-10–producing Tregs in the suppression of allergic pathology. These include Tr1 clones derived by antigen stimulation in the presence of IL-10, which could prevent Th2 sensitization and IgE production if adoptively transferred prior to sensitization (45). Tolerance to inhaled antigen challenge can also be induced by high-dose intranasal antigen delivery to the respiratory tract in mice. This tolerance was shown to be dependent on IL-10 production by mature dendritic cells, which in turn induced development of IL-10–producing regulatory CD4+ T cells (46, 47). In addition, prior exposure to heat-killed *Mycobacterium vaccae* reduced both Th2 airway inflammation and AHR to subsequent allergen challenge, and this was mediated by CD4+CD45RBlo T cells acting through both IL-10 and TGF-β (48). IL-10–producing Tregs have also been derived by antigen stimulation of either mouse or human naive CD4+ T cells in the presence of immunosuppressive drugs, dexamethasone, and vitamin D3 (49), although these have not been tested in mouse models of allergic disease. A further possible means of inducing regulation was demonstrated by Hoyne and colleagues. House dust mite–pulsed antigen-presenting cells that had been engineered to overexpress the Notch ligand Jagged-1 were injected into mice and shown to induce CD4+ T cells that could prevent house dust mite–induced airway pathology in an allergen challenge model (50). The Notch signaling system is involved in cell fate decision (including in the immune system) and may influence whether a proinflammatory or anti-inflammatory response is mounted.

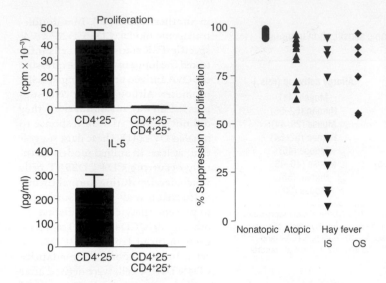

Figure 1
Peripheral blood CD4+CD25+ T cells from atopic volunteers show reduced suppressive ability in allergen-stimulated T cell cultures. CD4+CD25− T cells were separated from peripheral blood by immunomagnetic separation, then cultured with allergen extracts either alone or mixed with CD4+CD25+ T cells. Proliferation was assessed by incorporation of tritiated thymidine (shown as cpm), and IL-5 was measured in supernatants at day 6 of cultures by Luminex bead array. Data shown are means and standard errors for cpm and IL-5 from 9 separate nonatopic donors, showing almost complete suppression of responses of CD4+CD25− T cells when CD4+CD25+ T cells were added. When these data were expressed as percentage suppression (reduction in counts in the mixed culture compared with those in CD4+CD25− T cells alone), suppression was significantly less when cells were obtained from atopic donors or volunteers with hay fever studied in or out of season (IS or OS). Suppression out of season was significantly more than that seen in season but still significantly less than in nonatopic controls. Figure adapted from *Lancet* (54), with permission from Elsevier.

Thus murine models of allergic disease demonstrate that Tregs have potential in preventing allergic sensitization and that such cells can be induced in vivo. However, which regulatory pathways act in allergic disease and how these cells suppress Th2 responses clearly remains controversial.

Tregs and allergic responses in humans
There have been few reports of activity of human Tregs on Th2 cells in vitro. Human thymocyte Th2 clones were relatively resistant to suppression of allergen-induced activation by autologous CD4+CD25+ thymocytes (51). However, human peripheral blood CD4+CD25+ T cells suppressed proliferation and Th2-type cytokine production by CD4+CD25− T cells stimulated in vitro with allergen (52, 53). Thus Tregs may act to prevent inappropriate Th2 responses to environmental allergens. One possible reason for development of atopic sensitization would therefore be a deficiency or failure of such regulation. Although initial reports found no clear difference in the suppressive ability of CD4+CD25+ T cells between atopic allergic and nonatopic control subjects (52, 53), a more recent study showed that the suppressive ability of CD4+CD25+ T cells from atopic subjects in cocultures with allergen-stimulated autologous CD4+CD25− T cells was significantly less than that seen for cells from nonatopics (54) (Figure 1). There was no difference between atopic and nonatopic donors in the

ability of their CD4+CD25+ T cells to suppress CD4+CD25− T cells activated via anti-CD3 and anti-CD28 monoclonal antibodies, suggesting that atopy is not associated with a generalized defect in regulatory capacity of CD4+CD25+ T cells (54). One possibility was that the CD4+CD25+ T cells isolated from the atopic subjects contained fewer Tregs and more activated effector T cells, since CD25 is also expressed by recently activated or memory effector T cells. However, no differences in other phenotypic markers were detected to suggest an excess of activated T cells, and differences in CD4+CD25+ T cell suppression of grass pollen–stimulated cultures between patients with hay fever and nonatopic controls persisted even outside the pollen season (Figure 1). Human CD4+CD25+ T cells were confirmed to express mRNA for the Foxp3 transcription factor, which is consistent with a role for Foxp3 in the regulatory phenotype of CD4+CD25+ T cells in humans as well as mice (54). These findings show that human CD4+CD25+ T cells can indeed suppress T cell activation by allergens and that this process may be deficient in atopic subjects.

In addition to CD4+CD25+ T cells, IL-10–producing Tregs have been implicated in the prevention of atopic sensitization. Akdis et. al. showed an increased frequency of allergen-specific IL-10–producing CD4+ T cells in the blood of nonatopic individuals when compared to atopics and, furthermore, that these IL-10–producing T cells could specifically inhibit allergen-activated IL-4–producing T cells. Suppression was reversed by blocking antibodies to IL-10 and TGF-β (55). The cells were isolated after allergen stimulation and were CD25+. How these cells relate to the CD4+CD25+ T cells studied by others remains unclear; it is of note that suppression by induced CD4+CD25+ T cells (derived from coculture of CD25− T cells with CD4+CD25+ T cells in vitro) was dependent on IL-10, in contrast to CD4+CD25+ T cells freshly isolated from blood, in which suppression was not reversed by blocking antibodies to IL-10 or IL-10 receptor (54, 56).

Development of allergic sensitization: suppression versus activation of Th2 responses
The factors determining the balance between a suppressive or regulatory response to allergen and activation of a Th2 response either during maturation of the immune system or after allergen sensitization are uncertain. One hypothesis for the increasing prevalence of Th2-associated allergic disease, the hygiene hypothesis, is that Western lifestyles reduce contact with environmental microorganisms. Such microorganisms are thought to modify antigen-presenting cells (in part via toll-like receptors [TLRs]) so that they induce either immune deviation in favor of Th1 responses or the development of Treg responses to allergens. Susceptibility to such factors may explain genetic associations of atopic disease with polymorphisms such as those described for the LPS receptor CD14 and TLR2 (57–59). Different concentrations of LPS could favor either Th1 or Th2 responses, and activation of TLR4 via LPS overcame regulation by CD4+CD25+ T cells (60–62). Route, dose, and timing of allergen exposure, in addition to coexposure to pathogens or environmental bacteria, are all probably important in allergic sensitization. However, it is of note that allergic sensitization can occur in adulthood, so it is not the prerogative of the developing immune system. Both prior inhaled allergen exposure and exposure to mycobacterial antigens were shown to suppress

subsequent allergic sensitization in mice through mechanisms that include induction of IL-10–producing Tregs (46–48). Such mechanisms may explain the apparent protective effect of exposure to cats during early childhood on the later development of atopic sensitization to cat allergens. High-dose exposure to cat allergen has been associated with a dominant IL-10 response in peripheral blood mononuclear cells cultured with cat allergen and elevated serum concentrations of cat allergen-specific IgG4; this has been termed a "modified Th2" response (63). Whether this relates to the development of Tregs is at present unclear. In contrast, house dust mite exposure showed no such protective effect. It will clearly be of interest to determine whether this high-exposure tolerance is akin to high-dose tolerance in mouse models, which is dependent on IL-10–producing Tregs, and whether the normal nonatopic state results from low-dose chronic tolerance as described in mouse models dependent on surface TGF-β^+ Tregs (44). Recent data suggest that both IL-10–producing T cells and CD4$^+$CD25$^+$ T cells may contribute to the development of tolerance since children lose atopic sensitization to cows' milk allergens (64, 65).

Further understanding of the maturation of regulatory responses will be required if preventive vaccination strategies are to be developed for allergic disease. The data from animal models of allergic airway sensitization suggest that naturally occurring CD4$^+$CD25$^+$ T cells derived from the thymus may not be sufficient to prevent airway inflammation and that more than one type of Treg may be required.

Influence of allergy treatment on Treg populations

Corticosteroids. How might Tregs be induced by therapy to control or prevent allergic disease? Corticosteroids are widely used as topical or systemic anti-inflammatory treatment for allergic disease and inhibit T cell activation and Th2 cytokine expression both in vitro and in vivo. In addition, corticosteroids increase IL-10 production by T cells and macrophages. In vitro exposure of human CD4$^+$CD25$^+$ T cells to corticosteroids have been shown to increase their suppressive capacity in subsequent allergen-stimulated cultures, in part through an increase in IL-10 production (66). In addition, corticosteroids with vitamin D3 could induce an IL-10–producing Treg phenotype in vitro in both mouse and human CD4$^+$ T cells, and these IL-10–producing Tregs inhibited experimental allergic encephalomyelitis (49) and suppressed human allergen-specific Th2 cells in vitro (Catherine Hawrylowicz, King's College, London, United Kingdom; personal communication). Whether such strategies can be modified for in vivo use in treating human allergic disease remains to be determined. Current clinical experience indicates that, although corticosteroids can control allergic inflammation effectively, symptoms recur if they are discontinued. The development and duration of regulatory memory is poorly understood, but current steroid regimens do not appear to alter underlying sensitization. In contrast, allergen immunotherapy modulates Th2 responses to allergen, and this effect can persist for at least 3–4 years following discontinuation of therapy (67).

Allergen immunotherapy. Allergen immunotherapy involves the subcutaneous injection of incrementally increasing doses of allergen in order to suppress symptoms on subsequent reexposure to that allergen (68). In patients with severe seasonal pollinosis, grass pollen vaccination is highly effective (67–70). In children, 3 years of house dust mite extract immunotherapy prevented the onset of new allergen sensitivities (71) and resulted in a 2- to 3-fold reduction in the risk of developing physician-diagnosed asthma (72).

Some (73, 74), but not all (75) studies have described decreased peripheral blood T cell responsiveness to allergen and/or immune deviation in favor of Th1 responses (76, 77). In some early reports, evidence of the induction of regulatory responses, for example, increased expression of IL-10, was not sought. Increased proportions of CD4$^+$CD25$^+$ cells were found in grass pollen–stimulated peripheral blood mononuclear cell cultures following immunotherapy (78); these could represent a Treg population which might act to suppress Th2 T cells. In support of this hypothesis, grass pollen stimulation of PBMCs led to increased IL-10 production, which was colocalized to CD4$^+$CD25$^+$ T cells. These findings are consistent with previous observations in venom-sensitive patients, which also suggest a role for IL-10 in suppression of allergen-specific Th2 responses (73, 79). A report of immunotherapy in patients with house dust mite allergy demonstrated increased IL-10$^+$ CD4$^+$CD25$^+$ cells in peripheral blood after 70 days of treatment compared with before therapy (80). Whether these CD4$^+$CD25$^+$ T cells arise from the expansion of naturally occurring CD4$^+$CD25$^+$ T cells or are analogous to IL-10–dependent CD4$^+$CD25$^+$ T cells induced in vitro (56) remains to determined.

Effective suppression of allergen-responsive T cells in target organs (such as the lung) requires trafficking of Tregs. Studies of in vivo suppression in animal models of inflammatory bowel disease suggest a role for locally produced IL-10 or TGF-β (17). Allergen immunotherapy treatment is associated with reduced local nasal tissue infiltration by eosinophils, basophils, and Th2 T cells after allergen challenge together with local increases in T cells producing IL-2 and IFN-γ (Th1 cells), as detected by in situ hybridization (75, 81–83). In the same nasal biopsies, obtained during the peak pollen season, there were increases in the number of IL-10$^+$ and TGF-β^+ cells, detectable at both mRNA and protein levels (84). These increases in IL-10 and TGF-β were not detectable outside the pollen season, suggesting that local mucosal contact with allergen was necessary for these effects. Neither were these cells detectable in normal nonatopic subjects, which suggested that these IL-10 and TGF-β responses were not a feature of the normal mucosal response to pollen nor due to a toxic or other effect of pollen per se, in the absence of allergic sensitization. Colocalization studies confirmed that 15–20% of these nasal IL-10– and TGF-β–producing cells were T cells, raising the possibility of an immunotherapy-induced, local adaptive mucosal T regulatory response following natural allergen exposure. Clearly the development of specific markers for Tregs is an urgent requirement for their tissue localization or isolation. However, IL-10 and TGF-β also colocalized to macrophages (40%) and alternative sources, including B cells and dendritic cells. It is therefore possible that local increases in IL-10 and TGF-β from these innate immune cells may also result from the observed local increases in IFN-γ. Whether due to an immunotherapy-induced, allergen-specific T regulatory response (for example, production of IL-10 and/or TGF-β) and/or Th1 immune deviation (Th2 to Th1 shift in cytokine expression patterns), the production of these inhibitory cytokines within the nasal mucosa has the potential to suppress local allergen-specific Th2 cell responses and redirect antibody class switching in favor of IgG4 (IL-10 isotype switch factor), IgG (IFN-γ isotype switch factor), and IgA (TGF-β isotype switch factor).

In agreement with previous studies, our observations showed 40- to 100-fold increases in serum allergen-specific total IgG and IgG4 subclass antibodies and blunting of seasonal increases in IgE (84). It is possible that such antibody changes are important for

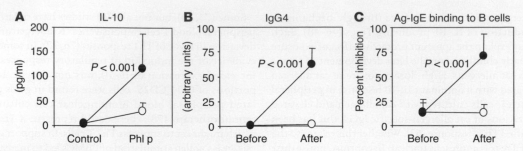

Figure 2

Immunological changes after allergen immunotherapy. Following 2-year grass pollen immunotherapy (closed circles), there were significant increases in (**A**) allergen-stimulated PBMC production of IL-10 (78); (**B**) serum concentrations of grass pollen (Phleum P5) allergen-specific IgG4 (84); and (**C**) serum inhibitory activity for allergen-IgE binding to B cells (88) compared with controls (open circles). These changes were accompanied by a reduction in symptoms and rescue medication use during the pollen season (83); inhibition of the allergen (grass pollen) induced late cutaneous response.

clinical efficacy of immunotherapy. For example, allergen–IgE-IgG complexes may coaggregate inhibitory Fc-γRIIb receptors with high-affinity FcεRI IgE receptors (85), resulting in the inhibition of FcεRI signalling (86, 87). Alternatively, they may inhibit IgE-facilitated allergen presentation to T cells (88), a rate-limiting step in allergen-specific Th2 T cell–driven allergic responses (89). Increases in peripheral IL-10 production, serum allergen-specific IgG4, and IgG-dependent serum-inhibitory activity for allergen-IgE binding to B cells after immunotherapy are shown in Figure 2. A hypothesis concerning the putative mechanism of allergen-injection immunotherapy is summarized in Figure 3.

Modifications of allergen immunotherapy

Despite the impressive efficacy of allergen-injection immunotherapy with whole allergen extracts, its widespread usage is confined to specialist centers in view of the risk of occasional IgE-mediated adverse events, which include systemic anaphylaxis. A number of strategies aim to modify immunotherapy for allergic disease in order to separate allergenicity (IgE cross-linking) from immunogenicity (induction of protective, non-IgE immunity).

Approaches have included the generation of allergoids through disruption of surface B cell epitopes of the protein. This has been achieved with amino–cross-linking agents such as glutaraldehyde

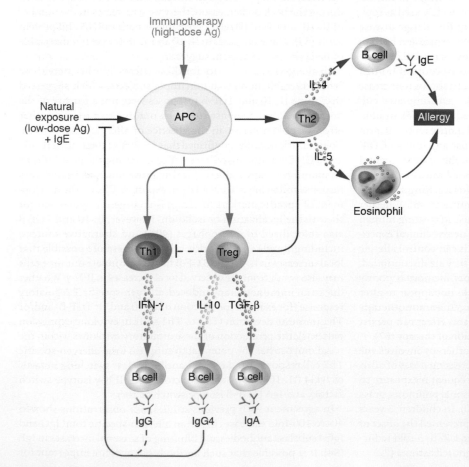

Figure 3

Potential mechanisms of conventional allergen immunotherapy. High-dose allergen exposure during immunotherapy results in both immune deviation of Th2 responses in favor of a Th0/Th1 response and in the generation of IL-10– and TGF-β–producing CD4+CD25+ T cells, possibly Tregs. IFN-γ–induced activation of bystander macrophages and/or other cells represents an alternative source of these inhibitory cytokines. During subsequent natural environmental exposure to allergens, the activation and/or maintenance of the usual atopic Th2 T cell response is inhibited. Additionally, these cytokines induce preferential switching of B cell responses in favor of IgG and IgG4 antibodies (and possibly IgA antibodies under the influence of TGF-β). IgG may also inhibit IgE-facilitated allergen binding to antigen-presenting cells with subsequent downregulation of IgE-dependent Th2 T lymphocyte responses. Blue arrows represent immune response pathway to natural exposure (low-doses Ag and IgE); green arrows represent immune response pathway to immunotherapy (high-dose Ag); red blocked lines represent inhibition (high-dose Ag); dotted lines represent possible means of action not yet proven.

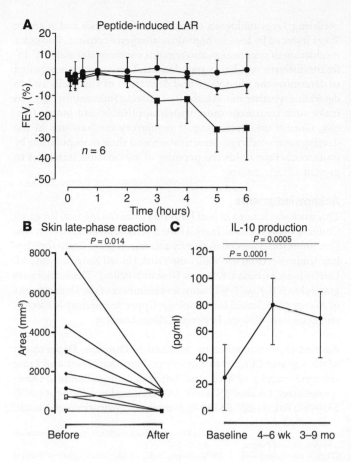

A Peptide-induced LAR

B Skin late-phase reaction   **C** IL-10 production

Figure 4
Peptide immunotherapy is associated with allergen-specific hyporesponsiveness and the induction of IL-10. (**A**) Cat-allergic asthmatic subjects (n = 6) were challenged intradermally with a mixture of 12 peptides (5 µg each) from the sequence of Fel d 1 or vehicle alone. Lung function was measured by spirometry for 6 hours. Challenges were separated by 14 days or more. Challenge with vehicle (circles) did not significantly modify forced expiratory volume in one second (FEV_1). Initial peptide challenge (squares) resulted in an isolated LAR that significantly differed from baseline (P = 0.02 area under the curve; AUC). A second challenge (triangles) with the same dose of peptide was associated with an attenuated or absent LAR. Values are mean of 6 individuals with standard error. (**B**) Cat-allergic asthmatic subjects underwent intradermal allergen challenge (volar aspect of the forearm) before and after administration of a mixture of 11 peptides in incremental divided doses. Peptide treatment significantly reduced the magnitude of the cutaneous late-phase reaction. (**C**) Treatment of 16 cat-allergic asthmatic individuals with a mixture of 12 peptides (incremental divided doses) resulted in elevated production of IL-10 by peripheral blood mononuclear cells at both 4–6 weeks and 3–9 months after the completion of treatment. Data presented as median and interquartile range.

(or formalin). In general, allergoids have demonstrated efficacy and safety profiles similar to those described for unmodified allergen extracts (90). In other studies, the introduction of point mutations or deletions has been used to reduce the number of conformational B cell epitopes displayed by an allergen (91). More recently, allergen has been coupled to synthetic bacterial DNA sequences (92). In the case of DNA-allergen conjugates, the proposed mechanisms of action are 2-fold: steric hindrance of the interaction between allergen and IgE together with modulation of T cell responses toward a Th1 phenotype through activation of the TLR-9 molecule by bacterial DNA CpG motifs (93). Clinical studies evaluating the safety and efficacy of genetically engineered allergens (trimers and fragments of the major birch pollen allergen Bet v 1) and allergen-DNA conjugates (for example, immunostimulatory sequence–conjugated ragweed allergen) have thus far only been reported in abstract form.

During the last decade, synthetic peptides derived from the amino acid sequences of allergens have been evaluated for the treatment of allergic individuals. The majority of data has been obtained from the use of peptides derived from the major cat allergen Fel d 1, although more recently, studies have been performed with peptides from the bee venom allergen phospholipase A_2 (PLA_2). The potential of peptide immunotherapy to reduce IgE-mediated allergenicity while maintaining immunogenicity has been examined in a number of clinical studies in recent years.

Peptide immunotherapy

Peptide therapy has been used in several animal models of human disease in an attempt to modulate T cell responses (94–99). Peptides from a variety of antigens have been administered intra-

nasally, orally, or by injection, leading to reduced proinflammatory responses accompanied by the induction of immunoregulatory cytokines such as IL-10 and/or TGF-β. Evidence for the functional existence of Tregs has been obtained in some murine models. For example, intramolecular suppression was demonstrated following intranasal administration of house dust mite peptides (100). Amelioration of disease was achieved following adoptive transfer of cells (101). Recently, oral administration of peptides to subjects with rheumatoid arthritis (phase I, open study) was associated with reductions in the PBMC production of IL-2, IFN-γ, and TNF-α as well as an increase in the production of IL-10, and an increase in the expression of mRNA encoding *Foxp3* (102).

Short peptides (16/17 residues) representing T cell epitopes of the major cat allergen Fel d 1 did not trigger basophil histamine release or early asthmatic reactions upon intradermal injection or inhaled challenge in cat allergic asthmatic subjects (103–106). However, intradermal injection of allergen peptides did, in some individuals, lead to the induction of isolated late asthmatic reactions (LARs), which were MHC class II–restricted (103). Subsequent injection of peptides did not reproduce the response, implying the induction of tolerance (Figure 4). Development of peptide therapy based on these observations has enabled the safe delivery of molar doses of peptides far in excess of those possible with intact allergen molecules (103, 105).

The mechanism of abrogation of LAR and reduction of late-phase skin reactions to allergen challenge was investigated by analysis of in vitro responses of PBMCs to allergen before and after peptide therapy (106). Supernatants harvested from cultures of PBMCs of cat-allergic subjects receiving Fel d 1 peptides or bee venom–allergic subjects receiving PLA_2 peptides showed reductions in proliferation of CD4+ T cells and both Th1 and Th2 cytokine production, with a significant increase in IL-10 production. Similar results were reported by Müller and colleagues following PLA_2 peptide therapy (107).

Preliminary data from clinical studies with Fel d 1 peptides supports a central role for Tregs in the downregulation of allergic responses. An antigen-specific regulatory activity was present within the CD4+ peripheral T cell population after Fel d 1 peptide therapy (108). Comparison of peripheral blood CD4+CD25+ T cell suppression of cat allergen–stimulated CD4$^+$CD25$^-$ T cells before and after peptide immunotherapy did not show any change in the suppressive ability of the CD4$^+$CD25$^+$ T cell subset despite significant reduction of CD4$^+$ T cell responses to allergen in this double-blind study (109). Thus peptide therapy may induce a regulatory subset separate from CD4$^+$CD25$^+$ T cells, in other words, an adaptive rather than a naturally occurring regulatory population. However, analysis of 24-hour biopsy samples taken from allergen-induced cutaneous late-phase reactions to whole cat dander allergen extract showed increases in the number of CD4$^+$CD25$^+$ cells, increases in the number of CD4$^+$IFN-γ^+ cells, and increases in TGF-β expression following peptide therapy (110). Thus evidence for both immune deviation (increased IFN-γ expression) and an indication of regulation (TGF-β) were found in this study.

Conclusion

In summary, current evidence suggests that human CD4$^+$CD25$^+$ T cells and IL-10–producing Tregs have the capacity to suppress Th2 responses to allergen and that this process may be defective in those who develop allergic sensitization. Allergen immunotherapy modifies T cell responses to allergen and may do so through induction of IL-10–producing Tregs, although other mechanisms, including Th1 immune deviation, are probably involved. Controversy remains about the nature and interrelation of different Treg populations; it is likely that CD4$^+$CD25$^+$ T cells isolated from human adult blood represent a mixed population of naturally occurring Tregs analogous to those studied in mice and adaptive Tregs induced by low- or high-dose allergen exposure. The exact mechanism of suppression also remains controversial and may differ for different regulatory populations. Further work is required to determine the phenotype and derivation of such cells and to determine whether modification of existing immunotherapy can make such treatment more widely applicable and more effective. Greater understanding of regulatory mechanisms in the development of allergic sensitization and their manipulation by immunotherapy holds the promise of vaccination strategies to prevent allergic disease.

Acknowledgments

This work was funded in part by grants from the Medical Research Council, the Wellcome Trust, United Kingdom, and Asthma UK. D.S. Robinson is funded in part by a Research Leave Grant for Clinical Academics from the Wellcome Trust, United Kingdom, and M. Larché is an Asthma UK Senior Research Fellow. The authors are grateful to A.B. Kay, T. Williams, and members of the Departments of Allergy and Clinical Immunology, Upper Respiratory Medicine, and Leukocyte Biology, Imperial College London.

Address correspondence to: Stephen R. Durham, Department of Allergy and Clinical Immunology, National Heart and Lung Institute, Faculty of Medicine, Imperial College London, Dovehouse Street, London SW3 6LY, United Kingdom. Phone: 44-207-3518992; Fax: 44-207-3518949; E-mail: s.durham@imperial.ac.uk.

1. Kay, A.B. 2001. Allergy and allergic diseases. *N. Engl. J. Med.* **344**:30–37.
2. Lloyd, C.M., Gonzalo, J.A., Coyle, A.J., and Gutierrez-Ramos, J.C. 2001. Mouse models of allergic airway disease. *Adv. Immunol.* **77**:263–295.
3. Robinson, D.S. 2000. Th-2 cytokines in allergic disease. *Br. Med. Bull.* **56**:956–968.
4. Baecher-Allan, C., and Hafler, D.A. 2004. Suppressor T cells in human diseases. *J. Exp. Med.* **200**:273–276.
5. Sakaguchi, S., Sakaguchi, N., Asano, M., Itoh, M., and Toda, M. 1995. Immunologic self-tolerance maintained by activated T cells expressing IL-2 receptor alpha-chains (CD25). Breakdown of a single mechanism of self-tolerance causes various autoimmune diseases. *J. Immunol.* **155**:1151–1164.
6. Sakaguchi, S. 2000. Regulatory T cells: key controllers of immunologic self-tolerance. *Cell.* **101**:455–458.
7. Shevach, E.M. 2002. CD4+ CD25+ suppressor T cells: more questions than answers. *Nat. Rev. Immunol.* **2**:389–400.
8. Levings, M.K., Sangregorio, R., and Roncarolo, M.-G. 2001. Human CD25$^+$CD4$^+$ T regulatory cells suppress naive and memory T cell proliferation and can be expanded in vitro without loss of function. *J. Exp. Med.* **193**:1295–1302.
9. Jonuleit, H., et al. 2001. Identification and functional characterization of human CD4$^+$CD25$^+$ T cells with regulatory properties isolated from peripheral blood. *J. Exp. Med.* **193**:1285–1294.
10. Dieckmann, D., Plottner, H., Berchtold, S., Berger, S., and Schuler, G. 2001. Ex vivo isolation and characterization of CD4$^+$CD25$^+$ T cells with regulatory properties from human blood. *J. Exp. Med.* **193**:1303–1310.
11. Taams, L.S., et al. 2001. Human anergic/suppressive CD4(+)CD25(+) T cells: a highly differentiated and apoptosis-prone population. *Eur. J. Immunol.* **31**:1122–1131.
12. Ng, W.F., et al. 2001. Human CD4(+)CD25(+) cells: a naturally occurring population of regulatory T cells. *Blood.* **98**:2736–2744.
13. Stephens, L.A., Mottet, C., Mason, D., and Powrie, F. 2001. Human CD4(+)CD25(+) thymocytes and peripheral T cells have immune suppressive activity in vitro. *Eur. J. Immunol.* **31**:1247–1254.
14. Bluestone, J.A., and Abbas, A.K. 2003. Natural versus adaptive regulatory T cells. *Nat. Rev. Immunol.* **3**:253–257.
15. Stephens, L.A., and Mason, D. 2000. CD25 is a marker for CD4+ thymocytes that prevent autoimmune diabetes in rats, but peripheral T cells with this function are found in both CD25+ and CD25- subpopulations. *J. Immunol.* **165**:3105–3110.
16. Chen, Y., Kuchroo, V.K., Inobe, J., Hafler, D.A., and Weiner, H.L. 1994. Regulatory T cell clones induced by oral tolerance: suppression of autoimmune encephalomyelitis. *Science.* **265**:1237–1240.
17. Maloy, K.J., and Powrie, F. 2001. Regulatory T cells in the control of immune pathology. *Nat. Immunol.* **2**:816–822.
18. Baecher-Allan, C., Brown, J.A., Freeman, G.J., and Hafler, D.A. 2001. CD4+CD25high regulatory cells in human peripheral blood. *J. Immunol.* **167**:1245–1253.
19. Takahashi, T., et al. 2000. Immunologic self-tolerance maintained by CD25(+)CD4(+) regulatory T cells constitutively expressing cytotoxic T lymphocyte-associated antigen 4. *J. Exp. Med.* **192**:303–310.
20. Read, S., Malmstrom, V., and Powrie, F. 2000. Cytotoxic T lymphocyte-associated antigen 4 plays an essential role in the function of CD25(+)CD4(+) regulatory cells that control intestinal inflammation. *J. Exp. Med.* **192**:295–302.
21. Nakamura, K., Kitani, A., and Strober, W. 2001. Cell contact-dependent immunosuppression by CD4(+)CD25(+) regulatory T cells is mediated by cell surface-bound transforming growth factor beta. *J. Exp. Med.* **194**:629–644.
22. Shimizu, J., Yamazaki, S., Takahashi, T., Ishida, Y., and Sakaguchi, S. 2002. Stimulation of CD25(+)CD4(+) regulatory T cells through GITR breaks immunological self-tolerance. *Nat. Immunol.* **3**:135–142.
23. McHugh, R.S., et al. 2002. CD4(+)CD25(+) immunoregulatory T cells: gene expression analysis reveals a functional role for the glucocorticoid-induced TNF receptor. *Immunity.* **16**:311–323.
24. Piccirillo, C.A., et al. 2002. CD4(+)CD25(+) regulatory T cells can mediate suppressor function in the absence of transforming growth factor beta1 production and responsiveness. *J. Exp. Med.* **196**:237–246.
25. Thornton, A.M., and Shevach, E.M. 1998. CD4+CD25+ immunoregulatory T cells suppress polyclonal T cell activation in vitro by inhibiting interleukin 2 production. *J. Exp. Med.* **188**:287–296.
26. Hori, S., Nomura, T., and Sakaguchi, S. 2003. Control of regulatory T cell development by the transcription factor Foxp3. *Science.* **299**:1057–1061.
27. Fontenot, J.D., Gavin, M.A., and Rudensky, A.Y. 2003. Foxp3 programs the development and function of CD4+CD25+ regulatory T cells. *Nat. Immunol.* **4**:330–336.
28. Khattri, R., Cox, T., Yasayko, S.A., and Ramsdell, F. 2003. An essential role for Scurfin in CD4+CD25+ T regulatory cells. *Nat. Immunol.* **4**:337–342.
29. Chen, W., et al. 2003. Conversion of peripheral CD4+CD25- naive T cells to CD4+CD25+ regulatory T cells by TGF-beta induction of transcription factor Foxp3. *J. Exp. Med.* **198**:1875–1886.
30. Wildin, R.S., et al. 2001. X-linked neonatal diabetes mellitus, enteropathy and endocrinopathy syndrome is the human equivalent of mouse scurfy. *Nat. Genet.* **27**:18–20.
31. O'Garra, A., and Vieira, P. 2003. Twenty-first century Foxp3. *Nat. Immunol.* **4**:304–306.
32. Vieira, P.L., et al. 2004. IL-10-secreting regulatory T cells do not express Foxp3 but have comparable regulatory function to naturally occurring CD4+CD25+ regulatory T cells. *J. Immunol.* **172**:5986–5993.
33. Cohn, L., Homer, R.J., Niu, N., and Bottomly, K. 1999. T helper 1 cells and interferon gamma regulate

allergic airway inflammation and mucus production. *J. Exp. Med.* **190**:1309–1318.

34. Szabo, S.J., et al. 2002. Distinct effects of T-bet in TH1 lineage commitment and IFN-gamma production in CD4 and CD8 T cells. *Science.* **295**:338–342.

35. Finotto S., et al. 2002. Development of spontaneous airway changes consistent with human asthma in mice lacking T-bet. *Science.* **295**:336–338.

36. Umetsu, D.T., Akbari, O., and Dekruyff, R.H. 2003. Regulatory T cells control the development of allergic disease and asthma. *J. Allergy Clin. Immunol.* **112**:480–487.

37. Stassen, M., et al. 2004. Differential regulatory capacity of CD25+ T regulatory cells and preactivated CD25+ T regulatory cells on development, functional activation, and proliferation of Th2 cells. *J. Immunol.* **173**:267–274.

38. Curotto de Lafaille, M.A., et al. 2001. Hyper immunoglobulin E response in mice with monoclonal populations of B and T lymphocytes. *J. Exp. Med.* **194**:1349–1359.

39. Suto, A., et al. 2001. Role of CD4(+) CD25(+) regulatory T cells in T helper 2 cell-mediated allergic inflammation in the airways. *Am. J. Respir. Crit. Care Med.* **164**:680–687.

40. Jaffar, Z., Sivakuru, T., and Roberts, K. 2004. CD4+CD25+ T cells regulate airway eosinophilic inflammation by modulating the Th2 cell phenotype. *J. Immunol.* **172**:3842–3849.

41. Hadeiba, H., and Locksley, R.M. 2003. Lung CD25 CD4 regulatory T cells suppress type 2 immune responses but not bronchial hyperreactivity. *J. Immunol.* **170**:5502–5510.

42. McMenamin, C., Pimm, C., McKersey, M., and Holt, P.G. 1994. Regulation of IgE responses to inhaled antigen in mice by antigen-specific gamma delta T cells. *Science.* **265**:1869–1871.

43. Hurst, S.D., Seymour, B.W., Muchamuel, T., Kurup, V.P., and Coffman, R.L. 2001. Modulation of inhaled antigen-induced IgE tolerance by ongoing Th2 responses in the lung. *J. Immunol.* **166**:4922–4930.

44. Ostroukhova, M., et al. 2004. Tolerance induced by inhaled antigen involves CD4+ T cells expressing membrane-bound TGF-β and FOXP3. *J. Clin. Invest.* **114**:28–38. doi:10.1172/JCI200420509.

45. Cottrez, F., Hurst, S.D., Coffman, R.L., and Groux, H. 2000. T regulatory cells 1 inhibit a Th2-specific response in vivo. *J. Immunol.* **165**:4848–4853.

46. Akbari, O., et al. 2002. Antigen-specific regulatory T cells develop via the ICOS-ICOS-ligand pathway and inhibit allergen-induced airway hyperreactivity. *Nat. Med.* **8**:1024–1032.

47. Akbari, O., DeKruyff, R.H., and Umetsu, D.T. 2001. Pulmonary dendritic cells producing IL-10 mediate tolerance induced by respiratory exposure to antigen. *Nat. Immunol.* **2**:725–731.

48. Zuany-Amorim, C., et al. 2002. Suppression of airway eosinophilia by killed Mycobacterium vaccae-induced allergen specific regulatory T-cells. *Nat. Med.* **8**:625–629.

49. Barrat, F.J., et al. 2002. In vitro generation of interleukin 10-producing regulatory CD4(+) T cells is induced by immunosuppressive drugs and inhibited by T helper type 1 (Th1)- and Th2-inducing cytokines. *J. Exp. Med.* **195**:603–616.

50. Hoyne, G.F., Dallman, M.J., Champion, B.R., and Lamb, J.R. 2001. Notch signalling in the regulation of peripheral immunity. *Immunol. Rev.* **182**:215–227.

51. Cosmi, L., et al. 2004. Th2 cells are less susceptible than Th1 cells to the suppressive activity of CD25+ regulatory thymocytes because of their responsiveness to different cytokines. *Blood.* **103**:3117–3121.

52. Tiemessen, M.M., et al. 2002. CD4 CD25 regulatory T cells are not functionally impaired in adult patients with IgE-mediated cow's milk allergy. *J. Allergy Clin. Immunol.* **110**:934–936.

53. Bellinghausen, I., Klostermann, B., Knop, J., and Saloga, J. 2003. Human CD4+CD25+ T cells derived from the majority of atopic donors are able to suppress TH1 and TH2 cytokine production. *J. Allergy Clin. Immunol.* **111**:862–868.

54. Ling, E.M., et al. 2004. Relation of CD4+CD25+ regulatory T-cell suppression of allergen-driven T-cell activation to atopic status and expression of allergic disease. *Lancet.* **363**:608–615.

55. Akdis, M., et al. 2004. Immune responses in healthy and allergic individuals are characterized by a fine balance between allergen-specific T regulatory 1 and T helper 2 cells. *J. Exp. Med.* **199**:1567–1575.

56. Dieckmann, D., Bruett, C.H., Ploettner, H., Lutz, M.B., and Schuler, G. 2002. Human CD4(+)CD25(+) regulatory, contact-dependent T cells induce interleukin 10-producing, contact-independent type 1-like regulatory T cells. *J. Exp. Med.* **196**:247–253.

57. Umetsu, D.T., McIntire, J.J., Akbari, O., Macaubas, C., and DeKruyff, R.H. 2002. Asthma: an epidemic of dysregulated immunity. *Nat. Immunol.* **3**:715–720.

58. Baldini, M., et al. 1999. A polymorphism* in the 5′ flanking region of the CD14 gene is associated with circulating soluble CD14 levels and with total serum immunoglobulin E. *Am. J. Respir. Cell Mol. Biol.* **20**:976–983.

59. Eder, W., et al. 2004. Toll-like receptor 2 as a major gene for asthma in children of European farmers. *J. Allergy. Clin. Immunol.* **113**:482–488.

60. Eisenbarth, S.C., et al. 2002. Lipopolysaccharide-enhanced, toll-like receptor 4-dependent T helper cell type 2 responses to inhaled antigen. *J. Exp. Med.* **196**:1645–1651.

61. Pasare, C., and Medzhitov, R. 2003. Toll-like receptors: balancing host resistance with immune tolerance. *Curr. Opin. Immunol.* **15**:677–682.

62. Pasare, C., and Medzhitov, R. 2003. Toll pathway-dependent blockade of CD4+CD25+ T cell-mediated suppression by dendritic cells. *Science.* **299**:1033–1036.

63. Platts-Mills, T., Vaughan, J., Squillace, S., Woodfolk, J., and Sporik, R. 2001. Sensitisation, asthma, and a modified Th2 response in children exposed to cat allergen: a population-based cross-sectional study. *Lancet.* **357**:752–756.

64. Karlsson, M.R., Rugtveit, J., and Brandtzaeg, P. 2004. Allergen-responsive CD4+CD25+ regulatory T cells in children who have outgrown cow's milk allergy. *J. Exp. Med.* **199**:1679–1688.

65. Tiemessen, M.M., et al. 2004. Cow's milk-specific T-cell reactivity of children with and without persistent cow's milk allergy: key role for IL-10. *J. Allergy Clin. Immunol.* **113**:932–939.

66. Robinson, D.S., and Nguyen, X.D. 2004. Fluticasone propionate increases suppression of allergen-driven T cell proliferation by CD4+CD25+ T cells. *J. Allergy Clin. Immunol.* **114**:296–301.

67. Durham, S.R., et al. 1999. Long-term clinical efficacy of grass-pollen immunotherapy. *N. Engl. J. Med.* **341**:468–475.

68. Till, S.J., and Durham, S.R. 2004. Immunological responses to allergen immunotherapy. *Clin. Allergy Immunol.* **18**:85–104.

69. Varney, V.A., et al. 1991. Usefulness of immunotherapy in patients with severe summer hay fever uncontrolled by antiallergic drugs. *BMJ.* **302**:265–269.

70. Bousquet, J., et al. 1998. Allergen immunotherapy: therapeutic vaccines for allergic diseases. World Health Organization. American Academy of Allergy, Asthma and Immunology. *Ann. Allergy Asthma Immunol.* **81**:401–405.

71. Pajno, G.B., Barberio, G., De Luca, F., Morabito, L., and Parmiani, S. 2001. Prevention of new sensitizations in asthmatic children monosensitized to house dust mite by specific immunotherapy. A six-year follow-up study. *Clin. Exp. Allergy.*

31:1392–1397.

72. Moller, C., et al. 2002. Pollen immunotherapy reduces the development of asthma in children with seasonal rhinoconjunctivitis (the PAT-study). *J. Allergy. Clin. Immunol.* **109**:251–256.

73. Akdis, M., et al. 1996. Epitope-specific T-cell tolerance to phospholipase A2 in bee venom immunotherapy and recovery by IL-2 and IL-15 in vitro. *J. Clin. Invest.* **98**:1676–1683.

74. Till, S., et al. 1997. IL-5 production by allergen-stimulated T-cells following grass pollen immunotherapy for seasonal allergic rhinitis. *Clin. Exp. Immunol.* **110**:114–121.

75. Wachholz, P., et al. 2002. Grass pollen immunotherapy for hayfever is associated with increases in local nasal mucosal but not peripheral Th1/Th2 ratios. *Immunology.* **105**:56–62.

76. Ebner, C., et al. 1997. Immunological changes during specific immunotherapy of grass pollen allergy: reduced lymphoproliferative responses to allergen and shift from TH2 to TH1 in T-cell clones specific for Phl p 1, a major grass pollen allergen. *Clin. Exp. Allergy.* **27**:1007–1015.

77. Jutel, M., et al. 1995. Bee venom immunotherapy results in decrease of IL-4 and IL-5 and increase of IFN-gamma secretion in specific allergen-stimulated T-cell cultures. *J. Immunol.* **154**:4187–4194.

78. Francis, J.N., Till, S.J., and Durham, S.R. 2003. Induction of IL-10+CD4+CD25+ T cells by grass pollen immunotherapy. *J. Allergy Clin. Immunol.* **111**:1255–1261.

79. Bellinghausen, I., et al. 1997. Insect venom immunotherapy induces interleukin-10 production and a Th2- to-Th1 shift, and changes surface marker expression in venom-allergic subjects. *Eur. J. Immunol.* **27**:1131–1139.

80. Jutel, M., et al. 2003. IL-10 and TGF-beta cooperate in the regulatory T cell response to mucosal allergens in normal immunity and specific immunotherapy. *Eur. J. Immunol.* **33**:1205–1214.

81. Varney, V.A., et al. 1993. Influence of grass pollen immunotherapy on cellular infiltration and cytokine mRNA expression during allergen-induced late-phase cutaneous responses. *J. Clin. Invest.* **92**:644–651.

82. Durham, S.R., et al. 1996. Grass pollen immunotherapy inhibits allergen-induced infiltration of CD4+ T lymphocytes and eosinophils in the nasal mucosa and increases the number of cells expressing messenger RNA for interferon-gamma. *J. Allergy Clin. Immunol.* **97**:1356–1365.

83. Wilson, D.R., et al. 2001. Grass pollen immunotherapy inhibits seasonal increases in basophils and eosinophils in the nasal epithelium. *Clin. Exp. Allergy.* **31**:1705–1713.

84. Nouri-Aria, K., et al. 2004. Grass pollen immunotherapy induces mucosal and peripheral IL-10 responses and blocking IgG activity. *J. Immunol.* **172**:3252–3259.

85. Daeron, M., et al. 1995. The same tyrosine-based inhibition motif, in the intracytoplasmic domain of Fc gamma RIIB, regulates negatively BCR-, TCR-, and FcR-dependent cell activation. *Immunity.* **3**:635–646.

86. Malbec, O., et al. 1998. Fc epsilon receptor I-associated lyn-dependent phosphorylation of Fc gamma receptor IIB during negative regulation of mast cell activation. *J. Immunol.* **160**:1647–1658.

87. Zhu, D., Kepley, C.L., Zhang, M., Zhang, K., and Saxon, A. 2002. A novel human immunoglobulin Fc gamma Fc epsilon bifunctional fusion protein inhibits Fc epsilon RI-mediated degranulation. *Nat. Med.* **8**:518–521.

88. Wachholz, P.A., Soni, N.K., Till, S.J., and Durham, S.R. 2003. Inhibition of allergen-IgE binding to B cells by IgG antibodies after grass pollen immunotherapy. *J. Allergy Clin. Immunol.* **112**:915–922.

89. Van Neerven, R.J., et al. 1999. Blocking antibodies

induced by specific allergy vaccination prevent the activation of CD4+ T cells by inhibiting serum-IgE-facilitated allergen presentation. *J. Immunol.* **163**:2944–2952.

90. Norman, P.S., Lichtenstein, L.M., and Marsh, D.G. 1981. Studies on allergoids from naturally occurring allergens. IV. Efficacy and safety of long-term allergoid treatment of ragweed hay fever. *J. Allergy Clin. Immunol.* **68**:460–470.

91. Pauli, G., et al. 2000. Comparison of genetically engineered hypoallergenic rBet v 1 derivatives with rBet v 1 wild-type by skin prick and intradermal testing: results obtained in a French population. *Clin. Exp. Allergy.* **30**:1076–1084.

92. Tighe, H., et al. 2000. Conjugation of immunostimulatory DNA to the short ragweed allergen amb a 1 enhances its immunogenicity and reduces its allergenicity. *J. Allergy Clin. Immunol.* **106**:124–134.

93. Rachmilewitz, D., et al. 2004. Toll-like receptor 9 signaling mediates the anti-inflammatory effects of probiotics in murine experimental colitis. *Gastroenterology.* **126**:520–528.

94. Gaur, A., Wiers, B., Liu, A., Rothbard, J., and Fathman, C.G. 1992. Amelioration of autoimmune encephalomyelitis by myelin basic protein synthetic peptide-induced anergy. *Science.* **258**:1491–1494.

95. Clayton, J.P., et al. 1989. Peptide-specific prevention of experimental allergic encephalomyelitis. Neonatal tolerance induced to the dominant T cell determinant of myelin basic protein. *J. Exp. Med.* **169**:1681–1691.

96. Daniel, D., and Wegmann, D.R. 1996. Protection of non-obese diabetic mice from diabetes by intranasal or subcutaneous administration of insulin peptide

B-(9-23). *Proc. Natl. Acad. Sci. U. S. A.* **93**:956–960.

97. Metzler, B., and Wraith, D.C. 1993. Inhibition of experimental autoimmune encephalomyelitis by inhalation but not oral administration of the encephalitogenic peptide: influence of MHC binding affinity. *Int. Immunol.* **5**:1159–1165.

98. Burkhart, C., Liu, G.Y., Anderton, S.M., Metzler, B., and Wraith, D.C. 1999. Peptide-induced T cell regulation of experimental autoimmune encephalomyelitis: a role for IL-10. *Int. Immunol.* **11**:1625–1634.

99. Prakken, B.J., et al. 1997. Peptide-induced nasal tolerance for a mycobacterial heat shock protein 60 T cell epitope in rats suppresses both adjuvant arthritis and nonmicrobially induced experimental arthritis. *Proc. Natl. Acad. Sci. U. S. A.* **94**:3284–3289.

100. Hoyne, G.F., O'Hehir, R.E., Wraith, D.C., Thomas, W.R., and Lamb, J.R. 1993. Inhibition of T cell and antibody responses to house dust mite allergen by inhalation of the dominant T cell epitope in naive and sensitized mice. *J. Exp. Med.* **178**:1783–1788.

101. Sundstedt, A., O'Neill, E.J., Nicolson, K.S., and Wraith, D.C. 2003. Role for IL-10 in suppression mediated by peptide-induced regulatory T cells in vivo. *J. Immunol.* **170**:1240–1248.

102. Prakken, B.J., et al. 2004. Epitope-specific immunotherapy induces immune deviation of proinflammatory T cells in rheumatoid arthritis. *Proc. Natl. Acad. Sci. U. S. A.* **101**:4228–4233.

103. Haselden, B.M., Kay, A.B., and Larche, M. 1999. Immunoglobulin E-independent major histocompatibility complex-restricted T cell peptide epitope-induced late asthmatic reactions. *J. Exp. Med.* **189**:1885–1894.

104. Ali, F.R., Oldfield, W.L., Higashi, N., Larche, M., and

Kay, A.B. 2004. Late asthmatic reactions induced by inhalation of allergen-derived T cell peptides. *Am. J. Respir. Crit. Care Med.* **169**:20–26.

105. Oldfield, W.L., Larche, M., and Kay, A.B. 2002. Effect of T-cell peptides derived from Fel d 1 on allergic reactions and cytokine production in patients sensitive to cats: a randomised controlled trial. *Lancet.* **360**:47–53.

106. Oldfield, W.L., Kay, A.B., and Larche, M. 2001. Allergen-derived T cell peptide-induced late asthmatic reactions precede the induction of antigen-specific hyporesponsiveness in atopic allergic asthmatic subjects. *J. Immunol.* **167**:1734–1739.

107. Muller, U., et al. 1998. Successful immunotherapy with T-cell epitope peptides of bee venom phospholipase A2 induces specific T-cell anergy in patients allergic to bee venom. *J. Allergy Clin. Immunol.* **101**:747–754.

108. Verhoef, A., Alexander, C., Kay, A.B., and Larché, M. 2004. Allergen-based peptide immunotherapy is associated with functional modifications in allergen-specific T-cell subpopulations [abstract]. *J. Allergy Clin. Immunol.* **113**(Suppl.):918.

109. Smith, T.R.F., Alexander, C., Kay, A.B., Larché, M., and Robinson, D.S. 2004. Cat allergen peptide immunotherapy reduces CD4⁺ T cell responses to cat allergen but does not alter suppression by CD4⁺CD25⁺ T cells. *Allergy.* **59**:1097–1101.

110. Alexander, C., Ying, S., Kay, A.B., and Larché, M. 2004. Fel d 1-derived T cell peptide therapy induces recruitment of activated (CD4+CD25+;CD4+IFN-gamma+) Th1 cells to sites of allergen-induced late-phase skin reactions in cat-allergic subjects. *Clin. Exp. Allergy.* In press.

Tregs and transplantation tolerance

Patrick T. Walsh, Devon K. Taylor, and Laurence A. Turka

Department of Medicine, University of Pennsylvania, Philadelphia, Pennsylvania, USA.

The induction and maintenance of immune tolerance to transplanted tissues constitute an active process involving multiple mechanisms that work cooperatively to prevent graft rejection. These mechanisms are similar to inherent tolerance toward self antigens and have a requirement for active immunoregulation, largely T cell mediated, that promotes specific unresponsiveness to donor alloantigens. This review outlines our current understanding of the Treg subsets that contribute to allotolerance and the mechanisms by which these cells exert their effects as well as their potential for therapy.

Introduction

Peripheral tolerance to self antigens is maintained by a dynamic process involving several different mechanisms that restrict the development of a potentially destructive autoaggressive T cell response. These mechanisms include T cell depletion through activation-induced cell death, "ignorance" of self antigens (meaning the apparent absence of antigen recognition), and the induction of T cell anergy (1). While these mechanisms are clearly important in the maintenance of self tolerance, they are by themselves not sufficient, as there is also a need for active suppression of autoreactive T cells by Tregs (2). Although initial characterization of these Treg subsets defined their role in the maintenance of tolerance to self, it is now clear that such regulatory cells play an important role in suppressing immune responses directed against alloantigens expressed on transplanted organs and tissues (3).

Overview of graft rejection and tolerance

Graft rejection occurs as a consequence of polymorphisms in histocompatibility genes, primarily those located within the MHC (4). T cells respond to foreign (allogeneic) MHC molecules in the same fashion as to any foreign protein: they secrete cytokines, divide, and differentiate (5). This generates a large population of activated effector cells, primarily T cells and macrophages, which are the primary mediators of graft destruction.

Alloresponsive T cells can recognize antigens present in transplanted tissues by 1 of 2 distinct pathways. In the direct pathway, the responding T cells recognize intact allogeneic MHC molecules on the surface of donor-derived APCs, whereas in the indirect pathway, recipient APCs process donor-derived allo-MHC molecules into peptides and then present those peptides to T cells on self-MHC molecules. It is generally accepted that the direct pathway predominates in the immediate aftermath of transplantation, when graft-resident APCs (passenger leukocytes) migrate to the surrounding lymphoid tissue, where they stimulate alloresponsive T cells. As donor-derived APCs are relatively short lived, the indirect pathway of allorecognition is generally believed to predominate as the alloresponse progresses (6).

Experimental methods to induce transplantation tolerance are typically divided into 2 categories. "Central" tolerance refers (in most instances) to the use of bone marrow transplantation as a means to induce hematopoietic chimerism (7). This results in the coexistence of donor- and recipient-derived lymphoid and myeloid cells. As a result, developing T cells that are donor reactive are deleted before they can exit the thymus, in the same manner as self-reactive T cells (8). "Peripheral" tolerance refers to the use of antibodies (or occasionally pharmacologic agents) that block or modulate T cell activation or growth factor receptor pathways in mature T cells. In most instances, this has the net result of promoting apoptosis among the T cells that are responding to alloantigens (9).

An important characteristic of alloimmune responses is the high frequency of T cells that are able to recognize and respond to alloantigens (primarily the products of genes encoded within the MHC) (10). Because of this, and based on data from studies on rodent models, many investigators believe that it is necessary to achieve large-scale deletion of alloreactive T cells in order to create transplantation tolerance (9). Both central and peripheral tolerance strategies achieve this during the early "induction" phase of therapy (i.e., the first 1–2 weeks after transplantation). In the case of central tolerance, this alone appears to be sufficient, as newly developing T cells with potential anti-donor reactivity will be eliminated within the thymus following encounter with donor-derived cells (7, 8). However, in the case of peripheral tolerance strategies, a large body of data derived from experimental animals suggests that following depletion, the "maintenance" phase of tolerance requires Tregs that can act on both any remaining alloresponsive T cells and on new thymic emigrants (Figure 1).

Multiple types of Tregs

Studies of Tregs in transplantation have identified multiple populations of cells with different cell-surface phenotypes and, to some extent, with different mechanisms of action (11). One population is a naturally occurring subset of CD4+ T cells that arises during T cell development in the thymus and is best defined by constitutive expression the α chain of the IL-2 receptor, CD25 (2). A second population consists of induced Treg subsets that may arise during the course of a normal immune response (presumably to help terminate the response when the pathogen is eliminated and prevent secondary autoimmunity). These "induced" Tregs, while largely contained within the CD4+ compartment, are distinct from their naturally occurring CD4+CD25+ counterparts. In addition, CD8+ Tregs, TCR+CD4-CD8- T cells, and NK Tregs have also been reported to play a role in different models of transplantation tolerance (12–15).

Naturally occurring Tregs

Characterization. The study of Tregs was historically crippled by the lack of reliable cellular or molecular markers that are necessary to identify these cells. The absence of such tools led to

Conflict of interest: The authors have declared that no conflict of interest exists.

Citation for this article: *J. Clin. Invest.* **114**:1398–1403 (2004). doi:10.1172/JCI200423238.

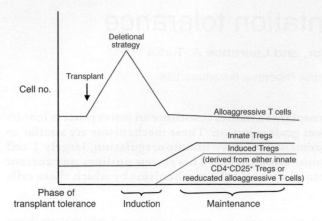

Figure 1
Altering the balance between alloaggressive and Treg subsets. Deletional strategies employed at or around the time of transplant reduce the number of potentially graft destructive T cells and facilitate the action of Treg subsets. During the maintenance phase of tolerance, these Tregs, either naturally occurring or induced, can thus act more efficiently on a greatly reduced number of effector T cells. Cell number, as denoted on the y axis, represents an illustration as to how the relative ratio of effector versus Treg subsets alters during the establishment of transplant tolerance and is not meant for comparison between groups.

the question of whether Tregs exist. This changed, however, with the discovery that the molecular marker CD25, previously thought to be expressed only on recently activated T cells, was also expressed on a subset of resting CD4+ T cells with regulatory function (1). Naturally occurring CD4+CD25+ Tregs develop in the thymus. These Tregs constitute approximately 5–10% of mature CD4+CD8- thymocytes and about 10% of peripheral CD4+ T cells (16, 17). Other cell-surface markers, such as CD45RB, CTLA-4, glucocorticoid-induced TNF receptor family–related receptor (GITR or TNFRSF18), CD122, CD103 ($\alpha_E\beta_7$ integrin), CD134 (OX40), and CD62L (L-selectin), whose relative expression levels can be used to define and isolate CD4+CD25+ Tregs have also been identified (3). However, as with CD25, none of these molecules alone represents a definitive marker for naturally occurring Tregs, as they are also expressed on other CD4+ T cell subsets, particularly activated T cells. Recently, the forkhead/winged helix transcription factor Foxp3 was shown to be uniquely expressed by naturally occurring CD4+CD25+ Tregs, and it is thought to act as a master switch controlling Treg differentiation (18, 19). However, the intracellular location of Foxp3 places an obvious limitation on its use in identifying and studying Tregs. Therefore, efforts continue to find a definitive cell-surface marker for Tregs.

Mechanism of action. Although the exact mechanism by which these cells exert their immunosuppressive effect remains elusive, CD4+CD25+ Tregs are known to suppress effector T cell proliferation in vitro through a cell contact–dependent mechanism that is largely cytokine independent (20–22). In particular, these studies implicate a role for accessory molecules such CTLA-4 and GITR expressed on the surface of Tregs (23, 24). This is in contrast to in vivo models, where blockade of both IL-10 and TGF-β has been reported to abrogate Treg-mediated unresponsiveness to alloantigens (25, 26). These apparent discrepancies could be explained by a requirement for cell contact with a third

cell, such as an APC, and subsequent elaboration of cytokines that may directly suppress other cells or recruit them to become regulators (see also below).

Role in transplantation. A role for naturally occurring CD4+CD25+ Tregs in the development of transplantation tolerance was first indicated by their ability to suppress graft verses host disease in murine models of allogeneic bone marrow transplantation. While transfer of allogeneic CD4+CD25- naive or effector T cells normally leads to graft versus host disease, cotransfer of purified CD4+CD25+ Tregs along with the CD4+CD25- T cells significantly delayed disease onset (27). Other groups have confirmed these findings, although it is not clear whether the activity of the cotransferred Tregs is amplified by or dependent upon other cell types, such as APCs or other T cells in vivo, or whether the Tregs themselves are sufficient to suppress alloresponsive T cells (28). In solid organ and tissue transplantation, cotransfer of CD4+CD25+ T cells into T cell–deficient mice along with naive CD4+CD25- cells can block the ability of the latter cell subset to reject minor or MHC-mismatched allogeneic skin grafts (25, 29). Collectively, these studies indicate that naturally occurring Tregs can play a role in achieving transplantation tolerance.

Inducible CD4+ Tregs
Characterization. The naturally occurring population of Tregs described above has inherent suppressive capabilities. However, populations of T cells whose immunosuppressive activity is induced/acquired in the periphery have also been identified. There are primarily 2 populations of these inducible Tregs important for transplantation tolerance: Th3 cells and Tr1 cells. Th3 cells were first identified because of their role, through the secretion of TGF-β, in the development of immune tolerance following the ingestion of antigens (termed oral tolerance) (30). Tr1 cells are similar to Th3 cells, but they secrete large amounts of IL-10 and were first characterized on the basis of their role in preventing autoimmune colitis (31).

There are a number of differences between naturally occurring Tregs and those induced in the periphery. First, CD4+CD25+ Tregs undergo development in the thymus, while there is no evidence to suggest thymic development of either Th3 or Tr1 cells. Instead, induced Tregs depend on peripheral factors such as the maturity or type of the stimulating APC and the availability of cytokines such as TGF-β (32). Second, in contrast to CD4+CD25+ Tregs, which exert their suppressive function through a cell contact–dependent and cytokine-independent mechanism, both Th3 and Tr1 cells appear to function independently of cell-to-cell contact and suppress immune responses through the secretion of immunosuppressive cytokines, such as IL-10 and TGF-β (33). Finally, the ability of Tr1 cells to home to anatomic sites differs from that of CD4+CD25+ T cells. Tr1 cells tend to migrate toward sites of inflammation, while naturally occurring CD4+CD25+ T cells are predominantly found in lymphoid organs (34). Agreement on this point is not universal, however, as Graca et al. recently demonstrated the existence of CD4+CD25+ Tregs within tolerated allografts (35).

At present, induced Treg subsets are largely identified on the basis of their secretion of immunosuppressive cytokines. As with CD4+CD25+ Tregs, there is no specific cell-surface marker to distinguish them from other T cell subsets. While Foxp3 is expressed by CD4+CD25+ Tregs, it is not yet clear whether it regulates the development of either Th3 or Tr1 suppressor T cell subsets (36).

However, TGF-β, a cytokine mediator of the effects of Th3 cells, has been shown recently to convert nonregulatory CD4+CD25− T cells into regulatory CD4+CD25+ T cells, in conjunction with induction of *Foxp3* expression (37).

Role in transplantation. In the context of allograft transplantation, the induction of a regulatory T cell phenotype in otherwise alloresponsive T cells has been proposed as a major contributing factor for the maintenance of tolerance achieved through selected strategies (38). Indeed it has been reported that repetitive stimulation of naive T cells with immature allogeneic DCs results in the development of a suppressive phenotype by responding T cells (39). The maturation status and types of stimulating DCs present in the grafted tissue is undoubtedly a critical factor in determining the outcome of an alloimmune response. Phenotypically, immature DCs do not stimulate optimal effector T cell responses, due to low expression of T cell costimulatory factors and proinflammatory cytokines. In fact, such cells are often able to induce a Treg phenotype in responding T cells (5). Beyond their maturational state, however, it is also important to consider the multiplicity of existing DC subtypes, as a number of recent reports demonstrate that particular DC subsets can induce a Treg phenotype (e.g., Th3 or Tr1 cells) irrespective of their maturation state (40, 41).

While induced Tregs represent a subset distinct from their naturally occurring CD4+CD25+ counterparts, there is considerable evidence indicating that CD4+CD25+ T cells play an important role in the "development" of these cells, promoting otherwise potentially graft-destructive effector T cells to adopt a Tr1 suppressor phenotype (42, 43). At present however, the mechanism for this activity is not known and could involve either direct cell-cell interaction, involvement of a third cell (such as an APC), soluble mediators, or some combination of the three. As noted above, it has recently been demonstrated that nonregulatory T cells may also convert to a CD4+CD25+ suppressor phenotype, under the influence of TGF-β (37).

Interestingly, TGF-β has been found in tolerated grafts, which suggests that induced Tregs may develop and exert their influence directly at the site of the graft (26). Karim et al. have also shown that CD4+CD25+ Tregs can develop from CD25− precursors in thymectomized mice (44) and that these Tregs can suppress skin allograft rejection. These data suggest that inducible Treg subsets can prolong allograft survival without newly formed innate Tregs entering the periphery. Although appropriate strategies were employed to deplete innate CD4+CD25+ Tregs, one cannot completely exclude the possibility that residual nondepleted cells contributed to tolerance. A role for CD4+CD25+ T cells in the induction of a regulatory phenotype in otherwise nonsuppressive T cells provides an attractive hypothesis bringing together the observations of numerous groups concerning the respective roles of both innate and induced Treg subsets in promoting transplantation tolerance (34, 45). This suggests a model in which the 2 subsets act in a cooperative fashion to suppress potentially inflammatory immune responses directed toward transplanted tissues. The ability of these cells to induce regulatory function in other populations would also explain a paradox that has been raised regarding the potency of CD4+CD25+ Tregs. In vitro, meaningful suppression of activated T cells by CD4+CD25+ Tregs generally requires at least a 1:3 ratio of Tregs to effectors; lower ratios yield little suppression (46). However, the frequency of CD4+CD25+ Tregs in vivo is only approximately

10% that of CD4+ T cells, and approximately 3% of all T cells (16). Thus, some combination of selective homing and/or induction of suppressive function in other cells must be occurring in vivo.

Other Treg types

As mentioned above, Treg subsets have also been described outside the CD4+ compartment. CD8+ Tregs generated through oral exposure to alloantigen have been observed in tolerated allografts. The precise mechanism of regulation by this subset is unclear but may be associated with increased IL-4 production (13). Similarly, Seino et al. demonstrated a requirement for NK T cells in acquiring long-term cardiac allograft acceptance after costimulatory blockade (15). Additionally, there is evidence for a TCR+CD4−CD8− Treg subset that can mediate acceptance of skin allografts, possibly by inducing the deletion of alloreactive CD8+ T cells (14). The full significance of these Treg subsets outside the CD4+ compartment with regard to transplantation tolerance, however, remains unknown given the still-early stages of investigation on these cell types.

Indirect allorecognition, linked suppression, and infectious tolerance

A great deal of work on the role of Tregs in transplantation tolerance has addressed their mode of allorecognition. As discussed above, it is believed that indirect allorecognition predominates with increasing duration of engraftment. This finding, coupled with observations from a number of groups that there is a requirement for the continuous presence of antigen in order to maintain transplant tolerance (47, 48), suggests that it is the indirect pathway that is the predominant mode of allorecognition by Tregs (49). Yamada et al. explored the relative contribution of the direct and indirect pathways of antigen presentation to transplant tolerance by using either donor mice deficient in MHC class II molecules (in which recognition is indirect only) or recipients that lack MHC class II molecules in the periphery but possess CD4+ T cells as a consequence of transgenic expression of MHC II on thymic epithelium (in which recognition is direct only). Using agents that block T cell costimulatory pathways, the authors found that tolerance could be easily achieved in the absence of direct recognition but that recipients lacking indirect allorecognition pathways were notably difficult to tolerize (50).

If Tregs recognize alloantigens via indirect allorecognition, one might ask how cells that recognize only a small subset of graft-derived antigens can block the response to all graft-expressed antigens. Indeed, even before the clear identification of CD4+CD25+ Tregs, these questions were already being investigated. For example, more than a decade ago, transplantation tolerance in a rat model was achieved through the oral administration of multiple MHC-derived peptide alloantigens (51). Later studies implicated a role for immunoregulatory T cells in contributing to tolerance in this model (52). Niimi and colleagues further extended these observations with the finding that allograft tolerance could be achieved through oral administration of a single alloantigen present in the graft (53). The tolerizing potential of a single alloantigen, which can subsequently dominantly confer nonresponsiveness against all other antigens present within the graft, has been termed "linked suppression" and is dependent on the action of Tregs (38). Linked suppression occurs when a potentially alloreactive T cell comes under the tolerizing influence of a Treg, such as a CD4+CD25+ Treg, as both cells recognize their respective alloantigens presented by the same APC (33, 54, 55). In effect this leads to a "reeducation"

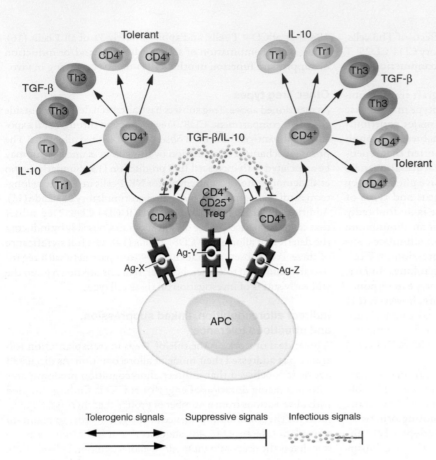

Figure 2
Infectious tolerance and linked suppression induced by CD4+CD25+ Tregs. CD4+CD25+ Treg cells can suppress alloreactive CD4+ T cells either directly via cell contact or secretion of IL-10 and TGF-β or alternatively by influencing the stimulating APC. Linked suppression arises when tolerance generated against a specific antigen (Ag-Y) leads to tolerance against unrelated or third-party antigens (Ag-X and Ag-Z), providing that these unrelated antigens are expressed on the surface of the same APC. The secretion of IL-10 and TGF-β by Tregs has been implicated in this process, which is thought to reinforce the infectious nature of transplant tolerance.

of the potentially destructive alloresponsive T cell with the induction of a Treg phenotype (Figure 2).

Importantly, the reeducated alloaggressive cell, which becomes instead a "secondary" regulatory cell, can in turn induce other naive CD4+ T cells to adopt a regulatory phenotype, thus propagating the tolerant state. Such mechanisms are thought to explain some seminal observations with regard to suppressive T cells and transplantation tolerance made almost 30 years ago (56). However, it was not until recently that the term "infectious tolerance" was coined, by Waldmann and colleagues, to describe the transferable nature of allograft tolerance from one recipient to another (reviewed in ref. 38). Tolerance achieved after a short course of nondepleting CD4 and CD8 antibodies prior to transplantation of minor histocompatibility–mismatched skin grafts was associated with the development of a regulatory CD4+ T cell phenotype within the recipient (57). Transfer of CD4+ T cells from these tolerized mice could prevent graft rejection in naive recipients. Most strikingly, the regulatory cells from the tolerized mice induced new Tregs in the naive-transplanted recipients, and these secondary Tregs could do the same if transferred to "tertiary" recipients, which illustrates the infectious nature of the process. Later studies by a number of groups confirmed these observations using more stringent models of transplant tolerance, such as cardiac allograft across major histocompatibility barriers, leading to a general acceptance of the existence of this phenomenon (47, 58). The generation of these Treg subsets has also been implicated in tolerance achieved through numerous strategies including costimulatory blockade and treatment with cyclosporine (59–61). The precise mechanisms by which infectious tolerance is mediated by CD4+ suppressor T cells remain

elusive, and it is not possible to assign functions to specific Treg subsets, although evidence indicates that immunosuppressive cytokines such as IL-10 and TGF-β play an important role (62).

Potential for utilizing Tregs in transplantation
One can consider 2 separate issues regarding CD4+CD25+ Tregs in transplantation: What is their role in models of tolerance? and What is their potential as a therapeutic tool? At present, there are almost no data in humans, although the issue has been examined extensively in animal models. Many studies establish the ability of Tregs to prevent graft rejection and facilitate tolerance in manipulated situations, such as using immune reconstituted immunodeficient hosts (reviewed in ref. 6). Because of confounding effects of homeostatic proliferation in T cell–deficient animals (63), this does not provide definitive proof of their role in tolerance in a normal host. However, such a role is strongly suggested by other studies, such as those showing Tregs in tolerated grafts (35). Moreover, both sets of studies clearly show the potential to use Tregs as a deliberate therapeutic tool. This might be achieved either by ex vivo expansion and activation of Tregs followed by infusion or by manipulating the immune response in vivo in a manner that promotes Treg development (64). Recent advances regarding the role of TGF-β, IL-10, and Foxp3 in the development of subsets of Tregs may provide a means to achieve these goals.

Concluding remarks
The large number of alloreactive cells present even in naive hosts is believed to underlie the requirement for deletion of these cells in the induction of tolerance across MHC-mismatched

barriers by strategies such as costimulatory blockade that target mature peripheral T cells (65, 66). From these investigations, we and others theorized that such a reduction in the uniquely high numbers of potentially graft destructive T cells would facilitate immunoregulation by Treg subsets, thus promoting and maintaining a tolerant state (9). Subsequently, this hypothesis has been strengthened by other groups who have demonstrated roles for both naturally occurring and induced Treg subsets in the development of allotolerance achieved through costimulatory blockade (27, 54). As a result, it is being increasingly proposed that clinical transplantation tolerance protocols that target alloreactive T lymphocytes for deletion need to specifically spare Tregs (1). Along those lines, Strom and colleagues recently demonstrated that selective lysis of nonregulatory CD25+ alloreactive T cells and persistence of Treg cells could be achieved through administration of an agonistic IL-2–Fc receptor fusion protein. Indeed, administration of this fusion protein in combination with selective blockade of IL-15 signaling, which is important for effector T cell proliferation and memory generation, resulted in graft acceptance in a very stringent transplantation model (67). Taken together, these reports strongly support the idea that deletion and regulation play complementary roles in the development of transplantation tolerance. This has raised the potential for ex vivo expansion/promotion of regulatory cells, which could then be used as adoptive immunotherapy in transplantation. Determining whether such approaches are feasible will undoubtedly be the focus of study for many years to come. It may be premature to predict whether or not they will succeed, but it is not premature to state that Tregs in transplantation have finally "arrived."

Address correspondence to: Laurence A. Turka, University of Pennsylvania, 700 CRB, 415 Curie Boulevard, Philadelphia, Pennsylvania 19104-6144, USA. Phone: (215) 898-1018; Fax: (215) 573-2880; E-mail: turka@mail.med.upenn.edu.

1. Lechler, R.I., Garden, O.A., and Turka, L.A. 2003. The complementary roles of deletion and regulation in transplantation tolerance. *Nat. Rev. Immunol.* **3**:147–158.

2. Sakaguchi, S., Sakaguchi, N., Asano, M., Itoh, M., and Toda, M. 1995. Immunologic self-tolerance maintained by activated T cells expressing IL-2 receptor alpha-chains (CD25). Breakdown of a single mechanism of self-tolerance causes various autoimmune diseases. *J. Immunol.* **155**:1151–1164.

3. Wood, K.J., and Sakaguchi, S. 2003. Regulatory T cells in transplantation tolerance. *Nat. Rev. Immunol.* **3**:199–210.

4. Rogers, N.J., and Lechler, R.I. 2001. Allorecognition. *Am. J. Transplant.* **1**:97–102.

5. Walsh, P.T., Strom, T.B., and Turka, L.A. 2004. Routes to transplant tolerance versus rejection; the role of cytokines. *Immunity.* **20**:121–131.

6. Chiffoleau, E., Walsh, P.T., and Turka, L. 2003. Apoptosis and transplantation tolerance. *Immunol. Rev.* **193**:124–145.

7. Nikolic, B., and Sykes, M. 1997. Mixed hematopoietic chimerism and transplantation tolerance. *Immunol. Res.* **16**:217–228.

8. Sykes, M. 2001. Mixed chimerism and transplant tolerance. *Immunity.* **14**:417–424.

9. Li, X.C., Strom, T.B., Turka, L.A., and Wells, A.D. 2001. T cell death and transplantation tolerance. *Immunity.* **14**:407–416.

10. Suchin, E.J., et al. 2001. Quantifying the frequency of alloreactive T cells in vivo: new answers to an old question. *J. Immunol.* **166**:973–981.

11. Jonuleit, H., and Schmitt, E. 2003. The regulatory T cell family: distinct subsets and their interrelations. *J. Immunol.* **171**:6323–6327.

12. Gilliet, M., and Liu, Y.J. 2002. Generation of human CD8 T regulatory cells by CD40 ligand-activated plasmacytoid dendritic cells. *J. Exp. Med.* **195**:695–704.

13. Zhou, J., Carr, R.I., Liwski, R.S., Stadnyk, A.W., and Lee, T.D. 2001. Oral exposure to alloantigen generates intragraft CD8+ regulatory cells. *J. Immunol.* **167**:107–113.

14. Zhang, Z.X., Yang, L., Young, K.J., DuTemple, B., and Zhang, L. 2000. Identification of a previously unknown antigen-specific regulatory T cell and its mechanism of suppression. *Nat. Med.* **6**:782–789.

15. Seino, K.I., et al. 2001. Requirement for natural killer T (NKT) cells in the induction of allograft tolerance. *Proc. Natl. Acad. Sci. U. S. A.* **98**:2577–2581.

16. Itoh, M., et al. 1999. Thymus and autoimmunity: production of CD25+CD4+ naturally anergic and suppressive T cells as a key function of the thymus in maintaining immunologic self-tolerance. *J. Immunol.* **162**:5317–5326.

17. Jordan, M.S., et al. 2001. Thymic selection of CD4+CD25+ regulatory T cells induced by an agonist self-peptide. *Nat. Immunol.* **2**:301–306.

18. Hori, S., Nomura, T., and Sakaguchi, S. 2003. Control of regulatory T cell development by the transcription factor Foxp3. *Science.* **299**:1057–1061.

19. Fontenot, J.D., Gavin, M.A., and Rudensky, A.Y. 2003. Foxp3 programs the development and function of CD4+CD25+ regulatory T cells. *Nat. Immunol.* **4**:330–336.

20. Takahashi, T., et al. 1998. Immunologic self-tolerance maintained by CD25+CD4+ naturally anergic and suppressive T cells: induction of autoimmune disease by breaking their anergic/suppressive state. *Int. Immunol.* **10**:1969–1980.

21. Suri-Payer, E., and Cantor, H. 2001. Differential cytokine requirements for regulation of autoimmune gastritis and colitis by CD4(+)CD25(+) T cells. *J. Autoimmun.* **16**:115–123.

22. Piccirillo, C.A., et al. 2002. CD4(+)CD25(+) regulatory T cells can mediate suppressor function in the absence of transforming growth factor beta1 production and responsiveness. *J. Exp. Med.* **196**:237–246.

23. Takahashi, T., et al. 2000. Immunologic self-tolerance maintained by CD25(+)CD4(+) regulatory T cells constitutively expressing cytotoxic T lymphocyte-associated antigen 4. *J. Exp. Med.* **192**:303–310.

24. Shimizu, J., Yamazaki, S., Takahashi, T., Ishida, Y., and Sakaguchi, S. 2002. Stimulation of CD25(+)CD4(+) regulatory T cells through GITR breaks immunological self-tolerance. *Nat. Immunol.* **3**:135–142.

25. Hara, M., et al. 2001. IL-10 is required for regulatory T cells to mediate tolerance to alloantigens in vivo. *J. Immunol.* **166**:3789–3796.

26. Josien, R., et al. 1998. A critical role for transforming growth factor-β in donor transfusion-induced allograft tolerance. *J. Clin. Invest.* **102**:1920–1926.

27. Taylor, P.A., Noelle, R.J., and Blazar, B.R. 2001. CD4(+)CD25(+) immune regulatory cells are required for induction of tolerance to alloantigen via costimulatory blockade. *J. Exp. Med.* **193**:1311–1318.

28. Hoffmann, P., Ermann, J., Edinger, M., Fathman, C.G., and Strober, S. 2002. Donor-type CD4(+)CD25(+) regulatory T cells suppress lethal acute graft-versus-host disease after allogeneic bone marrow transplantation. *J. Exp. Med.* **196**:389–399.

29. Graca, L., et al. 2002. Both CD4(+)CD25(+) and CD4(+)CD25(-) regulatory cells mediate dominant transplantation tolerance. *J. Immunol.* **168**:5558–5565.

30. Chen, Y., Kuchroo, V.K., Inobe, J., Hafler, D.A., and Weiner, H.L. 1994. Regulatory T cell clones induced by oral tolerance: suppression of autoimmune encephalomyelitis. *Science.* **265**:1237–1240.

31. Groux, H., et al. 1997. A CD4+ T-cell subset inhibits antigen-specific T-cell responses and prevents colitis. *Nature.* **389**:737–742.

32. Steinman, R.M., Hawiger, D., and Nussenzweig, M.C. 2003. Tolerogenic dendritic cells. *Annu. Rev. Immunol.* **21**:685–711.

33. Stassen, M., Schmitt, E., and Jonuleit, H. 2004. Human CD(4+)CD(25+) regulatory T cells and infectious tolerance. *Transplantation.* **77**(1 Suppl.):S23–S25.

34. Cottrez, F., and Groux, H. 2004. Specialization in tolerance: innate CD(4+)CD(25+) versus acquired TR1 and TH3 regulatory T cells. *Transplantation.* **77**(1 Suppl.):S12–S15.

35. Graca, L., Cobbold, S.P., and Waldmann, H. 2002. Identification of regulatory T cells in tolerated allografts. *J. Exp. Med.* **195**:1641–1646.

36. Sakaguchi, S. 2003. The origin of FOXP3-expressing CD4+ regulatory T cells: thymus or periphery. *J. Clin. Invest.* **112**:1310–1312. doi:10.1172/JCI200320274.

37. Chen, W., et al. 2003. Conversion of peripheral CD4+CD25- naive T cells to CD4+CD25+ regulatory T cells by TGF-beta induction of transcription factor Foxp3. *J. Exp. Med.* **198**:1875–1886.

38. Waldmann, H., and Cobbold, S. 2001. Regulating the immune response to transplants: a role for CD4+ regulatory cells? *Immunity.* **14**:399–406.

39. Jonuleit, H., Schmitt, E., Schuler, G., Knop, J., and Enk, A.H. 2000. Induction of interleukin 10-producing, nonproliferating CD4(+) T cells with regulatory properties by repetitive stimulation with allogeneic immature human dendritic cells. *J. Exp. Med.* **192**:1213–1222.

40. Wakkach, A., et al. 2003. Characterization of dendritic cells that induce tolerance and T regulatory 1 cell differentiation in vivo. *Immunity.* **18**:605–617.

41. Lavelle, E.C., et al. 2003. Cholera toxin promotes the induction of regulatory T cells specific for bystander antigens by modulating dendritic cell activation. *J. Immunol.* **171**:2384–2392.

42. Jonuleit, H., et al. 2002. Infectious tolerance: human CD25(+) regulatory T cells convey suppressor activity to conventional CD4(+) T helper cells. *J. Exp. Med.* **196**:255–260.

43. Dieckmann, D., Bruett, C.H., Ploettner, H., Lutz, M.B., and Schuler, G. 2002. Human CD4(+)CD25(+) regulatory, contact-dependent T cells induce interleukin 10-producing, contact-independent type 1-like regulatory T cells [corrected]. *J. Exp. Med.* **196**:247–253.

44. Karim, M., Kingsley, C.I., Bushell, A.R., Sawitzki, B.S., and Wood, K.J. 2004. Alloantigen-induced CD25+CD4+ regulatory T cells can develop in vivo

from CD25-CD4+ precursors in a thymus-independent process. *J. Immunol.* **172**:923–928.

45. Waldmann, H., et al. 2004. Regulatory T cells and organ transplantation. *Semin. Immunol.* **16**:119–126.

46. Kuniyasu, Y., et al. 2000. Naturally anergic and suppressive CD25(+)CD4(+) T cells as a functionally and phenotypically distinct immunoregulatory T cell subpopulation. *Int. Immunol.* **12**:1145–1155.

47. Chen, Z.K., Cobbold, S.P., Waldmann, H., and Metcalfe, S. 1996. Amplification of natural regulatory immune mechanisms for transplantation tolerance. *Transplantation.* **62**:1200–1206.

48. Onodera, K., Volk, H.D., Ritter, T., and Kupiec-Weglinski, J.W. 1998. Thymus requirement and antigen dependency in the "infectious" tolerance pathway in transplant recipients. *J. Immunol.* **160**:5765–5772.

49. Scully, R., Qin, S., Cobbold, S., and Waldmann, H. 1994. Mechanisms in CD4 antibody-mediated transplantation tolerance: kinetics of induction, antigen dependency and role of regulatory T cells. *Eur. J. Immunol.* **24**:2383–2392.

50. Yamada, A., et al. 2001. Recipient MHC class II expression is required to achieve long-term survival of murine cardiac allografts after costimulatory blockade. *J. Immunol.* **167**:5522–5526.

51. Sayegh, M.H., Khoury, S.J., Hancock, W.W., Weiner, H.L., and Carpenter, C.B. 1992. Induction of immunity and oral tolerance with polymorphic class II

major histocompatibility complex allopeptides in the rat. *Proc. Natl. Acad. Sci. U. S. A.* **89**:7762–7766.

52. Hancock, W.W., Sayegh, M.H., Kwok, C.A., Weiner, H.L., and Carpenter, C.B. 1993. Oral, but not intravenous, alloantigen prevents accelerated allograft rejection by selective intragraft Th2 cell activation. *Transplantation.* **55**:1112–1118.

53. Niimi, M., Shirasugi, N., Ikeda, Y., and Wood, K.J. 2001. Oral antigen induces allograft survival by linked suppression via the indirect pathway. *Transplant Proc.* **33**:81.

54. Honey, K., Cobbold, S.P., and Waldmann, H. 1999. CD40 ligand blockade induces CD4+ T cell tolerance and linked suppression. *J. Immunol.* **163**:4805–4810.

55. Davies, J.D., Leong, L.Y., Mellor, A., Cobbold, S.P., and Waldmann, H. 1996. T cell suppression in transplantation tolerance through linked recognition. *J. Immunol.* **156**:3602–3607.

56. Kilshaw, P.J., Brent, L., and Pinto, M. 1975. Suppressor T cells in mice made unresponsive to skin allografts. *Nature.* **255**:489–491.

57. Qin, S., et al. 1993. "Infectious" transplantation tolerance. *Science.* **259**:974–977.

58. Yin, D., and Fathman, C.G. 1995. CD4-positive suppressor cells block allotransplant rejection. *J. Immunol.* **154**:6339–6345.

59. Larsen, C.P., et al. 1996. Long-term acceptance of skin and cardiac allografts after blocking CD40 and CD28 pathways. *Nature.* **381**:434–438.

60. Graca, L., Honey, K., Adams, E., Cobbold, S.P., and Waldmann, H. 2000. Cutting edge: anti-CD154 therapeutic antibodies induce infectious transplantation tolerance. *J. Immunol.* **165**:4783–4786.

61. Hall, B.M., Jelbart, M.E., Gurley, K.E., and Dorsch, S.E. 1985. Specific unresponsiveness in rats with prolonged cardiac allograft survival after treatment with cyclosporine. Mediation of specific suppression by T helper/inducer cells. *J. Exp. Med.* **162**:1683–1694.

62. Cobbold, S., and Waldmann, H. 1998. Infectious tolerance. *Curr. Opin. Immunol.* **10**:518–524.

63. Goldrath, A.W., Bogatzki, L.Y., and Bevan, M.J. 2000. Naive T cells transiently acquire a memory-like phenotype during homeostasis-driven proliferation. *J. Exp. Med.* **192**:557–564.

64. Waldmann, H., and Cobbold, S. 2004. Exploiting tolerance processes in transplantation. *Science.* **305**:209–212.

65. Wells, A.D., et al. 1999. Requirement for T-cell apoptosis in the induction of peripheral transplantation tolerance. *Nat. Med.* **5**:1303–1307.

66. Li, Y., et al. 1999. Blocking both signal 1 and signal 2 of T-cell activation prevents apoptosis of alloreactive T cells and induction of peripheral allograft tolerance. *Nat. Med.* **5**:1298–1302.

67. Zheng, X.X., et al. 2003. Favorably tipping the balance between cytopathic and regulatory T cells to create transplantation tolerance. *Immunity.* **19**:503–514.

C-reactive protein: a critical update

Mark B. Pepys and Gideon M. Hirschfield

Centre for Amyloidosis and Acute Phase Proteins, Department of Medicine, Royal Free and University College Medical School, London, United Kingdom.

Introduction

In the mid 1990s, immunoassays for C-reactive protein (CRP), with greater sensitivity than those previously in routine use, revealed that increased CRP values, even within the range previously considered normal, strongly predict future coronary events. These findings triggered widespread interest, especially, remarkably, in the US, where the clinical use of CRP measurement had been largely ignored for about 30 years. CRP production is part of the nonspecific acute-phase response to most forms of inflammation, infection, and tissue damage and was therefore considered not to provide clinically useful information. Indeed, CRP values can never be diagnostic on their own and can only be interpreted at the bedside, in full knowledge of all other clinical and pathological results. However, they can then contribute powerfully to management, just as universal recording of the patient's temperature, an equally nonspecific parameter, is of great clinical utility.

The present torrent of studies of CRP in cardiovascular disease and associated conditions is facilitated by the ready commercial availability of automated CRP assays and of CRP itself as a research reagent. However, unlike the earlier rejection in the US of CRP as an empirical test because of its perceived lack of specificity, the current enthusiasm over CRP in cardiovascular disease is widely characterized by failure to recognize appropriately the nonspecific nature of the acute-phase response, and by lack of critical biological judgment. Quality control of the source, purity, and structural and functional integrity of the CRP, and the relevance of experimental design before ascribing pathophysiological functions, are also often ignored.

This article provides information about CRP as a protein and an acute-phase reactant, and a knowledge based framework for interpretation and analysis of clinical observations of CRP in relation to cardiovascular and other diseases. We also review the properties of CRP, its possible role in pathogenesis of disease, and our own observations that identify it as a possible therapeutic target.

The acute-phase response

CRP, named for its capacity to precipitate the somatic C-polysaccharide of *Streptococcus pneumoniae*, was the first acute-phase protein to be described and is an exquisitely sensitive systemic marker of inflammation and tissue damage (1). The acute-phase response comprises the nonspecific physiological and biochemical responses of endothermic animals to most forms of tissue damage, infection, inflammation, and malignant neoplasia. In particular, the synthesis of a number of proteins is rapidly upregulated, principally in hepatocytes, under the control of cytokines originating at the site of pathology. Other acute-phase proteins include proteinase inhibitors and coagulation, complement, and transport proteins, but the only molecule that displays sensitivity, response speed, and dynamic range comparable to those of CRP is serum amyloid A protein (SAA) (Table 1) (1).

Circulating CRP concentration

In healthy young adult volunteer blood donors, the median concentration of CRP is 0.8 mg/l, the 90th centile is 3.0 mg/l, and the 99th centile is 10 mg/l (2), but, following an acute-phase stimulus, values may increase from less than 50 µg/l to more than 500 mg/l, that is, 10,000-fold. Plasma CRP is produced only by hepatocytes, predominantly under transcriptional control by the cytokine IL-6, although other sites of local CRP synthesis and possibly secretion have been suggested. De-novo hepatic synthesis starts very rapidly after a single stimulus, serum concentrations rising above 5 mg/l by about 6 hours and peaking around 48 hours. The plasma half-life of CRP is about 19 hours and is constant under all conditions of health and disease, so that the sole determinant of circulating CRP concentration is the synthesis rate (3), which thus directly reflects the intensity of the pathological process(es) stimulating CRP production. When the stimulus for increased production completely ceases, the circulating CRP concentration falls rapidly, at almost the rate of plasma CRP clearance. In unselected general populations of ostensibly healthy subjects, the median CRP value is slightly higher than among blood donors and tends to increase with age, presumably reflecting the increasing incidence of subclinical pathologies (4). However, surprisingly in view of the sensitivity, speed, and range of the CRP response, subjects in the general population tend to have stable CRP concentrations characteristic for each individual, apart from occasional spikes presumably related to minor or subclinical infections, inflammation, or trauma. There is no significant seasonal variation in base-line CRP concentration, and, remarkably, the self correlation coefficient of measurements repeated years apart is about 0.5, which is comparable to that of cholesterol. Twin studies show a highly significant hereditable component in base-line CRP values that is independent of age and BMI. Associations between CRP production and genetic polymorphisms in IL-1 and IL-6 have been suggested, and a polymorphic GT repeat in the intron of the *CRP* gene is reportedly associated with differences in base-line CRP concentrations in normal individuals and in patients with systemic lupus erythematosus (5), and also with susceptibility to invasive pneumococcal disease. If such polymorphisms, particularly in the *CRP* gene itself, can be shown to reliably correlate with base-line CRP concentrations, and/or CRP production in the acute-phase response, and also with clinical outcome, the case for a pathogenetic and/or host-defense role of CRP in inflammatory disease will be strengthened.

In most, though not all, diseases (Table 2), the circulating value of CRP reflects ongoing inflammation and/or tissue damage much more accurately than do other laboratory parameters of the acute-phase response, such as plasma viscosity and the erythrocyte sedimentation rate. Importantly, acute-phase CRP values show no diurnal variation and are unaffected by eating. Liver failure impairs

Nonstandard abbreviations used: C-reactive protein (CRP); serum amyloid A protein (SAA); serum amyloid P component (SAP).

Conflict of interest: The authors have declared that no conflict of interest exists.

Citation for this article: *J. Clin. Invest.* **111**:1805–1812 (2003). doi:10.1172/JCI200318921.

Table 1

Changes in concentrations of plasma proteins in the acute-phase response

	Increased	Decreased
Proteinase inhibitors	α_1-Antitrypsin	Inter-α-antitrypsin
	α_1-Antichymotrypsin	
Coagulation proteins	Fibrinogen	
	Prothrombin	
	Factor VIII	
	Plasminogen	
Complement proteins	C1s	Properdin
	C2	
	B	
	C3	
	C4	
	C5	
	C1 inhibitor	
Transport proteins	Haptoglobin	
	Hemopexin	
	Ceruloplasmin	
Miscellaneous	CRP	Albumin
	SAA	Transthyretin
	Fibronectin	HDL
	α_1-acid glycoprotein	LDL
	Gc globulin	

CRP production, but no other intercurrent pathologies and very few drugs reduce CRP values unless they also affect the underlying pathology providing the acute-phase stimulus. The CRP concentration is thus a very useful nonspecific biochemical marker of inflammation, measurement of which contributes importantly to (a) screening for organic disease, (b) monitoring of the response to treatment of inflammation and infection, and (c) detection of intercurrent infection in immunocompromised individuals, and in the few specific diseases characterized by modest or absent acute-phase responses (Table 3) (1). It is not known why systemic lupus erythematosus and the other conditions listed with it in Table 2 fail to elicit major CRP production despite evident inflammation and tissue damage, nor why the CRP responses to intercurrent infection are apparently intact in patients with such conditions.

Structure and phylogeny of CRP

CRP belongs to the pentraxin family of calcium-dependent ligand-binding plasma proteins, the other member of which in humans is serum amyloid P component (SAP). The human CRP molecule (M_r-115,135) is composed of five identical nonglycosylated polypeptide subunits (Mr 23,027), each containing 206 amino acid residues. The protomers are noncovalently associated in an annular configuration with cyclic pentameric symmetry (6) (Figure 1). Each protomer has the characteristic "lectin fold," composed of a two-layered β sheet with flattened jellyroll topology. The ligand-binding site, composed of loops with two calcium ions bound 4 Å apart by protein side-chains, is located on the concave face. The other face carries a single α helix (Figure 1). The pentraxin family, named for its electron micrographic appearance from the Greek *penta* (five) *ragos* (berries), is highly conserved in evolution, with homologous proteins throughout the vertebrates and even in the phylogenetically distant arachnid, *Limulus polyphemus*, the horseshoe crab. SAP, named for its universal presence in amyloid deposits, is a constitu-

tive, non–acute-phase plasma glycoprotein in humans and all other species studied, except the mouse, in which it is the major acute-phase protein. In contrast, mouse CRP is a trace protein whose concentration increases only modestly, to a maximum of about 2 mg/l, during the acute-phase response. No mouse CRP knockout, to our knowledge, has yet been made, and in-vivo work on CRP function has largely been confined to passive administration of exogenous, heterologous CRP or to mice transgenic for rabbit or human CRP. These artifactual heterologous systems may not provide physiologically relevant information. Despite the evolutionary conservation of sequence, subunit organization, and protein fold, there are considerable variations between CRPs of different species with respect to fine ligand-binding specificity, presence and nature of glycosylation, protomer assembly, capacity to precipitate and aggregate ligands, baseline circulating concentrations, behavior as acute-phase proteins, and capacity to activate autologous complement (7–9). Indeed, only human CRP has been rigorously shown to activate complement in isologous serum. These differences command extreme caution in extrapolating from animal models to humans.

Ligand binding and biological role of CRP

Human CRP binds with highest affinity to phosphocholine residues, but it also binds to a variety of other autologous and extrinsic ligands, and it aggregates or precipitates the cellular, particulate, or molecular structures bearing these ligands. Autologous ligands include native and modified plasma lipoproteins (10), damaged cell membranes (11), a number of different phospholipids and related compounds, small nuclear ribonucleoprotein particles (12), and apoptotic cells (13). Extrinsic ligands include many glycan, phospholipid, and other constituents of microorganisms, such as capsular and somatic components of bacteria, fungi, and parasites, as well as plant products. When aggregated or bound to macromolecular ligands, human CRP is recognized by C1q and potently activates the classical complement pathway, engaging C3, the main adhesion molecule of the complement system, and the terminal membrane attack complex, C5–C9 (14, 15). Bound CRP may also provide secondary binding sites for factor H and thereby regulate alternative-pathway amplification and C5 convertases.

The secondary effects of CRP that follow ligand binding resemble some of the key properties of antibodies, suggesting that under various circumstances CRP may contribute to host defense against infection, function as a proinflammatory mediator, and participate in physiological and pathophysiological handling of autologous constituents. Evidence of CRP functioning in these various roles is available from experimental animal models, but there is no rigorous information from physiological isologous systems. The absence of any known deficiency or protein polymorphism of human CRP, and the phylogenetic conservation of CRP structure and its ligand-binding specificity for phosphocholine and related substances, suggest that this protein must have had survival value. Microbial infection is a major driving force of change during evolution, and CRP has many features compatible with a role in innate immunity. In addition, the impaired CRP response in active systemic lupus (1) and the marked spontaneous antinuclear autoimmunity of SAP knockout mice (16) are compatible with the possibility that pentraxins function to prevent autoimmunity.

Phosphocholine is a component of many prokaryotes and is almost universally present in eukaryotes (17), and a substantial proportion of germline-encoded, highly conserved natural antibodies resemble CRP in specifically recognizing phosphocholine.

Table 2

CRP responses in disease

Major CRP acute-phase response

Infections	Bacterial
	Systemic/Severe fungal, mycobacterial, viral
Allergic complications of infection	Rheumatic fever
	Erythema nodosum
Inflammatory disease	Rheumatoid arthritis
	Juvenile chronic arthritis
	Ankylosing spondylitis
	Psoriatic arthritis
	Systemic vasculitis
	Polymyalgia rheumatica
	Reiter disease
	Crohn disease
	Familial Mediterranean fever
Necrosis	Myocardial infarction
	Tumor embolization
	Acute pancreatitis
Trauma	Surgery
	Burns
	Fractures
Malignancy	Lymphoma
	Carcinoma
	Sarcoma

Modest or absent CRP acute-phase response

Systemic lupus erythematosus
Scleroderma
Dermatomyositis
Ulcerative colitis
Leukemia
Graft-versus-host disease

The capacity to bind these residues may thus be important for both host defense and handling of autologous constituents (1) including necrotic (18) and apoptotic cells (13, 19). Activation of complement by human CRP may then opsonize and enhance phagocytosis of these various ligands but could also mediate pro-inflammatory pathophysiological effects (1). Intriguingly, the spectrum of autologous ligands recognized by CRP overlaps that of anti-phospholipid autoantibodies that are associated with premature cardiovascular disease in autoimmune syndromes.

Some functions that have been claimed for CRP seem inherently unlikely. For example, it is improbable that a plasma protein with a dynamic range of 10,000-fold within hours would function like a cytokine or be a fine modulator of sophisticated cellular or physiological systems. Another implausible speculation concerns disso-ciated denatured CRP subunits, so-called modified or neo-CRP, for which various biological effects have been reported in-vitro. Native CRP is actually very stable, and release of separate protomers requires exposure of the protein to harsh denaturing conditions. There is no compelling evidence for the persistence of denatured CRP in-vivo, and rapid complete catabolism of such material would be expected.

CRP and cardiovascular disease

Earlier work suggested a prognostic association between increased CRP production and outcome after acute myocardial infarction (20) and in acute coronary syndromes (21). However, our original

study with high sensitivity measurements of CRP in patients with severe unstable angina (22), and the European Concerted Action on Thrombosis and Disabilities Angina Pectoris Study of outpatients with both stable and unstable angina (23, 24), first drew attention to the predictive significance of such measurements for future coronary events. Analysis of CRP values in stored sera from large epidemiological studies rapidly followed, and the subsequent availability of routine high-sensitivity assays for CRP has enabled a flood of studies demonstrating a predictive relationship between increased CRP production and future atherothrombotic events, including coronary events, stroke, and progression of peripheral arterial disease (25–30). Meta-analysis of all published studies up to the year 2000, comprising a total of 1,953 coronary events, showed a relative risk of 2.0 for a future coronary event in subjects with a single initial base-line CRP value in the upper third compared with those in the lower third of the distribution in the general population (29) (see "CRP and atherothrombotic events"). There have been further studies comprising 2,648 coronary events up to the year 2002, and the meta-analysis of these, together with the Reykjavik Icelandic Heart Study of about 19,000 individuals including 2,459 coronary-event patients who underwent follow-up for almost 20 years, is currently in preparation and will provide the most robust estimate to date of the relative risk predicted by CRP.

The recent emphasis in cardiovascular medicine on "high-sensitivity" or "highly sensitive" CRP, abbreviated as so-called hs-CRP, seems to have created a false impression in some quarters that this is somehow a different analyte from "conventional" CRP. This is incorrect. The "high sensitivity" refers simply to the lower detection limit of the assay procedures being used. The actual CRP analyte, the plasma protein that is being measured, is the same regardless of the assay range. Very sensitive CRP assays have been reported from research laboratories for 30 years (1). The new development is the introduction of commercial and automated routine CRP immunoassay systems with greater sensitivity than before.

Circulating CRP values correlate closely with other markers of inflammation, some of which show similar, albeit generally less significant, predictive associations with coronary events (31, 32). The attention focused on CRP reflects in part the fact that it is an exceptionally stable analyte in serum or plasma and that immunoassays for it are robust, well standardized, reproducible, and readily available. Furthermore, the intrinsic biological properties of CRP as an acute-phase reactant are, as explained above, especially favorable for its use as a sensitive quantitative systemic readout of the acute-phase response. In contrast, none of the other systemic markers of inflammation, whether upstream cytokine mediators, other sensitive acute-phase proteins such as SAA, negative acute-phase proteins such as albumin, or cruder multifactorial measures such as erythrocyte sedimentation rate or polymorph count, has such robust and desirable characteristics. The inherent properties of CRP and its behavior may sufficiently explain why it provides closer associations and better predictions than other markers of inflammation. However, CRP may also have specific associations with cardiovascular disease, as discussed below.

Atherosclerosis and inflammation

The mechanisms responsible for the low-grade upregulation of CRP production that predicts coronary events in general populations (27–29) are unknown (see "CRP and atherothrombotic events"). The causes of the often more substantial CRP values associated with poor prognosis in severe unstable angina (22) or

Table 3

Routine clinical uses of CRP measurement

Screening test for organic disease

Assessment of disease activity in inflammatory conditions

Juvenile chronic (rheumatoid) arthritis
Rheumatoid arthritis
Ankylosing spondylitis
Reiter disease
Psoriatic arthropathy
Vasculitides   Behçet syndrome
            Wegener granulomatosis
            Polyarteritis nodosa
            Polymyalgia rheumatica
Crohn disease
Rheumatic fever
Familial fevers including familial Mediterranean fever
Acute pancreatitis

Diagnosis and management of infection

Bacterial endocarditis
Neonatal septicemia and meningitis
Intercurrent infection in systemic lupus erythematosus
Intercurrent infection in leukemia and its treatment
Postoperative complications including infection
and thromboembolism

Differential diagnosis/classification of inflammatory disease

Systemic lupus erythematosus vs. rheumatoid arthritis
Crohn disease vs. ulcerative colitis

after angioplasty are also obscure. Atherosclerosis, and the evolution of plaque instability underlying atherothrombotic events, are inflammatory processes. It has thus been widely assumed, without direct supporting evidence, that the relevant acute-phase stimuli arise from inflammation within atheromatous lesions themselves and reflect their extent and/or severity. This is certainly possible, although there are conflicting observations regarding the association between CRP values and various indirect measures of atheroma burden. However, chronic systemic, nonvascular inflammation is known to be proatherogenic in general, and acute systemic inflammatory episodes are strongly

associated with atherothrombotic events. The increased production of CRP that predicts atherothrombotic events may therefore reflect inflammation elsewhere in the body, although there is no correlation with serological evidence of the various chronic microbial infections, such as *Chlamydia-pneumoniae* and *Helicobacter-pylori*, that have been putatively linked with coronary heart disease (29). Another possibility is that individuals vary in their sensitivity to the general background of intercurrent low-grade acute-phase stimuli to which everybody is exposed, and that those who are higher "CRP responders," through genetic and/or acquired mechanisms, are also more susceptible to progression and complications of atherosclerosis, regardless of whether there is a causal relationship.

There is a strong positive association between base-line CRP concentration and BMI (32), and weight loss lowers the CRP value. Raised base-line CRP values are also associated with many features of the insulin resistance or metabolic syndrome (33, 34), up to and including frank diabetes mellitus (35). This may reflect, in part, the fact that adipocytes are the source of a substantial portion of base-line IL-6 production (36) and perhaps also synthesize and secrete some of the base-line CRP itself. More generally, these associations raise the possibility that aspects of the inflammatory-marker profile associated with increased atherothrombotic risk in the population at large may not be triggered by inflammation or tissue damage in the classical sense. Rather, they may reflect a particular metabolic state that happens also to be proatherogenic and/or to predispose to atherothrombotic events. Indeed, CRP production predicts the development of type 2 diabetes independently of traditional risk factors (37). In insulin-resistant obese individuals, elevated CRP values fall in parallel with improvements in insulin resistance that are associated with weight loss, but the association between CRP and insulin resistance is independent of body mass (38). Other potentially important physiological rather than pathophysiological influences are suggested by the finding that oral contraceptive use (39) and systemic, but not transdermal, postmenopausal hormone replacement therapy (40, 41) are also associated with significantly raised base-line CRP concentrations without any sign of tissue-damaging inflammation. Similarly, physical exercise (42) and moderate alcohol consumption (43) are both associated with lowering of base-line CRP concentration. In contrast, the positive association

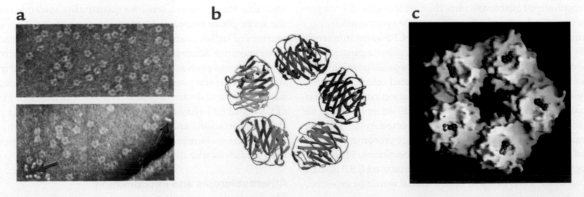

Figure 1

Molecular structure and morphology of human CRP. (a) Negatively stained electron micrograph showing the typical pentameric disc-like structure face-on and side-on (arrows). (b) Ribbon diagram of the crystal structure, showing the lectin fold and the two calcium atoms (spheres) in the ligand-binding site of each protomer (6). (c) Space-filling model of the CRP molecule, showing a single phosphocholine molecule located in the ligand-binding site of each protomer (6).

CRP and atherothrombotic events
Raised base-line CRP values predict coronary events, stroke, and progression of peripheral disease
Relative risk of coronary event is 2.0 for single base-line CRP concentration >2.4 mg/l versus <1 mg/l
What is driving this increased CRP production?
- Inflammation in atherosclerotic plaques?
- Total atherosclerotic burden?
- Inflammation elsewhere in the body?
- Metabolic status rather than inflammation?
- Individual higher CRP responsiveness to prevalent endogenous or environmental acute-phase stimuli?
CRP is not uniquely associated with cardiovascular disease; other systemic markers of inflammation show similar associations, though less marked ones than that of CRP

of CRP values with other classical cardiovascular disease risk factors, such as periodontal disease and smoking (32), seems more clearly related to local nonarterial inflammation.

HMG CoA-reductase inhibitors, the drugs known as statins, reproducibly reduce CRP values, independently of their effects on lipid profiles (44). It is not known whether this reflects direct effects on hepatocytes, anti-inflammatory effects within atherosclerotic plaques, and/or anti-inflammatory effects elsewhere. However, recent studies suggest that statins reduce the risk of future cardiovascular events to the same extent in patients with raised LDL cholesterol values and in those with normal LDL but with base-line CRP concentrations above the median (45). If these observations are confirmed, measurement of CRP may become an indication for prophylactic antiatherosclerotic therapy in otherwise apparently low-risk individuals and populations.

However, it is critically important to recognize that the CRP response is nonspecific and is triggered by many disorders unrelated to cardiovascular disease (Table 2). In using CRP for assessment of cardiovascular risk, it is therefore essential to clearly establish true base-line CRP values that are not distorted by either trivial or serious intercurrent pathologies. If the initial CRP result is in the low risk range, less than 1 mg/l, a single measurement is sufficient, but if it is in the higher-risk range, greater than about 2.5 mg/l, two or more serial samples taken at intervals of 1 week or more should be retested until a stable base-line value is seen. If the CRP value persistently remains above 10 mg/l, indicating the presence of a significant acute-phase response, a full history and physical examination of the patient is indicated, ideally together with relevant investigations, to determine the cause and alleviate it if possible. Interestingly, chronic inflammatory conditions, such as rheumatoid arthritis and hemodialysis for end-stage renal failure, that are characterized by persistently elevated CRP concentrations in some individuals, are associated with premature cardiovascular disease.

CRP and pathogenesis of atherosclerosis?

Binding of CRP to lipids, especially lecithin (phosphatidyl choline), and to plasma lipoproteins has been known for over 60 years, but the first suggestion of a possible relationship to atherosclerosis came when we demonstrated that aggregated, but not native, nonaggregated, CRP selectively bound only LDL and some VLDL from whole serum (10, 46). However, native CRP does bind to oxidized LDL (19) and to partly degraded LDL, as found in atheromatous plaques, and then activates complement (47). Furthermore, CRP is present in most such plaques examined ex-vivo (48, 49). This CRP could promote complement activation and thus inflammation in the plaques, and there is experimental evidence supporting a possible role of complement in atherogenesis (see "Possible specific associations of CRP with cardiovascular disease"). Addition of CRP to LDL in cell culture systems has been reported to stimulate formation of foam cells, which are a typical feature of atherosclerotic plaques (50). It is not known whether this reflects opsonization of the LDL particles by CRP or an effect of CRP on the phagocytic cells themselves. It has been claimed that CRP is recognized by a subset of cellular Fcγ receptors and could thereby directly opsonize its ligands and/or engage multiple processes of inflammation. However, robustly controlled studies, using recombinant and highly purified human CRP and avoiding use of whole IgG anti-CRP antibodies, do not confirm such interactions with human cells (51, 52).

CRP has also been reported to stimulate tissue factor production by peripheral blood monocytes in-vitro and could thereby have important procoagulant effects (53). Other studies show a variety of effects in cell cultures exposed to CRP, including, for example, increased expression of adhesion molecules, and modulation of NO synthesis. However, these experiments have generally used commercially sourced CRP of incompletely defined provenance and purity, and there have been few robust specificity controls. The findings must be treated with caution until the purity

Possible specific associations of CRP with cardiovascular disease
CRP binds selectively to LDL, especially "damaged" LDL
CRP is deposited in most atherosclerotic plaques
Aggregated and/or ligand-complexed CRP activates complement and can be proinflammatory
CRP is co-deposited with activated complement in all acute myocardial infarction lesions
Human CRP and complement increase final myocardial infarction size in experimental models
CRP may therefore be a therapeutic target

and structural and functional integrity of the CRP have been rigorously established and the specificity of the observed effects confirmed, for example by comparison with appropriate control proteins and by the use of specific CRP absorbents, ligands, antibodies, and inhibitors of binding.

CRP and myocardial infarction

Tissue necrosis is a potent acute-phase stimulus, and, following myocardial infarction, there is a major CRP response, the magnitude of which reflects the extent of myocardial necrosis (20). Furthermore, the peak CRP values at around 48-hours after the onset powerfully predict outcome after myocardial infarction (20, 54–56). Importantly, CRP is co-deposited with activated complement within all acute myocardial infarcts (57, 58), and compelling experimental evidence now suggests that the CRP response not only reflects tissue damage in this context but may also contribute significantly to the severity of ischemic myocardial injury (59).

We specifically investigated in-vivo the complement-dependent proinflammatory role of human CRP in rats, taking advantage of the fact that rat CRP does not activate rat complement while human CRP potently activates both rat and human complement (8). When rats that had undergone coronary artery ligation received daily injections of pure human CRP, they became sicker than similarly operated rats that received buffer alone or the closely related pentraxin human SAP, which does not activate complement (59). Injection of human CRP into unoperated rats had no adverse effects. Some of the coronary artery–ligated rats treated with human CRP died, and those that survived to day 5, when all animals were killed, had infarcts 40% larger than those of buffer- or SAP-treated controls (59). This dramatic enhancement of infarct size by human CRP was completely abrogated by in-vivo complement depletion of the rats using cobra venom factor, and hence it was absolutely complement dependent (59).

CRP: a target for therapy in human disease?

We have long speculated that CRP may have significant proinflammatory effects, and that, by binding to ligands exposed on cells or other autologous structures as a result of infection, inflammation, ischemia, and other pathologies, and triggering complement activation, it may exacerbate tissue damage, leading to more severe disease (1). The rat myocardial infarction model provided the first direct evidence of these processes in-vivo (59), but they are not necessarily confined to cardiovascular disease. The excellent correlation of circulating CRP concentrations with the severity, extent, and progression of many different pathologies, and the prognostic significance of these associations, are consistent with CRP not just being a marker of disease but also contributing to pathogenesis. A definitive way to test this concept will be the use of novel drugs that specifically block CRP binding and its proinflammatory effects in-vivo (60). If these compounds are effective, they may find very broad applicability. Such drugs would be a powerful tool for determining whether increased CRP production merely reflects atherosclerosis or does indeed participate in its pathogenesis and complications, and they could also have cardioprotective effects in acute myocardial infarction. Knowledge of the structure and function of CRP — including its three-dimensional structure alone and complexed with ligands (6) — coupled with experience in developing an inhibitor of the related protein SAP (61) establishes an excellent platform for drug design.

Acknowledgments

G.M. Hirschfield is a Medical Research Council Clinical Training Fellow. The work of the Centre for Amyloidosis and Acute Phase Proteins is supported by grants from the Medical Research Council (United Kingdom), the Wellcome Trust, and the Wolfson Foundation, and by National Health Service Research and Development Funds. We thank Beth Jones for expert assistance in preparing the manuscript.

Address correspondence to: Mark B. Pepys, Department of Medicine, Centre for Amyloidosis and Acute Phase Proteins, Royal Free and University College Medical School, Rowland Hill Street, London NW3 2PF, United Kingdom. Phone: 44-20-7433-2801; Fax: 44-20-7433-2803; E-mail: m.pepys@rfc.ucl.ac.uk.

1. Pepys, M.B., and Baltz, M.L. 1983. Acute phase proteins with special reference to C-reactive protein and related proteins (pentaxins) and serum amyloid A protein. *Adv. Immunol.* **34**:141–212.
2. Shine, B., de Beer, F.C., and Pepys, M.B. 1981. Solid phase radioimmunoassays for C-reactive protein. *Clin. Chim. Acta.* **117**:13–23.
3. Vigushin, D.M., Pepys, M.B., and Hawkins, P.N. 1993. Metabolic and scintigraphic studies of radioiodinated human C-reactive protein in health and disease. *J. Clin. Invest.* **91**:1351–1357.
4. Hutchinson, W.L., et al. 2000. Immunoradiometric assay of circulating C-reactive protein: age-related values in the adult general population. *Clin. Chem.* **46**:934–938.
5. Szalai, A.J., McCrory, M.A., Cooper, G.S., Wu, J., and Kimberly, R.P. 2002. Association between baseline levels of C-reactive protein (CRP) and a dinucleotide repeat polymorphism in the intron of the CRP gene. *Genes Immun.* **3**:14–19.
6. Thompson, D., Pepys, M.B., and Wood, S.P. 1999. The physiological structure of human C-reactive protein and its complex with phosphocholine. *Structure.* **7**:169–177.
7. Oliveira, E.B., Gotschlich, E.C., and Liu, T.-Y. 1980. Comparative studies on the binding properties of human and rabbit C-reactive proteins. *J. Immunol.* **124**:1396–1402.
8. de Beer, F.C., et al. 1982. Isolation and characteri-

sation of C-reactive protein and serum amyloid P component in the rat. *Immunology.* **45**:55–70.
9. Baltz, M.L., et al. 1982. Phylogenetic aspects of C-reactive protein and related proteins. *Ann. N. Y. Acad. Sci.* **389**:49–75.
10. Pepys, M.B., Rowe, I.F., and Baltz, M.L. 1985. C-reactive protein: binding to lipids and lipoproteins. *Int. Rev. Exp. Pathol.* **27**:83–111.
11. Volanakis, J.E., and Wirtz, K.W.A. 1979. Interaction of C-reactive protein with artificial phosphatidylcholine bilayers. *Nature.* **281**:155–157.
12. Du Clos, T.W. 1989. C-reactive protein reacts with the U1 small nuclear ribonucleoprotein. *J. Immunol.* **143**:2553–2559.
13. Gershov, D., Kim, S., Brot, N., and Elkon, K.B. 2000. C-reactive protein binds to apoptotic cells, protects the cells from assembly of the terminal complement components, and sustains an antiinflammatory innate immune response: implications for systemic autoimmunity. *J. Exp. Med.* **192**:1353–1363.
14. Volanakis, J.E. 1982. Complement activation by C-reactive protein complexes. *Ann. N. Y. Acad. Sci.* **389**:235–250.
15. Mold, C., Gewurz, H., and Du Clos, T.W. 1999. Regulation of complement activation by C-reactive protein. *Immunopharmacology.* **42**:23–30.
16. Bickerstaff, M.C.M., et al. 1999. Serum amyloid P component controls chromatin degradation and prevents antinuclear autoimmunity. *Nat. Med.*

5:694–697.
17. Harnett, W., and Harnett, M.M. 1999. Phosphorylcholine: friend or foe of the immune system? *Immunol. Today.* **20**:125–129.
18. Kushner, I., and Kaplan, M.H. 1961. Studies of acute phase protein. I. An immunohistochemical method for the localization of Cx-reactive protein in rabbits. Association with necrosis in local inflammatory lesions. *J. Exp. Med.* **114**:961–973.
19. Chang, M.K., Binder, C.J., Torzewski, M., and Witztum, J.L. 2002. C-reactive protein binds to both oxidized LDL and apoptotic cells through recognition of a common ligand: phosphorylcholine of oxidized phospholipids. *Proc. Natl. Acad. Sci. U. S. A.* **99**:13043–13048.
20. de Beer, F.C., et al. 1982. Measurement of serum C-reactive protein concentration in myocardial ischaemia and infarction. *Br. Heart J.* **47**:239–243.
21. Berk, B.C., Weintraub, W.S., and Alexander, R.W. 1990. Elevation of C-reactive protein in "active" coronary artery disease. *Am. J. Cardiol.* **65**:168–172.
22. Liuzzo, G., et al. 1994. The prognostic value of C-reactive protein and serum amyloid A protein in severe unstable angina. *N. Engl. J. Med.* **331**:417–424.
23. Thompson, S.G., Kienast, J., Pyke, S.D.M., Haverkate, F., and van de Loo, J.C.W. 1995. Hemostatic factors and the risk of myocardial infarction or sudden death in patients with angina pectoris. *N. Engl. J. Med.* **332**:635–641.

24. Haverkate, F., Thompson, S.G., Pyke, S.D.M., Gallimore, J.R., and Pepys, M.B. 1997. Production of C-reactive protein and risk of coronary events in stable and unstable angina. *Lancet.* **349**:462–466.

25. Kuller, L.H., Tracy, R.P., Shaten, J., and Meilahn, E.N. 1996. Relation of C-reactive protein and coronary heart-disease in the MRFIT nested case control study. *Am. J. Epidemiol.* **144**:537–547.

26. Tracy, R.P., et al. 1996. C-reactive protein and incidence of cardiovascular disease in older women: the rural health promotion project and the cardiovascular health study. *Circulation.* **93**:622.

27. Ridker, P.M., Cushman, M., Stampfer, M.J., Tracy, R.P., and Hennekens, C.H. 1997. Inflammation, aspirin, and the risk of cardiovascular disease in apparently healthy men. *N. Engl. J. Med.* **336**:973–979.

28. Koenig, W., et al. 1999. C-reactive protein, a sensitive marker of inflammation, predicts future risk of coronary heart disease in initially healthy middle-aged men: results from the MONICA (Monitoring Trends and Determinants in Cardiovascular Disease) Augsburg Cohort Study, 1984 to 1992. *Circulation.* **99**:237–242.

29. Danesh, J., et al. 2000. Low grade inflammation and coronary heart disease: prospective study and updated meta-analyses. *BMJ.* **321**:199–204.

30. Ridker, P.M., Hennekens, C.H., Buring, J.E., and Rifai, N. 2000. C-reactive protein and other markers of inflammation in the prediction of cardiovascular disease in women. *N. Engl. J. Med.* **342**:836–843.

31. Danesh, J., Collins, R., Appleby, P., and Peto, R. 1998. Association of fibrinogen, C-reactive protein, albumin, or leukocyte count with coronary heart disease. *J. Am. Coll. Cardiol.* **279**:1477–1482.

32. Danesh, J., et al. 1999. Risk factors for coronary heart disease and acute-phase proteins. A population-based study. *Eur. Heart J.* **20**:954–959.

33. Fröhlich, M., et al. 2000. Association between C-reactive protein and features of the metabolic syndrome: a population-based study. *Diabetes Care.* **23**:1835–1839.

34. Chambers, J.C., et al. 2001. C-reactive protein, insulin resistance, central obesity, and coronary heart disease risk in Indian Asians from the United Kingdom compared with European whites. *Circulation.* **104**:145–150.

35. Ford, E.S. 1999. Body mass index, diabetes, and C-reactive protein among U.S. adults. *Diabetes Care.* **22**:1971–1977.

36. Yudkin, J.S., Stehouwer, C.D.A., Emeis, J.J., and Coppack, S.W. 1999. C-reactive protein in healthy subjects: associations with obesity, insulin resistance, and endothelial dysfunction. A potential role for cytokines originating from adipose tissue? *Arterioscler. Thromb. Vasc. Biol.* **19**:972–978.

37. Freeman, D.J., et al. 2002. C-reactive protein is an independent predictor of risk for the development of diabetes in the West of Scotland Coronary Prevention Study. *Diabetes.* **51**:1596–1600.

38. McLaughlin, T., et al. 2002. Differentiation between obesity and insulin resistance in the association with C-reactive protein. *Circulation.* **106**:2908–2912.

39. Fröhlich, M., et al. 1999. Oral contraceptive use is associated with a systemic acute phase response. *Fibrinolysis and Proteolysis.* **13**:239–244.

40. Cushman, M., et al. 1999. Effect of postmenopausal hormones on inflammation-sensitive proteins: the Postmenopausal Estrogen/Progestin Interventions (PEPI) Study. *Circulation.* **100**:717–722.

41. Ridker, P.M., Hennekens, C.H., Rifai, N., Buring, J.E., and Manson, J.E. 1999. Hormone replacement therapy and increased plasma concentration of C-reactive protein. *Circulation.* **100**:713–716.

42. Ford, E.S. 2002. Does exercise reduce inflammation? Physical activity and C-reactive protein among U.S. adults. *Epidemiology.* **13**:561–568.

43. Imhof, A., et al. 2001. Effect of alcohol consumption on systemic markers of inflammation. *Lancet.* **357**:763–767.

44. Ridker, P.M., Rifai, N., Pfeffer, M.A., Sacks, F., and Braunwald, E. 1999. Long-term effects of pravastatin on plasma concentration of C-reactive protein. *Circulation.* **100**:230–235.

45. Ridker, P.M., et al. 2001. Measurement of C-reactive protein for the targeting of statin therapy in the primary prevention of acute coronary events. *N. Engl. J. Med.* **344**:1959–1965.

46. de Beer, F.C., et al. 1982. Low density and very low density lipoproteins are selectively bound by aggregated C-reactive protein. *J. Exp. Med.* **156**:230–242.

47. Bhakdi, S., Torzewski, M., Klouche, M., and Hemmes, M. 1999. Complement and atherogenesis. Binding of CRP to degraded, nonoxidized LDL enhances complement activation. *Arterioscler. Thromb. Vasc. Biol.* **19**:2348–2354.

48. Torzewski, J., et al. 1998. C-reactive protein frequently colocalizes with the terminal complement complex in the intima of early atherosclerotic lesions of human coronary arteries. *Arterioscler. Thromb. Vasc. Biol.* **18**:1386–1392.

49. Zhang, Y.X., Cliff, W.J., Schoefl, G.I., and Higgins, G. 1999. Coronary C-reactive protein distribution: its relation to development of atherosclerosis. *Atherosclerosis.* **145**:375–379.

50. Zwaka, T.P., Hombach, V., and Torzewski, J. 2001. C-reactive protein-mediated low density lipoprotein uptake by macrophages: implications for atherosclerosis. *Circulation.* **103**:1194–1197.

51. Hundt, M., Zielinska-Skowronek, M., and Schmidt, R.E. 2001. Lack of specific receptors for C-reactive protein on white blood cells. *Eur. J. Immunol.* **31**:3475–3483.

52. Saeland, E., et al. 2001. Human C-reactive protein does not bind to FcgammaRIIa on phagocytic cells. *J. Clin. Invest.* **107**:641–643.

53. Cermak, J., et al. 1993. C-reactive protein induces human peripheral blood monocytes to synthesize tissue factor. *Blood.* **82**:513–520.

54. Pietilä, K.O., Harmoinen, A.P., Jokiniitty, J., and Pasternack, A.I. 1996. Serum C-reactive protein concentration in acute myocardial infarction and its relationship to mortality during 24 months of follow-up in patients under thrombolytic treatment. *Eur. Heart J.* **17**:1345–1349.

55. Ueda, S., et al. 1996. C-reactive protein as a predictor of cardiac rupture after acute myocardial infarction. *Am. Heart J.* **131**:857–860.

56. Anzai, T., et al. 1997. C-reactive protein as a predictor of infarct expansion and cardiac rupture after a first Q-wave acute myocardial infarction. *Circulation.* **96**:778–784.

57. Kushner, I., Rakita, L., and Kaplan, M.H. 1963. Studies of acute phase protein. II. Localization of Cx-reactive protein in heart in induced myocardial infarction in rabbits. *J. Clin. Invest.* **42**:286–292.

58. Lagrand, W.K., et al. 1997. C-reactive protein colocalizes with complement in human hearts during acute myocardial infarction. *Circulation.* **95**:97–103.

59. Griselli, M., et al. 1999. C-reactive protein and complement are important mediators of tissue damage in acute myocardial infarction. *J. Exp. Med.* **190**:1733–1739.

60. Pepys, M.B. 1999. The Lumleian Lecture. C-reactive protein and amyloidosis: from proteins to drugs? In *Horizons in medicine.* Volume 10. G. Williams, editor. Royal College of Physicians of London. London, United Kingdom. 397–414.

61. Pepys, M.B., et al. 2002. Targeted pharmacological depletion of serum amyloid P component for treatment of human amyloidosis. *Nature.* **417**:254–259.

New insights into atopic dermatitis

Donald Y.M. Leung,[1] Mark Boguniewicz,[1] Michael D. Howell,[1] Ichiro Nomura,[1] and Qutayba A. Hamid[2]

[1]Division of Pediatric Allergy-Immunology, National Jewish Medical and Research Center, Department of Pediatrics, University of Colorado Health Sciences Center, Denver, Colorado, USA. [2]Meakins-Christie Laboratories, Montreal, Quebec, Canada.

Atopic dermatitis is a chronic inflammatory skin disease associated with cutaneous hyperreactivity to environmental triggers and is often the first step in the atopic march that results in asthma and allergic rhinitis. The clinical phenotype that characterizes atopic dermatitis is the product of interactions between susceptibility genes, the environment, defective skin barrier function, and immunologic responses. This review summarizes recent progress in our understanding of the pathophysiology of atopic dermatitis and the implications for new management strategies.

Historical perspective

Atopic dermatitis (AD) is a chronic inflammatory skin disease associated with cutaneous hyperreactivity to environmental triggers that are innocuous to normal nonatopic individuals (1). Although written descriptions of AD date back to the early 1800s, an objective laboratory test does not exist for AD. The diagnosis of AD is based on the following constellation of clinical findings: pruritus, facial and extensor eczema in infants and children, flexural eczema in adults, and chronicity of the dermatitis.

AD usually presents during early infancy and childhood, but it can persist into or start in adulthood (2). The lifetime prevalence of AD is 10–20% in children and 1–3% in adults. Its prevalence has increased two- to threefold during the past three decades in industrialized countries but remains much lower in countries with predominantly rural or agricultural areas. Wide variations in prevalence have been observed within countries inhabited by groups with similar genetic backgrounds, suggesting that environmental factors play a critical role in determining expression of AD.

A precise understanding of the mechanisms underlying AD is critical for development of more effective management strategies (Table 1). Various studies indicate that AD has a complex etiology, with activation of multiple immunologic and inflammatory pathways (3). The clinical phenotype that characterizes AD is the product of complex interactions among susceptibility genes, the host's environment, defects in skin barrier function, and systemic and local immunologic responses. An understanding of the relative role of these factors in the pathogenesis of AD has been made possible by a variety of approaches, including the analysis of cellular and cytokine gene expression in AD skin lesions in humans as well as gene knockout and transgenic mouse models of candidate genes in AD. The current review will summarize progress in our understanding of the pathophysiology of AD and implications for therapy.

Atopy as a systemic disease

Several observations suggest that AD is the cutaneous manifestation of a systemic disorder that also gives rise to asthma, food

allergy, and allergic rhinitis (1, 2). These conditions are all characterized by elevated serum IgE levels and peripheral eosinophilia. AD is often the initial step in the so-called "atopic march," which leads to asthma and allergic rhinitis in the majority of afflicted patients. In experimental models of AD, the induction of allergic skin inflammation by epicutaneous application of allergens has been found to augment the systemic allergic response and airway hyperreactivity characteristic of asthma (4).

At least two forms of AD have been delineated: an "extrinsic" form associated with IgE-mediated sensitization involving 70–80% of the patients, and an "intrinsic" form without IgE-mediated sensitization involving 20–30% of the patients (5). Both forms of AD have associated eosinophilia. In extrinsic AD, memory T cells expressing the skin homing receptor, cutaneous lymphocyte-associated antigen (CLA), produce increased levels of Th2 cytokines. These include IL-4 and IL-13, which are known to induce isotype switching to IgE synthesis, as well as IL-5, which plays an important role in eosinophil development and survival. These CLA⁺ T cells also produce abnormally low levels of IFN-γ, a Th1 cytokine known to inhibit Th2 cell function. Intrinsic AD is associated with less IL-4 and IL-13 production than extrinsic AD.

Immune responses in AD skin

Clinically unaffected skin in AD is not normal. It frequently manifests increased dryness and a greater irritant skin response than healthy controls. Unaffected AD skin contains a sparse perivascular T cell infiltrate not seen in normal healthy skin (see Figure 1). Analyses of biopsies from clinically unaffected skin of AD patients, as compared with normal nonatopic skin, demonstrate an increased number of Th2 cells expressing IL-4 and IL-13, but not IFN-γ, mRNA (6).

Acute eczematous skin lesions present clinically as intensely pruritic, erythematous papules associated with excoriation and serous exudation. In the dermis of acute lesions, there is a marked infiltration of CD4⁺ activated memory T cells. When compared to normal skin or uninvolved skin of AD patients, acute skin lesions have a significantly greater number of IL-4, IL-5, and IL-13 mRNA–expressing cells, but few IFN-γ or IL-12 mRNA–expressing cells. APCs (e.g., Langerhans cells [LCs], inflammatory dendritic epidermal cells [IDECs], and macrophages) in lesional and, to a lesser extent, in nonlesional skin bear IgE molecules (7).

Chronic lichenified skin lesions have undergone tissue remodeling (Figure 2) due to chronic inflammation and are characterized by thickened plaques with increased markings (lichenification) and dry, fibrotic papules. There is an increased

Nonstandard abbreviations used: atopic dermatitis (AD); cutaneous lymphocyte-associated antigen (CLA); Fcε receptor I (FcεRI); glucocorticoid receptor (GCR); house dust mite (HDM); inflammatory dendritic epidermal cell (IDEC); Langerhans cell (LC); macrophage inflammatory protein (MIP); monocyte chemotactic protein (MCP).

Conflict of interest: D.Y.M. Leung is a consultant, grant recipient, and member of the Speaker's Bureau for Novartis, Fujisawa, and Glaxo/SKB. The remaining authors have declared that no conflict of interest exists.

Citation for this article: *J. Clin. Invest.* **113**:651–657 (2004). doi:10.1172/JCI200421060.

Table 1

Important concepts in the pathobiology of AD

Cutaneous hyperreactivity to environmental triggers
Skin barrier dysfunction
Immunologic triggers: food and inhalant allergens, microbial infection, autoantigens
Systemic Th2 response; first step in atopic march
Biphasic T cell response in the skin (Th2 cells in acute AD; Th1 cells in chronic AD)
Reduced skin innate immune response
Skin inflammatory response driven by chemokine expression
Skin remodeling in chronic AD
Key effector cells: T cells, FcεRI+/IgE+ DCs, keratinocytes

number of IgE-bearing LCs and IDECs in the epidermis, and macrophages dominate the dermal mononuclear cell infiltrate. Eosinophils also contribute to the inflammatory response, and T cells remain present, although in smaller numbers than seen in acute AD. Chronic AD skin lesions have significantly fewer IL-4 and IL-13 mRNA–expressing cells, but greater numbers of IL-5, GM-CSF, IL-12, and IFN-γ mRNA–expressing cells, than in acute AD. Recent studies suggest that collagen deposition during chronic AD is due to increased gene expression of the profibrotic cytokine, IL-11 (8).

The evolution of AD skin lesions is orchestrated by the local tissue expression of proinflammatory cytokines and chemokines.

Cytokines such as TNF-α and IL-1 from resident cells (keratinocytes, mast cells, and DCs) bind to receptors on vascular endothelium, activating cellular signaling including the NF-κB pathway and inducing expression of vascular endothelial cell adhesion molecules. These events initiate the process of tethering, activation, and adhesion to the endothelium followed by extravasation of inflammatory cells. Once the inflammatory cells have infiltrated into the tissue, they respond to chemotactic gradients established by chemoattractant cytokines and chemokines, which emanate from sites of injury or infection (9). These molecules play a central role in defining the nature of the inflammatory infiltrate in AD (10). IL-16, an LC-derived chemoattractant cytokine for CD4+ T cells, is increased in acute AD skin lesions. C-C chemokine ligand 27 is highly upregulated in AD and preferentially attracts CLA+ T cells into the skin. As compared to psoriasis, the C-C chemokines, RANTES, monocyte chemotactic protein-4, and eotaxin (Figure 3) are increased in AD skin lesions and likely contribute to the chemotaxis of C-C chemokine receptor 3–expressing (CCR3-expressing) eosinophils, macrophages, and Th2 lymphocytes into AD skin. Selective recruitment of CCR4-expressing Th2 cells into AD skin may also be mediated by macrophage-derived chemokine and thymus and activation–regulated cytokine, which are increased in AD. Persistent skin inflammation in chronic lesions may be due to elevated IL-5 and GM-CSF expression in the skin leading to enhanced survival of eosinophils and monocyte-macrophages as well as LCs. In addition, extracellular matrix molecules deposited into chronic skin lesions have been found to enhance the survival of memory T cells (11).

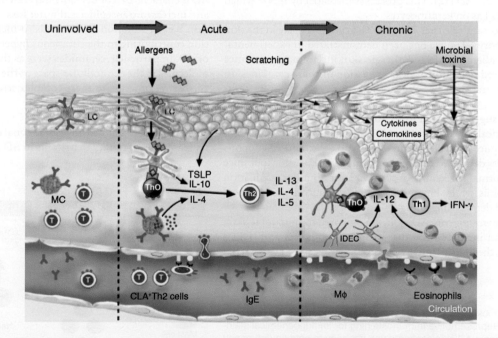

Figure 1

Immunologic pathways in AD. Th2 cells circulating in the peripheral blood of AD patients result in elevated serum IgE and eosinophils. These T cells express the skin homing receptor, CLA, and recirculate through unaffected AD skin where they can engage allergen-triggered IgE+ LCs and mast cells (MCs) that contribute to Th2 cell development. Skin injury by environmental allergens, scratching, or microbial toxins activates keratinocytes to release proinflammatory cytokines and chemokines that induce the expression of adhesion molecules on vascular endothelium and facilitate the extravasation of inflammatory cells into the skin. Keratinocyte-derived thymic stromal lymphopoietin (TSLP) and DC-derived IL-10 also enhance Th2 cell differentiation. AD inflammation is associated with increased Th2 cells in acute skin lesions, but chronic AD results in the infiltration of inflammatory IDECs, macrophages (Mφ), and eosinophils. IL-12 production by these various cell types results in the switch to a Th1-type cytokine milieu associated with increased IFN-γ expression. Figure modified with permission from *The Journal of Allergy and Clinical Immunology* (35).

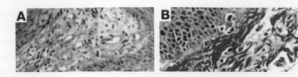

Figure 2
Skin remodeling in AD. Van Gieson staining (original magnification, ×400) from acute AD (A) and chronic AD (B) showing extensive fibrosis in chronic, as compared to acute, AD lesions. Figures reproduced with permission from *The Journal of Allergy and Clinical Immunology* (8).

Key effector cells in AD skin

T cells. The key role of immune effector T cells in AD is supported by the observation that individuals with primary T cell immunodeficiency disorders frequently have elevated serum IgE levels and eczematous skin lesions which clear following successful bone marrow transplantation. Furthermore, in animal models of AD, the eczematous rash does not occur in the absence of T cells. Use of the atopy patch test technique as a model for the induction of eczema in patients with AD has shown that house dust mite (HDM) allergen–induced skin lesions display two phases: an initial phase with predominantly IL-4 producing Th2 cells and a subsequent phase after 24 to 48 hours characterized by IFN-γ producing Th1 cells (1). This switch is thought to be initiated by the local production of IL-12 from infiltrating eosinophils and/or IDECs (3). Activated T cells expressing Fas ligand have also been shown to induce keratinocyte apoptosis contributing to the spongiosis found in acute AD (12). This process is mediated by IFN-γ, which upregulates Fas on keratinocytes.

The important role that Th1 and Th2 cytokines play in the skin inflammatory response has been demonstrated in experimental models of allergen-induced allergic skin inflammation in mice with targeted deletions or overexpression of these cytokines. In this regard, transgenic mice genetically engineered to overexpress IL-4 in their skin develop inflammatory pruritic skin lesions similar to AD, suggesting that local skin expression of Th2 cytokines plays a critical role in AD (13). Allergen-sensitized skin from IL-5 knockout mice has been found to have no detectable eosinophils and exhibits decreased thickening; skin from IL-4 knockout mice displays normal thickening of the skin layers, but has a reduction in eosinophils, while skin of IFN-γ knockout mice is characterized by reduced dermal thickening (14).

Antigen-presenting cells. AD skin, as compared to nonatopic skin, contains an increased number of IgE-bearing LCs and IDECs expressing the high affinity IgE receptor Fcε receptor I (FcεRI) (7).

Figure 3
Role of chemokines in AD. (A) Chemokines activated in AD versus psoriasis. (B) Representative immunostaining for immunochemical staining of a chronic AD skin section using antibody to eotaxin (left panel; original magnification, ×400), in situ hybridization of chronic AD skin for MCP-4 using a complementary RNA radio-labeled probe. The image of mRNA-cRNA complex was developed with autoradiography. Dark field illustration shows positive signal in the epidermis of inflammatory cells (middle panel; original magnification, ×400). Immunostaining of psoriasis skin section with antibody to eotaxin shows weak staining (right panel; original magnification, ×400). TARC, thymus and activation–regulated cytokine; PARC, pulmonary and activation–regulated chemokine; GROβ, growth-related β.

The increased expression of FcεRI on DCs in atopic skin is due to enhanced expression of the FcεRIγ chain and is preserved by increased IgE levels (15). LCs and IDECs play an important role in allergen presentation to Th2 and Th1 cells, respectively (3). FcεRI-bound IgE on LCs facilitates the capture and internalization of allergens prior to their processing and antigen presentation to T cells in atopic skin. FcεRI+/IgE+ LCs may also migrate to the lymph nodes and stimulate naive T cells to expand the pool of Th2 cells. The clinical importance of these cells is supported by the observation that, using an experimental model of aeroallergen-induced patch test reactions on atopic skin, the presence of FcεRI+/IgE+ LC is required to provoke eczematous skin lesions. IL-10 expression by APCs and T cells has recently been demonstrated to play a key role in Th2 cell development in an animal model of AD (16).

Keratinocytes. Epidermal keratinocytes from AD patients produce a unique profile of chemokines and cytokines following mechanical stimulation, e.g., scratching, or exposure to proinflammatory cytokines, including abnormally high levels of RANTES following stimulation with TNF-α and IFN-γ. They are also an important source of thymic stromal lymphopoietin, which activates DCs to prime naive Th cells to produce IL-4 and IL-13. These observations may explain the link between scratching and the triggering of Th2-mediated skin inflammation in AD. As discussed below, keratinocytes in AD are also deficient in their ability to synthesize antimicrobial peptides needed for innate immune responses against microbes (17, 18).

Skin barrier dysfunction

AD is characterized by dry skin and even involves nonlesional skin and increased transepidermal water loss. This impairment of the skin barrier function in AD leads to increased antigen absorption contributing to the cutaneous hyperreactivity characteristic of AD. In particular, ceramides serve as the major water-retaining molecules in the extracellular space of the cornified envelope, and the barrier function of these complex structures is provided by a matrix of structural proteins that are bound to ceramides (19). A reduced content of ceramides has been reported in the cornified envelope of both lesional and nonlesional skin in AD patients. The increased susceptibility to irritants in AD may therefore represent

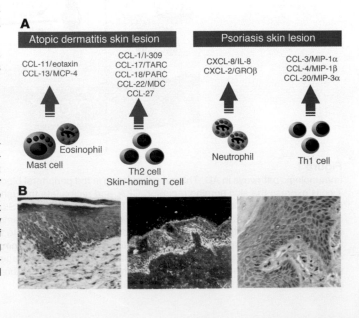

a primary defect of epidermal differentiation compounded by the presence of inflammation-induced skin damage.

Genetics

AD is a genetically complex disease that has a high familial occurrence. Twin studies of AD have shown concordance rates of 0.72–0.86 in monozygotic, and 0.21–0.23 in dizygotic, twin pairs, indicating that genetic factors play an important role in the development of this disease (20). Relevant to AD as part of a systemic atopic disorder, candidate genes involving IgE and Th2 cytokines have been identified. There has been particular focus on chromosome 5q31-33, as it contains a clustered family of Th2 cytokine genes, i.e., IL-3, IL-4, IL-5, IL-13, and GM-CSF (21). Case-control comparisons have suggested a genotypic association between the T allele of the -590C/T polymorphism of the IL-4 gene promoter region and AD. The fact that this allele is associated with increased IL-4 gene promoter activity suggests that it may increase allergic responses in AD. Similarly, IL-13 coding region variants, a gain-of-function polymorphism in the α subunit of the IL-4 receptor (located on chromosome 16q12), and a functional mutation in the promoter region of RANTES (located on chromosome 17q11) have been reported in AD. There have been controversial linkage findings between AD and markers on chromosome 11q13, including the gene encoding for the β-chain of the high affinity receptor for IgE (FcεRIβ). Most of these studies included patients with elevated IgE levels, i.e., patients with extrinsic AD. Thus, studies of so-called "pure" or intrinsic AD are needed to identify AD genes unrelated to IgE biology. Importantly, linkages to AD have generally not corresponded to loci associated with asthma, suggesting the presence of separate or additional genes in AD that may not act through atopic mechanisms.

Genome screens have been performed to identify susceptibility loci for AD. One screen in families of German and Scandinavian children found a linkage for AD on chromosome 3q21 (20). This region encodes the costimulatory molecules CD80 and CD86 and therefore may modulate T cell responses. A second screen reported linkage of AD to loci on chromosomes 1q21, 17q25, and 20p. Interestingly, these same regions are known to contain psoriasis susceptibility genes, which suggests common candidate genes involved in the control of skin inflammation. Although AD and psoriasis are distinct skin diseases, both conditions involve dry, scaly skin and disrupted epidermal differentiation.

A Glu420Lys polymorphism variant in the SPINK5 gene, which encodes serine proteinase inhibitor, Kazal type 5, has shown significant association with AD in two independent cohorts (22). This gene is expressed in the outermost layers of the skin and has been implicated in Netherton disease, an autosomal recessive disorder characterized by ichthyosis and atopy. It may have a protective role against allergens that are serine proteinases.

The rapid rise in prevalence of atopic diseases has been ascribed to the so-called hygiene hypothesis. According to this theory, early infections or exposure to microbial-derived material, such as LPS, early in life prevent the development of Th2-driven allergic disease. Thus, polymorphisms of genes involved in the recognition of microbial material may alter the balance between Th1- and Th2-driven immune responses and could change an individual's susceptibility to development of allergic diseases. Indeed, polymorphisms of a number of innate immunity genes such as CD14 and Toll-like receptors have been associated with the development of allergy. Recently, a polymorphism (G2722C) that results in functional impairment of caspase recruitment domain–containing protein 15, an intracellular receptor for LPS involved in NF-κB activation, has been associated with a twofold increased risk for development of AD (23). Thus, not only reduced microbial exposure in the environment, but also impaired molecular recognition of microbial molecules, may give rise to enhanced Th2 responses.

Factors contributing to flares of AD

General. The skin represents the interface between the body and the surrounding environment. It is therefore subjected to numerous insults, which act as triggers of inflammation. These include irritants as well as allergens. These factors trigger the scratching that ensues to induce and sustain the inflammatory cascade initiated by release of proinflammatory cytokines from atopic keratinocytes. Stress itself has been found to induce immunologic changes and combined with scratching will trigger the exacerbation of AD.

Allergens. Placebo-controlled, food challenge studies have demonstrated that food allergens can induce eczematoid skin rashes in nearly 40% of children with moderate to severe AD. In a subset of these patients, urticarial reactions, or noncutaneous symptoms, are elicited, which can trigger the itch-scratch cycle that flares this skin condition. Children with food allergies generally have positive immediate skin tests or serum IgE directed to various foods, particularly eggs, milk, wheat, soy, and peanuts. Importantly, food allergen–specific T cells have been cloned from the skin lesions of patients with AD, providing direct evidence that foods can contribute to skin inflammation. In mouse models of AD, oral sensitization with foods results in the elicitation of eczematous skin lesions on repeat oral food challenges.

After the age of 3, children frequently outgrow food allergy but may become sensitized to inhalant allergens. Pruritus and skin lesions can develop after intranasal or bronchial inhalation challenge with aeroallergens in sensitized AD patients. Epicutaneous application of aeroallergens (e.g., HDMs, weeds, animal danders, and molds) by atopy patch test on uninvolved skin of AD patients elicits eczematoid reactions in 30–50% of patients with AD. A combination of effective HDM-reduction measures has been reported to improve AD. The degree of IgE sensitization to aeroallergens is directly associated with the severity of AD. The isolation from AD skin lesions and allergen patch test sites of T cells that selectively respond to *Dermatophagoides pteronyssinus* and other aeroallergens supports the concept that immune responses in AD skin can be elicited by environmental aeroallergens.

Autoallergens. Patients with severe AD have been reported to generate IgE antibodies directed against human proteins. The autoallergens identified to date have been intracellular proteins, which can be detected in IgE immune complexes of AD sera. These data suggest that while IgE immune responses are initiated by environmental allergens, allergic inflammation can be maintained by the release of human proteins derived from damaged skin in chronic AD.

Microbes. Most patients with AD are colonized with *Staphylococcus aureus* and suffer relapses of their skin disease following overgrowth of this organism (24). The importance of *S. aureus* is supported by the observation that, in AD patients with secondary infection, treatment with a combination of antistaphylococcal antibiotics and topical corticosteroids results in greater clinical improvement than treatment with topical corticosteroids alone. One strategy by which *S. aureus* exacerbates AD is by secreting toxins called superantigens, which stimulate activation of T cells and macrophages. Most AD patients make specific IgE antibodies directed against staphylococcal superantigens (25), and these IgE antisuperantigens correlate with skin disease severity. Superantigens also induce corticosteroid resistance, suggesting that several mechanisms exist by which superantigens increase AD severity.

Increased binding of *S. aureus* to skin is driven by underlying AD skin inflammation. This is clinically supported by studies demonstrating that treatment with topical corticosteroids or tacrolimus reduces *S. aureus* counts on atopic skin. In experimental animal models, *S. aureus* binding was significantly greater at skin sites with Th2- as compared to Th1-mediated skin inflammation due to IL-4–induced expression of fibronectin. AD skin has also been found to be deficient in antimicrobial peptides needed for host defense against bacteria, fungi, and viruses (17, 18). Thus, once *S. aureus* binds to AD skin, inadequate host defense allows bacteria to colonize and grow. The lack of skin innate immune responses may predispose these patients to infection as well to fungi and viruses. Patients with AD have an increased propensity toward disseminated infections with herpes simplex or vaccinia virus. As such, smallpox vaccination is contraindicated in patients with AD unless there is imminent danger of exposure to smallpox.

Translating the lessons of molecular immunology into current management of AD

Successful management of AD requires a multipronged approach. This includes the avoidance of irritants and specific immunologic stimuli, including foods and aeroallergens, which can either induce the dermatitis or trigger the itch-scratch cycle that results in AD (1). Skin hydration and use of emollients to repair the impaired skin barrier function is a key part of management. Addition of a "ceramide-dominant" emollient to standard therapy results in both clinical improvement and decreased transepidermal water loss and improvement of stratum corneum integrity in children with "stubborn-to-recalcitrant" AD (26). Although viral or fungal infection can trigger AD, *S. aureus* colonization or infection is the most common cause of increased AD severity. In such patients a course of antibiotics, in combination with anti-inflammatory therapy, will lead to better control of skin disease. The key to successful long-term management of AD is the introduction of effective anti-inflammatory therapy.

Topical corticosteroids. This class of drugs is the mainstay of anti-inflammatory treatment, showing efficacy in the control of both acute and chronic skin inflammation. Corticosteroids mediate their anti-inflammatory effects through a cytoplasmic glucocorticoid receptor (GCR) in target cells. Upon ligand binding, the corticosteroid/GCR complex translocates into the nucleus where it mediates its anti-inflammatory effects via two major actions. First, the GCR complex can induce gene transcription by binding of GCR dimers to glucocorticoid response elements (a process known as transactivation) in the promoter regions of target genes. Transactivation is mainly responsible for the unwanted side effects of glucocorticoids. The major mechanism by which GCR mediates its anti-inflammatory effects is via a process called transrepression, which is independent of GCR DNA binding. In this process, ligand-bound GCR binds to various transcription factors, including activator protein-1 and NF-κB, via protein-protein interactions to inhibit the transcriptional activity of various proinflammatory genes encoding proinflammatory proteins such as cytokines (IL-1, IL-2, IL-3, IL-4, IL-5, IL-6, IL-11, IL-13, TNF-α, and GM-CSF), chemokines (IL-8, RANTES, macrophage inflammatory protein–1α [MIP-1α], monocyte chemotactic protein-1 [MCP-1], MCP-3, MCP-4, and eotaxin), and adhesion molecules (ICAM-1, VCAM-1, and E-selectin).

Due to concerns about potential side effects associated with chronic use, topical corticosteroids have not been used for maintenance therapy, especially on nonlesional skin, in AD. However, normal-appearing skin in AD is associated with subclinical inflammation, suggesting that maintenance anti-inflammatory therapy may be required to prevent relapse. Indeed, several studies with fluticasone propionate have shown that once control of clinical disease is achieved, long-term control can be maintained with twice weekly therapy (27).

Topical calcineurin inhibitors. Topical FK506 and pimecrolimus have recently been FDA-approved for the treatment of AD. Both agents act by binding with high affinity to the 12 kDa macrophilin and inhibit the phosphatase activity of the calcium-dependent serine/threonine phosphatase, calcineurin. In the presence of these calcineurin inhibitors, the transcription factor, nuclear factor of activated T cell protein (NF-ATp), is not dephosphorylated and therefore cannot translocate into the nucleus to activate transcription of various Th1 and Th2 cytokine genes. Tacrolimus and pimecrolimus inhibit the activation of a number of key effector cells involved in AD, including T cells and mast cells.

Short-term, multicenter, blinded, vehicle-controlled trials, in both adults and children, have shown both topical tacrolimus and pimecrolimus to be effective. In a minority of patients, stinging and local irritation have been reported with this new class of drugs, but overall they have been found to be quite safe. Long-term studies with both drugs have been performed in adults and children, with demonstrated sustained efficacy and no significant side effects. Unlike topical glucocorticoids, topical calcineurin inhibitors are not atrophogenic and have been used safely for facial and eyelid eczema. Several studies with pimecrolimus cream have also found that introducing the medication at the earliest

---

Table 2

Evolving therapeutic targets in AD

Anti-allergic approaches
　Anti-IgE
　Allergen-selective immunotherapy
　Immunization with CpG motifs

T cell targets
　Probiotics (e.g., *Lactobacillus rhamnosus* strain GG)
　Reduction of Th2 cytokine responses (anti–IL-4, soluble IL-4
　　receptor, anti–IL-13,
　　antisense oligonucleotide approaches)
　Mycobacterium vaccae vaccination
　Inhibition of T cell activation; e.g., alefacept (blocks T cell
　　CD2–leukocyte functional antigen-1 [LFA-1] and –LFA-3 APC
　　interactions), efalizumab (blocks LFA-1–ICAM-1 and –ICAM-2
　　APC interactions)

Anti-inflammatory agents
　Oral pimecrolimus
　TNF inhibitors

Antimicrobial approaches augmenting T cell responses
　Antimicrobial peptides

Blockade of inflammatory cell recruitment
　Chemokine antagonists (C-C chemokine receptor-4, cutaneous
　　lymphocyte–associated antigen)
　CLA inhibitors

CpG, cytosine phosphate guanine.

signs of clinical disease results in significantly less need for corticosteroid "rescue" therapy.

The approval of topical calcineurin inhibitors for the treatment of AD represents a significant advance in our management options for this disease. The distinction between pimecrolimus and tacrolimus is that pimecrolimus is a cream that is somewhat weaker than tacrolimus but less irritating. Tacrolimus is currently marketed as an ointment that is more potent but also more irritating. Importantly, there are situations in which topical calcineurin inhibitors may be advantageous over topical corticosteroids and may be useful as first-line therapy. These would include treatment of patients who are poorly responsive to topical steroids or have steroid phobia, and treatment of face and neck dermatitis where ineffective, low-potency topical corticosteroids are usually used due to fears of steroid-induced skin atrophy. The potential use of topical calcineurin inhibitors as maintenance therapy is also intriguing for prevention of AD flares and progression of the atopic march (28, 29). However, although systemic absorption of these compounds is low, there is a need for careful surveillance to rule out the possibility that skin cancers and increased viral skin infections will appear when such agents are used long-term.

Alternative anti-inflammatory approaches. UV light therapy can be a useful treatment modality for chronic recalcitrant AD. The photoimmunologic effects target key cells in atopic inflammation, such as LCs and keratinocytes, interfering with cytokine production and decreasing the expression of activation markers such as HLA-DR and IL-2 receptor on CLA$^+$ T cells. Unlike traditional UVA-UVB phototherapy, which appears less effective for acute exacerbations and acts primarily in the epidermis, high-dose UVA1 therapy has been shown to significantly decrease dermal IgE-binding cells, including mast cells, LCs, and DCs. UVA1 also inhibits LC migration out of the epidermis.

Cyclosporin A is a potent systemic calcineurin inhibitor. A number of studies have demonstrated its efficacy in both children and adults with severe, refractory AD, although toxicity, primarily renal, limits its chronic use. Preliminary data with oral pimecrolimus suggest that it may have a greater safety margin than either systemic cyclosporin or tacrolimus, which could make treatment more acceptable to some patients and reach inflammatory targets not accessible to topical medications (30). Antimetabolites, including mycophenolate mofetil, a purine biosynthesis inhibitor, methotrexate, and azathiaprine, have also been utilized for recalcitrant AD, although the potential for systemic toxicities restricts their use and requires close monitoring.

New approaches to modulate atopic inflammation

Although a number of anecdotal and case reports suggest clinical benefit from allergen-specific desensitization in AD, double-blinded, controlled trials have failed to show consistent efficacy of immunotherapy compared to placebo in the treatment of AD. Recently, omalizumab, a humanized IgG1 monoclonal antibody against IgE that recognizes and masks an epitope in the CH3 region of IgE responsible for binding to the high-affinity FcεR on mast cells and basophils, has been shown to be effective in treatment of allergic asthma and allergic rhinitis. Thus, it could potentially neutralize the effects of IgE in AD. However, the high serum IgE levels seen in AD may limit the usefulness of this antibody, although it may have a role in food-induced AD. In this respect, treatment of a population of peanut-allergic patients with anti-IgE significantly increased their threshold of sensitivity to peanuts

on oral food challenge, suggesting protection against unintended ingestion of the food allergen (31).

Insights into the role that certain cells and cytokines play in AD create opportunities for the development of targeted therapy. However, given the complexity of the biological processes involved, none of the tested compounds to date has proven to be the magic bullet. Some promising agents have been limited by associated toxicity. In this respect, several studies with recombinant human IFN-γ demonstrated clinical efficacy that correlated with a decrease in blood eosinophilia, even with long-term therapy. Early treatment with microbial probiotics may be more beneficial by boosting Th1 immune responses in AD (32). Additional approaches include cytokine modulation (e.g., soluble IL-4 receptor; anti–IL-5 monoclonal antibody; TNF inhibitors), blockade of inflammatory cell recruitment (chemokine receptor antagonists, CLA inhibitors, ref. 33), inhibition of T cell activation (alefacept, efalizumab, ref. 34), and use of synthetic antimicrobial peptides (Table 2).

Future directions

Our challenge for the future will be the development of more effective and safer drugs in the treatment of AD. Given the complexity of immune pathways that lead to AD, it is possible that more selective anti-inflammatory or immunomodulatory agents will be less effective. Thus, it will be important to better characterize the key immune pathways leading to the different phenotypes of AD, as medications may vary in their effectiveness for treatment of different forms of AD. For example, it is important to resolve whether the immune pathways leading to so-called intrinsic (or pure) vs. extrinsic (IgE-mediated) AD are fundamentally different. The role that IgE plays in allergic disease is still debated. Animal models of allergy can occur in IgE-knockout mice. This molecule may primarily act to facilitate allergen processing for Th2 cell activation and immediate reaction at extremely low environmental concentrations of allergens. Immune responses to allergen may occur in the absence of IgE in intrinsic AD but require higher allergen concentrations for T cell activation in the absence of immediate reaction. Indeed, atopy patch tests can be positive in a subset of AD patients lacking IgE responses to the relevant allergen. The relative role of microbes and autoantigens in the initiation and progression of AD also requires further clarification.

New advances are likely to require better definitions for the various clinical phenotypes of AD, including identification of the susceptibility genes leading to the different forms of AD and delineation of the relative role of immunoregulatory abnormalities and structural skin barrier defects underlying AD skin. New treatment paradigms are needed for preventing disease progression of AD to more severe forms of this skin disease and halting the so-called atopic march, which results in development of asthma. The factors that determine the chronicity, skin remodeling, and natural history of this disease remain poorly characterized. An understanding of the genes responsible for individual variation in response to therapy will be tied to the development of pharmacogenetics and the targeting of effective therapies to the different phenotypes of AD.

Address correspondence to: Donald Y.M. Leung, National Jewish Medical and Research Center, 1400 Jackson Street, Room K926, Denver, Colorado 80206, USA. Phone: (303) 398-1379; Fax: (303) 270-2182; E-mail: leungd@njc.org.

1. Leung, D.Y., and Bieber, T. 2003. Atopic dermatitis. *Lancet.* **361**:151–160.

2. Spergel, J.M., and Paller, A.S. 2003. Atopic dermatitis and the atopic march. *J. Allergy Clin. Immunol.* **112**:S128–S139.

3. Novak, N., Bieber, T., and Leung, D.Y.M. 2003. Immune mechanisms leading to atopic dermatitis. *J. Allergy Clin. Immunol.* **112**:S128–S139.

4. Spergel, J.M., et al. 1998. Epicutaneous sensitization with protein antigen induces localized allergic dermatitis and hyperresponsiveness to methacholine after single exposure to aerosolized antigen in mice. *J. Clin. Invest.* **101**:1614–1622.

5. Novak, N., and Bieber, T. 2003. Allergic and nonallergic forms of atopic diseases. *J. Allergy Clin. Immunol.* **112**:252–262.

6. Hamid, Q., Boguniewicz, M., and Leung, D.Y. 1994. Differential in situ cytokine gene expression in acute versus chronic atopic dermatitis. *J. Clin. Invest.* **94**:870–876.

7. Novak, N., Kraft, S., and Bieber, T. 2003. Unraveling the mission of FcepsilonRI on antigen-presenting cells. *J. Allergy Clin. Immunol.* **111**:38–44.

8. Toda, M., et al. 2003. Polarized in vivo expression of IL-11 and IL-17 between acute and chronic skin lesions. *J. Allergy Clin. Immunol.* **111**:875–881.

9. Ono, S.J., et al. 2003. Chemokines: roles in leukocyte development, trafficking, and effector function. *J. Allergy Clin. Immunol.* **111**:1185–1199.

10. Nomura, I., et al. 2003. Distinct patterns of gene expression in the skin lesions of atopic dermatitis and psoriasis: a gene microarray analysis. *J. Allergy Clin. Immunol.* **112**:1195–1202.

11. Akdis, M., et al. 2003. T helper (Th) 2 predominance in atopic diseases is due to preferential apoptosis of circulating memory/effector Th1 cells. *FASEB J.* **17**:1026–1035.

12. Trautmann, A., et al. 2000. T cell-mediated Fas-induced keratinocyte apoptosis plays a key pathogenetic role in eczematous dermatitis. *J. Clin. Invest.* **106**:25–35.

13. Chan, L.S., Robinson, N., and Xu, L. 2001. Expression of interleukin-4 in the epidermis of transgenic mice results in a pruritic inflammatory skin disease: an experimental animal model to study atopic dermatitis. *J. Invest. Dermatol.* **117**:977–983.

14. Spergel, J.M., Mizoguchi, E., Oettgen, H., Bhan, A.K., and Geha, R.S. 1999. Roles of TH1 and TH2 cytokines in a murine model of allergic dermatitis. *J. Clin. Invest.* **103**:1103–1111.

15. Novak, N., et al. 2003. Evidence for a differential expression of the FcepsilonRIgamma chain in dendritic cells of atopic and nonatopic donors. *J. Clin. Invest.* **111**:1047–1056. doi:10.1172/JCI200315932.

16. Laouini, D., et al. 2003. IL-10 is critical for Th2 responses in a murine model of allergic dermatitis. *J. Clin. Invest.* **112**:1058–1066. doi:10.1172/JCI200318246.

17. Ong, P.Y., et al. 2002. Endogenous antimicrobial peptides and skin infections in atopic dermatitis. *N. Engl. J. Med.* **347**:1151–1160.

18. Nomura, I., et al. 2003. Cytokine milieu of atopic dermatitis, as compared to psoriasis, skin prevents induction of innate immune response genes. *J. Immunol.* **171**:3262–3269.

19. Sator, P.G., Schmidt, J.B., and Honigsmann, H. 2003. Comparison of epidermal hydration and skin surface lipids in healthy individuals and in patients with atopic dermatitis. *J. Am. Acad. Dermatol.* **48**:352–358.

20. Cookson, W.O., and Moffatt, M.F. 2002. The genetics of atopic dermatitis. *Curr. Opin. Allergy Clin. Immunol.* **2**:383–387.

21. Liu, X., et al. 2003. Associations between total serum IgE levels and the 6 potentially functional variants within the genes IL4, IL13, and IL4RA in German children: the German Multicenter Atopy Study. *J. Allergy Clin. Immunol.* **112**:382–388.

22. Kato, A., et al. 2003. Association of SPINK5 gene polymorphisms with atopic dermatitis in the Japanese population. *Br. J. Dermatol.* **148**:665–669.

23. Kabesch, M., et al. 2003. Association between polymorphisms in caspase recruitment domain containing protein 15 and allergy in two German populations. *J. Allergy Clin. Immunol.* **111**:813–817.

24. Leung, D.Y. 2003. Infection in atopic dermatitis. *Curr. Opin. Pediatr.* **15**:399–404.

25. Leung, D.Y., et al. 1993. Presence of IgE antibodies to staphylococcal exotoxins on the skin of patients with atopic dermatitis. Evidence for a new group of allergens. *J. Clin. Invest.* **92**:1374–1380.

26. Chamlin, S.L., et al. 2002. Ceramide-dominant barrier repair lipids alleviate childhood atopic dermatitis: changes in barrier function provide a sensitive indicator of disease activity. *J. Am. Acad. Dermatol.* **47**:198–208.

27. Berth-Jones, J., et al. 2003. Twice weekly fluticasone propionate added to emollient maintenance treatment to reduce risk of relapse in atopic dermatitis: randomised, double blind, parallel group study. *BMJ.* **326**:1367.

28. Kapp, A., et al. 2002. Long-term management of atopic dermatitis in infants with topical pimecrolimus, a nonsteroid anti-inflammatory drug. *J. Allergy Clin. Immunol.* **110**:277–284.

29. Boguniewicz, M., Eichenfield, L.F., and Hultsch, T. 2003. Current management of atopic dermatitis and interruption of the atopic march. *J. Allergy Clin. Immunol.* **112**:S140–S150.

30. Rappersberger, K., et al. 2002. Pimecrolimus identifies a common genomic anti-inflammatory profile, is clinically highly effective in psoriasis and is well tolerated. *J. Invest. Dermatol.* **119**:876–887.

31. Leung, D.Y., et al. 2003. Effect of anti-IgE therapy in patients with peanut allergy. *N. Engl. J. Med.* **348**:986–993.

32. Kalliomaki, M., Salminen, S., Poussa, T., Arvilommi, H., and Isolauri, E. 2003. Probiotics and prevention of atopic disease: 4-year follow-up of a randomised placebo-controlled trial. *Lancet.* **361**:1869–1871.

33. Dimitroff, C.J., Kupper, T.S., and Sackstein, R. 2003. Prevention of leukocyte migration to inflamed skin with a novel fluorosugar modifier of cutaneous lymphocyte-associated antigen. *J. Clin. Invest.* **112**:1008–1018. doi:10.1172/JCI200319220.

34. Kupper, T.S. 2003. Immunologic targets in psoriasis. *N. Engl. J. Med.* **349**:1987–1990.

35. Leung, D.Y. 2000. Atopic dermatitis: new insights and opportunities for therapeutic intervention. *J. Allergy Clin. Immunol.* **105**:860–876.

Recent insights into the immunopathogenesis of psoriasis provide new therapeutic opportunities

Brian J. Nickoloff[1] and Frank O. Nestle[2]

[1]Skin Disease Research Laboratory and Cardinal Bernardin Cancer Center, Loyola University of Chicago, Medical Center, Maywood, Illinois, USA.
[2]University of Zurich, Zurich, Switzerland.

Chronic and excessive inflammation in skin and joints causes significant morbidity in psoriasis patients. As a prevalent T lymphocyte–mediated disorder, psoriasis, as well as the side effects associated with its treatment, affects patients globally. In this review, recent progress is discussed in the areas of genetics, the immunological synapse, the untangling of the cytokine web and signaling pathways, xenotransplantation models, and the growing use of selectively targeted therapies. Since psoriasis is currently incurable, new management strategies are proposed to replace previous serendipitous approaches. Such strategic transition from serendipity to the use of novel selective agents aimed at defined targets in psoriatic lesions is moving rapidly from research benches to the bedsides of patients with this chronic and debilitating disease.

Historical perspective, and clinical and histological features of psoriasis

Psoriasis (OMIM 177900) is an ancient and universal inflammatory disease, initially described at the beginning of medicine in the *Corpus Hippocraticum*. Hippocrates (460–377 BCE) used the term *psora*, meaning "to itch." While the cause of psoriasis remains unknown, it appears to result from a combination of genetic and environmental factors. It is frequently inherited and passed from one generation to the next, but not following a classical autosomal mendelian profile. While it may have originally been confused with leprosy (*lepra*, "to scale"), it is generally easy to recognize psoriasis when it presents in one of three typical presentations: guttate, pustular, and plaque-stage. Figure 1 provides a clinical view of untreated chronic stationary plaques distributed on the lower back.

Approximately 2–3% of the population worldwide is afflicted by psoriasis. Several authors, including John Updike and Vladimir Nabokov, as well as British experimental dramatist Dennis Potter (*The singing detective*), carefully recorded the suffering that psoriasis, and side effects of its treatment, had on their lives. Not only can skin lesions be pruritic and disfiguring; in 10–30% of patients there can also be nail dystrophy accompanied by psoriatic arthritis. The inflammation in the joints is similar in some ways to rheumatoid arthritis (RA), although in psoriasis it is a seronegative arthritis (no rheumatoid factor is present in the blood). Thus, the various clinical manifestations of psoriasis make it more than a dermatological nuisance, as it interferes with many normal daily activities, such as use of hands, walking, sleeping, and sexual activity. At least 30% of patients contemplate suicide, which places psoriasis on par with other major medical diseases such as depression, heart disease, and diabetes

(1). The most frequent extracutaneous medical problem associated with psoriasis (besides arthritis of small joints) is the inflammatory bowel disorder Crohn disease (2).

Psoriasis can begin at any age, although epidemiological studies demonstrate that it most commonly appears for the first time between the ages of 15 and 25 years (3). It is a lifelong inflammatory disease with spontaneous remissions and exacerbations. In the plaque stage, lesions are characterized as being symmetrically distributed, and well demarcated from adjacent symptomless skin, with erythema topped by white-silvery scale (Figure 1). Many triggering factors initiate or exacerbate psoriasis, including bacterial pharyngitis, stress, HIV-1, and various medications (e.g., lithium and β-blockers). Microscopically, plaque-stage lesions reveal significantly thickened skin with confluent parakeratotic scale (Figure 1; insets compare symptomless skin with psoriatic plaques), loss of the granular cell layer, and increased number of epidermal cell layers with mitotic figures in basal-layer keratinocytes. Rete ridges, representing downward extensions of epidermis, are elongated, and the papillary dermal blood vessels appear tortuous and dilated. The inflammatory cell infiltrate may contain neutrophil collections in the epidermis, but a more consistent finding is the presence of T cells in the dermis and epidermis accompanied by increased numbers of dermal DCs, macrophages, and mast cells.

The history of the treatment of psoriasis is also of interest, as it reflects not only the uncertainty regarding its pathogenesis, but also the limited options available to clinicians in the past. Translational arcs of discovery for therapeutic agents used in psoriasis in the past, present, and future are displayed in Figure 2. Note that in the past, physicians used many different compounds and serendipitously found several that helped treat psoriasis, including arsenic (Fowler's solution) and ammoniated mercury. In many instances, neither the specific target nor the mechanism of action for the treatment was known. However, current treatments include a new wave of selective therapeutic agents that emerged following introduction of cyclosporin A into dermatology clinics, reflecting our ever increasing understanding of psoriasis pathophysiology (4). The next section presents a brief overview of psoriasis pathophysiology, followed by detailed analyses of genetic, immunological, and animal-model

Nonstandard abbreviations used: Food and Drug Administration (FDA); high-mobility group B1 (HMGB1); NK type T cell (NKT cell); programmed cell death 1 (gene) (*PDCD1*); regulatory single-nucleotide polymorphism (rSNP); rheumatoid arthritis (RA); receptor for advanced-glycation end products (RAGE); runt-related transcription factor (RUNX); single-nucleotide polymorphism (SNP); T cell receptor (TCR); very late antigen (VLA).

Conflict of interest: The authors have declared that no conflict of interest exists.

Citation for this article: *J. Clin. Invest.* **113**:1664–1675 (2004). doi:10.1172/JCI200422147.

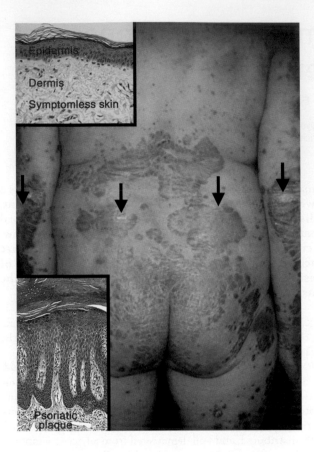

Figure 1
Clinical and histological appearance of stable chronic psoriatic plaques. Note the well-demarcated erythematous plaques covered by white-silvery scale (arrows), widely distributed on the lower back and extremities. Prepsoriatic symptomless skin is clearly demarcated from lesional skin. Psoriasis often develops at skin sites where minor trauma may occur, such as elbows and buttocks. Induction of psoriatic lesions by trauma is referred to as the Köbner phenomenon. Insets: Histological appearance of prepsoriatic (symptomless) skin (upper left panel), with unremarkable epidermis and dermis, and rare mononuclear cells present predominantly in the dermis. By contrast, a chronic psoriatic plaque (lower left panel) reveals markedly thickened skin due primarily to accumulation of scale and elongation of rete ridges. In addition, there is loss of the granular cell layer, increased layers of epidermal keratinocytes, and an influx of lymphocytes, DCs, and macrophages into the dermis, accompanied by the presence of dilated and tortuous blood vessels.

developments; we then delve into new therapeutic advances. It should be noted up front that the current listing of psoriasis with other putative autoimmune diseases such as multiple sclerosis is largely based on essentially negative results, reflecting the failure of investigators to consistently isolate a specific infectious agent or identify circulating autoantibodies. As will become clear in the following sections, psoriasis challenges investigators to understand the interplay between genetic-susceptibility alleles and immunological and environmental factors that contribute to the chronic inflammatory process.

Overview: pathophysiology of psoriasis

As mentioned earlier, the cause of psoriasis remains unknown, and there is no cure. Several hypotheses have been advanced and models proposed over the years concerning its pathogenesis. In this section, a brief overview of psoriasis pathophysiology is presented. As one considers hypotheses for psoriasis, it is important to recognize at least three unique and characteristic features of this enigmatic disease: (a) Psoriatic plaques represent highly localized sites of dysregulated growth and inflammation, yet these sites almost never develop into, or harbor, malignant clones of keratinocytes, melanocytes, or T cells (5); (b) Despite significantly altered barrier function due to aberrant epidermal cell differentiation, psoriatic plaques are highly resistant to bacterial, viral, and fungal infections (6); (c) Psoriatic plaques, either spontaneously or after various treatments, can revert back to symptomless or apparently healthy skin, with little or no trace of pre-existing disease activity.

In theories and models advanced to explain psoriatic plaque development, just about every cell type in the lesion has been given

center stage. Thus, clinicians and investigators initially focused on the epidermal keratinocyte and viewed psoriatic lesion formation from an "outside-in" perspective (7). Not surprisingly, early therapeutic strategies targeted the hyperplastic epidermis following the lead of oncologists to arrest keratinocyte growth using antiproliferative agents such as arsenic and, later, methotrexate. Other theories implicated psoriatic fibroblasts (8); neutrophils; mast cells (9); nerve cell endings (10); endothelial cells (11); T lymphocytes (12–14), and, specifically, clonal expansion of T cells (15, 16); and DCs (13). The successful use of cyclosporin, which acts to block cytokine release, resulted in a paradigm shift away from epidermal keratinocytes to various immunocyte populations including T cell subsets and dendritic APCs.

One unifying hypothesis of psoriasis pathophysiology, which assimilated the confederacy of cell types and plethora of soluble mediators, was the cytokine network model (17). In this model, either an exogenously derived stimulus such as trauma, or an endogenous stimulus such as HIV-1, neuropeptides, or ingested medications, was portrayed as triggering a plexus of cellular events by inciting a cascade of cytokines. Initially, this model featured primary pathogenic roles for TNF-α derived from dendritic APCs and keratinocytes, and IFN-γ produced by activated Th1-type lymphocytes. As presented later, considerable experimental data and therapeutic responses in human subjects support the cytokine network theory of psoriasis. While this theory explains the maintenance of psoriatic plaques via establishment of a vicious cycle, precisely how T cells are activated in the genesis of psoriatic lesions is unknown. It is also unknown whether the inciting antigen is self-derived (thereby qualifying psoriasis as an autoimmune disease) or is of non-self origin.

Figure 3 presents a working model for the immunopathogenesis of psoriasis, with emphasis on homeostatic versus pathological trafficking patterns of immunocytes. Symptomless skin (left panel), also known as prepsoriatic or uninvolved skin, is not entirely normal, as numerous molecular differences between symptomless skin and healthy skin from non-psoriatic patients are emerging from global gene expression profiling studies (18). Known mediators regulating trafficking of T cells and DCs in symptomless skin are portrayed, including those chemokines and chemokine receptors that facilitate trafficking to regional lymph nodes. The middle panel of Figure 3 shows acute lesions after a stimulus. A memory T cell has become activated upon interaction with a dendritic APC. The figure emphasizes costimulatory pathways involving CD2:LFA-3 and LFA-1:ICAM-1. As described

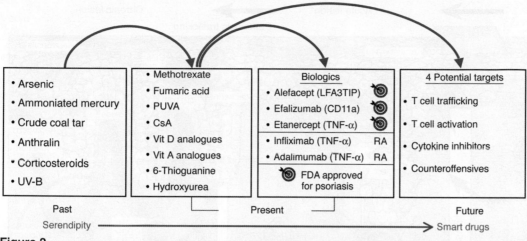

Figure 2
Translational arcs of discovery in psoriasis. In the past, therapeutic agents used by dermatologists were generally discovered serendipitously. Current therapeutic agents are characterized by more specific targeting of defined molecules in the pathological pathways, including, most recently, the use of biologics, which target T cells (alefacept, efalizumab, and etanercept; approved by the US Food and Drug Administration [FDA] for the treatment of psoriasis), and TNF inhibitors (etanercept, infliximab, and adalimumab; approved by the FDA for the treatment of RA). Several agents listed (Past column), including corticosteroids and UV-B light, are still currently used by some dermatologists to treat psoriasis patients. Future drug development should provide additional smart drugs that target specific molecular mediators implicated in the immunopathogenesis of psoriasis. Note: Biologics are defined as therapeutic agents produced by organisms through the use of recombinant biotechnology. CsA, cyclosporin A; PUVA, psoralen and UV-A light therapy; Vit, vitamin.

harboring disease-susceptibility alleles with single-nucleotide polymorphism (SNP) analysis. By using genome-wide scans, investigators have mapped (with varying degrees of confidence) at least six different susceptibility loci, designated *PSORS1–PSORS6* (23). A list of potentially important loci, including genetic loci that overlap with other chronic inflammatory diseases such as eczema and RA, is provided in Table 1. Several other psoriasis-susceptibility loci have been mapped, including *PSORS7* (1p) and *PSORS9* (4q31), and additional studies are ongoing by many laboratories (24, 25) and the International Psoriasis Genetics Consortium (26).

in detail in a later section, these molecules participate in formation of an immunological synapse to enhance APC:T cell interactions. Once resident (and recruited) T cells become activated, a complex network of cytokines, chemokines, and growth factors is created, and the envisioned vicious cycle facilitates further T cell and DC activation. Ultimately, a chronic plaque is produced (right panel), characterized by intraepidermal CD8+ T cells and neutrophils, and accompanied by epidermal thickening, scale production, and an angiogenic tissue reaction completely remodeling the involved or lesional skin site. We return to Figure 3 in later sections with more detailed molecular analysis to complement the aforementioned emphasis on cellular components and trafficking patterns that contribute to the pathophysiology of psoriasis. In the next section, a review of the genetic approach to understanding psoriasis is presented.

Genetic studies of psoriasis
Based on pioneering work by Gunnar Lomholt, who explored the relative roles of the environment and heredity in residents of the Faroe Islands (19), and Farber and Nall, who later studied concordance rates in monozygotic twins (~70%) and documented kindreds with multiple afflicted family members (20), a search for the genetic basis of psoriasis began. One of the earliest candidate genes for predisposition to psoriasis was the HLA class I allele, specifically HLA-Cw6 (21). A dosage effect of HLA-Cw*0602 has been observed, where heterozygotes have a relative risk of developing psoriasis of 8.9, compared with 23.1 in homozygous individuals. Also, homozygous individuals experience an earlier onset, but not necessarily a more severe disease course (22). Rapid advancements in human genetic studies have led investigators beyond classical mendelian approaches to use other research methods, including linkage analysis and association-based fine mapping of haplotypes

The major genetic determinant for psoriasis is within the *PSORS1* region of the MHC on chromosome 6p21, as reported by several independent groups, accounting for 30–50% genetic susceptibility (23). Unfortunately, despite intense study, the definitive gene in this region has not been unequivocally identified. The two most likely candidate loci are HLA-Cw*0602 and the *corneodesmosin* gene (27), which encodes an adhesive protein expressed by keratinocytes and is important in terminal differentiation of the epidermis, although interpretation is complicated by the extraordinary linkage disequilibrium observed around the MHC. Thus, it is possible that genes such as HLA-Cw6 are in linkage disequilibrium with another gene, or block of genes, at a distinct locus such as killer Ig-like receptor (KIR), as recently reported by Martin et al. (28). KIRs may be important in psoriasis (and psoriatic arthritis), since they are expressed on NK cells, as well as NK type T cells (NKT cells), which have been observed in lesional skin and have thus given rise to the consideration of innate immunity in psoriatic lesions (29).

The second most well-characterized disease-susceptibility locus (*PSORS2*) resides within 17q24–q25. Linkage of psoriasis

Table 1
Potential psoriasis-susceptibility loci

Locus name	Approximate location	Association with other inflammatory diseases
PSORS1	6p21	Asthma
PSORS2	17q25	Eczema
PSORS3	4q34	-
PSORS4	1q21	Eczema
PSORS5	3q21	RA
PSORS6	19p13	-

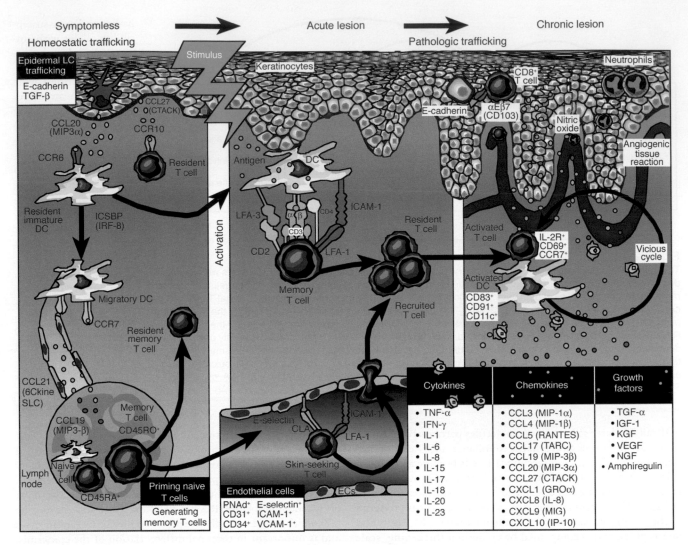

Figure 3
Working model for immunopathogenesis of psoriasis. Multiple stages are proposed for trafficking patterns of immunocytes, involving signals in which symptomless skin is converted into a psoriatic plaque. Symptomless skin is endowed with a confederacy of bone marrow–derived cells, and continuous leukocyte migration between skin and lymph nodes provides immunological vigilance to monitor invading pathogenic organisms. Known mediators of homeostatic trafficking for Langerhans cells (LCs), resident DCs, and T cells are portrayed (left panel). Ideal therapeutic agents for psoriasis should not perturb this physiological process. Following a stimulus, an acute psoriatic lesion forms in which DCs and T cells become activated with formation of an immunological synapse. No consistent antigen has been identified (middle panel). Stimuli may include a danger signal, either an extrinsic, pathogen-associated signal (e.g., pathogen-associated molecules that bind to pattern-recognition Toll-like receptors) or an intrinsic signal derived from within the body (e.g., heat shock proteins that bind to receptors, HIV-1, and ingested medications such as lithium or β-blockers). Once dendritic APCs and T cells become activated, they release cytokines, chemokines, and growth factors that trigger keratinocyte proliferation, altered differentiation, and an angiogenic tissue response. A vicious cycle of continuous T cell and DC activation can be envisioned within the chronic psoriatic plaque (right panel). A list of relevant cytokines, chemokines, and growth factors that likely conspire with resident and recruited cells to create and sustain psoriatic plaques is provided (lower right panel). Key inflammatory events include intraepidermal trafficking by CD8+ T cells and neutrophils.

to this locus has been identified by independent family sets (30, 31). Two candidate genes in this region are *SLC9A3R1* and *NAT9*, which are discussed in the following section (32). To understand how these genotyping results may be linked to the immune system and thereby contribute to the immunopathogenesis of psoriasis, it is helpful to review the role of the immunological synapse in psoriasis and autoimmunity, as described in the next section. It should be noted that the field of human molecular genetics and autoimmunity is progressing rapidly, and the ini-

tial view of candidate genes is bewildering, yet intriguing. Next, we provide a working model of the immune synapse of T cells in psoriasis, integrating recent reports describing T cell signaling pathways in autoimmunity.

The immunological synapse and psoriasis
In mature lymphocytes, T cell receptor (TCR) signaling is mediated by formation of a multimolecular complex at the T cell–APC interface, referred to as the immunological synapse (33).

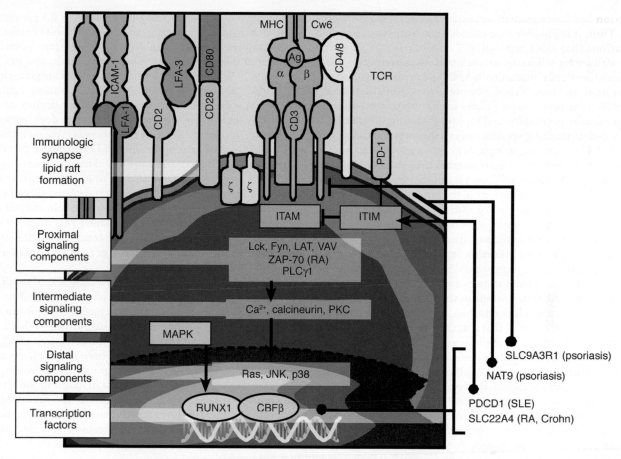

Figure 4

Immune synapse–related signaling pathway and autoimmune diseases. A possible role for genetic mutations involved in several autoimmune diseases is highlighted, including ZAP-70 (RA); SLC9A3R1 and NAT9 (psoriasis); PDCD1 (systemic lupus erythematosus [SLE]); and SLC22A4 (RA and Crohn disease). T lymphocytes contain both surface receptors and intracellular signaling components that become activated when the TCR is engaged by interactions with an appropriate APC in which antigen is presented in the context of MHC molecules. The APC is displayed in the upper portion of this schematic view. The TCR mediates signaling in conjunction with molecules located in the plasma membrane, including CD3 and ζ-chains, which contribute to the formation of an immunological synapse and a lipid raft, leading to activation of proximal, intermediate, and distal signaling components. Several but not all components are portrayed in this figure. Ultimately, transcription factors become activated and bind to respective promoter regions to either enhance or suppress target gene expression. Potential mechanisms linking the genetic findings to functional components of the immunological synapse are shown by solid bold lines.

The immunological synapse is a discrete cluster of molecules formed between a T cell and an APC that facilitates immune cell interactions. Two major technologies have been used to define events in the immunological synapse: live-cell fluorescent imaging of T cells in contact with supporting plasma bilayers, and real-time confocal imaging of the T cell–APC interface. Key molecular components include the TCR and, surrounding it, a ring of adhesion molecules, such as LFA-1, which can bind to ICAM-1 expressed by the adjacent cell, e.g., a keratinocyte or APC. The LFA-1 component of the synapse is especially important in psoriasis, as a therapeutic agent (anti–LFA-1 antibody; efalizumab) that blocks this adhesive interaction has recently been approved by the US Food and Drug Administration (FDA) for the treatment of psoriasis (34).

While the immunological synapse controls T cell activation, additional contributory molecules, including other adhesion molecules and costimulatory molecules, also influence T cell responsiveness. The cell surface molecular pairs CD2:LFA-3 (35)

and CD28:CD80/CD86 (B7.1/B7.2) are examples. Many of these molecules have also been the subjects of clinical trials in psoriatic patients, in which inhibitory reagents have been designed to block T cell activation. For example, an LFA-3 Ig fusion protein (alefacept) has been found to reduce psoriatic lesions (36), and a different fusion protein, CTLA4Ig, which blocks CD28:CD80/CD86 interactions, also improved psoriatic lesions (37). Attempts to modulate T cells directly in psoriasis include the use of antibodies directed against CD3 (38) and CD4 (39), and a fusion protein containing the cell-binding domain of diphtheria toxin (i.e., denileukin diftitox) has been produced to target and kill activated T cells bearing receptors for IL-2 (40).

For several reasons, the signaling molecules linked to the immunological synapse have become the focus of geneticists interested in defining the molecular basis of autoimmunity (Figure 4). It has been suggested that alterations in TCR signaling, which is critical for central tolerance (elimination of autoreactive T cells), may cause a breakdown in thymic negative selection and hence mediate appearance of autoreactive T cells (41). Thus, it is possible that psoriasis patients have inherited mutations that block removal of T cells bearing TCRs with a high affinity for self-antigens, such that upon activation, these autoreactive T cells interact with APCs and keratinocytes to create psoriatic plaques. It should be noted that the field involving the molecular basis that regulates formation of the mature T cell repertoire is complex and rapidly changing, so caution is warranted in ascribing specific roles to the immunological synapse and TCR signaling as regards negative selection in human subjects. Nonetheless, to gain insights into this possible scenario, it is useful to review progress in other autoimmune diseases, beginning with mutations in the proximal components of the TCR signaling pathway (e.g., ZAP-70). Recently, Sakaguchi et al. (42) reported a strain of mice that spontaneously develop a chronic systemic inflammatory disease resembling RA. CD4$^+$ T cells bearing a point mutation in a gene encoding an SH2 domain of ZAP-70, a key upstream signal-transducing molecule, could transfer the disease state in a variety of different mouse strains. Clinical manifestations in this model were also influenced by environmental conditions, including a susceptibility locus in the MHC complex that is analogous to that for psoriasis (*PSORS1* and *PSORS2*). In preliminary studies, this group also identified heterozygous mutations in the immunoreceptor tyrosine-based activating motif (ITAM) region of the TCR-ζ chain in 2.5% of RA patients, which is relevant because of the physical association between ITAM and ZAP-70.

Three recent reports also merit consideration, as they involve polymorphisms in regulatory (promoter) regions of genes that impact proper immune synapse function with emergence of autoreactive T cells (43). In Nordic multicase families with systemic lupus erythematosus (SLE), the strongest candidate for disease association was found to be within the programmed cell death 1 gene (*PDCD1*) (44). The associated allele of the SNP altered a binding site for runt-related transcription factor 1 (RUNX1). RUNX1 (also known as AML1) belongs to a family of transcription factors that can either inhibit or promote expression of a variety of genes. The transactivating potential of RUNX1 can be enhanced by phosphorylation events mediated by the MAPK pathway. Since *PDCD1* encodes a protein (PD-1) with an immunoreceptor tyrosine-based inhibitory motif (ITIM), which could thereby inhibit autoreactive T cells (i.e., participate in peripheral tolerance), deletion of RUNX1-binding sites may contribute to aberrant regulation of PDCD1, leading to dysregulated self-tolerance and persistent lymphocyte activation.

Besides identification of disease-associated regulatory single-nucleotide polymorphisms (rSNPs) in RUNX-binding sites for lupus, other groups identified rSNPs in psoriasis and RA patients (31, 45). In psoriatic patients, the rSNPs of interest may be regulating expression of two separate genes — *SLC9A3R1* and *NAT9* (32). SLC9A3R1 (solute carrier family 9, isoform 3 regulating factor 1) is a PD-2 domain–containing phosphoprotein that associates with members of the ezrin-radixin-moesin family. It can negatively regulate immune synapse function, influencing negative selection during development (central tolerance) and thus leading to the emergence of autoreactive T cells (46), or alter local immune responses in the skin with inappropriate or dysregulated T cell activation. Similarly, NAT9, a new member of the *N*-acetyltransferase family, alters glycosylation patterns of various immunoregulatory proteins, including MHC class I, and

components of the immunological synapse. In RA patients, the rSNP may regulate expression of *SLC22A4* (solute carrier family 22, isoform 4), which is an organic cationic transporter. The location of this gene is 5q31, a locus also implicated in Crohn disease (47). While this locus contains a cytokine gene cluster, SLC22A4 is not a cytokine, and its specific immunoregulatory role or impact on inflammatory processes is currently unclear. While this specific organic cation transporter does not appear to regulate uptake of L-arginine (48), there is growing interest in the importance of cationic amino acid transporter expression in skin diseases. If SLC22A4 were found to influence L-arginine levels, this could, in turn, modulate extracellular levels of nitric oxide and thereby alter TCR-ζ chain expression (49), providing a further link between the genetics of autoimmunity and the immunological synapse. Before finishing this brief review of ionic transporters, it should be noted that a different family of cation/chloride cotransporters has also been implicated in psoriasis (PSORS5) (50).

As can be appreciated from these new clinical trial results and DNA sequencing/genotyping data, advances related to non–antigen-specific (i.e., costimulatory) pathways have outpaced attempts to define or identify a precise antigen in this T cell–dependent skin disease. Unfortunately, no consistent antigen has been identified, nor has a specific gene been validated, as a definitive cause of psoriasis. Perhaps more gratifying for investigators have been the consensus and progress in defining the cytokine network operative in psoriatic lesions. Indeed, as described next, delineation of cytokine cascades facilitated transition in the clinical arena from serendipity to more selective and highly targeted therapeutics developed specifically for psoriasis.

Molecular effector pathways in psoriasis

Figure 3 provides our working hypothesis for the immunopathogenesis of psoriasis. This model contains elements from earlier proposals (13, 17). Multiple signaling pathways are envisioned to contribute to the pathological process whereby symptomless skin is converted to psoriatic plaques. An initial activating signal is portrayed as perturbing epidermal keratinocytes (13). Mild skin trauma, such as a cut or abrasion, in which epidermal keratinocytes are damaged can trigger psoriasis (this is known as the Köbner phenomenon) and a change in the epidermal maturation pathway. Exactly what happens in keratinocytes to create a "danger signal" is unclear (51), but several possible molecular events may occur to activate resting dendritic APCs (52), followed by delivery of an antigenic signal to T cells (so-called signal no. 1). These possible events include perturbation of the barrier function of skin with release of pre-formed or rapidly produced cytokines such as IL-1 and TNF-α (53); exposure of DCs constitutively expressing MHC class II molecules such as HLA-DR in the epidermis and/or dermis to bacterial products from skin flora (54), including superantigens (55); release of heat shock proteins from epidermal keratinocytes that could bind to CD91 expressed by dendritic APCs (56); exposure of keratinocytes and dendritic APCs to glycolipids that bind to CD1d (57, 58); and engagement of Toll-like receptors on dendritic APCs by unspecified molecular determinants (59).

The next step involves delivery of additional costimulatory signals (signal no. 2), which is likely to involve CD28 and B7 family members including B7.1 (CD80) and B7.2 (CD86) based on in vitro and in vivo studies (37, 60). New B7-related family members (and their ligands) have been identified, and it will be interesting to determine whether they also play a role in the pathogenesis of psoriasis. Once T cells and dendritic APCs are fully activated, they can create a "cytokine storm" composed of numerous cytokines, chemokines, and growth factors, as summarized in the lower right panel of Figure 3. A vicious cycle can then be envisioned in which keratinocytes, endothelial cells, neutrophils, and immunocytes in the vicinity become activated and conspire in the creation of a psoriatic plaque. Macrophages are also present, and it remains to be determined whether they are functioning to enhance the inflammatory response or to limit local immune reactions.

One of the first questions addressed by investigators was whether a polarized Th1- versus Th2-type cytokine-production profile would apply to human diseases. Initial analysis of psoriatic plaques with rather limited cytokine analysis revealed a Th1-type profile (61, 62). IFN-γ, TNF-α, and IL-12, but not IL-4, IL-5, or IL-10, were documented within psoriatic plaques at the mRNA and protein levels. As investigators probed the interactions among the Th1-type cytokines in vitro, it became apparent that there was synergy between IFN-γ and TNF-α with regard to production of adhesion molecules such as ICAM-1, and chemotactic polypeptides such as IL-8 or monocyte chemotactic activating factor-1 (MCAF-1). The challenging task of integrating the in vitro and in vivo findings into a coherent cytokine network that had spatial and temporal validity was first undertaken in 1991 (17). This initial portrayal of the cytokine network in psoriasis placed TNF-α at center stage as a key primary cytokine involved in the induction and maintenance of plaques. Polymorphisms in genes regulating cytokine production have been identified for TNF-α and other inflammatory mediators (63, 64). Given pleiotropic effects by which TNF-α influences a wide variety of cell types in both skin and joints of psoriatic patients, several pharmaceutical companies began targeting this cytokine, using two approaches. However, before we enter the therapeutic arena, additional details of the cytokine network are presented.

During the past decade, numerous reports have provided additional molecular details concerning the cytokine network in psoriatic plaques including chemokines, growth factors, and signal transduction pathways. With the advent of high-throughput cDNA-based microarrays, this list has grown exponentially (65, 66). Space constraints do not permit a complete summary of all of these mediators; however, some of the most prominent components are listed in Figure 3. Besides these mediators, consideration of high-mobility group B1 (HMGB1) protein binding to its receptor, receptor for advanced-glycation end products (RAGE) (65), is warranted, since HMGB1 influences cytokines such as TNF-α. In other chronic inflammatory diseases, HMGB1 is also considered an important regulator (67). Production of cytokines derived from APCs includes TNF-α (68) and IL-23 (69, 70). T cells are the likely source for IFN-γ, IL-15 (71, 72), and IL-17 (73), whereas keratinocytes can produce IL-1, IL-6, and IL-8, as well as IL-18 (74) and IL-20 (75). Besides these cytokines, numerous chemokines and chemokine receptors are present in psoriatic plaques. When chemokines bind to their respective receptors, they activate the cells, which may be important not only for recruitment into the skin, but also for their local release of cytokines and growth factors.

Chemokines and chemokine receptors of interest in the immunopathogenesis of psoriasis include TARC (CCL17), MIG (CXCL9), IP10 (CXCL10), MDC (CCL22), and RANTES (CCL5), as recently reviewed by Krueger (4), as well as CXCR2, CXCR3, CCR4, CCL27-CCR10, MIP3α (CCL20), MIP3β (CCL19), and CCR6. In addition, nitric oxide is present, which may contribute to an angiogenic tissue reaction, accompanied by many growth factors present at elevated levels within psoriatic plaques, including TGF-α, IGF-1, keratinocyte growth factor (KGF), VEGF, nerve growth factor (NGF), amphiregulin, and IL-20 (4). Given the plethora of these cytokines, chemokines, and growth factors, it should not be surprising that the end result is a thick, erythematous scaly plaque. In general, activated CD4+ T cells are primarily located in the dermis and CD8+ T cells in the psoriatic epidermis, accompanied by tangled collections of dendritic APCs predominantly located in the dermis.

Besides those molecules listed, potentially important signaling pathways including NF-κB, STAT-1, STAT-3, and IFN-α–inducible proteins are also now on the radar screen of psoriasis research (65, 66). Not only can high-throughput cDNA-based technology be used to identify specific transcripts that are elevated in psoriatic plaques; sequential analysis of plaques following treatment can be completed to explore the pharmacogenomics of psoriasis (65). Knowing which transcripts are reduced as lesions improve provides potential new therapeutic targets for future clinical trials, as detailed in the last section of this review. In the next section, additional lessons regarding the immunopathogenesis of psoriasis, derived from two different animal models, are reviewed. Besides xenogeneic animal models, transgenic murine models are important, as exemplified by a recent report in which genetic inactivation of a chemokine receptor 2 unexpectedly provoked a severe disease state similar to human RA, rather than the predicted reduction in inflammation (76).

Lessons learned from xenogeneic animal models

Up until the introduction of cyclosporin A, the prevailing dogma in dermatology was that psoriasis fundamentally represented a disease of epidermal keratinocytes (77). Early clinical evidence implicating bone marrow derived immunocytes as primary pathogenic cells emanated from bone marrow transplantations. Formal proof that psoriasis could be transferred by immunocytes required use of animal models in which human skin was engrafted onto SCID mice (78, 79). SCID mice are a strain with introduced mutations that render them genetically incapable of producing either T cells or B cells but retain their capacity to produce neutrophils and NK cells. Hence, they cannot reject human skin grafts and became widely used for organ-engraftment studies. It should be noted that psoriasis appears to be a uniquely human disease, as it has never been observed in primates or other animals. Moreover, given the likelihood that psoriasis is polygenic, attempts to develop transgenic rodent models with single-gene mutations or deletions have failed to generate skin lesions with all the relevant clinical, histological, and immunophenotypic abnormalities.

The ability to create bona fide psoriatic plaques using human skin grafted onto SCID mice opened up new avenues of investigation. Two important sets of observations will be highlighted. First, these new in vivo–based tools allowed researchers to determine whether pathogenic T cell subsets included CD4+ or CD8+ T cells, and whether other T cell subsets, such as NKT cells, may be relevant to psoriasis. The initial observation in which prepsoriatic skin was converted to psoriatic plaques using the SCID-Hu model suggested that CD4+ T cells, but not CD8+ T cells, were

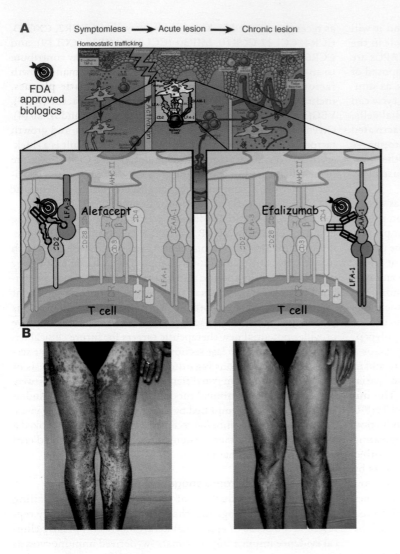

Figure 5
(**A**) T cell–targeted therapies in psoriasis. Two FDA-approved biologics (alefacept and efalizumab) are portrayed; the molecular target is identified on the T cell surface (upper panels). Note that the site of action of these agents is depicted in the skin, but the therapeutic efficacy may include other anatomical sites such as lymph nodes or other secondary lymphoid tissues. Alefacept targets the CD2:LFA-3 ligand/receptor interactions, whereas efalizumab targets the LFA-1:ICAM-1 ligand/receptor pair of surface molecules expressed by T cells and APCs, respectively. (**B**) The clinical response of a patient with severe psoriasis (left) to efalizumab administered for 2 months. Note the almost complete clearing of lesions on the lower extremities and hand (right).

capable of triggering psoriasis (80). A second group, using a pure murine inflammatory skin model supported this conclusion (81). Results implicating CD4⁺ T cells were unexpected, as the genetic data overwhelmingly implicated HLA-Cw6 (21, 23), an MHC class I molecule usually linked to CD8⁺ T cells rather than CD4⁺ T cells. This surprising result led to a search for other surface receptors on CD4⁺ T cells that can recognize MHC class I molecules such as molecules shared by NK cells (so-called NKT cells). Indeed, two independent groups have now found evidence for a possible role of NKT cells in the immunopathogenesis of psoriasis (82, 83). However, the precise identity of the actual pathogenic T cell is not currently known. Intraepidermal CD8⁺ T cells, which are activated and present in psoriatic plaques, might actually be the true pathogenic T cell, particularly since HLA-Cw6 remains the strongest genetic candidate locus for psoriasis susceptibility.

The second set of observations related to use of the SCID-Hu model has shown that the model is useful as a pharmacologically validated experimental tool for preclinical screening of promising therapeutic agents. Numerous reports have now described the favorable responses of engrafted psoriatic plaques to a wide range of agents targeting a variety of putative molecular mediators (72, 84–86). Not only can the SCID-Hu model assist the pharmaceutical and biotechnology indus-

tries, but response of grafted human skin to the targeted pathways will also provide valuable new insights into the relevance of specific molecules in maintaining the psoriatic phenotype. Such new insights are unlikely to be achieved by examination of various cell types only in tissue culture, or by the use of transgenic rodents.

Besides the SCID-Hu xenogeneic model, a new and different approach has recently been described in which psoriatic plaques are spontaneously created. This animal model uses AGR129 mice (which are deficient in type I and type II IFN receptors) and has recombination-activating gene 2 knocked out (87). When symptomless skin is engrafted onto AGR129 mice, plaques spontaneously develop without exogenous delivery of CD4⁺ T cells. Moreover, when human skin is serially examined after transplantation, resident human T cells undergo local proliferation with production of TNF-α. Such T cell proliferation and cytokine release are crucial for development of a psoriatic phenotype, since selective agents used to block T cell proliferation (anti–human CD3 antibody) or TNF inhibitors (infliximab or etanercept) prevent conversion of prepsoriatic skin to psoriatic plaques.

Taken together, these results warrant a paradigm shift in which resident T cells and APCs are viewed as necessary and sufficient for induction of psoriasis. The precise mechanism underlying resident T cell proliferation and cytokine release is unclear, but AGR129 mice differ from SCID mice in that they lack NK cells. Thus, AGR129 mice may serve as a host that cannot reject human CD4$^+$ T cells or DCs in the graft by virtue of the absence of NK cells. Furthermore, the lack of IFN receptors in AGR129 mice may facilitate a cytokine milieu (including IL-7 or IL-23) conducive to activation and proliferation of dormant pathogenic T cells residing in the prepsoriatic skin. Before conclusion of this section, three additional points should be emphasized. First, the AGR129 results support a key pathogenic role for TNF-α in the creation of plaques. Second, these results nicely complement earlier results using SCID mice in which engraftment of psoriatic plaques revealed a sustained phenotype after transplantation, indicating that immunocytes contained within the plaque were necessary and sufficient for maintenance of the chronic inflammatory pathological process. Third, to our knowledge, the AGR129 mouse model represents the first spontaneous animal model for a human autoimmune disease process, and it will facilitate development and testing of agents that could be used to prevent, rather than simply reverse, psoriatic lesions.

However, there are limitations in using xenograft models. First, engrafted skin has been disconnected from its lymphatic drainage, and, upon engraftment onto mice, no further human bone marrow–derived cells can participate in lesion formation; this precludes studies of trafficking between circulating cells and skin. Second, the models are cumbersome and require availability of patients willing to donate keratome samples from their skin. Third, the surgical procedure used to procure tissue, as well as the transplantation process, may produce wound-healing responses that complicate interpretation of the results. Next, a review of exciting new clinical developments that improve psoriatic skin lesions and psoriatic arthritis is presented.

Immunological targeted therapy in psoriasis

Because of space limitations, this section is not a complete review of all agents used or tested for management of psoriatic patients; it focuses instead on T cell–targeted treatments recently approved by the FDA, and those agents that target cytokines. Figure 5 profiles two different T cell–targeted therapies approved by the FDA. Alefacept is an LFA-3 Ig fusion protein that interferes with CD2: LFA-3 interactions (upper left panel). Efalizumab is an anti–LFA-1 antibody that interferes with LFA-1:ICAM-1 interactions, which also improves psoriatic lesions (upper right and lower panels). Besides these T cell–targeted therapies, three main cytokine-targeted approaches have been successfully used in the treatment of RA (Figure 2), psoriatic arthritis (4), and psoriatic skin lesions.

In the first approach, small molecules were developed that prevent cytokine release from immunocytes. Initial success was obtained using cyclosporin A, which is a calcineurin inhibitor that blocks signal transduction pathways that lead to cytokine release. Another calcineurin inhibitor is tacrolimus, which is 50–100 times more potent than cyclosporin A (88). Other immunosuppressant drugs that block cytokine release and have been found to be effective in psoriasis include sirolimus (rapamycin) (89) and pimecrolimus (ASM981) (90). These agents are derived from various soil fungi, contrasting sharply with a new wave of biological agents produced using recombinant DNA technology.

The other two strategies, based on use of biotechnology, attempt to neutralize cytokines once they are produced. Several agents target TNF-α, although many other approaches that target different cytokines such as IFN-γ and the p40 subunit shared by IL-12 and IL-23 are under active study. One strategy, using recombinant

biotechnology to neutralize TNF-α, is to produce highly specific mAb's that bind TNF-α, thereby preventing it from interacting with its surface receptors. Examples of such neutralizing antibodies against TNF-α include both chimeric antibodies (proteins containing a predominantly human backbone sequence combined with murine amino acid sequences), such as infliximab (91), and the fully humanized antibody adalimumab (92). The second strategy produces soluble receptors lacking signaling domains so as to bind and thereby neutralize TNF-α. This strategy derived from consideration of how viruses escape immunosurveillance using soluble receptors. Examples of soluble receptors are etanercept, which binds TNF-α and improves psoriatic arthritis and skin lesions (93, 94); and onercept, which improves Crohn disease (95).

Not only do many of the aforementioned treatments targeting TNF-α clearly benefit the majority of patients with psoriatic skin lesions, but they also can improve the inflammatory process in psoriatic joints, as well as in patients with RA and inflammatory bowel diseases such as Crohn disease. However, not all of these drugs work equally well in all patient groups, nor does the same drug necessarily improve both skin and joint manifestations equally. Also, since these drugs require injections for systemic delivery, there are injection-site reactions and concerns about potential infectious complications and oncological issues that will require long-term follow-up and clinical vigilance.

Finally, as a derivative of the second biotechnology approach, i.e., counterattack of existing cytokines, investigators are also producing recombinant human Th2 cytokines, and administering them to try to nullify the Th1-type cytokines or facilitate selective cytokine skewing of pathogenic T cells. Examples include delivery of IL-10 (96), IL-11 (65, 97), or IL-4 (98). While anti-inflammatory effects of IL-10 and IL-11 were relatively modest compared with those of the TNF-α inhibitors, clearly the entire field is rapidly moving from nonspecific treatments such as arsenic, corticosteroids, methotrexate, and UV light to more selective treatments aimed at defined pathogenic targets for patients with psoriasis and other chronic inflammatory diseases.

Future directions

There is no question that advances in understanding the cellular immunology and biology of psoriasis, when coupled with the biotechnology revolution and rapid advances derived from human genetic studies of autoimmunity, have enhanced insights into the cause and treatment of psoriasis. However, we still lack a cure for this common and enigmatic disease, and we have not unequivocally identified genes or antigens responsible for its occurrence worldwide. Moreover, many patients still experience significant side effects from currently available treatments, including development of skin cancers in nonlesional skin (with psoralen and UV-A light therapy), abnormal renal function (with cyclosporin A), and liver abnormalities (with methotrexate), to name a few.

As should be evident from this review, these are exciting times for clinicians and scientists interested in chronic inflammatory diseases such as psoriasis. Figure 6 emphasizes that there are still new innovations worth considering in psoriasis research. Future potential targets include T cell trafficking, T cell activation, cytokine inhibitors, and counteroffensive strategies. Briefly, trafficking targets may include very late antigen-1 (VLA-1, or $\alpha_1\beta_1$-integrin) expressed by T cells, which mediates binding to ECM molecules including collagen I and IV. Antibody against VLA-1 inhibits cutaneous hypersensitivity and arthritis in animal models (99). Another

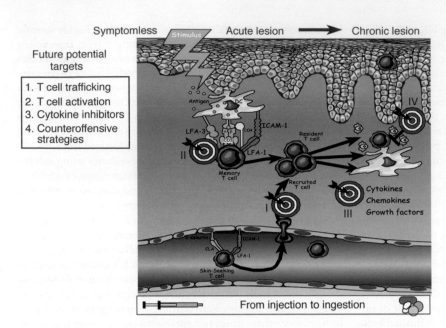

Figure 6
Innovations in psoriasis therapy. Future potential targets in the treatment of psoriasis are depicted. These therapeutic opportunities can be considered in the context of four broad areas: (I) T cell trafficking, (II) T cell activation, (III) cytokine inhibitors, and (IV) counteroffensive strategies. To enhance patient compliance and safety, anti-inflammatory and next-generation immunosuppressants ideally would be able to be ingested rather than injected.

trafficking-related target is VLA-4, which recognizes VCAM-1, and antibodies against VLA-4 improve multiple sclerosis and Crohn disease (100). T cell activation may also be impacted by these antibodies against surface molecules, as well as by other agents, that target various components of the immunological synapse depicted in Figure 4. Given the remarkable success of several different agents that target TNF-α in psoriasis, it is likely that additional reagents aimed at other cytokines, as described below, will surface in dermatology clinics. In the category of counteroffensive strategies, it may be possible in the near future to enhance relative levels for regulatory T cell subsets to negate the pathogenic T cell subsets.

Future challenges are severalfold and include the following: First, carefully monitor patients on currently approved new biologics and determine the long-term sequelae of chronic inflammatory inhibitors as regards potential risks of developing infections and neoplasms. Second, continue to define and refine the cytokine network working both upstream and downstream from TNF-α. In other words, identify what specific triggers account for the initial production of TNF-α. Such studies are likely to consider both innate and adaptive immunity. Cytokines that may impact TNF-α production include HMGB1, IL-15, and IL-23, to name a few. Third, elucidate the genetic basis by which this disease is transmitted from generation to generation. Our colleagues investigating Crohn disease have provided an encouraging precedent in which specific mutations in the *NOD2* gene are implicated in a subset of patients with this chronic inflammatory disease (101, 102). At least one group has identified a mutation in the *NOD2* gene (also known as *CARD15*) that confers susceptibility to psoriatic arthritis (103). Originally, it was believed that NOD2 was primarily expressed in monocytes and macrophages, but more recent studies indicate that NOD2 is also present in specialized epithelial cells known as Paneth cells, which are most numerous in the terminal ileum (104). Such new insights require reconsideration of an important role for epithelial cells in the immunopathogenesis of psoriasis. It would be ironic if the quest to identify the key cell in psoriasis, begun with a focus on keratinocytes, were concluded by studies aimed at keratinocytes once again.

Regarding the genetics of psoriasis, it remains to be determined whether final elucidation will reveal a rare gene variant for a common disease (as seen in Crohn disease), or whether this apparently complex genetic trait is created by multiple common low-risk susceptibility genes. It is intriguing to consider the possibility that mutations responsible for psoriasis may impact negative selection, thereby facilitating emergence of autoreactive T cells. Since only 5% of developing thymocytes emerge as mature T cells, ridding the immune system of autoreactive T cells may be imperfect, and tantalizing clues are beginning to point to this possibility for psoriasis and other autoimmune diseases, as highlighted in this review (Figure 4).

The fourth and final challenge for the future, based on the AGR129 mouse xenograft studies, is to focus on other resident cells besides keratinocytes, particularly localized subsets of immunocytes. Similarly, new clinical studies point to consideration of topical approaches targeting resident skin cells that may be associated with less systemic adverse events if the psoriatic patient does not have concomitant joint involvement. Besides novel transdermal delivery systems, the next generation of targeted therapy could include use of statins that interfere with LFA-1 (105), or novel small molecules that can interfere with inflammatory responses mediated by p38 signaling (106), cytokine receptors (107), or rationally designed dominant negative TNF-α variants (108).

Ultimately, once specific targets are validated with injectable reagents, such as the biological reagents, it will be possible to design orally available drugs to enhance patient compliance, safety, and medical costs. For example, orally ingestible statins that interfere with LFA-1 signaling should improve psoriasis, given the clinical efficacy of the anti–LFA-1 antibody (efalizumab). Perhaps lessons learned from studies in psoriasis and the skin immune system will also pay dividends for patients with other cytokine-mediated chronic inflammatory autoimmune diseases, including RA, lupus erythematosus, and inflammatory bowel disorders.

Acknowledgments

The authors appreciate the assistance of Stephanie Hiffman for manuscript and figure preparation. Brian Bonish created original artwork. The NIH has provided 13 years of continuous funding to Brian J. Nickoloff to support laboratory studies of psoriasis

(NIH R01 AR40065). June Robinson provided a clinical photograph (Figure 1), and James Krueger provided before/after clinical photographs (Figure 5). We also thank Pamela Pollock, Barbara Osborne, Kenneth Gordon, Jonathan Barker, and Jonathan Curry for reviewing the manuscript.

The authors apologize for omitting numerous important references because of space constraints. However, a list of over 200 relevant citations, arranged alphabetically by author, is available at http://www.jci.org/cgi/content/full/113/12/1664/DC1 to properly credit the other research groups responsible for data and concepts mentioned in the review.

Address correspondence to: Brian J. Nickoloff, Skin Disease Research Laboratory, Cardinal Bernardin Cancer Center, Loyola University of Chicago, 2160 S. First Avenue, Building 112, Room 301, Maywood, Illinois 60153, USA. Phone: (708) 327-3241; Fax: (708) 327-3239; E-mail: bnickol@lumc.edu.

1. Krueger, G., et al. 2001. The impact of psoriasis on quality of life: results of a 1998 National Psoriasis Foundation patient-membership survey. *Arch. Dermatol.* **137**:280–284.

2. Bhalerao, J., and Bowcock, A.M. 1998. The genetics of psoriasis: a complex disorder of the skin and immune system. *Hum. Mol. Genet.* **7**:1537–1545.

3. Henseler, T., and Christophers, E. 1985. Psoriasis of early and late onset: characterization of two types of psoriasis vulgaris. *J. Am. Acad. Dermatol.* **13**:450–456.

4. Krueger, J.G. 2002. The immunologic basis for the treatment of psoriasis with new biologic agents. *J. Am. Acad. Dermatol.* **46**:1–23; quiz 23–26.

5. Nickoloff, B.J. 2001. Creation of psoriatic plaques: the ultimate tumor suppressor pathway. A new model for an ancient T-cell-mediated skin disease. Viewpoint. *J. Cutan. Pathol.* **28**:57–64.

6. Henseler, T., and Christophers, E. 1995. Disease concomitance in psoriasis. *J. Am. Acad. Dermatol.* **32**:982–986.

7. Barker, J.N. 1991. The pathophysiology of psoriasis. *Lancet.* **338**:227–230.

8. Saiag, P., Coulomb, B., Lebreton, C., Bell, E., and Dubertret, L. 1985. Psoriatic fibroblasts induce hyperproliferation of normal keratinocytes in a skin equivalent model in vitro. *Science.* **230**:669–672.

9. Ackermann, L., et al. 1999. Mast cells in psoriatic skin are strongly positive for interferon-gamma. *Br. J. Dermatol.* **140**:624–633.

10. Raychaudhuri, S.P., Rein, G., and Farber, E.M. 1995. Neuropathogenesis and neuropharmacology of psoriasis. *Int. J. Dermatol.* **34**:685–693.

11. Lowe, P.M., et al. 1995. The endothelium in psoriasis. *Br. J. Dermatol.* **132**:497–505.

12. Gottlieb, A.B. 1988. Immunologic mechanisms in psoriasis. *J. Am. Acad. Dermatol.* **18**:1376–1380.

13. Nickoloff, B.J. 1999. The immunologic and genetic basis of psoriasis. *Arch. Dermatol.* **135**:1104–1110.

14. Valdimarsson, H., Baker, B.S., Jonsdottir, I., Powles, A., and Fry, L. 1995. Psoriasis: a T-cell-mediated autoimmune disease induced by streptococcal superantigens? *Immunol. Today.* **16**:145–149.

15. Chang, J.C., et al. 1997. Persistence of T-cell clones in psoriatic lesions. *Arch. Dermatol.* **133**:703–708.

16. Vollmer, S., Menssen, A., and Prinz, J.C. 2001. Dominant lesional T cell receptor rearrangements persist in relapsing psoriasis but are absent from non-lesional skin: evidence for a stable antigen-specific pathogenic T cell response in psoriasis vulgaris. *J. Invest. Dermatol.* **117**:1296–1301.

17. Nickoloff, B.J. 1991. The cytokine network in psoriasis. *Arch. Dermatol.* **127**:871–884.

18. Zhou, X., et al. 2003. Novel mechanisms of T-cell and dendritic cell activation revealed by profiling of psoriasis on the 63,100-element oligonucleotide array. *Physiol. Genomics.* **13**:69–78.

19. Lomholt, G. 1963. *Psoriasis: prevalence, spontaneous course, and genetics; a census study on the prevalence of skin diseases on the Faroe Islands.* Copenhagen, Denmark. G.E.C.-Gad. 295 pp.

20. Farber, E.M., and Nall, M.L. 1974. The natural history of psoriasis in 5,600 patients. *Dermatologica.* **148**:1–18.

21. Tiilikainen, A., Lassus, A., Karvonen, J., Vartiainen, P., and Julin, M. 1980. Psoriasis and HLA-Cw6. *Br. J. Dermatol.* **102**:179–184.

22. Gudjonsson, J.E., et al. 2003. Psoriasis patients who are homozygous for the HLA-Cw*0602 allele have a 2.5-fold increased risk of developing psoriasis compared with Cw6 heterozygotes. *Br. J. Dermatol.* **148**:233–235.

23. Capon, F., Munro, M., Barker, J., and Trembath, R. 2002. Searching for the major histocompatibility complex psoriasis susceptibility gene. *J. Invest. Dermatol.* **118**:745–751.

24. Nair, R.P., et al. 1997. Evidence for two psoriasis susceptibility loci (HLA and 17q) and two novel candidate regions (16q and 20p) by genome-wide scan. *Hum. Mol. Genet.* **6**:1349–1356.

25. Trembath, R.C., et al. 1997. Identification of a major susceptibility locus on chromosome 6p and evidence for further disease loci revealed by a two stage genome-wide search in psoriasis. *Hum. Mol. Genet.* **6**:813–820.

26. International Psoriasis Genetics Consortium. 2003. The International Psoriasis Genetics Study: assessing linkage to 14 candidate susceptibility loci in a cohort of 942 affected sib pairs. *Am. J. Hum. Genet.* **73**:430–437.

27. Ishihara, M., et al. 1996. Genetic polymorphisms in the keratin-like S gene within the human major histocompatibility complex and association analysis on the susceptibility to psoriasis vulgaris. *Tissue Antigens.* **48**:182–186.

28. Martin, M.P., et al. 2002. Cutting edge: susceptibility to psoriatic arthritis: influence of activating killer Ig-like receptor genes in the absence of specific HLA-C alleles. *J. Immunol.* **169**:2818–2822.

29. Nickoloff, B.J. 1999. Skin innate immune system in psoriasis: friend or foe? *J. Clin. Invest.* **104**:1161–1164.

30. Tomfohrde, J., et al. 1994. Gene for familial psoriasis susceptibility mapped to the distal end of human chromosome 17q. *Science.* **264**:1141–1145.

31. Speckman, R.A., et al. 2003. Novel immunoglobulin superfamily gene cluster, mapping to a region of human chromosome 17q25, linked to psoriasis susceptibility. *Hum. Genet.* **112**:34–41.

32. Helms, C., et al. 2003. A putative RUNX1 binding site variant between SLC9A3R1 and NAT9 is associated with susceptibility to psoriasis. *Nat. Genet.* **35**:349–356.

33. Bromley, S.K., et al. 2001. The immunological synapse. *Annu. Rev. Immunol.* **19**:375–396.

34. Lebwohl, M., et al. 2003. A novel targeted T-cell modulator, efalizumab, for plaque psoriasis. *N. Engl. J. Med.* **349**:2004–2013.

35. Nickoloff, B.J., et al. 1993. Activated keratinocytes present bacterial-derived superantigens to T lymphocytes: relevance to psoriasis. *J. Dermatol. Sci.* **6**:127–133.

36. Ellis, C.N., and Krueger, G.G. 2001. Treatment of chronic plaque psoriasis by selective targeting of memory effector T lymphocytes. *N. Engl. J. Med.* **345**:248–255.

37. Abrams, J.R., et al. 2000. Blockade of T lymphocyte costimulation with cytotoxic T lymphocyte-associated antigen 4-immunoglobulin (CTLA4Ig) reverses the cellular pathology of psoriatic plaques, including the activation of keratinocytes, dendritic cells, and endothelial cells. *J. Exp. Med.* **192**:681–694.

38. Weinshenker, B.G., Bass, B.H., Ebers, G.C., and Rice, G.P. 1989. Remission of psoriatic lesions with muromonab-CD3 (orthoclone OKT3) treatment. *J. Am. Acad. Dermatol.* **20**:1132–1133.

39. Gottlieb, A.B., et al. 2000. Anti-CD4 monoclonal antibody treatment of moderate to severe psoriasis vulgaris: results of a pilot, multicenter, multiple-dose, placebo-controlled study. *J. Am. Acad. Dermatol.* **43**:595–604.

40. Gottlieb, S.L., et al. 1995. Response of psoriasis to a lymphocyte-selective toxin (DAB389IL-2) suggests a primary immune, but not keratinocyte, pathogenic basis. *Nat. Med.* **1**:442–447.

41. Werlen, G., Hausmann, B., Naeher, D., and Palmer, E. 2003. Signaling life and death in the thymus: timing is everything. *Science.* **299**:1859–1863.

42. Sakaguchi, N., et al. 2003. Altered thymic T-cell selection due to a mutation of the ZAP-70 gene causes autoimmune arthritis in mice. *Nature.* **426**:454–460.

43. Alarcon-Riquelme, M.E. 2003. A RUNX trio with a taste for autoimmunity. *Nat. Genet.* **35**:299–300.

44. Prokunina, L., et al. 2002. A regulatory polymorphism in PDCD1 is associated with susceptibility to systemic lupus erythematosus in humans. *Nat. Genet.* **32**:666–669.

45. Tokuhiro, S., et al. 2003. An intronic SNP in a RUNX1 binding site of SLC22A4, encoding an organic cation transporter, is associated with rheumatoid arthritis. *Nat. Genet.* **35**:341–348.

46. Itoh, K., et al. 2002. Cutting edge: negative regulation of immune synapse formation by anchoring lipid raft to cytoskeleton through Cbp-EBP50-ERM assembly. *J. Immunol.* **168**:541–544.

47. Rioux, J.D., et al. 2001. Genetic variation in the 5q31 cytokine gene cluster confers susceptibility to Crohn disease. *Nat. Genet.* **29**:223–228.

48. Yabuuchi, H., et al. 1999. Novel membrane transporter OCTN1 mediates multispecific, bidirectional, and pH-dependent transport of organic cations. *J. Pharmacol. Exp. Ther.* **289**:768–773.

49. Rodriguez, P.C., et al. 2003. L-arginine consumption by macrophages modulates the expression of CD3 zeta chain in T lymphocytes. *J. Immunol.* **171**:1232–1239.

50. Hewett, D., et al. 2002. Identification of a psoriasis susceptibility candidate gene by linkage disequilibrium mapping with a localized single nucleotide polymorphism map. *Genomics.* **79**:305–314.

51. Matzinger, P. 2002. The danger model: a renewed sense of self. *Science.* **296**:301–305.

52. Soumelis, V., et al. 2002. Human epithelial cells trigger dendritic cell mediated allergic inflammation by producing TSLP. *Nat. Immunol.* **3**:673–680.

53. Nickoloff, B.J., and Naidu, Y. 1994. Perturbation of epidermal barrier function correlates with initiation of cytokine cascade in human skin. *J. Am. Acad. Dermatol.* **30**:535–546.

54. Baker, B.S., et al. 2001. Skin T cell proliferative response to M protein and other cell wall and membrane proteins of group A streptococci in chronic plaque psoriasis. *Clin. Exp. Immunol.* **124**:516–521.

55. Travers, J.B., et al. 1999. Epidermal HLA-DR and the enhancement of cutaneous reactivity to superantigenic toxins in psoriasis. *J. Clin. Invest.* **104**:1181–1189.

56. Curry, J.L., Qin, J.Z., Robinson, J., and Nickoloff, B.J. 2003. Reactivity of resident immunocytes in normal and prepsoriatic skin using an ex vivo skin-explant model system. *Arch. Pathol. Lab. Med.* **127**:289–296.

57. Bonish, B., et al. 2000. Overexpression of CD1d by keratinocytes in psoriasis and CD1d-dependent IFN-gamma production by NK-T cells. *J. Immunol.* **165**:4076–4085.

58. Gerlini, G., et al. 2001. CD1d is expressed on dermal dendritic cells and monocyte-derived dendritic cells. *J. Invest. Dermatol.* **117**:576–582.

59. Curry, J.L., et al. 2003. Innate immune-related receptors in normal and psoriatic skin. *Arch. Pathol. Lab. Med.* **127**:178–186.

60. Nestle, F.O., Turka, L.A., and Nickoloff, B.J. 1994. Characterization of dermal dendritic cells in psoriasis. Autostimulation of T lymphocytes and induction of Th1 type cytokines. *J. Clin. Invest.* **94**:202–209.

61. Uyemura, K., Yamamura, M., Fivenson, D.F., Modlin, R.L., and Nickoloff, B.J. 1993. The cytokine network in lesional and lesion-free psoriatic skin is characterized by a T-helper type 1 cell-mediated response. *J. Invest. Dermatol.* **101**:701–705.

62. Schlaak, J.F., et al. 1994. T cells involved in psoriasis vulgaris belong to the Th1 subset. *J. Invest. Dermatol.* **102**:145–149.

63. Reich, K., et al. 2002. Promoter polymorphisms of the genes encoding tumor necrosis factor-alpha and interleukin-1beta are associated with different subtypes of psoriasis characterized by early and late disease onset. *J. Invest. Dermatol.* **118**:155–163.

64. Asadullah, K., et al. 2001. Interleukin-10 promoter polymorphism in psoriasis. *J. Invest. Dermatol.* **116**:975–978.

65. Oestreicher, J.L., et al. 2001. Molecular classification of psoriasis disease-associated genes through pharmacogenomic expression profiling. *Pharmacogenomics J.* **1**.272–287.

66. Bowcock, A.M., et al. 2001. Insights into psoriasis and other inflammatory diseases from large-scale gene expression studies. *Hum. Mol. Genet.* **10**:1793–1805.

67. Nathan, C. 2002. Points of control in inflammation. *Nature.* **420**:846–852.

68. Nickoloff, B.J., et al. 1991. Cellular localization of interleukin-8 and its inducer, tumor necrosis factor-alpha in psoriasis. *Am. J. Pathol.* **138**:129–140.

69. Aggarwal, S., Ghilardi, N., Xie, M.H., de Sauvage, F.J., and Gurney, A.L. 2003. Interleukin-23 promotes a distinct CD4 T cell activation state characterized by the production of interleukin-17. *J. Biol. Chem.* **278**:1910–1914.

70. Lee, E., et al. 2004. Increased expression of interleukin 23 p19 and p40 in lesional skin of patients with psoriasis vulgaris. *J. Exp. Med.* **199**:125–130.

71. Ruckert, R., et al. 2000. Inhibition of keratinocyte apoptosis by IL-15: a new parameter in the pathogenesis of psoriasis? *J. Immunol.* **165**:2240–2250.

72. Villadsen, L.S., et al. 2003. Resolution of psoriasis upon blockade of IL-15 biological activity in a xenograft mouse model. *J. Clin. Invest.* **112**:1571–1580. doi:10.1172/JCI200318986.

73. Albanesi, C., et al. 2000. Interleukin-17 is produced by both Th1 and Th2 lymphocytes, and modulates interferon-gamma- and interleukin-4-induced activation of human keratinocytes. *J. Invest. Dermatol.* **115**:81–87.

74. Ohta, Y., Hamada, Y., and Katsuoka, K. 2001. Expression of IL-18 in psoriasis. *Arch. Dermatol. Res.* **293**:334–342.

75. Blumberg, H., et al. 2001. Interleukin 20: discovery, receptor identification, and role in epidermal function. *Cell.* **104**:9–19.

76. Quinones, M.P., et al. 2004. Experimental arthritis in CC chemokine receptor 2–null mice closely mimics severe human rheumatoid arthritis. *J. Clin. Invest.* **113**:856–866. doi:10.1172/JCI200420126.

77. Krueger, G.G., Bergstresser, P.R., Lowe, N.J., Voorhees, J.J., and Weinstein, G.D. 1984. Psoriasis. *J. Am. Acad. Dermatol.* **11**:937–947.

78. Wrone-Smith, T., and Nickoloff, B.J. 1996. Dermal injection of immunocytes induces psoriasis. *J. Clin. Invest.* **98**:1878–1887.

79. Boehncke, W.H., Dressel, D., Zollner, T.M., and Kaufmann, R. 1996. Pulling the trigger on psoriasis. *Nature.* **379**:777.

80. Nickoloff, B.J., and Wrone-Smith, T. 1999. Injection of pre-psoriatic skin with CD4+ T cells induces psoriasis. *Am. J. Pathol.* **155**:145–158.

81. Kess, D., et al. 2003. CD4(+) T cell-associated pathophysiology critically depends on CD18 gene dose effects in a murine model of psoriasis. *J. Immunol.* **171**:5697–5706.

82. Nickoloff, B.J., Bonish, B., Huang, B.B., and Porcelli, S.A. 2000. Characterization of a T cell line bearing natural killer receptors and capable of creating psoriasis in a SCID mouse model system. *J. Dermatol. Sci.* **24**:212–225.

83. Gilhar, A., et al. 2002. Psoriasis is mediated by a cutaneous defect triggered by activated immunocytes: induction of psoriasis by cells with natural killer receptors. *J. Invest. Dermatol.* **119**:384–391.

84. Zeigler, M., et al. 2001. Anti-CD11a ameliorates disease in the human psoriatic skin-SCID mouse transplant model: comparison of antibody to CD11a with Cyclosporin A and clobetasol propionate. *Lab. Invest.* **81**:1253–1261.

85. Schon, M.P., et al. 2002. Efomycine M, a new specific inhibitor of selectin, impairs leukocyte adhesion and alleviates cutaneous inflammation. *Nat. Med.* **8**:366–372.

86. Zollner, T.M., et al. 2002. Proteasome inhibition reduces superantigen-mediated T cell activation and the severity of psoriasis in a SCID-hu model. *J. Clin. Invest.* **109**:671–679. doi:10.1172/JCI200212736.

87. Boyman, O., et al. 2004. Spontaneous development of psoriasis in a new animal model shows an essential role for resident T cells and tumor necrosis factor-{alpha}. *J. Exp. Med.* **199**:731–736.

88. Reynolds, N.J., and Al-Daraji, W.I. 2002. Calcineurin inhibitors and sirolimus: mechanisms of action and applications in dermatology. *Clin. Exp. Dermatol.* **27**:555–561.

89. Reitamo, S., et al. 2001. Efficacy of sirolimus (rapamycin) administered concomitantly with a subtherapeutic dose of cyclosporin in the treatment of severe psoriasis: a randomized controlled trial. *Br. J. Dermatol.* **145**:438–445.

90. Rappersberger, K., et al. 2002. Pimecrolimus identifies a common genomic anti-inflammatory profile, is clinically highly effective in psoriasis and is well tolerated. *J. Invest. Dermatol.* **119**:876–887.

91. Antoni, C., and Manger, B. 2002. Infliximab for psoriasis and psoriatic arthritis. *Clin. Exp. Rheumatol.* **20**(Suppl. 28):S122–S125.

92. Scheinfeld, N. 2003. Adalimumab (HUMIRA): a review. *J. Drugs Dermatol.* **2**:375–377.

93. Mease, P.J., et al. 2000. Etanercept in the treatment of psoriatic arthritis and psoriasis: a randomized trial. *Lancet.* **356**:385–390.

94. Leonardi, C.L., et al. 2003. Etanercept as monotherapy in patients with psoriasis. *N. Engl. J. Med.* **349**:2014–2022.

95. Rutgeerts, P., et al. 2003. Treatment of active Crohn's disease with onercept (recombinant human soluble p55 tumour necrosis factor receptor): results of a randomized, open-label, pilot study. *Aliment. Pharmacol. Ther.* **17**:185–192.

96. Asadullah, K., et al. 1998. IL-10 is a key cytokine in psoriasis. Proof of principle by IL-10 therapy: a new therapeutic approach. *J. Clin. Invest.* **101**:783–794.

97. Trepicchio, W.L., et al. 1999. Interleukin-11 therapy selectively downregulates type I cytokine proinflammatory pathways in psoriasis lesions. *J. Clin. Invest.* **104**:1527–1537.

98. Ghoreschi, K., et al. 2003. Interleukin-4 therapy of psoriasis induces Th2 responses and improves human autoimmune disease. *Nat. Med.* **9**:40–46.

99. de Fougerolles, A.R., et al. 2000. Regulation of inflammation by collagen-binding integrins alpha-1beta1 and alpha2beta1 in models of hypersensitivity and arthritis. *J. Clin. Invest.* **105**:721–729.

100. von Andrian, U.H., and Engelhardt, B. 2003. Alpha4 integrins as therapeutic targets in autoimmune disease. *N. Engl. J. Med.* **348**:68–72.

101. Hugot, J.P., et al. 2001 Association of NOD2 leucine-rich repeat variants with susceptibility to Crohn's disease. *Nature.* **411**:599–603.

102. Ogura, Y., et al. 2001. A frameshift mutation in NOD2 associated with susceptibility to Crohn's disease. *Nature.* **411**:603–606.

103. Rahman, P., et al. 2003. CARD15: a pleiotropic autoimmune gene that confers susceptibility to psoriatic arthritis. *Am. J. Hum. Genet.* **73**:677–681.

104. Lala, S., et al. 2003. Crohn's disease and the NOD2 gene: a role for paneth cells. *Gastroenterology.* **125**:47–57.

105. Weitz-Schmidt, G., et al. 2001. Statins selectively inhibit leukocyte function antigen-1 by binding to a novel regulatory integrin site. *Nat. Med.* **7**:687–692.

106. Kumar, S., Boehm, J., and Lee, J.C. 2003. p38 MAP kinases: key signalling molecules as therapeutic targets for inflammatory diseases. *Nat. Rev. Drug Discov.* **2**:717–726.

107. Changelian, P.S., et al. 2003. Prevention of organ allograft rejection by a specific Janus kinase 3 inhibitor. *Science.* **302**:875–878.

108. Steed, P.M., et al. 2003. Inactivation of TNF signaling by rationally designed dominant-negative TNF variants. *Science.* **301**:1895–1898.

Exploiting dendritic cells to improve vaccine efficacy

Ralph M. Steinman[1] and Melissa Pope[2]

[1]Laboratory of Cellular Physiology and Immunology, Rockefeller University, New York, New York, USA.
[2]Center for Biomedical Research, Population Council, New York, New York, USA

The challenges to vaccine biology are dramatized by the current situation with an AIDS vaccine (see Letvin, this Perspective series, ref. 1). For years, data have been available on the HIV-1 genome and its proteins, as well as numerous antigens recognized by the immune system. Still, this information has not been readily translated into candidate vaccines that induce the broad and long-lasting T cell–mediated immunity thought to be necessary to protect people from acquiring AIDS (2–5). Vaccines are also lacking for many other serious infections in which T cell–mediated immunity should be protective. These include pathogens whose genomic sequences and antigenic proteins are well characterized: tuberculosis, malaria, and the herpes simplex, papilloma, Epstein-Barr, and hepatitis C viruses. In essence, the identification of foreign antigens is necessary but not sufficient for producing vaccines that are effective in the T cell sphere. Better vaccine delivery and vaccine adjuvants, or enhancers of immunity, are required (6, 7).

We propose that dendritic cell (DC) physiology should be considered and exploited in meeting each of the challenges in vaccine biology that lie ahead (see Table 1). DCs act as nature's adjuvants for regulating antigen-specific immunity. As antigen-presenting cells, DCs capture antigens, process them into peptides, and present them on products of the MHC to T cells. DCs are both efficient and specialized in antigen presentation, and they control the magnitude, quality, and memory of the ensuing immune response. DCs have been used successfully as cellular adjuvants in mice to elicit protective T cell–mediated immunity against pathogens and tumors (8, 9). These cells are now being used to prime and expand T cells specific for human cancers (refs. 10–12; see also Yu and Restifo, this Perspective series, ref. 13). The responding T cells include helper cells, especially Th1 CD4⁺ cells, which produce IFN-γ; and killer cells, especially CD8⁺ cytolytic T lymphocytes (CTLs), which exocytose granules rich in perforin and granzyme. New information indicates that DCs control responses by other classes of lymphocytes (B, NK, and NKT cells) and elicit T cell memory, a critical goal of vaccination.

Developing the capacity to harness DCs for vaccination seems particularly urgent in confronting infectious agents that, like HIV-1, pose unusual demands with respect to safety; the time-honored approach of microbial attenuation is now being set aside as vaccine biologists turn to defined antigens, poorly replicating vectors, and DNA. Although these vaccines introduce foreign microbial products, they often generate weak immunity, especially T cell–mediated immunity. Consequently, greater emphasis on underlying immunologic processes is needed, notably the strong adjuvant roles of DCs. Interestingly, as we discuss below, even the classical vaccine approach of microbial attenuation, used successfully for smallpox and measles, may have unknowingly exploited the adjuvant roles of DCs.

DCs as natural adjuvants

In vitro studies. DCs are potent stimulators of T cell responses and T-dependent antibody formation in tissue culture. Relatively few DCs and relatively low doses of antigen are required to elicit high levels of lymphocyte proliferation and differentiation. Initially, because DCs had to be isolated directly from lymphoid tissues (or, in the case of humans, from blood), the scarcity of these cells imposed a serious limitation on DC research. Typically, DCs make up less than 1% of a given cell population — a figure that is somewhat misleading since the frequency of these cells is at least 100 times greater than that of T cells specific for any given antigen. Moreover, DCs are extensively ramified in regions of the lymph nodes through which T cells recirculate (Figure 1).

Most investigators now study DCs produced in much larger numbers from either CD34⁺ proliferating progenitors or CD14⁺ nonproliferating monocytic precursors. These DCs are charged or "pulsed" with antigens, which they efficiently process and display as MHC-peptide complexes. Antigen-pulsed DCs can be placed into culture with lymphocytes, whereupon T cells begin to proliferate and to produce lymphokines and various cytotoxic molecules. Primary responses to microbial antigens can be difficult to achieve in short-term (1 week) culture, because the initial number of antigen-responsive cells is so low (<1 in 10⁵ lymphocytes), but mature DCs rapidly induce recall responses to many antigens, including those from HIV-1 and influenza. These viral antigens are presented to primed CD4⁺ and CD8⁺ T cells even when delivered to the DCs in poorly replicating vectors and as ultraviolet light and chemically inactivated virions. The potency of DCs in stimulating T cells in vitro reflects both their specialized ability to capture and present antigens and the effects of other molecules, not present in MHC complexes, that enhance T cell binding and stimulation.

In vivo evidence for DCs as strong adjuvants. Vaccination with DCs leads to protective immunity against infections and tumors (8, 9) and, in the case of certain self antigens, autoimmunity. DCs can be exposed to an antigen either in vivo, by introducing the antigen directly, or ex vivo, by pulsing the cells with antigen while they are in culture and administering them to genetically matched animals. After antigenic proteins are given to mice, DCs are found to be the main cells capturing antigen in an immunogenic form. When mice are challenged with microbes, DCs also are the principal cells producing the key protective cytokine IL-12. Ex vivo–activated DCs can prime recipient animals in an antigen-specific manner, allowing them to respond to an antigenic challenge within a week. These DCs migrate to the recipients' lymph nodes and lodge in the T cell areas, sites through which lymphocytes enter the lymph nodes via high endothelial venules (Figure 1). This

Citation for this article: *J. Clin. Invest.* **109**:1519–1526 (2002). doi:10.1172/JCI200215962.

movement positions the DCs in a seemingly ideal niche to select antigen-specific T cells when the latter percolate through the node. Such selection can be observed directly in situ: Following activation in contact with DCs, the T cells leave the lymph node, freeing them to fight infections and tumors. Some also become memory T cells, a response whose mechanism remains to be unraveled.

For purposes of vaccine design, it may be most straightforward to target antigens selectively to DCs in situ. This has been achieved through the DEC-205 receptor (CD205) (14), which mediates antigen uptake and processing in DCs. Crucially, induction of immunity also requires a stimulus that matures the DEC-205+ DCs. Antigen presentation sets the stage for antigen-specific T cell recognition, but maturation controls the T cell response. Therefore, vaccines must not only contain the requisite antigens to initiate protective immunity but also provide stimuli to promote DC maturation.

Exploiting the adjuvant roles of DCs

To date, the role of DCs in vaccine efficacy has been studied in detail only in mice receiving DNA vaccinations. Nevertheless, it is already evident that DCs have several features that could be modulated using appropriately designed vaccines to generate stronger T cell–mediated immunity. Below, we consider three aspects of DC biology that are of particular interest in vaccine development: antigen presentation, DC maturation, and DC mobilization.

Antigen handling and presentation. Vaccine antigens are not presented directly to the immune system but must first be captured, processed, and bound to antigen-presenting molecules, typically those of the class I or class II MHC. Humoral immunity depends on the fact that B cells and their antibody products react directly with native antigens on pathogens or their toxins, thus neutralizing the pathogen or toxin extracellularly, prior to binding and/or entry into cells. In contrast, T cell–mediated immunity to intracellular infections requires recognition of fragmented antigens produced within infected targets. The fragments are typically peptides that bind to highly polymorphic class I and class II products of the MHC and are then displayed on the cell surface as MHC-peptide complexes (15). Other, less polymorphic antigen-presenting molecules have been found, including the CD1 family, which is responsible for the presentation of microbial glycolipids (16), and the so-called MHC class Ib products, which present formylated bacterial peptides (15).

Despite the fact that DCs can capture and present to T cells even nonspecific, soluble proteins that prove to be poorly immunogenic, DC targeting clearly offers a valuable strategy for vaccination. Quantitative efficiency is one significant benefit of this approach — a peptide sequence within a protein delivered specifically to DCs is 100–1,000 times more efficient than a peptide given in a nonspecific adjuvant like CFA (14). A second benefit relates to the quality of the antigen processing (17). For example, by targeting select antigen uptake receptors on DCs, the vaccine can access their more efficient antigen processing and presentation pathways, particularly the exogenous pathway discussed below, which allows proteins and poorly replicating vaccines to load both MHC class I and class II molecules, as well as CD1.

Antigen presentation on MHC class I products, including the exogenous pathway. The presentation of vaccine antigens on MHC class I is needed to activate CD8+ CTLs, which kill infected targets early in the microbial life cycle, thereby blocking replication and spread of the pathogen. The classical, "endogenous" pathway for present-

ing peptides on MHC class I products begins when DCs or other cells are productively infected, as when DCs present antigens from influenza and recombinant vaccinia virus. Following endogenous synthesis within DCs, microbial proteins are clipped by the proteasome, and peptide fragments are transported via transporters for antigen presentation (TAPs) into the rough endoplasmic reticulum (15). There, the resulting peptides are affixed to the peptide-binding grooves of newly formed MHC class I products, and the MHC-peptide complexes exit via the Golgi apparatus to reach the surface for presentation to antigen receptors on T cells (15).

DCs are proving to be quite specialized in their capacities to form MHC class I–peptide complexes, which go beyond the classical endogenous pathway summarized above. One specialty is to present viruses that have been inactivated by ultraviolet light, heat, or chemical treatment, responses not seen with most other cell types. The inactive viruses retain their capacity to fuse with the plasma or endocytic vacuole membrane, thereby delivering some virion proteins into the cytoplasm. Subsequent efficient processing of incoming virions, or processing of newly synthesized proteins produced at low levels, may explain the capacity of DCs to present inactivated but fusogenic viruses, but this possibility needs further study.

Another specialty of DCs is to bring about what is termed "exogenous presentation" or "cross-presentation." These pathways act, respectively, on proteins derived from immune complexes or inactivated microbes, or on antigens originally synthesized in other cells, which then "cross" to the MHC products of DCs. In all such cases, antigens depend on selective endocytic uptake receptors to gain access to the cytoplasm of DCs, after which they likely engage

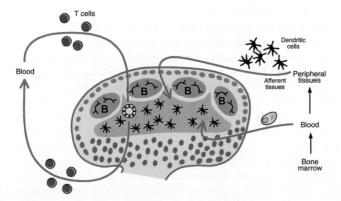

Figure 1
Lymphocyte and DC circulations. Naive lymphocytes circulate from blood via high endothelial venules into lymphoid tissues. B cells then move into follicles while T cells percolate through T cell areas, both eventually leaving the node via efferent lymphatics to return to the blood. Upon antigen recognition, some activated B and T cells, as well as DCs and follicular dendritic cells (a distinct cell type that, unlike DCs, retains native antigens as immune complexes), congregate in the follicles to generate the germinal center reaction for antibody formation. Other activated B and T cells return to inflammatory sites via the blood or become memory cells. Some of the latter are termed "effector memory" cells, because they can rapidly produce cytokines and are positioned in peripheral tissues. In parallel to the circulation of lymphocytes, DCs move from blood to tissues and then into afferent lymphatics, which bring DCs into the T cell areas where they eventually die. The plasmacytoid subset of DCs enters the T cell areas directly from blood; their subsequent fate is unclear.

Table 1

Challenges in vaccine biology requiring improved control of antigen presentation

Enhance antibody and T cell–mediated immune memory, especially in humans
Improve the quality of the T cell response, e.g., CD4+ Th1 helper and CD8+ killer cells
Achieve mucosal immunity, particularly for sexually transmitted diseases
Design therapeutic as well as preventive vaccines, e.g., against HIV-1 and cancer
Identify vaccines that dampen immunity, e.g., in autoimmune disease

the known DC systems that allow for protein ubiquitination, proteasomal cleavage of antigens, and TAP-mediated peptide transport. Thus, endocytosed antigens can gain access to the cytoplasm without the need for a viral envelope to mediate delivery. The exogenous pathway allows DCs to present many forms of nonreplicating antigens on MHC class I and thereby to elicit CD8+ CTLs. Active infection and biosynthesis do not need to take place in the DCs (18, 19). A good example is vaccinia virus: This prototype for successful vaccines is actually presented, at least in mice, almost entirely by the exogenous or cross-presenting routes (20).

Several DC receptors lead to MHC class I–peptide complex formation via the exogenous pathway. These include the FcγR, which binds immune complexes and antibody-coated tumor cells; the integrin $\alpha_v\beta_5$ and the phosphatidylserine receptor, which bind dying cells; and various receptors for heat shock proteins. Subsequent delivery of antigen into the cytosol is postulated to require a transporter that allows macromolecules to escape the endocytic vacuole. Once in the cytoplasm, proteins may be subject to the newly recognized heightened capacity of maturing DCs to polyubiquitinylate proteins. Ubiquitin conjugation marks the proteins for efficient proteasomal processing. It is anticipated that additional DC specializations will be found for increasing their efficiency in MHC I–peptide complex formation.

The exogenous and cross-presentation pathways via DCs constitute important routes to natural immunity in many infectious diseases, because DCs can capture and present antigens from immune complexes or dying infected cells to elicit CD8+ T cell immunity (18, 19). These pathways also substantially change how one approaches the design of vaccines for cell-mediated immunity. Nonreplicating vaccines, such as protein subunits and chemically inactivated vaccines, are generally thought to be unable to elicit CD8+ CTLs, which may be critical for defense against certain chronic intracellular infections and tumors. Subunit and inactive vaccines lose efficacy, it is thought, because they do not lead to the new intracellular synthesis of proteins required for processing in the classical endogenous pathway to MHC class I. However, the barrier to developing vaccines that engage the class I MHC seems no longer insurmountable if immunologists can learn to exploit the exogenous pathway in DCs.

It should be noted that, although many investigators use the terms "exogenous pathway" and "cross-presentation" to refer exclusively to presentation on MHC class I, DCs simultaneously present exogenous proteins and cellular antigens on MHC class I and II. Thus, as considered below, CD4+ helper T cells can be engaged to amplify the CD8+ T cell–mediated, MHC class I–restricted responses initiated by DC presentation.

Antigen presentation on MHC class II products. The MHC class II pathway, which forms MHC-peptide complexes to be recognized by CD4+ helper T cells, is particularly efficient in DCs. To illustrate, when a protein is delivered to DCs from dead cells, the for-

mation of MHC II–peptide complexes is actually many thousand times more efficient than when preprocessed peptides are applied. DCs have many candidate receptors for dying cells (21), but active receptors in vivo remain to be identified. Conversely, several DC-restricted uptake receptors are known (Figure 2) for which natural ligands remain to be identified. One example is the DEC-205 (CD205) uptake receptor, which traffics in a distinct way through DCs and greatly enhances antigen presentation relative to other adsorptive endocytic receptors. DEC-205 can recycle through the acidic late endosomal/lysosomal vacuoles in maturing DCs, compartments that are enriched for MHC class II molecules and proteinases like the cathepsins that mediate antigen processing and MHC class II–peptide complex formation.

These complexes, once formed, are transported to the DC surface within distinctive nonlysosomal transport vesicles. The vesicles contain both the MHC-peptide complexes, recognized by the T cell receptor, and the CD86 molecules, required to costimulate T cell growth. Upon arrival at the DC surface, the processed antigen and CD86 remain coclustered in aggregates that contain so-called tetraspannin membrane proteins. This situation seems ideal to set up immunologic synapses between DCs and the T cells that they activate. At this final mature stage, the DCs silence transcription of MHC class II products (whose genes are activated by the transcriptional activator CIITA) and shut down much of their endocytic activity, while actively presenting antigens captured in the periphery or vaccine site at lymphoid tissues (Figure 1).

Antigen presentation on CD1 glycolipid-binding molecules

DCs express the known members of the CD1 family of antigen-presenting molecules, but individual CD1 molecules can be restricted to subsets of DCs. CD1a is typically found on epidermal Langerhans cells in skin, while CD1b and c are expressed on dermal DCs. CD1 molecules present microbial glycolipids (22), but in addition, CD1d on monocyte-derived DCs presents the drug α galactosylceramide. The CD1d-restricted cells are called NKT cells. Following recognition of glycosphingolipid on CD1d, NKT cells orchestrate the production of large amounts of cytokines from several cell types and have the capacity to act as adjuvants for T cell–mediated immunity (23). Interestingly, none of the CD1 molecules have been found on the plasmacytoid subset of DCs discussed below.

DC maturation. In the absence of a perturbation such as infection or vaccination, most DCs remain at an immature stage of differentiation. To exploit DCs in vaccine design, the vaccine must not only provide protective antigens that are captured by DCs; it must also induce DC maturation.

Immature DCs can capture antigens, but they must differentiate or mature to become strong inducers of immunity. DC maturation is the control point that determines whether an antigen is to become an immunogen, and it can take place not just as a response to microbial infections, but also in other forms of strong T cell–mediated immunity such as transplantation reactions, contact allergy, and autoimmunity. There are two well-studied classes of maturation stimuli. One class is provided when the microbe or vaccine signals DCs through toll-like receptors (TLRs); a second class is provided by lymphocytes and other cells (either T, B, NK, NKT, platelets, or mast cells) that deliver TNF-type signals to the DCs.

Many defined microbial products initiate DC maturation through TLR signaling (24, 25). Cytokine production is triggered quickly, as is also the case with many other cell types. However, DCs can produce particularly high levels of immune-enhancing cytokines like IL-12, IFN-α, and even, in some situations, IL-2. Over longer periods, DCs mature to become strong adjuvants for T cell immunity. Expression of specific TLRs can be high in DCs, particularly TLR9, which responds to microbial DNA (26), and TLR3, a receptor for double-stranded RNA. TLRs can respond to particular small molecules, like specific CpG deoxyoligonucleotide sequences, or to complex microbial macromolecules like DNA. As discussed below, distinct DC subsets express different complements of TLRs. In terms of signal transduction, TLRs use the MyD88 adaptor protein to trigger cytokine release from different cell types (24). However, DC maturation through certain TLRs is also influenced by a MyD88-independent mechanism (24, 25) that will be important to identify and manipulate.

TNF family members that stimulate DCs include TNF itself, Fas ligand (FasL), CD40 ligand (CD40L), and TRANCE (RANKL). These molecules are expressed in a membrane-bound form by activated T cells and signal the corresponding activating subclass of TNF receptors (TNF-Rs). When microbial or vaccine stimuli mature DCs, CD40 and TRANCE receptor (TRANCE-R) are induced. As a result, once antigen-capturing DCs reach the lymph node (Figure 1), control of DC function can switch from the microbe to the T cell. Possibly, different maturation stimuli (TLR signaling via the microbe, for instance, rather than TNF-R signaling via the T cell) have different consequences for DCs. Full expression of some DC functions, such as IL-12 production, may also require concerted signaling by both of these receptor types.

DC maturation is an intricate differentiation process whose different components may be under separate control. Antigen processing and presentation are regulated at several levels, notably through the control of intracellular proteinase activity. Thus, maturation diminishes the level of the cysteine protease inhibitor cystatin C within the endocytic system, permitting increased catabolism of the invariant chain by cathepsin S, and promoting the binding of antigenic peptides to MHC class II molecules. The expression of CD40 and other T cell interaction molecules is also enhanced by maturation. Signaling through CD40, induced by CD40L on activated T cells, mast cells, and platelets, leads to the production of DC cytokines and chemokines and enhances DC migration and survival. Maturing DCs alter their expression of the costimulatory molecules CD80 and CD86 and of TNF family members, all of which can influence the extent and quality of the immune response. Maturing DCs also reshape their repertoire of chemokine receptors (27). Mature DCs lose CCR5 and CCR2, which respond to chemokines in an inflammatory site, but gain CCR7, which responds to chemokines in the lymphatic vessels and lymphoid organs. Maturation nicely illustrates the importance of taking a DC perspective in vaccine design. By targeting a vaccine to immature DCs and also maturing the cells, one implements a large spectrum of features (from antigen handling to proper positioning in vivo to optimal control of the magnitude and quality of the immune response) conducive to strong antigen-specific immunity.

Many existing vaccines may induce DC maturation, although their mechanisms can be quite complex. Some current vaccine vectors — recombinant yeast vaccines and DNA vaccines (28) among them — induce DCs to become strong stimulators of immunity, probably by directly stimulating TLRs. The attenuated smallpox and measles vaccines appear to mature DCs in a different manner. These organisms are both infectious and cytotoxic for DCs and yield infected dead cells that can then be processed efficiently through the exogenous pathway in other DCs. Furthermore, perhaps through the release of heat shock proteins, dying cells can mature the antigen-capturing DCs. Thus, vaccines may produce stronger immunity when they initially kill some DCs.

Effects of DC mobilization. DC mobilization entails both an increase in the population of these cells and a change in their migratory properties (27, 29). DC numbers can be increased tenfold using cytokines like G-CSF and flt-3L, while DC differentiation from nonproliferating precursors can be influenced by other hematopoietins (GM-CSF, IL-4) and IFNs. The requisite chemokine receptors for DCs to traffic into a vaccination site may vary, with CCR6 likely responding to macrophage inflammatory protein 3α (MIP-3α) at mucosal surfaces, and CCR5 and CCR2 responding to MIP-1s and monocyte chemoattractant proteins in other interstitial compartments. For vaccines administered into the skin and muscles, migration to lymph nodes requires afferent lymphatics (30, 31), but the DC-lymphatic interaction is still poorly understood.

DCs must also migrate in a directed way to the T cell areas, responses that are influenced by cysteinyl leukotrienes and transporters of the multidrug resistance family, as well as the distinct

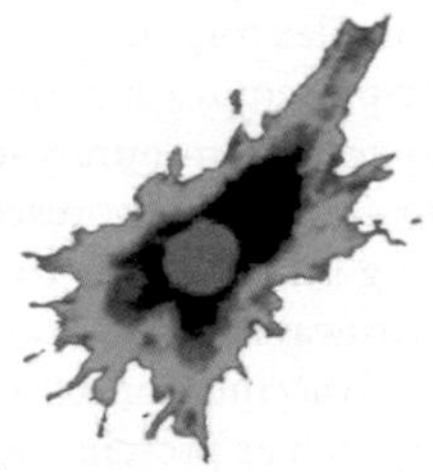

Antigen uptake receptors
DEC-205, MMR
Langerin, BDCA-2
DC-SIGN, ASGP-R
FCg-R, HSP-R, $\alpha_v\beta_5$

Maturation receptors
TLRs
TNF-Rs

T cell adhesion &
costimulatory molecules
DC-SIGN
CD86 + MHC clusters

Exogenous pathway
Dead cells
Immune complexes

Regulation of antigen processing
Proteolysis
MHC-peptide transport

Figure 2

Some specializations of DCs for vaccine capture, MHC-peptide complex formation, and T cell stimulation. DCs express many adsorptive uptake receptors whose natural ligands are generally not yet known. For this reason, anti-receptor antibodies are often used experimentally as surrogate antigens. Several receptors are type II transmembrane proteins with a single external C-type lectin domain found on distinct DC subsets: DC-SIGN and the asialoglycoprotein receptor on monocyte-derived DCs, BDCA-2 on plasmacytoid cells, and Langerin on Langerhans cells. MMR and DEC-205 are type I proteins with eight to ten contiguous C-type lectin domains; these receptors can also be expressed on certain endothelia and epithelia. Other receptors, such as FcγR, are not DC-restricted but function selectively in DCs to mediate the exogenous pathway for presentation on MHC class I products. Beyond antigen capture, DCs (or particular DC subsets) express high levels of select TLRs and thereby mature in response to specific microbial stimuli. During maturation, DCs produce and export high levels of several costimulatory molecules for T cell growth and differentiation. DC maturation regulates many of the elements involved in antigen capture and processing.

TREM-2 signaling molecule, each of which acts on the CCR7 lymph node homing receptor on DCs. Once in the T cell area, DCs are short-lived, apparently dying within a few days. Their lifespans can be prolonged through membrane bound TNFs on the T cell, e.g., CD40L and TRANCE (RANKL).

In summary, we propose that vaccine efficacy or immunogenicity can be improved by altering DC functions at three levels: by enhancing vaccine capture and processing, by promoting DC maturation, and by increasing DC numbers by stimulating DC replication, survival, and migration, to the lymph nodes.

Other endpoints in the immune response that can be achieved via DCs

Improved cell-mediated responses and T cell memory. Antigen-primed DCs rapidly prime an individual to form IFN-γ–producing or Th1CD4+ helper cells. When DC maturation is blocked, IL-4– and IL-5–producing Th2 helper cells seem to be induced, resulting in less efficient T cell–mediated immunity and memory, as well as the production of undesirable antibody subclasses — notably IgEs, which mediate allergy. Th1 helpers are especially critical in activating macrophages to resist intracellular bacteria and protozoa, and they are also the most effective form of helper for CD8+ CTL resistance to experimental viral infections and tumors. CD4+ Th1 cells additionally render DCs resistant to killing by CD8+ CTLs and directly lyse important MHC class II–expressing infected cells through a FasL-dependent (rather than perforin/granzyme–dependent) mechanism. The induction of Th1 cells is often ascribed to IL-12, but mature DCs can lose high-level IL-12 production while maintaining their ability to induce strong CD4+ Th1 and CD8+ CTL responses in vivo, possibly through other cytokines or special B7 and TNF family members.

When DCs directly stimulate CD8+ CTLs in humans, the T cells can kill targets in the presence of lower doses of peptide; in this way, the functional affinity of the CD8+ T cell is improved. A recent intriguing mechanism for this is that antigen-reactive T cells somehow remove MHC-peptide from the DCs, favoring selection of the more competitive, higher affinity T cells. Importantly, DCs induce T cell memory for both high-affinity CD8+ and Th1 CD4+ responses.

Generation of antibody-forming B cells. Classically, DCs enhance antibody formation by promoting the formation of antigen-specific CD4+ helper T cells, which induce antigen-specific B cells to proliferate and make antibody. In situ, IFN-α enhances T-dependent antibody formation, isotype switching, and memory. To obtain this result, DCs are the only cells that need to express the requisite type I IFN receptors. DCs can have direct effects on B cells that greatly enhance Ig secretion and isotype switching, including the production of the IgA subclass of antibodies, which contribute to mucosal immunity. Recently, DCs have been pulsed ex vivo with cell wall constituents from *Streptococcus pneumoniae*. When the DCs are reinfused into genetically matched mice, strong T-dependent B cell responses are induced to microbial proteins and capsular polysaccharides. This requires IL-6 production by the DCs and quite possibly a *ménage à trois* wherein B and T cells respond to native and processed antigens being presented on the same DC or DC subset. Consequently, if vaccines are delivered to the appropriate DCs, combined B cell and T cell immunity can ensue, an important consideration in the context of HIV-1 and other chronic infections.

Implementation of mucosal immunity. Vaccines are lacking for many sexually transmitted diseases. Strong T cell immunity and IgA antibodies may be required to provide protection from HIV-1, Epstein-Barr, herpes simplex, and human papilloma viruses. DCs are located beneath the antigen-transporting epithelium (M cells) (32) of mucosal lymphoid organs, and they may extend their processes through standard, lining epithelia to capture antigens. Access of vaccines to mucosal DCs should prove valuable for inducing mucosal immunity. However, maturation is again likely to be needed. Some mucosal DCs in the steady state produce IL-10 and may induce regulatory or immunosuppressive T cells, as discussed below. The latter would compromise vaccine efficacy.

Newly appreciated features of DCs relevant to vaccination

DC-induced tolerance. The immune-enhancing or adjuvant roles of DCs are exerted in two phases. In the first immediate or innate phase, DCs capture the antigen, begin to mature in response to stimuli, particularly microbial components, and produce cytokines and chemokines that mobilize and differentiate other cells, including NK cells. In the slower (adaptive) phase, DCs stimulate several components of the T cell response: clonal expansion; differentiation, specifically into Th1 helper or killer cells; and memory. Much of current effort by DC biologists focuses on a different effect of these cells, namely their capacity, when in the immature state, to induce antigen-specific unresponsiveness or tolerance following antigen capture.

DCs in the steady state are immature and can silence immunity in an antigen-specific manner through two recently identified mechanisms (33). Here, the "steady state" refers to the absence of acute inflammation and infection, with the latter providing stimuli that mature DCs via TLRs and TNF-Rs. One tolerance mechanism is exerted by DCs bearing the DEC-205 receptor in the lymph nodes. When antigen is targeted to these DCs in the absence of a maturation stimulus, interacting T cells proliferate but are soon deleted. Effector functions (IFN-γ secretion) and memory therefore do not develop, and the animal becomes tolerant to rechallenge with the peptide in a strong adjuvant (14). In contrast, chronic stimulation of DCs may predispose to the development of autoimmunity, including lupus erythematosus. A second tolerance mechanism involves the induction by certain immature DCs of IL-10–producing T cells, which in turn can silence other effector T cells. The tolerogenic roles of DCs could compromise vaccine efficacy. Conversely, the capacity to induce regulatory T cells may be useful in the design of a new class of vaccines for suppressing immunity in autoimmune diseases, allergy, and transplantation.

Contributions of DC subsets to innate immunity. There are many different forms of DCs in situ. These comprise the Langerhans cells in the skin and other epithelia and various DC precursors in blood, including monocytes and plasmacytoid DCs. Many DCs move from blood to tissues to lymph and then to the lymph node, but plasmacytoid DCs can move from blood directly into the lymph node via high endothelial venules, presumably by virtue of their CD62L expression (Figure 1). A hallmark of the plasmacytoid DC is the capacity to produce prodigious levels of IFN-α upon challenge with many viruses.

The raison d'être for these DC subsets has been a mystery, but some recent findings provide new perspectives. DC subsets vary in their expression of TLRs and therefore respond to different microbial stimuli. For example, CD11c+ CD14− cells in blood are the principal expressers of TLR3, a receptor for double-stranded viral RNA, whereas plasmacytoid cells are the main cell in blood expressing TLR9, the receptor for bacterial DNA and specific CpG deoxyoligonucleotides (26). Furthermore, distinct DC subsets can produce large amounts of individual cytokines (IL-12, TNF, IFN-α) in response to TLR signaling. This, in turn, may influence the kinds of lymphocytes (Th1 helpers, CD8+ CTLs) that are expanded by the antigen-presenting DCs. Importantly, because DC subsets also express different endocytic receptors, it may be possible to design selective vaccines targeting Langerhans cells, plasmacytoid DCs, or monocyte-derived DCs (Figure 2). As immune responses to specific pathogen-associated antigens are elucidated, it may prove important to target vaccines through the appropriate DC subset to take advantage of their distinct pathways for antigen uptake, maturation, and cytokine release.

DCs in the response to DNA vaccines

DNA vaccines are now showing promise in priming for resistance to simian immunodeficiency virus. At face value, the efficacy of DNA vaccines is perplexing, since the vaccine proteins are expressed primarily within skin or muscle cells, which are weak antigen-presenting cells for inducing immunity. However, successful DNA vaccination may involve DCs (28). A few DCs may be directly transfected with the vaccine and be responsible for immune priming. DCs may also capture antigens expressed in other cells that die following transfection and protein expression. Importantly, DNA itself activates many other DCs that are not transfected, most likely via TLR9 (26). Unfortunately, the requisite DNA receptors (and their CpG deoxyoligonucleotide mimics) may be expressed in different DC types in primates and in rodents. In human blood, for example, TLR9 is primarily expressed by the plasmacytoid DC, while, in mice, other subsets of DCs respond to DNA and CpG oligonucleotides. Therefore it remains challenging for DNA vaccines to direct the maturation of human DCs. This maturation can be regarded as a universal platform for vaccine efficacy.

The boosting of DNA-vaccinated individuals with viral vectors may also exploit DCs — again through several mechanisms. The vectors may directly infect DCs or be presented by the exogenous pathway. Some viral vectors may kill infected DCs, leading to uptake and maturation by other DCs as discussed above. In spite of these possibilities, there are few studies of DC function in the setting of DNA prime-viral vector boost vaccination in animals and, as a result, little direct information on whether DCs could be exploited to a greater extent.

Ex vivo–derived DCs in cancer vaccines. A new field has emerged in the setting of cancer immunotherapy (see Yu and Restifo, this Perspective series, ref. 13). Human DCs are generated ex vivo from progenitors, charged with antigens from the tumor, and then reinfused to boost a patient's immunity in an antigen-specific manner (10–12). Beyond the goal of developing new therapeutic vaccines that prevent the initial development or recurrence of tumors, the ex vivo DC approach provides an opportunity to investigate many pertinent features of human DCs as natural adjuvants. For example, the way to load DCs with a large spectrum of antigens can be monitored and optimized, the functions of distinct DC subsets can be assessed, and the maturation status of the DCs can be manipulated.

The immunologic impact of DC interactions with specific pathogens. Despite the evident promise of DCs for vaccination, it is important not to overlook the immune-evasive capacities of individual pathogens, many of which can directly disrupt components of DC function. HIV-1 and measles can be cytopathic, particularly follow-

ing syncytium formation between DCs and T cells. Herpes simplex, cytomegalovirus, lymphocytic choriomeningitis virus, and tumors can block DC functions, including maturation.

The premise of vaccine design is that strong immunity will be able to block the pathogen before it can significantly compromise immunity, even at the level of DCs. Nevertheless, two disquieting aspects of this interaction have come to light, especially in the setting of HIV-1. First, DCs can act like a Trojan horse to transport HIV-1 to T cell sites for replication. First, The DC-restricted lectin DC-SIGN, which is normally used by DCs to bind ICAM-3 on resting T cells, also binds HIV-1 and enhances its infectivity for T cells following initial uptake into DCs (34). Second, the large amounts of HIV-1 produced during chronic HIV-1 infection are proposed to exploit the normal tolerizing role of immature DCs (33), and to induce T cell deletion and regulatory T cell formation. HIV-1 virions bind several receptors on the tolerizing, immature form of DCs, including DC-SIGN, CD4, CCR5. In the difficult case of HIV-1, DCs therefore occupy both fronts of vaccine biology, guiding the offense of the pathogen and bolstering the defense of the host. Thus, pathogen interactions with DCs can present a formidable challenge to the development of safe vaccines that target these cells.

Conclusions

Vaccine design extends beyond the identification of antigens. It needs to harness the immunologic mechanisms that lead to strong and lasting immunity. In many instances, particularly the induction of T cell–mediated immunity, these mechanisms are controlled by antigen-presenting DCs, potent stimulators of specific T cell immunity, in tissue culture, in model organisms, and in humans. DCs in essence act as nature's adjuvants to generate immune resistance.

We have outlined three areas of DC biology that might be exploited to improve vaccine efficacy. First, DCs have select receptors for enhancing antigen uptake and processing; these could be targeted to improve the presentation of vaccine antigens to both CD4+ Th1 helper cells and CD8+ CTLs. Second, DCs undergo a process of terminal differentiation or maturation, typically in response to signaling via TLRs; vaccine-based stimulation of DC maturation is required in addition to antigen capture for DCs to elicit strong T cell immunity. Third, the numbers and migration of DCs in situ can be controlled to improve the selection of particular antigen-responsive lymphocytes. Several newly recognized DC functions pertinent to vaccine design are emerging. DCs or certain DC subsets exert innate functions, such as the production of large amounts of immune-enhancing cytokines. DCs also influence antibody production, control mucosal immunity, and, in the absence of maturation, induce antigen-specific silencing or tolerance. Ironically, because many of the pathogens for which vaccines are desired, especially HIV-1, have the capacity to exploit these cells during their replication or as means to evade immune defenses, DCs contribute to the pathogenesis and protective fronts of vaccine biology.

Acknowledgments

The authors thank Carol Moberg, John Mascola, and Robert Seder for help with the manuscript. Ralph Steinman is supported by NIH grants AI-13013, AI-40045, AI-40874, and CA-84512, and by Direct Effect. Melissa Pope is supported by NIH grants AI-40877, AI-47681, AI-52060, HD-41757, and HD-41752, and by The Rockefeller Foundation. Melissa Pope is an Elizabeth Glaser Scientist, supported by the Elizabeth Glaser Pediatric AIDS Foundation. Our website contains an earlier version of this article with an expanded reference list that extends beyond the background reading below.

Address correspondence to: Melissa Pope, Center for Biomedical Research, Population Council, 1230 York Avenue, New York, New York 10021, USA. Phone: (212) 327-7794; Fax: (212) 327-7764; E-mail: mpope@popcbr.rockefeller.edu.

1. Letvin, N.L. 2002. Strategies for an HIV vaccine. *J. Clin. Invest.* In press.
2. Seder, R.A., and Hill, A.V. 2000. Vaccines against intracellular infections requiring cellular immunity. *Nature.* **406**:793–798.
3. Mascola, J.R., and Nabel, G.J. 2001. Vaccines for the prevention of HIV-1 disease. *Curr. Opin. Immunol.* **13**:489–495.
4. Barouch, D.H., Letvin, N.L. 2001. CD8+ cytotoxic T lymphocyte responses to lentiviruses and herpesviruses. *Curr. Opin. Immunol.* **13**:479–482.
5. Berzofsky, J.A., Ehlers, J.D., and Belyakov, I.M. 2001. Strategies for designing and optimizing new generation vaccines. *Nat. Rev. Immunol.* **1**:209–219.
6. Singh, M., and O'Hagan, D. 1999. Advances in vaccine adjuvants. *Nat. Biotechnol.* **17**:1075–1081.
7. Schijns, V.E. 2000. Immunological concepts of vaccine adjuvant activity. *Curr. Opin. Immunol.* **12**:456–463.
8. Pulendran, B., Palucka, K., and Banchereau, J. 2001. Sensing pathogens and tuning immune responses. *Science.* **293**:253–256.
9. Banchereau, J., and Steinman, R.M. 1998. Dendritic cells and the control of immunity. *Nature.* **392**:245–252.
10. Fong, L., and Engleman, E.G. 2000. Dendritic cells in cancer immunotherapy. *Annu. Rev. Immunol.* **18**:245–273.
11. Nestle, F.O., Banchereau, J., and Harr, D. 2001. Dendritic cells: on the move from bench to bedside. *Nat. Med.* **7**:761–765.
12. Steinman, R.M., and Dhodapkar, M. 2001. Active immunization against cancer with dendritic cells: the near future. *Int. J. Cancer.* **94**:459–473.
13. Yu, Z., and Restifo, N.P. 2002. Cancer vaccines: progress reveals new complexities. *J. Clin. Invest.* In press.
14. Hawiger, D., et al. 2001. Dendritic cells induce peripheral T cell unresponsiveness under steady state conditions in vivo. *J. Exp. Med.* **194**:769–780.
15. Pamer, E., and Cresswell, P. 1998. Mechanisms of MHC class I–restricted antigen processing. *Annu. Rev. Immunol.* **16**:323 358.
16. Matsuda, J.L., and Kronenberg, M. 2001. Presentation of self and microbial lipids by CD1 molecules. *Curr. Opin. Immunol.* **13**:19–25.
17. Mellman, I., and Steinman, R.M. 2001. Dendritic cells: specialized and regulated antigen processing machines. *Cell.* **106**:255–258.
18. den Haan, J.M.M., and Bevan, M. 2001. Antigen presentation to CD8+ T cells: cross-priming in infectious diseases. *Curr. Opin. Immunol.* **13**:437–441.
19. Heath, W.R., Carbone, F.R. 2001. Cross-presentation in viral immunity and self tolerance. *Nat. Rev. Immunol.* **1**:126–134.
20. Sigal, L.J., Crotty, S., Andino, R., and Rock, K.L. 1999. Cytotoxic T-cell immunity to virus-infected non-haematopoietic cells requires presentation of exogenous antigen. *Nature.* **398**:77–80.
21. Somersan, S., and Bhardwaj, N. 2001. Tethering and tickling: a new role for the phosphatidylserine receptor. *J. Cell Biol.* **155**:501–504.
22. Porcelli, S.A., and Modlin, R.L. 1999. The CD1 system: antigen-presenting molecules for T cell recognition of lipids and glycolipids. *Annu. Rev. Immunol.* **17**:297–329.
23. Gonzalez-Aseguinolaza, G., et al. 2002. NKT cell ligand α-galactosylceramide enhances protective immunity induced by malaria vaccines. *J. Exp. Med.* **195**:617–624.
24. Medzhitov, R. 2001. Toll-like receptors and innate immunity. *Nat. Rev. Immunol.* **1**:135–145.
25. Akira, S., Takeda, K., and Kaisho, T. 2001. Toll-like receptors: critical proteins linking innate and acquired immunity. *Nat. Immunol.* **2**:675–680.
26. Hemmi, H., et al. 2000. A Toll-like receptor recognizes bacterial DNA. *Nature.* **408**:740–745.
27. Sallusto, F., and Lanzavecchia, A. 1999. Mobilizing dendritic cells for tolerance, priming, and chronic inflammation. *J. Exp. Med.* **189**:611–614.
28. Steinman, R.M., Bona, C., and Inaba, K. 2002. Dendritic cells: important adjuvants during DNA vaccination. In *DNA vaccines.* H. Ertl, editor. Landes Bioscience. Georgetown, Texas, USA. In press.
29. Lanzavecchia, A., and Sallusto, F. 2001. The instructive role of dendritic cells on T cell responses: lineages, plasticity and kinetics. *Curr. Opin. Immunol.* **13**:291–298.
30. Romani, N., et al. 2001. Migration of dendritic cells into lymphatics. The Langerhans cell example: routes, regulation, and relevance. *Int. Rev. Cytol.* **207**:237–270.
31. Matsuno, K., and Ezaki, T. 2000. Dendritic cell dynamics in the liver and hepatic lymph. *Int. Rev. Cytol.* **197**:83–136.
32. Neutra, M.R., Mantis, N.J., and Kraehenbuhl, J.P. 2001. Collaboration of epithelial cells with organized mucosal lymphoid tissues. *Nat. Immunol.* **2**:1004–1009.
33. Steinman, R.M., and Nussenzweig, M.C. 2002. Avoiding horror autotoxicus: the importance of dendritic cells in peripheral T cell tolerance. *Proc. Natl. Acad. Sci. USA.* **99**:351–358.
34. Kwon, D.S., Gregario, G., Bitton, N., Hendrickson, W.A., and Littman, D.R. 2002. DC-SIGN-mediated internalization of HIV is required for *trans*-enhancement of T cell infection. *Immunity.* **16**:135–144.

Part 4
Infectious Diseases

The enigma of sepsis

Niels C. Riedemann, Ren-Feng Guo, and Peter A. Ward

Department of Pathology, University of Michigan Medical School, Ann Arbor, Michigan, USA.

Sepsis remains a serious cause of morbidity and mortality, and the pathophysiology of the disease is not clear. The definition of the clinical manifestations of sepsis is ever evolving. This review discusses the search for effective therapeutic interventions, hurdles in translational sepsis research, and new therapies in development in current clinical trials.

Historical perspective and definition

Sepsis affects approximately 700,000 people annually and accounts for about 210,000 deaths per year in the US. According to recent reports, the incidence is rising at rates between 1.5% and 8% per year (1, 2), despite technical developments in intensive care units (ICUs) and advanced supportive treatment. Septic patients are generally hospitalized for extended periods, rarely leaving the ICU before 2–3 weeks. Accordingly, sepsis represents a major burden to the US health care system, with costs estimated to be approximately $16.7 billion per year (1). The question of why the incidence is rising has been extensively discussed, but a final answer has not yet been found. Interestingly, the spectrum of responsible microorganisms appears to have shifted from predominantly Gram-negative bacteria in the late 1970s and 1980s to predominantly Gram-positive bacteria at present (2). In addition, the rate of fungal infections is reported to have increased more than 200% during the same period (2). As difficult as the treatment of septic patients and the improvement of their survival rates have been, equally difficult has been the development of a clear clinical definition of sepsis. The lack of consensus regarding a definition has resulted in great difficulties in making meaningful comparisons of study results. Often, the lack of a precise definition of sepsis has also compromised the reproducibility of results of phase I clinical trials since the patient populations being studied were nonhomogeneous. An analysis of studies in recent years has demonstrated that the risk of death (and therefore the severity of disease) significantly correlates with the effectiveness of anti-inflammatory treatment (3). This finding implies a need for a better clinical definition of the disease in order for patients to be enrolled in treatment groups with well-defined entry criteria. In other words, the exact clinical status of each patient needs to be established.

For nearly a century, sepsis has been defined as the systemic host response to an infection, and even though many subclassifications have been made over the years, there has been little modification of this definition. Originally, sepsis was believed to be associated with the presence of bacteria in the blood (bacteremia), and the terms "sepsis" and "septicemia" were frequently interchanged in the clinical setting. In 1989, Bone et al. established a simple definition for sepsis syndrome, which was based on specific clinical symptoms

and included a known source of infection (4). The clinical signs of sepsis, however, were frequently presented by patients not characterized by measurable levels of bacteria in the blood, or by those with acute pancreatitis (5) or trauma (6). This discrepancy was first taken into account at a Consensus Conference held by the Society of Critical Care Medicine and the American College of Chest Physicians in 1992, when the term "systemic inflammatory response syndrome" (SIRS) was established (Table 1), for which no definable presence of bacterial infection was required (7, 8). In addition, the terms "severe sepsis" and "septic shock" were introduced to differentiate between different stages of disease. The criteria for the various clinical definitions shown in Table 1 have been well established in clinical use.

However, discussion at the most recent consensus conference led to the addition of several new criteria for the diagnosis of SIRS and sepsis (Table 2) (9). This was due in part to the documentation of additional clinical symptoms and laboratory findings that are frequently present in patients with acute systemic response to infection or other insults. At this same consensus conference, the definitions of SIRS, sepsis, severe sepsis, and septic shock, including multiple organ dysfunction syndrome established in 1992 were endorsed. It can be argued that the earlier definitions did not facilitate the diagnosis of precise stages of disease or allow for accurate prognosis of the host response to infection. Therefore, a new staging system has been developed, similar to that used to stage tumor progression. The PIRO staging system facilitates a more accurate characterization of the stage of sepsis disease and the associated risks and prognosis. We endorse the introduction of the PIRO system for diagnosing and tracking sepsis progression, which characterizes the disease based on predisposition (especially to genetic factors), the insult infection (especially the type of infection, source, etc.), the response of the host system (SIRS, septic shock, etc.; specific markers like IL-6, protein C, TNF, etc.), and organ dysfunction (9). This classification would take into account the clinical status of the patient in addition to biochemical analyses. This should ultimately permit more precise classification of sepsis-related disorders and might be particularly helpful in more closely defining entry criteria for clinical trials of sepsis therapies.

Early experimental basis for clinical anti-inflammatory therapeutic interventions

The idea that sepsis was caused by an overwhelming reaction of the patient to invading microorganisms was probably at least partially based on the observation that, on many occasions, no clinical evidence for infection (e.g., positive bacterial blood cultures) was found in patients with septic symptoms. In 1972, Lewis Thomas noted in *The New England Journal of Medicine* that "it is our response that makes the disease" and that the patient was, therefore, more

Nonstandard abbreviations used: intensive care unit (ICU); systemic inflammatory response syndrome (SIRS); predisposition, insult, response, organ dysfunction (PIRO); cecal ligation and puncture (CLP); activated protein C (APC); plasminogen activator inhibitor-1 (PAI-1); monocyte chemoattractant protein-1 (MCP-1); high-mobility group B-1 protein (HMGB1); macrophage migration inhibitory factor (MIF); C5a receptor (C5aR); tissue factor plasminogen inhibitor (TFPI).

Conflict of interest: The authors have declared that no conflict of interest exists.

Citation for this article: *J. Clin. Invest.* **112**:460–467 (2003). doi:10.1172/JCI200319523.

Table 1

Clinical definition of sepsis

SIRS	Temperature >38.3°C or <36°C
	Heart rate >90 beats/min
	Respiratory rate >20 breaths/min or PaCO$_2$ <32 mmHg
	White blood cell count >12 × 10^9/l or <4 × 10^9/l, or >10% immature band forms
Sepsis	Systemic response to infection, manifested by two or more of the conditions mentioned under SIRS (SIRS + evidence of infection)
Severe sepsis	Sepsis associated with organ dysfunction, hypoperfusion, or hypotension including lactic acidosis, oliguria, or acute alteration in mental state
Septic shock	Sepsis-induced hypotension (e.g., systolic blood pressure <90 mmHg or a reduction of >40 mmHg from base line) despite adequate fluid resuscitation, along with the presence of perfusion abnormalities that may include lactic acidosis, oliguria, or an acute alteration in mental state. Vasopressor- or inotropic-treated patients may not be hypotensive at the time of measurement
MODS	The presence of altered organ function in an acutely ill patient such that homeostasis cannot be maintained without intervention

PaCO$_2$, arterial partial pressure of carbon dioxide; MODS, multiple organ dysfunction syndrome.

endangered by this response than by the invading microorganisms (10). According to this view, interventions designed to attenuate immune and inflammatory responses might be clinically useful. In the 1960s, the first clinical trials featuring suppression of the immune and inflammatory responses were conducted, in which septic patients were treated with supraphysiological doses of glucocorticoids (11) (Table 3). However, these studies were unsuccessful, despite the fact that more recent studies have suggested benefits from low-dose glucocorticoid treatment (see "Failure of clinical trials in sepsis").

As mentioned above, the predominant source of infection in septic patients before the late 1980s was Gram-negative bacteria. LPS, the main component of the Gram-negative bacterial cell wall, was known to stimulate release of inflammatory mediators from various cell types and induce acute infectious symptoms when injected into animals. In 1969, Davis and colleagues found that infusion of IgGs improved survival in an experimental setting after endotoxin infusion (12). Based on these and other findings, LPS blockade became a target for clinical intervention (13).

Given the frequency of Gram-negative infections in septic patients, it was assumed that large amounts of circulating LPS must be present. Based on this assumption, animal models of sepsis were established mainly in rodents, in which large doses of LPS had been administered. In contrast to the responses observed following bacterial infection (e.g., bacteria delivered by intraperitoneal or i.v. injection, or released into the peritoneal cavity in the cecal ligation and puncture [CLP] model), LPS infusion models often did not mimic the changes observed during sepsis (Figure 1). This fact became apparent in the case of TNF-α, a potent proinflammatory mediator. In addition, infusion of TNF-α into animals induced the symptoms characteristic of sepsis (14), while passive immunization with anti–TNF-α was protective (15, 16). High levels of TNF-α have been found in the serum of humans following i.v. injection of LPS (15). In 1985, Beutler and colleagues found that passive immunization against TNF-α protected mice from lethal endotoxic shock (16). These results were confirmed when Tracey et al. found that TNF-α blockade was beneficial in animal models of shock following infusion of endotoxin or *Escherichia coli* (17, 18). But subsequent clinical trials failed to demonstrate the utility of this therapy in septic humans (19) and in CLP mice (20). Why?

One possible explanation is that the results obtained from studies performed in LPS infusion animal models did not accurately reflect clinical developments in human sepsis. Serum TNF-α levels after infusion of LPS into mice were later described to be more than 200-fold higher than in CLP animals (20). In a similar study, CLP mice treated with anti–TNF α antibodies showed not improved survival but, rather, a tendency toward worsened outcomes (21). Interestingly, the circulating LPS levels in the CLP sepsis model (which more accurately reflects the dynamics of sepsis occurring in humans) were found to be very low, and extreme elevations of TNF-α levels were not observed in rodent LPS infusion models (21). As indicated above, TNF-α levels observed in CLP models are generally very low and not comparable to TNF-α levels found after LPS infusion. In addition, LPS levels in septic patients are also reported to be low, and while in some cases of sepsis in humans (e.g., meningococcal sepsis in infants) elevated serum levels of TNF-α have been found in up to 90% of patients (22), several other clinical studies in septic patients reported only minimally elevated or undetectable levels of TNF-α (23). The failure of anti–TNF-α and anti-LPS interventions in septic patients can be seen as an example of how conclusions based on animal models may not hold true in humans, or may not be applicable to human sepsis because of incorrect assumptions underlying the animal models (e.g., that LPS is a major initiator of sepsis and is present in the serum at high levels during sepsis) (Figure 1). Currently, there is general agreement among researchers in the field that LPS injection may serve as a model for endotoxic shock but not for sepsis.

Failure of clinical trials in sepsis

The nearly 40-year history of clinical trials of anti-inflammatory strategies for the treatment of sepsis might be referred to as a "graveyard" for pharmaceutical companies, since almost none of these strategies has resulted in significantly improved survival of patients. In Table 3 the anti-inflammatory strategies used in clinical trials are summarized. The various strategies are discussed in detail elsewhere (23, 24). Often, promising initial results from small uncontrolled phase I or II clinical trials subsequently could not be confirmed in larger randomized clinical trials. The possible reasons are numerous. Some of these failures resulted from the use of

Table 2
Extended criteria for the diagnosis of sepsis

General variables	Fever or hypothermia (temperature >38.3°C or <36°C)
	Heart rate >90 beats/min or >2 SD above the normal value for age
	Tachypnea
	Altered mental state
	Significant edema or positive fluid balance (>20 ml/kg over 24 h)
	Hyperglycemia (plasma glucose >120 mg/dl or >7.7 mmol/l)
	in the absence of diabetes
Inflammatory variables	Leukocytosis (wbc count >12,000/μl)
	Leukopenia (wbc count <4,000/μl)
	Normal wbc count with >10% immature forms
	Plasma C-reactive protein level >2 SD above the normal value
	Plasma procalcitonin level >2 SD above the normal value
Hemodynamic variables	Arterial hypotension (SBP <90 mmHg, MAP <70, or an SBP decrease
	>40 mmHg in adults or <2 SD below normal for age)
	S_vO_2 >70%
	Cardiac index >3.5 l/min × M^{-23}
Organ dysfunction	Arterial hypoxemia (PaO_2/FiO_2 <300)
	Acute oliguria (urine output <0.5 ml/kg/h or 45 mmol/l for at least 2 h)
	Creatinine increase >0.5 mg/dl
	Coagulation abnormalities (INR >1.5 or aPTT >60 s)
	Ileus (absent bowel sounds)
	Thrombocytopenia (platelet count <100,000/μl)
	Hyperbilirubinemia (plasma total bilirubin >4 mg/dl or >70 mmol/l)
Tissue perfusion	Hyperlactatemia (>1 mmol/l)
	Decreased capillary refill

wbc, white blood cell; SBP, systolic blood pressure; MAP, mean arterial pressure; S_vO_2, mixed venous oxygen saturation; PaO_2, arterial partial pressure of oxygen; FiO_2, fraction of inspired oxygen; INR, international normalized ratio; aPTT, activated partial thromboplastin time.

unsuitable animal models (LPS infusion) and wrong assumptions, as discussed previously. In the case of glucocorticoids, follow-up studies in the 1990s suggested that lower doses of glucocorticoids might be beneficial in septic patients. In 2001, a large multicenter phase III clinical trial was completed, which suggested that 7-day treatment with low doses of hydrocortisone and fludrocortisone significantly reduced the risk of death in patients with septic shock who also demonstrated relative adrenal insufficiency (25). Such benefits, however, were not reported in patients who lacked evidence of adrenal insufficiency. Thus, in this case, the dose of the therapeutic intervention as well as the patient group appeared to determine failure or success.

Retrospectively, a crucial problem in most of the clinical trials investigating anti-inflammatory agents appears to be the non-homogeneity of the patient population enrolled, which partially stems from an inability to more effectively classify the immune status of the septic patient. The recent finding that the success of anti-inflammatory treatments in animals and humans with sepsis closely correlates with the severity of disease (3) may explain why some trials report benefits (especially those with patients in severe sepsis or septic shock) while others do not.

The immune system in a septic individual undergoes substantial modifications during sepsis. Experimental data support the theory that an early intense proinflammatory response of the immune system after infection or trauma can cause harm or set the stage for subsequent organ damage, but it is also well documented that, during sepsis, the innate immune system frequently loses the ability to effectively kill invading microorganisms (26). Depending on the ability of the immune system to respond to infection, an anti-

inflammatory strategy may not be helpful and could even be harmful, as in the clinical trial in which a TNF-α antagonist was reported to increase mortality (27).

The new PIRO scoring system can be seen as an attempt to stage a septic patient, with the ultimate goal of treating sepsis according to the patient's immunological status. A substantial hurdle will be development of diagnostic tools that facilitate an efficient and accurate determination of this status.

The mysterious success of activated protein C

In November 2001, drotrecogin alfa (Xigris), a recombinant form of activated protein C (APC), was approved by the US Food and Drug Administration for the treatment of patients with severe sepsis. The original large multicenter randomized trial, known as Protein C World Wide Evaluation in Severe Sepsis (PROWESS), that evaluated the utility of this drug against sepsis (27–29) revealed an overall reduction of 19.4% in the relative risk of death and a reduction of 6.1% in the absolute risk of death (30.8% mortality at 28 days in the placebo group vs. 24.7% with drotrecogin alfa). A detailed analysis revealed that only patients with a high risk of death (APACHE II scores ≥ 25) showed significantly reduced mortality, by 13% (44% mortality with placebo vs. 31% with drotrecogin alfa). Concerns emerged regarding infrequent bleeding complications (intracerebral hemorrhage) that occurred when the recommended doses of APC were used. Warnings were added to the labels for drotrecogin alfa, stating that its efficacy in septic infants, and its safety for severely septic patients receiving concomitant heparin, were unknown. The recently started Administration of Drotrecogin Alfa in Early Severe Sepsis trial (ADDRESS) will be

Table 3
Targets in clinical trials for the treatment of sepsis

Immune modulation	Glucocorticoids (inhibition of overactivation of the immune and inflammatory systems)
	IVIG (improvement of host defense)
Endotoxin (LPS)	Anti-endotoxin antibodies:
	polyclonal human antiserum
	human monoclonal anti–lipid A (HA-1A)
	murine monoclonal anti–lipid A (E5)
	human mAb's
	Bactericidal/permeability-increasing protein
	LPS elimination (hemofiltration)
TNF-α	TNF-α antibodies:
	murine monoclonal antibodies
	F(ab)$_2$ anti-TNFα
	Soluble TNF receptors (TNF inhibition)
IL-1	IL-1 receptor antagonist
PAF	Phospholipase A2 antagonist (reduction of PAF)
	PAF antagonist
	PAF-acetylhydrolase (PAF inactivation)
Bradykinin	Bradykinin antagonist
Arachidonic acid metabolites	Prostaglandin (PG) E1 and liposomes containing PGE1 (anti-inflammatory)
	Thromboxane inhibitors (anti-inflammatory)
	Ketoconazole (thromboxane synthetase inhibition)
	Ibuprofen (COX inhibition)
Reactive oxygen species	N-acetylcysteine (restoration of cellular antioxidant potential)
	Selenium has been used to bolster selenium-dependent glutathione peroxidase, which is a scavenger for O_2^-
NO	L-NAME (NOS inhibition)
	L-NMMA (iNOS inhibition)
	Methylene blue (guanylyl cyclase inhibition)
	PHP (NO scavenger)
Phosphodiesterase	Pentoxifylline (phosphodiesterase inhibition, cAMP increase)
Neutrophil activity	IFN-γ (reactivation of neutrophil immune functions)
	G-CSF, GM-CSF (increase of immune-competent blood cells)
	PGG-glucan (increase of phagocytosis and bacterial killing in neutrophils)
Complement system	C1 inhibitor (inhibition of classical and lectin pathway activation)1
Coagulation	Antithrombin III (inhibition of thrombin, factors IXa, Xa, XIa, and XIIa)
	TFPI (inhibition of factors X and IX)
	APC (inactivation of factors Va and VIIIa)

IVIG, intravenous IgG; PAF, platelet-activating factor; PGE1, prostaglandin E1; L-NAME, *N*-nitro-L-arginine methyl ester; L-NMMA, *N*(G)-monomethyl-L-arginine ; PHP, pyridoxylated hemoglobin polyoxyethylene conjugate; PGG, poly-(1-6)-β-glucotriosyl-(1-3)-β-glucopyranose; APC, activated protein C.

the largest clinical sepsis trial in history, aiming to enroll 11,000 patients to determine the potential benefits of this drug in septic patients with a lower risk of death and with dysfunction involving only one major organ (lungs, liver, or kidneys).

Thrombin, after binding to thrombomodulin on the cell surface of endothelial cells, can activate plasma protein C, which, together with its cofactor protein S, then functions as a proteolytic inhibitor of the clotting factors Va and VIIIa, thereby acting as an anticoagulant. In addition, APC is a powerful anti-inflammatory molecule capable of inhibiting cytokine production (TNF-α, IL-1, and IL-6) in monocytes and reducing adhesive interactions between neutrophils and endothelial cells (30). APC also indirectly increases the fibrinolytic response by inhibiting plasminogen activator inhibitor-1 (PAI-1). In humans with sepsis, plasma levels of proteins C and S fall significantly, and thrombomodulin present on endothelial cells is downregulated, leading to reduced APC generation. A recent study demonstrated that APC employs the endothelial cell protein C receptor as a coreceptor for cleavage of protease-activated receptor-1 (PAR-1) on endothelial cells (30). The study suggests that APC may play a role in cell signaling

and gene transcription of protective genes through selective PAR-1 activation on endothelial cells and related induction of monocyte chemoattractant protein-1 (MCP-1), which protects cells. Treatment of sepsis with APC is associated with a risk of significant complications (e.g., intracerebral bleeding) and is therefore the subject of intensive debate (30–32).

The mechanisms by which APC treatment improves survival in humans cannot be explained at this time. So far, significant reductions in mortality have been observed only in patients with severe sepsis, who seem to be more likely to benefit from anti-inflammatory strategies. Given this fact, one might speculate that the anti-inflammatory potential of APC provides protection in severely septic patients, especially since other anticoagulant strategies, such as antithrombin III and tissue factor plasminogen inhibitor (TFPI), have not resulted in significant benefits (ref. 33 and Table 3).

New strategies raise new hope
The success of APC provides new hope that other new therapeutic strategies for the treatment of sepsis may be introduced into the clinic in the near future. The main target groups for anti-inflam-

matory treatment, according to the lessons learned, will likely be patients with severe sepsis or septic shock. Some new developments in the area of sepsis research appear to be specifically promising.

Apoptosis inhibition

Programmed cell death (apoptosis) is differentiated from cell necrosis by typical features, such as DNA fragmentation, condensation of chromatin, membrane blebbing, and cell shrinkage. Several proinflammatory mediators (such as TNF-α) produced during experimental sepsis are known inducers of apoptosis involving various cell types. During sepsis, lymphocytes have been shown to undergo rapid apoptosis, whereas other cells, like neutrophils, demonstrate delayed apoptosis (34, 35). Hotchkiss and colleagues recently demonstrated that the prevention of lymphocyte cell death during sepsis could improve outcome (35–37). Caspase

inhibitors were found to be an effective tool to protect animals from death during pneumonia-induced sepsis (38). These findings led to the hypothesis that immunodepression resulting from the loss of lymphocytes could represent a central pathogenic event in sepsis. Further studies have suggested that inhibition of intestinal epithelial apoptosis (by selective Bcl-2 overexpression) significantly improved outcomes in two different models of sepsis in mice (38, 39). Recently it was demonstrated that adoptive transfer of apoptotic splenocytes in septic mice worsened the outcome, while transfer of necrotic splenocytes had protective effects (40). The underlying mechanisms of these studies are not yet understood. Treatment of septic patients with apoptosis inhibitors has to overcome several problems, including the selectivity of such inhibitors and potential uncontrolled cell growth. However, the concept of regulated apoptosis warrants further investigation.

Assumptions

1. Gram-negative bacteria are cause of sepsis infection
2. Bacteria causing disease shed LPS
3. High levels of serum LPS achieved in septic patients

Possible reasons for failure

1. Incorrect assumptions based on initiating factors of disease
2. Incomplete clinical observations

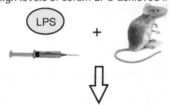

Observations in animals

1. High serum levels of TNF-α achieved following LPS infusion

1. Unsuitable animal model, not translatable to humans
2. High serum levels of TNF-α not achieved in humans during sepsis

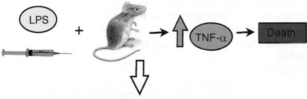

Outcomes of intervention

1. Treatment with anti–TNF-α antibodies increases survival

1. Unable to block all TNF-α
2. Results incorrect

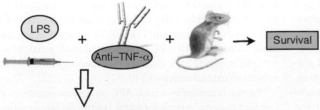

Clinical trials in humans with sepsis

1. Anti–TNF-α antibodies not protective

1. Anti-TNF-α antibodies not protective
2. Study design insufficient
3. Sepsis definition insufficient
4. Drug not working (not tested, etc.)
5. Wrong dose, time point, etc.
6. Wrong approach (see example provided)

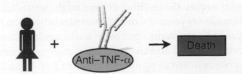

Figure 1
Possible reasons for failure in sepsis trials. The flow diagram reflects various stages in the development of sepsis therapies that precede clinical trials in sepsis patients. At each stage, possible reasons for failure of the strategy are listed. The murine LPS infusion model illustrates that anti–TNF-α antibody treatment can successfully increase survival rates of septic animals; however, this same therapy proved unsuccessful in clinical trials in humans with sepsis.

High-mobility group B-1 protein

Kevin Tracey and his group have recently demonstrated that high-mobility group B-1 protein (HMGB1) acts as a late mediator in LPS-induced endotoxicity (41, 42). HMGB1 was originally described as a nuclear binding protein, facilitating gene transcription by stabilizing nucleosome formation. HMGB1 can bind to the cellular receptor for advanced glycation end products (RAGE), facilitating activation of the transcription factor NF-κB and MAPK (43), and thereby inducing generation of proinflammatory mediators in monocytes (43). In endothelial cells, HMGB1 induces VCAM-1, ICAM-1, and RAGE expression, as well as secretion of TNF-α, IL-8, MCP-1, PAI-1, and tissue plasminogen activator (44). A recent study found that HMGB1 increases the permeability in enterocytic monolayers and bacterial translocation to lymph nodes in mice in vivo (45). These findings demonstrate the proinflammatory potential of HMGB1 and its linkage to the coagulation system. Recently, ethyl pyruvate, which inhibits HMGB1 production in vivo, was found to improve survival in a sepsis model in mice when administered 24 hours after the onset of sepsis (46). This finding makes HMGB1 a potential therapeutic target for the treatment of sepsis.

Macrophage migration inhibitory factor

Blockade of macrophage migration inhibitory factor (MIF) or targeted disruption of the MIF gene significantly improved survival in a model of septic shock in mice (47, 48). Earlier studies found that administration of MIF induced greater lethality after LPS infusion (47, 48) and that MIF knockout mice demonstrated improved survival after endotoxin challenge and showed reduced levels of TNF-α (48).

MIF was originally described as a T cell cytokine and later found to be produced by other cells such as pituitary gland cells (49), eosinophils (50), tubular epithelial cells in kidney (51), epithelial cells in lung (52), and macrophages (53), in which glucocorticoids acted as strong inducers of MIF (54). MIF is capable of inducing production of various proinflammatory mediators in macrophages and other cells and thereby can override the glucocorticoid-induced inhibition of proinflammatory-cytokine production (54). Septic patients show elevated serum levels of MIF (55, 56), as do patients with adult respiratory distress syndrome. In both cases, elevated levels of MIF are correlated with poor prognosis and death. Accordingly, MIF is an interesting target for therapeutic intervention in septic patients.

C5a and C5a receptor

The various complement pathways are activated on bacterial surfaces and by bacterial products such as LPS, by acute-phase proteins (e.g., C-reactive protein), by immune complexes, and by many other stimuli. Complement activation then results in assembly of the terminal membrane attack complex C5b-9, which forms pores in invading microorganisms and leads to their lysis. The anaphylatoxin C5a is cleaved from the complement protein C5 and exerts numerous proinflammatory effects, such as chemotactic responses of neutrophils (57), release of granular enzymes from phagocytic cells (58), neutrophil production of superoxide anion (59), vasodilatation, increased vascular permeability (60), and induction of thymocyte apoptosis during sepsis (61, 62).

Blockade of C5a generation with antibodies during the onset of sepsis in rodents has been shown to greatly improve survival (63). Similar findings were made when the C5a receptor (C5aR) was blocked, either with antibodies or with a small molecular inhibitor (64, 65). Earlier experimental studies in monkeys have suggested that antibody blockade of C5a attenuated E. coli–induced septic shock and adult respiratory distress syndrome (66, 67). In humans with sepsis, C5a was elevated and associated with significantly reduced survival rates together with multiorgan failure, when compared with that in less severely septic patients and survivors (68–70). The mechanisms by which C5a exerts its harmful effects during sepsis are yet to be investigated in greater detail, but recent data suggest that generation of C5a during sepsis significantly compromises innate immune functions of blood neutrophils (71, 72), their ability to express a respiratory burst, and their ability to generate cytokines (73). In addition, C5a generation during sepsis appears to have procoagulant effects (74). The concept of blockade of C5a/C5aR during sepsis therefore has therapeutic potential, especially in the context of prevention of development of sepsis.

Conclusion

The definition of sepsis appears to be very complicated because of the nonhomogenous nature of the patient populations studied and of the underlying conditions related to sepsis. The difficulties in precise clinical classification of septic patients could at least partially explain the failure of so many clinical trials that have used anti-inflammatory strategies. Therapeutic interventions are currently considered more likely to be successful in patients who exhibit severe sepsis or septic shock. Recent clinical studies suggested that early aggressive-volume resuscitation is beneficial in patients with severe sepsis or septic shock (75). Intensive insulin therapy for critically ill diabetic patients has reduced the frequency of occurrence of sepsis by 46% and lowered the mortality in patients with bacteremia (76). Significant improvements in supportive treatment in ICUs (e.g., more specific antibiotic treatment, improved mechanical ventilation, improved monitoring of circulation, etc.) are mainly responsible for improvements in survival statistics of septic patients. Still, the incidence of sepsis is rising; at this point, this cannot easily be explained. Accordingly, other strategies that more specifically target pathophysiological disorders in septic patients must be pursued to counteract these developments. With the approval of APC, new hope has been raised regarding the success of such strategies in general, including some of the emerging new strategies that have been discussed here. The development of new diagnostic tools that allow more accurate determination of the immune/inflammatory status of a septic patient may significantly contribute to the success of strategies that are based on defined and specific disorders in the context of sepsis.

Address correspondence to: Peter A. Ward, Department of Pathology, University of Michigan Medical School, 1301 Catherine Road, Ann Arbor, Michigan 48109-0602, USA. Phone: (734) 763-6384; Fax: (734) 763-4782; E-mail: pward@umich.edu.

1. Angus, D.C., et al. 2001. Epidemiology of severe sepsis in the United States: analysis of incidence, outcome, and associated costs of care. *Crit. Care Med.* **29**:1303–1310.

2. Martin, G.S., Mannino, D.M., Eaton, S., and Moss, M. 2003. The epidemiology of sepsis in the United States from 1979 through 2000. *N. Engl. J. Med.* **348**:1546–1554.

3. Eichacker, P.Q., et al. 2002. Risk and the efficacy of antiinflammatory agents: retrospective and confirmatory studies of sepsis. *Am. J. Respir. Crit. Care Med.* **166**:1197–1205.

4. Bone, R.C., et al. 1989. Sepsis syndrome: a valid clinical entity. Methylprednisolone Severe Sepsis Study Group. *Crit. Care Med.* **17**:389–393.

5. Goris, R.J., te Boekhorst, T.P., Nuytinck, J.K., and Gimbráere, J.S. 1985. Multiple-organ failure. Generalized autodestructive inflammation? *Arch. Surg.* **120**:1109–1115.

6. Nuytinck, H.K., Offermans, X.J., Kubat, K., and Goris, J.A. 1988. Whole-body inflammation in

trauma patients. An autopsy study. *Arch. Surg.* **123**:1519–1524.

7. Bone, R.C., et al. 1992. Definitions for sepsis and organ failure and guidelines for the use of innovative therapies in sepsis. The ACCP/SCCM Consensus Conference Committee. American College of Chest Physicians/Society of Critical Care Medicine. *Chest.* **101**:1644–1655.

8. Bone, R.C., Sprung, C.L., and Sibbald, W.J. 1992. Definitions for sepsis and organ failure. *Crit. Care Med.* **20**:724–726.

9. Levy, M.M., et al. 2003. 2001 SCCM/ESICM/ACCP/ATS/SIS International Sepsis Definitions Conference. *Crit. Care Med.* **31**:1250–1256.

10. Thomas, L. 1972. Germs. *N. Eng. J. Med.* **287**:553–555.

11. Bennett, I.L., et al. 1963. The effectiveness of hydrocortisone in the management of severe infection. *JAMA.* **183**:462–465.

12. Davis, C.E., Brown, K.R., Douglas, H., Tate, W.J., 3rd, and Braude, A.I. 1969. Prevention of death from endotoxin with antisera. I. The risk of fatal anaphylaxis to endotoxin. *J. Immunol.* **102**:563–572.

13. Cohen, J. 1999. Adjunctive therapy in sepsis: a critical analysis of the clinical trial programme. *Br. Med. Bull.* **55**:212–225.

14. Michie, H.R., et al. 1988. Tumor necrosis factor and endotoxin induce similar metabolic responses in human beings. *Surgery.* **104**:280–286.

15. Michie, H.R., et al. 1988. Detection of circulating tumor necrosis factor after endotoxin administration. *N. Engl. J. Med.* **318**:1481–1486.

16. Beutler, B., Milsark, I.W., and Cerami, A.C. 1985. Passive immunization against cachectin/tumor necrosis factor protects mice from lethal effect of endotoxin. *Science.* **229**:869–871.

17. Tracey, K.J., et al. 1986. Shock and tissue injury induced by recombinant human cachectin. *Science.* **234**:470–474.

18. Tracey, K.J., et al. 1987. Anti-cachectin/TNF monoclonal antibodies prevent septic shock during lethal bacteraemia. *Nature.* **330**:662–664.

19. Reinhart, K., and Karzai, W. 2001. Anti-tumor necrosis factor therapy in sepsis: update on clinical trials and lessons learned. *Crit. Care Med.* **29**(Suppl.):S121–S125.

20. Eskandari, M.K., et al. 1992. Anti-tumor necrosis factor antibody therapy fails to prevent lethality after cecal ligation and puncture or endotoxemia. *J. Immunol.* **148**:2724–2730.

21. Newcomb, D., Bolgos, G., Green, L., and Remick, D.G. 1998. Antibiotic treatment influences outcome in murine sepsis: mediators of increased morbidity. *Shock.* **10**:110–117.

22. Girardin, E., Grau, G.E., Dayer, J.M., Roux-Lombard, P., and Lambert, P.H. 1988. Tumor necrosis factor and interleukin-1 in the serum of children with severe infectious purpura. *N. Engl. J. Med.* **319**:397–400.

23. Hotchkiss, R.S., and Karl, I.E. 2003. The pathophysiology and treatment of sepsis. *N. Engl. J. Med.* **348**:138–150.

24. Riedemann, N.C., and Ward, P.A. 2003. Anti-inflammatory strategies for the treatment of sepsis. *Expert Opin. Biol. Ther.* **3**:339–350.

25. Annane, D., et al. 2002. Effect of treatment with low doses of hydrocortisone and fludrocortisone on mortality in patients with septic shock. *JAMA.* **288**:862–871.

26. Riedemann, N.C., Guo, R.F., and Ward, P.A. 2003. Novel strategies for the treatment of sepsis. *Nat. Med.* **9**:517–524.

27. Fisher, C.J., Jr., et al. 1996. Treatment of septic shock with the tumor necrosis factor receptor:Fc fusion protein. The Soluble TNF Receptor Sepsis Study Group. *N. Engl. J. Med.* **334**:1697–1702.

28. Bernard, G.R., et al. 2001. Efficacy and safety of recombinant human activated protein C for severe sepsis. *N. Engl. J. Med.* **344**:699–709.

29. Healy, D.P. 2002. New and emerging therapies for

sepsis. *Ann. Pharmacother.* **36**:648–654.

30. Riewald, M., Petrovan, R.J., Donner, A., Mueller, B.M., and Ruf, W. 2002. Activation of endothelial cell protease activated receptor 1 by the protein C pathway. *Science.* **296**:1880–1882.

31. Siegel, J.P. 2002. Assessing the use of activated protein C in the treatment of severe sepsis. *N. Engl. J. Med.* **347**:1030–1034.

32. Manns, B.J., Lee, H., Doig, C.J., Johnson, D., and Donaldson, C. 2002. An economic evaluation of activated protein C treatment for severe sepsis. *N. Engl. J. Med.* **347**:993–1000.

33. Warren, B.L., et al. 2001. Caring for the critically ill patient. High-dose antithrombin III in severe sepsis: a randomized controlled trial. *JAMA.* **286**:1869–1878.

34. Oberholzer, C., Oberholzer, A., Clare-Salzler, M., and Moldawer, L.L. 2001. Apoptosis in sepsis: a new target for therapeutic exploration. *FASEB J.* **15**:879–892.

35. Hotchkiss, R.S., et al. 1999. Overexpression of Bcl-2 in transgenic mice decreases apoptosis and improves survival in sepsis. *J. Immunol.* **162**:4148–4156.

36. Hotchkiss, R.S., et al. 1999. Prevention of lymphocyte cell death in sepsis improves survival in mice. *Proc. Natl. Acad. Sci. U. S. A.* **96**:14541–14546.

37. Hotchkiss, R.S., et al. 2000. Caspase inhibitors improve survival in sepsis: a critical role of the lymphocyte. *Nat. Immunol.* **1**:496–501.

38. Coopersmith, C.M., et al. 2002. Inhibition of intestinal epithelial apoptosis and survival in a murine model of pneumonia-induced sepsis. *JAMA.* **287**:1716–1721.

39. Coopersmith, C.M., et al. 2002. Overexpression of Bcl-2 in the intestinal epithelium improves survival in septic mice. *Crit. Care Med.* **30**:195–201.

40. Hotchkiss, R.S., et al. 2003. Adoptive transfer of apoptotic splenocytes worsens survival, whereas adoptive transfer of necrotic splenocytes improves survival in sepsis. *Proc. Natl. Acad. Sci. U. S. A.* **100**:6724–6729.

41. Wang, H., et al. 1999. HMG-1 as a late mediator of endotoxin lethality in mice. *Science.* **285**:248–251.

42. Wang, H., Yang, H., Czura, C.J., Sama, A.E., and Tracey, K.J. 2001. HMGB1 as a late mediator of lethal systemic inflammation. *Am. J. Respir. Crit. Care Med.* **164**:1768–1773.

43. Andersson, U., et al. 2000. High mobility group 1 protein (HMG-1) stimulates proinflammatory cytokine synthesis in human monocytes. *J. Exp. Med.* **192**:565–570.

44. Fiuza, C., et al. 2002. Inflammatory promoting activity of HMGB1 on human microvascular endothelial cells. *Blood.* **101**:2652–2660.

45. Sappington, P.L., et al. 2002. HMGB1 B box increases the permeability of Caco-2 enterocytic monolayers and impairs intestinal barrier function in mice. *Gastroenterology.* **123**:790–802.

46. Ulloa, L., et al. 2002. Ethyl pyruvate prevents lethality in mice with established lethal sepsis and systemic inflammation. *Proc. Natl. Acad. Sci. U. S. A.* **99**:12351–12356.

47. Calandra, T., et al. 2000. Protection from septic shock by neutralization of macrophage migration inhibitory factor. *Nat. Med.* **6**:164–170.

48. Bozza, M., et al. 1999. Targeted disruption of migration inhibitory factor gene reveals its critical role in sepsis. *J. Exp. Med.* **189**:341–346.

49. Bernhagen, J., et al. 1993. MIF is a pituitary-derived cytokine that potentiates lethal endotoxaemia. *Nature.* **365**:756–759.

50. Rossi, A.G., et al. 1998. Human circulating eosinophils secrete macrophage migration inhibitory factor (MIF). Potential role in asthma. *J. Clin. Invest.* **101**:2869–2874.

51. Rice, E.K., et al. 2003. Induction of MIF synthesis and secretion by tubular epithelial cells: a novel action of angiotensin II. *Kidney Int.* **63**:1265–1275.

52. Arndt, U., et al. 2002. Release of macrophage migration inhibitory factor and CXCL8/interleukin-8 from lung epithelial cells rendered necrotic by

influenza A virus infection. *J. Virol.* **76**:9298–9306.

53. Calandra, T., Bernhagen, J., Mitchell, R.A., and Bucala, R. 1994. The macrophage is an important and previously unrecognized source of macrophage migration inhibitory factor. *J. Exp. Med.* **179**:1895–1902.

54. Calandra, T., et al. 1995. MIF as a glucocorticoid-induced modulator of cytokine production. *Nature.* **377**:68–71.

55. Lehmann, L.E., et al. 2001. Plasma levels of macrophage migration inhibitory factor are elevated in patients with severe sepsis. *Intensive Care Med.* **27**:1412–1415.

56. Gando, S., et al. 2001. Macrophage migration inhibitory factor is a critical mediator of systemic inflammatory response syndrome. *Intensive Care Med.* **27**:1187–1193.

57. Shin, H.S., Snyderman, R., Friedman, E., Mellors, A., and Mayer, M.M. 1968. Chemotactic and anaphylatoxic fragment cleaved from the fifth component of guinea pig complement. *Science.* **162**:361–363.

58. Goldstein, I.M., and Weissmann, G. 1974. Generation of C5-derived lysosomal enzyme-releasing activity (C5a) by lysates of leukocyte lysosomes. *J. Immunol.* **113**:1583–1588.

59. Sacks, T., Moldow, C.F., Craddock, P.R., Bowers, T.K., and Jacob, H.S. 1978. Oxygen radicals mediate endothelial cell damage by complement-stimulated granulocytes. An in vitro model of immune vascular damage. *J. Clin. Invest.* **61**:1161–1167.

60. Schumacher, W.A., Fantone, J.C., Kunkel, S.E., Webb, R.C., and Lucchesi, B.R. 1991. The anaphylatoxins C3a and C5a are vasodilators in the canine coronary vasculature in vitro and in vivo. *Agents Actions.* **34**:345–349.

61. Guo, R.F., et al. 2000. Protective effects of anti-C5a in sepsis-induced thymocyte apoptosis. *J. Clin. Invest.* **106**:1271–1280.

62. Riedemann, N.C., et al. 2002. C5a receptor and thymocyte apoptosis in sepsis. *FASEB J.* **16**:887–888.

63. Czermak, B.J., et al. 1999. Protective effects of C5a blockade in sepsis. *Nat. Med.* **5**:788–792.

64. Huber-Lang, M.S., et al. 2002. Protection of innate immunity by C5aR antagonist in septic mice. *FASEB J.* **16**:1567–1574.

65. Riedemann, N.C., et al. 2002. Increased C5a receptor expression in sepsis. *J. Clin. Invest.* **110**:101–108. doi:10.1172/JCI200215409.

66. Hangen, D.H., et al. 1989. Complement levels in septic primates treated with anti-C5a antibodies. *J. Surg. Res.* **46**:195–199.

67. Stevens, J.H., et al. 1986. Effects of anti-C5a antibodies on the adult respiratory distress syndrome in septic primates. *J. Clin. Invest.* **77**:1812–1816.

68. Nakae, H., et al. 1994. Serum complement levels and severity of sepsis. *Res. Commun. Chem. Pathol. Pharmacol.* **84**:189–195.

69. Nakae, H., Endo, S., Inada, K., and Yoshida, M. 1996. Chronological changes in the complement system in sepsis. *Surg. Today.* **26**:225–229.

70. Bengtson, A., and Heideman, M. 1988. Anaphylatoxin formation in sepsis. *Arch. Surg.* **123**:645–649.

71. Huber-Lang, M.S., et al. 2002. Complement-induced impairment of innate immunity during sepsis. *J. Immunol.* **169**:3223–3231.

72. Guo, R.F., et al. 2003. Neutrophil C5a receptor and outcome in sepsis. *FASEB J.* In press.

73. Riedemann, N.C., et al. 2003. Regulation by C5a of neutrophil activation during sepsis. *Immunity.* In press.

74. Laudes, I.J., et al. 2002. Anti-c5a ameliorates coagulation/fibrinolytic protein changes in a rat model of sepsis. *Am. J. Pathol.* **160**:1867–1875.

75. Rivers, E., et al. 2001. Early goal-directed therapy in the treatment of severe sepsis and septic shock. *N. Engl. J. Med.* **345**:1368–1377.

76. van den Berghe, G., et al. 2001. Intensive insulin therapy in the critically ill patients. *N. Engl. J. Med.* **345**:1359–1367.

Antimicrobial resistance: the example of *Staphylococcus aureus*

Franklin D. Lowy

Division of Infectious Diseases, Departments of Medicine and Pathology, Columbia University, College of Physicians and Surgeons, New York, New York, USA.

In the early 1970s, physicians were finally forced to abandon their belief that, given the vast array of effective antimicrobial agents, virtually all bacterial infections were treatable. Their optimism was shaken by the emergence of resistance to multiple antibiotics among such pathogens as *Staphylococcus aureus*, *Streptococcus pneumoniae*, *Pseudomonas aeruginosa*, and *Mycobacterium tuberculosis*. The evolution of increasingly antimicrobial-resistant bacterial species stems from a multitude of factors that includes the widespread and sometimes inappropriate use of antimicrobials, the extensive use of these agents as growth enhancers in animal feed, and, with the increase in regional and international travel, the relative ease with which antimicrobial-resistant bacteria cross geographic barriers (1–3).

The irony of this trend toward progressively more resistant bacteria is that it coincides with a period of dramatically increased understanding of the molecular mechanisms of antimicrobial resistance. Unfortunately, while this insight has resulted in the identification of novel drug targets, it has not yet resulted in effective new chemotherapeutic agents. This paradox stands in sharp contrast to the dramatic progress made in antiviral (notably antiretroviral) therapy in the past ten years, where a number of newly discovered molecular targets have resulted in clinically effective therapeutic agents.

Nowhere has this issue been of greater concern than with the Gram-positive bacteria pneumococci, enterococci, and staphylococci. Multidrug resistance is now the norm among these pathogens. *S. aureus* is perhaps the pathogen of greatest concern because of its intrinsic virulence, its ability to cause a diverse array of life-threatening infections, and its capacity to adapt to different environmental conditions (4, 5). The mortality of *S. aureus* bacteremia remains approximately 20–40% despite the availability of effective antimicrobials (6). *S. aureus* is now the leading overall cause of nosocomial infections and, as more patients are treated outside the hospital setting, is an increasing concern in the community (7, 8).

S. aureus isolates from intensive care units across the country and from blood culture isolates worldwide are increasingly resistant to a greater number of antimicrobial agents (4, 8). Inevitably this has left fewer effective bactericidal antibiotics to treat these often life-threatening infections (Figure 1). As rapidly as new antibiotics are introduced, staphylococci have developed efficient mechanisms to neutralize them (Table 1).

Nonstandard abbreviations used: methicillin-resistant *Staphylococcus aureus* (MRSA); staphylococcal cassette chromosome *mec* (SCC*mec*); methicillin-susceptible *S. aureus* (MSSA); penicillin-binding protein (PBP); quinolone resistance–determining region (QRDR); minimal inhibitory concentration (MIC); vancomycin intermediate-resistant *S. aureus* (VISA); vancomycin-resistant *S. aureus* (VRSA).

Conflict of interest: The author has received research support from GlaxoSmithKline and is involved in ongoing clinical trials with Cubist Pharmaceuticals Inc.

Citation for this article: *J. Clin. Invest.* **111**:1265–1273 (2003). doi:10.1172/JCI200318535.

Recent reports of *S. aureus* isolates with intermediate or complete resistance to vancomycin portend a chemotherapeutic era in which effective bactericidal antibiotics against this organism may no longer be readily available (9, 10). This review will focus on the emergence of antimicrobial resistance in *S. aureus*. It will review the historical evolution of resistant strains, their spread, the molecular mechanisms of resistance for selected antibiotics, and progress toward the development of alternative drug targets or novel approaches for therapeutic or prophylactic intervention.

Penicillin resistance

History and epidemiology. The mortality of patients with *S. aureus* bacteremia in the pre-antibiotic era exceeded 80%, and over 70% developed metastatic infections (11). The introduction of penicillin in the early 1940s dramatically improved the prognosis of patients with staphylococcal infection. However, as early as 1942, penicillin-resistant staphylococci were recognized, first in hospitals and subsequently in the community (12). By the late 1960s, more than 80% of both community- and hospital-acquired staphylococcal isolates were resistant to penicillin. This pattern of resistance, first emerging in hospitals and then spreading to the community, is now a well-established pattern that recurs with each new wave of antimicrobial resistance (13).

Kirby first demonstrated that penicillin was inactivated by penicillin-resistant strains of *S. aureus* (14). Bondi and Dietz (15) subsequently identified the specific role of penicillinase. More than 90% of staphylococcal isolates now produce penicillinase, regardless of the clinical setting. The gene for β-lactamase is part of a transposable element located on a large plasmid, often with additional antimicrobial resistance genes (e.g., gentamicin and erythromycin). Spread of penicillin resistance primarily occurs by spread of resistant strains.

Mechanisms of resistance. Staphylococcal resistance to penicillin is mediated by *blaZ*, the gene that encodes β-lactamase (Figure 2a). This predominantly extracellular enzyme, synthesized when staphylococci are exposed to β-lactam antibiotics, hydrolyzes the β-lactam ring, rendering the β-lactam inactive. *blaZ* is under the control of two adjacent regulatory genes, the antirepressor *blaR1* and the repressor *blaI* (16). Recent studies have demonstrated that the signaling pathway responsible for β-lactamase synthesis requires sequential cleavage of the regulatory proteins BlaR1 and BlaI. Following exposure to β-lactams, BlaR1, a transmembrane sensor-transducer, cleaves itself (17, 18). Zhang et al. (18) hypothesize that the cleaved protein functions as a protease that cleaves the repressor BlaI, directly or indirectly (an additional protein, BlaR2, may be involved in this pathway) and allows *blaZ* to synthesize enzyme.

Methicillin resistance

History and epidemiology. Methicillin, introduced in 1961, was the first of the semisynthetic penicillinase-resistant penicillins.

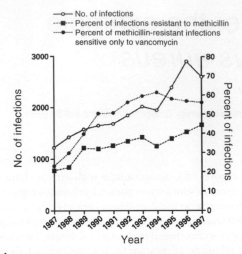

Figure 1
S. aureus infections in intensive care units in the National Nosoco-mial Infections Surveillance System. Data include the total number of infections from 1987 to 1997. Isolates were tested for sensitivity to the following antimicrobial agents: gentamicin, tobramycin, amikacin, cip-rofloxacin, clindamycin, erythromycin, chloramphenicol, trimethoprim-sulfamethoxazole, and vancomycin. Some hospitals did not test for all of the antibiotics. Reproduced with permission from *New England Journal of Medicine* (4).

Its introduction was rapidly folloawed by reports of methicillin-resistant isolates (19). For clinicians, the spread of these methi-cillin-resistant strains has been a critical one. The therapeutic outcome of infections that result from methicillin-resistant *S. aureus* (MRSA) is worse than the outcome of those that result from methicillin-sensitive strains (20). The difference has been ascribed to the underlying medical problems of the often sicker and older MRSA-infected patients as well as the less effective bac-tericidal drugs available to treat these infections, rather than to enhanced virulence of the MRSA strains.

First reported in a British hospital, MRSA clones rapidly spread across international borders. Waves of clonal dissemination with different dominant phage types (e.g., 83 complex) were reported in the 1960s and were responsible for a large proportion of cases (21, 22). Once identified in a new setting, these unique MRSA clones rapidly spread, often becoming the resident clones and accounting for an increasing percentage of nosocomial infections (23, 24). Like the penicillin-resistant strains, the MRSA isolates also frequently carried resistance genes to other antimicrobial agents (25).

The spread of methicillin-resistant clones is reminiscent of the emergence of penicillin resistance in the 1940s. First detected in hospitals in the 1960s, methicillin resistance is now increasingly recognized in the community (13). While many of these infections occurred in patients with some antecedent hospital experience, recently there has been an increasing number of subjects with no prior hospital exposure. These community-based infections have been reported in patients from both rural and urban settings (26–29). Of concern is the high mortality associated with some of these community-acquired MRSA infections. In one instance, clonally related MRSA strains caused the deaths of four otherwise healthy children (28). The empiric selection of β-lactams as ini-tial therapy may have contributed to the increased morbidity in these infections. In addition, the presence of virulence genes such

as the enterotoxins or the Panton-Valentine leukocidin may also have contributed to patient morbidity in some of the community-acquired infections (30, 31).

The *mecA* gene (the gene responsible for methicillin resistance) is part of a mobile genetic element found in all MRSA strains. Kata-yama et al. (32) demonstrated that *mecA* is part of a genomic island designated staphylococcal cassette chromosome *mec* (SCC*mec*). To date, four different SCC*mec* elements varying in size from 21 to 67 kb have been characterized (33). In contrast to the numer-ous different strains of methicillin-susceptible *S. aureus* (MSSA) that cause infections, only a limited number of clones are respon-sible for the epidemic spread of MRSA. This distinction reflects the genetic constraints of horizontal transfer of the *mec* element from related staphylococcal species into *S. aureus*. The frequency of this transfer event is subject to some debate, but it is clearly uncommon (34–37). The *mec* element SCC*mec* has been identified in several different MSSA genetic backgrounds. Using the DNA fingerprint of the recipient MSSA in combination with the genetic organization of the particular SCC*mec*, an MRSA biotype as well as an evolutionary profile of a particular clone can be established (33, 35–38). Oliveira et al. identified a putative pathway for the evolu-tion of several pandemic clones in Europe (38). By comparing the genetic backgrounds of epidemic MSSA clones, these investiga-tors also identified the likely ancestral MSSA recipients of the *mec* element. These studies suggest that the emergence of "epidemic" MRSA clones was in part the result of the successful horizontal transfer of the *mec* gene into an ecologically fit and transmission-efficient MSSA clone (37, 39).

The recent upsurge of community-acquired MRSA infections reported in patients from different countries was associated with the detection of a unique SCC*mec*, type IV (40). This element, smaller than the other elements, appears more genetically mobile and does not, at present, carry additional antimicrobial resistance genes. It also appears to occur in a more diverse range of MSSA genetic backgrounds, suggesting that it has been heterologously transferred more readily from other staphylococcal species (41).

Mechanisms of resistance. Methicillin resistance requires the pres-ence of the chromosomally localized *mecA* gene (16, 42) (Figure 2b). *mecA* is responsible for synthesis of penicillin-binding pro-tein 2a (PBP2a; also called PBP2′) a 78-kDa protein (43–45). PBPs are membrane-bound enzymes that catalyze the transpeptida-tion reaction that is necessary for cross-linkage of peptidoglycan chains (46). Their activity is similar to that of serine proteases, from which they appear to have evolved. PBP2a substitutes for the other PBPs and, because of its low affinity for all β-lactam antibiotics, enables staphylococci to survive exposure to high concentrations of these agents. Thus, resistance to methicillin confers resistance to all β-lactam agents, including cephalospo-rins. Recent studies determined the crystal structure of a soluble derivative of PBP2a. PBP2a differs from other PBPs in that its active site blocks binding of all β-lactams but allows the trans-peptidation reaction to proceed (47).

Phenotypic expression of methicillin resistance is variable, and each MRSA strain has a characteristic profile of the proportion of bacterial cells that grow at specific concentrations of methicillin (48). Expression of resistance in some MRSA strains is regulated by homologues of the regulatory genes for *blaZ*. These genes, *mecI* and *mecR1*, regulate the *mecA* response to β-lactam antibiotics in a fashion similar to that of the regulation of *blaZ* by the genes *blaR1* and *blaI* upon exposure to penicillin (Figure 2). In fact, the

Table 1

Mechanisms of *S. aureus* resistance to antimicrobials[A]

Antibiotic	Resistance gene(s)	Gene product(s)	Mechanism(s) of resistance	Location(s)
β-Lactams	1) *blaZ*	1) β-Lactamase	1) Enzymatic hydrolysis of β-lactam nucleus	1) Pl:Tn
	2) *mecA*	2) PBP2a	2) Reduced affinity for PBP	2) C:SCC*mec*
Glycopeptides	1) Unknown (VISA)	1) Altered peptidoglycan	1) Trapping of vancomycin in the cell wall	1) C
	2)	2) D-Ala-D-Lac	2) Synthesis of dipeptide with reduced affinity for vancomycin	2) Pl:Tn
Quinolones	1) *parC*	1) ParC (or GrlA) component of topoisomerase IV	1,2) Mutations in the QRDR region, reducing affinity of enzyme-DNA complex for quinolones	1) C
	2) *gyrA* or *gyrB*	2) GyrA or GyrB components of gyrase		2) C
Aminoglycosides (e.g., gentamicin)	Aminoglycoside-modifying enzymes (e.g., *aac*, *aph*)	Acetyltransferase, phosphotransferase	Acetylating and/or phosphorylating enzymes modify aminoglycosides	Pl, Pl:Tn
Trimethoprim-sulfamethoxazole (TMP-SMZ)	1) Sulfonamide: *sulA*	1) Dihydropteroate synthase	1) Overproduction of *p*-aminobenzoic acid by enzyme	1) C
	2) TMP: *dfrB*	2) Dihydrofolate reductase (DHFR)	2) Reduced affinity for DHFR	2) C
Oxazolidinones	*rrn*	23S rRNA	Mutations in domain V of 23S rRNA component of the 50S ribosome. Interferes with ribosomal binding	C
Quinupristin-dalfopristin (Q-D)	1) Q: *ermA*, *ermB*, *ermC*	1) Ribosomal methylases	1) Reduce binding to the 23S ribosomal subunit	1) Pl, C
	2) D: *vat*, *vatB*	2) Acetyltransferases	2) Enzymatic modification of dalfopristin	2) Pl

[A]Examples of several of the *S. aureus* mechanisms of resistance to selected antibiotics (77, 95–97). Pl, plasmid; C, chromosome; Tn, transposon; QRDR, quinolone resistance–determining region.

DNA sequences bound by the repressor genes to achieve inhibition of gene activation are identical (49). The sequence homology of *mecI-mecR1* with the *blaR1-blaI* regulatory genes results in the induction of *mecA* expression from this leaky alternative system. Deletions or mutations in *mecI* or the promoter region of *mecA* result in constitutive expression rather than variable expression of *mec* (50). Rosato et al. (51) have recently found that either *mecI* or *blaI* must be functional in all MRSA, and they suggest that this may be a protective mechanism preventing overproduction of a toxic protein. An additional series of genes, the *fem* genes (factor essential for resistance to methicillin resistance, play a role in cross-linking peptidoglycan strands and also contribute to the heterogeneity of expression of methicillin resistance (52).

As noted earlier, the *mecA* gene is invariably part of a larger unique, mobile genetic element, SCC*mec*. These islands may also contain additional genes for antimicrobial resistance and insertion sequences, as well as genes of uncertain function. The four SCC*mec*'s contain two recombinases, *ccrA* and *ccrB* from the invertase/resolvase family, that are responsible for site-specific integration and excision from the chromosome at *attBscc*, a part of an open reading frame of unknown function near the origin of replication (32, 53). The genetic mechanisms responsible for the transfer of these large mobile elements are uncertain.

Since no homologue of *mecA* exists in methicillin-susceptible staphylococci, it has been assumed that *mecA* was acquired from one of several coagulase-negative staphylococcal species (54). Couto et al. (55) identified a *mecA* gene in a methicillin-sensitive *Staphylococcus sciuri* with 88% homology on the amino acid level to MRSA. Transduction of the *S. sciuri mecA* into an MSSA resulted

in increased resistance to methicillin coupled with the detection of PBP2a (56). These studies therefore suggest one possible source of the *mecA* element in *S. aureus*. Hiramatsu and associates (41, 57) have speculated that the simultaneous detection of the new type IV SCC*mec* in different geographic regions of the world potentially reflects its enhanced mobility and multiple simultaneous transmissions from another coagulase-negative staphylococcus.

Quinolone resistance

History and epidemiology. Fluoroquinolones were initially introduced for the treatment of Gram-negative bacterial infections in the 1980s. However, because of their Gram-positive bacterial spectrum, they have also been used to treat bacterial infections caused by pneumococci and staphylococci. Quinolone resistance among *S. aureus* emerged quickly, more prominently among the methicillin-resistant strains. As a result, the ability to use fluoroquinolones as antistaphylococcal agents was dramatically reduced. The reasons for the disparity in rates of quinolone resistance between MSSA and MRSA strains are uncertain. One contributing factor is likely antibiotic selective pressure, especially in the hospital setting, resulting in the selection and spread of the more antibiotic-resistant MRSA strains.

Fluoroquinolone resistance develops as a result of spontaneous chromosomal mutations in the target of the antibiotic, topoisomerase IV or DNA gyrase, or by the induction of a multidrug efflux pump. When quinolones are used to treat infections caused by other bacterial pathogens, subjects colonized with *S. aureus* (e.g., on their skin or mucosal surfaces) are likely exposed to subtherapeutic antibiotic concentrations and are therefore at risk of becoming

a

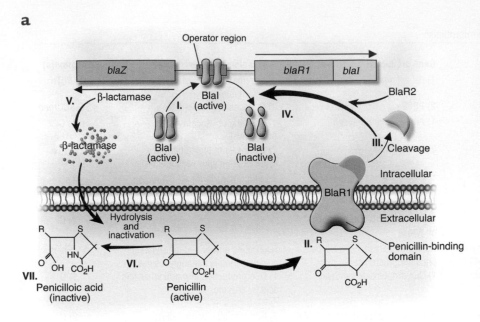

b

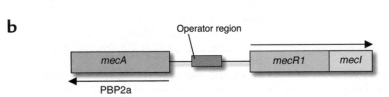

Figure 2
(**a**) Induction of staphylococcal β-lactamase synthesis in the presence of the β-lactam antibiotic penicillin. I. The DNA-binding protein BlaI binds to the operator region, thus repressing RNA transcription from both *blaZ* and *blaR1-blaI*. In the absence of penicillin, β-lactamase is expressed at low levels. II. Binding of penicillin to the transmembrane sensor-transducer BlaR1 stimulates BlaR1 autocatalytic activation. III–IV. Active BlaR1 either directly or indirectly (via a second protein, BlaR2) cleaves BlaI into inactive fragments, allowing transcription of both *blaZ* and *blaR1-blaI* to commence. V–VII. β-Lactamase, the extracellular enzyme encoded by *blaZ* (V), hydrolyzes the β-lactam ring of penicillin (VI), thereby rendering it inactive (VII). (**b**) Mechanism of *S. aureus* resistance to methicillin. Synthesis of PBP2a proceeds in a fashion similar to that described for β-lactamase. Exposure of MecR1 to a β-lactam antibiotic induces MecR1 synthesis. MecR1 inactivates MecI, allowing synthesis of PBP2a. MecI and BlaI have coregulatory effects on the expression of PBP2a and β-lactamase.

colonized with resistant mutants (58). These resident, resistant strains then become the reservoir for future infections. Høiby et al. (59) demonstrated that ciprofloxacin therapy rapidly increased the proportion of coagulase-negative staphylococcal strains colonizing the nares and skin that were resistant to both ciprofloxacin and methicillin. Since *S. aureus* is also a part of our commensal flora, a similar selection process is likely to occur.

Mechanisms of resistance. Resistance to quinolones results from the stepwise acquisition of chromosomal mutations. The confluence of high bacterial density, the likely preexistence of resistant subpopulations, and the sometimes limited quinolone concentrations achieved at sites of staphylococcal infections creates an environment that fosters selection of resistant mutants (58). The quinolones act on DNA gyrase, which relieves DNA supercoiling, and topoisomerase IV, which separates concatenated DNA strands. Amino acid changes in critical regions of the enzyme-DNA complex (quinolone resistance–determining region [QRDR]) reduce quinolone affinity for both of its targets. The ParC subunit (GrlA in *S. aureus*) of topoisomerase IV and the GyrA subunit in gyrase are the most common sites of resistance mutations; topoisomerase IV mutations are the most critical, since they are the primary drug targets in staphylococci (58, 60). Single amino acid mutations are sometimes sufficient to confer clinical resistance, but for the more active fluoroquinolones additional mutations appear necessary. Resistance mutations can accumulate in the QRDR sites, increasing the levels of resistance. It is fairly common for both targets to have resistance mutations. An additional mechanism of resistance in *S. aureus* is induction of the NorA multidrug resistance efflux pump. Increased expression of this pump in *S. aureus* can result in low-level quinolone resistance (61). In an interesting linkage of

virulence to antimicrobial resistance, a recent study showed that exposure of a quinolone-resistant isolate to a quinolone increased the organism's expression of fibronectin-binding protein, a surface protein that mediates adherence to tissue surfaces (62).

While newer fluoroquinolones (8-methoxyquinolones such as moxifloxacin) retain in vitro activity against ciprofloxacin-resistant staphylococci and appear less likely to select for resistant mutants (63), it is not clear that this translates into therapeutic efficacy. Entenza et al. (64) reported that while moxifloxacin retained a low minimal inhibitory concentration (MIC) against ciprofloxacin-resistant strains of *S. aureus*, treatment with this newer quinolone failed in an experimental model of endocarditis. This suggests a potential disparity between in vitro susceptibility testing and in vivo therapeutic efficacy.

Vancomycin resistance

History and epidemiology. The dramatic increase in use of vancomycin to treat infections caused by methicillin-resistant staphylococci (both coagulase-positive and -negative), *Clostridium difficile*, and enterococcal infections preceded the emergence of vancomycin-resistant staphylococci (65). Staphylococcal resistance to vancomycin in a clinical isolate was first reported in a strain of *Staphylococcus haemolyticus* (66). In 1997, the first report of vancomycin intermediate-resistant *S. aureus* (VISA) came from Japan, and additional cases were subsequently reported from other countries (9, 67). The VISA isolates were all MRSA and were not clonal. Many of the patients had received vancomycin therapy and had MRSA infections (68).

Two recent reports of infections caused by vancomycin-resistant *S. aureus* (VRSA) are of great concern because they reflect both

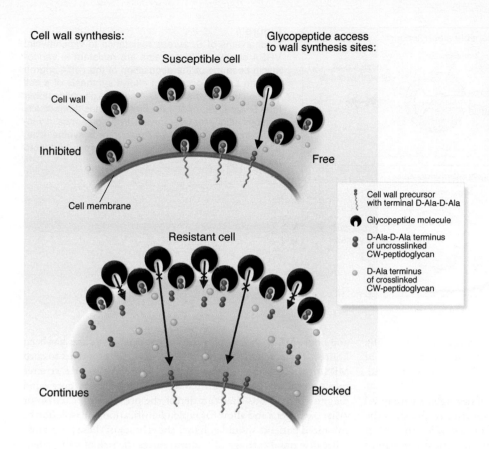

Cell wall synthesis:

Susceptible cell

Cell wall

Inhibited

Cell membrane

Glycopeptide access
to wall synthesis sites:

Free

Resistant cell

Continues   Blocked

Cell wall precursor
with terminal D-Ala-D-Ala

Glycopeptide molecule

D-Ala-D-Ala terminus
of uncrosslinked
CW-peptidoglycan

D-Ala terminus
of crosslinked
CW-peptidoglycan

Figure 3
Mechanisms of *S. aureus* resistance to vancomycin: VISA strains. VISA strains appear to be selected from isolates that are heterogeneously resistant to vancomycin. These VISA strains synthesize additional quantities of peptidoglycan with an increased number of D-Ala-D-Ala residues that bind vancomycin, preventing the molecule from getting to its bacterial target. Adapted from ref. 98.

complete resistance and a different mechanism for dissemination. In contrast to the chromosomally mediated resistance for VISA strains, the VRSA strains acquire resistance by conjugal transfer of the *vanA* operon from an *Enterococcus faecalis*, raising the specter of a far more efficient means for dissemination of the resistance gene among strains of staphylococci (10, 69).

Mechanisms of resistance. Two forms of *S. aureus* resistance to vancomycin have now been identified (70). One form has been identified in the VISA strains, which have MICs to vancomycin of 8–16 µg/ml (9). A pre-VISA stage of resistance, heterogeneously resistant, has also been identified. The heteroresistant strains remain susceptible to vancomycin but contain resistant subpopulations. It is hypothesized that, on exposure to vancomycin, the VISA isolates are selected from the vancomycin-resistant subpopulations (71). The reduced susceptibility to vancomycin appears to result from changes in peptidoglycan biosynthesis (Figure 3). The VISA strains are notable for the additional quantities of synthesized peptidoglycan that result in irregularly shaped, thickened cell walls. There is also decreased cross-linking of peptidoglycan strands, which leads to the exposure of more D-Ala-D-Ala residues (72, 73). The altered cross-linking results from reduced amounts of L-glutamine that are available for amidation of D-glutamate in the pentapeptide bridge (70). As a result there are more D-Ala-D-Ala residues available to bind and trap vancomycin (Figure 3). The bound vancomycin then acts as a further impediment to drug molecules reaching their target on the cytoplasmic membrane. HPLC provided further proof of this novel resistance mechanism by showing that large quantities of vancomycin become trapped in the abnormal peptidoglycan (74). The molecular mechanisms for these alterations in peptidoglycan biosynthesis are unexplained.

The second form of vancomycin resistance has resulted from the probable conjugal transfer of the *vanA* operon from a vancomycin-resistant *E. faecalis*. Showsh et al. (75) reported that the enterococcal plasmid containing *vanA* also encodes a sex pheromone that is synthesized by *S. aureus*, suggesting a potential facilitator of conjugal transfer. These VRSA isolates demonstrate complete vancomycin resistance, with MICs of ≥128 µg/ml. Resistance in these isolates is caused by alteration of the terminal peptide to D-Ala-D-Lac instead of D-Ala-D-Ala (Figure 4). Synthesis of D-Ala-D-Lac occurs only with exposure to low concentrations of vancomycin. As a result, the additional biosynthetic demands are limited and the VRSA strain is ecologically fit (76). This ecological fitness, the possibility that this plasmid exchange will occur more frequently (due to the ever increasing likelihood of patients being colonized with both MRSA and vancomycin-resistant enterococci), and the resistance of these strains to both β-lactams and glycopeptides all increase the likelihood that VRSA strains will rapidly become more prevalent.

Prospects for the future: new antimicrobials, new approaches

Currently available agents. Quinupristin-dalfopristin and linezolid are two of the newer antimicrobial agents currently available with activity against drug-resistant staphylococci (including most VISA and VRSA strains in vitro). Both agents are protein synthesis inhibitors with a Gram-positive spectrum. Quinupristin-dalfopristin retains bactericidal antistaphylococcal activity if the strain is susceptible to erythromycin and lincosamide. Linezolid is bacteriostatic. Cross-resistance has not been noted for linezolid, but at least one clinical isolate has developed resistance during therapy (77). Daptomycin, a novel bactericidal agent that damages the

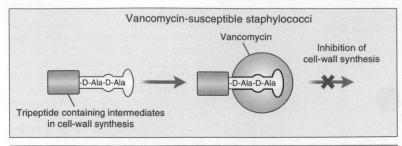

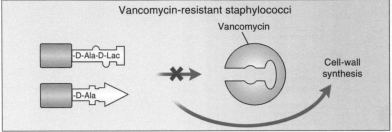

Figure 4
Mechanisms of *S. aureus* resistance to vancomycin: VRSA strains. VRSA strains are resistant to vancomycin because of the acquisition of the *vanA* operon from an enterococcus that allows synthesis of a cell wall precursor that ends in D-Ala-D-Lac dipeptide rather than D-Ala-D-Ala. The new dipeptide has dramatically reduced affinity for vancomycin. In the presence of vancomycin, the novel cell wall precursor is synthesized, allowing continued peptidoglycan assembly. Adapted from ref. 99.

cytoplasmic membrane, is currently undergoing clinical trials (78). Other agents with antistaphylococcal activity in varying stages of development include modified glycopeptides, carbapenems, oxazolidinones, quinolones, and tetracyclines.

Novel approaches and targets. The supply of new agents with novel mechanisms of action is limited, however, and emphasizes the need for the development of new drug targets (79, 80). Unfortunately, an increasing number of pharmaceutical companies have either eliminated or dramatically reduced their anti-infectives units. This results partly from financial considerations but also from frustration that target-based biochemical screening has failed to develop any clinically useful products. The failure has been attributed, in part, to the realization that target-based strategies do not take into account the "intrinsic" mechanisms of bacterial resistance (e.g., biofilms, multidrug efflux pumps) that contribute to in vivo bacterial resistance (79).

Despite these developments, a number of interesting models for identification of new drug targets have emerged. One approach has been to integrate genomic information on potential drug targets with high-throughput screening followed by chemical modification and efficacy animal testing. There has been a renewed interest in characterization of essential components of critical biosynthetic pathways (e.g., peptidoglycan assembly or fatty acid biosynthesis) as potential targets. Several different techniques, including in vivo expression technology, signature-tagged mutagenesis, and recognition of expressed *S. aureus* antigens, have been used to identify potential targets that are expressed during infection (80–82). Analysis of the crystal structure of drug targets (e.g., modifications of β-lactams that attack the active site of PBP2a) and synthesis of carbohydrate-modified compounds (glycopeptide analogues with altered carbohydrates) are increasingly being used to develop alternative agents (47, 83). Modification of *S. aureus* genes associated with virulence reduces infectivity (84, 85). Whether these genes can be successfully used as potential targets is uncertain.

Possible preventive measures. Prevention of *S. aureus* infections has to date been limited to the application of infection control measures. In some countries, such as The Netherlands and Denmark, where strict isolation policies have prevailed, these precautions have been effective in preventing dissemination of MRSA, while in the USA

and England, the success of infection control procedures has been limited. The potential of strict infection control programs to curb MRSA transmission suggests that, given recent trends, stricter infection control guidelines are warranted (86). Newer, more rapid diagnostic methods that can detect the presence of *S. aureus* or other pathogens and allow for rapid identification and isolation of colonized patients should enhance the efficacy of these programs.

Because nasal carriage of *S. aureus* raises the risk of subsequent infection, efforts have been directed to the elimination of carriage using topical antimicrobials (87–89). These approaches have been variably successful. More recently, the potential use of novel agents for this purpose, such as endopeptidase, lysostaphin, or phage lytic enzymes has also been considered (90–92).

Finally, several *S. aureus* vaccine candidates are under investigation. A capsular polysaccharide protein conjugate vaccine underwent a clinical trial with hemodialysis patients, with encouraging but inconclusive results (93). Other candidate vaccines directed at *S. aureus* virulence determinants such as the surface adhesins or enterotoxins are in varying stages of development (94).

The difficult therapeutic problem of multidrug-resistant *S. aureus* is just one example of the diminishing efficacy of antimicrobial agents for the treatment of bacterial infections. This trend is particularly alarming for *S. aureus* because of the severity and diversity of disease caused by this uniquely versatile pathogen. While effective antistaphylococcal agents still exist, their shelf-life is likely to be increasingly limited. Novel approaches to therapy and prevention will become more and more important, especially with the diminishing availability of new "wonder drugs."

Acknowledgments

This work was supported by NIH grants DA-09656, DA-11868, and DA-15018. Abigail Zuger's critical review of this manuscript is gratefully acknowledged.

Address correspondence to: Franklin D. Lowy, Division of Infectious Diseases, Departments of Medicine and Pathology, Columbia University, College of Physicians and Surgeons, 630 West 168th Street, New York, New York 10032, USA. Phone: (212) 305-5787; Fax: (212) 305-5794; E-mail: fl189@columbia.edu.

1. Cohen, M.L. 1992. Epidemiology of drug resistance: implications for a post-antimicrobial era. *Science*. **257**:1050–1055.

2. Tomasz, A. 1994. Multiple-antibiotic-resistant pathogenic bacteria. A report on the Rockefeller University Workshop. *N. Engl. J. Med.* **330**:1247–1251.

3. Swartz, M.N. 1997. Use of antimicrobial agents and drug resistance. *N. Engl. J. Med.* **337**:491–492.

4. Lowy, F.D. 1998. *Staphylococcus aureus* infections. *N. Engl. J. Med.* **339**:520–532.

5. Waldvogel, F.A. 2000. *Staphylococcus aureus* (including staphylococcal toxic shock). In *Principles and practice of infectious diseases*. G.L. Mandell, J.E. Bennett, and R. Dolin, editors. Churchill Livingstone. Philadelphia, Pennsylvania, USA. 2069–2092.

6. Mylotte, J.M., McDermott, C., and Spooner, J.A. 1987. Prospective study of 114 consecutive episodes of *Staphylococcus aureus* bacteremia. *Rev. Infect. Dis.* **9**:891–907.

7. 2001. CDC NNIS System: National Nosocomial Infections Surveillance (NNIS) system report, data summary from January 1992–April 2001, issued August 2001. *Am. J. Infect. Control.* **29**:400–421.

8. Diekema, D.J., et al. 2001. Survey of infections due to Staphylococcus species: frequency of occurrence and antimicrobial susceptibility of isolates collected in the United States, Canada, Latin America, Europe, and the Western Pacific region for the SENTRY Antimicrobial Surveillance Program, 1997–1999. *Clin. Infect. Dis.* **32**(Suppl. 2):S114–S132.

9. Hiramatsu, K., et al. 1997. Methicillin-resistant *Staphylococcus aureus* clinical strain with reduced vancomycin susceptibility. *J. Antimicrob. Chemother.* **40**:135–136.

10. 2002. *Staphylococcus aureus* resistant to vancomycin. United States, 2002. *MMWR.* **51**:565–567.

11. Skinner, D., and Keefer, C.S. 1941. Significance of bacteremia caused by *Staphylococcus aureus*. *Arch. Intern. Med.* **68**:851–875.

12. Rammelkamp, C.H., and Maxon, T. 1942. Resistance of *Staphylococcus aureus* to the action of penicillin. *Proc. Royal Soc. Exper. Biol. Med.* **51**:386–389.

13. Chambers, H.F. 2001. The changing epidemiology of *Staphylococcus aureus*? *Emerg. Infect. Dis.* **7**:178–182.

14. Kirby, W.M.M. 1944. Extraction of a higly potent penicillin inactivator from penicillin resistant staphylococci. *Science*. **99**:452–453.

15. Bondi, J.A., and Dietz, C.C. 1945. Penicillin resistant staphylococci. *Proc. Royal Soc. Exper. Biol. Med.* **60**:55–58.

16. Kernodle, D.S. 2000. Mechanisms of resistance to β-lactam antibiotics. In *Gram-positive pathogens*. V.A. Fischetti, R.P. Novick, J.J. Ferretti, D.A. Portnoy, and J.I. Rood, editors. American Society for Microbiology. Washington, DC, USA. 609–620.

17. Gregory, P.D., Lewis, R.A., Curnock, S.P., and Dyke, K.G. 1997. Studies of the repressor (BlaI) of beta-lactamase synthesis in *Staphylococcus aureus*. *Mol. Microbiol.* **24**:1025–1037.

18. Zhang, H.Z., Hackbarth, C.J., Chansky, K.M., and Chambers, H.F. 2001. A proteolytic transmembrane signaling pathway and resistance to beta-lactams in staphylococci. *Science*. **291**:1962–1965.

19. Jevons, M.P. 1961. "Celbenin"-resistant staphylococci. *Br. Med. J.* **1**:124–125.

20. Cosgrove, S.E., et al. 2003. Comparison of mortality associated with methicillin-resistant and methicillin-susceptible *Staphylococcus aureus* bacteremia: a meta-analysis. *Clin. Infect. Dis.* **36**:53–59.

21. Jessen, O., Rosendal, K., Bulow, P., Faber, V., and Eriksen, K.R. 1969. Changing staphylococci and staphylococcal infections. A ten-year study of bacteria and cases of bacteremia. *N. Engl. J. Med.* **281**:627–635.

22. Parker, M.T., and Hewitt, J.H. 1970. Methicillin resistance in *Staphylococcus aureus*. *Lancet*. **1**:800–804.

23. Panlilio, A.L., et al. 1992. Methicillin-resistant *Staphylococcus aureus* in U.S. hospitals, 1975–1991. *Infect. Control Hosp. Epidemiol.* **13**:582–586.

24. Couto, I., et al. 1995. Unusually large number of methicillin-resistant *Staphylococcus aureus* clones in a Portuguese hospital. *J. Clin. Microbiol.* **33**:2032–2035.

25. Lyon, B.R., Iuorio, J.L., May, J.W., and Skurray, R.A. 1984. Molecular epidemiology of multiresistant *Staphylococcus aureus* in Australian hospitals. *J. Med. Microbiol.* **17**:79–89.

26. Moreno, F., Crisp, C., Jorgensen, J.H., and Patterson, J.E. 1995. Methicillin-resistant *Staphylococcus aureus* as a community organism. *Clin. Infect. Dis.* **21**:1308–1312.

27. Herold, B.C., et al. 1998. Community-acquired methicillin-resistant *Staphylococcus aureus* in children with no identified predisposing risk. *J. Am. Med. Assoc.* **279**:593–598.

28. 1999. Four pediatric deaths from community-acquired methicillin-resistant *Staphylococcus aureus*: Minnesota and North Dakota, 1997–1999. *MMWR.* **48**:707–710.

29. Groom, A.V., et al. 2001. Community-acquired methicillin-resistant *Staphylococcus aureus* in a rural American Indian community. *J. Am. Med. Assoc.* **286**:1201–1205.

30. Baba, T., et al. 2002. Genome and virulence determinants of high virulence community-acquired MRSA. *Lancet*. **359**:1819–1827.

31. Fey, P.D., et al. 2003. Comparative molecular analysis of community- or hospital-acquired methicillin-resistant *Staphylococcus aureus*. *Antimicrob. Agents Chemother.* **47**:196–203.

32. Katayama, Y., Ito, T., and Hiramatsu, K. 2000. A new class of genetic element, staphylococcus cassette chromosome mec, encodes methicillin resistance in *Staphylococcus aureus*. *Antimicrob. Agents Chemother.* **44**:1549–1555.

33. Hiramatsu, K., Cui, L., Kuroda, M., and Ito, T. 2001. The emergence and evolution of methicillin-resistant *Staphylococcus aureus*. *Trends Microbiol.* **9**:486–493.

34. Kreiswirth, B., et al. 1993. Evidence for a clonal origin of methicillin resistance in *Staphylococcus aureus*. *Science*. **259**:227–230.

35. Hiramatsu, K. 1995. Molecular evolution of MRSA. *Microbiol. Immunol.* **39**:531–543.

36. Fitzgerald, J.R., Sturdevant, D.E., Mackie, S.M., Gill, S.R., and Musser, J.M. 2001. Evolutionary genomics of *Staphylococcus aureus*: insights into the origin of methicillin resistant strains and the toxic shock syndrome epidemic. *Proc. Natl. Acad. Sci. U. S. A.* **98**:8821–8826.

37. Enright, M.C., et al. 2002. The evolutionary history of methicillin-resistant *Staphylococcus aureus* (MRSA). *Proc. Natl. Acad. Sci. U. S. A.* **99**:7687–7692.

38. Oliveira, D.C., Tomasz, A., and de Lencastre, H. 2001. The evolution of pandemic clones of methicillin-resistant *Staphylococcus aureus*: identification of two ancestral genetic backgrounds and the associated mec elements. *Microb. Drug Resist.* **7**:349–361.

39. Crisostomo, M.I., et al. 2001. The evolution of methicillin resistance in *Staphylococcus aureus*: similarity of genetic backgrounds in historically early methicillin-susceptible and -resistant isolates and contemporary epidemic clones. *Proc. Natl. Acad. Sci. U. S. A.* **98**:9865–9870.

40. Ma, X.X., et al. 2002. Novel type of staphylococcal cassette chromosome mec identified in community-acquired methicillin-resistant *Staphylococcus aureus* strains. *Antimicrob. Agents Chemother.* **46**:1147–1152.

41. Okuma, K., et al. 2002. Dissemination of new methicillin-resistant *Staphylococcus aureus* clones in the community. *J. Clin. Microbiol.* **40**:4289–4294.

42. Chambers, H.F. 1997. Methicillin resistance in staphylococci: molecular and biochemical basis and clinical implications. *Clin. Microbiol. Rev.* **10**:781–791.

43. Hartman, B.J., and Tomasz, A. 1984. Low-affinity penicillin-binding protein associated with beta-lactam resistance in *Staphylococcus aureus*. *J. Bacteriol.*

158:513–516.

44. Utsui, Y., and Yokota, T. 1985. Role of an altered penicillin-binding protein in methicillin- and cephem-resistant *Staphylococcus aureus*. *Antimicrob. Agents Chemother.* **28**:397–403.

45. Song, M.D., Wachi, M., Doi, M., Ishino, F., and Matsuhashi, M. 1987. Evolution of an inducible penicillin-target protein in methicillin-resistant *Staphylococcus aureus* by gene fusion. *FEBS Lett.* **221**:167–171.

46. Ghuysen, J.M. 1994. Molecular structures of penicillin-binding proteins and beta-lactamases. *Trends Microbiol.* **2**:372–380.

47. Lim, D., and Strynadka, N.C. 2002. Structural basis for the beta lactam resistance of PBP2a from methicillin-resistant *Staphylococcus aureus*. *Nat. Struct. Biol.* **9**:870–876.

48. Tomasz, A., Nachman, S., and Leaf, H. 1991. Stable classes of phenotypic expression in methicillin-resistant clinical isolates of staphylococci. *Antimicrob. Agents Chemother.* **35**:124–129.

49. Archer, G.L., and Bosilevac, J.M. 2001. Microbiology. Signaling antibiotic resistance in staphylococci. *Science*. **291**:1915–1916.

50. Niemeyer, D.M., Pucci, M.J., Thanassi, J.A., Sharma, V.K., and Archer, G.L. 1996. Role of mecA transcriptional regulation in the phenotypic expression of methicillin resistance in *Staphylococcus aureus*. *J. Bacteriol.* **178**:5464–5471.

51. Rosato, A.E., et al. 2003. mecA-blaZ corepressors in clinical *Staphylococcus aureus* isolates. *Antimicrob. Agents Chemother.* **47**:1460–1463.

52. Berger-Bachi, B. 1994. Expression of resistance to methicillin. *Trends Microbiol.* **2**:389–393.

53. Ito, T., Katayama, Y., and Hiramatsu, K. 1999. Cloning and nucleotide sequence determination of the entire mec DNA of pre-methicillin-resistant *Staphylococcus aureus* N315. *Antimicrob. Agents Chemother.* **43**:1449–1458.

54. Archer, G.L., and Niemeyer, D.M. 1994. Origin and evolution of DNA associated with resistance to methicillin in staphylococci. *Trends Microbiol.* **2**:343–347.

55. Couto, I., et al. 1996. Ubiquitous presence of a mecA homologue in natural isolates of *Staphylococcus sciuri*. *Microb. Drug Resist.* **2**:377–391.

56. Couto, I., Wu, S.W., Tomasz, A., and de Lencastre, H. 2003. Development of methicillin resistance in clinical isolates of *Staphylococcus sciuri* by transcriptional activation of the mecA homologue native to the species. *J. Bacteriol.* **185**:645–653.

57. Hiramatsu, K., et al. 2002. New trends in *Staphylococcus aureus* infections: glycopeptide resistance in hospital and methicillin resistance in the community. *Curr. Opin. Infect. Dis.* **15**:407–413.

58. Hooper, D.C. 2002. Fluoroquinolone resistance among Gram-positive cocci. *Lancet Infect. Dis.* **2**:530–538.

59. Høiby, N., et al. 1997. Excretion of ciprofloxacin in sweat and multiresistant Staphylococcus epidermidis. *Lancet*. **349**:167–169.

60. Ng, E.Y., Trucksis, M., and Hooper, D.C. 1996. Quinolone resistance mutations in topoisomerase IV: relationship to the flqA locus and genetic evidence that topoisomerase IV is the primary target and DNA gyrase is the secondary target of fluoroquinolones in *Staphylococcus aureus*. *Antimicrob. Agents Chemother.* **40**:1881–1888.

61. Ng, E.Y., Trucksis, M., and Hooper, D.C. 1994. Quinolone resistance mediated by norA: physiologic characterization and relationship to flqB, a quinolone resistance locus on the *Staphylococcus aureus* chromosome. *Antimicrob. Agents Chemother.* **38**:1345–1355.

62. Bisognano, C., Vaudaux, P., Rohner, P., Lew, D.P., and Hooper, D.C. 2000. Induction of fibronectin-binding proteins and increased adhesion of quinolone-resistant *Staphylococcus aureus* by subinhibitory levels of ciprofloxacin. *Antimicrob. Agents Chemother.* **44**:1428–1437.

63. Ince, D., Zhang, X., and Hooper, D.C. 2003. Activity of and resistance to moxifloxacin in *Staphylococcus aureus*. *Antimicrob. Agents Chemother.* **47**:1410–1415.

64. Entenza, J.M., Que, Y.A., Vouillamoz, J., Glauser, M.P., and Moreillon, P. 2001. Efficacies of moxifloxacin, ciprofloxacin, and vancomycin against experimental endocarditis due to methicillin-resistant *Staphylococcus aureus* expressing various degrees of ciprofloxacin resistance. *Antimicrob. Agents Chemother.* **45**:3076–3083.

65. Kirst, H.A., Thompson, D.G., and Nicas, T.I. 1998. Historical yearly usage of vancomycin. *Antimicrob. Agents Chemother.* **42**:1303–1304.

66. Schwalbe, R.S., Stapleton, J.T., and Gilligan, P.H. 1987. Emergence of vancomycin resistance in coagulase-negative staphylococci. *N. Engl. J. Med.* **316**:927–931.

67. Smith, T.L., et al. 1999. Emergence of vancomycin resistance in *Staphylococcus aureus*. Glycopeptide-Intermediate *Staphylococcus aureus* Working Group. *N. Engl. J. Med.* **340**:493–501.

68. Fridkin, S.K., et al. 2003. Epidemiological and microbiological characterization of infections caused by *Staphylococcus aureus* with reduced susceptibility to vancomycin, United States, 1997–2001. *Clin. Infect. Dis.* **36**:429–439.

69. 2002. Vancomycin-resistant *Staphylococcus aureus*: Pennsylvania, 2002. *MMWR.* **51**:902.

70. Walsh, T.R., and Howe, R.A. 2002. The prevalence and mechanisms of vancomycin resistance in *Staphylococcus aureus*. *Annu. Rev. Microbiol.* **56**:657–675.

71. Hiramatsu, K., et al. 1997. Dissemination in Japanese hospitals of strains of *Staphylococcus aureus* heterogeneously resistant to vancomycin. *Lancet.* **350**:1670–1673.

72. Hanaki, H., et al. 1998. Increase in glutamine-non-amidated muropeptides in the peptidoglycan of vancomycin-resistant *Staphylococcus aureus* strain Mu50. *J. Antimicrob. Chemother.* **42**:315–320.

73. Hanaki, H., et al. 1998. Activated cell-wall synthesis is associated with vancomycin resistance in methicillin-resistant *Staphylococcus aureus* clinical strains Mu3 and Mu50. *J. Antimicrob. Chemother.* **42**:199–209.

74. Sieradzki, K., Roberts, R.B., Haber, S.W., and Tomasz, A. 1999. The development of vancomycin resistance in a patient with methicillin-resistant *Staphylococcus aureus* infection. *N. Engl. J. Med.* **340**:517–523.

75. Showsh, S.A., De Boever, E.H., and Clewell, D.B. 2001. Vancomycin resistance plasmid in *Enterococcus faecalis* that encodes sensitivity to a sex pheromone also produced by *Staphylococcus aureus*. *Antimicrob. Agents Chemother.* **45**:2177–2178.

76. Gonzalez-Zorn, B., and Courvalin, P. 2003. vanA-mediated high level glycopeptide resistance in MRSA. *Lancet Infect. Dis.* **3**:67–68.

77. Tsiodras, S., et al. 2001. Linezolid resistance in a clinical isolate of *Staphylococcus aureus*. *Lancet.* **358**:207–208.

78. Fuchs, P.C., Barry, A.L., and Brown, S.D. 2002. In vitro bactericidal activity of daptomycin against staphylococci. *J. Antimicrob. Chemother.* **49**:467–470.

79. Projan, S.J., and Youngman, P.J. 2002. Antimicrobials: new solutions badly needed. *Curr. Opin. Microbiol.* **5**:463–465.

80. Projan, S.J. 2002. New (and not so new) antibacterial targets: from where and when will the novel drugs come? *Curr. Opin. Pharmacol.* **2**:513–522.

81. McDevitt, D., and Rosenberg, M. 2001. Exploiting genomics to discover new antibiotics. *Trends Microbiol.* **9**:611–617.

82. Etz, H., et al. 2002. Identification of in vivo expressed vaccine candidate antigens from *Staphylococcus aureus*. *Proc. Natl. Acad. Sci. U. S. A.* **99**:6573–6578.

83. Ge, M., et al. 1999. Vancomycin derivatives that inhibit peptidoglycan biosynthesis without binding D-Ala-D-Ala. *Science.* **284**:507–511.

84. Cheung, A.L., et al. 1994. Diminished virulence of a sar-/agr- mutant of *Staphylococcus aureus* in the rabbit model of endocarditis. *J. Clin. Invest.* **94**:1815–1822.

85. Mazmanian, S.K., Liu, G., Jensen, E.R., Lenoy, E., and Schneewind, O. 2000. *Staphylococcus aureus* sortase mutants defective in the display of surface proteins and in the pathogenesis of animal infections. *Proc. Natl. Acad. Sci. U. S. A.* **97**:5510–5515.

86. Verhoef, J., et al. 1999. A Dutch approach to methicillin-resistant *Staphylococcus aureus*. *Eur. J. Clin. Microbiol. Infect. Dis.* **18**:461–466.

87. Yu, V.L., et al. 1986. *Staphylococcus aureus* nasal carriage and infection in patients on hemodialysis. Efficacy of antibiotic prophylaxis. *N. Engl. J. Med.* **315**:91–96.

88. Kluytmans, J., van Belkum, A., and Verbrugh, H. 1997. Nasal carriage of *Staphylococcus aureus*: epidemiology, underlying mechanisms, and associated risks. *Clin. Microbiol. Rev.* **10**:505–520.

89. Perl, T.M., et al. 2002. Intranasal mupirocin to prevent postoperative *Staphylococcus aureus* infections. *N. Engl. J. Med.* **346**:1871–1877.

90. Peacock, S.J., de Silva, I., and Lowy, F.D. 2001. What determines nasal carriage of *Staphylococcus aureus*? *Trends Microbiol.* **9**:605–610.

91. Fischetti, V.A. 2001. Phage antibacterials make a comeback. *Nat. Biotechnol.* **19**:734–735.

92. Climo, M.W., Patron, R.L., Goldstein, B.P., and Archer, G.L. 1998. Lysostaphin treatment of experimental methicillin-resistant *Staphylococcus aureus* aortic valve endocarditis. *Antimicrob. Agents Chemother.* **42**:1355–1360.

93. Shinefield, H., et al. 2002. Use of a *Staphylococcus aureus* conjugate vaccine in patients receiving hemodialysis. *N. Engl. J. Med.* **346**:491–496.

94. Michie, C.A. 2002. Staphylococcal vaccines. *Trends Immunol.* **23**:461–463.

95. Lyon, B.R., and Skurray, R. 1987. Antimicrobial resistance of *Staphylococcus aureus*: genetic basis. *Microbiol. Rev.* **51**:88–134.

96. Lina, G., et al. 1999. Distribution of genes encoding resistance to macrolides, lincosamides, and streptogramins among staphylococci. *Antimicrob. Agents Chemother.* **43**:1062–1066.

97. Allignet, J., Aubert, S., Morvan, A., and el Solh, N. 1996. Distribution of genes encoding resistance to streptogramin A and related compounds among staphylococci resistant to these antibiotics. *Antimicrob. Agents Chemother.* **40**:2523–2528.

98. Sieradzki, K., Pinho, M.G., and Tomasz, A. 1999. Inactivated pbp4 in highly glycopeptide-resistant laboratory mutants of *Staphylococcus aureus*. *J. Biol. Chem.* **274**:18942–18946.

99. Murray, B.E. 2000. Vancomycin-resistant enterococcal infections. *N. Engl. J. Med.* **342**:710–721.

Bacterial infectious disease control by vaccine development

Roy Curtiss III

Department of Biology, Washington University, St. Louis, Missouri, USA.

There is a need to develop vaccines with the potential for global use against bacterial infectious diseases. These should be inexpensive to produce, stable in the absence of refrigeration, safe and efficacious, and able to be given orally or intranasally rather than by injection. Here, I will focus on efforts to develop live attenuated bacterial vaccines and the use of recombinant attenuated bacteria to induce protective immunity against heterologous bacterial pathogens. I will not discuss bacterial vaccine vectors expressing viral, parasite, or fungal antigens, since lack of posttranslational modification and improper folding generally lead to induction of immune responses that are not protective, although induction of cell-mediated immunity can often be protective. Also omitted will be a discussion of the use of attenuated bacterial vaccine vectors for the delivery of DNA vaccines, wherein problems of processing of viral, parasite, and fungal antigens in bacteria are rectified by the synthesis and modification of these antigens within the eukaryotic cell. Lastly, although I acknowledge their value, especially in the developed world, I will not dwell on progress in the development of subunit and conjugate vaccines. Studies of such vaccines, however, often provide valuable information on identification and characterization of protective antigens that can be expressed in live recombinant attenuated bacterial vaccines of the kind I consider below.

Attributes of safe and effective live bacterial vaccines

The bacterial vaccine or antigen delivery vector. Attenuation should be sufficient to decrease if not eliminate induction of undesirable disease symptoms. In this regard, the nutritional status and health of the population to be vaccinated should be considered. The attenuation should be an inherent property of the bacterial vaccine and not be dependent on fully functional host defenses and immune response capabilities. The attenuation should not be reversible by diet or by host modification of diet constituents, including by host-resident microbial flora. The attenuation should not lead to the development of a persistent carrier state for the vaccine. However, it is conceivable that there may be situations in which a persisting live vaccine could be beneficial. The attenuated vaccine should be sufficiently invasive and persistent to stimulate both strong primary and lasting memory immune responses.

A vaccine should also be designed to minimize tissue damage that is not needed to induce an effective immune response. For example, vaccines for enteric pathogens must access the gut-associated lymphoid tissue (GALT) via invasion and transcytosis through M cells. However, attachment to and invasion into enterocytes can lead to undesirable diarrheal episodes that do not contribute to the desired immune response. As even attenuated vaccine strains may cause disease in a few unlucky individuals, safety considerations dictate that any live bacterial vaccine should be susceptible to all clinically useful antibiotics. Lastly, the attenuated vaccine should possess some containment features to reduce its shedding and/or survival in nature to preclude vaccination of individuals who did not elect to be vaccinated. However, persistence of a live vaccine with potential for individual-to-individual spread could have a positive public health benefit, as is probably the case for vaccination against polio.

The plasmid vector component. Stability of protective antigen expression in vivo is essential for recombinant vaccine efficacy. Insertion of genes into the chromosome can increase stability, but the level of antigen expression is generally too low to stimulate an adequate immune response. Since the level of protein synthesis in bacteria is very much dependent upon gene copy number, antigen production can be vastly increased by use of multicopy plasmid vectors. In this case, the use of a balanced-lethal host-vector system wherein the plasmid possesses a gene complementing a chromosomal deletion mutation of a vital gene, such as for cell wall synthesis or DNA stability or replication, ensures that the plasmid is maintained (1). An alternate approach to ensuring retention of the plasmid vector is the "Hok-Sok" strategy, which also results in bacterial cell death if the plasmid is lost (2). When a plasmid vector with high copy number is used, the level of expression of the gene encoding the vector-selective marker can be far in excess of that necessary for maintenance of the vector. In these instances, overexpression of such a gene product further attenuates the vaccine, presumably due to the added energy drain on the recombinant vaccine. To address this problem, the selective marker gene can be designed to include a ribosome-binding recognition sequence but no promoter (3). As a safety consideration, it is desirable that plasmid vectors possess some containment features to minimize the possibility of transfer to and maintenance in other bacterial species. The vector should therefore be nonconjugative, should preferably be nonmobilizable, should possess a narrow replicon host range, and should not specify resistance to any antibiotic.

In general, the level of antigen synthesis can be lower when a Th1 cell-mediated immunity-type (CMI-type) response is desired but must be much higher if Th2 mucosal and systemic antibody responses are necessary. Promoters driving the expression of foreign antigens have often been constitutive (4), but constant antigen synthesis may decrease vaccine fitness to result in increased attenuation and decreased immunogenicity. Improved results have frequently been observed by use of the *nirB* (5), *pagC* (6), and *dmsA* (7) *Salmonella* promoters that initiate transcription at higher levels in vivo. Antigen expression can also be delayed until after the vaccine has invaded lymphoid tissues by use of a regulatory cascade with *araC* P$_{BAD}$-controlled expression (8) of a repressor whose

Nonstandard abbreviations used: gut-associated lymphoid tissue (GALT); Bacillus Calmette-Guérin (BCG).

Conflict of interest: No conflict of interest has been declared.

Citation for this article: *J. Clin. Invest.* **110**:1061–1066 (2002). doi:10.1172/JCI200216941.

synthesis is dependent on the presence of arabinose during in vitro cultivation. In this case, the antigen gene is coupled to a promoter with transcription repressed by the *araC* P_{BAD}–regulated repressor. Since there is no free arabinose in vertebrate tissues, the foreign antigen gene is eventually de-repressed in vivo as the repressor is diluted out as a consequence of vaccine cell division.

The immune response to a protective antigen can be influenced by the location of the antigen. Although good immune responses have been observed for antigens retained in the cytoplasm or secreted into the periplasm of Gram-negative vaccines (9), placement on the surface or secretion into the supernatant fluid making use of heterologous (10) or homologous (11) secretion mechanisms can further enhance the level and type of immune response induced. In addition, by employing the type III secretion apparatus of *Salmonella* and *Yersinia*, antigens with T cell epitopes can be delivered into the cytoplasm of antigen-presenting cells within the immunized eukaryotic host, resulting in a CD8-restricted CTL response (12).

The host-vector combination. The synthesis of pathogen antigens by recombinant attenuated host-vector antigen delivery systems can contribute to further attenuation to reduce immunogenicity (13), another important consideration in vaccine design. To enhance the likelihood of sufficient colonization and persistence in lymphoid tissues to stimulate protective immunity, the growth properties of the vaccine construct should match those of the host-vector control not expressing the antigen. The recombinant system should also be designed to maximize the immune response to the foreign antigen(s) and minimize the competing immune responses to bacterial vector antigens.

Attenuated bacterial vaccines

Mycobacterium bovis Bacillus Calmette-Guérin (BCG), a vaccine to prevent *Mycobacterium tuberculosis* infection, was the first and continues to be the most widely used live bacterial vaccine in the world. Nevertheless, *M. tuberculosis* infects one-third of the human population and causes 2 million deaths each year. The strains of BCG used as vaccines have recently been carefully characterized (14) and shown to lack expression of certain antigens present in *M. tuberculosis* that might be important for protective immunity. The means of achieving attenuation by repetitive culture passage likely led to a number of genetic changes that not only contribute to attenuation but also reduce immunogenicity. Since *M. tuberculosis* infection often results in a carrier state with reactivation many years later, it is critical, in the development of new live, attenuated *M. tuberculosis* vaccine candidates, that one preclude the creation of a vaccine that can be as persistent in a hidden form as *M. tuberculosis* is.

Salmonella typhi Ty21a is a widely used licensed live vaccine that is exceedingly safe but requires three to four immunizations to induce long-term (at least 6–7 years) protective immunity in two-thirds of individuals immunized (15). Newer, more immunogenic, live *S. typhi* vaccines are currently being clinically evaluated with the goal of developing an attenuated *S. typhi* vaccine that can induce protective immunity in a high percentage of individuals with a single dose. This effort has been hampered by the fact that most means of attenuation, with one exception (*phoPQ*), that were effective in attenuating *Salmonella typhimurium* and rendering it highly immunogenic in mice fail to attenuate *S. typhi* sufficiently in humans (16, 17). For example, *S. typhi* strains with double defined deletion mutations in *aro* genes that possess requirements for a number of vitamin cofactors theoretically unavailable in vertebrates. Such strains should therefore be unable to grow for more than seven to ten generations in vivo. Nevertheless, positive blood cultures with fevers are observed when high doses of the double *aro* vaccine are administered to human volunteers (18). This unacceptable level of reactogenicity has been corrected by introduction of an additional mutation, *htrA*, eliminating a heat shock protease (19), which confers additional attenuation without sacrificing too much immunogenicity (but see ref. 20). Two such double *aro htrA* vaccine candidates are being clinically evaluated (21). Ty800, which is very likely the best-attenuated *S. typhi* vaccine candidate, has a defined deletion mutation of the two-component virulence-regulating genes, *phoPQ*. In human trials, Ty800 induced high-level immune responses with a single dose and without adverse side effects, except for diarrhea in a small number of individuals (22).

The live attenuated *Bacillus anthracis* strain Sterne, licensed as a veterinary vaccine, was used for human immunization in the former Soviet Union but was too reactogenic (23). New, improved, live attenuated *B. anthracis* vaccines have given promising results in inducing protective immunity in guinea pigs (24, 25) but have yet to undergo clinical evaluation in humans.

Development of a live attenuated *Shigella* vaccine that is sufficiently attenuated to be nonreactogenic yet adequately invasive to be highly immunogenic has been attempted for over 30 years. Due to substantial progress in understanding the molecular genetic basis of virulence of *Shigella* since the early 1980s, there are now two candidate vaccines with defined attenuating deletion mutations that induce protective immunity to *Shigella* challenge in humans without severe associated diarrheal symptoms (26, 27). Additional clinical trials on these vaccines are likely. Nevertheless, most vaccine development work has been with specific *Shigella flexneri* serotypes (of which there are 12), and serotype-specific immunity is important. In addition, severe dysentery is caused by *Shigella dysenteriae* and less severe but more prevalent dysentery by *S. sonnei* and *S. boydii*. Development of effective vaccines to prevent the global *Shigella* infections of 1 million or more each year, caused by some 38 serotypes, will therefore not be easily achieved.

Two live attenuated *Vibrio cholera* vaccine candidates have been constructed and clinically evaluated, and one, CVD 103-HgR, is now commercially available (28). Both of these vaccines possess defined deletion mutations that preclude the ability to express cholera toxin as well as some of the other accessory toxins that contribute somewhat to the diarrhea caused by *V. cholera* infection. Although much effort has gone into the design of these strains, it is difficult to see how these noninvasive candidate vaccines will induce a long-term mucosal or systemic immune memory. The very fact that the vaccine strains must persist in the intestinal tract to stimulate an immune response, and that they do so by the expression of toxin-coregulated pili, indicates that some diarrhea will result from immunization, since nontoxigenic strains of bacteria that possess the ability to express the pilus adhesins necessary for colonization can still cause significant diarrhea and sometimes death in animals (29).

Attenuated recombinant bacterial host-vector systems

Strains of *S. typhimurium* and *S. typhi* were first used as recombinant vectors for antigen delivery (4, 9), and BCG was later developed as an antigen delivery vector (30). Most recently, attenuated derivatives of *S. flexneri*, *Listeria monocytogenes*, and *B. anthracis* are being evaluated as recombinant antigen delivery vaccines.

Over 15 years of research, with a substantial and ever-increasing pace of published information, explores the use of attenuated *S. typhimurium* and *S. typhi* antigen delivery vaccines. Although recombinant attenuated *S. typhimurium* strains expressing a diversity of antigens from various types of pathogens have often induced protective immunity in mice and a number of agriculturally important animals, recombinant attenuated *S. typhi* Ty2 strains evaluated in human volunteers have induced diminished immune responses to the *Salmonella* vector and insignificant or nonexistent immune responses to expressed antigens from several different pathogens (31–33). Although there are several possible explanations, one significant problem may be the choice of the *S. typhi* antigen delivery strain, Ty2. We have demonstrated in *S. typhimurium* that chromosomal RpoS-regulated genes are necessary for invasion into and colonization of M cells overlying the GALT (34). In accord with this, *rpoS* mutants of *S. typhimurium*, with or without the virulence plasmid that possesses RpoS-regulated genes that play no role in colonization of the GALT, exhibit diminished immunogenicity, an observation confirmed by others (35). We demonstrated the same defect in dependence on RpoS-regulated chromosomal genes for effective colonization of lymphoid tissues, presumably the nasal-associated lymphoid tissue and bronchial-associated lymphoid tissue, following intranasal immunization in mice. Unfortunately, *S. typhi* Ty2, which was isolated in 1918 and maintained in the laboratory ever since, possesses an *rpoS* allelic defect that is also present in the not-so-immunogenic licensed *S. typhi* Ty21a vaccine. It is known that RpoS⁺ *S. typhi* strains are more virulent for human volunteers than are the *S. typhi* Ty2 derivatives attenuated by the same mutations (18). Use of RpoS⁺ *S. typhi* antigen delivery vectors will therefore necessitate degrees of attenuation in excess of what is necessary to attenuate *S. typhi* Ty2 strains for humans, but these means of attenuation must not compromise immunogenicity.

Despite these setbacks, results with recombinant attenuated *S. typhimurium* vaccines in animals suggest that recombinant attenuated *S. typhi* vaccines could be developed to protect against infections due to Hepatitis B virus (HBV), papillomaviruses, *Streptococcus pneumoniae*, *Helicobacter pylori*, *Yersinia pestis*, human enterotoxigenic and *Escherichia coli* strains, *M. tuberculosis*, *L. monocytogenes*, *B. anthracis*, *Clostridium tetani*, and *Clostridium difficile*. In addition, some progress is being made in developing vaccines to protect against infections due to *Chlamydia*, *Rickettsia*, *Francisella tularensis*, *Neisseria meningitidis*, *Brucella*, and *Porphyromonas*. One of the more tantalizing challenges is to develop a recombinant attenuated multivalent vaccine that would substitute for the injectable DPT vaccines (36). Fragment C of *C. tetani* has been expressed in attenuated *Salmonella* in mice, using a codon-optimized sequence and induced protective immunity to *C. tetani* toxin challenge. It is the only antigen delivered by an attenuated *S. typhi* Ty2 strain to induce a modest level of immunity in human volunteers (37). Research on expression of the *Corynebacterium diphtheriae* CRM 197 in *S. typhimurium* and *S. typhi* looks encouraging (38). A number of *Bordetella pertussis* protective antigens have been expressed in recombinant attenuated *S. typhimurium*, and significant immune responses in mice have been observed. Certainly, a multivalent recombinant attenuated *S. typhi* expressing all of these antigens would result in a vaccine that could be administered orally or intranasally and would do much to reduce mortality due to these pathogens in the developing world.

Since phagocytic cells do not readily lyse mycobacterial cells, BCG or other attenuated mycobacterial strains do not seem promising as antigen delivery vectors unless the expressed antigens can be secreted from the cells. In the case of protective antigens from *Borrelia burgdorferi*, this approach has successfully induced protection in mice, but it failed to induce antibody responses to the *Borrelia* OspA antigen in human volunteers (39). Secretion of *M. tuberculosis* antigens by BCG to induce protection against *M. tuberculosis* challenge in guinea pigs has also been achieved (40). Use of recombinant attenuated *L. monocytogenes* (41), *B. anthracis* (42), and *S. flexneri* (43) as antigen delivery vectors is hampered by lack of a suitable animal host for developmental research, the reliance on results with injected candidate recombinant vaccines, and/or the need to develop strains that are attenuated, nonreactogenic, and highly immunogenic in human volunteers.

Antigen selection

Identification of protective antigens of a pathogen to express in a recombinant attenuated bacterial vaccine is not an easy task. Presence of antibodies in a host surviving infection can offer clues. However, a high-level antibody response does not always correlate with protection. Of course, if one can obtain an mAb to an antigen that can passively protect against the pathogen, that is a good sign that that antigen is likely to induce a protective response. Similarly, the identification of antigen-specific T cell populations whose passive transfer can induce protective immunity will also identify candidate antigens for vaccine development. Evidence that inability to express an antigen due to mutation is associated with avirulence is another useful criteria for antigen selection especially if the antigen is surface-localized or secreted.

Many protective antigens gain access to antigen-presenting pathways, and one can employ genomics mining with algorithms to identify subsets of proteins that are secreted by various secretion pathways or are surface-localized. For example, antigens delivered by the type III secretion system of various pathogens such as *Salmonella* and *Yersinia* are delivered to the cytoplasm of host cells to result in CD8-restricted CTL responses (12). This result may also be true for proteins secreted by other pathogens such as *Mycobacterium* species, if these proteins are secreted into the cytoplasm of cells within the infected individual. Also, if a surface antigen from one pathogen is found to induce protective immunity, one can search by computer for homologs in other pathogens that can be evaluated for their ability to induce protective immune responses.

There is a possibility that given protective antigen is synthesized only in vivo, with the gene encoding it responding to some stress or signal. Such antigens are not expressed during in vitro cultivation. Genetic screens such as In Vivo Expression Technology (IVET) (44), Signature-Tagged Mutagenesis (STM) (45), and Selective Capture of Transcribed Sequences (SCOTS) (46) can be useful in identifying in vivo–expressed genes encoding proteins essential for survival of the pathogen in vivo. Proteins constitutively synthesized by recombinant clones of genes uniquely expressed in vivo can then be used to recognize either antibodies or lymphocyte populations responding to that antigen. Proteomic analysis, quantitating proteins synthesized when the pathogen is in an in vivo compartment or in cells in culture, can identify proteins synthesized in the greatest amounts. Such proteins have a higher likelihood of inducing immune responses may or may not be protective.

Biotechnology and vaccine design

Most articles describing design of recombinant attenuated vaccines omit mention of biotechnological means to improve the performance of the recombinant attenuated antigen delivery vaccine. One significant exception is the consideration that has been given to the promoter sequence used to drive expression of a plasmid-encoded protective antigen. In this regard, use of the *nirB* promoter, which functions optimally under the anaerobic conditions encountered in vivo, has given very promising results with recombinant attenuated *Salmonella* vaccine constructs (5). Often ignored among vaccine properties, codon usage patterns have species-specific differences, and codons in genes specifying an antigen from a heterologous species often need to be optimized for high-level expression in the vaccine strain. It may also be desirable to add DNA sequences specifying 5' and 3' untranslated mRNA sequences that create stem-loop structures and to remove potential targets within the coding mRNA sequences to minimize mRNA degradation by the mRNA degradosome (47). Deletion of the gene for poly(A) polymerase I can also increase mRNA stability (48) to possibly improve the productive fitness of the recombinant vaccine strain.

The stability of the synthesized antigen in the bacterial antigen delivery strain must be considered, in order to preclude toxicity of breakdown products as well as, after delivery to the immunized host, to enhance induction of the desired type of immune response. Thus, T cell antigens, which must be degraded by the proteosome prior to antigen presentation, might benefit from changes that decrease their stability. Conversely, antigens to which a humoral immune response is desired should be engineered for greater stability in the immunized host. Attention to the N-end rule for protein degradation (49) and the presence of PEST sequences (50) and other protease cleavage sites in the antigen can help in design considerations. If the foreign antigen contains cysteine residues, one must be concerned about anomalous protein folding, which can be rectified by deletion of Cys-encoding sequences, changing of the environment of the antigen delivery strain with regard to Eh potential, inclusion of a chaperone for the expressed antigen, or modification of the *dsb* system for isomerization of proteins with disulfide bonds (51). Another means to enhance immunogenicity of an expressed protective antigen might be inclusion of sequences leading to lipidation, since lipoproteins can be taken up more efficiently by antigen-presenting cells to induce an array of desired immune responses (52). Since expression of proteins with extensive hydrophobic domains very often results in toxicity to the antigen delivery vector, deleting of these sequences should be considered, since immune responses to them are seldom protective. Fusion of proteins to facilitate the secretion of protective antigens (3, 11), or their delivery to the cytoplasm by type III secretion systems to stimulate T cell immunity (12), has been considered in several cases. Such fusions can also be used to stabilize proteins or to enhance induction of immune responses by the fusion partner encoding a strong or universal epitope. Coexpression of cytokines along with antigenic sequences might be helpful as well, since many of these factors can be synthesized and secreted in a bioactive form from recombinant attenuated bacterial antigen delivery strains.

Evaluation in animals prior to human testing

Attaining the ultimate goal of vaccines that are safe and efficacious in humans can be both advanced and hindered by studies in animal systems. For example, the use of *S. typhimurium* in mice as a surrogate for *S. typhi* in humans has many real advantages. On the other hand, the disease caused by *S. typhimurium* in mice is not exactly the same as that caused by *S. typhi* infection in humans. To make matters worse, most scientists rely on mice that are inbred rather than outbred and that possess unique susceptibility to *S. typhimurium* infection. Similarly, use of outbred mice to evaluate candidate recombinant attenuated *S. typhi* vaccines is likely very misleading. Evaluation of *S. typhi* attenuation has often employed intraperitoneal inoculation of outbred mice in the presence of hog gastric mucin (53). This assay seems to have worked very well for attenuated strains with *aro* mutations that preclude growth of the *S. typhi* but would lead one to believe that bacteria with *phoPQ* deletions are as virulent as wild-type bacteria, since their growth is not impeded and the mice succumb to endotoxic shock after growth of the *S. typhi* strain (54). It is also well known that *S. typhi* is unable to survive in murine macrophages of diverse types, and therefore the immunogenicity of *S. typhi* constructs after intranasal inoculation into mice is probably no different from that achieved by a diversity of pathogens that are unable to cause lethal infection in mice regardless of whether attenuating mutations are present. The development of strains of mice with enhanced susceptibility to *S. typhi* might overcome these problems. In this regard, the recent generation of transgenic mice with a receptor allowing *L. monocytogenes* infection via oral inoculation (55) might provide a better model for development of recombinant attenuated *L. monocytogenes* vaccines for humans. Work toward development of improved *M. tuberculosis*–derived recombinant vaccines necessitates the discovery of attenuated *M. tuberculosis* strains unable to establish latent infections. Nevertheless, results from tests of vaccine candidates in mice or guinea pigs probably will not be reflective of responses in humans. Therefore, more studies on these various vaccine candidates must be conducted with human volunteers.

Conclusion

Progress in the development of attenuated bacterial vaccines to control diseases caused by those bacterial pathogens in humans is encouraging. The ultimate benefit, however, is likely to be achieved by the use of some of these attenuated bacterial vaccines, or modified derivatives thereof, to express antigens from other pathogens to induce protective immunity to those pathogens. Use of live recombinant attenuated bacterial antigen delivery systems would be akin to immunization of the individual with a vaccine factory, since production of the protective antigen and its presentation to the immunized individual occur in vivo after immunization. Such vaccines are inexpensive to manufacture and can be lyophilized to be reconstituted at the time of use, thus likely avoiding the necessity of refrigeration. Their oral or intranasal administration eliminates the need for reliance on syringes with their inherent added cost and association with accessory unwanted hazards. Potential problems associated with repeated use of the same or similar antigen delivery vectors for the delivery of different protective antigens require further careful study and might necessitate modification of surface antigen attributes on the antigen delivery vectors.

Acknowledgments

I thank Josephine E. Clark-Curtiss, Richard Groger, Ho Young Kang, and Joel Dworkin for helpful suggestions to improve the manuscript. I thank Erika Arch, Wendy Bollen, and Xin Zhang for help with reference collection and manuscript preparation. Research has been supported by grants from the NIH, the US Department of Agriculture, and Bristol-Myers Squibb Co.

Address correspondence to: Roy Curtiss III, Department of Biology, Washington University, 1 Brookings Drive, Campus Box 1137, St. Louis, Missouri 63130-4899, USA. Phone: (314) 935-6819; Fax: (314) 935-7246; E-mail: rcurtiss@biology.wustl.edu.

1. Nakayama, K., Kelly, S.M., and Curtiss, R., III. 1988. Construction of an Asd⁺ expression-cloning vector: stable maintenance and high level expression of cloned genes in a *Salmonella* vaccine strain. *Biotechnology (NY)*. **6**:693–697.

2. Galen, J.E., et al. 1999. Optimizing of plasmid maintenance in the attenuated live vector vaccine strain *Salmonella typhi* CVD 908-*htrA*. *Infect. Immun.* **67**:6424–6433.

3. Kang, H.Y., Srinivasan, J., and Curtiss, R., III. 2002. Immune responses to recombinant pneumococcal PspA antigen delivered by live attenuated *Salmonella enterica* serovar Typhimurium vaccine. *Infect. Immun.* **70**:1739–1749.

4. Curtiss, R., III. 1990. Attenuated *Salmonella* strains as live vectors for the expression of foreign antigens. In *New generation vaccines*. G.C. Woodrow and M.M. Levine, editors. Marcel Dekker Inc. New York, New York, USA. 161–188.

5. Roberts, M., Li, J., Bacon, A., and Chatfield, S. 1998. Oral vaccination against tetanus: comparison of the immunogenicities of *Salmonella* strains expressing fragment C from the *nirB* and *htrA* promoters. *Infect. Immun.* **66**:3080–3087.

6. Dunstan, S.J., Simmons, C.P., and Strugnell, R.A. 1999. Use of in vivo-regulated promoters to deliver antigens from attenuated *Salmonella enterica* var. Typhimurium. *Infect. Immun.* **67**:5133–5144.

7. Orr, N., Galen, J.E., and Levine, M.M. 2001. Novel use of anaerobically induced promoter, *dmsA*, for controlled expression fragment C of tetanus toxin in live attenuated *Salmonella enterica* serovar typhi strain CVD 908-*htrA*. *Vaccine*. **19**:1694–1700.

8. Guzman, L.-M., Belin, D., Carson, M.J., and Beckwith, J. 1995. Tight regulation, modulation, and high-level expression by vectors containing the arabinose P$_{BAD}$ promoter. *J. Bacteriol.* **177**:4121–4130.

9. Roberts, M., Chatfield, S.N., and Dougan, G. 1994. *Salmonella* as carriers of heterologous antigens. In *Novel delivery systems for oral vaccines*. D.T. O'Hagan, editor. CRC Press. Boca Raton, Florida, USA. 27–58.

10. Hess, J., et al. 2000. Protection against murine tuberculosis by an attenuated recombinant *Salmonella typhimurium* vaccine strain that secretes the 30-kDa antigen of *Mycobacterium bovis* BCG. *FEMS Immunol. Med. Microbiol.* **27**:283–289.

11. Hess, J., et al. 2000. Secretion of different listeriolysin cognates by recombinant attenuated *Salmonella typhimurium*: superior efficacy of haemolytic over non-haemolytic constructs after oral vaccination. *Microbes Infect.* **2**:1799–1806.

12. Rüssmann, H., et al. 1998. Delivery of epitopes by the *Salmonella* Type III secretion system for vaccine development. *Science*. **281**:565–568.

13. Galen, J.E., and Levine, M.M. 2001. Can a 'flawless' live vector vaccine strain be engineered? *Trends Microbiol.* **9**:372–376.

14. Behr, M.A., et al. 1999. Comparative genomics of BCG vaccines by whole-genome DNA microarray. *Science*. **284**:1520–1523.

15. Levine, M.M., et al. 1999. Duration of efficacy Ty21a, attenuated *Salmonella typhi* live oral vaccine. *Vaccine*. **17**(Suppl. 2):S22–S27.

16. Hone, D., et al. 1988. A *galE via* (Vi antigen-negative) mutant of *Salmonella typhi* Ty2 retains virulence in humans. *Infect. Immun.* **56**:1326–1333.

17. Bumann, D., Hueck, C., Aebischer, T., and Meyer, T.F. 2000. Recombinant live *Salmonella* spp. for human vaccination against heterologous pathogens. *FEMS Immunol. Med. Microbiol.* **27**:357–364.

18. Tacket, C.O., et al. 1992. Comparison of the safety and immunogenicity of ΔaroC ΔaroD and Δcya Δcrp Salmonella typhi strains in adult volunteers. *Infect. Immun.* **60**:536–541.

19. Chatfield, S.N., et al. 1992. Evaluation of *Salmonella typhimurium* strains harboring defined mutations in *htrA* and *aroA* in the murine salmonellosis model. *Microb. Pathog.* **12**:145–151.

20. Roberts, M., Chatfield, S., Pickard, D., Li, J., and Bacon, A. 2000. Comparison of abilities of *Salmonella enterica* serovar Typhimurium *aroA aroD* and *aroA htrA* mutants to act as live vectors. *Infect. Immun.* **68**:6041–6043.

21. Tacket, C.O., et al. 2000. Phase 2 clinical trial of attenuated *Salmonella enterica* serovar Typhi oral live vector vaccine CVD 908-*htrA* in U.S. volunteers. *Infect. Immun.* **68**:1196–1201.

22. Hohmann, E.L., Oletta, C.A., Killeen, K.P., and Miller, S.I. 1999. phoP/phoQ-deleted *Salmonella typhi* (Ty800) is a safe and immunogenic single-dose typhoid fever vaccine in volunteers. *J. Infect. Dis.* **173**:1408–1414.

23. Shlyakhov, E.N., and Rubinstein, E. 1993. Human live anthrax vaccine in the former USSR. *Vaccine*. **12**:727–730.

24. Barnard, J.P., and Friedlander, A.M. 1999. Vaccination against anthrax with attenuated recombinant strains of *Bacillus anthracis* that produce protective antigen. *Infect. Immun.* **67**:562–567.

25. Cohen, S., et al. 2000. Attenuated nontoxinogenic and nonencapsulated recombinant *Bacillus anthracis* spore vaccines protect against anthrax. *Infect. Immun.* **68**:4549–4558.

26. Coster, T.S., et al. 1999. Vaccination against shigellosis with attenuated *Shigella flexneri* 2a strain SC602. *Infect. Immun.* **67**:3437–3443.

27. Kotloff, K.L., et al. 2000. *Shigella flexneri* 2a strain CVD 1207, with specific deletions in *virG, sen, set* and *guaBA*, is highly attenuated in humans. *Infect. Immun.* **68**:1034–1039.

28. Ryan, E.T., and Calderwood, S.B. 2000. Cholera vaccines. *Clin. Infect. Dis.* **31**:561–565.

29. Svanborg, C., Orskov, F., and Orskov, I. 1994. Fimbriae and disease. In *Fimbriae adhesion, genetics, biogenesis, and vaccines*. P. Klemm, editor. CRC Press. Boca Raton, Florida, USA. 239–254.

30. Ohara, N., and Yamada, T. 2001. Recombinant BCG vaccines. *Vaccine*. **19**:4089–4098.

31. Nardelli-Haefliger, D., et al. 1996. Oral and rectal immunization of adult female volunteers with a recombinant attenuated *Salmonella typhi* vaccine strain. *Infect. Immun.* **64**:5219–5224.

32. Tacket, C.O., et al. 1997. Safety and immunogenicity in humans of an attenuated *Salmonella typhi* vaccine vector strain expressing plasmid encoded hepatitis B antigens stabilized by the Asd-balanced lethal vector system. *Infect. Immun.* **65**:3381–3385.

33. DiPetrillo, M.D., Tibbetts, T., Kleanthous, H., Killeen, K.P., and Hohmann, E.L. 1999. Safety and immunogenicity of phoP/phoQ-deleted *Salmonella typhi* expressing *Helicobacter pylori* urease in adult volunteers. *Vaccine*. **18**:449–459.

34. Nickerson, C.A., and Curtiss, R., III. 1997. Role of sigma factor RpoS in initial stages of *Salmonella typhimurium* infection. *Infect. Immun.* **65**:1814–1823.

35. Coynault, C., and Norel, F. 1999. Comparison of the abilities of *Salmonella typhimurium* rpoS, aroA and rpoS aroA strains to elicit humoral immune responses in BALB/c mice and to cause lethal infection in athymic BALB/c mice. *Microb. Pathog.* **26**:299–305.

36. Gómez-Duarte, O.G., et al. 1995. Expression of fragment C of tetanus toxin fused to a carboxyl-terminal fragment of diphtheria toxin in *Salmonella typhi* CVD 908 vaccine strain. *Vaccine*. **13**:1596–1602.

37. Tacket, C.O., et al. 2000. Safety and immune responses to attenuated *Salmonella enterica* serovar Typhi oral live vector vaccines expressing tetanus toxin fragment C. *Clin. Immunol.* **97**:146–153.

38. Orr, N., Galen, J.E., and Levine, M.M. 1999. Expression and immunogenicity of a mutant diphtheria toxin molecule, CRM₁₉₇, and its fragments in *Salmonella typhi* vaccine strain CVD 908-*htrA*. *Infect. Immun.* **67**:4290–4294.

39. Edelman, R., et al. 1998. Safety and immunogenicity of recombinant Bacille Calmette-Guérin (rBCG) expressing *Borrelia burgdorferi* outer surface protein A (OspA) lipoprotein in adult volunteers: a candidate Lyme disease vaccine. *Vaccine*. **17**:904–914.

40. Horwitz, M.A., Harth, G., Dilon, B.J., and Maslesa-Galic, S. 2000. Recombinant Bacillus Calmette-Guérin (BCG) vaccines expressing the *Mycobacterium tuberculosis* 30-kDa major secretory protein induce greater protective immunity against tuberculosis than conventional BCG vaccines in a highly susceptible animal model. *Proc. Natl. Acad. Sci. USA*. **97**:13853–13858.

41. Gunn, G.R., et al. 2001. Two *Listeria monocytogenes* vaccine vectors that express different molecular forms for human papilloma virus-16 (HPV-16) E7 induce qualitatively different T cell immunity that correlates with their ability to induce regression of established tumors immortalized by HPV-16. *J. Immunol.* **167**:6471–6479.

42. Mesnage, S., Weber-Levy, M., Haustant, M., Mock, M., and Fouet, A. 1999. Cell surface-exposed tetanus toxin fragment C produced by recombinant *Bacillus anthracis* protects against tetanus toxin. *Infect. Immun.* **67**:4847–4850.

43. Altboum, Z., Barry, E.M., Losonsky, G., Galen, J.E., and Levine, M.M. 2001. Attenuated *Shigella flexneri* 2a ΔguaBA strain CVD 1204 expressing enterotoxigenic *Escherichia coli* (ETEC) CS2 and CS3 fimbriae as a live mucosal vaccine against *Shigella* and ETEC infection. *Infect. Immun.* **69**:3150–3158.

44. Mahan, M.J., Slauch, J.M., and Mekalanos, J.J. 1993. Selection of bacterial virulence genes that are specifically induced in host tissues. *Science*. **259**:686–688.

45. Hensel, M., et al. 1995. Simultaneous identification of bacterial virulence genes by negative selection. *Science*. **269**:400–403.

46. Graham, J.E., and Clark-Curtiss, J.E. 1999. Identification of *Mycobacterium tuberculosis* RNAs synthesized in response to phagocytosis by human macrophages by selective capture of transcribed sequences (SCOTS). *Proc. Natl. Acad. Sci. USA*. **96**:11554–11559.

47. Liou, G.-G., Jane, W.-N., Cohen, S.N., Lin, N. S., and Lin-Chao, S. 2001. RNA degradosomes exist *in vivo* in *Escherichia coli* as multicomponent complexes associated with the cytoplasmic membrane via the N-terminal region of ribonuclease E. *Proc. Natl. Acad. Sci. USA*. **98**:63–68.

48. O'Hara, E.B., et al. 1995. Polyadenylylation helps regulate mRNA decay in *Escherichia coli*. *Proc. Natl. Acad. Sci. USA*. **92**:1807–1811.

49. Varshavsky, A. 1996. The N-end rule: functions, mysteries, uses. *Proc. Natl. Acad. Sci. USA*. **93**:12142–12149.

50. Rogers, S., Wells, R., and Rechsteiner, M. 1986. Amino acid sequences common to rapidly degraded proteins: the PEST hypothesis. *Science*. **234**:364–368.

51. Ritz, D., and Beckwith, J. 2001. Roles of thiol-redox pathways in bacteria. *Annu. Rev. Microbiol.* **55**:21–48.

52. Baier, W., Masihi, N., Huber, M., Hoffmann, P., and Bessler, W.G. 2000. Lipopeptides as immunoadjuvants and immunostimulants in mucosal immunization. *Immunobiology*. **201**:391–405.

53. Powell, C.J., DeSett, C.R., Lowenthal, J.P., and Berman, S. 1980. The effect of adding iron to mucin on the enhancement of virulence for mice of *Salmonella typhi* strain TY 2. *J. Biol. Stand.* **8**:79–85.

54. Baker, S.J., Daniels, C., and Morona, R. 1997. PhoP/Q regulated genes in *Salmonella typhi* identification of melittin sensitive mutants. *Microb. Pathog.* **22**:165–179.

55. Lecuit, M., et al. 2001. A transgenic model for Listeriosis: role of internalin in crossing the intestinal barrier. *Science*. **292**:1665–1667.

Development of an improved vaccine for anthrax

Stephen H. Leppla,[1] John B. Robbins,[2] Rachel Schneerson,[2] and Joseph Shiloach[3]

[1]National Institute of Dental and Craniofacial Research, [2]National Institute of Child Health and Human Development, and [3]National Institute of Diabetes and Digestive and Kidney Diseases, NIH, Bethesda, Maryland, USA.

Bacillus anthracis are aerobic or facultatively anaerobic Gram-positive, nonmotile rods measuring 1.0 μm wide by 3.0–5.0 μm long. Under adverse conditions, *B. anthracis* form highly resistant endospores (Figure 1). These are found in soil at sites where infected animals previously died. Interest in the pathogenesis, immunity, and vaccine development for anthrax was heightened by the deliberate contamination of mail with *B. anthracis* spores soon after the September 11 attacks. At that time, the only US-licensed human vaccine (anthrax vaccine adsorbed, or AVA) was not available because the manufacturer, BioPort Corp., had not received FDA certification of its new manufacturing process.

Although Pasteur had already demonstrated protection of sheep by injection of heat-attenuated *B. anthracis* cultures in 1881, our current knowledge of immunity to anthrax in humans remains limited. Widespread vaccination of domesticated animals with attenuated strains such as the Sterne strain began in the 1930s and has virtually abolished anthrax in industrialized countries. In the US, the licensed human vaccine (AVA, newly renamed BioThrax) is an aluminum hydroxide–adsorbed, formalin-treated culture supernatant of a toxigenic, noncapsulated, nonproteolytic *B. anthracis* strain, V770-NP1-R, derived from the Sterne strain (1). AVA was developed in the early 1950s, when purified components of *B. anthracis* were not available. Its only demonstrable protective component is the protective antigen (PA) protein (2). A similar culture supernatant–derived human vaccine is produced in the United Kingdom.

Data from a 1950s trial of wool-sorters immunized with a vaccine similar to AVA, coupled with long experience with AVA and the United Kingdom vaccine, have shown that a critical level of serum antibodies to the *B. anthracis* PA confers immunity to anthrax (3, 4). As early as 1959, a British Ministry of Labour report noted that, following the introduction of regular immunization the previous year, the staff of the Government Wool Disinfection Station in Liverpool were free of the disease "despite the high risk to which they are exposed" (5). AVA also protects laboratory animals and cattle from both cutaneous and inhalational challenge with *B. anthracis* (1, 6, 7).

Although safe and efficacious (8), AVA has limitations that justify the widespread interest in developing improved vaccines consisting solely of well-characterized components. First, standardization of AVA is based on the manufacturing process and a potency assay involving protection of guinea pigs challenged intracutaneously with *B. anthracis* spores (7, 9). PA is not measured in the vaccine, and there is no standardized assay of PA antibodies in animals or humans vaccinated with AVA. These factors probably explain why it has been difficult to maintain consistency of AVA. Second, this vaccine contains other cellular elements that probably contribute to the relatively high rate of local and systemic reactions (8). Finally, the schedule of AVA administration (subcutaneous injections at 0, 2, and 4 weeks and 6, 12, and 18 months with subsequent yearly boosters) is probably not optimal. This schedule, introduced in the 1950s, was designed for rapid induction of immunity (10), but it was recently shown that increasing the interval between the first two injections enhances the level of AVA-induced antibodies to PA (11). Moreover, there is no experimental support for including the injections given at 6, 12, and 18 months.

B. anthracis as a human pathogen

Humans contract anthrax almost exclusively from contact with, ingestion of, or inhalation of *B. anthracis* spores. Cutaneous anthrax results from a break in the skin and has a mortality rate of about 20% in untreated cases. Incubation is usually 2–3 days, although it can occur within 12 hours and as late as 2 weeks. A small papule appears, followed by a surrounding ring of vesicles about 24 hours later. The lesions ulcerate and become black and edematous. In pulmonary anthrax, inhaled spores are carried by macrophages from the lungs to adjacent lymph nodes. The spores germinate, multiply, and cause septicemia. Primary inflammation of the lungs (pneumonia) may not be detectable. Because the signs and symptoms of infection are mimicked by administration of anthrax toxin to monkeys, anthrax can be considered a toxin-mediated disease.

The virulence of *B. anthracis* results from the action of materials that are expressed from genes on two large plasmids, pX01 and pX02 (12, 13). pX01 encodes the proteins that make up the anthrax toxin. The massive edema and organ failure seen in anthrax patients are caused largely by the action of three individually nontoxic proteins: protective antigen (PA, 83 kDa), edema factor (EF, adenylate cyclase, 89 kDa), and lethal factor (LF, zinc protease, 90 kDa) (14). The latter two combine with the PA to form edema toxin and lethal toxin, respectively. PA, EF, and LF fit the A-B toxin model proposed by Gill (15). Thus, following its interaction with host cells, PA (the "B" subunit) is activated by the cellular protease furin, causing the release of a 20-kDa N-terminal domain (16). The remaining 63-kDa polypeptide creates a heptameric structure that constitutes a channel in the host cell membrane (17) through which LF and EF (each of which represents an alternative "A" subunit in this model) are translocated to the cytosol. The unregulated adenylate cyclase activity of EF leads to production of unphysiologically high concentrations of cAMP, one consequence of which is incapacitation of phagocytic cells (14). LF cleaves several mitogen-activated protein kinase kinases, thereby blocking signal transduction pathways by which host immune cells normally respond to pathogens (18, 19).

Plasmid pX02 encodes the poly-γ-linked D-glutamic acid (PGA) capsule, demonstrable by a Quellung (antibody-induced swelling) reaction (20). Strains lacking pX02 are avirulent. PGA confers virulence to *B.-anthracis* by its antiphagocytic activity, in the same way that capsular polysaccharides confer virulence to the pneumococci.

Citation for this article: *J. Clin. Invest.* **110**:141–144 (2002). doi:10.1172/JCI200216204.

Figure 1
A spore (left) and vegetative cells and a chain of vegetative rod cells of *B. anthracis*. Electron micrograph courtesy of the Centers for Disease Control and Prevention.

The essential role of anti-toxin antibodies in immunity to *B. anthracis*

Experience with AVA and PA (native or recombinant) in animals challenged by inhalation or by intracutaneous injection is extensive and points strongly to PA as an essential component of living, inactivated, or protein-based vaccines for anthrax. Thus, strains of *B. anthracis* cured of pX01, and therefore lacking the ability to express the components of anthrax toxin, are not virulent and do not confer immunity to animals (21) (with the exception of mice, as discussed below). Moreover, PA IgG antibodies, either actively induced or passively administered as polyclonal or monoclonal proteins, confer protection to mice, rabbits, rats, guinea pigs, and monkeys challenged with *B. anthracis* either intracutaneously or by aerosol. Pitt et al. (22) recently reported an excellent correlation between the level of AVA-induced antibodies and immunity to inhalational anthrax in rabbits. Conversely, antibodies to EF or LF alone do not confer strong immunity to infection, although antibodies to LF induced by a DNA vaccine protect against toxin challenge (23) and merit further study for their potential to act synergistically with antibodies to PA.

In certain other bacterial infections, antitoxins (serum-neutralizing antibodies) can exert antibacterial activity. For example, the toxin-mediated respiratory diseases caused by *Corynebacterium diphtheriae* and *Bordetella pertussis* are prevented by immunization with their respective toxoids (24). Both pathogens are almost always confined to the epithelial surface of the respiratory tract, where their secreted toxins cause local inflammation and inactivate phagocytic cells. Vaccine-induced antitoxins confer immunity to these diseases by permitting phagocytosis of *C. diphtheria* and *B. pertussis*. In the case of anthrax, there is the additional effect that antibodies protect mononuclear cells from toxin action, preventing the release of intracellular cytokines, lymphokines, and other agents that may contribute to the fever, shock, and multiple organ failure that characterize anthrax (25). The antibacterial effect of vaccine-induced antitoxin, therefore,

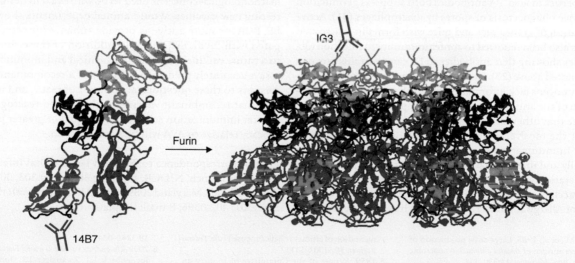

Figure 2
PA regions targeted by neutralizing antibodies. PA monomer (left) is cleaved by cellular furin, releasing domain 1a (amino acids 1–167, colored gray), and allowing the remaining domains to assemble into a heptameric channel (right). Domains remaining in the heptamer are 1b (yellow), 2 (red), 3 (blue), and 4 (green). Indicated in black within domain 4 is the "small loop," amino acids 680–692, which is involved in binding to the PA receptor. The neutralizing mouse mAb 14B7 reacts with this region, preventing PA binding to the receptor. In the heptamer, the "top" of domain 1a contains a surface on which LF and EF bind to initiate internalization. Neutralizing mouse mAb 1G3 binds to this same surface, competing with EF and LF, and thereby blocking their action. This figure was created with Protein Explorer (35), available at http://www.umass.edu/microbio/chime/explorer, using existing PA monomer and heptamer structure files contained in an Atlas of Macromolecules available at http://www.umass.edu/microbio/chime/explorer/index2.htm.

is indirect and is mediated by protection of lymphoid cells against the actions of anthrax toxin, thus permitting phagocytosis and killing of *B. anthracis*.

Toxin-neutralizing mouse mAb's identify two critical sites on PA (Figure 2); a site on domain 4 that binds the cellular receptor, and a site on the PA 63-kDa heptamer to which the LF and EF bind (26). It is probable that vaccine-induced immunity depends on antibodies to these two sites, and possibly to other sites. The key role of the receptor-binding site is shown by the ability of domain 4 alone to induce antibodies that protect mice from infection with *B. anthracis* (27), and by the potent toxin-neutralizing ability of affinity-enhanced recombinant antibodies directed to an epitope that includes amino acids 680–692 within domain 4 (28).

The potential role of antibodies to capsular polypeptide, spores, and LF

PGA is a poor immunogen, probably because of its resistance to proteolysis in antigen-presenting cells and its simple, repeating structure, which makes it a T cell–independent antigen (29). Thus, there has not yet been a definitive test of whether anti-capsular antibodies contribute to immunity to *B. anthracis*. The species of animals used to evaluate PGA is critical. Encapsulated *B. anthracis* strains are quite virulent in mice, regardless of whether they produce toxin (21). PGA, therefore, rather than the toxin, is the major virulence factor in mice, and vaccines based on PA show reduced efficacy in this species (30). For this reason, it may be easier to demonstrate a role for anti-PGA antibodies in protective immunity using mice, once effective strategies for inducing such antibodies have been established. However, it will be difficult to extrapolate conclusions about capsular antibodies to human anthrax infection, given that the contributions of PGA and PA to pathogenesis differ substantially between mice and humans.

Unexpectedly, antibodies to PA have been found to have sporicidal properties in vivo. PA antibodies both suppress germination and enhance phagocytosis of spores by macrophages (31). Active immunization of guinea pigs and mice with formalin-inactivated spores has also been reported to confer immunity to infection (30, 32). Studies showing that antibodies to LF can neutralize toxin in vivo were noted above (23). However, these studies on other candidate immunogens lack information about the specificity and concentration of the antibodies mediating the protection observed.

Evidence that other antigens can contribute to immunity suggests that the most effective vaccines would contain multiple antigens. Immunization trials with these additional antigens, individually and in combination, could lead to highly efficacious third-generation vaccines. However, in the short term, improved anthrax vaccines will consist primarily of PA. Thus, the National Institute of Allergy and Infectious Diseases has an accelerated program for vaccine development that seeks to make 25 million doses of a recombinant PA vaccine available within two years (33).

How will an investigational anthrax vaccine be standardized?

Limited clinical data with AVA and substantial animal experimentation indicate that a critical level of serum anti-PA antibodies confer immunity to both cutaneous and inhalational anthrax (8, 34). An improved anthrax vaccine, therefore, could be a single-component, purified protein that elicits concentrations of PA antibodies comparable to those induced by AVA. However, there are a number of uncertainties that may complicate the seemingly simple transition to a recombinant PA vaccine. First, it is unclear what concentration of serum PA antibodies in humans confers immunity to anthrax. Data for the efficacy of AVA are limited to one trial and long experience without a vaccine failure. For this reason, it is not obvious what point in the AVA injection schedule, or what resulting level of protective anti-PA antibodies, should be chosen as the standard for comparison with new vaccine candidates. Second, the level of Ab's required to protect people from the effects of a bioterrorist attack is uncertain, because the number of spores inhaled under those circumstances might greatly exceed that encountered by the previously studied population of wool-sorters. Would a vaccine need to protect against 5 LD_{50} or 5000 LD_{50}? (We presume that exposure would be to aerosolized spores from an "anthrax bomb," but it is possible that *B. anthracis* could be added to drinking water or food. This latter route would not be as immediately dangerous, but the spores would not be totally inactivated by boiling and could pose a continual threat). Third, selection of appropriate doses and schedules for pediatric vaccination also requires study. Would the schedule used for primary immunization with the diphtheria and tetanus toxoids be satisfactory? Finally, several methodological concerns need to be addressed in designing and testing new vaccines. Would animal experiments showing that LF, PGA, or spore antigens provide enhanced protection compared with PA alone justify the addition of these components to a future vaccine? Can physicochemical and immunochemical assays accurately predict the efficacy of a recombinant vaccine? Answers to these questions are not out of reach, and it is probable that recombinant vaccines with reduced reactogenicity, a shorter immunization schedule, and equal or greater protective efficacy relative to AVA will be available soon.

Address correspondence to: Stephen Leppla, Oral Infection and Immunity Branch, NIDCR, Building 30, Room 303, 30 Convent Drive, Bethesda, Maryland 20892-4350, USA. Phone: (301) 594-2865; Fax: (301) 402-0396; E-mail: Leppla@nih.gov.

1. Puziss, M., et al. 1963. Large-scale production of protective antigen of *Bacillus anthracis* in anaerobic cultures. *Appl. Microbiol.* **11**:330–334.
2. Turnbull, P.C., Broster, M.G., Carman, J.A., Manchee, R.J., and Melling, J. 1986. Development of antibodies to protective antigen and lethal factor components of anthrax toxin in humans and guinea pigs and their relevance to protective immunity. *Infect. Immun.* **52**:356–363.
3. Brachman, P.S., et al. 1962. Field evaluation of a human anthrax vaccine. *Am. J. Public Health.* **52**:632–645.
4. Food and Drug Administration. 1985. Biological products: bacterial vaccines and toxoids; implementation of efficacy review. Proposed rule. *Federal Register.* **50**:51002–51117.
5. 1959. Report of the Committee of Inquiry on Anthrax. Ministry of Labour (United Kingdom). Her Majesty's Stationery Office. London, United Kingdom. 83 pp.
6. Wright, G.G., Green, T.W., and Kanode, R.G. 1954. Studies on immunity in anthrax. V. Immunizing activity of alum-precipitated protective antigen. *J. Immunol.* **73**:387–391.
7. Fellows, P.F., et al. 2001. Efficacy of a human anthrax vaccine in guinea pigs, rabbits, and rhesus macaques against challenge by *Bacillus anthracis* isolates of diverse geographical origin. *Vaccine.* **19**:3241–3247.
8. 2002. *The anthrax vaccine. Is it safe? Does it work?* L.M. Joellenbeck, L.L. Zwanziger, J.S. Durch, and B.L. Strom, editors. National Academy Press. Washington, DC, USA. 265 pp. Online at www.nap.edu.
9. 1994. Anthrax Vaccine Adsorbed (AVA). *Code of Federal Regulations.* 21. Parts 600–799, chapter 1(4-1-94), subpart C.
10. Puziss, M., and Wright, G.G. 1963. Studies on immunity in anthrax. X. Gel-adsorbed protective antigen for immunization of man. *J. Bacteriol.* **85**:230–236.
11. Pittman, P.R., et al. 2000. Anthrax vaccine: increasing intervals between the first two doses enhances

antibody response in humans. *Vaccine.* **19**:213–216.

12. Green, B.D., Battisti, L., Koehler, T.M., Thorne, C.B., and Ivins, B.E. 1985. Demonstration of a capsule plasmid in *Bacillus anthracis. Infect. Immun.* **49**:291–297.

13. Uchida, I., Hashimoto, K., and Terakado, N. 1986. Virulence and immunogenicity in experimental animals of *Bacillus anthracis* strains harbouring or lacking 110 MDa and 60 MDa plasmids. *J. Gen. Microbiol.* **132**:557–559.

14. Leppla, S.H. 1999. The bifactorial *Bacillus anthracis* lethal and oedema toxins. In *Comprehensive sourcebook of bacterial protein toxins.* J.E. Alouf and J.H. Freer, editors. Academic Press. London, United Kingdom. 243–263.

15. Gill, D.M. 1978. Seven toxin peptides that cross cell membranes. In *Bacterial toxins and cell membranes.* J. Jeljaszewicz and T. Wadstrom, editors. Academic Press Inc. New York, New York, USA. 291–332.

16. Klimpel, K.R., Molloy, S.S., Thomas, G., and Leppla, S.H. 1992. Anthrax toxin protective antigen is activated by a cell surface protease with the sequence specificity and catalytic properties of furin. *Proc. Natl. Acad. Sci. USA.* **89**:10277–10281.

17. Singh, Y., Klimpel, K.R., Goel, S., Swain, P.K., and Leppla, S.H. 1999. Oligomerization of anthrax toxin protective antigen and binding of lethal factor during endocytic uptake into mammalian cells. *Infect. Immun.* **67**:1853–1859.

18. Duesbery, N.S., et al. 1998. Proteolytic inactivation of MAP-kinase-kinase by anthrax lethal factor. *Science.* **280**:734–737.

19. Erwin, J.L., et al. 2001. Macrophage-derived cell lines do not express proinflammatory cytokines after exposure to *Bacillus anthracis* lethal toxin. *Infect. Immun.* **69**:1175–1177.

20. Makino, S., Uchida, I., Terakado, N., Sasakawa, C., and Yoshikawa, M. 1989. Molecular characterization and protein analysis of the cap region, which is essential for encapsulation in *Bacillus anthracis. J. Bacteriol.* **171**:722–730.

21. Welkos, S.L., Vietri, N.J., and Gibbs, P.H. 1993. Non-toxigenic derivatives of the Ames strain of *Bacillus anthracis* are fully virulent for mice: role of plasmid pX02 and chromosome in strain-dependent virulence. *Microb. Pathog.* **14**:381–388.

22. Pirr, M.I., et al. 2001. In vitro correlate of immunity in a rabbit model of inhalational anthrax. *Vaccine.* **19**:4768–4773.

23. Price, B.M., et al. 2001. Protection against anthrax lethal toxin challenge by genetic immunization with a plasmid encoding the lethal factor protein. *Infect. Immun.* **69**:4509–4515.

24. Schneerson, R., Robbins, J.B., Taranger, J., Lagergard, T., and Trollfors, B. 1996. A toxoid vaccine for pertussis as well as diphtheria? Lessons to be relearned. *Lancet.* **348**:1289–1292.

25. Hanna, P.C., Acosta, D., and Collier, R.J. 1993. On the role of macrophages in anthrax. *Proc. Natl. Acad. Sci. USA.* **90**:10198–10201.

26. Little, S.F., et al. 1996. Characterization of lethal factor binding and cell receptor binding domains of protective antigen of *Bacillus anthracis* using monoclonal antibodies. *Microbiology.* **142**:707–715.

27. Flick-Smith, H.C., et al. 2002. A recombinant carboxy-terminal domain of the protective antigen of *Bacillus anthracis* protects mice against anthrax infection. *Infect. Immun.* **70**:1653–1656.

28. Maynard, J.A., et al. 2002. Protection against anthrax toxin by recombinant antibody fragments correlates with antigen affinity. *Nat. Biotechnol.* **20**:597–601.

29. Goodman, J.W., and Nitecki, D.E. 1966. Immunochemical studies on the poly-gamma-D-glutamyl capsule of *Bacillus anthracis.* I. Characterization of the polypeptide and of the specificity of its reaction with rabbit antisera. *Biochemistry.* **5**:657–665.

30. Brossier, F., Levy, M., and Mock, M. 2002. Anthrax spores make an essential contribution to vaccine efficacy. *Infect. Immun.* **70**:661–664.

31. Welkos, S., Little, S., Friedlander, A., Fritz, D., and Fellows, P. 2001. The role of antibodies to *Bacillus anthracis* and anthrax toxin components in inhibiting the early stages of infection by anthrax spores. *Microbiology.* **147**:1677–1685.

32. Cohen, S., et al. 2000. Attenuated nontoxinogenic and nonencapsulated recombinant *Bacillus anthracis* spore vaccines protect against anthrax. *Infect. Immun.* **68**:4549–4558.

33. 2002. Development and testing of vaccines against anthrax. Request for proposal. National Institute of Allergy and Infectious Diseases, NIH. http://www.niaid.nih.gov/contract/archive/RFP0226.pdf.

34. Friedlander, A.M., Pittman, P.R., and Parker, G.W. 1999. Anthrax vaccine: evidence for safety and efficacy against inhalational anthrax. *JAMA.* **282**:2104–2106.

35. Martz, E. 2002. Protein Explorer: easy yet powerful macromolecular visualization. *Trends Biochem. Sci.* **27**:107–109.

Helicobacter pylori persistence: biology and disease

Martin J. Blaser[1] and John C. Atherton[2]

[1]Department of Medicine and Department of Microbiology, New York University School of Medicine, and New York Harbor Veterans Affairs Medical Center, New York, New York, USA. [2]Wolfson Digestive Diseases Centre and Institute of Infection, Immunity and Inflammation, University of Nottingham, Nottingham, United Kingdom.

Helicobacter pylori are bacteria that have coevolved with humans to be transmitted from person to person and to persistently colonize the stomach. Their population structure is a model for the ecology of the indigenous microbiota. A well-choreographed equilibrium between bacterial effectors and host responses permits microbial persistence and health of the host but confers risk of serious diseases, including peptic ulceration and gastric neoplasia.

The twin hallmarks of the interaction between *Helicobacter pylori* and humans are its persistence during the life of the host, and the host's responses to its continuing presence. This conflict appears paradoxical, but both the microbes and the host adapt to the other in the form of a long-standing dynamic equilibrium (1, S1). Our understanding of the phenomena underlying these interactions is growing. The relationships are important, both because of the major role of *H. pylori* in promoting risk of peptic ulcer disease (2) and non-cardia adenocarcinoma of the stomach (3), and because of the emerging evidence that gastric *H. pylori* colonization has a protective role in relation to severe gastro-esophageal reflux disease and its sequelae, Barrett esophagus and adenocarcinoma of the esophagus (reviewed in ref. 4). New studies suggest other important impacts of *H. pylori* colonization on human physiology (5, 6).

We now present a general model for this host-microbial interaction and then turn to examples of specific operating mechanisms. Although *H. pylori* is unique in colonizing the human stomach, the principles governing the interaction are paradigms for understanding both commensalism and long-term parasitism. Such insights aid our understanding of disease processes as diverse as chronic inflammation, oncogenesis, and hormonal dysregulation and may be relevant to modern epidemic problems such as obesity and diabetes.

A general model of host-microbial persistence

Much evidence indicates that *Helicobacter* species are the indigenous biota of mammalian stomachs, and that *H. pylori* is the human-specific inhabitant (Figure 1a), having been present for at least tens of thousands of years, and probably for considerably longer (7–9). Therefore, coevolution of microbe and host might be expected, and for *H. pylori*, substantial evidence supports this notion (10), with important implications.

Microbial colonization of a host locale affects the surrounding tissue through the occupation of niches, utilization of resources, and excretion, all of which may be considered as signals to the host (Figure 1b). The host also signals the microbe in the form

of pressure, temperature, and chemical milieu (including host-defense molecules). Although these signals could be uncoordinated, coevolution implies linkage, in which the signals of one party affect the signals of the other (Figure 1c). Microbial persistence requires equilibrium, which only can occur when negative feedback is present (1, S1). This simple model forms the basis for understanding *H. pylori* persistence, and microbial persistence in general. If the microbial population includes differing strains, as clearly occurs for *H. pylori* (11, S2, S3), then the host signals are selective forces (Figure 1c), as it is this selected microbial population rather than the individual cell that is the host-signaling entity (Figure 1d). Many bacterial populations are not entirely clonal, reflecting both point mutations and recombination; *H. pylori* is a particularly extreme example with both a high mutation rate and a very high recombinational frequency (12, 13). Thus, each host is colonized not by a single clone but rather by a cloud of usually closely related organisms (11), resembling the "quasispecies" observed with persistent RNA viruses, such as hepatitis C and HIV. This microbial variation affects the signals to the host; for example, within an *H. pylori* population, individual cells may or may not express specific host-interaction molecules (e.g., CagA) that affect host biology in a directed manner. Consequent host "signals," ranging from increased nutrient supply through immune effectors to changes in the gastric microenvironment, are differentially selective for specific *H. pylori* genes. Thus, each host is colonized by a fluid bacterial gene pool, with genotype dominance determined by selection (Figure 1e). In sum, concepts of such highly plastic populations subject to host-specific selection provide models to explain the facility of *H. pylori* to persist, the presence of different strains as well as variants of these strains in individual hosts, and the ability of *H. pylori* to colonize essentially all humans (Figure 1f), despite our heterogeneity.

H. pylori adaptations that facilitate persistence

H. pylori mechanisms to increase diversity. The remarkable diversity of *H. pylori* (12, S4) may be viewed as evidence of a versatile population, able to maximize resource utilization in a variety of niches and microniches and to avoid host constraints. Such constraints include not only host immunity but also developmental changes in the gastric epithelium, acidity, and nutrient availability. Generation of diversity typically involves endogenous (point) mutations, and recombination; *H. pylori* has mechanisms for each (Table 1). Mutation rates are not constant in bacterial populations but sub-

Nonstandard abbreviations used: C-terminal Src kinase (Csk); G protein–coupled receptor kinase-interactor 1 (Git1); Toll-like receptor (TLR).

Conflict of interest: As co-discoverer of *cagA* and *vacA*, Martin Blaser may benefit from commercial exploitation of the intellectual property held by Vanderbilt University.

Citation for this article: *J. Clin. Invest.* **113**:321–333 (2004). doi:10.1172/JCI200420925.

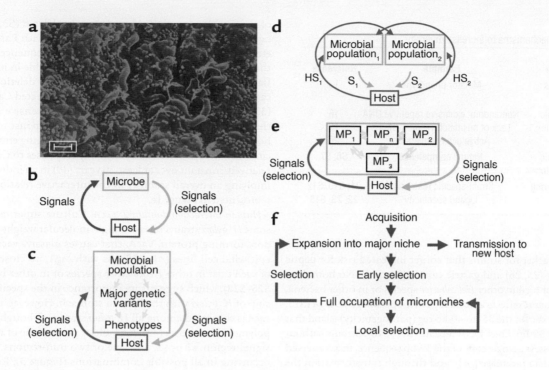

Figure 1

Models of the cross-signaling between obligate parasites and their hosts. (**a**) *H. pylori* in the gastric (antral) niche, inhabiting the luminal mucus layer overlaying the epithelium. Scale bar: 1 μm. (**b**) Model 1: A coevolved microbe sends chemical and physical (contact) signals to its host. The host's signals, including motion, temperature, and chemicals (including host-defense molecules), affect microbial growth. In a local niche, equilibrium can be reached if there is negative feedback in the cross-signaling (1, S1). The microbial signals may alter selection of growing host-cell populations (e.g., by antiapoptotic mechanisms). (**c**) Model 2: The microbial population includes genetic variants. Now, the host signals select for different bacterial phenotypes, and thus genotypes. (**d**) Model 3: Due to ongoing variation and selection, each microbe becomes a microbial population of related variants. The signals of each variant affect not only the local host signals, but also those affecting other microbial populations. This is represented by two distinct microbial populations, with each signaling the host (S_1, S_2) and inducing specific host signals (HS_1, HS_2). Increasing numbers of populations markedly and nonlinearly increase the complexity. (**e**) Model 4: The microbial populations are not clonal but can exchange genetic information; host selection, based on microbial genes or gene fragments rather than cells, determines the (dynamic) equilibria between the microbial populations (MP_n). For the naturally competent, extensively recombining *H. pylori* cells, model 4 best reflects the fluidity of population structures during persistent colonization (11, S2, S3). (**f**) Schematic of adaptation to individual hosts. After *H. pylori* acquisition and expansion into a major niche, early selection allows occupation of microniches, where further local selection determines predominant populations. Both local selection and global selection determine overall population structure, and the probability of transfer of particular genotypes to new hosts. Local resource differences and barriers to diffusion of bacterial cells allow establishment of distinct subpopulations. External fluctuations, such as incomplete antibiotic treatment, can markedly change genotype distribution.

ject to environmental signals, and within large populations a small proportion of cells arise that have heightened mutation rates (S5). For *H. pylori*, most strains would be considered to have this hypermutator phenotype (13), which favors the emergence of variants after selective pressure. A good example of efficient adaptive point mutation by *H. pylori* is the rapid development in the bacterial population of high-level resistance to commonly used antibiotics, such as the macrolide clarithromycin (14).

H. pylori cells are also highly competent for uptake of DNA from other *H. pylori* strains (S6, S7, S8). Analysis of *H. pylori* sequences shows strong evidence of recombination between strains, to the degree that clonal lineages are largely obscured (12, 15). Substantial intragenomic recombination occurs, based largely on the presence of repetitive DNA sequences (16, 17). Repetitive DNA allows for high-frequency deletion and duplication, including slipped-strand mispairing (18, S9). However, because *H. pylori* is naturally competent, any genetic element that is lost may be regained from either unaffected sectors of the population of that strain or from another strain (17, S10). A lack of mismatch-repair systems (19)

may increase the frequency of random variation, but it also may facilitate gene conversion, which minimizes genomic diversity of those alleles present in multiple copies (20). Thus, *H. pylori* can maximize diversity of sequences under strong selective pressure, while maintaining alleles that are critical for its lifestyle.

Introduction of a new *H. pylori* strain into an already colonized host increases the total diversity in the population, since each resembles a quasispecies; however, transformation of one strain by the other reduces diversity. All *H. pylori* strains contain multiple restriction-modification systems, but rarely do any two have the same complement (21, S11). Thus, there are restriction barriers to transformation, a property that may maximize coexistence of parallel gene pools, by slowing genetic exchange (S12). Local selection also may add to genetic diversity within an individual stomach; separate gastric microniches are likely colonized by subpopulations that have particular attributes to maximize fitness, for example, ligand specificity for local receptors (22, 23, S13).

Human interaction domain 1: cag island. In 1989, a strain-specific *H. pylori* gene, *cagA*, was identified (24), which now has been recog-

Table 1

H. pylori mechanisms to increase diversity

Mechanism	Example	Reference
Endogenous mutation	Mutator phenotype	13
Intragenomic recombination	Nonrandom, extensive repetitive DNA	16
	Lack of mismatch-repair systems	19
	Active gene conversion	20
Intergenomic recombination	Natural competence	S6, S7
Niche sectoring	Strain-specific restriction	21, S10, S12
	Ligand specificity	22, 23, S13

nized as a marker for strains that confer increased risk for peptic ulcer disease (25, 26) and gastric cancer (27, S14). No homologs are known for *cagA* in other *Helicobacter* species or in other bacteria, suggesting that it reflects a human gastric-specific gene. *cagA* (S15, S16) is a marker for the 35- to 40-kb *cag* (pathogenicity) island that is flanked by 39-bp direct DNA repeats (S17, S18). Strains without the island possess a single copy of the 39-bp sequence, in a conserved gene (glutamate racemase [*glr*]), and through transformation the entire island may be restored or lost (28). *H. pylori* strains with partial *cag* islands also have been identified, and variation in island size and genotype within individual hosts is well described (S3, S19, S20).

The island contains genes encoding a type IV secretion system, which in other bacteria inject macromolecules (i.e., DNA and proteins, such as pertussis toxin) into host cells (29). One substrate for the type IV system in *H. pylori* is the *cagA* product (30, 31, S21–S23), which is injected into epithelial cells, both in vitro (30, 31, S22, S23) and in vivo (32) (Figure 2). In many strains, the CagA protein contains tyrosine-phosphorylation sites (30, 31, 33, S21–S23) that are recognized by the host cell Src kinase (34). Once phosphorylated, CagA interacts with SHP-2, a tyrosine phosphatase (35), which affects spreading, migration, and adhesion of epithelial cells (32). This phenomenon can be assessed in vitro by a change in epithelial cell morphology to the scattered, or "hummingbird," phenotype (31).

The injected CagA protein also interacts with Grb2 and activates the Ras/MEK/ERK pathway, leading to the phenotypes of cell scattering (in AGS cells) and proliferation (in MDCK cells) (36). Tyrosine-phosphorylated CagA binds and activates C-terminal Src kinase (Csk) via its SH2 domain, which in turn inactivates the Src family of protein-tyrosine kinases. Since this signaling may induce apoptosis, the Csk pathway may attenuate the other CagA interactions (37). By inactivating Src, tyrosine-phosphorylated CagA induces dephosphorylation of cortactin, which then colocalizes with filamentous actin (F-actin), in the tip and base of hummingbird protrusions (38). Thus, the *H. pylori* CagA protein interacts with several of the major signal-transduction pathways present in epithelial cells. *H. pylori* cells with the *cag* island deleted have remarkably little interaction with AGS cells in tissue culture (39); conversely, the *cag* apparatus promotes antiapoptotic pathways, which may aid persistence by slowing turnover of the epithelial cells to which they are attached.

There is extensive *H. pylori* variation in this major interactive modality; clonality and the lack thereof each imply important, albeit different, selective pressures. In individual strains, parts

of the *cag* island, including *cagA*, may be deleted (S3, S19, S20); *cagA* itself shows phylogeographic variation with Eastern, Western, and hybrid genotypes (S24). The DNA sequences encoding the tyrosine-phosphorylation motifs are variable in number and flanked by repetitive elements, allowing their deletion or duplication, which affects the phenotype of the injected CagA protein (33). Thus, *H. pylori* populations possess extensive repertoires that permit variation of Cag phenotypes in response to particular hosts, microniches within these hosts, or changing environmental circumstances. Nevertheless, antibody responses to CagA remain relatively constant over at least 20 years (40) in an individual host, implying an overall stability in the interactive relationship, best represented in Figure 1e.

Human interaction domain 2: vacA. Culture supernatants from some *H. pylori* strains release a high–molecular weight multimeric pore-forming protein, VacA, that causes massive vacuolation in epithelial cell lines (41, S25). As with *cagA*, no close homologs of *vacA* exist in other *Helicobacter* species or in other bacteria (42, S26–S28), which suggests its importance in the specific relationship of *H. pylori* with the human stomach. However, unlike *cagA*, *vacA* is conserved among all *H. pylori* strains, although significant polymorphism exists (43). *vacA* alleles possess one of two types of signal region, s1 or s2, and one of two mid-regions, m1 or m2, occurring in all possible combinations (Figure 3). Research has focused on the most interactive (vacuolating) type, s1/m1.

VacA has several specific effects that may contribute to *H. pylori* persistence in the gastric niche. Firstly, it forms pores in epithelial cell membranes, allowing egress of anions and urea (44, 45, S29, S30). This is important since urea hydrolysis, catalyzed by *H. pylori* urease, protects against gastric acidity (S31). VacA also induces loosening of epithelial tight junctions, potentially allowing nutrients to cross the mucosal barrier to *H. pylori*'s gastric luminal niche (46, S32). Recent work in vitro suggests that VacA may help *H. pylori* persistence by specific immune suppression. VacA blocks phagosome maturation in macrophages (47), selectively inhibits antigen presentation in T cells (48, S33), blocks T cell proliferation, and downregulates Th1 effects by interacting with calcineurin to block signaling (49). Besides these actions that may benefit *H. pylori* persistence, VacA also has direct cell-damaging effects in vitro, inducing cytoskeletal changes, apoptosis, and suppression of epithelial cell proliferation and migration (50–52, S34, S35), as well as cell vacuolation. Whether these effects are germane in vivo is unknown, but cell damage could aid nutrient delivery from the gastric mucosa.

Which in vitro effects of VacA are most important for *H. pylori* persistence in vivo is unclear, and animal models have not clarified this. In piglets, gerbils, and mice, VacA-null strains persist without apparent disadvantage (S36–S38), although in competition experiments in mice, VacA-null mutants colonize less well than their VacA⁺ wild-type parents (S38). However, animal models have proved useful for characterizing the damaging effects of VacA. Although VacA is not necessary for gastric ulcer formation, in *H. pylori*–colonized Mongolian gerbils its presence increases the risk (53). Mice administered VacA orally develop gastric ulcers (54, S28), but mice deficient in the protein tyrosine phosphatase receptor type Z, polypeptide 1 (*Ptprz⁻/⁻* mice) do not (54). Ptprz is one of several putative VacA cellular receptors, and VacA-induced activation increases tyrosine phosphorylation of G protein–coupled receptor kinase-interactor 1 (Git1), leading ultimately to epithelial cell detachment (54). VacA also may have important effects on

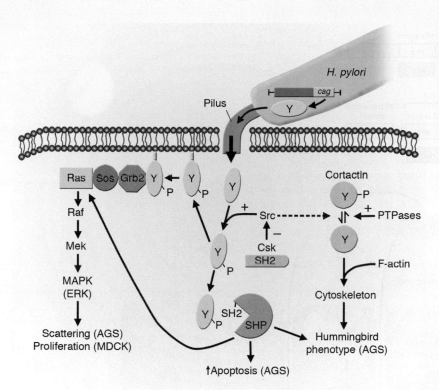

Figure 2
CagA interaction with epithelial cells. *H. pylori* cells
with intact *cag* islands, including an active type IV
secretion system, possess a pilus composed of
CagY protein. The *cagA* product is injected into the
cytoplasm of the host cell, where tyrosine (Y) residues near its COOH-terminus are phosphorylated.
Phosphotyrosine-CagA interacts with several major
signal-transduction pathways in the host cell (40,
S113), affecting phenotypes including cell morphology, proliferation, and apoptosis (see text). ERK,
extracellular signal–regulated kinase; PTPase, protein-tyrosine phosphatase; P, phosphate.

nonepithelial cells: in rats, only VacA⁺ strains induce macromolecular leakage from the gastric microcirculation (S39).

H. pylori strains with different forms of *vacA* exhibit varied phenotypes and have particular associations with gastro-duodenal
diseases. The *vacA* signal region encodes the signal peptide and the
N-terminus of the processed VacA toxin: type s1 VacA is fully active,
but type s2 has a short N-terminus extension that blocks vacuole
formation (55, 56) and attenuates pore formation in eukaryotic
membranes (S40). *vacA* s2 strains are rarely isolated from patients
with peptic ulcers or gastric adenocarcinoma (43, 37, S41, S42).
The *vacA* mid-region encodes part of the toxin cell binding domain.
s1/m2 forms of VacA bind to and vacuolate a narrower range of
cells than s1/m1 forms and induce less damage, yet they also act
as efficient membrane pores and increase paracellular permeability
(56, S30, S32, S43). *vacA* s1/m1 strains are most closely associated
with gastric carcinoma (58, 59, S44). Natural persistence of distinct
polymorphic forms of *vacA* in diverse human populations implies
that each offers a survival advantage. That particular forms of VacA
potentially induce different levels of *H. pylori*–host interaction fits
well with the general model in which less interactive strains cause
diminished tissue injury and disease, and highly interactive strains,
while benefiting from their interaction, are more likely to affect
their niche and thus injure their host.

Interaction between the interacting domains. The *cag* island and *vacA*
are far apart on the *H. pylori* chromosome (19, 60), yet there is a
strong statistical linkage between the s1 genotype of *vacA* and the
presence of the *cag* island (43, S24); similarly, the s2 genotype is
associated with lack of the *cag* island (43). Although these phenomena may reflect founder effects, the panmixis of *H. pylori* (12)
suggests selection for the skewed relationships. This is not a simple interdependence of function: in a *cag⁺/vacA* s1/m1 strain, *cag*
mutagenesis does not abolish vacuolating cytotoxin activity, nor
does disruption of *vacA* abolish the *cag⁺* phenotype (29, 31, 39, S45,

S46). However, there are subtle quantitative effects: *vacA* disruption slightly reduces early tyrosine phosphorylation of CagA during epithelial cell interaction, whereas, in contrast, *cag* disruption
slightly increases VacA-induced vacuolation (61). One contributor
to these effects may be the colocalization on lipid rafts of tyrosine-
phosphorylated Git1/Cat1 (substrate molecules of the VacA receptor phosphatase Ptprz) and tyrosine-phosphorylated CagA (61).

The minor interactions between CagA-induced and VacA-induced
effects on epithelial cells are unlikely to explain their close linkage
in *H. pylori*. An alternative hypothesis is that VacA selects for a functional *cag* island, since VacA-induced immune suppression might
not permit adequate nutrition of the *H. pylori* population (Figure 1).
The effect of *cag⁺* strains in weakening epithelial tight junctions (62)
may enhance nutrient flow to the bacteria and allow better VacA
delivery to the mucosa. For the less potent s2 strains, this selection
would be less substantial and might be counterbalanced by the
phenotypic costs of maintaining the *cag* island. Whatever the true
selective benefits VacA and *cag* offer each other, in conditions with
multiple strains colonizing individual hosts, each major genotypic
combination (*cag⁺/vacA* s1 or *cag⁻/vacA* s2) could occupy a relatively
exclusive physiological niche. Recombination events mixing the
loci would be selected against. With diminishing multiplicities of
H. pylori colonization (7), the strain (and genotype) diversity within a
host would be reduced, diminishing the resource base for the overall
population. Thus, once *H. pylori* transmission within a human population declines, the decline may accelerate because of diminished
vitality of the colonizing bacterial populations in individual hosts.

Immune evasion and manipulation by H. pylori. If a microbe is to persist in a vertebrate host, its biggest challenge is to avoid clearance
by the immune system. Transient *H. pylori* colonization has been
documented in both primates and humans (63, 64, S47), implying
that persistence does not inevitably follow acquisition. The race
between *H. pylori* adaptation to a specific host (Figure 1) and the

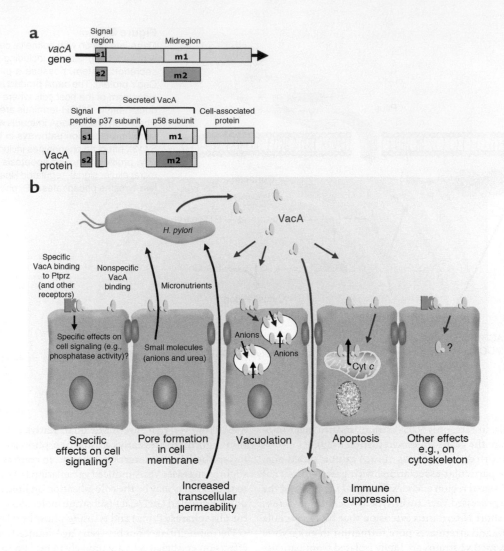

Figure 3

VacA polymorphism and function. (**a**) VacA polymorphism. The gene, *vacA*, is a polymorphic mosaic with two possible signal regions, s1 and s2, and two possible mid-regions, m1 and m2. The translated protein is an autotransporter with N- and C-terminal processing during bacterial secretion. The s1 signal region is fully active, but the s2 region encodes a protein with a different signal-peptide cleavage site, resulting in a short N-terminal extension to the mature toxin that blocks vacuolating activity and attenuates pore-forming activity. The mid-region encodes a cell-binding site, but the m2 type binds to and vacuolates fewer cell lines in vitro. (**b**) VacA biological activities. Secreted VacA forms monomers and oligomers; the monomeric form binds to epithelial cells both nonspecifically and through specific receptor binding, for example, to Ptprz, which may modulate cell signaling. Membrane-bound VacA forms pores. Following VacA endocytosis, large vacuoles form, but, although marked in vitro, these are rarely seen in vivo. VacA also induces apoptosis, in part by forming pores in mitochondrial membranes, allowing cytochrome *c* (Cyt *c*) egress. Although the presence of cytoplasmic VacA is implied rather than demonstrated, yeast two-hybrid experiments show potential for specific binding to cytosolic targets including cytoskeletal proteins, consistent with observed cytoskeletal effects. Finally, VacA has suppressive effects on immune cell function. The relative importance of these effects on *H. pylori* persistence and host interaction remains unclear.

development of effective immunity also implies the feasibility of vaccine development. However, usually, following *H. pylori* acquisition, there is rapid host recognition in the form of both innate and acquired immune responses, including generation of specific local and systemic antibodies (65, S48–S51). Once chronicity is established, the immune stimulation appears remarkably constant; for example, antibody titers remain stable for over 20 years (40), consistent with a model of dynamic equilibrium (Figure 1). The ubiquity and duration of host recognition of *H. pylori* and yet the lifelong colonization by the bacterium demonstrate the effectiveness of *H. pylori*'s strategies to evade host immunity. The important first step is to survive without tissue invasion (Table 2), and

the bulk of *H. pylori* cells, if not all of them (Figure 1a), reside in the gastric lumen (66, S52) beyond the reach of most host immune recognition and effector mechanisms (S48, S52, S53). However, even in this niche, some *H. pylori* cells establish intimate contact with the surface epithelium (S52, S54), some *H. pylori* proteins cross the epithelial barrier (67), and both innate and acquired immune systems are activated (65, S48–S50). Although it is not able to completely avoid immune activation, *H. pylori* has evolved mechanisms to reduce recognition by immune sensors, downregulate activation of immune cells, and escape immune effectors.

Innate immune system recognition of microorganisms involves Toll-like receptors (TLRs) that discriminate pathogen-associated

Table 2
Immune evasion by *H. pylori*

Mechanism	Example	Reference
Sequestration	Lumenal colonization	66, S52
Minimization of innate immunity	Low toxicity of LPS	S58–S60
	DNA methylation	S11
Downregulation of immune effectors	VacA suppression of T cell responses	48, 49
	Interference with phagocytosis	76
Mimicry of host antigens	Lewis expression	74
Antigenic variation	CagY	75

molecular patterns (S55). TLR stimulation triggers proinflammatory signaling through NF-κB activation, and *H. pylori* has evolved to minimize such stimulation. TLR5 recognizes bacterial flagella such as those of *Salmonella typhimurium* but is not stimulated by *H. pylori* flagella (S56). TLR9 recognizes the largely unmethylated DNA of most bacteria (S57), but the highly methylated *H. pylori* DNA likely minimizes recognition (S11). *H. pylori* LPS is anergic compared with that of other enteric bacteria because of lipid A core modifications (S58–S61), and while it stimulates macrophage TLR4 (68, S61), it does not stimulate gastric epithelial TLR4 (69). Although *H. pylori* is relatively camouflaged from innate immune sensors on cell surfaces, *cag*+ strains do stimulate NF-κB activation in epithelial cells (70, S62), apparently through recognition by Nod1 (S63), an innate intracellular pathogen-recognition molecule that detects soluble components of bacterial peptidoglycan (71). How such components are delivered to the epithelial cytoplasm by the *cag*-encoded type IV secretion system remains unclear, but the resultant NF-κB–induced proinflammatory cytokine expression is an important and continuing stimulus to inflammatory cell infiltration and thus to pathogenesis (65, S49).

H. pylori also activates the acquired immune system, as indicated by both humoral and cellular recognition of its antigens (72, S48, S50, S53), although it has evolved to substantially downregulate and avoid acquired immune effectors. Recognition by the acquired immune system requires antigen presentation, and *H. pylori* interferes with both uptake and processing of antigens, partially through a VacA effect (48). *H. pylori* also suppresses T cell proliferation and activation and induces selective T cell apoptosis, again partially through specific VacA effects on signaling (49, 73, S64–S66). It evades host adaptive responses by mimicry of the gastric epithelial fucosylated (Lewis) antigens (74, S67), and by antigenic variation of surface proteins including a critical pilus molecule, CagY (75). As this high-frequency antigenic variation occurs through mutation (usually slipped-strand mispairing) and intragenomic recombination between homologous sequences (19, 23, S9, S68), these genetic mechanisms are important contributors to immune evasion. Finally, *H. pylori* can also suppress less specific immune mechanisms such as phagocytosis (47, 76). The relative contributions of the different host manipulation and evasion strategies to *H. pylori* persistence are not established, possibly differing in individual hosts, but the existence of these varied mechanisms implies that immune surveillance of the gastric lumen is powerful, and that bacterial survival requires its subversion.

Host responses to *H. pylori* and their role in disease

The immune response to H. pylori and its importance in pathogenesis. Despite the mechanisms *H. pylori* has evolved to avoid and downregulate host immune responses, substantial immune activation occurs following *H. pylori* infection. This is manifested by continuous epithelial cell cytokine signaling and gastric mucosal infiltration by neutrophils, macrophages, and lymphocytes, all of which are more pronounced in colonization with a *cag*+ strain (25, 65, 77, S69). There is a pronounced specific acquired immune response, including generation of antibodies and effector T cells, and although this includes both a Th1 and a Th2 component, mucosal cytokine profiles imply Th1 predominance (72, S50). This is unusual for extracellular, toxin-producing bacteria, which usually are met by B cell activation and high-level antibody production (Th2 responses). However, studies in mice suggest that the predominant Th1 response is appropriate to control *H. pylori*: *Helicobacter* colonization density is lower in mice with predominant Th1 responses, whether genetically programmed or manipulated by experimental helminth infection (78, 79, S70, S71).

Despite its apparent propriety, the immune response, and in particular its Th1 component, is a major factor in *H. pylori*–associated pathogenesis (78, 80, 81, S70). Mice with a predominant Th1 response develop more gastric inflammation during *Helicobacter* colonization than those with a Th2 response (78, 79, 81, S70, S71). Experiments that use T cell transfer between mice show that these effects are dependent on Th1 cells (78). Gastric inflammation and atrophic changes are abrogated in the absence of the key Th1 cytokine IFN-γ (81, S70) and are induced by IFN-γ infusion, even without *Helicobacter* (S72). In humans, peptic ulceration is rare during immune suppression with cyclosporin A (S73) and pregnancy (S74), a Th2-predominant state. One hypothesis is that the relative sparsity of *H. pylori*–associated disease in Africa despite high *H. pylori* prevalence (the "African enigma") may be due to predominant Th2 responses to *H. pylori* among black Africans. These responses may be induced by endemic helminth infection (79) or may reflect a genetic predisposition selected by malaria (82).

The importance of heterogeneity in immune responses among human populations and individuals is further demonstrated by the contribution of cytokine polymorphisms to disease risk. Polymorphisms that increase the IL-1β response to *H. pylori* are associated with an increased risk of developing gastric atrophy, hypochlorhydria, and adenocarcinoma (83–85, S14, S75). Polymorphisms in TNF-α and IL-10 genes have a similar, but less pronounced, association (S14, S76). Thus the degree of activation of the immune response, which underlies *H. pylori*–associated pathology, is dependent on both *H. pylori* strain determinants and host genetic factors; the combined effect of these on disease outcome appears synergistic (S14), as predicted by the equilibrium model (Figure 1).

Effect of H. pylori–induced inflammation on acid homeostasis and its importance in upper-gastrointestinal diseases. *H. pylori*–induced proinflammatory cytokine expression and inflammation affect various cell types in the stomach that are important in acid homeostasis, including somatostatin-producing D cells, gastrin-producing G cells, and acid-producing parietal cells (86, 87, S77–S79). *H. pylori* gastritis causes a reduction in somatostatin levels (87, S77, S80) and, since somatostatin negatively regulates gastrin, hypergastrinemia (88). Gastrin is a specific growth factor for *H. pylori* (89), so this potentially creates a positive-feedback loop. Gastrin expression may be enhanced by a direct stimulant effect of *H. pylori*–induced

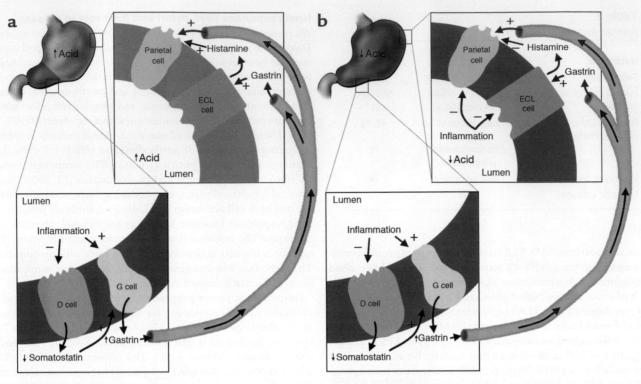

Figure 4
Relation of topography of inflammation to gastric physiology and clinical outcome. (a) *H. pylori*–induced antral-predominant inflammation. Antral inflammation results in hypergastrinemia, which stimulates a physiologically intact corpus to secrete acid, increasing risk for duodenal ulceration. (b) *H. pylori*–induced pan-gastritis. The inflammatory process suppresses corpus acid production, despite the gastrin stimulus from the antrum. Hypochlorhydria enhances risk of gastric ulcer and adenocarcinoma but conversely decreases risk for severe gastro-esophageal reflux disease and its sequelae. ECL, enterochromaffin-like.

proinflammatory cytokines on G cells (S78). Removal of *H. pylori* reverses these effects (S81, S82) (Figure 4).

The effects of gastrin levels on acid homeostasis and disease depend crucially on the topographic distribution in the stomach of the *H. pylori*–induced inflammation (Figure 4). In antral-predominant gastritis, the enterochromaffin-like and parietal cells in the gastric corpus are relatively uninvolved; thus, high gastrin levels lead to greater acid secretion (90, S83, S84). Persistently increased gastrin levels also increase parietal cell mass, enhancing these effects (90, 91). The increased acid load delivered to the duodenum induces gastric metaplasia, a protective phenotypic change. *H. pylori* cannot colonize the normal duodenum but colonizes gastric metaplasia, with resultant inflammation and ulceration (92–94, S85). When the inflammation involves the corpus (pan-gastritis or corpus-predominant gastritis), *H. pylori*–induced inflammatory mediators suppress acid production both indirectly, by inhibiting enterochromaffin-like cell histamine production, and directly by inhibiting parietal cell function (86, S79). Reduced acid secretion further augments gastrin levels, which, while ineffective in raising acid production from the inflamed gastric corpus, provide an ongoing proliferative stimulus to gastric epithelial cells. Continuing proliferation and inflammation affect epithelial cell cycle characteristics (95, 96, S86–S88) and lead to progressive loss of gastric glands. Such atrophic changes markedly increase risk of gastric ulceration and non-cardia gastric adenocarcinoma (4, 97, S89) but, because acid production is lowered, are protective against duodenal ulcer-

ation, and probably against acid-induced complications of gastroesophageal reflux (98, 99).

The topographic distribution of gastritis during chronic *H. pylori* colonization is at least partly host specific; for example, polymorphisms leading to high IL-1β production are associated with pan-gastritis with its accompanying reduced acid production and gastric atrophy (84). However, environmental factors likely also play a crucial role; duodenal ulceration (and so presumably antral-predominant gastritis) is largely a 20th-century disease (100, S90) associated with socioeconomic development. Prior to 20th-century increases in duodenal ulcer incidence, the predominant gastritis pattern was probably that found commonly today in developing countries: pan-gastritis and progressive atrophy. As humans have coevolved with *H. pylori* over at least thousands of years (8, 9) and our genes cannot have evolved appreciably over the last century, unknown environmental influences such as older age at acquisition, reduced number of colonizing strains, changed proportion of strains preadapted by passage through family members, reduction in other microorganisms colonizing the stomach, and improved nutritional status must be responsible for this change. In even more recent times, absence of *H. pylori* from late 20th- and 21st-century stomachs in developed countries, perhaps for the first time in our evolutionary history, may have had further effects on human acid homeostasis and health. As the predominant historical result of colonization was pan-gastritis and reduced acid production, absence of *H. pylori* would be expected to increase mean acid production in the general population, and we speculate that

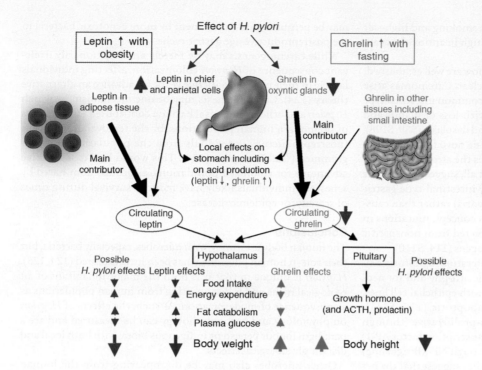

Figure 5
The described effects of *H. pylori* on leptin and ghrelin, and postulated subsequent effects on satiety, energy expenditure, weight, and height. Although leptin and ghrelin have other important paracrine, auto-crine, and endocrine effects, here we concentrate on actions that affect body habitus. The observed effects of *H. pylori* on leptin and ghrelin are based on observational and interventional (*H. pylori* eradication) human studies. Other observational human studies support the portrayed effects of *H. pylori* on weight and height.

this has contributed to the observed rise in acid-associated complications of gastro-esophageal reflux (severe reflux esophagitis, Barrett esophagus, and esophageal adenocarcinoma) in the late 20th century (101). In support of this, patients with severe or complicated acid-esophageal diseases have a reduced *H. pylori* prevalence, particularly of *cag*⁺ strains (101–103, S91), and a low prevalence of gastric atrophy (98, 99). Consequently, the current iatrogenic *H. pylori* removal may have important costs as well as benefits.

The ubiquity of *H. pylori* in unacculturated human populations has led to speculation that colonization also may be beneficial to the pre-reproductive host (S92, S93). We speculate that over the long course of human evolution, adult stomachs were mainly atrophic, and antral-predominant gastritis and hyperchlorhydria were largely conditions of childhood; children would be most likely to benefit from *H. pylori* colonization, by an enhanced acid barrier protecting against diarrheal pathogens (104). If so, there would be strong selection for maintenance of *H. pylori* in populations with poor sanitation; with improvement, such selection would be progressively lost.

Effects of H. pylori on leptin and ghrelin, hormones involved in appetite and satiety. Recently, gastric *H. pylori* colonization has been shown to affect expression of leptin and ghrelin, hormones that control appetite and satiety (5, 6, S94) (Figure 5). Leptin is secreted from adipose tissue and from the stomach (S95, S96); gastric leptin is produced by chief and parietal cells, and released in response to meals and associated hormonal stimuli (105, S96, S97). Leptin signals satiety to the hypothalamus, causing reduced food intake, increased energy expenditure, reduced gastrin and acid secretion, and increased gastric mucosal cell proliferation (106, S98, S99). Ghrelin, produced in oxyntic glands, is released during fasting, and suppressed by feeding and leptin (107, 108). In rats, ghrelin stimulates food intake, reduces energy expenditure, and increases acid secretion (107, S100).

Gastric leptin levels are higher in *H. pylori*–colonized than in noncolonized adults, and eradication leads to their reduction,

although serum levels may not be affected (5, S94). Evidence conflicts as to whether serum ghrelin levels are higher in *H. pylori*–negative persons (109, S101), but they increase after *H. pylori* eradication (6). In an animal model, immunity to *Helicobacter* is associated with upregulation of adipocyte genes, including adipsin, resistin, and adiponectin (110). Inquiry in this field is at an early stage, but if early findings are confirmed, the implications may be important. Weight gain after *H. pylori* eradication is common (5), and these changes in hormonal levels may contribute (Figure 5). Similarly, obesity is increasing in developed countries, as *H. pylori* prevalence is falling. In developing countries, most children acquire *H. pylori* by age 5, and nearly all by age 10, whereas progressively fewer children in developed countries are becoming colonized (111, S102). Whether *H. pylori* genes represent microbial contributions to the complement of "thrifty" (calorie-conserving) genes of humans, and whether *H. pylori* disappearance plays a role in childhood (and adult) adiposity, remain to be determined.

Chronic effects of H. pylori on the gastric epithelium and carcinogenesis. Persistent microbial colonization resulting in inflammation and cell damage is an important cause of carcinogenesis. Examples include *H. pylori* and distal gastric adenocarcinoma, *Schistosoma haematobium* and bladder carcinoma, and hepatitis B virus and hepatoma (112, S103–S105). In evolutionary terms, such cancers are probably neutral; their expression is mostly modern, possibly because of increased human lifespan, and they may be regarded as a cost of chronic colonization, which for gastric cancer occurs in about 1–3% of *H. pylori*–colonized persons.

H. pylori–induced gastric carcinogenesis is more likely when the interaction between *H. pylori* and the host is more interactive; inflammation is more intense and the effects on epithelial cells are more damaging (S106). This may reflect colonization by more interactive *H. pylori* strains: *cag*⁺ strains induce more inflammation and cell cycle effects (4), and *vacA* s1/m1 strains cause more direct epithelial damage (27, 58, 59, S105). Host cytokine polymorphisms enhance the inflammatory response (83, 84, S14, S75).

Damaging environmental factors, such as smoking and high-salt diets, further increase risk, whereas diets high in antioxidants are protective (113, S89, S107).

Although risk factors for gastric cancer now are well established, the mechanism of carcinogenesis is less clear. Carcinomas arise in stomachs with pan-gastritis; the more common intestinal type occurs following progressive atrophy (with loss of glands and hypochlorhydria), intestinal metaplasia, and dysplasia (S89, S108), whereas the diffuse type (S109) may arise de novo from *H. pylori*–colonized mucosa (112). *H. pylori* colonizes the atrophic stomach poorly, and intestinal metaplasia hardly at all, suggesting that the bacteria may create the environment for intestinal-type gastric carcinogenesis (atrophy and hypochlorhydria) rather than causing the cancer directly. In support of this concept, mutations in gastric carcinoma appear random, as expected from nonspecific DNA damage from environmental carcinogens (114, S110).

Disturbance of the epithelial cell proliferation/apoptosis balance is considered a risk factor for gastric atrophy and for neoplastic transformation. When cocultured with epithelial cell lines, *H. pylori* are antiproliferative and proapoptotic (115, S111), although *cag* signaling is essentially pro-proliferative (through MAPK signaling and expression of the transcription factor AP-1) (116, 117) and pro- and antiapoptotic (through NF-κB signaling) (70, 118). Animal models and human studies suggest that the net effect of *H. pylori* colonization is pro-proliferative and proapoptotic (95, 96, 119, 120, S87, S88). Pro-proliferative signaling increases cell replication and the chance of mutation, whereas apoptosis may be protective by inducing death of DNA-damaged cells. However, the consequences of both effects on the epithelial stem cell compartment is likely to be pro-proliferative (to replace apoptotic cells), potentially predisposing to senescence and atrophy or increased mutation and diffuse-type malignant transformation. Stem cell proliferation also may potentially arise more directly from *H. pylori*–induced hypergastrinemia, since gastrin is pro-proliferative for gastrointestinal epithelia; in *H. pylori*–infected gerbils, epithelial proliferation correlates well with serum gastrin levels (96).

Ultimately, carcinogenesis requires DNA damage, which can be induced directly by *H. pylori* products or indirectly through oxygen free radicals released by neutrophils (118, 121, 122). Gastric ascorbic acid, which neutralizes free radicals, is reduced in *H. pylori*–positive stomachs (S112), and *H. pylori* can also directly interfere with the epithelial mismatch-repair system (122). In the atrophic stomach, *H. pylori* colonization is sparse, but atrophy is associated with continuing epithelial proliferation and an inflammatory cell infiltrate. Reactive oxygen species survive longer in the low-acid environment of the atrophic stomach, and ascorbic acid concentrations remain low (123); colonization by oral and intestinal bacteria, which themselves can release reactive oxygen and nitrogen species, may occur. By leading to gastric atrophy, *H. pylori*

may be permitting its replacement by more genotoxic bacteria in the postreproductive-age gastric niche.

While carcinogenesis may be merely an evolutionarily irrelevant consequence of *H. pylori* colonization, affecting individuals largely in their postreproductive years, we advance an alternative theory (124). Carcinogenesis may be one mechanism by which *H. pylori* and other commensal bacteria contributed to the fitness of premodern human populations, by the removal of senescent (postreproductive) individuals from the population in a programmed ("safe") manner (124). This would lead to a selective advantage for colonized populations, as groups dominated by senescent individuals likely have reduced survival during times of scarcity, or epidemic disease.

Conclusions

The human body is teeming with microbes, especially bacteria, but their role in human physiology has been little explored (124, 125). *H. pylori* is unique in that it is both the major inhabitant of an ecological niche and is disappearing from human populations as a consequence of modernization. As such, the effects of *H. pylori* on physiology and pathophysiology can be measured and are a paradigm for our persistent indigenous biota, with both local and distant physiological effects.

Other microbes also may be disappearing from the human "microbiome" (125). We cannot yet ascertain this phenomenon because of the complexity of the indigenous flora, but parallels likely are present. Could extinction of *H. pylori* and other coevolved microbes have affected our physiological signaling? If so, could this in part be responsible for diseases that have increased in modern times, such as gastro-esophageal reflux, obesity, diabetes, asthma, and several malignancies?

Acknowledgments

This work was supported in part by the NIH (R01GM63270, R01DK53707, R01CA97946, and R21DK063603), the Medical Research Service of the Department of Veterans Affairs, a Cancer Research United Kingdom grant, and the award of a Medical Research Council (United Kingdom) Senior Clinical Fellowship to John C. Atherton.

Note: Due to space constraints, a number of important references could not be included in this reference list. Interested readers can find a supplementary reference list at http://www.jci.org/cgi/content/full/113/3/321/DC1.

Address correspondence to: Martin J. Blaser, New York University School of Medicine, 550 First Avenue, OBV-606, New York, New York 10016, USA. Phone: (212) 263-6394; Fax: (212) 263-3969; E-mail: martin.blaser@med.nyu.edu.

1. Blaser, M.J., and Kirschner, D. 1999. Dynamics of *Helicobacter pylori* colonization in relation to the host response. *Proc. Natl. Acad. Sci. U. S. A.* **96**:8359–8364.
2. Nomura, A.M.Y., et al. 2002. *Helicobacter pylori cagA* seropositivity and gastric carcinoma risk in a Japanese American population. *J. Infect. Dis.* **186**:1138–1144.
3. *Helicobacter* and Cancer Collaborative Group. 2001. Gastric cancer and *Helicobacter pylori*: a combined analysis of twelve case-control studies nested within prospective cohorts. *Gut.* **49**:347–353.
4. Peek, R.M., and Blaser, M.J. 2002. *Helicobacter pylori* and gastrointestinal tract adenocarcinomas. *Nat.*

Rev. Cancer. **2**:28–37.
5. Azuma, T., et al. 2001. Gastric leptin and *Helicobacter pylori* infection. *Gut.* **49**:324–329.
6. Nwokolo, C.U., Freshwater, D.A., O'Hare, P., and Randeva, H.S. 2003. Plasma ghrelin following cure of *Helicobacter pylori*. *Gut.* **52**:637–640.
7. Blaser, M.J. 1998. Helicobacters are indigenous to the human stomach: duodenal ulceration is due to changes in gastric microecology in the modern era. *Gut.* **43**:721–727.
8. Ghose, C., et al. 2002. East Asian genotypes of *Helicobacter pylori*: strains in Amerindians provide evidence for its ancient human carriage. *Proc. Natl.*

Acad. Sci. U. S. A. **99**:15107–15111.
9. Falush, D., et al. 2003. Traces of human migration in *Helicobacter pylori* populations. *Science.* **299**:1582–1585.
10. Blaser, M.J., and Berg, D.E. 2001. *Helicobacter pylori* genetic diversity and risk of human disease. *J. Clin. Invest.* **107**:767–773.
11. Kuipers, E.J., et al. 2000. Quasispecies development of *Helicobacter pylori* observed in paired isolates obtained years apart in the same host. *J. Infect. Dis.* **181**:273–282.
12. Suerbaum, S., et al. 1998. Free recombination with *Helicobacter pylori*. *Proc. Natl. Acad. Sci. U. S. A.*

95:12619–12624.

13. Bjorkholm, B., et al. 2001. Mutation frequency and biological cost of antibiotic resistance in *Helicobacter pylori*. *Proc. Natl. Acad. Sci. U. S. A.* **98**:14607–14612.

14. Sjolund, M., Wreiber, K., Andersson, D.I., Blaser, M.J., and Engstrand, L. 2003. Long-term persistence of resistant Enterococcus species after antibiotics to eradicate *Helicobacter pylori*. *Ann. Intern. Med.* **139**:483–487.

15. Falush, D., et al. 2001. Recombination and mutation during long-term gastric colonization by *Helicobacter pylori*: estimates of clock rates, recombination size, and minimal age. *Proc. Natl. Acad. Sci. U. S. A.* **98**:15056–15061.

16. Aras, R.A., Kang, J., Tschumi, A., Harasaki, Y., and Blaser, M.J. 2003. Extensive repetitive DNA facilitates prokaryotic genome plasticity. *Proc. Natl. Acad. Sci. U. S. A.* **100**:13579–13584.

17. Aras, R.A., Takata, T., Ando, T., Van der Ende, A., and Blaser, M.J. 2001. Regulation of the HpyII restriction-modification system of *Helicobacter pylori* by gene deletion and horizontal reconstitution. *Mol. Microbiol.* **42**:369–382.

18. Wang, G., Ge, Z., Rasko, D.A., and Taylor, D.E. 2000. Lewis antigens in *Helicobacter pylori*: biosynthesis and phase variation. *Mol. Microbiol.* **36**:1187–1196.

19. Tomb, J.F., et al. 1997. The complete genome sequence of the gastric pathogen *Helicobacter pylori*. *Nature.* **388**:539–547.

20. Pride, D.T., and Blaser, M.J. 2002. Concerted evolution between duplicated genetic elements in *Helicobacter pylori*. *J. Mol. Biol.* **316**:629–642.

21. Xu, Q., Morgan, R.D., Roberts, R.J., and Blaser, M.J. 2000. Identification of type II restriction and modification systems in *Helicobacter pylori* reveals their substantial diversity among strains. *Proc. Natl. Acad. Sci. U. S. A.* **97**:9671–9676.

22. Ilver, D., et al. 1998. *Helicobacter pylori* adhesion binding fucosylated histo-blood group antigens revealed by retagging. *Science.* **279**:373–377.

23. Pride, D.T., Meinersmann, R.J., and Blaser, M.J. 2001. Allelic variation within *Helicobacter pylori* babA and babB. *Infect. Immun.* **69**:1160–1171.

24. Cover, T.L., Dooley, C.P., and Blaser, M.J. 1990. Characterization and human serologic response to proteins in *Helicobacter pylori* broth culture supernatants with vacuolizing cytotoxin activity. *Infect. Immun.* **58**:603–610.

25. Crabtree, J.E., et al. 1991. Mucosal IgA recognition of *Helicobacter pylori* 120 kDa protein, peptic ulceration, and gastric pathology. *Lancet.* **338**:332–335.

26. Nomura, A.M.Y., Perez-Perez, G.I., Lee, J., Stemmerman, G., and Blaser, MJ. 2002. Relationship between *H. pylori* cagA status and risk of peptic ulcer disease. *Am. J. Epidemiol.* **155**:1054–1059.

27. Blaser, M.J., et al. 1995. Infection with *Helicobacter pylori* strains possessing *cagA* associated with an increased risk of developing adenocarcinoma of the stomach. *Cancer Res.* **55**:2111–2115.

28. Kersulyte, D., Chalkauskas, H., and Berg, D.E. 1999. Emergence of recombinant strains of *Helicobacter pylori* during human infection. *Mol. Microbiol.* **31**:31–43.

29. Tummuru, M.K.R., Sharma, S.A., and Blaser, M.J. 1995. *Helicobacter pylori* picB, a homologue of the *Bordetella pertussis* toxin secretion protein, is required for induction of IL-8 in gastric epithelial cells. *Mol. Microbiol.* **18**:867–876.

30. Odenbreit, S., et al. 2000. Translocation of *Helicobacter pylori* CagA into gastric epithelial cells by type IV secretion. *Science.* **287**:1497–1500.

31. Segal, E.D., Cha, J., Lo, J., Falkow, S., and Tompkins, L.S. 1999. Altered states: involvement of phosphorylated CagA in the induction of host cellular growth changes by *Helicobacter pylori*. *Proc. Natl. Acad. Sci. U. S. A.* **96**:14559–14564.

32. Yamazaki, S., et al. 2003. The CagA protein of *Helicobacter pylori* is translocated into epithelial cells

and binds to SHP-2 in human gastric mucosa. *J. Infect. Dis.* **187**:334–337.

33. Aras, R.A., et al. 2003. Natural variation in populations of persistently colonizing bacteria affect human host cell phenotype. *J. Infect. Dis.* **188**:486–496.

34. Selbach, M., Moese, S., Hauck, C.R., Meyer, T.F., and Backert, S. 2002. Src is the kinase of *Helicobacter pylori* CagA protein in vitro and in vivo. *J. Biol. Chem.* **277**:6775–6778.

35. Higashi, H., et al. 2002. SHP-2 tyrosine phosphatase as an intracellular target of *Helicobacter pylori* CagA protein. *Science.* **295**:683–686.

36. Mimuro, H., et al. 2002. Grb2 is a key mediator of *Helicobacter pylori* CagA protein activities. *Mol. Cell.* **10**:745–755.

37. Tsutsumi, R., Higashi, H., Higuchi, M., Okada, M., and Hatakeyama, M. 2003. Attenuation of *Helicobacter pylori* CagA x SHP-2 signaling by interaction between CagA and C-terminal Src kinase. *J. Biol. Chem.* **278**:3664–3670.

38. Selbach, M., et al. 2003. The *Helicobacter pylori* CagA protein induces cortactin dephosphorylation and actin rearrangement by c-Src inactivation. *EMBO J.* **22**:515–528.

39. Guillemin, K., Salama, N., Tompkins, L., and Falkow, S. 2002. Cag pathogenicity island-specific responses of gastric epithelial cells to *Helicobacter pylori* infection. *Proc. Natl. Acad. Sci. U. S. A.* **99**:15136–15141.

40. Perez-Perez, G.I., et al. 2002. Evidence that *cagA*+ *Helicobacter pylori* strains are disappearing more rapidly than *cagA*- strains. *Gut.* **50**:295–298.

41. Cover, T.L., and Blaser, M.J. 1992. Purification and characterization of the vacuolating toxin from *Helicobacter pylori*. *J. Biol. Chem.* **267**:10570–10575.

42. Cover, T.L., Tummuru, M.K.R., Cao, P., Thompson, S.A., and Blaser, M.J. 1994. Divergence of genetic sequences for the vacuolating cytotoxin among *Helicobacter pylori* strains. *J. Biol. Chem.* **269**:10566–10573.

43. Atherton, J., et al. 1995. Mosaicism in vacuolating cytotoxin alleles of *Helicobacter pylori*: association of specific *vacA* types with cytotoxin production and peptic ulceration. *J. Biol. Chem.* **270**:17771–17777.

44. Iwamoto, H., Czajkowsky, D.M., Cover, T.L., Szabo, G., and Shao, Z. 1999. VacA from *Helicobacter pylori*: a hexameric chloride channel. *FEBS Lett.* **450**:101–104.

45. Tombola, F., et al. 2001. The *Helicobacter pylori* VacA toxin is a urea permease that promotes urea diffusion across epithelia. *J. Clin. Invest.* **108**:929–937. doi:10.1172/JCI200113045.

46. Papini, E., et al. 1998. Selective increase of the permeability of polarized epithelial cell monolayers by *Helicobacter pylori* vacuolating toxin. *J. Clin. Invest.* **102**:813–820.

47. Zheng, P.Y., and Jones, N.L. 2003. *Helicobacter pylori* strains expressing the vacuolating cytotoxin interrupt phagosome maturation in macrophages by recruiting and retaining TACO (coronin 1) protein. *Cell. Microbiol.* **5**:25–40.

48. Molinari, M., et al. 1998. Selective inhibition of Ii-dependent antigen presentation by *Helicobacter pylori* toxin VacA. *J. Exp. Med.* **187**:135–140.

49. Gebert, B., Fischer, W., Weiss, E., Hoffmann, R., and Haas, R. 2003. Helicobacter pylori vacuolating cytotoxin inhibits T lymphocyte activation. *Science.* **301**:1099–1102.

50. Pai, R., Cover, T.L., and Tarnawski, A.S. 1999. *Helicobacter pylori* vacuolating cytotoxin (VacA) disorganizes the cytoskeletal architecture of gastric epithelial cells. *Biochem. Biophys. Res. Commun.* **262**:245–250.

51. Kuck, D., et al. 2001. Vacuolating cytotoxin of *Helicobacter pylori* induces apoptosis in the human gastric epithelial cell line AGS. *Infect. Immun.* **69**:5080–5087.

52. Cover, T.L., Krishna, U.S., Israel, D.A., and Peek,

R.M., Jr. 2003. Induction of gastric epithelial cell apoptosis by *Helicobacter pylori* vacuolating cytotoxin. *Cancer Res.* **63**:951–957.

53. Ogura, K., et al. 2000. Virulence factors of *Helicobacter pylori* responsible for gastric diseases in Mongolian gerbil. *J. Exp. Med.* **192**:1601–1610.

54. Fujikawa, A., et al. 2003. Mice deficient in protein tyrosine phosphatase receptor type Z are resistant to gastric ulcer induction by VacA of *Helicobacter pylori*. *Nat. Genet.* **33**:375–381.

55. Letley, D.P., and Atherton, J.C. 2000. Natural diversity in the N terminus of the mature vacuolating cytotoxin of *Helicobacter pylori* determines cytotoxin activity. *J. Bacteriol.* **182**:3278–3280.

56. Letley, D.P., Rhead, J.L., Twells, R.J., Dove, B., and Atherton, J.C. 2003. Determinants of non-toxicity in the gastric pathogen *Helicobacter pylori*. *J. Biol. Chem.* **278**:26734–26741.

57. Atherton, J.C., Peek, R.M., Tham, K.T., Cover, T.L., and Blaser, M.J. 1997. The clinical and pathological importance of heterogeneity in *vacA*, encoding the vacuolating cytotoxin of *Helicobacter pylori*. *Gastroenterology.* **112**:92–99.

58. Kidd, M., Lastovica, A.J., Atherton, J.C., and Louw, J.A. 1999. Heterogeneity in the *Helicobacter pylori* vacA and cagA genes: association with gastroduodenal disease in South Africa? *Gut.* **45**:499–502.

59. Miehlke, S., et al. 2000. The *Helicobacter pylori* vacA s1, m1 genotype and cagA is associated with gastric carcinoma in Germany. *Int. J. Cancer.* **87**:322–327.

60. Alm, R.A., et al. 1999. Genomic-sequence comparison of two unrelated isolates of the human gastric pathogen *Helicobacter pylori*. *Nature.* **397**:176–180.

61. Asahi, M. 2003. *Helicobacter pylori* CagA containing ITAM-like sequences localized to lipid rafts negatively regulates VacA-induced signaling in vivo. *Helicobacter.* **8**:1–14.

62. Amieva, M.R., et al. 2003. Disruption of the epithelial apical-junctional complex by *Helicobacter pylori* CagA. *Science.* **300**:1430–1434.

63. Dubois, A., et al. 1996. Transient and persistent experimental infection of non-human primates with *Helicobacter pylori*: implications for human disease. *Infect. Immun.* **64**:2885–2891.

64. Perez-Perez, G.I., et al. 2003. Transient and persistent *Helicobacter pylori* colonization in Native American children. *J. Clin. Microbiol.* **41**:2401–2407.

65. Peek, R.M., et al. 1995. Heightened inflammatory response and cytokine expression to *cagA*+ *Helicobacter pylori* strains. *Lab. Invest.* **73**:760–770.

66. Warren, J.R., and Marshall, B. 1983. Unidentified curved bacilli on gastric epithelium in active chronic gastritis. *Lancet.* **1**:1273–1275.

67. Mai, U.E.H., et al. 1992. Surface proteins from *Helicobacter pylori* exhibit chemotactic activity for human leucocytes and are present in gastric mucosa. *J. Exp. Med.* **175**:517–525.

68. Maeda, S., et al. 2001. Distinct mechanism of *Helicobacter pylori*-mediated NF-kappa B activation between gastric cancer cells and monocytic cells. *J. Biol. Chem.* **276**:44856–44864.

69. Backhed, F., et al. 2003. Gastric mucosal recognition of *Helicobacter pylori* is independent of Toll-like receptor 4. *J. Infect. Dis.* **187**:829–836.

70. Foryst-Ludwig, A., and Naumann, M. 2000. p21-activated kinase 1 activates the nuclear factor kappa B (NF-kappa B)-inducing kinase-Ikappa B kinases NF-kappa B pathway and proinflammatory cytokines in *Helicobacter pylori* infection. *J. Biol. Chem.* **275**:39779–39785.

71. Giardin, S.E., et al. 2003. Nod1 detects a unique muropeptide from Gram-negative bacterial peptidoglycan. *Science.* **300**:1584–1587.

72. Bamford, K.B., et al. 1998. Lymphocytes in the human gastric mucosa during *Helicobacter pylori* have a T helper cell 1 phenotype. *Gastroenterology.* **114**:482–492.

73. Wang, J., et al. 2001. Negative selection of T cells by

Helicobacter pylori as a model for bacterial strain selection by immune evasion. *J. Immunol.* **167**:926–934.

74. Wirth, H.P., Yang, M., Peek, R.M., Tham, K.T., and Blaser, M.J. 1997. *Helicobacter pylori* Lewis expression is related to the host Lewis phenotype. *Gastroenterology.* **113**:1091–1098.

75. Aras, R.A., et al. 2003. Plasticity of repetitive DNA sequences within a bacterial (type IV) secretion system component. *J. Exp. Med.* **198**:1349–1360.

76. Allen, L.A., Schlesinger, L.S., and Kang, B. 2000. Virulent strains of *Helicobacter pylori* demonstrate delayed phagocytosis and stimulate homotypic phagosome fusion in macrophages. *J. Exp. Med.* **191**:115–128.

77. Atherton, J.C., Tham, K.T., Peek, R.M., Cover, T.L., and Blaser, M.J. 1996. Density of *Helicobacter pylori* infection *in vivo* as assessed by quantitative culture and histology. *J. Infect. Dis.* **174**:552–556.

78. Mohammadi, M., et al. 1997. Murine CD4 T-cell response to Helicobacter infection: TH1 cells enhance gastritis and TH2 cells reduce bacterial load. *Gastroenterology.* **113**:1848–1857.

79. Fox, J.G., et al. 2000. Concurrent enteric helminth infection modulates inflammation and gastric immune responses and reduces helicobacter-induced gastric atrophy. *Nat. Med.* **6**:536–542.

80. Guiney, D.G., Hasegawa, P., and Cole, S.P. 2003. *Helicobacter pylori* preferentially induces interleukin 12 (IL-12) rather than IL-6 or IL-10 in human dendritic cells. *Infect. Immun.* **71**:4163–4166.

81. Smythies, L.E., et al. 2000. *Helicobacter pylori*-induced mucosal inflammation is Th1 mediated and exacerbated in IL-4, but not IFN-gamma, gene-deficient mice. *J. Immunol.* **165**:1022–1029.

82. Blaser, M.J. 1993. Malaria and the natural history of *Helicobacter pylori* infection. *Lancet.* **342**:551. (Letter)

83. Machado, J.C., et al. 2003. A proinflammatory genetic profile increases the risk for chronic atrophic gastritis and gastric carcinoma. *Gastroenterology.* **125**:364–371.

84. El-Omar, E., et al. 2000. Interleukin-1 polymorphisms associated with increased risk of gastric cancer. *Nature.* **404**:398–402.

85. El-Omar, E.M., et al. 2003. Increased risk of noncardia gastric cancer associated with proinflammatory cytokine gene polymorphisms. *Gastroenterology.* **124**:1193–1201.

86. Beales, I.L., and Calam, J. 1998. Interleukin 1 beta and tumour necrosis factor alpha inhibit acid secretion in cultured rabbit parietal cells by multiple pathways. *Gut.* **42**:227–234.

87. Moss, S.F., Legon, S., Bishop, A.E., Polak, J.M., and Calam, J. 1992. Effect of *Helicobacter pylori* on gastric somatostatin in duodenal ulcer disease. *Lancet.* **340**:930–932.

88. Levi, S., et al. 1989. Campylobacter pylori and duodenal ulcers: the gastrin link. *Lancet.* **1**:1167–1168.

89. Chowers, M.Y., et al. 2002. A defined human gastrin sequence stimulates the growth of *Helicobacter pylori*. *FEMS Microbiol. Lett.* **217**:231–236.

90. El-Omar, E.M., et al. 1995. *Helicobacter pylori* infection and abnormalities of acid secretion in patients with duodenal ulcer disease. *Gastroenterology.* **109**:681–691.

91. Gillen, D., et al. 1998. The acid response to gastrin distinguishes duodenal ulcer patients from *Helicobacter pylori*-infected healthy subjects. *Gastroenterology.* **114**:50–57.

92. Hamlet, A., and Olbe, L. 1996. The influence of *Helicobacter pylori* infection on postprandial duodenal acid load and duodenal bulb pH in humans. *Gastroenterology.* **111**:391–400.

93. Khulusi, S., et al. 1996. Pathogenesis of gastric metaplasia of the human duodenum: role of *Helicobacter pylori*, gastric acid, and ulceration. *Gastroenterology.* **110**:452–458.

94. Ohkusa, T., et al. 2003. *Helicobacter pylori* infection induces duodenitis and superficial duodenal ulcer in Mongolian gerbils. *Gut.* **52**:797–803.

95. Peek, R.M., Jr., et al. 1997. *Helicobacter pylori* cagA+ strains and dissociation of gastric epithelial cell proliferation from apoptosis. *J. Natl. Cancer Inst.* **89**:863–868.

96. Peek, R.M., Jr., et al. 2000. *Helicobacter pylori* alters gastric epithelial cell cycle events and gastrin secretion in Mongolian gerbils. *Gastroenterology.* **118**:48–59.

97. Kuipers, E.J., et al. 1995. Long-term sequelae of *Helicobacter pylori* gastritis. *Lancet.* **345**:1525–1528.

98. Yamaji, Y., et al. 2001. Inverse background of *Helicobacter pylori* antibody and pepsinogen in reflux oesophagitis compared with gastric cancer: analysis of 5732 Japanese subjects. *Gut.* **49**:335–340.

99. Koike, T., et al. 2001. *Helicobacter pylori* infection prevents erosive reflux oesophagitis by decreasing gastric acid secretion. *Gut.* **49**:330–334.

100. Baron, J.H., and Sonnenberg, A. 2001. Period- and cohort-age contours of deaths from gastric and duodenal ulcer in New York 1804-1998. *Am. J. Gastroenterol.* **96**:2887–2891.

101. El-Serag, H.B., and Sonnenberg, A. 1998. Opposing time trends of peptic ulcer and reflux disease. *Gut.* **43**:327–333.

102. Chow, W.H., et al. 1998. An inverse relation between cagA+ strains of *Helicobacter pylori* infection and risk of esophageal and gastric cardia adenocarcinoma. *Cancer Res.* **58**:588–590.

103. Vaezi, M.F., et al. CagA-positive strains of *Helicobacter pylori* may protect against Barrett's esophagus. *Am. J. Gastroenterol.* **95**:2206–2211.

104. Rothenbacher, D., Blaser, M.J., Bode, G., and Brenner, H. 2000. An inverse relationship between gastric colonization by *Helicobacter pylori* and diarrheal illnesses in children: results of a population-based cross-sectional study. *J. Infect. Dis.* **182**:1446–1449.

105. Sobhani, I., et al. 2002. Vagal stimulation rapidly increases leptin secretion in human stomach. *Gastroenterology.* **122**:259–263.

106. Goiot, H., et al. 2001. Antral mucosa expresses functional leptin receptors coupled to STAT-3 signaling, which is involved in the control of gastric secretions in the rat. *Gastroenterology.* **121**:1417–1427.

107. Asakawa, A., et al. 2001. Ghrelin is an appetite-stimulatory signal from stomach with structural resemblance to motilin. *Gastroenterology.* **120**:337–345.

108. Lee, H.M., et al. 2002. Ghrelin, a new gastrointestinal endocrine peptide that stimulates insulin secretion: enteric distribution, ontogeny, influence of endocrine, and dietary manipulations. *Endocrinology.* **143**:185–190.

109. Nishi, Y., et al. 2003. The relationship between ghrelin and *Helicobacter pylori* infection. *Helicobacter.* **8**:402. (Abstr.)

110. Mueller, A., et al. 2003. Protective immunity against *Helicobacter* is characterized by a unique transcriptional signature. *Proc. Natl. Acad. Sci. U. S. A.* **100**:12289–12294.

111. Parsonnet, J. 1995. The incidence of *Helicobacter pylori* infection. *Aliment. Pharmacol. Ther.* **9**:45–51.

112. Nomura, A., et al. 1991. *Helicobacter pylori* infection and gastric carcinoma among Japanese Americans in Hawaii. *N. Engl. J. Med.* **325**:1132–1136.

113. Hansson, L.E., et al. 1993. *Helicobacter pylori* infection: independent risk indicator of gastric adenocarcinoma. *Gastroenterology.* **105**:1098–1103.

114. Simpson, A.J., Caballero, O.L., and Pena, S.D. 2001. Microsatellite instability as a tool for the classification of gastric cancer. *Trends Mol. Med.* **7**:76–80.

115. Wagner, S., et al. 1997. Regulation of gastric epithelial cell growth by *Helicobacter pylori*: evidence for a major role of apoptosis. *Gastroenterology.* **113**:1836–1847.

116. Meyer-ter-Vehn, T., et al. 2000. *Helicobacter pylori* activates mitogen-activated protein kinase cascades and induces expression of the proto-oncogenes c-fos and c-jun. *J. Biol. Chem.* **275**:16064–16072.

117. Naumann, M., et al. 1999. Activation of activator protein 1 and stress response kinases in epithelial cells colonized by *Helicobacter pylori* encoding the cag pathogenicity island. *J. Biol. Chem.* **274**:31655–31662.

118. Maeda, S., et al. 2002. Analysis of apoptotic and antiapoptotic signaling pathways induced by *Helicobacter pylori*. *Mol. Pathol.* **55**:286–293.

119. Fan, X.J., et al. 1998. The effect of class II major histocompatibility complex expression on adherence of *Helicobacter pylori* and induction of apoptosis in gastric epithelial cells: a mechanism for T helper cell type 1-mediated damage. *J. Exp. Med.* **187**:1659–1669.

120. Rudi, J., et al. 1998. Involvement of the CD95 (APO-1/Fas) receptor and ligand system in *Helicobacter pylori*-induced gastric epithelial apoptosis. *J. Clin. Invest.* **102**:1506–1514.

121. Smoot, D.T., et al. 2000. Influence of *Helicobacter pylori* on reactive oxygen-induced gastric epithelial cell injury. *Carcinogenesis.* **21**:2091–2095.

122. Kim, J.J., et al. 2002. *Helicobacter pylori* impairs DNA mismatch repair in gastric epithelial cells. *Gastroenterology.* **23**:542–553.

123. Ruiz, B., et al. 1994. Vitamin C concentration in gastric juice before and after anti-*Helicobacter pylori* treatment. *Am. J. Gastroenterol.* **89**:533–539.

124. Blaser, M.J. 1997. The ecology of *Helicobacter pylori* in the human stomach. *J. Clin. Invest.* **100**:759–762.

125. Lederberg, J. 2000. Infectious history. *Science.* **288**:287–293.

The emergence of Lyme disease

Allen C. Steere,[1] Jenifer Coburn,[2] and Lisa Glickstein[1]

[1]Center for Immunology and Inflammatory Diseases, Division of Rheumatology, Allergy and Immunology, Massachusetts General Hospital, Harvard Medical School, Boston, Massachusetts, USA. [2]Division of Geographic Medicine and Infectious Diseases, Tufts-New England Medical Center, Tufts University School of Medicine, Boston, Massachusetts, USA.

Since its identification nearly 30 years ago, Lyme disease has continued to spread, and there have been increasing numbers of cases in the northeastern and north central US. The Lyme disease agent, *Borrelia burgdorferi*, causes infection by migration through tissues, adhesion to host cells, and evasion of immune clearance. Both innate and adaptive immune responses, especially macrophage- and antibody-mediated killing, are required for optimal control of the infection and spirochetal eradication. Ecological conditions favorable to the disease, and the challenge of prevention, predict that Lyme disease will be a continuing public health concern.

In the late 20th century, Lyme disease, or Lyme borreliosis, was recognized as an important emerging infection (1). It is now the most commonly reported arthropod-borne illness in the US and Europe and is also found in Asia (2). Since surveillance for Lyme disease was begun in the US by the Centers for Disease Control and Prevention, the number of reported cases has increased steadily, and in the year 2000, more than 18,000 cases were reported (3).

Lyme disease was recognized as a separate entity in 1976 because of geographic clustering of children in the Lyme, Connecticut, area who were thought to have juvenile rheumatoid arthritis (4, S1). It then became apparent that Lyme arthritis was a late manifestation of an apparently tick-transmitted, multisystem disease, of which some manifestations had been recognized previously in Europe and America (S2–S6). In 1981, Burgdorfer and colleagues discovered a previously unidentified spirochetal bacterium, called *Borrelia burgdorferi*, in a nymphal *Ixodes scapularis* (also called *Ixodes dammini*) tick (S7). This spirochete was then cultured from patients with early Lyme disease, and patients' immune responses were linked conclusively with that organism, proving the spirochetal etiology of the infection (S8, S9).

Based on genotyping of isolates from ticks, animals, and humans, the formerly designated *B. burgdorferi* has now been subdivided into multiple *Borrelia* species, including three that cause human infection. In the US, the sole cause is *B. burgdorferi* (S10). Although all three species are found in Europe, most of the disease there is due to *Borrelia afzelii* or *Borrelia garinii*, and only these two species seem to be responsible for the illness in Asia (S11, S12). During the 20th century, conditions evolved in the northeastern US that were especially favorable for enzootic *B. burgdorferi* infection (5). In this setting, Lyme disease continues to flourish and spread.

Biology of *B. burgdorferi*

The agents of Lyme borreliosis belong to the eubacterial phylum of spirochetes, which are vigorously motile, corkscrew-shaped bacteria (Figure 1). The spirochetal cell wall consists of a cytoplasmic membrane surrounded by peptidoglycan and flagella and then by a loosely associated outer membrane. The *B. burgdorferi* (strain B31) genome has been completely sequenced. It has a small linear chromosome that is just under one megabase (6), and nine circular and 12 linear plasmids that constitute 40% of its DNA (7). Some of these plasmids are indispensable and could be thought of as mini-chromosomes. Although it has been difficult to manipulate the *B. burgdorferi* genome, progress has recently been made using modified selectable markers and shuttle vectors (8, S13–S18).

The most remarkable aspect of the *B. burgdorferi* genome is the large number of sequences encoding predicted or known lipoproteins, including outer-surface proteins (Osp's) A through F (6). Lipoproteins are found in the outer leaflet of the cytoplasmic membrane, and in both the inner and the outer leaflets of the outer membrane. Some of these proteins are differentially expressed, and one surface-exposed lipoprotein, called VlsE, undergoes extensive antigenic variation (9). In contrast, the genome encodes very few proteins with recognizable biosynthetic activity, and therefore, the organism depends on the host for most of its nutritional requirements. A very unusual feature of *B. burgdorferi* is that it does not require iron, at least for growth in vitro (10). This may allow the spirochete to circumvent the usual host defense of limiting the availability of iron. Finally, the *B. burgdorferi* genome encodes no recognizable toxins. Instead, this extracellular pathogen causes infection by migration through tissues, adhesion to host cells, and evasion of immune clearance.

Enzootic cycles of *B. burgdorferi* infection

The genus *Borrelia* currently includes three pathogenic species that cause Lyme borreliosis (S10–S12) and eight closely related species that rarely if ever cause human infection (S19–S25). These spirochetes live in nature in enzootic cycles involving ticks of the *Ixodes ricinus* complex (also called the *Ixodes persulcatus* complex) and a wide range of animal hosts (Table 1) (11). These enzootic cycles have evolved somewhat differently in different locations (S26). The important vectors of the three pathogenic species of human Lyme borreliosis are the deer tick, *I. scapularis*, in the northeastern and north central US; *Ixodes pacificus* in the western US; the sheep tick, *I. ricinus*, in Europe; and the taiga tick, *I. persulcatus*, in Asia.

In the northeastern US from Maine to Maryland and in the north central states of Wisconsin and Minnesota, a highly efficient, horizontal cycle of *B. burgdorferi* transmission occurs among larval and nymphal *I. scapularis* ticks and certain rodents, particularly white-footed mice and chipmunks (12, S27). This cycle results in high rates of infection among rodents and nymphal ticks and many new cases of human Lyme disease during the late spring and early summer months (Figure 2). White-tailed deer,

Nonstandard abbreviations used: outer-surface protein (Osp); erythema migrans (EM); linear plasmid (lp); lymphocyte function–associated antigen 1 (LFA-1).

Conflict of interest: The authors have declared that no conflict of interest exists.

Citation for this article: *J. Clin. Invest.* **113**:1093–1101 (2004). doi:10.1172/JCI200421681.

297

Figure 1
A scanning electron micrograph of *B. burgdorferi* spirochetes in the midgut of a nymphal *I. scapularis* tick. The picture is a kind gift of Willy Burgdorfer.

which are not involved in the life cycle of the spirochete, are the preferred host of adult *I. scapularis*, and they seem to be critical for the survival of the ticks (S28).

The vector ecology of *B. burgdorferi* is quite different on the West Coast in northern California, where the frequency of Lyme disease is low. There, two intersecting cycles are necessary for disease transmission, one involving the dusky-footed wood rat and *Ixodes spinipalpis* (also called *Ixodes neotomae*) ticks, which do not bite humans and which maintain the cycle in nature, and the other involving wood rats and *I. pacificus* ticks, which are less often infected but do bite humans (S29). Similarly, in Colorado, wood rats and *I. spinipalpis* ticks may be infected with *Borrelia bissettii*, one of the non-pathogenic species, in a cycle that is not known to cause human infection (13). In the southeastern US, nymphal *I. scapularis* feed primarily on lizards, which are resistant to *B. burgdorferi* infection because of complement-mediated killing of the spirochete (14). Therefore, Lyme disease is rare in that part of the country.

In Europe, there is still debate about the preferred animal hosts of *I. ricinus*. These ticks feed on more than 300 animal species, including large and small mammals, birds, and reptiles (15). In Asia, immature *I. persulcatus* commonly feed on voles, shrews, and birds, and adult ticks feed on virtually all larger animals, including hares, deer, and cattle (16). Because the *Borrelia* species differ in their resistance to complement-mediated killing, small rodents are important reservoirs for *B. afzelii*, *B. bissettii*, and *Borrelia japonica*, whereas birds are strongly associated with *B. garinii*, *Borrelia valaisiana*, and *Borrelia turdi* (17).

Emergence of human Lyme borreliosis

The earliest known American cases of Lyme disease occurred in Cape Cod in the 1960s (S30). However, *B. burgdorferi* DNA has been identified by PCR in museum specimens of ticks and mice from Long Island dating from the late 19th and early 20th centuries (S31), and the infection has probably been present in North America for millennia (2). During the European colonization of North America, woodland in New England was cleared for farming, and deer were hunted almost to extinction (5). However, during the 20th century, conditions improved in the northeastern US for the ecology of Lyme disease. As farmland reverted to woodland, deer proliferated, white-footed mice were plentiful, and the deer tick

thrived. Soil moisture and land cover, as found near rivers and along the coast, were favorable for tick survival (18). Finally, these areas became heavily populated with both humans and deer, as more rural wooded areas became wooded suburbs in which deer were without predators and hunting was prohibited.

During the past 40 years, the infection has continued to spread in the northeastern US (19); it has caused focal outbreaks in some coastal areas (S30, S32, S33), and it now affects suburban locations near Boston, New York, Philadelphia, and Baltimore, the most heavily populated parts of the country (S34). In the year 2000, the overall incidence of reported cases in Connecticut, the state with the highest reported frequency of Lyme disease, was 111 per 100,000 residents (3). However, most of the cases still clustered in foci, particularly in two counties in the southeastern part of the state where the original epidemiologic investigation took place in the town of Lyme (S1). In a large, two-year vaccine trial, such high-risk areas had a yearly incidence of the disease of greater than 1 per 100 participants, and the frequency of seropositivity to *B. burgdorferi* at study entry was as high as 5 per 100 participants (20).

As in America, the European agents of Lyme borreliosis have probably been present there for many thousands of years. They are now known to be widely established in Europe's remaining forested areas (15). The highest reported frequencies of the disease are in middle Europe, particularly in Germany, Austria, Slovenia, and Sweden (2). In 1995, the yearly incidence of the disease in Slovenia and Austria was estimated to be 120–130 cases per 100,000 residents (2), similar to the frequency in Connecticut.

Clinical manifestations and disease pathogenesis

To maintain its complex enzootic cycle, *B. burgdorferi* must adapt to markedly different environments, the tick and the mammalian or avian host. The spirochete survives in a dormant state in the nymphal tick midgut during the fall, winter, and early spring, where it expresses primarily OspA (21). When the tick feeds in the

Table 1
The genospecies of *Borrelia burgdorferi* and their tick vectors and locations

	Principal tick vector	Location
Three pathogenic species		
Borrelia burgdorferi	Ixodes scapularis	Northeastern and north central US
	Ixodes pacificus	Western US
	Ixodes ricinus	Europe
Borrelia garinii	Ixodes ricinus	Europe
	Ixodes persulcatus	Asia
Borrelia afzelii	Ixodes ricinus	Europe
	Ixodes persulcatus	Asia
Eight minimally pathogenic or nonpathogenic species		
Borrelia andersonii	Ixodes dentatus	Eastern US
Borrelia bissettii	Ixodes spinipalpis	Western US
	Ixodes pacificus	
Borrelia valaisiana	Ixodes ricinus	Europe and Asia
Borrelia lusitaniae	Ixodes ricinus	Europe
Borrelia japonica	Ixodes ovatus	Japan
Borrelia tanukii	Ixodes tanukii	Japan
Borrelia turdae	Ixodes turdus	Japan
Borrelia sinica	Ixodes persulcatus	China

Figure 2
The enzootic cycle of *B. burgdorferi* infection in the northeastern US and intersection with human Lyme disease. *I. scapularis* ticks feed once during each of the three stages of their usual 2-year life cycle. Typically, larval ticks take one blood meal in the late summer (**A**), nymphs feed during the following late spring and early summer (**B**), and adults feed during the fall (**C**), after which the female tick lays eggs (**D**) that hatch the next summer (**E**). It is critical that the tick feeds on the same host species in both of its immature stages (larval and nymphal), because the life cycle of the spirochete (wavy red line) depends on horizontal transmission: in the early summer, from infected nymphs to certain rodents, particularly mice or chipmunks (**B**); and in the late summer, from infected rodents to larvae (**A**), which then molt to become infected nymphs that begin the cycle again in the following year. Therefore, *B. burgdorferi* spends much of its natural cycle in a dormant state in the midgut of the tick. During the summer months, after transmission to rodents, the spirochete must evade the immune response long enough to be transferred to feeding larval ticks. Although the tick may attach to humans at all three stages, it is primarily the tiny nymphal tick (~1 mm) that transmits the infection (**F**). This stage of the tick life cycle has a peak period of questing in the weeks surrounding the summer solstice. Humans are an incidental host and are not involved at all in the life cycle of the spirochete.

late spring or early summer, the expression of a number of spirochetal proteins is altered (S35). For example, OspA is downregulated, and OspC is upregulated (21). OspC expression is required for infection of the mammalian host (22, 23). In addition, the spirochete binds mammalian plasminogen and its activators, present in the blood meal, which facilitates spreading of the organism within the tick (24). Within the salivary gland, OspC expression predominates, but some organisms express only OspE and OspF; OspA and OspB are absent (25).

After transmission of the spirochete, human Lyme disease generally occurs in stages, with remissions and exacerbations and different clinical manifestations at each stage (4). Early infection consists of stage 1, localized infection of the skin, followed within days or weeks by stage 2, disseminated infection, and months to years later by stage 3, persistent infection. However, the infection is variable; some patients have only localized infection of the skin, while others have only later manifestations of the illness, such as arthritis. Moreover, there are regional variations, primarily between the illness found in America and that found in Europe and Asia (1).

Localized infection
After an incubation period of 3–32 days, a slowly expanding skin lesion, called erythema migrans (EM), forms at the site of the tick bite in 70–80% of cases (26, 27). In the US, the skin lesion is frequently accompanied by flu-like symptoms, such as malaise and fatigue, headache, arthralgias, myalgias, and fever, and by signs that suggest dissemination of the spirochete (28). In about 18% of cases (27), these symptoms are the presenting manifestation of the illness (29). In contrast, EM in Europe is more often an indolent, localized infection, and spirochetal dissemination is less common (30).

In addition to the Lyme disease agent, *I. scapularis* ticks in the US and *I. ricinus* ticks in Europe may transmit *Babesia microti* (a red-blood-cell parasite) or *Anaplasma phagocytophilum* (formerly referred to as "the agent of human granulocytic ehrlichiosis") (31–33). In a recent prospective study in the US, 4% of patients with culture-proven EM had coinfection with one of these other two tick-borne agents (34). Although these two infections are usually asymptomatic, coinfection may lead to more severe, acute flu-like illness (35).

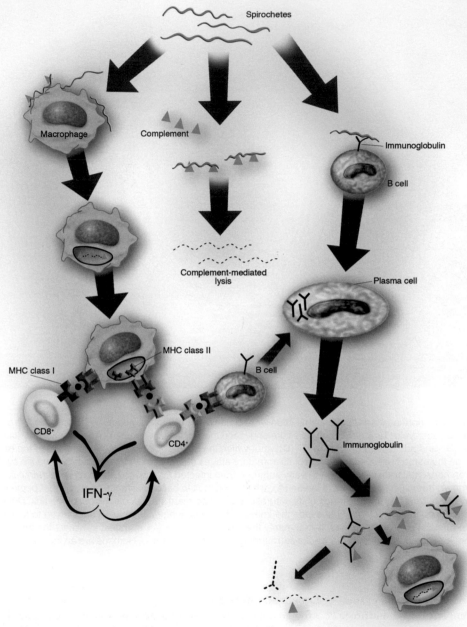

Figure 3
Host mechanisms of spirochetal killing. Complement-mediated lysis of the organism may be the first line of host defense. Spirochetal lipoproteins and other spirochetal signals activate macrophages, leading to the production of strong proinflammatory cytokines, especially TNF-α and IL-1β. Macrophages engulf spirochetes and degrade them in intracellular compartments. Spirochetal lipoproteins, which are B cell mitogens, also stimulate adaptive T cell–independent B cell responses. Humoral immune responses to nonlipidated spirochetal proteins are more likely to be T cell dependent. The primary role of *B. burgdorferi*–specific CD4+ Th1 cells is to prime T cell–dependent B cell responses, and antigen-specific CD8+ T cells may be a significant source of IFN-γ. Antibody-mediated spirochetal killing occurs by complement fixation and opsonization.

by Ken Beauchamp *J. Clin. Invest.*

In most patients, immune cells first encounter *B. burgdorferi* at the site of the tick bite. Depending on the *Borrelia* species and the host, complement-mediated lysis of the organism may be the first line of host defense (Figure 3) (36). On histologic examination, the resulting EM skin lesions consist of mild to marked perivascular infiltrates of lymphocytes, DCs, macrophages, and small numbers of plasma cells (37). As a part of the innate immune response, macrophages engulf and kill spirochetes (38–41). Inflammatory cells within the lesion produce primarily proinflammatory cytokines, including TNF-α and IFN-γ (37, 42). *B. burgdorferi*-stimulated PBMCs from patients with EM produce Th1 proinflammatory cytokines, especially IFN-γ (43). Within days after disease onset, most patients have an IgM antibody response to OspC or the 41-kDa flagellar protein of the spirochete (44). Thus, both innate and adaptive cellular elements are mobilized to fight the infection.

Disseminated infection

Within days to weeks after disease onset, *B. burgdorferi* often disseminates widely. During this period, the spirochete has been recovered from blood and cerebrospinal fluid (S7, S36, S37), and it has been seen in small numbers in specimens of myocardium, retina, muscle, bone, spleen, liver, meninges, and brain (45). Possible clinical manifestations include secondary annular skin lesions, acute lymphocytic meningitis, cranial neuropathy, radiculoneuritis, atrioventricular nodal block, migratory musculoskeletal pain in joints, bursae, tendon, muscle, or bone, and, rarely, eye manifestations (reviewed in ref. 4). Less often, spirochetal dissemination is asymptomatic.

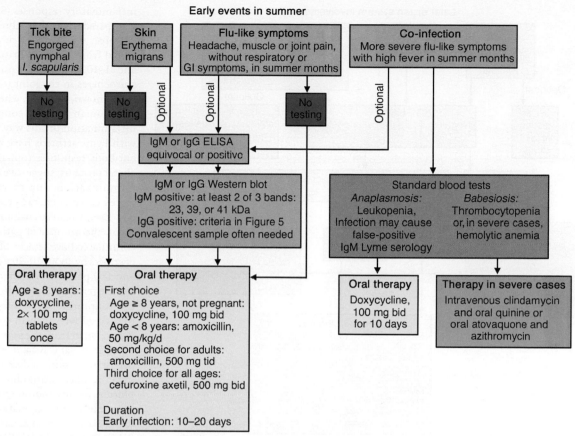

Early events in summer

Figure 4
An algorithm for the diagnosis and treatment of the early events surrounding Lyme disease in the summer months. Serologic testing for Lyme disease has limited utility during the first 1 or 2 weeks of infection, and early treatment, without serologic testing, is recommended. If serologic testing is done, acute and convalescent samples should be obtained. GI, gastrointestinal.

To disseminate, *B. burgdorferi* binds certain host proteins and adheres to integrins, proteoglycans, or glycoproteins on host cells or tissue matrices. As in the tick, spreading of the spirochete through tissue matrices may be facilitated by the binding of plasminogen and its activators to the surface of the organism (24). A 47-kDa spirochetal protein (BBK32) binds fibronectin, an ECM protein (46). The sequences of OspC vary considerably among strains, and only a few sequences are associated with disseminated infection (47), probably because they bind as-yet unidentified host structures. A 66-kDa outer-surface protein of the spirochete binds the fibrinogen receptor ($\alpha_{IIb}\beta_3$) and the vitronectin receptor ($\alpha_v\beta_3$) (48), which may allow the organism to establish an initial foothold and disseminate in the vasculature. A 26-kDa *Borrelia* glycosaminoglycan-binding (GAG-binding) protein, Bgp, binds to the GAG side chains of heparan sulfate on endothelial cells, and to both heparan sulfate and dermatan sulfate on neuronal cells (49, 50). Finally, spirochetal decorin-binding proteins A and B (DbpA and DbpB) bind decorin, a proteoglycan that associates with collagen (51). This may explain the alignment of spirochetes with collagen fibrils in the ECM of the heart, nervous system, or joints (45).

Despite an active immune response, *B. burgdorferi* may survive during dissemination by changing or minimizing antigenic expression of surface proteins and by inhibiting certain critical host immune responses. Two linear plasmids (lp's) seem to be essential, including lp25, which encodes a nicotinamidase (52), and

lp28-1, which encodes the VlsE lipoprotein (53), the protein that undergoes antigenic variation. In addition, the spirochete has a number of families of highly homologous, differentially expressed lipoproteins, including the OspE/F paralogs, which further contribute to antigenic diversity (54). *B. burgdorferi* may downregulate lipoproteins because of host immune pressure (54, 55). For example, in a mouse model, the development of antibody to OspC, a prominent early response, induces downregulation of OspC; and therefore, this antibody response does not completely clear the infection (56). Finally, *B. afzelii* and, to a lesser degree, *B. burgdorferi* have complement regulator-acquiring surface proteins that bind complement factor H and factor H-like protein 1 (57). These complement factors inactivate C3b, which protects the organism from complement-mediated killing (57, S38–S41). In contrast, *B. garinii* is efficiently killed by complement (S42).

As shown definitively in mouse models, both innate and adaptive immune responses are required for optimal control of disseminated infection (Figure 3). *B. burgdorferi* lipoproteins, which are B cell mitogens (S39), stimulate adaptive T cell–independent B cell responses (58, S43, S44). For example, antibody responses to OspC kill spirochetes (59). In addition, humoral immune responses to nonlipidated spirochetal proteins, which are more likely to be T cell–dependent, aid in spirochetal killing (60, 61). The primary role of *B. burgdorferi*-specific Th1 cells is to prime these T cell–dependent B cell responses (62). The combination of these responses

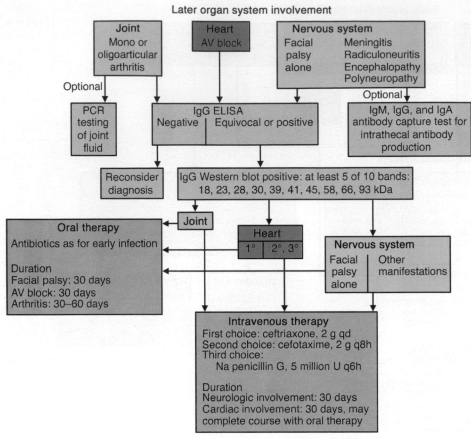

Figure 5

An algorithm for the diagnosis and treatment of late organ-system involvement in Lyme disease. By the time that organ-system involvement is present, which is at least several weeks after the onset of infection, almost all patients have a positive IgG response to *B. burgdorferi*. Depending on the manifestation, treatment with either oral or intravenous antibiotic therapy is recommended.

inflammatory responses (S49, S50). Compared with other inbred strains of mice, C57BL/6 mice are protected from severe arthritis by IL-6 and IL-10, despite large numbers of spirochetes in the joint (67, 68). It is unknown, however, whether certain human patients control joint inflammation in this way. Patients with Lyme arthritis have very high antibody responses to many spirochetal proteins, suggestive of hyperimmunization due to recurrent waves of spirochetal growth (63, 64). Even without antibiotic treatment, the number of patients who continue to have attacks of arthritis decreases by about 10–20% each year, and few patients have had attacks for longer than 5 years (65). Thus, these immune mechanisms seem to succeed eventually in eradicating *B. burgdorferi* from the joint.

In Europe and Asia, *B. afzelii* may persist in the skin for decades, resulting in acrodermatitis chronica atrophicans, a skin condition that occurs primarily on sun-exposed surfaces of distal extremities in elderly women (S51). Compared with EM lesions, infiltrates of T cells and macrophages in acrodermatitis lesions had a restricted cytokine profile, lacking IFN-γ production (37). Consistent with this finding, ultraviolet B irradiation of *B. burgdorferi*–infected C3H mice decreased the Th1 response (69). Thus, spirochetal persistence in acrodermatitis skin lesions may involve both spirochetal factors and an ineffective local immune response.

B. garinii, which is also found only in Europe and Asia, appears to be the most neurotropic of the three *Borrelia* species. It may cause an exceptionally wide range of neurologic abnormalities (70), including borrelial encephalomyelitis (S52), a multiple sclerosis–like illness. In the US, a rare, late neurologic syndrome has been described, called Lyme encephalopathy or polyneuropathy, which is manifested primarily by subtle cognitive disturbances, spinal radicular pain, or distal paresthesias (71, S53). With each of these three late neurologic complications, the possible duration of spirochetal persistence and the pathogenetic mechanisms are unknown.

Putative postinfectious syndromes

Treatment-resistant Lyme arthritis. About 10% of patients with Lyme arthritis have persistent joint inflammation for months or even several years after standard courses of antibiotic treatment (72), a complication rarely noted in Europe (S54). Although *B. burgdorferi* DNA can often be detected by PCR in the joint fluid of these patients prior to antibiotic treatment (73, S55), PCR results are usually negative after antibiotic treatment (73), suggesting that joint swelling may persist after complete or nearly complete eradication of the spirochete from the joint with antibiotic therapy.

leads to the production of antibodies against many components of the organism (63, 64), which promote spirochetal killing by complement fixation and opsonization (S45). Within several weeks to months, these antibody responses, in conjunction with innate immune mechanisms, control widely disseminated infection even without antibiotic treatment, and generalized symptoms resolve.

Persistent infection

After weeks of disseminated infection, the Lyme disease agents may still survive in localized niches for several years. By this time, systemic symptoms are minimal or absent altogether. Although each of the three pathogenic species may spread to the joints, nervous system, or other skin sites, they seem to vary in the frequency of dissemination to these sites and in their ability to persist there. *B. burgdorferi*, the sole cause of the infection in the US, seems to be the most arthritogenic. Months after the onset of illness, about 60% of untreated patients with this infection experience intermittent attacks of arthritis, primarily of the large joints, especially the knee (65).

As shown in a mouse model, neutrophil extravasation into the infected joint is a key initial step in the development of joint inflammation (66). In the human infection, CD4+ Th cells are of the proinflammatory Th1 subset (S46, S47), and *B. burgdorferi*-specific CD8+ T cells are found as well (S48). Within the joint, *B. burgdorferi*-specific γδ T cells may aid in the regulation of these

To explain this course, it has been hypothesized that these patients may have persistent infection or infection-induced autoimmunity (74). In support of the persistent-infection hypothesis, ex vivo–infected synovial cells contained *B. burgdorferi* in the cytosol (75), a site that might be protected from antibiotics. However, spirochetes have not been seen in intracellular locations in situ in human or mouse synovia (45). Moreover, PCR results for *B. burgdorferi* DNA were negative in synovial tissue in all 26 patients with treatment-resistant arthritis who underwent arthroscopic synovectomy a median of 7 months after the completion of antibiotic therapy (76). This methodology may be insufficient to identify rare spirochetes and would not detect retained spirochetal antigens.

In support of the autoimmunity hypothesis, treatment-resistant Lyme arthritis is associated with HLA-DRB1*0401, 0101, and other related alleles (77), and with cellular and humoral immune responses to OspA of *B. burgdorferi* (78, S56–S58). In an epitope-mapping study, 15 of 16 treatment-resistant patients had T cell reactivity with the OspA$_{165-173}$ epitope, the immunodominant epitope presented by the DRB1*0401 or 0101 molecule, compared with only one of five treatment-responsive patients (78). One homolog of this epitope, human lymphocyte function–associated antigen 1$\alpha_{L332-340}$ (LFA-1$\alpha_{L332-340}$), acted as a weak, partial agonist for OspA$_{165-173}$-reactive T cells in DRB1*0401-positive patients (79, 80), but the LFA-1 peptide did not bind the 0101 molecule (77), which suggests that the LFA-1 peptide is unlikely to be a relevant autoantigen. Although the pathogenesis of this syndrome is incompletely delineated, future technologies may allow the identification of spirochetal components or a relevant autoantigen, or both, in the synovia of these patients.

Post–Lyme disease syndrome. A small percentage of patients with well-documented Lyme disease may develop disabling musculoskeletal pain, neurocognitive symptoms, or fatigue along with or soon after symptoms of the infection (S59–S62). This post–Lyme disease syndrome, or chronic Lyme disease (the terms are used interchangeably), which is similar to chronic fatigue syndrome or fibromyalgia, persists for months or years after standard antibiotic treatment of the infection. In a study of such patients who then received intravenous ceftriaxone for 30 days followed by oral doxycycline for 60 days, or intravenous or oral placebo preparations for the same duration, no significant differences were found between the groups in the percentage of patients who said that their symptoms had improved, gotten worse, or stayed the same (81). Therefore, it is hypothesized that *B. burgdorferi* may trigger immunologic or neurohormonal processes in the brain that cause persistent pain, neurocognitive, or fatigue symptoms, despite spirochetal killing with antibiotic therapy (82). Among *B. burgdorferi*–infected patients, a prior history of depression or anxiety seems to be a risk factor for the development of chronic Lyme disease (83).

A counterculture has emerged regarding chronic Lyme disease (84). In contrast with the findings of evidence-based medicine, some people believe that the tests for Lyme disease are often inaccurately negative, and that antibiotic therapy is necessary for months or years to suppress the symptoms of this often incurable illness. A number of investigators at academic medical centers have reported series of patients referred for chronic Lyme disease in which the majority of patients had pain or fatigue syndromes with little or no evidence of past or present *B. burgdorferi* infection (85–87). Prolonged antibiotic therapy may be harmful. In studies of patients with unsubstantiated Lyme disease, minor side effects were common (86), prolonged ceftriaxone therapy sometimes

resulted in biliary complications (88), and in one reported case, the prolonged administration of cefotaxime resulted in death (89). Furthermore, prolonged use of antibiotics was recently associated with an increased risk of breast cancer (90). Although antibiotic use may not be causally related to cancer, this observation reinforces the advisability of prudent use of antibiotics.

Diagnosis and treatment

Algorithms for the diagnosis and treatment of early or late Lyme disease are presented in Figures 4 and 5. Except in those with active EM, the diagnosis is usually based on the recognition of a characteristic clinical picture (S63) and a positive antibody response to *B. burgdorferi* by whole-cell sonicate ELISA and Western blot, interpreted according to the criteria of the Centers for Disease Control and Prevention (S64). Evidence-based treatment recommendations are incorporated from those presented by the Infectious Disease Society of America (91).

Every summer, the lay public and physicians in endemic areas deal with the early events surrounding Lyme disease, including tick bites, early infection, and coinfection (Figure 4). Since 24–72 hours of tick attachment is necessary before transmission of the spirochete occurs, removal of the tick within 24 hours of attachment is usually sufficient to prevent Lyme disease (54, S65, S66). If an engorged nymphal *I. scapularis* tick is found, a single, 200-mg dose of doxycycline usually prevents the infection (92). Serologic tests are insensitive during the first 1 or 2 weeks of infection and depend largely on detection of a positive IgM response, which may still represent a false-positive response (63, S67). Because of these limitations, treatment is recommended for 10 to 20 days, without serologic testing, for presumed EM, most commonly with doxycycline in adults or amoxicillin in children (93, S68). If serologic testing is done, both acute and convalescent samples should be obtained, since most patients have a positive IgM or IgG response by convalescence at the conclusion of antibiotic treatment, and the demonstration of seroconversion provides better serologic support for the diagnosis. Reinfection may occur in patients who are treated with antibiotics early in the illness (94).

Flu-like illness during summer is a more difficult issue, since most cases are not caused by *B. burgdorferi* infection. However, if a patient from a highly endemic area has a febrile illness with headache and joint or muscle pain, without respiratory or gastrointestinal symptoms, in the weeks surrounding the summer solstice, antibiotic treatment may be indicated (Figure 4). For such patients, a second-generation serologic test, an IgG ELISA that uses a peptide in the sixth invariant region of the VlsE lipoprotein of *B. burgdorferi*, may be valuable, since this test typically becomes positive before five IgG bands are present on Western blot (29, 95). Although both babesiosis and anaplasmosis are usually asymptomatic, coinfection should be considered in a patient with more severe flu-like symptoms, including high fever, particularly if the patient is very young or old or asplenic (Figure 4). Fortunately, Lyme disease and anaplasmosis can both be treated with doxycycline. For severe cases of babesiosis, intravenous clindamycin and oral quinine, or oral atovaquone and azithromycin, may be effective (S69).

By the time that organ-system involvement is present in Lyme disease, which is at least several weeks after the onset of infection, almost all patients have a positive IgG response to *B. burgdorferi* (63) (Figure 5). Objective neurologic abnormalities require treatment with intravenous antibiotic therapy, usually intravenous ceftriaxone (S70), with the possible exception of facial palsy alone, without other neurologic manifestations. Lyme arthritis may be treated with either oral or intravenous therapy (72, S70), but oral therapy is easier to administer, is associated with fewer side effects, and is considerably less expensive (S71). Reinfection has not been reported in patients with the expanded immune response associated with Lyme arthritis.

After antibiotic treatment, antibody titers fall slowly, but IgG and even IgM responses may persist for years (96), as may the IgG response to the VlsE peptide (97). Moreover, asymptomatic IgG seroconversion to *B. burgdorferi* occurs in about 7% of patients in the US (98). If patients with asymptomatic seroconversion or past infection have symptoms caused by another illness, the danger is to attribute them incorrectly to Lyme disease, and therefore, the clinical picture must always be considered with the serologic result.

If patients with Lyme arthritis have persistent joint inflammation after 2 months of oral antibiotics or 1 month of intravenous antibiotics and the results of PCR testing are negative, we treat them with nonsteroidal anti-inflammatory agents, disease-modifying antirheumatic drugs, or arthroscopic synovectomy. In those with post–Lyme disease syndrome, we follow the guidelines for treating chronic fatigue syndrome or fibromyalgia (82).

Prevention
Ecological conditions favorable to Lyme disease, the steady increase in the number of cases, and the challenge of prevention predict that the infection will be a continuing public health concern. Personal protection measures, including protective clothing, repellents or acaricides, tick checks, and landscape modifications in or near residential areas, may be helpful (99). However, these measures are difficult to perform regularly throughout the summer. Attempts to control the infection on a larger scale by the eradication of deer or widespread use of acaricides, which may be effective, have had limited public acceptance (99). New methods of tick control, including host-targeted acaricides against rodents and deer, are being developed and may provide help in the future.

In the 1990s, recombinant OspA vaccines were developed and shown to be safe and effective for the prevention of Lyme disease in the US (20, S72, S73). Although one of the vaccines was licensed commercially, its acceptance by the public and by physicians was also limited, and it was withdrawn by the manufacturer in 2002 (100). Some of the reasons why its acceptance was limited included the low risk of Lyme disease in most parts of the country, the need for booster injections every year or every other year, and the relatively high cost of this preventive approach compared with antibiotic treatment of early infection (S74, S75). In addition, there was a theoretical, though never proven, concern that in rare cases, vaccination might trigger autoimmune arthritis.

For now, control of Lyme disease depends primarily on public and physician education about personal protection measures, signs and symptoms of the disease, and appropriate antibiotic therapy (99). However, if the risk of the infection continues to increase or if public perceptions change, vaccine development may again become a priority. Experience gained in the last ten years has proven the feasibility of vaccination for the prevention of this complex, tick-transmitted infection.

Acknowledgments
This work was supported in part by a grant from the NIH (AR20358), a Centers for Disease Control and Prevention Cooperative Agreement (CCU110291), the English, Bonter, Mitchell Foundation, the Lyme/Arthritis Research Fund, and the Eshe Fund.

Due to space constraints, a number of important references could not be included in this article. Interested readers can find a supplementary reading list at http://www.jci.org/cgi/content/full/ 113/8/1093/DC1.

Address correspondence to: Allen C. Steere, Massachusetts General Hospital, 55 Fruit Street, CNY 149/8301, Boston, Massachusetts 02114, USA. Phone: (617) 726-1527; Fax: (617) 726-1544; E-mail: asteere@partners.org.

1. Steere, A.C. 2001. Lyme disease. *N. Engl. J. Med.* **345**:115–125.
2. Dennis, D.T., and Hayes, E.B. 2002. Epidemiology of Lyme Borreliosis. In *Lyme borreliosis: biology, epidemiology and control.* O. Kahl, J.S. Gray, R.S. Lane, and G. Stanek, editors. CABI Publishing. Oxford, United Kingdom. 251–280.
3. 2002. Lyme disease: United States, 2000. *Morb. Mortal. Wkly. Rep.* **51**:29–31.
4. Steere, A.C. 1989. Lyme disease. *N. Engl. J. Med.* **321**:586–596.
5. Spielman, A. 1994. The emergence of Lyme disease and human babesiosis in a changing environment. *Ann. N. Y. Acad. Sci.* **740**:146–156.
6. Fraser, C.M., et al. 1997. Genomic sequence of a Lyme disease spirochete, *Borrelia burgdorferi. Nature.* **390**:580–586.
7. Casjens, S., et al. 2000. A bacterial genome in flux: the twelve linear and nine circular extrachromosomal DNAs in an infectious isolate of the Lyme disease spirochete *Borrelia burgdorferi. Mol. Microbiol.* **35**:490–516.
8. Cabello, F.C., Sartakova, M.L., and Dobrikova, E.Y. 2001. Genetic manipulation of spirochetes. Light at the end of the tunnel. *Trends Microbiol.* **9**:245–248.
9. Zhang, J.-R., and Norris, S.J. 1998. Genetic variation of the *Borrelia burgdorferi* gene vlsE involves cassette-specific, segmental gene conversation. *Infect. Immun.* **66**:3698–3704.
10. Posey, J.E., and Gherardini, F.C. 2000. Lack of a

role for iron in the Lyme disease pathogen. *Science.* **288**:1651–1653.
11. Xu, G., Fang, Q.Q., Keirans, J.E., and Durden, L.A. 2003. Molecular phylogenetic analyses indicate that the *Ixodes ricinus* complex is a paraphyletic group. *J. Parasitol.* **89**:452–457.
12. LoGiudice, K., Ostfeld, R.S., Schmidt, K.A., and Keesing, F. 2003. The ecology of infectious disease: effects of host diversity and community composition on Lyme disease risk. *Proc. Natl. Acad. Sci. U. S. A.* **100**:567–571.
13. Schneider, B.S., Zeidner, N.S., Burkot, T.R., Maupin, G.O., and Piesman, J. 2000. *Borrelia* isolates in northern Colorado identifies *Borrelia bissettii. J. Clin. Microbiol.* **38**:3103–3105.
14. Kuo, M.M., Lane, R.S., and Giclas, P.C. 2000. A comparative study of mammalian and reptilian alternative pathway of complement-mediated killing of the Lyme disease spirochete *Borrelia burgdorferi. J. Parasitol.* **86**:1223–1228.
15. Gern, L., and Pierre-Francois, H. 2002. Ecology of *Borrelia burgdorferi* sensu lato in Europe. In *Lyme borreliosis: biology, epidemiology and control.* O. Kahl, J.S. Gray, R.S. Lane, and G. Stanek, editor. CABI Publishing. Oxford, United Kingdom. 149–174.
16. Korenberg, E.I., Gorelova, N.B., and Kovalevskii, Y.V. 2002. Ecology of *Borrelia burgdorferi* sensu lato in Russia. In *Lyme borreliosis: biology, epidemiology and control.* O. Kahl, J.S. Gray, R.S. Lane, and G. Stanek, editors. CABI Publishing. Oxford, United King-

dom. 175–200.
17. Kurtenbach, K., et al. 2002. Host association of *Borrelia burgdorferi* sensu lato: the key role of host complement. *Trends Microbiol.* **10**:74–79.
18. Guerra, M., et al. 2002. Predicting the risk of Lyme disease: habitat suitability for *Ixodes scapularis* in the north central United States. *Emerg. Infect. Dis.* **8**:289–297.
19. Glavanakov, S., et al. 2001. Lyme disease in New York State: spatial pattern at a regional scale. *Am. J. Trop. Med. Hyg.* **65**:538–545.
20. Steere, A.C., et al. 1998. Vaccination against Lyme disease with recombinant *Borrelia burgdorferi* outer-surface lipoprotein A with adjuvant. *N. Engl. J. Med.* **339**:209–215.
21. Schwan, T.G., and Piesman, J. 2000. Temporal changes in outer surface proteins A and C of the Lyme disease-associated spirochete, *Borrelia burgdorferi*, during the chain of infection in ticks and mice. *J. Clin. Microbiol.* **38**:382–388.
22. Pal, U., et al. 2004. OspC facilitates *Borrelia burgdorferi* invasion of *Ixodes scapularis* salivary glands. *J. Clin. Invest.* **113**:220–230. doi:10.1172/JCI200419894.
23. Grimm, D., et al. 2004. Outer-surface protein C of the Lyme disease spirochete: a protein induced in ticks for infection of mammals. *Proc. Natl. Acad. Sci. U. S. A.* **101**:3142–3147.
24. Coleman, J.L., et al. 1997. Plasminogen is required for efficient dissemination of *B. burgdorferi* in ticks

and for enhancement of spirochetemia in mice. *Cell.* **89**:1111–1119.

25. Hefty, P.S., et al. 2002. OspE-related, OspF-related, and Elp lipoproteins are immunogenic in baboons experimentally infected with *Borrelia burgdorferi* and in human Lyme disease patients. *J. Clin. Microbiol.* **40**:4256–4265.

26. Smith, R.P., et al. 2002. Clinical characteristics and treatment outcome of early Lyme disease in patients with microbiologically confirmed erythema migrans. *Ann. Intern. Med.* **136**:421–428.

27. Steere, A.C., and Sikand, V.K. 2003. The presenting manifestations of Lyme disease and the outcomes of treatment. *N. Engl. J. Med.* **348**:2472–2474.

28. Steere, A.C., et al. 1983. The early clinical manifestations of Lyme disease. *Ann. Intern. Med.* **99**:76–82.

29. Steere, A.C., et al. 2003. Systemic symptoms without erythema migrans as the presenting picture of early Lyme disease. *Am. J. Med.* **114**:58–62.

30. Strle, F., et al. 1999. Comparison of culture-confirmed erythema migrans caused by *Borrelia burgdorferi* sensu stricto in New York state and by *Borrelia afzelii* in Slovenia. *Ann. Intern. Med.* **130**:32–36.

31. Krause, P.J., et al. 2002. Disease-specific diagnosis of coinfecting tickborne zoonoses: babesiosis, human granulocytic ehrlichiosis, and Lyme disease. *Clin. Infect. Dis.* **34**:1184–1191.

32. Hunfeld, K.P., et al. 2002. Seroprevalence of Babesia infections in humans exposed to ticks in midwestern Germany. *J. Clin. Microbiol.* **40**:2431–2436.

33. Bjoersdorff, A., Wittesjo, B., Berglun, J., Massung, R.F., and Eliasson, I. 2002. Human granulocytic ehrlichiosis as a common cause of tick-associated fever in Southeast Sweden: report from a prospective clinical study. *Scand J Infect Dis.* **34**:187–191.

34. Steere, A.C., et al. 2003. Prospective study of coinfection in patients with erythema migrans. *Clin. Infect. Dis.* **36**:1078–1081.

35. Krause, P.J., et al. 1996. Concurrent Lyme disease and babesiosis: evidence for increased severity and duration of illness. *JAMA.* **275**:1657–1660.

36. Breitner-Ruddock, S., Wurzner, R., Schulze, J., and Brade, V. 1997. Heterogeneity in the complement-dependent bacteriolysis within the species of *Borrelia burgdorferi. Med. Microbiol. Immunol. (Berl.)* **185**:253–260.

37. Mullegger, R.R., et al. 2000. Differential expression of cytokine mRNA in skin specimens from patients with erythema migrans or acrodermatitis chronica atrophicans. *J. Invest. Dermatol.* **115**:1115–1123.

38. Wooten, R.M., et al. 2002. Toll-like receptor 2 is required for innate, but not acquired, host defense to *Borrelia burgdorferi. J. Immunol.* **168**:348–355.

39. Talkington, J., and Nickell, S.P. 2001. Role of Fc gamma receptors in triggering host cell activation and cytokine release by *Borrelia burgdorferi. Infect. Immun.* **69**:413–419.

40. Modolell, M., Schaible, U.E., Rittig, M., and Simon, M.M. 1994. Killing of *Borrelia burgdorferi* by macrophages is dependent on oxygen radicals and nitric oxide and can be enhanced by antibodies to outer surface proteins of the spirochete. *Immunol. Lett.* **40**:139–146.

41. Montgomery, R.R., Lusitani, D., de Boisfleury Chevance, A., and Malawista, S.E. 2002. Human phagocytic cells in the early innate immune response to *Borrelia burgdorferi. J. Infect. Dis.* **185**:1773–1779.

42. Salazar, J.C., et al. 2003. Coevolution of markers of innate and adaptive immunity in skin and peripheral blood of patients with erythema migrans. *J. Immunol.* **171**:2660–2670.

43. Glickstein, L., et al. 2003. Inflammatory cytokine production predominates in early Lyme disease in patients with erythema migrans. *Infect. Immun.* **71**:6051–6053.

44. Vaz, A., et al. 2001. Cellular and humoral immune responses to *Borrelia burgdorferi* antigens in patients with culture-positive early Lyme disease. *Infect.*

Immun. **69**:7437–7444.

45. Duray, P.H., and Steere, A.C. 1988. Clinical pathologic correlations of Lyme disease by stage. *Ann. N. Y. Acad. Sci.* **539**:65–79.

46. Probert, W.S., and Johnson, B.J.B. 1998. Identification of a 47kDa fibronectin-binding protein expressed by *Borrelia burgdorferi* isolate B31. *Mol. Microbiol.* **30**:1003–1015.

47. Seinost, G., et al. 1999. Four clones of *Borrelia burgdorferi* sensu stricto cause invasive infection in humans. *Infect. Immun.* **67**:3518–3524.

48. Coburn, J., Chege, W., Magoun, L., Bodary, S.C., and Leong, J.M. 1999. Characterization of a candidate *Borrelia burgdorferi* β$_3$-chain integrin ligand identified using a phage display library. *Mol. Microbiol.* **34**:926–940.

49. Parveen, N., and Leong, J.M. 2000. Identification of a candidate glycosaminoglycan-binding adhesin of the Lyme disease spirochete *Borrelia burgdorferi. Mol. Microbiol.* **35**:1220–1234.

50. Leong, J.M., et al. 1998. Different classes of proteoglycans contribute to the attachment of *Borrelia burgdorferi* to cultured endothelial and brain cells. *Infect. Immun.* **66**:994–999.

51. Guo, B.P., Brown, E.L., Dorward, D.W., Rosenberg, L.C., and Hook, M. 1998. Decorin-binding adhesins from *Borrelia burgdorferi. Mol. Microbiol.* **30**:711–723.

52. Purser, J.E., et al. 2003. A plasmid-encoded nicotinamidase (PncA) is essential for infectivity of *Borrelia burgdorferi* in a mammalian host. *Mol. Microbiol.* **48**:753–764.

53. Labandeira-Rey, M., and Skare, J.T. 2001. Decreased infectivity in *Borrelia burgdorferi* strain B31 is associated with loss of linear plasmid 25 or 28-1. *Infect. Immun.* **69**:446–455.

54. Hefty, P.S., Jolliff, S.E., Caimano, M.J., Wikel, S.K., and Akins, D.R. 2002. Changes in temporal and spatial patterns of outer surface lipoprotein expression generate population heterogeneity and antigenic diversity in the Lyme disease spirochete, *Borrelia burgdorferi. Infect. Immun.* **70**:3468–3478.

55. Liang, F.T., Nelson, F.K., and Fikrig, E. 2002. Molecular adaptation of *Borrelia burgdorferi* in the murine host. *J. Exp. Med.* **196**:275–280.

56. Liang, F.T., Jacobs, M.B., Bowers, L.C., and Philipp, M.T. 2002. An immune evasion mechanism for spirochetal persistence in Lyme borreliosis. *J. Exp. Med.* **195**:415–422.

57. Kraiczy, P., et al. 2004. Complement resistance of *Borrelia burgdorferi* correlates with the expression of BbCRASP-1, a novel linear plasmid-encoded surface protein that interacts with human factor H and FHL-1 and is unrelated to Erp proteins. *J. Biol. Chem.* **279**:2421–2429.

58. McKisic, M.D., and Barthold, S.W. 2000. T-cell-independent responses to *Borrelia burgdorferi* are critical for protective immunity and resolution of Lyme disease. *Infect. Immun.* **68**:5190–5197.

59. Rousselle, J.C., et al. 1998. Borreliacidal antibody production against outer surface protein C of *Borrelia burgdorferi. J. Infect. Dis.* **178**:733–741.

60. Fikrig, E., et al. 1997. *Borrelia burgdorferi* P35 and P37 proteins, expressed in vivo, elicit protective immunity. *Immunity.* **6**:531–537.

61. Hanson, M.S., et al. 1998. Active and passive immunity against *Borrelia burgdorferi* decorin binding protein A (DbpA) protects against infection. *Infect. Immun.* **66**:2143–2153.

62. Keane-Myers, A., and Nickell, S.P. 1995. T cell subset-dependent modulation of immunity to *Borrelia burgdorferi* in mice. *J. Immunol.* **154**:1770–1776.

63. Dressler, F., Whalen, J.A., Reinhardt, B.N., and Steere, A.C. 1993. Western blotting in the serodiagnosis of Lyme disease. *J. Infect. Dis.* **167**:392–400.

64. Akin, E., McHugh, G.L., Flavell, R.A., Fikrig, E., and Steere, A.C. 1999. The immunoglobin (IgG) antibody response to OspA and OspB correlates with severe and prolonged Lyme arthritis and the

IgG response to P35 correlates with mild and brief arthritis. *Infect. Immun.* **67**:173–181.

65. Steere, A.C., Schoen, R.T., and Taylor, E. 1987. The clinical evolution of Lyme arthritis. *Ann. Intern. Med.* **107**:725–731.

66. Brown, C.R., Blaho, V.A., and Loiacono, C.M. 2003. Susceptibility to experimental Lyme arthritis correlates with KC and monocyte chemoattractant protein-1 production in joints and requires neutrophil recruitment via CXCR2. *J. Immunol.* **171**:893–901.

67. Anguita, J., et al. 1998. *Borrelia burgdorferi*-infected, interleukin-6-deficient mice have decreased Th2 responses and increased Lyme arthritis. *J. Infect. Dis.* **178**:1512–1515.

68. Brown, J.P., Zachary, J.F., Teuscher, C., Weis, J.J., and Wooten, R.M. 1999. Dual role of interleukin-10 in murine Lyme disease: regulation of arthritis severity and host defense. *Infect. Immun.* **67**:5142–5150.

69. Brown, E.L., et al. 1995. Modulation of immunity to *Borrelia burgdorferi* by ultraviolet irradiation: differential effect on Th1 and Th2 immune responses. *Eur. J. Immunol.* **25**:3017–3022.

70. Oschmann, P., et al. 1998. Stages and syndromes of neuroborreliosis. *J. Neurol.* **245**:262–272.

71. Logigian, E.L., Kaplan, R.F., and Steere, A.C. 1990. Chronic neurologic manifestations of Lyme disease. *N. Engl. J. Med.* **323**:1438–1444.

72. Steere, A.C., et al. 1994. Treatment of Lyme arthritis. *Arthritis Rheum.* **37**:878–888.

73. Nocton, J.J., et al. 1994. Detection of *Borrelia burgdorferi* DNA by polymerase chain reaction in synovial fluid in Lyme arthritis. *N. Engl. J. Med.* **330**:229–234.

74. Steere, A.C., and Glickstein, L. 2004. Elucidation of Lyme arthritis. *Nat. Rev. Immunol.* **4**:143–152.

75. Girschick, H.J., Huppertz, H.I., Rüssmann, H., Krenn, V., and Karch, H. 1996. Intracellular persistence of *Borrelia burgdorferi* in human synovial cells. *Rheumatol. Int.* **16**:125–132.

76. Carlson, D., et al. 1999. Lack of *Borrelia burgdorferi* DNA in synovial samples in patients with antibiotic treatment-resistant Lyme arthritis. *Arthritis Rheum.* **42**:2705–2709.

77. Steere, A.C., et al. 2003. Binding of outer surface protein A and human lymphocyte function-associated antigen 1 peptides to HLA-DR molecules associated with antibiotic treatment-resistant Lyme arthritis. *Arthritis Rheum.* **48**:534–540.

78. Chen, J., et al. 1999. Association of antibiotic treatment-resistant Lyme arthritis with T cell responses to dominant epitopes of outer-surface protein A (OspA) of *Borrelia burgdorferi. Arthritis Rheum.* **42**:1813–1822.

79. Gross, D.M., et al. 1998. Identification of LFA-1 as a candidate autoantigen in treatment-resistant Lyme arthritis. *Science.* **281**:703–706.

80. Trollmo, C., Meyer, A.L., Steere, A.C., Hafler, D.A., and Huber, B.T. 2001. Molecular mimicry in Lyme arthritis demonstrated at the single cell level: LFA-1α(L) is a partial agonist for outer surface protein A-reactive T cells. *J. Immunol.* **166**:5286–5291.

81. Klempner, M.S., et al. 2001. Two controlled trials of antibiotic treatment in patients with persistent symptoms and a history of Lyme disease. *N. Engl. J. Med.* **345**:85–92.

82. Steere, A.C. 2002. A 58-year-old man with a diagnosis of chronic Lyme disease. *JAMA.* **288**:1002–1010.

83. Solomon, S.P., Hilton, E., Weinschel, B.S., Pollack, S., and Grolnick, E. 1998. Psychological factors in the prediction of Lyme disease course. *Arthritis Care Res.* **11**:419–426.

84. Sigal, L.H., and Hassett, A.L. 2002. Contributions of societal and geographical environments to "chronic Lyme disease": the psychopathogenesis and aporology of a new "medically unexplained symptoms" syndrome. *Environ. Health Perspect.* **110**(Suppl. 4):607–611.

85. Steere, A.C., Taylor, E., McHugh, G.L., and Logigian, E.L. 1993. The overdiagnosis of Lyme disease.

JAMA. **269**:1812–1816.

86. Reid, M.C., Schoen, R.T., Evans, J., Rosenberg, J.C., and Horwitz, R.I. 1998. The consequences of overdiagnosis and overtreatment of Lyme disease: an observational study. *Ann. Intern. Med.* **128**:354–362.

87. Qureshi, M.Z., New, D., Zulqarni, N.J., and Nachman, S. 2002. Overdiagnosis and overtreatment of Lyme disease in children. *Pediatr. Infect. Dis. J.* **21**:12–14.

88. Ettestad, P.J., et al. 1995. Biliary complications in the treatment of unsubstantiated Lyme disease. *J. Infect. Dis.* **171**:356–361.

89. Patel, R., Grogg, K.L., Edwards, W.D., Wright, A.J., and Schwenk, N.M. 2000. Death from inappropriate therapy for Lyme disease. *Clin. Infect. Dis.* **31**:1107–1109.

90. Velicer, C.M., et al. 2004. Antibiotic use in relation to the risk of breast cancer. *JAMA.* **291**:827–835.

91. Wormser, G.P., et al. 2000. Practice guidelines for the treatment of Lyme disease. The Infectious Diseases Society of America. *Clin. Infect. Dis.* **31**(Suppl. 1):S1–S14.

92. Nadelman, R.B., et al. 2001. Prophylaxis with single-dose doxycycline for the prevention of Lyme disease after an *Ixodes scapularis* tick bite. *N. Engl. J. Med.* **345**:79–84.

93. Wormser, G.P., et al. 2003. Duration of antibiotic therapy for early Lyme disease. A randomized, double-blind, placebo-controlled trial. *Ann. Intern. Med.* **138**:697–704.

94. Bennet, L., and Berglund, J. 2002. Reinfection with Lyme borreliosis: a retrospective follow-up study in southern Sweden. *Scand. J. Infect. Dis.* **34**:183–186.

95. Bacon, R.M., et al. 2003. Serodiagnosis of Lyme disease by kinetic enzyme-linked immunosorbent assay using recombinant VlsE1 or peptide antigens of *Borrelia burgdorferi* compared with 2-tiered testing using whole-cell lysates. *J. Infect. Dis.* **187**:1187–1199.

96. Kalish, R.A., et al. 2001. Persistence of immunoglobulin M or immunoglobulin G antibody responses to *Borrelia burgdorferi* 10-20 years after active Lyme disease. *Clin. Infect. Dis.* **33**:780–785.

97. Peltomaa, M., McHugh, G., and Steere, A.C. 2003. Persistence of antibody response to the VlsE sixth invariant region (IR6) peptide of *Borrelia burgdorferi* after successful antibiotic treatment of Lyme disease. *J. Infect. Dis.* **187**:1178–1186.

98. Steere, A.C., Sikand, V.J., Schoen, R.T., and Nowakowski, J. 2003. Asymptomatic infection with *Borrelia burgdorferi*. *Clin. Infect. Dis.* **37**:528–532.

99. Hayes, E.B., and Piesman, J. 2003. How can we prevent Lyme disease? *N. Engl. J. Med.* **348**:2424–2430.

100. Hanson, M.S., and Edelman, R. 2003. Progress and controversy surrounding vaccines against Lyme disease. *Expert Rev. Vaccines.* **2**:683–703.

The application of biofilm science to the study and control of chronic bacterial infections

William Costerton,[1] Richard Veeh,[1] Mark Shirtliff,[2] Mark Pasmore,[1] Christopher Post,[3] and Garth Ehrlich[3,4]

[1]Center for Biofilm Engineering, Montana State University, Bozeman, Montana, USA. [2]Department of Biomedical Sciences, Dental School, University of Maryland–Baltimore, Baltimore, Maryland, USA. [3]Center for Genomic Sciences, Allegheny Singer Research Institute, Pittsburgh, Pennsylvania, USA. [4]Department of Microbiology and Immunology, Drexel University College of Medicine, Pittsburgh, Pennsylvania, USA.

Unequivocal direct observations have established that the bacteria that cause device-related and other chronic infections grow in matrix-enclosed biofilms. The diagnostic and therapeutic strategies that have served us so well in the partial eradication of acute epidemic bacterial diseases have not yielded accurate data or favorable outcomes when applied to these biofilm diseases. We discuss the potential benefits of the application of the new methods and concepts developed by biofilm science and engineering to the clinical management of infectious diseases.

Introduction

Clinicians who deal with device-related and other chronic bacterial infections have gradually defined a new category of infectious disease that differs radically from the acute epidemic bacterial diseases that predominated until the middle of the last century (1). The happy fact that acute epidemic diseases are largely in the past is a consequence of progress in microbiology, and can be directly attributed to the vaccines and antibiotics that were developed as we came to understand planktonic bacteria. The free-floating bacteria that caused diphtheria and cholera were accurately modeled in test tube cultures, their pathogenic mechanisms (e.g., toxins) became epitopes for vaccines, and antibiotics that killed them were derived from nature or from directed synthesis. These specialized pathogens caused disease in perfectly healthy individuals, they ran their courses quickly, and they retreated to different animal populations or to natural reservoirs when the population under attack became immune.

As we began to gain control over epidemic diseases, another type of disease came to the fore. These diseases are much less aggressive than acute infections, they often persist for months or years, and they progress through periods of quiescence that alternate with periods of acute exacerbation. When the sophisticated tools of microbiology were brought to bear on these chronic diseases, many anomalies emerged. The pathogens were common environmental organisms, with which individuals had daily contact and against which they often had adequate immunity, and their pathogenic mechanisms were often diffuse and poorly defined. When they were grown in conventional lab cultures, these environmental organisms appeared to be sensitive to conventional antibiotics, but these same antibiotics failed to resolve the bacterial infections, although they gave some relief during acute exacerbations. Even more puzzling was the observation that, in many cases, it was not possible at all to recover any bacteria by traditional culture mechanisms. This led many

investigators to posit that these chronic disease states were sterile inflammatory conditions that persisted after the eradication of all microorganisms. However, the application of molecular diagnostics demonstrated unequivocally that bacteria were both present and metabolically active, even when no bacteria were recovered by plating. In some cases (e.g., otitis media, cholesteatoma, tonsillitis), affected individuals were cured by surgical treatment or by growth-related anatomical changes, while many other victims were relegated to intermittent antibiotic therapy for the remainder of their lives (cystic fibrosis [CF], prostatitis). When we looked at the infecting bacteria as they grew in affected tissues, we noted that they actually lived in matrix-enclosed communities that closely resembled the biofilms that are the predominant form of growth of bacteria in industrial and environmental ecosystems. The simple fact that the organisms that cause device-related and other chronic infections grow in biofilms (1) goes some distance toward explaining the perceived anomalies of these diseases, and offers a measure of hope that they can eventually be controlled.

Chronic bacterial disease: a clinical entity. When the model used to analyze a natural process is incorrect, our attempts to understand and manipulate the process will fail, many honest and conscientious people will be frustrated, and the reputations of whole research groups will be damaged. In the case of chronic bacterial diseases, diagnostic microbiology labs reported that cultures of *Pseudomonas aeruginosa* from CF patients were sensitive to antibiotics (e.g., cloxacillin), but pulmonary clinicians saw little improvement when these antibiotics were used. The sera of CF patients contained very large amounts of specific antibodies against *Pseudomonas*, but the disease persisted, and the use of anti-*Pseudomonas* vaccines resulted in the deaths of some patients. Middle ear specimens from children with chronic otitis media with effusion (COME) yielded negative bacterial cultures, so that a host-sustained inflammatory etiology was suspected, but the factors driving the inflammation could not be identified and serology did not confirm the persistent involvement of viruses. Patients with raging febrile prostate infections yielded expressed prostatic secretion (EPS) samples that produced cultures negative for bacteria, and material recovered from osteomyelitis debridations with frank pus yielded only a few colonies of skin and environmental organisms. All was shadows and fog, and the reputations of the

Nonstandard abbreviations used: cystic fibrosis (CF); chronic otitis media with effusion (COME); expressed prostatic secretion (EPS); acyl homoserine lactone (AHL); polymorphonuclear neutrophil (PMN); intrauterine contraceptive device (IUD).

Conflict of interest: The authors have declared that no conflict of interest exists.

Citation for this article: *J. Clin. Invest.* **112**:1466–1477 (2003). doi:10.1172/JCI200320365.

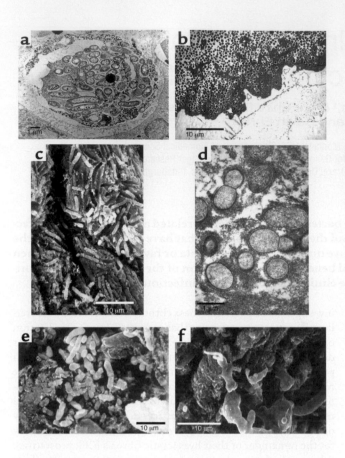

Figure 1

Electron micrographs of pathogenic bacterial biofilms from a variety of bacterial infections. (**a**) Transmission Electron Micrograph (TEM) of a section of lung tissue taken (postmortem) from a CF patient. The matrix-enclosed micro-colony of *P. aeruginosa* cells is surrounded by a prominent electron-dense "crust" of material that reacted very strongly with antibodies directed against IgG. Image published with permission from *Infection and Immunity* (59). (**b**) TEM of a section from the affected bone of a patient with very long-term osteomyelitis that had been treated with antibiotics (for four decades) and several debridations. Note the large number of Gram-positive cells (*S. aureus* was cultured) and the dehydrated remnants of the fibrous matrix of this massive biofilm. (**c**) Scanning electron micrograph (SEM) of a struvite crystal from the hilus of the kidney of a patient with acute pyleonephritis, who was affected by "staghorn" calculi. Cells of the infecting agent (*P. vulgaris*) have formed a biofilm whose matrix has become infused with struvite to produce a "petrified" biofilm. (**d**) TEM of a section from a vegetation formed on the endocardium of a rabbit in an animal model of native-valve endocarditis. Cells of the infecting agent (viridans group streptococci) are seen to have formed this macroscopic biofilm and to have produced very large amounts of fibrous matrix material. (**e**) SEM of tissue from a culture-negative prostatitis, showing the presence of rod-shaped bacterial cells. (**f**) SEM of tissue from the prostatitis patient in **e**, which had been reacted with the patient's serum, so that the matrix material of this well-developed biofilm was protected from dehydration, and is shown at its full extent.

microbiology units of many hospitals plummeted from the high levels they had attained earlier.

When our research group was unable to imagine why bacteria associated with chronic infections would grow as planktonic cells when biofilms predominated in virtually all industrial and environmental ecosystems, we undertook a series of simple morphological examinations. The bacteria in a very large number of device-related infections were seen to grow in biofilms (2–4), and these infections were recalcitrant to antibiotic therapy and insensitive to host defense mechanisms. The cells of *P. aeruginosa* in the sputum and in the lungs (postmortem) of CF patients were seen to grow in biofilms in which the cells were surrounded by very large expanses of matrix material (Figure 1a) (5). The cells of the pathogens that caused osteomyelitis in patients and in lab animals were seen to grow in enormous biofilms in which millions of bacterial cells (Figure 1b) were embedded in thick matrix material (6). Cells of *Proteus vulgaris* that caused chronic pyelonephritis, with complications involving the formation of "staghorn" calculi, were seen to grow in biofilms whose matrix had been "petrified" by crystalline struvite (Figure 1c). Vegetations in endocarditis caused by *Candida* species were composed of yeast and hyphal elements embedded in matrix material, and vegetations in experimental and clinical native-valve endocarditis were composed of streptococcal cells in equally extensive fibrous matrices (Figure 1d). The prostate of a patient whose EPS samples had yielded no bacteria was seen to be colonized by bacterial cells (Figure 1e) whose matrix material could be visualized especially well when it was reacted with the patient's serum to yield the classic quellung reaction (Figure 1f). In COME luxuriant biofilms were demonstrated to be growing directly on the

mucosal surface, and vital dye imaging showed that the embedded bacteria were all viable in spite of antibiotic treatment and loss of culturability. In some infections the biofilms were composed almost exclusively of bacterial cells and matrix material, while the microbial communities in other infections also contained fragmented platelets and host molecules (e.g., fibrin), so that the bacteria were widely dispersed. However, biofilms were found in all of the chronic infections examined in the 12 years during which this morphological series was pursued.

Morphological data are unequivocal in that it is highly unlikely that bacteria do not exist in a tissue in which we have found the distinctive features of prokaryotic cells, and these cells are integrated into their matrices and into the structure of the tissue itself. However, morphological data tell us little about the species of bacteria that are present in these chronic infections, or about their viability and their phenotypic pattern of gene expression at the time that the sample was obtained. These data are provided by modern techniques for the analysis of nucleic acids, and the Center for Genomic Sciences (Pittsburgh, Pennsylvania, USA) has established that living bacteria are present in otitis media and in otorrhea with effusion (7). Curtis Nickel's group has used nucleic acid analysis to establish that bacteria are present in EPS samples from prostatitis patients and from some individuals without overt symptoms (8). Similar analysis of sputum from CF patients has shown that cells of *P. aeruginosa* are present and that they express a phenotype different from that seen in planktonic cells in culture. Finally, Pradeep Singh and colleagues have analyzed the ratio between two types of acyl homoserine lactone (AHL) signals to show that the cells of *P. aeruginosa* in the CF lung do indeed grow in the biofilm phenotype (9). At some time in recent history

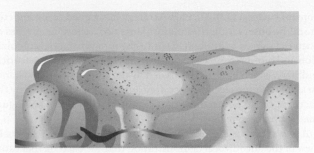

Figure 2
Diagrammatic representation of the tower- and mushroom-like micro-colonies that constitute the structural elements of biofilms. Sessile cells constitute only approximately 15% of the volume of their matrix-enclosed communities, so the micro-colonies are viscoelastic and deformable in high shear. Well-developed water channels conduct water in convective flow and deliver nutrients to most parts of the community. Figure is reproduced with permission from the American Society for Microbiology (60).

the conceptual balance tipped in favor of a biofilm etiology, at a different time in the case of each chronic disease, and the general notion that device-related and other chronic bacterial infections are caused by biofilms is now widely accepted (1).

Biofilm science as it applies to chronic infectious diseases. Biofilm science and biofilm engineering have been active fields of study since these sessile communities were first described and named in 1978 (10). Biofilms were first visualized as matrix-enclosed aggregates of bacteria, very similar to the planktonic cells of the species concerned, that were simply immobilized on surfaces or at interfaces in the ecosystems in which they were known to predominate. Later, direct examination of metabolic processes in natural biofilms showed that these sessile populations are much more active than their planktonic counterparts (11). Direct structural examination of biofilms showed that their component micro-colonies (Figure 2), which are composed of cells (±15% by volume) embedded in matrix material (±85% by volume), are bisected by ramifying water channels that carry bulk fluid into the community by convective flow (12). Direct measurement of several parameters by the use of microelectrodes established that biofilms are structurally and metabolically heterogeneous and that aerobic and anaerobic processes occur simultaneously in different parts of the multicellular community. The structural sophistication of biofilms suggested that these communities must be regulated by signals analogous to the hormones and pheromones that regulate multicellular eukaryotic communities, and the first of these regulatory signals were identified in 1998 (13). As we began to study the patterns of gene expression in biofilms, we found that these patterns produce a distinct biofilm phenotype, which differs very profoundly from that of the biofilms' planktonic counterparts (14).

Biofilms as self-assembling multicellular communities. When planktonic bacteria encounter a surface or an interface, they adhere to that interface in a reversible fashion while they "explore" the locale to ascertain whether it offers nutrient or other advantages. If the "decision" favors permanent settlement, the adherent cells upregulate the genes (15, 16) involved in matrix production (within as few as 12 minutes), and the process of biofilm formation begins. Kolter et al. (17) and others have described

biofilm formation as a developmental sequence that varies to some extent between species but generally results in the formation of a mature community of tower- and mushroom-shaped micro-colonies (Figure 2). Both biofilm formation and biofilm detachment are under the control of chemical signals of the same type that regulates quorum sensing (18, 19), and these regulatory molecules guide the formation of slime-enclosed micro-colonies and water channels. The cells are remarkably evenly distributed in the biofilm matrix, suggesting that some matrix component may dictate their precise location (20); Stoodley et al. have shown that the community as a whole has material properties similar to those of a viscous fluid. The micro-colonies are deformable, they oscillate in high-shear systems, and they break and detach as biofilm fragments if their tensile strength is exceeded by the shear forces (21). Biofilms also show "creeping" activity, in high shear, and whole biofilms can be seen moving across surfaces, with the development of transitory waves and areas of enhanced detachment. These mechanical properties of biofilms are documented in video sequences that are available on the Center for Biofilm Engineering website (http://www.erc.montana.edu). Biofilms form in many high-shear environments, like the vegetations that form on native heart valves, so these material properties are germane and can be used to predict when and where fragments will detach and where they will end up in a flowing system.

Biofilms as protected sessile communities. Stoodley et al. (20) have made the point that microbial biofilms constitute the most "defensive" life strategy that can be adopted by prokaryotic cells. In very hostile environments, in which many locations are too hot or too acid or too dry, the stationary mode of growth is inherently defensive, because bacterial cells are not swept into areas where they will be killed. These stationary sessile communities, which seem to have predominated in primitive earth, were attacked by bacteriophages and by amoebae, and their collective growth in matrix-enclosed micro-colonies gave complete protection from these predators. One of the most important protections afforded by the biofilm mode of growth is protection from drying and from ultraviolet light. These sessile communities are especially favored, and very visible, in the intermittently wet splash zones of freshwater and marine ecosystems, where these communities allow prokaryotic cells to survive daylong exposures to drying and intense sunlight and to revive when they are rehydrated. Experiments conducted in military defense establishments have shown that planktonic cells of lab-adapted strains of bacteria survive for very short lengths of time when they are released from high-flying airplanes, whereas cells in biofilm fragments survive long enough to reach the ground. In their 2002 review, Stoodley et al. (20) made the intriguing suggestion that bacteria may have developed their biofilm phenotype early in the evolutionary process, when survival in a hostile environment was a *sine qua non*. They suggest that the planktonic phenotype, with its genetically "expensive" and very sophisticated chemotactic and motility mechanisms, may have developed later, for the purposes of dissemination and the colonization of new habitats.

The biofilm as a distinct, signal-controlled phenotype. The inherent resistance of biofilm bacteria to antibiotics (22) and to virtually all antibacterial agents was an early indication that sessile cells differ radically from their planktonic counterparts. When it was established that this resistance was not the result of diffusion limitation (23) but was predicated on some change in cellular characteristics (24), the research community missed the central point and wandered clueless for more than a decade. Then various groups began to use the tools of modern molecular biology, like mRNA analysis by gene arrays (25–27) and proteomic analysis of gene products (28–30), and they showed that cells in biofilms express a radically different set of genes. This virtual "bombshell" of the concept of a distinct biofilm phenotype has changed most of the intellectual constructs in the biofilm field. The genes expressed in a biofilm differ from those expressed in the corresponding planktonic cell more than they do from the genes expressed in a spore or in the biofilm cells of another species (28). There is no single biofilm phenotype, but gene expression in sessile communities goes through a whole spectrum of changes as the community matures (20), and the planktonic phenotype begins to emerge as the biofilm begins to shed mobile cells. Now that we know that bacteria adopt a radically different phenotype when they adhere to a surface or interface and initiate biofilm formation, we understand such mysteries as the resistance of biofilms to antibiotics in terms of the expression of different sets of genes. Many other characteristics of sessile bacteria can now be explained in the same way, and matrix-assisted laser desorption-ionization time-of-flight (MALDITOF) and gene chip analyses of the genes expressed in biofilms are already painting a picture of considerable variety and specialization in these multicellular communities. In the microbiological community, the notions are taking shape that biofilms predominate in most ecosystems and that these sessile cells express genes we have never seen expressed and reach levels of interaction and community behavior that we have never imagined. Now that the phenotypes of planktonic and biofilm cells have been shown to be so profoundly different, pure cultures of planktonic bacteria seem to be a very poor model for the study of these organisms, which actually grow in mixed-species sessile communities in natural and medical systems (31).

The simple observations that microbial biofilms are composed of structured micro-colonies and that water channels are created and maintained suggested that some type of cell-cell signal system must be operative in these multicellular communities. We speculated that these signals would resemble the hormones that dictate the structure of more complex eukaryotic organisms and communities, and we visualized a type of "embryology" of biofilms. We demonstrated that the quorum-sensing molecules that had been discovered in planktonic bacteria (18) also control biofilm development (13) and many other bacterial behaviors, and we now speculate that these signal systems may have evolved to control biofilms, especially in the light organ of the squid. The AHL signals of Gram-negative bacteria and the two-component peptide signals of Gram-positive bacteria (32) have been shown to affect biofilm formation, and the newly discovered autoinducer II signal (33) appears to have a similar effect. Analysis of signal-negative mutants suggests that all aspects of bacterial behavior are regulated by an interactive network of signals, with large inputs from environmental factors, so that biofilms can now be seen as integrated communities that are very reactive to their environments. This demonstration of cross-talk among bacteria of the same and of different species presents us with the Rosetta stone of the bacterial language that controls biofilm formation, species interactions, growth rates, toxin production, invasive properties, and many other behaviors of great interest to humankind. We estimate that dozens of new signal systems will be discovered in the next decade, and that control measures based on signal manipulation will largely replace bactericidal agents in human attempts to control bacteria (34). This new strategy has gained some impetus from the observation that many plants use natural analogues of bacterial signals to inhibit (35) or to encourage bacterial colonization and biofilm formation on their surfaces. The pharmaceutical industry, which has been bedeviled by the emergence of resistance to all of their most carefully crafted antibiotics, is intrigued by the fact that resistance to these natural antibiofilm signal analogues has not yet been recorded.

A re-examination of chronic bacterial diseases using biofilm concepts. In view of the unequivocal demonstration that the bacteria that cause device-related and other chronic bacterial infections grow in biofilms, it seems useful to survey this large number of infections for new insights in terms of this new microbiological concept.

What new biofilm-specific methods tell us about biofilm bacteria. One of the most time-honored of microbiological conventions is the detection of bacteria by recovery methods, which usually involve mechanical removal of the organisms (often by swabbing) and their propagation in liquid or on solid media. We recently conducted a survey (36) of the bacterial colonization of the vaginas of 3,000 human subjects using standard swabbing and culture (plating) techniques, and concluded (as most similar studies have) that 10.8% of individuals carried *Staphylococcus aureus*. We then examined subsets of these individuals, using PCR to identify *S. aureus* by its DNA and using FISH probes to identify cells of this species by their 16S RNA content, and we found that 100% of these individuals carried this organism. We then examined whether individuals who yielded positive data in swab and plate tests carried more *S. aureus* as detected by PCR and FISH, and found there was no correlation (36). If certain bacteria are present on a tissue or an inert surface, the swab may or may not pick them up, they may be present in huge aggregates of hundreds of cells that will yield only one colony on plating, and they may be expressing the set of genes that constitute the biofilm phenotype and be unable to grow in the culture conditions provided. Swabbing and plating are techniques that date from the middle 1800s, but they are still used to screen key medical personnel for nasal carriage of *S. aureus* as well as to detect bacteria in suspected infections and to identify the pathogens concerned. It may be time for us to adopt more modern and accurate methods.

The biofilm concept of chronic infections offers at least a partial explanation for a phenomenon that has troubled clinicians for some time, that blood cultures from patients who show many signs of overt bacterial infection are often negative. If we remember that the infecting organisms exist both as biofilms and as individual planktonic cells that trigger most of the overt symptoms of infection, we must expect the planktonic cells to be killed by circulating antibiotics and activated phagocytes. Because the planktonic cells are killed and the biofilm cells are not released, except as multicellular matrix-enclosed biofilm fragments, no colonies develop on the plates, and negative reports are returned. The same problem occurs in middle ear (37) and prostate (8) infections, in that the reservoirs of biofilms in the tissues shed few fragments, and the planktonic cells that are detached from these sessile communities are killed by antibiotics and phagocytes before they can be recovered and grown. Until these biofilm bacteria were detected by molecular methods such as PCR and FISH probing, their bacterial etiology was in question and bizarre viral and/or immunological etiologies were being proposed.

We have found bacterial biofilms in a series of orthopedic implant patients who had been diagnosed as having "aseptic loosening"

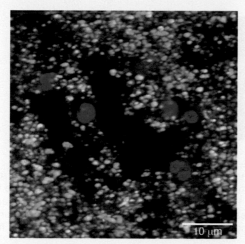

Figure 3
Confocal scanning-laser micrograph showing the invasion of a biofilm by PMNs. The PMNs (large red nuclei) have entered the biofilm via the open water channels and have invaded short distances (1–5 μm) into the biofilm. The bacterial cells have been stained with the live/dead BacLight stain (BacLight Bacterial Viability kit; Molecular Probes, Eugene, Oregon, USA) and living bacterial cells (green) are seen in very close proximity (<1 μm) to PMNs. We conclude that PMNs invade biofilms but are virtually inactive in killing sessile cells and resolving biofilm infections. Reproduced with permission from *Infection and Immunity* (40).

of hip prostheses (G. Maale, personal communication), because no bacteria had ever been cultured from blood, from tissues, or from the prostheses themselves. What had previously been seen as a mechanical problem became an infection problem, and replacement of the prostheses has been accomplished very successfully with the use of aggressive perioperative antibiotic therapy. When early detection of biofilm infections is particularly important, as in the case of vascular grafts that may fail catastrophically if they develop biofilm infections, new ELISA tests to detect antibodies against biofilm-specific epitopes common to all staphylococcal species are proving to be particularly useful. Incipient biofilms stimulate antibodies against biofilm within 10 days of the initial colonization of these vascular grafts, but staphylococcal cells are virtually never recovered by conventional culture techniques.

When bacteria are isolated from patients with biofilm diseases, they are plated to produce single colonies, and material from these single colonies is used to inoculate liquid media for their subsequent propagation. When cells of *P. aeruginosa* are first cultured in liquid media, they form macroscopic biofilms on the surfaces of the culture vessel, especially at the air-water interface. When planktonic cells are withdrawn from the culture to inoculate subsequent liquid media, the biofilms are left behind, and the next culture shows less "scum" formation on surfaces. After cultures have been passaged several times in liquid media, we have all seen that the cells of *P. aeruginosa* grow as an evenly turbid suspension in the liquid medium, and thus we have transformed a biofilm-forming pathogen into a planktonic lab-adapted strain. These lab-adapted cultures are really not good models for the study of diseases that have been shown unequivocally to be caused by bacteria growing in biofilms (5).

One of the most spectacular cases in which modern methods have been used to resolve long-standing anomalies is in the area

of the resistance of biofilms to host defenses. Ward et al. (38) showed that preformed biofilms could be introduced into the peritonea of rabbits that had both bactericidal and opsonizing antibodies as well as normal phagocytic cells, and that the sessile communities persisted for weeks and even months. At first, this anomaly was explained in terms of exclusion of both antibodies and phagocytes by the biofilm matrix (39), but more recently Leid et al. (40) have shown that activated polymorphonuclear neutrophils (PMNs) are attracted to biofilms and may penetrate these sessile communities. Figure 3 shows PMNs that have entered into the water channels of a biofilm (see movie at http://www.erc.montana.edu), and have penetrated 5–10 μm into individual micro-colonies. However, although the membranes of these phagocytes remain intact, they seem to be "paralyzed" in that they have not internalized any bacteria and that bacteria that are only 3–5 μm from these normally very aggressive leucocytes are alive (Figure 3, green) in this viability stain. This resistance to normal host defense factors may have been acquired when bacteria formed biofilms in the primitive earth (20) for defense against bacteriophages and against free-living amoebae.

The potential effect of the biofilm concept on the management of chronic infections. If we (the scientific microbiology community and the clinical infectious disease community) visualize the bacteria that cause device-related and other chronic infections as planktonic cells swimming or floating at or near the surfaces of the affected biomaterials and tissues, we will persist in the control strategies that have failed us so often in the past. If we grasp the concept that these bacteria are embedded in matrix material, that they have adopted a distinct biofilm phenotype, and that they have formed interactive communities, then we can bring all of the power of biofilm science to bear on chronic infections. If we acknowledge that the surface layers of the skin are colonized by living biofilms of *S. epidermidis*, even after the bacteria and fungi on the skin surface have been killed by surgical preparations, we will not allow devices to touch the cut edges of the skin when they are installed. If we realize that all vascular catheters are heavily colonized by biofilms if they have been in place for more than 1 week, we will not replace them over wire catheter replacement guides (J wires), because the wire will displace biofilm fragments and the new catheter will be inoculated as it slides over the wire. When we visualize macroscopic bacterial biofilms (vegetations) on natural or mechanical heart valves, we will realize that these large micro-colonies will form living bacterial emboli in the nearest capillary beds if they are detached by surgical interventions. We will treat acute exacerbations of device-related infections with antibiotics, and we will expect to see the resolution of many overt symptoms, but we will not expect to kill all the sessile cells in the biofilm on the device. If we realize that we are confronted by classic biofilm infections, we will probably remove devices (with their adherent biofilms) sooner rather than later, and we will use aggressive antibiotic therapy to prevent the recolonization of the replacement devices.

When we acknowledge that all device-related and other chronic bacterial infections are caused by bacteria living in biofilms, we will use biofilm-specific methods in diagnosis and research-based therapy. Because biofilm science has discredited "swab and plate" census methods, we will use direct microscopic methods to map and speciate the natural bacterial populations of human tissues, including the middle ears of children and the prostate glands of older men. These direct methods usually begin with the simple

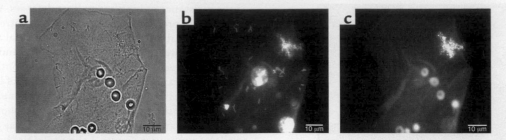

Figure 4

Fluorescence micrographs of epithelial cells recovered from the vaginas of human volunteers. (**a**) A single epithelial cell with highly refractile blood cells (the volunteer was menstruating) and colonized by rod-shaped and coccoid bacteria. (**b**) A similar cell reacted with the fluorescent eubacterial domain (EU 338) probe, whose base sequence reacts with the 16S RNA of all eubacteria, including both lactobacilli and staphylococci. (**c**) The same cell as in **a** reacted with a fluorescent probe that reacts only with the 16S RNA of *S. aureus*. This organism was found in the vaginal flora of all volunteers tested, and this result was confirmed by PCR. Figures reproduced with permission from the *Journal of Infectious Diseases* (36).

visualization of the tissue surface (Figure 4a) in situ or ex situ; the adherent bacteria can usually be resolved by simple phase-contrast microscopy. Then all of the colonizing bacteria can be stained with fluorescein or with the eubacterial domain (known as EUB 338) 16S RNA probe (Figure 4b), and living bacteria can be differentiated from dead cells by the use of the BacLight live/dead stain. After this overall census, we can identify and locate the cells of species of interest (Figure 4c) using specific 16S RNA probes in a FISH reaction that causes only the cells of the species for which the probes were designed to fluoresce.

If we know what natural and pathogenic bacterial populations are present on human tissues in many organ systems and we can determine the species identity and viability of these colonizing organisms, we can place diagnosis and treatment on a solid logical base. If we can see cells of *P. aeruginosa* growing in biofilms in the lungs of CF patients and we can detect very high titers of antibodies against *Pseudomonas* in the sera of these patients, we will be unlikely to continue with unsuitable immunization experiments (41). If we can see bacteria growing in biofilms in the middle ears of all children while we note symptoms of infection in only a subset that does not differ in the nature or extent of colonization, we will probably examine the host response to this colonization. If we examine the host response to this and other microbial colonization of tissues, we may be led to manipulate the Th1 and Th2 immune responses, instead of trying (unsuccessfully) to use antibiotics to kill all of the biofilm bacteria on a naturally colonized tissue surface.

When we examine the proteome of cells in biofilms (14) or analyze the biofilm phenotype by using DNA arrays (26), we find that sessile cells express many genes that are never expressed in planktonic cells, and vice versa. Now that we know that the inherent antibiotic resistance of biofilms does not result from diffusion limitation, and that all presently available antibiotics were selected for their ability to kill planktonic cells, we are looking for new biofilm-specific antibiotic targets. Ideally, these new agents will kill planktonic and sessile cells with equal efficacy, and their effects may be enhanced by the specific blockage of the *fmtC* gene, which mediates antibiotic resistance in biofilm cells of all species of *Staphylococcus*. Because biofilm science has discovered that many aspects of biofilm behavior are controlled by cell-cell signals, specific signal analogues have been developed to shut off

toxin production and to block biofilm formation and "lock" bacteria in the planktonic phenotype. One of these biofilm-control signal analogues (the RNAIII-inhibiting peptide analogue of the RNAIII-activating peptide signal) has been shown to block biofilm formation by all species of *Staphylococcus* and thus facilitate the killing of cells of these species even in the presence of a biomaterial (34). Biofilm engineering has also contributed new technologies of potential interest in the control of biofilm infections, in that biofilms have been shown to be much more susceptible to conventional antibiotics in direct current electric fields (42) or when treated with ultrasonic radiation (43). As biofilm science and engineering continue their exponential growth and biofilm problems in industry are solved, these concepts and technologies will be made available to medicine as long as we hold fast to the biofilm concept of chronic infection.

The role of biofilm fragments in the initiation of infections. Because the majority of bacteria in virtually all natural ecosystems grow in biofilms (44), microbial challenge to humans often comes from this source. We know that natural aquatic biofilms recruit and retain pathogens, such as *Salmonella* and *Escherichia coli*, and that immunocompetent humans can be infected from these sources. Immunocompromised individuals may be especially susceptible to the nontubercular *Mycobacterium avium* complex organisms that are a major component of many natural aquatic biofilms. Horizontal gene transfer is facilitated in biofilms (45), and G. Ehrlich and colleagues have recently suggested (46, 47) that these sessile communities may play a major role in the pathogenicity of bacterial species (e.g., *Vibrio cholerae*) that alternate between human hosts and natural reservoirs. This group has offered evidence that species of bacteria are composed of multiple strains, each of which contains a unique distribution of contingency genes from a population-based supragenome that is much larger than the genome of any single pathogen. During periods of stress, bacteria upregulate autocompetence and autotransformation systems within biofilms to promote the reassortment of genes that will result in the creation of some strains that may have a selective advantage under prevailing environmental conditions. They further suggest that natural marine and freshwater biofilms constitute a staging area in which the small DNA sequences can be cobbled together to recreate the pathogenicity "islands" necessary for the organisms' attack on human populations. It is clear that all examinations of communi-

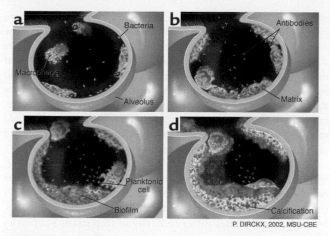

P. DIRCKX, 2002, MSU-CBE

Figure 5
Diagrammatic representation of the defense strategies of the lung. **(a)** The surface of the alveolar epithelium is "patrolled" by PMNs and macrophages, which phagocytose incoming planktonic bacteria quickly and easily. **(b)** The alveolar phagocytes are unable to engulf bacteria in matrix-enclosed biofilm fragments, even when these invaders are reacted with specific antibodies. **(c)** Biofilm fragments grow and burgeon in the colonized lung, and release occasional planktonic cells that react with antibodies and are phagocytosed. **(d)** The mature biofilm reaches a "standoff" with the immune system, and parts of the microbial community become calcified to form a long-term pulmonary nidus.

cable enteric diseases must use biofilm models and that investigators of these systems should subscribe to the biofilm concept.

Although the digestive system and the integument may be colonized by bacteria and fungi from environmental biofilms, the human organ system that is by far the most susceptible to attack from environmental biofilms is the pulmonary system. The trachea and the lungs are well designed to resist colonization by planktonic bacteria, and animal experiments have shown the clearance of as many as 1×10^6 bacterial cells in as little as 20 minutes, provided the challenging cells are single and unaggregated (48). Experiments in the same animal species using the same bacterial species (5) have shown that the lungs of normal animals are not able to clear bacteria that are introduced in the form of biofilm fragments or of cells enclosed in artificial matrices (e.g., as agar beads). When biofilm fragments or agar beads containing bacteria are introduced into the lung, these aggregates resist phagocytosis by resident phagocytes and killing by both innate and acquired immune factors, and they persist for weeks and even months (48). Figure 1a illustrates the mode of growth of the micro-colonies that comprise the *Pseudomonas* biofilm in the lung in animal models of CF, and Figure 5 shows how this matrix-enclosed population persists despite the immune reactions of the host. Woods and his colleagues proposed that the lungs may be colonized by the detachment of biofilm fragments from the oropharynx, which becomes overgrown by *P. aeruginosa* during periods of stress (49), and that these fragments cannot be cleared by pulmonary defense mechanisms. These data from animal experiments appear to have been confirmed by clinical examination of CF patients, and they raise the very grim specter of the inevitable colonization of the lungs with biofilm fragments when endotracheal tubes become colonized by mixed-species biofilms. Examination of endotracheal tubes that have been used for assist-

ed ventilation has shown massive aggregations of mixed-species biofilms (Figure 6); simple detachment of fragments could lead to chronic colonization of the lungs. In studies of intubated patients in intensive care units, we noted that the biofilms on endotracheal tubes often contained bacteria from the digestive tract when nasogastric tubes were also in use, and that these organisms were often found in the lungs of patients who had died after assisted ventilation. It is well documented that biofilm fragments are aspirated into the lung, and that these matrix-enclosed organisms cannot be cleared by pulmonary defense mechanisms and develop into micro-colonies that persist for months and may give rise to disseminated infection.

If the pulmonary system is in fact susceptible to colonization by aspirated biofilm fragments that cannot be cleared and may act as foci for chronic and/or acute bacterial infection, then other environmental organisms should be able to invade mammalian lungs. For this reason, it is germane to examine a trio of environmental bacteria that are "card-carrying" members of natural ecosystems and that share with *P. aeruginosa* the invidious ability to colonize mammalian lungs and cause serious diseases in immunocompetent mammals. *Pasteurella haemolytica* is a component of the normal oropharyngeal flora of cattle that proliferates when the animals are shipped to feedlot operations, and the aspiration of fragments of these biofilms into the lungs of these stressed animals results in respiratory infections that kill as many as 2% of these calves (50, 51). *Legionella pneumophila* are the predominant inhabitants of warm littoral zones in freshwater lakes, where they grow in association with green algae (*Fisheria* sp.) and avoid predation by free-living amoebae by forming biofilms and secreting antiphagocytic factors. As humans devised air conditioners and elaborate domestic hot water systems, *L. pneumophila* found an alternate home in hotels and hospitals, and emerged to kill several

Figure 6
SEM of a large multispecies biofilm aggregate that formed on the lumenal surface of an endotracheal tube used to ventilate a patient in a Systems Failure Intensive Care Unit. These uvula-shaped aggregates have a rubbery consistency, and they routinely break off of these surfaces and are aspirated into patients' lungs.

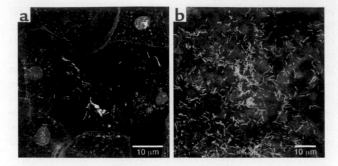

Figure 7
Confocal scanning-laser micrographs of tissue recovered from the vaginal epithelium of volunteers and stained with fluorescein. (**a**) The orange nuclei and green cytoplasm of the human cells are clearly seen, as are the bacterial cells in a biofilm aggregate partly detached from human cell surfaces. (**b**) The green rod-shaped cells of the vaginal biofilms are seen, with the lighter green matrix material that surrounds them, in aggregates at least 30 μm "tall." The extent and thickness of the vaginal biofilm are indicated by the fact that the orange tissue nuclei appear to be "buried" by this beneficial microbial population.

dozen elderly gentlemen in a notorious hotel in Philadelphia, as well as many other people in various hospitals and office buildings. An organism that lives in warm water, forms biofilms, and resists phagocytosis by amoeboid enemies survives very well in the condensate trays of air conditioners or in human lungs (52). The remarkable thing is not the pathogenicity of an environmental organism with no previous history of attacking humans, but the chilling realization that thousands of people must have died of legionellosis before the disease was defined by the Centers for Disease Control. If an individual dies of idiopathic pneumonia, this death causes very little interest, but the cause of this death may have been the mobilization of an acute respiratory infection from niduses of biofilm infection that were acquired by the aspiration of biofilm fragments. The sources of these fragments may be environmental, as in the case of cooling towers that cause seroconversion to *L. pneumophila* in people working downwind of the towers, or they may be very focal, as in the case of dental hygienists who breath aerosols for long periods of time every working day.

Perhaps the best example of a pulmonary disease that is caused by biofilms is meloidosis, which affects people in Southeast Asia and Australia, and caused pulmonary infections in US soldiers in Vietnam. The causative pathogen is *Burkholderia pseudomallei*, which is a natural component of the freshwater ecosystems in the area; humans make contact with biofilms formed by these organisms when they work in rice paddies or swim in local rivers (53). Rates of seroconversion indicate that more than 80% of the people in northeast Thailand have been exposed to this pathogen, presumably by aspiration of fragments of its exuberant biofilms, and we have visualized a population with multiple micro-colonies in their lungs (53). When we set up animal experiments by introducing agar beads containing these organisms into the lower left lobes of the lungs of guinea pigs, we induced an asymptomatic chronic infection that persisted for months (54). However, when we stressed these chronically infected animals with steroid injections (54), planktonic bacteria were released from the biofilm microcolonies, and the animals rapidly succumbed to the resultant acute infections and bacteremias. Several hundreds of people die

of acute meloidosis in northeast Thailand when they are stressed by seasonal starvation cycles, so the response to stress seen in the animal model appears to operate in human populations. These data add to the Damoclean image of the pulmonary health of modern humans, because bacteria from biofilm niduses acquired by the aspiration of biofilm fragments from many sources may be mobilized in times of stress and may cause acute pneumonias. So, many among us may carry the seeds of fatal pneumonia (the "old man's friend") in their lungs as they enjoy their air-conditioned offices and their homes with spas and hot tubs.

Biofilms may also be involved in the dissemination of disease within the body of an infected individual. When we speak of the hematogenous spread of infection, we must now specify whether the infectious units are planktonic cells or biofilm fragments, because these entities differ radically in important properties such as their antibiotic resistance and their adhesion characteristics. Planktonic cells are shed from virtually all mature biofilms, they are generally susceptible to antibiotics, and they adhere to certain tissues and to inert surfaces with considerable avidity. For this reason it is logical to use prophylactic antibiotic therapy to prevent the colonization of recently installed medical devices by planktonic bacteria introduced into the bloodstream by routine tooth brushing or by dental manipulations. On the other hand, many of the cells that detach from biofilms growing on native heart valves (resulting in endocarditis) or vascular catheters are in the form of matrix-enclosed biofilm fragments (21) that are very resistant to antibiotics, and they usually circulate until they "jam" in a capillary bed. For this reason, low-dose antibiotic therapy does not prevent the dissemination of bacteria in these biofilm diseases, and the best way of preventing this process is very aggressive high-dose treatment of native valve endocarditis and rapid removal of colonized vascular devices. We have developed animal models of dissemination in biofilm diseases, and we have been amazed at how well sheep lungs can tolerate the small hemorrhages that result from hundreds of biofilm fragments lodging in capillary beds. However, it is clear that one or two large biofilm fragments can cause profound damage if they detach and find their way to critical loci in the lungs or in the brain.

The role of commensal biofilms in protection from infection. One of the conceptual areas in which microbial ecology can assist medical practitioners most effectively and immediately is the area of natural microbial biofilms on the surfaces of normal human tissues. These tissues are among the most attractive surfaces in nature because they are homeostatic and rich in simple nutrients, and bacteria have adapted to virtually all types of mammalian tissues with considerable success. Perhaps the least welcoming tissue is dry skin, but cells of *S. epidermidis* grow in prolific biofilms between the squamous cells of the outer three to ten layers of this stratified epithelium and colonize hair follicles and sebaceous glands very successfully. Secretory epithelia like those of the eye, the vagina, and the mouth attract many species of bacteria. The eye limits colonization by producing antibacterial factors (surfactants and defensins); the vagina develops an exclusive acid-dominated environment favoring *Lactobacilli* (Figure 4, a–c, and Figure 7); and the mouth is colonized by a remarkable series of multi-species biofilms on different surfaces. Mucus-secreting tissues, like those of the trachea and the intestine, are covered by "mucus blankets," and this viscous material is often moved across the surface of the tissues by such forces as ciliary beating or by peristalsis. Bacteria have great difficulty in accessing mucus-covered tissues,

Figure 8

SEM of the lumenal surface of the intestine of a mouse. The dehydration-condensed residue of the intestinal biofilm is seen to occupy much of the tissue surface, and to be composed of a rich mixture of bacterial and protozoan species. A large *Giardia* cell is seen to be attached to the surface (P) of the intestinal epithelium, while a detached cell of the same species shows its well-developed sucker structure, and the microvillar surface of opposite side shows the scars of previous protozoan attachment. Attachment plays a large role in the microbial ecology of the intestine, because the intestinal mucus exerts powerful shear forces that tend to remove loosely attached organisms.

especially if the mucus blanket is 200–250 μm in thickness and is moving at a considerable speed across the tissue surface. For this reason, commensal organisms often use "hold fast" mechanisms and flagella that operate well in viscous fluids just to stay in the tissue surface ecosystem, and successful pathogens often find ways to disrupt the mucus blanket and gain access to the tissue surface. The outer surfaces of "mucus blankets" constitute a rich, if somewhat ephemeral, bacterial habitat, and many species proliferate on these surfaces (Figure 8).

Many organ systems in the human body must, in order to fulfill their normal functions, make contact with the bacteria-laden environment at their distal extremes, while maintaining strict asepsis in their proximal "core" organs. A simple example is the biliary system, which must deliver bile to the intestine while maintaining a bacteria-free situation in the liver, which is extraordinarily sensitive to bacterial infection (55). In extensive microbiological studies of the biliary system in the cat, we discovered that the common bile duct is not colonized by gut bacteria, but that these organisms enter the duct if the valve-like sphincter of Oddi is compromised or if a biliary stent is inserted (55). Planktonic bacteria make periodic excursions from the gut to the gall bladder, but in the absence of an inert surface on which to form a biofilm, they cannot colonize the tissues in the presence of bile salts, and they are washed back to the intestine. We speculated that most cholangitis results from vascular challenge, not from traffic in the bile duct, and we showed that the introduction of as few as 1×10^5 cells into the posterior vena cava was sufficient to overwhelm the defenses of the liver and cause acute hepatic disease (56). A more complex example of an organ system that must maintain strict sterility in core organs, while allowing microbiological traffic in more distal organs, is the female reproductive tract. The peritoneum, the ovaries, and the fallopian tubes must be protected from bacterial colonization, and the fallopian tubes are particularly sensitive to scarring (with resultant sterility) if they are exposed to bacteria. In contrast, the vagina and the uterus are exposed to bacteria from environmental and intestinal sources, and they must accommodate the passage of sperm while providing a sterile environment for the implanta-

tion of the egg and the development of the fetus. The fact that this system has functioned so well, even in primitive mammals and in humans without modern housing and bathing facilities, is evidence for the efficacy of its microbiological "design."

The partnership between the vaginal tissues and species-specific strains of *Lactobacillus* is exemplary, in that the lactic acid–rich environment of this organ is credited with enabling internal fertilization in the evolution of mammals from amphibians. It seems impossible that these lactic acid–producing organisms dominate the surfaces of these tissues by first access, and we have proposed that they are attracted to this niche by animal-bacterial signals like those that dictate the colonization of the digestive tissue of newborn ruminants (57). We have shown that mammalian sperm cotransport bacteria when they traverse the vagina and enter the uterus, and we speculate that the acidic environment of the vagina may slow growth and biofilm formation in non-*Lactobacillus* species. Direct examination of the surfaces of the endometrium in human uteri removed because of fibroid development has shown the presence of bacterial biofilms that cover as much as one third of the surface of this organ. Moreover, in people using many types of intrauterine contraceptive devices (IUDs), these plastic and metal devices were very extensively colonized by very thick bacterial biofilms (58). The use of IUDs did not make people permanently sterile, and normal uteri are colonized by bacterial biofilms, so we conclude that neither the biofilms on the IUD nor those on the endometrium regularly shed planktonic cells that infect the fallopian tubes. We therefore speculate that the host defenses of the endometrium must be sufficiently efficient to kill or engulf all planktonic cells that are released from these biofilms before they can infect "upstream" organs.

The methods available to modern microbial ecologists (Figures 3, 4, and 7) allow these intrepid scientists to map the microbial colonization of tissues within organ systems, to identify and locate all species of interest, and to assess viability and biofilm formation. A full knowledge of the extent and nature of bacterial biofilms on these tissues will allow us to determine the role that they play in normal tissue functions and in resistance to disease. Happily, these methods can also be used to determine the invidious effects of broad-spectrum antibiotic therapy and of other common medical procedures, and we may be able to avoid damage to natural biofilm populations, on which we depend for much of our protection from infectious disease.

Epilogue

The bacteria that colonize the tissues of the human body are no different from those that colonize the surfaces in all other aquatic ecosystems, in that they form biofilms in preference to growing as planktonic cells. If the tissue in question is exposed to the environment and is "designed" to be colonized by bacteria, these commensal biofilm populations may be beneficial, in that they preclude colonization by pathogens. If biofilms form on the surfaces of tissues or of natural (e.g., teeth) or artificial and extraneous inert materials in the body, they may grow slowly and be well controlled, in that the planktonic cells that are released may be killed by normal defense mechanisms. In these cases, the colonizing biofilms may be essentially nonpathogenic but capable of causing inflammation and bacterial dissemination if they become more extensive or if the host becomes compromised. Once the balance between colonization and infection has been tipped in favor of overt infection, biofilms constitute a peculiar problem that characterizes

65% of infections treated by physicians in the developed world (1). Although antibiotic therapy and activated host defenses can kill derived planktonic cells and often obviate symptoms, they cannot kill the biofilm cells that constitute the niduses of these chronic infections. For this reason, colonized medical devices must often be removed in order to effect a cure, and patients with devices and tissues that cannot be removed must reconcile themselves to intermittent antibiotic therapy for the remainder of their lives. The most immediate hope in this dismal prognosis is the pace at which biofilm science discovers new agents that preclude biofilm formation and "lock" potential pathogens in the planktonic mode of growth, in which they can be killed by antibiotics and host defense factors.

Address correspondence to: William Costerton, Center for Biofilm Engineering, 366 EPS Building, P.O. Box 173980, Montana State University, Bozeman, Montana 59717-3980, USA. Phone: (406) 994-4770; Fax: (406) 994-6098; E-mail: bill_c@erc.montana.edu.

1. Costerton, J.W., Stewart, P.S., and Greenberg, E.P. 1999. Bacterial biofilms: A common cause of persistent infections. *Science.* **284**:1318–1322.

2. Marrie, T.J., Nelligan, J., and Costerton, J.W. 1982. A scanning and transmission electron microscopic study of an infected endocardial pacemaker lead. *Circulation.* **66**:1339–1341.

3. Marrie, T.J., and Costerton, J.W. 1984. Scanning and transmission electron microscopy of *in situ* bacterial colonization of intravenous and intraarterial catheters. *J. Clin. Microbiol.* **19**:687–693.

4. Khoury, A.E., Lam, K., Ellis, B., and Costerton, J.W. 1992. Prevention and control of bacterial infections associated with medical devices. *ASAIO Transactions.* **38**:M174–M178.

5. Lam, J., Chan, R., Lam, K., and Costerton, J.W. 1980. Production of mucoid microcolonies by *Pseudomonas aeruginosa* within infected lungs in cystic fibrosis. *Infect. Immun.* **28**:546–556.

6. Lambe, D.W., Jr., Ferguson, K.P., Mayberry-Carson, K.J., Tober-Meyer, B., and Costerton, J.W. 1991. Foreign-body-associated experimental osteomyelitis induced with *Bacteroides fragilis* and *Staphylococcus epidermidis* in rabbits. *Clin. Ortho.* **266**:285–294.

7. Rayner, M.G., et al. 1998. Evidence of bacterial metabolic activity in culture-negative otitis media with effusion. *JAMA.* **279**:296–299.

8. Nickel, J.C., Costerton, J.W., McLean, R.J.C., and Olson, M. 1994. Bacterial biofilms: Influence on the pathogenesis, diagnosis and treatment of urinary-tract infections. *J. Antimicrob. Chemother.* **33**:31–41.

9. Singh, P.K., et al. 2000. Quorum-sensing signals indicate that cystic fibrosis lungs are infected with bacterial biofilms. *Nature.* **407**:762–764.

10. Costerton, J.W., Geesey, G.G., and Cheng, G.K. 1978. How bacteria stick. *Sci. Am.* **238**:86–95.

11. Wyndham, R.C., and Costerton, J.W. 1981. Heterotrophic potentials and hydrocarbon degradation potentials of sediment microorganisms within the Athabasca oil sands deposit. *Appl. Environ. Microbiol.* **41**:783–790.

12. de Beer, D., Stoodley, P., and Lewandowski, Z. 1994. Liquid flow in heterogeneous biofilms. *Biotechnol. Bioeng.* **44**:636–641.

13. Davies, D.G., et al. 1998. The involvement of cell-to-cell signals in the development of a bacterial biofilm. *Science.* **280**:295–298.

14. Sauer, K., Camper, A.K., Ehrlich, G.D., Costerton, J.W., and Davies, D.G. 2002. *Pseudomonas aeruginosa* displays multiple phenotypes during development as a biofilm. *J. Bacteriol.* **184**:1140–1154.

15. Davies, D.G., Chakrabarty, A.M., and Geesey, G.G. 1993. Exopolysaccharide production in biofilms: Substratum activation of alginate gene expression by *Pseudomonas aeruginosa. Appl. Envir. Microbiol.* **59**:1181–1186.

16. Davies, D.G., and Geesey, G.G. 1995. Regulation of the alginate biosynthesis gene *algC* in *Pseudomonas aeruginosa* during biofilm development in continuous culture. *Appl. Environ. Microbiol.* **61**:860–867.

17. O'Toole, G.A., Kaplan, H.B., and Kolter, R. 2000. Biofilm formation as microbial development. *Annu. Rev. Microbiol.* **54**:49–79.

18. Fuqua, W.C., Winans, E.P., and Greenberg, E.P. 1994. Quorum sensing in bacteria: The Lux R-Lux I family of cell density-responsive transcriptional regulators. *J. Bacteriol.* **176**:269–275.

19. Fuqua, W.C., and Greenberg, E.P. 2002. Listening in on bacteria: acyl-homoserine lactone signaling. *Nat. Rev. Mol. Cell Biol.* **3**:685–695.

20. Stoodley, P., Sauer, K., Davies, D.G., and Costerton, J.W. 2002. Biofilms as complex differentiated communities. *Annu. Rev. Microbiol.* **56**:187–209.

21. Stoodley, P., et al. 2001. Growth and detachment of cell clusters from mature mixed species biofilms. *Appl. Environ. Microbiol.* **67**:5608–5613.

22. Nickel, J.C., Ruseska, I., Wright, J.B., and Costerton, J.W. 1985. Tobramycin resistance of cells of *Pseudomonas aeruginosa* growing as a biofilm on urinary catheter material. *Antimicrob. Agents Chemother.* **27**:619–624.

23. Suci, P.A., Mittelman, M.W., Yu, F.P., and Geesey, G.G. 1994. Investigation of ciprofloxacin penetration into *Pseudomonas aeruginosa* biofilms. *Antimicrob. Agents Chemother.* **38**:2125–2133.

24. Stewart, P.S. 1996. Theoretical aspects of antibiotic diffusion into microbial biofilms. *Antimicrob. Agents Chemother.* **40**:2517–2522.

25. Schoolnik, G.K., et al. 2001. Whole genome DNA microarray expression analysis of biofilm development by *Vibrio cholerae* O1 E1 Tor. *Methods Enzymol.* **336**:3–18.

26. Wagner, V.E., Bushnell, D., Passador, L., Brooks, A.I., and Iglewski, B.H. 2003. Microarray analysis of *Pseudomonas aeruginosa* quorum-sensing reulons: effects of growth phase and environment. *J. Bacteriol.* **185**:2080–2095.

27. Shirtliff, M.E., Leid, J.G., and Costerton, J.W. 2003. Basic science in musculoskeletal infections. In *Musculoskeletal Infections.* J.T. Mader and J.H. Calhoun, editors. Marcel Dekker Inc. New York, New York, USA. 1–61.

28. Sauer, K., Camper, A.K., Ehrlich, G.D., Costerton, J.W., and Davies, D.G. 2002. *Pseudomonas aeruginosa* displays multiple phenotypes during development as a biofilm. *J. Bacteriol.* **184**:1140–1154.

29. Tremoulet, F., Duche, O., Namane, A., Martinie, B., and Labadie, J.C. 2002. A proteomic study of *Escherichia coli* O157:H7 NCTC 12900 cultivated in biofilm or in planktonic growth mode. *FEMS Microbiol. Lett.* **215**:7–14.

30. Miller, B.S., and Diaz-Torres, M.R. 1999. Proteome analysis of biofilms: growth of *Bacillus subtilis* on solid medium as model. *Methods Enzymol.* **310**:433–441.

31. Costerton, J.W., Lewandowski, Z., Caldwell, D.E., Korber, D.R., and Lappin-Scott, H.M. 1995. Microbial biofilms. *Ann. Rev. Micro.* **49**:711–745.

32. Novick, R.P., et al. 1993. Synthesis of staphylococcal virulence factors is controlled by a regulatory RNA molecule. *EMBO J.* **12**:3967–3975.

33. Xavier, K.B., and Bassler, B.L. 2003. LuxS quorum sensing: more than just a numbers game. *Curr. Opin. Microbiol.* **6**:191–197.

34. Balaban, N., et al. 2003. Use of the quorum-sensing inhibitor RNAIII-inhibiting peptide to prevent biofilm formation in vivo by drug-resistant *Staphylococcus epidermidis. J. Infect. Dis.* **187**:625–630.

35. de Nys, R., et al. 1995. Broad spectrum effects of secondary metabolites from the red alga *Delisea pulchra* in antifouling assays. *Biofouling.* **8**:259–271.

36. Veeh, R.H., et al. 2003. Detection of *Staphylococcus aureus* biofilm on tampons and menses components. *J. Infect. Dis.* **188**:519–530.

37. Post, J.C. 2001. Direct evidence of bacterial biofilms in otitis media. *Laryngoscope.* **111**:2083–2094.

38. Ward, K.H., Olson, M.E., Lam, K., and Costerton, J.W. 1992. Mechanism of persistent infection associated with peritoneal implants. *J. Med. Micro.* **36**:406–413.

39. Jensen, E.T., Kharazmi, A., Lam, K., Costerton, J.W., and Hoiby, N. 1990. Human polymorphonuclear leukocyte response to *Pseudomonas aeruginosa* grown in biofilms. *Infect. Immun.* **58**:2383–2385.

40. Leid, J.G., Shirtliff, M.E., Costerton, J.W., and Stoodley, P. 2002. Human leukocytes adhere to, penetrate, and respond to *Staphylococcus aureus* biofilms. *Infect. Immun.* **70**:6339–6345.

41. Costerton, J.W., and Anwar, H. 1994. *Pseudomonas aeruginosa*: The microbe and pathogen. In *Pseudomonas aeruginosa infections and treatment.* A.L. Baltch and R.P. Smith, editors. Marcel Dekker Inc. New York, New York, USA. 1–20.

42. Costerton, J.W., Ellis, B., Lam, K., Johnson, F., and Khoury, A.E. 1994. Mechanism of electrical enhancement of efficacy of antibiotics in killing biofilm bacteria. *Antimicrob. Agents Chemother.* **38**:2803–2809.

43. Rediske, A.M., Hymas, W.C., Wilkinson, R., and Pitt, W.G. 1998. Ultrasonic enhancement of antibiotic action on several species of bacteria. *J. Gen. Appl. Microbiol.* **44**:283 288.

44. Costerton, J.W., et al. 1987. Bacterial biofilms in nature and disease. *Annu. Rev. Microbiol.* **41**:435–464.

45. Ghigo, J.-M. 2001. Natural conjugative plasmids induce biofilm development. *Nature.* **412**:442–445.

46. Shen, K., Wang, X., Post, J.C., and Ehrlich, G.D. 2003. Molecular and translational research approaches for the study of bacterial pathogenesis in otitis media. In *Evidence-based otitis media.* 2nd edition. R. Rosenfeld and C.D. Bluestone, editors. B.C. Decker Inc. Hamilton, Ontario, Canada. 91–119.

47. Ehrlich, G.D., Hu, Z.E., and Post, J.C. Role of biofilms in infectious diseases. ASM Press. Washington, DC, USA. In press.

48. Woods, D.E., Bass, J.A., and Johanson, W.G., Jr. 1980. Role of adherence in the pathogenesis of *Pseudomonas aeruginosa* lung infection in cystic fibrosis patients. *Infect. Immun.* **30**:694–701.

49. Woods, D.E., Straus, D.C., Johanson, W.G., and Bass, J.A. 1981. Role of fibronectin in the prevention of the adherence of *Pseudomonas aeruginosa* to buccal cells. *J. Infect. Dis.* **143**:784–790.

50. Morck, D.W., et al. 1987. Electron microscopic description of glycocalyx and fimbriae on the surface of *Pasturella haemolytica. Can. J. Vet. Res.* **51**:83–88.

51. Morck, D.W., et al. 1990. A guinea pig model of bovine pneumonic *Pasteurellosis. Can. J. Vet. Res.* **54**:139–145.

52. Wright, J.B., Athar, M.A., van Olm, T.M., Wootliff, J.S., and Costerton, J.W. 1989. Atypical legionellosis: Isolation of *Legionella pneumophila* serogroup 1 from a patient with aspiration pneumonia. *J. Hosp. Inf.* **13**:187–190.

53. Vorachit, M., Lam, K., Jayanetra, P., and Costerton, J.W. 1995. Electron microscopy study of the mode of growth of *Pseudomonas pseudomallei* in vitro and in vivo. *J. Trop. Med. Hyg.* **98**:379–391.

54. Vorachit, M., Lam, K., Jayanetra, P., and Costerton, J.W. 1995. The study of the pathogenicity of *Burkholderia pseudomallei*–a guinea pig model. *J. Infect. Dis. Antimicrob. Agents.* **12**:115–121.

55. Sung, J.Y., et al. 1992. Ascending infection of the biliary tract after surgical sphincterotomy and biliary stenting. *J. Gastroenterol. Hepatol.* **7**:240–245.

56. Sung, J.Y., et al. 1991. Bacterial invasion of the biliary system by way of the portal-venous system. *Hepatology.* **14**:313–317.

57. Cheng, K.-J., and Costerton, J.W. 1981. Adherent rumen bacteria: Their role in the digestion of plant material, urea, and epithelial cells. In *Digestive physiology and metabolism in ruminants.* Y. Ruchebusch and P. Thivend, editors. MTP Press. Lancaster, United Kingdom. 227–250.

58. Marrie, T.J., and Costerton, J.W. 1983. A scanning and transmission electron microscopy study of the surfaces of intrauterine contraceptive devices. *Am. J. Obstet. Gynecol.* **146**:384–394.

59. Lam, J.S., et al. 1983. Immunogenicity of *Pseudomonas aeruginosa* outer membrane antigens examined by crossed immunoelectrophoresis. *Infect. Immun.* **42**:88–89.

60. Costerton, J.W., and Stewart, P.S. 2000. Biofilms and device-related infections. In *Persistent bacterial infections.* J.P. Nataro, M.J. Blaser, and S. Cunningham-Rundels, editors. American Society for Microbiology. Washington, DC, USA. 423–437.

Quorum sensing in *Staphylococcus* infections

Jeremy M. Yarwood[1] and Patrick M. Schlievert[2]

[1]Department of Microbiology, University of Iowa Roy J. and Lucille A. Carver College of Medicine, Iowa City, Iowa, USA.
[2]Department of Microbiology, University of Minnesota Medical School, Minneapolis, Minnesota, USA.

Quorum sensing via the accessory gene regulator (*agr*) system has been assigned a central role in the pathogenesis of staphylococci, particularly *Staphylococcus aureus*. While the control of virulence gene expression in vitro by *agr* has been relatively straightforward to describe, regulation of both the quorum response itself and virulence genes in vivo is considerably more complex. The quorum response is highly dependent upon the environment in which the organism is grown and is strongly influenced by additional regulators that respond to signals other than cell density. There is increasing evidence that the *agr* phenotype may influence the behavior and pathogenesis of biofilm-associated *S. aureus* and *S. epidermidis* and may contribute to the chronic nature of some biofilm-associated infections.

Staphylococcus aureus and *Staphylococcus epidermidis* are Gram-positive cocci that normally colonize the epithelial surfaces of large numbers of humans (reviewed in ref. 1). *S. epidermidis* is considered part of the of the normal human microbial flora, while *S. aureus* is usually regarded as a transient member. Colonization by either species usually does not lead to adverse events. However, when these organisms or their extracellular products are allowed to breach the epithelial layer, serious disease can result (1). *S. aureus* has many cell surface virulence factors (such as protein A and clumping factor) and secreted exotoxins and enzymes that allow strains to cause a myriad of infections. These diseases range from relatively benign furuncles and subcutaneous abscesses to scalded skin syndrome, sepsis, necrotizing pneumonia, and toxic shock syndrome (TSS). While no single cell surface virulence factor has been shown to be uniquely required for mucous membrane attachment, once colonization occurs, numerous secreted exotoxins, including the pyrogenic toxin superantigens and exfoliative toxins, definitively cause serious human disease. Other secreted exotoxins, such as the four hemolysins (α, β, δ, and γ) and Panton-Valentine leukocidin have also been suggested to contribute to significant illnesses. *S. epidermidis* does not possess the array of extracellular toxins that *S. aureus* does, and its primary virulence factor is considered to be its ability to form biofilms (2, 3).

The accessory gene regulator quorum sensing system

To improve their ability to cause this variety of human disease and to occupy numerous niches within the host, staphylococci have developed quorum-sensing systems that enable cell-to-cell communication and regulation of numerous colonization and virulence factors. The staphylococcal accessory gene regulator (*agr*) quorum-sensing system decreases the expression of several cell surface proteins and increases the expression of many secreted virulence factors in the transition from late-exponential growth to stationary phase in vitro (4, 5). Expression of *agr* was found to contribute to staphylococcal pathogenesis in several infection models, including murine subcutaneous abscesses (6) and arthritis (7), as well as rabbit endocarditis (8). Expression of *agr* also appears to be involved in the invasion and apoptosis of epithelial cells (9). Interestingly, different *agr* groups, as defined by their production and recognition of distinct secreted signals, are associated predominantly with certain diseases (5). The reasons for this association between *agr* group and infection type are not yet clear, but a better understanding of this phenomenon may contribute to our understanding of the epidemiology of staphylococcal diseases.

Two primary transcripts, RNAII and RNAIII, are generated by the *agr* locus and originate from the P2 and P3 promoters, respectively (Figure 1). The P2 operon encodes four proteins that generate the *agr*-sensing mechanism. AgrB is a transmembrane protein that appears to be involved in (a) processing of the *agrD* product into an octapeptide; (b) secretion of the autoinducing peptide (AIP) signal; and (c) modification of the AIP by the formation of a cyclic thiolactone bond between an internal cysteine and the carboxyl terminus. AgrA and AgrC form a two-component regulatory system in which the transmembrane component, AgrC (histidine kinase), binds the extracellular AIP and in turn modulates the activity of AgrA, the response regulator. Through an as-yet-undefined mechanism, AgrA activity then leads to greatly increased P2 and P3 transcription in the late-log phase of growth, when the concentration of the signal in the medium is high. Sequence variation in *agrB*, *agrD*, and *agrC* has led to the identification of at least four *S. aureus agr* specificity groups in which AIP produced by one group inhibits *agr* expression in other groups (5).

Increased transcription of the P3 operon results in dramatically increased levels of intracellular RNAIII. RNAIII encodes the toxin δ-hemolysin (via *hld*) but, more importantly, increases the transcription (and in some cases, translation) of several secreted virulence factors, including TSS toxin-1 and other hemolysins. However, some toxins, such as enterotoxins A and K, made typically in low concentrations and during exponential phase, are not regulated by RNAIII. Other secreted toxins, such as enterotoxins B, C, and D, are only partially upregulated by RNAIII and can be made in high concentrations independently of *agr*. At the same time, the expression of several cell surface virulence factors is decreased. It is easy to imagine the role that such coordination of virulence gene expression might play in certain infections, such as the formation of a walled-off furuncle. Initially, the staphylococci, present in small numbers, express their cell surface virulence factors in order to evade the host immune system. For example, protein A binds the Fc portion of IgG, and clumping factor may help to form the walled-off infection site. When this site becomes depleted of nutri-

---

Nonstandard abbreviations used: toxic shock syndrome (TSS); accessory gene regulator (*agr*); autoinducing peptide (AIP); staphylococcal accessory regulator (SarA); repressor of toxins (Rot); polysaccharide intercellular adhesin (PIA).

Conflict of interest: The authors have declared that no conflict of interest exists.

Citation for this article: *J. Clin. Invest.* **112**:1620–1625 (2003). doi:10.1172/JCI200320442.

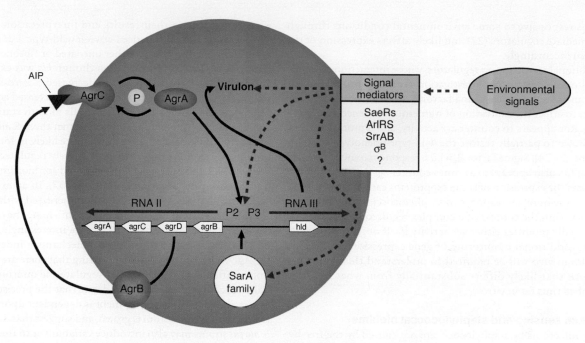

Figure 1
The accessory gene regulator (*agr*) system in *Staphylococcus*. The P2 operon encodes (via RNAII) the signaling mechanism, whereas the transcript of the P3 operon, RNAIII, acts as the effector molecule of the *agr* locus. Additional regulators of the quorum response and virulence genes, described in the text, are listed. Potential regulatory pathways used by environmental signals are indicated by the dashed red lines.

ents due to increased bacterial numbers, the organisms increase secreted factor production, allowing the organisms to escape the walled off site and spread through the host tissues.

The potential role of *agr*-mediated quorum sensing is not as clear in other host environments. We and other researchers have shown through in vivo human and animal studies that *agr* appears to be unnecessary in certain infections for the expression of secreted virulence factors (5, 10–12). For example, *agr* expression in a rabbit abscess model was decreased at the same time that the animals developed TSS through exotoxin production (10). In addition, an *agr* mutant was just as effective at causing TSS as the isogenic wild-type organism. These studies do not rule out roles for *agr* in other aspects of these diseases, as will be discussed later in this review. But the studies do suggest that additional regulatory mechanisms are integral in regulation of both the quorum response and overall virulence of staphylococci.

Additional regulators of the quorum response
The *agr* quorum-sensing system has historically been assigned a central role in the model of *S. aureus* pathogenesis. Thus, studies of other known regulators of staphylococcal virulence have usually examined their interaction with the *agr* system. These additional regulators allow the organism to respond to environmental signals in addition to bacterial cell density, and sometimes counter *agr* activity (reviewed in ref. 5).

Among these regulators are the two-component systems (a family that includes AgrAC) that allow staphylococci to sense and respond to various environmental stimuli. SaeRS was the second two-component system involved in global regulation of virulence factors to be identified (13, 14) after AgrAC. *sae* mutants produce substantially less hemolysin and coagulase but have no effect on the production of RNAIII. However, the expression of at least one

sae transcript was decreased in an *agr* mutant, suggesting that the *sae* system acts downstream of *agr* (15). It has been proposed, but not confirmed, that the *sae* locus responds to several environmental stimuli, including high salt, low pH, glucose, and subinhibitory concentrations of antibiotics (5).

ArlRS comprises a third two-component system that appears to counter *agr* autoinduction by repressing production of hemolysins and exoenzymes (16, 17). Expression of *arlRS* was itself reduced in an *agr* mutant. An *arlS* mutant was enhanced for biofilm formation, despite the increased expression of *agr* and the presumed downregulation of surface-associated adhesion factors. ArlRS also appears to regulate autolytic activity as well as the multidrug efflux pump NorA of *S. aureus*. These data suggest that under certain conditions, *agr* activity may influence the resistance of *S. aureus* to antibiotics.

A fourth two-component system, SrrAB, recently identified by our group (18) and others (19), was found to inhibit RNAIII expression and may itself be repressed by *agr*. Mutants of *srrAB* are unable to grow normally under anaerobic conditions (18, 19), and expression of *srrAB* was shown to regulate genes involved in energy metabolism (19). The signal for the system may be menaquinone, an intermediate in the oxidative respiratory pathway. Thus, SrrAB may be one link between energy metabolism in the cell and the quorum response.

The second major family of regulators of staphylococcal virulence are the DNA-binding proteins, including staphylococcal accessory regulator (SarA) and its homologs (reviewed in ref. 20). SarA, transcribed from three promoters within the same locus, was reported to be required for full *agr* transcription (21). In several reports, SarA has been shown to affect the expression of a wide array of virulence genes, sometimes acting independently of *agr* to decrease the expression of several exoproteins.

SarA is responsive to some environmental conditions through intermediate regulators (22) and likely affects expression of the *agr* locus accordingly.

Additional transcription regulators, whose interactions with the *agr* system deserve further investigation, are the repressor of toxins (Rot) and the alternative sigma factor B. Both of these have been shown to affect the expression of numerous virulence-associated genes. Rot appears to counter *agr* activity, and a mutation in *rot* was shown to partially restore the wild-type phenotype of an *agr* mutant (23, 24). Sigma factor B, which responds to environmental stress (25), also appears to at times counter *agr* activity in that it increases the expression of some exoproteins early in growth (26).

Thus, the quorum response in staphylococci during infection occurs within the context of a complex regulatory network that continually modifies either *agr* activity itself or its downstream effects. Additional monitoring of gene expression and protein profiles in vivo will be required to understand this regulatory network that likely differs substantially from what has been described thus far in vitro.

Quorum sensing and staphylococcal biofilms

Many infections by staphylococci are not caused by the free-living organism but rather by groups of interacting cells termed biofilms. Bacterial biofilms are broadly defined as a community of cells attached to either an abiotic or a biotic surface, are encased in a self-produced matrix, and generally exhibit an altered growth and gene expression profile compared with that of planktonic, or free-living, bacteria (reviewed in ref. 27). Biofilm-associated infections have special clinical relevance, as they are generally resistant to antibiotic therapy and clearance by host defenses. In staphylococcal infections, these diseases include endocarditis (8, 28), osteomyelitis (29, 30), implanted device–related infections (27), and even some skin infections (31).

Two stages of staphylococcal biofilm formation have been described (reviewed in ref. 3). The first stage involves attachment of cells to a surface. This stage of biofilm formation is likely to be mediated in part by cell wall–associated adhesins, including the microbial surface components recognizing adhesive matrix molecules (MSCRAMMs). The second stage of biofilm development includes cell multiplication and formation of a mature, multi-layered, structured community. This stage is associated with production of extracellular factors, including the polysaccharide intercellular adhesin (PIA) component of the extracellular matrix. Detachment of cells from the established biofilm may then allow staphylococci to spread and colonize new sites.

One of the most intriguing areas of investigation is determining what impact quorum sensing has on the growth, development, and pathogenesis of staphylococcal biofilms. There is mounting evidence that the *agr* phenotype and expression patterns may influence several aspects of biofilm behavior, including attachment of cells to surfaces, biofilm dispersal, and even the chronic nature of many biofilm-associated infections. Indeed, many of the products involved in biofilm development, including α-toxin, surface-associated adhesins, δ-hemolysin, and the autolysin AtlE (in *S. epidermidis*), are regulated by the *agr* system, at least in vitro. Furthermore, quorum sensing has been shown to be involved in biofilm development of several Gram-positive and Gram-negative bacteria, including *Streptococcus mutans* (32) and *Pseudomonas aeruginosa* (33).

The limited number of studies addressing the quorum response and staphylococcal biofilms appear at first glance to be somewhat conflicting in their results and interpretation. Pratten et al. (34) found little difference between wild-type *S. aureus* and an *agr* mutant in adherence to either uncoated or fibronectin-coated glass under flow conditions, even though *hld* was expressed. In another study, RNAIII expression decreased *S. aureus* adherence to fibrinogen under static conditions, but increased adherence to fibronectin and human endothelial cells in both static and flow conditions (35). Vuong et al. (36) found that those *S. aureus* strains with a nonfunctional *agr* were much more likely to form biofilms under static conditions. α-toxin, positively regulated by the *agr* system, was recently shown to be required for biofilm formation under both static and flow conditions (37). In an experimental endocarditis study, RNAIII expression increased with increasing *S. aureus* densities in vegetations (38), confirming the cell density–dependent expression of RNAIII in vivo. Interestingly, expression of RNAIII also occurred through a mechanism independent of the AgrAC signaling system, suggesting that there are additional, unidentified in vivo signals that regulate the quorum response. Taken together, these studies indicate that the precise role of *agr* expression in biofilm development is dependent upon the conditions in which the biofilm is grown, and suggest that differences in *S. aureus* strains may also introduce variability into the results.

Most infections of indwelling medical devices are caused by *S. epidermidis*, an organism with few exotoxins and for which the ability to form biofilms is considered the primary virulence factor (2). Recently, Vuong et al. (39) found that disruption of the *agr* locus in *S. epidermidis* resulted in increased attachment of the bacteria to polystyrene, increased biofilm formation, and higher expression of AtlE, which enhances attachment to abiotic surfaces. They also confirmed that the clinical isolate *S. epidermidis* O-47, the strain of choice for studying biofilm formation in *S. epidermidis*, was an *agr* mutant. Interestingly, *agr* did not regulate PIA expression.

Even in conditions in which *agr* does not appear to contribute to biofilm growth or development, it may still affect the virulence of biofilm-associated bacteria. We have observed expression of the *agr* system in conditions in which *agr* did not appear to affect biofilm growth or structure (40). It has been proposed that the production of δ-hemolysin, a molecule with surfactant properties that is encoded by the *agr* locus, may contribute to the detachment of cells from both *S. aureus* and *S. epidermidis* biofilms (36, 39). Should this in fact be the case, it has important clinical implications. Cells expressing *agr* and actively detaching from the biofilm not only may establish additional infection sites in the host but also may contribute to the toxemia associated with acute staphylococcal infections. These cells are likely to express secreted virulence factors, including the superantigenic toxins of *S. aureus*. On the other hand, cells that remain in the biofilm and do not express *agr* may contribute to persistent, low-level infections, particularly in the case of *S. epidermidis*.

Evidence of selection for an *agr*-negative phenotype in chronic infections is emerging. Schwan et al. (41) studied chronic wound infections using a murine abscess model. After establishing infections with a hemolytic (thus with a presumably functional *agr* locus) *S. aureus* strain, the number of nonhemolytic bacteria recovered from the wounds increased over time. The authors suggest that several of the nonhemolytic isolates had mutations in the *agr* locus, although this was not directly shown. In mixed-strain infection experiments using normal, hyperhemolytic, and nonhemolytic strains to inoculate the mouse, the population of hyperhemolytic isolates declined (44.0–9.3%) after 7 days, while the nonhemolytic group (presumably *agr* defective) increased (23.7–61.0%) over the same period of time. Conversely, in both this study and several previous ones, infections established with single strains resulted in decreased cell numbers recovered from the infection of the *agr* mutant compared with those of the wild-type strain. Thus, in the mixed-strain infection experiments, functions performed by the wild-type strain may assist the nonhemolytic group in establishing infection. Also consistent with a selection for an *agr*-negative phenotype, expression of virulence factors in epidemic methicillin-resistant strains is shifted away from extracellular toxins and enzymes toward expression of surface proteins and colonization factors (42). Furthermore, *agr* mutants can frequently be found in isolates from clinical settings (36, 39) and arise spontaneously in vitro (43). When Vuong et al. (36) examined the correlation between a functional *agr* system and the ability of *S. aureus* clinical isolates to adhere to polystyrene under static conditions, they found that only 6% of the isolates with a functional *agr* system formed a biofilm in these conditions, compared with 78% of the *agr*-defective isolates. Failure of the strains with functional *agr* loci to form a biofilm was thought to be due in part to the surfactant properties of the δ-hemolysin produced by these strains. These studies are all consistent with the idea that whereas secreted virulence factors may be important during the acute phase of infection, loss of *agr* function may enhance the long-term survival of staphylococci in the host and contribute to persistent (often biofilm-associated) infections. The enhanced survival of *agr* mutants might be due in part to the decreased production of immunostimulatory factors, such as superantigens, and increased expression of immune-evading factors, such as protein A.

It has been proposed that use of *agr*-inhibiting substances, such as AIP produced by staphylococci belonging to a different *agr* group, might be beneficial in the treatment of acute staphylococcal infection (5). Such treatments may decrease the production of extracellular virulence factors normally upregulated by *agr* expression. However, because loss of *agr* activity normally correlates with increased expression of adhesin factors and decreased expression of potential dispersion factors, inhibition of *agr* activity may instead result in the conversion of an acute infection into a chronic one, particularly in biofilm-associated infections. Indeed, inhibition of *agr* activity increases attachment of both *S. aureus* and

S. epidermidis to polystyrene (36, 39) and enhances biofilm formation. Furthermore, in certain animal models of infection, expression of *agr* does not significantly affect the expression of virulence factors, as would be expected from in vitro data (10–12). Understanding these caveats, additional investigation into *agr* inhibition is still warranted, given that inhibition of *agr* activity, and thus extracellular toxin and enzyme production, may be beneficial in some acute infection types.

Taken as a whole, the literature suggests that the role of quorum sensing in staphylococcal infections may not always be immediately obvious, as it varies with infection type, host environment, and even time. For example, in models in which *agr* does not appear to strongly influence virulence, its expression (or lack thereof) may contribute to biofilm formation. Conversely, in models in which *agr* expression does not appear to affect biofilm formation, it may still regulate virulence factor production. This makes the development of relevant in vitro and in vivo models a challenging proposition. Yet it is critical that further investigation of this area take place. The expression of the *agr* quorum-sensing system, already known to affect virulence factor production, may also affect everything from antibiotic resistance to energy metabolism through its interaction with other staphylococcal gene regulators. Furthermore, the *agr* phenotype and expression patterns in both *S. aureus* and *S. epidermidis* biofilms may influence when these infections become chronic or acute. Future studies should include sampling of staphylococcal infections for *agr* phenotype, evaluation of *agr* expression patterns in vivo using animal models of infection, and further investigation of *agr* function in biofilm growth and development.

Acknowledgments
J.M. Yarwood is supported by a Ruth M. Kirchstein National Research Service Award from the National Institutes of Health (GM069110). We thank Debra Murray for critical reading of the manuscript.

Address correspondence to: Jeremy M. Yarwood, Department of Microbiology, University of Iowa Roy J. and Lucille A. Carver College of Medicine, 540 Eckstein Medical Research Building, 500 Newton Road, Iowa City, Iowa 52242, USA. Phone: (319) 335-7996; Fax: (319) 335-7949; E-mail: jeremy-yarwood@uiowa.edu.

1. Tenover, F.C., and Gaynes, R.P. 2000. The epidemiology of *Staphylococcus aureus* infections. In *Gram-positive pathogens*. V.A. Fischetti, R.P. Novick, J.J. Ferretti, D.A. Portnoy, and J.I. Rood, editors. ASM Press. Washington, DC, USA. 414–421.
2. Raad, I., Alrahwan, A., and Rolston, K. 1998. *Staphylococcus epidermidis*: emerging resistance and need for alternative agents. *Clin. Infect. Dis.* 26:1182–1187.
3. Gotz, F. 2002. *Staphylococcus* and biofilms. *Mol. Microbiol.* 43:1367–1378.
4. Vuong, C., Gotz, F., and Otto, M. 2000. Construction and characterization of an *agr* deletion mutant of *Staphylococcus epidermidis. Infect. Immun.* 68:1048–1053.
5. Novick, R.P. 2003. Autoinduction and signal transduction in the regulation of staphylococcal virulence. *Mol. Microbiol.* 48:1429–1449.
6. Bunce, C., Wheeler, L., Reed, G., Musser, J., and Barg, N. 1992. Murine model of cutaneous infection with gram-positive cocci. *Infect. Immun.* 60:2636–2640.
7. Abdelnour, A., Arvidson, S., Bremell, T., Ryden, C., and Tarkowski, A. 1993. The accessory gene regulator (*agr*) controls *Staphylococcus aureus* virulence in a murine arthritis model. *Infect. Immun.* 61:3879–3885.
8. Cheung, A.L., et al. 1994. Diminished virulence of a *sar/agr* mutant of *Staphylococcus aureus* in the rabbit model of endocarditis. *J. Clin. Invest.* 94:1815–1822.
9. Wesson, C.A., et al. 1998. *Staphylococcus aureus* Agr and Sar global regulators influence internalization and induction of apoptosis. *Infect. Immun.* 66:5238–5243.
10. Yarwood, J.M., McCormick, J.K., Paustian, M.L., Kapur, V., and Schlievert, P.M. 2002. Repression of the *Staphylococcus aureus* accessory gene regulator in serum and in vivo. *J. Bacteriol.* 184:1095–1101.
11. Goerke, C., et al. 2000. Direct quantitative transcript analysis of the *agr* regulon of *Staphylococcus aureus* during human infection in comparison to the expression profile *in vitro. Infect. Immun.* 68:1304–1311.
12. Goerke, C., Fluckiger, U., Steinhuber, A., Zimmerli, W., and Wolz, C. 2001. Impact of the regulatory loci *agr, sarA* and *sae* of *Staphylococcus aureus* on the induction of α-toxin during device-related infection resolved by direct quantitative transcript analysis. *Mol. Microbiol.* 40:1439–1447.
13. Giraudo, A.T., Cheung, A.L., and Nagel, R. 1997. The *sae* locus of *Staphylococcus aureus* controls exoprotein synthesis at the transcriptional level. *Arch. Microbiol.* 168:53–58.
14. Giraudo, A.T., Calzolari, A., Cataldi, A.A., Bogni, C., and Nagel, R. 1999. The *sae* locus of *Staphylococcus aureus* encodes a two-component regulatory system. *FEMS Microbiol. Lett.* 177:15–22.
15. Giraudo, A.T., Mansilla, C., Chan, A., Raspanti, C., and Nagel, R. 2003. Studies on the expression of regulatory locus *sae* in *Staphylococcus aureus. Curr. Microbiol.* 46:246–250.
16. Fournier, B., and Hooper, D.C. 2000. A new two-component regulatory system involved in adhesion, autolysis, and extracellular proteolytic activity of *Staphylococcus aureus. J. Bacteriol.* 182:3955–3964.
17. Fournier, B., Klier, A., and Rapoport, G. 2001. The two-component system ArlS-ArlR is a regulator of virulence gene expression in *Staphylococcus aureus. Mol. Microbiol.* 41:247–261.
18. Yarwood, J.M., McCormick, J.K., and Schlievert, P.M. 2001. Identification of a novel two-component regulatory system that acts in global regulation of virulence factors of *Staphylococcus aureus. J. Bacteriol.* 183:1113–1123.

19. Throup, J.P., et al. 2001. The *srhSR* gene pair from *Staphylococcus aureus*: genomic and proteomic approaches to the identification and characterization of gene function. *Biochemistry.* **40**:10392–10401.

20. Cheung, A.L., and Zhang, G. 2002. Global regulation of virulence determinants in *Staphylococcus aureus* by the SarA protein family. *Front. Biosci.* **7**:1825–1842.

21. Cheung, A.L., Bayer, M.G., and Heinrichs, J.H. 1997. *sar* genetic determinants necessary for transcription of RNAII and RNAIII in the *agr* locus of *Staphylococcus aureus. J. Bacteriol.* **179**:3963–3971.

22. Deora, R., Tseng, T., and Misra, T. 1997. Alternative transcription factor σ^sb of *Staphylococcus aureus*: characterization and role in transcription of the global regulatory locus *sar. J. Bacteriol.* **179**:6355–6359.

23. McNamara, P.J., Milligan-Monroe, K.C., Khalili, S., and Proctor, R.A. 2000. Identification, cloning, and initial characterization of *rot*, a locus encoding a regulator of virulence factor expression in *Staphylococcus aureus. J. Bacteriol.* **182**:3197–3203.

24. Said-Salim, B., et al. 2003. Global regulation of *Staphylococcus aureus* genes by Rot. *J. Bacteriol.* **185**:610–619.

25. Chan, P.F., Foster, S.J., Ingham, E., and Clements, M.O. 1998. The *Staphylococcus aureus* alternative sigma factor σ^b controls the environmental stress response but not starvation survival or pathogenicity in a mouse abscess model. *J. Bacteriol.* **180**:6082–6089.

26. Nicholas, R.O., et al. 1999. Isolation and characterization of a *sigB* deletion mutant of *Staphylococcus aureus. Infect. Immun.* **67**:3667–3669.

27. Donlan, R.M., and Costerton, J.W. 2002. Biofilms: survival mechanisms of clinically relevant microorganisms. *Clin. Microbiol. Rev.* **15**:167–193.

28. van Wamel, W., et al. 2002. Regulation of *Staphylococcus aureus* type 5 capsular polysaccharides by *agr* and *sarA* in vitro and in an experimental endocarditis model. *Microb. Pathog.* **33**:73–79.

29. Gillaspy, A., et al. 1995. Role of the accessory gene regulator (*agr*) in pathogenesis of staphylococcal osteomyelitis. *Infect. Immun.* **63**:3373–3380.

30. Blevins, J.S., et al. 2003. Role of *sarA* in the pathogenesis of *Staphylococcus aureus* musculoskeletal infection. *Infect. Immun.* **71**:516–523.

31. Akiyama, H., et al. 2003. Confocal laser scanning microscopic observation of glycocalyx production by *Staphylococcus aureus* in skin lesions of bullous impetigo, atopic dermatitis and pemphigus foliaceus. *Brit. J. Dermatol.* **148**:526–532.

32. Li, Y.H., et al. 2002. A quorum-sensing signaling system essential for genetic competence in *Streptococcus mutans* is involved in biofilm formation. *J. Bacteriol.* **184**:2699–2708.

33. Davies, D.G., et al. 1998. The involvement of cell-to-cell signals in the development of a bacterial biofilm. *Science.* **280**:295–298.

34. Pratten, J., Foster, S.J., Chan, P.F., Wilson, M., and Nair, S.P. 2001. *Staphylococcus aureus* accessory regulators: expression within biofilms and effect on adhesion. *Microbes Infect.* **3**:633–637.

35. Shenkman, B., et al. 2002. Role of *agr* (RNAIII) in *Staphylococcus aureus* adherence to fibrinogen, fibronectin, platelets and endothelial cells under static and flow conditions. *J Med. Microbiol.* **51**:747–754.

36. Vuong, C., Saenz, H.L., Gotz, F., and Otto, M. 2000. Impact of the *agr* quorum-sensing system on adherence to polystyrene in *Staphylococcus aureus. J. Infect. Dis.* **182**:1688–1693.

37. Caiazza, N.C., and O'Toole, G.A. 2003. Alpha-toxin is required for biofilm formation by *Staphylococcus aureus. J. Bacteriol.* **185**:3214–3217.

38. Xiong, Y.Q., et al. 2002. Activation and transcriptional interaction between *agr* RNAII and RNAIII in *Staphylococcus aureus* in vitro and in an experimental endocarditis model. *J. Infect. Dis.* **186**:668–677.

39. Vuong, C., Gerke, C., Somerville, G.A., Fischer, E.R., and Otto, M. 2003. Quorum-sensing control of biofilm factors in *Staphylococcus epidermidis. J. Infect. Dis.* **188**:706–718.

40. Yarwood, J.M., Bartels, D.J., Volper, E.M., and Greenberg, E.P. 2003. Expression of the accesory gene regulator quorum sensing system in biofilms of *Staphylococcus aureus. ASM Conference on Biofilms 2003.* Victoria, British Columbia, Canada. 39. (Abstr.)

41. Schwan, W.R., Langhorne, M.H., Ritchie, H.D., and Stover, C.K. 2003. Loss of hemolysin expression in *Staphylococcus aureus agr* mutants correlates with selective survival during mixed infections in murine abscesses and wounds. *FEMS Immunol. Med. Microbiol.* **38**:23–28.

42. Papakyriacou, H., Vaz, D., Simor, A., Louie, M., and McGavin, M.J. 2000. Molecular analysis of the accessory gene regulator (*agr*) locus and balance of virulence factor expression in epidemic methicillin-resistant *Staphylococcus aureus. J. Infect. Dis.* **181**:990–1000.

43. Somerville, G.A., et al. 2002. In vitro serial passage of *Staphylococcus aureus*: changes in physiology, virulence factor production, and *agr* nucleotide sequence. *J. Bacteriol.* **184**:1430–1437.

Quorum sensing and biofilm formation in streptococcal infections

Dennis G. Cvitkovitch, Yung-Hua Li, and Richard P. Ellen

Dental Research Institute, University of Toronto, Toronto, Ontario, Canada.

Members of the bacterial genus *Streptococcus* are responsible for causing a wide variety of infections in humans. Many Streptococci use quorum-sensing systems to regulate several physiological properties, including the ability to incorporate foreign DNA, tolerate acid, form biofilms, and become virulent. These quorum-sensing systems are primarily made of small soluble signal peptides that are detected by neighboring cells via a histidine kinase/response regulator pair.

Streptococcal infections

Members of the genus *Streptococcus* (i.e., streptococci) are ubiquitous parasites of humans. Some are part of the indigenous microflora that are involved in opportunistic infections such as dental caries and others are exogenous pathogens that cause infections ranging from mild respiratory or skin diseases to life-threatening conditions such as pneumonia, septic shock, and necrotizing fasciitis. Contemporary research has found that many species of streptococci and other Gram positive bacteria have evolved similar peptide pheromone quorum-sensing systems that probably help them adapt to and survive host-imposed fluctuations in their local environment and coincidently regulate the expression of virulence factors that promote their pathogenicity. This review focuses on streptococcal peptide signaling pathways that are population density–dependent and that impact on vital survival and virulence traits.

Biofilms

Biofilms are dense aggregates of surface-adherent microorganisms embedded in an exopolysaccharide matrix. The study of bacteria residing in biofilms as an interactive community rather than free-living planktonic cells has recently gained a great deal of attention. This has arisen, in part, because of the estimate by the Centers for Disease Control and Prevention that 65% of human bacterial infections involve biofilms. Many species of streptococci are known to form biofilms; however, the relationship between the pathogenic state and the biofilm mode of growth has been most clearly established with the oral streptococci, which are known to initiate dental caries when the bacteria are living in the biofilm environment of dental plaque.

Dental plaque

The human oral cavity is a complex ecosystem that supports an extremely diverse microflora consisting of about 500 species of microorganisms (1). Numerous physical and nutritional interactions between oral bacteria contribute to a complex biofilm community (Figure 1) (2). Streptococci, including *Streptococcus mutans*, are ubiquitous in the oral microbiota of humans. *S. mutans* is considered to be a principal etiological agent of dental caries, where it can cause dissolution of tooth enamel by acid end-products resulting from carbohydrate metabolism. The tooth surface is an indispensable natural habitat for *S. mutans* (3). *S. mutans'* dental biofilm tropism most likely reflects its evolution of glucan synthesis and binding functions as well as its relative aciduricity. Its competitiveness for this ecological niche may also relate to cell density–dependent regulation of its acid tolerance response (ATR), natural genetic competence, and bacteriocin activity. Since *S. mutans* has evolved to depend on a biofilm lifestyle for survival and persistence in the oral cavity combined with its role as an opportunistic pathogen, it has become the best-studied example of a biofilm-forming, disease-causing *Streptococcus* (4). Biofilm-like populations of pathogenic streptococci may also reach higher densities in confined areas like heart valves, prosthetic devices, sinuses, tonsillar crypts, terminal respiratory passages, and in infectious skin lesions.

Quorum sensing in streptococci

Many bacteria, including streptococci, are known to regulate diverse physiological processes through a mechanism called quorum sensing. In Gram positive bacteria, quorum-sensing systems generally consist of three components, a signal peptide and a two-component regulatory system (also called two-component signal transduction system or TCSTS) that has a membrane-bound histidine kinase sensor and an intracellular response regulator (5). Quorum sensing in Gram positive bacteria has been found to regulate a number of physiological activities, including competence development in *Streptococccus gordonii*, *S. pneumoniae*, and *S. mutans* (6), sporulation in *Bacillus subtilus* (7), antibiotic biosynthesis in *Lactococcus lactis* (8), and induction of virulence factors in *Staphylococcus aureus* (8).

Many of the genes involved in competence induction and the transformation process of streptococci, which is very similar to that described for the mitis group of *Streptococcus* (including *S. pneumoniae*), have been identified. Induction of genetic competence in these streptococci is mediated by quorum sensing, which depends on a competence stimulating peptide (CSP) signaling system illustrated in Figure 2.

Evidence for the existence of other CSP-modulated pathways

It has been recently recognized that the quorum-sensing signal in *S. pneumoniae* initiates competence through the activity of a global transcription modulator, ComX, which acts as an alternate sigma factor during the development of genetic competence (9). Transcription of the *comX* gene is regulated by ComE, the response

Nonstandard abbreviations used: acid tolerance response (ATR); two-component signal transduction system (TCSTS); competence-stimulating peptide (CSP); extracellular induction component (EIC).

Conflict of interest: The authors have declared that no conflict of interest exists.

Citation for this article: *J. Clin. Invest.* 112:1626–1632 (2003). doi:10.1172/JCI200320430.

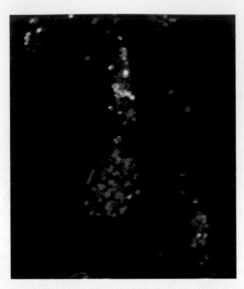

Figure 1
Eight-hour-old dental plaque was visualized using confocal scanning laser microscopy. Antibodies against *Streptococcus gordonii* DL1 and anti-receptor polysaccharides (RPS), commonly found on *Streptococcus oralis*, were used for indirect immunofluorescence along with the stain Syto 59. At least two staining types are seen within the aggregates: antibody-reactive (anti-DL1 [green], anti-RPS [purple]) cells found in direct association with antibody-unreactive (blue Syto 59–stained) cells. Single colonies containing all three staining types (anti-DL1, anti-RPS, and antibody unreactive) were frequently seen. Reproduced with permission from *The Journal of Bacteriology* (2).

regulator of the quorum-sensing system (10). The ComX sigma factor then initiates the transcription of competence-specific operons involved in DNA uptake and recombination by recognizing a *com*-box (also referred to as *cin*-box) consensus sequence (TAC-GAATA) in their promoter regions (9). *S. mutans* and *S. pyogenes* also have *comX* genes present in their genomes (9, 11), and inactivation of one copy of the *S. mutans comX* renders the cell transformation deficient (12). The presence of the conserved *com*-box consensus sequence in the promoter region of *S. mutans* late-competence genes such as *comFA, celA*, and *cglA* supports the hypothesis that transcriptional regulation of these genes is also mediated by the ComX sigma factor (13). There is much accumulating evidence that many genes not involved in competence are under the control of the CSP-ComX system. This includes the observations that many "other" *S. pneumoniae* genes contain the *com*-box sequence and are activated by CSP (14, 15). These genes likely encode products that aid in the cell's adaptation to a high cell density, and it is likely that they (or their homologs in other streptococci) contribute to the biofilm phenotype.

LuxS quorum sensing
Recently, the *luxS* gene has been identified in streptococci (16, 17). The LuxS protein is required for the biosynthesis of the type 2 autoinducer, AI-2, which is involved in quorum sensing in a wide range of bacterial species. Mutation of *luxS* in *S. mutans* caused a defect in biofilm formation, while disruption of this gene in *S. pneumoniae* resulted in reduced virulence in mouse infections. It has been suggested that the AI-2 pathway is a very good target for chemotherapeutic control of bacterial virulence.

Quorum sensing in *S. mutans* biofilms
Horizontal gene transfer through genetic transformation has been observed in many natural ecosystems, and recent studies suggest that growth of bacteria in biofilms may facilitate horizontal gene transfer among bacterial species via either transformation or conjugation (18, 19). Natural genetic transformation has been extensively studied in streptococci, but these studies relied exclusively on bacteria grown in fluid cultures, where they would become transiently competent after reaching a critical cell density. Since biofilms are more representative of bacterial growth in natural environments and *S. mutans* is an organism that relies on a biofilm lifestyle, we set forth to investigate the ability of this bacterium to transport and integrate exogenous DNA when living in its biofilm state. To facilitate assays for genetic transformation of biofilm-grown cells, we utilized a chemostat-based continuous flow biofilm system, which allows observation of physiological activities of a bacterial population under controlled growth conditions (20). Using this system, we have demonstrated that growth rates, culture pH, and biofilm age are several important factors that influence competence development of *S. mutans* growing in biofilms.

The most fascinating finding using this system was that *S. mutans* cells growing in biofilms were able to incorporate foreign DNA much more efficiently than their free-living counterparts. The transformation frequencies of biofilm-grown cells of *S. mutans* strains tested were about 10- to 600-fold higher than those of the planktonic cells (20). To our best knowledge, this is the first report to provide direct evidence that biofilm-grown bacteria can be efficiently induced to become genetically competent for transformation. In addition, *S. mutans* grown in a biofilm appeared to maintain a subpopulation of cells that were constantly competent for taking up DNA from the environment. This static state of competence is in contrast to that observed in fluid where most streptococci, including *S. mutans,* enter a transient physiological state that usually lasts for only 15–30 minutes during their growth cycle. The evidence from our study and another recent study that demonstrated genetic exchange in *S. pneumoniae* biofilms (21) clearly suggests that biofilm-growth mode in transformable streptococci appears to favor the induction and maintenance of genetic competence. It also appears that the biofilm environment provides conditions for optimal function of streptococcal cell-cell peptide signaling systems to activate genetic competence and facilitate genetic exchange.

Quorum sensing in biofilm formation
The discovery that the *S. mutans* quorum-sensing system functions optimally in growing biofilms led us to investigate other roles of this system in biofilm formation and biofilm physiology. Dental plaque is a complex biofilm community that harbors the most diverse resident microflora associated with humans. Bacteria in dental bio films, including *S. mutans*, are frequently exposed to various stresses, such as extreme nutrient shortage or excess, low pH, high osmolarity, oxidation, and consumption of antimicrobial agents or antibiotics by the host. Formation of a biofilm is considered an important mechanism used by a bacterium for adaptation to this environment (22). Although adaptation to environmental stress by genetic transformation is believed to occur very infrequently, such a rare event can be highly significant if the transforming DNA, such as an antibiotic resistance gene or a virulence factor, provides a selective advantage to the recipient cells (23). In addition to providing the community with an abundant extracellular gene pool, the biofilm environment facilitates the

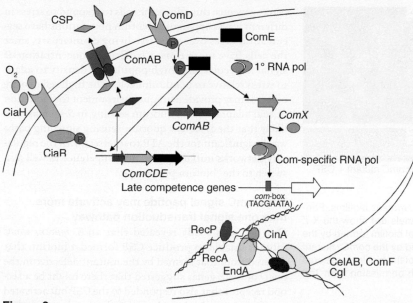

Figure 2
In *Streptococcus pneumoniae*, the induction of genetic competence (and potentially viru-lence) is regulated by a CSP–mediated quorum-sensing system (5, 9). Quorum sensing involves the expression of early gene products encoded by two genetic loci, *comAB* and *comCDE*. Genes in the operon, *comAB*, encode an ATP-binding cassette transporter (ComA) and an accessory protein to ComA (ComB). These secretory proteins are involved in the processing and export of the CSP. The loci, *comCDE*, respectively encode the precursor to the CSP, a histidine kinase that acts as a CSP receptor, and a response regulator that activates both *comAB* and *comCDE* operons. A second two-component regulatory system, CiaH-CiaR, affects the development of competence by negatively regu-lating *comCDE* expression. Quorum-sensing signals initiate competence through activity of ComX, a global transcription modulator, which was shown to act as an alternate sigma factor (40). This sigma factor initiates the transcription of competence-specific operons involved in DNA uptake and recombination by recognizing a *com*-box (also referred to as *cin*-box) consensus sequence (TACGAATA) in their promoter regions (9, 41, 42). Sev-eral of these late competence-specific operons include *cilA* (*ssb2*), *cilB* (*dal*, like *dprA* in *Haemophilus influenzae*), *cilC* (like *comC* in *Bacillus subtilis*), *cilD* (or *cglABCDE*), *cilE* (or *celAB*), *coi*, *cinA-recA*, *cfl* (like *comF* in *B. subtilis*), and *dpnA* of the DpnII restriction system. pol, polymerase; P, phosphate.

bacteria with a localized neighborhood where cell-cell signaling mechanisms likely abound.

Recent studies using genetic dissection of biofilm development have revealed that the formation of biofilms involves multiple, convergent signaling pathways and a genetic program for the tran-sition from planktonic growth state to the biofilm mode of growth (24). In Gram negative bacteria such as *Pseudomonas aeruginosa*, cell-cell signaling through quorum sensing has been found to play an important role in biofilm differentiation (25). The first evidence to suggest that quorum-sensing systems might influence the struc-ture of Gram positive biofilms came from a recent study of *S. gor-donii* where a biofilm-defective mutant was found to have a trans-poson insertion in the *comD* gene encoding the histidine kinase sensor protein of the TCSTS required for genetic competence (26). This implied that biofilm formation by *S. gordonii* involved cell-cell communication through quorum sensing. To test if the *S. mutans* ComCDE quorum-sensing system was involved in bio-film formation, we examined the ability of *S. mutans* mutants defective in various components of the system to form biofilms. We found that inactivation of any one of the genes encoding the components of the quorum-sensing signaling system results in

the formation of an abnormal biofilm (Figure 3) (12). Particularly, the *comC* mutant (unable to produce or secrete the signal peptide) formed a biofilm that lacked the wild-type architecture, whereas the *comD* and *comE* mutants defective in sensing and responding to the signal peptide formed biofilms with a reduced biomass. The architectural change in the *comC* mutant bio-films may be associated with a defect in cell sep-aration with mutations in this gene resulting in the formation of large aggregates or "weblike" biofilms that were easily removed from the sur-face. The observation that the mutants unable to produce or secrete the CSP formed biofilms that differed from those formed by the mutant defective in the *comD* or *comE* genes suggested that *S. mutans* has multiple CSP receptors. Although we have clearly demonstrated that the ComCDE quorum-sensing system is direct-ly connected to the ability of *S. mutans* to form biofilms, the molecular and biochemical mech-anisms involved in expression of the "wild-type biofilm phenotype" remain to be investigated.

CSP modulation of the ATR
The pH levels in dental biofilms are highly vari-able and frequently shift from above pH 7.0 in the resting pH state to as low as pH 3.0 during the ingestion of dietary carbohydrates by the host. Thus, pH exerts a significant ecological pressure on *S. mutans*, and its ability to tolerate and grow in low pH environments is crucial to its survival and eventual dominance in dental plaque, leading to caries (27). Considerable evidence has shown that *S. mutans* has evolved a number of sophisticated mechanisms to survive these pH changes includ-ing induction of an ATR in which exposure of *S. mutans* cells to a mild or moderately acidic pH (5.0–6.0) results in enhanced survival of a signifi-cant proportion of the cell population in a lower pH of 3.0–3.5 (28). This ATR involves a number of de novo proteins that appear to be important for adaptation to an acidic environment (29). Although many of the molecular mechanisms of the ATR in *S. mutans* remain unclear, this "signal pH" that results in synthesis of protective pro-teins appears to be important for induction of the ATR.

It is widely accepted that bacteria living in biofilms are more resistant to mechanical, physical, and chemical stresses. Since *S. mutans* normally resides in high cell density biofilms, the ability to withstand acid in this physiological state is likely an important adaptive response. We therefore addressed the question of whether acid adaptation involved cell density–dependent events or cell-cell signaling in biofilms. Changes in external pH can significantly influence many physiological parameters, such as energy coupling, ion transport, proton movement, and export of metabolic products, thereby triggering numerous secondary signals. During growth at pH 5.0, *Escherichia coli* can signal stress tolerance to other unadapted cells by secreting a proteinlike molecule, termed extracellular induc-tion component (EIC) (30). Although the signal molecule remains unidentified, induction of stress (including acid) adaptation in *E. coli* presumably involves cell-cell communication. Since *S. mutans*

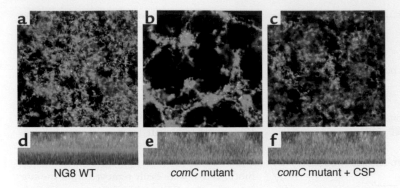

NG8 WT   comC mutant   comC mutant + CSP

Figure 3
Scanning confocal laser microscope images of 16 h S. mutans biofilms. Panels **a–c** show the X-Y planes (top view), while the panels **d–f** show the X-Z planes (side view). The panels illustrate (**a, d**) a normal biofilm formed by the parent strain NG8 and (**b, e**) an aberrant biofilm formed by the comC⁻ mutant defective in making a CSP. The wild-type appearance of the biofilm is restored by addition of synthetic CSP (**c, f**). Reproduced with permission from *The Journal of Bacteriology* (12).

normally encounters acid while living in dense biofilm communities, we proposed that quorum sensing at high cell densities of S. mutans might facilitate its survival against low pH challenges. Testing this hypothesis has led us to find that the ATR interfaces with the density-dependent signaling pathway that also initiates genetic competence. We have demonstrated that mutants defective in the comC, D, or E genes have a diminished log-phase ATR and the neutralized culture filtrates prepared from acid-adapted wild-type cells also induce a partial log-phase ATR in cells that have never encountered the signal pH (31). S. mutans grown at high cell density established adaptation to the signal pH more rapidly than cells grown at low density. Similarly, S. mutans cells grown in a high cell-density biofilm were more resistant to the killing pH than planktonic-phase cells. In fact, S. mutans cells grown in biofilms not only survived better than the planktonic cells but also were capable of growth at the lower pH following a glucose pulse. Based on the evidence obtained from this work, we propose that S. mutans, upon exposure to low pH in a growing culture, releases extracellular signal molecules, one of which is the CSP, to enhance induction of acid adaptation in the population. It is likely that optimal induction of acid adaptation in a population of S. mutans requires a coordinated activity through mechanisms involving both low pH induction and cell density–dependent intercellular signals.

Biofilms likely provide bacterial cells with a unique environment to fully express their adaptive survival mechanisms. Because of three-dimensional structures, high cell density, and diffusion barriers, bacterial cells at different locations within a biofilm may not sense the same degree of extracellular stress simultaneously. The cells that first sense a pH stress may rapidly process the information and pass their "secondary signal" to the other members of the population through cell-cell signaling systems to initiate a coordinated protective response against potentially lethal forces, like acid. Unlike some planktonic cells that need to reach a critical concentration of signal molecules and cell density, biofilms can allow signal molecules to accumulate rapidly in the local environment to initiate coordinated activities far more quickly (25). In addition, physiological states of bacterial cells living in a biofilm, in terms of growth rate, growth phase, or metabolic activities, are

heterogeneous; this allows the cells to respond to stress in different ways. Apparently, biofilm populations have several advantages over their free-living counterparts since the cells have more time, a sufficient concentration of signal molecules, and high population density to adapt to stress relative to planktonic cells. The high cell density biofilms may provide a unique environment for induction of acid adaptation via quorum sensing in S. mutans. It is likely that the S. mutans quorum-sensing signaling pathway is significant for the ATR to intersect with the regulatory networks initiating genetic competence as well as a switch to the "biofilm phenotype."

The ComC signal peptide may activate more than one signal transduction pathway

Our previous study revealed that an S. mutans comC mutant unable to produce CSP formed a biofilm that differed from that formed by the mutant defective in the comD or comE genes suggested that there might be a second receptor that also responded to the CSP but activated a different pathway in order to invoke the phenotype (12, 20). Based on the available information, we have proposed a "two-receptor" cell-cell signaling model to illustrate how the quorum-sensing system in S. mutans functions to regulate genetic competence, biofilm formation, and the ATR (12). The principle of this model is that the signal peptide (CSP) encoded by comC can simultaneously interact with two cognate receptors, one encoded by comD and another encoded by an unknown gene. These receptors likely transfer the input signal through two different pathways. Although the genes encoding the components involved in the second transduction pathway remain unknown, the work in our lab has recently characterized another TCSTS, named HK/RR11, which is also demonstrated to involve biofilm formation and acid resistance in S. mutans (32).

One of the defects observed in HK/RR11 mutant biofilms was the development of a spongelike architecture composed of cells in very long chains, a feature that we previously observed with the biofilm formed by the comC mutant unable to produce CSP. Since there is no putative substrate, signal or function assigned for the HK/RR11 transduction system, we suspected that the TCSTS encoded by hk/rr11 may activate a second pathway to respond to the CSP. To test this hypothesis, we added CSP to the HK/RR11 mutant cultures to assess the effect on chain formation by the hk11 and rr11 mutant biofilm cells. The results revealed that addition of CSP to the mutant cultures had no observable impact on the length of cell chains comprising the mutant biofilms. This result was consistent with HK11 acting as a CSP receptor but provided no direct evidence to conclusively assign a role to HK11 as a CSP receptor. A study for characterizing the relationship between the comC-encoded signal peptide and the HK/RR11 signal transduction pathway is now underway.

Evaluation of CSP analogs to act as inhibitors of biofilm formation

Analogs of quorum-sensing peptides can competitively inhibit the activity of the peptide-mediated phenotype. It has been demonstrated that analogous signal peptides from *Staphylococcus epidermidis* can inhibit the agr signaling system of *Staphylococcus aureus*, thereby modulating S. aureus virulence (33). Since peptide signaling–sensing systems are very similar among Gram positive bacteria (including the *S. mutans* CSP system) analogs of the *S. mutans* CSP may be able to interfere with the quorum-sensing process, leading to inhibition of induction of the "biofilm phenotype."

Mucosal pathogens

Quorum sensing among mucosal pathogens may not require high cell density. In contrast to the very dense dental plaques colonized by *S. mutans* and several other oral streptococci, pathogenic streptococci that infect mucosal tissues colonize environments where bacterial microcolonies are usually less dense due to bathing effects of secretions and desquamation of epithelium. Yet, several of these streptococcal and other Gram positive species have evolved peptide pheromone signaling pathways that have autoregulating functions analogous to the quorum-sensing systems that we described above for *S. mutans*. The finding that initiation of genetic competence in *S. pneumoniae* is sensitive to relatively low levels of CSP at about 10^7 bacteria/ml (34) suggests that pathogenic streptococci of mucosal surfaces may respond to gradients of cell densities that occur frequently during natural infections, raising the question of whether quorum sensing affects pathways by which pathogenic streptococci grow to predominance over normally protective indigenous species. It has been recently established that mutants of *S. pneumoniae* defective in the quorum-sensing competence induction pathway have diminished virulence in a murine model relative to the parent strain (35). Differential fluorescence induction also showed that several *Com* genes were induced during infection in mice (36). Clearly, the role of quorum sensing during infection by *S. pneumoniae* warrants further examination.

Bacteriocins

Bacteriocins are antimicrobial peptides that are generated by some bacteria and target others that are sensitive. The signaling networks that regulate bacteriocin production and transport as well as immunity to bacteriocins involve peptide pheromone sensing pathways that are very similar to those involved in genetic competence (8). Operons encoding bacteriocin peptide precursors, cognate peptide processing transporters, and two component regulatory sensors and response regulators are known to be under the control of cell density. For example, the virulence-related *blp* and *com* regulons of *S. pneumoniae* have many analogous features including highly similar signal peptides and peptide secretion systems (15). Moreover, activation of the competence and bacteriocin peptide pheromone response regulators may, in turn, activate downstream pathways that intersect in a common regulatory gene. One such example is the previously mentioned *comX*, a transcriptional regulator of so-called "late" competence genes (9). The genome of M1 *S. pyogenes* strain SF370 contains *comX* and a complete bacteriocin *salA1* locus; yet, it lacks the genes *comABC* that encode the CSP and its secretion apparatus (11). This implies that *S. pyogenes comX* (and downstream pathways) may be activated in response to environmental signals other than CSPs. *S. pyogenes* is a pathogen with a host range limited to humans and a unique set of virulence properties not shared by other streptococci. It is possible that it has evolved a sophisticated bacteriocin peptide regulatory system to help its population competitively emerge on mucosal surfaces, yet protect its genome from contamination with foreign DNA by deleting (or never evolving) genes for the cell density–dependent CSP. Recently, Jenkinson's lab reported that the sensory and immunity pathways for the pheromone antibiotics SalA1 of *S. pyogenes* and SalA of the indigenous oral species *Streptococcus salivarius* are conserved, closely related, and

cross-sensitive (37). Such novel findings suggest that determinants of population dynamics on bathed mucosal surfaces probably encompass subtle combinations of complex environmental sensing systems that are not limited to cell density, bacteriocin production, and competence stimulation. It is also possible that *S. pyogenes* utilizes a bacteriocin-like hemolysin as a quorum-sensing molecule. Recently, *Enterococcus faecalis* was shown to regulate gene expression by such a mechanism involving its cytolysin, a molecule with both antimicrobial and hemolytic activity (38). *S. pyogenes* has a hemolysin that is genetically very similar to quorum-sensing autoinduction operons found in streptococci (39).

Fluctuations in total bacterial burden and population density also determine the pathogenicity of bacterial microcolonies and biofilms. For example, the ability to suppress competing streptococci through bacteriocin activity may be considered a virulence property if it allows the pathogenic species to emerge sufficiently to damage the host. Evidence is growing that cell density–dependent gene regulation also affects the expression of many other virulence-related proteins of pathogenic streptococci. For example, virulence factor expression and regulation are, in part, determined by growth phase and activation of two-component regulatory sensors of environmental signals, which are characteristics of quorum sensing in Gram positive bacteria (38). Thus, the extracellular concentration of autoregulating peptides appears to serve as one of several environmental conditions that regulate virulence genes through signal transduction pathways that are initiated via two-component regulatory systems. In *S. pyogenes*, in addition to *sagA*, there appear to be several coordinated regulatory loci such as *mga*, *csrRS*, *fasBCA*, and *rgg* that affect the expression of genes for its numerous and diverse virulence factors.

Such a pattern of global regulation is reminiscent of the induction of virulence genes via quorum-sensing pathways in biofilms of *Pseudomonas aeruginosa*. Therefore, it is likely that the genomes of streptococci have evolved density-dependent regulons to control expression of downstream genes that affect bacterial survival in response to changing environmental conditions on mucosal and tooth surfaces, and the selective survival of the pathogenic streptococcal species leads to clinical infection. Although the key molecules of the quorum-sensing pathways of pathogenic streptococci such as *S. pneumoniae*, *S. pyogenes*, and *S. mutans* are distinct from those first reported for Gram negative bacteria, they seem to serve an analogous function, to modulate physiologic homeostasis and adaptation to environmental conditions in response to fluctuations in population density. The use of this information to exploit these pathways to control streptococcal infections is now being implemented and in the near future we may see a more selective and targeted approach to controlling persistent biofilm-dwelling bacteria.

Acknowledgments

Special thanks to Paul Kolenbrander and Rob Palmer from the National Institute of Dental and Craniofacial Research (NIDCR) for providing Figure 1. Work conducted by the authors was generously funded by the Canadian Institutes of Health Research, the NIDCR, and the Canada Foundation for Innovation.

Address correspondence to: Dennis Cvitkovitch, Room 449A, Dental Research Institute, University of Toronto, 124 Edward Street, Toronto, Ontario M5G 1G6, Canada. Phone: (416) 979-4917 ext. 4592; Fax: (416) 979-4936; E-mail: dennis.cvitkovitch@utoronto.ca.

1. Kroes, I., Lepp, P.W., and Relman, D.A. 1999. Bacterial diversity within the human subgingival crevice. *Proc. Natl. Acad. Sci. U. S. A.* **96**:14547–14552.

2. Palmer, R.J., Jr., et al. 2003. Coaggregation-mediated interactions of streptococci and actinomyces detected in initial human dental plaque. *J. Bacteriol.* **185**:3400–3409.

3. Carlsson, J., Soderholm, G., and Almfeldt, I. 1969. Prevalence of *Streptococcus sanguis* and *Streptococcus mutans* in the mouth of persons wearing full-dentures. *Arch. Oral Biol.* **14**:243–249.

4. Burne, R.A. 1998. Oral streptococci . . . products of their environment. *J. Dent. Res.* **77**:445–452.

5. Kleerebezem, M., et al. 1997. Quorum sensing by peptide pheromones and two-component signal-transduction systems in Gram-positive bacteria. *Mol. Microbiol.* **24**:895–904.

6. Cvitkovitch, D.G. 2001. Genetic competence and transformation in oral streptococci. *Crit. Rev. Oral Biol. Med.* **12**:217–243.

7. Lazazzera, B.A. 2000. Quorum sensing and starvation: signals for entry into stationary phase. *Curr. Opin. Microbiol.* **3**:177–182.

8. Kleerebezem, M., and Quadri, L.E. 2001. Peptide pheromone-dependent regulation of antimicrobial peptide production in Gram-positive bacteria: a case of multicellular behavior. *Peptides.* **22**:1579–1596.

9. Lee, M.S., and Morrison, D.A. 1999. Identification of a new regulator in *Streptococcus pneumoniae* linking quorum sensing to competence for genetic transformation. *J. Bacteriol.* 1999. **181**:5004–5016.

10. Ween, O., Gaustad, O., and Havarstein, L.S. 1999. Identification of DNA binding sites for ComE, a key regulator of natural competence in *Streptococcus pneumoniae. Mol. Microbiol.* **33**:817–827.

11. Ferretti, J.J., et al. 2001. Complete genome sequence of an M1 strain of *Streptococcus pyogenes. Proc. Natl. Acad. Sci. U. S. A.* **98**:4658–4663.

12. Li, Y.H., et al. 2002. A quorum-sensing signaling system essential for genetic competence in *Streptococcus mutans* is involved in biofilm formation. *J. Bacteriol.* **184**:2699–2708.

13. Lee, J.H., et al. 2001. Genetic transformation in *Streptococcus mutans*: identification of competence genes by functional genomic analysis. *Abstracts of the 101st Annual General Meeting of the American Society for Microbiology.* Orlando, Florida, USA. (Abstr. I-76)

14. Peterson, S., et al. 2000. Gene expression analysis of the *Streptococcus pneumoniae* competence regulons by use of DNA microarrays. *J. Bacteriol.* **182**:6192–6202.

15. de Saizieu, A., et al. 2000. Microarray-based identification of a novel *Streptococcus pneumoniae* regulon controlled by an autoinduced peptide. *J. Bacteriol.* **182**:4696–4703.

16. Merritt, J., et al. 2003. Mutation of luxS affects biofilm formation in *Streptococcus mutans. Infect. Immun.* **71**:1972–1979.

17. Stroeher, U.H., et al. 2003. Mutation of luxS of *Streptococcus pneumoniae* affects virulence in a mouse model. *Infect. Immun.* **71**:3206–3212.

18. Christensen, B.B., et al. 1998. Establishment of new genetic traits in a microbial biofilm community. *Appl. Environ. Microbiol.* **64**: 2247–2255.

19. Hausner, M., and Wuertz, S. 1999. High rates of conjugation in bacterial biofilms as determined by quantitative in situ analysis. *Appl. Environ. Microbiol.* **65**:3710–3713.

20. Li, Y.H., et al. 2001. Natural genetic transformation of *Streptococcus mutans* growing in biofilms. *J. Bacteriol.* **183**:897–908.

21. Waite, R.D., Struthers, J.K., and Dowson, C.G. 2001. Spontaneous sequence duplication within an open reading frame of the pneumococcal type 3 capsule locus causes high-frequency phase variation. *Mol. Microbiol.* **183**:1223–1232.

22. Donlan, R.M., and Costerton, J.W. 2002. Biofilms: survival mechanisms of clinically relevant microorganisms. *Clin. Microbiol. Rev.* **15**:167–193.

23. Dowson, C.G., et al. 1997. Horizontal gene transfer and the evolution of resistance and virulence determinants in *Streptococcus. Soc. Appl. Bacteriol. Symp. Ser.* **26**:42S–51S.

24. O'Toole, G.A., and Kolter, R. 1998. Initiation of biofilm formation in *Pseudomonas fluorescens* WCS365 proceeds via multiple, convergent signalling pathways: a genetic analysis. *Mol. Microbiol.* **28**:449–461.

25. Parsek, M.R., and Greenberg, E.P. 2000. Acyl-homoserine lactone quorum sensing in gram-negative bacteria: a signaling mechanism involved in associations with higher organisms. *Proc. Natl. Acad. Sci. U. S. A.* **97**:8789–8793.

26. Loo, C.Y., Corliss, D.A., and Ganeshkumar, N. 2000. *Streptococcus gordonii* biofilm formation: identification of genes that code for biofilm phenotypes. *J. Bacteriol.* **182**:1374–1382.

27. Kuramitsu, H.K. 1993. Virulence factors of mutans streptococci: role of molecular genetics. *Crit. Rev. Oral Biol. Med.* **4**:159–176.

28. Svensater, G., et al. 1997. Acid tolerance response and survival by oral bacteria. *Oral Microbiol. Immunol.* **12**:266–273.

29. Wilkins, J.C., Homer, K.A., and Beighton, D. 2002. Analysis of *Streptococcus mutans* proteins modulated by culture under acidic conditions. *Appl. Environ.*

Microbiol. **68**:2382–2390.

30. Rowbury, R.J., and Goodson, M. 2001. Extracellular sensing and signalling pheromones switch-on thermotolerance and other stress responses in *Escherichia coli. Sci. Prog.* **84**:205–233.

31. Li, Y.H., et al. 2001. Cell density modulates acid adaptation in *Streptococcus mutans*: implications for survival in biofilms. *J. Bacteriol.* **183**:6875–6884.

32. Li, Y.H., et al. 2002. A novel two-component regulatory system involved in biofilm formation and acid resistance in *Streptococcus mutans. J. Bacteriol.* **184**:6333–6342.

33. Otto, M., et al. 1999. Inhibition of virulence factor expression in *Staphylococcus aureus* by the *Staphylococcus epidermidis agr* pheromone and derivatives. *FEBS Lett.* **450**:257–262.

34. Steinmoen, H., Knutsen, E., and Havarstein, L.S. 2002. Induction of natural competence in *Streptococcus* pneumoniae triggers lysis and DNA release from a subfraction of the cell population. *Proc. Natl. Acad. Sci. U. S. A.* **99**:7681–7686.

35. Lau, G.W., et al. 2001. A functional genomic analysis of type 3 *Streptococcus pneumoniae* virulence. *Mol. Microbiol.* **40**:555–571.

36. Bartilson, M., et al. 2001. Differential fluorescence induction reveals *Streptococcus pneumoniae* loci regulated by competence stimulatory peptide. *Mol. Microbiol.* **39**:126–135.

37. Upton, M., et al. 2001. Intra- and interspecies signaling between *Streptococcus salivarius* and *Streptococcus pyogenes* mediated by SalA and SalA1 antibiotic peptides. *J. Bacteriol.* **183**:3931–3938.

38. Haas, W., Shepard, B.D., and Gilmore, M.S. 2002. Two-component regulator of *Enterococcus faecalis* cytolysin responds to quorum-sensing autoinduction. *Nature.* **415**:84–87.

39. Nizet, V., et al. 2000. Genetic locus for streptolysin S production by group A streptococcus. *Infect. Immun.* **68**:4245–4254.

40. Luo, P., and Morrison, D.A. 2003. Transient association of an alternative sigma factor, ComX, with RNA polymerase during the period of competence for genetic transformation in *Streptococcus pneumoniae. J. Bacteriol.* **185**:349–358.

41. Pestova, E.V., and Morrison, D.A. 1998. Isolation and characterization of three *Streptococcus pneumoniae* transformation-specific loci by use of a lacZ reporter insertion vector. *J. Bacteriol.* **180**:2701–2710.

42. Campbell, E.A., Choi, S.Y., and Masure, H.R. 1998. A competence regulon in *Streptococcus pneumoniae* revealed by genomic analysis. *Mol. Microbiol.* **27**:929–939.

Pharmacological inhibition of quorum sensing for the treatment of chronic bacterial infections

Morten Hentzer and Michael Givskov

Center for Biomedical Microbiology, BioCentrum-DTU, Technical University of Denmark (DTU), Lyngby, Denmark.

Traditional treatment of infectious diseases is based on compounds that aim to kill or inhibit bacterial growth. A major concern with this approach is the frequently observed development of resistance to antimicrobial compounds. The discovery of bacterial-communication systems (quorum-sensing systems), which orchestrate important temporal events during the infection process, has afforded a novel opportunity to ameliorate bacterial infection by means other than growth inhibition. Compounds able to override bacterial signaling are present in nature. Herein we discuss the known signaling mechanisms and potential antipathogenic drugs that specifically target quorum-sensing systems in a manner unlikely to pose a selective pressure for the development of resistant mutants.

One of the greatest accomplishments of modern medicine has been the development of antimicrobial pharmaceuticals for the treatment of infectious diseases. Alexander Fleming discovered the first antibiotic, penicillin, in 1928, and after over half a century of intense research, most acute bacterial infections can be treated effectively with antibiotics. Conventional antibiotics possess broad-range efficacy via toxic or growth-inhibitory effects on target organisms. However, an increased frequency of bacterial mutations has resulted in a significantly increased incidence of antibiotic resistance. The horizontal spread of resistance genes to other bacteria of the same or different species has been shown to rapidly create bacterial populations with (a) an increased ability to degrade antibacterial compounds; (b) decreased permeability; (c) decreased affinity for the antibiotic; or, finally, (d) increased efflux of many different antibiotics (1, 2). The increasing occurrence of multiresistant pathogenic bacterial strains has gradually rendered traditional antimicrobial treatment ineffective. Today, a global concern has emerged that we are entering a post-antibiotic era with a reduced capability to combat microbes, and, hence, the development of novel therapeutic approaches to the treatment of bacterial infections constitutes a focal point of modern research. The alternative to antibiotic-mediated bacteria killing or growth inhibition is attenuation of bacterial virulence such that the organism fails to establish successful infection and, in consequence, is cleared by the host immune response. Compounds with such abilities are the result of rational drug design and are termed antipathogenic drugs as opposed to antibacterial drugs (i.e., most traditional antibiotics). Antipathogenic drugs target key regulatory bacterial systems that govern the expression of virulence factors.

In recent years, researchers have come to appreciate that, in nature, most bacteria form complex surface-attached (sessile) communities called biofilms. Bacteria present within biofilms have characteristics distinct from those of free-swimming (planktonic) bacteria of the same species, including a signifi-

cantly increased tolerance to antimicrobial therapies and the host immune response (3). In modern clinical microbiology, the establishment of bacterial biofilms is often considered a pathogenicity trait during chronic infections (4). Biofilm formation is an example of microbial community behavior. Both Gram-positive and Gram-negative bacteria have been found to coordinate this behavior through cell-to-cell communication mediated by small, diffusible signals. This phenomenon has been termed quorum sensing and is prevalent among both symbiotic and pathogenic bacteria associated with plants and animals. Many of the phenotypes regulated by cell-to-cell communication are involved in bacterial colonization and virulence.

Among the Gram-negative bacteria, the most well studied quorum-sensing system is the LuxR-LuxI homologous system and the cognate signal molecules: N-acyl-homoserine lactones (AHLs) (3). This quorum-sensing system is widespread among Gram-negative genera and is involved in the regulation of many host-associated phenotypes, including production of virulence factors (5–7) and secondary metabolites (8). Emerging evidence points to the involvement of quorum sensing in biofilm formation and surface motility in the opportunistic pathogens *Pseudomonas aeruginosa* (9), *Burkholderia cepacia* (10), and *Aeromonas hydrophila* (11). These observations suggest that quorum sensing serves to link biofilm formation with virulence factor production. Interestingly, AHL-based cross-talk has been demonstrated between *P. aeruginosa* and *B. cepacia* (12) and between *S. liquefaciens* and *P. aeruginosa* (13).

The observation that quorum sensing is linked to virulence factor production and biofilm formation suggests that many virulent Gram-negative organisms could potentially be rendered nonpathogenic by inhibition of their quorum-sensing systems. Research into quorum sensing, and inhibition thereof, may provide a means of treating many common and damaging chronic infections without the use of growth-inhibitory agents, such as antibiotics, preservatives, and disinfectants, that unavoidably select for resistant organisms.

AHL-mediated quorum sensing

Quorum sensing is a generic regulatory mechanism used by many Gram-negative bacteria and Gram-positive bacteria to perceive and respond to factors as varied as changing microbial population density and the expression of specific genes. The concentration of a signal molecule reflects the density of bacterial cells in a defined

Nonstandard abbreviations used: N-acyl-homoserine lactone (AHL); exopolymeric substance (EPS); autoinducer 2 (AI-2); S-adenosyl methionine (SAM); *Pseudomonas* quinolone signal (PQS).

Conflict of interest: Michael Givskov is the founder and vice president of QSI Pharma A/S (Lyngby, Denmark).

Citation for this article: *J. Clin. Invest.* **112**:1300–1307 (2003). doi:10.1172/JCI200320074.

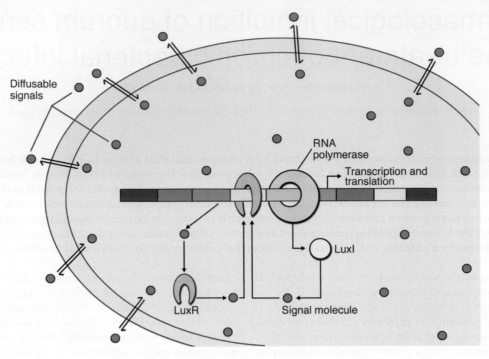

Figure 1
The archetypical Lux quorum sensor. The AHL signal (green circles) is synthesized by the *luxI* gene product LuxI (the synthase). At a certain threshold concentration, the AHL signal interacts with the receptor LuxR (encoded by *luxR*), which binds to the promoter sequence of the target genes (in this case, the *lux* operon) and in conjunction with the RNA polymerase promotes transcription.

environment, and the perception of a threshold level of that signal indicates that the population is "quorate," i.e., sufficiently dense to make a behavioral group–based decision. Quorum sensing is thought to afford pathogenic bacteria a mechanism to minimize host immune responses by delaying the production of tissue-damaging virulence factors until sufficient bacteria have amassed and are prepared to overwhelm host defense mechanisms and establish infection. In our laboratory, we also view quorum sensing as a mechanism by which bacteria expose part of their genetic repertoire for recognition by other organisms (prokaryotes as well as eukaryotes): a phenomenon referred to as cross-talk. One environment that contains a large number of bacteria in close proximity is the bacterial biofilm. Furthermore, the dense and diffusion-limited biofilm matrix seems to provide ideal conditions for accumulation of signal molecules and a protected environment for bacteria to induce quorum sensing–regulated virulence factors and launch an attack on the host.

AHL-mediated quorum-sensing systems are found in a large number of Gram-negative bacterial species belonging to the α, β, and γ subclasses of proteobacteria, including bacteria in the genera *Agrobacterium, Aeromonas, Burkholderia, Chromobacterium, Citrobacter, Enterobacter, Erwinia, Hafnia, Nitrosomonas, Obesumbacterium, Pantoea, Pseudomonas, Rahnella, Ralstonia, Rhodobacter, Rhizobium, Serratia, Vibrio, Xenorhabdus,* and *Yersinia* (reviewed in ref. 3). The quorum-sensing system consists, in brief, of a four-component circuit: an AHL signal molecule, a LuxI-type signal synthase, a LuxR-type signal receptor, and the target gene(s) (Figure 1). The AHL signal is synthesized at a low basal level by the AHL synthase. AHL signals diffuse out of the bacteria and into the surrounding environment. An increase in bacterial population density leads to an increase in

local AHL concentration, and, at a threshold concentration, this signal interacts with a cognate receptor (LuxR-type response regulator) that in turn is activated as a positive transcription factor and modulates the expression of quorum sensing–regulated genes. Often, the quorum sensor is subject to autoinduction, because the gene encoding the signal synthase is among the target genes; hence, a positive feedback regulatory loop is created. The autoinduction allows a rapid increase in signal production and dissemination, which in turn induces a quorum sensing–controlled phenotype throughout the bacterial population (see Figure 1).

Microbial biofilms and chronic infections

Biofilms are now considered ubiquitous in the natural world (14). Bacterial biofilms have been observed to be extremely heterogeneous, both structurally and with regard to the physiology of the bacterial cells within them. The prevailing conceptual model depicts bacterial biofilms as being made up of microcolonies, which serve as the basic unit of the greater biofilm structure. Microcolonies are hydrated structures consisting of bacterial cells enmeshed in a matrix of exopolymeric substances (EPSs). Bacteria may proliferate on the attachment surface, leading to microcolony expansion. Eventually, community growth becomes limited by substrate availability due to increased diffusion distances, and the biofilm reaches a steady state. Such mature biofilms often consist of "towers" and "mushrooms" of cells in an EPS matrix (Figure 2). Interstitial voids and channels separate the biofilm structures and facilitate a convective flow in order to transport nutrients to interior parts of the biofilm and remove waste products. The biofilm mode of growth has been shown to facilitate bacterial survival in a variety of environmental stresses, including antibiotics and disin-

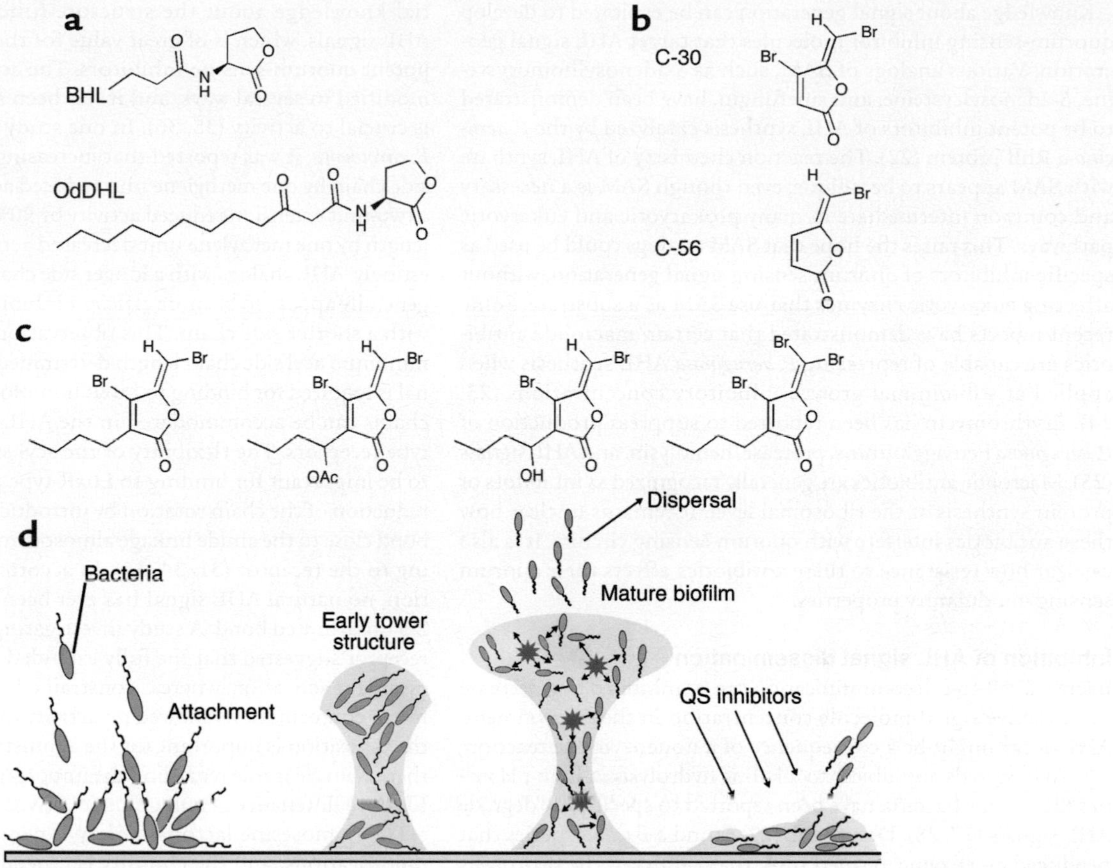

Figure 2
(a) Molecular structures of the two cognate signal molecules produced by *P. aeruginosa*, BHL ([*N*-butyryl]-L-homoserine lactone), and OdDHL (*N*-[3-oxo-dodecanoyl]-L-homoserine lactone). (b) Synthetic quorum-sensing (QS) inhibitors derived from (c) natural brominated furanone compounds isolated from *D. pulchra*. (d) Temporal biofilm development and dispersal. Stars represent QS.

fectants (4, 15). Biofilms have become evident in many, if not most, environmental, industrial, and medical bacteria-related problems. A recent public announcement from the NIH stated that more than 60% of all microbial infections involve biofilms (1).

Quorum sensing as a target for antimicrobial therapy

Given the many bacteria that employ quorum sensing in the control of virulence, quorum sensing constitutes a novel target for directed drug design (16). While AHL-mediated quorum-sensing systems are employed by Gram-negative bacteria, many Gram-positive bacteria, including *Bacillus subtilis*, *Streptococcus pneumoniae*, and *Staphylococcus aureus*, use small peptides or modified peptides for signaling. Autoinducer 2 (AI-2), a signaling molecule common to many diverse bacteria, has recently been described (17). Genome sequencing revealed the presence of *luxS* homologs (encoding the AI-2 signal synthase) in many pathogens, including *Escherichia coli*, *Helicobacter*, *Neisseria*, *Porphyromonas*, *Proteus*, *Salmonella*, *Enterococcus faecalis*, *Streptococcus pyogenes*, and *S. aureus* (18). In some bacteria, AI-2 signaling is required for virulence (19), but in other bacteria, it does not appear essential for bacterial virulence (20). Recently, Winzer and colleagues suggested that, in most bacteria, AI-2 is simply a metabolic by-product and, therefore, doubtful as a drug target (21).

Quorum sensing–inhibitory compounds might constitute a new generation of antimicrobial agents with applications

in many fields, including medicine (human and veterinary), agriculture, and aquaculture, and the associated commercial interests are substantial. Indeed, in recent years a number of biotechnology companies that aim specifically at developing anti-quorum-sensing and anti-biofilm drugs have emerged (QSI Pharma A/S, Lyngby, Denmark; Microbia, Cambridge, Massachusetts, USA; Quorex Pharmaceuticals Inc., Carlsbad, California, USA; and 4SC AG, Martinsried, Germany). Several strategies aiming at the interruption of bacterial quorum-sensing circuits are possible, including (a) inhibition of AHL signal generation, (b) inhibition of AHL signal dissemination, and (c) inhibition of AHL signal reception.

Inhibition of AHL signal generation

The vast majority of bacteria that produce AHL signals encode one or more genes homologous to *luxI* of *Vibrio fischeri* (see Figure 1). Expression of these genes in heterologous host backgrounds has demonstrated that the LuxI-type protein is required and sufficient for production of AHL signals. The catalysis of AHL synthesis has been studied in vitro for three LuxI family members. The reaction involves a sequentially ordered reaction mechanism that uses *S*-adenosyl methionine (SAM) as the amino donor for generation of the homoserine lactone ring moiety, and an appropriately charged acyl carrier protein (ACP) as the precursor for the acyl side chain of the AHL signal (22).

Knowledge about signal generation can be exploited to develop quorum-sensing inhibitor molecules that target AHL signal generation. Various analogs of SAM, such as S-adenosylhomocysteine, S-adenosylcysteine, and sinefungin, have been demonstrated to be potent inhibitors of AHL synthesis catalyzed by the *P. aeruginosa* RhII protein (22). The reaction chemistry of AHL synthase with SAM appears to be unique, even though SAM is a necessary and common intermediate in many prokaryotic and eukaryotic pathways. This raises the hope that SAM analogs could be used as specific inhibitors of quorum-sensing signal generation, without affecting eukaryotic enzymes that use SAM as a substrate. Some recent reports have demonstrated that certain macrolide antibiotics are capable of repressing *P. aeruginosa* AHL synthesis when applied at subliminal growth-inhibitory concentrations (23, 24). Erythromycin has been reported to suppress production of *P. aeruginosa* hemagglutinins, protease, hemolysin, and AHL signals (25). Macrolide antibiotics are generally recognized as inhibitors of protein synthesis at the ribosomal level. It remains unclear how these antibiotics interfere with quorum-sensing circuits. It is also unclear how resistance to these antibiotics affects their quorum sensing–modulatory properties.

Inhibition of AHL signal dissemination

Bacterial cell-to-cell communication can be inhibited by a decrease in the active signal-molecule concentration in the environment. AHL decay might be a consequence of a nonenzymatic reaction; e.g., AHL signals are subject to alkaline hydrolysis at high pH values (26). Some bacteria have been reported to specifically degrade AHL signals (27, 28). Dong et al. (27) found a *Bacillus* species that produced an enzyme, termed AiiA, that catalyzed the hydrolysis of AHL molecules. Expression of the *aiiA* gene in the plant pathogen *Erwinia carotovora* resulted in reduced release of AHL signals, decreased extracellular pectolytic enzyme activity, and attenuated soft rot disease symptoms in all plants tested (27). Moreover, transgenic plants expressing AiiA have been shown to be significantly less susceptible to infection by *E. carotovora* (29). In another study, a *Variovorax paradoxus* strain able to grow using 3-oxo-C6-N-homoserine lactone as the sole energy and nitrogen source was isolated from a soil sample (28). In our own laboratory, we have conducted similar screenings and found that bacteria able to degrade or metabolize AHL molecules can be isolated from environmental samples at a high frequency (M. Givskov et al., unpublished observations). The ecological significance of such AHL-degrading bacteria is not clear, but AHL-degrading enzymes are of great clinical interest for use in the prevention of diseases caused by quorum sensing–proficient bacterial populations.

Inhibition of AHL signal reception

Blocking of quorum-sensing signal transduction can be achieved by an antagonist molecule capable of competing or interfering with the native AHL signal for binding to the LuxR-type receptor. Competitive inhibitors would conceivably be structurally similar to the native AHL signal, in order to bind to and occupy the AHL-binding site but fail to activate the LuxR-type receptor. Noncompetitive inhibitors may show little or no structural similarity to AHL signals, as these molecules bind to different sites on the receptor protein.

Several reports describe the in vitro application of AHL analogs to achieve inhibition of the quorum-sensing circuits of various bacteria (30–34). These studies have generated substantial knowledge about the structure-function relationships of AHL signals, which is of great value for the continued search for potent quorum-sensing inhibitors. The acyl side chain has been modified in several ways, and it has been shown that the length is crucial to activity (35, 36). In one study of quorum sensing in *E. carotovora*, it was reported that increasing the length of the acyl side chain by one methylene unit reduced activity by 50%, whereas a two-unit extension reduced activity by 90%. Decreasing the chain length by one methylene unit decreased activity to 10% (35). Interestingly, AHL analogs with a longer side chain than the native AHL generally appear to be more efficient inhibitors than AHL analogs with a shorter side chain. This observation might suggest that a minimum acyl side chain length determined by the native AHL signal is required for binding to LuxR homologs and that longer acyl chains can be accommodated in the AHL-binding site of LuxR-type receptors. The flexibility of the acyl side chain also appears to be important for binding to LuxR-type proteins. For instance, reduction of the chain rotation by introduction of an unsaturated bond close to the amide linkage almost completely abolishes binding to the receptor (31, 34, 35). In accordance with this suggestion, no natural AHL signal has ever been reported to contain a 2,3 unsaturated bond. A study investigating the *P. aeruginosa* LasR receptor suggested that the fully extended chain geometry is necessary for activation, whereas constrained analogs locked into different conformations showed no activity (37). The substitution at the β-position is important for the agonistic activity of AHLs, but there is no clear rule regarding the importance of this substitution to the maintenance of antagonistic activity.

The homoserine lactone moiety is generally very sensitive to modifications, and the chirality is crucial to biological activity. Natural AHL signals are L-isomers, whereas D-isomers are generally devoid of biological activity (35). The acyl side chain appears essential for activity, as exemplified in *E. carotovora*, in which the unsubstituted homoserine lactone ring fails to activate the quorum-sensing system (35). Conversion of the homoserine lactone ring to a homoserine lactame ring results in a molecule without agonistic or antagonistic properties (30, 35). Interestingly, a change of the homoserine lactone structure to a homoserine thiolactone ring appears permissible in several quorum-sensing systems (30, 31, 35). A recent study showed that LasR and RhlR proteins responded differently to changes in the homoserine lactone moiety (38). This may indicate that the two *P. aeruginosa* AHL receptors differ significantly in their AHL-binding sites.

Quorum-sensing inhibitors expressed by higher organisms

A number of reports describe the ability of higher organisms to interfere with AHL-mediated quorum sensing. The best-characterized example is that of the Australian macroalga *Delisea pulchra*, described below. Recently, another example of eukaryotic interference with AHL-mediated signaling was provided by Teplitski et al. (39), who showed that several plants secrete substances that mimic bacterial AHL signal activities and affect quorum sensing–regulated behaviors in associated bacteria. Exudates from pea (*Pisum sativum*) were demonstrated to contain several separable activities that either stimulated or inhibited bacterial AHL-dependent phenotypes (39). Many plants and fungi have coevolved and established carefully regulated symbiotic associations with bacteria. Interestingly, many plant-associated proteobacteria possess AHL-mediated quorum-sensing systems (40). Importantly, both plants and

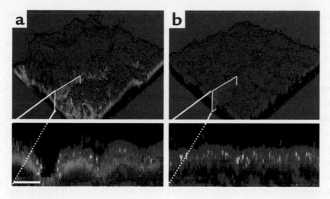

Figure 3
Furanone-treated *P. aeruginosa* biofilms are less tolerant to tobramycin. Scanning confocal laser photomicrographs of *P. aeruginosa* PAO1 biofilms grown in the absence (**a**) or the presence (**b**) of 10 μM C-30. After 3 days, the biofilms were exposed to 100 μg/ml tobramycin for 24 hours. Bacterial viability was assayed by staining using the LIVE/DEAD *Bac*Light Bacterial Viability Kit (Molecular Probes Inc., Eugene, Oregon, USA). Red areas are dead bacteria; green areas are live bacteria.

fungi are devoid of the active immune systems that are observed in mammals; rather, they rely on chemical defense systems to deal with bacteria in the environment. For these reasons, it might be expected that plants and fungi have evolved to produce chemical compounds to inhibit (or in other cases to stimulate) bacterial AHL-mediated communication. We have conducted screenings of plants (including some used in traditional herbal medicine) and fungal extracts for AHL-inhibitory activity (M. Givskov et al., unpublished results). A surprisingly large number of extracts contained quorum sensing–inhibitory activities. Not surprisingly, we found AHL-producing bacteria (which secrete hydrolytic exoenzymes) associated with these plants and their roots. We believe that the interplay of signals and signal inhibitors enables a stable coexistence of the eukaryotic host and the bacteria as long as the plant or root produces sufficient inhibitor to block the quorum-sensing systems of the colonizing organisms. Currently, work is in progress to characterize and isolate the pure compounds responsible for this quorum sensing–inhibitory activity.

Inhibition of quorum sensing by halogenated furanone compounds
The ability of bacteria to form biofilms is a major challenge for living organisms at risk of infection, such as humans, animals, and marine eukaryotes (41, 42). Marine plants are, in the absence of an advanced immune system, prone to disease (43, 44). Bacteria can be highly detrimental to marine algae and other eukaryotes (42). The Australian red macroalga *D. pulchra* produces a range of halogenated furanone compounds (45) that display antifouling and antimicrobial properties (46–48). This particular alga originally attracted the attention of marine biologists because it was devoid of surface colonization, i.e., biofouling, unlike other plants in the same environment. Biofouling is primarily caused by marine invertebrates and plants, but bacterial biofilms are believed to be the first colonizers of submerged surfaces, providing an initial conditioning biofilm to which other marine organisms may attach (49). Therefore, the abundance and composition of the bacterial community on the surface will significantly affect the subsequent

development of a macrofouling community (50, 51). Consequently, eukaryotes have developed chemical defense mechanisms (47, 52, 53) that, in several cases, include secondary metabolites that inhibit phenotypes relevant to bacterial colonization (54–56). Such secondary metabolites — furanones — are produced by the marine alga *D. pulchra* (54, 57, 58).

D. pulchra furanone compounds consist in general of a furan ring structure with a substituted acyl chain at the C-3 position and a bromine substitution at the C-4 position (see Figure 2). The substitution at the C-5 position may vary in terms of side chain structure. The natural furanone is halogenated at various positions by bromine, iodide, or chloride (45). *D. pulchra* produces at least 30 different species of halogenated furanone compounds, which are stored in specialized vesicles and are released at the surface of the thallus at a concentration ranging from 1 to 100 ng/cm². Field experiments have demonstrated that the surface concentration of furanones is inversely correlated to the degree of colonization by marine bacteria (55).

Givskov et al. (58) hypothesized that furanones of *D. pulchra* constitute a specific means of eukaryotic interference with bacterial signaling processes. An important discovery was the furanone-mediated displacement of radiolabeled AHL molecules from LuxR (59). This suggests that furanone compounds compete with the cognate AHL signal for the LuxR receptor site. Extensive experimental evidence in support of this model has accumulated during recent years. This includes the observations that furanones (a) repress AHL-dependent expression of *V. fischeri* bioluminescence (59); (b) inhibit AHL-controlled virulence factor production and pathogenesis in *P. aeruginosa* (60, 61); (c) inhibit quorum sensing–controlled luminescence and virulence of the black tiger prawn pathogen *Vibrio harveyi* (62); and, finally, (d) inhibit quorum sensing–controlled virulence of *E. carotovora* (63).

The natural furanone compounds have little or no effect on the quorum-sensing systems of *P. aeruginosa*. In collaboration with Staffan Kjelleberg's research group, we embarked on the process of drug development to find more potent quorum-sensing inhibitors. The natural furanone compounds were modified by chemical synthesis and screened for increased efficacy. Some derivatives of the *D. pulchra* furanone compounds were shown to repress quorum sensing in *P. aeruginosa* and reduce virulence factor expression (60, 61). Because synthetic compounds, which function well against planktonic cells, might be less efficient against biofilm bacteria, the efficacy of these quorum-sensing inhibitor compounds against bacterial biofilms was assayed. By means of AHL monitors built on the *P. aeruginosa* quorum sensors and the *lasB-gfp* target gene, the efficacy of these compounds was measured via GFP expression. The use of the GFP-based single-cell technology in combination with scanning confocal laser microscopy allowed estimation of furanone penetration and half-life and enabled us to identify synthetic compounds that not only inhibited the quorum sensors in the majority of the cells but also led to the formation of flat, undifferentiated biofilms that eventually detached (60). It is notable that the synthetic furanones, in concentrations that significantly lower quorum sensing–controlled gene expression in planktonic cells, were equally active against biofilm bacteria, despite the profoundly different modes of growth. In contrast, classical antibiotics used to treat *P. aeruginosa* infections, e.g., tobramycin and piperacillin, are required at concentrations 100- to 1,000-fold higher in order to kill biofilm bacteria than in order to kill their planktonic counterparts. In addition, we observed that furanone-treated biofilms were more susceptible to killing by tobramycin than their untreated counterparts (61) (Figure 3).

In a recently published report, we used DNA array technology to demonstrate that furanone compounds specifically repress expression of quorum sensing–controlled genes in *P. aeruginosa* (61). Microarray analysis of wild-type *P. aeruginosa* PAO1 showed that expression of 93 genes (1.7% of the *P. aeruginosa* genome) was affected by the addition of the furanone compound C-30. Overall, 85 genes (1.5%) were repressed and eight genes (0.1%) were activated in response to C-30. Genes encoding multidrug efflux pumps and transporters were predominant among the induced genes. The furanone-repressed genes included many previously known as quorum sensing–regulated genes, including numerous *P. aeruginosa* virulence factor genes such as *lasB* (encoding elastase), *lasA* (encoding LasA protease), *rhlAB* operon (regulating rhamnolipid production), *phzA-G* operon (encoding phenazine biosynthesis), *hcnABC* operon (regulating hydrogen cyanide production), and *chiC* (encoding chitinase). To determine whether the remaining furanone-repressed genes were in fact controlled by quorum sensing, parallel mapping of the quorum-sensing regulon was performed using a *lasI rhlI* double mutant grown with or without AHL signals. A comparative analysis showed that 80% of the furanone-repressed genes are indeed controlled by quorum sensing. Furthermore, furanone-repressed genes are not restricted to quorum sensing–regulated genes controlled by one of the two *P. aeruginosa* quorum-sensing circuits but may be controlled by either or both of the circuits (61). Microarray analysis demonstrated that expression of the *lasI/lasR* and *rhlI/rhlR* gene clusters, which encode the central components of the *P. aeruginosa* quorum-sensing system, was not notably affected by furanone treatment. This observation suggests that the furanone does not interfere with some of the regulatory systems controlling transcription of the *lasRI* and *rhlRI* genes, but rather that the furanone acts on these quorum-sensing regulators at the post-transcriptional level. Several genes involved in the biosynthesis of the *Pseudomonas* quinolone signal (PQS), including the *phnAB* operon and *pqsH* (64), were also repressed by the furanone compound. *Pseudomonas* quinolone signaling has been demonstrated, in concert with the AHL-based quorum-sensing systems, to be involved in the regulation of virulence factor production, in particular phenazine, pyocyanin, and hydrogen cyanide, and in autolysis of *P. aeruginosa* colonies (65).

A recent study has pointed at PQS signaling as an important regulatory function involved in *P. aeruginosa* adaptation and persistence in the cystic fibrotic lung environment (66). A pulmonary mouse infection model was used to study the effect of furanone compounds on the persistence of *P. aeruginosa* in chronic infections. Groups of mice infected with *P. aeruginosa* received subcutaneous furanone injections for 3 days, and this treatment was found to significantly reduce the bacterial load compared with that of the control group (67). Furthermore, the efficiency of bacterial clearing was positively correlated to the concentration of the furanone compound. The concentration used (as calculated by the whole-body concentration) was equal to or less than the concentrations required to inhibit expression of virulence factors in planktonic cultures and promote sloughing of in vitro biofilms.

Discussion

The purpose of research in this field has been to provide evidence that there are alternatives to the traditional mode of fighting bacterial infection. It is possible, from nature's own rich collection of chemical compounds, to generate powerful antipathogenic drugs that do not, per se, inhibit growth but instead interfere directly with microbial activity. The key to this concept is bacterial cell-to-cell communication. Knowledge of the molecular mechanisms underlying these signaling systems and their control of virulence, biofilm formation, and pathogenicity brings a completely new perspective to the potential control of microbial activity. Current halogenated furanones are too reactive, and therefore presumably too toxic, for the treatment of bacterial infections in humans. However, their proven ability to control *P. aeruginosa* infections in animal models is of considerable importance, since it demonstrates that quorum sensing is a useful and promising drug target in vivo. On the other hand, several obvious disadvantages are associated with AHL-based quorum-sensing antagonists. First, each antagonist has a narrow spectrum, and, therefore, specific antagonists have to be developed for each organism targeted. This might, however, prove advantageous in some scenarios. For instance, it would theoretically be possible to attenuate a single, pathogenic organism living in a mixed population of normal bacterial flora by a specific inhibitor while leaving the rest of the bacterial population unaffected. Second, the therapeutic use of AHL-based antagonists is complicated by the fact that some AHL signal molecules function as virulence factors per se, as they possess immunomodulatory activities and affect muscle tissue as well as tracheal gland cells.

The ability to control *P. aeruginosa* with antipathogenic drugs holds great promise that a whole range of opportunistic, pathogenic bacteria can be controlled by similar pharmaceuticals. *P. aeruginosa* is an attractive model organism for such studies, partly because of the recent development of suitable cDNA microarray technology by Affymetrix Inc. (Santa Clara, California, USA). We have used this technique to demonstrate the target specificity of certain first-generation antipathogenic drugs (61). We envision that this approach will be used in many primary research and pharmaceutical laboratories in the future quest for drugs that target specific cellular components or interactions. Given the large number of bacteria that employ quorum-sensing communication systems, chemical attenuation of unwanted bacterial activities rather than bactericidal or bacteriostatic strategies may find application in many different fields, e.g., in medicine, agriculture, and food technology. This new concept is highly attractive because it is unlikely to pose a selective pressure for the development of resistance. The present approach is therefore generic in nature and highly promising for defense against bacterial biofilms encountered in many infectious diseases, on medical implants, and in many industrial facilities and water pipelines.

Acknowledgments

The authors wish to acknowledge financial support from the Danish Technical Research Council, the Danish Medical Research Council, the Plasmid Foundation, the Villum Kann Rasmussen Foundation, and Cystic Fibrosis Foundation Therapeutics Inc.

Address correspondence to: Michael Givskov, Center for Biomedical Microbiology, BioCentrum, Building 301, Technical University of Denmark, DK-2800 Lyngby, Denmark. Phone: 45-45252769, or 45-21409867; Fax: 45-45932809; E-mail: immg@pop.dtu.dk.

1. Lewis, K. 2001. Riddle of biofilm resistance. *Antimicrob. Agents Chemother.* **45**:999–1007.

2. Hancock, R.E. 1998. Resistance mechanisms in *Pseudomonas aeruginosa* and other nonfermentative gram-negative bacteria. *Clin. Infect. Dis.* **27**(Suppl. 1):S93–S99.

3. Eberl, L. 1999. *N*-acyl homoserinelactone-mediated gene regulation in gram-negative bacteria. *Syst. Appl. Microbiol.* **22**:493–506.

4. Costerton, J.W., Stewart, P.S., and Greenberg, E.P. 1999. Bacterial biofilms: a common cause of persistent infections. *Science.* **284**:1318–1322.

5. Passador, L., Cook, J.M., Gambello, M.J., Rust, L., and Iglewski, B.H. 1993. Expression of *Pseudomonas aeruginosa* virulence genes requires cell-to-cell communication. *Science.* **260**:1127–1130.

6. von Bodman, S.B., and Farrand, S.K. 1995. Capsular polysaccharide biosynthesis and pathogenicity in *Erwinia stewartii* require induction by an *N*-acylhomoserine lactone autoinducer. *J. Bacteriol.* **177**:5000–5008.

7. Lewenza, S., Conway, B., Greenberg, E.P., and Sokol, P.A. 1999. Quorum sensing in *Burkholderia cepacia*: identification of the LuxRI homologs CepRI. *J. Bacteriol.* **181**:748–756.

8. Latifi, A., et al. 1995. Multiple homologues of LuxR and LuxI control expression of virulence determinants and secondary metabolites through quorum sensing in *Pseudomonas aeruginosa* PAO1. *Mol. Microbiol.* **17**:333–343.

9. Davies, D.G., et al. 1998. The involvement of cell-to-cell signals in the development of a bacterial biofilm. *Science.* **280**:295–298.

10. Huber, B., et al. 2001. The *cep* quorum-sensing system of *Burkholderia cepacia* H111 controls biofilm formation and swarming motility. *Microbiology.* **147**:2517–2528.

11. Lynch, M.J., et al. 2002. The regulation of biofilm development by quorum sensing in *Aeromonas hydrophila. Environ. Microbiol.* **4**:18–28.

12. Riedel, K., et al. 2001. *N*-acylhomoserine-lactone-mediated communication between *Pseudomonas aeruginosa* and *Burkholderia cepacia* in mixed biofilms. *Microbiology.* **147**:3249–3262.

13. Rasmussen, T.B., et al. 2000. How *Delisea pulchra* furanones affect quorum sensing and swarming motility in *Serratia liquefaciens* MG1. *Microbiology.* **146**:3237–3244.

14. Costerton, J.W., et al. 1987. Bacterial biofilms in nature and disease. *Annu. Rev. Microbiol.* **41**:435–464.

15. Xu, K.D., McFeters, G.A., and Stewart, P.S. 2000. Biofilm resistance to antimicrobial agents. *Microbiology.* **146**:547–549.

16. Hartman, G., and Wise, R. 1998. Quorum sensing: potential means of treating gram-negative infections? *Lancet.* **351**:848–849.

17. Chen, X., et al. 2002. Structural identification of a bacterial quorum-sensing signal containing boron. *Nature.* **415**:545–549.

18. Xavier, K.B., and Bassler, B.L. 2003. LuxS quorum sensing: more than just a numbers game. *Curr. Opin. Microbiol.* **6**:191–197.

19. Soo, Y.K., et al. 2003. Regulation of *Vibrio vulnificus* virulence by the LuxS quorum-sensing system. *Mol. Microbiol.* **48**:1647–1664.

20. Hubner, A., Revel, A.T., Nolen, D.M., Hagman, K.E., and Norgard, M.V. 2003. Expression of a *luxS* gene is not required for *Borrelia burgdorferi* infection of mice via needle inoculation. *Infect. Immun.* **71**:2892–2896.

21. Winzer, K., et al. 2002. LuxS: its role in central metabolism and the in vitro synthesis of 4-hydroxy-5-methyl-3(2H)-furanone. *Microbiology.* **148**:909–922.

22. Parsek, M.R., Val, D.L., Hanzelka, B.L., Cronan, J.E., Jr., and Greenberg, E.P. 1999. Acyl homoserine-lactone quorum-sensing signal generation. *Proc. Natl. Acad. Sci. U. S. A.* **96**:4360–4365.

23. Tateda, K., et al. 2001. Azithromycin inhibits quorum sensing in *Pseudomonas aeruginosa. Antimicrob. Agents Chemother.* **45**:1930–1933.

24. Pechere, J.C. 2001. Azithromycin reduces the production of virulence factors in *Pseudomonas aeruginosa* by inhibiting quorum sensing. *Jpn. J. Antibiot.* **54**:87–89.

25. Sofer, D., Gilboa-Garber, N., Belz, A., and Garber, N.C. 1999. 'Subinhibitory' erythromycin represses production of *Pseudomonas aeruginosa* lectins, autoinducer and virulence factors. *Chemotherapy.* **45**:335–341.

26. Yates, E.A., et al. 2002. N-acylhomoserine lactones undergo lactonolysis in a pH-, temperature-, and acyl chain length-dependent manner during growth of *Yersinia pseudotuberculosis* and *Pseudomonas aeruginosa. Infect. Immun.* **70**:5635–5646.

27. Dong, Y.H., Xu, J.L., Li, X.Z., and Zhang, L.H. 2000. AiiA, an enzyme that inactivates the acylhomoserine lactone quorum-sensing signal and attenuates the virulence of *Erwinia carotovora. Proc. Natl. Acad. Sci. U. S. A.* **97**:3526–3531.

28. Leadbetter, J.R., and Greenberg, E.P. 2000. Metabolism of acyl-homoserine lactone quorum-sensing signals by *Variovorax paradoxus. J. Bacteriol.* **182**:6921–6926.

29. Dong, Y.H., et al. 2001. Quenching quorum-sensing-dependent bacterial infection by an N-acyl homoserine lactonase. *Nature.* **411**:813–817.

30. Passador, L., et al. 1996. Functional analysis of the *Pseudomonas aeruginosa* autoinducer PAI. *J. Bacteriol.* **178**:5995–6000.

31. Schaefer, A.L., Hanzelka, B.L., Eberhard, A., and Greenberg, E.P. 1996. Quorum sensing in *Vibrio fischeri*: probing autoinducer-LuxR interactions with autoinducer analogs. *J. Bacteriol.* **178**:2897–2901.

32. Swift, S., et al. 1997. Quorum sensing in *Aeromonas hydrophila* and *Aeromonas salmonicida*: identification of the LuxRI homologs AhyRI and AsaRI and their cognate N-acylhomoserine lactone signal molecules. *J. Bacteriol.* **179**:5271–5281.

33. Swift, S., et al. 1999. Quorum sensing-dependent regulation and blockade of exoprotease production in *Aeromonas hydrophila. Infect. Immun.* **67**:5192–5199.

34. Zhu, J., et al. 1998. Analogs of the autoinducer 3-oxo-octanoyl-homoserine lactone strongly inhibit activity of the TraR protein of *Agrobacterium tumefaciens. J. Bacteriol.* **180**:5398–5405.

35. Chhabra, S.R., et al. 1993. Autoregulation of carbapenem biosynthesis in *Erwinia carotovora* by analogues of N-(3-oxohexanoyl)-L-homoserine lactone. *J. Antibiot. (Tokyo.)* **46**:441–454.

36. McClean, K.H., et al. 1997. Quorum sensing and *Chromobacterium violaceum*: exploitation of violacein production and inhibition for the detection of N-acylhomoserine lactones. *Microbiology.* **143**:3703–3711.

37. Kline, T., et al. 1999. Novel synthetic analogs of the *Pseudomonas* autoinducer. *Bioorg. Med. Chem. Lett.* **9**:3447–3452.

38. Smith, K.M., Bu, Y., and Suga, H. 2003. Induction and inhibition of *Pseudomonas aeruginosa* quorum sensing by synthetic autoinducer analogs. *Chem. Biol.* **10**:81–89.

39. Teplitski, M., Robinson, J.B., and Bauer, W.D. 2000. Plants secrete substances that mimic bacterial N-acyl homoserine lactone signal activities and affect population density-dependent behaviors in associated bacteria. *Mol. Plant Microbe Interact.* **13**:637–648.

40. Cha, C., Gao, P., Chen, Y.C., Shaw, P.D., and Farrand, S.K. 1998. Production of acyl-homoserine lactone quorum-sensing signals by gram-negative plant-associated bacteria. *Mol. Plant Microbe Interact.* **11**:1119–1129.

41. Kushmaro, A., Loya, Y., Fine, E., and Rosenberg, E. 1996. Bacterial infection and coral bleaching. *Nature.* **380**:396.

42. Littler, M.M., and Littler, D.S. 1995. Impact of CLOD pathogen on Pacific coral reefs. *Science.* **267**:1356–1360.

43. Correa, J.A. 1996. Diseases in seaweeds: an introduction. *Hydrobiologia.* **326**:87–88.

44. Fenical, W. 1997. New pharmaceuticals from marine organisms. *Trends Biotechnol.* **15**:339–341.

45. de Nys, R., Wright, A.D., König, G.M., and Sticher, O. 1993. New halogenated furanones from the marine alga *Delisea pulchra. Tetrahedron.* **49**:11213–11220.

46. de Nys, R., Steinberg, P., Rogers, C.N., Charlton, T.S., and Duncan, M.W. 1996. Quantitative variation of secondary metabolites in the sea hare *Aplysia parvula* and its host plant, *Delisea pulchra. Mar. Ecol. Prog. Ser.* **130**:135–146.

47. de Nys, R., et al. 1995. Broad spectrum effects of secondary metabolites from the red alga *Delisea pulchra* in antifouling assays. *Biofouling.* **8**:259–271.

48. Reichelt, J.L., and Borowitzka, M.A. 1984. Antimicrobial activity from marine algae: results of a large-scale screening programme. *Hydrobiology.* **116/117**:158–168.

49. Rice, S.A., Givskov, M., Steinberg, P., and Kjelleberg, S. 1999. Bacterial signals and antagonists: the interaction between bacteria and higher organisms. *J. Mol. Microbiol. Biotechnol.* **1**:23–31.

50. Belas, M.R. 2003. The swarming phenomenon of *Proteus mirabilis. ASM News.* **58**:15–22.

51. Henschel, J.R., and Cook, P.A. 1990. The development of a marine fouling community in relation to the primary film of microorganisms. *Biofouling.* **2**:1–11.

52. Davis, A.R., et al. 1989. Epibiosis of marine algae and benthic invertebrates: natural products chemistry and other mechanisms inhibiting settlement and overgrowth. *Bioorganic Marine Chemistry.* **3**:86–114.

53. Wahl, M. 2003. Marine epibiosis. Fouling and antifouling: some basic aspects. *Mar. Ecol. Prog. Ser.* **58**:175–189.

54. Kjelleberg, S., et al. 1997. Do marine natural products interfere with prokaryotic AHL regulatory systems? *Aquat. Microb. Ecol.* **13**:85–93.

55. Maximilien, R.R., et al. 1998. Chemical mediation of bacterial surface colonisation by secondary metabolites from the red alga *Delisea pulchra. Aquat. Microb. Ecol.* **15**:233–246.

56. Slattery, M., McClintoch, J.B., and Heine, J.N. 1995. Chemical defences in Antarctic soft corals: evidence for antifouling compounds. *J. Exp. Mar. Biol. Ecol.* **190**:61–77.

57. de Nys, R., et al. 1995. Broad spectrum effects of secondary metabolites from the red alga *Delisea pulchra* in antifouling assays. *Biofouling.* **8**:259–271.

58. Givskov, M., et al. 1996. Eukaryotic interference with homoserine lactone-mediated prokaryotic signalling. *J. Bacteriol.* **178**:6618–6622.

59. Manefield, M., et al. 1999. Evidence that halogenated furanones from *Delisea pulchra* inhibit acylated homoserine lactone (AHL)-mediated gene expression by displacing the AHL signal from its receptor protein. *Microbiology.* **145**:283–291.

60. Hentzer, M., et al. 2002. Inhibition of quorum sensing in *Pseudomonas aeruginosa* biofilm bacteria by a halogenated furanone compound. *Microbiology.* **148**:87–102.

61. Hentzer, M., et al. 2003. Attenuation of *Pseudomonas aeruginosa* virulence by quorum sensing inhibitors. *EMBO J.* **22**:3803–3815.

62. Manefield, M., Harris, L., Rice, S.A., de Nys, R., and Kjelleberg, S. 2000. Inhibition of luminescence and virulence in the black tiger prawn (*Penaeus monodon*) pathogen *Vibrio harveyi* by intercellular signal antagonists. *Appl. Environ. Microbiol.* **66**:2079–2084.

63. Manefield, M., Welch, M., Givskov, M., Salmond, G.P., and Kjelleberg, S. 2001. Halogenated fura-

nones from the red alga, *Delisea pulchra*, inhibit carbapenem antibiotic synthesis and exoenzyme virulence factor production in the phytopathogen *Erwinia carotovora*. *FEMS Microbiol. Lett.* **205**:131–138.

64. Gallagher, L.A., McKnight, S.L., Kuznetsova, M.S., Pesci, E.C., and Manoil, C. 2002. Functions

required for extracellular quinolone signaling by *Pseudomonas aeruginosa*. *J. Bacteriol.* **184**:6472–6480.

65. D'Argenio, D.A., Calfee, M.W., Rainey, P.B., and Pesci, E.C. 2002. Autolysis and autoaggregation in *Pseudomonas aeruginosa* colony morphology mutants. *J. Bacteriol.* **184**:6481–6489.

66. Guina, T., et al. 2003. Quantitative proteomic

analysis indicates increased synthesis of a quinolone by *Pseudomonas aeruginosa* isolates from cystic fibrosis airways. *Proc. Natl. Acad. Sci. U. S. A.* **100**:2771–2776.

67. Wu, H., et al. 2000. Detection of *N*-acylhomoserine lactones in lung tissues of mice infected with *Pseudomonas aeruginosa*. *Microbiology.* **146**:2481–2493.

Pseudomonas aeruginosa quorum sensing as a potential antimicrobial target

Roger S. Smith[1] and Barbara H. Iglewski[2]

[1]Department of Microbiology and Molecular Genetics, Harvard Medical School, Boston, Massachusetts, USA.
[2]Department of Microbiology and Immunology, University of Rochester School of Medicine and Dentistry, Rochester, New York, USA.

Pseudomonas aeruginosa has two complete quorum-sensing systems. Both of these systems have been shown to be important for Pseudomonas virulence in multiple models of infection. Thus, these systems provide unique targets for novel antimicrobial drugs.

Quorum sensing in *Pseudomonas aeruginosa*

Pseudomonas aeruginosa is a Gram-negative organism that is commonly found in soil and water. Although *P. aeruginosa* can survive under multiple harsh conditions, it is an opportunistic pathogen and is only able to infect hosts with defective immune system function, such as that observed in individuals with cystic fibrosis, burns, and HIV (1). To facilitate the establishment of infection, *P. aeruginosa* produces an impressive array of both cell-associated and extracellular virulence factors. Several of these virulence factors have been demonstrated to be regulated by quorum sensing (QS). QS is the mechanism whereby an individual bacterium produces small diffusible molecules that can be detected by surrounding organisms. In *P. aeruginosa*, and most Gram-negative bacteria, these signal molecules are acyl homoserine lactones (AHLs). Only when the concentration of AHLs in the environment increases, potentially because of increasing numbers of bacteria, are intracellular levels of AHLs sufficient to maximally induce the activation of transcriptional regulators. This mechanism of communication enables bacteria to act as a community in the coordinated regulation of gene expression. This regulated expression of virulence genes is thought to give the bacteria a selective advantage over host defenses and thus is important for the pathogenesis of the organism.

There are two QS systems in *P. aeruginosa*, which have been extensively studied. The *las* system consists of the LasR transcriptional regulator and the LasI synthase protein. LasI is essential for the production of the AHL signal molecule *N*-(3-oxododecanoyl)-L-homoserine lactone ($3O-C_{12}$-HSL) (2, 3). LasR requires $3O-C_{12}$-HSL in order to become an active transcription factor. It was recently demonstrated that, in the presence of $3O-C_{12}$-HSL, LasR forms multimers, and that only the multimeric form of this protein is able to bind DNA and regulate the transcription of multiple genes (Figure 1a) (4). A second QS system in *P. aeruginosa* consists of the RhlI and RhlR proteins. The RhlI synthase produces the AHL *N*-butyryl-L-homoserine lactone (C_4-HSL), and RhlR is the transcriptional regulator (5, 6). Only when RhlR is complexed with C_4-HSL does it regulate the expression of several genes. Both $3O-C_{12}$-HSL and C_4-HSL have been shown to freely diffuse out of bacterial cells; however, $3O-C_{12}$-HSL diffusion is significantly

slower than that of C_4-HSL. Removal of $3O-C_{12}$-HSL from bacteria is most efficiently accomplished via the MexAB-OprM efflux system (7). Recently, a third LuxR homologue termed QscR was identified, which has been shown to regulate the transcription of both *lasI* and *rhlI* (8). Although QscR exhibits significant homology to LasR and RhlR, it is currently unknown whether an AHL or similar molecule is needed to stimulate QscR function. Data indicate that *qscR* is important in regulating the production of several virulence factors but that this regulation may occur through control of the expression of both the *las* and the *rhl* systems.

Role of QS in the global regulation of *P. aeruginosa* genes

The importance of tight regulation of QS gene expression and AHL production has become evident with our increase in knowledge regarding QS-regulated genes. Several studies have identified numerous genes regulated by QS in *P. aeruginosa*, many of which are virulence factors (9, 10). With the sequencing of the *P. aeruginosa* genome and the availability of microarray technologies, a more comprehensive evaluation of QS regulation was recently undertaken. Three individual research groups have used microarray experiments to analyze the QS-regulated transcriptome of *P. aeruginosa* (11–13). All three studies used independently derived mutant *lasI/rhlI* strains of *P. aeruginosa* PAO1. Gene-expression levels were determined for this mutant bacterium when grown with or without exogenous $3O-C_{12}$-HSL and C_4-HSL. Schuster et al. (12) also used a *P. aeruginosa* PAO1 strain in which *lasR* and *rhlR* had been deleted. In all three studies, an overwhelmingly large number of genes were shown to be regulated by QS, with 3–7% of all *P. aeruginosa* open reading frames affected. Data from Hentzer et al. (13) represent only those genes that were induced fivefold or greater in comparison with control cultures, while the studies of Schuster et al. (12) and Wagner et al. (11) report all genes induced by QS. Although in many instances several genes were identified by only one of the three groups, a large number of genes were identified in two or more of the studies, and 97 induced genes were identified in all three studies (Figure 2).

Wagner et al. also identified 222 genes that showed statistically significant repression by the addition of AHLs to cultures (11), while Schuster et al. found only 38 such genes (12). Interestingly, there was little overlap between these two data sets. The disparity between these studies may in part be due to experimental design, including the choice of growth media, aeration of cultures, and the concentration of added AHLs. Not surprisingly, we found that the induction of various genes was greatly affected by the growth conditions used.

Nonstandard abbreviations used: quorum sensing (QS); acyl homoserine lactone (AHL); *N*-(3-oxododecanoyl)-L-homoserine lactone ($3O-C_{12}$-HSL); *N*-butyryl-L-homoserine lactone (C_4-HSL); acyl–acyl carrier protein (acyl-ACP).

Conflict of interest: The authors have declared that no conflict of interest exists.

Citation for this article: *J. Clin. Invest.* **112**:1460–1465 (2003). doi:10.1172/JCI200320364.

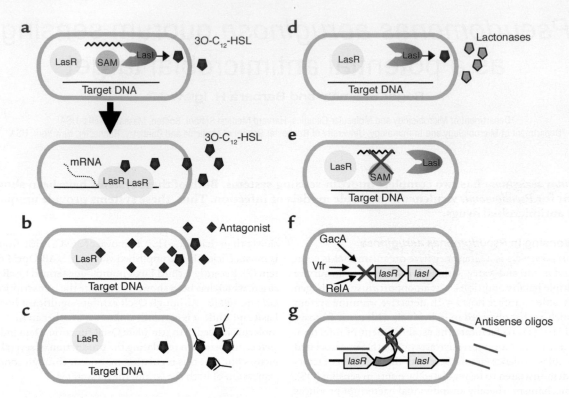

Figure 1
Potential QS targets for the inhibition of *P. aeruginosa* virulence. For simplicity, only the *las* QS system is shown; however, similar mechanisms could be used to inhibit the *rhl* system as well. (**a**) *P. aeruginosa* LasI synthase utilizes *S*-adenosyl methionine (SAM) and acyl-ACP to form 3O-C$_{12}$-HSL. As the density of the bacteria, and thus of 3O-C$_{12}$-HSL, increases, the 3O-C$_{12}$-HSL molecule binds to the LasR regulator, resulting in dimerization, DNA binding, and transcription of multiple genes. (**b**) Antagonistic analogues of cognate AHLs compete for binding to LasR but do not result in activation of the protein. (**c**) Specific antibodies bind to AHLs as they exit the bacteria, inhibiting their re-entry and thus inhibiting activation of LasR as well as their interaction with host cells. (**d**) Lactonases degrade AHLs as they leave the bacteria, thus inhibiting their activation of LasR and host cells. (**e**) Targeting the expression of LasI substrates would prevent the production of 3O-C$_{12}$-HSL, and thus QS activation. (**f**) Multiple factors have been shown to regulate *lasR* and *lasI*. Drugs that inhibit these factors would result in altered QS activation. (**g**) Specific antisense oligonucleotides (oligos) pair with *lasR* or *lasI* RNA and inhibit gene translation and thus protein production.

Changes in media composition and oxygen concentration appeared to have significant effects on the genes up- or downregulated by QS (11). Therefore, when evaluating variation in gene expression, it is important to consider the influence of growth medium, oxygen concentration, and other environmental conditions that may impact QS-regulated gene expression. Additionally, a similar more stringent statistical analysis of all three data sets may reveal additional overlap between experiments. Despite some differences in the modes of analysis of the QS-regulated transcriptome, a large group of genes was common to all three reports that appeared to be regulated by QS.

The majority of the genes regulated by QS were found to be hypothetical or of unknown function. However, the remaining genes encoded, for example, membrane proteins, putative enzymes, transcription factors, and secreted enzymes involved in a broad range of cellular functions, including two-component regulatory systems, energy metabolism, small-molecule transport, and export apparatus. In all three studies, a large number of the known genes were found to be probable virulence factors; however, several other identified genes may also fall into this category. To evaluate the interactions between QS factors and QS-regulated genes, an analysis was performed to identify QS-regulated genes that contained consensus DNA sequences for the binding of LasR and RhlR. Surprisingly, only approximately 7% of the QS-regulated genes pos-

sessed upstream DNA sequences homologous to known LasR- and RhlR-binding domains indicating that QS regulates many genes via indirect mechanisms (11, 12). This observation correlated with the demonstration that a large number of the QS-regulated genes were classified as transcriptional regulators or as members of a two-component regulatory system (11). By inducing these regulatory factors, QS may indirectly affect multiple mechanisms that regulate gene expression, and this may explain the low occurrence of specific DNA-binding domains for the QS regulators.

Overall, the data from these three studies demonstrate the global effects of QS in *P. aeruginosa*; however, the complete QS-regulated transcriptome may yet be incomplete. Although many genes were identified to be QS regulated, it is important to keep in mind that under different experimental conditions additional genes may be differentially regulated by QS.

Role of *P. aeruginosa* QS in pathogenesis
Considering that QS regulates such a wide range of factors that play such diverse roles in the function of *P. aeruginosa*, it is important to examine how QS affects the pathogenesis of this bacterium. Many studies examining QS during infection have used strains of *P. aeruginosa* that have deletions of one or more of the QS-related genes, in addition to wild-type strains. Studies of the role of

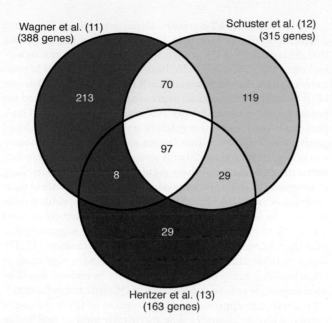

Wagner et al. (11)
(388 genes)

Schuster et al. (12)
(315 genes)

213

70

119

97

8

29

29

Hentzer et al. (13)
(163 genes)

Figure 2
Analyses of the QS-regulated transcriptome of *P. aeruginosa*. A comparison of QS-induced genes from microarray experiments performed by three different research groups is shown. The data reported by Hentzer et al. (13) represent only those genes that were induced five-fold or more. The studies by Wagner et al. (11) and Schuster et al. (12) include all induced genes. A total of 97 QS-regulated genes were common to all three studies.

P. aeruginosa QS in infection using a burnt-mouse model, a murine model of acute pneumonia, and a rat model of chronic lung infection have all demonstrated that deletions of one or more QS genes result in reduced *P. aeruginosa* virulence compared with wild-type *P. aeruginosa* (14–17). These data confirm that both the *las* and the *rhl* QS systems are important for *P. aeruginosa* to disseminate, which leads to septicemia, induces both acute and chronic lung infections, and causes pathology and mortality. QS has also been shown to be functional during *P. aeruginosa* infections in humans. In sputum samples from cystic fibrosis patients colonized with *P. aeruginosa*, levels of transcripts for QS genes were found to correlate with those of QS-regulated genes, indicating that QS was regulating their expression during infection (18, 19). Furthermore, AHLs were directly measured in the sputum of cystic fibrosis patients chronically colonized with *P. aeruginosa* (19). Collectively, these studies indicate that functional QS systems significantly affect the severity of both acute and chronic *P. aeruginosa* infections.

Additional studies have demonstrated that AHLs produced by *P. aeruginosa* are able to interact with eukaryotic cells and to stimulate the production of various factors that may affect the pathogenesis of this bacterium. In vitro experiments have shown that purified 3O-C$_{12}$-HSL stimulates the production of the inflammatory cytokine IL-8 from human lung bronchial epithelial cells (20, 21). It was subsequently demonstrated that 3O-C$_{12}$-HSL could stimulate a broad-spectrum response in vivo by inducing several inflammatory cytokines and chemokines (15). It has also been shown that 3O-C$_{12}$-HSL can inhibit the production of IL-12 and TNF-α from LPS-activated mouse peritoneal exudate cells or human PBMCs (22, 23). These data demonstrate that, under certain conditions, 3O-C$_{12}$-HSL also acts as an immunosuppressor.

In additional studies characterizing this response in leukocytes, it was observed that the structure of the AHL molecule was important for regulation of cytokine production. AHLs with a 3-oxo or 3-hydroxy substitution and an acyl chain of 12–14 carbons were the most active molecules in these experiments; however, AHLs with acyl side chains shorter than eight carbons were inactive (23).

These data demonstrate the importance of AHL production to *P. aeruginosa*. Not only are these molecules important for cell-to-cell communication between bacteria and the regulation of multiple bacterial factors, but they may also directly act as virulence factors. These data also suggest that AHL interactions with various host cells induce different responses. This diverse stimulation may result in different immune responses during various stages of *P. aeruginosa* infection. Therefore it is important to acknowledge that AHLs may have several different effects on the pathogenesis of *P. aeruginosa*.

Inhibition of QS as a therapeutic approach to *P. aeruginosa* infections

Based on the facts that QS regulates such an array of *P. aeruginosa* factors and that deletion of QS regulators attenuates *P. aeruginosa* virulence, it is conceivable that QS would be an ideal target for the inhibition of *Pseudomonas* infections. Multidrug-resistant *P. aeruginosa* are becoming more prevalent, and current antimicrobial treatments for cystic fibrosis are unable to eradicate *P. aeruginosa* infections. Therefore, alternative mechanisms for targeting *P. aeruginosa* have been the focus of much research. Therapeutics that target and inhibit QS in *P. aeruginosa* would attenuate the virulence of the bacterium and thus potentially assist the host immune response in clearing the infection.

Several components of the QS networks represent ideal targets for potential therapeutics: (a) LasR and RhlR activation (b) AHL formation and activity, and (c) *lasR/lasI* and *rhlR/rhlI* expression.

LasR and RhlR activation. The most promising mechanism for inhibiting LasR or RhlR activation is the use of AHL analogues that act as antagonists for 3O-C$_{12}$-HSL and C$_4$-HSL. These molecules would most likely be similar in structure to natural AHLs produced by *P. aeruginosa* and would compete for binding to LasR or RhlR proteins (Figure 1b). Structural variations of the 3O-C$_{12}$-HSL and C$_4$-HSL molecules have revealed epitopes that are important for the activation and inhibition of LasR activity (24, 25). A recent study used a reporter assay to identify a group of compounds containing a common aniline-ring structure with a hydrogen-bond acceptor that were able to compete with 3O-C$_{12}$-HSL and subsequently inhibit the activation of LasR and elastase production (25). These data demonstrate that use of a high-throughput assay for the screening of large libraries of AHL analogues may be fruitful in the identification of additional QS inhibitors. Although neither LasR nor RhlR has been purified and structurally analyzed, the structure of TraR, a homologous transcriptional regulator in *Agrobacterium tumefaciens*, has been extensively studied (26, 27). Data from the crystal structure of this protein have been immensely important in understanding how TraR and similar proteins may interact with AHLs and DNA. Similar studies in *P. aeruginosa* would be extremely useful in elucidating potential antagonists and also for enhancing our understanding of how cognate AHLs interact with LasR and RhlR.

Additional studies have identified a synthetic halogenated-furanone compound that is able to inhibit the production of many QS-induced factors (13). Microarray analysis of *P. aeruginosa* following exposure to this compound revealed that 80% of the genes repressed were QS regulated, thus revealing the specificity of the furanone for the QS system. In a mouse model of chronic *P. aeruginosa* lung infection, animals treated with 0.7 µg/g body weight of this furanone compound demonstrated a 3-log decrease in the number of bacteria that could be isolated from their lungs. These data are very exciting, as QS inhibitors have not previously been shown to be effective during in vivo infections. This compound and other similar analogues hold significant promise as potential treatments for patients fighting *P. aeruginosa* infection. The most significant advantage to this method of QS inhibition is in the small size and ease of delivery of these molecules.

AHL formation and activity. As previously discussed, these molecules diffuse or are actively pumped out of the bacterial cell. Therefore, once they appear in the extracellular environment, they are potential targets for destruction or inactivation. We have evaluated the potential of using AHL-specific antibodies to inhibit AHL activity. After AHLs diffuse out of the bacterial cell into the extracellular environment, they can potentially be bound by these antibodies and prevented from interacting with eukaryotic cells or from re-entering the bacterial cell and activating their cognate transcriptional regulators (Figure 1c). Preliminary experiments in our laboratory have demonstrated that mAb's made from a $3O$-C_{12}-HSL–protein conjugate were able to substantially inhibit $3O$-C_{12}-HSL activation of a *lasB* transcriptional reporter in *P. aeruginosa* as well as the production of IL-8 from epithelial cells (unpublished observations). Although additional studies are needed to further evaluate this approach, these data suggest that antibodies specific for AHLs inhibit QS and may be useful as therapeutics.

Numerous studies have recently identified lactonase enzymes, which are produced by several bacteria including *Bacillus* sp., *Variovorax paradoxus*, *Arthrobacter* sp., and *A. tumefaciens* (28–31). These enzymes cleave the lactone ring and thus produce nonfunctional molecules that are unable to activate their cognate transcriptional regulators (Figure 1d). When one of these lactonases was expressed in *P. aeruginosa*, there was a significant decrease in AHL production and virulence factor expression (32). Although these enzymes may not be ideal therapeutics, because of the difficulty in delivering active enzymes to the site of infection, they should prove to be strong tools for the study of QS. Identification of other bacterial strains that produce these enzymes may also provide insight into bacterial interactions in various environments.

Since the discovery of QS in *P. aeruginosa*, extensive research has evaluated the mechanisms involved in AHL production. LasI and RhlI synthase proteins utilize components of the amino acid and fatty acid biosynthesis pathways to produce AHLs. The homoserine lactone rings are typically derived from *S*-adenosyl methionine (SAM) and the acyl chains from the acyl-acyl carrier protein (acyl-ACP) pools in the bacteria (33, 34). Although LasI has a preference for 12-carbon acyl chains, it is able to utilize chains of varying lengths. Recent in vitro studies have shown that alterations in fatty acid biosynthesis that result in increased pools of short-chain acyl-ACPs saturate LasI and result in decreased production of the active AHL, $3O$-C_{12}-HSL (35). These data demonstrate that LasI is very sensitive to environmental conditions and that control of LasI substrates may have a significant effect on functional AHL production (Figure 1e). Therefore, drugs that target this pathway or inhibit LasI or RhlI functions would be ideal candidates to achieve inhibition of virulence factor production.

lasR/lasI and rhlR/rhlI expression. Given that QS regulates such a broad spectrum of factors in *P. aeruginosa*, it is not surprising that the regulation of QS genes is itself controlled. Several factors, including GacA, Vfr, and RelA, have been demonstrated to positively regulate the expression of LasR (Figure 1f) (1, 9). Deletion of *vfr* virtually eliminated all expression of *lasR* and reduced the production of virulence factors. Additionally, a deletion of GacA, part of a two-component regulatory system, resulted in reduced production of many virulence factors and a decrease in pathogenesis in a burnt-mouse model of *P. aeruginosa* infection. These data reveal the significance of these regulatory factors in controlling *P. aeruginosa* QS and virulence production. The importance of these other factors was also alluded to in transcriptional-profile studies in which the addition of exogenous AHLs to low-density cultures was unable to prematurely induce gene expression, suggesting that other factors need to be produced prior to QS induction (12, 13). Based on these data, various factors involved in the regulation of QS expression, many of which may not yet be known, would be ideal targets for potential disruption of QS virulence induction.

An alternative approach to the inhibition of QS is the use of antisense oligonucleotides that specifically bind to *lasR/lasI* or *rhlR/rhlI* transcripts and inhibit gene expression (Figure 1g). The use of antisense technologies to specifically inhibit translation of particular genes has been comprehensively studied in eukaryotic systems, and many of these technologies are currently being evaluated in clinical trials (36). However, the efficacy of these techniques for the inhibition of bacterial genes has not been extensively evaluated. Studies in *Staphylococcus aureus*, *Mycobacterium tuberculosis*, and *Escherichia coli* have successfully demonstrated that antisense oligonucleotides can specifically bind to target transcripts and inhibit gene expression (37–39). Although such an approach has multiple obstacles, such as cell wall permeability and efficacy of the mode of delivery, studies using modified oligonucleotides have shown significant promise (37, 38). The advantages of antisense technology are that antisense regulation is a common phenomenon in bacteria and that this mechanism of regulation is substantially different from most antimicrobials (39).

QS is an attractive therapeutic target because of the role that it plays in the global regulation of multiple *P. aeruginosa* factors and the importance of this role for the virulence of the organism in multiple different infections. As with any drug that is used to inhibit bacterial infections, there is a concern that resistant mutants may arise during the course of treatment. Inhibition of QS is not immune from such possibilities. It was previously demonstrated that *las*-deficient *P. aeruginosa* strains grown under selective pressure gave rise to spontaneous mutations that restored production of certain *las*-regulated virulence factors; however, such mutations could not be found when both *las* and *rhl* were deleted (1). These data demonstrate the importance of targeting both QS systems by therapeutics to effectively inhibit virulence factor production. Because QS can be inhibited in numerous ways, several different approaches may be used to achieve a more complete inhibition. Additionally, in concert with currently used antipseudomonal therapies, these inhibitors may be potent drugs for the eradication of *P. aeruginosa* infections.

Address correspondence to: Barbara H. Iglewski, Department of Microbiology and Immunology, University of Rochester School of Medicine and Dentistry, 601 Elmwood Avenue, Box 672, Rochester, New York 14642, USA. Phone: (585) 275-3402; Fax: (585) 473-9573; E-mail: bigl@mail.rochester.edu.

1. Van Delden, C., and Iglewski, B.H. 1998. Cell-to-cell signaling and *Pseudomonas aeruginosa* infections. *Emerg. Infect. Dis.* **4**:551–560.

2. Gambello, M.J., and Iglewski, B.H. 1991. Cloning and characterization of the *Pseudomonas aeruginosa* lasR gene, a transcriptional activator of elastase expression. *J. Bacteriol.* **173**:3000–3009.

3. Pearson, J.P., et al. 1994. Structure of the autoinducer required for expression of *Pseudomonas aeruginosa* virulence genes. *Proc. Natl. Acad. Sci. U. S. A.* **91**:197–201.

4. Kiratisin, P., Tucker, K.D., and Passador, L. 2002. LasR, a transcriptional activator of *Pseudomonas aeruginosa* virulence genes, functions as a multimer. *J. Bacteriol.* **184**:4912–4919.

5. Ochsner, U.A., Koch, A.K., Fiechter, A., and Reiser, J. 1994. Isolation and characterization of a regulatory gene affecting rhamnolipid biosurfactant synthesis in *Pseudomonas aeruginosa*. *J. Bacteriol.* **176**:2044–2054.

6. Pearson, J.P., Passador, L., Iglewski, B.H., and Greenberg, E.P. 1995. A second N-acylhomoserine lactone signal produced by *Pseudomonas aeruginosa*. *Proc. Natl. Acad. Sci. U. S. A.* **92**:1490–1494.

7. Pearson, J.P., Van Delden, C., and Iglewski, B.H. 1999. Active efflux and diffusion are involved in transport of *Pseudomonas aeruginosa* cell-to-cell signals. *J. Bacteriol.* **181**:1203–1210.

8. Chugani, S.A., et al. 2001. QscR, a modulator of quorum-sensing signal synthesis and virulence in *Pseudomonas aeruginosa*. *Proc. Natl. Acad. Sci. U. S. A.* **98**:2752–2757.

9. de Kievit, T.R., and Iglewski, B.H. 2000. Bacterial quorum sensing in pathogenic relationships. *Infect. Immun.* **68**:4839–4849.

10. Whiteley, M., Lee, K.M., and Greenberg, E.P. 1999. Identification of genes controlled by quorum sensing in *Pseudomonas aeruginosa*. *Proc. Natl. Acad. Sci. U. S. A.* **96**:13904–13909.

11. Wagner, V.E., Bushnell, D., Passador, L., Brooks, A., and Iglewski, B. 2003. Microarray analysis of *Pseudomonas aeruginosa* quorum-sensing regulons: effects of growth phase and environment. *J. Bacteriol.* **185**:2080–2095.

12. Schuster, M., Lostroh, P., Ogi, T., and Greenberg, E.P. 2003. Identification, timing, and signal specificity of *Pseudomonas aeruginosa* quorum-controlled genes: a transcriptome analysis. *J. Bacteriol.* **185**:2066–2079.

13. Hentzer, M., et al. 2003. Attenuation of *Pseudomonas aeruginosa* virulence by quorum sensing inhibitors. *EMBO J.* **22**:3803–3815.

14. Rumbaugh, K.P., Griswold, J.A., Iglewski, B.H., and Hamood, A.N. 1999. Contribution of quorum sensing to the virulence of *Pseudomonas aeruginosa* in burn wound infections. *Infect. Immun.* **67**:5854–5862.

15. Smith, R.S., Harris, S.G., Phipps, R., and Iglewski, B.H. 2002. The *Pseudomonas aeruginosa* quorum-sensing molecule N-(3-oxododecanoyl)homoserine lactone contributes to virulence and induces inflammation in vivo. *J. Bacteriol.* **184**:1132–1139.

16. Pearson, J.P., Feldman, M., Iglewski, B.H., and Prince, A. 2000. *Pseudomonas aeruginosa* cell-to-cell signaling is required for virulence in a model of acute pulmonary infection. *Infect. Immun.* **68**:4331–4334.

17. Wu, H., et al. 2001. *Pseudomonas aeruginosa* mutations in lasI and rhlI quorum sensing systems result in milder chronic lung infection. *Microbiology.* **147**:1105–1113.

18. Storey, D.G., Ujack, E.E., Rabin, H.R., and Mitchell, I. 1998. *Pseudomonas aeruginosa* lasR transcription correlates with the transcription of lasA, lasB, and toxA in chronic lung infections associated with cystic fibrosis. *Infect. Immun.* **66**:2521–2528.

19. Erickson, D.L., et al. 2002. *Pseudomonas aeruginosa* quorum-sensing systems may control virulence factor expression in the lungs of patients with cystic fibrosis. *Infect. Immun.* **70**:1783–1790.

20. DiMango, E., Zar, H.J., Bryan, R., and Prince, A. 1995. Diverse *Pseudomonas aeruginosa* gene products stimulate respiratory epithelial cells to produce interleukin-8. *J. Clin. Invest.* **96**:2204–2210.

21. Smith, R.S., et al. 2001. IL-8 production in human lung fibroblasts and epithelial cells activated by the *Pseudomonas* autoinducer N-3-oxododecanoyl homoserine lactone is transcriptionally regulated by NF-kappaB and activator protein-2. *J. Immunol.* **167**:366–374.

22. Telford, G., et al. 1998. The *Pseudomonas aeruginosa* quorum-sensing signal molecule N-(3-oxododecanoyl)-L-homoserine lactone has immunomodulatory activity. *Infect. Immun.* **66**:36–42.

23. Chhabra, S.R., et al. 2003. Synthetic analogues of the bacterial signal (quorum sensing) molecule N-(3-oxododecanoyl)-l-homoserine lactone as immune modulators. *J. Med. Chem.* **46**:97–104.

24. Passador, L., Tucker, K., Guertin, K., Kende, A., and Iglewski, B. 1996. Functional analysis of the *Pseudomonas aeruginosa* autoinducer PAI. *J. Bacteriol.* **178**:5595–6000.

25. Smith, K.M., Bu, Y., and Suga, H. 2003. Library screening for synthetic agonists and antagonists of a *Pseudomonas aeruginosa* autoinducer. *Chem. Biol.* **10**:563–571.

26. Zhang, R.G., et al. 2002. Structure of a bacterial quorum-sensing transcription factor complexed with pheromone and DNA. *Nature.* **417**:971–974.

27. Vannini, A., et al. 2002. The crystal structure of the quorum sensing protein TraR bound to its autoinducer and target DNA. *EMBO J.* **21**:4393–4401.

28. Dong, Y.-H., Gusti, A.R., Zhang, Q.-L., Xu, J., and Zhang, L.-H. 2002. Identification of quorum-quenching N-acyl homoserine lactonases from *Bacillus* species. *Appl. Environ. Microbiol.* **68**:1754–1759.

29. Leadbetter, J.R., and Greenberg, E.P. 2000. Metabolism of acyl-homoserine lactone quorum-sensing signals by *Variovorax paradoxus*. *J. Bacteriol.* **182**:6921–6926.

30. Park, S.Y., et al. 2003. AhlD, an N-acylhomoserine lactonase in *Arthrobacter sp.*, and predicted homologues in other bacteria. *Microbiology.* **149**:1541–1550.

31. Carlier, A., et al. 2003. The Ti plasmid of *Agrobacterium tumefaciens* harbors an attM-paralogous gene, aiiB, also encoding N-Acyl homoserine lactonase activity. *Appl. Environ. Microbiol.* **69**:4989–4993.

32. Reimmann, C., et al. 2002. Genetically programmed autoinducer destruction reduces virulence gene expression and swarming motility in *Pseudomonas aeruginosa* PAO1. *Microbiology.* **148**:923–932.

33. Parsek, M.R., Val, D.L., Hanzelka, B.L., Cronan, J.E., and Greenberg, E.P. 1999. Acyl homoserine-lactone quorum-sensing signal generation. *Proc. Natl. Acad. Sci. U. S. A.* **98**:4360–4365.

34. Hoang, T.T., and Schweizer, H.P. 1999. Characterization of the *Pseudomonas aeruginosa* enoyl-acyl carrier protein reductase: a target for triclosan and its role in acylated homoserine lactone synthesis. *J. Bacteriol.* **181**:5489–5497.

35. Hoang, T.T., Sullivan, S.A., Cusick, J.K., and Schweizer, H.P. 2002. β-Ketoacyl acyl carrier protein reductase (FabG) activity of the fatty acid biosynthetic pathway is a determining factor of 3-oxohomoserine lactone acyl chain lengths. *Microbiology.* **148**:3849–3856.

36. Kurreck, J. 2003. Antisense technologies. Improvement through novel chemical modifications. *Eur. J. Biochem.* **270**:1628–1644.

37. Harth, G., Zamecnik, P.C., Tang, J.-Y., Tabatadze, D., and Horwitz, M. 2000. Treatment of *Mycobacterium tuberculosis* with antisense oligonucleotides to glutamine synthetase mRNA inhibits glutamine synthetase activity, formation of the poly L-glutamate/glutamine cell wall structure, and bacterial replication. *Proc. Natl. Acad. Sci. U. S. A.* **97**:418–423.

38. Good, L., Awasthi, S.K., Dryselius, R., Larsson, O., and Nielsen, P. 2001. Bactericidal antisense effects of peptide-PNA conjugates. *Nat. Biotechnol.* **19**:360–364.

39. Ji, Y., Zhang, et al. 2001. Identification of critical *Staphylococcal* genes using conditional phenotypes generated by antisense RNA. *Science.* **293**:2266–2269.

Antimalarial drug resistance

Nicholas J. White[1,2]

[1]Faculty of Tropical Medicine, Mahidol University, Bangkok, Thailand. [2]Centre for Vaccinology and Tropical Medicine, Churchill Hospital, Oxford, United Kingdom.

Malaria, the most prevalent and most pernicious parasitic disease of humans, is estimated to kill between one and two million people, mainly children, each year. Resistance has emerged to all classes of antimalarial drugs except the artemisinins and is responsible for a recent increase in malaria-related mortality, particularly in Africa. The de novo emergence of resistance can be prevented by the use of antimalarial drug combinations. Artemisinin-derivative combinations are particularly effective, since they act rapidly and are well tolerated and highly effective. Widespread use of these drugs could roll back malaria.

Malaria is a hematoprotozoan parasitic infection transmitted by certain species of anopheline mosquitoes (Figure 1). Four species of plasmodium commonly infect humans, but one, *Plasmodium falciparum*, accounts for the majority of instances of morbidity and mortality. There has been a resurgence of interest in malaria in recent years as the immensity of the burden it imposes on poor countries in the tropics has become apparent, and as efforts at control have foundered after the failure of the global eradication campaign in the 1960s. Control has traditionally relied on two arms: control of the anopheline mosquito vector through removal of breeding sites, use of insecticides, and prevention of contact with humans (via the use of screens and bed nets, particularly ones that are impregnated with insecticides); and effective case management. A long-hoped-for third arm, an effective malaria vaccine, has not materialized and is not expected for another decade. Case management has relied largely on antimalarials (mainly chloroquine, and more recently sulfadoxine-pyrimethamine [SP]), which are inexpensive and widely available and are eliminated slowly from the body. Together with antipyretics, antimalarials are among the most commonly used medications in tropical areas of the world. Misuse is widespread. In many parts of the tropics, the majority of the population has detectable concentrations of chloroquine in the blood. The extensive deployment of these antimalarial drugs, in the past fifty years, has provided a tremendous selection pressure on human malaria parasites to evolve mechanisms of resistance (Table 1). The emergence of resistance, particularly in *P. falciparum*, has been a major contributor to the global resurgence of malaria in the last three decades (1). Resistance is the most likely explanation for a doubling of malaria-attributable child mortality in eastern and southern Africa (2).

P. falciparum is now highly resistant to chloroquine in most malaria-affected areas. Resistance to SP is also widespread and has developed much more rapidly. Resistance to mefloquine is confined only to those areas where it has been used widely (Thailand, Cambodia, and Vietnam) but has arisen within six years of systematic deployment (3). The epidemiology of resistance in *Plasmodium vivax* is less well studied; chloroquine resistance is serious only in parts of Indonesia, Papua New Guinea, and adjacent areas. SP resistance in *P. vivax* is more widespread.

Unfortunately, most malaria-affected countries have less than $10 per capita annually to spend on all aspects of health, and so for a disease that is one of the most common causes of fever, a treatment cost of more than 50 cents becomes prohibitive. As a result, the nationally recommended treatment in most countries is antimalarial drugs (i.e., chloroquine or SP), which are partially or completely ineffective. The effects of resistance on morbidity and mortality are usually underestimated (4, 5). Predicting the emergence and spread of resistance to current antimalarials and newly introduced compounds is necessary for planning malaria control and instituting strategies that might delay the emergence of resistance (6). Resistance has already developed to all the antimalarial drug classes with one notable exception — the artemisinins. These drugs are already an essential component of treatments for multidrug-resistant falciparum malaria (7). If we lose artemisinins to resistance, we may be faced with untreatable malaria. In this review, the emergence of resistance to current antimalarial drugs is considered in two parts: first, the initial genetic event that produces the resistant mutant, and second, the subsequent selection process in which the survival advantage in the presence of the antimalarial drug leads to preferential transmission and the spread of resistance (8).

Antimalarial drug resistance

Most symptomatic malaria infections are uncomplicated and manifest as fever, chills, malaise, often abdominal discomfort, and mild anemia. In falciparum malaria, the mortality associated with this presentation is approximately 0.1%, if effective drugs are readily available. In a small proportion of *P. falciparum* infections, untrammeled parasite multiplication leads to heavy parasite burdens, which produce vital-organ dysfunction with impairment of consciousness, acidosis, and more severe anemia. Seizures, hypoglycemia, and severe anemia are common manifestations of severe malaria in children, whereas jaundice, pulmonary edema, and acute renal failure are more common in adults. The mortality despite treatment rises to 15–20%. As death in severe malaria usually occurs within 48 hours of presentation, i.e., one asexual cycle of the blood-stage infection, it is mainly the current generation of *P. falciparum* malaria parasites (i.e., those parasites present when the patient presents to medical attention) that will determine whether the patient lives or dies, and so prevention of their maturation from the less pathogenic circulating ring stages (0–16 hours) to the more pathogenic sequestered stages is important. Stage specificity of drug action is therefore an important consideration. But in uncomplicated malaria, inhibition of parasite multiplication has greater importance, as this prevents the progression to severe disease and leads to resolu-

---

Nonstandard abbreviations used: intermittent presumptive treatment (IPT); minimum inhibitory concentration (MIC); parasite multiplication rate (PMR); sulfadoxine-pyrimethamine (SP).

Conflict of interest: The author has declared that no conflict of interest exists.

Citation for this article: *J. Clin. Invest.* **113**:1084–1092 (2004). doi:10.1172/JCI200421682.

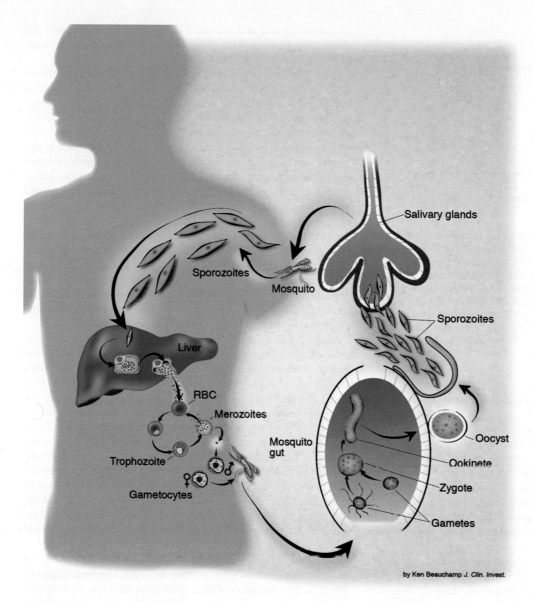

Figure 1
The life cycle of malaria parasites in the human host and anopheline mosquito vector.

tion of fever and other symptoms. Inhibition of parasite multiplication is a first-order process, which leads to a log-linear reduction in parasite numbers with time (9). Uninhibited blood-stage multiplication at 100% efficiency results in a parasite multiplication rate (PMR) equal to the median number of viable merozoites liberated by rupturing schizonts (5). In vivo efficiencies may exceed 50% in nonimmune patients, resulting in PMRs of approximately 10 per asexual cycle (10). Antimalarial drugs convert this positive value to a negative value, resulting in PMRs that range between 10^{-1} and 10^{-4} per cycle. These negative PMRs are also termed parasite killing rates or parasite reduction ratios (11). The higher values (i.e., lower killing rates) are obtained after therapy with drugs with relatively weak antimalarial activity, such as tetracyclines, and the highest values are obtained with artemisinin derivatives (Figures 2 and 3). Drug resistance to an anti-infective compound is defined by a right shift in the concentration-effect (dose-response) relationship (Figure 4). For uncomplicated malaria this refers to prevention of multiplica-

tion, and so for any given free plasma concentration of antimalarial drug there is less inhibition of parasite multiplication as resistance increases.

Genetic basis of antimalarial drug resistance

The genetic events that confer antimalarial drug resistance (while retaining parasite viability) are spontaneous and rare and are thought to be independent of the drug used. They are mutations in or changes in the copy number of genes encoding or relating to the drug's parasite target or influx/efflux pumps that affect intraparasitic concentrations of the drug (Table 1). A single genetic event may be all that is required, or multiple unlinked events may be necessary (epistasis). As the probability of multigenic resistance arising is the product of the individual component probabilities, this is a significantly rarer event. *P. falciparum* parasites from Southeast Asia have been shown to have an increased propensity to develop drug resistance (12).

Chloroquine resistance in *P. falciparum* may be multigenic and is initially conferred by mutations in a gene encoding a transporter (*Pf*CRT) (13). In the presence of *Pf*CRT mutations, mutations in a second transporter (*Pf*MDR1) modulate the level of resistance in vitro, but the role of *Pf*MDR1 mutations in determining the therapeutic response following chloroquine treatment remains unclear (13). At least one other as-yet unidentified gene is thought to be involved. Resistance to chloroquine in *P. falciparum* has arisen spontaneously less than ten times in the past fifty years (14). This suggests that the per-parasite probability of developing resistance de novo is on the order of 1 in 10^{20} parasite multiplications. The single point mutations in the gene encoding cytochrome *b* (*cytB*), which confer atovaquone resistance, or in the gene encoding dihydrofolate reductase (*dhfr*), which confer pyrimethamine resistance, have a per-parasite probability of arising de novo of approximately 1 in 10^{12} parasite multiplications (5). To put this in context, an adult with

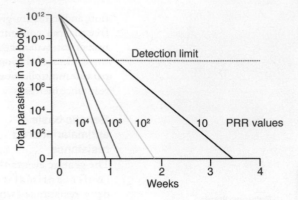

Figure 2
Pharmacodynamics: the parasite reductions produced by the different antimalarial drugs in vivo (in an adult patient with 2% parasitemia). Parasite reduction ratios (PRR; fractional reduction per asexual cycle) vary from less than 10 (antibiotics with antimalarial activity, antimalarials for which resistance is high grade) to 10,000 (artemisinin derivatives). Antimalarial drugs must be present at levels greater than the minimum inhibitory concentration (MIC) until eradication of the infection in nonimmune patients to ensure cure of the infection. Adapted with permission from from *Trends in Parasitology* (60).

approximately 2% parasitemia has 10^{12} parasites in his or her body. But in the laboratory, much higher mutation rates thane 1 in every 10^{12} are recorded (12).

Mutations may be associated with fitness disadvantages (i.e., in the absence of the drug they are less fit and multiply less well than their drug-sensitive counterparts). Another factor that may explain the discrepancy between in vitro and much lower apparent in vivo rates of spontaneous mutation is host immunity. Even a previously nonimmune individual develops a specific immune response to a malaria infection. This response is systematically evaded by the parasite population through programmed antigenic variation of the main red cell surface–expressed epitopes. In falciparum malaria, *P. falciparum* erythrocyte membrane protein 1 (*Pf*EMP1), which is encoded by the *var* multigene family, changes in 2–3% of parasites each asexual cycle (15). The untreated infection is characterized by successive waves of parasites, each comprising largely one antigenically distinct surface phenotype. It is likely that this specific immune response directed against the immunodominant surface antigens will reduce the probability of the usually single mutant parasite ever multiplying sufficiently to transmit as for *P. falciparum*; there is only a 2–3% chance that the genetic event causing resistance would arise in the antigenically variant subpopulation that will expand to reach transmissible densities.

The cause of chloroquine resistance in *P. vivax* has not been found. Resistance to mefloquine and other structurally related arylaminoalcohols in *P. falciparum* results from amplifications (i.e., duplications, not mutations) in *Pfmdr*, which encodes an energy-demanding p-glycoprotein pump (Pgh) (16–19). This is a more common genetic event. It is tempting to speculate that the relatively poor fidelity in mitotic duplication of this sequence has evolved to allow parasite populations to respond to environmental stresses, such as alterations in human diet. But the gene amplifications may well confer a fitness disadvantage to the parasite once the population stress has passed. The consequences of these various genetic events are reduced intracellular concentrations of the antimalarial quinolines (the relative importance

of reduced uptake and increased efflux remains unresolved). All these drugs interfere with the parasites' ability to detoxify heme liberated from hemoglobin.

For *P. falciparum* and *P. vivax*, resistance to antifols (pyrimethamine and cycloguanil) results from the sequential acquisition of mutations in *dhfr* (13). Each mutation confers a stepwise reduction in susceptibility. Resistance to the sulfonamides and sulfones, which are often administered in synergistic combination with antifols, also results from sequential acquisition of mutations in the gene *dhps*, which encodes the target enzyme dihydropteroate synthase (20). Resistance to atovaquone results from point mutations in the gene *cytB*, coding for cytochrome *b*. Atovaquone is deployed only in a fixed combination with proguanil (chloroguanide). In this combination, it is proguanil itself acting on the mitochondrial membrane, rather than the *dhfr*-inhibiting proguanil metabolite cycloguanil, that appears to be the important actor. Whether and how resistance develops to proguanil's mitochondrial action are not known (21). Although the target for the artemisinins has recently been identified (*Pf*ATPase6) (22), preliminary studies have not so far associated polymorphisms in the gene encoding this enzyme with reduced susceptibility to artemisinins (18).

Assuming an equal distribution of probabilities of spontaneous occurrence throughout the malaria parasites' life cycle, the genetic event resulting in resistance is likely to take place in only a single parasite at the peak of infection. These genetic events may result in moderate changes in drug susceptibility, such that the drug still remains effective (e.g., the serine-to-asparagine mutation at position 108 in *Pfdhfr* that confers pyrimethamine resistance), or, less commonly, very large reductions in susceptibility, such that achievable concentrations of the drug are completely ineffective (e.g., the mutations in *cytB* that confer atovaquone resistance) (16, 21, 23). It had been thought that resistance to some antimalarial compounds (notably pyrimethamine and SP) in human malaria parasites emerged relatively frequently. This suggested that prevention of the emergence of resistance would be very difficult, and control efforts would be better directed at limiting the subsequent spread of resistance. Recent remarkable molecular epidemiological studies in South America, southern Africa, and Southeast Asia have challenged this view. By examina-

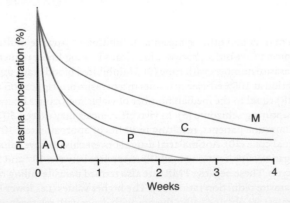

Figure 3
Pharmacokinetic properties of the generally available antimalarial drugs. The origin represents the maximum concentration (100%) achieved after a therapeutic dose. A, artemisinins; Q, quinine; P, pyrimethamine; C, chloroquine; M, mefloquine. Adapted with permission from from *Trends in Parasitology* (60).

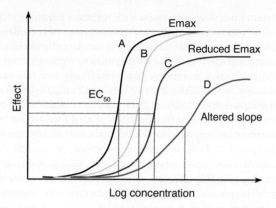

Figure 4
The dose-response curve in malaria. Increasing drug resistance leads to a rightward shift in the dose-response or concentration-effect relationship. The principal effect in uncomplicated malaria is parasite killing. This shift can be parallel, or the shape of the curve and the maximum effect can change. Adapted with permission from *Trends in Parasitology* (60).

tion of the sequence of the regions flanking the *Pfdhfr* gene, it has become apparent that, even for SP, multiple de novo emergence of resistance has not been a frequent event, and that, instead, a single parasite (with a mutation in *Pfdhfr* at positions 51, 59, and 108) has in recent years swept across each of these continents (24–26). The ability of these resistant organisms to spread has been phenomenal and may well relate to the apparent stimulation of gametocytogenesis that characterizes poor therapeutic responses to SP (27). Gametocyte carriage is considerably augmented following SP treatment of resistant infections. Studies to date do not suggest reduced infectivity for these gametocytes. There is a sigmoid relationship between gametocyte densities in blood and infectivity, which in volunteer studies was shown to saturate at gametocyte densities above 1,000 per microliter (a relatively high density in field observations). Thus it is the relative transmission advantage conferred by increased gametocyte carriage that drives the spread of resistance (5, 8).

Step 1: de novo selection of resistance
In experimental animal models, drug resistance mutations can be selected for, without mosquito passage (i.e., without meiotic recombination), by exposure of large numbers of malaria parasites (in vitro, in animals, or, in the past, in volunteers) to subtherapeutic antimalarial drug concentrations (28).

In order to assess the factors determining the emergence and spread of resistance, we need to consider the numbers of malaria parasites likely to be exposed to the drugs, both within an individual and in the entire human population. Fortunately this estimate of parasite numbers is much more precise than for almost any other human pathogen. Malaria parasites are eukaryotes. Meiosis occurs after a female anopheline mosquito has taken viable gametocytes in its blood meal. All the other 10^8–10^{13} cell divisions in the life cycle are mitotic. Nearly all these divisions take place in the bloodstream of the human host. Usually, less than ten sporozoite parasites are inoculated by an infected mosquito in order to establish malaria infection (29, 30) (Figure 1). These rapidly find their way to the liver. During *P. falciparum* infection, each infected hepatocyte liberates approximately 30,000 merozoites after 5–6 days of pre-erythrocytic schizogony. Thus approximately 100,000–300,000 merozoites are liberated into the bloodstream to begin the 48-hour asexual reproduction cycle. This is an important number, as it is the number of parasites that would encounter residual drug levels from a previous antimalarial treatment or drug levels during chemoprophylaxis (see below) (8). The density of parasites in the blood at which symptoms and fever occur (the pyrogenic density), and thus the stage at which appropriate antimalarial treatment could be given, vary considerably (31–33). In nonimmune people, nonspecific symptoms often occur a day or two before parasites are detectable on the blood smear (about 50 parasites per microliter of blood). This density corresponds to a total of between 10^8 and 10^9 asexual parasites in an adult with a red cell volume of about 2 l. In areas of moderate- or high-intensity transmission, parasitemias considerably higher than this level may be tolerated without symptoms, although densities over 10,000 per microliter (between 10^{10} and 10^{11} parasites in the body of an adult, and correspondingly less in children) are usually symptomatic, even in very high-transmission settings (34). Median or geometric mean parasite counts in malariometric surveys are usually below this value (i.e., most people with detectable parasitemias in these endemic areas are not obviously ill). It is estimated that approximately 300 million people in the world now have malaria parasites in their blood. Using current epidemiological data we have estimated that there must be less than 3×10^{16} malaria parasites in the world's asymptomatic carriers (Figure 5) (8).

Geometric mean or median admission parasitemias in clinical studies of falciparum malaria usually lie between 5,000 and 50,000 per microliter, with the lower figure coming from low-transmission settings, and the higher figure from high-transmission settings. Thus, if between one million and ten million people are symptomatic in any 2-day period (i.e., 180 million to 1800 million

Table 1
Factors determining the probability of selection of de novo antimalarial drug resistance

1. The frequency with which the resistance mechanism arises
2. The fitness cost to the parasite associated with the resistance mechanism
3. The number of parasites in the human host that are exposed to the drug
4. The concentrations of drug to which these parasites are exposed (i.e., the doses used and pharmacokinetic properties of the antimalarial drug or drugs)
5. The pharmacodynamic properties of the antimalarial drug or drugs
6. The degree of resistance (the shift in the concentration-effect relationship) that results from the genetic changes
7. The level of host defense (nonspecific and specific immunity)
8. The simultaneous presence of other antimalarial drugs or substances in the blood that will still kill the parasite if it develops resistance to one drug (i.e., the use of combinations)

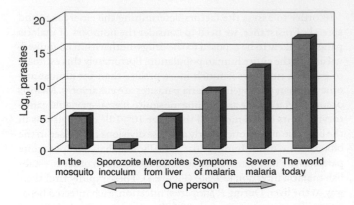

Figure 5
Total numbers of malaria parasites (log scale), from inoculation by
an anopheline mosquito, through the development of infection in the
human host, to the total estimated in the world today.

symptomatic infections per year), then, based on their age distribution and their blood volumes and parasitemias, these ill people
would contain between 5×10^{16} and 5×10^{17} malaria parasites (8).

Thus, on any day, although the majority of people infected with
malaria are asymptomatic, a significant proportion, and probably the majority, of malaria parasites in the world are in people
who are ill. It has been argued that, if the probability of a de novo
resistance mutation arising is distributed evenly among all these
parasites, then, because of their logarithmic distribution, those
patients with high parasitemias who survive their infection to
transmit viable gametocytes carry a significant proportion of
all the world's "potentially transmissible" malaria parasites (8).
They must therefore be an important potential source of resistance. Although mortality is increased in hyperparasitemic infections, thereby stopping transmission, if the patient survives, the
chance of *selection* and preferential survival of such a drug-resistant
mutant from this patient is greater than if a similar mutant arose
in a person with much lower parasitemia. This is because a hyperparasitemic patient has limited immunity, particularly against the
"strain" causing the infection (otherwise a high parasitemia would
not have developed). The immune response kills parasites irrespective of their sensitivity to antimalarial drugs. A hyperparasitemic
patient will receive antimalarial drug treatment, whereas nearly
all the asymptomatic patients will not, and the hyperparasitemic
patient may also be seriously ill, resulting in vomiting and malabsorption of antimalarial treatment. High parasite counts are associated with a higher chance of treatment failure than infections
with lower parasite numbers (35).

Step 2: the spread of resistance
In the emergence and spread of resistance to antimalarial drugs,
there are many parallels with antibiotic resistance (36, 37) — particularly antituberculous drug resistance, where, as for malaria,
transferable resistance genes are not involved in the emergence of
resistance. Resistance to one drug may be selected for by another
drug in which the mechanism of resistance is similar (a phenomenon known as cross-resistance). Antimalarial resistance in malaria parasites spreads because it confers a survival advantage in the
presence of the antimalarial and therefore results in a greater probability of transmission for resistant than for sensitive parasites.
Resistant infections are more likely to recrudesce, and eventually,

as resistance worsens, infections with resistant parasites respond
more slowly to treatment. Both increased rates of recrudescence
and slow initial responses to treatment increase the likelihood of
generating sufficient gametocyte densities to transmit, compared
with drug-sensitive infections. Mathematically, it is this ratio of
transmission probabilities in drug-resistant compared with drug-
sensitive infections that drives the spread of resistance. The recrudescence and subsequent transmission of an infection that generated resistant malaria parasites de novo are essential for resistance
to be propagated (5). If resistance is low grade (i.e., a small shift in
the concentration-effect relationship), or combination treatment
is given that is highly effective, then resistance may confer only a
very small increase in the treatment failure rate, and a correspondingly slow rate of spread. As resistance worsens, failure rates rise,
and the rate of spread accelerates. In the rare but important infection in which resistance arises de novo, killing of the transmissible
sexual stages (gametocytes) during the primary infection does *not*
affect resistance, because these gametocytes derive from drug-
sensitive parasites. Gametocytes carrying the resistance genes will
not reach transmissible densities until the resistant biomass has
expanded to a population size close to that necessary to produce
illness ($>10^7$ parasites) (38). Thus, to prevent spread of resistance,
gametocyte production from the subsequent recrudescent-resis-
tant infection must be prevented.

The central role of immunity in preventing the emergence and spread of resistance in high-transmission settings
Immunity to malaria is acquired slowly and imperfectly. A state of
sterile immunity against all infections is never attained. Malaria
parasites, like other successful parasites, have developed sophisticated immune-evasion strategies. In low-transmission areas, where
infections are acquired infrequently (e.g., less than three times a
year), the majority of malaria infections are symptomatic and
selection of resistance therefore takes place in the context of specific antimalarial treatment. Relatively large numbers of parasites
in an individual encounter antimalarial drugs. In higher-transmis-
sion areas the majority of infections are asymptomatic and these
are acquired repeatedly throughout life. Symptomatic and sometimes fatal disease occurs in the first years of life, but thereafter
malaria becomes increasingly likely to be asymptomatic. In areas of

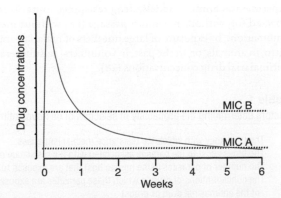

Figure 6
A slowly eliminated antimalarial such as chloroquine or piperaquine
presents a lengthy opportunity for the selection of resistance among
sensitive parasites (MIC A), but once resistance has become established (MIC B), the terminal elimination phase is no longer selective,
because the blood concentrations are no longer inhibitory.

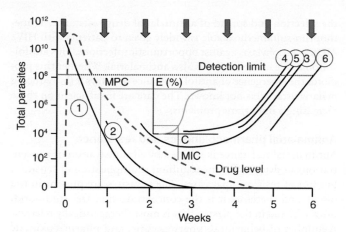

Figure 7

Opportunities for the de novo selection of antimalarial drug resistance in an area of high transmission (entomological inoculation rate 50 per year; each inoculation is depicted as a green arrow) in a young child treated for acute falciparum malaria with a slowly eliminated drug such as mefloquine (red dotted line). The initial infection (infection 1) is eliminated. The next infection acquired (infection 2) is also eliminated. Infections 3 and 4 are suppressed temporarily but eventually reach detectable densities. Infections 5 and 6 are under no selection pressure and also reach detectable densities. The inset shows the pharmacodynamic events, the relationship between concentration (C) and effect (E). When mefloquine levels fall below the minimum parasiticidal concentration (MPC) giving maximum parasite killing (E_{max}), then the rate of decline in parasitemia (PRR) falls until the PRR reaches 1. This results from an MIC of mefloquine and occurs in infection 3. Thereafter, parasitemia rises again and becomes detectable nearly 6 weeks after initial treatment.

higher malaria transmission, people still receive antimalarial treatments throughout their lives, as the term malaria is often used to describe any type of fever, and most treatment is given empirically without microscopy or dipstick confirmation. But these inappropriate treatments for other febrile illnesses are largely unrelated to the peaks of parasitemia, which reduces the individual probability of resistance selection (8). Host defense mechanisms contribute a major antiparasitic effect, with which any spontaneously generated drug-resistant mutant malaria parasite must contend. This

reduces significantly the survival probability of individual malaria parasites. Even if the resistant mutant does survive the initial drug treatment and multiplies, the chance that this will result in sufficient gametocytes for transmission is reduced as a result of both asexual-stage immunity (which reduces the multiplication rate and lowers the density at which the infection is controlled) and specific antigametocyte (transmission-blocking) immunity. Furthermore, other parasite genotypes are likely to be present, since infections are acquired continuously. These compete with the resistant parasites for red cells and increase the possibility of outbreeding of multigenic-resistance mechanisms or competition in the feeding anopheline mosquito. These factors, which reduce the probability of selecting for and transmitting resistance in high transmission areas, are balanced against the increased frequency of vector biting, and thus the increased probability that a feeding anopheline will encounter the resistance-bearing gametocytes. In some areas of the tropics, new malaria infections are acquired more than once each day. Even if the resistance-bearing parasites do establish themselves in the anopheline mosquito, they must still be transmitted to a susceptible recipient for resistance to spread. In areas where the majority of the population is immune, the individual probability of propagation is reduced, as inoculation in a subsequent mosquito-feeding event often does not result in an infection capable of being transmitted (i.e., an infection generating sufficient gametocytes for onward transmission).

In high-transmission areas, where malaria-associated illness and death are largely confined to young children, the chance of a drug encountering large numbers of parasites in a semi-immune host is confined to the first few years of life. The net result is considerable reduction in the probability of de novo selection and subsequent transmission of a resistant parasite mutant in high-transmission compared with low-transmission areas. Historically, chloroquine resistance emerged in low-transmission areas, and antifol resistance has increased more rapidly in low-transmission than in high-transmission areas.

Pregnancy

The control of malaria infections is impaired in pregnancy. In low-transmission settings, *P. falciparum* infections are more severe, but at all levels of transmission there is an associated reduction in birth weight of infants born to mothers with malaria (both with

Table 2

Antimalarial drug resistance; identified associations in the malaria parasite (gene products; mutation positions in corresponding resistant alleles)

Organism	Resistant to	Low- to intermediate-level resistance	High-level resistance
P. falciparum	Chloroquine	CRT; 76	CRT; 76 and other mutations, MDR; 86, and other undefined gene products
	Mefloquine, halofantrine, lumefantrine	MDR; amplification of wild-type allele	
	Pyrimethamine	DHFR; 108 then 51 and 59	DHFR; 108 + 51 + 59 + 164
	Cycloguanil, Chlorcycloguanil	DHFR; 16 + 108	DHFR; 108 + 51 + 59 + 164
	Atovaquone	No low-level resistance documented	Cytochrome *b*; 133 ± 280
	Sulfonamides and sulfones	DHPS; 436, 437, 540, 581, 613[A]	
	Artemisinin and derivatives	No resistance	No resistance
P. vivax	Pyrimethamine	DHFR; 117 + 58	DHFR; 117 + 58 + 59 + 61 + 13
	Chloroquine	Unknown	Unknown

CRT, chloroquine resistance transporter; MDR, multidrug resistance p-glycoprotein pump; DHFR, dihydrofolate reductase; DHPS, dihydropteroate synthase. Numbers refer to the positions of point mutations associated with resistance. [A]The relationship between mutations in *Pfdhps* and levels of resistance is still unclear.

P. falciparum and with *P. vivax* infection). For *P. falciparum* the adverse effects are greatest in primigravidae (39). The placenta is a site of *P. falciparum* sequestration and appears to be a "privileged" site for parasite multiplication, although exactly how this local immune paresis to malaria parasites operates is unclear. This has implications for the greater emergence and spread of resistance, which have not been evaluated. Responses to antimalarial drug treatment regimes in low-transmission settings are always worse in pregnant women compared with age-matched nonpregnant women from the same location (40). Treatment failures drive the development of resistance. The placenta may contain large numbers of parasites, thereby increasing the selection probability. These parasites are usually of a single surface-antigen phenotype (they bind to chondroitin sulphate A, and hyaluronic acid), suggesting expression of a single conserved *var* gene (41). After establishment of the infection in a pregnant woman who has not had malaria in pregnancy before (usually a primigravida in an endemic area), the infecting parasites are not apparently selected by the immune response to surface-expressed antigens, and so if a drug-resistant mutation arises it does not need to arise in a variant subpopulation to ensure its survival. Other factors also favor the emergence and spread of resistance. Antimalarial drug pharmacokinetics are usually altered, often with an expanded apparent volume of distribution (quinine, mefloquine, atovaquone, and proguanil), resulting in lower drug levels for any given dose. There are even data suggesting that pregnant women are more attractive to mosquitoes (42). It is widely recommended that pregnant women receive antimalarial prophylaxis, but the only drugs considered safe are chloroquine, which is ineffective against *P. falciparum* nearly everywhere, and proguanil, to which widespread resistance exists, and which has reduced biotransformation to the active antifol metabolite cycloguanil. Prophylaxis for pregnant women has given way to intermittent presumptive treatment (IPT) with SP, in which a treatment dose is given two or three times during the pregnancy, although SP is falling to resistance. Since in IPT the antimalarial drug is usually administered to healthy women, the biomass of parasites confronted by the drug is less than that present in symptomatic infections, but how much less has not been investigated. Taken together, these observations suggest that pregnant women could be an important contributor to antimalarial drug resistance.

HIV infection

There is now increasing evidence that there is an interaction between falciparum malaria and HIV infection. In settings of high malaria transmission, malaria is largely a problem of childhood, whereas HIV has higher mortality rates in infants and adults. But with the increasing availability of antiretroviral drugs, HIV-infected patients will live longer, and so the two infections will coincide more often. HIV coinfection in pregnancy is associated with greater reduction in birth weight than that associated with malaria infection alone (43). IPT with SP must be given monthly in order to achieve the same improvements in birth weight as 8–12 weekly administrations in HIV-negative pregnant women. Compared with HIV-negative nonimmune patients, more severe malaria is seen in HIV-infected nonimmune patients, and severely immunocompromised HIV-infected patients in high-transmission settings have higher parasite densities (44, 45). This suggests that the immunosuppression associated with HIV infection can affect the control of malaria-parasite numbers and would therefore compromise the effect of antimalarial immunity in reducing

the selection and spread of antimalarial drug resistance. Trimethoprim-sulfamethoxazole is widely given to patients with HIV/AIDS as prophylaxis against opportunistic infections. This antifol-sulfonamide combination is also antimalarial. Whether this promotes the emergence of antifol resistance or delays it (by reducing malaria attacks) is not known. The data are insufficient on these increasingly important problems.

Antimalarial pharmacokinetics and resistance

Antimalarial resistance is selected by administration of concentrations of drug sufficient to inhibit multiplication of sensitive, but not of resistant, parasites. The parasites are present in the blood, and therefore it is the concentration of free (unbound) drug achieved in the plasma that is most therapeutically relevant. A number of behavioral, pharmaceutic, and pharmacokinetic factors affect the probability of parasites encountering subtherapeutic levels of antimalarial agents. Several antimalarial drugs (notably lumefantrine, halofantrine, atovaquone, and, to a lesser extent, mefloquine) are lipophilic, hydrophobic, and quite variably absorbed (interindividual variation in bioavailability varies up to 20-fold) (46, 47). There is also large interindividual variability in distribution volumes. Together these result in considerable interindividual variations in blood concentration profiles (48). Since doses are chosen based on the therapeutic ratio — which defines the difference between a therapeutically effective dose and a dose capable of inducing adverse effects — poor oral bioavailability with a consequent wide range in blood levels will favor the emergence of resistance. Improving oral bioavailability thus reduces doses required to clear infection (and thus reduces costs) and should reduce the emergence and spread of resistance.

All people living in a high-transmission area have some malaria parasites in their blood all the time, and each person harbors many different parasite genotypes (although many are at densities below the level of PCR detection). Antimalarial treatment for symptomatic malaria exposes not only the parasites causing that infection to the drug, but also any newly acquired infections that emerge from the liver during the drug's elimination phase. The longer the terminal elimination half-life of the drug, the greater is the probability that any newly acquired parasite will encounter a partially effective (i.e., selective) drug concentration (49–51). The length of the terminal elimination half-life is therefore an important determinant of the propensity for an antimalarial drug to become ineffective because of the development of resistance, provided that the concentrations in the terminal phase traverse the steep part of the sigmoid concentration-effect relationship for the prevalent malaria parasites. This caveat is important since there has been a tendency to concentrate on the terminal-phase half-life as the main determinant of the rate of spread of resistance, but this is true only if this phase is "selective." For example, as chloroquine resistance increases, the chloroquine terminal elimination phase increasingly encompasses ineffective drug concentrations; it therefore no longer selects for higher levels of resistance (Figure 6). Some antimalarial drugs, e.g., the artemisinin derivatives, are never presented to infecting malaria parasites at intermediate selective drug concentrations, because they are eliminated completely within the 2-day life cycle of the asexual parasite. Other drugs (e.g., mefloquine, piperaquine, and chloroquine) have elimination half-lives of weeks or months. The prolonged presence of these drugs in the host's blood provides a lengthy exposure time in which resistant parasites may be selected. The probability of achieving a selective drug concentration in the plasma, and thus preferential survival of a resistant parasite during the elimination phase, depends on the degree of rightward shift in the concentration-effect relationship curve, its slope, and the duration of the elimination phase of the drug (5). The probability of subsequent transmission depends on the level of immunity; subsequent drug exposure; parasite multiplication capacity, which must take into account any fitness disadvantage conferred by the resistance mechanism; the reduction in antimalarial susceptibility, i.e., degree of resistance, conferred by the resistance mechanism; and intrahost competition from coexistent drug-sensitive parasites.

It has been suggested that the repeated exposure of parasite populations to residual drug concentrations of slowly eliminated drugs in areas of frequent infection is an important source of resistance (52). But this did not take into account the numbers of parasites involved (Figure 7). Selection of de novo resistance from an infection that emerges from the liver during the elimination phase of antimalarial treatment given to treat a previous infection (or during prophylaxis) would usually occur in the first generation of blood-stage malaria parasites — a total of approximately 10^5 parasites. This is because, for a resistant mutant to arise, and survive, from a larger number of parasites in generations subsequent to the first following hepatic schizogony, the antimalarial blood concentrations must have fallen below the minimum inhibitory concentration (MIC; the concentration associated with a multiplication rate of 1) for the sensitive parasites — otherwise their numbers would not have increased. If antimalarial concentrations exceed the sensitive parasites' MIC, then total parasite numbers will fall, and the chance of resistance selection in subsequent generations will fall in parallel. Assuming an equal probability of mutations arising among blood-stage parasites, the probability of resistance arising during the first asexual cycle following emergence from liver (10^5 parasites) is therefore between 1,000 and 10^7 times lower than in a symptomatic infection. To put this in context, if an individual acquired 20 symptomatic and potentially transmissible infections per year for fifty years, then the de novo selection probability from residual drug exposure to newly acquired infections in that half-century would be 1% of that in a single symptomatic infection of 10^{12} parasites. Taken together, the balance of evidence strongly favors acute symptomatic infection as the source of de novo antimalarial resistance. But the long elimination phase of some of the antimalarial drugs does provide a very efficient selective filter for resistant infections acquired from elsewhere, as it allows resistant infections to develop and then spread but suppresses sensitive infections. This selectively amplifies resistance. Thus, although it is a very unlikely source of de novo resistance, the duration of the antimalarial drug's elimination phase *is* an important determinant of the spread of antimalarial drug resistance (49–51). These calculations also suggest that antimalarial prophylaxis regimes, when adhered to, and mass treatment with effective drugs would not be major contributors to resistance (53), unlike mass continuous administration of subtherapeutic doses, as in table salt (the Pinotti method), which was disastrous (54).

Prevention of resistance by antimalarial combination therapy

The theory underlying combination drug treatment of tuberculosis, leprosy, and HIV infection is well known and is now generally accepted for malaria (5, 8, 55–58). If two drugs are used with different modes of action, and therefore different resistance mechanisms, then the per-parasite probability of developing resistance to both drugs is the product of their individual per-parasite probabilities. This is particularly powerful in malaria, because there are only about 10^{17} malaria parasites in the entire world. For example, if the per-parasite probabilities of developing resistance to drug A and drug B are both 1 in 10^{12}, then a simultaneously resistant mutant will arise spontaneously every 1 in 10^{24} parasites. As there is a cumulative total of less than 10^{20} malaria parasites in existence in one year, such a simultaneously resistant parasite would arise spontaneously roughly once every 10,000 years — provided the drugs always confronted the parasites in combination. Thus the lower the de novo per-parasite probability of developing resistance, the greater the delay in the emergence of resistance.

Stable, therapeutically significant resistance to the artemisinin derivatives has not yet been identified and cannot be induced yet in the laboratory, which suggests that it may be a very rare event. But it would be foolish to bank on its not happening, and should it arise, it would be a global disaster. For mutual protection against the emergence of drug resistance, these drugs should be used only in combination with other antimalarials.

Artemisinin derivatives are particularly effective in combinations because of their very high killing rates (parasite reduction ratios ~10,000-fold per cycle), lack of adverse effects, and absence of significant resistance (11). The ideal pharmacokinetic properties for an antimalarial drug have been greatly debated. From a resistance-prevention perspective, the combination partners should have similar pharmacokinetic properties. Rapid elimination ensures that the residual concentrations do not provide a selective filter for resistant parasites, but these drugs (if used alone) must be given for 7 days, and adherence to 7-day regimens is poor. Even 7-day regimens of artemisinin derivatives are associated with approximately 10% failure rates. In order to be highly effective in a 3-day regimen, terminal elimination half-lives of at least one drug component need to exceed 24 hours. Combinations of artemisinin derivatives (which are eliminated very rapidly) given for 3 days, with a slowly eliminated drug such as mefloquine (artemisinin combination treatment) provide complete protection for the artemisinin derivatives from selection of a de novo resistant mutant if adherence is good (i.e., no parasite is exposed to artemisinin during one asexual cycle without mefloquine being present). But this does leave the slowly eliminated "tail" of mefloquine unprotected by the artemisinin derivative. The residual number of parasites exposed to mefloquine alone, following two asexual cycles, is a tiny fraction (less than 0.00001%) of those present at the peak of the acute symptomatic infection. Furthermore, these residual parasites are exposed to relatively high levels of mefloquine, and, even if susceptibility was reduced, these levels are usually sufficient to eradicate infection. The long "tail" of the mefloquine elimination phase does, however, provide a selective filter for resistant parasites acquired from elsewhere and, as described earlier, contributes to the spread of resistance once it has developed. Yet on the northwestern border of Thailand, an area of low transmission where mefloquine resistance had developed already, systematic deployment of artesunate-mefloquine combination therapy was dramatically effective, both in stopping resistance and also in reducing the incidence of malaria (3, 59). This strategy would be expected to be effective at preventing the de novo emergence of resistance at higher levels of transmission, where high-biomass infections still constitute the major source of de novo resistance.

The main obstacles to the success of combination treatment in preventing the emergence of resistance will be incomplete coverage, or inadequate treatment, and, as for antituberculous drugs, use of one of the combination partners alone. Drugs of poor quality are common in tropical areas of the world, adherence to antimalarial treatment regimens is often incomplete, and antimalarials are available widely in the market place. Resistance to the artemisinins may not have happened yet. If it does, it will most likely arise in a hyperparasitemic patient who received an inadequate dose of a single antimalarial drug, not in combination with another suitable antimalarial agent. Irrespective of the epidemiological setting, ensuring that patients with high parasitemias receive a full course of adequate doses of artemisinin combination treatment would be an effective method of slowing the emergence of antimalarial drug resistance.

Acknowledgments

Nicholas J. White is a Wellcome Trust Principal Fellow.

Address correspondence to: Nicholas J. White, Faculty of Tropical Medicine, Mahidol University, 420/6 Rajvithi Road, Bangkok 10400, Thailand. Phone: 662-354-9172; Fax: 662-354-9169; E-mail: nickw@tropmedres.ac.

1. Marsh, K. 1998. Malaria disaster in Africa. *Lancet.* **352**:924–925.
2. Korenromp, E.L., Williams, B.G., Gouws, E., Dye, C., and Snow, R.W. 2003. Measurement of trends in childhood malaria mortality in Africa: an assessment of progress toward targets based on verbal autopsy. *Lancet Infect. Dis.* **3**:349–358.
3. Nosten, F., et al. 2000. Effects of artesunate-mefloquine combination on incidence of *Plasmodium falciparum* malaria and mefloquine resistance in western Thailand: a prospective study. *Lancet.* **356**:297–302.
4. Trape, J.F., et al. 1998. Impact of chloroquine resistance on malaria mortality. *C. R. Acad. Sci. III.* **321**:689–697.
5. White, N.J. 1999. Antimalarial drug resistance and combination chemotherapy. *Philos. Trans. R. Soc. Lond. B Biol. Sci.* **354**:739–749.
6. Hastings, I.M., and D'Alessandro, U. 2000. Modelling a predictable disaster: the rise and spread of drug-resistant malaria. *Parasitol. Today.* **16**:340–347.
7. WHO. 2001. Antimalarial drug combination therapy. Report of a technical consultation. WHO. Geneva, Switzerland. WHO/CDS/RBM 2001.35.
8. White, N.J., and Pongtavornpinyo, W. 2003. The de-novo selection of drug resistance in malaria parasites. *Philos. Trans. R. Soc. Lond. B Biol. Sci.* **270**:545–554.
9. Day, N.P.J., et al. 1996. Clearance kinetics of parasites and pigment-containing leukocytes in severe malaria. *Blood.* **88**:4696–4700.
10. Simpson, J.A., Aarons, L., Collins, W.E., Jeffery, G., and White, N.J. 2002. Population dynamics of the *Plasmodium falciparum* parasite within the adult human host in the absence of antimalarial drugs. *Parasitology.* **124**:247–263.
11. White, N.J. 1997. Assessment of the pharmacodynamic properties of antimalarial drugs in-vivo. *Antimicrob. Agents Chemother.* **41**:1413–1422.
12. Rathod, P.K., McErlean, T., and Lee, P.C. 1997. Variations in frequencies of drug resistance in *Plasmodium falciparum. Proc. Natl. Acad. Sci. U. S. A.* **94**:9389–9393.
13. Plowe, C.V. 2003. Monitoring antimalarial drug resistance: making the most of the tools at hand. *J. Exp. Biol.* **206**:3745–3752.
14. Su, X., Kirkman, L.A., Fujioka, H., and Wellems, T.E. 1997. Complex polymorphisms in an approximately 330 kDa protein are linked to chloroquine-resistant *P. falciparum* in Southeast Asia and Africa. *Cell.* **91**:593–603.
15. Kyes, S., Horrocks, P., and Newbold, C. 2001. Antigenic variation at the infected red cell surface in malaria. *Annu. Rev. Microbiol.* **55**:673–707.
16. Cowman, A.F., Galatis, D., and Thompson, J.K. 1994. Selection for mefloquine resistance in *Plasmodium falciparum* is linked to amplification of the *pfmdr1* gene and cross-resistance to halofantrine and quinine. *Proc. Natl. Acad. Sci. U. S. A.* **91**:1143–1147.
17. Reed, M.B., Saliba, K.J., Caruana, S.R., Kirk, K., and Cowman, A.F. 2000. *Pgh1* modulates sensitivity and resistance to multiple antimalarials in *Plasmodium*

falciparum. Nature. **403**:906–909.
18. Price, R.N., et al. 2004. Mefloquine resistance in *Plasmodium falciparum* results from increased *pfmdr1* gene copy number. *Lancet.* In press.
19. Triglia, T., Foote, S.J., Kemp, D.J., and Cowman, A.F. 1991. Amplification of the multidrug resistance gene *Pfmdr1* in *Plasmodium falciparum* has arisen as multiple independent events. *Mol. Cell. Biol.* **11**:5244–5250.
20. Alifrangis, M., et al. 2003. Prediction of *Plasmodium falciparum* resistance to sulfadoxine/pyrimethamine in vivo by mutations in the dihydrofolate reductase and dihydropteroate synthetase genes: a comparative study between sites of differing endemicity. *Am. J. Trop. Med. Hyg.* **69**:601–606.
21. Korsinczky, M., et al. 2000. Mutations in Plasmodium falciparum cytochrome b that are associated with atovaquone resistance are located at a putative drug-binding site. *Antimicrob. Agents Chemother.* **44**:2100–2108.
22. Eckstein-Ludwig, U., et al. 2003. Artemisinins target the SERCA of *Plasmodium falciparum. Nature.* **424**:957–961.
23. Looareesuwan, S., et al. 1996. Clinical studies of atovaquone, alone or in combination with other antimalarial drugs, for treatment of acute uncomplicated malaria in Thailand. *Am. J. Trop. Med. Hyg.* **54**:62–66.
24. Cortese, J.F., Caraballo, A., Contreras, C.E., and Plowe, C.V. 2002. Origin and dissemination of *Plasmodium falciparum* drug-resistance mutations in South America. *J. Infect. Dis.* **186**:999–1006.
25. Nair, S., et al. 2003. A selective sweep driven by pyrimethamine treatment in southeast Asian malaria parasites. *Mol. Biol. Evol.* **20**:1526–1536.
26. Roper, C., et al. 2003. Antifolate antimalarial resistance in southeast Africa: a population-based analysis. *Lancet.* **361**:1174–1181.
27. Bousema, J.T., et al. 2003. Treatment failure of pyrimethamine-sulphadoxine and induction of *Plasmodium falciparum* gametocytaemia in children in western Kenya. *Trop. Med. Int. Health.* **8**:427–430.
28. Peters, W. 1987. *Chemotherapy and drug resistance in malaria.* 2nd edition. Academic Press. London, United Kingdom. 1091 pp.
29. Rosenburg, R., and Wirtz, R.A. 1990. An estimation of the number of sporozoites ejected by a feeding mosquito. *Trans. R. Soc. Trop. Med. Hyg.* **84**:209–212.
30. Ponnudurai, T., et al. 1991. Feeding behaviour and sporozoite ejection by infected *Anopheles stephensi. Trans. R. Soc. Trop. Med. Hyg.* **85**:175–180.
31. James, S.P., Nichol, W.D., and Shute, P.G. 1932. A study of induced malignant tertian malaria. *Proc. R. Soc. Med.* **25**:1153–1186.
32. Fairley, N.H. 1947. Sidelights on malaria in man obtained by subinoculation experiments. *Trans. R. Soc. Trop. Med. Hyg.* **40**:521–676.
33. Kitchen, S.F. 1949. Symptomatology: general considerations and falciparum malaria. In *Malariology.* Volume 2. M.F. Boyd, editor. W.B. Saunders Co. Philadelphia, Pennsylvania, USA. 996–1017.
34. Smith, T., Schellenberg, J.A., and Hayes, R. 1994. Attributable fraction estimates and case defini-

tions for malaria in endemic areas. *Stat. Med.* **13**:2345–2358.
35. ter Kuile, F.O., et al. 1995. Predictors of mefloquine treatment failure: a prospective study in 1590 patients with uncomplicated falciparum malaria. *Trans. R. Soc. Trop. Med. Hyg.* **89**:660–664.
36. Austin, D.J., and Anderson, R.M. 1999. Studies of antibiotic resistance within the patient, hospitals and the community using simple mathematical models. *Philos. Trans. R. Soc. Lond. B Biol. Sci.* **354**:721–738.
37. Lipsitch, M., and Levin, B.R. 1997. The population dynamics of antimicrobial chemotherapy. *Antimicrob. Agents Chemother.* **41**:363–373.
38. Jeffery, G.M., and Eyles, D.E. 1955. Infectivity to mosquitoes of *Plasmodium falciparum* as related to gametocyte density and duration of infection. *Am. J. Trop. Med. Hyg.* **4**:781–789.
39. Shulman, C.E., and Dorman, E.K. 2003. Importance and prevention of malaria in pregnancy. *Trans. R. Soc. Trop. Med. Hyg.* **97**:30–35.
40. McGready, R., et al. 2001. Randomized comparison of quinine-clindamycin versus artesunate in the treatment of multi-drug resistant falciparum malaria in pregnancy. *Trans. R. Soc. Trop. Med. Hyg.* **95**:651–656.
41. Salanti, A., et al. 2003. Selective upregulation of a single distinctly structured var gene in chondroitin sulphate A-adhering *Plasmodium falciparum* involved in pregnancy-associated malaria. *Mol. Microbiol.* **49**:179–191.
42. Ansell, J., Hamilton, K.A., Pinder, M., Walraven, G.E., and Lindsay, S.W. 2002. Short-range attractiveness of pregnant women to *Anopheles gambiae* mosquitoes. *Trans. R. Soc. Trop. Med. Hyg.* **96**:113–116.
43. Ayisi, J.G., et al. 2003. The effect of dual infection with HIV and malaria on pregnancy outcome in western Kenya. *AIDS.* **17**:585–594.
44. Chirenda, J., Siziya, S., and Tshimanga, M. 2000. Association of HIV infection with the development of severe and complicated malaria cases at a rural hospital in Zimbabwe. *Cent. Afr. J. Med.* **46**:5–9.
45. Grimwade, K., et al. 2003. Childhood malaria in a region of unstable transmission and high human immunodeficiency virus prevalence. *Pediatr. Infect. Dis. J.* **22**:1057–1063.
46. White, N.J. 1992. Antimalarial pharmacokinetics and treatment regimens. *Br. J. Clin. Pharmacol.* **34**:1–10.
47. White, N.J., van Vugt, M., and Ezzet, F. 1999. Clinical pharmacokinetics and pharmacodynamics of artemether-lumefantrine. *Clin. Pharmacokinet.* **37**:105–125.
48. White, N.J. 2002. Plasmodium species (malaria). In *Antimicrobial therapy and vaccines.* 2nd edition. V. Yu, R. Weber, and D. Raoult, editors. Apple Trees Productions LLC. New York, New York, USA. 1609–1634.
49. Watkins, W.M., and Mosobo, M. 1993. Treatment of *Plasmodium falciparum* malaria with pyrimethamine-sulphadoxine: selective pressure for resistance is a function of long elimination half-life. *Trans. R. Soc. Trop. Med. Hyg.* **87**:75–78.

50. Watkins, W.M., Mberu, E.K., Winstanley, P.A., and Plowe, C.V. 1999. More on the efficacy of antifolate antimalarial combinations in Africa. *Parasitol. Today.* **15**:131–132.

51. Hastings, I., Watkins, W.M., and White, N.J. 2002. Pharmacokinetic parameters affecting the evolution of drug-resistance in malaria: the role of the terminal elimination half-life. *Philos. Trans. R. Soc. Lond. B Biol. Sci.* **357**:505–519.

52. Bloland, P.B., Ettling, M., and Meek, S. 2000. Combination therapy for malaria in Africa: hype and hope. *Bull. World Health Organ.* **78**:1378–1388.

53. von Seidlein, L., and Greenwood, B.M. 2003. Mass administrations of antimalarial drugs. *Trends Parasitol.* **19**:452–460.

54. Verdrager, J. 1986. Epidemiology of the emergence and spread of drug-resistant falciparum malaria in South-East Asia and Australasia. *J. Trop. Med. Hyg.* **8**:277–289.

55. Peters, W. 1969. Drug resistance: a perspective. *Trans. R. Soc. Trop. Med. Hyg.* **63**:25–45.

56. Peters, W. 1990. The prevention of antimalarial drug resistance. *Pharmacol. Ther.* **47**:499–508.

57. Chawira, A.N., Warhurst, D.C., Robinson, B.L., and Peters, W. 1987. The effect of combinations of qinghaosu (artemisinin) with standard antimalarial drugs in the suppressive treatment of malaria in mice. *Trans. R. Soc. Trop. Med. Hyg.* **81**:554–558.

58. Curtis, C.F., and Otoo, L.N. 1986. A simple model of the build-up of resistance to mixtures of antimalarial drugs. *Trans. R. Soc. Trop. Med. Hyg.* **80**:889–892.

59. Brockman, A., et al. 2000. *Plasmodium falciparum* antimalarial drug susceptibility on the northwestern border of Thailand during five years of extensive artesunate-mefloquine use. *Trans. R. Soc. Trop. Med. Hyg.* **94**:537–544.

60. White, N.J. 2002. The assessment of antimalarial drug efficacy. *Trends Parasitol.* **18**:458–464.

Human African trypanosomiasis of the CNS: current issues and challenges

Peter G.E. Kennedy

Department of Neurology, Division of Clinical Neurosciences, University of Glasgow, Institute of Neurological Sciences,
Southern General Hospital, Glasgow, Scotland, United Kingdom.

Human African trypanosomiasis (HAT), also known as sleeping sickness, is a major cause of mortality and morbidity in sub-Saharan Africa. Current therapy with melarsoprol for CNS HAT has unacceptable side-effects with an overall mortality of 5%. This review discusses the issues of diagnosis and staging of CNS disease, its neuropathogenesis, and the possibility of new therapies for treating late-stage disease.

Historical perspective

Human African trypanosomiasis (HAT), also known as sleeping sickness, comes in two variants: East African and West African. Caused by protozoan parasites of the genus *Trypanosoma*, it has emerged over the last few decades as a major threat to human health in Africa. While for centuries there was an awareness of the disease and of its propensity to induce a fatal sleep disorder, it was not until the period 1894–1910 that the cause of sleeping sickness in humans and cattle was discovered. Preeminent in this discovery was David Bruce, who, while working in Zululand on a wasting disease of cattle known as *nagana*, identified trypanosomes in the blood of affected cattle (1, 2). He then established experimentally that healthy game animals were host reservoirs of the disease, which was transmitted by the bite of the tsetse fly to domestic animals, which then became ill (2). In 1899 the causative parasite was identified as *Trypanosoma brucei*, and in 1902 Everett Dutton first identified, in a European patient, a subspecies of trypanosomes called *Trypanosoma brucei gambiense* (2) that is now recognized as the cause of West African sleeping sickness. In 1903, Aldo Castellani, working with Bruce, identified trypanosomes in the blood and cerebrospinal fluid (CSF) of a patient with sleeping sickness (2, 3), and in 1910 J.W.W. Stephens and H.B. Fantham first described *Trypanosoma brucei rhodesiense* (2), which is now recognized as causing East African sleeping sickness. There are characteristic differences between the biology and clinical features of *T.b. gambiense* disease and *T.b. rhodesiense* disease, probably due to a greater adaptation of the *gambiense* parasite to humans.

Currently, HAT occurs in 36 countries in sub-Saharan Africa; about 60 million people worldwide are at risk from developing the disease (4). The annual incidence of the disease is approximately 300,000 cases, and the area of Africa that is infested by the tsetse fly encompasses approximately ten million square kilometers (2, 5) — a third of the land mass of Africa (Figure 1). HAT was almost brought under control during the early 1950s (6), with a significant decrease in the number of newly registered cases from

1949 to 1965, but a variety of factors led to its recent reemergence. These include socio-economic unrest — especially war — causing disruption of disease surveillance and control, inadequate financial allocation of critical resources to the disease during peacetime, increasing parasite drug resistance, changes in climate and vegetation, the emergence of new virulent parasite strains, unpredicted population movements of animal reservoirs, and changes in host disease susceptibility (5, 6). Many of these factors may operate simultaneously, and there have been several significant epidemics and focal resurgences of the disease in various regions of Africa in recent years.

Trypanosome biology

In both types of HAT the disease is transmitted by the blood-sucking tsetse fly of the genus *Glossina*. Infected wild animals and domestic animals, such as cattle, are the reservoirs of parasites causing human disease. The flies become infective approximately 21 days after feeding on an infected animal host (5). The fly ingests the trypanosomes during a blood meal from the infected animal, following which the parasites undergo a series of morphological and biochemical changes in the fly's anterior midgut where the infection is initially established (7). Long slender parasitic forms produced in the midgut then move to the salivary glands to become epimastigotes, which then change into short stumpy infective metacyclic trypanosomes which enter via the wound of a bitten individual. A primary lesion known as a trypanosomal chancre usually develops 5–15 days later at the site of the bite, soon after which the trypanosomes invade the bloodstream, lymph nodes, and other tissues. HAT is invariably fatal if untreated. A fly remains infective for life, and human/fly contact is therefore a crucial component of the disease. The life cycle of the organism in the human and the tsetse fly is summarized in Figure 2.

Much work has been carried out on the sequencing and mapping of the trypanosome genome (8). Trypanosomes causing HAT are diploid and have a haploid nuclear DNA content of approximately 35 Mb (8, 9). Three classes of chromosome in *T. brucei* have been identified based on size, namely megabase (1–6 Mb), of which there are at least 11, intermediate (200–900 kb), and at least 100 minichromosomes (50–150 kb) (8, 9). The entire genome contains about 10,000 genes, 10% of which are thought to be variant surface glycoprotein (VSG) genes encoding variable surface glycoproteins (9). The VSG genes are of great pathogenic importance as they provide the molecular basis for the antigenic variation seen in trypanosome infection, and only one of them is expressed at any one time,

---

Nonstandard abbreviations used: human African trypanosomiasis (HAT); cerebrospinal fluid (CSF); variant surface glycoprotein (VSG); glycosylphosphatidylinositol (GPI); card agglutination trypanosomiasis test (CATT); white blood cell (WBC); electroencephalogram (EEG); difluoromethylornithine (DFMO); intramuscular (i.m.); post-treatment reactive encephalopathy (PTRE); Substance P (SP); macrophage inflammatory protein (MIP).

Conflict of interest: The author has declared that no conflict of interest exists.

Citation for this article: *J. Clin. Invest.* **113**:496–504 (2004). doi:10.1172/JCI200421052.

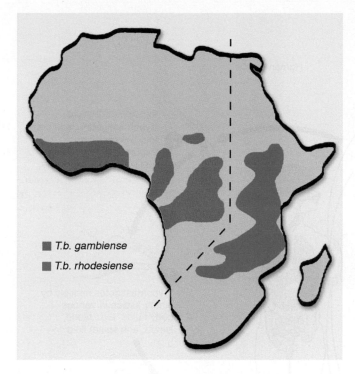

Figure 1
Diagrammatic representation of the distribution of the two types of human African trypanosomiasis in Africa. Figure modified with permission from Butterworth-Heinemann (5).

■ *T.b. gambiense*
■ *T.b. rhodesiense*

the rest being transcriptionally silent (9). The VSGs are distributed over the surface of the trypanosome and are anchored to the outer membrane by a glycosylphosphatidylinositol (GPI) anchor (9). Ten million copies of a single species of VSG cover the trypanosome surface at any one time. During infection of the host, a constant low frequency gene conversion process switches transcriptionally inactive basic copy VSG genes in and out of the expression site, and this antigenic variation allows the parasite to continuously evade the host's immune response. As a result, the parasite undergoes rapid multiplication in the blood of the host, producing waves of parasitaemia that characterize this disease (10).

Clinical features

There are two recognized stages in the clinical presentation of HAT, namely the early hemolymphatic stage, and the late encephalitic stage when the CNS is involved. However, the transition from the early to the late stage is not always distinct in *rhodesiense* infection (5). The tempo of the disease is usually acute in *rhodesiense* disease — CNS invasion by the parasite occurs early, within a few months after initial infection — whereas *gambiense* infection is usually a slower, chronic infection, with late CNS infection lasting months to years.

Early (hemolymphatic) stage. The onset is variable but usually occurs 1–3 weeks after the bite. Episodes of fever lasting 1–7 days occur together with generalized lymphadenopathy. The early symptoms tend to be non-specific: malaise, headache, arthralgia, generalized weakness, and weight loss (11). Multiple organs may then be infected (5, 12), including the spleen, liver, skin, cardiovascular system, endocrine system, and eyes. This involvement underlies the wide spectrum of systemic dysfunction that may occur (5).

Late (encephalitic) stage. The onset is insidious and the potential clinical phenotype is wide (5, 12). The broad neurologic spectrum has been detailed elsewhere (5), and the reported features can be grouped into general categories such as psychiatric, motor, and sensory abnormalities, and sleep disturbances. The mental disturbances may be subtle, and include irritability, lassitude, headache, apparent personality changes, and overt psychiatric presentations such as violence, hallucinations, suicidal tendencies, and mania (5, 12). Motor system involvement may include limb tremors, tongue and limb muscle fasciculation, limb hypertonia and pyramidal weakness, choreiform and athetoid movements, dysarthria, cerebellar ataxia, and polyneuritis (5, 13). Pout and palmar-mental reflexes may also be present. Sensory involvement may manifest as painful hyperaesthesia, pruritis, and also deep hyperaesthesia (Kerandel's sign), the latter being reported as particularly common in Europeans (12). The characteristic sleep disturbances include lassitude, distractibility, and spontaneous, uncontrollable urges to sleep, along with a reversal of the normal sleep-wake cycle in which daytime somnolence alternates with nocturnal insomnia. While these various features, including the sleep abnormalities, are typical of HAT, they are not individually diagnostic, since some of them may also be seen during other CNS infections. If untreated, the patient progresses to the final stage of the disease, which is characterized by seizures, severe somnolence, double incontinence, cerebral edema, coma, systemic organ failure, and inevitable death.

Disease diagnosis

The diagnosis of HAT is based on a combination of clinical and investigative data. A typical clinical presentation in the context of a geographical location where HAT is known to be endemic is clearly the key diagnostic clue. However, the non-specific nature of many of the clinical features makes it imperative to exclude other infections such as malaria, tuberculosis, HIV infection, leishmaniasis, toxoplasmosis, hookworm infection, typhoid, and viral encephalitis (5). A particular pitfall is that inappropriate antimalarial treatment may actually reduce the fever due to HAT, thus confounding and delaying the correct diagnosis (5), and these two conditions may also co-exist.

Specific diagnosis at the hemolymphatic stage ideally involves demonstration of the trypanosomes in the peripheral blood using stained thick and thin films (Figure 3), or in other infected tissues such as lymph node aspirates or occasionally bone marrow (5). While parasite detection in the blood is frequently successful in *rhodesiense* infection because of the permanent parasitaemia, this method is very difficult in *gambiense* infection, in which few parasites are present in the peripheral circulation other than at periods of cyclic parasitaemia, which reflects the chronicity of the disease. Therefore, serologic tests are of crucial importance in the diagnosis of *gambiense* infection. Currently the antibody-detecting card agglutination trypanosomiasis test (CATT) is in frequent use for serological *gambiense* diagnosis, being simple, easy to perform, and rapid (14).

The key issue in HAT diagnosis and therapeutic decision making is to distinguish reliably the late encephalitic stage of HAT from the early stage. Accurate staging of HAT is critical because failure to treat a patient with CNS involvement will lead inevitably to death from the disease, yet inappropriate CNS treatment in an early-

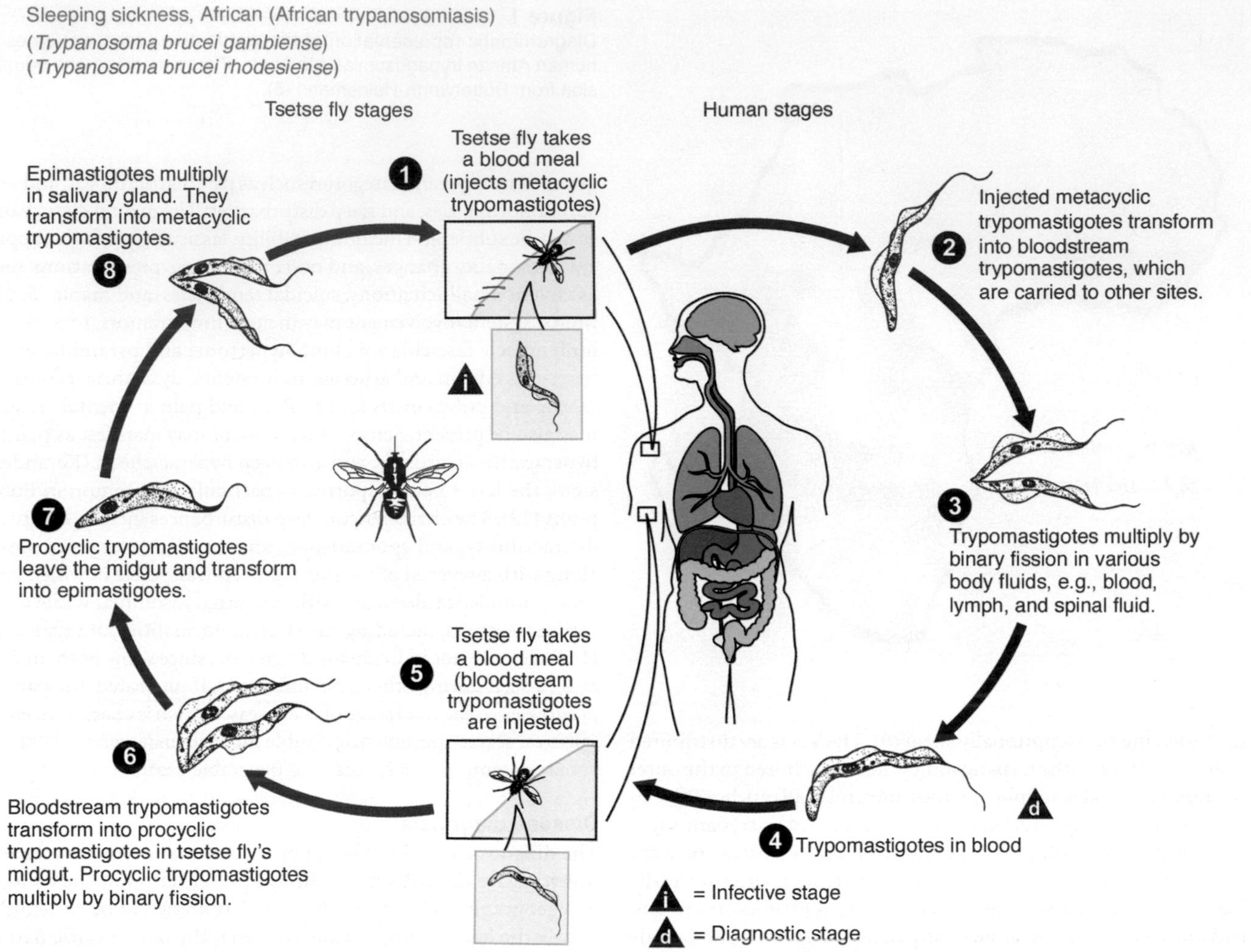

Sleeping sickness, African (African trypanosomiasis)
(*Trypanosoma brucei gambiense*)
(*Trypanosoma brucei rhodesiense*)

Figure 2
Diagrammatic representation of the life cycle of *Trypanosoma brucei* in the human and the tsetse fly. Image credit: Alexander J. da Silva and Melanie Moser, Centers for Disease Control Public Health Image Library.

stage patient carries a high risk of unnecessary drug toxicity (see below). In patients with suspected late-stage disease it is imperative to perform a lumbar puncture, which typically shows a lymphocytic pleiocytosis and raised protein level of 40–200 mg/100 ml (5). Further, all CATT-positive patients also need to undergo a lumbar puncture, as there are no reliable clinical suspicion criteria for early-stage disease. The WHO criteria for CNS involvement, and therefore for CNS drug treatment, are demonstration of the parasites in the CSF or a white blood cell (WBC) count of >5/µl (15). However, these criteria have been challenged by some investigators. Thus, in Angola and the Ivory Coast the criterion used for CNS involvement is 20 WBCs/µl in the CSF (16). It has also been pointed out that concentration techniques for trypanosome detection in the CSF vary, and that a CSF pleiocytosis may be non-specific (17). Recently it has been shown that detection of intrathecal IgM synthesis is a very sensitive marker for CNS involvement in sleeping sickness (18). The latex agglutination assay for CSF IgM quantitation can be applied in the field and has considerable promise for both staging CNS sleeping sickness and monitoring the development of treatment relapses (18).

CSF PCR to detect trypanosome DNA has also been used in the diagnosis of HAT, but considerable care must be used in the correct choice of primers, and problems with assay reproducibility have been documented (19). It has recently been reported that CSF PCR has a sensitivity rate of 96%, although its value for therapeutic decision making has been questioned (17). Therefore, PCR has not yet superseded serological diagnosis and, crucially, it is not readily available in field conditions. It has recently been suggested by Lejon et al. that the WHO criteria should be replaced by the presence of intrathecal IgM synthesis or the presence of >20 WBCs/µl, independent of the presence of trypanosomes in the CSF (16). This author regards the presence of trypanosomes in the CSF as compelling evidence of CNS involvement. However, patients with *gambiense* disease who have trypanosomes in the CSF and <20 WBCs/µl have been treated successfully with pentamidine (16), so perhaps one might speculate that there is a kind of 'intermediate stage' in which the trypanosomes can cross the blood-brain barrier without invading and damaging brain structures at that stage. Thus there are two critical, and not necessarily congruent, issues involved, one being the biological definition

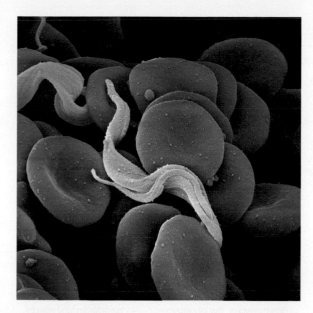

Figure 3
Colored scanning electron micrograph of *Trypanosoma brucei* in human blood. Image credit: Science Photo Library.

of CNS involvement, and the other being the ground for therapeutic choices. This lack of a universal consensus on the operational definition of late-stage HAT remains very problematic, but the clear requirement is to develop robust surrogate markers to guide therapeutic choices. These diagnostics need to be novel, simple, and affordable.

Electroencephalogram (EEG) and sophisticated neuroimaging are limited to specialist centers, but both have shown abnormalities in HAT. During the encephalitis stage the EEG shows non-specific abnormalities, which correlate with the severity of the disease. Changes include at least three different types of abnormal EEG patterns, which become normal after clinical improvement (5). Abnormalities reported on computed tomography scans and MRI are non-specific and not pathognomonic (Figure 4), but if available these tests should be carried out, partly to monitor the response to treatment, and also where the diagnosis is in doubt or where raised intracranial pressure is present. MRI of the brain may show diffuse asymmetric white matter abnormalities, diffuse hyperintensities in the basal ganglia, and ventricular enlargement (5, 20).

Overview of current treatment
The current treatment of HAT is based on four main drugs, namely suramin, pentamidine, melarsoprol, and eflornithine (difluoromethylornithine, or DFMO), with nifurtimox undergoing evaluation. Table 1 summarizes their disease spectrum, stage-specificity, route of administration, postulated mode of action, and main side-effects (21–26). It should be appreciated that most of these drugs were developed in the first half of the twentieth century, some of them would probably not pass current high safety standards (26), and there have been no new registered drugs for HAT since 1981. Early-stage disease is treated with i.v. suramin in *rhodesiense* disease and with intramuscular (i.m.) pentamidine in *gambiense* disease according to established treatment protocols. Treatment is effective and prevents disease progression.

The trivalent organic arsenical melarsoprol is the only effective drug for late-stage disease in both forms of HAT, as the drug crosses the blood-brain barrier (5, 27). Specific treatment regimes vary considerably among different centers and depending on whether the infection is due to *rhodesiense* or *gambiense*. Typically, a course of 3–4 i.v. doses are given daily over a week for a total period of 3–4 weeks (27). Ideally, patients are then followed up every 6 months with clinical evaluation and CSF examination for a total of 2 years, at which point a cure has been established if the CSF is normal. However, this policy is very difficult to carry out in routine practice in the field. Although about 80–90% of patients are cured with standard treatment regimes (5), there is evidence of increasing drug resistance, with treatment failure rates of 30% reported among patients in Northern Uganda (27, 28). But the major problem with melarsoprol treatment is that it is followed by a severe post-treatment reactive encephalopathy (PTRE) in up to 10% of cases, with a fatality rate of about 50% (29). Thus the overall mortality rate from melarsoprol therapy is 5%, which is unacceptably high (30). A prospective, randomized, non-blinded trial involving 598 patients with *gambiense* disease showed that the incidence of melarsoprol-induced encephalopathy and death was reduced in patients who were given concurrent administration of prednisolone and melarsoprol compared with melarsoprol therapy alone. However, this combined treatment regime was not associated with a reduction in the incidence of the other complications of PTRE or the relapse rate after melarsoprol therapy (31). Treatment of PTRE is focussed on treating seizures, general management of the comatose patient with i.v. hydration, antipyretics, steroids, and reduction of cerebral edema (5). After recovery, melarsoprol has to be restarted, possibly with a smaller initial dose, and the course then completed (5). A new, shorter treatment regime consisting of a 10-day course of daily melarsoprol injections was recently found

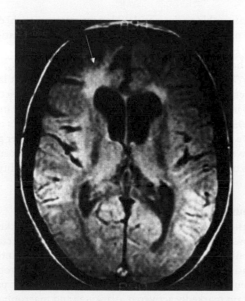

Figure 4
MRI scan (proton density) of a 13-year-old patient with CNS trypanosomiasis 3 years after successful completion of multiple treatment regimens for numerous relapses of the disease. Ventricular enlargement (especially of the frontal horns) is seen as well as diffuse white matter changes, which are prominent in the right frontal (see arrow) and periventricular regions. Reproduced with permission from Butterworth-Heinemann (5).

Table 1

Drugs currently used for the treatment of human African trypanosomiasis

Drug	Spectrum	Indication	Year of first use	Route of administration	Mode of action	Side-effects/ comments
Suramin	*T.b. rhodesiense*	Stage 1	Early 1920s	i.v.	Evidence for inhibition of acute phase protein stimulation, mediated by impairment of receptor function and/or signal transduction (21)	Anaphylactic shock, renal failure, skin lesions, neurologic effects
Pentamidine	*T.b. gambiense*	Stage 1	1940	i.m.	Evidence for interference with action and synthesis of polyamines and selective inhibition of parasite plasma membrane Ca^{2+}-Mg^{2+}-ATPase and Ca^{2+} transport (22)	Hypotension, hypoglycaemia, hyperglycaemia
Melarsoprol (Mel B)	*T.b. gambiense* *T.b. rhodesiense*	Stage 2	1949	i.v.	Acts on trypanothione, a trypanosome molecule which maintains an intracellular reducing environment (23)	PTRE, cardiac arrhythmias, dermatitis, agranulocytosis; increasing treatment failure
Eflornithine (DFMO)	*T.b. gambiense*	Stage 2	1981	i.v.	Irreversible inhibition of ornithine decarboxylase, the key enzyme in polyamine biosynthesis (24)	Bone marrow toxicity, gastrointestinal effects (oral drugs), alopecia, seizures; drug is expensive, oral bioavailability not high
Nifurtimox	*T.b. gambiense* *T.b. rhodesiense?*	Stage 2?	1977	Oral	Evidence for interference with trypanothione metabolism (25, 26)	Gastrointestinal and neurologic effects; toxicity poorly documented; registered for Chagas disease but not for HAT; no standard treatment protocols; possible role in combination therapy

Table reproduced in part, and modified with permission, from *Lancet Infect. Dis.* (27). i.m., intramuscular.

to be comparable to the standard longer treatment schedule over a period of 26 days, in terms of both cure and complication rates, and may be increasingly adopted in the future (32).

There has been great interest in developing safer drugs for late-stage HAT. DFMO has been used successfully to treat late-stage disease, especially melarsoprol-refractory *gambiense* infection (33), and also increasingly as first-line therapy, but is largely ineffective for *rhodesiense* infection. The problems with this drug's availability will be mentioned later. Nifurtimox is the only other potential alternative treatment for late-stage disease, but well-documented evidence of its efficacy and safety is lacking, and its utility is more likely to be in the context of combination therapy.

Neuropathogenesis

The pathologic substrate of late-stage sleeping sickness is a meningoencephalitis in which cellular proliferation occurs in the leptomeninges, and a diffuse perivascular white matter infiltration consisting of lymphocytes, plasma cells, and macrophages is prominent (5, 34). The perivascular cuffs and adjacent parenchyma contain markedly activated astrocytes and macrophages, and the white matter contains pathognomonic morular or Mott cells, which are thought to be modified plasma cells containing eosinophilic inclusions comprising IgM (34) (Figure 5). PTRE shows an exacerbation of these pathologic features.

Current understanding of the highly complex pathogenesis of sleeping sickness is based mainly on studies carried out either on patients' blood and CSF samples or in experimental animal models. In both cases, correlation of specific clinical features or stages with alterations of different biochemical or immunological parameters has often yielded interesting results, but caution must be used in assuming a cause-and-effect relationship between the investigation and the disease phenotype. Care must also be used in extrapolating results obtained in animal models to the human disease.

Alteration of cytokine levels has been detected in patients with CNS sleeping sickness. For example, significant elevations of IL-10 were detected in both the plasma and CSF in both early- and late-stage *rhodesiense* disease, and declined after treatment to the levels found in uninfected control persons (35). Total, but not free, plasma TNF-α levels were also higher in late-stage disease compared with levels obtained after treatment. However, the source of IL-10 elevation is unclear. Similar studies in patients with *gambiense* infection have also reported elevations of CSF IL-10 levels in late-stage disease, as well as a rise in IL-6 and IL-8 (36). Other abnormalities which have been reported in patients with CNS HAT include very raised CSF levels of prostaglandin D2 (37), which may be related to the marked somnolence, and raised blood and CSF endotoxin levels that may also contribute to the CNS pathology (38).

Several possible causes of PTRE have been suggested, including subcurative chemotherapy, abnormal immune responses to glial cell–attached antigens released from killed parasites following melarsoprol treatment, immune complex deposition, arsenical toxicity, and autoimmune mechanisms (39–42). PTRE has

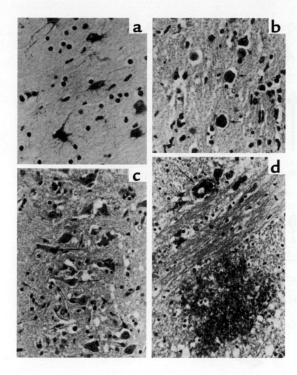

Figure 5
Neuropathology of CNS human African trypanosomiasis. (**a**) Late-stage disease in a patient who died 3–5 months after first injection of melarsoprol. Many large astrocytes are located in white matter. Stained for glial fibrillary acidic protein by immunoperoxidase. Original magnification, ×400. (**b**) Morular cells (indicated by arrows) observed in the brain of a patient with CNS trypanosomiasis who had not received melarsoprol. Morular cells are plasma cells filled with immunoglobulin. H&E stain. Original magnification, ×400. (**c**) PTRE in a patient dying 9 days after receiving melarsoprol. Ischaemic cell changes (indicated by arrows) are seen in neurons in the hippocampus. H&E stain. Original magnification, ×250. (**d**) PTRE with acute haemorrhagic leukoencephalopathy in a patient 9 days after receiving melarsoprol. There is fibrinoid necrosis in an arteriole (indicated by arrow) and focal haemorrhage in the pons. Martius scarlet blue stain. Original magnification, ×250. Reproduced with permission from *Neuropathol. Appl. Neurobiol.* (34).

RANTES, and MIP-1α produced by astrocytes, microglia, and T cells early in the CNS infection in a rat model (50). It should also be pointed out that in both human disease and animal models the cellular sources of these cytokines and neuropeptides are sometimes not known and are only inferred, with multiple stimuli for their secretion likely.

It is clear that macrophage activation by both parasite components and host-derived cytokines is central to HAT pathogenesis. Both VSG and GPI anchors are known to be potent macrophage activators (51, 52), as is IFN-γ, which itself may derive from several sources, including CD4+ and NK cells (53). A molecule called trypanosome-derived lymphocyte triggering factor has been described in mouse and rat models (54, 55). This molecule triggers the CD8+ T cell to produce IFN-γ, which both activates macrophages and apparently has growth-enhancing effects on trypanosomes (54, 56). The overall picture that is now emerging is a highly complex network of cytokine-brain interactions, with early astrocyte activation, macrophage activation, and, at least in animal models, an inflammatory cytokine response being prominent features (43, 52, 57, 58) (summarized schematically in Figure 6).

Prospects for CNS sleeping sickness

Advances in this field are likely to be made in several areas. The increasing use of animal models, including host and parasite gene knockouts, should help unravel the complex neuropathogenesis of sleeping sickness, in particular the role of specific neuropeptides and the importance of the balance of proinflammatory versus counterinflammatory cytokines. But such studies will need to be interpreted in the context of sophisticated analyses of the serum and CSF of human subjects.

Physicians need to reach a consensus as to what does and does not define late-stage HAT, and to be able to use reliable surrogate markers that will allow them to make rational therapeutic decisions, in particular when embarking on a treatment modality that currently has an overall mortality rate of 5%. There is a pressing need to developing a quick, easy to perform, reliable, and cheap diagnostic test that can be used in the field to diagnose and, crucially, to stage both *gambiense* and *rhodesiense* disease.

Control of sleeping sickness will require: (a) continuing and improved case surveillance with screening of humans in at-risk areas and also of domestic cattle; this will require both political will and stability and significantly increased funding to improve the screening infrastructure; (b) better treatment of human disease and of animal reservoirs; and (c) increased public health measures

been studied in a reproducible mouse model that mirrors many of the pathologic features of the disease in humans (43). Injection of *Trypanosoma brucei* into mice via the intraperitoneal route leads to a chronic infection in which the parasites are detectable in the CNS after 21 days. If the drug berenil (diminazene aceturate), which does not cross the blood-brain barrier and therefore clears the parasites from the extravascular compartment but not the CNS, is given 21–28 days after infection, the mice develop a severe post-treatment meningoencephalitis, which persists after the parasitemic phase is over. This condition shows strong pathologic similarity to PTRE in humans. A consistent observation in this model is that astrocytes are activated 14–21 days after infection and prior to the development of the inflammatory response (43, 44), and that transcripts for several cytokines such as TNF-α, IL-1, IL-4, IL-6 and IFN-γ can be detected in the brain at this time (44, 45). Early astrocyte activation is therefore likely to be of central importance in generating the CNS inflammatory response.

Different types of drug have been shown to modulate the inflammatory response in this mouse model. The trypanostatic drug DFMO has the ability to prevent the development of PTRE or ameliorate it once it is established in terms of greatly reducing both the neuropathology and the degree of astrocyte activation (46). The immunosuppressant drug azathiaprine can prevent but not cure PTRE (47), and the non-peptide Substance P (SP) antagonist RP-67,580 has been shown to significantly ameliorate both the neuroinflammatory reaction and the level of astrocyte activation (48). Although this showed that SP plays a role in generating the inflammatory response in this PTRE, recent evidence has shown that this is complex, since infected SP knockout mice show a novel phenotype in which the clinical and neuroinflammatory responses were dissociated with evidence of alternative tachykinin receptor usage (49). There is also evidence for the role of various chemokines such as macrophage inflammatory protein (MIP)-2,

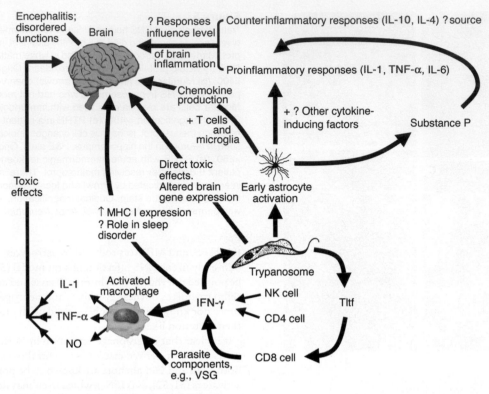

Figure 6

Schematic representation of possible immunopathologic pathways leading to brain dysfunction in late-stage human African trypanosomiasis. Concepts are based on a combination of human and animal model data and ideas, particularly from refs. 43, 48, 51, 53, and 56. Cytokines shown in red probably have important roles in neuropathogenesis. The schematic emphasizes the central importance of early astrocyte activation, cytokine responses, and macrophage activation. One should note that there are likely to be multiple factors acting together to produce brain damage and also multiple potential sources of different cytokines. Tltf, trypanosome-derived lymphocyte triggering factor.

to significantly decrease, and ultimately eradicate, human/tsetse fly contact through the use of, for example, increasingly sophisticated fly traps in infected areas, spraying of insecticides, and molecular genetic approaches such as the replacement of susceptible insect phenotypes with their engineered refractory counterparts to result in decreased HAT transmission (7).

The unacceptable toxicity of the currently available drugs for HAT underpins the urgency of developing more effective and safer drug regimes. A safe drug that is effective in the treatment of CNS HAT would dramatically change the control and management of sleeping sickness, as it would obviate the current difficulties of staging with CSF analysis. However, in reality no new drugs are likely to appear within the next 5 years, and even that may be overly optimistic. An effective oral drug is required for early-stage disease, but several recent candidate compounds have been abandoned because of unacceptable toxicity or lack of efficacy. The best, indeed only, candidate is DB 289, which is a diamidine derivative and the oral prodrug of an active form called DB 75 (27). A phase IIa clinical trial with DB 289 has just been completed, with good results in terms of safety and efficacy, but the drug will probably only be effective in early-stage disease (C. Burri, personal communication). A phase IIb multi-center, randomized, controlled trial of 80 patients with DB 289 is currently under way (C. Burri, personal communication). DFMO is effective for late-stage *gambiense* disease and is far less toxic than melarsoprol, but it became an orphan drug, as it was expensive and non-profitable

for pharmaceutical companies. Only as a result of the remarkable efforts of Médecins Sans Frontières, working with the WHO and the drug companies Aventis Pharma and Bristol-Myers, who had developed a renewed interest in this drug, is DFMO currently available for HAT treatment in Africa (59).

Another avenue of treatment is the use of combination therapy in order to increase efficacy, decrease toxicity, and delay the onset of drug resistance. Drug combinations also have the potential to solve the problems of complexity and high costs of current alternatives to melarsoprol. This approach can also be tested in the mouse model of HAT, which can provide valuable clues for novel treatment strategies (43). Current regimes of combination therapy which can be explored in humans with CNS HAT disease include DFMO/melarsoprol, melarsoprol/nifurtimox, and DFMO/nifurtimox. The latter regime, having shown lower toxicity in limited clinical studies, is currently under evaluation in a controlled clinical trial (G. Priotto, personal communication). A major hope for the future is that the morbidity and mortality from PTRE can be reduced from their current high level. Recent advances in our understanding of normal blood-brain barrier function and permeability have raised the possibility that existing or new trypanocidal drugs may be modified so as to cross the blood-brain barrier, thereby opening up a new therapeutic dimension for CNS sleeping sickness. A further approach is to modify dose regimes of currently available drugs, as has been the case with melarsoprol (32). More targeted approaches to treatment should also be possible, such as adjunct therapy of standard drug regimes with humanized neuropeptide antagonists to specifically block key components of the inflammatory response.

Acknowledgments

I wish to express my sincere gratitude to Jorge Atouguia, Els Torreele, Jeremy Sternberg, and Max Murray for their help with this article. Personal research described here was carried out with the financial support of the Wellcome Trust and the Sir Jules Thorne Charitable Trust.

Address correspondence to: Peter G.E. Kennedy, Department of Neurology, Division of Clinical Neurosciences, University of Glasgow, Institute of Neurological Sciences, Southern General Hospital, 1345 Govan Road, Glasgow G51 4TF, Scotland, United Kingdom. Phone: 44-141-201-2474; Fax: 44-141-201-2993; E-mail: P.G.Kennedy@clinmed.gla.ac.uk.

1. Vickerman, K. 1997. Landmarks in trypanosome research. In *Trypanosomiasis and leishmaniasis*. G. Hide, J.C. Mottram, G.H. Coombs, and P.H. Holmes, editors. Cab International. Oxford, United Kingdom. 1–37.
2. Williams, B.I. 1996. African trypanosomiasis. In *The Wellcome Trust illustrated history of tropical diseases*. F.E.A.G. Cox, editor. The Wellcome Trust. London, United Kingdom. 178–191.
3. Bentivoglio, M., Grassi-Zucconi, G., and Kristensson, K. 1994. From trypanosomes to the nervous system, from molecules to behavior: a survey, on the occasion of the 90th anniversary of Castellani's discovery of the parasites in sleeping sickness. *Ital. J. Neurol. Sci.* **15**:77–89.
4. 1986. Epidemiology and control of African trypanosomiasis. Report of a WHO expert committee. World Health Organization. Geneva, Switzerland. Technical Report Series, No. 739. 126 pp.
5. Atouguia, J.L.M., and Kennedy, P.G.E. 2000. Neurological aspects of human African trypanosomiasis. In *Infectious diseases of the nervous system*. L.E. Davis and P.G.E. Kennedy, editors. Butterworth-Heinemann. Oxford, United Kingdom. 321–372.
6. Kuzoe, F.A. 1993. Current situation of African trypanosomiasis. *Acta Trop.* **54**:153–162.
7. Aksoy, S. 2003. Control of tsetse flies and trypanosomes using molecular genetics. *Vet. Parasitol.* **115**:125–145.
8. El-Sayed, N.M.A., and Donelson, J.E. 1997. Sequencing and mapping the African trypanosome enome. In *Trypanosomiasis and leishmaniasis*. G. Hide, J.C. Mottram, G.H. Coombs, and P.H. Holmes, editors. Cab International. Oxford, United Kingdom. 51–55.
9. Donelson, J.E. 2002. Antigenic variation and the African trypanosome genome. *Acta Tropica*. **85**:391–404.
10. Barry, J.D. 1997. The biology of antigenic variation in African trypanosomes. In *Trypanosomiasis and leishmaniasis*. G. Hide, J.C. Mottram, G.H. Coombs, and P.H. Holmes, editors. Cab International. Oxford, United Kingdom. 89–107.
11. Apted, F.I.C. 1970. Clinical manifestations and diagnosis of sleeping sickness. In *The African trypanosomiasis*. H.W. Mulligan, editor. George Allen & Unwin. London, United Kingdom. 661–683.
12. Duggan, A.J., and Hutchington, M.P. 1966. Sleeping sickness in Europeans: a review of 109 cases. *J. Trop. Med. Hyg.* **69**:124–131.
13. Kristensson, K., Grassi-Zucconi, G., and Bentivoglio, M. 1995. Nervous system dysfunctions in African trypanosomiasis. In *Recent advances in tropical neurology*. F. Clifford Rose, editor. Elsevier Science BV. Oxford, United Kingdom. 165–174.
14. Truc, P., et al. 2002. Evaluation of the micro-CATT, CATT/Trypanosoma brucei gambiense, and LATEX/T.b. gambiense methods for serodiagnosis and surveillance of human African trypanosomiasis in West and Central Africa. *Bull. World Health Organ.* **80**:882–886.
15. 1998. Control and surveillance of African trypanosomiasis. Report of a WHO expert committee. World Health Organization. Geneva, Switzerland. Technical Report Series, No. 881. 114 pp.
16. Lejon, V., et al. 2003. Intrathecal immune response pattern for improved diagnosis of central nervous system involvement in trypanosomiasis. *J. Infect. Dis.* **187**:1475–1483.
17. Jamonneau, V., et al. 2003. Stage determination and therapeutic decision in human African trypanosomiasis: value of polymerase chain reaction and immunoglobulin M quantification on the cerebrospinal fluid of sleeping sickness patients in Côte d'Ivoire. *Trop. Med. Int. Health.* **8**:589–594.
18. Lejon, V., et al. 2002. IgM quantification in the cerebrospinal fluid of sleeping sickness patients by a latex card agglutination test. *Trop. Med. Int. Health.* **7**:685–692.
19. Solano, P., et al. 2002. Comparison of different DNA preparation protocols for PCR diagnosis of human African trypanosomosis in Côte d'Ivoire. *Acta Trop.* **82**:349–356.
20. Gill, D.S., Chatha, D.S., and del Carpio-O'Donovan, R. 2003. MR imaging findings in African trypansomiasis. *Am. J. Neuroradiol.* **24**:1383–1385.
21. Baumann, H., and Strassmann, G. 1993. Suramin inhibits the stimulation of acute phase plasma protein genes by IL-6-type cytokines in rat hepatoma cells. *J. Immunol.* **151**:1456–1462.
22. Benaim, G., Lopez-Estrano, C., Docampo, R., and Moreno, S.N. 1993. A calmodulin-stimulated Ca2+ pump in plasma-membrane vesicles from Trypanosoma brucei; selective inhibition by pentamidine. *Biochem. J.* **296**:759–763.
23. Fairlamb, A.H., Henderson, G.B., and Cerami, A. 1989. Trypanothione is the primary target for arsenical drugs against African trypanosomes. *Proc. Natl. Acad. Sci. U. S. A.* **86**:2607–2611.
24. Metcalf, B.W., et al. 1978. Catalytic irreversible inhibition of mammalian ornithine decarboxylase by substrate and product analogues. *J. Am. Chem. Soc.* **100**:2551–2553.
25. Henderson, G.B., et al. 1988. "Subversive" substrates for the enzyme trypanothione disulfide reductase: alternative approach to chemotherapy of Chagas disease. *Proc. Natl. Acad. Sci. U. S. A.* **85**:5374–5378.
26. Fairlamb, A.H. 1990. Future prospects for the chemotherapy of human trypanosomiasis. 1. Novel approaches to the chemotherapy of trypanosomiasis. *Trans. R. Soc. Trop. Med. Hyg.* **84**:613–617.
27. Legros, D., et al. 2002. Treatment of human African trypanosomiasis — present situation and needs for research and development. *Lancet Infect. Dis.* **2**:437–440.
28. Legros, D., Evans, S., Maiso, F., Enyaru, J.C.K., and Mbulamberi, D. 1994. Risk factors for treatment failure after melarsoprol for Trypanosoma brucei gambiense trypanosomiasis in Uganda. *Trans. R. Soc. Trop. Med. Hyg.* **93**:439–442.
29. Pepin, J., and Milord, F. 1994. The treatment of human African trypanosomiasis. *Adv. Parasitol.* **33**:1–47.
30. Pepin, J., et al. 1994. Gambiense trypanosomiasis: frequency of, and risk factors for, failure of melarsoprol therapy. *Trans. R. Soc. Trop. Med. Hyg.* **88**:447–452.
31. Pepin, J., et al. 1989. Trial of prednisolone for prevention of melarsoprol-induced encephalopathy in gambiense sleeping sickness. *Lancet.* **1**:1246–1249.
32. Burri, C., et al. 2000. Efficacy of new, concise schedule for melarsoprol in treatment of sleeping sickness caused by Trypanosoma brucei gambiense: a randomised trial. *Lancet.* **355**:1419–1425.
33. Burri, C., and Brun, R. 2003. Eflornithine for the treatment of human African trypanosomiasis. *Parasitol. Res.* **90**(Suppl. 1):S49–S52.
34. Adams, J.H., et al. 1986. Human African trypanosomiasis (T.b. gambiense): a study of 16 fatal cases of sleeping sickness with some observations on acute reactive arsenical encephalopathy. *Neuropathol. Appl. Neurobiol.* **12**:81–94.
35. MacLean, L., Odiit, M., and Sternberg, J.M. 2001. Nitric oxide and cytokine synthesis in human African trypanosomiasis. *J. Infect. Dis.* **184**:1086–1090.
36. Lejon, V., et al. 2002. Interleukin (IL)-6, IL-8 and IL-10 in serum and CSF of Trypanosoma brucei gambiense sleeping sickness patients before and after treatment. *Trans. R. Soc. Trop. Med. Hyg.* **96**:329–333.
37. Pentreath, V.W. 1995. Trypanosomiasis and the nervous system. Pathology and immunology. *Trans. R. Soc. Trop. Med. Hyg.* **89**:9–15.
38. Pentreath, V.W. 1989. Neurobiology of sleeping sickness. *Parasitol. Today.* **5**:215–218.
39. Pepin, J., and Milord, F. 1991. African trypanosomiasis and drug induced encephalopathy: risk factors and pathogenesis. *Trans. R. Soc. Trop. Med. Hyg.* **85**:222–224.
40. Lambert, P.H., Berney, M., and Kazyumba, G.L. 1981. Immune complexes in serum and cerebrospinal fluid in sleeping sickness. Correlation with polyclonal B-cell activation and with intracerebral immunoglobulin synthesis. *J. Clin. Invest.* **67**:77–85.
41. Hunter, C.A., Jennings, F.W., Adams, J.H., Murray, M., and Kennedy, P.G.E. 1992. Subcurative chemotherapy and fatal post-treatment reactive encephalopathies in African trypanosomiasis. *Lancet.* **339**:956–958.
42. Poltera, A.A. 1980. Immunopathological and chemotherapeutic studies in experimental trypanosomiasis with special reference to the heart and brain. *Trans. R. Soc. Trop. Med. Hyg.* **74**:706–715.
43. Kennedy, P.G.E. 1999. The pathogenesis and modulation of the post-treatment reactive encephalopathy in a mouse model of human African trypanosomiasis. *J. Neuroimmunol.* **100**:36–41.
44. Hunter, C.A., Jennings, F.W., Kennedy, P.G.E., and Murray, M. 1992. Astrocyte activation correlates with cytokine production in central nervous system of Trypanosoma brucei brucei-infected mice. *Lab. Invest.* **67**:635–642.
45. Hunter, C.A., Gow, J.W., Kennedy, P.G.E., Jennings, F.W., and Murray, M. 1991. Immunopathology of experimental African sleeping sickness: detection of cytokine mRNA in the brains of Trypanosoma brucei brucei-infected mice. *Infect. Immun.* **59**:4636–4640.
46. Jennings, F.W., et al. 1997. The role of the polyamine inhibitor eflornithine in the neuropathogenesis of experimental murine African trypanosomiasis. *Neuropathol. Appl. Neurobiol.* **23**:225–234.
47. Hunter, C.A., Jennings, F.W., Kennedy, P.G.E., and Murray, M. 1992. The use of azathioprine to ameliorate post-treatment encephalopathy associated with African trypanosomiasis. *Neuropathol. Appl. Neurobiol.* **18**:619–625.

48. Kennedy, P.G.E., et al. 1997. A substance P antagonist, RP-67,580, ameliorates a mouse meningoencephalitic response to Trypanosoma brucei brucei. *Proc. Natl. Acad. Sci. U. S. A.* **94**:4167–4170.

49. Kennedy, P.G.E., et al. 2003. Clinical and neuro-inflammatory responses to meningoencephalitis in Substance P receptor knockout mice. *Brain.* **16**:1683–1690.

50. Sharafeldin, A., Eltayeb, R., Pashendov, M., and Bakhiet, M. 2000. Chemokines are produced in the brain early during the course of experimental African trypanosomiasis. *J. Neuroimmunol.* **103**:165–170.

51. Paulnock, D.M., and Coller, S.P. 2001. Analysis of macrophage activation in African trypanosomiasis.

J. Leukoc. Biol. **69**:685–690.

52. Magez, S., et al. 1998. The glycosyl-inositol-phosphate and dimyristoylglycerol moieties of the glycosylphosphatidylinositol anchor of the trypanosome variant-specific surface glycoprotein are distinct macrophage-activating factors. *J. Immunol.* **160**:1949–1956.

53. Sternberg, J.M. 1998. Immunobiology of African trypanosomiasis. *Chem. Immunol.* **70**:186–199.

54. Olsson, T., et al. 1993. CD8 is critically involved in lymphocyte activation by a T. brucei brucei-released molecule. *Cell.* **72**:715–727.

55. Vaidya, T., et al. 1997. The gene for a T lymphocyte triggering factor from African trypanosomes.

J. Exp. Med. **186**:433–438.

56. Bentivoglio, M., Grassi-Zucconi, G., Olsson, T., and Kristensson, K. 1994. Trypanosoma brucei and the nervous system. *Trends Neurosci.* **17**:325–329.

57. Hunter, C.A., and Kennedy, P.G.E. 1992. Immuno-pathology in central nervous system human African trypanosomiasis. *J. Neuroimmunol.* **36**:91–95.

58. Schleifer, K.W., Filutowicz, H., Schopf, L.R., and Mansfield, J.M. 1993. Characterization of T helper cell responses to the trypanosome variant surface glycoprotein. *J. Immunol.* **150**:2910–2919.

59. Kennedy, P.G.E., Murray, M., Jennings, F., and Rodgers, J. 2002. Sleeping sickness: new drugs from old? *Lancet.* **359**:1695–1696.

Part 5

Viral Diseases

SARS coronavirus: a new challenge for prevention and therapy

Kathryn V. Holmes

University of Colorado Health Sciences Center, Department of Microbiology, Denver, Colorado, USA.

A new and deadly clinical syndrome now called severe acute respiratory syndrome (SARS) was brought to the attention of the WHO by Dr. Carlo Urbani and his colleagues in a Vietnamese hospital in February 2003 (1). The WHO, the medical staffs in hospitals where the disease had appeared, and local and regional governments, together with a dozen cooperating laboratories across the globe, immediately responded. They provided a provisional case definition to identify the extent and geographic distribution of the outbreak (2), laboratory investigations to identify the infectious agent, and travel advisories and quarantines to limit the spread of the disease (3, 4). This extraordinary and effective collaboration limited the potentially explosive spread of the outbreak, while initial case reports with clinical and epidemiological information were quickly posted on the Internet to help physicians identify additional cases of the new syndrome (2, 4–9). The press and scientific journals played valuable roles in rapidly distributing accurate information about SARS to the frightened public and making key scientific publications about SARS available via the Internet before they could appear in print. A stroke of good fortune in this crisis was the discovery that a novel virus could be readily isolated from patients' lungs and sputum and cultivated in a monkey kidney cell line (8, 10, 11). Laboratory investigations using electron microscopy, virus-discovery microarrays containing conserved nucleotide sequences characteristic of many virus families, randomly primed RT-PCR, and serological tests quickly identified the virus as a new coronavirus (8, 10, 11). Inoculation of monkeys with the SARS-associated coronavirus (SARS-CoV) caused interstitial pneumonia resembling SARS, and the virus was isolated from the nose and throat (12). No viral or bacterial copathogen was needed to induce the disease. These experiments fulfilled Koch's postulates and proved that SARS-CoV is the cause of SARS.

Lessons from the pathophysiology and epidemiology of known coronavirus diseases of humans and animals

Until SARS appeared, human coronaviruses were known as the cause of 15–30% of colds (13). Because there is no small-animal model for coronavirus colds, the pathophysiology of human coronavirus infection of the upper respiratory tract was studied in human volunteers (14, 15). Intranasal inoculation induces colds in a small percentage of volunteers, although virus replication in nasal epithelium is detected in most volunteers. Colds are generally mild, self-limited infections, and significant increases in neutralizing antibody titer are found in nasal secretions and serum after infection. Nevertheless, some unlucky individuals can be reinfected with the same coronavirus soon after recovery and get symptoms again. Coronavirus colds are more frequent in winter, and the two known human coronaviruses vary in prevalence from year to year. If SARS becomes established in humans, will it also have a seasonal incidence of clinical disease? Prospective studies of hospitalized patients showed that human respiratory coronaviruses only rarely cause lower respiratory tract infection, perhaps in part because they grow poorly at 37°C. Although coronavirus-like particles have been observed by electron microscopy in human feces, and serological studies of necrotizing enterocolitis in infants occasionally show rises in antibody titer to coronaviruses (16–18), infectious human coronaviruses have been, until SARS, extremely difficult to isolate from feces (19).

Coronaviruses cause economically important diseases of livestock, poultry, and laboratory rodents (20). Most coronaviruses of animals infect epithelial cells in the respiratory and/or enteric tracts, causing epizootics of respiratory diseases and/or gastroenteritis with short incubation periods (2–7 days), such as those found in SARS. In general, each coronavirus causes disease in only one animal species. In immunocompetent hosts, infection elicits neutralizing antibodies and cell-mediated immune responses that kill infected cells. In SARS patients, neutralizing antibodies are detected 2–3 weeks after the onset of disease, and 90% of patients recover without hospitalization (10). In animals, reinfection with coronaviruses is common, with or without disease symptoms. The duration of shedding of SARS-CoV from respiratory secretions of SARS patients appears to be quite variable. Some animals can shed infectious coronavirus persistently from the enteric tract for weeks or months without signs of disease, transmitting infectious virus to neonates and other susceptible animals. SARS-CoV has been detected in the feces of patients by RT-PCR and virus isolation (8, 11). Studies are being done to learn whether SARS-CoV is shed persistently from the respiratory and/or enteric tracts of some humans without signs of disease. Host factors such as age, strain or genotype, immune status, coinfection with other viruses, bacteria, or parasites, and stress affect susceptibility to coronavirus-induced diseases of animals, and the ability to spread virus to susceptible animals. It is important to learn what host factors and/or virus differences are responsible for the "super-spreader" phenomenon observed in SARS, in which a few patients infect many people through brief casual contact or possibly environmental contamination, even though most patients infect only people in close contact with them during the period of overt disease.

Several coronaviruses can cause fatal systemic diseases in animals, including feline infectious peritonitis virus (FIPV), hemagglutinating encephalomyelitis virus (HEV) of swine, and some strains of avian infectious bronchitis virus (IBV) and mouse hepa-

Nonstandard abbreviations used: severe acute respiratory syndrome (SARS); SARS coronavirus (SARS-CoV); feline infectious peritonitis virus (FIPV); hemagglutinating encephalomyelitis virus (HEV); infectious bronchitis virus (IBV); mouse hepatitis virus (MHV); transmissible gastroenteritis coronavirus (TGEV); porcine respiratory coronavirus (PRCoV); porcine epidemic diarrhea virus (PEDV).

Conflict of interest: The author has declared that no conflict of interest exists.

Citation for this article: *J. Clin. Invest.* **111**:1605–1609 (2003). doi:10.1172/JCI200318819.

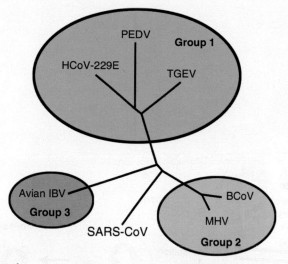

Figure 1
Phylogenetic analysis of coronaviruses, based on the polymerase gene, shows that SARS coronavirus is different from each of the three groups of the previously known coronaviruses. HCoV-229E, human coronavirus 229E; BCoV, bovine coronavirus. Adapted with permission from ref. 24.

titis virus (MHV). These coronaviruses can replicate in liver, lung, kidney, gut, spleen, brain, spinal cord, retina, and other tissues. SARS-CoV has been found in patients' lungs, feces, and kidney. Further studies with sensitive methods of detection will reveal which additional tissues may be infected with SARS-CoV. The pathophysiology of coronavirus diseases of animals has been studied extensively, but there is no coronavirus disease of animals that closely resembles SARS. Immunopathology plays a role in tissue damage in MHV and FIPV, and cytokines are responsible for some signs of disease. Significantly, in cats with persistent, inapparent infection with feline enterotropic coronavirus, virulent virus mutants can arise and cause fatal infectious peritonitis, a systemic disease (21). Are virulent mutants of SARS-CoV associated with the fatal cases? Comparison of the genomes of SARS-CoVs isolated from fatal versus milder cases will identify any virus mutations that may be associated with increased virulence.

In animals, coronaviruses cause enzootic or epizootic diseases. Four different coronaviruses infect pigs, and the epidemiology of these porcine diseases is informative. Transmissible gastroenteritis coronavirus (TGEV) can infect the enteric and respiratory tracts, causing severe diarrhea in suckling pigs, and milder or inapparent infection in adult pigs. Mutant TGEVs with spontaneous deletions of more than 200 amino acids in the viral spike glycoprotein, or several point mutations in the same region, have arisen separately in Europe and the US and are called porcine respiratory coronavirus (PRCoV). The mutant viruses cause mild respiratory disease and cannot infect the gut (22). PRCoV may serve as a natural vaccine to protect piglets from TGEV. A third porcine coronavirus, HEV, causes vomiting and wasting disease of piglets and can cause encephalomyelitis. A "new" porcine coronavirus, porcine epidemic diarrhea virus (PEDV), was first detected in European pigs during widespread outbreaks of fatal diarrhea of piglets during the 1980s (23). Serology suggests that, before this time, pigs had not been exposed to PEDV. How do such "new" coronaviruses emerge? What viral or host factors make them able to spread so effectively?

How did SARS-CoV suddenly appear in humans?

Human sera collected before the SARS outbreak do not contain antibodies directed against SARS-CoV (8, 10), suggesting that this virus is new to humans. Additional studies on human sera from the region where the outbreak began are needed to confirm this preliminary finding. Did SARS-CoV jump to humans by mutation of an animal coronavirus or by recombination between several known human or animal coronaviruses?

The complete 29,727-nucleotide sequence of the RNA genome of SARS-CoV (GenBank accession nos. AY274119 and AY278741) (24, 25) proves that it is a member of the *Coronaviridae* family and provides some insight into its possible origin. The SARS-CoV genome encodes all five of the coronavirus proteins needed for production of new virions. It contains the enormous (20-kb) gene that encodes the unique RNA-dependent RNA polymerase common to all coronaviruses. The order of the genes encoding the RNA polymerase and structural proteins is conserved in the genomes of all coronaviruses, including SARS-CoV. Interspersed between these genes are several nonconserved open reading frames encoding proteins that are not required for virus replication. The SARS genome, like that of other coronaviruses, contains several nonconserved, open reading frames that encode small nonstructural proteins with unknown functions.

The genes of SARS-CoV were compared with the corresponding genes of known coronaviruses of humans, pigs, cattle, dogs, cats, mice, rats, chickens, and turkeys. Each gene of SARS-CoV has only 70% or less identity with the corresponding gene of the known coronaviruses. Thus, SARS-CoV is only distantly related to the known coronaviruses of humans and animals. Phylogenetic analysis (Figure 1) suggests that SARS-CoV does not fit within any of the three groups that contain all other known coronaviruses (11, 24, 25). Its closest relatives are the murine, bovine, porcine, and human coronaviruses in group 2 and avian coronavirus IBV in group 1. These data show that SARS-CoV did not arise by mutation of human respiratory coronaviruses or by recombination between known coronaviruses. Instead, it is likely that SARS-CoV was enzootic in an unknown animal or bird species and had been genetically isolated there for a very long time before somehow suddenly emerging as a virulent virus of humans. Did this jump to humans occur only once because of an unlucky and unlikely combination of random mutations, or can SARS-CoV now infect both humans and its original host? Does the virus have the potential to jump repeatedly from its animal host to cause deadly outbreaks of human disease?

What features of SARS-CoV and its replication are potential targets for development of new antiviral drugs and vaccines?

Unfortunately, there are no approved antiviral drugs that are highly effective against coronaviruses. However, many steps unique to coronavirus replication could be targeted for development of antiviral drugs (Figure 2). Coronavirus infection begins with binding of the spike protein (S) on the viral envelope to a specific receptor on the cell membrane. Conformational changes are induced in S that probably lead to fusion of the viral envelope with the host cell membrane (23–25). Molecules that block binding to the receptor or inhibit the receptor-induced conformational change in S might block SARS-CoV infection (26–28). Inhibitors of HIV-1 entry and membrane fusion are good models for new drugs that target this first step in coronavirus infection.

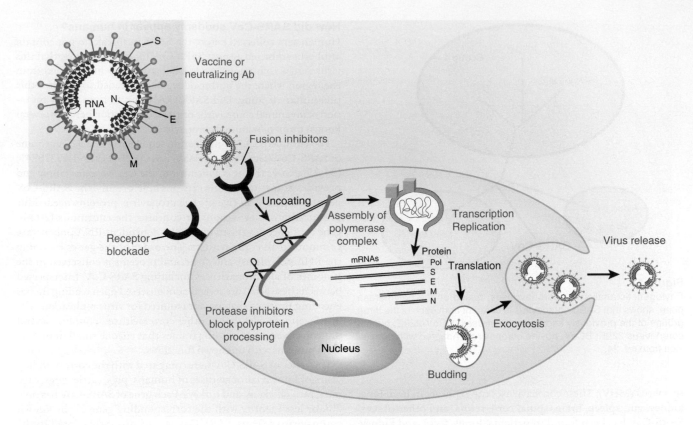

Figure 2
Steps in coronavirus replication that are potential targets for antiviral drugs and vaccines. The spike glycoprotein S is a good candidate for vaccines because neutralizing antibodies are directed against S. Blockade of the specific virus receptor on the surface of the host cell by monoclonal antibodies or other ligands can prevent virus entry. Receptor-induced conformational changes in the S protein can be blocked by peptides that inhibit membrane fusion and virus entry. The polyprotein of the replicase protein is cleaved into functional units by virus-encoded proteinases. Protease inhibitors may block replication. The polymerase functions in a unique membrane-bound complex in the cytoplasm, and the assembly and functions of this complex are potential drug targets. Viral mRNAs made by discontinuous transcription are shown in the cytoplasm with the protein that each encodes indicated at the right. The common 70 base long leader sequence on the 5′ end of each mRNA is shown in red. Budding and exocytosis are processes essential to virus replication that may be targets for development of antiviral drugs. M, membrane protein required for virus budding; S, viral spike glycoprotein that has receptor binding and membrane fusion activities; E, small membrane protein that plays a role in coronavirus assembly; N, nucleocapsid phosphoprotein associated with viral RNA inside the virion. Adapted with permission from ref. 35.

The large polyprotein encoded by the polymerase gene of coronaviruses must be proteolytically cleaved at specific sites by several virus-encoded proteases in order to have RNA polymerase activity (29–31). Protease inhibitors developed to treat other viral diseases as well as new protease inhibitors are being tested for the ability to inhibit cleavage of the SARS-CoV polymerase protein and viral RNA synthesis. Coronavirus RNA is synthesized in a virus-specific, flask-shaped cytoplasmic compartment bordered by a double membrane (32). Could the assembly or function of this unique organelle be inhibited?

The RNA genome of coronaviruses is transcribed discontinuously so that the complement of the 70-nucleotide leader sequence is joined to the 3′ ends of the subgenomic negative-strand RNAs that are templates for the nested set of subgenomic mRNAs (33, 34). Perhaps a small RNA or other inhibitor could be designed to block this unusual discontinuous RNA transcription. Alternatively, nucleoside inhibitors might be designed to block SARS-CoV replication specifically without damaging the cell.

Coronavirus structural proteins and newly synthesized RNA genomes assemble into virions by budding into pre-Golgi mem-

branes. Virus assembly is also a potential target for drug development. Coronaviruses are apparently released from living cells by exocytosis, so inhibitors of secretion should be tested for antiviral activity.

The spike glycoproteins on virions of several coronaviruses require cleavage by serine protease to activate viral infectivity, but it is not yet known whether this is also true for SARS-CoV. Inhibitors of serine proteases might block this late step in the coronavirus life cycle.

Fortunately, new antiviral drugs that will be developed to treat SARS-CoV may also be effective in the treatment of common colds and economically important coronavirus diseases of companion animals, livestock, and poultry. Antiviral drugs to treat other diseases of the respiratory tract, such as influenza, are most effective when used very soon after signs of disease appear. If this is also found to be true for antiviral drugs for SARS, then rapid viral diagnostic tests will be needed to differentiate SARS from other pulmonary infections soon after the onset of disease.

In about 10% of SARS patients, interstitial pneumonia is followed after 5–7 days by progressive diffuse alveolar damage, possibly due to immunopathology (5, 8–10). Corticosteroids have been used to try to reduce disease progression. When the pathophysiology of these severe cases of SARS is understood, more specific anti-inflammatory drugs may be found to prevent SARS-induced progressive tissue damage. If specific host factors are found to be associated with the most severe cases, it may be possible to modulate their expression or activity in order to prevent progression of the disease. If host factors play a role in SARS progression, a test to identify patients with the highest risk of severe SARS might guide decisions concerning potential benefits versus risks of the use of novel antiviral drugs and other treatments.

Passive immunization with convalescent serum has been tested as a way to treat SARS. A possible approach for prevention of SARS in people at high risk of exposure, such as health care workers, is administration of neutralizing mAb's against the spike protein of SARS-CoV, similar to the current use of a neutralizing mAb against respiratory syncytial virus to prevent lower respiratory tract disease in infants at high risk of complications. Such a neutralizing anti-SARS mAb might also be useful for treatment of SARS.

Control of SARS is most likely to be achieved by vaccination. Live attenuated vaccines prevent serious diseases caused by porcine and avian coronaviruses. It is likely that a similar live attenuated vaccine could be developed for SARS-CoV, especially since the virus can be grown to high titers in cell culture. It will be particularly important to test SARS vaccines for untoward effects, however, since several vaccines against feline coronavirus have caused antibody-dependent enhancement of disease when the vaccinated animals were subsequently infected with wild-type virus. A SARS vaccine would be used to protect health care workers and others at high risk in areas where the virus is circulating. However, because the incubation period of SARS is so short, the vaccine would have to be used prophylactically, and it would be unlikely to prevent disease when used after exposure to a SARS patient.

The SARS epidemic appears to be out of control now in some areas. New tests to identify SARS patients at the earliest stages of disease are expected to be widely available soon. These tests will guide quarantine decisions and other public health measures to limit the spread of infection. Nevertheless, it now appears likely that drugs and/or vaccines will be needed to control the epidemic. Development of effective drugs and vaccines for SARS is likely to take a long time. The world will anxiously watch the high-stakes race between the spread of the SARS epidemic and the development of effective SARS drugs and vaccines.

Address correspondence to: Kathryn V. Holmes, Department of Microbiology, Campus Box B-175, University of Colorado Health Sciences Center, 4200 East 9th Avenue, Denver, Colorado 80262, USA. Phone: (303) 315-7329; Fax: (303) 315-6785; E-mail: Kathryn.holmes@uchsc.edu.

1. 2003. World Health Organization. Severe acute respiratory syndrome (SARS): multi-country outbreak. http://www.who.int/csr/don/2003_03-16/en/.
2. 2003. Preliminary clinical description of severe acute respiratory syndrome. *MMWR Morb. Mortal. Wkly. Rep.* **52**:255–256.
3. 2003. WHO recommended measures for persons undertaking international travel from areas affected by severe acute respiratory syndrome (SARS). *Wkly. Epidemiol. Rec.* **78**:97–120.
4. Gerberding, J.L. 2003. Faster. But fast enough? Responding to the epidemic of severe acute respiratory syndrome. *N. Engl. J. Med.* In press.
5. Lee, N., et al. 2003. A major outbreak of severe acute respiratory syndrome in Hong Kong. *N. Engl. J. Med.* In press.
6. Poutanen, S.M., et al. 2003. Identification of severe acute respiratory syndrome in Canada. *N. Engl. J. Med.* In press.
7. Tsang, K.W., et al. 2003. A cluster of cases of severe acute respiratory syndrome in Hong Kong. *N. Engl. J. Med.* In press.
8. Peiris, J.S.M., et al. 2003. Coronavirus as a possible cause of severe acute respiratory syndrome. *Lancet.* **361**:1319–1325.
9. Chan-Yeung, M., and Yu, W.C. 2003. Outbreak of severe acute respiratory syndrome in Hong Kong Special Administrative Region: case report. *BMJ.* **326**:850–852.
10. Ksiazek, T.G., et al. 2003. A novel coronavirus associated with severe acute respiratory syndrome. *N. Engl. J. Med.* In press.
11. Drosten, C., et al. 2003. Identification of a novel coronavirus in patients with severe acute respiratory syndrome. *N. Engl. J. Med.* In press.
12. Fouchier, R.A.M., et al. 2003. Aetiology: Koch's postulates fulfilled for SARS virus. *Nature.* **423**:240.
13. Holmes, K.V. 2001. Coronaviruses. In *Fields' virology.* D. Knipe, et al., editors. Lippincott Williams & Wilkins. Philadelphia, Pennsylvania, USA. 1187–1203.
14. Bradburne, A.F., and Tyrrell, D.A.J. 1971. Coronaviruses of man. *Prog. Med. Virol.* **13**:373–403.
15. Chilvers, M.A., et al. 2001. The effects of coronavirus on human nasal ciliated respiratory epithelium. *Eur. Respir. J.* **18**:965–970.
16. Resta, S., Luby, J.P., Rosenfeld, C.R., and Siegel, J.D. 1985. Isolation and propagation of a human enteric coronavirus. *Science.* **229**:978–981.
17. Kapikian, A.Z. 1975. The coronaviruses. *Dev. Biol. Stand.* **28**:42–64.
18. Battaglia, M., Passarani, N., Di Matteo, A., and Gerna, G. 1987. Human enteric coronaviruses: further characterization and immunoblotting of viral proteins. *J. Infect. Dis.* **155**:140–143.
19. Macnaughton, M.R., and Davies, H.A. 1981. Human enteric coronaviruses: brief review. *Arch. Virol.* **70**:301–313.
20. Lai, M.M.C., and Holmes, K.V. 2001. Coronaviridae and their replication. In *Fields' virology.* D. Knipe, et al., editors. Lippincott Williams & Wilkins. Philadelphia, Pennsylvania, USA. 1163–1185.
21. Herrewegh, A.A., et al. 1997. Persistence and evolution of feline coronavirus in a closed cat-breeding colony. *Virology.* **234**:349–363.
22. Ballesteros, M.L., Sanchez, C.M., and Enjuanes, L. 1997. Two amino acid changes at the N-terminus of transmissible gastroenteritis coronavirus spike protein result in the loss of enteric tropism. *Virology.* **227**:378–388.
23. de Arriba, M.L., Carvajal, A., Pozo, J., and Rubio, P. 2002. Mucosal and systemic isotype-specific antibody responses and protection in conventional pigs exposed to virulent or attenuated porcine epidemic diarrhoea virus. *Vet. Immunol. Immunopathol.* **85**:85–97.
24. Rota, P.A., et al. 2003. Characterization of a novel coronavirus associated with severe acute respiratory syndrome. *Science.* doi:10.1126/science.1085952.
25. Marra, M.A., et al. 2003. The genome sequence of the SARS-associated coronavirus. *Science.* doi:10.1126/science.1085953.
26. Matsuyama, S., and Taguchi, F. 2002. Receptor-induced conformational changes of murine coronavirus spike protein. *J. Virol.* **76**:11819–11826.
27. Zelus, B.D., Schickli, J.H., Blau, D.M., Weiss, S.R., and Holmes, K.V. 2003. Conformational changes in the spike glycoprotein of murine coronavirus are induced at 37C either by soluble murine CEACAM1 receptors or by pH 8. *J. Virol.* **77**:830–840.
28. Lewicki, D.N., and Gallagher, T.M. 2002. Quaternary structure of coronavirus spikes in complex with carcinoembryonic antigen-related cell adhesion molecule cellular receptors. *J. Biol. Chem.* **277**:19727–19734.
29. Ziebuhr, J., and Siddell, S.G. 1999. Processing of the human coronavirus 229E replicase polyproteins by the virus-encoded 3C-like proteinase: identification of proteolytic products and cleavage sites common to pp1a and pp1ab. *J. Virol.* **73**:177–185.
30. Denison, M.R., et al. 1999. The putative helicase of the coronavirus mouse hepatitis virus is processed from the replicase gene polyprotein and localizes in complexes that are active in viral RNA synthesis. *J. Virol.* **73**:6862–6871.
31. Hegyi, A., and Ziebuhr, J. 2002. Conservation of substrate specificities among coronavirus main proteases. *J. Gen. Virol.* **83**:595–599.
32. Gosert, R., Kanjanahaluethai, A., Egger, D., Bienz, K., and Baker, S.C. 2002. RNA replication of mouse hepatitis virus takes place at double-membrane vesicles. *J. Virol.* **76**:3697–3708.
33. Sawicki, D., Wang, T., and Sawicki, S. 2001. The RNA structures engaged in replication and transcription of the A59 strain of mouse hepatitis virus. *J. Gen. Virol.* **82**:385–396.
34. Sethna, P.B., and Brian, D.A. 1997. Coronavirus genomic and subgenomic minus-strand RNAs copartition in membrane-protected replication complexes. *J. Virol.* **71**:7744–7749.
35. 1996. *Fundamental Virology.* B.N. Fields, D.M. Knipe, and P.M. Howley, editors. 3rd edition. Lippincott-Raven. Philadelphia, Pennsylvania, USA/New York, New York, USA. 544.

Acute HIV revisited: new opportunities for treatment and prevention

Christopher D. Pilcher,[1] Joseph J. Eron Jr.,[1] Shannon Galvin,[1] Cynthia Gay,[1] and Myron S. Cohen[1,2]

[1]Department of Medicine and [2]Department of Epidemiology, University of North Carolina at Chapel Hill, Chapel Hill, North Carolina, USA.

Inability to recognize incident infection has traditionally limited both scientific and public health approaches to HIV disease. Recently, some laboratories have begun adding HIV nucleic acid amplification testing to HIV diagnostic testing algorithms so that acute (antibody-negative) HIV infections can be routinely detected within the first 1–3 weeks of exposure. In this review article, we will highlight critical opportunities for HIV treatment and prevention that are presented by these diagnostic strategies.

Acute HIV infection

The natural history of HIV infection encompasses an acute/primary phase that lasts months, followed by an early/clinically latent phase that typically lasts 3–10 years, and ultimately by the immune collapse characterized by AIDS. "Acute" HIV infection best describes the interval during which HIV can be detected in blood serum and plasma before the formation of antibodies routinely used to diagnose infection. During this time, high levels of viremia and shedding at mucosal sites can be demonstrated, because HIV replication is unrestrained by immune responses. Approximately 30 days after infection, early virus-specific immune responses are mounted, with subsequent reduction of viremia. After 4–6 months, viral and host factors combine to determine a new pseudo–steady state of viremia (or virologic "set point") for each patient, heralding the beginning of the long clinical latency experienced by patients infected by HIV.

Recognition of acute HIV infection is important for several reasons. First, acute HIV provides a unique view of HIV transmission and pathogenesis, including early host-virus interactions that require further study. Second, prevention strategies directed at subjects with acute HIV infection may have great impact. Third, very early recognition may allow for HIV treatment that could alter the natural history of disease, or even eliminate infection.

Acute HIV revisited: new possibilities for early detection

Despite infection of nearly 60 million individuals worldwide with HIV, fewer than 1,000 cases have been diagnosed in the first month of infection, primarily because of a lack of a specific and recognizable acute retroviral syndrome. About half of people with acute HIV infection develop headache, fever, myalgias, anorexia, rash, and/or diarrhea (1–3) after an incubation period of around 14 days (1, 4–6). Symptoms are generally minor and last days to weeks (1). Genital or oral ulcers may be present, and coinfections with other sexually transmitted pathogens (e.g., herpes simplex virus, gonorrhea, syphilis, hepatitis viruses) are common (2, 7–9). The latter observation suggests common cotransmission of HIV and other sexually transmitted disease (STD) pathogens.

Since acute retroviral syndromes mimic many common febrile illnesses, including infectious mononucleosis, influenza, malaria, and rickettsial diseases (1, 10), the true diagnosis (acute HIV) is rarely considered at an initial patient encounter (1, 10–12). The diagnostic challenge in acute HIV infection is made more difficult by the fact that routine HIV antibody tests will typically remain negative for 1–2 weeks beyond the onset of acute retroviral symptoms (26–35 days following initial infection) (5, 13); additional virus-specific diagnostic tests (e.g., HIV p24 antigen ELISA and HIV nucleic acid amplification assays) are needed to detect HIV infection prior to the appearance of antibodies (Figure 1). While HIV nucleic acid amplification assays are now extremely sensitive and can reliably detect HIV by days 9–11 of infection (13, S1), they are vulnerable to false-positive rates as high as 1%. Such tests remain relatively expensive and have not traditionally been used for routine clinical HIV screening.

The blood-banking industry, however, combines antibody and HIV RNA testing to protect the blood supply. Blood banks submit all donations to routine antibody testing; antibody-positive donations are discarded first. They then combine aliquots from a number of antibody-negative donations, to create a specimen pool. The pools are then screened for HIV RNA. An individual donation is therefore only tested for HIV RNA if its pool first screens positive. In this way, only very few individual specimens are tested for HIV RNA — increasing throuput, decreasing costs, and dramatically reducing false-positive results (Figure 2; refs. 14, S2). Pooling strategies for HIV detection in clinical testing have recently been piloted in Swiss (15), Indian (16), and US (14) clinical-HIV-testing populations. The results suggest that HIV RNA testing — if based on pooling of antibody-negative specimens — can efficiently identify acute HIV infections in clinical testing. Especially in clinical-testing populations expected to contain low numbers of acute, antibody-negative infections, pooling affords remarkable increases in cost efficiency and predictive value. Clinical studies have clearly indicated that voluntary counseling and testing (VCT) enhanced to incorporate HIV RNA screening can be both clinically (14) and economically feasible (14, 16).

Furthermore, studies using various detection strategies (including individual tests for HIV RNA or p24 antigen, as well as pooling strategies) have revealed that acute HIV is not as rare in clinical-testing populations as is commonly assumed. Pincus and colleagues (12) found that five (1.0%) of 511 consecutive urgent-care-center attendees with "any viral symptoms" in Boston had acute HIV infection confirmed by seroconversion. Rosenberg and colleagues

---

Nonstandard abbreviations used: antiretroviral therapy (ART); CC chemokine receptor 5 (CCR5); CXC chemokine receptor 4 (CXCR4); sexually transmitted disease (STD); simian human immunodeficiency virus (SHIV); voluntary counseling and testing (VCT).

Conflict of interest: The authors have declared that no conflict of interest exists.

Citation for this article: *J. Clin. Invest.* **113**:937–945 (2004). doi:10.1172/JCI200421540.

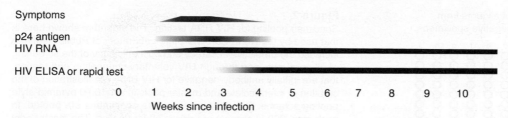

Symptoms

p24 antigen
HIV RNA

HIV ELISA or rapid test

Weeks since infection

Figure 1
Acute HIV diagnostic timeline. Symptoms, when present, typically occur around 2 weeks after infection. Viremia is detectable prior to symptoms in the form of HIV p24 antigen (detectable by ELISA) or HIV RNA (detectable by even more sensitive nucleic acid amplification). While viremia reaches extremely high levels in the month or two following infection, p24 antigen typically becomes undetectable shortly before seroconversion, because of the formation of early antibody-antigen complexes. A secondary antibody-negative, p24 antigen–negative period is sometimes observed.

examined specimens submitted for heterophile antibody testing for infectious mononucleosis in a university hospital and found that at least four (0.8%) of 536 sera examined were from acutely HIV-infected patients (11). Three (0.3%) of 1,000 emergency-department attendees with fever in North Carolina (17) and six (0.3%) of 2,300 consecutive, unselected emergency-department attendees in Baltimore (18) had antibody-negative acute HIV. In hyperendemic areas and among patients with STDs, acute HIV infections can be detected even more frequently: 58 (1.2%) of 4,999 patients at an Indian STD clinic had acute HIV (7), and 23 (2.5%) of 928 STD clinic attendees in Lilongwe, Malawi, were found to be acutely HIV-infected (2). While the overall percentage of individuals with acute HIV infection in these studies, in both low- and high-prevalence areas, may appear low, a striking finding was that HIV antibody tests have missed more than 3% of the detectable HIV infections in each testing population. In fact, in Pincus et al.'s and Rosenberg et al.'s studies, antibody-negative infections represented *more than a third* of all detectable infections among patients presenting for care with apparent viral symptoms. Together, these results indicate that adding HIV RNA screening to current diagnostic screens can commonly identify acutely infected patients and substantially improve the overall sensitivity of HIV testing for infection.

The North Carolina experience. In collaboration with the state of North Carolina, we have developed the Screening and Tracing Active Transmission (STAT) program, which incorporates HIV RNA screening into all publicly funded HIV testing performed for HIV VCT by the central State Laboratory of Public Health (approximately 120,000 specimens are processed per year). HIV RNA screening has directly increased the number of cases identified by 4% over antibody testing alone, false-positive results have been rare (less than one per 50,000 specimens), and the marginal cost of pooling/HIV RNA testing has remained low (approximately $2 per specimen) (3). Similar programs are now being launched in other states and abroad. It is therefore critical to revisit the opportunities that detection of acute HIV infection can provide for our understanding of HIV pathogenesis, and for our ability to help patients receive better clinical care and more effective prevention.

The biology of HIV expansion

Initial expansion of viral reservoirs. HIV's attachment to and entry into cells results from a complex interaction of the HIV envelope proteins gp120 and gp41 with cell surface receptors (CD4 and CC chemokine receptor 5 [CCR5] or CXC chemokine receptor 4 [CXCR4]) that are present on T cells and/or macrophages (19,

S3–S7). In acute HIV infection resulting from sexual exposure to HIV-1 (Figure 3), the virus successfully traverses the epithelial layer by one of several proposed mechanisms: transcytosis, epithelial cell capture, or exposure to subepithelial target cells through disruptions in mucosal integrity (reviewed in ref. 20). Mucosal DCs express C-type lectins (e.g., DC-SIGN), which can both specifically bind HIV and facilitate its internalization. HIV-1 that is bound and internalized by a DC may allow infection of nearby CD4+CCR5+ cells and/or efficient presentation of HIV-1 to susceptible target cells after DCs migrate to draining lymph nodes (21, S8). However, simian human immunodeficiency virus–infected (SHIV-infected) memory CD4+CCR5+ cells have been detected in mucosal tissues of macaques as early as 6 hours after exposure, which suggests the possibility of direct infection (22, S9, S10). Ultimately, HIV-infected CD4+ cells can be found in lymphoid tissue within 2 days of infection, and the magnitude of infection present in these tissues is sufficient to account for the subsequent rise in plasma viremia (23, S11).

Initial viremia expands exponentially, with a doubling time of approximately 0.3 days during the first 2–3 weeks of infection (13, S12–S14). Experimental primate models of mucosal HIV infection have demonstrated that maximal levels of viral expression occur in different tissue compartments sequentially over short periods of time (22, 24, 25). Viral loads in blood, genital secretions (26, 27), and other compartments peak at very high levels around 4 weeks after infection (27) and then decrease in association with the appearance of virus-specific CTL responses (4, 28).

The overall composition of the CD4+ cell population changes during the course of acute HIV infection. Provided the individual does not have another systemic infection, the CD4 cell population consists predominantly of resting (HLA-DR−CD38−) cells (S15–S19). Increasing HIV-1 replication during acute infection leads to a general activation of CD4+ and CD8+ cells, including HIV-1–specific CD4+ cells. Activated CD4+ cells are likely to be successfully infected with HIV-1 (29); it is therefore likely that the very cells necessary to orchestrate a robust immune response are targeted for destruction during acute infection.

Host-virus interactions: selection, compartmentalization, and diversification. The host-virus interaction shapes the population of HIV variants in systemic and tissue compartments. The number of replication events that occur during acute HIV infection is extremely large, and the opportunity for host factors to exert pressure on existing or newly mutated viral variants is substantial. Host factors exerting early selective pressure may include the cells initially infected, or microenvironments in the mucosa, submucosa, and lymph nodes, which present a variety of innate host defenses including IFNs and/or other molecules (30, 31, S20). Host selective pressures may either decrease or increase the diversity of the infecting-virus population. For example, most acute infections (97%) result from HIV variants that use CCR5, regardless of route of infection (32–35, S21–S25), which suggests some form of selection against variants that use CXCR4.

A   Pooling of aliquots from
90 HIV-seronegative specimens

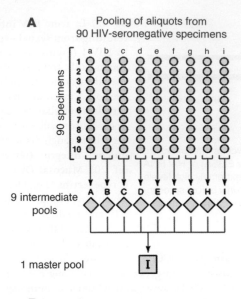

9 intermediate pools

1 master pool

B   Resolution testing of master pools

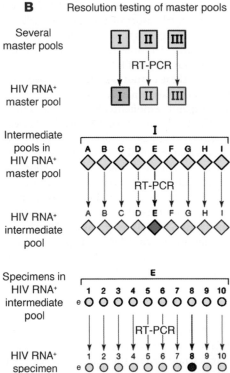

Several master pools

HIV RNA⁺ master pool

Intermediate pools in HIV RNA⁺ master pool

HIV RNA⁺ intermediate pool

Specimens in HIV RNA⁺ intermediate pool

HIV RNA⁺ specimen

Figure 2

Specimen pooling for HIV RNA testing. The algorithm shown here is that used by the North Carolina State Laboratory of Public Health to increase the predictive accuracy and cost efficiency of the screening of all sera, obtained through HIV voluntary counseling and testing, that are initially antibody-negative for HIV RNA. (**A**) Illustration of the creation of intermediate and master pools in a 1:10:90 pyramid-style pooling scheme. Only HIV-seronegative specimens are pooled. In each step, 200-μl aliquots are drawn off for pooling. The master pool ultimately contains sera from 90 antibody-negative individuals. (**B**) Illustration of the manner in which pools are screened using HIV RNA amplification testing. Positive results on a master pool trigger testing of intermediate pools and, in the final round of testing, individual specimens. When only a small number of specimens in a population are truly HIV RNA–positive, this procedure results in a dramatic reduction in the number of RNA tests used and virtually eliminates false-positive RNA test results in the final round of testing. Figure reprinted with permission from *Journal of the American Medical Association* (16).

ing a smaller, long-lived reservoir (37). These long-lived cells are invisible to the immune system and have a decay half-life on the order of 48 months (S26). Such slow decay is explainable either by persistence of infected cells or by continued, low-level viral replication (38, S29). Persistence of quiescent but infected cellular reservoirs makes HIV incurable with current interventions.

The precise dynamics of the establishment of this long-lived reservoir during acute HIV infection are unknown. Two clinical trials have quantified reservoirs of HIV in latently infected CD4⁺ cells in seronegative acutely HIV-infected patients treated with a standard duration of ART (39, 40). Each found that patients started on therapy while they were still HIV antibody–negative had lower levels of latently HIV-infected CD4⁺ cells, consistent with the hypothesis that these reservoirs are fully established only toward the end of the acute-infection interval.

The adaptive immune response. Infections with HIV-1 and with related viruses that infect nonhuman primates present a particularly complex interaction with the host immune system, as the primary target cells are critical CD4⁺ T cells, which are themselves essential to a robust pathogen-specific immune response (reviewed in ref. 41).

The initial HIV-1–specific immune response is characterized by the appearance of non-neutralizing antibodies detected by ELISA 2–4 weeks after infection (13, 42). These antibodies serve as the basis for the diagnosis of HIV. However, they have no effect on viral load and appear to exert little selective pressure (43). Neutralizing antibodies are detected 8 weeks or more after infection (43–45), well past the viral-load peak (S30–S33). Neutralizing antibodies seem to place strong immune pressure on the virus, as evidenced by escape variants that appear (44, 45) together with an explosion of viral diversity (43). While neutralizing antibodies exert strong selective pressure, they do not appear to have a sustained effect on HIV (or SHIV) RNA levels in plasma, which demonstrates the ease with which HIV-1 escapes from this immune response (45).

Retrovirus-specific CD8⁺ CTLs detectable early after infection are strongly associated with control of viremia in acute infection and established infection (4, 41, 46, 47, S34–S36). CTL escape variants can be identified as early as 4 weeks after acute SHIV infection in macaques (S37), and escape variants can establish HIV infection in vertical transmission (48, S38). The initial CTL response(s) may be directed against a few epitopes and subsequently broaden during prolonged antigen stimulation (S39). Epitopes targeted during acute infection often differ from those recognized during chronic infection

Establishment of infected cell pools in acute HIV infection. While exponential replication of HIV-1 with attendant CD4 cell activation and loss is the primary dynamic during very early infection, more subtle processes begin at the same time. Small populations of CD4⁺ cells with a resting memory cell phenotype begin to integrate HIV-1 DNA in the chromosome. Such cells are detected relatively early in HIV infection (36), and at least some cells appear to have a long half-life (S26–S28). Two reservoirs of latently infected cells are likely to exist from early in acute infection, the more abundant one having "pre-integration" HIV-1 and the other having integrated, replication-competent HIV. After initiation of antiretroviral therapy (ART), the two cell subpopulations can be distinguished. Cells with pre-integration HIV and a shorter half-life decay, reveal-

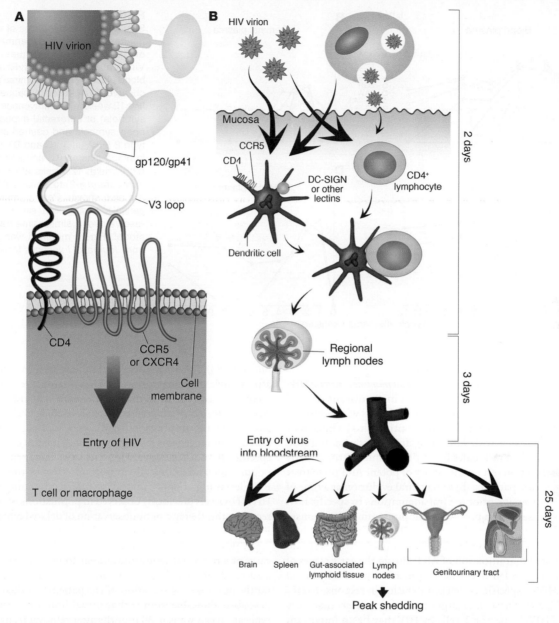

Figure 3
HIV transmission and the establishment of HIV reservoirs. (**A**) Interactions of HIV envelope glycoproteins, CD4, and CCR5 or CXCR4 coreceptors trigger fusion and entry of HIV. (**B**) Outline of the sequence and time course of events involved in viral dissemination. Figure adapted with permission from *New England Journal of Medicine* (98, 99).

(S39, S40). The importance of the CTL response in modulating HIV-1 viremia can be inferred by the association of specific HLA types with low HIV RNA levels and slowed disease progression (49–51; S41, S42); by the demonstration that specific HLA-restricted CTLs are associated with better control of SHIV and HIV replication (S43, S44), even in acute infection (S45); and by selection for viral escape (4, S37). However, all initial CTLs are not created equal. Some specific CTLs in the SHIV model select rapidly for escape variants, while others appear to offer little selective pressure (S46–S48). The avidity of the initial CTL responses may be extremely important, and the ease with which virus can escape highly avid responses may dictate the level of control. However, multiple previously unrecognized CTLs may be present (S39), and the sequence of events leading to effective

CTL responses during acute HIV infection is not clear. In particular, the potential effects of ART on developing cellular responses is a matter of controversy: in one study, patients in whom ART was initiated before seroconversion demonstrated narrower and lesser-magnitude CTL responses to HIV than did untreated patients (S49).

However, HIV-specific CD4+ responses may also be an essential component of the initial adaptive immune response (52, 53, S50) and, in acute infection, may be associated with robust CD8+ CTL responses (54, S49). In the study cited above, treatment during acute infection either before or after seroconversion led to higher CD4+ Th cell responses than treatment of chronically infected patients (S49). There is continued debate over the relative importance of quantitative HIV-1–specific CD4+ cell responses (S51).

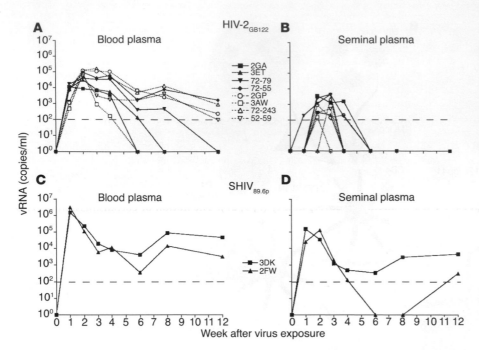

Figure 4
Similarity in the dynamics of viremia and genital shedding in experimentally HIV- or SHIV-infected macaques. (**A** and **B**) Virion-associated RNA (vRNA) levels in blood plasma (**A**) and seminal plasma (**B**), observed in pig-tailed macaques during the first 12 weeks after intravenous ($n = 4$; filled symbols) or intrarectal exposure ($n = 4$; open symbols and dashed lines) to HIV type 2 (HIV_{GB122}). (**C** and **D**) vRNA levels in blood plasma (**C**) and seminal plasma (**D**) through 12 weeks after intravenous exposure ($n = 2$) to simian/HIV ($SHIV_{89.6p}$) are also shown. Peak viremia and peak genital shedding occur simultaneously and resolve over a similar time frame in each model. Figure modified with permission from *Journal of Infectious Diseases* (82).

Acute HIV and clinical implications

A unique role for acute treatment: augmenting host immune control of HIV. Highly potent ART in current use results in significant reversal of disease progression in HIV-infected individuals with advanced disease. Prolonged treatment, however, continues to be limited by drug toxicity, problems with adherence, drug resistance, and cost. To date, the principal rationales for antiretroviral treatment early in acute HIV infection rest on the ability of potent ART (a) to treat some symptomatic patients; (b) to halt viral evolution at a time of minimal viral diversity, prior to viral adaptations to specific host immune responses; (c) to protect developing immune responses from the deleterious effects of sustained HIV viremia; (d) to reduce the viral set point; and (e) to limit the latent pool. Rosenberg and colleagues noted that initiation of ART during the early stages of acute HIV infection led to the development of unusually broad and strong HIV-1–specific Th cell proliferation in response to HIV antigens (52, 54). These data support the hypothesis that early targeting of HIV-1–specific T cells by HIV may be an important pathogenic mechanism allowing persistent, high-level viral replication in the host after resolution of acute infection (29) — and that early intervention with ART could boost initial host responses to infection. If true, this would suggest that treatment of acute HIV might serve as a method to augment host immunity and might thus obviate or delay the need for lifelong continuous ART.

Early observations supported a possible role of ART in the management of acute HIV infection. Anecdotes described patients that experienced prolonged periods of controlled viremia in conjunction with strong, HIV-specific cellular immune responses after discontinuing ART initiated during acute infection (55, S52). One of the first prospective studies on structured treatment interruption (STI) in acute HIV infection incorporated the concept of viral rebound as antigenic stimulus to enhance T cell responses. The protocol involved treatment interruption after a minimum of 8 months of undetectable viral load, with therapy reinitiated if the viral load increased to greater than 5,000 RNA copies per milliliter plasma for 3 consecutive weeks, or greater than 50,000 copies at

one time (54). Three patients achieved long-term viral suppression after a single treatment cessation. The remainder required re-treatment and a second interruption, after which all individuals experienced transiently suppressed plasma viremia, and two achieved sustained viral suppression. In all cases, interruptions were followed by enhancement in magnitude and breadth of cellular HIV-specific immune responses. This trial lacked a concurrent control population, and a clinically important effect of therapy on HIV RNA levels in patients can only be inferred by comparison with HIV RNA levels previously observed in populations of patients not receiving this therapy, or by observation of delayed rebound of HIV RNA in plasma after a period of relative suppression.

The principal adverse events in early studies have been isolated cases of a retroviral syndrome similar to primary HIV infection, associated with initial discontinuation of therapy (56, S53). Importantly, prolonged observation of the patients studied in the trial described above has shown that initial immune control of HIV replication can wane 6–30 months after removal from therapy, in association with viral rebound to more typical levels of viremia and concurrent evolution of escape mutations to HIV-specific CTL epitopes or neutralizing antibody (57, 58). These results emphasize that the benefits of acute treatment may not translate to permanent remission of HIV disease. Still, acute HIV remains the only setting in which the natural history of disease may be altered by the temporary use of ART.

The proving ground for curing HIV. The decay of HIV-infected cell populations in patients receiving potent ART resulted in initial enthusiasm for the "eradication hypothesis" (59). Since then, attempts to cure established HIV infection with antiviral agents have met with absolutely no success. As described above, a subpopulation of HIV-infected CD4+ lymphocytes avoids viral or immune cytolysis and enters the resting state (S27, S54). This reservoir, invisible to the immune system, appears to have a very low decay rate even in the setting of prolonged antiretroviral suppression (S26), though treatment during early infection may accelerate this decay (S28). Addition of cytoreductive chemotherapy (cyclophosphamide) to ART failed to reduce proviral DNA load in lymphatic tissue (S55). Several groups have studied use of global T cell activation agents to perturb latent reservoirs in aviremic, treated patients; IL-2, low-dose OKT3, and/or IFN-α–based regimens have shown evidence of decreasing latent reservoirs but have failed to eliminate latent infection (60, 61, S56). It is possible, however, that global T cell activation may induce viral replication and increase the number of susceptible uninfected target cells beyond a threshold that can be contained by ART. Newly described host mechanisms that regulate HIV gene expression involving histone deacetylase 1 (62, S57–S59) may present therapeutic targets for selective derepression of HIV gene expression, to allow outgrowth of quiescent HIV without the pitfalls of global T cell activation.

Regardless of the specific approach, acute HIV infection could be a critical proving ground for eradication strategies, for several reasons. First, it is possible that initiation of intensive ART in very early (antibody-negative) acute HIV can significantly limit the size of the initial latent pool (39, 40) — effectively lowering the bar for all strategies that target this reservoir. Second, patients treated in the first 3 weeks of infection may be able to marshal maximal HIV-specific immune responses to facilitate reservoir clearance (54, 63, 64). Third, in macaques, HIV infection can be prevented when ART is provided for 28 days within 72 hours after exposure (64); this suggests that nascent latent reservoirs can be cleared. Fourth, early acute HIV infection is the only setting in which clinical benefits have been associated with brief (<1 year) and interrupted courses of intensive ART (54).

Individual infectiousness, disease stage, and probability of sexual transmission. The main routes of HIV transmission are sexual, maternal-child, and parenteral. The majority of HIV transmission worldwide occurs through sexual contact (S60) with mucosal exposure to infected genital secretions. Regardless of the route of exposure, in every case transmission of HIV is determined by the infectiousness of the transmitting host and the susceptibility of the exposed person (66).

Determinants of host susceptibility may include CCR5/CXCR4 coreceptor polymorphisms (67, S61) and HLA types. Alterations in genital mucosa may also have a profound effect on HIV susceptibility. The presence of STDs is an important risk factor for acquiring HIV (66, 68, 69, S62). Both genital ulcerative diseases like herpes, syphilis, and chancroid and inflammatory diseases such as gonorrhea increase HIV acquisition (66, 68, 69, S62). Bacterial vaginosis in women has also been shown to be a risk factor for HIV acquisition (70); other factors such as vaginal thinning from estrogen deficiency (S63) and cervical ectopy (S64) affect HIV susceptibility as well. In men, changes in the penile epithelium that occur with circumcision can be protective (71).

Biological factors strongly influence infectiousness. High plasma viral loads correlate with risk of transmission by vertical (S65, S66) and sexual (72) routes. Conditions such as tuberculosis (S67) and malaria (73) that elevate plasma viral loads would therefore be expected to be associated with increased infectiousness. Genital viral load in semen (74–76, S68) or female genital fluids (77) can be increased to a level greater than that of blood plasma viral load by the inflammation caused by STDs (68, 69, 75, 78, 79, S62). In addition, nutritional deficiencies and hormonal influences in women (S69, S70) can increase shedding of HIV in the genital tract. Such transient elevations of genital or rectal shedding of HIV are considered periods of "hyperinfectiousness."

Disease stage is powerfully associated with individual infectiousness. Higher viral loads correlate with higher transmission probability in patients with advanced disease (80, S71). However,

the most striking elevations in HIV viremia (S72, S73) and genital-fluid shedding (27, 81, 82) occur early in acute infection. Based on a probabilistic model of the relationship between semen viral burden and probability of HIV transmission (83), increases in semen HIV load during acute HIV infection appear sufficient to account for an eight- to 20-fold increase in the odds of transmission per coital act (27). Other biological influences hypothesized to contribute to elevated individual infectiousness in acute infection include the relative homogeneity in *Env* (84), the near-universal R5 coreceptor usage (35), the absence of antibody in genital fluids (S74), and the frequent presence of other inflammatory or ulcerative STDs that might increase both shedding and partner susceptibility (7, 9, 27). The notion of cotransmission of STD pathogens suggests one location where acute HIV infection might be best detected: STD clinics. Recent studies have borne out that STD clinic attendees in the US (14), Malawi (2), and India (16) can have a surprisingly high prevalence of acute HIV infection, and Cameron et al. generated compelling evidence for cotransmission of herpes simplex virus and HIV in the late 1980s (85).

Role of acute HIV in epidemic spread. A high proximate risk of secondary HIV transmission has been clearly associated with stage of disease. Secondary infections (from an acutely infected index case to a susceptible partner) have been documented beginning as early as 7–14 days after initial infection, prior to the onset of acute retroviral symptoms in the acutely infected index case (S74). In one study examining sexual risk behavior during and after HIV seroconversion, Colfax and colleagues (86) documented alarming rates of partner change and sexual concurrency among acutely HIV-infected men who have sex with men in California: 39 (59%) of 66 acutely infected men in the study reported unprotected insertive anal intercourse with additional (HIV-susceptible) partners during the time of acute infection, with no apparent decrease in risk behavior associated with the immediate acute-infection period even in symptomatic patients.

Mathematical models (87, 88) have previously suggested that rapid global spread of HIV can be ascribed to hyperinfectiousness in the acute phase of HIV disease; empiric epidemiologic data now strongly support this hypothesis. Retrospective cohort studies of patients infected by sexual contact (89) or blood transfusion (S75) have confirmed significantly increased risk of infection among those exposed to index cases with acute or recent infection. Recently, Wawer and colleagues (80) examined the link between acute disease stage and transmission in a remarkable study of couples in Rakai, Uganda, who were screened for HIV infection and followed as part of an STD treatment trial. An estimated 48% of all HIV transmission events observed among all serodiscordant couples in the study occurred in situations where the first infected partner had acquired HIV during the preceding 5 months.

The potential magnitude of the contribution of acute hyperinfectiousness to transmission is revealed by two studies examining the phylogenetic relatedness of viral strains in newly infected patients in defined communities. Among 191 newly infected patients in the Swiss HIV Cohort, 29% shared viral sequences, time of infection, and geography with at least one other acutely infected patient (90). In a similar study in New York, New York, 51 (21%) of 241 patients were found to be clustered, and acute-to-acute transmission was confirmed by history in nine cases (S76). There are two principal alternative explanations for such extensive clustering. The first is efficient, rapid dissemination by acutely infected individuals, leading to serial infections in sexual networks. However, "core transmission" by a small unidentified subpopulation of HIV-positive men or women in the same population could also lead to the same outcome (91, S77). In either scenario, the ability to identify acutely HIV-infected men and women could open a unique window on active HIV transmission taking place in a community, and provide opportunities to interfere with the chain of HIV transmission.

Intervention: targeting prevention to networks with active disease transmission. Traditional surveillance systems, which are based on detection of undifferentiated, prevalent cases of HIV disease, can describe where an epidemic has been, but not where it is going. Because traditional surveillance cannot isolate the context or conditions of HIV incidence, the results cannot describe local forces (socioeconomic, cultural, and geographic) that drive HIV spread in communities (92). Incident STDs can be heavily concentrated in definable social networks and particular locales that are not immediately obvious (93, 94, S78, S79). Identifying local and regional patterns of HIV transmission can allow targeting of high-transmission areas for public health interventions (S60). Reliable information about the locations and correlates of incidence can furthermore ratify information gathered from other sources (e.g., behavioral surveys or cross-sectional serosurveys) and provide early indications of the effectiveness of interventions. The drive to identify regional variation in incidence rates has previously led to development of "detuned" HIV antibody assays that remain negative in some patients with recent infection and low-avidity antibodies; one such assay under evaluation by the Centers for Disease Control and Prevention has proven useful for estimating seroincidence in cross-sectional, population-based samples (95). Unfortunately, the current generation of detuned assays are unable to accurately classify individual patients as having truly recent infection.

One particularly promising approach to HIV control combines identification of acute disease with "network notification," linking case reporting to the tracing of contacts and offering partner counseling and referral services (96). Applying this strategy prospectively to incident HIV could harness classical infection-control techniques that have been successfully used for decades in tuberculosis- and syphilis-eradication campaigns. Modern epidemiologic techniques can supplement the information provided by STD contact tracing. For instance, spatial analysis based on geographic information systems can be applied to incidence data to define high-transmission areas for HIV (S80, S81), and modern social-network analysis (97, S82) can give a clear understanding of transmission patterns in communities. The effectiveness of any

such creative, proactive public health strategies for HIV prevention hinges critically on better understanding the effectiveness of specific interventions — such as partner counseling or ART (given as treatment, postexposure prophylaxis, or pre-exposure prophylaxis) — at reducing disease transmission.

Conclusions

While compelling evidence suggests the potential importance of detecting acute HIV infection, remarkably little attention has been devoted to this goal. This oversight results from a widespread belief that neither clinical nor serosurveillance methods have much chance of identifying patients with acute HIV infection. Not surprisingly, cogent treatment and prevention strategies for this group have not been developed. However, new methods in population-based serosurveillance and increased focus on some kinds of STD patients could allow identification of a large number of patients with acute HIV infection. The detection of acute disease will permit exciting research, clinical, and public health opportunities.

Acknowledgments

This work was supported by the NIH (RO1 MH68686, K23 AI01781, P30 AI50410, RR 00046, RO1 DK49381, [HPTN, std and pathogenesis training grant numbers]). The authors wish to thank Susan Fiscus, Ronald Swanstrom, William Miller, Charles Hicks, and Peter Leone for their scientific contributions targeting acute HIV. We acknowledge the invaluable collaboration of Todd McPherson, Leslie Wolf, and Lou Turner of the North Carolina State Laboratory of Public Health, as well as Evelyn Foust, Del Williams, Judy Owen O'Dowd, Rhonda Ashby, and Todd Vanhoy of North Carolina Department of Health and Human Services' HIV/STD Prevention and Care Branch. The UNC Retrovirology Core laboratory staff deserve special thanks for their dedication to innovation.

Due to space constraints, a number of important references could not be included in this article. Interested readers can find a supplementary reading list at http://www.jci.org/cgi/content/full/113/7/937/DC1.

Address correspondence to: Christopher D. Pilcher, CB#7215, 211A West Cameron Street, University of North Carolina at Chapel Hill, Chapel Hill, North Carolina 27599-7215, USA. Phone: (919) 843-2721; Fax: (919) 966-8928; E-mail: cpilcher@med.unc.edu.

1. Schacker, T., Collier, A.C., Hughes, J., Shea, T., and Corey, L. 1996. Clinical and epidemiologic features of primary HIV infection. *Ann. Intern. Med.* **125**:257–264.
2. Pilcher, C.D., et al. 2004. Frequent detection of acute primary HIV infection in men in Malawi. *AIDS.* **18**:1–8.
3. Pilcher, C.D., et al. 2004. The "Screening and Tracing Active Transmission" (STAT) program: real-time detection and monitoring of HIV incidence. *Program and abstracts: 11th Conference on Retroviruses and Opportunistic Infections.* San Francisco, California, USA. Abstr. 20.
4. Borrow, P., et al. 1997. Antiviral pressure exerted by HIV-1-specific cytotoxic T lymphocytes (CTLs) during primary infection demonstrated by rapid selection of CTL escape virus. *Nat. Med.* **2**:205–211.
5. Lindback, S., et al. 2000. Diagnosis of primary HIV-1 infection and duration of follow-up after HIV exposure. Karolinska Institute Primary HIV Infection Study Group. *AIDS.* **14**:2333–2339.

6. Pilcher, C.D., et al. 2001. Sexual transmission during the incubation period of primary HIV-1 infection. *JAMA.* **286**:1713–1714.
7. Bollinger, R.C., et al. 1997. Risk factors and clinical presentation of acute primary HIV infection in India. *JAMA.* **278**:2085–2089.
8. Daar, E.S., et al. 2001. Diagnosis of primary HIV-1 infection. *Ann. Intern. Med.* **134**:25–29.
9. Kinloch-de Loes, S., et al. 1993. Symptomatic primary infection due to human immunodeficiency virus type 1: review of 31 cases. *Clin. Infect. Dis.* **17**:59–65.
10. Weintrob, A.C., et al. 2003. Infrequent diagnosis of primary human immunodeficiency virus infection: missed opportunities in acute care settings. *Arch. Intern. Med.* **163**:2097–2100.
11. Rosenberg, E.S., Caliendo, A.M., and Walker, B.D. 1999. Acute HIV infection among patients tested for mononucleosis. *N. Engl. J. Med.* **340**:969.
12. Pincus, J.M., et al. 2003. Acute human immunodeficiency virus infection in patients presenting

to an urban urgent care center. *Clin. Infect. Dis.* **37**:1699–1704.
13. Fiebig, E.W., et al. 2003. Dynamics of HIV viremia and antibody seroconversion in plasma donors: implications for diagnosis and staging of primary HIV infection. *AIDS.* **17**:1871–1879.
14. Morandi, P.A., et al. 1998. Detection of human immunodeficiency virus type 1 (HIV-1) RNA in pools of sera negative for antibodies to HIV-1 and HIV-2. *J. Clin. Microbiol.* **36**:1534–1538.
15. Quinn, T.C., et al. 2000. Feasibility of pooling sera for HIV viral RNA to diagnose acute primary HIV-1 infection and estimate HIV incidence. *AIDS.* **14**:2751–2757.
16. Pilcher, C.D., et al. 2002. Real-time, universal screening for acute HIV infection in a routine HIV counseling and testing population. *JAMA.* **288**:216–221.
17. Weintrob, A.C., et al. 2003. Prevalence of undiagnosed HIV infection in febrile patients presenting to an emergency department in Southeastern

United States. *Program and abstracts, 2nd International AIDS Society Pathogenesis Meeting.* Paris, France. Abstr. 84.

18. Clark, S.J., et al. 1994. Unsuspected primary human immunodeficiency virus type 1 infection in seronegative emergency department patients. *J. Infect. Dis.* **170**:194–197.

19. Deng, H., et al. 1996. Identification of a major co-receptor for primary isolates of HIV-1. *Nature.* **381**:661–666.

20. Pope, M., and Haase, A.T. 2003. Transmission, acute HIV-1 infection and the quest for strategies to prevent infection. *Nat. Med.* **9**:847–852.

21. Geijtenbeek, T.B., et al. 2000. DC-SIGN, a dendritic cell-specific HIV-1-binding protein that enhances trans-infection of T cells. *Cell.* **100**:587–597.

22. Zhang, Z., et al. 1999. Sexual transmission and propagation of SIV and HIV in resting and activated CD4+ T cells. *Science.* **286**:1353–1357.

23. Schacker, T., et al. 2000. Rapid accumulation of human immunodeficiency virus (HIV) in lymphatic tissue reservoirs during acute and early HIV infection: implications for timing of antiretroviral therapy. *J. Infect. Dis.* **181**:354–357.

24. Spira, A.I., et al. 1996. Cellular targets of infection and route of viral dissemination after an intravaginal inoculation of simian immunodeficiency virus into rhesus macaques. *J. Exp. Med.* **183**:215–225.

25. Bogers, W.M., et al. 1998. Characteristics of primary infection of a European human immunodeficiency virus type 1 clade B isolate in chimpanzees. *J. Gen. Virol.* **79**:2895–2903.

26. Pilcher, C.D., et al. 2001. HIV in body fluids during primary HIV infection: implications for pathogenesis, treatment and public health. *AIDS.* **15**:837–845.

27. Pilcher, C.D., et al. 2004. Brief but efficient: acute HIV infection and the sexual transmission of HIV. *J. Infect. Dis.* In press.

28. Koup, R.A., et al. 1994. Temporal association of cellular immune responses with the initial control of viremia in primary human immunodeficiency virus type 1 syndrome. *J. Virol.* **68**:4650–4655.

29. Douek, D.C., et al. 2002. HIV preferentially infects HIV-specific CD4+ T cells. *Nature.* **417**:95–98.

30. Shugars, D.C., Alexander, A.L., Fu, K., and Freel, S.A. 1999. Endogenous salivary inhibitors of human immunodeficiency virus. *Arch. Oral Biol.* **44**:445–453.

31. Shugars, D.C., and Wahl, S.M. 1998. The role of the oral environment in HIV-1 transmission. *J. Am. Dent. Assoc.* **129**:851–858.

32. Roos, M.T., et al. 1992. Viral phenotype and immune response in primary human immunodeficiency virus type 1 infection. *J. Infect. Dis.* **165**:427–432.

33. Zhu, T., et al. 1993. Genotypic and phenotypic characterization of HIV-1 patients with primary infection. *Science.* **261**:1179–1181.

34. Wolinsky, S.M., et al. 1996. Adaptive evolution of human immunodeficiency virus-type 1 during the natural course of infection. *Science.* **272**:537–542.

35. Ritola, K., et al. 2004. HIV-1 V1/V2 and V3 env diversity during primary infection suggests a role for multiply infected cells in transmission. *Program and abstracts: 11th Conference on Retroviruses and Opportunistic Infections.* San Francisco, California, USA. Abstr. 386.

36. Chun, T.W., et al. 1995. In vivo fate of HIV-1-infected T cells: quantitative analysis of the transition to stable latency. *Nat. Med.* **1**:1284–1290.

37. Blankson, J.N., et al. 2000. Biphasic decay of latently infected CD4+ T cells in acute human immunodeficiency virus type 1 infection. *J. Infect. Dis.* **182**:1636–1642.

38. Zhang, L., et al. 1999. Quantifying residual HIV-1 replication in patients receiving combination antiretroviral therapy. *N. Engl. J. Med.* **340**:1605–1613.

39. Lori, F., et al. 1999. Treatment of human immunodeficiency virus infection with hydroxyurea,

didanosine, and a protease inhibitor before seroconversion is associated with normalized immune parameters and limited viral reservoir. *J. Infect. Dis.* **180**:1827–1832.

40. Lafeuillade, A., Poggi, C., Hittinger, G., Counillon, E., and Emilie, D. 2003. Predictors of plasma human immunodeficiency virus type 1 RNA control after discontinuation of highly active antiretroviral therapy initiated at acute infection combined with structured treatment interruptions and immune-based therapies. *J. Infect. Dis.* **188**:1426–1432.

41. Letvin, N.L., and Walker, B.D. 2003. Immunopathogenesis and immunotherapy in AIDS virus infections. *Nat. Med.* **9**:861–866.

42. Henrard, D.R., et al. 1995. Virologic and immunologic characterization of symptomatic and asymptomatic primary HIV-1 infection. *J. Acquir. Immune Defic. Syndr. Hum. Retrovirol.* **9**:305–310.

43. Rybarczyk, B.J., et al. 2004. Correlation between env V1/V2 region diversification and neutralizing antibodies during primary infection by simian immunodeficiency virus sm in rhesus macaques. *J. Virol.* In press.

44. Wei, X., et al. 2003. Antibody neutralization and escape by HIV-1. *Nature.* **422**:307–312.

45. Richman, D.D., Wrin, T., Little, S.J., and Petropoulos, C.J. 2003. Rapid evolution of the neutralizing antibody response to HIV type 1 infection. *Proc. Natl. Acad. Sci. U. S. A.* **100**:4144–4149.

46. Koup, R.A., et al. 1994. Temporal association of cellular immune responses with the initial control of viremia in primary human immunodeficiency virus type 1 syndrome. *J. Virol.* **68**:4650–4655.

47. Schmitz, J.E., et al. 1999. Control of viremia in simian immunodeficiency virus infection by CD8+ lymphocytes. *Science.* **283**:857–860.

48. Goulder, P.J., et al. 2001. Evolution and transmission of stable CTL escape mutations in HIV infection. *Nature.* **412**:334–338.

49. Malhotra, U., et al. 2001. Role for HLA class II molecules in HIV-1 suppression and cellular immunity following antiretroviral treatment. *J. Clin. Invest.* **107**:505–517.

50. Gao, X., et al. 2001. Effect of a single amino acid change in MHC class I molecules on the rate of progression to AIDS. *N. Engl. J. Med.* **344**:1668–1675.

51. Kaslow, R.A., et al. 1996. Influence of combinations of human major histocompatibility complex genes on the course of HIV-1 infection. *Nat. Med.* **2**:405–411.

52. Rosenberg, E.S., et al. 1997. Vigorous HIV-1-specific CD4+ T cell responses associated with control of viremia. *Science.* **278**:1447–1450.

53. Pitcher, C.J., et al. 1999. HIV-1-specific CD4+ T cells are detectable in most individuals with active HIV-1 infection, but decline with prolonged viral suppression. *Nat. Med.* **5**:518–525.

54. Rosenberg, E.S., et al. 2000. Immune control of HIV-1 after early treatment of acute infection. *Nature.* **407**:523–526.

55. Lisziewicz, J., et al. 1999. Control of HIV despite the discontinuation of antiretroviral therapy. *N. Engl. J. Med.* **340**:1683–1684.

56. Kilby, J., et al. 2000. Recurrence of the acute HIV syndrome after interruption of antiretroviral therapy in a patient with chronic HIV infection: a case report. *Ann. Intern. Med.* **133**:435–438.

57. Kaufmann, D., et al. 2004. Limited durability of immune control following treated acute HIV infection. *Program and abstracts: 11th Conference on Retroviruses and Opportunistic Infections.* San Francisco, California, USA. Abstr. 24.

58. Montefiori, D.C., et al. 2003. Viremia control despite escape from a rapid and potent autologous neutralizing antibody response after therapy cessation in an HIV-1-infected individual. *J. Immunol.* **170**:3906–3914.

59. Perelson, A.S., et al. 1997. Decay characteristics of

HIV-1-infected compartments during combination therapy. *Nature.* **387**:188–191.

60. Lafeuillade, A., et al. 2001. Impact of immune interventions on proviral HIV-1 DNA decay in patients receiving highly active antiretroviral therapy. *HIV Med.* **2**:189–194.

61. Kulkosky, J., et al. 2002. Intensification and stimulation therapy for human immunodeficiency virus type 1 reservoirs in infected persons receiving virally suppressive highly active antiretroviral therapy. *J. Infect. Dis.* **186**:1403–1411.

62. Coull, J.J., et al. 2002. Targeted derepression of the human immunodeficiency virus type 1 long terminal repeat by pyrrole-imidazole polyamides. *J. Virol.* **76**:12349–12354.

63. Oxenius, A., et al. 2000. Early highly active antiretroviral therapy for acute HIV-1 infection preserves immune function of CD8+ and CD4+ T lymphocytes. *Proc. Natl. Acad. Sci. U. S. A.* **97**:3382–3387.

64. Malhotra, U., et al. 2000. Effect of combination antiretroviral therapy on T-cell immunity in acute human immunodeficiency virus type 1 infection. *J. Infect. Dis.* **181**:121–131.

65. Otten, R.A., et al. 2000. Efficacy of postexposure prophylaxis after intravaginal exposure of pig-tailed macaques to a human-derived retrovirus (human immunodeficiency virus type 2). *J. Virol.* **74**:9771–9775.

66. Galvin, S.R., and Cohen, M.S. 2004. The role of sexually transmitted diseases in HIV transmission. *Nat. Rev. Microbiol.* **2**:33–42.

67. Hogan, C.M., and Hammer, S.M. 2001. Host determinants in HIV infection and disease. Part 2: genetic factors and implications for antiretroviral therapeutics. *Ann. Intern. Med.* **134**:978–996.

68. Fleming, D., and Wasserheit, J. 1999. From epidemiological synergy to public health policy and practice: the contribution of other sexually transmitted diseases to sexual transmission of HIV infection. *Sex. Transm. Infect.* **75**:3–17.

69. Rottingen, J.A., Cameron, D.W., and Garnett, G. 2001. A systematic review of the epidemiologic interactions between classic sexually transmitted diseases and HIV. *Sex. Transm. Dis.* **28**:579–597.

70. Taha, T.E. 1998. Bacterial vaginosis and disturbances of vaginal flora: association with increased acquisition of HIV. *AIDS.* **12**:1699–1706.

71. Gray, R.H. 2000. Male circumcision and HIV acquisition and transmission: cohort studies in Rakai, Uganda. *AIDS.* **14**:2371–2381.

72. Quinn, T.C. 2000. Viral load and heterosexual transmission of human immunodeficiency virus type 1. *N. Engl. J. Med.* **342**:921–929.

73. Hoffman, I.F. 1999. The effect of Plasmodium falciparum malaria on HIV-1 RNA blood plasma concentrations. *AIDS.* **13**:487–494.

74. Vernazza, P.L., et al. 1994. Detection and biologic characterization of infectious HIV-1 in semen of seropositive men. *AIDS.* **8**:1325–1329.

75. Dyer, J., et al. 1996. Quantitation of human immunodeficiency virus type 1 RNA in cell-free seminal plasma: comparison of NASBA with Amplicor reverse transcription-PCR amplification and correlation with quantitative culture. *J. Virol. Methods.* **60**:161–170.

76. Vernazza, P.L., et al. 1997. Quantification of HIV in semen: correlation with antiviral treatment and immune status. *AIDS.* **11**:987–993.

77. Hart, C. 1999. Correlation of human immunodeficiency virus type 1 RNA levels in blood and female genital tract. *J. Infect. Dis.* **179**:871–882.

78. Eron, J., Gilliam, B., Fiscus, S., Dyer, J., and Cohen, M. 1996. HIV-1 shedding and chlamydial urethritis. *JAMA.* **275**:36.

79. Cohen, M.S., et al. 1997. Reduction of concentration of HIV-1 in semen after treatment of urethritis: implications for prevention of sexual transmission of HIV-1. AIDSCAP Malawi Research Group.

Lancet. **349**:1868–1873.

80. Wawer, M.J., et al. 2003. HIV-1 transmission per coital act, by stage of HIV infection in the index partner, in discordant couples, Rakai, Uganda. *Program and abstracts: 10th Conference on Retroviruses and Opportunistic Infections.* Boston, Massachusetts, USA. Abstr. 40.

81. Dyer, J.R., et al. 1997. Shedding of HIV-1 in semen during primary infection. *AIDS.* **11**:543–545.

82. Pullium, J.K., et al. 2001. Pig-tailed macaques infected with human immunodeficiency virus (HIV) type 2GB122 or simian/HIV89.6p express virus in semen during primary infection: new model for genital tract shedding and transmission. *J. Infect. Dis.* **183**:1023–1030.

83. Chakraborty, H., et al. 2001. Viral burden in genital secretions determines male-to-female sexual transmission of HIV-1: a probabilistic empiric model. *AIDS.* **15**:621–627.

84. Zhu, T., et al. 1993. Genetic characterization of human immunodeficiency virus type 1 in blood and genital secretions: evidence for viral compartmentalization and selection during sexual transmission. *J. Virol.* **70**:3098–3107.

85. Cameron, D.W., et al. 1989. Female to male transmission of human immunodeficiency virus type 1:

risk factors for seroconversion in men. *Lancet.* **2**:403–407.

86. Colfax, G.N., et al. 2002. Sexual risk behaviors and implications for secondary HIV transmission during and after HIV seroconversion. *AIDS.* **16**:1529–1535.

87. Jacquez, J.A., Koopman, J.S., Simon, C.P., and Longini, I.M., Jr. 1994. Role of the primary infection in epidemics of HIV infection in gay cohorts. *J. Acquir. Immune Defic. Syndr.* **7**:1169–1184.

88. Koopman, J.S., et al. 1997. The role of early HIV infection in the spread of HIV through populations. *J. Acquir. Immune Defic. Syndr. Hum. Retrovirol.* **14**:249–258.

89. Leynaert, B., Downs, A.M., and de Vincenzi, I. 1998. Heterosexual transmission of human immunodeficiency virus: variability of infectivity throughout the course of infection. European Study Group on Heterosexual Transmission of HIV. *Am. J. Epidemiol.* **148**:88–96.

90. Yerly, S., et al. 2001. Acute HIV infection: impact on the spread of HIV and transmission of drug resistance. *AIDS.* **15**:2287–2292.

91. Clumeck, N., et al. 1989. A cluster of HIV infection among heterosexual people without apparent risk factors. *N. Engl. J. Med.* **321**:1460–1462.

92. Ruiz, M.S., et al., editors. 2000. *No time to lose: getting*

more from HIV prevention. National Academies Press. Washington, DC, USA. 227 pp.

93. Alvarez-Darder, C., Marrques, S., and Perea, E.J. 1985. Urban cluster of sexually transmitted diseases in the city of Seville, Spain. *Sex. Transm. Dis.* **12**:166–168.

94. Rothenberg, R.B. 1983. The geography of gonorrhea: empirical demonstration of core group transmission. *Am. J. Epidemiol.* **117**:688–694.

95. Janssen, R.S., et al. 1998. New testing strategy to detect early HIV-1 infection for use in incidence estimates and for clinical and prevention purposes [erratum 1999, **281**:1893]. *JAMA.* **280**:42–48.

96. Foust, E., et al. 2003. Partner counseling and referral services to identify persons with undiagnosed HIV: North Carolina, 2001. *MMWR Morb. Mortal. Wkly. Rep.* **52**:1181–1184.

97. Rothenberg, R.B., et al. 1998. Using social network and ethnographic tools to evaluate syphilis transmission. *Sex. Transm. Dis.* **25**:154–160.

98. Luster, A.D. 1998. Chemokines: chemotactic cytokines that mediate inflammation. *N. Engl. J. Med.* **338**:436–445.

99. Kahn, L.O., and Walker, B.D. 1998. Acute human immunodeficiency virus type 1 infection. *N. Engl. J. Med.* **339**:33–39.

Strategies for an HIV vaccine

Norman L. Letvin

Harvard Medical School, Beth Israel Deaconess Medical Center, Boston, Massachusetts, USA.

The development of an HIV vaccine poses an unprecedented challenge to the scientific community. The inexorable spread of HIV worldwide and the devastating clinical consequences of AIDS can only be contained by an effective vaccine. Yet, almost two decades after the first demonstration of HIV and its etiologic role in AIDS, this vaccine still is a goal, not a reality. In this Perspective, I consider the unique problems for vaccine development posed by the biology of HIV infection, and I summarize recent advances in our understanding of the immune control of HIV and the implications of these advances for vaccine development. After reviewing the sobering history of failed traditional vaccine strategies, I discuss the rationale for vaccines with novel designs and their experimental successes to date.

HIV biology

HIV offers a uniquely difficult target for vaccine development. The HIV isolates that infect humans and cause AIDS include a genetically diverse population of viruses (1). The HIV responsible for causing AIDS in much of West Africa is referred to as HIV-2; the HIV that causes AIDS throughout the rest of the world is referred to as HIV-1. HIV-2 and HIV-1 are so divergent in their genetic sequences that their envelope glycoproteins are often not immunologically cross-reactive. Moreover, the viruses of the HIV-1 group include disparate viral clades or subtypes that are clustered epidemiologically in distinct geographic regions. These various clades of HIV-1 isolates differ from one another so dramatically in their genetic sequences, and therefore their antigenic characteristics, that it has been suggested that different geographic regions of the world may actually require different vaccines. The parallel development of various region-specific HIV vaccines would clearly be both difficult and time-consuming.

Genetic diversity is also continuously generated in the course of an HIV infection in a single infected individual, as the inaccurate enzymatic machinery of this virus's replication results in ongoing production of mutant virions. This process engenders such a genetically heterogeneous population of virions that an antibody that can neutralize one HIV isolate may fail to neutralize another from the same individual. Such an extraordinary degree of genetic diversity among HIV isolates immeasurably complicates the process of HIV vaccine development.

Other aspects of the biology of HIV infections also have substantial implications for AIDS vaccine development. HIV is transmitted both venereally and hematogenously. Therefore, there is reason to suppose that an effective HIV vaccine must elicit both mucosal immunity, to contain sexually transmitted virus, and systemic immunity, to contain virus transmitted directly into the bloodstream. Furthermore, HIV is likely transmitted both as cell-free and as cell-associated virus. Therefore, more than a single type of immunity must be elicited by a vaccine if that vaccine is to be effective. Cell-free virions can be bound and neutralized by anti-

body, while cell-associated virus can be eliminated by cell-mediated immune responses. Finally, and most troubling for the prospects of developing an effective HIV vaccine, infection with this virus universally results in high levels of viral replication that persist in the face of seemingly robust anti-viral antibody and cell-mediated immune responses. Unlike most other viral infections in humans, replicating HIV is never fully cleared. Moreover, the level and persistence of viral replication are inexorably tied to the pathogenicity of the virus, with high persistent HIV replication being associated with rapid progression of clinical disease. The universal persistence of viral replication in spite of potent immune responses raises the specter that no vaccine-elicited immune response may be capable of fully eliminating or containing indefinitely the replication of HIV.

AIDS animal models

Appropriate animal models have been key to the elucidation of AIDS immunopathogenesis and the assessment of HIV vaccine strategies. Nonhuman primates represent a particularly powerful model for the study of AIDS. The HIV-1 and HIV-2 isolates that infect humans are members of a large family of lentiviruses that endemically infect nonhuman primate species of Africa. These nonhuman primate viruses are known as simian immunodeficiency viruses (SIVs). Interestingly, they cause no disease in their natural host species. However, some of these isolates cause AIDS when inoculated experimentally in Asian macaques (2). Moreover, chimeric viruses have been developed that express HIV-1 envelope glycoproteins on SIV backbones. Some of these simian human immunodeficiency viruses (SHIVs) can cause rapidly progressive AIDS-like disease in macaques (3).

AIDS immunopathogenesis

Many acute viral infections are cleared by neutralizing antibodies induced by the replicating viruses. These antibodies bind to viral particles and block the ability of the particles to attach to and subsequently infect cells. Our emerging understanding of antibody responses in HIV-infected individuals suggests that such immune responses are not likely to be critical in blocking HIV spread. While high-titer anti-HIV antibodies certainly develop in infected individuals, these antibodies display only weak HIV-neutralizing activity. Moreover, the partial containment of replicating HIV usually seen during the first weeks following initial infection precedes the development of antibodies that can neutralize the virus. These observations imply that the virus-specific antibody response does not play a critical role in either the chronic or early containment of HIV replication in the infected individual.

Nevertheless, there is reason to suppose that neutralizing antibodies will be important in the development of an effective HIV vaccine and that they can be elicited by immunization. A limited number of mAb's have been developed that neutralize diverse HIV-1 isolates, confirming that shared, neutralization-sensitive viral domains exist. Infection by cell-free virus can probably only be blocked immunologically by antibodies that target such viral epitopes. Moreover, studies in the SHIV/macaque model clearly

Citation for this article: *J. Clin. Invest.* **109**:15–20 (2002). doi:10.1172/ JCI200215985.

Table 1
Traditional designs for an HIV-1 vaccine

Design	Limitations
Live, attenuated virus	Pathogenicity in vaccinees
Inactivated viruses with adjuvants	Restricted specificity of neutralizing antibodies, absence of CTLs
Recombinant envelope protein	No neutralizing antibodies for patient isolates of HIV-1; absence of CTLs

indicate that passively administered neutralizing antibodies can prevent an AIDS virus infection if sufficiently high levels of circulating antibody are achieved (4). Thus, if such antibodies can be elicited through vaccination, they should be effective in blocking transmission of virus. It is problematic, however, that the neutralization-sensitive domains of the virus have proven poorly immunogenic. Configuring a subunit immunogen that can elicit an antibody response that neutralizes a diversity of HIV isolates stands as perhaps the greatest challenge facing HIV vaccine development at this time.

The immunologic mechanisms responsible for containing HIV replication are very different from those responsible for controlling many of the other viruses for which effective vaccines have been developed. These other viruses are contained primarily or solely by neutralizing antibody. HIV, on the other hand, appears to be controlled predominantly by cell-mediated immunity. Thus, soon after the indentification of HIV as the etiologic agent in AIDS, it was shown that CD8$^+$ lymphocytes can inhibit HIV replication in CD4$^+$ T cells in vitro (5). Later work showed that a virus-specific CD8$^+$ CTL response precedes the early, partial control of HIV replication in acutely infected individuals (6). Moreover, the clinical status of chronically HIV-infected individuals is associated with the levels of circulating virus-specific CD8$^+$ CTLs, high levels being predictive of a stable immunologic function (7). In fact, newly developed, highly quantitative assays for detecting these cell populations have recently demonstrated extremely high levels of HIV-specific CTLs in both acutely and chronically infected individuals (8). A definitive and direct demonstration of the importance of CD8$^+$ lymphocytes in the control of viral infection came from work in the SIV/macaque model, where monkeys depleted of CD8$^+$ lymphocytes by mAb infusion and then infected with SIV never controlled early viral replication. These animals went on to die with a rapidly progressive AIDS-like disease (9). Taken together, these findings make a strong argument for the importance of CD8$^+$ CTLs in controlling HIV replication, and they suggest that an effective HIV vaccine must elicit such an immune response.

Immune studies in normal mice and humans have clearly shown that CD8$^+$ CTLs function normally only when optimal CD4$^+$ T lymphocyte help is available to support CTL function. Hence, it is not surprising that control of HIV and stable clinical status have been shown to be associated with high levels of virus-specific CD4$^+$ T lymphocyte help (10). An effective HIV vaccine would therefore be expected to elicit virus-specific CD4$^+$ T lymphocyte help in addition to CD8$^+$ CTLs.

Traditional vaccine designs

The studies done to date to elucidate the replication of HIV and the immunopathogenesis of AIDS suggest that HIV is unique in its biology and may therefore not be amenable to control by immune responses elicited through traditional vaccine modalities. In fact, experiments in nonhuman primates and early-phase human studies bear out this supposition, providing convincing evidence that live attenuated virus vaccines, inactivated virus vaccines, and recombinant protein vaccines are all likely to be ineffective in preventing HIV infection and AIDS (Table 1).

Viruses can be attenuated in their in vivo pathogenicity by propagation in vitro, a process that generates limited numbers of mutations in the viruses. Because these viruses replicate in vivo and therefore elicit robust immune responses, infection with such pathogenically attenuated viruses represents an effective means of vaccinating humans to prevent measles, polio, and chicken pox. Preliminary studies in the SIV/macaque model suggested that viruses can be altered by molecular manipulation through the deletion of a limited amount of genetic material, and that such viruses become infectious but pathogenically attenuated. Moreover, these studies showed that prior infection with such attenuated viruses prevents subsequent infection with pathogenic wild-type virus (11). While reports of these findings raised hopes that a live attenuated HIV vaccine might be feasible, subsequent studies have shown that this approach to HIV vaccine design is flawed. Further work in the SIV/macaque model showed that newborn monkeys or adult monkeys infected for a long period of time with such vaccine strains of virus eventually develop AIDS and die (12). Similarly, a cluster of human infections has been described in which the HIV isolate was crippled in its replication competence and, accordingly, attenuated in its pathogenicity. However, as in the SIV/macaque studies, these infected individuals eventually went on to develop AIDS, albeit with a delay in time from infection to onset of disease. Such studies have suggested that it is unlikely to prove possible to uncouple the high level of replication of an attenuated AIDS virus needed to elicit protective immunity and the eventual pathogenicity of the virus. There is, therefore, little enthusiasm at this time for pursuing live attenuated virus strategies for an HIV vaccine.

While inactivated virus vaccine strategies have proven useful for preventing infections with influenza and polio virus in humans, they have been disappointing when assessed in the SIV/macaque model. This vaccine modality provided protection in monkeys that were challenged with SIV identical to the virus used in creating the vaccine, as well as in monkeys challenged at the time immunity was maximal (13). However, this vaccine protection proved neither broad nor robust. Thus, protection has not been demonstrated in this model when the strains of challenge virus and vaccine virus were even slightly disparate genetically or when the vaccinated monkeys were challenged even a few weeks after peak immunity was reached. Moreover, some studies suggested that the protection seen in this animal model may have reflected experimental artifact rather than virus-specific immunity (14). Nevertheless, inactivated virus immunogens were evaluated in limited early-phase human immunogenicity trials (15) with disappointing results. Such immunogens did not elicit antibody responses that neutralized HIV isolates, since the immunogen proved to have very little retained viral envelope glycoprotein. Moreover, since there was no synthesis of proteins in cells initiated by this immunogen, it did not induce CTL responses. Thus, as is the case for the live attenuated vaccine strategy, there seems to be little basis for pursuing further studies with inactivated HIV immunogens.

Finally, highly purified viral protein, produced through recombinant DNA technology, has been a highly effective immunogen for preventing hepatitis B virus infection in humans. Some years ago, this approach to HIV vaccine design was evaluated in nonhuman

Table 2

Novel designs for an HIV-1 vaccine

Design	Limitations
Plasmid DNA	Limited immunogenicity in humans
Live, recombinant vectors:	
Pox viruses	
Vaccinia	Dissemination in immunosuppressed vaccinees
MVA, NYVAC	Limited experience in humans
Canary pox	Limited immunogenicity in humans at achievable dosages
Gene-deleted adenovirus	Pre-existing immunity to adenovirus may limit immunogenicity
alphaviruses, adeno-associated virus	Limited experience in humans
Envelope subunit immunogens	No elicitation of neutralizing antibodies

primate models, using recombinant HIV envelope glycoprotein as an immunogen. Animals enjoyed only modest protection, and only when the challenge virus and immunizing envelope glycoprotein were identical in sequence (16). Nevertheless, early-phase human immunogenicity trials have been carried out. Antibody responses elicited in these studies were modest in titer and very limited in the spectrum of HIV isolates that they neutralized. Moreover, as would be expected with a subunit immunogen, these proteins did not elicit CTL responses. Phase III efficacy trials are underway at this time with envelope subunit immunogens in both the US and Southeast Asia, supported by private sector resources, but there is little optimism in the scientific community that these studies will demonstrate meaningful protection against HIV infection, since the immune responses crucial for HIV containment will not be elicited by this approach. Thus, the traditional approaches have proved disappointing in the effort to create an effective HIV vaccine.

Novel vaccine designs

Recognition of the limitations of these traditional immunization strategies for preventing HIV infections has inspired researchers to explore a plethora of novel vaccine designs. The most promising of these approaches involve the use of plasmid DNA immunogens and live, recombinant vectors (Table 2).

It was shown more than a decade ago that plasmids encoding proteins under the control of potent promoters can be immunogenic if inoculated intramuscularly in small laboratory animals. Subsequent work has shown that this immunogenicity can be substantially enhanced through the delivery of these plasmids formulated with particular adjuvants or cytokines. Such immunogens have proven particularly useful in eliciting cell-mediated immune responses (17). Moreover, a number of studies have suggested that plasmid DNA is quite effective as a priming or initial immunogen in a bimodal vaccine strategy (18). While this approach provides a safe means of eliciting CTL responses, preliminary data in large animal studies have raised questions as to whether sufficiently large inocula of plasmid DNA can be administered in humans to elicit useful immune responses. Nevertheless, a number of early-phase human studies are planned and ongoing to explore plasmid DNA immunogens as HIV vaccine candidates.

Live recombinant vectors are also being explored as tools for eliciting immune responses against HIV. Genes of HIV can be inserted by molecular approaches into live, replication-competent microorganisms. The resulting recombinant microorganisms then can serve to carry these genes. Upon infection with these recombinant microorganisms, immunity is elicited to the vector and to the product of the HIV gene carried by that vector. Such immunogens have proven particularly useful for eliciting CTLs, since the HIV proteins are produced intracellularly by the replicating vector and therefore enter the MHC class I processing pathway.

The microorganisms best studied as potential vaccine vectors are the pox family of viruses. The prototype member of this family is vaccinia, the replication-competent virus that served as the primary vaccine virus in the worldwide smallpox eradication campaign. Although studies in nonhuman primates have shown that recombinant vaccinia viruses can elicit potent CTL responses to HIV and SIV proteins (19), safety concerns have dampened enthusiasm for this vector as an HIV vaccine candidate. Vaccinia has been shown to disseminate in immunocompromised humans, sometimes causing a fatal encephalitis (20). Because an HIV vaccine is most desperately needed in regions of the world in which HIV infections are endemic, significant numbers of individuals already infected with HIV, and therefore immunosuppressed, would likely receive an AIDS vaccine during a campaign to immunize an entire population. A well-founded fear exists that a substantial number of those already HIV-infected individuals would develop a fatal vaccinia infection upon such a vaccination.

Other pox viruses have therefore received attention as potential HIV vaccine vectors. Perhaps the most interesting of these pox viruses is modified vaccinia Ankara (MVA). Generated from a parental vaccinia virus isolate by multiple in vitro passages, MVA carries a large number of deletions, leaving it infectious and immunogenic but highly attenuated in its pathogenicity. Nonhuman primate studies have shown that MVA elicits impressive immune responses, either as a stand alone immunogen or as a boosting immunogen following plasmid DNA priming. Another similarly generated, gene-deleted vaccinia virus, referred to as NYVAC, has been shown in nonhuman primate studies to elicit an immune response comparable to that elicited by MVA when used as a vaccine vector. Both MVA and NYVAC will soon be evaluated as HIV vaccine vectors in early-phase human testing.

The most extensively studied of the pox viruses as potential HIV vaccine vectors are the avian pox viruses. Canary pox undergoes an abortive cycle of replication in human cells but initiates the synthesis of viral proteins during that process. Presumably, this level of protein expression is sufficient for MHC class I processing of the expressed proteins to occur. Recombinant canary pox constructs have undergone extensive human testing in recent years (21). These studies have shown recombinant canary pox to be safe and immunogenic, eliciting antibody responses in 70% of vaccinees and CTL responses detectable at any single point in time following vaccination in approximately 30% of individuals. An efficacy trial in Southeast Asia is currently being considered for a recombinant canary pox immunogen.

Perhaps the most promising of the live recombinant vectors assessed to date as a potential HIV vaccine is the gene-deleted adenovirus that was developed as a vector for gene therapy. The serotype 5 adenovirus, made replication-incompetent by deletion or inactivation of the *E1* and *E3* genes, has demonstrated impressive immunogenicity in both murine and nonhuman primate studies

(22). These vectors have elicited both high-titer antibody and high-frequency CTL responses in these animal models. In fact, early-phase HIV immunogenicity trials with this vector are ongoing in humans. Preliminary findings in these trials have indicated that pre-existing antibody responses to adenovirus serotype 5 in humans who were previously infected with this common pathogen significantly dampen the in vivo expression, and therefore immunogenicity, of these vaccines, but a number of strategies are currently being considered to circumvent this problem. The immunogenicity of these recombinant vectors is most impressive if they are used to boost plasmid DNA–primed immune responses. Adequate immune responses to HIV may therefore be elicitable with this vector system in adenovirus serotype 5–immune humans by priming them with plasmid DNA vaccines before immunizing them with the recombinant adenovirus serotype 5 constructs. It may also be possible to construct analogous vaccines with comparable immunogenicity using unusual serotype adenoviruses, isolates to which most humans have not been previously exposed. Alternatively, vaccine constructs might be developed using adenovirus isolates from nonhuman primate species. Such viruses are not natural human pathogens and therefore have not infected humans. Vectors constructed from such viruses should, however, prove immunogenic in humans.

A number of other replication-competent viruses are also being explored as potential vectors for HIV vaccines. These include single-strand RNA alphaviruses (Venezuelan equine encephalitis virus and Semliki forest virus) and the parvovirus adeno-associated virus. The bacterial vector systems that are receiving attention at this time include the attenuated Mycobacterium Bacille Calmette-Guerin and some of the pathogenically attenuated enteric bacteria. All of these approaches will be tested in the near future for immunogenicity in human vaccine trials.

Developing an immunogen that elicits anti-HIV envelope antibody responses that can neutralize a diversity of HIV isolates has been one of the most intractable problems that have arisen in the process of HIV vaccine development. A number of novel strategies are being actively pursued in an effort to generate such an envelope immunogen. Assuming that genetically conserved, neutralizing epitopes of HIV envelope may be shielded from the immune system during the process of infection with the virus, investigators are assessing immunogens constructed by removing N-linked glycans and variable loop structures from the HIV envelope. Some have suggested that an effective vaccine might be created by constructing a subunit immunogen that conforms to the native folding of the HIV envelope glycoprotein. Since the envelope glycoproteins exist as oligomers on the native virus, attempts are being made to create a variety of stable oligomeric envelope proteins for evaluation as immunogens. Recent advances in our understanding of the process of HIV infection of a cell indicate that the envelope of the virus undergoes a series of stereotypic conformational changes during the process of viral fusion to the cell membrane. Attempts are being made to develop subunit immunogens that mimic such fusion intermediate forms of the virus envelope. Finally, since an effective HIV vaccine must elicit antibodies that bind to neutralizing determinants of a variety of HIV envelopes, some are arguing that polyvalent envelope mixtures should be assessed as potential immunogens. Incorporation of any such immunogen in a HIV vaccine would represent a dramatic departure from usual viral subunit vaccines.

Although the ultimate configuration of an effective HIV vaccine remains uncertain, there is a growing consensus that it will require more than a single vaccine modality. In a bimodal vaccine approach, a live recombinant vector or plasmid DNA could be used to elicit CTLs, while a subunit immunogen could be used to induce a neutralizing antibody response. Moreover, a growing body of evidence indicates that a high-frequency CTL response is best elicited by combining two complementing vaccine modalities. This has been most clearly demonstrated in studies combining plasmid DNA with live recombinant MVA or adenovirus immunizations. Thus, even a CTL-based vaccine is likely to make use of two distinct vaccine modalities as a prime/boost series of immunizations. Such an approach represents a radical change from traditional vaccine designs.

Vaccination to prevent clinical disease

We are accustomed to considering a particular viral vaccine effective if it blocks infections by that virus. A vaccine strategy that provides anything short of that ideal level of protection would be considered inadequate. There is, however, a shared conviction that for a vaccine to prevent HIV infections, the immunogen must elicit an antibody response that neutralizes a diversity of HIV isolates. Since we remain unable to elicit such an immune response with available vaccine prototypes, there has been a long-standing pessimism in the field of HIV research about the possibility of creating a useful HIV vaccine.

New findings in nonhuman primate studies have, however, suggested that available vaccine technologies may confer important benefits short of complete protection from infection. Monkeys vaccinated with a variety of novel immunogens, including plasmid DNA, cytokine-augmented plasmid DNA, recombinant MVA, recombinant gene-deleted adenovirus, and plasmid DNA followed by recombinant MVA or gene-deleted adenovirus (18, 22–25) show some improvement in their ability to control infection with highly pathogenic AIDS viruses. These vaccines did not elicit broadly neutralizing antibody responses and, accordingly, did not provide sterilizing immunity, but all of these vaccines elicited virus-specific CTL responses and conferred partial control of viral replication to vaccinated animals. Moreover, the level of viral replication in the animals correlated with the rapidity of disease progression. Those animals that demonstrated good containment of virus replication evidenced prolonged disease-free survival.

These findings suggest that available vaccine modalities that elicit potent CTL responses may be able to confer protection against the persistent high levels of viral replication routinely observed in HIV-infected individuals. Such vaccine-elicited immunity may slow the progression of disease in a vaccinated individual who subsequently is infected with HIV. A second potential benefit might also occur as a result of such a vaccine strategy. Vaccinated individuals who subsequently become infected with HIV may have low levels of replication-competent virus in their secretions. Since the likelihood of transmitting virus from an infected to an uninfected individual is correlated with the levels of virus in that infected individual's blood and secretions (26), such vaccinated and then infected individuals may be less likely to transmit virus than those who were infected without the benefit of prior vaccination. Thus, a CTL-inducing vaccine may ultimately slow the spread of HIV in the population. Vaccines that elicit immunity that attenuates disease and slows virus spread in a population may therefore be achievable. The potential benefits of such an outcome could be enormous, particularly in regions of the world where the absence of a medical infrastructure precludes the distribution of effective antiviral drugs to HIV-infected individuals.

However, a recent observation in nonhuman primates has suggested a likely limitation of a vaccine approach based solely on the elicitation of CTLs. In a monkey that was vaccinated and subsequently infected with an AIDS virus isolate, impressive early CTL control of replicating virus and clinical protection were lost as the virus accumulated mutations that allowed it to escape from CTL recognition (27). Viral escape from CTLs may therefore prove a general mechanism of vaccine failure in individuals receiving vaccines of this type. Nevertheless, such virologic events should occur infrequently if vaccine-elicited immune responses restrict viral replication to very low levels, since virus replicating at low levels would be expected to accumulate mutations relatively slowly.

Conclusions

Although studies in nonhuman primates have shown that neutralizing antibodies can protect against infection by an AIDS virus, vaccine strategies for eliciting such immune responses remain elusive. With accumulating evidence for the importance of CTLs in containing HIV spread in an infected individual, a number of vaccine strategies are being pursued for elicitation of these immune effector cells. These strategies include the use of plasmid DNA, live recombinant viral vectors, and combined modality or prime/boost approaches. Interestingly, while these CTL-inducing vaccines have not elicited immunity that prevents AIDS virus infections in monkeys, they have generated immunity that contributes to containment of virus replication subsequent to infection. This containment of virus replication results in prolonged disease-free survival in vaccinated and then challenged monkeys. Early-phase testing of these novel vaccine strategies is ongoing in human volunteer populations. The conviction is growing that an HIV vaccine that at least slows disease progression, if not one that prevents infection, is now possible.

Address correspondence to: Norman L. Letvin, Harvard Medical School, Beth Israel Deaconess Medical Center, 41 Avenue Louis Pasteur, RE-113, Boston, Massachusetts 02215, USA. Phone: (617) 667-2042; Fax: (617) 667-8210; E-mail: nletvin@caregroup.harvard.edu.

1. Malim, M.H., and Emerman, M. 2001. HIV-1 sequence variation: drift, shift, and attenuation. *Cell.* **104**:469–472.
2. Hirsch, V.M., and Johnson, P.R. 1994. Pathogenic diversity of simian immunodeficiency viruses. *Virus Res.* **32**:183–203.
3. Reimann, K.A., et al. 1996. A chimeric simian/human immunodeficiency virus expressing a primary patient human immunodeficiency virus type 1 isolate *env* causes an AIDS-like disease after in vivo passage in rhesus monkeys. *J. Virol.* **70**:6922–6928.
4. Mascola, J.R., et al. 2000. Protection of macaques against vaginal transmission of a pathogenic HIV-1/SIV chimeric virus by passive infusion of neutralizing antibodies. *Nat. Med.* **6**:207–210.
5. Walker, C.M., Moody, D.J., Stites, D.P., and Levy, J.A. 1986. CD8+ lymphocytes can control HIV infection in vitro by suppressing virus replication. *Science.* **234**:1563–1566.
6. Koup, R.A., et al. 1994. Temporal association of cellular immune responses with the initial control of viremia in primary human immunodeficiency virus type 1 syndrome. *J. Virol.* **68**:4650–4655.
7. Ogg, G.S., et al. 1998. Quantitation of HIV-1-specific cytotoxic T lymphocytes and plasma load of viral RNA. *Science.* **279**:2103–2106.
8. Altman, J.D., et al. 1996. Phenotypic analysis of antigen-specific T lymphocytes. *Science.* **274**:94–96.
9. Schmitz, J.E., et al. 1999. Control of viremia in simian immunodeficiency virus infection by CD8+ lymphocytes. *Science.* **283**:857–860.
10. Rosenberg, E.S., et al. 1997. Vigorous HIV-1-specific CD4+ T cell responses associated with con-

trol of viremia. *Science.* **278**:1447–1450.
11. Daniel, M.D., Kirchhoff, F., Czajak, S.C., Sehgal, P.K., and Desrosiers, R.C. 1992. Protective effects of a live attenuated SIV vaccine with a deletion in the *nef* gene. *Science.* **258**:1938–1941.
12. Baba, T.W., et al. 1995. Pathogenicity of live, attenuated SIV after mucosal infection of neonatal macaques. *Science.* **267**:1820–1825.
13. Murphey-Corb, M., et al. 1989. A formalin-inactivated whole SIV vaccine confers protection in macaques. *Science.* **246**:1293–1297.
14. Stott, E.J. 1991. Anti-cell antibody in macaques. *Nature.* **353**:393.
15. Levine, A.M., et al. 1996. Initial studies on active immunization of HIV-infected subjects using a gp120-depleted HIV-1 immunogen: long-term follow-up. *J Acquir Immune Defic. Syndr.* **11**:351–364.
16. Berman, P.W., et al. 1990. Protection of chimpanzees from infection by HIV-1 after vaccination with recombinant glycoprotein gp120 but not gp160. *Nature.* **345**:622–625.
17. Egan, M.A., et al. 2000. Simian immunodeficiency virus (SIV) *gag* DNA-vaccinated rhesus monkeys develop secondary cytotoxic T lymphocyte responses and control viral replication after pathogenic SIV infection. *J. Virol.* **74**:7485–7495.
18. Amara, R.R., et al. 2001. Control of a mucosal challenge and prevention of clinical AIDS in rhesus monkeys by a multiprotein DNA/MVA vaccine. *Science.* **292**:69–74.
19. Shen, L., et al. 1991. Recombinant virus vaccine-induced SIV-specific CD8+ cytotoxic T lymphocytes.

Science. **252**:440–443.
20. Redfield, R.R., et al. 1987. Disseminated vaccinia in a military recruit with human immunodeficiency virus (HIV) disease. *N. Engl. J. Med.* **316**:673–676.
21. Evans, T.G., et al. 1999. A canarypox vaccine expressing multiple human immunodeficiency virus type 1 genes given alone or with rgp120 elicits broad and durable CD8+ cytotoxic T lymphocyte responses in seronegative volunteers. *J. Infect. Dis.* **180**:290–298.
22. Shiver, J.W., et al. 2002. Replication-incompetent adenoviral vaccine vector elicits effective anti-immunodeficiency-virus immunity. *Nature.* **415**:331–335.
23. Ourmanov, I., et al. 2000. Comparative efficacy of recombinant modified vaccinia virus Ankara expressing simian immunodeficiency virus (SIV) Gag-Pol and/or Env in macaques challenged with pathogenic SIV. *J. Virol.* **74**:2740–2751.
24. Barouch, D.H., et al. 2000. Control of viremia and prevention of clinical AIDS in rhesus monkeys by cytokine-augmented DNA vaccination. *Science.* **290**:486–492.
25. Barouch, D.H., et al. 2001. Reduction of simian-human immunodeficiency virus 89.6P viremia in rhesus monkeys by recombinant modified vaccinia virus Ankara vaccination. *J. Virol.* **75**:5151–5158.
26. Quinn, T.C., et al. 2000. Viral load and heterosexual transmission of human immunodeficiency virus type 1. *N. Engl. J. Med.* **342**:921–929.
27. Barouch, D.H., et al. 2002. Eventual AIDS vaccine failure in a rhesus monkey by viral escape from cytotoxic T lymphocytes. *Nature.* **415**:335–339.

Dengue: defining protective versus pathologic immunity

Alan L. Rothman

Center for Infectious Disease and Vaccine Research, University of Massachusetts Medical School, Worcester, Massachusetts, USA.

Dengue is an expanding public health problem, and an effective vaccine remains elusive. This review discusses how the significant influence of sequential infection with different dengue virus serotypes on the severity of disease can be viewed in terms of beneficial and detrimental effects of heterologous immunity. A more complete understanding of these effects is likely to be critical for predicting optimal vaccine-induced immune responses.

Among scientists in the developed world, there has been a recent resurgence of interest in exotic diseases, with the aim of protecting the global population from emerging infectious disease threats. The renewed commitment of effort and resources has welcome implications for the developing world, whose population faces the greatest part of these threats. Among the biological threats considered most serious are the viral hemorrhagic fevers. Although this term usually brings to mind the deadly outbreaks of Ebola virus, in reality over 99% of the cases of viral hemorrhagic fever reported worldwide are related instead to dengue hemorrhagic fever (DHF) (1).

DHF is caused by the dengue viruses (DENVs), members of the *Flaviviridae* family of small enveloped viruses (2). The flaviviruses carry a single-stranded RNA genome of relatively simple organization. A single open reading frame in the RNA directs the synthesis of a long polyprotein that is processed by viral and host cell proteases to produce the ten viral proteins, including three structural proteins (core [C]; membrane [M], produced as a precursor protein; and envelope [E]) and seven nonstructural (NS) proteins (Figure 1). The DENV complex encompasses four closely related serotypes: DENV-1, DENV-2, DENV-3, and DENV-4. All four DENV serotypes are transmitted between humans in nature by mosquitoes of the genus *Aedes*, principally *Aedes aegypti*, which is highly domesticated and has a preference for biting humans.

It has been estimated that over 50 million DENV infections occur globally each year (3). Most of these infections are clinically inapparent. Among symptomatic cases, the majority of subjects experience uncomplicated dengue fever (DF), an acute febrile illness typically lasting 3–7 days, accompanied by headache, myalgias, and, less often, a maculopapular rash. The headache and myalgias may be quite debilitating, which originated the name "break-bone fever" that was recorded prior to the 1900s (4). Laboratory findings in patients with DF include leukopenia thrombocytopenia, and mild elevations in serum hepatic transaminases (5). Fatigue may be prolonged for months after resolution of fever, but patients eventually recover without sequelae. None of these features is sufficiently specific for accurate clinical diagnosis of DF. Laboratory support for detection of IgM antibody or virus (by RT-PCR or virus isolation) is therefore important for recognition of outbreaks.

DHF represents the severe clinical manifestation of DENV infection, which occurs in no more than 3% of infected individuals (6). Fever, headache, and myalgias are prominent symptoms in DHF, as they are in DF. DHF is distinguished from DF on clinical grounds, with the three primary criteria being the occurrence of a vascular permeability defect resulting in plasma leakage; multifactorial hemostatic abnormalities, including marked thrombocytopenia; and a bleeding diathesis (7). Plasma leakage is evidenced by hemoconcentration and/or effusions in the pleural or peritoneal spaces, which usually occur after 3–5 days of fever, near the time of defervescence. When severe, the plasma leakage leads to hypotension and circulatory collapse, and this is the most important concern for triage and therapy. Thrombocytopenia is quite severe in DHF, with a platelet count of less than 100,000 cells/mm^3 required to fulfill the case definition. The nadir of the platelet count is typically coincident with plasma leakage, and severe thrombocytopenia also occurs in a significant percentage of patients with DF; nevertheless, the decline in platelet count typically precedes plasma leakage, and this finding has been useful in deciding whether to hospitalize patients suspected to have DHF (8). Bleeding manifestations, for example, from the gastrointestinal tract, can be severe in patients with DHF but are not usually severe enough to require transfusion and are rarely the principal cause of hypotension. Despite the use of the term "hemorrhagic fever," many patients with DHF have minor hemorrhagic manifestations; often the occurrence of petechiae elicited by inflation of a blood pressure cuff on the arm for 5 minutes (the "tourniquet test") is the only clinical sign.

The emergence of DHF as a global infectious disease threat is striking in its persistence and magnitude. Southeast Asia and the western Pacific region, where DHF was first recognized as a clinical entity, have faced this health problem for almost 50 years. Nevertheless, during each decade the number of DHF cases, the number of countries affected, and the geographic distribution of DHF have all increased steadily (3). DHF is, in comparison, a new problem in the tropical Western hemisphere, but one that has rapidly expanded in scope. Although essentially absent from the area prior to 1981, DHF has during the past decade been seen throughout Latin America and the Caribbean. It is currently estimated that over 3 billion individuals live in areas at risk for DHF and that hundreds of thousands of cases and thousands of deaths due to DHF occur each year, representing an urgent health problem to be addressed.

Nonstandard abbreviations used: antibody-dependent enhancement of infection (ADE); core (C); dengue fever (DF); dengue hemorrhagic fever (DHF); dengue virus (DENV); envelope (E); membrane (M); nonstructural (NS).

Conflict of interest: The author has declared that no conflict of interest exists.

Citation for this article: *J. Clin. Invest.* **113**:946–951 (2004). doi:10.1172/JCI200421512.

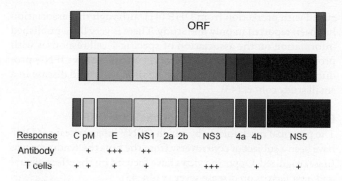

Response	C pM	E	NS1	2a 2b	NS3	4a 4b	NS5
Antibody	+	+++	++				
T cells	+ +	+	+		+++	+	+

Figure 1
Organization of the flavivirus genome and its resulting proteins and the location of the major targets of the immune response. The DENV genome is a single-stranded sense RNA with a single open reading frame (ORF; top). The ORF is translated as a single polyprotein (middle) that is cleaved by viral and host proteases to yield the ten viral proteins: the C protein; the M protein, which is synthesized as the larger precursor protein pre-M (pM); the major E glycoprotein; and seven NS proteins involved in viral replication among other functions (bottom). The strength of the antibody and T cell responses to individual viral proteins is indicated below.

Human activities that have led to enhanced mosquito breeding capacity, increased interaction between mosquitoes and humans, and increased dispersal of viruses in both mosquito and human hosts have contributed (and continue to contribute) to the emergence of DHF, as discussed in several recent reviews (9, 10). Poorly planned development has been a major factor, especially in urban centers of the developing world. The attendant inadequate sanitation and potable water supply, leading to an increase in mosquito breeding places, and the concentration of susceptible human hosts in close proximity to the mosquitoes also facilitate transmission. In the Western hemisphere, *A. aegypti* have also taken advantage of the lapse in mosquito control efforts that has occurred since the 1970s and have re-established essentially the full geographical range they commanded prior to the yellow fever eradication efforts of the mid-twentieth century.

At present, there is no specific therapy available for DHF. Appropriate fluid management to correct hypovolemia has been successful in reducing the mortality of DHF (7, 11, 12), but access to medical services remains problematic in many developing countries. Mosquito control, which is costly and often ineffective, remains the only method of preventing DHF currently available. Development of an effective vaccine against DENVs has therefore been considered a high priority by the WHO.

Background of dengue vaccine development

In principle, an effective vaccine against DENV is highly feasible. Unlike the situation with HIV and hepatitis C virus, DENV does not cause chronic infection. Viral replication is effectively controlled after a short (3- to 7-day) period of viremia, and individuals who have recovered from DENV infection are immune to rechallenge (13). Studies conducted in the 1940s demonstrated that a DENV strain derived by serial propagation in suckling mouse brain was attenuated in humans and could elicit protective immunity against DENV challenge (13).

There are a number of candidate DENV vaccines currently in development (Table 1) (14). Live attenuated strains of all four DENV serotypes have been derived by the traditional approach of serial propagation in primary dog or monkey kidney cells (15–17). This effort has been slowed by the lack of an animal model or in vitro markers of attenuation in humans. Improved molecular virology techniques and an improved understanding of the genomic structure of DENV have permitted a more rapid approach based on engineering of attenuating mutations into infectious cDNA clones of each of the four DENV serotypes, with the added theoretical advantage of a lower possibility of reversion to virulence (18–20). Another molecular approach being utilized is the creation of four separate infectious chimeric flaviviruses, each of which contains the pre-M and E genes of one of the four DENV serotypes in a single "backbone" containing the C and NS proteins of an attenuated flavivirus, either the yellow fever vaccine strain or an attenuated DENV strain (21, 22). DNA vaccines consisting of plasmids expressing one or a few proteins from each DENV serotype are in an earlier stage of development, as are subunit vaccines based on purified recombinant DENV proteins (23, 24). Several of these approaches have demonstrated protective efficacy in animal models of DENV infection, and a few have shown safety and immunogenicity in early phase clinical studies (14, 18, 25, 26).

Two main obstacles have delayed further clinical development of these candidate vaccines. First, an effective vaccine needs to prevent infection with all four DENV serotypes. Natural DENV infection induces long-lasting protective immunity only to the same serotype and only short-term (months) protection from infection with other serotypes (13). Although live attenuated and recombinant vaccines to all four DENV serotypes have been developed, incorporation of these into a tetravalent formulation that retains the immunogenicity of all four components has proven difficult, requiring the use of more complicated, multiple-dose immunization regimens (25).

The second and more significant obstacle is the current inability to predict whether candidate DENV vaccines will be at all effective in preventing DHF. DHF was unrecognized prior to

Table 1
Immunological considerations and current status in development of candidate dengue vaccines

	Live attenuated	Chimeric virus	Plasmid DNA	Subunit
No. of dengue antigens included	All (10)	2 (10 for viral "backbone" only)	1 to several	Usually 1
In vivo replication	Yes	Yes	No	No
Elicits T and B cell memory	Best	Best	Excellent	Fair: Th, B; Poor: CTL
Anticipated durability of immune response	Best	Best	Good	Usually poor
Protective in animal models	Yes	Yes	Yes	Yes
Status of vaccine development	Phase I/II	Phase I/II	Preclinical	Preclinical

the late 1950s. Therefore, the experimental challenge studies in humans conducted during and shortly after World War II did not attempt to test the efficacy of candidate vaccines against DHF. Subsequent studies of candidate vaccines have analyzed efficacy only in experimental animal models, none of which faithfully reproduce the DHF syndrome seen in humans. As a result, selection of the most promising DENV vaccine candidates will necessarily rely on comparing vaccine-induced immune responses to a profile of protective immunity developed from observation of natural DENV infections.

Immune responses to DENV

The principal targets of the immune response to DENV are illustrated in Figure 1. The E glycoprotein is the principal component of the external surface of the DENV virion (27) and is a dominant target of the response consisting of antibodies against DENV. Antibodies against E have been shown to inhibit viral binding to cells and to neutralize viral infectivity in vitro (28). Passive transfer of antibodies against E protected mice from DENV challenge (29). Antibodies against E show variable degrees of cross-reactivity among the DENV serotypes, although neutralization of infectivity by antibodies is usually more serotype specific than virion binding (28). Binding of antibody to virus at non-neutralizing epitopes, or at concentrations below the neutralization endpoint, can enhance the uptake of virions into monocytic cell lines and primary human monocytes in vitro through interaction with cell surface Ig receptors, a phenomenon known as antibody-dependent enhancement of infection (ADE) (30). These "enhancing" antibodies do not increase infection of human DCs, which are highly permissive to DENV infection, however (31, 32).

While not a component of the virion, the NS1 protein is also an important target of antibodies against DENV. NS1 is expressed on the surface of infected cells and is also secreted into the circulation as a soluble multimer (33). Antibodies against NS1 can trigger complement-mediated lysis of DENV-infected cells in vitro and have been shown to protect mice from DENV challenge (34). A monoclonal antibody directed against the pre-M protein has also been shown to protect mice from DENV challenge (35); however, the mechanism of this protection and its relevance to natural protective immunity are uncertain.

As with other viruses, the CD4+ and CD8+ T cell response to DENV is directed against multiple viral proteins (Figure 1), although the NS3 protein appears to be particularly immunogenic, with a preponderance of the T cell epitopes identified (36, 37). DENV-reactive T cells vary in their ability to recognize different DENV serotypes, depending upon the degree of homology at a given epitope. However, cross-reactivity with multiple serotypes is common, especially at epitopes in the more highly conserved NS proteins (38, 39). DENV-reactive CD4+ and CD8+ T cells predominantly produce high levels of IFN-γ as well as TNF-α, TNF-β, and chemokines including macrophage inhibitory protein-1β upon interaction with DENV-infected antigen presenting cells, and are efficient at lysis of DENV-infected cells in vitro (39, 40).

The extent to which these immune responses contribute to the long-term protective immunity afforded by natural primary DENV infection has not been fully defined. Symptomatic DENV infections, including cases of DHF, can occur despite the presence of antibodies capable of neutralizing in vitro infection of epithelial cell lines (41, 42). Neutralization of infection in monocytic cells (instead of enhancement of infection) may be more strongly asso-

ciated with protection from DHF (41), although this association has been reported in only one study. There is very little published information on the association of specific T cell responses with protection. However, broadly serotype–cross-reactive IFN-γ production in vitro showed a weak association with mild disease in a small study cohort (43).

Evidence of an immunological basis for DHF

The principal mechanisms by which DENV infection causes DHF have been a subject of controversy from the time the syndrome was first recognized. Opposing views have focused on the effect of viral and host factors on disease severity (44, 45).

Differences in virulence between naturally circulating DENV strains had been suspected based on differences in clinical profiles observed during isolated DENV outbreaks in Indonesia and the Pacific Islands (46, 47). More recently, studies in Peru and Sri Lanka have provided more convincing data, demonstrating the association of DHF with specific viral genotypes and not with others (48–50). Specific genetic determinants potentially explaining this association have been mapped to the E gene and to the 5′ and 3′ untranslated regions of the DENV genome (51–53). However, in the absence of a validated model of virulence in humans, it has not been possible to verify that these genetic elements are responsible for virulence or to confirm any proposed mechanism for virulence.

Key seroepidemiological studies by Halstead and colleagues in Thailand during the 1960s first suggested an association of increased risk for DHF with a secondary DENV infection; that is, a new DENV infection in an individual who had previously experienced one or more DENV infections (54, 55). This hypothesis has since received strong confirmation from the "experiment of nature" in a large outbreak of DHF associated with DENV-2 in Cuba (56), as well as several independent prospective cohort studies in Southeast Asia (57–59). These observations have provided the basis for exploration of the possibility of a pathologic role for the immune response to DENV in the development of DHF. Age, race/ethnicity, nutritional state, and underlying chronic diseases such as asthma have also been suggested to contribute to DHF risk, although the data supporting these associations are weaker and the mechanisms involved are poorly defined.

Antibodies against DENV can affect the course of disease through multiple mechanisms (Table 2). Passive transfer of antibody against DENV did increase viremia titers in nonhuman primates in one study (60), and recent studies have demonstrated a positive correlation between peak viremia titer and disease severity in humans (61, 62), supporting the idea of the potential in vivo importance of ADE. The occurrence of DHF during primary DENV infection in the first year of life in children born to DENV-immune mothers, and who therefore acquire antibody against DENV transplacentally, also supports the idea of an in vivo role for ADE (63). However, high viremia titers in older children and adults with primary DENV infections and in clinically mild secondary DENV infections indicate that other factors are involved (64, 65). Immune complex formation in vivo has been detected in association with complement activation in patients with severe disease (66). Cross-reactivity of antibodies against E with plasminogen has been associated with bleeding in acute DENV infection, although not with DHF (67).

Cytokine production and cytolysis by activated T lymphocytes are additional potential contributors to disease (Table 2). Elevations of circulating levels of activation markers including soluble

Table 2
Postulated pathologic mechanisms of DENV-reactive immune responses in the development of DHF

Mechanism	Postulated effect(s)	References
Antibodies		
Enhancement of infection	Increased cellular infection, increased viral burden	(45, 60, 84)
Immune complex formation	Complement activation	(66, 85, 86)
Cross-reactivity to coagulation and endothelial cell proteins	Bleeding, endothelial cell dysfunction	(67, 87, 88)
T lymphocytes		
Proinflammatory cytokine production	Increased vascular permeability	(61, 68, 69)
Lysis of bystander (uninfected) cells	Hepatic injury	(40)

TNF receptors, soluble IL-2 receptors, and soluble CD8 have been shown to correlate with disease severity (61, 68–71). Similar associations with disease severity have been shown for the expression of activation markers on circulating CD8+ T cells (72) and the expansion of DENV epitope–specific T cell populations (73, 74). Increased production of various cytokines, including IFN-γ, TNF-α, IL-10, and chemokines has also been detected in acute DENV infection (69–71, 75–77). Although elevations in both type 1 and type 2 cytokine levels are detected in DHF, the timing of their production appears to be an important factor with earlier induction of type 1 cytokines being associated with more severe disease (61). Analysis of T cell responses to DENV prior to secondary DENV infections showed an association between in vitro TNF-α responses to DENV antigens and more severe disease during the subsequent infection (43).

Heterologous immunity and the pathogenesis of DHF

Although similar response mechanisms appear to operate in both protective and pathological immunity to DENV, a principal immunological consideration is the heterologous nature of secondary DENV infection. The DENV causing secondary disease is always of a different serotype than the virus that induced immune responses during the earlier DENV infection, and therefore the antibodies and memory T lymphocytes induced by the primary DENV infection typically encounter antigens that differ (although some may be the same) in sequence from their original target antigen. The magnitude of these sequence differences is epitope dependent, but all sequence differences have the potential to affect the quality of the effector response and also modify the immunological repertoire. We speculate that the immune response to these heterologous sequences has the net effect of altering the balance between a protective and pathological outcome.

The effect of sequence differences at antibody-binding sites is to reduce the avidity of the interactions between the pre-existing antibodies and the new DENV serotype. These less avid interactions have a significant effect on the ability of the antibodies to neutralize virus infectivity. However, there is sufficient binding of antibody to the virion to trigger ADE. In addition, these low-avidity antibodies increase in titer in preference to new antibodies with high avidity for the new DENV serotype, because the pre-existing memory B cells and plasma cells are more rapidly activated than naive B cells. The pattern of antibody responses in convalescence, therefore, is influenced heavily by the serotype of the primary DENV infection, a phenomenon referred to as "original antigenic sin" (78).

The effect of sequence differences at T cell epitopes is more complex, consistent with the wider range of effector functions of T cells. Variant peptides, or altered peptide ligands, induce dif-

ferent activation signals in antigen-specific T lymphocytes and thereby modulate specific effector functions of CD4+ and CD8+ T cells (79, 80). In general, such partial T cell agonists have the greatest effect on the proliferation responses and less effect on cytotoxicity responses. Some partial T cell agonists fail to induce cytokine production, whereas others induce an altered cytokine response profile. We have observed just such effects in the in vitro responses of DENV-reactive T cells to peptides of heterologous DENV serotypes (81). Furthermore, a phenomenon analogous to "original antigenic sin" would be expected to occur in the T cell response to secondary DENV infections, in which expansion of the pre-existing lower avidity memory T cell populations takes precedence over the expansion of naive T cell populations with higher avidity for the new DENV serotype. Recent studies of the T cell repertoire after secondary DENV infection support this view (74).

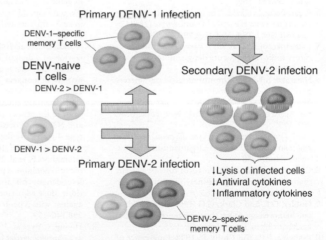

Figure 2
Proposed model of heterologous immunity in secondary dengue virus infections and its implications for the pathogenesis of dengue hemorrhagic fever. Primary DENV-2 infection and sequential DENV-1 and DENV-2 infections are compared for illustration purposes. The naive T cell repertoire (pale colors) likely contains some cells with higher avidity for DENV-1 than DENV-2 (red; DENV-1 > DENV-2) and other cells with higher avidity for DENV-2 than DENV-1 (blue; DENV-2 > DENV-1). During primary infection, T cell populations with higher avidity for the infecting serotype are preferentially expanded and enter the memory pool (shown as darker colors). When DENV-2 infection follows DENV-1 infection, the memory T cell populations with higher avidity for the earlier infection expand more rapidly than do naive T cell populations. Because these DENV-1–specific memory T cells have lower avidity for DENV-2, viral clearance mechanisms are suboptimal, whereas proinflammatory responses contribute to disease.

The potential importance of heterologous immunity in influencing the severity of DENV disease is bolstered by data from an animal model of sequential virus infections. In this experimental system, previous infection with one virus (e.g., lymphocytic choriomeningitis virus) altered the response to a subsequent vaccinia virus challenge, altering the hierarchy of epitope-specific T cell responses and resulting in lower virus titers but greater tissue inflammation (82, 83). A model for the effect of heterologous immunity in the pathogenesis of DHF is shown in Figure 2. We postulate that the low-avidity T cells that dominate the response to secondary DENV infections are less than optimally efficient at elimination of DENV-infected cells and that the pattern of cytokine production is altered such that proinflammatory cytokine production (e.g., TNF-α) is enhanced and antiviral cytokine production (e.g., IFN-γ) is reduced. These effects could combine to induce the plasma leakage symptomatic of DHF.

Responding to the challenge

To maximize protection and minimize pathologic heterologous immunity and achieve long-lasting immune responses, an optimal vaccine to prevent DENV infections including DHF should induce high-avidity antibodies and T lymphocytes against all four DENV serotypes. The theoretical advantages of live attenuated virus vaccines or plasmid DNA vaccines (Table 1) have placed them on a faster track toward clinical development. Tetravalent attenuated vaccines have been developed, but researchers have encountered problems in finding a formulation that balances the immunogenicity of all four viruses (25). Recombinant chimeric flavivirus vaccines show promise in induction of antibodies against E (21), but it will be important to determine whether the common epitopes in the NS proteins of the viral "backbone" will induce optimal T cell responses. Some findings suggest that individuals with mild and severe disease target different epitopes in their T cell response to DENV (73). Nucleotide sequences from different DENVs could theoretically be introduced into a single vaccine "backbone" to provide the most favorable targets for a high-avidity, cross-serotype immune response. Much additional research is needed to identify and characterize protective and pathological immune responses in order to make an effective DENV vaccine a reality.

Address correspondence to: Alan L. Rothman, Center for Infectious Disease and Vaccine Research, Room S5-326, University of Massachusetts Medical School, 55 Lake Avenue North, Worcester, Massachusetts 01655, USA. Phone: (508) 856-4182; Fax: (508) 856-4890; E-mail: alan.rothman@umassmed.edu.

1. Rigau-Perez, J.G., et al. 1998. Dengue and dengue haemorrhagic fever. *Lancet.* **352**:971–977.
2. Henchal, E.A., and Putnak, J.R. 1990. The dengue viruses. *Clin. Microbiol. Rev.* **3**:376–396.
3. Pinheiro, F.P., and Corber, S.J. 1997. Global situation of dengue and dengue haemorrhagic fever, and its emergence in the Americas. *World Health Stat. Q.* **50**:161–169.
4. Rigau-Perez, J.G. 1998. The early use of break-bone fever (*Quebranta huesos*, 1771) and *dengue* (1801) in Spanish. *Am. J. Trop. Med. Hyg.* **59**:272–274.
5. Kalayanarooj, S., et al. 1997. Early clinical and laboratory indicators of acute dengue illness. *J. Infect. Dis.* **176**:313–321.
6. Halstead, S.B. 1980. Immunological parameters of togavirus disease syndromes. In *The togaviruses. biology, structure, replication.* R.W. Schlesinger, editor. Academic Press. New York, New York, USA. 107–173.
7. 1997. *Dengue haemorrhagic fever: diagnosis, treatment and control.* World Health Organization. Geneva, Switzerland. 1–58.
8. Chin, C.K., et al. 1993. Protocol for out-patient management of dengue illness in young adults. *J. Trop. Med. Hyg.* **96**:259–263.
9. Gubler, D.J., and Clark, G.G. 1995. Dengue/dengue hemorrhagic fever: the emergence of a global health problem. *Emerg. Infect. Dis.* **1**:55.
10. Pinheiro, F.P., and Chuit, R. 1998. Emergence of dengue hemorrhagic fever in the Americas. *Infect. Med.* **15**:244–251.
11. Ngo, N.T., et al. 2001. Acute management of dengue shock syndrome: a randomized double-blind comparison of 4 intravenous fluid regimens in the first hour. *Clin. Infect. Dis.* **32**:204–213.
12. Harris, E., et al. 2003. Fluid intake and decreased risk for hospitalization for dengue fever, Nicaragua. *Emerg. Infect. Dis.* **9**:1003–1006.
13. Sabin, A.B. 1952. Research on dengue during World War II. *Am. J. Trop. Med. Hyg.* **1**:30–50.
14. Barrett, A.D. 2001. Current status of flavivirus vaccines. *Ann. N. Y. Acad. Sci.* **951**:262–271.
15. Bhamarapravati, N., and Sutee, Y. 2000. Live attenuated tetravalent dengue vaccine. *Vaccine.* **18**:44–47.
16. Eckels, K.H., et al. 2003. Modification of dengue virus strains by passage in primary dog kidney cells: preparation of candidate vaccines and immunization of monkeys. *Am. J. Trop. Med. Hyg.* **69**:12–16.
17. Innis, B.L., and Eckels, K.H. 2003. Progress in development of a live-attenuated, tetravalent dengue virus vaccine by the United States Army Medical Research and Materiel Command. *Am. J. Trop. Med. Hyg.* **69**:1–4.
18. Durbin, A.P., et al. 2001. Attenuation and immunogenicity in humans of a live dengue virus type-4 vaccine candidate with a 30 nucleotide deletion in its 3'-untranslated region. *Am. J. Trop. Med. Hyg.* **65**:405–413.
19. Markoff, L., et al. 2002. Derivation and characterization of a dengue type 1 host range-restricted mutant virus that is attenuated and highly immunogenic in monkeys. *J. Virol.* **76**:3318–3328.
20. Blaney, J.E., Jr., Manipon, G.G., Murphy, B.R., and Whitehead, S.S. 2003. Temperature sensitive mutations in the genes encoding the NS1, NS2A, NS3, and NS5 nonstructural proteins of dengue virus type 4 restrict replication in the brains of mice. *Arch. Virol.* **148**:999–1006.
21. Guirakhoo, F., et al. 2002. Viremia and immunogenicity in nonhuman primates of a tetravalent yellow fever-dengue chimeric vaccine: genetic reconstructions, dose adjustment, and antibody responses against wild-type dengue virus isolates. *Virology.* **298**:146–159.
22. Huang, C.Y., et al. 2003. Dengue 2 PDK-53 virus as a chimeric carrier for tetravalent dengue vaccine development. *J. Virol.* **77**:11436–11447.
23. Chang, G.J., Davis, B.S., Hunt, A.R., Holmes, D.A., and Kuno, G. 2001. Flavivirus DNA vaccines: current status and potential. *Ann. N. Y. Acad. Sci.* **951**:272–285.
24. Simmons, M., Murphy, G.S., Kochel, T., Raviprakash, K., and Hayes, C.G. 2001. Characterization of antibody responses to combinations of a dengue-2 DNA and dengue-2 recombinant subunit vaccine. *Am. J. Trop. Med. Hyg.* **65**:420–426.
25. Kanesa-thasan, N., et al. 2001. Safety and immunogenicity of attenuated dengue virus vaccines (Aventis Pasteur) in human volunteers. *Vaccine.* **19**:3179–3188.
26. Sabchareon, A., et al. 2004. Safety and immunogenicity of a three dose regimen of two tetravalent live-attenuated dengue vaccines in five- to twelve-year-old Thai children. *Pediatr. Infect. Dis. J.* **23**:99–109.
27. Kuhn, R.J., et al. 2002. Structure of dengue virus: implications for flavivirus organization, maturation, and fusion. *Cell.* **108**:717–725.
28. Roehrig, J.T., Bolin, R.A., and Kelly, R.G. 1998. Monoclonal antibody mapping of the envelope glycoprotein of the dengue 2 virus, Jamaica. *Virology.* **246**:317–328.
29. Kaufman, B.M., Summers, P.L., Dubois, D.R., and Eckels, K.H. 1987. Monoclonal antibodies against dengue 2 virus E-glycoprotein protect mice against lethal dengue infection. *Am. J. Trop. Med. Hyg.* **36**:427–434.
30. Morens, D.M., and Halstead, S.B. 1990. Measurement of antibody-dependent infection enhancement of four dengue virus serotypes by monoclonal and polyclonal antibodies. *J. Gen. Virol.* **71**:2909–2914.
31. Wu, S.J., et al. 2000. Human skin Langerhans cells are targets of dengue virus infection. *Nat. Med.* **6**:816–820.
32. Library, D.H., Pichyangkul, S., Ajariyakhajorn, C., Endy, T.P., and Ennis, F.A. 2001. Human dendritic cells are activated by dengue virus infection: enhancement by gamma interferon and implications for disease pathogenesis. *J. Virol.* **75**:3501–3508.
33. Young, P.R., Hilditch, P.A., Bletchly, C., and Halloran, W. 2000. An antigen capture enzyme-linked immunosorbent assay reveals high levels of the dengue virus protein NS1 in the sera of infected patients. *J. Clin. Microbiol.* **38**:1053–1057.
34. Schlesinger, J.J., Brandriss, M.W., and Walsh, E.E. 1987. Protection of mice against dengue 2 virus encephalitis by immunization with the dengue 2 virus non-structural glycoprotein NS1. *J. Gen. Virol.* **68**:853–857.
35. Kaufman, B.M., et al. 1989. Monoclonal antibodies for dengue virus prM glycoprotein protect mice against lethal dengue infection. *Am. J. Trop. Med. Hyg.* **41**:576–580.
36. Kurane, I., Brinton, M.A., Samson, A.L., and Ennis, F.A. 1991. Dengue virus-specific, human CD4$^+$ CD8$^-$ cytotoxic T-cell clones: multiple patterns of virus cross-reactivity recognized by NS3-specific T-cell clones. *J. Virol.* **65**:1823–1828.
37. Lobigs, M., Arthur, C.E., Mullbacher, A., and Blanden, R.V. 1994. The flavivirus nonstructural protein NS3 is a dominant source of cytotoxic T

cell peptide determinants. *Virology.* **202**:195–201.

38. Kurane, I., Zeng, L., Brinton, M.A., and Ennis, F.A. 1998. Definition of an epitope on NS3 recognized by human CD4⁺ cytotoxic T lymphocyte clones cross-reactive for dengue virus types 2, 3, and 4. *Virology.* **240**:169–174.

39. Zivny, J., et al. 1995. A single nine-amino acid peptide induces virus-specific, CD8⁺ human cytotoxic T lymphocyte clones of heterogenous serotype specificities. *J. Exp. Med.* **182**:853–863.

40. Gagnon, S.J., Ennis, F.A., and Rothman, A.L. 1999. Bystander target cell lysis and cytokine production by dengue virus-specific human CD4⁺ cytotoxic T lymphocyte clones. *J. Virol.* **73**:3623–3629.

41. Kliks, S.C., Nisalak, A., Brandt, W.E., Wahl, L., and Burke, D.S. 1989. Antibody-dependent enhancement of dengue virus growth in human monocytes as a risk factor for dengue hemorrhagic fever. *Am. J. Trop. Med. Hyg.* **40**:444–451.

42. Endy, T.P., et al. 2004. Relationship of preexisting dengue virus (DV) neutralizing antibody levels to viremia and severity of disease in a prospective cohort study of DV infection in Thailand. *J. Infect. Dis.* **189**:990–1000.

43. Mangada, M.M., et al. 2002. Dengue-specific T cell responses in peripheral blood mononuclear cells obtained prior to secondary dengue virus infections in Thai schoolchildren. *J. Infect. Dis.* **185**:1697–1703.

44. Rosen, L. 1977. The Emperor's New Clothes revisited, or reflections on the pathogenesis of dengue hemorrhagic fever. *Am. J. Trop. Med. Hyg.* **26**:337–343.

45. Halstead, S.B. 1989. Antibody, macrophages, dengue virus infection, shock, and hemorrhage: a pathogenetic cascade. *Rev. Infect. Dis.* **11**:S830–S839.

46. Gubler, D.J., Suharyono, W., Tan, R., Abidin, M., and Sie, A. 1981. Viraemia in patients with naturally acquired dengue infection. *Bull. World Health Organ.* **59**:623–630.

47. Gubler, D.J., Suharyono, W., Lubis, I., Eram, S., and Gunarso, S. 1981. Epidemic dengue 3 in central Java, associated with low viremia in man. *Am. J. Trop. Med. Hyg.* **30**:1094–1099.

48. Watts, D.M., et al. 1999. Failure of secondary infection with American genotype dengue 2 to cause dengue haemorrhagic fever. *Lancet.* **354**:1431–1434.

49. Messer, W.B., et al. 2002. Epidemiology of dengue in Sri Lanka before and after the emergence of epidemic dengue hemorrhagic fever. *Am. J. Trop. Med. Hyg.* **66**:765–773.

50. Messer, W.B., Gubler, D.J., Harris, E., Sivananthan, K., and de Silva, A.M. 2003. Emergence and global spread of a dengue serotype 3, subtype III virus. *Emerg. Infect. Dis.* **9**:800–809.

51. Leitmeyer, K.C., et al. 1999. Dengue virus structural differences that correlate with pathogenesis. *J. Virol.* **73**:4738–4747.

52. Pryor, M.J., et al. 2001. Replication of dengue virus type 2 in human monocyte-derived macrophages: comparisons of isolates and recombinant viruses with substitutions at amino acid 390 in the envelope glycoprotein. *Am. J. Trop. Med. Hyg.* **65**:427–434.

53. Cologna, R., and Rico-Hesse, R. 2003. American genotype structures decrease dengue virus output from human monocytes and dendritic cells. *J. Virol.* **77**:3929–3938.

54. Halstead, S.B., Nimmannitya, S., and Cohen, S.N.

1970. Observations related to pathogenesis of dengue hemorrhagic fever. IV. Relation of disease severity to antibody response and virus recovered. *Yale J. Biol. Med.* **42**:311–328.

55. Halstead, S.B. 1970. Observations related to pathogenesis of dengue hemorrhagic fever. VI. Hypotheses and discussion. *Yale J. Biol. Med.* **42**:350–362.

56. Guzman, M.G., et al. 1990. Dengue hemorrhagic fever in Cuba, 1981: a retrospective seroepidemiologic study. *Am. J. Trop. Med. Hyg.* **42**:179–184.

57. Sangkawibha, N., et al. 1984. Risk factors for dengue shock syndrome: a prospective epidemiologic study in Rayong, Thailand. I. The 1980 outbreak. *Am. J. Epidemiol.* **120**:653–669.

58. Burke, D.S., Nisalak, A., Johnson, D.E., and Scott, R.M. 1988. A prospective study of dengue infections in Bangkok. *Am. J. Trop. Med. Hyg.* **38**:172–180.

59. Thein, S., et al. 1997. Risk factors in dengue shock syndrome. *Am. J. Trop. Med. Hyg.* **56**:566–572.

60. Halstead, S.B. 1979. In vivo enhancement of dengue virus infection in rhesus monkeys by passively transferred antibody. *J. Infect. Dis.* **140**:527–533.

61. Libraty, D.H., et al. 2002. Differing influences of viral burden and immune activation on disease severity in secondary dengue 3 virus infections. *J. Infect. Dis.* **185**:1213–1221.

62. Murgue, B., Roche, C., Chungue, E., and Deparis, X. 2000. Prospective study of the duration and magnitude of viraemia in children hospitalised during the 1996-1997 dengue-2 outbreak in French Polynesia. *J. Med. Virol.* **60**:432–438.

63. Kliks, S.C., Nimmanitya, S., Nisalak, A., and Burke, D.S. 1988. Evidence that maternal dengue antibodies are important in the development of dengue hemorrhagic fever in infants. *Am. J. Trop. Med. Hyg.* **38**:411–419.

64. Sudiro, T.M., et al. 2001. Analysis of plasma viral RNA levels during acute dengue virus infection using quantitative competitor reverse transcription-polymerase chain reaction. *J. Med. Virol.* **63**:29–34.

65. Vaughn, D.W., et al. 1997. Dengue in the early febrile phase: viremia and antibody responses. *J. Infect. Dis.* **176**:322–330.

66. Theofilopoulos, A.N., Wilson, C.B., and Dixon, F.J. 1976. The Raji cell radioimmune assay for detecting immune complexes in human sera. *J. Clin. Invest.* **57**:169–182.

67. Chungue, E., et al. 1994. Correlation between detection of plasminogen cross-reactive antibodies and hemorrhage in dengue virus infection. *J. Infect. Dis.* **170**:1304–1307.

68. Kurane, I., et al. 1991. Activation of T lymphocytes in dengue virus infections. High levels of soluble interleukin 2 receptor, soluble CD4, soluble CD8, interleukin 2, and interferon-gamma in sera of children with dengue. *J. Clin. Invest.* **88**:1473–1480.

69. Green, S., et al. 1999. Early immune activation in acute dengue is related to development of plasma leakage and disease severity. *J. Infect. Dis.* **179**:755–762.

70. Hober, D., Delannoy, A.S., Benyoucef, S., De Groote, D., and Wattre, P. 1996. High levels of sTNFR p75 and TNF alpha in dengue-infected patients. *Microbiol. Immunol.* **40**:569–573.

71. Bethell, D.B., et al. 1998. Pathophysiologic and prognostic role of cytokines in dengue hemorrhagic fever. *J. Infect. Dis.* **177**:778–782.

72. Green, S., et al. 1999. Early CD69 expression on

peripheral blood lymphocytes from children with dengue hemorrhagic fever. *J. Infect. Dis.* **180**:1429–1435.

73. Zivna, I., et al. 2002. T cell responses to an HLA B*07-restricted epitope on the dengue NS3 protein correlate with disease severity. *J. Immunol.* **168**:5959–5965.

74. Mongkolsapaya, J., et al. 2003. Original antigenic sin and apoptosis in the pathogenesis of dengue hemorrhagic fever. *Nat. Med.* **9**:921–927.

75. Green, S., et al. 1999. Elevated plasma interleukin-10 levels in acute dengue correlate with disease severity. *J. Med. Virol.* **59**:329–334.

76. Mustafa, A.S., Elbishbishi, E.A., Agarwal, R., and Chaturvedi, U.C. 2001. Elevated levels of interleukin-13 and IL-18 in patients with dengue hemorrhagic fever. *FEMS Immunol. Med. Microbiol.* **30**:229–233.

77. Hober, D., et al. 1993. Serum levels of tumor necrosis factor-α (TNF-α), interleukin-6 (IL-6), and interleukin-1β (IL-1β) in dengue-infected patients. *Am. J. Trop. Med. Hyg.* **48**:324–331.

78. Halstead, S.B., Rojanasuphot, S., and Sangkawibha, N. 1983. Original antigenic sin in dengue. *Am. J. Trop. Med. Hyg.* **32**:154–156.

79. Evavold, B.D., Sloan-Lancaster, J., and Allen, P.M. 1993. Tickling the TCR: selective T-cell functions stimulated by altered peptide ligands. *Immunol. Today* **14**:602–609.

80. Sloan-Lancaster, J., and Allen, P.M. 1996. Altered peptide ligand-induced partial T cell activation: molecular mechanisms and role in T cell biology. *Annu. Rev. Immunol.* **14**:1–27.

81. Zivny, J., et al. 1999. Partial agonist effect influences the CTL response to a heterologous dengue virus serotype. *J. Immunol.* **163**:2754–2760.

82. Selin, L.K., Varga, S.M., Wong, I.C., and Welsh, R.M. 1998. Protective heterologous antiviral immunity and enhanced immunopathogenesis mediated by memory T cell populations. *J. Exp. Med.* **188**:1705–1715.

83. Chen, H.D., Fraire, A.E., Joris, I., Welsh, R.M., and Selin, L.K. 2003. Specific history of heterologous virus infections determines anti-viral immunity and immunopathology in the lung. *Am. J. Pathol.* **163**:1341–1355.

84. Morens, D.M. 1994. Antibody-dependent enhancement of infection and the pathogenesis of viral disease. *Clin. Infect. Dis.* **19**:500–512.

85. Bokisch, V.A., Top, F.H., Jr., Russell, P.K., Dixon, F.J., and Muller-Eberhard, H.J. 1973. The potential pathogenic role of complement in dengue hemorrhagic shock syndrome. *N. Engl. J. Med.* **289**:996–1000.

86. Malasit, P. 1987. Complement and dengue haemorrhagic fever/shock syndrome. *SE Asian J. Trop. Med. Pub. Health.* **18**:316–320.

87. Falconar, A.K.I. 1997. The dengue virus nonstructural-1 protein (NS1) generates antibodies to common epitopes on human blood clotting, integrin/ adhesion proteins and binds to human endothelial cells: potential implications in haemorrhagic fever pathogenesis. *Arch. Virol.* **142**:897–916.

88. Markoff, L.J., Innis, B.L., Houghten, R., and Henchal, L.S. 1991. Development of cross-reactive antibodies during the immune response to dengue virus infection. *J. Infect. Dis.* **164**:294–301.

West Nile virus: a growing concern?

L. Hannah Gould and Erol Fikrig

Department of Epidemiology and Public Health and Section of Rheumatology, and Department of Internal Medicine,
Yale University School of Medicine, New Haven, Connecticut, USA.

West Nile virus was first detected in North America in 1999 and has subsequently spread throughout the United States and Canada and into Mexico and the Caribbean. This review describes the epidemiology and ecology of West Nile virus in North America and the prospects for effective treatments and vaccines.

Introduction and epidemiology

West Nile virus (WNV), a member of the family *Flaviviridae*, genus *Flavivirus*, was first isolated from the serum of a febrile woman in 1937 in the West Nile district of Uganda (1). Following its original isolation, WNV was implicated as the cause of sporadic outbreaks of mild viral illness in Africa, the Middle East, western Asia, and Australia (Kunjin virus subtype) (2, 3). Since the 1990s, however, WNV outbreaks have also occurred in Europe and North America, and these recent outbreaks have been associated with higher rates of viral encephalitis and other neurological symptoms (2–4).

In August and September 1999, a cluster of encephalitis cases caused by WNV was identified in New York City (5, 6). Previously unidentified in North America, the WNV epidemic in 1999 was responsible for seven human fatalities, as well as the death of hundreds of birds and horses in New York, New Jersey, and Connecticut. Although the virus spread westward during the next two years, only modest disease activity was seen until 2002, when the number of cases increased dramatically, and by the end of 2003 the epizootic had spread to all but two of the lower 48 states (7, 8). The number of cases of WNV has also continued to rise (4); by mid-February 2004, 9175 human cases and 230 deaths were reported as a result of the 2003 outbreak (7, 9).

The range of WNV in the Western Hemisphere has continued to expand, and in addition to the United States, virus activity has been detected in Canada, Mexico, and the West Indies (Figure 1). In Canada, WNV was first isolated from birds in 2001 and humans in 2002, and by the end of 2003, disease activity had spread to a total of nine provinces and territories (10). In 2002, WNV-neutralizing antibodies were detected in samples from birds captured in Jamaica, the Dominican Republic, and Guadeloupe (11–13), and horses in Mexico (14), suggesting establishment of the virus in the neotropics. Although only one human case has been reported from the Caribbean (15), virus activity in birds and horses suggests that human cases may soon be regularly detected in these areas. In the United States, human infections occur from May to December, with 85% of cases occurring in August and September. In the neotropics, however, transmission is likely to occur year-round and WNV may pose a significant public health problem in future years. Additionally, because the Caribbean region serves as the wintering ground for many migratory birds, presence of the virus in this region is likely to contribute to the maintenance of virus in its avian reservoirs.

Nonstandard abbreviations used: cerebrospinal fluid (CSF); envelope (E); intravenous immunoglobulin (IVIG); Japanese Encephalitis virus (JEV); premembrane (prM); Saint Louis Encephalitis virus (SLEV); West Nile virus (WNV).

Conflict of interest: The authors have declared that no conflict of interest exists.

Citation for this article: *J. Clin. Invest.* **113**:1102–1107 (2004). doi:10.1172/JCI200421623.

WNV is classified into two lineages. Lineage 1 viruses are responsible for human disease while lineage 2 viruses are found primarily in enzootic cycles in Africa and typically do not cause severe, neurologic illness in humans, although significant outbreaks of lineage 2 WNV have occurred (16). The lineage 1 virus found in North America is almost genetically identical to a strain circulating in Israel from 1997–2000, suggesting a potential origin of the North American epidemic (6, 17, 18). Since its importation the virus has undergone only minor genetic evolution, suggesting a lack of strong selective pressure (19); however, the virus may change more rapidly as immune responses to WNV become more common in reservoir hosts. While it is still unknown how WNV first reached North America, a number of theories have been advanced, including introduction by a viremic human or bird, or the accidental importation of a WNV-infected mosquito. Because humans infected with WNV have only a low and transient viremia, it is unlikely that WNV was introduced by an infected person. Although it is possible that a migratory bird blown off its flight path introduced the virus, this is also doubtful, as the bird would need to have maintained a high viremia over a considerable period of time. Evidence of direct transmission within some birds suggests that the introduction of a bird with a low viremia could potentially have been sufficient to pass and amplify the virus for subsequent transmission from mosquito to humans.

Ecology

WNV is maintained in an enzootic bird-mosquito-bird cycle (Figure 2). In Africa, southern Europe, and western Asia, WNV has been isolated from mosquitoes of more than 40 species, primarily those in the genus *Culex* (3); in the United States, WNV has been isolated from mosquitoes belonging to 43 species since 1999 (9, 15). In the United States the majority of WNV isolates have been from *Culex* species, particularly *Cx. pipiens, Cx. restuans*, and *Cx. salinarius*; however, isolates have also been recovered from species in other genera, including *Aedes, Anopheles, Coquillettidia*, and *Ochlerotatus* (15, 20). *Culex* species are the most important maintenance vectors within the avian cycle, with other species serving as bridge vectors from birds to humans and horses in mid to late summer. While mosquitoes belonging to many species are competent vectors of the virus in the laboratory, their vector competence in humans is still unknown. In the northeastern and southern US transmission among birds is likely mediated by *Cx. pipiens*; however, the most important bridge vectors to humans are unidentified. The large outbreak of WNV in the western US, especially Colorado, in 2003 was likely due to *Cx. tarsalis*, an indiscriminant feeder on both birds and mammals. In regions where the same vector contributes to both the mammalian and the avian

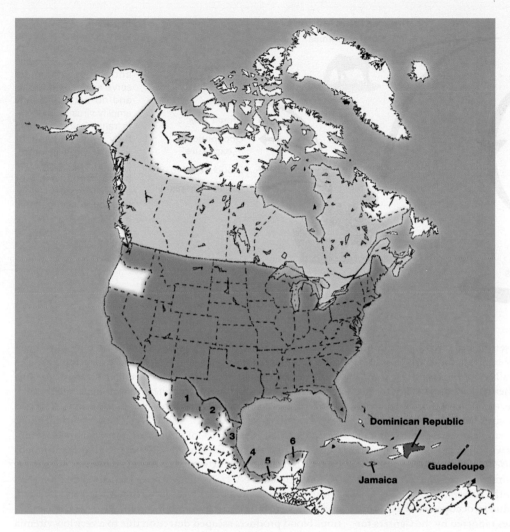

Figure 1
WNV activity in North America as of January 2004. Shading represents areas in which virus activity has been recorded. 1, Chihuahua; 2, Coahuila; 3, Tamaulipas; 4, Veracruz; 5, Tabasco; 6, Yucatan.

cycles, disease activity will continue to be more pronounced. WNV has also been isolated from both hard and soft ticks; however, the ability of ticks to successfully and significantly transmit the virus in nature is unclear (21, 22).

In the United States, Canada, and Israel, WNV is responsible for significant avian mortality. WNV has been isolated from 198 bird species in North America (23), and mortality may approach 100% in some species (24). Passerine birds, including crows, house sparrows, and blue jays, serve as the primary amplifying hosts of the virus, and develop a high-level viremia that lasts for several days (24). A study of WNV transmission in 25 species of birds found cloacal shedding of virus in 17 of 24 species and oral shedding in 12 of 14 species (24). In addition, contact transmission was identified in four species, and oral transmission in five species (24). Although the precise contribution of direct transmission to disease activity among birds has not been quantified, it has potentially significant implications for disease epidemiology, because even with effective vector control virus may be amplified and transmitted. Direct transmission may be aggravated in commercial settings because of cannibalism and feather-picking of sick birds (25).

In contrast to some bird species, in which high-level viremia is seen, humans and horses develop only low-level and transient viremia and are unlikely to contribute to virus amplification (26, 27).

WNV infection has also been demonstrated in a number of other wild vertebrate species, including wolves, bears, crocodiles, alligators, and bats (28–31), and in domestic animals, including cats and dogs (28, 32). Because of the low viremia in many of these species, it is similarly unlikely that they contribute to the amplification of the virus, but instead serve as dead-end hosts.

Routes of transmission

Most human infections of WNV are the result of transmission of the virus by infected mosquitoes; however, several novel modes of transmission are now recognized in the United States (Figure 2). Taken together, these new modes of transmission suggest a complex epidemiology of this disease, and have important implications for the control and containment of WNV infection. Because of the low incidence of WNV worldwide until recently, it is unclear whether these novel modes of transmission have always been present but rare or if the epidemiology of the virus is truly changing. As WNV continues to expand, an awareness of these additional, nonvector modes of transmission are crucial to effective surveillance and disease control.

The first case of intrauterine WNV transmission was reported in 2002. A pregnant woman in her second trimester was infected with WNV and subsequently transmitted the virus to her fetus in utero. The infant, delivered at term, had bilateral chorioretinitis,

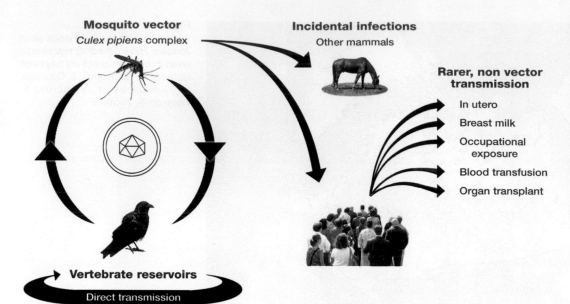

Figure 2
WNV is maintained in an enzootic mosquito-bird-mosquito cycle. Humans and other mammals serve as dead-end hosts and do not sufficiently amplify virus for mosquito transmission, although they may transmit or acquire virus in utero, through breast milk, via blood transfusion or organ transplantation, or through occupational exposure.

severe cerebral abnormalities, and serum and cerebrospinal fluid (CSF) WNV-specific IgM and WNV-neutralizing antibodies (33). Mother-to-child transmission of WNV via breast milk has also been identified; the lactating mother had acquired WNV from a blood transfusion (34). Intrauterine infection with either Japanese Encephalitis virus (JEV) or a dengue virus has been associated with spontaneous abortion and severe dengue infection in the infant (35, 36). Pregnant women in regions with high risk of WNV should take precautions to reduce their disease risk; however, given the low incidence of intrauterine transmission, routine screening of pregnant women is not indicated at the current time.

Several cases of WNV infection due to occupational exposure have been documented. In two cases reported by the Centers for Disease Control and Prevention in 2002, exposure of laboratory workers occurred through accidental inoculation (37). In one of these cases the infected individual possessed heterologous *Flavivirus* antibodies; however, these did not confer protection against infection with WNV, although the severity of the illness may have been reduced (37).

In late 2002, two cases of WNV were reported in workers at a turkey-breeding farm in Wisconsin (38). A survey of workers at this farm revealed a higher prevalence of WNV antibodies and incidence of febrile illness than in workers at several other local turkey farms and meat processing facilities or in the local citizens who lived near the affected farm. In addition, WNV seroprevalence was nearly 100% among birds at this facility. Because turkeys do not develop sufficient viremia to serve as amplifying hosts for WNV transmission by mosquitoes, it is likely that nonvector transmission was responsible for this outbreak (39). While the mode of transmission to the workers remains unclear, this outbreak was likely due to an occupational exposure, perhaps due to percutaneous injury or aerosol exposure to turkey feces. This finding has important implications for individuals who work in areas with high concentrations of potentially infected animals, and precautions beyond those recommended to prevent mosquito bites should be widely implemented.

In 2002, 23 people were reported to have acquired WNV from blood components collected from 16 WNV-viremic blood donors,

and as many as 500 WNV-positive donations may have been collected. In response to these cases, blood collection agencies implemented nucleic acid amplification tests to screen donations for WNV in 2003. Between late June and mid-September 2003, 0.05% of approximately 2.5 million blood donations screened tested positive for WNV and 489 WNV-viremic blood donors were identified. Nearly 90% of viremic donors remained asymptomatic and only two developed WNV-associated meningoencephalitis. Despite this extensive screening, two cases of transfusion-associated WNV transmission were identified in 2003 (40). The infectious donors in both cases were identified and retested, and both were found to have seroconverted. Whether these infectious blood products escaped detection due to a very low viremia or test efficacy is unknown (41). In the future, blood collection agencies should consider additional methods, such as inactivation of the pathogen in blood products, as the risk of WNV infection from blood transfusion remains.

In August 2002, WNV was identified in an organ donor and four transplant recipients. The organ donor had received blood components from 63 individuals, one of whom was subsequently identified as WNV-seropositive. Three of the organ recipients developed encephalitis and one developed a febrile illness (42). Several additional cases of WNV transmission by organ transplantation have subsequently been reported (43, 44).

Prevention of WNV transmission by organ transplantation or transfusion of blood components relies on the accurate exclusion of viremic donors. Nucleic acid tests were introduced in blood donor screening in 2003, and may prove useful for routine diagnosis in the future, especially in immunocompromised patients with impaired antibody development. As seen in 2003, screening of blood donors can significantly reduce the incidence of transfusion of contaminated blood; however, more sensitive tests are needed to identify potential donors with very low levels of viremia. Routine screening of the blood supply and of organ donors is probably too costly to be warranted; however, during the summer months in locations with a high level of virus activity, such measures should be encouraged and are necessary to reduce transmission to already immunocompromised individuals.

Virus detection

Laboratory findings show mostly normal or elevated leukocyte counts in peripheral blood samples (45). CSF samples of patients with neurologic manifestations show pleocytosis, usually with a predominance of lymphocytes and elevated protein levels (45–47). Computerized axial tomography scans of the brain tend to reveal no evidence of acute disease; magnetic resonance imaging findings from most patients are normal but can show lesions in the pons, basal ganglia, and thalamus as well as enhancement of the leptomeninges and/or the periventricular areas (46, 47).

The presence of IgM antibody in the serum or CSF is the most efficient and reliable indicator of infection. The IgM antibody-capture ELISA (MAC-ELISA) test is considered the most reliable (45, 46). IgM antibody does not readily cross the blood-brain barrier, so its presence in the CSF indicates infection of the CNS. More than 90% of patients with meningoencephalitis have IgM antibody in the CSF within 8 days of symptom onset (48).

Because of the close antigenic relationships among some of the flaviviruses, laboratory findings must be interpreted with care. Persons recently vaccinated against or infected with another *Flavivirus* may have IgM antibody to WNV. While tests exist that help distinguish infections, the persistence of IgM antibody for longer than 500 days in a small proportion of patients, the fact that most cases are asymptomatic, and the presence of IgM antibody may reflect a previous infection with an unrelated *Flavivirus* or vaccination against a *Flavivirus*, such as Yellow Fever virus (46, 49). Two newly developed and highly sensitive assays, based on nonstructural protein 5 (50) and a recombinant form of the envelope (E) protein (51), reliably distinguish between infections with WNV, dengue viruses, and SLEV, differentiate between immunity from vaccination and natural infection, indicate recent infection status, and may shorten testing time to less than three hours (50, 51).

Treatment and vaccine prospects

Most infections with WNV are clinically inapparent and go undetected. A serosurvey in 1999 in New York showed that only approximately 20% of infected persons developed fever caused by WNV, and of these, only about half visited a physician for their illness (52). Approximately 1 in 150 patients progress to severe neurologic illnesses (encephalitis, meningitis, acute flaccid paralysis), and while prevalence rates are fairly uniform across age-groups, rates of neurological disease increase substantially with age, as does the clinical:subclinical infection ratio (53, 54). Case-fatality rates among hospitalized patients have ranged from 4% (Romania, 1996) to 12% in the New York 1999 outbreak and as high as 14% in an outbreak in Israel in 2000. Advanced age is the main risk factor for death, with individuals more than 70 years of age at particularly high risk. Among such individuals, the case-fatality rate ranges from 15% to 29% (45). The reason for increased mortality in the elderly is not yet known but may be related to a decreased capacity of these individuals to develop a protective immune response to help control infection. Clearly, in mice, select populations of immune cells including B cells (55), γδ T cells (56), and CD8+ T cells (57) are important in both immunity against and the pathogenesis of WNV. Future studies in humans, guided in part by experimental work in animal models, will determine whether selected immune responses to WNV are decreased in the elderly, leading to new strategies to treat disease.

Currently, the only treatments for WNV infection are supportive. In vitro studies have found ribavirin and IFN-α2b to be effective against the virus; however, a patient treated with both agents did not improve, and there have been no controlled clinical trials of either agent (58–60). Because the induction of a WNV-specific, neutralizing response early in the course of murine WNV infection limits viremia and dissemination into the CNS, antibodies may be effective as both prophylaxis and therapy for WNV infection, particularly in the elderly, who have decreased antibody production and limited response following vaccination (61). In mice administered human gamma globulin prior to infection, protection against disease was afforded and clinical outcome improved even after the virus had spread to the CNS (62, 63). Most significantly, antibody-dependent enhancement of infection — a phenomenon that is associated with severe hemorrhagic syndrome in secondary infections with dengue viruses — was not observed (64).

Several human case reports indicate that treatment with intravenous immunoglobulin (IVIG) may aid in recovery from infection (65–67). A number of questions must be resolved to determine the efficacy of IVIG as therapy in humans. Of critical importance is the window in which passive therapy is effective. In animal models, the use of IVIG appears effective before or immediately after viral challenge; however, once cerebral infection occurs, use is limited (68). Because the precise timing of infection in a human being is usually undetermined and most individuals do not present to their clinician prior to severe illness, administration of antibodies may provide limited use therapeutically. As prophylaxis, however, antibodies could prove useful for individuals at high risk of infection (69). Hamsters administered immunoglobulin 24 hours prior to infection were completely protected from infection (70), indicating that passive immunization may serve as a vaccine strategy for short-term, immediate exposures in individuals at high risk. In addition, the dose of antibody required to boost the immune response must be addressed. Given the lack of current treatments for WNV and the paucity of case reports supporting the utility of antibody therapy, a controlled clinical trial of immunoglobulin therapy is warranted to determine dose, timing, and efficacy more completely.

A variety of WNV vaccine candidates are in various stages of testing. Because of the low incidence of disease in humans and the sporadic nature of most outbreaks, it is difficult to target human populations for vaccination and to assess the economic feasibility of a human vaccine. An equine vaccine has been in use since 2001, and was licensed by the United States Department of Agriculture in 2003. This vaccine is a formalin-inactivated WNV, and was found to prevent development of viremia in 94% of immunized horses (71). Killed vaccines, while safe, may have to be administered in multiple doses to elicit and sustain an immune response, and the manufacturer of the commercially available vaccine recommends annual revaccination. A live attenuated WNV/dengue virus serotype 4 chimera also produced a strong neutralizing antibody response and prevented viremia in monkeys challenged with WNV (72). Two live attenuated virus variants have been used successfully to prevent infection in geese (73).

Several vaccines have been developed based on WNV structural proteins. As the predominant surface glycoprotein and the primary target of WNV-neutralizing antibodies (74, 75), the E protein is the most likely candidate for a successful vaccine. An experimental chimeric WNV vaccine candidate containing the premembrane (prM) and E genes from wild-type New York 1999 virus within the Yellow Fever 17D vaccine virus (76, 77) elicits a strong and potentially long-lasting humoral immune response in a hamster model (78). Additional vaccine candidates include recombinant DNA vaccines expressing the prM and E (79) or capsid proteins (80), and a recombinant E protein subunit vaccine (81). While it is likely that several of these vaccines will be effective, the benefits and risks of vaccination remain to be determined. In general, live attenuated vaccines often have a higher perceived risk than recombinant antigens, and may not receive enthusiastic public support. As most people in areas of virus activity are exposed to mosquitoes, the vaccine could be made widely available; however, since the risk of acquiring WNV is relatively low, this does not seem warranted at the present time. Perhaps a vaccine that targets elderly and immunocompromised individuals in high-risk areas would be an appropriate initial strategy.

Several studies have assessed the efficacy of heterologous *Flavivirus* vaccines against WNV. Partial cross-protective immunity has been generated in hamsters and mice using JEV, Yellow Fever virus, Saint Louis Encephalitis virus (SLEV), and dengue virus vaccines (70, 82); however, a study of human volunteers showed no WNV-neutralizing antibodies after vaccination against either JEV or a dengue virus (83). While vaccination with JEV or Yellow Fever virus may not prevent infection with WNV, such vaccination may limit disease severity and progression. Vaccination with the WNV subtype Kunjin virus–based DNA vaccine protected mice against WNV infection, and may serve as an effective vaccination strategy because of the close antigenic relationship between the Kunjin subtype and the strain infecting North America (84). Likewise, it is theoretically possible that a WNV vaccine could afford partial protection against other flaviviral infections.

Future directions

WNV was initially diagnosed as SLEV when it appeared in New York in 1999. In North America these two viruses share similar mosquito vectors and avian hosts, and may share a common epidemiologic pattern. SLEV is the cause of sporadic outbreaks that are difficult to predict and which cause a highly variable number of cases, ranging from a handful to more than 2000 in the 1975 epidemic (46, 85). While there are many similarities between these two viruses, the sustained levels of WNV activity of the past few years combined with the complex epidemiology of transmission, the high levels of viremia seen in wild reservoirs, and the large number of species of mosquitoes that have been found infected with this virus, suggest that WNV will be a greater challenge for the public health system and clinicians than SLEV. As WNV continues to spread in the Western Hemisphere, it is likely that transmission will become a year-round, constant presence. The need for effective vaccines and treatments is increasing in importance as this disease continues to expand its range and increase in severity.

Address correspondence to: Erol Fikrig, Yale University School of Medicine, S525 TAC, 300 Cedar Street, New Haven, Connecticut 06520, USA. Phone: (203) 785-2453; Fax: (203) 785-7053; E-mail: erol.fikrig@yale.edu.

1. Smithburn, J.S., Hughes, T.P., Burke, A.W., and Paul, J.H. 1970. A neurotropic virus isolated from the blood of a native of Uganda. *Am. J. Trop. Med. Hyg.* **20**:471–492.

2. Monath, T.P. 1990. Flaviviruses. In *Virology.* B.N. Fields and D.M. Knipe, editors. Raven Press. New York, New York, USA. 763–814.

3. Hayes, C.G. 1989. West Nile fever. In *The arboviruses: epidemiology and ecology.* T.P. Monath, editor. CRC Press. Boca Raton, Florida, USA. 59–88.

4. Marfin, A.A., and Gubler, D.J. 2001. West Nile encephalitis: an emerging disease in the United States. *Clin. Infect. Dis.* **33**:1713–1719.

5. Anderson, J.F., et al. 1999. Isolation of West Nile virus from mosquitoes, crows, and a Cooper's hawk in Connecticut. *Science.* **286**:2331–2333.

6. Lanciotti, R.S., et al. 1999. Origin of the West Nile virus responsible for an outbreak of encephalitis in the northeastern United States. *Science.* **286**:2333–2337.

7. CDC. 2003. West Nile virus activity — United States, November 20–25, 2003. *Morb. Mortal. Wkly Rep.* **52**:1160.

8. CDC. 2002. Provisional surveillance summary of the West Nile virus epidemic — United States, January–November 2002. *Morb. Mortal. Wkly Rep.* **51**:1129–1133.

9. CDC. Division of vector-borne infectious diseases. West Nile virus. http://www.cdc.gov/ncidod/dvbid/westnile/.

10. Health Canada. West Nile virus. http://www.hc-sc.gc.ca/english/westnile/.

11. Dupuis, A.P., II, Marra, P.P., and Kramer, L.D. 2003. Serologic evidence of West Nile virus transmission, Jamaica, West Indies. *Emerg. Infect. Dis.* **9**:860–863.

12. Komar, O., et al. 2003. West Nile virus transmission in resident birds, Dominican Republic. *Emerg. Infect. Dis.* **9**:1299–1302.

13. Quirin, R., et al. 2004. West Nile Virus, Guadeloupe. *Emerg. Infect. Dis.* http://www.cdc.gov/ncidod/EID/vol10no4/03-0465.htm

14. Lorono-Pino, M.A., et al. 2003. Serologic evidence of West Nile virus infection in horses, Yucatan State, Mexico. *Emerg. Infect. Dis.* **9**:857–859.

15. 2002. West Nile Virus activity — United States, 2001. *Morb. Mortal. Wkly Rep.* **51**:497–501.

16. Jupp, P.G. 2001. The ecology of West Nile virus in South Africa and the occurrence of outbreaks in humans. *Ann. N. Y. Acad. Sci.* **951**:143–152.

17. Briese, T., Jia, X.Y., Huang, C., Grady, L.J., and Lipkin, W.I. 1999. Identification of a Kunjin/West Nile-like flavivirus in brains of patients with New York encephalitis. *Lancet.* **354**:1261–1262.

18. Jia, X.Y., et al. 1999. Genetic analysis of West Nile New York 1999 encephalitis virus. *Lancet.* **354**:1971–1972.

19. Beasley, D.W., et al. 2003. Limited evolution of West Nile virus has occurred during its southwesterly spread in the United States. *Virology.* **309**:190–195.

20. Turell, M.J., Sardelis, M.R., Dohm, D.J., and O'Guinn, M.L. 2001. Potential North American vectors of West Nile virus. *Ann. N. Y. Acad. Sci.* **951**:317–324.

21. Abbassy, M.M., Osman, M., and Marzouk, A.S. 1993. West Nile virus (Flaviviridae:Flavivirus) in experimentally infected Argas ticks (Acari:Argasidae). *Am. J. Trop. Med. Hyg.* **48**:726–737.

22. Anderson, J.F., Main, A.J., Andreadis, T.G., Wikel, S.K., and Vossbrinck, C.R. 2003. Transstadial transfer of West Nile virus by three species of ixodid ticks (Acari: Ixodidae). *J. Med. Entomol.* **40**:528–533.

23. Komar, N. 2003. West Nile virus: epidemiology and ecology in North America. *Adv. Virus Res.* **61**:185–234.

24. Komar, N., et al. 2003. Experimental infection of North American birds with the New York 1999 strain of West Nile virus. *Emerg. Infect. Dis.* **9**:311–322.

25. Banet-Noach, C., Simanov, L., and Malkinson, M. 2003. Direct (non-vector) transmission of West Nile virus in geese. *Avian Pathol.* **32**:489–494.

26. Southam, C.M., and Moore, A.E. 1954. Induced virus infections in man by the Egypt isolates of West Nile virus. *Am. J. Trop. Med. Hyg.* **3**:19–50.

27. Bunning, M.L., et al. 2001. Experimental infection of horses with West Nile virus and their potential to infect mosquitoes and serve as amplifying hosts. *Ann. N. Y. Acad. Sci.* **951**:338–339.

28. 2000. Update: West Nile virus activity — eastern United States, 2000. *Morb. Mortal. Wkly Rep.* **49**:1044–1047.

29. Farajollahi, A., et al. 2003. Serologic evidence of West Nile virus infection in black bears (Ursus americanus) from New Jersey. *J. Wildl. Dis.* **39**:894–896.

30. Steinman, A., et al. 2003. West Nile virus infection in crocodiles. *Emerg. Infect. Dis.* **9**:887–889.

31. Miller, D.L., et al. 2003. West Nile virus in farmed alligators. *Emerg. Infect. Dis.* **9**:794–799.

32. Blackburn, N.K., Reyers, F., Berry, W.L., and Shepherd, A.J. 1989. Susceptibility of dogs to West Nile virus: a survey and pathogenicity trial. *J. Comp. Pathol.* **100**:59–66.

33. 2002. Intrauterine West Nile virus infection — New York, 2002. *Morb. Mortal. Wkly Rep.* **51**:1135-1136.

34. 2002. Possible West Nile virus transmission to an infant through breast-feeding — Michigan, 2002. *Morb. Mortal. Wkly Rep.* **51**:877–878.

35. Chaturvedi, U.C., et al. 1980. Transplacental infection with Japanese encephalitis virus. *J. Infect. Dis.* **141**:712–715.

36. Chye, J.K., et al. 1997. Vertical transmission of dengue. *Clin. Infect. Dis.* **25**:1374–1377.

37. 2002. Laboratory-acquired West Nile virus infections — United States, 2002. *Morb. Mortal. Wkly Rep.* **51**:1133–1135.

38. 2003. West Nile virus infection among turkey breeder farm workers — Wisconsin, 2002. *Morb. Mortal. Wkly Rep.* **52**:1017–1019.

39. Swayne, D.E., Beck, J.R., and Zaki, S. 2000. Pathogenicity of West Nile virus for turkeys. *Avian Dis.* **44**:932–937.

40. 2003. Update: Detection of West Nile virus in blood donations — United States, 2003. *Morb. Mortal. Wkly Rep.* **52**:916–919.

41. 2003. Detection of West Nile virus in blood donations — United States, 2003. *Morb. Mortal. Wkly Rep.* **52**:769–772.

42. Iwamoto, M., et al. 2003. Transmission of West Nile virus from an organ donor to four transplant recipients. *N. Engl. J. Med.* **348**:2196–2203.

43. DeSalvo, D., et al. 2004. West Nile virus encephalitis in organ transplant recipients: another high-risk group for meningoencephalitis and death. *Transplantation.* **77**:466–469.

44. Kumar, D., Prasad, G.V., Zaltzman, J., Levy, G.A., and Humar, A. 2004. Community-acquired West Nile virus infection in solid-organ transplant recipients. *Transplantation.* **77**:399–402.

45. Petersen, L.R., and Marfin, A.A. 2002. West Nile virus: a primer for the clinician. *Ann. Intern. Med.* **137**:173–179.

46. Campbell, G.L., Marfin, A.A., Lanciotti, R.S., and Gubler, D.J. 2002. West Nile virus. *Lancet Infect. Dis.* **2**:519–529.

47. Sejvar, J.J., et al. 2003. Neurologic manifestations and outcome of West Nile virus infection. *JAMA.* **290**:511–515.

48. Petersen, L.R., Marfin, A.A., and Gubler, D.J. 2003. West Nile virus. *JAMA.* **290**:524–528.

49. Roehrig, J.T., et al. 2003. Persistence of virus-reactive serum immunoglobulin m antibody in confirmed West Nile virus encephalitis cases. *Emerg. Infect. Dis.* **9**:376–379.

50. Wong, S.J., et al. 2003. Immunoassay targeting nonstructural protein 5 to differentiate West Nile virus infection from dengue and St. Louis encephalitis virus infections and from flavivirus vaccination. *J. Clin. Microbiol.* **41**:4217–4223.

51. Wong, S.J., et al. 2004. Detection of human anti-flavivirus antibodies with a West Nile virus recombinant antigen microsphere immunoassay. *J. Clin. Microbiol.* **42**:65–72.

52. Mostashari, F., et al. 2001. Epidemic West Nile encephalitis, New York, 1999: results of a household-based seroepidemiological survey. *Lancet.* **358**:261–264.

53. Nash, D., et al. 2001. The outbreak of West Nile virus infection in the New York City area in 1999. *N. Engl. J. Med.* **344**:1807–1814.

54. Tsai, T.F., Popovici, F., Cernescu, C., Campbell, G.L., and Nedelcu, N.I. 1998. West Nile encephalitis epidemic in southeastern Romania. *Lancet.* **352**:767–771.

55. Diamond, M.S., Shrestha, B., Marri, A., Mahan, D., and Engle, M. 2003. B cells and antibody play critical roles in the immediate defense of disseminated infection by West Nile encephalitis virus. *J. Virol.* **77**:2578–2586.

56. Wang, T., et al. 2003. IFN-gamma-producing gamma delta T cells help control murine West Nile virus infection. *J. Immunol.* **171**:2524–2531.

57. Wang, Y., Lobigs, M., Lee, E., and Mullbacher, A. 2003. CD8+ T cells mediate recovery and immunopathology in West Nile virus encephalitis. *J. Virol.*

77:13323–13334.

58. Anderson, J.F., and Rahal, J.J. 2002. Efficacy of interferon alpha-2b and ribavirin against West Nile virus in vitro. *Emerg. Infect. Dis.* **8**:107–108.

59. Jordan, I., Briese, T., Fischer, N., Lau, J.Y., and Lipkin, W.I. 2000. Ribavirin inhibits West Nile virus replication and cytopathic effect in neural cells. *J. Infect. Dis.* **182**:1214–1217.

60. Weiss, D., et al. 2001. Clinical findings of West Nile virus infection in hospitalized patients, New York and New Jersey, 2000. *Emerg. Infect. Dis.* **7**:654–658.

61. Diamond, M.S., Shrestha, B., Mehlhop, E., Sitati, E., and Engle, M. 2003. Innate and adaptive immune responses determine protection against disseminated infection by West Nile encephalitis virus. *Viral Immunol.* **16**:259–278.

62. Ben-Nathan, D., et al. 2003. Prophylactic and therapeutic efficacy of human intravenous immunoglobulin in treating West Nile virus infection in mice. *J. Infect. Dis.* **188**:5–12.

63. Engle, M.J., and Diamond, M.S. 2003. Antibody prophylaxis and therapy against West Nile virus infection in wild-type and immunodeficient mice. *J. Virol.* **77**:12941–12949.

64. Morens, D.M. 1994. Antibody-dependent enhancement of infection and the pathogenesis of viral disease. *Clin. Infect. Dis.* **19**:500–512.

65. Shimoni, Z., Niven, M.J., Pitlick, S., and Bulvik, S. 2001. Treatment of West Nile virus encephalitis with intravenous immunoglobulin. *Emerg. Infect. Dis.* **7**:759.

66. Hamdan, A., et al. 2002. Possible benefit of intravenous immunoglobulin therapy in a lung transplant recipient with West Nile virus encephalitis. *Transpl. Infect. Dis.* **4**:160–162.

67. Haley, M., Retter, A.S., Fowler, D., Gea-Banacloche, J., and O'Grady, N.P. 2003. The role for intravenous immunoglobulin in the treatment of West Nile virus encephalitis. *Clin. Infect. Dis.* **37**:e88–e90.

68. Agrawal, A.G., and Petersen, L.R. 2003. Human immunoglobulin as a treatment for West Nile virus infection. *J. Infect. Dis.* **188**:1–4.

69. Roehrig, J.T., Staudinger, L.A., Hunt, A.R., Mathews, J.H., and Blair, C.D. 2001. Antibody prophylaxis and therapy for flavivirus encephalitis infections. *Ann. N. Y. Acad. Sci.* **951**:286–297.

70. Tesh, R.B., Travassos da Rosa, A.P., Guzman, H., Araujo, T.P., and Xiao, S.Y. 2002. Immunization with heterologous flaviviruses protective against fatal West Nile virus infection. *Emerg. Infect. Dis.* **8**:245–251.

71. Ng, T., et al. 2003. Equine vaccine for West Nile virus. *Dev. Biol.* **114**:221–227.

72. Pletnev, A.G., et al. 2003. Molecularly engineered live-attenuated chimeric West Nile/dengue virus vaccines protect rhesus monkeys from West Nile virus. *Virology.* **314**:190–195.

73. Lustig, S., et al. 2000. A live attenuated West Nile virus strain as a potential veterinary vaccine. *Viral Immunol.* **13**:401–410.

74. Chambers, T.J., Halevy, M., Nestorowicz, A., Rice, C.M., and Lustig, S. 1998. West Nile virus envelope proteins: nucleotide sequence analysis of strains differing in mouse neuroinvasiveness. *J. Gen. Virol.* **79**:2375–2380.

75. Konishi, E., Yamaoka, M., Khin-Sane, W., Kurane, I., and Mason, P.W. 1998. Induction of protective immunity against Japanese encephalitis in mice by immunization with a plasmid encoding Japanese encephalitis virus premembrane and envelope genes. *J. Virol.* **72**:4925–4930.

76. Arroyo, J., Miller, C.A., Catalan, J., and Monath, T.P. 2001. Yellow fever vector live-virus vaccines: West Nile virus vaccine development. *Trends Mol. Med.* **7**:350–354.

77. Monath, T.P., Arroyo, J., Miller, C., and Guirakhoo, F. 2001. West Nile virus vaccine. *Curr. Drug Targets Infect. Disord.* **1**:37–50.

78. Tesh, R.B., et al. 2002. Efficacy of killed virus vaccine, live attenuated chimeric virus vaccine, and passive immunization for prevention of West Nile virus encephalitis in hamster model. *Emerg. Infect. Dis.* **8**:1392–1397.

79. Davis, B.S., et al. 2001. West Nile virus recombinant DNA vaccine protects mouse and horse from virus challenge and expresses in vitro a noninfectious recombinant antigen that can be used in enzyme-linked immunosorbent assays. *J. Virol.* **75**:4040–4047.

80. Yang, J.S., et al. 2001. Induction of potent Th1-type immune responses from a novel DNA vaccine for West Nile virus New York isolate (WNV-NY1999). *J. Infect. Dis.* **184**:809–816. Epub 2001 Aug 2029.

81. Wang, T., et al. 2001. Immunization of mice against West Nile virus with recombinant envelope protein. *J. Immunol.* **167**:5273–5277.

82. Takasaki, T., et al. 2003. Partial protective effect of inactivated Japanese encephalitis vaccine on lethal West Nile virus infection in mice. *Vaccine.* **21**:4514–4518.

83. Kanesa-Thasan, N., Putnak, J.R., Mangiafico, J.A., Saluzzo, J.F., and Ludwig, G.V. 2002. Short report: absence of protective neutralizing antibodies to West Nile virus in subjects following vaccination with Japanese encephalitis or dengue vaccines. *Am. J. Trop. Med. Hyg.* **66**:115–116.

84. Hall, R.A., et al. 2003. DNA vaccine coding for the full-length infectious Kunjin virus RNA protects mice against the New York strain of West Nile virus. *Proc. Natl. Acad. Sci. U. S. A.* **100**:10460–10464.

85. Monath, T.P. 1980. Epidemiology. In *St. Louis Encephalitis.* T.P. Monath, editor. American Public Health Association. Washington, DC, USA. 239–312.

Progress on new vaccine strategies against chronic viral infections

Jay A. Berzofsky,[1] Jeffrey D. Ahlers,[2] John Janik,[3] John Morris,[3] SangKon Oh,[1] Masaki Terabe,[1] and Igor M. Belyakov[1]

[1]Molecular Immunogenetics and Vaccine Research Section, Vaccine Branch, The Center for Cancer Research, National Cancer Institute, [2]Division of AIDS, National Institute of Allergy and Infectious Diseases, and [3]Clinical Trials Team, Metabolism Branch, The Center for Cancer Research, National Cancer Institute, NIH, Bethesda, Maryland, USA.

Among the most cost-effective strategies for preventing viral infections, vaccines have proven effective primarily against viruses causing acute, self-limited infections. For these it has been sufficient for the vaccine to mimic the natural virus. However, viruses causing chronic infection do not elicit an immune response sufficient to clear the infection and, as a result, vaccines for these viruses must elicit more effective responses — quantitative and qualitative — than does the natural virus. Here we examine the immunologic and virologic basis for vaccines against three such viruses, HIV, hepatitis C virus, and human papillomavirus, and review progress in clinical trials to date. We also explore novel strategies for increasing the immunogenicity and efficacy of vaccines.

Vaccines have proven among the most cost-effective strategies for preventing infectious diseases — following only the provision of safe drinking water and sanitation. During the 20th century, vaccines for bacterial toxins and many common acute viral infections were developed and made widely available. Vaccines have changed the face of viral disease as much as antibiotics have affected the course of bacterial disease. They have been most successful in cases in which acute natural infection is self-limited and leads to long-lasting protective immunity if the patient survives the initial infection. In these cases, the best vaccine has usually been the one that most closely mimics the natural infection, such as a live, attenuated virus. Indeed, just this year, a new, live, attenuated influenza vaccine was licensed for intranasal aerosol administration (1).

However, development of a vaccine that is effective against viruses that cause chronic infections, such as HIV, hepatitis C virus (HCV), and human papillomavirus (HPV), may require consideration of a paradigm different from that described above. These viruses cause chronic infections with different frequencies; virtually 100% of cases of HIV infection, 55–85% of cases of HCV infection, and over 30% of cases of HPV result in chronic viral infection. In most of these cases, the immune response to the natural infection is not sufficient to eradicate the infection. Therefore, a vaccine that just mimics natural infection is not likely to be adequate to induce protection. Also, there is much concern about the use of live attenuated viruses for vaccination against these diseases. These viruses have evolved to escape or evade the immune system, not to act as an optimal vaccine. The challenge for the 21st century is to apply the latest fundamental knowledge in molecular biology, virology, and immunology to developing vaccines that are more effective at eliciting immunity than the natural infection and, consequently, effective against chronic viral and other infectious diseases in addition to cancer, which do not fit the classic paradigm.

Although advances in molecular biology have raised great hope for the development of new vaccine strategies and much effort has been invested in this endeavor, only one recombinant viral protein vaccine — a hepatitis B surface antigen vaccine — has been licensed to date, and that advance occurred about 17 years ago (2, 3). In the last 5–10 years, however, many new vaccine strategies have been designed based on substantial increases in fundamental knowledge of the immune system, and some of these vaccines have advanced to clinical trials. Most of these strategies are based on improved ways of inducing antibodies, which can prevent infection if present at high enough levels at the time of exposure, or inducing CTLs that can detect and destroy cells infected with virus and thereby control and ultimately clear infection. These CTLs can detect any viral protein made within the infected host cell even if it is not present on the cell surface. They are able to respond to peptide fragments of these proteins produced by proteasomal cleavage and transported to the endoplasmic reticulum. Here they bind newly synthesized class I MHC proteins, such as HLA-A, -B, and -C in humans, which carry the peptides to the cell surface and present them to T cells.

In addition to CD8[+] T cell responses, CD4[+] T cell responses have been found to be critical in the maintenance of adequate CD8[+] T cell function and control of viremia in both HIV and HCV infection (4–7). However, HIV-specific CD4[+] T cells may be preferentially infected and deleted by HIV (8), limiting the ability of vaccines to induce crucial T cell help after the early stages of infection. In addition, memory CD8[+] T cells have now been subdivided into effector memory T cells, which home to tissues, and central memory cells, which recirculate in the body (9–11). Chronic antigen stimulation during a persistent infection may inhibit the transition of memory CD8[+] T cells to central memory cells. However, central memory cells are more effective at protection because they are better able to proliferate when reexposed to antigen (12). Thus, chronic viral infection may perpetuate itself by preventing the development of the most effective form of T cell memory. Therefore the challenge for an effective vaccine is to induce long-lived central memory CD8[+] T cells as well as CD4[+] helper T cells. While space limitations preclude comprehensive coverage, this review article will attempt to highlight some of the exciting progress in vaccine development, primarily for three chronic infections on which much research has been focused: HIV, HCV, and HPV.

Nonstandard abbreviations used: canarypox virus (ALVAC); hepatitis C virus (HCV); highly active antiretroviral therapy (HAART); human papillomavirus (HPV); modified vaccinia Ankara (MVA); pan-DR epitope (PADRE); principal neutralizing determinant (PND); simian immunodeficiency virus (SIV); simian-human immunodeficiency virus strain 89.6 (SHIV 89.6P); virus-like particle (VLP).

Conflict of interest: The authors have declared that no conflict of interest exists.

Citation for this article: *J. Clin. Invest.* **114**:450–462 (2004). doi:10.1172/JCI200422674.

Table 1
Vaccine strategies for prevention of selected chronic viral infections

Virus	Desired immune response	Major strategies in development
HIV	Generation of CD4+ Th cells and CD8+ CTLs	Heterologous prime-boost
		Cytokine-adjuvanted DNA delivery
		Mucosal immunization
	Generation of broadly cross-reactive neutralizing antibody	Delivery of stabilized trimeric envelope protein
		Delivery of prefusion intermediate HIV envelope structures
		Delivery of modified and variable loop-deleted envelope proteins
		Delivery of multiclade and multiquasi species envelope immunogens
HCV	Generation of neutralizing antibody against HCV envelope proteins (E1 and E2)	Delivery of recombinant proteins (recombinant E1 vaccine is in clinical trial)
		Delivery of plasmid DNA and recombinant vectors (E1/E2)
	Generation of T cell–mediated immunity against HCV proteins	Delivery of plasmid DNA vaccine (E1/E2 and NS proteins)
		Delivery of synthetic peptide vaccines (Core, NS, and envelope proteins)
		Delivery of recombinant viruses (NS3, E2)
HPV	Prevention of HPV infection: generation of humoral mucosal immunity	Delivery of: L1 or L1/L2 virus–like particles (VLPs).
	Therapy of HPV-associated tumor: generation of cellular immunity directed against oncogene products E6/E7.	Delivery of vaccinia vector expressing E6/E7
		Delivery of chimeric VLPs (with E6/E7 peptides or L1/L2-E7 fusions)

NS, nonstructural.

One may ask how viruses that cause chronic infections differ from those that cause acute, self-limited infections. Each of these viral types has specific mechanisms of evading or attacking the immune system, but certain common features may be discerned. A number of factors probably play a role, including the size of the virus inoculum, the kinetics of viral replication, the viral genotype, host genetics, and the competence of the host immune system. For example, for hepatitis B virus (HBV), chronicity occurs in about 90% of cases of vertical (maternal) transmission when the immune system of the recipient is immature but in only about 10% of cases of horizontal transmission to immunocompetent adults (2). The common scenario for chronic viral infections, we suspect, is that the initial, acute infection does not usually cause very severe disease and does not provoke an immune response that is adequate to eradicate the infection, only one that can reduce the viral load. Consequently, a balance is struck between the immune system and the virus in which a steady-state level of virus is maintained within the host. The immune system keeps the virus partially in check, but the virus also evades or in some cases inhibits the immune system. Finally, disease is due primarily to the chronic insult of the lower level of viral infection over time, rather than to high viremia. When viremia does increase again, as occurs in the case of HIV, it is usually after the virus has taken its toll on the immune system as well as after viral mutations have allowed a transition to a more aggressive phenotype. In contrast, viruses that cause acute infections usually cause more severe disease initially and provoke an immune response that clears the infection. For example, in the case of hepatitis B and C, individuals who develop chronicity often experience an initially milder acute hepatitis syndrome and weaker cellular immune responses than those that clear the infection and do not develop chronic infection (13–15). The goal of vaccines in the case of viruses that cause chronic infections is therefore to tip this balance in favor of the immune response. Thus, the real hope for dealing with chronic viral infections is the development of a new generation vaccine, important to developing countries. Effective vaccination could be accomplished either prophylactically, by establishing an early immune advantage via the generation of pre-existing vaccine-induced immune memory, or therapeutically, by increasing the strength or quality of the immune response beyond that controlling the level of steady-state viral load during chronic infection. For HIV, HCV, and HPV, we will explore the approaches being undertaken to accomplish these goals (see Table 1).

Human immunodeficiency virus

The basis for current HIV vaccine strategies. There is little direct evidence for immune correlates of protection against HIV in humans since no individual has mounted an immune response capable of spontaneously clearing the infection, even though there are some long-term nonprogressors who have remained infected without developing AIDS. Nevertheless, there is much evidence in animal studies and indirect evidence in humans that CD4+ and CD8+ T cells, broadly neutralizing antibodies, and innate immunity all play an important role in the control of infection with HIV and its close cousin, simian immunodeficiency virus (SIV), in macaques.

Antibodies neutralizing AIDS viruses clearly play an important role in protection. Passive transfer of IgG1 monoclonal antibodies was shown to be sufficient to protect macaques against i.v. challenge or against mucosal transmission (16, 17). However, a high level of monoclonal antibody is required to achieve complete protection while partial protection could be achieved with a lower-antibody titer. Therefore early studies primarily focused on the HIV envelope protein gp160 as the primary target of neutralizing antibodies. However, while it was possible to achieve neutralizing antibodies against a specific virus strain grown in the laboratory, the difficulty of obtaining antibodies that neutralized a broad array of strains, particularly primary isolates, provided incentive both to devise novel approaches for the induction of the relevant antibodies and to target T cell immunity as an alternative strategy.

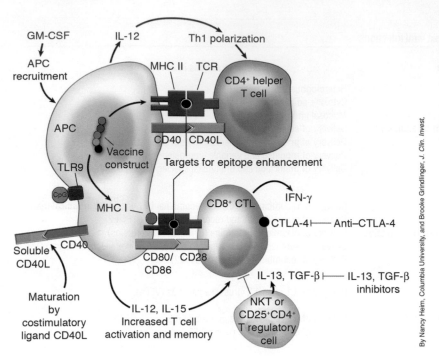

By Nancy Heim, Columbia University, and Brooke Grindlinger. *J. Clin. Invest.*

Figure 1
New strategies for second generation vaccines based on cellular immunity. CD4⁺
helper T cells mature and activate APCs through recognition of epitopes presented by class II MHC molecules (MHC II) and interaction of CD40 and CD40 ligand
(CD40L). The CD40-CD40L interaction causes the APC to upregulate expression of
costimulatory molecules such as CD80 and CD86 and to secrete cytokines IL-12 and
IL-15. The costimulatory molecules interact with CD28 on the CD8⁺ CTL to provide a
second CTL activation signal in addition to T cell receptor (TCR) recognition of an antigenic peptide presented by a class I MHC molecule (signal 1). IL-12 also contributes
to activating the CTL and polarizing the T helper cell to produce Th1 cytokines, such
as IFN-γ. IL-15 contributes to induction and maintenance of CTL memory and longevity. Regulatory T cells, including NK T cells and CD25⁺CD4⁺ T cells, can dampen or
inhibit the CTL response in order to prevent autoimmunity, but also reduce the immune
response to the vaccine. Various strategies may be employed to improve the natural T
cell response. Epitope enhancement of class I or class II MHC–binding peptides can
increase their affinity for the respective MHC molecules and their immunogenicity.
Incorporation of (a) cytokines such as IL-15 to recruit more memory CTLs; (b)
IL-12 to steer the T helper cell population towards a Th1 response; (c) GM-CSF
to recruit dendritic cells; or (d) costimulatory ligands such as CD40L to activate and induce
maturation of the dendritic cells recruited by the GM-CSF can synergistically amplify the
immune response. In addition, CpG-containing oligonucleotides can act through toll-like
receptor 9 (TLR 9) to activate dendritic cells. Increased levels of costimulatory molecules
can selectively induce higher-avidity CTLs that are more effective at clearing virus infections. Agents that block factors secreted or induced by regulatory T cells, such as IL-13
and TGF-β, can synergize with other strategies to allow the CTL response to reach its
full potential. Similarly, blockade of the inhibitory receptor, CTLA-4, on the T cell, can
increase the T cell response. See multiple references in the text for the other strategies.

A number of lines of evidence implicate CD8⁺ T lymphocytes
(especially CTLs) in controlling HIV or SIV infection (reviewed in
refs. 18–21). The acute viremia in both HIV and SIV was found to
decline concomitant with the rise of the CTL response and prior to
the appearance of neutralizing antibodies. Many HIV-infected long-
term nonprogressors have expressed a high level of HIV-specific
CTLs, and African sex workers that had been exposed to HIV but
remained uninfected possessed high CTL responses. However, the
most direct evidence comes from studies of HIV-infected chimpanzees and SIV-infected macaques, in which depletion of CD8⁺ T cells

in vivo led to increases in viral load that were later
reversed when the T cells reappeared (22–24).
For this reason, most strategies studied in nonhuman primates today are based on eliciting
an effective HIV- or SIV–specific CTL response.
Although CTLs can be elicited by peptides and
other constructs, the most straightforward
approaches have involved agents that induce
endogenous expression of the viral antigens
in a professional antigen-presenting cell, such
as a dendritic cell, because the most efficient
and natural way to load class I MHC molecules
with peptides for presentation to CD8⁺ T cells
is endogenous expression of the protein within
the cell. This can be accomplished with DNA
vaccines or with viral vectors, which can introduce the antigen gene into antigen-presenting
cells. A recent study indicates that a DNA vaccine augmented by IL-2 can induce a strong
HIV-specific CTL response and can control
pathogenic viral challenge and prevent AIDS in
rhesus macaques (25).

Compared to repeated DNA vaccination or
viral-vector–based vaccination alone, one of
the most effective strategies for eliciting HIV-
specific immunity is a heterologous prime and
boost regimen, which involves administration
of a DNA vaccine followed by a viral-vector vaccine, which induces a stronger CTL response
than can be achieved by priming and boosting
with the same agent (26), or a recombinant protein boost, which induces a significant level of
neutralizing antibodies (27). HIV DNA vaccines
were shown to be effectively boosted by a recombinant vaccinia virus, such as modified vaccinia
Ankara (MVA) (28) or replication-deficient
recombinant adenovirus (29), which induced
high frequencies of CD8⁺ CTLs and neutralizing antibodies and protected against viral challenge. MVA is a highly attenuated vaccinia virus
that has lost the ability to replicate in primate
cells and can be considered a safe vaccine. In a
study comparing SIV gag recombinant MVA
and recombinant adenoviral vectors in prime-
boost regimens in macaques, DNA prime–
recombinant adenovirus boosting was most
effective at eliciting long-lasting CD8⁺ IFN-γ
responses and protection against viral challenge (30). However, mutation of AIDS viruses,
leading to escape from immune control medi-
ated by CTLs (31), and the breadth of the protection against more
distant strains of challenge virus remain concerns. Indeed, several
of these studies used the highly pathogenic simian-human immunodeficiency virus strain 89.6 (SHIV 89.6P) as the challenge virus
as it has an atypically rapid disease course, but this virus may not
be representative. Protection against heterologous viral challenge
will be critical to demonstrating the breadth of vaccine efficacy.

These difficulties led to the development of new vaccine strategies against AIDS viruses as described below (reviewed in ref.
32), including but not limited to the following: (a) creation by

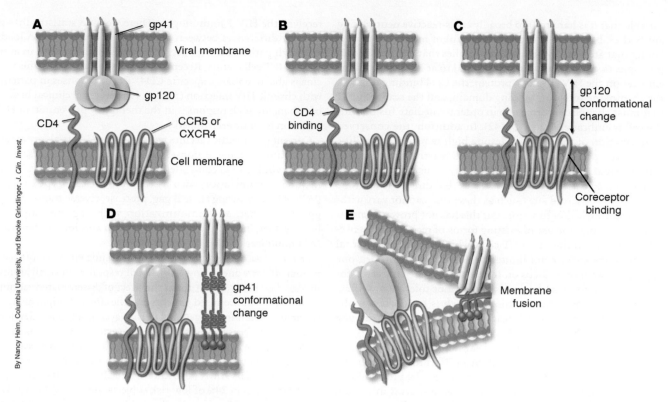

By Nancy Heim, Columbia University, and Brooke Grindlinger, *J. Clin. Invest.*

Figure 2
Interactions of HIV envelope glycoproteins, CD4, and chemokine receptors CCR5 or CXCR4 trigger fusion and entry of HIV. These interactions determine critical regions of the HIV envelope glycoprotein against which neutralizing antibodies could be raised. After the envelope protein interacts with CD4 on the target cell (**A** and **B**), it undergoes a conformational change allowing its interaction with a chemokine receptor (**C**). This second interaction induces a further conformational change in the gp41 portion of the envelope glycoprotein that mediates the fusion event (**D** and **E**). Blockade of any of these three steps can prevent viral entry.

sequence modification of enhanced epitopes that bind with higher affinity to MHC molecules; (b) targeted induction of mucosal immunity; (c) use of synergistic combinations of cytokines, chemokines, or costimulatory molecules to enhance the immune response; (d) relief of negative regulatory or suppressive mechanisms that inhibit the immune response; (e) use of dendritic cells as vaccine vehicles; (f) induction of alloimmunity for protection against HIV; and (g) formulation of the vaccine to incorporate agents inducing innate immunity (Figure 1).

Mucosal vaccination of mice with an HIV peptide induced systemic and mucosal CTL responses, while parental vaccine administration induced predominantly systemic CTL responses (33). Systemic CTLs were not sufficient to protect against mucosal virus transmission, so it is important that the CTLs be locally present in the mucosa (34). Mucosal delivery of the vaccine is usually the most effective route to induce mucosal immunity although, interestingly, transcutaneous immunization can also induce mucosal CTLs and may provide an alternative vaccination option (35). Moreover, mucosal vaccination against AIDS viruses, which induced CD8+ CTLs in the gut mucosa of immunized Rhesus macaques, more effectively cleared the major reservoir for SIV replication in the gut and thus reduced plasma viral load below the level of detection. The same vaccine administered subcutaneously was less effective, leaving residual viremia (36). Also, control of SIV infection was associated with mucosal CTLs (37). These findings make a strong argument for mucosal delivery of an AIDS

vaccine even if some partial protection against mucosal challenge can be observed with a systemic vaccine (38).

Cytokine and chemokine gene codelivery along with DNA-encoding immunogens can modulate the direction and magnitude of immune responses (39) and can improve vaccine efficacy compared to delivery of DNA alone (25). For example, RANTES coinjection induced high levels of CD8+ CTLs, as did a synergistic combination of GM-CSF and CD40 ligand when negative regulatory mechanisms were blocked (40).

Overall it is widely believed that induction of both antibodies and T cells will be needed for an effective AIDS vaccine. Several major strategies are being studied for development of a preventive vaccine against AIDS viruses (reviewed in ref. 41). Most of these approaches focus on the generation of either neutralizing antibodies or CTL responses, but ultimately some combination of these approaches may be needed.

Another major hurdle for HIV vaccine development is the extraordinary diversity of the virus and its ability to rapidly mutate within each infected individual. The genetic subtypes of HIV, clades A–E, are responsible for the main epidemics in different parts of the world: clade B is prevalent in North America and Europe; clades A, C, and D in Africa; and clades E and B in Thailand. In addition to the diversity of viral subtypes that must be targeted by a vaccine, the high level of mutability, due to the error-prone nature of reverse transcriptase, facilitates escape mutation. Some of the major neutralizing epitopes of HIV are so variable that it is hard to find broadly cross-reactive neutralizing antibodies, but the existence of a handful of monoclonal antibodies that are broadly neutralizing implies that it is possible, in principle, to do so. These antibodies bind to at least three different sites on the HIV envelope protein: the CD4 binding domain, the chemokine receptor–binding domain, and the stalk of gp41 that must change conformation in order to mediate fusion with the cell membrane (Figure 2) (42). In addition, the conserved conformation of the V3 loop, despite high sequence variability, has allowed production of broadly cross-reactive antibodies to this principal neutralizing region (43, 44). Thus, with respect to antibodies, a goal is to develop vaccines that direct the immune response to conserved sites such as these that cannot vary without a resulting loss of function, but this has not proven straightforward through the use of existing forms of recombinant envelope proteins. In the case of T cells, which can target internal viral proteins and are not limited to neutralizing epitopes, one approach has been to focus on sequences that are conserved for functional or structural reasons and cannot tolerate modification by escape mutations. This approach is supported by the finding that more cross-clade reactivity has been observed among T cells than antibodies (45, 46).

HIV-1 vaccine clinical trials. This year heralds the deployment worldwide of multiple vaccine trials in an effort to stem the ongoing HIV epidemic. Safety and immunogenicity data for multiple vaccines and immunization platforms currently moving into phase I/II studies will establish the criteria for phase III efficacy studies. A listing of ongoing preventive trials of HIV vaccines is available at the International AIDS Vaccine Initiative website (http://www.iavi.org/trialsdb/basicsearchform.asp), and the ongoing and planned protocols of the HIV Vaccine Trials Network in association with the National Institute of Allergy and Infectious Diseases can be found at http://chi.ucsf.edu/vaccines/.

Therapeutic vaccine trials. Early clinical studies tested the ability of HIV-1 clade B envelope protein (gp160 or gp120) vaccines or peptide vaccines based on the V3 loop of gp120 that had been identified as the principal neutralizing determinant (PND), in order to elicit neutralizing antibodies. Other vaccine trials in HIV-1 infected individuals initiated during the pre–highly active antiretroviral therapy (pre-HAART) era attempted to elicit both cellular and humoral immune responses using synthetic peptides containing promiscuous CD4+ T helper cell epitopes linked to the PND located at the crown of the V3 loop (47, 48). Various strategies (e.g., the use of adjuvants and multimeric and multivalent immunogens) were employed to increase vaccine immunogenicity and cross-reactivity. Although these approaches proved capable of eliciting high-titered type-specific neutralizing antibodies to tissue culture lab–adapted virus strains in addition to some lymphoproliferative responses in immunized patients, these antibodies failed to neutralize primary viral isolates (49–51).

An early approach targeting specific cellular immunity used gp120-depleted, whole, killed virus (52). The idea was to elicit cellular immunity to internal viral proteins while avoiding induction of so-called enhancing antibodies, specific for the envelope protein, that facilitate virus uptake by cells, as well as avoiding induction of other potential deleterious effects of the envelope protein. A large multicenter, double-blind, placebo-controlled, randomized trial was conducted on a whole inactivated Zairian HIV-1 isolate in incomplete Freund's adjuvant given intramuscularly at 12-week intervals. A total of 1262 subjects of 2527 HIV-1 infected patients received the HIV-1 immunogen. There were no statistically significant differences between groups in plasma HIV RNA loads although patients in the vaccine group had an increase in average CD4+ T cell counts. Recent studies have shown that this vaccine is able to enhance specific CD4+ T cell responses in patients with chronic HIV infection (53, 54). However, the clinical benefit of enhancing such responses in the therapy or prevention of HIV infection is yet undetermined.

Current therapeutic vaccines entering phase I/II trials are aimed at boosting immune responses in HIV-1–positive patients where plasma viral load is controlled by antiretroviral therapy. These include immunization with a recombinant canarypox vector (VCP1452) expressing clade B gag, protease, reverse transcriptase, gp120 and nef, and immunization with gag, pol, and nef lipopeptides, in addition to peptide-pulsed autologous dendritic cell immunization plus IL-2.

Preventive vaccine trials. Recombinant vaccinia virus vectors have proven effective at eliciting CD8+ T cell responses in small animal models; however, concern about the effect of disseminated vaccinia on immunocompromised patients and the effect of prior smallpox vaccination on immunogenicity of the vaccine led to the development of a number of live, attenuated pox-vectored HIV vaccines that do not replicate in human cells and can be administered repeatedly. In a study by Evans et al. (55), recombinant canarypox expressing only gp160 of HIV-1MN (also known as vCP125) elicited anti-HIV env CD8+ CTLs in 24% of low-risk subjects. Anti–HIV-1 env CTLs were detected in 12 subjects and anti–HIV-1 gag CTLs were detected in 7 of the 20 vaccine subjects receiving canarypox virus (known as ALVAC) expressing HIV-1 *env, gag,* and protease (vCP205) vaccine alone or with clade B HIV-1 strain SF-2 recombinant gp120 protein (called rgp120 SF) (56, 57). Coadministration with SF-2 rgp120 vaccine enhanced lymphocyte proliferation in response to HIV-1 envelope glycoprotein and broadened envelope-stimulated cytokine secretion. Belshe et al. (58) reported a cumulative positive response frequency of 33% for anti–HIV-1 env or gag CTLs among 170 subjects in a phase II trial of vCP205. The vaccines were safe, and all patients developed binding antibody to monomeric gp120; approximately 60% developed antibody to gag p24.

Overall, lymphoproliferative responses to gp120 varied among ALVAC vCP205 studies, with from 50–100% of vaccinated subjects demonstrating CD4+ T cell proliferation. Fifteen to twenty percent of vaccinees developed CD8+ CTL responses, mostly against the envelope protein, with cross-clade reactivity seen in some subjects (45, 59). The maximum positive CTL response (35/84; 42%) was observed after four immunizations (57). It is noteworthy that the first successful phase I vaccine study initiated in Africa involved vaccinating uninfected volunteers in Uganda with clade B ALVAC vCP205 (60). Future vaccine strategies involving variations of the canarypox vector currently being developed as combination vaccines include replacing the clade B *env* sequences in vCP205 with clade A or clade E *env* sequences or sequences from a primary clade B isolate and the addition of reverse transcriptase (RT) and nef epitope sequences.

Heterologous prime-boost strategies. DNA vaccines used alone have not proven as immunogenic in humans and nonhuman primates as in mice; however, strategies involving DNA priming and boosting with a viral vector are capable of eliciting potent CD8+ and modest CD4+ T cell and antibody responses in macaques (28, 29). In an effort to improve immunogenicity results obtained with ALVAC vectors, current studies employing multiple viral gene products in complex combinations of DNA and viral vectors such as MVA, attenuated Vaccinia Copenhagen strain with deletions in virulence genes (NYVAC), fowlpox, adenovirus, and Venezuelan equine encephalitis virus–like replicon particles are underway. Major considerations in the development of recombinant viral vectors are ways to circumvent preexisting immunity and the production of high-titered stable vectors. Current phase I placebo-controlled trials are aimed at defining optimum vaccination regimens for eliciting cellular immune responses by varying the dose (dose escalation), number of doses, intervals between doses, and routes of administration. Results of phase I adenovirus trials in humans show safe, strong, long-lasting CD8$^+$ T cell responses measured by IFN-γ ELISPOT. The University of New South Wales, Australia, has recently initiated a phase I trial of a prime-boost protocol using DNA and a fowlpox vector each expressing clade B gag-pol (without integrase), tat, nef, and gp160 env coding sequences. It is encouraging that recent clinical trials evaluating heterologous prime-boost regimens with HIV-1 and a recent malaria sporozoite protection trial with pre-erythrocytic *Plasmodium falciparum* immunogens have demonstrated the ability of these regimens to elicit strong IFN-γ–secreting CD8$^+$ T cell responses equivalent to or better than those achieved during natural infection (61). Current vaccines moving into clinical trial that incorporate multiple immunogenic viral gene products are designed to address the issues of HLA polymorphism and escape mutation, and to identify correlates of immune protection. Second-generation DNA and viral-vectored vaccines include multiclade gag, pol, and env and may include nef and the accessory gene products tat and vpu. In addition, several novel agents are currently in phase I trials. One examines modified HIV envelope immunogens (e.g., ΔCFI [cleavage site deletion (C), fusion peptide deletion (F), deletion of interspace between gp41 (N), and C heptad repeats (I)]) and clade A, B, and C DNA envelope immunogens developed by the Vaccine Research Center. A second examines a V2-deleted trimeric gp140 protein developed at Chiron Corp. Two other studies (at St. Jude's Children's Hospital, Memphis, Tennessee, USA; and University of Massachusetts Medical School, Worcester, Massachusetts, USA) test multiclade, multienvelope DNA, recombinant vaccinia virus with protein boost strategies in an effort to elicit broadly neutralizing antibody in addition to CD4$^+$ and CD8$^+$ T cell responses. In the Vaccine Research Center phase I DNA vaccine trial with a gag-pol-nef and multiclade env vaccine, CD4$^+$ T cell responses were more frequent than CD8$^+$ T cell responses and were primarily directed toward env but not the gag-pol-nef fusion protein. Coadministration of plasmid DNA expressing IL-2-Ig to enhance cellular immune responses is currently being tested with the Vaccine Research Center DNA vaccine (clade B gag-pol-nef and multiclade env) in a phase I trial. Cytokines IL-12 and IL-15, which have been shown to enhance induction of cellular immune responses and memory to vaccine antigens in small animal models, are scheduled for testing in human phase I HIV-1 vaccine trials in the near future.

Although cross-clade CD8$^+$ CTL responses have been reported with clade B immunogens (45, 46), the importance of clade diversity will only be definitively addressed in phase III trials that compare vaccine candidates in parallel trials in different geographic regions. A recent study of HIV-1 subtype C–specific immune responses during natural infection in individuals in Botswana emphasizes the need to match vaccine epitopes to immunodominant epitopes detected in the target population based upon HLA frequencies (62). Recently,

the first phase I HIV vaccine trial to be conducted simultaneously in Africa and the United States (sites in Gabarone, Botswana; Boston, Massachusetts, USA; and St. Louis, Missouri, USA) was initiated to test a DNA vaccine developed by Epimmune, composed of a promiscuous helper T cell epitope, pan-DR epitope (PADRE), and 21 specific epitopes optimized to elicit CD8$^+$ CTL responses in individuals expressing one of three HLA alleles: HLA-A2, HLA-A3, and HLA-B7. Results obtained from this polyepitope study regarding immunogenicity will provide information for future vaccine design since the immunogen is not specifically selected for epitopes or matched for HLA types prevalent in this African population. Results of a previous single phase I trial in Africa employing a DNA vaccine expressing clade A gag and a stretch of 25 CTL epitopes known to be expressed in the vaccinated population revealed modest CD8$^+$ T cell IFN-γ responses to gag, but no responses to the individual epitopes included in the vaccine were seen. After a single MVA boost (5×10^7 pfu) 6 months to 1 year later, CD8$^+$ IFN-γ ELISPOT responses were detected in 19 of 26 individuals. In addition, an increase in the breadth of responses was seen after boosting. A phase II trial of this vaccine is in progress (63).

Lessons for the future. Only one HIV vaccine construct has yet progressed through large-scale phase III studies testing efficacy. A phase III trial completed in early 2003 (known as VAX003) in the United States, Canada, and the Netherlands and another in Thailand completed in 2004 (known as VAX004), both testing bivalent formulations of gp120 protein subunit vaccine (AIDSVAX B/B and B/E; VaxGen Inc.) aimed at targeting neutralizing antibodies, failed to demonstrate efficacy. Although a difference in the infection rate of African-American placebo recipients (9/116; 7.8%) versus African-American vaccine recipients (2.6%; 6/233) was found, based upon the small number of infections, further analysis is necessary to determine the significance of these differences. This suggestive result in a retrospective stratification emphasizes the need to adequately power future phase III trials to address differences in immune responses based upon gender and ethnicity. Phylogenetic analysis representing the overall diversity of viral isolates from the complete VAX004 data set showed no differences in any treatment group based upon race, gender, or geography (64). Results from these trials were consistent with those of previous studies, in which monomeric gp120 was not proven effective at eliciting broadly cross-reactive neutralizing antibodies.

A critical component for future vaccine prime-boost regimens is the inclusion of an envelope immunogen capable of eliciting broadly cross-reactive neutralizing antibodies against primary HIV-1 isolates. Although HIV-1–infected individuals are capable of developing neutralizing antibodies to primary viral quasispecies, serial escape occurs; consequently, the neutralizing antibody response lags one step behind the evolution of the viral envelope (65–67). The majority of cross-reactive neutralizing antibodies directed against HIV-1 glycoproteins have been mapped to conserved regions within the CD4 binding site and CD4 inducible epitope, V2, V3, the carboxy-terminus of C5, the leucine zipper-like region of gp41, and the ELDKWAS motif in the transmembrane region of gp41. Monoclonal antibodies directed to these epitopes neutralize primary isolates from multiple clades to varying degrees. Monoclonal antibodies, 2F5 and 4E10, directed to membrane proximal domains in gp41 are the most potent, in that they cross-neutralize 67% and 100%, respectively, of all clade isolates tested (68). The inability of primary isolates to elicit cross-reactive, neutralizing antibody may be explained by the low immunogenicity of these epitopes, resulting from conformational dynamics within the viral envelope that maintain a structure that makes these sites inaccessible or only transiently exposed. The rational design of new envelope immunogens should focus on engineering structures that expose, and direct antibody responses to, conserved epitopes on native trimers that are recognized by broadly cross-reactive neutralizing antibodies, and that at the same time prevent induction of dominant non-neutralizing antibodies.

An effective HIV-1 vaccine will require both potent and durable cell-mediated immune responses as well as effective neutralizing antibody responses. In the coming years, phase III efficacy trials of vaccines of proven immunogenicity will determine the need to employ immunization strategies focusing on eliciting mucosal immune responses, as noted above, since delivery of current vaccines primarily targets induction of systemic responses. Another important issue is whether sterilizing immunity, in which a vaccine is completely successful in preventing infection and which is likely to require high titers of broadly neutralizing antibodies, is essential or whether a vaccine based only on inducing cellular immunity, which controls viral loads so as to both prevent disease in the individual and reduce the risk of transmission to others (69, 70), will be sufficient to contain this pandemic.

Hepatitis C virus

Most cases of acute viral hepatitis are caused by hepatitis A virus (HAV), HBV, HCV, and hepatitis D virus (HDV). Of those viruses, HBV and HCV are the most important causes of chronic infection and liver-related morbidity and mortality (71). For both HAV and HBV, effective vaccines are currently available and include inactivated whole virus for HAV and recombinant hepatitis B surface antigen for HBV (reviewed in ref. 72). In addition, a new combined HAV and HBV vaccine has recently been approved for use in individuals 18 years and older. Lack of a licensed vaccine for HDV is of less concern because HDV requires HBV for pathogenicity. Unlike HBV, HCV is an RNA virus that does not integrate into the host genome. HCV infection results in persistent infection in 55–85% of patients (reviewed in ref. 72). Chronic HCV and HBV infections are leading causes of liver cirrhosis and hepatocellular carcinoma worldwide. Although chronic HBV infection remains widespread despite the existing prophylactic vaccine and immunotherapeutic approaches are needed, this review will be confined to HCV.

Immune responses to HCV. Although serum antibodies to HCV are detected following virus infection, humoral immunity alone may not play a critical role in viral clearance during acute infection, perhaps because HCV is capable of rapid outgrowth of antibody escape mutants (73). In addition, no data are available to show that HCV-infected patients have long-lasting protective antibody responses. In contrast, cellular immunity does seem to play a role in the virological outcome during acute infection and persists for decades after viral clearance (5, 6, 13–15, 74, 75). A wide variety of vigorous CD4+ T cell responses persists for many years, and memory CD8+ T cells may also be maintained (74). However, these responses are significantly weaker among patients who later progress to chronic infection, suggesting that the intensity of cellular immunity in the early stage of infection is a critical factor in limiting the spread of HCV. The unfortunate consequence of a vigorous cellular immune response to acute hepatitis infection is liver damage, especially that resulting from CD8+ T cells killing infected hepatocytes. Indeed, the possibility has been raised that different subsets of CD8+ T cells

contribute to immunopathology and to viral clearance (6). Thus, the goal of a vaccine against HCV is to induce an initial immune response of sufficient strength and type to clear the infection without causing severe acute hepatitis. Host genetics, including HLA type, have also been shown to contribute to HCV clearance and chronicity (reviewed in ref. 76).

Vaccines for HCV. Development of a vaccine for HCV has been delayed by many impediments, including lack of a suitable small animal model, a high degree of genomic diversity, and inability to grow large amounts of virus in vitro. Recent studies suggest hope for the development of prophylactic and therapeutic vaccines against HCV infection.

An early prophylactic vaccine approach against HCV, targeting recombinant HCV envelope glycoprotein (E1/E2) in chimpanzees (77), was designed to induce neutralizing antibodies. Later, studies using recombinant E1/E2 protein and peptide vaccines showed that antibodies induced could neutralize low levels of homologous HCV challenge in nonhuman primates (78). Although the vaccine failed to protect against high-dose virus challenge, the reduction in risk of chronic infection is a great success because most morbidity and mortality of HCV is a consequence of chronic infection. A recombinant E1 protein is currently being evaluated in clinical trials as a therapeutic vaccine against HCV (reviewed in ref. 79). To achieve humoral immunity with both prophylactic and therapeutic vaccines, several strategies, including use of DNA plasmids, recombinant viruses or bacteria, and virus-like particles (VLPs), are under study. In particular, VLPs are attractive because the particulate multivalent structure is more immunogenic than soluble proteins.

The complementary approach of targeting T cell–mediated immunity has been given an impetus by studies of reinfection after recovery from HCV infection. Although HCV can cause more than one episode of acute hepatitis in the same individual under certain circumstances, such as in thalassemia patients (80), other studies in previously infected humans and chimpanzees indicate that the risk of a second infection becoming chronic is greatly reduced compared to that of a primary infection, and this protection correlates with both CD4+ and CD8+ T cell responses (75, 81–84). Indeed, the critical role of CD8+ T cell recruitment in this protective response was demonstrated by depletion of these cells in chimpanzees, which resulted in high viral loads until the CD8+ T cell population recovered, despite the persistence of a CD4+ T cell response (84). The role of T cells in these protective responses was supported by the results of several studies demonstrating the absence of detectable HCV envelope glycoprotein–specific antibodies in the protected animals (75, 83, 84). Thus, effort has been invested in defining HCV CTL epitopes and designing vaccine constructs (85–90). Such approaches to HCV vaccine development include the use of DNA plasmids (90, 91), recombinant viral vectors expressing HCV antigens (91–93), and HCV virus–like particles (94, 95). To improve on the ability of the wild-type viral sequence to induce T cell immunity, the amino acid sequence of epitopes has been modified to increase affinity for the HLA molecule to make the epitopes more potent vaccines (known as epitope enhancement) (88). This epitope-enhancement approach can be applied to any type of vaccine construct.

However, recent evidence suggests that CD8+ T cell responses alone are not sufficient for protection against HCV infection and that CD4+ T cells may also be critical. Depletion of the CD4+ T cell population from two immune chimpanzees before HCV reinfection led to viral persistence despite continued HCV-specific

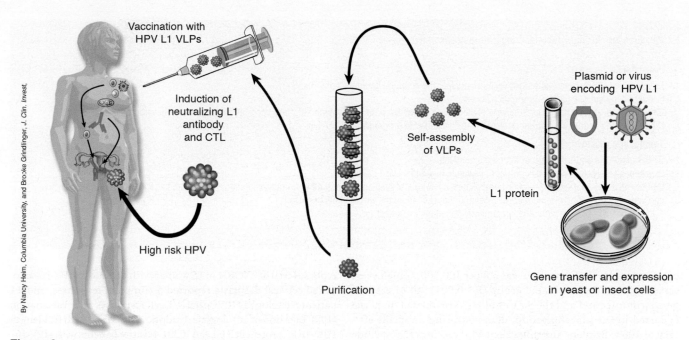

By Nancy Heim, Columbia University, and Brooke Grindlinger, *J. Clin. Invest.*

Figure 3
Vaccination against HPV infection using genotype-specific HPV L1 VLPs. Recombinant HPV-16 or HPV-18 L1 capsid protein made in yeast or baculovirus-infected insect cells self-assembles to form VLPs that are very potent at inducing neutralizing antibodies but are not infectious because they lack any viral nucleic acid. Such VLP vaccines show promise for prevention of HPV infection and HPV-associated cervical cancer. Depicted within the vaccinated subject are dendritic cells that present antigen to helper T cells (blue) and B cells (pink), which induces the B cells to become plasma cells (shown as ellipses). Plasma cells then generate antibodies (red) capable of neutralizing the virus.

CD8$^+$ T cell memory, and this inadequate control of viral load in the absence of CD4$^+$ T cell help led to the emergence of viral escape mutants (7). Likewise, in humans, clearance of acute HCV infection was associated with a strong CD4$^+$ T cell response, and absence or loss of this response was associated with viral persistence or recurrence (5, 6). Thus, the role of CD4$^+$ T cell help may parallel that described as occurring during HIV infection. Therefore, recent strategies for inducing both humoral and CD4$^+$ and CD8$^+$ cellular immunity could potentially provide more complete protective immunity against HCV infection. Some of these vaccines have induced CTLs in HLA transgenic or other mice sufficient to protect against a recombinant vaccinia virus expressing HCV antigens, used as a surrogate challenge virus in mice (91, 93). To induce Th1-type immunity and improve T cell–mediated immunity, inclusion of cytokines and other biological adjuvants may be necessary for both prophylactic and therapeutic HCV vaccines in the future.

Human papillomavirus
Persistent infection with oncogenic strains of HPV is the major cause of cervical cancer. HPV genotypes 16, 18, 31, 33, 45, and 56 account for more than 95% of cases (96, 97). HPV has also been implicated in cancers of the anus, penis, vulva, and oropharynx. As humoral immunity to HPV is genotype specific, effective vaccines must be polyvalent. Approximately 100 HPV genotypes have been identified, about 40 of which infect the genital tract, and of these, at least fifteen are believed to be oncogenic. More than 70% of women who become infected even with high-risk HPV genotypes will clear the infection within 2 years, and only a minority of the women with persistent infection will develop dysplasia and progress to cancer. Vaccine development has been slowed by the

large number of genotypes needing to be addressed and the lack of an efficient HPV culture system (98). Two HPV vaccine strategies are currently under study: (a) prophylactic vaccination to prevent primary infection, usually aimed at generating neutralizing antibodies to the L1 major capsid protein; and (b) therapeutic vaccines inducing CTLs, usually against the viral E6 and/or E7 oncoproteins expressed in HPV-associated dysplastic and cancerous lesions and responsible for malignant transformation.

Prophylactic vaccines targeting capsid proteins with antibodies. Prophylactic HPV vaccines have focused on the use of type-specific VLPs generated from recombinant L1 major capsid protein or L1/L2 capsids to induce neutralizing antibodies. Overexpressed L1 capsid protein with or without L2 will self-assemble to generate noninfectious, nononcogenic VLPs (99). In animal models, L1-VLP induced high titers of neutralizing antibodies and protected animals against HPV infection (98). Passive immunization of animals using serum from vaccinated animals confirmed the ability of neutralizing antibodies to prevent infection. L1 or L1/L2 VLP HPV vaccines are currently in clinical trials. Preclinical work includes the development of L1/L2-E2-E7 fusion protein VLPs and the incorporation of plasmid expression vectors into the VLPs (100).

Clinical trials of VLPs have focused on prevention of primary HPV infection (Figure 3). A series of trials showed that an HPV-16 L1-VLP vaccine was well tolerated and generated high levels of antibodies against HPV-16 as well as CD4$^+$ and CD8$^+$ T cell responses to L1 (97, 101). In a randomized, double-blind, placebo-controlled trial of an HPV-16 L1-VLP vaccine in 1533 women aged 16–23 designed to determine whether such a vaccine could prevent persistent HPV-16 infection, Koutsky et al. found that over 99% of the women receiving the vaccine developed antibodies against HPV (102). In the control arm, the rate

Vaccines for chronic viral infections: future needs and goals

Prophylactic vaccines:

Development of effective multivalent vaccines against the common genotypes of virus.
Reduction in cost of vaccines to broaden access.
Simplified storage, handling, and delivery of vaccines to allow easier implementation of mass vaccination programs in developing countries.
Definition and expansion of target populations for vaccination programs.

Therapeutic vaccines:

Earlier diagnosis and treatment of infected individuals.
Broadened viral protein/epitope content of vaccines to avoid escape mutations.
Use of enhanced epitopes, cytokines, chemokines, costimulatory molecules, and other immunostimulatory agents and agents that block suppressive pathways to boost the immune response beyond that elicited by the chronic infection itself.
Combination of vaccines with other treatments, including antiviral drugs.

of persistent HPV-16 infection was 3.8 per 100,000 woman-years versus none in the vaccinated group ($P < 0.001$). All 41 cases of persistent infection and the 9 cases of HPV-16–related dysplasia occurred in the placebo group, demonstrating a vaccine efficacy of 100%. Bivalent and polyvalent VLP vaccines that include HPV-16 and -18, or HPV-6, -11, -16, and -18 are under development by at least three groups (98), and phase III trials of these promising VLP vaccines are underway. The initial success of these VLP-based prophylactic vaccines for HPV infection offers hope that soon it will finally be possible to save millions of lives by widespread prevention of HPV-associated cervical cancer.

Therapeutic vaccines targeting HPV oncoproteins with CTLs. The major targets for preventing the progression of persistent HPV infection to cancer and for treating cancer are the E6 and E7 oncoproteins. In high-grade dysplasia and carcinoma, genotype-specific E6 and E7 are expressed in virtually all cells. Their expression is both necessary and sufficient for the maintenance of the transformed phenotype (103), ensuring retention of oncoprotein expression by the tumor. Vaccine development for late-stage HPV-associated disease has focused on generating cellular immunity against these antigens.

In patients with recurrent or advanced cervical cancer, a recombinant vaccinia virus encoding nonfunctioning HPV-16 and -18 E6/E7 fusion proteins (known as TA-HPV) produced no clinical responses; however, 2 out of 8 patients remained alive and tumor free 15 and 21 months after vaccination (104). One of three patients tested developed HPV-18 E6/E7-specific CTLs that were not detected prior to vaccination. Unfortunately, complicating the interpretation of this trial was the fact that these patients also received other treatments.

In a phase I/II study, 15 HLA-A*0201–positive cervical cancer patients were vaccinated with HLA-restricted E7 peptides and PADRE in Montanide ISA-51. Proliferative responses were elicited in 4 patients, but no clinical responses or CTLs were observed (105). In another phase I trial, 18 HLA-A*0201–positive women with HPV-16–associated high-grade genital dysplasia were vaccinated with HPV-16 E7 peptides in incomplete Freund's adjuvant (106). Three patients cleared their dysplasia, and 6 others achieved a partial response.

A phase II study of TA-GW, a recombinant HPV-6 L2E7 protein in alum, produced 5 complete responses in 27 men with genital warts (107). Vaccination using a polymer-encapsulated plasmid (ZYC101) expressing HLA-A*0201–restricted HPV-16 E7 epitopes fused with HLA DRA0101 in 12 men with high-grade anal dysplasia produced 3 partial responses, and 10 patients demonstrated antigen-specific IFN-γ–producing T cells for up to 6 months after vaccination (108).

A phase I trial of ZYC101 in 15 women with persistent HPV-16–associated cervical dysplasia reported 5 complete responses, and 11 patients developed HPV-specific T cell responses after vaccination (109). Injection of autologous dendritic cells pulsed with full-length HPV-16 E7 protein in 3 HLA-A*0201–positive patients with HPV-16–associated cervical cancers elicited CD8$^+$ CTLs against autologous tumor cells (110), and one patient had a complete response that lasted 23 months (111). A heat-shock protein-65–HPV-16 E7 fusion protein (HspE7) vaccine that may offer broader immunity against a number of HPV genotypes is currently in clinical trials (112).

There is real optimism for the development of an effective HPV vaccine in the near future. HPV L1–VLP vaccines appear safe and effective for the prevention of primary infection. Although more progress has been made in the development of prophylactic HPV vaccines, those aimed at the treatment of premalignant lesions and cancer are also promising, albeit still investigational. In the not too distant future, preventative vaccination strategies similar to those used for hepatitis B hold great promise for reducing the burden of HPV-associated cancer.

Future directions

Viruses have evolved to evade the immune system, not to induce an antiviral immune response. New strategies are being developed to improve vaccines so that they will generate immunity beyond that induced by the virus itself and these are needed, particularly in the fight against chronic viral infection. A number of novel strategies are being studied, including epitope enhancement to modify the amino acid sequence of individual epitopes in order to increase their affinity for MHC molecules; incorporation of cytokines, chemokines, and costimulatory molecules to increase and steer the immune response toward the desired type of immunity; blockade of pathways inhibiting immune responses; and development of approaches to increase CTL avidity (32, 113).

Epitope enhancement. As viral sequences may have been selected by immune pressure to differ from sequences that optimally bind to MHC molecules in order to allow viruses to evade immune elimination, modifying the sequence of weaker epitopes may make them more effective vaccines (32, 113). In our early studies of HIV epitopes, we found that the binding of a helper epitope to its class II MHC molecule could be improved by replacing peptide amino acid residues that created adverse interactions (114). Such modification also increased the peptide's efficacy in a vaccine to maximize the CTL response to an attached CTL epitope (115). Further, we found that the helper response was skewed toward a Th1 cytokine pattern with predominant IFN-γ production. The mechanism could be explained by a reciprocal interaction between helper T cells and dendritic cells in which the higher affinity peptide induced more CD40L expression on the surface of the helper T cells. These in turn induced more IL-12 production by the dendritic cells as well as more costimulatory molecule expression, which skewed the helper T cell phenotype to the Th1 type, made the dendritic cells more effective at activating CTL precursors, and improved protective efficacy (116). Several studies have described epitope enhancement of HIV peptides with low affinity for the most common human class I HLA molecule HLA-A*0201 (117–119), but when modified peptides are used, care must be taken to induce T cells that still respond well to the natural viral sequence (119). This approach has great potential for improving not only peptide vaccines, but any form of vaccine in which such T cell epitopes occur, including recombinant protein, DNA, and viral vector vaccines as well as attenuated viruses.

Use of cytokines, chemokines, and costimulatory molecules. Besides carrying immunogenic epitopes, viruses can also trigger the innate immune system, which alerts the body to danger and helps initiate adaptive immune responses. The signals that transmit these messages from the innate immune system are largely cytokines, chemokines, and costimulatory molecules. Incorporation of these into synthetic vaccines can make the vaccines as effective or more effective than live viruses at eliciting an immune response (reviewed in refs. 32, 120). In addition to individual cytokines, of which the most broadly applicable appears to be GM-CSF, synergies between cytokines such as GM-CSF and IL-12 or the two combined with TNF-α generate more potent immune responses (121–124). IL-15 expressed by a vaccine vector can selectively induce longer-lived memory CTLs (125). Cytokines also synergize with costimulatory molecules to improve the CTL response and antiviral protection (40). A triad of costimulatory molecules can greatly augment CTL responses (126). Chemokines can also be used as adjuvants to enhance immune responses (127, 128). These molecules plus other activators of the innate immune system, such as DNA oligonucleotides of high CpG content, which mimic bacterial DNA (129, 130), can potentiate vaccine efficacy by either triggering or mimicking the innate immune system. They can also steer the immune response toward more protective responses, such as Th1 cytokine production, rather than inhibitory responses. This approach is expected to be a critical component of second-generation vaccine strategies.

Blockade of negative regulatory pathways. Recent work suggests that negative regulation of the immune system is an important braking mechanism that must be overcome to maximize vaccine-induced immune responses (32, 131). CD4$^+$CD25$^+$ T regulatory cells have been found to inhibit other T cell responses (132, 133), and their elimination can improve immune responsiveness to a vaccine (134). Another regulatory T cell subset, CD4$^+$ NK T cells, expresses NK cell markers in addition to conventional T cell receptors and responds to glycolipids presented by CD1 (135). We have found that CD4$^+$ NK T cells can inhibit CTL-mediated tumor immunosurveillance (136–138). These cells act, at least in part, through production of IL-13 (136, 137) and induction of TGF-β (138). Elimination of NK T cells or blockade of IL-13 can increase vaccine-induced CTL responses and protection against an HIV-surrogate virus in a murine model (40). Finally, CTLA-4 has been found to be an inhibitory receptor that binds costimulatory molecules CD80 and CD86 but inhibits T cell responses rather than activating them (139, 140). Blockade of this molecule with antibodies can improve vaccine responses (134, 139), and such anti–CTLA-4 antibodies are now in clinical trials in conjunction with cancer vaccines but may be equally applicable to viruses causing chronic infections. During chronic viral infections, mechanisms that dampen the immune response may contribute to failure to eradicate the virus. Thus we propose that blockade of these regulatory mechanisms may be an important component of a second-generation vaccine strategy for chronic viral diseases.

Induction of high avidity CTLs. High-avidity CTLs are much more effective at eliminating viral infections than low-avidity CTLs (141–143). We recently reviewed the role of high-avidity CTLs in both virus infections and cancer (144). In vitro, high-avidity CTLs can be selectively grown by stimulation with very low concentrations of antigen. However, vaccine strategies to selectively induce high-avidity CTLs in vivo were lacking because very low concentrations of antigen induced no response. However, recently we found that augmentation of costimulation (signal 2) could compensate for a low level of antigen (signal 1) and allow induction of high-avidity CTLs (145). We also recently found that expression of IL-15 by a vaccine vector can select for higher-avidity CTLs that persist longer than low-avidity CTLs, promoting avidity maturation over time (S. Oh, L.P. Perera, D.S. Burke, T.A. Waldmann, and J.A. Berzofsky, unpublished data). We propose that use of such strategies may also be critical for designing the most effective vaccines capable of preventing or eradicating chronic viral infections.

For acute infectious diseases, vaccines have been the most cost-effective agents, saving many millions of lives. However, for chronic viral infections, parasitic and mycobacterial infections, and cancer, the traditional approaches may not be sufficient. Besides implementation of the strategies just described, there are other important goals that need to be attained (see *Vaccines for chronic viral infections: future needs and goals*). Some viruses, such as HIV, invade through mucosal surfaces and grow in mucosal sites, and for those, delivering vaccines by routes that induce mucosal immunity may be critical (32, 34, 36). Mucosal immunization can also provide another benefit in overcoming preexisting systemic immunity to the vaccine vector, such as vaccinia, because of the asymmetry between the mucosal and systemic compartments (146). Use of a DNA-prime and recombinant viral vector boost strategy may also circumvent some preexisting immunity to the viral vector (147). Successful therapeutic vaccination may also require combination with antiviral drug therapy (148). Recent understanding of the immune system has facilitated second-generation vaccine approaches that hold promise for preventing or controlling many of these diseases (149). Some of these new vaccine strategies are being translated into clinical trials, and a combination of these may be necessary to achieve protection against chronic viral infections.

Acknowledgments

We wish to thank Steve Feinstone, Jorge Flores, Allan Hildesheim, and Barbara Rehermann for critical reading of the manuscript and very helpful suggestions. We thank John Mascola from the Vaccine Research Center, National Institute for Allergy and Infectious Diseases, NIH, for suggestions and advice in the preparation of Figure 2.

Address correspondence to: Jay A. Berzofsky, Vaccine Branch, Center for Cancer Research, National Cancer Institute, National Institutes of Health, Building 10, Room 6B-12, 10 Center Drive (MSC#1578), Bethesda, Maryland 20892-1578, USA. Phone: (301) 496-6874; Fax: (301) 480-0681; E-mail: berzofsk@helix.nih.gov.

1. Belshe, R.B. 2004. Current status of live attenuated influenza virus vaccine in the US. *Virus Res.* **103**:177–185.

2. Hilleman, M.R. 2001. Overview of the pathogenesis, prophylaxis and therapeusis of viral hepatitis B, with focus on reduction to practical applications. *Vaccine.* **19**:1837–1848.

3. McAleer, W.J., et al. 1984. Human hepatitis B vaccine from recombinant yeast. *Nature.* **307**:178–180.

4. Rosenberg, E.S., et al. 1997. Vigorous HIV-1-specific CD4+ T cell responses associated with control of viremia. *Science.* **278**:1447–1450.

5. Gerlach, J.T., et al. 1999. Recurrence of hepatitis C virus after loss of virus-specific CD4(+) T-cell response in acute hepatitis C. *Gastroenterology.* **117**:933–941.

6. Thimme, R., et al. 2001. Determinants of viral clearance and persistence during acute hepatitis C virus infection. *J. Exp. Med.* **194**:1395–1406.

7. Grakoui, A., et al. 2003. HCV persistence and immune evasion in the absence of memory T cell help. *Science.* **302**:659–662.

8. Douek, D.C., et al. 2002. HIV preferentially infects HIV-specific CD4+ T cells. *Nature.* **417**:95–98.

9. Sallusto, F., Lenig, D., Forster, R., Lipp, M., and Lanzavecchia, A. 1999. Two subsets of memory T lymphocytes with distinct homing potentials and effector functions. *Nature.* **401**:708–712.

10. Masopust, D., Vezys, V., Marzo, A.L., and Lefrancois, L. 2001. Preferential localization of effector memory cells in nonlymphoid tissue. *Science.* **291**:2413–2417.

11. Seder, R.A., and Ahmed, R. 2003. Similarities and differences in CD4+ and CD8+ effector and memory T cell generation. *Nat. Immunol.* **4**:835–842.

12. Wherry, E.J., et al. 2003. Lineage relationship and protective immunity of memory CD8 T cell subsets. *Nat. Immunol.* **4**:225–234.

13. Missale, G., et al. 1996. Different clinical behaviors of acute hepatitis C virus infection are associated with different vigor of the anti-viral cell-mediated immune response. *J. Clin. Invest.* **98**:706–714.

14. Diepolder, H.M., et al. 1995. Possible mechanism involving T-lymphocyte response to non-structural protein 3 in viral clearance in acute hepatitis C virus infection. *Lancet.* **346**:1006–1007.

15. Gerlach, J.T., et al. 2003. Acute hepatitis C: high rate of both spontaneous and treatment-induced viral clearance. *Gastroenterology.* **125**:80–88.

16. Mascola, J.R., et al. 2000. Protection of macaques against vaginal transmission of a pathogenic HIV-1/SIV chimeric virus by passive infusion of neutralizing antibodies. *Nat. Med.* **6**:207–210.

17. Baba, T.W., et al. 2000. Human neutralizing monoclonal antibodies of the IgG1 subtype protect against mucosal simian-human immunodeficiency virus infection. *Nat. Med.* **6**:200–206.

18. Rowland-Jones, S., Tan, R., and McMichael, A. 1997. Role of cellular immunity in protection against HIV infection. *Adv. Immunol.* **65**:277–346.

19. Goulder, P.J.R., Rowland-Jones, S.L., McMichael, A.J., and Walker, B.D. 1999. Anti-HIV cellular immunity: recent advances towards vaccine design. *AIDS.* **13**(Suppl A):S121–S136.

20. Levy, J.A. 1993. Pathogenesis of human immunodeficiency virus infection [review]. *Microbiol. Rev.* **57**:183–289.

21. Letvin, N.L. 2002. Strategies for an HIV vaccine. *J. Clin. Invest.* **110**:15–20. doi:10.1172/JCI200215985.

22. Castro, B.A., Homsy, J., Lennette, E., Murthy, K.K., Eichberg, J.W., and Levy, J.A. 1992. HIV-1 expression in chimpanzees can be activated by CD8+ cell depletion or CMV infection. *Clin. Immunol. Immunopathol.* **65**:227–233.

23. Schmitz, J.E., et al. 1999. Control of viremia in simian immunodeficiency virus infection by CD8+ lymphocytes. *Science.* **283**:857–860.

24. Jin, X., et al. 1999. Dramatic rise in plasma viremia after CD8+ T cell depletion in simian immunodeficiency virus-infected macaques. *J. Exp. Med.* **189**:991–998.

25. Barouch, D.H., et al. 2000. Control of viremia and prevention of clinical AIDS in rhesus monkeys by cytokine-augmented DNA vaccination. *Science.* **290**:486–492.

26. Hanke, T., et al. 1998. Enhancement of MHC class I-restricted peptide-specific T cell induction by a DNA prime/MVA boost vaccination regime. *Vaccine.* **16**:439–445.

27. Otten, G., et al. 2003. Induction of broad and potent anti-human immunodeficiency virus immune responses in rhesus macaques by priming with a DNA vaccine and boosting with protein-adsorbed polylactide coglycolide microparticles. *J. Virol.* **77**:6087–6092.

28. Amara, R.R., et al. 2001. Control of a mucosal challenge and prevention of AIDS by a multiprotein DNA/MVA vaccine. *Science.* **292**:69–74.

29. Shiver, J.W., et al. 2002. Replication-incompetent adenoviral vaccine vector elicits effective anti-immunodeficiency-virus immunity. *Nature.* **415**:331–335.

30. Casimiro, D.R., et al. 2003. Comparative immunogenicity in rhesus monkeys of DNA plasmid, recombinant vaccinia virus, and replication-defective adenovirus vectors expressing a human immunodeficiency virus type 1 gag gene. *J. Virol.* **77**:6305–6313.

31. Barouch, D.H., et al. 2002. Eventual AIDS vaccine failure in a rhesus monkey by viral escape from cytotoxic T lymphocytes. *Nature.* **415**:335–339.

32. Berzofsky, J.A., Ahlers, J.D., and Belyakov, I.M. 2001. Strategies for designing and optimizing new generation vaccines. *Nat. Rev. Immunol.* **1**:209–219.

33. Belyakov, I.M., et al. 1998. Mucosal immunization with HIV-1 peptide vaccine induces mucosal and systemic cytotoxic T lymphocytes and protective immunity in mice against intrarectal recombinant HIV-vaccinia challenge. *Proc. Natl. Acad. Sci. U. S. A.* **95**:1709–1714.

34. Belyakov, I.M., et al. 1998. The importance of local mucosal HIV-specific CD8+ cytotoxic T lymphocytes for resistance to mucosal-viral transmission in mice and enhancement of resistance by local administration of IL-12. *J. Clin. Invest.* **102**:2072–2081.

35. Belyakov, I.M., Hammond, S.A., Ahlers, J.D., Glenn, G.M., and Berzofsky, J.A. 2004. Transcutaneous immunization induces mucosal CTL and protective immunity by migration of primed skin dendritic cells. *J. Clin. Invest.* **113**:998–1007. doi:10.1172/JCI200420261.

36. Belyakov, I.M., et al. 2001. Mucosal AIDS vaccine reduces disease and viral load in gut reservoir and blood after mucosal infection of macaques. *Nat. Med.* **7**:1320–1326.

37. Murphey-Corb, M., et al. 1999. Selective induction of protective MHC class I restricted CTL in the intestinal lamina propria of rhesus monkeys by transient SIV infection of the colonic mucosa. *J. Immunol.* **162**:540–549.

38. Belyakov, I.M., and Berzofsky, J.A. 2004. Immunobiology of mucosal HIV infection and the basis for development of a new generation of mucosal AIDS vaccines. *Immunity.* **20**:247–253.

39. Kim, J.J., Yang, J.S., Dentchev, T., Dang, K., and Weiner, D.B. 2000. Chemokine gene adjuvants can modulate immune responses induced by DNA vaccines. *J. Interferon Cytokine Res.* **20**:487–498.

40. Ahlers, J.D., et al. 2002. A push-pull approach to maximize vaccine efficacy: abrogating suppression with an IL-13 inhibitor while augmenting help with GM-CSF and CD40L. *Proc. Natl. Acad. Sci. U. S. A.* **99**:13020–13025.

41. McMichael, A., and Hanke, T. 2002. The quest for an AIDS vaccine: is the CD8+ T-cell approach feasible? [review]. *Nat. Rev. Immunol.* **2**:283–291.

42. Wyatt, R., et al. 1998. The antigenic structure of the HIV gp120 envelope glycoprotein. *Nature.* **393**:705–711.

43. Gorny, M.K., et al. 2002. Human monoclonal antibodies specific for conformation-sensitive epitopes of V3 neutralize human immunodeficiency virus type 1 primary isolates from various clades. *J. Virol.* **76**:9035–9045.

44. Sharon, M., et al. 2003. Alternative conformations of HIV-1 V3 loops mimic beta hairpins in chemokines, suggesting a mechanism for coreceptor selectivity. *Structure (Camb.).* **11**:225–236.

45. Ferrari, G., et al. 1997. Clade B-based HIV-1 vaccines elicit cross-clade cytotoxic T lymphocyte reactivities in uninfected volunteers. *Proc. Natl. Acad. Sci. U. S. A.* **94**:1396–1401.

46. Shiver, J. 2003. A non-replicating adenoviral vector as a potential HIV vaccine. *Res. Initiat. Treat. Action.* **8**:14–16.

47. Pinto, L.A., et al. 1999. HIV-specific immunity following immunization with HIV synthetic envelope peptides in asymptomatic HIV-infected patients. *AIDS.* **13**:2003–2012.

48. Bartlett, J.A., et al. 1998. Safety and immunogenicity of an HLA-based HIV envelope polyvalent synthetic peptide immunogen. *AIDS.* **12**:1291–1300.

49. Beddows, S., Lister, S., Cheingsong, R., Bruck, C., and Weber, J. 1999. Comparison of the antibody repertoire generated in healthy volunteers following immunization with a monomeric recombinant gp120 construct derived from a CCR5/CXCR4-using human immunodeficiency virus type 1 isolate with sera from naturally infected individuals. *J. Virol.* **73**:1740–1745.

50. Schooley, R.T., et al. 2000. Two double-blinded, randomized, comparative trials of 4 human immunodeficiency virus type 1 (HIV-1) envelope vaccines in HIV-1- infected individuals across a spectrum of disease severity: AIDS Clinical Trials Groups 209 and 214. *J. Infect. Dis.* **182**:1357–1364.

51. Lambert, J.S., et al. 2001. A phase I safety and immunogenicity trial of UBI microparticulate monovalent HIV-1 MN oral peptide immunogen with parenteral boost in HIV-1 seronegative human subjects. *Vaccine.* **19**:3033–3042.

52. Kahn, J.O., Cherng, D.W., Mayer, K., Murray, H., and Lagakos, S. 2000. Evaluation of HIV-1 immunogen, an immunologic modifier, administered to patients infected with HIV having 300 to 549 x 10(6)/L CD4 cell counts: A randomized controlled trial. *JAMA.* **284**:2193–2202.

53. Moss, R.B., et al. 2000. HIV-1-specific CD4 helper function in persons with chronic HIV-1 infection on antiviral drug therapy as measured by ELISPOT after treatment with an inactivated, gp120-depleted HIV-1 in incomplete Freund's adjuvant. *J. Acquir. Immune Defic. Syndr.* **24**:264–269.

54. Robbins, G.K., et al. 2003. Augmentation of HIV-1-specific T helper cell responses in chronic HIV-1 infection by therapeutic immunization. *AIDS.* **17**:1121–1126.

55. Evans, T.G., et al. 1999. A canarypox vaccine expressing multiple human immunodeficiency virus type 1 genes given alone or with rgp120 elicits broad and durable CD8+ cytotoxic T lymphocyte responses in seronegative volunteers. *J. Infect. Dis.* **180**:290–298.

56. Clements-Mann, M.L., et al. 1998. Immune responses to human immunodeficiency virus (HIV) type 1 induced by canarypox expressing HIV-1MN gp120, HIV-1SF2 recombinant gp120, or both vaccines in seronegative adults. *Infect. Dis.* **177**:1230–1246.

57. Gupta, K., et al. 2002. Safety and immunogenicity of a high-titered canarypox vaccine in combination with rgp120 in a diverse population of HIV-1-uninfected adults: AIDS Vaccine Evaluation Group Protocol 022A. *J. Acquir. Immune Defic. Syndr.*

29:254–261.

58. Belshe, R.B., et al. 2001. Safety and immunogenicity of a canarypox-vectored human immunodeficiency virus Type 1 vaccine with or without gp120: a phase 2 study in higher- and lower-risk volunteers. *J. Infect. Dis.* **183**:1343–1352.

59. Gorse, G.J., Patel, G.B., and Belshe, R.B. 2001. HIV type 1 vaccine-induced T cell memory and cytotoxic T lymphocyte responses in HIV type 1-uninfected volunteers. *AIDS Res. Hum. Retroviruses.* **17**:1175–1189.

60. Cao, H., et al. 2003. Immunogenicity of a recombinant human immunodeficiency virus (HIV)-canarypox vaccine in HIV-seronegative Ugandan volunteers: results of the HIV Network for Prevention Trials 007 Vaccine Study. *J. Infect. Dis.* **187**:887–895.

61. McConkey, S.J., et al. 2003. Enhanced T-cell immunogenicity of plasmid DNA vaccines boosted by recombinant modified vaccinia virus Ankara in humans. *Nat. Med.* **9**:729–735.

62. Novitsky, V., et al. 2002. Magnitude and frequency of cytotoxic T-lymphocyte responses: identification of immunodominant regions of human immunodeficiency virus type 1 subtype C. *J. Virol.* **76**:10155–10168.

63. Mwau, M., and McMichael, A.J. 2003. A review of vaccines for HIV prevention. *J. Gene Med.* **5**:3–10.

64. Jobes, D.V., et al. 2003. Molecular phylogenetic and epidemiologic study of HIV-1 infected individuals from a large-scale phase III vaccine efficacy trial (VAX004). Presented at: AIDS Vaccine 2003 Conference. September 18–21. New York, New York, USA.

65. Richman, D.D., Wrin, T., Little, S.J., and Petropoulos, C.J. 2003. Rapid evolution of the neutralizing antibody response to HIV type 1 infection. *Proc. Natl. Acad. Sci. U. S. A.* **100**:4144–4149.

66. Wei, X., et al. 2003. Antibody neutralization and escape by HIV-1. *Nature.* **422**:307–312.

67. Montefiori, D.C., et al. 2003. Viremia control despite escape from a rapid and potent autologous neutralizing antibody response after therapy cessation in an HIV-1-infected individual. *J. Immunol.* **170**:3906–3914.

68. Burton, D.R., et al. 2004. HIV vaccine design and the neutralizing antibody problem. *Nat. Immunol.* **5**:233–236.

69. Gray, R.H., et al. 2001. Probability of HIV-1 transmission per coital act in monogamous, heterosexual, HIV-1-discordant couples in Rakai, Uganda. *Lancet.* **357**:1149–1153.

70. Quinn, T.C., et al. 2000. Viral load and heterosexual transmission of human immunodeficiency virus type 1. Rakai Project Study Group. *N. Engl. J. Med.* **342**:921–929.

71. Duseja, A., et al. 2002. Hepatitis B and C virus — prevalence and prevention in health care workers. *Trop. Gastroenterol.* **23**:125–126.

72. Koff, R.S. 2003. Hepatitis vaccines: recent advances. *Int. J. Parasitol.* **33**:517–523.

73. Weiner, A.J., et al. 1992. Evidence for immune selection of hepatitis C virus (HCV) putative envelope glycoprotein variants: potential role in chronic HCV infections. *Proc. Natl. Acad. Sci. U. S. A.* **89**:3468–3472.

74. Takaki, A., et al. 2000. Cellular immune responses persist and humoral responses decrease two decades after recovery from a single-source outbreak of hepatitis C. *Nat. Med.* **6**:578–582.

75. Racanelli, V., and Rehermann, B. 2003. Hepatitis C virus infection: when silence is deception. *Trends Immunol.* **24**:456–464.

76. Isaguliants, M.G., and Ozeretskovskaya, N.N. 2003. Host background factors contributing to hepatitis C virus clearance. *Curr. Pharm. Biotechnol.* **4**:185–193.

77. Choo, Q.L., et al. 1994. Vaccination of chimpanzees against infection by the hepatitis C virus. *Proc. Natl. Acad. Sci. U. S. A.* **91**:1294–1298.

78. Esumi, M., et al. 1999. Experimental vaccine activities of recombinant E1 and E2 glycoproteins and hypervariable region 1 peptides of hepatitis C virus in chimpanzees. *Arch. Virol.* **144**:973–980.

79. Tan, S.L., Pause, A., Shi, Y., and Sonenberg, N. 2002. Hepatitis C therapeutics: current status and emerging strategies. *Nat. Rev. Drug Discov.* **1**:867–881.

80. Lai, M.E., et al. 1994. Hepatitis C virus in multiple episodes of acute hepatitis in polytransfused thalassaemic children. *Lancet.* **343**:388–390.

81. Major, M.E., et al. 2002. Previously infected and recovered chimpanzees exhibit rapid responses that control hepatitis C virus replication upon rechallenge. *J. Virol.* **76**:6586–6595.

82. Mehta, S.H., et al. 2002. Protection against persistence of hepatitis C. *Lancet.* **359**:1478–1483.

83. Nascimbeni, M., et al. 2003. Kinetics of CD4+ and CD8+ memory T-cell responses during hepatitis C virus rechallenge of previously recovered chimpanzees. *J. Virol.* **77**:4781–4793.

84. Shoukry, N.H., et al. 2003. Memory CD8+ T cells are required for protection from persistent hepatitis C virus infection. *J. Exp. Med.* **197**:1645–1655.

85. Shirai, M., et al. 1995. CTL responses of HLA-A2.1-transgenic mice specific for hepatitis C viral peptides predict epitopes for CTL of humans carrying HLA-A2.1. *J. Immunol.* **154**:2733–2742.

86. Cerny, A., et al. 1995. Cytotoxic T lymphocyte response to hepatitis C virus-derived peptides containing the HLA A2.1 binding motif. *J. Clin. Invest.* **95**:521–530.

87. Battegay, M., et al. 1995. Patients with chronic hepatitis C have circulating cytotoxic T cells which recognize hepatitis C virus-encoded peptides binding to HLA-A2.1 molecules. *J. Virol.* **69**:2462–2470.

88. Sarobe, P., et al. 1998. Enhanced in vitro potency and in vivo immunogenicity of a CTL epitope from hepatitis C virus core protein following amino acid replacement at secondary HLA-A2.1 binding positions. *J. Clin. Invest.* **102**:1239–1248.

89. Saito, T., et al. 1997. Plasmid DNA-based immunization for hepatitis C virus structural proteins, immune responses in mice. *Gastroenterology.* **112**:1321–1330.

90. Frelin, L., et al. 2003. Low dose and gene gun immunization with a hepatitis C virus nonstructural (NS) 3 DNA based vaccine containing NS4A inhibit NS3/4A-expressing tumors in vivo. *Gene Ther.* **10**:686–699.

91. Pancholi, P., Perkus, M., Tricoche, N., Liu, Q., and Prince, A.M. 2003. DNA immunization with hepatitis C virus (HCV) polycistronic genes or immunization by HCV DNA priming-recombinant canarypox virus boosting induces immune responses and protection from recombinant HCV-vaccinia virus infection in HLA-A2.1-transgenic mice. *J. Virol.* **77**:382–390.

92. Makimura, M., et al. 1996. Induction of antibodies against structural proteins of hepatitis C virus in mice using recombinant adenovirus. *Vaccine.* **14**:28–36.

93. Matsui, M., Moriya, O., and Akatsuka, T. 2003. Enhanced induction of hepatitis C virus-specific cytotoxic T lymphocytes and protective efficacy in mice by DNA vaccination followed by adenovirus boosting in combination with the interleukin-12 expression plasmid. *Vaccine.* **21**:1629–1639.

94. Lechmann, M., et al. 2001. Hepatitis C virus-like particles induce virus-specific humoral and cellular immune responses in mice. *Hepatology.* **34**:417–423.

95. Qiao, M., Murata, K., Davis, A.R., Jeong, S.-H., and Liang, T.J. 2003. Hepatitis C virus-like particles combined with novel adjuvant systems enhance virus-specific immune responses. *Hepatology.* **37**:52–59.

96. Walboomers, J.M., et al. 1999. Human papillomavirus is a necessary cause of invasive cervical cancer worldwide. *J. Pathol.* **189**:12–19.

97. Harro, C.D., et al. 2001. Safety and immunogenicity trial in adult volunteers of a human papillomavirus 16 L1 virus-like particle vaccine. *J. Natl. Cancer. Inst.* **93**:284–292.

98. Schiller, J.T., and Lowy, D.R. 2001. Papillomavirus-like particle based vaccines: cervical cancer and beyond. *Expert Opin. Biol. Ther.* **1**:571–581.

99. Kirnbauer, R., et al. 1993. Efficient self-assembly of human papillomavirus type 16 L1 and L1-L2 into virus-like particles. *J. Virol.* **67**:6929–6936.

100. Greenstone, H.L., et al. 1998. Chimeric papillomavirus virus-like particles elicit antitumor immunity against the E7 oncoprotein in an HPV16 tumor model. *Proc. Natl. Acad. Sci. U. S. A.* **95**:1800–1805.

101. Pinto, L.A., et al. 2003. Cellular immune responses to HPV-16 L1 in healthy volunteers immunized with recombinant HPV-16 L1 virus-like particles. *J. Infect. Dis.* **188**:327–338.

102. Koutsky, L.A., et al. 2002. A controlled trial of a human papillomavirus type 16 vaccine. *N. Engl. J. Med.* **347**:1645–1651.

103. Munger, K., Phelps, W.C., Bubb, V., Howley, P.M., and Schlegel, R. 1989. The E6 and E7 genes of the human papillomavirus type 16 together are necessary and sufficient for transformation of primary human keratinocytes. *J. Virol.* **63**:4417–4421.

104. Borysiewicz, L.K., et al. 1996. A recombinant vaccinia virus encoding human papillomavirus types 16 and 18, E6 and E7 proteins as immunotherapy for cervical cancer. *Lancet.* **347**:1523–1527.

105. van Driel, W.J., et al. 1999. Vaccination with HPV16 peptides of patients with advanced cervical carcinoma: clinical evaluation of a phase I-II trial. *Eur. J. Cancer.* **35**:946–952.

106. Muderspach, L., et al. 2000. A phase I trial of a human papillomavirus (HPV) peptide vaccine for women with high-grade cervical and vulvar intraepithelial neoplasia who are HPV 16 positive. *Clin. Cancer Res.* **6**:3406–3416.

107. Lacey, C.J., et al. 1999. Phase IIa safety and immunogenicity of a therapeutic vaccine, TA-GW, in persons with genital warts. *J. Infect. Dis.* **179**:612–618.

108. Klencke, B., et al. 2002. Encapsulated plasmid DNA treatment for human papillomavirus 16-associated anal dysplasia: a phase I study of ZYC101. *Clin. Cancer Res.* **8**:1028–1037.

109. Sheets, E.E., et al. 2003. Immunotherapy of human cervical high-grade cervical intraepithelial neoplasia with microparticle-delivered human papillomavirus 16 E7 plasmid DNA. *Am. J. Obstet. Gynecol.* **188**:916–926.

110. Santin, A.D., et al. 1999. Induction of human papillomavirus-specific CD4(+) and CD8(+) lymphocytes by E7-pulsed autologous dendritic cells in patients with human papillomavirus type 16- and 18-positive cervical cancer. *J. Virol.* **73**:5402–5410.

111. Santin, A.D., Bellone, S., Gokden, M., Cannon, M.J., and Parham, G.P. 2002. Vaccination with HPV-18 E7-pulsed dendritic cells in a patient with metastatic cervical cancer. *N. Engl. J. Med.* **346**:1752–1753.

112. Goldstone, S.E., Palefsky, J.M., Winnett, M.T., and Neefe, J.R. 2002. Activity of HspE7, a novel immunotherapy, in patients with anogenital warts. *Dis. Colon Rectum.* **45**:502–507.

113. Berzofsky, J.A., et al. 1999. Approaches to improve engineered vaccines for human immunodeficiency virus (HIV) and other viruses that cause chronic infections. *Immunol. Rev.* **170**:151–172.

114. Boehncke, W.-H., et al. 1993. The importance of dominant negative effects of amino acids side chain substitution in peptide-MHC molecule interactions and T cell recognition. *J. Immunol.* **150**:331–341.

115. Ahlers, J.D., Takeshita, T., Pendleton, C.D., and Berzofsky, J.A. 1997. Enhanced immunogenicity

of HIV-1 vaccine construct by modification of the native peptide sequence. *Proc. Natl. Acad. Sci. U. S. A.* **94**:10856–10861.

116. Ahlers, J.D., Belyakov, I.M., Thomas, E.K., and Berzofsky, J.A. 2001. High affinity T-helper epitope induces complementary helper and APC polarization, increased CTL and protection against viral infection. *J. Clin. Invest.* **108**:1677–1685. doi:10.1172/JCI200113463.

117. Pogue, R.R., Eron, J., Frelinger, J.A., and Matsui, M. 1995. Amino-terminal alteration of the HLA-A*0201-restricted human immunodeficiency virus pol peptide increases complex stability and in vitro immunogenicity. *Proc. Natl. Acad. Sci. U. S. A.* **92**:8166–8170.

118. Tourdot, S., et al. 2000. A general strategy to enhance immunogenicity of low-affinity HLA-A2.1- associated peptides: implication in the identification of cryptic tumor epitopes. *Eur. J. Immunol.* **30**:3411–3421.

119. Okazaki, T., Pendleton, D.C., Lemonnier, F., and Berzofsky, J.A. 2003. Epitope-enhanced conserved HIV-1 peptide protects HLA-A2-transgenic mice against virus expressing HIV-1 antigen. *J. Immunol.* **171**:2548–2555.

120. Ahlers, J.D., Belyakov, I.M., and Berzofsky, J.A. 2003. Cytokine, chemokine and costimulatory molecule modulation to enhance efficacy of HIV vaccines. *Curr. Mol. Med.* **3**:285–301.

121. Ahlers, J.D., Dunlop, N., Alling, D.W., Nara, P.L., and Berzofsky, J.A. 1997. Cytokine-in-adjuvant steering of the immune response phenotype to HIV-1 vaccine constructs: GM-CSF and TNFα synergize with IL-12 to enhance induction of CTL. *J. Immunol.* **158**:3947–3958.

122. Ahlers, J.D., Belyakov, I.M., Matsui, S., and Berzofsky, J.A. 2001. Mechanisms of cytokine synergy essential for vaccine protection against viral challenge. *Int. Immunol.* **13**:897–908.

123. Belyakov, I.M., Ahlers, J.D., Clements, J.D., Strober, W., and Berzofsky, J.A. 2000. Interplay of cytokines and adjuvants in the regulation of mucosal and systemic HIV-specific cytotoxic T lymphocytes. *J. Immunol.* **165**:6454–6462.

124. Iwasaki, A., Stiernholm, B.J.N., Chan, A.K., Berinstein, N.L., and Barber, B.H. 1997. Enhanced CTL responses mediated by plasmid DNA immunogens encoding costimulatory molecules and cytokines. *J. Immunol.* **158**:4591–4601.

125. Oh, S., Berzofsky, J.A., Burke, D.S., Waldmann, T.A., and Perera, L.P. 2003. Coadministration of HIV vaccine vectors with vaccinia viruses expressing IL-15 but not IL-2 induces long-lasting cellular immunity. *Proc. Natl. Acad. Sci. U. S. A.* **100**:3392–3397.

126. Hodge, J.W., et al. 1999. A triad of costimulatory molecules synergize to amplify T-cell activation. *Cancer Res.* **59**:5800–5807.

127. Biragyn, A., et al. 2002. DNA vaccines encoding human immunodeficiency virus-1 glycoprotein 120 fusions with proinflammatory chemoattractants induce systemic and mucosal immune responses. *Blood.* **100**:1153–1159.

128. Biragyn, A., Tani, K., Grimm, M.C., Weeks, S., and Kwak, L.W. 1999. Genetic fusion of chemokines to a self tumor antigen induces protective, T-cell dependent antitumor immunity. *Nat. Biotechnol.* **17**:253–258.

129. Klinman, D.M., Barnhart, K.M., and Conover, J. 1999. CpG motifs as immune adjuvants. *Vaccine.* **17**:19–25.

130. Horner, A.A., et al. 2001. Immunostimulatory DNA-based vaccines elicit multifaceted immune responses against HIV at systemic and mucosal sites. *J. Immunol.* **167**:1584–1591.

131. Terabe, M., and Berzofsky, J.A. 2004. Immunoregulatory T cells in tumor immunity. *Curr. Opin. Immunol.* **16**:157–162.

132. Shimizu, J., Yamazaki, S., and Sakaguchi, S. 1999. Induction of tumor immunity by removing CD25+ CD4+ T cells: a common basis between tumor immunity and autoimmunity. *J. Immunol.* **163**:5211–5218.

133. Suri-Payer, E., Amar, A.Z., Thornton, A.M., and Shevach, E.M. 1998. CD4+CD25+ T cells inhibit both the induction and effector function of autoreactive T cells and represent a unique lineage of immunoregulatory cells. *J. Immunol.* **160**:1212–1218.

134. Sutmuller, R.P.M., et al. 2001. Synergism of cytotoxic T lymphocyte-associated antigen 4 blockade and depletion of CD25+ regulatory T cells in antitumor therapy reveals alternative cytotoxic T lymphocyte responses. *J. Exp. Med.* **194**:823–832.

135. Kronenberg, M., and Gapin, L. 2002. The unconventional lifestyle of NKT cells. *Nat. Rev. Immunol.* **2**:557–568.

136. Terabe, M., et al. 2000. NKT cell-mediated repression of tumor immunosurveillance by IL-13 and the IL-4R-STAT6 pathway. *Nat. Immunol.* **1**:515–520.

137. Terabe, M., Park, J.M., and Berzofsky, J.A. 2003. Role of IL-13 in negative regulation of anti-tumor immunity. *Cancer Immunol. Immunother.* **53**:79–85.

138. Terabe, M., et al. 2003. Transforming growth factor-β production and myeloid cells are an effector mechanism through which CD1d-restricted T cells block cytotoxic T lymphocyte-mediated tumor immunosurveillance: abrogation prevents tumor recurrence. *J. Exp. Med.* **198**:1741–1752.

139. Egen, J.G., Kuhns, M.S., and Allison, J.P. 2002. CTLA-4: new insights into its biological function and use in tumor immunotherapy. *Nat. Immunol.* **3**:611–618.

140. Walunas, T.L., et al. 1994. CTLA-4 can function as a negative regulator of T cell activation. *Immunity.* **1**:405–413.

141. Alexander-Miller, M.A., Leggatt, G.R., and Berzofsky, J.A. 1996. Selective expansion of high or low avidity cytotoxic T lymphocytes and efficacy for adoptive immunotherapy. *Proc. Natl. Acad. Sci. U. S. A.* **93**:4102–4107.

142. Derby, M.A., Alexander-Miller, M.A., Tse, R., and Berzofsky, J.A. 2001. High avidity CTL exploit two complementary mechanisms to provide better protection against viral infection than low avidity CTL. *J. Immunol.* **166**:1690–1697.

143. Gallimore, A., Dumrese, T., Hengartner, H., Zinkernagel, R.M., and Rammensee, H.G. 1998. Protective immunity does not correlate with the hierarchy of virus-specific cytotoxic T cell responses to naturally processed peptides. *J. Exp. Med.* **187**:1647–1657.

144. Snyder, J.T., Alexander-Miller, M.A., Berzofsky, J.A., and Belyakov, I.M. 2003. Molecular mechanisms and biological significance of CTL avidity. *Curr. HIV Res.* **1**:287–294.

145. Oh, S., et al. 2003. Selective induction of high avidity CTL by altering the balance of signals from antigen presenting cells. *J. Immunol.* **170**:2523–2530.

146. Belyakov, I.M., Moss, B., Strober, W., and Berzofsky, J.A. 1999. Mucosal vaccination overcomes the barrier to recombinant vaccinia immunization caused by preexisting poxvirus immunity. *Proc. Natl. Acad. Sci. U. S. A.* **96**:4512–4517.

147. Yang, Z.Y., et al. 2003. Overcoming immunity to a viral vaccine by DNA priming before vector boosting. *J. Virol.* **77**:799–803.

148. Hel, Z., et al. 2000. Viremia control following antiretroviral treatment and therapeutic immunization during primary SIV251 infection of macaques. *Nat. Med.* **6**:1140–1146.

149. Berzofsky, J.A., et al. 2004. Progress on new vaccine strategies for the immunotherapy and prevention of cancer [review]. *J. Clin. Invest.* **113**:1515–1525. doi:10.1172/JCI200421926.

Entry of parainfluenza virus into cells as a target for interrupting childhood respiratory disease

Anne Moscona

Department of Pediatrics and Department of Microbiology and Immunology, Weill Medical College of Cornell University, New York, New York, USA.

Human parainfluenza viruses cause several serious respiratory diseases in children for which there is no effective prevention or therapy. Parainfluenza viruses initiate infection by binding to cell surface receptors and then, via coordinated action of the 2 viral surface glycoproteins, fuse directly with the cell membrane to release the viral replication machinery into the host cell's cytoplasm. During this process, the receptor-binding molecule must trigger the viral fusion protein to mediate fusion and entry of the virus into a cell. This review explores the binding and entry into cells of parainfluenza virus type 3, focusing on how the receptor-binding molecule triggers the fusion process. There are several steps during the process of binding, triggering, and fusion that are now understood at the molecular level, and each of these steps represents potential targets for interrupting infection.

The paramyxovirus family of viruses and the parainfluenza viruses

Viruses belonging to the paramyxovirus family, particularly respiratory syncytial virus (RSV), the recently identified human metapneumovirus (1), and the human parainfluenza viruses (HPIVs) types 1, 2, and 3, cause the majority of childhood cases of croup, bronchiolitis, and pneumonia worldwide (2). HPIV3 alone is responsible for approximately 11% of pediatric respiratory hospitalizations in the US (3, 4) and is the predominant cause of croup in young infants, while HPIV1 and -2 tend to infect older children and adolescents. While other causes of respiratory disease in children — influenza and measles — have yielded in part to vaccination programs and antiviral therapy, children are still virtually unaided in their battle against the major causes of croup and bronchiolitis. RSV has been extensively studied, and some effective strategies of prophylaxis have been developed (5), but for the parainfluenza viruses, there are no therapeutic weapons; advances in preventing and treating diseases caused by both groups of viruses, especially the parainfluenza viruses, are far behind those in combatting diseases caused by many more genetically complex pathogens.

The parainfluenza viruses replicate in the epithelium of the upper respiratory tract and spread from there to the lower respiratory tract. Epithelial cells of the small airways become infected, and this is followed by the appearance of inflammatory infiltrates. The relationship among the tissue damage caused by the virus, the immune responses that help to clear the virus, and the inflammatory responses that contribute to disease is still quite enigmatic. Both humoral and cellular components of the immune system appear to contribute to both protection and pathogenesis (6, 7). Infection with HPIV in immunocompromised children (e.g., transplant recipients) is associated with a range of disease, from mild upper-respiratory symptoms to severe disease requiring mechanical ventilation and leading to death (8).

The hurdle for developing modes of preventing and treating croup and bronchiolitis caused by parainfluenza has been in large part a result of the gaps in our understanding of fundamental processes of viral biology and of the interaction of these viruses with their hosts during pathogenesis. For example, an inactivated HPIV1, -2, -3 vaccine used in infants in the late 1960s was immunogenic but did not offer protection from infection (9, 10), which highlights the challenge of identifying which elements of the immune response confer protection from HPIVs. Primary infection with any HPIV does not confer permanent immunity against that virus, and repeated reinfection with the same agent within a year of the previous infection is common in young children. Immunity generated after the first infection is, however, often sufficient to restrict virus replication in the lower respiratory tract and prevent severe disease. Efforts are currently underway to develop live attenuated vaccines against HPIV1, -2, and -3, and an increased understanding of the molecular basis for attenuation of virulence may eventually lead to live HPIV vaccines that can be designed to be both attenuated and immunogenic and even to the development of combination respiratory virus vaccines (reviewed in ref. 11). Deeper understanding of the interplay among virus-mediated pathology, beneficial immune responses, and exaggerated or disease-enhancing inflammatory responses will be vital for developing safe and effective vaccine strategies.

Antiviral therapy for the parainfluenza viruses has not been explored but, in light of the complexities involved in vaccination, could be a principal weapon against these diseases. Several features of the viral life cycle make these viruses vulnerable to attack. HPIVs enter their target cell by binding to a receptor molecule and then fusing their viral envelope with the cell membrane to gain admittance to the cytoplasm. Binding, fusion, and entry are therefore critical stages at which we could interfere with the viral life cycle and prevent disease. A firm grasp of the molecular mechanisms of these events is the basis for understanding respiratory disease pathogenesis and developing potential approaches to prevention and treatment.

The parainfluenza virus life cycle

HPIVs are members of the Respirovirus and Rubulavirus genera within the Paramyxoviridae family. The viruses are roughly spherical in shape, approximately 150–400 nm in diameter, and have

Nonstandard abbreviations used: F protein, viral fusion protein; 4-GU-DANA, 4-guanidino-Neu5Ac2en; HN, hemagglutinin-neuraminidase; HPIV, human parainfluenza virus; HR, heptad repeat; HR-C, C-terminal HR; HR-N, N-terminal HR; NA, neuraminidase protein; NDV, Newcastle disease virus; RSV, respiratory syncytial virus.

Conflict of interest: The author has declared that no conflict of interest exists.

Citation for this article: J. Clin. Invest. **115**:1688–1698 (2005). doi:10.1172/JCI25669.

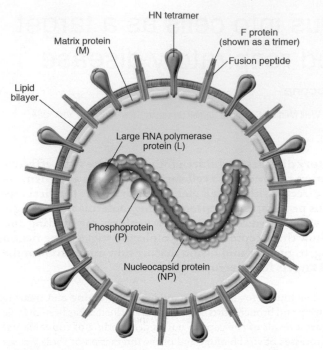

Figure 1

A schematic diagram of the parainfluenza virion. L, large RNA polymerase protein; M, matrix protein; NP, nucleocapsid protein; P, phosphoprotein. Modified with permission from *New England Journal of Medicine* (91).

an envelope composed of host cell lipids and viral glycoproteins derived from the plasma membrane of the host cell during viral budding. The HPIV genome is single-stranded, negative-sense RNA that must be transcribed into message-sense RNA before it can be translated into protein. Like all negative-stranded RNA viruses, the HPIVs encode and package an RNA-dependent RNA polymerase in the virion particles (12). The RNA genome is approximately 15,500 nucleotides in length and is encapsidated by the viral nucleocapsid protein, forming helical nucleocapsids (Figure 1) (13).

The first step in infection of a cell by all HPIVs is binding to the target cell, via interaction of the viral receptor-binding molecule (hemagglutinin-neuraminidase [HN]) with sialic acid–containing receptor molecules on the cell surface (Figure 2). The viral envelope then is thought to fuse directly with the plasma membrane of the cell, mediated by the viral fusion protein (F protein), releasing the nucleocapsid into the cytoplasm (14, 15). The nucleocapsid released into the cytoplasm after fusion contains the genome RNA in tight association with the viral nucleocapsid protein, and this RNA/protein complex is the template both for transcription and for replication of the genome RNA that is packaged into progeny virions. The 6 viral genes encode the 2 surface glycoproteins HN and F; the matrix protein, which is involved in assembly and budding; the RNA polymerase proteins and a protein that encapsidates the RNA; and, through alternative reading frames and/or RNA editing, 1 or more proteins that are expressed only in the infected cell and whose roles include evasion of the host immune response.

Virions are formed, according to the prevailing model for virion assembly, when newly assembled nucleocapsids containing the full-length viral RNA genome along with the polymerase proteins bud out through areas of the plasma membrane that contain the

F and HN proteins and the matrix protein. In polarized epithelial cells, the viruses bud from the apical surface of the cell. The matrix protein binds to the nucleocapsid and also interacts with the cytoplasmic tails of the HN and F proteins, in this way mediating the alignment of the nucleocapsid with the areas of the plasma membrane containing viral glycoproteins in order to set the scenario for budding (16). The neuraminidase or receptor-cleaving activity of the HN molecule cleaves sialic acid–containing receptor moieties that would attach the viral HN protein to the cell surface and allows the release of newly budded particles from the cell to begin a new round of infection (17, 18).

Role of the parainfluenza surface protein HN in receptor binding, receptor cleaving, and F protein activation to mediate fusion

The HN proteins of HPIVs are different from the receptor-binding glycoproteins of other members of the paramyxovirus family in that they possess both hemagglutinating (sialic acid–containing receptor-binding) *and* neuraminidase (sialic acid–containing receptor–cleaving) activities. The parainfluenza HN proteins are oriented such that their amino termini extend into the cytoplasm, while the C termini are extracellular (Figure 1). The HN protein is present on the cell surface and on the virion as a tetramer composed of disulfide-linked dimers (19). The molecule contains a cytoplasmic domain, a membrane-spanning region, a stalk region, and a globular head. Crystal structures of the HN protein of the avian paramyxovirus known as Newcastle disease virus (NDV) (20, 21) and more recently the HPIV3 HN protein (22) demonstrate that the globular head contains the primary sialic acid–binding site and the neuraminidase active site.

Far from simply serving to attach the virus to the surface of the cell and to release virus after replication, the interaction of the HN protein with its receptor is required for F protein-mediated membrane fusion during viral infection (23, 24). Studies of several related paramyxoviruses have revealed that, for most members of this family, the HN protein is essential to the F protein-mediated fusion process (25–27). While receptor binding is an important component of this process, attachment is not sufficient (23, 25, 28); many F proteins demonstrate a requirement for the presence of an HN protein from the same type of virus (the homotypic HN protein) in order to mediate fusion (25, 26). One proposed explanation for this requirement is that the interaction between the HN and F proteins may be type specific and/or that a specific relationship between the structures and/or activities of the 2 proteins is required in order to maintain function (29). This final key function of the HN protein — promotion of fusion — has become amenable to mechanistic study only recently; upon binding to its receptor, parainfluenza HN protein plays a critical role in activating or "triggering" the F protein to assume its fusion-ready conformation (30, 31). Since insertion of the fusion peptide region of the F protein into the target cell membrane after the activation step is the key event leading to membrane fusion, efficiency of F protein triggering by the HN protein is an important variable influencing the extent of fusion mediated by the F protein and thus the extent of viral entry.

Triggering of fusion during entry by enveloped viruses

Entry of all enveloped viruses into host cells requires fusion of the viral and cell membranes. The fusion protein that mediates these processes differs among the enveloped viruses, but thus far these have been mechanistically grouped into just 2 classes of proteins.

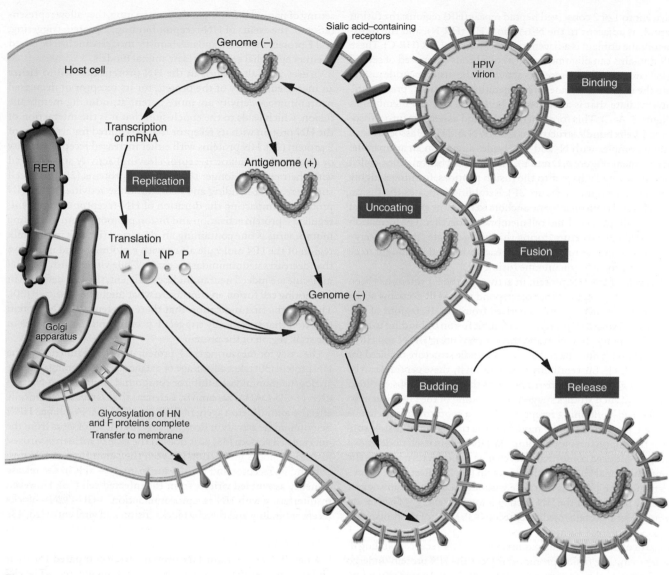

Figure 2
A schematic illustration of the parainfluenza viral life cycle. RER, rough endoplasmic reticulum.

The first, termed class I (reviewed in ref. 32), includes the paramyxovirus fusion proteins as well as the influenza hemagglutinin protein, the HIV gp120 fusion protein, and the Ebola virus fusion protein. Each is synthesized as a single polypeptide chain that forms trimers and is then cleaved by host proteases into 2 subunits, exposing a fusion peptide that will insert into the target cell membrane (33, 34) (reviewed in ref. 35). The trigger that initiates a series of conformational changes in the F protein leading to membrane merger differs depending on the pathway the virus uses to enter the cell and thus whether fusion needs to occur at the surface at neutral pH or in the endosome. The influenza HA protein has been the most extensively studied model for class I fusion (36), and the conformational change is triggered by the acidic pH of the endosome, which then allows the viral and endosomal membranes to fuse (35). Class II fusion proteins include the flavivirus dengue virus E protein (37), tick-borne encephalitis virus E protein (38), and togavirus Semliki Forest virus E1 protein (39), and despite pronounced differences in the structures of class I and class II fusion proteins, their transi-

tion to the post-fusion state proceeds through structures similar enough to suggest a common mechanism (40).

The paramyxovirus fusion process is thought to occur at the surface of the target cell at neutral pH, and activation of the F proteins occurs when the adjacent HN protein binds to the sialic acid–containing receptor, permitting fusion to occur. For HPIV3, the binding of HN protein to its receptor triggers the F protein to fuse with the target cell membrane, and alterations in the HN protein can alter its ability to trigger the F protein (30). The fusion peptides, which are buried within the F protein trimer, must be exposed in order to insert into the target cell membrane, and additional coreceptor binding events, for either the HN or F protein, have not been ruled out. New structural and experimental information about paramyxovirus F proteins has led to models for the structural transitions that occur during class I fusion (31, 32, 40–42) (Figure 3). The ectodomain of the membrane-anchored subunit of the F protein contains 2 hydrophobic domains, the fusion peptide and the transmembrane-spanning domain. Each of these domains is adjacent to 1 of 2 conserved heptad repeat (HR) regions: the fusion peptide is adjacent to the N-terminal HR (HR-N), and the transmembrane domain is adjacent to the C-terminal HR (HR-C). These HR domains can oligomerize into coiled coils composed of several α-helices. Once the F protein is activated, the fusion peptide inserts into the target membrane, first generating a transient "prehairpin" intermediate that is anchored to both viral and cell membranes (Figure 3, A–C). This form then refolds and assembles into a fusogenic 6-helix bundle structure as the HR-N and HR-C associate into a tight complex with N- and C-peptides aligned in an antiparallel arrangement (Figure 3, D and E). The resultant helical coiled-coiled rods are located adjacent to the fusion peptides, forming a highly stable 6-helix bundle (Figure 3F). Refolding relocates the fusion peptides and transmembrane anchors to the same end of the coiled coil, bringing the viral and cell membranes together. The formation of a coiled-coil structure during this step generates the free energy for the membranes to bend toward each other and is thought to be the driving force for membrane fusion (31).

The role of the HN protein in activating the F protein has been explored using peptides that correspond to the HR domains of the F protein. Synthetic peptides derived from the HR regions of several paramyxovirus F proteins can inhibit fusion by binding to their complementary HR region and thereby preventing HR-N and HR-C from refolding into the stable 6-helix bundle structure required for fusion (32, 41). Susceptibility to inhibition by these peptides can be used as a gauge of F protein's progress through the steps outlined above, and such studies suggest that binding of the HN protein to a sialic acid–containing receptor induces a conformational change in the F protein (31). One model for the triggering of fusion suggests that, upon receptor binding, the HN protein itself undergoes a receptor-induced conformational change, which in turn triggers the conformational change in the F protein (14, 24). Results of experiments using F proteins that fuse without the HN protein suggest that the presence of the HN protein lowers the energy of activation required for F protein–mediated fusion (43). Many of the fundamental aspects of this fusion activation process remain to be understood; in particular, why is there a requirement for HN-receptor binding in order to initiate fusion promotion? Does the HN protein undergo conformational change upon receptor binding? An understanding of how the HN protein carries out its F protein–triggering function is central to understanding paramyxovirus entry.

Use of variant HN proteins to scrutinize the HN-triggered F protein–mediated fusion process

We have focused on the process whereby HPIV3 HN protein triggers the F protein after receptor interaction in order to understand this step in the entry of HPIV3 into the host cell. Since the HN protein has 3 functions (receptor binding, neuraminidase, and F protein activation) that each impact the fusion potential of the virus, the field had been hindered by the inability to study just 1 function — triggering — independently of the influence of other HN protein activities. A strategy for quantitating F protein triggering was developed using a panel of mutant HN proteins (17, 18, 24, 30) in order to map the 3 functions of the HN protein to specific regions of the protein. One of the mutant HN proteins displayed a defect in triggering the F protein, which demonstrated for the first time that F protein triggering is a distinct function of the HN protein (30). In our assays, insertion of F protein into target rbc membranes — or fusion between cells — served as a surrogate for the first steps of viral entry into target cells; while these systems differ from the

setting of natural infection in the human lung, they allow representation of the events of HN-receptor binding, F protein triggering, and F protein insertion, and mechanistic models can then be tested in either epithelial cell cultures or animal models.

Viruses with alterations in the HN protein that led to either an increased affinity of the protein for its receptor or decreased neuraminidase activity are more potent at inducing membrane fusion, which leads to the conclusion that it is the interaction of the HN protein with its receptor that is required for activation of F protein (29). HN proteins with either increased receptor affinity or decreased enzymatic (receptor-cleaving) activity stay engaged with their receptor longer than WT HN proteins (29). Both cell surface receptor–binding and neuraminidase activities of the HN protein, by impacting the duration of HN-receptor engagement, regulate F protein activation and fusion promotion. An intriguing mutant virus is one containing an alteration (P111S) in the stalk region of the HN molecule along with a globular head mutation that decreases neuraminidase activity. While viruses with this HN molecule are indeed neuraminidase-deficient, they are defective in promoting cell fusion, in contrast to what might be predicted (30). This was the first indication that there is a separate HN protein function that specifically triggers F protein and that it resides in the stalk region of the protein.

The assay for measuring the F protein–triggering function of the HN protein (30) takes advantage of a sialic acid–containing receptor analog/neuraminidase inhibitor compound, 4-guanidino-Neu5Ac-2en (4-GU-DANA; zanamivir, Relenza) that serves as a clinically effective antiinfluenza agent (44). While 4-GU-DANA inhibits HPIV neuraminidase activity, it does not prevent release of virus from the infected cell surface (45) as it does in the case of influenza viruses; instead, it blocks interaction between the parainfluenza HN protein and its receptor and thus — surprisingly — aids in the release of newly assembled virions from the infected cell (18). However, by interfering with HN-receptor interaction, 4-GU-DANA blocks receptor binding and thereby blocks fusion and viral entry (18, 45). These findings have stimulated interest in designing binding/entry inhibitors for treatment of paramyxovirus infection.

When WT and variant HN proteins were compared for their ability to trigger WT F protein, several findings emerged. The HN protein derived from the neuraminidase-deficient/fusion-defective virus has 2 amino acid alterations (D216N, in the globular head, and P111S, in the stalk) and is slower at activating F protein than either the WT HN protein or the singly mutated D216N HN protein. Comparison with a singly mutated P111S HN protein revealed that this triggering delay is entirely attributable to the P111S mutation (30). F protein triggering was dramatically reduced by this change in the stalk region of the molecule, although there was no decrease in receptor binding avidity. Conducting the experiments at a temperature and pH not permissive of neuraminidase activity eliminated the effects of the HN protein's neuraminidase activity. As a result of the diminished triggering, cell fusion was also markedly reduced. The virus containing the doubly mutated D216N/P111S HN protein is the only paramyxovirus variant virus found to be specifically defective in the HN protein's F protein–triggering and fusion promotion function, and this was the first time that a fusion defect could be specifically attributed to the HN protein's triggering function, independent of the effects of the other 2 HN protein functions (30). The fact that 1 change in the HN protein led to a specific defect in fusion promotion showed that it is, indeed, the HN protein that activates the F protein.

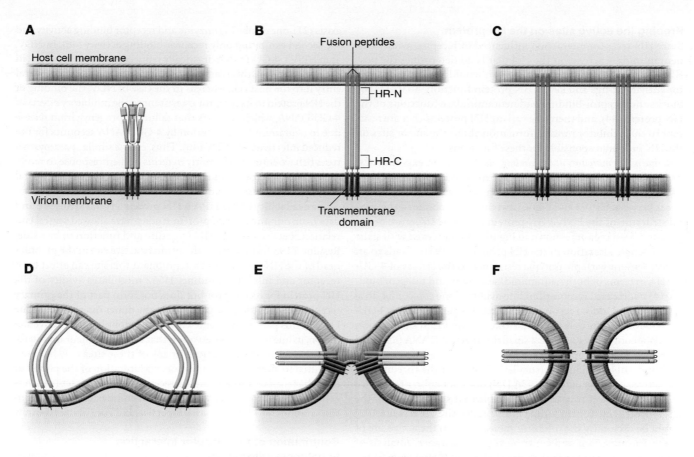

Figure 3
Model of class I fusion protein–mediated membrane fusion. (**A** and **B**) The trimeric paramyxovirus F protein contains 2 hydrophobic domains: the fusion peptide and the transmembrane-spanning domain. Each is adjacent to 1 of 2 HR regions, HR-N and HR-C. (**B**) The F protein binds to a receptor on the host cell membrane, causing a conformational change and subsequent insertion of the hydrophobic fusion peptide into the host cell membrane. (**C**) Multiple F protein trimers are believed to mediate the fusion process. Protein refolding occurs when the host and viral cell membranes bend toward each other (**D**), and formation of a hemifusion stalk allows the lipids on the outer part of the membranes to interact (**E**). (**F**) When protein refolding is completed, the fusion peptide and the transmembrane domain are antiparallel in the same membrane, which creates the most stable form of the fusion protein. Figure modified with permission from *Nature* (40).

Properties of the HN protein that modulate its ability to trigger the F protein

When known variant HN proteins are studied under experimental conditions that allow assessment of all 3 HN protein functions, it becomes evident that the balance among these properties determines entry (29). F protein triggering by the WT HN protein is dramatically reduced at a pH close to the optimum for neuraminidase (pH 5.7); target cell receptors are released from HN protein by neuraminidase, and little triggering occurs. For the neuraminidase-dead HN protein (D216N/P111S HN), however, the rate and extent of F protein triggering are *the same* at both pH 5.7 and pH 8.0 (a pH at which the neuraminidase is not active), which confirms that for the WT HN protein, it is the enhanced neuraminidase activity at low pH that diminishes F protein triggering. Comparison of the doubly mutated protein with the P111S HN (30) revealed the effect of the HN protein's neuraminidase activity: the P111S HN (with residual neuraminidase activity) releases reversibly bound receptors from the HN protein, and unlike the D216N/P111S HN, its triggering cannot "catch up" by remaining in longer contact with receptor. A pH conducive to increased neuraminidase activity (pH 5.7) completely abolishes triggering by the P111S HN. This comparison of

different HN proteins — with neuraminidase activity as the only variable — illustrates the key role of this enzyme in regulating F protein triggering: neuraminidase reduces the chance that the HN protein will remain in contact with target cell receptors and thus prevents the slowly triggering HN protein from performing.

To assess the impact of receptor avidity on triggering, a variant HN (T193A) with higher avidity for receptor than the WT HN, and with WT neuraminidase activity, was useful. For this variant, F protein triggering remained as high at pH 5.7 as at pH 8.0, and target cell receptor release remained as low at pH 5.7 as at pH 8.0, which suggests that higher receptor avidity counterbalances the effect of increased neuraminidase. Both neuraminidase activity and receptor-binding avidity impact receptor availability and thereby the efficiency of the third function, F protein triggering. Thus, while mutations in the stalk region (e.g., P111S) affect the triggering potential of the HN protein, expression of this potential is also modulated by alterations in the globular head that affect HN-receptor interaction. Triggering absolutely depends on HN-receptor interaction, and each of the 3 discrete properties of the HN protein independently affect the ability of the HN protein to complement the F protein in mediating fusion.

Probing the active sites on the HN protein

Since HN-receptor interaction influenced by receptor avidity and neuraminidase activity in the globular head determines the possible extent of triggering, this domain would be a prime target for antiviral drugs and key to HN protein function. 4-GU-DANA blocks the receptor-binding and neuraminidase functions of the HN protein (44), and therefore variant HN proteins that are resistant to this inhibitor reveal information about the site or sites on the HN protein responsible for these functions.

Resistance to inhibitors at the binding site. Just when experimental data had allowed us to generate predictions about the mechanism of 4-GU-DANA resistance in HPIV3 as well as potential differences between HPIV3 and NDV, Lawrence et al. obtained the crystal structure of the HPIV3 HN protein (22). The structure of the globular head region is shown in Figure 4A complexed with sialic acid. A single alteration in the HN protein — T193I — leads to an HPIV3 variant with phenotypic resistance to the effects of 4-GU-DANA in terms of both neuraminidase activity and receptor binding (45). Increased receptor-binding avidity alone can confer drug resistance and indeed accounts for part of the variant virus's 4-GU-DANA–resistant properties (45). However, the T193I substitution does not confer resistance to a smaller molecule, DANA (identical to 4-GU-DANA except for a smaller substituent group at C4). In addition, substitution of a (smaller) alanine for the threonine in the active site (to generate T193A HN) does not confer resistance to 4-GU-DANA. It thus seemed possible that substitution of the larger isoleucine for threonine in the active site (at residue 193) might be excluding the inhibitor molecule from the active site of the resistant variant and contributing to resistance. Analysis of the crystal structure indeed shows that a T193I alteration in the HPIV3 HN protein would likely place the side chain of the isoleucine in a conformation that could lead to steric clash between the isoleucine at 193 and the guanidinium moiety of the 4-GU-DANA. Figure 4B shows the active site region complexed with 4-GU-DANA and reveals the extension of the guanidine moiety into the pocket. The structure thus supports the notion that part of the resistance of the T193I HN variant to 4-GU-DANA indeed arises from a reduction in binding of 4-GU-DANA due to the bulk of the isoleucine side chain. This pathway whereby HPIV3 could develop resistance to such compounds is an issue that needs to be carefully considered in the design of antiviral analogs.

An alteration at the HN protein dimer interface that affects avidity. One HPIV3 variant HN protein (H552Q HN) is resistant to 4-GU-DANA solely due to its higher avidity for the receptor; this HN protein is like the WT HN protein in its neuraminidase activity and neuraminidase sensitivity to 4-GU-DANA. H552 lies at the HN dimeric interface (22) and does not appear to be involved in forming the primary receptor-binding site, which is consistent with the fact that the mutation has no effect on 4-GU-DANA binding affinity or on neuraminidase activity. How can the increased receptor-binding avidity of the H552Q variant (24, 45) be explained? Either the mutation causes an indirect conformational change at the binding site or H552 could represent part of a second receptor-binding site. The possible existence of a second binding site for the HPIV3 HN protein at or near the dimer interface is under study.

HN molecules from related paramyxoviruses differ in their response to receptor analog inhibitors. The 3D crystal structure of the HN protein of the avian paramyxovirus NDV (20) suggests that one site could carry out both binding and neuraminidase activities, but researchers have postulated that an additional HN receptor-binding site

exists (21), one exhibiting enzyme and receptor binding activity and the second exhibiting only receptor binding activity (46). In NDV, as in HPIV3, 4-GU-DANA drastically reduces infection. However, in the case of HPIV3, this is an effect mediated by the blocking of viral entry into the host cell, whereas in the case of NDV, the binding of the HN protein to its receptor is resistant to the inhibitory effects of 4-GU-DANA, which suggests that failure of progeny virion release due to neuraminidase inhibition by 4-GU-DANA accounts for the reduced infectivity for NDV (46). Thus, even 2 similar paramyxoviruses behave entirely differently in terms of their response to receptor analog inhibitors, a finding central to the discussion of antiviral approaches (discussed below).

The availability of the HPIV3 HN protein crystal structure and data from our mutant HN protein studies have allowed several correlations between HPIV3 HN structure and function to be made. Residue T193 forms part of the primary active site in the globular head of the HN protein, and alterations at this site can affect both receptor and inhibitor binding. H552 modulates avidity of the HN protein for its receptor but does not form part of the primary receptor-binding site, lying instead at the dimer interface. Residue D216 forms part of the framework of the globular head active site region, in line with the observation that it is key for neuraminidase activity (Figure 4A). Mutations at any of these sites may have the potential to alter fusion promotion, via alteration of the period of time that the HN protein and its receptor are in contact (which is essential for F protein activation), which emphasizes the relationship among the 3 HN protein properties that contribute to entry.

Contribution of HN-receptor interaction to pathogenesis in vivo

For HPIVs, the interplay among virus-mediated pathology, beneficial immune responses, and disease-enhancing inflammatory responses is not well understood, and it is likely that, as for RSV (47), in many cases, disease severity is increased and the pathology of clinical disease is actually caused by the inflammatory response rather than by the cytopathic effects of the virus (47). This fundamental concept is highlighted by the fact that virus titers in the infected host are generally waning by the time disease symptoms become apparent (2) and that virus titer does not correlate with the severity of lower-respiratory disease. A cotton rat model of disease has proven useful in initial analyses of the factors affecting the pathogenesis of HPIV3 in vivo. Experimental infection of the cotton rat leads to infection of bronchiolar epithelial cells and to bronchiolitis, mimicking human disease, which makes this a relevant model for HPIV3 lower-respiratory infection (48). In a study of cotton rats infected with WT HPIV3 and 3 of the variant HN viruses described above — HN T193A (high receptor avidity, globular head mutation), HN H552Q (high receptor avidity, dimer interface mutation), and HN D216N (low neuraminidase activity, globular head mutation) (49) — there was normal clearance of the variant viruses compared with WT viruses, the variant plaque morphology was preserved in vivo, and there was no reversion to WT phenotype in the infected animals. Quite surprisingly, each of the HN protein alterations led to striking differences in the ability of HPIV3 to cause extensive disease, and this effect was dissociated from effects on viral replication. The variants caused alveolitis and an interstitial infiltrate, while the WT virus only caused peribronchiolitis. The enhanced disease caused by the HN variants was manifested by greatly increased inflammatory cell infiltrate in the alveoli and interstitial spaces in the lung, characterized by notably

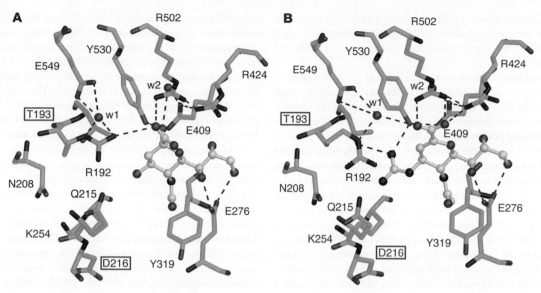

Figure 4
Images of the active site of the HPIV3 HN protein complexed with sialic acid (**A**) and 4-GU-DANA (zanamivir) (**B**) (both shown in yellow). Figure modified with permission from the *Journal of Molecular Biology* (22).

thickened alveolar walls and marked recruitment of inflammatory cells within the air spaces. These results suggest that these differences are indeed due to modulation of the inflammatory response through the different HN protein activity of the variants and are dissociated from viral replication or infectivity.

We hypothesize that changes in the HN protein that alter either the affinity of the HN protein for receptor or receptor-cleaving activity may alter the nature of the inflammatory response of the host. Using HN variants to dissect the etiology of enhanced disease, it may be possible to identify which component(s) of the immune system's response to HPIV3 contributes to disease. Indeed, preliminary experiments suggest that the enhanced pathology observed following infection of the cotton rat with HN-variant HPIVs correlates with specific alterations in the chemokine response to infection that are distinct for each variant HN protein (50). If further experiments support the finding that HN protein alterations specifically alter chemokine expression, this will provide information about the immune contribution to pathogenesis that can be used to develop therapies to modulate an overactive inflammatory response following HPIV3 infection.

For influenza, the severity of disease may be related to the ability of individual strains to induce proinflammatory cytokine expression (51, 52), and cytokine levels appear to correlate with severity of illness (53–55). In a mouse model of disease, it is the HA protein (receptor-binding protein) of the highly virulent 1918 influenza "Spanish flu" that confers the ability to cause severe disease; the disease (as in the cotton rat experiments described above) was widespread and involved recruitment of neutrophils to the alveoli, while viruses with WT HA protein led to only limited involvement of the alveoli, an effect not attributable to differences in replicative ability (56). These findings correlated with greatly enhanced cytokine production, which suggests that this specific HA protein is a critical determinant of macrophage activation and of production of neutrophil chemoattractants. These findings are reminiscent of the enhanced disease caused by the HPIV3 HN variants.

Strategies for blocking fusion and viral entry

Drawing on all that is known about entry into the cell by HPIV and other enveloped viruses, a number of potential strategies for influencing viral fusion become evident: first, blocking or perturbing F-triggering, and second, blocking HN protein–receptor binding. Both events would result in failure of the virus to enter the target cell. As mentioned earlier, peptides derived from the HR-N and HR-C regions of class I F proteins (called HR-N and HR-C peptides) can interfere with fusion intermediates of the F protein (41, 57–61). For example, the HIV envelope glycoprotein gp160 attaches to cellular receptors via its gp120 subunit and mediates fusion via its gp41 subunit; HIV peptides corresponding to the HR-C domain of gp41 are effective for treatment of HIV in humans, and T-20 was the first synthetic HR-C peptide approved for HIV treatment (62, 63). The C-terminus of the HR-N trimer contains a hydrophobic pocket that provides a potential binding site for small molecules that might interfere with the stability of the hairpin structure (64) and could provide advantages over peptides for clinical use. Intriguingly, a low-molecular-weight molecule that is highly effective in inhibiting RSV fusion was recently shown to bind within this hydrophobic pocket of HR-N, which suggests that indeed a small molecule that disrupts the hairpin can derail the fusion process (65); similar results have been obtained for the paramyxovirus simian virus 5, which suggests the general applicability of this approach (66). Inhibition of the F protein triggering process, by peptides or other small molecules that interact with the HR regions, is a promising area for development of antiviral therapies.

While F protein activation is key for entry, the correct timing of F protein activation is also essential; triggering must occur when the F protein is in contact with the target cell membrane. A fusion inhibitor effective against influenza was shown to prematurely trigger the conformational change in HA protein, rendering the virus incapable of fusion (67). A similar mechanism has been proposed for HIV (32, 68). We suggest (29) that correct timing of activation, which for HPIV3 must depend on the balance among the HN protein's receptor-binding, receptor-cleaving, and triggering activities, is critical to entry and represents a potential target for intervention. Swaying the balance of HN protein activities toward premature triggering of parainfluenza F protein may be a strategy for preventing entry.

Since HN-receptor interaction is the critical prelude to F protein triggering, it is an attractive step for blocking viral entry, and the recent availability of 3D structures for NDV and HPIV3 would seem to make it more feasible to design inhibitors that specifically fit into the binding pocket on the globular head of the HN protein. For HPIV3, sialic acid analogs such as 4-GU-DANA, while they do inhibit neuraminidase (18), counteract infection by inhibiting receptor binding. It is thus possible that for HPIV3, sialic acid analogs may be viable antiviral agents by functioning as binding/entry inhibitors. For NDV, however, recent results (see above) indicate the opposite (46); these compounds inhibit neuraminidase and reduce infection in culture but do not completely prevent binding or block viral entry. It is possible that sialic acid analogs that are specifically designed to inhibit the active site of NDV neuraminidase may inhibit virion release as they do in the case of influenza virus (69). These data encourage optimism that receptor blockade may be effective for treating some paramyxoviruses and indeed for HPIV3. The data also highlight the fact that the paramyxoviruses differ from each other in terms of the properties of the HN protein in ways that bear on antiviral development and that each pathogen must be considered individually.

Inhibition of neuraminidase: how HPIV is different from influenza and what can be learned from the influenza experience

While neuraminidase inhibition does not seem a promising strategy for interfering with HPIV infection, this strategy has met with success in treating influenza virus infections. For influenza, HA protein, which recognizes the sialic acid moiety on the cell surface receptor, mediates both receptor binding of the virus to the cell and fusion of the viral envelope with the endosomal membrane; the neuraminidase protein (NA) is necessary for promoting the release of newly formed virions from the cell surface because it removes receptors for the virus, preventing self-aggregation (70). While in the case of HPIV infection, 4-GU-DANA interferes with HN-receptor interaction and thus actually enhances virus release (18), in the case of influenza virus infection, the clinical effectiveness of this molecule has been attributed to its ability to halt spread of the virus when given early in infection (71, 72). We have found that sialic acid–based inhibitors of influenza virus NA can also exert a direct effect on the function of the other envelope protein, HA protein (44). Recent experiments in primary cultures of human airway epithelium cells demonstrated that oseltamivir (a sialic acid analog related to 4-GU-DANA, discussed below) interfered with influenza infection at the early stage of entry (73). Thus, while the effects of 4-GU-DANA on influenza virus have been ascribed purely to the prevention of viral release by neuraminidase inhibition, these results suggested that the antiviral mechanism of action of 4-GU-DANA might be broader and may extend to interfering with viral entry (44). It will be of great interest to determine whether, as is possible for HPIV, neuraminidase may play a role in early infection and whether inhibition at this stage of the viral life cycle contributes to clinical effectiveness.

The design of 4-GU-DANA as a sialic acid analog antiviral compound that mimics the virus's natural substrate proceeded directly from 3D structural studies of the influenza NA (69). Zanamivir is administered by oral inhalation, which delivers the drug directly to the respiratory tract epithelium, and is clinically effective if given early in infection, with remarkably few side effects (74). Shortly after the introduction of zanamivir in clinical practice, an orally available NA inhibitor, oseltamivir, was developed. The NA inhibitors as a class are effective against all NA subtypes and therefore against all strains of influenza (74–82), including the 2004 avian influenza H5N1 strains that are resistant to the M2 inhibitors (83, 84).

An important aspect of the utility of these compounds is that until recently, there seemed to be very little development of resistance to neuraminidase inhibitors (85). The structure-based design of the neuraminidase inhibitors contributed to the fact that it is unlikely for the viral neuraminidase to change in such a way as to confer resistance, while still maintaining function. The neuraminidase inhibitors must fit directly into the enzyme's active site pocket in order to block the enzyme's activity, and since zanamivir was designed to closely resemble the natural substrate, mutations that interfere with zanamivir binding rarely permit enzyme function. In vitro experiments (86) timed the emergence of mutations conferring resistance to several of the then-new neuraminidase inhibitors. The rapidity with which the virus developed resistance to each compound was directly related to how different the inhibitor molecule was from the structure of the natural substrate. This led to the idea that the closer the drug structure is to that of the natural substrate, the less likely it is that the neuraminidase can mutate and maintain function (87); oseltamivir has a variety of modifications from the natural substrate, and hence resistance was considered more likely to develop to this drug than to zanamivir.

It seems plausible now that both the optimistic predictions for zanamivir and the concerns raised about oseltamivir's design are being borne out in clinical practice. For zanamivir, no resistant virus has been isolated after treatment of immunocompetent people, but for oseltamivir, the frequency of post-treatment neuraminidase resistance is higher. While only about 0.4% of treated adults harbored influenza viruses with oseltamivir-resistant neuraminidases, this number rose to at least 4% for treated children. Recently, a small study of children in Japan (88) found that, out of 50 children treated with oseltamivir, 9 (18%) harbored viruses with drug-resistance mutations in the neuraminidase gene; the mutations were located where predicted from the in vitro studies discussed above. These mutations occurred far more frequently than has been previously observed (89), but it has not yet been established whether this is a general phenomenon, nor whether the oseltamivir-resistant viruses are transmissible or pathogenic (90).

This tale, both in the elements of resounding success and in the elements of increasing concern about the development of resistance (88), points to the great utility of structural and in vitro studies brought to bear on development of antiviral therapies. In order to continue to benefit from these potent antiviral compounds, it is critical to understand more about which features of neuraminidase inhibitors will discourage the emergence of resistance (90). By understanding the structural basis of resistance, it ought to be possible to design effective neuraminidase inhibitors that are less likely to select for resistant neuraminidase molecules. The same principles should hold true for designing molecules to interact with the sites on the parainfluenza HN protein that participate in receptor binding and F protein triggering.

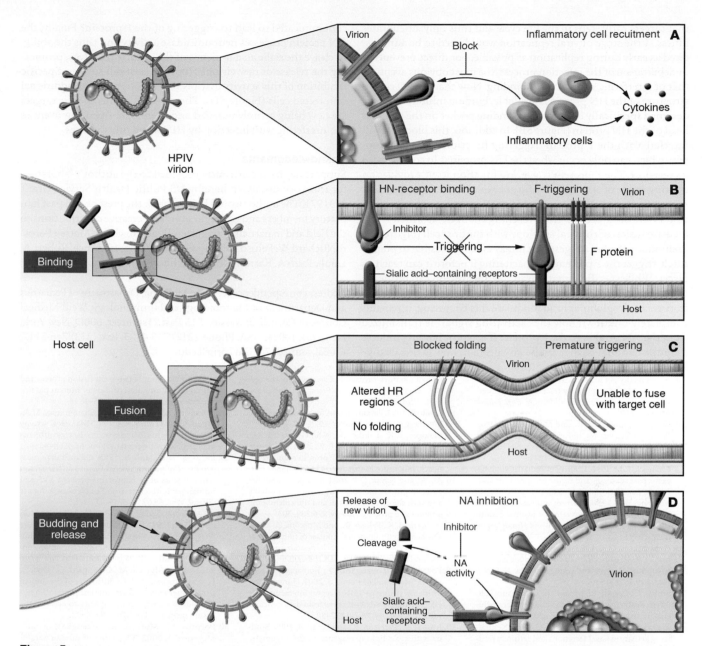

Figure 5
Sites or steps within the viral life cycle that represent potential targets for antiviral molecules. (**A**) Agents directed at blocking the ability of the HN protein to recruit inflammatory cells to the lung and subsequent cytokine expression may reduce the inflammatory response to the HN protein and ameliorate disease. (**B**) Molecules that fit into the binding pocket on the HN globular head may inhibit HN-receptor binding and the subsequent F protein triggering action of the HN protein stalk. On the left, the HN protein is shown bound with an inhibitor, precluding the scenario shown on the right, in which the HN protein's binding to the cell has led to its activation of F protein. (**C**) F protein peptides may be designed to prevent the refolding event that is essential to fusion during virus entry into the host cell. In addition, the F protein could perhaps be prematurely triggered and become incapacitated before it reaches the target host cell membrane. (**D**) Finally, the HN protein has NA activity and thus the ability to cleave the sialic acid moieties of the cellular receptors, promoting the release of new virions from the host cell surface. Specific inhibition of this activity may prevent virion entry into additional uninfected cells.

Using what we know to block parainfluenza virus pathogenesis, and learning more

The multiple roles of the parainfluenza envelope glycoproteins lend themselves to potential strategies for interfering with viral entry, pathogenesis, and disease (Figure 5). Since parainfluenza pathogenesis is likely due in large part to the exuberant inflammatory response to infection, the finding that specific alterations in the HN protein correlate with enhanced pathology, possibly due to the HN protein's role in induction of inflammatory responses, suggests approaches to modulate this inflammatory response and ameliorate disease (Figure 5A). Given the key role of the inflammatory response as well as the facts that viral replication in the respiratory tract peaks soon after disease onset and that viral titers do not correlate directly with disease severity, any antiviral strategy that targets early steps in the life cycle and thus only ameliorates illness at the stage of viral replication would need to be administered as early during replication as possible. For direct prevention or debilitation of the infection process, the most obvious avenue is that of interfering with receptor binding. Now that the 3D crystal structure of the HN protein is available, binding inhibitors can be designed specifically to fit into the binding pocket on the globular head of the HN protein (Figure 5B). In addition, this blockade will interfere with the F protein triggering function of the HN protein, which can only occur when the HN protein is in contact with its receptor. The F protein triggering function itself provides an exciting range of strategies for future exploration. First, peptides corresponding to the HR repeats of the F protein can be designed to prevent the F protein from reaching the state at which it can mediate fusion of the viral envelope with the host cell (Figure 5C). Preliminary studies suggest that it may even be possible to prematurely trigger the F protein, incapacitating it before it can reach its target. While specific mutations in the stalk region can influence the ability of the HN protein to trigger the F protein, and specific features of the globular head modulate this triggering, it remains completely unknown how the "activating signal" is transmitted from the HN protein to the F protein. If receptor binding induces a conformational change in the HN protein, how is this change communicated to lead to triggering of the F protein? Finally, the HN protein possesses neuraminidase activity and thus the ability to cleave the sialic acid moieties of the cellular receptors, promoting the release of new virions from the host cell surface. Specific inhibition of this activity may prevent virion entry into additional uninfected cells (Figure 5D). These potential therapeutic targets are now being actively pursued and promise to open new avenues for interfering with infection by HPIVs and other viruses.

Acknowledgments

Support has been generously provided to the author's laboratory for the work discussed here by US Public Health Service Grant AI31971. The author would like to thank the present and past laboratory members and colleagues for their creative contributions to the field and in particular to gratefully acknowledge Matteo Porotto, Richard W. Peluso, Olga Greengard, Gregory Prince, Robert A. Lamb, Ruth A. Karron, Kurt Hirschhorn, and Carol Heilman.

Address correspondence to: Anne Moscona, Department of Pediatrics and Department of Microbiology and Immunology, Weill Medical College of Cornell University, 515 East 71st Street, 600D, New York, New York 10021, USA. Phone: (212) 746-4523; Fax: (212) 746-8117; E-mail: anm2047@med.cornell.edu.

1. Williams, J.V., et al. 2004. Human metapneumovirus and lower respiratory tract disease in otherwise healthy infants and children. N. Engl. J. Med. 350:443–450.
2. Collins, P., Chanock, R., and McIntosh, K. 1996. Parainfluenza viruses. In Fields virology. B. Fields et al., editors. Lippincott-Raven Publishers. Philadelphia, Pennsylvania, USA. 1205–1241.
3. Chanock, R.M. 1990. Control of pediatric viral diseases: past successes and future prospects [review]. Pediatr. Res. 27(Suppl.):S39–S43.
4. Murphy, B.R. 1988. Current approaches to the development of vaccines effective against parainfluenza viruses [review]. Bull. World Health Organ. 66:391–397.
5. Groothuis, J., et al. 1993. Prophylactic administration of respiratory syncytial virus immune globulin to high-risk infants and young children. N. Engl. J. Med. 329:1524–1530.
6. Smith, C., Purcell, R., Bellanti, J., and Chanock, R. 1966. Protective effect of antibody to parainfluenza type 1 virus. N. Eng. J. Med. 275:1145–1152.
7. Tremonti, L., Lin, J., and Jackson, G. 1968. Neutralizing activity in nasal secretions and serum in resistance of volunteers to parainfluenza virus type 2. J. Immunol. 101:572–577.
8. Apalsch, A.M., Green, M., Ledesma-Medina, J., Nour, B., and Wald, E.R. 1995. Parainfluenza and influenza virus infections in pediatric organ transplant recipients. Clin. Infect. Dis. 20:394–399.
9. Chin, J., Magoffin, R., Shearer, L., Schieble, J., and Lennette, E.1969. Field evaluation of a respiratory syncytial virus vaccine and a trivalent parainfluenza virus vaccine in a pediatric population. Am. J. Epidemiol. 89:449–463.
10. Fulginiti, V., et al. 1969. Respiratory virus immunization. I. A field trial of two inactivated respiratory virus vaccines; an aqueous trivalent parainfluenza virus vaccine and an alum-precipitated respiratory syncytial virus vaccine. Am. J. Epidemiol. 89:435–448.
11. Durbin, A.P., and Karron, R.A. 2003. Progress in the development of respiratory syncytial virus and parainfluenza virus vaccines. Clin. Infect. Dis. 37:1668–1677.
12. Lamb, R.A., Mahy, B.W., and Choppin, P.W. 1976. The synthesis of sendai virus polypeptides in infected cells. Virology. 69:116–131.
13. Choppin, P.W., and Compans, R. 1975. Reproduction of paramyxoviruses. In Comprehensive virology. R.R. Wagner, editor. Plenum Press. New York, New York, USA. 95–178.
14. Lamb, R. 1993. Paramyxovirus fusion: a hypothesis for changes [review]. Virology. 197:1–11.
15. Plemper, R.K., Lakdawala, A.S., Gernert, K.M., Snyder, J.P., and Compans, R.W. 2003. Structural features of paramyxovirus F protein required for fusion initiation. Biochemistry. 42:6645–6655.
16. Ali, A., and Nayak, D.P. 2000. Assembly of Sendai virus: M protein interacts with F and HN proteins and with the cytoplasmic tail and transmembrane domain of F protein. Virology. 276:289–303.
17. Huberman, K., Peluso, R., and Moscona, A. 1995. The hemagglutinin-neuraminidase of human parainfluenza virus type 3: role of the neuraminidase in the viral life cycle. Virology. 214:294–300.
18. Porotto, M., Greengard, O., Poltoratskaia, N. Horga, M.-A., and Moscona, A. 2001. Human parainfluenza virus 3 HN-receptor interaction: the effect of 4-GU-DANA on a neuraminidase-deficient variant. J. Virology. 76:7481–7488.
19. Russell, R., Paterson, R., and Lamb, R. 1994. Studies with cross-linking reagents on the oligomeric form of the paramyxovirus fusion protein. Virology. 199:160–168.
20. Crennell, S., Takimoto, T., Portner, A., and Taylor, G. 2000. Crystal structure of the multifunctional paramyxovirus hemagglutinin-neuraminidase. Nat. Struct. Biol. 7:1068–1074.
21. Zaitsev, V., et al. 2004. Second sialic acid binding site in Newcastle disease virus hemagglutinin-neuraminidase: implications for fusion. J. Virol. 78:3733–3741.
22. Lawrence, M.C., et al. 2004. Structure of the haemagglutinin-neuraminidase from human parainfluenza virus type III. J. Mol. Biol. 335:1343–1357.
23. Moscona, A., and Peluso, R.W. 1991. Fusion properties of cells persistently infected with human parainfluenza virus type 3: participation of hemagglutinin-neuraminidase in membrane fusion. J. Virol. 65:2773–2777.
24. Moscona, A., and Peluso, R.W. 1993. Relative affinity of the human parainfluenza virus 3 hemagglutinin-neuraminidase for sialic acid correlates with virus-induced fusion activity. J. Virol. 67:6463–6468.
25. Hu, X., Ray, R., and Compans, R.W. 1992. Functional interactions between the fusion protein and hemagglutinin-neuraminidase of human parainfluenza viruses. J. Virol. 66:1528–1534.
26. Horvath, C.M., Paterson, R.G., Shaughnessy, M.A., Wood, R., and Lamb, R.A. 1992. Biological activity of paramyxovirus fusion proteins: factors influencing formation of syncytia. J. Virol. 66:4564–4569.
27. Bagai, S., and Lamb, R. 1995. Quantitative measurement of paramyxovirus fusion: differences in requirements of glycoproteins between SV5 and human parainfluenza virus 3 or Newcastle disease virus. J. Virol. 69:6712–6719.
28. Sergel, T., McGinnes, L.W., Peeples, M.E., and Morrison, T.G. 1993. The attachment function of the Newcastle disease virus hemagglutinin-neuraminidase protein can be separated from fusion promotion by mutation. Virology. 193:717–726.
29. Porotto, M., Murrell, M., Greengard, O., Doctor, L., and Moscona, A. 2005. Influence of the human parainfluenza virus 3 attachment protein's neuraminidase activity on its capacity to activate the fusion protein. J. Virol. 79:2383–2392.
30. Porotto, M., Murrell, M., Greengard, O., and Moscona, A. 2003. Triggering of human parainfluenza virus 3 fusion protein (F) by the hemagglutinin-neuraminidase (HN): an HN mutation diminishing the rate of F activation and fusion. J. Virol. 77:3647–3654.
31. Russell, C.J., Jardetzky, T.S., and Lamb, R.A. 2001. Membrane fusion machines of paramyxoviruses: capture of intermediates of fusion. EMBO J. 20:4024–4034.
32. Colman, P.M., and Lawrence, M.C. 2003. The structural biology of type I viral membrane fusion. Nat. Rev. Mol. Cell Biol. 4:309–319.
33. Scheid, A., and Choppin, P. 1974. Identification of biological activities of paramyxovirus glycoproteins. Activation of cell fusion, hemolysis and infectivity by proteolytic cleavage of an inactive protein of Sendai virus. Virology. 57:470–490.
34. Homma, M., and Ouchi, M. 1973. Trypsin action on the growth of sendai virus in tissue culture cells. Structural difference of Sendai viruses grown in eggs and tissue culture cells. J. Virol. 12:1457–1465.
35. Hernandez, L.D., Hoffman, L.R., Wolfsberg, T.G., and White, J.M. 1996. Virus-cell and cell-cell fusion. Annu. Rev. Cell Dev. Biol. 12:627–661.

36. Skehel, J.J., and Wiley, D.C. 2000. Receptor binding and membrane fusion in virus entry: the influenza hemagglutinin. *Annu. Rev. Biochem.* **69**:531–569.

37. Modis, Y., Ogata, S., Clements, D., and Harrison, S.C. 2004. Structure of the dengue virus envelope protein after membrane fusion. *Nature.* **427**:313–319.

38. Bressanelli, S., et al. 2004. Structure of a flavivirus envelope glycoprotein in its low-pH-induced membrane fusion conformation. *EMBO J.* **23**:728–738.

39. Gibbons, D.L., et al. 2004. Conformational change and protein-protein interactions of the fusion protein of Semliki Forest virus. *Nature.* **427**:320–325.

40. Jardetzky, T.S., and Lamb, R.A. 2004. Virology: a class act. *Nature.* **427**:307–308.

41. Baker, K.A., Dutch, R.E., Lamb, R.A., and Jardetzky, T.S. 1999. Structural basis for paramyxovirus-mediated membrane fusion. *Mol. Cell.* **3**:309–319.

42. Dutch, R.E., Jardetzky, T.S., and Lamb, R.A. 2000. Virus membrane fusion proteins: biological machines that undergo a metamorphosis. *Biosci. Rep.* **20**:597–612.

43. Paterson, R.G., Russell, C.J., and Lamb, R.A. 2000. Fusion protein of the paramyxovirus SV5: destabilizing and stabilizing mutants of fusion activation. *Virology.* **270**:17–30.

44. Greengard, O., Poltoratskaia, N., Leikina, E., Zimmerberg, J., and Moscona, A. 2000. The anti-influenza virus agent 4-GU-DANA (zanamivir) inhibits cell fusion mediated by human parainfluenza virus and influenza virus HA. *J. Virol.* **74**:11108–11114.

45. Murrell, M., Porotto, M., Weber, T., Greengard, O., and Moscona, A. 2003. Mutations in human parainfluenza virus type 3 HN causing increased receptor binding activity and resistance to the transition state sialic acid analog 4-GU-DANA (zanamivir). *J. Virol.* **77**:309–317.

46. Porotto, M., et al. 2004. Inhibition of parainfluenza type 3 and Newcastle disease virus hemagglutinin-neuraminidase receptor binding: effect of receptor avidity and steric hindrance at the inhibitor binding sites. *J. Virol.* **78**:13911–13919.

47. Openshaw, P.J., Culley, F.J., and Olszewska, W. 2001. Immunopathogenesis of vaccine-enhanced RSV disease. *Vaccine.* **20**(Suppl. 1):S27–S31.

48. Porter, D., Prince, G., Hemming, V., and Porter, H. 1991. Pathogenesis of human parainfluenza virus 3 infection in two species of cotton rat: Sigmodon hispidus develops bronchiolitis, while Sigmodon fulviventer develops interstitial pneumonia. *J. Virol.* **65**:103–111.

49. Prince, G.A., Ottolini, M.G., and Moscona, A. 2001. Contribution of the human parainfluenza virus type 3 HN-receptor interaction to pathogenesis in vivo. *J. Virol.* **75**:12446–12451.

50. Eichelberger, M., Bauchiero, S., Ottolini, M., Prince, G., and Moscona, A. 2003. Individual mutations that alter the hemagglutinin-neuraminidase molecule result in distinct alterations in the inflammatory response to human parainfluenza 3 viruses. In *Abstracts of papers presented at the 2003 Meeting on Molecular Approaches to Vaccine Design.* Cold Spring Harbor Laboratory. Cold Spring Harbor, New York, USA. 21.

51. Guan, Y., et al. 2004. H5N1 influenza: a protean pandemic threat. *Proc. Natl. Acad. Sci. U. S. A.* **101**:8156–8161.

52. Cheung, C.Y., et al. 2002. Induction of proinflammatory cytokines in human macrophages by influenza A (H5N1) viruses: a mechanism for the unusual severity of human disease? *Lancet.* **360**:1831–1837.

53. Hayden, F.G., et al. 1998. Local and systemic cytokine responses during experimental human influenza A virus infection. Relation to symptom formation and host defense. *J. Clin. Invest.* **101**:643–649.

54. Fritz, R.S., et al. 1999. Nasal cytokine and chemokine responses in experimental influenza A virus infection: results of a placebo-controlled trial of intravenous zanamivir treatment. *J. Infect. Dis.* **180**:586–593.

55. Kaiser, L., Fritz, R.S., Straus, S.E., Gubareva, L., and Hayden, F.G. 2001. Symptom pathogenesis during acute influenza: interleukin-6 and other cytokine responses. *J. Med. Virol.* **64**:262–268.

56. Kobasa, D., et al. 2004. Enhanced virulence of influenza A viruses with the haemagglutinin of the 1918 pandemic virus. *Nature.* **431**:703–707.

57. Rapaport, D., Ovadia, M., and Shai, Y. 1995. A synthetic peptide corresponding to a conserved heptad repeat domain is a potent inhibitor of Sendai virus-cell fusion: an emerging similarity with functional domains of other viruses. *EMBO J.* **14**:5524–5531.

58. Lambert, D.M., et al. 1996. Peptides from conserved regions of paramyxovirus fusion (F) proteins are potent inhibitors of viral fusion. *Proc. Natl. Acad. Sci. U. S. A.* **93**:2186–2191.

59. Yao, Q., and Compans, R.W. 1996. Peptides corresponding to the heptad repeat sequence of human parainfluenza virus fusion protein are potent inhibitors of virus infection. *Virology.* **223**:103–112.

60. Wild, C.T., Shugars, D.C., Greenwell, T.K., McDanal, C.B., and Matthews, T.J. 1994. Peptides corresponding to a predictive alpha-helical domain of human immunodeficiency virus type 1 gp41 are potent inhibitors of virus infection. *Proc. Natl. Acad. Sci. U. S. A.* **91**:9770–9774.

61. Lu, M., Blacklow, S.C., and Kim, P.S. 1995. A trimeric structural domain of the HIV-1 transmembrane glycoprotein. *Nat. Struct. Biol.* **2**:1075–1082.

62. Wild, C., Oas, T., McDanal, C., Bolognesi, D., and Matthews, T. 1992. A synthetic peptide inhibitor of human immunodeficiency virus replication: correlation between solution structure and viral inhibition. *Proc. Natl. Acad. Sci. U. S. A.* **89**:10537–10541.

63. Kilby, J.M., et al. 1998. Potent suppression of HIV-1 replication in humans by T-20, a peptide inhibitor of gp41-mediated virus entry. *Nat. Med.* **4**:1302–1307.

64. Chan, D.C., Chutkowski, C.T., and Kim, P.S. 1998. Evidence that a prominent cavity in the coiled coil of HIV type 1 gp41 is an attractive drug target. *Proc. Natl. Acad. Sci. U. S. A.* **95**:15613–15617.

65. Cianci, C., et al. 2004. Targeting a binding pocket within the trimer-of-hairpins: small-molecule inhibition of viral fusion. *Proc. Natl. Acad. Sci. U. S. A.* **101**:15046–15051.

66. Russell, C.J., Kantor, K.L., Jardetzky, T.S., and Lamb, R.A. 2003. A dual-functional paramyxovirus F protein regulatory switch segment: activation and membrane fusion. *J. Cell Biol.* **163**:363–374.

67. Hoffman, L.R., Kuntz, I.D., and White, J.M. 1997. Structure-based identification of an inducer of the low-pH conformational change in the influenza virus hemagglutinin: irreversible inhibition of infectivity. *J. Virol.* **71**:8808–8820.

68. Neurath, A.R., Strick, N., Jiang, S., Li, Y.Y., and Debnath, A.K. 2002. Anti-HIV-1 activity of cellulose acetate phthalate: synergy with soluble CD4 and induction of "dead-end" gp41 six-helix bundles. *BMC Infect. Dis.* [serial online]. **2**:6. http://www.biomedcentral.com/1471-2334/2/6.

69. von Itzstein, M., et al. 1993. Rational design of potent sialidase-based inhibitors of influenza virus replication. *Nature.* **363**:418–423.

70. Air, G.M., and Laver, W.G. 1989. The neuraminidase of influenza virus. *Proteins.* **6**:341–356.

71. Palese, P., and Compans, R.W. 1976. Inhibition of influenza virus replication in tissue culture by 2-deoxy-2,3-dehydro-N-trifluoroacetylneuraminic acid (FANA): mechanism of action. *J. Gen. Virol.* **33**:159–163.

72. Woods, J.M., et al. 1993. 4-Guanidino-2,4-dideoxy-2,3-dehydro-N-acetylneuraminic acid is a highly effective inhibitor both of the sialidase (neuraminidase) and of growth of a wide range of influenza A and B viruses in vitro. *Antimicrob. Agents Chemother.* **37**:1473–1479.

73. Matrosovich, M.N., Matrosovich, T.Y., Gray, T., Roberts, N.A., and Klenk, H.D. 2004. Neuraminidase is important for the initiation of influenza virus infection in human airway epithelium. *J. Virol.* **78**:12665–12667.

74. Hayden, F.G., et al. 1997. Efficacy and safety of the neuraminidase inhibitor zanamivir in the treatment of influenzavirus infections. GG167 Influenza Study Group. *N. Engl. J. Med.* **337**:874–880.

75. Hayden, F.G., et al. 1999. Use of the oral neuraminidase inhibitor oseltamivir in experimental human influenza: randomized controlled trials for prevention and treatment. *JAMA.* **282**:1240–1246.

76. Hayden, F.G., et al. 2000. Oral oseltamivir in human experimental influenza B infection. *Antivir. Ther. (Lond.).* **5**:205–213.

77. Hayden, F.G., et al. 2000. Inhaled zanamivir for the prevention of influenza in families. Zanamivir Family Study Group. *N. Engl. J. Med.* **343**:1282–1289.

78. Monto, A.S., et al. 1999. Zanamivir in the prevention of influenza among healthy adults: a randomized controlled trial. *JAMA.* **282**:31–35.

79. Welliver, R., et al. 2001. Effectiveness of oseltamivir in preventing influenza in household contacts: a randomized controlled trial. *JAMA.* **285**:748–754.

80. Tumpey, T.M., et al. 2002. Existing antivirals are effective against influenza viruses with genes from the 1918 pandemic virus. *Proc. Natl. Acad. Sci. U. S. A.* **99**:13849–13854.

81. Leneva, I.A., Goloubeva, O., Fenton, R.J., Tisdale, M., and Webster, R.G. 2001. Efficacy of zanamivir against avian influenza A viruses that possess genes encoding H5N1 internal proteins and are pathogenic in mammals. *Antimicrob. Agents Chemother.* **45**:1216–1224.

82. Leneva, I.A., Roberts, N., Govorkova, E.A., Goloubeva, O.G., and Webster, R.G. 2000. The neuraminidase inhibitor GS4104 (oseltamivir phosphate) is efficacious against A/Hong Kong/156/97 (H5N1) and A/Hong Kong/1074/99 (H9N2) influenza viruses. *Antiviral Res.* **48**:101–115.

83. Centers for Disease Control. 2004. Update on influenza A(H5N1) and SARS: interim recommendations for enhanced U.S. surveillance, testing, and infection controls. http://www.cdc.gov/flu/han020302.htm.

84. Commonwealth Scientific and Industrial Research Organization. 2004. CSIRO based drug effective against bird flu. http://www.csiro.au/index.asp?type=mediaRelease&docid=PrBirdFlu5&style=mediarelease.

85. McKimm-Breschkin, J., et al. 2003. Neuraminidase sequence analysis and susceptibilities of influenza virus clinical isolates to zanamivir and oseltamivir. *Antimicrob. Agents Chemother.* **47**:2264–2272.

86. McKimm-Breschkin, J., et al. 1998. Mutations in a conserved residue in the influenza virus neuraminidase active site decreases sensitivity to Neu5Ac 2en-derived inhibitors. *J. Virol.* **72**:2456–2462.

87. Varghese, J.N., et al. 1998. Drug design against a shifting target: a structural basis for resistance to inhibitors in a variant of influenza virus neuraminidase. *Structure.* **6**:735–746.

88. Kiso, M., et al. 2004. Resistant influenza A viruses in children treated with oseltamivir: descriptive study. *Lancet.* **364**:759–765.

89. Whitley, R.J., et al. 2001. Oral oseltamivir treatment of influenza in children. *Pediatr. Infect. Dis. J.* **20**:127–133.

90. Moscona, A. 2004. Oseltamivir-resistant influenza? *Lancet.* **364**:733–734.

91. Breese Hall, C. 2001. Respiratory syncytial virus and parainfluenzae virus. *N. Engl. J. Med.* **344**:1917–1928.

Influenza vaccines: present and future

Peter Palese and Adolfo García-Sastre

Department of Microbiology, Mount Sinai School of Medicine, New York, New York, USA.

The impact of influenza on individuals and on society remains underappreciated. Up to 20% of the population may develop influenza in any given year, and epidemics of influenza are responsible — on average — for 20,000 deaths per year in the US. In some epidemics, this number may climb to more than 40,000 deaths, with over 300,000 influenza-related hospitalizations in a single winter season (1). By far the most catastrophic impact of influenza during the past 100 years was the pandemic of 1918, which cost more than 500,000 lives in the United States (2) and lowered the average life expectancy by almost 10 years.

Epidemiology

The epidemiological success of influenza can be largely explained by the fact that the viruses undergo antigenic change. Strains belonging to two different types (A and B) of influenza virus circulate in the population and are associated with most cases of human influenza. Although their genes have less than 30% overall sequence identity, the two types share a common ancestor and both possess genomes consisting of eight RNAs of negative sense polarity. Hemagglutinin (HA) and neuraminidase (NA) decorate the surface of the lipid-containing virus particles, and these proteins are primarily responsible for the antigenic changes observed in influenza viruses.

New subtypes emerge, it is believed, when genes from animal influenza viruses are captured by the human virus via reassortment, which occurs when a host has been simultaneously infected by both virus types. Such an *antigenic shift* appears to present the human immune system with a novel antigenic experience, usually resulting in high morbidity and mortality. As shown in Figure 1, influenza A viruses of the H1N1 (hemagglutinin 1, neuraminidase 1) subtype circulated from 1918 until 1957. They were then replaced by viruses of the H2N2 subtype, which continued to circulate until 1968. Since 1968, H3N2 viruses have been found in the population. Because H1N1 viruses returned in 1977, two influenza A viruses are presently cocirculating. In addition to antigenic shift, slight changes in the surface proteins caused by point mutations (*antigenic drift*) also allow the viruses to evade the human system, but it may take 3–5 years for a virus of a given subtype to accumulate enough point mutations to cause disease when it reinfects a previously exposed person.

Because type B influenza viruses circulate almost exclusively in humans, these viruses cannot undergo reassortment with animal strains and thus are changed only by antigenic drift. The type C influenza virus represents a distant third in disease-causing potential and is probably of little public health concern.

The present

Inactivated influenza virus vaccines. The main option for reducing the impact of influenza is vaccination. In the US, only inactivated influenza virus vaccines are approved at this time. To be effective, current vaccines must contain an H1N1, an H3N2, and a B virus component (see Table 1). Over the past several years, at least one of the components in the formulation had to be changed due to antigenic drift of the strain circulating in the human population. To prepare vaccines, the viral strains are grown in embryonated eggs, and the virus is then purified and made noninfectious by chemical inactivation.

The effectiveness of the vaccines that are currently available for influenza depends primarily on the antigenic "match" of the circulating viruses with the strains used for vaccination, as well as on a subject's age and immune status. If efficacy is defined as preventing illness, studies show that current inactivated vaccines prevent illness in approximately 70–80% of healthy people under the age of 65, although this number may be far lower (30–40%) in the elderly population. On the other hand, if efficacy is defined as preventing death in the vaccinated individual, the current inactivated vaccines may be up to 80% effective, even in such a high-risk segment of the population as the elderly (3). Recommendations at present include vaccination of all people over the age of 50 and of high-risk groups and those who can transmit influenza to high-risk individuals. The potential benefits of vaccination in preventing disease, hospitalization, or even death greatly outweigh the possible risk of side effects or rare adverse reactions associated with the vaccine.

Antiviral agents. Any discussion of measures against influenza must consider the availability of effective antivirals. Four drugs are approved at present in the US: Amantadine and rimantadine are chemically related inhibitors of the ion-channel M2 protein involved in viral uncoating (4); and zanamivir and oseltamivir are NA inhibitors (5) and prevent the proper release of influenza virus particles from the cytoplasmic membrane. These four drugs are important adjuncts for any medical intervention against influenza (6–8), and three of them can be used in prophylaxis against the virus (zanamivir has not yet been approved). Although these drugs are generally effective, their widespread use is limited by concerns over side effects, patients' compliance, and the possible emergence of drug-resistant variants. Nevertheless, these agents could be of extraordinary value should a new pandemic strain emerge against which a vaccine has not yet been developed. There is good reason to predict that these drugs would be effective against an emerging virus subtype that threatens to start a new pandemic.

The future of influenza vaccine design

Cold-adapted influenza virus vaccine. In Russia, live, cold-adapted influenza virus vaccines have been administered to tens of millions of children with protective efficacy and without evidence of deleterious side effects. In addition, there is no sign of the spread of virulent revertants, either within Russia or globally (9). In the US, the development of a cold-adapted influenza virus vaccine for humans has been pursued for more than 20 years, but licensure has not yet been obtained. Maassab and colleagues (10) passaged influenza viruses in chicken kidney cells and in embryonated

Citation for this article: *J. Clin. Invest.* **110**:9–13 (2002). doi:10.1172/JCI200215999.

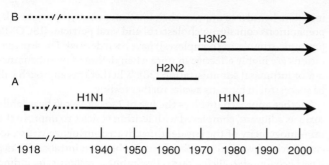

Figure 1
Epidemiology of human influenza A and B viruses. Three different influenza A virus hemagglutinin subtypes (H1, H2, and H3) and two neuraminidase subtypes (N1 and N2) have been identified in humans over the last century. Although no live 1918 virus (solid square) is available, viral RNAs have been sequenced following RT-PCR from formalin-fixed or frozen tissue samples of 1918 victims (35). Rescue of virus containing reconstructed 1918 genes allows study of virulence characteristics of this long-gone strain (36).

eggs to adapt them to growth at 25°C. The resulting cold-adapted master strains are temperature-sensitive and are well suited for use as live vaccines because their pathogenicity, in animals as well as humans, is strongly attenuated. The annually updated vaccine formulations could therefore be generated by making "6:2 reassortants" in which the two genes encoding major viral surface antigens (HA and NA) reflect the sequence found in the current strains, whereas the remaining six genes derive from the cold-adapted master strains.

Such live-virus vaccines can be administered by nasal spray — a distinct advantage over the more difficult and costly route of intramuscular injection using needles. Live viruses can also induce local neutralizing immunity and cell-mediated immune responses, which may be associated with a longer-lasting and more cross-protective immunity than is elicited by chemically inactivated virus preparations. Finally, overall protection may be better in certain age groups (children 6 months to 9 years), and there is also evidence of a drastic reduction in secondary bacterial infections causing otitis media, and thus in the need for use of antibiotics (11). The further use of live influenza virus vaccines will shed light on the benefits, potential risks, and economic and logistical consequences of this approach. Continuing surveillance and monitoring will be needed to safeguard against unexpected complications that might arise from the widespread use of cold-adapted vaccines.

Genetically engineered live influenza virus vaccines. The advent of techniques to engineer site-specific changes in the genomes of negative-strand RNA viruses (12, 13) has made it possible to consider new vaccine approaches. Specifically, it is now possible to tailor-make strains with unique properties that lead to attenuation. For example, exchanging the promoter region of the *NA* gene of an influenza A virus with that of an influenza B virus gene attenuates that strain in mice (14). Alternatively, engineered changes in the *PB2* gene led to a live influenza A virus candidate (15, 16) with interesting biological characteristics.

Live influenza virus vaccine candidates expressing altered NS1 genes. A further improvement of reverse genetics techniques now allows the rescue of influenza virus vaccine candidates from cells transfected with plasmids (Figure 2) (17, 18). This plasmid-only rescue system makes it possible to engineer deletions in the genomes of influenza viruses for improved stability.

We have shown that the NS1 protein of influenza viruses has IFN-antagonist activity (19). Following infection by a virus, the host usually mounts an antiviral IFN response. Many, if not all, viruses express an anti-IFN protein or an IFN-antagonist activity. In the case of influenza viruses, changing the NS1 protein can result in an altered virulence characteristic. Specifically, truncated NS1 proteins are responsible for increased attenuation of both influenza A and influenza B viruses in mice (20). In humans, a virus lacking the *NS1* gene may be too attenuated to be suitable for vaccines. However, we have found that viruses expressing the N-terminal 99 or 126 amino acids of the NS1 protein possess intermediate IFN-antagonist activity in mice. Such genetically engineered viruses may have optimal phenotypic characteristics for stimulating a robust immune response in humans, while at the same time being safely attenuated because they cannot completely overcome the IFN response by the host. It is thought that humans have redundant IFN genes in order to make the IFN response "fail-safe." Consequently, humans — even when immune-compromised — may respond effectively to a virus that has a reduced anti-IFN activity, and thus would not become ill. The fact that the IFN response should be higher in humans infected with NS1 mutants than in those infected with wild-type virus may lead to a vigorous (enhanced) humoral immune response. This is based on the finding that type I IFNs can potently enhance the primary antibody response to proteins and can act as adjuvants in mice (21). If this immune-stimulatory effect is also observed in humans inoculated with NS1-mutant viruses, lower amounts of virus may be used for vaccination. If the inoculum size could be reduced by a factor of 100 relative to the dosage needed for the cold-adapted vaccine, protection of large segments of the population in developing as well as developed countries would become feasible.

Table 1
Changing formulations of influenza virus vaccines

	H3N2	**H1N1**	**B**
1999–2000	A/Sydney/5/1997[A]	A/Beijing/262/1995	B/Beijing/184/1993
2000–2001	A/Moscow/10/1999	A/New Caledonia/20/1999	B/Beijing/184/1993
2001–2002	A/Moscow/10/1999	A/New Caledonia/20/1999	B/Sichuan/379/1999
2002–2003	A/Moscow/10/1999	A/New Caledonia/20/1999	B/Hong Kong/330/2001

Current vaccine formulations are tripartite, containing representative strains from both of the influenza A subtypes that are now circulating (H3N2 and H1N1), as well as an influenza B type. The recommendations for each season identify a specific strain, but antigenically equivalent strains or high-yield reassortants (37) with the respective HA and NA can be used. [A]The strain designation denotes: influenza virus type/location of isolation/isolate number/year of isolation.

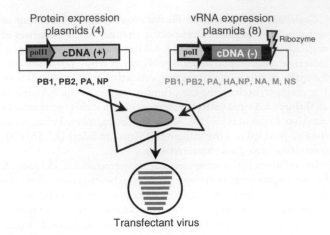

Figure 2
Plasmid-only rescue of infectious influenza virus. Twelve plasmids are introduced into mammalian cells: four plasmids lead to expression of the viral proteins required for viral RNA replication (PA, PB1, PB2, and NP), and the eight transcription plasmids express precise copies of the eight viral RNA segments (PA, PB1, PB2, HA, NP, NA, M, and NS). The resulting viral RNAs are replicated and transcribed by the reconstituted influenza virus RNA-dependent RNA polymerase. Recombinant infectious influenza virus is generated 48–72 hours after transfection of cells (17).

Use of replication-defective influenza viruses as vaccine candidates. Another promising approach for the development of live influenza virus vaccines is the construction of virus particles that undergo only a single cycle of replication. For example, infection of cells with a preparation of virus particles lacking the gene for the NEP (NS2) protein will express viral proteins but will not result in the formation of infectious particles (22). Thus, these preparations induce a protective antibody response and also stimulate a strong cell-mediated immune response without allowing the replication of infectious virus. Another way to attenuate the virus and generate a replication-defective strain is by the elimination of the *M2* gene. Such a deletion strain grows efficiently in tissue culture substrates but only poorly in mice and thus represents a potential live-virus vaccine candidate (23). Since it is believed that the analogous genes of influenza A and B viruses have identical functions, mutations or deletions found to be effective in influenza A viruses could also be tried in influenza B viruses.

DNA vaccination. DNA vaccination involves the administration — by injection or by topical application — of plasmid DNA encoding one or more of the influenza virus proteins. To date, reports on DNA vaccination against influenza have been limited to studies in animal models, including nonhuman primates (24), with most of the work being done in mice, chickens, and ferrets (25, 26). Although progress has been impressive, with protection shown against influenza challenge following DNA vaccination, it is likely that vaccination using DNA will be more appropriate for diseases such as AIDS, for which the use of attenuated vaccine strains may be difficult for safety reasons (see Letvin, this Perspective series, ref. 27). On the other hand, further improvements with DNA vaccination techniques may offer a universal approach to generating protective humoral and cell-mediated responses to a variety of foreign antigens, which may result in the development of effective vaccines against a wide variety of pathogens.

Novel adjuvant approaches. The killed or inactivated influenza virus vaccines presently in use are administered by intramuscular injection. In order to improve their immunogenicity, liposome-like preparations containing cholesterol and viral particles (ISCOMS [immune-stimulating complexes]) have been devised. These preparations are highly effective in mice when delivered by subcutaneous or intranasal administration (28). Whether this approach will be successful in humans awaits further testing.

Another approach involves the use of *Escherichia coli* heat-labile toxin as adjuvant, complexed with lecithin vesicles to improve the immunogenicity of the trivalent inactivated influenza virus vaccine. Such "virosomal" preparations can be given intranasally and have been found to elicit a protective immune response, including an influenza-specific cell-mediated immunity (29). Unfortunately, this vaccine, which was licensed for use in Switzerland, has been recalled since cases of Bell's palsy were reported in some vaccinees.

A universal vaccine against influenza? The possibility of developing a universal influenza vaccine has attracted the attention of many researchers, because the continuing antigenic change of influenza viruses necessitates reformulating the vaccine on a nearly annual basis. Although some components of the virus are more conserved than others, a convincing approach to a universal vaccine based on these conserved parts of the virus has yet to emerge, primarily because the most conserved components of the virus, the minor antigens, are less immunogenic and thus less likely to induce a protective response. Neirynck et al. (30) have attempted to generate a universal vaccine against influenza A virus by fusing the extracellular domain of the conserved M2 protein of an influenza A virus to the hepatitis B virus core protein. They found that intraperitoneal or intranasal administration in mice provided a high degree of protection against viral challenge, but it remains to be seen whether an immune response against a minor surface antigen like the M2 protein will be sufficient to provide full protection against influenza in the human population at large. Vaccines lacking HA and based solely on viral proteins like M2 (30) or NA (31) may not represent an improvement over currently approved vaccines unless the immune responses to these antigens can be strengthened, perhaps by adding appropriate cytokines to the viral protein preparations (32) or by developing acceptable adjuvants for use in humans (33).

As an alternative to targeting only the well-conserved minor antigens, it may be possible to construct a "generic" HA that could be used as an immunogen in vaccines, based on a framework of conserved amino acids. Alternatively, predicting the evolution of human influenza viruses (34) or at least identifying an antigenic trend could be of great benefit in determining the most appropriate strains for use in an influenza virus vaccine. It is even possible that genetic engineering could be used to construct synthetic strains that anticipate evolutionary trends, and that these could be used in future vaccine formulations. Such vaccine candidates may have a better fit with concurrently circulating strains and might not need to be so frequently changed.

Conclusion
Influenza remains a serious disease despite the availability of antivirals and inactivated trivalent vaccines, which are effective for most recipients. Although these modalities of medical intervention are helpful, new approaches are being developed. Specifically, cold-adapted live-virus vaccines, which have been used in millions of people outside the US, are now being considered for approval by the FDA. Other second-generation live-virus vaccines are being designed and tested in animals and are des-

tined to be studied in humans. Major improvements, based on novel adjuvants and recombinant DNA techniques, promise to change the landscape of vaccinology against influenza and many other infectious diseases.

Address correspondence to: Peter Palese, Department of Microbiology, Mount Sinai School of Medicine, One Gustave Levy Place, New York, New York 10029, USA. Phone: (212) 241-7318; Fax: (212) 722-3634; E-mail: peter.palese@mssm.edu.

1. 2001. Prevention and control of influenza. Recommendations of the Advisory Committee on Immunization Practices (ACIP). *Morb. Mortal. Wkly. Rep.* **50**:1–49.
2. Heilman, C., and La Montagne, J. 1990. Influenza: status and prospects for its prevention, therapy and control. *Pediatr. Clin. North Am.* **37**:669–688.
3. Patriarca, P., et al. 1985. Efficacy of influenza vaccine in nursing homes. Reduction in illness and complications during an influenza A (H3N2) epidemic. *JAMA.* **253**:1136–1139.
4. Hay, A., Wolstenholme, A., Skehel, J., and Smith, M. 1985. The molecular basis of the specific anti-influenza action of amantadine. *EMBO J.* **4**:3021–3024.
5. Palese, P., and Compans, R.W. 1976. Inhibition of influenza virus replication in tissue culture by 2-deoxy-2,3-dehydro-N-trifluoroacetylneuraminic acid (FANA): mechanism of action. *J. Gen. Virol.* **33**:159–163.
6. Tominack, R., and Hayden, F. 1987. Rimantadine hydrochloride and amantadine hydrochloride use in influenza A virus infections. *Infect. Dis. Clin. North Am.* **1**:459–478.
7. Treanor, J., et al. 2000. Efficacy and safety of the oral neuraminidase inhibitor oseltamivir in treating acute influenza: a randomized controlled trial. *JAMA.* **283**:1016–1024.
8. Monto, A., Moutl, A., and Sharp, S. 2000. Effect of zanamivir on duration and resolution of influenza symptoms. *Clin. Ther.* **22**:1294–1305.
9. Wareing, M., and Tannock, G. 2001. Live attenuated vaccines against influenza; an historical review. *Vaccine.* **19**:3320–3330.
10. Maassab, H., Heilman, C., and Herlocher, M. 1990. Cold-adapted influenza viruses for use as live vaccines for man. *Adv. Biotechnol. Processes.* **14**:203–242.
11. Nichol, K. 2001. Live attenuated influenza virus vaccines: new options for the prevention of influenza. *Vaccine.* **19**:4373–4377.
12. Enami, M., Luytjes, W., Krystal, M., and Palese, P. 1990. Introduction of site-specific mutations into the genome of influenza virus. *Proc. Natl. Acad. Sci. USA.* **87**:3802–3805.
13. García-Sastre, A. 1998. Negative-strand RNA viruses: applications to biotechnology. *Trends Biotechnol.*

16:230–235.
14. Muster, T., Subbarao, E.K., Enami, M., Murphy, B.R., and Palese, P. 1991. An influenza A virus containing influenza B virus 5′ and 3′ noncoding regions on the neuraminidase gene is attenuated in mice. *Proc. Natl. Acad. Sci. USA.* **88**:5177–5181.
15. Parkin, N., Chiu, P., and Coelingh, K. 1997. Genetically engineered live attenuated influenza A virus vaccine candidates. *J. Virol.* **71**:2772–2778.
16. Murphy, B., Park, E., Gottlieb, P., and Subbarao, K. 1997. An influenza A live attenuated reassortant virus possessing three temperature-sensitive mutations in the PB2 polymerase gene rapidly loses temperature sensitivity following replication in hamsters. *Vaccine.* **15**:1372–1378.
17. Fodor, E., et al. 1999. Rescue of influenza A virus from recombinant DNA. *J. Virol.* **73**:9679–9682.
18. Neumann, G., et al. 1999. Generation of influenza A viruses entirely from cloned cDNAs. *Proc. Natl. Acad. Sci. USA.* **96**:9345–9350.
19. García-Sastre, A., et al. 1998. Influenza A virus lacking the NS1 gene replicates in interferon-deficient systems. *Virology.* **252**:324–330.
20. Talon, J., et al. 2000. Influenza A and B viruses expressing altered NS1 proteins: a vaccine approach. *Proc. Natl. Acad. Sci. USA.* **97**:4309–4314.
21. LeBon, A., et al. 2001. Type I interferons potently enhance humoral immunity and can promote isotype switching by stimulating dendritic cells in vivo. *Immunity.* **14**:461–470.
22. Watanabe, T., Watanabe, S., Neumann, G., Kida, H., and Kawaoka, Y. 2002. Immunogenicity and protective efficacy of replication-incompetent influenza virus-like particles. *J. Virol.* **76**:767–773.
23. Watanabe, T., Watanabe, S., Ito, H., Kida, H., and Kawaoka, Y. 2001. Influenza A virus can undergo multiple cycles of replication without M2 ion channel activity. *J. Virol.* **75**:5656–5662.
24. Donnelly, J., et al. 1995. Preclinical efficacy of a prototype DNA vaccine: enhanced protection against antigenic drift in influenza virus. *Nat. Med.* **1**:583–587.
25. Ljungberg, K., et al. 2000. Effective construction of DNA vaccines against variable influenza genes by homologous recombination. *Virology.* **268**:244–250.

26. Kodihalli, S., Kobasa, D., and Webster, R. 2000. Strategies for inducing protection against avian influenza A virus subtypes with DNA vaccines. *Vaccine.* **18**:2592–2599.
27. Letvin, N.L. 2002. Strategies for an HIV vaccine. *J. Clin. Invest.* **110**:15–20.
28. Sambhara, S., et al. 2001. Heterosubtypic immunity against human influenza A viruses, including recently emerged avian H5 and H9 viruses, induced by FLU-ISCOM vaccine in mice requires both cytotoxic T-lymphocyte and macrophage function. *Cell. Immunol.* **211**:143–153.
29. Glueck, R. 2001. Review of intranasal influenza vaccine. *Adv. Drug Deliv. Rev.* **51**:203–211.
30. Neirynck, S., et al. 1999. A universal influenza A vaccine based on the extracellular domain of the M2 protein. *Nat. Med.* **5**:1157–1163.
31. Kilbourne, E., et al. 1995. Purified influenza A virus N2 neuraminidase vaccine is immunogenic and non toxic in humans. *Vaccine.* **13**:1799–1803.
32. Babai, I., et al. 2001. A novel liposomal influenza vaccine (INFLUSOME-VAC) containing hemagglutinin-neuraminidase and IL-2 or GM-CSF induces protective anti-neuraminidase antibodies cross-reacting with a wide spectrum of influenza A viral strains. *Vaccine.* **20**:505–515.
33. Illum, L., Jabbal-Gill, I., Hinchcliffe, M., Fisher, A., and Davis, S. 2001. Chitosan as a novel nasal delivery system for vaccines. *Adv. Drug Deliv. Rev.* **51**:81–96.
34. Bush, R., Bender, C., Subbarao, K., Cox, N., and Fitch, W. 1999. Predicting the evolution of human influenza A. *Science.* **286**:1866–1867.
35. Reid, A., Fanning, T., Hultin, J., and Taubenberger, J. 1999. Origin and evolution of the 1918 "Spanish" influenza virus hemagglutinin gene. *Proc. Natl. Acad. Sci. USA.* **96**:1651–1656.
36. Basler, C.F., et al. 2001. Sequence of the 1918 pandemic influenza virus nonstructural gene (NS) segment and characterization of recombinant viruses bearing the 1918 NS genes. *Proc. Natl. Acad. Sci. USA.* **98**:2746–2751.
37. Kilbourne, E. 1969. Future influenza vaccines and the use of genetic recombinants. *Bull. World Health Organ.* **41**:643–645.

Live-attenuated virus vaccines for respiratory syncytial and parainfluenza viruses: applications of reverse genetics

Brian R. Murphy and Peter L. Collins

Respiratory Viruses Section, Laboratory of Infectious Diseases, National Institute of Allergy and Infectious Diseases, NIH, Bethesda, Maryland, USA.

Respiratory syncytial virus (RSV) and the human parainfluenza viruses (PIVs) are enveloped nonsegmented negative-strand RNA viruses of the paramyxovirus family. RSV is the leading cause of severe viral respiratory disease in infants and children, and in the United States RSV is responsible for 73,000 to 126,000 annual hospitalizations of infants younger than 1 year of age (1). Two antigenic subgroups of RSV, designated A and B, have been distinguished on the basis of antigenic and sequence dimorphism that is most pronounced for the ectodomain of the G protein, one of the two major protective antigens. Although there is substantial cross-reactivity between the two subgroups, they are sufficiently distinct that an effective RSV vaccine likely will need to represent both.

The human PIV serotypes 1, 2, and 3 also cause severe respiratory tract disease that leads to hospitalization of infants and young children (2). PIV1, PIV2, and PIV3 are distinct serotypes that do not induce significant cross-neutralization or cross-protection. In a long term study of infants and children over a 20-year period, PIV1, PIV2, and PIV3 were identified as etiologic agents responsible for 6.0%, 3.2%, and 11.5%, respectively, of hospitalizations for pediatric respiratory tract disease (1). In total, the three PIVs accounted for only slightly less than the 23% of hospitalizations for pediatric respiratory tract disease caused by RSV. Therefore, there is a need for a vaccine to protect against these three serotypes of PIV and the two antigenic subgroups of RSV. An additional pediatric vaccine will likely be needed for the newly described but related human respiratory tract pathogen human metapneumovirus (3).

The molecular virology of RSV and the PIVs

The molecular virology of RSV and the PIVs is relatively well understood and forms a strong foundation on which to build a vaccine development program using reverse genetics. The genome of RSV (Figure 1a) is a single-stranded negative-sense RNA of 15.2 kb that encodes ten subgenomic mRNAs. Transcription of the viral genes is directed by short, conserved gene-start and gene-end cis-acting signals that flank each gene and are separated by intergenic regions of varying length. These mRNAs are translated into 11 known proteins: four nucleocapsid proteins, namely, nucleocapsid N protein, phosphoprotein P, large polymerase subunit L, and transcription elongation factor M2-1; three transmembrane envelope glycoproteins, namely, fusion F protein, attachment G protein, and small hydrophobic SH protein; two nonstructural proteins, NS1 and NS2; a matrix M protein; and an RNA regulatory factor, M2-2. The M2-1 and M2-2 proteins are expressed from separate open reading frames

(ORFs) in the M2 mRNA. The G and F glycoproteins are the major protective antigens and induce RSV-neutralizing antibodies and resistance to infection.

The genetic maps of the PIVs (shown for PIV3 and PIV1 in Figure 2, a and b, respectively) have some similarity to that of RSV, encoding N, P, M, F, and L proteins that are distantly related functional counterparts to the same proteins in RSV. On the other hand, the human PIVs lack NS1, NS2, SH, and M2 proteins, and the PIV P gene gives rise to additional, accessory proteins (designated C, C', Y1, Y2, V, D and X) that vary in occurrence among the different PIVs (1, 2). The two protective antigens of the PIVs are the HN (hemagglutinin-neuraminidase) attachment protein and the F protein.

We have focused our efforts on developing live-attenuated intranasally administered pediatric RSV and PIV vaccines. This strategy, which mimics natural infection, has the major advantage of inducing a balanced immune response that includes serum and mucosal virus-neutralizing antibodies as well as protective and regulatory components of cellular and innate immunity. In addition, mucosally administered vaccines partially escape the immunosuppressive effect of maternally derived RSV- and PIV-specific serum antibodies present in the young infant. Another important consideration is that vaccines based on inactivated virus or purified proteins have been associated with disease enhancement upon subsequent RSV infection, whereas live-attenuated RSV vaccines or natural infection induces protective immunity without disease enhancement (4). The attenuated viruses are developed using reverse genetics, whereby infectious virus is produced entirely from cDNA, and predetermined changes in the nucleotide sequence can thus be introduced into infectious virus via the cDNA intermediate. The work described here employed reverse genetics systems based on the RSV subgroup A strain A2 (5), PIV3 virus strain JS (6), and bovine PIV3 (BPIV3) (7). More recently, human PIV1 (8) and PIV2 (9) have also been recovered from cDNA and are now available for attenuation to yield new candidate vaccine strains.

The genetic basis of attenuation in biologically derived, live-attenuated virus vaccine candidates

It is very advantageous to know the specific mutations within a live-attenuated vaccine that confer the attenuation phenotype, as well as to know their relative contributions to that phenotype. Knowledge of the genetic basis of attenuation allows one to monitor the stability of the relevant mutations during all phases of vaccine manufacture and usage in humans. Knowledge of the number of attenuating mutations and their nature (for instance, whether they are point mutations or deletion mutations) also helps in interpreting data on the stability of the attenuation phe-

Citation for this article: J. Clin. Invest. 110:21–27 (2002). doi:10.1172/JCI200216077.

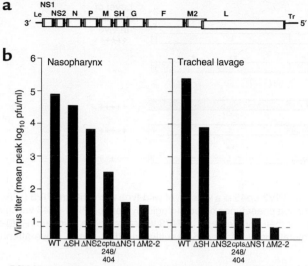

Figure 1

Map of the RSV genome and spectrum of attenuation exhibited by gene-deletion viruses. (**a**) Map of the RSV genome, a single negative-sense RNA of 15.2 kb. Each viral gene is represented by a box, and the gene-start and gene-end transcription signals that flank each gene are indicated by gray and black bars, respectively. The 3' and 5' ends of the genome consist of the extragenic leader (Le) and trailer (Tr) regions, respectively. The genes are separated by short intergenic regions except in the case of the *M2* and *L* genes, which overlap but are nonetheless transcribed into separate mRNAs, as are all of the other genes (1). The diagram is only approximately to scale. (**b**) Mean peak virus titers in the upper (nasopharynx) and lower (tracheal lavage) respiratory tract of chimpanzees that had been inoculated by the intranasal and intratracheal routes simultaneously with the indicated gene-deletion virus or, as controls, with wild-type RSV or with the rRSV*cpts*248/404 virus. This last virus serves as a reference point for a virus with mild residual virulence for seronegative infants (4).

notype following growth in vitro and in vivo. Furthermore, once attenuating mutations have been identified, they can be introduced into cDNA-derived viruses to generate improved candidate vaccine viruses.

Reverse genetics provides a powerful means to define the genetic basis of attenuation of existing biologically derived RSV and PIV vaccine candidates that had been attenuated by conventional methods, such as cold-passage or chemical mutagenesis (10, 11). As a first step, the complete consensus nucleotide sequence must be determined for each virus in question and compared with that of its wild-type parent. The identified mutations or sets of mutations must then be inserted into wild-type recombinant virus via reverse genetics to determine whether they confer an attenuation phenotype (10, 11). For example, a cold-passaged (*cp*) PIV3 (PIV3-*cp*45) that is a promising vaccine candidate has been found to contain 15 potentially significant nucleotide substitutions (Figure 2). The reintroduction of these 15 mutations into wild-type recombinant virus by reverse genetics showed that only six mutations, including both temperature-sensitive (*ts*) and non-*ts* types, contribute substantially to the attenuation phenotype. The presence of multiple *ts* and non-*ts* attenuating mutations provides a likely explanation for the high level of phenotypic stability previously observed for PIV3-*cp*45 following its replication in vivo (12), since

experience with live-attenuated influenza virus vaccines indicates that *ts* mutations are much less likely to revert in the presence of non-*ts* attenuating mutations (13).

In the case of RSV, the most promising live-attenuated viruses are based on a virus called *cp*RSV, which was generated by extensive passage of RSV at suboptimal temperatures. *cp*RSV is only moderately attenuated in chimpanzees and humans and has been found to contain five non-*ts* amino acid substitutions in three proteins (N, F, and L) that, as a set, confer the attenuation phenotype. *cp*RSV was subjected to chemical mutagenesis, and two promising *ts* mutants (RSV*cpts*248 and RSV*cpts*530) were identified and subjected to a second round of mutagenesis, yielding several mutants (including RSV*cpts*248/404 and RSV*cpts*530/1030) with a lower shut-off temperature. Interestingly, each chemical mutagenesis step introduced a single attenuating *ts* mutation (10). All but one of these attenuating *ts* mutations involved single nucleotide substitutions, resulting in single amino acid substitutions in the L polymerase protein. The exception was a strongly attenuating *ts* mutation in RSV*cpts*248/404 involving a single nucleotide substitution in the noncoding, *cis*-acting gene-start signal of the *M2* gene. None of these biologically derived RSV vaccine candidates proved to be sufficiently attenuated for use as a pediatric vaccine, but they represented a starting point for further attenuation by reverse genetics. Furthermore, in a number of cases, the amino acid substitution that conferred the attenuation phenotype could be designed to involve two nucleotide changes relative to the wild-type assignment, rather than the single nucleotide substitution found in the original biological mutants, which should greatly reduce the frequency of reversion and thus increase phenotypic stability.

Reverse genetics has also been used to determine the genetic basis of attenuation of a Jennerian PIV3 vaccine candidate, that is, a BPIV3 wild-type virus that is antigenically related to human PIV3 and is highly attenuated and phenotypically stable in humans due to a natural host range restriction (14, 15). Bailly et al. (16) constructed a chimeric recombinant human-bovine PIV3 virus (rPIV3-NB) in which the nucleoprotein (N) ORF of the human PIV3 was replaced by its counterpart from BPIV3. The replication of rPIV3-NB in rhesus monkeys was restricted to an extent similar to that of its BPIV3 parent, showing that the BPIV3 N protein helps determine the host range restriction (that is, the attenuation phenotype) of BPIV3 in primates (16). There are 79 differences out of a total of 515 amino acids between the N proteins of human PIV3 and BPIV3, many of which probably contribute to the host range attenuation phenotype of rPIV3-NB. Therefore, this strain should be stable even following extensive replication in vivo. rPIV3-NB, which combined the antigenic determinants of PIV3 with the host range restriction and attenuation phenotype of BPIV3, induced a level of resistance to PIV3 challenge in monkeys that was indistinguishable from that conferred by immunization with human PIV3. Thus, the process of analyzing the genetic basis of attenuation of BPIV3 for primates led to the development of the promising vaccine candidate rPIV3-NB and identified chimerization of human and bovine PIV3 viruses as a novel method to generate an attenuated PIV3 vaccine. In summary, reverse genetics has allowed us to explore the genetic basis of attenuation of at least ten biologically derived RSV and PIV3 vaccine candidates and has provided insight into the genetic basis of stability of the attenuation phenotype following replication of these vaccine viruses in vivo.

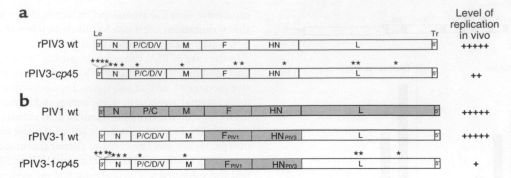

Figure 2
Structure of the PIV genome in wild-type, attenuated, and chimeric viruses. (**a**) rPIV3-*cp*45 is a version of wild-type recombinant PIV3 (rPIV3 wt) that contains 15 point mutations (asterisks) that confer the *ts* and attenuation phenotypes of the biologically derived PIV3-*cp*45 vaccine candidate virus. (**b**) The *HN* and *F* ORFs of PIV1 wt virus were substituted for their counterparts in PIV3 wt, creating the chimeric rPIV3-1 virus. This was then attenuated by the further introduction of 12 of the 15 mutations of PIV3-*cp*45, resulting in rPIV3-1*cp*45, a live-attenuated vaccine candidate for PIV1.

Developing a menu of attenuating viral mutations

Practical considerations dictate that a live-attenuated vaccine candidate must be not only satisfactorily attenuated but also capable of efficient replication in cultured cells suitable for large scale production and use in humans. This is particularly important because attenuated viruses often exhibit decreased infectivity for the host, necessitating the administration of a relatively high dose of the vaccine (17). Attenuating mutations with desired properties can be assembled into a large menu of mutations that can be combined to derive a vaccine strain that possesses the required balance between attenuation and immunogenicity.

In our studies, three major sources of attenuating mutations were used to compile menus for RSV and PIV3. The first source was existing, biologically derived attenuated vaccine candidate viruses, including multiply passaged viruses (such as PIV3-*cp*45 and *cp*RSV), mutants derived by chemical mutagenesis (such as the *cpts* RSV mutants), and host range–restricted viruses (such as BPIV3). These viruses replicate efficiently in cell culture and have been confirmed to be safe and attenuated in preclinical and clinical studies; hence they represent a valuable resource. In addition, the attenuating host range sequences in the BPIV3 N protein exhibit a high level of phenotypic stability. Attenuating *ts* mutations are desirable because they can be grown efficiently in vitro at permissive temperature and because different mutations can specify different levels of attenuation (18). For respiratory viruses such as RSV and the PIVs, attenuating *ts* mutations also are desirable because they specify restricted replication preferentially in the warmer lower respiratory tract, the site of the most severe disease. Six RSV and three PIV attenuating *ts* mutations have been identified by sequence analysis of biologically derived vaccine candidates and by reverse genetics (10, 11).

The second source of attenuating mutations is the generation of novel mutations not found among existing virus mutants or vaccine candidates. RSV and the PIVs encode several nonessential proteins whose major function is to facilitate virus replication in vivo, for instance, by antagonizing host defense factors. Deleting or silencing such accessory genes or ORFs is attractive because the resulting virus frequently retains replicative efficiency in cell culture. If these viral accessory factors enhance pathogenesis or downregulate host immune responses, their ablation might improve the safety and efficacy of the resulting vaccine (10). Such mutations can be made by deleting a complete transcriptional unit encoding an mRNA and protein, such as has been done for the *NS1*, *NS2*, and *SH* genes of RSV, or by deleting or silencing a single ORF contained within a complex mRNA encoding more than one protein, as has been done to eliminate expression of the M2-2 protein of RSV and the C, D, and V proteins of PIV3 (19, 20). Deletion of a complete gene is a desirable mutation because it would be expected to be more genetically and phenotypically stable than point mutations.

The ΔSH, ΔNS1, ΔNS2, and ΔM2-2 deletion mutants of RSV exhibit a spectrum of attenuation for the respiratory tract of chimpanzees (Figure 1b) (21, 22) and other experimental animals (23–25). In particular, the deletion mutant RSVΔNS1 is slightly more attenuated in the chimpanzee than is RSV*cpts*248/404, a biologically derived vaccine candidate that retains some mild residual virulence for the upper respiratory tract of the 2-month-old human infant (4, 22). Despite its highly restricted replication in chimpanzees, RSVΔNS1 induces a high level of resistance to RSV challenge and represents a very promising vaccine candidate.

Deletion mutants of the *D* and *V* ORFs of PIV3 appear to be useful additions to the menu of attenuating mutations, whereas deletion of the PIV3 *C* ORF produced a virus that appears to be overattenuated (19, 20). Reverse genetics also could be used to develop new point mutations not previously observed in biologically derived strains. For example, mutation could be introduced via the cDNA intermediate into individual proteins either randomly or by replacing single or paired charged residues with alanine residues. In this case, each mutation would have to be characterized with regard to viability, replication efficiency, and *ts* and attenuation phenotype of the recovered virus. Other novel attenuating mutations that can be generated using reverse genetics include large nucleotide insertions or changes in the gene order. However, these mutations typically reduce the efficiency of virus replication in vitro and thus would complicate the manufacture of vaccine virus (26, 27).

A third source of attenuating mutations involves existing attenuated mutants of other paramyxoviruses. Using existing sequence relatedness among paramyxoviruses as a guide, mutations that involve conserved residues or domains can be "transferred" from one paramyxovirus to another. For example, a point mutation (F521L) in the L polymerase of RSV*cpts*530 (a biologically derived live-attenuated RSV vaccine candidate mentioned above)

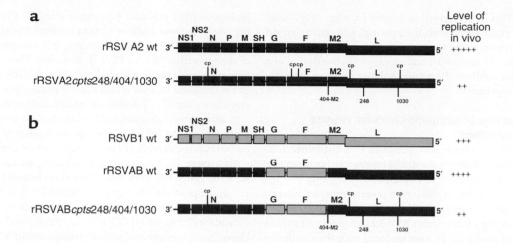

Figure 3

Use of reverse genetics to introduce new combinations of mutations into rRSV A2 of subgroup A and to expedite the development of a live-attenuated vaccine virus specific to RSV subgroup B. (**a**) Recombinant wild-type RSV (rRSV A2 wt) was used as a backbone for the introduction of the five point mutations of *cp*RSV (cp), two attenuating mutations from RSV*cpts*248/404 (248-L and 404-M2), and one attenuating mutation from RSV*cpts*530/1030 (1030) to create the new vaccine candidate rRSVA2*cpts*248/404/1030. (**b**) The genes encoding the F and G antigenic determinants of the subgroup B RSV B1 wt virus were substituted into the backbone of rRSV A2 wt to create the chimeric rRSV AB wt virus, which was then attenuated by the introduction of three mutations of *cp*RSV, two mutations from RSV*cpts*248/404, and one mutation from RSV*cpts*530/1030. The level of viral replication in the respiratory tract of rodents or chimpanzees is indicated on the right.

has been transferred using reverse genetics to the homologous position (residue 456) of the distantly related PIV3 *cp*45 (10). The acquisition of the F456L mutation by PIV3-*cp*45 resulted in a small, incremental increase in its level of attenuation in rhesus monkeys and chimpanzees (28). This virus is now available for evaluation should the ongoing clinical evaluation of PIV3-*cp*45 indicate that a small increase in attenuation would be desirable. In a second example, a point mutation (F170S) in the C protein of a multiply passaged attenuated mutant of Sendai virus (29) (a PIV1 of murine origin) has been introduced into the homologous position (residue 164) of the C protein of human PIV3 (19). The resulting recombinant PIV3, designated rF164S, replicates very efficiently in vitro but is attenuated in monkeys, especially for the upper respiratory tract. On this basis, the F164S mutation has been added to the menu of attenuating PIV3 mutations. The ability to import attenuating mutations from one virus to another greatly expands the menu of attenuating mutations that can be used to produce candidate vaccines.

An important purpose of compiling a list of attenuating mutations is that mutations from this list can be added to insufficiently attenuated viruses to augment both their level of attenuation and their phenotypic stability in vivo. This principle is illustrated by RSV vaccine candidates presently in clinical trials (30). The RSV 1030 attenuating *ts* mutation (Y1321N) in the L protein of RSV*cpts*530/1030 has been introduced into the insufficiently attenuated RSV*cpts*248/404 vaccine candidate (4) to yield rRSVA2*cpts*248/404/1030, which is more *ts* and more attenuated than its RSV*cpts*248/404 parent and represents a promising new RSV vaccine candidate (Figure 3). A further modification, namely deletion of the *SH* gene, created rA2*cpts*248/404/1030ΔSH, which is now in clinical trials in infants and children and appears to be more attenuated than RSV*cpts*248/404 (31). Thus, rRSVA2*cpts*248/404/1030ΔSH has mutations gathered using reverse genetics from five different RSV mutants: *cp*RSV, RSV*cpts*248, RSV*cpts*248/404, RSV*cpts*530/1030, and RSVΔSH. This illustrates the power of com-

bining mutations from a menu of attenuating mutations into a single mutant in response to information gathered from ongoing clinical trials, and it shows that knowledge of the individual properties of the mutations can guide the choice of which mutations to include in a new vaccine candidate.

Several factors need to be considered in compiling a menu of useful attenuating mutations for respiratory viruses. First, it is preferable to include mutations that yield various levels of attenuation, since it might be necessary to further restrict the level of replication of an incompletely attenuated candidate over a 10- to 1000-fold range. Second, the menu should include non-*ts* mutations, which can enhance the genetic stability of *ts* mutations present in a vaccine virus. Third, mutations or sequences that specify a high level of genetic and phenotypic stability should be included, such as deletion mutations or host range determinants (e.g., BPIV3 *N* ORF). Fourth, the menu should be reasonably large, since not all combinations of attenuating mutations are viable (30) and since the level of attenuation specified by a combination of mutations is not always the sum of the attenuation specified by the individual mutations. For example, the attenuating effect of an individual mutation can be masked when placed in the context of certain other attenuating mutations. Including such mutations in a single strain can be desirable, since it allows a greater number of attenuating mutations to be incorporated into a vaccine candidate, which would restrain phenotypic change should reversion occur at one or more sites. Fifth, as mentioned above, it is desirable, when possible, to generate amino acid point mutations using codons that differ from the wild-type assignment by more than one nucleotide, in order to reduce the frequency of reversion.

Trial and error is needed to identify compatible mutations and to confirm the level of attenuation of any particular combination. It might seem that the potential number of vaccine candidates would be unworkably large, but in practice the design of new viruses is guided and simplified by comparison to previous vaccine candidates that have undergone preclinical and clinical evaluation. This is illustrated in Figure 1a, where RSV deletion mutants are compared with the existing, well-characterized RSVcpts248/404 virus. Using these principles, a menu of attenuating mutations can be assembled and employed in the design of a vaccine virus that exhibits the proper balance between attenuation and immunogenicity.

The design and use of antigenic chimeric viruses to create new vaccines

As indicated above, human RSV exists as two antigenic subgroups A and B, both of which should be represented in a vaccine. To expedite the development of a vaccine for subgroup B — for which promising vaccine candidates, a reverse genetics system, and a menu of attenuating mutations were lacking — we have modified the existing recombinant RSV A2 (subgroup A) so that its G and F glycoprotein genes were replaced with their counterparts from the RSV subgroup B strain B1 (Figure 3b) (32). This replacement of the protective antigens was initially done for the wild-type recombinant RSV A2 virus to create an RSV AB wild-type chimeric virus (rRSVAB wt), and then for a series of RSV A2 derivatives that contain various combinations of RSV A2–derived attenuating mutations located in genes other than F and G, one of which is depicted in Figure 3b (32). The rRSVAB wt chimeric virus replicates in cell culture with an efficiency comparable to that of the RSV A2 and B1 wild-type parents. Likewise, the rRSV-ABcpts248/404/1030 chimeric recombinant containing ts mutations in the RSV A2 background exhibits a level of temperature sensitivity in vitro similar to that of the RSV A2 recombinant bearing the same mutations.

In chimpanzees, the replication of the rRSVAB wt chimera is intermediate between that of the RSV A2 and B1 wild-type viruses (Figure 3b) and is accompanied by moderate rhinorrhea. rRSV-ABcpts248/404/1030 is highly attenuated in both the upper and lower respiratory tract of chimpanzees and is immunogenic and protective against challenge with rRSVAB wt virus. Therefore, rRSVABcpts248/404/1030 chimeric virus represents a promising vaccine candidate for RSV subgroup B and should be evaluated in humans. Furthermore, these results suggest that additional attenuating mutations derived from RSV A2 can be inserted into the RSV A2 background of rRSVAB as necessary to modify the attenuation phenotype in a reasonably predictable manner to achieve an optimal balance between attenuation and immunogenicity in a virus bearing the major protective antigens of RSV subgroup B.

We took a similar approach to designing a live-attenuated vaccine for PIV1, for which promising vaccine candidates did not exist and which did not have a reverse genetics system available until very recently (8). Taking advantage of the existing PIV3 reverse genetics system, we replaced the HN and F ORFs of PIV3 with their PIV1 counterparts, creating a chimeric virus, rPIV3-1 wt, that replicated as efficiently as a wild-type virus in vitro and in rodents and monkeys (Figure 2b). In the same way, we introduced the PIV1 HN and F ORFs into a recombinant version of the well-characterized PIV3-cp45 vaccine candidate to create a live-attenuated chimeric PIV1 vaccine candidate, termed rPIV3-1cp45, that combines the attenuated backbone of PIV3-cp45 with the neutralization antigens of PIV1 (Figure 2b) (33). Three of the 15 mutations of PIV3-cp45 lie within the HN and F genes and thus were lost by the swap involving these genes. Nonetheless, the chimeric rPIV3-1cp45 virus was found to be somewhat more restricted in replication in hamsters than was rPIV3-cp45, indicating that the introduction of the het-

erologous PIV1 HN and F proteins into PIV3 had an attenuating effect that was additive to that conferred by the 12 PIV3-cp45 mutations. rPIV3-1cp45 is immunogenic and protective against challenge with wild-type PIV1 in hamsters. This virus shows sufficient promise that it should be evaluated further as a candidate live-attenuated vaccine strain for the prevention of severe lower respiratory tract PIV1 disease in infants and young children.

In another permutation of this strategy, we used reverse genetics to replace the HN and F glycoprotein genes of BPIV3, the Jennerian vaccine candidate for human PIV3 described above, with their human PIV3 counterparts, creating an attenuated chimeric virus called rB/HPIV3. Because this chimeric virus bears the HN and F protective antigens of human PIV3, it induces a higher level of serum antibody to HPIV3 than does the BPIV3 vaccine candidate (7, 34). rB/HPIV3, like the rPIV3-NB chimera described above, combines the antigenic determinants of human PIV3 with the host range restriction and attenuation phenotype of BPIV3, but in the case of rB/HPIV3 all of the genes other than the HN and F are of bovine PIV3 origin. It is very likely that more than one BPIV3 gene is involved in the host range restriction of replication of BPIV3 for primates, in which case rB/HPIV3 would be even more phenotypically stable than rPIV3-NB.

Although antigenic chimeric viruses provide a strategy for the rapid generation of new vaccine candidates, there might be a price to pay for their use. The epidemiology of RSV and PIV1, PIV2, and PIV3 suggests that it would be optimal to administer vaccines in a sequential schedule, and the use of antigenic chimeric viruses as vaccines must be compatible with this sequential schedule. RSV and PIV3 cause significant illness within the first 4 months of life, whereas most of the illness caused by PIV1 and PIV2 occurs after 6 months of age. A desirable immunization sequence employing live-attenuated RSV and PIV vaccines would be administration of RSV and PIV3 vaccines together as a combined vaccine given two or more times, with the first dose administered at or before 1 month of age, followed by a bivalent PIV1 and PIV2 vaccine at 4 and 6 months of age. Should cross-protection between a PIV3 vaccine and a recombinant PIV3-1 vaccine virus occur, it would complicate this type of sequential immunization scheme. We have found that prior infection with PIV3 moderately decreases the replication, immunogenicity, and efficacy of an rPIV3-1 vaccine candidate against PIV1 (35). This interference likely was mediated by resistance conferred by T cell immunity to the shared internal PIV3 proteins of the two viruses. In hamsters, this effect wanes within a period of 4 months. The magnitude and duration of cell-mediated immunity to PIV3 in primates and, in particular, young human vaccinees is unknown. Thus, a sequential immunization protocol using an rPIV3-1 antigenic chimeric vaccine in PIV3-immune persons might pose potential problems for the infectivity and immunogenicity of these vaccines. Importantly, this limitation does not apply to the RSV subgroup A vaccine virus and its A/B antigenic chimeric counterpart, since these two viruses would be given simultaneously. A successful vaccine against PIV1 and PIV2 might require the development of attenuated human PIV1 and PIV2 vaccine viruses using the recently described reverse genetics systems and employing the general principles outlined here.

Address correspondence to: Brian R. Murphy, NIH, Building 50, Room 6517, 50 South Drive MSC 8007, Bethesda, Maryland 20892-8007, USA. Phone: (301) 594-1616; Fax: (301) 480-5033; E-mail: bmurphy@niaid.nih.gov.

1. Collins, P.L., Chanock, R.M., and Murphy, B.R. 2001. Respiratory syncytial virus. In *Fields virology*. 4th edition. Volume 1. D.M. Knipe et al., editors. Lippincott Williams & Wilkins. Philadelphia, Pennsylvania, USA. 1443–1486.

2. Chanock, R.M., Murphy, B.R., and Collins, P.L. 2001. Parainfluenza viruses. In *Fields virology*. 4th edition. Volume 1. D.M. Knipe et al., editors. Lippincott Williams & Wilkins. Philadelphia, Pennsylvania, USA. 1341–1379.

3. van den Hoogen, B.G., et al. 2001. A newly discovered human pneumovirus isolated from young children with respiratory tract disease. *Nat. Med.* 7:719–724.

4. Wright, P.F., et al. 2000. Evaluation of a live, cold-passaged, temperature-sensitive, respiratory syncytial virus vaccine candidate in infancy. *J. Infect. Dis.* 182:1331–1342.

5. Collins, P.L., et al. 1995. Production of infectious human respiratory syncytial virus from cloned cDNA confirms an essential role for the transcription elongation factor from the 5′ proximal open reading frame of the M2 mRNA in gene expression and provides a capability for vaccine development. *Proc. Natl. Acad. Sci. USA.* 92:11563–11567.

6. Durbin, A.P., et al. 1997. Recovery of infectious human parainfluenza virus type 3 from cDNA. *Virology.* 235:323–332.

7. Schmidt, A.C., et al. 2000. Bovine parainfluenza virus type 3 (BPIV3) fusion and hemagglutinin-neuraminidase glycoproteins make an important contribution to the restricted replication of BPIV3 in primates. *J. Virol.* 74:8922–8929.

8. Newman, J.T., et al. 2002. Sequence analysis of the Washington/1964 strain of human parainfluenza virus type 1 (HPIV1) and recovery and characterization of wild type recombinant HPIV1 produced by reverse genetics. *Virus Genes.* 24:77–92.

9. Kawano, M., et al. 2001. Recovery of infectious human parainfluenza type 2 virus from cDNA clones and properties of the defective virus without V-specific cysteine-rich domain. *Virology.* 284:99–112.

10. Collins, P.L., et al. 1999. Rational design of live-attenuated recombinant vaccine virus for human respiratory syncytial virus by reverse genetics. *Adv. Virus Res.* 54:423–451.

11. Skiadopoulos, M.H., et al. 1999. Identification of mutations contributing to the temperature-sensitive, cold-adapted, and attenuation phenotypes of the live-attenuated cold-passage 45 (cp45) human parainfluenza virus 3 candidate vaccine. *J. Virol.* 73:1374–1381.

12. Karron, R.A., et al. 1995. A live human parainfluenza type 3 virus vaccine is attenuated and immunogenic in healthy infants and children. *J. Infect. Dis.* 172:1445–1450.

13. Murphy, B.R., Park, E.J., Gottlieb, P., and Subbarao, K. 1997. An influenza A live attenuated reassortant virus possessing three temperature-sensitive mutations in the PB2 polymerase gene rapidly loses temperature sensitivity following replication in hamsters. *Vaccine.* 15:1372–1378.

14. Karron, R.A., et al. 1995. A live attenuated bovine parainfluenza virus type 3 vaccine is safe, infectious, immunogenic, and phenotypically stable in infants and children. *J. Infect. Dis.* 171:1107–1114.

15. Karron, R.A., et al. 1996. Evaluation of a live attenuated bovine parainfluenza type 3 vaccine in two- to six-month-old infants. *Pediatr. Infect. Dis. J.* 15:650–654.

16. Bailly, J.E., et al. 2000. A recombinant human parainfluenza virus type 3 (PIV3) in which the nucleocapsid N protein has been replaced by that of bovine PIV3 is attenuated in primates. *J. Virol.* 74:3188–3195.

17. Murphy, B.R. 1993. Use of live attenuated cold-adapted influenza A reassortant virus vaccines in infants, children, young adults and elderly adults. *Infectious Diseases in Clinical Practice.* 2:174–181.

18. Richman, D.D., and Murphy, B.R. 1979. The association of the temperature-sensitive phenotype with viral attenuation in animals and humans: implications for the development and use of live virus vaccines. *Rev. Infect. Dis.* 1:413–433.

19. Durbin, A.P., McAuliffe, J.M., Collins, P.L., and Murphy, B.R. 1999. Mutations in the C, D, and V open reading frames of human parainfluenza virus type 3 attenuate replication in rodents and primates. *Virology.* 261:319–330.

20. Nagai, Y., and Kato, A. 1999. Paramyxovirus reverse genetics is coming of age. *Microbiol. Immunol.* 43:613–624.

21. Whitehead, S.S., et al. 1999. Recombinant respiratory syncytial virus bearing a deletion of either the NS2 or SH gene is attenuated in chimpanzees. *J. Virol.* 73:3438–3442.

22. Teng, M.N., et al. 2000. Recombinant respiratory syncytial virus that does not express the NS1 or M2-2 protein is highly attenuated and immunogenic in chimpanzees. *J. Virol.* 74:9317–9321.

23. Jin, H., et al. 1998. Recombinant human respiratory syncytial virus (RSV) from cDNA and construction of subgroup A and B chimeric RSV. *Virology.* 251:206–214.

24. Jin, H., Cheng, X., Zhou, H.Z., Li, S., and Seddiqui, A. 2000. Respiratory syncytial virus that lacks open reading frame 2 of the M2 gene (M2-2) has altered growth characteristics and is attenuated in rodents. *J. Virol.* 74:74–82.

25. Jin, H., et al. 2000. Recombinant respiratory syncytial viruses with deletions in the NS1, NS2, SH, and M2-2 genes are attenuated in vitro and in vivo. *Virology.* 273:210–218.

26. Skiadopoulos, M.H., Surman, S.R., Durbin, A.P., Collins, P.L., and Murphy, B.R. 2000. Long nucleotide insertions between the HN and L protein coding regions of human parainfluenza virus type 3 yield viruses with temperature-sensitive and attenuation phenotypes. *Virology.* 272:225–234.

27. Wertz, G.W., Perepelitsa, V.P., and Ball, L.A. 1998. Gene rearrangement attenuates expression and lethality of a nonsegmented negative strand RNA virus. *Proc. Natl. Acad. Sci. USA.* 95:3501–3506.

28. Skiadopoulos, M.H., et al. 1999. Attenuation of the recombinant human parainfluenza virus type 3 cp45 candidate vaccine virus is augmented by importation of the respiratory syncytial virus cpts530 L polymerase mutation. *Virology.* 260:125–135.

29. Itoh, M., Isegawa, Y., Hotta, H., and Homma, M. 1997. Isolation of an avirulent mutant of Sendai virus with two amino acid mutations from a highly virulent field strain through adaptation to LLC-MK2 cells. *J. Gen. Virol.* 78:3207–3215.

30. Whitehead, S.S., et al. 1999. Addition of a missense mutation present in the L gene of respiratory syncytial virus (RSV) cpts530/1030 to RSV vaccine candidate cpts248/404 increases its attenuation and temperature sensitivity. *J. Virol.* 73:871–877.

31. Karron, R.A., et al. 2001. Evaluation of live rRSV A2 vaccines in infants and children. In *Respiratory syncytial viruses after 45 years*. Instituto de Salud Carlos, Ministerio de Sanidad y Consumo. Segovia, Spain. p. 151. (Abstr.)

32. Whitehead, S.S., et al. 1999. Replacement of the F and G proteins of respiratory syncytial virus (RSV) subgroup A with those of subgroup B generates chimeric live attenuated RSV subgroup B vaccine candidates. *J. Virol.* 73:9773–9780.

33. Skiadopoulos, M.H., Tao, T., Surman, S.R., Collins, P.L., and Murphy, B.R. 1999. Generation of a parainfluenza virus type 1 vaccine candidate by replacing the HN and F glycoproteins of the live-attenuated PIV3 cp45 vaccine virus with their PIV1 counterparts. *Vaccine.* 18:503–510.

34. Haller, A.A., Miller, T., Mitiku, M., and Coelingh, K. 2000. Expression of the surface glycoproteins of human parainfluenza virus type 3 by bovine parainfluenza virus type 3, a novel attenuated virus vaccine vector. *J. Virol.* 74:11626–11635.

35. Tao, T., et al. 2000. A live attenuated recombinant chimeric parainfluenza virus (PIV) vaccine containing the hemagglutinin-neuraminidase and fusion glycoproteins of PIV1 and the remaining proteins from PIV3 induces resistance to PIV1 even in animals immune to PIV3. *Vaccine.* 18:1359–1366.

Herpes simplex viruses: is a vaccine tenable?

Richard J. Whitley[1] and Bernard Roizman[2]

[1]Department of Pediatrics, Microbiology and Medicine, University of Alabama at Birmingham, Birmingham, Alabama, USA.
[2]The Marjorie B. Kovler Viral Oncology Laboratories, The University of Chicago, Chicago, Illinois, USA

Human herpes simplex virus (HSV) infections have been documented since ancient Greek times. Greek scholars, notably Hippocrates, used the word "herpes," meaning to creep or crawl, to describe spreading cutaneous lesions. Over the ensuing centuries, the clinical manifestations were described, including a distinction between genital and orofacial herpes, accounting for the designation of HSV-1 and HSV-2. Importantly, these viruses have a unique propensity to establish latency and recur over time, in spite of specific host immune responses, making successful vaccine development a challenge. The spectrum of disease caused by HSV includes primary and recurrent infections of mucous membranes (e.g., gingivostomatitis, herpes labialis, and genital HSV infections), keratoconjunctivitis, neonatal HSV infection, visceral HSV infections of the immunocompromised host, encephalitis, Kaposi varicella-like eruption, and an association with erythema multiforme. As it relates to genital herpes, 22% of the general population is infected with HSV-2, and the seroprevalence is even higher: 30–50% in patients attending sexually transmitted diseases clinics. Here, we will focus on HSV vaccine development with specific reference to translational molecular biology.

Molecular virology relevant to vaccine development

A brief discussion of HSV's molecular biology will set the stage for a discussion of the current status of vaccines. HSVs are members of a family of viruses whose genomes consist of a single large double-stranded DNA molecule (1). The HSV virion consists of four components: (a) an electron-dense core containing viral DNA; (b) an icosadeltahedral capsid; (c) an amorphous, at-times eccentric layer of proteins, designated tegument, which surrounds the capsid; and (d) an envelope. The capsid consists of 162 capsomeres and is surrounded by the tightly adhering tegument. The envelope surrounds the capsid-tegument structure and consists of at least 11 glycosylated and several nonglycosylated viral proteins, lipids, and polyamines. Host immune responses are greatest against the structural proteins and the envelope glycoproteins. The DNA of HSV-1 and -2 consists of two covalently linked components, designated L (long) and S (short). Each component is composed of unique sequences (UL or US, respectively) flanked by relatively large inverted repeats. The inverted repeat sequences flanking UL are ab and $b'a'$, whereas those flanking US are $a'c'$ and ca. The two components can invert relative to one another to yield four populations of DNA molecules differing solely in the relative orientation of these DNA sequences (2). Knowledge of the structure of HSV allowed for the development of genetic engineering to generate candidate vaccines and gene therapy vectors.

HSV-1 — and, by extension, HSV-2 — are now thought to encode at least 84 different polypeptides (3). To initiate infection, HSV must attach to cell-surface receptors, fuse its envelope to the plasma membrane, and allow the de-enveloped capsid to be transported to the nuclear pores, where DNA is released into the nucleus. The key events in viral replication that occur in the nucleus include transcription, DNA synthesis, capsid assembly, DNA packaging, and envelopment. Viral surface glycoproteins, key targets of protective humoral responses, mediate attachment and penetration of the virus into cells. Currently at least 11 viral glycoproteins (designated gB, gC, gD, gE, gG, gH, gI, gJ, gK, gL, and gM) are known, and another is (gN) predicted. Two of the glycoproteins encoded by HSV-2, gB-2 and gD-2, have been tested in subunit vaccines, and a mutant lacking gH is a candidate for use as a disabled infectious single cycle (DISC) vaccine (4).

The synthesis of viral gene products, both RNA and proteins, takes place in three sequential waves. The α, or immediate early, proteins regulate the reproductive cycle of the virus or block antigenic peptide presentation on the infected cell surfaces. The β, or early, proteins, synthesized next, are responsible for viral nucleic acid metabolism and are the main target of antiviral chemotherapy. These gene products include the viral thymidine kinase and the viral DNA polymerase. Finally, the late, or γ, gene products are, for the most part, the structural components of the virion (1).

Several aspects of HSV protein function are relevant to the pathogenesis of human disease, and to prospects for prophylaxis and antiviral chemotherapy. First, most viral proteins examined to date play multiple roles and, in many instances, interact with diverse cellular proteins. They are also extensively modified posttranslationally, and evidence is mounting that the specific function expressed by each protein is determined at least in part by their modifications. Second, of the 84 genes, as many as 47 can be deleted singly or in small groups without affecting the ability of the virus to replicate — at least in some cultured cell types.

Most of these genes, although dispensable in vitro, are essential for efficient replication in experimental animal systems. Many of the dispensable viral proteins, as well as some of the essential ones, are designed to scavenge cells for useful cellular proteins capable of blocking any attempt of the host to inhibit viral replication. Thus, a tegument protein, designated virion host shutoff (vhs), induces degradation of cellular and viral RNA early in infection. The sheer volume of viral RNA transcripts accumulating in infected cells enables viral gene expression. The infected cell protein 27 (ICP27), which is made immediately after infection, is another example. This protein blocks splicing of cellular mRNA, thereby reinforcing the shutoff of cellular protein synthesis initiated by vhs. In addition, since viral entry into cells and expression of viral gene functions induce a stress response that can, if unchecked, activate proapoptotic pathways that modulate virus spread. At least three different viral proteins, glycoproteins D and J and the protein kinase US3, play a role in blocking programmed cell death induced by exogenous agents or by viral gene products.

Another viral gambit that nullifies a protective host response involves the HSV protein $\gamma_1 34.5$. Since open reading frames exist on both strands of HSV DNA, complementary transcripts abound in the infected cell, leading to activation of protein

Citation for this article: *J. Clin. Invest.* **109**:145–151 (2002). doi:10.1172/JCI200216126.

426

kinase R (PKR), phosphorylation of the α subunit of the translation initiation factor eIF2, and total shutoff of protein synthesis. Activation of PKR is an innate response to infection; most viruses have evolved one or multiple mechanisms to evade it. Protein $\gamma_1 34.5$ binds protein phosphatase 1 and redirects it to dephosphorylate eIF2α. The dephosphorylation of eIF2α is highly efficient and insures uninterrupted synthesis of viral proteins. Viral protein ICP47, made immediately after infection, blocks immune responses to HSV-infected cells by binding the antigenic peptide transporter TAP1/TAP2, thus preventing the presentation of viral epitopes on MHC class I proteins of infected cells. Interestingly, two viral products, vhs and $\gamma_1 34.5$, cooperate to block endocytosis and the transport of MHC class II to the cell surface. ICP0 has recently been shown to function as a ubiquitin ligase, which is now thought to specifically target for degradation proteins that are inimical to viral replication.

Finally, certain gene products that are important in vivo but dispensable in cell culture help ensure the proper cellular environment and mobilization of cellular resources necessary for efficient viral replication. Key proteins of this class may be found in the tegument or expressed among the α genes. These include the α-trans-inducing factor (αTIF), or VP16, which augments the basal level of expression of α genes; ICP4, which both promotes and represses viral gene expression; ICP22, which controls the expression of a subset of late genes; and ICP27, which shuttles RNA from nucleus to cytoplasm late in infection.

Thus, the immunity induced by an HSV vaccine must overcome the highly evolved capacity of the virus to hide the infected cell from the immune system and the very efficient machinery to produce virus progeny, allowing it to spread from cell to cell.

HSV involvement of neurologic tissues

Two unique biologic properties of HSVs profoundly influence human disease and will need to be targeted by a successful vaccine. First, these viruses can invade and replicate in the CNS. Deletion of virtually any of the genes dispensable for viral replication in cell culture reduces the capacity of the virus to invade and replicate in CNS. Mutations affecting neuroinvasiveness have also been mapped to glycoprotein genes. Since neuronal cells do not make cellular DNA, they lack the precursors for viral DNA synthesis that are also encoded by the viral genes dispensable for replication in cell culture. $\gamma_1 34.5$ is of particular interest in that mutants deleted in this gene are among the least virulent known, even though they multiply well in a variety of cells in culture (5). The molecular basis for the failure of $\gamma_1 34.5$ mutants to multiply in the CNS is unclear and may not be related to the protein's known ability to dephosphorylate eIF2α.

The second salient characteristic of the HSVs is that of latency. Following entry into and infection of nerve endings, both HSV-1 and HSV-2 are transported in a retrograde fashion to the nuclei of sensory ganglia, where they multiply in a small number of sensory neurons. In the vast majority of the infected neurons, the viral genome remains in an episomal state for the entire life of the individual. Reactivations can occur in spite of ongoing cell-mediated and humoral immune responses and may be induced by a variety of local or systemic stimuli, such as physical or emotional stress, fever, exposure to ultraviolet light, or tissue damage, as well as by immune suppression. Recurrent herpes labialis is three times more frequent in febrile patients than in nonfebrile controls. Latent virus can be retrieved from the trigeminal, sacral,

and vagal ganglia of humans either unilaterally or bilaterally (6). The recovery of virus by in vitro cultivation of trigeminal ganglia helps explain the observation of vesicles that recur at the same site in humans, usually the vermilion border of the lip.

Little is known regarding the mechanisms by which the virus establishes and maintains a latent state or becomes reactivated. Likely, establishment of latency is based on several events occurring simultaneously. These could well involve the expression of viral genes that block proteins from being made, the absence of factors that enable high-level expression of genes, or the induction of cellular factors that repress viral gene expression.

The rationale for HSV vaccines

An efficacious HSV vaccine is very much needed, as indicated best by disease burden. In the US alone, over 100 million individuals are infected by HSV-1, and at least 40 million to 60 million individuals have been infected by HSV-2. Annually, a minimum of 2500 cases of neonatal herpes and 3000 cases of herpes simplex encephalitis result in significant morbidity and mortality in spite of efficacious antiviral therapy. Furthermore, because HSV results in genital ulcerative disease, the risk of acquisition of HIV is significantly increased by a factor of three or more.

An ideal vaccine should induce immune responses adequate to prevent infection. If primary infection were prevented, the colonization of the sensory ganglia would not occur and, therefore, no source of virus for either subsequent recurrences or transmission would exist. No one knows whether these objectives can be met. Short of this ideal, a candidate vaccine might be considered successful if it (a) mitigates primary clinical episodes, (b) prevents colonization of the ganglia, (c) helps reduce the frequency or severity of recurrences, and (d) reduces viral shedding in actively infected or asymptomatic individuals. These goals must be approached in light of the age of the target population and the duration of the desired results. Arguably, fundamental to a successful vaccine is its ability to prevent person-to-person transmission. Prospective clinical trials, a time-consuming and expensive exercise, will be required to appropriately define the true utility of an HSV vaccine using appropriate markers of clinical efficacy. Toward this end, it is of the utmost importance to determine which aspects of the host response — humoral or cell-mediated immunity; local or systemic immunity; antibody-dependent or -independent cellular cytotoxicity — are protective against HSV infection (7).

Based upon human epidemiologic data, the rationale for an HSV vaccine is fourfold. First, exogenous reinfection is exceedingly uncommon in the immune-competent host (6). Second, many more individuals are infected by HSV than either shed viruses or experience recurrences. Third, prior HSV-1 infection ameliorates the frequency and severity of subsequent HSV-2 disease. Finally, transplacental antibodies significantly decrease the risk of infection in the exposed newborn at the time of delivery. Additional encouraging findings come from experiments in animal models. Thus, both humoral and cell-mediated immune responses have been effective in a variety of animal species in preventing HSV challenge. Passive antibody administration with either polyclonal or monoclonal antibodies specific for gB-2 and/or gD-2 protects against experimental challenge in murine and guinea pig models. Immunization with either or both of these glycoproteins results in immune responses that prevent viral disease. Lastly, the genetic engineering of HSV has led to replication-competent and DISC vaccines that are similarly efficacious in several animal models.

Taken together, these observations strongly suggest that an efficacious vaccine could be designed.

HSV vaccine development

Two of the more promising approaches to HSV vaccine development now being pursued build upon entirely different theoretical approaches. The first is based on the use of gB-2 and/or gD-2 as subunit vaccines in combination with an adjuvant. The second is a genetically engineered live, attenuated vaccine from which putative neurovirulence sequences have been removed. However, prior attempts at HSV vaccine development offer important lessons for the design of clinical and preclinical studies to evaluate vaccines of this kind.

Live and inactivated wild-type virus vaccines

Numerous clinicians attempted to alter the pattern of recurrences by inoculation of autologous virus, of virus from another infected individual, or, in one set of experiments, of virus recovered from an experimentally infected rabbit. The consequences were obvious, with lesions appearing at the site of inoculation in 40–80% of volunteers. In spite of the appearance of lesions, the evaluation of only a very limited number of patients, and the absence of important controls, efficacy was reported. In some cases, inoculation led to recurrences (8). Such live-virus vaccines were abandoned on the grounds that many patients did not develop lesions at the site of inoculation and, therefore, it was not perceived that the patient had an "adequate take."

Killed virus vaccines have been studied in a variety of animal models, often with good results. Unfortunately, most studies of HSV-infected individuals receiving these preparations failed to include an appropriate control group. Because patients have been reported to experience a 30–70% decrease in the frequency of recurrences as well as improvement in severity simply from having received placebo (9, 10), such controls are clearly needed to evaluate the protection afforded by the vaccine. The initial inactivated vaccines had the additional disadvantage that they were derived from phenol-treated infected animal tissues. Because of the possibility that administration of animal proteins might lead to demyelination, these vaccines did not attract much biomedical attention. More recently, ultraviolet light inactivation of purified virus derived from tissue culture has replaced phenol inactivation.

Over the past two decades, numerous studies have yielded conflicting evidence about the value of the killed-virus approach. Viral antigen obtained from amniotic or allantoic fluid, chorioallantoic membranes, chick cell cultures, sheep kidney cells, or rabbit kidney cells and inactivated by formalin, ultraviolet light, or heat led to a series of vaccine studies in thousands of patients (9, 11–13). Unfortunately, placebo-controlled studies of inactivated HSV vaccines are rare (10, 14), and the results have been widely discrepant, even between studies using the same vaccine. Few side effects are generally noted, although some authors have noted their concern that, in patients with keratitis, autoimmune phenomena might make the herpetic disease worse (14). With one exception, each of these studies reported significant improvement in as many as 60–80% of patients (12, 13). Interestingly, these studies show that, despite repeated inoculations, antibody titers (as measured by neutralization or complement fixation) remain little changed in most subjects. Some investigators have concluded that vaccination provided some initial benefit for patients with recurrent infection, but that long-term benefit could not be established. The only prospective study of prevention of HSV infections by vaccination was performed by Anderson et al. in children in an orphanage (15). In this study, ten children received vaccine and ten placebo; HSV stomatitis developed in an equal number of patients on follow-up.

Subunit vaccines

Subunit vaccines evolved out of attempts to remove viral DNA and eliminate the potential for cellular transformation, to enhance antigenic concentration in order to induce stronger immunity, and, finally, to exclude any possibility of residual live-virus contamination. The immunogenicity and efficacy of glycoprotein vaccines derived from the entire HSV envelope, free of viral DNA, have been demonstrated in animals (16–24). Neutralizing antibodies are detected in varying amounts, correlating with the degree of protection upon challenge. Because of the variety of challenge models and the use of different routes and dosages, interpretation of these results is extremely difficult. While there are conflicting animal model studies, in general, the subunit vaccines appear to elicit a degree of protection, as evidenced by amelioration of morbidity and reduction in mortality in the immunized animals. Nevertheless, several injections were required to induce protection, and adjuvant must be included as well. Protection of rodents is significantly more successful than that of higher primate species, perhaps reflecting the fact that HSV is not indigenous to rodent species. Such protection studies may therefore be totally irrelevant when evaluating human responses. Vaccination of primates, specifically rhesus monkeys, chimpanzees (17), and cebus monkeys, induces neutralizing antibodies, leading to an amnestic response following subsequent injection months later, but the significance of even these data for human experimentation remains unclear.

Several human subunit vaccine trials have now been completed. One of the earliest human vaccine experiments was with a Merck Sharp & Dohme (Rahway, New Jersey, USA) glycoprotein envelope vaccine (25), which was produced from purified envelope glycoproteins. In a phase IIA study, carried out in sexual partners of patients known to have genital herpes, the number of individuals developing herpetic infection was nearly equal in placebo and vaccine recipients; thus, vaccination failed to provide any benefit at all. More recent clinical trials have evaluated the Chiron Corporation (Emeryville, California, USA) gB-2 and gD-2 and SmithKline Beecham (King of Prussia, Pennsylvania, USA) gD-2 purified subunit vaccines in humans. These vaccines incorporate one or both glycoproteins, as well as adjuvants unique to each company. This work provided important lessons for vaccine development: Extensive rodent experiments employing the guinea pig and murine genital herpes models demonstrated that either combined gB-2 and gD-2 or gD-2 alone, with Freund's adjuvant, completely protected against both primary and spontaneous recurrent disease following intravaginal viral inoculation (22). However, because complete Freund's adjuvant is not acceptable for human administration, alternative adjuvants have been explored, including Chiron's MF-59 and a proprietary SmithKline Beecham adjuvant. Both afford a high level of protection from HSV disease (26). The quantity of neutralizing antibody elicited by immunization and the total HSV antibody titers were higher after vaccination, and the latter value correlated with protection from disease (22, 27).

Data from the largest published series of vaccinated individuals with the Chiron construct failed to demonstrate significant long-term prevention of infection in susceptible sexual partners, although initial benefit was apparent for the first 5 months. In this trial, there was a 50% reduction in the rate of infection during this short window. The overall efficacy of the vaccine was 9%, although the vaccine had no effect on the frequency of recurrences. Even when all variables that could have influenced acquisition of infection were controlled, the beneficial effect was maintained. This large multi-institutional trial was preceded by a carefully performed phase II study that suggested clinical benefit (28). No further vaccine studies are planned for this construct. Of note, the adjuvant for these studies was MF-59, a potent inducer of Th2 responses.

Another series of clinical trials on gD-2 from Glaxo SmithKline has been reported (29). Here, the adjuvant was alum plus monophosphoryl lipid A, a potent inducer of Th1 responses. In these studies, women who were seronegative for both HSV-1 and HSV-2 were protected from both disease (72% efficacy) and infection (43% efficacy). However, no significant clinical benefit could be demonstrated for individuals seropositive for HSV-1 (irrespective of sex) or for seronegative men. These results are being pursued in follow-up clinical trials.

Live vaccines revisited

Live vaccines, in general, are considered preferable to killed or subunit vaccines, because they are more likely to induce a broad range of immune responses to the expressed gene products, providing a higher level of protection, as with such viral pathogens as measles, mumps, and rubella. Furthermore, since these vaccines replicate in the recipient, the resulting immunity should be longer-lasting. Moreover, they usually require smaller doses of antigen and, therefore, should be more economical. In addition to the direct approach of using live HSV (typically an avirulent virus), it is now possible to engineer other viruses to display the HSV epitopes. Alternatively, HSV itself can be engineered to produce a vaccine strain that is protective without causing human disease. This latter approach is particularly attractive because nonpathogenic strains derived from serial passage in cell culture or animal hosts readily revert to pathogenicity. This lack of genetic stability is unacceptable for potential human vaccines (27).

Vaccinia virus has been proposed as a vector for delivering antigens to humans (30). The principle of inserting foreign genes into a vaccinia vector was exploited for the expression of the *gD* and *gB* genes of HSV (31–33). Unfortunately, vaccinia has been shown to cause vaccinia gangrenosum and disseminated vaccinia in individuals who were vaccinated to prevent smallpox. Moreover, immune memory in individuals who have previously received vaccinia may prevent recognition of any foreign gene insert. Adenoviruses and canary pox have also been proposed as alternative expression vectors, on the grounds that they might be safer than vaccinia (34).

Recombinant HSVs have also been constructed as prototypes of HSV vaccines, using a technology developed by Post, Roizman, and colleagues (35–37). These vaccines were designed to be attenuated, to protect against HSV-1 and HSV-2 infections, to provide serological markers of immunization different from those that would be seen following wild-type infections, and to serve as vectors to express immunogens of other human pathogens. To excise some of the genetic loci responsible for neurovirulence and to create convenient sites and space within the genome for insertion of other genes, the genome of HSV-1 strain HSV-1(F) was deleted in the domain of the viral thymidine kinase (*TK*) gene and in the inverted repeats flanking the junction region between the UL and US sequences. An HSV-2 DNA fragment encoding the HSV-2 glycoproteins D, J, G, and I was inserted in place of the internal

inverted repeat. These type 2 genes were included in the hope of broadening the spectrum of the immune response and to create a chimeric pattern of antibody specificities as a serological marker of vaccination. The resulting recombinant, designated R7017, lacks TK activity and should therefore be resistant to acyclovir. Another recombinant, designated R7020, was generated by inserting the *TK* gene next to the HSV-2 DNA fragment. Since this virus expresses TK, it is susceptible to antiviral chemotherapy with acyclovir.

In rodent models, these two constructs induce protective immunity but appear considerably attenuated in their pathogenicity and ability to establish latency. The recombinants did not regain virulence, nor did they change DNA restriction enzyme cleavage patterns when subjected to serial passages in the mouse brain (37). Surprisingly, the TK⁻ virus, R7017, was no more attenuated than the TK⁺ virus, R7020. These results have been corroborated by studies in owl monkeys (38): While 100 plaque-forming units (PFU) of wild-type viruses administered by peripheral routes were fatal to the monkeys, recombinants given by various routes in amounts at least 105-fold greater were innocuous or produced mild infections, even in the presence of immunosuppression by total lymphoid irradiation (37).

Unfortunately, human studies with this vaccine have been disappointing. The maximum dose of vaccine administered, 10^5 PFU, elicited only mild immunogenicity, even with the administration of two doses (38). In many respects, the R7020 construct appears overly attenuated. However, as noted below, this virus is now being studied for gene therapy of adenocarcinoma metastases from the colon to the liver. Regardless, these same principles of genetic engineering have been applied to a generation of newer constructs.

The recent identification of the neurovirulence gene $\gamma_1 34.5$ provided an important target for genetic engineering (5). An HSV-1 variant lacking this gene is more attenuated than the R7020 mutant. In contrast, the corresponding HSV-2 variant appears to be far more virulent, necessitating the deletion of additional genes — which, unfortunately, impairs its ability to replicate to levels required for commercial production. An additional problem is the isolation of second-site compensatory mutants that regain some but not all of the virulence of wild-type strains.

Alternative approaches

Two other approaches to live-virus HSV vaccines are worthy of brief note. First, the so-called DISC vaccines are replication-defective. One version lacks the essential gene *gH* and can therefore undergo only one round of replication. Alternative DISC vaccines have been generated by deletion of either the *ICP27* or the *ICP8* gene (39, 40). Although these candidate viruses are unable to spread, they should generate a broad range of immune responses, and indeed, these vaccines have proved immunogenic, stimulating both humoral and cell-mediated immune responses in both mice and guinea pigs (4). A second approach of note is the use of naked DNA, inoculated directly into muscle. The advantage of such an approach is that it allows for both MHC class I and II presentation, perhaps resulting in broader immune responses. The genes attracting the most interest for this application encode ICP27 and HSV-2 gD-2.

Gene therapy applications of replication-attenuated HSV

Genetically engineered HSV has mainly been assessed for the treatment of human glioblastoma multiforme. These constructs have included mutations in the viral genes for thymidine kinase, ribonucleotide reductase, and $\gamma_1 34.5$ (41–43), all of which are intended to improve the therapeutic index in the treatment of gliomas. While virtually any alteration of HSV ameliorates neurovirulence, only the deletions in the $\gamma_1 34.5$ gene consistently demonstrate safety and efficacy in animal models. Numerous glioma models, including mouse, rat, and human glioma cell lines and human glioma explants, have been used to demonstrate the tumoricidal effects of these agents in vitro and in vivo – notably increased survival and some tumor cures in intracranial implant models. These effects are reproducible in vivo in both immunodeficient and immunocompetent animals (41, 43–45). The time course of infection indicates impaired replication, with limited spread of virus to the brain (46). The use of two selected mutations, affecting $\gamma_1 34.5$ and ribonucleotide reductase, appears to avoid second-site mutations with reversion to wild-type phenotype (47). Furthermore, the retention of the native HSV TK allows for acyclovir susceptibility in the engineered strain (43). The safety of these constructs has been established in susceptible primates (47).

HSV can also be used as a vector to deliver therapeutic gene products to tumors (refs. 45, 48; for review see ref. 49). Indeed the HSV construct G207 (41) demonstrates an adequate safety profile in both cell culture and animal studies (50) and has proved efficacious in several tumor models in vivo (41–43). This candidate therapeutic is deleted in both copies of the $\gamma_1 34.5$ gene as well as in the gene for ribonucleotide reductase. Numerous other constructs have been developed, encoding cytokines and chemokines, enzymes, and receptors (51).

G207, a conditionally replicating HSV mutant, has been evaluated in a phase I safety trial for a subset of patients with unresponsive recurrent malignant gliomas (52). A total of 21 patients receiving escalating doses of G207 at five intratumoral sites showed no toxicity or adverse events that could be unequivocally ascribed to G207 administration, although adverse events were noted in several patients. Importantly, no patient developed herpes simplex encephalitis. Of five seronegative volunteers, two have survived more than 3 years with stable Karnofsky scores. These data provide the basis for phase IB and II clinical trials, recently approved by the Food and Drug Administration, for further dose escalation after tumor debridement or administration of concomitant radiotherapy.

The mutant HSV strain 1716, which is deleted in both copies of the $\gamma_1 34.5$ gene and expresses lacZ under the control of the latency-associated transcript, has been studied in Scotland in a similar population. In this trial, a total of nine patients were evaluated at one of three doses of virus, beginning at 1×10^3 and escalating by a factor of 10 to 1×10^5 (53). Again, there were no reports of significant adverse events directly attributable to virus administration. Four of the nine patients were alive 14–24 months after injection. Of note, the maximum amount of virus administered in this trial was four logs lower than that in the study performed in the US (52). These promising studies have led to phase II trials in both the US and the United Kingdom.

Although the two trials examined different genetically engineered constructs and doses of virus for administration, the remarkable demonstration of safety following intratumoral inoculation paves the way for the evaluation of genetically engineered HSV in phase II trials. Future studies are required to address the extent and magnitude of viral replication in the tumor and to characterize the host response in much more detail. Second-generation constructs that express the cytokine IL-12 are expected to induce an enhanced effect within the tumor bed and will be used in human investigations in the immediate future. Furthermore, a construct identical to R7020, called NV1020, is under investigation for the selective treatment of colorectal metastases to the liver. In a dose escalation study, this virus has been administered at up to 3×10^7 PFU without adverse events. Such studies not only lay the groundwork for the broader use of HSV and related vectors in cancer therapy but may well offer important lessons for the development of vaccines to prevent HSV infections.

Conclusion

Several creative potential prophylactic vaccine candidates have emerged and undergone extensive testing. The results in humans are mixed, but a hint of efficacy exists. We have learned, for example, that seronegative individuals at high risk for infection represent ideal candidates for participation in vaccine trials, while individuals with frequent recurrences are not significantly affected by vaccines. As a consequence, vaccination should be scheduled for a time prior to exposure of the offending pathogen. For a vaccine designed to prevent HSV-2 infections, this would be early in adolescence prior to the onset of sexual activity.

Adequate methodology has not uniformly been applied to clinical evaluations of HSV vaccines. Current and future studies, to benefit from this experience, should be double-blind and placebo-controlled and should involve a sufficient number of volunteers to provide adequate power for the necessary statistical analyses. Interim analyses, predicated on results obtained during the performance of the trial, will guarantee the ethical nature of the trial design. The value or failure of the current, seemingly promising, approaches to engineering HSV vaccines should be evident within the next several years.

Acknowledgments

Studies performed by the authors and reported herein were initiated and supported under a contract (NO1-AI-65306, NO1-AI-15113, NO1-AI-62554) with the Development and Applications Branch of the National Institute of Allergy and Infectious Diseases, a Program Project Grant, University of Alabama at Birmingham (PO1-AI-24009) (B. Roizman), and grants from the General Clinical Research Center Program (RR-032) (R. Whitley), the State of Alabama (R. Whitley), and the National Cancer Institute (CA-78766, CA-71933, CA-83939, CA-88860) (B. Roizman).

Address correspondence to: Richard J. Whitley, University of Alabama at Birmingham Children's Hospital, ACC 616, 1600 Seventh Avenue South, Birmingham, Alabama 35233, USA. Phone: (205) 934-5316; Fax: (205) 934-8559; E-mail: rwhitley@peds.uab.edu.

1. Roizman, B., and Pellett, P.E. 2001. Herpesviridae. In *Fields virology*. 4th edition. D.M. Knipe and R.M. Howley, editors. Lippincott Williams & Wilkins. Philadelphia, Pennsylvania, USA. 2381–2397.
2. Roizman, B., and Knipe, D.M. 2001. Herpes simplex viruses and their replication. In *Fields virology*. 4th edition. D.M. Knipe and R.M. Howley, editors. Lippincott Williams & Wilkins. Philadelphia, Pennsylvania, USA. 2399–2459.
3. Ward, P.L., and Roizman, B. 1994. Herpes simplex genes: the blueprint of a successful human pathogen. *Trends Genet.* **10**:267–274.
4. Boursnell, M.E., et al. 1997. A genetically inactivated herpes simplex virus type 2 (HSV-2) vaccine provides effective protection against primary and recurrent HSV-2 disease. *J. Infect. Dis.* **175**:16–25.
5. Chou, J., Kern, E.R., Whitley, R.J., and Roizman, B. 1990. Mapping of herpes simplex virus-1 neuro-virulence to gamma 134.5, a gene nonessential for growth in culture. *Science.* **250**:1262–1266.
6. Whitley, R.J. 2002. Herpes simplex vaccines. In *New generation vaccines*. M.M. Levine, J.B. Kaper, R. Rappuoli, M. Liu, and M. Good, editors. Marcel Dekker Inc. New York, New York, USA. In press.
7. Meignier, B., Martin, B., Whitley, R., and Roizman, B. 1990. In vivo behavior of genetically engineered herpes simplex viruses R7017 and R7020. II. Stud-

ies in immunocompetent and immunosuppressed owl monkeys (Aotus trivirgatus). *J. Infect. Dis.* **162**:313–321.

8. Lazar, M.P. 1956. Vaccination for recurrent herpes simplex infection: initiation of a new disease site following the use of unmodified material containing the live virus. *Arch. Dermatol.* **73**:70.

9. Jawetz, E., Allende, M.E., and Coleman, V.R. 1955. Studies on herpes simplex virus. VI. Observations on patients with recurrent herpetic lesions injected with herpes viruses or their antigens. *Am. J. Med. Sci.* **229**:477–485.

10. Kern, A.B., and Schiff, B.L. 1964. Vaccine therapy in recurrent herpes simplex. *Arch. Dermatol.* **89**:844–845.

11. Chapin, H.B., Wong, S.C., and Reapsome, J. 1962. The value of tissue culture vaccine in the prophylaxis of recurrent attacks of herpetic keratitis. *Am. J. Ophthalmol.* **54**:255–265.

12. Nasemann, T., and Schaeg, G. 1973. Herpes simplex virus, type II: microbiology and clinical experiences with attenuated vaccine. *Hautarzt.* **24**:133–139.

13. Dundarov, S., Andonov, P., Bakalov, B., Nechev, K., and Tomov, C. 1982. Immunotherapy with inactivated polyvalent herpes vaccines. *Dev. Biol. Stand.* **52**:351–358.

14. Soltz-Szots, J. 1971. Therapy of recurrent Herpes simplex. *Z. Haut. Geschlechtskr.* **46**:267–272.

15. Anderson, S.G., Hamilton J., and Williams, S. 1950. An attempt to vaccinate against herpes simplex. *Aust. J. Exp. Biol. Med. Sci.* **28**:579–584.

16. Mertz, G.J., et al. 1984. Herpes simplex virus type-2 glycoprotein-subunit vaccine: tolerance and humoral and cellular responses in humans. *J. Infect. Dis.* **150**:242–249.

17. Cappel, R., DeCuyper, F., and Rikaert, F. 1980. Efficacy of a nucleic acid free herpetic subunit vaccine. *Arch. Virol.* **65**:15–23.

18. Kitces, E.N., Morahan, P.O., Tew, J.G., and Murray, B.K. 1977. Protection from oral herpes simplex virus infection by a nucleic acid-free virus vaccine. *Infect. Immun.* **16**:955–960.

19. Dix, R.D., and Mills, J. 1985. Acute and latent herpes simplex virus neurological disease in mice immunized with purified virus-specific glycoproteins gB or gD. *J. Med. Virol.* **17**:9–18.

20. Lasky, L.A., Dowbenko, D., Simonsen, C.C., and Berman, P.W. 1984. Protection of mice from lethal herpes-simplex virus-infection by vaccination with a secreted form of cloned glycoprotein-D. *Bio-Technology.* **2**:527.

21. Meignier, B., Jourdier, T.M., Norrild, B., Pereira, L., and Roizman, B. 1987. Immunization of experimental animals with reconstituted glycoprotein mixtures of herpes simplex virus 1 and 2: protection against challenge with virulent virus. *J. Infect. Dis.* **155**:921–930.

22. Stanberry, L.R., Bernstein, D.I., Burke, R.L., Pachl, C., and Myers, M.G. 1987. Vaccination with recombinant herpes simplex virus glycoproteins: protection against initial and recurrent genital herpes. *J. Infect. Dis.* **155**:914–920.

23. Stanberry, L.R., Myers, M.G., Stephanopoulos, D.E., and Burke, R.L. 1989. Preinfection prophylaxis with herpes simplex virus glycoprotein immunogens: factors influencing efficacy. *J. Gen. Virol.* **70**:3177–3185.

24. Wachsman, M., et al. 1987. Protection of guinea pigs from primary and recurrent herpes simplex virus (HSV) type 2 cutaneous disease with vaccinia virus recombinants expressing HSV glycoprotein D. *J. Infect. Dis.* **155**:1188–1197.

25. Mertz, G.J., et al. 1990. Double-blind, placebo-controlled trial of a herpes simplex virus type 2 glycoprotein vaccine in persons at high risk for a genital herpes infection. *J. Infect. Dis.* **161**:653–660.

26. Burke, R.L., et al. 1989. Development of herpes simplex virus subunit vaccine. In *Vaccines 89: modern approaches to new vaccines including prevention of AIDS.* R.A. Lerner, H. Ginsberg, R.M. Chanock, and F. Brown, editors. Cold Spring Harbor Laboratory Press. Cold Spring Harbor, New York, USA. 377–382.

27. Kaerner, H.C., Schroder, C.H., Ott-Hartmann, A., Kummel, G., and Kirchner, H. 1983. Genetic variability of herpes simplex virus: development of a pathogenic variant during passaging of a non-pathogenic herpes simplex virus type 1 virus strain in mouse brain. *J. Virol.* **46**:83–93.

28. Straus, S.E., et al. 1994. Placebo-controlled trial of vaccination with recombinant glycoprotein D of herpes simplex virus type 2 for immunotherapy of genital herpes. *Lancet.* **343**:1460–1463.

29. Spruance, S.L., et al. 2000. Gender-specific efficacy of a prophylactic SBAS4-adjuvanted gD2 subunit vaccine against genital herpes disease (GHD): results of two clinical efficacy trials. SmithKline Beecham (SB) Herpes Vaccine Efficacy Study Group. Paper presented at: 40th Interscience Conference on Antimicrobial Agents and Chemotherapy; September 17–20, 2000; Toronto, Canada.

30. Smith, G.L., Mackett, M., and Moss, B. 1983. Infectious vaccinia virus recombinants that express hepatitis B virus surface antigen. *Nature.* **302**:490–495.

31. Paoletti, E., Lipinskas, B.R., Samsonoff, C., Mercer, S., and Panicali, D. 1984. Construction of live vaccines using genetically engineered poxviruses: biological activity of vaccinia virus recombinants expressing the hepatitis B virus surface antigen and the herpes simplex virus glycoprotein D. *Proc. Natl. Acad. Sci. USA.* **81**:193–197.

32. Cremer, K.J., Mackett, M., Wohlenberg, C., Notkins, A.L., and Moss, B. 1985. Vaccinia virus recombinant expressing herpes simplex virus type 1 glycoprotein D prevents latent herpes in mice. *Science.* **228**:737–740.

33. Cantin, E.M., et al. 1987. Expression of herpes simplex virus 1 glycoprotein D by a recombinant vaccinia virus and protection of mice against lethal herpes simplex virus infection. *Proc. Natl. Acad. Sci. USA.* **84**:5908–5912.

34. McDermott, M.R., Graham, F.L., Hanke, T., and Johnson, D.C. 1989. Protection of mice against lethal challenge with herpes simplex virus by vaccination with an adenovirus vector expressing HSV glycoprotein B. *Virology.* **169**:244–247.

35. Post, L.E., and Roizman, B. 1981. A generalized technique for deletion of specific genes in large genomes: alpha gene 22 of herpes simplex virus 1 is not essential for growth. *Cell.* **25**:227–232.

36. Mocarski, E.S., Post, L.E., and Roizman, B. 1980. Molecular engineering of the herpes simplex virus genome: insertion of a second L-S junction into the genome causes additional genome inversions. *Cell.* **22**:243–255.

37. Meignier, B., Longnecker, R., and Roizman, B. 1988. In vivo behavior of genetically engineered herpes simplex virus R7017 and R7020. Construction and evaluation in rodents. *J. Infect. Dis.* **158**:602–614.

38. Cadoz, M., et al. Phase 1 trial of R7020: a live attenuated recombinant herpes simplex (HSV) candidate vaccine. Paper presented at: 32nd Interscience Conference on Antimicrobial Agents and Chemotherapy; October 11–14, 1992; Anaheim, California, USA.

39. Da Costa, X.J., Jones, C.A., and Knipe, D.M. 1999. Immunization against genital herpes with a vaccine virus that has defects in productive and latent infection. *Proc. Natl. Acad. Sci. USA.* **96**:6994–6998.

40. de Bruyn Kops, A., Uprichard, S.L., Chen, M., and Knipe, D.M. 1998. Comparison of the intranuclear distributions of herpes simplex virus proteins involved in various viral functions. *Virology.* **252**:162–178.

41. Martuza, R.L., Malick, A., Markert, J.M., Ruffner, K.L., and Coen, D.M. 1991. Experimental therapy of human glioma by means of a genetically engineered virus mutant. *Science.* **252**:854–856.

42. Boviatsis, E.J., et al. 1994. Antitumor activity and reporter gene transfer into rat brain neoplasms inoculated with herpes simplex virus vectors defective in thymidine kinase or ribonucleotide reductase. *Gene Ther.* **1**:323–331.

43. Chambers, R., et al. 1995. Comparison of genetically engineered herpes simplex viruses for the treatment of brain tumors in a scid mouse model of human malignant glioma. *Proc. Natl. Acad. Sci. USA.* **92**:1411–1425.

44. Mineta, T., et al. 1994. CNS tumor therapy by attenuated herpes simplex viruses. *Gene Ther.* **1**(Suppl. 1):S78.

45. Parker, J., et al. 2000. Engineered herpes simplex virus expressing interleukin 12 in the treatment of experimental murine tumors. *Proc. Natl. Acad. Sci. USA.* **97**:2208–2213.

46. Boviatsis, E.J., et al. 1994. Gene transfer into experimental brain tumors mediated by adenovirus, herpes simplex virus, and retrovirus vectors. *Hum. Gene Ther.* **5**:183–191.

47. Yazaki, T., Manz, H.J., Rabkin, S.D., and Martuza, R.L. 1995. Treatment of human malignant meningiomas by G207, a replication-competent multimutated herpes simplex virus 1. *Cancer Res.* **55**:4752–4756.

48. Andreansky, S., et al. 1998. Treatment of intracranial gliomas in immunocompetent mice using herpes simplex viruses that express murine interleukins. *Gene Ther.* **5**:121–130.

49. Chung, S.-M., et al. 2002. The use of a genetically engineered herpes simplex virus (R7020) with ionizing radiation for experimental hepatoma. *Gene Ther.* **9**:75–80.

50. Hunter, W.D., et al. 1999. Attenuated, replication-competent herpes simplex virus type 1 mutant G207: safety evaluation of intracerebral injection in nonhuman primates. *J. Virol.* **73**:6319–6326.

51. Markert, J.M., Gillespie, Y.G., Weichselbaum, R.R., Roizman, B., and Whitley, R.J. 2000. Genetically engineered HSV in the treatment of glioma: a review. *Rev. Med. Virol.* **10**:17–30.

52. Markert, J.M., et al. 2000. Conditionally replicating herpes simplex virus mutant G207 for the treatment of malignant glioma: results of a phase I trial. *Gene Ther.* **7**:867–874.

53. Rampling, R., et al. 2000. Toxicity evaluation of replication-competent herpes simplex virus (ICP 34.5 null mutant 1716) in patients with recurrent malignant glioma. *Gene Ther.* **7**:859–866.

Prophylactic human papillomavirus vaccines

Douglas R. Lowy and John T. Schiller

Laboratory of Cellular Oncology, Center for Cancer Research, National Cancer Institute, NIH, Bethesda, Maryland, USA.

Human papillomavirus (HPV) infection causes virtually all cases of cervical cancer, the second most common cause of death from cancer among women worldwide. This Review examines prophylactic HPV subunit vaccines based on the ability of the viral L1 capsid protein to form virus-like particles (VLPs) that induce high levels of neutralizing antibodies. Following preclinical research by laboratories in the nonprofit sector, Merck and GlaxoSmithKline are developing commercial versions of the vaccine. Both vaccines target HPV16 and HPV18, which account for approximately 70% of cervical cancer. The Merck vaccine also targets HPV6 and HPV11, which account for approximately 90% of external genital warts. The vaccines have an excellent safety profile, are highly immunogenic, and have conferred complete type-specific protection against persistent infection and associated lesions in fully vaccinated women. Unresolved issues include the most critical groups to vaccinate and when the vaccine's cost may be low enough for widespread implementation in the developing world, where 80% of cervical cancer occurs.

Human papillomavirus infection and disease

Of the 10 million cases of cancer that develop annually throughout the world, more than 15% are estimated to be attributable to infectious agents (1). Infection by human papillomaviruses (HPVs) accounts for approximately 30% of these cancers (~5% of all cancers), with hepatitis B and C viruses and *Helicobacter pylori* together accounting for another 60% of cancers with an infectious etiology.

HPVs infect the stratified squamous epithelia of skin and mucous membranes, where they cause benign lesions, some of which have the potential to progress to invasive cancer. (2–4). HPVs are small, nonenveloped viruses whose approximately 8-kb circular genome encodes 2 structural proteins, L1 and L2, that form the viral capsid, plus several nonstructural proteins that are important for the virus life cycle but are not incorporated into virions. To establish infection, microtrauma or erosion of the overlying epithelial layers is thought to enable HPVs to infect cells of the basal epithelial layer, where the stem cells and other long-lived cells are found (Figure 1). HPV infections tend to last months or years because the viral genome successfully parasitizes these cells and because the virus evades the immune system by limiting most viral gene expression and viral replication to suprabasal cell layers. Most infections are self-limited, presumably because the host eventually mounts a successful immune response.

The benign lesions induced by HPVs include nongenital and anogenital skin warts, oral and laryngeal papillomas, and anogenital mucosal condylomata (Figure 2). Anogenital infections are almost always transmitted sexually. Long-term infection by a subset of HPVs can lead to malignant anogenital tumors, including cancers of the anus, penis, vulva, vagina, and cervix (5–7). A proportion of oral cancer is also attributable to HPV (8, 9). While HPV infection has been associated on limited occasions with esophageal cancer and skin cancer, a frequent causal link, although plausible, remains more tenuous (6, 10).

Among the cancers attributable to HPV infection, cervical cancer has received the most attention (11), as it accounts for about 10% of all cancers in women worldwide. Cervical cancer is the sec-

ond most common cause of death from cancer, after breast cancer, among women worldwide. The interval between the acquisition of HPV infection and malignant progression usually takes at least 10 years (Figure 3) and is frequently longer (7, 12). Cervical cancer is therefore very uncommon in women under 25; the incidence rises progressively for women over 25 and is highest for women over 40. About 80% of cervical cancers occur in less-developed countries, primarily because they lack sufficient resources for high-quality cervical cancer screening programs that detect cervical abnormalities via Pap smear testing or testing for the presence of cervical HPV DNA (13). Virtually all cases of cervical cancer are attributable to sexually acquired HPV infection (14), while infection by HPV accounts for a variable proportion of the other tumors in which the virus has been implicated etiologically. Of all cancer cases linked etiologically to HPV, cervical cancer accounts for about two-thirds of them. This cancer can result from infection by any 1 of about 15 oncogenic HPV types, but HPV16 and HPV18 predominate, accounting for about 50% and 20% of cervical cancer, respectively (15). These 2 types account for an even higher proportion of the other genital and mucosal cancers attributable to HPV infection (6).

Genital HPV infection is believed to be the most common sexually transmitted viral infection, with an estimated prevalence of about 20–40% among sexually active 20-year-old women, an estimated 3-year cumulative incidence of more than 40% in studies of college women in the United States, and an estimated lifetime risk for women of at least 75% for one or more genital HPV infections (16, 17). Most genital HPV infections are benign, subclinical, and self-limited, and a high proportion of infections associated with low-grade cervical dysplasias (Figure 2) also regress spontaneously (7, 17). By contrast, persistent cervical infection (often defined as an infection that is detected more than once in an interval of 6 months or longer) with an oncogenic HPV type, especially HPV16 and HPV18, is the most important risk factor for progression to high-grade dysplasia (Figure 2) (18), which is recognized as a precancerous lesion that should be treated to prevent the development of invasive cancer. Locally, ablative therapy is used successfully to treat high-grade dysplasia (13).

Prophylactic HPV vaccine: development and efficacy studies

Identification of a viral agent such as HPV as a cause of disease(s) implies that successful prophylactic or therapeutic intervention

Nonstandard abbreviations used: CIN, cervical intraepithelial neoplasia; HPV, human papillomavirus; HSV, herpes simplex virus; VLP, virus-like particle.

Conflict of interest: The authors, as employees of the National Cancer Institute, NIH, are inventors of the HPV VLP vaccine technology described in this Review. The technology has been licensed by the NIH to the 2 companies, Merck and GlaxoSmithKline, that are developing the commercial HPV vaccines described herein.

Citation for this article: *J. Clin. Invest.* **116**:1167–1173 (2006). doi:10.1172/JCI28607.

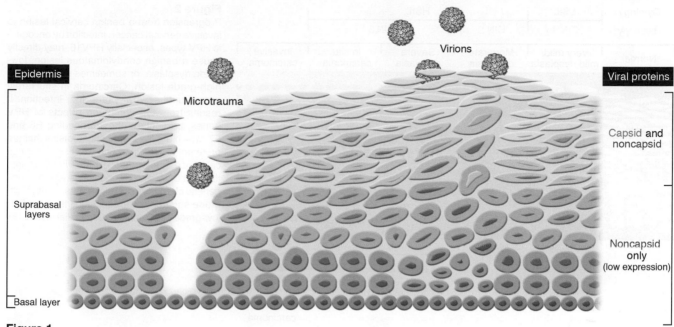

Figure 1

Papillomavirus life cycle. To establish infection, the virus must infect basal epithelial cells that are long lived or have stem cell–like properties. Microtrauma to the suprabasal epidermal cells probably enables the virus to infect the cell within the basal layer. The viral genome maintains itself as an episome in basal cells, where the viral genes are poorly expressed. Viral replication takes place in suprabasal layers and is tied to the epidermal differentiation process. The presence of the virus causes morphological abnormalities in the epithelium, including papillomatosis, parakeratosis, and koilocytosis. Progeny virus is released in desquamated cells.

against the viral agent should prevent the disease(s) it causes. A preexisting viral infection can theoretically be targeted by an antiviral or a therapeutic vaccine. Successful antivirals have been developed for the treatment of some viral infections, including diseases such as HIV and influenza (19), but not against viruses such as HPV. A therapeutic vaccine against HPV infection would be highly desirable to prevent the cancer-associated complications of HPV infection, which only develop after many years of infection. However, despite ongoing efforts to develop effective therapeutic vaccines against HPV and other viral infections, none has been shown to be highly effective clinically (20), probably because the vaccines have not yet adequately mimicked critical aspects of a curative immune response. On the other hand, prophylactic vaccines have been developed against a variety of human viral pathogens and are often a cost-effective approach to interfere with the diseases caused by these pathogens (21). To be widely implemented, a prophylactic vaccine generally needs to confer high-level protection for at least several years without boosting and to be particularly safe, as it is given to healthy individuals.

The recognition of HPV as the cause of cervical cancer and other diseases therefore implied that an effective HPV vaccine should be able to interfere with the benign and malignant conditions attributable to HPV infection. However, approved prophylactic vaccines have been directed against infectious agents that cause systemic disease, and efforts to develop vaccines against sexually transmitted agents such as HPV whose disease results from local infection had not proven successful. It is believed that neutralizing antibodies form the cornerstone of most prophylactic vaccines, and viruses that cause disease only after passing through the circulation are accessible to the neutralizing antibodies pres-

ent in the blood (22). Another limitation was that the presence of oncogenes in HPVs suggested that a subunit vaccine approach would theoretically be preferable to an inactivated vaccine or an attenuated live-virus vaccine, and it was unclear whether a subunit HPV vaccine would have the potential to be effective against the local infections caused by genital HPVs.

Despite these uncertainties, 2 pharmaceutical companies, Merck and GlaxoSmithKline, have recently reported a remarkable degree of protection by candidate prophylactic HPV vaccines (see below). The vaccines that both companies are developing are subunit virus-like particle (VLP) vaccines composed of a single viral protein, L1, which is the major structural (capsid) protein of the virus and contains the immunodominant neutralization epitopes of the virus. The vaccines are based primarily on preclinical research showing that (a) when expressed in cells, L1 has the intrinsic ability to self-assemble into VLPs (Figure 4) (23–26) that can induce high levels of neutralizing antibodies (23, 27, 28); (b) in animal models of animal papillomavirus infection, parenteral vaccination with L1 VLPs protects from high-dose challenge with homologous virus (29–31), while animals are not protected by systemic immunization with denatured L1 or L1 VLPs from a heterologous papillomavirus because L1 neutralization epitopes are conformationally dependent and predominantly type specific (29); and (c) protection can be passively transferred by immune IgG (29, 30). The VLPs from the Merck vaccine are produced in yeast (32–34), while the VLPs from the GlaxoSmithKline vaccine are produced in insect cells via recombinant baculovirus (35). Merck uses alum as an adjuvant in its vaccine, while GlaxoSmithKline uses AS04, a proprietary adjuvant composed of alum plus monophosphoryl lipid A (a detoxified form of lipopolysaccharide). Both vaccines use

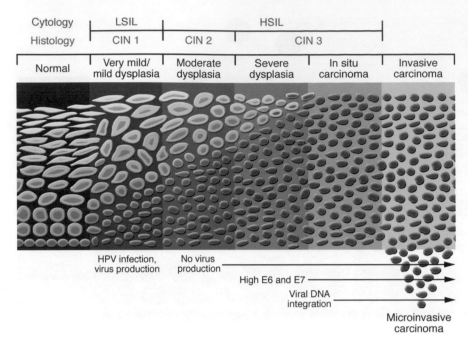

Figure 2

Progression from a benign cervical lesion to invasive cervical cancer. Infection by oncogenic HPV types, especially HPV16, may directly cause a benign condylomatous lesion, low-grade dysplasia, or sometimes even an early high-grade lesion. Carcinoma in situ rarely occurs until several years after infection. It results from the combined effects of HPV genes, particularly those encoding E6 and E7, which are the 2 viral oncoproteins that are preferentially retained and expressed in cervical cancers; integration of the viral DNA into the host DNA; and a series of genetic and epigenetic changes in cellular genes. HSIL, high-grade squamous intraepithelial lesion; LSIL, low-grade squamous intraepithelial lesion.

purified particles, which are given as 3 intramuscular injections over a 6-month period.

Of course, it would be desirable for an HPV vaccine to have the ability to prevent all cases of cervical cancer. However, although the 15 oncogenic types implicated in cervical cancer are more closely related to each other phylogenetically than they are to the HPVs that cause nongenital skin lesions (warts) (36), the immunodominant epitopes in L1 VLPs induce neutralizing antibodies that are predominantly type specific (37). It has therefore been necessary, at least for the first generation HPV vaccines, to focus on the HPV types found most frequently in cervical cancer. This consideration has led both companies to focus on HPV16 and HPV18, which, as noted above, account for about 70% of cases of cervical cancer. The Merck vaccine also targets HPV6 and HPV11, which together account for about 90% of external genital warts (38); the latter 2 types also infect the cervix, but are not implicated in cervical cancer. Thus, the GlaxoSmithKline vaccine that is currently in phase III trials is a bivalent vaccine composed of VLPs from HPV16 and HPV18, while the vaccine that Merck has used for its phase III trials is a quadrivalent vaccine that contains VLPs from HPV6, HPV11, HPV16, and HPV18.

When considering appropriate end points for determining vaccine efficacy, the most relevant end points recommended by an FDA vaccine advisory panel (39) were a reduction in the incidence of vaccine type–specific persistent infections and of associated moderate- and high-grade cervical dysplasias and carcinomas in situ, which together are referred to as cervical intraepithelial neoplasia 2+ (CIN2+; CIN is graded as CIN1, CIN2, and CIN3 for low-, moderate-, and high-grade dysplasia, respectively). HPV DNA testing results have been shown to be substantially more reproducible than the pathological diagnosis of dysplasia (40), but moderate- and high-grade dysplasias represent clinical end points that trigger therapeutic intervention. It is important to note that it would be unethical to use cervical cancer as a primary end point for vaccine efficacy trials, as cervical cancer screening can prevent the vast majority of cancer through the identification of precancers, which

are then treated. Also, the interval between infection and the development of invasive cancer usually takes more than 10 years (7, 12).

Following the observations that systemic vaccination with a monovalent HPV16 L1 VLP vaccine was safe and highly immunogenic (41), even without adjuvant, a Merck-sponsored proof-of-principle efficacy trial of an HPV16 L1 VLP vaccine reported that fully vaccinated women who were HPV negative throughout the vaccination period were completely protected against the development of persistent incident infection with HPV16 when followed for an average of 17 months (Table 1 and ref. 32). This vaccine cohort has now been followed for an average of 3.5 years, and the high level of protection was maintained throughout this period (Table 1 and ref. 34). After the initial peak, the levels of serum antibodies appeared to decline approximately 10-fold over the first 2 years following vaccination but then remained stable for the remainder of the follow-up period. If the level of serum antibodies represents a surrogate for protection against infection, the stability of the antibody titers would suggest that high-level protection may continue substantially beyond 3.5 years. Preliminary efficacy results from Merck's quadrivalent vaccine were also published and showed excellent protection against the viruses targeted by the vaccine (Table 1 and ref. 33). In unpublished studies, Merck has reported at scientific meetings on the clinical efficacy of the multinational phase III trial of their quadrivalent vaccine, with an average follow-up of 1.5 years (42, 43). When fully vaccinated women who remained negative for infection throughout the vaccination period were analyzed against the comparable placebo cohort, the vaccine was 100% effective in preventing CIN2+ associated with HPV16 or HPV18 and also in preventing external genital warts associated with HPV6 or HPV11. Even when the efficacy was compared starting 1 month after the first immunization, protection for these end points was greater than 90%. Thus systemic immunization with a subunit HPV vaccine can achieve a high degree of protection in women against benign and premalignant diseases induced by this sexually transmitted local infection of the genital mucosa or skin.

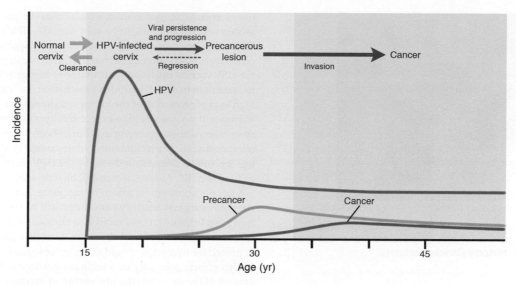

Figure 3
Relationship among incidences of cervical HPV infection, precancer, and cancer. The HPV curve emphasizes the high incidence of infection that develops soon after women initiate sexual activity and subsequent lower incidence because a high proportion of infections are self-limited. The precancer incidence curve follows several years behind the HPV incidence curve and is substantially lower than that of HPV incidence, as there is generally a delay between the acquisition of HPV infection and precancer development, and only a subset of infected women develop precancers. The cancer incidence curve follows several years behind the precancer curve, reflecting the relatively long interval between precancer and progression to invasive cancer. As women approach 40 years of age, the incidence of cancer begins to approach the incidence of precancer. Figure modified with permission from the *New England Journal of Medicine* (53).

The Merck-sponsored trials have not reported on the comparative incidence of infection by other HPV types not included in the vaccines. However, the proof-of-principle monovalent HPV16 trial has reported on the total number of patients with various grades of cervical dysplasia, and whether the dysplasias were associated with HPV16, for the fully vaccinated women who were HPV negative throughout the vaccination period (32). In contrast to the complete protection seen against HPV16-associated dysplasias, there was no difference between the placebo and vaccinated groups in the number of non–HPV16-associated dysplasias in the initial report on this cohort. In the 3.5-year follow-up report (34), there was also no evidence that non–HPV16-associated dysplasias in the vaccinated group were less frequent in number than those in the nonvaccinated group. These results strongly suggest that the protection conferred by Merck HPV16 monovalent vaccine was predominantly HPV type specific.

GlaxoSmithKline has also sponsored a proof-of-principle efficacy trial of an HPV16 and HPV18 bivalent vaccine similar to the one that is currently in phase III trials (Table 1 and ref. 35). Their placebo-controlled proof-of-principle trial had a smaller number of participants who were followed for an average of 18 months. In the group of women who were fully vaccinated and remained uninfected throughout the vaccination period, all subsequent persistent infections associated with HPV16 and HPV18 occurred in the placebo group, although there were only a total of 7 cases (5 with HPV16 and 2 with HPV18). When incident persistent infection was monitored starting 1 month after the first dose of vaccine, combined protection against HPV16 and HPV18 was still about 90%. Another potentially important result, which has been presented at meetings but not yet published, is that the vaccine has been associated with some cross-protection against HPV types closely related to HPV16 and HPV18, although this protection was less complete than that offered against HPV16 and HPV18 (44, 45). As Merck has not reported results analyzed in this manner, it is difficult to know whether the cross-protection represents an activity that may be greater in the GlaxoSmithKline vaccine. It will be important for the ongoing large-scale efficacy trials to analyze this parameter, as cross-protection against HPV types not in the vaccine could enhance its overall utility.

Vaccine implementation issues
The notable efficacy results (Table 1) have thus far been correlated with an excellent safety profile, which make it likely that one or both vaccines will be licensed in the near future. In fact, Merck applied to the FDA in December 2005 for a license to sell their quadrivalent vaccine. It is anticipated that application for licensure will also come from GlaxoSmithKline if their phase III trial results are similarly positive.

Several issues about the vaccine remain to be addressed. It will be important to confirm that the strong safety profile remains intact as more individuals receive the vaccine. As noted above, it remains to be determined how long the high level of type-specific protection is maintained, as this issue will have implications for whether and when booster injections might be advisable and will also contribute to the cost-effectiveness of the vaccine. It remains to be seen whether the cross-protection against HPV types not in the vaccine, as noted in the GlaxoSmithKline proof-of-principle trial, will be confirmed in the large-scale trials of this vaccine and whether the Merck vaccine possesses similar properties. A substantial degree of cross-protection could increase the potential impact of the vaccine by further reducing the incidence of serious genital HPV infections and by reducing the number of abnormal Pap tests and the cost of their follow-up. On the other hand, protection against heterologous HPV types is likely to wane more quickly than that

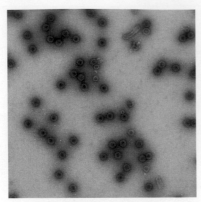

Figure 4
Electron micrograph of HPV16 L1 VLPs. Original magnification, ×14,500. Image courtesy of Yuk-Ying Susana Pang (Laboratory of Cellular Oncology, National Cancer Institute).

against the HPV types specifically targeted by the vaccine, which could also have implications for boosting.

Another unanswered question is whether vaccination might alter the natural history of prevalent HPV infection by reducing the incidence of persistent infection or cytological abnormalities. The large vaccine trials may have a sufficient number of prevalent infections attributable to the HPV types in the vaccine to address this question. If such a "therapeutic effect" were seen, it could provide an added rationale to vaccinate sexually active women who might have prevalent infection with one of the types in the vaccine. The most likely explanation for such an effect would be that the vaccine had reduced the efficiency of transmission of an early infection from one genital site to other genital sites, presumably via specific antibodies in the genital tract (46). It is also possible that the vaccine could have direct therapeutic effects against established lesions. However, this possibility seems less likely, as persistent infection is usually attributed to the presence of the viral genome in long-lived basal epithelial cells, which do not express L1. The experimental evidence also does not support this possibility: when the effect of the VLP vaccine was tested on established lesions in an animal model (bovine papillomavirus type 4 [BPV-4], which induced benign oral mucosal lesions), it did not induce their regression, although the vaccine was very effective when given prophylactically (31).

Although the immune response of men to the vaccine is similar to that of women (41), it is not yet known whether the vaccine will be protective in men. Many vaccines have comparable efficacy in males and females. However, a subunit vaccine for type 2 herpes simplex virus (HSV gD vaccine), another sexually transmitted viral infection, was found to be effec-

tive in women but not in men, which raises the possibility that an analogous difference might be seen with the HPV vaccine (47). As HSV infection is more likely to be mucosal in women and cutaneous in men, it was speculated that the difference in protection from the HSV vaccine might be attributable to higher antibody titers in mucosa than in skin. Even if this explanation is relevant to HSV, the high level of protection of the Merck vaccine against cutaneous genital warts in women would be expected to apply to men as well, since these warts appear on cornified skin in both sexes. It may also be relevant that the protection induced in women by the HSV vaccine was less robust than that induced by the HPV vaccine (47). Efficacy trials of the HPV vaccine in men will directly address this issue.

A key question will be whom to vaccinate. In the United States, the main national advisory committee will be the Advisory Committee on Immunization Practices at the Centers for Disease Control and Prevention. Until there are data that show the vaccine is protective in men, it would seem most logical to focus public health efforts primarily on vaccinating women. If the principal activity of the vaccine is the prevention of incident HPV infection, the greatest reduction in the number of infections would likely result from immunizing girls or women before they become sexually active. In the United States, this consideration would imply that pre- or young adolescent girls would be prime candidates for the vaccine. Of course, older girls and women with no prior sexual exposure should also achieve maximum benefit from the vaccine, which implies that "catch-up" vaccination for these groups should be seriously considered. Giving the vaccine to women who have had some prior sexual activity could also reduce their number of infections, although their degree of benefit from the vaccine would probably be inversely related to their degree of prior sexual activity. If the vaccines are shown to be highly effective in men, it will be logical to include them in vaccination programs in high-resource settings such as the United States, especially if the Merck vaccine is able to protect men against genital warts. From a public health perspective, however, the degree to which vaccination of males would contribute to herd immunity is unclear. In some models of sexually transmitted infections, if a high proportion of one gender is vaccinated and the vaccine is effective in preventing transmission, vaccinating the other gender achieves a relatively

Table 1
Proof-of-principle HPV VLP prophylactic efficacy trials

Study	Koutsky et al. (32)	Harper et al. (35)	Villa et al. (33)	Mao et al. (34)
HPV VLP type	16	16, 18	6, 11, 16, 18	16
Adjuvant	Alum	AS04	Alum	Alum
Sponsor	Merck	GSK	Merck	Merck
Trial site	United States	United States, Canada, Brazil	United States, European Union, Brazil	United States
Subject age	16–23	15–25	16–23	16–23
No. subjects (ATP)	1,533	721	468	1,505
Vaccination schedule (mo)	0, 2, 6	0, 1, 6	0, 2, 6	0, 2, 6
Follow-up (yr)	1.5	1.5	2.5	3.5
Persistent infections[A]	42/0 (100)	7/0 (100)	36/4[B] (90)	111/7[C] (94)
CIN1+[D]	9/0 (100)	6/0 (100)	3/0 (100)	24/0 (100)

Shown are according-to-protocol (ATP) analyses for the HPV types included in the vaccines. [A]Values are shown as number of controls versus number of vaccinees with persistent infections; values in parentheses indicate percent efficacy. [B]Ten of 36 controls and 3 of 4 vaccinees were HPV DNA positive only at the last visit. [C]Nineteen of 111 controls and 7 of 7 vaccinees were HPV DNA positive only at the last visit. [D]Values are shown as number of controls versus number of vaccinees that were CIN1+; values in parentheses indicate percent efficacy. GSK, GlaxoSmithKline.

small increase in herd immunity (48). In addition, more than 80% of cancers attributable to HPV infection occur in women (1). These considerations suggest that vaccination of women should probably have the highest priority in settings with limited resources.

Anticipated impact of vaccination

The previous section has noted several important unanswered questions that make it difficult to predict the actual impact of a vaccine on HPV infection and disease. However, it is possible to give some estimates if it is assumed that the vaccine will provide at least 90% type-specific efficacy, with boosters as necessary to ensure that this protection is of long duration. Under those circumstances, the greatest short-term impact in industrialized nations such as the United States would be a reduction in the overall number of CIN2+ cases to about one-third to one-half as many such lesions in vaccinated women compared with nonvaccinated women, given that HPV16 and HPV18 together account for 60–70% of such lesions. This protection would translate to a substantial reduction in medical and psychological morbidity and treatment together with a reduction in the related costs. The anticipated eventual reduction in the incidence of cervical cancers and its consequences would be anticipated to be at least as great. If the vaccine were widely administered to populations that historically are less likely to be screened regularly, the vaccine could prevent most of those serious infections that currently are not detected because of a lack of screening (49). The impact on subclinical and low-grade dysplasias would be expected to be more modest, as only a minority of these infections are attributable to HPV16 and HPV18 (or to HPV6/11/16/18, when considering the quadrivalent Merck vaccine infection) (50). Although these anticipated reductions in CIN2+ and invasive cervical cancer would be impressive, it must be noted that there would still be many serious

HPV infections against which the vaccine would not protect. For this reason, it will be essential to educate health care providers and patients about this limitation of the vaccine and to emphasize that it will be necessary for vaccinated women to follow current cervical cancer screening guidelines.

What about vaccination in developing countries, where 80% of cervical cancers occur, but where medical resources are relatively scarce? The vaccine probably has the potential to prevent several hundred thousand cancers annually, many of which affect relatively young women and therefore have an enormous impact on their life expectancies (11). However, vaccination in low-resource settings would probably be cost-effective only if the expense of vaccination were modest, especially in view of the long interval between infection and the development of invasive cancer. The history of hepatitis B vaccination suggests that it may take many years to achieve a cost-effective vaccination program in developing areas (51). Thus reducing the cost of vaccination would be a priority for the developing world. In the interim, a substantial reduction in the incidence of cervical cancer might be achievable in these settings, in less time than vaccination would lead to a reduction, by the alternate modality of a low-cost once- or twice-in-a-lifetime screen-and-treat approach to cervical cancer prevention (52, 53).

Acknowledgments

This work was supported by the Intramural Research Program of the NIH, National Cancer Institute, Center for Cancer Research.

Address correspondence to: Douglas R. Lowy, Laboratory of Cellular Oncology, Center for Cancer Research, National Cancer Institute, National Institutes of Health, Bethesda, Maryland 20892, USA. Phone: (301) 496-9513; Fax: (301) 480-5322; E-mail: drl@helix.nih.gov.

1. Pisani, P., Parkin, D.M., Munoz, N., and Ferlay, J. 1997. Cancer and infection: estimates of the attributable fraction in 1990. *Cancer Epidemiol. Biomarkers Prev.* **6**:387–400.
2. Lowy, D.R., and Howley, P.H. 2001. Papillomaviruses. In *Fields virology*. D.M. Knipe and P.H. Howley, editors. Lippincott, Williams, & Wilkins. Philadelphia, Pennsylvania, USA. 2231–2264.
3. Zur Hausen, H. 2002. Papillomaviruses and cancer: from basic studies to clinical application. *Nat. Rev. Cancer.* **2**:342–350.
4. Bosch, F.X., Schiffman, M., and Solomon, D., editors. 2003. Future directions in epidemiologic and preventive research on human papillomaviruses and cancer. *J. Natl. Cancer Inst. Monogr.* **31**:131.
5. Frisch, M. 2002. On the etiology of anal squamous carcinoma. *Dan. Med. Bull.* **49**:194–209.
6. Gillison, M.L., and Shah, K.V. 2003. Chapter 9: role of mucosal human papillomavirus in nongenital cancers. *J. Natl. Cancer Inst. Monogr.* **31**:57–65.
7. Schiffman, M., and Kjaer, S.K. 2003. Chapter 2: natural history of anogenital human papillomavirus infection and neoplasia. *J. Natl. Cancer Inst. Monogr.* **31**:14–19.
8. Gillison, M.L., and Lowy, D.R. 2004. A causal role for human papillomavirus in head and neck cancer. *Lancet.* **363**:1488–1489.
9. Szentirmay, Z., et al. 2005. Human papillomavirus in head and neck cancer: molecular biology and clinicopathological correlations. *Cancer Metastasis Rev.* **24**:19–34.
10. Pfister, H. 2003. Chapter 8: human papillomavirus and skin cancer. *J. Natl. Cancer Inst. Monogr.* **31**:52–56.
11. Yang, B.H., Bray, F.I., Parkin, D.M., Sellors, J.W., and Zhang, Z.F. 2004. Cervical cancer as a priority for prevention in different world regions: an evaluation using years of life lost. *Int. J. Cancer.* **109**:418–424.
12. Snijders, P.J., Steenbergen, R.D., Heideman, D.A., and Meijer, C.J. 2006. HPV-mediated cervical carcinogenesis: concepts and clinical implications. *J. Pathol.* **208**:152–164.
13. Wright, T.C., Jr., Cox, J.T., Massad, L.S., Twiggs, L.B., and Wilkinson, E.J. 2002. 2001 consensus guidelines for the management of women with cervical cytological abnormalities. *JAMA.* **287**:2120–2129.
14. Bosch, F.X., Lorincz, A., Munoz, N., Meijer, C.J., and Shah, K.V. 2002. The causal relation between human papillomavirus and cervical cancer. *J. Clin. Pathol.* **55**:244–265.
15. Munoz, N., et al. 2004. Against which human papillomavirus types shall we vaccinate and screen? The international perspective. *Int. J. Cancer.* **111**:278–285.
16. Cates, W., Jr. 1999. Estimates of the incidence and prevalence of sexually transmitted diseases in the United States. American Social Health Association Panel. *Sex. Transm. Dis.* **26**(4 Suppl.):S2–S7.
17. Baseman, J.G., and Koutsky, L.A. 2005. The epidemiology of human papillomavirus infections. *J. Clin. Virol.* **32**(Suppl. 1):S16–S24.
18. Khan, M.J., et al. 2005. The elevated 10-year risk of cervical precancer and cancer in women with human papillomavirus (HPV) type 16 or 18 and the possible utility of type-specific HPV testing in clinical practice. *J. Natl. Cancer Inst.* **97**:1072–1079.
19. Littler, E., and Oberg, B. 2005. Achievements and challenges in antiviral drug discovery. *Antivir. Chem. Chemother.* **16**:155–168.
20. Roden, R.B., Ling, M., and Wu, T.C. 2004. Vaccination to prevent and treat cervical cancer. *Hum. Pathol.* **35**:971–982.
21. Ehreth, J. 2005. The economics of vaccination from a global perspective: present and future. 2-3 December, 2004, Vaccines: all things considered, San Francisco, CA, USA. *Expert Rev. Vaccines.* **4**:19–21.
22. Robbins, J.B., Schneerson, R., and Szu, S.C. 1995. Hypothesis: serum IgG antibody is sufficient to confer protection against infectious diseases by inactivating the inoculum. *J. Infect. Dis.* **171**:1387–1398.
23. Kirnbauer, R., Booy, F., Cheng, N., Lowy, D.R., and Schiller, J.T. 1992. Papillomavirus L1 major capsid protein self-assembles into virus-like particles that are highly immunogenic. *Proc. Natl. Acad. Sci. U. S. A.* **89**:12180–12184.
24. Hagensee, M.E., Yaegashi, N., and Galloway, D.A. 1993. Self-assembly of human papillomavirus type 1 capsids by expression of the L1 protein alone or by coexpression of the L1 and L2 capsid proteins. *J. Virol.* **67**:315–322.
25. Rose, R.C., Bonnez, W., Reichman, R.C., and Garcea, R.L. 1993. Expression of human papillomavirus type 11 L1 protein in insect cells: in vivo and in vitro assembly of viruslike particles. *J. Virol.* **67**:1936–1944.
26. Kirnbauer, R., et al. 1993. Efficient self-assembly of human papillomavirus type 16 L1 and L1-L2 into virus-like particles. *J. Virol.* **67**:6929–6936.
27. Rose, R.C., Reichman, R.C., and Bonnez, W. 1994. Human papillomavirus (HPV) type 11 recombinant virus-like particles induce the formation of neutralizing antibodies and detect HPV-specific antibodies in human sera. *J. Gen. Virol.* **75**:2075–2079.
28. Roden, R.B.S., et al. 1996. *In vitro* generation and type-specific neutralization of a human papillomavirus type 16 virion pseudotype. *J. Virol.* **70**:5875–5883.

29. Breitburd, F., et al. 1995. Immunization with virus-like particles from cottontail rabbit papillomavirus (CRPV) can protect against experimental CRPV infection. *J. Virol.* **69**:3959–3963.

30. Suzich, J.A., et al. 1995. Systemic immunization with papillomavirus L1 protein completely prevents the development of viral mucosal papillomas. *Proc. Natl. Acad. Sci. U. S. A.* **92**:11553–11557.

31. Kirnbauer, R., et al. 1996. Virus-like particles of Bovine Papillomavirus type 4 in prophylactic and therapeutic immunization. *Virology.* **219**:37–44.

32. Koutsky, L.A., et al. 2002. A controlled trial of a human papillomavirus type 16 vaccine. *N. Engl. J. Med.* **347**:1645–1651.

33. Villa, L.L., et al. 2005. Prophylactic quadrivalent human papillomavirus (types 6, 11, 16, and 18) L1 virus-like particle vaccine in young women: a randomised double-blind placebo-controlled multicentre phase II efficacy trial. *Lancet Oncol.* **6**:271–278.

34. Mao, C., et al. 2006. Efficacy of human papillomavirus-16 vaccine to prevent cervical intraepithelial neoplasia: a randomized controlled trial. *Obstet. Gynecol.* **107**:18–27.

35. Harper, D.M., et al. 2004. Efficacy of a bivalent L1 virus-like particle vaccine in prevention of infection with human papillomavirus types 16 and 18 in young women: a randomised controlled trial. *Lancet.* **364**:1757–1765.

36. De Villiers, E.M., Fauquet, C., Broker, T.R., Bernard, H.U., and zur Hausen, H. 2004. Classification of papillomaviruses. *Virology.* **324**:17–27.

37. Roden, R.B.S., et al. 1996. Assessment of the serological relatedness of genital human papillomaviruses by hemagglutination inhibition. *J. Virol.* **70**:3298–3301.

38. Greer, C.E., et al. 1995. Human papillomavirus (HPV) type distribution and serological response to HPV 6 virus-like particle in patients with genital warts. *J. Clin. Microbiol.* **33**:2058–2063.

39. [No authors listed]. 2001. FDA vaccines and related biological products advisory committee. Center for Biologics Evaluation and Research meeting #88. November 28–29. Bethesda, Maryland, USA. http://www.fda.gov/ohrms/dockets/ac/cber01.htm#Vaccines.

40. Stoler, M.H. 2003. Human papillomavirus biology and cervical neoplasia: implications for diagnostic criteria and testing. *Arch. Pathol. Lab. Med.* **127**:935–939.

41. Harro, C.D., et al. 2001. Safety and immunogenicity trial in adult volunteers of a human papillomavirus 16 L1 virus-like particle vaccine. *J. Natl. Cancer Inst.* **93**:284–292.

42. Barr, E. 2005. Gardasil HPV vaccine. Presented at the Centers for Disease Control and Prevention, National Immunization Program meeting of the Advisory Committee on Immunization Practices. February 10–11. Atlanta, Georgia, USA.

43. Barr, E. 2006. Garadasil HPV vaccine. Presented at the Centers for Disease Control and Prevention, National Immunization Program meeting of the Advisory Committee on Immunization Practices. February 21–22. Atlanta, Georgia, USA.

44. Dubin, G. 2005. Enhanced immunogenicity of a candidate human papillomavirus (HPV) 16/18 L1 virus like particle (VLP) vaccine with novel AS04 adjuvant in preteens/adolescents. Presented at the 45th Annual International Conference on Antimicrobial Agents and Chemotherapy (ICAAC). December 16–19. Washington, DC, USA.

45. Dubin, G. 2006. Cervarix HPV vaccine. Presented at the Centers for Disease Control and Prevention, National Immunization Program meeting of the Advisory Committee on Immunization Practices. February 21–22. Atlanta, Georgia, USA.

46. Nardelli-Haefliger, D., et al. 2003. Specific antibody levels at the cervix during the menstrual cycle of women vaccinated with human papillomavirus 16 virus-like particles. *J. Natl. Cancer Inst.* **95**:1128–1137.

47. Stanberry, L.R., et al. 2002. Glycoprotein-D-adjuvant vaccine to prevent genital herpes. *N. Engl. J. Med.* **347**:1652–1661.

48. Garnett, G.P. 2005. Role of herd immunity in determining the effect of vaccines against sexually transmitted disease. *J. Infect. Dis.* **191**(Suppl. 1):S97–S106.

49. Freeman, H., and Wingrove, B. 2005. Excess cervical cancer mortality: a marker for low access to health care in poor communities. National Cancer Institute, Center to Reduce Cancer Health Disparities. Rockville, Maryland, USA. http://crchd.nci.nih.gov/initiatives/#Reducing. 79 pp.

50. Clifford, G.M., et al. 2005. Human papillomavirus genotype distribution in low-grade cervical lesions: comparison by geographic region and with cervical cancer. *Cancer Epidemiol. Biomarkers Prev.* **14**:1157–1164.

51. Ulmer, J.B., and Liu, M.A. 2002. Ethical issues for vaccines and immunization. *Nat. Rev. Immunol.* **2**:291–296.

52. Goldie, S.J., et al. 2005. Cost-effectiveness of cervical-cancer screening in five developing countries. *N. Engl. J. Med.* **353**:2158–2168.

53. Schiffman, M., and Castle, P.E. 2005. The promise of global cervical-cancer prevention. *N. Engl. J. Med.* **353**:2101–2104.

Part 6
Respiratory System

Disorders of lung matrix remodeling

Harold A. Chapman

Department of Medicine and Cardiovascular Research Institute, University of California at San Francisco,
San Francisco, California, USA.

A set of lung diseases share the tendency for the development of progressive fibrosis ultimately leading to respiratory failure. This review examines the common pathogenetic features of these disorders in light of recent observations in both humans and animal models of disease, which reveal important pathways of lung matrix remodeling.

Pulmonary fibrosis is a common consequence and often a central feature of many lung diseases. In some disorders fibrosis develops focally and to a limited degree. For example, in asthma and chronic obstructive pulmonary disease fibrotic changes occur around conducting airways where scarring may be important to the pathophysiology (1, 2). But fibrosis is not the dominant pathological feature of either disease. In contrast, a subset of lung diseases are particularly vexing because the degree of fibroproliferation and fibrosis is a dominant determinant of clinical outcome and yet for the most part current therapies are ineffective or only marginally effective. These diseases share in common the propensity for progressive fibrosis leading to respiratory failure and thus they are, in a sense, disorders of lung matrix remodeling. They also share common elements of pathobiology favoring matrix and architectural remodeling and disease progression, especially repeated epithelial cell death, and for all of these reasons are grouped together (Table 1). The disorders listed in Table 1 are mostly diseases of the modern era. This grouping underrepresents causes of severe fibrotic lung disease in regions of heavy dust exposure and endemic tuberculosis. But better control of infections, work conditions, longer life spans, and new medical technology such as mechanical ventilation and organ transplantation have resulted in disorders of lung matrix remodeling becoming more prevalent. A patient over the age of 50 presenting with several months of dyspnea without signs of infection, bilateral infiltrates on chest X-ray, and a restrictive pattern of pulmonary function abnormalities now has almost a 40% chance of having idiopathic pulmonary fibrosis (IPF) as the underlying disease (3).

Differences in the tempo and sites of disease progression among these disorders (Table 1) tend to obscure common clinical features. The adult respiratory distress syndrome (ARDS) is a dramatic example of acute, diffuse alveolar damage and non-cardiogenic pulmonary edema. But an important issue for long-term survival is the degree of parenchymal fibrosis and loss of lung function (4–6). This development during the acute or subacute setting, heralded by a progressive increase in pulmonary vascular resistance, leads to the hallmark clinical features of pulmonary fibrosis common to all of these disorders: persistent patchy radiographic infiltrates, reduced lung compliance indicative of restrictive lung disease, progressively impaired gas exchange, and ultimately cor pulmonale. The same applies to acute interstitial pneumonitis (AIP), which has a similar time course and pathology as ARDS but no known precipitating event (7). In stark contrast to the dramatic nature of ARDS and AIP is the prototypic fibrotic lung disease, IPF, a diagnosis more recently restricted to the histological pattern, usual interstitial pneumonitis (UIP) (8). In this disease, there is virtually no evidence of an acute process on plain chest radiographs. Fibrosis associated with well-defined collagen vascular diseases (CVDs) such as scleroderma fall somewhere between ARDS and IPF, with elements of acute and chronic inflammation and injury. Three of these disorders are strikingly bronchocentric: sarcoidosis, eosinophilic granuloma, and bronchiolitis obliterans (BO). Accordingly, their clinical features are a mixture of both restriction (from progressive fibrosis) and obstruction (from airway narrowing). Although respiratory failure is very uncommon in sarcoidosis and variable in eosinophilic granuloma, both disorders are found as indications for lung transplantation in the setting of end-stage fibrosis (9). Progressive intralumenal and airway matrix organization in BO is a major cause of disability and fatality following lung or bone marrow transplantation (10).

In spite of their clinical distinctions, the disorders of matrix remodeling share a common paradigm of disease progression: provisional matrices formed in the context of injury emit signals to activate an inflammatory response and epithelial cells, provoking ingrowth and/or expansion of connective tissue elements that lead to persistent and at times permanent matrix reordering. This is highlighted in Figure 1, which illustrates the loose matrices of lesional activity in UIP, BO, and cryptogenic organizing pneumonia (COP) being covered by epithelial cells and invaded by inflammatory cells and fibroblasts. Several comprehensive reviews focused on the distinguishing clinical, radiographic, and histological patterns among these diseases have been recently published (3, 8, 11). This review will focus on our current understanding of the pathobiology common to the progressively fibrotic lung diseases (Table 1). Understanding molecular mechanisms driving the fibrotic process for even one of these disorders may empower new efforts to monitor and treat the group of disorders where lung matrix remodeling dominates.

Nonstandard abbreviations used: idiopathic pulmonary fibrosis (IPF); adult respiratory distress syndrome (ARDS); acute interstitial pneumonitis (AIP); usual interstitial pneumonitis (UIP); collagen vascular disease (CVD); bronchiolitis obliterans (BO); cryptogenic organizing pneumonia (COP); bronchoalveolar lavage (BAL); alveolar macrophage (AM); activated protein C (APC); sonic hedgehog (SHH); epidermal growth factor (EGF); bone morphogenic protein (BMP); epithelial-mesenchymal transition (EMT); latency-associated peptide (LAP); thrombospondin-1 (TSP-1); latent TGF-binding protein (LTBP); plasminogen activator inhibitor-1 (PAI-1); protease-activated receptor-1 (PAR-1); connective tissue growth factor (CTGF); monocyte chemotactic protein-1 (MCP-1); surfactant protein C (SPC); single nucleotide polymorphism (SNP); nonspecific interstitial pneumonia (NSIP).

Conflict of interest: The author has declared that no conflict of interest exists.

Citation for this article: *J. Clin. Invest.* **113**:148–157 (2004). doi:10.1172/JCI200420729.

Historical perspectives and emergence of current concepts

Role of inflammation. A seminal set of findings in the mid-1970s and early 1980s, based on the introduction into clinical research

Table 1

Disorders of lung matrix remodeling

Idiopathic pulmonary fibrosis
Adult respiratory distress syndrome
Fibrosis with collagen vascular disease
Cryptogenic organizing pneumonia
Bronchiolitis obliterans, transplant associated
Sarcoidosis
Histiocytosis X (eosinophilic granuloma)
Hermansky-Pudlak syndrome

Disorders where fibrosis is frequently prominent but not usually a cause of death: asbestosis, hypersensitivity pneumonitis, drug-induced lung disease, localized fibrosis around airways in asthma and emphysema, and lung irradiation.

of bronchoalveolar lavage (BAL), revealed that patients with UIP or any of the related set of chronic fibrotic disorders (Table 1) have a persistent alveolitis (12, 13). This was found to be true whether or not any evidence of edema or inflammation was detectable on the plain radiograph, leading to the paradigm that early and persistent inflammation was the cause of injury and subsequent development of fibrosis. This idea was seemingly corroborated by then-emerging high resolution chest imaging techniques: changes consistent with edema and inflammation, i.e., hazy increased densities on the radiograph not obscuring the underlying architecture (termed "ground glass" changes), were found to be very common, even in UIP, and the extent of these changes tended to correlate with less fibrotic, early phases of disease (8). Because the alveolar space could be readily sampled in these patients, this paradigm also spawned many studies examining the levels of inflammatory mediators, biomarkers of injury, and cytokines thought to be relevant to matrix remodeling and fibrosis. Over time, several paradigms have emerged. One paradigm is that chemokines, cytokines, and other mediators found to be upregulated in the BAL of patients with fibrotic processes act at more than one point in the inflammatory response. Leukotrienes promote the inflammatory response and impair barrier function of the lung epithelium, perhaps important to host defense (14). Cysteinyl-leukotrienes such as LTC4 also stimulate fibroblast proliferation and collagen production, a potential mechanism for an exaggerated healing response (15). Other chemokines exhibit similar profiles (16). Conversely, a cytokine strongly linked to fibroblast activation and matrix production, TGF-β_1, also regulates lung permeability in response to acute injury (17). A second paradigm is that there are patterns of chemokine and/or cytokine expression in patient BAL which appear to correlate with progression of the fibrotic process. The group of profibrotic mediators shown in Table 2 are elevated in the BAL and tissues of several of the disorders listed in Table 1, though the majority of studies have been done with patients thought to have IPF (18–24). Because individual chemokines, cytokines, and leukotrienes act at multiple points and regulate each others' expression levels, determining the importance of any single mediator among these patients remains a challenge. Furthermore, key events in matrix remodeling take place at sites of cell-matrix and cell-cell interaction not necessarily reflected by soluble mediator levels. Perhaps for these reasons, serial measurements of chemokines and other soluble mediators (Table 2) have so far not proven valuable in predicting disease progression or treatment response. Recent results in experimental models, dis-

cussed further below, also call into question the degree to which inflammation per se actually drives fibrosis. Nonetheless, to date, efforts to control pulmonary fibrosis have mostly targeted proinflammatory mediators found in Table 2.

Importance of the provisional matrix. The recognition of an alveolitis in patients at risk of interstitial fibrosis led to careful pathological studies examining the relationship between alveolitis and interstitial fibrosis in diffuse alveolar damage, UIP, and other chronic fibrotic processes (25, 26). These studies indicated that a major pathway of interstitial fibrosis is organization of a provisional matrix that appears in the alveolus as a consequence of alveolar wall injury, its ultimate appearance as a thickened alveolar wall being due to incorporation of the alveolar fibrotic process by re-epithelialization (Figure 1, Figure 2, a and c). This is not to say that the interstitium itself is unaffected; indeed, expansion of mesenchymal elements (meaning interstitial connective tissue) occurs early and prominently in lung injury. But the lung appears particularly robust in its development and organization of fibrinous, provisional alveolar matrices. The structure of the lung suggests a reason for this propensity. The airway and alveolar compartment is virtually an open space surrounded by the entire blood volume and separated from it by microscopic epithelial and endothelial barriers. The lung defends this precarious situation not only by shunting blood flow away from areas of low oxygen level (e.g., areas poorly ventilated after edema or hemorrhage occurs), but also by expressing high levels of the key procoagulant, tissue factor, promoting coagulation along alveolar and airway surfaces (Figure 2a). Both alveolar macrophages (AMs) and epithelial cells constitutively express tissue factor, which is released in lipid vesicles into the surfactant-rich airway lining fluid (27–29). The appearance of proteinaceous exudates in alveoli as a result of barrier breakdown provides substrate to empower thrombin activation and fibrin formation. Further thrombin formation is then normally limited by local thrombin activation of activated protein C (APC) which then degrades a key procoagulant, Factor V.

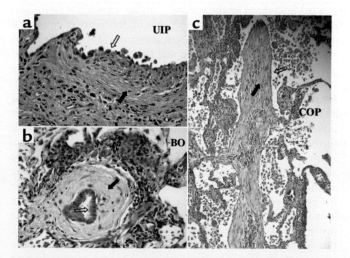

Figure 1

Photomicrographs of active matrix remodeling in: (**a**) UIP, (**b**) BO, and (**c**) COP. Filled arrows point to areas of active matrix remodeling in each disorder. Open arrows point to airway or alveolar epithelial cells overlying the remodeling matrix. Images were kindly provided by Kirk Jones, Department of Pathology, University of California, San Francisco, San Francisco, California, USA. Magnifications: (**a, b**) ×400; (**c**) ×100.

Table 2

Profibrotic mediators of human lung fibrosis

Chemokines
　　IL-1β
CXC ligand epithelial neutrophil activating protein-78
　　MCP-1[A]
Leukotrienes
Leukotrienes C4, E4[A]
Cytokines
　　TGF-α, TGF-β[A]
　　IGF-1
　　TNFα
Connective tissue growth factor
　　PDGF
　　Endothelin-1[A]
　　Thrombin[A]
Protease inhibitors
　　PAI-1[A]
　　APC inhibitor
Tissue inhibitor of metalloprotease-1

BAL fluids may also show relatively low levels of PGE2, IFN-γ, and certain chemokines thought to counteract profibrotic mediators. Replacement of these downregulated mediators is also a therapeutic strategy. [A]Mediators validated as profibrotic in the lung by loss-of-function studies in mice (see text).

Thrombin activity is blocked by complexing with plasma-derived antithrombin III. Concurrently, low but significant levels of the plasminogen activator urokinase are continuously released along alveolar surfaces to facilitate timely resolution of extensive fibrin deposition and associated provisional matrix proteins (30). This is important because the insoluble matrix which accumulates in alveolar spaces contains both chemotactic and growth factors to support an influx of fibroblasts and fibroproliferation. This reportedly occurs within days of plasma leakage into alveoli in ARDS (Figure 1) (31–33). The importance of this resolution pathway is highlighted by the spontaneous appearance of fibrin and lung fibrosis in mice as a consequence of deficiency of combined urokinase and tissue plasminogen activators (34). Thus, normal coordination of the activation and function of the cascades of coagulation and fibrinolysis facilitates resolution of proteinaceous exudates and/or blood from alveolar spaces (Figure 1, a and b). Parallel with resorption of proteinaceous deposits is the removal of other matrix components such as the hyaluronans through macrophage scavenger receptors, such as CD44, which otherwise further provokes a remodeling response (35).

Role of epithelial lining cells. The focus on the alveolus in the 1980s also led to greater attention to the biology of alveolar and airway epithelial cells in lung injury and repair. The epithelial barrier is a potent regulator of inflammation in the lung. In part this is because of the extensive network of dendritic cells interdigitated with epithelial cells along the entire conducting airway surfaces (36). But epithelial cells themselves have been demonstrated to release chemotactic factors and, along with macrophages, control the type of inflammatory cell influx into the lung. Also, alveolar epithelial cell death, mainly by apoptosis, is an early and consistent finding in these disorders, whether dramatic as in ARDS or very focal as in UIP or BO (37–39). Thus epithelial cells have emerged as a key site of initial injury as well as a major determinant of repair. The mechanisms of epithelial cell apoptosis in patients developing ARDS or UIP are

still unclear but recent evidence points to the Fas/Fas ligand pathway (40–42). The implication that epithelial apoptosis is an early event is an important concept both because it can explain much of the breakdown of barrier function and because activation of the Fas signaling cascade culminating in apoptosis can also lead to release of pro-inflammatory mediators and overt inflammation (43). IL-8 expression and neutrophilic inflammation occur as a consequence of Fas activation in macrophages (44). Apoptosis, inflammation, and matrix remodeling in the lung appear intricately linked.

Finally, key signaling pathways through which the epithelium and mesenchyme (embryonic connective tissue) communicate to effect lung development have recently been found to reappear during lung injury (45). Though this field is still maturing, the prospects for new mechanistic insights are promising. Capitalizing on clues from invertebrate models, studies of vertebrate lung development have identified several mediators of epithelial and mesenchymal crosstalk, including sonic hedgehog (SHH), epidermal growth factor (EGF), fibroblast growth factor 10, TGF-β₁/bone morphogenic protein, Wnt, fibronectin, and others (46–48). Each of these mediators has corresponding cellular receptors, which initiate signaling pathways. Integration of these signaling pathways orchestrates proliferation, migration, differentiation, and apoptosis of airway epithelial and mesenchymal cells during airway morphogenesis. At least two of these pathways, and likely others, are active during the response to lung injury in humans. SHH, a secreted ligand of embryonic epithelial cells critical to early airway formation, reappears in response to injury and promotes proliferation especially of neuroendocrine cells in the airway epithelium (49). SHH could also be expected to promote mesenchymal expansion, a point encouraging further investigation (50). Another pathway is signaling through β-catenin, a transcription factor connected to the cytoplasmic tails of epithelial adhesion proteins (cadherins) and implicated in promoting epithelial cell motility and matrix invasion (51). β-catenin was recently found by immunostaining in nuclei of hyperplastic epithelial cells and fibroblast-like cells within fibroblastic foci of UIP patients (52). Nuclear accumulation of β-catenin is a direct consequence of Wnt signaling, another critical signaling network for early lung development (53). A prominent downstream target of β-catenin signaling, MMP7 is highly expressed in lungs of UIP patients and has been shown to be an important protease in regulation of both epithelial cell apoptosis and epithelial cell migration (54, 55). Epithelial-mesenchymal transition (EMT), a consequence of β-catenin signaling, is a prominent pathway of matrix remodeling and fibrosis in injured kidney, liver, and breast tissues, where the appearance of matrix-secreting myofibroblasts derives in part from epithelial cells stimulated to undergo EMT through TGF-β₁, EGF, and Wnt/β-catenin signaling pathways (56, 57). The evidence for an active Wnt pathway in lung matrix remodeling is rather preliminary, but an attractive concept in that EMT could promote epithelial-mesenchymal interactions not only by movement of fibroblastoid epithelial cells into contact with resident fibroblasts but also by promoting provisional matrix invasion and organization by epithelial cells undergoing this putative phenotypic switch. The documented involvement of signaling pathways known to be prominent in pulmonary fibrosis in the development of EMT in other tissues invites studies to explore the importance of this pathway in disorders of lung matrix remodeling.

In summary, early and repeated epithelial cell stimulation and injury, and its consequent interplay with inflammatory and

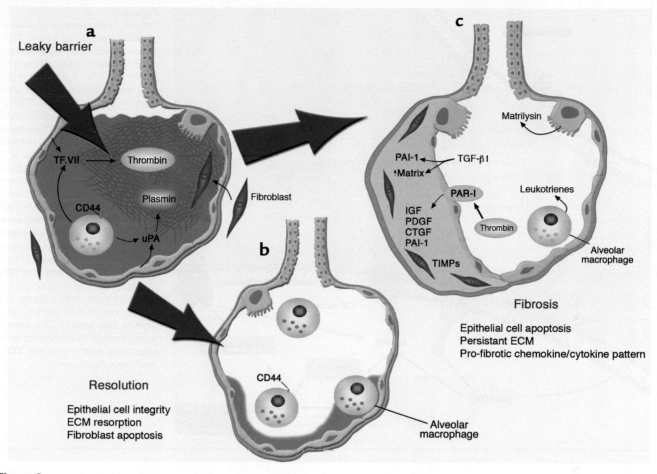

Figure 2
Determinants of lung response to breakdown in epithelial barrier function. (**a**) Coordinated processes of tissue factor/factor VII complexes initiating thrombin formation and plasminogen activation by urokinase (uPA). Plasmin facilitates initial exudate turnover. Ingrowth of fibroblasts and organization of exudates into a collagenous matrix occurs quickly. (**b**) Maintenance of epithelial integrity, resorption of ECMs by AMs, and fibroblast apoptosis favor resolution. (**c**) Epithelial cell apoptosis, activation of receptors of the profibrotic cytokines thrombin, leukotrienes, and TGF-β_1, and persistence of the ECM secondary to excess protease inhibitors (PAI-1 and tissue inhibitor of metalloprotease) with accumulation of fibroblast growth factors (IGF, PDGF, CTGF) favors fibrosis.

mesenchymal cells, has emerged as a focal point in disorders of lung matrix remodeling (Figure 2). Therapeutic intervention, however, requires validation of the suspected molecular events. In this regard, targeted gene deficiency in mice has proven invaluable.

Lessons learned from genetically altered mice
Studies of pulmonary fibrosis in mice have largely used a model of drug-induced lung injury, bleomycin-induced lung fibrosis. This model emanates from clinical observations that the use of bleomycin as an anticancer drug in humans is limited by development of widespread alveolar epithelial cell injury and development of pulmonary fibrosis (58). The mouse model has key elements of not only human bleomycin toxicity but also the prominent pathological features of the matrix remodeling disorders of unknown cause (Table 1): epithelial cell injury, inflammation and alveolar exudates, fibroblast activation and expansion with new collagen deposition, and thickened alveolar walls with reduced lung compliance. Also, like the human disorders, the intensity and progression of bleomycin injury is dependent on genetic background. Some mouse strains experience progressive fibrosis and in other

models the process wanes and even reverses (59). What has been learned from studies of this model?

TGF-β_1. The cytokine most consistently linked both experimentally and by association in human and animal studies with tissue fibrosis is TGF-β_1. Indeed, microarray analysis of whole lung mRNA during bleomycin-induced fibrosis shows most of the known TGF-β_1–inducible genes to be upregulated (60). Overexpression of active TGF-β_1 along alveolar surfaces of mice leads to a vigorous fibrotic response, and inhibition of TGF-β_1 by antibodies or decoys abrogates bleomycin-induced fibrosis (61). TGF-β_1's activation, signaling, and actions to promote fibrosis are complicated but important to consider in the context of current and future clinical interventions.

TGF-β_1 is synthesized as an inactive precursor, but its subsequent activation is rather complex. Initial cleavage of pro-TGF-β_1 results in an inactive complex of mature TGF-β_1 and its propiece (termed the latency-associated peptide, LAP). Full activation and competency for receptor engagement require an additional step to remove LAP. As indicated in Figure 3, there are at least three distinct mechanisms for TGF-β_1 activation, and studies in animals

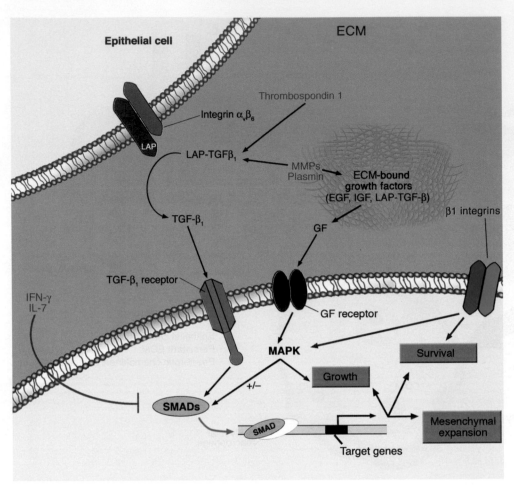

Figure 3
Pre- and postreceptor regulation of TGF-β1 signaling. Latent TGF-β1 can be activated by proteases (such asMMPs, plasmin) and by binding to TSP-1 or the integrin α$_v$β$_6$, expressed on epithelial cells. Signaling through the TGF-β1 receptor leads to phosphorylation of SMADs 2 and 3 and their translocation to the nucleus to mediate transcription of target genes. IFN-γ and IL-7 induce counter-regulatory SMAD 7. Cellular responses to TGF-β1 signaling are modulated by concurrent signaling through growth factor (GF) and adhesion receptors (β1 integrins). Integrated inputs from the pathways shown in the figure link deposition and turnover of ECM elements and GFs to expansion of mesenchymal elements such as fibroblasts and smooth muscle cells and the matrix proteins they secrete. Chemokines acting through G protein–coupled receptors also modulate TFβ1 responses (not shown).

suggest that all of them are potentially active in the lung. In the bleomycin-induced lung fibrosis model the key pathway of TGF-β activation appears to be through binding of LAP/ TGF-β1 to the epithelial cell integrin α$_v$/β$_6$ (illustrated in Figure 3) and its subsequent activation by a presumed conformational change mediated by the integrin and leading to presentation of active TGF-β1 to TGF-β1 receptors on adjacent cells (62). Importantly, LAP contains a typical integrin-binding sequence (arginine-glycine-aspartate). Munger and colleagues found that mice deficient in the beta6 integrin, expressed exclusively in epithelial cells, were virtually completely protected from bleomycin-induced lung fibrosis, implying the integrin pathway is dominant in vivo. TGF-β1 can also be activated by at least two additional pathways: binding of LAP/TGFβ to the matrix protein thrombospondin-1 (TSP-1), again releasing TGF-β1, and proteolytic cleavage of LAP by plasmin and certain metalloproteases (63, 64). Therefore regulation of integrin α$_v$/β$_6$ expression and activation in alveolar epithelial cells of TSP-1 deposition and turnover in provisional matrices during injury, and the proteolytic activities of plasmin and metalloproteases, may all influence the levels of active TGF-β1 in the lung. Yet another level of regulation is the spatial distribution of TGF-β1, regulated by a set of proteins capable of binding both TGF-β1 and the extracellular matrix (ECM): latent TGF-binding proteins (LTBPs) (65). The intricacy of regulation of TGF-β1 activation attests to the critical importance the level of active TGF-β1 has on lung biology.

As illustrated in Figure 3, the consequences of the binding of TGF-β1 to its receptor are also complex. Once binding to the serine-threonine kinase receptor occurs, a signaling complex is formed, leading to the phosphorylation of SMADs 2 and 3, the main cytoplasmic protein mediators of TGF-β1 signaling, and their subsequent trafficking as a complex to the nucleus where they bind a well-defined SMAD-response element (66). Because direct DNA binding by SMADs is relatively weak, the SMADs must organize a nuclear transcriptional complex consisting of coactivators and the SMADs, which together initiate new gene transcription. All steps in this process appear to be elaborately regulated, and thus the consequences of SMAD signaling depend on the context in which it occurs. For example, TGF-β1 signaling has different effects in embryonic and adult lung. Active TGF-β1 suppresses alveolar epithelial cell proliferation. In the embryonic lung, TGF-β1 overexpression leads to hypoplasia and lack of alveolar development (67). In the adult lung, TGF-β1 overexpression leads to progressive fibrosis (68).

One key step of SMAD regulation is the expression of inhibitory SMAD elements, mainly SMADs 6 and 7, which act by preventing productive complex formation between SMADs 2 and 3 (Figure 3) (66). Potent inducers of SMAD7 are IFN-γ and IL-7 (69). Induction of inhibitory SMADs by IFN-γ is important in the interplay between TGF-β1 and these cytokines in immune cell function. Although there are a number of rationales for using IFN-γ in the treatment of patients with IPF, blockade of TGF-β1 signaling may be the most compelling. Unfortunately, to date there is no information available as to whether lung TGF-β_1 signaling is altered in patients given IFN-γ infusions (70, 71).

TGF-β_1 signaling is also critically regulated by integration with other cell signaling pathways (highlighted in Figure 3). One key point of intersection among signaling intermediates is MAP kinase activation occurring following engagement of growth factors, integrins, and even chemokine receptors (72–76). In most cases, activated MAP kinases promote TGF-β_1's actions to enhance cell migration and mesenchymal expansion. Thus determinants of growth factor receptor activation, integrin signaling, and the pattern of cytokine/chemokine signaling all influence the response of cells to TGF-β_1. A future challenge is to determine critical points where therapeutic manipulation of active TGF-β_1 levels, or signaling pathways which influence the cells' response to TGF-β_1, will ameliorate excessive mesenchymal expansion without deleterious side effects.

Plasmin and thrombin. One gene highly induced in response to TGF-β_1 is the plasminogen activator inhibitor PAI-1. Mice deficient in PAI-1 are protected from, and mice overproducing PAI-1 are more susceptible to, bleomycin-induced lung fibrosis (77, 78). Further, inducible expression of the plasminogen activator urokinase in the alveolar compartment ameliorates fibrosis in this model (79). These findings demonstrate that the urokinase/plasmin/PAI-1 system is functionally important in lung matrix remodeling. These animal studies are consistent with many pathological and biochemical observations in humans with ARDS, UIP, scleroderma, and sarcoidosis, all of which indicate an activation of both procoagulant and antifibrinolytic pathways (Figure 2) in these disorders (28, 80, 81). Recent studies in mice, however, reveal that turnover of fibrin as the principal component of a provisional matrix does not explain the functional role of PAI-1 and plasmin in matrix remodeling: fibrinogen null mice develop bleomycin-induced fibrosis comparably to wild-type mice, and yet PAI-1 still promotes the process (82). Thus the urokinase/PAI-1/plasmin system must also act at other sites, possibly by affecting mediator turnover or matrix-integrin signaling (Figure 3).

Thrombin is also a key player in the response of the lung to injury, with evidence of increased thrombin activation consistently seen in the BAL of fibrosis patients. Apart from its initial role in fibrinous matrix formation (Figure 2a), thrombin signaling through its cognate receptor, especially protease-activated receptor-1 (PAR-1), evokes production of secondary profibrotic cytokines such as IL-1β and connective tissue growth factor (CTGF) (83, 84). Inhibition of thrombin activity suppresses fibrosis in the bleomycin-induced model of lung fibrosis in both rats and mice, supporting a mechanistic role for thrombin in tissue repair in the lung similar to that found at other sites of injury (85). Recently, enhanced expression of PAR-1 in the epithelial cells of patients with UIP was reported (86). Enhanced thrombin formation persists in the lung of patients with disordered remodeling not only because of enhanced tissue factor activity but also because patients with UIP, sarcoidosis, and CVDs have lower levels of APC in their lavage fluid and enhanced levels of an APC inhibitor (87). Coordinated expression of the APC inhibitor and PAR-1 by epithelial cells likely acts to promote wound repair, but in the lung this also appears to favor excessive matrix expansion (Figure 2c).

Lipid mediators and chemokines. Studies in mice with null mutations in lipid mediator synthesis point to this family of mediators as important to matrix remodeling. Mice deficient in cyclo-oxygenase–2, required for prostaglandin E2 (PGE2) synthesis, showed enhanced fibrosis in response to bleomycin challenge (88). Conversely, mice deficient in phospholipase A2 (required for synthesis of all lipid mediators) or 5-lipoxygenase synthetase (required for leukotriene synthesis) show attenuated fibrotic responses to bleomycin (89, 90). These findings support the idea that PGE2 could suppress, and leukotrienes enhance, matrix remodeling leading to fibrosis. This conclusion in the murine model mirrors earlier conclusions from studies of isolated fibroblasts and tissues from patients with IPF (91). The mechanism(s) underlying this set of data, however, is (are) uncertain and likely complex. It is possible that these mediators ultimately act at least in part through regulation of the TGF-β_1 signaling pathway (92).

Consistently, studies of null mutations in cytokine and chemokine pathways in mice and the bleomycin model show a lack of correlation between regulation of the inflammatory response and matrix remodeling. Mice deficient in the CC chemokine receptor 2, the receptor for monocyte chemotactic protein-1 (MCP-1), show an attenuated fibrotic response to bleomycin but no change in any indices of acute or chronic inflammation (21, 93). Similarly, mice deficient in platelet-activating factor show defective acute inflammatory responses to bleomycin injection but no change in the overall degree of fibrosis. Inflammation in the lungs of integrin $\beta6$-null mice was enhanced, even though fibrosis was blocked (62). These observations highlight again the distinctions between regulatory pathways of inflammation and the critical molecular events leading to matrix accumulation and progressive fibrosis. This conclusion is endorsed by the clinical experience that anti-inflammatory agents are relatively ineffective in blocking progressive fibrosis in most disorders of matrix remodeling. Validation in mice, as summarized above, by loss-of-function studies of several of the profibrotic mediators prominent in human disorders of matrix remodeling (indicated by note A in Table 2) offers new alternative approaches to this problem.

Disorders of matrix remodeling: lessons from human mutations

UIP. Although there is little risk to relatives for most patients with UIP, recent studies of small subgroups of patients with familial forms of pulmonary fibrosis are informative. Two families carrying separate mutations in the surfactant protein C (SPC) gene have been reported with progressive pneumonitis and lung fibrosis. SPC, a normal component of alveolar surfactant, is synthesized in a precursor form requiring C-terminal proteolytic processing for proper folding, assembly with lipid, and secretion. Nogee et al. reported childhood onset of interstitial pneumonitis and pulmonary fibrosis in an affected parent within a family carrying an SPC mutation (94). The mutation resulted in a truncated form of SPC that accumulated in the endoplasmic reticulum of type II alveolar cells. A survey of familial pulmonary fibrosis subjects identified at least one large kindred showing an autosomal dominant pattern of linkage to chromosome 6 in the region of the SPC gene (95). A single amino acid change, L188N, in the C-terminal SPC region was found in all six available family members with disease, but also in two unaffected obligate heterozygotes, and in none of four unaffected siblings or any of 88 unrelated control chromosomes. Expression of this mutant in murine alveolar cells showed lipid accumulation and toxicity, suggesting, as had the truncation mutant, that normal SPC function is critical to the health of type II cells. Studies of mice null for SPC also show pneumonitis and lung injury, confirming that loss of SPC function per se is sufficient to cause disease (96). Interestingly, the effects of SPC deficiency on lung injury were strongly influenced by the genetic background of the mice. This parallels the kindred with the L188N transversion, in which different subjects had markedly different histological patterns on lung biopsy, disease severity, and age of disease onset, implying that expression of the risk incurred by mutations in SPC is dependent on input from other genes.

Hermansky-Pudlak syndrome. This syndrome is a triad of albinism (of variable penetrance), platelet dysfunction with bleeding tendencies, and progressive lysosomal accumulation of ceroid lipids. The syndrome is due to an autosomal recessive mutation in any one of several proteins that were recently identified as forming a cytoplasmic protein complex involved in lysosome-organelle genesis. Thus the basic defect in these patients appears to be in lysosome formation and trafficking. A subset of these patients, especially those of Puerto Rican heritage, develop progressive pulmonary fibrosis, typically presenting in the mid-30s, and die of respiratory failure (97). The histological pattern, though not its distribution, is reminiscent of UIP. Recent studies identify lysosomal distortion and cellular toxicity of alveolar type II cells in Hermansky Pudlak patients (98). The type II cells show enlarged lysosomal granules engorged with surfactant. The studies point to injury of type II cells as an early and likely causal event in the initiation of the fibrotic process, reminiscent of the cellular changes seen in patients with mutations in the SPC gene.

Together these studies of patients with rare but defined mutations leading to progressive fibrosis confirm two main observations in the broad range of patients with disordered matrix remodeling: the importance of epithelial cells as a focal point of repeated injury and the strong influence of genetic background on disease progression. This underscores the importance of further detailed studies of genetic variation and disease progression, not only in UIP, but in all the disorders listed in Table 1. Reports of polymorphisms, mainly single nucleotide polymorphisms (SNPs), associated with many of these diseases are already beginning to appear in the literature, but much more experience and ultimately validation of hypotheses resulting from SNP analyses await.

Conclusion

From a clinical perspective, a crucial issue in managing patients with disorders of matrix remodeling is predicting which patients will develop more progressive disease. The current classification of interstitial pneumonias, reflecting radiographic patterns, histopathological evaluations, and pulmonary function testing, is focused on this issue. A radiographic and biopsy pattern of nonspecific interstitial pneumonia (NSIP), coupled with preserved (>50% predicted) and relatively stable vital capacity, conveys a much better prognosis than UIP even though it is quite uncertain whether NSIP is a disorder or only a pattern of persistent inflammation somewhere between resolution and progression (99, 100). In a similar vein, COP responds to corticosteroids, whereas UIP, diffuse alveolar damage, and BO do not. At the moment these important clinical distinctions have no molecular explanation. While all disorders of lung matrix remodeling appear to share common pathways of propagation, the different inciting events must be a determinant of progression. These are also not understood. And yet these issues are energized by the ongoing pace of elucidation of critical molecular pathways in epithelial injury, matrix remodeling, and signaling within the mesenchyme as reviewed here. These insights have already had medical impact. The focus of newer clinical trials, such as IFN-γ for patients with IPF (70), has turned away from agents that nonspecifically block inflammation and toward mediators proven by experimental models to be involved in fibroproliferation and matrix remodeling. The promise of genomic approaches at both the experimental and clinical levels augurs for new therapeutic approaches based on these insights in the not distant future. In the meantime, better understanding of the process will likely lead to new screening tools to establish risk and predict disease progression (101–103).

Acknowledgments

The author thanks his many colleagues at the University of California at San Francisco for helpful discussions regarding this manuscript. This work is supported in part by NIH grant HL-44712.

Address correspondence to Harold A. Chapman, Pulmonary and Critical Care Division, University of California at San Francisco, 513 Parnassus Avenue, San Francisco, California 94143-0130, USA. Phone: (415) 514-1210; Fax: (415) 502-4995; E-mail: halchap@itsa.ucsf.edu.

1. Jeffery, P.K. 2001. Remodeling in asthma and chronic obstructive lung disease. *Am. J. Respir. Crit. Care Med.* **164**:S28–S38.
2. Tiddens, H., Silverman, M., and Bush, A. 2000. The role of inflammation in airway disease: remodeling. *Am. J. Respir. Crit. Care Med.* **162**:S7–S10.
3. Green, F.H. 2002. Overview of pulmonary fibrosis. *Chest.* **122**:334S–339S.
4. Heffner, J.E., Brown, L.K., Barbieri, C.A., Harpel, K.S., and DeLeo, J. 1995. Prospective validation of an acute respiratory distress syndrome predictive score. *Am. J. Respir. Crit. Care Med.* **152**:1518–1526.
5. Martin, C., Papazian, L., Payan, M.J., Saux, P., and Gouin, F. 1995. Pulmonary fibrosis correlates with outcome in adult respiratory distress syndrome. A study in mechanically ventilated patients. *Chest.* **107**:196–200.
6. Tomashefski, J.F., Jr. 2000. Pulmonary pathology of acute respiratory distress syndrome. *Clin. Chest Med.* **21**:435–466.
7. Shimabukuro, D.W., Sawa, T., and Gropper, M.A. 2003. Injury and repair in lung and airways. *Crit. Care Med.* **31**:S524–S531.
8. 2002. American Thoracic Society and European Respiratory Society. American Thoracic Society/European Respiratory Society international

multidisciplinary consensus classification of the idiopathic interstitial pneumonias. This joint statement of the American Thoracic Society (ATS), and the European Respiratory Society (ERS) was adopted by the ATS board of directors, June 2001, and by the ERS Executive Committee, June 2001. *Am. J. Respir. Crit. Care Med.* **165**:277–304.
9. Sulica, R., Teirstein, A., and Padilla, M.L. 2001. Lung transplantation in interstitial lung disease. *Curr. Opin. Pulm. Med.* **7**:314–322.
10. Estenne, M., and Hertz, M.I. 2002. Bronchiolitis obliterans after human lung transplantation. *Am. J. Respir. Crit. Care Med.* **166**:440–444.
11. Pandit-Bhalla, M., Diethelm, L., Ovella, T., Sloop, G.D., and Valentine, V.G. 2003. Idiopathic interstitial pneumonias: an update. *J. Thorac. Imaging.* **18**:1–13.
12. Weinberger, S.E., et al. 1978. Bronchoalveolar lavage in interstitial lung disease. *Ann. Intern. Med.* **89**:459–466.
13. Keogh, B.A., and Crystal, R.G. 1982. Alveolitis: the key to the interstitial lung disorders. *Thorax.* **37**:1–10.
14. Peters-Golden, M. 2003. Arachidonic acid metabolites: potential mediators and therapeutic targets in fibrotic lung disease. In *Idiopathic pulmonary fibrosis.*

J.P. Lynch III, editor. Marcel Dekker Inc. New York, New York, USA. 421–452.
15. Phan, S.H., McGarry, B.M., Loeffler, K.M., and Kunkel, S.L. 1988. Binding of leukotriene C4 to rat lung fibroblasts and stimulation of collagen synthesis in vitro. *Biochemistry.* **27**:2846–2853.
16. Yamamoto, T., Eckes, B., Mauch, C., Hartmann, K., and Krieg, T. 2000. Monocyte chemoattractant protein-1 enhances gene expression and synthesis of matrix metalloproteinase-1 in human fibroblasts by an autocrine IL-1 alpha loop. *J. Immunol.* **164**:6174–6179.
17. Pittet, J.F., et al. 2001. TGF-beta is a critical mediator of acute lung injury. *J. Clin. Invest.* **107**:1537–1544.
18. Keane, M.P., et al. 2002. Imbalance in the expression of CXC chemokines correlates with bronchoalveolar lavage fluid angiogenic activity and procollagen levels in acute respiratory distress syndrome. *J. Immunol.* **169**:6515–6521.
19. Kunkel, S.L., Lukacs, N., and Strieter, R.M. 1995. Chemokines and their role in human disease. *Agents Actions Suppl.* **46**:11–22.
20. Smith, R.E., et al. 1995. A role for C-C chemokines in fibrotic lung disease. *J. Leukoc. Biol.* **57**:782–787.
21. Belperio, J.A., et al. 2001. Critical role for the chemokine MCP-1/CCR2 in the pathogenesis of

bronchiolitis obliterans syndrome. *J. Clin. Invest.* **108**:547–556. doi:10.1172/JCI200112214.

22. Krein, P.M., Sabatini, P.J., Tinmouth, W., Green, F.H., and Winston, B.W. 2003. Localization of insulin-like growth factor-I in lung tissues of patients with fibroproliferative acute respiratory distress syndrome. *Am. J. Respir. Crit. Care Med.* **167**:83–90.

23. Reichenberger, F., et al. 2001. Different expression of endothelin in the bronchoalveolar lavage in patients with pulmonary diseases. *Lung.* **179**:163–174.

24. Homma, S., et al. 1995. Localization of platelet-derived growth factor and insulin-like growth factor I in the fibrotic lung. *Am. J. Respir. Crit. Care Med.* **152**:2084–2089.

25. Kuhn, C., III, et al. 1989. An immunohistochemical study of architectural remodeling and connective tissue synthesis in pulmonary fibrosis. *Am. Rev. Respir. Dis.* **140**:1693–1703.

26. Fukuda, Y., et al. 1987. The role of intraalveolar fibrosis in the process of pulmonary structural remodeling in patients with diffuse alveolar damage. *Am. J. Pathol.* **126**:171–182.

27. Gross, T.J., Simon, R.H., and Sitrin, R.G. 1992. Tissue factor procoagulant expression by rat alveolar epithelial cells. *Am. J. Respir. Cell Mol. Biol.* **6**:397–403.

28. Idell, S. 2003. Coagulation, fibrinolysis, and fibrin deposition in acute lung injury. *Crit. Care Med.* **31**:S213–S220.

29. Chapman, H.A., Stahl, M., Allen, C.L., Yee, R., and Fair, D.S. 1988. Regulation of the procoagulant activity within the bronchoalveolar compartment of normal human lung. *Am. Rev. Respir. Dis.* **137**:1417–1425.

30. Marshall, B.C., et al. 1991. Alveolar epithelial cells express both plasminogen activator and tissue factor. Potential role in repair of lung injury. *Chest.* **99**:25S–27S.

31. Marshall, R.P., et al. 2000. Fibroproliferation occurs early in the acute respiratory distress syndrome and impacts on outcome. *Am. J. Respir. Crit. Care Med.* **162**:1783–1788.

32. Pugin, J., Verghese, G., Widmer, M.C., and Matthay, M.A. 1999. The alveolar space is the site of intense inflammatory and profibrotic reactions in the early phase of acute respiratory distress syndrome. *Crit. Care Med.* **27**:304–312.

33. Chesnutt, A.N., Matthay, M.A., Tibayan, F.A., and Clark, J.G. 1997. Early detection of type III procollagen peptide in acute lung injury. Pathogenetic and prognostic significance. *Am. J. Respir. Crit. Care Med.* **156**:840–845.

34. Carmeliet, P., et al. 1994. Physiological consequences of loss of plasminogen activator gene function in mice. *Nature.* **368**:419–424.

35. Teder, P., et al. 2002. Resolution of lung inflammation by CD44. *Science.* **296**:155–158.

36. Lambrecht, B.N., Prins, J.B., and Hoogsteden, H.C. 2001. Lung dendritic cells and host immunity to infection. *Eur. Respir. J.* **18**:692–704.

37. Martin, T.R., Nakamura, M., and Matute-Bello, G. 2003. The role of apoptosis in acute lung injury. *Crit. Care Med.* **31**:S184–S188.

38. Uhal, B.D. 2002. Apoptosis in lung fibrosis and repair. *Chest.* **122**:293S–298S.

39. Kuwano, K., et al. 1999. The involvement of Fas-Fas ligand pathway in fibrosing lung diseases. *Am. J. Respir. Cell Mol. Biol.* **20**:53–60.

40. Albertine, K.H., et al. 2002. Fas and fas ligand are up-regulated in pulmonary edema fluid and lung tissue of patients with acute lung injury and the acute respiratory distress syndrome. *Am. J. Pathol.* **161**:1783–1796.

41. Kuwano, K., et al. 2002. Increased circulating levels of soluble Fas ligand are correlated with disease activity in patients with fibrosing lung diseases. *Respirology.* **7**:15–21.

42. Wang, R., Zagariya, A., Ang, E., Ibarra-Sunga, O., and Uhal, B.D. 1999. Fas-induced apoptosis of alveolar epithelial cells requires ANG II generation and receptor interaction. *Am. J. Physiol.* **277**:L1245–L1250.

43. Chen, J.J., Sun, Y., and Nabel, G.J. 1998. Regulation of the proinflammatory effects of Fas ligand (CD95L). *Science.* **282**:1714–1717.

44. Park, D.R., et al. 2003. Fas (CD95) induces proinflammatory cytokine responses by human monocytes and monocyte-derived macrophages. *J. Immunol.* **170**:6209–6216.

45. Warburton, D., et al. 2001. Do lung remodeling, repair, and regeneration recapitulate respiratory ontogeny? *Am. J. Respir. Crit. Care Med.* **164**:S59–S62.

46. Warburton, D., Zhao, J., Berberich, M.A., and Bernfield, M. 1999. Molecular embryology of the lung: then, now, and in the future. *Am. J. Physiol.* **276**:L697–L704.

47. Costa, R.H., Kalinichenko, V.V., and Lim, L. 2001. Transcription factors in mouse lung development and function. *Am. J. Physiol. Lung Cell. Mol. Physiol.* **280**:L823–L838.

48. Sakai, T., Larsen, M., and Yamada, K.M. 2003. Fibronectin requirement in branching morphogenesis. *Nature.* **423**:876–881.

49. Watkins, D.N., et al. 2003. Hedgehog signalling within airway epithelial progenitors and in small-cell lung cancer. *Nature.* **422**:313–317.

50. Pepicelli, C.V., Lewis, P.M., and McMahon, A.P. 1998. Sonic hedgehog regulates branching morphogenesis in the mammalian lung. *Curr. Biol.* **8**:1083–1086.

51. Muller, T., Bain, G., Wang, X., and Papkoff, J. 2002. Regulation of epithelial cell migration and tumor formation by beta-catenin signaling. *Exp. Cell Res.* **280**:119–133.

52. Chilosi, M., et al. 2003. Aberrant Wnt/beta-catenin pathway activation in idiopathic pulmonary fibrosis. *Am. J. Pathol.* **162**:1495–1502.

53. Morrisey, E.E. 2003. Wnt signaling and pulmonary fibrosis. *Am. J. Pathol.* **162**:1393–1397.

54. Dunsmore, S.E., et al. 1998. Matrilysin expression and function in airway epithelium. *J. Clin. Invest.* **102**:1321–1331.

55. Zuo, F., et al. 2002. Gene expression analysis reveals matrilysin as a key regulator of pulmonary fibrosis in mice and humans. *Proc. Natl. Acad. Sci. U. S. A.* **99**:6292–6297.

56. Iwano, M., et al. 2002. Evidence that fibroblasts derive from epithelium during tissue fibrosis. *J. Clin. Invest.* **110**:341–350. doi:10.1172/JCI200215518.

57. Yang, J., and Liu, Y. 2001. Dissection of key events in tubular epithelial to myofibroblast transition and its implications in renal interstitial fibrosis. *Am. J. Pathol.* **159**:1465–1475.

58. Adamson, I.Y. 1984. Drug-induced pulmonary fibrosis. *Environ. Health Perspect.* **55**:25–36.

59. Haston, C.K., et al. 2002. Bleomycin hydrolase and a genetic locus within the MHC affect risk for pulmonary fibrosis in mice. *Hum. Mol. Genet.* **11**:1855–1863.

60. Kaminski, N., et al. 2000. Global analysis of gene expression in pulmonary fibrosis reveals distinct programs regulating lung inflammation and fibrosis. *Proc. Natl. Acad. Sci. U. S. A.* **97**:1778–1783.

61. Kelly, M., Kolb, M., Bonniaud, P., and Gauldie, J. 2003. Re-evaluation of fibrogenic cytokines in lung fibrosis. *Curr. Pharm. Des.* **9**:39–49.

62. Munger, J.S., et al. 1999. The integrin alpha v beta 6 binds and activates latent TGF beta 1: a mechanism for regulating pulmonary inflammation and fibrosis. *Cell.* **96**:319–328.

63. Murphy-Ullrich, J.E., and Poczatek, M. 2000. Activation of latent TGF-beta by thrombospondin-1: mechanisms and physiology. *Cytokine Growth Factor Rev.* **11**:59–69.

64. Mu, D., et al. 2002. The integrin alpha(v)beta8 mediates epithelial homeostasis through MT1-MMP-dependent activation of TGF-beta1. *J. Cell Biol.* **157**:493–507.

65. Dallas, S.L., Rosser, J.L., Mundy, G.R., and Bonewald, L.F. 2002. Proteolysis of latent transforming growth factor-beta (TGF-beta)-binding protein-1 by osteoclasts. A cellular mechanism for release of TGF-beta from bone matrix. *J. Biol. Chem.* **277**:21352–21360.

66. Massague, J., and Chen, Y.G. 2000. Controlling TGF-beta signaling. *Genes Dev.* **14**:627–644.

67. Zeng, X., Gray, M., Stahlman, M.T., and Whitsett, J.A. 2001. TGF-beta1 perturbs vascular development and inhibits epithelial differentiation in fetal lung in vivo. *Dev. Dyn.* **221**:289–301.

68. Gauldie, J., Sime, P.J., Xing, Z., Marr, B., and Tremblay, G.M. 1999. Transforming growth factor-beta gene transfer to the lung induces myofibroblast presence and pulmonary fibrosis. *Curr. Top. Pathol.* **93**:35–45.

69. Huang, M., et al. 2002. IL-7 inhibits fibroblast TGF-β production and signaling in pulmonary fibrosis. *J. Clin. Invest.* **109**:931–937. doi:10.1172/JCI200214685.

70. Ziesche, R., Hofbauer, E., Wittmann, K., Petkov, V., and Block, L.H. 1999. A preliminary study of long-term treatment with interferon gamma-1b and low-dose prednisolone in patients with idiopathic pulmonary fibrosis. *N. Engl. J. Med.* **341**:1264–1269.

71. Kalra, S., Utz, J.P., and Ryu, J.H. 2003. Interferon gamma-1b therapy for advanced idiopathic pulmonary fibrosis. *Mayo Clin. Proc.* **78**:1082–1087.

72. Thannickal, V.J., et al. 2003. Myofibroblast differentiation by transforming growth factor-beta1 is dependent on cell adhesion and integrin signaling via focal adhesion kinase. *J. Biol. Chem.* **278**:12384–12389.

73. Bhowmick, N.A., Zent, R., Ghiassi, M., McDonnell, M., and Moses, H.L. 2001. Integrin beta 1 signaling is necessary for transforming growth factor-beta activation of p38MAPK and epithelial plasticity. *J. Biol. Chem.* **276**:46707–46713.

74. Hayashida, T., Decaestecker, M., and Schnaper, H.W. 2003. Cross-talk between ERK MAP kinase and Smad signaling pathways enhances TGF-beta-dependent responses in human mesangial cells. *FASEB J.* **17**:1576–1578.

75. Chen, Y., et al. 2002. CTGF expression in mesangial cells: involvement of SMADs, MAP kinase, and PKC. *Kidney Int.* **62**:1149–1159.

76. Janda, E., et al. 2002. Ras and TGF[beta] cooperatively regulate epithelial cell plasticity and metastasis: dissection of Ras signaling pathways. *J. Cell Biol.* **156**:299–313.

77. Eitzman, D.T., et al. 1996. Bleomycin-induced pulmonary fibrosis in transgenic mice that either lack or overexpress the murine plasminogen activator inhibitor-1 gene. *J. Clin. Invest.* **97**:232–237.

78. Olman, M.A., Mackman, N., Gladson, C.L., Moser, K.M., and Loskutoff, D.J. 1995. Changes in procoagulant and fibrinolytic gene expression during bleomycin-induced lung injury in the mouse. *J. Clin. Invest.* **96**:1621–1630.

79. Sisson, T.H., et al. 2002. Inducible lung-specific urokinase expression reduces fibrosis and mortality after lung injury in mice. *Am. J. Physiol. Lung Cell. Mol. Physiol.* **283**:L1023–L1032.

80. Kotani, I., et al. 1995. Increased procoagulant and antifibrinolytic activities in the lungs with idiopathic pulmonary fibrosis. *Thromb. Res.* **77**:493–504.

81. Chapman, H.A., Allen, C.L., and Stone, O.L. 1986. Abnormalities in pathways of alveolar fibrin turnover among patients with interstitial lung disease. *Am Rev. Respir. Dis.* **133**:437–443.

82. Hattori, N., et al. 2000. Bleomycin-induced pulmonary fibrosis in fibrinogen-null mice. *J. Clin. Invest.* **106**:1341–1350.

83. Chambers, R.C., Leoni, P., Blanc-Brude, O.P., Wembridge, D.E., and Laurent, G.J. 2000. Thrombin is a potent inducer of connective tissue growth factor production via proteolytic activation of protease-

activated receptor-1. *J. Biol. Chem.* **275**:35584–35591.

84. Ruf, W., and Riewald, M. 2003. Tissue factor-dependent coagulation protease signaling in acute lung injury. *Crit. Care Med.* **31**:S231–S237.

85. Howell, D.C., et al. 2001. Direct thrombin inhibition reduces lung collagen, accumulation, and connective tissue growth factor mRNA levels in bleomycin-induced pulmonary fibrosis. *Am. J. Pathol.* **159**:1383–1395.

86. Howell, D.C., Laurent, G.J., and Chambers, R.C. 2002. Role of thrombin and its major cellular receptor, protease-activated receptor-1, in pulmonary fibrosis. *Biochem. Soc. Trans.* **30**:211–216.

87. Kobayashi, H., et al. 1998. Protein C anticoagulant system in patients with interstitial lung disease. *Am. J. Respir. Crit. Care Med.* **157**:1850–1854.

88. Keerthisingam, C.B., et al. 2001. Cyclooxygenase-2 deficiency results in a loss of the anti-proliferative response to transforming growth factor-beta in human fibrotic lung fibroblasts and promotes bleomycin-induced pulmonary fibrosis in mice. *Am. J. Pathol.* **158**:1411–1422.

89. Nagase, T., et al. 2002. A pivotal role of cytosolic phospholipase A(2) in bleomycin-induced pulmonary fibrosis. *Nat. Med.* **8**:480–484.

90. Peters-Golden, M., et al. 2002. Protection from pulmonary fibrosis in leukotriene-deficient mice. *Am. J. Respir. Crit. Care Med.* **165**:229–235.

91. Wilborn, J., et al. 1996. Constitutive activation of 5-lipoxygenase in the lungs of patients with idiopathic pulmonary fibrosis. *J. Clin. Invest.* **97**:1827–1836.

92. Ricupero, D.A., Rishikof, D.C., Kuang, P.P., Poliks, C.F., and Goldstein, R.H. 1999. Regulation of connective tissue growth factor expression by prostaglandin E(2). *Am. J. Physiol.* **277**:L1165–L1171.

93. Moore, B.B., et al. 2001. Protection from pulmonary fibrosis in the absence of CCR2 signaling. *J. Immunol.* **167**:4368–4377.

94. Nogee, L.M., et al. 2001. A mutation in the surfactant protein C gene associated with familial interstitial lung disease. *N. Engl. J. Med.* **344**:573–579.

95. Thomas, A.Q., et al. 2002. Heterozygosity for a surfactant protein C gene mutation associated with usual interstitial pneumonitis and cellular nonspecific interstitial pneumonitis in one kindred. *Am. J. Respir. Crit. Care Med.* **165**:1322–1328.

96. Glasser, S.W., et al. 2003. Pneumonitis and emphysema in sp-C gene targeted mice. *J. Biol. Chem.* **278**:14291–14298.

97. Brantly, M., et al. 2000. Pulmonary function and high-resolution CT findings in patients with an inherited form of pulmonary fibrosis, Hermansky-Pudlak syndrome, due to mutations in HPS-1. *Chest.* **117**:129–136.

98. Nakatani, Y., et al. 2000. Interstitial pneumonia in Hermansky-Pudlak syndrome: significance of florid foamy swelling/degeneration (giant lamellar body degeneration) of type-2 pneumocytes. *Virchows Arch.* **437**:304–313.

99. Latsi, P.I., et al. 2003. Fibrotic idiopathic interstitial pneumonia: the prognostic value of longitudinal functional trends. *Am. J. Respir. Crit. Care Med.* **168**:531–537.

100. Flaherty, K.R., et al. 2003. Fibroblastic foci in usual interstitial pneumonia: idiopathic versus collagen vascular disease. *Am. J. Respir. Crit. Care Med.* **167**:1410–1415.

101. Ware, L.B., Fang, X., and Matthay, M.A. 2003. Protein C and thrombomodulin in human acute lung injury. *Am. J. Physiol. Lung Cell. Mol. Physiol.* **285**:L514–L521.

102. Greene, K.E., et al. 2002. Serum surfactant proteins-A and -D as biomarkers in idiopathic pulmonary fibrosis. *Eur. Respir. J.* **19**:439–446.

103. Takahashi, H., et al. 2000. Serum surfactant proteins A and D as prognostic factors in idiopathic pulmonary fibrosis and their relationship to disease extent. *Am. J. Respir. Crit. Care Med.* **162**:1109–1114.

New insights into the pathogenesis of asthma

Jack A. Elias, Chun Geun Lee, Tao Zheng, Bing Ma, Robert J. Homer, and Zhou Zhu

Yale University School of Medicine, New Haven, Connecticut, USA.

Historical perspective

Asthma is a disease whose ability to cause episodic symptomatology has been appreciated since antiquity. Although the fine points of the definition can be debated, it is reasonable to think of asthma as a pulmonary disorder characterized by the generalized reversible obstruction of airflow and to define reversibility as a greater than 12% increase in the patient's forced expiratory volume in 1 second (FEV_1) that occurs either spontaneously or with therapy. Airway hyperresponsiveness, an exaggerated bronchospastic response to nonspecific agents such as methacholine and histamine or specific antigens, is the physiologic cornerstone of this disorder. A diagnosis of asthma is established based on a history of recurrent wheeze, cough, or shortness of breath, reversible airway obstruction demonstrated by pulmonary-function testing, and, in cases where questions exist, a methacholine challenge demonstrating airway hyperresponsiveness. It has long been assumed that patients with asthma experience intermittent attacks and have relatively normal lung function during intervening periods. More recent studies have demonstrated that asthma can cause progressive lung impairment and, in some patients, eventuate in partially reversible or irreversible airway obstruction.

Any discussion of asthma must take into account the recent increase in its prevalence. Since approximately 1980, the frequency of this disorder has almost doubled. As a result of this "epidemic," asthma now affects approximately 8–10% of the population in the US, is the leading cause of hospitalization among children less than 15 years of age, and costs society billions of dollars annually. This increase in prevalence is not simply due to diagnostic transference or increased diagnostic awareness, since asthma mortality rates have also increased during this interval.

An aerosol antigen challenge of an appropriately sensitized asthmatic patient can induce two types of airway responses. The early response is an acute bronchospastic event that occurs 15–30 minutes after exposure and resolves over time. The late-phase response peaks 4–6 hours after exposure and can cause prolonged symptomatology. Over the years, a variety of concepts of pathogenesis have been put forth in an attempt to explain one or both of these responses (Table 1). Early investigators postulated that there was an intrinsic airway smooth muscle abnormality at the root of the asthmatic diathesis. However, many studies with airway myocytes in culture have not corroborated this contention. This was followed by the contention that asthma is an autonomic dysfunction syndrome characterized by excess cholinergic and/or tachykinin pathway activity. This was never proven or disproven. Instead, IgE-mediated mast cell and/or basophil degranulation with the release of leukotrienes, histamine, prostaglandins, tryptase, cytokines (such as IL-4 and IL-5), and other mediators

Nonstandard abbreviations used: T regulatory [cell] (Tr); signal transducer and activator of transcription (STAT); chemokine receptor 2 (CCR2); histone deacetylase (HDAC).

Conflict of interest: The authors have declared that no conflict of interest exists.

Citation for this article: *J. Clin. Invest.* **111**:291–297 (2003). doi:10.1172/JCI200317748.

was appreciated to be a key event in the acute response. The prominent eosinophil-, macrophage-, and lymphocyte-rich inflammatory response in the airways of patients with asthma (Figure 1) and the efficacy of steroids in the majority of patients with asthma then led to the present-day concept that asthma is a chronic inflammatory disorder of the airway and that T cells are pivotal initiators and regulators of this response. Structural alterations including airway wall thickening, fibrosis in the lamina reticularis and adventitia of the airway, mucus metaplasia, myocyte hypertrophy and hyperplasia, and neovascularization are all readily appreciated in the asthmatic airway (Figure 1). This led to the hypothesis that the inflammatory response in the asthmatic airway causes these remodeling events, and to the belief that these events contribute to disease pathogenesis. Studies using new immunologic and molecular approaches have provided impressive insights into the nature of this inflammatory response and the relationship between this response and the remodeling and physiologic alterations characteristic of the disorder.

The Th1/Th2 paradigm

It has been known for over 50 years that people tend to mount antibody- or cell-mediated immune responses to specific antigens. A major advance in our knowledge of the mechanisms responsible for these divergent effects was achieved when it was discovered, initially in studies in mice, that the type of response that is seen is influenced greatly by the type of T cells that accumulate at the site of local antigen deposition. In the mouse, a number of functionally distinct $CD4^+$ T cells have been defined based on the profile of cytokines that they elaborate. Although the differentiation is not as clear in humans, similarly differentiated cells in humans have been described. Th1 and Th2 cells have been the topic of the most intense study, with the former elaborating IFN-γ, IL-2, and lymphotoxin, and the latter elaborating IL-4, IL-5, IL-9, IL-13, and IL-10. Significantly, less is known about T regulatory (Tr) cells and Th3 cells, which produce IL-10 and TGF-$β_1$, respectively (1, 2). As shown in Figure 2, Th1 and Th2 cells are formed from a common naive precursor T cell and differentiate into polarized populations based on signals from the local microenvironment. In the presence of $CD8α^+$ DCs and/or IL-12, IL-18, or IFN-γ, they differentiate into Th1 cells. This evolution is mediated by a mechanism that is dependent on signal transducer and activator of transcription-1 (STAT-1) and the T-bet transcription factor (1, 3). In the presence of $CD8α^-$ DCs and/or IL-4 (which can come from IgE-activated mast cells or DCs), Th2 cells are formed. This is a complex process that involves STAT-6–mediated signal transduction and the activation of a variety of transcription factors, including GATA-3, nuclear factor of activated T cells-c ($NFAT_c$), and c-maf (2, 4). Interestingly, Th1/Th2 counter-regulation has also been described, with each cell population able to inhibit and/or regulate the development and/or phenotype induced by the other. Th1 polarized responses play a key role in macrophage activation in delayed-type hypersensitivity reactions. They are a key feature in the pathogenesis of diseases like rheumatoid arthritis, sarcoidosis, and tuberculosis. In contrast, Th2-dominant responses stimulate antibody-mediated responses, activate mast cells, and elicit tissue eosinophilia. They play a key role in allergy

Table 1
Evolving concepts of asthma pathogenesis

1. A primary abnormality of airway-myocyte hyperresponsiveness
2. Autonomic dysfunction with exaggerated activity of cholinergic or tachykinin pathways
3. IgE-mediated mast cell/basophil degranulation
4. Complex T lymphocyte–mediated airway inflammation
5. Airway remodeling

and antiparasite responses. They are also the predominant responses in the asthmatic airway, where elevated levels of IL-4, IL-5, IL-13, and IL-9 have been detected by a variety of investigators (5).

Characterization of the chronic effector functions of Th2 cytokines

In contrast to the vast majority of injury and repair responses in the lung and other organs, asthmatic inflammation frequently starts in childhood and persists throughout the life of the afflicted individual. In addition, physicians treating asthmatics invariably find themselves attempting to deal with manifestations of established disease, often in the setting of a disease exacerbation. Surprisingly, the model systems that have been used most frequently in studies designed to understand asthma pathogenesis have not appropriately taken these issues into account. Instead, the most commonly employed modeling systems evaluate the acute responses that are elicited in the lung after normal animals are sensitized to and then challenged with an aeroallergen. The asthma-like inflammation and physiologic dysregulation that are seen in these models are an end result of the cellular and molecular events involved in sensitization, Th2 cell development, Th2 cytokine elaboration, and the activation of Th2 cytokine effector pathways. Interventions that inhibit any of these steps can appear to have a beneficial effect on the asthma-relevant readouts that are employed. However, since it is likely that antigen sensitization, Th2 cell development, and Th2 cytokine elaboration have already occurred in patients with established disease and/or a disease exacerbation, interventions at these sites will likely be less than useful therapeutically. In contrast, interventions that regulate Th2 cytokine effector pathways are attractive as therapies. Until recently, modeling systems that allowed the inflammatory and remodeling effects of chronically elaborated Th2 cytokines to be selectively evaluated did not exist. Overexpression-transgenic methodology has, however, powerfully addressed this issue.

The standard approaches used to generate overexpression-transgenic mice are illustrated in Figure 3. First, a DNA construct is prepared that contains the gene that the investigator wishes to express and a promoter to drive the expression of this gene in the desired organ and/or tissue (Figure 3a). If temporally regulated gene expression is desired, recent advances in transgenic methodology that involve the generation of double- and triple-transgenic animals allow the transgene to be selectively turned on or off at any time during the life of the animal (6, 7). When asthma-relevant questions are being asked, the Clara cell 10-kDa protein (CC10) promoter is used to target gene expression, because it is selectively expressed by the Clara cells that make up 40% of the epithelium of the murine airway. To generate transgenic mice, male and female mice are allowed to mate, and the fertilized eggs are washed out of the female's oviduct. The desired DNA construct is then directly microinjected into the pronuclei of these eggs, and the eggs are placed in the uterus of a

pseudopregnant mouse. A pseudopregnant mouse is a female that has been mated with a vasectomized male. She is behaving, from a hormonal perspective, as if she is pregnant and becomes pregnant when the fertilized eggs are deposited in her uterus. She subsequently carries to term, delivering a litter of pups, some of which have the transgene randomly integrated into their genome, others of which do not (Figure 3b). Transgene-positive and -negative mice can be differentiated by extracting DNA from tail biopsies from the pups and determining whether the transgene is present by Southern or PCR analysis. Thus, an outstanding experimental system is established where one can compare the phenotypes of mice born to the same mother, on the same day, that are exposed to the same environment and differ only in the one gene that was inserted. The power of this approach can be easily appreciated in studies designed to define the effects of IL-13 in the asthmatic airway (8).

IL-13 is the product of a gene on chromosome 5 at q31, a site that has been repeatedly implicated in genetic studies looking for the genes involved in the asthmatic diathesis. It was originally discovered as an IL-4–like molecule and was presumed to have an effector profile identical to that of IL-4. It has since become clear that IL-13 and IL-4 differ in their effector properties, with IL-4 and IL-13 playing more prominent roles in the initiation and the effector phases of Th2 inflammation, respectively.

The effector functions of IL-13 were defined and clarified using overexpression-transgenic modeling systems. These studies demonstrated that IL-13 is a potent inducer of an eosinophil-, macrophage-, and lymphocyte-rich inflammatory response, airway fibrosis, mucus metaplasia, and airway hyperresponsiveness (8) (Figure 4). These studies also demonstrated that other Th2 cytokines, such as IL-9, mediate their effects in the lung via their ability to induce IL-13 (9), suggesting that IL-13 may be a final common pathway for Th2-mediated inflammatory responses. Importantly, these transgenic systems were also manipulated to define the mechanisms by which IL-13 generates these critical tissue responses. This was done using standard methods of quantitating mRNA and gene chip methodology to define the genes that are regulated by IL-13 in lungs from transgenic mice. This was followed by a variety of manipulations that characterized the contributions of specific genes to the pathogenesis of the IL-13 phenotype. One such manipulation was the use of neutralizing antibodies against the gene products in question. Another was the breeding of the IL-13 transgenic mice with mice with null mutations of select-

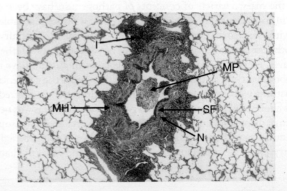

Figure 1
Inflammation and remodeling in the asthmatic airway. There is impressive inflammation (I), mucus plugging (MP), subepithelial fibrosis (SF), myocyte hypertrophy and hyperplasia (MH), and neovascularization (N) in this autopsy lung section from a teenage asthmatic individual.

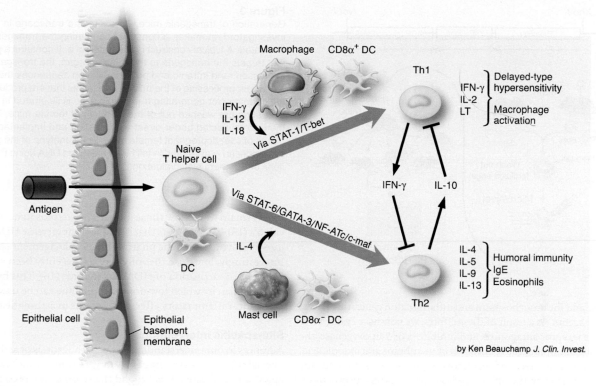

Figure 2

Development of Th1 and Th2 lymphocytes. Antigens enter through the endobronchial tree, cross the epithelial surface, and interact with naive Th cells and DCs. As a result of signals from the surrounding microenvironment, they differentiate into Th1 cells, which produce IFN-γ, IL-2, and lymphotoxin (LT), or Th2 cells, which produce IL-4, IL-5, IL-9, IL-13, and IL-10. Polarization into Th1 cells occurs via a STAT-1– and T-bet–dependent pathway under the influence of CD8α+ DCs and macrophage-derived cytokines such as IFN-γ, IL-12, and IL-18. Differentiation into Th2 cells occurs via a pathway that involves STAT-6, GATA-3, nuclear factor of activated T cells-c (NFATc), and c-maf under the influence of CD8α– DCs and IL-4, which may come from mast cells.

ed downstream genes, followed by characterization of the effects of the transgene in mice that were sufficient or deficient in the downstream gene in question (Figure 5). As can be seen in Figure 6, these studies have provided impressive insights into the mechanisms of IL-13–induced inflammation and tissue fibrosis. The inflammatory response is mediated by the ability of IL-13 to stimulate the elaboration of chemotactic cytokines called chemokines and proteolytic enzymes called matrix metalloproteinases (MMPs) (10, 11). These studies demonstrate that chemokine receptor 2 (CCR2), MMP-9, and MMP-12 play key roles in these responses (10, 11). They also demonstrate that the fibrotic response results from the ability of IL-13 to stimulate the production and activation of the fibrogenic cytokine TGF-β_1, and that TGF-β_1 is activated via an MMP-9– and plasmin-dependent pathway in this setting (12). These studies provide a road map that defines the pathways that IL-13 uses to generate tissue inflammatory and remodeling responses. They also highlight target genes against which therapies can be directed to control selected aspects of the IL-13–induced tissue response.

Contributions of new pathogenic insights to the therapy of asthma

For decades, the overt bronchospasm and impressive benefit that patients experienced with bronchodilators led physicians and patients to rely heavily on these agents to control asthmatic symptomatology. Although steroids were also well known for their ability to control asthmatic symptoms, their side effect profile caused them

to be used only in the most severe cases. The appreciation that asthma is an inflammatory disorder, the availability of effective aerosol steroid preparations, and the belief that chronic unchecked inflammation leads to airway remodeling all contributed to a change in this pattern of practice. This change is reflected in the treatment guidelines for asthma that have been promulgated in recent years by the NIH and other organizations (13). These guidelines stress the need for anti-inflammatory therapy for all but the mildest patients with infrequent symptoms. At present, steroids are the cornerstone of this anti-inflammatory intervention. However, recent advances in our understanding of asthma pathogenesis have provided insights into the mechanisms by which some of our present therapies alter airway inflammation and have provided the rationale for therapies directed against new selective inflammation-regulating targets that may have better side effect profiles than the current therapies. Lastly, our new insights raise the possibility of interventions that might actually prevent asthma in at-risk individuals. Each is reviewed below.

Inflammation regulating the effects of standard therapies

Over sixty years ago, slow-reacting substance of anaphylaxis (SRSA) was appreciated as a spasmogenic activity in lung effluent. Subsequent studies demonstrated that SRSA was mediated by leukotriene products of arachidonic acid metabolism and defined the pathways responsible for the generation of LTB$_4$ and the cystinyl leukotrienes (LTC$_4$, LTD$_4$, and LTE$_4$). They also demonstrated the striking accumulation of cystinyl leukotrienes in biologic fluids from patients with

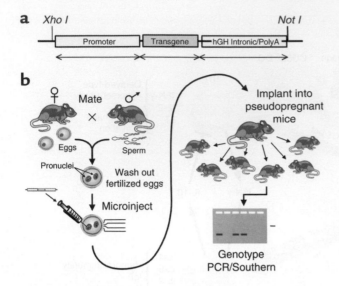

Figure 3
Generation of transgenic mice. To express a transgene in vivo, the investigator first makes a construct containing the transgene being evaluated. A typical construct is illustrated in **a**. It contains a promoter that targets the transgene to the desired organ, the transgene being expressed, and intronic and polyadenylation sequences that ensure the proper processing of the mRNA transcripts that are produced. The methodology for generating transgenic mice is illustrated in **b**. Fertilized eggs are washed out of the oviducts of female mice. They are then microinjected under direct visualization and implanted into the uterus of pseudopregnant female mice. The genotype of the pups that are produced is evaluated in tail biopsy–derived DNA using PCR reactions or Southern blot evaluations.

asthma, and the bronchospasm and inflammation generating effects of these agents. As a result of these efforts, we now have cystinyl leukotriene receptor antagonists and inhibitors of 5-lipoxygenase, the enzyme that initiates the breakdown of membrane arachidonic acid. These agents have received regulatory approval for asthma therapy and are the first new therapies to be licensed for asthma in over thirty years. Their ability to improve lung function and ameliorate aspects of asthmatic inflammation is well documented. Recent studies have also demonstrated that these agents may also control tissue fibrotic responses such as that in asthmatic airway remodeling (14).

Although theophyllines have been used to treat asthma for over seventy years, their use has declined in recent decades because of their low therapeutic index and the appreciation that their anti-inflammatory effects are modest compared with those of corticosteroids. However, theophyllines can diminish airway hyperresponsiveness. In addition, the administration of low-dose theophylline to patients on steroids gives a greater improvement in asthma control than can be achieved by a doubling of the steroid dose itself. Recent studies have demonstrated that these effects may be mediated by a novel mechanism relating to chromatin remodeling. In a quiescent state, the chromatin in genes is tightly wound around core histone proteins. During gene activation, histones are acetylated, which unwinds the chromatin, allowing transcription factors and RNA polymerase II to bind and increase gene transcription (15, 16). This process is mediated by histone acetyl transferase (HAT). Corticosteroids inhibit this process by recruiting histone deacetylase (HDAC), which suppresses the activity of HAT. This suppression leaves the chromatin densely wound and, as a result, decreases target gene transcription. Patients with asthma have decreased levels of HDAC at base line (15). In addition, although steroids increase the levels of HDAC in asthmatic patients, those

levels remain well below those induced by steroids in normal individuals (15). This suggests that therapies that increase HDAC activity will prove useful in the pharmacologic management of asthma. Interestingly, low-dose theophylline has recently been shown to augment the activation of HDAC by steroids (16). This highlights a mechanism by which low-dose theophylline can be used to augment anti-inflammatory effects of steroids in asthmatic tissues.

Site-specific interventions

Advances in our understanding of the pathogenesis of asthma have opened the door to a variety of new targets against which novel therapies can be directed. It is beyond the scope of the present article to describe all of the targets that are being studied. It is, however, convenient to categorize them as interventions that alter the development of Th2 immune responses and interventions that alter the effector functions of Th2 cytokines that have been elaborated. The generation of a Th2 response requires antigen presentation by APCs such as DCs, and T cell proliferation and Th2 differentiation. These events can be inhibited by blockade of APC function (e.g., of costimulatory molecules like B7.2 or the inducible costimulator ICOS) and by blockade of T cell proliferation and/or differentiation (e.g., block-

Figure 4
Demonstration of the effects of transgenic IL-13 on airway fibrosis and mucus metaplasia. (**a**) Trichrome stains are used to compare the amount of blue-staining collagen around airways from transgene-negative mice (left) and transgene-positive mice (right). (**b**) Alcian blue stains are used to demonstrate mucus accumulation in airways from transgene-negative mice (left) and transgene-positive mice (right). Mucus is blue in this evaluation.

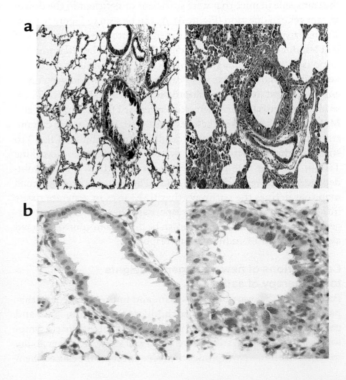

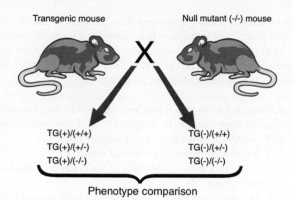

Figure 5

Use of null mutant (knockout) mice to define the pathways that transgenes use to generate disease-relevant phenotypes. In these experiments, transgenic mice with a disease-relevant phenotype (for example, fibrosis or inflammation) are mated with mice that have a null mutation of a downstream gene that is believed to play an important role in the generation of this phenotype. Transgene-positive (TG[+]) and transgene-negative (TG[–]) mice are generated that have normal downstream genes (+/+), are heterozygote knockout at the downstream gene in question (+/–), or are null-mutant for the downstream gene in question (–/–). The presence and intensity of the phenotypes of these mice are then compared. These comparisons allow an investigator to define the role(s) that this downstream gene plays in the generation of the pathologic response.

ade of IL-4 or GATA-3, or the administration of IFN-γ or IL-12). Examples of interventions that block Th2 effector pathways include therapies that block IL-5, IL-13, or IL-9, decrease eosinophil influx and effector function, and/or alter the production and/or effector pathways of inflammation-inducing chemokines. Treatment with a humanized antibody against IgE also alters effector pathways and has shown promise in preliminary investigations (17). Since IL-13 may be a final common pathway for Th2 cytokines (9), approaches that control IL-13 are particularly appealing (18). As illustrated in Figure 6, this includes therapies directed at the STAT-6 signal transducer that mediates the effects of IL-13, therapies directed against the multimeric IL-13 receptor system (IL-4 receptor α and IL-13 receptor α_1), and the administration of IL-13 receptor α_2, a decoy receptor protein that binds IL-13 but does not transduce a signal. It also includes therapies directed against the genes that are downstream of IL-13 and mediate its effects, for example, CCR2 (11).

Preventing asthma

Our present therapeutic approach to asthma focuses exclusively on symptom amelioration. However, our knowledge of the processes that regulate Th1 and Th2 immune responses has raised the possibility of designing interventions that can prevent asthma in at-risk individuals. Many of these approaches are based on in vitro studies demonstrating that Th1 cytokines (such as IFN-γ) can inhibit Th2 cell development, and in vivo observations demonstrating that Th1 responses can feed back to diminish Th2 tissue responses. A variety of approaches have been proposed to accomplish this, including the administration of IFN-γ or IFN-γ–inducing cytokines such as IL-12, and the administration of infectious agents or vaccinations that induce Th1 immunity, such as *Mycobacterium bovis* bacilli Calmette-Guérin (BCG). In addition, therapies have been proposed that entail the administration of

oligodeoxynucleotides that contain unmethylated CpG motifs. Unmethylated CpG DNA motifs are more common in bacterial than in mammalian DNA. The administration of these agents mimics infections and induces Th1 inflammation via a Toll-like receptor 9–dependent mechanism. The ultimate utility of each of these approaches is unclear. This is due, in part, to safety concerns because it is now known that Th1 responses can contribute to the initiation of Th2 responses, that exaggerated Th1 responses contribute to autoimmunity and pulmonary pathology, and that Th1-like signal transduction pathway activation may also contribute to the pathogenesis of asthma (19).

The lung is unique among mucosal compartments in that it is constantly exposed to airborne antigens. As a consequence, the respiratory immune system must differentiate harmless antigens and potentially harmful infectious agents. This is done, in part, by a system of barriers that ensures that harmless antigens induce tolerance and do not elicit sensitization while potentially harmful exposures illicit sensitization and immune system activation. These tolerogenic pathways are altered in patients with asthma. As a result, asthmatics manifest a heightened ability to sensitize and to mount Th2 immune responses to antigens (for example, ragweed) that would not elicit similar responses in normal individuals. The details of the tolerogenic pathways in normal people and their defects in asthmatics are poorly understood. There is evidence that tolerance is mediated by special regulatory T cell populations (Tr cells) that produce IL-10 or TGF-β₁ (2). Treatments that induce tolerance in asthmatics or people at risk for asthma are therefore of potential benefit in preventing and/or ameliorating asthmatic responses. Oral allergen immunotherapy and conventional allergen immunotherapy appear to work, in part, via the induction of tolerance. It is hoped that, over time, more effective tolerance-inducing therapies will be developed.

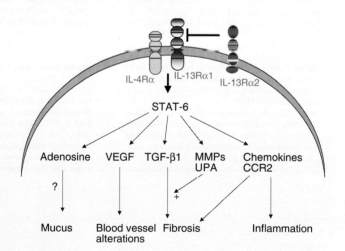

Figure 6

Mechanisms of IL-13–induced phenotype generation. IL-13 binds to the IL-13 receptor complex made up of IL-4 receptor α (IL-4Rα) and IL-13 receptor α_1 (IL-13Rα_1). IL-13 also binds to IL-13Rα_2, which is a decoy receptor that inhibits IL-13 responses. After binding to the IL-13 receptor complex, IL-13 activates STAT-6 signal transduction pathways. Pathways that involve chemokines, the chemokine receptor CCR2, MMPs, urinary plasminogen activator (UPA), TGF-β₁, VEGF, and/or adenosine are then activated, and inflammation, fibrosis, blood vessel alterations, and mucus responses are generated. Each of these pathways is a site against which therapeutic agents can be directed.

Major issues and future directions

Disorders such as asthma are believed to be the result of a dysregulated mucosal immune system and pathologic T cell responses in genetically susceptible individuals (1). Over the next five to ten years we will need to define the relationships between the inflammatory response, the structural alterations noted in the remodeled asthmatic airway, and asthmatic symptomatology, physiologic dysregulation, and disease progression. These investigations will allow us to determine which inflammatory and structural changes contribute to disease pathogenesis (and thus need to be suppressed) and which are more reasonably thought of as aspects of an appropriate healing response in the injured airway (and thus should not be inhibited). It will also be essential to define the genetic alterations that contribute to asthmatic susceptibility, the genetic alterations responsible for the person-to-person variability in asthma presentation and severity, the processes regulating tolerance in the asthmatic and normal lung, and the mechanisms responsible for the lifelong nature of asthmatic inflammation. It is conceivable that, at some time in the future, DNA samples from patients with asthma or people at risk for asthma will be assessed for the presence or absence of polymorphisms of specific genes. Based on these assays, physicians would know whether the patient would develop asthma and, if so, what the natural history of the disease would be and what therapies the patient would optimally respond to. From this perspective, it is easy to see how our knowledge of asthma pathogenesis can have impressive ramifications for our practice of medicine. It is important to point out, however, that it has taken us over a century to get where we are. One is humbled by the appreciation that in 1892 Sir William Osler, in *The Principles and Practice of Medicine*, said that

> Asthma is a term which has been applied to various conditions associated with dyspnea . . . Of the numerous theories the following are most important:
> 1. It is due to spasm of bronchial muscles.
> 2. The attack is due to swelling of the bronchial mucous membrane.
> 3. In many cases it is a special form of inflammation of the smaller bronchioles.

Hopefully, it will not take us another century to go from our understanding of the inflammatory nature of the asthmatic diathesis to effective interventions that control and even prevent this common and debilitating disorder.

Address correspondence to: Jack A. Elias, Yale University School of Medicine, Section of Pulmonary and Critical Care Medicine, 333 Cedar Street/105 LCI, PO Box 208057, New Haven, Connecticut 06520-8057, USA. Phone: (203) 785-4163; Fax: (203) 785-3826; E-mail: Jack.Elias@yale.edu.

1. Neurath, M.F., Finotto, S., and Glimcher, L.H. 2002. The role of Th1/Th2 polarization in mucosal immunity. *Nat. Med.* **8**:567–573.
2. Umetsu, D.T., McIntire, J.J., Akbari, O., Macaubas, C., and DeKruyff, R.H. 2002. Asthma: an epidemic of dysregulated immunity. *Nat. Immunol.* **3**:715–720.
3. Finotto, S., et al. 2002. Development of spontaneous airway changes consistent with human asthma in mice lacking T-bet. *Science.* **295**:336–338.
4. Ray, A., and Cohn, L. 1999. Th2 cells and GATA-3 in asthma: new insights into the regulation of airway inflammation. *J. Clin. Invest.* **104**:1001–1006.
5. Tournoy, K.G., Kips, J.C., and Pauwels, R.A. 2002. Is Th1 the solution for Th2 in asthma? *Clin. Exp. Allergy.* **32**:17–29.
6. Zheng, T., et al. 2000. Inducible targeting of IL-13 to the adult lung causes matrix metalloproteinase- and cathepsin-dependent emphysema. *J. Clin. Invest.* **106**:1081–1093.
7. Zhu, Z., Ma, B., Homer, R.J., Zheng, T., and Elias, J.A. 2001. Use of the tetracycline-controlled transcriptional silencer (tTS) to eliminate transgene leak in inducible overexpression transgenic mice. *J. Biol. Chem.* **276**:25222–25299.
8. Zhu, Z., et al. 1999. Pulmonary expression of interleukin-13 causes inflammation, mucus hypersecretion, subepithelial fibrosis, physiologic abnormalities and eotaxin production. *J. Clin. Invest.* **103**:779–788.
9. Temann, U.A., Ray, P., and Flavell, R.A. 2002. Pulmonary overexpression of IL-9 induces Th2 cytokine expression, leading to immune pathology. *J. Clin. Invest.* **109**:29–39. doi:10.1172/JCI200213696.
10. Lanone, S., et al. 2002. Overlapping and enzyme-specific contributions of matrix metalloproteinases-9 and -12 in IL-13–induced inflammation and remodeling. *J. Clin. Invest.* **110**:463–474. doi:10.1172/JCI200214136.
11. Zhu, Z., et al. 2002. IL-13-induced chemokine responses in the lung: role of CCR2 in the pathogenesis of IL-13-induced inflammation and remodeling. *J. Immunol.* **168**:2953–2962.
12. Lee, C.G., et al. 2001. Interleukin-13 induces tissue fibrosis by selectively stimulating and activating TGF-β$_1$. *J. Exp. Med.* **194**:809–821.
13. 1995. Global initiative for asthma: global strategy for asthma management and prevention. NHLBI/WHO workshop report. NIH, National Heart, Lung and Blood Institute. Bethesda, Maryland, USA. Publication number 95-3659.
14. Henderson, W.R., Jr., et al. 2002. A role for cysteinyl leukotrienes in airway remodeling in a mouse asthma model. *Am. J. Respir. Crit. Care Med.* **165**:108–116.
15. Ito, K., et al. 2002. Expression and activity of histone deacetylases in human asthmatic airways. *Am. J. Respir. Crit. Care Med.* **166**:392–396.
16. Ito, K., et al. 2002. A molecular mechanism of action of theophylline: induction of histone deacetylase activity to decrease inflammatory gene expression. *Proc. Natl. Acad. Sci. USA.* **99**:8921–8926.
17. Milgrom, H., et al. 1999. Treatment of allergic asthma with monoclonal anti-IgE antibody. rhuMAb-E25 Study Group. *N. Engl. J. Med.* **341**:1966–1973.
18. Donaldson, D.D., Elias, J.A., and Wills-Karp, M. 2001. IL-13 antagonism. In *Progress in respiratory research: new drugs in asthma, allergy and COPD.* Volume 31. T. Hansel and P. Barnes, editors. Karger Press. Basel, Switzerland. 260–263.
19. Sampath, D., Castro, M., Look, D.C., and Holtzman, M.J. 1999. Constitutive activation of an epithelial signal transducer and activator of transcription (STAT) pathway in asthma. *J. Clin. Invest.* **103**:1353–1361.

JAK-STAT signaling in asthma

Alessandra B. Pernis[1] and Paul B. Rothman[2]

[1]Department of Medicine, and [2]Department of Microbiology, Columbia University, New York, New York, USA.

The past two decades have witnessed a dramatic increase in the prevalence of asthma worldwide (1). Asthma is a chronic disease characterized by variable airway obstruction, airway hyperresponsiveness (AHR), and airway inflammation and remodeling. Histological studies show that airways of asthmatic patients contain a chronic inflammatory infiltrate composed of lymphocytes, eosinophils, and mast cells. This infiltrate is usually accompanied by desquamation of the bronchial epithelial layer, goblet cell hyperplasia, and thickening of the submucosa. In most cases, the asthmatic inflammatory process results from inappropriate immune responses to common environmental antigens in a genetically susceptible individual (2). These inappropriate immune responses are orchestrated by a subset of CD4+ T helper cells termed T helper 2 (Th2) cells.

Cytokines play a pivotal role in the development of asthma by regulating the expansion of Th2 cells and by mediating many of the Th2 effector functions that underlie the pathogenic events of an asthmatic response. Much effort has recently been placed in elucidating the pathways used by cytokines to mediate their actions. These studies have revealed that cytokine-mediated signals are primarily transduced by the Jak-Stat signaling cascade (3). In this review we will highlight the recent advances made in dissecting the roles of this signaling pathway in the pathogenesis of asthma.

Jak-Stat signaling in Th1 and Th2 differentiation

The two major subsets of CD4+ Th cells, termed Th1 and Th2, secrete mutually distinct profiles of cytokines and thereby coordinate different classes of immune response (4). Th1 cells secrete IL-2, IFN-γ, and TNF-β, whereas Th2 cells produce IL-4, IL-5, IL-6, IL-10, and IL-13. It is now well accepted that Th1 cells are critically involved in the generation of delayed-type hypersensitivity responses, whereas Th2 cells can direct B cells to mount strong humoral responses. Polarization of an immune response toward a Th2 phenotype, while extremely useful in the clearance of parasites, may prove harmful if directed against an otherwise innocuous environmental antigen, as occurs in the pathogenesis of allergic diseases like asthma.

The Th2 cytokines, primarily IL-4, IL-5, and IL-13, control all the major components that characterize an inflammatory asthmatic response, including IgE isotype switching, mucus production, and the recruitment and activation of eosinophils. The involvement of Th2 cells in the pathophysiology of asthma has been corroborated by studies in both humans and mice. The population of Th2 cells is notably expanded in the airways of asthmatic subjects, and presence of these cells correlates with AHR and airway eosinophilia (2). Work in murine models of AHR confirms this correlation and demonstrates that adoptively transferred antigen-specific Th2, but not Th1, cells can mediate airway eosinophilia, mucus hypersecretion, and AHR when recipient mice are exposed to inhaled antigen (5).

The cytokines IL-12 and IL-4 direct the differentiation of Th1 and Th2 cells, respectively, from naive T helper cells (6, 7). Genet-

ic studies have confirmed the physiologic importance of these cytokines in vivo. Mice deficient in either IL-12 or the IL-12 receptor (IL-12R) are unable to mount Th1 responses, while mice lacking IL-4 or the IL-4 receptor α (IL-4Rα) chain display defects in the generation of Th2 responses (8, 9). Indeed, IL-12 and IL-4 not only drive the expansion of their corresponding Th subset but simultaneously block the generation of the opposing subset. Given the importance of IL-4 and IL-12 in controlling Th differentiation, much effort has been placed over the past few years in dissecting the mechanisms by which these cytokines mediate their actions. These studies have revealed that, like most other cytokines, IL-4 and IL-12 activate the Jak-Stat signaling cascade discussed elsewhere in this Perspective series. In this signaling pathway, binding of a cytokine to its receptor leads to the activation of members of the JAK family of receptor-associated kinases. These kinases subsequently activate, via tyrosine phosphorylation, preexistent cytoplasmic factors termed Stats (signal transducer and activator of transcription). Tyrosine phosphorylation allows the Stat proteins to dimerize and translocate to the nucleus, where they mediate changes in gene expression by binding specific DNA elements.

Although both IL-4 and IL-12 follow this basic signaling framework, the two cytokines differ in the specific Jak and Stat components that they activate (10). IL-4 stimulates Jak1 and Jak3 to activate Stat6. In contrast, interaction of IL-12 with its receptor leads to the activation of Jak2 and Tyk2 and the subsequent phosphorylation of Stat4. Activation of Stat6 and Stat4 are thus critical events in the signaling cascades of IL-4 and IL-12, respectively. Given the pivotal roles of these two cytokines in skewing Th cells toward either a Th2 or a Th1 phenotype, it is not surprising that Stat6 and Stat4 control multiple aspects of the Th differentiation programs, as shown in Figure 1 and detailed below.

Stat6 in Th differentiation

The importance of Stat6 in Th2 cell differentiation has been confirmed by the generation of Stat6-deficient mice (11–13), which fail to mount Th2 responses either in vitro — in Th differentiation systems — or in vivo, upon infection with parasitic pathogens that elicit Th2 responses. Th2 memory cell development and survival are dependent on Stat6 (14, 15), whereas Stat6-independent pathways of IL-4 production seem to contribute only to the initial responses mounted by naive T cells, rather than to the later regulation of the Th2 population. Stat6 appears to be not only necessary but also sufficient to drive Th2 differentiation, since introduction of constitutively active forms of Stat6 results in the expression of Th2-type cytokines even when this form of Stat6 is expressed in cells already differentiating toward a Th1 phenotype (16, 17).

Stat6's role in Th2 differentiation is not confined to activating and silencing the expression of specific cytokines but extends to other aspects of the distinctive pattern of gene expression found in Th2 cells, including the heightened expression in these cells of the chemokine receptors CCR4 and CCR8 (18). Stat6-deficient T cells are significantly impaired in their ability to expand upon IL-4 stimulation, thus implicating this transcription factor in

Citation for this article: *J. Clin. Invest.* **109**:1279–1283 (2002). doi:10.1172/JCI200215786.

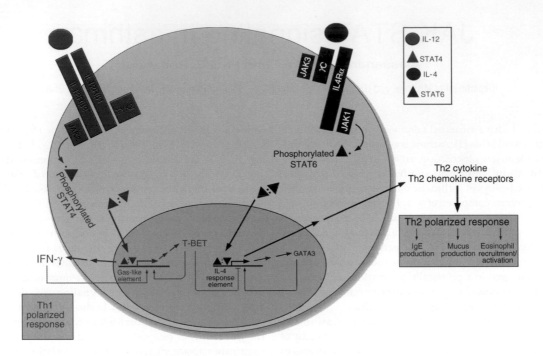

Figure 1
JAK-STAT signaling and the generation of Th1 and Th2 cells. Following antigen presentation, a naive CD4+ T cell will differentiate along the Th1 or Th2 pathway, depending on the nature of the cytokines it contacts. Signaling through the IL-12 receptor (red pathway) and its associated Jak and Stat proteins culminates in the expression of Th1-specific gene products, particularly the cytokine interferon-γ (IFN-γ) Conversely, IL-4 activates a distinct receptor complex, containing different Jaks and Stats, and favors the expression of Th2 cytokines and other gene products (black pathway). These pathways can regulate one another at the transcriptional level. In particular, the transcription factor GATA3, whose expression is induced by IL4 signaling, is essential for the transactivation of other Th2-associated genes, such as those for Th2 cytokines and chemokine receptors. Another transcription factor, T-BET, is induced by IFN-γ (and therefore indirectly by IL-12 stimulation) and also favors expression of this cytokine — thus establishing a positive feedback loop that supports Th1 polarization. In addition, T-BET silences expression of Th2-associated genes. Polarization toward the Th2 response leads to the expression of various cytokines — IL-4, IL-5, IL-6, IL-10, and IL-13 — that contribute to the pathologies seen in asthma.

IL-4–driven proliferation as well as differentiation (17, 19). This defect is due to a block in the progression from the G1 to the S phase of the cell cycle and correlates with an impaired ability of Stat6-deficient T cells to downregulate the cell cycle–dependent kinase inhibitor p27^{kip} in response to IL-4 stimulation. In parallel with its induction of Th2-specific gene expression programs, Stat6 appears to suppress Th1-specific pathways. Indeed, generation of mice deficient in both Stat4 and Stat6 has revealed that in the absence of Stat6, Stat4-deficient CD4$^+$ T cells can differentiate into Th1 cells (20). This effect is not detected in mice deficient for Stat4 alone, suggesting that the presence of Stat6 blocks the ability of Th cells to acquire the Th1 phenotype.

The mechanisms by which Stat6 controls Th2 differentiation are complex and involve subtype-specific induction of specific transcription factors, as well as changes in the chromatin structure and the pattern of cytosine methylation at the *IL4* locus. Th2-specific factors like GATA3 and c-maf synergize with NFAT and AP-1, which are more broadly expressed (6, 7), to activate the characteristic Th2 pattern of gene expression. While both the transcriptional and the epigenetic changes depend on Stat6, it remains unclear whether either of these effects is direct. Some reports also suggest that Stat6 can bind to specific sites within the *IL4* promoter and to a site in the 3′ untranslated region of *IL4* that may function as a Th1-specific silencer (21–23). The significance of Stat6 targeting to these regulatory regions for Th2 commitment has not been established.

Stat4 in Th differentiation
While activation of Stat6 in response to IL-4 is critical for the generation of Th2 responses, activation of a different Stat, Stat4, skews Th cells toward the Th1 phenotype following IL-12 stimulation. Phenotypic analysis of Stat4-deficient mice shows that activation of Stat4 is critical to this process. Stat4-deficient T cells are unable to produce high levels of the Th1 cytokine IFN-γ after exposure to IL-12 (20, 24). Stat4-independent pathways of Th1 differentiation also exist, but, as mentioned above, their effects are seen most clearly in the absence of Stat6 (20, 24). Recent studies suggest that Stat4 activation is particularly critical for the sustained, rather than the initial, production of Th1-type cytokines (15). Stat4 also mediates the downregulation of the cell-cycle inhibitor p27^{kip} in IL-12–treated Th1 cells (19, 25), allowing it (like Stat6) to control Th proliferation as well as differentiation.

As with Th2 cytokines, induction of Th1-specific cytokine gene expression requires epigenetic remodeling, as well as the induction of Th1-specific transcription factors like the recently described T-bet (26). The precise mechanisms employed by Stat4 to mediate its effects on Th1 differentiation have not been fully delineated. In particular, it is unclear whether Stat4 activation participates in the induction of the Th1-specific factor T-bet, whose deficiency in mice has recently been shown to lead to the spontaneous development of airway changes characteristic of human asthma (15, 25, 27, 28). Stat4 is reported to transactivate IFN-γ directly (29), although

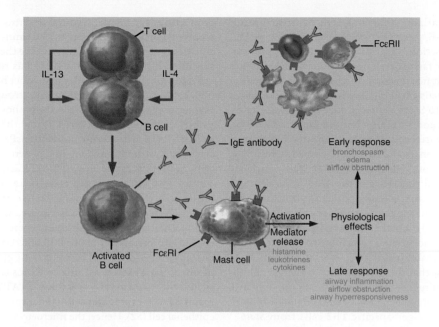

Figure 2
IgE in the pathogenesis of asthmatic responses. Production of Th2-type cytokines (IL-4 and IL-13) by T cells in response to antigens like airborne allergens will drive IgE synthesis by B cells. IgE can then bind to high-affinity IgE receptors on mast cells. Cross-linking of the bound IgE molecules upon reexposure to the antigen provokes mast cell degranulation with the subsequent release of a variety of mediators, which trigger both early and late inflammatory asthmatic responses.

additional factors are likely to be required for full expression of this cytokine (7). Activated Stat4 also blocks differentiation along the Th2 pathway by repressing expression of the Th2-specific factor Gata3 (30), establishing yet another parallel between the role of Stat4 in Th1 development and that of Stat6 in Th2 development. Further progress toward understanding the physiological basis of the actions of these two Stat proteins will require the identification of the key targets of these two transcription factors.

The Jak-Stat signaling cascade in IgE regulation
Levels of IgE correlate with the incidence of atopic asthma in humans. The recent success of anti-IgE therapy in the treatment of atopic asthmatics also highlights the importance of this class of antibodies in human asthma (31) (Figure 2). Work over the past ten years has detailed how cytokines (IL-4 and IL-13, in particular) induce the production of IgE and how other cytokines, such as IFN-γ, block this induction. IL-4 initiates signaling by oligomerizing the heterodimeric IL-4 receptor, which, in hematopoietic cells, is composed of the ligand-specific IL-4Rα chain and the common γ chain (γC) (32). This oligomerization initiates signaling by activating Jak1 and Jak3, which associate constitutively with the cytoplasmic tail of cytokine receptor subunits (Jak1 with IL-4Rα and Jak3 with γC). IL-13 likewise binds IL-4Rα and can activate signaling through Jak1, but this cytokine also binds a more specific receptor subunit, the IL-13Rα1 chain, which associates with the Jak-family kinase Tyk2.

After binding of either IL-4 or IL-13, the activated Jaks phosphorylate tyrosines within the cytoplasmic domain of the IL-4Rα, which act as docking sites for Stat6. Phosphorylated Stat6 homodimerizes, translocates to the nucleus, and activates transcription of genes involved in B cell differentiation, including the germline Iε and Iγ1 genes in mice and germline Iε and Iγ4 genes in humans (33). Induction of these germline promoters, and expres-

sion of the corresponding germline "sterile" transcripts, have been shown in mice to be essential for Ig class switching, recombination events that are required for the production of the various classes of secreted antibodies (33). Stat6 contributes to class switching to produce IgE and IgG1 (11–13). In mice, IFN-γ, which acts via Stat1 to activate the transcription of many early response genes, inhibits the transcription of germline Iε and Iγ1 and thereby blocks B cell production of IgE and IgG1 (34). This regulation appears to be mediated by the suppressor of cytokine signaling-1 (Socs-1), one of a new family of inhibitory Socs molecules defined by the presence of conserved SH2 domains and a novel motif termed a "Socs box" (35). These proteins are induced by cytokines and appear to function in a negative feedback loop. Socs-1 can bind to all the Jaks and inhibit Stat activation. The ability of IFN-γ to inhibit IL-4 signaling seems to be due to the induction of Socs-1 and the resulting suppression of Stat6 activity (36, 37).

A second inhibitor of Jak-Stat signaling implicated in atopic immune responses is Bcl-6, the product of a putative protooncogene that is rearranged or mutated in non-Hodgkin lymphomas. Normally, Bcl-6 functions as a transcriptional repressor, and it has been shown to bind to Stat6-binding sites, such as one found in the Iε promoter. Mice lacking Bcl-6 develop an inflammatory disease characterized by increased levels of IgE, Th2 cells, and mast cell infiltrates. Their B cells produce high levels of IgE (38, 39), a phenotype that requires Stat6 expression by these cells (40, 41). The importance of these regulators of Jak-Stat signaling in human asthma is still undefined.

Stat6 in asthma pathophysiology
In light of the extensive evidence that JAK-STAT signaling controls many of the physiologic events that are deregulated in asthma, several groups have pursued the role of this pathway in murine

models of pulmonary inflammation and AHR. Most such studies have focused on Stat6, given its involvement in directing Th2 responses and IgE production (42–46). Following allergen provocation Stat6 deficient-mice fail to develop a pulmonary Th2 response or AHR, and they exhibit no detectable increase in IgE production or in the number of mucus-containing cells. Because intravenous administration of IL-5 to Stat6-deficient mice restores the development of eosinophilia and AHR after antigen sensitization (46), it appears that one essential role of Stat6 in the development of AHR is to drive Th2 responses and IL-5 production. Interestingly, different groups have reported distinct effects of Stat6 deficiency on the inflammatory infiltrate. Mice of the BALB/c background show only a 50% decrease in the eosinophilic infiltrate, whereas antigen-induced eosinophilia is completely blocked in Stat6-deficient C57BL/6 mice (42, 43). Whether distinctions in sensitization protocols and/or strain background contribute to the differences reported remains to be established.

Recent reports suggest that the development of allergic pulmonary inflammation requires the activation of Stat6 not only in T cells, but also in the parenchymal cells of the lung (47). Indeed, adoptively transferred antigen-specific Th2 cells from $Stat6^{+/+}$ mice fail to mediate allergic inflammation in Stat6-deficient mice. These defects are believed to be due to impairments of these Th2 cells to traffic to the lung, possibly because of defects in the production of chemokines like eotaxin, which control recruitment of Th2 cells. Another feature of asthma that appears to require Stat6 is mucus production: $Stat6^{-/-}$ mice do not demonstrate goblet cell hyperplasia and have decreased mucus secretion in the ovalbumin-asthma challenge asthma model (42, 43, 48). Again, transfer of antigen-specific $Stat6^{+/+}$ T cells fails to complement this defect, thus implicating Stat6 within the lung as a key mediator of mucus

production (47). Similar findings in terms of AHR have also been reported. It will be interesting to determine whether cell type–specific transcription factors can influence the effects of Stat6 within specific cell types, given that IL-13 has recently been shown to activate different gene profiles in distinct human airway cell types (49). Taken together, all these studies strongly support the notion that Stat6 plays a central role in the pathogenesis of asthma.

Several groups have also investigated the expression and activation of STAT6 in asthmatic individuals. Peripheral blood lymphocytes from asthmatic and allergic patients do not display significant differences in the level of STAT6 activity relative to healthy controls (50), but these patients do have a higher density of STAT6-expressing cells in their airways (51). Intriguingly, the density of these STAT6-expressing cells is significantly higher in atopic than in nonatopic asthmatics, although there is no significant difference in expression of the other characteristic Th2 transcription factors GATA3 and c-maf between these groups. One recent study of subjects with severe asthma confirms that such patients show significantly elevated airway levels of STAT6 and also identifies the major STAT6-expressing cell type in this tissue as the bronchial epithelial cell (52). Hence, the microenvironment of asthmatic airways may contribute to deregulating STAT6 expression. Further studies will be needed to investigate the functional consequences of the enhanced expression of STAT6 in asthmatic patients and to determine whether deregulated Stat6 expression is indeed a critical distinction between atopic and nonatopic asthma.

Address correspondence to: Alessandra Pernis, Department of Medicine, Columbia University, 630 West 168th Street, New York, New York 10032, USA. Phone: (212) 305-3763; Fax: (212) 305-4478; E-mail: abp1@columbia.edu.

1. Beasley, R., Crane, J., Lai, C.K.W., and Pearce, N. 2000. Prevalence and etiology of asthma. *J. Allergy Clin. Immunol.* **105**:S466–S472.

2. Wills-Karp, M. 1999. Immunologic basis of antigen-induced airway hyperresponsiveness. *Annu. Rev. Immunol.* **17**:255–281.

3. Darnell, J.E. 1997. STATs and gene regulation. *Science.* **277**:1630–1635.

4. Seder, R., and Paul, W. 1994. Acquisition of lymphokine-producing phenotype by CD4+ T cells. *Annu. Rev. Immunol.* **12**:635–673.

5. Cohn, L., and Ray, A. 2000. T-helper type 2 cell–directed therapy for asthma. *Pharmacol. Ther.* **88**:187–196.

6. Rengarajan, J., Szabo, S.J., and Glimcher, L.H. 2000. Transcriptional regulation of Th1/Th2 polarization. *Immunol. Today.* **21**:479–483.

7. Murphy, K.M., et al. 2000. Signaling and transcription in T helper development. *Annu. Rev. Immunol.* **18**:451–494.

8. Kopf, M., et al. 1993. Disruption of the murine IL-4 gene blocks Th2 cytokine responses. *Nature.* **362**:245–248.

9. Noben-Trauth, N., et al. 1997. An interleukin 4 (IL-4)-independent pathway for CD4+ T cell IL-4 production is revealed in IL-4 receptor-deficient mice. *Proc. Natl. Acad. Sci. USA.* **94**:10838–10843.

10. Wurster, A.L., Tanaka, T., and Grusby, M.J. 2000. The biology of Stat4 and Stat6. *Oncogene.* **19**:2577–2584.

11. Kaplan, M.H., Schindler, U., Smiley, S.T., and Grusby, M.J. 1996. Stat6 is required for mediating responses to IL-4 and for the development of Th2 cells. *Immunity.* **4**:313–319.

12. Takeda, K., et al. 1996. Essential role of Stat6 in IL-4 signalling. *Nature.* **380**:627–630.

13. Shimoda, K., et al. 1996. Lack of IL-4-induced Th2

response and IgE class switching in mice with disrupted Stat6 gene. *Nature.* **380**:630–633.

14. Finkelman, F.D., et al. 2000. Stat6 regulation of in vivo IL-4 responses. *J. Immunol.* **164**:2303–2310.

15. Grogan, J.L., et al. 2001. Early transcription and silencing of cytokine genes underlie polarization of T helper cell subsets. *Immunity.* **14**:205–215.

16. Kurata, H., Lee, H.J., O'Garra, A., and Arai, N. 1999. Ectopic expression of activated Stat6 induces the expression of Th2-specific cytokines and transcription factors in developing Th1 cells. *Immunity.* **11**:677–688.

17. Zhu, J., Guo, L., Watson, C.J., Hu-Li, J., and Paul, W.E. 2001. Stat6 is necessary and sufficient for IL-4's role in Th2 differentiation and cell expansion. *J. Immunol.* **166**:7276–7281.

18. Zhang, S., Lukacs, N.W., Lawless, V.A., Kunkel, S.L., and Kaplan, M.H. 2000. Differential expression of chemokines in Th1 and Th2 cells is dependent on Stat6 but not on Stat4. *J. Immunol.* **165**:10–14.

19. Kaplan, M.H., Daniel, C., Schindler, U., and Grusby, M.J. 1998. Stat proteins control lymphocyte proliferation by regulating p27Kip1 expression. *Mol. Cell. Biol.* **18**:1996–2003.

20. Kaplan, M.H., Sun, Y.L., Hoey, T., and Grusby, M.J. 1996. Impaired IL-12 responses and enhanced development of Th2 cells in Stat4-deficient mice. *Nature.* **382**:174–177.

21. Curiel, R.E., et al. 1997. Identification of a Stat6-responsive element in the promoter of the human interleukin-4 gene. *Eur. J. Immunol.* **27**:1982–1987.

22. Georas, S.N., et al. 1998. Stat6 inhibits human interleukin-4 promoter activity in T cells. *Blood.* **92**:4529–4538.

23. Kubo, M., et al. 1997. T-cell subset-specific expression of the IL-4 gene is regulated by a silencer element and STAT6. *EMBO J.* **16**:4007–4020.

24. Thierfelder, W.E., et al. 1996. Requirement for Stat4 in interleukin-12-mediated responses of natural killer and T cells. *Nature.* **382**:171–174.

25. Mullen, A.C., et al. 2001. Role of T-bet in commitment of TH1 cells before IL-12-dependent selection. *Science.* **292**:1907–1910.

26. Szabo, S.J., et al. 2000. A novel transcription factor, T-bet, directs Th1 lineage commitment. *Cell.* **100**:655–669.

27. Lighvani, A.A., et al. 2001. T-bet is rapidly induced by interferon-gamma in lymphoid and myeloid cells. *Proc. Natl. Acad. Sci. USA.* **98**:15137–15142.

28. Finotto, S., et al. 2002. Development of spontaneous changes consistent with human asthma in mice lacking T-bet. *Science.* **295**:336–338.

29. Xu, X., Sun, Y.L., and Hoey, T. 1996. Cooperative DNA binding and sequence-selective recognition conferred by the STAT amino-terminal domain. *Science.* **273**:794–797.

30. Ouyang, W., et al. 1998. Inhibition of Th1 development mediated by GATA-3 through an IL-4-independent mechanism. *Immunity.* **9**:745–755.

31. Milgrom, H., et al. 1999. Treatment of allergic asthma with monoclonal anti-IgE antibody. rhuMAb-E25 Study Group. *N. Engl. J. Med.* **341**:1966–1973.

32. Jiang, H., Harris, M.B., and Rothman, P. 2000. IL-4/IL-13 signaling beyond JAK/STAT. *J. Allergy Clin. Immunol.* **105**:1063–1070.

33. Coffman, R.L., Lebman, D.A., and Rothman, P. 1993. Mechanism and regulation of immunoglobulin isotype switching. *Adv. Immunol.* **54**:229–270.

34. Linehan, L.A., Warren, W.D., Thompson, P.A., Grusby, M.J., and Berton, M.T. 1998. STAT6 is required for IL-4-induced germline Ig gene transcription and switch recombination. *J. Immunol.* **161**:302–310.

35. Chen, X.P., Losman, J.A., and Rothman, P. 2000.

SOCS proteins, regulators of intracellular signaling. *Immunity.* **13**:287–290.

36. Venkataraman, C., Leung, S., Salvekar, A., Mano, H., and Schindler, U. 1999. Repression of IL-4-induced gene expression by IFN-gamma requires Stat1 activation. *J. Immunol.* **162**:4053–4061.

37. Dickensheets, H.L., and Donnelly, R.P. 1999. Inhibition of IL-4-inducible gene expression in human monocytes by type I and type II interferons. *J. Leukoc. Biol.* **65**:307–312.

38. Dent, A.L., Shaffer, A.L., Yu, X., Allman, D., and Staudt, L.M. 1997. Control of inflammation, cytokine expression, and germinal center formation by BCL-6. *Science.* **276**:589–592.

39. Ye, B., et al. 1997. The BCL-6 proto-oncogene controls germinal-centre formation and Th2-type inflammation. *Nat. Genet.* **16**:161–170.

40. Harris, M.B., et al. 1999. Transcriptional repression of Stat6-dependent interleukin-4-induced genes by BCL-6: specific regulation of iepsilon transcription and immunoglobulin E switching. *Mol. Cell. Biol.* **19**:7264–7275.

41. Dent, A.L., Doherty, T.M., Paul, W.E., Sher, A., and Staudt, L.M. 1999. BCL-6-deficient mice reveal an IL-4-independent, STAT6-dependent pathway that controls susceptibility to infection by Leishmania major. *J. Immunol.* **163**:2098–2103.

42. Kuperman, D., Schofield, B., Wills-Karp, M., and Grusby, M.J. 1998. Signal transducer and activator of transcription factor 6 (Stat6)-deficient mice are protected from antigen-induced airway hyperresponsiveness and mucus production. *J. Exp. Med.* **187**:939–948.

43. Akimoto, T., et al. 1998. Abrogation of bronchial eosinophilic inflammation and airway hyperreactivity in signal transducers and activators of transcription (STAT)6-deficient mice. *J. Exp. Med.* **187**:1537–1542.

44. Miyata, S., et al. 1999. STAT6 deficiency in a mouse model of allergen-induced airways inflammation abolishes eosinophilia but induces infiltration of CD8+ T cells. *Clin. Exp. Allergy.* **29**:114–123.

45. Herrick, C.A., MacLeod, H., Glusac, E., Tigelaar, R.E., and Bottomly, K. 2000. Th2 responses induced by epicutaneous or inhalational protein exposure are differentially dependent on IL-4. *J. Clin. Invest.* **105**:765–775.

46. Tomkinson, A., et al. 1999. The failure of STAT6 deficient mice to develop eosinophilia and airway hyperresponsiveness is overcome by interleukin-5. *Am. J. Respir. Crit. Care Med.* **160**:1283–1291.

47. Mathew, A., et al. 2001. Signal transducer and activator of transcription 6 controls chemokine production and T helper cell type 2 cell trafficking in allergic pulmonary inflammation. *J. Exp. Med.* **193**:1087–1096.

48. Tomkinson, A., et al. 2001. Temporal association between airway hyperresponsiveness and airway eosinophilia in ovalbumin-sensitized mice. *Am. J. Respir. Crit. Care Med.* **163**:721–730.

49. Lee, J.H., et al. 2001. Interleukin-13 induces dramatically different transcriptional programs in three human airway cell types. *Am. J. Respir. Cell Mol. Biol.* **25**:474–485.

50. Miller, R.L., Eppinger, T.M., McConnell, D., Cunningham Rundles, C., and Rothman, P. 1998. Analysis of cytokine signaling in patients with extrinsic asthma and hyperimmunoglobulin E. *J. Allergy Clin. Immunol.* **102**:503–511.

51. Christodoulopoulos, P., et al. 2001. Th2 cytokine-associated transcription factors in atopic and nonatopic asthma: evidence for differential signal transducer and activator of transcription 6 expression. *J. Allergy Clin. Immunol.* **107**:586–591.

52. Mullings, R.E., et al. 2001. Signal transducer and activator of transcription 6 (STAT-6) expression and function in asthmatic bronchial epithelium. *J. Allergy Clin. Immunol.* **108**:832–838.

Rescuing protein conformation: prospects for pharmacological therapy in cystic fibrosis

Marina S. Gelman[1] and Ron R. Kopito[2]

[1]PPD Discovery Inc., Menlo Park, California, USA. [2]Department of Biological Sciences, Stanford University, Stanford, California, USA.

Cystic fibrosis (CF), a fatal autosomal recessive genetic disease that affects over 60,000 people worldwide, is caused by mutations in *CFTR* (1). This gene encodes the cystic fibrosis transmembrane conductance regulator protein, which functions as a Cl⁻ channel at the apical membranes of pulmonary epithelial cells (2). The CFTR channel is also found in certain other epithelia, such as the sweat ducts and part of the gastrointestinal tract, but lung pathology is by far the most prominent cause of clinical disease in *CFTR* homozygotes and compound heterozygotes. Precisely how the loss of functional, surface-expressed CFTR channels and the consequent decrease in Cl⁻ conductance lead to CF pathogenesis is controversial (3). Still, the recognition that the majority of cases of CF are the result of a defect in biogenesis or intracellular trafficking of the protein, and that the mutant protein retains at least partial function, has stimulated an intensive search for therapeutic strategies aimed at rescuing the function of the mutant CFTR. Herein, we discuss some of the results of this search and the prospects of such a protein-repair strategy in the context of CFTR biogenesis and intracellular trafficking.

CFTR biogenesis

CFTR is a polytopic integral membrane glycoprotein composed of 1,480 amino acids (Figure 1). Biogenesis of CFTR begins with the targeting of nascent chain-ribosome complexes to the endoplasmic reticulum (ER) membrane, followed by translocation and integration of transmembrane domains into the lipid bilayer (4) (Figure 2a, step 1). Conformational maturation of wild-type CFTR in the ER (step 2) is an inefficient process; approximately 75% of newly synthesized CFTR molecules are degraded by cytoplasmic proteasomes shortly after synthesis (step 3). Maturation of CFTR to post-ER compartments (step 4) can be readily detected as an approximately 20-kDa decrease in electrophoretic mobility. This decrease reflects conversion by enzymes in the Golgi apparatus of the two Asn-linked glycans in the fourth extracellular loop from immature, high-mannose forms into mature, complex oligosaccharides. Once delivered to the plasma membrane, CFTR is subject to rapid internalization to a pool of subapical vesicles (step 5) that can be recycled to the plasma membrane (step 6) or delivered to lysosomes for degradation (5) (step 8).

Over 1,200 mutations and sequence variants in the *CFTR* gene have been linked to CF (Cystic Fibrosis Mutation Data Base, http://www.genet.sickkids.on.ca/cftr/). These mutations have been grouped into four classes (2): class I mutations abrogate the synthesis of CFTR protein (Figure 2a, step 1), class II mutants are defective in protein trafficking (steps 2 and 4), class III mutations lead to the presence of unstable or nonfunctional protein at the plasma membrane (steps 5, 6, and 8), and class IV mutations interfere with channel activation and regulation by physiological agonists (step 7). Despite this large number of *CFTR* disease alleles, the vast majority (>90%) of CF patients of Northern European origin have at least one copy of a single mutant allele, ΔF508, which encodes a CFTR molecule lacking a phenylalanine at position 508 (1, 6) (Figure 1).

What's wrong with ΔF508 CFTR?

When expressed heterologously in cultured epithelial or nonepithelial cells, ΔF508 CFTR is found as an immature, core-glycosylated species localized by immunofluorescence microscopy to the ER membrane, whereas wild-type CFTR is predominantly found as a complex glycosylated species at the plasma membrane (4). CFTR immunoreactivity is restricted to internal membranes in sweat ducts from ΔF508 CFTR homozygotes (7), although recent studies suggest that the degree to which ΔF508 CFTR is detectable at the plasma membrane may be tissue-specific (8). These findings led to the initial assignment of ΔF508 CFTR as a class II, or trafficking, mutant (2).

Further study of the function and fate of this mutant protein indicates that the trafficking model is incomplete. As discussed below, folding defects in ΔF508 CFTR biosynthesis alter the protein's interactions with the quality control system in the early secretory pathway and also directly or indirectly affect its activity as an anion channel and its stability as a cell surface glycoprotein. Thus, ΔF508 should be regarded not as a simple class II mutant, but as a mixed mutant with the properties of classes II, III, and IV.

ΔF508 CFTR is folding-defective

Quality control mechanisms in the ER recognize misfolded or partially assembled membrane and secretory proteins and block their deployment to distal compartments, either by retaining them in this compartment through prolonged association with molecular chaperones or by targeting them for degradation (9). Pulse-chase analysis shows that nascent ΔF508 CFTR molecules are rapidly degraded by the ubiquitin-proteasome system without detectable lag following their release from the ribosome (4). Together with biochemical studies of the thermal stability and refolding properties of recombinant ΔF508 CFTR fragments, these studies have led to the conclusion that ΔF508 CFTR is fundamentally a folding mutant (10) and that its steady-state distribution in the ER reflects the operation of the quality control machinery in that organelle, rather than a trafficking defect per se (4). Simply increasing ΔF508 CFTR expression in heterologous cells (11, 12) leads to increased expression of functional ΔF508 CFTR molecules at the plasma membrane, indicating that

Nonstandard abbreviations used: cystic fibrosis (CF); endoplasmic reticulum (ER); trimethylamine-*N*-oxide (TMAO); 4-phenyl butyrate (4PB); 8-cyclopentenyl-1,3-dipropylxanthine (CPX); deoxyspergualin (DSG); thapsigargin (TG); doxorubicin (Dox).

Conflict of interest: The authors have declared that no conflict of interest exists.

Citation for this article: *J. Clin. Invest.* **110**:1591–1597 (2002). doi:10.1172/JCI200216786.

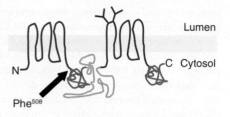

Figure 1
Cartoon representation of CFTR structure. Indicated are the transmembrane domains (blue), the two Asn-linked glycans (purple), the "R" domain (green), and the two nucleotide-binding domains (red). N, amino terminus; C, carboxy terminus.

the phenotype of this mutation is leaky. Although this leakiness could be a simple consequence of overwhelming the quality control machinery, it is more likely that deletion of Phe[508] reduces the efficiency with which CFTR folds, possibly due to kinetic partitioning of an intermediate in the early folding pathway (Figure 2, step 2). This view of ΔF508 CFTR as a folding mutant is also supported by the observation, discussed below, that the fractional folding of ΔF508 CFTR in vivo (4) (or of a fragment of the protein in vitro [ref. 13]) can be increased by reducing the temperature of incubation or by adding chemical chaperones.

ΔF508 CFTR gating shows diminished response to β-adrenergic agonists

Comprehensive biophysical analysis from many laboratories has established that, while the basic pore properties of ΔF508 CFTR are indistinguishable from those of wild-type CFTR, the kinetics of CFTR gating is significantly altered by this mutation (12, 14–16). The observed difference in macroscopic halide conductance properties can be explained by a three- to fourfold reduction in the channel's open probability, mainly due to increased closed times (12, 15). Whether this difference is a direct consequence of a primary structural change in CFTR resulting from deletion of Phe[508] or represents a secondary effect of altered interaction with kinases or phosphatases that control channel activation (17), it is clear that ΔF508 CFTR channel function in situ is severely reduced at any given agonist concentration (Figure 2, step 7).

Rescued ΔF508 CFTR is less stable than wild-type CFTR

Once delivered to the plasma membrane (Figure 2, step 4), CFTR is subject to endocytic recycling through a subapical vesicular compartment (5). The steady-state distribution of ΔF508 CFTR at the plasma membrane is determined by the relative rates of endocytosis (step 5) and exocytosis (step 6). The overall half-life of CFTR molecules in this recycling pool is determined by partitioning between exocytosis (step 6) and degradation by the lysosome (step 8). Estimates of the half-life of the wild-type plasma membrane/recycling pool of CFTR range from 18–48 hours in BHK cells (18–20) to 24–48 hours in the epithelial line LCPK1. In marked contrast, the half-life of ΔF508 CFTR is only about 4 hours, as determined in cultured cells exposed to chemical chaperones or held at a reduced temperature (treatments that allow the protein to accumulate on the surface prior to the assay) (18–20). Pulse-chase labeling studies indicate that this reduced half-life is a consequence of increased lysosomal degradation and not simply a redistribution between

surface and subapical recycling compartments. The finding that complex glycosylated ΔF508 CFTR molecules are 20 times less stable than wild-type CFTR at modestly elevated temperatures (40°C) but exhibit stability similar to that of wild-type CFTR at reduced temperatures suggests that the shorter half-life of plasma membrane ΔF508 CFTR is due to structural differences between ΔF508 and wild-type CFTR (19). Data from proteolytic mapping studies confirm this view.

Pharmacological rescue of ΔF508 CFTR trafficking

Because both the instability of the rescued ΔF508 CFTR at the plasma membrane and the protein's inefficient conformational maturation in the ER are phenotypes of Phe[508] deletion, it is possible that this mutant protein is prone to adopt a structurally similar, non-native conformation in both membranes. If so, this conformation may be detected by quality control machinery linked in the ER to ubiquitin-dependent proteolysis but linked in the distal secretory pathway to lysosomal degradation. This model would suggest that a small-molecule drug that facilitates ΔF508 CFTR folding in the ER might also be effective at stabilizing the protein in the plasma membrane. On the other hand, it is also possible that compounds that specifically suppress off-pathway conformers in the ER do not interact with the non-native conformation in the plasma membrane. In either case, these studies clearly underscore the need to consider ΔF508 CFTR stability at the plasma membrane in any pharmacological rescue therapy regimen.

Fortunately, there are promising drug candidates capable of increasing CFTR activation kinetics (see ref. 21 for a comprehensive recent review). These compounds, which include the alkylxanthines, flavones, and phosphatase inhibitors, appear to stimulate CFTR by a combination of amplifying the cAMP signal transduction/phosphorylation cascade and directly binding to CFTR. Below, we consider a subset of these compounds that have been reported to ameliorate the ΔF508 CFTR defect by increasing its delivery to or its stability in the plasma membrane.

Chemical and pharmacological chaperones

The term "chemical chaperone" loosely describes a family of low-molecular weight compounds including polyols (e.g., glycerol, sorbitol, and *myo*-inositol), amines (e.g., betaine and trimethylamine-N-oxide [TMAO]), and solvents such as DMSO and D_2O. These compounds have long been recognized to have protein-stabilizing properties in vitro, due largely to their ability to increase protein hydration (22). Endogenously produced compounds like *myo*-inositol and betaine serve as osmolytes, balancing osmotic forces in cells and organisms that are chronically exposed to osmotic stress.

Incubation of ΔF508 CFTR–expressing cells in high concentrations (>1 M) of glycerol or other chemical chaperones (e.g., TMAO or D_2O) increases the steady-state level of complex-glycosylated ΔF508 CFTR on immunoblotting and increases functional cAMP-activated channel activity, as seen in measurements of halide efflux and whole-cell patch clamp analysis (23). Pulse-chase analysis shows that the increase in steady-state expression of functional cell surface ΔF508 CFTR in the presence of glycerol is due to increased maturation of core-glycosylated nascent ΔF508 CFTR to a post-ER compartment and decreased degradation, most likely as a result of enhanced folding (24). Fischer and colleagues report significant increases in forskolin-activated intestinal Cl⁻ transepithelial transport in matched wild-type and ΔF508 CFTR (but not in CFTR-null) mice injected subcutaneously with TMAO, suggesting that a simi-

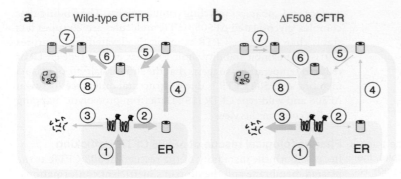

a Wild-type CFTR   **b** ΔF508 CFTR

Figure 2
Biogenesis and intracellular trafficking pathway of wild-type (**a**) and ΔF508 (**b**) CFTR. The width of the gray arrows is proportional to the relative flux through a particular branch of the pathway. Synthesis and cotranslational integration (step 1) in the ER membrane are followed by folding to a native conformation (step 2). About 25% of wild-type and more than 99% of ΔF508 CFTR molecules are degraded by a process that is mediated by cytoplasmic proteasomes (step 3). Native CFTR molecules (light blue cylinder) are delivered via the Golgi apparatus (not shown) to the plasma membrane (step 4), where they are subject to rapid endocytosis (step 5) to subapical vesicles (light blue lumen). CFTR is recycled to the plasma membrane (step 6), where it can be activated by cAMP-dependent kinases (step 7). Differences in the relative rates of recycling and degradation in lysosomes (pink lumen; step 8) are likely to account for the substantial differences in half-lives between wild-type and ΔF508 CFTR.

lar rescue of ΔF508 CFTR is feasible in vivo (25). This functional restoration was dependent upon the dose of TMAO; at a level near the LD_{50} for the compound, this treatment supported Cl⁻ transport levels less than one-sixth of those observed in mice expressing wild-type CFTR. These data, together with the demonstration that TMAO and glycerol increase folding of ΔF508 CFTR in cell culture, confirm the feasibility of pharmacological rescue of at least some misfolded CFTR mutants and should provide the impetus for drug-discovery screens for compounds that are active in rescuing ΔF508 CFTR folding at nontoxic pharmacological doses.

A related approach to treating CFTR misfolding involves the use of "pharmacological chaperones" (26), defined substrates or ligands of cell surface–borne channels and receptors. Mutant membrane proteins that have been studied in this regard include P-glycoprotein and the V2 vasopressin and δ-opioid receptors. The identification of pharmacological chaperones constitutes proof-of-principle demonstration that small, high-affinity compounds can influence the conformational maturation pathways of proteins and ameliorate their folding defects.

Upregulation of CFTR surface expression using butyrate

Butyrate is a short-chain fatty acid that has long been recognized to function at millimolar concentrations as a transcriptional activator, most likely due to its ability to influence chromatin structure via inhibition of histone deacetylase activity (27). The molecular targets of butyrate have not been identified. Possibly as a direct result of its effects on transcription, or in combination with other mechanisms, butyrate has been reported to modulate a wide spectrum of cellular events, including cellular differentiation and apoptosis (28).

Treatment of cells expressing ΔF508 CFTR under control of the *Metallothionein* (29) and cytomegalovirus (M. Gelman and R. Kopi-

to, unpublished observations) promoters with sodium butyrate stimulates expression of ΔF508 CFTR, with a concomitant increase in functional CFTR at the cell surface. Elevated levels of ΔF508 CFTR mRNA and protein synthesis following butyrate treatment suggest that the increase in functional ΔF508 CFTR is a consequence of overexpression, rather than a specific effect on folding, since no effect of butyrate on folding efficiency or degradation kinetics has been reported. The ΔF508 CFTR folding defect is leaky, a finding that is best explained by a kinetic model in which a fixed fraction of ΔF508 CFTR molecules are able to fold and, hence, to mature to the plasma membrane. Wild-type CFTR, which is also folded inefficiently, should likewise be upregulated by this treatment (4). An increase in surface ΔF508 CFTR can thus be achieved simply by increasing ΔF508 CFTR gene expression, independent of any effect on trafficking.

Whether butyrate and the butyrate analog 4-phenyl butyrate (4PB) influence ΔF508 CFTR folding or trafficking in addition to their effects on transcription is controversial, as is their efficacy as therapeutic agents. Rubenstein and colleagues tested the effects of 4PB on ΔF508 CFTR expression in primary nasal polyp cultures from CF patients with ΔF508 CFTR alleles or the immortalized CF cell line 1B3-1 (30). They observed increased forskolin-activated Cl⁻ conductance, concomitant with increased accumulation of the complex glycosylated form of CFTR, and they argued that 4PB alters the intracellular trafficking of ΔF508 CFTR. However, since they did not report in that study whether butyrate treatment affected ΔF508 CFTR transcription, and since data on ΔF508 CFTR folding efficiency and turnover are lacking, the observed increase in functional surface ΔF508 CFTR might result from enhanced transcription, rather than increased trafficking efficiency. Indeed, Heda and Marino (31) have found that butyrate treatment increased both the amount of functional cell surface ΔF508 CFTR and ΔF508 CFTR mRNA levels.

It has been argued that the effect of 4PB on ΔF508 CFTR trafficking is explained by decreased levels of heat shock cognate protein Hsc70 (32). While this model is not implausible, it remains to be shown whether Hsc70 is limiting for ΔF508 CFTR degradation. In light of the pleiotropic effects of butyrate on cell function (28), the contradictory data on the effect of butyrate on the inducible heat shock protein Hsp70 (32, 33), and the possibility that butyrate's effect on ΔF508 CFTR is simply a consequence of overexpression, it would be premature to conclude that butyrate or 4PB influences the folding or the stability of ΔF508 CFTR.

The utility of butyrates as potential therapeutics may be limited by their inhibitory effects on ion transport systems. Butyrates inhibit CFTR directly (34, 35) and may block CFTR function indirectly by inhibiting sodium pumps (36) or other unidentified transporters in the basolateral membrane (37), as well as Na-K-2Cl cotransporters (38), all of which are required for transepithelial salt transport. Given these concerns about the mechanisms and potential side effects of these drugs, additional caution is warranted before proceeding with further clinical trials.

CPX

The adenosine A1 receptor 8-cyclopentenyl-1,3-dipropylxanthine (CPX), an alkylxanthine now in phase 2 clinical trials for CF, has

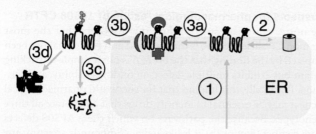

Figure 3
Following initial synthesis and membrane integration (step 1), nascent CFTR molecules are inefficiently folded (step 2) and are recognized as misfolded by one or more components of the quality control machinery (red; step 3a), which may interact with cytoplasmic, membrane, or lumenal portions of CFTR. This interaction is necessary to dislocate CFTR to the cytoplasmic ubiquitin (green spheres) conjugation machinery (step 3b), which then directs it to the proteasome for degradation (step 3c). Cytoplasmic aggregates form (step 3d) when the rate of production and dislocation of misfolded CFTR exceeds the capacity of the ubiquitin-proteasome system for degradation.

been reported to enhance Cl⁻ efflux from NIH3T3 cells expressing ΔF508 CFTR (39). CPX is also reported to correct the trafficking defect of ΔF508 CFTR (40, 41). The supporting data consist of immunoblot analyses, immunofluorescence microscopy, and laser scanning cytometry, which together show that CPX treatment leads to increased expression of total ΔF508 CFTR and a small concomitant increase in the steady-state levels of a band corresponding to the complex glycoform of CFTR (41). While these data are consistent with CPX improving folding or delivery of ΔF508 CFTR to the plasma membrane, they are readily explained by an increase in synthesis of this leaky mutant. In the absence of any data directly measuring the effect of the drug on the efficiency of ΔF508 CFTR folding or exit from the ER, the claim that CPX can repair or correct the trafficking of ΔF508 CFTR is unjustified.

Drugs that interfere with molecular chaperones and proteasomal degradation

The simple model presented in Figure 2 predicts that inhibition of ΔF508 CFTR degradation should, by mass action, increase production of folded ΔF508 CFTR molecules. However, acute or chronic exposure of ΔF508 CFTR–expressing cells to proteasome inhibitors leads to accumulation of detergent-insoluble, multiubiquitylated ΔF508 CFTR molecules, with no detectable increase in folded CFTR (4). These findings suggest that the substrate of proteasome action is not in direct equilibrium with intermediates on the folding pathway. These and subsequent studies on the degradation of misfolded ΔF508 CFTR and other proteins in the secretory pathway (see Kaufman, this Perspective series, ref. 42; see also ref. 4) reveal that degradation is a multistep process (Figure 3) that involves recognition (step 3a), ubiquitin conjugation and dislocation across the ER membrane (step 3b), and unfolding and degradation by the proteasome (step 3c). Proteasome inhibitors (inhibition of step 3c) cause proteins to aggregate in the cytoplasm (step 3d), interfering with the dislocation process (step 3b) (43). Given the strong aggregation-promoting effect of proteasome inhibitors and the accumulating evidence that protein aggregation is a toxic event linked to the pathogenesis of many neurodegenerative diseases (43), therapeutic strategies that target the proteasome or associated machinery should be approached with extreme cau-

tion, all the more so because many of these diseases take decades to manifest, so toxic side effects of proteasome inhibitors may fail to be detected by conventional toxicity assays.

A second class of compounds that have been proposed as having potential utility in rescuing ΔF508 CFTR misfolding are those which target molecular chaperones systems. Nascent CFTR molecules interact transiently with cytoplasmic members of the Hsp70 (44, 45), Hsp40 (45), and Hsp90 (46) families and the ER lumenal chaperone calnexin (47); and these chaperones may contribute to recognition of misfolded CFTR and ΔF508 CFTR. The immunosuppressant deoxyspergualin (DSG) binds Hsp70 and Hsp90 in vitro with micromolar affinity and appears to compete for binding of substrates to a subset of these chaperones (48). Treatment of ΔF508 CFTR–expressing epithelial cells with DSG increases forskolin-stimulated halide efflux and whole-cell currents to levels comparable to those achieved by low temperature incubation (49), suggesting that, at least at rather high concentrations (50 μg/ml), this drug can partially restore function to ΔF508 CFTR. However, there are no data to support the hypothesis that DSG influences the folding or delivery to the plasma membrane, directly or indirectly. Even though it appears to exhibit some specificity with respect to Hsp70 family members (50), the pleiotropic effects of interfering with both the Hsp70 and the Hsp90 chaperone systems could have indirect consequences on multiple cellular pathways, including those that influence CFTR activation or gating mechanisms. Moreover, although it is clear that DSG interacts with Hsp70 and Hsp90, there is no evidence that any of the biological effects of this drug are mediated by the drug's interaction with these chaperones; other targets of lower abundance and higher affinity may remain to be identified. Interestingly, treatment of CFTR-expressing cells with the benzoquinone ansamycins geldanamycin and herbimycin A — drugs that bind with high affinity to Hsp90-family chaperones — appears to perturb the interaction between Hsp90 and CFTR, suppressing its maturation and accelerating its degradation (46). Recent studies suggest that geldanamycin activates a general heat shock response, influencing other classes of molecular chaperones and components of the degradation apparatus, perhaps as an indirect consequence of its effects on Hsp90 (51).

A recent report by Egan et al. (52) suggests that thapsigargin (TG), an inhibitor of the ER calcium pump, can also increase expression of forskolin-stimulated Cl⁻ channels and immunochemically detectable ΔF508 CFTR molecules in the apical plasma membrane. This drug is proposed to enhance ΔF508 CFTR egress from the ER by interfering with its interaction with one or more Ca-dependent ER chaperones. Although TG treatment leads to an increase in the complex glycosylated form of ΔF508 CFTR, the data do not rule out the possibility that this drug directly or indirectly influences the partitioning of ΔF508 CFTR molecules through the endocytic pathway (Figure 2a, steps 7 and 8). Although the authors show that TG treatment does not appear to induce a general heat shock response, its well-established role as an inducer of the "unfolded protein response" (UPR; see Kaufman, this Perspective series, ref. 42; and ref. 53) suggests that its effect on ΔF508 CFTR might be nonspecific. It will be important to determine whether other UPR inducers such as tunicamycin have similar effects. Perturbation of cellular Ca homeostasis and ER chaperone function and the attendant impact on signal transduction, protein folding, and degradation may promote proteotoxicity over the long term, despite the reported lack of acute toxicity in ΔF508 CFTR mice (52).

Molecular chaperones play essential roles both in protein biogenesis, where they facilitate folding and suppress aggregation, and in protein degradation, where they contribute to recognition and destruction of misfolded molecules. The studies described here suggest that drugs that interfere with the action of broad classes of molecular chaperones can influence the functionality of ΔF508 CFTR molecules in model systems. Concerns over the utility of chaperone antagonists or agonists center on our lack of understanding of the role of specific chaperones in the intracellular fate of ΔF508 CFTR, and a recognition of the pleiotropic nature of chaperone action and of the potential for induction of proteotoxicity through aggregate formation.

Clearly, considerable basic investigation of these questions is required before this class of compounds can be considered as potential therapeutics. Perhaps ubiquitin E3 ligases or other relatively specific components of the quality control apparatus will offer more reasonable targets. Because E3s participate in the earliest stages of ΔF508 CFTR recognition (Figure 3, step 3a), identification of E3s in ΔF508 CFTR degradation should be viewed as an important goal.

Other compounds reported to influence ΔF508 CFTR trafficking

Substituted benzo[c]quinolizinium (MBP) compounds MBP-07 and MBP-91 have also been reported to activate wild-type and ΔF508 CFTR Cl conductance in cultured ΔF508 CFTR epithelial cells (54, 55). These compounds appear to act at the low millimolar range to increase CFTR Cl transport function, and limited biochemical data suggest that they can increase the amount of complex glycosylated ΔF508 CFTR and immunohistochemically detectable staining at the apical plasma membrane of nasal epithelial cells. Additional data will be needed to rule out the possibility that they simply increase ΔF508 CFTR transcription or stabilize the plasma membrane pool of ΔF508 CFTR. Moreover, the observation that these compounds increase intracellular ΔF508 CFTR staining in a juxtanuclear location (55) suggests that MBP-07 and MBP-91 may actually promote ΔF508 CFTR aggregation.

The anthracycline doxorubicin (Dox) was recently reported to increase functional cell surface expression of ΔF508 CFTR (56). Low concentrations (0.25 μM) of this drug promote increased halide efflux and short-circuit current in monolayers of T84 cells and appear to produce a modest increase in steady-state accumulation of complex glycosylated CFTR without apparent effect on CFTR mRNA levels. While these data suggest that Dox could increase the efficiency of CFTR folding or stabilize the plasma membrane pool, the effect observed was quite transient: Enhanced CFTR expression was observed only 24 hours after Dox treatment and was lost by 48 hours. It is also curious that Dox has no significant effect on the activity or processing of CFTR in another epithelial cell line, MDCK. Finally, the extremely modest effect of this compound on the cell surface expression of ΔF508 CFTR and the absence of evidence indicating functional activation of ΔF508 CFTR, together with the considerable toxicity of this anticancer drug, raise doubts about its utility as a drug.

Prospects for pharmacological rescue of ΔF508 CFTR

The initial optimism surrounding the discovery that the most common CF allele, ΔF508, is mislocalized and functional has been tempered by the finding that the protein is not a simple trafficking mutant but exhibits multiple defects in folding, stability, and activation. Practically, this implies that for successful pharmacological rescue it may be essential to identify drugs that ameliorate all three phenotypes. Because the pathways on which these ΔF508 defects lie are distinct (Figure 2b), it is likely that compounds affecting any single pathway will act synergistically. For example, because the cell surface half-life of ΔF508 CFTR is so significantly reduced, it is likely that any treatment that enhances ΔF508 CFTR folding will be greatly enhanced by even a modest prolongation of cell surface residence. In support of this view, Heda and Marino report that sodium butyrate, a transcriptional enhancer, acts synergistically with reduced temperature, a treatment that increases folding efficiency (31).

Recognition of the multiple ΔF508 CFTR phenotypes must be considered in the design of any rational high-throughput screen for pharmacological agents to rescue ΔF508 CFTR. Screening efforts currently underway are focused on cell-based reporters that can identify compounds that increase cAMP-dependent transmembrane halide movement (57). These assays have the advantage of identifying any compound that increases CFTR function, since they make no assumption about mechanism. On the other hand, small effects of any compound on a single pathway could be greatly enhanced if combined with a well-defined effector of another pathway. An alternative approach would be to customize screens to identify compounds that act on a specific phenotype — for example, to search for molecules that increase the amount of cell surface ΔF508 CFTR irrespective of function. Such a screen could identify molecules that enhance either folding or stability of the mutant protein, and it might yield either potential leads or compounds that could be useful in conjunction with functionally based screens.

In summary, there is reason to expect that pharmacological rescue of ΔF508 CFTR is an achievable goal. However, the compounds that have been thus far described to ameliorate the folding or stability defect are either unsuitable for pharmacological use because of inherent toxicity (e.g., chemical chaperones like TMAO and glycerol) or too poorly characterized at the cellular or molecular level to support the claims that they specifically rescue or correct the ΔF508 trafficking defect (e.g., CPX, 4PB, MBP, and Dox). The optimal strategy in the development of effective therapeutics for CF will be to simultaneously target the defects in folding, stability, and activation of ΔF508 CFTR. This goal will require a more complete understanding of the mechanism of action of currently available and newly discovered drugs.

Acknowledgments
The authors are grateful to G. Endemann for critical reading of the manuscript.

Address correspondence to: Ron R. Kopito, Department of Biological Sciences, Stanford University, Stanford, California 94305-5020, USA. Phone: (650) 723-7581; Fax: (650) 723-8475; E-mail: kopito@stanford.edu.

1. Kerem, B.-S., et al. 1989. Identification of the cystic fibrosis gene: genetic analysis. *Science.* 245:1073–1080.
2. Welsh, M.J., and Smith, A.E. 1993. Molecular mechanisms of CFTR chloride channel dysfunction in cystic fibrosis. *Cell.* 73:1251–1254.
3. Wine, J.J. 1999. The genesis of cystic fibrosis lung disease. *J. Clin. Invest.* 103:309–312.
4. Kopito, R.R. 1999. Biosynthesis and degradation of CFTR. *Physiol. Rev.* 79(Suppl.):S167–S173.
5. Kleizen, B., Braakman, I., and de Jonge, H.R. 2000. Regulated trafficking of the CFTR chloride channel. *Eur. J. Cell Biol.* 79:544–556.
6. Riordan, J.R., et al. 1989. Identification of the cystic fibrosis gene: cloning and characterization of

complementary DNA. *Science.* **245**:1066–1073.

7. Kartner, N., Augustinas, O., Jensen, T.J., Naismith, A.L., and Riordan, J.R. 1992. Mislocalization of delta F508 CFTR in cystic fibrosis sweat gland. *Nat. Genet.* **1**:321–327.

8. Kälin, N., Claaß, A., Sommer, M., Puchelle, E., and Tümmler, B. 1999. ΔF508 CFTR protein expression in tissues from patients with cystic fibrosis. *J. Clin. Invest.* **103**:1379–1389.

9. Ellgaard, L., Molinari, M., and Helenius, A. 1999. Setting the standards: quality control in the secretory pathway. *Science.* **286**:1882–1888.

10. Qu, B.H., Strickland, E., and Thomas, P.J. 1997. Cystic fibrosis: a disease of altered protein folding. *J. Bioenerg. Biomembr.* **29**:483–490.

11. Cheng, S.H., et al. 1995. Functional activation of the cystic fibrosis trafficking mutant DF508-CFTR by overexpression. *Am. J. Physiol.* **268**:L615–L624.

12. Dalemans, W., et al. 1991. Altered chloride ion channel kinetics associated with the DF508 cystic fibrosis mutaion. *Nature.* **354**:526–528.

13. Qu, B.H., and Thomas, P.J. 1996. Alteration of the cystic fibrosis transmembrane conductance regulator folding pathway. *J. Biol. Chem.* **271**:7261–7264.

14. Drumm, M.L., et al. 1991. Chloride conductance expressed by DF508 and other mutant CFTRs in *Xenopus* oocytes. *Science.* **254**:1797–1799.

15. Haws, C.M., et al. 1996. Delta F508-CFTR channels: kinetics, activation and potentiation by xanthines. *Am. J. Physiol.* **270**:C1544–C1555.

16. Denning, G.M., et al. 1992. Processing of mutant cystic fibrosis transmembrane conductance regulator is temperature-sensitive. *Nature.* **358**:761–764.

17. Li, C., et al. 1993. The cystic fibrosis mutation (delta F508) does not influence the chloride channel activity of CFTR. *Nat. Genet.* **3**:311–316.

18. Lukacs, G.L., et al. 1993. The delta F508 mutation decreases the stability of cystic fibrosis transmembrane conductance regulator in the plasma membrane. Determination of functional half-lives on transfected cells. *J. Biol. Chem.* **268**:21592–21598.

19. Sharma, M., Benharouga, M., Hu, W., and Lukacs, G.L. 2001. Conformational and temperature-sensitive stability defects of the delta F508 cystic fibrosis transmembrane conductance regulator in post-endoplasmic reticulum compartments. *J. Biol. Chem.* **276**:8942–8950.

20. Heda, G.D., Tanwani, M., and Marino, C.R. 2001. The Delta F508 mutation shortens the biochemical half-life of plasma membrane CFTR in polarized epithelial cells. *Am. J. Physiol. Cell Physiol.* **280**:C166–C174.

21. Roomans, G.M. 2001. Pharmacological treatment of the ion transport defect in cystic fibrosis. *Expert Opin. Investig. Drugs.* **10**:1–19.

22. Gekko, K., and Ito, H. 1990. Competing solvent effects of polyols and guanidine hydrochloride on protein stability. *J. Biochem. (Tokyo).* **107**:572–577.

23. Brown, C.R., Hong-Brown, L.Q., and Welch, W.J. 1997. Strategies for correcting the delta F508 CFTR protein-folding defect. *J. Bioenerg. Biomembr.* **29**:491–502.

24. Sato, S., Ward, C.L., Krouse, M.E., Wine, J.J., and Kopito, R.R. 1996. Glycerol reverses the misfolding phenotype of the most common cystic fibrosis mutation. *J. Biol. Chem.* **271**:635–638.

25. Fischer, H., et al. 2001. Partial restoration of defec-

tive chloride conductance in DeltaF508 CF mice by trimethylamine oxide. *Am. J. Physiol. Lung Cell. Mol. Physiol.* **281**:L52–L57.

26. Morello, J.P., Petaja-Repo, U.E., Bichet, D.G., and Bouvier, M. 2000. Pharmacological chaperones: a new twist on receptor folding. *Trends Pharmacol. Sci.* **21**:466–469.

27. Kruh, J. 1982. Effects of sodium butyrate, a new pharmacological agent, on cells in culture. *Mol. Cell. Biochem.* **42**:65–82.

28. Marks, P.A., Richon, V.M., and Rifkind, R.A. 2000. Histone deacetylase inhibitors: inducers of differentiation or apoptosis of transformed cells. *J. Natl. Cancer Inst.* **92**:1210–1216.

29. Cheng, S.H., et al. 1995. Functional activation of the cystic fibrosis trafficking mutant delta F508-CFTR by overexpression. *Am. J. Physiol.* **268**:L615–L624.

30. Rubenstein, R.C., Egan, M.E., and Zeitlin, P.L. 1997. In vitro pharmacologic restoration of CFTR-mediated chloride transport with sodium 4-phenylbutyrate in cystic fibrosis epithelial cells containing delta F508-CFTR. *J. Clin. Invest.* **100**:2457–2465.

31. Heda, G.D., and Marino, C.R. 2000. Surface expression of the cystic fibrosis transmembrane conductance regulator mutant DeltaF508 is markedly upregulated by combination treatment with sodium butyrate and low temperature. *Biochem. Biophys. Res. Commun.* **271**:659–664.

32. Rubenstein, R.C., and Zeitlin, P.L. 2000. Sodium 4-phenylbutyrate downregulates Hsc70: implications for intracellular trafficking of DeltaF508-CFTR. *Am. J. Physiol. Cell Physiol.* **278**:C259–C267.

33. Choo-Kang, L.R., and Zeitlin, P.L. 2001. Induction of HSP70 promotes DeltaF508 CFTR trafficking. *Am. J. Physiol. Lung Cell. Mol. Physiol.* **281**:L58–L68.

34. Linsdell, P. 2001. Direct block of the cystic fibrosis transmembrane conductance regulator Cl(–) channel by butyrate and phenylbutyrate. *Eur. J. Pharmacol.* **411**:255–260.

35. Loffing, J., Moyer, B.D., Reynolds, D., and Stanton, B.A. 1999. PBA increases CFTR expression but at high doses inhibits Cl(–) secretion in Calu-3 airway epithelial cells. *Am. J. Physiol.* **277**:L700–L708.

36. Moyer, B.D., Loffing-Cueni, D., Loffing, J., Reynolds, D., and Stanton, B.A. 1999. Butyrate increases apical membrane CFTR but reduces chloride secretion in MDCK cells. *Am. J. Physiol.* **277**:F271–F276.

37. Loffing-Cueni, D., et al. 2001. Trafficking of GFP-tagged DeltaF508-CFTR to the plasma membrane in a polarized epithelial cell line. *Am. J. Physiol. Cell Physiol.* **281**:C1889–C1897.

38. Matthews, J.B., et al. 1998. Na-K-2Cl cotransporter gene expression and function during enterocyte differentiation. Modulation of Cl⁻ secretory capacity by butyrate. *J. Clin. Invest.* **101**:2072–2079.

39. Guay-Broder, C., et al. 1995. A1 receptor antagonist 8-cyclopentyl-1,3-dipropylxanthine selectively activates chloride efflux from human epithelial and mouse fibroblast cell lines expressing the cystic fibrosis transmembrane regulator delta F508 mutation. *Biochemistry.* **34**:9079–9087.

40. Srivastava, M., Eidelman, O., and Pollard, H.B. 1999. Pharmacogenomics of the cystic fibrosis transmembrane conductance regulator (CFTR) and the cystic fibrosis drug CPX using genome microarray analysis. *Mol. Med.* **5**:753–767.

41. Eidelman, O., Zhang, J., Srivastava, M., and Pollard,

H.B. 2001. Cystic fibrosis and the use of pharmacogenomics to determine surrogate endpoints for drug discovery. *Am. J. Pharmacogenomics.* **1**:221–238.

42. Kaufman, R.J. 2002. Orchestrating the unfolded protein response in health and disease. *J. Clin. Invest.* **110**:1389–1398. doi:10.1172/JCI200216886.

43. Kopito, R.R. 2000. Aggresomes, inclusion bodies and protein aggregation. *Trends Cell Biol.* **10**:524–530.

44. Yang, Y., Janich, S., Cohn, J.A., and Wilson, J.M. 1993. The common variant of cystic fibrosis transmembrane conductance regulator is recognized by hsp70 and degraded in a pre-Golgi nonlysosomal compartment. *Proc. Natl. Acad. Sci. USA.* **90**:9480–9484.

45. Meacham, G.C., et al. 1999. The Hdj-2/Hsc70 chaperone pair facilitates early steps in CFTR biogenesis. *EMBO J.* **18**:1492–1505.

46. Loo, M.A., et al. 1998. Perturbation of Hsp90 interaction with nascent CFTR prevents its maturation and accelerates its degradation by the proteasome. *EMBO J.* **17**:6879–6887.

47. Pind, S., Riordan, J.R., and Williams, D.B. 1994. Participation of the endoplasmic reticulum chaperone calnexin (p88, IP90) in the biogenesis of the cystic fibrosis transmembrane conductance regulator. *J. Biol. Chem.* **269**:12784–12788.

48. Nadler, S.G., Tepper, M.A., Schacter, B., and Mazzucco, C.E. 1992. Interaction of the immunosuppressant deoxyspergualin with a member of the Hsp70 family of heat shock proteins. *Science.* **258**:484–486.

49. Jiang, C., et al. 1998. Partial restoration of cAMP-stimulated CFTR chloride channel activity in DeltaF508 cells by deoxyspergualin. *Am. J. Physiol.* **275**:C171–C178.

50. Brodsky, J.L. 1999. Selectivity of the molecular chaperone-specific immunosuppressive agent 15-deoxyspergualin: modulation of Hsc70 ATPase activity without compromising DnaJ chaperone interactions. *Biochem. Pharmacol.* **57**:877–880.

51. Sittler, A., et al. 2001. Geldanamycin activates a heat shock response and inhibits huntingtin aggregation in a cell culture model of Huntington's disease. *Hum. Mol. Genet.* **10**:1307–1315.

52. Egan, M.E., et al. 2002. Calcium-pump inhibitors induce functional surface expression of DeltaF508-CFTR protein in cystic fibrosis epithelial cells. *Nat. Med.* **8**:485–492.

53. Bertolotti, A., Zhang, Y., Hendershot, L.M., Harding, H.P., and Ron, D. 2000. Dynamic interaction of BiP and ER stress transducers in the unfolded-protein response. *Nat. Cell Biol.* **2**:326–332.

54. McPherson, M.A., et al. 2001. The CFTR-mediated protein secretion defect: pharmacological correction. *Pflugers Arch.* **443**(Suppl. 1):S121–S126.

55. Dormer, R.L., et al. 2001. Correction of delF508-CFTR activity with benzo(c)quinolizinium compounds through facilitation of its processing in cystic fibrosis airway cells. *J. Cell Sci.* **114**:4073–4081.

56. Maitra, R., Shaw, C.M., Stanton, B.A., and Hamilton, J.W. 2001. Increased functional cell surface expression of CFTR and DeltaF508-CFTR by the anthracycline doxorubicin. *Am. J. Physiol. Cell Physiol.* **280**:C1031–C1037.

57. Galietta, L.V., Jayaraman, S., and Verkman, A.S. 2001. Cell-based assay for high-throughput quantitative screening of CFTR chloride transport agonists. *Am. J. Physiol. Cell Physiol.* **281**:C1734–C1742.

Epithelial-mesenchymal transition and its implications for fibrosis

Raghu Kalluri[1] and Eric G. Neilson[2]

[1]Center for Matrix Biology, Beth Israel Deaconess Medical Center and Harvard Medical School, Boston, Massachusetts, USA.
[2]Departments of Medicine and Cell and Developmental Biology, Vanderbilt University School of Medicine, Nashville, Tennessee, USA.

Epithelial to mesenchymal transition (EMT) is a central mechanism for diversifying the cells found in complex tissues. This dynamic process helps organize the formation of the body plan, and while EMT is well studied in the context of embryonic development, it also plays a role in the genesis of fibroblasts during organ fibrosis in adult tissues. Emerging evidence from studies of renal fibrosis suggests that more than a third of all disease-related fibroblasts originate from tubular epithelia at the site of injury. This review highlights recent advances in the process of EMT signaling in health and disease and how it may be attenuated or reversed by selective cytokines and growth factors.

For one hundred and forty-five years, biologists have known that cells come from cells (1). This concept is so fundamental today that we accept it implicitly; cells either divide asymmetrically to preserve stem cell progenitors, partition into sister cells, differentiate along fate pathways, or undergo oncogenesis following the formation of normal tissues. Specification and diversification of cell lineages are initiated by genetic programs under the control of morphogenic cues (2–4). These lineages evolve in a hierarchical manner conforming to developmental boundaries and oscillating biological clocks until reaching terminal differentiation (5). Epithelia from metazoans are emblematic of this process, and at maturity cover outer surfaces (6, 7) or line hollow cavities formed by tubular structures in complex tissues (8–10). Since epithelia typically serve specialized functions (11–13), it is assumed that a state of terminal differentiation is necessary and protected once development is complete.

In recent years, however, this formidable notion has been challenged by observations that mature epithelia change their phenotype following morphogenic pressure from injured tissue. Since the phenomenon of epithelial plasticity was described before it had a firm biochemical basis (14, 15), one is confronted with a plethora of seemingly interchangeable vocabulary. Today, the terms "epithelial-mesenchymal transformation, interactions, or transition" are comingled inappropriately with the term "epithelial-mesenchymal transdifferentiation." "Transformation" classically describes the oncogenic conversion of epithelia. Likewise, the induction of bone marrow stem cells to form somatic cells probably should be considered differentiation rather than transdifferentiation (16). Epithelial-mesenchymal interaction refers to proximate paracrine cross-talk between tissue epithelia and stromal fibro blasts and is completely different from the concept of epithelial-mesenchymal transition (EMT). EMT is a variant of transdifferentiation and a well-recognized mechanism for dispersing cells in vertebrate embryos (17), forming fibroblasts in injured tissues (18, 19), or initiating metastases in epithelial cancer (20–23). We prefer the term "transition" to describe this conversion instead of "epithelial-mesenchymal transdifferentiation" because "transdifferentiation" classically refers to differentiated cells changing into other differentiated cells (24). Transdifferentiation has been observed in retinal pigmented cells that become lens epithelia (25, 26), in the conversion from white to brown adipocytes (27), endothelial cells that become vascular smooth muscle cells (28), lactotrophs that interconvert to somatotrophs in the pituitary (29), pancreatic acinar cells that become ductal epithelium (30, 31) or hepatocytes (32, 33), and hepatocytes that morph into pancreatic ductal cells (34). Although many investigators fail to make the distinction between transdifferentiation and transition, it may be time to do so. It is not yet clear whether the fibroblast transition of EMT is an expected middle phase of transdifferentiating epithelium or whether EMT producing fibroblasts is an arrested form of transdifferentiation (24). EMT of terminally differentiated epithelium, in its purest sense, produces a tissue fibroblast (19).

Developmental biologists have also known for decades that epiblasts undergo EMT to form primary mesenchyme in the creation of triploblastic germ layers (17, 35–37; Figure 1). In the mesoderm this is followed by mesenchymal-epithelial transitions to create secondary epithelium as part of somitogenesis (38, 39) and the further commitment and diversification of cells forming mesoendodermal structures (40–42). Secondary epithelium in mature or adult tissues can also undergo EMT following epithelial stress, such as inflammation (18, 19) or wounding (17, 43) that leads to fibroblast production and fibrogenesis. Epithelia forming tumors also use EMT when carcinomas become metastatic (23, 44, 45).

We review here recent observations regarding the mechanism of EMT in culture and during fibrogenesis, especially associated with kidney disease. The problem of tissue fibrosis is that epithelial units are overtaken by scarification and lose their morphogenic cues, leaving involved organs to fail. While traditional studies of fibrosis have focused on the production of extracellular matrix, recent information now suggests that epithelia contribute to the problem by creating new fibroblasts. Experiments demonstrating the reversibility of organ fibrosis also highlight the need to consider cellular mechanisms of fibrogenesis and the basic biology that will, one hopes, contribute new molecules as useful therapeutics.

Nonstandard abbreviations used: epithelial-mesenchymal transition (EMT); bone morphogenic protein 7 (BMP-7); integrin-linked kinase (ILK); lymphoid enhancer factor (LEF); glycogen synthase kinase (GSK); fibroblast-specific protein-1 (FSP1); α-smooth muscle actin (αSMA); tubular basement membrane (TBM); tissue plasminogen activator (tPA).

Conflict of interest: The authors have declared that no conflict of interest exists.

Citation for this article: *J. Clin. Invest.* **112**:1776–1784 (2003). doi:10.1172/JCI200320530.

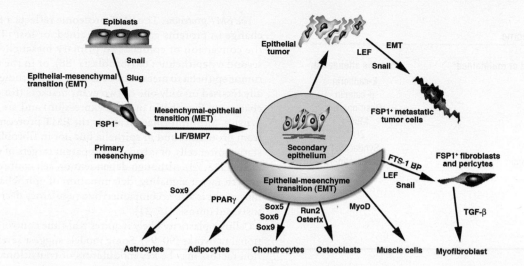

Figure 1

Primitive epithelia (epiblasts) form tropoblastic germ layers through EMT. The primary mesenchyme that migrates after EMT is reinduced to secondary epithelium by mesenchymal-epithelial transition. Secondary epithelia differentiate to form new epithelial tissues and undergo a second round of EMT to form the cells of connective tissue, including astrocytes, adipocytes chondrocytes, osteoblasts, muscle cells, and fibroblasts. Mature secondary epithelia that form epithelial organs can also transform into primary tumors that later undergo EMT to metastasize. These processes are regulated by morphogenic cues and a variety of transcription factors, and are potentially plastic in their adaptation to new biologic circumstances.

The mechanism of EMT

From a general perspective, EMT is about disaggregating epithelial units and reshaping epithelia for movement. Epithelium in transition lose polarity, adherens junctions, tight junctions, desmosomes, and cytokeratin intermediate filaments in order to rearrange their F-actin stress fibers and express filopodia and lamellopodia. This phenotypic conversion requires the molecular reprogramming of epithelium with new biochemical instructions. Much of this conversion has been studied fractionally, during experiments that expose new transduction and signaling pathways, in epithelia that transition in culture, and more recently in fibrogenic tissues. Below we describe an enlarging picture of EMT from a broad and increasingly complex literature.

Induction of EMT. EMT is easily engaged by a combination of cytokines associated with proteolytic digestion of basement membranes upon which epithelia reside. Metalloproteinases (46, 47) or membrane assembly inhibitors (48) initiate the process by dismantling the local basement membrane. Local expression of TGF-β, EGF, IGF-II, or FGF-2 facilitates EMT (Figure 2) by binding epithelial receptors with ligand-inducible intrinsic kinase activity (49–52). The TGF-β effect depends on β-integrin transduction (53), Smad3-dependent transcription (54), or Smad-independent p38MAP kinase activation and GTPase-mediated signaling (53, 55, 56). Depending on the tissue, all three isoforms of TGF-β may be involved sequentially (57–59). While TGF-β is considered prototypical in its induction of EMT (50, 60, 61), there is an increase in epithelial EGF receptors in the EMT microenvironment (62), and EGF can assist in completing the conversion (50). IGF-II also directs the redistribution of β-catenins from the cell surface to the nucleus and facilitates the intracellular degradation of E-cadherin (51), while FGF-2 and TGF-β are required for the expression of MMP-2 and MMP-9 to assist in basement membrane degradation (52). Combinations of cytokines are generally present in most areas of tissue injury, so it is difficult to assign priorities or hierarchy. Each moiety

may contribute a unique inducement to the transition. Furthermore, the role of HGF and FGFs depends on the timed expression and selective distribution of receptors (63, 64). While HGF action through its c-Met/Crk adaptor proteins (65) induces EMT during somitogenesis and endocardial cushion development (66, 67) and modulates the connectivity of intercellular junctions between polarized intestinal and kidney epithelium (68), in the fibrogenic kidney it has the opposite effect of protecting epithelium from EMT (69). In this regard, the expression of bone morphogenic protein 7 (BMP-7) also counterbalances EMT in the kidney (54). Like many biological systems, countervening processes that modulate EMT effector events are beginning to appear.

Engagement of protein kinases. Epithelial signaling that leads to EMT has been studied in a variety of cultured epithelia and seems to have broad generality across numerous phenotypes (Figure 2). As a result of ligand-inducible receptor kinase activation (14), there is a downstream engagement of GTPases from the Ras superfamily (70, 71) or commitment of SH2-SH3 protein domains of the nonreceptor tyrosine kinase (c-Src and Btk) pathways (14, 72) that shift the intracellular balance of small GTPases (Rho, Rac, and Cdc42). Raf/MAP kinases are subsequently activated with several interconnected consequences: engagement of the EMT transcriptome (73) followed by actin rearrangement of the cytoskeleton (70, 74). Integrin-linked kinase (ILK) activation by TGF-β–activated Smad proteins (75) or integrin signaling (76) enhances β-catenin/lymphoid enhancer factor (LEF) expression, suppressing E-cadherin. Activation of Src kinases favors the PI3K pathway, stabilization of β-catenin for nuclear import (77), protection from apoptosis, and disruption of β-integrin binding and E-cadherin complexes (21, 53, 78, 79). Many of these pathways collaboratively reinforce EMT.

The nuclear import of LEF proteins with Smad3 or β-catenin from the cytoplasm is beginning to look like one of several key molecular steps in EMT (73, 80). The phosphorylation of

Table 1
The EMT proteome

Proteins gained or maintained:	Proteins attenuated:
Snail	E-cadherin
Slug	β-catenin
Scratch	Desmoplakin
SIP1	Muc-1
E47	ZO-1
Ets	Syndecan-1
FTS binding protein	Cytokeratin-18
RhoB	
FSP1	
TGF-β	
FGF-1,-2,-8	
MMP-2	
MMP-9	
Vimentin	
αSMA	
Fibronectin	
Collagen type I	
Collagen type III	
Thrombospondin	
PAI-1	

The EMT proteome. The EMT proteome reflects a fundamental change in proteins gained, maintained, or lost (Table 1) with the conversion of epiblasts to primary mesenchyme (37, 87), secondary epithelium to fibroblasts (88), or in the transition of tumor epithelia to metastatic cells (89). Many studies have generally focused on only one or two event markers (for example, the changes in E-cadherin or Snail expression) and are not comprehensive. Most information about the EMT proteome is inferred from proteins found in epithelia but not in fibroblasts or metastatic cancer cells, or is based on apparent targets of transcription factors (15, 73). Although dependent on cell context and ease of growth factor signaling, delamination of epithelia to facilitate movement is also accompanied by a regulatory decrease in apoptosis and mitosis (15, 21).

Cellular plasticity likely requires real-time control by transcriptional networks (90–94). Some models suggest several transcription factors may be key modulators of transitional events. The Snail superfamily of zinc-finger proteins has two evolutionary branches, one for Scratch and the other for Snail and Slug (15). These proteins recognize an E-box binding motif on the promoter for E-cadherin (among others) in competition with the basic helix-loop-helix protein SIP1. Ras/MAPK activates Snail while TGF-β regulates Smad-dependent pathways to engage *SIP1* and *Snail* (15, 95). Subsequently E-cadherin, cytokeratin, muc-1, and desmoplakin are repressed, while fibroblast-specific protein-1 (FSP1), fibronectin, vimentin, and Rho are increased (15, 54). Repression of E-cadherin by Snail proteins frees up more cytoplasmic β-catenin, which, as mentioned above, is co-imported with LEF to the nucleus where its activation is strongly associated with EMT (73). Ets transcription factors regulate EMT in the heart (96, 97).

One of the more interesting proteins found in the EMT proteome is FSP1 (18), also known as S100A4 (98). Support for the notion that EMT is a major source of fibroblasts comes from experiments showing FSP1 expression in cultured epithelium during EMT following exposure to TGF-β and EGF (50), histologic evidence that epithelial units expressing FSP1 disaggregate as organ tissues devolve during the early stages of fibrogenesis (18), and direct observations of EMT in transgenic mice carrying marked epithelium (19). Dividing fibroblasts exposed to nucleoside analogues are also selectively eliminated in transgenic mice expressing thymidine kinase under control of the *FSP1* promoter (99). Members of the S100 superfamily have been implicated in cytoskeletal-membrane interactions, calcium signal transduction, and cellular growth and differentiation (100). In the presence of calcium, FSP1 dimerizes and binds the c-terminal of p53 in the cytoplasm. In this way, FSP1 may sequester p53 from the APC ubiquination pathway (101, 102), perhaps raising levels of free β-catenin. We suspect that FSP1 facilitates and may even maintain the EMT phenotype through this mechanism (Figure 2). While the precise function of FSP1 is not entirely clear, its interaction with cytoskeletal moieties and its early role in EMT suggest that FSP1 may fashion mesenchymal cell shape to enable motility (50) and induce angiogenesis (103). The expression of FSP1 indicates the potential presence of a molecular program determining fibroblast phenotype.

The promoter for *FSP1* is also part of a transcriptome that shares putative FTS-1/CArG box sites in the early promoter regions of a group of genes that would be expected in EMT-derived fibroblasts, including those encoding c-myc, c-Fos, H-ras, Slap, TGF-β, FGF-1, -2, and -8, FSP1, vimentin, α-smooth muscle actin (αSMA),

β-catenin by glycogen synthase kinase- (GSK-3β) allows it to form a complex with APC suppressor protein and Axin (81). p53 activation of APC-dependent pathways also forms a complex with β-catenin (82), and both of these pathways lead to direct loss of free β-catenin through ubiquination (83). If, however, phosphorylation by GSK-3β is inhibited (77), cytoplasmic β-catenin is stabilized by re-entering the E-cadherin complex (84) or binding to the B-box of LEF where together they move into the nucleus to engage the EMT transcriptome (85). Wnt-1 (81), IGF-II (51), Ras (77), and ILK (76) all stabilize cytoplasmic levels of β-catenin to facilitate EMT, perhaps by GSK-3β (or other kinase) inhibition. Smad3 activation by TGF-β family members can also activate LEF-1 in the absence of β-catenin (80). These latter findings suggest either a synergistic or independent control of LEF-1 by at least two EMT-linked signaling pathways. While levels of APC suppressor protein in epithelia may protect the state of terminal differentiation from EMT (73, 86), activation of Smad pathways may provide a countervailing leak in this stability. β-catenin and Smad3 also require the engagement of different transcriptional coactivators, depending on the promoter. These differences may regulate the selectivity or availability of the EMT transcriptome.

Recently, there also has been some attempt to distinguish true EMT from an epithelial phenocopy called "reversible scatter" (21, 73). Reversible scatter following cytokine stimulation looks like EMT because the cells assume a spindlelike shape and undergo a brief period of transcription. But because transcription is not sustained on withdrawal of the inducement and/or if the cells are protected from apoptosis, the epithelia return to their original state (21). TGF-β and Ras classically produce EMT, while EGF, HGF, and FGF favor scattering (21), but not in all cells (52, 68). A scatter effect may be facilitated by varying levels of cytoplasmic APC suppressor protein (73) or preferred activation of the PI3K pathway (21). Whether scatter reverses or goes on to EMT may really just be a timing issue in the continuum of transition, as it is not clear what biological function reversible scatter serves on its own.

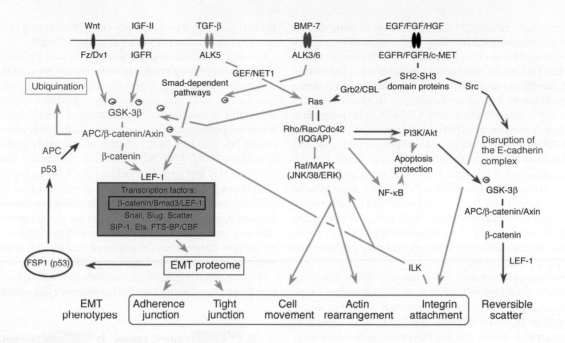

Figure 2
Epithelial plasticity can lead to classical EMT (loss of cell-cell and cell-substratum attachments, new actin rearrangements, and gain of mobility) or reversible scatter, which looks like EMT but is not enduring and can revert. These events are regulated by ligand-inducible intrinsic kinase receptors on the cell surface, which modulate small GTPases, Smads, PI3Ks, MAP kinases, and the availability of β-catenin to coactivate LEF in the nucleus. Free levels of β-catenin are regulated by E-cadherin or APC/β-catenin/Axin complexes, the latter of which shuttle β-catenin between ubiquination or utilization in adherens junctions. Activation of nuclear transcription provides new transcriptional regulators (Snail, SIP1, Ets, and FTS-BP/CarG box binding factor) of the EMT proteome. The EMT proteome comprises proteins listed in Table 1. The variability of receptors, kinases, and the emergence of combined preferences for signaling pathways determine the plasticity unique to each epithelium.

aggrecan, collagen types I and III, thrombospondin I, and matrix metalloproteinases 2 and 9 (19). One hypothesis is that the selective engagement of FTS-1/CArG box sites by a transcriptional complex of proteins may be one of several key regulators of EMT; preliminary evidence suggests a new Kruppel-like zinc finger protein called FTS/CArG box binding factor (FTS-BP/CBF) (refs. 104, 105; and our unpublished observations) may contribute.

GTPase modulation of cell shape and movement. Epithelia that undergo EMT during development, inflammation, or carcinogenesis become mesenchymal cells, fibroblasts, or metastatic tumor cells, respectively (19, 72, 106). This conversion of epithelia is dependent on molecular switches under the control of the Ras superfamily of small GTPases (71). Ras and Rho families of GTPases are activated by guanine nucleotide exchange factors (107) and deactivated by GTPase activating proteins (108). GTPases are a signaling link between cell surface receptor activation and the actin cytoskeleton; some GTPases can also cooperatively modulate EMT with cytokine pressure (21, 109). Three of the best-studied small GTPases are Rho, Rac, and Cdc42. The cross-talk between these members suggests they are activated independently or in series: Ras or Cdc42 can activate Rac, and Rac can inhibit or activate Rho (110–113). Rho helps reconfigure actin stress fibers and stimulates actin-myosin contraction in the cell body, Rac induces the assembly of actin surface protrusions called lamellopodia, and Cdc42 promotes the formation of actin-rich finger extensions called filopodia and modulates cellular asymmetry (71). Their differential activation and balance ensure not only epithelialization but also its dissolution. The cellular properties of contraction, migration, proliferation, and phagocytosis are also under GTPase control

(70). The cellular actions of these small GTPases engage downstream MAP kinases, alter gene transcription, and are integral to shaping cell phenotype during EMT.

Fibroblasts derive from a niche

Since the original observations of Cohnheim (114), investigators have debated the origin of tissue fibroblasts. Three notions persist regarding their lineage: The longest-held concept is that fibroblasts are simply residual embryonic mesenchymal cells left over from organogenesis. While this hypothesis explains the incorrect but often interchangeable substitution of the term "fibroblast" for "mesenchymal cell," the idea itself has no proof and is unlikely since primary mesenchymal cells do not express FSP1 (18). A second notion argues that fibroblasts emerge from the bloodstream after release from the bone marrow (115), and a third view suggests that fibroblasts derive locally in tissues following EMT (19). The second and third hypotheses are mechanistically identical; that is, all fibroblasts probably arise from EMT.

Interstitial fibroblasts appear after gastrulation (after E8.5 in mice) (18) and form as a result of EMT from secondary epithelium (19). Support for the notion that EMT is a major source of local fibroblasts comes from experiments described above. FSP1 is also expressed in some endosteal lining cells and marrow stromal cells (19), and about 14–15% of fibroblasts in fibrosing kidney are derived from marrow. Not much is known about the origin of endosteal lining cells (116, 117). Endosteal bone marrow lining cells precede the formation of the marrow cavity and its contents (117), and in some species are separated from medullary hematopoiesis by a marrow sac comprising a mixture of simple epithelium and/or condensed stromal-like cells (118, 119). This sac appears to have a structural and biochemical interface with endosteal lining cells in nodal regions of bone (118, 120), and the collective structure may be an EMT niche for osteogenic precursor cells, indifferent endosteum, fibroblasts, and marrow stromal cells. Recent evidence also suggests it can be a niche for hematopoietic stem cells (121). FSP1⁺ cells in the marrow are mostly CD34⁻. CD34⁻ progenitor cells cycle their expression of CD34 in the marrow (122–124); both CD34⁻ and CD34⁺ stromal cells circulate in peripheral blood, and, following bone marrow reconstitution, CD34 cells locate as bone-lining endosteum (116, 124). Since some marrow stromal cells can be released into the circulation (124), CD34⁻, FSP1⁺ bone marrow "fibroblasts" might derive from an endosteal EMT niche transitioning to marrow stromal cells (19), which then may evolve into circulating fibrocytes (115). The contribution of CD34⁻, Strol⁺, CD73⁺ mesenchymal stem cells in the accumulation of fibroblasts in this setting is yet unknown.

Most investigators accept with conventional wisdom that fibroblasts represent a cell type of limited diversity. Fibroblast shape, cytoskeletal structure, secretion of interstitial collagens, mobility, participation in tissue fibrosis, and behavior in culture all tend to support this belief. The EMT hypothesis, however, challenges this notion of homogeneity, as do observations that fibroblasts express subtle biochemical differences (125) and phenotypic variability (126–128), and respond differently to cytokines and matrix (129), depending on their tissue of origin. Recent evaluation of the transcriptome from a variety of fibroblasts suggests there is topographic differentiation perhaps based on a "Hox code" (130). Consequently, fibroblasts formed by EMT may differentially express a profile of genes, or a few residual receptors or signaling pathways representative of their previous life as mature epithelium, and, theoretically, can be as heterogeneous as the universe of epithelia.

EMT and fibrosis

The role of EMT during tissue injury leading to organ fibrosis (deposition of collagens, elastin, tenacin, and other matrix molecules) is becoming increasingly clear (Figure 3). A great bulk of such evidence exists for EMT associated with progressive kidney diseases (19), and isprobably true for the lung (131) and possibly the liver. Typical experimental models of kidney fibrosis in mice or rats include progressive glomerulonephritis from anti-glomerular basement membrane disease (132), Alport syndrome (133), or spontaneous lupus nephritis (134), and NOD or db/db nephritic mice (models for diabetic nephropathy) (135), all of which chronically progress at a slow pace, and unilateral ureteral obstruction (136), which progresses to end-stage quickly but leaves the contralateral kidney normal as a control. A number of studies demonstrating EMT during kidney fibrosis correlate with the expression of FSP1 (described above) (18, 50). FSP1 identifies tubular epithelial cells undergoing transition in damaged nephrons trapped by interstitial injury and tracks with increasing numbers of fibroblasts as fibrosis grows worse (50, 137). These FSP1⁺ epithelia traverse through damaged tubular basement membrane (TBM) and accumulate in the interstitium of the kidney (138) where they lose their epithelial markers as they gain a fibroblast phenotype (50, 137). Fibroblasts are not particularly abundant in normal kidneys as they are in lungs, lymphoid nodes, and spleens. When renal fibrogenesis sets in, about 36% of new fibroblasts come from local EMT, about 14–15% from the bone marrow, and

the rest from local proliferation (19). This finding reinforces the notion that fibrogenesis is a local epithelial event.

It is worth mentioning that fibroblasts have little in the way of other distinguishing anatomic features, and most of the proteins they express are not highly specific (18). Vimentin is not fibroblast-specific (50, 137), and type I collagen synthesis is generally only detectable in selected subpopulations of fibroblasts (139–142). Some subpopulations of fibroblasts during fibrogenesis express αSMA (46, 128, 143, 144), a marker of activated fibroblasts (myofibroblasts) (145, 146). While much of the fibrosis literature has relied on this less specific marker, αSMA is not expressed by all fibroblasts (137), suggesting that it does not define the universe of fibroblasts and potentially also identifies smooth muscle cells separated from local blood vessels during tissue injury (137, 147). The increased number of αSMA⁺ smooth muscle cells in fibrotic tissue may derive from delaminated endothelial cells following endothelial-mesenchymal transition (28).

Why are tubular epithelia susceptible to EMT? Injury to the kidney is associated with many inflammatory cells which can incite

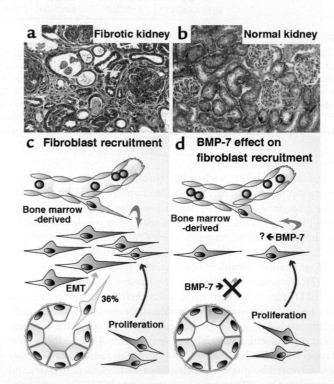

Figure 3
Origin of fibroblasts during kidney fibrosis. (**a**) Fibrotic kidney which displays accumulation of numerous fibroblasts (blue arrow), damaged kidney tubules (yellow arrow), and blood vessels (green arrow). (**b**) Normal kidney with proper tubular structures and very few fibroblasts. (**c**) Schematic illustration of three possible mechanisms via which fibroblasts can originate during kidney injury. Recent experiments suggest that approximately 14–15% of fibroblasts are from bone marrow, 36% can arise via local EMT involving tubular epithelial cells under inflammatory stress, and the rest are likely contributed by proliferation of fibroblasts from all sources. (**d**) Systemic treatment of mice with renal fibrosis using recombinant human BMP-7 results in reversal of renal disease due to severe decrease in EMT-derived fibroblasts and potentially bone marrow–derived fibroblasts. Such events likely have a cascade of beneficial effects that decreasing the overall number of fibroblasts in the kidney, and attenuating fibrosis.

EMT using growth factors such as TGF-β, EGF, and FGF-2 (52). Under the influence of such growth factors, resident fibroblasts and tubular epithelia induce basement membrane–degrading enzymes such as MMP-2 and MMP-9 (48). Degradation of TBM results in disruption of tubular nephrons, and delaminated epithelial cells either fall off into the tubular fluid or migrate towards the interstitium under the influence of increasing growth factor gradients and chemoattractants (47). This initial recruitment of tubular epithelial cells for EMT can be inhibited by blocking the expression of MMP-9 through the disruption of tissue plasminogen activator (tPA, an activator of MMP-9) (148). Other studies have also demonstrated that HGF can decrease levels of TGF-β, restore TGF-β–mediated loss of E-cadherin, and potentially decrease amounts of active MMP-9 (149). In this regard, ILK is now identified as a key mediator of TGF-β–induced EMT associated with tubular epithelial cells (75).

The relevance of TGF-β–induced EMT for progression of kidney fibrosis was recently addressed in studies using BMP-7 as an intracellular competitor of TGF-β signaling (54, 150, 151). BMP-7 is the endogenous antagonist of TGF-β–induced EMT in the kidney and elsewhere (54, 150, 151). BMP-7 reverses the decrease of E-cadherin caused by TGF-β (54). Restoration of E-cadherin by BMP-7 is mediated by its ALK3/6 receptors and Smad5. The capacity of BMP-7 to reverse TGF-β–induced EMT in culture is also observed in mouse models of kidney fibrosis. Systemic administration of recombinant BMP-7 in mice with kidney fibrosis following ureteral obstruction results in reversal of EMT and repair of damaged tubular structures with repopulation of healthy tubular epithelial cells (54, 152). This reversal is also associated with return of renal function, a significant decrease in FSP1+ interstitial fibroblasts and de novo activation of BMP-7 signaling (54). Renal protection from BMP-7 has also been observed in murine models of diabetic nephropathy (153), Alport syndrome, and lupus nephritis (150). Today, TGF-β signaling attenuated by BMP-7 is the closest paradigm in EMT arguing in favor of privileged pathways.

Summary

Progress in understanding EMT has been an exercise in coming to appreciate the level of complexity required for changing cellular identity. The mechanism of transition highlights an integration of nuclear regulation and network signaling with alterations in microenvironment to create a moving cell. Remarkably, differentiating epithelia make these transitions during development, and terminally differentiated epithelia use them for physiologic repair or to advance oncogenesis. EMT is a form of molecular exaptation, a mechanism of economy by which cells reuse known physiologic processes to provide new functions (154). With the foundation established by current studies, new questions regarding the definition and role of pericytes and myofibroblasts can be explored, and a framework for a better understanding of other transitions, like endothelial-mesenchymal transition, is possible. EMT also provides a mechanism for creating ancestral relationships between local cells and may be particularly important in tumor expansion. Lastly, fibroblasts may carry forward remnants of a unique epithelial signature. And if all fibroblasts or tumor cells which arise via EMT are not created equal, then therapies to combat fibrosis or metastatic disease may need more specificity. Nevertheless, the future holds great promise for EMT as a viable therapeutic target.

Acknowledgements

This work was supported by grants NIH/DK55001, and NIH/DK62987, The Espinosa Liver Fibrosis Fund, and research support from the Center for Matrix Biology at Beth Israel Deaconess Medical Center, all to R. Kalluri. E. Neilson is supported by NIH/DK46282 and Yamanouchi USA Foundation.

Address correspondence to: Raghu Kalluri, Center for Matrix Biology, Beth Israel Deaconess Medical Center, 330 Brookline Ave. (DANA 514), Boston, Massachusetts 02215, USA. Phone: (617) 667-0455; Fax: (617) 975-5663; E-mail: rKalluri@BIDMC.Harvard.edu.

1. Virchow, R. 1858. *Die Cellularpathologie in ihrer Begr9Fndung auf physiologische und pathologische Gewebelebre*. A. Hirschwald. Berlin, Germany. 456 pp.
2. Slack, J.M.W. 2000. Stem cells in epithelial tissues. *Science.* **287**:1431–1433.
3. Watt, F.M., and Hogan, B.L. 2000. Out of Eden: stem cells and their niches. *Science.* **287**:1427–1430.
4. Blau, H.M., Brazelton, T.R., and Weimann, J.M. 2001. The evolving concept of a stem cell: entity or function? *Cell.* **105**:829–841.
5. Irvine, K.D., and Rauskolb, C. 2001. Boundaries in development: formation and function. *Annu. Rev. Cell Dev. Biol.* **17**:189–214.
6. Sengel, P. 1990. Pattern formation in skin development. *Int. J. Dev. Biol.* **34**:33–50.
7. Erickson, C.A., and Reedy, M.V. 1998. Neural crest development: the interplay between morphogenesis and cell differentiation. *Curr. Top. Dev. Biol.* **40**:177–209.
8. Hogan, B.L., and Kolodziej, P.A. 2002. Organogenesis: molecular mechanisms of tubulogenesis. *Nat. Rev. Genet.* **3**:513–523.
9. Krasnow, M.A., and Nelson, W.J. 2002. Tube morphogenesis. *Trends Cell Biol.* **12**:351.
10. Lubarsky, B., and Krasnow, M.A. 2003. Tube morphogenesis: making and shaping biological tubes. *Cell.* **112**:19–28.
11. Gumbiner, B.M. 1992. Epithelial morphogenesis. *Cell.* **69**:385–387.
12. Yeaman, C., Grindstaff, K.K., Hansen, M.D., and Nelson, W.J. 1999. Cell polarity: versatile scaffolds keep things in place. *Curr. Biol.* **9**:R515–517.
13. Al-Awqati, Q., Vijayakumar, S., and Takito, J. 2003. Terminal differentiation of epithelia. *Biol. Chem.* **384**:1255–1258.
14. Boyer, B., Valles, A.M., and Edme, N. 2000. Induction and regulation of epithelial-mesenchymal transitions. *Biochem. Pharmacol.* **60**:1091–1099.
15. Nieto, M.A. 2002. The snail superfamily of zinc-finger transcription factors. *Nat. Rev. Mol. Cell Biol.* **3**:155–166.
16. Tsai, R.Y., Kittappa, R., and McKay, R.D. 2002. Plasticity, niches, and the use of stem cells. *Dev. Cell.* **2**:707–712.
17. Hay, E.D. 1995. An overview of epithelio-mesenchymal transformations. *Acta Anat.* **154**:8–20.
18. Strutz, F., et al. 1995. Identification and characterization of a fibroblast marker: FSP1. *J. Cell Biol.* **130**:393–405.
19. Iwano, M., et al. 2002. Evidence that fibroblasts derive from epithelium during tissue fibrosis. *J. Clin. Invest.* **110**:341–350. doi:10.1172/JCI200215518.
20. Kiemer, A.K., Takeuchi, K., and Quinlan, M.P. 2001. Identification of genes involved in epithelial-mesenchymal transition and tumor progression. *Oncogene.* **20**:6679–6688.
21. Janda, E., et al. 2002. Ras and TGF[beta] cooperatively regulate epithelial cell plasticity and metastasis: dissection of Ras signaling pathways. *J. Cell Biol.* **156**:299–313.
22. Vincent-Salomon, A., and Thiery, J.P. 2003. Host microenvironment in breast cancer development: epithelial-mesenchymal transition in breast cancer development. *Breast Cancer Res.* **5**:101–106.
23. Xue, C., Plieth, D., Venkov, C., Xu, C., and Neilson, E.G. 2003. The gatekeeper effect of epithelial-mesenchymal transition regulates the frequency of breast cancer metastasis. *Cancer Res.* **63**:3386–3394.
24. Slack, J.M., and Tosh, D. 2001. Transdifferentiation and metaplasia–switching cell types. *Curr. Opin. Genet. Dev.* **11**:581–586.
25. Eguchi, G., and Kodama, R. 1993. Transdifferentiation. *Curr. Opin. Cell Biol.* **5**:1023–1028.
26. Rio-Tsonis, K.D., and Tsonis, P.A. 2003. Eye regeneration at the molecular age. *Dev. Dyn.* **226**:211–224.
27. Cinti, S. 2002. Adipocyte differentiation and transdifferentiation: plasticity of the adipose organ. *J. Endocrinol. Invest.* **25**:823–835.
28. Frid, M.G., Kale, V.A., and Stenmark, K.R. 2002. Mature vascular endothelium can give rise to smooth muscle cells via endothelial-mesenchymal transdifferentiation: in vitro analysis. *Circ. Res.* **90**:1189–1196.
29. Vidal, S., Horvath, E., Kovacs, K., Lloyd, R.V., and Smyth, H.S. 2001. Reversible transdifferentiation: interconversion of somatotrophs and lactotrophs in pituitary hyperplasia. *Mod. Pathol.* **14**:20–28.
30. Rooman, I., Heremans, Y., Heimberg, H., and Bouwens, L. 2000. Modulation of rat pancreatic acino-ductal transdifferentiation and expression of PDX-1 in vitro. *Diabetologia.* **43**:907–914.
31. Hall, P.A., and Lemoine, N.R. 1992. Rapid acinar to ductal transdifferentiation in cultured human exocrine pancreas. *J. Pathol.* **166**:97–103.
32. Shen, C.N., Slack, J.M., and Tosh, D. 2000. Molecular basis of transdifferentiation of pancreas to liver. *Nat. Cell Biol.* **2**:879–887.
33. Shen, C.N., Horb, M.E., Slack, J.M., and Tosh, D. 2003. Transdifferentiation of pancreas to liver.

Mech. Dev. **120**:107–116.

34. Horb, M.E., Shen, C.N., Tosh, D., and Slack, J.M. 2003. Experimental conversion of liver to pancreas. *Curr. Biol.* **13**:105–115.

35. Tam, P.P., and Behringer, R.R. 1997. Mouse gastrulation: the formation of a mammalian body plan. *Mech. Dev.* **68**:3–25.

36. Narasimha, M., and Leptin, M. 2000. Cell movements during gastrulation: come in and be induced. *Trends Cell Biol.* **10**:169–172.

37. Carver, E.A., Jiang, R., Lan, Y., Oram, K.F., and Gridley, T. 2001. The mouse snail gene encodes a key regulator of the epithelial-mesenchymal transition. *Mol. Cell. Biol.* **21**:8184–8188.

38. Summerbell, D., and Rigby, P.W. 2000. Transcriptional regulation during somitogenesis. *Curr. Top. Dev. Biol.* **48**:301–318.

39. Pourquie, O. 2001. Vertebrate somitogenesis. *Annu. Rev. Cell Dev. Biol.* **17**:311–350.

40. Ekblom, P. 1989. Developmentally regulated conversion of mesenchyme to epithelium. *FASEB J.* **3**:2141–2150.

41. Birchmeier, W., and Birchmeier, C. 1994. Mesenchymal-epithelial transitions. *Bioessays.* **16**:305–307.

42. Barasch, J., et al. 1999. Mesenchymal to epithelial conversion in rat metanephros is induced by LIF. *Cell.* **99**:377–386.

43. Desmouliere, A. 1995. Factors influencing myofibroblast differentiation during wound healing and fibrosis. *Cell Biol. Int.* **19**:471–476.

44. Lochter, A. 1998. Plasticity of mammary epithelia during normal development and neoplastic progression. *Biochem. Cell Biol.* **76**:997–1008.

45. Guarino, M., Micheli, P., Pallotti, F., and Giordano, F. 1999. Pathological relevance of epithelial and mesenchymal phenotype plasticity. *Pathol. Res. Pract.* **195**:379–389.

46. Yang, J., and Liu, Y. 2001. Dissection of key events in tubular epithelial to myofibroblast transition and its implications in renal interstitial fibrosis. *Am. J. Pathol.* **159**:1465–1475.

47. Zeisberg, M., Maeshima, Y., Mosterman, B., and Kalluri, R. 2002. Renal fibrosis. Extracellular matrix microenvironment regulates migratory behavior of activated tubular epithelial cells. *Am. J. Pathol.* **160**:2001–2008.

48. Zeisberg, M., et al. 2001. Renal fibrosis: collagen composition and assembly regulates epithelial-mesenchymal transdifferentiation. *Am. J. Pathol.* **159**:1313–1321.

49. Fan, J.M., et al. 1999. Transforming growth factor-beta regulates tubular epithelial-myofibroblast transdifferentiation in vitro. *Kidney Int.* **56**:1455–1467.

50. Okada, H., Danoff, T.M., Kalluri, R., and Neilson, E.G. 1997. The early role of FSP1 in epithelial-mesenchymal transformation. *Am. J. Physiol.* **273**:563–574.

51. Morali, O.G., et al. 2001. IGF-II induces rapid beta-catenin relocation to the nucleus during epithelium to mesenchyme transition. *Oncogene.* **20**:4942–4950.

52. Strutz, F., et al. 2002. Role of basic fibroblast growth factor-2 in epithelial-mesenchymal transformation. *Kidney Int.* **61**:1714–1728.

53. Bhowmick, N.A., Zent, R., Ghiassi, M., McDonnell, M., and Moses, H.L. 2001. Integrin beta 1 signaling is necessary for transforming growth factor-beta activation of p38MAPK and epithelial plasticity. *J. Biol. Chem.* **276**:46707–46713.

54. Zeisberg, M., et al. 2003. BMP-7 counteracts TGF-beta1-induced epithelial-to-mesenchymal transition and reverses chronic renal injury. *Nat. Med.* **9**:964–968.

55. Boyer, A.S., Erickson, C.P., and Runyan, R.B. 1999. Epithelial-mesenchymal transformation in the embryonic heart is mediated through distinct pertussis toxin-sensitive and TGFβ signal transduction mechanisms. *Dev. Dyn.* **214**:81–91.

56. Yu, L., Hebert, M.C., and Zhang, Y.E. 2002. TGF-beta receptor-activated p38 MAP kinase mediates Smad-independent TGF-beta responses. *EMBO J.* **21**:3749–3759.

57. Boyer, A.S., et al. 1999. TGFbeta2 and TGFbeta3 have separate and sequential activities during epithelial-mesenchymal cell transformation in the embryonic heart. *Dev. Biol.* **208**:530–545.

58. Bhowmick, N.A., et al. 2001. Transforming growth factor-beta1 mediates epithelial to mesenchymal transdifferentiation through a RhoA-dependent mechanism. *Mol. Biol. Cell.* **12**:27–36.

59. Camenisch, T.D., et al. 2002. Temporal and distinct TGFbeta ligand requirements during mouse and avian endocardial cushion morphogenesis. *Dev. Biol.* **248**:170–181.

60. Miettinen, P.J., Ebner, R., Lopez, A.R., and Derynck, R. 1994. TGF-beta induced transdifferentiation of mammary epithelial cells to mesenchymal cells: involvement of type I receptors. *J. Cell Biol.* **127**:2021–2036.

61. Derynck, R., and Zhang, Y.E. 2003. Smad-dependent and Smad-independent pathways in TGF-beta family signalling. *Nature.* **425**:577–584.

62. Citterio, H.L., and Gaillard, D.A. 1994. Expression of transforming growth factor alpha (TGF alpha), epidermal growth factor receptor (EGF-R) and cell proliferation during human palatogenesis: an immunohistochemical study. *Int. J. Dev. Biol.* **38**:499–505.

63. Morabito, C.J., Dettman, R.W., Kattan, J., Collier, J.M., and Bristow, J. 2001. Positive and negative regulation of epicardial-mesenchymal transformation during avian heart development. *Dev. Biol.* **234**:204–215.

64. Thery, C., and Stern, C.D. 1996. Roles of kringle domain-containing serine proteases in epithelial-mesenchymal transitions during embryonic development. *Acta Anat. (Basel).* **156**:162–172.

65. Lamorte, L., Royal, I., Naujokas, M., and Park, M. 2002. Crk adapter proteins promote an epithelial-mesenchymal-like transition and are required for HGF-mediated cell spreading and breakdown of epithelial adherens junctions. *Mol. Biol. Cell.* **13**:1449–1461.

66. Song, W., Majka, S.M., and McGuire, P.G. 1999. Hepatocyte growth factor expression in the developing myocardium: evidence for a role in the regulation of the mesenchymal cell phenotype and urokinase expression. *Dev. Dyn.* **214**:92–100.

67. Thery, C., Sharpe, M.J., Batley, S.J., Stern, C.D., and Gherardi, E. 1995. Expression of HGF/SF, HGF1/MSP, and c-met suggests new functions during early chick development. *Dev. Genet.* **17**:90–101.

68. Nusrat, A., et al. 1994. Hepatocyte growth factor/scatter factor effects on epithelia. Regulation of intercellular junctions in transformed and non-transformed cell lines, basolateral polarization of c-met receptor in transformed and natural intestinal epithelia, and induction of rapid wound repair in a transformed model epithelium. *J. Clin. Invest.* **93**:2056–2065.

69. Mizuno, S., et al. 1998. Hepatocyte growth factor prevents renal fibrosis and dysfunction in a mouse model of chronic renal disease. *J. Clin. Invest.* **101**:1827–1834.

70. Etienne-Manneville, S., and Hall, A. 2002. Rho GTPases in cell biology. *Nature.* **420**:629–635.

71. Bar-Sagi, D., and Hall, A. 2000. Ras and Rho GTPases: a family reunion. *Cell.* **103**:227–238.

72. Savagner, P. 2001. Leaving the neighborhood: molecular mechanisms involved during epithelial-mesenchymal transition. *Bioessays.* **23**:912–923.

73. Kim, K., Lu, Z., and Hay, E.D. 2002. Direct evidence for a role of beta-catenin/LEF-1 signaling pathway in induction of EMT. *Cell Biol. Int.* **26**:463–476.

74. Hall, A. 1998. Rho GTPases and the actin cytoskeleton. *Science.* **279**:509–514.

75. Li, Y., Yang, J., Dai, C., Wu, C., and Liu, Y. 2003. Role for integrin-linked kinase in mediating tubular epithelial to mesenchymal transition and renal interstitial fibrogenesis. *J. Clin. Invest.* **112**:503–516. doi:10.1172/JCI200317913.

76. Novak, A., et al. 1998. Cell adhesion and the integrin-linked kinase regulate the LEF-1 and beta-catenin signaling pathways. *Proc. Natl. Acad. Sci. U. S. A.* **95**:4374–4379.

77. Espada, J., Perez-Moreno, M., Braga, V.M., Rodriguez-Viciana, P., and Cano, A. 1999. H-Ras activation promotes cytoplasmic accumulation and phosphoinositide 3-OH kinase association of beta-catenin in epidermal keratinocytes. *J. Cell Biol.* **146**:967–980.

78. Behrens, J., et al. 1993. Loss of epithelial differentiation and gain of invasiveness correlates with tyrosine phosphorylation of the E-cadherin/beta-catenin complex in cells transformed with a temperature-sensitive v-SRC gene. *J. Cell Biol.* **120**:757–766.

79. Smith, D.E., Franco del Amo, F., and Gridley, T. 1992. Isolation of Sna, a mouse gene homologous to the Drosophila genes snail and escargot: its expression pattern suggests multiple roles during postimplantation development. *Development.* **116**:1033–1039.

80. Attisano, L., and Wrana, J.L. 2002. Signal transduction by the TGF-beta superfamily. *Science.* **296**:1646–1647.

81. He, X. 2003. A wnt-wnt situation. *Dev. Cell.* **4**:791–797.

82. Sadot, E., Geiger, B., Oren, M., and Ben-Ze'ev, A. 2001. Down-regulation of beta-catenin by activated p53. *Mol. Cell. Biol.* **21**:6768–6781.

83. Aberle, H., Bauer, A., Stappert, J., Kispert, A., and Kemler, R. 1997. beta-catenin is a target for the ubiquitin-proteasome pathway. *EMBO J.* **16**:3797–3804.

84. von Kries, J.P., et al. 2000. Hot spots in beta-catenin for interactions with LEF-1, conductin and APC. *Nat. Struct. Biol.* **7**:800–807.

85. Kim, K., and Hay, E.D. 2001. New evidence that nuclear import of endogenous beta-catenin is LEF-1 dependent, while LEF-1 independent import of exogenous beta-catenin leads to nuclear abnormalities. *Cell Biol. Int.* **25**:1149–1161.

86. Midgley, C.A., et al. 1997. APC expression in normal human tissues. *J. Pathol.* **181**:426–433.

87. Ip, Y.T., and Gridley, T. 2002. Cell movements during gastrulation: snail dependent and independent pathways. *Curr. Opin. Genet. Dev.* **12**:423–429.

88. Zavadil, J., et al. 2001. Genetic programs of epithelial cell plasticity directed by transforming growth factor-beta. *Proc. Natl. Acad. Sci. U. S. A.* **98**:6686–6691.

89. Ramaswamy, S., Ross, K.N., Lander, E.S., and Golub, T.R. 2003. A molecular signature of metastasis in primary solid tumors. *Nat. Genet.* **33**:49–54.

90. Blau, H.M., and Blakely, B.T. 1999. Plasticity of cell fate: insights from heterokaryons. *Semin. Cell Dev. Biol.* **10**:267–272.

91. Emerson, B.M. 2002. Specificity of gene regulation. *Cell.* **109**:267–270.

92. Cremer, T., and Cremer, C. 2001. Chromosome territories, nuclear architecture and gene regulation in mammalian cells. *Nat. Rev. Genet.* **2**:292–301.

93. Mannervik, M., Nibu, Y., Zhang, H., and Levine, M. 1999. Transcriptional coregulators in development. *Science.* **284**:606–609.

94. Young, B.A., Gruber, T.M., and Gross, C.A. 2002. Views of transcription initiation. *Cell.* **109**:417–420.

95. Peinado, H., Quintanilla, M., and Cano, A. 2003. Transforming growth factor beta 1 induces snail transcription factor in epithelial cell lines. Mechanisms for epithelial-mesenchymal transitions. *J. Biol. Chem.* **278**:21113–21123.

96. Lie-Venema, H., et al. 2003. Ets-1 and Ets-2 transcription factors are essential for normal coronary and myocardial development in chicken embryos. *Circ. Res.* **92**:749–756.

97. Macias, D., Perez-Pomares, J.M., Garcia-Garrido, L., Carmona, R., and Munoz-Chapuli, R. 1998. Immunoreactivity of the ets-1 transcription factor correlates with areas of epithelial-mesenchymal transition in the developing avian heart. *Anat. Embryol. (Berl.).* **198**:307–315.

98. Ridinger, K., et al. 1998. Clustered organization of S100 genes in human and mouse. *Biochim. Biophys. Acta.* **1448**:254–263.

99. Iwano, M., et al. 2001. Conditional abatement of tissue fibrosis using nucleoside analogs to selectively corrupt DNA replication in transgenic fibroblasts. *Mol. Ther.* **3**:149–159.

100. Barraclough, R. 1998. Calcium-binding protein S100A4 in health and disease. *Biochim. Biophys. Acta.* **1448**:190–199.

101. Grigorian, M., et al. 2001. Tumor suppressor p53 protein is a new target for the metastasis-associated Mts1/S100A4 protein: functional consequences of their interaction. *J. Biol. Chem.* **276**:22699–22708.

102. Chen, H., et al. 2001. Binding to intracellular targets of the metastasis-inducing protein, S100A4 (p9Ka). *Biochem. Biophys. Res. Commun.* **286**:1212–1217.

103. Ambartsumian, N., et al. 2001. The metastasis-associated Mts1(S100A4) protein could act as an angiogenic factor. *Oncogene.* **20**:4685–4695.

104. Okada, H., et al. 1998. Novel cis-acting elements in the FSP1 gene regulate fibroblast-specific transcription. *Am. J. Physiol.* **275**:306–314.

105. Tian, Y.C., Fraser, D., Attisano, L., and Phillips, A.O. 2003. TGF-{beta}1-mediated alterations of renal proximal tubular epithelial cell phenotype. *Am. J. Physiol. Renal Physiol.* **285**:F130–F142.

106. Thiery, J.P., and Chopin, D. 1999. Epithelial cell plasticity in development and tumor progression. *Cancer Metastasis Rev.* **18**:31–42.

107. Quilliam, L.A., Rebhun, J.F., and Castro, A.F. 2002. A growing family of guanine nucleotide exchange factors is responsible for activation of Ras-family GTPases. *Prog. Nucleic Acid Res. Mol. Biol.* **71**:391–444.

108. Bernards, A. 2003. GAPs galore! A survey of putative Ras superfamily GTPase activating proteins in man and Drosophila. *Biochim. Biophys. Acta.* **1603**:47–82.

109. Masszi, A., et al. 2003. Central role for Rho in TGF-beta1-induced alpha-smooth muscle actin expression during epithelial-mesenchymal transition. *Am. J. Physiol. Renal Physiol.* **284**:F911–924.

110. Ridley, A.J., Paterson, H.F., Johnston, C.L., Diekmann, D., and Hall, A. 1992. The small GTP-binding protein rac regulates growth factor-induced membrane ruffling. *Cell.* **70**:401–410.

111. Nobes, C.D., and Hall, A. 1995. Rho, rac and cdc42 GTPases: regulators of actin structures, cell adhesion and motility. *Biochem. Soc. Trans.* **23**:456–459.

112. Sander, E.E., ten Klooster, J.P., van Delft, S., van der Kammen, R.A., and Collard, J.G. 1999. Rac downregulates Rho activity: reciprocal balance between both GTPases determines cellular morphology and migratory behavior. *J. Cell Biol.* **147**:1009–1022.

113. Zondag, G.C., et al. 2000. Oncogenic Ras downregulates Rac activity, which leads to increased Rho activity and epithelial-mesenchymal transition. *J. Cell Biol.* **149**:775–782.

114. Cohnheim, J. 1867. 86ber Entz9Fndung und Eiterung. *Virchows Arch.* **40**:1–79.

115. Abe, R., Donnelly, S.C., Peng, T., Bucala, R., and Metz, C.N. 2001. Peripheral blood fibrocytes: differentiation pathway and migration to wound sites. *J. Immunol.* **166**:7556–7562.

116. Huss, R. 2000. Isolation of primary and immortalized CD34-hematopoietic and mesenchymal stem cells from various sources. *Stem Cells.* **18**:1–9.

117. Bianco, P., and Robey, P.G. 2000. Marrow stromal cells. *J. Clin. Invest.* **105**:1663–1668.

118. Bi, L.X., Simmons, D.J., Hawkins, H.K., Cox, R.A., and Mainous, E.G. 2000. Comparative morphology of the marrow sac. *Anat. Rec.* **260**:410–415.

119. Simmons, D.J. 1996. The in vivo role of bone marrow fibroblast-like stromal cells. *Calcif. Tissue Int.* **58**:129–132.

120. Weiss, L., and Geduldig, U. 1991. Barrier cells: stromal regulation of hematopoiesis and blood cell release in normal and stressed murine bone marrow. *Blood.* **78**:975–990.

121. Zhang, J., et al. 2003. Identification of the haematopoietic stem cell niche and control of the niche size. *Nature.* **425**:836–841.

122. Krause, D.S., et al. 2001. Multi-organ, multi-lineage engraftment by a single bone marrow-derived stem cell. *Cell.* **105**:369–377.

123. Huss, R., Hong, D.S., McSweeney, P.A., Hoy, C.A., and Deeg, H.J. 1995. Differentiation of canine bone marrow cells with hemopoietic characteristics from an adherent stromal cell precursor. *Proc. Natl. Acad. Sci. U. S. A.* **92**:748–752.

124. Huss, R. 2000. Perspectives on the morphology and biology of CD34-negative stem cells. *J. Hematother. Stem Cell Res.* **9**:783–793.

125. Garrett, D.M., and Conrad, G.W. 1979. Fibroblast-like cells from embryonic chick cornea, heart, and skin are antigenically distinct. *Dev. Biol.* **70**:50–70.

126. Schor, S.L., and Schor, A.M. 1987. Clonal heterogeneity in fibroblast phenotype: implications for the control of epithelial-mesenchymal interactions. *Bioessays.* **7**:200–204.

127. M9Fller, G.A., and Rodemann, H.P. 1991. Characterization of human renal fibroblasts in health and disease. I. Immunophenotyping of cultured tubular epithelial cells and fibroblasts derived from kidneys with histologically proven interstitial fibrosis. *Am. J. Kidney Dis.* **17**:680–683.

128. Dugina, V., Alexandrova, A., Chaponnier, C., Vasiliev, J., and Gabbiani, G. 1998. Rat fibroblasts cultured from various organs exhibit differences in alpha-smooth muscle actin expression, cytoskeletal pattern, and adhesive structure organization. *Exp. Cell. Res.* **238**:481–490.

129. Alvarez, R.J., et al. 1992. Biosynthetic and proliferative characteristics of tubulointerstitial fibroblasts probed with paracrine cytokines. *Kidney Int.* **41**:14–23.

130. Chang, H.Y., et al. 2002. Diversity, topographic differentiation, and positional memory in human fibroblasts. *Proc. Natl. Acad. Sci. U. S. A.* **99**:12877–12882.

131. Chilosi, M., et al. 2003. Aberrant Wnt/beta-catenin pathway activation in idiopathic pulmonary fibrosis. *Am. J. Pathol.* **162**:1495–1502.

132. Kalluri, R., Danoff, T.M., Okada, H., and Neilson, E.G. 1997. Susceptibility to anti-glomerular basement membrane disease and Goodpasture syndrome is linked to MHC class II genes and the emergence of T cell-mediated immunity in mice. *J. Clin. Invest.* **100**:2263–2275.

133. Cosgrove, D., et al. 1996. Collagen COL4A3 knockout: a mouse model for autosomal Alport syndrome. *Genes Dev.* **10**:2981–2992.

134. Anders, H., and Schlondorff, D. 2000. Murine models of renal disease: possibilities and problems in studies using mutant mice. *Exp. Nephrol.* **8**:181–193.

135. Janssen, U., Phillips, A.O., and Floege, J. 1999. Rodent models of nephropathy associated with type II diabetes. *J. Nephrol.* **12**:159–172.

136. Diamond, J.R., Ricardo, S.D., and Klahr, S. 1998. Mechanisms of interstitial fibrosis in obstructive nephropathy. *Semin. Nephrol.* **18**:594–602.

137. Okada, H., et al. 2000. Progressive renal fibrosis in murine polycystic kidney disease: an immunohistochemical observation. *Kidney Int.* **58**:587–597.

138. Okada, H., Strutz, F., Danoff, T.M., Kalluri, R., and Neilson, E.G. 1996. Possible mechanisms of renal fibrosis. *Contrib. Nephrol.* **118**:147–154.

139. Breen, E., Falco, V.M., Absher, M., and Cutroneo, K.R. 1990. Subpopulations of rat lung fibroblasts with different amounts of type I and type III collagen mRNAs. *J. Biol. Chem.* **265**:6286–6290.

140. Goldring, S.R., Stephenson, M.L., Downie, E., Krane, S.M., and Korn, J.H. 1990. Heterogeneity in hormone responses and patterns of collagen synthesis in cloned dermal fibroblasts. *J. Clin. Invest.* **85**:798–803.

141. Jelaska, A., Strehlow, D., and Korn, J.H. 1999. Fibroblast heterogeneity in physiological conditions and fibrotic disease. *Springer Semin. Immunopathol.* **21**:385–395.

142. Rossert, R.A., Chen, S.S., Eberspaecher, H., Smith, C.N., and De Crombrugghe, B. 1996. Identification of a minimal sequence of the mouse pro-a1(I) collagen promoter that confers high-level osteoblast expression in transgenic mice and that binds a protein selectively present in osteoblasts. *Proc. Natl. Acad. Sci. U. S. A.* **93**:1027–1031.

143. Zeisberg, M., Strutz, F., and Muller, G.A. 2000. Role of fibroblast activation in inducing interstitial fibrosis. *J. Nephrol.* **13**(Suppl. 3):S111–S120.

144. Serini, G., and Gabbiani, G. 1999. Mechanisms of myofibroblast activity and phenotypic modulation. *Exp. Cell Res.* **250**:273–283.

145. Tang, W.W., et al. 1996. Platelet-derived growth factor-BB induces renal tubulointerstitial myofibroblast formation and tubulointerstitial fibrosis. *Am. J. Pathol.* **148**:1169–1180.

146. Ng, Y.Y., et al. 1998. Tubular epithelial-myofibroblast transdifferentiation in progressive tubulointerstitial fibrosis in 5/6 nephrectomized rats. *Kidney Int.* **54**:864–876.

147. Eyden, B. 2001. The myofibroblast: an assessment of controversial issues and a definition useful in diagnosis and research. *Ultrastruct. Pathol.* **25**:39–50.

148. Yang, J., et al. 2002. Disruption of tissue-type plasminogen activator gene in mice reduces renal interstitial fibrosis in obstructive nephropathy. *J. Clin. Invest.* **110**:1525–1538. doi:10.1172/JCI200216219.

149. Yang, J., and Liu, Y. 2002. Blockage of tubular epithelial to myofibroblast transition by hepatocyte growth factor prevents renal interstitial fibrosis. *J. Am. Soc. Nephrol.* **13**:96–107.

150. Zeisberg, M., et al. 2003. Bone morphogenic protein-7 inhibits progression of chronic renal fibrosis associated with two genetic mouse models. *Am. J. Physiol. Renal Physiol.* **285**:F1060–F1067.

151. Kalluri, R., and Zeisberg, M. 2003. Exploring the connection between chronic renal fibrosis and bone morphogenic protein-7. *Histol. Histopathol.* **18**:217–224.

152. Morrissey, J., et al. 2002. Bone morphogenetic protein-7 improves renal fibrosis and accelerates the return of renal function. *J. Am. Soc. Nephrol.* **13**(Suppl. 1):S14–S21.

153. Wang, S., et al. 2003. Bone morphogenic protein-7 (BMP-7), a novel therapy for diabetic nephropathy. *Kidney Int.* **63**:2037–2049.

154. Gould, S.J., and Vrba, E.S. 1982. Exaptation — a missing term in the science of form. *Paleobiology.* **8**:4–15.

Part 7
Cardiovascular System

Oxygen, oxidative stress, hypoxia, and heart failure

Frank J. Giordano

Department of Medicine, Yale University School of Medicine, New Haven, Connecticut, USA.

A constant supply of oxygen is indispensable for cardiac viability and function. However, the role of oxygen and oxygen-associated processes in the heart is complex, and they and can be either beneficial or contribute to cardiac dysfunction and death. As oxygen is a major determinant of cardiac gene expression, and a critical participant in the formation of ROS and numerous other cellular processes, consideration of its role in the heart is essential in understanding the pathogenesis of cardiac dysfunction.

The mammalian heart is an obligate aerobic organ. At a resting pulse rate, the heart consumes approximately 8–15 ml O_2/min/100 g tissue. This is significantly more than that consumed by the brain (approximately 3 ml O_2/min/100 g tissue) and can increase to more than 70 ml O_2/min/100 g myocardial tissue during vigorous exercise (1, 2). Mammalian heart muscle cannot produce enough energy under anaerobic conditions to maintain essential cellular processes; thus, a constant supply of oxygen is indispensable to sustain cardiac function and viability. The story of oxygen in the heart is complex, however, and goes well beyond its role in energy metabolism.

Oxygen is a major determinant of myocardial gene expression, and as myocardial O_2 levels decrease, either during isolated hypoxia or ischemia-associated hypoxia, gene expression patterns in the heart are significantly altered (3). Oxygen participates in the generation of NO, which plays a critical role in determining vascular tone, cardiac contractility, and a variety of additional parameters. Oxygen is also central in the generation of reactive oxygen species (ROS), which can participate as benevolent molecules in cell signaling processes or can induce irreversible cellular damage and death. Oxygen is thus both vital and deleterious (4).

The role of oxygen in myocardial energetics and metabolism

The heart can utilize a variety of metabolic fuels, including fatty acids, glucose, lactate, ketones, and amino acids. In the fed state, fatty acids are the preferred fuel, accounting for up to 90% of the total acetyl-CoA provided to cardiac mitochondria (5). Fatty acids are metabolized by β-oxidation, producing acetyl-CoA, NADH, and $FADH_2$. The acetyl-CoA enters the Krebs cycle, producing more NADH and $FADH_2$. Glucose is metabolized initially via the glycolytic pathway, producing a relatively small amount of ATP and also pyruvate, which enters the Krebs cycle, producing NADH and $FADH_2$. In the absence of oxygen, the total amount of energy produced by these processes is insufficient to meet cardiac needs. The cardiac energy requirement is met, however, by entry of the resultant NADH and $FADH_2$ into the electron transport chain, which gener-

ates ATP by oxidative phosphorylation in the mitochondria. Oxygen serves as the terminal electron acceptor in the electron transport chain, and in the absence of sufficient oxygen, electron transport ceases and cardiac energy demands are not met (Figure 1).

Generation and counterbalancing of ROS

ROS can be formed in the heart by a variety of mechanisms, including generation during oxidative phosphorylation in the mitochondria as a byproduct of normal cellular aerobic metabolism (4, 6). Thus, the major process from which the heart derives sufficient energy can also result in the production of ROS (6). Each oxygen atom contains 2 unpaired electrons in its outermost shell. Atoms or molecules with unpaired electrons are designated free radicals and are highly reactive entities that can readily participate in a variety of chemical/biochemical reactions. Molecular oxygen, O_2, is characterized as diradical, a property that allows liquid oxygen to be attracted to the poles of a magnet. This property also dictates that full reduction of oxygen to water as a terminal event in the electron transport chain requires 4 electrons. The sequential donation of electrons to oxygen during this process can generate ROS as intermediates, and "electron leakage" can also contribute to the formation of ROS (4, 7, 8). Donation of a single electron to molecular oxygen results in the formation of the superoxide radical ($O_2^{\cdot-}$). Donation of a second electron yields peroxide, which then undergoes protonation to yield hydrogen peroxide (H_2O_2). Donation of a third electron, such as occurs in the Fenton reaction ($Fe^{2+} + H_2O_2 \rightarrow Fe^{3+} + {\cdot}OH + OH^-$), results in production of the highly reactive hydroxyl radical (${\cdot}OH$). Finally, donation of a fourth electron yields water. Singlet oxygen ($^1O_2$), a very short-lived and reactive form of molecular oxygen in which the outer electrons are raised to a higher energy state, can be formed by a variety of mechanisms, including the Haber-Weiss reaction ($H_2O_2 + O_2^{\cdot-} \rightarrow {\cdot}OH + OH^- + {}^1O_2$) (9).

ROS can be formed in the heart, and other tissues, by several mechanisms; they can be produced by xanthine oxidase (XO), NAD(P)H oxidases, cytochrome P450; by autooxidation of catecholamines; and by uncoupling of NO synthase (NOS) (10–14). NO contains an unpaired electron, and under certain conditions can react with $O_2^{\cdot-}$ to form peroxynitrite ($ONOO^-$), a powerful oxidant. ROS formation in the heart can be induced by the action of cytokines and growth factors as well. Angiotensin II (ATII), PDGF, and TNF-α, for example, can induce H_2O_2 and $O_2^{\cdot-}$ formation via activation of the NAD(P)H oxidases (10, 15). This NAD(P)H-dependent pathway is best described in vascular smooth muscle cells but has also been documented in other cell types, including

Nonstandard abbreviations used: ARNT, aryl hydrocarbon nuclear translocase; ASK-1, apoptosis-signaling kinase 1; ATII, angiotensin II; HIF-1α, hypoxia-inducible factor 1α; MI, myocardial infarction; NOS, NO synthase; SOD, superoxide dismutase; VHL, von Hippel–Lindau protein; XO, xanthine oxidase.

Conflict of interest: The author has declared that no conflict of interest exists.

Citation for this article: *J. Clin. Invest.* **115**:500–508 (2005). doi:10.1172/JCI200524408.

A

Metabolism in cytosol

Metabolism within mitochondria

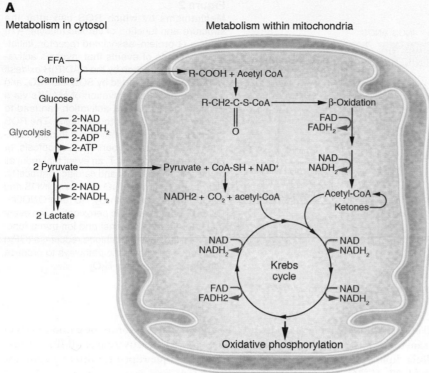

Figure 1
Role of oxygen in myocardial metabolism. (**A**) Schematic depiction of the pathways by which cardiac muscle utilizes various fuels, including fatty acids, glucose, lactate, and ketones. Glycolysis occurs in the cytosol and does not require oxygen. β-Oxidation of fatty acids, ketone metabolism, and the metabolism of glucose-derived intermediates all generate reduced flavoproteins ($NADH_2$ and $FADH_2$). (**B**) Schematic depiction of the process of oxidative phosphorylation in the mitochondria. Complexes 1–4 refer to specific electron transfer steps that occur in the mitochondria. A series of electron transfers among the flavoproteins ($FMNH_2$, $NADH_2$, $FADH_2$), iron-sulfur, coenzyme Q, and the cytochromes a–$c1$, results in accumulation of protons in the space between the inner and outer mitochondrial membranes. This proton gradient provides the energy for ATP production via complex 5. Sustaining this crucial process requires the continuous availability of oxygen as the terminal electron acceptor in the chain. $Fe^{2+}S$, reduced iron-sulfur; $Fe^{3+}S$, oxidized iron-sulfur; FMN, flavin mononucleotide; cyt, cytochrome; CoQ, coenzyme Q.

B

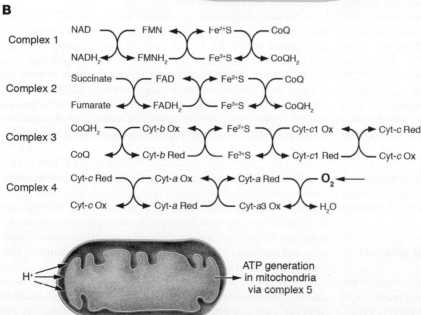

antioxidant and redox regulatory system that has been implicated in a wide variety of ROS-related processes (21). Thioredoxin and thioredoxin reductase can catalyze the regeneration of many antioxidant molecules, including ubiquinone (Q10), lipoic acid, and ascorbic acid, and as such constitute an important antioxidant defense against ROS. Deletion of thioredoxin reductase results in developmental heart abnormalities and in cardiac death secondary to a severe dilated cardiomyopathy (26).

Nonenzymatic mechanisms include intracellular antioxidants such as the vitamins E, C, and β-carotene (a precursor to vitamin A), ubiquinone, lipoic acid, and urate (21). They also include glutathione, which acts as a reducing substrate for the enzymatic activity of glutathione peroxidase.

The biological significance of ROS

ROS have an important role in several important biological processes, including the oxidative burst reaction essential to phagocytes (27). They are involved in a variety of cellular signaling pathways (28), acting in some instances as second messengers downstream of specific ligands, including TGF-β1, PDGF, ATII, FGF-2, endothelin, and others (14, 15, 29, 30). ROS are also involved in modulating the activity of specific transcription factors, including NF-κB and activator protein-1 (AP-1) (20, 31–34). NF-κB, for example, becomes more transcriptionally active in response to the contribution of ROS to the degradation of IκB, the inhibitory partner of NF-κB that sequesters it in the cytosol. Thus ROS can play an important role in modulating inflammation.

cardiomyocytes (16–20). A number of additional ligands have been associated with the induction of ROS, including several with particular relevance to the cardiovascular system (reviewed by Thannickal et al. in ref. 15).

There are several cellular mechanisms that counterbalance the production of ROS, including enzymatic and nonenzymatic pathways (21). Among the best-characterized enzymatic pathways are catalase and glutathione peroxidase, which coordinate the catalysis of H_2O_2 to water, and the superoxide dismutases (SODs), which facilitate the formation of H_2O_2 from O_2^{-} (22–25). Thioredoxin and thioredoxin reductase together form an additional enzymatic

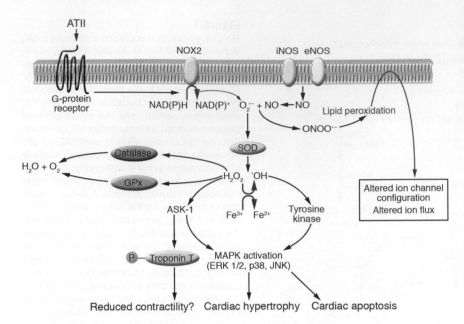

Figure 2
Mechanisms by which ROS can alter the structure and function of cardiac muscle. ATII binds a G-protein–associated receptor, initiating a cascade of events that involves activation of $O_2^{\cdot-}$ production by the NAD(P)H oxidase NOX2. $O_2^{\cdot-}$ is converted by SOD into H_2O_2 and $\cdot OH$ that mediates activation of MAPKs via a tyrosine kinase. MAPK activation can lead to cardiac hypertrophy or to apoptosis. The ROS that is generated can also signal through ASK-1 to induce cardiac hypertrophy, apoptosis, or phosphorylate troponin T, an event that reduces myofilament sensitivity and cardiac contractility. NO production by the NO synthases iNOS and eNOS can interact with $O_2^{\cdot-}$ to form $ONOO^{\cdot-}$. $ONOO^{\cdot-}$ can cause lipid peroxidation, an event that can alter ion channel and ion pump function. Catalase and glutathione reductase (GPx) are shown as enzymatic pathways to produce water and oxygen from H_2O_2.

Perhaps the most widely recognized biological effects of ROS, however, are those that occur when cellular antioxidant defenses are overwhelmed and ROS react directly with cellular lipids, proteins, and DNA, causing cell damage and death (4, 27, 35, 36). Lipid peroxidation, for example, is a well-characterized effect of ROS that results in damage to the cell membrane as well as to the membranes of cellular organelles (37, 38). ROS can contribute to mutagenesis of DNA by inducing strand breaks, purine oxidation, and protein-DNA cross-linking, and other ROS-mediated alterations in chromatin structure may significantly affect gene expression (39, 40). Modification of proteins by ROS can cause inactivation of critical enzymes and can induce denaturation that renders proteins nonfunctional (41, 42). General aging and age-related alteration in the cardiovascular system have been attributed to the long-term cumulative effects of ROS, although the relative contribution of ROS to the aging process remains the subject of debate (43, 44). Figure 2 depicts several pathways by which ROS can mediate biological effects germane to the cardiovascular system.

Actions of ROS relevant to the heart, and the potential role of ROS in heart failure

A significant number of in vitro and animal studies have demonstrated ROS activation in the cardiovascular system in response to various stressors and in the failing heart (6, 14, 20, 36, 45–47). Further, although the results are somewhat inconsistent, animal studies have also delineated that antioxidants and ROS defense pathways can ameliorate ROS-mediated cardiac abnormalities (26, 48–50).

Many ROS-mediated biological processes that are germane to the heart and to the genesis of heart failure have been described.

ROS in ischemic syndromes

Given that coronary artery disease (CAD) with consequent myocardial ischemia and necrosis is a leading cause of heart failure worldwide, it is important to note that ROS may play an important role in the genesis and progression of CAD (51, 52). ROS activity in the vessel wall, for example, is thought to contribute to the forma-

tion of oxidized LDL, a major contributor to the pathogenesis of atherosclerosis (53). ROS-associated activation of MMPs may play an important role in vessel plaque rupture, initiating coronary thrombosis and occlusion (54).

In the setting of acute myocardial infarction (MI), ROS are purported to play a significant role in tissue necrosis and reperfusion injury (55, 56). Transgenic overexpression of SOD has been shown to reduce infarct size in mice, which supports the contention that ROS are important mediators of myocardial damage in MI (50). However, attempts to target SOD to improve outcome after MI in other animal models have yielded mixed results, suggesting that ROS activation is only a single contributor to post-MI necrosis and reperfusion injury and/or that SOD alone is insufficient to neutralize the deleterious effects of ROS in this setting. ROS also play a significant role in the pathogenesis of myocardial stunning, which can complicate acute ischemic syndromes (57). XO expression is significantly increased in the setting of MI, which increases the probability that XO-induced ROS will be generated. Interestingly, administration of the XO inhibitor allopurinol in the setting of acute MI attenuates stunning and ameliorates the excitation-contraction uncoupling that occurs in stunned myocardium (58, 59).

At least 20% of patients who suffer MI go on to develop heart failure. The manner in which the ventricle heals and remodels after MI is a major determinant of eventual cardiac function and the progression to heart failure (60). ROS may contribute to the remodeling processes in a number of ways, including activating MMPs that participate in reconfiguration of the extracellular matrix; acting as signaling molecules in the development of compensatory hypertrophy; and contributing to myocyte loss via apoptosis or other cell death mechanisms (20, 61, 62). Recently it was shown that inhibition of XO with allopurinol after experimental MI in dogs diminished ROS production in the myocardium and attenuated maladaptive LV remodeling, leading to improved post-MI cardiac function (63). While this does not prove that ROS are a major clinical target for decreasing the progression to failure after MI, these findings are part of a growing list of data suggesting a major contribution of ROS to this process.

ROS in cardiac hypertrophy, apoptosis, and the transition to failure

Cardiac hypertrophy can be either compensatory and adaptive or a maladaptive precursor to cardiac failure. Mounting evidence has strongly implicated ROS signaling in the genesis of cardiac hypertrophy (64–68). Many extracellular factors are capable of inducing hypertrophy of cardiomyocytes, and many of the various downstream signaling pathways that mediate the hypertrophic growth response to these factors can be activated directly or indirectly by ROS. These include PKC; the MAPKs p38, JNK, apoptosis-signaling kinase 1 (ASK-1), and ERK1/2; PI3K; Akt; several tyrosine kinases (e.g., src and FAK); NF-κB; and calcineurin (20, 69). An example of purported direct activation is ROS-mediated activation of PKC via oxidation of cysteine residues (70). An example in which an extracellular signal induces cardiac hypertrophy via a ROS-dependent pathway is ATII-induced hypertrophy. The clinical role of ATII in the development and progression of heart failure is now well established, and blocking either ATII generation or ATII binding to the AT1 receptor prolongs life in specific patient cohorts. ATII induces cardiac hypertrophy via a G-protein–linked pathway that involves generation of ROS and ROS-associated activation of several downstream signals, including MAPKs (71). Interestingly, inhibition of ROS by antioxidants blocks ATII-mediated cardiac hypertrophy (71, 72).

As mentioned above, it has been shown that ATII induces ROS in large measure via NAD(P)H oxidases, although the mechanism by which ATII activates these oxidases has remained incompletely understood. In vitro and in vivo data suggest that the small G-protein Rho may be involved in this link. Inhibition of Rho with a dominant-negative construct in neonatal rat cardiomyocytes prevents ATII-mediated intracellular oxidation events, and inhibition of Rho activation by the HMG-CoA reductase inhibitor simvastatin blocked ATII-mediated increases in protein synthesis in these cells, leading to smaller cell sizes (20, 73, 74). Simvastatin administration in vivo also prevented cardiac hypertrophy in response to either ATII administration or pressure overload induced by aortic banding, which establishes that the effects of ROS on hypertrophy are not an artifactual in vitro finding (74).

Another mechanism by which ROS can induce cardiac hypertrophy is via transcription factor–mediated alterations in gene expression. For example, ATII stimulates ROS-mediated activation of the transcription factor NF-κB, which is purported to play an important role in the induction of cardiac hypertrophy. AP-1 activity is also purportedly regulated by ROS and is involved in the transcriptional expression of several genes involved in cardiac hypertrophy. In addition, it is possible that regulation of transcription by ROS may be far more widespread than anticipated and that ROS-mediated alterations in gene structure and function contribute significantly to the pathogenesis of disease in many organ systems, including the heart. Recently the roles of ROS in chromatin remodeling and in DNA damage have become more established, and consideration of the potential involvement of these effects in the failing heart is warranted.

Another MAPK family member linking ROS and hypertrophy is the redox-sensitive kinase ASK-1. ASK-1 is strongly activated by ROS and in turn activates MAPKs p38 and JNK. Deletion of ASK-1 attenuates p38 and JNK activation and the cardiac hypertrophic response to ATII (75). Expression of a dominant-negative ASK-1 attenuates NF-κB activation and inhibits cardiac hypertrophy in response to ROS-generating G-protein receptor agonists, which

demonstrates a role for ASK-1 in the link among ROS, NF-κB, and cardiac hypertrophy (31). ASK-1 also mediates TNF-α–induced apoptosis, which constitutes another link between ROS and an event (apoptosis) that contributes to the pathogenesis of heart disease (76). Overexpression of ASK-1 induces apoptosis in cardiomyocytes, and ASK-1–null mice demonstrate attenuated ventricular remodeling in response to pressure overload, a finding attributed in part to a reduction in cardiac apoptosis (77).

Cardiomyocyte apoptosis occurs in hypertrophied, ischemic, and failing hearts and may contribute to the development and progression of cardiac dysfunction and heart failure (47, 78). Experimental evidence suggests that ROS can mediate apoptosis by a variety of mechanisms, including direct mediation of genotoxicity (62, 68, 79). Interestingly, whether or not apoptosis is induced in cardiomyocytes by oxidative stress appears to be dependent upon the level of ROS produced (64). For instance, in adult cardiomyocytes, relatively low levels of H_2O_2 are associated with the activation of ERK1/2 MAPK and the stimulation of protein synthesis. Conversely, a higher level of H_2O_2, while still activating ERK1/2, also activates the JNK and p38 MAPKs and Akt and induces apoptosis (64). Another clinically relevant potential link between heart failure and ROS involves adrenergic signaling (14, 80). That β-adrenergic receptor (βAR) blockers prolong life and are beneficial in heart failure is well established. One mechanism may be the prevention of βAR-induced apoptosis, an event that appears to involve ROS activation (80).

ROS effects on ion channels and calcium flux

Ion flux is critical to normal cardiac function, and there is significant evidence that ROS alter ion channel flux and membrane ion pump function in a biologically important manner in heart muscle (81). General membrane damage secondary to ROS-mediated lipid peroxidation is one mechanism by which this can occur; however, more specific ROS-mediated effects also contribute. ROS can target L-type calcium channels on the sarcolemma and suppress the Ca^{2+} current (82). ROS depress the activity of the sarcoplasmic reticulum Ca^{2+} ATPase SERCA2, a membrane calcium pump that has been shown to play a crucial role in cardiac calcium handling and as a determinant of myocardial contractility (83). SERCA2 expression is concomitantly reduced in cardiomyocytes stimulated to hypertrophy via ROS-associated signaling pathways. ROS generation can also alter the function of cardiac sodium channels, potassium channels, and ion exchangers, such as the Na/Ca exchanger (84, 85).

In another ROS-mediated pathway that may lead to reduced contractility, ROS can decrease the calcium sensitivity of the myofilaments. Recently it was shown that the ROS-related kinase ASK-1 associates with and phosphorylates troponin T in vitro and in vivo and that this event diminishes contractility and alters calcium handling in cardiomyocytes (86). Whether this pathway contributes to human heart failure remains unknown. It has been postulated, however, that via mechanisms such as this ROS-mediated abnormality in excitation-contraction coupling, chronic exposure to ROS contributes to the progression of failure.

Lipotoxicity and ROS

Myocardial lipotoxicity refers to the accumulation of intramyocardial lipids concomitant with contractile dysfunction, often associated with myocyte death (87, 88). Recently it was shown that lipid accumulation is a significant feature of clinical heart failure (88). Lipid accumulation occurs when there is an imbalance between lipid uptake and β-oxidation, a circumstance that can occur via a variety of mechanisms. Lipid accumulation induces an increase in the PPARα, a nuclear receptor that alters gene expression in response to lipids. PPARα increases fatty acid oxidation, and increased expression of PPARα has been associated with the development of cardiac dysfunction, including diabetic cardiomyopathy (89). Although the mechanism by which this occurs remains unclear, β-oxidation of fatty acids generates ROS, and data suggests that ROS play a role in the pathogenesis of PPARα-associated cardiomyopathies and lipotoxicity (88).

Reactive nitrogen species, NO, and cardiac function

NO, by virtue of its unpaired outer shell electron, is a reactive molecule. In addition to the expansive array of crucially important biological processes in which it plays a role, NO is also a determinant of cardiac contractility. Initially NO was characterized as a contractility depressant, although the role of NO, and the NO-generating NO synthases, is now understood to be much more complex with regard to effects on cardiac contractility (90, 91). Although its general biology will not be reviewed here, NO does react and interact with ROS, and this crosstalk can also have significant effects on cardiac function (92).

NO can mediate the S-nitrosylation of proteins at specific cysteine residues (93). This process also occurs in the heart and has significant functional implications, especially with regard to calcium flux and excitation-contraction coupling (94, 95). S-nitrosylation is facilitated by $O_2^{\cdot-}$ when $O_2^{\cdot-}$ is present at "physiologic" levels. When levels of $O_2^{\cdot-}$ increase, however, it becomes inhibitory to normal S-nitrosylation. Increased $O_2^{\cdot-}$ levels also facilitate interaction of $O_2^{\cdot-}$ with NO to form deleterious reactive molecules, including peroxynitrite (96). Thus, at an optimal $NO/O_2^{\cdot-}$ stoichiometry, the crosstalk between these 2 reactive species facilitates essential cellular processes, a relationship termed nitroso-redox balance in a recent editorial addressing the results of the African American Heart Failure Trial (95). In that trial, combined therapy with hydralazine, a vasodilator that inhibits generation of $O_2^{\cdot-}$, and isosorbide dinitrate improved quality-of-life scores and decreased mortality by approximately 45% in African Americans with severe heart failure (97). A compelling argument has been made that the effectiveness of this therapy is due in part to restoration of nitroso-redox balance.

Do ROS play a role in clinical heart failure?

Several clinical observations support the hypothesis that ROS play a role in human heart failure, although to date the clinical data has been conflicting and less than compelling (98, 99). In a handful of defined cardiomyopathies, the contribution of ROS is well established, including alcohol-mediated and anthrocycline-induced cardiomyopathies (100). In other forms of human heart failure, the role of ROS has not been definitively established. One problem is that it has been difficult to determine ROS activity in vivo, and clinical studies have relied on indirect measures, using biochemical markers of ROS activity, including indices of lipid peroxidation. Further, the patients included in these studies often had confounding comorbidities that could induce alterations in these biochemical markers, independent of cardiac-specific ROS activity (99). Finally, the presence of ROS in the heart concomitant with heart failure does not prove a causal relationship, and clinical trials using antioxidant therapy have yielded mixed and often disappointing results. This may, however, reflect the difficulty in altering ROS-associated processes in vivo rather than an absence of their involvement in human cardiac failure or may indicate that the relationship between ROS and heart failure is too complex to be addressable by a single intervention (101). Representative examples of the conflicting clinical data follow.

In failing human hearts of patients with either ischemic or dilated cardiomyopathy, NAD(P)H oxidase–linked ROS activity was elevated and associated with increased rac-1 GTPase activity (98). Treatment of heart failure patients with statins decreased rac-1 activity in myocardium from these patients, possibly via statin effects on ROS activity. Treatment with the XO inhibitor allopurinol improved cardiac contractility and restored normal vasomotor reactivity (102, 103). Treatment with vitamin C inhibited endothelial cell apoptosis (104). Carvedilol has been shown in clinical trials to be superior to other β-blockers for the treatment of congestive heart failure, and one potential mechanism that has been put forth to explain this is that carvedilol has antioxidant effects (105). Conversely, large clinical trials of the antioxidant vitamins or precursors have not shown benefit in preventing cardiac morbidity or mortality (106–108). A larger clinical trial of a XO inhibitor to follow up on the earlier positive clinical results is planned, but at this time, the jury remains out regarding the therapeutic utility of antioxidant therapy for heart failure (109).

Oxygen as a determinant of myocardial gene expression and cardiac function

Myocardial oxygen consumption and availability must be matched to ensure normal cardiac function and viability. Although physiological mechanisms are in place to coordinate this balance dynamically in response to acute and subacute alterations in cardiac workload, alterations in gene expression patterns also play a critical role, generally as adaptive responses to cardiac stressors that alter either myocardial O_2 consumption (e.g., pressure overload) or O_2 delivery (e.g., coronary artery disease). Gene expression is adjusted to oxygen availability in the heart by several mechanisms, including regulation of gene transcription by the hypoxia-inducible basic helix-loop-helix transcription factor hypoxia-inducible factor 1α (HIF-1α) (3, 110). HIF-1α regulates the transcription of an extensive repertoire of genes, including many involved in angiogenesis and vascular remodeling, erythropoiesis, metabolism, apoptosis, control of ROS, vasomotor reactivity and vascular tone, and inflammation (111–113).

HIF-1α alters the transcription of these genes by dimerizing with the aryl hydrocarbon nuclear translocase (ARNT, or HIF-1β) and then binding to specific hypoxia response elements (HREs) in their regulatory regions. The general response to HIF is transcriptional activation, coordinated by recruitment of the p300 coactivator. ARNT is generally abundant; thus the availability of HIF-1α is rate limiting in formation of the HIF-ARNT dimer. The major mechanism coordinating the effects of HIF-1α on gene expression with oxygen availability involves the posttranslational regulation of HIF-1α abundance. HIF-1α is constitutively transcribed and translated in the heart and most other tissues, but under conditions in which oxygen is abundant, HIF undergoes prolyl-hydroxylation at the direction of specific cellular prolyl hydroxylases (114). This hydroxylation event produces a binding site recognized by the von Hippel–Lindau protein (VHL) (115). VHL, which is deficient in the von Hippel–Lindau syndrome, is a subunit in an E3 ubiquitin ligase complex that polyubiquitylates HIF-1α, thus targeting it for rapid destruction by the proteosome. When oxygen is less abun-

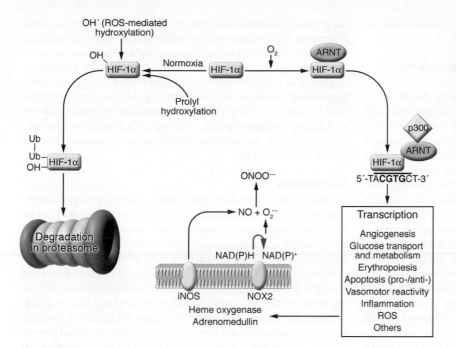

Figure 3
Transcriptional gene regulation by the hypoxia-inducible factor HIF-1α. HIF-1α protein undergoes rapid prolyl hydroxylation under normoxic conditions by specific cellular prolyl hydroxylases. Direct hydroxylation by ROS is a purported alternative pathway. Hydroxylated HIF interacts with the VHL, a critical member of an E3 ubiquitin ligase complex that polyubiquitylates HIF (Ub, ubiquitin). Polyubiquitylation targets HIF-1α for destruction by the proteosome. Under hypoxia ($\downarrow O_2$) hydroxylation does not occur and HIF-1α is stabilized. Heterodimerization with ARNT forms the active HIF complex that binds to a core hypoxia response element in a wide array of genes involved in a diversity of biological processes germane to cardiovascular function. Transcriptional activation of iNOS expression is shown as an example of how HIF-mediated gene expression can affect ROS generation by generating NO that interacts with $O_2^{\bullet-}$ to form $ONOO^{\bullet-}$. NOX2 is shown as a cellular source of $O_2^{\bullet-}$.

dant, prolyl-hydroxylation does not occur, and consequently VHL does not bind to and ubiquitylate HIF-1α. HIF-1α thus accumulates in the cell, associates with ARNT, binds the HRE in hypoxia-responsive genes, and alters the transcription of those genes. This post-translational level of control assures that HIF-mediated transcriptional responses to hypoxia occur rapidly. Control of HIF-1α expression at the mRNA level does occur but is much less important than this post-translational mechanism.

Cardiomyocyte-specific deletion of HIF-1α leads to altered expression of multiple genes in the normoxic heart, which establishes that HIF-1α plays an important role in coordinating gene expression with oxygen availability within the normal range of oxygen tensions encountered in the heart, independent of hypoxia or ischemia. This is borne out by the finding that expression of HIF-1α in cardiomyocytes is required to maintain normal myocardial metabolism, vascularity, calcium handling, and contractile function under normoxic conditions (3). Interestingly, it was recently shown that HIF-1α levels are also coordinated by a Fenton reaction in the endoplasmic reticulum, which provides an important link between ROS and oxygen-sensitive gene expression (116). Conversely, HIF-1α coordinates the expression of several genes that could have a significant impact on ROS and nitroso-redox balance (117, 118). For example, HIF regulates the expression of iNOS and, indirectly (via VEGF), the levels and activity of eNOS and thus has an effect on NO availability. HIF also has marked effects on the expression of genes involved in cardiac metabolism, including all of the glycolytic enzymes, a critical glucose transporter, and lactate dehydrogenase (3). These alterations in cardiac metabolism could have a significant impact on ROS generation. Evidence that hypoxia-inducible gene expression may play a role in control of oxidative stress includes the demonstration in a recent publication that deletion of HIF-2α (EPAS-1, an HIF family member that also dimerizes with ARNT) causes multiple organ pathology, including severe cardiomyopathy, associated with loss of ROS homeostasis (46). Figure 3 summarizes the mechanisms by which HIF levels are regulated by oxygen and the manner in which ROS and HIF may interact.

Hypoxia-inducible gene expression via HIF-1α is beneficially adaptive in the context of acute or subacute myocardial hypoxia or ischemia (119). Examples of HIF-mediated adaptations include increased expression of VEGF to promote angiogenesis; glucose transporter 1 to enhance glucose uptake; the glycolysis-associated enzymes to facilitate glucose metabolism; and erythropoietin to enhance hematopoiesis and increase oxygen carrying capacity. It remains an open question, however, whether chronic activation of HIF-1α, or other hypoxia-responsive pathways, is maladaptive.

Chronic ischemia can lead to progressive heart failure, independent of the acute loss of heart muscle by necrosis. Whether progressive loss of function in the setting of chronic ischemia is due to incremental loss of myocytes and decreased contractility resulting from an inadequate adaptive response to ischemia or is due to maladaptive effects of chronically activated hypoxia-ischemia response pathways in the heart, or both, remains unclear. HIF-1α levels have been shown to be elevated in the myocardium of patients with ischemic cardiomyopathy (120), and it is conceivable that chronic transcriptional activation of HIF-1α–responsive genes is deleterious. In addition to HIF-1α, there are at least 2 additional HIF subunits, HIF-2α (EPAS-1) and HIF-3α, that are regulated in a manner similar to HIF-1α and that can partner with ARNT. The relative roles of HIF-1α and HIF-2α remain unclear, and the coordination of their activities appears to be complex (121). As mentioned above, HIF-2α–null mice manifest a severe cardiomyopathy that appears to be related to ROS activation (46). Whether this phenotype is exclusively due to the loss of HIF-2α–mediated effects on gene expression or involves in some reciprocal manner alterations in HIF-1α–mediated activity is not clear from the literature. Interestingly, we have recently found that cardiac-specific deletion of VHL, an intervention that causes an increase in HIF-1α levels, causes severe dilated cardiomyopathy with pathologic changes consistent with those seen in ischemic cardiomyopathy (122). Whether this is due entirely to elevated HIF-1α levels in the absence of VHL or to loss of an HIF-independent role of VHL remains unclear, but the data from these studies strongly supports a critical role of the HIF pathway in regulating cardiac function.

Is there relationship between ventricular geometry, oxygen, ROS, and cardiac function?

Ventricular geometry is a major determinant of cardiac function and is also a critical determinant of myocardial oxygen consumption. Myocardial oxygen consumption is proportional to ventricular wall tension, and by Laplace's law, ventricular wall tension is proportional to $P \times r / 2\pi$ (where P is pressure, r is the radius of curvature of the ventricle, and π is ventricular wall thickness). Thus, at any given pressure and myocardial thickness, a larger ventricle will consume more oxygen per gram tissue than a smaller one. Therefore, irrespective of the etiology, a failing dilated heart requires more oxygen per gram tissue than a nonfailing smaller heart. The consequences of this remain unclear, especially in the setting of heart failure unrelated to coronary disease, but it is reasonable to postulate that this alteration in myocardial oxygen consumption leads to alterations in gene expression and possibly in ROS generation. Ventricular volume reduction by aneurysmectomy or partial left ventriculectomy can lead to improved cardiac function, although in the latter case there is no clear correlation with increased survival (123). The role of decreased wall stress, decreased myocardial oxygen consumption, and consequent alterations in the expression of hypoxia-responsive genes in the clinical response to these procedures is unclear but of significant interest. The Acorn device, a mesh that is wrapped around the heart to limit ventricular dilation, is under clinical investigation for the treatment of heart failure. Interestingly, it is postulated that one potential benefit of this approach will be to decrease ROS generation associated with cardiac dilation (124).

One additional, as-yet-unproven therapeutic approach to heart failure is the induction of angiogenesis with growth factors such as VEGF, even in the absence of coronary artery disease. This approach is based on the hypothesis that myocardial hypoxia occurs as a consequence of mismatch in the relationship between myocardial mass, myocardial oxygen demand, myocardial vascularity, and oxygen delivery and that this mismatch contributes to the genesis of heart failure. This can theoretically occur in a variety of settings, including in nonischemic dilated cardiomyopathy as a consequence of the Laplace relationship as describe above or in pathologic cardiac hypertrophy in which myocardial vascularity might not be sufficient for the increased myocardial muscle mass. Supporting this hypothesis is our finding that cardiomyocyte-specific deletion of VEGF in the mouse heart leads to hypovascularity and dilated cardiomyopathy, although this phenotype may also reflect defects in normal development of the myocardium in the absence of VEGF (125).

Conclusions

Oxygen, beyond its indispensable role in cardiac energy metabolism, plays a central role in other biological processes that can be determinants of cardiac function, including the generation of ROS and the determination of cardiac gene expression patterns. Although their role in the pathogenesis of clinical heart failure remains unclear, ROS have been implicated in most processes thought to have a significant effect on cardiac function, including hypertrophy, ion flux and calcium handling, EC coupling, extracellular matrix configuration, vasomotor function, metabolism, gene expression, and downstream signaling of several growth factors and cytokines. Clinical trials based on antioxidant therapies have been, however, generally disappointing. Whether this is a function of the particular antioxidants used is unclear, and planned trials with XO inhibitors and other alternative agents should help answer this question. The role of hypoxia-induced alterations in gene expression in the genesis of heart failure also remains unclear, although experimental data suggest that these changes in gene expression can be either adaptive or maladaptive, depending on context. Given the central role of oxygen in cardiovascular biology, further investigational focus on oxygen-related processes in the genesis of heart failure is warranted.

Address correspondence to: Frank J. Giordano, Department of Medicine, Yale University School of Medicine, BCMM 436C, 295 Congress Avenue, New Haven, Connecticut 06510, USA. Phone: (203) 785-7361; Fax: (203) 737-2290; E-mail: Frank.Giordano@yale.edu.

1. West, J.B. 1991. Cardiac energetics and myocardial oxygen consumption. *Physiologic basis of medical practice.* Williams and Wilkins. Baltimore, Maryland, USA. 250–260.
2. Braunwald, E. 2001. Coronary blood flow and myocardial ischemia. *Heart disease: a textbook of cardiovascular medicine.* W.B. Saunders Company. Philadelphia, Pennsylvania, USA. 1161–1183.
3. Huang, Y., et al. 2004. Cardiac myocyte-specific HIF-1alpha deletion alters vascularization, energy availability, calcium flux, and contractility in the normoxic heart. *FASEB J.* **18**:1138–1140.
4. Davies, K.J. 1995. Oxidative stress: the paradox of aerobic life. *Biochem. Soc. Symp.* **61**:1–31.
5. Jafri, M.S., Dudycha, S.J., and O'Rourke, B. 2001. Cardiac energy metabolism: models of cellular respiration. *Annu. Rev. Biomed. Eng.* **3**:57–81.
6. Ide, T., et al. 1999. Mitochondrial electron transport complex I is a potential source of oxygen free radicals in the failing myocardium. *Circ. Res.* **85**:357–363.
7. Miwa, S., and Brand, M.D. 2003. Mitochondrial matrix reactive oxygen species production is very sensitive to mild uncoupling. *Biochem. Soc. Trans.* **31**:1300–1301.
8. Genova, M.L., et al. 2003. Mitochondrial production of oxygen radical species and the role of Coenzyme Q as an antioxidant. *Exp. Biol. Med. (Maywood).* **228**:506–513.
9. Toufektsian, M.C., Boucher, F.R., Tanguy, S., Morel, S., and de Leiris, J.G. 2001. Cardiac toxicity of singlet oxygen: implication in reperfusion injury. *Antioxid. Redox. Signal.* **3**:63–69.
10. Seshiah, P.N., et al. 2002. Angiotensin II stimulation of NAD(P)H oxidase activity: upstream mediators. *Circ. Res.* **91**:406–413.
11. Xia, Y., Tsai, A.L., Berka, V., and Zweier, J.L. 1998. Superoxide generation from endothelial nitric-oxide synthase. A Ca2+/calmodulin-dependent and tetrahydrobiopterin regulatory process. *J. Biol. Chem.* **273**:25804–25808.
12. Xia, Y., Roman, L.J., Masters, B.S., and Zweier, J.L. 1998. Inducible nitric-oxide synthase generates superoxide from the reductase domain. *J. Biol. Chem.* **273**:22635–22639.
13. Griendling, K.K., Sorescu, D., and Ushio-Fukai, M. 2000. NAD(P)H oxidase: role in cardiovascular biology and disease. *Circ. Res.* **86**:494–501.
14. Sawyer, D.B., et al. 2002. Role of oxidative stress in myocardial hypertrophy and failure. *J. Mol. Cell. Cardiol.* **34**:379–388.
15. Thannickal, V.J., and Fanburg, B.L. 2000. Reactive oxygen species in cell signaling. *Am. J. Physiol. Lung Cell Mol. Physiol.* **279**:L1005–L1028.
16. Bendall, J.K., Cave, A.C., Heymes, C., Gall, N., and Shah, A.M. 2002. Pivotal role of a gp91(phox)-containing NADPH oxidase in angiotensin II-induced cardiac hypertrophy in mice. *Circulation.* **105**:293–296.
17. Wu, M.L., Chan, C.C., and Su, M.J. 2000. Possible mechanism(s) of arachidonic acid-induced intracellular acidosis in rat cardiac myocytes. *Circ. Res.* **86**:E55–E62.
18. Sauer, H., et al. 2004. Involvement of reactive oxygen species in cardiotrophin-1-induced proliferation of cardiomyocytes differentiated from murine embryonic stem cells. *Exp. Cell Res.* **294**:313–324.
19. Heymes, C., et al. 2003. Increased myocardial NADPH oxidase activity in human heart failure. *J. Am. Coll. Cardiol.* **41**:2164–2171.
20. Sabri, A., Hughie, H.H., and Lucchesi, P.A. 2003. Regulation of hypertrophic and apoptotic signaling pathways by reactive oxygen species in cardiac myocytes. *Antioxid. Redox Signal.* **5**:731–740.
21. Nordberg, J., and Arner, E.S. 2001. Reactive oxygen species, antioxidants, and the mammalian thioredoxin system. *Free Radic. Biol. Med.* **31**:1287–1312.
22. Kirkman, H.N., and Gaetani, G.F. 1984. Catalase: a tetrameric enzyme with four tightly bound molecules of NADPH. *Proc. Natl. Acad. Sci. U. S. A.*

81:4343–4347.

23. Kirkman, H.N., Rolfo, M., Ferraris, A.M., and Gaetani, G.F. 1999. Mechanisms of protection of catalase by NADPH. Kinetics and stoichiometry. *J. Biol. Chem.* **274**:13908–13914.

24. Ursini, F., et al. 1995. Diversity of glutathione peroxidases. *Methods Enzymol.* **252**:38–53.

25. de Haan, J.B., et al. 2004. Fibroblasts derived from Gpx1 knockout mice display senescent-like features and are susceptible to H2O2-mediated cell death. *Free. Radic. Biol. Med.* **36**:53–64.

26. Conrad, M., et al. 2004. Essential role for mitochondrial thioredoxin reductase in hematopoiesis, heart development, and heart function. *Mol. Cell. Biol.* **24**:9414–9423.

27. Hensley, K., Robinson, K.A., Gabbita, S.P., Salsman, S., and Floyd, R.A. 2000. Reactive oxygen species, cell signaling, and cell injury. *Free Radic. Biol. Med.* **28**:1456–1462.

28. Nishida, M., et al. 2000. G alpha(i) and G alpha(o) are target proteins of reactive oxygen species. *Nature.* **408**:492–495.

29. Griendling, K.K., and FitzGerald, G.A. 2003. Oxidative stress and cardiovascular injury: part I: basic mechanisms and in vivo monitoring of ROS. *Circulation.* **108**:1912–1916.

30. Machida, Y., et al. 2003. Overexpression of tumor necrosis factor-alpha increases production of hydroxyl radical in murine myocardium. *Am. J. Physiol. Heart Circ. Physiol.* **284**:H449–H455.

31. Hirotani, S., et al. 2002. Involvement of nuclear factor-kappaB and apoptosis signal-regulating kinase 1 in G-protein-coupled receptor agonist-induced cardiomyocyte hypertrophy. *Circulation.* **105**:509–515.

32. Hsu, T.C., Young, M.R., Cmarik, J., and Colburn, N.H. 2000. Activator protein 1 (AP-1)- and nuclear factor kappaB (NF-kappaB)-dependent transcriptional events in carcinogenesis. *Free Radic. Biol. Med.* **28**:1338–1348.

33. Turpaev, K.T. 2002. Reactive oxygen species and regulation of gene expression. *Biochemistry Mosc.* **67**:281–292.

34. Wu, H.M., Chi, K.H., and Lin, W.W. 2002. Proteasome inhibitors stimulate activator protein-1 pathway via reactive oxygen species production. *FEBS Lett.* **526**:101–105.

35. Hemnani, T., and Parihar, M.S. 1998. Reactive oxygen species and oxidative DNA damage. *Indian J. Physiol. Pharmacol.* **42**:440–452.

36. Suematsu, N., et al. 2003. Oxidative stress mediates tumor necrosis factor-alpha-induced mitochondrial DNA damage and dysfunction in cardiac myocytes. *Circulation.* **107**:1418–1423.

37. Rathore, N., John, S., Kale, M., and Bhatnagar, D. 1998. Lipid peroxidation and antioxidant enzymes in isoproterenol induced oxidative stress in rat tissues. *Pharmacol. Res.* **38**:297–303.

38. Thollon, C., Iliou, J.P., Cambarrat, C., Robin, F., and Vilaine, J.P. 1995. Nature of the cardiomyocyte injury induced by lipid hydroperoxides. *Cardiovasc. Res.* **30**:648–655.

39. Rahman, I. 2003. Oxidative stress, chromatin remodeling and gene transcription in inflammation and chronic lung diseases. *J. Biochem. Mol. Biol.* **36**:95–109.

40. Konat, G.W. 2003. H2O2-induced higher order chromatin degradation: a novel mechanism of oxidative genotoxicity. *J. Biosci.* **28**:57–60.

41. Lockwood, T.D. 2000. Redox control of protein degradation. *Antioxid. Redox Signal.* **2**:851–878.

42. Stadtman, E.R., and Levine, R.L. 2003. Free radical-mediated oxidation of free amino acids and amino acid residues in proteins. *Amino Acids.* **25**:207–218.

43. Sinclair, D.A. 2002. Paradigms and pitfalls of yeast longevity research. *Mech. Ageing Dev.* **123**:857–867.

44. Lakatta, E.G. 2003. Arterial and cardiac aging: major shareholders in cardiovascular disease enterprises: part III: cellular and molecular clues to heart and arterial aging. *Circulation.* **107**:490–497.

45. Wallace, D.C. 2001. Mouse models for mitochondrial disease. *Am. J. Med. Genet.* **106**:71–93.

46. Scortegagna, M., et al. 2003. Multiple organ pathology, metabolic abnormalities and impaired homeostasis of reactive oxygen species in Epas1-/- mice. *Nat. Genet.* **35**:331–340.

47. Cesselli, D., et al. 2001. Oxidative stress-mediated cardiac cell death is a major determinant of ventricular dysfunction and failure in dog dilated cardiomyopathy. *Circ. Res.* **89**:279–286.

48. Ho, Y.S., Magnenat, J.L., Gargano, M., and Cao, J. 1998. The nature of antioxidant defense mechanisms: a lesson from transgenic studies. *Environ. Health Perspect.* **106**(Suppl. 5):1219–1228.

49. Yen, H.C., Oberley, T.D., Vichitbandha, S., Ho, Y.S., and St. Clair, D.K. 1996. The protective role of manganese superoxide dismutase against adriamycin-induced acute cardiac toxicity in transgenic mice. *J. Clin. Invest.* **98**:1253–1260.

50. Chen, E.P., Bittner, H.B., Davis, R.D., Folz, R.J., and Van Trigt, P. 1996. Extracellular superoxide dismutase transgene overexpression preserves postischemic myocardial function in isolated murine hearts. *Circulation.* **94**:II412–II417.

51. Pennathur, S., Wagner, J.D., Leeuwenburgh, C., Litwak, K.N., and Heinecke, J.W. 2001. A hydroxyl radical-like species oxidizes cynomolgus monkey artery wall proteins in early diabetic vascular disease. *J. Clin. Invest.* **107**:853–860.

52. Khatri, J.J., et al. 2004. Vascular oxidant stress enhances progression and angiogenesis of experimental atheroma. *Circulation.* **109**:520–525.

53. Witztum, J.L., and Steinberg, D. 1991. Role of oxidized low density lipoprotein in atherogenesis. *J. Clin. Invest.* **88**:1785–1792.

54. Rajagopalan, S., Meng, X.P., Ramasamy, S., Harrison, D.G., and Galis, Z.S. 1996. Reactive oxygen species produced by macrophage-derived foam cells regulate the activity of vascular matrix metalloproteinases in vitro. Implications for atherosclerotic plaque stability. *J. Clin. Invest.* **98**:2572–2579.

55. Asimakis, G.K., Lick, S., and Patterson, C. 2002. Postischemic recovery of contractile function is impaired in SOD2(+/-) but not SOD1(+/-) mouse hearts. *Circulation.* **105**:981–986.

56. Yoshida, T., Maulik, N., Engelman, R.M., Ho, Y.S., and Das, D.K. 2000. Targeted disruption of the mouse Sod I gene makes the hearts vulnerable to ischemic reperfusion injury. *Circ. Res.* **86**:264–269.

57. Bolli, R. 1998. Causative role of oxyradicals in myocardial stunning: a proven hypothesis. A brief review of the evidence demonstrating a major role of reactive oxygen species in several forms of postischemic dysfunction. *Basic Res. Cardiol.* **93**:156–162.

58. Perez, N.G., Gao, W.D., and Marban, E. 1998. Novel myofilament Ca2+-sensitizing property of xanthine oxidase inhibitors. *Circ. Res.* **83**:423–430.

59. Charlat, M.I., et al. 1987. Evidence for a pathogenetic role of xanthine oxidase in the "stunned" myocardium. *Am. J. Physiol.* **252**:H566–H577.

60. Pfeffer, M.A., and Braunwald, E. 1990. Ventricular remodeling after myocardial infarction. Experimental observations and clinical implications. *Circulation.* **81**:1161–1172.

61. Mann, D.L., and Spinale, F.G. 1998. Activation of matrix metalloproteinases in the failing human heart: breaking the tie that binds. *Circulation.* **98**:1699–1702.

62. von Harsdorf, R., Li, P.F., and Dietz, R. 1999. Signaling pathways in reactive oxygen species-induced cardiomyocyte apoptosis. *Circulation.* **99**:2934–2941.

63. Engberding, N., et al. 2004. Allopurinol attenuates left ventricular remodeling and dysfunction after experimental myocardial infarction: a new action for an old drug? *Circulation.* **110**:2175–2179.

64. Kwon, S.H., Pimentel, D.R., Remondino, A., Sawyer, D.B., and Colucci, W.S. 2003. H(2)O(2) regulates cardiac myocyte phenotype via concentration-dependent activation of distinct kinase pathways. *J. Mol. Cell. Cardiol.* **35**:615–621.

65. Li, J.M., Gall, N.P., Grieve, D.J., Chen, M., and Shah, A.M. 2002. Activation of NADPH oxidase during progression of cardiac hypertrophy to failure. *Hypertension.* **40**:477–484.

66. Date, M.O., et al. 2002. The antioxidant N-2-mercaptopropionyl glycine attenuates left ventricular hypertrophy in in vivo murine pressure-overload model. *J. Am. Coll. Cardiol.* **39**:907–912.

67. Higuchi, Y., et al. 2002. Involvement of reactive oxygen species-mediated NF-kappa B activation in TNF alpha-induced cardiomyocyte hypertrophy. *J. Mol. Cell. Cardiol.* **34**:233–240.

68. Pimentel, D.R., et al. 2001. Reactive oxygen species mediate amplitude-dependent hypertrophic and apoptotic responses to mechanical stretch in cardiac myocytes. *Circ. Res.* **89**:453–460.

69. Ghosh, M.C., Wang, X., Li, S., and Klee, C. 2003. Regulation of calcineurin by oxidative stress. *Methods Enzymol.* **366**:289–304.

70. Gopalakrishna, R., Gundimeda, U., and Chen, Z.H. 1997. Cancer-preventive selenocompounds induce a specific redox modification of cysteine-rich regions in Ca(2+)-dependent isoenzymes of protein kinase C. *Arch. Biochem. Biophys.* **348**:25–36.

71. Nakamura, K., et al. 1998. Inhibitory effects of antioxidants on neonatal rat cardiac myocyte hypertrophy induced by tumor necrosis factor-alpha and angiotensin II. *Circulation.* **98**:794–799.

72. Delbosc, S., Cristol, J.P., Descomps, B., Mimran, A., and Jover, B. 2002. Simvastatin prevents angiotensin II-induced cardiac alteration and oxidative stress. *Hypertension.* **40**:142–147.

73. Sugden, P.H., and Clerk, A. 1998. "Stress-responsive" mitogen-activated protein kinases (c-Jun N-terminal kinases and p38 mitogen-activated protein kinases) in the myocardium. *Circ. Res.* **83**:345–352.

74. Takemoto, M., et al. 2001. Statins as antioxidant therapy for preventing cardiac myocyte hypertrophy. *J. Clin. Invest.* **108**:1429–1437. doi:10.1172/JCI200113350.

75. Izumiya, Y., et al. 2003. Apoptosis signal-regulating kinase 1 plays a pivotal role in angiotensin II-induced cardiac hypertrophy and remodeling. *Circ. Res.* **93**:871–883.

76. Gotoh, Y., and Cooper, J.A. 1998. Reactive oxygen species- and dimerization-induced activation of apoptosis signal-regulating kinase 1 in tumor necrosis factor-alpha signal transduction. *J. Biol. Chem.* **273**:17477–17482.

77. Yamaguchi, O., et al. 2003. Targeted deletion of apoptosis signal-regulating kinase 1 attenuates left ventricular remodeling. *Proc. Natl. Acad. Sci. U. S. A.* **100**:15883–15888.

78. Rayment, N.B., et al. 1999. Myocyte loss in chronic heart failure. *J. Pathol.* **188**:213–219.

79. Adeghate, E. 2004. Molecular and cellular basis of the aetiology and management of diabetic cardiomyopathy: a short review. *Mol. Cell. Biochem.* **261**:187–191.

80. Remondino, A., et al. 2003. Beta-adrenergic receptor-stimulated apoptosis in cardiac myocytes is mediated by reactive oxygen species/c-Jun NH2-terminal kinase-dependent activation of the mitochondrial pathway. *Circ. Res.* **92**:136–138.

81. Kourie, J.I. 1998. Interaction of reactive oxygen species with ion transport mechanisms. *Am. J. Physiol.* **275**:C1–C24.

82. Guerra, L., Cerbai, E., Gessi, S., Borea, P.A., and Mugelli, A. 1996. The effect of oxygen free radicals on calcium current and dihydropyridine binding sites in guinea-pig ventricular myocytes. *Br. J. Pharmacol.* **118**:1278–1284.

83. Kaplan, P., Babusikova, E., Lehotsky, J., and Dobrota, D. 2003. Free radical-induced protein

modification and inhibition of Ca2+-ATPase of cardiac sarcoplasmic reticulum. *Mol. Cell. Biochem.* **248**:41–47.

84. Nakaya, H., Takeda, Y., Tohse, N., and Kanno, M. 1992. Mechanism of the membrane depolarization induced by oxidative stress in guinea-pig ventricular cells. *J. Mol. Cell. Cardiol.* **24**:523–534.

85. Goldhaber, J.I. 1996. Free radicals enhance Na+/Ca2+ exchange in ventricular myocytes. *Am. J. Physiol.* **271**:H823–H833.

86. He, X., et al. 2003. ASK1 associates with troponin T and induces troponin T phosphorylation and contractile dysfunction in cardiomyocytes. *Am. J. Pathol.* **163**:243–251.

87. Unger, R.H. 2002. Lipotoxic diseases. *Annu. Rev. Med.* **53**:319–336.

88. Sharma, S., et al. 2004. Intramyocardial lipid accumulation in the failing human heart resembles the lipotoxic rat heart. *FASEB J.* **18**:1692–1700.

89. Finck, B.N., et al. 2003. A critical role for PPARalpha-mediated lipotoxicity in the pathogenesis of diabetic cardiomyopathy: modulation by dietary fat content. *Proc. Natl. Acad. Sci. U. S. A.* **100**:1226–1231.

90. Balligand, J.L., Kelly, R.A., Marsden, P.A., Smith, T.W., and Michel, T. 1993. Control of cardiac muscle cell function by an endogenous nitric oxide signaling system. *Proc. Natl. Acad. Sci. U. S. A.* **90**:347–351.

91. Bonaventura, J., and Gow, A. 2004. NO and superoxide: opposite ends of the seesaw in cardiac contractility. *Proc. Natl. Acad. Sci. U. S. A.* **101**:16403–16404.

92. Khan, S.A., et al. 2004. Neuronal nitric oxide synthase negatively regulates xanthine oxidoreductase inhibition of cardiac excitation-contraction coupling. *Proc. Natl. Acad. Sci. U. S. A.* **101**:15944–15948.

93. Gaston, B.M., Carver, J., Doctor, A., and Palmer, L.A. 2003. S-nitrosylation signaling in cell biology. *Mol. Interv.* **3**:253–263.

94. Paolocci, N., et al. 2000. cGMP-independent inotropic effects of nitric oxide and peroxynitrite donors: potential role for nitrosylation. *Am. J. Physiol. Heart Circ. Physiol.* **279**:H1982–H1988.

95. Hare, J.M. 2004. Nitroso-redox balance in the cardiovascular system. *N. Engl. J. Med.* **351**:2112–2114.

96. Ferdinandy, P., Danial, H., Ambrus, I., Rothery, R.A., and Schulz, R. 2000. Peroxynitrite is a major contributor to cytokine-induced myocardial contractile failure. *Circ. Res.* **87**:241–247.

97. Taylor, A.L., et al. 2004. Combination of isosorbide dinitrate and hydralazine in blacks with heart failure. *N. Engl. J. Med.* **351**:2049–2057.

98. Maack, C., et al. 2003. Oxygen free radical release in human failing myocardium is associated with increased activity of rac1-GTPase and represents a target for statin treatment. *Circulation.* **108**:1567–1574.

99. Mak, S., and Newton, G.E. 2001. The oxidative stress hypothesis of congestive heart failure: radical thoughts. *Chest.* **120**:2035–2046.

100. Jaatinen, P., Saukko, P., and Hervonen, A. 1993. Chronic ethanol exposure increases lipopigment accumulation in human heart. *Alcohol Alcohol.* **28**:559–569.

101. Mak, S., and Newton, G.E. 2004. Redox modulation of the inotropic response to dobutamine is impaired in patients with heart failure. *Am. J. Physiol. Heart Circ. Physiol.* **286**:H789–H795.

102. Cappola, T.P., et al. 2001. Allopurinol improves myocardial efficiency in patients with idiopathic dilated cardiomyopathy. *Circulation.* **104**:2407–2411.

103. Doehner, W., et al. 2002. Effects of xanthine oxidase inhibition with allopurinol on endothelial function and peripheral blood flow in hyperuricemic patients with chronic heart failure: results from 2 placebo-controlled studies. *Circulation.* **105**:2619–2624.

104. Rossig, L., et al. 2001. Vitamin C inhibits endothelial cell apoptosis in congestive heart failure. *Circulation.* **104**:2182–2187.

105. Packer, M., et al. 2001. Effect of carvedilol on survival in severe chronic heart failure. *N. Engl. J. Med.* **344**:1651–1658.

106. Hennekens, C.H., et al. 1996. Lack of effect of long-term supplementation with beta carotene on the incidence of malignant neoplasms and cardiovascular disease. *N. Engl. J. Med.* **334**:1145–1149.

107. Yusuf, S., Dagenais, G., Pogue, J., Bosch, J., and Sleight, P. 2000. Vitamin E supplementation and cardiovascular events in high-risk patients. The Heart Outcomes Prevention Evaluation Study Investigators. *N. Engl. J. Med.* **342**:154–160.

108. Lonn, E., et al. 2002. Effects of vitamin E on cardiovascular and microvascular outcomes in high-risk patients with diabetes: results of the HOPE study and MICRO-HOPE substudy. *Diabetes Care.* **25**:1919–1927.

109. Freudenberger, R.S., et al. 2004. Rationale, design and organisation of an efficacy and safety study of oxypurinol added to standard therapy in patients with NYHA class III - IV congestive heart failure. *Expert Opin. Investig. Drugs.* **13**:1509–1516.

110. Giordano, F.J., and Johnson, R.S. 2001. Angiogenesis: the role of the microenvironment in flipping the switch. *Curr. Opin. Genet. Dev.* **11**:35–40.

111. Ryan, H.E., Lo, J., and Johnson, R.S. 1998. HIF-1 alpha is required for solid tumor formation and embryonic vascularization. *EMBO J.* **17**:3005–3015.

112. Carmeliet, P., et al. 1998. Role of HIF-1alpha in hypoxia-mediated apoptosis, cell proliferation and tumour angiogenesis. *Nature.* **394**:485–490.

113. Iyer, N.V., et al. 1998. Cellular and developmental control of O2 homeostasis by hypoxia-inducible factor 1 alpha. *Genes Dev.* **12**:149–162.

114. Semenza, G.L. 2001. HIF-1, O(2), and the 3 PHDs: how animal cells signal hypoxia to the nucleus. *Cell.* **107**:1–3.

115. Kaelin, W.G., Jr. 2002. Molecular basis of the VHL hereditary cancer syndrome. *Nat. Rev. Cancer.* **2**:673–682.

116. Liu, Q., et al. 2004. A Fenton reaction at the endoplasmic reticulum is involved in the redox control of hypoxia-inducible gene expression. *Proc. Natl. Acad. Sci. U. S. A.* **101**:4302–4307.

117. Cormier-Regard, S., Nguyen, S.V., and Claycomb, W.C. 1998. Adrenomedullin gene expression is developmentally regulated and induced by hypoxia in rat ventricular cardiac myocytes. *J. Biol. Chem.* **273**:17787–17792.

118. Lee, P.J., et al. 1997. Hypoxia-inducible factor-1 mediates transcriptional activation of the heme oxygenase-1 gene in response to hypoxia. *J. Biol. Chem.* **272**:5375–5381.

119. Semenza, G.L. 2004. O2-regulated gene expression: transcriptional control of cardiorespiratory physiology by HIF-1. *J. Appl. Physiol.* **96**:1173–1177; discussion 1170–1172.

120. Lee, S.H., et al. 2000. Early expression of angiogenesis factors in acute myocardial ischemia and infarction. *N. Engl. J. Med.* **342**:626–633.

121. Park, S.K., et al. 2003. Hypoxia-induced gene expression occurs solely through the action of hypoxia-inducible factor 1alpha (HIF-1alpha): role of cytoplasmic trapping of HIF-2alpha. *Mol. Cell. Biol.* **23**:4959–4971.

122. Lei, L., et al. 2004. Malignant transformation of heart muscle in the absence of the von Hippel-Lindau protein. *Circulation.* **110**:III-45.

123. Batista, R.J., et al. 1997. Partial left ventriculectomy to treat end-stage heart disease. *Ann. Thorac. Surg.* **64**:634–638.

124. Gorman, R.C., Jackson, B.M., and Gorman, J.H. 2004. The potential role of ventricular compressive therapy. *Surg. Clin. North Am.* **84**:45–59.

125. Giordano, F.J., et al. 2001. A cardiac myocyte vascular endothelial growth factor paracrine pathway is required to maintain cardiac function. *Proc. Natl. Acad. Sci. U. S. A.* **98**:5780–5785.

NO/redox disequilibrium in the failing heart and cardiovascular system

Joshua M. Hare[1] and Jonathan S. Stamler[2]

[1]Johns Hopkins University School of Medicine, Baltimore, Maryland, USA. [2]Howard Hughes Medical Institute, Departments of Medicine and Biochemistry, Duke University School of Medicine, Durham, North Carolina, USA.

There is growing evidence that the altered production and/or spatiotemporal distribution of reactive oxygen and nitrogen species creates oxidative and/or nitrosative stresses in the failing heart and vascular tree, which contribute to the abnormal cardiac and vascular phenotypes that characterize the failing cardiovascular system. These derangements at the integrated system level can be interpreted at the cellular and molecular levels in terms of adverse effects on signaling elements in the heart, vasculature, and blood that subserve cardiac and vascular homeostasis.

Cellular damage versus malfunction in signaling

Altered cellular production of ROS and/or reactive nitrogen species (RNS) is a ubiquitous feature of human disease. Many years of study, beginning with the discovery of superoxide (1) and superoxide dismutase (2), provide a sound understanding of the diverse chemical mechanisms by which these agents damage lipids, DNA, and proteins, and a pathophysiologic context is imparted by analyses of diseased tissues, which show the chemical footprints of oxidative and nitrosative stress (3–6). Thus, it has been widely assumed that direct chemical (oxidative and nitrosative) injury is a principal factor in the damage or disruption of cellular and subcellular structure–function that typifies such pathologic situations. A major difficulty with this hypothesis, however, lies in understanding how the salient features of chemical injury, which are common across organ systems, underlie the diverse pathophysiology of chronic disease. Moreover, irreversible oxidative damage cannot be easily reconciled with the acute restoration of cardiovascular performance by certain classes of antioxidants (see ref. 7 for example). It thus remains unclear to what extent the damage caused by RNS and ROS contributes to disease pathogenesis.

It was appreciated early on that NO and/or related congeners have a function in signal transduction, and over the past decade, the idea that additional redox-active species may have signaling roles has been strengthened (8–13). Although the molecular mechanisms by which RNS and ROS modulate cellular signal transduction remain incompletely understood, there is a general consensus that cysteine residues are principal sites of redox regulation (10, 13, 14). Crosstalk between ROS- and RNS-regulated pathways may occur at both the chemical interaction level (15, 16) and through their coordinate effects on target proteins (17, 18); this is reflected in a growing awareness that RNS and ROS may subserve conjoint signaling roles. Analysis of several such examples (17–22) has shown that S-nitrosylation (the covalent attachment of NO

Nonstandard abbreviations used: cGMP, 3′,5′-cyclic guanosine monophosphate; E-C, excitation-contraction; GSNO, S-nitrosoglutathione; GSNOR, S-nitrosoglutathione reductase; Hb, hemoglobin; HF, heart failure; NOS, NO synthase; NOS1, neuronal NOS; PTP, protein tyrosine phosphatase; RNS, reactive nitrogen species; RyR, ryanodine receptor calcium–release channel; SNO, S-nitrosothiol; SNO-Hb, S-nitrosohemoglobin; SR, sarcoplasmic reticulum; XDH, xanthine dehydrogenase; XO, xanthine oxidase; XOR, xanthine oxidoreductase.

Conflict of interest: J.M. Hare serves as a consultant to Cardiome Pharma, and J.S. Stamler has a financial interest in Nitrox LLC.

Citation for this article: *J. Clin. Invest.* **115**:509–517 (2005). doi:10.1172/JCI200524459.

to cysteine thiol) may play a primary effector role while oxygen and other ROS may control the responsivity to S-nitrosylation (much as ROS control the strength of phosphorylation signaling by inhibiting phosphatases; ref. 11). This article addresses the physiologic and pathophysiologic consequences of the interplay between ROS and RNS in the cardiovascular system (Figure 1). We review work that directly implicates NO/redox-based signaling in the physiologic regulation of cardiac contractility and blood flow, and that consolidates the idea that malfunction in physiologic signaling plays a central role in the pathophysiology of heart failure (HF). It is further proposed that the impairments in NO-based signaling that contribute to the failing of the cardiovascular system can be ascribed to disequilibria between ROS and RNS in the heart, vasculature, and blood.

ROS/RNS axis

Molecular basis of interaction. ROS (e.g., O_2^-, H_2O_2, and $OH^·$), may contribute to cardiac injury (4, 23) both by oxidizing cellular constituents, including proteins critical for excitation-contraction (E-C) coupling, (24, 25) and by diminishing NO bioactivity (22). One way that ROS may diminish the effect of NO is by directly inactivating it. But the extent to which interactions between RNS and ROS constitute a pathophysiologic mechanism in vivo is unclear. A second way that ROS may affect NO responses is by oxidizing sites in proteins with which NO reacts (direct competition) or which otherwise influence NO binding (allosteric modulation) (Figure 2). Evidence suggests that this mode of ROS action may contribute to cardiac pathophysiology.

Signaling through cysteine thiols. In addition to binding to transition metal centers in proteins such as guanylate cyclase and cytochrome c, NO-based modification of proteins is subserved by the attachment of a NO group to the thiol side chain of the amino acid cysteine. Well over a hundred proteins of all classes are substrates for S-nitrosylation (for review see refs. 20, 26), and the ubiquitous effects exerted by NO in cells are conveyed in large part by this mechanism of protein regulation (20, 27–30). Although early ideas about the chemistry of NO raised questions about the feasibility of S-nitrosylation reactions in vivo, a substantial body of recent work, which demonstrates the presence of multiple S-nitrosylated proteins at basal conditions and increases in S-nitrosylation that are coupled to activation of all isoforms of NO synthase (NOS) (stimuli include calcium, growth factors, cytokines, hormones, and multiple ligands), has established the physiologic relevance of this

A

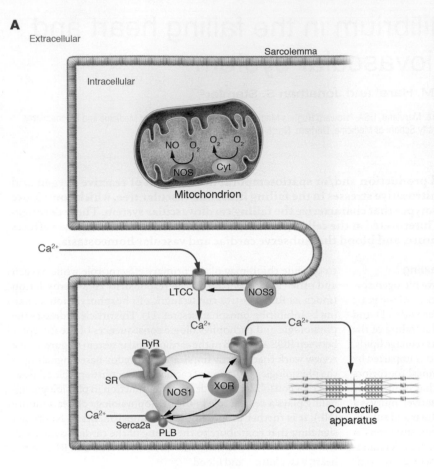

B

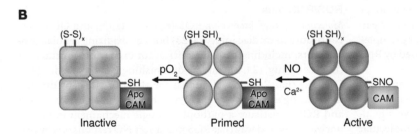

Figure 1
(**A**) Spatial localization of NOSs and oxidases in the cardiac myocyte. NOS3 localizes to the sarcolemmal caveolae (85, 97), where it participates in regulation of L-type Ca^{2+} channel (LTCC) currents, mediated either by cGMP formation (S64) or by S-nitrosylation of the LTCC (S13). NOS1 localizes to the SR (32), where it facilitates the SR Ca^{2+} cycle (97) by S-nitrosylation of the RyR and possibly the SR Ca^{2+} ATPase (SERCA2a) (17, 33). XOR also localizes to the SR in the cardiac myocyte; upregulation of protein or activity (caused by SR NOS1 deficiency) disrupts SNO regulation of the RyR (22). Other oxidases (e.g., NADPH oxidase) have been described in the cardiac myocyte (50), but precise signaling roles, identities, and/or subcellular localizations have not been elucidated. The mitochondria are an additional source of both O_2^- and NO, which may participate in control of mitochondrial respiration (S65–S67) and apoptosis (S41). Cyt, cytochrome; PLB, phospholamban. (**B**) Regulation of the RyR by S-nitrosylation. S-nitrosylation occurs at a single cysteine residue (1 of approximately 50 free thiols), which resides within a calmodulin-binding domain of the cardiac RyR (17). NO binding (shown for RyR1 of skeletal muscle) occurs in an oxygen-concentration–dependent manner and primes the channel for calmodulin regulation (18). Higher pO_2 oxidizes a small set of RyR-associated thiols that regulate the channel's responsiveness to NO. SH, reduced thiol; S-S, oxidized thiols; CAM, calmodulin; x refers to a small set between 1 and 3.

protein modification reaction (28–30). Moreover, recent studies in mice deficient in an S-nitrosothiol–metabolizing (SNO-metabolizing) enzyme (31) and analyses of S-nitrosylation in multiple cellular systems, including site-directed mutagenesis of cysteines in more than 20 proteins of different classes (20), has demonstrated that this NO-derived posttranslational modification serves as a major effector of NO bioactivity and an important mode of cellular signal transduction.

While the ubiquity of thiols across protein classes would appear to support a diversity of roles for S-nitrosylation, the conventional idea that NO acts promiscuously has posed a problem for understanding how specificity might be achieved. Recent literature, however, reveals a growing appreciation of the many determinants of RNS specificity (20), including colocalization of NOSs with their substrates (32, 33). In particular, spatial confinement of NO-signaling components (33, 34) and stimulus-coupled regulation of S-nitrosylation within the confines of signaling modules (26) contrast with ideas about free diffusion of NO and redox state alterations (affecting the cell or

subcellular compartment uniformly) and thus allow S-nitrosylation to serve as a prototype mechanism for redox-based cellular signal transduction. Compartmentalization of NOS isoforms is well exemplified in the cardiac myocyte and disrupted in HF.

Notably, ROS, in particular hydrogen peroxide (H_2O_2), share with NO/RNS a chemical reactivity toward thiols (13, 17, 35), and the concept of target-thiol specificity is reflected in the recent literature on both the differential reactivity of protein thiols toward H_2O_2 (13) and the colocalization of enzymatic sources of ROS (oxidases) with thiol-containing phosphatases (13, 36). However, while it has been convincingly demonstrated that inactivation of protein tyrosine phosphatases (PTPs) by endogenous ROS is required for full phosphorylation signaling by receptor tyrosine kinases (11–13, 36), the broader role of ROS in signal transduction remains to be elucidated, that is, the identity of additional proteins, sites of protein modification (by physiologic ROS), and nature of the in situ oxidations are still largely unknown. Furthermore, since ROS can easily upset the local and cellular redox milieu, the extent to

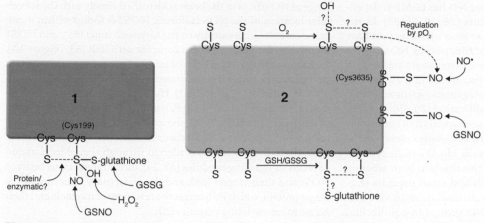

Figure 2
NO/redox-based signaling and nitrosative stress. Molecular recognition by cysteine-containing proteins is achieved either through the existence of single classes of thiols that are adapted to differentiate NO modification (S-nitrosylation) from oxidations (S-glutathionylation, S-S [intramolecular disulfide] and/or sulfur oxides [SO_x^-, where x is 1–3]) — exemplified in protein 1 — or through the presence of multiple classes of thiols, each adapted to recognize different redox-related molecules, including NO, GSNO, H_2O_2, O_2, and cellular redox potential (for protein 2, note that some classes of thiols may be functionally linked to others, exemplified in the pO_2-dependent oxidation of RyR thiols that promotes S-nitrosylation). In model 1, thiol oxidation would adversely impact nitrosylation signaling. In model 2, signal malfunction may result from altered amounts, timing, and/or the nature of RNS/ROS-based modifications. S, cysteine thiol; GSH/GSSG, glutathione/glutathione disulfide; pO_2, partial pressure of O_2.

which they elicit physiologic versus pathologic signaling is not always clear — the mitogenic activity that is potentiated by ROS inhibition of PTPs is a case in point. Resolution of this issue may be facilitated by experimental paradigms that replicate cellular O_2 tensions rather than the oxidative stress represented by ambient pO_2 (18, 37). One intriguing possibility consistent with available data is that ROS may operate to change the set points for initiation of phosphorylation and nitrosylation signaling (18, 20, 36) rather than signaling alone. Phosphorylation and nitrosylation would serve to propagate signals. Altered production of ROS, as will be described in the failing heart, would thus impair physiologic signal transduction.

Cardiovascular sources of O_2^-/ROS and NO/RNS

The major cardiovascular sources of ROS include the enzymes xanthine oxidoreductase (XOR) (38), NAD(P)H oxidase (multisubunit membrane complexes) (39), and NOS (40, 41) as well as the mitochondrial cytochromes (42) and hemoglobin (43, 44). NOSs and hemoglobin are also principal sources of RNS, including NO and SNOs (NO-modified cysteine thiols in amino acids, peptides, and proteins), which convey NO bioactivity. NO and SNOs can also be generated from nitrite in acidic compartments (10, 30), such as the lysosome or the inner membrane of the mitochondria, and from nitrite and nitrate by the actions of enzymes such as XOR (38).

Xanthine oxidoreductase. XOR, a prominent cardiovascular source of ROS, exhibits increased abundance and activity in HF (7, 45, 46). XOR is upregulated in the heart (7) and vasculature (46) of patients with HF and contributes to mechanoenergetic uncoupling (7) and vasoconstriction (46), respectively. A molybdenum-containing enzyme, XOR is expressed as a 150-kDa homodimer that produces superoxide or hydrogen peroxide as byproducts of the terminal steps of purine metabolism (ref. 1; for review see ref. 38). The enzyme has 2 forms: xanthine oxidase (XO) and xanthine

dehydrogenase (XDH) (47). XO is a variant of XDH, resulting from either irreversible proteolytic cleavage or reversible oxidation of sulfhydryl residues of XDH. Whereas XDH uses NAD^+ as a cofactor (reducing it to NADH), XO utilizes molecular oxygen (reducing it to O_2^-/H_2O_2) (reviewed in refs. 38, 47, 48). The production of $O_2^{\cdot-}$ by XO has potential pathophysiologic relevance. XO can also generate NO from nitrite and nitrate at low pO_2, but whether this is a physiologically relevant pathway in vivo is unclear (38, 49). Also unclear is whether XO-catalyzed production of NO is beneficial or deleterious in the heart.

NAD(P)H oxidase. NAD(P)H oxidases are multisubunit enzymes that catalyze single-electron reductions of O_2 using NAD(P)H as electron sources (39). The vascular enzymes are similar to the macrophage NAD(P)H oxidase and contain Nox family subunits, such as Nox1, Nox4, p22phox, and gp91phox. Like XOR, NAD(P)H oxidase subunits are increased in human HF myocytes (50) and in ischemia-reperfusion (51). There is crosstalk between NAD(P)H oxidase and XO. NAD(P)H oxidase activity maintains endothelial XO levels and participates in the conversion of XDH to XO (52). Mice deficient in gp91phox still exhibit NAD(P)H-dependent superoxide generation and develop pressure overload–induced hypertrophy (53), suggesting an alternative source of ROS. Interestingly, XO has a NADH oxidase activity, which can be inhibited by diphenyleneiodonium (DPI) (54) but not by allopurinol (54, 55). The idea that XOR may be the main pathophysiologic source of ROS generation is supported by the functional consequences of XO inhibition (XOI) in HF. Importantly, both XO (56, 57) and NADPH oxidases (58, 59) are inhibited by NO, providing a mechanism through which NOS activity regulates superoxide production and thereby maintains O_2^-/NO homeostasis.

NO synthases. NOSs oxidize the terminal guanidino nitrogen of L-arginine to form NO and the amino acid L-citrulline (60). The NOS isoforms, neuronal NOS (nNOS or NOS1), iNOS or calcium-independent NOS (NOS2), and endothelial NOS (eNOS or NOS3), play modulatory roles in most cells, tissues and organs, including the nervous, immune, respiratory, urologic, and cardiovascular systems that collectively have an impact on cardiovascular performance. NOS1 and NOS3 can be activated by calcium and calmodulin whereas NOS2 is effectively calcium-independent by virtue of its high basal Ca^{2+}/calmodulin affinity. NO activates soluble guanylate cyclase (S-GC) by binding to its heme prosthetic group (61). This activation leads to the production of 3′,5′-cyclic guanosine monophosphate (cGMP), which in turn activates protein kinase G and a cascade of biological signaling events (for reviews see refs. 62, 63).

While a role for cGMP in cardiac inotropy, lusitropy, and ion channel responsivity (64, 65) is reasonably well established, the extent to which natriuretic peptides versus NOS mediate these effects is less clear. It is also evident that NO has cGMP-independent effects in the heart and vasculature (26, 29, 66–69). In virtually all cases examined to date where sites of NO action have been identified in vivo, modulation of function by NO involves modification of 1 or very few cysteines, often contained within characteristic acid-base or hydrophobic motifs (20, 26), and as previously mentioned, site-directed mutagenesis experiments support the specificity of this cysteine modification (20, 69). Notably, the ryanodine receptor calcium–release channel (RyR) of cardiac myocytes is constitutively S-nitrosylated at a single cysteine in the rabbit heart (17), and NO circulates in the blood stream of mammals as a NO-derivative of single-reactive thiols in both albumin and hemoglobin (69, 70). Amino acids and small peptides (e.g., cysteine, glutathione) also undergo nitrosylation in vivo (15, 71), providing a reactive source of NO groups in equilibrium with SNO-proteins (72, 73). A putative motif for S-nitrosylation of proteins by S-nitrosoglutathione (GSNO) has been recently adduced (20). In addition, a GSNO reductase (GSNOR; formally alcohol dehydrogenase type III) has been discovered, which selectively metabolizes GSNO (31). Animals deficient in GSNOR show increased steady-state levels of circulating SNOs at basal conditions and widespread elevations of protein-SNO in tissues following challenge by cytokines (31, 74). These studies suggest that the role played by GSNOR is akin to that of phosphatases on the one hand (i.e., regulated denitrosylation) and of superoxide dismutase on the other (i.e., protection from nitrosative stress) and provide genetic evidence for the role of SNOs in physiology and disease.

Hemoglobin (SNO synthase and heme-oxidase). Hemoglobin (a tetramer comprised of 2 α and 2 β subunits) is the largest reservoir of both O_2 and NO in the body; O_2 is carried at hemes, and NO at both hemes and cysteine thiols (S-nitrosohemoglobin [SNO-Hb]) (75, 76). The O_2/NO-binding functions of hemoglobin (Hb) are governed, in significant part, by an equilibrium between 2 structures: deoxy (T) and oxy (R). An allosteric change from R to T structure lowers the affinity of hemes for O_2 and promotes the transfer of NO groups from SNO-Hb to acceptor thiols (75). rbcs thereby provide a NO vasodilator activity that may enable increases in tissue blood flow (O_2 delivery). Rebinding of O_2 to Hb in the lungs helps to regenerate *SNO-Hb* by promoting intramolecular NO transfer from heme to thiol (75, 77, 78). Hemoglobin also possesses an intrinsic heme-oxidase activity, which leads to production of superoxide (43, 44, 79). The release of superoxide by Hb (with concomitant heme oxidation) is favored in the T structure (43, 75, 80, 81). Thus, sustained or excessive desaturation of hemoglobin, characteristic of HF, would increase ROS production. In addition, low O_2 saturations may cause NO to migrate from the β chains (which dispense NO bioactivity) to the α chains (75, 82), which are impaired in their ability to support production of SNO (75). Indeed, it has recently been reported that HF is characterized by accumulations of heme-NO (relative to controls) that are correlated with venous desaturations (78). NO/redox disequilibrium in situ may impair rbc vasodilation and thereby contribute to tissue ischemia.

Cardiac NO signaling

Substrate specificity. Ion channels of most classes are targets for S-nitrosylation (26); within the heart (63), the plasmalemmal L-type calcium channel and the sarcoplasmic reticulum (SR) RyR are notable examples (17, 18, 66) (Figure 1). cGMP is also involved in eliciting NO effects on myocardial E-C coupling (67, 83, 84). The range of NO effects in the heart is identified closely with the subcellular location of the NOS isoforms. NOS3 is found within membrane caveolae in proximity to the L-type channel (85), and NOS1 localizes to the SR (32, 86) in a complex with RyR (33) (Figure 1A). As discussed below, NOS1 can be targeted to caveolae under certain circumstances, where it is associated with the sarcolemmal calmodulin-dependent Ca^{2+} pump (87). Hydrophobic compartmentalization of the target Cys within RyR, coupled with allosteric regulation of thiol reactivity by O_2/ROS (shown for RyR1), serves to direct NO to only 1 of approximately 50 channel thiols (66) (Figure 1B). More generally, NO specificity is conferred through spatial localization of NOSs to signaling modules (33, 34), direct interactions between NOSs and their targets (88), and the quaternary structure of the target protein, which influences the reactivity of its thiols and their access to nitrosylating reagents (20).

Functional specificity. There is substantial physiologic support obtained from studies in *NOS* deletion mice for the notion that NOS1 and NOS3 exert independent and, in some cases, opposite effects on cardiac contractility (22, 33, 34, 89–91) — actions that are well rationalized by the spatial localization of the NO-signaling module (Figure 1). Specifically, NOS3 exerts its effects on signal-transduction events occurring at the plasmalemmal membrane, inhibiting the L-type Ca^{2+} channel and in turn attenuating β-adrenergic myocardial contractility (33, 92, 93). In contrast, NOS1 exerts its effects in the SR, facilitating the Ca^{2+} cycling between SR and cytosol and therefore enhancing myocardial contractility stimulated by either catecholamines (33) or increasing heart rate (22, 34). Such opposite effects of SR NOS1 versus plasmalemmal NOS3 on myocardial contractility indicate that NOS-derived NO does not act as a freely diffusible messenger within the myocyte. Additional support for the paradigm of specificity through compartmentalization of NOS and its targets comes from studies in ischemic (94) or failing myocardium (95) where NOS1 "translocates" from the SR to the plasmalemmal caveolae; there NOS1 exerts "NOS3-like" effects, inhibiting β-adrenergic inotropy (94). Thus NOS isoform specificity derives directly from its location within the cell. NOS structural motifs and fatty acylation enable such spatial localization: NOS1 has a PDZ-domain motif, which facilitates key protein-protein interactions (e.g., syntrophin, ref. 96; plasma membrane calcium ATPase, ref. 87), and NOS3 is myristoylated and palmitoylated, which targets it to the membrane (97). Both proteins can interact with caveolin, and an interaction of NOS1 with XO has been detected in the SR (22). In addition to their location-specific effects on contractility, the NOS isoforms contribute independently to other cardiac phenotypes — notably cardiac hypertrophy. For example, a cumulative hypertrophic phenotype emerges, at both structural and genetic levels, when both NOS isoforms are absent from myocardium (33, 89, 90). The cardiovascular phenotypes of these mice have been previously reviewed (63, 98).

Terminating NO signals. Termination of NO-based signaling is partly governed by the actions of enzymes. cGMP is metabolized by phosphodiesterase-5 (PDE5), which is spatially localized in proximity to NOS (65); in myocytes, PDE5 is found at the cell membrane associated with caveolae. GSNOR regulates the levels of proteins-SNOs that are in equilibrium with GSNO by shifting the position of the equilibrium toward GSNO (31, 74). GSNOR, which shows activity in the endothelium and cardiac myocyte, degrades GSNO into glutathione disulphide (GSSG) and ammonia (NH_3) (74).

Heart failure

NO/redox imbalance. Oxidant-producing enzymes are upregulated in congestive HF (7, 50), and NO-producing enzymes — NOSs and XO — are altered in either their abundance or spatial localization (95). A relative NO deficiency may further promote oxidase activities (22, 59), which suggests that NO may be a global modulator of O_2^-/ROS production (22, 56, 57). In particular, vascular NADPH oxidases (99) are increased in abundance in the failing circulation, at least in part due to increased levels of angiotensin II, which suggests a link between neurohormonal activation and NO/redox disequilibrium (100, S1). Increases in superoxide may inactivate NO, reducing its control over the vascular oxidase (59). XO, which is produced in the liver, gut, and heart, is increased in abundance and activity throughout the cardiovascular system in HF (46), contributing to vasoconstriction and depressed cardiac function (7, 38, S2). In HF, increased XO activity is directly reflected in dysregulation of NO signaling (45). Furthermore, a hemoglobin oxidase activity may create a major oxidant burden in failing systems; venous desaturations characteristic of HF promote the T structure in Hb that supports the release of superoxide (43). The hypoxemia may be aggravated by a deficiency in rbc-SNO that could reduce tissue perfusion (75).

Dysregulated reactions with cysteine thiols. The relative flux of NO and O_2^- — a function of abundance and location of both NOSs and oxidases in the heart — determines the chemical fate of their interactions: $NO > O_2^-$ (characteristic of the physiologic situation) favors *S*-nitrosylation, whereas NO/O_2^- disequilibrium (characteristic of HF) favors oxidation reactions (15, 69). More specifically, at low physiologic levels, NO may act as an antioxidant (S3), abating fenton-type reactions, terminating radical chain reactions, and inhibiting peroxidases and oxidases (e.g., by nitrosylation of allosteric thiols; ref. 59) (S3, S4). Further, NO reactions with O_2^- at basal conditions produce nitrosating reagents that react preferential with thiols (69). Thus, controlled production of RNS and ROS not only preserves an antioxidant environment (15), but may also serve as a mechanism of channeling NO to cysteine substrates. Conversely, when NO and/or O_2^- are elevated, both the nature of target modification and the specificity of targeting are impaired (S5, S6) (Figures 1 and 3). In other words, superoxide/ROS production may facilitate protein *S*-nitrosylation at basal conditions but disrupts this signaling mechanism at higher concentrations (17, 18, 22). Interestingly, superoxide dismutase, which preserves physiologic superoxide levels, may also catalyze *S*-nitrosylation reactions (75, S7).

Nitrosative stress. In circumstances of acute ischemia, sepsis, or HF, iNOS (NOS2) abundance may increase, leading to nitrosative stress, a pathophysiologic situation characterized by accumulation of *S*-nitrosylated proteins to hazardous levels (amount and/or spatiotemporal distribution) (63). Nitrosative stress may be exacerbated in situ by oxidants (oxidative stress); stimuli that lead to iNOS induction may also upregulate oxidases, and concomitant elevations in NO/RNS and O_2^-/ROS may lead to formation of higher oxides (NO_x), including peroxynitrite (38). Increased amounts of SNO/NO_x favor polynitrosylation and oxidation of cysteine thiols as well as nitration of tyrosines in proteins (16, S8, S9). This situation is relevant to the failing heart, in which the loss of spatial confinement of NOS1, which redistributes from SR to the plasma membrane, may be a proximate cause of oxidative/nitrosative stress both by relieving the local (SR) control of XO (22, 56, 57, S10) and by altering NO/redox balance at the sarcolemma. At the molecular level, poly–*S*-nitrosylation, oxidation and nitration of the Ca^{2+} ATPase (S11) and RyR (17) may ensue with adverse effect on calcium homeostasis. Thus, a central pathophysiologic consequence of redox disequilibrium is the disruption of NO signaling by alteration of the occurrence or nature of the posttranslational modifications (S12).

Pathophysiologic impact of disruption in NO-based signaling

Cardiac function. To the extent that nitric oxide governs key processes in the heart, NO/redox disequilibrium may adversely affect cardiac performance (Figure 3). In the myocyte, ion channels regulate the calcium cycle responsible for normal systolic and diastolic function (63), and the L-type and RyR, which subserve calcium homeostasis, are well-characterized NO targets (17, S13) (Figure 1). Lines of evidence supporting this view include: (a) cardiac NO levels have been shown to change on millisecond timescales, commensurate with physiologic E-C coupling (S14); (b) NOS1 is localized to the SR (32) in proximity to the RyR (NOS1 and the RyR coimmunoprecipitate in mice, ref. 33; and humans, ref. 95), and NOS3 is in the immediate vicinity of the L-type channel (97, S15); (c) NOS1 potentiates both contractility and SR Ca^{2+} release during inotropic stimulation by β-adrenergic agonists (33) or by pacing (the force-frequency response) (22, 34); (d) RyR isolated from hearts is constitutively *S*-nitrosylated (1 NO/RyR subunit) (17); (e) in lipid bilayer experiments, NO activates RyR by *S*-nitrosylating a single thiol, recapitulating the activity of NOS1 in vivo; likewise, in patch clamp experiments, NO activates the L-type channel independently of cGMP (63); (f) functional impairments observed in the NOS1 (and NOS3) knockout mice establish that NO regulates E-C coupling. Thus, there is substantial in vitro biochemical and in vivo functional data supporting the view that NOS1 positively modulates myocardial E-C coupling through RyR *S*-nitrosylation.

Disruption of the normal E-C coupling by ROS and RNS has been demonstrated for both the skeletal (18) and cardiac (17) RyRs. In vitro, the oxidatively modified channel exhibits sustained activation — directionally similar to the response with physiologic NO — but is unresponsive to physiologic effectors (17, 18). Polynitrosylation of RyR also results in dysregulated channel activity (17). Irreversible activation of the RyR by oxidants leads to SR Ca^{2+} leak (S16), depressed SR stores (S17), and a classic HF phenotype (S18) reminiscent of that described for RyR hyperphosphorylation (S19) and altered FKBP12.6 (calstabin2) interactions (S20–S22).

An additional body of work with inhibitors of NOSs and oxidases supports the concept of a critical balance between NO/SNO and O_2^-/ROS in physiologic vs. pathophysiologic cardiac function. XO upregulation in the failing heart (7, S23) represents a cause of oxidative stress (38). Inhibition of XO increases myofilament Ca^{2+} responsiveness in postischemic stunning (S24), improves abnormal mechanoenergetics in the failing heart (45, S25), and ameliorates remodeling postmyocardial infarction (S26, S27). Notably, the beneficial effects of XO inhibition are dependent on NOS, that is, reducing O_2^- levels is beneficial only if NO is available (45). Conversely, while XO inhibition does not affect the normal heart, it reverses the mechanoenergetic uncoupling caused by NOS inhibition. These data are consistent with the finding that NOS1 deficiency results in increased XO activity and impaired contractility (22), and they are well rationalized by the observation that NOS1 and XO interact within the SR (22). Thus, NO derived from NOS1 tonically inhibits XO, and loss of inhibition results in a HF phenotype. Pessah and colleagues have recently expanded this paradigm

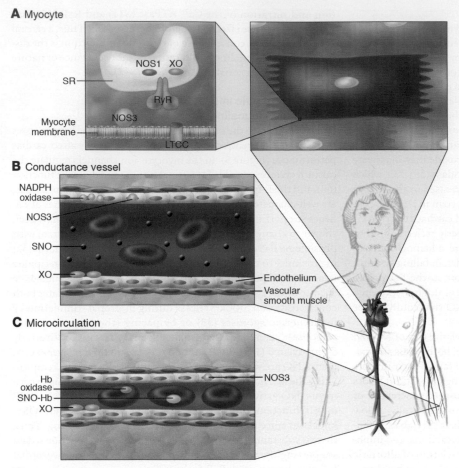

A Myocyte

SR
Myocyte membrane
NOS1 XO
RyR
NOS3
LTCC

B Conductance vessel

NADPH oxidase
NOS3
SNO
XO
Endothelium
Vascular smooth muscle

C Microcirculation

Hb oxidase
SNO-Hb
XO
NOS3

Figure 3

Consequences of NO/redox disequilibrium in the cardiovascular system — congestive HF phenotype. The balance between nitric oxide (NOS and hemoglobin–based activities) and superoxide/ROS production (oxidase activity) plays a pivotal role in cell/organ function at key sites in the cardiovascular system, including the heart (**A**), large- and medium-sized conductance blood vessels (**B**), and the microvasculature (**C**) (S12). At each of these sites, NO/redox disequilibrium is identified with dysregulated NO-based signaling. (**A**) In the cardiac myocyte, NO regulates receptor-mediated signal transduction, the calcium cycle, mitrochondrial respiration, and myofilament contractility. Loss of NOS in the cardiac SR (95) impairs NO signaling and creates oxidative stress (by relieving inhibition of the oxidase) (22). Upregulation of inducible NOS (NOS2) may further disrupt physiologic NO regulation by producing a nitrosative stress. The NO/redox disequilibrium that ensues in HF is characterized by disruption and/or impairment of the cardiac calcium cycle, mitochondrial respiration, and myofilament responsiveness to activator calcium. (**B**) In conductance vessels, vasoconstriction may result from diminished endothelial NOS activity and/or impaired delivery of plasma-borne NO bioactivity. A NO/redox disequilibrium is linked to increased expression or activity of both vascular NADPH oxidase (Nox4) (99) and circulating XO (46). (**C**) In the microvasculature, rbcs govern NO bioactivity. Lower venous O_2 saturation in HF may subserve a NO/redox disequilibrium by impairing NO release from rbcs (SNO-Hb) and promoting hemoglobin oxidase activity (44, 75). Impaired vasodilation by rbcs may exacerbate tissue ischemia.

by showing that NO also regulates an NADH oxidase that inhibits SR Ca^{2+} release and couples SR calcium flux with mitochondrial energetics (S28). Collectively, the data support the idea that ROS pathophysiology does not result from overt oxidative injury to the heart but rather from disruption of physiologic NO signaling that controls E-C coupling and energetics. Thus, the balance between NO and ROS can be understood to have a direct impact upon cardiac regulation through its effects on Ca^{2+} signaling.

Vascular function. NO/redox disequilibrium resulting from increased oxidase activity at the vascular wall may contribute to vasoconstriction, a central HF phenotype. Vascular ROS may originate from circulating XO (S2), NADPH oxidase (99), and/or NOS3 (if cofactors such as tetrahydrobiopterin are diminished) (40). The balance between NO and superoxide may be tipped in HF by a number of effectors, including endothelin, epinephrine, and angiotensin II, which stimulate ROS production, and by arginase (S29) and methylarginines (S30), which diminish NO production.

A function for SNOs in regulation of vasomotor tone has been established in GSNOR-deficient mice (31), and studies in humans demonstrate an association between abnormal SNO levels and adverse cardiovascular events (S31–S33). Notably, the turnover of circulating SNOs may be impaired in clinical situations characterized by oxidative stress (70). Thus, the increased oxidant production in HF may impair plasma-borne NO delivery.

Blood function. A deficit in tissue O_2 delivery is a key component of the HF syndrome. It has been recently appreciated that blood flow in the microcirculation is not linked primarily to eNOS activity or pO_2, but rather to the O_2 saturation of Hb (S34, S35), implicating a role for rbcs (75, 76). Impaired SNO-Hb function and/or vasodilation by rbcs has been observed in several disorders characterized by tissue O_2 deficits, including HF (69, 78, S36, S37). Impaired vasodilation by rbcs may disrupt the O_2 gradients (the normal decline in O_2 saturation as vessel size decreases) in failing tissues. Interestingly these O_2 gradients also regulate *S*-nitrosylation of RyR1, subserving increases in skeletal muscle force (37). Thus, low physiologic pO_2 may provide a concerted mechanism for increasing blood flow to actively contracting muscle though *S*-nitrosylation of RyR and hemoglobin. Conversely, disruption of NO signaling in the microcirculation may be a major contributor to vasoconstriction as well as to the uncoupling of oxygen delivery from the work of skeletal muscle (S38), both of which are characteristics of HF.

Dysfunction of rbcs in the failing circulation (78) may result from both sustained hypoxemia and NO/redox disequilibrium within the rbc. Hypoxemia will favor retention of NO on the hemes of T structured (deoxygenated) Hb, inhibiting SNO-Hb production in the lung (75, S36). Moreover, the oxidase activity of Hb would be increased by chronic venous and tissue hypoxia (43, 79). The consequent production of ROS by hypoxic rbcs may disrupt the delivery of NO to the vessel wall by oxidizing thiol acceptors (73, 79, S39). Thus, the lowered venous pO_2 in HF (78) may compromise the delivery of NO bioactivity that subserves regulation of blood flow.

Cardiac hypertrophy and apoptosis. Apoptosis and cardiac hypertrophy are implicated in cardiac remodeling in HF. NO may exert antiapoptotic effects by *S*-nitrosylating and thereby inhibiting caspases 3 and 9 (S40, S41), the kinase activities of both apoptosis signaling kinase-1 (S42) and c-Jun N-terminal kinase (S43), and the transcriptional activity of jun (S44). NO can also activate the oxidoreductase activity of endothelial thioredoxin by *S*-nitrosylation of an allosteric thiol, thereby helping to preserve the NO/redox equilibrium (S45). In addition, denitrosylation of IκB kinase allows NF-κB to translocate to the nucleus and induce protective genes (S46). The mechanisms by which the different NOSs may exert their influence in modulating apoptosis in the failing heart are not known.

On the other hand, oxidative and nitrosative stress may promote apoptosis (S47). For example, increased levels of SNO may activate apoptosis signaling kinase-1 (by relieving an inhibitory association with thioredoxin) while also blocking the binding of NF-κB to DNA (20). Thus, *S*-nitrosylation of NF-κB would prevent transcription of antiapoptotic genes. In addition, Fas-associated apoptosis is associated with denitrosylation of caspases 3 and 9 (S48). Oxidative stress (mediated by both NADPH oxidase, ref. S49; and XO, ref. S26) is not only linked to apoptosis (through multiple mechanisms) but also to cardiac hypertrophy (23, S49), consistent with the involvement of ROS in mitogenic signaling (12), whereas NO is reported to be antihypertrophic (S50). The precise molecular targets of NO/ROS in cardiac hypertrophy remain to be determined (S49).

Implications for HF therapeutics

Angiotensin converting enzyme inhibitors, adrenergic receptor antagonists, and statins. Insights into NO/redox-based signaling may offer new perspectives on the mechanism of action of existing therapies as well as guiding the development of novel therapeutic strategies for HF. Indeed, many currently used drug treatments influence NO/redox balance, including inhibitors of the renin-angiotensin aldosterone pathway (S51), the sympathetic nervous system, and the HMG-CoA reductase pathways (S52). Angiotensin-converting enzyme inhibitors stimulate NO production by increasing bradykinin formation and reduce O_2^-/ROS production by suppressing angiotensin II–stimulation of NADPH oxidase. The sympathetic nervous system is coupled intimately to NO/O_2^- at multiple levels, and the benefits of HMG-CoA reductases are mediated in significant part through increased expression of NOS (S53).

Hydralazine nitrate combinations. The combination of isosorbide dinitrate and hydralazine as a treatment for HF may be relevant to NO/redox homeostasis (S54–S56). It has been reported that both hydralazine (S57, S58) and NO (59) can inhibit the NADPH oxidase, which is upregulated in failing hearts, and that hydralazine might also inhibit the production of ROS accompanying the use of nitrates. Some of these ideas, however, may need to be revisited in light of the finding that the mitochondria (not the NADH oxidase) are the principle source of ROS in nitroglycerin tolerance (S59), and the extent to which the NADPH oxidase contributes to the oxidant burden in HF is still unknown. It also remains to be shown that antioxidant effects of hydralazine (S57) can be achieved at the concentrations that are employed clinically. Future studies are needed to determine if the remarkable benefits of this drug combination result primarily from the nitrate or hydralazine components or from both.

XO inhibitors. Allopurinol and oxypurinol were described as XO inhibitors in the 1950s by Gertrude Elion (S60, S61) and have enjoyed widespread clinical use in the treatment of a variety of conditions characterized by hyperuricemia, most notably gout and hematologic malignancies. As discussed above, XO is upregulated in HF, which leads to oxidative stress in both the heart (7) and vasculature (46, S62). Based on the findings that XO may serve as a primary source of the superoxide (38) implicated in myocyte apoptosis (4), endothelial dysfunction, and mechanoenergetic uncoupling (45, 90, S25), clinical trials of oxypurinol have been initiated in patients with symptomatic HF (S63).

rbc therapeutics. The understanding that impaired vasodilation by rbcs may represent a primary cause of tissue O_2 deficits that characterize HF (75) and, moreover, that rbcs are by far the largest source of oxidase activity in the cardiovascular system — yet one not previously considered — suggests that efforts directed at restoring the NO/redox homeostasis in red blood cells might have important implications for the treatment of the HF. Early efforts are underway to test these new concepts.

Conclusion

Insights into the integrated physiology of the cardiovascular system have been advanced by the understanding that *S*-nitrosylation is established as a route through which NO can modulate diverse cellular processes, including cardiac E-C coupling, endothelial/vascular function, and tissue oxygen delivery. This understanding supports a perspective in which oxidative and nitrosative stress disrupt physiologic signaling, and by inference, one in which malfunction in NO/redox-based signaling, rather than overt chemical injury, may subserve cardiovascular dysfunction, including structural alterations (e.g., cardiac hypertrophy and programmed cell death). The theme of "NO/redox disequilibrium" – reflecting increased oxidase activities in heart (XO), vasculature (NAD(P)H oxidase), and red blood cells (hemoglobin oxidase) — threads through our evolving understanding of a NO-deficient cardiovascular system. Viewed from this perspective, HF can be seen as a pathophysiologic state that is potentially amenable to therapeutic modulation through targeted restoration of NO/redox balance.

Acknowledgments

The authors are supported by a Paul Beeson Physician Faculty Scholar in Aging award (to J.M. Hare), the Donald W. Reynolds Foundation (to J.M. Hare), and by NIH grants HL-065455 (to J.M. Hare), PO1-HL75443 (to J.S. Stamler), and 5PO1-HL42444 (to J.S. Stamler).

Note: Due to space constraints, a number of important references could not be included. References S1–S67 are available online with this article; doi:10.1172/JCI200524459DS1.

Address correspondence to: Joshua M. Hare, The Johns Hopkins University School of Medicine, Department of Medicine, Cardiology Division, 720 Rutland Avenue, Ross 1059, Baltimore, Maryland 21205, USA. Phone: (410) 614-4161; Fax: (443) 287-7945; E-mail: jhare@mail.jhmi.edu. Or to: Jonathan S. Stamler, Duke University School of Medicine, MSRB Room 321, Durham, North Carolina 27710, USA. Phone: (919) 684-6933; Fax: (919) 684-6998; Email: STAML001@mc.duke.edu.

1. McCord, J.M., and Fridovitch, I. 1968. The reduction of cytochrome c by milk xanthine oxidase. *J. Biol. Chem.* **243**:5753–5760.

2. McCord, J.M., and Fridovitch, I. 1969. Superoxide dismutase. An enzymic function for erythrocuprein (hemocuprein). *J. Biol. Chem.* **244**:6049–6055.

3. Keith, M., et al. 1998. Increased oxidative stress in patients with congestive heart failure. *J. Am. Coll. Cardiol.* **31**:1352–1356.

4. Cesselli, D., et al. 2001. Oxidative stress-mediated cardiac cell death is a major determinant of ventricular dysfunction and failure in dog dilated cardiomyopathy. *Circ. Res.* **89**:279–286.

5. Dhalla, A.K., Hill, M.F., and Singal, P.K. 1996. Role of oxidative stress in transition of hypertrophy to heart failure. *J. Am. Coll. Cardiol.* **28**:506–514.

6. Vanderheyden, M., et al. 2004. Hernodynamic effects of inducible nitric oxide synthase and nitrotyrosine generation in heart failure. *J. Heart Lung Transplant.* **23**:723–728.

7. Cappola, T.P., et al. 2001. Allopurinol improves myocardial efficiency in patients with idiopathic dilated cardiomyopathy. *Circulation.* **104**:2407–2411.

8. Palmer, R.M.J., Ferrige, A.G., and Moncada, S. 1987. Nitric-oxide release accounts for the biological-activity of endothelium-derived relaxing factor. *Nature.* **327**:524–526.

9. Ignarro, L.J., Buga, G.M., Wood, K.S., Byrns, R.E., and Chaudhuri, G. 1987. Endothelium-derived relaxing factor produced and released from artery and vein is nitric-oxide. *Proc. Natl. Acad. Sci. U. S. A.* **84**:9265–9269.

10. Stamler, J.S., Singel, D.J., and Loscalzo, J. 1992. Biochemistry of nitric-oxide and its redox-activated forms. *Science.* **258**:1898–1902.

11. Sundaresan, M., Yu, Z.X., Ferrans, V.J., Irani, K., and Finkel, T. 1995. Requirement for generation of H2O2 for platelet-derived growth-factor signal-transduction. *Science.* **270**:296–299.

12. Irani, K., et al. 1997. Mitogenic signaling mediated by oxidants in Ras-transformed fibroblasts. *Science.* **275**:1649–1652.

13. Forman, H.J., Fukuto, J.M., and Torres, M. 2004. Redox signaling: thiol chemistry defines which reactive oxygen and nitrogen species can act as second messengers. *Am. J. Physiol. Cell Physiol.* **287**:C246–C256.

14. Stamler, J.S. 1994. Redox signaling: nitrosylation and related target interactions of nitric oxide. *Cell.* **78**:931–936.

15. Wink, D.A., et al. 1997. Superoxide modulates the oxidation and nitrosation of thiols by nitric oxide-derived reactive intermediates. *J. Biol. Chem.* **272**:11147–11151.

16. Reiter, C.D., Teng, R.J., and Beckman, J.S. 2000. Superoxide reacts with nitric oxide to nitrate tyrosine at physiological pH via peroxynitrite. *J. Biol. Chem.* **275**:32460–32466.

17. Xu, L., Eu, J.P., Meissner, G., and Stamler, J.S. 1998. Activation of the cardiac calcium release channel (ryanodine receptor) by poly-S-nitrosylation. *Science.* **279**:234–237.

18. Eu, J.P., Sun, J., Xu, L., Stamler, J.S., and Meissner, G. 2000. The skeletal muscle calcium release channel: Coupled O2 sensor and NO signaling functions. *Cell.* **102**:499–509.

19. Jia, L., Bonaventura, C., Bonaventura, J., and Stamler, J.S. 1996. S-nitrosohaemoglobin: A dynamic activity of blood involved in vascular control. *Nature.* **380**:221–226.

20. Hess, D.T., Matsumoto, A., Kim, S.O., Marshall, H., and Stamler, J.S. 2004. Protein S-nitrosylation: purview and parameters. *Nat. Rev. Mol. Cell. Biol.* **6**:150–166.

21. Abu-Soud, H., Rousseau, D.L., and Stuehr, D.J. 2005. Nitric oxide binding to the heme of neuronal nitric-oxide synthase links its activity to changes in oxygen tension. *J. Biol. Chem.* **271**:32515–32518.

22. Khan, S.A., et al. 2004. Neuronal nitric oxide synthase negatively regulates xanthine oxidoreductase inhibition of cardiac excitation-contraction coupling. *Proc. Natl. Acad. Sci. U. S. A.* **101**:15944–15948.

23. Siwik, D.A., et al. 1999. Inhibition of copper-zinc superoxide dismutase induces cell growth, hypertrophic phenotype, and apoptosis in neonatal rat cardiac myocytes in vitro. *Circ. Res.* **85**:147–153.

24. Chiamvimonvat, N., et al. 1995. Functional consequences of sulfhydryl modification in the pore-forming subunits of cardiovascular Ca^{2+} and Na$^+$ channels. *Circ. Res.* **76**:325–334.

25. Thomas, J.A., Poland, B., and Honzatko, R. 1995. Protein sulfhydryls and their role in the antioxidant function of protein S-thiolation. *Arch. Biochem. Biophys.* **319**:1–9.

26. Stamler, J.S., Lamas, S., and Fang, F.C. 2001. Nitrosylation. the prototypic redox-based signaling mechanism. *Cell.* **106**:675–683.

27. Gow, A.J., Buerk, D.G., and Ischiropoulos, H. 1997. A novel reaction mechanism for the formation of S-nitrosothiol in vivo. *J. Biol. Chem.* **272**:2841–2845.

28. Jaffrey, S.R., Erdjument-Bromage, H., Ferris, C.D., Tempst, P., and Snyder, S.H. 2001. Protein S-nitrosylation: a physiological signal for neuronal nitric oxide. *Nat. Cell Biol.* **3**:193–197.

29. Gow, A.J., et al. 2002. Basal and stimulated protein S-nitrosylation in multiple cell types and tissues. *J. Biol. Chem.* **277**:9637–9640.

30. Rhee, K.Y., Erdjument-Bromage, H., Tempst, P., and Nathan, C. 2005. S-nitroso proteome of Mycobacterium tuberculosis: enzymes of intermediary metabolism and antioxidant defense. *Proc. Natl. Acad. Sci. U. S. A.* **102**:467–472.

31. Liu, L., et al. 2004. Essential roles of S-nitrosothiols in vascular homeostasis and endotoxic shock. *Cell.* **116**:617–628.

32. Xu, K.Y., Huso, D.L., Dawson, T., Bredt, D.S., and Becker, L.C. 1999. NO synthase in cardiac sarcoplasmic reticulum. *Proc. Natl. Acad. Sci. U. S. A.* **96**:657–662.

33. Barouch, L.A., et al. 2002. Nitric oxide regulates the heart by spatial confinement of nitric oxide synthase isoforms. *Nature.* **416**:337–340.

34. Khan, S., et al. 2003. Nitric oxide regulation of myocardial contractility and calcium cycling: independent impact of neuronal and endothelial nitric oxide synthases. *Circ. Res.* **92**:1322–1329.

35. Stamler, J.S., and Hausladen, A. 1998. Oxidative modifications in nitrosative stress. *Nat. Struct. Biol.* **5**:247–249.

36. Kwon, J., et al. 2004. Reversible oxidation and inactivation of the tumor suppressor PTEN in cells stimulated with peptide growth factors. *Proc. Natl. Acad. Sci. U. S. A.* **101**:16419–16424.

37. Eu, J.P., et al. 2003. Concerted regulation of skeletal muscle contractility by oxygen tension and endogenous nitric oxide. *Proc. Natl. Acad. Sci. U. S. A.* **100**:15229–15234.

38. Berry, C.E., and Hare, J.M. 2004. Xanthine oxidoreductase in the cardiovascular system: molecular mechanisms and pathophysiologic implications. *J. Physiol.* **555**:589–606.

39. Griendling, K.K., Sorescu, D., and Ushio-Fukai, M. 2000. NAD(P)H oxidase: role in cardiovascular biology and disease. *Circ. Res.* **86**:494–501.

40. Landmesser, U., et al. 2003. Oxidation of tetrahydrobiopterin leads to uncoupling of endothelial cell nitric oxide synthase in hypertension. *J. Clin. Invest.* **111**:1201–1209. doi:10.1172/JCI200314172.

41. Kuzkaya, N., Weissmann, N., Harrison, D.G., and Dikalov, S. 2003. Interactions of peroxynitrite, tetrahydrobiopterin, ascorbic acid, and thiols — implications for uncoupling endothelial nitric-oxide synthase. *J. Biol. Chem.* **278**:22546–22554.

42. Ide, T., et al. 1999. Mitochondrial electron transport complex I is a potential source of oxygen free radicals in the failing myocardium. *Circ. Res.* **85**:357–363.

43. Balagopalakrishna, C., Manoharan, P.T., Abugo, O.O., and Rifkind, J.M. 1996. Production of superoxide from hemoglobin-bound oxygen under hypoxic conditions. *Biochemistry.* **35**:6393–6398.

44. Misra, H.P., and Fridovitch, I. 1972. The generation of superoxide radical during the autoxidation of hemoglobin. *J. Biol. Chem.* **247**:6960–6962.

45. Saavedra, W.F., et al. 2002. Imbalance between xanthine oxidase and nitric oxide synthase signaling pathways underlies mechanoenergetic uncoupling in the failing heart. *Circ. Res.* **90**:297–304.

46. Landmesser, U., et al. 2002. Vascular oxidative stress and endothelial dysfunction in patients with chronic heart failure - role of xanthine-oxidase and extracellular superoxide dismutase. *Circulation.* **106**:3073–3078.

47. McCord, J.M. 1985. Oxygen-derived free radicals in postischemic tissue injury. *N. Eng. J. Med.* **312**:159–163.

48. Saugstad, O.D. 1996. Role of xanthine oxidase and its inhibitor in hypoxia: reoxygenation injury. *Pediatrics.* **98**:103–107.

49. Webb, A., et al. 2004. Reduction of nitrite to nitric oxide during ischemia protects against myocardial ischemia-reperfusion damage. *Proc. Natl. Acad. Sci. U. S. A.* **101**:13683–13688.

50. Heymes, C., et al. 2003. Increased myocardial NADPH oxidase activity in human heart failure. *J. Am. Coll. Cardiol.* **41**:2164–2171.

51. Duilio, C., et al. 2001. Neutrophils are primary source of O2 radicals during reperfusion after prolonged myocardial ischemia. *Am. J. Physiol. Heart Circ. Physiol.* **280**:H2649–H2657.

52. McNally, J.S., et al. 2003. Role of xanthine oxidoreductase and NAD(P)H oxidase in endothelial superoxide production in response to oscillatory shear stress. *Am. J. Physiol. Heart Circ. Physiol.* **285**:H2290–H2297.

53. Maytin, M., et al. 2004. Pressure overload-induced myocardial hypertrophy in mice does not require gp91(phox). *Circulation.* **109**:1168–1171.

54. Sanders, S.A., Eisenthal, R., and Harrison, R. 1997. NADH oxidase activity of human xanthine oxidoreductase generation of superoxide anion. *Eur. J. Biochem.* **245**:541–548.

55. Zhang, Z., et al. 1998. A reappraisal of xanthine dehydrogenase and oxidase in hypoxic reperfusion injury: the role of NADH as an electron donor. *Free Radic. Res.* **28**:151–164.

56. Cote, C.G., Yu, F.S., Zulueta, J.J., Vosatka, R.J., and Hassoun, P.M. 1996. Regulation of intracellular xanthine oxidase by endothelial-derived nitric oxide. *Am. J. Physiol. Lung Cell. Mol. Physiol.* **15**:L869–L874.

57. Hassoun, P.M., Yu, F.S., Zulueta, J.J., White, A.C., and Lanzillo, J.J. 1995. Effect of nitric oxide and cell redox status on the regulation of endothelial cell xanthine dehydrogenase. *Am. J. Physiol.* **268**:L809–L817.

58. Shinyashiki, M., Pan, C.J.G., Lopez, B.E., and Fukuto, J.M. 2004. Inhibition of the yeast metal reductase heme protein Fre1 by nitric oxide (NO): A model for inhibition of NADPH oxidase by NO. *Free Radic. Biol. Med.* **37**:713–723.

59. Clancy, R.M., Leszczynskapiziak, J., and Abramson, S.B. 1992. Nitric-oxide, an endothelial-cell relaxation factor, inhibits neutrophil superoxide anion production via a direct action on the NADPH oxidase. *J. Clin. Invest.* **90**:1116–1121.

60. Michel, T., and Feron, O. 1997. Nitric oxide synthases: Which, where, how and why? *J. Clin. Invest.* **100**:2146–2152.

61. Wink, D.A., and Mitchell, J.B. 1998. Chemical biology of nitric oxide: insights into regulatory, cytotoxic, and cytoprotective mechanisms of nitric oxide. *Free Radic. Biol. Med.* **25**:434–456.

62. Schmidt, H.H.H.W., and Walter, U. 1994. NO at work. *Cell.* **78**:919–925.

63. Hare, J.M. 2003. Nitric oxide and excitation-contraction coupling. *J. Mol. Cell. Cardiol.* **35**:719–729.

64. Layland, J., Li, J.M., and Shah, A.M. 2002. Role of cyclic GMP-dependent protein kinase in the contractile response to exogenous nitric oxide in rat cardiac myocytes. *J. Physiol.* **540**:457–467.

65. Senzaki, H., et al. 2001. Cardiac phosphodiesterase 5 (cGMP-specific) modulates beta-adrenergic signaling in vivo and is down-regulated in heart failure. *FASEB J.* **15**:1718–1726.

66. Sun, J., Xin, C., Eu, J.P., Stamler, J.S., and Meissner, G. 2001. Cysteine-3635 is responsible for skeletal muscle ryanodine receptor modulation by NO. *Proc. Natl. Acad. Sci. U. S. A.* **98**:11158–11162.

67. Paolocci, N., et al. 2000. cGMP-independent inotropic effect of nitric oxide and peroxynitrite donors: Potential role for S-nitrosylation. *Am. J. Physiol.* **279**:H1982–H1988.

68. Bolotina, V.M., Najibi, S., Palacino, J.J., Pagano, P.J., and Cohen, R.A. 1994. Nitric-oxide directly activates calcium-dependent potassium channels in vascular smooth-muscle. *Nature.* **368**:850–853.

69. Foster, M.W., McMahon, T.J., and Stamler, J.S. 2003. S-nitrosylation in health and disease. *Trends Mol. Med.* **9**:160–168.

70. Foster, M.W., Pawloski, J.R., Singel, D., and Stamler, J.S. 2005. Role of circulating S-nitrosothiols in control of blood pressure. *Hypertension.* **45**:15–17.

71. Scharfstein, J.S., et al. 1994. In vivo transfer of nitric oxide between a plasma protein-bound reservoir and low molecular weight thiols. *J. Clin. Invest.* **94**:1432–1439.

72. Stamler, J.S., et al. 1992. Nitric oxide circulates in mammalian plasma primarily as an S-nitroso adduct of serum albumin. *Proc. Natl. Acad. Sci. U. S. A.* **89**:7674–7677.

73. Lipton, A.J., et al. 2001. S-nitrosothiols signal the ventilatory response to hypoxia. *Nature.* **413**:171–174.

74. Liu, L., et al. 2001. A metabolic enzyme for S-nitrosothiol conserved from bacteria to humans. *Nature.* **410**:490–494.

75. Singel, D.J., and Stamler, J.S. 2005. Chemical physiology of blood flow regulation by red blood cells: The role of nitric oxide and S-nitrosohemoglobin. *Annu. Rev. Physiol.* **67**:99–145.

76. Singel, D.J., and Stamler, J.S. 2004. Blood traffic control. *Nature.* **430**:297.

77. McMahon, T.J., et al. 2002. Nitric oxide in the human respiratory cycle. *Nat. Med.* **8**:711–717.

78. Datta, B., et al. 2004. Red blood cell nitric oxide as an endocrine vasoregulator - A potential role in congestive heart failure. *Circulation.* **109**:1339–1342.

79. Gow, A.J., and Stamler, J.S. 1998. Reactions between nitric oxide and haemoglobin under physiological conditions. *Nature.* **391**:169–173.

80. Wallace, W.J., Houtchens, R.A., Maxwell, J.C., and Caughey, W.S. 1982. Mechanism of autoxidation for hemoglobins and myoglobins - promotion of superoxide production by protons and anions. *J. Biol. Chem.* **257**:4966–4977.

81. Mansouri, A., and Winterhalter, K.H. 1974. Nonequivalence of chains in hemoglobin oxidation and oxygen binding. Effect of organic phosphates. *Biochemistry.* **13**:3311–3314.

82. Taketa, F., Antholine, W.E., and Chen, J.Y. 1978. Chain nonequivalence in binding of nitric-oxide to hemoglobin. *J. Biol. Chem.* **253**:5448–5451.

83. Vila-Petroff, M.G., Younes, A., Egan, J., Lakatta, E.G., and Sollott, S.J. 1999. Activation of distinct cAMP-dependent and cGMP-dependent pathways by nitric oxide in cardiac myocytes. *Circ. Res.* **84**:1020–1031.

84. Chesnais, J.M., Fischmeister, R., and Mery, P.F. 1999. Positive and negative inotropic effects of NO donors in atrial and ventricular fibres of the frog heart. *J. Physiol.* **518**:449–461.

85. Feron, O., Saldana, F., Michel, J.B., and Michel, T. 1998. The endothelial nitric-oxide synthase-caveolin regulatory cycle. *J. Biol. Chem.* **273**:3125–3128.

86. Sears, C., et al. 2003. Cardiac neuronal nitric oxide synthase isoform regulates myocardial contraction and calcium handling. *Circ. Res.* **92**:e52–e59.

87. Schuh, K., Uldrijan, S., Telkamp, M., Rothlein, N., and Neyses, L. 2001. The plasmamembrane calmodulin-dependent calcium pump: a major regulator of nitric oxide synthase I. *J. Cell Biol.* **155**:201–205.

88. Matsumoto, A., Comatas, K.E., Liu, L., and Stamler, J.S. 2003. Screening for nitric oxide-dependent protein-protein interactions. *Science.* **301**:657–661.

89. Barouch, L.A., et al. 2003. Combined loss of cardiac neuronal and endothelial nitric oxide synthase causes premature mortality and age-associated hypertropic cardiac remodeling in mice. *J. Mol. Cell. Cardiol.* **35**:637–644.

90. Cappola, T.P., et al. 2003. Deficiency of different nitric oxide synthase isoforms activates divergent transcriptional programs in cardiac hypertrophy. *Physiol. Genomics.* **14**:25–34.

91. Sears, C.E., et al. 2003. Myocardial NOS1 controls the lusitropic response to β-adrenergic stimulation in vivo and in vitro [abstract]. *Circulation.* **108**:IV-249.

92. Balligand, J.-L., et al. 1995. Nitric oxide-dependent parasympathetic signaling is due to activation of constitutive endothelial (type III) nitric oxide synthase in cardiac myocytes. *J. Biol. Chem.* **270**:14582–14586.

93. Gyurko, R., Kuhlencordt, P., Fishman, M.C., and Huang, P.L. 2000. Modulation of mouse cardiac function in vivo by eNOS and ANP. *Am. J. Physiol. Heart Circ. Physiol.* **278**:H971–H981.

94. Damy, T., et al. 2003. Up-regulation of cardiac nitric oxide synthase 1-derived nitric oxide after myocardial infarction in senescent rats. *FASEB J.* **17**:1934–1936.

95. Damy, T., et al. 2004. Increased neuronal nitric oxide synthase-derived NO production in the failing human heart. *Lancet.* **363**:1365–1367.

96. Brenman, J.E., et al. 1996. Interaction of nitric oxide synthase with the postsynaptic density protein PSD-95 and α1-syntrophin mediated by PDZ domains. *Cell.* **84**:757–767.

97. Feron, O., et al. 1998. Modulation of endothelial nitric-oxide synthase-caveolin interaction in cardiac myocytes. *J. Biol. Chem.* **273**:30249–30254.

98. Massion, P.B., Feron, O., Dessy, C., and Balligand, J.L. 2003. Nitric oxide and cardiac function - Ten years after, and continuing. *Circ. Res.* **93**:388–398.

99. Hilenski, L.L., Clempus, R.E., Quinn, M.T., Lambeth, J.D., and Griendling, K.K. 2004. Distinct subcellular localizations of Nox1 and Nox4 in vascular smooth muscle cells. *Arterioscler. Thromb. Vasc. Biol.* **24**:677–683.

100. Mollnau, H., et al. 2002. Effects of angiotensin II infusion on the expression and function of NAD(P)H oxidase and components of nitric oxide/cGMP signaling. *Circ. Res.* **90**:E58–E65.

Genetic causes of human heart failure

Hiroyuki Morita, Jonathan Seidman, and Christine E. Seidman

Department of Genetics, Harvard Medical School, Division of Cardiology, Brigham and Women's Hospital, and
Howard Hughes Medical Institute, Boston, Massachusetts, USA.

Factors that render patients with cardiovascular disease at high risk for heart failure remain incompletely defined. Recent insights into molecular genetic causes of myocardial diseases have highlighted the importance of single-gene defects in the pathogenesis of heart failure. Through analyses of the mechanisms by which a mutation selectively perturbs one component of cardiac physiology and triggers cell and molecular responses, studies of human gene mutations provide a window into the complex processes of cardiac remodeling and heart failure. Knowledge gleaned from these studies shows promise for defining novel therapeutic targets for genetic and acquired causes of heart failure.

Introduction

Despite considerable public awareness and technological advances that foster early diagnosis and aggressive therapeutic interventions for cardiovascular disease, heart failure, a pathophysiologic state in which blood delivery is inadequate for tissue requirements, remains a critical unsolved problem. Reaching almost epidemic proportions, heart failure currently affects 4.8 million Americans, and each year over 500,000 new cases are diagnosed. These statistics are associated with considerable social and economic costs. In 2003 heart failure contributed to over 280,000 deaths and accounted for 17.8 billion health care dollars (1).

Heart failure is a complex pathophysiologic state in which delivery of blood and nutrients is inadequate for tissue requirements. Heart failure almost universally arises in the context of antecedent cardiovascular disease: atherosclerosis, cardiomyopathy, myocarditis, congenital malformations, or valvular disease. Knowledge of why some patients with these conditions are at high risk for heart failure is incomplete, and deciphering critical mediators from nonspecific factors is often difficult in the context of underlying heart disease. The study of single-gene mutations that trigger heart failure provides an opportunity for defining important molecules involved in these processes. Although these monogenic disorders account for only a small subset of overall heart failure cases, insights into the responses triggered by gene mutations are likely to also be relevant to more common etiologies of heart failure.

The earliest clinical manifestation of an underlying genetic cause of heart failure is ventricular remodeling. One of 2 distinct morphologies occurs: left ventricular hypertrophy (increased wall thickness without chamber expansion) or dilation (normal or thinned walls with enlarged chamber volumes). Each is associated with specific hemodynamic changes. Systolic function is normal, but diastolic relaxation is impaired in hypertrophic remodeling; diminished systolic function characterizes dilated remodeling. Clinical recognition of these cardiac findings usually prompts diagnosis of hypertrophic cardiomyopathy (HCM) or dilated cardiomyopathy (DCM). These were once considered as specific diagnoses, but there is now considerable evidence that many different gene mutations can cause these pathologies (Figure 1), and with these discoveries has come recognition of distinct histopathologic features that further delineate several subtypes of remodeling. While compilation of the full repertoire of genes that remodel the heart remains a work in progress, the current compendia already suggest a multiplicity of pathways by which the human heart can fail. To facilitate a discussion, we have grouped known cardiomyopathy genes according to the probable functional consequences of mutations on force generation and transmission, metabolism, calcium homeostasis, or transcriptional control. This approach, while convenient for organization, is undoubtedly somewhat arbitrary; gene mutations in one functional category inevitably have an impact on multiple myocyte processes. Indeed, the eventual delineation of signals between functional groups may be critical to understanding nodal points that result in, or protect against, cardiac decompensation and heart failure development.

Force generation and propagation

Generation of contractile force by the sarcomere and its transmission to the extracellular matrix are the fundamental functions of heart cells. Inadequate performance in either component prompts cardiac remodeling (hypertrophy or dilation), produces symptoms, and leads to heart failure. Given the importance of these processes for normal heart function and overt clinical manifestations of deficits in either force generation or transmission, it is not surprising that more single-gene mutations have been identified in molecules involved in these critical processes than in those of other functional classes (Figure 2).

Sarcomere protein mutations. Human mutations in the genes encoding protein components of the sarcomere cause either HCM or DCM. While progression to heart failure occurs with both patterns of remodeling, the histopathology, hemodynamic profiles, and biophysical consequences of HCM or DCM mutations suggest that distinct molecular processes are involved.

Over 300 dominant mutations in genes encoding β-cardiac myosin heavy chain (*MYH7*), cardiac myosin-binding protein-C (*MYBPC3*), cardiac troponin T (*TNNT2*), cardiac troponin I (*TNNI3*), essential myosin light chain (*MYL3*), regulatory myosin light chain (*MYL2*), α-tropomyosin (*TPM1*), cardiac actin (*ACTC*), and titin (*TTN*) have been reported to cause HCM (Figure 2) (2, 3). Recent reports of comprehensive sequencing of sarcomere protein genes in diverse patient populations indicate that *MYBPC3* and *MYH7* mutations are most frequent (4, 5). Sarcomere gene mutations that

Nonstandard abbreviations used: ARVD, arrhythmogenic right ventricular cardiomyopathy; DCM, dilated cardiomyopathy; HCM, hypertrophic cardiomyopathy; HDAC, histone deacetylase; K_{ATP} channel, cardiac ATP-sensitive potassium channel; LAMP2, lysosome-associated membrane protein-2; MLP, muscle LIM protein; NFAT, nuclear factor of activated T cells; PLN, phospholamban; RXRα, retinoid X receptor α; RyR2, cardiac ryanodine receptor; SERCA2a, cardiac sarcoplasmic reticulum Ca^{2+}-ATPase pump; SR, sarcoplasmic reticulum; ZASP, Z-band alternatively spliced PDZ-motif protein.

Conflict of interest: The authors have declared that no conflict of interest exists.

Citation for this article: *J. Clin. Invest.* **115**:518–526 (2005).
doi:10.1172/JCI200524351.

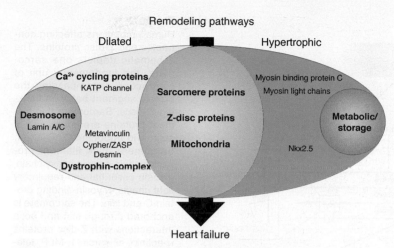

Figure 1
Human gene mutations can cause cardiac hypertrophy (blue), dilation (yellow), or both (green). In addition to these two patterns of remodeling, particular gene defects produce hypertrophic remodeling with glycogen accumulation (pink) or dilated remodeling with fibrofatty degeneration of the myocardium (orange). Sarcomere proteins denote β-myosin heavy chain, cardiac troponin T, cardiac troponin I, α-tropomyosin, cardiac actin, and titin. Metabolic/storage proteins denote AMP-activated protein kinase γ subunit, LAMP2, lysosomal acid α 1,4–glucosidase, and lysosomal hydrolase α-galactosidase A. Z-disc proteins denote MLP and telethonin. Dystrophin-complex proteins denote δ-sarcoglycan, β-sarcoglycan, and dystrophin. Ca^{2+} cycling proteins denote PLN and RyR2. Desmosome proteins denote plakoglobin, desmoplakin, and plakophilin-2.

cause HCM produce a shared histopathology with enlarged myocytes that are disorganized and die prematurely, which results in increased cardiac fibrosis. The severity and pattern of ventricular hypertrophy, age at onset of clinical manifestations, and progression to heart failure are in part dependent on the precise sarcomere protein gene mutation. For example, TNNT2 mutations are generally associated with a high incidence of sudden death despite only mild left ventricular hypertrophy (6, 7). While only a small subset (10–15%) of HCM patients develop heart failure, this end-stage phenotype has a markedly poor prognosis and often necessitates cardiac transplantation. Accelerated clinical deterioration has been observed with *MYH7* Arg719Trp, *TNNT2* Lys273Glu, *TNNI3* Lys-183del, and *TPM1* Glu180Val mutations (8–11).

Most HCM mutations encode defective polypeptides containing missense residues or small deletions; these are likely to be stably incorporated into cardiac myofilaments and to produce hypertrophy because normal sarcomere function is disturbed. Many HCM mutations in *MYBPC3* fall within carboxyl domains that interact with titin and myosin; however, the exact biophysical properties altered by these defects remain unknown (Figure 2). HCM mutations in myosin are found in virtually every functional domain, which suggests that the biophysical consequences of these defects may vary. Genetic engineering of some human myosin mutations into mice has indicated more consistent sequelae. Isolated single-mutant myosin molecules containing different HCM mutations had increased actin-activated ATPase activity and showed greater force production and faster actin-filament sliding, biophysical properties that may account for hyperdynamic contractile performance observed in HCM hearts and that suggest a mechanism for premature myocyte death in HCM (12–14). Uncoordinated contraction due to heterogeneity of mutant and normal sarcomere proteins, increased energy consumption, and changes

in Ca^{2+} homeostasis could diminish myocyte survival and trigger replacement fibrosis. With insidious myocyte loss and increased fibrosis, the HCM heart transitions from hypertrophy to failure.

Mice that are engineered to carry a sarcomere mutation replicate the genetics of human disease; heterozygous mutations cause HCM. One exception is a deletion of proximal myosin-binding protein-C sequences; heterozygous mutant mice exhibited normal heart structure while homozygous mutant mice developed hypertrophy (15). Remarkably, while most heterozygous mouse models with a mutation in myosin heavy chain, myosin-binding protein-C, or troponin T developed HCM (16–18), homozygous mutant mice (19, 20) developed DCM with fulminant heart failure and, in some cases, premature death. These mouse studies might indicate that HCM, DCM, and heart failure reflect gradations of a single molecular pathway. Alternatively, significant myocyte death caused by homozygous sarcomere mutations may result in heart failure. Human data suggest a more complicated scenario. The clinical phenotype of rare individuals who carry homozygous sarcomere mutations in either *MYH7* (21) or in *TNNT2* (22) is severe hypertrophy, not DCM. Furthermore, individuals with compound heterozygous sarcomere mutations exhibit HCM, not DCM. The absence of ventricular dilation in human hearts with 2 copies of mutant sarcomere proteins is consistent with distinct cellular signaling programs that remodel the heart into hypertrophic or dilated morphologies.

DCM sarcomere protein gene mutations affect distinct amino acids from HCM-causing mutations, although the proximity of altered residues is remarkable. The histopathology of sarcomere DCM mutations is quite different from those causing HCM, and is remarkably nonspecific. Degenerating myocytes with increased interstitial fibrosis are present, but myocyte disarray is notably absent. There are 2 mechanisms by which sarcomere mutations may cause DCM and heart failure: deficits of force production and deficits of force transmission. Diminished force may occur in myosin mutations (e.g., *MYH7* Ser532Pro) that alter actin-binding residues involved in initiating the power stroke of contraction. Impaired contractile force may also occur in DCM troponin mutations (*TNNT2* ΔLys210, ref. 23; and *TNNI3* Ala2Val, ref. 24) that alter residues implicated in tight binary troponin interactions. Because troponin molecules modulate calcium-stimulated actomyosin ATPase activity, these defects may cause inefficient ATP hydrolysis and therein decrease contractile power.

Other DCM sarcomere mutations are more likely to impair force transmission (Figure 2). For example, a myosin mutation (at residue 764) located within the flexible fulcrum that transmits movement from the head of myosin to the thick filament is likely to render ineffectual the force generated by actomyosin interactions (23). DCM *TPM1* mutations (25) are predicted to destabilize actin interactions and compromise force transmission to neighboring sarcomere. Likewise, *ACTC* mutations (26) that impair binding of actin to Z-disc may compromise force propagation. *TTN* mutations provide quintessential evidence that deficits in force transmission cause DCM and heart failure. By spanning the sarcomere from Z-disc to M-line, this giant muscle protein assembles contractile filaments and provides elasticity through serial spring elements. Titin interacts with α-actinin and telethonin (T-cap) at the Z-disc, with calpain3

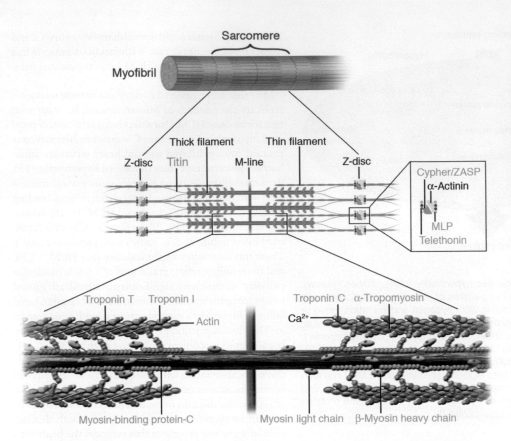

Figure 2
Human mutations affecting contractile and Z-disc proteins. The schematic depicts one sarcomere, the fundamental unit of contraction encompassing the protein segment between flanking Z discs. Sarcomere thin filament proteins are composed of actin and troponins C, T, and I. Sarcomere thick filament proteins include myosin heavy chain, myosin essential and regulatory light chains, myosin-binding protein-C and titin. The sarcomere is anchored through titin and actin interactions with Z disc proteins α-actinin, calsarcin-1, MLP, telethonin (T-cap), and ZASP. Human mutations (orange text) in contractile proteins and Z-disc proteins can cause HCM or DCM.

and obscurin at the I-band (the extensible thin filament regions flanking Z-discs), and with myosin-binding protein-C, calmodulin, and calpain3 at the M-line region. Human mutations identified in the Z-disc–I-band transition zone (27), in the telethonin and α-actinin–binding domain, and in the cardiac-specific N2B domain (an I-band subregion; ref. 28) each cause DCM and heart failure.

Intermediate filaments and dystrophin-associated glycoprotein mutations. Intermediate filaments function as cytoskeletal proteins linking the Z-disc to the sarcolemma. Desmin is a type III intermediate filament protein, which, when mutated, causes skeletal and cardiac muscle disease (Figure 3). The hearts of mice deficient in desmin (29) are more susceptible to mechanical stress, which is consistent with the function of intermediate proteins in force transmission.

Through dystrophin and actin interactions, the dystrophin-associated glycoprotein complex (composed of α- and β-dystroglycans, α-, β-, γ- and δ-sarcoglycans, caveolin-3, syntrophin, and dystrobrevin) provides stability to the sarcomere and transmits force to the extracellular matrix. Human mutations in these proteins cause muscular dystrophy with associated DCM and heart failure (Figure 3). Skeletal muscle manifestations can be minimal in female carriers of X-linked dystrophin defects, and some individuals present primarily with heart failure (30). In the mouse experiment, coxsackievirus B3–encoded protease2A, which can cleave dystrophin, was shown to produce sarcolemmal disruption and cause DCM, which suggests that dystrophin is also involved in the pathologic mechanism of DCM and heart failure that follow viral myocarditis (31).

While deficiencies of proteins that link the sarcomere to the extracellular matrix are likely to impair force transmission, recent studies of mice engineered to carry mutations in these molecules indicate other mechanisms for heart failure. A model of desmin-related cardiomyopathies (32) uncovered striking intracellular aggresomes,

electron dense accumulations of heat shock and chaperone protein, α-B-crystalline, desmin, and amyloid in association with sarcomeres. While particularly abundant in the amyloid heart, aggresomes were also found in some DCM and HCM specimens, which suggests that excessive degenerative processing induced by myocyte stress or gene mutation may be toxic to sarcomere function.

Analyses of δ-sarcoglycan null mice (33) also yielded unexpected disease mechanisms, primary coronary vasospasm and myocardial ischemia. Selective restoration of δ-sarcoglycan to the cardiac myocytes extinguished this pathology, thereby implicating chronic ischemia as a contributing factor to heart failure development in patients with sarcoglycan mutations.

Mutations in intercalated and Z-disc proteins. To generate contraction, one end of each actin thin filament must be immobilized. The Z-disc defines the lateral boundary of the sarcomere, where actin filaments, titin, and nebulette filaments are anchored. Metavinculin provides attachment of thin filaments to the plasma membrane and plays a key role in productive force transmission. Two metavinculin gene mutations cause DCM by disruption of disc structure and actin-filament organization (34).

Other Z-disc protein constituents may also function as mechano-stretch receptors (35). Critical components include α-actinin, which aligns actin and titin from neighboring sarcomeres and interacts with muscle LIM protein (MLP encoded by *CSRP3*), telethonin (encoded by *TCAP*), which interacts with titin and MLP to subserve overall sarcomere function, and Cypher/Z-band alternatively spliced PDZ-motif protein (Cypher/ZASP), a striated muscle-restricted protein that interacts with α-actinin–2 through a PDZ domain and couples to PKC-mediated signaling via its LIM domains (Figure 2). Mutations in these molecules cause either DCM (35, 36) or HCM (37, 38) and predispose the affected

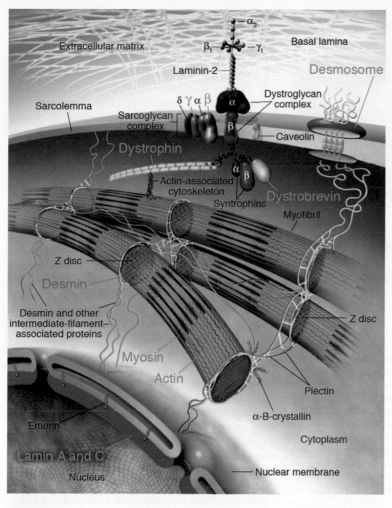

Figure 3
Human mutations (orange text) in components of myocyte cytoarchitecture cause DCM and heart failure. Force produced by sarcomeric actin-myosin interactions is propagated through the actin cytoskeleton and dystrophin to the dystrophin-associated glycoprotein complex (composed of α- and β-dystroglycans, α-, β-, γ- and δ-sarcoglycans, caveolin-3, syntrophin, and dystrobrevin). Desmosome proteins plakoglobin, desmoplakin, and plakophilin-2, provide functional and structural contacts between adjacent cells and are linked through intermediate filament proteins, including desmin, to the nuclear membrane, where lamin A/C is localized. Adapted from ref. 96.

individuals to heart failure. Genetically engineered mice with MLP deficiency (39) help to model the mechanism by which mutations in distinct proteins cause disease. Without MLP, telethonin is destabilized and gradually lost from the Z-disc; as a consequence, MLP-deficient cardiac papillary muscle shows an impairment in tension generation following the delivery of a 10% increase in passive stretch of the muscle and a loss of stretch-dependent induction of molecular markers (e.g., brain natriuretic peptide), which suggests that an MLP-telethonin–titin complex is an essential component of the cardiac muscle mechanical stretch sensor machinery. An important question is how signaling proteins (e.g., Cyper/ZASP) within the Z-disc translate mechanosensing into activation of survival or cell death pathways.

Lamin A/C mutations. The inner nuclear-membrane protein complex contains emerin and lamin A/C. Defects in emerin cause X-linked Emery-Dreifuss muscular dystrophy, joint con-

tractures, conduction system disease, and DCM. Dominant *lamin A/C* mutations exhibit a more cardiac-restricted phenotype with fibrofatty degeneration of the myocardium and conducting cells, although subclinical involvement of skeletal muscles and contractures are sometimes apparent. The remarkable electrophysiologic deficits (progressive atrioventricular block and atrial arrhythmias) observed in mutations of lamin A/C and emerin indicate the particular importance of these proteins in electrophysiologic cells. A recent study of lamin A/C mutant mice showed evidence of marked nuclear deformation, fragmentation of heterochromatin, and defects in mechanotransduction (40, 41), all of which likely contribute to reduced myocyte viability. The similarities of cardiac histopathology (fibrofatty degeneration) observed in mutations of the nuclear envelope and desmosomes raise the possibility that these structures may both function as important mechanosensors in myocytes (Figure 3).

Desmosome protein mutations. Arrhythmogenic right ventricular cardiomyopathy (ARVD) identifies an unusual group of cardiomyopathies characterized by progressive fibrofatty degeneration of the myocardium, electrical instability, and sudden death (42). While right ventricular dysplasia predominates, involvement of the left ventricle also occurs. Progressive myocardial dysfunction is seen late in the course of disease, often with right-sided heart failure. ARVD occurs in isolation or in the context of Naxos syndrome, an inherited syndrome characterized by prominent skin (palmar-plantar keratosis), hair, and cardiac manifestations. Mutations in protein components of the desmosomes (Figure 3) (plakoglobin, ref. 43; desmoplakin, refs. 44, 45; and plakophilin-2, ref. 46) and in the cardiac ryanodine receptor (RyR2) (ref. 47; discussed below) cause syndromic and nonsyndromic ARVD. Desmosomes are organized cell membrane structures that provide functional and structural contacts between adjacent cells and that may be involved in signaling processes. Whether mutations in the desmosomal proteins render cells of the heart (and skin) inappropriately sensitive to normal mechanical stress or cause dysplasia via another mechanism is unknown.

Energy production and regulation

Mitochondrial mutations. Five critical multiprotein complexes, located within the mitochondria, synthesize ATP by oxidative phosphorylation. While many of the protein components of these complexes are encoded by the nuclear genome, 13 are encoded by the mitochondrial genome. Unlike nuclear gene mutations, mitochondrial gene mutations exhibit matrilineal inheritance. In addition, the mitochondrial genome is present in multiple copies, and mutations are often heteroplasmic, affecting some but not all copies. These complexities, coupled with the dependence of virtually all tissues on mitochondrial-derived energy supplies, account for the considerable clinical diversity of mitochondrial gene mutations (Figure 4). While most defects cause either dilated or hypertrophic cardiac remodeling in the context of mitochondrial syndromes such as Kearns-Sayre syndrome, ocular myopathy, mitochondrial encephalomyopathy with lactic-acidosis and stroke-like episodes

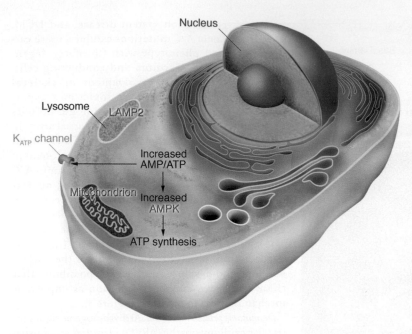

Figure 4

Human gene mutations affecting cardiac energetics and metabolism. Energy substrate utilization is directed by critical metabolic sensors in myocytes, including AMP-activated protein kinase (AMPK), which, in response to increased AMP/ATP levels, phosphorylates target proteins and thereby regulates glycogen and fatty acid metabolism, critical energy sources for the heart. Glycogen metabolism involves a large number of proteins including α-galactosidase A (mutated in Fabry disease) and LAMP2 (mutated in Danon disease). Glycogen and fatty acids are substrates for multiprotein complexes located within the mitochondria for the synthesis of ATP. K_{ATP} channels composed of an enzyme complex and a potassium pore participate in decoding metabolic signals to maximize cellular functions during stress adaptation. Human mutations (orange text) that cause cardiomyopathies have been identified in the regulatory SUR2A subunit of K_{ATP}, the γ2 subunit of AMPK, mitochondrial proteins, α-galactosidase A, and LAMP2.

(MELAS), and myoclonus epilepsy with ragged-red fibers (MERFF) (48), there is some evidence that particular mitochondrial mutations can produce predominant or exclusive cardiac disease (49, 50). An association between heteroplasmic mitochondrial mutations and DCM has been recognized (51).

Nuclear-encoded metabolic mutations. Nuclear gene mutations affecting key regulators of cardiac metabolism are emerging as recognized causes of hypertrophic cardiac remodeling and heart failure (Figure 4). Mutations in genes encoding the γ2 subunit of AMP-activated protein kinase (*PRKAG2*), α-galactosidase A (*GLA*), and lysosome-associated membrane protein-2 (*LAMP2*) can cause profound myocardial hypertrophy in association with electrophysiologic defects (52). AMP-activated protein kinase functions as a metabolic-stress sensor in all cells. This heterotrimeric enzyme complex becomes activated during energy-deficiency states (low ATP, high ADP) and modulates (by phosphorylation) a large number of proteins involved in cell metabolism and energy (53). Most *GLA* mutations can cause multisystem classic Fabry disease (angiokeratoma, corneal dystrophy, renal insufficiency, acroparesthesia, and cardiac hypertrophy), but some defects produce primarily cardiomyopathy. *LAMP2* mutations can also produce either multisystem Danon disease (with skeletal muscle, neurologic, and hepatic manifestations) or a more restricted cardiac phenotype.

Cardiac histopathology reveals that, unlike sarcomere gene mutations, which cause hypertrophic remodeling, the mutations in *PRKAG2*, *LAMP2*, and *GLA* accumulate glycogen in complexes with protein and/or lipids, thereby defining these pathologies as storage cardiomyopathies. Progression from hypertrophy to heart failure is particularly common and occurs earlier with *LAMP2* mutations than with other gene mutations that cause metabolic cardiomyopathies. Since both *GLA* and *LAMP2* are encoded on chromosome X, disease expression is more severe in men, but heterozygous mutations in women are not entirely benign, perhaps due to X-inactivation that equally extinguishes a normal or mutant allele. The cellular and molecular pathways that produce either profound hypertrophy or progression to heart failure from *PRKAG2*, *GLA*, or *LAMP2* mutations are incompletely understood. While accumulated byproducts are likely to produce toxicity, animal models indicate that mutant proteins cause far more profound consequences by changing cardiac metabolism and altering cell signaling. This is particularly evident in *PRKAG2* mutations that increase glucose uptake by stimulating translocation of the glucose transporter GLUT-4 to the plasma membrane, increase hexokinase activity, and alter expression of signaling cascades (54).

The cooccurrence of electrophysiologic defects in metabolic mutations raises the possibility that pathologic cardiac conduction and arrhythmias contribute to cardiac remodeling and heart failure in these gene mutations. One mechanism for electrophysiologic defects appears to be the direct consequence of storage: transgenic mice that express a human *PRKAG2* mutation (55) developed ventricular pre-excitation due to pathologic atrioventricular connections by glycogen-filled myocytes that ruptured the annulus fibrosis (the normal anatomic insulator which separates atrial and ventricular myocytes). A second and unknown mechanism may be that these gene defects are particularly deleterious to specialized cells of the conduction system. Little is known about the metabolism of these cells, although historical histopathologic data indicate glycogen to be particularly more abundant in the conduction system than in the working myocardium (56–58).

Ca²⁺ cycling

Considerable evidence indicates the presence of abnormalities in myocyte calcium homeostasis to be a prevalent and important mechanism for heart failure. Protein and RNA levels of key calcium modulators are altered in acquired and inherited forms of heart failure, and human mutations in molecules directly involved in calcium cycling have been found in several cardiomyopathies (Figure 5).

Calcium enters the myocyte through voltage-gated L-type Ca²⁺ channels; this triggers release of calcium from the sarcoplasmic reticulum (SR) via the RyR2. Emerging data define FK506-binding protein (FKBP12.6; calstabin2) as a critical stabilizer of RyR2 function (59), preventing aberrant calcium release during the relaxation phase of the cardiac cycle (Figure 5). Stimuli that phosphorylate RyR2 (such as exercise) by protein kinase A (PKA) dissociate calstabin2 from the receptor, thereby increasing calcium release and enhancing contractility. At low concentrations of intracellular calcium, troponin I and actin interactions block actomyosin ATPase activity; increasing levels foster calcium binding

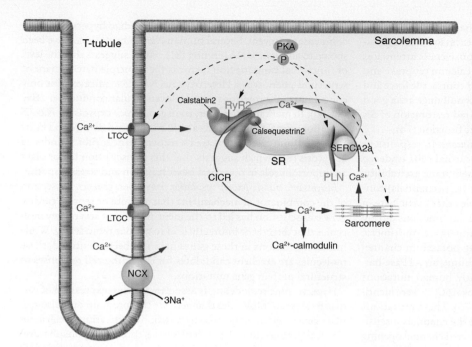

Figure 5
Human mutations affecting Ca²⁺ cycling proteins. Intracellular Ca²⁺ handling is the central coordinator of cardiac contraction and relaxation. Ca²⁺ entering through L-type channels (LTCC) triggers Ca²⁺ release (CICR) from the SR via the RyR2, and sarcomere contraction is initiated. Relaxation occurs with SR Ca²⁺ reuptake through the SERCA2a. Calstabin2 coordinates excitation and contraction by modulating RyR2 release of Ca²⁺. PLN, an SR transmembrane inhibitor of SERCA2a modulates Ca²⁺ reuptake. Dynamic regulation of these molecules is effected by PKA-mediated phosphorylation. Ca²⁺ may further function as a universal signaling molecule, stimulating Ca²⁺-calmodulin and other molecular cascades. Human mutations (orange text) in molecules involved in calcium cycling cause cardiac remodeling and heart failure. NCX, sodium/calcium exchanger.

to troponin C, which releases troponin I inhibition and stimulates contraction. Cardiac relaxation occurs when calcium dissociates from troponin C, and intracellular concentrations decline as calcium reuptake into the SR occurs through the cardiac sarcoplasmic reticulum Ca²⁺-ATPase pump (SERCA2a). Calcium reuptake into SR is regulated by phospholamban (PLN), an inhibitor of SERCA2a activity that when phosphorylated dissociates from SERCA2a and accelerates ventricular relaxation.

RyR2 mutations. While some mutations in the *RyR2* are reported to cause ARVD (47) (see discussion of desmosome mutations), defects in this calcium channel are more often associated with catecholaminergic polymorphic ventricular tachycardia (60, 61), a rare inherited arrhythmic disorder characterized by normal heart structure and sudden cardiac death during physical or emotional stress. Mutations in calsequestrin2, an SR calcium-binding protein that interacts with RyR2, also cause catecholaminergic polymorphic ventricular tachycardia (62, 63). Whether the effect of *calsequestrin2* mutations directly or indirectly alters RyR2 function is unknown (Figure 5).

While *RyR2* mutations affect residues in multiple functional domains of the calcium channel, those affecting residues involved in calstabin2-binding provide mechanistic insights into the substantial arrhythmias found in affected individuals. Mutations that impair calstabin2-binding may foster calcium leak from the SR and trigger depolarization. Diastolic calcium leak can also affect excitation-contraction coupling and impair systolic contractility.

Studies of mice deficient in *FKBP12.6* (64) confirmed the relevance of SR calcium leak from RyR2 to clinically important arrhythmias. RyR2 channel activity in *FKBP12.6*-null mice was significantly increased compared with that of wild-type mice, consistent with a diastolic Ca²⁺ leak. Mutant myocytes demonstrated delayed afterdepolarizations, and exercise-induced syncope, ventricular arrhythmias, and sudden death were observed in *FKBP12.6*-null mice.

Calcium dysregulation is also a component of hypertrophic remodeling that occurs in sarcomere gene mutations. Calcium cycling is abnormal early in the pathogenesis of murine HCM (65, 66): SR calcium stores are decreased and calcium-binding proteins and

RyR2 levels are diminished. Whether calcium changes contribute to ventricular arrhythmias in mouse and human HCM remains an intriguing question.

Related mechanisms may contribute to ventricular dysfunction and arrhythmias in acquired forms of heart failure, in which chronic phosphorylation of RyR2 reduces calstabin2 levels in the channel macromolecular complex and increases calcium loss from SR stores. These data indicate the potential benefit of therapeutics that improve calstabin2-mediated stabilization of RyR2 (67, 68); such agents may both improve ventricular contractility and suppress arrhythmias in heart failure.

PLN mutations. Rare human *PLN* mutations cause familial DCM and heart failure (69, 70). The pathogenetic mechanism of one mutation (*PLN* Arg9Cys) was elucidated through biochemical studies, which indicated unusual PKA interactions that inhibited phosphorylation of mutant and wild-type PLN. The functional consequence of the mutation was predicted to be constitutive inhibition of SERCA2a, a result confirmed in transgenic mice expressing mutant, but not wild-type, PLN protein. In mutant transgenic mice, calcium transients were markedly prolonged, myocyte relaxation was delayed, and these abnormalities were unresponsive to β-adrenergic stimulation. Profound biventricular cardiac dilation and heart failure developed in mutant mice, providing clear evidence of the detrimental effects of protracted SERCA2a inhibition due to excess PLN activity.

The biophysical consequences accounting for DCM in humans who are homozygous for a *PLN* null mutation (Leu39stop; ref. 70) are less clear. PLN-deficient mice show increased calcium reuptake into the SR and enhanced basal contractility (71). Indeed, these effects on calcium cycling appear to account for the mechanism by which *PLN* ablation rescues DCM in *MLP*-null mice (72). However, normal responsiveness to β-adrenergic stimulation is blunted in PLN-deficient myocytes, and cells are less able to recover from acidosis that accompanies vigorous contraction or pathologic states, such as ischemia (73). The collective lesson from human *PLN* mutations appears to be that too little or too much PLN activity is bad for long-term heart function.

Acquired causes of heart failure are also characterized by a relative decrease in SERCA2a function due to excessive PLN inhibition. Downregulation of β-adrenergic responsiveness attenuates PLN phosphorylation, which compromises calcium reuptake and depletes SR calcium levels, which may impair contractile force and enhance arrhythmias. Heterozygote *SERCA2* null mice are a good model of this phenotype and exhibit impaired restoration of SR calcium with deficits in systolic and diastolic function (74).

Cardiac ATP-sensitive potassium channel mutations. In response to stress such as hypoxia and ischemia, myocardial cells undergo considerable changes in metabolism and membrane excitability. Cardiac ATP-sensitive potassium channels (K_{ATP} channels) contain a potassium pore and an enzyme complex that participate in decoding metabolic signals to maximize cellular functions during stress adaptation (Figure 4) (75). K_{ATP} channels are multimeric proteins containing the inwardly rectifying potassium channel pore (Kir6.2) and the regulatory SUR2A subunit, an ATPase-harboring, ATP-binding cassette protein. Recently, human mutations in the regulatory SUR2A subunit (encoded by *ABCC9*) were identified as a cause of DCM and heart failure (76). These mutations reduced ATP hydrolytic activities, rendered the channels insensitive to ADP-induced conformations, and affected channel opening and closure. Since K_{ATP}-null mouse hearts have impaired response to stress and are susceptible to calcium overload (75), some of the pathophysiology of human K_{ATP} mutations (DCM and arrhythmias) may reflect calcium increases triggered by myocyte stress.

Transcriptional regulators

Investigation of the molecular controls of cardiac gene transcription has led to the identification of many key molecules that orchestrate physiologic expression of proteins involved in force production and transmission, metabolism, and calcium cycling. Given that mutation in the structural proteins involved in these complex processes is sufficient to cause cardiac remodeling, it is surprising that defects in transcriptional regulation of these same proteins have not also been identified as primary causes of heart failure. Several possible explanations may account for this. Transcription factor gene mutations may be lethal or may at least substantially impair reproductive fitness so as to be rapidly lost. The consequences of transcription factor gene mutations may be so pleiotropic that these cause systemic rather than single-organ disease. Changes in protein function (produced by a structural protein mutation) may be more potent for remodeling than changes in levels of structural protein (produced by transcription factor mutation). While many other explanations may be relevant, the few human defects discovered in transcriptional regulators that cause heart failure provide an important opportunity to understand molecular mechanisms for heart failure.

Nkx2.5 mutations. The homeodomain-containing transcription factor Nkx2.5, a vertebrate homolog of the Drosophila homeobox gene *tinman*, is one of the earliest markers of mesoderm. When *Nkx2.5* is deleted in the fly, cardiac development is lost (77). Targeted disruption of *Nkx2.5* in mice (*Nkx2.5⁻/⁻*) causes embryonic lethality due to the arrested looping morphogenesis of the heart tube and growth retardation (78, 79). Multiple human dominant *Nkx2.5* mutations have been identified as causing primarily structural malformations (atrial and ventricular septation defects) accompanied by atrioventricular conduction delay, although cardiac hypertrophic remodeling has also been observed (80). Although the mechanism for ventricular hypertrophy in humans with *Nkx2.5* mutations is not fully understood, the pathology is unlike that found

in HCM, which perhaps indicates that cardiac hypertrophy is a compensatory event. Several human *Nkx2.5* mutations have been shown to abrogate DNA binding (81), which suggests that the level of functional transcription factor is the principle determinant of structural phenotypes. Heterozygous *Nkx2.5⁺/⁻* mice exhibit only congenital malformations with atrioventricular conduction delay (82, 83). Remarkably, however, transgenic mice expressing *Nkx2.5* mutations develop profound cardiac conduction disease and heart failure (84) and exhibit increased sensitivity to doxorubicin-induced apoptosis (85), which suggests that this transcription factor plays an important role in postnatal heart function and stress response.

Insights into transcriptional regulation from mouse genetics. Dissection of the combinatorial mechanisms that activate or repress cardiac gene transcription has led to the identification of several key molecules that directly or indirectly lead to cardiac remodeling. While human mutations in these genes have not been identified, these molecules are excellent candidates for triggering cell responses to structural protein gene mutations.

Hypertrophic remodeling is associated with reexpression of cardiac fetal genes. Molecules that activate this program may also regulate genes that directly cause hypertrophy. Activation of calcineurin (Ca^{2+}/calmodulin-dependent serine/threonine phosphatase) results in dephosphorylation and nuclear translocation of nuclear factor of activated T cells 3 (NFAT3), which, in association with the zinc finger transcription factor GATA4, induces cardiac fetal gene expression. Transgenic mice that express activated calcineurin or NFAT3 in the heart develop profound hypertrophy and progressive decompensation to heart failure (86), responses that were prevented by pharmacologic inhibition of calcineurin. Although these data implicated NFAT signaling in hypertrophic heart failure, pharmacologic inhibition of this pathway fails to prevent hypertrophy caused by sarcomere gene mutations in mice and even accelerates disease progression to heart failure (65). Mice lacking calsarcin-1, which is localized with calcineurin to the Z-disc, showed an increase in Z-disc width, marked activation of the fetal gene program, and exaggerated hypertrophy in response to calcineurin activation or mechanical stress, which suggests that calsarcin-1 plays a critical role in linking mechanical stretch sensor machinery to the calcineurin-dependent hypertrophic pathway (87).

Histone deacetylases (HDACs) are emerging as important regulators of cardiac gene transcription. Class II HDACs (4/5/7/9) bind to the cardiac gene transcription factor MEF2 and inhibit MEF2-target gene expression. Stress-responsive HDAC kinases continue to be identified but may include an important calcium-responsive cardiac protein, calmodulin kinase. Kinase-induced phosphorylation of class II HDACs causes nuclear exit, thereby releasing MEF2 for association with histone acetyltransferase proteins (p300/CBP) and activation of hypertrophic genes. Mice deficient in HDAC9 are sensitized to hypertrophic signals and exhibit stress-dependent cardiac hypertrophy. The discovery that HDAC kinase is stimulated by calcineurin (88) implicates crosstalk between these hypertrophic signaling pathways.

Recent attention has also been focused on Hop, an atypical homeodomain-only protein that lacks DNA-binding activity. Hop is expressed in the developing heart, downstream of Nkx2-5. While its functions are not fully elucidated, Hop can repress serum response factor–mediated (SRF-mediated) transcription. Mice with *Hop* gene ablation have complex phenotypes. Approximately half of *Hop*-null embryos succumb during mid-gestation with poorly developed myocardium; some have myocardial rupture and pericardial effusion. Other *Hop*-null embryos survive to adulthood with apparently normal heart structure and function. Cardiac transgenic overexpression of epitope-tagged Hop causes hypertrophy, possibly by recruitment of class I HDACs that may inhibit anti-hypertrophic gene expression (89–92).

PPARα plays important roles in transcriptional control of metabolic genes, particularly those involved in cardiac fatty acid uptake and oxidation. Mice with cardiac-restricted overexpression of *PPARα* replicate the phenotype of diabetic cardiomyopathy: hypertrophy, fetal gene activation, and systolic ventricular dysfunction (93). Heterozygous *PPARγ*-deficient mice, when subjected to pressure overload, developed greater hypertrophic remodeling than wild-type controls, implicating the PPARγ-pathway as a protective mechanism for hypertrophy and heart failure (94).

Retinoid X receptor α (RXRα) is a retinoid-dependent transcriptional regulator that binds DNA as an RXR/retinoic acid receptor (RXR/RAR) heterodimer. *RXRα*-null mice die during embryogenesis with hypoplasia of the ventricular myocardium. In contrast, overexpression of RXRα in the heart does not rescue myocardial hypoplasia but causes DCM (95).

Integrating functional and molecular signals

Study of human gene mutations that cause HCM and DCM provides information about functional triggers of cardiac remodeling. In parallel with evolving information about molecular-signaling cascades that influence cardiac gene expression, there is considerable opportunity to define precise pathways that cause the heart to fail. To understand the integration of functional triggers with molecular responses, a comprehensive data set of the transcriptional and proteomic profiles associated with precise gene mutations is needed. Despite the plethora of information associated with such studies, bioinformatic assembly of data and deduction of pathways should be feasible and productive for defining shared or distinct responses to signals that cause cardiac remodeling and heart failure. Accrual of this data set in humans is a desirable goal, although confounding clinical variables and tissue acquisition pose considerable difficulties that can be more readily addressed by study of animal models with heart disease. With more knowledge about the pathways involved in HCM and DCM, strategies may emerge to attenuate hypertrophy, reduce myocyte death, and diminish myocardial fibrosis, processes that ultimately cause the heart to fail.

Acknowledgments

This work was supported by grants from the Howard Hughes Medical Institute and the NIH.

Address correspondence to: Christine E. Seidman, Department of Genetics, New Research Building (NRB) Room 256, Harvard Medical School, 77 Avenue Louis Pasteur, Boston, Massachusetts 02115, USA. Phone: (617) 432-7871; Fax: (617) 432-7832; E-mail: cseidman@genetics.med.harvard.edu.

1. American Heart Association. 2004. Heart disease and stroke statistics – 2004 update. American Heart Association. http://www.americanheart.org.
2. Seidman, J.G., and Seidman, C. 2001. The genetic basis for cardiomyopathy: from mutation identification to mechanistic paradigms. *Cell.* **104**:557–567.
3. CardioGenomics. Genomics of cardiovascular development, adaptation, and remodeling. NHLBI program for genomic applications. Harvard Medical School. http://cardiogenomics.med.harvard.edu.
4. Morita, H., et al. 2002. Molecular epidemiology of hypertrophic cardiomyopathy. *Cold Spring Harb. Symp. Quant. Biol.* **67**:383–388.
5. Richard, P., et al. 2003. Hypertrophic cardiomyopathy: distribution of disease genes, spectrum of mutations, and implications for a molecular diagnosis strategy. *Circulation.* **107**:2227–2232.
6. Watkins, H., et al. 1995. Mutations in the genes for cardiac troponin T and alpha-tropomyosin in hypertrophic cardiomyopathy. *N. Engl. J. Med.* **332**:1058–1064.
7. Moolman, J.C., et al. 1997. Sudden death due to troponin T mutations. *J. Am. Coll. Cardiol.* **29**:549–555.
8. Anan, R., et al. 1994. Prognostic implications of novel β cardiac myosin heavy chain gene mutations that cause familial hypertrophic cardiomyopathy. *J. Clin. Invest.* **93**:280–285.
9. Fujino, N., et al. 2002. A novel mutation Lys273Glu in the cardiac troponin T gene shows high degree of penetrance and transition from hypertrophic to dilated cardiomyopathy. *Am. J. Cardiol.* **89**:29–33.
10. Kokado, H., et al. 2000. Clinical features of hypertrophic cardiomyopathy caused by a Lys183 deletion mutation in the cardiac troponin I gene. *Circulation.* **102**:663–669.
11. Regitz-Zagrosek, V., Erdmann, J., Wellnhofer, E., Raible, J., and Fleck, E. 2000. Novel mutation in the alpha-tropomyosin gene and transition from hypertrophic to hypocontractile dilated cardiomyopathy. *Circulation.* **102**:E112–E116.
12. Tyska, M.J., et al. 2000. Single-molecule mechanics of R403Q cardiac myosin isolated from the mouse model of familial hypertrophic cardiomyopathy.

Circ. Res. **86**:737–744.
13. Palmer, B.M., et al. 2004. Differential cross-bridge kinetics of FHC myosin mutations R403Q and R453C in heterozygous mouse myocardium. *Am. J. Physiol. Heart Circ. Physiol.* **287**:H91–H99.
14. Palmer, B.M., et al. 2004. Effect of cardiac myosin binding protein-C on mechanoenergetics in mouse myocardium. *Circ. Res.* **94**:1615–1622.
15. Harris, S.P., et al. 2002. Hypertrophic cardiomyopathy in cardiac myosin binding protein-C knockout mice. *Circ. Res.* **90**:594–601.
16. Geisterfer-Lowrance, A.A., et al. 1996. A mouse model of familial hypertrophic cardiomyopathy. *Science.* **272**:731–734.
17. Carrier, L., et al. 2004. Asymmetric septal hypertrophy in heterozygous cMyBP-C null mice. *Cardiovasc. Res.* **63**:293–304.
18. Oberst, L., et al. 1998. Dominant-negative effect of a mutant cardiac troponin T on cardiac structure and function in transgenic mice. *J. Clin. Invest.* **102**:1498–1505.
19. Fatkin, D., et al. 1999. Neonatal cardiomyopathy in mice homozygous for the Arg403Gln mutation in the α cardiac myosin heavy chain gene. *J. Clin. Invest.* **103**:147–153.
20. McConnell, B.K., et al. 1999. Dilated cardiomyopathy in homozygous myosin-binding protein-C mutant mice. *J. Clin. Invest.* **104**:1235–1244.
21. Nishi, H., et al. 1994. Possible gene dose effect of a mutant cardiac beta-myosin heavy chain gene on the clinical expression of familial hypertrophic cardiomyopathy. *Biochem. Biophys. Res. Commun.* **200**:549–556.
22. Ho, C.Y., et al. 2000. Homozygous mutation in cardiac troponin T: implications for hypertrophic cardiomyopathy. *Circulation.* **102**:1950–1955.
23. Kamisago, M., et al. 2000. Mutations in sarcomere protein genes as a cause of dilated cardiomyopathy. *N. Engl. J. Med.* **343**:1688–1696.
24. Murphy, R.T., et al. 2004. Novel mutation in cardiac troponin I in recessive idiopathic dilated cardiomyopathy. *Lancet.* **363**:371–372.
25. Olson, T.M., Kishimoto, N.Y., Whitby, F.G., and

Michels, V.V. 2001. Mutations that alter the surface charge of alpha-tropomyosin are associated with dilated cardiomyopathy. *J. Mol. Cell. Cardiol.* **33**:723–732.
26. Olson, T.M., Michels, V.V., Thibodeau, S.N., Tai, Y.S., and Keating, M.T. 1998. Actin mutations in dilated cardiomyopathy, a heritable form of heart failure. *Science.* **280**:750–752.
27. Gerull, B., et al. 2002. Mutations of TTN, encoding the giant muscle filament titin, cause familial dilated cardiomyopathy. *Nat. Genet.* **30**:201–204.
28. Itoh-Satoh, M., et al. 2002. Titin mutations as the molecular basis for dilated cardiomyopathy. *Biochem. Biophys. Res. Commun.* **291**:385–393.
29. Milner, D.J., Weitzer, G., Tran, D., Bradley, A., and Capetanaki, Y. 1996. Disruption of muscle architecture and myocardial degeneration in mice lacking desmin. *J. Cell Biol.* **134**:1255–1270.
30. Franz, W.M., et al. 2000. Association of nonsense mutation of dystrophin gene with disruption of sarcoglycan complex in X-linked dilated cardiomyopathy. *Lancet.* **355**:1781–1785.
31. Badorff, C., et al. 1999. Enteroviral protease 2A cleaves dystrophin: evidence of cytoskeletal disruption in an acquired cardiomyopathy. *Nat. Med.* **5**:320–326.
32. Sanbe, A., et al. 2004. Desmin-related cardiomyopathy in transgenic mice: a cardiac amyloidosis. *Proc. Natl. Acad. Sci. U. S. A.* **101**:10132–10136.
33. Wheeler, M.T., et al. 2004. Smooth muscle cell-extrinsic vascular spasm arises from cardiomyocyte degeneration in sarcoglycan-deficient cardiomyopathy. *J. Clin. Invest.* **113**:668–675. doi:10.1172/JCI200420410.
34. Olson, T.M., et al. 2002. Metavinculin mutations alter actin interaction in dilated cardiomyopathy. *Circulation.* **105**:431–437.
35. Knoll, R., et al. 2002. The cardiac mechanical stretch sensor machinery involves a Z disc complex that is defective in a subset of human dilated cardiomyopathy. *Cell.* **111**:943–955.
36. Vatta, M., et al. 2003. Mutations in Cypher/ZASP in patients with dilated cardiomyopathy and left

ventricular non-compaction. *J. Am. Coll. Cardiol.* **42**:2014–2027.

37. Geier, C., et al. 2003. Mutations in the human muscle LIM protein gene in families with hypertrophic cardiomyopathy. *Circulation.* **107**:1390–1395.

38. Hayashi, T., et al. 2004. Tcap gene mutations in hypertrophic cardiomyopathy and dilated cardiomyopathy. *J. Am. Coll. Cardiol.* **44**:2192–2201.

39. Arber, S., et al. 1997. MLP-deficient mice exhibit a disruption of cardiac cytoarchitectural organization, dilated cardiomyopathy, and heart failure. *Cell.* **88**:393–403.

40. Nikolova, V., et al. 2004. Defects in nuclear structure and function promote dilated cardiomyopathy in lamin A/C-deficient mice. *J. Clin. Invest.* **113**:357–369. doi:10.1172/JCI200419448.

41. Lammerding, J., et al. 2004. Lamin A/C deficiency causes defective nuclear mechanics and mechanotransduction. *J. Clin. Invest.* **113**:370–378. doi:10.1172/JCI200419670.

42. McKenna, W.J., et al. 1994. Diagnosis of arrhythmogenic right ventricular dysplasia/cardiomyopathy. Task Force of the Working Group Myocardial and Pericardial Disease of the European Society of Cardiology and of the Scientific Council on Cardiomyopathies of the International Society and Federation of Cardiology. *Br. Heart J.* **71**:215–218.

43. McKoy, G., et al. 2000. Identification of a deletion in plakoglobin in arrhythmogenic right ventricular cardiomyopathy with palmoplantar keratoderma and woolly hair (Naxos disease). *Lancet.* **355**:2119–2124.

44. Norgett, E.E., et al. 2000. Recessive mutation in desmoplakin disrupts desmoplakin-intermediate filament interactions and causes dilated cardiomyopathy, woolly hair and keratoderma. *Hum. Mol. Genet.* **9**:2761–2766.

45. Rampazzo, A., et al. 2002. Mutation in human desmoplakin domain binding to plakoglobin causes a dominant form of arrhythmogenic right ventricular cardiomyopathy. *Am. J. Hum. Genet.* **71**:1200–1206.

46. Gerull, B., et al. 2004. Mutations in the desmosomal protein plakophilin-2 are common in arrhythmogenic right ventricular cardiomyopathy. *Nat. Genet.* **36**:1162–1164.

47. Tiso, N., et al. 2001. Identification of mutations in the cardiac ryanodine receptor gene in families affected with arrhythmogenic right ventricular cardiomyopathy type 2 (ARVD2). *Hum. Mol. Genet.* **10**:189–194.

48. Anan, R., et al. 1995. Cardiac involvement in mitochondrial diseases. A study on 17 patients with documented mitochondrial DNA defects. *Circulation.* **91**:955–961.

49. Santorelli, F.M., et al. 1996. Maternally inherited cardiomyopathy and hearing loss associated with a novel mutation in the mitochondrial tRNA(Lys) gene (G8363A). *Am. J. Hum. Genet.* **58**:933–939.

50. Santorelli, F.M., et al. 1999. Maternally inherited cardiomyopathy: an atypical presentation of the mtDNA 12S rRNA gene A1555G mutation. *Am. J. Hum. Genet.* **64**:295–300.

51. Arbustini, E., et al. 1998. Mitochondrial DNA mutations and mitochondrial abnormalities in dilated cardiomyopathy. *Am. J. Pathol.* **153**:1501–1510.

52. Arad, M., Seidman, J.G., and Seidman, C.E. 2002. Phenotypic diversity in hypertrophic cardiomyopathy. *Hum. Mol. Genet.* **11**:2499–2506.

53. Kemp, B.E., et al. 2003. AMP-activated protein kinase, super metabolic regulator. *Biochem. Soc. Trans.* **31**:162–168.

54. Tian, R., Musi, N., D'Agostino, J., Hirshman, M.F., and Goodyear, L.J. 2001. Increased adenosine monophosphate-activated protein kinase activity in rat hearts with pressure-overload hypertrophy. *Circulation.* **104**:1664–1669.

55. Arad, M., et al. 2003. Transgenic mice overexpressing mutant PRKAG2 define the cause of Wolff-Parkinson-White syndrome in glycogen storage cardiomyopathy. *Circulation.* **107**:2850–2856.

56. Henry, C.G., and Lowry, O.H. 1985. Enzymes and metabolites of glycogen metabolism in canine cardiac Purkinje fibers. *Am. J. Physiol.* **248**:H599–H605.

57. Kubler, W., Schomig, A., and Senges, J. 1985. The conduction and cardiac sympathetic systems: metabolic aspects. *J. Am. Coll. Cardiol.* **5**:157B–161B.

58. Gabrielli, F., Aita, M., Arturi, E., and Alcini, E. 1992. A comparative enzyme histochemical study of glucose metabolism in the conduction system of mammalian hearts. *Cell. Mol. Biol.* **38**:449–455.

59. Wehrens, X.H., et al. 2004. Protection from cardiac arrhythmia through ryanodine receptor-stabilizing protein calstabin2. *Science.* **304**:292–296.

60. Laitinen, P.J., et al. 2001. Mutations of the cardiac ryanodine receptor (RyR2) gene in familial polymorphic ventricular tachycardia. *Circulation.* **103**:485–490.

61. Priori, S.G., et al. 2001. Mutations in the cardiac ryanodine receptor gene (hRyR2) underlie catecholaminergic polymorphic ventricular tachycardia. *Circulation.* **103**:196–200.

62. Lahat, H., et al. 2001. A missense mutation in a highly conserved region of CASQ2 is associated with autosomal recessive catecholamine-induced polymorphic ventricular tachycardia in Bedouin families from Israel. *Am. J. Hum. Genet.* **69**:1378–1384.

63. Postma, A.V., et al. 2002. Absence of calsequestrin 2 causes severe forms of catecholaminergic polymorphic ventricular tachycardia. *Circ. Res.* **91**:e21–e26.

64. Wehrens, X.H., et al. 2003. FKBP12.6 deficiency and defective calcium release channel (ryanodine receptor) function linked to exercise-induced sudden cardiac death. *Cell.* **113**:829–840.

65. Fatkin, D., et al. 2000. An abnormal Ca^{2+} response in mutant sarcomere protein-mediated familial hypertrophic cardiomyopathy. *J. Clin. Invest.* **106**:1351–1359.

66. Semsarian, C., et al. 2002. The L-type calcium channel inhibitor diltiazem prevents cardiomyopathy in a mouse model. *J. Clin. Invest.* **109**:1013–1020. doi:10.1172/JCI200214677.

67. Doi, M., et al. 2002. Propranolol prevents the development of heart failure by restoring FKBP12.6-mediated stabilization of ryanodine receptor. *Circulation.* **105**:1374–1379.

68. Yano, M., et al. 2003. FKBP12.6-mediated stabilization of calcium-release channel (ryanodine receptor) as a novel therapeutic strategy against heart failure. *Circulation.* **107**:477–484.

69. Schmitt, J.P., et al. 2003. Dilated cardiomyopathy and heart failure caused by a mutation in phospholamban. *Science.* **299**:1410–1413.

70. Haghighi, K., et al. 2003. Human phospholamban null results in lethal dilated cardiomyopathy revealing a critical difference between mouse and human. *J. Clin. Invest.* **111**:869–876. doi:10.1172/JCI200317892.

71. Luo, W., et al. 1994. Targeted ablation of the phospholamban gene is associated with markedly enhanced myocardial contractility and loss of beta-agonist stimulation. *Circ. Res.* **75**:401–409.

72. Minamisawa, S., et al. 1999. Chronic phospholamban-sarcoplasmic reticulum calcium ATPase interaction is the critical calcium cycling defect in dilated cardiomyopathy. *Cell.* **99**:313–322.

73. DeSantiago, J., Maier, L.S., and Bers, D.M. 2004. Phospholamban is required for CaMKII-dependent recovery of Ca transients and SR Ca reuptake during acidosis in cardiac myocytes. *J. Mol. Cell. Cardiol.* **36**:67–74.

74. Periasamy, M., et al. 1999. Impaired cardiac performance in heterozygous mice with a null mutation in the sarco(endo)plasmic reticulum Ca2+-ATPase isoform 2 (SERCA2) gene. *J. Biol. Chem.* **274**:2556–2562.

75. Zingman, L.V., et al. 2002. Kir6.2 is required for adaptation to stress. *Proc. Natl. Acad. Sci. U. S. A.* **99**:13278–13283.

76. Bienengraeber, M., et al. 2004. ABCC9 mutations identified in human dilated cardiomyopathy disrupt catalytic KATP channel gating. *Nat. Genet.* **36**:382–387.

77. Komuro, I., and Izumo, S. 1993. Csx: a murine homeobox-containing gene specifically expressed in the developing heart. *Proc. Natl. Acad. Sci. U. S. A.* **90**:8145–8149.

78. Lyons, I., et al. 1995. Myogenic and morphogenetic defects in the heart tubes of murine embryos lacking the homeo box gene Nkx2-5. *Genes Dev.* **9**:1654–1666.

79. Tanaka, M., Chen, Z., Bartunkova, S., Yamasaki, N., and Izumo, S. 1999. The cardiac homeobox gene Csx/Nkx2.5 lies genetically upstream of multiple genes essential for heart development. *Development.* **126**:1269–1280.

80. Schott, J.J., et al. 1998. Congenital heart disease caused by mutations in the transcription factor NKX2-5. *Science.* **281**:108–111.

81. Kasahara, H., and Benson, D.W. 2004. Biochemical analyses of eight NKX2.5 homeodomain missense mutations causing atrioventricular block and cardiac anomalies. *Cardiovasc. Res.* **64**:40–51.

82. Biben, C., et al. 2000. Cardiac septal and valvular dysmorphogenesis in mice heterozygous for mutations in the homeobox gene Nkx2-5. *Circ. Res.* **87**:888–895.

83. Tanaka, M., et al. 2002. A mouse model of congenital heart disease: cardiac arrhythmias and atrial septal defect caused by haploinsufficiency of the cardiac transcription factor Csx/Nkx2.5. *Cold Spring Harb. Symp. Quant. Biol.* **67**:317–325.

84. Kasahara, H., et al. 2001. Progressive atrioventricular conduction defects and heart failure in mice expressing a mutant Csx/Nkx2.5 homeoprotein. *J. Clin. Invest.* **108**:189–201. doi:10.1172/JCI200112694.

85. Toko, H., et al. 2002. Csx/Nkx2-5 is required for homeostasis and survival of cardiac myocytes in the adult heart. *J. Biol. Chem.* **277**:24735–24743.

86. Molkentin, J.D., et al. 1998. A calcineurin-dependent transcriptional pathway for cardiac hypertrophy. *Cell.* **93**:215–228.

87. Frey, N., et al. 2004. Mice lacking calsarcin-1 are sensitized to calcineurin signaling and show accelerated cardiomyopathy in response to pathological biomechanical stress. *Nat. Med.* **10**:1336–1343.

88. Zhang, C.L., et al. 2002. Class II histone deacetylases act as signal-responsive repressors of cardiac hypertrophy. *Cell.* **110**:479–488.

89. Chen, F., et al. 2002. Hop is an unusual homeobox gene that modulates cardiac development. *Cell.* **110**:713–723.

90. Shin, C.H., et al. 2002. Modulation of cardiac growth and development by HOP, an unusual homeodomain protein. *Cell.* **110**:725–735.

91. Kook, H., et al. 2003. Cardiac hypertrophy and histone deacetylase–dependent transcriptional repression mediated by the atypical homeodomain protein Hop. *J. Clin. Invest.* **112**:863–871. doi:10.1172/JCI200319137.

92. McKinsey, T.A., and Olson, E.N. 2004. Cardiac histone acetylation--therapeutic opportunities abound. *Trends Genet.* **20**:206–213.

93. Finck, B.N., et al. 2002. The cardiac phenotype induced by PPARα overexpression mimics that caused by diabetes mellitus. *J. Clin. Invest.* **109**:121–130. doi:10.1172/JCI200214080.

94. Asakawa, M., et al. 2002. Peroxisome proliferator-activated receptor gamma plays a critical role in inhibition of cardiac hypertrophy in vitro and in vivo. *Circulation.* **105**:1240–1246.

95. Subbarayan, V., et al. 2000. RXRα overexpression in cardiomyocytes causes dilated cardiomyopathy but fails to rescue myocardial hypoplasia in RXRα-null fetuses. *J. Clin. Invest.* **105**:387–394.

96. Dalakas, M.C., et al. 2000. Desmin myopathy, a skeletal myopathy with cardiomyopathy caused by mutations in the desmin gene. *N. Engl. J. Med.* **342**:770–780.

Protein kinase cascades in the regulation of cardiac hypertrophy

Gerald W. Dorn II[1] and Thomas Force[2]

[1]Heart and Vascular Center of the University of Cincinnati, Cincinnati, Ohio, USA. [2]Molecular Cardiology Research Institute,
Tufts-New England Medical Center and Tufts University School of Medicine, Boston, Massachusetts, USA.

In broad terms, there are 3 types of cardiac hypertrophy: normal growth, growth induced by physical condition-ing (i.e., physiologic hypertrophy), and growth induced by pathologic stimuli. Recent evidence suggests that normal and exercise-induced cardiac growth are regulated in large part by the growth hormone/IGF axis via signaling through the PI3K/Akt pathway. In contrast, pathological or reactive cardiac growth is triggered by autocrine and paracrine neurohormonal factors released during biomechanical stress that signal through the Gq/phospholipase C pathway, leading to an increase in cytosolic calcium and activation of PKC. Here we review recent developments in the area of these cardiotrophic kinases, highlighting the utility of animal models that are helping to identify molecular targets in the human condition.

Introduction

In the 20 years since Paul Simpson initially demonstrated that neurohormonal stimulation of cultured neonatal cardiomyo-cytes results in cellular hypertrophy, characteristic changes in car-diac gene expression, and activation of specific kinase signaling pathways (1–3), protein kinases have attracted attention as can-didate mediators of the cardiac biomechanical stress and trophic responses. Various kinases are downstream effectors of neuro-hormone receptors that transduce signals from the sympathetic nervous and renin-angiotensin-aldosterone systems. Involvement of these pathways in the acute and chronic cardiac responses to hemodynamic overload or myocardial injury is incontrovertible, and targeting these events constitutes the rationale for current therapeutics aimed at blocking neurohormonal responses in con-gestive heart failure (4).

Epinephrine, norepinephrine, angiotensin II, and aldosterone have been identified as the most important neurohormones stimulating stress-mediated or reactive cardiac hypertrophy, i.e., pathological hypertrophy, and contributing to its progression to heart failure. In experimental models of heart failure and the human clinical syn-dromes, receptor antagonists or synthesis inhibitors for each can modulate the hypertrophy response and improve the prognosis (5). There are, however, important differences in the cardiac respons-es to these agents, and individual roles for catecholamines versus renin-angiotensin in the cardiac hypertrophy response needed to be defined. In addition, normal cardiac postnatal growth (also known as eutrophy) and adaptive growth in response to physical condi-tioning, i.e., physiological hypertrophy, appear to be stimulated not by neurohormones, but by peptide growth factors that may have therapeutic benefits, depending on method of delivery, duration,

and level of expression (6–9). Attempts to further define the signal-ing pathways for cardiac eutrophy, physiological hypertrophy, and pathological hypertrophy have employed a reductionist approach, delineating downstream signaling effectors of each receptor-hor-mone system and their specific manipulation in tissue culture or in physiologically stressed and genetically modified animal models. The accumulated data reveal that the multiple aspects of reactive cardiac hypertrophy may be beneficial or harmful, depending upon physiological context. Likewise, the molecular events that signal hypertrophy are more complex than initially anticipated, with many parallel and redundant transducer and effector pathways. Protein kinases and phosphatases, such as MAPKs, JAKs, cyclin-dependent kinase-9, calcium/calmodulin-dependent protein kinases, and calmodulin-dependent phosphatases, are among the best estab-lished mediators of hypertrophy, and have been the subject of recent surveys (10, 11). This review examines recent findings in 2 kinase signaling pathways that have been identified as critically important mediators of maladaptive and adaptive hypertrophy: the Gq/PKC and PI3K/Akt pathways, respectively. Particular attention is given to recently described genetically modified mouse models wherein the consequences of overexpressing, activating, ablating, or inhibiting a specific kinase on cardiac hypertrophy and contractile function in the intact cardiorenovascular system have been assessed.

Kinase signaling in adaptive hypertrophy

Adaptive cardiac growth occurs as a feature of normal postnatal cardiac eutrophy or as the physiological hypertrophy resulting from exercise conditioning (12). Maladaptive hypertrophy devel-ops in response to excess hemodynamic workload; if the inciting pathologic stimulus is not removed, reactive hypertrophy that is initially a functional, although not essential, compensation (13–15) inevitably undergoes ventricular remodeling/dilation, with func-tional decompensation and development of overt heart failure (16). A third form of hypertrophy, also maladaptive, is caused by genetic mutations affecting sarcomeric or cytoskeletal proteins or proteins involved in calcium homeostasis and is reviewed else-where (17). Thus, it is critical to define and distinguish among the pathways that regulate adaptive versus maladaptive hypertrophy in order to target the latter in human disease using novel pharma-cological or gene transfer approaches.

Nonstandard abbreviations used: ASK1, apoptosis signaling kinase 1; GH, growth hormone; GSK-3, glycogen synthase kinase–3; LVAD, left-ventricular assist device; MEKK1, MAPK/ERK kinase 1; mTOR, mammalian target of rapamycin; NFAT, nuclear factor of activated T cells; p110α, PI3K subgroup Iα; PDK1, 3-phosphoinosit-ide–dependent protein kinase-1; PLCβ, phospholipase Cβ; PTEN, phosphatase and tensin homolog on chromosome 10; RACK, receptor for activated C kinases; S6K1, S6 kinase 1; Tak1, TGFβ-activated kinase 1.

Conflict of interest: The authors have declared that no conflict of interest exists.

Citation for this article: *J. Clin. Invest.* **115**:527–537 (2005).
doi:10.1172/JCI200524178.

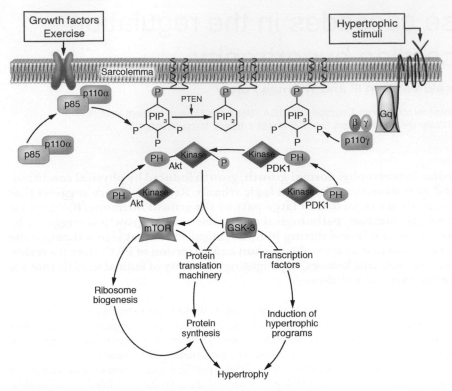

Figure 1
Mechanisms of activation of PI3K/Akt signaling in adaptive versus maladaptive hypertrophy. In adaptive hypertrophy, binding of growth factors to their cognate receptors triggers translocation of the PI3K isoform p110α to the cell membrane, a process triggered by the interaction of the p85 subunit of PI3K with specific tyrosine phosphorylated residues in the growth factor receptor. p110α then phosphorylates phosphatidylinositols in the membrane at the 3′ position of the inositol ring. The pleckstrin homology (PH) domains of both Akt and its activator, PDK1, associate with the 3′ phosphorylated lipids, allowing PDK1 to activate Akt. Full activation of Akt requires phosphorylation by a second kinase, PDK2 (not shown), that may be the DNA-dependent protein kinase (DNA-PK). Activation of Akt then leads to activation of mTOR, a central regulator of protein synthesis, via its effects on both ribosome biogenesis and activation of the protein translation machinery. Akt also phosphorylates and inhibits a kinase, GSK-3 (of which there are 2 isoforms, α and β). Since GSK-3 inhibits a key component of the protein translation machinery, as well as a number of transcription factors believed to play roles in the induction of the hypertrophic program of gene expression, inhibition of GSK-3 promotes both protein synthesis and gene transcription. Maladaptive hypertrophy, triggered by neurohormonal mediators and biomechanical stress, also activates Akt, but the mechanism involves activation of heterotrimeric G-protein–coupled receptors coupled to the G-protein family Gq/G11. The PI3K isoform p110γ associates with the βγ subunits of Gq and phosphorylates membrane phosphatidylinositols, which leads to the recruitment of PDK1 and Akt. Maladaptive hypertrophy also recruits alternative pathways to activation of mTOR and Akt. Also shown is the phosphatase, PTEN, which, by dephosphorylating the 3′ position of phosphatidylinositol trisphosphate (PIP₃), shuts off signaling down the pathway.

Cardiac eutrophy and physiological hypertrophy are largely mediated by signaling through the peptide growth factors: IGF-1 and growth hormone (GH), the latter acting predominantly via increased production of IGF-1 (18). When IGF-1, insulin, and other growth factors bind to their membrane tyrosine kinase receptors (Figure 1), a 110-kDa lipid kinase, PI3K subgroup Iα (hereafter referred to as p110α) is activated (19) and phosphorylates the membrane phospholipid phosphatidylinositol 4,5 bisphosphate at the 3′ position of the inositol ring. This leads to recruitment of the protein kinase Akt (also known as PKB) and its activator, 3-phosphoinositide–dependent protein kinase-1 (PDK1), to the cell membrane via interac-

tions between kinase pleckstrin homology domains and the 3′-phosphorylated lipid (Figure 1) (20). This enforced colocalization of Akt and PDK1 causes the latter to phosphorylate and activate the former.

Accumulated data suggest that PI3K/Akt signaling transduces adaptive cardiac hypertrophy. The whole-genome knockout of p110α was lethal at E9.5–E10.5 (showing a severe proliferative defect; ref. 21) and therefore was of limited usefulness for cardiac studies. However, a central role of the p110α pathway in IGF-1–induced growth and normal and exercise-induced hypertrophy was demonstrated utilizing mice expressing constitutively active or dominant-negative mutants of PI3K specifically in the heart (22). Strikingly, the adaptive hypertrophy seen with constitutive activation of cardiomyocyte PI3K did not transition into a maladaptive hypertrophy. In contrast, cardiac expression of a mutant dominant-negative p110α impaired normal eutrophic heart growth and prevented exercise-induced hypertrophy induced by swim training (23). It is important to note that p110α was not, however, necessary for the hypertrophic response to pressure overload (although it may be important in the maintenance of left-ventricular function in the setting of pressure overload; ref. 23). Further supporting a critical role for the PI3K/PDK1/Akt pathway in regulating normal heart growth is the finding that cardiac-specific ablation of PDK1 leads to reduced cardiac growth and a cardiomyopathic picture (24). Finally, cardiac-specific inactivation of phosphatase and tensin homolog on chromosome 10 (PTEN), a tumor-suppressor phosphatase that negatively regulates the PI3K/Akt pathway by dephosphorylating 3′-phosphorylated phosphoinositides, resulted in cardiac hypertrophy (25, 26).

As noted above, a major kinase effector of PI3K signaling is Akt. Of the 3 Akt genes, only Akt1 and Akt2 are highly expressed in the heart. Cardiac-specific overexpression of constitutively active Akt mutants stimulates heart growth that may (27) or may not (28, 29) culminate in LV decompensation, likely depending on the degree of overexpression. In addition, expression of Akt confers protection from ischemia-induced cell death and cardiac dysfunction (27, 29, 30). Consistent with the general trophic function of Akt, the Akt1 whole-genome–knockout mice weigh approximately 20% less than wild-type littermates and have a proportional reduction in size of all somatic tissues, including the heart (31). In contrast, Akt2-knockout mice have only a modest reduction in organ size. Thus, data from the available Akt-knockout models support a critical role specifically for Akt1 in normal growth of the heart.

Akt1/Akt2 double-knockout mice suffer from marked growth deficiency and a striking defect in cell proliferation. Investigating *Akt1+/−* and *Akt1−/−* mice for resistance to hypertrophy and confirming these findings in a conditional, cardiac-specific Akt1-knockout model (thereby increasing the likelihood that the observed phenotype is secondary to the deletion of *Akt1* rather than to the compensations for long-term, whole-body deletion of this essential kinase) will reevaluate long-standing concepts regarding a central role of Akt signaling in pathologic stress–induced hypertrophy and in the hypertrophic response to neurohormonal agonists (Figure 1).

Akt is at a signaling cascade branch point. While its effects on cell death/survival are directly mediated via phosphorylation of the FOXO family of transcription factors and other regulators of apoptosis (20), it is the 2 signaling branches downstream from Akt, not Akt itself, that largely determine the nature of a given hypertrophic response. One branch leads to mammalian target of rapamycin (mTOR) and the protein synthetic machinery, which is essential for all forms of hypertrophy (Figure 1 and see below). The other branch leads to glycogen synthase kinase–3 (GSK-3), which also regulates the general protein translational machinery (Figure 1) (32) as well as specific transcription factor targets implicated in both normal and pathologic cardiac growth. Of note, activity of both of these branches can also be regulated by stress-activated, Gq-dependent mechanisms that are independent of Akt (Figure 1) (32, 33), which likely explains in part the ability of the *Akt1−/−* mouse heart to hypertrophy in response to pathologic stress.

Kinase signaling in maladaptive hypertrophy

Gq/phospholipase C and cross-talk with PI3K/Akt

The heterotrimeric G-proteins Gq and G11 are functionally redundant transducers of phospholipase C signaling from prohypertrophic heptahelical receptors for angiotensin, endothelin, norepinephrine, and other neurohormones (34). PKC- and inositol 1,4,5-triphosphate–mediated (IP3-mediated) calcium release are considered to be the major effectors of Gq signaling (see below). However, PI3K-dependent signaling is also activated by this pathway but differs from physiological PI3K signaling in that the activated PI3K isoform (γ) is distinct from that activated by IGF-1 (α). The mechanism of its activation also differs (19, 34) (Figure 1): whereas p110α is activated via tyrosine phosphorylation by ligand-occupied growth factor receptors, p110γ is activated by recruitment to the sarcolemma by βγ subunits of activated Gq/11, providing access to membrane phosphoinositides. Strikingly, while p110α is required for normal or exercise-induced growth, but not pathologic stress-induced growth (23), p110γ is required for stress-induced hypertrophy, but not for normal growth (25, 35). Thus, PI3K signaling, including that of Akt and both arms of its downstream signaling pathways (mTOR and GSK-3), is activated in response to both physiologic and pathologic stimuli, and either branch downstream of Akt can regulate adaptive and maladaptive growth. It is therefore unlikely that this pathway is the sole determinant of adaptive versus maladaptive growth. However, it is possible that signal intensity or duration, which may differ between p110α and p110γ, helps to determine adaptive versus maladaptive growth. Although this hypothetical effect has not yet been critically examined, aortic-banded animals exhibit a sustained increase in the amount of the p110γ protein (35) that could lead to more prolonged activation than the typically brief exercise-induced activation of p110α. We believe that the major determinant of adaptive versus maladaptive growth is likely to be recruitment of

additional signaling pathways — the Gq/phospholipase Cβ/Ca2+ (Gq/PLCβ/Ca2+) module signaling to PKC and the calcineurin/nuclear factor of activated T cells (calcineurin/NFAT) pathway in response to pathologic (but not physiologic) stressors.

Gq/11 and their effectors

When activated by biomechanical stress/neurohormonal mediators, Gq and the functionally similar G11 activate PLCβ, which leads to an IP3-mediated increase in cytosolic [Ca2+] and generation of diacylglycerols. The sustained increase in [Ca2+] activates the protein phosphatase calcineurin and its target, the NFAT family of transcription factors, which are critical mediators of pathologic, but not physiologic, hypertrophy (Figure 2A). The other effector arm of the Gq/PLCβ signaling cascade is the PKC family of diverse kinases that share structural homology and activation by lipid products of phospholipase C or D activity (Figure 2B) (36). In the heart, the 4 most functionally significant PKC family members belong to the "conventional" group (PKCα and -β; calcium- and diacylglycerol-activated) and the "novel" group (PKCδ and -ε; diacylglycerol-activated with no requirement for calcium) (37). These PKC isoforms are activated by membrane receptors coupled to phospholipase C via Gq/G11 heterotrimeric G-proteins. Virtually every cardiomyocyte receptor that couples to Gq stimulates cardiac or cardiomyocyte hypertrophy, the most important of which are the α1-adrenergic receptors for norepinephrine and phenylephrine, the AT-1 receptor for angiotensin II, and the ET receptor for endothelin-1 (38). A critical role for Gq signaling in cardiomyocyte hypertrophy was first demonstrated when forced gain or loss of Gq function was observed to control hypertrophy of cultured neonatal cardiomyocytes (39). Subsequently, studies involving in vivo cardiac-specific transgenic overexpression, dominant inhibition, and gene ablation have proven that cardiomyocyte Gq signaling was both necessary for pressure overload hypertrophy (40, 41) and sufficient to produce pressure overload–like cardiac hypertrophy in the absence of hemodynamic stress (42).

Three features of Gq overexpression–induced hypertrophy are notable. First, despite an increase in cardiomyocyte cross-sectional area that recapitulates pressure overload hypertrophy, the ventricular geometry of Gq overexpressors exhibited eccentric hypertrophy (i.e., the ratio of ventricular dimension to wall thickness did not change), in contrast to the concentric hypertrophy of pressure overload (42, 43). This suggests that the determinants of organ-level ventricular modeling in pressure overload are distinct from those for an individual cardiomyocyte's growth. Second, whereas baseline ventricular systolic function was within normal limits, and hence there was no overt heart failure, the contractile function of individual ventricular myocytes was depressed. Neither the intact hearts nor the individual cardiomyocytes responded normally to β-adrenergic receptor stimulation (42, 43), which in the absence of sympathetic hyperactivity (not seen in these functionally compensated animals) indicates that contractile depression and β-adrenergic unresponsiveness can be intrinsic properties of hypertrophy and determinants of maladaption. Third, under specific forms of genetic, biochemical, or physiological stress, nonfailing Gq-overexpressing hearts rapidly failed due to induction of cardiomyocyte apoptosis (44, 45), which established a plausible cellular and molecular mechanism for the transition from hypertrophy to failure.

In a study identifying likely downstream mediators of Gq-stimulated hypertrophy, it was observed that PKCα was increased at both the protein and mRNA levels and that PKCε exhibited a change in

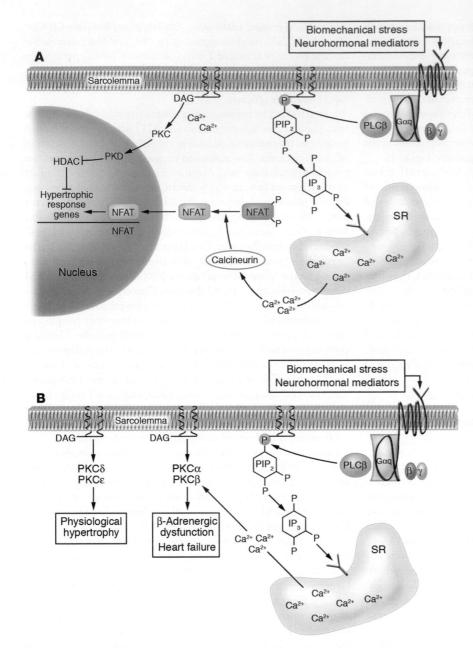

Figure 2
Gq/11-activated pathways in maladaptive hypertrophy. (**A**) Calcineurin/NFAT pathway. Hypertrophic stimuli, acting via the α subunit of Gq or G11, recruit PLCβ to the membrane, where it hydrolyses phosphatidylinositol 4,5 bisphosphate (PIP$_2$), releasing inositol 1,4,5-triphosphate (IP$_3$) and diacylglycerol (DAG). IP$_3$ binds to receptors in the sarcoplasmic reticulum (SR), releasing calcium. The increase in cytosolic [Ca^{2+}], together with calmodulin, activates the protein phosphatase calcineurin. Calcineurin dephosphorylates several residues in the amino-terminal region of the transcription factor NFAT, allowing it to translocate to the nucleus and activate transcription of hypertrophic response genes. (**B**) PKCs. Activation of PKC isoforms is accomplished by the IP$_3$-mediated release of calcium from the SR together with DAG (classical isoforms), whereas the so-called novel isoforms are activated by DAG alone. See text for details of the roles of the various PKC isoforms in hypertrophy. One PKC-regulated pathway not discussed is that leading to the inhibition of a subset of histone deacetylases (HDACs 5 and 9) that appear to specifically regulate cellular hypertrophy. In this pathway, one or more PKC isoforms activate another protein kinase, PKD, that then phosphorylates the HDAC, leading to its export from the nucleus and, thus, inactivation. This pathway is the subject of a review in this series (109).

both subcellular distribution and level of expression (42, 46). Since PKCε was known to be activated in other forms of maladaptive or stress-mediated hypertrophy (ref. 38; see below), it was proposed to be the mediator of Gq-stimulated hypertrophy. As discussed below, subsequent studies combining Gq overexpression with specific modulation of PKCε and -α, as well as individual overexpression of these PKC isoforms, have indicated otherwise.

PKC isoform regulation in hypertrophy and heart failure
Based on in vivo and tissue culture experiments using phorbol esters as general PKC agonists, PKCs have long been implicated in cell proliferation, survival, and programmed death (47). In cultured cardiomyocytes, PKCs regulate contractility and hypertrophy (48). However, there are at least 12 different isoforms of PKC, according to molecular cloning studies, and the multiplicity of family members produces varied cellular responses depending

upon isoform activity and physiological context. In cardiac tissue, PKC isoform expression differs with species, cell type, and developmental stage, with most adult mammalian myocardia expressing PKCα, -β1, -β2, -δ, -ε, and -λ/ζ (Figure 2B) (49, 50).

The activity of any given PKC isoform is dependent upon its expression level, its localization within the cell, and its phosphorylation state (51). Each of these factors is regulated in cardiac disease, although coexpression and parallel activation of multiple PKC isoforms, isoform interdependence and cross-talk, and overlapping isoform effects are potential confounders for measuring PKC signaling. Despite these complexities and the differences between experimental models and human syndromes, studies of myocardial hypertrophy or heart failure largely report similar overall findings: PKCα and PKCβ are upregulated, PKCε is either upregulated or preferentially activated, and levels of PKCδ and -λ/ζ do not change (50, 52, 53). However, this correlative approach does not distinguish primary pathological effects from secondary compensatory events, and simultaneous regulation of multiple PKC isoforms with different subcellular destinations, substrates, and cellular effects precluded assignment of individual pathological consequences based on associations alone. Accordingly, the field moved to generating PKC isoform overexpression and gene knockout models.

Genetic models of cardiac PKC isoform regulation: gene ablation studies

While gene knockout models have been highly informative in many instances, this has not generally been the case for myocardial PKC. To date, mice have been generated that are null for PKCα, -β, -δ, and -ε (54–57), i.e., each of the myocardial PKC isoforms reported to be regulated in hypertrophy or heart failure, plus other isoforms not expressed in the heart. These PKC isoform–null models are all notable for the absence of a significant baseline cardiac phenotype. Instead, the phenotypes have been endocrine, immunological, and neural (reviewed in ref. 58), and only under physiological stress such as ischemia-reperfusion or pressure overload have subtle cardiac phenotypes been provoked (54, 59, 60). Does the absence of a cardiac phenotype associated with PKC isoform gene ablation indicate that the postulated roles for these kinases, based upon their activation in heart disease and their effects on hypertrophy of cultured cardiomyocytes, were incorrect? Indeed, some subsequent studies have indicated that the initial conclusions regarding the mediation of cardiomyocyte growth by PKCα or PKCβ2 (based on viral or transgenic overexpression) were in error (see below). Paradoxically, it may be the gene ablation results that are most misleading, because the multiplicity of PKC isoforms in cardiomyocytes, with parallel activation and overlapping functions, results in opportunistic compensation of the null gene by related PKC isoforms. An example of this phenomenon is ablation of the Gq gene, which resulted only in a platelet defect (61), and ablation of functionally redundant G11, which caused no phenotype whatsoever (62). However, when the Gq- and G11-null mice were interbred (Gq/G11 double-knockout mice), the result was embryonic lethality and a hypoplastic heart (62), which demonstrates that each could substitute for the other during embryonic development. Likewise, we believe that cardiomyocyte-specific PKC isoform ablation, individually and in combination, will be required to unambiguously define the roles for myocardial-expressed PKCs in normal and pathological cardiac growth.

Genetic models of cardiac PKC isoform regulation: transgenic expression

In contrast to genome-wide gene ablation, overexpression of mutationally activated or wild-type PKCs, of dominant inhibitors of individual PKC isoforms, or of isoform-specific PKC translocation modifiers has been highly informative in defining the functions of specific isoforms in the heart. Clearly there are limitations to all transgenic expression approaches, which result in lack of complete specificity due to: (a) altered stoichiometry between overexpressed enzyme and endogenous substrate (elegantly demonstrated for high-level overexpression of PKCε, which promiscuously interacted with PKCβ anchoring proteins; ref. 63); (b) absence of normal regulation for mutant constitutively activated kinases; and (c) nonspecific interactions or incomplete suppression of the activity of transgenic dominant inhibitors. However, these limitations are largely avoided through the use of transgenesis to target expression of peptide activators or inhibitors of PKC isoform translocation to

cardiomyocytes. Indeed, PKC isoform translocation modulation maintains normal enzyme-substrate expression levels and only minimally affects basal activity, although it is clearly less specific and complete than gene ablation. This approach to modulating PKC isoform activity in an isoform- and tissue-specific manner is based on perturbing normal activation-mediated translocation of PKC isoforms to distinct subcellular compartments and binding to isoform-specific anchoring proteins, designated receptors for activated C kinases (RACKs) (64). Short peptides that mimic a PKC-RACK binding domain act as competitive inhibitors for PKC-RACK binding, thus preventing PKC translocation and inhibiting enzyme activity (48). Likewise, peptides that mimic a PKC pseudo-RACK site selectively bind to specific PKC isoforms and expose the RACK binding domain, thus promoting PKC isoform translocation and activation. As described below, such peptides have been transgenically expressed in the mouse heart, where they have been demonstrated to be highly specific in their isoform modulating effects and have provided insight into the consequences of chronically modulated PKCα, -δ, and -ε activity on cardiac hypertrophy and contractile function.

Effects of PKC isoforms on cardiac hypertrophy and contractility

PKCα. Although it is the most highly expressed of the myocardial PKC isoforms (65), PKCα is the least studied of the cardiac PKCs because, unlike PKCδ and -ε, it is not regulated in acute myocardial ischemia (48). Likewise, in contrast to PKCβ, PKCα is not regulated in diabetes (66). An initial comparative analysis of PKC isoforms using adenovirus-mediated transfection of wild-type or dominant inhibitory forms of PKCα, -β2, -δ, and -ε in neonatal rat cardiomyocytes suggested that only PKCα was sufficient to stimulate cell hypertrophy and only inhibition of PKCα inhibited agonist-mediated hypertrophy (67). The implication of this work was that PKCα is a key regulator of cardiomyocyte hypertrophic growth. However, an in vivo analysis of PKCα effects in the mouse heart utilizing gene ablation and transgenic overexpression revealed no effect of PKCα overexpression on cardiac growth and no effect of PKCα inhibition on the hypertrophic response to pressure overload (54). Instead, ablation of PKCα improved contractility, while overexpression diminished it.

The notion that PKCα is more important as a regulator of myocardial contractility than cardiac hypertrophy mirrors the findings from studies using in vivo translocation modulation (65). Here, RACK binding and pseudo-RACK peptides derived from PKCβ were transgenically expressed in the mouse heart. Because of sequence homology of these regions among all 4 conventional PKC isoforms (PKCα, -β1, -β2, and -γ), these peptides have the potential to regulate translocation and activation of each. However, the only measurable effects of the peptides was on the dominant myocardial conventional PKC isoform PKCα. Interestingly, cardiac mass was not altered with increased PKCα activity. However, chronic activation of PKCα diminished baseline ventricular ejection performance and, in combination with Gq-mediated hypertrophy (in which PKCα is transcriptionally upregulated; refs. 42, 46), caused a lethal cardiomyopathy. In contrast, chronic PKCα inhibition improved myocardial contractility and inhibited Gq-mediated cardiac hypertrophy (65). Thus, the results of studies showing gain-of-function by overexpression or translocation facilitation and loss-of-function by gene ablation or translocation inhibition agree: PKCα has minimal effects on cardiac hypertrophy but is a critical determinant of myocardial systolic function. Proposed mechanisms for PKCα-mediated contractile dysfunction include regulation of sarcoplasmic reticular ATPase-mediated calcium cycling through the phospholamban/protein phosphatase inhibitor-1 axis (54) and phosphorylation-mediated uncoupling of β-adrenergic receptors from adenylyl cyclase (65). The relative contribution to myocardial contractility of these 2 equally plausible mechanisms is not yet known.

PKCβ. PKCβ was the first PKC isoform to be studied using cardiac-targeted expression, in part because its activity and expression are increased in diabetes mellitus and human heart failure (66). However, there is little PKCβ in adult mouse myocardium (49, 65), which raises the possibility that potentially important functions of this isoform in human heart disease are performed by related isoforms in the mouse. The PKCβ-transgenic models may therefore represent ectopic expression rather than mimicking naturally occurring upregulation.

Two PKCβ-transgenic models were developed independently. A model expressing wild-type (i.e., nonactivated) PKCβ2 exhibited hypertrophy with contractile dysfunction and pathological gene expression (68). The other model used an inducible transgenic system to express mutationally activated PKCβ2 in either neonatal or adult mouse hearts (69). In adults, PKCβ2 expression resembled wild-type overexpression, with myocardial hypertrophy and ventricular dysfunction, but activated PKCβ overexpression in the neonate was lethal due to effects of unregulated PKCβ activity on postnatal cardiac eutrophic growth. Thus, in all 3 cases, the phenotypes support an important role for PKCβ in transducing myocardial hypertrophy. However, PKCβ-knockout mouse hearts were found to hypertrophy normally to phenylephrine or aortic banding (70). Thus, the role of PKCβ in myocardial hypertrophy signaling is unclear. PKCβ is sufficient to produce cardiac hypertrophy in mice but is not necessary for normal hypertrophy in response to α-adrenergic stimulation or pressure overload. Because of interspecies differences in PKC isoform expression, the relevance of these mouse studies to the human condition is a matter of dispute.

PKCδ. Although it has long been recognized as being activated in myocardial ischemia, relatively little is known about this PKC isoform in cardiac hypertrophy. A cardiac PKCδ–transgenic mouse model has not been described, and PKCδ-knockout mice have no basal cardiac phenotype (56, 60). Translocation modification has confirmed that PKCδ is a critical mediator of postischemic cardiomyocyte necrosis and contractile dysfunction in mice, rats, and pigs (71, 72).

Transgenic expression of PKCδ translocation activator or inhibitor peptides in mouse hearts revealed a prohypertrophic role for PKCδ in the nonstressed heart (71). As is the case with PKCε (see below), increasing basal PKCδ activation by 15–20% resulted in a form of normally functioning or adaptive hypertrophy. In contrast, high-level chronic expression of a PKCδ inhibitor caused a myofibrillar cardiomyopathy characterized by disruption of the cardiomyocyte cytoskeleton (73). These findings indicate that PKCδ can regulate normal cardiomyocyte growth, but we conclude that PKCδ is likely more important in cardiac ischemia than myocardial growth, consistent with it being a critical stress-response gene that can perform varied tasks depending upon physiological context.

PKCε. The best-characterized PKC isoform in cardiac hypertrophy is PKCε. Implicated in hypertrophic signaling because it is activated by mechanical stress as well as genetic (Gq) and physiological (pressure overload) hypertrophic stimuli (42, 74), PKCε was perceived to be a key mediator of maladaptive hypertrophy. However, both transgenic PKCε overexpression and translocation activation result in mice with normally functioning, mildly enlarged hearts, i.e., adaptive hypertrophy (75, 76). Thus, correlative studies seemed to conflict with genetic gain-of-function. Determining whether PKCε is a direct mediator of maladaptive hypertrophy or a compensatory event demanded a loss-of-function approach. As noted, the PKCε-knockout mouse exhibited no basal cardiac phenotype (57). However, mice with a high degree of PKCε translocation inhibition developed lethal perinatal heart failure with myocardial hypoplasia (76), consistent with an essential role for PKCε in normal eutrophic cardiac growth. Why then did the PKCε-knockout mouse not exhibit a similar hypoplastic phenotype? The knockout is genome-wide and exists from fertilization. Thus, a viable PKCε-knockout mouse would likely require compensatory signaling by another PKC isoform (perhaps PKCδ, since it has similar prohypertrophic effects; ref. 71) during critical stages of embryonic development. In contrast, as a consequence of the α–myosin heavy chain promoter, first expression of the inhibitory peptide in the neonatal period occurs after critical stages of embryonic development and in the context of normal expression of PKCε and related isoforms. In the absence of compensatory regulation, the phenotype was unmasked.

The effects of PKCε in hypertrophic syndromes were determined by breeding a low-expressing and phenotypically normal PKCε translocation inhibitor mouse with Gq mice. The consequence of "normalizing" PKCε activity in Gq-mediated hypertrophy was catastrophic, with ventricular dilation and lethal heart failure (77). Conversely, coexpression of Gq with the PKCε activator peptide, which exaggerated the biochemical "abnormality" of increased PKCε translocation, diminished cardiac hypertrophy and improved contractile function. These reciprocal phenotypes strongly support a role for PKCε in adaptive hypertrophy and indicate that its activation in pathological hypertrophy is a compensatory event.

Taken together, the analyses of Gq/PKC-mediated hypertrophy signaling have revealed previously unrecognized fundamental characteristics of reactive cardiac hypertrophy. It is indisputable that, as determined by Laplace's law ($T = pr/2h$), increased ventricular wall thickness (h) will lower wall tension (T), given constant pressure (p) and chamber radius (r). Paradoxically, there are now several examples of hypertrophy-deficient mouse models that fare well under increased loads, such as after transverse aortic banding (13). Indeed, Gq/G11-null mice and Gq-inhibited mice have diminished hypertrophy after induction of surgical pressure overload but maintain functional compensation (41, 42). In contrast, PKCε activity is salutary, whereas PKCα is detrimental in Gq hypertrophy, even though there are no effects on the extent of hypertrophy (65, 77). Thus, one must consider the quality or form of hypertrophy, and not only its quantity, in determining whether it is functionally compensatory.

Glycogen synthase kinase: a negative regulator of hypertrophy

GSK-3β, which was among the first negative regulators of cardiac hypertrophy to be identified, was found to block cardiomyocyte hypertrophy in response to ET-1, PE, isoproteronol, and Fas signaling (78–80). Subsequently, GSK-3β has been found to be a negative regulator of both normal (81) and pathologic stress–induced (isoproteronol infusion or pressure overload; ref. 82) growth in vivo (Figure 3). The expression of tetracycline-regulated GSK3-β in a mouse has further suggested a role for GSK-3β in regression of established pressure overload hypertrophy (83).

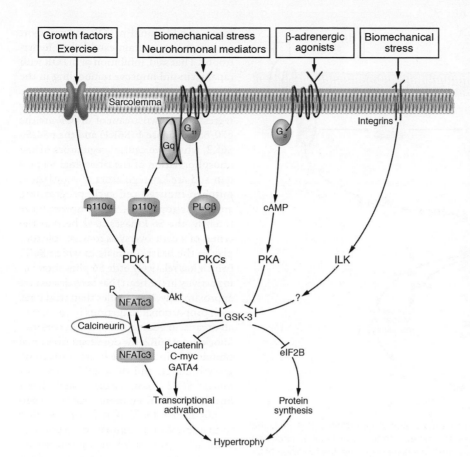

Figure 3
GSK-3 as a convergence point in hypertrophic signaling. Inhibition of GSK-3 appears to be a key element in both adaptive and maladaptive hypertrophy. Growth factors, acting via Akt; neurohormonal mediators, acting via both Akt and PKCs (particularly PKCα); β-adrenergic agonists, acting via PKA; and biomechanical stress, acting via several mechanisms, possibly involving the integrin-linked kinase (ILK) or an ILK-associated protein, all lead to the inactivation of GSK-3. Therefore, GSK-3 appears to serve as a convergence point, integrating inputs from many prohypertrophic signals. Inhibition of GSK-3 releases a number of transcription factors from tonic inhibition, and also releases eIF2B, allowing activation of the protein synthetic machinery. Thus GSK-3 affects both key components of the response, reprogramming of gene expression and activation of protein synthesis. Additional negative regulators of GSK-3 not shown include the serum and glucocorticoid–induced kinase (SGK) and, possibly, the ERK pathway target p90 ribosomal S6 kinase (RSK1).

GSK-3β is unlike most kinases in that it is negatively regulated by growth factors and hypertrophic agonists. It is "on" in the cell until it is turned "off" by these and other stimuli. GSK-3β negatively regulates most of its substrates. Thus, inhibition of GSK-3β in response to growth stimuli releases its substrates from tonic inhibition. For example, GSK-3β phosphorylates and negatively regulates the protein translation initiation factor eIF2B (32) (Figure 3). Overexpression of an eIF2B mutant that cannot be phosphorylated and inactivated by GSK-3β induces hypertrophy of cultured cardiomyocytes (79, 84). GSK-3β also inhibits the activity of a number of transcription factors directly implicated in cardiac growth, including c-Myc, GATA4, and β-catenin (85–87) and therefore may be particularly important in the reprogramming of gene expression that characterizes both adaptive and maladaptive hypertrophy (Figure 3). Finally, GSK-3β is a counter-regulator of calcineurin/NFAT signaling (79), phosphorylating NFAT amino-terminal residues that are dephosphorylated by calcineurin (Figure 3). This prevents nuclear translocation of the NFATs, thereby restricting access to target genes. Indeed, concomitant GSK-3β overexpression markedly reduced hypertrophy of calcineurin-transgenic mice (82).

A final important difference in signaling pathways activated in adaptive versus maladaptive hypertrophy is the strong recruitment of stress-activated MAPKs, p38 MAPKs and JNKs by the latter but only weak (or no) recruitment by the former. The role of these kinases in pathologic hypertrophy remains somewhat uncertain, but it seems that their major role is not in regulating growth directly, but rather in regulating matrix remodeling, direct and indirect contractile function (88), and the progression of left-ventricular dysfunction (89, 90).

Stress-activated MAPKs are the downstream kinases in a 3-tiered cascade in which a MAP3K activates a MAP2K (MEK), which then activates the MAPK. At the MAP3K level, several kinases have been implicated as regulators of hypertrophy in cultured cardiomyocytes or in transgenics (apoptosis signaling kinase 1 [Ask1], TGF-β–activated kinase 1 [Tak1], MAPK/ERK kinase kinase 1 [MEKK1]). However, when the more definitive studies in knockout mice in vivo have been done, results have often been confusing. For example, deletion of MEKK1, a kinase that is reasonably selective for the JNK pathway, blocked hypertrophy in the Gq-overexpressing mouse but did not reduce pressure overload hypertrophy (91). Similarly, deletion of Ask1, which is upstream of both JNKs and p38 MAPKs, blocked angiotensin II–induced hypertrophy but had little or no effect on hypertrophy in either banding or myocardial infarction models. There was, however, a significant reduction in apoptosis, which suggests that Ask1, as the name implies, may be a much more important regulator of cell death than hypertrophy. These disparate results may not be surprising given the growing consensus that the targets of these MAP3Ks, the stress-activated MAPKs, do not play a major role in regulating hypertrophy.

Regulators of protein synthesis in hypertrophic growth

Regulation of protein synthesis is critical in all forms of hypertrophy and has 2 components: global control of protein synthesis and regulation of translation of specific mRNAs. Global control itself has 2 components, ribosome biogenesis and activation of the translational machinery. Ribosome biogenesis involves the enhanced translation of mRNAs encoding ribosomal proteins. Translation

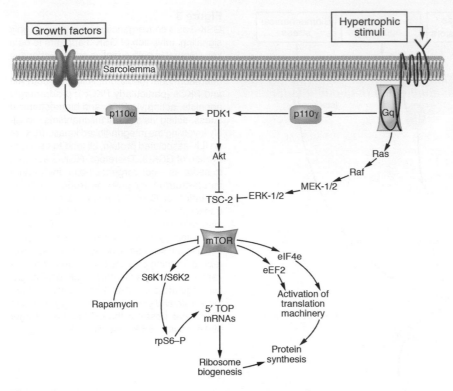

Figure 4
Regulation of protein translation in adaptive versus maladaptive hypertrophy. It is likely that all hypertrophic stimuli must activate mTOR and the general protein translational machinery in order to allow the full expression of the phenotype. This is mediated via the inhibition of the tuberous sclerosis gene product, tuberin (TSC2), mutations of which lead to benign hamarto-mas in various tissues including the heart. TSC2 can be phosphorylated and inhibited by Akt and, in some instances, via ERK-1/2 or an ERK target. The latter may be an Akt-independent mechanism of activation of mTOR that may be particularly relevant to pathologic stress–induced growth. Shown are the pathways to ribosome biogenesis as well as the regulators of the trans-lational machinery (initiation factors [IF] and elongation factors [EF]) regulated by mTOR in the heart. As noted in the text, recent surprising findings related to this pathway have included the limited role for S6K1 and S6K2 in both adaptive and, particularly, maladaptive hypertrophy and the identification of Akt1 as a possible antihypertrophic factor in pathologic hypertrophy but a prohypertrophic factor in physiologic hypertrophy.

ment of additional mTOR-independent pathways that sustain established hyper-trophy. That said, inhibition of mTOR with rapamycin did improve remodeling in the decompensated group (92).

What mTOR targets mediate hypertrophy regression? Downstream of mTOR are the p70/85 S6 kinase 1 (S6K1) and the p54/56 S6K2, which are central regulators of the phosphorylation of the ribosomal S6 pro-tein and are key regulators of translation, mitogen-induced cell cycle progression, and hypertrophy (93, 94). However, more recently, the S6 kinases have been at the center of a controversy as to what role they play in the heart. Animals in which S6K1 (which has relatively little S6 phosphorylat-ing activity in the heart) has been deleted are approximately 20% smaller than wild type, with proportional reductions in the size of all organs, including the heart. In contrast, knockout of S6K2, the dominant ribosomal protein S6 kinase in the heart, leads to no growth defect, and the double knockout (S6K1$^{-/-}$S6K2$^{-/-}$) is no smaller than the S6K1 knockout (95). Even more striking, hyper-trophy in the double-knockout is similar to that in wild-type animals in response to swim training or aortic banding, which dem-onstrates that, while S6K1 plays a modest role in normal heart growth, neither S6K1 nor S6K2 are necessary for exercise-induced hypertrophy, pathologic hypertrophy, or IGF-1/PI3K–dependent hypertrophy (92) (Figure 4). Thus, while mTOR dependent, these types of hypertrophy are not p70S6K dependent, and mTOR undoubtedly has other targets critical for the upregulation of general protein synthesis. Two such tar-gets are not part of the ribosome biogenesis pathway but are factors that regulate activation of the translational machinery, the translation initiation factor, eIF4E, and the transla-tion elongation factor, eEF2 (Figure 4). mTOR releases both eIF4E and eEF2 from repression by 4E-binding protein 1 and the eEF2 kinase, respectively, allowing translation to proceed.

Alterations of signaling in human heart failure

We have discussed the signaling pathways that regulate hypertrophy in the diseased hearts of experimental animals. But how do the sig-naling alterations seen in the hearts of these animals compare with signaling alterations in the hearts of patients with hypertrophy or heart failure? And is there any evidence that dysregulation of signal-ing pathways seen in these clinical scenarios are a cause of heart fail-ure (as opposed to a consequence of the heart failure), and, therefore, will manipulating their activity alter the progression of disease?

There are very limited data available on hearts with compen-sated hypertrophy, but it appears that calcineurin activity may be increased in these hearts (96). In addition, despite the fact that sev-eral studies have examined the signaling profile of hearts explant-ed from patients with extremely advanced failure (either going

of these mRNAs and those encoding some of the proteins that directly regulate translation is dependent upon the protein kinase mTOR. The primary activators of mTOR in mammalian cells are growth stimuli, such as peptide growth factors (GH, IGF) or neuro-hormonal hypertrophic agonists (angiotensin II, endothelin, nor-epinephrine). Representing what may be another key difference in signaling between these 2 classes of agents, peptide growth factors activate mTOR primarily via *p110α/Akt*, whereas neurohormonal mediators likely do so via p110γ, acting through ERKs (Figure 4). This may provide Akt-independent activation of mTOR in patho-logic, as opposed to physiologic, hypertrophy. mTOR occupies a central position in this schema, likely being a final common path-way through which all growth signals must pass to induce protein synthesis. Rapamycin, which inhibits mTOR, blunts the develop-ment of pressure overload hypertrophy induced by aortic band-ing and regresses banding-induced established hypertrophy (92). Interestingly, rapamycin is significantly more effective at regressing established hypertrophy in compensated hearts (i.e., without sig-nificant contractile dysfunction or remodeling) than in decompen-sated hearts, which suggests that decompensation leads to recruit-to transplant or undergoing left-ventricular assist device [LVAD] placement prior to transplant) (96–104), no clear consensus has emerged, except that in 3 studies examining tissues sampled from patients before and after LVAD placement, ERK activity decreased after LVAD placement, coincident with a decrease in cardiomyocyte size (i.e., regression of hypertrophy) (97, 100, 104). Where examined, calcineurin expression and activity were increased, though not to the same degree as seen in the hypertrophied hearts discussed above (96, 101), and there is a fairly consistent pattern of PKC isoform expression/activation, as reviewed above. Thus, not only can one not define a unique signaling profile of the failing heart at this time, it is also unclear whether individual signaling alterations are a cause or consequence of heart failure. Therefore, the effect, if any, of manipulating these pathways on the progression of heart disease is uncertain. Finally, and probably most importantly, the signaling profile of hearts with less advanced failure is unknown.

The lack of consensus may be due to any number of factors — patient variability and differences in medication, age, etc. However, it may be that progression of heart failure, especially late progression, may be due more to alterations in survival pathways, in energy production, in calcium homeostasis, and in β-adrenergic signaling than to alterations in the growth pathways responsible for the development of hypertrophy (105–108). In any case, the complexity of the signaling abnormalities and heterogeneity among patients with heart failure creates a great deal of uncertainty and leads to significant challenges for translational research in this area. Thus, it is not surprising that therapies targeting the β-adrenergic receptor kinase are being considered for patients with heart failure (thoroughly reviewed in ref. 107), but to our knowledge, no trials currently planned will target growth pathways, and probably none will until our field gains a better understanding of signaling alterations at earlier stages of the disease.

In summary, molecular and functional dissection of multiple components from the PI3K pathway has established a role for this signaling cascade in normal, exercise-induced, and reactive stress-mediated cardiac hypertrophy. This pathway, with its 2 branches, the mTOR and GSK-3 pathways, is a dominant determinant of cardiomyocyte and heart size in mammals. However, it appears that the primary determinant of whether hypertrophy will be adaptive or maladaptive is whether neurohormonal-stimulated/calcium-activated pathways, including calcineurin and PKCs, are recruited. In the case of PKCs, multiple isoforms that are differentially regulated and activated, and that are uniquely targeted to distinct subcellular locales, provide for specific but partially overlapping functional profiles. Whether these pathways, as opposed to pathways regulating apoptosis, oncosis, or expression/activity of various calcium-handling proteins, will prove to be viable targets in the hearts of patients with hypertrophy and advanced heart failure is a question that may require improved experimental models for resolution.

Acknowledgments

This work is supported by National Heart, Lung, and Blood Institute grants HL58010, HL59888, and HL52319 (to G.W. Dorn II); and HL69779, HL61688, and HL67371 (to T. Force).

Address correspondence to: Gerald W. Dorn II, University of Cincinnati Medical Center, 231 Albert Sabin Way, M.L. 0542, Cincinnati, Ohio 45267-0542, USA. Phone: (513) 558-3065; Fax: (513) 558-3060; E-mail: dorngw@ucmail.uc.edu. Or to: Thomas Force, Molecular Cardiology Research Institute, Tufts-New England Medical Center, 750 Washington Street, Box 8486, Boston, Massachusetts 02111, USA. Phone: (617) 636-0719; Fax: (617) 636-5649; E-mail: tforce@tufts-nemc.org.

1. Simpson, P. 1983. Norepinephrine-stimulated hypertrophy of cultured rat myocardial cells is an α 1 adrenergic response. *J. Clin. Invest.* **72**:732–738.
2. Sadoshima, J., and Izumo, S. 1993. Molecular characterization of angiotensin II–induced hypertrophy of cardiac myocytes and hyperplasia of cardiac fibroblasts. Critical role of the AT1 receptor subtype. *Circ. Res.* **73**:413–423.
3. Shubeita, H.E., et al. 1990. Endothelin induction of inositol phospholipid hydrolysis, sarcomere assembly, and cardiac gene expression in ventricular myocytes. A paracrine mechanism for myocardial cell hypertrophy. *J. Biol. Chem.* **265**:20555–20562.
4. Bristow, M.R. 2000. beta-adrenergic receptor blockade in chronic heart failure. *Circulation.* **101**:558–569.
5. Mehra, M.R., Uber, P.A., and Francis, G.S. 2003. Heart failure therapy at a crossroad: are there limits to the neurohormonal model? *J. Am. Coll. Cardiol.* **41**:1606–1610.
6. Duerr, R.L., et al. 1995. Insulin-like growth factor-1 enhances ventricular hypertrophy and function during the onset of experimental cardiac failure. *J. Clin. Invest.* **95**:619–627.
7. Colao, A. 2004. Cardiovascular effects of growth hormone treatment: potential risks and benefits. *Horm. Res.* **62**(Suppl. 3):42–50.
8. Rosenthal, N., and Musaro, A. 2002. Gene therapy for cardiac cachexia? *Int. J. Cardiol.* **85**:185–191.
9. Fazio, S., et al. 1996. A preliminary study of growth hormone in the treatment of dilated cardiomyopathy. *N. Engl. J. Med.* **334**:809–814.
10. Molkentin, J.D. 2004. Calcineurin-NFAT signaling regulates the cardiac hypertrophic response in coordination with the MAPKs. *Cardiovasc. Res.*

63:467–475.
11. Zhang, T., and Brown, J.H. 2004. Role of Ca2+/calmodulin-dependent protein kinase II in cardiac hypertrophy and heart failure. *Cardiovasc. Res.* **63**:476–486.
12. Scheuer, J., Malhotra, A., Hirsch, C., Capasso, J., and Schaible, T.F. 1982. Physiologic cardiac hypertrophy corrects contractile protein abnormalities associated with pathologic hypertrophy in rats. *J. Clin. Invest.* **70**:1300–1305.
13. Hill, J.A., et al. 2000. Cardiac hypertrophy is not a required compensatory response to short-term pressure overload. *Circulation.* **101**:2863–2869.
14. Esposito, G., et al. 2002. Genetic alterations that inhibit in vivo pressure-overload hypertrophy prevent cardiac dysfunction despite increased wall stress. *Circulation.* **105**:85–92.
15. Sano, M., and Schneider, M.D. 2002. Still stressed out but doing fine: normalization of wall stress is superfluous to maintaining cardiac function in chronic pressure overload. *Circulation.* **105**:8–10.
16. Grossman, W., Jones, D., and McLaurin, L.P. 1975. Wall stress and patterns of hypertrophy in the human left ventricle. *J. Clin. Invest.* **56**:56–64.
17. Seidman, J.G., and Seidman, C. 2001. The genetic basis for cardiomyopathy: from mutation identification to mechanistic paradigms. *Cell.* **104**:557–567.
18. Lupu, F., Terwilliger, J.D., Lee, K., Segre, G.V., and Efstratiadis, A. 2001. Roles of growth hormone and insulin-like growth factor 1 in mouse postnatal growth. *Dev. Biol.* **229**:141–162.
19. Oudit, G.Y., et al. 2004. The role of phosphoinositide-3 kinase and PTEN in cardiovascular physiology and disease. *J. Mol. Cell. Cardiol.* **37**:449–471.
20. Brazil, D.P., Yang, Z.Z., and Hemmings, B.A. 2004.

Advances in protein kinase B signalling: AKTion on multiple fronts. *Trends Biochem. Sci.* **29**:233–242.
21. Bi, L., Okabe, I., Bernard, D.J., Wynshaw-Boris, A., and Nussbaum, R.L. 1999. Proliferative defect and embryonic lethality in mice homozygous for a deletion in the p110alpha subunit of phosphoinositide 3-kinase. *J. Biol. Chem.* **274**:10963–10968.
22. Shioi, T., et al. 2000. The conserved phosphoinositide 3-kinase pathway determines heart size in mice. *EMBO J.* **19**:2537–2548.
23. McMullen, J.R., et al. 2003. Phosphoinositide 3-kinase(p110alpha) plays a critical role for the induction of physiological, but not pathological, cardiac hypertrophy. *Proc. Natl. Acad. Sci. U. S. A.* **100**:12355–12360.
24. Mora, A., et al. 2003. Deficiency of PDK1 in cardiac muscle results in heart failure and increased sensitivity to hypoxia. *EMBO J.* **22**:4666–4676.
25. Crackower, M.A., et al. 2002. Regulation of myocardial contractility and cell size by distinct PI3K-PTEN signaling pathways. *Cell.* **110**:737–749.
26. Schwartzbauer, G., and Robbins, J. 2001. The tumor suppressor gene PTEN can regulate cardiac hypertrophy and survival. *J. Biol. Chem.* **276**:35786–35793.
27. Shioi, T., et al. 2002. Akt/protein kinase B promotes organ growth in transgenic mice. *Mol. Cell. Biol.* **22**:2799–2809.
28. Condorelli, G., et al. 2002. Akt induces enhanced myocardial contractility and cell size in vivo in transgenic mice. *Proc. Natl. Acad. Sci. U. S. A.* **99**:12333–12338.
29. Matsui, T., et al. 2002. Phenotypic spectrum caused by transgenic overexpression of activated Akt in the heart. *J. Biol. Chem.* **277**:22896–22901.
30. Matsui, T., et al. 2001. Akt activation preserves cardiac

function and prevents injury after transient cardiac ischemia in vivo. *Circulation.* **104**:330–335.

31. Cho, H., Thorvaldsen, J.L., Chu, Q., Feng, F., and Birnbaum, M.J. 2001. Akt1/PKBalpha is required for normal growth but dispensable for maintenance of glucose homeostasis in mice. *J. Biol. Chem.* **276**:38349–38352.

32. Proud, C.G. 2004. Ras, PI3-kinase and mTOR signaling in cardiac hypertrophy. *Cardiovasc. Res.* **63**:403–413.

33. Doble, B.W., and Woodgett, J.R. 2003. GSK-3: tricks of the trade for a multi-tasking kinase. *J. Cell Sci.* **116**:1175–1186.

34. Rockman, H.A., Koch, W.J., and Lefkowitz, R.J. 2002. Seven-transmembrane-spanning receptors and heart function. *Nature.* **415**:206–212.

35. Patrucco, E., et al. 2004. PI3Kgamma modulates the cardiac response to chronic pressure overload by distinct kinase-dependent and -independent effects. *Cell.* **118**:375–387.

36. Nishizuka, Y. 1986. Studies and perspectives of protein kinase C. *Science.* **233**:305–312.

37. Naruse, K., and King, G.L. 2000. Protein kinase C and myocardial biology and function. *Circ. Res.* **86**:1104–1106.

38. Molkentin, J.D., and Dorn, G.W., II. 2001. Cytoplasmic signaling pathways that regulate cardiac hypertrophy. *Annu. Rev. Physiol.* **63**:391–426.

39. LaMorte, V.J., et al. 1994. Gq- and ras-dependent pathways mediate hypertrophy of neonatal rat ventricular myocytes following alpha 1-adrenergic stimulation. *J. Biol. Chem.* **269**:13490–13496.

40. Akhter, S.A., et al. 1998. Targeting the receptor-Gq interface to inhibit in vivo pressure overload myocardial hypertrophy. *Science.* **280**:574–577.

41. Wettschureck, N., et al. 2001. Absence of pressure overload induced myocardial hypertrophy after conditional inactivation of Galphaq/Galpha11 in cardiomyocytes. *Nat. Med.* **7**:1236–1240.

42. D'Angelo, D.D., et al. 1997. Transgenic Galphaq overexpression induces cardiac contractile failure in mice. *Proc. Natl. Acad. Sci. U. S. A.* **94**:8121–8126.

43. Sakata, Y., Hoit, B.D., Liggett, S.B., Walsh, R.A., and Dorn, G.W. 1998. Decompensation of pressure-overload hypertrophy in G alpha q-overexpressing mice. *Circulation.* **97**:1488–1495.

44. Adams, J.W., et al. 1998. Enhanced Galphaq signaling: a common pathway mediates cardiac hypertrophy and apoptotic heart failure. *Proc. Natl. Acad. Sci. U. S. A.* **95**:10140–10145.

45. Yussman, M.G., et al. 2002. Mitochondrial death protein Nix is induced in cardiac hypertrophy and triggers apoptotic cardiomyopathy. *Nat. Med.* **8**:725–730.

46. Dorn, G.W., Tepe, N.M., Wu, G., Yatani, A., and Liggett, S.B. 2000. Mechanisms of impaired beta-adrenergic receptor signaling in G(alphaq)-mediated cardiac hypertrophy and ventricular dysfunction. *Mol. Pharmacol.* **57**:278–287.

47. Murray, N.R., Thompson, L.J., and Fields, A.P. 1997. The role of protein kinase C in cellular proliferation and cell cycle control. In *Molecular biology intelligence unit.* P.J. Parker and L. Dekker, editors. Landes. Austin, Texas, USA. 97–120.

48. Dorn, G.W., and Mochly-Rosen, D. 2002. Intracellular transport mechanisms of signal transducers. *Annu. Rev. Physiol.* **64**:407–429.

49. Sabri, A., and Steinberg, S.F. 2003. Protein kinase C isoform-selective signals that lead to cardiac hypertrophy and the progression of heart failure. *Mol. Cell. Biochem.* **251**:97–101.

50. Bowling, N., et al. 1999. Increased protein kinase C activity and expression of Ca2+-sensitive isoforms in the failing human heart. *Circulation.* **99**:384–391.

51. Malhotra, A., Kang, B.P., Opawumi, D., Belizaire, W., and Meggs, L.G. 2001. Molecular biology of protein kinase C signaling in cardiac myocytes. *Mol. Cell. Biochem.* **225**:97–107.

52. Wang, J., Liu, X., Arneja, A.S., and Dhalla, N.S. 1999. Alterations in protein kinase A and protein kinase C levels in heart failure due to genetic cardiomyopathy. *Can. J. Cardiol.* **15**:683–690.

53. Wang, J., Liu, X., Sentex, E., Takeda, N., and Dhalla, N.S. 2003. Increased expression of protein kinase C isoforms in heart failure due to myocardial infarction. *Am. J. Physiol. Heart Circ. Physiol.* **284**:H2277–H2287.

54. Braz, J.C., et al. 2004. PKC-alpha regulates cardiac contractility and propensity toward heart failure. *Nat. Med.* **10**:248–254.

55. Leitges, M., et al. 1996. Immunodeficiency in protein kinase cbeta-deficient mice. *Science.* **273**:788–791.

56. Miyamoto, A., et al. 2002. Increased proliferation of B cells and auto-immunity in mice lacking protein kinase Cdelta. *Nature.* **416**:865–869.

57. Khasar, S.G., et al. 1999. A novel nociceptor signaling pathway revealed in protein kinase C epsilon mutant mice. *Neuron.* **24**:253–260.

58. Dempsey, E.C., et al. 2000. Protein kinase C isozymes and the regulation of diverse cell responses. *Am. J. Physiol. Lung Cell Mol. Physiol.* **279**:L429–L438.

59. Gray, M.O., et al. 2004. Preservation of base-line hemodynamic function and loss of inducible cardioprotection in adult mice lacking protein kinase C epsilon. *J. Biol. Chem.* **279**:3596–3604.

60. Mayr, M., et al. 2004. Ischemic preconditioning exaggerates cardiac damage in PKC-delta null mice. *Am. J. Physiol. Heart Circ. Physiol.* **287**:H946–H956.

61. Offermanns, S., Toombs, C.F., Hu, Y.H., and Simon, M.I. 1997. Defective platelet activation in G alpha(q)-deficient mice. *Nature.* **389**:183–186.

62. Offermanns, S., et al. 1998. Embryonic cardiomyocyte hypoplasia and craniofacial defects in G alpha q/G alpha 11-mutant mice. *EMBO J.* **17**:4304–4312.

63. Pass, J.M., et al. 2001. PKCepsilon activation induces dichotomous cardiac phenotypes and modulates PKCepsilon-RACK interactions and RACK expression. *Am. J. Physiol. Heart Circ. Physiol.* **280**:H946–H955.

64. Mochly-Rosen, D., Khaner, H., and Lopez, J. 1991. Identification of intracellular receptor proteins for activated protein kinase C. *Proc. Natl. Acad. Sci. U. S. A.* **88**:3997–4000.

65. Hahn, H.S., et al. 2003. Protein kinase Calpha negatively regulates systolic and diastolic function in pathological hypertrophy. *Circ. Res.* **93**:1111–1119.

66. He, Z., and King, G.L. 2004. Protein kinase Cbeta isoform inhibitors: a new treatment for diabetic cardiovascular diseases. *Circulation.* **110**:7–9.

67. Braz, J.C., Bueno, O.F., De Windt, L.J., and Molkentin, J.D. 2002. PKC alpha regulates the hypertrophic growth of cardiomyocytes through extracellular signal-regulated kinase1/2 (ERK1/2). *J. Cell Biol.* **156**:905–919.

68. Wakasaki, H., et al. 1997. Targeted overexpression of protein kinase C beta2 isoform in myocardium causes cardiomyopathy. *Proc. Natl. Acad. Sci. U. S. A.* **94**:9320–9325.

69. Bowman, J.C., et al. 1997. Expression of protein kinase C β in the heart causes hypertrophy in adult mice and sudden death in neonates. *J. Clin. Invest.* **100**:2189–2195.

70. Roman, B.B., Geenen, D.L., Leitges, M., and Buttrick, P.M. 2001. PKC-beta is not necessary for cardiac hypertrophy. *Am. J. Physiol. Heart Circ. Physiol.* **280**:H2264–H2270.

71. Chen, L., et al. 2001. Opposing cardioprotective actions and parallel hypertrophic effects of delta PKC and epsilon PKC. *Proc. Natl. Acad. Sci. U. S. A.* **98**:11114–11119.

72. Inagaki, K., et al. 2003. Inhibition of delta-protein kinase C protects against reperfusion injury of the ischemic heart in vivo. *Circulation.* **108**:2304–2307.

73. Hahn, H.S., et al. 2002. Ischemic protection and myofibrillar cardiomyopathy: dose-dependent effects of in vivo deltaPKC inhibition. *Circ. Res.* **91**:741–748.

74. Gu, X., and Bishop, S.P. 1994. Increased protein kinase C and isozyme redistribution in pressure-overload cardiac hypertrophy in the rat. *Circ. Res.* **75**:926–931.

75. Takeishi, Y., et al. 2000. Transgenic overexpression of constitutively active protein kinase C epsilon causes concentric cardiac hypertrophy. *Circ. Res.* **86**:1218–1223.

76. Mochly-Rosen, D., et al. 2000. Cardiotrophic effects of protein kinase C epsilon: analysis by in vivo modulation of PKCepsilon translocation. *Circ. Res.* **86**:1173–1179.

77. Wu, G., Toyokawa, T., Hahn, H., and Dorn, G.W. 2000. Epsilon protein kinase C in pathological myocardial hypertrophy. Analysis by combined transgenic expression of translocation modifiers and Galphaq. *J. Biol. Chem.* **275**:29927–29930.

78. Badorff, C., et al. 2002. Fas receptor signaling inhibits glycogen synthase kinase 3β and induces cardiac hypertrophy following pressure overload. *J. Clin. Invest.* **109**:373–381. doi:10.1172/JCI200213779.

79. Haq, S., et al. 2000. Glycogen synthase kinase-3beta is a negative regulator of cardiomyocyte hypertrophy. *J. Cell Biol.* **151**:117–130.

80. Morisco, C., et al. 2000. The Akt-glycogen synthase kinase 3beta pathway regulates transcription of atrial natriuretic factor induced by beta-adrenergic receptor stimulation in cardiac myocytes. *J. Biol. Chem.* **275**:14466–14475.

81. Michael, A., et al. 2004. Glycogen synthase kinase-3beta regulates growth, calcium homeostasis, and diastolic function in the heart. *J. Biol. Chem.* **279**:21383–21393.

82. Antos, C.L., et al. 2002. Activated glycogen synthase-3 beta suppresses cardiac hypertrophy in vivo. *Proc. Natl. Acad. Sci. U. S. A.* **99**:907–912.

83. Sanbe, A., et al. 2003. Reengineering inducible cardiac-specific transgenesis with an attenuated myosin heavy chain promoter. *Circ. Res.* **92**:609–616.

84. Hardt, S.E., Tomita, H., Katus, H.A., and Sadoshima, J. 2004. Phosphorylation of eukaryotic translation initiation factor 2Bepsilon by glycogen synthase kinase-3beta regulates beta-adrenergic cardiac myocyte hypertrophy. *Circ. Res.* **94**:926–935.

85. Haq, S., et al. 2003. Stabilization of beta-catenin by a Wnt-independent mechanism regulates cardiomyocyte growth. *Proc. Natl. Acad. Sci. U. S. A.* **100**:4610–4615.

86. Xiao, G., et al. 2001. Inducible activation of c-Myc in adult myocardium in vivo provokes cardiac myocyte hypertrophy and reactivation of DNA synthesis. *Circ. Res.* **89**:1122–1129.

87. Pikkarainen, S., Tokola, H., Kerkela, R., and Ruskoaho, H. 2004. GATA transcription factors in the developing and adult heart. *Cardiovasc. Res.* **63**:196–207.

88. Liao, P., et al. 2002. p38 Mitogen-activated protein kinase mediates a negative inotropic effect in cardiac myocytes. *Circ. Res.* **90**:190–196.

89. Liang, Q., and Molkentin, J.D. 2003. Redefining the roles of p38 and JNK signaling in cardiac hypertrophy: dichotomy between cultured myocytes and animal models. *J. Mol. Cell. Cardiol.* **35**:1385–1394.

90. Petrich, B.G., and Wang, Y. 2004. Stress-activated MAP kinases in cardiac remodeling and heart failure; new insights from transgenic studies. *Trends Cardiovasc. Med.* **14**:50–55.

91. Minamino, T., et al. 2002. MEKK1 is essential for cardiac hypertrophy and dysfunction induced by Gq. *Proc. Natl. Acad. Sci. U. S. A.* **99**:3866–3871.

92. McMullen, J.R., et al. 2004. Inhibition of mTOR signaling with rapamycin regresses established cardiac hypertrophy induced by pressure overload. *Circulation.* **109**:3050–3055.

93. Fingar, D.C., Salama, S., Tsou, C., Harlow, E., and Blenis, J. 2002. Mammalian cell size is controlled

by mTOR and its downstream targets S6K1 and 4EBP1/eIF4E. *Genes Dev.* **16**:1472–1487.

94. Fingar, D.C., et al. 2004. mTOR controls cell cycle progression through its cell growth effectors S6K1 and 4E-BP1/eukaryotic translation initiation factor 4E. *Mol. Cell. Biol.* **24**:200–216.

95. McMullen, J.R., et al. 2004. Deletion of ribosomal S6 kinases does not attenuate pathological, physiological, or insulin-like growth factor 1 receptor-phosphoinositide 3-kinase-induced cardiac hypertrophy. *Mol. Cell. Biol.* **24**:6231–6240.

96. Haq, S., et al. 2001. Differential activation of signal transduction pathways in human hearts with hypertrophy versus advanced heart failure. *Circulation.* **103**:670–677.

97. Baba, H.A., et al. 2003. Dynamic regulation of MEK/Erks and Akt/GSK-3beta in human end-stage heart failure after left ventricular mechanical support: myocardial mechanotransduction-sensitivity as a possible molecular mechanism. *Cardiovasc. Res.* **59**:390–399.

98. Communal, C., et al. 2002. Reciprocal modulation of mitogen-activated protein kinases and mitogen-activated protein kinase phosphatase 1 and 2 in failing human myocardium. *J. Card. Fail.* **8**:86–92.

99. Cook, S.A., Sugden, P.H., and Clerk, A. 1999. Activation of c-Jun N-terminal kinases and p38-mitogen-activated protein kinases in human heart failure secondary to ischaemic heart disease. *J. Mol. Cell. Cardiol.* **31**:1429–1434.

100. Flesch, M., et al. 2001. Differential regulation of mitogen-activated protein kinases in the failing human heart in response to mechanical unloading. *Circulation.* **104**:2273–2276.

101. Lim, H.W., and Molkentin, J.D. 1999. Calcineurin and human heart failure. *Nat. Med.* **5**:246–247.

102. Ng, D.C., Court, N.W., dos Remedios, C.G., and Bogoyevitch, M.A. 2003. Activation of signal transducer and activator of transcription (STAT) pathways in failing human hearts. *Cardiovasc. Res.* **57**:333–346.

103. Razeghi, P., et al. 2003. Mechanical unloading of the failing human heart fails to activate the protein kinase B/Akt/glycogen synthase kinase-3beta survival pathway. *Cardiology.* **100**:17–22.

104. Razeghi, P., and Taegtmeyer, H. 2004. Activity of the Akt/GSK-3beta pathway in the failing human heart before and after left ventricular assist device support. *Cardiovasc. Res.* **61**:196–197.

105. del Monte, F., et al. 1999. Restoration of contractile function in isolated cardiomyocytes from failing human hearts by gene transfer of SERCA2a. *Circulation.* **100**:2308–2311.

106. Hirota, H., et al. 1999. Loss of a gp130 cardiac muscle cell survival pathway is a critical event in the onset of heart failure during biomechanical stress. *Cell.* **97**:189–198.

107. Hata, J.A., Williams, M.L., and Koch, W.J. 2004. Genetic manipulation of myocardial beta-adrenergic receptor activation and desensitization. *J. Mol. Cell. Cardiol.* **37**:11–21.

108. Mani, K., and Kitsis, R.N. 2003. Myocyte apoptosis: programming ventricular remodeling. *J. Am. Coll. Cardiol.* **41**:761–764.

109. McKinsey, T.A., and Olson, E.N. 2005. Toward transcriptional therapies for the failing heart: chemical screens to modulate genes. *J. Clin. Invest.* **115**:538–546. doi:10.1172/JCI200524144.

Toward transcriptional therapies for the failing heart: chemical screens to modulate genes

Timothy A. McKinsey[1] and Eric N. Olson[2]

[1]Myogen Inc., Westminster, Colorado, USA. [2]Department of Molecular Biology, University of Texas Southwestern Medical Center, Dallas, Texas, USA.

In response to acute and chronic stresses, the heart frequently undergoes a remodeling process that is accompanied by myocyte hypertrophy, impaired contractility, and pump failure, often culminating in sudden death. The existence of redundant signaling pathways that trigger heart failure poses challenges for therapeutic intervention. Cardiac remodeling is associated with the activation of a pathological gene program that weakens cardiac performance. Thus, targeting the disease process at the level of gene expression represents a potentially powerful therapeutic approach. In this review, we describe strategies for normalizing gene expression in the failing heart with small molecules that control signal transduction pathways directed at transcription factors and associated chromatin-modifying enzymes.

Introduction

Heart failure, a complex disorder in which cardiac contractility is insufficient to meet the metabolic demands of the body, is the leading cause of death in the Western world. Approximately 5 million individuals in the United States (2–3% of the population) are afflicted with this syndrome, and the numbers are rising. Heart failure results from diverse acute and chronic insults, including coronary artery disease, myocardial infarction, hypertension, valve abnormalities, and inherited mutations in sarcomere and cytoskeletal proteins.

Currently, heart transplantation represents the most effective therapy for end-stage heart failure, but this approach obviously cannot reach the millions of affected individuals worldwide and is not suitable for patients with milder forms of the disease. Traditional therapies for heart failure have involved the use of multiple drugs to improve cardiac contractile function by modifying neurohumoral signaling (e.g., β blockers and angiotensin-converting enzyme inhibitors) or normalizing calcium handling by the cardiomyocyte (1). While such strategies promote short-term improvement in cardiac function, the 5-year mortality rate for heart failure patients remains close to 50%. Thus, there is a great need for the development of novel therapeutics, preferably new drugs, that will improve the quality of life and prolong survival of heart failure patients. An understanding of the mechanistic underpinnings of heart failure represents an essential step toward that goal.

Heart failure is frequently preceded by pathological enlargement of the heart due to hypertrophy of cardiac myocytes (2–5). Cardiac hypertrophy and failure are accompanied by the reprogramming of cardiac gene expression and the activation of "fetal" cardiac genes, which encode proteins involved in contraction, calcium handling, and metabolism (Figure 1) (6–9). Such transcriptional reprogramming has been shown to correlate with loss of cardiac function

and, conversely, improvement in cardiac function in response to drug therapy or implantation of a left ventricular assist device is accompanied by normalization of cardiac gene expression (10–12). Strategies to control cardiac gene expression, therefore, represent attractive, albeit challenging, approaches for heart failure therapy.

Pharmacological normalization of cardiac gene expression in the settings of hypertrophy and heart failure will require the identification of new drug targets that serve as nodal regulators to integrate and transmit stress signals to the genome of the cardiac myocyte. Transcription factors are generally considered to be poor drug targets due to their lack of enzymatic activity and inaccessibility in the nucleus. However, we and others have recently found that cardiac stress response pathways control cardiac gene expression by modulating the activities of chromatin-remodeling enzymes, which act as global regulators of the cardiac genome during pathological remodeling of the heart (13). Here we describe strategies for manipulating chromatin structure to alter cardiac gene expression in the settings of pathological hypertrophy and heart failure as a new means of "transcriptional therapy" for these disorders. We focus on pathways and mechanisms that govern the activity of the nuclear factor of activated T cells (NFAT) and myocyte enhancer factor–2 (MEF2) transcription factors, which integrate cardiac stress signals and play pivotal roles in transcriptional reprogramming of the hypertrophic and failing heart.

Transcriptional remodeling of the hypertrophic and failing heart

In response to acute and chronic insults, the adult heart undergoes distinct remodeling responses, which can take the form of ventricular wall thickening, accompanied by myocyte hypertrophy; or dilatation, accompanied by myocyte elongation (eccentric hypertrophy), serial assembly of sarcomeres, and myocyte apoptosis. While there may be salutary aspects of cardiac hypertrophy, for example, the normalization of ventricular wall stress, it is clear that prolonged hypertrophy in response to stress is deleterious and is a major predictor for heart failure and sudden death (2–5). On the other hand, physiological hypertrophy, as occurs in highly trained athletes or during normal postnatal development, represents a beneficial form of cardiac growth. A major challenge in designing potential therapies for cardiac hypertrophy and failure is to selectively target components of pathological signaling

---

Nonstandard abbreviations used: CaMK, calcium/calmodulin-dependent protein kinase; HAT, histone acetyltransferase; HDAC, histone deacetylase; 5-HT, 5-hydroxytryptamine; MCIP, modulatory calcineurin-interacting protein; MEF2, myocyte enhancer factor–2: MHC, myosin heavy chain; NAD, nicotinamide adenine dinucleotide; NFAT, nuclear factor of activated T cells; PAMH, pyridine activator of myocyte hypertrophy.

Conflict of interest: The authors own stock in Myogen Inc.

Citation for this article: *J. Clin. Invest.* **115**:538–546 (2005).
doi:10.1172/JCI200524144.

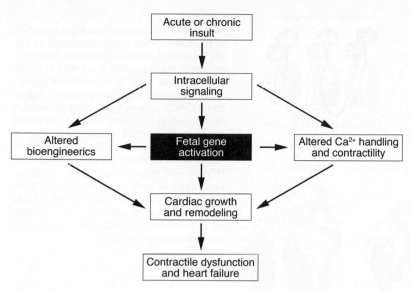

Figure 1
Abnormalities associated with cardiac remodeling during pathological hypertrophy and heart failure.

mechanisms without affecting mechanisms of physiological cardiac growth and function.

Heart failure is typically a disorder of pump function, although it can also arise from acute volume overload (acute aortic insufficiency), high-output disorders (thyroid hormone excess), and pericardial restriction. A hallmark of maladaptive cardiac growth and remodeling is the differential regulation of the 2 myosin heavy chain (MHC) isoforms, α and β, which has a profound effect on cardiac function (14). α-MHC, which is upregulated in the heart after birth, has high ATPase activity, whereas β-MHC has low ATPase activity. Pathological remodeling of the heart in rodent models is accompanied by upregulation of β-MHC expression and downregulation of α-MHC, with consequent reduction in myofibrillar ATPase activity and reduced shortening velocity of cardiac myofibers, leading to eventual contractile dysfunction. Remarkably, minor changes in α-MHC content of the heart can have a profound influence on cardiac performance (15).

Whereas β-MHC is the predominant MHC isoform in the adult rodent heart, the adult human heart contains primarily α-MHC, which has led some to doubt the relevance of MHC isoform switching in human heart failure. Nonetheless, there is compelling evidence indicating a role for changes in MHC isoform expression in the pathogenesis of heart failure in humans. Indeed, α-MHC mRNA and protein levels are markedly reduced in failing human hearts (16, 17), and improvement of left-ventricular ejection fraction through β blocker therapy is associated with normalization of α-MHC expression (10). Additionally, a mutation in the human *α-MHC* gene was identified in association with hypertrophic cardiomyopathy (18), which demonstrates that, despite its low abundance, α-MHC is critical for normal heart function and further validates the hypothesis that MHC isoform switching plays a significant role in human heart failure.

In addition to the α-MHC/β-MHC switch, the gene encoding the sarcoplasmic reticulum Ca²⁺ ATPase (SERCA) is commonly downregulated in the hypertrophic and failing heart, with consequent loss of efficient calcium cycling. Conversely, genes encoding the

natriuretic peptides atrial natriuretic factor (ANF) and B-type natriuretic peptide (BNP) are upregulated as a compensatory mechanism to promote natriuresis and suppress myocyte hypertrophy (14). During pathological remodeling, the heart also undergoes a shift in its mode of energy utilization from oxidative toward glycolytic, which may further contribute to cardiac demise (19). Modulation of RNA polymerase II activity by cyclin-dependent protein kinases has been shown to contribute to many of the above-mentioned processes in the diseased heart (20).

Signaling pathways leading to cardiac hypertrophy and failure

Diverse neurohumoral signals acting through numerous, interwoven signal transduction pathways lead to pathological cardiac hypertrophy and heart failure (6–9). Many such agonists act through cell surface receptors coupled with Gαq to mobilize intracellular calcium, with consequent activation of downstream kinases and the calcium- and calmodulin-dependent phosphatase calcineurin. MAPK signaling pathways are also interconnected at multiple levels with calcium-dependent kinases and calcineurin (21). The details of these pathways have been reviewed elsewhere (6–9, 22). β-Adrenergic agonists also influence cardiac growth and function through the generation of cAMP, which activates PKA and other downstream effectors (23). While we focus in this review on the roles of transcriptional regulators as mediators of cardiac stress-response pathways, it is important to emphasize that such signaling pathways also target a variety of substrates in the cardiomyocyte, including components of the contractile apparatus, calcium channels, and their regulatory proteins.

Control of gene expression by histone acetylation/deacetylation

In eukaryotes, histone-dependent packaging of genomic DNA into chromatin is a central mechanism for gene regulation. The basic unit of chromatin, the nucleosome, comprises DNA wrapped around a histone octamer. Nucleosomes interact to create a highly compact structure that limits access of transcriptional machinery to genomic DNA, thereby repressing gene expression (24).

Histone acetyltransferases (HATs) and histone deacetylases (HDACs) act in an opposing manner to control the acetylation state of nucleosomal histones (Figure 2A). Acetylation of the conserved amino-terminal histone tails by HATs is thought to relax nucleosomal structure by weakening the interaction of the positively charged histone tails with the negatively charged phosphate backbone of DNA, allowing access of transcriptional activators and gene induction. Deacetylation of nucleosomal histones by HDACs results in transcriptional repression.

There are multiple mammalian HDACs, which fall into 3 classes on the basis of structural and biochemical characteristics (25). Class I HDACs consist primarily of a catalytic domain, while class II HDACs contain amino-terminal extensions of approximately 500 residues that harbor binding sites for other transcriptional regulators. The amino-terminal extensions of class II HDACs also contain conserved phosphorylation sites that serve as targets for signaling pathways, thereby connecting signals from outside the cell to the genome. Class III HDACs are

A

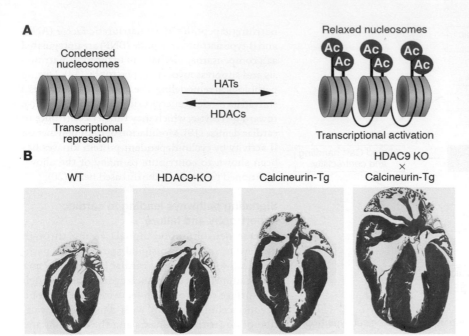

B

Figure 2
Model of HDAC function and hypersensitivity of HDAC9-knockout mice to cardiac stress. (**A**) Schematic of chromatin structure and the actions of HATs and HDACs. HATs acetylate (Ac) histones, causing relaxation of nucleosomal structure and transcriptional activation. HDACs oppose the actions of HATs by deacetylating histones, causing chromatin condensation and transcriptional repression. (**B**) Histological sections of adult mouse hearts of the indicated genotypes are shown. HDAC9-null mice do not display a cardiac phenotype at early age. The cardiac calcineurin transgene (Calcineurin-Tg) induces dramatic cardiac growth, which is exacerbated in the absence of HDAC9. Adapted with permission from *Cell* (26).

unique in that they require nicotinamide adenine dinucleotide (NAD) for catalytic activity.

Control of cardiac growth by class II HDACs

Evidence supporting a role for HDACs in the control of stress-induced cardiac remodeling initially came from studies of class II HDACs. Forced overexpression of class II HDACs 5 or 9 in cardiac myocytes prevents hypertrophy in response to diverse agonists (26–28). More important, mice in which the gene encoding either HDAC5 or HDAC9 has been disrupted by homologous recombination are hypersensitive to pathological signals, developing cardiomegaly and eventual cardiac failure in response to stresses such as pressure overload or introduction of the calcineurin transgene (Figure 2B) (26–29).

Abnormal cardiac growth in HDAC-knockout animals correlates with superactivation of the MEF2 transcription factor (26), which suggests a causal relationship between MEF2 activity and the development of cardiac pathology. Indeed, prior studies established that MEF2 factors selectively associate with class II HDACs via an 18-amino-acid motif present only in these HDACs. Class II HDACs form a complex with MEF2 on gene regulatory elements, resulting in repression of genes harboring MEF2 binding sites (30).

Posttranslational regulation of class II HDACs

Class II HDAC levels do not appear to change in stressed myocardium (26, 29). Instead, these HDACs are shuttled from the nucleus to the cytoplasm in response to stress (Figure 3), which provides a posttranslational mechanism to override HDAC-mediated repression of cardiac growth (27, 28, 31). This redistribution of HDACs frees MEF2 (and other transcription factors) to associate with HATs (32, 33), resulting in increased local histone acetylation and activation of downstream genes that promote cell growth.

Nucleocytoplasmic shuttling of class II HDACs is dependent on phosphorylation of 2 serine-containing motifs found exclusively in these HDACs (34–36). When phosphorylated, these motifs associate with a chaperone protein, termed 14-3-3, which results in the unmasking of a class II HDAC nuclear export sequence (NES) (35–38). The NES is subsequently bound by the CRM1 nuclear export receptor, which escorts class II HDACs from the nucleus to the cytoplasm (Figure 4).

Figure 3
Agonist-dependent nuclear export of HDAC5 correlates with cardiomyocyte hypertrophy. Stimulation of cardiomyocytes with neurohumoral agonists evokes a hypertrophic response characterized by fetal gene activation, sarcomere assembly, and hypertrophy. The upper panels show the subcellular distribution of HDAC5 fused to GFP. In unstressed myocytes, HDAC5-GFP is localized to the nucleus, whereas stimulation with a hypertrophic agonist causes it to redistribute to the cytoplasm. A small molecular inhibitor of the signaling pathway from the cell surface to HDAC5 prevents nuclear export of HDACs and blocks hypertrophy, as detected by staining for sarcomeric α-actinin.

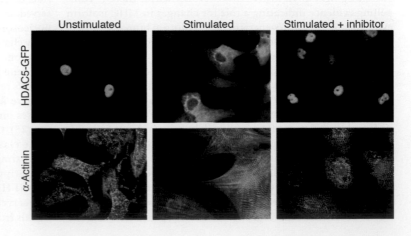

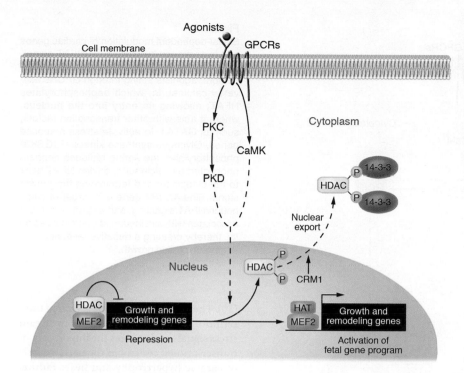

Figure 4
Signal-dependent modulation of cardiac genes
and hypertrophy by class II HDACs. MEF2
recruits class II HDACs to target genes, which
results in transcriptional repression due to chromatin condensation. Stimulation of cardiomyocytes with neurohumoral agonists acting through
G-protein coupled receptors (GPCRs) activates
kinase pathways that culminate with the phosphorylation of class II HDACs and their export
to the cytoplasm as a complex with 14-3-3
proteins. The nuclear export protein CRM1 is
required for HDAC nuclear export. The release
of class II HDACs from MEF2 allows for the
association of HATs with MEF2 and conse-
quentially chromatin relaxation and transcriptional activation of fetal cardiac genes.

Signal-dependent neutralization of class II HDACs is a key step in the control of stress-mediated cardiac growth. Thus, there is interest in identifying signaling molecules that govern posttranslational modification and trafficking of these transcriptional regulators. The availability of small molecules that specifically sustain the repressive function of class II HDACs by inhibiting such signaling factors should provide a novel means to control pathological cardiac remodeling. Initial studies demonstrated that calcium/calmodulin-dependent protein kinase (CaMK) is a potent class II HDAC kinase (34). However, an important issue was whether CaMK is the sole kinase responsible for regulating HDAC nuclear export in the heart or whether multiple kinases might converge on the regulatory HDAC phosphorylation sites, and thus different HDAC kinases might be activated in response to different stimuli.

We recently showed that signaling via PKC leads to the phosphorylation of the same sites in HDAC5 that are phosphorylated by CaMK (28). The PKC family includes at least 12 different isoforms, many, but not all, of which are expressed at appreciable levels in the myocardium (39). Direct activation of PKC by phorbol ester is sufficient to induce nuclear export of HDAC5, and hypertrophic agonists stimulate nuclear export of HDAC5 in cardiac myocytes through a signaling pathway that depends on PKC activation (28).

PKCs are unable to directly phosphorylate HDACs and instead modulate HDAC phosphorylation via the downstream effector kinase PKD. PKD, which is phosphorylated and activated by PKC, physically associates with HDAC5 and promotes phosphorylation of the 14-3-3 binding sites on the protein, which results in HDAC nuclear export. Importantly, small molecule inhibitors that target PKC and PKD, but not CaMK, abolish agonist-mediated nuclear export of HDAC5 cardiac myocytes, which suggests a predominant role for this pathway in the control of HDAC5 in the heart.

PKD inhibitors
Numerous kinase inhibitors are in development for a variety of clinical indications (40). Our data suggest that PKD inhibitors

could be used to block cardiac remodeling in settings where class II HDACs are subject to signal-dependent neutralization. To date, PKD-specific inhibitors have not been identified. However, a heterocyclic derivative of the general serine/threonine kinase inhibitor staurosporine, termed Gö-6976, has been shown to function as a potent PKD inhibitor (41). Although Gö-6976 blocks the activity of several kinases in addition to PKD, it may be possible to synthesize analogs with enhanced specificity.

PKD was recently shown to have the capacity to phosphorylate cardiac troponin I (cTnI), resulting in enhanced contractility (42). Whether PKD is a bona fide TnI kinase, and the implications of this finding on the potential for PKD inhibitors as heart failure therapeutics awaits further investigation. It should also be noted that PKD is expressed in many cell types in addition to cardiomyocytes, and has been implicated in processes as diverse as Golgi transport, proliferation, and apoptosis (43). Clearly, PKD inhibitors may have effects, both positive and negative, in the heart and in other tissues. In vivo proof-of-concept testing with PKD inhibitors awaits the identification of novel bioavailable small molecules with potent and selective activity against this kinase. High-throughput screening and medicinal chemistry approaches may reveal such compounds.

A potential complication of therapeutically targeting a single kinase such as PKD to control class II HDACs in the heart is the possible existence of redundant kinases able to bypass the blockade of PKD. An approach to circumvent this problem may involve inhibiting events that occur following HDAC phosphorylation, such as dissociation of phospho-HDAC from MEF2 or HDAC shuttling to the cytosol.

MEF2/HDAC — additional considerations
It is important to emphasize that HDACs and HDAC kinases do not function in isolation to control MEF2 activity. Indeed, MEF2 is regulated by a plethora of signaling networks that include MAPKs, as well as the calcineurin phosphatase (30, 44). One mechanism

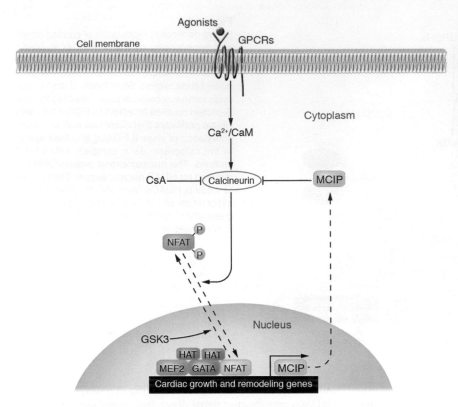

Figure 5
Signal-dependent modulation of cardiac genes and hypertrophy by calcineurin/NFAT signaling. Stimulation of cardiomyocytes with neurohumoral agonists acting through GPCRs activates calcineurin, which dephosphorylates NFAT, allowing its entry into the nucleus, where it acts with other transcription factors, such as GATA4, to activate stress-response genes. Glycogen synthase kinase–3 (GSK3) phosphorylates the serine residues dephosphorylated by calcineurin, driving NFAT back to the cytoplasm and terminating the growth signal. The *MCIP1* gene is a target of calcineurin/NFAT signaling, and its protein product associates with calcineurin to restrain its activity, thereby creating a negative feedback loop to govern cardiac growth.

for calcineurin-dependent activation of MEF2 involves NFAT transcription factors, which, upon dephosphorylation by calcineurin, translocate to the nucleus and facilitate recruitment of HATs to MEF2-response elements (see below) (45).

A challenge in developing inhibitors of MEF2 (or other components of signaling pathways) will be to ensure that sufficient MEF2 activity is maintained for homeostatic control in the heart. Indeed, ablation of MEF2 in the heart can lead to dilated cardiomyopathy in mice and has been linked to increased risk of myocardial infarction in humans (46, 47). These phenotypes may be related to the ability of MEF2 to regulate expression of the PPARγ coactivator-1α (PGC-1) transcription factor, a master regulator of mitochondrial biogenesis (48). In what may be a related finding, cardiomyopathy induced by cyclin-dependent kinase 9 signaling was recently shown to be associated with repression of MEF2-mediated PGC-1 expression (49).

Control of cardiac growth by the calcineurin/NFAT axis
Kinase cascades can certainly regulate cardiac gene expression through negative effects on HDACs. However, protein phosphatases likely play an equally important role in the regulation of chromatin structure during cardiac remodeling. In this regard, the calcium- and calmodulin-dependent protein phosphatase calcineurin is activated in response to cardiac stress signaling, and its activation has been shown to be sufficient and, in many cases, necessary for pathological cardiac hypertrophy (50, 51). Calcineurin dephosphorylates members of the NFAT family of transcription factors, which enables them to translocate into the nucleus, where they activate transcription in cooperation with other transcription factors, including MEF2 and GATA4 (Figure 5).

Misexpression of a mutant form of NFAT, which lacks the regulatory phosphorylation sites and therefore mimics the activated form of the protein, is sufficient to induce cardiac hypertrophy in vivo

(50). The ability of this single activated transcription factor to substitute for upstream stress signals underscores the importance of transcriptional control in the pathogenesis of cardiac hypertrophy and heart failure. The prohypertrophic action of calcineurin can be suppressed by the activity of glycogen synthase kinase–3 (GSK3) (52–54), which phosphorylates the calcineurin substrate sites in NFAT proteins, thereby driving NFAT from the nucleus to the cytoplasm and preventing hypertrophy.

NFAT factors activate gene expression, in part, by recruiting HATs to gene regulatory elements harboring NFAT and MEF2 binding sites. Thus, calcineurin inhibitors should block the ability of NFAT to trigger chromatin-remodeling events that stimulate expression of genes required for pathological cardiac growth. Consistent with this hypothesis, inhibition of calcineurin signaling with cyclosporin A or FK-506 can prevent cardiac hypertrophy in animal models (50, 51, 55). However, since these compounds also suppress T cell function and have other side effects, they will not likely be suitable for long-term treatment of patients with heart failure.

There has been interest in developing small molecules that specifically suppress calcineurin/NFAT signaling in the heart and thereby circumvent the undesirable immunosuppressive effects associated with global calcineurin inhibition. In this regard, calcineurin signaling is inhibited by a family of muscle-enriched proteins, referred to as modulatory calcineurin-interacting proteins (MCIPs) (56), which associate with the calcineurin catalytic subunit and restrain its activity. Thus, it may be possible to selectively suppress calcineurin in the heart by enhancing the expression of MCIP or its association with calcineurin.

MCIP1 expression is activated by calcineurin through a series of NFAT binding sites in the promoter of the *MCIP1* gene, which provides a calcineurin-dependent negative feedback loop whereby the heart can restrain potentially deleterious calcineurin signaling (Figure 5) (57). Consistent with the notion that MCIP1 acts as an endogenous suppressor of calcineurin, MCIP1-knockout mice are hypersensitive to calcineurin signaling and rapidly develop fatal cardiac hypertrophy in response to calcineurin activation (Figure 6) (58). Conversely, overexpression of MCIP in the heart protects against diverse stress signals and blocks hypertrophy (Figure 6)

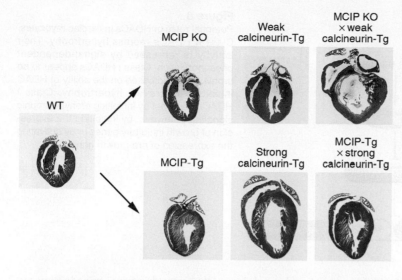

WT

MCIP KO

Weak
calcineurin-Tg

MCIP KO
× weak
calcineurin-Tg

MCIP-Tg

Strong
calcineurin-Tg

MCIP-Tg
× strong
calcineurin-Tg

Figure 6
Modulation of cardiac calcineurin signaling by MCIP expression. The upper set of panels shows the effect of calcineurin signaling in MCIP1-knockout mice. In the absence of MCIP, the heart is sensitized to calcineurin signaling and rapidly undergoes fatal hypertrophy. Because of the heightened sensitivity of MCIP1-knockout mice to calcineurin, only a weak calcineurin transgene can be expressed in the absence of MCIP1. A stronger calcineurin transgene, as used in the lower panels, results in heart failure and death within the first few days after birth. The lower set of panels shows the effect of calcineurin signaling in mice that overexpress MCIP1 in the heart. Overexpression of MCIP1 prevents pathological hypertrophy. Adapted with permission from *Proceedings of the National Academy of Sciences of the United States of America* (58, 59).

(59, 60). MCIP overexpression has also been shown to sustain cardiac function and enhance survival following myocardial infarction in mice (61). Importantly, blockade to hypertrophy by MCIP overexpression does not result in cardiac decompensation, which supports the notion that strategies to prevent pathological hypertrophy will not necessarily be counterproductive (55, 59, 60). MCIP also plays a permissive role in calcineurin activation, the basis of which is incompletely understood (58). However, the inhibitory activity of MCIP toward calcineurin is dominant under conditions of MCIP overexpression.

A chemical screen for modulators of the calcineurin/NFAT axis

Based on the apparent cardioprotective functions of MCIP1, a high-throughput screen for small molecular activators of MCIP1 expression in cardiac myocytes was designed (27). In principle, such activators could induce MCIP1 expression through stimulation of the calcineurin/NFAT signaling pathway, in which case they would likely be prohypertrophic, or they could act through other mechanisms. In a study using regulatory DNA sequences from the *MCIP1* gene linked to a luciferase reporter, several novel small molecules capable of inducing luciferase expression were identified (27). One such activator, pyridine activator of myocyte hypertrophy (PAMH), dramatically stimulated MCIP1 expression.

Scrutiny of the structure of PAMH revealed an embedded pharmacophore resembling serotonin (5-hydroxytryptamine [5-HT]) (Figure 7). Indeed, 5-HT receptor antagonists abolished PAMH activity on cardiomyocytes, and direct ligand binding assays showed that PAMH binds the 5-HT$_{2B}$ receptor with an affinity of approximately 60 nM (27).

Through analysis of the effects of a series of small molecule inhibitors on PAMH activity, PAMH was found to induce MCIP1 expression by activating calcineurin with consequent nuclear import of NFAT and by promoting the phosphorylation-dependent nuclear export of class II HDACs. Thus, although PAMH potently upregulates MCIP expression, it does so through mechanisms that are predicted to exacerbate pathological remodeling of the heart. Indeed, PAMH was found to dramatically stimulate hypertrophy of cultured myocytes. Given this shortcoming, additional screens were devised to identify small molecules that elevate cardiac MCIP protein levels through calcineurin-independent mechanisms.

Although PAMH is unlikely to progress as a development candidate for heart failure, it has altered the paradigm for cardiac calcineurin signaling and has uncovered novel possibilities for manipulation of calcium signaling in the heart. Because PAMH is able to mimic 5-HT, we screened structurally related compounds to find one that might also modulate calcineurin signaling via the 5-HT$_{2B}$ receptor and discovered another novel 4-amino pyridine, A-PAMH, which antagonizes the activity of PAMH (27). Intriguingly, A-PAMH also blocks the hypertrophic effects of PE, which acts through the α-adrenergic receptor. We do not currently understand the molecular basis for this antagonism, which could reflect crosstalk between the 5-HT$_{2A/B}$ and α-adrenergic receptors, competition between A-PAMH and PE for binding to the α-adrenergic receptor, or another mechanism.

Consistent with the finding that signaling through 5-HT$_{2A/B}$ receptors can drive cardiac growth, overexpression of the 5-HT$_{2B}$ receptor in the heart leads to cardiac hypertrophy (62), and 5-HT$_{2B}$ receptor–knockout mice die during mid-gestation from a failure of ventricular growth (63). We propose a model in which 5-HT$_{2A/B}$ signaling promotes cardiac hypertrophy by stimulating calcineurin/NFAT signaling, with consequent recruitment of HATs to regulatory regions of NFAT target genes. Since PAMH also stimulates nuclear export of class II HDACs, it is likely that MEF2 plays a role in the mechanism by which 5-HT$_{2A/B}$ signaling triggers cardiac remodeling.

Normalization of cardiac gene expression with HDAC inhibitors

In addition to indirectly regulating chromatin structure via small molecules that target signaling pathways, it may also be possible

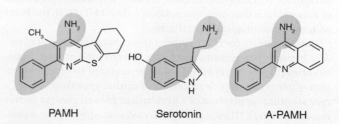

PAMH   Serotonin   A-PAMH

Figure 7
Structures of serotonin, PAMH, and A-PAMH.

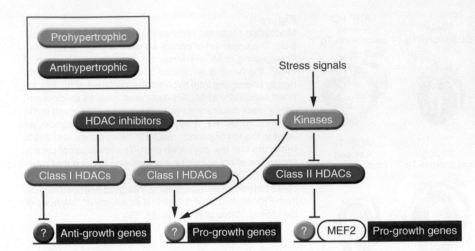

Figure 8

Potential roles of HDACs in cardiac myocytes. Class II HDACs repress hypertrophy. Their activity is repressed by signal-dependent phosphorylation. Class I HDACs appear to be prohypertrophic based on the ability of HDAC inhibitors to prevent hypertrophy. Class I HDACs could act by inhibiting prohypertrophic signaling pathways, by inhibiting the expression of growth inhibitory genes or by activating the expression of pro-growth genes.

to therapeutically control cardiac remodeling by directly inhibiting chromatin-modifying enzymes. In this regard, synthetic and naturally occurring small molecule inhibitors of HATs have been described (64, 65). Several classes of enzymatic inhibitors of HDACs, including hydroxamic acids, short chain fatty acids, benzamides, cyclic tetrapeptides, and bicyclic depsipeptides, have also been identified (66).

The effect of HAT inhibitors on cardiac remodeling remains unknown. However, given the ability of the p300 HAT to stimulate dilated cardiomyopathy in mice (67), this class of compounds may prove beneficial in the context of heart failure. Surprisingly, HDAC inhibitors have been found to potently repress agonist-dependent cardiac hypertrophy in a manner that correlates with increased histone acetylation (68, 69). This finding was paradoxical, since class II HDACs block cardiac hypertrophy, and HDAC inhibitors would logically be predicted to neutralize this repressive function.

How might the paradoxical effects of class II HDACs and HDAC inhibitors on cardiac growth be explained? The demonstration that class II HDAC catalytic activity is not required to repress the hypertrophic program suggests that these HDACs act through a mechanism independent of HDAC activity to repress hypertrophy (26). The antihypertrophic action of class II HDACs likely lies in their capacity to prohibit binding of HATs to MEF2 or other transcription factors and to repress transcription via HDAC-independent mechanisms (13).

Based on the results with HDAC inhibitors, one or more HDACs may play a positive or permissive role in the control of cardiac hypertrophy. Given the antihypertrophic function of class II HDACs, a class I HDAC(s) likely fulfills this role. Of note, the HDAC inhibitors used in studies of cardiac hypertrophy do not antagonize NAD-dependent class III HDACs. A testable prediction of this model is that overexpression of a class I HDAC in the heart will cause pathological hypertrophy.

The mechanism for HDAC inhibitor–mediated repression of cardiac hypertrophy remains unknown. We envision at least 3 possibilities (Figure 8). (a) HDACs may be required to block expression of genes that encode repressors of cardiac growth. (b) HDACs may stimulate expression of a pro-cardiac growth gene(s). In this regard, although HDACs are typically associated with gene repression, there is an increasing number of examples in which HDACs have been linked to gene induction (70). (c) HDACs may directly

or indirectly modulate prohypertrophic signal transduction cascades. Indeed, several nonhistone targets for HATs/HDACs have been identified, including factors implicated in cardiac remodeling, such as tubulin and the GATA transcription factor (67, 71).

Potential for HDAC inhibitors in heart failure therapy

The fortuitous discovery that HDAC inhibitors repress cardiac hypertrophy and normalize cardiac gene expression in the face of stress may ultimately affect the treatment of heart failure in humans. Importantly, HDAC inhibition in vitro and in vivo results in downregulation β-MHC expression, with a concomitant increase in the levels of α-MHC (68, 72). Thus, we predict that HDAC inhibitors will not only antagonize deleterious cardiac growth, but will also increase myofibrillar ATPase activity and improve contractility in the failing heart.

Advancement of an HDAC inhibitor into the clinic for the treatment of heart failure awaits rigorous preclinical testing in animal models of pathological cardiac remodeling. Nonetheless, enthusiasm for this novel therapeutic approach is buoyed by successes with HDAC inhibitors in other disease models as well as in humans. Indeed, at least 9 independent approaches for HDAC inhibition are currently being tested in human clinical trials for cancer, and therapeutic benefit and tolerability have been observed (66).

Looking to the future

The primary focus of this review has been on pathways and interventions that affect cardiomyocyte hypertrophy as an endpoint. While cardiac hypertrophy is clearly an important prognostic indicator of poor clinical outcome (2–5), it is but one component of the remodeling process, and deleterious remodeling can occur in the absence of cardiac cell growth. A common feature of remodeling, regardless of etiology, is fetal cardiac gene induction, which is thought to contribute to cardiac demise through the dysregulation of genes encoding proteins that control cardiac contractility (e.g., MHC). The types of transcriptional therapies proposed here would be designed to block not hypertrophy per se, but rather the pathological gene program that underlies the remodeling process.

Despite the success of lipid-lowering drugs for the prevention of coronary artery disease and myocardial infarction, and several decades of effort toward modifying cardiac function through traditional drug targets, heart failure remains the most common cause of disease-related death in the Western world, and all projections for the future point to a continued increase in its prevalence. Thus, major advances toward the reduction of heart failure incidence will require new therapeutic approaches. We propose that pathological cardiac remodeling in the settings of hypertrophy and heart failure can be viewed, at least in part, as a transcriptional disorder and that normalization of the gene expression pattern of the cardiac myocyte through transcriptional therapies represents a promising and largely unexploited approach for cardiac therapy. Recent advances in understanding stress-response pathways responsible for heart failure and mechanisms of cardiac gene expression, combined with new technologies for high throughput screening for novel small molecular modifiers of cellular functions, offer promising and untapped opportunities for success in failure.

Acknowledgments

Work in Eric Olson's laboratory was supported by the NIH, the Donald W. Reynolds Cardiovascular Clinical Research Center, and the Robert A. Welch Foundation. We thank A. Tizenor, L. Melvin, E. Bush, and L. Castonguary for assistance with graphics, and are grateful to R. Gorczynski, L. Melvin, N. Pagratis, and E. Bush for invaluable scientific input.

Address correspondence to: Timothy A. McKinsey, Myogen Inc., Westminster, Colorado 80021, USA. Phone: (303) 533-1736; Fax: (303) 410-6669; E-mail: timothy.mckinsey@myogen.com. Or to: Eric N. Olson, Department of Molecular Biology, University of Texas Southwestern Medical Center, Dallas, Texas 75390, USA. Phone: (214) 648-1187; Fax: (214) 648-1196; E-mail: Eric.Olson@utsouthwestern.edu.

1. Linseman, J.V., and Bristow, M.R. 2003. Drug therapy and heart failure prevention. Circulation. 107:1234–1236.
2. Vakili, B.A., Okin, P.M., and Devereux, R.B. 2001. Prognostic implications of left ventricular hypertrophy. Am. Heart J. 141:334–341.
3. Okin, P.M., et al.; for the LIFE Study Investigators. 2004. Regression of electrocardiographic left ventricular hypertrophy during antihypertensive treatment and the prediction of major cardiovascular events. JAMA. 292:2343–2349.
4. Devereux, R.B., et al. 2004. Prognostic significance of left ventricular mass change during treatment of hypertension. JAMA. 292:2350–2356.
5. Gardin, J.M., and Lauer, M.S. 2004. Left ventricular hypertrophy: the next treatable, silent killer? JAMA. 292:2396–2398.
6. Olson, E.N., and Schneider, M.D. 2003. Sizing up the heart: development redux in disease. Genes Dev. 17:1937–1956.
7. Chien, K.R. 1999. Stress pathways and heart failure. Cell. 98:555–558.
8. Marks, A.R. 2003. A guide for the perplexed: towards an understanding of the molecular basis of heart failure. Circulation. 107:1456–1459.
9. Molkentin, J.D., and Dorn, I.G., 2nd. 2001. Cytoplasmic signaling pathways that regulate cardiac hypertrophy. Annu. Rev. Physiol. 63:391–426.
10. Abraham, W.T., et al. 2002. Coordinate changes in Myosin heavy chain isoform gene expression are selectively associated with alterations in dilated cardiomyopathy phenotype. Mol. Med. 8:750–760.
11. Lowes, B.D., et al. 2002. Myocardial gene expression in dilated cardiomyopathy treated with beta-blocking agents. N. Engl. J. Med. 346:1357–1365.
12. Blaxall, B.C., Tschannen-Moran, B.M., Milano, C.A., and Koch, W.J. 2003. Differential gene expression and genomic patient stratification following left ventricular assist device support. J. Am. Coll. Cardiol. 41:1096–1106.
13. McKinsey, T.A., and Olson, E.N. 2004. Cardiac histone acetylation--therapeutic opportunities abound. Trends Genet. 20:206–213.
14. Braunwald, E., and Bristow, M.R. 2000. Congestive heart failure: fifty years of progress. Circulation. 102(Suppl. 4):IV14–IV23.
15. Herron, T.J., and McDonald, K.S. 2002. Small amounts of alpha-myosin heavy chain isoform expression significantly increase power output of rat cardiac myocyte fragments. Circ. Res. 90:1150–1152.
16. Nakao, K., Minobe, W., Roden, R., Bristow, M.R., and Leinwand, L.A. 1997. Myosin heavy chain gene expression in human heart failure. J. Clin. Invest. 100:2362–2370.
17. Miyata, S., Minobe, W., Bristow, M.R., and Leinwand, L.A. 2000. Myosin heavy chain isoform expression in the failing and nonfailing human heart. Circ. Res. 86:386–390.
18. Niimura, H., et al. 2002. Sarcomere protein gene mutations in hypertrophic cardiomyopathy of the elderly. Circulation. 105:446–451.
19. Ashrafian, H., Redwood, C., Blair, E., and Watkins, H. 2003. Hypertrophic cardiomyopathy: a paradigm for myocardial energy depletion. Trends Genet. 19:263–268.
20. Sano, M., et al. 2002. Activation and function of cyclin T-Cdk9 (positive transcription elongation factor-b) in cardiac muscle-cell hypertrophy. Nat. Med. 8:1310–1317.
21. Sugden, P.H., and Clerk, A. 1998. "Stress-responsive" mitogen-activated protein kinases (c-Jun N-terminal kinases and p38 mitogen-activated protein kinases) in the myocardium. Circ. Res. 83:345–352.
22. Frey, N., and Olson, E.N. 2003. Cardiac hypertrophy: the good, the bad, and the ugly Annu. Rev. Physiol. 65:45–79.
23. Rockman, H.A., Koch, W.J., and Lefkowitz, R.J. 2002. Seven-transmembrane-spanning receptors and heart function. Nature. 415:206–212.
24. Fischle, W., Wang, Y., and Allis, C.D. 2003. Histone and chromatin cross-talk. Curr. Opin. Cell Biol. 15:172–183.
25. Verdin, E., Dequiedt, F., and Kasler, H.G. 2003. Class II histone deacetylases: versatile regulators. Trends Genet. 19:286–293.
26. Zhang, C.L., et al. 2002. Class II histone deacetylases act as signal-responsive repressors of cardiac hypertrophy. Cell. 110:479–488.
27. Bush, E., et al. 2004. A small molecular activator of cardiac hypertrophy uncovered in a chemical screen for modifiers of the calcineurin signaling pathway. Proc. Natl. Acad. Sci. U. S. A. 101:2870–2875.
28. Vega, R.B., et al. 2004. Protein kinases C and D mediate agonist-dependent cardiac hypertrophy through nuclear export of histone deacetylase 5. Mol. Cell. Biol. 24:8374–8385.
29. Chang, S., et al. 2004. Histone deacetylases 5 and 9 govern responsiveness of the heart to a subset of stress signals and play redundant roles in heart. Mol. Cell. Biol. 24:8467–8476.
30. McKinsey, T.A., Zhang, C.L., and Olson, E.N. 2002. MEF2: a calcium-dependent regulator of cell division, differentiation and death. Trends Biochem. Sci. 27:40–47.
31. Harrison, B.C., et al. 2004. The CRM1 nuclear export receptor controls of pathological cardiac gene expression. Mol. Cell. Biol. 24:10636–10649.
32. Youn, H.D., Grozinger, C.M., and Liu, J.O. 2000. Calcium regulates transcriptional repression of myocyte enhancer factor 2 by histone deacetylase 4. J. Biol. Chem. 275:22563–22567.
33. Han, A., et al. 2003. Sequence-specific recruitment of transcriptional co-repressor Cabin1 by myocyte enhancer factor-2. Nature. 422:730–734.
34. McKinsey, T.A., Zhang, C.L., Lu, J., and Olson, E.N. 2000. Signal-dependent nuclear export of a histone deacetylase regulates muscle differentiation. Nature. 408:106–111.
35. McKinsey, T.A., Zhang, C.L., and Olson, E.N. 2000. Activation of the myocyte enhancer factor-2 transcription factor by calcium/calmodulin-dependent protein kinase-stimulated binding of 14-3-3 to histone deacetylase 5. Proc. Natl. Acad. Sci. U. S. A. 97:14400–14405.
36. Grozinger, C.M., and Schreiber, S.L. 2000. Regulation of histone deacetylase 4 and 5 and transcriptional activity by 14-3-3-dependent cellular localization. Proc. Natl. Acad. Sci. U. S. A. 97:7835–7840.
37. McKinsey, T.A., Zhang, C.L., and Olson, E.N. 2001. Identification of a signal-responsive nuclear export sequence in class II histone deacetylases. Mol. Cell. Biol. 21:6312–6321.
38. Wang, A.H., and Yang, X.J. 2001. Histone deacetylase 4 possesses intrinsic nuclear import and export signals. Mol. Cell. Biol. 21:5992–6005.
39. Das, D.K. 2003. Protein kinase C isozymes signaling in the heart. J. Mol. Cell. Cardiol. 35:887–889.
40. Force, T., Kuida, K., Namchuk, M., Parang, K., and Kyriakis, J.M. 2004. Inhibitors of protein kinase signaling pathways: emerging therapies for cardiovascular disease. Circulation. 109:1196–1205.
41. Gschwendt, M., et al. 1996. Inhibition of protein kinase C mu by various inhibitors. Differentiation from protein kinase c isoenzymes. FEBS Lett. 392:77–80.
42. Haworth, R.S., et al. 2004. Protein kinase D is a novel mediator of cardiac troponin I phosphorylation and regulates myofilament function. Circ Res. 95:1091–1099.
43. Van Lint, J., et al. 2002. Protein kinase D: an intracellular traffic regulator on the move. Trends Cell Biol. 12:193–200.
44. Han, J., and Molkentin, J.D. 2000. Regulation of MEF2 by p38 MAPK and its implication in cardiomyocyte biology. Trends Cardiovasc. Med. 10:19–22.
45. Youn, H.D., Chatila, T.A., and Liu, J.O. 2000. Integration of calcineurin and MEF2 signals by the coactivator p300 during T-cell apoptosis. EMBO J. 19:4323–4231.
46. Naya, F.J., et al. 2002. Mitochondrial deficiency and cardiac sudden death in mice lacking the MEF2A transcription factor. Nat. Med. 8:1303–1309.
47. Wang, L., Fan, C., Topol, S.E., Topol, E.J., and Wang, Q. 2003. Mutation of MEF2A in an inherited disorder with features of coronary artery disease. Science. 302:1578–1581.
48. Czubryt, M.P., McAnally, J., Fishman, G.I., and Olson, E.N. 2003. Regulation of peroxisome proliferator-activated receptor gamma coactivator 1 alpha (PGC-1 alpha) and mitochondrial function by MEF2 and HDAC5. Proc. Natl. Acad. Sci. U. S. A. 100:1711–1716.
49. Sano, M., et al. 2004. Activation of cardiac Cdk9 represses PGC-1 and confers a predisposition to heart failure. EMBO J. 23:3559–3569.

50. Molkentin, J.D., et al. 1998. A calcineurin-dependent transcriptional pathway for cardiac hypertrophy. *Cell.* **93**:215–228.

51. Leinwand, L.A. 2001. Calcineurin inhibition and cardiac hypertrophy: a matter of balance. *Proc. Natl. Acad. Sci. U. S. A.* **98**:2947–2949.

52. Antos, C.L., et al. 2002. Activated glycogen synthase kinase-3 beta suppresses cardiac hypertrophy in vivo. *Proc. Natl. Acad. Sci. U. S. A.* **99**:907–912.

53. Haq, S., et al. 2000. Glycogen synthase kinase-3beta is a negative regulator of cardiomyocyte hypertrophy. *J. Cell Biol.* **151**:117–130.

54. Morisco, C., et al. 2000. The Akt-glycogen synthase kinase 3beta pathway regulates transcription of atrial natriuretic factor induced by beta-adrenergic receptor stimulation in cardiac myocytes. *J. Biol. Chem.* **275**:14466–14475.

55. Frey, N., Katus, H.A., Olson, E.N., and Hill, J.A. 2004. Hypertrophy of the heart: a new therapeutic target? *Circulation.* **109**:1580–1589.

56. Vega, R.B, Bassel-Duby, R., and Olson, E.N. 2003. Control of cardiac growth and function by calcineurin signaling. *J. Biol. Chem.* **278**:36981–36984.

57. Yang, J., et al. 2000. Independent signals control expression of the calcineurin inhibitory proteins MCIP1 and MCIP2 in striated muscles. *Circ. Res.* **87**:E61–E68.

58. Vega, R.B., et al. 2003. Dual roles of modulatory calcineurin-interacting protein 1 in cardiac hypertrophy. *Proc. Natl. Acad. Sci. U. S. A.* **100**:669–674.

59. Rothermel, B.A., et al. 2001. Myocyte-enriched calcineurin-interacting protein, MCIP1, inhibits cardiac hypertrophy in vivo. *Proc. Natl. Acad. Sci. U. S. A.* **98**:3328–3333.

60. Hill, J.A., et al. 2002. Targeted inhibition of calcineurin in pressure-overload cardiac hypertrophy. Preservation of systolic function. *J. Biol. Chem.* **277**:10251–10255.

61. van Rooij, E., et al. 2004. MCIP1 overexpression suppresses left ventricular remodeling and sustains cardiac function after myocardial infarction. *Circ. Res.* **94**:e18–e26.

62. Nebigil, C.G., et al. 2003. Overexpression of the serotonin 5-HT2B receptor in heart leads to abnormal mitochondrial function and cardiac hypertrophy. *Circulation.* **107**:3223–3229.

63. Nebigil, C.G., et al. 2001. Ablation of serotonin 5-HT(2B) receptors in mice leads to abnormal cardiac structure and function. *Circulation.* **103**:2973–2979.

64. Lau, O.D., et al. 2000. HATs off: selective synthetic inhibitors of the histone acetyltransferases p300 and PCAF. *Mol. Cell.* **5**:589–595.

65. Balasubramanyam, K., Swaminathan, V., Ranganathan, A., and Kundu, T.K. 2003. Small molecule modulators of histone acetyltransferase p300. *J. Biol. Chem.* **278**:19134–19140

66. Piekarz, R., and Bates, S. 2004. A review of depsipeptide and other histone deacetylase inhibitors in clinical trials. *Curr. Pharm. Des.* **10**:2289–2298.

67. Yanazume, T., et al. 2003. Cardiac p300 is involved in myocyte growth with decompensated heart failure. *Mol. Cell. Biol.* **23**:3593–3606.

68. Antos, C.L., et al. 2003. Dose-dependent blockade to cardiomyocyte hypertrophy by histone deacetylase inhibitors. *J. Biol. Chem.* **278**:28930–28937.

69. Kook, H., et al. 2003. Cardiac hypertrophy and histone deacetylase–dependent transcriptional repression mediated by the atypical homeodomain protein Hop. *J. Clin. Invest.* **112**:863–871. doi:10.1172/JCI200319137

70. Chang, H.M., et al. 2004. Induction of interferon-stimulated gene expression and antiviral responses require protein deacetylase activity. *Proc. Natl. Acad. Sci. U. S. A.* **101**:9578–9583.

71. Westermann, S., and Weber, K. 2003. Post-translational modifications regulate microtubule function. *Nat. Rev. Mol. Cell Biol.* **4**:938–947.

72. Davis, F.J., Pillai, J.B., Gupta, M., and Gupta, M.P. 2005. Concurrent opposite effects of an inhibitor of histone deacetylases, Trichostatin-A, on the expression of cardiac {alpha}-myosin heavy chain and tubulins: implication for gain in cardiac muscle contractility. *Am. J. Physiol. Heart Circ. Physiol.* **288**:H1477–H1490.

Mitochondrial energy metabolism in heart failure: a question of balance

Janice M. Huss[1,2] and Daniel P. Kelly[1,2,3,4]

[1]Center for Cardiovascular Research and [2]Department of Medicine, [3]Department of Molecular Biology & Pharmacology, and
[4]Department of Pediatrics, Washington University School of Medicine, St. Louis, Missouri, USA.

The mitochondrion serves a critical role as a platform for energy transduction, signaling, and cell death pathways relevant to common diseases of the myocardium such as heart failure. This review focuses on the molecular regulatory events and downstream effector pathways involved in mitochondrial energy metabolic derangements known to occur during the development of heart failure.

Introduction

All cellular processes are driven by ATP-dependent pathways. The heart has perpetually high energy demands related to the maintenance of specialized cellular processes, including ion transport, sarcomeric function, and intracellular Ca^{2+} homeostasis. Myocardial workload (energy demand) and energy substrate availability (supply) are in continual flux, yet the heart has a limited capacity for substrate storage. Thus, ATP-generating pathways must respond proportionately to dynamic fluctuations in physiological demands and fuel delivery. The time frame of such metabolic regulatory responses ranges from seconds to minutes (acute) or hours to days (chronic) and involves regulation at multiple levels, including allosteric control of enzyme activity via metabolic intermediates, signal transduction events, and the regulation of genes encoding rate-limiting enzymes and proteins.

Metabolic regulation is inextricably linked with cardiac function. This metabolism-function relationship is relevant to diseases that lead to cardiac hypertrophy and heart failure. The progression to heart failure of any cause is associated with a gradual but progressive decline in the activity of mitochondrial respiratory pathways leading to diminished capacity for ATP production. Reduced capacity for energy transduction leads to secondary dysregulation of cellular processes critical for cardiac pump function, including Ca^{2+} handling and contractile function, which results in a downward spiral of increased energy demand and diminished function. Evidence has emerged that energy deficiency can be a cause and an effect of heart failure. Cardiac metabolic regulatory events may also be adaptive in certain disease states. For example, in the ischemic heart, a reduction in mitochondrial oxidative capacity serves to reduce oxygen consumption in the context of limited O_2 availability. Many of the metabolic regulatory events that dictate fuel selection and capacity for ATP production in the normal and failing heart occur at the level of gene expression. The consequence of specific metabolic gene regulatory events as adaptive ver-

sus maladaptive in the context of myocardial disease is an area of intense investigation.

The purposes of this review are to provide a brief overview of cardiac energy metabolic pathways with emphasis on the mitochondrion, to describe the gene regulatory circuitry involved in the regulation of cardiac mitochondrial energy metabolism in the normal heart, to summarize the current knowledge about how this metabolic regulatory network is altered during the development of heart failure, and to review the evidence that links altered energy metabolism to the development of heart failure. Emphasis will be given to gene regulatory mechanisms and upstream signaling pathways.

Cardiac energy metabolic pathways

Oxidation of fatty acids (FAs) and glucose in mitochondria accounts for the vast majority of ATP generation in the healthy adult heart (1, 2). FAs are the preferred substrate in the adult myocardium, supplying about 70% of total ATP (3–5). FAs derived from circulating triglyceride-rich lipoproteins and albumin-bound nonesterified FAs are oxidized in the mitochondrial matrix by the process of FA β-oxidation (FAO), whereas pyruvate derived from glucose and lactate is oxidized by the pyruvate-dehydrogenase (PDH) complex, localized within the inner mitochondrial membrane (Figure 1). Acetyl-CoA, derived from both pathways, enters the tricarboxylic acid (TCA) cycle. Reduced flavin adenine dinucleotide ($FADH_2$) and NADH are generated via substrate flux through the β-oxidation spiral and the TCA cycle, respectively. The reducing equivalents enter the electron transport chain, producing an electrochemical gradient across the mitochondrial membrane that drives ATP synthesis in the presence of molecular oxygen (oxidative phosphorylation).

Mitochondrial enzymes are encoded by both nuclear and mitochondrial genes (reviewed in ref. 6). All of the enzymes of β-oxidation and the TCA cycle, and most of the subunits of electron transport/oxidative phosphorylation, are encoded by nuclear genes. The mitochondrial genome is comprised of 1 circular double-stranded chromosome that encodes 13 electron transport chain subunits within complexes I, III, and IV (7). Since mitochondrial number and function require both nuclear and mitochondrial-encoded genes, coordinated mechanisms exist to regulate the 2 genomes and determine overall cardiac oxidative capacity. In addition, distinct pathways exist to coordinately regulate nuclear genes encoding component mitochondrial pathways.

Nonstandard abbreviations used: Ant, adenine nucleotide translocator; BAT, brown adipose tissue; ERR, estrogen-related receptor; FA, fatty acid; $FADH_2$, reduced flavin adenine dinucleotide; FAO, FA β-oxidation; NRF, nuclear respiratory factor; PDH, pyruvate dehydrogenase; PET, positron emission tomography; PGC-1, PPARγ coactivator-1; PRC, PGC-1–related coactivator; RXRα, retinoid X receptor α; TCA, tricarboxylic acid; Tfam, mitochondrial transcription factor A.

Conflict of interest: The authors have declared that no conflict of interest exists.

Citation for this article: *J. Clin. Invest.* **115**:547–555 (2005).
doi:10.1172/JCI200524405.

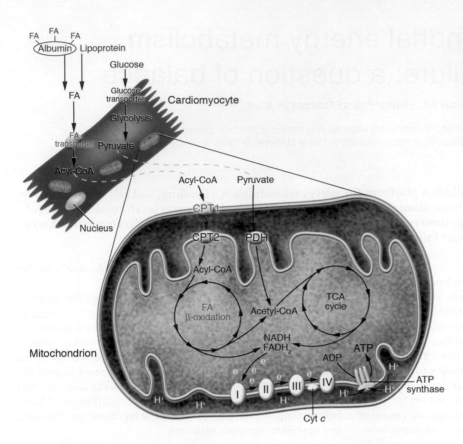

Figure 1
Pathways involved in cardiac energy metabolism. FA and glucose oxidation are the main ATP-generating pathways in the adult mammalian heart. Acetyl-CoA derived from FA and glucose oxidation is further oxidized in the TCA cycle to generate NADH and FADH₂, which enter the electron transport/oxidative phosphorylation pathway and drive ATP synthesis. Genes encoding enzymes involved at multiple steps of these metabolic pathways (i.e., uptake, esterification, mitochondrial transport, and oxidation) are transcriptionally regulated by PGC-1α with its nuclear receptor partners, including PPARs and ERRs (blue text). Glucose uptake/oxidation and electron transport/oxidative phosphorylation pathways are also regulated by PGC-1α via other transcription factors, such as MEF-2 and NRF-1. Cyt *c*, cytochrome *c*.

The transcriptional network regulating cardiac mitochondrial biogenesis and respiratory function

PGC-1α: an inducible integrator of transcriptional circuits regulating mitochondrial biogenesis and function. The PPARγ coactivator-1 (PGC-1) family of transcriptional coactivators is involved in regulating mitochondrial metabolism and biogenesis. PGC-1α was the first member discovered through its functional interaction with the nuclear receptor PPARγ in brown adipose tissue (BAT), a mitochondrial-rich tissue involved in thermogenesis (8). Two PGC-1α–related coactivators, PGC-1β (also called PERC) and PGC-1–related coactivator (PRC), have since been identified (9–11). PRC is ubiquitously expressed and coactivates transcription factors involved in mitochondrial biogenesis; however, to date there is no direct empirical evidence that PRC drives or is necessary for mitochondrial biogenesis (9, 12). PGC-1α and PGC-1β share some regulatory overlap. Both are preferentially expressed in tissues with high oxidative capacity, such as heart, slow-twitch skeletal muscle, and BAT, where they serve critical roles in the regulation of mitochondrial functional capacity (8, 10, 13–15). PGC-1α regulates additional metabolic pathways, including hepatic gluconeogenesis and skeletal muscle glucose uptake (16–18). Based on its tissue expression pattern, PGC-1β probably plays a role in regulating energy metabolism in the heart, although its precise role in the normal heart and in mediating alterations in energy metabolism observed in heart failure has not been investigated.

PGC-1α is distinct from other PGC-1 family members, indeed from most coactivators, in its broad responsiveness to developmental alterations in energy metabolism and physiological and pathological cues at the level of expression and transactivation. In the heart, PGC-1α expression increases at birth coincident with an increase in cardiac oxidative capacity and a perinatal shift from

reliance on glucose metabolism to the oxidation of fats for energy (19). PGC-1α is induced by physiological stimuli that increase ATP demand and stimulate mitochondrial oxidation, including cold exposure, fasting, and exercise (8, 19–23). Activation of the PGC-1α regulatory cascade increases cardiac mitochondrial oxidative capacity in the heart. In cardiac myocytes in culture, PGC-1α increases mitochondrial number, upregulates expression of mitochondrial enzymes, and increases rates of FA oxidation and coupled respiration (19, 24). A mouse cardiac-specific conditional transgenic expression system was used to demonstrate that PGC-1α triggers mitochondrial biogenesis in vivo in a developmental stage–dependent manner: permissive during the neonatal period but less so in the adult heart (25). In summary, PGC-1α is an inducible coactivator that coordinately regulates cardiac fuel selection and mitochondrial ATP-producing capacity.

How does PGC-1α exert its pleiotropic effects on mitochondrial biogenesis and respiratory function? Recent work by several laboratories has provided insight into the downstream transcriptional regulatory circuits through which PGC-1α mediates its effects (Figure 2) (6). PGC-1α activates expression of nuclear respiratory factor-1 (NRF-1) and NRF-2 and directly coactivates NRF-1 on its target gene promoters (13). NRF-1 and NRF-2 regulate expression of mitochondrial transcription factor A (Tfam), a nuclear-encoded transcription factor that binds regulatory sites on mitochondrial DNA and is essential for replication, maintenance, and transcription of the mitochondrial genome (26–28). Furthermore, NRF-1 and NRF-2 regulate the expression of nuclear genes encoding respiratory chain subunits and other proteins required for mitochondrial function (29, 30).

PGC-1α coactivates the PPAR and ERR nuclear receptors, critical regulators of myocardial FA utilization. PGC-1α regulates genes involved in the cellular uptake and mitochondrial oxidation of FAs through

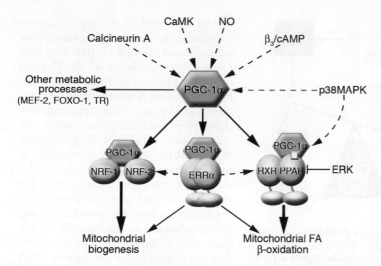

Figure 2
PGC-1α is an integrator of the transcriptional network regulating mitochondrial biogenesis and function. Numerous signaling pathways, including Ca²⁺-dependent, NO, MAPK, and β-adrenergic pathways (β₃/cAMP), activate the PGC-1α directly by increasing either PGC-1α expression or activity. Additionally, the p38MAPK pathway selectively activates PPARα, which may bring about synergistic activation in the presence of PGC-1α, whereas ERK-MAPK has the opposite effect. These signaling pathways transduce physiological stimuli, such as stress, fasting, and exercise, to the PGC-1α pathway. PGC-1α, in turn, coactivates transcriptional partners, including NRF-1 and -2, ERRα, and PPARα, which regulate mitochondrial biogenesis and FA-oxidation pathways. Dashed lines indicate activation mediated by signal transduction pathways in contrast to the coactivation by PGC-1α, which is denoted by solid lines. The arrows from ERRα to the NRFs and the PPAR complex indicate that ERRα activates these pathways at the level of expression.

direct coactivation of PPARs and estrogen-related receptors (ERRs) (Figure 2) (31–34). PPARs are FA-activated members of the nuclear receptor superfamily of transcription factors that serve as central regulators of cardiac FA metabolism. There are 3 PPAR isoforms (α, β/δ, and γ); PPARα and PPARβ are the primary regulators of FA metabolism in the heart. PPARs function by binding as obligate heterodimers with the retinoid X receptor α (RXRα) and recruiting coactivators, including PGC-1α, in response to direct binding and activation by FAs and their derivatives, which serve as ligands. PPARα regulates genes involved in virtually every step of cardiac FA utilization (reviewed in ref. 35). PPARα-null mice have reduced cardiac expression of genes involved in the cellular uptake, mitochondrial transport, and mitochondrial (and peroxisomal) oxidation of FAs (36–38). Myocardial FA uptake and oxidation rates are decreased in these mice, while glucose oxidation rates are increased (39). PPARα-null mice have also provided evidence that PPARα serves an important homeostatic function in the context of physiological and dietary stressors. When subjected to fasting or pharmacological inhibition of FAO, PPARα⁻/⁻ mice develop massive cardiac myocyte lipid accumulation and hepatic steatosis (40, 41). These results have defined PPARα as an important regulator of myocardial energy substrate preference and FAO capacity within the PGC-1α transcriptional regulatory cascade.

In contrast to the biological function of PPARα, that of PPARβ (also known as PPARδ) in the heart has not been extensively studied. However, recent evidence indicates that cardiac PPARβ target genes significantly overlap with those of PPARα. PPARβ selective ligands induce mitochondrial FAO enzyme genes and increase palmitate oxidation rates in cardiac myocytes from both wild-type and PPARα⁻/⁻ mice (42, 43). Consistent with findings from activation studies, cardiac-specific deletion of the PPARβ gene results in reduced expression of FAO enzyme genes and diminished palmitate oxidation rates, which is similar to what occurs in PPARα-null mice (44). In contrast to PPARα⁻/⁻ mice, however, cardiac-specific PPARβ-null mice do not exhibit a fasting-induced phenotype. Rather, PPARβ-deficient mice develop a cardiomyopathy under basal conditions. Collectively, these results suggest that, although PPARα and PPARβ drive similar gene targets, PPARβ probably serves to regulate basal metabolism whereas PPARα is perhaps more important in the response to physiological conditions that increase FA delivery.

ERRs are a subfamily of orphan nuclear receptors for which a role in cardiac energy metabolism has recently been described. There are 3 members of the ERR family: ERRα, ERRβ, and ERRγ (45–47). Cardiac ERRα expression increases dramatically following birth, coincident with the switch to FAs as an energy substrate and the upregulation of PPARα and PGC-1α and of enzymes involved in FA uptake and mitochondrial oxidation (33, 48). Both ERRα and ERRγ are directly coactivated by PGC-1α while ERRα is further regulated by PGC-1α at the gene expression level (33, 34, 49, 50). Recent evidence suggests that ERRα cross-regulates other transcription factors within the PGC-1α network. In cardiac myocytes, ERRα activates expression of many known PPARα and NRF target genes involved in cellular FA utilization and mitochondrial electron transport/oxidative phosphorylation, respectively (24). This apparently overlapping function was recently shown to occur, in part, via direct transactivation of the *PPARα* and *Gabpa* (NRF-2 subunit) gene promoters. (Figure 2) (24, 51). In addition, ERRα cooperates with NRF-1 and NRF-2 to regulate mitochondrial biogenesis and expression of oxidative phosphorylation enzyme genes (51, 52). Collectively, these results support a role for ERRα as an amplifier of cardiac oxidative energy metabolism downstream of PGC-1α (Figure 2).

Signaling pathways trigger changes in mitochondrial function through the PGC-1α regulatory cascade. How do PGC-1α and its transcription factor partners receive input from physiological and pathological stimuli? Insights into the physiological responsiveness of the PGC-1α pathway come from identification of signal transduction pathways that modulate the activity of PGC-1α and its downstream partners (Figure 2). PGC-1α is upregulated in response to β-adrenergic signaling, consistent with the involvement of this pathway in thermogenesis (8, 53). The stress-activated MAPK p38 activates PGC-1α by both increasing PGC-1α protein stability and promoting dissociation of a repressor (21, 54, 55). p38 also increases mitochondrial FAO through selective activation of the PGC-1α partner, PPARα (56). Conversely, the ERK-MAPK pathway inactivates the PPARα/RXRα complex via direct phosphorylation (57). Therefore, distinct limbs of the MAPK pathway exert opposing regulatory influences on the PGC-1α cascade. Recently, NO has emerged as a novel signaling molecule proposed to integrate pathways involved in regulating mitochondrial biogenesis. NO induces mitochondrial prolif-

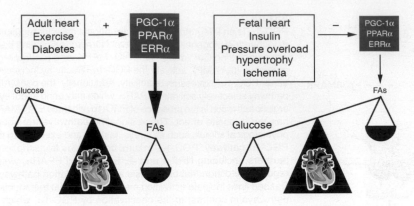

Figure 3
Cardiac energy substrate selection is a dynamic balance influenced by developmental, physiological, and pathological cues. In the fetal heart, glucose oxidation is favored, whereas FA oxidation serves as the major ATP-generating pathway in the adult myocardium. Significant shifts in substrate preference occur in response to dietary (insulin) and physiological (exercise) stimuli. Certain pathophysiological contexts, such as hypertrophy and ischemia, drive metabolism toward glucose utilization, whereas in uncontrolled diabetes, the heart utilizes FAs almost exclusively. In some cases, as in early response to pressure overload–induced hypertrophy, these metabolic shifts are thought to be protective. Alterations in activity or expression of nuclear receptors (PPARs and ERRs) and PGC-1α mediate these shifts in energy substrate utilization.

eration and increases coupled respiration and ATP content in numerous cell types via a cGMP-dependent pathway (58, 59). The effects of NO/cGMP appear to be mediated, at least in part, by transcriptional activation of PGC-1α expression levels (58).

Recent evidence has also implicated Ca^{2+} signaling in the control of the PGC-1α/PPARα regulatory pathway. Transgenic mice overexpressing constitutively active calcineurin (Cn) or Ca^{2+}/calmodulin-dependent kinase (CaMK) in skeletal muscle displayed increased PGC-1α expression and expansion of slow-twitch oxidative fibers and mitochondrial biogenesis (22, 60, 61). Activation of PGC-1α by Cn and CaMK is mediated transcriptionally through myocyte enhancer factor-2 (MEF-2) and cAMP response element–binding protein (CREB), respectively (62). In the heart, exercise training activates the Cn and CaMK pathways coincident with maintained or enhanced mitochondrial function; however, paradoxically, overexpression of constitutively activated forms of Cn or CaMK results in reduced mitochondrial oxidation and heart failure (63–65). Recent studies have shown that Cn and CaMK both activate PGC-1α expression in cardiac myocytes but have distinct effects on downstream targets of the PGC-1α pathway (66). CaMK activates genes involved in glucose oxidation and mitochondrial electron transport while Cn selectively activates mitochondrial FAO enzyme genes. The selective effects of Cn are thought to be mediated, in part, through direct transcriptional activation of the *PPARα* gene. Consistent with the gain-of-function results, reduction in Cn catalytic activity through cardiac-specific deletion of the CnB regulatory subunit results in reduced PGC-1α and PPARα expression and downregulation of FAO enzymes (66).

Perturbations in PGC-1/PPAR signaling in the hypertrophied and failing heart

Numerous studies have demonstrated altered cardiac substrate preference in the hypertrophied and failing heart. However, the direction of the "fuel shifts" varies with the etiology and severity of ventricular dysfunction. Studies using animal models of ven-

tricular hypertrophy due to hypertension or imposed pressure overload have consistently demonstrated a myocardial shift from FAO toward glucose oxidation (67–72). Changes in gene expression in the hypertrophied heart, including downregulation of mitochondrial FAO enzymes, are consistent with the observed metabolic alterations (73, 74).

The results of studies of energy substrate shifts in the failing heart, particularly in humans, have not led to a clear conclusion. Some investigators have shown that expression of FAO genes and corresponding enzymatic activities are reduced in the failing rodent and human heart (70, 73, 75). Consistent with these findings, recent cardiac positron emission tomography (PET) studies in humans have shown a shift away from FA utilization with hypertensive cardiac hypertrophy and idiopathic cardiomyopathy (76, 77). However, others have demonstrated the opposite metabolic profile or no change in substrate utilization in humans and animal models of heart failure of mild-to-moderate severity (78–80). Animal models of failure induced by pacing or myocardial infarction show that earlier stages are not associated with a switch from FAs to glucose as the primary energy substrate. Dogs with moderate severity coronary microembolism–induced heart failure exhibited FA and glucose uptake patterns that were indistinguishable from those of controls (81). These apparent discrepancies have been attributed to both the severity of failure and temporal differences during the progressive remodeling that characterizes the transition to heart failure. The specific etiology of the myocardial disease may also play an important role.

Significant progress has been made in delineating the gene regulatory events driving the reduction of myocardial FA utilization in the hypertrophied, failing, and hypoxic heart. Given the importance of PPARα for the transcriptional control of cardiac lipid metabolism, this nuclear receptor has served as an important focus for such studies. In rodent models of pressure overload hypertrophy, expression of PPARα and PGC-1α is reduced in the hypertrophied and failing heart and in hypertrophied cardiac myocytes in culture (82–87). Levels of the PPAR partner, RXRα, are also reduced in a canine pacing-induced model of heart failure and by hypoxia in cardiac myocytes (88, 89). Interestingly, a reciprocal induction of transcription factors, including chicken ovalbumin upstream promoter-transcription factor (COUP-TF) and Sp1 and Sp3, which repress promoter activity of key FAO enzyme genes, is observed in hypertrophy and heart failure models and is thought to contribute to the reduction in myocardial FAO (82, 90). Moreover, PPARα activity is inhibited posttranslationally in the hypertrophied cardiac myocyte in culture through ERK-MAPK–dependent phosphorylation (83). The results of these studies suggest that one key mechanism involved in the energy substrate switches in the hypertrophied and failing heart involves deactivation of the PGC-1α/PPARα complex at both transcriptional and posttranscriptional levels (Figure 3).

In contrast to heart failure related to pressure overload or ischemia, the cardiomyopathy that develops in the context of insulin resistance and frank diabetes is associated with increased cardiac reliance on FAs as the primary energy substrate (Figure 3) (91–94). This fuel switch is linked to the combined effects of myocyte insulin resistance and high-circulating free FAs. We have found that, consistent with this metabolic profile, the expression and activity of PPARα, PGC-1α, and the enzymes of mitochondrial FAO are induced in both insulin-deficient and insulin-resistant forms of diabetes in mouse models (95, 96). The mechanism involved in the activation of PPARα signaling in the diabetic heart is unknown but likely involves increased cellular import of FAs, which serve as ligands for this nuclear receptor. In contrast, other studies have demonstrated a reduction in PPARα transcript levels in diabetic myocardium (97, 98). This apparent discrepancy could reflect temporal-dependent regulatory events during progression of diabetic myocardial disease. Future studies aimed at defining the precise time-course of PPARα-driven regulatory events during the development of diabetic cardiomyopathy are needed.

Do derangements in mitochondrial energy metabolism cause heart failure?

Myocardial fuel shifts as a cause of heart failure. Do deactivation of PGC-1α/ PPARα and the switch away from FAO in the hypertrophied heart contribute to the pathological remodeling that leads to heart failure? Similarly, does chronic activation of PPARα and myocardial FAO in the diabetic heart lead to cardiac dysfunction? Alternatively, do these myocardial substrate shifts serve adaptive functions in the diseased heart? The answers to these questions have not been resolved, but evidence provided largely by observations in rare genetic human diseases and genetically modified mice offers some insight. However, perhaps the strongest evidence supporting deleterious effects of reduced capacity for cardiac FA utilization comes from the observations of the cardiac phenotype of human genetic defects in FAO pathway enzymes. Children with deficiencies in enzymes involved in mitochondrial long-chain FAO often develop a stress-induced cardiomyopathy associated with myocardial lipid accumulation (99). In addition, mouse models in which the FAO enzymes, very-long-chain acyl-CoA dehydrogenase (VLCAD) or long-chain acyl-CoA dehydrogenase (LCAD), have been disrupted exhibit cardiomyopathies similar to that observed in humans (100, 101). Furthermore, the cardiomyopathic phenotype of mice with cardiac-specific deletion of the PPARβ gene also supports the conclusion that a reduction in capacity to oxidize FAs in the heart has deleterious effects (44). The PPARβ$^{-/-}$ hearts accumulate lipids in association with the development of cardiac hypertrophy, which ultimately leads to dilated cardiomyopathy, myocyte apoptosis, and death.

One proposed mechanism for cardiac dysfunction in these models is excess intracellular lipid accumulation resulting in myocyte dysfunction or death, termed "lipotoxicity" (reviewed in ref. 102). The heart and other high-energy flux organs are adapted to closely match energy substrate import and utilization, not storage. A mismatch can result from increased lipid delivery, such as occurs in obesity or diabetes, or impaired FA oxidation, as in the aforementioned models. The resulting derangements in cellular lipid homeostasis can lead to accumulation of lipid intermediates such as acyl-CoA thioesters, acylcarnitines, ceramides, and triglycerides, molecules that could confer cellular toxicity. Transgenic models in which proteins involved in FA uptake/delivery are overexpressed also exhibit lipid accumulation and systolic dysfunction (103, 104). These hearts display evidence of myocyte dropout due to activation of apoptotic pathways. Indeed, direct treatment with saturated long-chain FAs has been shown to trigger apoptosis in cardiac myocytes as well as in other cell types in

culture via mechanisms that may involve generation of reactive oxygen species (105–107). Collectively, observations in mice and humans indicate that severe reduction in mitochondrial FAO capacity sets the stage for cardiac lipotoxic effects related to lipid intermediates that accumulate in the context of impaired catabolism.

Despite compelling evidence for the deleterious effects of reduced mitochondrial FAO in genetic models and human deficiency states, there is some data to suggest that a shift from FAO to glucose utilization in the hypertrophied heart may be adaptive, at least in the short term. Taegtmeyer and coworkers have shown that reactivation of FAO in a rat ventricular-pressure overload model causes ventricular dysfunction (108). In addition, despite reduced cardiac FAO rates, PPARα-null mice do not exhibit overt ventricular dysfunction (39). It is likely that the degree and duration of the pathophysiological stimulus as well as the systemic metabolic state (e.g., levels of circulating lipids) ultimately determine whether alterations in FAO capacity contribute to the pathogenesis of heart failure.

As described above, the insulin resistant and diabetic heart is characterized by increased FAO rates due, perhaps in part, to chronic activation of the PPARα gene regulatory pathway. Studies of mice genetically modified to mimic the metabolic derangements of the diabetic heart have provided evidence that chronically increased reliance on FAs for energy leads to pathological signatures of the diabetic heart. Mice with cardiac-specific overexpression of PPARα (*MHC-PPARα* mice) exhibit increased expression of PPARα target genes involved in cellular FA import and peroxisomal and mitochondrial FAO, coincident with lipid accumulation and increased rates of FAO (96, 109). Interestingly, myocardial glucose uptake and oxidation rates are reciprocally decreased in the *MHC-PPARα* mouse, a metabolic phenotype that mimics the diabetic heart. These results demonstrate that a primary drive on the PPARα gene regulatory pathway triggers cross-talk suppression of glucose utilization pathways. Importantly, the metabolic derangements of *MHC-PPARα* mice are associated with ventricular diastolic/systolic dysfunction at baseline, which becomes more severe in the context of increased delivery of FAs to the heart such as occurs with high-fat feeding or insulinopenia (95). Thus, restricting the heart to reliance on FAs to the virtual exclusion of glucose oxidation leads to development of cardiomyopathy (109).

The metabolic dysregulation in the *MHC-PPARα* heart is associated with neutral lipid accumulation and increased production of reactive species consistent with the importance of oxidative stress and mitochondrial dysfunction in diabetic cardiomyopathy (95). Direct characterization of mitochondria from hearts in type 1 diabetic models has revealed evidence of damaged mitochondria and impaired mitochondrial respiration, presumably due to oxidative stress (110). In the same type 1 model, overexpression of antioxidant proteins such as metallothionein and catalase reduced reactive oxygen species and rescued cardiac contractility (111–113). A similar effect was observed with catalase overexpression in a mouse model of type 2 diabetes (112). The observations from elevated FA flux and FAO-deficient states, which indicate that either a chronic increase (diabetes) or decrease (pressure overload hypertrophy) in myocardial FAO can lead to heart failure, emphasize the importance of substrate flexibility for normal cardiac function (Figure 3).

Derangements in mitochondrial ATP generation. Evidence for a link between mitochondrial respiratory dysfunction and heart failure is compelling. The cardiomyopathic phenotype of humans with mitochondrial genome defects underscores the importance of high-capacity mitochondrial ATP production for normal striated-muscle function. Mutations in both nuclear- and mito-chondrial-encoded genes account for heritable respiratory chain defects (for an extensive review of this topic, see refs. 114–117). Respiratory chain defects typically present as multisystem dysfunction disproportionately affecting organs with high ATP demand, such as the heart, skeletal muscle, and the central nervous system. Cardiomyopathy may develop during childhood or at later ages. Mouse models of mitochondrial dysfunction have also provided important information about the role of mitochondrial proteins in regulating mitochondrial number and function as well as the way in which altered mitochondrial energetics contribute to the development of heart failure. Tfam is a nuclear-encoded mitochondrial transcription factor necessary for mitochondrial biogenesis and gene expression. Cardiac-specific Tfam−/− mice exhibit reduced respiratory capacity and mitochondrial DNA in the heart before birth coincident with a high neonatal mortality (118). Surviving animals develop hypertrophy, progressing to dilated cardiomyopathy and conduction abnormalities, and die by 4 months of age. Interestingly, cardiac dysfunction is accompanied by a metabolic shift from FAO to glucose oxidation (119).

Adenine nucleotide translocators (Ants) are mitochondrial-membrane proteins involved in the transport of cytoplasmic ADP in exchange for mitochondrial ATP, so mitochondria deficient in Ant have reduced capacity for substrate level phosphorylation. Mice targeted for the Ant1 gene, the cardiac/skeletal muscle expressed isoform, exhibit reduced ADP-dependent respiration rates in heart and skeletal muscle in spite of increased mitochondrial number (120, 121). The mitochondrial proliferation is likely a compensatory response to cellular energy deficiency but is thought to contribute to progressive cardiac hypertrophy that develops in these mice (121, 122). ATP deficiency is thought to be a primary cause for the observed pathophysiology, but these models also support a role for mitochondrial-derived ROS in mediating cellular damage (123). Although the precise cellular insults are still unknown, these genetic models demonstrate that mitochondrial dysfunction is sufficient for the development of heart failure.

Several recently developed mouse models have identified exciting potential links between PGC-1α–mediated control of mitochondrial function and the development of heart failure. Sano et al. recently found that overexpression of cyclin T/Cdk9, an RNA polymerase kinase, triggers cardiac hypertrophy (124). Further studies revealed that in the context of ventricular pressure overload, mice overexpressing Cdk9 in the heart develop a fulminant apoptotic cardiomyopathy (125). Gene expression–profiling studies demonstrated that Cdk9 suppresses expression of PGC-1α and its downstream targets involved in mitochondrial respiratory function. Rescue of PGC-1α expression in cardiac myocytes in culture prevented Cdk9-triggered apoptosis. In a separate mouse model, chronic activation of PGC-1α in the heart, as occurs in diabetes, led to ventricular dysfunction. In a study using an inducible, cardiac-specific transgenic system, chronic overexpression of PGC-1α protein in the adult mouse heart caused mitochondrial ultrastructural abnormalities and reduced myofibrillar density, which led to cardiomyopathy and diastolic dysfunction (25). Interestingly, the mitochondrial proliferative response is reversible and the cardiomyopathy rescued upon cessation of transgene expression. The basis for this reversible cardiomyopathy is unknown but could involve the accumulation of reactive intermediates or abnormalities in ATP generation.

Taken together, these recent findings suggest that, in the context of a mechanical stress such as pressure overload, reduced levels of PGC-1α predispose the heart to pathological remodeling related to mitochondrial dysfunction and apoptosis. Conversely, chronic activation of PGC-1α and its downstream targets, such as PPARα, that mimic the diabetic state also leads to ventricular dysfunction through mechanisms that are reversible.

Despite strong evidence for a link between mitochondrial dysfunction and heart failure in genetic models, the role of altered mitochondrial ATP generation in the pathogenesis of acquired forms of heart failure is less clear. Phosphocreatine (pCR) serves as the main energy store in myocardium (for review, see ref. S1; see Supplemental References; supplemental material available online with this article; doi:10.1172/JCI200524405DS1). Early studies revealed that myocardial pCR/ATP ratios are reduced in the hypertrophied and failing heart while absolute ATP levels are only detectably reduced in heart failure (S2–S4). However, the concentration of ATP measured in the failing myocardium is still above the Km for most cellular ATPases. Consistent with a gradual fall in intracellular high-energy phosphates, the "energy sensor kinase," AMP-activated protein kinase (AMPK), is upregulated and activated during hypertrophy (S5). These results suggest that myocardial energy reserves are disproportionately reduced during hypertrophy and heart failure progression compared to absolute ATP levels. In dog models of pacing-induced heart failure, ATP concentrations fall gradually coincident with a reduction in mitochondrial respiration rates during the progression to heart failure (S6, S7). These functional changes are associated with evidence of structural abnormalities in mitochondria from these hearts (S8). Reduced mitochondrial oxidative capacity has also been observed in rodent heart failure models (87). Further studies will be necessary to accurately delineate the temporal pattern of alterations in bioenergetics during the development of heart failure. Murine genetic loss-of-function and pharmacological rescue strategies in larger animal models should serve as useful experimental strategies to determine cause-and-effect relationships.

Metabolic modulators as a new treatment strategy for heart failure? A question of balance

In summary, evidence is emerging to support the concept that alterations in myocardial fuel selection and energetics are linked to the development and progression of heart failure. Accordingly, metabolic pathways involved in cardiac FA and glucose utilization or ATP generation are attractive targets of novel therapeutic strategies aimed at the prevention or early treatment of heart failure. Indeed, specific activators for each of the PPARs have been developed and are currently used for treatment of hyperlipidemia (e.g., fibrates) and diabetes (thiazolidinediones). Activation of the PPAR pathway in heart or extracardiac tissues, such as adipose or liver, could theoretically reduce cardiac lipotoxicity by reducing lipid delivery or increasing mitochondrial oxidation. However, a strong word of caution is necessary. As described above, it is now clear that the degree of metabolic modulation is an important determinant of whether shifts in energy substrate utilization serve adaptive or maladaptive functions in the context of disease states that predispose patients to heart failure. Conversely, chronic activation of PPARα could lead to deleterious effects, particularly in the context of diabetes, hyperlipidemic states, or the ischemic heart, given the potential for increased mitochondrial oxidative flux, which could generate ROS and increase oxygen consumption. It is likely that the response will be disease-specific and dependent on the systemic metabolic phenotype. Studies in large animals and humans will be particularly informative. Such studies will require rigorous metabolic phenotyping approaches with imaging modalities such as PET and spectroscopy to define baseline myocardial substrate utilization and energetic profiles, respectively. Similarly, both metabolic and functional endpoints will be necessary to assess response to therapy. If metabolic modulator therapy proves useful, the standard diagnostic approach to the heart failure patient will likely include metabolic imaging or surrogate metabolic biomarkers to guide therapeutic decisions and assess the patient's response. In this regard, an interdisciplinary approach involving cardiologists, endocrinologists, and radiologists might be envisioned.

Acknowledgments

This work was supported by NIH grants R01-DK45416, R01-HL58493, and P01-HL57278 and Digestive Diseases Core Center grant P30-DK52574 (to D.P. Kelly). J.M. Huss is supported by NIH grant K01-DK063051 and Washington University School of Medicine Diabetes Research Training Center grant P60-DK20579. The authors thank Adam Wende for valuable contributions to the figures and Mary Wingate for expert assistance in preparing this manuscript.

Address correspondence to: Daniel P. Kelly, Center for Cardiovascular Research, Washington University School of Medicine, St. Louis, Missouri 63110, USA. Phone: (314) 362-8908; Fax: (314) 362-0186; E-mail: dkelly@im.wustl.edu.

1. Stanley, W.C., and Chandler, M.P. 2002. Energy metabolism in the normal and failing heart: potential for therapeutic interventions. *Heart Fail. Rev.* **7**:115–130.

2. Taegtmeyer, H. 1994. Energy metabolism of the heart: from basic concepts to clinical applications. *Curr. Probl. Cardiol.* **19**:59–113.

3. Bing, R.J., Siegel, A., Ungar, I., and Gilbert, M. 1954. Metabolism of the human heart. II. Studies on fat, ketone and amino acid metabolism. *Am. J. Med.* **16**:504–515.

4. Shipp, J.C., Opie, L.H., and Challoner, D. 1961. Fatty acid and glucose metabolism in the perfused heart. *Nature.* **189**:1018–1019.

5. Wisnecki, J.A., Gertz, E.Q., Neese, R.A., and Mayr, M. 1987. Myocardial metabolism of free fatty acids: studies with ^{14}C labelled substrates in humans. *J. Clin. Invest.* **79**:359–366.

6. Kelly, D.P., and Scarpulla, R.C. 2004. Transcriptional regulatory circuits controlling mitochondrial biogenesis and function. *Genes Dev.* **18**:357–368.

7. Anderson, S., et al. 1981. Sequence and organization of the human mitochondrial genome. *Nature.* **290**:457–465.

8. Puigserver, P., et al. 1998. A cold-inducible coactivator of nuclear receptors linked to adaptive thermogenesis. *Cell.* **92**:829–839.

9. Andersson, U., and Scarpulla, R.C. 2001. PGC-1-related coactivator, a novel, serum-inducible coactivator of nuclear respiratory factor-1-dependent transcription in mammalian cells. *Mol. Cell. Biol.* **21**:3738–3749.

10. Lin, J., Puigserver, P., Donovan, J., Tarr, P., and Spiegelman, B.M. 2002. Peroxisome proliferator-activated receptor γ coactivator 1β (PGC-1β), a novel PGC-1-related transcription coactivator associated with host cell factor. *J. Biol. Chem.* **277**:1645–1648.

11. Kressler, D., Schreiber, S.N., Knutti, D., and Kralli, A. 2002. The PGC-1-related protein PERC is a selective coactivator of estrogen receptor alpha. *J. Biol. Chem.* **277**:13918–13925.

12. Savagner, F., et al. 2003. PGC-1-related coactivator and targets are upregulated in thyroid oncocytoma. *Biochem. Biophys. Res. Commun.* **310**:779–784.

13. Wu, Z., et al. 1999. Mechanisms controlling mitochondrial biogenesis and respiration through the thermogenic coactivator PGC-1. *Cell.* **98**:115–124.

14. Kamei, Y., et al. 2003. PPARγ coactivator 1β/ERR ligand 1 is an ERR protein ligand, whose expression induces a high-energy expenditure and antagonizes obesity. *Proc. Natl. Acad. Sci. U. S. A.* **100**:12378–12383.

15. St-Pierre, J., et al. 2003. Bioenergetic analysis of peroxisome proliferator-activated receptor γ coactivators 1α and 1β (PGC-1α and PGC-1β) in muscle cells. *J. Biol. Chem.* **278**:26597–26603.

16. Herzig, S., et al. 2001. CREB regulates hepatic gluconeogenesis through the coactivator PGC-1. *Nature.* **413**:179–183.

17. Rhee, J., et al. 2003. Regulation of hepatic fasting response by PPARγ coactivator-1α (PGC-1α): requirement for hepatocyte nuclear factor 4α in gluconeogenesis. *Proc. Natl. Acad. Sci. U. S. A.* **100**:4012–4017.

18. Michael, L.F., et al. 2001. Restoration of insulin-sensitive glucose transporter (GLUT4) gene expression in muscle cells by the transcriptional coactivator PGC-1. *Proc. Natl. Acad. Sci. U. S. A.* **98**:3820–3825.

19. Lehman, J.J., et al. 2000. PPARγ coactivator-1 (PGC-1) promotes cardiac mitochondrial biogenesis. *J. Clin. Invest.* **106**:847–856.

20. Baar, K., et al. 2002. Adaptations of skeletal muscle to exercise: rapid increase in the transcriptional coactivator PGC-1α. *FASEB J.* **16**:1879–1886.

21. Puigserver, P., et al. 2001. Cytokine stimulation of energy expenditure through p38 MAP kinase activation of PPARγ coactivator-1. *Mol. Cell.* **8**:971–982.

22. Wu, H., et al. 2002. Regulation of mitochondrial biogenesis in skeletal muscle by CaMK. *Science.* **296**:349–352.

23. Goto, M., et al. 2000. cDNA cloning and mRNA analysis of PGC-1 in epitrochlearis muscle in swimming-exercised rats. *Biochem. Biophys. Res. Commun.* **274**:350–354.

24. Huss, J.M., Pinéda Torra, I., Staels, B., Giguère, V., and Kelly, D.P. 2004. ERRα directs PPAR α signaling in the transcriptional control of energy metabolism in cardiac and skeletal muscle. *Mol. Cell. Biol.* **24**:9079–9091.

25. Russell, L.K., et al. 2004. Cardiac-specific induction of the transcriptional coactivator peroxisome proliferator-actived receptor γ coactivator-1α promotes mitochondrial biogenesis and reversible cardiomyopathy in a developmental stage-dependent manner. *Circ. Res.* **94**:525–533.

26. Fisher, R.P., Lisowsky, T., Parisi, M.A., and Clayton, D.A. 1992. DNA wrapping and bending by a mitochondrial high mobility group-like transcriptional activator protein. *J. Biol. Chem.* **267**:3358–3367.

27. Garesse, R., and Vallejo, C.G. 2001. Animal mitochondrial biogenesis and function: a regulatory cross-talk between two genomes. *Gene.* **263**:1–16.

28. Larsson, N.-G., et al. 1998. Mitochondrial transcription factor A is necessary for mtDNA maintenance and embryogenesis in mice. *Nat. Genet.* **18**:231–236.

29. Scarpulla, R.C. 2002. Nuclear activators and coactivators in mammalian mitochondrial biogenesis. *Biochim. Biophys. Acta.* **1576**:1–14.

30. Virbasius, C.A., Virbasius, J.V., and Scarpulla, R.C. 1993. NRF-1, an activator involved in nuclear-mitochondrial interactions, utilizes a new DNA-binding domain conserved in a family of developmental regulators. *Genes Dev.* **7**:2431–2445.

31. Vega, R.B., Huss, J.M., and Kelly, D.P. 2000. The coactivator PGC-1 cooperates with peroxisome proliferator-activated receptor α in transcriptional control of nuclear genes encoding mitochondrial fatty acid oxidation enzymes. *Mol. Cell. Biol.* **20**:1868–1876.

32. Dressel, U., et al. 2003. The peroxisome proliferator-activated receptor β/δ agonist, GW501516, regulates the expression of genes involved in lipid catabolism and energy uncoupling in skeletal muscle cells. *Mol. Endocrinol.* **17**:2477–2493.

33. Huss, J.M., Kopp, R.P., and Kelly, D.P. 2002. PGC-1α coactivates the cardiac-enriched nuclear receptors estrogen-related receptor-α and -γ. *J. Biol. Chem.* **277**:40265–40274.

34. Schreiber, S.N., Knutti, D., Brogli, K., Uhlmann, T., and Kralli, A. 2003. The transcriptional coactivator PGC-1 regulates the expression and activity of the orphan nuclear receptor estrogen-related receptor α (ERRα). *J. Biol. Chem.* **278**:9013–9018.

35. Desvergne, B., and Wahli, W. 1999. Peroxisome proliferator-activated receptors: nuclear control of metabolism. *Endocr. Rev.* **20**:649–688.

36. Lee, S.S.T., et al. 1995. Targeted disruption of the α isoform of the peroxisome proliferator-activated receptor gene in mice results in abolishment of the pleiotropic effects of peroxisome proliferators. *Mol. Cell. Biol.* **15**:3012–3022.

37. Watanabe, K., et al. 2000. Constitutive regulation of cardiac fatty acid metabolism through peroxisome proliferator-activated receptor α associated with age-dependent cardiac toxicity. *J. Biol. Chem.* **275**:22293–22299.

38. Djouadi, F., et al. 1999. The role of the peroxisome proliferator-activated receptor α (PPARα) in the control of cardiac lipid metabolism. *Prostaglandins Leukot. Essent. Fatty Acids.* **60**:339–343.

39. Campbell, F.M., et al. 2002. A role for PPARα in the control of cardiac malonyl-CoA levels: reduced fatty acid oxidation rates and increased glucose oxidation rates in the hearts of mice lacking PPARα are associated with higher concentrations of malonyl-CoA and reduced expression of malonyl-CoA decarboxylase. *J. Biol. Chem.* **277**:4098–4103.

40. Kersten, S., et al. 1999. Peroxisome proliferator-activated receptor α mediates the adaptive response to fasting. *J. Clin. Invest.* **103**:1489–1498.

41. Leone, T.C., Weinheimer, C.J., and Kelly, D.P. 1999. A critical role for the peroxisome proliferator-activated receptor alpha (PPARα) in the cellular fasting response: the PPARα-null mouse as a model of fatty acid oxidation disorders. *Proc. Natl. Acad. Sci. U. S. A.* **96**:7473–7478.

42. Gilde, A.J., et al. 2003. PPARα and PPARβ/δ, but not PPARγ, modulate the expression of genes involved in cardiac lipid metabolism. *Circ. Res.* **92**:518–524.

43. Cheng, L., et al. 2004. Peroxisome proliferator-activated receptor δ activates fatty acid oxidation in cultured neonatal and adult cardiomyocytes. *Biochem. Biophys. Res. Commun.* **313**:277–286.

44. Cheng, L., et al. 2004. Cardiomyocyte-restricted peroxisome proliferator-activated receptor-δ dele-

tion perturbs myocardial fatty acid oxidation and leads to cardiomyopathy. *Nat. Med.* **10**:1245–1250.

45. Giguère, V., Yang, N., Segui, P., and Evans, R.M. 1988. Identification of a new class of steroid hormone receptors. *Nature.* **331**:91–94.

46. Heard, D.J., Norbu, P.L., Holloway, J., and Vissing, H. 2000. Human ERRγ, a third member of the estrogen receptor-related receptor (ERR) subfamily of orphan nuclear receptors: tissue-specific isoforms are expressed during development and in the adult. *Mol. Endocrinol.* **14**:383–392.

47. Hong, H., Yang, L., and Stallcup, M.R. 1999. Hormone-independent transcriptional activation and coactivator binding by novel orphan nuclear receptor ERR3. *J. Biol. Chem.* **274**:22618–22626.

48. Sladek, R., Bader, J.-A., and Giguère, V. 1997. The orphan nuclear receptor estrogen-related receptor α is a transcriptional regulator of the human medium-chain acyl coenzyme a dehydrogenase gene. *Mol. Cell. Biol.* **17**:5400–5409.

49. Hentschke, M., Susens, U., and Borgmeyer, U. 2002. PGC-1 and PERC, coactivators of the estrogen receptor-related receptor gamma. *Biochem. Biophys. Res. Commun.* **299**:872–879.

50. Laganiere, J., et al. 2004. A polymorphic autoregulatory hormone response element in the human estrogen-related receptor α (ERRα) promoter dictates peroxisome proliferator-activated receptor γ coactivator-1α control of ERRα expression. *J. Biol. Chem.* **274**:18504–18510.

51. Mootha, V.K., et al. 2004. ERRα and Gabpa/b specify PGC-1α-dependent oxidative phosphorylation gene expression that is altered in diabetic muscle. *Proc. Natl. Acad. Sci. U. S. A.* **101**:6570–6575.

52. Schreiber, S.N., et al. 2004. The estrogen-related receptor alpha (ERRα) functions in PPARγ coactivator 1α (PGC-1α) - induced mitochondrial biogenesis. *Proc. Natl. Acad. Sci. U. S. A.* **101**:6472–6477.

53. Boss, O., et al. 1999. Role of the β₃-adrenergic receptor and/or putative β₄-adrenergic receptor on the expression of uncoupling proteins and peroxisome proliferator-activated receptor-γ coactivator-1. *Biochem. Biophys. Res. Commun.* **261**:870–876.

54. Knutti, D., Kressler, D., and Kralli, A. 2001. Regulation of the transcriptional coactivator PGC-1 via MAPK-sensitive interaction with a corepressor. *Proc. Natl. Acad. Sci. U. S. A.* **98**:9713–9718.

55. Fan, M., et al. 2004. Suppression of mitochondrial respiration through recruitment of p160 myb binding protein to PGC-1alpha: modulation by p38 MAPK. *Genes Dev.* **18**:278–289.

56. Barger, P.M., Browning, A.C., Garner, A.N., and Kelly, D.P. 2001. p38 MAP kinase activates PPARα: a potential role in the cardiac metabolic stress response. *J. Biol. Chem.* **276**:44495–44501.

57. Ballal, K., Sekiguchi, K., Nanda, S., and Barger, P. 2004. ERK MAPK regulation of cardiac retinoid X receptor is involved in impaired fatty acid β-oxidation during hypertrophic growth [abstract]. *Circulation.* **110**(Suppl.):946.

58. Nisoli, E., et al. 2003. Mitochondrial biogenesis in mammals: the role of endogenous nitric oxide. *Science.* **299**:896–899.

59. Nisoli, E., Clementi, E., Moncada, S., and Carruba, M.O. 2004. Mitochondrial biogenesis as a cellular signaling framework. *Biochem. Pharmacol.* **67**:1–15.

60. Naya, F.J., et al. 2000. Stimulation of slow skeletal muscle fiber gene expression by calcineurin in vivo. *J. Biol. Chem.* **275**:4545–4548.

61. Ryder, J.W., Bassel-Duby, R., Olson, E.N., and Zierath, J.R. 2003. Skeletal muscle reprogramming by activation of calcineurin improves insulin action on metabolic pathways. *J. Biol. Chem.* **278**:44298–44304.

62. Handschin, C., Rhee, J., Lin, J., Tam, P.T., and Spiegelman, B.M. 2003. An autoregulatory loop controls peroxisome proliferator-activated receptor γ coactivator 1α expression in muscle. *Proc. Natl. Acad. Sci. U. S. A.* **100**:7111–7116.

63. Eto, Y., et al. 2000. Calcineurin is activated in rat hearts with physiological left ventricular hypertrophy induced by voluntary exercise training. *Circulation.* **101**:2134–2137.

64. Molkentin, J.D., et al. 1998. A calcineurin-dependent transcriptional pathway for cardiac hypertrophy. *Cell.* **93**:215–228.

65. Passier, R., et al. 2000. CaM kinase signaling induces cardiac hypertrophy and activates the MEF2 transcription factor in vivo. *J. Clin. Invest.* **105**:1395–1406.

66. Schaeffer, P.J., et al. 2004. Calcineurin and calcium/calmodulin-dependent protein kinase activate distinct metabolic gene regulatory programs in cardiac muscle. *J. Biol. Chem.* **279**:39593–39603.

67. Bishop, S.P., and Altschuld, R.A. 1970. Increased glycolytic metabolism in cardiac hypertrophy and congestive failure. *Am. J. Physiol.* **218**:153–159.

68. Allard, M.F., Schonekess, B.O., Henning, S.L., English, D.R., and Lopaschuk, G.D. 1994. Contribution of oxidative metabolism and glycolysis to ATP production in hypertrophied hearts. *Am. J. Physiol.* **267**:H742–H750.

69. Christe, M.D., and Rodgers, R.L. 1994. Altered glucose and fatty acid oxidation in hearts of the spontaneously hypertensive rat. *J. Mol. Cell. Cardiol.* **26**:1371–1375.

70. Taegtmeyer, H., and Overturf, M.L. 1988. Effects of moderate hypertension on cardiac function and metabolism in the rabbit. *Hypertension.* **11**:416–426.

71. Massie, B.M., et al. 1995. Myocardial high-energy phosphate and substrate metabolism in swine with moderate left ventricular hypertrophy. *Circulation.* **91**:1814–1823.

72. Sambandam, N., Lopaschuk, G.D., Brownsey, R.W., and Allard, M.F. 2002. Energy metabolism in the hypertrophied heart. *Heart Fail. Rev.* **7**:161–173.

73. Sack, M.N., et al. 1996. Fatty acid oxidation enzyme gene expression is downregulated in the failing heart. *Circulation.* **94**:2837–2842.

74. Razeghi, P., et al. 2001. Metabolic gene expression in fetal and failing human heart. *Circulation.* **104**:2923–2931.

75. Razeghi, P., Young, M.E., Abbasi, S., and Taegtmeyer, H. 2001. Hypoxia in vivo decreases peroxisome proliferator-activated receptor alpha-regulated gene expression in rat heart. *Biochem. Biophys. Res. Commun.* **287**:5–10.

76. Davila-Roman, V.G., et al. 2002. Altered myocardial fatty acid and glucose metabolism in idiopathic dilated cardiomyopathy. *J. Am. Coll. Cardiol.* **40**:271–277.

77. de las Fuentes, L., et al. 2003. Myocardial fatty acid metabolism: independent predictor of left ventricular mass in hypertension and in left ventricular dysfunction. *Hypertension.* **41**:83–87.

78. Recchia, F.A., et al. 1998. Reduced nitric oxide production and altered myocardial metabolism during the decompensation of pacing-induced heart failure in the conscious dog. *Circ. Res.* **83**:969–979.

79. Paolisso, G., et al. 1994. Total-body and myocardial substrate oxidation in congestive heart failure. *Metabolism.* **43**:174–179.

80. Wallhaus, T.R., et al. 2001. Myocardial free fatty acid and glucose use after carvedilol treatment in patients with congestive heart failure. *Circulation.* **103**:2441–2446.

81. Chandler, M.P., et al. 2004. Moderate severity heart failure does not involve a downregulation of myocardial fatty acid oxidation. *Am. J. Physiol. Heart Circ. Physiol.* **287**:H1538–H1543.

82. Sack, M.N., Disch, D.L., Rockman, H.A., and Kelly, D.P. 1997. A role for Sp and nuclear receptor transcription factors in a cardiac hypertrophic growth program. *Proc. Natl. Acad. Sci. U. S. A.* **94**:6438–6443.

83. Barger, P.M., Brandt, J.M., Leone, T.C., Weinheimer, C.J., and Kelly, D.P. 2000. Deactivation of peroxisome proliferator-activated receptor-α during cardiac hypertrophic growth. *J. Clin. Invest.* **105**:1723–1730.

84. Lehman, J.J., and Kelly, D.P. 2002. Transcriptional activation of energy metabolic switches in the developing and hypertrophied heart. *Clin. Exp. Pharmacol. Physiol.* **29**:339–345.

85. Barger, P.M., and Kelly, D.P. 2000. PPAR signaling in the control of cardiac energy metabolism. *Trends Cardiovasc. Med.* **10**:238–245.

86. Kanda, H., Nohara, R., Hasegawa, K., Kishimoto, C., and Sasayama, S. 2000. A nuclear complex containing PPARα/RXRα is markedly downregulated in the hypertrophied rat left ventricular myocardium with normal systolic function. *Heart Vessels.* **15**:191–196.

87. Garnier, A., et al. 2003. Depressed mitochondrial transcription factors and oxidative capacity in rat failing cardiac and skeletal muscles. *J. Physiol.* **551**:491–501.

88. Huss, J.M., Levy, F.H., and Kelly, D.P. 2001. Hypoxia inhibits the PPARα/RXR gene regulatory pathway in cardiac myocytes. *J. Biol. Chem.* **276**:27605–27612.

89. Osorio, J.C., et al. 2002. Impaired myocardial fatty acid oxidation and reduced protein expression of retinoid X receptor-alpha in pacing-induced heart failure. *Circulation.* **106**:606–612.

90. Tian, R. 2003. Transcriptional regulation of energy substrate metabolism in normal and hypertrophied heart. *Curr. Hypertens. Rep.* **5**:454–458.

91. Wall, S.R., and Lopaschuk, G.D. 1989. Glucose oxidation rates in fatty acid perfused isolated working hearts from diabetic rat. *Biochim. Biophys. Acta.* **1006**:97–103.

92. Saddik, M., and Lopaschuk, G.D. 1994. Triacylglycerol turnover in isolated working hearts of acutely diabetic rats. *Can. J. Physiol. Pharmacol.* **72**:1110–1119.

93. Belke, D.D., Larsen, T.S., Gibbs, E.M., and Severson, D.L. 2000. Altered metabolism causes cardiac dysfunction in perfused hearts from diabetic (db/db) mice. *Am. J. Physiol.* **279**:E1104–E1113.

94. Neitzel, A.S., Carley, A.N., and Severson, D.L. 2003. Chylomicron and palmitate metabolism by perfused hearts from diabetic mice. *Am. J. Physiol. Endocrinol. Metab.* **284**:E357–E365.

95. Finck, B., et al. 2003. A critical role for PPARα-mediated lipotoxicity in the pathogenesis of diabetic cardiomyopathy: modulation of phenotype by dietary fat content. *Proc. Natl. Acad. Sci. U. S. A.* **100**:1226–1231.

96. Finck, B., et al. 2002. The cardiac phenotype induced by PPARα overexpression mimics that caused by diabetes mellitus. *J. Clin. Invest.* **109**:121–130. doi:10.1172/JCI200214080.

97. Depre, C., et al. 2000. Streptozotocin-induced changes in cardiac gene expression in the absence of severe contractile dysfunction. *J. Mol. Cell. Cardiol.* **32**:985–996.

98. Young, M.E., et al. 2001. Uncoupling protein 3 transcription is regulated by peroxisome proliferator-activated receptor α in the adult rodent heart. *FASEB J.* **15**:833–845.

99. Kelly, D.P., and Strauss, A.W. 1994. Inherited cardiomyopathies. *N. Engl. J. Med.* **330**:913–919.

100. Exil, V.J., et al. 2003. Very-long-chain Acyl-coenzyme A dehydrogenase deficiency in mice. *Circ. Res.* **93**:448–455.

101. Kurtz, D.M., et al. 1998. Targeted disruption of mouse long-chain acyl-CoA dehydrogenase gene reveals crucial roles for fatty acid oxidation. *Proc. Natl. Acad. Sci. U. S. A.* **95**:15592–15597.

102. Schaffer, J.E. 2003. Lipotoxicity: when tissues overeat. *Curr. Opin. Lipidol.* **14**:281–287.

103. Chiu, H.-C., et al. 2001. A novel mouse model of lipotoxic cardiomyopathy. *J. Clin. Invest.* **107**:813–822.

104. Yagyu, H., et al. 2003. Lipoprotein lipase (LpL) on the surface of cardiomyocytes increases lipid uptake and produces a cardiomyopathy. *J. Clin. Invest.* **111**:419–426. doi:10.1172/JCI200316751.

105. Listenberger, L.L., Ory, D.S., and Schaffer, J.E. 2001.

Palmitate-induced apoptosis can occur through a ceramide-independent pathway. *J. Biol. Chem.* **276**:14890–14895.

106. deVries, J.E., et al. 1997. Saturated but not mono-unsaturated fatty acids induce apoptotic cell death in neonatal rat ventricular myocytes. *J. Lipid Res.* **38**:1384–1394.

107. Kong, J.Y., and Rabkin, S.W. 2002. Palmitate-induced cardiac apoptosis is mediated through CPT-1 but not influenced by glucose and insulin. *Am. J. Physiol. Heart Circ. Physiol.* **282**:717–725.

108. Young, M.E., Laws, F.A., Goodwin, G.W., and Taegtmeyer, H. 2001. Reactivation of peroxisome proliferator activated receptor alpha is associated with contractile dysfunction in hypertrophied rat heart. *J. Biol. Chem.* **276**:44390–44395.

109. Hopkins, T.A., et al. 2003. Control of cardiac pyruvate dehydrogenase activity in peroxisome proliferator-activated receptor-α transgenic mice. *Am. J. Physiol. Heart Circ. Physiol.* **285**:H270–H276.

110. Shen, X., et al. 2004. Cardiac mitochondrial damage and biogenesis in a chronic model of type I diabetes. *Am. J. Physiol. Endocrinol. Metab.* **287**:896–905.

111. Ye, G., Metreveli, N.S., Ren, J., and Epstein, P.N. 2003. Metallothionein prevents diabetes-induced deficits in cardiomyocytes by inhibiting reactive oxygen species production. *Diabetes.* **52**:777–783.

112. Ye, G., et al. 2004. Catalase protect cardiomyocyte function in models of type 1 and type 2 diabetes. *Diabetes.* **53**:1336–1343.

113. Liang, Q., et al. 2002. Overexpression of metallothionein reduces diabetic cardiomyopathy. *Diabetes.* **51**:174–181.

114. Wallace, D.C. 1999. Mitochondrial diseases in man and mouse. *Science.* **283**:1482–1488.

115. DiMauro, S., and Schon, E.A. 2001. Mitochondrial DNA mutations in human disease. *Am. J. Med. Genet.* **106**:18–26.

116. Larsson, N.G., and Oldfors, A. 2001. Mitochondrial myopathies. *Acta Physiol. Scand.* **171**.385–393.

117. Kirby, D.M., et al. 2004. NDUFS6 mutations are a novel cause of lethal neonatal mitochondrial comlex I deficiency. *J. Clin. Invest.* **114**:837–845. doi:10.1172/JCI200420683.

118. Li, H., et al. 2000. Genetic modification of survival in tissue-specific knockout mice with mitochondrial cardiomyopathy. *Proc. Natl. Acad. Sci. U. S. A.* **97**:3467–3472.

119. Hansson, A., et al. 2004. A switch in metabolism precedes increased mitochondrial biogenesis in respiratory chain-deficient hearts. *Proc. Natl. Acad. Sci. U. S. A.* **101**:3136–3141.

120. Stepien, G., Torroni, A., Chung, A.B., Hodge, J.A., and Wallace, D.C. 1992. Differential expression of adenine nucleotide translocator isoforms in mammalian tissues and during muscle cell differentiation. *J. Biol. Chem.* **267**:14592–14597.

121. Graham, B.H., et al. 1997. A mouse model for mitochondrial myopathy and cardiomyopathy resulting from a deficiency in the heart/muscle isoform of the adenine nucleotide translocator. *Nat. Genet.* **16**:226–234.

122. Murdock, D.G., Boone, B.E., Esposito, L.A., and Wallace, D.C. 1999. Up-regulation of nuclear and mitochondrial genes in the skeletal muscle of mice lacking the heart/muscle isoform of the adenine nucleotide translocator. *J. Biol. Chem.* **274**:14429–14433.

123. Esposito, L.A., Melov, S., Panov, A., Cottrell, B.A., and Wallace, D.C. 1999. Mitochondrial disease in mouse results in increased oxidative stress. *Proc. Natl. Acad. Sci. U. S. A.* **96**:4820–4825.

124. Sano, M., et al. 2002. Activation and function of cyclin T-Cdk9 (positive transcription elongation factor-b) in cardiac muscle-cell hypertrophy. *Nat. Med.* **8**:1310–1317.

125. Sano, M., et al. 2004. Activation of cardiac Cdk9 represses PGC-1 and confers a predisposition to heart failure. *EMBO J.* **23**:3559–3569.

Death begets failure in the heart

Roger S.-Y. Foo,[1] Kartik Mani,[1] and Richard N. Kitsis[1,2]

[1]Departments of Medicine and Cell Biology, Cardiovascular Research Center, and [2]Cancer Center,
Albert Einstein College of Medicine, Bronx, New York, USA.

Recently, low — but abnormal — rates of cardiomyocyte apoptosis have been observed in failing human hearts. Genetic and pharmacological studies suggest that this cell death is causally linked to heart failure in rodent models. Herein, we review these data and discuss potential therapeutic implications.

Introduction

Heart failure is a heterogeneous syndrome that can result from primary cardiomyopathies or, more commonly, myocardial infarction, hypertension, and valvular heart disease, among other disorders. The prevalence of heart failure has increased dramatically as modern therapies have reduced the mortality of acute myocardial infarction. However, current treatments for heart failure are woefully inadequate, and the availability of hearts for transplantation is severely limited. Even those therapies that successfully target biologically relevant pathways (e.g., β-adrenergic receptor blockers, angiotensin-converting enzyme inhibitors) become less effective with time (1). These limitations underscore the need for understanding the biology of heart failure at the most fundamental level.

The underlying cause of heart failure has remained an enigma since this syndrome was first described by Richard Lower in 1669 (2). Multiple mechanisms have been proposed, including desensitization of β-adrenergic receptor signaling (3), dysregulation of excitation-contraction coupling (4, 5), alterations in cytoskeletal proteins (6), myosin isoform switches (7), and dysfunctional energy utilization (8, 9), all topics covered elsewhere in this series. Indeed, these mechanisms have been implicated in the progressive loss of contractile function in heart failure. On the other hand, there is a longstanding notion that heart failure involves not only myocyte contractile dysfunction, but also cell "drop-out." The loss of myocytes would be predicted to decrease contractility and promote slippage of muscle bundles, wall thinning, and dilatation — the archetypical changes observed in heart failure.

In this review, we first discuss the evidence that cardiomyocyte death plays a mechanistic role in heart failure and go on to consider how this mechanism may provide a target for novel therapies.

Central apoptosis pathways

Cell death can occur by apoptosis, necrosis, or perhaps autophagy. A framework has been described for apoptosis, a highly regulated cell suicide process that is hard wired into all metazoan cells (10). Much less is known about necrosis, although recent work suggests that this form of cell death may be actively regulated and is not necessarily accidental (11, 12). Autophagy is an important recycling process in which macromolecules are degraded in the lysosome so that their components can be used as energy sub-strates by the cell (13). Whether autophagy functions as a death process remains controversial (11, 14).

Apoptosis is mediated by 2 evolutionarily conserved central death pathways: the extrinsic pathway, which utilizes cell surface death receptors; and the intrinsic pathway, involving mitochondria and the ER (Figure 1) (10). In the extrinsic pathway, death ligands (e.g., FasL) initiate apoptosis by binding their cognate receptors (15). This stimulates the recruitment of the adaptor protein Fas-associated via death domain (FADD), which then recruits procaspase-8 into the death-inducing signaling complex (DISC) (16–18). Procaspase-8 is activated by dimerization in this complex and subsequently cleaves and activates procaspase-3 and other downstream procaspases (19).

In contrast to the extrinsic pathway that mediates a specialized subset of death signals, the intrinsic pathway transduces a wide variety of extracellular and intracellular stimuli including loss of survival/trophic factors, toxins, radiation, hypoxia, oxidative stress, ischemia-reperfusion, and DNA damage. Although a myriad of peripheral pathways connect these signals with the central death machinery, each ultimately feeds into a variety of proapoptotic Bcl-2 proteins that possess only Bcl-2 homology domain 3 (BH3-only proteins) and the proapoptotic multidomain Bcl-2 proteins Bax and Bak (20). These proteins undergo activation through a diversity of mechanisms to trigger the release of mitochondrial apoptogens, such as cytochrome c, into the cytoplasm (21–27). Once in the cytoplasm, cytochrome c binds Apaf-1 along with dATP. This stimulates Apaf-1 to homo-oligomerize and recruit procaspase-9 into the multiprotein complex called the apoptosome (28–39). Within the apoptosome, procaspase-9 is activated by dimerization, after which it cleaves and activates downstream procaspases. Bid, a BH3-only protein, unites the extrinsic and intrinsic pathways: following cleavage by caspase-8, Bid's C-terminal portion translocates to the mitochondria and triggers apoptogen release (40–42).

The extrinsic and intrinsic pathways are held in check by a variety of endogenous inhibitors of apoptosis. FLICE-like (Fas-associated death domain protein-like-interleukin-1–converting enzyme–like) inhibitory protein (FLIP), whose expression is highly enriched in striated muscle, binds to and inhibits procaspase-8 in the DISC (43). Antiapoptotic Bcl-2 proteins, such as Bcl-2 and Bcl-x_L, inhibit mitochondrial apoptogen release through biochemical mechanisms that are still incompletely understood (10). Ku-70 and humanin bind Bax and block its conformational activation and translocation to the mitochondria (44, 45). X-linked inhibitor of apoptosis (XIAP) and related proteins that contain baculovirus inhibitor of apoptosis repeats bind to and inhibit already activated caspases-9, -3, and -7, as well as interfering with procaspase-9 dimerization and activation (46–50). Each of these inhibitors act on circumscribed portions of either the extrinsic or intrinsic pathway. In contrast,

Nonstandard abbreviations used: ARC, apoptosis repressor with a CARD (caspase recruitment domain); BH3, Bcl-2 homology domain 3; DISC, death-inducing signaling complex; FADD, Fas-associated via death domain; Gαq, α subunit of Gq; IP3R, inositol 1,4,5-triphosphate receptor; XIAP, X-linked inhibitor of apoptosis.

Conflict of interest: The authors have declared that no conflict of interest exists.

Citation for this article: *J. Clin. Invest.* **115**:565–571 (2005).
doi:10.1172/JCI200524569.

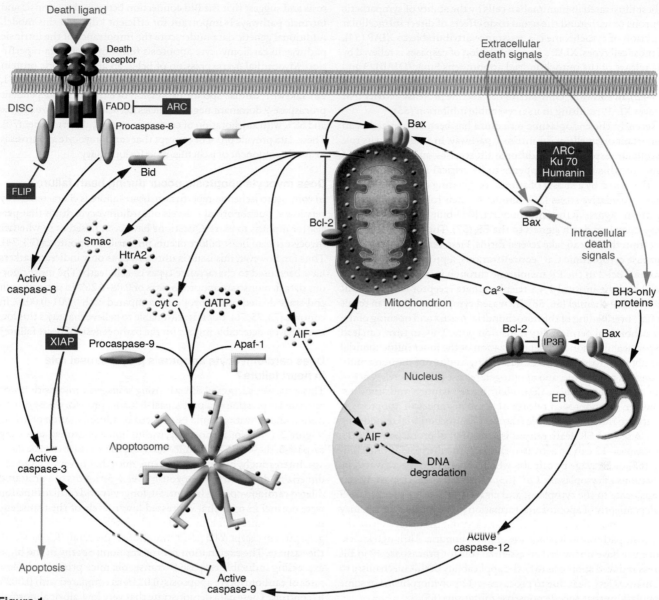

Figure 1

In the extrinsic pathway, ligand binding induces death receptors to recruit FADD, which recruits procaspase-8. Within this complex (DISC), procaspase-8 dimerizes and activates. Caspase-8 proteolytically activates procaspase-3, which cleaves cellular proteins causing cell death. In the intrinsic pathway, extracellular and intracellular stimuli signal to the mitochondria through a variety of BH3-only proteins (e.g., Bid) and through Bax, which translocates to and inserts into the outer mitochondrial membrane. Bax and Bak (not shown) stimulate the release of cytochrome *c* (cyt *c*) and other apoptogens. Once released, cytochrome *c* binds Apaf-1 along with dATP. This triggers oligomerization of Apaf-1 and recruitment of procaspase-9. Within this complex (apoptosome), procaspase-9 dimerizes and activates. Caspase-9 proteolytically activates procaspase-3. Caspase-8 also cleaves Bid, and its C-terminal fragment translocates to the mitochondria, where it activates Bax and Bak stimulating apoptogen release. FLIP binds and inhibits procaspase-8 in the DISC. ARC binds Fas, FADD, and procaspase-8, inhibiting DISC assembly. ARC, Ku-70, and Humanin bind Bax and inhibit its conformational activation and translocation. Bcl-2 and Bcl-x_L (not shown) inhibit mitochondrial apoptogen release. XIAP binds and inhibits activated caspase-9 and -3. Once cytoplasmic, the mitochondrial apoptogens Smac and HtrA2 bind XIAP, displacing and disinhibiting caspases. HtrA2 also has a serine protease activity that cleaves XIAP (not shown). AIF translocates to the nucleus and, in conjunction with a presumed endonuclease, mediates large-scale DNA fragmentation. A subset of intrinsic pathway death signals stimulate the ER, possibly through BH3-only proteins. This causes the release of intraluminal Ca^{2+} into the cytoplasm, which is mediated by Bax and Bcl-2 through the IP3R. Ca^{2+} translocates to the mitochondrial matrix and stimulates opening of the mitochondrial permeability transition pore in the inner membrane (not shown), which indirectly results in apoptogen release. The ER pathway may also activate procaspase-12, which can cleave and activate procaspase-9 independently of apoptosome formation (see text).

apoptosis repressor with a CARD (ARC), which is expressed preferentially in striated muscle and some neurons, antagonizes both central apoptosis pathways (51). The extrinsic pathway is inhibited by ARC's direct interactions with Fas, FADD, and procaspase-8,

which prevent DISC assembly, while the intrinsic pathway is inhibited by direct binding and inhibition of Bax (51–53).

Efficient cell killing usually requires neutralization of inhibitory pathways as well as activation of effector mechanisms. This is best illustrated in mammalian cells by the ability of sympathetic neurons to withstand the usual toxic effects of direct intracellular injection of cytochrome c, a resistance attributable to XIAP (54). In most cell types, XIAP's tonic inhibition of caspases is relieved by the release of the mitochondrial apoptogens Smac/DIABLO and Omi/HtrA2, which directly bind XIAP, thereby displacing caspases (23–25). In addition, Omi/HtrA2 has a serine protease activity that cleaves XIAP, resulting in its irreversible inhibition (55).

Recently, the endoplasmic reticulum has been recognized as an important organelle in the intrinsic pathway. In addition to its role in cellular responses to traditional ER stresses, such as misfolded proteins, this organelle appears to be critical in mediating cell death elicited by a subset of stimuli originating outside of the ER, such as oxidative stress (56). Similar to their roles in transducing upstream signals to the mitochondria, BH3-only proteins appear to relay upstream death signals to the ER (57). The death signal output from the ER can take several forms. First, certain death stimuli increase cytoplasmic Ca^{2+} concentrations, a process controlled by Bax and Bcl-2 in the ER membrane through interactions between Bcl-2 and the inositol 1,4,5-trisphosphate receptor (IP3R), an ER Ca^{2+} release channel (56, 58). Increased cytoplasmic Ca^{2+} can result in Ca^{2+} overloading of the mitochondrial matrix and opening of the mitochondrial permeability transition pore. This, in turn, can lead to permeabilization and depolarization of the inner mitochondrial membrane, gross mitochondrial swelling, rupture of the outer mitochondrial membrane, and apoptogen release (59). The relevance of this mechanism may be dependent on cell context and stimulus, however, as apoptogen release in some instances of apoptosis is limited to much more subtle features of mitochondrial remodeling (60). A second ER death output is activation of procaspase-12 (61). Procaspase-12 can be activated through the intrinsic ER machinery (62) or cleavage by calpain, which can, in turn, be activated by elevations in cytoplasmic Ca^{2+} (63). Once activated, caspase-12 can translocate to the cytoplasm and cleave and activate procaspase-9 independently of apoptosome formation (64). This mechanism may provide a means by which the ER can carry out the distal steps in the intrinsic pathway independently of mitochondria. Although knockout mice have uncovered an essential role for procaspase-12 in ER stress-induced apoptosis (61), the applicability of this mechanism to humans is less clear due to procaspase-12 polymorphisms in some populations that encode nonsense mutations (65).

Lessons from myocardial ischemia-reperfusion

The significance of these apoptotic pathways in cardiomyocytes is most clearly revealed in the context of ischemia-reperfusion, a model characterized by a robust burst of apoptosis over a limited time frame (66, 67). Ischemia-reperfusion activates both the extrinsic and intrinsic pathways. Moreover, a variety of genetic studies indicate that both pathways play critical roles in the genesis of ischemia-reperfusion–induced myocardial infarction. Thus, a loss of function mutation in the death receptor Fas (*lpr* mouse) results in a 64% decrease in cardiomyocyte apoptosis and a 63% decrease in infarct size following ischemia-reperfusion in vivo (68). The source of death ligands during ischemia-reperfusion may be the heart cells (myocytes or nonmyocytes) themselves, as transudates from isolated perfused wild-type hearts contain FasL, TNF-α, and TNF-related apoptosis-inducing ligand (TRAIL) in the reperfusion phase (69). Similarly, infarct size following ischemia-reperfusion is reduced by 53% in mice lacking Bid, and this is accompanied by reduced cardiac dysfunction (70). These data implicate the extrinsic pathway in cardiomyocyte apop-

tosis and suggest that the Bid connection between the extrinsic and intrinsic pathways is important for efficient killing in this model. Additional genetic data underscore the importance of the intrinsic pathway in cardiomyocyte apoptosis following ischemia-reperfusion. Myocardial overexpression of Bcl-2, an antiapoptotic protein that inhibits cytochrome c release, reduces infarct size by 48–64% (71, 72). Likewise, cardiac-specific expression of either of 2 independent procaspase-9 dominant negative alleles decreases infarct size by 53% and 68%, with amelioration of cardiac functional abnormalities (70). These data provide proof of concept that cardiomyocyte apoptosis is a causal component of ischemia-reperfusion injury.

Does myocyte apoptosis occur during heart failure?

In contrast to ischemia-reperfusion, heart failure is characterized by very low — but abnormal — levels of cardiomyocyte death that persist for months to years. Questions have been raised as to whether myocyte loss in heart failure occurs primarily by apoptosis (73, 74). Thus far, however, this issue is unresolved, as only indirect markers have been used to characterize types of cell death. The most rigorous data demonstrate apoptosis rates of 0.08–0.25% in patients with end-stage dilated cardiomyopathy compared with 0.001–0.002% in controls (73, 75, 76). But is it reasonable to believe that rates this low could have a detectable impact on the pathogenesis of heart failure?

Does cardiomyocyte apoptosis play a causal role in heart failure?

This issue was addressed directly using transgenic mice with heart-restricted expression of a procaspase-8 fusion protein, whose dimerization and activation could be induced by administration of a drug (Figure 2) (77). Not surprisingly, within hours of acutely activating caspase-8, these transgenic mice died due to extensive cardiac damage. Interestingly, however, transgenic mice that never received the dimerizing drug died spontaneously over 2–6 months of a profound dilated cardiomyopathy. In contrast, longevity and cardiac function were normal in mice that expressed lower levels of the transgene protein and mice that expressed similar levels of an identical transgene protein except for a point mutation in the catalytic cysteine of the caspase. The explanation for the cardiomyopathy in the high-expressing inducible caspase-8–transgenic mice proved to be low rates of cardiomyocyte apoptosis: 0.023% as compared with 0.002% in controls. These data demonstrate that very low, albeit abnormal, rates of cardiomyocyte apoptosis are sufficient to cause lethal dilated cardiomyopathy. Given that the rates of cardiomyocyte apoptosis in patients with end-stage dilated cardiomyopathy are 5- to 10-fold higher than those in this transgenic model (73, 75, 76), myocyte apoptosis may also play a causal role in human heart failure. To test whether apoptosis is required for the development of cardiomyopathy in this model, a broad-spectrum caspase inhibitor was administered systemically starting before cardiac decompensation. Caspase inhibition abrogated cardiac dilatation and markedly ameliorated contractile dysfunction. These experiments provide direct evidence that low levels of cardiomyocyte apoptosis may be a causal component of heart failure.

Support for the importance of cardiomyocyte apoptosis in heart failure is also provided by another, more physiological model (Figure 2). Gq transduces humoral (e.g., angiotensin II) and mechanical stimuli that are important in cardiac hypertrophy (78). Myocardial overexpression of the α subunit of Gq (Gαq) bypasses the need for stimulus and elicits cardiac hypertrophy and dilated cardiomyopathy (79–81). This phenotype is accompanied by cardiomyocyte

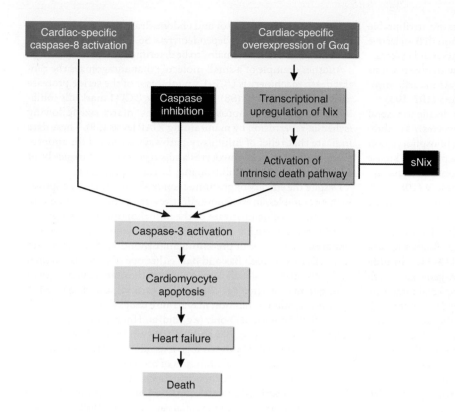

Figure 2
Cardiomyocyte apoptosis plays a mechanistic role in murine heart failure models. Cardiac-restricted expression of a conditional caspase-8 transgene indicates that low, but abnormal, levels of cardiomyocyte apoptosis (0.023% vs. 0.002% in controls) are sufficient to cause a lethal dilated cardiomyopathy. Similarly, cardiac-specific expression of Gαq induces the transcriptional activation of the BH3-only–like protein Nix, which results in cardiomyocyte apoptosis and dilated cardiomyopathy. The latter is severely exacerbated and lethal during pregnancy. Caspase inhibitors or expression of sNix, an antiapoptotic Nix splice variant, ameliorate cardiomyopathy in these models. In addition, caspase inhibition and sNix decrease mortality in the Gαq peripartum cardiomyopathy model.

apoptosis, which is mediated by the transcriptional induction of Nix/BNip3L, a BH3-only–like protein (82). In addition to this baseline cardiomyopathy, 30% of pregnant Gαq-transgenic mice exhibit fulminant lethal heart failure accompanied by further increases in apoptosis. Inhibition of apoptosis, either through expression of sNix, an antiapoptotic Nix splice variant (82), or administration of a broad-spectrum caspase inhibitor (83), markedly decreases the severity and mortality of the Gαq peripartum cardiomyopathy.

Hemodynamic overload is a unifying stimulus in multiple forms of heart failure. Several studies provide evidence that inhibition of cardiomyocyte apoptosis lessens the severity of cardiomyopathy resulting from hemodynamic overload. Overexpression of Bcl-2 (84) or ARC (85) inhibits cardiomyopathy induced by prior myocardial infarction. Conversely, loss of a basal survival mechanism can exacerbate cardiomyopathy induced by hemodynamic overload. This is illustrated by a study on mice with ventricular-restricted deletion of gp130 (86), a component of the receptor for several prosurvival cytokines that induces the STAT-dependent transcription of Bcl-x_L (87). While these mice are normal at baseline, they develop lethal heart failure following aortic constriction. Taken together, these and other studies (88, 89) demonstrate a mechanistic role for cardiomyocyte apoptosis in heart failure.

What pathways mediate cardiomyocyte apoptosis in the failing heart?

Although the fundamental blueprint for cell death signaling has been highly conserved over more than a billion years of evolution, the precise molecular events — especially upstream ones — are often specific to cell type and/or death stimulus. Relevant stimuli in heart failure probably include stretch (90), ROS (91), β₁-adrenergic agonists (92, 93), angiotensin II (94), proinflammatory cytokines (95), cytoskeletal abnormalities (96), and drugs such as anthracy-

clines (97), although the relative importance of these stimuli in the most common instances of heart failure has not yet been established. Likewise, with the exception of the angiotensin II → Gαq → Nix and gp130 → STAT → Bcl-x_L pathways noted above, little is known about the connections between the myriad of upstream signaling pathways and the central death machinery. Moreover, the pathophysiological significance of some upstream events is poorly understood. For example, strong evidence implicates telomere dysfunction as a cause for myocyte apoptosis in heart failure (98–100). However, it remains unclear whether telomere uncapping represents an abortive attempt by cardiomyocytes to reenter the cell cycle or a response to the activation of multiple stress pathways.

Even the importance of each of the central death pathways in heart failure needs to be better defined. For example, while both intrinsic and extrinsic death pathways play significant roles in ischemia-reperfusion injury, little is known about the importance of the extrinsic pathway in heart failure. Similarly, although the ER pathway is known to be activated in failing human hearts (101), the extent to which it plays a mechanistic role in cardiomyocyte demise remains to be determined. The study of mice harboring mutations in various components of the central death machinery, which has been so informative with respect to understanding ischemia-reperfusion injury, should be useful in addressing some of these questions.

Potential therapeutic targets and future questions

We have described experimental evidence showing that cardiomyocyte apoptosis is a causal component of ischemia-reperfusion injury and heart failure in rodent models. If this paradigm extends to humans, it follows that prevention or inhibition of myocyte death may provide a novel therapeutic approach to these most common and lethal heart syndromes. How might inhibition of cardiomyocyte apoptosis be achieved in a clinical setting?

We will limit our comments to small molecule therapies, which, at present, are the most practical options. In general, receptors and enzymes constitute the molecules most amenable to pharmacological manipulation. Several receptor/enzyme pathways influence survival in cardiomyocytes. In fact, β-adrenergic receptor antagonists and inhibitors of the angiotensin II axis (angiotensin-converting enzyme inhibitors and angiotensin II type 1 receptor antagonists) are already mainstays in the treatment of heart failure. The extent to which the salutary effects of these agents are attributable to their inhibition of apoptosis is not known. Although β_1-selective adrenergic receptor blockers are used primarily to avoid extracardiac side effects, this approach is also consistent with observations that β_1-adrenergic receptors activate death pathways in cardiomyocytes, which are opposed by β_2-adrenergic receptors (102, 103).

The serine-threonine kinase Akt is a central molecule in several receptor-mediated survival pathways in cardiomyocytes. Akt inhibits apoptotic signaling at multiple levels including phosphorylation of the BH3-only protein Bad (104, 105), IKKβ (a proximal kinase that activates NF-κB signaling) (106), the proapoptotic transcription factor Foxo3a (107), and perhaps procaspase-9 (108, 109). Akt can be activated in cardiomyocytes by receptor tyrosine kinase ligands, such as IGF-1 (110, 111), and by exogenously administered thymosin $\beta 4$ (112). IGF-1 and Akt have been demonstrated to reduce cardiomyocyte apoptosis and infarct size following ischemia-reperfusion and prolonged ischemia (111, 113–117). In addition, cardiac-restricted expression of an IGF-1 transgene has beneficial effects on cardiac remodeling following myocardial infarction (114) and in a genetic model of cardiomyopathy (118), although interpretation of these studies is complicated by the fact that myocardial IGF-1 expression was present in the mice beginning in the fetus, which resulted in baseline increases in the number of cardiomyocytes (119). Interestingly, growth hormone, which increases serum IGF-1 levels, was shown in a small clinical study to improve clinical and functional parameters in patients with idiopathic dilated cardiomyopathy (120). Although questions remain concerning the deleterious effects of Akt on the heart (121), the data discussed here raise the possibility that IGF-1 may ameliorate heart failure through inhibition of cardiomyocyte apoptosis. Similarly, activation of the gp130 → STAT → Bcl-x_L survival pathway with ligands such as cardiotrophin-1 (CT-1) may provide an alternative means of achieving this end (122).

Although manipulation of receptor-mediated upstream pathways is attractive because of their accessibility, this approach can be confounded by redundancy and complicated by undesirable pleiotropic effects. These issues provide a strong rationale for interventions that focus directly on the central death pathways. Where in the central death pathways would inhibition of cardiomyocyte apoptosis be most efficiently achieved? Both central death pathways converge on caspases. Broad-spectrum caspase inhibitors reduce infarct size by 21–52% and decrease cardiac dysfunction following ischemia-reperfusion (123–125, S1). Caspase inhibitors have also been shown to ameliorate cardiac dysfunction and/or inhibit mortality in the caspase-8 and Gαq models of dilated cardiomyopathy discussed above (77, 83). These effects of caspase inhibition correlate with inhibition of cardiomyocyte death, which suggests that the benefit of these agents results from their antiapoptotic properties. Recently, however, caspases have been demonstrated to cleave cardiac contractile proteins and at least 1 cardiac transcription factor (S2–S4). This raises the possibility that caspase inhibitors may also preserve cardiac function independently of their effects on cell death.

The efficacy of caspase inhibitors in these settings poses important questions. Caspase activation in cardiomyocytes occurs after mitochondrial damage and the release of apoptogens. Moreover, as in many systems, cytochrome c release in the myocardium is unaffected by caspase inhibition (S5). Given these mitochondrial abnormalities, it is curious that caspase inhibitors exert such marked improvement on disease pathogenesis. Whether their beneficial effects result from inhibiting upstream caspases, the release of other apoptogens (such

as apoptosis-inducing factor and endonuclease G, whose translocation appears to be caspase dependent; refs. S6, S7), or other yet-to-be-described mechanisms remains to be determined.

Another example of a small molecule that antagonizes the central death machinery is UCF-1, an inhibitor of the serine protease activity of Omi/HtrA2 (S8). Unexpectedly, UCF-1 markedly inhibits cardiomyocyte apoptosis and reduces infarct size following ischemia-reperfusion by maintaining XIAP levels (S9). These data indicate that relief of inhibitory pathways is critical for apoptosis to proceed in cardiomyocytes and suggest that strategies built around maintaining inhibition may be cardioprotective.

Despite the efficacy of postmitochondrial inhibition of apoptosis in these examples, an even more effective strategy may be to intercept apoptotic signaling upstream of the mitochondria. While a premitochondrial approach might be limited by redundancy, it offers the important advantage of preventing mitochondrial dysfunction. In fact, cells lacking both Bax and Bak, either one of which is required for upstream death signals to gain access to the mitochondria, exhibit long-term protection against multiple noxious stimuli (S10, S11). Premitochondrial inhibition may require drugs that interfere with interactions between BH3-only (e.g., Bid) and multidomain proapoptotic Bcl-2 proteins (e.g., Bax and Bak) (Figure 1). Another attractive possibility is provided by the endogenous inhibitor ARC. ARC would provide premitochondrial inhibition of both extrinsic and intrinsic pathways and preserve mitochondrial function (51, S12). In addition, the relatively restricted expression pattern of endogenous ARC (52) may offer a means to avoid potential carcinogenic effects of diffuse long-term inhibition of apoptosis. A major outstanding issue is how best to exploit endogenous ARC as a therapeutic agent.

Conclusions

Our understanding of the significance of myocyte loss during heart failure has increased substantially since this phenomenon was initially observed by cardiac pathologists. Molecular and genetic studies demonstrate clearly that cardiomyocyte apoptosis is a critical process in the pathogenesis of heart failure in rodent models. If this paradigm extends to humans, apoptosis will be a logical target for novel therapies. Despite success in establishing a mechanistic link between apoptosis and heart failure, our knowledge of the precise molecular mechanisms that regulate cell death specifically in this syndrome remains rudimentary. An understanding of these mechanisms will be indispensable for the rational design of future antiapoptotic therapies.

Acknowledgments

We are indebted to Thierry H. LeJemtel and James Scheuer for critical reading of the manuscript. This work was supported by NIH R01 grants HL60665, HL61550, and HL73732 (to R.N. Kitsis) and a Wellcome Trust Advanced Fellowship (to R.S.-Y. Foo). R.N. Kitsis was also supported by The Dr. Gerald and Myra Dorros Chair in Cardiovascular Disease of the Albert Einstein College of Medicine and the Monique Weill-Caulier Career Scientist Award.

Due to space constraints, a number of important references could not be included in this article. References S1–S12 are available online with this article; doi:10.1172/JCI200524569DS1.

Address correspondence to Richard N. Kitsis, Albert Einstein College of Medicine, 1300 Morris Park Avenue, Bronx, New York 10461, USA. Phone: (718) 430-2609; Fax: (718) 430-8989; E-mail: kitsis@aecom.yu.edu.

1. Mann, D.L. 1999. Mechanisms and models in heart failure: a combinatorial approach. *Circulation.* **100**:999–1008.

2. Lower, R. 1669. Tractatus de corde. Item de motu et colore sanguinis et chyli in eum transitu. Jo. Redmayne for Jacob Allestry. London, United Kingdom.

3. Lefkowitz, R.J., Rockman, H.A., and Koch, W.J. 2000. Catecholamines, cardiac beta-adrenergic receptors, and heart failure. *Circulation.* **101**:1634–1637.

4. Marks, A.R. 2002. Ryanodine receptors, FKBP12, and heart failure. *Front. Biosci.* **7**:d970–d977.

5. Luo, W., et al. 1994. Targeted ablation of the phospholamban gene is associated with markedly enhanced myocardial contractility and loss of beta-agonist stimulation. *Circ. Res.* **75**:401–409.

6. Chien, K.R. 1999. Stress pathways and heart failure. *Cell.* **98**:555–558.

7. Nakao, K., Minobe, W., Roden, R., Bristow, M.R., and Leinwand, L.A. 1997. Myosin heavy chain gene expression in human heart failure. *J. Clin. Invest.* **100**:2362–2370.

8. Taegtmeyer, H. 2002. Switching metabolic genes to build a better heart. *Circulation.* **106**:2043–2045.

9. Kelly, D.P., and Strauss, A.W. 1994. Inherited cardiomyopathies. *N. Engl. J. Med.* **330**:913–919.

10. Danial, N.N., and Korsmeyer, S.J. 2004. Cell death: critical control points. *Cell.* **116**:205–219.

11. Yuan, J., Lipinski, M., and Degterev, A. 2003. Diversity in the mechanisms of neuronal cell death. *Neuron.* **40**:401–413.

12. Zong, W.X., Ditsworth, D., Bauer, D.E., Wang, Z.Q., and Thompson, C.B. 2004. Alkylating DNA damage stimulates a regulated form of necrotic cell death. *Genes Dev.* **18**:1272–1282.

13. Klionsky, D.J., and Emr, S.D. 2000. Autophagy as a regulated pathway of cellular degradation. *Science.* **290**:1717–1721.

14. Shintani, T., and Klionsky, D.J. 2004. Autophagy in health and disease: a double-edged sword. *Science.* **306**:990–995.

15. Ashkenazi, A., and Dixit, V.M. 1998. Death receptors: signaling and modulation. *Science.* **281**:1305–1308.

16. Kischkel, F.C., et al. 1995. Cytotoxicity-dependent APO-1 (Fas/CD95)-associated proteins form a death-inducing signaling complex (DISC) with the receptor. *EMBO J.* **14**:5579–5588.

17. Boldin, M.P., Goncharov, T.M., Goltsev, Y.V., and Wallach, D. 1996. Involvement of MACH, a novel MORT1/FADD-interacting protease, in Fas/APO-1- and TNF receptor-induced cell death. *Cell.* **85**:803–815.

18. Muzio, M., et al. 1996. FLICE, a novel FADD-homologous ICE/CED-3-like protease, is recruited to the CD95 (Fas/APO-1) death-inducing signaling complex. *Cell.* **85**:817–827.

19. Boatright, K.M., et al. 2003. A unified model for apical caspase activation. *Mol. Cell.* **11**:529–541.

20. Crow, M.T., Mani, K., Nam, Y.J., and Kitsis, R.N. 2004. The mitochondrial death pathway and cardiac myocyte apoptosis. *Circ. Res.* **95**:957–970.

21. Liu, X., Kim, C.N., Yang, J., Jemmerson, R., and Wang, X. 1996. Induction of apoptotic program in cell-free extracts: requirement for dATP and cytochrome c. *Cell.* **86**:147–157.

22. Kluck, R.M., Bossy-Wetzel, E., Green, D.R., and Newmeyer, D.D. 1997. The release of cytochrome c from mitochondria: a primary site for Bcl-2 regulation of apoptosis. *Science.* **275**:1132–1136.

23. Du, C., Fang, M., Li, Y., Li, L., and Wang, X. 2000. Smac, a mitochondrial protein that promotes cytochrome c-dependent caspase activation by eliminating IAP inhibition. *Cell.* **102**:33–42.

24. Verhagen, A.M., et al. 2000. Identification of DIABLO, a mammalian protein that promotes apoptosis by binding to and antagonizing IAP proteins. *Cell.* **102**:43–53.

25. Suzuki, Y., et al. 2001. A serine protease, HtrA2, is released from the mitochondria and interacts with XIAP, inducing cell death. *Mol. Cell.* **8**:613–621.

26. Susin, S.A., et al. 1999. Molecular characterization of mitochondrial apoptosis-inducing factor. *Nature.* **397**:441–446.

27. Li, L.Y., Luo, X., and Wang, X. 2001. Endonuclease G is an apoptotic DNase when released from mitochondria. *Nature.* **412**:95–99.

28. Zou, H., Henzel, W.J., Liu, X., Lutschg, A., and Wang, X. 1997. Apaf-1, a human protein homologous to C. elegans CED-4, participates in cytochrome c-dependent activation of caspase-3. *Cell.* **90**:405–413.

29. Li, P., et al. 1997. Cytochrome c and dATP-dependent formation of Apaf-1/caspase-9 complex initiates an apoptotic protease cascade. *Cell.* **91**:479–489.

30. Srinivasula, S.M., Ahmad, M., Fernandes-Alnemri, T., and Alnemri, E.S. 1998. Autoactivation of pro-caspase-9 by Apaf-1-mediated oligomerization. *Mol. Cell.* **1**:949–957.

31. Hu, Y., Ding, L., Spencer, D.M., and Nunez, G. 1998. WD-40 repeat region regulates Apaf-1 self-association and procaspase-9 activation. *J. Biol. Chem.* **273**:33489–33494.

32. Saleh, A., Srinivasula, S.M., Acharya, S., Fishel, R., and Alnemri, E.S. 1999. Cytochrome c and dATP-mediated oligomerization of Apaf-1 is a prerequisite for procaspase-9 activation. *J. Biol. Chem.* **274**:17941–17945.

33. Zou, H., Li, Y., Liu, X., and Wang, X. 1999. An APAF-1.cytochrome c multimeric complex is a functional apoptosome that activates procaspase-9. *J. Biol. Chem.* **274**:11549–11556.

34. Qin, H., et al. 1999. Structural basis of procaspase-9 recruitment by the apoptotic protease-activating factor 1. *Nature.* **399**:549–557.

35. Vaughn, D.E., Rodriguez, J., Lazebnik, Y., and Joshua-Tor, L. 1999. Crystal structure of Apaf-1 caspase recruitment domain: an alpha-helical Greek key fold for apoptotic signaling. *J. Mol. Biol.* **293**:439–447.

36. Zhou, P., Chou, J., Olea, R.S., Yuan, J., and Wagner, G. 1999. Solution structure of Apaf-1 CARD and its interaction with caspase-9 CARD: a structural basis for specific adaptor/caspase interaction. *Proc. Natl. Acad. Sci. U. S. A.* **96**:11265–11270.

37. Day, C.L., Dupont, C., Lackmann, M., Vaux, D.L., and Hinds, M.G. 1999. Solution structure and mutagenesis of the caspase recruitment domain (CARD) from Apaf-1. *Cell Death Differ.* **6**:1125–1132.

38. Jiang, X., and Wang, X. 2000. Cytochrome c promotes caspase-9 activation by inducing nucleotide binding to Apaf-1. *J. Biol. Chem.* **275**:31199–31203.

39. Acehan, D., et al. 2002. Three-dimensional structure of the apoptosome: implications for assembly, procaspase-9 binding, and activation. *Mol. Cell.* **9**:423–432.

40. Luo, X., Budihardjo, I., Zou, H., Slaughter, C., and Wang, X. 1998. Bid, a Bcl2 interacting protein, mediates cytochrome c release from mitochondria in response to activation of cell surface death receptors. *Cell.* **94**:481–490.

41. Li, H., Zhu, H., Xu, C.J., and Yuan, J. 1998. Cleavage of BID by caspase 8 mediates the mitochondrial damage in the Fas pathway of apoptosis. *Cell.* **94**:491–501.

42. Gross, A., et al. 1999. Caspase cleaved BID targets mitochondria and is required for cytochrome c release, while BCL-XL prevents this release but not tumor necrosis factor-R1/Fas death. *J. Biol. Chem.* **274**:1156–1163.

43. Peter, M.E. 2004. The flip side of FLIP. *Biochem. J.* **382**:e1–e3.

44. Sawada, M., et al. 2003. Ku70 suppresses the apoptotic translocation of Bax to mitochondria. *Nat. Cell Biol.* **5**:320–329.

45. Guo, B., et al. 2003. Humanin peptide suppresses apoptosis by interfering with Bax activation. *Nature.* **423**:456–461.

46. Sun, C., et al. 2000. NMR structure and mutagenesis of the third Bir domain of the inhibitor of apoptosis protein XIAP. *J. Biol. Chem.* **275**:33777–33781.

47. Chai, J., et al. 2001. Structural basis of caspase-7 inhibition by XIAP. *Cell.* **104**:769–780.

48. Huang, Y., et al. 2001. Structural basis of caspase inhibition by XIAP: differential roles of the linker versus the BIR domain. *Cell.* **104**:781–790.

49. Riedl, S.J., et al. 2001. Structural basis for the inhibition of caspase-3 by XIAP. *Cell.* **104**:791–800.

50. Shiozaki, E.N., et al. 2003. Mechanism of XIAP-mediated inhibition of caspase-9. *Mol. Cell.* **11**:519–527.

51. Nam, Y.J., et al. 2004. Inhibition of both the extrinsic and intrinsic death pathways through nonhomotypic death-fold interactions. *Mol. Cell.* **15**:901–912.

52. Koseki, T., Inohara, N., Chen, S., and Nunez, G. 1998. ARC, an inhibitor of apoptosis expressed in skeletal muscle and heart that interacts selectively with caspases. *Proc. Natl. Acad. Sci. U. S. A.* **95**:5156–5160.

53. Gustafsson, A.B., Tsai, J.G., Logue, S.E., Crow, M.T., and Gottlieb, R.A. 2004. Apoptosis repressor with caspase recruitment domain protects against cell death by interfering with Bax activation. *J. Biol. Chem.* **279**:21233–21238.

54. Potts, P.R., Singh, S., Knezek, M., Thompson, C.B., and Deshmukh, M. 2003. Critical function of endogenous XIAP in regulating caspase activation during sympathetic neuronal apoptosis. *J. Cell Biol.* **163**:789–799.

55. Yang, Q.H., Church-Hajduk, R., Ren, J., Newton, M.L., and Du, C. 2003. Omi/HtrA2 catalytic cleavage of inhibitor of apoptosis (IAP) irreversibly inactivates IAPs and facilitates caspase activity in apoptosis. *Genes Dev.* **17**:1487–1496.

56. Scorrano, L., et al. 2003. BAX and BAK regulation of endoplasmic reticulum Ca2+: a control point for apoptosis. *Science.* **300**:135–139.

57. Morishima, N., Nakanishi, K., Tsuchiya, K., Shibata, T., and Seiwa, E. 2004. Translocation of Bim to the endoplasmic reticulum (ER) mediates ER stress signaling for activation of caspase-12 during ER stress-induced apoptosis. *J. Biol. Chem.* **279**:50375–50381.

58. Oakes, S.A., et al. 2005. Proapoptotic BAX and BAK regulate the type 1 inositol trisphosphate receptor and calcium leak from the endoplasmic reticulum. *Proc. Natl. Acad. Sci. U. S. A.* **102**:105–110.

59. Halestrap, A.P., McStay, G.P., and Clarke, S.J. 2002. The permeability transition pore complex: another view. *Biochimie.* **84**:153–166.

60. Scorrano, L., et al. 2002. A distinct pathway remodels mitochondrial cristae and mobilizes cytochrome c during apoptosis. *Dev. Cell.* **2**:55–67.

61. Nakagawa, T., et al. 2000. Caspase-12 mediates endoplasmic-reticulum-specific apoptosis and cytotoxicity by amyloid-beta. *Nature.* **403**:98–103.

62. Szegezdi, E., Fitzgerald, U., and Samali, A. 2003. Caspase-12 and ER-stress-mediated apoptosis: the story so far. *Ann. N. Y. Acad. Sci.* **1010**:186–194.

63. Nakagawa, T., and Yuan, J. 2000. Cross-talk between two cysteine protease families. Activation of caspase-12 by calpain in apoptosis. *J. Cell Biol.* **150**:887–894.

64. Morishima, N., Nakanishi, K., Takenouchi, H., Shibata, T., and Yasuhiko, Y. 2002. An endoplasmic reticulum stress-specific caspase cascade in apoptosis. Cytochrome c-independent activation of caspase-9 by caspase-12. *J. Biol. Chem.* **277**:34287–34294.

65. Saleh, M., et al. 2004. Differential modulation of endotoxin responsiveness by human caspase-12 polymorphisms. *Nature.* **429**:75–79.

66. Gottlieb, R.A., Burleson, K.O., Kloner, R.A., Babior, B.M., and Engler, R.L. 1994. Reperfusion injury induces apoptosis in rabbit cardiomyocytes. *J. Clin. Invest.* **94**:1621–1628.

67. Fliss, H., and Gattinger, D. 1996. Apoptosis in ischemic and reperfused rat myocardium. *Circ. Res.* **79**:949–956.

68. Lee, P., et al. 2003. Fas pathway is a critical media-

tor of cardiac myocyte death and MI during ischemia-reperfusion in vivo. *Am. J. Physiol. Heart Circ. Physiol.* **284**:H456–H463.

69. Jeremias, I., et al. 2000. Involvement of CD95/Apo1/Fas in cell death after myocardial ischemia. *Circulation.* **102**:915–920.

70. Peng, C.-F., et al. 2001. Multiple independent mutations in apoptotic signaling pathways markedly decrease infarct size due to myocardial ischemia-reperfusion [abstract]. *Circulation.* **104**(Suppl. II):II-187.

71. Brocheriou, V., et al. 2000. Cardiac functional improvement by a human Bcl-2 transgene in a mouse model of ischemia/reperfusion injury. *J. Gene Med.* **2**:326–333.

72. Chen, Z., Chua, C.C., Ho, Y.S., Hamdy, R.C., and Chua, B.H. 2001. Overexpression of Bcl-2 attenuates apoptosis and protects against myocardial I/R injury in transgenic mice. *Am. J. Physiol. Heart Circ. Physiol.* **280**:H2313–H2320.

73. Guerra, S., et al. 1999. Myocyte death in the failing human heart is gender dependent. *Circ. Res.* **85**:856–866.

74. Hein, S., et al. 2003. Progression from compensated hypertrophy to failure in the pressure-overloaded human heart: structural deterioration and compensatory mechanisms. *Circulation.* **107**:984–991.

75. Olivetti, G., et al. 1997. Apoptosis in the failing human heart. *N. Engl. J. Med.* **336**:1131–1141.

76. Saraste, A., et al. 1999. Cardiomyocyte apoptosis and progression of heart failure to transplantation. *Eur. J. Clin. Invest.* **29**:380–386.

77. Wencker, D., et al. 2003. A mechanistic role for cardiac myocyte apoptosis in heart failure. *Cell.* **111**:1497–1504.

78. Dorn, G.W., 2nd, and Brown, J.H. 1999. Gq signaling in cardiac adaptation and maladaptation. *Trends Cardiovasc. Med.* **9**:26–34.

79. D'Angelo, D.D., et al. 1997. Transgenic Galphaq overexpression induces cardiac contractile failure in mice. *Proc. Natl. Acad. Sci. U. S. A.* **94**:8121–8126.

80. Adams, J.W., et al. 1998. Enhanced Galphaq signaling: a common pathway mediates cardiac hypertrophy and apoptotic heart failure. *Proc. Natl. Acad. Sci. U. S. A.* **95**:10140–10145.

81. Sakata, Y., Hoit, B.D., Liggett, S.B., Walsh, R.A., and Dorn, G.W., 2nd. 1998. Decompensation of pressure-overload hypertrophy in G alpha q-overexpressing mice. *Circulation.* **97**:1488–1495.

82. Yussman, M.G., et al. 2002. Mitochondrial death protein Nix is induced in cardiac hypertrophy and triggers apoptotic cardiomyopathy. *Nat. Med.* **8**:725–730.

83. Hayakawa, Y., et al. 2003. Inhibition of cardiac myocyte apoptosis improves cardiac function and abolishes mortality in the peripartum cardiomyopathy of Galpha(q) transgenic mice. *Circulation.* **108**:3036–3041.

84. Chatterjee, S., et al. 2002. Viral gene transfer of the antiapoptotic factor Bcl-2 protects against chronic postischemic heart failure. *Circulation.* **106**:I212–I217.

85. Chatterjee, S., et al. 2003. Blocking the development of postischemic cardiomyopathy with viral gene transfer of the apoptosis repressor with caspase recruitment domain. *J. Thorac. Cardiovasc. Surg.* **125**:1461–1469.

86. Hirota, H., et al. 1999. Loss of a gp130 cardiac muscle cell survival pathway is a critical event in the onset of heart failure during biomechanical stress. *Cell.* **97**:189–198.

87. Fujio, Y., Kunisada, K., Hirota, H., Yamauchi-Takihara, K., and Kishimoto, T. 1997. Signals through gp130 upregulate bcl-x gene expression via STAT1-binding cis-element in cardiac myocytes. *Cell.* **99**:2898–2905.

88. Xing, H., Zhang, S., Weinheimer, C., Kovacs, A., and Muslin, A.J. 2000. 14-3-3 proteins block apoptosis and differentially regulate MAPK cascades. *EMBO J.* **19**:349–358.

89. Zhang, D., et al. 2000. TAK1 is activated in the myocardium after pressure overload and is sufficient to provoke heart failure in transgenic mice. *Nat. Med.* **6**:556–563.

90. Cheng, W., et al. 1995. Stretch-induced programmed myocyte cell death. *Cell.* **96**:2247–2259.

91. Ide, T., et al. 2000. Direct evidence for increased hydroxyl radicals originating from superoxide in the failing myocardium. *Circ. Res.* **86**:152–157.

92. Communal, C., Singh, K., Pimentel, D.R., and Colucci, W.S. 1998. Norepinephrine stimulates apoptosis in adult rat ventricular myocytes by activation of the beta-adrenergic pathway. *Circulation.* **98**:1329–1334.

93. Shizukuda, Y., et al. 1998. beta-adrenergic stimulation causes cardiocyte apoptosis: influence of tachycardia and hypertrophy. *Am. J. Physiol.* **275**:H961–H968.

94. Cigola, E., Kajstura, J., Li, B., Meggs, L.G., and Anversa, P. 1997. Angiotensin II activates programmed myocyte cell death in vitro. *Exp. Cell Res.* **231**:363–371.

95. Mann, D.L. 1999. Inflammatory mediators in heart failure: homogeneity through heterogeneity. *Lancet.* **353**:1812–1813.

96. Nikolova, V., et al. 2004. Defects in nuclear structure and function promote dilated cardiomyopathy in lamin A/C-deficient mice. *Cell.* **113**:357–369.

97. Wang, L., Ma, W., Markovich, R., Chen, J.W., and Wang, P.H. 1998. Regulation of cardiomyocyte apoptotic signaling by insulin-like growth factor I. *Circ. Res.* **83**:516–522.

98. Oh, H., et al. 2001. Telomerase reverse transcriptase promotes cardiac muscle cell proliferation, hypertrophy, and survival. *Proc. Natl. Acad. Sci. U. S. A.* **98**:10308–10313.

99. Leri, A., et al. 2003. Ablation of telomerase and telomere loss leads to cardiac dilatation and heart failure associated with p53 upregulation. *EMBO J.* **22**:131–139.

100. Oh, H., et al. 2003. Telomere attrition and Chk2 activation in human heart failure. *Proc. Natl. Acad. Sci. U. S. A.* **100**:5378–5383.

101. Okada, K., et al. 2004. Prolonged endoplasmic reticulum stress in hypertrophic and failing heart after aortic constriction: possible contribution of endoplasmic reticulum stress to cardiac myocyte apoptosis. *Circulation.* **110**:705–712.

102. Chesley, A., et al. 2000. The beta(2)-adrenergic receptor delivers an antiapoptotic signal to cardiac myocytes through G(i)-dependent coupling to phosphatidylinositol 3'-kinase. *Circ. Res.* **87**:1172–1179.

103. Xiao, R.P. 2001. Beta-adrenergic signaling in the heart: dual coupling of the beta2- adrenergic receptor to G(s) and G(i) proteins. *Sci. STKE.* **2001**:RE15.

104. Datta, S.R., et al. 1997. Akt phosphorylation of BAD couples survival signals to the cell-intrinsic death machinery. *Cell.* **91**:231–241.

105. Mehrhof, F.B., et al. 2001. In cardiomyocyte hypoxia, insulin-like growth factor-I-induced antiapoptotic signaling requires phosphatidylinositol-3-OH-kinase-dependent and mitogen-activated protein kinase-dependent activation of the transcription factor cAMP response element-binding protein. *Circulation.* **104**:2088–2094.

106. Romashkova, J.A., and Makarov, S.S. 1999. NF-kappaB is a target of AKT in anti-apoptotic PDGF signalling. *Nature.* **401**:86–90.

107. Brunet, A., et al. 1999. Akt promotes cell survival by phosphorylating and inhibiting a Forkhead transcription factor. *Cell.* **96**:857–868.

108. Cardone, M.H., et al. 1998. Regulation of cell death protease caspase-9 by phosphorylation. *Science.* **282**:1318–1321.

109. Fujita, E., et al. 1999. Akt phosphorylation site found in human caspase-9 is absent in mouse caspase-9. *Biochem. Biophys. Res. Commun.* **264**:550–555.

110. Matsui, T., et al. 1999. Adenoviral gene transfer of activated phosphatidylinositol 3'-kinase and Akt inhibits apoptosis of hypoxic cardiomyocytes in vitro. *Circulation.* **100**:2373–2379.

111. Fujio, Y., Nguyen, T., Wencker, D., Kitsis, R.N., and Walsh, K. 2000. Akt promotes survival of cardiomyocytes in vitro and protects against ischemia-reperfusion injury in mouse heart. *Circulation.* **101**:660–667.

112. Bock-Marquette, I., Saxena, A., White, M.D., Dimaio, J.M., and Srivastava, D. 2004. Thymosin beta4 activates integrin-linked kinase and promotes cardiac cell migration, survival and cardiac repair. *Nature.* **432**:466–472.

113. Buerke, M., et al. 1995. Cardioprotective effect of insulin-like growth factor I in myocardial ischemia followed by reperfusion. *Proc. Natl. Acad. Sci. U. S. A.* **92**:8031–8035.

114. Li, Q., et al. 1997. Overexpression of insulin-like growth factor-1 in mice protects from myocyte death after infarction, attenuating ventricular dilation, wall stress, and cardiac hypertrophy. *Cell.* **100**:1991–1999.

115. Miao, W., Luo, Z., Kitsis, R.N., and Walsh, K. 2000. Intracoronary, adenovirus-mediated Akt gene transfer in heart limits infarct size following ischemia-reperfusion injury in vivo. *J. Mol. Cell. Cardiol.* **32**:2397–2402.

116. Matsui, T., et al. 2001. Akt activation preserves cardiac function and prevents injury after transient cardiac ischemia in vivo. *Circulation.* **104**:330–335.

117. Chao, W., et al. 2003. Strategic advantages of insulin-like growth factor-I expression for cardioprotection. *J. Gene Med.* **5**:277–286.

118. Welch, S., et al. 2002. Cardiac-specific IGF-1 expression attenuates dilated cardiomyopathy in tropomodulin-overexpressing transgenic mice. *Circ. Res.* **90**:641–648.

119. Reiss, K., et al. 1996. Overexpression of insulin-like growth factor-1 in the heart is coupled with myocyte proliferation in transgenic mice. *Proc. Natl. Acad. Sci. U. S. A.* **93**:8630–8635.

120. Fazio, S., et al. 1996. A preliminary study of growth hormone in the treatment of dilated cardiomyopathy. *N. Engl. J. Med.* **334**:809–814.

121. Matsui, T., et al. 2002. Phenotypic spectrum caused by transgenic overexpression of activated Akt in the heart. *J. Biol. Chem.* **277**:22896–22901.

122. Pennica, D., Wood, W.I., and Chien, K.R. 1996. Cardiotrophin-1: a multifunctional cytokine that signals via LIF receptor-gp 130 dependent pathways. *Cytokine Growth Factor Rev.* **7**:81–91.

123. Yaoita, H., Ogawa, K., Maehara, K., and Maruyama, Y. 1998. Attenuation of ischemia/reperfusion injury in rats by a caspase inhibitor. *Circulation.* **97**:276–281.

124. Holly, T.A., et al. 1999. Caspase inhibition reduces myocyte cell death induced by myocardial ischemia and reperfusion in vivo. *J. Mol. Cell. Cardiol.* **31**:1709–1715.

125. Huang, J.Q., Radinovic, S., Rezaiefar, P., and Black, S.C. 2000. In vivo myocardial infarct size reduction by a caspase inhibitor administered after the onset of ischemia. *Eur. J. Pharmacol.* **402**:139–142.

Unchain my heart: the scientific foundations of cardiac repair

Stefanie Dimmeler,[1] Andreas M. Zeiher,[1] and Michael D. Schneider[2]

[1]Department of Molecular Cardiology, Department of Internal Medicine IV, University of Frankfurt, Frankfurt am Main, Germany.
[2]Center for Cardiovascular Development, Departments of Medicine, Molecular & Cellular Biology, and Molecular Physiology & Biophysics, Baylor College of Medicine, Houston, Texas, USA.

In humans, the biological limitations to cardiac regenerative growth create both a clinical imperative — to offset cell death in acute ischemic injury and chronic heart failure — and a clinical opportunity; that is, for using cells, genes, and proteins to rescue cardiac muscle cell number or in other ways promote more efficacious cardiac repair. Recent experimental studies and early-phase clinical trials lend credence to the visionary goal of enhancing cardiac repair as an achievable therapeutic target.

Heart failure — a severe deficiency in ventricular pump function — arises through a finite number of terminal effector mechanisms, regardless of the cause. These include: defects intrinsic to cardiac muscle cells' contractility due to altered expression or operation of calcium-cycling proteins, components of the sarcomere, and enzymes for cardiac energy production; defects extrinsic to cardiac muscle cells, such as interstitial fibrosis, affecting organ-level compliance; and myocyte loss, unmatched by myocyte replacement. Cardiac regeneration is robust for certain organisms such as the newt and zebrafish, in which total replacement can transpire even for an amputated limb, fin, or tail, via production of an undifferentiated cell mass called the blastema (1). Such a degree of restorative growth might also be dependent on the retention of proliferative potential in a subset of adult cardiomyocytes (2) and is impossible in mammals under normal, unassisted biological circumstances. Several complementary strategies can be foreseen as potentially aiding this process: overriding cell-cycle checkpoints that constrain the reactive proliferation of ventricular myocytes (3); supplementing the cytoprotective mechanisms that occur naturally, or inhibiting pro-death pathways (4, 5); supplementing the angiogenic mechanisms that occur naturally using defined growth factors or vessel-forming cells (6, 7); or providing exogenous cells as a surrogate or precursor for cardiac muscle itself (8–10).

Among these conceptual possibilities, cell implantation in various forms has been the first strategy to be translated from bench to bedside. The possibility of tissue repair by autologous adult progenitor cells — suggested by the auspicious findings in experimental studies of various cell sources — immediately captured the attention of clinicians confronted with the disabling, life-threatening circumstance of patients who suffer from heart failure in acute or chronic ischemic heart disease. The promise of cellular cardiomyogenesis and neovascularization, individually or in tandem, offered altogether novel opportunities for treatment, tailored to the underlying pathobiology.

Within just the past 3 years, more than a half-dozen early clinical studies have been published, ranging from case reports to formal trials, deploying a range of differing cell-based therapies with the shared objective of improving cardiac repair (11–16). Clinical follow-up for as long as a year is now available for some patients (17, 18). Despite their different strategies and cells, and lack of double-blinded controls, these small initial human trials in general point to a functional improvement; yet key questions remain open. Understanding better just why and how grafting works will be essential, alongside needed empirical trials, to engineer the soundest future for regenerative therapy in human heart disease.

The inauguration of human cardiac repair

Ventricular dysfunction is the sine qua non of heart failure. Conventional palliative medical management does not correct, or attempt to correct, underlying defects in cardiac muscle cell number. Moreover, while indispensable to current treatment of end-stage heart failure, more aggressive interventions such as cardiac transplantation and the use of mechanical LV assist devices (either as a bridge to transplantation or as "destination" therapy) are thwarted by comorbidity or finite effectiveness, inciting the search for more advanced methods (19, 20). This need for causally directed treatment — to complement mere support of the remaining healthy myocardium — has prompted the translation of a decade's experimental studies into clinical pilot trials studying cell-based myocardial regeneration and repair.

Cell types

Currently, a variety of adult progenitor cells are undergoing clinical evaluation — all autologous, so that tissue rejection is obviated (Figure 1). The first clinically relevant cells to be proposed as a surrogate for cardiomyocytes were skeletal muscle myoblasts — undifferentiated proliferation-competent cells that serve as precursors to skeletal muscle (21–23). For clinical use, autologous human myoblasts are isolated from skeletal muscle biopsies, propagated and expanded ex vivo for a few days or weeks, then injected directly into the ventricular wall (11, 24, 25).

Bone marrow is, at present, the most frequent source of cells used for clinical cardiac repair (12–14, 16). Bone marrow contains a complex assortment of progenitor cells, including HSCs; so-called side population (SP) cells, defined by their ability to expel a Hoechst dye, which account for most if not all long-term self-renewal (26, 27) and reconstitute the full panoply of hematopoietic lineages

Nonstandard abbreviations used: BMP, bone morphogenetic protein; EGFP, enhanced green fluorescent protein; EPC, endothelial progenitor cell; HMGB1, high mobility group box protein 1; MAPC, multipotential adult progenitor cell; MHC, myosin heavy chain; MSC, mesenchymal stem cell; SDF-1, stromal cell–derived factor-1; SP, side population; TERT, telomerase reverse transcriptase.

Conflict of interest: The authors have declared that no conflict of interest exists.

Citation for this article: *J. Clin. Invest.* **115**:572–583 (2005). doi:10.1172/JCI200524283.

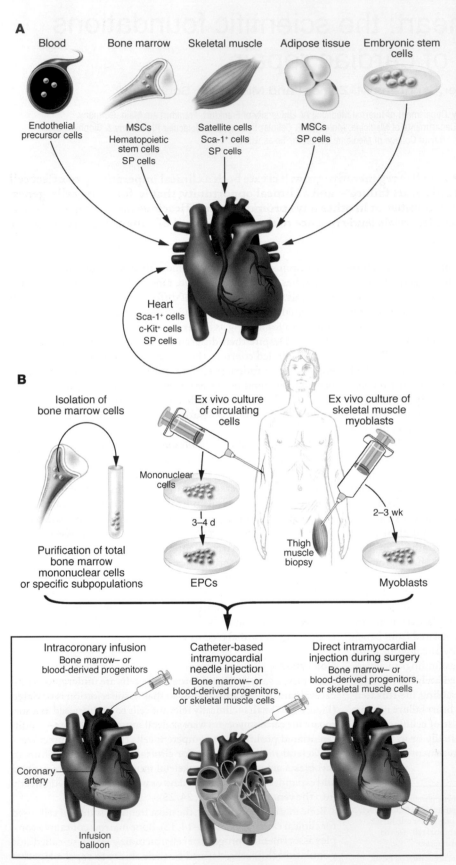

Figure 1

Sources of cells for cardiac repair, and routes of their administration. (**A**) Cells in current human trials include skeletal muscle myoblasts, unfractionated bone marrow, and circulating (endothelial) progenitor cells. Cells in preclinical studies include bone marrow MSCs, multipotent cells from other sources, and novel progenitor or stem cells discovered in the adult myocardium; see text for details. (**B**) Existing trials use intracoronary delivery routes (over-the-wire balloon catheters), intramuscular delivery via catheters (e.g., the NOGA system for electromechanical mapping), or direct injection during cardiac surgery. Not represented here are the theoretical potential for systemic delivery, suggested by the homing of some cell types to infarcted myocardium (39), and strategies to mobilize endogenous cells from other tissue sites to the heart.

after single-cell grafting (28); mesenchymal stem cells or stromal cells (29); and multi-potential adult progenitor cells (MAPCs), a subset of mesenchymal stem cells (MSCs) (30). Bone marrow is aspirated under local or general anesthesia, the entire mononuclear cell fraction is obtained (a heterogeneous mix of the above-mentioned cells), or specific subpopulations are purified, and isolated cells are injected into the heart without need of further ex vivo expansion. Expansion in cell culture could be desirable or essential, though, if defined but minute subpopulations prove to be advantageous (30).

Last, peripheral blood–derived progenitor cells are used both for clinical cardiac repair (13) and for neovascularization in peripheral arterial occlusive disease (7). These circulating cells (endothelial progenitor cells [EPCs]) are bone marrow derived (31), and, historically, therapeutic angiogenesis is the objective of virtually all clinical studies using bone marrow or its circulating derivatives for ischemic myocardium. For clinical use, EPCs are isolated from mononuclear blood cells and selected ex vivo by culturing in "endothelium-specific" medium for 3 days, prior to reinjection into the heart. The added hypothesis that such cells also might trans-differentiate to create new cardiomyocytes (32) is unrelated to the clinical studies' origin and dispensable to their rationale. Critiques of the clinical studies — where based on the absence or paucity of myocyte formation in mice (33–35) — raise useful questions as to the mechanisms for success (as measured to date by improvements in ventricular pump function), but overlook this most salient point, namely, the actual rationale.

Table 1

Clinical trials of intracoronary progenitor cells for acute myocardial infarction

Study	N	Days after MI	Cell type	Cell preparation (volume/purification/culture)	Mean cell no. ($\times 10^6$)	Safety	Myocardial function
Strauer et al. (12)	10	8	BMCs	40 ml/Ficoll/overnight (Teflon)	28	+	Regional contractility ↑ (LVA); endsystolic volume ↓ (LVA); perfusion ↑ (scintigraphy)
TOPCARE-AMI (13, 17, 122)	59	4.9	CPCs	250 ml/blood/3 days	16	+	Global contractility ↑ (LVA/MRI); end systolic volume ↓ (LVA/MRI); viability ↑ (MRI);
			BMCs	50 ml/Ficoll/none	213		flow reserve ↑ (Doppler); similar results for both cell types
BOOST (16)	30 vs. 30 randomized controls	4.8	BMCs	150 ml/gelatin-polysuccinate sedimentation/none	2,460	+	Global contractility ↑ (MRI)
Fernández-Avilés et al. (123)	20	13.5	BMCs	50 ml/Ficoll/overnight (Teflon)	78	+	Global contractility ↑ (MRI); end systolic volume ↓ (MRI)

↑ indicates increase; ↓ indicates decrease; + denotes lack of adverse events. BMC, bone marrow–derived cell; CPC, circulating progenitor cell (EPCs); LVA, LV cineangiography; MI, myocardial infarction.

On the horizon of being tested for potential clinical application are other progenitor/stem cell populations: fat tissue–derived multipotent stem cells (36); multipotential cells from bone marrow or skeletal muscle (minuscule subpopulations, distinct from the unfractionated bone marrow and the myoblasts used in current trials) (30, 37); somatic stem cells from placental cord blood (38); and cardiac-resident progenitor cells that have a heightened predisposition to adopt the cardiac muscle fate (39–43). In each of these newer cases, techniques to isolate and purify the numerically minor population of potent cells will need to be optimized for clinical use, and enabling data from mammals larger than the mouse will surely be warranted.

Routes of application

Thus far, progenitor cells for cardiac repair have been delivered in 3 ways: via an intracoronary arterial route or by injection of the ventricular wall via a percutaneous endocardial or surgical epicardial approach (Figure 1). The advantage of intracoronary infusion — using standard balloon catheters — is that cells can travel directly into myocardial regions in which nutrient blood flow and oxygen supply are preserved, which hence ensures a favorable environment for cells' survival, a prerequisite for stable engraftment. Conversely, homing of intra-arterially applied progenitor cells requires migration out of the vessel into the surrounding tissue, so that unperfused regions of the myocardium are targeted far less efficiently, if at all. Moreover, whereas bone marrow–derived and blood-derived progenitor cells are known to extravasate and migrate to ischemic areas (44), skeletal myoblasts do not and furthermore may even obstruct the microcirculation after intra-arterial administration, leading to embolic myocardial damage.

By contrast, direct delivery of progenitor cells into scar tissue or areas of hibernating myocardium by catheter-based needle injection, direct injection during open-heart surgery, and minimally invasive thoracoscopic procedures are not limited by cell uptake from the circulation or by embolic risk. An offsetting consideration is the risk of ventricular perforation, which may limit the use of direct needle injection into freshly infarcted hearts. In addition, it is hard to envisage that progenitor cells injected into uni-

formly necrotic tissue — lacking the syncytium of live muscle cells that may furnish instructive signals and lacking blood flow for the delivery of oxygen and nutrients — would receive the necessary cues and environment to engraft and differentiate. Most cells, if injected directly, simply die (45). For this reason, electromechanical mapping of viable but "hibernating" myocardium may be useful to pinpoint the preferred regions for injection (14). Finally, in diffuse diseases such as dilated nonischemic cardiomyopathy, focal deposits of directly injected cells might be poorly matched to the underlying anatomy and physiology.

Thus, it already appears likely that patients' individual pathobiology — the specific underpinnings of their heart failure — will ultimately influence, if not dictate, the source and route chosen among potential progenitor cell therapies. Given such variations in the underlying clinical context, it is not yet possible on the basis of existing pilot clinical trials, whose design and findings are detailed below, to assert an "optimal" cell type or "best" mode of delivery.

Initial clinical results, 2000–2004

Tables 1 and 2 summarize the results of all clinical trials studying cell-based myocardial repair published to date. It is vital to distinguish between those investigations performed on patients with acute myocardial infarction (Table 1) and those on patients with chronic heart failure due to prior myocardial infarction (Table 2), not only because of the different cell types and modes of delivery used, but also because fundamentally different pathophysiological processes are targeted. For example, in patients with acute myocardial infarction, progenitor cell transplantation is predicted to significantly modify postinfarction LV remodeling through enhanced neovascularization and reduced cardiomyocyte apoptosis, irrespective of long-term engraftment and transdifferentiation. Conversely, the former 2 mechanisms acting alone may have little or no benefit in patients with long-established scars, apart from the functional rescue of hibernating myocytes.

Given that myocardial ischemia acutely and potently upregulates the chemoattractants for neoangiogenesis (46), it was logical to test clinically an intracoronary infusion of bone marrow- or blood-derived progenitor cells in patients with acute myocardial infarc-

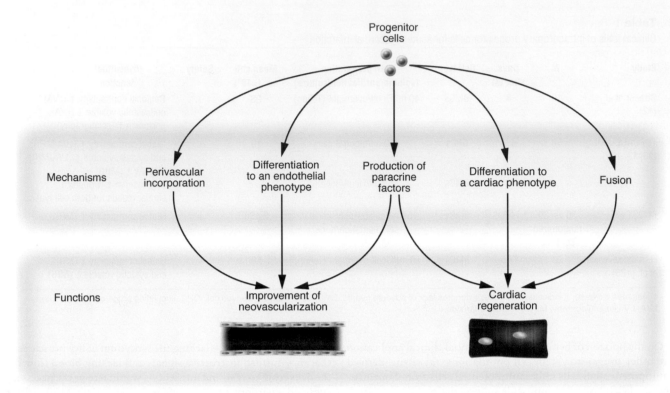

Figure 2
Mechanisms of action. Progenitor cells may improve functional recovery of infarcted or failing myocardium by various potential mechanisms, including direct or indirect improvement of neovascularization. Paracrine factors released by progenitor cells may inhibit cardiac apoptosis, affect remodeling, or enhance endogenous repair (e.g., by tissue-resident progenitor cells). Differentiation into cardiomyocytes may contribute to cardiac regeneration. The extent to which these different mechanisms are active may critically depend on the cell type and setting, such as acute or chronic injury.

tion (Table 1 and Figure 2). All the published trials reported nearly identical results — a 7–9% improvement in global LV ejection fraction, significantly reduced end-systolic LV volumes, and improved perfusion in the infarcted area — 4–6 months after cell transplantation. Notably, in the only prospectively randomized trial so far (BOOST), global LV function was significantly improved in the cell-treated group compared with nontreated control subjects (16). Moreover, recent data from the TOPCARE-AMI trial, in which magnetic resonance was used as the optimal measure of LV function and mass, demonstrate that the improvement of LV function and absence of reactive hypertrophy are preserved even after 1 year, which suggests a sustained benefit on LV remodeling (17).

What technical variables might influence outcome? Surprisingly, none of the studies showed an association between cell numbers infused and extent of functional improvement over the ranges tested. Moreover, although the volumes of bone marrow harvested, the cell isolation procedures, and the number of cells infused varied considerably among trials, the reported improvements were nearly identical. Thus, it not yet possible to extract any critical elements of design that would maximize future studies' chance of success. Importantly, however, in all 4 trials — comprising a total of more than 100 patients who received intracoronary progenitor cell transplants — the observed complications did not exceed those expected in patients with acute myocardial infarction. Specifically, no arrhythmic complications resulted from delivery of bone marrow–derived progenitor cells, whether at surgery or by percutaneous approaches.

By comparison, in patients with chronic ischemic heart disease and old myocardial infarction, the initial attempts at cell-based myocardial repair were more heterogeneous in outcome, likely owing in part to the more heterogeneous populations treated (Table 2). The first such trial used skeletal muscle–derived progenitor cells, directly injected into the scarred region of the LV during open heart surgery for coronary artery bypass grafting (11). Global and regional LV function were significantly and persistently improved, although concomitant revascularization complicates the assessment of benefit. Indeed, in patients not undergoing simultaneous revascularization, transcatheter injection of myoblasts into the scar that had resulted from myocardial infarction 5–6 years earlier reduced the symptoms of heart failure, but without objective evidence of improved global LV function (15). Unfortunately, even though previous extensive animal experiments provided no hint of an arrhythmogenic risk (21, 23), the enthusiasm for injecting myoblasts into scar tissue for cardiac repair has been dampened by the fact that patients receiving this treatment have experienced life-threatening arrhythmias (47). Mechanistically, this phenomenon may relate to the lack of electrical coupling of skeletal muscle to neighboring cardiomyocytes (48) or, alternatively, be contingent on coupling by the few hybrid cells formed by fusion with adjacent cardiomyocytes, which generate spatially heterogeneous calcium transients (49). Therefore, currently, implantation of skeletal myoblasts requires the placement as well of an implantable cardioverter/defibrillator, as a mandatory adjunct to therapy (47). Importantly, in the 1 small, nonrandomized trial using bone marrow–derived progenitor cells for chronic ischemic heart failure, injection sites were chosen by electromechanical mapping of the LV endocardial surface to find areas of myocardial hibernation: significant increases in global LV ejection fraction resulted, with decreased end-systolic volumes and improved exercise capacity (14). While this functional improvement might be secondary to an improved blood supply to hibernating cardiomyocytes, it is also conceivable that an area of hibernating myocardium may provide a more favorable microenvironment for injected cells' survival and engraftment than a cell-depleted scar.

Lessons from clinical pilot trials: more questions than answers

Thus, patients in 3 disparate clinical scenarios — days, months, or years following infarction — have been subjected to cell transplantation for cardiac repair. In the case of acute myocardial infarction, the established safety and suggestive efficacy of intracoronary progenitor cell transplantation provide a cogent rationale for larger, randomized, double-blind trials and for expanding such studies from Europe to the United States. In the case of chronic ischemic heart failure, an additional question is whether identifying hibernating myocardium to direct cell therapy is essential to an effective outcome. It must also be shown whether delivery of skeletal myoblasts to established scar tissue late in the disease improves clinical outcome, once patients are protected against potential arrhythmias by an implantable defibrillator. Trials of each kind are currently ongoing and their results anticipated eagerly. Ultimately, it must be proven that cellular therapy aimed at cardiac repair not only improves pump function but also reduces mortality, morbidity, or both.

Beyond safety, beyond efficacy, what else do we need to know clinically? For ischemic disease, the technical armament is in hand for treating patients' hearts with progenitor cells but still at a very early stage — rudimentary experimental knowledge is being applied in the clinical arena, yet a variety of pivotal but straightforward utilitarian questions still remain unanswered (optimal patient selection, usefulness of repeated treatments). Nonischemic heart disease has yet to be addressed at all.

More complex and challenging is a series of pathobiological concerns, which have sent the scientific community from bedside to bench and back again. Certain patients' cells may be unsatisfactory,

in their naive and unmanipulated state, which is now prompting systematic dissection of each step in progenitor cell function, from recruitment to plasticity. This task, in turn, is complicated by the fact that we do not yet understand the mechanisms underlying cell-based cardiac repair. For instance, the efficacy of skeletal muscle myoblasts (23) provided the impetus for human trials of skeletal muscle cells, but in the rabbit, diastolic functions are improved even by injected fibroblasts (50) and systolic performance improved to the same degree with bone marrow–derived cells as with skeletal muscle ones (51). Along with the issue of skeletal muscle cells' electrical isolation from host myocardium, this prompts the question of how mechanical improvements arise even in this ostensibly straightforward instance. Another reason to consider potential indirect mechanisms is that studies have called into question the extent to which bone marrow–derived cells implanted in the heart form cardiomyocytes (33–35). The majority of this review is, therefore, devoted to the biological horizons — namely mobilization, homing, neoangiogenesis, and cardiac differentiation (Figures 3 and 4) — and to evolving new insights that may enable cell therapy for cardiac repair to surpass the present state of the art.

Cell mobilization

The first hints that cytokine-induced mobilization may be a way to enhance cardiac repair came as an extrapolation of findings of results from efforts to increase EPC levels for neovascularization in another context — hind limb ischemia. Indeed, VEGF (52) and GM-CSF (53) were found to augment EPC levels and improve neovascularization, and subsequent studies documented EPC mobilization by numerous other proangiogenic growth factors — stromal cell–derived factor-1 (SDF-1), angiopoietin-1, placental growth factor, and erythropoietin (54–56). A wide array of interventions even more accessible clinically than growth factor administration enhance the number of circulating EPCs in adults, including treatment with HMG CoA reductase inhibitors (statins) and estrogens as well as exercise (57–59). Most studies confirmed an improvement in endothelial regeneration or neovascularization by mobilizing agents. However, such functional improvements may not rely entirely on EPC mobilization but may also — at least in part — be explained by direct proangiogenic or antiapoptotic effects. Hence, as discussed as a recurring theme in this review, the existence of known (and potential unknown) pleiotropic modes of action complicates the interpretation of regenerative therapies, even in cases where the beneficial effect is clear-cut and assured (Figure 3).

Table 2

Clinical trials of catheter-based progenitor cell delivery for chronic coronary heart disease

Study	N	Delivery technique	Cell type	Cell preparation (volume/purification/ culture)	Mean cell no. (×10⁶)	Safety	Myocardial function
Tse et al. (25)	8	Endocardial injection, guided by electromechanical mapping	BMCs	40 ml/Ficoll/none	Not reported	+	Wall motion and thickening ↑; hypoperfusion ↓
Fuchs et al. (124)	10	Endocardial injection, guided by electromechanical mapping	BMCs	Filtered/none	78.3	+	Angina score ↓; stress-induced ischemia ↓
Perin et al. (14, 18)	14	Endocardial injection, guided by electromechanical mapping	BMCs	50 ml/Ficoll/none	30	+	Global contractility ↓; endsystolic volumes ↓; reversible perfusion defects ↓; exercise capacity ↑

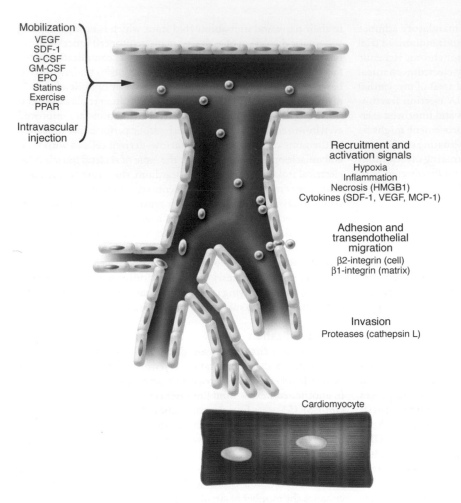

Mobilization
VEGF
SDF-1
G-CSF
GM-CSF
EPO
Statins
Exercise
PPAR

Intravascular injection

Recruitment and activation signals
Hypoxia
Inflammation
Necrosis (HMGB1)
Cytokines (SDF-1, VEGF, MCP-1)

Adhesion and transendothelial migration
β2-integrin (cell)
β1-integrin (matrix)

Invasion
Proteases (cathepsin L)

Cardiomyocyte

Figure 3
Mobilization and homing. After intravascular delivery or mobilization from bone marrow, progenitor cells are targeted to the sites of injury by multiple signals. Homing is mediated by a multistep process including the initial adhesion, transmigration, and invasion. Molecular mechanisms contributing to these individual steps are indicated but likely vary depending on the cell type and model. EPO, erythropoietin; MCP-1, monocyte chemoattractant protein–1.

tor cells were predominantly trapped by the spleen when given to athymic nude rats (44), and cardiac regeneration elicited by treatment with G-CSF plus SCF was documented only for animals lacking a spleen (60). The use of leukocyte-mobilizing cytokines might be most worthwhile combined with selective enhancements of progenitor cell homing or as a prelude to isolating cells for local delivery (63).

Cell homing
Defining the events in progenitor cell homing may enable better targeting of cells, most obviously when cells are mobilized from the bone marrow into the bloodstream. Later steps in homing, though, are instrumental to the impact even of progenitor cells infused locally into coronary arteries. Homing is a multistep cascade including the initial adhesion to activated endothelium or exposed matrix, transmigration through the endothelium, and, finally, migration and invasion of the target

A shift in emphasis from the heart's vessels to the heart itself was prompted by the report that bone marrow–derived cells can differentiate into cardiomyocytes when injected into injured myocardium and regenerate the heart effectively (32). Based on this discovery, hematopoietic stem cell–mobilizing factors — G-CSF and SCF (Kit ligand) — were used to improve cardiac regeneration experimentally (60), which quickly led to the initiation of clinical trials studying the ability of G-CSF to mobilize stem/progenitor cells in patients with coronary artery disease. This cytokine is used routinely in the treatment of humans, e.g., to help in harvesting cells for bone marrow transplantation. Although results from these first small trials do not permit any conclusion of efficacy, the safety of G-CSF in acute myocardial infarction has already come into question (61). The observed increase in restenosis may be partially explained by the study design (which precluded the standard clinical practice of promptly stenting the obstructed vessel), but the rise in leukocyte number to leukemic levels may be directly responsible, via plaque growth or destabilization. Adverse vascular events have also been attributed to G-CSF in patients with intractable angina who were not candidates for revascularization and even in patients without cardiac disease (62). In the future, it may be preferable to use strategies that augment circulating progenitor cells without causing massive inflammation.

A second open question regarding systemic mobilization is whether enough progenitor cells will home where needed, to the sites of cardiac injury (63). Systemically administered human progeni-

tissue (Figure 3). The capacity to migrate and invade may be pivotal to functional integration even when cells are injected intramuscularly. Particularly in patients who lack the endogenous stimuli incited by acute ischemic injury, the enhancement of local homing signals or cells' ability to respond may be of critical importance.

While homing of hematopoietic progenitor cells to bone marrow has been studied extensively (64), the mechanisms for progenitor cells' homing to sites of tissue injury are only understood rudimentarily. SDF-1 appears to be one key factor that regulates trafficking of stem and progenitor cells to ischemic tissue (65), and local delivery of SDF-1 can enhance EPC recruitment and neovascularization (63, 66). Cell necrosis causes the release of a chromatin-binding protein, high mobility group box protein 1 (HMGB1), whose release acts as an extracellular "danger signal" and may stimulate progenitor cells' homing (67). Extracellular HMGB1 attracts mesoangioblasts in vitro and in vivo and likely plays a role in muscle regeneration (68). HMGB1 interacts with the receptor for advanced glycation end products, as well as Toll-like receptors 2 and 4 (69). The exact mechanisms that mediate cell attraction by HMGB1 are not yet clear and may involve additional receptors that are as yet unidentified.

Adhesion and transmigration of stem and progenitor cells are mediated by integrins. Indeed, integrin-dependent adhesion of EPCs is one effect of SDF-1 (70). Particularly, β2 integrins were found to be essential for homing and improvement of neovascularization mediated by EPCs after hind limb ischemia (71). In a study of tumor angiogenesis focused on in vivo homing by embryonic EPCs from cord blood, the circulating cells arrested within tumor microvessels, extravasated into the interstitium, and incorporated into neovessels (72). The cells' initial physical arrest was suggested to be mediated by E- and P-selectin and P-selectin glycoprotein ligand-1 (72). However, this study was performed with embryonic endothelial progenitor cells in a tumor model of neovascularization — distinct from therapeutic angiogenesis in both respects. Many of the chemokines and adhesion molecules induced by cardiac ischemic injury are familiar players in other disorders (73), but different cell types may use different mechanisms for homing, and ischemia may differ from other attractants. Thus, a molecular dissection is essential to define the multiple steps of progenitor cell homing to and invasion of the myocardium, especially for those cells in current use for clinical cardiac repair and for other, novel, auspicious cells now in preclinical studies (39).

Neoangiogenesis

To date, there is no direct clinical evidence that cellular cardiomyogenesis in fact occurs in the human heart after transplantation of progenitor cells. Angiogenesis, improvements in scar tissue, and cytoprotection must be considered, along with transdifferentiation, as among the most important possible consequences of cell-based therapies for cardiac repair (Figure 3). Of these, most obviously, progenitor cells may improve neovascularization, which in turn would augment oxygen supply. Progenitor cells are expected to be of most benefit to cardiac regeneration or performance when used to treat jeopardized or hibernating cardiomyocytes. Neovascularization, in turn, can be mediated by the physical incorporation of progenitor cells into new capillaries (74, 75) or, in some settings, perivascular cells (76). Incorporated progenitor cells of most if not all types may release growth factors that promote angiogenesis by acting on mature endothelial cells (77). The extent to which progenitor cells contribute to vasculogenesis by becoming physical elements of newly formed vessels versus acting through secreted factors may plausibly depend on the circumstances of cell type and cardiac injury. However, human bone marrow–derived angioblasts exert both types of effect (78).

Cells engineered to overexpress angiogenic factors might enhance both their own survival and that of the recipient myocardium. As exemplified by the finding that myoblast-based delivery of VEGF led to an improved treatment outcome compared with direct viral gene transfer (79), cell therapy can be envisioned as a platform for secreted proteins' local production in the injured myocardium. More simply, hypoxic "preconditioning" enhances the ability of EPCs to rescue hind limb ischemia (80), conceivably through activation of the angiogenic gene program: in differentiating embryonic stem cells, formation of hemangioblasts (the common precursor of endothelial and hematopoietic progenitor cells) is activated by oxygen deprivation, requires hypoxia-inducible factor, and is regulated by VEGF plus other local signals (81).

Cardiac myogenesis by noncardiac cells

The possibility of cardiomyocyte formation by multipotent progenitor/stem cells first was raised by pioneering studies in which embryonic stem cells were grafted into mouse myocardium (82), the one cell type, along with germ cells, for which totipotency is assured. The formation of functional cardiomyocytes by mouse and human embryonic stem cells is proven according to many criteria (83, 84). Although religious, ethical, and political objections

to using human embryonic stem cells have received justifiable attention (85), human embryonic stem cells, being allogeneic, also pose the clinical challenge of immunological barriers (86), which are obviated in all forms of autologous cell therapy.

At times equalling the intensity, if not the religious character, of the dispute over human embryonic stem cells is the debate regarding the extent — and even the existence — of the plasticity of progenitor cells (87–90). Challenging but appropriate concerns have been raised regarding the nature of proof. Is the appearance of multiple lineages due merely to a mixed assortment of starting cells? For which cell types, settings, and means of administration does fusion of donor and host cells create the appearance of transdifferentiation or multiple potentials? Is plasticity in propagated cells merely acquired in culture and not a reflection of native biology? The latter question is especially thorny, since cloning cells to homogeneity is a useful response to the concern that the starting population is mixed. Finally, developmental potential during normal maturation and in healthy adults may differ from plasticity after injury, and it is necessary therefore to test the latter explicitly. It simply makes the most sense to study cells' potential contributions to cardiac regeneration in settings, like infarction, where regeneration is most needed, rather than study plasticity just in normal hearts. The raging of this generic controversy about the plasticity of adult stem cells provides a partial explanation for the hostile climate in which claims of cardiac myogenesis by adult progenitor and stem cells have been received.

As proof of the concept that components of the heart can arise from a bone marrow source, myocardial ischemia/reperfusion injury was induced in mice following bone marrow reconstitution by treatment with ROSA26 SP cells (CD34$^{-/low}$Kit$^+$Sca-1$^+$) that were genetically labeled to express β-gal ubiquitously (75). Donor-derived myocytes were detected in the peri-infarct region, albeit at a prevalence of only 0.02%; endothelial cells derived from bone marrow were seen far more frequently (3.3%), as expected from the known origin of circulating EPCs. Recruitment of circulating bone marrow–derived cells to form cardiomyocytes also was shown in dystrophic mdx mice (91). In a particularly interesting study (32), female mice were subjected to permanent coronary artery occlusion, and viable myocardium bordering the infarction was injected with Lin$^-$Kit$^+$ cells from the bone marrow of male transgenic mice widely expressing enhanced GFP (EGFP). These donor cells lack the surface markers of mature hematopoietic lineages but express the SCF receptor indicative of HSCs and some other primitive cell types. Extensive restorative growth was reported and donor origin shown by costaining for cardiac myosin plus the Y chromosome or green fluorescent protein. In addition, some EPFP$^+$ cells stained for an endothelial or smooth muscle marker and were, morphologically, organized into donor-derived vessels. Encouragingly, ventricular pump function was improved.

Subsequently, these results were challenged by researchers using seemingly identical conditions as well as complementary ones (33–35). In one example (34), the peri-infarct zone was injected with Lin$^-$Kit$^+$ cells from bone marrow of transgenic mice bearing an α-myosin heavy chain driven (αMHC-driven; cardiomyocyte-specific) transgene encoding nuclear-localized β-gal. No nuclei stained blue, nor was regeneration seen by alternative criteria, e.g., ectopic staining for sarcomeric MHC in the area of necrosis. Other bone marrow–derived populations (Lin$^-$Kit$^-$Sca$^+$ cells from the αMHC-nLacZ mice; Lin$^-$Kit$^+$ cells from αMHC- or βAct-EGFP mice) likewise failed to differentiate as donor cells (33–35). βAct-EGFP is expressed in the donor cells even when undifferentiated, and its use as a constitutive marker confirms that donor cells were successfully injected and retained in those studies where little or no cardiogenic differentiation was seen.

Supporting the hypothesis of progenitor cell plasticity, though, wild-type bone marrow cells from β*Act-EGFP* donors give rise, rarely, to cardiac muscle fibers and, more often, to small, presumptively developing myocytes, when transplanted into the bone marrow of *Sod1* mutant mice (92). Independent studies have shown that bone marrow–derived progenitor cells can be reprogrammed to express cardiac marker genes in vitro, including bone marrow stromal cells (93), EPCs grown together with cardiomyocytes (94), and bone marrow cells grown in the presence of ostensibly angiogenic factors (95). Hence, there is little room to doubt that bone marrow cells and their derivatives can acquire a cardiomyogenic phenotype, but the extent to which such differentiation occurs in vivo for the cell types in clinical use still needs to be better defined.

Apart from technical disparities in the end points, differences in the starting cells must be considered, too, as numerically minor components could be the responsible ones — e.g., MAPCs and SP cells. Determining the number of donor MSCs might be of particular value: these are relatively abundant in bone marrow, can express c-Kit, and have clearer potential than hematopoietic cells do to form cardiomyocytes upon injection into the heart (96–98). In a side-by-side comparison to define the subpopulations from which cardiomyocytes were formed (99), mice received bone marrow transplantation through the use of whole bone marrow, single long-term repopulating HSCs ("Tip"-SP Lin⁻CD34⁻c-Kit⁺Sca-1⁺ cells; ref. 28), or a clonal MSC line, all expressing EGFP. Recipient mice were subjected to coronary artery ligation, with or without G-CSF to help mobilize engrafted cells. Whole bone marrow generated dozens of EGFP⁺ α-actinin⁺ cells per heart, or thousands if G-CSF had been used. Although single HSCs effectively reconstituted the bone marrow and the peripheral blood nucleated cells, purified HSCs provided little or no contribution to bone marrow MSCs, mesenchymal lineages in culture, or cardiomyocytes after infarction. Conversely, bone marrow transplantation with MSCs yielded hundreds of EGFP⁺ α-actinin⁺ cells. A plausible interpretation is that the MSCs in bone marrow — not HSCs — chiefly contribute to the creation of new cardiomyocytes following infarction. To date, human trials studying cardiac repair all have employed unfractionated bone marrow cells (presumably including bone marrow MSCs).

Cardiac myogenesis by adult cardiac progenitor cells

The quest for novel heart-forming cells in adult myocardium can be traced to several instigating rationales: the inability of skeletal myocytes to transdifferentiate; challenges to claims of bone marrow–derived cells' far-ranging plasticity; and an emerging countermodel of tissue-resident progenitor cells, sharing some signatures of "stemness" (100) yet predisposed to differentiate into lineages of the organ in which they reside.

In adults, the unique self-renewal potential of progenitor/stem cells — along with tumor cells and germ cells — is associated with telomerase reverse transcriptase (TERT), an RNA-dependent DNA polymerase that maintains the lariat-like loop (telomere) that caps chromosome ends (101). Telomerase expression and activity are markedly downregulated in the mouse heart soon after birth. A small number of cardiac cells express the surface marker stem cell antigen–1 (Sca-1), and this population, unlike Sca-1⁻ cells, contains telomerase at levels akin to those in newborn hearts (39). Cardiac-resident Sca-1⁺ cells lack the HSC markers CD45 and CD34 (also a marker of EPCs), lack hematopoietic transcription factors (Lmo2, Gata2, Tal), and thus are readily distinguished from bone marrow HSCs. Cardiac Sca-1⁺ cells also lack transcripts for cardiac structural genes but, intriguingly, express the majority of known cardiogenic transcription factors (*Mef2c*, *Gata4*, *Srf*, and *Tead1/TEF-1*, excepting a handful of others), congruent with the postulated properties of cardiac progenitor cells as undifferentiated yet predisposed to become cardiomyocytes.

In culture, cardiac Sca-1⁺ cells express *Nkx2.5*, after which cardiac structural genes are activated, if treated with 5′-azacytidine to relax condensed chromatin (ref. 39; and see ref. 93). As evidence that differentiation was specific, and not a nondescript response to the DNA methylation inhibitor, inducing αMHC required the signal transduction pathway for bone morphogenetic proteins (BMPs), as differentiation was blocked by ex vivo deletion of the type IA BMP receptor (39). Adult cardiac Sca-1⁺ cells form beating cardiomyocytes with spontaneous calcium transients and may possess multipotency (41). However, multipotency needs to be proven in the progeny of single cells.

When given intravenously to mice just after ischemia-reperfusion injury, cardiac Sca-1⁺ cells home selectively to injured myocardium, interdigitate with surrounding host myocytes, and demonstrate robust differentiation in situ, constituting approximately 15% of the myocytes in the infarct "border zone" 2 weeks after grafting (39). Roughly 5% of donor-derived myocytes were still proliferating, as indicated by serine 10 phosphorylation of histone H3, a marker of Cdc2 activity. Given the possible existence of cell fusion as a confounding factor (102), a pair of genetic tags was utilized: αMHC-Cre, the cardiomyocyte–specific expression of the DNA recombinase used widely for tissue-restricted knockout mutations, plus *R26R*, a ubiquitously transcribed but latent gene for β-gal (behind a Cre-deletable "stop" signal). αMHC-Cre is inactive in cardiac Sca-1⁺ cells in their initial undifferentiated state. Hence, the induction of Cre protein fulfills 2 roles — as a marker of donor cell identity and of differentiation in situ. Cre⁺LacZ⁻ cells are de novo differentiation products, whereas Cre⁺LacZ⁺ ones denote differentiation associated with fusion (though not indicating which came first). The two modes of differentiation were equally prevalent. The extent of cell fusion in other models of cardiac repair must be determined empirically.

An intriguing subgroup within the Sca-1⁺ fraction consists of SP cells, discussed earlier in connection with bone marrow. SP cells have been found in other tissues, through use of the same flow cytometry test for dye efflux as is used for the SP cells in bone marrow itself. While many related transport proteins may provide a molecular marker for SP cells of the heart, Abcg2 seems to best serve this function in development and disease (42). Genes whose expression is enriched in cardiac SP cells include *Ly6a* (encoding Sca-1), genes for the cardiogenic transcription factors MEF2A and MEF2C, and genes suggesting that SP cells share an origin or developmental pathway with endothelial and hematopoietic cells.

Although cardiac Sca-1⁺ cells were typically negative for c-Kit (39, 41), this second marker also holds importance in heart. Prompted by the successful use of Kit⁺ bone marrow cells in cardiac repair (32), cardiac-resident Kit⁺ cells were sought (40): Lin⁻ c-Kit⁺ cells were found in adult rat myocardium with a prevalence of approximately 1 per 10⁴ mature myocytes, frequently in clusters, with varying expression of cardiogenic transcription factors and rarer expression of cardiac-restricted sarcomeric proteins. As with cardiac Sca-1+ cells (39), cardiac c-Kit+ cells lack CD45 and CD34 (40). Cardiac c-Kit+ cells were self-renewing (propagated for months), clonogenic (expanded in culture after plating 1 cell per well), and multipotential (generating cardiomyocytes, smooth muscle cells, and endothelial cells) (40). When injected into the border of new infarcts, cardiac c-Kit+ cells led to bands of regenerating myocardium in the region of necrosis, contributed to endothelium and vascular smooth muscle, and improved pump function and chamber geometry (40), much as the authors had reported using bone marrow c-Kit+ cells (32). Results with clonally derived cardiac c-Kit+ cells were equivalent to those with the initial cardiac c-Kit+ population, which suggests that an expandable source from within the heart might be applied to cardiac repair.

Where and how do cardiac progenitor or stem cells arise in the heart? Various models can be considered, ranging from persistence as undifferentiated remnants of heart-forming tissue in the early embryo, to a hematogenous origin (from bone marrow or even sites of earlier hematopoiesis, losing HSC markers in the process), to mechanisms involving ingrowth of the developing coronary vasculature. As an example of the first mechanism, postnatal cardioblasts numbering just a few hundred per heart have been identified on the basis of persistent expression of a LIM-homeodomain transcription factor, Isl1, especially in the atria, right ventricle, and outflow tract — regions where Isl1 is most prevalently expressed during cardiac organogenesis (43). By contrast, for cardiac Sca-1+ cells, the third potential mechanism may be favored, given the cells' striking similarities to the mesoangioblast, which include surface labeling, microarray findings, and earliest sites of marker expression (39, 103). How many cardiomyocytes, if any, are generated in the normal heart after birth by these new routes to heart muscle cell formation? What is the contribution of these pathways, unassisted, to cardiomyocyte formation in disease, as a reserve for the replacement of dead and dying cells? To answer these questions, the genetic strategy of fate mapping is likely relevant — indelibly tagging cells with an irreversible marker of their status — using, for example, a Sca-1–driven gene for Cre recombinase plus a Cre-dependent reporter to permanently label the progeny of Sca-1+ cells, even once Sca-1 is no longer expressed.

Cell augmentation for cardiac repair

The findings of unexpected persistence of cardiopoietic cells in adult hearts and effective cardiac repair by the noncardiac cells in current trials raise a number of fascinating questions. Although cell-based therapy has only taken its first steps into the clinic, limiting factors have already arisen as targets for future improvement. Ultimately, engineered cells may supercede naive cells.

First, how can one make delivered cells more durable, considering the adverse environment? Most stem cells share the property of stress resistance (100), but even stem cells die in the absence of blood flow. This concern is especially apt when progenitor cells are directly delivered into unperfused necrotic myocardium. Targeting injections to the margin of injury where oxygen supply persists is immediately workable (14), and biological means to augment the cells' survival are on the horizon, including the use of angiogenic factors, as discussed (79), or modification of MSCs with the anti-apoptotic gene Akt (96).

Second, how can one restore progenitor cells to normal, where they are deficient in number or function? The functional capacity of bone marrow–derived cells is defective in patients with heart failure (104), and risk factors for atherosclerosis correlate inversely with the numbers and function of circulating EPCs (105, 106). Stem cell defects also occur as a consequence of aging (107, 108). Potential remedies include statins, which not only reduce risk factor load by lipid-lowering but also delay the onset of cellular senescence, telomere "uncapping," and DNA damage signals in human EPCs (109, 110). A Myb-like telomere-capping protein, telomere repeat-binding factor–2, is downregulated by cell stress cascades (4) and mediates the beneficial effect of statins on EPC function (110). Direct interference with telomere-based aging and death signals might also be achieved through the forcible expression of TERT (4, 111, 112). Antioxidants or growth factors also might be used to promote telomerase activity (108, 113). TERT and other cell cycle activators may be useful in prolonging donor cell cycling either ex vivo or, especially if tightly controlled, following engraftment. Interestingly, the nuclear export of cyclin D1 and cyclin D3 limits these proteins' impact on cardiac regeneration when overexpressed in their wild-type form, whereas cyclin D2 is not excluded from the nucleus and led to marked infarct

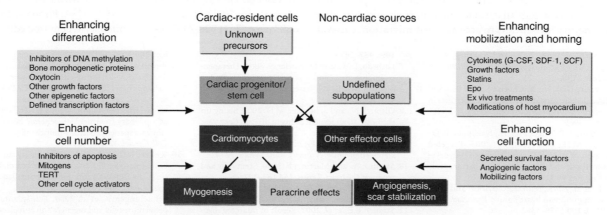

Figure 4
Current challenges for cell-based therapy in cardiac repair include identifying the origins of the novel cardiac progenitor and stem cells found within the heart, pinpointing the biologically active cells from bone marrow and other mixed populations, optimizing cell mobilization and homing, augmenting grafted cells' survival, defining the cues for cardiac differentiation, promoting donor cell proliferation ex vivo (or, if safe, in vivo), and exploiting cell therapy as a platform for secretory signals.

regression (3). In this regard, it should prove instructive to understand better the molecular mechanisms that lock the cell cycle in terminally differentiated mammalian cardiomyocytes, including tumor suppressor proteins of the retinoblastoma family (114), and those that maintain a competence for cardiac proliferation in model organisms with enhanced capacity for cardiac repair (1, 2). The utility of zebrafish arises not only from the accessibility of its development outside the mother, but also (unlike chicks and frogs) from the fact that genome-wide mutagenesis is workable, while daunting, and several genes needed to drive heart repair have already been identified. These include *Mps1*, encoding a mitotic checkpoint kinase that is induced in the highly proliferative cells of the regeneration blastema (1). Blastemal regeneration, including that of the zebrafish heart, is also thought to entail the transient induction of Notch pathway components and Msx transcription factors (115, 116).

Third, what drives the cardiac fate in adult heart-forming cells? While some mechanisms for cardiac specification in injured adult hearts might differ from those in early embryos, better knowledge of the "wiring diagram" for commitment to the cardiac fate is expected to yield useful clues to enhance the differentiation process in susceptible cell types and even extend the range of donor cells that form heart muscle well (8, 117). Evidence for the importance of oxytocin in both early heart and adult cardiac Sca-1+ cells is a recent example (41, 118). Multiple autocrine or paracrine factors, from host myocardium or donor cells, provide essential instructions to enter the cardiac muscle lineage. Blocking TGF-β or its relative, BMPs, prevents embryonic stem cells from becoming cardiomyocytes during coculture with ventricular muscle cells and following injection of the heart (119). Conversely, PDGF-AB enhances the induction of cardiac genes and beating cell aggregates in cultured bone marrow cells and increases the number of intra-scar cardiomyocyte islands when coinjected with bone marrow cells (95). As-yet-undefined local factors mediate the inductive effect of cardiomyocytes on cocultured EPCs (94) and isl1+ cardiac cells (43). Notably, azacytidine was key to activating the cardiomyocyte program in immortalized bone marrow stromal cells (93) and in cardiac Sca-1+ cells (39), and other epigenetic modifiers can be foreseen. High-throughput chemical and genetic screens will likely identify novel activators of cardiac myogenesis.

In summary, the existence of cardiac and noncardiac progenitor cells that are efficacious in cardiac repair next should stimulate vigorous inquiry into means to activate their migration, survival, growth, and differentiation (Figure 4). These are questions of great importance, whether one is envisioning the manipulation of cells ex vivo for subsequent administration or contemplating, instead, the activation of latent cells within the injured heart.

Finally, another approach to cardiac repair might be deemed "cell-free therapy. It remains formally unproven whether cardiac myogenesis is instrumental to improved cardiac function in the cell-based therapies discussed here, and other modes of action can be considered, beyond just angiogenesis, rescue of hibernating myocytes, and prevention of ventricular thinning. Potential paracrine actions of cell therapies might include immunomodulatory effects, as has been shown for mesenchymal stem cells (120), and other signals to modulate scar healing and remodeling. Additionally, paracrine mechanisms might suppress host cell death directly, both in acute infarction and in chronic heart failure, regardless of cause. A remarkable illustration of this principle is the discovery that thymosin β4 — a secreted protein associated with the early heart on the one hand and with wound healing on the other — promotes cardiac cell migration, survival, and repair, working through the Akt survival pathway (121). Perhaps a primary role of cell therapy — at least for some modalities — is paracrine or chemoattractive. To what extent defined signals can bypass the need for cells in cardiac repair is unknown. At least for the moment, our cells likely know more than we do.

Acknowledgments

We thank K. Chien, J. Edelberg, L. Field, K. Fukuda, D. Losordo, B. Martin, and S. Rafii for suggestions. The authors are members of the Fondation Leducq Transatlantic Network of Excellence for Cardiac Regeneration. S. Dimmeler and A. Zeiher are supported by the Deutsche Forschungsemeinschaft and the EU European Vascular Genomics Network. M. Schneider is supported by the NIH, the Donald W. Reynolds Foundation, and the M.D. Anderson Foundation Professorship.

Address correspondence to: Stefanie Dimmeler, Molecular Cardiology, Department of Internal Medicine IV, University of Frankfurt, Theodor Stern-Kai 7, 60590 Frankfurt, Germany. Phone: 49-69-6301-6667; Fax: 49-69-6301-7113; E-mail: Dimmeler@em.uni-frankfurt.de. Or to: Michael D. Schneider, Center for Cardiovascular Development, Baylor College of Medicine, One Baylor Plaza, Room 506D, Houston, Texas 77030, USA. Phone: (713) 798-6683; Fax: (713) 798-7437; E-mail: michaels@bcm.tmc.edu.

1. Poss, K.D., Wilson, L.G., and Keating, M.T. 2002. Heart regeneration in zebrafish. *Science.* **298**:2188–2190.
2. Bettencourt-Dias, M., Mittnacht, S., and Brockes, J.P. 2003. Heterogeneous proliferative potential in regenerative adult newt cardiomyocytes. *J. Cell Sci.* **116**:4001–4009.
3. Pasumarthi, K.B., Nakajima, H., Nakajima, H.O., Soonpaa, M.H., and Field, L.J. 2005. Targeted expression of cyclin D2 results in cardiomyocyte DNA synthesis and infarct regression in transgenic mice. *Circ. Res.* **96**:110–118.
4. Oh, H., et al. 2003. Telomere attrition and chk2 activation in human heart failure. *Proc. Natl. Acad. Sci. U. S. A.* **100**:5378–5383.
5. Foo, R.S.-Y., Mani, K., and Kitsis, R.N. 2005. Death begets failure in the heart. *J. Clin. Invest.* **115**:565–571. doi:10.1172/JCI200524569.
6. Losordo, D.W., and Dimmeler, S. 2004. Therapeutic angiogenesis and vasculogenesis for ischemic disease: part I: angiogenic cytokines. *Circulation.*

109:2487–2491.
7. Losordo, D.W., and Dimmeler, S. 2004. Therapeutic angiogenesis and vasculogenesis for ischemic disease: part II: cell-based therapies. *Circulation.* **109**:2692–2697.
8. Olson, E.N., and Schneider, M.D. 2003. Sizing up the heart: development redux in disease. *Genes Dev.* **17**:1937–1956.
9. Rosenthal, N. 2003. Prometheus's vulture and the stem-cell promise. *N. Engl. J. Med.* **349**:267–274.
10. Mathur, A., and Martin, J.F. 2004. Stem cells and repair of the heart. *Lancet.* **364**:183–192.
11. Menasché, P., et al. 2001. Myoblast transplantation for heart failure. *Lancet.* **357**:279–280.
12. Strauer, B.E., et al. 2002. Repair of infarcted myocardium by autologous intracoronary mononuclear bone marrow cell transplantation in humans. *Circulation.* **106**:1913–1918.
13. Assmus, B., et al. 2002. Transplantation of progenitor cells and regeneration enhancement in acute myocardial infarction (TOPCARE-AMI). *Circula-*

tion. **106**:3009–3017.
14. Perin, E.C., et al. 2003. Transendocardial, autologous bone marrow cell transplantation for severe, chronic ischemic heart failure. *Circulation.* **107**:2294–2302.
15. Smits, P.C., et al. 2003. Catheter-based intramyocardial injection of autologous skeletal myoblasts as a primary treatment of ischemic heart failure: clinical experience with six-month follow-up. *J. Am. Coll. Cardiol.* **42**:2063–2069.
16. Wollert, K.C., et al. 2004. Intracoronary autologous bone-marrow cell transfer after myocardial infarction: the BOOST randomised controlled clinical trial. *Lancet.* **364**:141–148.
17. Schachinger, V., et al. 2004. Transplantation of progenitor cells and regeneration enhancement in acute myocardial infarction final one-year results of the TOPCARE-AMI trial. *J. Am. Coll. Cardiol.* **44**:1690–1699.
18. Perin, E.C., et al. 2004. Improved exercise capacity and ischemia 6 and 12 months after transendocardial injection of autologous bone marrow

mononuclear cells for ischemic cardiomyopathy. *Circulation.* **110**:II213–II218.

19. Reinlib, L., and Field, L. 2000. Cell transplantation as future therapy for cardiovascular disease? A workshop of the national heart, lung, and blood institute. *Circulation.* **101**:182–187.

20. Reinlib, L., and Abraham, W. 2003. Recovery from heart failure with circulatory assist: a working group of the national, heart, lung, and blood institute. *J. Card. Fail.* **9**:459–463.

21. Koh, G.Y., Klug, M.G., Soonpaa, M.H., and Field, L.J. 1993. Differentiation and long-term survival of c2c12 myoblast grafts in heart. *J. Clin. Invest.* **92**:1548–1554.

22. Chiu, R.C., Zibaitis, A., and Kao, R.L. 1995. Cellular cardiomyoplasty: myocardial regeneration with satellite cell implantation. *Ann. Thorac. Surg.* **60**:12–18.

23. Taylor, D.A., et al. 1998. Regenerating functional myocardium: improved performance after skeletal myoblast transplantation. *Nat. Med.* **4**:929–933.

24. Hagege, A.A., et al. 2003. Viability and differentiation of autologous skeletal myoblast grafts in ischaemic cardiomyopathy. *Lancet.* **361**:491–492.

25. Tse, H.F., et al. 2003. Angiogenesis in ischaemic myocardium by intramyocardial autologous bone marrow mononuclear cell implantation. *Lancet.* **361**:47–49.

26. Goodell, M.A., Brose, K., Paradis, G., Conner, A.S., and Mulligan, R.C. 1996. Isolation and functional properties of murine hematopoietic stem cells that are replicating in vivo. *J. Exp. Med.* **183**:1797–1806.

27. Goodell, M.A., et al. 1997. Dye efflux studies suggest that hematopoietic stem cells expressing low or undetectable levels of cd34 antigen exist in multiple species. *Nat. Med.* **3**:1337–1345.

28. Matsuzaki, Y., Kinjo, K., Mulligan, R.C., and Okano, H. 2004. Unexpectedly efficient homing capacity of purified murine hematopoietic stem cells. *Immunity.* **20**:87–93.

29. Pittenger, M.F., and Martin, B.J. 2004. Mesenchymal stem cells and their potential as cardiac therapeutics. *Circ. Res.* **95**:9–20.

30. Jiang, Y., et al. 2002. Pluripotency of mesenchymal stem cells derived from adult marrow. *Nature.* **418**:41–49.

31. Asahara, T., et al. 1999. Bone marrow origin of endothelial progenitor cells responsible for postnatal vasculogenesis in physiological and pathological neovascularization. *Circ. Res.* **85**:221–228.

32. Orlic, D., et al. 2001. Bone marrow cells regenerate infarcted myocardium. *Nature.* **410**:701–705.

33. Balsam, L.B., et al. 2004. Haematopoietic stem cells adopt mature haematopoietic fates in ischaemic myocardium. *Nature.* **428**:668–673.

34. Murry, C.E., et al. 2004. Haematopoietic stem cells do not transdifferentiate into cardiac myocytes in myocardial infarcts. *Nature.* **428**:664–668.

35. Nygren, J.M., et al. 2004. Bone marrow-derived hematopoietic cells generate cardiomyocytes at a low frequency through cell fusion, but not transdifferentiation. *Nat. Med.* **10**:494–501.

36. Planat-Benard, V., et al. 2004. Spontaneous cardiomyocyte differentiation from adipose tissue stroma cells. *Circ. Res.* **94**:223–229.

37. Jiang, Y., et al. 2002. Multipotent progenitor cells can be isolated from postnatal murine bone marrow, muscle, and brain. *Exp. Hematol.* **30**:896–904.

38. Kogler, G., et al. 2004. A new human somatic stem cell from placental cord blood with intrinsic pluripotent differentiation potential. *J. Exp. Med.* **200**:123–135.

39. Oh, H., et al. 2003. Cardiac progenitor cells from adult myocardium: homing, differentiation, and fusion after infarction. *Proc. Natl. Acad. Sci. U. S. A.* **100**:12313–12318.

40. Beltrami, A.P., et al. 2003. Adult cardiac stem cells are multipotent and support myocardial regeneration. *Cell.* **114**:763–776.

41. Matsuura, K., et al. 2004. Adult cardiac sca-1-positive cells differentiate into beating cardiomyocytes. *J. Biol. Chem.* **279**:11384–11391.

42. Martin, C.M., et al. 2003. Persistent expression of the atp-cassette transporter, abcg2, identifies cardiac stem cells in the adult heart. In *From stem cells to therapy.* Keystone Symposia. Steamboat Springs, Colorado, USA. 75.

43. Laugwitz, K.-L., et al. 2005. Post-natal isl1+ cardioblasts enter fully differentiated cardiomyocyte lineages. *Nature.* In press.

44. Aicher, A., et al. 2003. Assessment of the tissue distribution of transplanted human endothelial progenitor cells by radioactive labeling. *Circulation.* **107**:2134–2139.

45. Beauchamp, J.R., Morgan, J.E., Pagel, C.N., and Partridge, T.A. 1999. Dynamics of myoblast transplantation reveal a discrete minority of precursors with stem cell-like properties as the myogenic source. *J. Cell Biol.* **144**:1113–1122.

46. Lee, S.H., et al. 2000. Early expression of angiogenesis factors in acute myocardial ischemia and infarction. *N. Engl. J. Med.* **342**:626–633.

47. Menasché, P., et al. 2003. Autologous skeletal myoblast transplantation for severe postinfarction left ventricular dysfunction. *J. Am. Coll. Cardiol.* **41**:1078–1083.

48. Leobon, B., et al. 2003. Myoblasts transplanted into rat infarcted myocardium are functionally isolated from their host. *Proc. Natl. Acad. Sci. U. S. A.* **100**:7808–7811.

49. Rubart, M., et al. 2003. Physiological coupling of donor and host cardiomyocytes after cellular transplantation. *Circ. Res.* **92**:1217–1224.

50. Hutcheson, K.A., et al. 2000. Comparison of benefits on myocardial performance of cellular cardiomyoplasty with skeletal myoblasts and fibroblasts. *Cell. Transplant.* **9**:359–368.

51. Thompson, R.B., et al. 2003. Comparison of intracardiac cell transplantation: autologous skeletal myoblasts versus bone marrow cells. *Circulation.* **108**(Suppl. 1):II264–II271.

52. Asahara, T., et al. 1999. Vegf contributes to postnatal neovascularization by mobilizing bone marrow-derived endothelial progenitor cells. *EMBO J.* **18**:3964–3972.

53. Takahashi, T., et al. 1999. Ischemia- and cytokine-induced mobilization of bone marrow-derived endothelial progenitor cells for neovascularization. *Nat. Med.* **5**:434–438.

54. Hattori, K., et al. 2001. Plasma elevation of stromal cell-derived factor-1 induces mobilization of mature and immature hematopoietic progenitor and stem cells. *Blood.* **97**:3354–3360.

55. Hattori, K., et al. 2002. Placental growth factor reconstitutes hematopoiesis by recruiting vegfr1(+) stem cells from bone-marrow microenvironment. *Nat. Med.* **8**:841–849.

56. Heeschen, C., et al. 2003. Erythropoietin is a potent physiologic stimulus for endothelial progenitor cell mobilization. *Blood.* **102**:1340–1346.

57. Dimmeler, S., et al. 2001. HMG-CoA reductase inhibitors (statins) increase endothelial progenitor cells via the PI 3-kinase/Akt pathway. *J. Clin. Invest.* **108**:391–397. doi:10.1172/JCI200113152.

58. Laufs, U., et al. 2004. Physical training increases endothelial progenitor cells, inhibits neointima formation, and enhances angiogenesis. *Circulation.* **109**:220–226.

59. Iwakura, A., et al. 2003. Estrogen-mediated, endothelial nitric oxide synthase-dependent mobilization of bone marrow-derived endothelial progenitor cells contributes to reendothelialization after arterial injury. *Circulation.* **108**:3115–3121.

60. Orlic, D., et al. 2001. Mobilized bone marrow cells repair the infarcted heart, improving function and survival. *Proc. Natl. Acad. Sci. U. S. A.* **98**:10344–10349.

61. Kang, H.J., et al. 2004. Effects of intracoronary infusion of peripheral blood stem-cells mobilised with granulocyte-colony stimulating factor on left ventricular systolic function and restenosis after coronary stenting in myocardial infarction: the magic cell randomised clinical trial. *Lancet.* **363**:751–756.

62. Matsubara, H. 2004. Risk to the coronary arteries of intracoronary stem cell infusion and g-csf cytokine therapy. *Lancet.* **363**:746–747.

63. Askari, A.T., et al. 2003. Effect of stromal-cell-derived factor 1 on stem-cell homing and tissue regeneration in ischaemic cardiomyopathy. *Lancet.* **362**:697–703.

64. Papayannopoulou, T. 2003. Bone marrow homing: the players, the playfield, and their evolving roles. *Curr. Opin. Hematol.* **10**:214–219.

65. Ceradini, D.J., et al. 2004. Progenitor cell trafficking is regulated by hypoxic gradients through hif-1 induction of sdf-1. *Nat. Med.* **10**:858–864.

66. Yamaguchi, J., et al. 2003. Stromal cell-derived factor-1 effects on ex vivo expanded endothelial progenitor cell recruitment for ischemic neovascularization. *Circulation.* **107**:1322–1328.

67. Scaffidi, P., Misteli, T., and Bianchi, M.E. 2002. Release of chromatin protein hmgb1 by necrotic cells triggers inflammation. *Nature.* **418**:191–195.

68. Palumbo, R., et al. 2004. Extracellular hmgb1, a signal of tissue damage, induces mesoangioblast migration and proliferation. *J. Cell Biol.* **164**:441–449.

69. Bianchi, M.E., and Manfredi, A. 2004. Chromatin and cell death. *Biochim. Biophys. Acta.* **1677**:181–186.

70. De Falco, E., et al. 2004. Sdf-1 involvement in endothelial phenotype and ischemia-induced recruitment of bone marrow progenitor cells. *Blood.* **104**:3472–3482.

71. Chavakis, E., et al. 2005. Role of b2-integrins for homing and neovascularization capacity of endothelial progenitor cells. *J. Exp. Med.* **201**:63–72.

72. Vajkoczy, P., et al. 2003. Multistep recruitment of microvascular recruitment of ex vivo-expanded embryonic endothelial progenitor cells during tumor angiogenesis. *J. Exp. Med.* **197**:1755–1765.

73. Dewald, O., et al. 2004. Of mice and dogs: species-specific differences in the inflammatory response following myocardial infarction. *Am. J. Pathol.* **164**:665–677.

74. Urbich, C., and Dimmeler, S. 2004. Endothelial progenitor cells: characterization and role in vascular biology. *Circ. Res.* **95**:343–353.

75. Jackson, K.A., et al. 2001. Regeneration of ischemic cardiac muscle and vascular endothelium by adult stem cells. *J. Clin. Invest.* **107**:1395–1402.

76. De Palma, M., Venneri, M.A., Roca, C., and Naldini, L. 2003. Targeting exogenous genes to tumor angiogenesis by transplantation of genetically modified hematopoietic stem cells. *Nat. Med.* **9**:789–795.

77. Fuchs, S., et al. 2001. Transendocardial delivery of autologous bone marrow enhances collateral perfusion and regional function in pigs with chronic experimental myocardial ischemia. *J. Am. Coll. Cardiol.* **37**:1726–1732.

78. Kocher, A.A., et al. 2001. Neovascularization of ischemic myocardium by human bone-marrow-derived angioblasts prevents cardiomyocyte apoptosis, reduces remodeling and improves cardiac function. *Nat. Med.* **7**:430–436.

79. Askari, A., et al. 2004. Cellular, but not direct, adenoviral delivery of vascular endothelial growth factor results in improved left ventricular function and neovascularization in dilated ischemic cardiomyopathy. *J. Am. Coll. Cardiol.* **43**:1908–1914.

80. Akita, T., et al. 2003. Hypoxic preconditioning augments efficacy of human endothelial progenitor cells for therapeutic neovascularization. *Lab. Invest.* **83**:65–73.

81. Ramirez-Bergeron, D.L., et al. 2004. Hypoxia affects mesoderm and enhances hemangioblast specification during early development. *Development.* **131**:4623–4634.

82. Klug, M.G., Soonpaa, M.H., Koh, G.Y., and Field, L.J. 1996. Genetically selected cardiomyocytes from differentiating embryonic stem cells form stable intracardiac grafts. *J. Clin. Invest.* **98**:216–224.

83. Kehat, I., et al. 2001. Human embryonic stem cells can differentiate into myocytes with structural and functional properties of cardiomyocytes. *J. Clin. Invest.* **108**:407–414. doi:10.1172/JCI200112131.

84. Xu, C., Police, S., Rao, N., and Carpenter, M.K. 2002. Characterization and enrichment of cardiomyocytes derived from human embryonic stem cells. *Circ. Res.* **91**:501–508.

85. Blackburn, E. 2004. Bioethics and the political distortion of biomedical science. *N. Engl. J. Med.* **350**:1379–1380.

86. Fairchild, P.J., Cartland, S., Nolan, K.F., and Waldmann, H. 2004. Embryonic stem cells and the challenge of transplantation tolerance. *Trends Immunol.* **25**:465–470.

87. Anderson, D.J., Gage, F.H., and Weissman, I.L. 2001. Can stem cells cross lineage boundaries? *Nat. Med.* **7**:393–395.

88. Verfaillie, C.M. 2002. Adult stem cells: assessing the case for pluripotency. *Trends Cell Biol.* **12**:502–508.

89. Pomerantz, J., and Blau, H.M. 2004. Nuclear reprogramming: a key to stem cell function in regenerative medicine. *Nat. Cell Biol.* **6**:810–816.

90. Wagers, A.J., Sherwood, R.I., Christensen, J.L., and Weissman, I.L. 2002. Little evidence for developmental plasticity of adult hematopoietic stem cells. *Science.* **297**:2256–2259.

91. Bittner, R.E., et al. 1999. Recruitment of bone-marrow-derived cells by skeletal and cardiac muscle in adult dystrophic mdx mice. *Anat. Embryol.* **199**:391–396.

92. Corti, S., et al. 2004. Wild-type bone marrow cells ameliorate the phenotype of sod1-g93a als mice and contribute to cns, heart and skeletal muscle tissues. *Brain.* **127**:2518–2532.

93. Makino, S., et al. 1999. Cardiomyocytes can be generated from marrow stromal cells in vitro. *J. Clin. Invest.* **103**:697–705.

94. Badorff, C., et al. 2003. Transdifferentiation of blood-derived human adult endothelial progenitor cells into functionally active cardiomyocytes. *Circulation.* **107**:1024–1032.

95. Xaymardan, M., et al. 2004. Platelet-derived growth factor-ab promotes the generation of adult bone marrow-derived cardiac myocytes. *Circ. Res.* **94**:E39–E45.

96. Mangi, A.A., et al. 2003. Mesenchymal stem cells modified with akt prevent remodeling and restore performance of infarcted hearts. *Nat. Med.* **9**:1195–1201.

97. Toma, C., Pittenger, M.F., Cahill, K.S., Byrne, B.J., and Kessler, P.D. 2002. Human mesenchymal stem cells differentiate to a cardiomyocyte phenotype in the adult murine heart. *Circulation.* **105**:93–98.

98. Saito, T., Kuang, J.Q., Lin, C.C., and Chiu, R.C. 2003. Transcoronary implantation of bone marrow stromal cells ameliorates cardiac function after myocardial infarction. *J. Thorac. Cardiovasc. Surg.* **126**:114–123.

99. Kawada, H., et al. 2004. Nonhematopoietic mesenchymal stem cells can be mobilized and differentiate into cardiomyocytes after myocardial infarction. *Blood.* **104**:3581–3587.

100. Ramalho-Santos, M., Yoon, S.J., Matsuzaki, Y., Mulligan, R.C., and Melton, D.A. 2002. "stemness": transcriptional profiling of embryonic and adult stem cells. *Science.* **298**:597–600.

101. Harrington, L. 2004. Does the reservoir for self-renewal stem from the ends? *Oncogene.* **23**:7283–7289.

102. Alvarez-Dolado, M., et al. 2003. Fusion of bone-marrow-derived cells with Purkinje neurons, cardiomyocytes and hepatocytes. *Nature.* **425**:968–973.

103. Minasi, M.G., et al. 2002. The meso-angioblast: a multipotent, self-renewing cell that originates from the dorsal aorta and differentiates into most mesodermal tissues. *Development.* **129**:2773–2783.

104. Heeschen, C., et al. 2004. Profoundly reduced neovascularization capacity of bone marrow mononuclear cells derived from patients with chronic ischemic heart disease. *Circulation.* **109**:1615–1622.

105. Vasa, M., et al. 2001. Number and migratory activity of circulating endothelial progenitor cells inversely correlate with risk factors for coronary artery disease. *Circ. Res.* **89**:E1–E7.

106. Hill, J.M., et al. 2003. Circulating endothelial progenitor cells, vascular function, and cardiovascular risk. *N. Engl. J. Med.* **348**:593–600.

107. Edelberg, J.M., Tang, L., Hattori, K., Lyden, D., and Rafii, S. 2002. Young adult bone marrow-derived endothelial precursor cells restore aging-impaired cardiac angiogenic function. *Circ. Res.* **90**:E89–E93.

108. Torella, D., et al. 2004. Cardiac stem cell and myocyte aging, heart failure, and insulin-like growth factor-1 overexpression. *Circ. Res.* **94**:514–524.

109. Assmus, B., et al. 2003. Hmg-coa reductase inhibitors reduce senescence and increase proliferation of endothelial progenitor cells via regulation of cell cycle regulatory genes. *Circ. Res.* **92**:1049–1055.

110. Spyridopoulos, I., et al. 2004. Statins enhance migratory capacity by upregulation of the telomere repeat-binding factor trf2 in endothelial progeni-

tor cells. *Circulation.* **110**:3136–3142.

111. Murasawa, S., et al. 2002. Constitutive human telomerase reverse transcriptase expression enhances regenerative properties of endothelial progenitor cells. *Circulation.* **106**:1133–1139.

112. Oh, H., et al. 2001. Telomerase reverse transcriptase promotes cardiac muscle cell proliferation, hypertrophy, and survival. *Proc. Natl. Acad. Sci. U. S. A.* **98**:10308–10313.

113. Haendeler, J., et al. 2004. Antioxidants inhibit nuclear export of telomerase reverse transcriptase and delay replicative senescence of endothelial cells. *Circ. Res.* **94**:768–775.

114. MacLellan, W.R., et al. 2005. Overlapping roles of pocket proteins in the myocardium are unmasked by germline deletion of p130 plus heart-specific deletion of rb. *Mol. Cell. Biol.* In press.

115. Raya, A., et al. 2003. Activation of notch signaling pathway precedes heart regeneration in zebrafish. *Proc. Natl. Acad. Sci. U. S. A.* **100**(Suppl. 1):11889–11895.

116. Keating, M.T. 2004. Genetic approaches to disease and regeneration. *Philos. Trans. R. Soc. Lond., B, Biol. Sci.* **359**:795–798.

117. Foley, A., and Mercola, M. 2004. Heart induction: embryology to cardiomyocyte regeneration. *Trends Cardiovasc. Med.* **14**:121–125.

118. Jankowski, M., et al. 2004. Oxytocin in cardiac ontogeny. *Proc. Natl. Acad. Sci. U. S. A.* **101**:13074–13079.

119. Behfar, A., et al. 2002. Stem cell differentiation requires a paracrine pathway in the heart. *FASEB J.* **16**:1558–1566.

120. Tse, W.T., Pendleton, J.D., Beyer, W.M., Egalka, M.C., and Guinan, E.C. 2003. Suppression of allogeneic T-cell proliferation by human marrow stromal cells: implications in transplantation. *Transplantation.* **75**:389–397.

121. Bock-Marquette, I., et al. 2004. Thymosin b4 activates integrin-linked kinase and promotes cardiac cell migration, survival, and cardiac repair. *Nature.* **432**:466–472.

122. Britten, M.B., et al. 2003. Infarct remodeling after intracoronary progenitor cell treatment in patients with acute myocardial infarction (TOPCARE-AMI): mechanistic insights from serial contrast-enhanced magnetic resonance imaging. *Circulation.* **108**:2212–2218.

123. Fernandez-Aviles, F., et al. 2004. Experimental and clinical regenerative capability of human bone marrow cells after myocardial infarction. *Circ. Res.* **95**:742–748.

124. Fuchs, S., et al. 2003. Catheter-based autologous bone marrow myocardial injection in no-option patients with advanced coronary artery disease: a feasibility study. *J. Am. Coll. Cardiol.* **41**:1721–1724.

Biological basis for the cardiovascular consequences of COX-2 inhibition: therapeutic challenges and opportunities

Tilo Grosser, Susanne Fries, and Garret A. FitzGerald

Institute for Translational Medicine and Therapeutics and Department of Pharmacology, University of Pennsylvania, Philadelphia, Pennsylvania, USA.

Inhibitors selective for prostaglandin G/H synthase-2 (PGHS-2) (known colloquially as COX-2) were designed to minimize gastrointestinal complications of traditional NSAIDs — adverse effects attributed to suppression of COX-1–derived PGE_2 and prostacyclin (PGI_2). Evidence from 2 randomized controlled-outcome trials (RCTs) of 2 structurally distinct selective inhibitors of COX-2 supports this hypothesis. However, 5 RCTs of 3 structurally distinct inhibitors also indicate that such compounds elevate the risk of myocardial infarction and stroke. The clinical information is biologically plausible, as it is compatible with evidence that inhibition of COX-2–derived PGI_2 removes a protective constraint on thrombogenesis, hypertension, and atherogenesis in vivo. However, the concept of simply tipping a "balance" between COX-2–derived PGI_2 and COX-1–derived platelet thromboxane is misplaced. Among the questions that remain to be addressed are the following: (a) whether this hazard extends to all or some of the traditional NSAIDs; (b) whether adjuvant therapies, such as low-dose aspirin, will mitigate the hazard and if so, at what cost; (c) whether COX-2 inhibitors result in cardiovascular risk transformation during chronic dosing; and (d) how we might identify individuals most likely to benefit or suffer from such drugs in the future.

One should not increase, beyond what is necessary, the number of entities required to explain anything.
— Occam's razor

Arachidonic acid (AA) is subject to metabolism by prostaglandin G/H synthase (PGHS; commonly known as COX) enzymes, lipoxygenases, and epoxygenases to form a mesmerizing array of biologically active products. The COX enzymes are bisfunctional proteins, possessing both COX and hydroperoxidase (HOX) activities, catalyzing the biotransformation of AA into the PG endoperoxide intermediates PGG_2 and PGH_2. These are, in turn, acted on by isomerases and synthases to form the PGs and thromboxane A_2 (TxA_2) (1–3). All of these products activate G protein–coupled receptors; the phenotypes resulting from deletion of these receptors has informed considerably our understanding of prostanoid biology (4). NSAIDs, which include both traditional NSAIDs (tNSAIDs) and selective inhibitors of COX-2 and which are among the most commonly used drugs (5), relieve pain and inflammation by suppressing the COX function of PGHS and the consequent

formation of PGE_2 (6) and prostacyclin (PGI_2) (7), but perhaps also of other prostanoids. The failure of NSAIDs to inhibit PGHS HOX–dependent free radical formation may contribute to the failure of these drugs to modify disease progression in arthritis (8).

Several groups made observations (9–12) that predicted the discovery of a second COX enzyme (13–15). Unlike COX-1, which appeared to be expressed constitutively in most tissues, COX-2 was subject to rapid induction by inflammatory cytokines and mitogens and was speculated to account largely if not exclusively for PG formation in inflammation and cancer. Drug screening in cellular and biological systems identified compounds selective for COX-2 (16, 17). The elucidation of the COX structures subsequently explained this capability (18, 19). The COX enzymes are remarkably similar, both sharing a hydrophobic tunnel that affords access of the lipid substrate to the active site, deep within the proteins. However, the COX-2 tunnel is more accommodating and includes a side pocket not present in COX-1 (Figure 1). This affords both broader substrate recognition and a structural explanation for the ability to detect drugs selective for COX-2 in pharmacological screens (20). The attraction in developing such compounds was that they might not induce the commonest complication of tNSAIDs — gastrointestinal (GI) intolerance — which the "COX-2 hypothesis" attributed entirely to inhibition of COX-1–derived protective PGE_2 and PGI_2 by gastroduodenal epithelium and platelet COX-1–derived TxA_2 (21). Subsequently, the simplicity of this concept has been challenged by increasing evidence supporting the importance of COX-2 in resolution of mucosal inflammation and in ulcer healing (22). Despite this and epidemiological evidence suggesting that the incidence of severe tNSAID-related GI adverse effects was in decline (23), the race for the approval of drugs designed as COX-2–specific inhibitors gained momentum. The conventional bases for approval by the FDA of the first 3 of these drugs — celecoxib, rofecoxib, and valdecoxib —were relatively small: clinical studies consisted mostly of hundreds of volunteers and were short-term (mostly less than 6 months) studies in which endoscopic visualization of drug-induced

Nonstandard abbreviations used: AA, arachidonic acid; APC, Adenoma Prevention with Celecoxib; APPROVe, Adenomatous Polyp Prevention on Vioxx; CABG, coronary artery bypass grafting; CLASS, Celecoxib Long-term Arthritis Safety Study; EDGE, Etoricoxib versus Diclofenac Sodium Gastrointestinal Evaluation; EMEA, European Medicines Agency; GI, gastrointestinal; HOX, hydroperoxidase; IP, PGI_2 receptor; MEDAL, Multinational Etoricoxib and Diclofenac Arthritis Long-term; PGHS, prostaglandin G/H synthase; PGI_2, prostacyclin; PGIM, PGI_2 metabolite; RCT, randomized controlled-outcome trial; TARGET, Therapeutic Arthritis Research and Gastrointestinal Event Trial; tNSAID, traditional NSAID; TxA_2, thromboxane A_2; VIGOR, Vioxx Gastrointestinal Outcome Research.

Conflict of interest: G.A. FitzGerald receives financial support for investigator-initiated research from Bayer, Merck, and Boehringer Ingelheim, all of which manufacture drugs that target COXs. G.A. FitzGerald is a member of the Steering Committee of the Multinational Etoricoxib and Diclofenac Arthritis Long-term (MEDAL) Study Program. This author also serves as a consultant for Johnson & Johnson, Bayer, Merck, GlaxoSmithKline, Novartis, Boehringer Ingelheim, and NiCox.

Citation for this article: *J. Clin. Invest.* **116**:4–15 (2006). doi:10.1172/JCI27291.

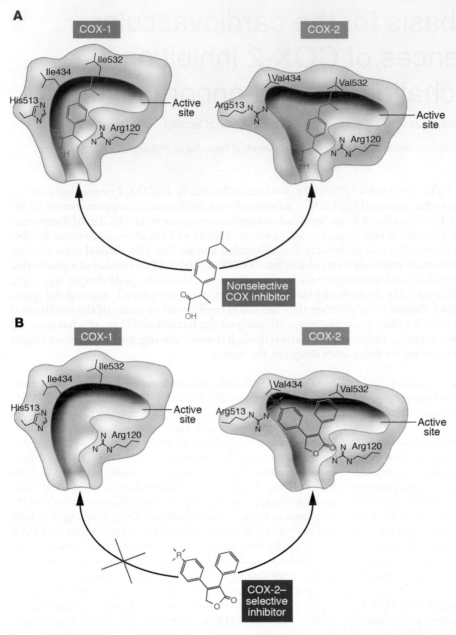

Figure 1
Schematic depiction of the structural differences between the substrate-binding channels of COX-1 and COX-2 that allowed the design of selective inhibitors. The amino acid residues Val434, Arg513, and Val523 form a side pocket in COX-2 that is absent in COX-1. (**A**) Nonselective inhibitors have access to the binding channels of both isoforms. (**B**) The more voluminous residues in COX-1, Ile434, His513, and Ile532, obstruct access of the bulky side chains of COX-2 inhibitors. Figure modified with permission from *Nature* from protein structures reported in refs. 18 and 20.

ulceration was compared among the coxib, a tNSAID, and placebo. The superiority of the COX-2 inhibitors over their tNSAID comparators in these studies was striking (24–28).

Mechanistic basis for a cardiovascular hazard resulting from inhibition of COX-2

During the course of drug development, we found that both celecoxib and rofecoxib suppressed urinary 2, 3-dinor 6-keto $PGF_{1\alpha}$, a stable PGI_2 metabolite (PGIM), to a degree comparable

to that attained by treatment with structurally distinct tNSAIDs (29, 30). While the latter drugs inhibited platelet aggregation ex vivo transiently at the time of peak action, the coxibs had no such effect, compatible with the absence of COX-2 from mature human platelets (31). Unlike the tNSAID comparators in these studies — ibuprofen and indomethacin — neither celecoxib nor rofecoxib inhibited COX-1–derived TxA_2 coincident with its impact on PGI_2. Thus, the cardiovascular effects of TxA_2 would be expected to be exaggerated. However, as PGI_2 was known to act as a general restraint on *any* recognized stimulus to platelet activation, it was not suggested that upsetting a notional "balance" between the 2 prostanoids was likely to be the mechanism of drug action. Correspondingly, variation in other endogenous mediators, such as NO, would be expected to modulate the impact of COX-2 inhibition on cardiovascular function. Given that similar observations were made with both celecoxib and rofecoxib, it appeared that this effect was mechanism based, rather than an off-target effect restricted to 1 compound. Despite the presence of only COX-1 in endothelial cells under static conditions in vitro (32), PGI_2 — a dominant product of endothelium (33) — appeared largely to derive from COX-2 under physiological conditions in humans. Prior findings of Topper and Gimbrone (34) were invoked to explain these observations. They had found that subjection of endothelial cells in culture to laminar shear upregulated COX-2 expression. Thus, induction of COX-2 was likely to have occurred in response to blood flow under physiological conditions in vivo. Studies that led to approval of the coxibs were too short and too small in subject number to have excluded a risk of myocardial infarction or stroke attributable to this hypothesis.

Insight into the consequences of suppressing PGI_2 in vivo were limited at the time; deletion of the PGI_2 receptor (IP) augmented the response to an exogenous thrombogenic stimulus in mice (7). However, it was claimed that redundancy with other antithrombotic systems, particularly the elaboration of NO, would annul the impact of PGI_2 suppression in vivo. When deletion of the IP was shown to restrain the effect of endogenous TxA_2 on platelet activation and vascular proliferation in response to injury (35), it was questioned whether loss of 2 copies of the IP would mimic the substantial but incomplete suppression of PGIM attained in humans (S1). Subsequent experiments revealed that IP deletion predisposed to thrombosis in a gene dose–dependent fashion; the effect of a COX-2 inhibitor in thrombosis models was intermediate between $IP^{+/-}$ and $IP^{-/-}$ mice

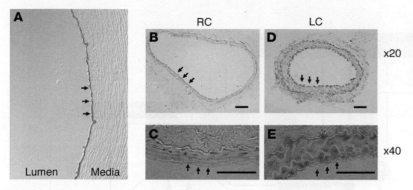

Figure 2
Expression of *COX-2* mRNA in the endothelium (arrows) of human (**A**) and COX-2 protein in murine (**B–E**) arteries. (**A**) In situ detection of *COX-2* mRNA in the endothelium of a human umbilical artery. Image kindly provided by James N. Topper, Frazier Healthcare Ventures, Palo Alto, California, USA. (**B–E**) Immunostaining shows COX-2 upregulation 4 weeks after left common carotid artery ligation in mice. Scale bars: 50 μm. Baseline COX-2 expression (brown staining) is evident in the intima in cross sections of the right common, unligated carotid artery, which served as a control. Magnification, ×20 (**B**); ×40 (**C**). Flow reduction induced further COX-2 expression in the intimal layer and marked endothelial expression as shown in **D** and **E**. Magnification, ×20 (**D**); ×40 (**E**). LC, left common carotid artery; RC, right common carotid artery. **B–E** are reproduced here with permission from *Circulation Research* (55).

(Y. Cheng, personal communication). It has also been suggested that the presence of PGIM does not reflect endothelial biosynthesis of PGI_2 and that COX-2 is undetectable in endothelial cells ex vivo (S2). While one can never attribute with certainty a tissue of origin to a metabolite measured in urine (36) or plasma (37), studies in vitro had indicated that endothelium is the major tissue source of PGI_2 (33) and local vascular stimulation and short-term systemic administration of PGI_2 is reflected by readily detectable alterations in urinary PGIM levels (38, 39). Furthermore, expression of *COX-2* is evident in human endothelial cells ex vivo (Figure 2A), and expression of endothelial COX-2 may be modulated in a flow-dependent manner in mice (Figure 2, B–E). Indeed, the human cDNAs for COX-2 were originally cloned from unstimulated endothelial cells (40, 41), reflective of constitutive expression of the enzyme. Aside from physiological conditions, one would expect that vascular stimulation by the products of platelet activation and by inflammatory cytokines might upregulate endothelial and vascular smooth muscle cell expression of COX-2, as occurs in atherosclerotic lesions (42, 43). Indeed, excretion of both PGIM and the TxA_2 metabolite 2, 3-dinor TxB_2, are together increased in patients with severe atherosclerosis (43, 44, S3). The failure of some studies to report COX-2 expression in endothelial cells ex vivo may reflect the particular experimental circumstances and/or discordance between the offset kinetics of flow-induced gene expression and the time of sample preparation.

Subsequent work expanded our understanding of the effects of PGI_2 on cardiovascular biology (Figure 3). Celecoxib suppressed PGI_2-dependent vascular bioactivity and undermined the antithrombotic effect of aspirin in dogs (45). Selective COX-2 inhibition suppressed PGI_2 and predisposed to platelet activation and arterial thrombosis under conditions of hypoxia-induced pulmonary hypertension in rodents (46). Similarly, selective COX-2 inhibition suppressed PGI_2 and enhanced platelet–vessel wall interactions in vivo and platelet adhesion to hamster cheek pouch arterioles (47). PGI_2 was shown in endothelial cells to stimulate substantially thrombomodulin (48). Thus, removal of this natural constraint to

thrombin activation would interact with augmented platelet activation to promote assembly of the prothrombinase complex and consequent thrombosis, perhaps particularly in the microvasculature. Reperfusion injury of the myocardium is augmented in mice lacking the IP (49), suggesting a limit to the benefit of such therapeutic strategies in patients who suffered a thrombosis, and COX-2–dependent PGI_2 was shown to afford protection against oxidant injury to cardiomyocytes in vivo (50).

Aside from effects most relevant to acute human syndromes of thrombotic vascular occlusion, suppression of COX-2 may also predispose to a more gradual elevation of cardiovascular risk during prolonged dosing with inhibitors. Deletion of the IP was found to promote initiation and early progression of atherosclerosis in mice genetically predisposed to hyperlipidemia (51, 52). This appears to reflect removal of a constraint to the activation of both neutrophils and platelets, their interaction with the vessel wall, and the resultant oxidant stress. Additionally, COX-2–dependent PGI_2 formation appears to contribute to the atheroprotection afforded by estrogen and mediated via its ER-α receptor in vivo (52). Deletion of the IP (53, 54), like COX-2 inhibition (37), elevates blood pressure (54) and augments the pressor response to dietary sodium (53, 54). Finally, both deletion of the IP and inhibition of COX-2 modulate vascular remodeling induced by hemodynamic stress — such as hypertension — in vivo (55). Additional effects on AA metabolism resulting from COX-2 inhibition, including a failure to metabolize the vasoconstrictor 20-hydroxyeicosatetraenoic acid (20-HETE) to a vasodilator product (56) and augmented metabolism via lipoxygenase and cytochrome P450 enzymes, may impact blood pressure regulation. However, evidence that such effects contribute to the renovascular effects of COX-2 inhibitors remains to be provided in vivo. COX-2 is also the dominant source of PGE_2 and PGD_2 biosynthesis under physiological conditions in humans (S4, S5). Deletion of the EP2 receptor for PGE_2, like deletion of the IP, results in salt-sensitive hypertension (57), and both PGE_2 (via the IP) and PGD_2 inhibit platelet activation, at least in vitro (S6, S7). The extent to which inhibition of these other PGs might contribute to a cardiovascular hazard of COX-2 inhibitors remains to be established. In the interim, one might speculate how the disparate effects of PGI_2 suppression on atherogenesis, blood pressure, and the remodeling response might converge, over time, to result in transformation of cardiovascular risk in patients initially at low risk when exposed to chronic COX-2 inhibition (58).

In summary, a substantial body of evidence has accumulated that 1 mechanism, suppression of COX-2–dependent PGI_2 formation, can both augment the response to thrombotic and hypertensive stimuli and initiate and accelerate atherogenesis.

Concordance of clinical experience with mechanism-based predictions

Given the assumption that this mechanism explains the observed cardiovascular complications of COX-2 inhibitors, how would such a hazard be expected to become clinically manifest?

First, the actual degree of selectivity attained at the vascular interface in vivo would be an important variable. Although assays in whole blood in vitro suggest a clear segregation between the degree of selectivity attained by the drugs under consideration, there are

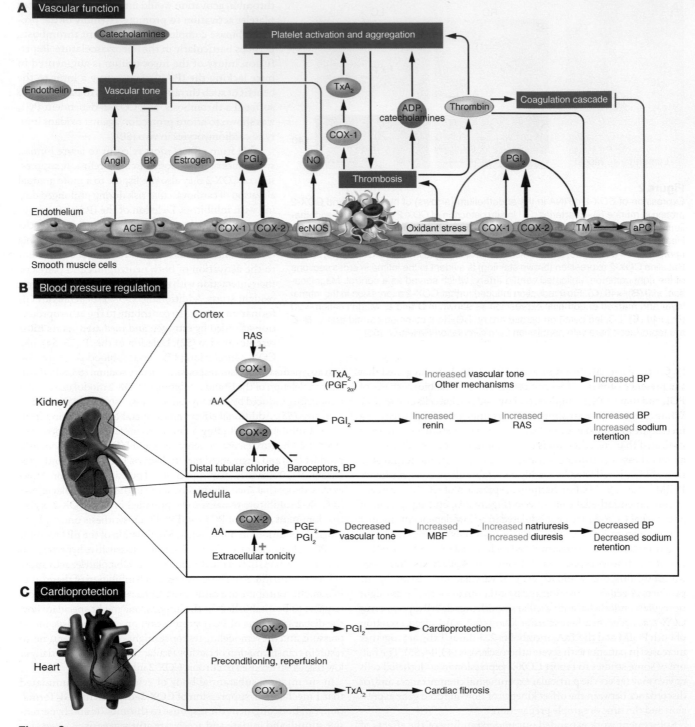

Figure 3
Roles of the COX isozymes in cardiovascular (**A** and **C**) and renal (**B**) biology. ACE, angiotensin-converting enzyme; ADP, adenosine diphosphate; aPC, activated protein C; BK, bradykinin; ecNOS, endothelial cell NOS; MBF, medullary blood flow; RAS, renin-angiotensin system; TM, thrombomodulin.

substantial interindividual differences in drug response (59) and consequent overlap in the degree of selectivity attained in vivo. Selectivity for COX-2 can be viewed as a continuous variable within the class of NSAIDs. Indeed, some tNSAIDs — diclofenac, nimesulide, meloxicam, and nabumetone — express average selectivity for COX-2

similar to that of celecoxib in human whole blood in vitro (28). Sufficient concentration of any selective COX-2 inhibitor becomes nonselective as it begins to inhibit COX-1, at least in vitro (28, 60). Second, the more prolonged the drug exposure (determined by dose, duration of action, and duration of treatment), the more likely an

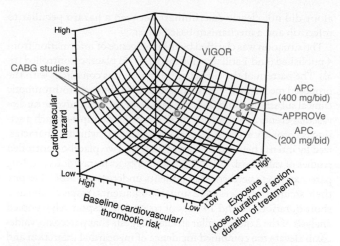

Figure 4
Illustration of the expected interaction of baseline cardiovascular and thrombotic risk with components of drug exposure including dose, duration of action, and duration of treatment with a selective inhibitor of COX-2. The approximate relationship of cardiovascular hazard detected in controlled studies within this interaction are indicated (not to scale). APC study, ref. 81; APPROVe study, ref. 72; CABG studies, parecoxib/valdecoxib after bypass surgery, refs. 77, 78; VIGOR study, ref. 62.

adverse consequence. Third, concordant administration of low-dose aspirin, which favors inhibition of COX-1 (61), would be expected to mitigate but not abolish the hazard. The degree and duration of simultaneous inhibition of the 2 COX enzymes would also be expected to influence the existence of a cardiovascular hazard from tNSAIDs (see below). Finally, $IP^{-/-}$ mice are more responsive to thrombogenic stimuli; they do not develop spontaneous thrombosis (7). Thus, a clinical or genetic predisposition to thrombosis would favor emergence of a drug-related cardiovascular event.

Aside from the question posed by clinical pharmacology, the first evidence consistent with the hypothetical cardiovascular hazard emerged in the Vioxx Gastrointestinal Outcome Research (VIGOR) study (62), in which a 2-fold divergence in the incidence of serious GI adverse events between rofecoxib and the tNSAID naproxen coincided with a 5-fold divergence in the incidence of myocardial infarction (20 versus 4 events). This study was conduct-

Figure 5
The spectrum of selectivity for COX inhibition. (**A**) The relative affinities of tNSAIDs and coxibs (open circles) for COX-1 and COX-2. The concentrations required to inhibit COX-1 and COX-2 by 50% (IC50) have been measured using whole-blood assays of COX-1 and COX-2 activity in vitro. The diagonal line indicates equivalent COX-1 and COX-2 inhibition. Drugs plotted below the line (orange) are more potent inhibitors of COX-2 than drugs plotted above the line (green). The distance to the line is a measure of selectivity. Note the log scale. For example, lumiracoxib is the compound with the highest degree of selectivity for COX-2 as its distance to the line is the largest. Celecoxib and diclofenac have similar degrees of selectivity for COX-2, as their distances to the line are similar; however, diclofenac is active at lower concentrations and thus located more to the left. Figure modified with permission from *The New England Journal of Medicine* (28). (**B**) Implication of the relative degrees of selectivity. Increasing degrees of selectivity for COX-2 are associated with augmented cardiovascular risk while increasing degrees of selectivity for COX-1 are associated with augmented GI risk. The relative size of the circles indicates approximately the variation in sample sizes among the trials.

ed with a high dose (50 mg/day) of rofecoxib in patients in whom low-dose aspirin was precluded. Most of the patients suffered from RA, a disease associated with an odds ratio of a myocardial infarction roughly 50% higher than in patients with osteoarthritis or no arthritis (63). These results generated considerable controversy; some researchers claimed that rofecoxib was neutral and that the result reflected a cardioprotective effect of naproxen, based on its extended duration of action (64), permitting this mixed inhibitor of COX-1 and COX-2 to behave like aspirin.

The corresponding outcomes study of celecoxib (Celecoxib Long-term Arthritis Safety Study [CLASS]) was published in a highly unorthodox manner (65). Partial presentation of the data seemed to suggest that high-dose (800 mg/day) celecoxib had caused fewer GI adverse effects than its tNSAID comparators; however, this turned out not to be the case when the full data set was revealed (66). This study, conducted with, on average, a shorter-lived, less selective COX-2 inhibitor than rofecoxib, also demonstrated no difference in the incidence of cardiovascular events. Around 20% of the patients took aspirin, and much was made of the apparent divergent incidence of GI adverse effects on ibuprofen versus celecoxib in a post hoc analysis of nonaspirin users. Perhaps aspirin had masked the GI advantage of celecoxib. However, if so, it may also have masked the cardiovascular hazard. A similar underpowered and retrospective analysis suggests that cardiovascular events occurred more often with celecoxib than with ibuprofen in nonaspirin users. Interestingly, the incidence of both GI and cardiovascular events on diclofenac and celecoxib appeared to be similar (67).

In summary, the number of events reported in the VIGOR study was small. However, if the estimate of the difference between the 2 treatment groups was reliable, this was larger than might be expected from an "aspirin-like" effect of naproxen; clearly it was

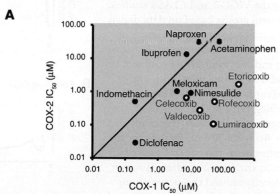

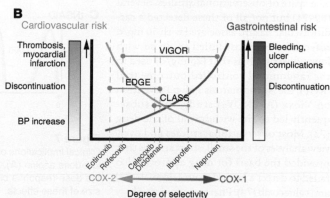

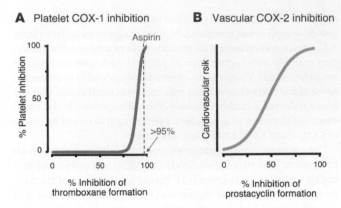

Figure 6
Discordant dose-response relationships for inhibition of platelet COX-1 (**A**) and vascular COX-2 (**B**). Derived from data reported in ref. 28.

compatible with the coincidence of a cardiovascular hazard from rofecoxib and some protection from naproxen.

The traditional approach to drug safety is to rely upon pharmaco-epidemiology. However, this is an insensitive detector system when the need is to identify a small absolute increase — maybe 1–2% in retrospect — in the absolute incidence of a problem that occurs commonly in the age group under study. In addition to these limitations, epidemiological (observational) studies are subject to many sources of bias, and in this particular case, the common use of prescription databases was also potentially confounded by unrecorded over-the-counter use of tNSAIDs and aspirin.

Further attention to the prospect of a cardiovascular hazard from the COX-2 inhibitors (coxibs) was prompted initially by a comparison of trial data for both drugs, including the VIGOR and CLASS studies, with a control group, based on data drawn from the placebo groups of 4 primary prevention trials of low-dose aspirin (68). However, this indirect analysis was subject to considerable methodological criticism. The estimated cardiovascular event rates in 2 of the placebo groups lay below while 2 were above those calculated for rofecoxib and celecoxib. Roughly 70% of the data for the pooled estimate in the control group was drawn from the first 2 studies. However, the controversy around this paper prompted a spate of observational studies. Several (69–71) but not all of these detected a cardiovascular hazard associated with 50 mg/d rofecoxib, but most failed to do so with lower doses such as that (25 mg/d) used in the randomized controlled-outcome trial (RCT), the Adenomatous Polyp Prevention on Vioxx (APPROVe) study, that subsequently led to the withdrawal of the drug (72). Most observational studies and overview analyses of the small, short studies that provided the basis for drug approval also failed to detect a hazard from celecoxib (73) and valdecoxib (74). Pharmacoepidemiology

alone did not clearly discriminate between a hazard peculiar to rofecoxib and a mechanism-based effect.

The situation was clarified by the emergence of information from 4 published (and 1 still unpublished; ref. 75) placebo-controlled trials. The pattern of the clinical information was consistent with the proposed mechanism. For reasons discussed above, a prothrombotic clinical substrate would favor the rapid emergence of adverse cardiovascular events in a relatively small study. An example of such a setting is coronary artery bypass grafting (CABG), which is characterized by intense hemostatic activation (76). Two placebo-controlled studies of valdecoxib (77, 78), anteceded by its intravenous prodrug parecoxib, were performed in patients undergoing CABG. Despite their small study sizes (462 and 1636 patients, respectively) and short duration (10 and 14 days of treatment, respectively), pooled analysis of the 2 quite similar studies suggests that parecoxib/valdecoxib elevate the combined incidence of myocardial infarction and stroke by 3-fold in this population (79). Although the patients were prescribed aspirin, the timing of its administration relative to the incidence of the vascular events is unclear. CABG is also a setting of apparent "aspirin resistance" (80). These studies are compatible with the rapid emergence of a cardiovascular hazard based on suppression of COX-2–derived PGI_2 in a population with preexisting, intense hemostatic activation. Similarly, one would anticipate that a less pronounced prothrombotic substrate, such as the patients with RA in the 9-month VIGOR trial, might reveal a hazard more gradually. The rapidity with which a cardiovascular risk might become manifest would reflect in part the intensity of a genetic or environmental predisposition to thrombosis (Figure 4).

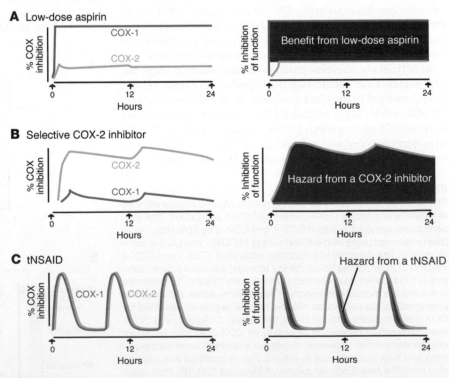

Figure 7
Clinical implications of differences in the dose-response relationships for COX-1 and COX-2 of low-dose aspirin (**A**), a selective inhibitor of COX-2 (**B**), and a tNSAID (**C**). The area between the dose-response curves would correspond to benefit (**A**) and hazard (**B** and **C**) and to the size of these effects.

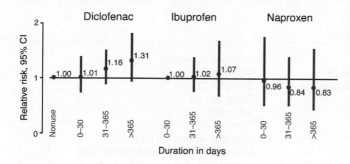

Figure 8
Duration of use of tNSAIDs and individual tNSAIDs among current users (use within a month) and risk of myocardial infarction. Redrawn with permission from *BMC Medicine* (106). CI, confidence interval; nonuse, reference group with relative risk of 1.00.

APPROVe (72) and Adenoma Prevention with Celecoxib (APC), 2 studies in patients with colonic adenomata, presumed initially to be at low risk of cardiovascular events, revealed the gradual emergence of a cardiovascular risk attributable respectively to rofecoxib (72) and celecoxib (81) after dosing for more than 1 year (Figure 4). Supportive of this being a true drug-related effect, the hazard in patients taking celecoxib 200 mg/bid and 400 mg/bid appeared to be dose related (81).

Several comparative studies of COX-2 inhibitors and tNSAIDs failed to detect a discriminant incidence of cardiovascular events. However, in each case, these studies were substantially underpowered to exclude this possibility. These include the Therapeutic Arthritis Research and Gastrointestinal Event Trial (TARGET) (82), which comprised two 1-year long, comparator studies of lumiracoxib, 1 with ibuprofen, 1 with naproxen. While cardiovascular events tended to be higher in the lumiracoxib group, the study included patients mostly at low risk, and the power of the comparisons was undermined (83). Furthermore, TARGET was not designed to establish noninferiority of cardiovascular risk among the treatment groups; thus, it had no predefined upper confidence interval for relative risk (84) and used an intention-to-treat analysis. While it has been suggested that the pharmacokinetics of lumiracoxib favor a transient exposure in the vascular compartment with prolonged availability in the joint space (85), 400 mg/d of lumiracoxib exceeds considerably the dose necessary to inhibit COX-2 at the time of peak drug action. Given at this dose, it has a prolonged systemic pharmacodynamic half-life, depressing PGIM excretion to a similar extent and for a similar duration as rofecoxib (ref. 86 and P. Patrignani, personal communication). Unpublished studies also substantially underpowered to exclude a cardiovascular hazard include the prematurely and unconventionally terminated Alzheimer's Disease Anti-inflammatory Prevention (ADAPT) study of celecoxib, naproxen, and placebo in Alzheimer disease and the multinational, placebo-controlled evaluation of celecoxib (400 mg/d) in chemoprevention of colonic adenomata (http://www.clinicaltrials.gov/ct/show/NCT00087256).

In summary, while the number of cardiovascular events in *all* of the relevant individual RCTs addressing this issue is small, the currently available clinical evidence is remarkably compatible with a unitary mechanism for which there is comprehensive biological plausibility, attained in vivo. The clinically concordant evidence includes the following: (a) the easiest detection of a signal in epidemiological studies for a long-lived compound with a high degree of selectivity for COX-2, rofecoxib, given at a high dose (50 mg/d); (b) the rapid

emergence of a signal in 2 relatively small RCTs of valdecoxib in a setting of intense hemostatic activation and likely aspirin resistance; (c) the intermediate time to detection of a hazard in RA patients in the VIGOR study in whom hemostatic activation and risk of thrombosis is considerably less than in those individuals that have undergone CABG but exceeds that in patients without arthritis; (d) the similarity of the overview analyses of etoricoxib versus naproxen to what was observed in VIGOR (67) and evidence in trials to date (87) consistent with a cardiovascular hazard from this drug; (e) the delayed emergence of a hazard in 2 RCTs of prolonged treatment with rofecoxib and celecoxib, which is compatible with risk transformation in patients initially at low risk of cardiovascular disease; and (f) the evidence of hazard involving 3 structurally distinct selective COX-2 inhibitors — belying the notion that this is an off-target effect of rofecoxib. Finally, the issue of a mitigating effect of low-dose aspirin has not been addressed in the RCTs. This seems biologically plausible, as COX-1 knockdown in mice, which genetically mimics the impact of low-dose aspirin (88), attenuates the prothrombotic and hypertensive effect of COX-2 inhibition (Y. Cheng, personal communication). However, aspirin use was only prespecified in one of the RCTs in humans: TARGET. As mentioned, this was underpowered to address the cardiovascular question. However, the available evidence is compatible with risk attenuation; the relative risk of myocardial infarction was reduced from 2.37 in nonusers to 1.36 in those patients taking aspirin when lumiracoxib was compared with naproxen (83). However, this might also reflect a differential capacity of naproxen versus lumiracoxib to interact with and undermine the antiplatelet effect of low-dose aspirin (see below).

Cardiovascular risk and the tNSAIDs
Given the worldwide withdrawal of rofecoxib, the withdrawal of valdecoxib from the US, Australian, and European markets, and the substantial decline in the number of prescriptions for celecoxib, the safety of tNSAIDs has attracted substantial attention. Unfortunately, we do not have placebo-controlled RCTs addressing the cardiovascular safety of tNSAIDs, only observational studies, information from basic and human pharmacology, and the previously discussed tNSAID comparator RCTs. Aside from its relevance to clinical decision making, variability among the tNSAID comparators may be relevant to the heterogeneity of outcome among RCTs of COX-2–selective drugs.

A discussion of this information must be tempered by the reminder that all of these drugs, including those designed to be selective for COX-2, are NSAIDs and that the degree of selectivity among NSAIDs is best viewed as a continuous variable, given the substantial interindividual differences in response to drug administration (59). Thus, while valdecoxib is more selective for COX-2 than celecoxib in vitro, each individual will have idiosyncratic factors that modulate his or her dose-response relationship; there may well be some patients in whom celecoxib is the more selective inhibitor in vivo. It became fashionable after the VIGOR study to compare naproxen with "nonnaproxen" tNSAIDs. This approach should be abandoned, given the likely heterogeneity among the latter group with respect to cardiovascular risk.

Several tNSAIDs resemble, even in vitro, the selectivity profile of celecoxib (Figure 5A). These include diclofenac, the most commonly consumed tNSAID worldwide, and meloxicam, a marked beneficiary of the recent shift in NSAID prescriptions since the withdrawal of rofecoxib and valdecoxib in the US. Besides this pattern of selectivity, additional data substantiate the likelihood that diclofenac resembles

Table 1

Suggested consideration for preferred treatment options (level of evidence)[A]

COX inhibitors with proven cardioprotective efficacy	Low-dose aspirin (1a)
COX inhibitors with potential cardioprotective efficacy, variable among individuals	Naproxen (3a)
COX inhibitors with the potential to offset the cardioprotective effect of low-dose aspirin	Ibuprofen (3a)
	Flubiprofen (5)
	Indomethacin (5)
	Naproxen (5)
COX -2 inhibitors with proven gastroprotective efficacy	Rofecoxib (withdrawn) (1b)
	Lumiracoxib (FDA approval pending) (1b)
Treatment options for chronic treatment of patients with low cardiovascular and low GI risk	Naproxen (2b, 2a)
	Ibuprofen (2b, 2a)
Treatment options for chronic treatment of patients with low cardiovascular and high GI risk	Naproxen + proton pump inhibitor (2b, 2a)
	Ibuprofen + proton pump inhibitor (2b, 2a)
	Diclofenac + proton pump inhibitor (2b, 2a)
	Possibly celecoxib (although GI advantage vs. tNSAID not proven) (3, 2)
Treatment options for chronic treatment of patients with high cardiovascular and low GI risk	Naproxen + Clopidogrel (to avoid potential interaction with low-dose aspirin; however, the GI toxicity of this combination is likely to be at least that of tNSAID + low-dose aspirin and may warrant addition of a proton pump inhibitor) (5)
	Ibuprofen + clopidogrel (see comment above) (5)
Treatment options for chronic treatment of patients with high cardiovascular and high GI risk	Naproxen + proton pump inhibitor + clopidogrel (5)
	Ibuprofen + proton pump inhibitor + clopidogrel (5)

[A]The levels of evidence are based on the scoring system of the Oxford Centre for Evidence-Based Medicine (http://www.cebm.net/levels_of_evidence.asp): 1a, systematic reviews (with homogeneity) of RCTs; 1b, individual RCTs (with narrow confidence interval); 1c, all or none RCTs; 2a, systematic reviews (with homogeneity) of cohort studies; 2b, individual cohort study or low-quality RCTs; 2c, outcomes research, ecological studies; 3a, systematic review (with homogeneity) of case-control studies; 3b, individual case-control study; 4, case series (and poor quality cohort and case-control studies); 5, expert opinion based on physiology, bench research, or first principles.

celecoxib in practice. Neither cardiovascular nor GI outcomes differed between high-dose diclofenac and celecoxib in the CLASS study, and this pattern was sustained in the post hoc analysis in nonusers of aspirin. Secondly, NSAIDs that inhibit COX-1, such as ibuprofen, may interact pharmacodynamically to undermine the cardioprotective effects of low-dose aspirin (89). This interaction does not occur with drugs selective for COX-2, as it is not extant in platelets. Diclofenac, just like rofecoxib and celecoxib, but unlike ibuprofen or naproxen, does not subserve this interaction (89–91). Interestingly, a small epidemiological study of survivors of myocardial infarction suggested that concurrent ibuprofen but not diclofenac undermined the efficacy of aspirin in preventing a second myocardial infarction (92). It is unknown whether differences in the degree of selectivity attained on average by diclofenac versus purpose built COX-2 inhibitors will translate into differences in clinical outcomes. However, the Multinational Etoricoxib and Diclofenac Arthritis Long-term (MEDAL) study program (93, 94) consists of 3 randomized, double-blind trials in osteoarthritis and RA patients comparing etoricoxib to diclofenac (150 mg/d). Roughly 35,000 patients will be randomized, and the primary analysis is a noninferiority comparison of confirmed cardiovascular events, defined as an upper bound of the 95% confidence interval less than 1.30 (93, 94). The MEDAL program incorporates 2 smaller studies, the Etoricoxib versus Diclofenac Sodium Gastrointestinal Evaluation (EDGE) trials, designed primarily to evaluate GI outcomes respectively in patients with osteoarthritis (EDGE) and RA (EDGE 2). Etoricoxib was used at 90 mg/d in both EDGE studies but at 60 mg/d in the later stages of the MEDAL program. Interestingly, while GI and cardiovascular outcomes did not differ between the treatment groups in EDGE, discontinuations due to GI intolerance were more common with diclofenac while discontinuations due to hypertension were more common on etoricoxib (93, 94). Assimilation of the information from these comparator trials may begin to define the functional implications of progressive selectivity for COX-2 among the NSAIDs (Figure 5B). Less information is available concerning the other tNSAIDs with selectivity similar to diclofenac. Overview analyses suggest that serious GI toxicity is dose related with meloxicam, and at 15 mg/d this was similar to diclofenac. However, this impression was based largely on studies of fewer than 60 days duration (95), and there have been no adequately sized RCTs to address the GI toxicity of this compound (96). Similarly, we have no outcome data on cardiovascular events caused by meloxicam. Even less information is available with respect to nimesulide (97) and nabumetone (98).

A second group of tNSAIDs includes ibuprofen, flubiprofen, and indomethacin. These drugs favor somewhat inhibition of COX-1 over COX-2 in vitro and inhibit both enzymes reversibly during the dosing interval. Pharmacodynamic studies have raised the possibility that such drugs interact with and undermine the cardioprotective effect of aspirin (89). While observational studies support (92, 99, 100) and fail to support (101, 102) this hypothesis, the issue has not been addressed in an adequately powered RCT. Provocatively, concurrent high-dose ibuprofen appeared to undermine the benefit of aspirin in the TARGET trial (82); however, the number of events was too small to address the issue with confidence.

Most observational studies of tNSAID use last less than 1 year and most (102) but not all (103, 104) are consistent with no increased risk of cardiovascular events on ibuprofen. However, it is theoretically possible that a cardiovascular hazard from ibuprofen, albeit considerably less pronounced than for selective inhibitors of COX-2, may exist. First, this may derive from discordant rates of offset of inhibition of the 2 COX enzymes in vivo. This has never been documented, but should inhibition of COX-1 wane faster than that of COX-2, the drug would be selective for the latter enzyme for some portion of the dosing interval. A second possibility pertains even if the inhibition time profiles of the 2 enzymes are dynamically aligned during the dosing interval. We had previously shown a highly non-linear relationship in humans between the degree of inhibition of platelet COX-1 and platelet TxA_2–dependent function (105); one must inhibit the capacity of the enzyme by greater than 95% before impacting on platelet activation in vivo (Figure 6A). Studies in mice, which reveal that deletion of the IP has a gene dose–dependent effect on thrombosis and vascular function (Y. Cheng, personal communication), are consistent with a more linear relationship between inhibition of COX-2 and the functional consequences of suppressing COX-2–derived PGI_2 formation (Figure 6B). Discordance between these 2 curves might result in a "window of hazard," despite dynamic alignment of enzyme inhibition during the dosing interval (Figure 7A). Should such a window exist, it would be a smaller aperture than that for the risk from sustained selective inhibition of COX-2 throughout the dosing interval or, indeed, for the sustained benefit from inhibiting platelet COX-1 by low-dose aspirin (Figure 7, B and C). Thus, one would have to conduct much larger and/or longer trials to detect a hazard from ibuprofen than was necessary to detect a cardiovascular hazard from the coxibs or a benefit from low-dose aspirin. Interestingly, a recent epidemiological analysis of long-term use of NSAIDs suggests heterogeneity of effect consistent with their pharmacology (106). There is an apparent time-dependent emergence of a hazard with diclofenac; the data are also consistent with the possibility of a small hazard from ibuprofen only emerging upon prolonged dosing, if at all, while naproxen (see below) seems neutral or somewhat protective (Figure 8).

Naproxen has attracted particular attention because of the outcome of the VIGOR trial (62) and evidence of a prolonged pharmacokinetic half-life, at least in some individuals (64, 107). It is assumed that drugs that act like ibuprofen do not afford cardioprotection because they act reversibly and only sustain inhibition of platelet COX-1 in the functionally relevant zone transiently in the dosing interval (89). Epidemiological analyses of naproxen have suggested that it might have a dilute aspirin effect (103) consistent with an extended half-life and sustained platelet inhibition in some but not all individuals (64, 107). Indeed, this benefit of naproxen might be further undermined by irregular compliance outside the rigors of an RCT. We do not have evidence from cardiovascular outcome studies for naproxen and can only speculate as to how it may have contributed to the result reported in the VIGOR study. Recently, naproxen was shown to interact with aspirin in a manner similar to ibuprofen (91).

Aside from their putative effects on thrombosis, the potential of all NSAIDs, including those selective for COX-2, to raise blood pressure is well recognized. This may reflect diverse effects on salt and water handling and vascular reactivity, which have been discussed in detail elsewhere (108). The propensity of COX-2 deletion to elevate blood pressure is dependent on genetic background in mice (109), and it seems likely that genetic modifiers condition the existence and magnitude of this sporadic response to NSAID intake in humans. Despite the suggestion from experiments in mice (110) that hypertension might result most commonly from inhibition of COX-2 and the selectivity with which that is achieved, this has never been addressed directly by studies in humans, although an overview analysis consistent with that hypothesis has been published recently (111). Hypertension, reported as a serious adverse event, related to dose with rofecoxib and celecoxib and was more common with the more selective and longer-lived drug. Other factors, such as a documented COX-independent effect on vascular function (112) and drug potency, may explain perceived frequency of hypertension in patients taking indomethacin. Given COX inhibition by acetaminophen, reports of hypertension on this drug (113) are unsurprising. The commonest daily dose, 1000 mg, results in approximately 50% inhibition of both COX-1 and COX-2 (89), and an observational study suggests that higher doses, which may attain complete inhibition, result in a GI adverse event profile as in the tNSAIDs (114). It is unknown whether other aspects of acetaminophen action may modulate the impact of COX inhibition on cardiovascular function.

In summary, we lack information from placebo-controlled RCTs on the cardiovascular effects of the tNSAIDs. Thus, while a small but absolute risk of cardiovascular events is established for rofecoxib, valdecoxib, and celecoxib, we have no evidence of comparable quality for the tNSAIDs. Presently, it seems plausible to think of them in several clusters: (a) drugs such as diclofenac and meloxicam that are likely to resemble celecoxib with a small, but absolute risk; (b) drugs such as ibuprofen which, in themselves, may be neutral but may undermine the effectiveness of aspirin; (c) naproxen, which may afford protection in some individuals but which may also interact with aspirin; and finally, (d) a heterogeneous group of drugs such as indomethacin and acetaminophen, which may possess off-target cardiovascular effects that compound their profile. Clearly, the assumptions that underlie this classification can only be tested by RCTs. In the interim, however, it may afford a reasonable basis for therapeutic decision making (Table 1).

Some lessons learned and outstanding questions
An ambivalent legacy surrounds the aggressive strategy, heavily reliant upon direct-to-consumer marketing, that rendered the selective COX-2 inhibitors "blockbuster drugs." Ironically, the rationale for their development supported a niche concept — patients who had GI intolerance for tNSAIDs. After the drug withdrawals, it has been estimated that less than 5% of the patients previously taking coxibs had been at high risk of serious GI adverse effects from tNSAIDs (115).

An interdisciplinary approach to drug surveillance. The questions raised by mechanistic studies in humans performed before the first selective inhibitors of COX-2 were approved failed to prompt further studies to address the hypothetical mechanism of a cardiovascular hazard by the manufacturers. When such proof of concept did emerge, the data failed to inform substantially the interpretation of the pharmacoepidemiology. However, these different silos of information, mechanistic studies in humans, proof-of-principal studies in mice, and observational studies, finally afforded a powerful context within which to interpret the RCTs. In the future, we need to develop a more integrated, translational approach to information on drug safety, continuously refining our perception, perhaps exploiting formal Bayesian decision-making strategies as applied commonly in other fields. Post-marketing surveillance or pharmacovigilance might be strengthened considerably by the integration of large-scale databases from third-party payers, provisional periods of drug approval, and access to individual data from industry-sponsored clinical trials for independent analysis. However, such developments must be integrated within a surveillance system that prompts rapid performance of mechanistic studies to address hypotheses of concern, even when they emerge after drug approval.

An individualized approach to defining efficacy and risk. Even though the development strategy of the coxibs largely bypassed their comparative efficacy with tNSAIDs, it is often claimed that these drugs work uniquely in some patients; however, the evidence is strictly anecdotal. It is possible to design studies to determine if there is variability between individuals in the efficacy of NSAIDs and, if so, to provide an explanation. The pharmacological response to administration of distinct selective COX-2 inhibitors is strikingly variable among individuals, in part due to genetic sources of variance (59); factors such as body mass, age, and sex may also be of relevance. Similarly, we can easily reduce risk by avoiding these drugs in patients with a high-to-moderate risk of cardiovascular disease. It would also be judicious to exclude from therapy with a selective COX-2 inhibitor patients with recognized prothrombotic environmental exposure (e.g., anovulants) or genetic (e.g., factor V Leiden) variants (116). However, given the biological plausibility of the time course of the results of the APPROVe and APC studies, we need to address seriously the possibility that prolonged treatment with coxib-like drugs may predispose gradually to an emerging hazard in those previously at low risk of cardiovascular disease. Does extended therapy with COX-2 inhibitors result in accumulation of atherosclerotic plaque burden in humans? If so, does some combination of biomarkers of drug exposure, mechanism-based risk transformation, and atherogenesis combine with physiological responses (such as the rise in blood pressure) and genetic variants to predict the small number of the individuals treated who progress to clinical events? Such information would then enhance the design of RCTs so that they might provide information of value to individual therapeutic decisions in the future. Simplistic approaches to trial design (117) are likely to be inconclusive. TARGET was announced as a trial in patients at high cardiovascular risk that would provide a definitive answer to the cardiovascular question (118). In fact, too few patients at high risk were recruited for the trial to be powered to exclude a cardiovascular risk from lumiracoxib (82). However, given present evidence, performance of an RCT involving a selective COX-2 inhibitor in high-risk cardiovascular patients (117) is, at the least, ethically questionable.

A regulatory approach that synthesizes available information in a manner most pertinent to clinical decision making. Both the FDA and the European Medicines Agency (EMEA) (http://www.emea.eu.int/) concluded that rofecoxib, valdecoxib, and celecoxib conveyed a small but absolute hazard of myocardial infarction and stroke. Rofecoxib and valdecoxib have been withdrawn from the market in the US, Europe, and other jurisdictions. Both agencies agreed that more information was desirable concerning tNSAIDs but reacted in a distinct, but important way. The FDA applied a "black box" warning to celecoxib, which remained on the market, but also to the tNSAIDs (119). The EMEA, in contrast, imposed restrictions on celecoxib (and on etoricoxib, which is on the market in some European countries) but con-

cluded that there was no evidence to prompt a change in their advice about tNSAIDs (120). Adding a "black box" to the label of tNSAIDs is likely to mitigate the competitive damage to celecoxib and to diminish the hazard of litigation for all the relevant manufacturers by signaling risk to the consumer. However, an indiscriminate approach to warning about all remaining NSAIDs does not reflect the varied quality of the available evidence and is as practically valuable to patients and their doctors as the prior absence of explicit warning about the cardiovascular safety of any of these drugs.

Conclusions

Just as low-dose aspirin is effective in the secondary prevention of myocardial infarction and stroke and causes a small but definite risk of serious GI adverse effects (121), so selective inhibitors of COX-2 relieve pain and inflammation and convey a small but definite risk of myocardial infarction and stroke. While the preferential inhibition of COX-1 is *sufficient* to explain both the cardiovascular efficacy and adverse GI events observed with low-dose aspirin, so inhibition of COX-2 is *sufficient* to explain the antinflammatory efficacy and cardiovascular adverse events observed with the coxibs. Despite this, a plethora of additional actions of aspirin have been claimed to explain its action over the past 2 decades (122). The majority of these observations have been made in vitro and/or are of uncertain relevance to aspirin action at therapeutically tolerated doses in vivo. Similarly, a variety of alternative explanations for the cardiovascular effects of COX-2 inhibitors, again based largely on conjecture, in vitro data, and/or drug concentrations unlikely ever to be attained therapeutically have begun to emerge. Given the experience of aspirin, it might be wise to use the razor of William of Occam, the most celebrated proponent of the medieval principle of parsimony, to "shave off" unnecessary concepts, variables, and constructs. One should always choose the simplest explanation of a phenomenon, one that requires the fewest leaps of logic (123).

Acknowledgments

The authors are supported by grants from the NIH (MO 1RR00040, HL 54500, HL 62250, and HL70128) and the American Heart Association (National Scientist Development grant 0430148N to T. Grosser; Pennsylvania-Delaware Affiliate postdoctoral fellowship to S. Fries). Garret FitzGerald is the Elmer Bobst Professor of Pharmacology.

Note: References S1–S7 are available online with this article; doi:10.1172/JCI27291DS1.

Address correspondence to: G.A. FitzGerald, School of Medicine, Institute for Translational Medicine and Therapeutics, 153 Johnson Pavilion, University of Pennsylvania, Philadelphia, Pennsylvania 19104, USA. Phone: (215) 898-1184; Fax: (215) 573-9135; E-mail: garret@spirit.gcrc.upenn.edu.

1. Smyth, E., Burke, A., and FitzGerald, G.A. 2005. Lipid-derived autacoids. In *Goodman & Gilman's The pharmacological basis of therapeutics.* McGraw-Hill. New York, New York, USA. 653–670.
2. Funk, C.D. 2005. Leukotriene modifiers as potential therapeutics for cardiovascular disease. *Nat. Rev. Drug Discov.* 4:664–672.
3. Smith, W.L., DeWitt, D.L., and Garavito, R.M. 2002. Cyclooxygenases: structural, cellular, and molecular biology. *Annu. Rev. Biochem.* 69:145–182.
4. Narumiya, S., and FitzGerald, G.A. 2001. Genetic and pharmacological analysis of prostanoid receptor

function. *J. Clin. Invest.* 108:25–30. doi:10.1172/JCI200113455.
5. Burke, A., Smyth, E., and FitzGerald, G.A. 2005. Analgesic-anti-pyretic and anti-inflammatory agents and drugs employed in the treatment of gout. In *Goodman & Gilman's The pharmacological basis of therapeutics.* McGraw-Hill. New York, New York, USA. 673–715.
6. Goulet, J.L., et al. 2004. E-prostanoid-3 receptors mediate the proinflammatory actions of prostaglandin E$_2$ in acute cutaneous inflammation. *J. Immunol.* 173:1321–1326.

7. Murata, T., et al. 1997. Altered pain perception and inflammatory response in mice lacking prostacyclin receptor. *Nature.* 388:678–682.
8. Fajardo, M., and Di Caesare, P.E. 2005. Disease-modifying therapies for osteoarthritis: current status. *Drugs Aging.* 22:141–161.
9. Rosen, G.D., Birkenmeier, T.M., Raz, A., and Holzman, M.J. 1989. Identification of a cyclooxygenase-related gene and its potential role in prostaglandin formation. *Biochem. Biophys. Res. Commun.* 164:1358–1365.
10. Hla, T., and Bailey, J.M. 1989. Differential recovery

of prostacyclin synthesis in cultured vascular endothelial vs. smooth muscle cells after inactivation of cyclooxygenase with aspirin. *Prostaglandins Leukot. Essent. Fatty Acids.* **36**:175–184.

11. Sebalt, R.J., Sheller, J.R., Oates, J.A., Roberts, L.J., II, and FitzGerald, G.A. 1990. Inhibition of eicosanoid biosynthesis by glucocorticoids in humans. *Proc. Natl. Acad. Sci. U. S. A.* **87**:6974–6978.

12. Masferrer, J.L., Zweifel, B.S., Seibert, K., and Needleman, P. 1990. Selective regulation of cellular cyclooxygenase by dexamethasone and endotoxin in mice. *J. Clin. Invest.* **86**:1375–1379.

13. Kujubu, D., and Herschman, H. 1992. Dexamethasone inhibits mitogen induction of the TIS10 prostaglandin synthase/cyclooxygenase gene. *J. Biol. Chem.* **267**:7991–7994.

14. O'Banion, M.K., Winn, V.D., and Young, D. 1992. cDNA cloning and functional activity of a glucocorticoid-regulated inflammatory cyclooxygenase. *Proc. Natl. Acad. Sci. U. S. A.* **89**:4888–4892.

15. Sirois, J., Simmons, D.L., and Richards, J.S. 1992. Hormonal regulation of messenger ribonucleic acid encoding a novel isoform of prostaglandin endoperoxide H synthase in rat preovulatory follicles. Induction in vivo and in vitro. *J. Biol. Chem.* **267**:11586–11592.

16. Patrignani, P., et al. 1997. Differential inhibition of human prostaglandin endoperoxide synthase-1 and -2 by nonsteroidal anti-inflammatory drugs. *J. Physiol. Pharmacol.* **48**:623–631.

17. Warner, T.D., et al. 1999. Nonsteroid drug selectivities for cyclo-oxygenase-1 rather than cyclooxygenase-2 are associated with human gastrointestinal toxicity: a full in vitro analysis. *Proc. Natl. Acad. Sci. U. S. A.* **96**:7563–7568.

18. Picot, D., Loll, P.J., and Garavito, M. 1994. The X-ray crystal structure of the membrane protein prostaglandin H2 synthase-1. *Nature.* **367**:243–249.

19. Luong, C., et al. 1996. Flexibility of the NSAID binding site in the structure of human cyclooxygenase-2. *Nat. Struct. Biol.* **3**:927–933.

20. Kurumbail, R.G., et al. 1996. Structural basis for selective inhibition of cyclooxygenase-2 by anti-inflammatory agents. *Nature.* **384**:644–648.

21. Seibert, K., et al. 1995. Mediation of inflammation by cyclooxygenase-2. *Agents Actions.* **46**:41–50.

22. Devchand, P.L., and Wallace, J. 2005. Emerging roles for cyclooxygenase-2 in gastrointestinal mucosal defense. *Br. J. Pharmacol.* **145**:275–282.

23. Fries, J.F., et al. 2004. The rise and decline of nonsteroidal antiinflammatory drug-associated gastropathy in rheumatoid arthritis. *Arthritis Rheum.* **50**:2433–2440.

24. Simon, L.S., et al. 1999. Anti-inflammatory and upper gastrointestinal effects of celecoxib in rheumatoid arthritis: a randomized controlled trial. *JAMA.* **282**:1921–1928.

25. Lanza, F.L., et al. 1999. Specific inhibition of cyclooxygenase-2 with MK-0966 is associated with less gastroduodenal damage than either aspirin or ibuprofen. *Aliment. Pharmacol. Ther.* **13**:761–767.

26. Langman, M.J., et al. 1999. Adverse upper gastrointestinal effects of rofecoxib compared with NSAIDs. *JAMA.* **282**:1929–1933.

27. Sikes, D.H., et al. 2002. Incidence of gastroduodenal ulcers associated with valdecoxib compared with that of ibuprofen and diclofenac in patients with osteoarthritis. *Eur. J. Gastroenterol. Hepatol.* **14**:1101–1111.

28. FitzGerald, G.A., and Patrono, C. 2001. The coxibs, selective inhibitors of cyclooxygenase-2. *N. Engl. J. Med.* **345**:433–442.

29. McAdam, B.F., et al. 1999. Systemic biosynthesis of prostacyclin by cyclooxygenase (COX)-2: the human pharmacology of a selective inhibitor of COX-2. *Proc. Natl. Acad. Sci. U. S. A.* **96**:272–277.

30. Catella-Lawson, F., et al. 1999. Effects of specific inhibition of cyclooxygenase-2 on sodium balance, hemo-

dynamics, and vasoactive eicosanoids. *J. Pharmacol. Exp. Ther.* **289**:735–741.

31. Patrignani, P., et al. 1999. COX-2 is not involved in thromboxane biosynthesis by activated human platelets. *J. Physiol. Pharmacol.* **50**:661–667.

32. Creminon, C., et al. 1995. Differential measurement of constitutive (COX-1) and inducible (COX-2) cyclooxygenase expression in human umbilical vein endothelial cells using specific immunometric enzyme immunoassays. *Biochim. Biophys. Acta.* **1254**:341–348.

33. Moncada, S., Higgs, J., and Vane, J.R. 1977. Human arterial and venous tissues generate prostacyclin (prostaglandin x), a potent inhibitor of platelet aggregation. *Lancet.* **1**:18–20.

34. Topper, J.N., Cai, J., Falb, D., and Gimbrone, M.A., Jr. 1996. Identification of vascular endothelial genes differentially responsive to fluid mechanical stimuli: cyclooxygenase-2, manganese superoxide dismutase, and endothelial cell nitric oxide synthase are selectively up-regulated by steady laminar shear stress. *Proc. Natl. Acad. Sci. U. S. A.* **93**:10417–10422.

35. Cheng, Y., et al. 2002. Role of prostacyclin in the cardiovascular response to thromboxane A2. *Science.* **296**:539–541.

36. Catella, F., and FitzGerald, G.A. 1990. Measurement of thromboxane metabolites by gas chromatography-mass spectrometry. *Methods Enzymol.* **187**:42–50.

37. FitzGerald, G.A., Pedersen, A.K., and Patrono, C. 1983. Analysis of prostacyclin and thromboxane biosynthesis in cardiovascular disease. *Circulation.* **67**:1174–1177.

38. FitzGerald, G.A., Brash, A.R., Falardeau, P., and Oates, J.A. 1981. Estimated rate of prostacyclin secretion into the circulation of normal man. *J. Clin. Invest.* **68**:1272–1276.

39. Roy, L., Knapp, H., Robertson, R., and FitzGerald, G.A. 1985. Endogenous biosynthesis of prostacyclin during cardiac catheterization and angiography in man. *Circulation.* **71**:434–440.

40. Hla, T., and Nielsen, K. 1992. Human cyclooxygenase-2 cDNA. *Proc. Natl. Acad. Sci. U. S. A.* **89**:7384–7388.

41. Jones, D.A., Carlton, D.P., McIntyre, T.M., Zimmerman, G.A., and Prescott, S. 1993. Molecular cloning of human prostaglandin endoperoxide synthase type II and demonstration of expression in response to cytokines. *J. Biol. Chem.* **268**:9049–9054.

42. Schonebeck, U., Sukhova, G.K., Graber, P., Coulter, S., and Libby, P. 1999. Augmented expression of cyclooxygenase-2 in human atherosclerotic lesions. *Am. J. Pathol.* **155**:1281–1291.

43. Belton, O., Byrne, D., Kearney, D., Leahey, A., and Fitzgerald, D.J. 2000. Cyclooxygenase-1 and -2-dependent prostacyclin formation in patients with atherosclerosis. *Circulation.* **102**:840–845.

44. FitzGerald, G.A., Smith, B., Pedersen, A.K., and Brash, A.R. 1984. Increased prostacyclin biosynthesis in patients with severe atherosclerosis and platelet activation. *N. Engl. J. Med.* **310**:1065–1068.

45. Hennan, J.K., et al. 2001. Effects of selective cyclooxygenase-2 inhibition on vascular responses and thrombosis in canine coronary arteries. *Circulation.* **104**:820–825.

46. Pigeon, G.P., et al. 2004. Intravascular thrombosis after hypoxia-induced pulmonary hypertension: regulation by cyclooxygenase-2. *Circulation.* **110**:2701–2707.

47. Buerkle, M.A., et al. 2004. Selective inhibition of cyclooxygenase-2 enhances platelet adhesion in hamster arterioles in vivo. *Circulation.* **110**:2053–2059.

48. Rabausch, K., et al. 2005. Regulation of thrombomodulin expression in human vascular smooth muscle cells by COX-2-derived prostaglandins. *Circ. Res.* **96**:e1–e6.

49. Xiao, C.Y., et al. 2001. Roles of prostaglandin I2 and thromboxane A2 in cardiac ischemia-reperfusion

injury: a study using mice lacking their respective receptors. *Circulation.* **104**:2210–2215.

50. Dowd, N.P., Scully, M., Adderley, S.R., Cunningham, A.J., and Fitzgerald, D.J. 2001. Inhibition of cyclooxygenase-2 aggravates doxorubicin-mediated cardiac injury in vivo. *J. Clin. Invest.* **108**:585–590. doi:10.1172/JCI200111334.

51. Kobayashi, T., et al. 2004. Roles of thromboxane A2 and prostacyclin in the development of atherosclerosis in apoE-deficient mice. *J. Clin. Invest.* **114**:784–794. doi:10.1172/JCI200421446.

52. Egan, K., et al. 2004. COX-2-derived prostacyclin confers atheroprotection on female mice. *Science.* **306**:1954–1957.

53. Watanabe, H., et al. 2005. Effects of salt loading on blood pressure in mice lacking the prostanoid receptor gene. *Circ. J.* **69**:124–126.

54. Francois, H., et al. 2005. Hypertension in prostacyclin receptor deficient mice. *Cell. Metab.* **2**:201–207.

55. Rudic, R.D., et al. 2005. COX-2-derived prostacyclin modulates vascular remodeling. *Circ. Res.* **96**:1240–1247.

56. Cheng, M.K., McGiff, J.C., and Carroll, M.A. 2003. Renal arterial 20-hydroxyeicosatetraenoic acid levels: regulation by cyclooxygenase. *Am. J. Physiol. Renal Physiol.* **284**:F474–F479.

57. Kennedy, C.R., et al. 1999. Salt-sensitive hypertension and reduced fertility in mice lacking the prostaglandin EP2 receptor. *Nat. Med.* **5**:217–220.

58. FitzGerald, G.A. 2004. Coxibs and cardiovascular disease. *N. Engl. J. Med.* **351**:1709–1711.

59. Fries, S., Grosser, T., Lawson, J., DeMarco, S., and FitzGerald, G.A. 2005. Marked interindividual variability in the response to selective inhibitors of cyclooxygenase-2. *Gastroenterology.* Online publication ahead of print.

60. Brune, K., and Hinz, B. 2004. Selective cyclooxygenase-2 inhibitors: similarities and differences. *Scand. J. Rheumatol.* **33**:1–6.

61. McAdam, B., et al. 2000. Effect of regulated expression of human cyclooxygenase isoforms on eicosanoid and isoeicosanoid production in inflammation. *J. Clin. Invest.* **105**:1473–1482.

62. Bombardier, C., et al. 2000. Comparison of upper gastrointestinal toxicity of rofecoxib and naproxen in patients with rheumatoid arthritis. VIGOR Study Group. *N. Engl. J. Med.* **343**:1520–1528.

63. Watson, D.J., Rhoades, T., and Guess, H.A. 2003. All-cause mortality and vascular events among patients with rheumatoid arthritis, osteoarthritis, or no arthritis in the UK General Practice Research Database. *J. Rheumatol.* **30**:1196–1202.

64. VanHecken, A., et al. 2000. Comparative inhibitory activity of rofecoxib, meloxicam, diclofenac, ibuprofen and naproxen on COX-2 versus COX-1 in healthy volunteers. *J. Clin. Pharmacol.* **40**:1109–1120.

65. Silverstein, F.E., et al. 2000. Gastrointestinal toxicity with celecoxib vs nonsteroidal anti-inflammatory drugs for osteoarthritis and rheumatoid arthritis: the CLASS Study: a randomized controlled trial. *JAMA.* **284**:1247–1255.

66. Juni, P., Rutjes, M.A., and Dieppe, P.A. 2002. Are selective COX 2 inhibitors superior to traditional non steroidal anti-inflammatory drugs? *BMJ.* **324**:1287–1288.

67. FitzGerald, G.A. 2003. COX-2 and beyond: Approaches to prostaglandin inhibition in human disease. *Nat. Rev. Drug Discov.* **2**:879–890.

68. Mukherjee, D., Nissen, S.E., and Topol, E.J. 2001. Risk of cardiovascular events associated with selective COX-2 inhibitors. *JAMA.* **286**:954–959.

69. Ray, W.A., et al. 2002. COX-2 selective non-steroidal anti-inflammatory drugs and risk of serious coronary heart disease. *Lancet.* **360**:1071–1073.

70. Solomon, D.H., et al. 2004. Relationship between selective cyclooxygenase-2 inhibitors and acute myocardial infarction in older adults. *Circulation.* **109**:2068–2073.

71. Solomon, D.H., Schneeweiss, S., Levin, R., and Avorn, J. 2004. Relationship between COX-2 specific inhibitors and hypertension. *Hypertension.* **44**:140–145.

72. Bresalier, R.S., et al. 2005. Cardiovascular events associated with rofecoxib in a colorectal adenoma chemoprevention trial. *N. Engl. J. Med.* **352**:1092–1102.

73. Kimmel, S., et al. 2005. Patients exposed to rofecoxib and celecoxib have different odds of nonfatal myocardial infarction. *Ann. Intern. Med.* **142**:157–164.

74. White, W.B., Strand, V., Roberts, R., and Whelton, A. 2004. Effects of the cyclooxygenase-2 specific inhibitor valdecoxib versus nonsteroidal antiinflammatory agents and placebo on cardiovascular thrombotic events in patients with arthritis. *Am. J. Ther.* **11**:244–250.

75. Anonymous. 2005. Petition to remove the COX-2 inhibitors celecoxib (CELEBREX) and valdecoxib (BEXTRA) from the market. *Public Citizen.* http://www.citizen.org/publications/release.cfm?ID=7358.

76. Cannata, A., Biglioli, P., Tremoli, E., and Parolari, A. 2004. Biological effects of coronary surgery: role of surgical trauma and CPB [letter]. *Eur. J. Cardiothorac. Surg.* **26**:664.

77. Ott, E., et al. 2003. Efficacy and safety of the cyclooxygenase-2 inhibitors parecoxib and valdecoxib in patients undergoing coronary artery bypass surgery. *J. Thorac. Cardiovasc. Surg.* **125**:1481–1492.

78. Nussmeier, N.A., et al. 2005. Complications of the COX-2 inhibitors parecoxib and valdecoxib after cardiac surgery. *N. Engl. J. Med.* **352**:1081–1091.

79. Furberg, C., Psaty, B., and FitzGerald, G.A. 2005. Parecoxib, valdecoxib, and cardiovascular risk [editorial]. *Circulation.* **111**:249.

80. Zimmermann, N., et al. 2003. Functional and biochemical evaluation of platelet aspirin resistance after coronary artery bypass surgery. *Circulation.* **108**:542–547.

81. Solomon, S.D., et al. 2005. Cardiovascular risk associated with celecoxib in a clinical trial for colorectal adenoma prevention. *N. Engl. J. Med.* **352**:1071–1080.

82. Schnitzer, T., et al. 2004. Comparison of lumiracoxib with naproxen and ibuprofen in the Therapeutic Arthritis Research and Gastrointestinal Event Trial (TARGET), reduction in ulcer complications: randomised controlled trial. *Lancet.* **364**:665–674.

83. Topol, E.J., and Falk, G.W. 2004. A coxib a day won't keep the doctor away. *Lancet.* **364**:639–640.

84. Jones, B., Jarvis, P., Lewis, J.A., and Ebbutt, A.F. 1996. Trials to assess equivalence: the importance of rigorous methods. *BMJ.* **313**:36–39.

85. Scott, G., et al. 2004. Pharmacokinetics of lumiracoxib in plasma and synovial fluid. *Clin. Pharmacokinet.* **43**:467–478.

86. Tacconelli, S., Capone, M.L., and Patrignani, P. 2004. Clinical pharmacology of novel selective COX-2 inhibitors. *Curr. Pharm. Des.* **10**:589–601.

87. Aldington, S., Shirtcliffe, P., Weatherall, M., and Beasley, R. 2005. Systematic review and meta-analysis of the risk of major cardiovascular events with etoricoxib therapy. *N. Z. Med. J.* **118**:U1684.

88. Yu, Y., et al. 2005. Differential impact of prostaglandin H synthase 1 knockdown on platelets and parturition. *J. Clin. Invest.* **115**:986–995. doi:10.1172/JCI200523683.

89. Catella-Lawson, F., et al. 2001. Cyclooxygenase inhibitors and the antiplatelet effects of aspirin. *N. Engl. J. Med.* **345**:1809–1817.

90. Wilner, K.D., et al. 2002. Celecoxib does not affect the antiplatelet activity of aspirin in healthy volunteers. *J. Clin. Pharmacol.* **42**:1027–1030.

91. Capone, M.L., et al. 2005. Pharmacodynamic interaction of naproxen with low-dose aspirin in healthy subjects. *J. Am. Coll. Cardiol.* **45**:1295–1301.

92. MacDonald, T.M., and Wei, L. 2003. Effect of ibuprofen on cardioprotective effect of aspirin. *Lancet.* **361**:573–574.

93. Cannon, C., Curtis, S., Bolognese, J.A., and Laine, L. 2005. Cardiovascular outcomes with a COX-2 selective inhibitor vs. a traditional NSAID in patients with osteoarthritis and rheumatoid arthritis: clinical trial design and patient demographics of the Multinational Etoricoxib and Diclofenac Arthritis Long-term (MEDAL) Study Program. *Am. J. Cardiol.* In press.

94. US Food and Drug Administration. 2005. Briefing package for NDA 21-389 Etoricoxib. http://www.fda.gov/ohrms/dockets/ac/05/briefing/2005-4090B1_31_AA-FDA-Tab-T.pdf.

95. Singh, G., Lanes, S., and Triadafilopoulos, G. 2004. Risk of serious upper gastrointestinal and cardiovascular thromboembolic complications with meloxicam. *Am. J. Med.* **117**:100–106.

96. Mickelwright, R., et al. 2003. Review article: NSAIDs, gastroprotection and cyclo-oxygenase-II-selective inhibitors. *Aliment. Pharmacol. Ther.* **17**:321–332.

97. Panara, M.R., et al. 1998. Effects of nimesulide on constitutive and inducible prostanoid biosynthesis in human beings. *Clin. Pharmacol. Ther.* **63**:672–681.

98. Hedner, T., et al. 2004. Nabumetone: therapeutic use and safety profile in the management of osteoarthritis and rheumatoid arthritis. *Drugs.* **64**:2315–2343.

99. Kurth, T., et al. 2003. Inhibition of clinical benefits of aspirin on first myocardial infarction by nonsteroidal antiinflammatory drugs. *Circulation.* **108**:1191–1195.

100. Hudson, M., Baron, M., Rahme, E., and Pilote, L. 2005. Ibuprofen may abrogate the benefits of aspirin when used for secondary prevention of myocardial infarction. *J. Rheumatol.* **32**:1589–1593.

101. Curtis, J.P., and Krumholz, H. 2004. The case for an adverse interaction between aspirin and non-steroidal anti-inflammatory drugs: is it time to believe the hype? *J. Am. Coll. Cardiol.* **43**:991–993.

102. Garcia Rodriguez, L., et al. 2004. Nonsteroidal antiinflammatory drugs and the risk of myocardial infarction in the general population. *Circulation.* **109**:3000–3006.

103. Graham, D.J., et al. 2005. Risk of acute myocardial infarction and sudden cardiac death in patients treated with cyclo-oxygenase 2 selective and non-selective non-steroidal anti-inflammatory drugs: nested case-control study. *Lancet.* **365**:475–481.

104. Hippisley-Cox, J., and Coupland, C. 2005. Risk of myocardial infarction in patients taking cyclo-oxygenase-2 inhibitors or conventional non-steroidal anti-inflammatory drugs: population based nested case-control analysis. *BMJ.* **330**:1366–1372.

105. Reilly, I.A., and FitzGerald, G.A. 1987. Inhibition of thromboxane formation in vivo and ex vivo: implications for therapy with platelet inhibitory drugs. *Blood.* **69**:180–186.

106. Garcia Rodriguez, L.A., and Gonzalez-Perez, A. 2005. Long-term use of traditional non-steroidal anti-inflammatory drugs and the risk of myocardial infarction in the general population. *BMC Med.* **3**:17.

107. Capone, M.L., et al. 2004. Clinical pharmacology of platelet, monocyte, and vascular cyclooxygenase inhibition by naproxen and low-dose aspirin in healthy subjects. *Circulation.* **109**:1468–1471.

108. Fries, S., Grosser, T., and FitzGerald, G.A. 2005. Cyclooxygenase inhibition and blood pressure regulation. In *Hypertension.* E. Mohler, editor. BC Decker. Hamilton, Ontario, Canada. In press.

109. Yang, T., et al. 2004. Influence of genetic background and gender on hypertension and renal failure in COX-2-deficient mice. *Am. J. Physiol. Renal Physiol.* **288**:F1125–F1132.

110. Qi, Z., et al. 2002. Opposite effects of cyclooxygenase-1 and -2 activity on the pressor response to angiotensin II. *J. Clin. Invest.* **110**:61–69. doi:10.1172/JCI200214752.

111. Aw, T.J., Haas, S.J., Liew, D., and Krum, H. 2005. Meta-analysis of cyclooxygenase-2 inhibitors and their effects on blood pressure. *Arch. Intern. Med.* **165**:490–496.

112. Elund, A. et al. 1985. Coronary flow regulation in patients with ischemic heart disease: release of purines and prostacyclin and the effect of inhibitors of prostaglandin formation. *Circulation.* **71**:1113–1120.

113. Forman, J.P., Stampfer, M.J., and Curhan, G.C. 2005. Non-narcotic analgesic dose and risk of incident hypertension in US women. *Hypertension.* **46**:500–507.

114. García Rodríguez, L.A., and Hernández-Díaz, S. 2001. Relative risk of upper gastrointestinal complications among users of acetaminophen and nonsteroidal anti-inflammatory drugs. *Epidemiology.* **12**:570–576.

115. Dai, C., Stafford, R.S., and Alexander, G.C. 2005. National trends in cyclooxygenase-2 inhibitor use since market release: nonselective diffusion of a selectively cost-effective innovation. *Arch. Intern. Med.* **165**:171–177.

116. Westgate, E., and FitzGerald, G.A. 2005. Pulmonary embolism in a woman taking oral contraceptives and valdecoxib. *PLoS Med.* **2**:e197.

117. Herman, M., Krum, H., and Ruschitzka, F. 2005. To the heart of the matter: coxibs, smoking, and cardiovascular risk. *Circulation.* **112**:941–945.

118. Hawkey, C.J., et al. 2004. Therapeutic arthritis research and gastrointestinal event trial of lumiracoxib - study design and patient demographics. *Aliment. Pharmacol. Ther.* **20**:51–63.

119. Anonymous. 2004. Public health advisory: non-steriodal anti-inflammatory drug products (NSAIDS). US Food and Drug Administration. http://www.fda.gov/cder/drug/advisory/nsaids.htm.

120. Anonymous. 2005. EMEA press release. European Medicines Agency. http://www.emea.eu.int/pdfs/human/press/pr/24732305en.pdf.

121. Patrono, C., García Rodríguez, L.A., Landolfi, R., and Baigent, C. 2005. Low dose aspirin for the prevention of atherothrombosis. *N. Engl. J. Med.* **353**:49–59.

122. Tegeder, I., Pfeilschifter, J., and Geisslinger, G. 2001. Cyclooxygenase-independent actions of cyclooxygenase inhibitors. *FASEB J.* **15**:2057–2072.

123. Beckett, D. 1994. Biography. http://wotug.kent.ac.uk/parallel/www/occam/occam-bio.html.

Sex is a potent modifier
of the cardiovascular system

Leslie A. Leinwand

Department of Molecular, Cellular, and Developmental Biology, University of Colorado, Boulder, Colorado, USA.

Despite the growing number of reports in the literature identifying sex-related differences in cardiac function in both rodents and humans, the underlying mechanisms have yet to be determined. Here, variables of experimental studies such as diet, animal model utilized, and age, in addition to sex hormones and other factors that may play a role in sex-related variations in cardiac responses to various pathophysiological conditions are discussed, suggesting that current approaches used in the study of cardiac disease require reevaluation.

There are numerous health problems that are affected by gender. Women are more susceptible than men to depression, osteoporosis, asthma, lung cancer due to smoking, and autoimmune disease (1). Gender effects in disease are complex, however, as exemplified by the observation that while the incidence of melanoma is slightly higher among women than men, mortality from melanoma is higher in men. Not all medical problems show gender dimorphism. For example, males do not differ from females in terms of their responses to infection (1). When it comes to heart disease, generally, of those individuals diagnosed with heart disease, women fare much better than men. Little is known about the basis for this difference in cardiovascular disease. Much focus has been placed on the potential cardioprotective role of estrogen. However, the recent finding that estrogen replacement in postmenopausal women actually increased heart disease has challenged this view (2).

Deriving a coherent view of potential players in sex-dependent differences in the heart requires analysis of both the clinical literature and the literature on sex-dependent differences in the cardiovascular system of laboratory animals. There is a major limiting factor in comparing rodent laboratory studies: the vast majority have been carried out on males only. However, this limitation appears to be changing. In the past year, a number of studies have appeared in which both males and females have been analyzed. An article in a recent issue of the *JCI* by O'Connell et al. (3) provided a glimpse into what promises to be a new way of thinking about genetic sex. In this report, the authors found that the α_1-adrenergic receptors are critical in determining heart size and the ability of the heart to respond to both pathologic and physiologic stimuli, but only in male animals. Intriguingly, this sex-related difference did not disappear following ovariectomy of females. This article will discuss the fact that sex is an extremely potent modifier of the myocardium and will identify which pathways have been implicated in some of these differences. I hope it will become clear that future research should focus on the mechanisms by which both sex and diet can modify cardiovascular phenotypes.

Nonstandard abbreviations used: hypertrophic cardiomyopathy (HCM); alcohol dehydrogenase (ADH); myosin heavy chain (α-MyHC); estrogen receptor (ER).

Conflict of interest: The author has declared that no conflict of interest exists.

Citation for this article: *J. Clin. Invest.* **112**:302–307 (2003). doi:10.1172/JCI200319429.

Sex-related differences in normal male and female hearts in human cardiac disease

What are the sexually dimorphic cardiovascular differences to which I refer? Before puberty, there are no statistically significant sex-related differences in heart size (4). After puberty, absolute heart mass is greater in men than in women by about 15–30% (5). Both males and females start out with the same number of cardiac myocytes and the size of their myocytes is initially the same. Therefore, it appears that male myocytes undergo a greater degree of hypertrophy during normal postnatal growth than their female counterparts. In terms of normal function, echocardiography has demonstrated better diastolic function in young women compared to age-matched men, and with aging, men have a decrease in systolic function that is not observed in women (6).

In the most common forms of cardiovascular disease, premenopausal women show a much better prognosis than men. Women show supranormal contractility with increased wall thickness compared to men with similar degrees of aortic stenosis, hypertension, aging, and hypertrophic cardiomyopathy (HCM). Men have poor contractility, chamber dilation, and wall thinning (7–14). Is the more favorable prognosis for females due to the cardioprotective properties of estrogen? It is probably not that simple. I say this because of the now famous study that showed that hormone replacement therapy increased cardiovascular events in postmenopausal women (2, 15). Further, it is not always the case that women fare better than men in cardiovascular disease. In cases of idiopathic dilated cardiomyopathy, females have a significantly poorer prognosis than males (16). Women are also more sensitive to alcohol-induced cardiac disease (17). In addition, the sex-related differences reverse with age when examining mortality due to cardiovascular disease. More males (39%) than females die from heart disease between the ages of 45 and 64. However, after age 65 the death rate due to heart disease in women exceeds that in men by 22% (18). The combination of these sex-related differences observed in the progression of cardiovascular disease and the increase in mortality in postmenopausal women at the very least deserves more research. Currently, due to the limitations of human studies, the cellular and molecular mechanisms of these differences have yet to be characterized. Therefore, the questions remain: what is the basis for the better prognosis for premenopausal females in heart failure, aortic stenosis, hypertension, and HCM; and, what is the mechanism whereby postmenopausal women have increased mortality compared to their male counterparts?

Rodent studies on sex-related differences in the cardiovascular system

What can we learn from animal studies about sex-related differences in heart disease? The literature regarding studies in animals falls into two general categories: (i) studies that deliberately address sex-related differences in the cardiovascular system; and

Table 1

Summary of differences observed in cardiovascular function between males and females

Cardiovascular component or characteristic	Model system	Observed effect		Reference
		Males	Females	
Aortic constriction	Rat	↓ Systolic function ↑ Diastolic pressure	↑ Concentric remodeling	(51)
Hypertension	Rat	Left ventricular chamber ↑ ↓ Fractional shortening	Left ventricular chamber significantly ↑	(52) (53)
Angiotensin-converting enzyme	Mouse	↑ Relative to females Orchiectomy ↓ ACE	↓ Relative to males Ovariectomy ↑ ACE	(54)
α-MyHC	Human	↑ α-MyHC in left ventricle	↓ α-MyHC in left ventricle, relative to males	(55)
K⁺ current	Mouse	↑ Relative to females ↓ Kv1.5 channel	↓ Relative to males ↑ Action potential	(56)
Repolarization reserves	Mouse	↑ Relative to females	↓ Relative to males	(57)

(ii) studies in which the inactivation or manipulation of genes has resulted in the discovery of a sexually dimorphic phenotype. Most of the studies in the first category have involved subjecting wild-type rodents of both sexes to some pathologic or physiologic stimulus and observing how the heart responds (Table 1). These studies have generally been consistent with observations in clinical populations. Studies in the second category are beginning to provide the groundwork for understanding the basis for sex-related differences noted in the first category (Table 2). However, when we examine the animal model literature, there are two important things to note: Firstly, the majority of the literature is devoted to the study of male animals. Most of the basis for this (although not explicitly stated) is that female animals would need to be studied only at a fixed point in the estrous cycle and this would be impractical. A systematic analysis is warranted in order to analyze the car-

diac responsiveness of male and female mice to a variety of stimuli and investigate how diet might modify cardiovascular phenotypes (see Phytoestrogens and heart disease).

On the surface however, rodent studies do not always appear to be consistent. It is important to ask whether the inconsistencies are telling us something meaningful. If the experimental rodent literature is examined closely, some of the inconsistencies can be explained because the studies themselves are not comparable. For example, different results have been obtained in comparing contractile performance in male and female rats. One study in papillary muscles from 6-month old rats found that males had significantly slower rates of contraction and relaxation than females (19). A similar study using the isolated working heart preparation revealed increased cardiac output in males (20). A third study using younger papillary muscles found

Table 2

Summary of male- and female-related differences in genetically manipulated mice

Cardiovascular component or characteristic	Observed effect		Reference
	Males	Females	
α₁ₐ and α₁ᵦ-adrenergic receptor null	Small heart; ↓ exercise performance	↓ Aortic restriction response; heart normal following ovariectomy	(3)
Constitutively active Akt	Heart size ↑ compared to wild type	Heart size ↑ compared to wild type; heart size ↑ compared to transgenic males	(44)
Mutant α-MyHC	Cardiac chamber dilation; heart failure	Hypertrophy; preserved function	(58)
Mutant cardiac troponin T (R92Q)	Heart size ↓ compared to wild type	No change in heart size when compared to wild type	(59) (L. Leinwand, unpublished observations)
Superinhibition of phospholamban	Progress to dilated cardiomyopathy at 6 months of age	Hypertrophy; normal cardiac function	(60)
Phospholamban overexpression	Hypertrophy; mortality: 15 months of age	Delayed hypertrophy; mortality: 22 months of age despite similar dysfunction in males	(61)
ADH overexpression	No phenotype	Myocytes more sensitive to ethanol; depression of contractions	(62)
TNF-α overexpression	Cardiac chamber dilation; Ca²⁺ handling is low	Hypertrophy; normal Ca²⁺	(63)
Lipoprotein lipase I overexpression/PPAR-null	Die at 4 months of age	Live	(64)
Relaxin null	Cardiomyopathy	No altered phenotype	(65)
β₂-adrenergic receptor overexpression	↑ Contractility; ↑ ischemia reperfusion injury	↑ Contractility; normal response to injury	(66)

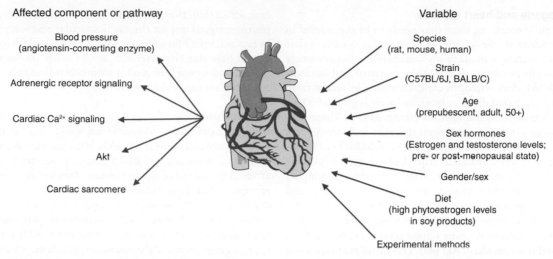

Figure 1
Pathways and factors affecting the cardiovascular system. The figure indicates how numerous variables, along with factors known to be important in the cardiovascular system, can converge.

no difference between the sexes until animals were 6 months of age (21). These alleged inconsistencies suggest that it is not possible to compare apples and oranges. It seems clear that the following parameters need to be assessed when trying to compare one study to another: (i) what are the ages of the animals; (ii) what species and what strain were used in the studies; and (iii) how, in the case of function, were the measurements made? For example, is it possible to freely compare cardiac function as assessed by echocardiography with that assessed in an isolated working heart preparation?

In the case of genetically manipulated rodents, most studies have been done in mice, narrowing one of these variables — species. Strain is still an issue, but most investigators now backcross their strains onto the C57B1/6 background. The age of the animals is still a major issue for several reasons. Phenotypes of genetically manipulated mice have been shown to vary over time (as mentioned above with wild-type rats). One striking example describes mice overexpressing the β_2-adrenergic receptor (β_2AR). At 2–4 months of age, male β_2AR mice exhibited enhanced cardiac contractility. Between 8 and 12 months of age the mice developed severe fibrosis and contractile dysfunction (22). A second, and perhaps more important, issue in correlating rodent and human studies, has to do with extrapolating the time frame of rodent studies to humans. Many of the sex-related differences seen in humans do not emerge until well into adulthood, which in a mouse is roughly equivalent to at least 10–12 months of age. When reviewing the literature, it becomes apparent that many studies have been conducted on rats and mice of 4–6 months of age, so conclusions of a lack of sex-related differences in cardiac phenotypes may be premature. That said, now that more studies are including animals of both sexes, it is clear that sex plays a major role in cardiovascular function in rodents. Figure 1 shows a number of pathways implicated in these sex-related differences. As is obvious from the summary of recent studies that have documented sex-related differences in cardiovascular function of genetically manipulated animals (Table 2), in general, females have blunted cardiac responses compared to their male counterparts (or no response in some cases) following manipulation of a

number of genes. They also have blunted responses to the inactivation of certain genes. The cardiovascular phenotype observed in females is also blunted or normal in cases of overexpression/activation or expression of a mutation. An augmented response in females has been reported in the case of constitutive activation of Akt, and a worse outcome for females was noted in mice overexpressing alcohol dehydrogenase (ADH).

Sex hormones
Although the role of estrogen in providing cardioprotection is no longer so clear, this is not to say that sex hormones do not have an effect on the cardiovascular system. Functional estrogen receptors (ERs) (α and β) have been demonstrated in ventricular myocardium of both males and females (23–25). Both receptors have a similar affinity for 17β-estradiol. Estrogen binding has genomic effects as ERs are ligand-activated transcription factors and can activate transcription of a number of genes whose promoter regions contain tandem estrogen response elements (AGGTCANNNTGACCT)(26). Estrogen also has nongenomic effects. Estrogen has been shown to rapidly decrease calcium current through the L-type Ca^{2+} channel in isolated cardiac myocytes (27). The effect of estrogen on Ca^{2+} could be an important regulatory mechanism in cardiac muscle given the central role of intracellular Ca^{2+} in sarcomere function. Androgen receptors have also been demonstrated in the myocardium of multiple species, including mice (28). In addition, testosterone has been observed to transcriptionally regulate cardiac gene expression (29, 30). Gonadectomy is a well-established method to evaluate sex hormone effects on end-organ physiology in animals (31–35) and has been used to evaluate changes in cardiac myosin heavy chain (MyHC) isoform expression (3, 33–35). In gonadectomized rats subjected to swimming or renovascular hypertension, gender-appropriate hormone replacement has been shown to provide a beneficial effect on myosin isoform expression, which is considered a marker for pathologic versus physiologic hypertrophy (32). In the recent article by O'Connell et al. (3) as described above, a very interesting sex-specific effect was seen in mice null for both the α_{1A} and α_{1B}-adrenergic receptors.

Phytoestrogens and heart disease

Another recently emerging issue that needs to be considered in evaluating rodent studies of the cardiovascular system is that standard rodent chow is made of soy and therefore has extremely high levels of the potent phytoestrogens genistein and daidzein (see refs. 36–38). Phytoestrogen concentrations in serum from rodents fed a soy-based diet can be exceedingly high (30,000- to 60,000-fold higher than endogenous estrogen) (36). These phytoestrogens can bind to the estrogen receptor and have potent effects (39). For example, genistein has a tenfold higher affinity for ERβ and can transcriptionally stimulate estrogen responsive genes (40, 41). For example, male mice fed genistein have a higher degree of DNA methylation than control mice (39). Genistein and other phytoestrogens have been shown to exert potent physiologic effects (39, 42). Genistein, in fact, is routinely used in the laboratory setting to inhibit tyrosine kinase receptors and has been shown to inhibit activation of Akt (43). This latter property could be very important in the setting of the heart since Akt activation has been shown to cause significant cardiac hypertrophy (44). Soy-based diets can have an antihypertensive effect on spontaneously hypertensive rats (45). Male rats on a casein versus soy diet have been shown to exhibit less exercise-induced MyHC degradation in skeletal muscle (46). Diets containing phytoestrogen also appear to play a cardioprotective role in ischemia-reperfusion injury in female rats (47). Many human food sources such as soy and soy products have a high phytoestrogen content, so there may also be relevance to the human condition. Soy milk can lower blood pressure in men and women (48) and soy protein supplementation can reduce hypertension in perimenopausal women (49). The Food and Drug Administration recently approved the claim that foods with more than 6.25 grams of soy protein per serving can reduce the risk of heart disease because of its hypocholesterol-emic effect (50). How does a diet with supraphysiologic levels of phytoestrogens impact the cardiovascular phenotypes reported in rats and mice? Since most people in the world have diets that do not have this characteristic, are we really able to extrapolate rodent findings in humans? It seems clear that this is an area of research that needs to be pursued.

Conclusions

Being male or female is a variable that should be dealt with in both basic science and clinical research. It is clear that the response of humans and animals to various disease states can be profoundly affected by the sex of the individual. Further, it is important to recognize that these differences are unlikely to be due solely to sex hormones. For example, some genes on the X chromosome are expressed at higher levels in females than in males, despite the process of X chromosome inactivation. Additionally, males express genes on the Y chromosome, which are clearly not present in females. Given the different outcomes exhibited by males and females in disease, it should be possible to determine which pathways contribute to these differences. There are now many candidate pathways to probe such as calmodulin-dependent protein kinase, Akt, glycogen synthase kinase 3β, and myocyte enhancer factor 2 signaling. The other major issue that needs to be addressed is the evaluation of the results of studies conducted on animals fed a soy diet. I predict that diet will be a strong confounding factor in many of these studies and have an effect on sexually dimorphic cardiovascular phenotypes.

Address correspondence to: Leslie A. Leinwand, Department of Molecular, Cellular, and Developmental Biology, 347 UCB, University of Colorado, Boulder, Colorado 80309, USA. Phone: (303) 492-7606; Fax: (303) 492-8907; E-mail: Leslie.Leinwand@colorado.edu.

1. Committee on Understanding the Biology of Sex and Gender Differences. 2001. *Exploring the biological contributions to human health: does sex matter?* The National Academies Press. Washington, D.C., USA. 288 pp.
2. Rossouw, J.E., et al. 2002. Risks and benefits of estrogen plus progestin in healthy postmenopausal women: principal results from the Women's Health Initiative randomized control trial. *JAMA.* **288**:321–333.
3. O'Connell, T.D., et al. 2003. The $\alpha_{1A/C}$- and α_{1B}-adrenergic receptors are required for physiological cardiac hypertrophy in the double-knockout mouse. *J. Clin. Invest.* **111**:1783–1791. doi:10.1172/JCI200316100.
4. Malcolm, D.D., Burns, T.L., Mahoney, L.T., and Lauer, R.M. 1993. Factors affecting left ventricular mass in childhood: the Muscatine Study. *Pediatrics.* **92**:703–709.
5. de Simone, G., et al. 1995. Gender differences in left ventricular growth. *Hypertension.* **26**:979–983.
6. Grandi, A.M., et al. 1992. Influence of age and sex on left ventricular anatomy and function in normals. *Cardiology.* **81**:8–13.
7. Aronow, W.S., Ahn, C., and Kronzon, I. 1999. Comparison of incidences of congestive heart failure in older African-Americans, Hispanics, and whites. *Am. J. Cardiol.* **84**:611–612.
8. De Maria, R., et al. 1993. Comparison of clinical findings in idiopathic dilated cardiomyopathy in women versus men. The Italian Multicenter Cardiomyopathy Study Group (SPIC). *Am. J. Cardiol.* **72**:580–585.
9. Adams, K.F., Jr., et al. 1999. Gender differences in survival in advanced heart failure. Insights from the FIRST study. *Circulation.* **99**:1816–1821.
10. Douglas, P.S., et al. 1995. Gender differences in left ventricle geometry and function in patients undergoing balloon dilatation of the aortic valve for isolated aortic stenosis. NHLBI Balloon Valvuloplasty Registry. *Br. Heart J.* **73**:548–554.
11. Aurigemma, G.P., Silver, K.H., McLaughlin, M., Mauser, J., and Gaasch, W.H. 1994. Impact of chamber geometry and gender on left ventricular systolic function in patients > 60 years of age with aortic stenosis. *Am. J. Cardiol.* **74**:794–798.
12. Aurigemma, G.P., and Gaasch, W.H. 1995. Gender differences in older patients with pressure-overload hypertrophy of the left ventricle. *Cardiology.* **86**:310–317.
13. Kimmelstiel, C.D., and Konstam, M.A. 1995. Heart failure in women. *Cardiology.* **86**:304–309.
14. Legget, M.E., et al. 1996. Gender differences in left ventricular function at rest and with exercise in asymptomatic aortic stenosis. *Am. Heart J.* **131**:94–100.
15. Grady, D., et al. 2000. Postmenopausal hormone therapy increases risk for venous thromboembolic disease. The Heart and Estrogen/Progestin Replacement Study. *Ann. Intern. Med.* **132**:689–696.
16. Mohan, S.B., Parker, M., Wehbi, M., and Douglass, P. 2002. Idiopathic dilated cardiomyopathy: a common but mystifying cause of heart failure. *Cleve. Clin. J. Med.* **69**:481–487.
17. Fernandez-Sola, J., and Nicolas-Arfelis, J.M. 2002. Gender differences in alcoholic cardiomyopathy. *J. Gend. Specif. Med.* **5**:41–47.
18. National Center for Health Statistics. 1997. Statistical abstract of the United States. 117th edition. US Bureau of the Census. Washington, DC, USA. 83 pp. http://www.census.gov/prod/www/statistical-abstract-us.html.
19. Capasso, J.M., Remily, R.M., Smith, R.H., and Sonnenblick, E.H. 1983. Sex differences in myocardial contractility in the rat. *Basic Res. Cardiol.* **78**:156–171.
20. Schaible, T.F., and Scheuer, J. 1984. Comparison of heart function in male and female rats. *Basic Res. Cardiol.* **79**:402–412.
21. Leblanc, N., Chartier, D., Gosselin, H., and Rouleau, J.L. 1998. Age and gender differences in excitation-contraction coupling of the rat ventricle. *J. Physiol.* **511**:533–548.
22. Freeman, K., et al. 2001. Alterations in cardiac adrenergic signaling and calcium cycling differentially affect the progression of cardiomyopathy. *J. Clin. Invest.* **107**:967–974.
23. Pelzer, T., Shamim, A., Wolfges, S., Schumann, M., and Neyses, L. 1997. Modulation of cardiac hypertrophy by estrogens. *Adv. Exp. Med. Biol.* **432**:83–89.
24. Grohe, C., et al. 1994. Functional estrogen receptors in myocardial and myogenic cells. *Circulation.* **90**:I538. (Abstr.)
25. Grohe, C., et al. 1996. Modulation of hypertensive heart disease by estrogen. *Steroids.* **61**:201–204.
26. Klein-Hitpass, L., Schorpp, M., Wagner, U., and Ryffel, G.U. 1986. An estrogen-responsive element derived from the 5′ flanking region of the Xenopus vitellogenin A2 gene functions in transfected human cells. *Cell.* **46**:1053–1061.
27. Grohe, C., et al. 1997. Cardiac myocytes and fibroblasts contain functional estrogen receptors. *FEBS Lett.* **416**:107–112.
28. Marsh, J.D., et al. 1998. Androgen receptors mediate

hypertrophy in cardiac myocytes. *Circulation.* **98**:256–261.

29. Koenig, H., Goldstone, A., and Lu, C.Y. 1982. Testosterone-mediated sexual dimorphism of the rodent heart. Ventricular lysosomes, mitochondria, and cell growth are modulated by androgens. *Circ. Res.* **50**:782–787.

30. Morano, I., et al. 1990. Regulation of myosin heavy chain expression in the hearts of hypertensive rats by testosterone. *Circ. Res.* **66**:1585–1590.

31. Malhotra, A., Buttrick, P., and Scheuer, J. 1990. Effects of sex hormones on development of physiological and pathological cardiac hypertrophy in male and female rats. *Am. J. Physiol.* **259**:H866–H871.

32. Morris, G.S., et al. 1998. Ovariectomy fails to modify the cardiac myosin isoenzyme profile of adult rats. *Horm. Metab. Res.* **30**:84–87.

33. Scheuer, J., Malhotra, A., Schaible, T.F., and Capasso, J. 1987. Effects of gonadectomy and hormonal replacement on rat hearts. *Circ. Res.* **61**:12–19.

34. Schaible, T.F., Malhotra, A., Ciambrone, G., and Scheuer, J. 1984. The effects of gonadectomy on left ventricular function and cardiac contractile proteins in male and female rats. *Circ. Res.* **54**:38–49.

35. Sharkey, L.C., et al. 1998. Effect of ovariectomy in heart failure-prone SHHF/Mcc-facp rats. *Am. J. Physiol.* **275**:R1968–R1976.

36. Thigpen, J.E., et al. 1999. Phytoestrogen content of purified, open- and closed-formula laboratory animal diets. *Lab Anim. Sci.* **49**:530–536.

37. Degen, G.H., Janning, P., Diel, P., and Bolt, H.M. 2002. Estrogenic isoflavones in rodent diets. *Toxicol. Lett.* **128**:145–157.

38. Boettger-Tong, H., et al. 1998. A case of a laboratory animal feed with high estrogenic activity and its impact on in vivo responses to exogenously administered estrogens. *Environ. Health Perspect.* **106**:369–373.

39. Day, J.K., et al. 2002. Genistein alters methylation patterns in mice. *J. Nutr.* **132**:2419S–2423S.

40. Barkhem, T., et al. 1998. Differential response of estrogen receptor alpha and estrogen receptor beta to partial estrogen agonists/antagonists. *Mol. Pharmacol.* **54**:105–112.

41. Kuiper, G.G., et al. 1998. Interaction of estrogenic chemicals and phytoestrogens with estrogen receptor beta. *Endocrinology.* **139**:4252–4263.

42. Chinni, S.R., Alhasan, S.A., Multani, A.S., Pathak, S., and Sarkar, F.H. 2003. Pleotropic effects of genistein on MCF-7 breast cancer cells. *Int. J. Mol. Med.* **12**:29–34.

43. Matsui, T., et al. 2002. Phenotypic spectrum caused by transgenic overexpression of activated Akt in the heart. *J. Biol. Chem.* **277**:22896–22901.

44. Shioi, T., et al. 2002. Akt/protein kinase B promotes organ growth in transgenic mice. *Mol. Cell Biol.* **22**:2799–2809.

45. Martin, D.S., Breitkopf, N.P., Eyster, K.M., and Williams, J.L. 2001. Dietary soy exerts an antihypertensive effect in spontaneously hypertensive female rats. *Am. J. Physiol. Regul. Integr. Comp. Physiol.* **281**:R553–R560.

46. Nikawa, T., et al. 2002. Effects of a soy protein diet on exercise-induced muscle protein catabolism in rats. *Nutrition.* **18**:490–495.

47. Zhai, P., et al. 2001. Effects of dietary phytoestrogen on global myocardial ischemia-reperfusion injury in isolated female rat hearts. *Am. J. Physiol. Heart Circ. Physiol.* **281**:H1223–H1232.

48. Rivas, M., et al. 2002. Soy milk lowers blood pressure in men and women with mild to moderate essential hypertension. *J. Nutr.* **132**:1900–1902.

49. Washburn, S., Burke, G.L., Morgan, T., and Anthony, M. 1999. Effect of soy protein supplementation on serum lipoproteins, blood pressure, and menopausal symptoms in perimenopausal women. *Menopause.* **6**:7–13.

50. Food and Drug Administration Talk Paper. October 20, 1999. http://www.fda.gov/bbs/topics/answers/ans00980.html.

51. Weinberg, E.O., et al. 1999. Gender differences in molecular remodeling in pressure overload hypertrophy. *J. Am. Coll. Cardiol.* **34**:264–273.

52. Tamura, T., Said, S., and Gerdes, A.M. 1999. Gender-related differences in myocyte remodeling in progression to heart failure. *Hypertension.* **33**:676–680.

53. Wallen, W.J., Cserti, C., Belanger, M.P., and Wittnich, C. 2000. Gender-differences in myocardial adaptation to afterload in normotensive and hypertensive rats. *Hypertension.* **36**:774–779.

54. Freshour, J.R., Chase, S.E., and Vikstrom, K.L. 2002. Gender differences in cardiac ACE expression are normalized in androgen-deprived male mice. *Am. J. Physiol. Heart Circ. Physiol.* **283**:H1997–H2003.

55. Miyata, S., Minobe, W., Bristow, M.R., and Leinwand, L.A. 2000. Myosin heavy chain isoform expression in the failing and nonfailing human heart. *Circ. Res.* **86**:386–390.

56. Trepanier-Boulay, V., St-Michel, C., Tremblay, A., and Fiset, C. 2001. Gender-based differences in cardiac repolarization in mouse ventricle. *Circ. Res.* **89**:437–444.

57. Wu, Y., and Anderson, M.E. 2002. Reduced repolarization reserve in ventricular myocytes from female mice. *Cardiovasc. Res.* **53**:763–769.

58. Freeman, K., et al. 2001. Progression from hypertrophic to dilated cardiomyopathy in mice that express a mutant myosin transgene. *Am. J. Physiol. Heart Circ. Physiol.* **280**:151–159.

59. Tardiff, J.C., et al. 1999. Cardiac troponin T mutations result in allele-specific phenotypes in a mouse model for hypertrophic cardiomyopathy. *J. Clin. Invest.* **104**:469–481.

60. Haghighi, K., et al. 2001. Superinhibition of sarcoplasmic reticulum function by phospholamban induces cardiac contractile failure. *J. Biol. Chem.* **276**:24145–24152.

61. Dash, R., et al. 2003. Differential regulation of p38 mitogen-activated protein kinase mediates gender-dependent catecholamine-induced hypertrophy. *Cardiovasc. Res.* **57**:704–714.

62. Duan, J., et al. 2003. Influence of gender on ethanol-induced ventricular myocyte contractile depression in transgenic mice with cardiac overexpression of alcohol dehydrogenase. *Comp. Biochem. Physiol. Mol. Integr. Physiol.* **134**:607–614.

63. Janczewski, A.M., et al. 2003. Morphological and functional changes in cardiac myocytes isolated from mice overexpressing TNF-α. *Am. J. Physiol. Heart Circ. Physiol.* **284**:H960–H969.

64. Nohammer, C., et al. 2003. Myocardial dysfunction and male mortality in peroxisome proliferator-activated receptor alpha knockout mice overexpressing lipoprotein lipase in muscle. *Lab Invest.* **83**:259–269.

65. Du, X.J., et al. 2003. Increased myocardial collagen and ventricular diastolic dysfunction in relaxin deficient mice: a gender-specific phenotype. *Cardiovasc. Res.* **57**:395–404.

66. Cross, H.R., Murphy, E., Koch, W.J., and Steenbergen, C. 2002. Male and female mice overexpressing the beta(2)-adrenergic receptor exhibit differences in ischemia/reperfusion injury: role of nitric oxide. *Cardiovasc. Res.* **53**:662–671.

Cholesterol in health and disease

Ira Tabas

Department of Medicine, and Department of Anatomy and Cell Biology, Columbia University, New York, New York, USA.

Imagine if an excess amount of a critical, life-sustaining molecule like ATP were, by a perverse series of events involving the lifestyle of modern humans, causally related to a major human disease. The thought of "ATP" being synonymous with poor health and poor living in the minds of the lay public and press, even of health care providers, would be difficult for any self-respecting scientist to accept. So one might view the biomedical history of cholesterol — indeed, this history might be seen as even stranger than the hypothetical ATP scenario, given that evolution has devoted close to 100 genes to the synthesis, transport, metabolism, and regulation of cholesterol. This structurally fascinating lipid is utterly essential to the proper functioning of cells and organisms. Cholesterol, cholesterol metabolites, and immediate biosynthetic precursors of cholesterol play essential roles in cellular membrane physiology, dietary nutrient absorption, reproductive biology, stress responses, salt and water balance, and calcium metabolism. Indeed, many of the articles in this Perspective series are devoted to the normal physiology of cholesterol. Still, there is little doubt that the disease process responsible for the leading cause of death in the industrialized world — atherosclerosis — is a disorder in which an excess of cholesterol is a major culprit. How did evolution come up with a molecule that is critical for so many aspects of normal physiology, and what went "wrong"?

The physiology of cholesterol

Cholesterol and cellular membranes: an evolutionary perspective. As organisms became more complex, cells required membranes that would provide the proper conformational environment for a wide variety of integral membrane proteins, such as channels, transporters, and enzymes. Moreover, the cells of these advanced organisms required sophisticated signaling machinery, and this machinery would have to be organized as multiprotein complexes in focal, nonhomogenous areas of cellular membranes. Put simply, these requirements were met by focally increasing the "stiffness" or viscosity of phospholipid bilayers. In theory, this goal could be achieved by increasing the degree of saturation of the fatty acyl moieties of membrane phospholipids. However, unsaturated fatty acids in these phospholipids, particularly in the *sn*-2 position, are needed for a wide variety of cellular signaling functions, and so an exogenous "stiffening factor" was needed. This factor would have to be able to pack tightly with the long saturated and unsaturated fatty acyl chains of membrane phospholipids through van der Waals interactions. This would require a long, flat, and properly shaped hydrophobic molecule, accompanied by additional features (see below) to help further stabilize this interaction.

To meet these needs, nature began with molecules, probably originating from prebiotic anaerobic times, that were formed by the sequential condensation of the two-carbon acetate molecule. Prokaryotes evolved enzymes to synthesize more complex linear molecules from these precursors (e.g., carotenoids), which were able to meet the membrane-organizing requirements of most of these primitive organisms. These primitive molecules, however, lacked the proper shape, rigidity, and amphipathic nature to properly organize the membranes of more advanced organisms and, interestingly, two species of bacteria (see below) (1). At some point in the evolution of the earliest eukaryotes or their nearest Archaeal relatives, the largest of these linear molecules was cyclized to form steroids, planar structures containing four rings (2). This shape change conferred rigidity and altered the molecule's ability to pack within the fatty acyl tails of neighboring phospholipids. Additional features were added that helped stabilize these lateral interactions, including the introduction of a C-C double bond (3) and of an –OH group, which converted the steroid to a sterol; the precise role of the latter modification is controversial (1).

These molecular transformations correspond to the initial conversion of acetate into acyclic, apolar squalene, followed evolutionarily by the oxygenation and cyclization of squalene to lanosterol (Figure 1). Lanosterol is then converted to cholesterol by oxidative demethylation. As referred to above, two species of bacteria, *Methylococcus capsulatus* and *Mycoplasma* species, require sterols for growth; while *Mycoplasma* obtains these sterols from the cells of higher organisms, *M. capsulatus* actually synthesizes sterols (1). In both cases, however, the structural requirements of these sterols are more lax than those described above for eukaryotes (1). Interestingly, the cyclization of squalene and the conversion of lanosterol to cholesterol require molecular oxygen and could therefore occur only after the evolution of aerobic cells. However, in a fascinating "evolutionary preview" of some of these features, acidophilic bacteria evolved enzymes to anaerobically synthesize cyclic membrane-organizing lipids called hopanoids from squalene (Figure 1). Haines has speculated that hopanoids fill a unique role in acidophile membrane structure, namely, the prevention of inward leakage of protons (4).

The structural requirements outlined above demand precision. For example, a molecule almost identical to cholesterol but with a 3α-hydroxyl group instead of a 3β-hydroxyl group ("epicholesterol") cannot function properly in biological membranes (5). Moreover, the conversion of lanosterol to cholesterol involves a complex series of 18 enzymatic reactions, even though lanosterol is already a cyclic 3β alcohol. In fact, lanosterol and other methylated derivatives cannot substitute for cholesterol in mammalian cell mutants that are auxotrophic for sterols (6). Why? The three methyl groups that are removed from lanosterol to form cholesterol are all on the so-called α-face of the molecule and thus protrude from this otherwise flat surface (Figure 2a). This is particularly true of the axial 14α-methyl group of lanosterol. Bloch and others have suggested that the removal of these three methyl groups allows proper fitting of the ring structure to the fatty acyl chains of membrane phospholipids, particularly those that are saturated (1, 7) (Figure 2b). According to this model, cholesterol optimally binds natural membrane phospholipids, which usually contain a saturated fatty acid at *sn*-1 and an

Conflict of interest: No conflict of interest has been declared.

Citation for this article: *J. Clin. Invest.* **110**:583–590 (2002). doi:10.1172/JCI200216381.

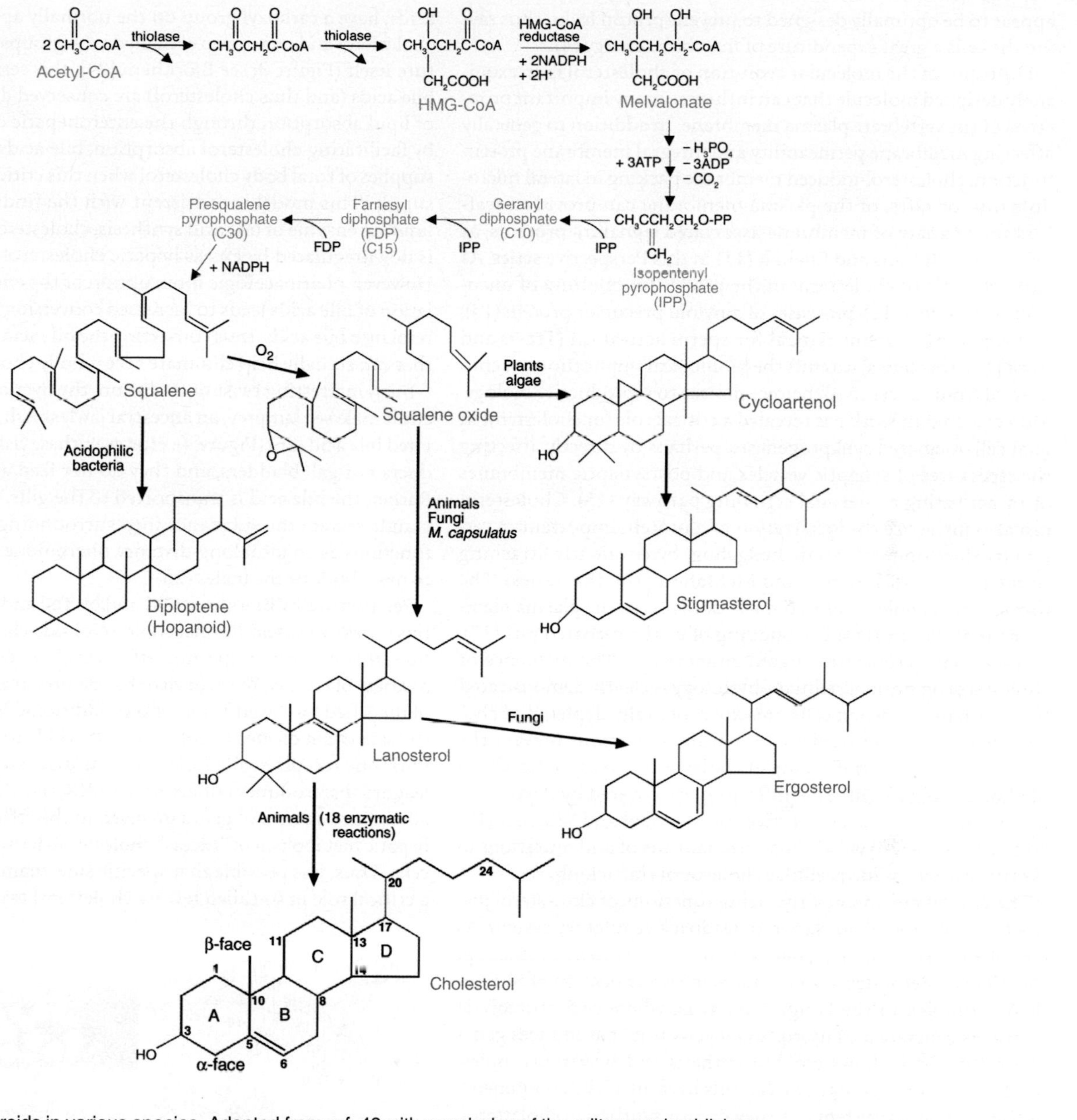

Figure 1
Biosynthesis of steroids in various species. Adapted from ref. 43 with permission of the editors and publisher.

unsaturated fatty acid at *sn*-2. In particular, its flat α-face is ideally suited to interact with the *sn*-1 saturated fatty acyl group, whereas the methylated β-face is optimal for interaction with the more flexible unsaturated fatty acyl chains in the *sn*-2 position of an adjacent phospholipid molecule. The α-face of cholesterol, but not lanosterol, has the additional feature of a C-C double bond, which further enhances interaction with neighboring phospholipids (3). In this context, Mouritsen and colleagues (8) have used deuterium nuclear magnetic resonance spectroscopy on model membranes to demonstrate the advantage of cholesterol over lanosterol in the formation of liquid-ordered membrane domains, or "rafts" (see below).

Finally, the side chain of cholesterol is saturated and relatively unsubstituted with extra methyl groups, resulting in a hydrophobic structure that is usually linear and aligned with the plane of

the steroid nucleus (Figure 2a). This design allows optimal van der Waals interactions with phospholipid fatty acyl groups while avoiding disruption of the bilayer (1, 7). Interestingly, plants evolved a different series of enzymatic reactions in which squalene is converted to cycloartenol instead of lanosterol, followed by further modifications that result in the synthesis of phytosterols (Figure 1). Phytosterols are characterized by double bonds and additional methyl and/or ethyl substitutions in the side chain. These bulky side chains, which are often aligned out of the plane of the steroid nucleus, alter critical properties of the membranes of higher animal cells (9). Indeed, higher animals have evolved specific, energy-requiring molecules to excrete such sterols following intestinal absorption (10). Plant cell plasma membranes, however, encounter a much higher proton gradient than most animal cell plasma membranes, and the bulky side chains of phytosterols appear to be optimally designed to prevent proton leaks, thus saving the cells a great expenditure of metabolic energy (4).

The result of the molecular evolution of cholesterol is an exquisitely designed molecule that can influence many important properties of the vertebrate plasma membrane. In addition to generally affecting membrane permeability and integral membrane protein function, cholesterol-induced membrane packing in lateral microdomains, or rafts, of the plasma membrane can provide a scaffold for a variety of membrane-associated signaling proteins, as reviewed by Simons and Ehehalt (11) in this Perspective series. As such, the role of cholesterol in the proper functioning of oncogenic G proteins (12), proteases of amyloid precursor protein (13), and signaling proteins critical for sperm activation (Travis and Kopf [14], this series) extends the biomedical implications of cholesterol from cancer to Alzheimer disease to reproductive biology. Moreover, recent work has revealed a critical role for cholesterol in glial cell–mediated synaptogenesis, perhaps by directly affecting the structures of synaptic vesicles and postsynaptic membranes or by activating a neuronal signaling pathway (15). Cholesterol can also influence the localization of a protein important in vertebrate development, Sonic hedgehog, by covalently attaching to the protein itself (Jeong and McMahon [16], this series). The formation of cholesterol-rich microdomains in the plasma membrane is also critical for the budding of clathrin-coated pits (17), a key step in receptor-mediated endocytosis. The influence of cholesterol on normal cellular physiology is clearly demonstrated by what happens when cells are experimentally depleted of cholesterol or when, by accidents of nature, mutations prevent the proper synthesis or trafficking of cholesterol. The specific effects of cholesterol biosynthetic mutations are covered by Porter (18) in this series, and the Perspectives by Maxfield and Wüstner (19) and by Jefcoate (20) will discuss mechanisms of and mutations in specific aspects of intracellular cholesterol trafficking.

The critical membrane-organizing functions of cholesterol justify the evolution of an elaborate feedback regulatory system. As described by Horton, Goldstein, and Brown in a recent Spotlight in the *JCI* (21), sterol starvation induces the translocation of SREBP-SCAP complex to the Golgi apparatus, where two proteolytic cleavages generate a transcriptional activator that induces genes controlling both cholesterol biosynthesis and exogenous cholesterol uptake. Interestingly, insect cells have all of the components of this regulatory system, yet they do not synthesize cholesterol, and their ability to process SREBP is not affected by exogenous sterols. Rather, the system is used in phosphatidylethanolamine-mediated feedback regulation of the biosynthesis of fatty acids and phospholipids, which critically affect membrane structure in these cells (22). Because phatidylethanolamine and cholesterol share the ability to alter membrane structure and thus to alter SCAP conformation and SREBP processing (22), vertebrate evolution appears to have usurped a pre-existing system that was devoted to membrane lipid homeostasis even before cholesterol emerged as the key molecular organizer of cell membranes.

Functions of cholesterol metabolites and immediate biosynthetic precursors of cholesterol. As if the membrane effects of cholesterol were not enough, nature has used the cholesterol backbone as a precursor for a variety of biologically active molecules (Figures 3 and 4). In one pathway, at least 14 different liver enzymes add polarity to the ring structure and side chain of cholesterol to create ideal micellar solubilization agents for the absorption of dietary fats and fat-soluble vitamins (23). Specifically, these solubilizing agents, or bile

acids, have a carboxyl group on the normally apolar side chain of cholesterol and one to two extra hydroxyl groups on the ring structure itself (Figure 3; see Björkhem [24], this series). Importantly, bile acids (and thus cholesterol) are conserved during the process of lipid absorption through the enterohepatic circulation. Thus, by facilitating cholesterol absorption, bile acids ensure adequate supplies of total body cholesterol when this critical lipid is in short supply. This model is consistent with the finding that the rate-limiting enzyme of bile acid synthesis, cholesterol 7α-hydroxylase, is downregulated by excess hepatic cholesterol in primates (25). However, pharmacologic interruption of the enterohepatic circulation of bile acids leads to increased conversion of hepatic cholesterol into bile acids, thus converting the bile acid pathway into one that can actually help eliminate excess body cholesterol (23).

In an interesting twist of evolution, the liver of the spermiating adult male sea lamprey, an ancestral jawless fish, synthesizes a sulfated bile acid (26) (Figure 4). However, these fish have neither bile ducts nor gall bladders, and they do not feed while spermiating. Rather, the bile acid is transported to the gills, where specialized glands secrete the substance into surrounding waters. There, it functions as an ideal long-distance pheromone to attract females to nests built by the males (26).

Perspectives by Björkhem (24) and by Tall and colleagues (27) in this series discuss additional cholesterol side-chain oxidation reactions that may have important effects in cholesterol metabolism in a variety of tissues. Work in vitro has demonstrated that cholesterol modified by the addition of one additional hydroxyl group on the side chain creates a molecule that is able to activate a nuclear hormone receptor called LXR (28). As discussed by Tall and colleagues, heterodimers of activated LXR and RXR transcriptionally induce a series of genes involved in the efflux, transport, and hepatic metabolism of "excess" cholesterol derived from peripheral cells. Thus, it is possible that specific side-chain hydroxylases play a critical role in so-called reverse cholesterol transport. Oxidation

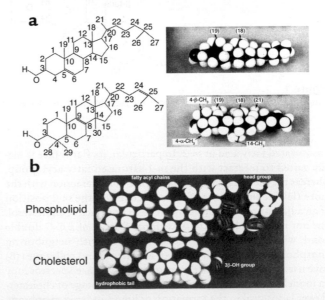

Figure 2
(**a**) Structural and space-filling models of cholesterol and lanosterol. Reproduced from ref. 8 with permission of the author and journal. (**b**) Space-filling model of cholesterol interacting with a phospholipid molecule. Reproduced from ref. 7 with permission of the author and journal.

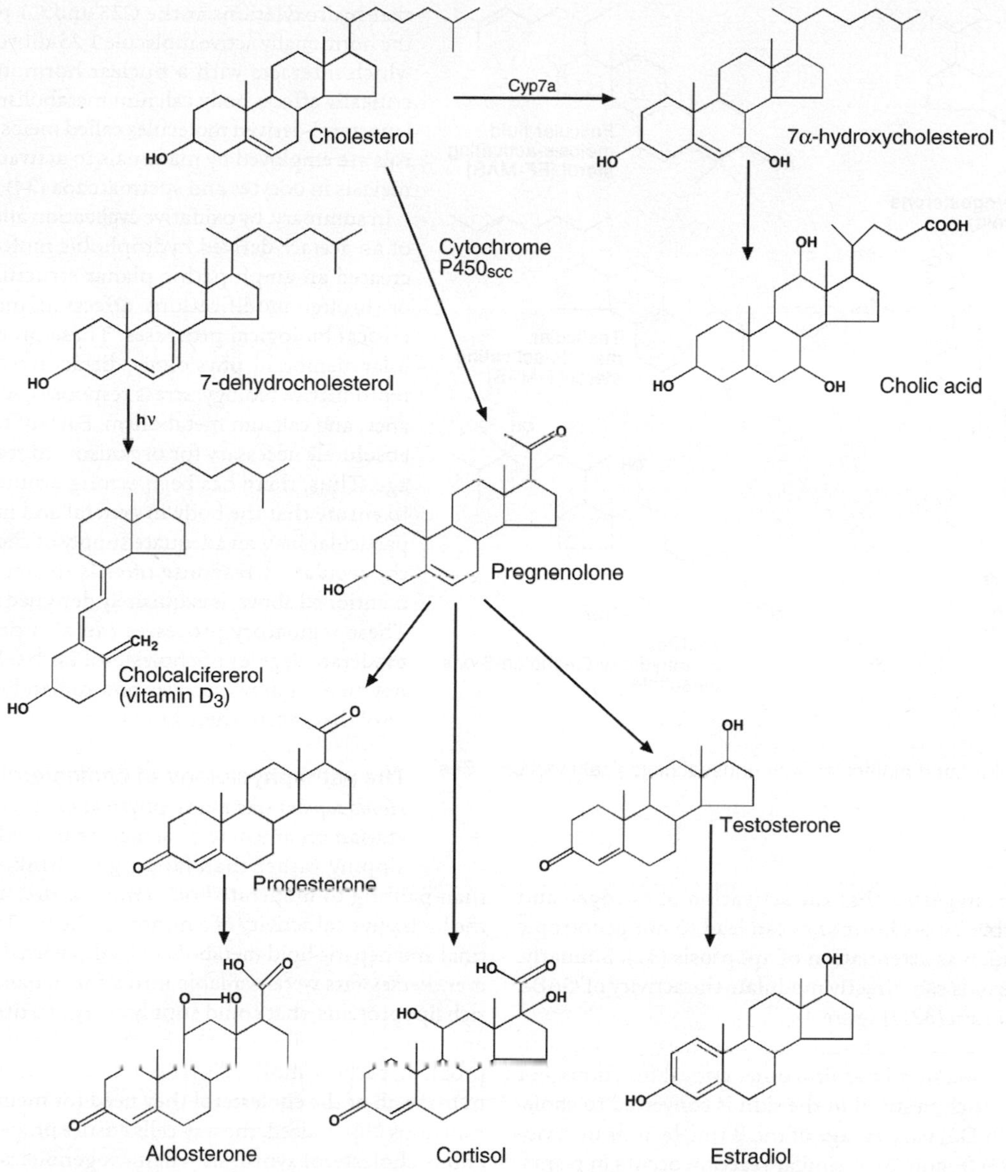

Figure 3
Biosynthesis of cholesterol-derived hormones. Adapted from ref. 43 with permission of the editors and publisher.

of the side chain to a hydroxyl or carboxylic acid group may also directly promote the efflux of sterols from cells by increasing their aqueous solubility (Björkhem [24], this series).

Another scenario involves the oxidative cleavage of the side chain of cholesterol to a C20 ketone moiety (Figure 3). This reaction is catalyzed by a specific cytochrome P450 enzyme in the mitochondria of steroidogenic cells (see Jefcoate [20], this series) and possibly brain cells. The resulting molecule, Δ-5-pregnenolone, is a precursor for a large number of new molecules that have the proper shape to bind and activate nuclear hormone receptors involved in a variety of critical metabolic processes (29) (Figure 3). In certain cases, the C20 ketone moiety of pregnenolone is retained, and the 3β-hydroxyl group is oxidized to a ketone (e.g., progesterone, cortisol, and aldosterone) (Figure 3). In other cases, the side chain is oxidatively cleaved further to yield a C17 hydroxyl group (e.g.,

testosterone and estradiol), and in one of these molecules (estradiol) the A ring is reduced to a phenol (Figure 3). Different degrees of ring demethylations, oxidations, and reductions add further variety to the shape of the steroid nucleus and thus confer their specificity for individual hormone receptors (29).

In the classic pathway defined for steroid hormone signaling, the hormone-receptor complex enters the nucleus and transcriptionally activates genes critically involved in reproductive biology (e.g., progesterone, estradiol, and testosterone), the stress response (e.g., cortisol), intravascular salt and volume homeostasis (e.g., aldosterone), and possibly higher brain functions (e.g., sex hormones) (29). An evolutionary preview of this paradigm is found in insects, where exogenously derived sterols are oxidized to form ecdysteroids (Figure 4), which control metamorphosis by interacting with nuclear hormone receptors (30). Interestingly,

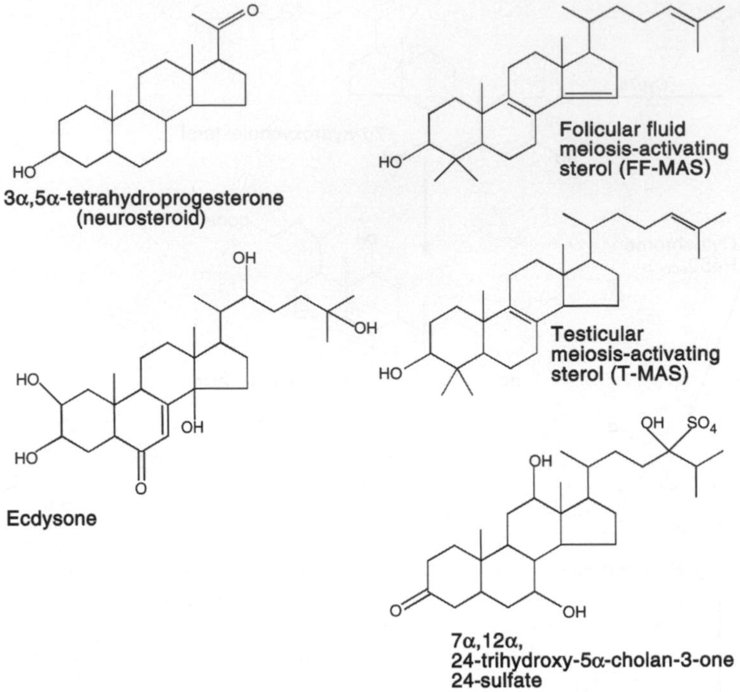

3α,5α-tetrahydroprogesterone
(neurosteroid)

Ecdysone

Folicular fluid
meiosis-activating
sterol (FF-MAS)

Testicular
meiosis-activating
sterol (T-MAS)

7α,12α,
24-trihydroxy-5α-cholan-3-one
24-sulfate

Figure 4
Structure of sterol-related molecules with unusual biological functions. See text for details.

ther hydroxylations at the C25 and C1 positions lead to the hormonally active molecule 1,25-dihydroxy-vitamin D, which interacts with a nuclear hormone receptor that critically affects body calcium metabolism (33). Likewise, lanosterol-derived molecules called meiosis-activating sterols are employed by mammals to activate resumption of meiosis in oocytes and spermatozoa (34) (Figure 4).

In summary, by oxidative cyclization and demethylation of an acetate-derived hydrophobic molecule, nature has created an amphipathic planar structure that, directly or through modifications, affects an incredible array of critical biological processes. These processes affect cellular membrane physiology, dietary nutrient absorption, reproductive biology, stress responses, salt and water balance, and calcium metabolism. Each of these processes is absolutely necessary for organisms to reach reproductive age. Thus, there has been strong evolutionary pressure to ensure that the body in general and individual cells in particular have an adequate supply of cholesterol. Indeed, the regulatory response of cells to sterol starvation, as mentioned above, is exquisitely designed for this purpose. These regulatory processes can also protect cells from moderate degrees of cholesterol excess. What nature did not plan for, however, was how to handle levels of cholesterol that exceed these limits.

The pathophysiology of cholesterol

Homo sapiens has many physical characteristics of a vegetarian organism (e.g., structure of teeth and intestines, sipping rather than lapping to drink, sweating rather than panting to moderate body temperature) and evolved with the high physical activity of a hunter-gatherer (35). As such, intestinal and hepatic lipid metabolism likely evolved to convert whatever dietary fats were available into a form, namely, triglyceride-rich lipoproteins, that could supply energy to tissues. The limited amount of cholesterol in the hunter-gatherer's diet was not a problem, because most cells have a more than adequate capacity to make all of the cholesterol they need for membrane structural functions (36). Indeed, the way cells ensure proper levels of endogenous cholesterol synthesis when exogenous sources are low is part of the exquisite story of cellular regulation that was described

recent data have suggested that the activation of estrogen and androgen receptors by sex hormones can lead to nongenotropic effects in cells, such as attenuation of apoptosis (31). Similarly, certain neurosteroids can directly modulate the activity of GABA receptors in the brain (32) (Figure 4).

Finally, nature has also taken advantage of some of the immediate precursors of cholesterol to fulfill other critical functions. For example, 7-dehydrocholesterol in the skin is converted to cholecalciferol (vitamin D3) via cleavage of the B ring by near ultraviolet irradiation (33) (Figure 3). A similar reaction occurs in plants, where ergosterol is converted to ergocalciferol (vitamin D2). Fur-

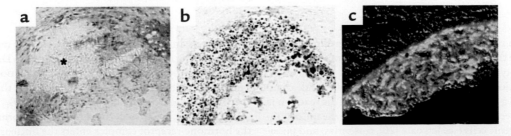

Figure 5
Cholesterol in atherosclerosis. The images show proximal aorta atherosclerotic lesions from the atherosclerosis-susceptible apoE knockout mice fed a cholesterol- and saturated fat–enriched diet for 25 weeks. (**a** and **b**) Adjacent sections from the same lesion stained with hematoxylin (**a**) or oil red O (**b**). The asterisk in **a** marks an area filled with cholesterol crystals. The bright red-orange staining in **b** shows areas rich in neutral lipids, most likely cholesteryl ester inclusions in macrophage foam cells and cholesteryl ester-rich emulsions in extracellular regions of the lesion. Areas of the lesion that stain poorly with oil red O are often rich in free (i.e., unesterified) cholesterol. (**c**) Fluorescence microscopy image of another atherosclerotic lesion stained with the fluorescent dye filipin (shown as blue in this image), which binds specifically to areas enriched in free cholesterol. The bright filipin staining in almost the entire diseased intima indicates that advanced atherosclerotic lesions in this model are very rich in free cholesterol.

recently in the *JCI* by Horton, Goldstein, and Brown (21). Those cells that use cholesterol as a precursor to other molecules, such as steroidogenic cells (steroid hormones) and hepatocytes (bile acids), may need an exogenous supply under certain conditions, but this could easily be met by the internalization of the small amount of cholesterol left in the remnants of triglyceride-rich lipoproteins or in the LDL-like and HDL-like by-products of triglyceride-rich lipoprotein metabolism.

Intruding on this serene picture of metabolic harmony was the development of the superior intellect of modern humans, which enabled our species to secure the higher energy content of animal-derived foods at a much lower cost to metabolic energy. For reasons that are still not clear, the ingestion of saturated fats often causes a dramatic rise in plasma cholesterol levels, including those in remnant lipoproteins and LDL. The increase in plasma cholesterol is further exacerbated by obesity, insulin resistance, and perturbations of glucose and fatty acid metabolism, all of which accompany sedentary lifestyles.

Even with these problems, the degree of plasma cholesterol elevation that occurs at or before reproductive age usually poses no immediate threat. For cells that internalize some of this excess cholesterol, there are multiple protective mechanisms, including downregulation of endogenous cholesterol biosynthesis and of LDL receptors, intracellular esterification of the excess cholesterol by acyl-coenzyme A:cholesterol acyltransferase, and efflux of cholesterol by multiple pathways (37). For tissues, the adverse reactions that occur in response to the accumulation of large amounts of extracellular cholesterol-rich lipoproteins take years to develop.

In post-reproductive life, however, the pressures of evolution have little impact, and the toll for elevated cholesterol is eventually paid. Clearly, the highest toll is associated with atherosclerotic vascular disease, although other consequences of persistently elevated body cholesterol include cholesterol gallstones, liver dysfunction, cholesterol crystal emboli, and dermatological abnormalities (e.g., xanthomas). In the case of vascular disease, the persistently high levels of circulating cholesterol-rich lipoproteins lead gradually but inexorably to their accrual in the subendothelial space of certain segments of midsize and small arteries (38) (Figure 5). This is "foreign" material, and a tenet of human physiology that is critical for survival through the reproductive years holds that foreign material must be attacked. Thus, macrophages, T cells, and their inflammatory cytokines enter the affected areas of the arterial wall in response to excess cholesterol-rich lipoproteins, but what they encounter is no ordinary bacteria or virus. Rather, cellular reactions to cholesterol, products of cholesterol oxidation, and other lipids that are packaged with cholesterol-carrying lipoproteins (e.g., unsaturated fatty acids) lead to a series of responses that promote the atherogenic response (38–41). Cholesterol accumulates in atherosclerotic lesions in large amounts in cells and in the extracellular space, both in esterified and in unesterified (or "free") forms (Figure 5). Free cholesterol and certain oxidized forms of cholesterol are particularly toxic to lesional cells, and death of lesional macrophages exposed to excess free cholesterol or oxys-

terols is likely an important cause of lesional necrosis (ref. 37; see also Perspectives by Tabas [42] and by Björkhem [24], this series). The end result is a festering inflammatory "abscess," which all too often leads to destruction of the arterial wall, acute thrombotic vascular occlusion, and infarction of critical tissues supplied by the affected vessel (40, 41). Thus, a molecule essential to life is associated with the leading cause of death in the industrialized world.

In this series

The panoply of functions of cholesterol, its metabolites, and its immediate biosynthetic precursors prohibits complete coverage of all of these areas. Rather, the goal of this Perspective series is to provide a taste of interesting topics related to both the physiological and the pathophysiological roles of cholesterol. The function of cholesterol in biological membranes will be covered by several articles. Simons and Ehehalt (11) will review the role of cholesterol-mediated membrane rafts in intracellular signaling, and Maxfield and Wüstner (19) will cover various aspects related to normal and perturbed trafficking of cholesterol among cellular membranes. Travis and Kopf (14) will examine a more focused area related to cholesterol and membrane function, namely, the fascinating role of acrosomal membrane cholesterol efflux in sperm capacitation. Jefcoate (20) will deal with how cholesterol is transported to the inner mitochondrial membrane, which is critical for the conversion of cholesterol to steroid hormones. Mitochondria can also convert cholesterol into oxysterols, whose cellular and physiological functions will be reviewed by Björkhem (24). As discussed above, certain proteins become adducted with cholesterol, and Jeong and McMahon (16) will discuss how cholesterol modification affects the activity of Sonic hedgehog as a morphogen. One demonstration of the critical role played by cholesterol in normal physiology is the devastating effects of genetic mutations in sterol biosynthesis, a topic covered by Porter (18). Finally, the series includes two articles examining various aspects of the cholesterol-loaded macrophage, which is a critical cellular component of atherosclerotic lesions. Tall and colleagues (27) will review pathways of cholesterol efflux from cholesterol-loaded macrophages, a key protective process that undoubtedly has important therapeutic implications. Tabas (42) will examine what happens when this and other protective functions fail, as they all too often do in atherosclerosis, by reviewing pathways by which the accumulation of unesterified cholesterol in macrophages leads to the death of these cells.

Acknowledgments

I gratefully acknowledge the comments and suggestions of Donald M. Small, Dennis Vance, Jean Vance, Thomas Haines, and John Ashkenas.

Address correspondence to: Ira Tabas, Department of Medicine, Columbia University, 630 West 168th Street, New York, New York 10032, USA. Phone: (212) 305-9430; Fax: (212) 305-4834; E-mail: iat1@columbia.edu.

1. Bloch, K.E. 1979. Speculations on the evolution of sterol structure and function. *CRC Crit. Rev. Biochem.* **7**:1–5.
2. Cavalier-Smith, T. 2002. The neomuran origin of archaebacteria, the negibacterial root of the universal tree and bacterial megaclassification. *Int. J. Syst. Evol. Microbiol.* **52**:7–76.
3. Ranadive, G.N., and Lala, A.K. 1987. Sterol-phos-pholipid interaction in model membranes: role of C5-C6 double bond in cholesterol. *Biochemistry.* **26**:2426–2431.
4. Haines, T.H. 2001. Do sterols reduce proton and sodium leaks through lipid bilayers? *Prog. Lipid Res.* **40**:299–324.
5. Esfahani, M., Scerbo, L., and Devlin, T.M. 1984. A requirement for cholesterol and its structural features for a human macrophage-like cell line. *J. Cell. Biochem.* **25**:87–97.
6. Vance, D.E., and Van den Bosch, H. 2000. Cholesterol in the year 2000. *Biochim. Biophys. Acta.* **1529**:1–8.
7. Yeagle, P.L. 1985. Cholesterol and the cell membrane. *Biochim. Biophys. Acta.* **822**:267–287.
8. Miao, L., et al. 2002. From lanosterol to cholesterol: structural evolution and differential effects

on lipid bilayers. *Biophys. J.* **82**:1429–1444.

9. Awad, A.B., and Fink, C.S. 2000. Phytosterols as anticancer dietary components: evidence and mechanism of action. *J. Nutr.* **130**:2127–2130.

10. Berge, K.E., et al. 2000. Accumulation of dietary cholesterol in sitosterolemia caused by mutations in adjacent ABC transporters. *Science.* **290**:1771–1775.

11. Simons, K., and Ehehalt, R. 2002. Cholesterol, lipid rafts, and disease. *J. Clin. Invest.* In press.

12. Prior, I.A., and Hancock, J.F. 2001. Compartmentalization of Ras proteins. *J. Cell Sci.* **114**:1603–1608.

13. Simons, M., Keller, P., Dichgans, J., and Schulz, J.B. 2001. Cholesterol and Alzheimer's disease: is there a link? *Neurology.* **57**:1089–1093.

14. Travis, A.J., and Kopf, G.S. 2002. The role of cholesterol efflux in regulating the fertilization potential of mammalian spermatozoa. *J. Clin. Invest.* In press.

15. Mauch, D.H., et al. 2001. CNS synaptogenesis promoted by glia-derived cholesterol. *Science.* **294**:1354–1357.

16. Jeong, J., and McMahon, A.P. 2002. Cholesterol modification of Hedgehog family proteins. *J. Clin. Invest.* In press.

17. Subtil, A., et al. 1999. Acute cholesterol depletion inhibits clathrin-coated pit budding. *Proc. Natl. Acad. Sci. USA.* **96**:6775–6780.

18. Porter, F.D. 2002. Malformation syndromes due to inborn errors of cholesterol synthesis. *J. Clin. Invest.* In press.

19. Maxfield, F.R., and Wüstner, D. 2002. Intracellular cholesterol transport. *J. Clin. Invest.* In press.

20. Jefcoate, C. 2002. High-flux mitochondrial cholesterol trafficking, a specialized function of the adrenal cortex. *J. Clin. Invest.* In press.

21. Horton, J.D., Goldstein, J.L., and Brown, M.S. 2002. SREBPs: activators of the complete program of cholesterol and fatty acid synthesis in the liver. *J. Clin. Invest.* **109**:1125–1131. doi:10.1172/JCI200215593.

22. Dobrosotskaya, I.Y., Seegmiller, A.C., Brown, M.S., Goldstein, J.L., and Rawson, R.B. 2002. Regulation of SREBP processing and membrane lipid production by phospholipids in Drosophila. *Science.* **296**:879–883.

23. Russell, D.W., and Setchell, K.D. 1992. Bile acid biosynthesis. *Biochemistry.* **31**:4737–4749.

24. Björkhem, I. 2002. Do oxysterols control cholesterol homeostasis? *J. Clin. Invest.* In press.

25. Rudel, L., Deckelman, C., Wilson, M., Scobey, M., and Anderson, R. 1994. Dietary cholesterol and downregulation of cholesterol 7 alpha-hydroxylase and cholesterol absorption in African green monkeys. *J. Clin. Invest.* **93**:2463–2472.

26. Li, W., et al. 2002. Bile acid secreted by male sea lamprey that acts as a sex pheromone. *Science.* **296**:138–141.

27. Tall, A.R., Costet, P., and Wang, N. 2002. Regulation and mechanisms of macrophage cholesterol efflux. *J. Clin. Invest.* In press.

28. Peet, D.J., Janowski, B.A., and Mangelsdorf, D.J. 1998. The LXRs: a new class of oxysterol receptors. *Curr. Opin. Genet. Dev.* **8**:571–575.

29. Bondy, P.K. 1985. Disorders of the adrenal cortex. In *Textbook of endocrinology.* J.D. Wilson and D.W. Foster, editors. W.B. Saunders Co. Philadelphia, Pennsylvania, USA. 816–890.

30. Gilbert, L.I., Rybczynski, R., and Warren, J.T. 2002. Control and biochemical nature of the ecdysteroidogenic pathway. *Annu. Rev. Entomol.* **47**:883–916.

31. Kousteni, S., et al. 2001. Nongenotropic, sex-nonspecific signaling through the estrogen or androgen receptors: dissociation from transcriptional activity. *Cell.* **104**:719–730.

32. Stoffel-Wagner, B. 2001. Neurosteroid metabolism in the human brain. *Eur. J. Endocrinol.* **145**:669–679.

33. Aurbach, G.D., Marx, S.J., and Spiegel, A.M. 1985. Parathyroid hormone, calcitonin, and calciferols. In *Textbook of endocrinology.* J.D. Wilson and D.W. Foster, editors. W.B. Saunders Co. Philadelphia, Pennsylvania, USA. 1137–1217.

34. Byskov, A.G., Andersen, C.Y., Leonardsen, L., and Baltsen, M. 1999. Meiosis activating sterols (MAS) and fertility in mammals and man. *J. Exp. Zool.* **285**:237–242.

35. Macko, S.A., Lubec, G., Teschler-Nicola, M., Andrusevich, V., and Engel, M.H. 1999. The Ice Man's diet as reflected by the stable nitrogen and carbon isotopic composition of his hair. *FASEB J.* **13**:559–562.

36. Dietschy, J.M., Turley, S.D., and Spady, D.K. 1993. Role of the liver in the maintenance of cholesterol and low density lipoprotein homeostasis in different animal species, including humans. *J. Lipid Res.* **34**:1637–1659.

37. Tabas, I. 1997. Free cholesterol-induced cytotoxicity. A possible contributing factor to macrophage foam cell necrosis in advanced atherosclerotic lesions. *Trends Cardiovasc. Med.* **7**:256–263.

38. Williams, K.J., and Tabas, I. 1995. The response-to-retention hypothesis of early atherogenesis. *Arterioscler. Thromb. Vasc. Biol.* **15**:551–561.

39. Ross, R. 1999. Atherosclerosis: an inflammatory disease. *N. Engl. J. Med.* **340**:115–126.

40. Lusis, A.J. 2000. Atherosclerosis. *Nature.* **407**:233–241.

41. Glass, C.K., and Witztum, J.L. 2001. Atherosclerosis. the road ahead. *Cell.* **104**:503–516.

42. Tabas, I. 2002. Consequences of cellular cholesterol accumulation: basic concepts and physiological implications. *J. Clin. Invest.* In press.

43. Gurr, M.I., and Harwood, J.L. 1991. Metabolism of structural lipids. In *Lipid biochemistry.* M.I. Gurr and J.L. Harwood, editors. Chapman & Hall. New York, New York, USA. 297–337.

Cholesterol, lipid rafts, and disease

Kai Simons and Robert Ehehalt

Max Planck Institute of Molecular Cell Biology and Genetics, Dresden, Germany.

Lipid rafts are dynamic assemblies of proteins and lipids that float freely within the liquid-disordered bilayer of cellular membranes but can also cluster to form larger, ordered platforms. Rafts are receiving increasing attention as devices that regulate membrane function in eukaryotic cells. In this Perspective, we briefly summarize the structure and regulation of lipid rafts before turning to their evident medical importance. Here, we will give some examples of how rafts contribute to our understanding of the pathogenesis of different diseases. For more information on rafts, the interested reader is referred to recent reviews (1, 2).

Composition of lipid rafts

Lipid rafts have changed our view of membrane organization. Rafts are small platforms, composed of sphingolipids and cholesterol in the outer exoplasmic leaflet, connected to phospholipids and cholesterol in the inner cytoplasmic leaflet of the lipid bilayer. These assemblies are fluid but more ordered and tightly packed than the surrounding bilayer. The difference in packing is due to the saturation of the hydrocarbon chains in raft sphingolipids and phospholipids as compared with the unsaturated state of fatty acids of phospholipids in the liquid-disordered phase (3). Thus, the presence of liquid-ordered microdomains in cells transforms the classical membrane fluid mosaic model of Singer and Nicholson into a more complex system, where proteins and lipid rafts diffuse laterally within a two-dimensional liquid.

Membrane proteins are assigned to three categories: those that are mainly found in the rafts, those that are present in the liquid-disordered phase, and those that represent an intermediate state, moving in and out of rafts. Constitutive raft residents include glycophosphatidylinositol-anchored (GPI-anchored) proteins; doubly acylated proteins, such as tyrosine kinases of the Src family, Gα subunits of heterotrimeric G proteins, and endothelial nitric oxide synthase (eNOS); cholesterol-linked and palmitate-anchored proteins like Hedgehog (see Jeong and McMahon, this Perspective series, ref. 4); and transmembrane proteins, particularly palmitoylated proteins such as influenza virus hemagglutinin and β-secretase (BACE) (1). Some membrane proteins are regulated raft residents and have a weak affinity for rafts in the unliganded state. After binding to a ligand, they undergo a conformational change and/or become oligomerized. When proteins oligomerize, they increase their raft affinity (5). A peripheral membrane protein, such as a nonreceptor tyrosine kinase, can be reversibly palmitoylated and can lose its raft association after depalmitoylation (6). By these means, the partitioning of proteins in and out of rafts can be tightly regulated.

Cholesterol and raft biogenesis

Cholesterol is thought to serve as a spacer between the hydrocarbon chains of the sphingolipids and to function as a dynamic glue that keeps the raft assembly together (1). Cholesterol partitions between the raft and the nonraft phase, having higher affinity to raft sphingolipids than to unsaturated phospholipids. Removal of raft cholesterol leads to dissociation of most proteins from rafts and renders them nonfunctional.

Association with detergent-resistant membranes (DRMs) is a useful criterion to estimate whether a protein associates with lipid rafts (2). After solubilization of membranes or cells with Triton X-100 or with CHAPS at 4°C, raft-associated lipids and proteins remain insoluble and can then be floated to low density by sucrose gradient centrifugation. If cholesterol is extracted by methyl-β-cyclodextrin or complexed by saponin, the raft proteins usually, but not always, become detergent-soluble.

Lipid rafts are first assembled in the Golgi complex in mammalian cells (3). Cholesterol is synthesized in the endoplasmic reticulum (ER), as is ceramide, the hydrophobic backbone of sphingolipids. However, most of the sphingolipid head groups are attached to ceramide in the Golgi complex, where raft assembly takes place (7). There is an increasing concentration of cholesterol and sphingolipids from the ER to the plasma membrane. This increase seems to be achieved by excluding lipid rafts from the retrograde traffic between the Golgi complex and the ER (8). Thus, lipid rafts are moved forward from the Golgi complex to the plasma membrane, where they concentrate but also spread into the endocytic recycling pathways (9). Cholesterol and sphingolipid concentrations are tightly regulated and limit the supply of lipid rafts to organelles supplied by the Golgi apparatus.

The cell pays a price for using cholesterol as a spacer for keeping rafts together. Cholesterol is toxic, and its cellular levels are kept in tight control by an intricate network of transcriptional regulation of cholesterol biosynthesis and cellular uptake as well as by deposition of cholesterol into fat droplets in an esterified form, and by cellular efflux (ref. 10; see also Tall et al. [ref. 11] and Tabas [ref. 12], this Perspective series). Disturbance of these tightly regulated processes leads to a variety of diseases of lipid metabolism, as shown in Table 1.

Regulation of raft size

One important issue in raft function is their size. There is consensus that rafts are too small to be resolved by light microscopy. Pralle et al. (13) employed photonic force microscopy to measure the size of lipid rafts and found that rafts in the plasma membrane of fibroblasts diffuse as assemblies of 50 nm diameter, corresponding to a surface area covered by about 3,000 sphingolipids. Based on data from cultured baby hamster kidney (BHK) cells, whose lipid composition and organelle surface area have been examined in detail, it appears that an individual cell has a surface area of approximately 2,000 μm^2. The lipid composition of the cell plasma membrane contains 26% phosphatidylcholine, 24% sphingomyelin, and 12% glycosphingolipids. Due to the

---

Nonstandard abbreviations used: glycophosphatidylinositol-anchored (GPI-anchored); endothelial nitric oxide synthase (eNOS); β-secretase (BACE); endoplasmic reticulum (ER); baby hamster kidney (BHK); detergent-resistant membranes (DRMs); Alzheimer disease (AD); amyloid-β-peptide (Aβ).

Conflict of interest: No conflict of interest has been declared.

Citation for this article: *J. Clin. Invest.* **110**:597–603 (2002). doi:10.1172/JCI200216390.

Table 1

Diseases for which rafts and raft proteins are targets

Alzheimer disease
Parkinson disease
Muscular dystrophy
Polyneuropathies, demyelinating diseases
Autoimmune disease, chronic inflammation, vaccine response
B cell response
T cell response
Asthma and allergic response
Neoplasia
Atherosclerosis
Hypertension, hemodynamic regulation
Diabetes
Hyperparathyroidism
Osteoarthritis
Gastrointestinal ulceration
Paroxysmal nocturnal hemoglobinuria
Lysosomal storage disease
Niemann-Pick disease
Tay-Sachs disease, morbus Fabry, metachromatic leukodystrophy
Pilzaeus-Merzbacher disease
Postsqualene cholesterol biosynthesis disorders
Pore-forming toxins (gas gangrene)
Sepsis, septic shock

Bacterial infections

    Escherichia coli
    Mycobacteria tuberculosis and *bovis*
    Campylobacter jejuni
    Vibrio cholerae
    Clostridium difficile (pseudomembranous colitis)
    Clostridium tetani
    Salmonella, Shigella

Viral infections

    Influenza virus
    HIV-1
    Measles virus
    Respiratory syncytial cell virus
    Filoviridae (Ebolavirus, Marburgvirus)
    Papillomaviridae and polyomaviridae
    Epstein-Barr virus
    Echovirus 1

Other pathogens

    Plasmodium (malaria)
    Trypanosoma (sleeping sickness)
    Leishmania
    Prions (Creutzfeldt-Jakob disease, Kuru, Gerstmann-Sträussler-Scheinker syndrome)
    Toxoplasma gondii

See supplemental reading list (www.jci.org/cgi/content/full/110/5/597/DC1) for detailed references.

asymmetric nature of the lipid organization in the plasma membrane, most of the sphingolipids occupy the outer leaflet of the bilayer, while less than half of the phosphatidylcholine has been estimated to be in this leaflet (14).

Assuming that most of the sphingolipid is raft-associated, rafts would cover more than half of the cell surface. The density of membrane proteins has been estimated to be around 20,000 molecules per μm^2. Thus, the plasma membrane would accordingly contain about 40×10^6 protein molecules. The number of 50-nm rafts would be about 10^6, and if the density of proteins is the same in rafts as in the surrounding bilayer, each raft would carry about 20 protein molecules. If BHK cells are representative, it follows that the density of rafts floating in the fibroblast plasma membrane is high. If 20×10^6 raft protein molecules were distributed more or less randomly, each raft would likely contain a different subset of proteins. A kinase attached to the cytosolic leaflet of a raft is, therefore, unlikely to meet its substrate in the same individual raft.

The small size of an individual raft may be important for keeping raft-borne signaling proteins in the "off" state. Accordingly, for activation to occur, many rafts have to cluster together, forming a larger platform, where the protein participants in a signal transduction process can meet, undisturbed by what happens outside the platform (Figure 1). Thus, rafts are small, and, when activated, they cluster to form larger platforms in which functionally related proteins can interact. One way to analyze raft association and clustering is to patch raft and nonraft components on the surface of living cells by specific antibodies (5, 15). If two raft components are cross-linked by antibodies, they will form overlapping patches in the plasma membrane. However, patching of a raft protein and a nonraft marker such as the transferrin receptor leads to the formation of segregated patches. In general, co-patching of two raft components is dependent on the simultaneous addition of both antibodies to the cells. If antibodies are added sequentially, segregated patches predominate. Notably, the patching behavior is cholesterol-dependent. As a consequence of the small size and the heterogeneous composition of individual rafts, these structures must be clustered in specific ways if signaling is to ensue.

One example of such a raft clustering process encountered in daily clinical practice is the IgE signaling during the allergic immune response (16, 17). The allergen that elicits the allergic reaction by stimulating the degranulation of a mast or basophilic cell is multivalent, binding several IgE antibody molecules. Cross-linking of two or more IgE receptors [Fc(ε)RI] increases their association with rafts, as measured by increased detergent resistance. Within the rafts, cross-linked Fc(ε)RI becomes tyrosine-phosphorylated by raft-associated Lyn, a double-acylated Src-related kinase. The Fc(ε)RI phosphorylation recruits Syk-related kinases, which are activated and lead to binding and scaffolding of downstream signaling molecules and, finally, to the formation of a signaling platform. This structure includes the raft protein LAT (linker of activation of T cells), which guides the clustering of additional rafts into the expanding platform (18). Signaling leads to calcium mobilization, which triggers the release of preformed mediators such as histamine from the intracellular stores.

The more participants are collected into the raft platform, the higher the signaling response. Uncontrolled amplification of the signaling cascade by raft clustering might trigger hyperactivation, with life-threatening consequences such as Quinke edema and allergic shock. The whole signaling assembly can be dissociated by dephosphorylation or downregulated by internalization of the components by endocytosis (19). Thus, in IgE signaling, lipid rafts serve to increase the efficiency by concentrating the participating proteins into fluid microdomains and limiting their lateral diffusion so that proteins remain at the site of signaling. Even a small change of partitioning into lipid rafts can, through amplification, initiate a signaling cascade or prompt a deleterious overshoot, as occurs in allergic reactions (20).

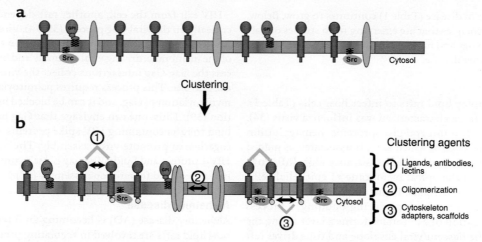

Figure 1

Mechanisms of raft clustering. (**a**) Rafts (red) are small at the plasma membrane, containing only a subset of proteins. (**b**) Raft size is increased by clustering, leading to a new mixture of molecules. This clustering can be triggered (1) at the extracellular side by ligands, antibodies, or lectins, (2) within the membrane by oligomerization, or (3) by cytosolic agents (cytoskeletal elements, adapters, scaffolds). Raft clustering occurs at the plasma membrane as well as intracellularly, e.g., in endosomal lumen. Ligand binding or oligomerization can alter the partitioning of proteins in and out of rafts. Increased raft affinity of a given protein and its activation within rafts (e.g., phosphorylation by Src-family kinases [yellow]) can initiate a cascade of events, leading to further increase of raft size by clustering.

Another clinically relevant example of raft clustering is the pathogenic mechanisms of pore-forming toxins, which are secreted by *Clostridium*, *Streptococcus*, and *Aeromonas* species, among other bacteria (21). These toxins may cause diseases ranging from mild cellulitis to gaseous gangrene and pseudomembranous colitis. Best studied is the toxin aerolysin from the marine bacterium *Aeromonas hydrophila* (22). Aerolysin is secreted and binds to a GPI-anchored raft protein on the surface of the host cell. The toxin is incorporated into the membrane after proteolysis and then heptamerizes in a raft-dependent manner to form a raft-associated channel through which small molecules and ions flow to trigger the pathogenic changes. The oligomerization of aerolysin can be triggered in solution but occurs at more than 10^3-fold lower toxin concentration at the surface of the living cell (22). This enormous increase in efficiency is due to activation by raft binding and by concentration into raft clusters, which is driven by the oligomerization of the toxin. Again, a small change can lead to a huge effect by amplification of raft clustering.

Caveolae and lipid rafts

One source of confusion in the raft field has been the interrelationship between caveolae and rafts. Fortunately, this issue has now been cleared up by the analysis of mice deficient in caveolins (23–25). Caveolae are small surface invaginations seen in many cell types (26). Some cells, such as adipocytes, endothelial cells, and smooth muscle cells, have plasma membranes with numerous caveolae, which have been postulated to be formed by clustering of rafts on the cell surface.

Caveolar invagination is driven by polymerization of caveolins, of which there are three: caveolin-1, -2, and -3. All caveolins are raft-associated, and mice deficient in caveolins lack caveolae (24). The mice are superficially healthy at birth, but detailed analysis shows that they have exercise intolerance accompanied by severe lung abnormalities, endothelial cell proliferations, and fibrosis, as well as dysregulation of the vascular tone; this last effect is secondary to eNOS activation and disturbance of calcium signaling. Caveolin-1–

deficient animals also develop a lung pathology resembling the end-state of a range of parenchymal lung disorders, including industrial lung diseases, sarcoidosis and connective tissue diseases, and idiopathic pulmonary fibrosis and chronic hypersensitivity pneumonitis (reviewed in ref. 27). Mutations in the human gene for the muscle-specific caveolin-3 lead to an autosomal dominant form of limb-girdle muscular dystrophy. In this disease, there is severe impairment of caveola formation in the striated muscle fibers, combined with a disorganization of the T-tubule system (28). The analysis of the caveolin-deficient mice has shown that most lipid raft functions are normal. Caveolae thus appear to have relatively narrow, tissue-specific functions.

Like rafts, caveolae are dynamic structures. Caveolae usually remain attached to the cell surface, but they can be endocytosed, for instance when they encounter Simian virus-40 (29). The ligand-triggered endocytosis of the virus is regulated by phosphorylation. After binding of the virus, it takes 20 minutes or more before caveolae pinch off from the plasma membrane and move into the cell. Video microscopy demonstrates that caveolae devoid of virus particles do not internalize. Cross-linking of raft proteins and lipids on the cell surface may result in movement into caveolae, where the cross-linked complexes become trapped. One physiological example is the inactivation of eNOS; after activation, this enzyme moves into caveolae, where the enzyme is downregulated (30). Normally, caveolae may primarily serve to store and downregulate raft proteins or act as reservoirs of rafts. Why some cells, such as adipocytes, have plasma membranes almost covered by caveolae remains an enigma.

Lipid rafts play a central role in many cellular processes, including membrane sorting and trafficking, cell polarization, and signal transduction processes that have been best studied in T cells (15), B cells (31), and, as mentioned, the allergic response (16, 17). Ceramide/sphingomyelin signaling, which regulates cell growth, survival, and death, also involves raft clustering (32). Several groups of pathogens, bacteria, prions, viruses, and parasites hijack lipid rafts for their purposes (33). The already impressive catalog of raft involvement in disease (Table 1) continues to grow. Below, we illustrate the principles that are emerging from studies on raft functions in physiology and human disease by focusing on viruses, prions, and Alzheimer disease.

Viral infection

Several viruses employ lipid rafts to infect host cells (Table 1). The first example to be characterized was influenza virus (34). The virus contains two integral glycoproteins, hemagglutinin and neuraminidase, both of which are raft-associated as judged by cholesterol-dependent detergent resistance (35). Influenza virus buds out from the apical membrane of epithelial cells, which is enriched in raft lipids. Influenza virus preferentially includes raft lipids in its envelope during budding, a process in which polymerization of M proteins forms a layer facing the cytosolic side of the nascent viral envelope and thus drives raft clustering (35). HIV-1, which likewise incorporates host raft lipids and proteins into its envelope, employs rafts for at least four key events in its life cycle: passage across a new host's mucosa, viral entry into immune cells, signaling of changes in host cell functions as well as viral exit from cells, and dispersion through the host's vascular system.

In the first process, breaching a host animal's mucosa, HIV binds to the glycosphingolipid galactosylceramide at the apical surface of mucosal epithelial cells and then transcytoses across the epithelium to be released on the basolateral side. Disrupting raft association blocks viral transcytosis (36).

During infection of target cells, the viral envelope components, as well as the internal Gag protein (which is essential for assembly of the viral envelope; ref. 37), are all initially associated with rafts — defined operationally as DRMs. Indeed, viral glycoproteins can co-patch with known raft-associated proteins on the surface of living cells after cross-linking with specific antibodies (38). Interestingly, the virus receptors on the host cell surface are also raft-associated. Manes et al. (38) found that the HIV glycoprotein gp120 co-patches with the cell surface receptor CD4 and with the co-receptors, the chemokine receptors CCR5 and CXLR4. CD4, CCR5, and CXLR4 are found in DRMs. Binding of the virus to its surface receptors, first to CD4 and then to the chemokine receptor, seems to lead to raft clustering and lateral assembly of a protein complex in the membrane to initiate fusion of the virus envelope with the cell membrane. Both cholesterol and specific glycosphingolipid species serve as crucial elements in organizing the fusion complex (39, 40).

An increasing body of data shows that viruses, bacteria, and parasites prepare their way into the host cell by changing the cellular state of signaling. This is also the case during HIV infection. Nef, an early HIV gene product, promotes infectivity of the virus via lipid rafts (41); infection with HIV-1 virions lacking Nef does not progress to AIDS (42). The Nef protein is a peripheral, myristoylated membrane protein with a proline-rich repeat that can bind to raft-associated nonreceptor tyrosine kinases of the Src family. It associates with DRMs and seems to prime the host cells for HIV infection by lowering the threshold necessary for T cell activation (41). Resting T cells do not support a productive HIV infection, but Nef activates T cells by increasing IL-2 secretion and obviates the need for costimulatory signals. By clustering lipid rafts carrying relevant host cell surface proteins, Nef oligomerization may aid in organizing the T cell signaling complex and the HIV budding site (41, 43).

HIV exit from the cell, another raft-dependent step, depends critically on the viral Gag protein (39, 44). Viruses contain 1,200–1,500 Gag molecules, which multimerize on the cytosolic leaflet of the membrane, driving viral assembly and budding. In this process the Gag-Gag interactions collect the virus spike proteins to the bud site. This process requires palmitoylation of gp120 and myristoylation of Gag, and it can be blocked by cholesterol depletion (39). Thus, one can envisage that Gag proteins specifically bind to rafts containing HIV spike proteins, which cluster rafts together to promote virus assembly. The interaction between HIV-1 protein and lipid rafts may cause a conformational change in Gag required for envelope assembly (37).

Alzheimer disease

Alzheimer disease (AD) is becoming an interesting example of how lipid rafts are involved in regulating protein trafficking and processing. Formation of senile plaques containing the amyloid-β-peptide (Aβ) is a hallmark of AD. Aβ, a fragment derived from the large type I transmembrane protein APP, the amyloid precursor protein (45), is cleaved sequentially by enzymes termed β-secretase (BACE) and γ-secretase.

BACE is a novel aspartyl-protease that sets the pace of Aβ generation. BACE cleaves APP in its luminal domain, generating a secreted ectodomain. The resulting 10-kDa C-terminal fragment is subsequently clipped by γ-secretase, which acts at the transmembrane domain of APP to release Aβ. A third enzymatic activity, the α-secretase, cleaves APP in the middle of the Aβ region, generating the α fragment (a secreted ectodomain), as well as a short C-terminal stub that is also cleaved by γ-secretase. Cleavage of α cuts APP within the Aβ region, yielding products that are non-amyloidogenic. Importantly, α-cleavage directly competes with β-cleavage for their common substrate APP. Therefore, a key issue for understanding how the disease-promoting peptide Aβ is generated is the question of how access of these enzymes to APP is regulated.

There is growing evidence that cholesterol plays a central role in regulating α- and β-cleavage (46). First, the ε4 allele of apoE, a carrier that plays a major role in cholesterol delivery within the CNS, represents a major risk factor for AD. Levels of total cholesterol and LDL in serum apparently correlate with the amount of Aβ in AD brains, and there is epidemiological evidence that elevated cholesterol level during mid-life increases the risk of developing AD (47). In addition, elevated dietary cholesterol uptake also increases amyloid plaque formation in transgenic mice and rabbits. Recently, two independent retrospective studies reported a strong decrease in the incidence of AD and dementia in patients treated with 3-hydroxy-3-methylglutaryl-CoA–reductive (HMG-CoA–reductase) inhibitors (48, 49).

Most importantly, it has been shown that cholesterol depletion inhibits β-cleavage and Aβ formation in neurons and other cells, while at the same time promoting α-cleavage (46). We hypothesize that APP is present in two cellular pools, one associated with lipid rafts, in which Aβ is generated, and another outside of rafts, where α-cleavage takes place. This model would explain how the same protein can be processed in mutually exclusive ways. Cholesterol depletion would shift the partitioning of APP from lipid rafts to the surrounding lipid bilayer, and the extent of this shift would determine the magnitude of BACE processing. Because cell surface rafts are small and contain only a subset of proteins, the likelihood that APP and BACE are in the same rafts is low. Therefore, β-cleavage is predicted to be limited on the cell surface, although it may occur after endocytosis, when internalized raft membrane can cluster and become redistributed.

Recent evidence from our laboratory shows that APP and BACE co-patch with one another following antibody cross-linking, to form a complex that also contains GPI-anchored raft markers but excludes the non–raft-associated transferrin receptor. A fraction of APP and BACE is found in DRMs, especially following antibody cross-linking. Remarkably, Aβ generation is strongly reduced after expression of a mutant form of dynamin or after activation of the Rab5 GTPase–activating protein, both of which treatments inhibit endocytosis. Conversely, Aβ production is strongly stimulated after cross-linking, presumably because clustering brings together surface rafts containing APP and BACE. As would be predicted, the effect of clustering is not sensitive to blockade by inhibition of endocytosis (R. Ehehalt et al., unpublished data).

Even after processing, interactions between Aβ and rafts may affect Alzheimer pathogenesis. Thus, raft-derived Aβ has been shown to promote fibrillogenesis of soluble Aβ, leading Mizuno et al. (50) to suggest that raft binding causes a conformational change that promotes amyloid plaque formation. Cholesterol depletion reduces the seeding properties of Aβ. Another raft component, the ganglioside GM1, is known to bind Aβ and perhaps change the latter's conformation (51). All these studies suggest a critical and decisive role for lipid rafts in Aβ generation.

Prion diseases

A conformational change resulting in amyloid formation is also involved in the pathogenesis of prion disease. Prion diseases are thought be promoted by an abnormal form (PrPsc) of a host-encoded protein (PrPc). PrPsc can interact with its normal counterpart PrPc and change the conformation of PrPc so that the protein turns into PrPsc. PrPsc then self-aggregates in the brain, and these aggregates are thought to cause the disorders manifested in humans as Creutzfeldt-Jakob disease, Kuru, or Gerstmann-Sträussler-Scheinker syndrome (52). The mechanism by which PrPc is converted to PrPsc is not known, but several lines of evidence suggest that lipid rafts are involved (53, 54).

PrP is a GPI-anchored protein. Both PrPc and PrPsc are associated with DRMs in a cholesterol-dependent manner. Cholesterol depletion of cells leads to decreased formation of PrPsc from PrPc. The GPI anchor is required for conversion. When the GPI anchor is exchanged with a transmembrane domain, conversion to abnormal proteins is blocked. In vitro, the conversion of PrPc to PrPsc, as monitored by PrP protease resistance, occurs when microsomes containing PrPsc are fused with DRMs containing PrP (53). Extraction with detergent leads to raft clustering in DRMs. Fusion of microsomes with DRMs was necessary in this experiment because simply mixing the membranes did not lead to measurable generation of new PrPsc. On the other hand, releasing PrP ectodomains from PrPsc by phospholipase C treatment also stimulated conversion of PrP to PrPsc in this system. Baron et al. (53) hypothesize that membrane components exchange between apposed cells; a possible mechanism for such an exchange is that the cells release membrane vesicles containing PrPsc that fuse with neighboring cells. Indeed, a similar process has been found to mediate transfer of the raft-associated chemokine receptor CCR5 (55). Alternatively, GPI-anchored PrPsc could be released as such from one cell and move across the extracellular aqueous phase to be inserted into another cell. Recently, it was shown that direct cell-cell contact is required for transfer of PrPsc infectivity in cell culture (56).

How lipid rafts promote abnormal prion conversion is not clear. No in vitro generated PrPsc has yet been shown to be infectious. Again, however, because of the small size of individual rafts, plasma membrane PrPc and PrPsc would in all likelihood be in separate rafts and therefore would not meet each other. Endocytosis has also been shown to play a role for prion conversion, as is the case for BACE cleavage of APP. We hypothesize that rafts containing PrPc and PrPsc become clustered after endocytosis. It is also possible that the protein factor X, postulated to mediate conversion, is involved in raft clustering after endocytosis. In support of this scenario is the finding that a pH of 6–7, which prevails in endosomes, promotes conversion. If PrPc and PrPsc were clustered into the same raft platform after endocytosis, an increase of interaction efficiency would result and lead to amplification of conversion.

Obviously more work is required to sort out this remarkable process, but since raft lipids have proved to be required for prion conversion to proceed efficiently, it may be that these structures serve not only to concentrate the components but also to directly regulate the conformational changes that yield infectious prions.

Perspectives

Research on lipid rafts is now entering an exciting phase. From what is already known, rafts represent versatile devices for compartmentalizing cellular membrane processes. In the nonactivated state they float freely, carrying a few passenger proteins, but, when activated, they coalesce to form larger platforms where proteins meet to perform functions in signaling, processing, and transport. Clustering of rafts is key to understanding raft function in human physiology as well as in development of human disease.

The biophysics of raft assembly is still poorly understood. How is the inner leaflet composed and linked to outer leaflet of raft assemblies? What limits the size of rafts in nonactivated and activated states? What role do the different constituents play? There is evidence for specific functions for different glycosphingolipids and phosphoinositides in rafts, but little is known about how these lipids provide specificity to membrane function in linking to proteins. What determines raft association of proteins and lipids? How is raft scaffolding accomplished? What is the role of the actin cytoskeleton, and how is the concentration of raft lipids in membranes regulated? A great deal is known about cholesterol regulation, but how are the different sphingolipid levels adjusted? Novel mass spectroscopic methodology is already available to analyze both proteins and lipids in membranes, but the field will still require additional tools to elucidate how lipids and proteins interact in cell membranes.

Acknowledgments

We thank T. Kurzchalia, W. Stremmel, and the members of the Simons laboratory for critical reading of this manuscript. The authors were supported by the Deutsche Forschungsgemeinschaft Schwerpunktprogramm SPP 1085, Zelluläre Mechanismen der Alzheimer Erkrankung.

Address correspondence to: Kai Simons, Max Planck Institute of Molecular Cell Biology and Genetics, Pfotenhauerstrasse 108, D-01307 Dresden, Germany. Phone: 49-351-2102800; Fax: 49-351-2102900; E-mail: simons@mpi-cbg.de.

Robert Ehehalt's present address is: University of Heidelberg, Department of Internal Medicine IV, Heidelberg, Germany.

1. Simons, K., and Toomre, D. 2000. Lipid rafts and signal transduction. *Nat. Rev. Mol. Cell Biol.* **1**:31–39.
2. London, E., and Brown, D.A. 2000. Insolubility of lipids in triton X-100: physical origin and relationship to sphingolipid/cholesterol membrane domains (rafts). *Biochim. Biophys. Acta.* **1508**:182–195.
3. Brown, D.A., and London, E. 1998. Functions of lipid rafts in biological membranes. *Annu. Rev. Cell Dev. Biol.* **14**:111–136.
4. Jeong, J., and McMahon, A.P. 2002. Cholesterol modification of Hedgehog family proteins. *J. Clin. Invest.* **110**:591–596. doi:10.1172/JCI200216506.
5. Harder, T., et al. 1998. Lipid domain structure of the plasma membrane revealed by patching of membrane components. *J. Cell Biol.* **141**:929–942.
6. Zacharias, D.A., et al. 2002. Partitioning of lipid-modified monomeric GFPs into membrane microdomains of live cells. *Science.* **296**:913–916.
7. van Meer, G. 1989. Lipid traffic in animal cells. *Annu. Rev. Cell Biol.* **5**:247–275.
8. Brugger, B., et al. 2000. Evidence for segregation of sphingomyelin and cholesterol during formation of COPI-coated vesicles. *J. Cell Biol.* **151**:507–518.
9. Mukherjee, S., and Maxfield, F.R. 2000. Role of membrane organization and membrane domains in endocytic lipid trafficking. *Traffic.* **1**:203–211.
10. Simons, K., and Ikonen, E. 2000. How cells handle cholesterol. *Science.* **290**:1721–1726.
11. Tall, A.R., Costet, P., and Wang, N. 2002. Regulation and mechanisms of macrophage cholesterol efflux. *J. Clin. Invest.* In press.
12. Tabas, I. 2002. Consequences of cellular cholesterol accumulation: basic concepts and physiological implications. *J. Clin. Invest.* In press.
13. Pralle, A., et al. 2000. Sphingolipid-cholesterol rafts diffuse as small entities in the plasma membrane of mammalian cells. *J. Cell Biol.* **148**:997–1008.
14. Allan, D. 1996. Mapping the lipid distribution in the membranes of BHK cells (mini-review). *Mol. Membr. Biol.* **13**:81–84.
15. Janes, P.W., et al. 2000. The role of lipid rafts in T cell antigen receptor (TCR) signaling. *Semin. Immunol.* **12**:23–34.
16. Sheets, E.D., Holowka, D., and Baird, B. 1999. Membrane organization in immunoglobulin E receptor signaling. *Curr. Opin. Chem. Biol.* **3**:95–99.
17. Holowka, D., and Baird, B. 2001. Fc(epsilon)RI as a paradigm for a lipid raft-dependent receptor in hematopoietic cells. *Semin. Immunol.* **13**:99–105.
18. Rivera, J., et al. 2001. A perspective: regulation of IgE receptor-mediated mast cell responses by a LAT-organized plasma membrane-localized signaling complex. *Int. Arch. Allergy Immunol.* **124**:137–141.
19. Xu, K., et al. 1998. Stimulated release of fluorescently labeled IgE fragments that efficiently accumulate in secretory granules after endocytosis in RBL-2H3 mast cells. *J. Cell Sci.* **111**:2385–2396.
20. Kholodenko, B.N., Hoek, J.B., and Westerhoff, H.V. 2000. Why cytoplasmic signaling proteins should be recruited to cell membranes. *Trends Cell Biol.* **10**:173–178.
21. Lesieur, C., et al. 1997. Membrane insertion: the strategies of toxins (review). *Mol. Membr. Biol.* **14**:45–64.
22. Abrami, L., and van Der Goot, F.G. 1999. Plasma membrane microdomains act as concentration platforms to facilitate intoxication by aerolysin. *J. Cell Biol.* **147**:175–184.
23. Razani, B., et al. 2002. Caveolin-2-deficient mice show evidence of severe pulmonary dysfunction without disruption of caveolae. *Mol. Cell. Biol.* **22**:2329–2344.
24. Drab, M., et al. 2001. Loss of caveolae, vascular dysfunction, and pulmonary defects in caveolin-1 gene-disrupted mice. *Science.* **293**:2449–2452.
25. Galbiati, F., et al. 2001. Caveolin-3 null mice show a loss of caveolae, changes in the microdomain distribution of the dystrophin-glycoprotein complex, and t-tubule abnormalities. *J. Biol. Chem.* **276**:21425–21433.
26. Kurzchalia, T.V., and Parton, R.G. 1999. Membrane microdomains and caveolae. *Curr. Opin. Cell Biol.* **11**:424–431.
27. Razani, B., and Lisanti, M.P. 2001. Caveolin-deficient mice: insights into caveolar function and human disease. *J. Clin. Invest.* **108**:1553–1561. doi:10.1172/JCI200114611.
28. Minetti, C., et al. 2002. Impairment of caveolae formation and T-system disorganization in human muscular dystrophy with caveolin-3 deficiency. *Am. J. Pathol.* **160**:265–270.
29. Pelkmans, L., and Helenius, A. 2002. Endocytosis via caveolae. *Traffic.* **3**:311–320.
30. Sowa, G., Pypaert, M., and Sessa, W.C. 2001. Distinction between signaling mechanisms in lipid rafts vs. caveolae. *Proc. Natl. Acad. Sci. USA.* **98**:14072–14077.
31. Cherukuri, A., Dykstra, M., and Pierce, S.K. 2001. Floating the raft hypothesis: lipid rafts play a role in immune cell activation. *Immunity.* **14**:657–660.
32. Kolesnick, R. 2002. The therapeutic potential of modulating the ceramide/sphingomyelin pathway. *J. Clin. Invest.* **110**:3–8. doi:10.1172/JCI200216127.
33. van der Goot, F.G., and Harder, T. 2001. Raft membrane domains: from a liquid-ordered membrane phase to a site of pathogen attack. *Semin. Immunol.* **13**:89–97.
34. Scheiffele, P., et al. 1999. Influenza viruses select ordered lipid domains during budding from the plasma membrane. *J. Biol. Chem.* **274**:2038–2044.
35. Zhang, J., Pekosz, A., and Lamb, R.A. 2000. Influenza virus assembly and lipid raft microdomains: a role for the cytoplasmic tails of the spike glycoproteins. *J. Virol.* **74**:4634–4644.
36. Alfsen, A., et al. 2001. Secretory IgA specific for a conserved epitope on gp41 envelope glycoprotein inhibits epithelial transcytosis of HIV-1. *J. Immunol.* **166**:6257–6265.
37. Campbell, S.M., Crowe, S.M., and Mak, J. 2001. Lipid rafts and HIV-1: from viral entry to assembly of progeny virions. *J. Clin. Virol.* **22**:217–227.
38. Manes, S., et al. 2000. Membrane raft microdomains mediate lateral assemblies required for HIV-1 infection. *EMBO Rep.* **1**:190–196.
39. Ono, A., and Freed, E.O. 2001. Plasma membrane rafts play a critical role in HIV-1 assembly and release. *Proc. Natl. Acad. Sci. USA.* **98**:13925–13930.
40. Hug, P., et al. 2000. Glycosphingolipids promote entry of a broad range of human immunodeficiency virus type 1 isolates into cell lines expressing CD4, CXCR4, and/or CCR5. *J. Virol.* **74**:6377–6385.
41. Zheng, Y.H., et al. 2001. Nef increases infectivity of HIV via lipid rafts. *Curr. Biol.* **11**:875–879.
42. Kirchhoff, F., et al. 1995. Brief report: absence of intact nef sequences in a long-term survivor with nonprogressive HIV-1 infection. *N. Engl. J. Med.* **332**:228–232.
43. Wang, J.K., et al. 2000. The Nef protein of HIV-1 associates with rafts and primes T cells for activation. *Proc. Natl. Acad. Sci. USA.* **97**:394–399.
44. Lindwasser, O.W., and Resh, M.D. 2001. Multimerization of human immunodeficiency virus type 1 Gag promotes its localization to barges, raft-like membrane microdomains. *J. Virol.* **75**:7913–7924.
45. Selkoe, D.J. 2001. Alzheimer's disease: genes, proteins, and therapy. *Physiol. Rev.* **81**:741–766.
46. Simons, M., et al. 2001. Cholesterol and Alzheimer's disease: is there a link? *Neurology.* **57**:1089–1093.
47. Kivipelto, M., et al. 2001. Midlife vascular risk factors and Alzheimer's disease in later life: longitudinal, population based study. *BMJ.* **322**:1447–1451.
48. Jick, H., et al. 2000. Statins and the risk of dementia. *Lancet.* **356**:1627–1631.
49. Wolozin, B., et al. 2000. Decreased prevalence of Alzheimer disease associated with 3-hydroxy-3-methyglutaryl coenzyme A reductase inhibitors. *Arch. Neurol.* **57**:1439–1443.
50. Mizuno, T., et al. 1999. Cholesterol-dependent generation of a seeding amyloid beta-protein in cell culture. *J. Biol. Chem.* **274**:15110–15114.
51. Choo-Smith, L.P., et al. 1997. Acceleration of amyloid fibril formation by specific binding of Abeta-(1-40) peptide to ganglioside-containing membrane vesicles. *J. Biol. Chem.* **272**:22987–22990.
52. Caughey, B. 2000. Transmissible spongiform encephalopathies, amyloidoses and yeast prions: common threads? *Nat. Med.* **6**:751–754.
53. Baron, G.S., et al. 2002. Conversion of raft associated prion protein to the protease-resistant state requires insertion of PrP-res (PrP(Sc)) into contiguous membranes. *EMBO J.* **21**:1031–1040.
54. Taraboulos, A., et al. 1995. Cholesterol depletion and modification of COOH-terminal targeting sequence of the prion protein inhibit formation of the scrapie isoform. *J. Cell Biol.* **129**:121–132.
55. Mack, M., et al. 2000. Transfer of the chemokine receptor CCR5 between cells by membrane-derived microparticles: a mechanism for cellular human immunodeficiency virus 1 infection. *Nat. Med.* **6**:769–775.
56. Kanu, N., et al. 2002. Transfer of scrapie prion infectivity by cell contact in culture. *Curr. Biol.* **12**:523–530.

Malformation syndromes due to inborn errors of cholesterol synthesis

Forbes D. Porter

Heritable Disorders Branch, National Institute of Child Health and Human Development, NIH, Bethesda, Maryland, USA.

Cholesterol has long been known to function as both a structural lipid and a precursor molecule for bile acid and steroid hormone synthesis. In addition, this ubiquitous lipid is now known to contribute fundamentally to the development and function of the CNS and the bones, and, as detailed in other articles in this Perspective series, it plays major roles in signal transduction, sperm development, and embryonic morphogenesis. Over the past decade, the identification of multiple congenital anomaly/mental retardation syndromes due to inborn errors of cholesterol synthesis has underscored the importance of cholesterol synthesis in normal development.

The prototypical example of a human malformation syndrome that results from a defect in cholesterol synthesis is the RSH/Smith-Lemli-Opitz syndrome (SLOS). To date, five additional human syndromes resulting from impaired cholesterol synthesis have been described. These include desmosterolosis, X-linked dominant chondrodysplasia punctata type 2 (CDPX2), CHILD syndrome (congenital hemidysplasia with ichthyosiform erythroderma/nevus and limb defects), Greenberg dysplasia, and, most recently, Antley-Bixler syndrome. Natural mouse mutations corresponding to CDPX2 (tattered) and CHILD syndrome (bare patches and striated) have been identified, and mouse models corresponding to SLOS and lathosterolosis have been produced by gene disruption. Identification of the biochemical defects present in these disorders has given insight into the role that cholesterol plays in normal embryonic development, has provided the initial step in understanding the pathophysiological processes underlying these malformation syndromes, and has given rise to treatment protocols for patients with SLOS.

Cholesterol is synthesized from lanosterol, the first sterol in the cholesterol synthesis pathway, via a series of enzymatic reactions shown in Figure 1. These include the demethylation at C4α, C4β, and C14, which converts the C30 molecule lanosterol to C27 cholesterol; isomerization of the $\Delta^{8(9)}$ double bond to a Δ^7 double bond; desaturation to form a Δ^5 double bond; and finally, reduction of Δ^{14}, Δ^{24}, and Δ^7 double bonds (Figure 1). Analyses of human and murine syndromes resulting from cholesterol synthetic defects have helped illuminate both the normal functions of cholesterol and a range of normal and teratogenic functions of the various precursor sterols that accumulate in these disorders.

Nonstandard abbreviations used: Smith-Lemli-Opitz syndrome (SLOS); chondrodysplasia punctata type 2 (CDPX2); congenital hemidysplasia with ichthyosiform erythroderma/nevus and limb defects syndrome (CHILD syndrome); emopamil-binding protein (EBP); hydrops–ectopic calcification–moth-eaten skeletal dysplasia (HEM dysplasia); Sonic hedgehog (Shh); Indian hedgehog (Ihh); Desert hedgehog (Dhh); Alzheimer disease (AD); selective Alzheimer disease indicator 1 (*seladin-1*); cerebral spinal fluid (CSF); liquid gas chromatography-mass spectrometry (LGC-MS); Autism Diagnostic Interview-Revised (ADI-R).

Conflict of interest: No conflict of interest has been declared.

Citation for this article: *J. Clin. Invest.* **110**:715–724 (2002). doi:10.1172/JCI200216386.

Here, I consider the clinical, molecular, biochemical, and developmental aspects of these disorders, focusing on the five malformation syndromes presented in Table 1.

Inborn errors of cholesterol metabolism

Smith-Lemli-Opitz syndrome. The Smith-Lemli-Opitz syndrome (SLOS; MIM no. 270400; see refs. 1–3 for detailed reviews), the prototypical example of a human malformation syndrome resulting from an inborn error of cholesterol synthesis, was first described in 1964. Smith, Lemli, and Opitz initially described three male patients with similar facial features, mental retardation, microcephaly, developmental delay, and hypospadias, and they designated the novel disorder as the RSH syndrome, referring to the names of the first three patients. In various populations, the clinical incidence of this condition has been reported to range from 1 in 10,000 to 1 in 60,000, although SLOS is thought to be more common in individuals of Northern European descent.

SLOS, along with a severe variant (type II SLOS, or Rutledge lethal multiple congenital anomaly syndrome; MIM no. 268670), is now known to represent a single, clinically heterogeneous genetic disorder. Infants at the severe end of the SLOS phenotypic spectrum often die due to multiple major congenital anomalies. Conversely, individuals at the mild end of the spectrum show only minor physical stigmata, coupled with behavioral problems that can include aspects of autism (4) and self-injurious behavior. Although near-normal intelligence has been reported, moderate to severe mental retardation is typical. The classical SLOS face is distinctive (Figure 2, a–c): Typical craniofacial features include microcephaly, metopic prominence, ptosis, a small upturned nose, cleft palate, broad alveolar ridges, and micrognathia, with cataracts developing pre- or postnatally. Limb abnormalities are frequent in SLOS, and syndactyly of the second and third toes is the most frequent single physical finding (Figure 2e). Because the SLOS phenotypic spectrum is so broad, the presence of second-third toe syndactyly in a child with significant mental or behavioral problems should prompt consideration of SLOS. Other limb anomalies include short and proximally placed thumbs, single palmar creases, and postaxial polydactyly of either the upper or lower extremities (Figure 2d). Congenital heart defects are common. Slow growth and poor weight gain are typical. Most SLOS infants are poor feeders, and gastrostomy tube placement is required in many cases. Gastrointestinal anomalies include colonic aganglionosis, pyloric stenosis, and malrotation. Genital malformations are common in male patients. Hypospadius is a typical finding, and more severely affected patients may have ambiguous genitalia (Figure 2f).

In 1993, Irons and coworkers (5) reported decreased serum levels of cholesterol and elevated levels of 7-dehydrocholesterol (7-DHC) in two SLOS patients. The conversion of 7-DHC to cholesterol, the last enzymatic reaction in the Kandutsch-Russel cholesterol synthetic pathway, is catalyzed by the 3β-hydroxysterol Δ^7-reductase

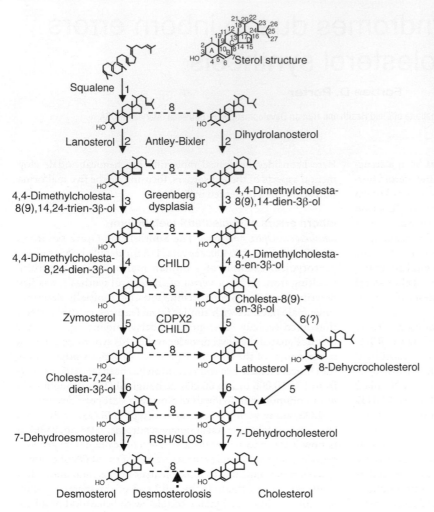

Figure 1
Cholesterol synthesis pathway and human malformation syndromes. The sterol ring structure and carbon position numbering are shown at the top of the figure. Human malformation syndromes are in boldface type. Cholesterol is synthesized from squalene in a series of enzymatic reactions denoted by numbers: 1, squalene monooxygenase and squalene cyclase; 2, lanosterol 14-α-demethylase; 3, 3β-hydroxysterol Δ14-reductase; 4, C4 demethylation complex (C4-sterol methyloxidase, C4-sterol decarboxylase [NSDHL], and 3-ketoreductase); 5, 3β-hydroxysterol Δ8,Δ7-isomerase; 6, lathosterol 5-desaturase; 7, 3β-hydroxysterol Δ7-reductase; 8, 3β-hydroxysterol Δ24-reductase.

alleles, this finding has been used to predict carrier frequencies of 1 in 30 and disease incidences on the order of 1 in 1,590 to 1 in 13,500 (7). This predicted disease incidence is much higher than clinical data would indicate, suggesting that a substantial number of severely affected fetuses with two severe alleles may be lost prenatally. An alternative explanation, that not all SLOS patients are currently being ascertained, could be addressed in a newborn screening trial. Indeed, given the availability of a therapeutic intervention, such a screen should be considered.

The high carrier frequency of the IVS8-1G→C mutation has led to speculation that heterozygotes are at a competitive advantage. Decreased cardiac or thromboembolic disease has been proposed as a possible advantage but seems unlikely, since these disorders usually affect individuals after their reproductive years. However, since many viruses rely on a membrane fusion event to infect cells, it is plausible that the presence of low levels of 7-DHC in the plasma membrane could protect the host from some viral diseases, a possibility that has yet to be tested in mouse models or human cell lines. In addition, since 7-DHC in the skin is the precursor for vitamin D synthesis, increased 7-DHC levels in heterozygotes could be adaptive in a Northern European population at risk of vitamin D–deficient rickets.

Dietary cholesterol supplementation has been attempted in SLOS, but the efficacy of this approach is probably limited by inefficient cholesterol transport across the blood-brain barrier and by the inability to reverse prior developmental effects of cholesterol deprivation. Biochemically, cholesterol supplementation leads to an improved plasma cholesterol/total sterol ratio, and prolonged therapy can even decrease plasma 7-DHC levels. Observational studies have reported multiple benefits of dietary cholesterol supplementation, including improved nutrition and growth, improved muscle tone and strength, decreased photosensitivity, decreased irritability, decreased tactile defensiveness, increased sociability and alertness, decreased self-injurious behavior, and decreased aggressiveness. To date, a controlled, blinded trial of dietary cholesterol supplementation confirming these observations has not been published. In a preliminary statement on one blinded trial, Kelley (3) reported that he found no difference between supplementation with 50 mg/kg/d and with 150 mg/kg/d. Since 50 mg/kg/d exceeds daily cholesterol

(Figure 1). In 1998, three groups, including ours (6), independently cloned the human 3β-hydroxysterol Δ7-reductase gene (*DHCR7*), which encodes this enzyme, and identified *DHCR7* mutations in SLOS patients. The DHCR7 protein is predicted to be a 475–amino acid polypeptide integral membrane protein with up to nine transmembrane domains. To date, 79 SLOS disease alleles have been identified in *DHCR7*, the most frequent of which is IVS8-1G→C (32%). This splice acceptor mutation, which results in the inclusion of 134 bp of intronic sequence in the DHCR7 mRNA, is a null allele, and IVS8-1G→C homozygotes typically have a severe phenotype. Other relatively common alleles include T93M (9%), W151X (7%), V326L (6%), R404C (5%), and R352W (3%). Establishing a genotype/phenotype correlation for SLOS has been confounded by the large number of different alleles and the fact that most patients are compound heterozygotes. The observation that patients with the same *DHCR7* genotype may have markedly different phenotypic severity suggests that other genetic, developmental, or maternal factors, perhaps affecting cholesterol biosynthesis or homeostasis, significantly influence a given patient's phenotype.

The IVS8-1G→C allele has been reported to have a carrier frequency of 1.06% (16 in 1,503) in an Oregon population (7). A 1% carrier frequency for the IVS8-1G→C mutation predicts an IVS8-1G→C homozygosity incidence of 1 in 40,000. Since the IVS8-1G→C mutation accounts for a third of the ascertained mutant *DHCR7*

Table 1

Human malformation syndromes due to inborn errors of cholesterol synthesis

	SLOS	Desmosterolosis	CDPX2	CHILD	Greenberg dysplasia	Antley-Bixler syndrome
Inheritance	AR	AR	XLD	XLD	AR	AR
Enzyme defect	3β-Hydroxysterol Δ⁷-reductase	3β-Hydroxysterol Δ²⁴-reductase	3β-Hydroxysterol Δ⁸Δ⁷-isomerase	3β-Hydroxysterol 4-decarboxylase	3β-Hydroxysterol Δ¹⁴-reductase	Lanosterol 14-α-demethylase
Chromosomal location	11q12-13	1p31.1-p33	Xp11.22-11.23	Xq28	NR	NR
Gene symbol	DHCR7	DHCR24	EBP	NSDHL	NR	NR
Mouse model	Dhcr7⁻ᐟ⁻	NR	Td	Bpa, Str	NR	NR

AR, autosomal recessive; XLD, X-Linked dominant; NR, not reported, *Td*, tattered; *Bpa*, bare patches; *Str*, striated.

needs in children, the clinical effect may already be maximal at this dose. Although dietary cholesterol supplementation has no effect on fixed developmental malformations, it does appear to improve the overall health status of these patients and to lessen the behavioral problems associated with this disorder.

Other treatment modalities have been considered as well. Bile acid supplementation has not shown a clear benefit. Fresh frozen plasma, a source of lipoproteins containing cholesterol, has been used for prenatal therapy and appears to be efficacious in the acute management of ill SLOS patients. Inhibition of HMG-CoA reductase, the rate-limiting enzyme in cholesterol synthesis, has been suggested as a means to reduce any toxic effects of 7-DHC. However, in a teratogenic rat model of SLOS, one inhibitor of this enzyme decreased cholesterol levels without decreasing 7 DHC levels (8). This animal study, combined with prior evidence that these drugs can precipitate an acute metabolic crisis in mevalonic aciduria patients and with anecdotal reports of poor outcomes in SLOS patients treated with HMG-CoA reductase inhibitors, led to the concern that inhibition of endogenous cholesterol synthesis in SLOS could be detrimental.

Nevertheless, one small trial of simvastatin therapy in SLOS has been reported. Jira et al. (9) treated two SLOS patients with simvastatin, and they reported decreased dehydrocholesterol levels, improved serum dehydrocholesterol/cholesterol ratios in serum and cerebral spinal fluid (CSF), and improved growth. Curiously, they also observed a paradoxical increase in plasma cholesterol levels. This surprising finding might be explained by upregulation of the mutant DHCR7. Because statins induce transcription of HMG-CoA reductase and many genes involved in cholesterol synthesis are coordinately regulated, simvastatin treatment may well have induced the synthesis of the 3β-hydroxysterol Δ⁷-reductase protein in these patients. If a mutant allele encodes a protein with residual enzymatic function, such an increase in protein levels could result in increased conversion of 7-DHC to cholesterol. Conversely, simvastatin upregulation of a null allele would clearly not increase the conversion of 7-DHC to cholesterol, but it could inhibit the endogenous sterol synthesis. Since 7-DHC appears to substitute for cholesterol in some cellular functions, the net effect of this situation may be detrimental.

Defining the mechanism of action of the HMG-CoA reductase inhibitors in SLOS will be important for safely designing a clinical trial. If upregulation of a partially functioning enzyme is the mechanism of the observed effect, initial trials should be limited to patients who can be demonstrated to have significant residual enzymatic function. Since simvastatin crosses the blood-brain barrier, it may directly affect the biochemical defect in the CNS and thus, perhaps, the behavioral phenotype of the disease.

Several pharmacological inhibitors of 3β-hydroxysterol Δ⁷-reductase have been used in studies of SLOS teratogenesis (reviewed in ref. 10). However, these studies are confounded by the effects of these drugs on maternal cholesterol synthesis and on other enzymes involved in cholesterol biosynthesis. Wassif et al. (10) produced a genetic mouse model of SLOS by disruption of *Dhcr7*. As found in human patients, *Dhcr7⁻ᐟ⁻* pups have markedly reduced tissue cholesterol levels and increased 7-DHC levels. Phenotypic overlap between this mouse model and the human syndrome includes intrauterine growth retardation; variable craniofacial malformations, including cleft palate; poor feeding with an abnormal suck; and neurological abnormalities, including apparent hypotonia, apparent weakness, and decreased movement.

Desmosterolosis. As shown in Figure 1, two major pathways of cholesterol synthesis operate in parallel. The first uses 7-DHC as

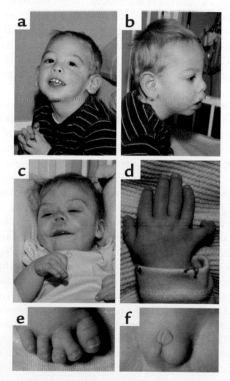

Figure 2

Phenotypic features of SLOS. (**a–c**) Facial appearance including microcephaly, ptosis, a small upturned nose, and micrognathia. (**d**) Hand malformations including postaxial polydactyly and a short thumb. (**e**) Second-third toe syndactyly. (**f**) Ambiguous genitalia in a 46, XY patient.

the immediate precursor of cholesterol. In the second, the enzyme 3β-hydroxysterol Δ²⁴-reductase reduces the Δ²⁴ double bond in desmosterol to yield cholesterol. Desmosterolosis (MIM no. 602398; see refs. 3, 11 for extensive reviews), an autosomal recessive malformation syndrome resulting from a deficiency of this enzyme, was the second human malformation syndrome shown to arise from an inborn error of cholesterol synthesis.

Since only two cases have been identified, the desmosterolosis phenotype has yet to be fully delineated. The initial patient, identified by Clayton et al. (12), was an infant who died soon after birth. This patient had macrocephaly, craniofacial malformations including cleft palate and a thick alveolar ridge, a congenital heart malformation, hypoplastic lungs, renal hypoplasia, ambiguous genitalia, short limbs, and osteosclerosis. The second patient was a 4-year-old male with marked developmental delay, microcephaly, dysmorphic facial features, and limb malformations (13). Sterol analysis by liquid gas chromatography-mass spectrometry (LGC-MS) showed marked elevations of desmosterol in serum and tissues from both of these patients.

The human 3β-hydroxysterol Δ²⁴-reductase gene (*DHCR24*), identified and cloned by Waterham et al. (13), encodes a 516–amino acid polypeptide with one predicted transmembrane domain. The first patient had one missense mutation, Y471S, on one allele, and two missense mutations, N294T and K306N, on the second allele. The second patient was homozygous for an E191K mutation. All four missense mutations independently decreased DHCR24 activity, and the combination of N294T with K306N was more severe than either mutation by itself.

X-linked dominant chondrodysplasia punctata type 2. Chondrodysplasia punctata is a defect of endochondral bone formation characterized by abnormal foci of calcification in cartilaginous elements. In x-rays, chondrodysplasia punctata appears as epiphyseal stippling. X-linked dominant chondroplasia punctata type 2 (CDPX2, also known as Conradi-Hünermann-Happle syndrome or Happle syndrome; MIM no. 302960) was first delineated by Happle (14) in 1979. This X-linked dominant disorder is typically lethal in hemizygous males. Heterozygous females present at birth with generalized congenital ichthyosiform erythroderma, a scaly hyperkeratotic rash in linear and blotchy patterns following the lines of Blaschko, consistent with the effects of X-inactivation. Older patients suffer from ichthyosis, atrophoderma, layered splitting of the nails, trichorrhexis nodosa, and patchy alopecia. Skeletal abnormalities include chondrodysplasia punctata, asymmetrical rhizomelic limb shortening, scoliosis, and hexadactyly. Craniofacial anomalies include an asymmetrical head shape, due to unilateral hypoplasia, and a flat nasal bridge. Asymmetrical and sectorial cataracts are found in CDPX2. Like the distribution of the dermatological findings, the asymmetrical nature of the craniofacial malformations and the cataracts is likely due to differential X-inactivation.

Because chondrodysplasia punctata has been described in SLOS, Kelley et al. (15) investigated the possibility that abnormal sterol synthesis could cause other genetic disorders with chondrodysplasia punctata. This group identified five CDPX2 patients with increased tissue 8-dehydrocholesterol and cholesta-8(9)-en-3β-ol, consistent with a deficiency of 3β-hydroxysterol Δ⁸,Δ⁷-isomerase activity (Figure 1). This enzyme was first identified as a high-affinity emopamil-binding protein (EBP) (16), a putative four–transmembrane domain integral membrane protein of 230 amino acids. Mutations of the *EBP* gene in CDPX2 patients, first reported by Derry et al. (17) and Braverman et al. (18), include a preponderance of presumed null

alleles, suggesting that mild missense alleles either are less penetrant or yield a different phenotypic presentation.

In the tattered mouse, a missense mutation (G107R) of *Ebp* confers an X-linked male-lethality trait (17). Heterozygous females are growth-retarded and have a "tattered" appearance due to hyperkeratotic patches of alopecia. The hemizygous male phenotype includes intrauterine growth retardation, micrognathia, short limbs, retarded ossification, and intestinal agenesis.

CHILD syndrome. CHILD syndrome (MIM # 308050), an X-linked dominant disorder which consists of congenital hemidysplasia with ichthyosiform erythroderma/nevus and limb defects, overlaps both phenotypically and genetically with CDPX2. This disorder is distinguished by the striking unilateral presentation of ichthyosiform and inflammatory nevus in heterozygous females, although contralateral skin involvement can be present. In 1980, Happle et al. (19) reviewed the CHILD phenotype in 20 cases and found that ipsilateral limb defects range from phalangeal hypoplasia to limb truncation, and ipsilateral skeletal defects may include hypoplasia of the calvaria, mandible, scapula, vertebrae, or ribs. Chondrodysplasia punctata was noted in five of these 20 cases. Congenital heart disease has also been described, as well as ipsilateral renal agenesis and lung hypoplasia. Unlike CDPX2, cataracts are not reported in CHILD syndrome, and the patient's right side is more often affected than the left. Defects in cholesterol synthesis in patients with CHILD syndrome were simultaneously reported by two groups in 2000. König et al. (20) reported mutations in the NADPH sterol dehydrogenase–like protein gene (*NSDHL*) in six patients, and Grange et al. (21) reported mutation of the *EBP* gene in a single patient. The NSDHL protein likely functions in concert with a C4-sterol methyloxidase and a 3-ketoreductase to remove the C4 methyl groups of 4,4-dimethylcholesta-8(9)-en-3β-ol or 4,4-dimethylcholesta-8,24-dien-3β-ol to yield cholesta-8(9)-en-3β-ol or zymosterol, respectively (22). This enzymatic reaction immediately precedes the reaction catalyzed by EBP (Figure 1).

Although the diagnosis of the patient with the *EBP* mutation has been debated, CHILD syndrome as a clinical diagnosis is distinguished by its predominantly unilateral presentation. This presentation may reflect a disturbance of laterality determination induced by the presence of abnormal sterol metabolites. The function of Hedgehog proteins is affected by abnormal sterol metabolites, and both Sonic hedgehog (Shh) and Indian hedgehog (Ihh) are involved in left-right axis determination (see ref. 23). Although an epigenetic effect of random X-inactivation could lead to a unilateral presentation, this mechanism would not explain why all patients with *NSDHL* mutations appear to present with CHILD syndrome, whereas only some patients with *EBP* mutations present with a CHILD phenotype. However, because of the sequential nature of these two inborn errors of cholesterol synthesis, the elevated 4,4-dimethylcholesta-8-en-3β-ol and 4,4-dimethylcholesta-8,24-dien-3β-ol levels that are characteristic of *NSDHL* mutations could also occur as a result of mutations in *EBP*. A sterol-induced effect on left-right axis determination, perhaps specific to one of these metabolites, needs to be considered. Accumulation of different precursor sterols in patients with *NSDHL* or *EBP* mutations may explain the differences in the dermatological findings between these two syndromes and the development of cataracts in CDPX2 but not in CHILD syndrome.

Mutations of the *Nsdhl* are found in the bare patches (*Bpa*) and striated (*Str*) mice. *Bpa* and *Str* are allelic phenotypes with a hypomorphic presentation in *Str*. Heterozygous *Bpa* females demonstrate growth retardation and have patchy hyperkeratotic skin lesions, short limbs, chondrodysplasia punctata, and asymmetrical cataracts. The *Bpa* mutation is a preimplantation lethal for hemizygous male embryos. Liu et al. (24) showed that *Bpa* and *Str* mice carry mutations of *Nsdhl* and sterol profiles consistent with impaired Nsdhl function. The striking unilateral presentation of CHILD syndrome is not observed in *Bpa*, *Str*, or tattered (*Td*) mice (22).

Greenberg dysplasia. Greenberg dysplasia (MIM no. 215140), also called hydrops–ectopic calcification–moth-eaten skeletal dysplasia (HEM dysplasia), is a rare autosomal recessive skeletal dysplasia first described by Greenberg et al. (25). The phenotypic findings of this lethal disorder can include dysmorphic facial features, hydrops fetalis, cystic hygroma, incomplete lung lobation, pulmonary hypoplasia, extramedullary hematopoiesis, intestinal malrotation, polydactyly, and very short limbs. Radiological findings include a distinctive "moth-eaten" appearance of the long bones, platyspondyly with abnormal ossification centers, ectopic ossification of both the ribs and the pelvis, and deficient ossification of the skull. Histological characterization showed marked disorganization of cartilage and bone, with absence of cartilage column formation, nodular calcifications in cartilage, and islands of cartilage surrounded by bone.

Based on the observation that the ossification abnormalities found in HEM dysplasia are similar to the ossification abnormalities found in CDPX2, Kelley (3) analyzed sterols from cartilage obtained from four fetuses with HEM dysplasia. He found increased levels of cholesta-8,14-dien-3β-ol and cholesta-8,14,24-trien-3β-ol, consistent with a deficiency of 3β-hydroxysterol Δ^{14}-reductase activity. To date, the genetic cause of HEM dysplasia has not been identified.

Antley-Bixler syndrome. Antley-Bixler syndrome (MIM no. 207410) is one of a large number of craniosynostosis syndromes. Reviewing phenotypic findings in 22 cases, Bottero et al. (26) noted a variety of other morphogenetic defects, including brachycephaly, proptosis, midface hypoplasia with choanal atresia or stenosis, dysplastic ears, radiohumeral synostosis, femoral bowing, multiple joint contractures and long bone fractures, as well as genital anomalies in most subjects, and imperforate anus and congenital heart defects in occasional cases.

Antley-Bixler syndrome appears to affect a heterogenous group of patients with multiple genetic and teratogenic etiologies. Mutations of the FGF receptor 2 gene (*FGFR2*) have been reported in a number of patients with an Antley-Bixler–like phenotype (27). Mutations of FGF receptor gene family members (*FGFR1*, *FGFR2*, and *FGFR3*) are common in craniosynostosis syndromes. Reardon et al. (27) identified *FGFR2* mutations in seven of sixteen Antley-Bixler patients and described various abnormalities of steroid biogenesis in a different subset of these patients. Although only one of the patients exhibited both an abnormality in steroid biogenesis and an *FGFR2* mutation, this group has proposed that some cases of Antley-Bixler syndrome might be digenic in origin, with abnormal steroid biogenesis or an alteration of the steroid environment potentiating the effects of a hypomorphic *FGFR2* mutation.

Reardon et al. (27) also reviewed the role of fluconazole in the genesis of teratogenic cases of Antley-Bixler syndrome. This synthetic triazole antifungal agent appears to be a teratogen, causing a malformation syndrome with a phenotype similar to that of Antley-Bixler syndrome. Fluconazole inhibits the cytochrome P450 enzyme lanosterol 14-α-demethylase, which demethylates lanosterol. Based on this association with lanosterol 14-α-demethylase inhibition, Kelley et al. (28) hypothesized that autosomal recessive cases of Antley-Bixler syndrome might be due to a deficiency *CYP51*, which encodes this enzyme, and they showed that lymphoblasts from one Antley-Bixler patient who did not have a mutation of *FGFR2* exhibited markedly increased levels of lanosterol and dihydrolanosterol. While the subject studied appeared to be free of pathogenic mutations in *CYP51*, these data support the idea that loss of lanosterol 14-α-demethylase activity during embryonic development underlies at least some cases of autosomal recessive Antley-Bixler syndrome.

Sterols in development

The identification of human and murine malformation syndromes due to inborn errors of cholesterol synthesis has led to a better appreciation of the multiple functions of sterols during development. To understand the pathophysiological processes underlying the various malformations and clinical problems found in this group of malformation syndromes, one needs to consider the consequences of both the loss of cholesterol or its products and the teratogenic effects of accumulating precursor sterols.

Because some cholesterol precursors are biologically active, abnormally low or high concentrations of specific sterols may have functional consequences. For instance, 7-DHC is the precursor for vitamin D synthesis. The B-ring of 7-DHC is photolyzed by ultraviolet light to form cholecalciferol (vitamin D_3). Since serum vitamin D levels are elevated in patients with SLOS (our unpublished data), calcium homeostasis may be altered. Likewise, although functions for these metabolites have yet to be been determined, both 7-DHC and 8-DHC are naturally present at high concentrations in the rat epididymis (29). In addition, 4,4-dimethyl-5α-cholesta-8,24-dien-3β-ol and 4,4-dimethyl-5α-cholesta-8,14,24-trien-3β-ol are meiosis-activating sterols that have been shown to activate the LXRα nuclear receptor (30). Altered gene regulation by these nuclear receptor transcription factors could clearly have profound consequences on development. Another metabolite, desmosterol, is synthesized in the CNS and accumulates just prior to the onset of myelination (31). It is not known whether this accumulation of desmosterol is necessary for the normal myelination process or merely reflects limited DHCR24 activity in this tissue. Desmosterol is also a major sterol in the testes and spermatozoa, and its concentration in the flagella has led to the hypothesis that it increases membrane fluidity to allow tail movement in sperm motility (32). Thus, in addition to being a precursor for cholesterol, desmosterol likely has independent functions. It is likely that the accumulation of bioactive cholesterol precursors contributes to the developmental malformations found in the inborn errors of cholesterol synthesis.

Low levels of cholesterol or the substitution of other sterols for cholesterol may also have detrimental effects during development. In particular, the accumulation of 7- and 8-DHC in SLOS can give rise to aberrant bile acids (33) and steroid hormones (34) whose biological effects have not been defined. 7-DHC itself may impair the function of proteins with sterol-sensing domains, such as HMG-CoA reductase, the Niemann-Pick type C protein (NPC1; see Maxfield and Wüstner, this Perspective series, ref. 35), the sterol regulatory element binding protein cleavage–activating protein (SCAP; ref. 36), and two proteins, Patched and Dispatched, that participate in Hedgehog signaling during development (see Jeong and McMahon, this series, ref. 37). Fitzky et al. (38) have shown that 7-DHC inhibits cholesterol synthesis by accelerating the degradation of HMG-CoA reductase protein, and Wassif et al. (39) have shown that 7-DHC impairs LDL cholesterol metabolism in a manner similar to that seen in Neimann-Pick type C.

In addition to its role as a precursor molecule, cholesterol is a structural lipid in cellular membranes. With their elevated 7-DHC levels, SLOS fibroblasts are reported to have increased membrane fluidity (40). Altered 7-DHC/cholesterol ratios may also affect cellular interactions and morphology during development (41), perhaps by perturbing signal transduction in lipid rafts, ordered lipid microenvironments in the plasma membrane consisting of cholesterol and sphingolipids (see Simons and Ehehalt, this Perspective series, ref. 42; and ref. 43). Indeed, incorporation of 7-DHC changes the structural properties of lipid rafts (44) and may affect local protein structures or interactions. The effects of other cholesterol synthetic intermediates on the structural properties of lipid rafts have not yet been thoroughly investigated in vitro or in the various mouse models of cholesterol biosynthesis disorders.

The requirement for cholesterol in the maturation of the Hedgehog family morphogens has now been studied extensively in vertebrate and invertebrate systems (see Jeong and McMahon, this Perspective series, ref. 37). There are three vertebrate Hedgehog family members. Shh functions in the patterning of the neural tube and the limbs, Ihh is important in endochondral bone formation, and Desert hedgehog (Dhh) is necessary for normal testicular development (23). In the course of autocatalytic cleavage of the Hedgehog protein, a molecule of cholesterol is added covalently to the protein's amino-terminal signaling domain (45), thus attaching these proteins to the plasma membrane and altering the secretion and movement of these signaling proteins during development (23). Disturbance of Hedgehog signaling likely underlies some of the developmental abnormalities found in the cholesterol synthetic disorders. Shh signaling appears to be impaired in SLOS, as well as in model systems treated with inhibitors of postsqualene cholesterol biosynthesis. Indeed, pharmacologic inhibition of 3β-hydroxysterol Δ^7-reductase activity is teratogenic, causing holoprosencephaly, a phenotype seen in $Shh^{-/-}$ mouse embryos and in individuals with either SHH mutations (46) or severe SLOS (1, 41). The inhibitor AY9944 has been shown to perturb the normal expression of Shh, as well as of the developmentally important transcription factors Otx2 and Pax6 in the ventral neural tube (47). Because Hedgehog and the Hedgehog receptor complex of Patched and Smoothened localize to lipid raft domains (48, 49), altered physiochemical properties of lipid rafts incorporating sterols other than cholesterol (44) may represent a mechanism by which Hedgehog signaling could be impaired in the cholesterol synthetic defects.

Common clinical findings in the various malformation syndromes that occur in inborn errors of cholesterol synthesis include abnormalities of bone formation and limb patterning (11). Thus, rhizomesomelic limbs have been reported in SLOS, desmosterolosis, Greenberg dysplasia, and CDPX2. Postaxial polydactyly is found in SLOS, CDPX2, and Greenberg dysplasia. Chondrodysplasia punctata has been reported in SLOS, CDPX2, and CHILD syndrome. Other abnormalities, such as osteosclerosis in desmosterolosis, limb aplasia in CHILD syndrome, and the dramatic "moth-eaten" appearance of the skeleton in Greenberg dysplasia, are unique to a given syndrome. Disturbance of SHH, which plays a major role in both proximal-distal and preaxial-postaxial patterning of the limb, and of IHH, which is involved in the normal progression of cartilage to bone during endochondral bone formation (23), likely underlies some of the shared developmental abnormalities. However, the unique aspects of some of the skeletal defects argue for specific teratogenic effects of the various precursor sterols during skeletal development.

Sterols in CNS function and disease

Sterols and sterol synthetic enzymes appear to play a major role in CNS function. The mental retardation and behavioral phenotype found in SLOS may be due to developmental abnormalities, or to disturbance of CNS function arising from a deficit of cholesterol in this tissue, as suggested by the apparently beneficial effect of dietary cholesterol supplementation on behavioral aspects of this disorder. Understanding the pathophysiological processes underlying the clinical problems found in the inborn errors of cholesterol synthesis may offer insights into pathophysiological processes in more common disorders of the CNS.

Epidemiological studies have suggested an association between decreased serum cholesterol levels and problems such as depression, suicide, and increased impulsivity and aggression. It has been postulated that decreased cholesterol levels contribute to the serotonergic abnormalities found in these disorders (50). Indeed, older children and adults with SLOS frequently have clinical depression (11), and both impulsivity and aggression are part of the SLOS behavioral phenotype (4), consistent with impaired serotonergic function. Alternatively, the Hedgehog signaling pathway is involved in the generation of serotonergic neurons in the developing brain, so an abnormal sterol environment during development could disturb normal development of this neuronal population. Further study of existing mouse models of SLOS and the development of viable hypomorphic mouse models will allow some of these questions to be addressed.

Dietary cholesterol supplementation may also help prevent the development of autistic behavior in SLOS. In a retrospective study, Tierney et al. (4) found that the age of initiation of dietary cholesterol supplementation appeared to influence the development of autistic behavioral characteristics. Of nine SLOS patients who were started on dietary cholesterol supplementation prior to 5 years of age, two (22%) met the Autism Diagnostic Interview-Revised (ADI-R) criteria for autism. In contrast, seven of eight (88%) SLOS patients started on dietary cholesterol supplementation after 5 years of age met these criteria. In a pilot study by Tierney and Kelley (reported in ref. 3), cholesterol therapy reduced the mean ADI-R social domain score of five 4- to 5-year-old SLOS children from 18.2 ± 2.8 to 6.8 ± 1.8, where a normal score is 0 and scores above 10 are judged to be consistent with autism. Although SLOS patients likely represent a small subset of patients with autism, the characterization of this aspect of the SLOS behavioral phenotype may provide broader insights into the roles of sterols in the biology of autism.

The effect of dietary cholesterol supplementation suggests that the behavioral phenotype of SLOS is, at least in part, due to a biochemical disturbance, rather than to irreversible developmental problems arising during embryogenesis. Because the blood-brain barrier excludes significant transfer of dietary cholesterol to the CNS, it may be that the beneficial effects are mediated by neurosteroids, molecules that appear to modulate the activity of a number of neurotransmitter receptors, including NMDA glutamate, GABA$_A$, and cholinergic receptors (51, 52). Analogs of pregnenolone, pregnanetriol, DHEA, and androstenediol derived from 7-DHC have been identified (34, 53), but it is not yet known whether any of these compounds can either substitute for or antagonize the effects of their normal, 7-hydroxy analogs.

The σ-receptors are a class of receptors thought to mediate immunosuppressive, antipsychotic, and neuroprotective effects of a large number of drugs and neurosteroids (51, 52, 54). Three σ-receptor family members have been described, and at least two of these may have role in cholesterol synthesis (54). At the amino acid level, the guinea pig σ_1-receptor is 30% identical and 69% similar to the yeast ERG2 protein, a sterol Δ^8,Δ^7-isomerase involved in ergosterol synthesis. ERG2 has a predicted secondary structure similar to that of the σ_1-receptor and binds numerous σ_1 ligands. Although the human gene product fails to complement *ERG2* mutant yeast, the σ_1-receptor may function as a sterol isomerase in mammalian neurons, as suggested by Gibbs and Farb (51). Remarkably, the mammalian 3β-hydroxysterol Δ^8,Δ^7-isomerase protein (first identified as a high-affinity EBP, as noted above) is also a member of the σ-receptor family, although it is structurally distinct from the σ_1-receptor. This parallel is very intriguing and suggests that sterol isomerase function and σ-receptor function may be somehow linked. A third member of the σ-receptor family, σ_2, has yet to be purified or cloned. It will be interesting to determine whether this protein is structurally related to a sterol isomerase or to other enzymes in cholesterol synthesis.

Poor feeding, a major clinical problem encountered in SLOS infants, is also observed in *Dhcr7*$^{-/-}$ pups (10). These pups die during the first day of life due to an abnormal suck and failure to feed. Neurophysiological testing shows a decreased NMDA receptor response of cortical neurons to glutamate stimulation. Given both that pharmacological blockade of NMDA receptor function in newborn mice impairs suckling and that disruption of the *NMDAε2* subunit gene results in an uncoordinated suck similar to what we observe in *Dhcr7*$^{-/-}$ newborns, we speculate that impaired NMDA receptor function underlies the poor feeding and abnormal suck observed in *Dhcr7*$^{-/-}$ pups. Impaired NMDA function in the *Dhcr7*$^{-/-}$ mice could have several possible explanations. Decreased cholesterol synthesis by glial cells could disrupt normal synaptogenesis (55). Alternatively, the substitution of 7-DHC and 7-dehydrodesmosterol for cholesterol and desmosterol, respectively, could directly inhibit the NMDA receptor or impair translocation of the receptor subunits to the synaptosomal membrane. NMDA function could also be impaired indirectly by an effect on neurosteroid function, since NMDA glutamate receptor function is increased by neurosteroids, and decreased cholesterol levels might reduce the production of neurosteroids. Increased 7-DHC levels might also inhibit neurosteroid formation or lead to synthesis of an inhibitory analog. The finding of decreased NMDA glutamate receptor function in *Dhcr7*$^{-/-}$ cortical neurons may provide insight into some of the neurological and feeding problems found in SLOS infants. Further characterization of neurotransmitter function in these model systems should help clarify the pathophysiological basis of the behavioral and mental disturbances found in this human syndrome and may suggest therapies to ameliorate these problems.

One other link that has been proposed between cholesterol and neurological dysfunction concerns Alzheimer disease (AD). This link was suggested in part by epidemiological data showing that elevated serum cholesterol levels are a risk factor for later development of AD. In addition, animal studies indicate that dietary cholesterol supplementation causes β-amyloid accumulation, that diet-induced hypercholesterolemia can accelerate the development of AD-like pathology, and that cholesterol levels may directly affect the activity of β- and γ-secretases, processing enzymes that act on the amyloid precursor protein (reviewed in ref. 56; see also Simons, this Perspective series, ref. 42). Further evidence comes from genetic associations of *APOE*, which encodes a lipoprotein involved in cholesterol transport in the brain, since carriers of the ε4 allele of this gene are at markedly increased risk of developing AD (57). Also consistent with either a primary or a secondary disturbance of CNS cholesterol homeostasis in these patients, levels of 24S-hydroxycholesterol, which plays a major role in CNS cholesterol homeostasis, are elevated in the cerebrospinal fluid of AD patients (58). Recent epidemiological studies have shown that the prevalence of AD is decreased in patients on HMG-CoA inhibitors (59). Fassbender et al. (60), likewise, have recently reported that that simvastatin reduces β-amyloid peptide accumulation in vitro and in vivo.

A number of observations link AD to inborn errors of cholesterol synthesis. First, the 3β-hydroxysterol Δ^7-reductase inhibitor BM15.766, which has been used to produce teratogenic SLOS rodent models, reduces plaque formation in a transgenic mouse model of AD (61). Since the number of adult patients with SLOS is small, and no pathological analysis has been reported from older patients, it is not known whether SLOS patients or *DHCR7* heterozygotes are at increased or decreased risk for AD. Second, *DHCR24* is identical to a previously identified gene named *seladin-1* (selective Alzheimer disease indicator 1), whose expression is decreased in neurons from patients with AD. When overexpressed in cell culture, this gene product appears to have a protective effect against β-amyloid toxicity (62). Sterol profiles, including that of desmosterol, have not been well characterized in AD, but the recognition that seladin-1 is an enzyme involved in cholesterol synthesis suggests that desmosterol or other Δ^{24}-sterols contribute to the pathogenesis of AD.

Summary

The inborn errors of cholesterol synthesis are an interesting group of human malformation syndromes for both basic and clinical science. Cholesterol and precursor sterols are essential for many normal cellular and developmental processes. The importance of sterols in normal embryonic development has been underscored by the discovery that inborn errors of cholesterol synthesis cause human and murine malformation syndromes. Defining the clinical problems found in this group of disorders and studying the developmental perturbations caused by inborn errors of cholesterol synthesis continues to yield insights into the function of cholesterol and its precursors during normal development and to suggest potential therapeutic interventions.

Acknowledgments

I would like to thank Richard Kelley with sharing a copy of his Antley-Bixler manuscript with me prior to its publication. I would like to acknowledge the assistance of Lina Correa-Cerro, Diana Cozma, Brooke Wright, and Christopher Wassif in the preparation of this review. Finally, I would like to express my appreciation to the families of my patients.

Address correspondence to: Forbes D. Porter, Heritable Disorders Branch, NICHD, NIH, Building 10, Room 9S241, 10 Center Drive, Bethesda, Maryland 20892, USA. Phone: (301) 435-4432; Fax: (301) 480-5791; E-mail: fdporter@helix.nih.gov.

1. Kelley, R.I., and Hennekam, R.C. 2000. The Smith-Lemli-Opitz syndrome. *J. Med. Genet.* **37**:321–335.
2. Porter, F.D. 2000. RSH/Smith-Lemli-Opitz syndrome: a multiple congenital anomaly/mental retardation syndrome due to an inborn error of cholesterol biosynthesis. *Mol. Genet. Metab.* **71**:163–174.
3. Kelley, R.I. 2000. Inborn errors of cholesterol biosynthesis. *Adv. Pediatr.* **47**:1–53.
4. Tierney, E., et al. 2001. Behavior phenotype in the RSH/Smith-Lemli-Opitz syndrome. *Am. J. Med. Genet.* **98**:191–200.

5. Irons, M., Elias, E.R., Salen, G., Tint, G.S., and Batta, A.K. 1993. Defective cholesterol biosynthesis in Smith-Lemli-Opitz syndrome. *Lancet.* **341**:1414.

6. Wassif, C.A., et al. 1998. Mutations in the human sterol delta7-reductase gene at 11q12-13 cause Smith-Lemli-Opitz syndrome. *Am. J. Hum. Genet.* **63**:55–62.

7. Battaile, K.P., Battaile, B.C., Merkens, L.S., Maslen, C.L., and Steiner, R.D. 2001. Carrier frequency of the common mutation IVS8-1G>C in DHCR7 and estimate of the expected incidence of Smith-Lemli-Opitz syndrome. *Mol. Genet. Metab.* **72**:67–71.

8. Xu, G., et al. 1995. Treatment of the cholesterol biosynthetic defect in Smith-Lemli-Opitz syndrome reproduced in rats by BM 15.766. *Gastroenterology.* **109**:1301–1307.

9. Jira, P.E., et al. 2000. Simvastatin. A new therapeutic approach for Smith-Lemli-Opitz syndrome. *J. Lipid Res.* **41**:1339–1346.

10. Wassif, C.A., et al. 2001. Biochemical, phenotypic and neurophysiological characterization of a genetic mouse model of RSH/Smith-Lemli-Opitz syndrome. *Hum. Mol. Genet.* **10**:555–564.

11. Nwokoro, N.A., Wassif, C.A., and Porter, F.D. 2001. Genetic disorders of cholesterol biosynthesis in mice and humans. *Mol. Genet. Metab.* **74**:105–119.

12. Clayton, P., Mills, K., Keeling, J., and FitzPatrick, D. 1996. Desmosterolosis: a new inborn error of cholesterol biosynthesis. *Lancet.* **348**:404.

13. Waterham, H.R., et al. 2001. Mutations in the 3beta-hydroxysterol Delta24-reductase gene cause desmosterolosis, an autosomal recessive disorder of cholesterol biosynthesis. *Am. J. Hum. Genet.* **69**:685–694.

14. Happle, R. 1979. X-linked dominant chondrodysplasia punctata. Review of literature and report of a case. *Hum. Genet.* **53**:65–73.

15. Kelley, R.I., et al. 1999. Abnormal sterol metabolism in patients with Conradi-Hunermann-Happle syndrome and sporadic lethal chondrodysplasia punctata. *Am. J. Med. Genet.* **83**:213–219.

16. Moebius, F.F., et al. 1994. Purification and aminoterminal sequencing of the high affinity phenylalkylamine Ca2+ antagonist binding protein from guinea pig liver endoplasmic reticulum. *J. Biol. Chem.* **269**:29314–29320.

17. Derry, J.M., et al. 1999. Mutations in a delta 8-delta 7 sterol isomerase in the tattered mouse and X-linked dominant chondrodysplasia punctata. *Nat. Genet.* **22**:286–290.

18. Braverman, N., et al. 1999. Mutations in the gene encoding 3 beta-hydroxysteroid-delta 8, delta 7-isomerase cause X-linked dominant Conradi-Hunermann syndrome. *Nat. Genet.* **22**:291–294.

19. Happle, R., Koch, H., and Lenz, W. 1980. The CHILD syndrome. Congenital hemidysplasia with ichthyosiform erythroderma and limb defects. *Eur. J. Pediatr.* **134**:27–33.

20. König, A., Happle, R., Bornholdt, D., Engel, H., and Grzeschik, K.H. 2000. Mutations in the NSDHL gene, encoding a 3beta-hydroxysteroid dehydrogenase, cause CHILD syndrome. *Am. J. Med. Genet.* **90**:339–346.

21. Grange, D.K., Kratz, L.E., Braverman, N.E., and Kelley, R.I. 2000. CHILD syndrome caused by deficiency of 3beta-hydroxysteroid-delta8, delta7-isomerase. *Am. J. Med. Genet.* **90**:328–335.

22. Herman, G.E. 2000. X-Linked dominant disorders of cholesterol biosynthesis in man and mouse. *Biochim. Biophys. Acta.* **1529**:357–373.

23. Ingham, P.W. 2001. Hedgehog signaling: a tale of two lipids. *Science.* **294**:1879–1881.

24. Liu, X.Y., et al. 1999. The gene mutated in bare patch-es and striated mice encodes a novel 3beta-hydroxysteroid dehydrogenase. *Nat. Genet.* **22**:182–187.

25. Greenberg, C.R., et al. 1988. A new autosomal recessive lethal chondrodystrophy with congenital hydrops. *Am. J. Med. Genet.* **29**:623–632.

26. Bottero, L., Cinalli, G., Labrune, P., Lajeunie, E., and Renier, D. 1997. Antley-Bixler syndrome. Description of two new cases and a review of the literature. *Childs Nerv. Syst.* **13**:275–280.

27. Reardon, W., et al. 2000. Evidence for digenic inheritance in some cases of Antley-Bixler syndrome? *J. Med. Genet.* **37**:26–32.

28. Kelley, R.I., et al. 2002. Abnormal sterol metabolism in a patient with Antley-Bixler syndrome and ambiguous genitalia. *Am. J. Med. Genet.* **110**:95–102.

29. Lindenthal, B., et al. 2001. Neutral sterols of rat epididymis. High concentrations of dehydrocholesterols in rat caput epididymidis. *J. Lipid Res.* **42**:1089–1095.

30. Janowski, B.A., Willy, P.J., Devi, T.R., Falck, J.R., and Mangelsdorf, D.J. 1996. An oxysterol signalling pathway mediated by the nuclear receptor LXR alpha. *Nature.* **383**:728–731.

31. Hinse, C.H., and Shah, S.N. 1971. The desmosterol reductase activity of rat brain during development. *J. Neurochem.* **18**:1989–1998.

32. Connor, W.E., Lin, D.S., Wolf, D.P., and Alexander, M. 1998. Uneven distribution of desmosterol and docosahexaenoic acid in the heads and tails of monkey sperm. *J. Lipid Res.* **39**:1404–1411.

33. Honda, A., et al. 1999. Bile acid synthesis in the Smith-Lemli-Opitz syndrome: effects of dehydrocholesterols on cholesterol 7alpha-hydroxylase and 27-hydroxylase activities in rat liver. *J. Lipid Res.* **40**:1520–1528.

34. Shackleton, C.H., Roitman, E., and Kelley, R. 1999. Neonatal urinary steroids in Smith-Lemli-Opitz syndrome associated with 7-dehydrocholesterol reductase deficiency. *Steroids.* **64**:481–490.

35. Maxfield, F.R., and Wüstner, D. 2002. Intracellular cholesterol transport. *J. Clin. Invest.* In press. doi:10.1172/JCI200216500.

36. Horton, J.D., Goldstein, J.L., and Brown, M.S. 2002. SREBPs: activators of the complete program of cholesterol and fatty acid synthesis in the liver. *J. Clin. Invest.* **109**:1125–1131. doi:10.1172/JCI200215593.

37. Jeong, J., and McMahon, A.P. 2002. Cholesterol modification of Hedgehog family proteins. *J. Clin. Invest.* **110**:591–596. doi:10.1172/JCI200216506.

38. Fitzky, B.U., et al. 2001. 7-Dehydrocholesterol-dependent proteolysis of HMG-CoA reductase suppresses sterol biosynthesis in a mouse model of Smith-Lemli-Opitz/RSH syndrome. *J. Clin. Invest.* **108**:905–915. doi:10.1172/JCI200212103.

39. Wassif, C.A., et al. 2002. Cholesterol storage defect in RSH/Smith-Lemli-Opitz syndrome fibroblasts. *Mol. Genet. Metab.* **75**:325–334.

40. Tulenko, T.N., LaBelle, E., Boesze-Battaglia, K., Mason, R.P., and Tint, G.S. 1998. A membrane bilayer defect in the Smith-Lemli-Opitz syndrome. *FASEB J.* **12**:A827. (Abstr.)

41. Dehart, D.B., Lanoue, L., Tint, G.S., and Sulik, K.K. 1997. Pathogenesis of malformations in a rodent model for Smith-Lemli-Opitz syndrome. *Am. J. Med. Genet.* **68**:328–337.

42. Simons, K., and Ehehalt, R. 2002. Cholesterol, lipid rafts, and disease. *J. Clin. Invest.* **110**:597–603. doi:10.1172/JCI200216390.

43. Simons, K., and Toomre, D. 2000. Lipid rafts and signal transduction. *Nat. Rev. Mol. Cell Biol.* **1**:31–39.

44. Xu, X., et al. 2001. Effect of the structure of natural sterols and sphingolipids on the formation of ordered sphingolipid/sterol domains (rafts). Comparison of cholesterol to plant, fungal, and disease-associated sterols and comparison of sphingomyelin, cerebrosides, and ceramide. *J. Biol. Chem.* **276**:33540–33546.

45. Porter, J.A., Young, K.E., and Beachy, P.A. 1996. Cholesterol modification of hedgehog signaling proteins in animal development. *Science.* **274**:255–259.

46. Roessler, E., et al. 1996. Mutations in the human Sonic Hedgehog gene cause holoprosencephaly. *Nat. Genet.* **14**:357–360.

47. Gofflot, F., Kolf-Clauw, M., Clotman, F., Roux, C., and Picard, J.J. 1999. Absence of ventral cell populations in the developing brain in a rat model of the Smith-Lemli-Opitz syndrome. *Am. J. Med. Genet.* **87**:207–216.

48. Rietveld, A., Neutz, S., Simons, K., and Eaton, S. 1999. Association of sterol- and glycosylphosphatidylinositol-linked proteins with Drosophila raft lipid microdomains. *J. Biol. Chem.* **274**:12049–12054.

49. Karpen, H.E., et al. 2001. The sonic hedgehog receptor patched associates with caveolin-1 in cholesterol-rich microdomains of the plasma membrane. *J. Biol. Chem.* **276**:19503–19511.

50. Brunner, J., Parhofer, K.G., Schwandt, P., and Bronisch, T. 2002. Cholesterol, essential fatty acids, and suicide. *Pharmacopsychiatry.* **35**:1–5.

51. Gibbs, T.T., and Farb, D.H. 2000. Dueling enigmas: neurosteroids and sigma receptors in the limelight. *Sci. STKE* 2000:PE1. http://stke.sciencemag.org/cgi/reprint/sigtrans;2000/60/pe1.pdf.

52. Maurice, T., Urani, A., Phan, V.L., and Romieu, P. 2001. The inter-action between neuroactive steroids and the sigma1 receptor function: behavioral consequences and therapeutic opportunities. *Brain Res. Brain Res. Rev.* **37**:116–132.

53. Shackleton, C., Roitman, E., Guo, L., Wilson, W.K., and Porter, F.D. 2002. Identification of 7(8) and 8(9) unsaturated adrenal steroid metabolites produced by patients with 7-dehydrosterol-7-reductase deficiency (Smith-Lemli-Opitz syndrome). *J. Steroid Biochem. Mol. Biol.* In press.

54. Moebius, F.F., Striessnig, J., and Glossmann, H. 1997. The mysteries of sigma receptors: new family members reveal a role in cholesterol synthesis. *Trends Pharmacol. Sci.* **18**:67–70.

55. Mauch, D.H., et al. 2001. CNS synaptogenesis promoted by glia-derived cholesterol. *Science.* **294**:1354–1357.

56. Hartmann, T. 2001. Cholesterol, A beta and Alzheimer's disease. *Trends Neurosci.* **24**(Suppl.):S45–S48.

57. Eichner, J.E., et al. 2002. Apolipoprotein E polymorphism and cardiovascular disease: a HuGE review. *Am. J. Epidemiol.* **155**:487–495.

58. Papassotiropoulos, A., et al. 2002. 24S-hydroxycholesterol in cerebrospinal fluid is elevated in early stages of dementia. *J. Psychiatr. Res.* **36**:27–32.

59. Wolozin, B., Kellman, W., Ruosseau, P., Celesia, G.G., and Siegel, G. 2000. Decreased prevalence of Alzheimer disease associated with 3-hydroxy-3-methyglutaryl coenzyme A reductase inhibitors. *Arch. Neurol.* **57**:1439–1443.

60. Fassbender, K., et al. 2001. Simvastatin strongly reduces levels of Alzheimer's disease beta-amyloid peptides Abeta 42 and Abeta 40 in vitro and in vivo. *Proc. Natl. Acad. Sci. USA.* **98**:5856–5861.

61. Refolo, L.M., et al. 2001. A cholesterol-lowering drug reduces beta-amyloid pathology in a transgenic mouse model of Alzheimer's disease. *Neurobiol. Dis.* **8**:890–899.

62. Greeve, I., et al. 2000. The human DIMINUTO/DWARF1 homolog seladin-1 confers resistance to Alzheimer's disease-associated neurodegeneration and oxidative stress. *J. Neurosci.* **20**:7345–7352.

Intracellular cholesterol transport

Frederick R. Maxfield and Daniel Wüstner

Department of Biochemistry, Weill Medical College of Cornell University, New York, New York, USA.

The correct intracellular distribution of cholesterol among cellular membranes is essential for many biological functions of mammalian cells, including signal transduction and membrane traffic. Intracellular trafficking plays a major role in the proper disposition of internalized cholesterol and in the regulation of cholesterol efflux. Despite the importance of the transport and distribution of cholesterol within cells for normal physiology and in pathological conditions, many fundamental aspects of intracellular cholesterol movement are not well understood. For instance, the relative roles of vesicular and nonvesicular transport have not been fully determined, and the asymmetric distribution of cholesterol between the two leaflets of biological membranes is poorly characterized in many cases. Also, while it is clear that small, cholesterol-enriched microdomains (often referred to as rafts; see Simons and Ehehalt, this Perspective series, ref. 1) occur in many biological membranes, the composition, size, and dynamics of these microdomains remain uncertain. Here, we will discuss the current understanding of intracellular cholesterol transport.

The intracellular distribution of cholesterol

The membrane organelles of mammalian cells maintain distinct protein and lipid compositions that are essential for their proper function (Figure 1). In the biosynthetic secretory pathway, cholesterol is low in the endoplasmic reticulum (ER), but its level increases through the Golgi apparatus, with the highest levels in the plasma membrane (2). In the endocytic pathway, the endocytic recycling compartment (ERC), which contains recycling membrane proteins and lipids (3), also contains high levels of cholesterol (4). The cholesterol content of late endosomes and lysosomes is not well documented, but under normal conditions it appears to be lower than in the ERC (2, 4). In polarized epithelial cells, the apical membrane is enriched in cholesterol and sphingolipids relative to the basolateral membrane (5).

Although the ER is the site of cholesterol synthesis, the concentration of cholesterol in the ER is very low, comprising only about 0.5–1% of cellular cholesterol (6) even though the surface area of the ER exceeds that of the plasma membrane in many cells. Many aspects of cholesterol regulation are under tight feedback control and are sensitive to the concentration of cholesterol in the ER. For example, an increase in the cholesterol content of the ER accelerates the degradation of 3-hydroxy-3-methylglutaryl-coenzyme A reductase, the key regulated enzyme for cholesterol synthesis (7). The sterol response element–binding protein (SREBP) cleavage-

activating protein (SCAP) responds to reduction in cholesterol in the ER membrane by activating proteolytic cleavage and translocation of a fragment of SREBP into the nucleus, where it alters the transcription of several genes involved in cholesterol regulation (8). Thus, the sterol concentration in the ER membrane is crucial for cellular cholesterol homeostasis.

The cholesterol content of the Golgi apparatus is intermediate between those of the ER and the plasma membrane, but the measured cholesterol content of the Golgi depends on the method used for purification of these membranes (9). Electron microscopic filipin binding studies show an increased cholesterol content from the cis- to the trans-Golgi (7). It has been proposed that rafts enriched in sphingomyelin, glycosphingolipids, and cholesterol form in the Golgi apparatus and that these raft domains are selectively transported to the apical domain from the trans-Golgi (5, 10). The cholesterol in the ERC is similarly important for correct membrane trafficking. Reductions in cellular cholesterol have been shown to alter the recycling of glycosyl-phosphatidylinositol-anchored (GPI-anchored) proteins (11).

The plasma membrane is estimated to contain about 60–80% of total cellular cholesterol (2), and it has been estimated that cholesterol is as much as 30–40% of the lipid molecules in the plasma membrane. Despite some uncertainty in these numbers, it is clear from all studies that the plasma membrane is highly enriched in cholesterol relative to other cellular membranes.

Cholesterol-rich microdomains

Cholesterol and lipids are not homogeneously distributed within biological membrane bilayers. Cholesterol- and sphingolipid-enriched microdomains, which are resistant to solubilization at low temperature by non-ionic detergents such as Triton X-100, have been proposed to play an important role in cholesterol transport (5, 10, 12). The properties of lipids in such detergent-resistant membranes (DRMs) are similar to those in a liquid-ordered (Lo) phase that has been characterized in model membranes (13). Lo domains exhibit high lateral mobility of the lipids and tight packing in the hydrophobic core of the membrane. Rafts have not been observed directly by optical microscopy in living cells, presumably because their dimensions are below the resolution limit. The high lateral mobility within Lo domains and the small size of rafts in vivo (5) ensure that molecules will encounter the raft boundaries frequently and that individual molecules leave raftlike regions within seconds.

A high fraction of the plasma membrane of many types of cells is found in raftlike domains; about 70–80% of the surface area of several cell types has been shown to be resistant to solubilization by cold Triton X-100 (14). Fluorescence polarization studies also indicate that nearly half of the plasma membrane is in ordered domains at 37°C, consistent with the high concentration of cholesterol and sphingolipids in the plasma membrane. In polarized epithelial cells, sphingolipids and cholesterol are especially enriched in the apical membrane domains, which may exist almost entirely in an ordered state (5).

Nonstandard abbreviations used: endoplasmic reticulum (ER); endocytic recycling compartment (ERC); sterol response element–binding protein (SREBP); SREBP cleavage–activating protein (SCAP); detergent-resistant membrane (DRM); liquid-ordered (Lo); dehydroergosterol (DHE); ultraviolet (UV); horseradish peroxidase (HRP); steroidogenic acute regulatory protein (StAR); StAR-related lipid transfer (START); trans-Golgi network (TGN); Niemann-Pick C1 (NPC1).

Conflict of interest: No conflict of interest has been declared.

Citation for this article: J. Clin. Invest. **110**:891–898 (2002). doi:10.1172/JCI200216500.

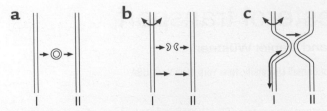

Figure 1
Basic mechanisms of cholesterol transport between two membranes.
(**a**) Vesicular transport. This process requires ATP but does not
require a change in the transversal distribution of cholesterol in the
donor membrane. (**b**) Diffusion through the cytoplasm either bound
to a carrier protein (upper arrows) or by free diffusion (lower arrows).
Cholesterol in the donor membrane must desorb from the cytoplasmic
leaflet, so the transbilayer distribution of cholesterol can affect this
process. (**c**) Transport across membrane contacts. Adjacent donor
and acceptor membranes come into close contact, resulting in cholesterol shuttling across the intermembrane space. This process requires
cholesterol in the cytosolic leaflet of the donor membrane and may be
facilitated by transport proteins.

The quantitative distribution of cholesterol between ordered
and disordered domains in the plasma membrane is not known.
Given the high overall concentration of cholesterol within the
plasma membrane, only a small degree of local enrichment may
be possible without disrupting the bilayer structure. Rather, the
plasma membrane may be maintained at a composition where
small changes in cholesterol content can cause large changes in
membrane fluidity, as has been seen in model membranes (14).

Most likely, many types of membrane microdomains coexist in
cells (14). Caveolae are one type of specialized, raftlike domain.
They are associated with caveolins, and they have a characteristic
flask shape with a diameter of about 60 nm. In most cells, caveolae
cover a few percent of the plasma membrane, so they represent
a small fraction of the plasma membrane DRMs. Under certain
circumstances, caveolae can pinch off from the plasma membrane (15), but under standard tissue culture conditions, plasma
membrane caveolae do not exchange rapidly with internal pools
of caveolin (16). Caveolins can bind cholesterol (17), but, as with
other raftlike domains, the relative enrichment of cholesterol in
caveolae is uncertain.

In addition to the lateral separation of membranes into various
domains, the two leaflets of biological membranes have distinct
compositions. Mechanisms for maintaining an asymmetric distribution of cholesterol are not well understood, but a likely mechanism is that the cholesterol distribution is largely determined by the
other lipids in the membrane. In general, the transbilayer distribution of cholesterol in biological membranes other than the plasma
membrane is not well characterized, and even in the plasma membrane there is some uncertainty. Some studies have indicated that
sterol is predominantly in the outer leaflet of the plasma membrane
(18), which would be consistent with a preferential association
with sphingomyelin (13). However, several other studies have indicated that cholesterol is enriched in the cytoplasmic leaflet (2, 19),
although the molecular basis for such an enrichment is unknown.

Methods for studying cholesterol transport and distribution

The uncertainty regarding cholesterol distribution in the inner
and outer leaflets of the plasma membrane is one example of the

more general difficulty of quantifying this and many other lipid
species in intact biological membranes. In this case, since no direct
measure of cholesterol is feasible, transbilayer distributions of cholesterol are determined by quenching emission from fluorescent
sterols using membrane impermeant quenchers. This approach is
not ideal, in part because there may be quantitative differences in
the properties of the fluorescent cholesterol analogs as compared
with cholesterol (see below).

Perhaps the most serious difficulty for the quantitative analysis of cholesterol in membranes is that, although cholesterol is
poorly soluble in water, it can spontaneously desorb from membranes at an appreciable rate. Most often it will return to the same
membrane, but it can also bind to whatever other hydrophobic
binding sites are available. For this reason, a significant fraction
of membrane cholesterol can redistribute among isolated organelles (20). In cells, soluble proteins can bind cholesterol, sometimes with high specificity and affinity (21, 22). Such proteins can
also mediate transfer of cholesterol between membranes in vitro
after cell disruption (22).

The transport and distribution of newly synthesized cholesterol
can be determined by introducing ³H-acetate into living cells and
measuring the amount of ³H cholesterol in isolated membranes
at different times. Radiolabeled cholesterol and cholesterol esters
can be delivered by lipoproteins, and labeled cholesterol can also
be delivered via specific cyclodextrin carriers, such as methyl-β-
cyclodextrin. Total cholesterol can be measured by direct chemical
methods such as gas chromatography–mass spectrometry or by
indirect methods such as assays based on cholesterol oxidase (2,
23). In order for any of these methods of measuring cholesterol
transport and distribution to be used, the various organelles of
interest must be purified. It is generally quite difficult to obtain
highly purified membrane fractions, so the possibility of effects
from contaminating membranes must be considered. In addition,
lengthy purification protocols may increase the risk of cholesterol
transfer. These methods are most useful when organelles can be
easily separated, as with the ER and the plasma membrane, but
they can be very difficult to interpret when organelles such as
endosomes and Golgi membranes are considered.

One of the most widely used tools for studying intracellular
cholesterol distribution is the fluorescent detergent filipin, which
binds selectively to cholesterol (and not to cholesterol esters) (24).
Filipin staining can be used to detect cholesterol in various membrane organelles in intact cells. Although filipin is a relatively weak
fluorophore, it is detected easily by cooled charge-coupled device
cameras. Generally, filipin has been used for qualitative analyses of
cholesterol distribution, since its fluorescence intensity is not necessarily linearly related to cholesterol content. For example, there
are differences in the accessibility of cholesterol in different pools
(25). Another limitation is that cholesterol might redistribute
during long incubations with filipin. Thus, quantification of the
intracellular cholesterol distribution from experiments using filipin is not possible, although cholesterol distributions observed with
filipin are generally consistent with the distributions obtained by
other methods. When filipin binds to cholesterol in membranes, it
produces a characteristic bump in the membrane that can be seen
in freeze-fracture micrographs. As with fluorescence studies, electron microscopy of filipin has been useful for qualitative analysis
of cholesterol distribution.

Several fluorescent derivatives of cholesterol have been used in
fluorescence microscopy studies. A major problem with fluorescent

adducts of cholesterol is that the fluorophore can greatly change the properties of the cholesterol so that its distribution among cellular membranes is drastically altered (9). A few sterols are intrinsically fluorescent and have lateral and transbilayer distribution properties in membranes that are fairly similar to cholesterol. One of these, dehydroergosterol (DHE), is a naturally occurring sterol produced by yeast (26). Although it is a rapidly photobleached, weak fluorophore with emission in the near ultraviolet (UV), DHE can be incorporated into cellular membranes in sufficient concentrations to be observed with UV-sensitive charge-coupled device cameras (4, 9, 27). DHE has also been used in a multiphoton imaging study (26). There are significant issues that need to be considered in validating DHE or other intrinsically fluorescent sterols as cholesterol analogs. DHE distributes similarly to cholesterol in cells (4, 9, 27), but there are quantitative differences, for example, in its rate of desorption from model lipid membranes (28). Thus, there may be quantitative differences in the transport of DHE as compared with cholesterol. The fluorescence properties of DHE are also sensitive to the local environment, so relative fluorescence intensities in various membranes may not reflect concentration with complete accuracy. Nonetheless, these reagents offer the significant advantage that they can be used to directly observe sterol redistribution in living cells. DHE can be delivered to the plasma membrane via a methyl-β-cyclodextrin carrier in pulses as short as 1 minute, and redistribution can then be observed and quantified in pulse-chase studies (4, 27). The photobleaching of DHE can also be used to advantage by photobleaching one region of a cell and then measuring the rate at which the fluorescence returns as unbleached DHE exchanges into the bleached area (4, 9, 27).

Some specialized techniques, such as the cholesterol oxidase assay (23), have been developed to quantify cholesterol in the plasma membrane. This assay will overestimate the amount of cholesterol on the plasma membrane if the enzyme gains access to intracellular compartments (e.g., by endocytosis in living cells or by membrane breakage) or if cholesterol moves to the plasma membrane during the assay. However, a modification of this approach, designed to minimize these problems (2), provides estimates of the fraction of cellular cholesterol at the plasma membrane (about 70%) that are in general agreement with other methods. Another useful method to quantify cholesterol delivery to the plasma membrane is extraction by extracellular cyclodextrin (29, 30). This selective cholesterol acceptor can remove at least a pool of cholesterol from the membrane within a minute or so. In such efflux experiments, cells are equilibrated with ³H-cholesterol before adding extracellular cyclodextrins. The extraction of labeled cholesterol is typically biphasic, with a slow component of extraction requiring tens of minutes (30). It is likely that a major portion of the slow phase is delivery from internal organelles (4), but it is possible that there is also a slowly extractable pool in the plasma membrane (30). A slowly extractable plasma membrane pool would lead to uncertainty in estimates of the fraction of cholesterol at the plasma membrane by this method.

An enzymatic density shift method that has been used to determine whether proteins are in the same compartment has been adapted for determining the relative amount of ³H-cholesterol in organelles (4). When horseradish peroxidase (HRP) catalyzes peroxidation of diaminobenzidine, an insoluble polymer is formed within the organelle that creates a very large increase in its density. The organelles containing the HRP-catalyzed reaction product can be separated rapidly from the other cellular components by centrifugation. HRP can be delivered to endosomes by coupling to a protein such as transferrin. According to studies using this procedure, about 35% of cellular cholesterol is found in the early endosomes, including the ERC (4). HRP can also be targeted to organelles on the biosynthetic secretory pathway by expressing constructs encoding chimeric proteins containing HRP and an ER or Golgi protein (31). This method is subject to concerns about redistribution of cholesterol during the assay, but it provides a reasonably rapid and easy way to obtain relatively clean isolation of one organelle.

While many methods have been developed to measure cholesterol distribution in cells, all of them are subject to various degrees of uncertainty in their interpretation. It is therefore necessary to compare results obtained by several different methods in order to get a reliable analysis of intracellular cholesterol distribution.

Mechanisms for intracellular sterol transport

Another great challenge in studying intracellular cholesterol transport is that several fundamentally different mechanisms to move sterols operate simultaneously in live cells (Figure 1). The combined action of these pathways can make it hard to obtain a clear overall understanding of cholesterol transport.

Cholesterol can be incorporated into transport vesicles or tubules that carry membrane constituents from one organelle to another. Pharmacological inhibition can be used to test the importance of various membrane trafficking steps. For example, the involvement of the Golgi apparatus can be tested using brefeldin A, which causes Golgi membranes to become fused into the ER (32). Microtubule-mediated movement of vesicles can be tested using nocodazole or other agents that disrupt microtubules (32). Expression of dominant inhibitory proteins can also be used to assess the role of vesicle traffic. For example, export of cholesterol or DHE from the ERC (Figure 2) requires ATP and can be blocked by expression of a mutant form of an ERC-associated protein, Rme-1 (4).

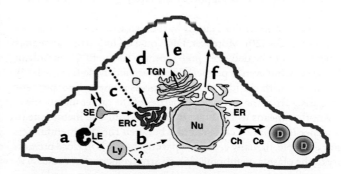

Figure 2
Cholesterol transport in nonpolarized cells. LDL carrying cholesterol and CE (esterified cholesterol) is transported (**a**) from sorting endosomes (SE) to late endosomes (LE) and lysosomes (Ly), from which cholesterol can efflux and reach the plasma membrane or the ER, where it gets esterified (**b**). Efflux from LE and Ly is poorly characterized, as indicated by dashed lines. Cholesterol can move from the plasma membrane to the ERC by a nonvesicular, ATP-independent process (**c**). In contrast, recycling of cholesterol occurs almost exclusively in vesicles also carrying other recycling markers (**d**). De novo synthesized cholesterol is mostly transported from the ER directly to the plasma membrane, bypassing the Golgi apparatus (**f**), but some follows the biosynthetic secretory pathway from the ER to the TGN (**e**). Excess cholesterol (Ch) in the ER becomes esterified (CE) and stored in cytoplasmic lipid droplets (D).

Since cholesterol can desorb from membranes at a significant rate and cells have many possible cholesterol carriers in the cytoplasm, carrier-mediated diffusion can play an important role in transport of cholesterol among cellular membranes (Figure 1). Nonvesicular transport is important for sterol transport from the plasma membrane to the ERC (4) (Figure 2) and for delivery of cholesterol to the inner mitochondrial membrane in steroidogenic cells (22). The cholesterol carriers could consist of a large number of cytosolic proteins each with low affinity and specificity for cholesterol. Alternatively, the carriers could be specialized cholesterol transport proteins. A family of high-affinity lipid and sterol carriers has been identified of which one of the prototypes is the steroidogenic acute regulatory protein (StAR/StarD1) (21), whose lipid- or sterol-binding domains are called StAR-related lipid transfer (START) domains. StAR/StarD1 has been implicated in the delivery of cholesterol to mitochondrial cytochrome P450 in steroidogenic cells (ref. 22; see also Jefcoate, this Perspective series, ref. 33). There are now several family members known, some of which have been shown to bind cholesterol with high affinity. The expression of one of these, StarD4, is regulated by cellular sterol levels (34). Interestingly, another family member, MLN64, has a transmembrane domain that localizes it to late endosomes and a START domain that binds cholesterol (35). It is not known whether the main function of these START domain proteins is sterol and lipid transport or whether they are primarily regulatory proteins.

Diffusible sterol-binding proteins provide a rapid mechanism for shuttling cholesterol among membranes, but the basis of specificity in membrane targeting is not known. One possibility is that these factors are targeted to certain compartments by binding to proteins or lipids that are enriched in those compartments. While it seems likely that specific targeting mechanisms exist, an alternate model holds that diffusible carriers distribute sterol among all possible target membranes, with the relative enrichment in various organelles determined by the ability of the membranes to serve as sterol acceptors. In support of the latter model, the initial rate of cholesterol transfer among organelles isolated from fibroblasts appears to be determined largely by the characteristics of the acceptor membrane (20). Moreover, DHE on a cyclodextrin carrier has been found to be delivered preferentially to the ERC not only in living cells, but even in permeabilized, formaldehyde-fixed cells (4), consistent with the idea that this process is energy-independent and arises from the intrinsic properties of the target membranes.

Another means to facilitate transport from one membrane to another is to have a close contact between the two membranes. For example, in many cell types, part of the ER is in close proximity to the plasma membrane (36), an arrangement that could facilitate rapid exchange of cholesterol between these two membranes, perhaps with the assistance of transfer proteins. Similarly, three-dimensional reconstructions of the Golgi apparatus have revealed extensive areas of close apposition between the *trans*-Golgi and ER (37). The cholesterol-esterifying enzyme ACAT is enriched in the parts of the ER near to the ERC and the *trans*-Golgi (38), perhaps allowing efficient delivery of cholesterol to ACAT from these cholesterol-rich membranes.

Caveolae and caveolin have been proposed to play an important role in intracellular cholesterol transport. In cultured fibroblasts, reduction in caveolin expression by antisense DNA suppresses cholesterol efflux, whereas transfection with caveolin cDNA stimulates this process (39). The mechanisms by which caveolin could affect cholesterol transport remain uncertain, although caveolin can bind cholesterol directly (17). It has been proposed that caveolin forms a complex with chaperone proteins that deliver cholesterol from the ER to the plasma membrane, bypassing the Golgi apparatus (40). Palmitoylation of caveolin-1 may be required for rapid ($t_{1/2}$ = 10 minutes) transfer of cholesterol from the ER to caveolae via this chaperone complex (40). However, some studies have failed to find evidence that this cholesterol chaperone complex participates in surface delivery of newly synthesized cholesterol (32), and several caveats need to be resolved about the contribution of caveolin to cholesterol transport. Thus, at least in cultured cells, the rate of caveolin flux between the cell interior and the plasma membrane is typically quite low (16). Moreover, mice lacking expression of both caveolin-1 and caveolin-2 (41, 42) appear to be normal with regard to their cholesterol levels and regulation, although a full study of cholesterol transport in these mice has not yet been reported.

A more indirect role for caveolins and other related proteins has been proposed in which increased accumulation of caveolin-1 in the ER promotes its interaction with newly formed lipid droplets (43). It is suggested that the membrane topology of caveolin, which has two cytoplasmic domains flanking a central region that penetrates but does not cross the bilayer, might allow it to associate preferentially with the membrane deformations that appear during the formation of a lipid droplet. Interestingly, a family of plant proteins associated with lipid droplets, the oleosins, have a topology similar to caveolins. Hence, caveolins and related proteins may alter the membrane bilayer structure, promoting membrane curvature, and perhaps facilitating the transbilayer redistribution of cholesterol.

General intracellular transport pathways

Transport to and from the ER. Since the ER is the site of cholesterol synthesis but maintains a low steady-state level of cholesterol, efficient transport mechanisms must exist to export cholesterol from the ER (Figure 2). Cholesterol export from the ER requires metabolic energy (2). As measured by extracting cholesterol upon arrival at the plasma membrane with cyclodextrin, the kinetics of its transport appears to be biphasic (29), with a first phase, corresponding to about half of the total, reaching the plasma membrane in 60 minutes, and the remaining material being released in a slow phase over 5–6 hours (29).

The vesicle-mediated protein secretory pathway through the Golgi apparatus could, in principle, provide a parallel route for newly synthesized cholesterol to the plasma membrane. Biochemical studies have demonstrated that this is not a major pathway of cholesterol transport from the ER to the plasma membrane. Brefeldin A treatment can quantitatively suppress cell surface transport of vesicular stomatitis virus G protein without blocking cholesterol transport (44). Still, although this vesicular route is not a major pathway of cholesterol export, a recent study found that some cholesterol exported from the ER becomes confined to DRMs before reaching the plasma membrane and that transport of cholesterol to the plasma membrane can be partly inhibited by brefeldin A (32). The passage of some cholesterol through the Golgi apparatus might be important for raft-dependent sorting in the *trans*-Golgi network (TGN) of polarized epithelia (5). In addition, energy depletion or treatments that interfere with vesicle traffic have been found to reduce the esterification of β-VLDL–derived cholesterol in macrophages at a stage after release of cholesterol from late endosomes and lysosomes (45). However, treatment of

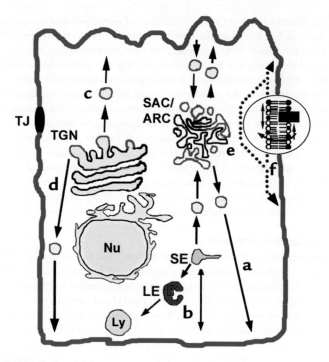

Figure 3
Cholesterol transport in polarized cells. Polarized cells form distinct apical (red) and basolateral (blue) membrane compartments, which are
separated by tight junctions (TJ). Proteins and lipids are sorted along
the biosynthetic and endocytic pathways indicated by blue (basolateral) and red (apical) vesicles. Plasma membrane cholesterol is transported in vesicles between the basolateral and apical membrane via a
subapical compartment or apical recycling compartment (SAC/ARC).
Recycling to the basolateral membrane can also occur from this compartment (**a**). LDL cholesterol has the same fate as in nonpolarized
cells (**b**). A fraction of de novo synthesized cholesterol is transported
along the biosynthetic pathway as in nonpolarized cells. In the TGN,
cholesterol might form microdomains or rafts along with sphingolipids
that segregate from the remaining TGN membrane and carry apically
destined proteins and lipids to the apical membrane shown in red (**c**).
Basolaterally destined vesicles (blue) bud off of the TGN but should
contain less cholesterol (**d**). Plasma membrane cholesterol can shuttle
rapidly between the plasma membrane domains by nonvesicular transport. This process involves fast transbilayer migration of cholesterol
to circumvent the lateral diffusion barrier created by the TJ in the exoplasmic leaflet of the plasma membrane. Transport through the cytoplasm bound to a protein carrier (**e**), and/or diffusion along the inner
monolayer (**f**) result in rapid exchange of cholesterol between the apical
and basolateral plasma membrane domain. CE formation occurs as in
nonpolarized cells but is omitted for clarity.

cells with bacterial sphingomyelinase, which releases cholesterol
from the plasma membrane, leads to ATP-independent cholesterol
esterification and is accompanied by an ATP-independent vesiculation of the plasma membrane (45, 46). It thus appears that both
vesicular and nonvesicular mechanisms operate throughout the
cholesterol transport process.

Export from late endosomes and lysosomes. Cholesterol esters in LDL
and other lipoproteins are hydrolyzed to free cholesterol in late
endosomes and lysosomes. The cholesterol is exported from the
lysosomes by a mechanism that is not fully characterized (Figure 2).
Transport from lysosomes to the ER or the plasma membrane can be
inhibited by progesterone or hydrophobic amines, such as U18666A,

imipramine, sphinganine, and stearylamine (2), but this inhibition
does not seem to be at the step of export from the late endosomes
and lysosomes, since LDL-derived cholesterol appears at the plasma
membrane with similar kinetics in treated and untreated cells (47).
After efflux from the late endosomes and lysosomes, cholesterol can
be delivered to various other organelles. It has been estimated that in
fibroblasts about 30% of the LDL-derived cholesterol is delivered to
the ER without transit through the plasma membrane (2, 29).

Niemann-Pick C disease is an inherited recessive disorder
characterized by accumulation of cholesterol and other lipids in
organelles that share many but not all characteristics of late endosomes (7, 48). The Niemann-Pick C1 (NPC1) protein is a multiple-
membrane-spanning protein that does not transport cholesterol
directly but that can facilitate the transbilayer transport of some
hydrophobic molecules (49). It is unclear at what step or steps
the NPC1 protein acts, but, as with progesterone or hydrophobic
amines, it appears that initial export from late endosomes is not
affected; cholesterol taken into the cells as an LDL-associated ester
exits the late endosomes and appears at the plasma membrane as
free cholesterol at the same rate in wild-type and *NPC1* mutant
cells (47, 50). Nevertheless, cholesterol ultimately accumulates in
storage organelles that resemble late endosomes and are enriched
in lysobisphosphatidic acid (48) and other lipids (7). The basis for
this cholesterol and lipid accumulation in the NPC1 mutants or in
cells treated with hydrophobic amines remains uncertain.

Efflux from the plasma membrane. There has been considerable
progress in the past few years in understanding the mechanisms
of cholesterol efflux to extracellular acceptors (see Tall et al., this
Perspective series, ref. 51). The protein that is defective in Tangier
disease, ABCA1, plays a key, but indirect, role in cholesterol efflux
to HDL acceptors (12). It appears that the specific role of ABCA1
is to facilitate phospholipid transfer to apoA-I. Cholesterol can
then be effluxed to the phospholipid-loaded apoA-I by a mechanism that does not require ABCA1 (52). The role of membrane
specializations in cholesterol efflux is an active area of investigation. Caveolae have been described as important sites for efflux,
but it is not clear whether these have been distinguished from
other, more abundant types of raftlike membranes (12). Interestingly, cells overexpressing ABCA1 carry plasma membrane protrusions (53) that may play a role in facilitating phospholipid
and cholesterol transfer.

Intracellular transport of cholesterol
in specialized cells

Hepatic cholesterol transport. Chylomicron remnants and other types
of lipoproteins are taken up by hepatocytes using a variety of
receptors (54). After hydrolysis of cholesterol esters, free cholesterol is released into the cell where it can be shuttled directly to the
canalicular membrane for biliary secretion, used for synthesis of
bile salts, or re-esterified and used for assembly of VLDLs in the ER
(55). Assembled VLDL is secreted at the basolateral membrane of
hepatocytes and carries cholesterol to peripheral tissues (56).

The intracellular trafficking pathways of cholesterol in
hepatocytes and other epithelia are poorly defined. In addition to
the pathways available in nonpolarized cells, epithelial cells maintain distinct lipid and protein compositions in their apical and
their basolateral plasma membranes, requiring specialized sorting
in the biosynthetic secretory pathway and in the endocytic recycling pathways (3, 57) (Figure 3). Rapid, nonvesicular exchange
of DHE between the apical and basolateral plasma membrane

domains has been observed in the polarized hepatoma cell line HepG2, which forms an apical vacuole resembling the biliary canaliculus (27). DHE is transported in vesicles to a subapical compartment or apical recycling compartment (SAC/ARC), suggesting that cholesterol derived from the plasma membrane of hepatocytic cells has rapid access to various intracellular compartments.

Cholesterol transport in macrophages. Since loading of cholesterol esters in cytoplasmic droplets of macrophages is an early step in formation of atherosclerotic lesions, the intracellular transport of cholesterol in these cells is of great interest (58). Many of the transport pathways are similar to those found in other cells. In macrophages, the intracellular fate of cholesterol derived from lipoproteins depends on the mechanism underlying its internalization. LDL is internalized by the LDL receptor and delivered to late endosomes, where cholesterol esters are hydrolyzed; homeostatic downregulation of LDL receptors limits the total cholesterol delivery from this source, and cholesterol ester lipid droplets do not accumulate in these cells. In contrast, cholesterol delivered via β-VLDL, which also binds to the LDL receptor, becomes esterified by ACAT and stored in cytoplasmic droplets, perhaps because a greater load of cholesterol is delivered by the larger β-VLDL particles. Oxidized or acetylated LDLs enter via scavenger receptors and also stimulate formation of cholesterol ester droplets, although the oxidized LDL is much less effective in activating this process (59). LDL aggregated and retained on the ECM does not become internalized by receptor-mediated endocytosis but remains in prolonged extracellular contact with macrophages; in this case, delivery of the associated cholesterol precedes internalization and degradation of the proteins (60). This interaction, which resembles the situation in atherosclerotic lesions, also leads to significant stimulation of cholesterol esterification (58).

The extent of cholesterol esterification by macrophages depends in a nonlinear fashion on the amount of cholesterol loading (61), such that much more esterification occurs once a threshold loading value is exceeded. The basis for this nonlinear response is not fully understood, but one component of it may be more efficient delivery of cholesterol to ACAT in the ER when free cholesterol is above the threshold. A similar nonlinear relationship between plasma membrane cholesterol elevation and levels of cholesterol in the ER has been found in fibroblasts (6), and such a relationship would be consistent with the effects seen in macrophages as cholesterol is elevated.

Cholesterol transport in steroidogenic cells. Cholesterol is the precursor for the synthesis of steroid hormones in mitochondria of gonadal and adrenal tissue. The synthesis is initiated by conversion of cholesterol to pregnenolone via the C27 cholesterol side chain cleavage cytochrome P450, located in the inner mitochondrial membrane. Delivery of cholesterol is rate-limiting to this process (21). Cholesterol destined for steroid hormone synthesis is mainly derived from the plasma membrane and from lipid droplets. How

cholesterol reaches the outer surface of mitochondria is not yet defined, but nonvesicular transport mechanisms via carrier proteins or via membrane contacts between mitochondria and lipid droplets are likely to be important (62, 63). It is generally assumed that the rate-limiting step in steroidogenesis is the translocation of cholesterol from the outer to the inner mitochondrial membrane, and several proteins have been proposed to mediate this transport step, including the peripheral-type benzodiazepine receptor, which binds cholesterol and other compounds known to stimulate steroidogenesis (64). StAR/StarD1 is thought to act selectively at the outer mitochondria surface to mediate import of cholesterol (refs. 65, 66; see also Jefcoate, this Perspective series, ref. 33, for an alternative view), but the exact mechanism of StAR-induced cholesterol import into mitochondria is not known.

Concluding remarks

The intracellular distribution of unesterified cholesterol among various cellular organelles is determined by multiple mechanisms including transbilayer flipping, stabilization in membrane microdomains by interactions with lipids (and perhaps proteins), enrichment or exclusion from transport vesicles as they form, and binding to diffusible carrier proteins in the cytoplasm and within intracellular organelles. Some of these transport mechanisms involve passive movement of cholesterol down a free-energy gradient, but others require metabolic energy to overcome kinetic barriers or to move cholesterol up a gradient. Cholesterol transport from one organelle to another can be accomplished by combinations of these mechanisms, which operate in parallel within the cell. The mechanisms and regulation of the resulting, complex pathways are still only partially understood. Further progress will require better understanding of transport pathways, including a kinetic and morphological description of the movement of cholesterol, as well as a more complete description of the proteins that control the key events in the various pathways.

Acknowledgments

We are grateful to Timothy McGraw and Mingming Hao for reading the manuscript. This work was supported by grants from the NIH (DK-27083) and the Ara Parseghian Medical Research Foundation. D. Wüstner is supported by a postdoctoral fellowship from the Charles H. Revson Foundation. Space limitations precluded referencing a number of important studies. Additional references may be found in the suggested reading list available at www.jci.org/cgi/content/full/110/07/891/DC1.

Address correspondence to: Frederick R. Maxfield, Department of Biochemistry, Room E-215, Weill Medical College of Cornell University, 1300 York Avenue, New York, New York 10021, USA. Phone: (212) 747-6405; Fax: (212) 746-8875; E-mail:frmaxfie@med.cornell.edu.

1. Simons, K., and Ehehalt, R. 2002. Cholesterol, lipid rafts, and disease. *J. Clin. Invest.* **110**:597–603. doi:10.1172/JCI200216390.
2. Liscum, L., and Munn, N.J. 1999. Intracellular cholesterol transport. *Biochim. Biophys. Acta.* **1438**:19–37.
3. Mukherjee, S., Ghosh, R.N., and Maxfield, F.R. 1997. Endocytosis. *Physiol. Rev.* **77**:759–803.
4. Hao, M., et al. 2002. Vesicular and non-vesicular sterol transport in living cells. The endocytic recycling compartment is a major sterol storage organelle. *J. Biol. Chem.* **277**:609–617.
5. Simons, K., and Ikonen, E. 1997. Functional rafts

in cell membranes. *Nature.* **387**:569–572.
6. Lange, Y., Ye, J., Rigney, M., and Steck, T.L. 1999. Regulation of endoplasmic reticulum cholesterol by plasma membrane cholesterol. *J. Lipid. Res.* **40**:2264–2270.
7. Blanchette-Mackie, E. 2000. Intracellular cholesterol trafficking: role of the NPC1 protein. *Biochim. Biophys. Acta.* **1486**:171–183.
8. Brown, M., and Goldstein, J. 1999. A proteolytic pathway that controls the cholesterol content of membranes, cells, and blood. *Proc. Natl. Acad. Sci. USA.* **96**:11041–11048.

9. Mukherjee, S., Zha, X., Tabas, I., and Maxfield, F. 1998. Cholesterol distribution in living cells: fluorescence imaging using dehydroergosterol as a fluorescent cholesterol analog. *Biophys. J.* **75**:1915–1925.
10. Ikonen, E. 2001. Roles of lipid rafts in membrane transport. *Curr. Opin. Cell Biol.* **13**:470–477.
11. Mayor, S., Sabharanjak, S., and Maxfield, F. 1998. Cholesterol-dependent retention of GPI-anchored proteins in endosomes. *EMBO J.* **17**:4626–4638.
12. Fielding, C., and Fielding, P. 2001. Cellular cholesterol efflux. *Biochim. Biophys. Acta.* **1533**:175–189.
13. Brown, D., and London, E. 2000. Structure and

function of sphingolipid- and cholesterol-rich membrane rafts. *J. Biol. Chem.* **275**:17221–17224.

14. Maxfield, F.R. 2002. Plasma membrane microdomains. *Curr. Opin. Cell Biol.* **14**:483–487.

15. Pelkmans, L., Puntener, D., and Helenius, A. 2002. Local actin polymerization and dynamin recruitment in SV40-induced internalization of caveolae. *Science.* **296**:535–539.

16. Thomsen, P., Roepstorff, K., Stahlhut, M., van Deurs, B. 2002. Caveolae are highly immobile plasma membrane microdomains, which are not involved in constitutive endocytic trafficking. *Mol. Biol. Cell.* **13**:238–250.

17. Ikonen, E., and Parton, R. 2000. Caveolins and cellular cholesterol balance. *Traffic.* **1**:212–217.

18. Boesze-Battaglia, K., Clayton, S., and Schimmel, R. 1996. Cholesterol redistribution within human platelet plasma membrane: evidence for a stimulus-dependent event. *Biochemistry.* **35**:6664–6673.

19. Schroeder, F., et al. 2001. Recent advances in membrane microdomains: rafts, caveolae, and intracellular cholesterol trafficking. *Exp. Biol. Med.* **226**:873–890.

20. Frolov, A., Woodford, J.K., Murphy, E.J., Billheimer, J.T., and Schroeder, F. 1996. Spontaneous and protein-mediated sterol transfer between intracellular membranes. *J. Biol. Chem.* **271**:16075–16083.

21. Stocco, D. 2001. StAR protein and the regulation of steroid hormone biosynthesis. *Annu. Rev. Physiol.* **63**:193–213.

22. Strauss, J., et al. 1999. The steroidogenic acute regulatory protein (StAR): a window into the complexities of intracellular cholesterol trafficking. *Recent Prog. Horm. Res.* **54**:369–394.

23. Lange, Y. 1992. Tracking cell cholesterol with cholesterol oxidase. *J. Lipid Res.* **33**:315–321.

24. Schroeder, F., Holland, J., and Bieber, L. 1971. Fluorometric evidence for the binding of cholesterol to the filipin complex. *J. Antibiot. (Tokyo).* **24**:846–849.

25. Steer, C.J., Bisher, M., Blumenthal, R., and Steven, A.C. 1984. Detection of membrane cholesterol by filipin in isolated rat liver coated vesicles is dependent upon removal of the clathrin coat. *J. Cell Biol.* **99**:315–319.

26. Frolov, A., et al. 2000. High density lipoprotein-mediated cholesterol uptake and targeting to lipid droplets in intact L-cell fibroblasts. A single- and multiphoton fluorescence approach. *J. Biol. Chem.* **275**:12769–12780.

27. Wüstner, D., Herrmann, A., Hao, M., and Maxfield, F.R. 2002. Rapid nonvesicular transport of sterol between the plasma membrane domains of polarized hepatic cells. *J. Biol. Chem.* **277**:30325–30336.

28. Ohvo-Rekila, H., Akerlund, B., and Slotte, J. 2000. Cyclodextrin-catalyzed extraction of fluorescent sterols from monolayer membranes and small unilamellar vesicles. *Chem. Phys. Lipids.* **105**:167–178.

29. Neufeld, E.B., et al. 1996. Intracellular trafficking of cholesterol monitored with a cyclodextrin. *J. Biol. Chem.* **271**:21604–21613.

30. Haynes, M., Phillips, M., and Rothblat, G. 2000. Efflux of cholesterol from different cellular pools. *Biochemistry.* **39**:4508–4517.

31. Connolly, C., Futter, C., Gibson, A., Hopkins, C., and Cutler, D. 1994. Transport into and out of the Golgi complex studied by transfecting cells with cDNAs encoding horseradish peroxidase. *J. Cell Biol.* **127**:641–652.

32. Heino, S., et al. 2000. Dissecting the role of the Golgi complex and lipid rafts in biosynthetic transport of cholesterol to the cell surface. *Proc. Natl. Acad. Sci. USA.* **97**:8375–8380.

33. Jefcoate, C. 2002. High-flux mitochondrial cholesterol trafficking, a specialized function of the adrenal cortex. *J. Clin. Invest.* **110**:881–890. doi:10.1172/JCI200216771.

34. Soccio, R., et al. 2002. The cholesterol-regulated StarD4 gene encodes a StAR-related lipid transfer protein with two closely related homologues, StarD5 and StarD6. *Proc. Natl. Acad. Sci. USA.* **99**:6943–6948.

35. Alpy, F., et al. 2001. The steroidogenic acute regulatory protein homolog MLN64, a late endosomal cholesterol-binding protein. *J. Biol. Chem.* **276**:4261–4269.

36. Putney, J. 1999. "Kissin' cousins": intimate plasma membrane-ER interactions underlie capacitative calcium entry. *Cell.* **99**:5–8.

37. Ladinsky, M., Mastronarde, D., McIntosh, J., Howell, K., and Staehelin, L. 1999. Golgi structure in three dimensions: functional insights from the normal rat kidney cell. *J. Cell Biol.* **144**:1135–1149.

38. Khelef, N., et al. 2000. Enrichment of acyl coenzyme A:cholesterol O-acyltransferase near trans-golgi network and endocytic recycling compartment. *Arterioscler. Thromb. Vasc. Biol.* **20**:1769–1776.

39. Fielding, C., Bist, A., and Fielding, P. 1999. Intracellular cholesterol transport in synchronized human skin fibroblasts. *Biochemistry.* **38**:2506–2513.

40. Uittenbogaard, A., and Smart, E. 2000. Palmitoylation of caveolin-1 is required for cholesterol binding, chaperone complex formation, and rapid transport of cholesterol to caveolae. *J. Biol. Chem.* **275**:25595–25599.

41. Drab, M., et al. 2001. Loss of caveolae, vascular dysfunction, and pulmonary defects in caveolin-1 gene-disrupted mice. *Science.* **293**:2449–2452.

42. Razani, B., et al. 2002. Caveolin-1-deficient mice are lean, resistant to diet-induced obesity, and show hypertriglyceridemia with adipocyte abnormalities. *J. Biol. Chem.* **277**:8635–8647.

43. Ostermeyer, A., et al. 2001. Accumulation of caveolin in the endoplasmic reticulum redirects the protein to lipid storage droplets. *J. Cell Biol.* **152**:1071–1078.

44. Urbani, L., and Simoni, R.D. 1990. Cholesterol and vesicular stomatitis virus G protein take separate routes from the endoplasmic reticulum to the plasma membrane. *J. Biol. Chem.* **265**:1919–1923.

45. Skiba, P.J., Zha, X., Maxfield, F.R., Schissel, S.L., and Tabas, I. 1996. The distal pathway of lipoprotein-induced cholesterol esterification, but not sphingomyelinase-induced cholesterol esterification, is energy-dependent. *J. Biol. Chem.* **271**:13392–13400.

46. Zha, X., et al. 1998. Sphingomyelinase treatment induces ATP-independent endocytosis. *J. Cell Biol.* **140**:39–47.

47. Lange, Y., Ye, J., Rigney, M., and Steck, T. 2000. Cholesterol movement in Niemann-Pick type C cells and in cells treated with amphiphiles. *J. Biol. Chem.* **275**:17468–17475.

48. Kobayashi, T., et al. 1999. Late endosomal membranes rich in lysobisphosphatidic acid regulate cholesterol transport. *Nat. Cell Biol.* **1**:113–118.

49. Davies, J., Chen, F., and Ioannou, Y. 2000. Transmembrane molecular pump activity of Niemann-Pick C1 protein. *Science.* **290**:2295–2298.

50. Cruz, J., Sugii, S., Yu, C., and Chang, T. 2000. Role of Niemann-Pick type C1 protein in intracellular trafficking of low density lipoprotein-derived cholesterol. *J. Biol. Chem.* **275**:4013–4021.

51. Tall, A.R., Costet, P., and Wang, N. 2002. Regulation and mechanisms of macrophage cholesterol efflux. *J. Clin. Invest.* **110**:899–904. doi:10.1172/JCI200216391.

52. Wang, N., Silver, D., Thiele, C., and Tall, A. 2001. ATP-binding cassette transporter A1 (ABCA1) functions as a cholesterol efflux regulatory protein. *J. Biol. Chem.* **276**:23742–23747.

53. Wang, N., Silver, D., Costet, P., and Tall, A. 2000. Specific binding of ApoA-I, enhanced cholesterol efflux, and altered plasma membrane morphology in cells expressing ABC1. *J. Biol. Chem.* **275**:33053–33058.

54. Havel, R. 1995. Chylomicron remnants: hepatic receptors and metabolism. *Curr. Opin. Lipidol.* **6**:312–316.

55. Vance, J., and Vance, D. 1990. Lipoprotein assembly and secretion by hepatocytes. *Annu. Rev. Nutr.* **10**:337–356.

56. Kang, S., and Davis, R. 2000. Cholesterol and hepatic lipoprotein assembly and secretion. *Biochim. Biophys. Acta.* **1529**:223–230.

57. Zegers, M., and Hoekstra, D. 1998. Mechanisms and functional features of polarized membrane traffic in epithelial and hepatic cells. *Biochem. J.* **336**:257–269.

58. Tabas, I. 2000. Cholesterol and phospholipid metabolism in macrophages. *Biochim. Biophys. Acta.* **1529**:164–174.

59. Tabas, I. 1999. Nonoxidative modifications of lipoproteins in atherogenesis. *Annu. Rev. Nutr.* **19**:123–139.

60. Sakr, S., et al. 2001. The uptake and degradation of matrix-bound lipoproteins by macrophages require an intact actin cytoskeleton, Rho family GTPases, and myosin ATPase activity. *J. Biol. Chem.* **276**:37649–37658.

61. Xu, X.X., and Tabas, I. 1991. Lipoproteins activate acyl-coenzyme A:cholesterol acyltransferase in macrophages only after cellular cholesterol pools are expanded to a critical threshold level. *J. Biol. Chem.* **266**:17040–17048.

62. Kallen, C., et al. 1998. Steroidogenic acute regulatory protein (StAR) is a sterol transfer protein. *J. Biol. Chem.* **273**:26285–26288.

63. Hall, P., and Almahbobi, G. 1997. Roles of microfilaments and intermediate filaments in adrenal steroidogenesis. *Microsc. Res. Tech.* **36**:463–479.

64. Papadopoulos, V., et al. 1997. Peripheral benzodiazepine receptor in cholesterol transport and steroidogenesis. *Steroids.* **62**:21–28.

65. Arakane, F., et al. 1998. The mechanism of action of steroidogenic acute regulatory protein (StAR). StAR acts on the outside of mitochondria to stimulate steroidogenesis. *J. Biol. Chem.* **273**:16339–16345.

66. Bose, H., Lingappa, V., and Miller, W. 2002. Rapid regulation of steroidogenesis by mitochondrial protein import. *Nature.* **417**:87–91.

Regulation and mechanisms of macrophage cholesterol efflux

Alan R. Tall, Philippe Costet, and Nan Wang

Division of Molecular Medicine, Department of Medicine, Columbia University, New York, New York, USA.

The accumulation of cholesterol in macrophage foam cells results from the uptake of retained and modified apoB lipoproteins (1) and is a central event in atherogenesis. Although it has long been known that HDL and its apolipoproteins can stimulate cholesterol efflux from macrophage foam cells (2, 3) and reduce atherogenesis in animal models (4, 5), this area of research has been enlivened by the elucidation of the genetic defect in Tangier disease, a condition in which there is a defect in apolipoprotein-mediated cholesterol efflux (6) and in which the hallmark pathology is the accumulation of macrophage foam cells in various tissues. Tangier disease is caused by mutations in the ATP-binding cassette transporter A1 (ABCA1). This cell surface transporter facilitates the efflux of phospholipids and cholesterol onto lipid-poor apolipoproteins, initiating the formation of HDL particles.

This Perspective will focus on recent studies on the regulation and role of ABCA1 in cellular cholesterol efflux, lipoprotein physiology, and atherogenesis, attempting to place this work in the broader context of studies on cellular cholesterol efflux, centripetal cholesterol transport, and ABC transporter structure and function.

Expression and regulation of ABCA1

The ABCA1 mRNA is widely expressed and is found in the liver, kidney, adrenal gland, intestine, and CNS (7). It is also prominent in the macrophage foam cells of atherosclerotic lesions. While basal levels of ABCA1 mRNA and protein in macrophages are low, both are induced with cholesterol loading and can be reversed by HDL-mediated cholesterol efflux (8). The mechanism of cholesterol induction involves oxysterol-dependent activation of the LXR/RXR transcription factors, which act on a DR4 site (direct repeat nuclear receptor-binding site with a spacing of four nucleotides) in the proximal promoter of the ABCA1 gene. ABCA1 is one of a battery of LXR/RXR target genes involved in the regulation of centripetal cholesterol transport, cholesterol excretion, and fatty acid biosynthesis (Figure 1). Thus, LXR/RXR transcription factors regulate genes mediating macrophage cholesterol efflux (APOE, ABCA1, and possibly ABCG1), transport (LPL, CETP, and several genes encoding isoforms of apoC), conversion of cholesterol into bile acids (CYP7A), and metabolism and excretion into bile or intestinal lumen (ABCG5 and ABCG8). Small-molecule activators of LXR/RXR might therefore have a favorable effect on cholesterol homeostasis.

Since LXR is activated by specific oxysterols but apparently not by cholesterol itself, the induction of ABCA1 expression following cellular uptake of cholesterol-rich lipoproteins requires the conversion of cholesterol into specific oxysterols, such as 22-OH cholesterol, 27-OH cholesterol, or 24(S),25-epoxycholesterol. Cyp27, a mitochondrial enzyme that converts cholesterol into 27-OH cholesterol, is mutated in cerebrotendinous xanthomatosis (CTX), and CTX fibroblasts fail to upregulate ABCA1 in response to cholesterol loading (ref. 9; for more discussion of the potential significance of oxysterols in cholesterol homeostasis, see Björkhem, this Perspective series, ref. 10). Since CTX patients develop xanthomatosis and premature atherosclerosis, macrophage foam cell formation in CTX may partly reflect a defect in ABCA1-mediated cholesterol efflux.

In addition to positive sterol regulation, ABCA1 is subject to negative regulation by non-sterol mevalonate products. Thus, geranylgeranylpyrophosphate inhibits ABCA1 gene expression by different mechanisms, acting through the DR4 element in the ABCA1 promoter (11). cAMP induces the ABCA1 mRNA in macrophages and some other cells by an unknown mechanism. A phosphodiesterase 4 inhibitor has been found to increase ABCA1 mRNA and cellular cholesterol efflux, suggesting a possible therapeutic approach to increasing ABCA1 expression (12). IFN-γ, conversely, decreases ABCA1 mRNA levels in cholesterol-loaded macrophages (13). Since IFN-γ is proatherogenic, this finding provides a potential link between the inflammatory aspect of atherosclerosis and a defect in foam cell cholesterol efflux.

Early reports suggested that PPARγ activators increase ABCA1 transcription and cholesterol efflux by sequentially upregulating LXRα and ABCA1. These findings have been widely cited as an explanation for antiatherogenic effects of PPARγ activators. However, although the LXRα gene contains an authentic PPARγ response element (14), some recent reports have failed to confirm that PPARγ activators, such as troglitazone or rosiglitazone, induce ABCA1 mRNA and cholesterol efflux in macrophages (15, 16). Perhaps the availability of LXR/RXR ligands, rather than the abundance of LXRs themselves, is usually rate-limiting for ABCA1 expression in differentiated cells. Nevertheless, PPARδ activators appear to induce ABCA1 expression and cholesterol efflux moderately, and they can increase HDL levels in an obese-monkey model (17).

There is also emerging evidence that ABCA1 is regulated at a posttranscriptional level: in addition to acting as inhibitors of LXR-stimulated gene expression (18), polyunsaturated fatty acids increase the already rapid turnover of ABCA1 in macrophages and can diminish cell surface expression of this protein (19). Since FFA levels may be increased in diabetes, this response suggests a mechanism that could produce a defect in cellular cholesterol efflux in poorly controlled diabetes. Lipid-poor apolipoproteins, conversely, can stabilize ABCA1 protein (20).

Pathways mediating cellular cholesterol efflux

Three distinct pathways of cellular cholesterol efflux involving HDL and its apolipoproteins have been described. First, plasma

---

Nonstandard abbreviations used: ATP-binding cassette (ABC); ATP-binding cassette transporter A1 (ABCA1); cerebrotendinous xanthomatosis (CTX); scavenger receptor B-I (SR-BI); knockout (KO).

Conflict of interest: No conflict of interest has been declared.

Citation for this article: *J. Clin. Invest.* **110**:899–904 (2002). doi:10.1172/JCI200216391.

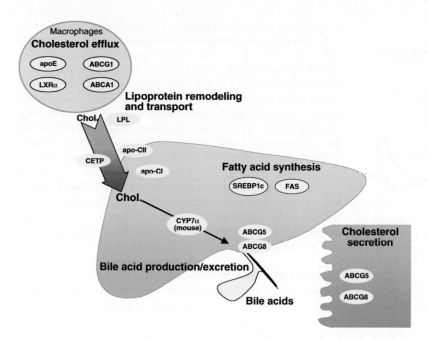

Figure 1
LXR target genes and their products. LXR regulates cholesterol (Chol) efflux, centripetal cholesterol transport, cholesterol excretion, and fatty acid biosynthesis by inducing multiple genes involved in lipid metabolism. Two products of these genes, ABCA1 and apoE, promote macrophage cholesterol efflux. Another LXR target, macrophage LPL, is activated by apoC-II, which is also regulated by LXR. Cholesteryl ester transfer protein (CETP) promotes transfer of HDL cholesteryl ester to VLDLs, which are taken up by the liver. In the liver, Cyp7α (cytochrome P450 7α) promotes cholesterol excretion by mediating conversion of cholesterol into bile acids. ABCG5 and ABCG8 facilitate hepatic and intestinal cholesterol excretion. LXR activates SREBP1c, regulating the expression of multiple lipogenic proteins, including FAS (fatty acid synthesis), which is also a direct target of LXR.

HDL particles can promote cholesterol efflux by a process of passive aqueous diffusion (21). Free cholesterol molecules spontaneously desorb from the plasma membrane, diffuse through the aqueous phase, and incorporate into HDL particles by collision. Second, scavenger receptor B-I (SR-BI), an HDL receptor that mediates the selective uptake of HDL cholesteryl esters into cells (22), also facilitates the efflux of cholesterol from cells to HDLs (23). Cholesterol efflux is blocked by antibodies that inhibit binding of HDLs to SR-BI (24), suggesting that the mechanism involves a direct interaction between HDL particles and the receptor and is therefore distinct from the aqueous diffusion mechanism. Finally, ABCA1 mediates the active efflux of phospholipid and cholesterol from cells to lipid-poor apolipoproteins, such as apoA-I. ApoA-I binds and cross links to ABCA1, suggesting a direct interaction that leads to lipid efflux (25).

Passive aqueous diffusion and SR-BI–facilitated efflux involve bidirectional exchange of cholesterol between cells and HDLs. In contrast, ABCA1/apolipoprotein–mediated cholesterol and phospholipid efflux is a unidirectional net transfer process (25). The specificity for cholesterol acceptors in the three pathways is different. Passive efflux is driven by the phospholipid content of lipoprotein acceptors (26). SR-BI can bind both apolipoproteins and HDL particles, but binding affinity is greatest for large, spherical HDL particles, suggesting that these may be the major substrate for SR-BI in vivo (27). In contrast, ABCA1 binds and cross-links lipid-poor apoA-I, while showing minor interaction with smaller HDL$_3$ and no interaction with larger HDL$_2$ subspecies (28). Thus, ABCA1 shows limited interaction with the major HDL species isolated from plasma, emphasizing the importance of a small pool of lipid-poor apolipoproteins either secreted by cells or generated by lipid exchange and lipolysis of HDL particles in the bloodstream. ABCA1 probably mediates the rapid efflux of cellular cholesterol in response to pre-β HDL, a minor but metabolically significant fraction of plasma HDLs comprising free apoA-I and apoA-I associated with a small number of phospholipid and cholesterol molecules (29). Following addition of lipid to apoA-I by ABCA1, there is further growth of HDL particles as a result of phospho-

lipid transfer by phospholipid transfer protein, cholesterol esterification by LCAT, and acquisition of apoE molecules. The large, apoE-rich HDLs are probably the optimal substrates for SR-BI. In summary, ABCA1 may play the major role in initiating cholesterol efflux from macrophages and other cells to lipid-poor apolipoproteins, while SR-BI principally clears cholesterol and cholesteryl ester from large HDLs in the liver and steroidogenic tissues.

ABC transporter–mediated lipid efflux

ABC transporter proteins comprise a large family of membrane proteins involved in the cellular export or import of a wide variety of different substances, including ions, lipids, cyclic nucleotides, peptides, and proteins. These transporters have been widely implicated in disease processes, such as Stargardt macular degeneration, cholestasis of pregnancy, and cystic fibrosis. In general, ABC transporters consist of two membrane-spanning domains forming a translocation pathway for a particular substrate, and two attached, cytoplasmic ABCs that link the transport process to the hydrolysis of ATP. ABCA1 conforms to this general model, and its likely membrane topology is shown in Figure 2. Similar to its closest neighbor (ABCR or ABCA4, the defective molecule in Stargardt retinal degeneration) (30), ABCA1 is likely to comprise two sets of six transmembrane domains, with two large extracellular loops joined by a disulfide bridge, and to have two paired cytoplasmic Walker and ABC signature motifs (Figure 2).

In the presence of apoA-I, ABCA1 promotes cellular efflux of phosphatidylcholine, sphingomyelin, and cholesterol. Because ABCA1 interacts with different lipid-poor apolipoproteins, including apoA-I, apoA-II, apoE, and likely apoA-IV, it appears to recognize a general feature in the apolipoprotein structure, such as a cluster of amphipathic helices (31). ABCA1 cross-links to apoA-I (25, 32), suggesting a direct interaction, likely involving the first and fourth extracellular loop (33) (Figure 2), as well as other regions of the molecule. Mutation of the cytoplasmic ATP-binding Walker motif results in defective lipid efflux and abolishes binding and cross-linking of apoA-I (28). One interpretation is that apoA-I binding to ABCA1 is linked to binding of

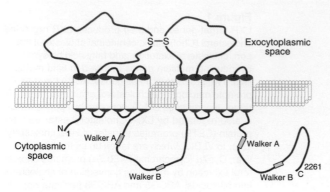

Figure 2
Schematic model of ABCA1. The model is based on studies by Bungert et al. (30) on ABCA4 membrane topology and by Fitzgerald et al. (33, 49) on ABCA1 membrane topology. ABCA4 is the closest relative of ABCA1.

ATP to cytoplasmic domains and a conformational change that fosters translocation of phospholipid and access to apoA-I (28). Mutation of the Walker motif would prevent the formation of this transition state causing defective phospholipid translocation and binding of apoA-I. Another interpretation is that the binding site consists primarily of phospholipids translocated by ABCA1 and lipid efflux is not dependent on a direct protein-protein interaction between apoA-I and ABCA1 (34). A recent study using mutagenesis analysis of several Tangier disease missense mutations in the extracellular loop of ABCA1 showed that binding of apoA-I is necessary but not sufficient for lipid efflux (33). The identification of a mutant with increased cross-linking of apoA-I but defective lipid efflux provides strong evidence that the binding site does not consist of translocated lipid (33).

Lipid rafts and ABCA1-mediated cholesterol efflux
Although the primary action of ABCA1 may be to act as a phospholipid translocase, different views have emerged on how ABCA1 mediates cholesterol efflux. One view is that ABCA1 promotes assembly of a cholesterol/phospholipid/apoA-I complex in a single step (Figure 3a). A variant of this model is that ABCA1 translocates phospholipid and cholesterol, giving rise to an excess of these lipids in the external membrane hemileaflet; apoA-I binds ABCA1 and absorbs the excess lipid, forming an HDL particle. A second view is that ABCA1 acts in a two-step process, initially giving rise to a phospholipid/apoA-I particle that then promotes cholesterol efflux from distinct cholesterol-enriched membrane microdomains, such as cholesterol- and sphingolipid-rich rafts or caveolae (28) (Figure 3c). A third model, blending features of each of the two earlier models, holds that ABCA1 occupies a border region between liquid and cholesterol-rich liquid–ordered membrane microdomains. Thus, although phospholipid and cholesterol efflux are normally coupled, this model (Figure 3b) predicts that the two processes might be dissociated when cholesterol-enriched rafts are far apart or depleted of cholesterol.

Although ABCA1 is localized on the cell surface, it is also found in intracellular organelles, including the Golgi apparatus, early and late endosomes, and lysosomes. Intracellular ABCA1 could be involved in lipid efflux, or it might represent nascent ABCA1 or molecules that have been targeted for degradation. However, energy-dependent trafficking of lipid vesicles between intracellular

sites and the cell surface appears to be involved in ABCA1-mediated lipid efflux, as disruption of Golgi membranes by brefeldin (35), and inhibitors of calcium-dependent retroendocytosis (36), inhibit apoA-I–mediated lipid efflux. Moreover, macrophages from Niemann-Pick C1–deficient mice, which are defective in cholesterol release from late endosomes and lysosomes, have a severe reduction in apoA-I–mediated cholesterol but not in phospholipid efflux (37). Similarly, defective cholesterol efflux from late endosomes can result from accumulation of sphingomyelin in Niemann-Pick diseases types A and B, resulting in a defect in cholesterol efflux (38). These studies suggest that cholesterol deposited in late endosomes or lysosomes by modified LDL acts as a preferential source of cholesterol for ABCA1-mediated efflux, perhaps because cholesterol-enriched vesicles are routed to plasma membrane cholesterol-rich rafts that donate cholesterol for ABCA1-mediated efflux.

ABCA1, LXRs, and atherogenesis
Studies in ABCA1 knockout and ABCA1 transgenic mice. ABCA1 knockout (KO) mice accumulate foam cells in their lungs but do not develop spontaneous atherosclerosis. Furthermore, even when crossed onto apoE- or LDL receptor–deficient atherogenic backgrounds, the *ABCA1* mutation causes no significant difference in the extent of atherosclerotic lesions, compared with controls (39). However, ABCA1 KO mice have reduced levels of VLDL/LDL cholesterol and apoB in these backgrounds, reminiscent of the reductions in apoB lipoproteins in Tangier disease. Thus, a decrease in the atherogenic stimulus provided by apoB lipoproteins may offset the defect in cholesterol efflux that occurs in ABCA1-deficient macrophages. Consistent with this idea, bone marrow transplantation from ABCA1 KO mice into apoE KO animals results in an approximately 50% increase in lesion area (39, 40). Although apoA-I and apoE plasma levels are slightly reduced (39, 40), overall HDL cholesterol levels are unchanged in these animals. The simplest

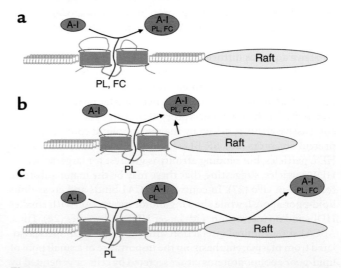

Figure 3
Models for ABCA1-mediated lipid efflux. (**a**) ABCA1 promotes phospholipid (PL) and cholesterol (FC) efflux from a membrane domain in a single step. (**b**) ABCA1, located in a border region between liquid and cholesterol-rich liquid–ordered domains (rafts), promotes phospholipid and cholesterol efflux to apoA-I. (**c**) ABCA1 first promotes phospholipid efflux to apoA-I to form intermediate complexes, which then remove cholesterol from rafts.

interpretation of these studies is that defective cholesterol efflux from foam cells can exacerbate atherosclerosis.

Of direct relevance to the idea of upregulating ABCA1 expression as a therapeutic modality are the results of overexpressing ABCA1. Here the effects on plasma lipoproteins and atherogenesis have been less consistent in different models. An initial study with BAC transgenic mice demonstrated increased macrophage and hepatic ABCA1 expression and cholesterol efflux, but no change in plasma HDL levels (41). The lack of change in plasma HDL may have reflected lower expression levels, since two subsequent studies, using either BAC transgenes containing the *ABCA1* gene, or the *APOE* promoter-enhancer driving the ABCA1 cDNA, found increased expression of ABCA1 in liver and macrophages, as well as increased HDL levels. ABCA1 transgenic mice in the C57BL/6 background had diminished atherosclerosis, with considerable overlap in the distribution of lesion areas between the groups (42). Moreover, crossing this transgene onto the apoE KO background increased atherosclerosis relative to the parental background. In contrast, ABCA1 transgenic mice produced from a BAC clone have diminished atherosclerosis in the apoE KO background (43). A possible explanation for the differing results is that in the negative study (43), the *APOE* promoter was used to drive ABCA1 expression in the apoE KO background. Since apoE could be an important ligand for ABCA1, expression of ABCA1 specifically in cells lacking apoE might limit the ability of ABCA1 to promote cholesterol efflux and to reverse foam cell formation. Further studies using tissue-specific overexpression of ABCA1 in liver or macrophages, and different promoters to drive transgenes, might help to clarify these issues.

Antiatherogenic effects of LXR/RXR targets in macrophages. ApoE and LDLR KO mice treated with small-molecule activators of LXR or RXR show a decrease in atherosclerosis (16, 44). Treatment of macrophages with LXR or RXR activator increases expression of ABCA1 mRNA and promotes cholesterol efflux to apoA-I (16). Moreover, apoE KO mice treated with LXR activator show increased expression of ABCA1 and ABCG1 mRNAs in lesions, presumably in macrophages (44). Remarkably, while short-term treatment with LXR activators increases plasma triglycerides and HDL in hamsters, chronic treatment in mice leads to no major changes in plasma lipoprotein levels, suggesting an important but poorly understood adaptive change. These findings suggest that increased macrophage cholesterol efflux, even in the absence of plasma lipoprotein changes, is likely to be antiatherogenic. Alternatively, there may be other LXR target genes expressed in the liver, intestine, or elsewhere, with a beneficial effect on atherogenesis, possibly including *ABCG5* or *ABCG8*.

Human studies. In case-control studies, Schaefer et al. (45) found a two- to threefold increase in risk for atherosclerotic cardiovascular disease in subjects with Tangier disease and obligate *ABCA1* heterozygotes, relative to age- and sex-matched controls from the Framingham study. Recently, van Dam et al. (46) studied carotid intima-media thickness in subjects with defined heterozygous mutations of *ABCA1*. These subjects have isolated low HDL levels, decreased on average by about 43%, but normal VLDL and LDL. In these subjects, the efficiency of apoA-I–mediated cholesterol efflux from individuals' fibroblasts correlates with plasma HDL

levels. These authors also found that carotid intima-media thickness is increased in subjects with *ABCA1* mutations compared with controls. Although these studies are suggestive of increased risk in humans with ABCA1 deficiency, the sample sizes are small. Furthermore, the control subjects (being, on average, 5 years younger than ABCA1-deficient individuals) were not perfectly matched to cases, although the conclusions appeared to hold even after age adjustment (46). Thus, while this correlation is promising, definitive evidence relating ABCA1 deficiency to atherogenesis in humans is still lacking.

Macrophage cholesterol efflux as a strategy for treating atherosclerosis

In principle, the results of recent mouse atherosclerosis studies represent a conceptual watershed for the development of treatment strategies for atherosclerosis. Together with earlier studies on macrophage-specific apoE expression (47), this work demonstrates the feasibility of upregulating macrophage genes that promote cholesterol efflux as an antiatherogenic treatment, and they indicate that small-molecule activators of LXR or RXR may offer a practical approach to achieving this goal. Although LXR activation is associated with upregulation of SREBP1c target genes (48), resulting in troublesome fatty liver and hypertriglyceridemia, the effects on plasma triglycerides appear to be short-lived (44). However, the issue of fatty liver, which is now thought to be a precursor of more serious liver pathology, remains a stumbling block. The various mechanisms that control ABCA1-mediated lipid efflux might also be therapeutically manipulated. In principle, the most successful strategies could involve macrophage-specific upregulation of cholesterol efflux, or perhaps upregulation of ABCG5 and ABCG8 to promote cholesterol excretion from the body.

Conclusions

Cellular cholesterol efflux is mediated by both active and passive processes and is regulated on multiple levels. The discovery that Tangier disease is caused by mutations in *ABCA1* has provided a molecular key to understanding the mechanisms and regulation of active cellular phospholipid and cholesterol efflux. ABCA1 interacts with lipid-free apolipoproteins, promoting phospholipid and cholesterol efflux from cells and giving rise to HDL particles. Based on analogy with other ABC transporters, ABCA1 may act as a phospholipid translocase facilitating phospholipid binding to apoA-I. Cholesterol may also be translocated with phospholipid, or it could be derived from membrane rafts in a two-step process. ABCA1 gene expression is upregulated in cholesterol-loaded cells as a result of activation of LXR/RXR–mediated gene transcription. LXR and RXR coordinately induce a battery of genes mediating cellular cholesterol efflux, centripetal cholesterol transport, and cholesterol excretion in bile. Small-molecule activators of LXR/RXR or other stimulators of macrophage or intestinal cholesterol efflux hold great promise as future treatments for atherosclerosis.

Address correspondence to: Alan R. Tall, Department of Medicine, Division of Molecular Medicine, Columbia University, 630 West 168th Street 8–101, New York, New York 10032, USA. Phone: (212) 305-9418; Fax: (212) 305-5052; E-mail: art1@columbia.edu.

1. Williams, K.J., and Tabas, I. 1998. The response-to-retention hypothesis of atherogenesis reinforced. *Curr. Opin. Lipidol.* **9**:471–474.
2. Oram, J.F., Albers, J.J., Cheung, M.C., and Bierman,

E.L. 1981. The effects of subfractions of high density lipoprotein on cholesterol efflux from cultured fibroblasts. Regulation of low density lipoprotein receptor activity. *J. Biol. Chem.* **256**:8348–8356.

3. Hara, H., and Yokoyama, S. 1991. Interaction of free apolipoproteins with macrophages. Formation of high density lipoprotein-like lipoproteins and reduction of cellular cholesterol. *J. Biol. Chem.*

266:3080–3086.

4. Rubin, E.M., Krauss, R.M., Spangler, E.A., Verstuyft, J.G., and Clift, S.M. 1991. Inhibition of early atherogenesis in transgenic mice by human apolipoprotein AI. *Nature.* **353**:265–267.

5. Plump, A.S., Scott, C.J., and Breslow, J.L. 1994. Human apolipoprotein A-I gene expression increases high density lipoprotein and suppresses atherosclerosis in the apolipoprotein E-deficient mouse. *Proc. Natl. Acad. Sci. USA.* **91**:9607–9611.

6. Francis, G.A., Knopp, R.H., and Oram, J.F. 1995. Defective removal of cellular cholesterol and phospholipids by apolipoprotein A-I in Tangier disease. *J. Clin. Invest.* **96**:78–87.

7. Lawn, R.M., Wade, D.P., Couse, T.L., and Wilcox, J.N. 2001. Localization of human ATP-binding cassette transporter 1 (ABC1) in normal and atherosclerotic tissues. *Arterioscler. Thromb. Vasc. Biol.* **21**:378–385.

8. Langmann, T., et al. 1999. Molecular cloning of the human ATP-binding cassette transporter 1 (hABC1): evidence for sterol-dependent regulation in macrophages. *Biochem. Biophys. Res. Commun.* **257**:29–33.

9. Fu, X., et al. 2001. 27-Hydroxycholesterol is an endogenous ligand for liver X receptor in cholesterol-loaded cells. *J. Biol. Chem.* **276**:38378–38387.

10. Björkhem, I. 2002. Do oxysterols control cholesterol homeostasis? *J. Clin. Invest.* **110**:725–730. doi:10.1172/JCI200216388.

11. Gan, X., et al. 2001. Dual mechanisms of ABCA1 regulation by geranylgeranyl pyrophosphate. *J. Biol. Chem.* **276**:48702–48708.

12. Lin, G., and Bornfeldt, K.E. 2002. Cyclic AMP-specific phosphodiesterase 4 inhibitors promote ABCA1 expression and cholesterol efflux. *Biochem. Biophys. Res. Commun.* **290**:663–669.

13. Panousis, C.G., and Zuckerman, S.H. 2000. Interferon-gamma induces downregulation of Tangier disease gene (ATP-binding-cassette transporter 1) in macrophage-derived foam cells. *Arterioscler. Thromb. Vasc. Biol.* **20**:1565–1571.

14. Chawla, A., et al. 2001. A PPAR gamma-LXR-ABCA1 pathway in macrophages is involved in cholesterol efflux and atherogenesis. *Mol. Cell.* **7**:161–171.

15. Akiyama, T.E., et al. 2002. Conditional disruption of the peroxisome proliferator-activated receptor gamma gene in mice results in lowered expression of ABCA1, ABCG1, and apoE in macrophages and reduced cholesterol efflux. *Mol. Cell. Biol.* **22**:2607–2619.

16. Claudel, T., et al. 2001. Reduction of atherosclerosis in apolipoprotein E knockout mice by activation of the retinoid X receptor. *Proc. Natl. Acad. Sci. USA.* **98**:2610–2615.

17. Oliver, W.R., Jr., et al. 2001. A selective peroxisome proliferator-activated receptor delta agonist promotes reverse cholesterol transport. *Proc. Natl. Acad. Sci. USA.* **98**:5306–5311.

18. Ou, J., et al. 2001. Unsaturated fatty acids inhibit transcription of the sterol regulatory element-binding protein-1c (SREBP-1c) gene by antagonizing ligand-dependent activation of the LXR. *Proc. Natl. Acad. Sci. USA.* **98**:6027–6032.

19. Wang, Y., and Oram, J.F. 2002. Unsaturated fatty acids inhibit cholesterol efflux from macrophages by increasing degradation of ATP-binding cassette transporter A1. *J. Biol. Chem.* **277**:5692–5697.

20. Arakawa, R., and Yokoyama, S. 2002. Helical apolipoproteins stabilize ATP-binding cassette transporter A1 by protecting it from thiol protease-mediated degradation. *J. Biol. Chem.* **277**:22426–22429.

21. Rothblat, G.H., et al. 1999. Cell cholesterol efflux: integration of old and new observations provides new insights. *J. Lipid Res.* **40**:781–796.

22. Acton, S., et al. 1996. Identification of scavenger receptor SR-BI as a high density lipoprotein receptor. *Science.* **271**:518–520.

23. Ji, Y., et al. 1997. Scavenger receptor BI promotes high density lipoprotein-mediated cellular cholesterol efflux. *J. Biol. Chem.* **272**:20982–20985.

24. Gu, X., Kozarsky, K., and Krieger, M. 2000. Scavenger receptor class B, type I-mediated [3H]cholesterol efflux to high and low density lipoproteins is dependent on lipoprotein binding to the receptor. *J. Biol. Chem.* **275**:29993–30001.

25. Wang, N., Silver, D.L., Costet, P., and Tall, A.R. 2000. Specific binding of ApoA-I, enhanced cholesterol efflux, and altered plasma membrane morphology in cells expressing ABC1. *J. Biol. Chem.* **275**:33053–33058.

26. Phillips, M.C., Johnson, W.J., and Rothblat, G.H. 1987. Mechanisms and consequences of cellular cholesterol exchange and transfer. *Biochim. Biophys. Acta.* **906**:223–276.

27. Liadaki, K.N., et al. 2000. Binding of high density lipoprotein (HDL) and discoidal reconstituted HDL to the HDL receptor scavenger receptor class B type I. Effect of lipid association and APOA-I mutations on receptor binding. *J. Biol. Chem.* **275**:21262–21271.

28. Wang, N., Silver, D.L., Thiele, C., and Tall, A.R. 2001. ATP-binding cassette transporter A1 (ABCA1) functions as a cholesterol efflux regulatory protein. *J. Biol. Chem.* **276**:23742–23747.

29. Fielding, C.J., and Fielding, P.E. 2001. Cellular cholesterol efflux. *Biochim. Biophys. Acta.* **1533**:175–189.

30. Bungert, S., Molday, L.L., and Molday, R.S. 2001. Membrane topology of the ATP binding cassette transporter ABCR and its relationship to ABC1 and related ABCA transporters: identification of N-linked glycosylation sites. *J. Biol. Chem.* **276**:23539–23546.

31. Remaley, A.T., et al. 2001. Apolipoprotein specificity for lipid efflux by the human ABCAI transporter. *Biochem. Biophys. Res. Commun.* **280**:818–823.

32. Oram, J.F., Lawn, R.M., Garvin, M.R., and Wade, D.P. 2000. ABCA1 is the cAMP-inducible apolipoprotein receptor that mediates cholesterol secretion from macrophages. *J. Biol. Chem.* **275**:34508–34511.

33. Fitzgerald, M.L., et al. 2002. Naturally occurring mutations in the largest extracellular loops of ABCA1 can disrupt its direct interaction with apolipoprotein A-I. *J. Biol. Chem.* **277**:33178–33187.

34. Chambenoit, O., et al. 2001. Specific docking of apolipoprotein A-I at the cell surface requires a functional ABCA1 transporter. *J. Biol. Chem.* **276**:9955–9960.

35. Neufeld, E.B., et al. 2001. Cellular localization and trafficking of the human ABCA1 transporter. *J. Biol. Chem.* **276**:27584–27590.

36. Takahashi, Y., and Smith, J.D. 1999. Cholesterol efflux to apolipoprotein AI involves endocytosis and resecretion in a calcium-dependent pathway. *Proc. Natl. Acad. Sci. USA.* **96**:11358–11363.

37. Chen, W., et al. 2001. Preferential ATP-binding cassette transporter A1-mediated cholesterol efflux from late endosomes/lysosomes. *J. Biol. Chem.* **276**:43564–43569.

38. Leventhal, A.R., Chen, W., Tall, A.R., and Tabas, I. 2001. Acid sphingomyelinase-deficient macrophages have defective cholesterol trafficking and efflux. *J. Biol. Chem.* **276**:44976–44983.

39. Aiello, R.J., et al. 2002. Increased atherosclerosis in hyperlipidemic mice with inactivation of ABCA1 in macrophages. *Arterioscler. Thromb. Vasc. Biol.* **22**:630–637.

40. van Eck, M., et al. 2002. Leukocyte ABCA1 controls susceptibility to atherosclerosis and macrophage recruitment into tissues. *Proc. Natl. Acad. Sci. USA.* **99**:6298–6303.

41. Cavelier, L.B., et al. 2001. Regulation and activity of the human ABCA1 gene in transgenic mice. *J. Biol. Chem.* **276**:18046–18051.

42. Joyce, C.W., et al. 2002. The ATP binding cassette transporter A1 (ABCA1) modulates the development of aortic atherosclerosis in C57BL/6 and apoE-knockout mice. *Proc. Natl. Acad. Sci. USA.* **99**:407–412.

43. Singaraja, R.R., et al. 2002. Increased ABCA1 activity protects against atherosclerosis. *J. Clin. Invest.* **110**:35–42. doi:10.1172/JCI200215748.

44. Joseph, S.B., et al. 2002. Synthetic LXR ligand inhibits the development of atherosclerosis in mice. *Proc. Natl. Acad. Sci. USA.* **99**:7604–7609.

45. Schaefer, E.J., Zech, L.A., Schwartz, D.E., and Brewer, H.B., Jr. 1980. Coronary heart disease prevalence and other clinical features in familial high-density lipoprotein deficiency (Tangier disease). *Ann. Intern. Med.* **93**:261–266.

46. van Dam, M.J., et al. 2002. Association between increased arterial-wall thickness and impairment in ABCA1-driven cholesterol efflux: an observational study. *Lancet.* **359**:37–42.

47. Bellosta, S., et al. 1995. Macrophage-specific expression of human apolipoprotein E reduces atherosclerosis in hypercholesterolemic apolipoprotein E-null mice. *J. Clin. Invest.* **96**:2170–2179.

48. Repa, J.J., et al. 2000. Regulation of mouse sterol regulatory element-binding protein-1c gene (SREBP-1c) by oxysterol receptors, LXRalpha and LXRbeta. *Genes Dev.* **14**:2819–2830.

49. Fitzgerald, M.L., et al. 2001. ATP-binding cassette transporter A1 contains an NH2-terminal signal anchor sequence that translocates the protein's first hydrophilic domain to the exoplasmic space. *J. Biol. Chem.* **276**:15137–15145.

Consequences of cellular cholesterol accumulation: basic concepts and physiological implications

Ira Tabas

Department of Medicine, and Department of Anatomy and Cell Biology, Columbia University, New York, New York, USA.

The cells of most organs and tissues satisfy their requirements for membrane cholesterol via endogenous cholesterol biosynthesis (1). Many cell types, however, have acquired mechanisms to internalize exogenous sources of cholesterol, usually in the form of plasma-derived lipoproteins (2). Examples include steroid-synthesizing cells, hepatocytes, and macrophages and smooth muscle cells in atherosclerotic lesions, often referred to as foam cells. In the case of steroidogenic cells, the internalization of lipoprotein-cholesterol represents a physiological process that provides cells with precursor cholesterol stores, to be used for "acute" steroid hormone production (3). Hepatocyte lipoprotein uptake mediates the clearance of various classes of plasma lipoproteins (1), which can lead to whole-body elimination of excess diet-derived cholesterol in the bile, a process known as reverse cholesterol transport (Tall, this Perspective series, ref. 4). The uptake of arterial-wall lipoproteins by macrophages and smooth muscle cells may be a type of physiological scavenging response that initially helps rid the endothelium of potentially harmful lipoprotein material (5). As will be discussed below, however, this cellular process eventually contributes to the progression and complications of atherosclerotic vascular disease.

Cells that rely totally or mostly on endogenous cholesterol synthesis cannot accumulate excess endogenous cholesterol because of homeostatic regulation at multiple steps in the cholesterol biosynthetic pathway (6). Cells that internalize exogenous cholesterol also repress endogenous cholesterol biosynthesis and LDL receptor expression in response to cholesterol loading. Furthermore, these cells have evolved other mechanisms to prevent the accumulation of excess unesterified, or "free," cholesterol (FC). One mechanism is cholesterol esterification, which is mediated by the microsomal enzyme acyl-coenzyme A:cholesterol acyltransferase (ACAT) (7) (Figure 1). The two forms, ACAT-1 and ACAT-2, differ in their sites of expression, with macrophages and most other cell types expressing ACAT-1. In humans, intestinal epithelial cells, but not hepatocytes, selectively express ACAT-2; mice, in contrast, express ACAT-2 in both of these cell types (7). Another important protective mechanism against FC accumulation is cellular efflux of cholesterol and certain cholesterol-derived oxysterols (Tall, this Perspective series, ref. 4; Björkhem, this series, ref. 8) (Figure 1). In addition, some of the physiological pathways described above, such as steroid and bile acid biosynthesis, may help limit

Nonstandard abbreviations used: free cholesterol (FC); acyl-coenzyme A:cholesterol acyltransferase (ACAT); cholesteryl ester (CE); Niemann-Pick C protein 1 (npc1); endoplasmic reticulum (ER); phosphatidylcholine (PC); CTP:phosphocholine cytidylyltransferase (CT).

Conflict of interest: No conflict of interest has been declared.

Citation for this article: *J. Clin. Invest.* **110**:905–911 (2002). doi:10.1172/JCI200216452.

the accumulation of intracellular FC in steroidogenic cells and hepatocytes, respectively (2).

The evolution of multiple pathways to prevent intracellular FC accumulation reflects the toxic effects of excess cellular FC (9). In this article, I first discuss possible mechanisms of FC-induced cytotoxicity. I then turn to the development of advanced atherosclerotic lesions, a critical pathophysiological scenario in which macrophages accumulate excess FC. This scenario provides a valuable model to study mechanisms of FC accumulation, the adaptive processes that cells employ to help protect themselves from excess FC, and pathways and consequences of FC-induced cell death. The conversion of the lesional macrophage from a cholesteryl ester–laden (CE-laden) foam cell into an FC-loaded cell may, I argue, represent a critical turning point in the progression of the atherosclerotic lesion.

Mechanisms of FC-induced cytotoxicity

A physiological FC/phospholipid ratio in cellular membranes is necessary to maintain proper membrane fluidity, or more precisely, a proper range of membrane fluidities (10). The degree of saturation of the fatty acyl moieties of membrane phospholipids is the major determinant of the fluidity of lateral membrane domains, which consist of well-packed, detergent-resistant liquid-ordered rafts and more fluid, detergent-soluble liquid-crystalline regions (Simons and Ehehalt, this Perspective series, ref. 11; Maxfield and Wüstner, this series, ref. 12). Nonetheless, the interaction of the hydrophobic rings of cholesterol with these fatty acyl chains has important effects (Tabas, Series Introduction, ref. 13). In particular, the ability of cholesterol to pack tightly with saturated fatty acyl groups of membrane phospholipids is critical for the formation of liquid-ordered rafts (10). Thus, cholesterol depletion causes marked disruption of these rafts. However, when the FC/phospholipid ratio rises above a physiological level, the liquid-ordered rafts may become too rigid, and the liquid-crystalline domains may begin to lose their fluidity. These events adversely affect certain integral membrane proteins that require conformational freedom for proper function and that can be inhibited by a high FC/phospholipid ratio (14). Such proteins include plasma membrane constituents like Na^+-K^+ ATPase, adenylate cyclase, alkaline phosphatase, rhodopsin, and transporters for glucose, organic anions, and thymidine. Similar observations have been made with proteins residing in internal membranes, such as the Na^+-Ca^{2+} transporter in the sarcoplasmic reticulum of cardiac muscle, the ATP-ADP transporter in the inner mitochondrial membrane, and UDP-glucuronyltransferase in liver microsomes (14). Interestingly, inhibition of ACAT activity in Chinese hamster ovary cells transfected with amyloid precursor protein blocks the generation of amyloid

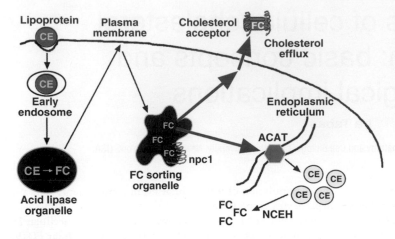

Figure 1

Mechanisms that protect cells from excess accumulation of FC. Lipoprotein-derived cholesterol is distributed to peripheral cellular sites from a putative FC-sorting organelle, which may be a type of late endosome. The npc1 protein is depicted in this organelle as one of the molecules that are known to influence cholesterol trafficking. The cholesterol trafficking itineraries depicted here include transport to ACAT in the endoplasmic reticulum, leading to cholesterol esterification, and to sites of cholesterol efflux in the plasma membrane, leading to removal of cholesterol if appropriate extracellular cholesterol acceptors are present. Not depicted here are those pathways that downregulate the LDL receptor and cholesterol biosynthetic enzymes and those pathways in specialized cells that lead to the metabolism of cholesterol to other molecules, like bile acids and steroid hormones. As described in the text, FC accumulation can occur via inhibition of cholesterol transport to ACAT or to the plasma membrane, or by direct inhibition of ACAT or cholesterol efflux molecules. An increase in neutral cholesteryl esterase (NCEH) activity in the absence of compensatory re-esterification or efflux of cholesterol could also lead to FC accumulation.

β-peptide (15). Although the mechanism of this effect is not yet known, one possibility is that the conformation of the amyloid precursor or the β-peptide–generating proteases is altered by an increase in the local FC/phospholipid ratio (Simons and Ehehalt, this series, ref. 11). High FC levels might therefore be proposed to kill cells in part by inhibiting one or more integral membrane proteins whose function is blocked or altered under conditions of high membrane rigidity. This and several other models for FC-induced cell death, discussed below, are summarized in Table 1.

Excess membrane cholesterol may also disrupt the function of signaling proteins that reside in membrane domains. For example, when human neutrophils with a normal plasma membrane cholesterol content are stimulated to migrate by exposure to chemokines, the actin-signaling protein Rac is recruited to detergent-sensitive non-raft membrane domains in leading lamella. However, when the plasma membrane of these cells are overloaded with cholesterol, Rac is recruited to the entire circumference of the plasma membrane, lamellar extension is nonvectorial, and neutrophil migration does not occur (L.M. Pierini and F.R. Maxfield, personal communication). One interpretation of these data is that excess plasma membrane cholesterol disrupts the function of certain signaling molecules that normally reside in non-raft domains. Experiments in vitro with model membranes suggest that, in membranes already enriched in sphingolipids (like those that exist in several types of epithelial cells), increasing the FC concentration even modestly above

the physiological concentration can actually suppress the formation of membrane domains (16).

Other mechanisms of cellular toxicity associated with FC accumulation include intracellular cholesterol crystallization, oxysterol formation (Björkhem, this series, ref. 8), and triggering of apoptotic signaling pathways (9). Needle-shaped cholesterol crystals form when the FC/phospholipid ratio reaches a very high level. Although typically seen in extracellular regions of advanced atherosclerotic lesions, intracellular cholesterol crystals have been observed both in cultured macrophages overloaded with cholesterol and in foam cells isolated directly from human coronary atherosclerotic lesions (17, 18). Intracellular cholesterol crystals can probably damage cells by physically disrupting the integrity of intracellular structures. Excess intracellular FC accumulation can also promote the oxidation of cholesterol to oxysterols, some of which may be cytotoxic (19). Finally, FC overloading of macrophages can trigger a series of apoptotic pathways (20–22), as discussed below.

Accumulation of FC in macrophages during atherogenesis

The hallmark of the early atherosclerotic lesion is the CE-laden macrophage foam cell (23). Remarkably, in the face of extremely high levels of CE, these early lesional cells maintain an FC content not markedly different from non–foam cell macrophages. However, lipid assays of lesional material from various stages of human and animal atheromata reveal a steady increase in FC content and a steady decrease in CE content as the lesions become more advanced (24). While a portion of this trend reflects extracellular events, analysis of foam cells isolated from advanced lesions clearly demonstrates an increase in macrophage FC content (9). As discussed below, cultured macrophage models of FC accumulation have revealed fascinating cellular responses to FC accumulation, as well as end-stage consequences of FC loading that are relevant to the progression and complications of atherosclerosis.

Mechanisms of FC accumulation

Macrophages are normally protected from the accumulation of excess FC by ACAT-1–mediated esterification and by cholesterol efflux (2) (Figure 1). In addition, the hydrolysis of stored CE by neutral CE hydrolase does not usually exceed a cell's capacity to export or re-esterify this pool of cholesterol. Thus, the progressive accumulation of FC by lesional macrophages might be explained by failure of one or more of these protective mechanisms as the atherosclerotic lesion progresses.

Both cholesterol efflux and cholesterol esterification would be adversely affected by disruption of intracellular cholesterol transport. Cholesterol that enters cells via internalization of lipoproteins is initially targeted to an organelle, probably a type of late endosome, that then distributes this cholesterol to the plasma membrane for efflux and the endoplasmic reticulum (ER) for esterification (2, 25, 26). Peripheral cholesterol transport is an energy-dependent vesicular process and involves molecules such as Niemann-Pick C protein 1 (npc1) and npc2 (HE1) and the lipids lysobisphosphatidic acid and sphingomyelin (12, 25). Disturbance of cellular membrane vesiculation, depletion of cellular ATP stores, or disruption of the function of these molecules could block FC efflux and esterification, thus favoring FC accumulation. Indeed, peripheral FC trafficking is inhibited in cultured macrophages incubated with oxidized LDL, a form of modified LDL found in atherosclerotic lesions. This response may be related to inhibition by oxysterols of lysosomal sphingomyelinase (27), leading to sphingomyelin accumulation. Excess intracellular sphingomyelin, in turn, disrupts normal peripheral cholesterol distribution by an unknown mechanism (28). Interestingly, experiments in cultured cells have revealed that lysosomal FC accumulation itself inhibits lysosomal sphingomyelinase activity (29). Thus, if cellular cholesterol influx begins to exceed the capacity of lysosomes to transport FC, an initially modest accumulation of lysosomal FC may be amplified by subsequent inhibition of lysosomal sphingomyelinase. This mechanism may help explain the finding that cultured macrophages that phagocytose large quantities of CE droplets accumulate FC in lysosomes (30).

Direct disruption of ACAT-1 or cellular efflux pathways would also promote FC accumulation. Efflux might be compromised by inhibiting ABCA1, SR-BI, or other plasma membrane proteins that mediate cholesterol efflux. Indeed, recent data from our laboratory suggest that excess FC accumulation in macrophages compromises the function of ABCA1, which may then amplify further FC accumulation (31). Efflux may also be blocked if access is lost to plasma-derived cholesterol acceptors like HDL and apoA-1, as may occur for macrophages buried deep in the intima of atherosclerotic lesions. In addition, if cellular cholesterol influx exceeds the capacity of ACAT-1, FC may begin to accumulate in the ER membrane domains where this enzyme resides. Because these membranes are normally cholesterol-poor (32), an FC-induced change in the physical properties of these membranes might compromise ACAT-1 activity (33). According to this model, FC accumulation would be amplified by progressive cholesterol accumulation in ER membranes, leading to further ACAT-1 dysfunction. Finally, efflux pathways may also be inhibited by conversion of cholesterol into oxysterols. While certain oxysterols actually promote sterol removal (e.g., 27-hydroxycholesterol in macrophages and bile acids in hepatocytes), other oxysterols (e.g., 25-hydroxycholesterol and 7-ketocholesterol) exacerbate cholesterol accumulation by inhibiting FC efflux or ACAT-1–mediated cholesterol esterification (19).

Recent experiments in cultured macrophages suggest that neutral CE hydrolase activity can further complicate the problem of cellular FC homeostasis in advancing atherogenic lesions. When these cells are incubated for a short period with atherogenic lipoproteins, the CE-rich lipid droplets formed have a high CE content and essentially normal FC content, as expected. After subsequent culture in the absence of lipoproteins, however, CE content drops and FC content rises correspondingly. Collections of FC can be seen in the cytoplasm of these cells by filipin staining, reflecting the ability of neutral CE hydrolase to act on preformed CE droplets and to generate excess FC under at least some physiological conditions (34). Whether this enzyme is also active when cells are continuously exposed to atherogenic lipoproteins remains to be determined. Nonetheless,

the action of this hydrolase is likely to contribute to FC accumulation in lesional macrophage foam cells whose FC efflux or re-esterification by ACAT-1 is compromised.

Adaptive processes

As shown in Figure 2, the responses of macrophages to FC loading can be divided into two phases — an initial adaptive stage in which phospholipid synthesis increases to offset the harmful effects of increasing FC, and a later stage when these defenses are overcome, leading to cell death. As investigators continue to explore the mechanisms of FC accumulation in vivo, a convenient cell-culture model has enabled the study of both of these stages (9, 35). In this model, macrophages are treated with one of several specific inhibitors of ACAT-1, either during or after incubation with atherogenic lipoproteins. Importantly, the perinuclear distribution of filipin-positive FC in the cultured cells mimics the distribution of FC observed in macrophages isolated from atherosclerotic lesions. Moreover, the various effects observed in this model are not seen with ACAT inhibitors in the absence of lipoproteins and are thus dependent on FC loading per se.

One of the earliest responses of macrophages to FC loading is an increase in phospholipid biosynthesis (35). While this response is observed with sphingomyelin and several other classes of phospholipids, the major species produced is phosphatidylcholine (PC). The mechanism involves posttranslational activation of the rate-limiting enzyme in PC biosynthesis, CTP:phosphocholine cytidylyltransferase (CT). Exactly how FC accumulation leads to CT activation is not known, but the process requires dephosphorylation of CT and possibly one or more CT regulatory proteins. The two-fold increment in PC mass in FC-loaded macrophages is remarkable, and it is reflected by the appearance of phospholipid-containing whorl-like membrane structures in the cells. Importantly, such structures have also been seen in lesional macrophages (36). This observation, together with other reports showing an increase in phospholipid biosynthesis in the cells of atherosclerotic lesions in vivo (37), suggests that the responses of FC-loaded cultured macrophages indeed mimic physiological events.

The central issue to arise from these observations concerned the functional role of this phospholipid response. The hypothesis that increasing PC represents an adaptive response, preventing FC-mediated cytotoxicity by keeping the FC/phospholipid ratio in check, received direct support from studies conducted with peritoneal macrophages from CT-deficient mice (38). There are two genes that encode CT, CCTα and CCTβ. Macrophages from mice with homozygous disruption of the CCTα gene have intact CTβ activity and thus retain approximately 20% of their total CT activity and PC biosynthetic capacity. Importantly, these cells appear healthy when grown under normal culture

Table 1

Potential mechanisms of FC-induced cytotoxicity

Event	Consequence
Loss of membrane fluidity	Dysfunction of integral membrane proteins
Disruption of membrane domains	Disruption of signaling events
Induction of apoptosis	Caspase-mediated cell death
Intracellular cholesterol crystallization	Organelle disruption
Formation of toxic oxysterols	? Oxidative damage
? Alteration of gene expression	? Change in balance of survival vs. death proteins

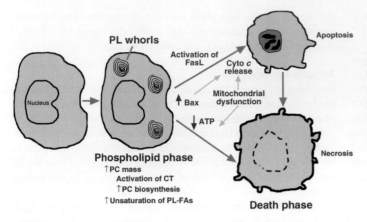

Figure 2

Sequential responses of cultured macrophages to FC loading. In the initial, adaptive phase, CT is activated, leading to increased PC biosynthesis and PC mass and resulting in a normalization of the FC/phospholipid ratio. In addition, there is an increase in the degree of unsaturation of phospholipid fatty acids (FAs). With continued FC loading, death ensues by both apoptotic and necrotic processes. Apoptosis involves both activation of Fas ligand and release of cytochrome c (cyto c) from mitochondria, which may be related to the increased levels of mitochondria-associated Bax in FC-loaded macrophages. FC-induced mitochondrial dysfunction also leads to depletion of cellular ATP stores, which could trigger cellular necrosis.

conditions and are essentially indistinguishable from wild-type macrophages. However, when subjected to an FC load, CTα-deficient macrophages cannot mount a substantial PC response, and they succumb to the toxic effects of FC much earlier than do wild-type macrophages. Thus, the ability of FC-loaded macrophages to activate CT and increase PC biosynthesis helps protect them from FC-induced cytotoxicity.

Ikonen and coworkers (39) recently reported another phospholipid response to FC loading, namely, an increase in plasma membrane phospholipid species with polyunsaturated acyl chains. These studies were conducted with human fibroblasts that had been loaded with FC either by incubation with serum in the face of a mutation that blocks cholesterol esterification or, in wild-type fibroblasts, by acute plasma membrane loading. In both cases, there was a significant increase in the content of polyunsaturated fatty acids in several classes of phospholipids. Although the mechanism of these fatty acid alterations is not known, these changes may represent another adaptive effect to FC loading, because membranes rich in phospholipids with polyunsaturated acyl chains are more resistant to the stiffening effects of cholesterol (40).

FC-induced cell death

Although cells are clearly capable of adaptive responses to FC loading, these mechanisms eventually fail with prolonged internalization of cholesterol, leading to cell death (9, 41) (Figure 2). The basis of adaptive failure is not known, although a decrease in CT activity has been observed before the onset of cellular toxicity (42). CT, like ACAT, is associated with normally cholesterol-poor cellular membranes and thus may become dysfunctional when its lipid microenvironment becomes too rigid.

By morphological criteria, cytotoxic FC-loaded macrophages show signs of both necrosis (e.g., disrupted cellular membranes) and apoptosis (e.g., condensed nuclei) (9, 41). By biochemical

criteria, apoptosis-associated caspases and their signaling pathways are activated in a portion of the cells (below). In all likelihood, as depicted in Figure 3, some cells in a population of FC-loaded macrophages become acutely necrotic due to direct and acute disruptive effects on membrane proteins, whereas others undergo a programmed apoptotic response. Moreover, cells that initially undergo an apoptotic program can subsequently demonstrate morphological signs of necrosis (so-called aponecrosis), perhaps as a result of chronic ATP depletion or failure of neighboring cells to phagocytose the apoptotic bodies (43).

In a typical cell-culture model of FC-loaded macrophages, approximately 25–30% of the cells show the apoptosis-associated hallmarks of phosphatidylserine externalization and DNA fragmentation, which can be completely prevented by inhibitors of effector caspases (21). Interestingly, partial inhibition of apoptosis is also observed when the Fas receptor signaling pathway is disrupted either by genetic mutations in Fas or its ligand or by use of a blocking anti–Fas ligand antibody. FC loading results in posttranslational activation of cell-surface Fas ligand, either by inducing a conformational change in the molecule or by stimulating its transport from intracellular stores to the plasma membrane (21).

Widespread mitochondrial dysfunction, indicated by a decrease in the mitochondrial transmembrane potential, is also observed in FC-loaded macrophages (22). Furthermore, FC-loaded macrophages show evidence of mitochondrial cytochrome c release and caspase-9 activation. Thus, in addition to involvement of the Fas pathway, a classic mitochondrial pathway of apoptosis is activated in FC-loaded macrophages. Importantly, these events are not mediated by oxysterols, because oxysterol-induced mitochondrial dysfunction and apoptosis are inhibited by the antioxidant glutathione, whereas the death pathways described above are resistant to this compound. The mechanisms by which FC overloading triggers these events is not known, but it appears to be independent of direct effects of FC on the plasma membrane (P.M. Yao and I. Tabas, unpublished data). Intracellular and mitochondrial levels of the proapoptotic protein Bax are increased in FC-loaded macrophages (22), but direct proof for the involvement of Bax is lacking. Other possibilities include direct toxic effects of FC on mitochondrial membranes and activation of a proapoptotic signaling pathway.

An important concept to arise from these cell-culture studies is the importance of intracellular cholesterol trafficking in FC-induced death. Rothblat and colleagues (20) were the first to show that amphipathic amines, which markedly inhibit peripheral cholesterol trafficking, protect FC-loaded macrophages from death. These results suggest that, whether FC triggers death by direct membrane effects or by activation of death-promoting molecules or both, FC must be able to traffic to peripheral sites in the cell to effect cell killing. Our recent data indicate that, whereas FC trafficking to the plasma membrane may be responsible for necrotic-type death, excess FC accumulation in the plasma membrane cannot explain apoptotic death. Moreover, FC-induced apoptosis is blocked by very subtle disruptions of intracellular trafficking, such as occur in macrophages from heterozygous NPC mice or in macrophages treated with very-low-dose amphipathic amines (ref. 44; and P.M. Yao and I. Tabas, unpublished data). Thus, the peripheral apoptosis-sensing mechanisms appear to be quite sensitive to FC. Clearly, an important

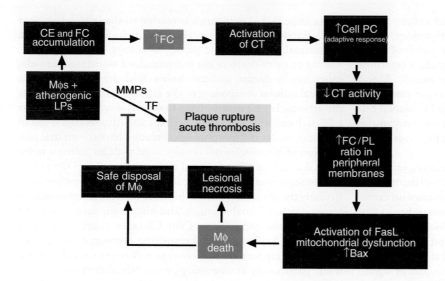

Figure 3
Hypothetical model relating FC loading of lesional macrophages (Mφs) to acute events in advanced atheromata. Progressive FC loading of lesional macrophages leads to a series of phospholipid-related adaptive responses, as described in the text. These adaptive responses eventually fail, leading to macrophage death (see Figure 2). On the one hand, macrophage death may contribute to plaque instability by promoting lesional necrosis. On the other hand, safe disposal of apoptotic macrophages could decrease the number of macrophages that secrete matrix metalloproteinases (MMPs), tissue factor (TF), and inflammatory cytokines. Because these molecules are thought to contribute to plaque rupture and acute thrombosis, macrophage death in this particular context might be protective.

goal of further research in this area is to define the molecular mechanisms that underlie the relationship between intracellular cholesterol trafficking and apoptosis in FC-loaded cells.

Relevance to atherosclerotic vascular disease

The presence of apoptotic and necrotic macrophages in atherosclerotic lesions has been well documented in many human and animal studies (45, 46). Among the potential causes of lesional macrophage death, FC-induced toxicity needs serious consideration, because macrophages from advanced lesions are known to be loaded with FC, a potent inducer of macrophage death (9). The functional significance of the cell death pathways depicted in Figure 3 remains uncertain. On the one hand, assuming harmless disposal of apoptotic bodies by neighboring phagocytes, macrophage apoptosis may limit the number of intimal cells in a physiologically "safe" manner that avoids inducing local inflammation. On the other hand, death of macrophages by either necrosis or apoptosis might lead to release of cellular proteases, inflammatory cytokines, and prothrombotic molecules, which could contribute to plaque instability, plaque rupture, and acute thrombotic vascular occlusion. Indeed, necrotic areas of advanced atherosclerotic lesions are known to be associated with death of macrophages, and ruptured plaques from human lesions have been shown to be enriched in apoptotic macrophages (46).

Two mouse models have begun to shed light on the in vivo consequences of FC-induced macrophage death. In the first model, LDL receptor knockout mice were reconstituted with ACAT-1–deficient macrophages by bone marrow transplantation (47). Compared with control mice reconstituted with ACAT-1–positive macrophages, the atherosclerotic lesions of the experimental mice were larger and had increased FC content and more apoptotic macrophages. Thus, in this model, FC-induced macrophage death promotes lesion development.

The second model was designed to address the effect of preventing macrophage death in an otherwise atherosclerosis-prone mouse. By crossing a disrupted allele of *Npc1*, in heterozygous form, into an apoE knockout background, our group produced animals with advanced atherosclerotic lesions whose macrophages were relatively resistant to FC-induced death. Control mice with normal npc1 function had large acellular areas filled with FC and

macrophage debris, but not with CE. In contrast, the lesions of the mice with partial npc1 deficiency had a greater content of CE-rich macrophages (44). Because npc1 deficiency does not protect macrophages from other inducers of apoptosis, these data support the hypothesis that FC-induced death, requiring intact FC trafficking, is an important cause of lesional macrophage death in vivo. Moreover, the lesions in the death-resistant model appear more stable, but future studies with rupture-susceptible mice will be needed to substantiate this point.

In light of these recent data, the therapeutic value of ACAT inhibitors may require some further scrutiny. These agents, which are intended to block the accumulation of CE in vascular macrophages, have been proposed for the prevention or treatment of atherosclerotic vascular disease (48). Indeed, ACAT inhibitors have been shown to prevent atherosclerosis in several animal models, and one such inhibitor is currently undergoing trials in humans. The site of drug action may be key to explaining the beneficial effects of these drugs despite the apparent risk that they could further the progression of atherosclerotic lesions. First, even for ACAT-1 inhibitors, which suppress macrophage-associated enzymatic activity in vitro, the drug's ability to enter the lesion may be limited. Second, moderate suppression of ACAT activity in these cells can probably be offset by cholesterol efflux, as long as the cells have access to cholesterol acceptors such as HDL or apoA-1. ACAT-2 inhibitors, conversely, should have no direct effect on lesional macrophages and could be strongly beneficial because of their ability to suppress lipoprotein production by the intestine and — in mice — the liver (49). For these reasons, ACAT inhibitors, whether directed at one or both of these isoforms, may be less likely to precipitate macrophage death and lesional necrosis than work with ACAT-1–deficient cells might suggest.

Concluding remarks and future directions

The study of cellular cholesterol excess provides an opportunity to address a variety of important topics ranging from biophysical chemistry to intracellular signaling pathways to mechanisms of clinical disease. To begin, a proper understanding of the FC-loaded cell requires an appreciation of membrane lipid phase behavior. One must then investigate the alterations of specific enzymes and other proteins to elucidate the mechanisms of specific consequences of FC loading on cellular physiology. These alterations may result either from a direct consequence of membrane alterations or from the activation of signaling reactions that themselves are triggered by membrane alterations. Moreover, effects of FC or FC metabolites on gene expression must also be considered. Finally, the biology of the FC-loaded cell must be placed in the context of the whole tissue and organism, as demonstrated by the potential effects of the FC-loaded macrophage on atherosclerosis.

In each area, much remains to be done. While the effects of high levels of FC on the physical properties of model membranes have been studied widely in vitro, biophysical studies on excess FC in membranes of living cells have suffered from technological difficulties. New advances in fluorescence microscopy, including the use of domain-specific probes (Maxfield and Wüstner, this series, ref. 12), are beginning to close this critical gap. The consequences of cellular FC excess on specific molecules or signaling pathways are gradually entering the scientific literature, but insight into the mechanisms linking these consequences with changes in FC-induced alterations in membrane structure and possibly gene expression is lacking. For example, much has been published on the effects of cellular cholesterol depletion on raft structure and function (Simons and Ehehalt, this series, ref. 11), but very little work has been published about the effects of cellular cholesterol excess on raft biology.

The impetus for such work is related to the issues discussed immediately above, namely, the role of the FC-loaded cell in organismal physiology and pathophysiology. In particular, the leading cause of death in the industrialized world is atherosclerotic vascular disease, and the cholesterol-loaded macrophage is a critical cellular component of the atherosclerotic lesion. While the atherosclerotic macrophage foam cell is typically viewed as a CE-rich cell with normal or perhaps slightly elevated levels of FC, biochemical and morphological studies have shown that progression of atherosclerosis is associated with an increase in FC and a decrease in CE in lesional macrophages. Given the potential consequences of FC loading for macrophage physiology, particularly the potential relationships between FC-induced macrophage death, lesional necrosis, and plaque rupture, one might argue that the conversion of the CE-laden macrophage into an FC-loaded cell is a critical transition point in atherogenesis (Figure 3). Support for this concept will require much further work, particularly in vivo using genetically altered mice. Nonetheless, given the current evidence and the importance of atherosclerosis, the rationale for gaining a thorough understanding of the biology of the FC-loaded cell is clear.

Address correspondence to: Ira Tabas, Department of Medicine, Columbia University, 630 West 168th Street, New York, New York 10032, USA. Phone: (212) 305-9430; Fax: (212) 305-4834; E-mail: iat1@columbia.edu.

1. Dietschy, J.M., Turley, S.D., and Spady, D.K. 1993. Role of the liver in the maintenance of cholesterol and low density lipoprotein homeostasis in different animal species, including humans. *J. Lipid Res.* **34**:1637–1659.

2. Tabas, I., and Kreiger, M. 1999. Lipoprotein receptors and cellular cholesterol metabolism in health and disease. In *Molecular basis of heart disease.* K.R. Chien, editor. W.B. Saunders Co. New York, New York, USA. 428–457.

3. Lin, D., et al. 1995. Role of steroidogenic acute regulatory protein in adrenal and gonadal steroidogenesis. *Science.* **267**:1828–1831.

4. Tall, A.R., Costet, P., and Wang, N. 2002. Regulation and mechanisms of macrophage cholesterol efflux. *J. Clin. Invest.* **110**:899–904. doi:10.1172/JCI200216391.

5. Glass, C.K., and Witztum, J.L. 2001. Atherosclerosis: the road ahead. *Cell.* **104**:503–516.

6. Horton, J.D., Goldstein, J.L., and Brown, M.S. 2002. SREBPs: activators of the complete program of cholesterol and fatty acid synthesis in the liver. *J. Clin. Invest.* **109**:1125–1131. doi:10.1172/JCI200215593.

7. Chang, T.Y., et al. 2001. Roles of acyl-coenzyme A:cholesterol acyltransferase-1 and -2. *Curr. Opin. Lipidol.* **12**:289–296.

8. Björkhem, I. 2002. Do oxysterols control cholesterol homeostasis? *J. Clin. Invest.* **110**:725–730. doi:10.1172/JCI200216388.

9. Tabas, I. 1997. Free cholesterol-induced cytotoxicity. A possible contributing factor to macrophage foam cell necrosis in advanced atherosclerotic lesions. *Trends Cardiovasc. Med.* **7**:256–263.

10. Simons, K., and Ikonen, E. 2000. How cells handle cholesterol. *Science.* **290**:1721–1726.

11. Simons, K., and Ehehalt, R. 2002. Cholesterol, lipid rafts, and disease. *J. Clin. Invest.* **110**:597–603. doi:10.1172/JCI200216390.

12. Maxfield, F.R., and Wüstner, D. 2002. Intracellular cholesterol transport. *J. Clin. Invest.* **110**:891–898. doi:10.1172/JCI200216500.

13. Tabas, I. 2002. Cholesterol in health and disease.

J. Clin. Invest. **110**:583–590. doi:10.1172/JCI200216381.

14. Yeagle, P.L. 1991. Modulation of membrane function by cholesterol. *Biochimie.* **73**:1303–1310.

15. Puglielli, L., et al. 2001. Acyl-coenzyme A:cholesterol acyltransferase modulates the generation of the amyloid beta-peptide. *Nat. Cell Biol.* **3**:905–912.

16. Milhiet, P.E., Giocondi, M.C., and Le Grimellec, C. 2002. Cholesterol is not crucial for the existence of microdomains in kidney brush-border membrane models. *J. Biol. Chem.* **277**:875–878.

17. Kellner-Weibel, G., et al. 1999. Crystallization of free cholesterol in model macrophage foam cells. *Arterioscler. Thromb. Vasc. Biol.* **19**:1891–1898.

18. Lupu, F., Danaricu, I., and Simionescu, N. 1987. Development of intracellular lipid deposits in the lipid-laden cells of atherosclerotic lesions. A cytochemical and ultrastructural study. *Atherosclerosis.* **67**:127–142.

19. Brown, A.J., and Jessup, W. 1999. Oxysterols and atherosclerosis. *Atherosclerosis.* **142**:1–28.

20. Kellner-Weibel, G., et al. 1998. Effects of intracellular free cholesterol accumulation on macrophage viability: a model for foam cell death. *Arterioscler. Thromb. Vasc. Biol.* **18**:423–431.

21. Yao, P.M., and Tabas, I. 2000. Free cholesterol loading of macrophages induces apoptosis involving the Fas pathway. *J. Biol. Chem.* **275**:23807–23813.

22. Yao, P.M., and Tabas, I. 2001. Free cholesterol loading of macrophages is associated with widespread mitochondrial dysfunction and activation of the mitochondrial apoptosis pathway. *J. Biol. Chem.* **276**:42468–42476.

23. Ross, R. 1995. Cell biology of atherosclerosis. *Annu. Rev. Physiol.* **57**:791–804.

24. Katz, S.S., Shipley, G.G., and Small, D.M. 1976. Physical chemistry of the lipids of human atherosclerotic lesions. Demonstration of a lesion intermediate between fatty streaks and advanced plaques. *J. Clin. Invest.* **58**:200–211.

25. Liscum, L., and Munn, N.J. 1999. Intracellular cholesterol transport. *Biochim. Biophys. Acta.* **1438**:19–37.

26. Lange, Y. 1998. Intracellular cholesterol movement and homeostasis. In *Intracellular cholesterol trafficking.* T.Y. Chang and D.A. Freeman, editors. Kluwer Academic Publishers. Boston, Massachusetts, USA. 15–27.

27. Maor, I., Mandel, H., and Aviram, M. 1995. Macrophage uptake of oxidized LDL inhibits lysosomal sphingomyelinase, thus causing the accumulation of unesterified cholesterol-sphingomyelin-rich particles in the lysosomes. A possible role for 7-ketocholesterol. *Arterioscler. Thromb. Vasc. Biol.* **15**:1378–1387.

28. Leventhal, A.R., Chen, W., Tall, A.R., and Tabas, I. 2001. Acid sphingomyelinase-deficient macrophages have defective cholesterol trafficking and efflux. *J. Biol. Chem.* **276**:44976–44983.

29. Reagan, J.W., Jr., Hubbert, M.L., and Shelness, G.S. 2000. Posttranslational regulation of acid sphingomyelinase in niemann-pick type C1 fibroblasts and free cholesterol-enriched chinese hamster ovary cells. *J. Biol. Chem.* **275**:38104–38110.

30. Tangirala, R.K., Mahlberg, F.H., Glick, J.M., Jerome, W.G., and Rothblat, G.H. 1993. Lysosomal accumulation of unesterified cholesterol in model macrophage foam cells. *J. Biol. Chem.* **268**:9653–9660.

31. Feng, B., and Tabas, I. 2002. ABCA1-mediated cholesterol efflux is defective in free cholesterol-loaded macrophages. Mechanism involves enhanced ABCA1 degradation in a process requiring full npc1 activity. *J. Biol. Chem.* In press.

32. Chang, T.Y., Chang, C.C.Y., and Cheng, D. 1997. Acyl-coenzyme A:cholesterol acyltransferase. *Annu. Rev. Biochem.* **66**:613–638.

33. Moynault, A., Luciani, M.F., and Chimini, G. 1998. ABC1, the mammalian homologue of the engulfment gene ced-7, is required during phagocytosis of both necrotic and apoptotic cells. *Biochem. Soc. Trans.* **26**:629–635.

34. Mori, M., et al. 2001. Foam cell formation containing lipid droplets enriched with free cholesterol by hyperlipidemic serum. *J. Lipid Res.* **42**:1771–1781.

35. Tabas, I. 2000. Cholesterol and phospholipid metabolism in macrophages. *Biochim. Biophys. Acta.*

1529:164–174.

36. Shio, H., Haley, N.J., and Fowler, S. 1979. Characterization of lipid-laden aortic cells from cholesterol-fed rabbits. III. Intracellular localization of cholesterol and cholesteryl ester. *Lab. Invest.* **41**:160–167.

37. Zilversmit, D.B., Shore, M.L., and Ackerman, R.F. 1954. The origin of aortic phospholipid in rabbit atheromatosis. *Circulation.* **9**:581–585.

38. Zhang, D., et al. 2000. Macrophages deficient in CTP:Phosphocholine cytidylyltransferase-alpha are viable under normal culture conditions but are highly susceptible to free cholesterol-induced death. Molecular genetic evidence that the induction of phosphatidylcholine biosynthesis in free cholesterol-loaded macrophages is an adaptive response. *J. Biol. Chem.* **275**:35368–35376.

39. Blom, T.S., et al. 2001. Mass spectrometric analysis reveals an increase in plasma membrane polyunsaturated phospholipid species upon cellular cholesterol loading. *Biochemistry.* **40**:14635–14644.

40. Huster, D., Arnold, K., and Gawrisch, K. 1998. Influence of docosahexaenoic acid and cholesterol on lateral lipid organization in phospholipid mixtures. *Biochemistry.* **37**:17299–17308.

41. Warner, G.J., Stoudt, G., Bamberger, M., Johnson, W.J., and Rothblat, G.H. 1995. Cell toxicity induced by inhibition of acyl coenzyme A: cholesterol acyltransferase and accumulation of unesterified cholesterol. *J. Biol. Chem.* **270**:5772–5778.

42. Tabas, I., Marathe, S., Keesler, G.A., Beatini, N., and Shiratori, Y. 1996. Evidence that the initial up-regulation of phosphatidylcholine biosynthesis in free cholesterol-loaded macrophages is an adaptive response that prevents cholesterol-induced cellular necrosis. Proposed role of an eventual failure of this response in foam cell necrosis in advanced atherosclerosis. *J. Biol. Chem.* **271**:22773–22781.

43. Formigli, L., et al. 2000. Aponecrosis: morphological and biochemical exploration of a syncretic process of cell death sharing apoptosis and necrosis. *J. Cell. Physiol.* **182**:41–49.

44. Zhang, D., et al. 2001. Niemann-Pick C heterozygosity confers marked protection from macrophage death and lesional necrosis. *Circulation.* **104**:II-45. (Abstr.)

45. Kockx, M.M. 1998. Apoptosis in the atherosclerotic plaque: quantitative and qualitative aspects. *Arterioscler. Thromb. Vasc. Biol.* **18**:1519–1522.

46. Mitchinson, M.J., Hardwick, S.J., and Bennett, M.R. 1996. Cell death in atherosclerotic plaques. *Curr. Opin. Lipidol.* **7**:324–329.

47. Fazio, S., et al. 2001. Increased atherosclerosis in LDL receptor null mice lacking ACAT1 in macrophages. *J. Clin. Invest.* **107**:163–171.

48. Brown, W.V. 2001. Therapies on the horizon for cholesterol reduction. *Clin. Cardiol.* **24**:III24–III27.

49. Buhman, K.K., et al. 2000. Resistance to diet-induced hypercholesterolemia and gallstone formation in ACAT2-deficient mice. *Nat. Med.* **6**:1341–1347.

Monogenic hypercholesterolemia: new insights in pathogenesis and treatment

Daniel J. Rader,[1] Jonathan Cohen,[2,3] and Helen H. Hobbs[3,4]

[1]Department of Medicine and Center for Experimental Therapeutics, University of Pennsylvania School of Medicine, Philadelphia, Pennsylvania, USA.
[2]Center for Human Nutrition, and [3]Eugene McDermott Center for Human Growth and Development and the Department of Molecular Genetics, University of Texas Southwestern Medical Center, Dallas, Texas, USA. [4]Howard Hughes Medical Institute, Dallas, Texas, USA.

Introduction

The careful clinical characterization of patients with genetic forms of severe hypercholesterolemia has played a critical role in the historic linkage of hypercholesterolemia to atherosclerosis. Elucidation of gene defects that cause severe hypercholesterolemia has provided molecular entrées into the biosynthetic and regulatory pathways that produce and eliminate cholesterol and has led to the development of potent pharmacological agents that dramatically reduce circulating levels of cholesterol. The last decade of the twentieth century culminated in the demonstration that pharmacological reductions in plasma cholesterol levels result in fewer cardiovascular events and reduce total mortality.

This review will summarize recent developments in our understanding of the molecular pathogenesis and treatment of monogenic forms of severe hypercholesterolemia, and some implications that these findings have for the management of common forms of hypercholesterolemia.

General overview of LDL metabolism. Cholesterol is a rigid, hydrophobic molecule that confers structural integrity to plasma membranes of vertebrate cells. Excess cellular cholesterol is esterified with fatty acids to form cholesteryl esters, which are either stored as lipid droplets in cells or packaged with other apolipoproteins to form VLDL in the liver and and chylomicrons in the intestine (Figure 1). The two major cholesterol-carrying lipoproteins in humans are LDL and HDL. Approximately 70% of circulating cholesterol is transported as LDL.

LDL is formed in the circulation from VLDL (Figure 1). The triglycerides and phospholipids of circulating VLDL are hydrolyzed by lipases anchored to vascular endothelial surfaces, forming cholesterol-enriched VLDL remnant particles. Approximately half of the VLDL remnants are cleared from the circulation by LDL receptor–mediated (LDLR-mediated) endocytosis in the liver, and the remainder undergoes further processing to produce LDL. Most LDL is removed from the circulation after binding to the hepatic LDLR via apoB-100.

Plasma levels of LDL-cholesterol (LDL-C) are directly related to the incidence of coronary events and cardiovascular deaths. Approximately 50% of the interindividual variation in plasma levels of LDL-C is attributable to genetic variation (1). The major portion of this genetic variation is polygenic, reflecting the cumulative effects of multiple sequence variants in any given individu-

Nonstandard abbreviations used: LDL receptor (LDLR); LDL-cholesterol (LDL-C); familial hypercholesterolemia (FH); familial defective apoB-100 (FDB); autosomal recessive hypercholesterolemia (ARH).

Conflict of interest: The authors have declared that no conflict of interest exists.

Citation for this article: *J. Clin. Invest.* **111**:1795–1803 (2003). doi:10.1172/JCI200318925.

al. A subset of patients with very high plasma LDL-C levels have monogenic forms of hyper-cholesterolemia, which are associated with the deposition of cholesterol in tissues, producing xanthomas and coronary atherosclerosis.

The clinical features, diagnosis, and pathophysiology of the known mendelian disorders of severe hypercholesterolemia will be serially reviewed (Table 1). This will be followed by a discussion of how insights gleaned from the study of these disorders may be extended to the treatment of hypercholesterolemia in the general population.

Familial hypercholesterolemia

Historical perspective. Familial hypercholesterolemia (FH), the most common and most severe form of monogenic hypercholesterolemia, was the first genetic disease of lipid metabolism to be clinically and molecularly characterized (2). The disease has an autosomal codominant pattern of inheritance and is caused by mutations in the *LDLR* gene; individuals with two mutated LDLR alleles (FH homozygotes) are much more severely affected than those with one mutant allele (FH heterozygotes). The plasma levels of LDL-C are uniformly very high in FH homozygotes, irrespective of diet, medications, or lifestyle. For example, FH homozygotes living in China, where the dietary intake of cholesterol and saturated fat is low, have plasma LDL-C levels similar to those of FH homozygotes living in Western countries (3).

FH homozygotes develop cutaneous (planar) xanthomas and coronary atherosclerosis in childhood (2). Atherosclerosis develops initially in the aortic root, causing supravalvular aortic stenosis, and then extends into the coronary ostia. The severity of atherosclerosis is proportional to the extent and duration of elevated plasma LDL-C levels (calculated as the cholesterol-year score) (4). If the LDL-C level is not effectively reduced, FH homozygotes die prematurely of atherosclerotic cardiovascular disease. Optimization of other cardiovascular risk factors has little impact on the clinical course of the disease.

Patients with homozygous FH are classified into one of two major groups based on the amount of LDLR activity measured in their skin fibroblasts: patients with less than 2% of normal LDLR activity (receptor-negative), and patients with 2–25% of normal LDLR activity (receptor-defective) (2). In general, plasma levels of LDL-C are inversely related to the level of residual LDLR activity. Untreated, receptor-negative patients with homozygous FH rarely survive beyond the second decade; receptor-defective patients have a better prognosis but, with few exceptions, develop clinically significant atherosclerotic vascular disease by age 30, and often sooner (2).

The plasma levels of LDL-C in FH heterozygotes are lower (elevated two- to threefold) and much more dependent on other

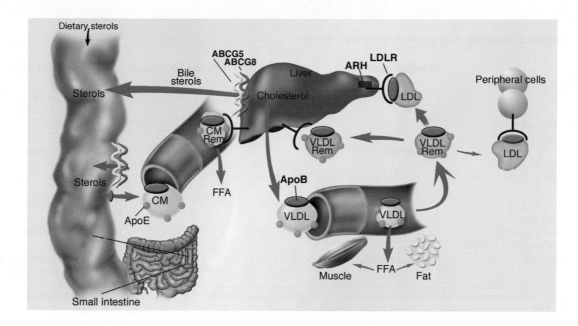

Figure 1

Overview of LDL metabolism in humans. Dietary cholesterol and triglycerides are packaged with apolipoproteins in the enterocytes of the small intestine, secreted into the lymphatic system as chylomicrons (CM). As chylomicrons circulate, the core triglycerides are hydrolyzed by lipoprotein lipase, resulting in the formation of chylomicron remnants (CM Rem), which are rapidly removed by the liver. Dietary cholesterol has four possible fates once it reaches the liver: it can be esterified and stored as cholesteryl esters in hepatocytes; packaged into VLDL particles and secreted into the plasma; secreted directly into the bile; or converted into bile acids and secreted into the bile. VLDL particles secreted into the plasma undergo lipolysis to form VLDL remnants (VLDL Rem). Approximately 50% of VLDL remnants are removed by the liver via the LDLR, and the remainder mature into LDL, the major cholesterol transport particle in the blood. An estimated 70% of circulating LDL is cleared by LDLR in the liver. ABCG5 and ABCG8 are located predominantly in the enterocytes of the duodenum and jejunum, the sites of uptake of dietary sterols, and in hepatocytes, where they participate in sterol trafficking into bile. Mutations in either transporter cause an increase in delivery of dietary sterols to the liver and a decrease in secretion of sterols into the bile. Autosomal recessive hypercholesterolemia (ARH) is a putative adaptor protein that is involved in the mechanics of LDLR-mediated endocytosis. ABCG: ATP-binding cassette, family G.

genetic and environmental factors than are those in FH homozygotes. Although the nature of the molecular defect has some impact on the severity of hypercholesterolemia, FH heterozygotes with the same LDLR mutation can have widely different plasma levels of LDL-C (2). The clinical prognosis of FH heterozygotes is related not only to the magnitude of the elevation in plasma LDL-C but also to the presence of other coronary risk factors (5).

New insights into pathogenesis. Despite our detailed knowledge of the molecular biology of the LDLR, fundamental questions regarding how the receptor delivers its cargo in cells without being degraded itself have only recently been elucidated. The crystal structure of the extracellular domain of the protein has provided a compelling model of how the receptor binds LDL with high affinity at the cell membrane and then releases it in the appropriate intracellular compartment (6). The extracellular domain consists of a ligand-binding domain, an EGF precursor homology domain, which contains a six-bladed β-propeller flanked by cysteine-rich EGF repeats (7), and an O-linked sugar-rich domain. When the LDLR is on the cell surface, the extracellular domain is extended, exposing the ligand-binding domain to LDL. After the receptor binds LDL, the receptor-ligand complex is internalized and delivered to endosomes. In the acidic environment of the endosome, the LDLR folds back on itself, bringing the β-propeller region of the EGF precursor domain into close apposition to the ligand-binding domain, thus displacing LDL. The β-propeller appears to function as a pseudosubstrate for the ligand-binding domain, per-

mitting release of the lipoprotein in the endosome and recycling of the LDLR to the cell surface. Naturally occurring LDLR mutations in humans that disrupt recycling of the receptor are located in residues critical to the structure of the EGF precursor domain (6, 7).

LDL turnover studies documented the key role of hepatic LDLRs in LDL catabolism (2). More recent studies indicate that the LDLR may also regulate the rate of entrance of VLDL into the circulation. Mice overexpressing the lipogenic transcription factor SREBP-1a have increased hepatic cholesterol and triglyceride synthesis, and hepatic steatosis, but normal plasma lipid levels. When LDLR expression is abolished in these mice, they become profoundly hyperlipidemic due to an increase in secretion of apoB-containing lipoproteins (8). Evidence from studies in cultured mouse hepatocytes suggests that the LDLR may also restrict hepatic apoB-100 secretion by promoting its intracellular degradation (9). Thus, the LDLR may limit the number of circulating triglyceride-rich particles by reducing their secretion and by promoting their recapture before they enter the circulation from the liver. The importance of the LDLR in limiting VLDL secretion in vivo in humans remains controversial, because isotopic studies cannot readily determine the proportion of newly formed VLDL that enters the systemic circulation.

Diagnosis. In general, the diagnosis of FH is straightforward and is based on a family history of hypercholesterolemia and premature coronary atherosclerosis, the lipid profile, and the presence of xanthomas. Heterozygous FH occurs in approximately 1 in 500

Table 1
Major monogenic diseases that cause severe hypercholesterolemia

Disease	Defective gene	Prevalence	Plasma LDL-C level	Metabolic defect
Autosomal dominant				
FH	*LDLR*			↓LDL clearance (1°)
				↑LDL production (2°)
Heterozygous FH		1 in 500	+++	
Homozygous FH		1 in 1 × 10⁶	+++++	
FDB	*APOB*			↓LDL clearance
Heterozygous FDB		1:1,000	++	
Homozygous FDB		1 in 4 × 10⁶	+++	
FH3	*PCSK9*			
Heterozygous FH3		<1 in 2500	+++	Unkown
Autosomal recessive				
Recessive autosomal	*ARH*	<1 in 5 × 10⁶	++++	↓LDL clearance
Hypercholesterolemia (ARH)				
Sitosterolemia	*ABCG5* or *ABCG8*	<1 in 5 × 10⁶	− to +++++	↓Cholesterol excretion (1°)
				↓LDL clearance (2°)

The number of + signs indicates the relative plasma level of LDL-C. 1°, primary major mechanism; 2°, secondary mechanism. FH, familial hypercholesterolemia; FDB, familial defective apoB-100; LDLR, LDL receptor.

persons worldwide but has a much higher incidence in certain populations, such as the Afrikaners, Christian Lebanese, Finns, and French-Canadians, due to founder effects (2). Over 900 mutations in the LDLR gene cause FH (10). Most mutations are unique, making the molecular diagnosis difficult, except in patients from populations where a limited number of mutations predominate. However, to date, there is no evidence that molecular diagnosis of the disease has important therapeutic implications.

Treatment. Heterozygous FH patients are responsive to statins, which inhibit HMG-CoA reductase and result in upregulation of the normal LDLR allele. Combination therapy is frequently required to achieve desired LDL-C levels (Table 2). Historically, bile acid sequestrants or nicotinic acid have been used for this purpose. Stanol esters, which decrease cholesterol absorption by displacing cholesterol from mixed micelles, are also effective in combination therapy (11). Recently, ezetimibe, a drug that specifically inhibits cholesterol absorption, became available (12). Ezetimibe binds to the microvilli of jejunal enterocytes and interferes with the enterohepatic circulation of cholesterol, much as bile acid sequestrants interrupt the enterohepatic circulation of bile acids. The drug is effective at very low concentrations and so presumably works by inhibiting a putative cholesterol transporter. Ezetimibe is more effective than dietary cholesterol restriction in lowering plasma cholesterol levels because it decreases the uptake of both dietary and biliary cholesterol. The reduced flux of cholesterol from the gut to the liver leads to a compensatory increase in hepatic LDLR, resulting in an approximately 20% reduction in LDL-C. The discovery of ezetimibe has rejuvenated the study of intestinal cholesterol absorption.

HMG-CoA reductase inhibitors and ezetimibe have only modest effects on plasma levels of LDL-C in FH homozygotes, even when administered at high doses (13). While some FH homozygotes with receptor-defective mutations may retain sufficient LDLR activity to respond to these potent lipid-lowering agents, drug therapy alone is never adequate treatment for these patients. The current treatment of choice for homozygous FH (and for heterozygotes whose plasma LDL-C remains elevated with drug therapy) is LDL apheresis. This process, in which the LDL par-

ticles are selectively removed from the circulation through extracorporeal binding to either dextran sulphate or heparin, can promote regression of xanthomas and may slow the progression of atherosclerosis (14). However, the procedure is time consuming and expensive and must be performed every 1–2 weeks. Although it retards the development of atherosclerosis, it does not prevent it, because of the recurrent hypercholesterolemia between procedures. Therefore, new therapies are urgently needed to treat the hypercholesterolemia of individuals suffering from homozygous FH. Inhibition of microsomal transfer protein (MTP), which is required for synthesis of apoB-containing lipoproteins, reduces plasma cholesterol levels in LDLR-deficient rabbits (15) and may be effective in homozygous FH, though the development of hepatic steatosis may be dose limiting. Liver-directed gene transfer of the LDLR theoretically is attractive but awaits the development of better and safer vectors.

Familial defective apoB-100

Historical perspective. A subset of individuals with a clinical presentation similar to FH and reduced rates of LDL catabolism were found to have normal LDLR activity. When LDL from these patients was infused into normal subjects, the heterologous LDL was cleared at a reduced rate compared with normal autologous LDL (16). The disease, familial defective apoB-100 (FDB), results from a missense mutation (Arg3500Gln) in the LDLR-binding domain of apoB-100 (17). Other less frequent mutations in *APOB* can also cause this disease. FDB occurs with a frequency of about 1 in 1,000 in Central Europe but is much less common in other populations (18). Like FH, FDB is characterized by elevated plasma LDL-C levels with normal triglycerides, tendon xanthomas, and premature atherosclerosis. The mean concentration of LDL-C is about 100 mg/dl higher in patients with FDB than in age matched controls, but the levels vary over a wide range. Over 25% of Europeans with FDB have plasma LDL-C levels below the 95th percentile of the population, which is a significantly higher percentage than is seen for FH (19).

FDB homozygotes have levels of plasma LDL-C comparable to those in FH heterozygotes rather than those in FH homozygotes

Table 2
Major LDL-C–lowering therapies for severe hypercholesterolemia

LDL-C–lowering therapy	Major effect	LDL-lowering response		
		Low (<10%)	Intermediate (10–25%)	High (>25%)
Statins	↑LDLR activity	Homozygous FH[A] Sitosterolemia		Heterozygous FH FDB ARH
Resins	↑LDLR activity	Homozygous FH[A]	Heterozygous FH FDB ARH	Sitosterolemia
Nicotinic acid	↓VLDL synthesis	Homozygous FH[A] Sitosterolemia		Heterozygous FH ARH FDB
Ezetimibe	↓Cholesterol absorption	Homozygous FH[A]	Heterozygous FH ARH (?) FDB (?)	Sitosterolemia
Stanol esters	↓Cholesterol absorption	Homozygous FH[A]	Heterozygous FH ARH (?) FDB (?)	
Low-fat, low-cholesterol diet	↑LDLR activity	Homozygous FH[A]	Heterozygous FH FDB	Sitosterolemia
LDL apheresis	Removes LDL			Homozygous FH Heterozygous FH[B]

[A]FH homozygotes with no LDLR function. FH homozygotes who have residual LDLR function can have intermediate responses to the lipid-lowering agents listed. [B]Plasmapheresis is used in FH heterozygotes who do not reach the target LDL-C on maximum tolerated doses of lipid-lowering agents. ?, no data available.

(18). The increased risk of atherosclerosis in FDB homozygotes confirms that LDL itself is atherogenic, since there is no accumulation of remnant particles in these patients. The development of coronary disease is slower, however, in FDB homozygotes than in FH homozygotes, since most FDB homozygotes identified to date are between 40 and 60 years of age (18).

New insights into pathogenesis. LDL from FDB homozygotes binds the LDLR with approximately 10% of normal affinity and is removed from the circulation at one-third to one-quarter of the normal rate (20). As a consequence of the delayed clearance of the defective LDL, the circulating LDL in FDB heterozygotes contains significantly more particles bearing the mutant apoB-100 than particles bearing the normal apoB-100 (20). The results of kinetic studies using stable isotopes are consistent with the idea that the lower plasma LDL concentrations in FDB than in FH result from reduced rates of LDL production (21). Patients with FDB are able to clear LDL precursor particles, since remnant uptake is mediated by apoE rather than apoB-100 (Figure 1).

Diagnosis. FDB cannot be clinically distinguished from heterozygous FH, although patients with FDB tend to have lower levels of plasma LDL-C and fewer xanthomas (18, 19). The single base pair substitution in the apoB-100 gene responsible for the most common missense mutation can be easily detected using a variety of molecular techniques (18), although absence of the sequence variation does not rule out the diagnosis of FDB, since some patients have other mutations in *APOB*. The clinical management of FDB and heterozygous FH is similar; thus, it is not necessary to establish the molecular diagnosis of FDB.

Treatment. It had been anticipated that patients with FDB might be less responsive to statin therapy than patients with FH, but this is not the case. FDB heterozygotes and homozygotes are responsive to the LDL-lowering effects of statins, presumably because of increased clearance of VLDL remnants (Table 2).

Other autosomal dominant forms of hypercholesterolemia

Recently the molecular defect responsible for an additional form of autosomal dominant hypercholesteremic has been identified. These patients are clinically indistinguishable from patients with heterozygous FH and FDB (except in one family in which the hypercholesterolemic family members had a lower body mass index [ref. 22]; however, the disease does not segregate with either *LDLR* or *APOB* (22, 23). The disease causing gene *PCSK9* encodes neural apoptosis-regulated convertase 1 (NARC-1), a member of the proteinase K family of subtilases (24, 25). Very little is known about this protein and elucidation of its function should provide interesting new insights into the control of plasma LDL-C.

Autosomal recessive forms of hypercholesterolemia

Historical perspective. In 1973, Khachadurian described an unusual Lebanese family in which all four offspring had clinical features of homozygous FH, including severe hypercholesterolemia and large tendon xanthomas, and yet LDLR function in their cultured fibroblasts was near normal (26). Subsequent characterization of Sardinian patients with a similar phenotype revealed that LDL clearance rates were as low as in patients with homozygous FH, and yet LDLR function in cultured fibroblasts was largely preserved (27). The hypercholesterolemia failed to segregate with either the *LDLR* or the *APOB* gene; the new disorder was named autosomal recessive hypercholesterolemia (ARH).

New insights into pathogenesis. ARH is caused by mutations in the *ARH* gene, which encodes a novel adaptor protein (28). To date, ten ARH mutations have been described (29, 30). All ten mutations interrupt the reading frame, precluding synthesis of a full-length protein. Although LDLR function is preserved in ARH fibroblasts, it is defective in transformed lymphocytes from these patients (31). The amount of LDLR is normal, but the distribution of immunodetectable receptor protein is significantly altered in ARH lymphocytes, with most of the LDLR residing on the plasma membrane (31). Cell surface LDL binding is increased, but LDL degradation is markedly reduced, indicating that ARH is involved in the internalization of the LDLR-LDL complex.

ARH contains an approximately 130-residue phosphotyrosine-binding domain, which is present in several adaptor proteins that bind a sequence motif (NPXY) in the cytoplasmic tails of a variety of cell surface receptors, including the LDLR (32). Brown, Goldstein, and colleagues showed that the integrity of the FDN-PXY sequence is required for internalization of the LDLR (33). ARH binds to this motif in a sequence-specific manner (34). This sequence also binds inositol phospholipids, which may anchor the protein to the plasma membrane (35). The C-terminal portion of the protein contains a canonical clathrin box consensus sequence (LLDLE in the human sequence) that binds the heavy chain of clathrin, and a highly conserved 27–amino acid sequence that binds the β_2-adaptin subunit of AP-2, a structural component of clathrin-coated pits (34). Taken together, these data suggest that ARH functions as a modular adaptor protein linking the LDLR to the endocytic machinery of the coated pit.

The exact role of ARH in LDLR function is not known. ARH may be required to chaperone LDLRs to coated pits, or simply to anchor receptors in the pits during internalization. A naturally occurring mutation in the Y of the NPXY motif (Y807C) that prevents LDLR clustering in coated pits and, consequently, LDLR internalization, also prevents ARH binding to the cytoplasmic tail in vitro (34, 36). This finding suggests that ARH may play a role in LDLR clustering. Direct observations of LDLR trafficking in ARH-deficient cells will be required to determine whether this is indeed the case.

Why is ARH required for LDLR function in some cell types, such as hepatocytes, but not in others, such as fibroblasts? ARH is one of a large family of adapter proteins, others of which have been documented to associate with the LDLR tail (37). Perhaps one of these other family members can effectively substitute for ARH in fibroblasts, but not in hepatocytes. Alternatively, the process of LDLR internalization may differ in hepatocytes and fibroblasts. Unlike in fibroblasts, where LDLRs are clustered in coated pits, in hepatocytes LDLRs are dispersed on the plasma membrane (38). ARH may perform a specific function in hepatocytes that is not required in fibroblasts.

ARH appears to be a near perfect phenocopy of FH (39), which is consistent with the possibility that all clinical sequelae of ARH mutations result from defective LDLR activity. ARH may be involved in other receptor pathways, but no phenotypes indicating defective function of other NPXY-containing proteins have been reported in subjects with ARH.

Diagnosis. Plasma levels of LDL-C in ARH patients tend to be intermediate between those in FH heterozygotes and those in FH homozygotes (39). The onset of clinically significant coronary atherosclerotic disease is later in ARH patients than in patients with homozygous FH. Despite having lower plasma levels of cholesterol than FH homozygotes, patients with ARH often have large, bulky xanthomas (26, 39).

Treatment. Subjects with ARH respond to lipid-lowering medications with more substantial plasma cholesterol reductions than patients with homozygous FH; in particular, statins produce striking reductions in plasma cholesterol in some ARH patients (39) (Table 2). The increased LDLR expression induced by statins may compensate to some degree for the defective hepatic LDLR function in ARH. Most ARH subjects do not reach optimal plasma cholesterol levels on lipid-lowering medications alone and are maintained on LDL apheresis (39).

Sitosterolemia

Historical perspective. In 1974, Bhattacharyya and Connor described two normocholesterolemic siblings with large xanthomas, elevated plasma levels of the major plant sterol sitosterol, and normolipidemic parents (40). In normal individuals, cholesterol constitutes more than 99% of circulating sterols; noncholesterol sterols, such as sitosterol, are present in only trace amounts. In sitosterolemia, plasma levels of sitosterol are elevated more than 50-fold, and approximately 15% of circulating and tissue sterols are derived from plants and shellfish (41).

Although the index cases of sitosterolemia were not hypercholesterolemic, most patients identified subsequently have had elevated plasma levels of LDL-C, especially in childhood. Children with sitosterolemia can have plasma LDL-C levels as high as those seen in FH homozygotes (>500 mg/dl) and also often develop planar xanthomas. Like patients with FH, these patients can develop aortic stenosis and premature coronary atherosclerotic disease. A distinctive clinical feature of sitosterolemia is the occurrence of low-level hemolysis, presumably due to the incorporation of plant sterols into red blood membranes.

Sitosterol, though differing from cholesterol by only minor modifications of the side chain, has a very different metabolic fate (41). Both sitosterol and cholesterol are taken up into enterocytes in the proximal small intestine, and between 20% and 80% of dietary cholesterol is incorporated into chylomicrons. In contrast, less than 5% of dietary sitosterol is absorbed. The small amount of sitosterol that is transported to the liver is preferentially secreted into the bile (42). Consequently, plasma levels of plant sterols in normal individuals are very low. Patients with sitosterolemia have increased fractional absorption of dietary sterols and a defect in the ability to secrete sterols into the bile, resulting in the accumulation of both animal and plant sterols in the blood and body tissues (41). The plasma cholesterol level in patients with sitosterolemia is extremely responsive to dietary cholesterol restriction, which decreases the input of sterols into the body, and to bile acid resin therapy, which promotes the excretion of cholesterol by increasing bile acid synthesis.

Patients with sitosterolemia have strikingly low rates of cholesterol synthesis (41) and yet accumulate cholesterol in selected tissues, even when they are not hypercholesterolemic. The sterol composition of xanthomas and atherosclerotic lesions mirrors that of the plasma, so cholesterol is the predominant component. Although about 60% of the plant sterols in plasma are esterified, tissue noncholesterol sterols are present in the free form (40).

New insights into pathogenesis. Whereas the primary metabolic defect in FH, FDB, and ARH is in the receptor-mediated uptake of circulating LDL, sitosterolemia results from a defect in sterol efflux from cells. Sitosterolemia is caused by mutations in either of two adjacent genes that encode ABC half-transporters, ABCG5 and ABCG8 (43, 44). More than 25 different mutations cause sitosterolemia (43–45). Sitosterolemic patients invariably have two mutant alleles of *ABCG5* or two mutant alleles of *ABCG8* alleles; none of the patients identified has one mutation in *ABCG5* and one in *ABCG8*. No obvious clinical differences in disease manifestation are apparent between patients with mutations in *ABCG5* or in *ABCG8*. Interestingly, most Caucasian patients have mutations in *ABCG8*, whereas all Japanese individuals with sitosterolemia identified to date have mutations in *ABCG5* (45).

ABCG5 and *ABCG8* are expressed almost exclusively in hepatocytes and enterocytes and are coordinately upregulated by cholesterol feeding in mice (43), thus limiting the absorption and promoting the elimination of dietary sterols. The responses of *ABCG5* and *ABCG8* to cholesterol feeding are dependent on liver X receptor α (LXRα), a nuclear hormone receptor that orchestrates the regulation of several genes involved in the trafficking of cholesterol from tissues to the liver (*ABCA1, APOE, CETP*, and others) (16). The crucial role of *ABCG5* and *ABCG8* in cholesterol homeostasis has been revealed by genetic manipulation of the expression of the two genes in mice. Mice expressing no ABCG5 or ABCG8 provide an excellent animal model of sitosterolemia; the mice are sitosterolemic, have an increased fractional absorption of dietary sitosterol, and have markedly reduced levels of biliary cholesterol (47). Whereas the plasma cholesterol level increases only marginally in response to increases in dietary cholesterol in wild-type mice, plasma and liver cholesterol levels both increase dramatically in mice lacking ABCG5 and ABCG8.

Diagnosis. The diagnosis of sitosterolemia should be considered in individuals with xanthomatosis and hypercholesterolemia whose parents are normocholesterolemic. The disease should be suspected in hypercholesterolemic patients who have a greater than usual response to dietary cholesterol restriction or to bile acid resins (Table 2). The diagnosis of sitosterolemia is made by extraction of lipids from the plasma and fractionation of the sterols using gas-liquid chromatography.

Treatment. Sitosterolemic patients respond poorly to HMG-CoA reductase inhibitors but are unusually responsive to dietary cholesterol restriction and bile acid–binding resins (Table 2). Recently, ezetimibe, when combined with a low-sterol diet, was shown to be particularly effective at lowering plasma levels of plant sterols as well as cholesterol in sitosterolemic patients (48). This suggests that sitosterol and cholesterol enter the enterocyte by the same pathway. Whether ezetimibe will reduce the cardiovascular risk associated with sitosterolemia remains unknown. The optimal treatment for sitosterolemia is dietary sterol restriction, ezetimibe, and a bile acid sequestrant to limit the accumulation of both plant and animal sterols.

Other recessive forms of hypercholesterolemia

Cholesterol 7α-hydroxylase deficiency. Recently, Pullinger et al. described three hypercholesterolemic siblings (two brothers and a sister) with genetic deficiency of cholesterol 7α-hydroxylase (*CYP7A1*), the first enzyme in the classical pathway for bile acid biosynthesis (49). The three affected siblings were homozygous for a frameshift mutation that results in premature termination of translation and a nonfunctional enzyme. The two affected brothers were hypertriglyceridemic as well as hypercholesterolemic, but the levels of remnant particles in these patients were not determined. Both brothers were reportedly resistant to statin therapy, although plasma cholesterol levels normalized on a regimen of atorvastatin and niacin. The bile acid content of a 24-hour stool sample from a single proband was reduced by a striking 94%, which is consistent with a dramatic reduction in bile acid synthesis.

7α-Hydroxylase deficiency presumably causes hypercholesterolemia by reducing hepatic LDLR activity. Evaluation of additional individuals with *CYP7A1* deficiency will be necessary to confirm that the hypercholesterolemia results from the genetic defect in *CYP7A1* and that the characteristic lipoprotein profile of this disorder is an isolated elevation in LDL-C. However, the finding that three siblings homozygous for the same mutation were all hypercholesterolemic suggests that disposal of cholesterol via the bile acid biosynthetic pathway is essential for the maintenance of normal plasma cholesterol levels in humans.

Thus, both major pathways for excretion of cholesterol into bile appear to be required for normal cholesterol homeostasis: cholesterol that accumulates because of defects in the ABCG5/ABCG8 transporter cannot simply be quantitatively excreted by conversion to bile acids, and direct secretion of cholesterol by ABCG5/ABCG8 cannot compensate for reduced bile acid synthesis.

Implications for the treatment of hypercholesterolemia

Insights derived from the study of patients with well-characterized monogenic disorders have implications for the management of hypercholesterolemia in the general population. The robust relationship between FH and premature coronary atherosclerotic disease provides the cornerstone supporting the primary role of elevations in plasma LDL-C levels in coronary atherosclerosis. Despite the inevitable development of clinically significant coronary atherosclerosis in FH homozygotes, the age of onset and severity of the disease vary widely, even among patients with identical LDLR mutations (50). These findings reflect the complexity of the atherosclerotic lesion and highlight the difficulties associated with predicting the clinical course of individual patients. Although the identification of DNA sequence variations associated with coronary atherosclerosis may provide windows into new therapeutic targets, it is highly unlikely that assaying single or even hundreds of DNA sequence polymorphisms will provide useful predictive information regarding disease risk (51). Changes in modifiable risk factors such as diet, cessation of smoking, weight and blood pressure control, and, of course, cholesterol reduction will have substantially more impact on the incidence and progression of atherosclerosis.

Large-scale clinical trials provide convincing evidence that substantial LDL lowering reduces cardiovascular morbidity and mortality, even in subjects who do not have elevated LDL-C levels (52). Since an ever increasing proportion of the population would benefit from LDL-lowering therapy, can we develop ways to predict which agents will be most effective in a given individual? The statins, our most effective LDL-lowering agents, do not lower plasma cholesterol levels in patients with sitosterolemia, suggesting that these drugs may be less effective in patients who have increased cholesterol absorption and reduced hepatic cholesterol synthesis. Miettinen and colleagues have provided evidence that plasma levels of plant sterols may predict the relative responsiveness of individuals to two of the major classes of hypolipidemic agents: those that interfere with cholesterol synthesis (statins) and those that interfere with cholesterol absorption (ezetimibe and stanol esters) (53). Stanol esters are most effective in individuals with increased cholesterol absorption; conversely, statins may be less efficacious in these individuals, since the increased influx of dietary cholesterol to the liver is associated with lower rates of hepatic cholesterol biosynthesis (53). The interindividual differences in the response to various lipid-lowering agents illustrate the genetic heterogeneity underlying polygenic hypercholesterolemia. A better understanding of the molecular mechanisms causing these genetic differences may contribute to the development of new, more individualized therapies for the treatment of hypercholesterolemia.

Acknowledgments

D.J. Rader is a recipient of a Burroughs Wellcome Fund Clinical Scientist Award in Translational Research and a Doris Duke Distinguished Clinical Scientist Award. H.H. Hobbs is an Investigator of the Howard Hughes Medical Institute. This work was supported by grants from the NIH (HL-20948, HL-72304, HL-53917, HL-55323, HL-59407, HL-70128, and RR-00040), the W.M. Keck Foundation, and the D.W. Reynolds Cardiovascular Clinical Research Center. We wish to thank Jay Horton and Scott Grundy for helpful discussions.

Address correspondence to: Helen H. Hobbs, Department of Molecular Genetics, University of Texas Southwestern Medical Center, 5323 Harry Hines Boulevard, Dallas, Texas 75390, USA. Phone: (214) 648-6724; Fax: (214) 648-7539; E-mail: Helen. Hobbs@UTSouthwestern.edu.

1. Heller, D.A., de Faire, U., Pedersen, N.L., Dahlen, G., and McClearn, G.E. 1993. Genetic and environmental influences on serum lipid levels in twins. *N. Engl. J. Med.* **328**:1150–1156.

2. Goldstein, J., Hobbs, H., and Brown, M. 2001. Familial hypercholesterolemia. In *The metabolic and molecular bases of inherited disease*. C. Scriver, A. Beaudet, W. Sly, and D. Valle, editors. McGraw-Hill. New York, New York, USA. 2863–2913.

3. Sun, X.M., et al. 1994. Familial hypercholesterolemia in China. Identification of mutations in the LDL-receptor gene that result in a receptor-negative phenotype. *Arterioscler. Thromb.* **14**:85–94.

4. Hoeg, J.M., Feuerstein, I.M., and Tucker, E.E. 1994. Detection and quantitation of calcific atherosclerosis by ultrafast computed tomography in children and young adults with homozygous familial hypercholesterolemia. *Arterioscler. Thromb.* **14**:1066–1074.

5. Hill, J.S., Hayden, M.R., Frohlich, J., and Pritchard, P.H. 1991. Genetic and environmental factors affecting the incidence of coronary artery disease in heterozygous familial hypercholesterolemia. *Arterioscler. Thromb.* **11**:290–297.

6. Rudenko, G., et al. 2002. Structure of the LDL receptor extracellular domain at endosomal pH. *Science.* **298**:2353–2358.

7. Jeon, H., et al. 2001. Implications for familial hypercholesterolemia from the structure of the LDL receptor YWTD-EGF domain pair. *Nat. Struct. Biol.* **8**:499–504.

8. Horton, J.D., Shimano, H., Hamilton, R.L., Brown, M.S., and Goldstein, J.L. 1999. Disruption of LDL receptor gene in transgenic SREBP-1a mice unmasks hyperlipidemia resulting from production of lipid-rich VLDL. *J. Clin. Invest.* **103**:1067–1076.

9. Gillian-Daniel, D.L., Bates, P.W., Tebon, A., and Attie, A.D. 2002. Endoplasmic reticulum localization of the low density lipoprotein receptor mediates presecretory degradation of apolipoprotein B. *Proc. Natl. Acad. Sci. U. S. A.* **99**:4337–4342.

10. The low density lipoprotein receptor (LDLR) gene in familial hypercholesterolemia. www.Ucl.ac.uk/fh.

11. Miettinen, T.A., and Gylling, H. 1999. Regulation of cholesterol metabolism by dietary plant sterols. *Curr. Opin. Lipidol.* **10**:9–14.

12. Van Heek, M., et al. 1997. In vivo metabolism-based discovery of a potent cholesterol absorption inhibitor, SCH58235, in the rat and rhesus monkey through the identification of the active metabolites of SCH48461. *J. Pharmacol. Exp. Ther.* **283**:157–163.

13. Gagne, C., Gaudet, D., and Bruckert, E. 2002. Efficacy and safety of ezetimibe coadministered with atorvastatin or simvastatin in patients with homozygous familial hypercholesterolemia. *Circulation.* **105**:2469–2475.

14. Thompson, G.R. 2003. LDL apheresis. *Atherosclerosis.* **167**:1–13.

15. Wetterau, J.R., et al. 1998. An MTP inhibitor that normalizes atherogenic lipoprotein levels in WHHL rabbits. *Science.* **282**:751–754.

16. Vega, G.L., and Grundy, S.M. 1986. In vivo evidence for reduced binding of low density lipoproteins to receptors as a cause of primary moderate hypercholesterolemia. *J. Clin. Invest.* **78**:1410–1414.

17. Soria, L.F., et al. 1989. Association between a specific apolipoprotein B mutation and familial defective apolipoprotein B-100. *Proc. Natl. Acad. Sci. U. S. A.* **86**:587–591.

18. Myant, N.B. 1993. Familial defective apolipoprotein B-100: a review, including some comparisons with familial hypercholesterolaemia. *Atherosclerosis.* **104**:1–18.

19. Hansen, P.S. 1998. Familial defective apolipoprotein B-100. *Dan. Med. Bull.* **45**:370–382.

20. Innerarity, T.L., et al. 1990. Familial defective apolipoprotein B-100: a mutation of apolipoprotein B that causes hypercholesterolemia. *J. Lipid Res.* **31**:1337–1349.

21. Schaefer, J.R., et al. 1997. Homozygous familial defective apolipoprotein B-100. Enhanced removal of apolipoprotein E-containing VLDLs and decreased production of LDLs. *Arterioscler. Thromb. Vasc. Biol.* **17**:348–353.

22. Haddad, L., et al. 1999. Evidence for a third genetic locus causing familial hypercholesterolemia. A non-LDLR, non-APOB kindred. *J. Lipid. Res.* **40**:1113–1122.

23. Varret, M., et al. 1999. A third major locus for autosomal dominant hypercholesterolemia maps to 1p34.1-p32. *Am. J. Hum. Genet.* **64**:1378–1387.

24. Abifadel, M., et al. 2003. Mutations in PCSK9 cause autosomal dominant hypercholesterolemia. *Nat. Genet.* In press.

25. Seidah, N.G., et al. 2003. The secretory proprotein convertase neural apoptosis-regulated convertase 1 (NARC-1): liver regeneration and neuronal differentiation. *Proc. Natl. Acad. Sci. U. S. A.* **100**:928–933.

26. Khachadurian, A.K., and Uthman, S.M. 1973. Experiences with the homozygous cases of familial hypercholesterolemia. A report of 52 patients. *Nutr. Metab.* **15**:132–140.

27. Zuliani, G, et al. 1995. Severe hypercholesterolaemia: unusual inheritance in an Italian pedigree. *Eur. J. Clin. Invest.* **25**:322–331.

28. Garcia, C.K., et al. 2001. Autosomal recessive hypercholesterolemia caused by mutations in a putative LDL receptor adaptor protein. *Science.* **292**:1394–1398.

29. Cohen, J.C., Kimmel, M., Polanski, A., and Hobbs, H.H. 2003. Molecular mechanisms of autosomal recessive hypercholesterolemia. *Curr. Opin. Lipidol.* **14**:121–127.

30. Eden, E.R., et al. 2002. Restoration of LDL receptor function in cells from patients with autosomal recessive hypercholesterolemia by retroviral expression of *ARH1*. *J. Clin. Invest.* **110**:1695–1702. doi:10.1172/JCI200216445.

31. Norman, D., et al. 1999. Characterization of a novel cellular defect in patients with phenotypic homozygous familial hypercholesterolemia. *J. Clin. Invest.* **104**:619–628.

32. Forman-Kay, J.D., and Pawson, T. 1999. Diversity in protein recognition by PTB domains. *Curr. Opin. Struct. Biol.* **9**:690–695.

33. Davis, C.G., van Driel, I.R., Russell, D.W., Brown, M.S., and Goldstein, J.L. 1987. The low density lipoprotein receptor. Identification of amino acids in cytoplasmic domain required for rapid endocytosis. *J. Biol. Chem.* **262**:4075–4082.

34. He, G., Gupta, S., Michaely, P., Hobbs, H.H., and Cohen, J.C. 2002. ARH is a modular adaptor protein that interacts with the LDL receptor, clathrin and AP-2. *J. Biol. Chem.* **277**:44044–44049.

35. Mishra, S.K., Watkins, S.C., and Traub, L.M. 2002. The autosomal recessive hypercholesterolemia (ARH) protein interfaces directly with the clathrin-coat machinery. *Proc. Natl. Acad. Sci. U. S. A.* **99**:16099–16104.

36. Davis, C.G., et al. 1986. The J.D. mutation in familial hypercholesterolemia: amino acid substitution in cytoplasmic domain impedes internalization of LDL receptors. *Cell.* **45**:15–24.

37. Gotthardt, M., et al. 2000. Interactions of the low-density lipoprotein receptor gene family with cytosolic adaptor and scaffold proteins suggest diverse biological functions in cellular communication and signal transduction. *J. Biol. Chem.* **275**:25616–25624.

38. Pathak, R.K., et al. 1990. Tissue-specific sorting of the human LDL receptor in polarized epithelia of transgenic mice. *J. Cell Biol.* **111**:347–359.

39. Arca, M., et al. 2002. Autosomal recessive hypercholesterolaemia in Sardinia, Italy, and mutations in ARH: a clinical and molecular genetic analysis. *Lancet.* **359**:841–847.

40. Bhattacharyya, A.K., and Connor, W.E. 1974. Beta-sitosterolemia and xanthomatosis. A newly described lipid storage disease in two sisters. *J. Clin. Invest.* **53**:1033–1043.

41. Salen, G., et al. 1992. Sitosterolemia. *J. Lipid Res.* **33**:945–955.

42. Salen, G., Ahrens, E.H., Jr., and Grundy, S.M. 1970. Metabolism of *beta*-sitosterol in man. *J. Clin. Invest.* **49**:952–967.

43. Berge, K.E., et al. 2000. Accumulation of dietary cholesterol in sitosterolemia caused by mutations in adjacent ABC transporters. *Science.* **290**:1771–1775.

44. Lee, M.H., et al. 2001. Identification of a gene, *ABCG5*, important in the regulation of dietary cholesterol absorption. *Nat. Genet.* **27**:79–83.

45. Lu, K., et al. 2001. Two genes that map to the *STSL* locus cause sitosterolemia: genomic structure and spectrum of mutations involving sterolin-1 and sterolin-2, encoded by *ABCG5* and *ABCG8*, respectively. *Am. J. Hum. Genet.* **69**:278–290.

46. Repa, J.J., et al. 2002. Regulation of ATP-binding cassette sterol transporters ABCG5 and ABCG8 by the liver X receptors alpha and beta. *J. Biol. Chem.* **277**:18793–18800.

47. Yu, L., et al. 2002. Disruption of Abcg5 and Abcg8 in mice reveals their crucial role in biliary cholesterol secretion. *Proc. Natl. Acad. Sci. U. S. A.* **99**:16237–16242.

48. Salen, G., et al. 2002. Ezetimibe is an effective treatment for homozygous sitosterolemia. *Circulation Supplement.* ll-185.

49. Pullinger, C.R., et al. 2002. Human cholesterol 7α-hydroxylase (CYP7A1) deficiency has a hypercholesterolemic phenotype. *J. Clin Invest.* **110**:109–117. doi:10.1172/JCI200215387.

50. Hobbs, H.H., Brown, M.S., Russell, D.W., Davignon, J., and Goldstein, J.L. 1987. Deletion in the gene for the low-density-lipoprotein receptor in a majority of French Canadians with familial hypercholesterolemia. *N. Engl. J. Med.* **317**:734–737.

51. Hegele, R.A. 2002. Environmental modulation of atherosclerosis end points in familial hypercholesterolemia. *Atheroscler. Suppl.* **2**:5–7.

52. 2002. MRC/BHF Heart Protection Study of cholesterol lowering with simvastatin in 20,536 high-risk individuals: a randomised placebo-controlled trial. *Lancet.* **360**:7–22.

53. Gylling, H., and Miettinen, T.A. 2002. Baseline intestinal absorption and synthesis of cholesterol regulate its response to hypolipidaemic treatments in coronary patients. *Atherosclerosis.* **160**:477–481.

Isoprenoids as mediators of the biological effects of statins

James K. Liao

Cardiovascular Division, Department of Medicine, Brigham and Women's Hospital and Harvard Medical School, Boston, Massachusetts, USA.

The 3-hydroxy-3-methylglutaryl coenzyme A (HMG-CoA) reductase inhibitors, or statins, are potent inhibitors of cholesterol biosynthesis. Several large clinical trials have demonstrated the benefits of cholesterol lowering with these agents in the primary and secondary prevention of coronary heart disease. The overall clinical benefits observed with statin therapy, however, appear to be greater than what might be expected from changes in lipid profile alone, suggesting that the beneficial effects of statins may extend beyond their effects on serum cholesterol levels.

Recent experimental and clinical evidence indicates that some of the cholesterol-independent, or so-called pleiotropic, effects of statins involve improving or restoring endothelial function, enhancing the stability of atherosclerotic plaques, decreasing oxidative stress and inflammation, and inhibiting the thrombogenic response in the vascular wall. Many of these cholesterol-independent effects reflect statins' ability to block the synthesis of important isoprenoid intermediates, which serve as lipid attachments for a variety of intracellular signaling molecules. In particular, the inhibition of small GTP-binding proteins Rho, Ras, and Rac, whose proper membrane localization and function are dependent upon isoprenylation, may play an important role in mediating the biological effects of statins.

Pharmacological properties of statins

Statins bind to HMG-CoA reductase at nanomolar concentrations, leading to competitive displacement of the natural substrate, HMG CoA, which binds at micromolar concentrations (1). In addition, inhibition of cholesterol biosynthesis is accompanied by an increase in hepatic LDL receptor, which promotes uptake and clearance of cholesterol from the bloodstream. While all statins inhibit hepatic HMG-CoA reductase to varying degrees, important structural differences exist among the statins that distinguish their lipophilicity, half-life, and potency (2). For example, one of the more potent newer statins, rosuvastatin, is relatively hydrophilic and has a greater number of bonding interactions with the catalytic site of HMG-CoA reductase compared with mevastatin, fluvastatin, simvastatin, cerivastatin, and atorvastatin (1, 3).

The lipophilic statins would be expected to penetrate cell membranes more effectively than the more hydrophilic statins, causing more side effects but, at the same time, eliciting more pleiotropic effects. However, the observation that hydrophilic statins have pleiotropic effects similar to those of lipophilic statins calls into question whether there are really any cholesterol-independent effects of statins. Indeed, recent evidence suggests that some of the cholesterol-independent effects of these agents may be mediated by inhibition of hepatic HMG-CoA

reductase, leading to subsequent reduction in circulating isoprenoid levels (4). This hypothesis may help explain why hydrophilic statins such as pravastatin and rosuvastatin are still able to exert cholesterol-independent benefits on the vascular wall without directly entering vascular wall cells. In this respect, the word "pleiotropic" probably does not reflect the hepatic versus nonhepatic effects of these agents.

Clinical trials with statins

Because serum cholesterol level is strongly associated with coronary heart disease (5), it has been generally assumed that cholesterol reduction by statins is the predominant, if not the only, mechanism underlying their beneficial effects in cardiovascular diseases. However, subgroup analysis of large clinical trials such as the 4S, WOSCOP, CARE, and HPS suggests that the clinical benefits of statins are not associated with base-line cholesterol levels or the degree of cholesterol reduction (6–9). Furthermore, in angiographic trials, clinical improvements with statins far exceed changes in the size of atherosclerotic lesions (10). It is quite likely that cholesterol lowering in these long-term trials stabilized atherosclerotic plaques and made them less prone to rupture. However, in the Myocardial Ischemia Reduction with Aggressive Cholesterol Lowering (MIRACL) trial, statins reduced recurrent ischemic events within 16 weeks following acute coronary ischemia (11). Although the serum LDL-cholesterol was decreased by 40%, this time frame was probably far too rapid for appreciable changes in lesion size and plaque stability to occur as a consequence of cholesterol reduction.

An intriguing but perplexing result of large clinical trials with statins is the reduction in ischemic stroke (12). Although myocardial infarction is closely associated with serum cholesterol levels, neither the Framingham Heart Study nor the Multiple Risk Factor Intervention Trial (MRFIT) demonstrated significant correlation between ischemic stroke and serum cholesterol levels (13, 14). Thus, the findings of these large statin trials raise the interesting question of how statins could reduce ischemic stroke when ischemic stroke and cholesterol are unrelated. It appears likely that some of the beneficial effects of statins in ischemic stroke are attributable to the pleiotropic effects of statins on endothelial function and fibrinolytic pathways.

Statins and isoprenylated proteins

By inhibiting L-mevalonic acid synthesis, statins also prevent the synthesis of other important isoprenoid intermediates of the cholesterol biosynthetic pathway, such as farnesylpyrophosphate (FPP) and geranylgeranylpyrophosphate (GGPP) (15) (Figure 1). These intermediates serve as important lipid attachments for the posttranslational modification of a variety of cell-signaling proteins. Protein isoprenylation permits the covalent attachment, subcellular localization, and intracellular trafficking of

Citation for this article: *J. Clin. Invest.* **110**:285–288 (2002). doi:10.1172/JCI200216421.

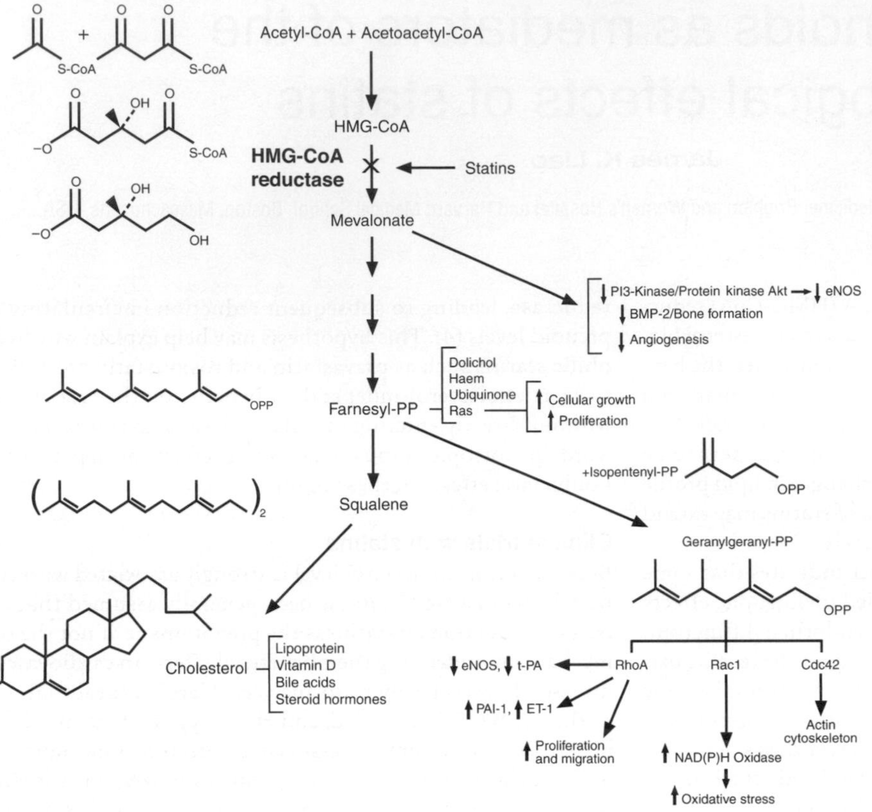

Figure 1

Biological actions of isoprenoids and cholesterol. This diagram of the cholesterol biosynthesis pathway shows the effects of inhibition of HMG-CoA reductase by statins. Decrease in isoprenylation of signaling molecules such as Ras, Rho, and Rac leads to modulation of various signaling pathways. BMP-2, bone morphogenetic protein-2; eNOS, endothelial nitric oxide synthase; t-PA, tissue-type plasminogen activator; ET-1, endothelin-1; PAI-1, plasminogen activator inhibitor-1.

membrane-associated proteins (16). Members of the Ras and Rho GTPase family are major substrates for posttranslational modification by isoprenylation and may be important targets for inhibition by statins. Indeed, statins induce changes in the actin cytoskeleton and assembly of focal adhesion complexes by inhibiting RhoA and Rac1 isoprenylation (Figure 2).

Besides altering the actin cytoskeleton, inhibition of RhoA by statins increases endothelial nitric oxide synthase (eNOS) expression and decreases severity of cerebral ischemia in a mouse model of ischemic stroke (17, 18). Similarly, statins also increase the expression of tissue-type plasminogen activator (19) and inhibit the expression of plasminogen activator inhibitor-1 (19) and endothelin-1 by mechanisms involving inhibition of geranylgeranylation (20). Because Ras and Rho also regulate the cell cycle, they are, in addition, likely targets for the direct antiproliferative effects of statins. Indeed, statins inhibit vascular smooth muscle cell proliferation in transplant-associated arteriosclerosis (21) and may have clinical benefits in inhibiting certain breast cancers (22). Finally, inhibition of Rac1 geranylgeranylation and Rac1-mediated NAD(P)H oxidase activity by statins attenuates angiotensin II–induced reactive oxygen species production in vascular smooth muscle cells and cardiac myocytes (23, 24) (Figure 3). These cholesterol-independent antioxidant effects of statins lead to the inhibition of hypertrophic responses in these tissues.

Statins and cardiovascular diseases

Plaque rupture is a major cause of acute coronary syndromes (25). Lipid lowering by statins contributes to plaque stability by reducing plaque size or by modifying the physiochemical properties of the lipid core. However, since the changes in plaque size associated with lipid lowering tend to occur over extended time and to be quite minimal as assessed by angiography, it appears that the clinical benefits from statins must have another explanation. Most likely, these benefits arise from a combined reduction in lipids and macrophage accumulation in atherosclerotic lesions and inhibition of matrix metalloproteinases and tissue factor production by activated macrophages (26, 27).

Recently, statins have been found to increase the number of circulating endothelial progenitor cells (EPCs), which may give rise to neovascularization in ischemic tissues (28). Indeed, statin therapy induces angiogenesis by promoting the proliferation, migration, and survival of circulating EPCs via the phosphatidylinositol (PI) 3-kinase/Akt pathway (29). In patients with angiographically documented, stable coronary artery disease, statins augment the number of circulating EPCs and enhanced functional activity (30). These findings agree with earlier data showing that statin therapy rapidly activates PI 3-kinase/Akt and eNOS, inhibits apoptosis,

Figure 2

Actin cytoskeletal effects of statins. Phalloidin staining of human endothelial cells shows the effects of the statin simvastatin (10 μM) on actin stress fibers and focal adhesion complexes (green) with and without L-mevalonate (L-Mev, 200 μM).

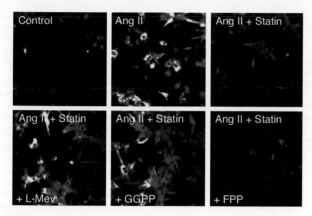

Figure 3
Antioxidant effects of statins. Intracellular oxidation (red) as determined by 2',7'-dichlorofluoroscein staining of rat cardiomyocytes treated with angiotensin II (Ang II, 10 nM), with and without simvastatin (statin, 10 μM), L-mevalonate (L-Mev, 200 μM), GGPP (100 μM), or FPP (100 μM).

and accelerates vascular structure formation (31). Interestingly, these angiogenic effects occur rapidly at very low concentrations of statins and are cholesterol-independent.

Are clinical benefits of statin therapy due entirely to cholesterol lowering?

Many clinicians, especially lipidologists, find it difficult to embrace the concept of statin pleiotropy for a number of reasons. First, patients receiving statin therapy invariably will have reduced lipid levels, and it is often difficult to separate the lipid-lowering from the non–lipid-lowering effects of statins in clinical trials. Second, many effects of statins, such as improvement in endothelial function, decreased inflammation, increased plaque stability, and reduced thrombogenic response, could all be accounted for, to some extent, by lipid lowering. Third, the concentrations used to demonstrate the biological effects of statins in cell culture and animal experiments, especially with regard to inhibition of Rho geranylgeranylation (but not PI 3-kinase/Akt activation), appear to be much higher than those prescribed clinically. Finally, both hydrophilic and lipophilic statins, which inhibit hepatic HMG-CoA reductase, appear to exert similar cholesterol-independent effects, despite the relative impermeability of hydrophilic statins in vascular tissues. Thus, it appears that statins are very potent cholesterol-lowering

agents and that reduction in cholesterol levels by statins contributes to many of their clinical benefits.

The evidence for cholesterol-independent effects of statins in humans, however, stems mostly from the rapidity of statin action in clinical trials (i.e., sometimes within days) and from evidence for clinical benefits that are not related to base-line cholesterol levels or the degree of cholesterol reduction. Furthermore, statins appear to exert clinical benefits beyond cardiovascular disease, including a reduction in the risk of dementia (32), Alzheimer disease (33), ischemic stroke (12), osteoporosis (34), and possibly breast cancer (22). Indeed, there is a growing body of biological, epidemiological, and limited but nonrandomized clinical evidence indicating that lowering serum cholesterol by statins may retard the pathogenesis of Alzheimer disease (35). Because neurons receive only small amounts of exogenous cholesterol, statins that reduce endogenous isoprenoid and cholesterol synthesis may inhibit the formation of Aβ-amyloid peptide by removing amyloid precursor protein from cholesterol- and sphingolipid-enriched membrane microdomains (36). However, in a recent prospective study, lipid and lipoprotein levels were not associated with the development of Alzheimer disease (37). These interesting observations suggest that the cellular or non-cholesterol-lowering effects of statins may be more important in influencing the progression of Alzheimer disease.

For osteoporosis, ischemic stroke, and other conditions for which statins appear to be beneficial, there is no clear association between cholesterol levels and risk of disease. Is it possible, then, that in normocholesterolemic individuals or in patients with ischemic stroke, plasma cholesterol, like L-mevalonate, is merely a marker of statins' inhibitory effect on HMG-CoA reductase, rather than the cause of the disease? Perhaps in patient populations where cholesterol is not an overt risk factor, other factors such as inflammation, which is also reduced by statin therapy, may be a more appropriate marker of statin efficacy than serum cholesterol levels (38). These uncertainties beg for further randomized clinical trials that would allow the cholesterol-dependent and -independent effects of statins to be evaluated separately. Only then will one be able to determine conclusively whether real clinical benefits of statin therapy beyond lipid lowering exist.

Address correspondence to: James K. Liao, Vascular Medicine Research, Brigham and Women's Hospital, 65 Landsdowne Street, Room 275, Cambridge, Massachusetts 02139, USA. Phone: (617) 768-8424; Fax: (617) 768-8425; E-mail: jliao@rics.bwh.harvard.edu.

1. Istvan, E.S., and Deisenhofer, J. 2001. Structural mechanism for statin inhibition of HMG-CoA reductase. *Science.* **292**:1160–1164.
2. Illingworth, D.R., and Tobert, J.A. 2001. HMG-CoA reductase inhibitors. *Adv. Protein Chem.* **56**:77–114.
3. McTaggart, F., et al. 2001. Preclinical and clinical pharmacology of Rosuvastatin, a new 3-hydroxy-3-methylglutaryl coenzyme A reductase inhibitor. *Am. J. Cardiol.* **87**:28B–32B.
4. Corsini, A., et al. 1999. New insights into the pharmacodynamic and pharmacokinetic properties of statins. *Pharmacol. Ther.* **84**:413–428.
5. Klag, M.J., et al. 1993. Serum cholesterol in young men and subsequent cardiovascular disease. *N. Engl. J. Med.* **328**:313–318.
6. 1994. Randomised trial of cholesterol lowering in 4444 patients with coronary heart disease: the Scandinavian Simvastatin Survival Study (4S).

Lancet. **344**:1383–1389.
7. Shepherd, J., et al. 1995. Prevention of coronary heart disease with pravastatin in men with hypercholesterolemia. West of Scotland Coronary Prevention Study Group. *N. Engl. J. Med.* **333**:1301–1307.
8. Sacks, F.M., et al. 1996. The effect of pravastatin on coronary events after myocardial infarction in patients with average cholesterol levels. Cholesterol and Recurrent Events Trial investigators. *N. Engl. J. Med.* **335**:1001–1009.
9. Collins, R., Peto, R., and Armitage, J. 2002. The MRC/BHF Heart Protection Study: preliminary results. *Int. J. Clin. Pract.* **56**:53–56.
10. Brown, B.G., Zhao, X.Q., Sacco, D.E., and Albers, J.J. 1993. Lipid lowering and plaque regression. New insights into prevention of plaque disruption and clinical events in coronary disease. *Circulation.* **87**:1781–1791.

11. Schwartz, G.G., et al. 2001. Effects of atorvastatin on early recurrent ischemic events in acute coronary syndromes. The MIRACL study: a randomized controlled trial. *JAMA.* **285**:1711–1718.
12. Crouse, J.R., Byington, R.P., and Furberg, C.D. 1998. HMG-CoA reductase inhibitor therapy and stroke risk reduction: an analysis of clinical trials data. *Atherosclerosis.* **138**:11–24.
13. Kannel, W.B., Castelli, W.P., Gordon, T., and McNamara, P.M. 1971. Serum cholesterol, lipoproteins, and the risk of coronary heart disease. The Framingham study. *Ann. Intern. Med.* **74**:1–12.
14. 1982. Multiple risk factor intervention trial: risk factor changes and mortality results. Multiple Risk Factor Intervention Trial Research Group. *JAMA.* **248**:1465–1477.
15. Goldstein, J.L., and Brown, M.S. 1990. Regulation of the mevalonate pathway. *Nature.* **343**:425–430.

16. Van Aelst, L., and D'Souza-Schorey, C. 1997. Rho GTPases and signaling networks. *Genes Dev.* **11**:2295–2322.

17. Endres, M., et al. 1998. Stroke protection by 3-hydroxy-3-methylglutaryl (HMG)-CoA reductase inhibitors mediated by endothelial nitric oxide synthase. *Proc. Natl. Acad. Sci. USA.* **95**:8880–8885.

18. Laufs, U., et al. 2000. Neuroprotection mediated by changes in the endothelial actin cytoskeleton. *J. Clin. Invest.* **106**:15–24.

19. Essig, M., et al. 1998. 3-Hydroxy-3-methylglutaryl coenzyme A reductase inhibitors increase fibrinolytic activity in rat aortic endothelial cells. Role of geranylgeranylation and Rho proteins. *Circ. Res.* **83**:683–690.

20. Hernandez-Perera, O., Perez-Sala, D., Soria, E., and Lamas, S. 2000. Involvement of rho GTPases in the transcriptional inhibition of preproendothelin-1 gene expression by simvastatin in vascular endothelial cells. *Circ. Res.* **87**:616–622.

21. Kobashigawa, J.A., et al. 1995. Effect of pravastatin on outcomes after cardiac transplantation. *N. Engl. J. Med.* **333**:621–627.

22. Denoyelle, C., et al. 2001. Cerivastatin, an inhibitor of HMG-CoA reductase, inhibits the signaling pathways involved in the invasiveness and metastatic properties of highly invasive breast cancer cell lines: an in vitro study. *Carcinogenesis.* **22**:1139–1148.

23. Wassmann, S., et al. 2001. Inhibition of geranylgeranylation reduces angiotensin II-mediated free radical production in vascular smooth muscle cells: involvement of angiotensin AT1 receptor expression and Rac1 GTPase. *Mol. Pharmacol.* **59**:646–654.

24. Takemoto, M., et al. 2001. Statins as antioxidant therapy for preventing cardiac myocyte hypertrophy. *J. Clin. Invest.* **108**:1429–1437. doi:10.1172/JCI200113350.

25. Libby, P. 1995. Molecular bases of the acute coronary syndromes. *Circulation.* **91**:2844–2850.

26. Aikawa, M., et al. 2001. An HMG-CoA reductase inhibitor, cerivastatin, suppresses growth of macrophages expressing matrix metalloproteinases and tissue factor in vivo and in vitro. *Circulation.* **103**:276–283.

27. Crisby, M., et al. 2001. Pravastatin treatment increases collagen content and decreases lipid content, inflammation, metalloproteinases, and cell death in human carotid plaques: implications for plaque stabilization. *Circulation.* **103**:926–933.

28. Llevadot, J., et al. 2001. HMG-CoA reductase inhibitor mobilizes bone marrow–derived endothelial progenitor cells. *J. Clin. Invest.* **108**:399–405. doi:1172/JCI200113131.

29. Dimmeler, S., et al. 2001. HMG-CoA reductase inhibitors (statins) increase endothelial progenitor cells via the PI 3-kinase/Akt pathway. *J. Clin. Invest.* **108**:391–397. doi:1172/JCI200113152.

30. Vasa, M., et al. 2001. Increase in circulating endothelial progenitor cells by statin therapy in patients with stable coronary artery disease. *Circulation.* **103**:2885–2890.

31. Kureishi, Y., et al. 2000. The HMG-CoA reductase inhibitor simvastatin activates the protein kinase Akt and promotes angiogenesis in normo-cholesterolemic animals. *Nat. Med.* **6**:1004–1010.

32. Jick, H., Zornberg, G.L., Jick, S.S., Seshadri, S., and Drachman, D.A. 2000. Statins and the risk of dementia. *Lancet.* **356**:1627–1631.

33. Wolozin, B., Kellman, W., Ruosseau, P., Celesia, G.G., and Siegel, G. 2000. Decreased prevalence of Alzheimer disease associated with 3-hydroxy-3-methyglutaryl coenzyme A reductase inhibitors. *Arch. Neurol.* **57**:1439–1443.

34. Chan, K.A., et al. 2000. Inhibitors of hydroxymethylglutaryl-coenzyme A reductase and risk of fracture among older women. *Lancet.* **355**:2185–2188.

35. Scott, H.D., and Laake, K. 2001. Statins for the prevention of Alzheimer's disease. *Cochrane Database Syst. Rev.* CD003160.

36. Simons, M., Keller, P., Dichgans, J., and Schulz, J.B. 2001. Cholesterol and Alzheimer's disease: is there a link? *Neurology.* **57**:1089–1093.

37. Moroney, J.T., et al. 1999. Low-density lipoprotein cholesterol and the risk of dementia with stroke. *JAMA.* **282**:254–260.

38. Ridker, P.M., et al. 2001. Measurement of C-reactive protein for the targeting of statin therapy in the primary prevention of acute coronary events. *N. Engl. J. Med.* **344**:1959–1965.

The channelopathies: novel insights into molecular and genetic mechanisms of human disease

Robert S. Kass

Department of Pharmacology, Columbia University Medical Center, New York, New York, USA.

Ion channels are pore-forming proteins that provide pathways for the controlled movement of ions into or out of cells. Ionic movement across cell membranes is critical for essential and physiological processes ranging from control of the strength and duration of the heartbeat to the regulation of insulin secretion in pancreatic β cells. Diseases caused by mutations in genes that encode ion channel subunits or regulatory proteins are referred to as channelopathies. As might be expected based on the diverse roles of ion channels, channelopathies range from inherited cardiac arrhythmias, to muscle disorders, to forms of diabetes. This series of reviews examines the roles of ion channels in health and disease.

Introduction

Ion channels are a diverse group of pore-forming proteins that cross the lipid membrane of cells and selectively conduct ions across this barrier. Ion channels coordinate electrical signals in most tissues and are thus involved in every heartbeat, every movement, and every thought and perception. They have evolved to selectively provide pathways for ions to move down their electrochemical gradients across cell membranes and either depolarize cells, by moving positively charged ions in, or repolarize cells, by moving positively charged ions out.

During the past 50 years, our understanding of the roles and molecular structure of ion channels has grown at a rapid pace and has bridged fundamental basic research with advances in clinical medicine. The link between basic science and clinical medicine has been the discovery of human diseases linked to mutations in genes coding for ion channel subunits or proteins that regulate them: the channelopathies.

Ion channels: from squid giant axons to atomic structure

Before 1982, insight into mechanisms underlying the ionic basis of electrical activity in excitable cells was limited to model systems. Our understanding of these mechanisms was based largely on the beautiful work of Hodgkin, Huxley, and Cole, among others, who unraveled the ionic basis of nerve excitation in the squid giant axon (1–4) and subsequently showed that similar mechanisms were responsible for excitation and contraction of amphibian skeletal muscle (5). Using mammalian preparations, other groups soon demonstrated similar, though more complex, mechanisms underlying electrical activity in the heart (6). However, the link between mechanisms in these model systems and human physiology remained indirect. This changed dramatically in 1982 when the first ion channel, acetylcholine receptor α-subunit, was cloned (7, 8). Molecular biology provided the techniques to identify genes encoding ion channels, and, as a result, a plethora of channels has been discovered to be critical to the physiological function of vir-

tually every tissue, controlling such diverse functions as hearing and insulin secretion. The combination of genetic identification of multiple channel genes and the development of patch-clamp electrophysiological procedures by Neher and Sakmann (9) made it possible to analyze in great detail the functional properties of ion channels in small cells, eliminating the restriction to model systems and extending understanding of the roles of molecular structures in the control of channel function.

The crystal structure of a bacterial potassium channel was solved in 1998 (10), revealing, at the atomic level, the structural basis of fundamental mechanisms of this class of ion channels. Insight into channel structure clarified the manner in which the channels open and close; the structural basis for selection of ions that can pass through the open channel pore; and the mechanism by which the channel proteins sense changes in transmembrane voltage that control the open or closed conformational states of the channel (Figure 1) (11). Thus investigations of ion channel proteins bridge fundamental physics with function of biologically critical proteins. But the link to human disease has come from clinical investigations of congenital disorders and the discoveries that defects in genes coding for ion channels or ion channel regulatory subunits cause diverse disease states. The number of diseases linked to these mutations is so large that the term "channelopathy" has been introduced to define this class of disease (Table 1).

In this issue of the *JCI*, we have assembled a series of review articles to provide the most up-to-date information on channelopathies and to assess how investigations of these disorders have increased our understanding of the mechanisms of human disease. Rather than serving as an extensive catalog of all channelopathies, these papers impart both a historical perspective and an opportunity to understand how mutations that cause common changes in ion channel function may be associated with diverse diseases depending on the tissues in which the channels are expressed.

Sodium channels

The channelopathies are introduced by A.L. George Jr. with a broad discussion of the impact of mutations of voltage-gated sodium channels on human disease (12). Mutations in genes coding for sodium channels are now known to cause nearly 20 disorders ranging from skeletal muscle defects to dysfunction of the nervous system (12).

Nonstandard abbreviations used: LQTS, long QT syndrome.

Conflict of interest: The author has declared that no conflict of interest exists.

Citation for this article: *J. Clin. Invest.* **115**:1986–1989 (2005).
doi:10.1172/JCI26011.

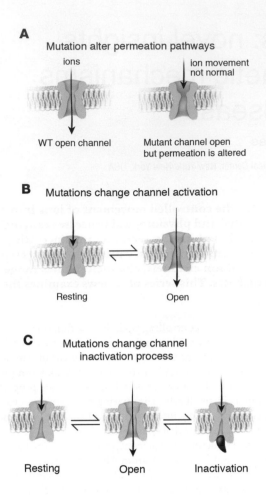

A Mutation alter permeation pathways

ions

ion movement
not normal

WT open channel

Mutant channel open
but permeation is altered

B Mutations change channel activation

Resting   Open

C Mutations change channel
inactivation process

Resting   Open   Inactivation

Figure 1
Inherited mutations alter ion channel function and structure and cause human disease. Mutations may alter the permeation pathway (**A**) to inhibit the movement of ions through an open channel pore and may also alter ion channel gating by changing either the process by which channels open (activate) (**B**) or the process by which they inactivate (**C**). Transitions from the open to the inactivated state reduce the number of channels that are available to conduct ions. Mutations that destabilize the inactivated, nonconducting state of the channel are gain-of-function mutations and are common to diverse diseases, including LQTS, certain forms of epilepsy, and muscle disorders such as hyperkalemic paralysis.

Investigation into the functional consequences of these mutations in heterologous expression systems has revealed novel mechanisms underlying disease processes (13–15), mutation-specific therapeutic approaches to disease management (16, 17), and, importantly, common functional mechanisms underlying such diverse diseases as cardiac arrhythmias and seizure disorders (12, 18).

Sodium channels open in response to membrane depolarization and, while open, provide the pathway for the inward movement of positively charged sodium ions. This results in a regenerative process that is necessary for conduction of electrical impulses in nerve and muscle. Within milliseconds of opening, most voltage-gated sodium channels enter a nonconducting inactivated state. The rapid transition from open to inactivated states is necessary to control durations of electrical impulses in all tissues. Most inherited sodium channel mutations that are associated with human disease alter the inactivation process and hence alter the essential control of electrical-impulse duration that is effected by transitions into the inactivated state. Analysis of skeletal muscle disorders identified some of the first channelopathies associated with mutations that alter sodium channel inactivation. In this review series, Jurkat-Rott and Lehmann-Horn describe these and other skeletal muscle channelopathies (19). These authors provide an overview of the roles of sodium channels in muscle physiology and explain how mutations that alter channel inactivation may result in the clinical pathologies and myotonias. They note that in Na⁺ channel myotonia and paramyotonia, mutation-induced gating defects of the Na⁺ channels destabilize the inactivated state of

the channel such that inactivation may be slowed or incomplete. This results in hyperexcitability, repetitive muscle stimulation, and fatigue (19). A critical discussion of muscle channelopathies that may be caused by mutations in other ion channel genes, by antibodies directed against ion channel proteins, or by changes of cell homeostasis leading to aberrant splicing of ion channel RNA or to disturbances of modification and localization of channel proteins is also presented (19).

The importance of mutation-altered sodium channel inactivation to human disease is further explored by Meisler and Kearney (20), who present a detailed discussion of the roles of sodium channel mutations in epilepsy and other seizure disorders. They point out that since the first mutations of the neuronal sodium channel *SCN1A* were identified 5 years ago, more than 150 mutations have been described in patients with epilepsy, and they suggest that sodium channel mutations may be significant factors in the etiology of numerous neurological diseases and may contribute to psychiatric disorders as well. Interestingly, mutations related to epilepsy predominantly affect sodium channel inactivation. In these cases, defective inactivation of the channels could contribute to hyperexcitability. Thus similar channel mutations may have diverse and markedly different clinical consequences depending on the tissue in which the channel is expressed.

Long QT syndrome

Long QT syndrome (LQTS), a rare genetic disorder associated with life-threatening arrhythmias, is next reviewed briefly by Moss and Kass (21). LQTS investigations have provided fundamental new insight into the molecular and genetic basis of electrical signaling in the heart. To date, mutations in at least 8 different genes have been associated with LQTS. The most prevalent LQTS genes code for ion channel subunits, but regulatory proteins are also associated with this disorder. Though LQTS is caused by a diverse group of genes, the common clinical phenotype is delayed ventricular repolarization, which, in most cases, is caused by mutation-induced loss of potassium channel function or a gain of sodium channel function. The mutation-altered channel function that underlies LQTS is remarkably similar to altered channel function in epilepsy and skeletal muscle disorders, and this similarity provides insight into the physiologically essential balance of ion channel function in diverse cells and tissue.

An important contribution of LQTS investigations discussed by Moss and Kass has been the discovery not only of mutation-specific risk factors for different genetic variants of LQTS (22, 23) but of mutation-specific therapeutic strategies to manage the disorder (16, 24). Because the clinical phenotype, prolonged QT interval on the ECG, is determined by routine noninvasive procedures, LQTS

Table 1

Selected channelopathies reviewed in this series

Protein	Gene	Disease	Functional defect	Reference
Na$_v$1.1	SCN1A	Generalized epilepsy with febrile seizures plus (GEFS+)	Hyperexcitability	12, 20
Na$_v$1.2	SCN2A	Generalized epilepsy with febrile and afebrile seizures	Hyperexcitability	12, 20
Na$_v$1.4	SCN4A	Paramyotonia congenita, potassium-aggravated myotonia, hyperkalemic periodic paralysis	Hyperexcitability	12, 19, 20
Na$_v$1.5	SCN5A	LQTS/Brugada syndrome	Heart action potential	21
SCN1B	SCN1B	Generalized epilepsy with febrile seizures plus (GEFS+)	Hyperexcitability	12, 20
KCNQ1	KCNQ1	Autosomal-dominant LQTS with deafness	Heart action potential/inner ear K$^+$ secretion	21
		Autosomal-recessive LQTS	Heart action potential	
KCNH2	KCNH2	LQTS	Heart action potential	21
Kir2.1	KCNJ2	LQTS with dysmorphic features	Heart action potential	21
HERG	KCNH2	Congenital and acquired LQTS	Heart action potential and excessive responses to drugs	21, 26
Ankyrin-B	ANKB	LQTS	Heart action potential	21
Ca$_v$1.2	CACNA2	Timothy syndrome	Multisystem disorders	21
Kir6.2	KCNJ11	Persistent hyperinsulinemic hypoglycemia of infancy	Insulin hypersecretion	29
		Diabetes mellitus	Insulin hyposecretion	
SUR1	SUR1	Persistent hyperinsulinemic hypoglycemia of infancy	Insulin hyposecretion	29
SUR2	SUR2	Dilated cardiomyopathy	Metabolic signaling	29
KCNE1	KCNE1	Autosomal-dominant LQTS with deafness	Heart action potential	21
		Autosomal-dominant LQTS	Heart action potential	
KCNE2	KCNE2	LQTS	Heart action potential	21
CFTR	ABCC7	Cystic fibrosis	Epithelial transport defect	28
ClC-1	CLCN1	Myotonia (autosomal-recessive or -dominant)	Defective muscle repolarization	19, 28
ClC-5	CLCN5	Dent disease	Defective endosome acidification	28
ClC-7	CLCN7	Osteopetrosis (recessive or dominant)	Defective bone resorption	28
ClC-Kb	CLCNKB	Bartter syndrome type III	Renal salt loss	28
RyR1	RyR1	Central core disease, malignant hyperthermia	Abnormal muscle activity	19, 27
RyR2	RyR2	Catecholaminergic polymorphic tachycardia	Exercise-related cardiac arrhythmias	27

has led to fundamentally novel understanding of the processes that control and regulate electrical activity in the heart.

Drug-induced QT prolongation defines "acquired" LQTS (25). Though, by definition, this disorder is not due to mutation-induced QT prolongation due directly to changes in ion channel gene products, considerable evidence indicates that, in many cases, variations in ion channel genes that are not sufficiently severe to be classified as mutations, but instead are referred to as polymorphisms, may contribute to unique drug responses in carriers of these gene variants. Drug-induced QT prolongation, so-called acquired LQTS, is reviewed by Roden and Viswanathan (26).

Ion channels in cellular organelles and nonexcitable cells

Ion channels expressed in plasma membranes affect cellular excitability in nerve and muscle, but expression of ion channels in nonexcitable cells and membranes of intracellular organelles such as the sarcoplasmic reticulum, lysosomes, and mitochondria confers a broad range of functional roles to these important proteins. Mutations in channels responsible for the control of intracellular calcium release from the principal intracellular calcium store in muscle, the sarcoplasmic reticulum, have only recently been discovered to underlie a class of intracellular calcium–dependent arrhythmias and disorders. Priori and Napolitano review the current knowledge about the mutations of the gene encoding the cardiac ryanodine receptor (RyR2) that cause cardiac arrhythmias (27). These authors discuss the similarities between the mutations identified in genes

coding for calcium release channels (ryanodine receptors). In the heart, mutations in the ryanodine receptor RyR2 cause cardiac arrhythmias. In skeletal muscle, mutations in the ryanodine receptor RyR1 cause malignant hyperthermia and central core disease.

Finally, reviews by Jentsch et al. (28) and Ashcroft (29) concentrate on channelopathies in classically nonexcitable tissue. Jentsch et al. (28) focus on diseases related to transepithelial transport and on disorders involving vesicular Cl$^-$ channels. As is pointed out in their review, the transport of anions across cellular membranes is crucial for multiple physiological functions that include the control of electrical excitability of muscle and nerve, transport of salt and water across epithelia, and the regulation of cell volume or the acidification and ionic homeostasis of intracellular organelles. Consequently, mutations in Cl$^-$ channels are associated with diverse human disease.

ATP-sensitive potassium (K$_{ATP}$) channels, so named because they are inhibited by intracellular ATP, couple cell metabolism to electrical activity of the plasma membrane. When intracellular ATP concentrations fall, these channels open and hyperpolarize cells. Conversely, when intracellular ATP concentrations are elevated, K$_{ATP}$ channels close, resulting in membrane depolarization. In pancreatic β cells, membrane depolarization induced by closure of these channels results in activation of voltage-dependent calcium channels and, consequently, glucose-dependent insulin secretion. Thus, K$_{ATP}$ channels serve as targets for drugs that are used to treat and control type 2 diabetes, such as the sulphonylureas. The review by Ashcroft (29) highlights insulin secretory disorders, such as congenital hyperinsulinemia and neonatal diabetes, that result from mutations in K_{ATP} channel genes and considers whether defective regulation of K_{ATP} channel activity contributes to the etiology of type 2 diabetes.

Summary

The aim of this special *JCI* series on channelopathies is to review selected disorders caused by mutations that result in defective ion channel function, regulation, or expression. Together, this collection of articles illustrates the impact that investigation of these diseases has had on our understanding of the mechanistic basis of human disease, revealing critical roles of ion channel function in physiological processes as diverse as muscle activation and regulation of insulin secretion. Remarkable conservation of the essential function of ion channels emerges from this series of review articles in this issue of the *JCI* — function that, when disrupted in neurons, may cause epilepsy, but, when disrupted in the heart, causes fatal cardiac arrhythmias. The lessons from these studies are dramatic and clearly call for additional collaborative investigation in which clinical and preclinical scientists, working together, continue to unravel the molecular basis of diverse human disorders.

Address correspondence to: Robert S. Kass, Department of Pharmacology, Columbia University Medical Center, 630 West 168th Street, PH 7W318, New York, New York 10032, USA. Phone: (212) 305-7444; Fax: (212) 305-8780; E-mail: rsk20@columbia.edu.

1. Hodgkin, A.L., and Huxley, A.F. 1952. A quantitative description of membrane current and its application to conduction and excitation in nerve. *J. Physiol.* **117**:500–544.
2. Hodgkin, A.L., and Rushton, W.A.H. 1946. The electrical constants of a crustacean nerve fiber. *Proc. R. Soc. Lond. B.* **B133**:444–479.
3. Hodgkin, A.L., and Huxley, A.F. 1952. Propagation of electrical signals along giant nerve fibers. *Proc. R. Soc. Lond. B Biol. Sci.* **140**:177–183.
4. Cole, K.S. 1979. Mostly membranes (Kenneth S. Cole). *Annu. Rev. Physiol.* **41**:1–24.
5. Hodgkin, A.L., and Horowicz, P. 1959. Movements of Na and K in single muscle fibres. *J. Physiol.* **145**:405–432.
6. Weidmann, S. 1952. The electrical constants of Purkinje fibres. *J. Physiol.* **118**:348–360.
7. Noda, M., et al. 1982. Primary structure of alpha-subunit precursor of Torpedo californica acetylcholine receptor deduced from cDNA sequence. *Nature.* **299**:793–797.
8. Giraudat, J., Devillers-Thiery, A., Auffray, C., Rougeon, F., and Changeux, J.P. 1982. Identification of a cDNA clone coding for the acetylcholine binding subunit of Torpedo marmorata acetylcholine receptor. *EMBO J.* **1**:713–717.
9. Neher, E., and Sakmann, B. 1976. Single-channel currents recorded from membrane of denervated frog muscle fibres. *Nature.* **260**:799–802.
10. Doyle, D.A., et al. 1998. The structure of the potassium channel: molecular basis of K+ conduction and selectivity. *Science.* **280**:69–77.
11. MacKinnon, R. 2003. Potassium channels. *FEBS Lett.* **555**:62–65.
12. George, A.L., Jr. 2005. Inherited disorders of voltage-gated sodium channels. *J. Clin. Invest.* **115**:1990–1999. doi:10.1172/JCI25505.
13. Clancy, C.E., Tateyama, M., Liu, H., Wehrens, X.H., and Kass, R.S. 2003. Non-equilibrium gating in cardiac Na+ channels: an original mechanism of arrhythmia. *Circulation.* **107**:2233–2237.
14. Clancy, C.E., and Kass, R.S. 2005. Inherited and acquired vulnerability to ventricular arrhythmias: cardiac Na+ and K+ channels. *Physiol. Rev.* **85**:33–47.
15. Nuyens, D., et al. 2001. Abrupt rate accelerations or premature beats cause life-threatening arrhythmias in mice with long-QT3 syndrome. *Nat. Med.* **7**:1021–1027.
16. Benhorin, J., et al. 2000. Effects of flecainide in patients with new SCN5A mutation: mutation-specific therapy for long-QT syndrome? *Circulation.* **101**:1698–1706.
17. Abriel, H., Wehrens, X.H., Benhorin, J., Kerem, B., and Kass, R.S. 2000. Molecular pharmacology of the sodium channel mutation D1790G linked to the long-QT syndrome. *Circulation.* **102**:921–925.
18. Lossin, C., Wang, D.W., Rhodes, T.H., Vanoye, C.G., and George, A.L., Jr. 2002. Molecular basis of an inherited epilepsy. *Neuron.* **34**:877–884.
19. Jurkat-Rott, K., and Lehmann-Horn, F. 2005. Muscle channelopathies and critical points in functional and genetic studies. *J. Clin. Invest.* **115**:2000–2009. doi:10.1172/JCI25525.
20. Meisler, M.H., and Kearney, J.A. 2005. Sodium channel mutations in epilepsy and other neurological disorders. *J. Clin. Invest.* **115**:2010–2017.

doi:10.1172/JCI25466.
21. Moss, A.J., and Kass, R.S. 2005. Long QT syndrome: from channels to cardiac arrhythmias. *J. Clin. Invest.* **115**:2018–2024. doi:10.1172/JCI25537.
22. Moss, A.J., et al. 2002. Increased risk of arrhythmic events in long-QT syndrome with mutations in the pore region of the human ether-a-go-go-related gene potassium channel. *Circulation.* **105**:794–799.
23. Schwartz, P.J., et al. 2001. Genotype-phenotype correlation in the long-QT syndrome: gene-specific triggers for life-threatening arrhythmias. *Circulation.* **103**:89–95.
24. Moss, A.J., et al. 2000. Effectiveness and limitations of beta-blocker therapy in congenital long-QT syndrome. *Circulation.* **101**:616–623.
25. Fenichel, R.R., et al. 2004. Drug-induced torsades de pointes and implications for drug development. *J. Cardiovasc. Electrophysiol.* **15**:475–495.
26. Roden, D.M., and Viswanathan, P.C. 2005. Genetics of acquired long QT syndrome. *J. Clin. Invest.* **115**:2025–2032. doi:10.1172/JCI25539.
27. Priori, S.G., and Napolitano, C. 2005. Cardiac and skeletal muscle disorders caused by mutations in the intracellular Ca^{2+} release channels. *J. Clin. Invest.* **115**:2033–2038. doi:10.1172/JCI25664.
28. Jentsch, T.J., Maritzen, T., and Zdebik, A.A. 2005. Chloride channel diseases resulting from impaired transepithelial transport or vesicular function. *J. Clin. Invest.* **115**:2039–2046. doi:10.1172/JCI25470.
29. Ashcroft, F.M. 2005. ATP-sensitive potassium channelopathies: focus on insulin secretion. *J. Clin. Invest.* **115**:2047–2058. doi:10.1172/JCI25495.

Mechanisms of sudden cardiac death

Michael Rubart[1,2] and Douglas P. Zipes[1]

[1]Krannert Institute of Cardiology and [2]Wells Center for Pediatric Research, Indiana University School of Medicine, Indianapolis, Indiana, USA.

Despite recent advances in preventing sudden cardiac death (SCD) due to cardiac arrhythmia, its incidence in the population at large has remained unacceptably high. Better understanding of the interaction among various functional, structural, and genetic factors underlying the susceptibility to, and initiation of, fatal arrhythmias is a major goal and will provide new tools for the prediction, prevention, and therapy of SCD. Here, we review the role of aberrant intracellular Ca^{2+} handling, ionic imbalances associated with acute myocardial ischemia, neurohumoral changes, and genetic predisposition in the pathogenesis of SCD due to cardiac arrhythmia. Therapeutic measures to prevent SCD are also discussed.

Sudden cardiac death (SCD) from any cause claims 300,000–400,000 lives a year in the United States. The most common sequence of events leading to SCD appears to be the degeneration of ventricular tachycardia (VT; abnormal acceleration of ventricular rate) into ventricular fibrillation (VF), during which disorganized contractions of the ventricles fail to eject blood effectively, often followed by asystole or pulseless electrical activity. Preexisting coronary artery disease and its consequences (acute myocardial ischemia, scarring from previous myocardial infarction, heart failure) are manifest in 80% of SCD victims. Dilated nonischemic and hypertrophic cardiomyopathies account for the second largest number of SCDs, whereas other cardiac disorders, including congenital heart defects and the known genetically determined ion channel anomalies, account for 5–10% of SCDs (1).

While the implantable cardioverter defibrillator (ICD) improves survival in high-risk patients (2), standard antiarrhythmic drug therapy has failed to reduce, and in some instances has increased, the incidence of SCD (3). In fact, the greatest reduction in cardiovascular mortality (including SCD) in patients with clinically manifest heart disease has resulted from the use of beta blockers (4) and non-antiarrhythmic drugs, i.e., those without major direct electrophysiological action in cardiac muscle or the specialized conduction system, such as angiotensin-converting enzyme (ACE) inhibitors, angiotensin receptor–blocking agents, lipid-lowering agents, spironolactone, thrombolytic and antithrombotic agents, and perhaps magnesium and omega-3 fatty acids (for review, see ref. 5). These drugs most likely exert their antiarrhythmic potential indirectly by inhibiting or delaying adverse functional and structural remodeling in the diseased heart, i.e., by affecting "upstream events" that contribute to the development of electrophysiological instability.

Nonstandard abbreviations used: ACE, angiotensin-converting enzyme; $[Ca^{2+}]_i$, intracellular Ca^{2+} concentration; CASQ2, calsequestrin 2; CPVT, catecholaminergic polymorphic VT; DAD, delayed after-depolarization; EAD, early after-depolarization; FKBP12.6, 12.6-kDa FK506-binding protein; $I_{Ca,L}$, L-type Ca^{2+} current; ICD, implantable cardioverter defibrillator; $I_{Na/Ca}$, Na^+/Ca^{2+} exchange current; IP$_3$, inositol-1,4,5 triphosphate; I_{to}, transient outward K^+ current; LQTS, long-QT syndrome; $[Na^+]_i$, intracellular Na^+ concentration; RAS, renin-angiotensin system; RyR2, ryanodine receptor 2; SCD, sudden cardiac death; SR, sarcoplasmic reticulum; VF, ventricular fibrillation; V_m, membrane potential; VT, ventricular tachycardia.

Conflict of interest: The authors have declared that no conflict of interest exists.

Citation for this article: *J. Clin. Invest.* 115:2305–2315 (2005). doi:10.1172/JCI26381.

Clinical trials, in general, have failed to define SCD risk markers for specific individuals in the larger general population, where the relative risk of SCD is low but the absolute number of deaths is high. Risk markers include abnormalities in cardiovascular function (left-ventricular ejection fraction), electrocardiographic variables (e.g., late potentials, T-wave alternans, QRS duration, dispersion of repolarization), results of electrophysiological testing (programmed electrical stimulation), and measures of cardiac autonomic function (heart rate at rest and during exercise, heart rate variability, and baroreflex sensitivity), as well as ambient ventricular arrhythmias (premature ventricular depolarizations, nonsustained and sustained VT).

Acquired functional and structural changes occurring in the diseased heart as well as genetic factors (e.g., mutations of ion channel–encoding genes, polymorphism of coagulation factors or β-adrenergic receptors) may contribute to an increased risk of dying suddenly, but these factors alone cannot explain the apparent randomness of the occurrence of fatal arrhythmias. The nature of the *immediate precipitating event* that triggers the fatal ventricular tachyarrhythmia at a specific time in an otherwise stable patient remains as the major unanswered question. Better understanding of the interaction among various functional, structural, and genetic factors that are thought to underlie the susceptibility to, and initiation of, fatal arrhythmias is the major goal of future research and should provide new tools for their prediction, prevention, and therapy. This review will highlight the role of some of these factors, with a focus on altered intracellular Ca^{2+} dynamics, acute myocardial ischemia, neurohumoral changes, and, briefly, genetic predisposition. Additional mechanisms as well as therapeutic and diagnostic approaches are discussed in other recently published review articles (6–9).

The normal electrocardiogram and its relation to the transmembrane action potential and activities of cardiac ion channels

Both the anatomy and physiology of the specialized conduction system and working cardiac muscle determine how the heart is activated. One cycle of a normal ECG is shown in Figure 1A. The P wave is produced by the electrical activation of the atria, and the PR interval represents the duration of conduction from the sinus node, through the atria to the ventricles; the QRS complex is generated by electrical activation of the ventricles; and the ST-T wave reflects recovery. During a normal cardiac cycle, electrical excitation starts in the sinus node, then moves to the atria, crossing the atrioventricular node and His bundle, and finally reaches the ventricles. The QRS complex reflects the sum of the spatial patterns of ventricular activation. Ventricular excitation rapidly

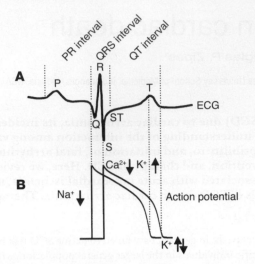

Figure 1
Temporal relationship between ECG and single cardiomyocyte action potential. (**A**) The waves and intervals of a normal ECG. (**B**) Schematic representation of a ventricular action potential and its major underlying ionic currents. The downward arrow indicates influx; the upward arrow, efflux.

spreads along the specialized intracardiac conduction system (His-Purkinje system) in the endocardium, resulting in depolarization of most of the endocardial surfaces of both ventricles within several milliseconds. The activation front then moves from endocardium to epicardium, starting at the Purkinje–muscle cell junctions and proceeding by cell-to-cell conduction through the muscle toward the epicardium. Ventricular recovery, like activation, occurs in a stereotypical geometrical pattern. Differences in recovery timing occur both across the ventricular wall and between regions of the left and right ventricles and give rise to the characteristic morphology of the ST-T wave in the electrocardiogram.

Transmembrane ionic currents are ultimately responsible for the electrical activity recorded in the ECG. Shown in Figure 1B are the major ion fluxes during a normal ventricular action potential. Initial Na^+ influx through open Na^+ channels causes rapid membrane depolarization, which in turn activates Ca^{2+} channels. The ensuing inflow of Ca^{2+} causes Ca^{2+} release channels in the sarcoplasmic reticulum (SR) membrane to open, increasing the cytosolic calcium level and causing contraction. Finally, an increase in K^+ efflux through activated K^+ channels restores the resting membrane potential (V_m; repolarization), and removal of Ca^{2+} from the cytosol deactivates contractile proteins, thereby relaxing the cardiac muscle.

The emerging role of altered intracellular Ca^{2+} dynamics in cardiac arrhythmogenesis

Alterations in intracellular calcium homeostasis play an important role in the development of ventricular tachyarrhythmias in the failing heart (10), as well as in some inherited syndromes leading to SCD. Con-

genital Ca^{2+} handling anomalies in the heart include defective function of the ryanodine receptor 2 (RyR2) in catecholaminergic polymorphic VT (CPVT) (11); a loss-of-function mutation in ankyrin-B causing type 4 long-QT cardiac arrhythmias (12); and a missense mutation of calsequestrin 2 (CASQ2) causing stress-induced polymorphic VT (13). Disruption of the gene encoding the 12.6-kDa FK506-binding protein (FKBP12.6), which reversibly associates with and modulates the activity of RyR2, causes similar alterations in intracellular Ca^{2+} signaling in cardiomyocytes of male and female mice but results in hypertrophy in male mice only (14). It is currently not known whether similar gender differences exist for the susceptibility to arrhythmias in hearts exhibiting inherited disorders of Ca^{2+} handling proteins.

Cardiac arrhythmogenesis associated with RyR2 dysfunction in CPVT and heart failure. RyR2 is the major Ca^{2+} release channel required for excitation-contraction coupling in cardiomyocytes. Depolar-

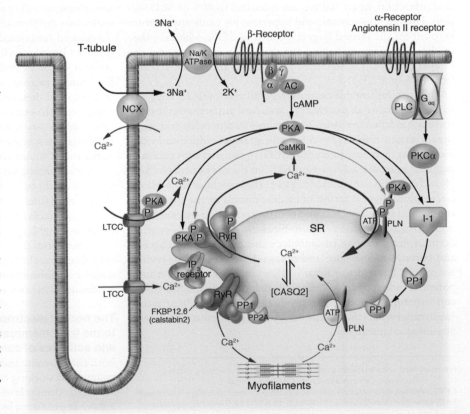

Figure 2
Schematic illustration of intracellular Ca^{2+} cycling and associated second messenger pathways in cardiomyocytes (figure modified from ref. 84). AC, adenylyl cyclase; α, G protein subunit α; α-receptor, α-adrenergic receptor; β, G protein subunit β; β-receptor, β-adrenergic receptor; γ, G protein subunit γ; LTCC, L-type Ca^{2+} channel; CAMKII, Ca^{2+}-calmodulin kinase II; I-1, inhibitor 1; NCX, Na^+/Ca^{2+} exchanger; P, phosphate group; PLC, phospholipase C; PLN, phospholamban; PP1, protein phosphatase 1; PP2A, protein phosphatase 2A; SERCA2a, SR Ca^{2+}-ATPase isoform 2a; T-tubule, transverse tubule.

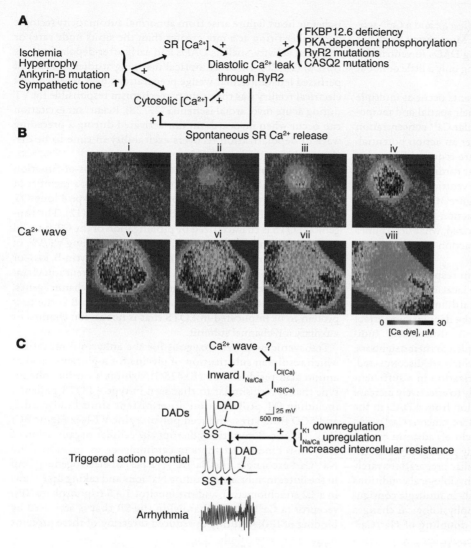

A

Ischemia
Hypertrophy
Ankyrin-B mutation
Sympathetic tone ↑

SR [Ca²⁺] ↑

Diastolic Ca²⁺ leak
through RyR2

{ FKBP12.6 deficiency
PKA-dependent phosphorylation
RyR2 mutations
CASQ2 mutations

Cytosolic [Ca²⁺] ↑

Spontaneous SR Ca²⁺ release

B

Ca²⁺ wave

i   ii   iii   iv

v   vi   vii   viii

0 ▮▮▮ 30
[Ca dye], µM

C

Ca²⁺ wave ?

Inward I$_{Na/Ca}$ I$_{Cl(Ca)}$ I$_{NS(Ca)}$

DADs   DAD

S S ↓   { I$_{K1}$ downregulation
I$_{Na/Ca}$ upregulation
Increased intercellular resistance

Triggered action potential   DAD

S S ↑↑

Arrhythmia

25 mV
500 ms

Figure 3
Proposed scheme of events leading to DADs and triggered tachyarrhythmia. **(A)** Congenital (e.g., ankyrin-B mutation) and/or acquired factors (e.g., ischemia, hypertrophy, increased sympathetic tone) will cause a diastolic Ca²⁺ leak through RyR2, resulting in localized and transient increases in [Ca²⁺]$_i$ in cardiomyocytes. **(B)** Representative series of images showing changes in [Ca²⁺]$_i$ during a Ca²⁺ wave in a single cardiomyocyte loaded with a Ca²⁺-sensitive fluorescent dye. Images were obtained at 117-ms intervals. Focally elevated Ca²⁺ (ii) diffuses to adjacent junctional SR, where it initiates more Ca²⁺ release events, resulting in a propagating Ca²⁺ wave (iii–viii). Reproduced with permission from *Biophysical Journal* (85). **(C)** The Ca²⁺ wave, through activation of Ca²⁺-sensitive inward currents, will depolarize the cardiomyocyte (DAD). In cardiomyocytes, the inward I$_{Na/Ca}$ is the major candidate for the transient inward current underlying DADs, although the role of the Ca²⁺-activated Cl⁻ current [I$_{Cl(Ca)}$] and a Ca²⁺-sensitive nonspecific cation current [I$_{NS(Ca)}$] cannot be excluded. If of sufficient magnitude, the DAD will depolarize the cardiomyocyte above threshold resulting in a single or repetitive premature heartbeat (red arrows), which can trigger an arrhythmia. Downregulation of the inward rectifier potassium current (I$_{K1}$), upregulation of I$_{Na/Ca}$, or a slight increase in intercellular electrical resistance can promote the generation of DAD-triggered action potentials. S, stimulus. Modified with permission from *Circulation Research* (26) and *Nature* (12).

izization of the cardiomyocyte cell membrane causes Ca²⁺ influx through activated voltage-dependent L-type Ca²⁺ channels, which in turn initiates Ca²⁺ release via RyR2s from the SR (Figure 2), known as Ca²⁺-induced Ca²⁺ release (15). RyR2-mediated Ca²⁺ release activates contractile proteins, which results in cardiac contraction during systole. During diastole, cytosolic Ca²⁺ re-sequesters into the SR via the phospholamban-regulated SR Ca²⁺-ATPase isoform 2a (SERCA2a); binding of FKBP12.6 (also known as calstabin2) to the RyR2 complex maintains the channel in a closed state to prevent leakage of SR Ca²⁺ into the cytoplasm. Phosphorylation of RyR2s by PKA during adrenergic stimulation (e.g., exercise) dissociates FKBP12.6 from the RyR2 channel complex, which results in increased RyR2 open probability, defined as the fraction of time the channel is in the open, i.e., conducting, state. RyR2 mutations in patients with CPVT reduce the affinity of the SR Ca²⁺ release channel for FKBP12.6, which causes Ca²⁺ to leak out of the SR during diastole (11). The electrocardiographic phenotype of clinical CPVT is mimicked in transgenic mice deficient in FKBP12.6 (16). Increasing the binding affinity of FKBP12.6 for the RyR2 complex prevents exercise- and catecholamine-induced polymorphic VT and SCD in FKBP12.6-deficient mice (16). In heart failure, the RyR2 channel–FKBP12.6 complex has been shown to be depleted of FKBP12.6 due to chronic PKA-mediated hyperphosphorylation, which results in an abnormal increase in RyR2 open probability during diastole (10).

Whereas these observations provide compelling evidence that abnormalities in intracellular Ca²⁺ homeostasis play a primary role in causing SCD due to cardiac arrhythmias, the mechanism by which spontaneous Ca²⁺ release from the SR can trigger fatal arrhythmias has remained elusive.

Congenital and/or acquired factors can cause a diastolic Ca²⁺ leak through RyR2 (Figure 3A), giving rise to localized increases in cytosolic calcium (Figure 3B, ii) in a single cardiomyocyte loaded with a Ca²⁺-sensitive fluorescent dye. The focally elevated Ca²⁺ then diffuses to adjacent junctional SR to activate Ca²⁺ release channels and induce release of Ca²⁺ that can then propagate throughout the cell. This regenerative propagating process has previously been termed a Ca²⁺ wave (17). Diastolic Ca²⁺ waves depolarize the cardiomyocyte membrane, triggering a delayed after-depolarization (DAD) via transient activation of a Ca²⁺-dependent inward current (18) (Figure 3C). Candidates for the transient inward current are the electrogenic Na⁺/Ca²⁺ exchange current (I$_{Na/Ca}$) operating in its forward mode (extruding Ca²⁺ and taking up Na⁺ in a 1:3 stoichiometry); the Ca^{2+}-activated Cl^- current ($I_{Cl(Ca)}$); and a Ca^{2+}-activated nonselective cation current ($I_{NS(Ca)}$). More recent work has supported a major role of $I_{Na/Ca}$ in causing DADs in ventricular myocytes, with $I_{Cl(Ca)}$ and $I_{NS(Ca)}$ contributing only a little or not at all to DAD initiation (18).

Only when spontaneous Ca^{2+} release events occur at multiple sites synchronously within the cell will their spatial and temporal summation result in a rise in intracellular Ca^{2+} concentration ($[Ca^{2+}]_i$) of sufficient magnitude to trigger an action potential. Increases in $[Ca^{2+}]_i$ exceeding 1 µmol/l are required for action potential activation in isolated ventricular cardiomyocytes (18), which can then propagate throughout the ventricles to induce an extra heartbeat, leading to VT and VF (Figure 3C). While factors determining whether a DAD triggers an action potential in isolated cardiomyocytes have been characterized, what determines whether a DAD produces a propagating action potential in the intact myocardium is less well known.

In the intact heart, the change in V_m in response to a given increase in inward current activated by the local increase in $[Ca^{2+}]_i$ is usually less than that in an isolated cardiomyocyte because more passive outward current opposes the depolarizing effect of locally activated Na^+ channel current in cardiac tissue. Individual cardiomyocytes are electrically coupled to their neighbors, as opposed to isolated cardiomyocytes, which are disconnected. Therefore, to induce membrane depolarization in a sufficient number of cardiomyocytes simultaneously to effectively increase the current load, action potential initiation from a DAD in the whole heart would probably require that spontaneous Ca^{2+} waves occur almost synchronously in multiple, closely adjacent cardiomyocytes or that a single Ca^{2+} wave rapidly propagate across cell borders. However, Ca^{2+} waves in intact cardiac preparations rarely spread to their neighboring cells under physiological conditions and do not appear to occur synchronously in multiple contiguous cardiomyocytes (17, 19, 20). Electrophysiological changes occurring in the failing heart, such as doubling of Na^+/Ca^{2+} exchanger expression and reduction in repolarizing K^+ currents (18), could facilitate action potential initiation from a DAD (6). Thus, for any given increase in $[Ca^{2+}]_i$, the inward current carried by the Na^+/Ca^{2+} exchanger will be doubled, and the reduction of outwardly directed K^+ currents will enhance the depolarizing effect of a given $I_{Na/Ca}$. Furthermore, reduced intercellular electrical coupling in nonischemic failing myocardium (21) and redistribution of gap junctions between cardiomyocytes in the epicardial border zone of chronic infarcts (22) may blunt electrotonic interactions between DAD-generating cardiomyocytes and their neighbors, reducing the inward current required to depolarize the cardiomyocytes above threshold. However, an increase in intercellular resistance may prevent the propagation of triggered action potentials. Finally, increased adrenergic activation in failing myocardium leads to SR Ca^{2+} overload, resulting in enhanced activity of RyR2 channels during diastole (23) and the enhanced probability of Ca^{2+} wave occurrence.

Collectively, these results strongly suggest that defective RyR2 gating and associated aberrant Ca^{2+} release from the SR during diastole may constitute the molecular trigger for arrhythmia initiation in the failing heart (Figure 3).

Three-dimensional mapping studies in intact hearts support the hypothesis that spontaneous SR Ca^{2+} release events act as focal triggers of arrhythmias by showing that nearly all VT in nonischemic heart failure and approximately 50% of those in ischemic heart failure arise from abnormal automaticity (ectopic pacemaker firing at a rate greater than the sinus node rate) or triggered activity, such as DADs and early after-depolarizations (EADs) (24–26). However, optical mapping studies in isolated perfused left-ventricular wedge preparations have revealed that electrical reentry was the primary mechanism responsible for VT during acute myocardial ischemia (27, 28). Reentrant excitation can occur when myocardium not activated during a preceding wave of depolarization recovers excitability in time to be discharged before the impulse dies out.

Ankyrin-B mutation in long-QT4 syndrome. A loss-of-function mutation in ankyrin-B (also known as ankyrin 2), a member of a family of membrane adaptor proteins, causes type 4 long-QT syndrome (LQTS) cardiac arrhythmias and SCD (12). The congenital LQTS is characterized by prolongation of the QT interval (see Figure 1), recurrent syncope, and life-threatening VT/VF. In addition to the mutation in the gene encoding ankyrin-B, loss- or gain-of-function mutations in a number of different ion channel–encoding genes, including cardiac Na^+ and K^+ channel genes, have been identified in inherited LQTS. Ankyrin-B is the first protein to be implicated in LQTS that is not an ion channel or auxiliary ion channel subunit.

Transgenic mice heterozygous for the ankyrin-B mutation, which results in substitution of glycine for a glutamic acid at amino acid residue 1425 (E1425G), exhibit a cardiac phenotype that is very similar to that seen in type 4 LQTS patients, including QT prolongation, intermittent sinus bradycardia, and SCD from stress-induced polymorphic VT (see Figure 3C) (12). Ankyrin-B mutations disrupt the cellular organization of ankyrin-B binding partners in cardiomyocytes, including the Na^+/Ca^{2+} exchanger, the Na^+/K^+ ATPase (an electrogenic pump in the outer membrane extruding Na^+ ions and taking up K^+ ions in a 3:2 stoichiometry), and the inositol-1,4,5 triphosphate (IP_3) receptor (a Ca^{2+} release channel in the SR that is activated by binding of IP_3), resulting in reduced targeting of these proteins to the transverse tubules (Figure 2) as well as reducing their overall expression levels. Peak amplitudes of depolarization-induced $[Ca^{2+}]_i$ transients are markedly increased in single cardiomyocytes isolated from heterozygous ankyrin-B mutant ($AnkB^{+/-}$) mice. It is conceivable that reduced activity of the Na^+/K^+ ATPase increases intracellular Na^+ concentration ($[Na^+]_i$), which in turn inhibits Ca^{2+} extrusion by the Na^+/Ca^{2+} exchanger into the extracellular space (see Figure 2). As Na^+/Ca^{2+} exchanger expression is reduced, total cellular Ca^{2+} content becomes elevated. Cellular Ca^{2+} overload can lead to increased incidence of DAD-generating Ca^{2+} waves. Transmembrane potential recordings in single $AnkB^{+/-}$ cardiomyocytes during β-adrenergic receptor activation revealed extrasystoles arising from DADs as well as from EADs (12), which suggests that these mechanisms are responsible for at least some of the VT/VF in LQTS patients.

Whereas it is generally accepted that DADs are initiated by spontaneous Ca^{2+} release from the SR, the involvement of SR Ca^{2+} release in the generation of EADs occurring in the plateau phase of the action potential is controversial. Some EADs are attributable to recovery from inactivation and reactivation of voltage-dependent L-type Ca^{2+} channels during prolonged action potentials, while EADs occurring late during repolarization may arise from spontaneous SR Ca^{2+} release and transient activation of a depolarizing inward current. The role of Ca^{2+} handling in the initiation of EADs and polymorphic VT (torsade de pointes [TdP]) has recently been studied by the simultaneously mapping of V_m and $[Ca^{2+}]_i$ in an animal model of type 2 LQTS (29). It was shown that at the sites of origin of EADs, increases in $[Ca^{2+}]_i$ preceded membrane depolarization, whereas away from these sites, increases in $[Ca^{2+}]_i$ coincided with or followed membrane depolarization. Conceivably, spontaneous Ca^{2+} release from internal stores activates inward currents to generate an EAD and a propagated action potential. Ischemia can enhance the development of EADs under some circumstances (30).

CASQ2 mutation in CPVT. A missense mutation in the gene encoding CASQ2, a Ca^{2+}-binding protein in the cardiac luminal SR, causes the recessive form of CPVT. Sequence analyses of the *CASQ2* genes of CPVT patients revealed substitution of histidine for an aspartic acid at amino acid residue 307 (D307H). Simultaneous recordings of V_m and $[Ca^{2+}]_i$ in single cardiomyocytes with adenovirus-mediated expression of *CASQ2* carrying the CPVT-linked mutation D307H ($CASQ2^{D307H}$) revealed Ca^{2+} waves and associated DADs during rapid pacing and β-adrenergic receptor stimulation (13). Primary alterations in cellular Ca^{2+} homeostasis, similar to those underlying CPVT and long-QT4 arrhythmias, could constitute the trigger for these arrhythmogenic events.

SR Ca^{2+} cycling in VF and electrical alternans. At normal heart rates, $[Ca^{2+}]_i$ reliably and consistently tracks changes in V_m. Mapping studies have demonstrated that SR Ca^{2+} cycling may exhibit intrinsic dynamics and may become independent of changes in V_m during VF. Non–voltage-gated SR Ca^{2+} release events, through modulation of Ca^{2+}-sensitive membrane currents, then cause local alterations in action potential duration and myocardial refractoriness, which in turn promote the maintenance of fibrillatory activity (31). Thus, spatial heterogeneity of $[Ca^{2+}]_i$ signaling may act not only as a trigger but also as a stabilizer of VT/VF.

Beat-to-beat variations in T-wave morphology and/or polarity (T-wave alternans), a marker of increased electrical instability and fibrillation (32), arise from alternation in action potential duration. Simultaneous measurements of $[Ca^{2+}]_i$ and V_m in single cardiomyocytes as well as in whole hearts during rapid pacing provide evidence that primary alternans in the amplitude and duration of $[Ca^{2+}]_i$ transients can drive action potential duration to alternate secondarily, by influencing several key membrane currents that are sensitive to changes in $[Ca^{2+}]_i$ (33–35).

Collectively, these observations support the concept that SR Ca^{2+} cycling plays an important role in contributing to action potential dynamics during VT/VF.

Intracellular calcium regulates second messenger pathways and ion channel gene transcription. Intracellular Ca^{2+} ions influence the activity of ion channels and/or transporters via direct interaction with the ion-conducting molecule and indirectly via modulation of Ca^{2+}-sensitive intracellular signaling pathways. Ca^{2+} can directly bind to a subunit of the voltage-dependent Na^+ channel, which results in altered Na^+ channel gating (36). Ca^{2+} also activates Ca^{2+}-calmodulin kinase II (CaMKII) (see Figure 2), which in turn targets key molecules that control intracellular Ca^{2+} homeostasis in cardiomyocytes, and increased CaMKII activity may trigger arrhythmia-initiating after-depolarizations by activating L-type Ca^{2+} channels or increasing inward current carried by the Na^+/Ca^{2+} exchanger (37, 38). Determining whether alterations in the sensitivity of the target molecules to changes in intracellular Ca^{2+} levels and/or alterations in intracellular second messenger signaling cascades contribute to arrhythmogenesis in conditions of abnormal intracellular Ca^{2+} handling will be important.

Collectively, these studies highlight the fundamental importance of investigating the spatial and temporal relationship of membrane voltage and intracellular Ca^{2+} dynamics in the whole heart. Techniques for simultaneously mapping transmembrane voltage and $[Ca^{2+}]_i$ with high temporal resolution in the intact heart have recently been developed (29) and will provide novel insights into the factors that facilitate the development of DAD- or EAD-related non-reentrant arrhythmias as well as the role of $[Ca^{2+}]_i$ in reentrant arrhythmias in animal models of SCD.

In addition to the transient effects of increased $[Ca^{2+}]_i$ on cardiomyocyte excitability, chronic changes in $[Ca^{2+}]_i$ in the regulation of ion channel expression in cardiomyocytes can effect lasting alterations in electrical properties (39). For example, long-term changes in intracellular Ca^{2+} handling in transgenic mice overexpressing SERCA1a caused downregulation of K^+ channel expression in the absence of cardiac hypertrophy and failure, leading to action potential prolongation (39). Determining the molecular underpinnings of these remodeling processes will potentially create novel targets for antiarrhythmic prevention.

Acute myocardial ischemia

Ionic imbalances during acute myocardial ischemia. Many SCDs from VT/VF occur during acute myocardial ischemia (40). Abrupt cessation of myocardial blood flow causes redistribution of a number of ions, including H^+, Na^+, Ca^{2+}, and K^+, across the cardiomyocyte membrane, an event that has profound electrophysiological consequences through its influence on the activity of a variety of ion channels and transporters. However, the details of their interactions vis-à-vis initiation of the triggering premature beat and maintenance of arrhythmias have remained largely elusive.

Net cellular K^+ loss and subsequent extracellular K^+ accumulation during acute myocardial hypoxia and/or ischemia causes sustained membrane depolarization that leads to a slowing of conduction and altered refractoriness, which in combination with other factors promote VT/VF and SCD (27, 28). Despite the fact that K^+ loss is an important pathogenic factor in ischemia-associated arrhythmogenesis, the underlying mechanisms are not completely understood. Net K^+ loss in part reflects passive intracellular Na^+ gain during myocardial hypoxia/ischemia (41). Net intracellular Na^+ gain implies that the Na^+/K^+ ATPase is unable to compensate for the amount of Na^+ ions entering the cardiomyocytes through the major Na^+ influx pathways, including voltage-gated Na^+ channels, Na^+/Ca^{2+} exchanger, Na^+/H^+ exchanger, and potentially plasmalemmal connexin-43 hemichannels (42) (Figure 4). It is unknown whether the influx/efflux mismatch results from a primary increase in Na^+ uptake rate beyond the maximal Na^+/K^+ ATPase pump capacity, primary inhibition of the pump, or both. Ischemia-induced decrease in intracellular ATP content may inhibit Na^+/K^+ ATPase. Increased intracellular proton generation increases Na^+ influx via increased activity of the Na^+/H^+ exchanger. Build-up of the ischemic metabolite lysophosphatidylcholine enhances Na^+ influx through voltage-dependent, tetrodotoxin-sensitive Na^+ channels (43). Major K^+ efflux pathways during hypoxia/ischemia involve ATP-dependent K^+ channels, inward rectifier K^+ channels, as well as other voltage-gated K^+ channels, and potentially connexin-43 hemichannels (42). Net cellular K^+ loss is thought to maintain electroneutrality of charge movement and cellular osmolarity. Increases in $[Na^+]_i$ may have detrimental electrophysiological effects. $[Na^+]_i$ regulates cardiac $[Ca^{2+}]_i$.

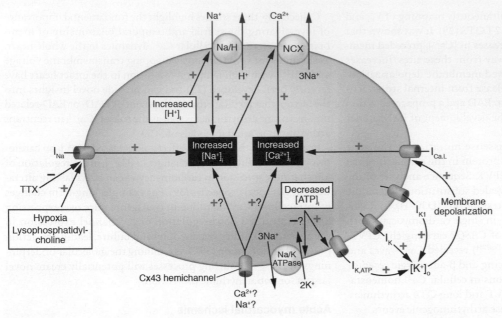

Figure 4

Proposed scheme of events leading to transmembrane ionic imbalances during myocardial ischemia. Net intracellular Na^+ gain due to a mismatch of Na^+ influx and efflux will cause net cellular K^+ loss, extracellular K^+ accumulation, and an increase in intracellular Ca^{2+} due to activation of the Na^+/Ca^{2+} exchanger operating in the reverse mode. Cellular Ca^{2+} overload will cause triggered arrhythmias by the mechanisms illustrated in Figure 3. Lysophosphatidylcholine is a product of diacyl phospholipid catabolism generated by the enzyme phospholipase A2 during ischemia. Question marks indicate that the pathway/mechanism is hypothetical. $[ATP]_i$, intracellular concentration of adenosine triphosphate; Cx43, connexin-43; $[H^+]_i$, intracellular proton concentration; I_K, delayed rectifier K^+ current; $I_{K,ATP}$, ATP-sensitive potassium current; I_{K1}, inward rectifier K^+ current; I_{Na}, fast Na^+ current; Na/H, sodium-hydrogen exchanger; $[K^+]_o$, extracellular K^+ concentration; TTX, tetrodotoxin (a specific blocker of the fast sodium current).

An increase in $[Na^+]_i$ will result in an increase in $[Ca^{2+}]_i$, due to an increase in Ca^{2+} influx via the Na^+/Ca^{2+} exchanger operating in the reverse mode and depolarization-activated L-type Ca^{2+} channels, and possibly direct Ca^{2+} influx through connexin-43 hemichannels (42) (Figure 4).

Role of intracellular calcium in ischemia-associated arrhythmias. Cellular Ca^{2+} overload secondary to Na^+ overload can increase the probability that DAD-generating Ca^{2+} waves will be induced. While the relationship between the amount of Ca^{2+} released from internal stores during a wave and the magnitude of the associated change in V_m has been characterized in isolated cardiomyocytes under normal conditions, it has not been rigorously quantified in partially depolarized cardiomyocytes typically found in the ischemic myocardium. It has also remained unknown whether spontaneous Ca^{2+} release through ryanodine-insensitive Ca^{2+} release channels (e.g., IP_3 receptors; see Figure 2) contributes to the induction of Ca^{2+} waves in the setting of acute ischemia and which membrane conductances are involved in the generation of the transient inward current underlying a DAD.

Even more important are the factors that determine whether a DAD results in a propagating action potential in the setting of acute myocardial ischemia. Although spontaneous Ca^{2+} waves occur in buffer-perfused isolated heart preparations under physiological conditions as well as following injury (17, 19, 20, 44), it is not known whether Ca^{2+} waves trigger arrhythmias under ischemic conditions. Conceivably, an increase in the electrical resistance between cardiomyocytes in the acutely ischemic myocardium

(45) initiates DAD-related extra beats because reduced intercellular coupling would diminish the opposing, i.e., repolarizing, effect of passive outward current generated by neighboring cardiomyocytes. It is also possible that ischemia induces cellular changes that cause either more Ca^{2+} to be released from internal stores or greater depolarizations for a given increase in $[Ca^{2+}]_i$, resulting in an increased propensity for triggered arrhythmias. Because intracellular Na^+ gain, through its secondary effects on cardiac $[Ca^{2+}]_i$ and extracellular $[K^+]$, appears to act as a primary arrhythmogenic factor in the setting of acute ischemia, it will be important to define the spatio-temporal relationships between V_m and intracellular concentrations of Na^+ and Ca^{2+} (and extracellular $[K^+]$) in animal models of acute ischemia–induced cardiac arrhythmias and SCD by simultaneously mapping transmembrane voltage and ion concentrations.

Although VT/VF occurring during reperfusion can be reproducibly induced in the experimental setting (46), the role of spontaneous reperfusion in triggering SCD from cardiac arrhythmias remains to be defined. The pathogenic mechanism underlying reperfusion arrhythmias is largely unknown, but an increase in $[Na^+]_i$ during ischemia and/or early reperfusion appears to play a major role (47). Intracellular Na^+ accumulation can then give rise to $[Ca^{2+}]_i$ overload, which in turn increases the propensity for DAD-related arrhythmias by the mechanisms described above.

Whereas there is experimental evidence for the role of spontaneous increases in $[Ca^{2+}]_i$ in inducing EADs and related ventricular arrhythmias in the nonischemic setting (29), the involvement of EADs in ischemia/reperfusion–induced arrhythmias has yet to be demonstrated experimentally. Other factors predisposing to EADs may have to be present, including action potential prolongation (30, 48) or slight reduction of the electrical conductivity between cells (45). Since failing as well as hypertrophic myocardium exhibit both delayed repolarization and increased electrical resistance (21, 22), acute ischemia occurring in the electrically remodeled failing heart may indeed give rise to EAD-related ventricular tachyarrhythmias.

Relationship between V_m and SR Ca^{2+} cycling in the acutely ischemic myocardium. A study utilizing simultaneous recordings of changes in membrane voltage and intracellular calcium recently showed that acute ischemia exerts contrasting effects on the kinetics of V_m and $[Ca^{2+}]_i$ transients (49). Ischemia was associated with marked shortening of action potential duration but significant prolongation of the $[Ca^{2+}]_i$ transient. Changes in action

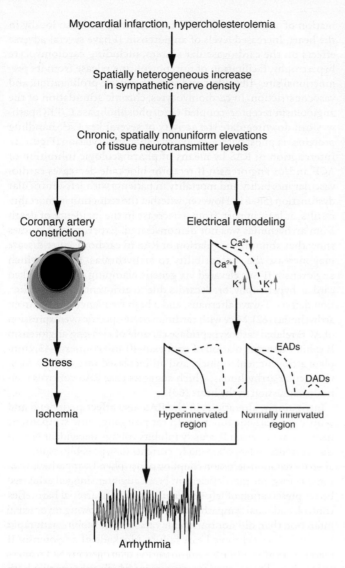

Myocardial infarction, hypercholesterolemia

Spatially heterogeneous increase
in sympathetic nerve density

Chronic, spatially nonuniform elevations
of tissue neurotransmitter levels

Coronary artery constriction

Electrical remodeling

Stress

Ischemia

EADs

DADs

Hyperinnervated region

Normally innervated region

Arrhythmia

Figure 5
Factors contributing to arrhythmogenesis in hearts with heterogeneous sympathetic innervation. Myocardial injury (e.g., myocardial infarction) or chronic hypercholesterolemia (51) will cause a spatially uneven increase in sympathetic nerve density in the heart, resulting in regional variations in release and, consequently, variations in tissue levels of sympathetic neurotransmitters. Chronic, nonuniform elevations of neurotransmitters, through alterations in the expression of L-type Ca^{2+} channels and K^+ channels, create spatial dispersion of action potential duration. Action potential prolongation and augmented Ca^{2+} influx through L-type Ca^{2+} channels combine to increase the susceptibility to EAD- and/or DAD-triggered activity in hyperinnervated regions. If the triggered beat propagates throughout the rest of the heart, the preexisting spatial dispersion of action potential duration and, thus, myocardial refractoriness facilitate the initiation of tachyarrhythmias. Locally elevated levels of neuropeptide Y and norepinephrine may increase coronary artery tone, thereby critically reducing the coronary perfusion reserve under conditions of increased oxygen demand (e.g., physical and/or emotional stress) and causing regional ischemia, which contributes to the development of an arrhythmia.

potential duration via augmentation of K^+ outward currents. The observation that pharmacological beta blockade confers a survival benefit on patients after myocardial infarction suggests an important role of the autonomic nervous system in the pathogenesis of SCD from VT/VF. Further evidence for a neural component in SCD comes from 2 recent experimental studies. In the first study, chronic infusion of nerve growth factor (NGF) to the left stellate ganglion in dogs with chronic myocardial infarction and complete atrioventricular block caused spatially heterogeneous sympathetic cardiac hyperinnervation (nerve sprouting) and dramatically increased the incidence of SCD from VT/VF (50). In a second study, Liu et al. (51) reported that a high-cholesterol diet resulted in myocardial hypertrophy and cardiac sympathetic hyperinnervation in rabbits in the absence of coronary artery stenoses and infarction. Furthermore, they found a marked increase in the incidence of VF, which was associated with enhanced dispersion of repolarization, prolongation of action potential duration and the QT interval, and increased L-type Ca^{2+} current ($I_{Ca,L}$) density (defined as the peak $I_{Ca,L}$ amplitude normalized to the cell surface area). Although results of these studies do not constitute an ultimate proof, they strongly suggest a causal relationship between altered autonomic innervation and SCD due to VT/VF. Intriguingly, explanted human hearts from transplant recipients with a history of arrhythmias exhibited a significantly higher and also more heterogeneous density of sympathetic nerve fibers than those from patients without arrhythmias (52), which indicates that heterogeneous overgrowth of cardiac nerve may occur in human hearts with myocardial infarctions even in the absence of exogenous NGF. Whether neural remodeling also involved parasympathetic nerve fibers in the heart was not examined in these studies.

The precise mechanism by which sympathetic hyperinnervation promotes cardiac arrhythmia is speculative at present. The ultimate manifestation of the arrhythmia is probably the end result of a variety of interacting factors, as schematically illustrated in Figure 5. Increased density of sympathetic nerve endings could augment release of and consequently result in higher than normal tissue concentrations of sympathetic neurotransmitters (e.g., norepinephrine, neuropeptide Y) during sympathetic excitation. This autonomic remodeling is associated with

potential duration were spatially more heterogeneous than was lengthening of $[Ca^{2+}]_i$ transient duration. Moreover, occurrence of $[Ca^{2+}]_i$ transient alternans was not consistently associated with alternans of the action potential duration. Thus, in contrast to similar optical mapping studies performed under nonischemic conditions, this study deemphasizes the primary role of alternation of $[Ca^{2+}]_i$ transient magnitude and/or duration in determining action potential duration and action potential duration alternans in the setting of acute ischemia.

Neurohumoral changes
Role of cardiac sympathetic innervation. Activation of cardiac β-adrenergic receptors by the neurotransmitters epinephrine and norepinephrine has several physiological effects, including increasing systolic contractility and diastolic relaxation rate as well as accelerating heart rate and atrioventricular conduction. At the cardiomyocyte level, stimulation of β-adrenergic receptors alters the activity of a number of ion channels and transporters via activation of the G protein/adenylyl cyclase/cAMP/PKA pathway (see Figure 2). This in turn results in an increase in both peak amplitude and rate of decline of the $[Ca^{2+}]_i$ transient via stimulation of key Ca^{2+} handling proteins as well as shortening of action

spatially heterogeneous electrical remodeling of the cardiomyocytes, including an increase in $I_{Ca,L}$ density (51) and decreases in K$^+$ current densities (53), resulting in action potential prolongation in hyperinnervated regions (Figure 5). Along with electrical remodeling of a number of ion channels and transporters in the surviving myocardium along the infarct border (i.e., the infarct borer zone), acute release of sympathetic neurotransmitters, through their effects on Ca^{2+}, K$^+$, and Cl$^-$ channels and Ca^{2+} transporters and enzymes, is likely to accentuate the preexisting heterogeneity of excitability and refractoriness, which in turn is likely to contribute to arrhythmia susceptibility in these models. Norepinephrine- and/or neuropeptide Y–induced arterial constriction can induce myocardial ischemia in hyperinnervated regions during emotional and/or physical stress, further promoting vulnerability to arrhythmic events.

Myocardial ischemia and infarction are known to result in injury of sympathetic nerves and sympathetic denervation of noninfarcted myocardium in areas distal to the infarct (54). Sympathetic denervation potentiates adrenergic responsiveness to catecholamines, a process termed denervation supersensitivity (55). Diffusion of neurotransmitters from normal regions to neighboring denervated areas during sympathetic excitation is likely to create steep local gradients of refractoriness and excitability, further contributing to the arrhythmogenic substrate. Furthermore, superimposed upon prolongation of action potential duration and increased $I_{Ca,L}$ density, sympathetic stimulation is likely to lead to intracellular Ca^{2+} overload, which in turn could give rise to triggered activity, which may underlie the spontaneous occurrence of VF in these animal models.

Future studies on the mechanisms by which neural remodeling contributes to arrhythmia susceptibility will have to address: (a) the molecular and cellular mechanisms regulating sympathetic (and parasympathetic) nerve processing in the normal and diseased heart; (b) the effects (both short- and long-term) of noncholinergic, nonadrenergic neurotransmitters, including neuropeptide Y, calcitonin gene–related product, vasoactive intestinal peptide, and substance P, on cardiomyocyte electrophysiology and Ca^{2+} handling; and (c) the second messenger pathways involved. All 4 peptides have profound physiological effects. Neuropeptide Y, for example, is a potent vasoconstrictor (56) and also appears to be an essential in vivo regulator of cardiac $I_{Ca,L}$ during postnatal development (57). Other questions include the following: what are the altered electrical properties of pre-/postganglionic sympathetic/parasympathetic neurons in the diseased state that result in altered excitation-secretion coupling? what are the mechanisms underlying sympathetically induced ventricular tachyarrhythmias in animal models of SCD? does sympathetic nerve stimulation acutely exaggerate any preexisting electrical heterogeneity in the diseased heart? and, finally, does sympathetic stimulation increase the incidence of abnormalities in intracellular Ca^{2+} handling in the whole heart (Ca^{2+} waves, beat-to-beat variations in Ca^{2+} amplitude and/or duration [alternans]) that may act as triggers of ventricular arrhythmias? To answer the latter question, optical mapping techniques will be required to simultaneously measure membrane voltage and [Ca^{2+}]$_i$ in whole hearts during autonomic stimulation.

Role of the renin-angiotensin system

Angiotensin II. The renin-angiotensin system (RAS) is an enzymatic cascade that controls conversion of angiotensinogen into angiotensin II. The final step in this process is mediated by ACE. Formation of angiotensin II occurs systemically and also locally in the heart. Increased levels of angiotensin II have several adverse effects on the cardiovascular system, including cardiomyocyte hypertrophy, facilitation of norepinephrine release from its prejunctional sites, fibroblast and smooth muscle proliferation, and vasoconstriction. In cardiomyocytes, chronic stimulation of the angiotensin receptor–coupled Gαq/phospholipase C/PKC pathway can downregulate the activity of several key Ca^{2+}-handling proteins via phosphatase-mediated dephosphorylation (Figure 2). Interruption of RAS by means of pharmacologic inhibition of ACE and/or angiotensin II receptor blockade decreases cardiovascular morbidity and mortality in patients with left-ventricular dysfunction (58–61). However, whether the reduction in mortality results, at least in part, from a decrease in the incidence of death from arrhythmias was not demonstrated. Experimental studies show that abnormal regulation of RAS in cardiovascular disease may increase the susceptibility to arrhythmias. Mice in which angiotensin II was elevated via genetic clamping exhibit marked cardiac hypertrophy, bradycardia due to atrioventricular conduction defects, T-wave alternans, and a high frequency of SCD from arrhythmias (62). Mice with cardiomyocyte-specific overexpression of ACE-related carboxypeptidase capable of cleaving angiotensin II exhibit downregulation of connexin-40 and connexin-43, complete atrioventricular block, and an increased rate of SCD from ventricular arrhythmias, which suggests that RAS controls connexin expression in the heart (63).

Intriguingly, alterations in the RAS also affect peripheral and central sympathetic function. In rat postganglionic sympathetic neurons, angiotensin II acutely inhibits voltage-dependent N-type Ca^{2+} currents and repolarizing K$^+$ currents (64), possibly resulting in altered excitation-secretion coupling of peripheral sympathetic neurons. Transgenic rats deficient in brain angiotensinogen exhibited better preservation of left-ventricular function, arterial baroreflex control, and renal sympathetic nerve activity following myocardial infarction than did nontransgenic rats (65). In rabbits with rapid pacing–induced chronic heart failure, blockade of angiotensin II receptors combined with administration of exogenous NO reduces elevated renal sympathetic nerve activity, which suggests that both loss of NO and an increase in angiotensin II are necessary for sympathoexcitation in chronic heart failure (66). It is still unknown whether preservation of normal sympathetic function translates into a reduction in the incidence of cardiac arrhythmias.

Effect of angiotensin II on ion channels and transporters. Angiotensin II has also been shown to directly modulate ion channels and transporters. Inhibition by angiotensin II of voltage-dependent, repolarizing K$^+$ currents in arterial smooth muscle causes vasoconstriction (67, 68), possibly leading to reduced myocardial blood flow in conditions associated with elevated angiotensin II. In cardiomyocytes, enhancement of $I_{Ca,L}$; reduction of the delayed rectifier K$^+$ current, I_K, and the transient outward K$^+$ current, I_{to}; and electrogenic Na/K ATPase activity act synergistically to prolong action potential duration and to increase intracellular Ca^{2+} load, thereby promoting arrhythmia susceptibility. ACE inhibitors, as well as angiotensin receptor blockers, reduce SCD in heart failure and postinfarction patients (58–61). Passive stretch elicits an outwardly rectifying Cl$^-$ current in ventricular myocytes by a mechanism that involves release of angiotensin II and activation of angiotensin II type 1 receptor in an autocrine/paracrine loop (69).

Aldosterone. Angiotensin II also enhances release of aldosterone from the adrenal cortex. Aldosterone has potent sodium-retaining properties and can exert adverse cardiovascular effects, including myocardial hypertrophy and fibrosis. Like angiotensin II, aldosterone regulates cardiac ion channels. Chronic exposure of rat ventricular myocytes to aldosterone reduces and increases, respectively, I_{to} and $I_{Ca,L}$ (70). The temporal relationship of the changes in current densities and the effect of blockers of the $I_{Ca,L}$ are consistent with the Ca^{2+}-dependent reduction of I_{to} expression. In a post–myocardial infarction model of heart failure in rats, aldosterone antagonism had a number of effects, including inhibiting fibrosis, reducing myocardial norepinephrine content, and increasing the VF threshold (71). Clinically, treatment with the aldosterone receptor antagonists spironolactone or eplerenone has been demonstrated to reduce SCD in heart failure patients (72, 73).

Collectively, these observations suggest that elevated angiotensin II and/or aldosterone levels exert adverse electrophysiological effects that may contribute to the development of a proarrhythmic substrate in conditions that are typically associated with abnormal activation of the RAS (e.g., heart failure). Antagonizing their effects could reduce the extent of adverse electrical remodeling and has been shown to prevent SCD. Because RAS signaling also has profound effects on heart structure, including induction of hypertrophy and interstitial fibrosis, it will be important to determine whether antiarrhythmic effects of RAS signaling antagonists derive indirectly from preservation of the structural integrity of the heart, directly from modulation of ion channels and transporters, or both.

Genetic predisposition

In addition to well-known heritable genetic disorders such as LQTS, genetic variations exist that determine an individual's risk for fatal arrhythmias. In the Paris Prospective study, a history of SCD in 1 parent was shown to increase the risk of fatal arrhythmia in the offspring by 80%; a history of SCD in both parents leads to an approximate 9-fold increase in the risk of SCD for the offspring (74). In another population-based study, the rate of SCD in first-degree relatives of SCD victims was 50% greater than the rate in control subjects and, importantly, was independent of other risk factors such as diabetes, hypertension, or cigarette smoking (75). It is very likely that a number of genes contribute to the phenotypic manifestation of SCD. Contributors may include known or unknown genes associated with myocardial ischemia, neurohumoral signaling, cardiomyocyte Ca^{2+} handling, and cardiac electrical properties. Once genetic factors have been identified by genomic screening, the major challenge will be to determine whether a specific gene variant constitutes a specific marker of the risk for SCD independently of the underlying structural heart disease and what role this variant plays in the pathogenesis of fatal arrhythmias.

Prevention of SCD from arrhythmia

Chronic beta blockade improves survival in patients after myocardial infarction (4, 76), reducing the incidence of SCD, particularly in patients with impaired left-ventricular function. The use of some compounds, such as encainide, flecainide, and D-sotalol, that block specific cardiac ion channels, i.e., "classic" antiarrhythmic drugs, has been shown to be associated with an adverse outcome in patients with chronic myocardial infarction (3, 77). Amiodarone, an antiarrhythmic agent, which at clinically relevant concentrations blocks a variety of cardiac ion channels and transporters, including sodium channels, L-type calcium channels, several types of potassium channels, and the sodium-calcium exchanger, and which inhibits α- and β-adrenergic receptors (78), also does not improve survival (79–81). Therefore, prophylactic use of specific ion channel blockers does not reduce mortality or may even increase it. In contrast, ICDs confer survival benefit compared with drugs in high-risk populations (2). ICD therapy is now standard care for patients who have survived life-threatening arrhythmias (2).

Patients with heart failure commonly have regions of delayed myocardial activation and contraction, which leads to cardiac dyssynchrony. Cardiac resynchronization, which entails placement of a right-atrial, a right-ventricular, and a left-ventricular lead, restores atrio-biventricular synchrony resulting in improved left-ventricular function and (without concomitant ICD therapy) can improve survival in patients with heart failure (ejection fraction <35%) and cardiac dyssynchrony (82).

Conclusion

The purpose of this review was to consider some of the many events that can transform an electrically stable heart into one that is unstable, in order to answer the clinical question of why a particular patient died at a particular time on a particular day and not at another time. This is one of the most important (and vexing) questions in clinical cardiology today, as it entails exploration of the proximate precipitators of SCD (83) in order to help identify the individual at risk. Because of space constraints, we have only discussed several of the many factors that may be important. The list is long and can be approached by considering fundamental molecular mechanisms, cellular changes, whole heart electrophysiology, and clinical risk factors. Nevertheless, it is an area richly deserving intense investigation if we are ever going to reduce the horrendous toll claimed by SCD.

Acknowledgments

This work was supported by grants from the Hermann C. Krannert Fund and the NIH (HL075165 to M. Rubart).

Address correspondence to: Michael Rubart, Herman B. Wells Center for Pediatric Research, Indiana University School of Medicine, 1044 West Walnut Street, Indianapolis, Indiana 46202-5225, USA. Phone: (317) 274-2207; Fax: (317) 278-5413; E-mail: mrubartv@iupui.edu.

1. Huikuri, H.V., Castellanos, A., and Myerburg, R.J. 2001. Sudden death due to cardiac arrhythmias. *N. Engl. J. Med.* **345**:1473–1482.
2. Yadav, A.V., Das, M., and Zipes, D.P. 2005. Selection of patients for ICDs: 'where are we in 2005?'. *ACC Curr. J. Review.* **14**:33–37.
3. Echt, D.S., et al. 1991. Mortality and morbidity in patients receiving encainide, flecainide, or placebo: the Cardiac Arrhythmia Suppression Trial. *N. Engl. J. Med.* **324**:781–788.
4. Gottlieb, S.S., McCarter, R.J., and Vogel, R.A. 1998. Effect of beta-blockade on mortality among high-risk and low-risk patients after myocardial infarction. *N. Engl. J. Med.* **339**:489–497.
5. Alberte, C., and Zipes, D.P. 2003. Use of nonantiarrhythmic drugs for prevention of sudden cardiac death. *J. Cardiovasc. Electrophysiol.* **14**:S87–S95.
6. Tomaselli, G.F., and Zipes, D.P. 2004. What causes sudden death in heart failure? *Circ. Res.* **95**:754–763.
7. Arking, D.E., Chugh, S.S., Chakravarti, A., and Spooner, P.M. 2004. Genomics in sudden cardiac death. *Circ. Res.* **94**:712–723.
8. Chen, P.S., et al. 2003. A tale of two fibrillations. *Circulation.* **108**:2298–2303.
9. Huikuri, H.V., et al. 2003. Prediction of sudden cardiac death: appraisal of the studies and methods assessing the risk of sudden arrhythmic death. *Circulation.* **108**:110–115.
10. Marx, S.O., et al. 2000. PKA phosphorylation dissociates FKBP12.6 from the calcium release channel (ryanodine receptor): defective regulation in failing hearts. *Cell.* **101**:365–376.
11. Wehrens, X.H.T., et al. 2003. FKBP12.6 deficiency and defective calcium release channel (ryanodine receptor) function linked to exercise-induced sudden cardiac death. *Cell.* **113**:829–840.

12. Mohler, P.J., et al. 2003. Ankyrin-B mutation causes type 4 long-QT cardiac arrhythmia and sudden cardiac death. *Nature.* **421**:634–639.

13. Viatchenko-Karpinski, S., et al. 2004. Abnormal calcium signaling and sudden cardiac death associated with mutation of calsequestrin. *Circ. Res.* **94**:471–477.

14. Xin, H.B., et al. 2002. Oestrogen protects FKBP12.6 null mice from cardiac hypertrophy. *Nature.* **416**:334–338.

15. Fabiato, A., and Fabiato, F. 1975. Contractions induced by a calcium-triggered release of calcium from the sarcoplasmic reticulum of single skinned cardiac cells. *J. Physiol.* **249**:469–495.

16. Wehrens, X.H.T., et al. 2004. Protection from cardiac arrhythmia through ryanodine receptor-stabilizing protein calstabin2. *Science.* **304**:292–296.

17. Wier, W.G., ter Keurs, H.E., Marban, E., Gao, W.D., and Balke, C.W. 1997. Ca²⁺ 'sparks' and waves in intact ventricular muscle resolved by confocal imaging. *Circ. Res.* **81**:462–469.

18. Schlotthauer, K., and Bers, D.M. 2000. Sarcoplasmic reticulum Ca(2+) release causes myocyte depolarization. Underlying mechanism and threshold for triggered action potentials. *Circ. Res.* **87**:774–780.

19. Baader, A.P., Buchler, L., Bircher-Lehmann, L., and Kleber, A.G. 2001. Real time, confocal imaging of Ca²⁺ waves in arterially perfused rat hearts. *Cardiovasc. Res.* **53**:105–115.

20. Tanaka, H., Oyamada, M., Tsujii, E., Nakajo, T., and Takamatsu, T. 2002. Excitation-dependent intracellular Ca²⁺ waves at the border zone of the cryo-injured rat heart revealed by real-time confocal microscopy. *J. Mol. Cell. Cardiol.* **34**:1501–1512.

21. Ai, X., and Pogwizd, S.M. 2005. Connexin 43 downregulation and dephosphorylation in nonischemic heart failure is associated with enhanced colocalized protein phosphatase type 2A. *Circ. Res.* **96**:54–63.

22. Yao, J.A., et al. 2003. Remodeling of gap junctional channel function in epicardial border zone of healing canine infarcts. *Circ. Res.* **92**:437–443.

23. Pogwizd, S.M., Schlotthauer, K., Li, L., Yuan, W., and Bers, D.M. 2001. Arrhythmogenesis and contractile dysfunction in heart failure: roles of sodium-calcium exchange, inward rectifier potassium current, and residual beta-adrenergic responsiveness. *Circ. Res.* **88**:1159–1167.

24. Pogwizd, S.M. 1995. Nonreentrant mechanism underlying spontaneous ventricular arrhythmias in a model of nonischemic heart failure in rabbits. *Circulation.* **92**:1034–1048.

25. Pogwizd, S.M., McKenzie, J.P., and Cain, M.E. 1998. Mechanisms underlying spontaneous and induced ventricular arrhythmias in patients with idiopathic dilated cardiomyopathy. *Circulation.* **98**:2404–2414.

26. Qin, D., et al. 1996. Cellular and ionic basis of arrhythmias in postinfarction remodeled ventricular myocardium. *Circ. Res.* **79**:461–473.

27. Wu, J., and Zipes, D.P. 2001. Transmural reentry during global acute ischemia and reperfusion in canine ventricular muscle. *Am. J. Physiol.* **280**:H2717–H2725.

28. Takahashi, T., et al. 2004. Optical mapping of the functional circuit of ventricular tachycardia in acute myocardial infarction. *Heart Rhythm.* **4**:451–459.

29. Choi, B., Burton, F., and Salama, G. 2002. Cytosolic Ca²⁺ triggers early afterdepolarizations and torsade de pointes in rabbit hearts with type 2 long QT syndrome. *J. Physiol.* **543**:615–631.

30. Ueda, N., Zipes, D.P., and Wu, J. 2004. Prior ischemia enhances arrhythmogenicity in isolated canine ventricular wedge model of long QT 3. *Cardiovasc. Res.* **63**:69–76.

31. Omichi, C., et al. 2004. Intracellular Ca dynamics in ventricular fibrillation. *Am. J. Physiol. Heart Circ. Physiol.* **286**:H1836–H1844.

32. Nearing, B.D., Huang, A.H., and Verrier, R.L. 1991. Dynamic tracking of cardiac vulnerability by complex demodulation of the T wave. *Science.* **252**:437–440.

33. Pruvot, E.J., Katra, R.P., Rosenbaum, D.S., and Laurita, K.R. 2004. Role of calcium cycling versus restitution in the mechanism of repolarization alternans. *Circ. Res.* **94**:1083–1090.

34. Goldhaber, J.I., et al. 2005. Action potential duration restitution and alternans in rabbit ventricular myocytes: the key role of intracellular calcium cycling. *Circ. Res.* **96**:459–466.

35. Chudin, E., Goldhaber, J., Garfinkel, A., Weiss, J., and Kogan, B. 1999. Intracellular Ca(2+) dynamics and the stability of ventricular tachycardia. *Biophys. J.* **77**:2930–2941.

36. Wingo, T.L., et al. 2004. An EF-hand in the sodium channel couples intracellular calcium to cardiac excitability. *Nat. Struct. Mol. Biol.* **11**:219–225.

37. Wu, Y., Roden, D.M., and Anderson, M.E. 1999. Calmodulin kinase inhibition prevents development of the arrhythmogenic transient inward current. *Circ. Res.* **84**:906–912.

38. Wu, Y., et al. 2002. Calmodulin kinase II and arrhythmias in a mouse model of cardiac hypertrophy. *Circulation.* **106**:1288–1293.

39. Xu, Y., et al. 2005. The effect of intracellular Ca²⁺ on cardiac K⁺ channel expression and activity: novel insights from genetically altered mice. *J. Physiol.* **562**:745–758.

40. Zipes, D.P., and Wellens, H.J.J. 1998. Sudden cardiac death. *Circulation.* **98**:2334–2351.

41. Shivkumar, K., et al.1997. Mechanism of hypoxic K loss in rabbit ventricle. *J. Clin. Invest.* **100**:1782–1788.

42. John, S.A., Kondo, R., Wang, S.Y., Goldhaber, J.I., and Weiss, J.N. 1999. Connexin-43 hemichannels opened by metabolic inhibition. *J. Biol. Chem.* **274**:236–240.

43. Yan, G.X., Park, T.H., and Corr, P.B. 1995. Activation of thrombin receptor increases intracellular Na⁺ during myocardial ischemia. *Am. J. Physiol.* **268**:H1740–H1748.

44. Kaneko, T., Tanaka, H., Oyamada, M., Kawata, S., and Takamatsu, T. 2000. Three distinct types of Ca⁽²⁺⁾ waves in Langendorff-perfused rat heart revealed by real-time confocal microscopy. *Circ. Res.* **86**:1093–1099.

45. Verkerk, A.O., Veldkamp, M.W., Coronel, R., Wilders, R., and van Ginneken, A.C. 2001. Effects of cell-to-cell uncoupling and catecholamines on Purkinje and ventricular action potentials: implications for phase-1b arrhythmias. *Cardiovasc. Res.* **51**:30–40.

46. Woodcock, E.A., Arthur, J.F., Harrison, S.N., Gao, X., and Du, X. 2001. Reperfusion-induced Ins(1,4,5)P3 generation and arrhythmogenesis require activation of the Na+/Ca2+ exchanger. *J. Mol. Cell. Cardiol.* **33**:1861–1869.

47. Imahashi, K., et al. 1999. Intracellular sodium accumulation during ischemia as the substrate for reperfusion injury. *Circ. Res.* **84**:1401–1406.

48. Ueda, N., Zipes, D.P., and Wu, J. 2004. Functional and transmural modulation of M cell behavior in canine ventricular wall. *Am. J. Physiol. Heart Circ. Physiol.* **287**:H2569–H2575.

49. Lakireddy, V., et al. 2005. Contrasting effects of ischemia on the kinetics of membrane voltage and intracellular calcium transient underlie electrical alternans. *Am. J. Physiol. Heart Circ. Physiol.* **288**:H400–H407.

50. Cao, J.M., et al. 2000. Nerve sprouting and sudden cardiac death. *Circ. Res.* **86**:816–821.

51. Liu, Y., et al. 2003. Sympathetic nerve sprouting, electrical remodeling, and increased vulnerability to ventricular fibrillation in hypercholesterolemic rabbits. *Circ. Res.* **92**:1145–1152.

52. Cao, J.M., et al. 2000. Relationship between regional cardiac hyperinnervation and ventricular arrhythmia. *Circulation.* **101**:1960–1969.

53. Heath, B.M., et al. 1998. Overexpression of nerve growth factor in the heart alters ion channel activity and β-adrenergic signaling in an adult transgenic mouse. *J. Physiol.* **512**:779–791.

54. Barber, M.J., Mueller, T.M., Henry, D.P., Felten, S.Y., and Zipes, D.P. 1983. Transmural myocardial infarction in the dog produces sympathectomy in noninfarcted myocardium. *Circulation.* **67**:787–796.

55. Warner, M.R., Wisler, P.L., Hodges, T.D., Watanabe, A.M., and Zipes, D.P. 1993. Mechanisms of denervation supersensitivity in regionally denervated canine hearts. *Am. J. Physiol.* **264**:H815–H820.

56. Komaru, T., et al. 1990. Neuropeptide Y modulates vasoconstriction in coronary microvessels in the beating canine heart. *Circ. Res.* **67**:1142–1151.

57. Protas, L., et al. 2003. Neuropeptide Y is an essential in vivo developmental regulator of cardiac ICa,L. *Circ. Res.* **93**:972–979.

58. Pfeffer, M.A., et al. 2003. Valsartan, captopril, or both in myocardial infarction complicated by heart failure, left ventricular dysfunction, or both. *N. Engl. J. Med.* **349**:1893–1906.

59. Pfeffer, M.A., et al. 1992. Effect of captopril on mortality and morbidity in patients with left ventricular dysfunction after myocardial infarction. Results of the survival and ventricular enlargement trial. The SAVE Investigators. *N. Engl. J. Med.* **327**:669–677.

60. The SOLVD Investigators. 1992. Effect of enalapril on mortality and the development of heart failure in asymptomatic patients with reduced left ventricular ejection fractions. *N. Engl. J. Med.* **327**:685–691.

61. Cohn, J.N., Tognoni, G., and Valsartan Heart Failure Trial Investigators. 2001. A randomized trial of the angiotensin-receptor blocker valsartan in chronic heart failure. *N. Engl. J. Med.* **345**:1667–1775.

62. Caron, K.M., et al. 2004. Cardiac hypertrophy and sudden death in mice with a genetically clamped renin transgene. *Proc. Nat. Acad. Sci. U. S. A.* **101**:3106–3111.

63. Donoghue, M., et al. 2003. Heart block, ventricular tachycardia, and sudden death in ACE2 transgenic mice with downregulated connexins. *J. Mol. Cell. Cardiol.* **35**:1043–1053.

64. Shapiro, M.S., Wollmuth, L.P., and Hille, B. 1994. Angiotensin II inhibits calcium and M current in rat sympathetic neurons via G proteins. *Neuron.* **12**:1319–1329.

65. Wang, H., Huang, B.S., Ganten, D., and Leenen, F.H.H. 2004. Prevention of sympathetic and cardiac dysfunction after myocardial infarction in transgenic rats deficient in brain angiotensinogen. *Circ. Res.* **94**:843–849.

66. Liu, J., and Zucker, I.H. 1999. Regulation of sympathetic nerve activity in heart failure. *Circ. Res.* **84**:417–423.

67. Toro, L., Amador, M., and Stefani, E. 1990. ANG II inhibits calcium-activated potassium channels from coronary smooth muscle in lipid bilayers. *Am. J. Physiol.* **258**:H912–H915.

68. Gelband, C.H., and Hume, J.R. 1995. [Ca²⁺]ᵢ inhibition of K⁺ channels in canine renal artery. Novel mechanism for agonist-induced membrane depolarization. *Circ. Res.* **77**:121–130.

69. Browe, D.M., and Baumgarten, C.M. 2004. Angiotensin II (AT1) receptors and NADPH oxidase regulate Cl⁻ current elicited by β1 integrin stretch in rabbit ventricular myocytes. *J. Gen. Physiol.* **14**:273–287.

70. Benitah, J.P., Perrier, E., Gomez, A.M., and Vassort, G. 2001. Effects of aldosterone on transient outward K⁺ current density in rat ventricular myocytes. *J. Physiol.* **537**:151–160.

71. Cittadini, A., et al. 2003. Aldosterone receptor blockade improves left ventricular remodeling and increases ventricular fibrillation threshold in experimental heart failure. *Cardiovasc. Res.* **58**:555–564.

72. Pitt, B., et al. 2003. The effect of spironolactone on morbidity and mortality in patients with severe heart failure. *N. Engl. J. Med.* **341**:709–717.

73. Pitt, B., et al. 2003. Eplerenone, a selective aldosterone blocker, in patients with left ventricular dysfunction after myocardial infarction. *N. Engl. J. Med.* **348**:1309–1321.

74. Jouven, X., Desnos, M., Guerot, C., and Ducimetiere, P. 1999. Predicting sudden death in the population: the Paris Prospective Study I. *Circulation.* **99**:1978–1983.

75. Friedlander, Y., et al. 1998. Family history as a risk factor for primary cardiac arrest. *Circulation.* **97**:155–160.

76. Yusuf, S., Peto, R., Lewis, J., Collins, R., and Sleight, P. 1985. Beta blockade during and after myocardial infarction: an overview of the randomized trials. *Prog. Cardiovasc. Dis.* **27**:335–371.

77. Waldo, A.L., et al. 1996. Effect of d-sotalol on mortality in patients with left ventricular dysfunction after recent and remote myocardial infarction. The SWORD Investigators. Survival With Oral d-Sotalol. *Lancet.* **348**:7–12.

78. Watanabe, Y., and Kimura, J. 2000. Inhibitory effect of amiodarone on Na^+/Ca^{2+} exchange in guinea-pig cardiomyocytes. *Br. J. Pharmacol.* **131**:80–84.

79. Julian, D.G., et al. 1997. Randomised trial of effect of amiodarone on mortality in patients with left-ventricular dysfunction after recent myocardial infarction: EMIAT. European Myocardial Infarct Amiodarone Trial Investigators. *Lancet.* **349**:667–674.

80. Cairns, J.A., Connolly, S.J., Roberts, R., and Gent, M. 1997. Randomised trial of outcome after myocardial infarction in patients with frequent or repetitive ventricular premature depolarisations: CAMIAT. Canadian Amiodarone Myocardial Infarction Arrhythmia Trial Investigators. *Lancet.* **349**:675–682.

81. Bardy, G.H., et al. 2005. Amiodarone or an implantable cardioverter-defibrillator for congestive heart failure. *N. Engl. J. Med.* **352**:225–237.

82. Cleland, J.G., et al. 2005. The effect of cardiac resynchronization on morbidity and mortality in heart failure. *N. Engl. J. Med.* **352**:1539–1549.

83. Zipes, D.P. 2003. Less heart is more. *Circulation.* **107**:2531–2532.

84. Yano, M., Ikeda, Y., and Matsuzaki, M. 2005. Altered intracellular Ca^{2+} handling in heart failure. *J. Clin. Invest.* **115**:556–564. doi:10.1172/JCI200524159.

85. Subramanian, S., Viatchenko-Karpinski, S., Lukyanenko, V., Gyorke, S., and Wiesner, T.F. 2001. Underlying mechanisms of symmetric calcium wave propagation in rat ventricular myocytes. *Biophys. J.* **80**:1–11.

Long QT syndrome: from channels to cardiac arrhythmias

Arthur J. Moss[1] and Robert S. Kass[2]

[1]Heart Research Follow-up Program, Department of Medicine, University of Rochester School of Medicine and Dentistry, Rochester, New York, USA.
[2]Department of Pharmacology, Columbia University Medical Center, New York, New York, USA.

Long QT syndrome, a rare genetic disorder associated with life-threatening arrhythmias, has provided a wealth of information about fundamental mechanisms underlying human cardiac electrophysiology that has come about because of truly collaborative interactions between clinical and basic scientists. Our understanding of the mechanisms that control the critical plateau and repolarization phases of the human ventricular action potential has been raised to new levels through these studies, which have clarified the manner in which both potassium and sodium channels regulate this critical period of electrical activity.

Background

The common form of long QT syndrome (LQTS), Romano-Ward syndrome (RWS), is a heterogeneous, autosomal dominant, genetic disease caused by mutations of ion channel genes involving the cell membranes of the cardiac myocytes. These channelopathies are associated with delayed ventricular repolarization and are clinically manifest by passing-out spells (syncope; see Glossary) and sudden death from ventricular arrhythmias, notably torsade de pointes (1). Clinically, LQTS is identified by abnormal QT interval prolongation on the ECG. The QT interval prolongation may arise from either a decrease in repolarizing cardiac membrane currents or an increase in depolarizing cardiac currents late in the cardiac cycle. Most commonly, QT interval prolongation is produced by delayed repolarization due to reductions in either the rapidly or the slowly activating delayed repolarizing cardiac potassium (K^+) current, I_{Kr} or I_{Ks} (2). Less commonly, QT interval prolongation results from prolonged depolarization due to a small persistent inward "leak" in cardiac sodium (Na^+) current, I_{Na} (3) (Figure 1A).

Patients with LQTS are usually identified by QT interval prolongation on the ECG during clinical evaluation of unexplained syncope, as part of a family study when 1 family member has been identified with the syndrome, or in the investigation of patients with congenital neural deafness. The first family with LQTS was reported in 1957 and was thought to be an autosomal recessive disorder (4), but in 1997 it was shown to result from a dominant, homozygous mutation involving the KvLQT1 gene (5), now called the KCNQ1 gene. The more common autosomal dominant RWS was described in 1963–1964, and over 300 different mutations involving 7 different genes (LQT1–LQT7) have now been reported (6). Most of the clinical information currently available regarding LQTS relates to RWS. There is considerable variability in the clinical presentation of LQTS, due to the different genotypes, different mutations, variable penetrance of the mutations, and possibly

genetic and environmental modifying factors. Clinical criteria have been developed to determine the probability of having LQTS, and genotype screening of individuals suspected of having LQTS and of members of known LQTS families has progressively increased the number of subjects with genetically confirmed LQTS. The genes associated with LQTS have been numerically ordered by the chronology of their discovery (LQT1 through LQT7), with 95% of the known mutations located in the first 3 of the 7 identified LQTS genes (Table 1). All of the LQT genes except LQT4 (ANKB) code for ion channels. LQT4 codes for a protein called ankyrin-B that anchors ion channels to specific domains in the plasma membrane. LQT7 is an ion channel gene, and mutations involving this gene result in a multisystem disease (Andersen-Tawil syndrome) that includes modest QT interval prolongation secondary to reduction in 1 of the potassium repolarization currents (Kir2.1). Prophylactic and preventive therapy for LQTS is directed toward reduction in the incidence of syncope and sudden death and has involved left cervicothoracic sympathetic ganglionectomy, β-blockers, pacemakers, implanted defibrillators, and gene/mutation–specific pharmacologic therapy (7).

Gene-specific triggers of cardiac arrhythmias

The discovery that distinct LQTS variants were associated with genes coding for different ion channel subunits has had a major impact on the diagnosis and analysis of LQTS patients. It is clear that there are distinct risk factors associated with the different LQTS genotypes, and this must be taken into account during patient evaluation and diagnosis. The greatest difference in risk factors becomes apparent when LQT3 syndrome patients (*SCN5A* mutations) are compared with patients with LQT1 syndrome (*KCNQ1* mutations) or LQT2 syndrome (*KCNH2* mutations). The potential for understanding a mechanistic basis for arrhythmia risk was realized soon after the first genetic information relating mutations in genes coding for distinct ion channels to LQTS became available (8, 9) and is still the focus of extensive investigation. This genotype-phenotype association was confirmed in an extensive study in which the risk of cardiac events was studied in LQTS-genotyped patients. This investigation, which focused on patients with KCNQ1 (LQT1), KCNH2 (LQT2), and SCN5A (LQT3) mutations, reported clear differences in arrhythmic risk that was correlated with gene-specific mutations. In the

---

Nonstandard abbreviations used: β-AR, β-adrenergic receptor; ATS, Andersen-Tawil syndrome; I_{Kr}, rapidly activating delayed repolarizing cardiac potassium current; I_{Ks}, slowly activating delayed repolarizing cardiac potassium current; I_{Na}, cardiac sodium current; LQTS, long QT syndrome; QTc, heart rate–corrected QT (interval); RWS, Romano-Ward syndrome; TS, Timothy syndrome.

Conflict of interest: The authors have declared that no conflict of interest exists.

Citation for this article: *J. Clin. Invest.* 115:2018–2024 (2005).
doi:10.1172/JCI25537.

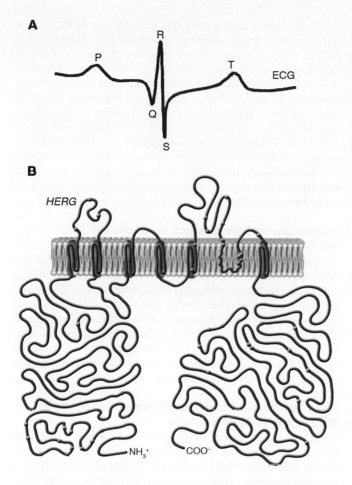

A

P   R   T   ECG
Q
S

B

HERG

NH₃⁺   COO⁻

Figure 1
(**A**) An illustrative example of a single cardiac cycle detected as spatial and temporal electrical gradients on the ECG. The P wave is generated by the spread of excitation through the atria. The QRS complex represents ventricular activation and is followed by the T wave, which reflects ventricular repolarization gradients. (**B**) Schematic representation of the *KCNH2* (HERG) potassium channel α subunit, involving the N-terminal part (NH₃⁺), 6 membrane-spanning segments with the pore region located from segment S5 to segment S6, and the C-terminal portion (COO⁻). Mutation locations are indicated by blue dots. Fourteen different mutations were located in 13 locations within the pore region. Reproduced with permission from *Circulation* (28).

case of SCN5A mutation carriers (LQT3), risk of cardiac events was greatest during rest (bradycardia), when sympathetic nerve activity is expected to be low. In contrast, cardiac events in LQT2 syndrome patients were associated with arousal and/or conditions in which patients were startled, whereas LQT1 syndrome patients were found to be at greatest risk of experiencing cardiac events during exercise or conditions associated with elevated sympathetic nerve activity (10).

Subsequently, additional evidence has linked dysfunctional regulation of the QT interval during exercise to LQT1 mutations (11–13). The contrast between LQT1 and LQT3 patients in the role of adrenergic input and/or heart rate in arrhythmia risk is clear and has raised the possibility of distinct therapeutic strategies in the management of patients with these LQTS variants. In fact, β-blocker therapy has been shown to be most effective in preventing recurrence of cardiac events and lowering the death rate in LQT1 and LQT2 syndrome patients but is much less effective in the treatment of LQT3 syndrome patients (14, 15). β-Blockers have minimal effects on the heart rate–corrected QT (QTc) interval but are associated with a significant reduction in cardiac events in LQTS patients, probably because these drugs modulate the stimulation of β-adrenergic receptors (β-ARs) and hence the regulation of downstream signaling targets during periods of elevated sympathetic nerve activity. Clinical data for genotyped patients provide strong support for the hypothesis that the effectiveness of β-blocking drugs depends critically on the genetic basis of the disease; for example, recent data provide evidence that there is still a high rate

of cardiac events in LQT2 and LQT3 patients treated with β-blocking drugs (15). Consequently, even β-blockers do not provide absolute protection against fatal cardiac arrhythmias.

The sodium channel and mutation-specific pharmacologic therapy

The SCN5A gene encodes the α subunit of the major cardiac sodium channel (16), and various LQT3 mutations can result in different functional alterations in this channel with similar degrees of QT interval prolongation and cardiac arrhythmias (17, 18). For example, the 9-bp deletion with loss of 3 amino acids (ΔKPQ) in the linker between the third and fourth domains of the α unit of the sodium channel and 3 missense mutations in this gene (N1325S, R1623Q, and R1644H) all promote sustained and inappropriate sodium entry into the myocardial cell during the plateau phase of the action potential, resulting in prolonged ventricular repolarization and the LQTS phenotype. The functional consequences of the D1790G missense mutation are quite different in that this mutation does not promote sustained inward sodium current but, rather, causes a negative shift in steady-state inactivation with a similar LQTS phenotype (19). It should be noted that other mutations in the SCN5A gene can result in Brugada syndrome and conduction system disorders without QT interval prolongation. At least 1 mutation (1795insD) has been shown to have a dual effect, with inappropriate sodium entry at slow heart rates (LQTS ECG pattern) and reduced sodium entry at fast heart rates (Brugada syndrome ECG pattern) (20).

Mutation-specific pharmacologic therapy has been reported in 2 specific SCN5A mutations associated with LQTS. In 1995, Schwartz et al. reported that a single oral dose of the sodium channel blocker mexiletine administered to 7 LQT3 patients with the ΔKPQ deletion produced significant shortening of the QTc interval within 4 hours (21). Similar shortening of the QTc interval in LQT3 patients with the ΔKPQ deletion has been reported with lidocaine and tocainide (22). Preliminary clinical experience with flecainide revealed normalization of the QTc interval with low doses of this drug in patients with the ΔKPQ deletion (23). In 2000, Benhorin et al. reported the effectiveness of open-label oral flecainide in shortening the QTc interval in 8 asymptomatic subjects with the D1790G mutation (24).

The SCN5A-D1790G mutation changes the sodium channel interaction with flecainide. This mutation confers a high sensitivity to use-dependent blockade by flecainide, due in large part to the marked slowing of the repriming of the mutant channels in the presence of the drug (19). Flecainide's tonic block is not affected by the D1790G mutation. These flecainide effects are different from those occurring with the ΔKPQ mutant channels (25).

Table 1

Nomenclature, gene names, and proteins associated with LQTS

Disease	Gene (historical name)	Protein
LQT1	*KCNQ1* (KVLQT1)	$I_{Ks}K^+$ channel α subunit
LQT2	*KCNH2* (HERG)	$I_{Kr}K^+$ channel α subunit
LQT3	SCN5A	$I_{Na}Na^+$ channel α subunit
LQT4	ANKB	Ankyrin-B
LQT5	*KCNE1* (minK)	$I_{Ks}K^+$ channel β subunit
LQT6	*KCNE2* (MiRP1)	$I_{Kr}K^+$ channel β subunit
LQT7	KCNJ2	$I_{Kr2.1}K^+$ channel α subunit
LQT8	*CACNA1*	Cav1.2 Calcium channel α subunit

These flecainide findings in patients with the ΔKPQ and D1790G mutations provide encouraging evidence in support of mutation-specific pharmacologic therapy for 2 specific forms of the LQT3 disorder. Larger clinical trials with flecainide in patients with these 2 mutations are needed before this therapy can be recommended as safe and effective for patients with these genetic disorders.

Clinical relevance of mutations in regions of KCNH2

The KCNH2 gene encodes the ion channel involved in the rapid component of the delayed rectifier repolarization current (I_{Kr}), and mutations in this gene are responsible for the LQT2 form of LQTS (26). Mutations in KCNH2 are associated with diminution in the repolarizing I_{Kr} current with resultant prolongation of ventricular repolarization and lengthening of the QT interval. During the 1990s, it was appreciated that several drugs, such as terfenadine and cisapride, caused QT interval prolongation by reducing I_{Kr} current through the pore region of the KCNH2 channel (27). These findings raised the question of whether mutations in the pore region of the KCNH2 channel would be associated with a more virulent form of LQT2 than mutations in non-pore regions.

In a report from the International LQTS Registry, 44 different KCNH2 mutations were identified in 201 subjects, with 14 mutations in 13 locations in the pore region (amino acid residues 550–650) (Figure 1B) (28). Thirty-five subjects had mutations in the pore region and 166 in non-pore regions. When birth was used as the time origin with follow-up through age 40, subjects with pore mutations had more severe clinical manifestations of the genetic disorder and experienced a higher frequency of arrhythmia-related cardiac events at an earlier age than did subjects with non-pore mutations. The cumulative probability of a first cardiac event before β-blockers were initiated in subjects with pore mutations and non-pore mutations in the KCNH2 channel is shown in Figure 2, with a hazard ratio in the range of 11 (P < 0.0001) at an adjusted QTc interval of 0.50 seconds. This study involved a limited number of different KCNH2 mutations and only a small number of subjects with each mutation. Missense mutations made up 94% of the pore mutations, and thus it was not possible to evaluate risk by the mutation type within the pore region.

These findings indicate that mutations in different regions of the KCNH2 potassium channel can be associated with different levels of risk for cardiac arrhythmias in LQT2. An important question is whether similar region-related risk phenomena exist in the other LQTS channels. Two studies evaluated the clinical risk of mutations located in different regions of the KCNQ1 (LQT1) gene and reported contradictory findings. One study found no significant differences in clinical presentation, ECG parameters,

and cardiac events among 294 LQT1 patients with KCNQ1 mutations located in the pre-pore region including the N-terminus (1 to 278), the pore region (279 to 354), and the post-pore region including the C-terminus (>354) (29). In contrast, another study of 66 LQT1 patients found that mutations in the transmembrane portion of KCNQ1 were associated with a higher risk of LQTS-related cardiac events and had greater sensitivity to sympathetic stimulation than mutations located in the C-terminal region (30). These different findings in the 2 LQT1 studies may reflect, in part, population-related genetic heterogeneity, since 1 population was almost entirely white and the subjects in the other study were Japanese. Much larger homogeneous populations need to be studied to resolve this issue.

Insights gained from interactions of clinical and basic scientists

Though a rare congenital disorder, LQTS has provided a wealth of information about fundamental mechanisms underlying human cardiac electrophysiology that has come about because of truly collaborative interactions between clinical and basic scientists. Our understanding of the mechanisms that control the critical plateau and repolarization phases of the human ventricular action potential (Figure 3) has been raised to new levels through these studies with clarification about the manner in which both potassium and sodium channels regulate this critical period of electrical activity.

Potassium channel currents and the action potential plateau: the delayed rectifiers

It had been known since 1969 that potassium currents with unique kinetic and voltage-dependent properties were important to the cardiac action potential plateau (31, 32). Because of the unique voltage dependence, these currents were referred to as delayed rectifiers. In a pivotal study, Sanguinetti and Jurkiewicz used a pharmacologic analysis to demonstrate 2 distinct components of the delayed rectifier potassium current in the heart: I_{Kr} and I_{Ks} (33). The I_{Ks} component had previously been shown to be under control

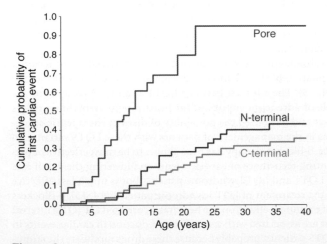

Figure 2

Kaplan-Meier cumulative probability of first cardiac events from birth through age 40 years for subjects with mutations in pore (*n* = 34), N-terminal (*n* = 54), and C-terminal (*n* = 91) regions of the *KCNH2* (HERG) channel. The curves are significantly different (P < 0.0001, log-rank), mainly because of the high first-event rate in subjects with pore mutations. Reproduced with permission from *Circulation* (28).

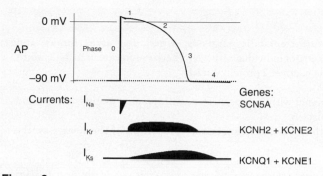

Figure 3
Three major cardiac ion channel currents (I_{Na}, I_{Kr}, and I_{Ks}) and respective genes responsible for generation of portions of the ventricular action potential (AP).

of the sympathetic nervous system, providing an increase in repolarization currents in the face of β-AR agonists in cellular models (34), but the molecular identity and the relevance to human electrophysiology were not only unclear, but controversial. The clear clinical importance and the genetic basis of these potassium currents were revealed through LQTS investigations.

The first report linking potassium channel dysfunction to LQTS revealed the molecular identity of 1 of the delayed rectifier channels and confirmed the pharmacologic evidence for independent channels underlying these currents (35). This ground breaking study, a collaborative effort between clinical and basic scientists, revealed that KCNH2 encodes the α (pore-forming) subunit of the I_{Kr} channel and that the rectifying properties of this channel, identified previously by pharmacologic dissection, are indigenous to the channel protein. This work not only provided the first clear evidence for a role of this channel in the congenital LQTS but also laid the base line for future studies that would show that it is the KCNH2 channel that underlies almost all cases of acquired LQTS (36).

The next breakthrough came in 1996 when a collaborative clinical/basic effort discovered that LQTS variant 1 (LQT1) was caused by mutations in a gene (KCNQ1) coding for an unusual potassium channel subunit that could be studied in heterologous expression systems (37). Basic studies almost immediately revealed the functional role of the KCNQ1 gene product: the α (pore-forming) subunit of the I_{Ks} channel (38, 39). Furthermore, these studies indicated that a previously reported but as-yet poorly understood gene (KCNE1) formed a key regulatory subunit of this important channel. Mutations in KCNE1 have subsequently been linked to LQT5 (40). Now the molecular identity of the 2 cardiac delayed rectifiers had been established.

As is summarized above, clinical studies had provided convincing evidence linking sympathetic nerve activity and arrhythmia susceptibility in LQTS patients, particularly in patients harboring LQT1 mutations. These data and previous basic reports of the robust sensitivity of the slow delayed rectifier component, I_{Ks}, to β-AR agonists (34) motivated investigation of the molecular links between KCNE1 channels and β-AR stimulation, which revealed, for the first time, that the KCNQ1/KCNE1 channel is part of a macromolecular signaling complex in the human heart (41). The channel complexes with an adaptor protein called Yotiao that in turn directly binds key enzymes in the β-AR signaling cascade (protein kinase A and protein phosphatase 1) and recruits them to form a local signaling environment to control the phosphorylation

state of the channel. Mutations in either KCNQ1 (41) or KCNE1 (42) can disrupt this regulation and create heterogeneity in the cellular response to β-AR stimulation, a novel mechanism that may contribute to the triggering of some arrhythmias in LQT1 and LQT5 (43). This fundamental work is a direct consequence of collaborative efforts between basic and clinical investigators.

Trafficking and its pharmacologic rescue as a mechanism in LQT2

LQT2 mutations reduce repolarizing current through HERG channels, contributing to delayed repolarization of the ventricular action potential and the resulting QT interval. Though all LQT2 mutations lead to the same clinical phenotype, prolonged QT intervals, multiple mechanisms at the level of the channel protein, including altered gating kinetics and dominant-negative effects via subunit assembly, may underlie this condition.

January and colleagues first reported that mutation-altered protein trafficking could also cause LQT2 (44). Trafficking defects reduce the number of functional channels expressed in the sarcolemmal membrane, contributing to a reduction in repolarizing current, but cause no alteration in the basic biophysical properties of those channels that are functional. Importantly, and uniquely, January and colleagues also made the important discovery that mutation-induced trafficking defects could be restored by temperature and incubation with pharmacologic agents that block HERG channels (45). These observations have subsequently been confirmed by multiple groups and for several distinct LQT2 mutations (46–48), and, most recently, experimental work has suggested that these LQT2 mutations may cause subtle protein misfolding and retention in the endoplasmic reticulum and that drug binding might act as a chaperone to restore maturation efficiency and hence a higher level of sarcolemmal channel protein (49).

Insights into cardiac sodium channels

The report that mutations in SCN5A, the gene coding for the α subunit of the major cardiac sodium channel, were associated with LQTS (17) was surprising, because this channel is associated most frequently with impulse conduction and hence the QRS but not the QT waveforms of the ECG. Sodium channels are voltage-gated channels that rapidly enter a nonconducting inactivated state during sustained depolarization such as the cardiac action potential plateau. Importantly, the first SCN5A mutation, the ΔKPQ mutation, physically disrupted a cytoplasmic peptide linker in the channel protein that, in basic biochemical and biophysical studies, had been shown to be a critical determinant of sodium channel inactivation: the inactivation gate (50, 51). This peptide links 2 domains (III and IV) of the channel and physically moves to occlude the channel pore upon depolarization. Once again, the combination of basic and clinical investigation has led to a clear understanding of the molecular basis of this key physiological parameter in the human heart. Further, the demonstration that small changes in sodium channel inactivation such at those changes that occur in LQT3 mutations can have life-threatening consequences confirms predictions made more than 50 years ago when it was shown that the cardiac action potential plateau was an exquisitely sensitive period of electrical activity that could adapt, with little energy expenditure, to small changes in ionic currents (52).

Subsequent investigations of LQT3 mutations have revealed that not only is the domain III/IV intracellular linker key to inactivation and maintenance of the action potential plateau (and hence QT interval), but also the channel carboxyterminal (C-T) domain is essential in this process. In fact, evidence generated by basic studies of LQT3 mutations has now shown that the C-T domain forms a molecular complex with the inactivation gate that is necessary to stabilize the nonconducting state of the channel during prolonged depolarization (53), and, further, that this interaction may be modulated by cell calcium and calmodulin (54). Thus there is the intriguing possibility that dysfunction in sodium channel inactivation that can lead to QT interval prolongation may occur in other forms of human cardiovascular pathology when regulation of cellular calcium concentration may be disrupted. Thus, once again, collaboration between clinical and basic scientists investigating a rare inherited disorder has resulted in fundamentally novel insight into cardiovascular physiology, which may have wide-ranging impact on our understanding of human cardiovascular disease.

LQTS variants and multisystem disorders

Andersen-Tawil syndrome (ATS) is a skeletal muscle and cardiac muscle disease. Clinical phenotypes of patients with ATS include periodic paralysis, developmental dysmorphisms, and prolonged QT intervals. Mutations in KCNJ2 were found to underlie this multisystem disorder (55–57). KCNJ2 encodes the inward rectifier K^+ channel α subunit Kir2.1, which is expressed in skeletal and cardiac muscle. Interestingly, like mutations in HERG described above, at least a subset of ATS KCNJ2 mutations have been shown to cause trafficking defects, and hence loss of functional channel activity, of inwardly rectifying potassium channels (58). These genetic and clinical data have confirmed the key role of this inwardly rectifying potassium channel not only in skeletal muscle physiology, but also in control of the duration of the ventricular action potential, a role not often associated with this ionic pathway.

Most recently, a second multiorgan syndrome, Timothy syndrome (TS), has been reported. Like ATS, clinical phenotypes of patients with TS include a wide variety of organ dysfunction, including lethal arrhythmias, webbing of fingers and toes, autism, immune deficiency, intermittent hypoglycemia, cognitive abnormalities, and markedly prolonged QT intervals on the ECG (59).

Importantly, this multisystem dysfunction was found to be caused by mutations in a splice variant of the gene (CACNA1I) coding for the α subunit (Cav1.2) of the L-type calcium channel. The Cav1.2 channel is a principal pathway for calcium entry in cells in which it is expressed, and, in the heart, its activation and inactivation are critical regulators of action potential waveforms and activation of contraction. Hence it is not surprising that mutations of Cav1.2 cause marked clinical dysfunction. Like LQT3 mutations, the TS mutations disrupt inactivation of Cav1.2 channels, but unlike voltage-gated sodium channels, inactivation of Cav1.2 channels is regulated by both voltage and calcium entry (60), and the TS mutations almost completely abolish the voltage-dependent inactivation component. Thus, as has been the case with all other LQTS-related mutations, the combination of clinical and basic investigations of TS patients and mutations will provide the clearest insight to date into the functional roles of these biophysical channel properties in the human heart.

LQT4 is caused by mutations in an adaptor protein

In 1995, Schott et al. reported a study of a 65-member family in which LQTS was also associated with marked sinus bradycardia (61). Genetic linkage was observed for a region of chromosome 4 located in the interval 4q25–q27, a region that did not contain candidate ion channel genes, and the identification of the genetic basis of variant LQT4 remained elusive. A clue to the basis of this variant emerged from a study by Chauhan et al. in which sodium channel activity in mice lacking ankyrin-B — not an ion channel protein, but an adaptor protein that associated with cytoskeletal interactions — was altered (62). The major form of ankyrin-B in cardiac cells is 220 kDa. Ankyrins typically contain 3 functional domains that consist of the membrane-binding domain, the spectrin-binding domain, and the regulatory domain. Ankyrins bind to multiple proteins that can contribute (directly or indirectly) to cardiac electrical activity. These include the anion exchanger (Cl^-/HCO_3^- exchanger), the Na^+,K^+–ATPase, voltage-sensitive sodium channels, the Na^+/Ca^{2+} exchanger (NCX, or I_{Na-Ca}), and calcium-release channels including those mediated by ryanodine or inositol trisphosphate (IP_3) receptors (63). The identify of the LQT4 gene as the gene coding for ankyrin-B was confirmed by Mohler et al., who reported a loss of function in the ankyrin-B mutation (E1425G) that causes dominantly inherited type 4 long QT cardiac arrhythmia in humans (64). Physiological insight into the mechanistic basis for LQT4 arrhythmias came from genetically altered mice heterozygous for a null mutation in ankyrin-B. Because of the multiple proteins that associate with ankyrin-B, the results indicated that these arrhythmias may develop because of a combined disruption in the cellular organization of the sodium pump, the sodium/calcium exchanger, and inositol-1,4,5-trisphosphate receptors. Interestingly, the ankyrin-B mutation also leads to altered Ca^{2+} signaling in adult cardiomyocytes, which plays a major role in the triggering of arrhythmias in this mouse model (64). Recently, further ankyrin-B mutations have been reported to be causally linked to LQTS (65). The mechanistic basis of LQT4-associated arrhythmias is complex, but, as is the case for TS, it will provide important insight into the pivotal role played by the regulation of intracellular calcium in the genesis and prevention of cardiac arrhythmias in humans.

Conclusions

We have made considerable progress in understanding the importance of ion channel structure to human physiology since the first ion channel was cloned in 1982. We now have a better understanding of the molecular genetics, ion channel structures, and cellular electrophysiology that contribute to the genesis of cardiac arrhythmias. Much of this improved insight has come directly from investigations of LQTS and other inherited arrhythmias and is being translated into more effective and more rational therapy for patients with electrical disorders of the cardiac rhythm. Much remains to be accomplished, and this will be done through continued collaboration of basic and clinical scientists based on the foundations laid by studies of LQTS.

Address correspondence to: Arthur J. Moss, Box 653, University of Rochester Medical Center, Rochester, New York 14642, USA. Phone: (585) 275-5391; Fax: (585) 273-5283; E-mail: heartajm@heart.rochester.edu.

1. Moss, A.J., et al. 1991. The long QT syndrome: prospecitve longitudinal study of 328 families. *Circulation.* **84**:1136–1144.
2. Sanguinetti, M.C., and Zou, A. 1997. Molecular physiology of cardiac delayed rectifier K+ channels [review]. *Heart Vessels.* **12**(Suppl.):170–172.
3. Bennett, P.B., Yazawa, K., Makita, N., and George, A.L. 1995. Molecular mechanism for an inherited cardiac arrhythmia. *Nature.* **376**:683–685.
4. Jervell, A., and Lange-Nielsen, F. 1957. Congenital deaf mutism, functional heart disease with prolongation of the Q-T interval and sudden death. *Am. Heart J.* **54**:59–68.
5. Splawski, I., Timothy, K.W., Vincent, G.M., Atkinson, D.L., and Keating, M.T. 1997. Molecular basis of the long-QT syndrome associated with deafness. *N. Engl. J. Med.* **336**:1562–1567.
6. Splawski, I., et al. 2000. Spectrum of mutations in long-QT syndrome genes: KVLQT1, HERG, SCN5A, KCNE1, and KCNE2. *Circulation.* **102**:1178–1185.
7. Moss, A.J. 2003. Long QT syndrome. *JAMA.* **289**:2041–2044.
8. Priori, S.G., Napolitano, C., Paganini, V., Cantu, F., and Schwartz, P.J. 1997. Molecular biology of the long QT syndrome: impact on management. *Pacing Clin. Electrophysiol.* **20**:2052–2057.
9. Zareba, W., et al. 1998. Influence of genotype on the clinical course of the long-QT syndrome. International Long-QT Syndrome Registry Research Group. *N. Engl. J. Med.* **339**:960–965.
10. Schwartz, P.J., et al. 2001. Genotype-phenotype correlation in the long-QT syndrome: gene-specific triggers for life-threatening arrhythmias. *Circulation.* **103**:89–95.
11. Ackerman, M.J., et al. 2002. Epinephrine-induced QT interval prolongation: a gene-specific paradoxical response in congenital long QT syndrome. *Mayo Clin. Proc.* **77**:413–421.
12. Takenaka, K., et al. 2003. Exercise stress test amplifies genotype-phenotype correlation in the LQT1 and LQT2 forms of the long-QT syndrome. *Circulation.* **107**:838–844.
13. Paavonen, K.J., et al. 2001. Response of the QT interval to mental and physical stress in types LQT1 and LQT2 of the long QT syndrome. *Heart.* **86**:39–44.
14. Moss, A.J., et al. 2000. Effectiveness and limitations of beta-blocker therapy in congenital long-QT syndrome. *Circulation.* **101**:616–623.
15. Priori, S.G., et al. 2004. Association of long QT syndrome loci and cardiac events among patients treated with beta-blockers. *JAMA.* **292**:1341–1344.
16. George, A.L., et al. 1995. Assignment of the human

heart tetrodotoxin-resistant voltage-gated Na+ channel alpha-subunit gene (SCN5A) to band 3P21. *Cytogenet. Cell Genet.* **68**:67–70.
17. Wang, Q., et al. 1995. SCN5A mutations associated with an inherited cardiac arrhythmia, long QT syndrome. *Cell.* **80**:805–811.
18. Wang, Q., et al. 1995. Cardiac sodium channel mutations in patients with long QT syndrome, an inherited cardiac arrhythmia. *Hum. Mol. Genet.* **4**:1603–1607.
19. Abriel, H., Wehrens, X.H., Benhorin, J., Kerem, B., and Kass, R.S. 2000. Molecular pharmacology of the sodium channel mutation D1790G linked to the long-QT syndrome. *Circulation.* **102**:921–925.
20. Veldkamp, M.W., et al. 2000. Two distinct congenital arrhythmias evoked by a multidysfunctional Na(+) channel. *Circ. Res.* **86**:E91–E97.
21. Schwartz, P.J., et al. 1995. Long QT syndrome patients with mutations of the SCN5A and HERG genes have differential responses to NA+ channel blockade and to increases in heart rate: implications for gene-specific therapy. *Circulation.* **92**:3381–3386.
22. Rosero, S.Z., Zareba, W., Robinson, J.L., and Moss, A. 1997. Gene-specific therapy for long QT syndrome: QT shortening with lidocaine and tocainide in patients with mutation of the sodium channel gene. *Ann. Noninvasive Electrocardiol.* **2**:274–278.
23. Windle, J.R., Geletka, R.C., Moss, A.J., Zareba, W., and Atkins, D.L. 2001. Normalization of ventricular repolarization with flecainide in long QT syndrome patients with SCN5A:DeltaKPQ mutation. *Ann. Noninvasive Electrocardiol.* **6**:153–158.
24. Benhorin, J., et al. 2000. Effects of flecainide in patients with new SCN5A mutation: mutation-specific therapy for long-QT syndrome? *Circulation.* **101**:1698–1706.
25. Nagatomo, T., January, C.T., and Makielski, J.C. 2000. Preferential block of late sodium current in the LQT3 DeltaKPQ mutant by the class I(C) antiarrhythmic flecainide. *Mol. Pharmacol.* **57**:101–107.
26. January, C.T., Gong, Q., and Zhou, Z. 2000. Long QT syndrome: cellular basis and arrhythmia mechanism in LQT2. *J. Cardiovasc. Electrophysiol.* **11**:1413–1418.
27. Sanguinetti, M.C., Curran, M.E., Spector, P.S., and Keating, M.T. 1996. Spectrum of HERG K channel dysfunction in an inherited cardiac arrhythmia. *Proc. Natl. Acad. Sci. U. S. A.* **93**:2208–2212.
28. Moss, A.J., et al. 2002. Increased risk of arrhythmic events in long-QT syndrome with mutations in the pore region of the human ether-a-go-go-related gene potassium channel. *Circulation.* **105**:794–799.
29. Zareba, W., et al. 2003. Location of mutation in

the KCNQ1 and phenotypic presentation of long QT syndrome. *J. Cardiovasc. Electrophysiol.* **14**:1149–1153.
30. Shimizu, W., et al. 2004. Mutation site-specific differences in arrhythmic risk and sensitivity to sympathetic stimulation in the LQT1 form of congenital long QT syndrome: multicenter study in Japan. *J. Am. Coll. Cardiol.* **44**:117–125.
31. Noble, D., and Tsien, R. 1968. The kinetics and rectifier properties of the slow potassium current in cardiac Purkinje fibres. *J. Physiol.* **195**:185–214.
32. Noble, D., and Tsien, R.W. 1969. Outward membrane currents activated in the plateau range of potentials in cardiac Purkinje fibres. *J. Physiol.* **200**:205–231.
33. Sanguinetti, M.C., and Jurkiewicz, N.K. 1990. Two components of cardiac delayed rectifier K+ current. Differential sensitivity to block by class III antiarrhythmic agents. *J. Gen. Physiol.* **96**:195–215.
34. Kass, R.S., and Wiegers, S.E. 1982. The ionic basis of concentration-related effects of noradrenaline on the action potential of calf cardiac purkinje fibres. *J. Physiol.* **322**:541–558.
35. Sanguinetti, M.C., Jiang, C., Curran, M.E., and Keating, M.T. 1995. A mechanistic link between an inherited and an acquired cardiac arrhythmia: HERG encodes the IKr potassium channel. *Cell.* **81**:299–307.
36. Mitcheson, J.S., Chen, J., Lin, M., Culberson, C., and Sanguinetti, M.C. 2000. A structural basis for drug-induced long QT syndrome. *Proc. Natl. Acad. Sci. U. S. A.* **97**:12329–12333.
37. Wang, Q., et al. 1996. Positional cloning of a novel potassium channel gene: KVLQT1 mutations cause cardiac arrhythmias. *Nat. Genet.* **12**:17–23.
38. Sanguinetti, M.C., et al. 1996. Coassembly of KvLQT1 and minK(ISK) proteins to form cardiac IKS potassium channel. *Nature.* **384**:80–83.
39. Barhanin, J., et al. 1996. K(V)LQT1 and lsK (minK) proteins associate to form the I(Ks) cardiac potassium current. *Nature.* **384**:78–80.
40. Splawski, I., Tristani-Firouzi, M., Lehmann, M.H., Sanguinetti, M.C., and Keating, M.T. 1997. Mutations in the hminK gene cause long QT syndrome and suppress IKs function. *Nat. Genet.* **17**:338–340.
41. Marx, S.O., et al. 2002. Requirement of a macromolecular signaling complex for beta adrenergic receptor modulation of the KCNQ1-KCNE1 potassium channel. *Science.* **295**:496–499.
42. Kurokawa, J., Chen, L., and Kass, R.S. 2003. Requirement of subunit expression for cAMP-mediated regulation of a heart potassium channel. *Proc. Natl. Acad. Sci. U. S. A.* **100**:2122–2127.
43. Kass, R.S., and Moss, A.J. 2003. Long QT syn-

drome: novel insights into the mechanisms of cardiac arrhythmias. *J. Clin. Invest.* **112**:810–815. doi:10.1172/JCI200319844.

44. Furutani, M., et al. 1999. Novel mechanism associated with an inherited cardiac arrhythmia: defective protein trafficking by the mutant HERG (G601S) potassium channel. *Circulation.* **99**:2290–2294.

45. Zhou, Z., Gong, Q., and January, C.T. 1999. Correction of defective protein trafficking of a mutant HERG potassium channel in human long QT syndrome. Pharmacological and temperature effects. *J. Biol. Chem.* **274**:31123–31126.

46. Ficker, E., Obejero-Paz, C.A., Zhao, S., and Brown, A.M. 2002. The binding site for channel blockers that rescue misprocessed human long QT syndrome type 2 ether-a-gogo-related gene (HERG) mutations. *J. Biol. Chem.* **277**:4989–4998.

47. Paulussen, A., et al. 2002. A novel mutation (T65P) in the PAS domain of the human potassium channel HERG results in the long QT syndrome by trafficking deficiency. *J. Biol. Chem.* **277**:48610–48616.

48. Rajamani, S., Anderson, C.L., Anson, B.D., and January, C.T. 2002. Pharmacological rescue of human K(+) channel long-QT2 mutations: human ether-a-go-go-related gene rescue without block. *Circulation.* **105**:2830–2835.

49. Gong, Q., Anderson, C.L., January, C.T., and Zhou, Z. 2004. Pharmacological rescue of trafficking defective HERG channels formed by coassembly of wild-type and long QT mutant N470D subunits. *Am. J. Physiol. Heart Circ. Physiol.* **287**:H652–H658.

50. Stuhmer, W., et al. 1989. Structural parts involved in activation and inactivation of the sodium channel. *Nature.* **339**:597–603.

51. Catterall, W.A. 1995. Structure and function of voltage-gated ion channels [review]. *Annu. Rev. Biochem.* **64**:493–531.

52. Weidmann, S. 1952. The electrical constants of Purkinje fibres. *J. Physiol.* **118**:348–360.

53. Motoike, H.K., et al. 2004. The Na+ channel inactivation gate is a molecular complex: a novel role of the COOH-terminal domain. *J. Gen. Physiol.* **123**:155–165.

54. Kim, J., et al. 2004. Calmodulin mediates Ca2+ sensitivity of sodium channels. *J. Biol. Chem.* **279**:45004–45012.

55. Donaldson, M.R., Yoon, G., Fu, Y.H., and Ptacek, L.J. 2004. Andersen-Tawil syndrome: a model of clinical variability, pleiotropy, and genetic heterogeneity. *Ann. Med.* **36**(Suppl. 1):92–97.

56. Andelfinger, G., et al. 2002. KCNJ2 mutation results in Andersen syndrome with sex-specific cardiac and skeletal muscle phenotypes. *Am. J. Hum. Genet.* **71**:663–668.

57. Tristani-Firouzi, M., et al. 2002. Functional and clinical characterization of KCNJ2 mutations associated with LQT7 (Andersen syndrome). *J. Clin. Invest.* **110**:381–388. doi:10.1172/JCI200215183.

58. Bendahhou, S., et al. 2003. Defective potassium channel Kir2.1 trafficking underlies Andersen-Tawil syndrome. *J. Biol. Chem.* **278**:51779–51785.

59. Splawski, I., et al. 2004. Ca(V)1.2 calcium channel dysfunction causes a multisystem disorder including arrhythmia and autism. *Cell.* **119**:19–31.

60. Kass, R.S., and Sanguinetti, M.C. 1984. Calcium channel inactivation in the cardiac Purkinje fiber. Evidence for voltage- and calcium-mediated mechanisms. *J. Gen. Physiol.* **84**:705–726.

61. Schott, J.J., et al. 1995. Mapping of a gene for long QT syndrome to chromosome 4q25-27. *Am. J. Hum. Genet.* **57**:1114–1122.

62. Chauhan, V.S., Tuvia, S., Buhusi, M., Bennett, V., and Grant, A.O. 2000. Abnormal cardiac Na(+) channel properties and QT heart rate adaptation in neonatal ankyrin(B) knockout mice. *Circ. Res.* **86**:441–447.

63. Yong, S., Tian, X., and Wang, Q. 2003. LQT4 gene: the "missing" ankyrin. *Mol. Interv.* **3**:131–136.

64. Mohler, P.J., et al. 2003. Ankyrin-B mutation causes type 4 long-QT cardiac arrhythmia and sudden cardiac death. *Nature.* **421**:634–639.

65. Mohler, P.J., et al. 2004. A cardiac arrhythmia syndrome caused by loss of ankyrin-B function. *Proc. Natl. Acad. Sci. U. S. A.* **101**:9137–9142.

Genetics of acquired long QT syndrome

Dan M. Roden[1] and Prakash C. Viswanathan[2]

[1]Departments of Medicine and Pharmacology and [2]Department of Anesthesiology, Vanderbilt University School of Medicine, Nashville, Tennessee, USA.

The QT interval is the electrocardiographic manifestation of ventricular repolarization, is variable under physiologic conditions, and is measurably prolonged by many drugs. Rarely, however, individuals with normal baseline intervals may display exaggerated QT interval prolongation, and the potentially fatal polymorphic ventricular tachycardia torsade de pointes, with drugs or other environmental stressors such as heart block or heart failure. This review summarizes the molecular and cellular mechanisms underlying this acquired or drug-induced form of long QT syndrome, describes approaches to the analysis of a role for DNA variants in the mediation of individual susceptibility, and proposes that these concepts may be generalizable to common acquired arrhythmias.

Introduction

The QT interval on the surface ECG is a representation of repolarization time in the ventricle. QT intervals in humans vary as a function of age, sex, heart rate, heart disease, and drugs and are generally less than 480 ms. "Acquired long QT syndrome" describes not one end of a physiologic spectrum, but rather pathologic QT interval prolongation, generally to greater than 550–600 ms, upon exposure to an environmental stressor and reversion back to normal following withdrawal of the stressor. When QT intervals are markedly prolonged in this fashion, the polymorphic ventricular tachycardia torsade de pointes becomes a real risk; torsade de pointes can be self-limited or can degenerate to fatal arrhythmias such as ventricular fibrillation. It is the potential for torsade de pointes and sudden death that has generated such attention to acquired long QT syndrome (1, 2). As discussed below, the principles elucidated in studies of drug-induced long QT syndrome likely apply broadly to more common arrhythmia phenotypes.

This review focuses on drug therapy, the most common cause of acquired long QT syndrome. Acquired QT interval prolongation and torsade de pointes can occur in other settings, such as heart block (3) and, rarely, acute myocardial infarction (4) (Table 1 and Figure 1). In addition, even minor degrees of QT interval prolongation have been associated with increased mortality in many settings, notably convalescence from acute myocardial infarction (5), advancing age (6), and heart failure (7–9). The extent to which the mechanisms described below apply to these settings is uncertain, although overlap seems likely.

Most recognized cases of drug-induced long QT syndrome arise during therapy with QT interval–prolonging antiarrhythmics, as listed in Table 1. For some of these, such as quinidine and dofetilide, estimates of incidence range as high as 3–5%, although patients at especially high or low risk can be identified on clinical grounds (10). Treatment with drugs not intended for cardiovascular therapy has also been associated with drug-induced QT interval prolongation and arrhythmias, although the frequency of the problem appears to be much smaller. Nevertheless, because this rare adverse effect can be fatal, its recognition after the marketing of a drug can profoundly affect the perception of risk versus benefit that goes into the approval or prescription of the drug. Indeed, QT interval prolongation, with the potential for fatal arrhythmias, has been the single most common cause of withdrawal or relabeling of marketed drugs in the last decade (2); examples have occurred in multiple drug classes and include antihistamines (terfenadine and astemizole), gastrointestinal agents (cisapride), antipsychotics (sertindole), and urologic agents (terodiline).

Mechanisms underlying QT interval prolongation and torsade de pointes

First principles. QT interval prolongation on the surface ECG represents prolongation of action potentials in at least some regions of the ventricle (Figure 2). First principles in cellular electrophysiology dictate that such action potential prolongation, in turn, must reflect either a decrease in outward, repolarizing currents (flowing primarily through potassium channels) or an increase in plateau inward current (flowing primarily through calcium and sodium channels). Importantly, and as discussed further by Moss and Kass in an accompanying article in this series (11), mutations in genes encoding potassium, sodium, and calcium channels (as well as the structural protein ankyrin-B) have been linked to the congenital form of long QT syndrome, a disease with features — including torsade de pointes — in common with the acquired syndrome. As predicted, the potassium channel mutations result in decreased outward currents, and the calcium and sodium channel mutations result in increased plateau inward current.

HERG/I_{Kr} blockade. While mutations in any 1 of at least 8 genes can cause congenital QT interval prolongation, drugs that produce acquired long QT syndrome almost inevitably target a specific potassium current, the rapid component of the delayed rectifier, termed I_{Kr} (12). I_{Kr} is generated by expression of the human ether-à-go-go–related gene (*HERG*, also known as *KCNH2*), mutated in the LQT2 form of the congenital syndrome (13, 14). In heterologous systems, expression of *HERG* is sufficient to generate I_{Kr}; expression of other genes, such as *KCR1* (15) or *KCNE2* (16), generates proteins that appear to modulate I_{Kr} function in these systems, although their role in cardiomyocytes is less well established.

The very interesting question of why *HERG* channels are so readily blocked by a wide range of drugs to produce acquired long QT syndrome, while other potassium channels seem much less susceptible, has been addressed by Sanguinetti and colleagues (17–19) (Figure 3). A common drug-binding site in the channel is located on the intracellular face on the pore region, as in many other ion channels. Two key structural features inferred in the *HERG* pore, absent in other potassium channels, appear to underlie the "promiscuity"

Nonstandard abbreviations used: CAST, Cardiac Arrhythmia Suppression Trial; EAD, early afterdepolarization; *HERG*, human ether-à-go-go–related gene; I_{Kr}, the rapid component of the delayed rectifier; I_{Ks}, the slow component of the delayed rectifier.

Conflict of interest: The authors have declared that no conflict of interest exists.

Citation for this article: *J. Clin. Invest.* **115**:2025–2032 (2005). doi:10.1172/JCI25539.

Table 1

Causes of acquired long QT syndrome

Drugs
    Drugs that frequently cause torsade de pointes
        Disopyramide
        Dofetilide
        Ibutilide
        Procainamide
        Quinidine
        Sotalol
    Drugs clearly associated with torsade de pointes
    but with low incidence[A]
        Amiodarone
        Arsenic trioxide
        Erythromycin
        Droperidol
        Haloperidol
        Thioridazine
        Methadone
Heart block
Hypokalemia, hypomagnesemia[B]
Acute myocardial infarction[B]
Subarachnoid hemorrhage and other CNS injury[B]
Liquid protein diets and other forms of starvation[B]

[A]There are case reports of torsade de pointes with many other drugs. Lists that give some sense of overall risk with a particular drug are maintained at http://www.torsades.org. [B]These are rare causes of torsade de pointes, but may increase risk when other causes are present.

of the channel's vulnerability to blockade by drugs. The first is the presence of multiple aromatic residues oriented to face the permeation pore; these provide high-affinity binding sites for a wide range of compounds. Other binding sites have been identified within the channel pore that may modulate on-and-off rates. The second key feature is absence of a pair of proline residues in the S6 helix that forms part of the pore. As a result, the S6 helix is not kinked in the *HERG* channel, and thus it is hypothesized that access to the binding site is less restricted than in other channels, allowing access to the blocking site by a wide range of drugs. Further understanding of drug binding might allow prediction of structures unlikely to bind to the channel, or structures that will unbind so quickly as to not produce the tonic blockade required to prolong QT intervals.

Why is potassium channel blockade arrhythmogenic? When preparations from the canine conduction system (Purkinje fibers) are exposed to conditions that promote torsade de pointes clinically (slow rates, low extracellular potassium, or drugs), action potentials not only lengthen but also develop distinctive morphologic abnormalities termed early afterdepolarizations (EADs) and triggered upstrokes (20–23) (Figure 4). These findings suggest that triggered upstrokes arising from EADs are 1 potential initiating mechanism for the arrhythmia. Importantly, EADs and triggered activity arise only indirectly from potassium channel inhibition; it seems likely that action potential prolongation by I_{Kr} blockade enables activation or reactivation of arrhythmogenic inward currents that underlie EADs and triggering. These may include calcium channels or the sodium-calcium exchanger (24, 25). Some studies also suggest a facilitatory role for intracellular calcium overload (26, 27), which would agree with the finding that heart failure appears to increase incidence of the arrhythmia (28, 29).

An increasingly well recognized feature of ventricular repolarization is that normal action potential durations and configurations vary across the ventricular wall (30, 31). It is this heterogeneity that results in a distinct positive T wave on the surface ECG (Figure 2). A key transmural difference is that action potential durations are longest in the midmyocardium (the "M cell" layer), and shorter in epicardial and endocardial regions. Two sets of experimental data, not necessarily mutually exclusive, have been advanced to explain this difference: M cells have been reported to display a decrease in a second, slow component of the delayed rectifier potassium current (32), termed I_{Ks}, as well as an increase in current flowing through sodium channels during the plateau ("late" sodium current) (33). The molecular basis for these changes has not been elucidated.

The effect of blocking I_{Kr} has been studied both in experimental systems in which all cell types are represented (such as the canine left ventricular perfused "wedge") and using computer models in which the effects of individual membrane currents on action potentials can be computed (34, 35). In both the in vitro and the in silico work, I_{Kr} blockade in M cells produces striking action potential prolongation, similar to that produced in Purkinje fibers, while epicardial and endocardial cells show much smaller changes (Figure 2). This exaggeration of physiologic heterogeneities of action potential duration in turn increases the susceptibility to transmural reentry, a likely mechanism underlying torsade de pointes (36, 37). Interestingly, this transmural dispersion mechanism has also been identified in congenital long QT syndrome, in other congenital diseases such as Brugada syndrome and the recently described short QT syndrome, and in heart failure (38–40). An important clinical implication of the recognition of the role of heterogeneity of repolarization in arrhythmia susceptibility has been that the measurement of the QT interval alone may not be an especially good guide to torsade de pointes risk. Instead, other indices of repolarization, such as T wave morphology or T peak and T end time, may be more sensitive indicators of this dispersion and hence the arrhythmogenic substrate (30, 31, 41, 42); however, these measures remain to be validated.

Animal models for the study of torsade de pointes. One animal model for the study of torsade de pointes is anesthetized rabbits, in which drugs producing the arrhythmia in patients regularly produce torsade de pointes but only if infused after pretreatment with the α-blocker methoxamine (43, 44). A second is dogs in which the atrioventricular node has been destroyed to create complete heart block (45, 46). After atrioventricular nodal ablation, QT intervals progressively prolong over weeks, and torsade de pointes is then readily induced by drug infusion. A likely mechanism in both situations is inhibition of I_{Ks} and perhaps other repolarizing currents to enhance the pharmacologic effect of I_{Kr} blockade. In the dog model, striking action potential lability on drug exposure separates arrhythmia-prone from arrhythmia-resistant animals (47), again suggesting that indices of repolarization beyond simple measurement of the QT interval may be useful in gauging risk.

Identifying patients at risk

Clinical features. The typical pause dependence of the arrhythmia is illustrated in Figure 1 (48–51). There also appears to be an increase in underlying heart rate before the onset of an episode of arrhythmia (52), suggesting a role for adrenergic activation. The arrhythmia is treated by withdrawal of offending agents, correction of hypokalemia to greater than 4–4.5 mEq/l (53), and empiric magnesium regardless of the serum magnesium (54); if the arrhythmia

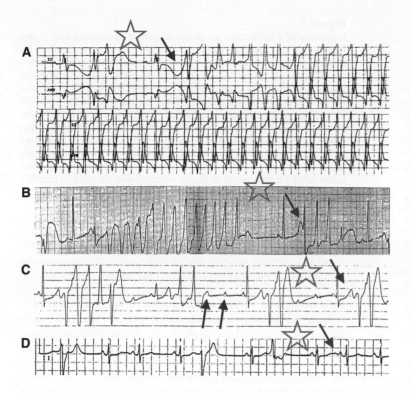

Figure 1
Examples of acquired long QT syndrome. A common feature is a pause (often after an ectopic beat), indicated by a star, with deranged repolarization in the following cycle (red arrows). (**A**) Continuous recording from a 79-year-old man with advanced heart disease treated with the antiarrhythmic dofetilide. The abnormal QT interval is followed by 7 beats of polymorphic ventricular tachycardia (torsade de pointes). In this patient, torsade de pointes then precipitated sustained monomorphic ventricular tachycardia, due to underlying heart disease. (**B**) Torsade de pointes in a patient treated with the antipsychotic haloperidol. (**C**) Torsade de pointes in a patient with complete heart block. The blue arrows indicate nonconducted atrial depolarizations. (**D**) Markedly abnormal postpause repolarization in a patient with advanced heart failure. Such disordered repolarization may represent increased risk for torsade de pointes (7).

recurs, temporary pacing or isoproterenol to prevent the pauses preceding the arrhythmia is used.

Over the past 2 decades, several clinical features have been consistently identified in multiple series of drug-induced torsade de pointes (48–51). These are listed in Table 2, and their identification at the clinical level has enabled interesting and important mechanistic research at the molecular level. For example, hypokalemia is a very common feature among patients with drug-induced torsade de pointes, and lowering of extracellular potassium decreases I_{Kr}, an effect that likely contributes to QT interval prolongation in hypokalemic patients (55, 56). However, this effect on I_{Kr} is unexpected, since simple electrochemical considerations predict an increase in outward potassium current with lowering of extracellular potassium. Two explanations have been advanced to explain this paradoxical behavior.

One is that sodium and potassium compete for access to extracellular binding sites on the channel and sodium is a potent blocker of the current (57). As a result, when extracellular potassium is lowered, the inhibitory effect of sodium on the current becomes more apparent. The second explanation involves the very rapid inactivation that I_{Kr} undergoes after opening during depolarizing pulses (55). Lowering of extracellular potassium enhances this fast inactivation, so with hypokalemia more channels are in the inactivated state and fewer in the open state during depolarizing pulses. This very rapid inactivation also explains why the HERG channel, which generates I_{Kr}, plays such a key role in repolarization (Figure 2). During the plateau, most HERG channels are in the inactivated state. As repolarization is initiated at the beginning of phase 3 of the action potential, channels recover from inactivation and enter the open state before closing. Thus, as the action potential starts its repolarizing trajectory, I_{Kr} increases (reflecting more channels in the open state), thereby further accelerating repolarization. As discussed above, this is a major protective mechanism against arrhythmias, since it prevents the development of arrhythmogenic inward currents during the end of the action potential.

Another twist on hypokalemia as a risk factor has been the observation that drug blockade is actually enhanced at low levels of extracellular potassium (58, 59). Thus, hypokalemia potentiates

Figure 2
Computed action potentials, using the Luo-Rudy simulation (94) modified to include a transient outward current. This simulation incorporates physiologically realistic numerical models of individual ion currents and other electrogenic events (e.g., exchangers) and thereby allows in silico prediction of the effects of lesions in individual components on the whole physiologic system. **A** and **B** each show (from top to bottom) epicardial action potential, I_{Kr} and I_{Ks} during the epicardial action potential, midmyocardial action potential, I_{Kr} and I_{Ks} during the midmyocardial action potential, and an ECG signal computed from a 1-dimensional fiber consisting of endocardial, midmyocardial, and epicardial cells connected through resistive gap junctions (95). (**A**) Control. The numbered phases of the action potential are shown on the epicardial signal. Note the increase in I_{Kr} at the beginning of phase 3; as discussed in the text, this serves to enhance repolarization. The dotted lines indicate the ends of repolarization in the epicardial and midmyocardial cells and correspond roughly to the peak and end of the T wave, respectively. (**B**) 75% I_{Kr} blockade. Note that action potentials at both sites are prolonged, and the difference between them is exaggerated. The T wave abnormality in the computed ECG also reflects formation of EADs in endocardial cells (not shown).

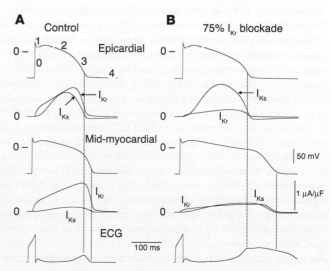

A

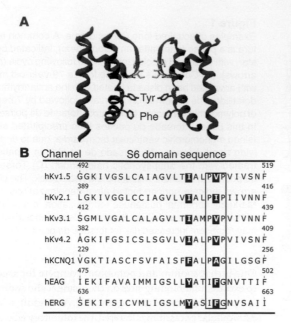

B

Channel	S6 domain sequence

hKv1.5 GGKIVGSLCAIAGVLTIALPVPVIVSNF

hKv2.1 LGKIVGGLCCIAGVLVIALPIPIIVNNF

hKv3.1 SGMLVGALCALAGVLTIAMPVPVIVNNF

hKv4.2 AGKIFGSICSLSGVLVIALPVPVIVSNF

hKCNQ1 VGKTIASCFSVFAISFFALPAGILGSGF

hEAG IEKIFAVAIMMIGSLLYATIFGNVTTIF

hERG SEKIFSICVMLIGSLMYASIFGNVSAII

torsade de pointes risk through at least 2 mechanisms: (a) a decrease in the repolarizing current itself, and (b) potentiation of drug blockade of residual current. The mechanisms underlying other factors listed in Table 2, such as increased risk immediately following cardioversion of atrial fibrillation (60), have not yet been elucidated at the molecular level.

Reduced repolarization reserve. Figure 5 summarizes the way in which blockade of a single channel, encoded by *HERG*, can culminate in sudden death due to ventricular fibrillation; the key intermediate steps are action potential prolongation, EADs, QT interval prolongation, and torsade de pointes. A unifying framework for approaching these underlying mechanisms, and thereby understanding variability in response to HERG channel blockade and, in particular, why only very few patients exposed to HERG blockers die suddenly, has been the concept of reduced repolarization reserve (61). The starting point for this concept is that cardiac repolarization is determined by net outward current over time, itself a function not only of I_{Kr} and I_{Ks}, but also of other inward and outward currents during the plateau of the action potential. A defect in any 1 of these mechanisms may, therefore, remain subclinical if other pathways to normal repolarization are intact. The animal models discussed above are one example. Another is the phenomenon of "exposure" of subclinical congenital long QT syndrome due to mutations in the genes encoding I_{Ks} (62–64). Such cases suggest that mutations reducing this repolarizing current may be tolerated because of a robust I_{Kr}. However, administration of an I_{Kr}-blocking drug to such patients may then expose the defect in repolarization and result in marked QT interval prolongation and torsade de pointes (Figure 4). It seems likely that this framework can be used to analyze the role of other less well understood risk factors, such as female sex, heart failure, or left ventricular hypertrophy. In each, it seems likely that a subclinical defect in repolarization is exposed by inhibition of I_{Kr}. This framework is actually a specific example of the more general concept that systems controlling many physiologic processes, such as blood pressure, xenobiotic elimination, and protection from cancer, are usually highly redundant. A single lesion in such a system thus often

Figure 3
Hypothesized molecular structure of the drug-binding site in the HERG channel. (**A**) The orientation of the channel pore, lined by S6 helices, is shown; drug access is via the intracellular face of the channel. Portions of 2 of the 4 subunits of the homotetrameric channel are shown, and the other 2 are omitted for clarity. The aromatic residues (tyrosine [Tyr] and phenylalanine [Phe]) that face the pore are thought to be high-affinity drug-binding sites. (**B**) Sequence comparisons between HERG and other potassium channels. With the exception of the closely related hEAG channel, the others have 1 or 2 prolines in S6 and 0 or 1 aromatic residues. As discussed in the text, these 2 features appear to determine the ease with which the HERG channel is blocked by a wide range of drugs. Adapted with permission from the *Journal of Biological Chemistry* (19).

remains clinically inapparent, and multiple lesions may be required to actually develop an overt phenotype, such as hypertension, an unusual drug reaction due to decreased clearance, or cancer.

Role of genetic variants in acquired long QT syndrome

QT interval prolongation, with the exception of that induced by quinidine, is increased at high plasma concentrations. Hence, genetic variants that impair elimination of an I_{Kr}-blocking drug may increase risk for torsade de pointes. The antipsychotic agent thioridazine is eliminated by the cytochrome P450 CYP2D6, which is functionally absent because of loss-of-function variants in the gene in approximately 7% of white and black individuals; the current FDA labeling warns of increased torsade de pointes risk with thioridazine in the poor-metabolizer group.

Subclinical long QT syndrome. The identification of congenital long QT syndrome disease genes has led to screening of large kindreds, and the recognition of incomplete penetrance, i.e., mutation carriers with normal ECGs. The identification of such individuals after an episode of torsade de pointes argues that mutations not generating baseline QT interval prolongation may nevertheless still con-

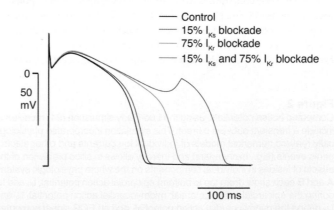

— Control
— 15% I_{Ks} blockade
— 75% I_{Kr} blockade
— 15% I_{Ks} and 75% I_{Kr} blockade

0
50
mV

100 ms

Figure 4
Luo-Rudy simulations showing the concept of repolarization reserve. The blue line shows the effect of reducing I_{Ks} by 15%, as might be expected in a subtle congenital long QT syndrome mutation. The green line shows the expected prolongation of the control action potential resulting from 75% I_{Kr} blockade. The red line shows the effect of the same degree of drug blockade applied to the simulation with 15% I_{Ks} blockade. Not only is there marked exaggeration of action potential prolongation, but an EAD with a triggered upstroke is also generated; L-type calcium current generates the upstroke in this model.

Table 2

Risk factors for torsade de pointes in the presence of a culprit drug[A]

Female sex
Hypokalemia
Hypomagnesemia
Bradycardia
Heart failure
Recent conversion from atrial fibrillation
    Ion channel variants
    Subclinical or unrecognized congenital long QT syndrome
    Polymorphisms in congenital long QT syndrome disease genes
    Other polymorphisms
High drug concentrations[B]
Left ventricular hypertrophy
Rapid i.v. drug bolus

[A]Evidence for most of these is derived from nonrandomized case series and/or studies in animal models. [B]Quinidine is an exception.

fer risk on drug exposure. Analyses of probands with drug-induced long QT syndrome have identified the subclinical congenital syndrome in a minority (less than 10%); mutations have been reported in *KCNQ1*, encoding the pore-forming subunit underlying I_{Ks}; *HERG* itself; the K⁺ channel subunit genes *KCNE1* and *KCNE2*; and *SCN5A*, encoding the cardiac sodium channel (64, 65).

Polymorphisms in congenital long QT syndrome disease genes. These analyses have also identified polymorphisms in long QT syndrome ion channel genes, some of which may be overrepresented in patients with drug-induced or other arrhythmias. One striking example is S1103Y in the cardiac sodium channel gene (66). When 23 black patients with a range of arrhythmias, including drug-induced long QT syndrome, were compared with black controls, this variant was overrepresented in the patients, with a minor allele frequency of 13%. In vitro studies identified a subtle gating defect that increased the risk of EADs with I_{Kr} blockade in computed (in silico) action potentials. The S1103Y variant has not been identified in other ethnic groups, except for a single report in a white family, in which it was implicated as the disease-causing mutation in manifest congenital long QT syndrome (67). These studies point to the increasingly well-recognized role of ethnicity in polymorphism frequencies and in modulation of important physiologic and drug-response phenotypes (68). Thus, any study examining the genetic determinants of these endpoints must include a consideration of ethnicity.

Another lesson in this regard was the *KCNE2* variant, which results in Q9E. This was initially described as a mutation in a black woman with drug-induced long QT syndrome, because it was absent in

more than 1,000 normal controls. However, this turns out to be a relatively common polymorphism, occurring in 3.2% of black people (69). Other rare polymorphisms with minor allele frequencies of 1–2% that have been implicated in drug-induced torsade de pointes include D85N (KCNE1) and T8A (KCNE2) (65, 70, 71).

Extending the list of candidate variants. These studies identified DNA variants — mutations and polymorphisms — associated with drug-induced long QT syndrome by testing the hypothesis that variants in the congenital long QT syndrome disease genes might contribute to risk in the drug-induced form. Patients with the target phenotype (drug-induced torsade de pointes) were screened for variants in these genes; the frequency of these variants was then determined in control populations, and the function of the variants was determined by in vitro heterologous expression of variant ion channels. An alternate paradigm may now be emerging, driven both by an increasing appreciation of the large number of genes that determine normal cardiac electrophysiology, and by improvements in high-throughput genetic technologies. Rather than confining the list of candidate genes to those with mutations causing congenital arrhythmia syndromes, the alternate approach generates a list of many dozens or more based on a current understanding of normal cardiac electrophysiology. High-throughput screening of these genes is then undertaken to identify common or functionally important polymorphisms, and their frequency is then compared in patients with the target phenotype and controls. Using this approach, we have identified I447V in *KCR1* as a potential modulator of the risk of drug-induced torsade de pointes; the valine variant occurred in 1.1% of patients, compared with 7% in controls, suggesting that the presence of valine in this position protects against drug-induced torsade de pointes (15).

Genome science and arrhythmia phenotypes. While this new approach appears to be an appealing way of integrating contemporary genomics into arrhythmia science, there are a number of obstacles. The first is the very large number of candidate genes and the correspondingly huge number of polymorphisms that have already been described in these candidates. The second is the problem of false positives, highlighted by reports that most association studies are not reproducible (72, 73). One way to bolster an association between a genetic variant and a clinical phenotype is to describe modified biology conferred by the variant that could explain the phenotype. Thus, the argument that a coding region polymorphism in an ion channel gene contributes to variability in an arrhythmia phenotype can be bolstered by demonstration that the polymorphism produces altered channel function. In addition, it is possible to then use computer simulations (e.g., Figures 2 and 4) to predict how such variant channel function might modify action potentials and arrhythmia susceptibility (35, 66, 70, 74). In this way, an association

Figure 5

Mechanisms of sudden death with HERG blockade. Drug blockade of the HERG channel (left) produces prolongation (blue) and an EAD (red) in the cardiac action potential. These changes, which are heterogeneous across the ventricular wall, generate QT interval prolongation and, through mechanisms described further in the text, torsade de pointes (right; upper panel). In this example, the arrhythmia degenerates to ventricular fibrillation (VF).

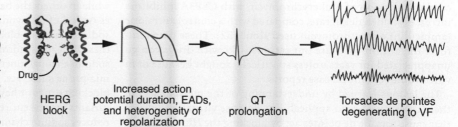

Drug

HERG block → Increased action potential duration, EADs, and heterogeneity of repolarization → QT prolongation → Torsades de pointes degenerating to VF

study is supported by an argument of biological plausibility. Unfortunately, in many instances, methods have not yet been developed to demonstrate that a DNA variant alters function or expression of the encoded protein, or that such changes alter the behavior of a complex system like the action potential.

Although there are millions of single-nucleotide and other polymorphisms in the human genome, it is apparent that large haplotype blocks display high linkage disequilibrium (75). Therefore, the number of polymorphisms to be analyzed in an association paradigm can be reduced by study of "haplotype tagging" polymorphisms. Indeed, a common haplotype blockade in the cardiac sodium channel — a key determinant of conduction velocity in the heart — has been associated with variability in the QRS duration (an index of conduction velocity in the ventricle) in a normal population (76). This haplotype blockade is in a noncoding region that includes the core promoter, and so the effect, if reproduced, seems likely attributable to variable sodium channel transcription as a contributor to interindividual variability in conduction velocity. As with the problem of predicting QT changes caused by a drug in an individual patient, the consequences of such differences may be minimal among healthy subjects but may variably engender important differences among individuals — in this case in the critically slow conduction that underlies many forms of reentry — on exposure to further stressors, such as sodium channel–blocking drugs and/or ischemia. Thus, it is even conceivable that the adverse effects of sodium channel blockers (including increased mortality due to arrhythmias, demonstrated by the Cardiac Arrhythmia Suppression Trial [CAST; ref. 77]) might reflect, in part, such genetically determined variable susceptibility to arrhythmias.

Implications for development of new drugs

Lessons learned after high-profile drug withdrawals, notably of terfenadine and cisapride, have important implications for development of new drugs. Both agents were developed before the molecular details of torsade de pointes outlined above were known. Although both turn out to be potent I_{Kr} blockers, torsade de pointes was actually quite rare, because both drugs undergo near-complete presystemic biotransformation to noncardioactive metabolites by members of the CYP3A family of cytochrome P450s (78–80); another key pharmacokinetic feature that the drugs share is that they lack robust alternate elimination mechanisms. It was largely in patients with impairment of this presystemic metabolism — due to overdose, liver disease, or concomitant therapy with potent CYP3A inhibitors such as ketoconazole or erythromycin — that the drugs accumulated in the systemic circulation and caused torsade de pointes.

Erythromycin itself is another example of a drug that may cause torsade de pointes rarely. Erythromycin is a weak I_{Kr} blocker that is also metabolized by CYP3A; the drug has been reported to cause torsade de pointes when used i.v. at high doses (81), and recent pharmacoepidemiologic data suggest that coadministration of oral erythromycin with CYP3A inhibitors increases sudden death rate, compared with a control antibiotic (ampicillin) or erythromycin used alone (82). These data reinforce the notion that rare but serious risks with drugs can go unappreciated for years unless specifically sought because of in vitro studies or clinical case reports.

The lessons learned by understanding of the mechanisms in these cases have broad applicability. Because virtually all drugs that cause torsade de pointes act by blocking the *HERG* channel, it has become standard practice to screen new drug candidates for this activity prior to clinical trials. However, because the channel is so readily blocked, many candidates blockade *HERG* channels, so a common question is whether this property is important for a potential new drug's safety profile (83–85). To answer this question requires several other pieces of information. The first is the potency of I_{Kr} blockade compared with that of activity at the target molecular site of action; the smaller the margin between the two, the more likely that HERG blockade could occur at clinically relevant doses. A second is the disposition kinetics of the drug: is it likely that some patients could generate very high plasma concentrations (of parent drug, or perhaps of active metabolites) that would place them at high risk? Such aberrant drug responses can arise because of drug interactions or because of a genetically based absence of a pathway for drug elimination. In either case, a marker of a high-risk situation is the presence of only a single pathway for drug elimination. A third piece of information is whether the drug candidate exerts other electrophysiologic effects that could potentiate, or blunt, action potential prolongation and thus torsade de pointes risk. Amiodarone and verapamil both block I_{Kr}; however, amiodarone causes torsade de pointes only rarely (86), while verapamil has actually been used to treat the arrhythmia (87). The drugs' effects on other channels, notably inward current via calcium channels, likely blunt action potential prolongation and afterdepolarizations due to I_{Kr} blockade. The antianginal agent ranolazine similarly blocks I_{Kr} but does not produce an arrhythmogenic phenotype in the wedge preparation; this has been attributed to blockade of plateau sodium current (88). Further clinical experience will be required to assess the clinical effects of this agent. Finally, a risk for torsade de pointes may be acceptable for a serious medical condition for which alternate therapies are not available and for which treatment in monitored conditions or for short periods of time is the norm; by contrast, even a tiny risk might be unacceptable for a drug being developed for long-term outpatient therapy of a nuisance symptom, and for which other therapies are available. Data attesting to the lack of torsade de pointes in pre-marketing clinical trials, which generally include no more than several thousand patients, cannot rule out serious risk, as the terfenadine, cisapride, and erythromycin examples show. Screening for HERG activity has presented a major headache for the pharmaceutical industry but has correspondingly reduced the likelihood that new drugs will unexpectedly cause torsade de pointes.

Other acquired arrhythmia syndromes

The concept that analysis of monogenic disease genes can be used as a starting point for analysis of more common clinical phenotypes can be extended from the long QT syndromes to other arrhythmias. Brugada syndrome is due to loss-of-function mutations in *SCN5A* in approximately 20% of patients (89). Some of these patients have a manifest electrocardiographic phenotype, while in others the baseline ECG is normal and the ECG changes typical of the syndrome are only exposed by challenge with a sodium channel–blocking drug. Taken together, the results of these Brugada syndrome studies and of the CAST point to loss of sodium channel function as a potential contributor to an arrhythmia-prone substrate. Thus, patients who take sodium channel–blocking drugs or have subclinical reduction-of-function *SCN5A* variants may be entirely asymptomatic until a further insult that reduces sodium channel function (e.g., transient myocardial ischemia) occurs, increasing their risk of fatal ventricular fibrillation.

The logic extends to very common arrhythmia phenotypes. Sudden cardiac death due to ventricular fibrillation affects 400,000–500,000 Americans each year, is the cause of death in more than 25% of adults, and is the first symptom of heart disease in over 50% of victims. Analyses from a number of databases indicate that a family history of sudden cardiac death increases risk in the proband; this suggests that genetic factors contribute to risk (90, 91). One obvious candidate gene is *SCN5A*, but many others could be inferred based on an increasingly sophisticated understanding of cardiac myocyte physiology. Similarly, atrial fibrillation affects millions of Americans and is generally thought to be a disease of aging. However, atrial fibrillation can occur in youth and midlife, and in these situations it is generally unassociated with any other disease (lone atrial fibrillation) or associated with hypertension, which is often mild. As with sudden death, family studies support a role for a genetic contributor to risk (92, 93). Therefore, atrial fibrillation may in fact be largely a genetic disease, but with incomplete penetrance; i.e., a genetic predisposition combined with as-yet unidentified environmental factors may be sufficient to elicit the arrhythmia in susceptible individuals.

Conclusions

Studies of both rare and common arrhythmias are thus converging on a common model in which genetic makeup interacts with environmental stressors to generate specific clinical phenotypes. In some cases, the phenotype may be manifest without additional provokers. The best examples are patients with monogenic arrhythmia syndromes such as congenital long QT syndrome, or full-blown Brugada syndrome. At the other end of this spectrum are common phenotypes such as sudden death and atrial fibrillation. Studies of drug-induced long QT syndrome can thus be viewed not simply as an interesting exercise in understanding a relatively uncommon adverse drug interaction, but as laying the foundation for a new paradigm in understanding the role of genetic factors in mediating common arrhythmia phenotypes.

Acknowledgments

The authors' work is supported in part by grants from the US Public Health Service (HL46681, HL49989, and HL65962). D.M. Roden is the holder of the William Stokes Chair in Experimental Therapeutics, a gift from the Dai-ichi Corp.

Address correspondence to: Dan M. Roden, Oates Institute for Experimental Therapeutics, Vanderbilt University School of Medicine, 532 Medical Research Building I, Nashville, Tennessee 37232, USA. Phone: (615) 322-0067; Fax: (615) 343-4522; E-mail: dan.roden@vanderbilt.edu.

1. Viskin, S., Justo, D., Halkin, A., and Zeltser, D. 2003. Long QT syndrome caused by noncardiac drugs. *Prog. Cardiovasc. Dis.* **45**:415–427.

2. Roden, D.M. 2004. Drug-induced prolongation of the QT interval. *N. Engl. J. Med.* **350**:1013–1022.

3. Dessertenne, F. 1966. La tachycardie ventriculaire à deux foyers opposés variables. *Arch. Mal. Coeur Vaiss.* **59**:263–272.

4. Halkin, A., et al. 2001. Pause-dependent torsade de pointes following acute myocardial infarction. A variant of the acquired long QT syndrome. *J. Am. Coll. Cardiol.* **38**:1168–1174.

5. Schwartz, P.J., and Wolf, S. 1978. QT interval prolongation as a predictor of sudden death in patients with myocardial infarction. *Circulation.* **56**:1074–1077.

6. de Bruyne, M.C., et al. 1999. Prolonged QT interval predicts cardiac and all-cause mortality in the elderly. The Rotterdam Study. *Eur. Heart J.* **20**:278–284.

7. Tomaselli, G.F., et al. 1994. Sudden cardiac death in heart failure. The role of abnormal repolarization. *Circulation.* **90**:2534–2539.

8. Spargias, K.S., et al. 1999. QT dispersion as a predictor of long-term mortality in patients with acute myocardial infarction and clinical evidence of heart failure. *Eur. Heart J.* **20**:1158–1165.

9. Barr, C.J., Naas, A., Freeman, M., Lang, C.C., and Struthers, A.D. 1994. QT dispersion and sudden unexpected death in chronic heart failure. *Lancet.* **343**:327–329.

10. Zeltser, D., et al. 2003. Torsade de pointes due to noncardiac drugs: most patients have easily identifiable risk factors. *Medicine (Baltimore).* **82**:282–290.

11. Moss, A.J., and Kass, R.S. 2005. Long QT syndrome: from channels to cardiac arrhythmias. *J. Clin. Invest.* **115**:2018–2024. doi:10.1172/JCI25537.

12. Sanguinetti, M.C., and Bennett, P.B. 2003. Antiarrhythmic drug target choices and screening. *Circ. Res.* **96**:491–499.

13. Sanguinetti, M.C., Jiang, C., Curran, M.E., and Keating, M.T. 1995. A mechanistic link between an inherited and an acquired cardiac arrhythmia: *HERG* encodes the I$_{Kr}$ potassium channel. *Cell.* **81**:299–307.

14. Curran, M.E., et al. 1995. A molecular basis for cardiac arrhythmia: *HERG* mutations cause long QT syndrome. *Cell.* **80**:795–803.

15. Petersen, C.I., et al. 2004. In vivo identification of ether-a-go-go related gene-interacting proteins in Caenorhabditis elegans that affect cardiac arrhythmias in humans. *Proc. Natl. Acad. Sci. U. S. A.* **101**:11773–11778.

16. Abbott, G.W., et al. 1999. MiRP1 forms I$_{Kr}$ potassium channels with HERG and is associated with cardiac arrhythmia. *Cell.* **97**:175–187.

17. Mitcheson, J.S., Chen, J., Lin, M., Culberson, C., and Sanguinetti, M.C. 2000. A structural basis for drug-induced long QT syndrome. *Proc. Natl. Acad. Sci. U. S. A.* **97**:12329–12333.

18. Sanchez-Chapula, J.A., Ferrer, T., Navarro-Polanco, R.A., and Sanguinetti, M.C. 2003. Voltage-dependent profile of human ether-a-go-go-related gene channel block is influenced by a single residue in the S6 transmembrane domain. *Mol. Pharmacol.* **63**:1051–1058.

19. Fernandez, D., Ghanta, A., Kauffman, G.W., and Sanguinetti, M.C. 2004. Physicochemical features of the HERG channel drug binding site. *J. Biol. Chem.* **279**:10120–10127.

20. Brachmann, J., Scherlag, B.J., Rosenshtraukh, L.V., and Lazzara, R. 1983. Bradycardia-dependent triggered activity: relevance to drug-induced multiform ventricular tachycardia. *Circulation.* **68**:846–856.

21. Strauss, H.C., Bigger, J.T., and Hoffman, B.F. 1970. Electrophysiological and beta-receptor blocking effects of MJ 1999 on dog and rabbit cardiac tissue. *Circ. Res.* **26**:661–678.

22. Dangman, K.H., and Hoffman, B.F. 1981. In vivo and in vitro antiarrhythmic and arrhythmogenic effects of N-acetyl procainamide. *J. Pharmacol. Exp. Ther.* **217**:851–862.

23. Roden, D.M., and Hoffman, B.F. 1985. Action potential prolongation and induction of abnormal automaticity by low quinidine concentrations in canine Purkinje fibers. Relationship to potassium and cycle length. *Circ. Res.* **56**:857–867.

24. Szabo, B., Sweidan, R., Rajagopalan, C.B., and Lazzara, R. 1994. Role of Na$^+$:Ca^{2+} exchange current in Cs$^+$-induced early afterdepolarizations in Purkinje fibers. *J. Cardiovasc. Electrophysiol.* **5**:933–944.

25. Nattel, S., and Quantz, M.A. 1988. Pharmacological response of quinidine induced early afterdepolarisations in canine cardiac Purkinje fibres: insights into underlying ionic mechanisms. *Cardiovasc. Res.*

22:808–817.

26. Burashnikov, A., and Antzelevitch, C. 1998. Acceleration-induced action potential prolongation and early afterdepolarizations. *J. Cardiovasc. Electrophysiol.* **9**:934–948.

27. Wu, Y., Roden, D.M., and Anderson, M.E. 1999. Calmodulin kinase inhibition prevents development of the arrhythmogenic transient inward current. *Circ. Res.* **84**:906–912.

28. Kober, L., et al. 2000. Effect of dofetilide in patients with recent myocardial infarction and left-ventricular dysfunction: a randomised trial. *Lancet.* **356**:2052–2058.

29. Torp-Pedersen, C., et al. 1999. Dofetilide in patients with congestive heart failure and left ventricular dysfunction. Danish Investigations of Arrhythmia and Mortality on Dofetilide Study Group. *N. Engl. J. Med.* **341**:857–865.

30. Antzelevitch, C. 2001. Transmural dispersion of repolarization and the T wave. *Cardiovasc. Res.* **50**:426–431.

31. Belardinelli, L., Antzelevitch, C., and Vos, M.A. 2003. Assessing predictors of drug-induced torsade de pointes. *Trends Pharmacol. Sci.* **24**:619–625.

32. Liu, D.W., and Antzelevitch, C. 1995. Characteristics of the delayed rectifier current (I$_{Kr}$ and I$_{Ks}$) in canine ventricular epicardial, midmyocardial, and endocardial myocytes: a weaker I$_{Ks}$ contributes to the longer action potential of the M cell. *Circ. Res.* **76**:351–365.

33. Zygmunt, A.C., Eddlestone, G.T., Thomas, G.P., Nesterenko, V.V., and Antzelevitch, C. 2001. Larger late sodium conductance in M cells contributes to electrical heterogeneity in canine ventricle. *Am. J. Physiol. Heart Circ. Physiol.* **281**:H689–H697.

34. Shimizu, W., and Antzelevitch, C. 2000. Differential effects of beta-adrenergic agonists and antagonists in LQT1, LQT2 and LQT3 models of the long QT syndrome. *J. Am. Coll. Cardiol.* **35**:778–786.

35. Viswanathan, P.C., Shaw, R.M., and Rudy, Y. 1999. Effects of IKr and IKs heterogeneity on action potential duration and its rate dependence: a simulation study. *Circulation.* **99**:2466–2474.

36. Yan, G.X., and Antzelevitch, C. 1998. Cellular basis for the normal T wave and the electrocardiographic manifestations of the long-QT syndrome. *Circulation.* **98**:1928–1936.

37. Akar, F.G., Yan, G.X., Antzelevitch, C., and Rosenbaum, D.S. 2002. Unique topographical distribution of M cells underlies reentrant mechanism of torsade de pointes in the long-QT syndrome. *Circulation.* **105**:1247–1253.

38. Extramiana, F., and Antzelevitch, C. 2004. Amplified transmural dispersion of repolarization as the basis for arrhythmogenesis in a canine ventricular-wedge model of short-QT syndrome. *Circulation.* **110**:3661–3666.

39. Yan, G.X., and Antzelevitch, C. 1999. Cellular basis for the Brugada syndrome and other mechanisms of arrhythmogenesis associated with ST-segment elevation. *Circulation.* **100**:1660–1666.

40. Akar, F.G., and Rosenbaum, D.S. 2003. Transmural electrophysiological heterogeneities underlying arrhythmogenesis in heart failure. *Circ. Res.* **93**:638–645.

41. Antzelevitch, C. 2004. Arrhythmogenic mechanisms of QT prolonging drugs: is QT prolongation really the problem? *J. Electrocardiol.* **37**(Suppl.):15–24.

42. Malik, M., and Camm, A.J. 2001. Evaluation of drug-induced QT interval prolongation: implications for drug approval and labelling. *Drug Saf.* **24**:323–351.

43. Carlsson, L., Almgren, O., and Duker, G. 1990. QTU-prolongation and torsades de pointes induced by putative class III antiarrhythmic agents in the rabbit: etiology and interventions. *J. Cardiovasc. Pharmacol.* **16**:276–285.

44. Carlsson, L., Abrahamsson, C., Andersson, B., Duker, G., and Schiller-Linhardt, G. 1993. Proarrhythmic effects of the class III agent almokalant: importance of infusion rate, QT dispersion, and early afterdepolarisations. *Cardiovasc. Res.* **27**:2186–2193.

45. Chezalviel-Guilbert, F., et al. 1995. Proarrhythmic effects of a quinidine analog in dogs with chronic A-V block. *Fundam. Clin. Pharmacol.* **9**:240–247.

46. Volders, P.G., et al. 1998. Cellular basis of biventricular hypertrophy and arrhythmogenesis in dogs with chronic complete atrioventricular block and acquired torsade de pointes. *Circulation.* **98**:1136–1147.

47. Thomsen, M.B., et al. 2004. Increased short-term variability of repolarization predicts d-sotalol-induced torsades de pointes in dogs. *Circulation.* **110**:2453–2459.

48. Jackman, W.M., et al. 1988. The long QT syndromes: a critical review, new clinical observations and a unifying hypothesis. *Prog. Cardiovasc. Dis.* **31**:115–172.

49. Kay, G.N., Plumb, V.J., Arciniegas, J.G., Henthorn, R.W., and Waldo, A.L. 1983. Torsades de pointes: the long-short initiating sequence and other clinical features: observations in 32 patients. *J. Am. Coll. Cardiol.* **2**:806–817.

50. Roden, D.M., Woosley, R.L., and Primm, R.K. 1986. Incidence and clinical features of the quinidine-associated long QT syndrome: implications for patient care. *Am. Heart J.* **111**:1088–1093.

51. Makkar, R.R., Fromm, B.S., Steinman, R.T., Meissner, M.D., and Lehmann, M.H. 1993. Female gender as a risk factor for torsades de pointes associated with cardiovascular drugs. *JAMA.* **270**:2590–2597.

52. Locati, E.H., Maison-Blanche, P., Dejode, P., Cauchemez, B., and Coumel, P. 1995. Spontaneous sequences of onset of torsade de pointes in patients with acquired prolonged repolarization: quantitative analysis of Holter recordings. *J. Am. Coll. Cardiol.* **25**:1564–1575.

53. Choy, A.M., et al. 1997. Normalization of acquired QT prolongation in humans by intravenous potassium. *Circulation.* **96**:2149–2154.

54. Tzivoni, D., et al. 1988. Treatment of torsade de pointes with magnesium sulfate. *Circulation.* **77**:392–397.

55. Yang, T., Snyders, D.J., and Roden, D.M. 1997. Rapid inactivation determines the rectification and $[K^+]_o$ dependence of the rapid component of the delayed rectifier K^+ current in cardiac cells. *Circ. Res.* **80**:782–789.

56. Wang, S., Liu, S., Morales, M.J., Strauss, H.C., and Rasmusson, R.L. 1997. A quantitative analysis of the activation and inactivation kinetics of HERG expressed in Xenopus oocytes. *J. Physiol.* **502**:45–60.

57. Numaguchi, H., Johnson, J.P., Jr., Petersen, C.I., and Balser, J.R. 2000. A sensitive mechanism for cation modulation of potassium current. *Nat. Neurosci.* **3**:429–430.

58. Yang, T., and Roden, D.M. 1996. Extracellular potassium modulation of drug block of I_{Kr}: implications for torsades de pointes and reverse use-dependence. *Circulation.* **93**:407–411.

59. Wang, S., Morales, M.J., Liu, S., Strauss, H.C., and Rasmusson, R.L. 1997. Modulation of HERG affinity for E-4031 by [K+]o and C-type inactivation. *FEBS Lett.* **417**:43–47.

60. Choy, A.M.J., Darbar, D., Dell'Orto, S., and Roden, D.M. 1999. Increased sensitivity to QT prolonging drug therapy immediately after cardioversion to sinus rhythm. *J. Am. Coll. Cardiol.* **34**:396–401.

61. Roden, D.M. 1998. Taking the idio out of idiosyncratic: predicting torsades de pointes. *Pacing Clin. Electrophysiol.* **21**:1029–1034.

62. Donger, C., et al. 1997. KVLQT1 C-terminal missense mutation causes a forme fruste long-QT syndrome. *Circulation.* **96**:2778–2781.

63. Napolitano, C., et al. 2000. Evidence for a cardiac ion channel mutation underlying drug-induced QT prolongation and life-threatening arrhythmias. *J. Cardiovasc. Electrophysiol.* **11**:691–696.

64. Yang, P., et al. 2002. Allelic variants in long QT disease genes in patients with drug-associated torsades de pointes. *Circulation.* **105**:1943–1948.

65. Paulussen, A.D., et al. 2004. Genetic variations of KCNQ1, KCNH2, SCN5A, KCNE1, and KCNE2 in drug-induced long QT syndrome patients. *J. Mol. Med.* **82**:182–188.

66. Splawski, I., et al. 2002. Variant of SCN5A sodium channel implicated in risk of cardiac arrhythmia. *Science.* **297**:1333–1336.

67. Chen, S., et al. 2002. SNP S1103Y in the cardiac sodium channel gene SCN5A is associated with cardiac arrhythmias and sudden death in a white family. *J. Med. Genet.* **39**:913–915.

68. Ackerman, M.J., et al. 2003. Ethnic differences in cardiac potassium channel variants: implications for genetic susceptibility to sudden cardiac death and genetic testing for congenital long QT syndrome. *Mayo Clin. Proc.* **78**:1479–1487.

69. Pharmacogenetics of Arrhythmia Therapy study group. 2005. KCNE2 genomic screening. http://www.pharmgkb.org/views/index.jsp?objId=69386963&objCls=VariantPosition&view=AlleleFrequencyInSampleSets.

70. Wei, J., et al. 1999. KCNE1 polymorphism confers risk of drug-induced long QT syndrome by altering kinetic properties of IKs potassium channels [abstract]. *Circulation.* **100**:I-495.

71. Sesti, F., et al. 2000. A common polymorphism associated with antibiotic-induced cardiac arrhythmia. *Proc. Natl. Acad. Sci. U. S. A.* **97**:10613–10618.

72. Hirschhorn, J.N., Lohmueller, K., Byrne, E., and Hirschhorn, K. 2002. A comprehensive review of genetic association studies. *Genet. Med.* **4**:45–61.

73. Ioannidis, J.P., Ntzani, E.E., Trikalinos, T.A., and Contopoulos-Ioannidis, D.G. 2001. Replication validity of genetic association studies. *Nat. Genet.* **29**:306–309.

74. Viswanathan, P., and Rudy, Y. 1999. Pause induced early afterdepolarizations in the long QT syndrome: a simulation study. *Cardiovasc. Res.* **42**:530–542.

75. The International HapMap Consortium. 2003. The International HapMap Project. *Nature.* **426**:789–796.

76. Pfeufer, A., et al. 2004. A common haplotype in the 5' region of the SCN5A gene is strongly associated with ventricular conduction impairment [abstract]. *Circulation.* **110**:III-2293.

77. The CAST Investigators. 1989. Increased mortality due to encainide or flecainide in a randomized trial of arrhythmia suppression after myocardial infarction. *N. Engl. J. Med.* **321**:406–412.

78. Woosley, R.L., Chen, Y., Freiman, J.P., and Gillis, R.A. 1993. Mechanism of the cardiotoxic actions of terfenadine. *JAMA.* **269**:1532–1536.

79. Wysowski, D.K., and Bacsanyi, J. 1996. Cisapride and fatal arrhythmia. *N. Engl. J. Med.* **335**:290–291.

80. Rampe, D., Roy, M.L., Dennis, A., and Brown, A.M. 1997. A mechanism for the proarrhythmic effects of cisapride (Propulsid): high affinity blockade of the human cardiac potassium channel HERG. *FEBS Lett.* **417**:28–32.

81. Nattel, S., Ranger, S., Talajic, M., Lemery, R., and Roy, D. 1990. Erythromycin-induced long QT syndrome: concordance with quinidine and underlying cellular electrophysiologic mechanism. *Am. J. Med.* **89**:235–238.

82. Ray, W.A., et al. 2004. Oral erythromycin and the risk of sudden death from cardiac causes. *N. Engl. J. Med.* **351**:1089–1096.

83. Fenichel, R.R., et al. 2004. Drug-induced torsades de pointes and implications for drug development. *J. Cardiovasc. Electrophysiol.* **15**:475–495.

84. Haverkamp, W., et al. 2000. The potential for QT prolongation and proarrhythmia by non-antiarrhythmic drugs: clinical and regulatory implications. Report on a policy conference of the European Society of Cardiology. *Eur. Heart J.* **21**:1216–1231.

85. Anderson, M.E., Al Khatib, S.M., Roden, D.M., and Califf, R.M. 2002. Cardiac repolarization: current knowledge, critical gaps, and new approaches to drug development and patient management. *Am. Heart J.* **144**:769–781.

86. Lazzara, R. 1989. Amiodarone and torsades de pointes. *Ann. Int. Med.* **111**:549–551.

87. Shimizu, W., et al. 1995. Effects of verapamil and propranolol on early afterdepolarizations and ventricular arrhythmias induced by epinephrine in congenital long QT syndrome. *J. Am. Coll. Cardiol.* **26**:1299–1309.

88. Antzelevitch, C., et al. 2004. Electrophysiological effects of ranolazine, a novel antianginal agent with antiarrhythmic properties. *Circulation.* **110**:904–910.

89. Antzelevitch, C., et al. 2005. Brugada syndrome: report of the second consensus conference. Endorsed by the Heart Rhythm Society and the European Heart Rhythm Association. *Circulation.* **111**:659–670.

90. Jouven, X., Desnos, M., Guerot, C., and Ducimetiere, P. 1999. Predicting sudden death in the population: the Paris Prospective Study I. *Circulation.* **99**:1978–1983.

91. Friedlander, Y., et al. 1998. Family history as a risk factor for primary cardiac arrest. *Circulation.* **97**:155–160.

92. Fox, C.S., et al. 2004. Parental atrial fibrillation as a risk factor for atrial fibrillation in offspring. *JAMA.* **291**:2851–2855.

93. Darbar, D., et al. 2003. Familial atrial fibrillation is a genetically heterogeneous disorder. *J. Am. Coll. Cardiol.* **41**:2185–2192.

94. Luo, C.H., and Rudy, Y. 1991. A model of the ventricular cardiac action potential. Depolarization, repolarization, and their interaction. *Circ. Res.* **68**:1501–1526.

95. Gima, K., and Rudy, Y. 2002. Ionic current basis of electrocardiographic waveforms: a model study. *Circ. Res.* **90**:889–896.

Muscle channelopathies and critical points in functional and genetic studies

Karin Jurkat-Rott and Frank Lehmann-Horn

Department of Applied Physiology, Ulm University, Ulm, Germany.

Muscle channelopathies are caused by mutations in ion channel genes, by antibodies directed against ion channel proteins, or by changes of cell homeostasis leading to aberrant splicing of ion channel RNA or to disturbances of modification and localization of channel proteins. As ion channels constitute one of the only protein families that allow functional examination on the molecular level, expression studies of putative mutations have become standard in confirming that the mutations cause disease. Functional changes may not necessarily prove disease causality of a putative mutation but could be brought about by a polymorphism instead. These problems are addressed, and a more critical evaluation of the underlying genetic data is proposed.

Introduction

Skeletal muscle was the first tissue in which hereditary diseases caused by ion channel defects, the myotonias and periodic paralyses, were described (1). It is now recognized that malignant hyperthermia (MH), central core disease, and the congenital myasthenic syndromes (CMSs) as well as the antibody-mediated myasthenia gravis should be included in the classification of muscle channelopathies (2, 3). Aberrant ion channel splicing due to mutations in other genes leads to muscle channelopathies, such as in the myotonic dystrophies (4, 5); and diseases caused by defects in proteins associated with trafficking, targeting, and clustering of ion channels are also ion channelopathies, e.g., disturbed clustering of acetylcholine receptors (AChRs) at the neuromuscular junction by rapsyn mutations (6). This review focuses on hereditary ion channelopathies and discusses mutation patterns, functional consequences, and possible interpretations of the findings. A brief overview of muscle physiology is provided to review the significance of the channels for muscle function.

Muscle physiology. Motoneuron activity is transferred to skeletal muscle at the neuromuscular junction, generating an endplate potential that depends on acetylcholine (ACh) release from the nerve terminal and its reaction with the subsynaptic nicotinic AChR, a pentameric ligand-gated ion channel (7). Normally, an endplate potential is large enough to induce a sarcolemmal action potential that propagates from the endplate to the tendon and along the transverse tubular system. This membrane region projects deeply into the cell to ensure even distribution of the impulse (Figure 1). The upstroke of the action potential is mediated by opening of the voltage-gated $Na_v1.4$ Na^+ channels (encoded by *SCN4A*), which elicit a Na^+ inward current with rapid activation kinetics. Repolarization of the membrane by fast Na^+ channel inactivation is supported by opening of delayed rectifier K^+ channels that mediate an outward K^+ current. Buffering of afterpotentials is achieved by a high Cl^- conductance near the resting potential, resulting from the homodimeric Cl^- channel ClC-1, encoded by

CLCN1 (2). At specialized junctions in the transverse tubular system, the signal is transmitted from the tubular membrane to the sarcoplasmic reticulum (SR), causing the release of Ca^{2+} ions into the myoplasm, which activate the contractile apparatus (8). This process is called excitation-contraction coupling. Two Ca^{2+} channel complexes are chiefly involved in this process, the voltage-gated pentameric $Ca_v1.1$ Ca^{2+} channel (also called the dihydropyridine receptor) located in the transverse tubular system, encoded by *CACNA1S*, and the homotetrameric ryanodine receptor, ryanodine receptor type 1 (RyR1), of the SR (9).

Muscle channelopathies due to altered membrane excitability

Most muscle channelopathies have similar clinical presentation: typically the symptoms occur as episodic attacks lasting from minutes to days that show spontaneous and complete remission, onset in the first or second decade of life, and — for unknown reasons — amelioration at the age of 40 or 50. Frequently, the attacks can be provoked by exercise, rest following physical activity, hormones, mental stress, or certain types of food and drugs.

CMSs: hypo- or hyperexcitable neuromuscular junctions

CMSs are a heterogeneous group of inherited disorders characterized by defective transmission of neuromuscular excitation resulting in muscle fatigue (10). Weakness is usually evident at birth or within the first year or 2 of life and is characterized by feeding difficulties, ptosis, impaired eye movements, and delayed motor milestones. In some cases, strength improves during adolescence and does not exhibit a progressive course. Reflexes are usually brisk, and muscle wasting does not occur. CMSs can lead to congenital arthrogryposis multiplex involving reduced fetal movement and multiple joint contractures in the neonate (11). Electromyography in CMS patients reveals a characteristic decrement of compound action potential amplitude on repetitive stimulation, and single-fiber recordings show an increased variability in the synaptic transmission time ("jitter") and transmission blocks (12).

Presynaptic, synaptic, and postsynaptic loss-of-function proteins. CMSs result from defects in presynaptic, synaptic, and postsynaptic proteins. Presynaptic defects reduce ACh release and resynthesis due to mutations in the choline acetyltransferase. Synaptic CMSs are caused by acetylcholinesterase (AChE) deficiency (13) due to mutations in the collagenic tail subunit (ColQ) that mediates AChE

Nonstandard abbreviations used: ACh, acetylcholine; AChR, ACh receptor; CMS, congenital myasthenic syndrome; MH, malignant hyperthermia; PP, periodic paralysis; RyR1, ryanodine receptor type 1; SR, sarcoplasmic reticulum.

Conflict of interest: The authors have declared that no conflict of interest exists.

Citation for this article: *J. Clin. Invest.* **115**:2000–2009 (2005). doi:10.1172/JCI25525.

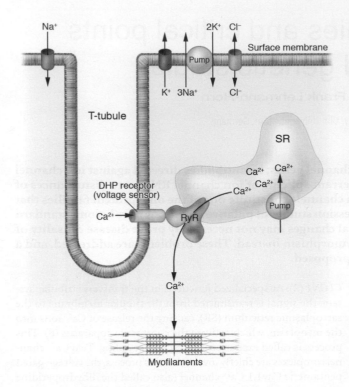

Figure 1

Excitation-contraction coupling of skeletal muscle. A muscle fiber is excited via the nerve by an endplate potential and generates an action potential, which spreads out along the surface membrane and the transverse tubular system into the deeper parts of the muscle fiber. The dihydropyridine (DHP) receptor senses the membrane depolarization, alters its conformation, and activates the ryanodine receptor, which releases Ca^{2+} from the SR, a Ca^{2+} store. Ca^{2+} binds to troponin and activates the so-called contractile machinery.

insertion into the synaptic basal lamina (10). Postsynaptic CMSs are caused by dominant or recessive mutations in 1 of the nicotinic AChR subunits (14) (Figure 2), or in proteins anchoring AChRs into the membrane, such as the rapsyn mutations (Table 1). Loss-of-function mutations of AChR subunits lead to compensatory expression of fetal δ subunits, yielding AChR complexes that differ functionally from the adult type.

Kinetic gain- and loss-of-function nicotinic AChR mutations. Rarely, postsynaptic CMSs are caused by mutations at different sites and different functional domains that alter the kinetic channel properties. These kinetic mutations result in the slow- or fast-channel syndromes. The low-affinity fast-channel syndrome is caused by loss-of-function mutations that have effects similar to those of AChR deficiency but is much rarer. Mutations at different sites lead to fewer and shorter channel activations. In contrast to all the CMSs described above, the slow-channel syndrome presents in childhood, adolescence, or adult life with upper-limb predominance and contractures, does not respond to anticholinesterase, and is progressive. CMS patients with a slow-channel syndrome show increased synaptic response to ACh with characteristic repetitive discharges in response to a single supramaximal stimulus. The syndrome results from gain-of-function mutations in the ion-conducting pore M2 (Figure 2). The leaky AChRs exert an excitotoxic effect and cause endplate myopathy via focal caspase activation (15–18).

Myotonia: plasmalemmal hyperexcitability due to mutant Na+ or Cl− channels

Muscle stiffness, termed myotonia, is ameliorated by exercise — the "warm-up phenomenon" — and can be associated with transient weakness during strenuous muscle activity. On the contrary, paradoxical myotonia (also called paramyotonia) worsens with cold and after exercise. Both myotonia and paramyotonia derive from uncontrolled repetitive action potentials of the sarcolemma following an initial voluntary activation. This may be noted as a myotonic burst

in the electromyogram. The involuntary electrical activity prevents the muscle from immediate relaxation after contraction, and the patients subsequently experience this as muscle stiffness.

Chloride channel myotonias: Thomsen and Becker. Dominant Thomsen myotonia and recessive Becker myotonia are caused by missense and nonsense mutations in the homodimeric Cl− channel encoded by *CLCN1* (19) (Figure 3). Functionally, the dominant mutants exert a dominant-negative effect on the dimeric channel complex as shown by coexpression studies, meaning that mutant/mutant and mutant/WT complexes are dysfunctional (20). The most common feature of the resulting Cl− currents is a shift of the activation threshold toward more positive membrane potentials almost out of the physiological range (21–23). As a consequence of this, the Cl− conductance is drastically reduced in the vicinity of the resting membrane potential. The recessive mutants that do not functionally hinder the associated subunit supply the explanation of why 2 mutant alleles are required to reduce Cl− conductance sufficiently for myotonia to develop in Becker myotonia.

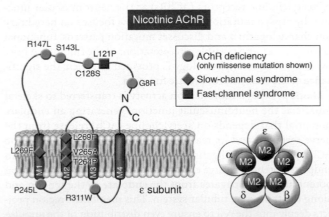

Figure 2

Muscle endplate nicotinic AChR. The nicotinic AChR of skeletal muscle is a pentameric channel complex consisting of 2 α subunits and 1 β, 1 γ, and 1 δ subunit in fetal and denervated muscle, and 2 α subunits and 1 β, 1 δ, and 1 ε subunit in adult muscle. All subunits have a similar structure with 4 transmembrane segments, M1 to M4. They form a channel complex with each subunit contributing equally to the ion-conducting central pore formed by the M2 segments. The pore is permeable to cations. The binding site for ACh is located in the long extracellular loop of the α subunit. The 3 main conformational states of the ligand-gated channels are closed, open, and desensitized. Binding of the transmitter opens the channel from the closed state, and, during constant presence of the transmitter, desensitization occurs. Only after removal of the transmitter, the channel can recover from desensitization and subsequently will be available for another opening. Mutations associated with subtypes of CMSs are indicated by conventional 1-letter abbreviations for the replaced amino acids.

Table 1

Hereditary muscle channelopathies

Gene	Locus	Channel protein	Disease	Heredity	Effect
SCN4A	17q23.1-25.3	Na$_v$1.4 Na$^+$ channel α subunit	Hyperkalemic periodic paralysis, paramyotonia congenita, K$^+$-aggravated myotonia	Dominant	Gain
			Hypokalemic periodic paralysis type 2	Dominant	Loss
CACNA1S	1q31-32	L-type Ca^{2+} channel α1 subunit, DHP receptor	Hypokalemic periodic paralysis type 1	Dominant	Unclear
			MH type 5	Dominant	Unclear
RyR1	19q13.1	RyR1, Ca^{2+} release channel	MH	Dominant	Gain
			Central core disease	Dominant or recessive	Gain
KCNJ2	17q23-24	Kir2.1 K$^+$ channel α subunit	Andersen syndrome	Dominant	Loss
KCNQ2	20q13.3	Kv7.2 K$^+$ channel α subunit	Neuromyotonia with benign neonatal familial convulsions	Dominant	Loss
CLCN1	7q32-qter	ClC-1 voltage-gated Cl$^-$ channel	Thomsen myotonia	Dominant	Loss
			Becker myotonia	Recessive	Loss
		Altered splicing of ClC-1	Myotonic dystrophy type 1	Dominant	Loss
			Myotonic dystrophy type 2	Dominant	Loss
CHRNA1	2q24-32	nAChR α1 subunit	CMS	Dominant or recessive	Gain or loss
CHRNB1	17p12-11	nAChR β1 subunit	CMS	Dominant or recessive	Gain or loss
CHRND	2q33-34	nAChR δ subunit	CMS	Dominant or recessive	Gain or loss
CHRNE	17p13-p12	nAChR ε1 subunit	CMS	Dominant or recessive	Gain or loss
RAPSN	11p11	Rapsyn, nAChR-associated	CMS	Recessive	Loss

DHP, dihydropyridine; nAChR, nicotinic AChR.

Chloride channel myotonia in myotonic dystrophies

Myotonic dystrophy, the most common inherited muscle disorder in adults, is a progressive multisystemic disease characterized by muscle wasting, myotonia, subcapsular cataracts, cardiac conduction defects, gonadal atrophy, hearing deficiencies, and cognitive deficits. There are 2 clinically distinguished types: type 1, with the classical phenotype caused by an expansion of an unstable CTG trinucleotide repeat in the 3′ untranslated region of the myotonic dystrophy protein kinase (DMPK) gene on chromosome 19q13.3 (24), and type 2, with a more proximal pattern of weakness (25) caused by an expansion of a CCTG tetranucleotide repeat in intron 1 of the ZNF9 gene coding for a zinc finger protein (26). The pathogenesis of the myotonia is based on an alternative splicing of the ClC-1 RNA, leading to loss of function of the channel protein (4, 5).

Sodium channel myotonia and paramyotonia

In Na$^+$ channel myotonia and paramyotonia, there is a gating defect of the Na$^+$ channels that destabilizes the inactivated state such that channel inactivation may be slowed or incomplete (27, 28). This results in an increased tendency of the muscle fibers to depolarize, which generates repetitive action potentials (myotonia). The mutant channels confer a dominant gain of function on the channel as well as on cell excitability (Figure 4).

One hot spot for the paramyotonia mutations is a special voltage-sensing transmembrane region that couples channel inactivation to channel activation (29); another hot spot is an intracellular protein loop containing the inactivation particle (30). The K$^+$-aggravated myotonia mutations are found in intracellular regions of the protein, potentially interfering with the channel inactivation process (28, 31). Corresponding to the severity of the disruption of the inactivation gate structure on the protein level, there are 3 clinical severities to be distinguished: (a) myotonia fluctuans, where patients may not be aware of their disorder; (b) myotonia

responsive to acetazolamide with a Thomsen-like clinical phenotype; and (c) myotonia permanens, where continuous electrical myotonia leads to a generalized muscle hypertrophy including facial and neck muscles, suggestive of facial dysmorphia. In all 3 types, body exertion or administration of depolarizing agents may result in a severe or even life-threatening myotonic crisis (32).

Periodic paralysis: plasmalemmal hypoexcitability due to mutant Na$^+$ or Ca^{2+} channels

Symptoms occur episodically with varying intervals of normal muscle function and excitation because ion channel defects are usually well compensated and an additional trigger is often required for muscle inexcitability. Two dominant episodic types of weakness with or without myotonia are distinguished by the serum K$^+$ level during the attacks of tetraplegia: hyperkalemic and hypokalemic periodic paralysis (PP). Intake of K$^+$ and intake of glucose have opposite effects in the 2 disorders: while K$^+$ triggers a hyperkalemic attack and glucose is a remedy, glucose provokes hypokalemic attacks, which are ameliorated by K$^+$ intake. Because of additional release of K$^+$ from hyperkalemic PP muscle and uptake of K$^+$ into hypokalemic PP muscle, serum K$^+$ disturbance can be so severe during a paralytic attack that cardiac complications arise. During an attack, death can also occur due to respiratory insufficiency (2).

Sodium channel PP with myotonia. Most Na$_v$1.4 mutations that cause hyperkalemic PP are situated at inner parts of the transmembrane segments or in intracellular protein loops (Figure 4) and affect structures that form the docking site for the fast-inactivation particle (33, 34). Thereby, they impair fast channel inactivation and lead to a persistent Na$^+$ current (35). At the beginning of an attack, the sustained inward current is associated with a mild membrane depolarization and leads to myotonia. The progressing attack is characterized by membrane inexcitability and muscle weakness, since the penetrated Na$^+$ ions go along with a more

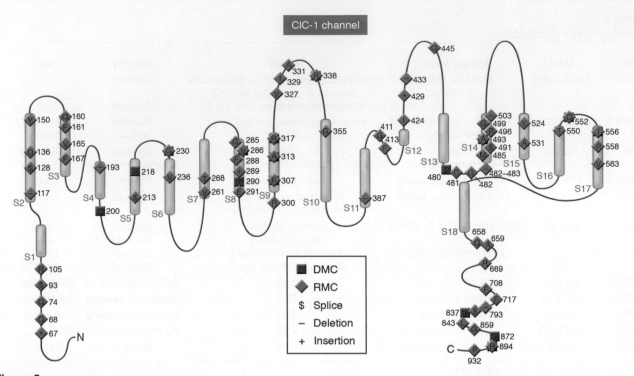

Figure 3
ClC-1, the major chloride channel of skeletal muscle. A membrane topology model of the ClC-1 monomer is shown. X-ray measurements and cryo-electron microscopy have elucidated the structure of the channel (104, 105) and confirmed the conclusions derived from electrophysiological results. The functional channel is an antiparallel assembled homodimer. It possesses 2 independent ion-conducting pores, each with a fast-opening mechanism of its own, 2 selectivity filters, and 2 voltage sensors (106). The channel is functional without any other subunits. Symbols are used for the mutations leading to either dominant myotonia congenita (DMC) or recessive myotonia congenita (RMC). The amino acids at which substitutions occur are indicated by 1-letter abbreviations and numbered according to the protein sequence. Adapted with permission from *Nature* (104).

severe sustained membrane depolarization that inactivates most Na⁺ channels. Depending on the location of the underlying mutation, symptoms typical of hyperkalemic PP, K⁺-aggravated myotonia, and paramyotonia congenita can overlap in a given patient (36). Sodium channel inhibitors such as mexiletine and flecainide are highly effective in preventing sodium channel myotonia and weakness in paramyotonia patients but not in patients with hyperkalemic PP. Particularly, mutant sodium channels that exhibit an enhanced closed-state inactivation offer a pharmacogenetic strategy for mutation-specific treatment (37).

Na⁺ and Ca²⁺ channel PP without myotonia. In contrast to the gain-of-function changes associated with hyperkalemic PP, hypokalemic PP is associated with a loss-of-function defect of 2 different ion channel types: Cav1.1 (hypokalemic PP type 1) and $Na_v1.4$ (clinically indistinguishable hypokalemic PP type 2) (38–40). The mutations are located exclusively in the voltage-sensing S4 segment of domain 2 of $Na_v1.4$ and domain 2 or 4 of Cav1.1 (Figure 4). Functionally, the inactivated state is stabilized in the Na⁺ channel mutants (39, 41, 42), while the channel availability is reduced for the Ca²⁺ channel mutants (43, 44). It is still unclear how the loss-of-function mutations of these 2 cation channels can produce the long-lasting depolarization that leads to the weakness seen in patients (45, 46). The attacks of weakness drastically reduce the patients' ability to perform activities of daily living. For many, loss of their jobs and social relationships is more distressing than the physical handicap.

K⁺ channel PP with cardiac arrhythmia. Patients with Andersen syndrome may experience a life-threatening ventricular arrhythmia independent of their PP, and long QT syndrome is the primary cardiac manifestation (47, 48). The syndrome is characterized by the highly variable clinical triad of PP, ventricular ectopy, and potential dysmorphic features (49, 50). The paralytic attack may be hyperkalemic or hypokalemic, and, accordingly, the response to oral K⁺ is unpredictable. Mutations of the Kir2.1 K⁺ channel, an inward rectifier expressed in skeletal and cardiac muscle, are causative of the disorder (51). Kir2.1 channels are essential for maintaining the highly negative resting membrane potential of muscle fibers and accelerating the repolarization phase of the cardiac action potential. The mutations mediate loss of channel function by haploinsufficiency or by dominant-negative effects on the WT allele (52) and may lead to long-lasting depolarization and membrane inexcitability.

Muscle channelopathies due to an altered excitation-contraction coupling

Muscle contractures, i.e., electrically silent contractions due to intracellular Ca²⁺ exceeding the mechanical threshold, as well as flaccid weakness are characteristic features of disturbed muscle excitation-contraction coupling. Two allelic forms are well studied: MH and central core disease.

Malignant hyperthermia

Susceptibility to MH is an autosomal dominant predisposition to respond abnormally when exposed to volatile anesthetics, depolarizing muscle relaxants, or extreme physical activity in hot environ-

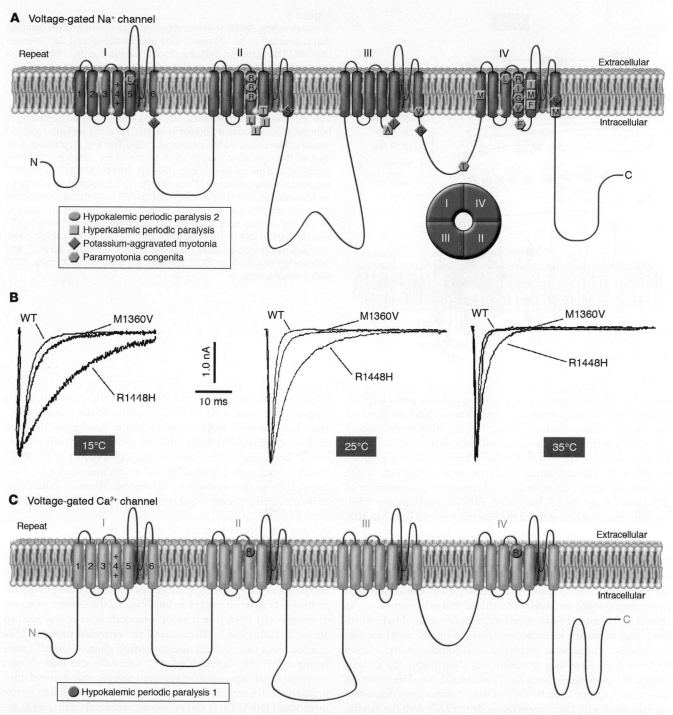

Figure 4

Voltage-gated Na+ and Ca2+ channels: structure and function. The α subunit consists of 4 highly homologous domains, I–IV, with 6 transmembrane segments each (S1–S6). The S5–S6 loops and the S6 transmembrane segments form the ion-selective pore, and the S4 segments contain positively charged residues that confer voltage dependence to the protein. The S4 segments are thought to move outward upon depolarization, thereby inducing channel opening. The repeats are connected by intracellular loops; in the Na+ channel (**A**), the III–IV linker contains the supposed inactivation particle, whereas the slowly activating and inactivating L-type Ca2+ channel does not possess a fast-inactivation gate (**C**). When inserted in the membrane, the 4 repeats of the protein fold to generate a central pore. Mutations associated with the various diseases are indicated. (**B**) Activation, inactivation, and recovery from the fast-inactivated to the resting state are voltage- and time-dependent processes. Compared is the fast inactivation of WT and 2 mutant skeletal muscle Na+ channels expressed in human embryonic kidney cells: R1448H, a cold-sensitive mutation causing paramyotonia congenita, and M1360V, a temperature-insensitive mutation causing hyperkalemic periodic paralysis. The whole-cell current responses to a depolarization from –100 mV to 0 mV were superimposed at 25°C and 35°C. Adapted with permission from the *Journal of Physiology* (107).

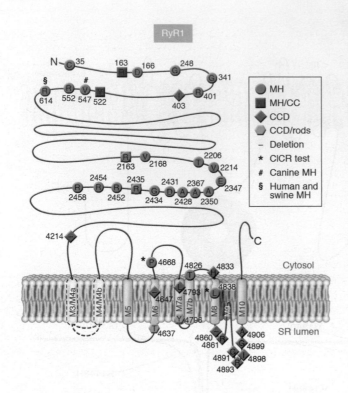

Figure 5
Skeletal muscle RyR1. RyR1 forms a homotetrameric protein complex that is situated in the SR membrane and functions as a Ca^{2+} release channel. The cytosolic part, the "foot," bridges the gap between the transverse tubular system and the SR. It contains binding sites for various activating ligands, like Ca^{2+} (μM), ATP (nM), calmodulin (nM), caffeine (mM), and ryanodine (nM), and inactivating ligands, like dantrolene (>10 μM), Ca^{2+} (>10 μM), ryanodine (>100 μM), and Mg^{2+} (μM). The transmembrane segments, M3–M10, are numbered according to both the earlier model of Zorzato et al. (108) and the recently modified model of MacLennan and colleagues (109). The first 2 cylinders, with dashed lines, indicate the tentative nature of the composition of the first predicted helical hairpin loop (M3–M4 or M4a–M4b). The long M7 sequence is designated as M7a and M7b. The proposed selectivity filter between M8 and M10 is designated as M9 even though it is clearly not a transmembrane sequence. Mutations causing susceptibility to MH and/or central core disease are indicated. Susceptibility to MH was defined by use of the in vitro contracture test or, in a single case, by the Japanese Ca-induced Ca release test (CICR test). MH/CC, MH with some central cores; CCD, central core disease; CCD/rods, CCD with nemaline rods.

ments (53). During exposure to triggering agents, a pathologically high increase in myoplasmic Ca^{2+} concentration leads to increased muscle metabolism and heat production, resulting in muscle contractures, hyperthermia associated with metabolic acidosis, hyperkalemia, and hypoxia. The metabolic alterations usually progress rapidly, and, without immediate treatment, up to 70% of the patients die. Early administration of dantrolene, an inhibitor of Ca^{2+} release from the SR, has successfully aborted numerous fulminant crises and has reduced the mortality rate to less than 10%.

In most families, mutations can be found in the gene encoding the skeletal muscle ryanodine receptor, RyR1 (Figure 1 and Figure 5). This Ca^{2+} channel is not voltage-dependent on its own but exists under the control of Cav1.1. MH mutations are usually situated in the cytosolic part of the protein and show gain-of-function effects: they increase RyR1 sensitivity to caffeine and other activators, as shown in functional tests of excised muscle, isolated native proteins, and ryanodine receptors expressed in muscle and nonmuscle cells (54). For another MH locus on chromosome 1q31-32, an R1086H disease-causing mutation was identified in the skeletal muscle L-type calcium channel α1 subunit (55, 56). The mutation is located in an intracellular loop of the protein, whose functional significance for EC coupling is under debate (57). Although mutations in the same gene cause hypokalemic PP type 1, this disorder is not thought to be associated with MH susceptibility (58, 59).

Central core disease
RyR1 mutations in the SR-luminal region cause central core disease, a congenital myopathy clinically characterized by muscle hypotrophy and weakness and a floppy-infant syndrome, often alongside other skeletal abnormalities such as hip displacement and scoliosis (60). Pathognomonic is the abundance of central cores devoid of oxidative enzyme activity along the predominant type 1 muscle fibers. Some mutations decrease the open probability of the RyR1 channel so that it loses the ability to release Ca^{2+}

in response to the altered conformation of the dihydropyridine receptor that is induced by depolarization of the plasma membrane (61). However, RyR1 retains the ability to influence the open probability of the dihydropyridine receptor, with which it interacts. Other mutations increase the open probability of the RyR1 channel, leading to depleted SR Ca^{2+} stores and weakness. Both dominant and rare recessive mutations have been described, the latter transiently presenting as multi-minicore disease (62).

In vitro functional studies of channel mutants
As illustrated above, functional expression of mutations has contributed to the understanding of the molecular pathogenesis of several muscle channelopathies. However, there are undeniable problems of interpreting changes in function brought about by mutants in in vitro expression systems. The overexpression of introduced DNA in a heterologous cell system may lead to an unphysiological localization of the encoded protein. This can lead to a false conclusion regarding channel significance (compare ref. 63 with refs. 64–67). Secondly, the cells chosen for functional expression may have endogenous channel subunits that can potentially interact with or be upregulated by the introduced DNA. These can generate currents that may be falsely assumed to appear due to the introduced DNA (compare ref. 68 with refs. 69, 70). Thirdly, heterologous expression systems may secondarily modify the channels chemically, which may lead to misinterpretation of the functional significance of the channel subunits of $Na_v1.4$, for example, which exhibits more rapid inactivation kinetics when expressed in human embryonic cell lines than in *Xenopus* oocytes. Originally, this finding was attributed to the lack of expression of the accessory β subunit in the oocytes; however, the rapid kinetics of the channel were also found when $Na_v1.4$ was expressed in cells without endogenous β subunits. Therefore, posttranslational modifications and association of sodium channels with other membrane proteins such

as cytoskeletal components are now considered responsible for the differences in kinetics (2).

The function of ion channels is highly dependent on the expression system used. The functional significance implied by these experiments may not necessarily be valid for the physiological situation in vivo. Additionally, a functional change may not necessarily prove that a naturally occurring amino acid substitution causes a disease. The functional change could be brought about by a polymorphism instead. In spite of inherent difficulties in obtaining such findings, several of these "functional polymorphisms" have been described. For example, S906T in the skeletal muscle Na$_v$1.4 sodium channel was found to segregate perfectly with PP in several large pedigrees and to alter entry into and recovery from slow channel inactivation. However, it occurs in 5% of the population without association with any disease (71). In the PP families mentioned, S906T turned out just to be linked to another change that is causing PP and that was identified much later.

Functional polymorphisms in cardiac Kv7.1 and Kv11.1 K$^+$ channels occur in approximately 11–30% of the population (72–74), and those in cardiac Na$_v$1.5 sodium channels have been detected in approximately 20% of the population (75, 76). These polymorphisms have been suggested to mediate susceptibility to life-threatening arrhythmia caused by elongated QT intervals in the ECG, even though the prevalence of the polymorphisms is several hundred times higher than that of the long QT syndrome.

The question arises of whether it is justified to consider up to 30% of the population to be at risk for long QT syndrome and what consequences such a high long QT prevalence should have. Given the variability with which humans are prone to polymorphisms, it is not surprising that these polymorphisms must be associated with changes in function that can generate disease-susceptible, disease-protective, or otherwise disadvantageous or advantageous features (such as intelligence or attractiveness). Therefore, it is not at all clear how to interpret changes of channel function brought about by naturally occurring amino acid changes when studying in vitro expression systems.

In contrast to polymorphisms that may lead to functional changes, mutations in ion channels may cause changes of function that are irrelevant to disease. For example, familial hypokalemic PP type 1 still baffles scientists even though the genetic cause was identified 10 years ago (38) and the functional defects of the mutant Ca^{2+} channel have been described in various expression systems (43, 44, 77, 78). Nevertheless, how a channel that is primarily involved in muscle excitation-contraction coupling can elicit the long-lasting membrane depolarization shown to be the cause of the paralysis (45, 46) is still not understood. Another striking example is that opposite changes of function that have been described for putative mutations of the same channel can cause the same clinical phenotype. This unexpected observation contradicts the idea that a singular pathogenetic mechanism can be deduced from similar functional defects observed in vitro. For example, several missense mutations in the voltage-gated neuronal Na$^+$ channels Na$_v$1.1 and Na$_v$1.2 are thought to cause a dominant monogenic form of epilepsy. When expressed in mammalian cells or in transgenic mice, some of the mutations enhanced channel inactivation and reduced membrane excitability, while others destabilized the inactivated channel state and increased cell excitability (79–82). The situation for familial hemiplegic migraine, a rare subtype of migraine with aura caused by Cav2.1 Ca^{2+} channel mutations, is similar: the mutations lead to either reduced or increased Ca^{2+} influx into the cytoplasm so that both gain and loss of channel mechanisms can cause the same phenotype (83–87).

Epidemiology and genetic linkage studies of channel mutants

Frequency of the putative mutation in a control population

Because of the shortcomings of the interpretation of functional studies, the genetic screening of large and adequately matched control populations for absence of the putative mutations is important to prove disease causality. Two reports have proposed the typing of 150–200 controls (300–400 chromosomes) for putative mutations with a prevalence of 1% by power analysis (88, 89). A more general algorithm that recommends exclusion of the putative mutation in ethnically matched control chromosomes has recently been proposed (90). For a proposed maximally tolerable error of 1% and a mutation present on 1% of tested patient chromosomes, the equation advises to test 460 control chromosomes (230 control individuals) (Figure 6).

Therefore, the common laboratory practice of excluding a novel mutation in approximately 100 healthy controls is insufficient. An example is an R83H substitution in a K$^+$ channel β subunit, MiRP2, suggested to cause PP because it showed a loss of function in vitro and was found in 2 of 100 of such patients but in none of 120 unaffected controls (91). In later studies, the substitution was identified in 1 of 104 and 1 of 138 patients, but also in 8 of 506 and 3 of 321 controls (90, 92). When these results are taken together, the substitution is present in 1.17% of patients

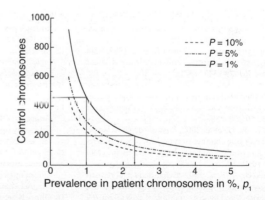

Figure 6
Proposed number of control chromosomes. A statistical algorithm helps to calculate the number of controls required to minimize the error. Let the prevalence of a mutation in patient chromosomes be p_1 and the prevalence in control chromosomes be p_0. Then the probability of an arbitrary control chromosome not carrying the mutation is $(1 - p_0)$. Because the world control population is large, the probability P of arbitrarily choosing n chromosomes thereof without the mutation may be approximated by $P = (1 - p_0)^n$. The null hypothesis would be that the mutation frequency is equal in patient and control chromosomes, i.e., $p_0 = p_1$ and $P = (1 - p_1)^n$. The number of control chromosomes to be tested can be calculated by resolution of the equation for the number $n = \ln(P)/\ln(1 - p_1)$. When the error probability P is set at 1%, the number of required control chromosomes is $n = -4.6/\ln(1 - p_1)$ and $n = 460$ for the example of $p_1 = 1\%$. The curve demonstrates that 100 control individuals (200 chromosomes) would be adequate for a p_1 of 2.5%, a prevalence that is much higher than that of the most frequent monogenic disorder. Adapted with permission from *Neurology* (90).

and in 1.16% of healthy controls, which does not support disease causality. Even though the difference between defining a putative mutation as truly disease-causing and defining it as a functional polymorphism may seem only marginal on a scientific level, this difference has drastic consequences for an affected carrier whose diagnosis is made or confirmed by the finding and who is being medically treated. This problem will increasingly need to be addressed in future studies when the number of known mutations and putatively associated phenotypes continues to increase.

Genetic linkage studies within families

Genetic linkage analyses were very successful in finding the gene loci in MH and also in hypokalemic PP (39, 93–95). In contrast to these genome-wide analyses in which large pedigrees were studied, a number of relatively small MH families for which linkage to the ryanodine receptor gene on chromosome 19 was excluded were tested in a candidate-gene approach. The first alternative locus was assigned to chromosome 17q11.2-q24, which suggested SCN4A, the gene encoding the skeletal muscle sodium channel, as candidate gene (96, 97). Apparently some of the families had a muscle Na$^+$ channelopathy (98, 99), which could explain the anesthesia-related events as exaggerated myotonic reactions. The generalized muscle spasms and resulting systemic alterations are usually triggered by succinylcholine in patients with a Na$^+$ channelopathy and can resemble MH; as further evidence for the myotonic origin of the crises, susceptibility to MH was excluded in such patients by the European in vitro contracture test (32, 58, 100–102). Suggestions of 4 further MH loci, made each in a single pedigree, still await confirmation.

Phylogenetic analysis of genes for conserved sequences and structural regions

Good conservation of amino acid residues is no guarantee that all changes of such a residue would lead to disease. For example, the polymorphism W118G in ClC-1 (103) concerns an amino acid that is highly conserved in several members of the ClC-channel family, such as ClC-2, ClC-7, ClC-Ka, and ClC-Kb. In contrast, several of the known disease-causing ClC-1 mutations such as F413C or Q552R affect residues are not equally well conserved in these channels.

These examples of questionable interpretation of genetic or functional data emphasize the importance of combining as many of the above criteria as possible to be able to make a reliable decision regarding whether a given variant may be deleterious or not. Genetic animal models or gene expression profiling may clarify these areas in the future.

Acknowledgments

This work was supported by the German Research Foundation (Deutsche Forschungsgemeinschaft, JU470/1) and by the European Commission's Human Potential Programme under contract HPRN-CT-2002-00331, EC coupling in striated muscle.

Address correspondence to: Karin Jurkat-Rott and Frank Lehmann-Horn, Department of Applied Physiology, Ulm University, Albert-Einstein-Allee 11, 89069 Ulm, Germany. Phone: 49-731-50-23250; Fax: 49-731-50-23260; E-mail: frank.lehmann-horn@medizin.uni-ulm.de (F. Lehmann-Horn). Phone: 49-731-50-23065; Fax: 49-731-50-23260; E-mail: karin.jurkat-rott@medizin.uni-ulm.de (K. Jurkat-Rott).

1. Hoffman, E.P., Lehmann-Horn, F., and Rüdel, R. 1995. Overexcited or inactive: ion channels in muscle diseases. Cell. 80:681–686.
2. Lehmann-Horn, F., and Jurkat-Rott, K. 1999. Voltage-gated ion channels and hereditary disease. Physiol. Rev. 79:1317–1371.
3. Vincent, A., and Mills, K. 1999. Genetic and antibody-mediated channelopathies at the neuromuscular junction. Electroencephalogr. Clin. Neurophysiol. Suppl. 50:250–258.
4. Charlet-B., N., et al. 2002. Loss of the muscle-specific chloride channel in type 1 myotonic dystrophy due to misregulated alternative splicing. Mol. Cell. 10:45–53.
5. Mankodi, A., et al. 2002. Expanded CUG repeats trigger aberrant splicing of ClC-1 chloride channel pre-mRNA and hyperexcitability of skeletal muscle in myotonic dystrophy. Mol. Cell. 10:35–44.
6. Ohno, K., et al. 2002. Rapsyn mutations in humans cause endplate acetylcholine-receptor deficiency and myasthenic syndrome. Am. J. Hum. Genet. 70:875–885.
7. Lindstrom, J.M. 2003. Nicotinic acetylcholine receptors of muscles and nerves: comparison of their structures, functional roles, and vulnerability to pathology. Ann. N. Y. Acad. Sci. 998:41–52.
8. Leong, P., and MacLennan, D.H. 1998. A 37-amino acid sequence in the skeletal muscle ryanodine receptor interacts with the cytoplasmic loop between domains II and III in the skeletal muscle dihydropyridine receptor. J. Biol. Chem. 273:7791–7794.
9. Meissner, G. 2002. Regulation of mammalian ryanodine receptors. Front. Biosci. 7:2072–2080.
10. Engel, A.G., Ohno, K., Shen, X.M., and Sine, S.M. 2003. Congenital myasthenic syndromes: multiple molecular targets at the neuromuscular junction. Ann. N. Y. Acad. Sci. 998:138–160.
11. Brownlow, S., et al. 2001. Acetylcholine receptor delta subunit mutations underlie a fast-channel myasthenic syndrome and arthrogryposis multiplex congenita. J. Clin. Invest. 108:125–130. doi:10.1172/JCI200112935.
12. Kullmann, D.M., and Hanna, M.G. 2002. Neurological disorders caused by inherited ion-channel mutations. Lancet Neurol. 1:157–166.
13. Hutchinson, D., et al. 1993. Congenital endplate acetylcholinesterase deficiency. Brain. 116:633–653.
14. Engel, A.G., Ohno, K., Bouzat, C., Sine, S.M., and Griggs, R.G. 1996. Endplate acetylcholine receptor deficiency due to nonsense mutations in the epsilon subunit. Ann. Neurol. 40:810–817.
15. Engel, A.G., et al. 1982. A newly recognised congenital myasthenic syndrome attributed to a prolonged open time of the acetylcholine-induced ion channel. Ann. Neurol. 11:553–569.
16. Ohno, K., et al. 1995. Congenital myasthenic syndrome caused by prolonged acetylcholine receptor channel openings due to a mutation in the M2 domain of the ε subunit. Proc. Natl. Acad. Sci. U. S. A. 92:758–762.
17. Milone, M., et al. 1997. Slow-channel myasthenic syndrome caused by enhanced activation, desensitization, and agonist binding affinity attributable to mutation in the M2 domain of the acetylcholine receptor alpha subunit. J. Neurosci. 17:5651–5665.
18. Vohra, B.P., et al. 2004. Focal caspase activation underlies the endplate myopathy in slow-channel syndrome. Ann. Neurol. 55:347–352.
19. Koch, M.C., et al. 1992. The skeletal muscle chloride channel in dominant and recessive human myotonia. Science. 257:797–800.
20. Saviane, C., Conti, F., and Pusch, M. 1999. The muscle chloride channel ClC-1 has a double-barreled appearance that is differentially affected in dominant and recessive myotonia. J. Gen. Physiol. 113:457–468.
21. Pusch, M., Steinmeyer, K., Koch, M.C., and Jentsch, T.J. 1995. Mutations in dominant human myotonia congenita drastically alter the voltage dependence of the ClC-1 chloride channel. Neuron. 15:1455–1463.
22. Wagner, S., et al. 1998. The dominant chloride channel mutant G200R causing fluctuating myotonia: clinical findings, electrophysiology and channel pathology. Muscle Nerve. 21:1122–1128.
23. Simpson, B.J., et al. 2004. Characterization of three myotonia-associated mutations of the CLCN1 chloride channel gene via heterologous expression [abstract]. Hum. Mutat. 24:185.
24. Harley, H.G., et al. 1992. Expansion of an unstable DNA region and phenotypic variation in myotonic dystrophy. Nature. 355:545–546.
25. Ricker, K., et al. 1994. Proximal myotonic myopathy: a new dominant disorder with myotonia, muscle weakness, and cataracts. Neurology. 44:1448–1452.
26. Liquori, C.L., et al. 2001. Myotonic dystrophy type 2 caused by a CCTG expansion in intron 1 of ZNF9. Science. 293:864–867.
27. Lehmann-Horn, F., Rüdel, R., and Ricker, K. 1987. Membrane defects in paramyotonia congenita (Eulenburg). Muscle Nerve. 10:633–641.
28. Lerche, H., et al. 1993. Human sodium channel myotonia: slowed channel inactivation due to substitutions for a glycine within the III/IV linker. J. Physiol. 470:13–22.
29. Chahine, M., et al. 1994. Sodium channel mutations in paramyotonia congenita uncouple inactivation from activation. Neuron. 12:281–294.
30. Bouhours, M., et al. 2004. Functional characterization and cold sensitivity of T1313A, a new mutation of the skeletal muscle sodium channel causing paramyotonia congenita in humans. J. Physiol. 554:635–647.
31. Heine, R., Pika, U., and Lehmann-Horn, F. 1993. A novel SCN4A mutation causing myotonia aggravated by cold and potassium. Hum. Mol. Genet. 2:1349–1353.

32. Ricker, K., Moxley, R.T., 3rd, Heine, R., and Lehmann-Horn, F. 1994. Myotonia fluctuans. A third type of muscle sodium channel disease. *Arch. Neurol.* **51**:1095–1102.

33. Rojas, C.V., et al. 1991. A Met-to-Val mutation in the skeletal muscle sodium channel α-subunit in hyperkalemic periodic paralysis. *Nature.* **354**:387–389.

34. Bendahhou, S., Cummins, T.R., Kula, R.W., Fu, Y.H., and Ptácek, L.J. 2002. Impairment of slow inactivation as a common mechanism for periodic paralysis in DIIS4-S5. *Neurology.* **58**:1266–1272.

35. Lehmann-Horn, F., et al. 1987. Adynamia episodica hereditaria with myotonia: a non-inactivating sodium current and the effect of extracellular pH. *Muscle Nerve.* **10**:363–374.

36. Kim, J., et al. 2001. Phenotypic variation of a Thr704Met mutation in skeletal sodium channel gene in a family with paralysis periodica paramyotonica. *J. Neurol. Neurosurg. Psychiatr.* **70**:618–623.

37. Mohammadi, B., et al. 2005. Preferred mexiletine block of human sodium channels with IVS4 mutations and its pH-dependence. *Pharmacogenet. Genomics.* **15**:235–244.

38. Jurkat-Rott, K., et al. 1994. A calcium channel mutation causing hypokalemic periodic paralysis. *Hum. Mol. Genet.* **3**:1415–1419.

39. Jurkat-Rott, K., et al. 2000. Voltage sensor sodium channel mutations cause hypokalemic periodic paralysis type 2 by enhanced inactivation and reduced current. *Proc. Natl. Acad. Sci. U. S. A.* **97**:9549–9554.

40. Vicart, S., et al. 2004. New mutations of SCN4A cause a potassium-sensitive normokalemic periodic paralysis. *Neurology.* **63**:2120–2127.

41. Struyk, A.F., Scoggan, K.A., Bulman, D.E., and Cannon, S.C. 2000. The human skeletal muscle Na channel mutation R669H associated with hypokalemic periodic paralysis enhances slow inactivation. *J. Neurosci.* **20**:8610–8617.

42. Kuzmenkin, A., et al. 2002. Enhanced inactivation and pH sensitivity of Na+ channel mutations causing hypokalemic periodic paralysis type II. *Brain.* **125**:835–843.

43. Jurkat-Rott, K., et al. 1998. Calcium currents and transients of native and heterologously expressed mutant skeletal muscle DHP receptor α1 subunits (R528H). *FEBS Lett.* **423**:198–204.

44. Morrill, J.A., and Cannon, S.C. 1999. Effects of mutations causing hypokalaemic periodic paralysis on the skeletal muscle L-Type Ca²⁺ channel expressed in Xenopus laevis oocytes. *J. Physiol.* **520**:321–336.

45. Rüdel, R., Lehmann-Horn, F., Ricker, K., and Küther, G. 1984. Hypokalemic periodic paralysis: in vitro investigation of muscle fiber membrane parameters. *Muscle Nerve.* **7**:110–120.

46. Ruff, R.L. 1999. Insulin acts in hypokalemic periodic paralysis by reducing inward rectifier K+ current. *Neurology.* **53**:1556–1563.

47. Andelfinger, G., et al. 2002. KCNJ2 mutation results in Andersen syndrome with sex-specific cardiac and skeletal muscle phenotypes. *Am. J. Hum. Genet.* **71**:663–668.

48. Tristani-Firouzi, M., et al. 2002. Functional and clinical characterization of KCNJ2 mutations associated with LQT7 (Andersen syndrome). *J. Clin. Invest.* **110**:381–388. doi:10.1172/JCI200215183.

49. Tawil, R., et al. 1994. Andersen's syndrome: potassium-sensitive periodic paralysis, ventricular ectopy, and dysmorphic features. *Ann. Neurol.* **35**:326–330.

50. Sansone, V., et al. 1997. Andersen's syndrome: a distinct periodic paralysis. *Ann. Neurol.* **42**:305–312.

51. Plaster, N.M., et al. 2001. Mutations in Kir2.1 cause the developmental and episodic electrical phenotypes of Andersen's syndrome. *Cell.* **105**:511–519.

52. Bendahhou, S., et al. 2003. Defective potassium channel Kir2.1 trafficking underlies Andersen-Tawil syndrome. *J. Biol. Chem.* **278**:51779–51785.

53. Gronert, G.A., Thompson, R.L., and Onofrio, B.M. 1980. Human malignant hyperthermia: awake episodes and correction by dantrolene. *Anesth. Analg.* **59**:377–378.

54. Tong, J., et al. 1997. Caffeine and halothane sensitivity of intracellular Ca²⁺ release is altered by 15 calcium release channel (ryanodine receptor) mutations associated with malignant hyperthermia and/or central core disease. *J. Biol. Chem.* **272**:26332–26339.

55. Monnier, N., Procaccio, V., Stieglitz, P., and Lunardi, J. 1997. Malignant-hyperthermia susceptibility is associated with a mutation of the α1-subunit of the human dihydropyridine sensitive L-type voltage-dependent calcium-channel receptor in skeletal muscle. *Am. J. Hum. Genet.* **60**:1316–1325.

56. Stewart, S.L., Hogan, K., Rosenberg, H., and Fletcher, J.E. 2001. Identification of the Arg1086His mutation in the alpha subunit of the voltage-dependent calcium channel (CACNA1S) in a North American family with malignant hyperthermia. *Clin. Genet.* **59**:178–184.

57. Weiss, R.G., et al. 2004. Functional analysis of the R1086H malignant hyperthermia mutation in the DHPR reveals an unexpected influence of the III-IV loop on skeletal muscle EC coupling. *Am. J. Physiol. Cell Physiol.* **287**:C1094–C1102.

58. Lehmann-Horn, F., and Iaizzo, P.A. 1990. Are myotonias and periodic paralyses associated with susceptibility to malignant hyperthermia? *Br. J. Anaesth.* **65**:692–697.

59. Marchant, C.L., Ellis, F.R., Halsall, P.J., Hopkins, P.M., and Robinson, R.L. 2004. Mutation analysis of two patients with hypokalemic periodic paralysis and suspected malignant hyperthermia *Muscle Nerve.* **30**:114–117.

60. Lynch, P.J., et al. 1999. A mutation in the transmembrane/luminal domain of the ryanodine receptor is associated with abnormal Ca²⁺ release channel function and severe central core disease. *Proc. Natl. Acad. Sci. U. S. A.* **96**:4164–4169.

61. Avila, G., O'Brien, J.J., and Dirksen, R.T. 2001. Excitation-contraction uncoupling by a human central core disease mutation in the ryanodine receptor. *Proc. Natl. Acad. Sci. U. S. A.* **98**:4215–4220.

62. Ferreiro, A., et al. 2002. A recessive form of central core disease, transiently presenting as multi-minicore disease, is associated with a homozygous mutation in the ryanodine receptor type 1 gene. *Ann. Neurol.* **51**:750–759.

63. Duan, D., Winter, C., Cowley, S., Hume, J.R., and Horowitz, B. 1997. Molecular identification of a volume-regulated chloride channel. *Nature.* **390**:417–421.

64. Li, X., Shimada, K., Showalter, L.A., and Weinman, S.A. 2000. Biophysical properties of ClC-3 differentiate it from swelling-activated chloride channels in Chinese hamster ovary-K1 cells. *J. Biol. Chem.* **275**:35994–35998.

65. Stobrawa, S.M., et al. 2001. Disruption of ClC-3, a chloride channel expressed on synaptic vesicles, leads to a loss of the hippocampus. *Neuron.* **29**:185–196.

66. Weylandt, K.H., et al. 2001. Human ClC-3 is not the swelling-activated chloride channel involved in cell volume regulation. *J. Biol. Chem.* **276**:17461–17467.

67. Arreola, J., et al. 2002. Secretion and cell volume regulation by salivary acinar cells from mice lacking expression of the Clcn3 Cl- channel gene. *J. Physiol.* **535**:207–216.

68. Wang, K.W., and Goldstein, S.A. 1995. Subunit composition of minK potassium channels. *Neuron.* **14**:1303–1309.

69. Barhanin, J., et al. 1996. K(V)LQT1 and lsK (minK) proteins associate to form the I(Ks) cardiac potassium current. *Nature.* **7**:78–80.

70. Sanguinetti, M.C., et al. 1996. Coassembly of K(V)LQT1 and minK (IsK) proteins to form cardiac I(Ks) potassium channel. *Nature.* **384**:80–83.

71. Kuzmenkin, A., Jurkat-Rott, K., Lehmann-Horn, F., and Mitrovic, N. 2003. Impaired slow inactivation due to a benign polymorphism and substitutions of Ser-906 in the II-III loop of the human Na_v1.4 channel. *Pflügers Arch.* **447**:71–77.

72. Kubota, T., et al. 2001. Evidence for a single nucleotide polymorphism in the KCNQ1 potassium channel that underlies susceptibility to life-threatening arrhythmias. *J. Cardiovasc. Electrophysiol.* **12**:1223–1229.

73. Paavonen, K.J., et al. 2003. Functional characterization of the common amino acid 897 polymorphism of the cardiac potassium channel KCNH2 (HERG). *Cardiovasc. Res.* **59**:603–611.

74. Bezzina, C.R., et al. 2003. A common polymorphism in KCNH2 (HERG) hastens cardiac repolarization. *Cardiovasc. Res.* **59**:27–36.

75. Viswanathan, P.C., Benson, D.W., and Balser, J.R. 2003. A common SCN5A polymorphism modulates the biophysical effects of an SCN5A mutation. *J. Clin. Invest.* **111**:341–346. doi:10.1172/JCI200316879.

76. Ye, B., Valdivia, C.R., Ackerman, M.J., and Makielski, J.C. 2003. A common human SCN5A polymorphism modifies expression of an arrhythmia causing mutation. *Physiol. Genomics.* **12**:187–193.

77. Lapie, P., Goudet, C., Nargeot, J., Fontaine, B., and Lory, P. 1996. Electrophysiological properties of the hypokalaemic periodic paralysis mutation (R528H) of the skeletal muscle α1s subunit as expressed in mouse L cells. *FEBS Lett.* **382**:244–248.

78. Lerche, H., Klugbauer, N., Lehmann-Horn, F., Hofmann, F., and Melzer, W. 1996. Expression and functional characterization of the cardiac L-type calcium channel carrying a skeletal muscle DHP-receptor mutation causing hypokalaemic periodic paralysis. *Pflügers Arch.* **431**:461–463.

79. Alekov, A.K., Rahman, M.M., Mitrovic, N., Lehmann-Horn, F., and Lerche, H. 2000. A sodium channel mutation causing epilepsy in man exhibits subtle defects in fast inactivation and activation in vitro. *J. Physiol.* **529**:533–539.

80. Kearney, J.A., et al. 2001. A gain-of-function mutation in the sodium channel gene Scn2a results in seizures and behavioral abnormalities. *Neuroscience.* **102**:307–317.

81. Lossin, C., Wang, D.W., Rhodes, T.H., Vanoye, C.G., and George, A.L., Jr. 2002. Molecular basis of an inherited epilepsy. *Neuron.* **34**:877–884.

82. Spampanato, J., Escayg, A., Meisler, M.H., and Goldin, A.L. 2003. Generalized epilepsy with febrile seizures plus type 2 mutation W1204R alters voltage-dependent gating of Nav1.1 sodium channels. *Neuroscience.* **116**:37–48.

83. Hans, M., et al. 1999. Functional consequences of mutations in the human alpha1A calcium channel subunit linked to familial hemiplegic migraine. *J. Neurosci.* **19**:1610–1619.

84. Kraus, R.L., Sinnegger, M.J., Glossmann, H., Hering, S., and Striessnig, J. 1998. Familial hemiplegic migraine mutations change alpha1A Ca2+ channel kinetics. *J. Biol. Chem.* **273**:5586–5590.

85. Kraus, R.L., et al. 2000. Three new familial hemiplegic migraine mutants affect P/Q-type Ca²⁺ channel kinetics. *J. Biol. Chem.* **275**:9293–9243.

86. Piedras-Renteria, E.S., et al. 2001. Increased expression of alpha 1A Ca2+ channel currents arising from expanded trinucleotide repeats in spinocerebellar ataxia type 6. *J. Neurosci.* **21**:9185–9193.

87. Urbano, F.J., et al. 2003. Altered properties of quantal neurotransmitter release at endplates of mice lacking P/Q-type Ca2+ channels. *Proc. Natl. Acad. Sci. U. S. A.* **100**:3491–3496.

88. Marchuk, D.A., et al. 1998. Laboratory approaches toward gene identification. In *Approaches to gene mapping in complex human diseases.* J.L. Haines and M.A. Pericak-Vance, editors. Wiley-Liss. New York,

New York, USA. 371–372.

89. Collins, J.S., and Schwartz, C.E. 2002. Detecting polymorphisms and mutations in candidate genes. *Am. J. Hum. Genet.* **71**:1251–1252.

90. Jurkat-Rott, K., and Lehmann-Horn, F. 2004. Periodic paralysis mutation MiRP2-R83H in controls: interpretations and general recommendation. *Neurology.* **62**:1012–1015.

91. Abbott, G.W., et al. 2001. MiRP2 forms potassium channels in skeletal muscle with Kv3.4 and is associated with periodic paralysis. *Cell.* **104**:217–231.

92. Sternberg, D., Tabti, N., Fournier, E., Hainque, B., and Fontaine, B. 2003. Lack of association of the potassium channel-associated peptide MiRP2-R83H variant with periodic paralysis. *Neurology.* **61**:857–859.

93. McCarthy, T.V., et al. 1990. Localization of the malignant hyperthermia susceptibility locus to human chromosome 19q11.2-13.2. *Nature.* **343**:562–563.

94. MacLennan, D.H., et al. 1990. Ryanodine receptor gene is a candidate for predisposition to malignant hyperthermia. *Nature.* **343**:559–561.

95. Fontaine, B., et al. 1994. Mapping of the hypokalaemic periodic paralysis (HypoPP) locus to chromosome 1q31-32 in three European families. *Nat. Genet.* **6**:267–272.

96. Levitt, R.C., et al. 1992. Evidence for the localization of a malignant hyperthermia susceptibility locus (MHS2) to human chromosome 17q. *Genomics.* **14**:562–566.

97. Olckers, A., et al. 1992. Adult muscle sodium channel alpha-subunit is a gene candidate for malignant hyperthermia susceptibility. *Genomics.* **14**:829–831.

98. Vita, G.M., et al. 1995. Masseter muscle rigidity associated with glycine1306-to-alanine mutation in the adult muscle sodium channel alpha-subunit gene. *Anesthesiology.* **82**:1097–1103.

99. Moslehi, R., Langlois, S., Yam, I., and Friedman, J.M. 1998. Linkage of malignant hyperthermia and hyperkalemic periodic paralysis to the adult skeletal muscle sodium channel (SCN4A) gene in a large pedigree. *Am. J. Med. Genet.* **76**:21–27.

100. Ricker, K., Lehmann-Horn, F., and Moxley, R.T., III. 1990. Myotonia fluctuans. *Arch. Neurol.* **47**:268–272.

101. Allen, G.C. 1993. Paramyotonia and MH. *Can. J. Anaesth.* **40**:580–581.

102. Iaizzo, P., and Lehmann-Horn, F. 1995. Anesthetic complications in muscle disorders. *Anesthesiology.* **82**:1093–1096.

103. Lehmann-Horn, F., Mailänder, V., Heine, R., and George, A.L. 1995. Myotonia levior is a chloride channel disorder. *Hum. Mol. Genet.* **4**:1397–1402.

104. Dutzler, R., Campbell, E.B., Cadene, M., Chait, B.T., and MacKinnon, R. 2002. X-ray structure of a ClC chloride channel at 3.0 Å reveals the molecular basis of anion selectivity. *Nature.* **415**:287–294.

105. Mindell, J.A., Maduke, M., Miller, C., and Grigorieff, N. 2001. Projection structure of a ClC-type chloride channel at 6.5 Å resolution. *Nature.* **409**:219–223.

106. Fahlke, C., Rhodes, T.H., Desai, R.R., and George, A.L., Jr. 1998. Pore stoichiometry of a voltage-gated chloride channel. *Nature.* **394**:687–690.

107. Mohammadi, B., Mitrovic, N., Lehmann-Horn, F., Dengler, R., and Bufler, J. 2003. Mechanisms of cold sensitivity of paramyotonia congenita mutation R1448H and overlap syndrome mutation M1360V. *J. Physiol.* **547**:691–698.

108. Zorzato, F., et al. 1990. Molecular cloning of cDNA encoding human and rabbit forms of the Ca2+ release channel (ryanodine receptor) of skeletal muscle sarcoplasmic reticulum. *J. Biol. Chem.* **265**:2244–2256.

109. Du, G.G., Sandhu, B., Khanna, V.K., Guo, X.H., and MacLennan, D.H. 2002. Topology of the Ca2+ release channel of skeletal muscle sarcoplasmic reticulum (RyR1). *Proc. Natl. Acad. Sci. U. S. A.* **99**:16725–16730.

Inherited disorders of voltage-gated sodium channels

Alfred L. George Jr.

Division of Genetic Medicine, Departments of Medicine and Pharmacology, Vanderbilt University, Nashville, Tennessee, USA.

A variety of inherited human disorders affecting skeletal muscle contraction, heart rhythm, and nervous system function have been traced to mutations in genes encoding voltage-gated sodium channels. Clinical severity among these conditions ranges from mild or even latent disease to life-threatening or incapacitating conditions. The sodium channelopathies were among the first recognized ion channel diseases and continue to attract widespread clinical and scientific interest. An expanding knowledge base has substantially advanced our understanding of structure-function and genotype-phenotype relationships for voltage-gated sodium channels and provided new insights into the pathophysiological basis for common diseases such as cardiac arrhythmias and epilepsy.

Introduction

Voltage-gated sodium channels (Na$_V$Chs) are important for the generation and propagation of signals in electrically excitable tissues like muscle, the heart, and nerve. Activation of Na$_V$Chs in these tissues causes the initial upstroke of the compound action potential, which in turn triggers other physiological events leading to muscular contraction and neuronal firing. Na$_V$Chs are also important targets for local anesthetics, anticonvulsants, and antiarrhythmic agents.

The essential nature of Na$_V$Chs is emphasized by the existence of inherited disorders (sodium "channelopathies") caused by mutations in genes that encode these vital proteins. Nearly 20 disorders affecting skeletal muscle contraction, cardiac rhythm, or neuronal function and ranging in severity from mild or latent disease to life-threatening or incapacitating conditions have been linked to mutations in human Na$_V$Ch genes (Table 1). Most sodium channelopathies are dominantly inherited, but some are transmitted by recessive inheritance or appear sporadic. Additionally, certain pharmacogenetic syndromes have been traced to variants in Na$_V$Ch genes. The clinical manifestations of these disorders depend primarily on the expression pattern of the mutant gene at the tissue level and the biophysical character of Na$_V$Ch dysfunction at the molecular level.

This review will cover the current state of knowledge of human sodium channelopathies and illustrate important links among clinical, genetic, and pathophysiological features of the major syndromes with the corresponding biophysical properties of mutant Na$_V$Chs. An initial brief overview of the structure and function of Na$_V$Chs will provide essential background information needed to understand the nuances of these relationships. This will be followed by a review of the major syndromes, organized by affected tissue. Emphasis will be placed on relating clinical phenotypes to patterns of channel dysfunction that underlie pathophysiology of these conditions.

Nonstandard abbreviations used: GEFS+, generalized epilepsy with febrile seizures plus; ICEGTC, intractable childhood epilepsy with frequent generalized tonic-clonic seizures; LQTS, long QT syndrome; Na$_V$Ch, voltage-gated sodium channel; SIDS, sudden infant death syndrome; SMEB, borderline severe myoclonic epilepsy of infancy; SMEI, severe myoclonic epilepsy of infancy; SUDS, sudden unexplained death syndrome.

Conflict of interest: The author has declared that no conflict of interest exists.

Citation for this article: *J. Clin. Invest.* **115**:1990–1999 (2005).
doi:10.1172/JCI25505.

Structure and function of Na$_V$Chs

Sodium channels are heteromultimeric, integral membrane proteins belonging to a superfamily of ion channels that are gated (opened and closed) by changes in membrane potential (1, 2). Sodium channel proteins from mammalian brain, muscle, and myocardium consist of a single large (approximately 260 kDa) pore-forming α subunit complexed with 1 or 2 smaller accessory β subunits (Figure 1). Nine genes (*SCN1A*, *SCN2A*, etc.) encoding distinct α subunit isoforms and 4 β subunit genes (*SCN1B*, *SCN2B*, etc.) have been identified in the human genome. Many isoforms are expressed in the central and peripheral nervous system (3), while skeletal muscle and cardiac muscle express more restricted Na$_V$Ch repertoires (4–9). The α subunits are constructed with a 4-fold symmetry consisting of structurally homologous domains (D1–D4) each containing 6 membrane-spanning segments (S1–S6) and a region (S5–S6 pore loop) controlling ion selectivity and permeation (Figure 1). The S4 segment, which functions as a voltage sensor (10, 11), is amphipathic with multiple basic amino acids (arginine or lysine) at every third position surrounded by hydrophobic residues. Each domain resembles an entire voltage-gated potassium channel subunit as well as a primitive bacterial Na$_V$Ch (12).

Na$_V$Chs switch between 3 functional states depending on the membrane potential (Figure 2) (13). In excitable membranes, a sudden membrane depolarization causes a rapid rise in local Na$^+$ permeability due to the opening (*activation*) of Na$_V$Chs from their resting closed state. For this to occur, voltage sensors (the 4 S4 segments) within the Na$_V$Ch protein must move in an outward direction, propelled by the change in membrane potential, and then translate this conformational energy to other structures (most likely S6 segments) that swing out of the way of incoming Na$^+$ ions. This increase in Na$^+$ permeability causes the sudden membrane depolarization that characterizes the initial phase of an action potential. Normally, activation of Na$_V$Chs is transient owing to *inactivation*, another gating process mediated by structures located on the cytoplasmic face of the channel protein (mainly the D3–D4 linker). Na$_V$Chs cannot reopen until the membrane is repolarized and they undergo *recovery from inactivation*. Membrane repolarization is achieved by fast inactivation of Na$_V$Chs and is augmented by activation of voltage-gated potassium channels. During recovery from inactivation, Na$_V$Chs may undergo *deactivation*, the transition from the open to the closed state (14). Activation, inactivation, and recovery from inactivation occur within a few milliseconds. In

Table 1

Inherited disorders of Na$_V$Chs

Muscle sodium channelopathies (*SCN4A*)
    Hyperkalemic periodic paralysis
    Paramyotonia congenita
    Potassium-aggravated myotonia
    Painful congenital myotonia
    Myasthenic syndrome
    Hypokalemic periodic paralysis type 2
    Malignant hyperthermia susceptibility
Cardiac sodium channelopathies (*SCN5A*)
    Congenital long QT syndrome (Romano-Ward)
    Idiopathic ventricular fibrillation (Brugada syndrome)
    Isolated cardiac conduction system disease
    Atrial standstill
    Congenital sick sinus syndrome
    Sudden infant death syndrome
    Dilated cardiomyopathy, conduction disorder, arrhythmia
Brain sodium channelopathies (*SCN1A, SCN2A, SCN1B*)
    Generalized epilepsy with febrile seizures plus
    Severe myoclonic epilepsy of infancy (Dravet syndrome)
    Intractable childhood epilepsy with frequent generalized tonic-clonic seizures
    Benign familial neonatal-infantile seizures
Peripheral nerve sodium channelopathies (*SCN9A*)
    Familial primary erythermalgia

addition to these rapid gating transitions, Na$_V$Chs are also susceptible to closing by slower inactivating processes (*slow inactivation*) if the membrane remains depolarized for a longer time (15). These slower events may contribute to determining the availability of active channels under various physiological conditions.

Muscle sodium channelopathies

Disturbances in the function of muscle Na$_V$Chs can affect the ability of skeletal muscle to contract or relax. Two symptoms are characteristic of muscle membrane (sarcolemma) Na$_V$Ch dysfunction, myotonia and periodic paralysis (16). Myotonia is characterized by delayed relaxation of muscle following a sudden forceful contraction and is associated with repetitive action potential generation, a manifestation of sarcolemmal hyperexcitability. By contrast, periodic paralysis represents a transient state of hypoexcitability or inexcitability in which action potentials cannot be generated or propagated.

Periodic paralysis and myotonia. Periodic paralysis is characterized by episodic weakness or paralysis of voluntary muscles occurring with normal neuromuscular transmission and in the absence of motor neuron disease. Patients with familial periodic paralysis present typically in childhood (17). Attacks of weakness are often associated with changes in the serum potassium (K$^+$) concentration as a result of abrupt redistribution of intracellular and extracellular K$^+$. This clinical epiphenomenon forms the basis for classifying periodic paralysis as hypokalemic, hyperkalemic, or normokalemic. In paramyotonia congenita, the dominant symptom is cold-induced muscle stiffness and weakness (17, 18). Potassium-aggravated myotonia is characterized by myotonia without weakness and worsening symptoms following K$^+$ ingestion (19). In general, these disorders are not associated with disabling muscular dystrophy, although chronic weakness may develop in some individuals with long-standing hyperkalemic periodic paralysis (20).

In vitro electrophysiological studies determined that both myotonia and periodic paralysis are associated with abnormal muscle cell membrane sodium conductance (21), and these findings pointed to *SCN4A* as the most plausible candidate gene. Genetic linkage studies confirmed this hypothesis (22–24). Hyperkalemic periodic paralysis, paramyotonia congenita, and potassium-aggravated myotonia are all associated with missense mutations in *SCN4A*. There are 2 predominant mutations associated with hyperkalemic periodic paralysis (T704M and M1592V), and these occur independently in unrelated kindreds (20, 25). Allelic diversity is greater for paramyotonia congenita and potassium-aggravated myotonia (26–32). In addition, approximately 15% of patients with genotype-defined hypokalemic periodic paralysis carry *SCN4A* mutations (33). Patients with *SCN4A* mutations may present rarely with life-threatening myotonic reactions upon exposure to succinylcholine resembling the syndrome of malignant hyperthermia susceptibility (34, 35). In 1 report, congenital myasthenia has been linked to *SCN4A* mutations (36).

Characterization of SCN4A mutations and pathophysiology. Using heterologously expressed recombinant Na$_V$Chs, several laboratories have characterized the biophysical properties of many mutations associated with either periodic paralysis or various myotonic disorders. These studies demonstrated that variable defects in the rate or extent of inactivation occur in virtually all cases. Mutations associated with hyperkalemic periodic paralysis exhibit incomplete inactivation leading to a small level (1–2% of peak current) of persistent Na$^+$ current that is predicted to cause sustained muscle fiber depolarization (Figure 3) (37, 38). Sustained depolarization will cause the majority of Na$_V$Chs (mutant and wild type) to become inactivated, and this explains conduction failure and electrical inexcitability observed in skeletal muscle during an attack of periodic paralysis (39, 40). By this mechanism, mutant Na$_V$Chs exert an indirect dominant-negative effect on normal channels. In addition, some, but not all, mutations associated with hyperkalemic periodic paralysis have impaired slow inactivation (41), and this may contribute to sustaining the effect of persistent Na$^+$ current (42).

SCN4A mutations in the myotonic disorders slow the rate of inactivation, speed recovery from inactivation, and slow deactivation (30, 43–47). These biophysical defects are predicted to lengthen the duration of muscle action potentials (48). Prolongation of action potentials along T-tubule membranes will exaggerate the local rise in extracellular K$^+$ concentration by efflux through persistently activated potassium channels. Extracellular K$^+$ in T-tubules exerts a depolarizing effect on the resting membrane potential, increasing the probability of an aberrant afterdepolarization. A large afterdepolarization can trigger spontaneous action potentials in adjacent surface membranes, which in turn cause persistent muscle contraction and delayed relaxation, the physiological hallmarks of myotonia (Figure 4) (49).

Treatment strategies for muscle sodium channelopathies. Pharmacological treatment for periodic paralysis with carbonic anhydrase inhibitors is often successful, but the mechanism of action is poorly understood (50, 51). Certain local anesthetic/antiarrhythmic agents have antimyotonic activity and are sometimes useful treatments for nondystrophic myotonias (52, 53). These drugs are effective because of their ability to interrupt rapidly conducted

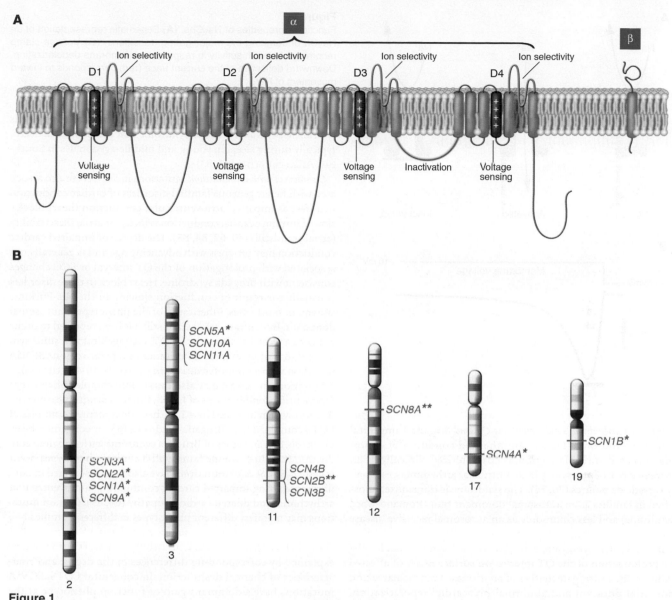

Figure 1
Structure and genomic location of human Na$_V$Chs. (**A**) Simple model representing transmembrane topology of α and β Na$_V$Ch subunits. Structural domains mediating key functional properties are labeled. (**B**) Chromosomal location of human genes encoding α (red) and β (blue) subunits across the genome. An asterisk next to the gene name indicates association with an inherited human disease. A double asterisk indicates association with murine phenotypes.

trains of action potentials through their use-dependent Na$_V$Ch-blocking action. Mexiletine is the most commonly used antimyotonic agent, and there have been in vitro studies demonstrating its effectiveness (54), but there have been no clinical trials comparing this agent with either placebo or other treatments. A more potent Na$_V$Ch blocker, flecainide, may also have utility in severe forms of myotonia that are resistant to mexiletine (55). The efficacy of flecainide for treating myotonia associated with certain *SCN4A* mutations may be greatest when there is a depolarizing shift of the steady-state fast inactivation curve for the mutant channel, whereas mutations that induce hyperpolarizing shifts in this curve are predicted to have greater sensitivity to mexiletine (56). Long-term treatment of myotonia with Na$_V$Ch blockers is often limited by drug side effects.

Cardiac sodium channelopathies
In the heart, Na$_V$Chs are essential for the orderly progression of action potentials from the sinoatrial node, through the atria, across the atrioventricular node, along the specialized conduction system of the ventricles (His-Purkinje system), and ultimately throughout the myocardium to stimulate rhythmic contraction. Mutations in *SCN5A*, the gene encoding the principal Na$_V$Ch α subunit expressed in the human heart, cause inherited susceptibility to ventricular arrhythmia (congenital long QT syndrome, idiopathic ventricular fibrillation) (57–59), impaired cardiac conduction (60), or both (61–65). *SCN5A* mutations may also manifest as drug-induced arrhythmias (66), sudden infant death syndrome (SIDS) (67, 68), and other forms of arrhythmia susceptibility (69).

A

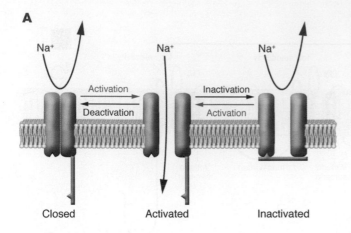

Closed   Activated   Inactivated

B

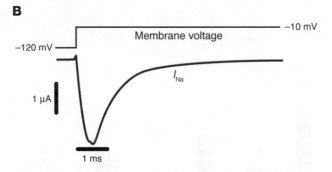

Inherited arrhythmia syndromes: long QT and Brugada. Congenital long QT syndrome (LQTS), an inherited condition of abnormal myocardial repolarization, is characterized clinically by an increased risk of potentially fatal ventricular arrhythmias, especially torsade de pointes (70, 71). The syndrome is transmitted most often in families as an autosomal dominant trait (Romano-Ward syndrome) and less commonly as an autosomal recessive disease combined with congenital deafness (Jervell and Lange-Nielsen syndrome). The syndrome derives its name from the characteristic prolongation of the QT interval on surface ECGs of affected individuals, a surrogate marker of an increased ventricular action potential duration and abnormal myocardial repolarization. Approximately 10% of LQTS cases are caused by *SCN5A* mutations, whereas the majority of Romano-Ward subjects harbor mutations in 2 cardiac potassium channel genes (*KCNQ1* and *HERG*) (72, 73). Triggering factors associated with arrhythmic events are different among genetic subsets of LQTS. *SCN5A* mutations often produce distinct clinical features including bradycardia, and a tendency for cardiac events to occur during sleep or rest (74, 75).

Mutations in *SCN5A* have also been associated with idiopathic ventricular fibrillation, including Brugada syndrome (59, 76) and sudden unexplained death syndrome (SUDS) (77, 78). Individuals with Brugada syndrome have an increased risk for potentially lethal ventricular arrhythmias (polymorphic ventricular tachycardia or fibrillation) without concomitant ischemia, electrolyte abnormalities, or structural heart disease. Individuals with the disease often exhibit a characteristic ECG pattern consisting of ST elevation in the right precordial leads, apparent right bundle branch block, but normal QT intervals (79). Administration of Na$_V$Ch-blocking agents (i.e., procainamide, flecainide, ajmaline) may expose this ECG pattern in latent cases (80). Inheritance is autosomal dominant with incomplete penetrance and a male

Figure 2
Functional properties of Na$_V$Chs. (**A**) Schematic representation of an Na$_V$Ch undergoing the major gating transitions. (**B**) Voltage-clamp recording of Na$_V$Ch activity in response to membrane depolarization. Downward deflection of the current trace (red) corresponds to inward movement of Na$^+$.

predominance. A family history of unexplained sudden death is typical. SUDS is a very similar syndrome that causes sudden death, typically during sleep, in young and middle-aged males in Southeast Asian countries (81–83).

Disorders of cardiac conduction. Mutations in *SCN5A* are also associated with heterogeneous familial disorders of cardiac conduction manifest as impaired atrioventricular conduction (heart block), slowed intramyocardial conduction velocity, or atrial inexcitability (atrial standstill) (60, 62, 84, 85). The degree of impaired cardiac conduction may progress with advancing age and is generally not associated with prolongation of the QT interval or ECG changes consistent with Brugada syndrome. Heart block in these disorders is usually the result of conduction slowing in the His-Purkinje system. In most cases, inheritance of the phenotype is autosomal dominant. By contrast, atrial standstill has been reported to occur either as a recessive disorder of *SCN5A* (congenital sick sinus syndrome) (85) or by digenic inheritance of a heterozygous *SCN5A* mutation with a promoter variant in the connexin-40 gene (84).

Mutations in *SCN5A* may also cause more complex phenotypes representing combinations of LQTS, Brugada syndrome, and conduction system disease. There have been documented examples of LQTS combined with Brugada syndrome (63) or congenital heart block (86, 87), and cases of Brugada syndrome with impaired conduction (88). In 1 unique family, all 3 clinical phenotypes occur together (65). *SCN5A* mutations have also been discovered in families segregating impaired cardiac conduction, supraventricular arrhythmia, and dilated cardiomyopathy (64, 89). Certain mutations may manifest different phenotypes in different families.

Characterization of SCN5A mutations and arrhythmogenesis. The clinical heterogeneity associated with *SCN5A* mutations is partly explained by corresponding differences in the degree and characteristics of channel dysfunction. In congenital LQTS, *SCN5A* mutations have a dominant gain-of-function phenotype at the

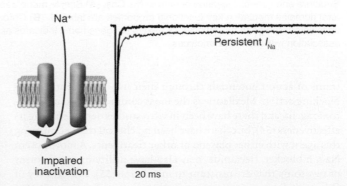

Figure 3
A common form of defective inactivation exhibited by mutant Na$_V$Chs associated with hyperkalemic periodic paralysis, long QT syndrome, and inherited epilepsy. The defect is caused by incomplete closure of the inactivation gate (left panel) resulting in an increased level of persistent current (right panel, red trace) as compared with Na$_V$Chs with normal inactivation (black trace).

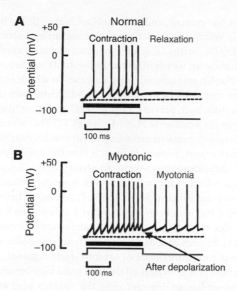

Figure 4

Differences between normal and myotonic muscle action potentials. (**A**) Generation of action potential spikes during electrical stimulation (horizontal blue line and square wave) of a normal muscle fiber. Contraction occurs during action potential firing, followed by muscle relaxation when stimulation ceases. (**B**) Action potentials in myotonic muscle during and immediately after electrical stimulation. An afterdepolarization triggers spontaneous action potentials that fire after termination of the electrical stimulus (myotonic activity).

molecular level. Specifically, most mutant cardiac Na_VChs associated with LQTS exhibit a characteristic impairment of inactivation, leading to persistent inward Na^+ current during prolonged membrane depolarizations (Figure 3) (90–92). A general slowing of inactivation may be present in mutations associated with severe LQTS (93), while some mutations alter voltage-dependence of activation and inactivation but do not have measurable non-inactivating current (94). Persistent Na^+ current during the cardiac action potential explains abnormal myocardial repolarization in LQTS (95). By contrast with nerve and muscle, cardiac action potentials last several hundred milliseconds because of a prolonged depolarization phase (plateau), the result of opposing inward (mainly Na^+ and Ca^{2+}) and outward (K^+) ionic currents. Repolarization occurs when net outward current exceeds net inward current. Non-inactivating behavior of mutant cardiac Na_VChs will shift this balance toward inward current and delay onset of repolarization, thus lengthening the action potential duration and the corresponding QT interval (Figure 5). Delayed repolarization predisposes to ventricular arrhythmias by exaggerating the dispersion of refractoriness throughout the myocardium and increasing the probability of early afterdepolarization, a phenomenon caused largely by reactivation of calcium channels during the action potential plateau (96). Both of these phenomena create conditions that allow electrical signals from depolarized regions of the heart to prematurely re-excite adjacent myocardium that has already repolarized, the basis for a reentrant arrhythmia. Additional proof of the role of cardiac Na_VCh mutations in LQTS has come from studies of mice heterozygous for a prototypic LQTS *SCN5A* mutation (delKPQ). These mice have

spontaneous life-threatening ventricular arrhythmias and a persistent Na^+ current in cardiac myocytes (97). *SCN5A* mutations associated with SIDS also exhibit this biophysical phenotype; this suggests a pathophysiological relationship with LQTS (67, 68).

The proposed cellular basis of Brugada syndrome involves a primary reduction in myocardial sodium current that exaggerates differences in action potential duration between the inner (endocardium) and outer (epicardium) layers of ventricular muscle (96, 98, 99). These differences exist initially because of an unequal distribution of potassium channels responsible for the transient outward current (I_{TO}), a repolarizing current more prominent in the epicardial layer that contributes to the characteristic spike and dome shape of the cardiac action potential. Reduced myocardial Na^+ current will cause disproportionate shortening of epicardial action potentials because of unopposed I_{TO}, leading to an exaggerated transmural voltage gradient, dispersion of repolarization, and a substrate promoting reentrant arrhythmias (Figure 6). This hypothesis has been validated using animal models and computational methods. The theory helps explain the characteristic ECG pattern observed in Brugada syndrome and the effects of Na_VCh-blocking agents to aggravate the phenotype.

Consistent with reduced sodium current as the primary pathophysiological event in Brugada syndrome, many *SCN5A* mutations associated with this disease cause frameshift errors, splice site defects, or premature stop codons (59, 100) that are predicted to produce nonfunctional channels. Furthermore, some missense mutations have also been demonstrated to be nonfunctional because of either impaired protein trafficking to the cell membrane or presumed disruption of Na^+ conductance through the channel (101–104). However, other missense mutations associated with Brugada syndrome are functional but have biophysical defects predicted to reduce channel availability, such as altered voltage-dependence of activation, more rapid fast inactivation, and enhanced slow inactivation (105–107).

Pathophysiology of SCN5A dysfunction in cardiac conduction disorders. Defects in cardiac Na_VCh function due to mutations associated

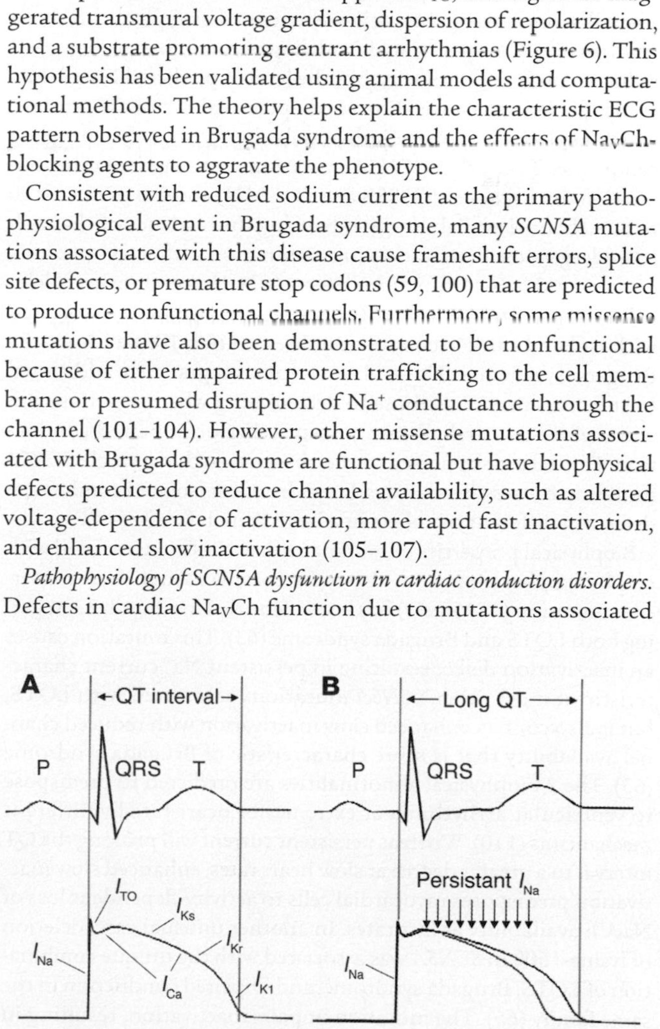

Figure 5

Electrophysiological basis for LQTS. (**A**) Relationship of surface ECG (top) with a representative cardiac action potential (bottom). The QT interval approximates the action potential duration. Individual ionic currents responsible for different phases of the action potential are labeled. (**B**) Prolongation of the QT interval and corresponding abnormal cardiac action potential (blue) resulting from persistent sodium current. I_{Ca}, calcium current; I_{K1}, inward rectifier current; I_{Kr}, rapid component of delayed rectifier current; I_{Ks}, slow component of delayed rectifier current; I_{Na}, sodium current; I_{TO}, transient outward current.

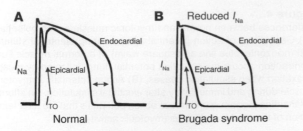

Figure 6
Electrophysiological basis for Brugada syndrome. (**A**) Comparison of endocardial and epicardial action potentials in normal heart. The epicardial action potential is shorter because of large transient outward current. (**B**) Endocardial and epicardial action potentials in Brugada syndrome. Reduced sodium current causes disproportionate shortening of epicardial action potentials with resulting exaggeration of the transmural voltage gradient (horizontal double arrow).

with disorders of cardiac conduction exhibit more complex biophysical properties (61, 62). Mutations causing isolated conduction defects have generally been observed to cause reduced Na$_V$Ch availability as a consequence of mixed gating disturbances. In the case of a Dutch family segregating a specific missense allele (G514C), the mutation causes unequal depolarizing shifts in the voltage-dependence of activation and inactivation such that a smaller number of channels are activated at typical threshold voltages (61). Computational modeling of these changes supports reduced conduction velocity, but the level of predicted Na$_V$Ch loss is insufficient to cause shortened epicardial action potentials, which explains why these individuals do not manifest Brugada syndrome. Two other *SCN5A* mutations causing isolated conduction disturbances (G298S and D1595N) are also predicted to reduce channel availability by enhancing the tendency of channels to undergo slow inactivation in combination with a complex mix of gain- and loss-of-function defects (62). However, other alleles exhibiting complete loss of function have also been associated with isolated cardiac conduction disease (108, 109) without the Brugada syndrome. These observations suggest that additional host factors may contribute to determining whether a mutation will manifest as arrhythmia susceptibility or impaired conduction. This idea is supported by the observation that a single *SCN5A* mutation causes either Brugada syndrome or isolated conduction defects in different members of a large French family (88).

Biophysical properties of mutant cardiac Na$_V$Chs associated with combined phenotypes are also more complex. An in-frame insertion mutation (1795insD) has been identified in a family segregating both LQTS and Brugada syndrome (63). This mutation causes an inactivation defect resulting in persistent Na$^+$ current characteristic of most other *SCN5A* mutations associated with LQTS, but it also confers enhanced slow inactivation with reduced channel availability that is more characteristic of Brugada syndrome (63). The 2 biophysical abnormalities are predicted to predispose to ventricular arrhythmia at extremes of heart rate by different mechanisms (110). Whereas persistent current will prolong the QT interval to a greater degree at slow heart rates, enhanced slow inactivation predisposes myocardial cells to activity-dependent loss of Na$_V$Ch availability at fast rates. In another unusual case, deletion of lysine-1500 in *SCN5A* was associated with the unique combination of LQTS, Brugada syndrome, and impaired conduction in the same family (65). The mutation impairs inactivation, resulting in

a persistent Na$^+$ current, and reduces Na$_V$Ch availability by opposing shifts in voltage-dependence of inactivation and activation.

Unlike LQTS, Brugada syndrome, and isolated cardiac conduction disease, in which affected individuals are heterozygous for single Na$_V$Ch mutations, there are cases in which individuals with severe impairments in cardiac conduction have inherited mutations from both parents. Lupoglazoff et al. described a child homozygous for a missense *SCN5A* allele (V1777M) who exhibited LQTS with rate-dependent atrioventricular conduction block (86). In a separate report, probands from 3 families exhibited perinatal sinus bradycardia progressing to atrial standstill (congenital sick sinus syndrome) and were found to have compound heterozygosity for mutations in *SCN5A* (85). Compound heterozygosity in *SCN5A* has also been observed in 2 infants with neonatal wide complex tachycardia and a generalized cardiac conduction defect (111). In each case of compound heterozygosity, individuals inherited 1 nonfunctional or severely dysfunctional mutation from 1 parent and a second allele with mild biophysical defects from the other parent. Interestingly, the parents who were carriers of single mutations were asymptomatic, which suggests that they had subclinical disease or other host factors affording protection. These unusually severe examples of *SCN5A*-linked cardiac conduction disorders illustrate the clinical consequence of nearly complete loss of Na$_V$Ch function. Complete absence of the murine *Scn5a* locus results in embryonic lethality (112), and it is likely that homozygous deletion or inactivation of human *SCN5A* is also not compatible with life.

Treatment strategies for cardiac sodium channelopathies. Specific therapeutic options for *SCN5A*-linked disorders are limited. β-Adrenergic blockers remain the first line of therapy in LQTS albeit this treatment strategy may be less efficacious in the setting of *SCN5A* mutations (113). Clinical and in vitro evidence suggests that mexiletine may counteract the aberrant persistent Na$^+$ current and shorten the QT interval (114, 115) in *SCN5A* mutation carriers, although there are no data indicating an improvement in mortality. Mexiletine has also been demonstrated to rescue trafficking defective *SCN5A* mutants in vitro (116). Flecainide has also been observed to shorten QT intervals in the setting of certain *SCN5A* mutations (117, 118), but some have raised concern over the safety of this therapeutic strategy (119). Class III–type antiarrhythmic agents (quinidine, sotalol) may be beneficial in Brugada syndrome (120, 121). Device therapy (implantable defibrillator for LQTS and Brugada syndrome; pacemaker for conduction disorders) is also an important treatment option.

Neuronal sodium channelopathies

Neuronal Na$_V$Chs are critical for the generation and propagation of action potentials in the central and peripheral nervous system. Most of the 13 genes encoding Na$_V$Ch α or β subunits are expressed in the brain, peripheral nerves, or both (1). In addition to their critical physiological function, neuronal Na$_V$Chs serve as important pharmacological targets for anticonvulsants and local anesthetic agents (122, 123). Their roles in genetic disorders including a variety of inherited epilepsy syndromes and a rare painful neuropathy have been revealed during the past 7 years.

Sodium channels and inherited epilepsies. Genetic defects in genes encoding 2 pore-forming α subunits (*SCN1A* and *SCN2A*) and the accessory β$_1$ subunit (*SCN1B*) are responsible for a group of epilepsy syndromes with overlapping clinical characteristics but divergent clinical severity (124–129). Generalized epilepsy with febrile seizures plus (GEFS+) is usually a benign disorder characterized by the frequent occurrence of febrile seizures in early childhood that persist beyond age 6 years, and epilepsy later in life associated with afebrile seizures with multiple clinical phenotypes (absence, myoclonic, atonic, myoclonic-astatic). Mutations in 3 neuronal Na$_V$Ch genes (SCN1A, SCN1B, and SCN2A) and a GABA receptor subunit (GABRG2) may independently cause GEFS+ or very similar disorders (130, 131). Mutations in SCN2A have also been associated with benign familial neonatal-infantile seizures (BFNIS), a seizure disorder of infancy that remits by age 12 months with no long-term neurological sequelae (129, 132). Interestingly, despite expression of SCN1A and SCN1B in the heart (9), there are no apparent cardiac manifestations associated with these disorders.

By contrast, severe myoclonic epilepsy of infancy (SMEI) and related syndromes have severe neurological sequelae. The diagnosis of SMEI is based on several clinical features, including (a) appearance of seizures, typically generalized tonic-clonic, during the first year of life, (b) impaired psychomotor development following onset of seizures, (c) occurrence of myoclonic seizures, (d) ataxia, and (e) poor response to antiepileptic drugs (133). Two designations, borderline SMEI (SMEB) (133, 134) and intractable childhood epilepsy with frequent generalized tonic-clonic seizures (ICEGTC) (128), have been assigned to patients with a condition resembling SMEI but in whom myoclonic seizures are absent and less severe psychomotor impairment is evident. SCN1A mutations have been identified in probands affected by all of these conditions.

More than 100 SCN1A mutations have been identified, with missense mutations being most common in GEFS+ (125, 135–139) and more deleterious alleles (nonsense, frameshift) representing the majority of SMEI mutations (126, 140, 141). Only missense mutations in SCN1A have been reported for patients diagnosed with either ICEGTC or SMEB. There are rare reports of families segregating both GEFS+ and either SMEI or ICEGTC (128). The overlapping phenotypes and molecular genetic etiologies among the SCN1A-linked epilepsies suggest that they represent a continuum of clinical disorders (142).

Sodium channel dysfunction and epileptogenesis. The first human Na$_V$Ch mutation associated with an inherited epilepsy (GEFS+) was discovered in SCN1B encoding the β1 accessory subunit (124). However, mutations in this gene have very rarely been associated with inherited epilepsy. Only 2 SCN1B mutations have been described to date, including a missense allele (C121W) and a 5-amino acid deletion (del70–74) (124, 143). Both mutations occur in an extracellular Ig-fold domain of the β1 subunit that is important for functional modulation of Na$_V$Ch α subunits (144, 145) and mediates protein-protein interactions critical for Na$_V$Ch subcellular localization in neurons (146). The C121W mutation disrupts a conserved disulfide bridge in this domain, and functional expression studies demonstrated a failure of the mutant to normally modulate the functional properties of recombinant brain Na$_V$Chs (124, 147). These findings and the observed seizure disorder in mice with targeted deletion of murine β1 subunit indicate that SCN1B loss of function explains the epilepsy phenotype (148). Functional characterization of the SCN1B deletion allele has not been reported.

Expression studies of α subunit mutations have demonstrated a wide range of functional disturbances. Early findings indicated that SCN1A mutations causing GEFS+ promote a gain of function, while mutations associated with SMEI are predicted to disable channel function. Two studies have demonstrated that increased persistent Na$^+$ current is caused by several GEFS+ mutations (149, 150). This behavior is reminiscent of the channel dysfunction associated with 2 other human sodium channelopathies discussed above, hyperkalemic periodic paralysis and LQTS (Figure 3). Noninactivating Na$^+$ current may facilitate neuronal hyperexcitability by reducing the threshold for action potential firing. However, not all GEFS+ mutations exhibit increased persistent current. For example, a shift in the voltage-dependence of inactivation to more depolarized potentials has been observed for 2 other GEFS+ mutations (T875M and D1866Y). This functional change is predicted to increase channel availability at voltages near the resting membrane potential and is sufficient to enhance excitability in a simple computational model of a neuronal action potential (150). This may be an oversimplification, as T875M also exhibits enhanced slow inactivation, which is predicted to decrease channel availability. For D1866Y, the changed voltage-dependence of inactivation was attributed to decreased modulation by the β1 subunit, a novel epilepsy-associated mechanism. Other GEFS+ mutations have been described that are nonfunctional (V1353L, A1685V) or exhibit depolarizing shifts in voltage-dependence of activation (I1656M, R1657C) predicted to reduce channel activity (151). These findings indicate that more than 1 biophysical mechanism accounts for seizure susceptibility in GEFS+.

Most SCN1A mutations associated with SMEI are predicted to produce nonfunctional channels by introducing premature termination or frameshifts into the coding sequence. This observation led to the notion that SMEI stems from SCN1A haploinsufficiency. Consistent with this idea was the finding that some missense mutations associated with SMEI are nonfunctional (151, 152). However, a simple dichotomy of gain versus loss of function to explain clinical differences between GEFS+ and SMEI is not consistent with recent observations. As mentioned above, some GEFS+ mutations exhibit loss-of-function characteristics. More recently, 2 SMEI missense alleles (R1648C and F1661S) were demonstrated to encode functional channels that exhibit a mixed pattern of biophysical defects consistent with either gain (persistent Na$^+$ current) or loss (reduced channel density, altered voltage-dependence of activation and inactivation) of function (152). The precise cellular mechanism by which this constellation of biophysical disturbances leads to epilepsy is uncertain and motivates further experiments in animal models to help determine the impact of Na$_V$Ch mutations.

SCN9A and painful inherited neuropathy. Mutations in another neuronal Na$_V$Ch gene, SCN9A, encoding an α subunit isoform expressed in sensory and sympathetic neurons, have been discovered in patients with familial primary erythermalgia, a rare autosomal dominant disorder characterized by recurrent episodes of severe pain, redness, and warmth in the distal extremities. Two missense SCN9A mutations were recently identified in Chinese patients (153). Both mutations cause a hyperpolarizing shift in the voltage-dependence of channel activation and slow the rate of deactivation (154). This combination of biophysical defects is predicted to confer hyperexcitability on peripheral sensory and sympathetic neurons, accounting for the episodic pain and vasomotor symptoms characteristic of the disease. Consistent with overactive Na$_V$Chs are anecdotal reports of improved symptoms during treatment with local anesthetic agents (i.e., lidocaine, bupivacaine) or mexiletine (155–157).

Summary and future challenges

Na$_V$Chs are important from many perspectives. Their recognized importance in the physiology and pharmacology of nerve, muscle, and heart is now further emphasized by their role in inherited disorders affecting these tissues. The sodium channelopathies provide outstanding illustrations of the delicate balances that maintain normal operation of critical physiological events such as muscle contraction and conduction of electrical signals.

Despite the extensive array of disorders listed in Table 1, it is likely that other inherited or pharmacogenetic disorders are caused by mutations or polymorphisms in NaVCh genes. Only 6 of the 13 known genes encoding NaVCh subunits have been linked to human disease. However, spontaneous or engineered disruption of 2 other genes (*Scn8a* and *Scn2b*) causes neurological phenotypes in mice (158–160), suggesting that other human sodium channelopathies might exist. Establishing new geno-type-phenotype relationships, exploring pathophysiology, and developing new treatment strategies remain exciting challenges for the future.

Acknowledgments

The author is supported by grants from the NIH (NS32387 and HL68880) and is the recipient of a Javits Neuroscience Award from the National Institute of Neurological Disorders and Stroke.

Address correspondence to: Alfred L. George Jr., Division of Genetic Medicine, 529 Light Hall, Vanderbilt University, Nashville, Tennessee 37232-0275, USA. Phone: (615) 936-2660; Fax: (615) 936-2661; E-mail: al.george@vanderbilt.edu.

1. Catterall, W.A. 1992. Cellular and molecular biology of voltage-gated sodium channels [review]. *Physiol. Rev.* **72**(Suppl. 4):S15–S48.
2. Catterall, W.A. 2000. From ionic currents to molecular mechanisms: the structure and function of voltage-gated sodium channels. *Neuron.* **26**:13–25.
3. Whitaker, W.R., et al. 2000. Distribution of voltage-gated sodium channel alpha-subunit and beta-subunit mRNAs in human hippocampal formation, cortex, and cerebellum. *J. Comp. Neurol.* **422**:123–139.
4. Trimmer, J.S., et al. 1989. Primary structure and functional expression of a mammalian skeletal muscle sodium channel. *Neuron.* **3**:33–49.
5. George, A.L., Komisarof, J., Kallen, R.G., and Barchi, R.L. 1992. Primary structure of the adult human skeletal muscle voltage-dependent sodium channel. *Ann. Neurol.* **31**:131–137.
6. Rogart, R.B., Cribbs, L.L., Muglia, L.K., Kephart, D.D., and Kaiser, M.W. 1989. Molecular cloning of a putative tetrodotoxin-resistant rat heart Na⁺ channel isoform. *Proc. Natl. Acad. Sci. U. S. A.* **86**:8170–8174.
7. Gellens, M.E., et al. 1992. Primary structure and functional expression of the human cardiac tetrodotoxin-insensitive voltage-dependent sodium channel. *Proc. Natl. Acad. Sci. U. S. A.* **89**:554–558.
8. Maier, S.K., et al. 2003. An unexpected requirement for brain-type sodium channels for control of heart rate in the mouse sinoatrial node. *Proc. Natl. Acad. Sci. U. S. A.* **100**:3507–3512.
9. Maier, S.K.G., et al. 2002. An unexpected role for brain-type sodium channels in coupling of cell surface depolarization to contraction in the heart. *Proc. Natl. Acad. Sci. U. S. A.* **99**:4073–4078.
10. Stühmer, W., et al. 1989. Structural parts involved in activation and inactivation of the sodium channel. *Nature.* **339**:597–603.
11. Yang, N.B., George, A.L., Jr., and Horn, R. 1996. Molecular basis of charge movement in voltage-gated sodium channels. *Neuron.* **16**:113–122.
12. Ren, D., et al. 2001. A prokaryotic voltage-gated sodium channel. *Science.* **294**:2372–2375.
13. Hodgkin, A.L., and Huxley, A.F. 1952. Currents carried by sodium and potassium ion through the membrane of the giant axon of *Loligo. J. Physiol.* **116**:449–472.
14. Kuo, C.C., and Bean, B.P. 1994. Na⁺ channels must deactivate to recover from inactivation. *Neuron.* **12**:819–829.
15. Vilin, Y.Y., and Ruben, P.C. 2001. Slow inactivation in voltage-gated sodium channels: molecular substrates and contributions to channelopathies. *Cell Biochem. Biophys.* **35**:171–190.
16. Cannon, S.C., and George, A.L. 2002. Pathophysiology of myotonia and periodic paralysis. In *Diseases of the nervous system: clinical neuroscience and therapeutic principles.* A.K. Asbury, G.M. Mckhann, W.I. McDonald, P.J. Goadsby, and J.C. McArthur, editors. Cambridge University Press. Cambridge, United Kingdom. 1183–1206.
17. Griggs, R.C. 1977. The myotonic disorders and the periodic paralyses. *Adv. Neurol.* **17**:143–159.
18. Streib, E.W. 1987. Differential diagnosis of myotonic syndromes. *Muscle Nerve.* **10**:603–615.
19. Rüdel, R., Ricker, K., and Lehmann-Horn, F. 1993. Genotype-phenotype correlations in human skeletal muscle sodium channel diseases. *Arch. Neurol.* **50**:1241–1248.
20. Ptacek, L.J., et al. 1991. Identification of a mutation in the gene causing hyperkalemic periodic paralysis. *Cell.* **67**:1021–1027.
21. Rüdel, R., and Lehmann-Horn, F. 1985. Membrane changes in cells from myotonia patients. *Physiol. Rev.* **65**:310–356.
22. Fontaine, B., et al. 1990. Hyperkalemic periodic paralysis and the adult muscle sodium channel alpha subunit gene. *Science.* **250**:1000–1002.
23. Ptacek, L.J., et al. 1991. Paramyotonia congenita and hyperkalemic periodic paralysis map to the same sodium channel gene locus. *Am. J. Hum. Genet.* **49**:851–854.
24. Ebers, G.C., et al. 1991. Paramyotonia congenita and non-myotonic hyperkalemic periodic paralysis are linked to the adult muscle sodium channel gene. *Ann. Neurol.* **30**:810–816.
25. Rojas, C.V., et al. 1991. A Met-to-Val mutation in the skeletal muscle Na⁺ channel α-subunit in hyperkalemic periodic paralysis. *Nature.* **354**:387–389.
26. McClatchey, A.I., et al. 1992. Temperature-sensitive mutations in the III-IV cytoplasmic loop region of the skeletal muscle sodium channel gene in paramyotonia congenita. *Cell.* **68**:769–774.
27. Ptacek, L.J., et al. 1992. Mutations in an S4 segment of the adult skeletal muscle sodium channel gene cause paramyotonia congenita. *Neuron.* **8**:891–897.
28. Ptacek, L.J., et al. 1993. Sodium channel mutations in paramyotonia congenita and hyperkalemic periodic paralysis. *Ann. Neurol.* **33**:300–307.
29. Ptacek, L.J., et al. 1994. Sodium channel mutations in acetazolamide-responsive myotonia congenita, paramyotonia congenita, and hyperkalemic periodic paralysis. *Neurology.* **44**:1500–1503.
30. Lerche, H., et al. 1993. Human sodium channel myotonia: slowed channel inactivation due to substitutions for a glycine within the III/IV linker. *J. Physiol.* **470**:13–22.
31. Lerche, H., Mitrovic, N., Dubowitz, V., and Lehmann-Horn, F. 1996. Paramyotonia congenita: the R1448P Na⁺ channel mutation in adult human skeletal muscle. *Ann. Neurol.* **39**:599–608.
32. Ricker, K., Moxley, R.T., III, Heine, R., and Lehmann-Horn, F. 1994. Myotonia fluctuans. A third type of muscle sodium channel disease. *Arch. Neurol.* **51**:1095–1102.
33. Miller, T.M., et al. 2004. Correlating phenotype and genotype in the periodic paralyses. *Neurology.* **63**:1647–1655.
34. Vita, G.M., et al. 1995. Masseter muscle rigidity associated with glycine-1306 to alanine mutation in the adult muscle sodium channel α-subunit gene. *Anesthesiology.* **82**:1097–1103.
35. Bendahhou, S., et al. 2000. A double mutation in families with periodic paralysis defines new aspects of sodium channel slow inactivation. *J. Clin. Invest.* **106**:431–438.
36. Tsujino, A., et al. 2003. Myasthenic syndrome caused by mutation of the SCN4A sodium channel. *Proc. Natl. Acad. Sci. U. S. A.* **100**:7377–7382.
37. Cannon, S.C., Brown, R.H., and Corey, D.P. 1991. A sodium channel defect in hyperkalemic periodic paralysis: potassium induced failure of inactivation. *Neuron.* **6**:619–626.
38. Cannon, S.C., and Strittmatter, S.M. 1993. Functional expression of sodium channel mutations identified in families with periodic paralysis. *Neuron.* **10**:317–326.
39. Lehmann-Horn, F., et al. 1983. Two cases of adynamia episodica hereditaria: in vitro investigation of muscle cell membrane and contraction parameters. *Muscle Nerve.* **6**:113–121.
40. Rüdel, R., Lehmann-Horn, F., Ricker, K., and Kuther, G. 1984. Hypokalemic periodic paralysis: in vitro investigation of muscle fiber membrane parameters. *Muscle Nerve.* **7**:110–120.
41. Cummins, T.R., and Sigworth, F.J. 1996. Impaired slow inactivation in mutant sodium channels. *Biophys. J.* **71**:227–236.
42. Ruff, R.L. 1994. Slow Na⁺ sodium channel inactivation must be disrupted to evoke prolonged depolarization-induced paralysis. *Biophys. J.* **66**:542–545.
43. Yang, N., et al. 1994. Sodium channel mutations in paramyotonia congenita exhibit similar biophysical phenotypes *in vitro. Proc. Natl. Acad. Sci. U. S. A.* **91**:12785–12789.
44. Mitrovic, N., George, A.L., Jr., Rudel, R., Lehmann-Horn, F., and Lerche, H. 1999. Mutant channels contribute <50% to Na⁺ current in paramyotonia congenita muscle. *Brain.* **122**:1085–1092.
45. Mitrovic, N., et al. 1994. K⁺-aggravated myotonia: destabilization of the inactivated state of the human muscle Na+ channel by the V1589M mutation. *J. Physiol.* **478**:395–402.
46. Mitrovic, N., et al. 1995. Different effects on gating of three myotonia-causing mutations in the inactivation gate of the human muscle sodium channel. *J. Physiol.* **487**:107–114.
47. Hayward, L.J., Brown, R.H., Jr., and Cannon, S.C. 1996. Inactivation defects caused by myotonia-associated mutations in the sodium channel III-IV linker. *J. Gen. Physiol.* **107**:559–576.
48. Cannon, S.C. 1997. From mutation to myotonia in sodium channel disorders. *Neuromuscul. Disord.* **7**:241–249.
49. Adrian, R.H., and Bryant, S.H. 1974. On the repetitive discharge in myotonic muscle fibres. *J. Physiol.* **240**:505–515.
50. Moxley, R.T., III. 2000. Channelopathies. *Curr. Treat. Options Neurol.* **2**:31–47.

51. Meola, G., and Sansone, V. 2000. Therapy in myotonic disorders and in muscle channelopathies [review]. *Neurol. Sci.* **21**(Suppl. 5):S953–S961.

52. Jackson, C.E., Barohn, R.J., and Ptacek, L.J. 1994. Paramyotonia congenita: abnormal short exercise test, and improvement after mexiletine therapy. *Muscle Nerve.* **17**:763–768.

53. Kwiecinski, H., Ryniewicz, B., and Ostrzycki, A. 1992. Treatment of myotonia with antiarrhythmic drugs. *Acta Neurol. Scand.* **86**:371–375.

54. Takahashi, M.P., and Cannon, S.C. 2001. Mexiletine block of disease-associated mutations in S6 segments of the human skeletal muscle Na⁺ channel. *J. Physiol.* **537**:701–714.

55. Rosenfeld, J., Sloan-Brown, K., and George, A.L., Jr. 1997. A novel muscle sodium channel mutation causes painful congenital myotonia. *Ann. Neurol.* **42**:811–814.

56. Desaphy, J.F., De Luca, A., Didonna, M.P., George, A.L., and Conte, C.D. 2003. Different flecainide sensitivity of hNaV1.4 channels and myotonic mutants explained by state-dependent block. *J. Physiol.* **554**:321–334.

57. Wang, Q., et al. 1995. *SCN5A* mutations associated with an inherited cardiac arrhythmia, long QT syndrome. *Cell.* **80**:805–811.

58. Wang, Q., et al. 1995. Cardiac sodium channel mutations in patients with long QT syndrome, an inherited cardiac arrhythmia. *Hum. Mol. Genet.* **4**:1603–1607.

59. Chen, Q., et al. 1998. Genetic basis and molecular mechanism for idiopathic ventricular fibrillation. *Nature.* **392**:293–296.

60. Schott, J.J., et al. 1999. Cardiac conduction defects associate with mutations in SCN5A. *Nat. Genet.* **23**:20–21.

61. Tan, H.L., et al. 2001. A sodium-channel mutation causes isolated cardiac conduction disease. *Nature.* **409**:1043–1047.

62. Wang, D.W., Viswanathan, P.C., Balser, J.R., George, A.L., Jr., and Benson, D.W. 2002. Clinical, genetic, and biophysical characterization of SCN5A mutations associated with atrioventricular conduction block. *Circulation.* **105**:341–346.

63. Bezzina, C., et al. 1999. A single Na⁺ channel mutation causing both long-QT and Brugada syndromes. *Circ. Res.* **85**:1206–1213.

64. McNair, W.P., et al. 2004. SCN5A mutation associated with dilated cardiomyopathy, conduction disorder, and arrhythmia. *Circulation.* **110**:2163–2167.

65. Grant, A.O., et al. 2002. Long QT syndrome, Brugada syndrome, and conduction system disease are linked to a single sodium channel mutation. *J. Clin. Invest.* **110**:1201–1209. doi:10.1172/JCI200215570.

66. Makita, N., et al. 2002. Drug-induced long-QT syndrome associated with a subclinical SCN5A mutation. *Circulation.* **106**:1269–1274.

67. Schwartz, P.J., et al. 2000. A molecular link between the sudden infant death syndrome and the long-QT syndrome. *N. Engl. J. Med.* **343**:262–267.

68. Ackerman, M.J., et al. 2001. Postmortem molecular analysis of SCN5A defects in sudden infant death syndrome. *JAMA.* **286**:2264–2269.

69. Splawski, I., et al. 2002. Variant of SCN5A sodium channel implicated in risk of cardiac arrhythmia. *Science.* **297**:1333–1336.

70. Keating, M.T. 1996. The long QT syndrome. A review of recent molecular genetic and physiologic discoveries [review]. *Medicine (Baltimore).* **75**:1–5.

71. Vincent, G.M. 1998. The molecular genetics of the long QT syndrome: genes causing fainting and sudden death. *Annu. Rev. Med.* **49**:263–274.

72. Curran, M.E., et al. 1995. A molecular basis for cardiac arrhythmia: *HERG* mutations cause long QT syndrome. *Cell.* **80**:795–803.

73. Wang, Q., et al. 1996. Positional cloning of a novel potassium channel gene: KVLQT1 mutations cause cardiac arrhythmias. *Nat. Genet.* **12**:17–23.

74. Schwartz, P.J., et al. 2001. Genotype-phenotype correlation in the long-QT syndrome: gene-specific triggers for life-threatening arrhythmias. *Circulation.* **103**:89–95.

75. Schwartz, P.J., et al. 1995. Long QT syndrome patients with mutations of the *SCN5A* and *HERG* genes have differential responses to Na⁺ channel blockade and to increases in heart rate. Implications for gene-specific therapy. *Circulation.* **92**:3381–3386.

76. Akai, J., et al. 2000. A novel SCN5A mutation associated with idiopathic ventricular fibrillation without typical ECG findings of Brugada syndrome. *FEBS Lett.* **479**:29–34.

77. Vatta, M., et al. 2002. Genetic and biophysical basis of sudden unexplained nocturnal death syndrome (SUNDS), a disease allelic to Brugada syndrome. *Hum. Mol. Genet.* **11**:337–345.

78. Sangwatanaroj, S., Yanatasneejit, P., Sunsaneewitayakul, B., and Sitthisook, S. 2002. Linkage analyses and SCN5A mutations screening in five sudden unexplained death syndrome (Lai-tai) families. *J. Med. Assoc. Thai.* **85**(Suppl. 1):S54–S61.

79. Brugada, J., and Brugada, P. 1997. Further characterization of the syndrome of right bundle branch block, ST segment elevation, and sudden cardiac death. *J. Cardiovasc. Electrophysiol.* **8**:325–331.

80. Brugada, R., et al. 2000. Sodium channel blockers identify risk for sudden death in patients with ST-segment elevation and right bundle branch block but structurally normal hearts. *Circulation.* **101**:510–515.

81. Baron, R.C., et al. 1983. Sudden death among Southeast Asian refugees. An unexplained nocturnal phenomenon. *JAMA.* **250**:2947–2951.

82. Nademanee, K., et al. 1997. Arrhythmogenic marker for the sudden unexplained death syndrome in Thai men. *Circulation.* **96**:2595–2600.

83. Sangwatanaroj, S., Ngamchareon, C., and Prechawat, S. 2001. Pattern of inheritance in three sudden unexplained death syndrome ("Lai-tai") families. *J. Med. Assoc. Thai.* **84**(Suppl. 1):S443–S451.

84. Groenewegen, W.A., et al. 2003. A cardiac sodium channel mutation cosegregates with a rare connexin40 genotype in familial atrial standstill. *Circ. Res.* **92**:14–22.

85. Benson, D.W., et al. 2003. Congenital sick sinus syndrome caused by recessive mutations in the cardiac sodium channel gene (SCN5A). *J. Clin. Invest.* **112**:1019–1028. doi:10.1172/JCI200318062.

86. Lupoglazoff, J.M., et al. 2001. Homozygous SCN5A mutation in long-QT syndrome with functional two-to-one atrioventricular block. *Circ. Res.* **89**:E16–E21.

87. Chang, C.C., et al. 2004. A novel SCN5A mutation manifests as a malignant form of long QT syndrome with perinatal onset of tachycardia/bradycardia. *Cardiovasc. Res.* **64**:268–278.

88. Kyndt, F., et al. 2001. Novel SCN5A mutation leading either to isolated cardiac conduction defect or Brugada syndrome in a large French family. *Circulation.* **104**:3081–3086.

89. Olson, T.M., et al. 2005. Sodium channel mutations and susceptibility to heart failure and atrial fibrillation. *JAMA.* **293**:447–454.

90. Bennett, P.B., Yazawa, K., Makita, N., and George, A.L., Jr. 1995. Molecular mechanism for an inherited cardiac arrhythmia. *Nature.* **376**:683–685.

91. Dumaine, R., et al. 1996. Multiple mechanisms of Na⁺ channel-linked long-QT syndrome. *Circ. Res.* **78**:916–924.

92. Wang, D.W., Yazawa, K., George, A.L., Jr., and Bennett, P.B. 1996. Characterization of human cardiac Na⁺ channel mutations in the congenital long QT syndrome. *Proc. Natl. Acad. Sci. U. S. A.* **93**:13200–13205.

93. Kambouris, N.G., et al. 1998. Phenotypic characterization of a novel long-QT syndrome mutation (R1623Q) in the cardiac sodium channel. *Circulation.* **97**:640–644.

94. Abriel, H., et al. 2001. Novel arrhythmogenic mechanism revealed by a long-QT syndrome mutation in the cardiac Na⁺ channel. *Circ. Res.* **88**:740–745.

95. Clancy, C.E., and Rudy, Y. 1999. Linking a genetic defect to its cellular phenotype in a cardiac arrhythmia. *Nature.* **400**:566–569.

96. Antzelevitch, C., Yan, G.X., and Shimizu, W. 1999. Transmural dispersion of repolarization and arrhythmogenicity: the Brugada syndrome versus the long QT syndrome [review]. *J. Electrocardiol.* **32**(Suppl.):158–165.

97. Nuyens, D., et al. 2001. Abrupt rate accelerations or premature beats cause life-threatening arrhythmias in mice with long-QT3 syndrome. *Nat. Med.* **7**:1021–1027.

98. Yan, G.X., and Antzelevitch, C. 1999. Cellular basis for the Brugada syndrome and other mechanisms of arrhythmogenesis associated with ST-segment elevation. *Circulation.* **100**:1660–1666.

99. Antzelevitch, C. 1999. Ion channels and ventricular arrhythmias: cellular and ionic mechanisms underlying the Brugada syndrome. *Curr. Opin. Cardiol.* **14**:274–279.

100. Schulze-Bahr, E., et al. 2003. Sodium channel gene (SCN5A) mutations in 44 index patients with Brugada syndrome: different incidences in familial and sporadic disease. *Hum. Mutat.* **21**:651–652.

101. Baroudi, G., Acharfi, S., Larouche, C., and Chahine, M. 2002. Expression and intracellular localization of an SCN5A double mutant R1232W/T1620M implicated in Brugada syndrome. *Circ. Res.* **90**:E11–E16.

102. Baroudi, G., et al. 2001. Novel mechanism for Brugada syndrome: defective surface localization of an SCN5A mutant (R1432G). *Circ. Res.* **88**:E78–E83.

103. Baroudi, G., Napolitano, C., Priori, S.G., Del Bufalo, A., and Chahine, M. 2004. Loss of function associated with novel mutations of the SCN5A gene in patients with Brugada syndrome. *Can. J. Cardiol.* **20**:425–430.

104. Valdivia, C.R., et al. 2004. A trafficking defective, Brugada syndrome-causing SCN5A mutation rescued by drugs. *Cardiovasc. Res.* **62**:53–62.

105. Dumaine, R., et al. 1999. Ionic mechanisms responsible for the electrocardiographic phenotype of the Brugada syndrome are temperature dependent. *Circ. Res.* **85**:803–809.

106. Wang, D.W., Makita, N., Kitabatake, A., Balser, J.R., and George, A.L., Jr. 2000. Enhanced Na⁺ channel intermediate inactivation in Brugada syndrome. *Circ. Res.* **87**:E37–E43.

107. Rook, M.B., et al. 1999. Human SCN5A gene mutations alter cardiac sodium channel kinetics and are associated with the Brugada syndrome. *Cardiovasc. Res.* **44**:507–517.

108. Herfst, L.J., et al. 2003. Na⁺ channel mutation leading to loss of function and non-progressive cardiac conduction defects. *J. Mol. Cell. Cardiol.* **35**:549–557.

109. Probst, V., et al. 2003. Haploinsufficiency in combination with aging causes SCN5A-linked hereditary Lenegre disease. *J. Am. Coll. Cardiol.* **41**:643–652.

110. Clancy, C.E., and Rudy, Y. 2002. Na(+) channel mutation that causes both Brugada and long-QT syndrome phenotypes: a simulation study of mechanism. *Circulation.* **105**:1208–1213.

111. Bezzina, C.R., et al. 2003. Compound heterozygosity for mutations (W156X and R225W) in SCN5A associated with severe cardiac conduction disturbances and degenerative changes in the conduction system. *Circ. Res.* **92**:159–168.

112. Papadatos, G.A., et al. 2002. Slowed conduction and ventricular tachycardia after targeted disruption of the cardiac sodium channel gene Scn5a. *Proc. Natl. Acad. Sci. U. S. A.* **99**:6210–6215.

113. Priori, S.G., et al. 2004. Association of long QT syndrome loci and cardiac events among patients treated with beta-blockers. *JAMA.* **292**:1341–1344.

114. Schwartz, P.J., et al. 1995. Long QT syndrome

patients with mutations of the *SCN5A* and *HERG* genes have differential responses to Na⁺ channel blockade and to increases in heart rate. Implications for gene-specific therapy. *Circulation.* **92**:3381–3386.

115. Wang, D.W., Yazawa, K., Makita, N., George, A.L., Jr., and Bennett, P.B. 1997. Pharmacological targeting of long QT mutant sodium channels. *J. Clin. Invest.* **99**:1714–1720.

116. Valdivia, C.R., et al. 2002. A novel SCN5A arrhythmia mutation, M1766L, with expression defect rescued by mexiletine. *Cardiovasc. Res.* **55**:279–289.

117. Windle, J.R., Geletka, R.C., Moss, A.J., Zareba, W., and Atkins, D.L. 2001. Normalization of ventricular repolarization with flecainide in long QT syndrome patients with SCN5A:DeltaKPQ mutation. *Ann. Noninvasive Electrocardiol.* **6**:153–158.

118. Abriel, H., Wehrens, X.H., Benhorin, J., Kerem, B., and Kass, R.S. 2000. Molecular pharmacology of the sodium channel mutation D1790G linked to the long-QT syndrome. *Circulation.* **102**:921–925.

119. Priori, S.G., et al. 2000. The elusive link between LQT3 and Brugada syndrome: the role of flecainide challenge. *Circulation.* **102**:945–947.

120. Belhassen, B., Glick, A., and Viskin, S. 2004. Efficacy of quinidine in high-risk patients with Brugada syndrome. *Circulation.* **110**:1731–1737.

121. Glatter, K.A., et al. 2004. Effectiveness of sotalol treatment in symptomatic Brugada syndrome. *Am. J. Cardiol.* **93**:1320–1322.

122. Macdonald, R.L., and Greenfield, L.J., Jr. 1997. Mechanisms of action of new antiepileptic drugs. *Curr. Opin. Neurol.* **10**:121–128.

123. Catterall, W.A. 1999. Molecular properties of brain sodium channels: an important target for anticonvulsant drugs. *Adv. Neurol.* **79**:441–456.

124. Wallace, R.H., et al. 1998. Febrile seizures and generalized epilepsy associated with a mutation in the Na⁺-channel β1 subunit gene SCN1B. *Nat. Genet.* **19**:366–370.

125. Escayg, A., et al. 2000. Mutations of SCN1A, encoding a neuronal sodium channel, in two families with GEFS+2. *Nat. Genet.* **24**:343–345.

126. Claes, L., et al. 2001. De novo mutations in the sodium-channel gene SCN1A cause severe myoclonic epilepsy of infancy. *Am. J. Hum. Genet.* **68**:1327–1332.

127. Sugawara, T., et al. 2001. A missense mutation of the Na⁺ channel α$_{II}$ subunit gene Na$_v$1.2 in a patient with febrile and afebrile seizures causes channel dysfunction. *Proc. Natl. Acad. Sci. U. S. A.* **98**:6384–6389.

128. Fujiwara, T., et al. 2003. Mutations of sodium channel alpha subunit type 1 (SCN1A) in intractable childhood epilepsies with frequent generalized tonic-clonic seizures. *Brain.* **126**:531–546.

129. Heron, S.E., et al. 2002. Sodium-channel defects in benign familial neonatal-infantile seizures. *Lancet.* **360**:851–852.

130. Baulac, S., et al. 2001. First genetic evidence of GABA$_A$ receptor dysfunction in epilepsy: a mutation in the γ2-subunit gene. *Nat. Genet.* **28**:46–48.

131. Wallace, R.H., et al. 2001. Mutant GABA$_A$ receptor γ2-subunit in childhood absence epilepsy and febrile seizures. *Nat. Genet.* **28**:49–52.

132. Berkovic, S.F., et al. 2004. Benign familial neonatal-infantile seizures: characterization of a new sodium channelopathy. *Ann. Neurol.* **55**:550–557.

133. Ohmori, I., et al. 2003. Is phenotype difference in severe myoclonic epilepsy in infancy related to SCN1A mutations? *Brain Dev.* **25**:488–493.

134. Fukuma, G., et al. 2004. Mutations of neuronal voltage-gated Na⁺ channel α1 subunit gene SCN1A in core severe myoclonic epilepsy in infancy (SMEI) and in borderline SMEI (SMEB). *Epilepsia.* **45**:140–148.

135. Escayg, A., et al. 2001. A novel *SCN1A* mutation associated with generalized epilepsy with febrile seizures plus — and prevalence of variants in patients with epilepsy. *Am. J. Hum. Genet.* **68**:866–873.

136. Wallace, R.H., et al. 2001. Neuronal sodium-channel α1-subunit mutations in generalized epilepsy with febrile seizures plus. *Am. J. Hum. Genet.* **68**:859–865.

137. Abou-Khalil, B., et al. 2001. Partial epilepsy and generalized epilepsy with febrile seizures plus and a novel *SCN1A* mutation. *Neurology.* **57**:2265–2272.

138. Sugawara, T., et al. 2001. Na$_v$1.1 mutations cause febrile seizures associated with afebrile partial seizures. *Neurology.* **57**:703–705.

139. Ito, M., et al. 2002. Autosomal dominant epilepsy with febrile seizures plus with missense mutations of the Na⁺-channel α1 subunit gene, *SCN1A.* *Epilepsy Res.* **48**:15–23.

140. Ohmori, I., Ouchida, M., Ohtsuka, Y., Oka, E., and Shimizu, K. 2002. Significant correlation of the SCN1A mutations and severe myoclonic epilepsy in infancy. *Biochem. Biophys. Res. Commun.* **295**:17–23.

141. Sugawara, T., et al. 2002. Frequent mutations of SCN1A in severe myoclonic epilepsy in infancy. *Neurology.* **58**:1122–1124.

142. Singh, R., et al. 2001. Severe myoclonic epilepsy of infancy: extended spectrum of GEFS+? *Epilepsia.* **42**:837–844.

143. Audenaert, D., et al. 2003. A deletion in SCN1B is associated with febrile seizures and early-onset absence epilepsy. *Neurology.* **61**:854–856.

144. Makita, N., Bennett, P.B., and George, A.L., Jr. 1996. Molecular determinants of β$_1$ subunit-induced gating modulation in voltage-dependent Na⁺ channels. *J. Neurosci.* **16**:7117–7127.

145. McCormick, K.A., et al. 1998. Molecular determinants of Na⁺ channel function in the extracellular domain of the β1 subunit. *J. Biol. Chem.* **273**:3954–3962.

146. McEwen, D.P., Meadows, L.S., Chen, C., Thyagarajan, V., and Isom, L.L. 2004. Sodium channel β1 subunit-mediated modulation of Na$_v$1.2 currents and cell surface density is dependent on interac-tions with contactin and ankyrin. *J. Biol. Chem.* **279**:16044–16049.

147. Meadows, L.S., et al. 2002. Functional and biochemical analysis of a sodium channel β1 subunit mutation responsible for generalized epilepsy with febrile seizures plus type 1. *J. Neurosci.* **22**:10699–10709.

148. Chen, C., et al. 2004. Mice lacking sodium channel β1 subunits display defects in neuronal excitability, sodium channel expression, and nodal architecture. *J. Neurosci.* **24**:4030–4042.

149. Lossin, C., Wang, D.W., Rhodes, T.H., Vanoye, C.G., and George, A.L., Jr. 2002. Molecular basis of an inherited epilepsy. *Neuron.* **34**:877–884.

150. Spampanato, J., et al. 2004. A novel epilepsy mutation in the sodium channel SCN1A identifies a cytoplasmic domain for beta subunit interaction. *J. Neurosci.* **24**:10022–10034.

151. Lossin, C., et al. 2003. Epilepsy-associated dysfunction in the voltage-gated neuronal sodium channel SCN1A. *J. Neurosci.* **23**:11289–11295.

152. Rhodes, T.H., Lossin, C., Vanoye, C.G., Wang, D.W., and George, A.L., Jr. 2004. Noninactivating voltage-gated sodium channels in severe myoclonic epilepsy of infancy. *Proc. Natl. Acad. Sci. U. S. A.* **101**:11147–11152.

153. Yang, Y., et al. 2004. Mutations in SCN9A, encoding a sodium channel alpha subunit, in patients with primary erythermalgia. *J. Med. Genet.* **41**:171–174.

154. Cummins, T.R., Dib-Hajj, S.D., and Waxman, S.G. 2004. Electrophysiological properties of mutant Na$_v$1.7 sodium channels in a painful inherited neuropathy. *J. Neurosci.* **24**:8232–8236.

155. Jang, H.S., et al. 2004. A case of primary erythromelalgia improved by mexiletine. *Br. J. Dermatol.* **151**:708–710.

156. Legroux-Crespel, E., et al. 2003. Treatment of familial erythermalgia with the association of lidocaine and mexiletine. *Ann. Dermatol. Venereol.* **130**:429–433.

157. Stricker, L.J., and Green, C.R. 2001. Resolution of refractory symptoms of secondary erythermalgia with intermittent epidural bupivacaine. *Reg. Anesth. Pain Med.* **26**:488–490.

158. Buchner, D.A., Seburn, K.L., Frankel, W.N., and Meisler, M.H. 2004. Three ENU-induced neurological mutations in the pore loop of sodium channel Scn8a Na$_v$1.6 and a genetically linked retinal mutation, rd13. *Mamm. Genome.* **15**:344–351.

159. Kohrman, D.C., Smith, M.R., Goldin, A.L., Harris, J., and Meisler, M.H. 1996. A missense mutation in the sodium channel Scn8a is responsible for cerebellar ataxia in the mouse mutant *jolting.* *J. Neurosci.* **16**:5993–5999.

160. Chen, C., et al. 2002. Reduced sodium channel density, altered voltage dependence of inactivation, and increased susceptibility to seizures in mice lacking sodium channel β2-subunits. *Proc. Natl. Acad. Sci. U. S. A.* **99**:17072–17077.

Cardiac and skeletal muscle disorders caused by mutations in the intracellular Ca²⁺ release channels

Silvia G. Priori[1,2] and Carlo Napolitano[1]

[1]Molecular Cardiology, Istituto di Ricovero e Cura a Carattere Scientifico Fondazione Maugeri, Pavia, Italy.
[2]Department of Cardiology, University of Pavia, Pavia, Italy.

Here we review the current knowledge about the mutations of the gene encoding the cardiac ryanodine receptor (*RyR2*) that cause cardiac arrhythmias. Similarities between the mutations identified in the *RyR2* gene and those found in the gene *RyR1* that cause malignant hyperthermia and central core disease are discussed. In vitro functional characterization of *RyR1* and *RyR2* mutants is reviewed, with a focus on the contribution that in vitro expression studies have made to our understanding of related human diseases.

Introduction

Rapid mobilization of calcium from the sarcoplasmic reticulum (SR) into the cytosol triggers activation of contractile elements, and it is therefore a fundamental process in the physiology of heart and muscles. The channels that regulate the duration and amplitude of calcium efflux from the SR are the ryanodine receptors (RyRs). Three subtypes of these proteins exist: RyR1 is mainly expressed in skeletal muscle, RyR2 is highly represented in cardiac tissue, and RyR3 is preferentially expressed in the brain. In this review we will focus on the genetic diseases that affect structure and function of the cardiac isoform of the ryanodine receptor, RyR2, causing development of ventricular arrhythmias and sudden cardiac death. Since genetic disorders affecting the skeletal muscle isoform (1) of the ryanodine receptor, RyR1, were described 10 years before those affecting RyR2 (2), and since observations initially made about mutations in the *RyR1* gene can be extended to those identified in the *RyR2* gene, we will discuss selected aspects of RyR1 diseases in light of their similarities with RyR2 diseases.

The cardiac ryanodine receptor and intracellular calcium release

Structure of the cardiac ryanodine receptor, RyR2

The cardiac ryanodine receptor is a protein composed of 4,967 amino acids with a molecular weight of 565 kDa (3) that is encoded by the large *RyR2* gene located on chromosome 1 (1q42-q43) (4). This is a large gene spanning 790 kb of genomic DNA with 105 exons (5). The functional channel is a tetramer that derives from the coassembly of 4 of the subunits encoded by the *RyR2* gene, resulting in a 2.2-megadalton complex (Figure 1). Each RyR2 monomer is composed of 2 major structural regions. The carboxy terminus (spanning approximately from aa 4,500 to the end of the protein) is the pore-forming region, which is composed of an even, but yet undefined, number of transmembrane segments (4–10 for each monomer, most likely 6) (6, 7). This portion of the protein, despite being relatively small, is the most important, because it contains sequences that control RyR2 localization and oligomerization (8) and is sufficient to form a functional Ca²⁺ release channel. The second structural region, which includes nine-tenths of the protein, corresponds to the large cytoplasmic domain that performs regulatory functions.

Function of the cardiac ryanodine receptor, RyR2

During the plateau phase of the action potential, the opening of L-type voltage-dependent calcium channels (dihydropyridine receptors) creates a small influx of calcium ions that triggers a massive release of calcium from the SR by opening the RyR2 channels: this process is called calcium-induced calcium release and is the pivotal mechanism that controls coordination between electrical and mechanical functions in cardiac myocytes (9). In the ventricles of mammals, the ryanodine receptors are located in the proximity of the surface membrane and of the T tubules; this location is strategic to ensure a prompt interaction of RyR2 with L-type calcium channels (10). Thanks to this architecture, for every L-type calcium channel that opens, 4–6 ryanodine receptors are synchronously activated to generate a Ca²⁺ spark (11). A cluster of approximately 100 ryanodine receptor channels and 25 L-type calcium channels creates a functional unit called a junction, or couplon (12). There are approximately 10,000 couplons in a ventricular myocyte that are simultaneously activated by Ca²⁺ influx for every heartbeat (13). The coordination in opening and closing (14) is critical for timing of intracellular calcium release to the systolic phase of the cardiac cycle, which thus ensures functional silence during diastole.

RyR2 and its macromolecular complex

The ryanodine receptor interacts with a multitude of accessory proteins to form a macromolecular calcium release complex. Junctin, triadin, and calsequestrin are the most important proteins that interact with RyR2 at the intra-SR level. It has been proposed that other proteins located in the SR, such as sarcalumenin and calreticulin, may also interact with RyR2 and be part of its macromolecular complex. Most of the regulatory functions are accomplished by proteins binding to the cytosolic portion of RyR2, such as calmodulin, protein phosphatase 1, protein phosphatase 2A,

---

Nonstandard abbreviations used: ARVD2, arrhythmogenic right ventricular cardiomyopathy type 2; CCD, central core disease; CPVT, catecholaminergic polymorphic ventricular tachycardia; DAD, delayed afterdepolarization; MH, malignant hyperthermia; PKA, protein kinase A; RyR, ryanodine receptor; SR, sarcoplasmic reticulum; VT, ventricular tachycardia.

Conflict of interest: The authors have declared that no conflict of interest exists.

Citation for this article: *J. Clin. Invest.* **115**:2033–2038 (2005).
doi:10.1172/JCI25664.

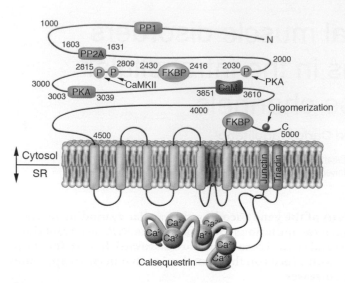

Figure 1
Schematic showing the predicted structure of the cardiac ryanodine receptor, RyR2, including the sites of interaction with ancillary proteins and the phosphorylation sites. Calsequestrin, junctin, and triadin, proteins interacting with ryanodine receptor in the SR, are also depicted. PP, protein phosphatase; P, phosphorylation sites; CaM, calmodulin; CaMKII, calmodulin-dependent protein kinase II. Adapted with permission from the *Journal of Molecular and Cellular Cardiology* (13).

protein kinase A (PKA), calmodulin-dependent protein kinase II, calcineurin, and PKC isoforms α and β (15). Among the cytosolic ligands, the peptidyl-prolyl isomerase FKBP12.6 (calstabin 2) has been extensively studied. It is a 12.6-kDa protein that binds to a specific region of RyR2 that, according to some authors, encompasses aa 2,416–2,430; it should be noted, however, that this binding site has been disputed (16, 17). FKBP12.6 stabilizes the closed state of the RyR2 channel, thus preventing aberrant calcium leakage (18, 19). The binding of FKBP12.6 to RyR2 is thought to be modulated by PKA-mediated phosphorylation that decreases the binding affinity of RyR2 for FKBP12.6, thus increasing the open probability of the channel and amplifying its response to calcium-dependent activation (18, 19). However, the role of this molecule in PKA phosphorylation has also been questioned, at least with regard to serine 2,030 (a major RyR2 phosphorylation site) (Figure 1) (20).

Ryanodine receptors and human diseases

Mutations in RyR1, central core disease, and malignant hyperthermia
The first evidence that human genetic diseases may be associated with mutations of ryanodine receptors came from studies performed on the skeletal muscle isoform, RyR1. Predisposition to 2 diseases, malignant hyperthermia (MH) and central core disease (CCD), was associated with *RyR1* mutations (21, 22). MH is characterized by acute hyperpyrexia that develops during or immediately after general anaesthesia. The disease shows paroxysmal manifestations that occur in carriers of a mutation upon exposure to volatile anaesthetics; in this setting, therefore, the mutations create a substrate that becomes apparent only in the presence of the specific triggering agent: i.e., it is needed, but not sufficient, to cause the phenotype. CCD is a rare neuromuscular disorder that causes

weakness in the muscles of the legs and, less frequently, in other muscles. Its name derives from the presence of abnormal "cores" identified at histological examination in type I muscle fibers that also present sarcomeric disorganization, lack of mitochondria, and lack of oxidative activity.

Molecular genetics and functional characterization of RyR1 mutants. Most of the mutations identified in RyR1 cluster in 3 regions: the amino terminus (including approximately the region between aa 35 and aa 614), a central region (spanning aa 1,787–2,458), and the carboxy terminus (extending from aa 4,796 to aa 4,973). It should be noted, however, that as more families with RyR1 mutations are being identified, the boundaries of these regions are becoming less well defined. It is interesting to observe that in the last 3 years, 9 distinct mutations have been reported in the region between aa 3,527 and aa 4,651 — a portion of the gene that was until then devoid of mutations. In analogy with what has happened with most genetic diseases, the spreading of genetic analysis in families with CCD and MH is broadening the spectrum of the clinical phenotypes and is disclosing unexpected findings. For example, despite that both CCD and MH are inherited as autosomal dominant conditions, CCD families with homozygous mutations have been reported (23), and 1 family with compound heterozygous mutations has been identified presenting with an overlapping phenotype of susceptibility to MH combined with a mild form of CCD (24).

In a few instances, *RyR1* mutations have been identified in patients with phenotypes unrelated to MH and CCD: for example, a proline-to-serine substitution at position 3,527 (25) and a splice error (26) have been associated with multi-minicore disease, a congenital myopathy transmitted as a recessive trait that manifests with hypotonia and generalized weakness in axial and proximal limb muscles. It seems premature to decide, based on these sporadic reports, whether the latter disease is allelic to CCD and MH; it seems more appropriate to wait until more genotyping data become available. A similar conservative approach is recommended for discussion of RyR2 mutations identified in a few families diagnosed with an atypical form of right ventricular cardiomyopathy, arrhythmogenic right ventricular cardiomyopathy type 2 (ARVD2) (see below).

Functional characterization of mutant proteins is a very important step in the understanding of the pathophysiological mechanisms of a genetic disease. In the next few paragraphs, we will try to summarize the key results of the in vitro studies of RyR1 mutants and will highlight data that clarify how mutations in the same gene may cause 2 apparently unrelated diseases such as CCD and MH.

Most of the *RyR1* mutations characterized in different in vitro systems lead to a gain of function. Richter et al. (27) showed that the G2434R mutation identified in patients with MH enhances the sensitivity of RyR1 to Ca^{2+} and caffeine. Similar findings were reported by Yang et al. (28) when they expressed 6 common *RyR1* mutants found in MH patients (R163C, G341R, R614C, R2163C, V2168M, and R2458H) in myotubes differentiated from cells lacking endogenous *RyR1*, and by Wehner et al. (29) when they characterized RyR1 mutants in primary human myotubes cultured from carriers of *RyR1* mutations. Overall, these studies suggest that the leading pathophysiological substrate for MH is RyR1, which, although it functions normally at resting conditions, presents enhanced sensitivity to activators and reduced sensitivity to endogenous inhibitors (Ca^{2+} and Mg^{2+}). Abnormal response to activators may therefore precipitate the clinical manifestations.

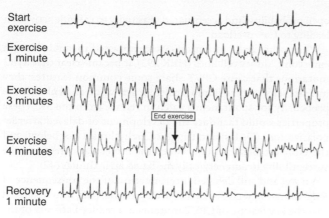

Figure 2
Exercise stress test in a patient with polymorphic VT and RyR2 mutation. Ventricular arrhythmias are observed with a progressive worsening during exercise. Typical bidirectional VT develops after 3 minutes of exercise with a sinus heart rate of approximately 120 bpm. Arrhythmias rapidly recede during recovery.

Different properties of RyR1 mutants have been highlighted in the study of defects identified in CCD patients, and at least 3 other abnormalities have been reported: (a) mutations occurring in exons 101 and 102 were shown to make RyR1 leaky in basal conditions without altering the response to pharmacological modulators with reduced intracellular calcium stores (30); (b) uncoupling of sarcolemmal excitation (voltage-gated Ca^{2+} channel) from SR Ca^{2+} release was reported for some CCD mutations (31, 32); and (c) mutations located in the amino terminus of the RyR1 protein have been reported to share the common characteristic of increasing Ca^{2+} release (leak) at negative (resting) membrane potentials (32).

In summary, some hypotheses have been advanced to explain the different phenotypes associated with *RyR1* mutations. DNA defects found in MH patients appear to produce channels that have a normal behavior at rest but overreact to exogenous activators. On the contrary, the majority of CCD mutants produce a channel that is leaky at base line, depletes the SR, and may also cause electrical uncoupling between RyR1 and the voltage-gated Ca^{2+} channels. To add to this interpretation, Dirksen and Avila (33) suggested that, since the degree of Ca^{2+} leakage and EC uncoupling induced by different mutations is variable, the combination of these effects may represent a critical modulator of the clinical phenotype.

Mutations in RyR2 and catecholaminergic polymorphic ventricular tachycardia

Mutations in the RyR2 gene have been identified in families and in sporadic patients affected by catecholaminergic polymorphic ventricular tachycardia (CPVT). This disease was initially described in 1978 (34) and more extensively characterized in 1995 (35). The name derives from the clinical evidence that these patients develop a typical polymorphic or bidirectional ventricular tachycardia (VT) that leads to syncopal episodes during catecholamine release in the setting of emotion or exercise (Figure 2). In most instances, CPVT is inherited as an autosomal dominant disease, and approximately 30% of patients have a family history of juvenile sudden death and/or stress-related syncope (35). The first clinical manifestations of the disease appear during childhood, although late-onset forms have been described (36).

Linkage studies localized the CPVT gene on the long arm of chromosome 1, 1q42-q43 (37). Thus the human cardiac ryanodine receptor gene (*RyR2*) that mapped in that region became a plausible candidate gene for CPVT. We performed screening of the coding region of this gene in DNA obtained from CPVT patients and were able to identify pathogenetic mutations (2). An uncommon autosomal recessive variant of CPVT has been identified and linked to chromosome 1p13-p21 (38); shortly after the publication of linkage data, mutations were found in the gene *Casq2* that encodes for calsequestrin, thus confirming the view that CPVT is a disease caused by intracellular calcium handling abnormalities (39, 40).

Treatment of CPVT patients. Antiadrenergic treatment with β-blockers is the mainstay of therapy for CPVT patients, and it is indicated in all patients with clinical diagnosis (35, 36). No data are available to define whether asymptomatic carriers of RyR2 mutations should also be treated with prophylactic β-blockers. These drugs reduce the number of runs of polymorphic VT and their rate and duration in the majority of patients, but they usually fail to completely suppress ventricular tachyarrhythmias. In those patients in which β-blockers do not prevent the recurrence of sustained VT, it is recommended to consider the implantation of a cardioverter defibrillator (41).

Interestingly, a single group of investigators claimed the identification of RyR2 mutations in families diagnosed with an atypical form of right ventricular cardiomyopathy, ARVD2 (5), that combines modest structural abnormalities of the right ventricle with typical stress-induced polymorphic VTs. The presence of RyR2 mutations in patients with the clinical diagnosis right ventricular cardiomyopathy has not been confirmed by other groups, and, at the present time, whether mutations in RyR2 cause ARVD2 remains a matter of debate that awaits further data.

Role of genetic testing in CPVT. Approximately 40% of CPVT patients carry a mutation on the RyR2 gene; as a consequence, it is recommended that all patients with this clinical diagnosis undergo genetic analysis. Because DNA analysis can be extended to all family members to identify those who are asymptomatic but genetically affected, it is important to know whether the clinically affected proband carries the mutation. Silent-mutation carriers should receive prophylactic treatment with β-blockers with the objective of preventing sudden cardiac death (35, 36). At the present time, no genotype-phenotype correlations have been established in CPVT, and it will be important to collect data on larger numbers of genotyped individuals with a long clinical follow-up to determine whether it is possible to define risk of cardiac events or response to therapy based on the genetic defect (42, 43).

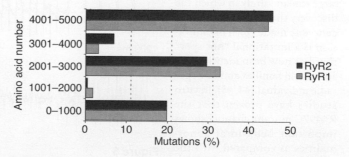

Figure 3
Bar graph showing the similar clustering of mutations in RyR1 (including both MH and CCD phenotypes) and RyR2 (both CPVT and ARVD2).

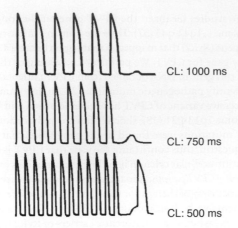

Figure 4
Representation of a cardiac action potential, showing DADs. DAD amplitude increases and DAD coupling interval decreases upon cycle length (CL) reduction. With further increase in frequency of stimulation, an automatic beat is generated from the DAD.

Molecular genetics and functional characterization of RyR2 mutants. All RyR2 mutations identified so far in CPVT patients are single-base-pair substitutions that lead to the replacement of a highly conserved amino acid. This is at variance with RyR1 mutations found in MH and CCD patients, where splice errors and deletions have also been reported. The distribution of mutations on the predicted topology of the proteins, however, is identical for RyR1 and RyR2. Most RyR2 mutations are clustered in the amino terminus, in the FKBP12.6-binding domains, and in the transmembrane domains of the protein (Figure 3). Very few RyR2 mutations have been functionally characterized, so it is still difficult to define whether functional differences exist among mutations and whether they account for the severity of the phenotype (33). Furthermore, some of the first mutations identified have been expressed by different groups yielding different results.

The first *RyR2* mutation to be functionally characterized was the mouse homolog of the mutation (R4497C) identified in the large Italian family in which the discovery that *RyR2* is the CPVT gene was made (2). This mutation is a mutational "hot spot" that has now been identified in additional families and in 1 sporadic individual (44, 45). In vitro studies have proven that the R4497C mutant channel shows important behavioral abnormalities as compared with the wild-type channel (46): R4497C-RyR2 presents enhanced resting activity, increased single-channel

open probability, and accentuated sensitivity to caffeine activation leading to Ca²⁺ overload.

When other *RyR2* mutations were expressed in stable human embryonic kidney cell lines (HEK 293), it became clear that RyR2 mutations that cause CPVT share some common features: they increase sensitivity to luminal calcium activation and enhance the propensity for spontaneous calcium release from the SR. These properties would facilitate the development of delayed afterdepolarizations (DADs) and triggered arrhythmias (see below) and therefore establish a link between the molecular defect and the susceptibility to adrenergically mediated arrhythmias (47).

A novel and still debated interpretation of the consequence of *RyR2* mutation was provided by Wehrens et al. (48), who suggested that the key feature of CPVT mutants is a reduced affinity of the channel for the binding of the regulatory protein FKBP12.6, leading to a channel that releases calcium during diastole. George et al. (49) did not confirm the data of Wehrens et al.: they found that mutant channels had a normal RyR2/FKBP12.6 interaction but an abnormally augmented response to isoproterenol and forskolin, resulting in excessive calcium release. The same group recently reported another interesting observation (50) when they expressed mutations identified in ARVD2 patients (5). Interestingly, 1 of the 4 mutants studied (50) was not associated with the conventional gain of function (increased response to RyR2 agonists that promote calcium release from the SR, such as caffeine) but rather induced a loss of function. In a speculative editorial, Gomez and Richard (51) put forward a highly provocative hypothesis by suggesting that a loss-of-function mutation would reduce Ca²⁺ release from RyR2 and that this could in turn slow the inactivation of the current conducted by the sarcolemmal calcium-dependent voltage channel and prolong cardiac action potential. In this view, arrhythmias triggered by loss-of-function mutations in RyR2 could still

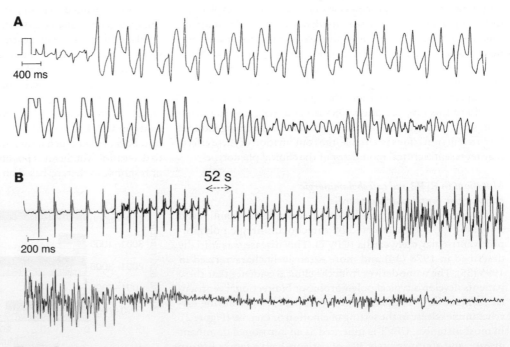

Figure 5
Bidirectional VT and ventricular fibrillation in CPVT. (**A**) ECG recording of bidirectional VT degenerating into ventricular fibrillation in a CPVT patient. (**B**) ECG recording of a bidirectional VT degenerating into ventricular fibrillation in a RyR2⁺/RyRR4496C mouse.

be triggered, but by early afterdepolarizations (EADs) rather than by DADs. Albeit appealing, this hypothesis is in contrast with common knowledge that EAD-induced arrhythmias occur in the setting of bradycardia; EADs therefore could hardly account for adrenergically mediated arrhythmias that originate during sinus tachycardia. A further improvement in the understanding of the pathophysiology of RyR2-related arrhythmias should now be provided by knock-in animal models.

Overall, functional characterization of mutants performed so far seems to agree on the presence of an increased calcium release response upon adrenergic stimulation and on a lower threshold for calcium spilling from the SR. All these dysfunctions are likely to promote the development of DADs and triggered arrhythmias. Less agreement exists on the mechanisms that mediate these abnormal responses, and, specifically, the role of FKBP12.6 needs to be further elucidated.

From functional characterization of mutants to electrophysiological mechanisms of CPVT. Even before the identification of the gene responsible for CPVT, the hypothesis that arrhythmias were initiated by DADs had been advanced based on the observation that the bidirectional VT observed in CPVT patients closely resembles digitalis-induced arrhythmias. A lot is known about the electrophysiological mechanisms underlying digitalis-induced arrhythmias: sound experimental data have demonstrated that in the presence of digitalis-induced intracellular calcium overload the Na^+/Ca^{2+} exchanger gets activated in the attempt to reduce calcium and exchanges 1 Ca^{2+} for 3 Na^+, thus generating a net inward current (the so-called transient inward current). This depolarizing current promotes diastolic depolarizations (called delayed afterdepolarizations, or DADs) (52) that may reach the threshold for I_{Na} activation and elicit premature action potentials. This mechanism for arrhythmia initiation is called triggered activity. Several in vitro studies have demonstrated that β-adrenergic stimulation induces DADs both in Purkinje fibers and in cardiac myocytes (52, 53). Based on the above-mentioned information, it becomes appealing to speculate that RyR2 mutations identified in CPVT patients promote the development of calcium overload in concomitance with sympathetic stimulation and the activation of the Na^+/Ca^{2+} exchanger, thus leading to the development of DADs and triggered arrhythmias (Figure 4). This hypothesis still awaits conclusive experimental support: the availability of the recently described knock-in mouse carrier of the R4496C RyR2 mutation (54), in which bidirectional VTs can be induced (Figure 5), will allow demonstration of whether RyR2 mutations lead to the development of DADs as the initiating mechanism for cardiac arrhythmias.

Conclusions

In the last few years, advancements in understanding the uncommon inherited arrhythmogenic disorder CPVT have given momentum to the study of the regulation of intracellular calcium handling.

The discovery of DNA mutations that modify the intracellular calcium release channel ryanodine receptor has highlighted the pivotal role of RyR2 in the control of intracellular calcium and has brought to general appreciation the link between RyR2 and electrophysiological properties of the cardiac myocytes. These observations are of major importance because they may lead to innovative curative approaches that may apply not only to patients with CPVT, but also to individuals with acquired arrhythmias associated with dysfunctional intracellular calcium handling, such as heart failure patients. As it did for long QT syndrome, collaboration between clinicians, molecular biologists, and biophysicists has, for CPVT as well, opened new and important areas of research that through the advancement of basic science will eventually develop new hope for the cure of human diseases.

Acknowledgments

We wish to acknowledge the helpful discussion and the constructive insights provided by Anthony F. Lai (Cardiff University, Cardiff, United Kingdom) during the preparation of this article.

Address correspondence to: Silvia G. Priori, Molecular Cardiology, Maugeri Foundation, University of Pavia, Via Ferrata 8, 27100 Pavia, Italy. Phone: 39-0382-592051; Fax: 39-0382-592059; E-mail: spriori@fsm.it.

1. Gillard, E.F., et al. 1992. Polymorphisms and deduced amino acid substitutions in the coding sequence of the ryanodine receptor (RYR1) gene in individuals with malignant hyperthermia. *Genomics.* **13**:1247–1254.

2. Priori, S.G., et al. 2001. Mutations in the cardiac ryanodine receptor gene (hRyR2) underlie catecholaminergic polymorphic ventricular tachycardia. *Circulation.* **103**:196–200.

3. Tunwell, R.E., et al. 1996. The human cardiac muscle ryanodine receptor-calcium release channel: identification, primary structure and topological analysis. *Biochem. J.* **318**:477–487.

4. Otsu, K., et al. 1993. Chromosome mapping of five human cardiac and skeletal muscle sarcoplasmic reticulum protein genes. *Genomics.* **17**:507–509.

5. Tiso, N., et al. 2001. Identification of mutations in the cardiac ryanodine receptor gene in families affected with arrhythmogenic right ventricular cardiomyopathy type 2 (ARVD2). *Hum. Mol. Genet.* **10**:189–194.

6. Du, G.G., Sandhu, B., Khanna, V.K., Guo, X.H., and MacLennan, D.H. 2002. Topology of the Ca2+ release channel of skeletal muscle sarcoplasmic reticulum (RyR1). *Proc. Natl. Acad. Sci. U. S. A.* **99**:16725–16730.

7. Ma, J., Hayek, S.M., and Bhat, M.B. 2004. Membrane topology and membrane retention of the ryanodine receptor calcium release channel. *Cell Biochem. Biophys.* **40**:207–224.

8. Stewart, R., Zissimopoulos, S., and Lai, F.A. 2003. Oligomerization of the cardiac ryanodine receptor C-terminal tail. *Biochem. J.* **376**:795–799.

9. Fabiato, A., and Fabiato, F. 1978. Calcium-induced release of calcium from the sarcoplasmic reticulum of skinned cells from adult human, dog, cat, rabbit, rat, and frog hearts and from fetal and new-born rat ventricles. *Ann. N. Y. Acad. Sci.* **307**:491–522.

10. Carl, S.L., et al. 1995. Immunolocalization of sarcolemmal dihydropyridine receptor and sarcoplasmic reticular triadin and ryanodine receptor in rabbit ventricle and atrium. *J. Cell Biol.* **129**:672–682.

11. Wang, S.Q., Song, L.S., Lakatta, E.G., and Cheng, H. 2001. Ca2+ signalling between single L-type Ca2+ channels and ryanodine receptors in heart cells. *Nature.* **410**:592–596.

12. Franzini-Armstrong, C., Protasi, F., and Ramesh, V. 1999. Shape, size, and distribution of Ca(2+) release units and couplons in skeletal and cardiac muscles. *Biophys. J.* **77**:1528–1539.

13. Bers, D.M. 2004. Macromolecular complexes regulating cardiac ryanodine receptor function. *J. Mol. Cell. Cardiol.* **37**:417–429.

14. Marx, S.O., et al. 2001. Coupled gating between cardiac calcium release channels (ryanodine receptors). *Circ. Res.* **88**:1151–1158.

15. Meissner, G. 2002. Regulation of mammalian ryanodine receptors. *Front. Biosci.* **7**:d2072–d2080.

16. Masumiya, H., Wang, R., Zhang, J., Xiao, B., and Chen, S.R. 2003. Localization of the 12.6-kDa FK506-binding protein (FKBP12.6) binding site to the NH2-terminal domain of the cardiac Ca2+ release channel (ryanodine receptor). *J. Biol. Chem.* **278**:3786–3792.

17. Zissimopoulos, S., and Lai, F.A. 2002. Evidence for a FKBP12.6 binding site at the C-terminus of the cardiac ryanodine receptor [abstract]. *Biophys. J.* **82**:59.

18. Marx, S.O., et al. 2000. PKA phosphorylation dissociates FKBP12.6 from the calcium release channel (ryanodine receptor): defective regulation in failing hearts. *Cell.* **101**:365–376.

19. Lehnart, S.E., Wehrens, X.H., and Marks, A.R. 2004. Calstabin deficiency, ryanodine receptors, and sudden cardiac death. *Biochem. Biophys. Res. Commun.* **322**:1267–1279.

20. Xiao, B., et al. 2005. Characterization of a novel PKA phosphorylation site, serine-2030, reveals no PKA hyperphosphorylation of the cardiac ryanodine receptor in canine heart failure. *Circ. Res.* **96**:847–855.

21. Loke, J., and MacLennan, D.H. 1998. Malignant hyperthermia and central core disease: disorders of Ca2+ release channels. *Am. J. Med.* **104**:470–486.

22. Jurkat-Rott, K., McCarthy, T., and Lehmann-Horn, F. 2000. Genetics and pathogenesis of malignant hyperthermia. *Muscle Nerve.* **23**:4–17.

23. Jungbluth, H., et al. 2002. Autosomal recessive inheritance of RYR1 mutations in a congenital myopathy with cores. *Neurology.* **59**:284–287.

24. Frank, J.P., Harati, Y., Butler, I.J., Nelson, T.E., and Scott, C.I. 1980. Central core disease and malignant hyperthermia syndrome. *Ann. Neurol.* **7**:11–17.

25. Davis, M.R., et al. 2003. Principal mutation hotspot for central core disease and related myopathies in the C-terminal transmembrane region of the RYR1 gene. *Neuromuscul. Disord.* **13**:151–157.

26. Monnier, N., et al. 2003. A homozygous splicing mutation causing a depletion of skeletal muscle RYR1 is associated with multi-minicore disease congenital myopathy with ophthalmoplegia. *Hum. Mol. Genet.* **12**:1171–1178.

27. Richter, M., Schleithoff, L., Deufel, T., Lehmann-Horn, F., and Herrmann-Frank, A. 1997. Functional characterization of a distinct ryanodine receptor mutation in human malignant hyperthermia-susceptible muscle. *J. Biol. Chem.* **272**:5256–5260.

28. Yang, T., Ta, T.A., Pessah, I.N., and Allen, P.D. 2003. Functional defects in six ryanodine receptor isoform-1 (RyR1) mutations associated with malignant hyperthermia and their impact on skeletal excitation-contraction coupling. *J. Biol. Chem.* **278**:25722–25730.

29. Wehner, M., Rueffert, H., Koenig, F., and Olthoff, D. 2004. Functional characterization of malignant hyperthermia-associated RyR1 mutations in exon 44, using the human myotube model. *Neuromuscul. Disord.* **14**:429–437.

30. Tilgen, N., et al. 2001. Identification of four novel mutations in the C-terminal membrane spanning domain of the ryanodine receptor 1: association with central core disease and alteration of calcium homeostasis. *Hum. Mol. Genet.* **10**:2879–2887.

31. Avila, G., and Dirksen, R.T. 2001. Functional effects of central core disease mutations in the cytoplasmic region of the skeletal muscle ryanodine receptor. *J. Gen. Physiol.* **118**:277–290.

32. Avila, G., O'Brien, J.J., and Dirksen, R.T. 2001. Excitation–contraction uncoupling by a human central core disease mutation in the ryanodine receptor. *Proc. Natl. Acad. Sci. U. S. A.* **98**:4215–4220.

33. Dirksen, R.T., and Avila, G. 2002. Altered ryanodine receptor function in central core disease: leaky or uncoupled Ca(2+) release channels [review]? *Trends Cardiovasc. Med.* **12**:189–197.

34. Coumel, P., Fidelle, J., Lucet, V., Attuel, P., and Bouvrain, Y. 1978. Catecholaminergic-induced severe ventricular arrhythmias with Adams-Stokes syndrome in children: report of four cases. *Br. Heart J.* **40**:28–37.

35. Leenhardt, A., et al. 1995. Catecholaminergic polymorphic ventricular tachycardia in children. A 7-year follow-up of 21 patients. *Circulation.* **91**:1512–1519.

36. Priori, S.G., et al. 2002. Clinical and molecular characterization of patients with catecholaminergic polymorphic ventricular tachycardia. *Circulation.* **106**:69–74.

37. Swan, H., et al. 1999. Arrhythmic disorder mapped to chromosome 1q42-q43 causes malignant polymorphic ventricular tachycardia in structurally normal hearts. *J. Am. Coll. Cardiol.* **34**:2035–2042.

38. Lahat, H., et al. 2001. Autosomal recessive catecholamine- or exercise-induced polymorphic ventricular tachycardia. *Circulation.* **103**:2822–2827.

39. Lahat, H., et al. 2001. A missense mutation in a highly conserved region of CASQ2 is associated with autosomal recessive catecholamine-induced polymorphic ventricular tachycardia in Bedouin families from Israel. *Am. J. Hum. Genet.* **69**:1378–1384.

40. Postma, A.V., et al. 2002. Absence of calsequestrin 2 causes severe forms of catecholaminergic polymorphic ventricular tachycardia. *Circ. Res.* **91**:e21–e26.

41. Cerrone, M., et al. 2004. Clinical and molecular characterization of a large cohort of patients affected with catecholaminergic polymorphic ventricular tachycardia [abstract]. *Circulation.* **110**(Suppl. 3):552.

42. Priori, S.G., et al. 2004. Association of long QT syndrome loci and cardiac events among patients treated with beta-blockers. *JAMA.* **292**:1341–1344.

43. Priori, S.G., et al. 2003. Risk stratification in the long-QT syndrome. *N. Engl. J. Med.* **348**:1866–1874.

44. Choi, G., et al. 2004. Spectrum and frequency of cardiac channel defects in swimming-triggered arrhythmia syndromes. *Circulation.* **110**:2119–2124.

45. Tester, D.J., Spoon, D.B., Valdivia, H.H., Makielski, J.C., and Ackerman, M.J. 2004. Targeted mutational analysis of the RyR2-encoded cardiac ryanodine receptor in sudden unexplained death: a molecular autopsy of 49 medical examiner/coroner's cases. *Mayo Clin. Proc.* **79**:1380–1384.

46. Jiang, D., Xiao, B., Zhang, L., and Chen, S.R. 2002. Enhanced basal activity of a cardiac Ca2+ release channel (ryanodine receptor) mutant associated with ventricular tachycardia and sudden death. *Circ. Res.* **91**:218–225.

47. Jiang, D., et al. 2004. RyR2 mutations linked to ventricular tachycardia and sudden death reduce the threshold for store-overload-induced Ca2+ release (SOICR). *Proc. Natl. Acad. Sci. U. S. A.* **101**:13062–13067.

48. Wehrens, X.H., et al. 2003. FKBP12.6 deficiency and defective calcium release channel (ryanodine receptor) function linked to exercise-induced sudden cardiac death. *Cell.* **113**:829–840.

49. George, C.H., Higgs, G.V., and Lai, F.A. 2003. Ryanodine receptor mutations associated with stress-induced ventricular tachycardia mediate increased calcium release in stimulated cardiomyocytes. *Circ. Res.* **93**:531–540.

50. Lowri, T.N., George, C.H., and Anthony, L.F. 2004. Functional heterogeneity of ryanodine receptor mutations associated with sudden cardiac death. *Cardiovasc. Res.* **64**:52–60.

51. Gomez, A.M., and Richard, S. 2004. Mutant cardiac ryanodine receptors and ventricular arrhythmias: is 'gain-of-function' obligatory? *Cardiovasc. Res.* **64**:3–5.

52. Kieval, R.S., Butler, V.P., Jr., Derguini, F., Bruening, R.C., and Rosen, M.R. 1988. Cellular electrophysiologic effects of vertebrate digitalis-like substances. *J. Am. Coll. Cardiol.* **11**:637–643.

53. Priori, S.G., and Corr, P.B. 1990. Mechanisms underlying early and delayed afterdepolarizations induced by catecholamines. *Am. J. Physiol.* **258**:H1796–H1805.

54. Cerrone, M., et al. 2005. Bidirectional ventricular tachycardia and fibrillation elicited in a knock-in mouse model carrier of a mutation in the cardiac ryanodine receptor (RyR2). *Circ. Res.* **96**:e77–e82.

Chloride channel diseases resulting from impaired transepithelial transport or vesicular function

Thomas J. Jentsch, Tanja Maritzen, and Anselm A. Zdebik

Zentrum für Molekulare Neurobiologie Hamburg, Universität Hamburg, Hamburg, Germany.

The transport of anions across cellular membranes is crucial for various functions, including the control of electrical excitability of muscle and nerve, transport of salt and water across epithelia, and the regulation of cell volume or the acidification and ionic homeostasis of intracellular organelles. Given this broad range of functions, it is perhaps not surprising that mutations in Cl⁻ channels lead to a large spectrum of diseases. These diverse pathologies include the muscle disorder myotonia, cystic fibrosis, renal salt loss in Bartter syndrome, kidney stones, deafness, and the bone disease osteopetrosis. This review will focus on diseases related to transepithelial transport and on disorders involving vesicular Cl⁻ channels.

Defects in transepithelial Cl⁻ transport

As Cl⁻ channels allow only for the passive, diffusional flux of Cl⁻ down its electrochemical gradient, the difference between cytoplasmic and extracellular Cl⁻ concentration ([Cl⁻]), together with the membrane voltage, determines whether the opening of a Cl⁻ channel will lead to an influx or efflux of this ion. Whereas in adult neurons intracellular [Cl⁻] ([Cl⁻]$_i$) is mostly below its equilibrium (predominantly because of the activity of the neuronal potassium chloride cotransporter KCC2; ref. 1), [Cl⁻]$_i$ in epithelial cells is often in the 30–40 mM range and thus above equilibrium at a voltage (–50 mV) typical for these cells. Several cotransporters may contribute to the relatively high cytoplasmic [Cl⁻]$_i$ in epithelia, most prominently NaK2Cl cotransporters. Opening of epithelial Cl⁻ channels will therefore lead mostly to an efflux of Cl⁻, and the localization of these channels will determine the transport direction (Figure 1). Hence, apical Cl⁻ channels are involved in Cl⁻ secretion (as shown in Figure 1B), whereas basolateral Cl⁻ channels play a role in Cl⁻ (and salt) (re)absorption (Figure 1D).

CFTR and cystic fibrosis

Cystic fibrosis is the most common and best-known genetic disease involving a defect in transepithelial Cl⁻ transport. It affects several epithelial organs, i.e., the lungs, pancreas, and intestine, among others. The most serious cystic fibrosis symptoms are generally observed in the lungs, where the fluid covering the airway epithelia becomes viscous and susceptible to bacterial infection.

CFTR, the cystic fibrosis transmembrane conductance regulator (2), functions as a cAMP- and ATP-regulated Cl⁻ channel. This discovery came as a surprise, because CFTR belongs to the gene family of ABC transporters, which normally function as transport ATPases, but not as ion channels. However, the line separating ion channels from transporters may be thin, as recently demonstrated (3) by the 2Cl⁻/H⁺ exchange activity of a bacterial homolog of mammalian Cl⁻ channels of the CLC gene family. In addition to the well-established function of CFTR as a Cl⁻ channel, many regulatory roles have been ascribed to it, but some of these remain controversial (4–6). CFTR

may negatively regulate the epithelial Na⁺ channel ENaC, the activation of which may contribute to the cystic fibrosis lung phenotype (4, 7). However, the purported interaction of both channels was questioned by others (8). Recent data indicate that CFTR may directly activate the anion exchangers Slc26a3 (DRA, downregulated in adenoma) and Slc26a6 (PAT-1) (9). Certain CFTR mutations identified in patients may impair the Cl⁻/HCO₃⁻ exchange activity of these transporters (9) and thus could result in decreased pancreatic HCO₃⁻ secretion as is often observed in cystic fibrosis.

The role of CFTR in lung physiology is complex, and the cystic fibrosis lung pathology is not reproduced in mouse models. By contrast, the role of CFTR in colonic Cl⁻ secretion is well understood (Figure 1B). In the colon, CFTR expression seems to be limited to crypts (10), which are the site of Cl⁻ secretion (11). Like in other epithelia and consistent with a secretory role, CFTR is present in the apical membrane of the crypt cells. The opening of CFTR Cl⁻ channels, which is triggered by a rise in cAMP, leads to a passive efflux of Cl⁻ because [Cl⁻]$_i$ is elevated above equilibrium by the activity of basolateral NKCC1 NaK2Cl cotransporters. K⁺ ions that are taken up together with Cl⁻ in a stoichiometrically coupled process need to be recycled through basolateral K⁺ channels that are most likely heteromers of KCNQ1 and KCNE3 subunits (12), which are coexpressed in crypt cells (12, 13). As predicted by this model (Figure 1B), the disruption of CFTR in cystic fibrosis impairs colonic Cl⁻ and fluid secretion. This results in thick feces (meconium ileus) in a subset of infants suffering from cystic fibrosis. Similar intestinal problems are the major phenotype observed in CFTR mouse models (14). Conversely, inadequately strong activation of CFTR by the drastic increase in cAMP that is elicited by cholera toxin leads to severe diarrhea, which can be alleviated in animal models by blocking of the basolateral K⁺ conductance (12, 15).

CFTR and ClC-2: a lack of synergism

The cystic fibrosis phenotype may be modulated by the genetic background both in humans (16, 17) and in mice (18, 19). This has fueled speculations that other Cl⁻ channels that are coexpressed in the same apical membranes as CFTR may partially compensate for its loss. In addition to putative Ca²⁺-activated Cl⁻ channels (18, 20), the ubiquitously expressed ClC-2 Cl⁻ channel has been widely speculated to play such a role (21–23). However, mice in which both CFTR and ClC-2 were disrupted survived slightly better than CFTR KO mice and displayed neither lung nor pancreatic pathology (24).

Nonstandard abbreviations used: CFTR, cystic fibrosis transmembrane conductance regulator; [Cl⁻], Cl⁻ concentration; [Cl⁻]$_i$, intracellular [Cl⁻]; DRA, downregulated in adenoma; 1,25(OH)₂-VitD₃, 1,25-dihydroxyvitamin D₃; PT, proximal tubule.

Conflict of interest: The authors have declared that no conflict of interest exists.

Citation for this article: *J. Clin. Invest.* **115**:2039–2046 (2005).
doi:10.1172/JCI25470.

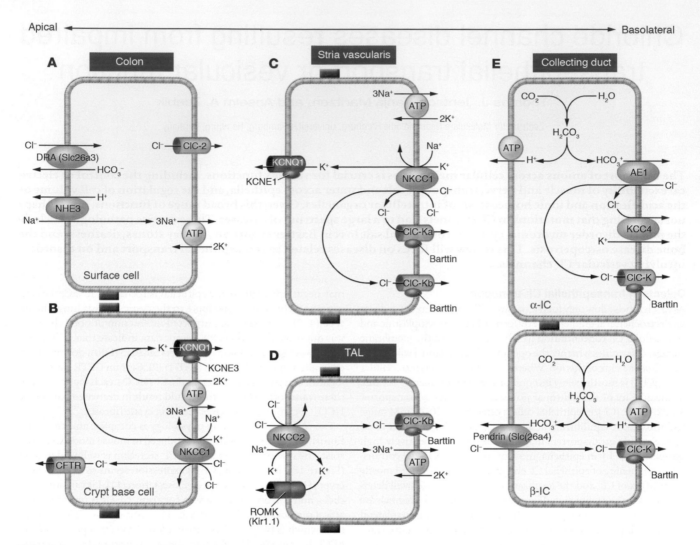

Figure 1

Diverse roles of Cl⁻ channels in transepithelial transport. In colonic epithelia, cells at the luminal surface (**A**) express a Cl⁻/HCO₃⁻ exchanger (which may be electrogenic) and the Na⁺/H⁺ exchanger NHE3 in their apical membrane, allowing for net NaCl reabsorption. Chloride probably crosses the basolateral membrane through ClC-2. Cells at the crypt base (**B**) secrete chloride, which is taken up by basolateral NKCC1, through apical CFTR channels. KCNQ1/KCNE3 heteromeric K⁺ channels are needed for K⁺ recycling. (**C**) Model for K⁺ secretion in the stria vascularis of the cochlea. K⁺ is taken up by the basolateral isoform of the NKCC cotransporter, NKCC1, and the Na,K-ATPase. Chloride is recycled by basolateral ClC-Ka and ClC-Kb/barttin channels. (**D**) Model for NaCl reabsorption in the thick ascending limb of Henle (TAL). NaCl is taken up by the apical NKCC2 transporter that needs the apical ROMK channel for K⁺ recycling. Cl⁻ leaves the cell through basolateral ClC-Kb/barttin channels. (**E**) Model for intercalated cells of the collecting duct. α-Intercalated cells (α-IC) secrete protons using a proton ATPase, while basolateral transport of acid equivalents is via the anion exchanger AE1. It is proposed that both KCC4 cotransporters (65) and ClC-K/barttin channels recycle Cl⁻. It is unknown whether ClC-K/barttin is involved in Cl⁻ reabsorption in β-intercalated cells as shown below.

Instead, they showed a superimposition of the intestinal phenotype of CFTR KO mice (14) with the retinal and testicular degeneration observed in ClC-2 KO mice (25). The degenerative phenotype in ClC-2 KO mice is thought to be due to defective Cl⁻ recycling for Cl⁻/HCO₃⁻ exchangers that may be needed to regulate the pH in the narrow extracellular clefts of these tissues, which depend heavily on lactate transport (25).

The improved survival of double-KO over CFTR KO mice (24) might be explained by an increased Cl⁻ secretion (or less Cl⁻ absorption) in ClC-2 KO colon, compatible with a basolateral instead of an apical localization of ClC-2 (24) (Figure 1A). ClC-2 has been variably reported to be present in apical (26, 27) or basolateral (28, 29)

membranes of lung and intestinal epithelia. In KO controlled immunocytochemistry, ClC-2 is exclusively detectable in basolateral membranes of surface epithelia of mouse colon, suggesting that it may be involved in Cl⁻ absorption. It is not significantly expressed in deeper sites of crypts, where CFTR is expressed (10). These findings, and in particular the CFTR/ClC-2 double-KO mice (24), should put an end to speculations that ClC-2 activation may ameliorate the cystic fibrosis phenotype.

Both ClC-2 (30) and CFTR (31) are also expressed in the kidney, but their intrarenal expression pattern is poorly defined. No renal phenotype has been described for the ClC-2 KO mouse (25). Neither is there an overt renal pathology in cystic fibrosis (32), although CFTR may interact with ROMK (Kir1.1) K⁺ channels and change their drug-sensitivity (33, 34) and was suggested to regulate ENaC in the kidney as well (35).

ClC-K/barttin: basolateral chloride channels in kidney and inner ear epithelia

The importance of the Cl⁻ channels ClC-Ka/barttin (ClC-K1/barttin in rodents) and ClC-Kb/barttin (ClC-K2/barttin in rodents) for the kidney, by contrast, is evident from their mutations in human genetic disease (36–38) and from a mouse model (39). Human loss-of-function mutations in ClC-Kb lead to severe renal salt loss in Bartter syndrome type III (36), whereas the disruption of ClC-K1 in mice causes a defect in urinary concentration (39). Human mutations in the common accessory β subunit barttin lead to Bartter syndrome type IV, which combines severe renal salt loss with congenital deafness (40).

Like ClC-2, ClC-Ka and ClC-Kb are members of the CLC family of Cl⁻ channels (41). They are 90% identical in their primary structure. Their genes are separated by just a few kilobases of DNA, indicating recent gene duplication. Both proteins are expressed in the kidney and the inner ear, as revealed by immunocytochemistry (42, 43) and the transgenic expression of a reporter gene driven by the ClC-Kb promoter (44, 45). This showed that ClC-Ka is expressed in the thin limb of Henle loop of the nephron, whereas ClC-Kb is present in the thick ascending limb of Henle loop and in the distal convoluted tubule, as well as in acid-transporting intercalated cells of the collecting duct. Also in the inner ear, ClC Ka and ClC Kb are expressed in epithelial cells. They are found in marginal cells of the stria vascularis and in dark cells of the vestibular organ, both of which probably coexpress these isoforms.

The high degree of amino acid identity between ClC-Ka and ClC-Kb makes it difficult to generate isoform-specific antibodies, resulting in some uncertainty whether, for instance, the thick ascending limb or intercalated cells express both isoforms. The exclusive expression of ClC-K1 in the thin limb of Henle loop in the kidney, however, has been ascertained by the absence of staining in this segment in ClC-K1 KO mice (39). Immunocytochemistry suggests that ClC-K1 and ClC-K2 are expressed in basolateral membranes of renal (42, 43) and cochlear (37) epithelia. The only exception might be the thin limb of Henle loop, where ClC-K1 was reported to be present in both apical and basolateral membranes by 1 group (42) but was found only in basolateral membranes by others (43). The β subunit barttin was colocalized with ClC-K in every tissue examined, e.g., in the thick ascending limb, the distal convoluted tubule, and intercalated cells of the kidney, as well as in inner ear epithelia (37). As yet, no cell or tissue has been identified that expresses barttin without a ClC-K protein or vice versa.

Barttin, a rather small protein with 2 predicted transmembrane domains (40), is necessary for the functional expression of ClC-K currents (37). The only and enigmatic exception seems to be rat ClC-K1, which yields plasma membrane currents also without barttin (46, 47). Coexpressing ClC-K1 with barttin, however, strongly increased currents (37) and diminished their sensitivity to extracellular Ca²⁺ (48). Barttin is necessary for the transport of ClC-K α subunits to the plasma membrane; this readily explains the stimulation of currents seen in coexpression (37). The cytoplasmic C-terminus of barttin contains a motif (PPYVRL) that is a potential site (PPY) for binding of WW domain–containing ubiquitin ligases, or that may serve as a tyrosine-based endocytosis signal (YVRL). Indeed, mutating the tyrosine led to a twofold increase in surface

expression and currents (37). This is compatible with either hypothesis, as PY-dependent ubiquitination may serve as a signal for endocytosis as described for the epithelial Na⁺ channel ENaC (49, 50) or ClC-5 (51). It was reported that the ubiquitin ligase Nedd4-2 may mediate this effect (52). However, in contrast to findings for ClC-5 (51) and ENaC (50), the expression of inactive forms of the WW domain–containing ubiquitin ligase did not increase ClC-K/barttin currents (52). Taking into account the poor consensus sequence of the PY motif of barttin for WW domain binding, the reported interaction with Nedd4 should be viewed with caution.

ClC-K/barttin in salt reabsorption. The pathology of Bartter syndrome is easily understood in terms of a transport model for the thick ascending limb (Figure 1D), a nephron segment resorbing large amounts of NaCl. Powered by the Na⁺ gradient that is generated by the basolateral Na,K-ATPase, the apical NaK2Cl cotransporter NKCC2 transports Cl⁻ and K⁺ into the cytoplasm. This raises [Cl⁻]ᵢ above its electrochemical equilibrium, allowing for its diffusional, passive exit through basolateral ClC-Kb/barttin Cl⁻ channels. Na⁺ ions are transported across the basolateral membrane via the ATPase, while K⁺ is recycled over the apical membrane via ROMK (Kir1.1) K⁺ channels. This model is impressively supported by genetic evidence: mutations in *NKCC2* underlie Bartter syndrome I (53); in *ROMK*, Bartter II (54); in *ClC-Kb*, Bartter III (36); and, finally, in *barttin*, Bartter IV (40).

ClC-K/barttin in inner ear K⁺ secretion. It is instructive to compare the renal thick ascending limb transport model (Figure 1D) with that of marginal cells of the cochlear stria vascularis (Figure 1C). The epithelium of the stria secretes K⁺ into the fluid of the scala media of the cochlea. The exceptionally high K⁺ concentration (150 mM) and positive potential (+100 mV) of this compartment are needed to provide the driving force for the depolarizing influx of K⁺ through apical mechanosensitive ion channels of sensory hair cells and are hence essential for hearing (55). In marginal cells, the secretory, basolateral NaK2Cl-cotransporter isoform NKCC1, together with the Na,K-ATPase, raises cellular [K⁺]. Potassium then leaves the cell through apical K⁺ channels that are assembled from KCNQ1 and KCNE1 subunits. The Na⁺ that is taken up together with K⁺ is extruded via the ATPase, while cotransported Cl⁻ ions must be recycled across the basolateral membrane. This occurs through basolateral ClC-Ka/barttin and ClC-Kb/barttin Cl⁻ channels, which are believed to be coexpressed in the stria (37). Again, this model is strongly supported by genetic evidence: mutations in either *KCNQ1* (56) or *KCNE1* (57) lead to deafness in Jervell and Lange-Nielsen syndrome. Disruption of *NKCC1* in mice also causes deafness (58), as do mutations in *barttin* in Bartter syndrome type IV (37, 40). *ClC-Kb* mutations in Bartter III cause salt loss without deafness, because ClC-Kb/barttin channels are rate-limiting in the thick ascending limb of the loop of Henle, but not in the stria, where their function can be replaced by ClC-Ka/barttin. Only the disruption of the common β subunit barttin results in a chloride-recycling defect that lowers strial K⁺ secretion to pathogenic levels (37). This model has now been supported by the identification of a single family in which the disruption of both ClC-Ka *and* ClC-Kb results in a pathology indistinguishable from Bartter IV (59).

ClC-K/barttin: other potential roles. The function of ClC-K/barttin in intercalated cells of the collecting duct, where it is expressed in both acid-secreting α and base-secreting β cells (37), is less well understood. In α cells, it might serve to recycle Cl⁻ for the basolateral Cl⁻/HCO₃⁻ exchanger AE1 (Figure 1E). This role is also performed by the K-Cl cotransporter KCC4, the disruption of which leads to renal tubular acidosis in mice (60). In the presence of the

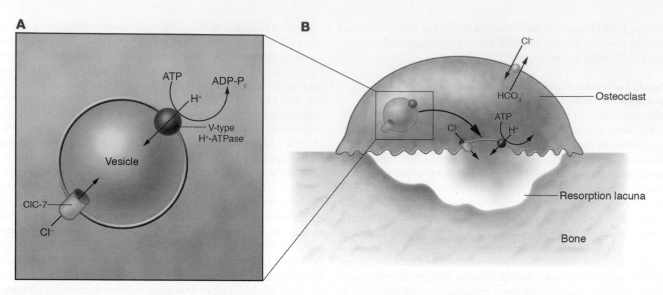

Figure 2

General concept of vesicular acidification exemplified by ClC-7. (**A**) Vesicles of the endosomal and lysosomal pathway are acidified by
H⁺-ATPases. Their current is neutralized by Cl⁻ channels. In their absence, efficient proton pumping is prevented. (**B**) Model for the resorption
lacuna acidification in osteoclasts. The H⁺-ATPase and ClC-7 are trafficked to the "ruffled border" membrane of osteoclasts. The acidification of
the resorption lacuna is required for dissolving the mineral phase of bone, as well as for the enzymatic degradation of the organic bone matrix
by lysosomal enzymes. It depends on the presence of both ClC-7 and the H⁺-ATPase in the ruffled border.

severe disturbance of renal salt handling in Bartter patients, an
additional distal acidification defect may not be easily detectable.

Interestingly, there is a common polymorphism in the human
gene encoding *ClC-Kb*. It exchanges a threonine for serine at position 481 before helix P. T481S mutant channels show dramatically increased currents when expressed in oocytes (61). If currents
through ClC-Kb/barttin were rate-limiting for NaCl reabsorption
in the thick ascending limb or distal convoluted tubule, this polymorphism could lead to increased salt reabsorption and possibly
hypertension. Indeed, a statistical association of the T481S variant
with hypertension was described (62). However, as with other association studies, firm conclusions may only be drawn after these
results have been replicated in other cohorts (63). A recent study
did not find a correlation between the T481S polymorphism and
high blood pressure in the Japanese population, in which, however,
the frequency of this polymorphism was only about 3% (64).

Human mutations in *ClC-Ka* alone have not yet been reported.
The targeted disruption of the mouse homolog *ClC-K1* leads to a
diabetes insipidus–like phenotype (39). This mouse model demonstrates that the large Cl⁻ conductance in the thin limb, which is
part of the countercurrent system, is essential for solute accumulation in the inner renal medulla (65).

Vesicular chloride channels

Ion channels are present not only in the plasma membrane, but
also in membranes of intracellular organelles like vesicles of the
endocytotic or secretory pathways, synaptic vesicles, the endoplasmic and sarcoplasmic reticulum, or mitochondria. With the
exception of intracellular Ca²⁺ channels or anion channels of
mitochondria, they have received much less attention than plasma
membrane channels. This is probably in part due to the experimental difficulties involved in their study. Recent genetic evidence
on the role of vesicular CLC gene family Cl⁻ channels, however, has
revealed their great physiological importance.

While ClC-1, -2, -Ka, and -Kb form the branch of CLC channels
that reside predominantly in plasma membranes, ClC-3, -4, -5, -6,
and -7 are located mainly in vesicles of the endocytotic and lysosomal pathway. As a word of caution, one should acknowledge the possibility that some of these proteins might not be channels, but Cl⁻/H⁺
exchangers similar to the bacterial homolog ClC-e1 (3). Within this
second group, ClC-3, -4, and -5 form a separate homology branch of
closely related proteins displaying 80% sequence identity.

Most intracellular organelles on which CLC channels have been
found are acidified by vesicular H⁺-ATPases. H⁺ transport unbalanced
by a parallel Cl⁻ current would generate a lumen-positive potential
ultimately preventing further acidification. Hence, vesicular CLC
proteins are thought to facilitate vesicular acidification (Figure 2A).

ClC-5, endocytosis, and Dent disease

Within this branch, the function of ClC-5 is best understood, as
mutations in its gene cause a renal disorder, Dent disease (66). This
rare X-linked disease is characterized by low–molecular weight
proteinuria that is in most cases accompanied by hypercalciuria,
nephrolithiasis, nephrocalcinosis, and sometimes renal failure
(67). The link between chloride channel function and the complex
disease pathology has been established mainly through the use of
KO mouse models (68–70).

The kidney is the major site of ClC-5 expression, followed by the
intestine (71, 72). ClC-5 is most prominently expressed in the proximal tubule (PT) and in intercalated cells of the collecting duct, with
significantly lower levels also being present in other segments like
the thick ascending limb (68–70). While the function of ClC-5 in
acid-transporting intercalated cells is not yet clear, its role in PTs
has been resolved in considerable detail. The PT is responsible for
the endocytotic uptake of low–molecular weight proteins that have
passed the glomerular filter. Immunofluorescence revealed a rim
stained for ClC-5 right underneath the brush border of the proximal tubular cells that colocalized with the V-type ATPase (68, 70).

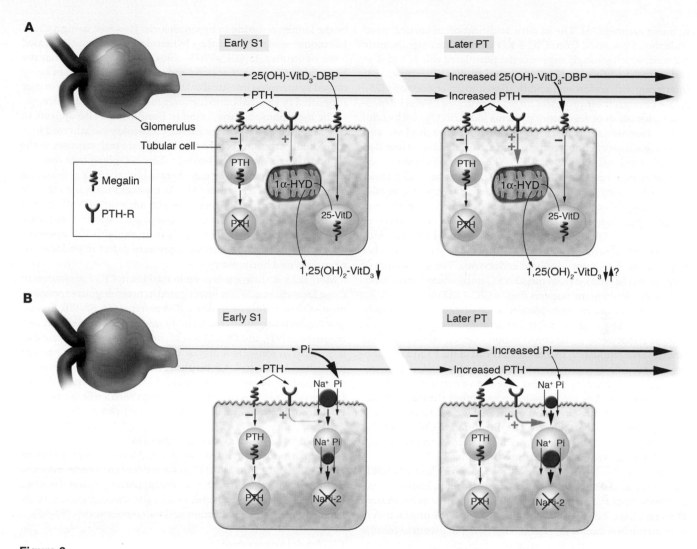

Figure 3

Model to explain hypercalciuria and hyperphosphaturia in Dent disease. (**A**) Alterations in vitamin D metabolism. Parathyroid hormone (PTH) is filtered into the primary urine from which it is normally cleared by megalin-mediated endocytosis and subsequent degradation. The impaired endocytosis due to a disruption of ClC-5 results in an increased luminal PTH concentration that leads to an enhanced activation of luminal PTH receptors (PTH-R). This stimulates the transcription of the mitochondrial enzyme 1α-hydroxylase (1α-HYD) that catalyzes the conversion of the vitamin D precursor 25(OH)-VitD$_3$ into the active metabolite 1,25(OH)$_2$-VitD$_3$. Increased enzyme activity would be expected to lead to an increased production of 1,25(OH)$_2$-VitD$_3$ that in turn would indirectly cause hypercalciuria by stimulating intestinal Ca^{2+} reabsorption. However, 25(OH)-VitD$_3$ (bound to its binding protein DBP) is mainly taken up apically by megalin- and ClC-5–dependent endocytosis. Hence, the endocytosis defect in Dent disease leads to a decreased availability of the substrate for 1α-HYD. Thus there is a delicate balance between enzyme activation and precursor scarcity that can turn toward decreased as well as increased production of 1,25(OH)$_2$-VitD$_3$. Furthermore, the active hormone is also lost into the urine. This may account for the variability of hypercalciuria observed in Dent disease patients as well as in ClC-5 KO mouse models. (**B**) Mechanism causing phosphaturia. The apical Na phosphate cotransporter NaPi-2a is regulated by PTH, which causes its endocytosis and degradation. The increased stimulation of apical PTH receptors that is due to the increased luminal PTH concentration caused by an impaired endocytosis of PTH in the absence of ClC-5 leads to less NaPi-2a in the apical membrane, resulting in a urinary loss of phosphate.

Immune electron microscopy confirmed its presence in this subapical region that is packed with endocytotic vesicles (68). When the uptake of labeled protein into proximal tubular cells was analyzed, ClC-5 colocalized with the endocytosed material only at early time points, arguing for its presence on early endosomes (68). This subcellular localization of ClC-5 points to an involvement in early endocytosis, which is obviously consistent with the proteinuria observed in Dent disease. Endosomes are acidified by a V-type ATPase that needs a counter-ion flux for effective operation (Figure 2A). Otherwise the electrogenic influx of protons into the vesicle would soon create a membrane potential that would hinder further proton pumping and thus prevent the creation of a large pH gradient. This counter-ion flux was suggested to be chloride. On the basis of these data, it was suggested that ClC-5, because it facilitated endosome acidification, was essential for renal endocytosis (68).

This hypothesis was fully confirmed by ClC-5 KO mouse models (68–70). The elimination of ClC-5 reproduced the low–molecular weight proteinuria observed in patients. The loss of ClC-5 drastically reduced apical fluid-phase and receptor-mediated endocytosis, as well as the retrieval of apical membrane proteins, in a cell-autonomous manner (73). The in vitro acidification of cortical renal endosomes prepared from ClC-5 KO animals was significantly reduced, which strongly supports the postulated role of ClC-5 in endosomal acidification (73, 74). Several studies have shown that interfering with endosomal acidification impairs endocytosis (75). A link between these processes might be provided by a pH-dependent association of regulatory proteins like Arf6 (76) with endosomes. However, this aspect has not been fully resolved yet, and other regulators of endocytosis such as rab GTPases were found to be unchanged in ClC-5 KO mice (77). The expression of the endocytotic receptor megalin was decreased in a cell-autonomous manner in PT cells lacking ClC-5; this suggests a defect in recycling megalin back to the surface (73). This was later confirmed by an electron microscopic study that also revealed a reduction of the coreceptor cubilin (77). The decrease of these endocytotic receptors, which show broad substrate specificity, is expected to lead to a further impairment of receptor-mediated endocytosis. The importance of megalin can be gleaned from megalin KO mice, whose renal phenotype resembles in many respects that of ClC-5 KO mice (78).

Hypercalciuria and hyperphosphaturia: a consequence of impaired endocytosis. It is likely that the most important function of proximal tubular endocytosis is the conservation of essential vitamins like vitamin D or retinol (79) that are reabsorbed in the PT in a megalin-dependent process together with their respective binding proteins. Whereas these binding proteins, like the vast majority of other endocytosed proteins, are degraded in lysosomes, the attached vitamins are recycled into the blood or are even processed to the active hormone in the PT (vitamin D). The disruption of ClC-5, like that of megalin (80, 81), led to a massive loss of retinol and vitamin D, as well as their binding proteins, into the urine (73, 74).

Both Dent disease patients and ClC-5 KO mice lose vitamin D and its metabolites into the urine. The situation is, however, complex because of the influence of parathyroid hormone on vitamin D metabolism. Parathyroid hormone enhances the production of 1,25-dihydroxyvitamin D_3 [$1,25(OH)_2$-$VitD_3$] in proximal tubular cells by stimulating the transcription of the enzyme 1α-hydroxylase that converts the inactive precursor $25(OH)$-$VitD_3$ into the active hormone. Being a small-peptide hormone, parathyroid hormone is freely filtered into the urine from which it is normally endocytosed by proximal tubular cells in a megalin-dependent manner (82). As predicted, parathyroid hormone concentration is elevated in the urine of ClC-5 KO mice (73) and of patients with Dent disease (83). This implies that a disruption of ClC-5 causes an increase in the concentration of parathyroid hormone along the length of the PT (73). In proximal tubular cells, parathyroid hormone receptors are found not only in basolateral but also in apical membranes. The increased stimulation of apical receptors by the elevated parathyroid hormone concentration then stimulates the transcription of 1α-hydroxylase and results in an increased ratio of serum $1,25(OH)_2$-$VitD_3$ to $25(OH)$-$VitD_3$ in the KO (73, 74) (Figure 3A). This, however, does not necessarily raise the absolute serum concentration of the active hormone $1,25(OH)_2$-$VitD_3$, because the lack of ClC-5 severely reduces the uptake of the precursor $25(OH)$-$VitD_3$ into proximal tubular cells. It seems that the balance between these 2 effects can turn in either direction, possibly depending on dietary or genetic factors. In most patients, serum vitamin D is slightly increased (67, 84), whereas it is consistently decreased in our KO mouse model (73). Elevated levels of $1,25(OH)_2$-$VitD_3$ are expected to stimulate the intestinal absorption of calcium, which may then be excreted in increased amounts

by the kidney, resulting in hypercalciuria. However, using a ClC-5 KO mouse model that displays hypercalciuria (85) and increased levels of serum $1,25(OH)_2$-$VitD_3$, Silva et al. (86) suggest that the hypercalciuria is rather of bone and renal origin instead of being caused by increased intestinal calcium absorption. This was in part supported by elevated bone-turnover markers in the KO mice.

The hyperphosphaturia found in Dent disease also appears to be a secondary effect of the increased urinary parathyroid hormone concentration (Figure 3B). Phosphate reabsorption in the PT occurs mainly through NaPi-2a. This Na$^+$ phosphate cotransporter is downregulated by parathyroid hormone via endocytosis and lysosomal degradation (87). As expected from the increase in luminal parathyroid hormone, the amount of NaPi-2a was decreased in ClC-5 KO mice, and the protein was mainly found in intracellular vesicles (73). This fully explains the observed hyperphosphaturia in terms of a primary defect in endocytosis of parathyroid hormone.

More than 30 different human mutations in ClC-5 are known to cause Dent disease. When investigated in heterologous expression, most of them reduce or abolish ClC-5 currents (66, 88–90). Recently, it has been suggested that Dent disease might be genetically heterogeneous (91), and *OCRL1* was identified as a second gene that is mutated in some Dent disease patients (92). This gene encodes phosphatidylinositol 4,5-bisphosphate-(PIP_2)-5-phosphatase, which had been known to be mutated in the multisystem disease Lowe syndrome. This finding is consistent with a role of phosphatidylinositol metabolites in endocytotic trafficking.

ClC-3: disruption leads to neurodegeneration
No human disease with mutations in ClC-3 has been reported so far, but in mice, the disruption of this channel leads to a severe neurodegeneration with a dramatic loss of the hippocampus and the retina (93). ClC-3 is located on synaptic vesicles and endosomes (93). Its disruption impairs the acidification of these compartments (93–95).

ClC-7: role in osteopetrosis
A KO mouse model led to the discovery that another vesicular Cl$^-$ channel, ClC-7, is mutated in human osteopetrosis (96). ClC-7 displays a very broad tissue distribution and is expressed in late endosomes and lysosomes. Its disruption in mice led to severely sick animals that died about 6 weeks after birth and displayed the typical hallmarks of osteopetrosis, as well as a rapidly progressing retinal degeneration (96). ClC-7 KO mice also display neurodegeneration in the CNS (97). In agreement with the lysosomal localization of ClC-7, it displays characteristics typical for lysosomal storage diseases. Electron microscopy revealed storage material reminiscent of neuronal ceroid lipofuscinosis. Interestingly, storage material was also observed in the PT, where protein turnover is high (97). Skeletal abnormalities included the loss of bone marrow cavities that were instead filled by bone material, as well as a failure of teeth to erupt. ClC-7 is highly expressed in osteoclasts (96), the cells involved in bone degradation. It localizes to the acid-secreting ruffled border, which is formed by the exocytotic insertion of H$^+$-ATPase–containing vesicles of late endosomal/lysosomal origin. It was suggested that ClC-7 is co-inserted with the H$^+$-ATPase into this membrane and serves, like ClC-5 in endosomes, as a shunt for the acidification of the resorption lacuna (Figure 2B). This acidification is crucial for the chemical dissolution of inorganic bone material, as well as for the activity of cosecreted lysosomal enzymes that degrade the organic bone matrix.

In a culture system, ClC-7 KO osteoclasts still attached to ivory but failed to acidify the resorption lacuna and were unable to degrade the bone surrogate (96).

The mouse pathology suggested that ClC-7 might also underlie recessive malignant infantile human osteopetrosis, and indeed ClC-7 mutations were identified in such patients (96). By now, about 30 human ClC-7 mutations are known to cause human osteopetrosis. Interestingly, this includes mutations found in autosomal dominant osteopetrosis of the Albers-Schönberg type (98). These mutations are present in a heterozygous state and presumably exert a dominant-negative effect on the coexpressed product of the normal allele. The situation is thus similar to findings with the skeletal muscle chloride channel ClC-1, mutations in which can cause recessive or dominant myotonia (99, 100). Because of the dimeric structure of CLC channels, about 25% of normal channel function should be left upon a 1:1 coexpression of dominant-negative and WT alleles. Hence, osteopetrosis in Albers-Schönberg disease is less severe, needs several years to decades to develop, and is usually not associated with blindness.

Conclusions

Our understanding of the physiological functions of chloride channels has been greatly advanced by mouse models and human diseases. The resulting pathologies have also yielded novel insights into vesicular chloride channels, disruption of which yields pathologies as diverse as proteinuria and osteopetrosis. This finally puts vesicular channels, which have received much less attention than plasma membrane channels, into the limelight.

Note added in proof. ClC-4 and ClC-5 were recently shown to be electrogenic Cl^-/H^+ antiporters rather than channels (101, 102). This very likely also applies to the highly homologous ClC-3. Such an exchange activity is also compatible with their role in vesicular acidification.

Address correspondence to: Thomas J. Jentsch, Zentrum für Molekulare Neurobiologie Hamburg, ZMNH, Universität Hamburg, Falkenried 94, D-20252 Hamburg, Germany. Phone: 49-40-42803-4741; Fax: 49-40-42803-4839; E-mail: Jentsch@zmnh.uni-hamburg.de.

1. Hübner, C., et al. 2001. Disruption of KCC2 reveals an essential role of K-Cl cotransport already in early synaptic inhibition. *Neuron.* **30**:515–524.
2. Riordan, J.R., et al. 1989. Identification of the cystic fibrosis gene: cloning and characterization of complementary DNA. *Science.* **245**:1066–1073.
3. Accardi, A., and Miller, C. 2004. Secondary active transport mediated by a prokaryotic homologue of ClC Cl- channels. *Nature.* **427**:803–807.
4. Stutts, M.J., et al. 1995. CFTR as a cAMP-dependent regulator of sodium channels. *Science.* **269**:847–850.
5. Reddy, M.M., Light, M.J., and Quinton, P.M. 1999. Activation of the epithelial Na+ channel (ENaC) requires CFTR Cl- channel function. *Nature.* **402**:301–304.
6. Nagel, G., Szellas, T., Riordan, J.R., Friedrich, T., and Hartung, K. 2001. Non-specific activation of the epithelial sodium channel by the CFTR chloride channel. *EMBO Rep.* **2**:249–254.
7. Mall, M., Grubb, B.R., Harkema, J.R., O'Neal, W.K., and Boucher, R.C. 2004. Increased airway epithelial Na+ absorption produces cystic fibrosis-like lung disease in mice. *Nat. Med.* **10**:487–493.
8. Nagel, G., Szellas, T., Riordan, J.R., Friedrich, T., and Hartung, K. 2001. Non-specific activation of the epithelial sodium channel by the CFTR chloride channel. *EMBO Rep.* **2**:249–254.
9. Ko, S.B., et al. 2004. Gating of CFTR by the STAS domain of SLC26 transporters. *Nat. Cell Biol.* **6**:343–350.
10. Strong, T.V., Boehm, K., and Collins, F.S. 1994. Localization of cystic fibrosis transmembrane conductance regulator mRNA in the human gastrointestinal tract by in situ hybridization. *J. Clin. Invest.* **93**:347–354.
11. Welsh, M.J., Smith, P.L., Fromm, M., and Frizzell, R.A. 1982. Crypts are the site of intestinal fluid and electrolyte secretion. *Science.* **218**:1219–1221.
12. Schroeder, B.C., et al. 2000. A constitutively open potassium channel formed by KCNQ1 and KCNE3. *Nature.* **403**:196–199.
13. Dedek, K., and Waldegger, S. 2001. Colocalization of KCNQ1/KCNE channel subunits in the mouse gastrointestinal tract. *Pflügers Arch.* **442**:896–902.
14. Clarke, L.L., et al. 1992. Defective epithelial chloride transport in a gene-targeted mouse model of cystic fibrosis. *Science.* **257**:1125–1128.
15. Rufo, P.A., et al. 1997. The antifungal antibiotic, clotrimazole, inhibits chloride secretion by human intestinal T84 cells via blockade of distinct basolateral K+ conductances. Demonstration of efficacy in intact rabbit colon and in an in vivo mouse model of cholera. *J. Clin. Invest.* **100**:3111–3120.
16. Salvatore, F., Scudiero, O., and Castaldo, G. 2002. Genotype-phenotype correlation in cystic fibrosis: the role of modifier genes. *Am. J. Med. Genet.* **111**:88–95.
17. Bronsveld, I., et al. 2001. Chloride conductance and genetic background modulate the cystic fibrosis phenotype of Delta F508 homozygous twins and siblings. *J. Clin. Invest.* **108**:1705–1715. doi:10.1172/JCI200112108.
18. Rozmahel, R., et al. 1996. Modulation of disease severity in cystic fibrosis transmembrane conductance regulator deficient mice by a secondary genetic factor. *Nat. Genet.* **12**:280–287.
19. Gyömörey, K., Rozmahel, R., and Bear, C.E. 2000. Amelioration of intestinal disease severity in cystic fibrosis mice is associated with improved chloride secretory capacity. *Pediatr. Res.* **48**:731–734.
20. Ritzka, M., et al. 2004. The CLCA gene locus as a modulator of the gastrointestinal basic defect in cystic fibrosis. *Hum. Genet.* **115**:483–491.
21. Gyömörey, K., Yeger, H., Ackerley, C., Garami, E., and Bear, C.E. 2000. Expression of the chloride channel ClC-2 in the murine small intestine epithelium. *Am. J. Physiol. Cell Physiol.* **279**:C1787–C1794.
22. Schwiebert, E.M., et al. 1998. Analysis of ClC-2 channels as an alternative pathway for chloride conduction in cystic fibrosis airway cells. *Proc. Natl. Acad. Sci. U. S. A.* **95**:3879–3884.
23. Blaisdell, C.J., Pellettieri, J.P., Loughlin, C.E., Chu, S., and Zeitlin, P.L. 1999. Keratinocyte growth factor stimulates ClC-2 expression in primary fetal rat distal lung epithelial cells. *Am. J. Respir. Cell Mol. Biol.* **20**:842–847.
24. Zdebik, A.A., Cuffe, J., Bertog, M., Korbmacher, C., and Jentsch, T.J. 2004. Additional disruption of the ClC-2 Cl- channel does not exacerbate the cystic fibrosis phenotype of CFTR mouse models. *J. Biol. Chem.* **279**:22276–22283.
25. Bösl, M.R., et al. 2001. Male germ cells and photoreceptors, both depending on close cell-cell interactions, degenerate upon ClC-2 Cl--channel disruption. *EMBO J.* **20**:1289–1299.
26. Mohammad-Panah, R., et al. 2001. ClC-2 contributes to native chloride secretion by a human intestinal cell line, Caco-2. *J. Biol. Chem.* **276**:8306–8313.
27. Murray, C.B., et al. 1995. CIC-2: a developmentally dependent chloride channel expressed in the fetal lung and downregulated after birth. *Am. J. Respir. Cell Mol. Biol.* **12**:597–604.
28. Lipecka, J., et al. 2002. Distribution of ClC-2 chloride channel in rat and human epithelial tissues. *Am. J. Physiol. Cell Physiol.* **282**:C805–C816.
29. Catalán, M., Niemeyer, M.I., Cid, L.P., and Sepúlveda, F.V. 2004. Basolateral ClC-2 chloride channels in surface colon epithelium: regulation by a direct effect of intracellular chloride. *Gastroenterology.* **126**:1104–1114.
30. Thiemann, A., Gründer, S., Pusch, M., and Jentsch, T.J. 1992. A chloride channel widely expressed in epithelial and non-epithelial cells. *Nature.* **356**:57–60.
31. Crawford, I., et al. 1991. Immunocytochemical localization of the cystic fibrosis gene product CFTR. *Proc. Natl. Acad. Sci. U. S. A.* **88**:9262–9266.
32. Gibney, E.M., and Goldfarb, D.S. 2003. The association of nephrolithiasis with cystic fibrosis. *Am. J. Kidney Dis.* **42**:1–11.
33. McNicholas, C.M., et al. 1996. Sensitivity of a renal K+ channel (ROMK2) to the inhibitory sulfonylurea compound glibenclamide is enhanced by coexpression with the ATP-binding cassette transporter cystic fibrosis transmembrane regulator. *Proc. Natl. Acad. Sci. U. S. A.* **93**:8083–8088.
34. Yoo, D., et al. 2004. Assembly and trafficking of a multiprotein ROMK (Kir 1.1) channel complex by PDZ interactions. *J. Biol. Chem.* **279**:6863–6873.
35. Kibble, J.D., Neal, A.M., Colledge, W.H., Green, R., and Taylor, C.J. 2000. Evidence for cystic fibrosis transmembrane conductance regulator-dependent sodium reabsorption in kidney, using Cftr(tm2cam) mice. *J. Physiol.* **526**:27–34.
36. Simon, D.B., et al. 1997. Mutations in the chloride channel gene, *CLCNKB*, cause Bartter's syndrome type III. *Nat. Genet.* **17**:171–178.
37. Estévez, R., et al. 2001. Barttin is a Cl--channel β-subunit crucial for renal Cl--reabsorption and inner ear K+-secretion. *Nature.* **414**:558–561.
38. Birkenhäger, R., et al. 2001. Mutation of *BSND* causes Bartter syndrome with sensorineural deafness and kidney failure. *Nat. Genet.* **29**:310–314.
39. Matsumura, Y., et al. 1999. Overt nephrogenic diabetes insipidus in mice lacking the CLC-K1 chloride channel. *Nat. Genet.* **21**:95–98.
40. Birkenhäger, R., et al. 2001. Mutation of BSND causes Bartter syndrome with sensorineural deafness and kidney failure. *Nat. Genet.* **29**:310–314.
41. Jentsch, T.J., Stein, V., Weinreich, F., and Zdebik, A.A. 2002. Molecular structure and physiological function of chloride channels. *Physiol. Rev.* **82**:503–568.
42. Uchida, S., et al. 1995. Localization and functional characterization of rat kidney-specific chloride

channel, ClC-K1. *J. Clin. Invest.* **95**:104–113.

43. Vandewalle, A., et al. 1997. Localization and induction by dehydration of ClC-K chloride channels in the rat kidney. *Am. J. Physiol.* **272**:F678–F688.

44. Kobayashi, K., Uchida, S., Okamura, H.O., Marumo, F., and Sasaki, S. 2002. Human CLC-KB gene promoter drives the EGFP expression in the specific distal nephron segments and inner ear. *J. Am. Soc. Nephrol.* **13**:1992–1998.

45. Maehara, H., et al. 2003. Expression of CLC-KB gene promoter in the mouse cochlea. *Neuroreport.* **14**:1571–1573.

46. Uchida, S., et al. 1993. Molecular cloning of a chloride channel that is regulated by dehydration and expressed predominantly in kidney medulla [erratum 1994, **269**:19192]. *J. Biol. Chem.* **268**:3821–3824.

47. Waldegger, S., and Jentsch, T.J. 2000. Functional and structural analysis of ClC-K chloride channels involved in renal disease. *J. Biol. Chem.* **275**:24527–24533.

48. Waldegger, S., et al. 2002. Barttin increases surface expression and changes current properties of ClC-K channels. *Pflügers Arch.* **444**:411–418.

49. Staub, O., et al. 1996. WW domains of Nedd4 bind to the proline-rich PY motifs in the epithelial Na⁺ channel deleted in Liddle's syndrome. *EMBO J.* **15**:2371–2380.

50. Abriel, H., et al. 1999. Defective regulation of the epithelial Na⁺ channel by Nedd4 in Liddle's syndrome. *J. Clin. Invest.* **103**:667–673.

51. Schwake, M., Friedrich, T., and Jentsch, T.J. 2001. An internalization signal in ClC-5, an endosomal Cl⁻-channel mutated in Dent's disease. *J. Biol. Chem.* **276**:12049–12054.

52. Embark, H.M., et al. 2004. Regulation of CLC-Ka/barttin by the ubiquitin ligase Nedd4-2 and the serum- and glucocorticoid-dependent kinases. *Kidney Int.* **66**:1918–1925.

53. Simon, D.B., et al. 1996. Bartter's syndrome, hypokalaemic alkalosis with hypercalciuria, is caused by mutations in the Na-K-2Cl cotransporter NKCC2. *Nat. Genet.* **13**:183–188.

54. Simon, D.B., et al. 1996. Genetic heterogeneity of Bartter's syndrome revealed by mutations in the K⁺ channel, ROMK. *Nat. Genet.* **14**:152–156.

55. Jentsch, T.J. 2000. Neuronal KCNQ channel: physiology and role in disease. *Nat. Rev. Neurosci.* **1**:21–30.

56. Neyroud, N., et al. 1997. A novel mutation in the potassium channel gene *KVLQT1* causes the Jervell and Lange-Nielsen cardioauditory syndrome. *Nat. Genet.* **15**:186–189.

57. Schulze-Bahr, E., et al. 1997. *KCNE1* mutations cause Jervell and Lange-Nielsen syndrome. *Nat. Genet.* **17**:267–268.

58. Delpire, E., Lu, J., England, R., Dull, C., and Thorne, T. 1999. Deafness and imbalance associated with inactivation of the secretory Na-K-2Cl co-transporter. *Nat. Genet.* **22**:192–195.

59. Schlingmann, K.P., et al. 2004. Salt wasting and deafness resulting from mutations in two chloride channels. *N. Engl. J. Med.* **350**:1314–1319.

60. Boettger, T., et al. 2002. Deafness and renal tubular acidosis in mice lacking the K-Cl cotransporter Kcc4. *Nature.* **416**:874–878.

61. Jeck, N., Waldegger, P., Doroszewicz, J., Seyberth, H., and Waldegger, S. 2004. A common sequence variation of the *CLCNKB* gene strongly activates ClC-Kb chloride channel activity. *Kidney Int.* **65**:190–197.

62. Jeck, N., et al. 2004. Activating mutation of the renal epithelial chloride channel ClC-Kb predisposing to hypertension. *Hypertension.* **43**:1175–1181.

63. Geller, D.S. 2004. A genetic predisposition to

hypertension? *Hypertension.* **44**:27–28.

64. Kokubo, Y., et al. 2005. Association analysis between hypertension and *CYBA, CLCNKB,* and *KCNMB1* functional polymorphisms in the Japanese population. *Circ. J.* **69**:138–142.

65. Akizuki, N., Uchida, S., Sasaki, S., and Marumo, F. 2001. Impaired solute accumulation in inner medulla of *Clcnk1⁻/⁻* mice kidney. *Am. J. Physiol.* **280**:F79–F87.

66. Lloyd, S.E., et al. 1996. A common molecular basis for three inherited kidney stone diseases. *Nature.* **379**:445–449.

67. Wrong, O.M., Norden, A.G., and Feest, T.G. 1994. Dent's disease: a familial proximal renal tubular syndrome with low-molecular-weight proteinuria, hypercalciuria, nephrocalcinosis, metabolic bone disease, progressive renal failure and a marked male predominance. *QJM.* **87**:473–493.

68. Günther, W., Lüchow, A., Cluzeaud, F., Vandewalle, A., and Jentsch, T.J. 1998. ClC-5, the chloride channel mutated in Dent's disease, colocalizes with the proton pump in endocytotically active kidney cells. *Proc. Natl. Acad. Sci. U. S. A.* **95**:8075–8080.

69. Devuyst, O., Christie, P.T., Courtoy, P.J., Beauwens, R., and Thakker, R.V. 1999. Intra-renal and subcellular distribution of the human chloride channel, CLC-5, reveals a pathophysiological basis for Dent's disease. *Hum. Mol. Genet.* **8**:247–257.

70. Sakamoto, H., et al. 1999. Cellular and subcellular immunolocalization of ClC-5 channel in mouse kidney: colocalization with H⁺-ATPase. *Am. J. Physiol.* **277**:F957–F965.

71. Fisher, S.E., et al. 1994. Isolation and partial characterization of a chloride channel gene which is expressed in kidney and is a candidate for Dent's disease (an X-linked hereditary nephrolithiasis). *Hum. Mol. Genet.* **3**:2053–2059.

72. Steinmeyer, K., Schwappach, B., Bens, M., Vandewalle, A., and Jentsch, T.J. 1995. Cloning and functional expression of rat CLC-5, a chloride channel related to kidney disease. *J. Biol. Chem.* **270**:31172–31177.

73. Piwon, N., Günther, W., Schwake, R., Bösl, M.R., and Jentsch, T.J. 2000. ClC-5 Cl⁻-channel disruption impairs endocytosis in a mouse model for Dent's disease. *Nature.* **408**:369–373.

74. Günther, W., Piwon, N., and Jentsch, T.J. 2003. The ClC-5 chloride channel knock-out mouse: an animal model for Dent's disease. *Pflügers Arch.* **445**:456–462.

75. Gekle, M., Mildenberger, S., Freudinger, R., and Silbernagl, S. 1995. Endosomal alkalinization reduces Jmax and Km of albumin receptor-mediated endocytosis in OK cells. *Am. J. Physiol.* **268**:F899–F906.

76. Maranda, B., et al. 2001. Intra-endosomal pH-sensitive recruitment of the Arf-nucleotide exchange factor ARNO and Arf6 from cytoplasm to proximal tubule endosomes. *J. Biol. Chem.* **276**:18540–18550.

77. Christensen, E.I., et al. 2003. Loss of chloride channel ClC-5 impairs endocytosis by defective trafficking of megalin and cubilin in kidney proximal tubules. *Proc. Natl. Acad. Sci. U. S. A.* **100**:8472–8477.

78. Leheste, J.R., et al. 1999. Megalin knockout mice as an animal model of low molecular weight proteinuria. *Am. J. Pathol.* **155**:1361–1370.

79. Cutillas, P.R., et al. 2004. The urinary proteome in Fanconi syndrome implies specificity in the reabsorption of proteins by renal proximal tubule cells. *Am. J. Physiol. Renal Physiol.* **287**:F353–F364.

80. Nykjaer, A., et al. 1999. An endocytic pathway essential for renal uptake and activation of the steroid 25-(OH) vitamin D3. *Cell.* **96**:507–515.

81. Christensen, E.I., et al. 1999. Evidence for an essential role of megalin in transepithelial transport of

retinol. *J. Am. Soc. Nephrol.* **10**:685–695.

82. Hilpert, J., et al. 1999. Megalin antagonizes activation of the parathyroid hormone receptor. *J. Biol. Chem.* **274**:5620–5625.

83. Norden, A.G., et al. 2001. Glomerular protein sieving and implications for renal failure in Fanconi syndrome. *Kidney Int.* **60**:1885–1892.

84. Scheinman, S.J. 1998. X-linked hypercalciuric nephrolithiasis: clinical syndromes and chloride channel mutations. *Kidney Int.* **53**:3–17.

85. Wang, S.S., et al. 2000. Mice lacking renal chloride channel, CLC-5, are a model for Dent's disease, a nephrolithiasis disorder associated with defective receptor-mediated endocytosis. *Hum. Mol. Genet.* **9**:2937–2945.

86. Silva, I.V., et al. 2003. The ClC-5 knockout mouse model of Dent's disease has renal hypercalciuria and increased bone turnover. *J. Bone Miner. Res.* **18**:615–623.

87. Murer, H., et al. 1999. Posttranscriptional regulation of the proximal tubule NaPi-II transporter in response to PTH and dietary Pi. *Am. J. Physiol.* **277**:F676–F684.

88. Lloyd, S.E., et al. 1997. Characterisation of renal chloride channel, CLCN5, mutations in hypercalciuric nephrolithiasis (kidney stones) disorders. *Hum. Mol. Genet.* **6**:1233–1239.

89. Morimoto, T., et al. 1998. Mutations in *CLCN5* chloride channel in Japanese patients with low molecular weight proteinuria. *J. Am. Soc. Nephrol.* **9**:811–818.

90. Igarashi, T., et al. 1998. Functional characterization of renal chloride channel, CLCN5, mutations associated with Dent's_Japan disease. *Kidney Int.* **54**:1850–1856.

91. Hoopes, R.R., Jr., et al. 2004. Evidence for genetic heterogeneity in Dent's disease. *Kidney Int.* **65**:1615–1620.

92. Hoopes, R.R., Jr., et al. 2005. Dent Disease with mutations in OCRL1. *Am. J. Hum. Genet.* **76**:260–267.

93. Stobrawa, S.M., et al. 2001. Disruption of ClC-3, a chloride channel expressed on synaptic vesicles, leads to a loss of the hippocampus. *Neuron.* **29**:185–196.

94. Yoshikawa, M., et al. 2002. CLC-3 deficiency leads to phenotypes similar to human neuronal ceroid lipofuscinosis. *Genes Cells.* **7**:597–605.

95. Hara-Chikuma, M., et al. 2005. ClC-3 chloride channels facilitate endosomal acidification and chloride accumulation. *J. Biol. Chem.* **280**:1241–1247.

96. Kornak, U., et al. 2001. Loss of the ClC-7 chloride channel leads to osteopetrosis in mice and man. *Cell.* **104**:205–215.

97. Kasper, D., et al. 2005. Loss of the chloride channel ClC-7 leads to lysosomal storage disease and neurodegeneration. *EMBO J.* **24**:1079–1091.

98. Cleiren, E., et al. 2001. Albers-Schönberg disease (autosomal dominant osteopetrosis, type II) results from mutations in the *ClCN7* chloride channel gene. *Hum. Mol. Genet.* **10**:2861–2867.

99. Koch, M.C., et al. 1992. The skeletal muscle chloride channel in dominant and recessive human myotonia. *Science.* **257**:797–800.

100. Pusch, M., Steinmeyer, K., Koch, M.C., and Jentsch, T.J. 1995. Mutations in dominant human myotonia congenita drastically alter the voltage dependence of the ClC-1 chloride channel. *Neuron.* **15**:1455–1463.

101. Scheel, O., Zdebik, A.A., Lourdel, S., and Jentsch, T.J. 2005. Voltage-dependent electrogenic chloride-proton exchange by endosomal CLC proteins. *Nature.* In press.

102. Picollo, A., and Pusch, M. 2005. Chloride/proton antiporter activity of mammalian CLC proteins ClC-4 and ClC-5. *Nature.* In press.

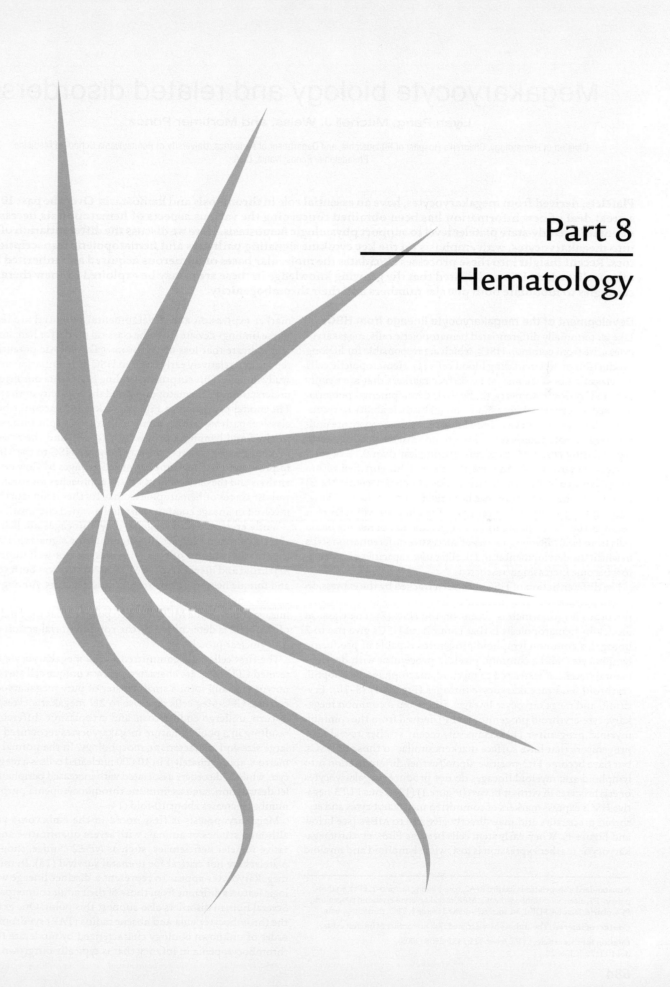

Part 8
Hematology

Megakaryocyte biology and related disorders

Liyan Pang, Mitchell J. Weiss, and Mortimer Poncz

Division of Hematology, Children's Hospital of Philadelphia, and Department of Pediatrics, University of Pennsylvania School of Medicine, Philadelphia, Pennsylvania, USA.

Platelets, derived from megakaryocytes, have an essential role in thrombosis and hemostasis. Over the past 10 years, a great deal of new information has been obtained concerning the various aspects of hematopoiesis necessary to maintain a steady-state platelet level to support physiologic hemostasis. Here we discuss the differentiation of HSCs into megakaryocytes, with emphasis on the key cytokine signaling pathways and hematopoietic transcription factors. Recent insight into these processes elucidates the molecular bases of numerous acquired and inherited hematologic disorders. It is anticipated that the growing knowledge in these areas may be exploited for new therapeutic strategies to modulate both platelet numbers and their thrombogenicity.

Development of the megakaryocyte lineage from HSCs

Like all terminally differentiated hematopoietic cells, megakaryocytes arise from common HSCs, which are responsible for lifelong production of all circulating blood cells (1). Hematopoietic cells are classified by 3 means: (a) by surface markers that are mainly detected by flow cytometry, (b) by their developmental potential assessed ex vivo in colony assays, and (c) by their ability to reconstitute host animals in vivo. Individual cells that reconstitute multilineage hematopoiesis for at least 6 months, termed long-term repopulating HSCs, are rare, constituting less than 0.1% of total nucleated marrow cells. In mice, these are highly enriched within a population of cells with surface markers Lin⁻Sca-1⁺c-kit^high (2–5) (Figure 1). This population has been referred to as the LSK long-term HSC population, but for ease of reading we will refer to it below as the "HSC" population. The production of mature blood cells from HSCs involves a series of successive differentiation steps in which the developmental and proliferative capacities of progenitors become increasingly restricted.

The differentiation of HSCs has been tracked by the expression of cell surface markers including the tyrosine kinase cytokine receptor Flt3 (6), which is absent on the HSC (7). The classical model for hematopoiesis is that committed HSCs give rise to 2 lineages, a common lymphoid progenitor capable of producing lymphocytes, and a common myeloid progenitor with developmental potential restricted to myeloid, macrophage, eosinophil, erythroid, and megakaryocyte lineages (Figure 1) (8–10). Erythroid and megakaryocyte lineages arise from a common megakaryocyte-erythroid progenitor (MEP) derived from the common myeloid progenitor (11). However, recent studies reveal that progenitors that have surface markers similar to those of HSCs, but have become Flt3-positive, upon further differentiation into lymphoid and myeloid lineages do not produce megakaryocytes or erythrocytes in vitro or in vivo (Figure 1) (12). Thus FLT3-negative HSCs express markers of committed megakaryocytes and erythroid precursors and may directly give rise to MEPs (see below and Figure 1). When early stem cells become Flt3⁺, erythromegakaryocytic marker expression is lost, while lymphoid and myeloid marker expression and developmental potential are retained. These findings deviate from the classical model for hematopoiesis and indicate that loss of erythromegakaryocytic potential may represent a relatively early event in HSC differentiation in certain study models. This surprising finding highlights our incomplete understanding of hematopoiesis and the plasticity of the process. The model presented in Figure 1 takes into account both the classical pathway, which predicts that HSCs split into common myeloid and lymphoid progenitors (8–10), and the newer findings suggesting a direct pathway from the HSC to the MEP (12). Improvement in fine mapping of cell lineages by flow cytometry analysis and the development of new approaches for studies of the earliest stages of hematopoiesis may further delineate the steps involved in lineage commitment under varied circumstances.

While erythroid and megakaryocyte lineages are believed to share a common MEP (8–10) (Figure 1), the signals that regulate the final separation of these lineages are not well understood. Erythroid and megakaryocytic precursors express both common and unique hematopoietic transcription factors. Among the latter, no single unique factors have been identified to determine lineage choice of the MEP. It is also possible that the final lineage of the MEP is determined by the combinatorial action of multiple nuclear proteins.

The first cells fully committed to the megakaryocyte lineage, termed CFU-Meg, are characterized by a unique cell surface phenotype (13) and form a small cluster of pure megakaryocytes in culture. CFU-Meg cells give rise to 2N megakaryocytes, which, in turn, undergo endomitosis and cytoplasmic differentiation, resulting in a pool of mature megakaryocytes recognized by their large size and characteristic morphology. In the normal human marrow, approximately 1 in 10,000 nucleated cells is a megakaryocyte, while in disorders associated with increased peripheral platelet destruction, such as immune thrombocytopenia purpura, the number increases about 10-fold (14).

Megakaryopoiesis is first noted in the embryonic yolk sac, although studies of animals with severe quantitative and qualitative platelet deficiencies, such as *NF-E2*⁻/⁻ mice, showed that platelets are not critical for prenatal survival (15). In mice, fetal megakaryocytes appear to represent a distinct lineage with biologic features different from those of their adult counterparts (16). Several human disorders also support this point. One example is the thrombocytopenia and absent radius (TAR) syndrome, a disorder of unknown etiology characterized by moderate to severe thrombocytopenia in infancy that is typically outgrown in early

Nonstandard abbreviations used: CBFA2, core-binding factor α-2; EPO, erythropoietin; ET, essential thrombocythemia; MEP, megakaryocyte-erythroid progenitor; PF4, platelet factor 4; SDF-1, stromal cell–derived factor-1; TPO, thrombopoietin.

Conflict of interest: The authors have declared that no conflict of interest exists.

Citation for this article: *J. Clin. Invest.* **115**:3332–3338 (2005). doi:10.1172/JCI26720.

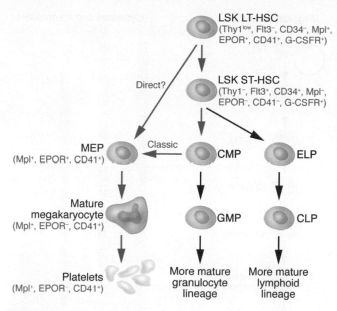

Figure 1
Megakaryopoiesis pathways. The figure extends from the HSC to platelets and offers a combination of the more "classical" pathway, leading to the common megakaryocyte-erythroid progenitor (MEP), and a proposed "direct" route from the HSC. Pathways leading to platelet production are indicated by blue arrows and other pathways by gray arrows. Surface markers of importance are noted in parentheses in red. LT-HSC, long-term HSC; ST-HSC, short-term HSC; Thy1, thymus 1 ("low" indicates low surface antigen and "–" indicates none detectable); Flt3, FMS-like tyrosine kinase 3; EPOR, erythropoietin receptor; CD41, glycoprotein IIb/IIIa or $\alpha_{IIb}\beta_3$ integrin receptor; G-CSFR, G-CSF receptor; CMP, common myeloid progenitor; ELP, early lymphoid progenitor; GMP, granulocyte/monocyte progenitor; CLP, common lymphoid progenitor.

childhood (17). Another example is transient myeloproliferative disorder and acute megakaryoblastic leukemia in Down syndrome, which develop nearly exclusively in the neonatal period and the first years of life, respectively (18). Interestingly, recent studies demonstrate that fetal megakaryocyte progenitors are uniquely sensitive to mutations in the transcription factor GATA-1, which accompany these disorders (19). Perhaps another example of the distinct nature of fetal/infant megakaryopoiesis is the well-known propensity of severely ill neonates to develop prolonged thrombocytopenia with slow marrow recovery of platelet production (20).

Unique aspects of megakaryocyte maturation
The hallmark of megakaryocyte development is the formation of a large cell (~50–100 μm diameter) containing a single, large, multilobulated, polyploid nucleus (21). Eventually, each megakaryocyte releases approximately 10^4 platelets (22). Unlike other cells, megakaryocytes undergo an endomitotic cell cycle during which they replicate DNA but do not undergo anaphase or cytokinesis; as a result, they acquire a DNA content of up to 256N per cell (23). The mechanisms regulating endomitosis are not fully understood. Clearly cyclins are involved, though a combined knockout of cyclins D1, D2, and D3, while specifically affecting hematopoiesis and causing late midgestation fetal loss due in part to anemia, was not noted to affect megakaryopoiesis (24). On the other hand, the cyclin E–null mouse clearly had a defect in megakaryopoiesis and

in development of trophoblasts, another cell line dependent on endomitosis (25). Other studies on chromosomal passenger proteins Aurora-B, survivin, and inner centromere protein showed normal levels overall in megakaryocytes (26), although one report suggests that survivin and Aurora-B may be mislocalized or absent during an important phase of endomitosis (27). The biologic importance of endoreduplication is unclear in terms of its necessity for cell size and for platelet release.

Cellular maturation of megakaryocytes is distinguished by accumulation of characteristic surface markers including $GP_{Ib}\alpha,\beta/GPIX/GPV$ receptors, a cytoplasmic demarcation system believed to participate in platelet formation, distinctive platelet organelles such as the α- and dense granules, and organelle granular proteins that participate in platelet function, such as platelet factor 4 (PF4) and vWF (28). Of note, the extent of polyploidization is not closely synchronized with cellular maturation, so that different degrees of ploidy are present at each stage. For this reason it has been hard to distinguish "early-onset" megakaryocyte-specific genes from "late-onset" ones (29).

Despite their close relationship in hematopoietic phylogeny and numerous common hematopoietic transcription factors, erythroid and megakaryocyte lineages do not share many specific proteins or organelles. However, it is interesting to note that both lineages circulate in anucleate forms. The platelet equivalent in fish is called a thrombocyte and is a circulating nucleated diploid cell (30, 31). The purpose of platelet and erythrocyte enucleation in mammals is unclear. One possibility is that loss of nuclei increases flexibility and distensibility of circulating cells, optimizing delivery of specialized functions within small-caliber capillary beds.

Cytokines involved in megakaryopoiesis
In humans, homeostatic mechanisms regulate the normal platelet count within an approximately 3-fold range (150×10^3 to 450×10^3 per cubic micrometer). Disorders that consume platelets increase their production. Numerous hematopoietic growth factors regulate different aspects of megakaryocyte biology (Figure 2). Certain cytokines, including GM-CSF, IL-3, IL-6, IL-11, IL-12, and erythropoietin (EPO), stimulate proliferation of megakaryocytic progenitors (32). Other cytokines, including IL-1α and leukemia inhibitory factor (LIF), modulate megakaryocyte maturation and platelet release (32, 33). Many of these cytokines have broad effects on all hematopoietic lineages. Presently, the multilineage cytokine IL-11 (Neumega) is the only clinically approved drug for treating thrombocytopenia (34).

More than 10 years ago, a more potent and relatively specific megakaryocyte/platelet cytokine, termed Mpl ligand or thrombopoietin (TPO), was identified. This cytokine is discussed below, along with 2 chemokines, stromal cell–derived factor-1 (SDF-1; CXCL12) and PF4 (CXCL4), that have important effects on megakaryopoiesis and platelet production. A detailed Review by Kenneth Kaushansky that focuses on the role of TPO and its receptor in thrombopoiesis is part of this series on the biology of megakaryocytes and platelets (35).

TPO. Mpl is a GP130 family member previously identified as important for megakaryocyte formation in vitro (36). This pivotal observation led to identification of the Mpl ligand, termed TPO, which was determined to markedly stimulate megakaryocyte production (37–40). TPO is highly homologous to EPO in its N-terminal half, reflecting a close evolutionary relationship between their respective receptor signaling pathways. Abrogation of either Mpl

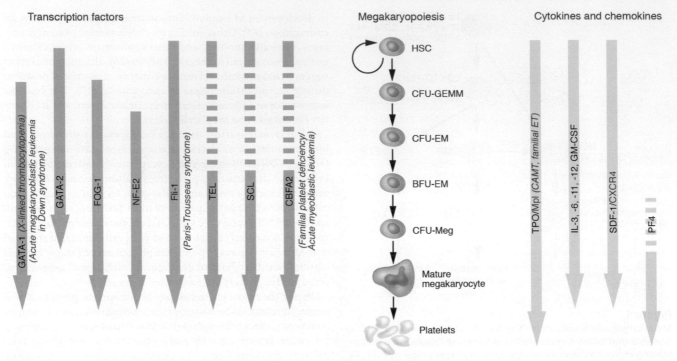

Figure 2
Regulation of megakaryopoiesis by cytokines, chemokines, and transcription factors. In the middle panel, a scheme based on the classical pathway of megakaryopoiesis is shown. Cytokines and chemokines that influence that process are shown on the right side as green arrows to indicate the approximate level of development at which they have their influence. Open white areas in arrows indicate levels at which the cytokine is not known to act. Blue text refers to cytokine receptors of significance in megakaryopoiesis. Transcription factors that affect megakaryopoiesis are shown on the left side, and the lilac-colored arrows indicate the approximate point of their influence. Open white areas in arrows indicate levels at which the transcription factor is not known to act. Clinically relevant diseases linked to defects of these regulators are noted in red, italicized text. CFU-GEMM, CFU, granulocyte, erythrocyte, macrophage, megakaryocyte; CFU-EM, CFU, erythrocyte, megakaryocyte; BFU-EM, burst-forming unit, erythrocyte, megakaryocyte; CFU-Meg, CFU, megakaryocyte.

or TPO in mice decreases megakaryocyte numbers in the marrow and circulating platelets by approximately 85% (41–43). Clearly the TPO:Mpl axis is important but not essential for megakaryopoiesis. Furthermore, studies of Mpl knockout animals showed that the TPO:Mpl axis functions in early hematopoietic progenitors, including HSCs (44, 45) (Figure 1). Thus, the TPO:Mpl axis appears to be important for hematopoiesis in general and megakaryopoiesis specifically. The discovery of TPO has contributed greatly to platelet biology, because it permits relatively large and pure cultures of megakaryocytes to be generated in vitro. TPO remains under development as a potential clinical thrombopoietic and/or hematopoietic agent and as a drug to stimulate ex vivo expansion of HSCs (46).

In adults, humoral regulation of thrombopoiesis differs from that of erythropoiesis, where the kidneys produce EPO in response to tissue hypoxia. In contrast, TPO is produced constitutively, and its circulating levels are regulated by its end product, platelets. Circulating TPO is believed to control endogenous megakaryocyte numbers and platelet count. In the steady state, TPO is synthesized predominantly and constitutively in the liver (47). Mpl receptors on circulating platelets absorb TPO to negatively regulate its availability for stimulating hematopoietic progenitor cells in the marrow (48). TPO is also produced by bone marrow stromal cells (49). The relative impact of circulating versus paracrine TPO production on platelet numbers is unclear. It is possible that these different modes of production satisfy distinct hematopoietic pools in different niches.

Defects in TPO:Mpl signaling occur in several human disorders. For example, Mpl mutations, mostly causing frameshifts and early termination, occur in congenital amegakaryocytic thrombocytopenia, a rare disorder of life-threatening thrombocytopenia and megakaryocyte deficiency in infancy (50–52) (Figure 2). Given the role of Mpl in HSC development, it is also possible that congenital amegakaryocytic thrombocytopenia patients are at risk for developing more diffuse hematopoietic defects, including aplastic anemia (53). Activating mutations in the *TPO* gene promoter (54) and the Mpl protein (55) occur in a subset of patients with familial essential thrombocythemia (ET) (Figure 2), a disorder characterized by increased numbers of hyperaggregable platelets. In contrast, the majority of patients with the more common acquired adult, myeloproliferative form of ET harbor somatic activating mutations in the JAK2 gene (56). These mechanistic differences could explain why familial ET carries an excellent long-term prognosis that differs from the high incidence of leukemic transformation in acquired ET (57).

SDF-1. SDF-1 enhances both megakaryopoiesis and homing of HSCs to the bone marrow during fetal development (58). SDF-1 stimulates megakaryopoiesis via TPO-independent CXCR4 receptor pathways by enhancing the chemotactic activity of their progenitors (59, 60). This activity of SDF-1 may be important for the movement of megakaryocyte progenitors from the proliferative "osteoblastic niche" to the "vascular niche" for platelet formation (61). Indeed, both in *TPO*–/– and in *Mpl*–/– mice, infusions of SDF-1 can rescue platelet production (61). There may be clinical utility for

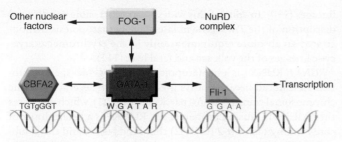

Figure 3
Assembly of transcription factor complex at a megakaryocyte-specific gene. A schematic representation of the proximal promoter region of a hypothetical megakaryocyte-specific gene modeled after the *Itga2b* gene. GATA-1 is shown in the middle binding to its known consensus sequence (69) and interacting with FOG-1, Fli-1, and CBFA2 of the RUNX complex (each interaction is indicated by a 2-headed arrow). Fli-1 and CBFA2 bind to their adjacent cognate on the DNA (89, 122), while FOG-1 is recruited through its interactions with the N-terminal zinc finger of GATA-1. In turn, FOG-1 recruits the NuRD complex and other nuclear factors.

SDF-1 infusions to improve platelet production. For example, HIV may cause thrombocytopenia by infecting megakaryocyte precursors through interactions with their CXCR4 receptors (62). Thus, SDF-1 mimetic drugs may improve HIV-related thrombocytopenia by competing with the virus for its megakaryocyte receptor (63).

PF4 and other chemokines. PF4 is an α-granule protein that inhibits megakaryocyte development and maturation in vitro (64), as other CXC and CC subfamily chemokines have subsequently been shown to do (65). The in vitro findings for PF4 are corroborated by altered platelet counts in *PF4*[−/−] mice and transgenic PF4-overexpressing mice (66). Platelet α-granules contain large stores not only of the platelet-specific chemokines PF4 and the closely related protein platelet basic protein (PBP; CXCL7), but also of other chemokines, especially RANTES (CCL5) and ENA-78 (CXCL5) (67, 68). They negatively regulate megakaryopoiesis and are mild platelet agonists through cognate receptors on developing megakaryocytes. These are weak agonists of platelet activation and also may be important in linking thrombosis and inflammation. Release of α-granular contents in the marrow could affect platelet numbers in pathologic states. For example, the discharge of chemokines during chemotherapy or radiation therapy may contribute to thrombocytopenia that occurs during these treatment modalities. In this case, strategies to inhibit this process could be used preemptively to prevent thrombocytopenia.

Transcription factors involved in megakaryopoiesis

Megakaryopoiesis is regulated by multiple cytokines influencing the survival and proliferation of increasingly committed progenitors as they transition from one hematopoietic niche to another in an organized fashion (61). During this process, a series of transcription factors coordinately regulate the chromatin organization of megakaryocyte-specific genes and prime them for expression en route to platelet formation. Some of these events are beginning to be understood, especially the hematopoietic-specific transcription factor complexes involved in terminal differentiation. Numerous nuclear proteins with important roles in megakaryocyte formation, growth regulation, and platelet release have been identified, mainly through loss-of-function studies in mice and analysis of human diseases. Important examples are discussed below.

GATA-1/FOG-1 complex. GATA-1 was first isolated as an essential 2–zinc finger, erythroid transcription factor that binds the DNA sequence WGATAR (69) (Figure 3). GATA-1 is also expressed and of functional consequence in megakaryocytes, mast cells, and eosinophils (70, 71). Transient expression reporter gene studies of megakaryocyte-specific proximal promoters defined several functionally important GATA-binding sites (72–74). A point mutation in a GATA-binding site of the *GP1bb* proximal promoter region causes a form of Bernard-Soulier syndrome (74).

While targeted disruption of the GATA-1 gene in mice causes embryonic lethality due to anemia (75), a megakaryocyte-specific knock down of GATA-1 expression results in significant thrombocytopenia and increased numbers of immature and dysmorphic megakaryocytes (76). GATA-2 is a closely related transcription factor and is also hematopoietic-specific, but it is expressed earlier and participates in maintenance of HSCs and multipotential progenitors (77). GATA-1 and GATA-2 are believed to have both overlapping and unique functions (69, 78). Continued GATA-2 expression during early megakaryopoiesis may explain the partial ability for platelet formation in the GATA-1 knockdown mouse (Figure 2).

FOG-1 (Friend of GATA-1, or Zfpm1) is a 9–zinc finger, hematopoietic-specific transcription factor isolated as a GATA-1 binding partner (79) (Figure 3). Targeted disruption of the FOG-1 gene markedly inhibits erythroid development, causing embryonic death from severe anemia. Detailed study of these animals and *FOG-1*[−/−] ES cells also demonstrated an early block to megakaryocytic development with no identifiable precursors. FOG-1 does not appear to bind DNA directly but, rather, associates with target genes indirectly through interactions with GATA proteins (Figure 3). FOG-1 is the only protein of its kind expressed in the erythro-megakaryocytic lineages. The severity of the megakaryocyte defect in the *FOG-1*[−/−] mouse suggests that most or all critical GATA-1– and GATA-2–related activities require interactions with FOG-1. In support of this, GATA-1 and FOG-1 synergistically enhance the expression of the megakaryocyte-specific α_{IIb} gene (80, 81). These observations have been expanded to show that direct contact is needed between the N-terminal zinc finger of GATA-1 and FOG-1. This synergism applies to multiple megakaryocyte-specific genes and also involves a specific Ets family transcription factor (see below). The N-terminus of FOG-1 plays a unique, nonredundant role in megakaryocyte-specific expression (82), possibly through its ability to recruit the corepressor complex NuRD (83). The clinical importance of the GATA-1/FOG-1 interaction in megakaryopoiesis is demonstrated by the identification of patients with X-linked thrombocytopenia and variable anemia who have GATA-1 mutations that impair FOG-1 binding (84–86) (Figure 2).

An important role for GATA-1 in regulating the maturation and proliferation of megakaryocyte progenitors is further evidenced by the recent discovery of acquired somatic mutations associated with both megakaryoblastic leukemia and transient myeloproliferative syndrome in infants with Down syndrome (87, 88). These mutations occur in the first coding exon, causing early termination and production of a truncated protein, termed GATA-1[Short], via translation initiation from a downstream internal methionine. This mutant form of GATA-1 may act as a dominant oncogene by specifically stimulating the proliferation of fetal megakaryocyte progenitors (19).

Fli-1 and TEL. The proximal promoters of many megakaryocyte-specific genes contain tandem-binding sites for GATA and Ets proteins, suggesting functional interactions between these 2 classes of transcription factors (71–73) (Figure 3). The Ets family is diverse with at least 30 members, all sharing an Ets-binding domain that recognizes a GGAA core sequence (89). Numerous Ets members are present in primary megakaryocytes and/or megakaryocytic cell lines (90–92). While a number of reports suggest a function for Ets-1 in megakaryopoiesis (93), the clearest story of a common transcriptional regulator appears to be that for Fli-1, an Ets transcription factor initially recognized to be important for T cell differentiation (94) and early hematopoiesis/vasculogenesis (95) (Figure 2). GATA-1/FOG-1 synergy for many megakaryocyte-specific genes appears to involve Fli-1 (81), and Fli-1 binds to the proximal promoter of these genes in vivo. The molecular basis of this synergy is still not fully known, but Fli-1 does bind GATA-1 (96) (Figure 3). In undifferentiated hematopoietic cell lines, overexpressed Fli-1 can induce megakaryocytic features (91). Moreover, *Fli1* gene–disrupted mice either have abnormal megakaryocytes with thrombocytopenia (97) or fail to develop recognizable megakaryocytes (98), depending on the size of the *Fli-1* gene deletion. Fli-1 expression also inhibits erythroid differentiation (99). Thus, Fli-1 may be a lineage-determining factor for megakaryocyte development. Hemizygous deficiency of Fli-1 expression causes thrombocytopenia associated with abnormal megakaryocytes in patients with Paris-Trousseau syndrome (100).

TEL, or ETV6, another Ets protein in the pointed domain subfamily, is closely related to Fli-1 and may also function in megakaryocytopoiesis (101). The pointed domain of TEL is a short N-terminal domain involved in self-oligomerization. Like Fli-1, TEL is important in early HSCs (102) (Figure 2), and TEL overexpression can drive megakaryocyte differentiation of hematopoietic cell lines (92). Remarkably, conditional disruption of the *TEL* gene demonstrated a unique, nonredundant role for TEL in megakaryopoiesis (103). Specifically, loss of TEL in the erythromegakaryocyte lineage results in large, highly proliferative megakaryocytes and mild thrombocytopenia. This phenotype resembles that of GATA-1 deficiency and also has overlapping features with the *NF-E2* knockout phenotype described below.

NF-E2. NF-E2 is a hematopoietic-specific transcription factor consisting of a tissue-specific p45 leucine zipper–containing subunit dimerized with a ubiquitous p18 subunit (104). In vitro studies suggested an important role for NF-E2 erythroid gene expression. Surprisingly, *Nfe2*-null mice do not develop anemia but, rather, exhibit severe thrombocytopenia with a marrow containing excessive immature, dysplastic megakaryocytes (15) (Figure 2). The molecular basis for this effect on megakaryocyte differentiation and platelet release has yet to be resolved. It may be that intracellular signaling pathways related to Rab27b (105) or cytoskeletal proteins (106) are underexpressed in the absence of p45, causing impaired proplatelet formation and platelet release.

SCL. SCL, or TAL1, is a basic helix-loop-helix transcription factor, initially identified in human T cell leukemias with multilineage characteristics (107), and has also been implicated in the earliest stages of hematopoiesis/vasculogenesis in the mouse embryo (108, 109) (Figure 2). LacZ knock-in studies suggest that SCL is expressed in myeloid, lymphoid, erythroid, and megakaryocytic

lineages (110). In spite of this widespread expression, conditional disruption of the *TAL1* gene in late-stage hematopoiesis demonstrated an absolute requirement only in the erythromegakaryocytic lineages of the yolk sac and fetal liver (111).

RUNX1. RUNX1 is a hematopoietic/vasculogenic–specific protein first noted because of its involvement in several leukemic chromosomal translocations, particularly t(8;21), which generates the AML1-ETO fusion protein (112). RUNX1 is a heterodimer of core-binding factor α-2 (CBFA2) that binds DNA and the β subunit CBFB, which does not directly bind DNA (113). RUNX1 was initially thought to be solely involved in myeloid differentiation (114), but studies on targeting of the *CBFA2* gene demonstrated an essential role for RUNX1 in early hematopoiesis and vasculogenesis (115, 116). A unique role for RUNX1 in adult megakaryopoiesis was established when a rare, dominantly inherited thrombocytopenia associated with an increased risk of developing acute myeloblastic leukemia was shown to be due to haploinsufficiency of CBFA2 (117, 118) (Figure 2). A role for CBFA2 in adult megakaryopoiesis was also confirmed in mice (119). RUNX1 appears to interact with GATA-1 (120) (Figure 3), and overexpression of RUNX1 can drive hematopoietic cell lines into a megakaryocytic phenotype (121), suggesting a role in lineage determination.

Conclusions

Mechanisms that regulate formation of the erythromegakaryocytic precursor and its commitment to unilineage megakaryocyte development are active areas of investigation. The discovery of new cytokines and transcription factors associated with megakaryopoiesis has enhanced our understanding of normal platelet development and human thrombocytopenias. One current challenge is to better define the developmental pathways though which MEPs and megakaryocytes arise from HSCs. For example, a recent finding that the onset of Flt3 receptor expression in early hematopoiesis coincides with loss of erythromegakaryocytic capacity suggests novel pathways for MEP formation.

Additionally, comparative studies of developmental hematopoiesis in embryos and adults should extend our understanding of normal and pathologic megakaryopoiesis at distinct developmental stages. In addition, recent studies support a central role for multifactor transcriptional complexes containing GATA-1, FOG-1, and Fli-1 in terminal megakaryocytic differentiation. Formation and regulation of these complexes may illustrate a mechanism that initiates megakaryocyte commitment from the MEP. Further studies into all of these various areas of megakaryopoiesis promises to provide new insights into numerous hematopoietic disorders and may also have broader clinical applications by elucidating novel strategies to regulate platelet count and/or platelet thrombogenicity.

Address correspondence to: Mortimer Poncz, Children's Hospital of Philadelphia, 34th Street and Civic Center Boulevard, Abramson Research Center Room 317, Philadelphia, Pennsylvania 19104, USA. Phone: (215) 590-3574; Fax: (267) 426-5467; E-mail: poncz@email.chop.edu.

1. Ogawa, M. 1993. Differentiation and proliferation of hematopoietic stem cells. *Blood.* **81**:2844–2853.
2. Spangrude, G.J., Heimfeld, S., and Weissman, I.L. 1988. Purification and characterization of mouse hematopoietic stem cells. *Science.* **241**:58–62.
3. Ikuta, K., and Weissman, I.L. 1992. Evidence that hematopoietic stem cells express mouse c-kit but

do not depend on steel factor for their generation. *Proc. Natl. Acad. Sci. U. S. A.* **89**:1502–1506.
4. Li, C.L., and Johnson, G.R. 1995. Murine hematopoietic stem and progenitor cells. I. Enrichment and biologic characterization. *Blood.* **85**:1472–1479.
5. Weissman, I.L., Anderson, D.J., and Gage, F. 2001. Stem and progenitor cells: origins, phenotypes,

lineage commitments and transdifferentiations. *Annu. Rev. Cell Dev. Biol.* **17**:387–403.
6. Rosnet, O., Marchetto, S., deLapeyriere, O., and Birnbaum, D. 1991. Murine Flt3, a gene encoding a novel tyrosine kinase receptor of the PDGFR/CSF1R family. *Oncogene.* **6**:1641–1650.
7. Adolfsson, J., et al. 2001. Upregulation of Flt3

expression within the bone marrow Lin(-)Sca1(+) c-kit(+) stem cell compartment is accompanied by loss of self-renewal capacity. *Immunity.* **15**:659–669.

8. Kanz, L., Straub, G., Bross, K.G., and Fauser, A.A. 1982. Identification of human megakaryocytes derived from pure megakaryocytic colonies (CFU-M), megakaryocytic-erythroid colonies (CFU-M/E), and mixed hemopoietic colonies (CFU-GEMM) by antibodies against platelet associated antigens. *Blut.* **45**:267–274.

9. Nakahata, T., Gross, A.J., and Ogawa, M. 1982. A stochastic model of self-renewal and commitment to differentiation of the primitive hemopoietic stem cells in culture. *J. Cell. Physiol.* **113**:455–458.

10. Akashi, K., Traver, D., Miyamoto, T., and Weissman, I.L. 2000. A clonogenic common myeloid progenitor that gives rise to all myeloid lineages. *Nature.* **404**:193–197.

11. Debili, N. 1996. Characterization of a bipotent erythro-megakaryocytic progenitor in human bone marrow. *Blood.* **88**:1284–1296.

12. Adolfsson, J., et al. 2005. Identification of Flt3+ lympho-myeloid stem cells lacking erythro-megakaryocytic potential: a revised road map for adult blood lineage commitment. *Cell.* **121**:295–306.

13. Nakorn, T.N., Miyamoto, T., and Weissman, I.L. 2003. Characterization of mouse clonogenic megakaryocyte progenitors. *Proc. Natl. Acad. Sci. U. S. A.* **100**:205–210.

14. Branehog, I., Ridell, B., Swolin, B., and Weinfeld, A. 1975. Megakaryocyte quantifications in relation to thrombokinetics in primary thrombocythaemia and allied diseases. *Scand. J. Haematol.* **15**:321–332.

15. Shivdasani, R.A., et al. 1995. Transcription factor NF-E2 is required for platelet formation independent of the actions of thrombopoietin/MGDF in megakaryocyte development. *Cell.* **81**:695–704.

16. Palis, J., and Koniski, A. 2004. Analysis of hematopoietic progenitors in the mouse embryo. *Methods Mol. Med.* **105**:289–302.

17. al-Jefri, A.H., Dror, Y., Bussel, J.B., and Freedman, M.H. 2000. Thrombocytopenia with absent radii: frequency of marrow megakaryocyte progenitors, proliferative characteristics, and megakaryocyte growth and development factor responsiveness. *Pediatr. Hematol. Oncol.* **17**:299–306.

18. Crispino, J.D. 2005. GATA1 mutations in Down syndrome: implications for biology and diagnosis of children with transient myeloproliferative disorder and acute megakaryoblastic leukemia. *Pediatr. Blood Cancer.* **44**:40–44.

19. Li, Z., et al. 2005. Developmental stage-selective effect of somatically mutated leukemogenic transcription factor GATA1. *Nat. Genet.* **37**:613–619.

20. Sola, M.C. 2004. Evaluation and treatment of severe and prolonged thrombocytopenia in neonates. *Clin. Perinatol.* **31**:1–14.

21. Cajano, A., and Polosa, P. 1950. Contribution to the study of the morphology of megakaryocytes and blood platelets with Feulgen's test. *Haematologica.* **34**:1113–11121.

22. Long, M.W. 1998. Megakaryocyte differentiation events. *Semin. Hematol.* **35**:192–199.

23. Odell, T.T., Jr., Jackson, C.W., and Gosslee, D.G. 1965. Maturation of rat megakaryocytes studied by microspectrophotometric measurement of DNA. *Proc. Soc. Exp. Biol. Med.* **119**:1194–1199.

24. Kozar, K. 2004. Mouse development and cell proliferation in the absence of D-cyclins. *Cell.* **118**:477–491.

25. Geng, Y. 2003. Cyclin E ablation in the mouse. *Cell.* **114**:431–443.

26. Geddis, A.E., and Kaushansky, K. 2004. Megakaryocytes express functional Aurora-B kinase in endomitosis. *Blood.* **104**:1017–1024.

27. Zhang, Y., et al. 2004. Aberrant quantity and localization of Aurora-B/AIM-1 and survivin during megakaryocyte polyploidization and the consequences of Aurora-B/AIM-1-deregulated expression. *Blood.* **103**:3717–3726.

28. Schmitt, A., Guichard, J., Masse, J.M., Debili, N., and Cramer, E.M. 2001. Of mice and men: comparison of the ultrastructure of megakaryocytes and platelets. *Exp. Hematol.* **29**:1295–1302.

29. Lefebvre, P., Winter, J.N., Meng, Y., and Cohen, I. 2000. Ex vivo expansion of early and late megakaryocyte progenitors. *J. Hematother. Stem Cell Res.* **9**:913–921.

30. Jagadeeswaran, P., Sheehan, J.P., Craig, F.E., and Troyer, D. 1999. Identification and characterization of zebrafish thrombocytes. *Br. J. Haematol.* **107**:731 738.

31. Gregory, M., and Jagadeeswaran, P. 2002. Selective labeling of zebrafish thrombocytes: quantitation of thrombocyte function and detection during development. *Blood Cells Mol. Dis.* **28**:418–427.

32. Gordon, M.S., and Hoffman, R. 1992. Growth factors affecting human thrombocytopoiesis: potential agents for the treatment of thrombocytopenia. *Blood.* **80**:302–307.

33. Vainchenker, W., Debili, N., Mouthon, M.A., and Wendling, F. 1995. Megakaryocytopoiesis: cellular aspects and regulation. *Crit. Rev. Oncol. Hematol.* **20**:165–192.

34. Orazi, A., et al. 1996. Effects of recombinant human interleukin-11 (Neumega rhIL-11 growth factor) on megakaryocytopoiesis in human bone marrow. *Exp. Hematol.* **24**:1289–1297.

35. Kaushansky, K. 2005. The molecular mechanisms that control thrombopoiesis. *J. Clin. Invest.* **115**:3339–3347. doi:10.1172/JCI26674.

36. Methia, N., Louache, F., Vainchenker, W., and Wendling, F. 1993. Oligodeoxynucleotides antisense to the proto-oncogene c-mpl specifically inhibit in vitro megakaryocytopoiesis. *Blood.* **82**:1395–1401.

37. Bartley, T.D., et al. 1994. Identification and cloning of a megakaryocyte growth and development factor that is a ligand for the cytokine receptor Mpl. *Cell.* **77**:1117–1124.

38. Lok, S., et al. 1994. Cloning and expression of murine thrombopoietin cDNA and stimulation of platelet production in vivo. *Nature.* **369**:565–568.

39. Kaushansky, K., et al. 1994. Promotion of megakaryocyte progenitor expansion and differentiation by the c-Mpl ligand thrombopoietin. *Nature.* **369**:568–571.

40. de Sauvage, F.J., et al. 1994. Stimulation of megakaryocytopoiesis and thrombopoiesis by the c-Mpl ligand. *Nature.* **369**:533–538.

41. Gurney, A.L., Carver-Moore, K., de Sauvage, F.J., and Moore, M.W. 1994. Thrombocytopenia in c-mpl-deficient mice. *Science.* **265**:1445–1447.

42. Alexander, W.S., Roberts, A.W., Nicola, N.A., Li, R., and Metcalf, D. 1996. Deficiencies in progenitor cells of multiple hematopoietic lineages and defective megakaryocytopoiesis in mice lacking the thrombopoietic receptor c-Mpl. *Blood.* **87**:2162–2170.

43. Murone, M., Carpenter, D.A., and de Sauvage, F.J. 1998. Hematopoietic deficiencies in c-mpl and TPO knockout mice. *Stem Cells.* **16**:1–6.

44. Debili, N., et al. 1995. The Mpl-ligand or thrombopoietin or megakaryocyte growth and differentiative factor has both direct proliferative and differentiative activities on human megakaryocyte progenitors. *Blood.* **86**:2516–2525.

45. Young, J.C., et al. 1996. Thrombopoietin stimulates megakaryocytopoiesis, myelopoiesis, and expansion of CD34+ progenitor cells from single CD34+Thy-1+Lin- primitive progenitor cells. *Blood.* **88**:1619–1631.

46. Basser, R. 2002. The impact of thrombopoietin on clinical practice. *Curr. Pharm. Des.* **8**:369–377.

47. Jelkmann, W. 2001. The role of the liver in the production of thrombopoietin compared with erythropoietin. *Eur. J. Gastroenterol. Hepatol.* **13**:791–801.

48. Kaushansky, K. 1997. Thrombopoietin: understanding and manipulating platelet production. *Annu. Rev. Med.* **48**:1–11.

49. Guerriero, A., et al. 1997. Thrombopoietin is synthesized by bone marrow stromal cells. *Blood.* **90**:3444–3455.

50. Ihara, K., et al. 1999. Identification of mutations in the c-mpl gene in congenital amegakaryocytic thrombocytopenia. *Proc. Natl. Acad. Sci. U. S. A.* **96**:3132–3136.

51. van den Oudenrijn, S., et al. 2000. Mutations in the thrombopoietin receptor, Mpl, in children with congenital amegakaryocytic thrombocytopenia. *Br. J. Haematol.* **110**:441–448.

52. Ballmaier, M., et al. 2001. c-mpl mutations are the cause of congenital amegakaryocytic thrombocytopenia. *Blood.* **97**:139–146.

53. Ballmaier, M., Germeshausen, M., Krukemeier, S., and Welte, K. 2003. Thrombopoietin is essential for the maintenance of normal hematopoiesis in humans: development of aplastic anemia in patients with congenital amegakaryocytic thrombocytopenia. *Ann. N. Y. Acad. Sci.* **996**:17–25.

54. Ghilardi, N., and Skoda, R.C. 1999. A single-base deletion in the thrombopoietin (TPO) gene causes familial essential thrombocythemia through a mechanism of more efficient translation of TPO mRNA. *Blood.* **94**:1480–1482.

55. Ding, J., et al. 2004. Familial essential thrombocythemia associated with a dominant-positive activating mutation of the c-MPL gene, which encodes for the receptor for thrombopoietin. *Blood.* **103**:4198–4200.

56. Kaushansky, K. 2005. On the molecular origins of the chronic myeloproliferative disorders: it all makes sense. *Blood.* **105**:4187–4190.

57. Barbui, T. 2004. The leukemia controversy in myeloproliferative disorders: is it a natural progression of disease, a secondary sequela of therapy, or a combination of both? *Semin. Hematol.* **41**:15–17.

58. Wang, J.F., Liu, Z.Y., and Groopman, J.E. 1998. The alpha-chemokine receptor CXCR4 is expressed on the megakaryocytic lineage from progenitor to platelets and modulates migration and adhesion. *Blood.* **92**:756–764.

59. Majka, M., et al. 2000. Stromal-derived factor 1 and thrombopoietin regulate distinct aspects of human megakaryopoiesis. *Blood.* **96**:4142–4151.

60. Kowalska, M.A., et al. 1999. Megakaryocyte precursors, megakaryocytes and platelets express the HIV co-receptor CXCR4 on their surface: determination of response to stromal-derived factor-1 by megakaryocytes and platelets. *Br. J. Haematol.* **104**:220–229.

61. Avecilla, S.T., et al. 2004. Chemokine-mediated interaction of hematopoietic progenitors with the bone marrow vascular niche is required for thrombopoiesis. *Nat. Med.* **10**:64–71.

62. Lee, B., Ratajczak, J., Doms, R.W., Gewirtz, A.M., and Ratajczak, M.Z. 1999. Coreceptor/chemokine receptor expression on human hematopoietic cells: biological implications for human immunodeficiency virus-type 1 infection. *Blood.* **93**:1145–1156.

63. Seibert, C., and Sakmar, T.P. 2004. Small-molecule antagonists of CCR5 and CXCR4: a promising new class of anti-HIV-1 drugs. *Curr. Pharm. Des.* **10**:2041–2062.

64. Gewirtz, A.M., Calabretta, B., Rucinski, B., Niewiarowski, S., and Xu, W.Y. 1989. Inhibition of human megakaryocytopoiesis in vitro by platelet factor 4 (PF4) and a synthetic COOH-terminal PF4 peptide. *J. Clin. Invest.* **83**:1477–1486.

65. Gewirtz, A.M., et al. 1995. Chemokine regulation of human megakaryocytopoiesis. *Blood.* **86**:2559–2567.

66. Eslin, D.E., et al. 2004. Transgenic mice studies demonstrate a role for platelet factor 4 in thrombosis: dissociation between anticoagulant and antithrombotic effect of heparin. *Blood.* **104**:3173–3180.

67. Kowalska, M.A., et al. 2000. Stromal cell-derived factor-1 and macrophage-derived chemokine: 2 chemokines that activate platelets. *Blood.* **96**:50–57.

68. Clemetson, K.J., et al. 2000. Functional expression of CCR1, CCR3, CCR4, and CXCR4 chemokine receptors on human platelets. *Blood.* **96**:4046–4054.

69. Weiss, M.J., and Orkin, S.H. 1995. GATA transcription factors: key regulators of hematopoiesis. *Exp. Hematol.* **23**:99–107.

70. Zon, L.I., et al. 1993. Expression of mRNA for the GATA-binding proteins in human eosinophils and basophils: potential role in gene transcription. *Blood.* **81**:3234–3241.

71. Lemarchandel, V., Ghysdael, J., Mignotte, V., Rahuel, C., and Romeo, P.H. 1993. GATA and Ets cis-acting sequences mediate megakaryocyte-specific expression. *Mol. Cell. Biol.* **13**:668–676.

72. Martin, F., Prandini, M.H., Thevenon, D., Marguerie, G., and Uzan, G. 1993. The transcription factor GATA-1 regulates the promoter activity of the platelet glycoprotein IIb gene. *J. Biol. Chem.* **268**:21606–21612.

73. Deveaux, S., et al. 1996. Analysis of the thrombopoietin receptor (MPL) promoter implicates GATA and Ets proteins in the coregulation of megakaryocyte-specific genes. *Blood.* **87**:4678–4685.

74. Ludlow, L.B., et al. 1996. Identification of a mutation in a GATA binding site of the platelet glycoprotein Ibbeta promoter resulting in the Bernard-Soulier syndrome. *J. Biol. Chem.* **271**:22076–22080.

75. Fujiwara, Y., Browne, C.P., Cunniff, K., Goff, S.C., and Orkin, S.H. 1996. Arrested development of embryonic red cell precursors in mouse embryos lacking transcription factor GATA-1. *Proc. Natl. Acad. Sci. U. S. A.* **93**:12355–12358.

76. Shivdasani, R.A., Fujiwara, Y., McDevitt, M.A., and Orkin, S.H. 1997. A lineage-selective knockout establishes the critical role of transcription factor GATA-1 in megakaryocyte growth and platelet development. *EMBO J.* **16**:3965–3973.

77. Tsai, F.Y., et al. 1994. An early haematopoietic defect in mice lacking the transcription factor GATA-2. *Nature.* **371**:221–226.

78. Fujiwara, Y., Chang, A.N., Williams, A.M., and Orkin, S.H. 2004. Functional overlap of GATA-1 and GATA-2 in primitive hematopoietic development. *Blood.* **103**:583–585.

79. Tsang, A.P., et al. 1997. FOG, a multitype zinc finger protein, acts as a cofactor for transcription factor GATA-1 in erythroid and megakaryocytic differentiation. *Cell.* **90**:109–119.

80. Gaines, P., Geiger, J.N., Knudsen, G., Seshasayee, D., and Wojchowski, D.M. 2000. GATA-1- and FOG-dependent activation of megakaryocytic alpha IIB gene expression. *J. Biol. Chem.* **275**:34114–34121.

81. Wang, X., et al. 2002. Control of megakaryocyte-specific gene expression by GATA-1 and FOG-1: role of Ets transcription factors. *EMBO J.* **21**:5225–5234.

82. Cantor, A.B., Katz, S.G., and Orkin, S.H. 2002. Distinct domains of the GATA-1 cofactor FOG-1 differentially influence erythroid versus megakaryocytic maturation. *Mol. Cell. Biol.* **22**:4268–4279.

83. Hong, W., et al. 2005. FOG-1 recruits the NuRD repressor complex to mediate transcriptional repression by GATA-1. *EMBO J.* **24**:2367–2378.

84. Nichols, K.E., et al. 2000. Familial dyserythropoietic anaemia and thrombocytopenia due to an inherited mutation in GATA1. *Nat. Genet.* **24**:266–270.

85. Yu, C., et al. 2002. X-linked thrombocytopenia with thalassemia from a mutation in the amino finger of GATA-1 affecting DNA binding rather than FOG-1 interaction. *Blood.* **100**:2040–2045.

86. Mehaffey, M.G., Newton, A.L., Gandhi, M.J., Crossley, M., and Drachman, J.G. 2001. X-linked thrombocytopenia caused by a novel mutation of GATA-1. *Blood.* **98**:2681–2688.

87. Wechsler, J., et al. 2002. Acquired mutations in GATA1 in the megakaryoblastic leukemia of Down syndrome. *Nat. Genet.* **32**:148–152.

88. Greene, M.E., et al. 2003. Mutations in GATA1 in both transient myeloproliferative disorder and acute megakaryoblastic leukemia of Down syndrome. *Blood Cells Mol. Dis.* **31**:351–356.

89. Oikawa, T., and Yamada, T. 2003. Molecular biology of the Ets family of transcription factors. *Gene.* **303**:11–34.

90. Terui, K., et al. 2000. Expression of transcription factors during megakaryocytic differentiation of CD34+ cells from human cord blood induced by thrombopoietin. *Tohoku J. Exp. Med.* **192**:259–273.

91. Athanasiou, M., et al. 1996. Increased expression of the ETS-related transcription factor FLI-1/ERGB correlates with and can induce the megakaryocytic phenotype. *Cell Growth Differ.* **7**:1525–1534.

92. Sakurai, T., et al. 2003. Effects of overexpression of the Ets family transcription factor TEL on cell growth and differentiation of K562 cells. *Int. J. Oncol.* **22**:1327–1333.

93. Jackers, P., Szalai, G., Moussa, O., and Watson, D.K. 2004. Ets-dependent regulation of target gene expression during megakaryopoiesis. *J. Biol. Chem.* **279**:52183–52190.

94. Anderson, M.K., Hernandez-Hoyos, G., Diamond, R.A., and Rothenberg, E.V. 1999. Precise developmental regulation of Ets family transcription factors during specification and commitment to the T cell lineage. *Development.* **126**:3131–3148.

95. Brown, L.A., et al. 2000. Insights into early vasculogenesis revealed by expression of the ETS-domain transcription factor Fli-1 in wild-type and mutant zebrafish embryos. *Mech. Dev.* **90**:237–252.

96. Eisbacher, M., et al. 2003. Protein-protein interaction between Fli-1 and GATA-1 mediates synergistic expression of megakaryocyte-specific genes through cooperative DNA binding. *Mol. Cell. Biol.* **23**:3427–3441.

97. Hart, A., et al. 2000. Fli-1 is required for murine vascular and megakaryocytic development and is hemizygously deleted in patients with thrombocytopenia. *Immunity.* **13**:167–177.

98. Kawada, H., et al. 2001. Defective megakaryopoiesis and abnormal erythroid development in Fli-1 gene-targeted mice. *Int. J. Hematol.* **73**:463–468.

99. Athanasiou, M., Mavrothalassitis, G., Sun-Hoffman, L., and Blair, D.G. 2000. FLI-1 is a suppressor of erythroid differentiation in human hematopoietic cells. *Leukemia.* **14**:439–445.

100. Raslova, H., et al. 2004. FLI1 monoallelic expression combined with its hemizygous loss underlies Paris-Trousseau/Jacobsen thrombopenia. *J. Clin. Invest.* **114**:77–84. doi:10.1172/JCI200421179.

101. Mackereth, C.D., et al. 2004. Diversity in structure and function of the Ets family PNT domains. *J. Mol. Biol.* **342**:1249–1264.

102. Wang, L.C., et al. 1998. The TEL/ETV6 gene is required specifically for hematopoiesis in the bone marrow. *Genes Dev.* **12**:2392–2402.

103. Hock, H., et al. 2004. Tel/Etv6 is an essential and selective regulator of adult hematopoietic stem cell survival. *Genes Dev.* **18**:2336–2341.

104. Andrews, N.C., Erdjument-Bromage, H., Davidson, M.B., Tempst, P., and Orkin, S.H. 1993. Erythroid transcription factor NF-E2 is a haematopoietic-specific basic-leucine zipper protein. *Nature.*

362:722–728.

105. Tiwari, S., et al. 2003. A role for Rab27b in NF-E2-dependent pathways of platelet formation. *Blood.* **102**:3970–3979.

106. Lecine, P., Italiano, J.E., Jr., Kim, S.W., Villeval, J.L., and Shivdasani, R.A. 2000. Hematopoietic-specific beta 1 tubulin participates in a pathway of platelet biogenesis dependent on the transcription factor NF-E2. *Blood.* **96**:1366–1373.

107. Aplan, P.D., Lombardi, D.P., and Kirsch, I.R. 1991. Structural characterization of SIL, a gene frequently disrupted in T-cell acute lymphoblastic leukemia. *Mol. Cell. Biol.* **11**:5462–5469.

108. Shivdasani, R.A., Mayer, E.L., and Orkin, S.H. 1995. Absence of blood formation in mice lacking the T-cell leukaemia oncoprotein tal-1/SCL. *Nature.* **373**:432–434.

109. Visvader, J.E., Fujiwara, Y., and Orkin, S.H. 1998. Unsuspected role for the T-cell leukemia protein SCL/tal-1 in vascular development. *Genes Dev.* **12**:473–479.

110. Elefanty, A.G., et al. 1998. Characterization of hematopoietic progenitor cells that express the transcription factor SCL, using a lacZ "knock-in" strategy. *Proc. Natl. Acad. Sci. U. S. A.* **95**:11897–11902.

111. Schlaeger, T.M., Mikkola, H.K., Gekas, C., Helgadottir, H.B., and Orkin, S.H. 2005. Tie2Cre-mediated gene ablation defines the stem-cell leukemia gene (SCL/tal1)-dependent window during hematopoietic stem-cell development. *Blood.* **105**:3871–3874.

112. Nucifora, G., and Rowley, J.D. 1995. AML1 and the 8;21 and 3;21 translocations in acute and chronic myeloid leukemia. *Blood.* **86**:1–14.

113. Speck, N.A., et al. 1999. Core-binding factor: a central player in hematopoiesis and leukemia. *Cancer Res.* **59**(Suppl. 7):1789s–1793s.

114. Zhang, D.E., et al. 1996. Function of PU.1 (Spi-1), C/EBP, and AML1 in early myelopoiesis: regulation of multiple myeloid CSF receptor promoters. *Curr. Top. Microbiol. Immunol.* **211**:137–147.

115. Takakura, N., et al. 2000. A role for hematopoietic stem cells in promoting angiogenesis. *Cell.* **102**:199–209.

116. North, T., et al. 1999. Cbfa2 is required for the formation of intra-aortic hematopoietic clusters. *Development.* **126**:2563–2575.

117. Song, W.J., et al. 1999. Haploinsufficiency of CBFA2 causes familial thrombocytopenia with propensity to develop acute myelogenous leukaemia. *Nat. Genet.* **23**:166–175.

118. Michaud, J., et al. 2002. In vitro analyses of known and novel RUNX1/AML1 mutations in dominant familial platelet disorder with predisposition to acute myelogenous leukemia: implications for mechanisms of pathogenesis. *Blood.* **99**:1364–1372.

119. Ichikawa, M., et al. 2004. AML-1 is required for megakaryocytic maturation and lymphocytic differentiation, but not for maintenance of hematopoietic stem cells in adult hematopoiesis. *Nat. Med.* **10**:299–304.

120. Elagib, K.E., et al. 2003. RUNX1 and GATA-1 coexpression and cooperation in megakaryocytic differentiation. *Blood.* **101**:4333–4341.

121. Niitsu, N., et al. 1997. AML1a but not AML1b inhibits erythroid differentiation induced by sodium butyrate and enhances the megakaryocytic differentiation of K562 leukemia cells. *Cell Growth Differ.* **8**:319–326.

122. Westendorf, J.J., and Hiebert, S.W. 1999. Mammalian runt-domain proteins and their roles in hematopoiesis, osteogenesis, and leukemia [review]. *J. Cell. Biochem.* **75**(Suppl.):51–58.

The molecular mechanisms that control thrombopoiesis

Kenneth Kaushansky

Department of Medicine, Division of Hematology/Oncology, University of California, San Diego, San Diego, California, USA.

Our understanding of thrombopoiesis — the formation of blood platelets — has improved greatly in the last decade, with the cloning and characterization of thrombopoietin, the primary regulator of this process. Thrombopoietin affects nearly all aspects of platelet production, from self-renewal and expansion of HSCs, through stimulation of the proliferation of megakaryocyte progenitor cells, to support of the maturation of these cells into platelet-producing cells. The molecular and cellular mechanisms through which thrombopoietin affects platelet production provide new insights into the interplay between intrinsic and extrinsic influences on hematopoiesis and highlight new opportunities to translate basic biology into clinical advances.

Overview of platelets and thrombopoiesis

An adequate supply of platelets is essential to repair the minute vascular damage that occurs with daily life, and to initiate thrombus formation in the event of overt vascular injury. Accumulating evidence also indicates vital roles for platelets in wound repair, the innate immune response, and metastatic tumor cell biology. The average platelet count in humans ranges from 150×10^9 to 400×10^9 per liter, although the level for any individual is maintained within fairly narrow limits from day to day. While 150×10^9 to 400×10^9 per liter is considered "normal," the values derived from the mean ± 2 SDs of a group of "healthy" individuals, epidemiological evidence indicates that individuals who display platelet counts in the highest quartile of the normal range have a 2-fold increased risk of adverse cardiovascular events (1), and, in both experimental animal models of metastatic cancer and patients with tumors, higher platelet levels carry an unfavorable prognosis (2).

With a lifespan of approximately 10 days, a blood volume of 5 liters, and one-third of platelets pooled in the spleen, the average adult must produce each day approximately 1×10^{11} platelets to maintain a normal platelet count under steady-state conditions, a level of production that can increase more than 10-fold under conditions of increased demand. The primary regulator of platelet production is thrombopoietin, an acidic glycoprotein produced primarily in the liver, kidney, and BM. The biochemistry and structure-activity relationships of thrombopoietin have been carefully evaluated, as have the binding sites to its receptor, the product of the cellular protooncogene c-Mpl (3, 4). This Review will focus on the regulation of platelet production, how thrombopoietin stimulates thrombopoiesis under normal and pathological conditions, and the molecular mechanisms through which the hormone produces its biological effects. In addition to providing an understanding of normal physiology, the discovery of the biological effects of thrombopoietin and its receptor provides a platform on which to understand a number of clinical disorders of hematopoiesis and serves to suggest novel therapeutic approaches to several diseases.

Nonstandard abbreviations used: ET, essential thrombocythemia; GSK3β, glycogen synthase kinase-3β; IMF, idiopathic myelofibrosis; ORF, open reading frame; PV, polycythemia vera; SDF-1, stromal cell–derived factor-1.

Conflict of interest: The author has declared that no conflict of interest exists.

Citation for this article: J. Clin. Invest. 115:3339–3347 (2005). doi:10.1172/JCI26674.

Discovery of thrombopoietin and the c-Mpl receptor

Evidence for a humoral regulator of thrombopoiesis was first provided in the late 1950s when plasma from bleeding or thrombocytopenic rats was found to induce thrombocytosis when transfused into secondary animals. Based on ongoing work in erythropoiesis, the term "thrombopoietin" was first coined in 1958 to describe the humoral substance responsible for increasing platelet production (5). In the late 1960s, in vivo assays were developed to detect thrombopoietin (6) but were cumbersome, hampering its purification. In the 1980s, in vitro megakaryocyte colony–forming assays were developed, allowing the identification of megakaryocyte colony–stimulating factors, but whether the identified substances, such as IL-3, GM-CSF, IL-6, or IL-11, were the same as thrombopoietin remained controversial (7).

Occasionally in science, findings from one discipline spur an entirely distinct field of investigation. The description of the murine myeloproliferative leukemia virus in 1986 (8) and its corresponding oncogene (v-Mpl) and cellular protooncogene (c-Mpl) in the early 1990s (9, 10) had such an effect on the search for thrombopoietin. Based on a number of shared primary and secondary structural features, c-Mpl was identified as an orphan member of the hematopoietic cytokine receptor family of proteins, leading to the cloning of its ligand, thrombopoietin, 2 years later (reviewed in ref. 11). Initial experiments with the recombinant protein established thrombopoietin as the primary physiological regulator of thrombopoiesis, as its levels were inversely related to platelet or megakaryocyte mass and its infusion massively increased platelet production (12).

Regulation of thrombopoietin production

Investigators have found that blood and marrow levels of thrombopoietin are inversely related to platelet count. Patients with aplastic anemia or thrombocytopenia secondary to myelosuppressive therapy display high levels of the hormone (13, 14). However, there are some notable exceptions to this relationship. The first is seen in states of platelet destruction, where levels of the hormone are not as high as would be anticipated from the degree of thrombocytopenia (15, 16). Such instances, seen in most patients with idiopathic thrombocytopenic purpura, are characterized by megakaryocyte hypertrophy, which likely contributes to thrombopoietin regulation (17). A second instance in which thrombopoietin levels are not accurately predicted by blood platelet count is in patients with inflam-

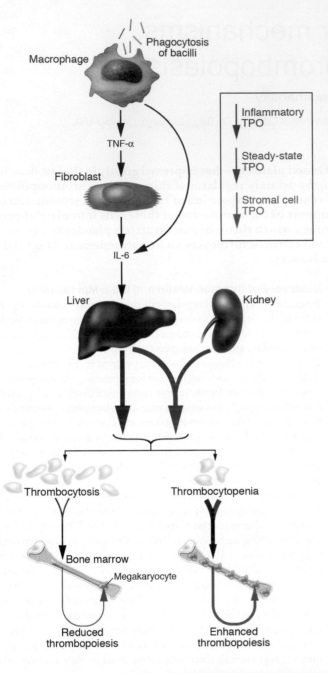

Figure 1
The regulation of thrombopoietin levels. A steady-state amount of hepatic thrombopoietin (TPO) is regulated by platelet c-Mpl receptor–mediated uptake and destruction of the hormone. Hepatic production of the hormone is depicted. Upon binding to platelet c-Mpl receptors, the hormone is removed from the circulation and destroyed, which reduces blood levels. In the presence of inflammation, IL-6 is released from macrophages and, through TNF-α stimulation, from fibroblasts and circulates to the liver to enhance thrombopoietin production. Thrombocytopenia also leads to enhanced marrow stromal cell production of thrombopoietin, although the molecular mediator(s) of this effect is not yet completely understood.

endothelial cell types proliferate or migrate in response to the hormone (25), transplantation studies have shown that endothelial cell c-Mpl does not materially affect thrombopoietin levels despite a 100-fold more expansive cell surface (and predicted greater c-Mpl mass) than that displayed by the totality of megakaryocytes and platelets (26). Therefore, the mere presence of c-Mpl does not guarantee that it is involved in regulating thrombopoietin blood levels.

A second exception to the relatively simple platelet-adsorption-and-destruction model of platelet homeostasis is illustrated by the physiological response to severe thrombocytopenia; studies in both mice and humans show that while marrow stromal cells display very little thrombopoietin mRNA under normal conditions, transcripts for the cytokine greatly increase in the presence of thrombocytopenia (Figure 1) (27). The precise humoral or cellular mediators of this effect are under intense study. For example, in one report, the platelet α-granule proteins PDGF and FGF-2 increased, but platelet factor 4, thrombospondin, and TGF-β decreased thrombopoietin production from primary human BM stromal cells (28). However, others have reported that HGF is responsible for thrombopoietin production from hepatocytes (29).

A third mechanism of thrombopoietin regulation occurs in states of reactive thrombocytosis, where hormone concentrations are higher than that predicted by the degree of thrombocytosis. For example, inflammatory stimuli affect thrombopoietin production, with the acute-phase response mediator IL-6 increasing thrombopoietin transcription from the liver (30). These in vitro effects are also seen in vivo; administration of IL-6 to mice or cancer patients increases thrombopoietin-specific mRNA in the liver and levels of the hormone in the blood. Since an anti-thrombopoietin antibody neutralizes the thrombopoietic effects of administered IL-6 (30), it is now clear that thrombopoietin is the final mediator of inflammation-induced thrombocytosis.

Additional modes of thrombopoietic regulation

In addition to thrombopoietin, additional factors likely influence thrombopoiesis, as the genetic elimination of thrombopoietin or its receptor leads to profound but not absolute thrombocytopenia (the platelet counts in these settings are about 10% of a normal level). In order to determine whether any of the known hematopoietic cytokines contributes to the residual thrombopoiesis in the *c-Mpl*–null state, such mice have been crossed with other cytokine- or cytokine receptor–deficient animals; from these studies it is clear that IL-3 (31, 32), IL-6, IL-11, and LIF (33) are not basal, physiological mediators of thrombopoiesis. However, the chemokine stromal cell–derived factor-1 (SDF-1) exerts numerous influences on megakaryopoiesis, and studies indicate that it may be responsible for thrombopoiesis not related to thrombopoietin.

matory, reactive thrombocytosis, where levels of the hormone are higher than expected (18–21). However, an important unanswered question is whether alterations in thrombopoietin production explain the thrombocytosis associated with iron deficiency.

A major component of thrombopoietin regulation is achieved by receptor-mediated uptake and destruction (Figure 1), a mechanism of hematopoietic growth factor regulation first established for M-CSF (22). Platelets bear high-affinity thrombopoietin receptors that remove the hormone from solution (23), thereby establishing an autoregulatory loop; as platelet counts rise, they remove more of the hormone from the circulation, driving levels down, whereas in thrombocytopenic states there are less platelets to adsorb thrombopoietin, allowing levels to rise and drive increased thrombopoiesis. However, not all thrombopoietin receptors contribute to this effect; while endothelial cells display c-Mpl receptors (24) and some

For example, SDF-1 acts alone and in synergy with thrombopoietin to enhance megakaryocyte colony formation in serum-free culture (34). The chemokine also affects the motility of megakaryocytes, driving their migration toward stromal cells (35), with which they productively interact in an integrin $\alpha_4\beta_1$–dependent manner (36). The administration of SDF-1, along with FGF-4, can nearly normalize the platelet count of c-Mpl–deficient mice and can enhance platelet recovery following myelosuppression (37). Thus, accumulating evidence points to SDF-1 and marrow stromal cells as important influences on thrombopoiesis, but whether levels of the chemokine or surface expression of integrins can be modulated in response to thrombocytopenia remains unknown.

Hematopoietic and other activities of thrombopoietin

Although thrombopoietin was initially postulated to exclusively promote the maturation of megakaryocytes and their fragmentation into platelets, its biological effects are more wide ranging than was initially thought. Thrombopoietin supports the survival and expansion of HSCs and all types of progenitor cells that display megakaryocyte potential, promotes the maturation of megakaryocytes, and enhances the platelet response to activating events.

Thrombopoietin is the most potent single stimulus of the growth of hematopoietic progenitor cells committed to the megakaryocyte lineage. It also acts in synergy with other hematopoietic cytokines, including SCF, IL-11, and erythropoietin, to promote progenitor cell proliferation (38). In suspension culture, thrombopoietin stimulates the formation of large, highly polyploid megakaryocytes that can form proplatelet processes that then fragment into immature and mature platelets (39). The hormone also affects mature platelets, reducing the level of ADP, collagen, or thrombin needed to induce aggregation (40, 41), and enhances platelet adhesion to fibrinogen, fibronectin, and vWF in parallel plate perfusion chamber assays (42).

In addition to its stimulation of most, if not all, aspects of megakaryopoiesis, thrombopoietin displays profound and nonredundant effects on HSCs. Using highly purified marrow-derived cells, 2 groups showed that, by itself, the hormone affects the survival of HSCs and works in synergy with IL-3 or SCF to promote proliferation in vitro in both murine and human cells (43, 44). Subsequent in vivo studies confirmed the physiological relevance of these findings; genetic elimination of thrombopoietin or its receptor in mice, or congenital absence of c-Mpl in children, leads to profound thrombocytopenia, and to equivalent reductions in the levels of HSCs and progenitor cells of all hematopoietic lineages (45–48).

While thrombopoietin clearly plays an important role in the maintenance of HSC numbers, it is not responsible for determination of the lineage fate of these cells. The distribution of hematopoietic progenitor cell types that emerge from cultures of HSCs is identical whether thrombopoietin is present or not (43). Rather, it is likely that the balance of transcription factors present in the HSC is responsible for commitment to one or another hematopoietic lineage. Recently, based on commitment to either the erythroid or the myeloid lineage, a model of mutual transcription factor antagonism has developed that can explain much of the stochastic commitment to one or another cell lineage (49, 50). For megakaryocytes, c-Myb seems particularly important; hypomorphic alleles of the gene result in megakaryocytic expansion (51), and 2 mutant c-Myb alleles, identified in a chemical mutagenesis screen, have been shown to skew lineage commitment toward megakaryopoiesis and alter cell responsiveness to hematopoietic growth factors (52). The transcription factors known to affect megakaryocyte commitment have been recently reviewed (53) and include GATA1, AML1/RUNX1, FOG1, FLI1, MYB, and NF-E2.

Mpl receptor expression, regulation, and signaling

The type I hematopoietic growth factor receptor family, of which c-Mpl is a member, consists of more than 20 molecules that bear 1 or 2 cytokine receptor motifs, an approximately 200–amino acid module containing 4 spatially conserved Cys residues, 14 β-sheets, and a juxtamembrane Trp-Ser-Xaa-Trp-Ser sequence (54). In addition to the cytokine receptor motif(s), type I receptors contain a 20- to 25-residue transmembrane domain and a 70- to 500–amino acid intracellular domain containing short sequences that bind intracellular kinases and other signal-transducing molecules. The thrombopoietin receptor is expressed primarily in hematopoietic tissues, specifically in megakaryocytes, their precursors, and their progeny. For the most part, c-Mpl is constitutively expressed in these tissues, although receptor display is modulated by thrombopoietin binding and receptor internalization. A second potential level of c-Mpl regulation exists; multiple spliceoforms of the receptor have been described that vary in their biological activity, and 1 form can alter receptor catabolism. Although the proportion of the various isoforms of the receptor differs in different tissues, they have not yet been shown to exert a regulatory effect.

The c-Mpl gene contains 12 exons and is organized like other members of the hematopoietic cytokine receptor family (55). A site for initiation of c-Mpl transcription resides 13 nucleotides upstream of the translation initiation codon, and although the promoter lacks conventional TATA and CAAT motifs, the 5′ flanking sequence contains consensus binding sequences for Ets and GATA transcription factors, proteins vital for the regulation of many megakaryocyte-specific genes. Analysis of c-Mpl transcripts has identified several alternately spliced forms, including extracellular domain deletions (56), an alternate intracellular domain (the K isoform; ref. 8), and a prematurely truncated isoform containing a unique carboxyl terminus (the Mpl-tr isoform; ref. 57). While potentially acting as a dominant-negative form of the receptor, and differentially expressed in certain cell types, the K isoform does not affect thrombopoietin signaling, as it does not interact with the wild-type receptor (58). However, Mpl-tr may play a physiological role, as it is the only isoform expressed in both human and murine cells. Of note, expression levels of c-Mpl are low, with only 25–100 surface receptors present per platelet (59, 60). The origin of the poor expression of c-Mpl appears to be related to the c-Mpl-tr isoform, as its coexpression with full-length c-Mpl leads to rapid degradation of the latter (57). However, whether this physiology is reflected in thrombopoietin signal regulation is, at present, only speculative.

Another aspect of c-Mpl regulation under intense study is its expression on hematopoietic cells of patients with myeloproliferative disorders (MPDs). While easily detectable on normal marrow megakaryocytes and platelets, the receptor is decreased on cells from patients with polycythemia vera and other myeloproliferative diseases (61, 62). While the molecular basis for this is not understood, it could be related to the hypersensitivity to cytokines and signaling abnormalities seen in these disorders. Another clue to this finding may lie in 2 recent observations, that coexpression of the signaling kinase JAK2 is vital for hematopoietic cytokine receptor expression (63), and that the activity of this kinase is altered in a substantial number of patients with MPDs.

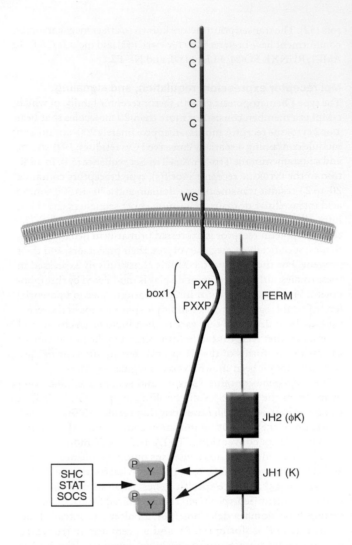

Figure 2

Hematopoietic cytokine receptor architecture and mechanism of initial signaling. A stylized hematopoietic cytokine receptor is shown, depicting the 1 or 2 cytokine receptor motifs (C, Cys; WS, Trp-Ser-Xaa-Trp-Ser), the transmembrane domain, and the box1 sequence to which JAK kinases bind. Also shown are the 3 major domains of JAK kinases, the FERM domain, which binds to box1, and the kinase JH1 and regulatory JH2 domains. Finally, upon JAK activation, the site of receptor tyrosine phosphorylation is shown, which then serves as a docking site for STATs and adapter proteins (SHC or SHP2).

to JH1, its active site is altered and inactivated and is thus termed the pseudokinase φK domain. The function of JH2 was identified by differential expression studies; the JH1 domain is an active kinase when expressed alone, whereas the activity of a JH1/JH2 polypeptide is greatly blunted (65). Thus, the JH2 domain regulates the kinase activity of JH1, a physiology put into structural terms by homology modeling of JH1/JH2; the JH2 domain interacts with the inactive, but not the active, conformation of the activation loop of JH1 (Figure 4; ref. 66), in a region of JH2 shown to be vital for kinase regulatory activity.

Once c-Mpl is activated by thrombopoietin engagement, its multiple effects on HSCs, megakaryocytes, and platelets are mediated by a series of biochemical signaling events. Thrombopoietin activates both JAK2 and TYK2 in c-Mpl–bearing cell lines, although only JAK2 is essential for signaling and is the predominant isoform activated in primary megakaryocytes (67). By generating a complex composed of the phosphatase SHP2, a scaffolding Gab/IRS protein, and the p85 regulatory subunit of PI3K, thrombopoietin stimulation of megakaryocytes and their precursors activates PI3K and its immediate downstream effector Akt (PKB) (Figure 3) (68, 69). Blocking this pathway inhibits thrombopoietin-induced cell survival and proliferation (70). In the mature platelet, the hormone enhances α-granule secretion and aggregation induced by thrombin in a PI3K-dependent fashion (71). The pathways downstream of Akt in megakaryocytes and platelets are under study and include the transcription factor FOXO3a, the cell cycle inhibitor p27, and glycogen synthase kinase-3β (GSK3β). In addition to PI3K, thrombopoietin stimulates 2 of the MAPK pathways (Figure 3), p42/p44 ERK1 and ERK2 (72) and p38 MAPK (73), events mediated by receptor phosphorylation, binding and phosphorylation of Grb2, SHC, and SOS, and exchange of GDP for GTP on Ras (74). The functional consequences of these events include induction of the transcription factor HoxB4 and expansion of HSCs mediated by p38 MAPK (73); translocation of the transcription factor HoxA9 from cytoplasm to nucleus, which also favorably affects HSC expansion (75); the ERK1/2–induced proliferation and polyploidization of megakaryocytes (76); and augmented thrombin-induced liberation of phospholipase A₂ and platelet activation (77).

Implications of thrombopoietin/Mpl signaling in clinical disorders of hematopoiesis

Two primary reasons to generate a detailed map of the signaling circuitry used by hematopoietic growth factors are to understand disorders of hematopoietic growth and to intervene in these processes for therapeutic benefit (either enhancing or blunting signaling). A number of human diseases in which blood cell production is altered can now be understood as disorders of growth factor signaling.

Upon binding, cognate ligand hematopoietic cytokine receptors such as c-Mpl are activated to transmit numerous biochemical signals. The molecular details of this process are now well understood based on studies of the erythropoietin receptor (EpoR). The EpoR exists in a homodimeric state in the absence of ligand, in a conformation that holds the cytoplasmic domains 73 Å apart (64). Upon ligand binding, receptor conformation shifts, bringing the cytoplasmic domains within 39 Å of one another. Additional studies indicate that the membrane-proximal box1 and box2 cytoplasmic domains constitutively bind JAK family kinases, even in an inactive state. Upon ligand binding, the closer juxtaposition of the 2 tethered kinases allows their cross-activation, initiating signal transduction. The active JAK kinase then phosphorylates (a) tyrosine residues within the receptor itself; (b) molecules that promote cell survival and proliferation, including the STATs, PI3K, and the MAPKs; and (c) those that limit cell signaling, including the SHP1 and SHIP1 phosphatases and SOCSs (Figures 2 and 3).

Additional insights into how the JAK kinases are regulated come from domain analysis of the proteins. All 4 members of the family (JAK1, JAK2, JAK3, and TYK2) display 3 major domains, JH1 (JAK homology 1), JH2, and FERM (four-point-one, ezrin, radixin, moesin), the latter responsible for binding to the cytoplasmic domain of the cytokine receptors (Figure 2). The JH1 domain carries the kinase activity of JAKs, and while JH2 bears significant homology

One of the most common abnormalities of the blood count is elevated platelet levels, and the most common cause of thrombo-

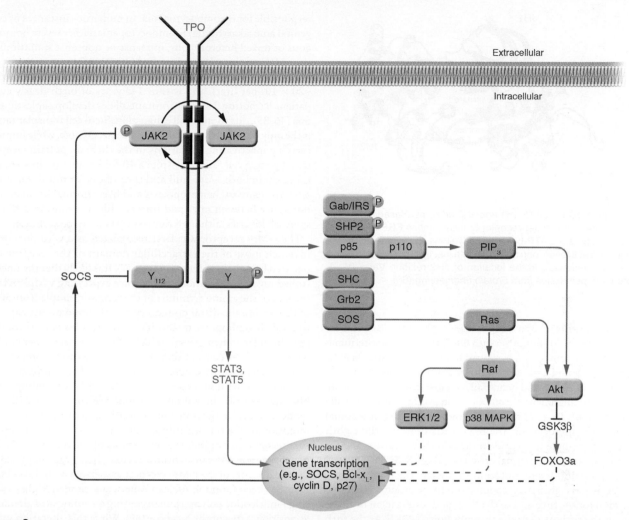

Figure 3

Signaling pathways activated by thrombopoietin. A stylized drawing of c-Mpl is shown in the activated (phosphorylated) form. Once phosphorylated, Tyr$_{112}$ serves as a docking site for STAT3 and STAT5, both activated by thrombopoietin in megakaryocytes, which leads to production of Bcl-x$_L$, among other antiapoptotic and pro-proliferative signaling molecules. The same site also serves to recruit SHC, which in turn recruits Grb2 and SOS (the latter a guanine nucleotide exchange factor for Ras), exchanging GTP for GDP, and thereby activating Ras. In succession, a MAPKKK (MAPK kinase kinase, e.g., Raf), a MAPKK (MAPK kinase), and the MAPK ERK1/2 or p38 MAPK are recruited and activated. As shown, Raf activation also contributes to PI3K activation. At a site proximal to Tyr$_{112}$, a complex containing the phosphatase SHP2, the adapter protein Gab1, and the regulatory subunit of PI3K (p85) forms upon phosphorylation by JAK2, which recruits the kinase subunit of PI3K (p110), leading to phosphorylation of cell membrane–bound phosphoinositol$_{4,5}$ biphosphate (PIP$_2$) and thus generating phosphoinositol$_{3,4,5}$ triphosphate (PIP$_3$). PIP$_3$ then recruits pleckstrin homology domain–containing proteins, including the Ser/Thr protein tyrosine kinase Akt. Once activated at the cell membrane, Akt phosphorylates (and inactivates) GSK3β, which also promotes cell proliferation. Akt also phosphorylates the transcription factor FOXO3a, leading to its nuclear exit and thus precluding its induction of the cell cycle inhibitor p27. Inhibition of cell signaling is also initiated by JAK activation; shown in red is the transcriptional regulation of SOCS proteins by STATs, and their subsequent blockade of signaling by preclusion of signaling molecule docking to P-Tyr residues of the receptor or their JAK-induced phosphorylation.

cytosis is reaction to an inflammatory insult (20). A number of cytokines are released from activated monocytes and macrophages, including IL-1, TNF-α, and IL-6. While IL-1 and TNF-α have mixed effects on hematopoiesis, in vitro IL-6 appears to act as a megakaryocyte maturation factor (78). Intravenous infusion of IL-6, in mice and during clinical trials in humans, leads to modest thrombocytosis (79, 80). While this initially suggested that inflammatory thrombocytosis is due to IL-6, several investigators have more recently shown that IL-6 induces hepatic thrombopoietin production and that the thrombocytosis associated with IL-6 is eliminated by blockade of thrombopoietin action (Figure 1) (30).

Much less common than inflammatory thrombocytosis is familial thrombocytosis. At least 4 different genetic alterations of the thrombopoietin gene have been shown to enhance production of the hormone and drive polyclonal thrombocytosis. The basis of all these cases is enhanced translation efficiency of the thrombopoietin transcript.

Unlike most structural genes, in which protein translation initiates from the first AUG codon in the transcript, the open reading frame (ORF) encoding thrombopoietin begins with the eighth AUG present in the mRNA (Figure 5); the previous 7 ORFs encode short, apparently functionless polypeptides, if they are translated

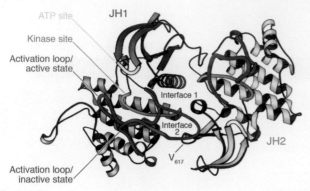

Figure 4
A molecular model of JAK2 JH1 and JH2 domains. Based on the tertiary structure of the dimer receptor tyrosine kinase FGF receptor-4, the model depicts the ATP-binding site (yellow), the kinase active site (orange), the activation loop of JH1 in both inactive (purple) and active (red) conformations, and the location of JH2 residue Val_{617} (V_{617}). Adapted with permission from *Protein Engineering* (66).

at all. Since the thrombopoietin initiation codon (AUG8) is embedded within the out-of-frame, seventh ORF, and since post-termination ribosomes cannot scan backward to initiate at upstream AUG codons (81), should a ribosome initiate at ORF7, it will terminate downstream of AUG8 and hence fail to translate thrombopoietin. Because of this alignment, thrombopoietin production is normally very inefficient. In 4 separate pedigrees displaying autosomal dominant thrombocytosis, the region surrounding the eighth AUG carries a single-nucleotide mutation that greatly enhances thrombopoietin translational efficiency (Figure 5; ref. 82). In 2 cases, a splice donor site mutation eliminates the exon carrying the seventh and eighth AUG codons, and the translation termination codon for the fifth and sixth ORFs, and results in an in-frame fusion of the fifth AUG to the thrombopoietin ORF. As the fifth AUG is a highly efficient initiation codon, thrombopoietin production rises, enhancing thrombopoiesis. The other 2 mutations within exon 3 of the thrombopoietin gene eliminate the reason for the unfavorable translation efficiency of the gene, the embedding of AUG8 within the seventh ORF (Figure 5), again resulting in greatly enhanced thrombopoietin production.

Disorders of the c-Mpl receptor

Thrombopoietin is the primary regulator of platelet production; thus, abnormalities of the hormone or its receptor might also be

responsible for thrombocytopenia. In numerous instances of congenital amegakaryocytic thrombocytopenia, either severe homozygous or mixed heterozygous, missense or nonsense mutations of the *c-Mpl* gene have been identified (46, 83, 84). Loss of the receptor leads to severe congenital thrombocytopenia (platelet counts $\sim 20 \times 10^9$ per liter), and within 1–3 years of birth nearly every patient harboring 2 severely mutant alleles develops aplastic anemia (46, 85), due to stem cell exhaustion. Stem cell transplantation is the only known treatment for the disease. Of note, while humans carrying inactivating mutations in the thrombopoietin receptor develop stem cell failure, despite a 10-fold reduction in stem cell numbers in both *c-Mpl*–null and thrombopoietin-null mice, the animals maintain hematopoiesis and live a normal lifespan. The difference between mice and humans likely resides in different stem cell kinetics, although this issue has yet to be addressed.

The c-Mpl receptor was first recognized as a viral oncogene, in which most of the extracellular domain of the receptor was replaced by a viral *gag* gene sequence (7). It is likely that the uncontrolled myeloproliferation seen in mice expressing v-Mpl reflects the loss of the amino terminus of the receptor; simple truncation of the membrane-distal domain of c-Mpl eliminates its capacity to bind thrombopoietin, and its expression leads to thrombopoietin-independent growth of cells (86). This and other studies have spurred the concept that the membrane-distal domain puts a brake on constitutive signaling by the membrane-proximal and transmembrane domains, a block relieved by thrombopoietin binding. Several clinical observations are consistent with this model of c-Mpl activation. For example, an activating Ser-to-Asn mutation of the transmembrane domain of the c-Mpl receptor, first recognized by random mutagenesis of the receptor (87), has now been found to cause thrombocytosis in a large family (88). A second disorder of the c-Mpl receptor associated with thrombocytosis has been found in African Americans, termed $Mpl_{Baltimore}$, a single-nucleotide polymorphism causing an amino acid alteration at position Asn_{34} of the receptor (89). While this alteration in the membrane-distal extracellular domain of c-Mpl associated with thrombocytosis also supports the developing model of c-Mpl activation, it remains possible that $Mpl_{Baltimore}$ leads to thrombocytosis through another mechanism, or that the single-nucleotide poly-

Figure 5
Genetic alterations in thrombopoietin that lead to enhanced translation efficiency. The normal thrombopoietin mRNA (light blue) is spliced from 7 exons, of which 3 are shown (**A**). The numbered initiation codons found in the primary thrombopoietin transcript are shown as within their corresponding ORFs (e.g., the thrombopoietin ORF is dark blue and initiates from AUG8). The sites of mutation that lead to enhanced translation of the thrombopoietin transcript do so (**B**) by eliminating exon 3 by altered splicing (ΔE3) to create a new thrombopoietin ORF initiated by a highly efficient initiation codon (AUG5); (**C**) by nonsense mutation, prematurely truncating ORF7, which embeds the normal thrombopoietin ORF; or (**D**) by shifting the efficiently initiated ORF7 (by a single-nucleotide insertion) to now include the thrombopoietin polypeptide. Adapted with permission from *Blood* (82).

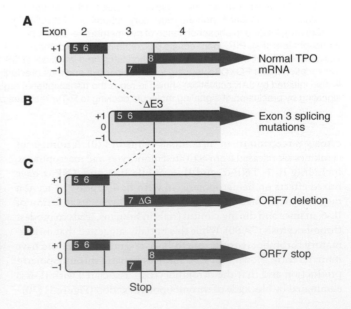

morphism lies in linkage disequilibrium with a distinct, causative, non–c-Mpl coding mutation in these individuals.

Disorders of hematopoietic cell signal transduction

The classic chronic MPDs polycythemia vera (PV), idiopathic myelofibrosis (IMF), and essential thrombocythemia (ET) share several clinical features, including overproduction of 1 or more hematopoietic cell lineages, a propensity for pathological hemorrhage or thrombosis, an excess or abnormality of megakaryocytes, which elaborate cytokines responsible for myelofibrosis, and a modestly elevated risk of progression to leukemia, a risk greatly hastened by exposure to alkylating agents (90). Several molecular features of the chronic MPDs suggest that they represent disorders of hematopoietic cell signaling. First, overexpression of numerous growth factor signaling mediators leads to chronic MPDs in mice — for example, unregulated expression of c-Mpl (91) or thrombopoietin (92), loss of either of the signal-inhibitory phosphatases SHP1 (93) and SHIP1 (94), conditional loss of the Ras regulator NF-1 (95), or expression of an activated form of K-Ras (96). Second, hematopoietic cells from patients with chronic MPDs express constitutively activated signaling molecules, including STAT3 (97), Bcl-x$_L$ (98), and Akt (99). Third, hematopoietic progenitor cells are hypersensitive to several different hematopoietic growth factors in PV, IMF, and ET (100–102), including erythropoietin and thrombopoietin; this implies a resetting of the molecular pathways that transduce growth factor signals. Based on these and other observations, several groups hypothesized that abnormalities in JAK2 kinase might underlie the chronic MPDs, and multiple groups reported this to be the case; 65–97% of patients with PV, 35–57% of patients with IMF, and 23–57% of patients with ET carry 1 or 2 mutant JAK2 alleles (overall, two-thirds are heterozygous and one-third homozygous, the latter because of mitotic recombination). Remarkably, every patient studied displays the same acquired mutation, Val-to-Phe substitution of amino acid 617, which leads to constitutive JAK2 activation in vitro, and polycythemia when introduced into hematopoietic cells in vivo (102–107). This region of the pseudokinase domain of JAK2 is necessary for proper JAK2 regulation (65) and, in a molecular model, interacts with the activation loop of the JH1 kinase domain of the molecule (Figure 4) (66). While of great interest, the finding of an activated signaling kinase in patients with chronic myeloproliferative diseases has also raised several questions: why is overexpression of the kinase necessary to produce cytokine hypersensitivity in vitro and polycythemia in vivo (since levels equivalent to wild-type JAK2 do not suffice); why are 3 distinct clinical disorders associated with the same mutation; are second "hits" necessary to generate disease; and what, if any, hematopoietic cytokine receptor does the mutant JAK2 kinase interact with to produce its effects? Regardless of these questions, it is almost certain that this discovery in chronic myeloproliferative diseases will yield new insights in patients with PV, IMF, and ET and spur research into identifying a therapeutic agent that can inhibit the mutant, but not the wild-type, form of JAK2 kinase.

Future research directions

Our understanding of the molecular basis of thrombopoiesis has progressed substantially in the past 100 years, beginning with James Homer Wright, who in 1906 provided evidence that megakaryocytes give rise to blood platelets (108). Until this time, little attention had been paid to platelets, then referred to as the "dust of the blood." But despite great progress, many questions remain: What is the mechanistic basis for proplatelet formation, a massive cytoplasmic reorganization of actin and tubulin that, upon fragmentation, generates platelets, and what, if any, are the exogenous signals that trigger the process? What is the reason for megakaryocyte polyploidy, and what are the mechanisms by which these cells uncouple DNA synthesis and cell division, one of the most closely guarded links in cell physiology? What is being sensed in marrow stromal cells that alter their production of thrombopoietin in thrombocytopenia, and what are the signals that accomplish this? How can an alteration in the JAK2 kinase lead to PV in some patients and ET in others? And how are stem cell decisions of lineage commitment orchestrated? Research in thrombopoiesis is presently in a logarithmic growth phase. The years to come will provide many new insights into how platelets develop and how that process is regulated.

Address correspondence to: Kenneth Kaushansky, Department of Medicine, Division of Hematology/Oncology, University of California, San Diego, 402 Dickinson Street, Suite 380, San Diego, California 92103-8811, USA. Phone: (619) 543-2259; Fax: (619) 543-3931; E-mail: kkaushansky@ucsd.edu.

1. Thaulow, E., Erikssen, J., Sandvik, L., Stormorken, H., and Cohn, P.F. 1991. Blood platelet count and function are related to total and cardiovascular death in apparently healthy men. *Circulation.* **84**:613–617.

2. Gupta, G.P., and Massague, J. 2004. Platelets and metastasis revisited: a novel fatty link. *J. Clin. Invest.* **114**:1691–1693. doi:10.1172/JCI200423823.

3. Kato, T., et al. 1998. Native thrombopoietin: structure and function. *Stem Cells.* **16**:322–328.

4. Jagerschmidt, A., et al. 1998. Human thrombopoietin structure-function relationships: identification of functionally important residues. *Biochem. J.* **333**:729–734.

5. Kelemen, E., Cserhati, I., and Tanos, B. 1958. Demonstration and some properties of human thrombopoietin in thrombocythemic sera. *Acta. Haematol.* **20**:350–355.

6. McDonald, T.P. 1976. A comparison of platelet size, platelet count, and platelet 35S incorporation as assays for thrombopoietin. *Br. J. Haematol.* **34**:257–267.

7. Williams, N. 1991. Is thrombopoietin interleukin 6? *Exp. Hematol.* **19**:714–718.

8. Wendling, F., Varlet, P., Charon, M., and Tambourin, P. 1986. MPLV: a retrovirus complex inducing an acute myeloproliferative leukemic disorder in adult mice. *Virology.* **149**:242–246.

9. Souyri, M., Vigon, I., Penciolelli, J.-F., Tambourin, P., and Wendling, F. 1990. A putative truncated cytokine receptor gene transduced by the myeloproliferative leukemia virus immortalizes hematopoietic progenitors. *Cell.* **63**:1137–1147.

10. Vigon, I., et al. 1992. Molecular cloning and characterization of *MPL*, the human homolog of the v-*mpl* oncogene: identification of a member of the hematopoietic growth factor receptor superfamily. *Proc. Natl. Acad. Sci. U. S. A.* **89**:5640–5644.

11. Kaushansky, K. 1995. Thrombopoietin: the primary regulator of platelet production. *Blood.* **86**:419–431.

12. Kaushansky, K. 1998. Thrombopoietin. *N. Engl. J. Med.* **339**:746–754.

13. Shinjo, K., et al. 1998. Serum thrombopoietin levels in patients correlate inversely with platelet counts during chemotherapy-induced thrombocytopenia. *Leukemia.* **12**:295–300.

14. Engel, C., Loeffler, M., Franke, H., and Schmitz, S. 1999. Endogenous thrombopoietin serum levels during multicycle chemotherapy. *Br. J. Haematol.* **105**:832–838.

15. Kosugi, S., et al. 1996. Circulating thrombopoietin level in chronic immune thrombocytopenic purpura. *Br. J. Haematol.* **93**:704–706.

16. Ichikawa, N., et al. 1996. Regulation of serum thrombopoietin levels by platelets and megakaryocytes in patients with aplastic anaemia and idiopathic thrombocytopenic purpura. *Thromb. Haemost.* **76**:156–160.

17. Nagasawa, T., et al. 1998. Serum thrombopoietin level is mainly regulated by megakaryocyte mass rather than platelet mass in human subjects. *Br. J. Haematol.* **101**:242–244.

18. Cerutti, A., Custodi, P., Duranti, M., Noris, P., and Balduini, C.L. 1997. Thrombopoietin levels in patients with primary and reactive thrombocytosis. *Br. J. Haematol.* **99**:281–284.

19. Hsu, H.C., et al. 1999. Circulating levels of thrombopoietic and inflammatory cytokines in patients with clonal and reactive thrombocytosis. *J. Lab. Clin. Med.* **134**:392–397.

20. Griesshammer, M., et al. 1999. Aetiology and clinical significance of thrombocytosis: analysis of 732

patients with an elevated platelet count. *J. Intern. Med.* **245**:295–300.

21. Wolber, E.M., Fandrey, J., Frackowski, U., and Jelkmann, W. 2001. Hepatic thrombopoietin mRNA is increased in acute inflammation. *Thromb. Haemost.* **86**:1421–1424.

22. Bartocci, A., et al. 1987. Macrophages specifically regulate the concentration of their own growth factor in the circulation. *Proc. Natl. Acad. Sci. U. S. A.* **84**:6179–6183.

23. Kuter, D.J., and Rosenberg, R.D. 1995. The reciprocal relationship of thrombopoietin (c-Mpl ligand) to changes in the platelet mass during busulfan-induced thrombocytopenia in the rabbit. *Blood.* **85**:2720–2730.

24. Methia, N., Louache, F., Vainchenker, W., and Wendling, F. 1993. Oligodeoxynucleotides antisense to the proto-oncogene c-mpl specifically inhibit in vitro megakaryocytopoiesis. *Blood.* **82**:1395–1401.

25. Cardier, J.E., and Dempsey, J. 1998. Thrombopoietin and its receptor, c-mpl, are constitutively expressed by mouse liver endothelial cells: evidence of thrombopoietin as a growth factor for liver endothelial cells. *Blood.* **91**:923–929.

26. Geddis, A.E., Fox, N.E., and Kaushansky, K. The Mpl receptor expressed on endothelial cells does not contribute significantly to the regulation of circulating thrombopoietin (TPO) levels. *Exp. Hematol.* In press.

27. Sungaran, R., Markovic, B., and Chong, B.H. 1997. Localization and regulation of thrombopoietin mRNA expression in human kidney, liver, bone marrow and spleen using in situ hybridization. *Blood.* **89**:101–107.

28. Sungaran, R., et al. 2000. The role of platelet alpha-granular proteins in the regulation of thrombopoietin messenger RNA expression in human bone marrow stromal cells. *Blood.* **95**:3094–3101.

29. Yamashita, K., et al. 2000. Hepatocyte growth factor/scatter factor enhances the thrombopoietin mRNA expression in rat hepatocytes and cirrhotic rat livers. *J. Gastroenterol. Hepatol.* **15**:83–90.

30. Kaser, A., et al. 2001. Interleukin-6 stimulates thrombopoiesis through thrombopoietin: role in inflammatory thrombocytosis. *Blood.* **98**:2720–2725.

31. Gainsford, T., et al. 1998. Cytokine production and function in c-mpl-deficient mice: no physiologic role for interleukin-3 in residual megakaryocyte and platelet production. *Blood.* **91**:2745–2752.

32. Chen, Q., Solar, G., Eaton, D.L., and de Sauvage, F.J. 1998. IL-3 does not contribute to platelet production in c-Mpl-deficient mice. *Stem Cells.* **16**(Suppl. 2):31–36.

33. Gainsford, T., et al. 2000. The residual megakaryocyte and platelet production in c-mpl-deficient mice is not dependent on the actions of interleukin-6, interleukin-11, or leukemia inhibitory factor. *Blood.* **95**:528–534.

34. Hodohara, K., Fujii, N., Yamamoto, N., and Kaushansky, K. 2000. Stromal cell derived factor 1 acts synergistically with thrombopoietin to enhance the development of megakaryocytic progenitor cells. *Blood.* **95**:769–775.

35. Hamada, T., et al. 1998. Transendothelial migration of megakaryocytes in response to stromal cell-derived factor 1 (SDF-1) enhances platelet formation. *J. Exp. Med.* **188**:539–548.

36. Fox, N.E., and Kaushansky, K. 2005. Engagement of integrin α4β1 enhances thrombopoietin induced megakaryopoiesis. *Exp. Hematol.* **33**:94–99.

37. Avecilla, S.T., et al. 2004. Chemokine-mediated interaction of hematopoietic progenitors with the bone marrow vascular niche is required for thrombopoiesis. *Nat. Med.* **10**:64–71.

38. Broudy, V.C., Lin, N.L., and Kaushansky, K. 1995. Thrombopoietin (c-mpl ligand) acts synergistically with erythropoietin, stem cell factor, and IL-11 to enhance murine megakaryocyte colony growth

and increases megakaryocyte ploidy in vitro. *Blood.* **85**:1719–1726.

39. Kaushansky, K., et al. 1995. Thrombopoietin, the Mpl-ligand, is essential for full megakaryocyte development. *Proc. Natl. Acad. Sci. U. S. A.* **92**:3234–3238.

40. Chen, J., Herceg-Harjacek, L., Groopman, J.E., and Grabarek, J. 1995. Regulation of platelet activation in vitro by the c-Mpl ligand, thrombopoietin. *Blood.* **86**:4054–4062.

41. Oda, A., et al. 1996. Thrombopoietin primes human platelet aggregation induced by shear stress and multiple agonists. *Blood.* **87**:4664–4670.

42. Van Os, E., et al. 2003. Thrombopoietin increases platelet adhesion under flow and decreases rolling. *Br. J. Haematol.* **121**:482–490.

43. Sitnicka, E., et al. 1996. The effect of thrombopoietin on the proliferation and differentiation of murine hematopoietic stem cells. *Blood.* **87**:4998–5005.

44. Kobayashi, M., Laver, J.H., Kato, T., Miyazaki, H., and Ogawa, M. 1996. Thrombopoietin supports proliferation of human primitive hematopoietic cells in synergy with steel factor and/or interleukin-3. *Blood.* **88**:429–436.

45. Carver-Moore, K., et al. 1996. Low levels of erythroid and myeloid progenitors in thrombopoietin- and c-mpl-deficient mice. *Blood.* **88**:803–808.

46. Ballmaier, M., et al. 2001. c-mpl mutations are the cause of congenital amegakaryocytic thrombocytopenia. *Blood.* **97**:139–146.

47. Solar, G.P., et al. 1998. Role of c-mpl in early hematopoiesis. *Blood.* **92**:4–10.

48. Kimura, S., Roberts, A.W., Metcalf, D., and Alexander, W.S. 1998. Hematopoietic stem cell deficiencies in mice lacking c-Mpl, the receptor for thrombopoietin. *Proc. Natl. Acad. Sci. U. S. A.* **95**:1195–1200.

49. Zhang, P., et al. 1999. Negative cross-talk between hematopoietic regulators: GATA proteins repress PU.1. *Proc. Natl. Acad. Sci. U. S. A.* **96**:8705–8710.

50. Nerlov, C., Querfurth, E., Kulessa, H., and Graf, T. 2000. GATA-1 interacts with the myeloid PU.1 transcription factor and represses PU.1-dependent transcription. *Blood.* **95**:2543–2551.

51. Emambokus, N., et al. 2003. Progression through key stages of haematopoiesis is dependent on distinct threshold levels of c-Myb. *EMBO J.* **22**:4478–4488.

52. Metcalf, D., et al. 2005. Anomalous megakaryocytopoiesis in mice with mutations in the c-Myb gene. *Blood.* **105**:3480–3487.

53. Schulze, H., and Shivdasani, R.A. 2004. Molecular mechanisms of megakaryocyte differentiation. *Semin. Thromb. Hemost.* **30**:389–398.

54. Ihle, J.N. 1995. Cytokine receptor signaling. *Nature.* **377**:591–594.

55. Alexander, W.S., and Dunn, A.R. 1995. Structure and transcription of the genomic locus encoding murine c-Mpl, a receptor for thrombopoietin. *Oncogene.* **10**:795–803.

56. Li, J., Sabath, D.F., and Kuter, D.J. 2000. Cloning and functional characterization of a novel c-mpl variant expressed in human CD34 cells and platelets. *Cytokine.* **12**:835–844.

57. Coers, J., Ranft, C., and Skoda, R.C. 2004. A truncated isoform of c-Mpl with an essential C-terminal peptide targets the full-length receptor for degradation. *J. Biol. Chem.* **279**:36397–36404.

58. Millot, G.A., et al. 2002. MplK, a natural variant of the thrombopoietin receptor with a truncated cytoplasmic domain, binds thrombopoietin but does not interfere with thrombopoietin-mediated cell growth. *Exp. Hematol.* **30**:166–175.

59. Broudy, V.C., Lin, N.L., Sabath, D.F., Papayannopoulou, T., and Kaushansky, K. 1997. Human platelets display high affinity receptors for thrombopoietin. *Blood.* **89**:1896–1904.

60. Li, J., Xia, Y., and Kuter, D.J. 1999. Interaction of thrombopoietin with the platelet c-mpl receptor in plasma: binding, internalization, stability and

pharmacokinetics. *Br. J. Haematol.* **106**:345–356.

61. Moliterno, A.R., Hankins, W.D., and Spivak, J.L. 1998. Impaired expression of the thrombopoietin receptor by platelets from patients with polycythemia vera. *N. Engl. J. Med.* **338**:572–580.

62. Yoon, S.Y., Li, C.Y., and Tefferi, A. 2000. Megakaryocyte c-Mpl expression in chronic myeloproliferative disorders and the myelodysplastic syndrome: immunoperoxidase staining patterns and clinical correlates. *Eur. J. Haematol.* **65**:170–174.

63. Huang, L.J., Constantinescu, S.N., and Lodish, H.F. 2001. The N-terminal domain of Janus kinase 2 is required for Golgi processing and cell surface expression of erythropoietin receptor. *Mol. Cell.* **8**:1327–1338.

64. Livnah, O., et al. 1999. Crystallographic evidence for preformed dimers of erythropoietin receptor before ligand activation. *Science.* **283**:987–993.

65. Saharinen, P., Vihinen, M., and Silvennoinen, O. 2003. Autoinhibition of Jak2 tyrosine kinase is dependent on specific regions in its pseudokinase domain. *Mol. Biol. Cell.* **14**:1448–1459.

66. Lindauer, K., Loerting, T., Liedl, K.R., and Kroemer, R.T. 2001. Prediction of the structure of human Janus kinase 2 (JAK2) comprising the two carboxy-terminal domains reveals a mechanism for autoregulation. *Protein Eng.* **14**:27–37.

67. Drachman, J.G., Millett, K.M., and Kaushansky, K. 1999. Mpl signal transduction requires functional JAK2, not TYK2. *J. Biol. Chem.* **274**:13480–13484.

68. Miyakawa, Y., Rojnuckarin, P., Habib, T., and Kaushansky, K. 2001. Thrombopoietin induces PI3K and SHP2 activation through Gab and IRS proteins in BaF3 cells and primary murine megakaryocytes. *J. Biol. Chem.* **276**:2494–2502.

69. Bouscary, D., et al. 2001. Role of Gab proteins in phosphatidylinositol 3-kinase activation by thrombopoietin (Tpo). *Oncogene.* **20**:2197–2204.

70. Geddis, A., Fox, N., and Kaushansky, K. 2001. Phosphatidylinositol 3-kinase (PI3K) is necessary but not sufficient for thrombopoietin-induced proliferation in engineered Mpl-bearing cell lines as well as in primary megakaryocytic progenitors. *J. Biol. Chem.* **276**:34473–34479.

71. Kojima, H., et al. 2001. Role of phosphatidylinositol-3 kinase and its association with Gab1 in thrombopoietin-mediated up-regulation of platelet function. *Exp. Hematol.* **29**:616–622.

72. Drachman, J.D., Sabath, D.F., Fox, N.E., and Kaushansky, K. 1997. Thrombopoietin signal transduction in purified murine megakaryocytes. *Blood.* **89**:483–492.

73. Kirito, K., Fox, N.E., and Kaushansky, K. 2003. Thrombopoietin stimulates expression of HoxB4: an explanation for the favorable effects of TPO on hematopoietic stem cells. *Blood.* **102**:3172–3178.

74. Rojnuckarin, P., et al. 2001. The roles of PI3K and PKCζ for thrombopoietin-induced MAP kinase activation in primary murine megakaryocytes. *J. Biol. Chem.* **276**:41014–41022.

75. Kirito, K., Fox, N.E., and Kaushansky, K. 2004. Thrombopoietin (TPO) induces the nuclear translocation of HoxA9 in hematopoietic stem cells (HSC): a potential explanation for the favorable effects of TPO on HSCs. *Mol. Cell. Biol.* **24**:6751–6762.

76. Rojnuckarin, P., Drachman, J.G., and Kaushansky, K. 1999. Thrombopoietin-induced activation of the mitogen activated protein kinase pathway in normal megakaryocytes: role in endomitosis. *Blood.* **94**:1273–1282.

77. van Willigen, G., Gorter, G., and Akkerman, J.W. 2000. Thrombopoietin increases platelet sensitivity to alpha-thrombin via activation of the ERK2-cPLA2 pathway. *Thromb. Haemost.* **83**:610–616.

78. Ishibashi, T., et al. 1989. Human interleukin 6 is a direct promoter of maturation of megakaryocytes in vitro. *Proc. Natl. Acad. Sci. U. S. A.* **86**:5953–5957.

79. Laterveer, L., van Damme, J., Willemze, R., and

Fibbe, W.E. 1993. Continuous infusion of interleukin-6 in sublethally irradiated mice accelerates platelet reconstitution and the recovery of myeloid but not of megakaryocytic progenitor cells in bone marrow. *Exp. Hematol.* **21**:1621–1627.

80. Sosman, J.A., et al. 1997. Concurrent phase I trials of intravenous interleukin 6 in solid tumor patients: reversible dose-limiting neurological toxicity. *Clin. Cancer Res.* **3**:39–46.

81. Kozak, M. 2001. Constraints on reinitiation of translation in mammals. *Nucleic Acids Res.* **29**:5226–5232.

82. Cazzola, M., and Skoda, R.C. 2000. Translational pathophysiology: a novel molecular mechanism of human disease. *Blood.* **95**:3280–3288.

83. Ihara, K., et al. 1999. Identification of mutations in the c-mpl gene in congenital amegakaryocytic thrombocytopenia. *Proc. Natl. Acad. Sci. U. S. A.* **96**:3132–3136.

84. van den Oudenrijn, S., et al. 2000. Mutations in the thrombopoietin receptor, Mpl, in children with congenital amegakaryocytic thrombocytopenia. *Br. J. Haematol.* **110**:441–448.

85. Ballmaier, M., Germeshausen, M., Krukemeier, S., and Welte, K. 2003. Thrombopoietin is essential for the maintenance of normal hematopoiesis in humans: development of aplastic anemia in patients with congenital amegakaryocytic thrombocytopenia. *Ann. N. Y. Acad. Sci.* **996**:17–25.

86. Sabath, D.F., Kaushansky, K., and Broudy, V.C. 1999. Deletion of the membrane-distal cytokine receptor homology domain of MPL results in constitutive cell growth and loss of thrombopoietin binding. *Blood.* **94**:365–367.

87. Onishi, M., et al. 1996. Identification of an oncogenic form of the thrombopoietin receptor using retrovirus-mediated gene transfer. *Blood.* **88**:1399–1406.

88. Ding, J., et al. 2004. Familial essential thrombocythemia associated with a dominant-positive activating mutation of the c-MPL gene, which encodes for the receptor for thrombopoietin. *Blood.* **103**:4198–4200.

89. Moliterno, A.R., et al. 2004. Mpl Baltimore: a thrombopoietin receptor polymorphism associated with thrombocytosis. *Proc. Natl. Acad. Sci. U. S. A.* **101**:11444–11447.

90. Spivak, J.L., et al. 2003. Chronic myeloproliferative disorders. *Hematology (Am. Soc. Hematol. Educ. Program).* **2003**:200–224.

91. Cocault, L., et al. 1996. Ectopic expression of murine TPO receptor (c-mpl) in mice is pathogenic and induces erythroblastic proliferation. *Blood.* **88**:1656–1665.

92. Yan, X.-Q., et al. 1996. A model of myelofibrosis and osteosclerosis in mice induced by overexpressing thrombopoietin (mpl ligand): reversal of disease by bone marrow transplant. *Blood.* **88**:402–409.

93. Tsui, H.W., Siminovitch, K.A., de Souza, L., and Tsui, F.W. 1993. Motheaten and viable motheaten mice have mutations in the haematopoietic cell phosphatase gene. *Nat. Genet.* **4**:124–129.

94. Helgason, C.D., et al. 1998. Targeted disruption of SHIP leads to hemopoietic perturbations, lung pathology, and a shortened life span. *Genes Dev.* **12**:1610 1620.

95. Gitler, A.D., et al. 2004. Tie2-Cre-induced inactivation of a conditional mutant Nf1 allele in mouse results in a myeloproliferative disorder that models juvenile myelomonocytic leukemia. *Pediatr. Res.* **55**:581–584.

96. Chan, I.T., et al. 2004. Conditional expression of oncogenic K-ras from its endogenous promoter induces a myeloproliferative disease. *J. Clin. Invest.* **113**:528–538. doi:10.1172/JCI200420476.

97. Roder, S., Steimle, C., Meinhardt, G., and Pahl, H.L. 2001. STAT3 is constitutively active in some patients with Polycythemia rubra vera. *Exp. Hematol.* **29**:694–702.

98. Silva, M., et al. 1998. Expression of Bcl-x in erythroid precursors from patients with polycythemia vera. *N. Engl. J. Med.* **338**:564–571.

99. Dai, C., Chung, I.J., and Krantz, S.B. 2005. Increased erythropoiesis in polycythemia vera is associated with increased erythroid progenitor proliferation and increased phosphorylation of Akt/PKB. *Exp. Hematol.* **33**:152–158.

100. Zanjani, E.D., Lutton, J.D., Hoffman, R., and Wasserman, L.R. 1997. Erythroid colony formation by polycythemia vera bone marrow in vitro. Dependence on erythropoietin. *J. Clin. Invest.* **59**:841–848.

101. Dai, C.H., et al. 1992. Polycythemia vera. II. Hypersensitivity of bone marrow erythroid, granulocyte-macrophage, and megakaryocyte progenitor cells to interleukin-3 and granulocyte-macrophage colony-stimulating factor. *Blood.* **80**:891–899.

102. Axelrad, A.A., Eskinazi, D., Correa, P.N., and Amato, D. 2000. Hypersensitivity of circulating progenitor cells to megakaryocyte growth and development factor (PEG-rHu MGDF) in essential thrombocythemia. *Blood.* **96**:3310–3321.

103. Baxter, E.J., et al. 2005. Acquired mutation of the tyrosine kinase JAK2 in human myeloproliferative diseases. *Lancet.* **365**:1054–1061.

104. Levine, R.L., et al. 2005. Activating mutation in the tyrosine kinase JAK2 in polycythemia vera, essential thrombocythemia, and myeloid metaplasia with myelofibrosis. *Cancer Cell.* **7**:387–397.

105. James, C., et al. 2005. A unique clonal *JAK2* mutation leading to constitutive signalling causes polycythaemia vera. *Nature.* **434**:1144–1148.

106. Kralovics, R., et al. 2005. A gain-of-function mutation of JAK2 in myeloproliferative disorders. *N. Engl. J. Med.* **352**:1779–1790.

107. Zhao, R., et al. 2005. Identification of an acquired JAK2 mutation in polycythemia vera. *J. Biol. Chem.* **280**:22788–22792.

108. Wright, J.H. 1906. The origin and nature of the blood plates. *Boston Med. Surg. J.* **23**:643–645.

Thrombus formation in vivo

Bruce Furie[1,2] and Barbara C. Furie[1,2]

[1]Division of Hemostasis and Thrombosis, Center for Vascular Biology Research, Beth Israel Deaconess Medical Center, Boston, Massachusetts, USA.
[2]Department of Medicine, Harvard Medical School, Boston, Massachusetts, USA.

To examine thrombus formation in a living mouse, new technologies involving intravital videomicroscopy have been applied to the analysis of vascular windows to directly visualize arterioles and venules. After vessel wall injury in the microcirculation, thrombus development can be imaged in real time. These systems have been used to explore the role of platelets, blood coagulation proteins, endothelium, and the vessel wall during thrombus formation. The study of biochemistry and cell biology in a living animal offers new understanding of physiology and pathology in complex biologic systems.

History of the study of hemostasis and thrombosis

The study of blood coagulation and thrombosis has moved through many stages. The classic studies of thrombus morphology and pathology were followed by a period in which blood coagulation proteins were identified based on the characterization of patients with hereditary clotting disorders. The fundamentals of platelet physiology, including platelet adhesion and aggregation, were defined by turbidimetric assay reporting platelet-platelet interaction and the adherence of platelets to glass beads, which reflects the ability of platelets to stick to surfaces that mimic the injured vessel wall. Only 40 years ago were the techniques of protein biochemistry applied to the study of hemostasis and thrombosis. Proteins were purified, their amino acid sequences determined, their genes isolated, and their interaction with other components analyzed in vitro. Protein domain structures were defined and their functional and structural relationship to other protein families determined. Site-specific mutagenesis and definition of the molecular basis of hemophilia identified functionally critical amino acids. From the analysis of the many enzymes, cofactors, structural proteins, ligands, and receptors involved in hemostasis and thrombosis, descriptive models emerged that explained the in vitro experimental results and the in vivo clinical observations in humans (1). These models have changed over the years, with continued refinement based on new information and fresh thinking (2–9). Still, many questions remain. We have learned what reactions and interactions *can* happen in vitro, but we need to understand what *does* happen in vivo. We are entering a new phase of discovery — the study of thrombosis in living animals to understand the mechanisms involved in this complex system.

Mechanisms of thrombus induction in animal models

Although the pathogenesis of thrombus formation can be both an acute and a chronic process in the natural condition, direct experimental observation of this process in animal models requires artificial methods. These methods take many forms, each with advantages and disadvantages, as they relate to the physiologic mechanisms of thrombus formation. With photochemical injury, dye (e.g., rose bengal) is infused into the circulation. Photo-excitation leads to oxidative injury of the vessel wall and subsequent thrombus formation (10). Mechanical (11) or electrical trauma (12) directly injures the endothelium and leads to thrombus formation. Vessel ligation causing stasis also initiates thrombus formation (13). Ferric chloride, introduced to initiate arterial thrombosis in small-animal models (14), generally initiates severe endothelial damage and vessel occlusion, monitored by the decrease in temperature distal to the developing thrombus. This model system has been widely used (15–17), modified with a Doppler flow probe to monitor vessel occlusion or with direct blood vessel visualization by intravital microscopy. Laser-induced injury causes heat damage to a limited region of the endothelium, with little morphologic change to the vessel wall (18). Because endothelial damage is limited and does not involve denudation of the endothelium, in contrast to ferric chloride–induced injury, the laser-induced injury appears to be a model of thrombosis more akin to the injury caused by inflammation than to that caused by trauma. Using the laser, both temporal and spatial resolution for thrombus generation is obtained, since the location and the precise time of injury are operator-controlled. The laser pulse, applied through the microscope optics, is targeted at the vessel wall. Upon injury, thrombus formation initiates with the rapid accumulation of platelets and the expression of tissue factor at the thrombus–vessel wall interface.

Intravital microscopy of thrombus formation

Intravital microscopy was introduced to study leukocyte interaction with the vessel wall in a living animal (19, 20). Since then, many investigators have captured in vivo images in real time using analog videomicroscopy (21, 22). Oude Egbrink et al. were among the first to combine experimental thrombosis, induced by mechanical puncture with glass micropipettes, with intravital videomicroscopy in a living animal (23). Others applied experimental thrombosis with in vivo microscopic analysis (24, 25), adopting the methods used for leukocyte rolling. Photochemical injury of microvessels in the mouse ear allowed analysis of the kinetics of platelet accumulation and vessel occlusion, leading to the observation that hirudin inhibited thrombus formation and promoted vessel recanalization (26). Denis et al. demonstrated that genetically altered mice lacking vWF showed defects in platelet accumulation following ferric chloride injury (16).

The discovery that P-selectin is an adhesion molecule that binds platelets to leukocytes (27) and the identification of the P-selectin counterreceptor PSGL-1 (28) prompted us to generate a PSGL-1 knockout mouse to better understand the biology of PSGL-1 (29). Methods of analog intravital microscopy to examine leukocyte rolling in living mice had been developed by

Conflict of interest: The authors have declared that no conflict of interest exists.

Citation for this article: *J. Clin. Invest.* **115**:3355–3362 (2005).
doi:10.1172/JCI26987.

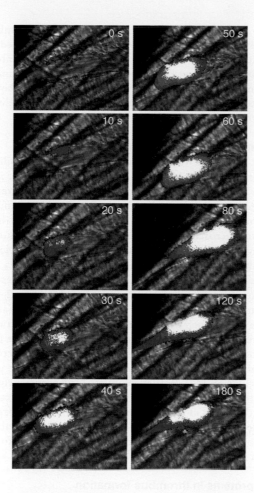

Figure 1
Birth of a thrombus. Intravital wide-field imaging of platelet, tissue factor, and fibrin deposition in the developing thrombus of a living WT mouse following endothelial injury. Blood flow is from right to left. Platelets, tissue factor, and fibrin were labeled using fluorescently tagged antibodies directed at CD41, tissue factor, and human fibrin, respectively. These components were imaged in 3 separate fluorescence channels. A black and white brightfield image indicates the histologic context of the composite image. To simplify analysis of the composite image, the dynamic range of the intensity of each pseudocolor was minimized. Red, platelets; green, tissue factor; blue, fibrin; yellow, platelets plus tissue factor; turquoise, tissue factor plus fibrin; magenta, platelets plus fibrin; white, platelets plus fibrin plus tissue factor.

description of in vivo thrombus formation remains incomplete, but the emerging concepts will be reviewed.

Initiation of thrombus formation

Laser injury initiates the expression of tissue factor activity on or near the vessel wall. Platelets rapidly accumulate on the endothelium, and fibrin can be observed at the platelet thrombus–vessel wall interface, at the leading edge of the thrombus, and, over seconds, within and throughout the thrombus (Figure 1). In contrast to classical models of thrombus formation in which it has been thought that the platelet thrombus forms and is then stabilized by the subsequent formation of a fibrin clot, these experiments demonstrate that platelet thrombus formation and fibrin clot formation overlap temporally and occur nearly simultaneously. Similar results are obtained with other forms of injury, including those that lead to thrombus formation through subendothelial matrix exposure.

Platelet adhesion

vWF plays an important role in thrombus formation. vWF-null mice displayed prolonged bleeding times and spontaneous bleeding in about 10% of neonates (16). These results simulate the phenotype of severe von Willebrand disease in humans. Via intravital microscopy using the ferric chloride injury model, significant impairment of platelet–vessel wall interaction was observed, leading to defective thrombus formation. Also, in the complete absence of vWF, platelet accumulation on the vessel wall was significantly decreased but not absent, and thrombi formed in some vessels. These results argue both for the importance of vWF in platelet deposition on the vessel wall and for a vWF-independent mechanism for arterial thrombus formation.

Fibrinogen is thought to play an important role in platelet–vessel wall interaction, particularly at the low shear rates characteristic of the venous circulation. To evaluate the contribution of fibrinogen to arterial thrombus formation, Ni et al., using fibrinogen-null mice, demonstrated that platelet deposition and the onset of thrombus formation were the same as in WT mice (33). However, thrombi in fibrinogen-null mice were unstable and embolized as they became larger. Mice deficient in both vWF and fibrinogen formed thrombi that proved unstable. Given the accepted tenet that these 2 adhesive proteins are critical for platelet–vessel wall interaction, the fact that thrombi formed in doubly deficient mice argues for additional adhesive molecules that participate in the platelet–vessel wall synapse. A role for fibronectin in this process remains plausible, since fibronectin

other laboratories (30). However, the ability to directly observe complex events in a living animal by optical microscopy offered new opportunities to understand thrombosis. The combination of available knockout mice, progress in optical spectroscopy, and advances in computer software and hardware supported the possibility of developing novel intravital imaging technology to study thrombus formation. An imaging system specifically designed for visualizing events in the vasculature of a living mouse required real-time imaging of thrombus formation using high image acquisition rates of 10–20 images per second in up to 3 fluorescent channels as well as a bright-field channel to appreciate the fluorescent image in a histologic context. The system needed the flexibility of confocal imaging to support 3D image reconstruction and a digital capture system for quantitative image analysis. A system that meets these criteria has been developed: an intravital microscopy system that supports high-speed wide-field and confocal imaging of the microcirculation of a living mouse (31, 32). In most experiments, vessel wall injury that initiates thrombosis is induced using a laser focused on the endothelium via the microscope optics. This technology has been applied to the study of thrombus formation in a living mouse, in order to reexamine many of the accepted tenets of thrombus formation and to determine whether or not the predictions of in vitro or ex vivo experiments are valid for understanding the in vivo system. This work, coupled with the in vivo experiments of others, promises to provide more detailed descriptions of the mechanisms involved in thrombus formation. The current deficiency delays thrombus formation in vivo through putative diminished platelet-platelet interaction (34). It seems that there are multiple participants in platelet–vessel wall interaction, some of which have yet to be directly implicated.

Platelet activation

The tissue factor pathway dominates platelet activation in the laser-induced thrombosis model. PAR4, the sole signaling protease-activated thrombin receptor on mouse platelets, is critical for thrombin-mediated platelet activation (35). Laser-induced injury in mice lacking PAR4 leads to a small, rapid accumulation of platelets (36). This initial thrombus is unstable and subsequently reduces in size. This initial phase of platelet accumulation may be mediated by vWF or another adhesive molecule. After 3–4 minutes, thrombus size in the PAR4-null mice is less that 10% of that in WT mice. During the first several minutes, there is no evidence of platelet activation, as monitored by the surface expression of P-selectin. However, although these thrombi in PAR4-null mice contain markedly reduced numbers of activated platelets, fibrin generation is normal. These results emphasize the importance of thrombin generation and PAR4 for platelet activation and thrombus formation in this model. A major unanswered question is the cellular or subcellular localization of the membrane surface that supports thrombin generation. Are the few activated platelets sufficient to supply the necessary membrane surfaces, or do other cells and cell-derived membranes support thrombin generation? Thrombin generation appears critically important to laser-induced thrombus formation. For example, in the absence of exposed collagen, laser-induced thrombus formation was normal in FcRγ mice lacking the collagen receptor glycoprotein VI (37). If there is collagen exposure, it is below the sensitivity of this imaging system. Thus, in the laser-induced model of thrombosis, platelets are likely activated by a mechanism independent of interaction of collagen with platelet collagen receptors.

Collagen exposure can play an important role in platelet activation during hemostasis, and the importance of this role varies with the type of vascular injury. Ferric chloride induces an oxidative injury exposing the subendothelial matrix (33), and type I collagen can be stained with anti-collagen antibodies in the blood in vivo after ferric chloride injury (37). Following ferric chloride injury, FcRγ-null mice lacking expression of glycoprotein VI failed to generate platelet thrombi in vivo, in contrast to WT mice. Furthermore, blocking antibodies against glycoprotein VI infused into WT mice also inhibited platelet thrombus formation. These results demonstrate the requirement for glycoprotein VI in collagen-mediated platelet activation in vivo.

Platelet activation can be monitored by P-selectin expression on the platelet surface. This marker of platelet activation correlates directly with exocytosis of α-granules. Although in vitro studies of P-selectin have established the kinetics of P-selectin expression following activation with various agonists, the kinetics of P-selectin expression in vivo has not been previously characterized. To this end, P-selectin expression on platelets incorporated into the developing thrombus was studied in vivo after laser-induced injury (38). P-selectin first appears on platelets at the vessel wall–thrombus interface. Over 3–4 minutes, a wave of P-selectin expression can be monitored through the thrombus from the vessel wall to the luminal surface of the thrombus. However, only when the P-selectin density on the luminal surface is sufficiently high to support leukocyte rolling on the arterial thrombus can any leukocyte-thrombus interaction be observed. Indeed, direct observation of leukocyte rolling on the developing thrombus cannot be appreciated until after about 3 minutes from the initiation of thrombus formation. These results emphasize a delayed role for leukocyte-thrombus interaction and tissue factor delivery via leukocytes during thrombus formation.

Calcium mobilization in vivo

Platelet activation leads to intracellular calcium mobilization, where increased calcium serves as a second messenger in initiating numerous signaling pathways. These calcium transients have been extensively studied in vitro using flow chambers to monitor platelet adherence (39–41). High-speed wide-field intravital microscopy has been used to image this calcium spike in living mice during thrombus formation (42). Platelets were isolated from a donor mouse and loaded with fura-2, a fluorochrome that is sensitive to calcium concentration. These platelets were infused into a recipient mouse, and the fura-2–labeled platelets were analyzed for significant changes in intracellular calcium during their circulation in blood, their interaction with the thrombus, and their incorporation into a stable platelet thrombus. These studies have revealed that platelet activation, as monitored by calcium mobilization, does not take place in the circulation (42). Rather, platelets bind transiently to the developing thrombus. After a short period, these platelets either undergo calcium mobilization and become stably incorporated into the thrombus or they disengage from the thrombus and float downstream. Calcium mobilization is required for stable platelet incorporation into the developing thrombus.

Other proteins in thrombus formation

The tissue factor–factor VIIa complex initiates thrombin generation and fibrin formation, and deficiency of any of the proteins within this pathway (e.g., factor IX, factor VIII, factor X, factor V, and prothrombin) decreases thrombin generation and thus thrombus formation. However, factor XII–null mice are characterized by defective arterial thrombus formation in vivo (43), as are factor XI–null mice (44). A role for factor XII in normal hemostasis has been long dismissed, since patients with factor XII deficiency have no bleeding phenotype. Recently a role for both factor XII and factor XI in thrombus formation has been demonstrated, although the actual mechanism for the participation of the intrinsic pathway in blood coagulation in vivo remains speculative. This raises the intriguing possibility that factor XII and factor XI are important for thrombosis but not hemostasis (43).

PECAM-1 is a cell adhesion molecule found on endothelial cells and platelets. PECAM-1–null mice formed larger arterial thrombi more rapidly than WT mice in the laser-induced thrombosis model (45). Using chimeric mice prepared by reciprocal bone marrow transplantation, platelet PECAM-1 was shown to be the critical component. These results suggest that PECAM-1 plays a role in negative regulation of thrombus formation.

Gas6, a γ-carboxyglutamic acid–containing membrane protein homologous to protein S, is present on the platelet membrane and binds to several receptor tyrosine kinases, including Axl. Gas6 amplified platelet aggregation and secretion responses to platelet agonists (46). Deficiency of Gas6, either in a Gas6−/− mouse or using blocking antibodies against Gas6, protected mice from fatal thrombosis but did not impair normal hemostasis. It would appear

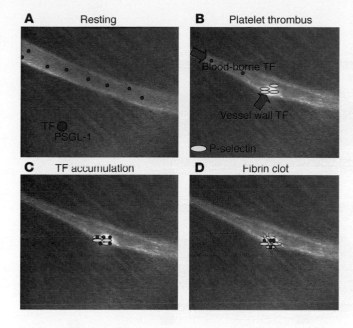

Figure 2
Model of P-selectin/PSGL-1–mediated tissue factor accumulation during thrombus formation. (**A**) Leukocyte microparticles (red) circulate constitutively in the blood under resting conditions. These microparticles express tissue factor (TF) and PSGL-1 on their surface. (**B**) Vessel wall tissue factor is expressed in response to laser-induced injury, leading to platelet activation and the subsequent expression of P-selectin on the stimulated platelets incorporated into the developing thrombus (initiation phase). (**C**) Blood-borne tissue factor associated with microparticles accumulates on the platelet thrombus through the binding of platelet P-selectin and microparticle PSGL-1. (**D**) Concentration of blood-borne tissue factor into the thrombus initiates thrombin generation and fibrin clot propagation within the thrombus (propagation phase).

The SLAM family of adhesion receptors, a subset of the CD2 Ig superfamily, is expressed on platelets (51). SLAM phosphorylation occurs during platelet aggregation. SLAM-deficient platelets showed defective aggregation, and SLAM-null mice were characterized in vivo by delayed thrombus formation but normal tail bleeding times. SLAM may play a secondary role in the formation of the platelet-platelet synapse that is otherwise dominated by $\alpha_{IIb}\beta_3$.

Tissue factor–bearing microparticles

P-selectin functions as an adhesion molecule (27) but was subsequently shown to have a role in fibrin formation (52). In a baboon arteriovenous shunt model of thrombosis, blocking antibodies against P-selectin not only inhibited leukocyte accumulation in the developing thrombus but also decreased fibrin formation. The molecular and cellular basis for this experimental observation was not clear at the time, but it was thought that leukocytes might generate tissue factor upon stimulation more rapidly in vivo than in the in vitro systems used to explore de novo tissue factor biosynthesis in stimulated cells (53, 54). Although a relationship of P-selectin to fibrin formation was secure, the basis for the inhibition of fibrin formation by anti–P-selectin antibodies remained unknown.

Using genetically altered mice and digital intravital microscopy imaging, this question was revisited (55). Tissue factor antigen and fibrin were observed throughout the thrombus generated in WT mice (Figure 1), a result that confirmed earlier in vitro experiments (56). However, minimal tissue factor antigen or fibrin was observed in thrombi generated in either P-selectin–null mice or PSGL-1–null mice (55). These results were similar to those obtained in the baboon thrombosis model using anti–P-selectin antibodies to block P-selectin action. We hypothesized that tissue factor and PSGL-1 must be physically coupled. Although this is true of activated monocytes, where tissue factor and PSGL-1 reside on the plasma membrane (57), there is no evidence that such monocytes circulate constitutively in blood (58). Furthermore, leukocytes do not interact with developing thrombi as rapidly as fibrin deposition begins (38). Rather, leukocyte microparticles might provide the basis for this observation. Leukocyte microparticles, first identified in 1994 (59), could express both tissue factor and PSGL-1 if derived from monocytes. Indeed, a population of microparticles exists in the circulation that is positive for both tissue factor antigen and PSGL-1 antigen. Using a monocyte-like cell line, fluorescently labeled microparticles were generated and infused into mice. During thrombus formation, microparticles accumulated

that Gas6 plays a role in amplifying signaling events induced by agonists and does not directly participate in platelet-platelet synapse formation of significant affinity.

CD40L, a transmembrane platelet granule protein, is expressed on the plasma membrane of activated platelets, where it can interact with CD40 that is widely distributed on vascular cells. Mice deficient in CD40L but not CD40 showed an in vivo defect in thrombus formation initiated by ferric chloride (47). These CD40L$^{-/-}$ mice showed delayed arterial occlusion and thrombus instability. CD40L appears to be an $\alpha_{IIb}\beta_3$ ligand required for stable formation of arterial thrombi.

Human platelets express 2 Eph kinases, Eph4 and EphB1, and at least 1 ligand, ephrinB1. During $\alpha_{IIb}\beta_3$-mediated platelet aggregation, Eph ephrin interactions on adjacent platelet surfaces contribute to high-affinity platelet-platelet contact (48). These interactions favor thrombus growth and stability, sustain contact-facilitated signaling via complex formation, and promote clot retraction. Inhibition of Eph-ephrin interaction, evaluated in vitro in a flow chamber, showed a 40% decrease in mean thrombus volume (48).

Outside-in $\alpha_{IIb}\beta_3$ signaling is required for normal platelet thrombus formation and is triggered by c-Src activation via PTP-1B (49). Studies of PTP-1B–deficient mouse platelets in vitro indicate that PTP-1B is required for fibrinogen-dependent platelet spreading and clot retraction. Thrombus formation in vivo is reduced in PTP-1B–null mice, a manifestation of ineffective calcium mobilization during platelet activation. PTP-1B is a positive regulator for the initiation of outside-in $\alpha_{IIb}\beta_3$ signaling.

CD39, the vascular ATP diphosphohydrolase, is largely expressed on endothelial cells. This enzyme converts ATP and ADP to AMP. Tail bleeding times were prolonged and platelet thrombus formation in vivo was delayed in CD39-null mice subjected to ferric chloride injury (50). These results appear consistent with the importance of ADP release during platelet activation for activation of adjacent platelets. However, interpretation of these results is complicated by the desensitization of the P2Y1 receptor on CD39-null platelets.

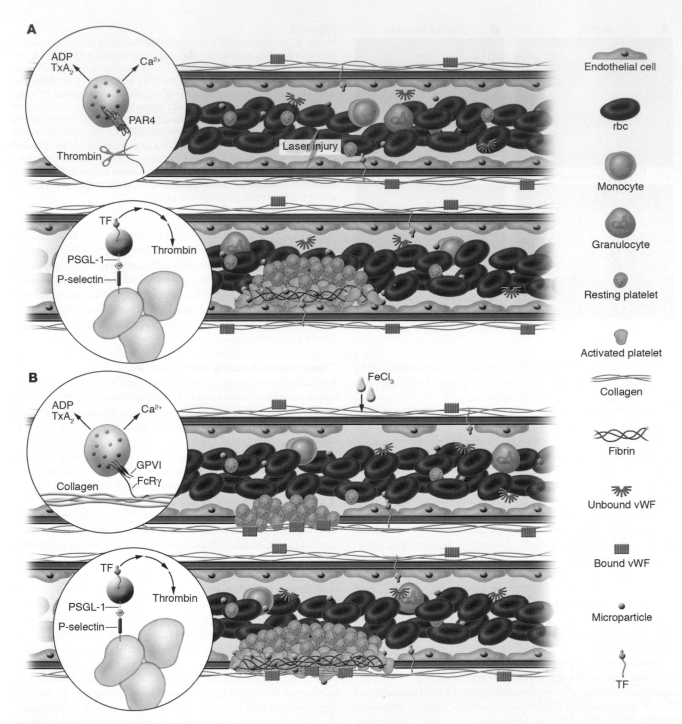

Figure 3

Experimental models of thrombosis. Platelets, red blood cells, monocytes, and granulocytes circulate in blood whereas endothelial cells line the vessel wall. Plasma proteins, including vWF, fibrinogen and other coagulation proteins, and microparticles are also present in the circulation. (**A**) Upon laser-induced injury of the vessel wall, vWF mediates the interaction of platelets with the endothelium. Tissue factor in the vessel wall leads to thrombin generation. Thrombin activates mouse platelets via the PAR4 receptor (inset). Activated platelets undergo calcium mobilization and the release of ADP and thromboxane A_2 (TxA$_2$) to accelerate platelet recruitment and activation and the formation of a platelet thrombus. These platelets express P-selectin, and leukocyte microparticles expressing PSGL-1 and tissue factor accumulate in the thrombus through the interaction of P-selectin with PSGL-1 (inset). The concentration of tissue factor initiates coagulation, the generation of more thrombin, and the propagation of a fibrin clot. (**B**) Upon vessel wall oxidative injury with ferric chloride, the endothelium is denuded and the subendothelial matrix exposed. Platelets interact with the matrix via GPIb-V-IX and $\alpha_{IIb}\beta_3$ on the platelet membrane and collagen and vWF in the matrix. Glycoprotein VI (GPVI) binding to collagen is required for platelet activation, and activated platelets undergo calcium mobilization and the release of ADP and thromboxane A_2 (inset) to accelerate platelet recruitment and activation and the formation of a thrombus. These platelets express P-selectin, and microparticles expressing PSGL-1 and tissue factor accumulate in the thrombus through the interaction of P-selectin with PSGL-1 (inset). The concentration of tissue factor leads to coagulation, the generation of more thrombin, and the propagation of a fibrin clot.

in the thrombi of WT mice. In contrast, no accumulation was observed in P-selectin–null mice.

These results are consistent with a model in which circulating microparticles expressing tissue factor and PSGL-1 accumulate in the developing thrombus via the interaction of P-selectin with PSGL-1. This delivers and concentrates tissue factor in the thrombus, leading to a critical concentration that can initiate blood coagulation (Figure 2). Numerous groups have reported tissue factor antigen in platelet-poor plasma, with levels varying from 100 to 150 pg/ml. However, Butenas et al. have recently reopened this issue (58). They report no detectable tissue factor activity in whole blood, no tissue factor antigen associated with unstimulated mononuclear cells in whole blood, and a level of tissue factor activity that cannot exceed 20 fM, equivalent to about 1 pg/ml, and is more likely lower. Since these authors demonstrate that 1 pg/ml of active tissue factor rapidly clots whole blood, it would seem that blood tissue factor concentration is much lower than 1 pg/ml and that a manyfold concentration of tissue factor within the thrombus is a critical component for the initiation of blood coagulation. Alternatively, an inactive form of tissue factor may undergo some form of activation to its biologically functional form.

Tissue factor resides in 3 distinct compartments: (a) the surface of extravascular cells, (b) the vessel wall, and (c) blood microparticles. Upon stimulation, both endothelial cells and monocytes have the capacity to express tissue factor. To determine whether tissue factor associated with blood microparticles contributes to fibrin formation during thrombosis in vivo, 1 strain of chimeric mice in which tissue factor was associated with the vessel wall but not the blood microparticles and another strain of chimeric mice in which tissue factor was associated with the blood microparticles but not the vessel wall were prepared (60). Such mice were generated by bone marrow transplantation of WT mice, with normal levels of tissue factor in both the vessel wall and blood microparticles, and low–tissue factor mice, with about 1% of the normal level of tissue factor (61). Chimeras generated by transplantation of low–tissue factor bone marrow into WT mice showed platelet thrombi containing markedly reduced tissue factor and fibrin (60). Conversely, chimeras generated by transplantation of WT bone marrow into low–tissue factor mice rescued tissue factor accumulation and fibrin generation in the platelet thrombus. These results emphasize that within the context of this in vivo model, fibrin propagation is dependent on tissue factor derived from blood microparticles. Both vessel wall tissue factor and microparticle tissue factor appear to contribute to thrombus formation. In thrombosis models where there is no vessel wall tissue factor (56), where vessel wall injury causes vessel wall tissue factor to predominate (62), or where there is no blood flow and thus the deposition of microparticles is eliminated (62), the balance between the contribution of vessel wall tissue factor and that of microparticle tissue factor can be altered, giving varying results. Likely, different pathologies associated with thrombosis may also differentially impact on the contributions of tissue factor from the vessel wall and from blood microparticles.

Animal models and their relevance to human disease
The laser-injury thrombosis model has numerous advantages for the study of thrombosis in vivo. This model permits the examination of thrombus formation in a living animal that is not anticoagulated and that has an intact vessel wall, all circulating cellular elements, and all circulating plasma proteins. Second, the precise location of the injury is known — within a micron or two — and the exact time of injury is known — within a second or two. This temporal and spatial resolution is in contrast to the ferric chloride model, where oxidative injury is generalized. Third, the laser injury is a heat injury, which does not induce morphologic changes to the vessel wall if the appropriate energy level is used. However, if excessive energy is used, tissue disruption and hemorrhage are observed.

Nonetheless, all of the models of thrombus formation are just models. For example, laser-induced injury, like mechanical disruption, electrical stimulation, chemical oxidation, or stasis, is nonphysiologic. The microcirculation of the cremaster muscle offers an ideal transparent vascular window for optical microscopy. However, atherothrombosis and peripheral arterial thrombosis are diseases of large arteries that are too thick to study by the current methods. Furthermore, the shear rates and flow dynamics within the microcirculation are different from those in large vessels. Lastly, mice are poor animal models for human atherosclerosis, although thrombosis may be more parallel for comparing with the human system. Insofar as in vitro studies have allowed the construction of an understanding of thrombosis, vascular injury models, albeit not perfect, are yet another step closer to studying thrombosis in the real thing: human arteries.

Summary
The ability to study the biochemistry and cell biology of complex systems in living animals has allowed reexamination of the tenets and models proposed to provide a foundation for the understanding of thrombosis. Although many of the constructs developed from in vitro studies appear reasonable, at least as a first approximation, intravital studies of thrombus formation have established some important features of the thrombotic process (Figure 3). First, it is now clear that platelet activation and platelet thrombus formation are intertwined with thrombin generation and fibrin clot propagation. These pathways are temporally and spatially integrated. Second, platelet accumulation, originally thought to be exclusively dependent on vWF, involves multiple proteins, including vWF, fibrinogen, and possibly fibronectin. Third, the platelet-platelet synapse, classically described as glycoprotein IIb/IIIa interaction with fibrinogen, appears to involve numerous adhesion molecules besides glycoprotein IIb/IIIa. Fourth, blood-borne tissue factor is initially delivered to the developing thrombus in a process dependent on P-selectin and PSGL-1; leukocytes bearing tissue factor either do not have a role or have a role later in thrombus formation. Fifth, laser-induced vessel wall injury activates the tissue factor pathway to thrombin generation. This thrombosis model, which has features similar to those characterized by inflammation, does not involve the subendothelial matrix and specifically does not involve collagen. In contrast, ferric chloride injury leads to collagen exposure in the subendothelial matrix; collagen triggers platelet activation in a mechanism mediated by glycoprotein VI. Sixth, intracellular calcium mobilization is necessary for stable platelet interaction with the thrombus. Seventh, thrombus generation is a highly complex process requiring many components, both structural and regulatory. Although there is some redundancy, elimination of any of these multiple components disrupts thrombus formation. This observation provides opportunities for targets for novel antithrombotics but also indi-

cates how empirical it will be to identify the optimal targets. Lastly, the dogma that activated platelets provide the membrane surface for thrombin generation needs to be questioned and the critical membrane surfaces in vivo determined.

Acknowledgments
We thank Klaus Ley for teaching us his cremaster model for the study of the microcirculation by intravital microscopy. Francis Castellino and Elliot Rosen demonstrated the laser-injury thrombosis model. Shinya Inoué introduced us to high-speed confocal microscopy. We are also indebted to the faculty of the Optical Microscopy course at the Marine Biological Laboratory, who were open-minded about the possibilities of this type of intravital microscopy and who gave us many ideas in the design of the architecture of this system. Nigel Mackman and Shaun Coughlin provided genetically altered mice as well as critical insight and analyses. Finally, past and present members of this laboratory have made considerable contributions to the development of the instrumentation, experimental design, and analysis and to the continued evolution of this form of intravital microscopy.

Address correspondence to: Bruce Furie, Division of Hemostasis and Thrombosis, Beth Israel Deaconess Medical Center, Research East 319, 330 Brookline Avenue, Boston, Massachusetts 02215, USA. Phone: (617) 667-0620; Fax: (617) 975-5505; E-mail: bfurie@bidmc.harvard.edu.

1. Furie, B., and Furie, B.C. 1988. The molecular basis of blood coagulation. *Cell.* **53**:505–518.
2. Goto, S., Ikeda, Y., Saldivar, E., and Ruggeri, Z.M. 1998. Distinct mechanisms of platelet aggregation as a consequence of different shearing flow conditions. *J. Clin. Invest.* **101**:479–486.
3. Ruggeri, Z.M. 2002. Platelets in atherothrombosis. *Nat. Med.* **8**:1227–1234.
4. Goto, S. 2004. Understanding the mechanism of platelet thrombus formation under blood flow conditions and the effect of new antiplatelet agents. *Curr. Vasc. Pharmacol.* **2**:23–32.
5. Andrews, R.K., and Berndt, M.C. 2004. Platelet physiology and thrombosis. *Thromb. Res.* **114**:447–453.
6. Gibbins, J.M. 2004. Platelet adhesion signalling and the regulation of thrombus formation. *J. Cell Sci.* **117**:3415–3425.
7. Furie, B., and Furie, B.C. 2004. Role of platelet P-selectin and microparticle PSGL-1 in thrombus formation. *Trends Mol. Med.* **10**:171–178.
8. Huo, Y., and Ley, K.F. 2004. Role of platelets in the development of atherosclerosis. *Trends Cardiovasc. Med.* **14**:18–22.
9. Robbie, L., and Libby, P. 2001. Inflammation and atherothrombosis. *Ann. N. Y. Acad. Sci.* **947**:167–179; discussion 179–180.
10. Watson, B.D., Dietrich, W.D., Busto, R., Wachtel, M.S., and Ginsberg, M.D. 1985. Induction of reproducible brain infarction by photochemically initiated thrombosis. *Ann. Neurol.* **17**:497–504.
11. Le Menn, R., Bara, L., and Samama, M. 1981. Ultrastructure of a model of thrombogenesis induced by mechanical injury. *J. Submicrosc. Cytol.* **13**:537–549.
12. Carmeliet, P., et al. 1997. Vascular wound healing and neointima formation induced by perivascular electric injury in mice. *Am. J. Pathol.* **150**:761–776.
13. Gitel, S.N., and Wessler, S. 1983. Dose-dependent antithrombotic effect of warfarin in rabbits. *Blood.* **61**:435–438.
14. Kurz, K.D., Main, B.W., and Sandusky, G.E. 1990. Rat model of arterial thrombosis induced by ferric chloride. *Thromb. Res.* **60**:269–280.
15. Farrehi, P.M., Ozaki, C.K., Carmeliet, P., and Fay, W.P. 1998. Regulation of arterial thrombolysis by plasminogen activator inhibitor-1 in mice. *Circulation.* **97**:1002–1008.
16. Denis, C., et al. 1998. A mouse model of severe von Willebrand disease: defects in hemostasis and thrombosis. *Proc. Natl. Acad. Sci. U. S. A.* **95**:9524–9529.
17. Fay, W.P., Parker, A.C., Ansari, M.N., Zheng, X., and Ginsburg, D. 1999. Vitronectin inhibits the thrombotic response to arterial injury in mice. *Blood.* **93**:1825–1830.
18. Rosen, E.D., et al. 2001. Laser-induced noninvasive vascular injury models in mice generate platelet- and coagulation-dependent thrombi. *Am. J. Pathol.* **158**:1613–1622.
19. Atherton, A., and Born, G.V. 1972. Quantitative investigations of the adhesiveness of circulating polymorphonuclear leucocytes to blood vessel walls. *J. Physiol.* **222**:447–474.
20. Schmid-Schonbein, G.W., Usami, S., Skalak, R., and Chien, S. 1980. The interaction of leukocytes and erythrocytes in capillary and postcapillary vessels. *Microvasc. Res.* **19**:45–70.
21. Tangelder, G.J., and Arfors, K.E. 1991. Inhibition of leukocyte rolling in venules by protamine and sulfated polysaccharides. *Blood.* **77**:1565–1571.
22. Sriramarao, P., Languino, L.R., and Altieri, D.C. 1996. Fibrinogen mediates leukocyte-endothelium bridging in vivo at low shear forces. *Blood.* **88**:3416–3423.
23. oude Egbrink, M.G., Tangelder, G.J., Slaaf, D.W., and Reneman, R.S. 1988. Thromboembolic reaction following wall puncture in arterioles and venules of the rabbit mesentery. *Thromb. Haemost.* **59**:23–28.
24. Roesken, F., et al. 1997. A new model for quantitative in vivo microscopic analysis of thrombus formation and vascular recanalisation: the ear of the hairless (hr/hr) mouse. *Thromb. Haemost.* **78**:1408–1414.
25. Rucker, M., Roesken, F., Vollmar, B., and Menger, M.D. 1998. A novel approach for comparative study of periosteum, muscle, subcutis, and skin microcirculation by intravital fluorescence microscopy. *Microvasc. Res.* **56**:30–42.
26. Roesken, F., Vollmar, B., Rucker, M., Seiffge, D., and Menger, M.D. 1998. In vivo analysis of antithrombotic effectiveness of recombinant hirudin on microvascular thrombus formation and recanalization. *J. Vasc. Surg.* **28**:498–505.
27. Larsen, E., et al. 1989. PADGEM protein: a receptor that mediates the interaction of activated platelets with neutrophils and monocytes. *Cell.* **59**:305–312.
28. Sako, D., et al. 1993. Expression cloning of a functional glycoprotein ligand for P-selectin. *Cell.* **75**:1179–1186.
29. Yang, J., et al. 1999. Targeted gene disruption demonstrates that P-selectin glycoprotein ligand 1 (PSGL-1) is required for P-selectin-mediated but not E-selectin-mediated neutrophil rolling and migration. *J. Exp. Med.* **190**:1769–1782.
30. Ley, K., et al. 1995. Sequential contribution of L- and P-selectin to leukocyte rolling in vivo. *J. Exp. Med.* **181**:669–675.
31. Falati, S., Gross, P., Merrill-Skoloff, G., Furie, B.C., and Furie, B. 2002. Real-time in vivo imaging of platelets, tissue factor and fibrin during arterial thrombus formation in the mouse. *Nat. Med.* **8**:1175–1181.
32. Celi, A., et al. 2003. Thrombus formation: direct real time observation and digital analysis of thrombus assembly in a living mouse by confocal and widefield intravital microscopy. *J. Thromb. Haemost.* **1**:60–68.
33. Ni, H., et al. 2000. Persistence of platelet thrombus formation in arterioles of mice lacking both von Willebrand factor and fibrinogen. *J. Clin. Invest.* **106**:385–392.
34. Ni, H., et al. 2003. Plasma fibronectin promotes thrombus growth and stability in injured arterioles. *Proc. Natl. Acad. Sci. U. S. A.* **100**:2415–2419.
35. Kahn, M.L., et al. 1998. A dual thrombin receptor system for platelet activation. *Nature.* **394**:690–694.
36. Vandendries, E., Hamilton, J.R., Coughlin, S.R., Furie, B.C., and Furie, B. 2004. Protease-activator receptor 4 is required for maximal thrombus growth, but not for fibrin generation in thrombi after laser injury [abstract]. *Blood.* **104**(Suppl.):624.
37. Dubois, C., Panicot-Dubois, L., Furie, B., and Furie, B.C. 2004. Importance of GPVI in platelet activation and thrombus formation in vivo [abstract]. *Blood.* **104**(Suppl.):842.
38. Gross, P., Furie, B.C., Merrill-Skoloff, G., Chou, J., and Furie, B. 2005. Leukocyte versus microparticle-mediated tissue factor transfer during arteriolar thrombus development. *J. Leukoc. Biol.* doi:10.1189/jlb.0405193.
39. Mazzucato, M., Pradella, P., Cozzi, M.R., De Marco, L., and Ruggeri, Z.M. 2002. Sequential cytoplasmic calcium signals in a 2-stage platelet activation process induced by the glycoprotein Ibalpha mechanoreceptor. *Blood.* **100**:2793–2800.
40. Mazzucato, M., Cozzi, M.R., Pradella, P., Ruggeri, Z.M., and De Marco, L. 2004. Distinct roles of ADP receptors in von Willebrand factor-mediated platelet signaling and activation under high flow. *Blood.* **104**:3221–3227.
41. Nesbitt, W.S., et al. 2003. Intercellular calcium communication regulates platelet aggregation and thrombus growth. *J. Cell Biol.* **160**:1151–1161.
42. Dubois, C., Panicot-Dubois, L., Furie, B.C., and Furie, B. 2004. Direct real time visualization of platelet calcium signaling in vivo: role of platelet activation and thrombus formation in a living mouse [abstract]. *Blood.* **104**(Suppl.):325.
43. Renne, T., et al. 2005. Defective thrombus formation in mice lacking coagulation factor XII. *J. Exp. Med.* **202**:271–281.
44. Wang, X., et al. 2005. Effects of factor IX or factor XI deficiency on ferric chloride-induced carotid artery occlusion in mice. *J. Thromb. Haemost.* **3**:695–702.
45. Falati, S., et al. 2005. Platelet PECAM-1 inhibits thrombus formation in vivo. *Blood.* doi:10.1182/blood-2005-04-1512.
46. Angellillo-Scherrer, A., et al. 2001. Deficiency or inhibition of Gas6 causes platelet dysfunction and protects mice against thrombosis. *Nat. Med.* **7**:215–221.
47. Andre, P., et al. 2002. CD40L stabilizes arterial thrombi by a beta3 integrin–dependent mechanism. *Nat. Med.* **8**:247–252.
48. Prevost, N., et al. 2005. Eph kinases and ephrins support thrombus growth and stability by regulating integrin outside-in signaling in platelets. *Proc.*

Natl. Acad. Sci. U. S. A. **102**:9820–9825.

49. Arias-Salgado, E.G., et al. 2005. Protein tyrosine phosphatase PTP-1B is an essential positive regulator of platelet thrombus formation and outside-in IIb 3 signaling. *J. Cell Biol.* **170**:837–845.

50. Enjyoji, K., et al. 1999. Targeted disruption of cd39/ATP diphosphohydrolase results in disordered hemostasis and thromboregulation. *Nat. Med.* **5**:1010–1017.

51. Nanda, N., et al. 2005. Platelet aggregation induces platelet aggregate stability via SLAM family receptor signaling. *Blood.* doi:10.1182/blood-2005-01-0333.

52. Palabrica, T., et al. 1992. Leukocyte accumulation promoting fibrin deposition is mediated in vivo by P-selectin on adherent platelets. *Nature.* **359**:848–851.

53. Semeraro, N., et al. 1983. Direct induction of tissue factor synthesis by endotoxin in human macrophages from diverse anatomical sites. *Immunology.* **50**:529–535.

54. Celi, A., et al. 1994. P-selectin induces the expression of tissue factor on monocytes. *Proc. Natl. Acad. Sci. U. S. A.* **91**:8767–8771.

55. Falati, S., et al. 2003. Accumulation of tissue factor into developing thrombi in vivo is dependent upon microparticle P-selectin glycoprotein ligand 1 and platelet P-selectin. *J. Exp. Med.* **197**:1585–1598.

56. Giesen, P.L., et al. 1999. Blood-borne tissue factor: another view of thrombosis. *Proc. Natl. Acad. Sci. U. S. A.* **96**:2311–2315.

57. Moore, K., et al. 1992. Identification of a specific glycoprotein ligand for P-selectin (CD62) on myeloid cells. *J. Cell Biol.* **118**:445–456.

58. Butenas, S., Bouchard, B.A., Brummel-Ziedins, K.E., Parhami-Seren, B., and Mann, K.G. 2005. Tissue factor activity in whole blood. *Blood.* **105**:2764–2770.

59. Satta, N., et al. 1994. Monocyte vesiculation is a possible mechanism for dissemination of membrane-associated procoagulant activities and adhesion molecules after stimulation by lipopolysaccharide. *J. Immunol.* **153**:3245–3255.

60. Chou, J., et al. 2004. Hematopoietic cell-derived microparticle tissue factor contributes to fibrin formation during thrombus propagation. *Blood.* **104**:3190–3197.

61. Parry, G.C., Erlich, J.H., Carmeliet, P., Luther, T., and Mackman, N. 1998. Low levels of tissue factor are compatible with development and hemostasis in mice. *J. Clin. Invest.* **101**:560–569.

62. Day, S.M., et al. 2005. Macrovascular thrombosis is driven by tissue factor derived primarily from the blood vessel wall. *Blood.* **105**:192–198.

Platelet genomics and proteomics
in human health and disease

Iain C. Macaulay,[1] Philippa Carr,[1] Arief Gusnanto,[2] Willem H. Ouwehand,[1,3]
Des Fitzgerald,[4] and Nicholas A. Watkins[1,3]

[1]Department of Haematology, University of Cambridge, Cambridge, United Kingdom. [2]Medical Research Council Biostatistics Unit,
Institute of Public Health, Cambridge, United Kingdom. [3]National Blood Service Cambridge, Cambridge, United Kingdom.
[4]Molecular Medicine Laboratory, Conway Institute of Biomolecular and Biomedical Research, University College Dublin, Dublin, Ireland.

Proteomic and genomic technologies provide powerful tools for characterizing the multitude of events that occur in the anucleate platelet. These technologies are beginning to define the complete platelet transcriptome and proteome as well as the protein-protein interactions critical for platelet function. The integration of these results provides the opportunity to identify those proteins involved in discrete facets of platelet function. Here we summarize the findings of platelet proteome and transcriptome studies and their application to diseases of platelet function.

Introduction

Normal hemostasis balances between pro- and antithrombotic behaviors, with platelets playing a pivotal role. Platelets are involved in maintaining vascular integrity by sensing and responding to endothelial damage. Platelets also have additional roles in wound healing and repair as well as activation of inflammatory and immune responses (1). The central role of the platelet in maintaining hemostasis raises the possibility that small genotypic variations may have dramatic phenotypic effects. Investigation of genetic variants that modulate platelet function will increase understanding of platelet biology and identify risk predictors for diseases involving platelets.

Several diseases resulting from disorders in platelet function are well characterized. In most cases these are rare, monogenic disorders, associated with well-defined phenotypes. Examples include Glanzmann thrombasthenia and Bernard-Soulier syndrome (2). The monogenic nature of these diseases means the underlying molecular defects can be identified via a reductionist approach focused on a single gene or its products. The study of rare allelic variants of these genes has contributed significantly to understanding platelet biology.

It is clear that platelets play a significant role in common diseases — notably in atherothrombosis and coronary artery disease (CAD). Atherothrombosis and CAD are the outcome of a complex interaction between genes and the environment, but it is clear that variations in platelet activity modulate thrombus formation. For example, large-scale clinical trials of oral "antiplatelet" drugs demonstrate that minor variations in platelet function can dramatically increase the risk of myocardial infarction (3). Evidence suggests that single nucleotide polymorphisms are associated with changes in platelet function (4, 5); however, conflicting results have been obtained when these polymorphisms are tested for disease association (6). It is more realistic to consider multiple contributing factors with the subtle effects of numerous genetic polymorphisms

combining to create a range of platelet response within the normal population. To understand how variation at the gene transcription, translation, and protein levels perturbs platelet function from a normal to a prothrombotic phenotype, a more holistic approach is required. Recent technological and methodological advances now make this possible.

Genomics without a genome: the platelet transcriptome

Platelets are produced in the bone marrow from megakaryocytes as cytoplasmic fragments without genomic DNA (7). This renders them incapable of transcription of nuclear material, and platelets were thought to have no synthetic capacity. However, platelets retain a small but functionally significant amount of megakaryocyte-derived RNA as well as the proteins and molecular machinery necessary for translation. Furthermore, platelets can respond to physiological stimuli using biosynthetic processes that are regulated at the level of protein translation (8, 9), demonstrating a functional role for platelet mRNA (10, 11).

Initial studies using platelet-derived mRNA in the 1980s involved the construction of cDNA libraries (12) that identified many transcripts present in platelets. The development of PCR and its application to platelet biology facilitated the characterization of platelet transcripts and was instrumental in defining many monogenic platelet disorders and the human platelet antigens (2, 13, 14). Recently, transcription profiling methods, such as serial analysis of gene expression (SAGE) and microarray technology, have been applied to more fully characterize transcripts in platelets (15–18).

Microarray technology, the fundamentals of which are described in Figure 1 and reviewed elsewhere (19), represents a rapid, semiquantitative system for gene expression profiling. The number of published microarray studies using platelet-derived mRNA is limited, but those that have been performed suggest that between 15% and 32% of genes studied are present in platelets (16–18) (Table 1). Several well-characterized platelet genes encode the most abundant transcripts (e.g., *glycoprotein Ibα, glycoprotein IIb*, and *platelet factor 4*). Numerous transcripts from genes involved in cytoskeletal organization were also identified (16–18). Extrapolating from the published studies, and assuming that the arrays have an unbiased coverage of the approximately 22,000 genes in the human genome (20), it could be estimated that as many as 3,000–6,000 transcripts are present in platelets. This figure is sur-

Nonstandard abbreviations used: 2D-GE, 2-dimensional gel electrophoresis; DRM, detergent-resistant membrane; LC, liquid chromatography; MALDI, matrix-assisted laser-desorption ionization; MS, mass spectrometry; MS/MS, tandem MS; PEAR1, platelet endothelial aggregation receptor 1; PV, polycythemia vera; SAGE, serial analysis of gene expression; TOF, time-of-flight.

Conflict of interest: The authors have declared that no conflict of interest exists.

Citation for this article: *J. Clin. Invest.* **115**:3370–3377 (2005).
doi:10.1172/JCI26885.

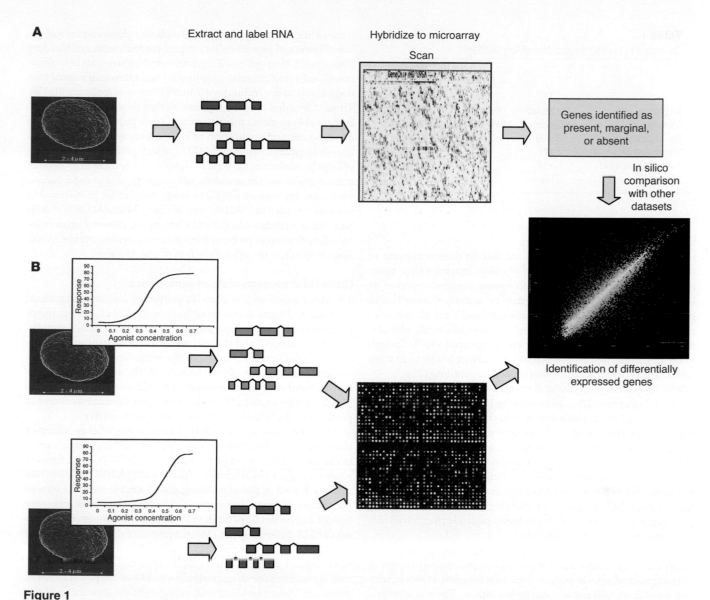

Figure 1

Platelet transcript profiling by microarray. Generally, 1 of 2 approaches can be used to identify expressed transcripts in platelets. (**A**) With single-channel oligonucleotide arrays, mRNA is isolated from platelets, labeled, and hybridized to the microarray. The array is then processed and scanned, and genes are identified as "present," "marginal," or "absent." Comparisons between samples can then be made in silico to identify differentially expressed genes. (**B**) In a 2-channel experiment, RNA from 2 individuals is isolated, each sample is labeled with a different fluorescent dye, and the 2 are compared directly on a microarray. In this case, the individuals have a different dose-response curve for a platelet agonist. Differentially expressed genes can then be directly identified.

prisingly high considering the limited amount of mRNA in platelets. Estimates suggest that each platelet contains just 0.002 fg mRNA (~12,500-fold less than a nucleated cell [ref. 21]). The low level of RNA increases the potential for interference from contaminating cells, emphasizing the need to obtain pure platelets prior to RNA isolation. A limited overlap with the leukocyte transcriptome (20% of the top 50) has been observed, suggesting that 25% of the detected genes are restricted to platelets (17).

The results of microarray analysis of the platelet transcriptome suggest a significant ontological bias toward metabolism, receptors, and signaling activities, categories that are concordant with known platelet function. Genes involved in the immune response were also present, consistent with the role of the plate-

let in inflammation and immunity (18). In addition, a number of highly expressed transcripts previously unknown in the platelet, such as *neurogranin* (a PKC substrate) and *clusterin* (a complement lysis inhibitor), were identified (17). Recently, reduced platelet expression of *clusterin* mRNA was demonstrated in patients with systemic lupus erythematosus (22); however, the significance of this remains to be determined.

A number of unexpected transcripts were identified in the published platelet microarray studies. In each study, erythroid transcripts were observed; this might suggest reticulocyte contamination, but residual expression from precursor cells remains a possibility. A number of histone transcripts were also present; however, the significance of this remains unclear.

Table 1

Summary of platelet transcript profiling studies

	Study		
	1	**2**	**3**
Microarray	HG-U95Av2	Pan 10K	HG-U95Av2
Manufacturer	Affymetrix	MWG Biotech	Affymetrix
No. of genes assayed	12,599	9,850	12,599
No. of positive genes	2,147	1,526	3,978–4,022
Percentage positive	13–17	15.5	31
Reference	(17)	(18)	(16)

SAGE in the platelet

Despite the lack of genomic DNA, the platelet does not appear to be entirely transcriptionally silent. Platelets contain a large number of mitochondria, each of which contains several copies of its own 16-kb circular genome that may be actively transcribed in platelets. The extent of this continued mitochondrial transcription was dramatically demonstrated by Gnatenko et al., who performed SAGE analysis in parallel with microarrays (17). Though lower in throughput than microarray analysis, SAGE is an open and quantitative strategy for transcriptome profiling (23).

When applied to the platelet, 89% of 2,033 SAGE tags were mitochondrial in origin (17). The remaining 11% corresponded to 126 unique genes, half of which were not on the microarray (17). Given this excess of mitochondrial transcripts, more than 300,000 SAGE tags may be required to completely characterize the platelet transcriptome. Subtractive SAGE is an alternative technique that may permit full characterization of the nonmitochondrial transcripts; however, with full genome expression arrays, the development of high-throughput sequencing technologies, and unbiased tiling path expression arrays (24–26), it seems possible that SAGE may be surpassed in future studies (11).

Studies of the platelet transcriptome, as well as that of the megakaryocyte (27, 28), have thus far given a comprehensive overview of the categories of genes important in platelet function. However, transcriptome analysis is not without its limitations. These studies produce long lists of transcripts that are classified as present or provide information on their relative expression levels (16–18) but give no information about localization, interactions, posttranslational modifications, or activation state of gene products. Thus there are several levels of cell function to which transcriptome analysis is oblivious. An integrated approach that identifies novel transcripts and confirms protein expression and functional significance is required.

Proteomics

While the transcriptome of a cell defines those genes that are expressed, it is the encoded proteins that perform the majority of cellular functions. The human genome encodes an estimated 22,000 genes, but the number of functional proteins that can be generated through alternative splicing and posttranslational modifications is estimated to be at least 50 times higher (29). The characterization of a cell's proteome represents a far more complex task than definition of the transcriptome.

Two-dimensional gel electrophoresis (2D-GE) (30), which separates proteins by both size and charge, is the basis of many current proteomic techniques but has been used for many years to study platelet biology (31, 32). Initial studies using reducing and nonreducing gels

resulted in the naming of many of the platelet glycoproteins and the identification of protein defects in patients with inherited bleeding disorders (2). Non–gel-based separation techniques such as multidimensional liquid chromatography (LC) and Multidimensional Protein Identification Technology (MudPIT) are now replacing 2D-GE. These techniques have the advantages that they can be automated and are able to detect membrane and basic proteins (33, 34). The development and application of electrospray and matrix-assisted laser-desorption ionization (MALDI), which permits the ionization of large biomolecules (35), led to significant advances in proteomic science. These ionization techniques can be applied in combination with mass spectrometry (MS) to study the platelet proteome. MS instruments, such as MALDI–time-of-flight–MS (MALDI-TOF-MS) and complex tandem MS (MS/MS) machines, allow the unambiguous identification of proteins from mixtures, permitting the identification of hundreds rather than tens of platelet proteins.

Global and focused platelet proteomics

The description of the platelet proteome has used 2 general approaches (Figure 2) that involve either the global cataloguing of proteins present in resting platelets (16, 36) or the characterization of "subproteomes," and changes within them in response to stimulation (37–43). Initial studies of the cytosolic platelet proteome using 2D-GE over a wide pI range (pI 3–10) identified 186 proteins, a number of which were phosphorylated in resting platelets (44). This represented a 10-fold increase over the number found in previous studies that had not used mass spectrometry.

More recent studies have shifted away from global profiling to the analysis of subfractions of the proteome and the identification of changes induced upon platelet activation (37–43). These focused studies allow the identification of many more platelet proteins than can be achieved by global profiling, giving a more complete view of the platelet proteome. For example, a study of the platelet proteome using a narrow, acidic pI range (pI 4–5) identified the protein products of 123 different genes (36). Extension of the same study into the pI 5–11 range identified an additional 760 protein features representing 311 different genes (38). Despite significant effort, this only represented the characterization of 54% of the pI 5–11 2D-GE proteome. These combined studies highlight the abundance of signaling (24%) and cytoskeletal proteins (15%) present in the platelet (38), mirroring the results obtained with microarrays and also identify several proteins not previously reported in platelets, such as the adapter molecule SH3GL1, an SH3 domain–containing Grb2-like protein, and RSU1, a regulator of G protein signaling. In addition, 15 predicted proteins were identified, demonstrating the utility of this approach in the identification of novel platelet proteins. Most recently, novel non–gel-based protein separation techniques (45, 46) have been applied to the study of the whole platelet proteome (47). This study identified 641 platelet proteins, 404 of which were novel or not previously described in the platelet; thus, this represents the largest catalogue of platelet proteins to date.

The platelet proteome can be broken down further by subcellular prefractionation prior to LC or 2D-GE (Figure 2), allowing the detection of low-abundance proteins that are masked in whole-cell analysis. These studies have focused on cytoskeleton-associated (42) and detergent-resistant membrane–associated (DRM-associated) proteins (48). Following platelet activation, several proteins, including tropomyosin, myosin, and caldesmon, translocate to the cytoskeleton through interactions with F-actin. A study by Gevaert et al. identified additional proteins that translocate to the cytoskeleton

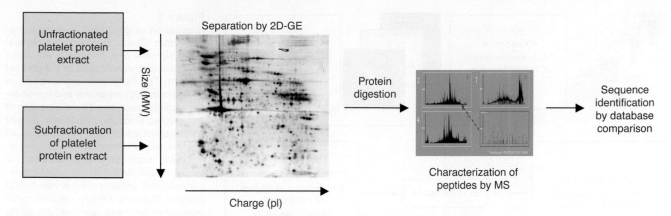

Figure 2
Analyzing the platelet proteome. 2D-GE has been extensively applied to the characterization of the platelet proteome. Published studies have analyzed the platelet proteome as a whole (36, 38, 44) or in subfractions (16, 40, 42). Proteins are separated on the gel by charge (pI) and molecular weight (MW). The protein spots are then digested with a proteolytic enzyme, typically trypsin, and the resultant peptides are characterized by mass spectrometry to generate partial sequence information for the protein. This information can then be used to identify the protein by comparison of the sequence with databases of known protein and nucleic acid sequences.

upon activation, including a number of actin-related proteins, heat shock proteins, and coronin-like protein (42). DRMs are sphingolipid- and cholesterol-rich domains in the plasma membrane (49) where receptors and downstream signaling molecules are highly concentrated. Several receptors on the surface of platelets show enhanced signal transduction when associated with DRMs (50). It is anticipated that the proteomic analysis of DRMs will identify proteins that are recruited to the rafts upon platelet activation (48).

Functional platelet proteomics
The platelet can rapidly respond to a variety of agonists, and the effect of these on the platelet proteome has been studied. Like many other cells, platelets secrete proteins from preformed storage granules in response to stimuli (51). Analysis of the secretome from thrombin-activated platelets has identified over 300 proteins that are secreted upon activation (41). A number of novel proteins, including secretogranin III, a monocyte chemoattractant precursor, were identified that may represent a group of proteins that mediate functions secondary to blood clot formation. In addition, a number of the secreted proteins have been identified in atherosclerotic lesions, suggesting a potential role in atherothrombosis (41).

Posttranslational modifications
Many signaling pathways in platelets are regulated by protein phosphorylation (52), and recent proteomic studies have focused on the identification of proteins that are differentially phosphorylated upon platelet activation (37, 40). Early studies identified phosphorylated proteins in thrombin-activated platelets (43, 44, 53), and enrichment for phosphotyrosine-containing proteins by immunoprecipitation detected 67 proteins that were differentially phosphorylated between resting and thrombin-activated platelets. Through the application of MALDI-TOF, the identity of 10 of these, including FAK and SYK, was confirmed (40). In a separate study that compared the unfractionated platelet proteomes from resting platelets and platelets activated by thrombin receptor–activating peptide (TRAP), 62 differentially regulated protein features were identified (37). Characterization of the phosphorylated targets in TRAP-activated platelets led to the

identification of a number of novel platelet proteins, such as the adapter protein downstream of tyrosine kinase-2 (DOK2), which is phosphorylated downstream of inside-out signaling through $\alpha_{IIb}\beta_3$ and also after glycoprotein VI activation. Additionally, activation-dependent phosphorylation of RGS10 and RGS18, 2 regulators of G protein signaling, was demonstrated (37). These proteins were not thought to undergo any posttranslational phosphorylation.

Tyrosine phosphorylation is just one posttranslational modification that is important for platelet function. Glycosylation, another key modification, has been studied in different cell types (54, 55), and the application of glycoproteomic analysis to the platelet will be of interest.

Other future directions include the application of isotype-coded affinity tags, which allows quantitative differential proteomics (56) that will aid the identification of proteins expressed differentially between phenotypically different platelets.

Integration of transcriptome and proteome data
The integration of proteomic and genomic studies is critical, but integrating data from different platforms is challenging. The correlation of protein and transcript levels is weak in nucleated cells (57) and poorly defined in platelets. McRedmond et al. performed a qualitative correlation of their platelet transcript profile with the proteome by identifying a confirmed set of 82 proteins secreted by thrombin-activated platelets (16). Seventy of these 82 proteins were represented on the microarray, and messages corresponding to 69% were detected. In the same study, a comparison of the transcript profile with those found in previously published platelet proteomics studies showed similar levels of correlation with proteins in the pI range 4–5 (68%) and tyrosine-phosphorylated platelet proteins (69%) (16). Thus, while it appears that transcript and protein abundance may correlate poorly, presence of transcript appears to be associated with presence of protein.

There is a clear discrepancy between the numbers of expressed genes identified by transcriptomics and confirmation of the protein in platelets. Given the diversity of proteins generated by posttranslational modification, coupled with the lack of tran-

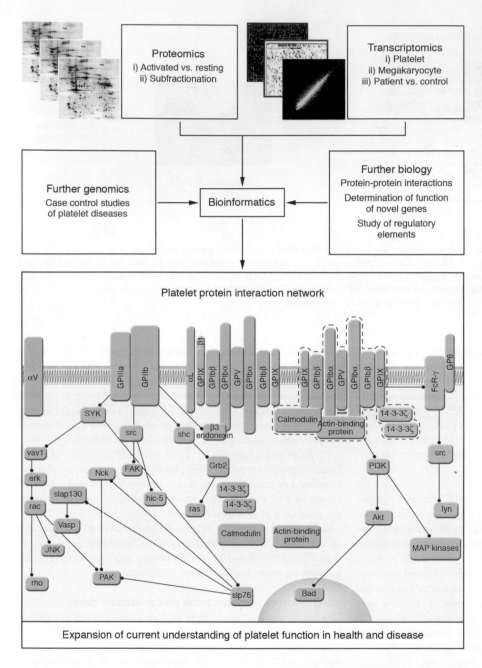

Figure 3
Integration of proteome and transcriptome data. Global studies of the platelet transcriptome and proteome will allow complete characterization of the platelet system. The functional significance of changes in both proteome and transcriptome can be considered by comparison of healthy and diseased individuals, as well as those individuals with extreme platelet phenotypes. Further genomics, involving case-control genotyping studies, will permit determination of the genetic components that underlie normal variation in platelet function. Finally, bioinformatic integration of these approaches, along with the study of protein function using model organisms or in vitro examination of protein interactions, will expand upon current models of platelet function in health and disease. This map was generated using CellDesigner version 2.0 (http://www.systems-biology.org/index.html).

scription in the platelet, one might expect more proteins than transcripts in the platelet. However, current experimental evidence runs contrary to this, with tens or hundreds of proteins being identified in platelet proteomic studies but thousands of transcripts being detected. To a large extent, this can be explained by the limitations of proteomic analysis; however, the application of novel techniques should reduce this discrepancy by increasing the number of identified proteins.

The recent study of Nanda et al. has applied both oligonucleotide microarrays and phosphoproteome analysis to identify and characterize a novel platelet transmembrane receptor (58). Using bioinformatic analysis of the platelet transcript profile, proteins with predicted transmembrane domains were identified. In a parallel proteomic study, phosphoproteins were purified from aggregated platelets by affinity chromatography, separated using 1-dimensional gel electrophoresis, and identified using LC/MS/MS. Platelet endothelial aggregation receptor 1 (PEAR1), which signals secondary to platelet-platelet interactions, was identified using both approaches. This study highlights how proteomics and transcriptomics can be combined to identify novel proteins with a role in platelet function. A combined analysis of published transcriptomic and proteomic datasets could lead to the identification of additional novel proteins present in platelets. However, such an analysis is not trivial, as data can be in different formats and access to raw data can be restricted. The submission of data to public databases at the time of publication will facilitate the integration of datasets and ensure that standardized documentation detailing the experimental approach is available. In the case of microarray data, public databases exist and there are clear guidelines, published by the Microarray Gene Expression Data

Society (www.mged.org), about the minimal data required for submission (59). For proteomic data, the guidelines and databases are currently under development.

The collation of proteomic and transcriptomic data will also allow a more complete characterization of the platelet through the identification of novel regulatory mechanisms and the construction of protein interaction networks (Figure 3). The cataloguing of large numbers of genes and proteins associated with the platelet lineage may also permit in silico study of regulatory elements responsible for megakaryocyte maturation and platelet formation. Software such as TFBScluster (60), which has been used to identify cis-regulatory sequences controlling blood and endothelial development (61), could also be applied to megakaryocyte-restricted genes to identify specific transcription factor signatures. Furthermore, novel bioinformatics approaches that visualize complex biochemical networks will expand current models of hemostasis to include proteins and processes identified in transcriptomic and proteomic studies (62).

Platelet proteomics and genomics in complex disease

Given the availability of whole-genome transcriptome platforms, robust RNA amplification techniques (15), and novel proteomics technologies, platelet biologists are now in a position to begin characterizing the full complement of transcripts and functionally relevant protein fractions in an individual's platelets. This allows a shift in focus away from determining the components of the platelet system to asking fundamental questions about the mechanisms that determine interindividual variation in platelet function. One of the difficulties in analyzing transcriptomic and proteomic data is the sheer volume of data generated from a single experiment. Comparative studies reduce the complexity of results by allowing the researcher to focus in on differentially expressed genes and proteins. Such studies, which are in their infancy, require large populations of individuals with well-defined clinical phenotypes and standardized protocols. The importance of this is highlighted by a comparison of 2 published microarray studies (16, 17), which used identical microarray platforms. While the most abundant transcripts in both experiments appear similar, there is an almost 2-fold difference in the number of transcripts identified as "present" (Table 1) because of differences in sample preparation and analysis parameters. In the study by McRedmond et al. (16), the analysis accounted for the low signal intensities obtained when platelet RNA was analyzed using microarrays.

The application of microarrays to the study of complex disease requires many statistical considerations (reviewed in Chui and Churchill, ref. 63). The development and application of sound statistical methods for the analysis of microarray data is critical to obtaining meaningful results (64, 65). Many microarray-based studies are underpowered and lack sufficient technical and biological replication to infer statistically significant datasets.

Genomics and proteomics in hematological disease

Transcript profiling is being used to investigate a number of platelet disorders, although no published studies have focused directly on the platelet. Comparative transcript profiling of neutrophil RNA from gray platelet syndrome (GPS) sufferers and healthy individuals showed that genes involved in the biosynthesis of cytoskeleton proteins are upregulated in GPS (66). However, the significance of these findings remains to be elucidated. Microarrays have also been applied to study myeloproliferative disorders such

as polycythemia vera (PV) and essential thrombocythemia (ET) (67–70). These acquired disorders are clonal HSC malignancies characterized by hypersensitivity to numerous cytokines. Transcript profiling of megakaryocytes differentiated in vitro from HSCs suggests that megakaryocytes from ET patients are more resistant to apoptosis. In this study (68), proapoptotic genes, such as Bcl-2–associated X protein (BAX) and Bcl-2/adenovirus E1B 19-kDa–interacting protein 3 (BNIP3), were downregulated in patients. Similarly, transcript profiling of samples from PV patients has identified a number of genes that are over- or underexpressed in these patients relative to controls. Transcript profiling of granulocyte RNA identified 253 genes that were upregulated and 391 that were downregulated relative to controls. In a similar study (70), antiapoptotic and survival factors were upregulated in granulocytes from PV patients, suggesting that these factors may promote cell survival. Recently, an acquired mutation in the JH2 pseudokinase domain of the JAK2 gene has been identified in more than 80% of PV patients (71–73). This discovery was made using 3 different but complementary techniques, none of which used transcriptome or proteome technologies. It is highly unlikely that the 1849G→T mutation, encoding a V617F substitution, would have been identified directly by transcript profiling; however, the effect of a constitutively active kinase that is directly linked to cytokine signaling should be detectable by expression profiling. It would be interesting to reconsider the transcript profiling results in light of the identification of the JAK2 mutation. That the studies identified a signature rather than the cause of the disease highlights the fact that the identification of differential expression is not enough. To be truly informative, such studies must consider not only changes in transcript abundance, but also the underlying causes of these changes. It is more likely that complex diseases will be characterized by the integration of transcriptomic, proteomic, and genomic studies.

Conclusions

Advances in technology and methodology of high-throughput experimental approaches, such as genomics, proteomics, and bioinformatics, provide an opportunity to describe the whole platelet system in health and disease. These technologies currently provide a limited catalogue of transcripts and proteins, but the application of additional experimental approaches, including the use of model organisms, may eventually allow the construction of a platelet "molecular interaction map." The ability to construct such maps is already contributing to our understanding of basic biology in model organisms (74–76), and such approaches have recently been applied to mammalian systems (77). This integrated understanding of platelet biology will allow a more complete characterization of the role of platelets in complex diseases and provide unique phenotypes for the discovery of genetic traits. The catalogues of genes that are being developed for the platelet and the megakaryocyte are the first step toward the identification of genetic variants that play a role in determining both interindividual variation in platelet response and disease risk. These insights will be gained from candidate gene and whole-genome genotyping studies of individuals with well-characterized platelet phenotypes.

While the identification of a "molecular signature" of prothrombotic platelet phenotypes, analogous to those identified by the application of proteomic and genomic studies of various cancers (78, 79), may seem fanciful, it would be careless to dismiss such a possibility. It is inevitable that the cost of these technologies will decrease as they become more widely applied. This will allow larger transcriptome and proteome studies to be carried out, which will, in turn, permit a detailed characterization of the variation of platelet gene expression within a large population of individuals, and thus investigation of the association, if any, between platelet phenotype and gene expression.

The development of diagnostic or therapeutic tools based, either directly or indirectly, on the application of proteomic and genomic analysis of platelets is still a long way off. However, these technologies are being applied more and more widely in the field of platelet biology. The extent to which such approaches will affect either the diagnosis or the treatment of platelet-associated diseases remains to be seen; however, it seems their impact on our understanding of basic platelet biology will be profound.

Address correspondence to: Nicholas A. Watkins, Department of Haematology, University of Cambridge, Long Road, Cambridge CB2 2PT, United Kingdom. Phone: 44-1223-548101; Fax: 44-1223-548136; E-mail: naw23@cam.ac.uk.

1. Weyrich, A.S., and Zimmerman, G.A. 2004. Platelets: signaling cells in the immune continuum. *Trends Immunol.* **25**:489–495.
2. Nurden, A.T., and Nurden, P., editors. 2002. *Inherited disorders of platelet function.* Academic Press. London, United Kingdom. 681–700.
3. Shields, D.C., et al. 2002. The contribution of genetic factors to thrombotic and bleeding outcomes in coronary patients randomised to IIb/IIIa antagonists. *Pharmacogenomics J.* **2**:182–190.
4. Hetherington, S.L., et al. 2005. Dimorphism in the P2Y1 ADP receptor gene is associated with increased platelet activation response to ADP. *Arterioscler. Thromb. Vasc. Biol.* **25**:252–257.
5. Joutsi-Korhonen, L., et al. 2003. The low-frequency allele of the platelet collagen signalling receptor glycoprotein VI is associated with reduced functional responses and expression. *Blood.* **101**:4372–4379.
6. Casas, J.P., Hingorani, A.D., Bautista, L.E., and Sharma, P. 2004. Meta-analysis of genetic studies in ischemic stroke: thirty-two genes involving approximately 18,000 cases and 58,000 controls. *Arch. Neurol.* **61**:1652–1661.
7. Italiano, J.E., Jr., and Shivdasani, R.A. 2003. Megakaryocytes and beyond: the birth of platelets. *J. Thromb. Haemost.* **1**:1174–1182.
8. Kieffer, N., Guichard, J., Farcet, J.P., Vainchenker, W., and Breton-Gorius, J. 1987. Biosynthesis of major platelet proteins in human blood platelets. *Eur. J. Biochem.* **164**:189–195.
9. Booyse, F.M., and Rafelson, M.E., Jr. 1968. Studies on human platelets. I. Synthesis of platelet protein in a cell-free system. *Biochim. Biophys. Acta.* **166**:689–697.
10. Weyrich, A.S., et al. 1998. Signal-dependent translation of a regulatory protein, Bcl-3, in activated human platelets. *Proc. Natl. Acad. Sci. U. S. A.* **95**:5556–5561.
11. Denis, M.M., et al. 2005. Escaping the nuclear confines: signal-dependent pre-mRNA splicing in anucleate platelets. *Cell.* **122**:379–391.
12. Wicki, A.N., et al. 1989. Isolation and characterization of human blood platelet mRNA and construction of a cDNA library in lambda gt11. Confirmation of the platelet derivation by identification of GPIb coding mRNA and cloning of a GPIb coding cDNA insert. *Thromb. Haemost.* **61**:448–453.
13. Metcalfe, P., et al. 2003. Nomenclature of human platelet antigens. *Vox Sang.* **85**:240–245.
14. Newman, P.J., et al. 1988. Enzymatic amplification of platelet-specific messenger RNA using the polymerase chain reaction. *J. Clin. Invest.* **82**:739–743.
15. Rox, J.M., et al. 2004. Gene expression analysis in platelets from a single donor: evaluation of a PCR-based amplification technique. *Clin. Chem.* **50**:2271–2278.
16. McRedmond, J.P., et al. 2004. Integration of proteomics and genomics in platelets: a profile of platelet proteins and platelet-specific genes. *Mol. Cell. Proteomics.* **3**:133–144.
17. Gnatenko, D.V., et al. 2003. Transcript profiling of human platelets using microarray and serial analysis of gene expression. *Blood.* **101**:2285–2293.
18. Bugert, P., Dugrillon, A., Gunaydin, A., Eichler, H., and Kluter, H. 2003. Messenger RNA profiling of human platelets by microarray hybridization.

19. Holloway, A.J., van Laar, R.K., Tothill, R.W., and Bowtell, D.D. 2002. Options available—from start to finish—for obtaining data from DNA microarrays II. *Nat. Genet.* **32**(Suppl.):481–489.
20. International Human Genome Sequencing Consortium. 2004. Finishing the euchromatic sequence of the human genome. *Nature.* **431**:931–945.
21. Fink, L., et al. 2003. Characterization of platelet-specific mRNA by real-time PCR after laser-assisted microdissection. *Thromb. Haemost.* **90**:749–756.
22. Wang, L., et al. 2004. Transcriptional down-regulation of the platelet ADP receptor P2Y(12) and clusterin in patients with systemic lupus erythematosus. *J. Thromb. Haemost.* **2**:1436–1442.
23. Velculescu, V.E., Zhang, L., Vogelstein, B., and Kinzler, K.W. 1995. Serial analysis of gene expression. *Science.* **270**:484–487.
24. Reinartz, J., et al. 2002. Massively parallel signature sequencing (MPSS) as a tool for in-depth quantitative gene expression profiling in all organisms. *Brief Funct. Genomic. Proteomic.* **1**:95–104.
25. Brenner, S., et al. 2000. Gene expression analysis by massively parallel signature sequencing (MPSS) on microbead arrays. *Nat. Biotechnol.* **18**:630–634.
26. Bertone, P., et al. 2004. Global identification of human transcribed sequences with genome tiling arrays. *Science.* **306**:2242–2246.
27. Kim, J.A., et al. 2002. Gene expression profile of megakaryocytes from human cord blood CD34(+) cells ex vivo expanded by thrombopoietin. *Stem Cells.* **20**:402–416.
28. Shim, M.H., Hoover, A., Blake, N., Drachman, J.G., and Reems, J.A. 2004. Gene expression profile of primary human CD34+CD38lo cells differentiating along the megakaryocyte lineage. *Exp. Hematol.* **32**:638–648.
29. Hochstrasser, D.F., Sanchez, J.C., and Appel, R.D. 2002. Proteomics and its trends facing nature's complexity. *Proteomics.* **2**:807–812.
30. Gorg, A., et al. 2000. The current state of two-dimensional electrophoresis with immobilized pH gradients. *Electrophoresis.* **21**:1037–1053.
31. Jenkins, C.S., et al. 1976. Platelet membrane glycoproteins implicated in ristocetin-induced aggregation. Studies of the proteins on platelets from patients with Bernard-Soulier syndrome and von Willebrand's disease. *J. Clin. Invest.* **57**:112–124.
32. Gravel, P., et al. 1995. Human blood platelet protein map established by two-dimensional polyacrylamide gel electrophoresis. *Electrophoresis.* **16**:1152–1159.
33. Washburn, M.P., Wolters, D., and Yates, J.R., 3rd. 2001. Large-scale analysis of the yeast proteome by multidimensional protein identification technology. *Nat. Biotechnol.* **19**:242–247.
34. Link, A.J., et al. 1999. Direct analysis of protein complexes using mass spectrometry. *Nat. Biotechnol.* **17**:676–682.
35. Pandey, A., and Mann, M. 2000. Proteomics to study genes and genomes. *Nature.* **405**:837–846.
36. O'Neill, E.E., et al. 2002. Towards complete analysis of the platelet proteome. *Proteomics.* **2**:288–305.
37. Garcia, A., et al. 2004. Differential proteome analysis of TRAP-activated platelets: involvement

Thromb. Haemost. **90**:738–748.

of DOK-2 and phosphorylation of RGS proteins. *Blood.* **103**:2088–2095.
38. Garcia, A., et al. 2004. Extensive analysis of the human platelet proteome by two-dimensional gel electrophoresis and mass spectrometry. *Proteomics.* **4**:656–668.
39. Marcus, K., and Meyer, H.E. 2004. Two-dimensional polyacrylamide gel electrophoresis for platelet proteomics. *Methods Mol. Biol.* **273**:421–434.
40. Maguire, P.B., et al. 2002. Identification of the phosphotyrosine proteome from thrombin activated platelets. *Proteomics.* **2**:642–648.
41. Coppinger, J.A., et al. 2004. Characterization of the proteins released from activated platelets leads to localization of novel platelet proteins in human atherosclerotic lesions. *Blood.* **103**:2096–2104.
42. Gevaert, K., Eggermont, L., Demol, H., and Vandekerckhove, J. 2000. A fast and convenient MALDI-MS based proteomic approach: identification of components scaffolded by the actin cytoskeleton of activated human thrombocytes. *J. Biotechnol.* **78**:259–269.
43. Immler, D., et al. 1998. Identification of phosphorylated proteins from thrombin-activated human platelets isolated by two-dimensional gel electrophoresis by electrospray ionization-tandem mass spectrometry (ESI-MS/MS) and liquid chromatography-electrospray ionization-mass spectrometry (LC-ESI-MS). *Electrophoresis.* **19**:1015–1023.
44. Marcus, K., Immler, D., Sternberger, J., and Meyer, H.E. 2000. Identification of platelet proteins separated by two-dimensional gel electrophoresis and analyzed by matrix assisted laser desorption/ionization-time of flight-mass spectrometry and detection of tyrosine-phosphorylated proteins. *Electrophoresis.* **21**:2622–2636.
45. Gevaert, K., et al. 2003. Exploring proteomes and analyzing protein processing by mass spectrometric identification of sorted N-terminal peptides. *Nat. Biotechnol.* **21**:566–569.
46. Gevaert, K., et al. 2004. Reversible labeling of cysteine-containing peptides allows their specific chromatographic isolation for non-gel proteome studies. *Proteomics.* **4**:897–908.
47. Martens, L., et al. 2005. The human platelet proteome mapped by peptide-centric proteomics: a functional protein profile. *Proteomics.* **5**:3193–3204.
48. Maguire, P.B., Foy, M., and Fitzgerald, D.J. 2005. Using proteomics to identify potential therapeutic targets in platelets. *Biochem. Soc. Trans.* **33**:409–412.
49. Simons, K., and Ikonen, E. 1997. Functional rafts in cell membranes. *Nature.* **387**:569–572.
50. Wonerow, P., et al. 2002. Differential role of glycolipid-enriched membrane domains in glycoprotein VI- and integrin-mediated phospholipase Cgamma2 regulation in platelets. *Biochem. J.* **364**:755–765.
51. Reed, G.L., editor. 2002. *Platelet secretion.* Academic Press. London, United Kingdom. 181–195.
52. Santos, M.T., et al. 2000. Participation of tyrosine phosphorylation in cytoskeletal reorganization, alpha(IIb)beta(3) integrin receptor activation, and aspirin-insensitive mechanisms of thrombin-stimulated human platelets. *Circulation.* **102**:1924–1930.
53. Marcus, K., Moebius, J., and Meyer, H.E. 2003. Differential analysis of phosphorylated proteins in

resting and thrombin-stimulated human platelets. *Anal. Bioanal. Chem.* **376**:973–993.

54. Zhang, H., Li, X.J., Martin, D.B., and Aebersold, R. 2003. Identification and quantification of N-linked glycoproteins using hydrazide chemistry, stable isotope labeling and mass spectrometry. *Nat. Biotechnol.* **21**:660–666.

55. Kaji, H., et al. 2003. Lectin affinity capture, isotope-coded tagging and mass spectrometry to identify N-linked glycoproteins. *Nat. Biotechnol.* **21**:667–672.

56. Gygi, S.P., et al. 1999. Quantitative analysis of complex protein mixtures using isotope-coded affinity tags. *Nat. Biotechnol.* **17**:994–999.

57. Hack, C.J. 2004. Integrated transcriptome and proteome data: the challenges ahead. *Brief Funct. Genomic. Proteomic.* **3**:212–219.

58. Nanda, N., et al. 2005. Platelet-endothelial aggregation receptor 1 (PEAR1), a novel epidermal growth factor repeat-containing transmembrane receptor, participates in platelet contact-induced activation. *J. Biol. Chem.* **280**:24680–24689.

59. Brazma, A., et al. 2001. Minimum information about a microarray experiment (MIAME): toward standards for microarray data. *Nat. Genet.* **29**:365–371.

60. Donaldson, I.J., Chapman, M., and Gottgens, B. 2005. TFBScluster: a resource for the characterization of transcriptional regulatory networks. *Bioinformatics.* **21**:3058–3059.

61. Donaldson, I.J., et al. 2005. Genome-wide identification of cis-regulatory sequences controlling blood and endothelial development. *Hum. Mol. Genet.* **14**:595–601.

62. Joshi-Tope, G., et al. 2005. Reactome: a knowledgebase of biological pathways. *Nucleic Acids Res.* **33**:D428–D432.

63. Cui, X., and Churchill, G.A. 2003. Statistical tests for differential expression in cDNA microarray experiments. *Genome Biol.* **4**:210.

64. Culhane, A.C., Perriere, G., and Higgins, D.G. 2003. Cross-platform comparison and visualisation of gene expression data using co-inertia analysis. *BMC Bioinformatics.* **4**:59.

65. Culhane, A.C., Thioulouse, J., Perriere, G., and Higgins, D.G. 2005. MADE4: an R package for multivariate analysis of gene expression data. *Bioinformatics.* **21**:2789–2790.

66. Hyman, T., et al. 2003. Use of a cDNA microarray to determine molecular mechanisms involved in grey platelet syndrome. *Br. J. Haematol.* **122**:142–149.

67. Goerttler, P.S., et al. 2005. Gene expression profiling in polycythaemia vera: overexpression of transcription factor NF-E2. *Br. J. Haematol.* **129**:138–150.

68. Tenedini, E., et al. 2004. Gene expression profiling of normal and malignant CD34-derived megakaryocytic cells. *Blood.* **104**:3126–3135.

69. Pellagatti, A., et al. 2004. Gene expression profiling in the myelodysplastic syndromes using cDNA microarray technology. *Br. J. Haematol.* **125**:576–583.

70. Pellagatti, A., et al. 2003. Gene expression profiling in polycythemia vera using cDNA microarray technology. *Cancer Res.* **63**:3940–3944.

71. Baxter, E.J., et al. 2005. Acquired mutation of the tyrosine kinase JAK2 in human myeloproliferative disorders. *Lancet.* **365**:1054–1061.

72. James, C., et al. 2005. A unique clonal JAK2 mutation leading to constitutive signalling causes polycythaemia vera. *Nature.* **434**:1144–1148.

73. Kralovics, R., et al. 2005. A gain-of-function mutation of JAK2 in myeloproliferative disorders. *N. Engl. J. Med.* **352**:1779–1790.

74. Tong, A.H., et al. 2004. Global mapping of the yeast genetic interaction network. *Science.* **303**:808–813.

75. Li, S., et al. 2004. A map of the interactome network of the metazoan C. elegans. *Science.* **303**:540–543.

76. Giot, L., et al. 2003. A protein interaction map of Drosophila melanogaster. *Science.* **302**:1727–1736.

77. Barrios-Rodiles, M., et al. 2005. High-throughput mapping of a dynamic signaling network in mammalian cells. *Science.* **307**:1621–1625.

78. van 't Veer, L.J., et al. 2002. Gene expression profiling predicts clinical outcome of breast cancer. *Nature.* **415**:530–536.

79. van de Vijver, M.J., et al. 2002. A gene-expression signature as a predictor of survival in breast cancer. *N. Engl. J. Med.* **347**:1999–2009.

The biogenesis of platelets
from megakaryocyte proplatelets

Sunita R. Patel, John H. Hartwig, and Joseph E. Italiano Jr.

Hematology Division, Department of Medicine, Brigham and Women's Hospital, Boston, Massachusetts, USA.

Platelets are formed and released into the bloodstream by precursor cells called megakaryocytes that reside within the bone marrow. The production of platelets by megakaryocytes requires an intricate series of remodeling events that result in the release of thousands of platelets from a single megakaryocyte. Abnormalities in this process can result in clinically significant disorders. Thrombocytopenia (platelet counts less than 150,000/μl) can lead to inadequate clot formation and increased risk of bleeding, while thrombocythemia (platelet counts greater than 600,000/μl) can heighten the risk for thrombotic events, including stroke, peripheral ischemia, and myocardial infarction. This Review will describe the process of platelet assembly in detail and discuss several disorders that affect platelet production.

Platelet formation

Megakaryocyte development. Megakaryocytes are rare myeloid cells (constituting less than 1% of these cells) that reside primarily in the bone marrow (1) but are also found in the lung and peripheral blood. In early development, before the marrow cavities have enlarged sufficiently to support blood cell development, megakaryopoiesis occurs within the fetal liver and yolk sac. Megakaryocytes arise from pluripotent HSCs that develop into 2 types of precursors, burst-forming cells and colony-forming cells, both of which express the CD34 antigen (2). Development of both cell types continues along an increasingly restricted lineage culminating in the formation of megakaryocyte precursors that develop into megakaryocytes (1). Thrombopoietin (TPO), the primary regulator of thrombopoiesis, is currently the only known cytokine required for megakaryocytes to maintain a constant platelet mass (3). TPO is thought to act in conjunction with other factors, including IL-3, IL-6, and IL-11, although these cytokines are not essential for megakaryocyte maturation (4).

Megakaryocytes tailor their cytoplasm and membrane systems for platelet biogenesis. Before a megakaryocyte has the capacity to release platelets, it enlarges considerably to an approximate diameter of 100 μm and fills with high concentrations of ribosomes that facilitate the production of platelet-specific proteins (5). Cellular enlargement is mediated by multiple rounds of endomitosis, a process that amplifies the DNA by as much as 64-fold (6–9). TPO, which binds to the c-Mpl receptor, promotes megakaryocyte endomitosis. During endomitosis, chromosomes replicate and the nuclear envelope breaks down. Although interconnected mitotic spindles assemble, the normal mitotic cycle is arrested during anaphase B. The spindles fail to separate, and both telophase and cytokinesis are bypassed. Nuclear envelope reformation (10, 11) results in a polyploid, multilobed nucleus with DNA contents ranging from 4N up to 128N within each megakaryocyte (12).

In addition to expansion of DNA, megakaryocytes experience significant maturation as internal membrane systems, granules, and organelles are assembled in bulk during their development. In particular, there is the formation of an expansive and interconnected membranous network of cisternae and tubules, called the demarcation membrane system (DMS), which was originally thought to divide the megakaryocyte cytoplasm into small fields where individual platelets would assemble and subsequently release (13). DMS membranes have continuity with the plasma membrane (14, 15) and are now thought to function primarily as a membrane reservoir for the formation of proplatelets, the precursors of platelets. A dense tubular network (16) and the open canalicular system, a channeled system for granule release, are also formed before the assembly of proplatelets begins. Specific proteins associated with platelets, such as vWF and fibrinogen receptors, are synthesized and sent to the megakaryocyte surface, while others are packaged into secretory granules with such factors as vWF, which is loaded into α-granules (17). Still other proteins, such as fibrinogen, are collected from plasma through endocytosis and/or pinocytosis by megakaryocytes and are selectively placed in platelet-specific granules (17, 18). Also assembled during megakaryocyte maturation are mitochondria and dense granules, which, like α-granules, derive from Golgi complexes. Thus, as terminally differentiated megakaryocytes complete maturation, they are fully equipped with the elements and machinery required for the major task of platelet biogenesis.

The flow model of platelet formation. Despite the identification of platelets over 120 years ago, there is still little consensus on many of the mechanisms involved in platelet biogenesis. However, recent evidence supports a modified flow model of platelet assembly. In this model, platelets are assembled along essential intermediate pseudopodial extensions, called proplatelets, generated by the outflow and evagination of the extensive internal membrane system of the mature megakaryocyte (19). In 1906, Wright introduced the initial concept that platelets arise from megakaryocyte extensions when he described the detachment of platelets from megakaryocyte pseudopods (20). Almost a century later, studies on megakaryocytes producing platelets in vitro have revealed the details of platelet assembly and have led us back to the classical proplatelet theory of platelet release in which platelets fragment from the ends of megakaryocyte extensions (21–23). The discovery and cloning of TPO in 1994 and its receptor, c-Mpl, have allowed major advances in the study of thrombopoiesis (24). TPO has facilitated the development of in vitro megakaryocyte culture systems through which the pro-

Nonstandard abbreviations used: DMS, demarcation membrane system; GP, glycoprotein; PKCα, protein kinase Cα; TPO, thrombopoietin.

Conflict of interest: The authors have declared that no conflict of interest exists.

Citation for this article: *J. Clin. Invest.* **115**:3348–3354 (2005).
doi:10.1172/JCI26891.

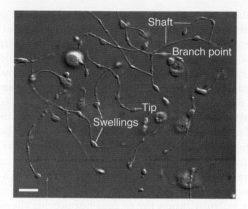

Figure 1
Anatomy of a proplatelet. Differential interference contrast image of proplatelets on a mouse megakaryocyte in vitro. Some of the hallmark features of proplatelets, including the tip, swellings, shafts, and a branch point, are indicated. Scale bar, 5 μm.

cess of platelet formation can be directly visualized and analyzed (25–29). These systems have successfully reconstituted the transition of terminally differentiated megakaryocytes into fully functional platelets. Megakaryocytes cultured in the presence of TPO extend numerous proplatelets, consistent with the flow model. The proplatelets generated in the in vitro systems are structurally similar to those seen in vivo extending into bone marrow sinusoids and within the bloodstream (30–32). Significantly, platelets released from megakaryocytes in vitro are structurally and functionally similar to those found in vivo (28, 29). They are discs, 2–3 μm in diameter, that contain a marginal microtubule band, and change shape in response to platelet agonists, including thrombin.

Transcriptional control of platelet formation. A complete understanding of platelet formation will rely heavily on the identification of cellular controls active at each step of the elaborate process described above. To date, only a few transcription and signaling factors have been implicated in platelet generation, in part because of the rarity of megakaryocytes in bone marrow.

GATA-1 and FOG (friend of GATA) are 2 transcription factors with major roles in thrombopoiesis. GATA-1 acts early in megakaryocyte development, where it is involved in lineage commitment of megakaryocytes, as well as erythrocytes, from their respective progenitor cells (33). GATA-1 also functions later in megakaryocyte development, controlling proliferation. Mutant mice that fail to accumulate GATA-1 within their megakaryocytes exhibit thrombocytopenia and possess an increased number of immature megakaryocytes within their bone marrow. These megakaryocytes exhibit small size, underdeveloped DMSs, reduced platelet-specific granule content, decreased polyploidization, and an excess of rough endoplasmic reticulum (34). In humans, missense mutations in GATA-1 that disrupt its interaction with FOG-1 lead to thrombocytopenia and abnormal bone marrow megakaryocytes. A truncated version of GATA-1, expressed in transient myeloproliferative disorder and acute megakaryoblastic leukemia, is able to interact with FOG-1 but lacks an N-terminal activation domain, resulting in decreased transcriptional activation (35). Thus GATA-1 and FOG-1 appear to play critical roles in megakaryocyte maturation.

The transcription factor NF-E2 has been identified as a major regulator of platelet biosynthesis. NF-E2 is a heterodimer of p45

and p18 subunits that assumes the basic leucine zipper motif (36). NF-E2 null mice experience lethal thrombocytopenia and die from hemorrhage since they lack circulating platelets (37). Megakaryocytes from NF-E2–null mice fail to undergo proplatelet formation, although megakaryocyte maturation appears intact (38). Interestingly, megakaryocytes from NF-E2 knockout mice lack β_1-tubulin, the major β-tubulin isoform expressed in megakaryocytes and platelets (39). NF-E2 has also been shown to interact with the promoter for Rab27b, a small GTPase identified in platelets. Rab27b expression is high in terminally differentiated megakaryocytes, and its inhibition in megakaryocytes results in attenuated proplatelet production, which suggests a role for Rab27b in proplatelet formation (40). NF-E2 may also affect thromboxane synthase (41) and caspase-12 (42), both of which are reduced in NF-E2–null megakaryocytes.

Overview of proplatelet formation. The assembly of platelets from megakaryocytes involves an elaborate dance that converts the cytoplasm into 100- to 500-μm-long branched proplatelets on which the individual platelets develop. The proplatelet and platelet formation process generally commences from a single site on the megakaryocyte where 1 or more broad pseudopodia form. Over a period of 4–10 hours, the pseudopodial processes continue to elongate and become tapered into proplatelets with an average diameter of 2–4 μm. Proplatelets are randomly decorated with multiple bulges or swellings, each similar in size to a platelet, which gives them the appearance of beads connected by thin cytoplasmic strings (Figure 1). The generation of additional proplatelets continues at or near the original site of proplatelet formation and spreads in a wavelike fashion throughout the remainder of the cell until the megakaryocyte cytoplasm is entirely transformed into an extensive and complex network of interconnected proplatelets (27, 36). The multilobed nucleus of the megakaryocyte cell body is compressed into a central mass with little cytoplasm and is eventually extruded and degraded. Platelet-sized swellings also develop at the proplatelet ends and are the primary sites of platelet assembly and release, as opposed to the swellings along the length of the proplatelet shaft (Figure 1). The precise events involved in platelet release from proplatelet ends have not been identified.

Microtubule organization in proplatelets. Microtubules, hollow polymers assembled from αβ-tubulin dimers, are the major structural component of the engine that drives the elongation of proplatelets. It is well known that microtubule bundles assembled by the megakaryocytes fill proplatelet processes (27, 43, 44), as shown in Figure 2. Likewise, when megakaryocytes are retrovirally directed to express GFP-tagged β_1-tubulin, fluorescent microtubules densely fill their proplatelets (45). The microtubules arrays are essential for proplatelet formation (26). Proplatelets fail to form in megakaryocytes treated with agents that inhibit microtubule assembly (1–10 μm nocodazole) (26, 27, 46). Transgenic mice lacking β_1-tubulin, the most abundant platelet β-tubulin isoform, assemble microtubules poorly, develop thrombocytopenia (15–30% reduction in platelet count), and have spherocytic circulating platelets, a consequence of defective marginal band formation (44). The marginal bands of β_1-tubulin–null platelets are composed of a reduced microtubule mass and have only 2–3 coilings instead of the normal 8–12. The importance of β_1-tubulin in platelet-shape maintenance is supported by a recent study that identified the first β_1-tubulin variation in humans, where a double-nucleotide mutation results in the substitution of a highly conserved glutamine with a proline (Q43P) (47). In heterozygous individuals carrying the Q43P mutation,

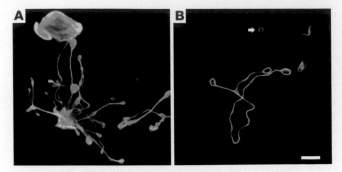

Figure 2
Localization of microtubules within proplatelets. (**A**) Immunofluorescence studies on murine megakaryocytes grown in culture and labeled with β_1-tubulin antibodies indicate that microtubules line the entire length of proplatelets, including shafts and the tip. (**B**) Immunofluorescence studies further show that microtubule coils similar to those seen in mature platelets occur in both proplatelets and released platelet-sized particles (arrow). Scale bar, 5 μm.

β_1-tubulin expression was reduced in platelets, which were enlarged and spherocytic because of defects in the microtubule marginal band. Additionally, the Q43P mutation is thought to occur quite frequently in the normal population (~10%) and impart a protective effect against cardiovascular disease in men (47).

Electron microscopy studies of megakaryocytes undergoing proplatelet formation have provided insights into how microtubule reorganization powers proplatelet growth (27). Just before proplatelet formation, microtubules consolidate in a mass just beneath the cortical plasma membrane. These microtubules align into bundles and fill the cortex of the first blunt process extended by megakaryocytes, signaling the beginning of proplatelet development. The microtubules merge into thick linear bundles that fill the proplatelet shafts when the proplatelets lengthen and taper. At the free proplatelet end, the microtubule bundles form loops, which reenter the proplatelet shaft. This process gives rise to the bulbous tips of the proplatelets, each measuring 3–5 μm in diameter. These studies also were the first to recognize that the platelet-sized swellings that occur along proplatelet shafts are not nascent platelets but are instead points where the microtubule bundles of the shaft diverge for a short distance and then reconvene to locally thicken the proplatelet shaft. Coiling of microtubules, the signature of circulating platelets, occurs only at the proplatelet ends and not within the platelet-sized swelling positioned along the proplatelet shaft. Therefore, the primary site of platelet assembly is at the end of each proplatelet.

Proplatelet elongation. Proplatelets grow from the megakaryocyte cell body at an average rate of 0.85 μm/min, in good agreement with the 4–10 hours required to convert the entire megakaryocyte cytoplasm into proplatelets with average lengths of 250–500 μm (48). Microtubule assembly dynamics within megakaryocytes and proplatelet formation are complex, and their exact relationship to growth, other than supplying microtubule mass, is unclear. In recent studies, EB3, a protein that binds the plus end of microtubules, fused to GFP was expressed in murine megakaryocytes and used as a marker of microtubule plus-end dynamics. Immature megakaryocytes without proplatelets employ a centrosomal-coupled microtubule nucleation/assembly reaction, which appears as a prominent starburst pattern when visualized with EB3-GFP.

Microtubules assemble only from the centrosomes and grow outward to the cell cortex, where they turn and run in parallel with the cell edges (49). However, just before proplatelet production, centrosomal assembly ceases and microtubules begin to collect in the cell cortex. Once proplatelet extension begins, microtubule nucleation and growth occur continuously throughout the entire proplatelet, including the shaft, swellings, and tip. The EB3-GFP studies also revealed that microtubules polymerize in both directions in proplatelets, e.g., toward both the tips and the cell body (49). This demonstrates that the microtubules composing the bundles have a mixed polarity. The rates of microtubule polymerization are approximately 10-fold faster than the proplatelet growth rate.

Although microtubules are continuously polymerizing in proplatelets, polymerization per se does not provide the forces for elongation. Proplatelets continue to elongate at normal rates even when microtubule polymerization is inhibited by drugs that block net microtubule assembly, which suggests another mechanism for proplatelet elongation (49). Consistent with this idea, proplatelets possess an inherent microtubule sliding mechanism, similar to the extension of a fire engine ladder. Dynein, a minus-end microtubule molecular motor protein, localizes along the microtubules of the proplatelet and appears to directly contribute to microtubule sliding, since inhibition of dynein, through disassembly of the dynactin complex, prevents proplatelet formation (49). Microtubule sliding can also be reactivated in detergent-permeabilized proplatelets. ATP, known to support the enzymatic activity of microtubule-based molecular motors, activates proplatelet elongation in permeabilized proplatelets (48) that contain both dynein and its regulatory complex, dynactin. Thus, dynein-facilitated microtubule sliding appears to be the key event in driving proplatelet elongation.

Platelet amplification. Each megakaryocyte has been estimated to generate and release thousands of platelets (50–52). If platelet formation is restricted to a relatively limited number of proplatelet ends, platelets would have to form and release on a minute time scale. (The average megakaryocyte has approximately 5–10 original proplatelets. If 1,000 platelets are constructed, then each end would have to produce 100–200 platelets over a 4-hour time course, equivalent to 25–50 platelets per hour.) Analysis of time-lapsed video microscopy of proplatelet development from megakaryocytes grown in vitro, however, has revealed that ends are amplified in an elaborate process that repeatedly bends and bifurcates the proplatelet shaft to form new ends (27). End amplification initiates when a proplatelet shaft is bent into a sharp kink, which then folds back on itself, forming a loop in the microtubule bundle. The new loop eventually elongates, forming a new proplatelet shaft branching from the side of the original proplatelet. Loops lead the proplatelet tip and define the site where nascent platelets will assemble and where platelet-specific contents are trafficked (Figure 3). In marked contrast to the microtubule-based motor that elongates proplatelets, actin-based force is used to bend the proplatelet in end amplification. Megakaryocytes treated with one of the actin toxins cytochalasin and latrunculin can extend long proplatelets but fail to branch and are decorated with few swellings along their length (27). Despite extensive characterization of actin filament dynamics during platelet activation, how actin participates in this reaction and the cytoplasmic signals that regulate bending have yet to be determined. Immunofluorescence and electron microscopy of megakaryocytes undergoing proplatelet formation indicate that actin filaments are distributed throughout the proplatelet and are particularly abundant within swellings and at proplatelet branch

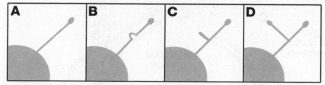

Figure 3
Proplatelet amplification. Megakaryocytes increase their proplatelet number through formation of branched extensions off of existing proplatelets. Initially, the shaft of the parent proplatelet (**A**) is sharply bent (**B**). This bend then folds back on itself to form a loop (**C**). The loop elongates to form a new proplatelet with a novel tip (**D**).

points (27, 53). Studies also indicate that protein kinase Cα (PKCα) associates with aggregated actin filaments in megakaryocytes undergoing proplatelet formation, and inhibition of PKCα or integrin signaling pathways prevents actin filament aggregation and proplatelet formation in megakaryocytes (53). However, the role of actin filament dynamics in platelet biogenesis remains unclear.

Organelle transport in proplatelets. Nascent platelets forming at the proplatelet tips must be loaded with their contents of organelles and platelet-specific granules. This process occurs along the shafts of proplatelets, as organelles and granules travel in a discontinuous fashion from cell body to proplatelet. Despite bidirectional movement along shafts, the particles are eventually captured at the proplatelet tip (54). Immunofluorescence and electron microscopy studies indicate that organelles are in direct contact with microtubules (48), and their movement appears to be independent of actin. Of the 2 major microtubule motors, kinesin and dynein, only the plus end–directed kinesin is situated in a pattern similar to that of organelles and granules and is likely responsible for transporting these elements along microtubules (54). It appears that a 2-fold mechanism of organelle and granule movement occurs in platelet assembly. First, organelles and granules travel along microtubules, and second, the microtubules themselves can slide bidirectionally in relation to other motile filaments to indirectly move organelles along proplatelets in a "piggyback" fashion.

Proplatelet release. In vivo, proplatelets extend into bone marrow vascular sinusoids, where they may be released and enter the bloodstream. The actual events surrounding platelet release in vivo have not been identified because of the rarity of megakaryocytes within the bone marrow. The events leading up to platelet release within cultured murine megakaryocytes have been documented. After complete conversion of the megakaryocyte cytoplasm into a network of proplatelets, a retraction event occurs, which releases individual proplatelets from the proplatelet mass (27). Proplatelets are released as chains of platelet-sized particles, with the most commonly released structure resembling a barbell, a narrow shaft connecting 2 teardrop-shaped tips. Electron microscopic analysis of these barbell structures illustrates linear microtubule bundles within the shafts and a coil of microtubules, reminiscent of the mature platelet marginal band, within the ends. Individual platelet-sized particles are also released from proplatelet-producing megakaryocytes or released proplatelets. Although the actual release event has yet to be captured, the platelet-sized particle is thought be liberated as the proplatelet shaft increasingly narrows. See Figure 4 for an overview of platelet formation.

Apoptosis in platelet biogenesis. The process of platelet assembly in megakaryocytes exhibits some characteristics associated with apoptosis, including cytoskeletal reorganization, membrane con-

densation, and ruffling. These similarities have led to further investigations aimed at determining whether apoptosis is a major force driving proplatelet formation and platelet release. Apoptosis, programmed cell death, is responsible for destruction of the nucleus in senescent megakaryocytes (55). However, it is thought that a specialized apoptotic process may lead to platelet assembly and release. Apoptosis has been described in megakaryocytes (56) and found to be more prominent in mature megakaryocytes as opposed to immature cells (57, 58). A number of apoptotic factors, both proapoptotic and antiapoptotic, have been identified in megakaryocytes (reviewed in ref. 59). Apoptosis-inhibitory proteins such as Bcl-2 and Bcl-x_L are expressed in early megakaryocytes. When overexpressed in megakaryocytes, both factors inhibit proplatelet formation (60, 61). Bcl-2 is absent in mature blood platelets, and Bcl-x_L is absent from senescent megakaryocytes (62), consistent with a role for apoptosis in mature megakaryocytes. Proapoptotic factors, including caspases and NO, are also expressed in megakaryocytes. Evidence indicating a role for caspases in platelet assembly is strong. Caspase activation has been established as a requirement for proplatelet formation. Caspase-3 and caspase-9 are active in mature megakaryocytes, and inhibition of these caspases blocks proplatelet formation (60). NO has been implicated in the release of platelet-sized particles from the megakaryocytic cell line Meg-01 and may work in conjunction with TPO to augment platelet release (63, 64). Other proapoptotic factors expressed in megakaryocytes and thought to be involved in platelet production include TGF-β_1 and SMAD proteins (65). Of interest is the distinct accumulation of apoptotic factors in mature megakaryocytes and mature platelets (66). For instance, caspase-3 and caspase-9 are active in terminally differentiated megakaryocytes. However, only caspase-3 is abundant in platelets (67), while caspase-9 is absent (66). Similarly, caspase-12, found in megakaryocytes, is absent in platelets (42). These data support differential mechanisms for programmed death in platelets and megakaryocytes and suggest the selective delivery and restriction of apoptotic factors to nascent platelets during proplatelet-based platelet assembly.

Disorders of platelet production

A diversity of factors can contribute to anomalous platelet counts; one of these is inappropriate platelet production. Disorders of inappropriate platelet production are grouped into 2 major categories, inherited and noninherited disorders. Inherited platelet disorders occur in individuals harboring genetic mutations within genes that are active during the process of platelet biogenesis. Identification of the precise genetic lesions that give rise to such disorders gives further insight into the mechanisms of platelet formation. These rare disorders, including the 3 major disorders described below, can result in severe thrombocytopenia and increased bleeding times.

Bernard-Soulier syndrome. Bernard-Soulier syndrome is an autosomal dominant disorder characterized by macrothrombocytopenia (thrombocytopenia with increased platelet volume), increased bleeding time, and impaired platelet agglutination (68). The underlying cause is absent or deficient expression of the glycoprotein (GP) Ib/IX/V complex, which forms the vWF receptor on the platelet surface (69). Binding of vWF to the GPIb/IX/V complex is an essential step in hemostasis. A host of mutations within the genes that encode GPIbα, GPIbβ, or GPIX have been identified and linked to Bernard-Soulier syndrome. How these mutations translate into macrothrombocytopenia, however, is still unknown.

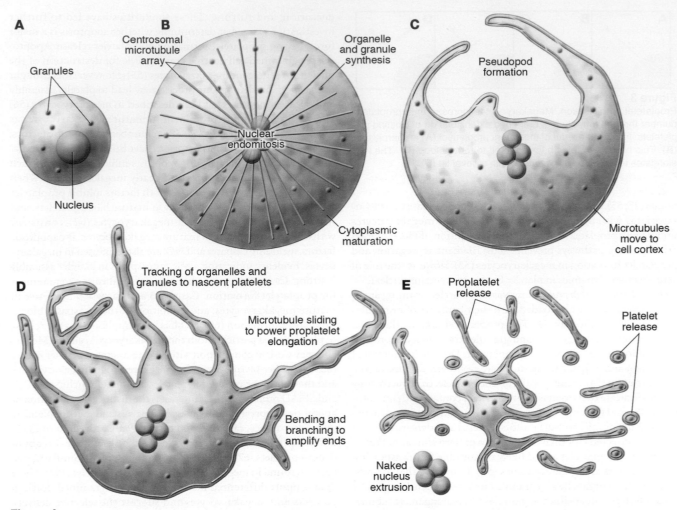

Figure 4
Overview of megakaryocyte production of platelets. As megakaryocytes transition from immature cells (**A**) to released platelets (**E**), a systematic series of events occurs. (**B**) The cells first undergo nuclear endomitosis, organelle synthesis, and dramatic cytoplasmic maturation and expansion, while a microtubule array, emanating from centrosomes, is established. (**C**) Prior to the onset of proplatelet formation, centrosomes disassemble and microtubules translocate to the cell cortex. Proplatelet formation commences with the development of thick pseudopods. (**D**) Sliding of overlapping microtubules drives proplatelet elongation as organelles are tracked into proplatelet ends, where nascent platelets assemble. Proplatelet formation continues to expand throughout the cell while bending and branching amplify existing proplatelet ends. (**E**) The entire megakaryocyte cytoplasm is converted into a mass of proplatelets, which are released from the cell. The nucleus is eventually extruded from the mass of proplatelets, and individual platelets are released from proplatelet ends.

Results from studies on GPIbα knockout mice, a model of Bernard-Soulier syndrome, indicate that the megakaryocyte DMS is disordered (70). These megakaryocytes went on to produce aberrantly large megakaryocyte fragments and proplatelets with reduced internal membrane content (71). Since the GPIb/IX/V complex is linked to the membrane skeleton, it is thought that the genetic defects of Bernard-Soulier syndrome may alter normal cytoskeletal dynamics during platelet formation (69).

MYH9-related disorders. MYH9-related disorders are characterized by macrothrombocytopenia. Patients with these disorders, including May-Hegglin anomaly, Sebastian syndrome, Fechtner syndrome, Alport syndrome, and Epstein syndrome, may also experience hearing loss, cataract, nephritis, and/or granulocyte inclusion body formation (69). The genetic defects in each of these illnesses occur within the *MYH9* gene, which encodes nonmuscle myosin heavy chain IIA, the sole myosin isoform expressed in platelets and

neutrophils. Myosin molecules are hexameric protein complexes composed of 2 heavy chains that dimerize and 2 pairs of myosin light chains. Each heavy chain consists of a globular head containing ATPase activity and actin-binding sites, and an α-helical tail or rod. Although numerous mutations have been identified throughout the *MYH9* gene, mutations are most commonly seen to occur in exons encoding the α-helical tail, a region that promotes myosin dimerization and filament formation. Studies examining the 4 most common myosin rod mutations identified in MYH9-related disorders indicate that the mutations result in diminished myosin dimer and filament formation in vitro (72). Moreover, studies completed on 2 MYH9 mutations that reside within the myosin head domain reveal that the defects dramatically decrease both the MgATPase activity and actin filament translocation of myosin in vitro (73).

The thrombocytopenia that occurs in MYH9-related disorders is thought to be a result of defective platelet production, since both megakaryocyte numbers and platelet clearance are normal (74). Myosin has been theorized to function in platelet formation, specifically during the bending and branching process, where an actomyosin interaction could possibly provide the forces required to contort the proplatelet shaft (27). Also, recent studies on platelets carrying MYH9 defects show that a greater amount of myosin is associated with the actin cytoskeleton in resting platelets and that, upon activation, MYH9-mutated platelets have altered cytoskeletal dynamics (75). Defects in surface expression of the GPIb/IX/V complex in platelets carrying MYH9 mutations have also been reported (76).

Gray platelet syndrome. Gray platelet syndrome is another autosomal dominant disease that presents with macrothrombocytopenia. These large platelets appear gray due to a reduction in α-granule content (77). α-granules normally contain a number of proteins including von Willebrand factor, and fibrinogen. In Gray platelet syndrome, platelets inadequately package these proteins within α-granules (78). As a consequence, a number of clot promoting factors fail to be released upon platelet activation, which increases the risk of bleeding. Although the precise genetic defects responsible for gray platelet syndrome are unknown, evidence indicates that cytoskeletal defects can result in poor α-granule packaging (79).

Conclusions

The transition from megakaryocyte to platelets is a complex process. Although the basic mechanisms of platelet production have been investigated, elucidating the specific molecular controls and cellular events involved in platelet formation and release is an unfinished task. Major issues still to be addressed include (a) determination of which factors induce proplatelet formation in mature megakaryocytes, (b) identification of the mechanism of platelet release from proplatelets, and (c) understanding of how the cytoskeleton drives the events that result in proplatelet production. Further examination of genetic defects that result in platelet disorders, in addition to continued molecular, cellular, and biochemical studies of megakaryocytes as they transition into platelets, will provide a clearer understanding of these processes. This knowledge may aid in the future ex vivo expansion of platelets or in vivo therapies aimed at enhancing platelet production in patients with thrombocytopenia.

Acknowledgments

This work was supported by NIH grant HL68130 (to J.E. Italiano Jr.). J.E. Italiano Jr. is an American Society of Hematology Scholar. S.R. Patel was supported by NIH training grant HL066978-04 and an NIH/National Heart, Lung, and Blood Institute postdoctoral fellowship award (HL082133).

Address correspondence to: Joseph E. Italiano Jr., Brigham and Women's Hospital, Division of Hematology, 1 Blackfan Circle, 6th Floor, Boston, Massachusetts 02115, USA. Phone: (617) 355-9007; Fax: (617) 355-9016; E-mail: jitaliano@rics.bwh.harvard.edu.

1. Ogawa, D. 1993. Differentiation and proliferation of hematopoietic stem cells. *Blood.* **81**:2844–2853.
2. Briddell, R., Brandt, J., Stravena, J., Srour, E., and Hoffman, R. 1989. Characterization of the human burst-forming unit-megakaryocyte. *Blood.* **59**:145–151.
3. Kaushansky, K. 2005. The molecular mechanisms that control thrombopoiesis. *J. Clin. Invest.* **115**:3339–3347. doi:10.1172/JCI26674.
4. Kaushansky, K., and Drachman, J.G. 2002. The molecular and cellular biology of thrombopoietin: the primary regulator of platelet production. *Oncogene.* **21**:3359–3367.
5. Long, M., Williams, N., and Ebbe, S. 1982. Immature megakaryocytes in the mouse: physical characteristics, cell cycle status, and in vitro responsiveness to thrombopoietic stimulatory factor. *Blood.* **59**:569–575.
6. Odell, T.T., Jr., Jackson, C.W., and Reiter, R.S. 1968. Generation cycle of rat megakaryocytes. *Exp. Cell Res.* **53**:321–328.
7. Ebbe, S. 1976. Biology of megakaryocytes. *Prog. Hemost. Thromb.* **3**:211–229.
8. Ebbe, S., and Stohlman, F. 1965. Megakaryocytopoiesis in the rat. *Blood.* **26**:20–34.
9. Ravid, K., Zimmet, J.M., and Jones, M.R. 2002. Roads to polyploidy: the megakaryocyte example. *J. Cell. Physiol.* **190**:7–20.
10. Nagata, N., Muro, Y., and Todokoro, K. 1997. Thrombopoietin-induced polyploidization of bone marrow megakaryocytes is due to a unique regulatory mechanism in late mitosis. *J. Cell Biol.* **139**:449–457.
11. Vitrat, N., et al. 1998. Endomitosis of human megakaryocytes is due to abortive mitosis. *Blood.* **91**:3711–3723.
12. Odell, T., Jackson, C., and Friday, T. 1970. Megakaryocytopoiesis in rats with special reference to polyploidy. *Blood.* **35**:775–782.
13. Yamada, F. 1957. The fine structure of the megakaryocyte in the mouse spleen. *Acta Anat. (Basel).* **29**:267–290.
14. Behnke, O. 1968. An electron microscope study of megakaryocytes of rat bone marrow. I. The development of the demarcation membrane system and the platelet surface coat. *J. Ultrastruct. Res.* **24**:412–433.
15. Nakao, K., and Angrist, A. 1968. Membrane surface specialization of blood platelet and megakaryocyte. *Nature.* **217**:960–961.
16. Gerrard, J., White, J., Rao, G., and Townsend, D. 1976. Localization of platelet prostaglandin production in the platelet dense tubular system. *Am. J. Pathol.* **101**:353–364.
17. Heijnen, H.F., et al. 1998. Multivesicular bodies are an intermediate stage in the formation of platelet alpha granules. *Blood.* **91**:2313–2325.
18. Handagama, P.J., George, M., Shuman, R., McEver, R., and Bainton, D.F. 1987. Incorporation of circulating protein into megakaryocyte and platelet granules. *Proc. Natl. Acad. Sci. U. S. A.* **84**:861–865.
19. Radley, J.M., and Haller, J.C. 1982. The demarcation membrane system of the megakaryocyte: a misnomer? *Blood.* **60**:213–219.
20. Wright, J. 1906. The origin and nature of blood platelets. *Boston Med. Surg. J.* **154**:643–645.
21. Becker, R.P., and DeBruyn, P.P. 1976. The transmural passage of blood cells into myeloid sinusoids and the entry of platelets into the sinusoidal circulation: a scanning electron microscopic investigation. *Am. J. Anat.* **145**:183–205.
22. Thiery, J., and Bessis, M. 1956. Genesis of blood platelets from the megakaryocytes in living cells [In French]. *C. R. Hebd. Seances Acad. Sci.* **242**:290–292.
23. Behnke, O. 1969. An electron microscope study of the rat megakaryocyte. II. Some aspects of platelet release and microtubules. *J. Ultrastruct. Res.* **26**:111–129.
24. Kaushansky, K. 2003. Thrombopoietin: a tool for understanding thrombopoiesis. *J. Thromb. Haemost.* **1**:1578–1592.
25. Leven, R.M., and Nachmias, V.T. 1982. Cultured megakaryocytes: changes in the cytoskeleton after ADP-induced spreading. *J. Cell Biol.* **92**:313–323.
26. Tablin, F., Castro, M., and Leven, R.M. 1990. Blood platelet formation in vitro. The role of the cytoskeleton in megakaryocyte fragmentation. *J. Cell Sci.* **97**:59–70.
27. Italiano, J.E., Lecine, P., Shivdasani, R., and Hartwig, J.H. 1999. Blood platelets are assembled principally at the ends of proplatelet processes produced by differentiated megakaryocytes. *J. Cell Biol.* **147**:1299–1312.
28. Choi, E., Nichol, J.L., Hokom, M.M., Hornkohl, A.C., and Hunt, P. 1995. Platelets generated in vitro from proplatelet-displaying human megakaryocytes are functional. *Blood.* **85**:402–413.
29. Cramer, E., et al. 1997. Ultrastructure of platelet formation by human megakaryocytes cultured with Mpl ligand. *Blood.* **89**:2336–2346.
30. Tavassoli, M., and Aoki, M. 1989. Localization of megakaryocytes in the bone marrow. *Blood Cells.* **15**:3–14.
31. Scurfield, G., and Radley, J.M. 1981. Aspects of platelet formation and release. *Am. J. Hematol.* **10**:285–296.
32. Lichtman, M., Chamberlain, J., Simon, W., and Santillo, P. 1978. Parasinusoidal location of megakaryocytes in marrow: a determinant of platelet release. *Am. J. Hematol.* **4**:303–312.
33. Crispino, J.D. 2005. GATA-1 in normal and malignant hematopoiesis. *Semin. Cell Dev. Biol.* **16**:137–147.
34. Shivdasani, R.A., Fujiwara, Y., McDevitt, M.A., and Orkin, S.H. 1997. A lineage-selective knockout establishes the critical role of transcription factor GATA-1 in megakaryocyte growth and platelet development. *EMBO J.* **16**:3965–3973.
35. Mehaffey, M.G., Newton, A.L., Gandhi, M.J., Crossley, M., and Drachman, J.G. 2001. X-linked thrombocytopenia caused by a novel mutation of GATA-1. *Blood.* **98**:2681–2688.
36. Andrews, N.H., Erdjument-Bromage, H., Davidson, M., Tempest, P., and Orkin, S. 1993. Erythroid transcription factor NF-E2 is a hematopoietic-specific basic-leucine zipper protein. *Nature.* **362**:722–728.
37. Shivdasani, R.A., et al. 1995. Transcription factor NF-E2 is required for platelet formation independent of the actions of thrombopoietin/MGDF in megakaryocytes. *Cell.* **81**:695–704.

38. Lecine, P., et al. 1998. Mice lacking transcription factor NF-E2 provide in vivo validation of the proplatelet model of thrombopoiesis and show a platelet production defect that is intrinsic to megakaryocytes. *Blood.* **92**:1608–1616.

39. Lecine, P., Italiano, J.E., Kim, S., Villeval, J., and Shivdasani, R. 2000. Hematopoietic-specific beta1 tubulin participates in a pathway of platelet biogenesis dependent on the transcription factor NF-E2. *Blood.* **96**:1366–1373.

40. Tiwari, S., et al. 2003. A role for Rab27b in NF-E2-dependent pathways of platelet formation. *Blood.* **102**:3970–3979.

41. Deveaux, S., et al. 1997. p45 NF-E2 regulates expression of thromboxane synthase in megakaryocytes. *EMBO J.* **16**:5654–5661.

42. Kerrigan, S.W., Gaur, M., Murphy, R.P., Shattil, S.J., and Leavitt, A.D. 2004. Caspase 12: a developmental link between G-protein-coupled receptors and integrin alphaIIbeta3 activation. *Blood.* **104**:1327–1334.

43. Topp, K.S., Tablin, F., and Levin, J. 1990. Culture of isolated bovine megakaryocytes on reconstituted basement membrane matrix leads to proplatelet process formation. *Blood.* **76**:912–924.

44. Schwer, H.P., Lecine, P., Tiwari, S., Italiano, J.E., and Hartwig, J. 2001. A lineage-restricted and divergent beta-tubulin isoform is essential for the biogenesis, structure and function of blood platelets. *Curr. Biol.* **11**:579–589.

45. Schulze, H., et al. 2004. Interactions between the megakaryocyte/platelet-specific beta 1 tubulin and the secretory leukocyte protease inhibitor SLPI suggest a role for regulated proteolysis in platelet functions. *Blood.* **104**:3949–3957.

46. Handagama, P.J., Feldman, B.F., Jain, N.C., Farver, T.B., and Kono, C.S. 1987. In vitro platelet release by rat megakaryocytes: effect of metabolic inhibitors and cytoskeletal disrupting agents. *Am. J. Vet. Res.* **48**:1142–1146.

47. Freson, K., et al. 2005. The *TUBB1* Q43P functional polymorphism reduces the risk of cardiovascular disease in men by modulating platelet function and structure. *Blood.* **106**:2356–2362.

48. Italiano, J.E., Shivdasani, R.A., and Hartwig, J.H. 2002. Cytoskeletal mechanics of platelet formation [abstract]. *Blood.* **100**:132a.

49. Patel, S.R., et al. 2005. Differential roles of microtubule assembly and sliding in proplatelet formation by megakaryocytes. *Blood.* doi:10.1182/blood-2005-06-2204.

50. Trowbridge, E., et al. 1984. The origin of platelet count and volume. *Clin. Phys. Physiol. Meas.* **5**:145–156.

51. Harker, L., and Finch, C. 1969. Thrombokinetics in man. *J. Clin. Invest.* **48**:963–974.

52. Kaufman, R., Airo, R., Pollack, S., and Crosby, W. 1965. Circulating megakaryocytes and platelet release in the lung. *Blood.* **26**:720–731.

53. Rojnuckarin, P., and Kaushansky, K. 2001. Actin reorganization and proplatelet formation in murine megakaryocytes: the role of protein kinase C alpha. *Blood.* **97**:154–161.

54. Richardson, J., Shivdasani, R., Boers, C., Hartwig, J., and Italiano, J.E. 2005. Mechanisms of organelle transport and capture along proplatelets during platelet production. *Blood.* doi:10.1182/blood-2005-06-2206.

55. Gordge, M.P. 2005. Megakaryocyte apoptosis: sorting out the signals. *Br. J. Pharmacol.* **145**:271–273.

56. Radley, J.M., and Haller, J.C. 1983. Fate of senescent megakaryocytes in the bone marrow. *Br. J. Haematol.* **53**:277–287.

57. Falcieri, E., et al. 2000. Ultrastructural characterization of maturation, platelet release and senescence of human cultured megakaryocytes. *Anat. Rec.* **258**:90–99.

58. Zauli, G., et al. 1997. In vitro senescence and apoptotic cell death of human megakaryocytes. *Blood.* **90**:2234–2243.

59. Kaluzhny, Y., and Ravid, K. 2004. Role of apoptotic processes in platelet biogenesis. *Acta Haematol.* **111**:67–77.

60. De Botton, S., et al. 2002. Platelet formation is the consequence of caspase activation within megakaryocytes. *Blood.* **100**:1310–1317.

61. Kaluzhny, Y., et al. 2002. BclxL overexpression in megakaryocytes leads to impaired platelet fragmentation. *Blood.* **100**:1670–1678.

62. Sanz, C., et al. 2001. Antiapoptotic protein Bcl-xL is up-regulated during megakaryocytic differentiation of CD34(+) progenitors but is absent from senescent megakaryocytes. *Exp. Hematol.* **29**:728–735.

63. Battinelli, E., and Loscalzo, J. 2000. Nitric oxide induces apoptosis in megakaryocytic cell lines. *Blood.* **95**:3451–3459.

64. Battinelli, E., Willoughby, S.R., Foxall, T., Valeri, C.R., and Loscalzo, J. 2001. Induction of platelet formation from megakaryocytoid cells by nitric oxide. *Proc. Natl. Acad. Sci. U. S. A.* **98**:14458–14463.

65. Kim, J.A., et al. 2002. Gene expression profile of megakaryocytes from human cord blood CD34+ cells ex vivo expanded by thrombopoietin. *Stem Cells.* **20**:402–416.

66. Clarke, M.C., Savill, J., Jones, D.B., Nobel, B.S., and Brown, S.B. 2003. Compartmentalized megakaryocyte death generates function platelets committed to caspase-independent death. *J. Cell Biol.* **160**:577–587.

67. Brown, S.B., Clarke, M.C., Magowan, L., Sanderson, H., and Savill, J. 2000. Constitutive death of platelets leading to scavenger receptor-mediated phagocytosis. A caspase-independent cell clearance program. *J. Biol. Chem.* **275**:5987–5996.

68. Bernard, J., and Soulier, J.P. 1948. Sur une nouvelle variet e de dystrophie thrombocytarie-hemorragipare congenitale. *Sem. Hop. Paris.* **24**:3217–3223.

69. Balduini, C.L., and Savoia, A. 2002. Inherited thrombocytopenias: from genes to therapy. *Haematologica.* **87**:860–880.

70. Ware, J., Russel, S., and Ruggeri, Z.M. 2000. Generation and rescue of a murine model of platelet dysfunction: the Bernard-Soulier syndrome. *Proc. Natl. Acad. Sci. U. S. A.* **97**:2803–2808.

71. Poujol, C., Ware, J., Nieswandt, B., Nurden, A.T., and Nurden, P. 2002. Absence of GPIbalpha is responsible for aberrant membrane development during megakaryocyte maturation: ultrastructural study using a transgenic model. *Exp. Hematol.* **30**:352–360.

72. Franke, J.D., Dong, F., Rickoll, W.L., Kelley, M.J., and Kiehart, D.P. 2004. Rod mutations associated with MYH9-related disorders disrupt non-muscle myosin-IIA assembly. *Blood.* **105**:161–169.

73. Hu, A., Wang, F., and Sellers, J.R. 2002. Mutations in human nonmuscle myosin IIA found in patients with May-Hegglin anomaly and Fechtner syndrome result in impaired enzymatic function. *J. Biol. Chem.* **277**:46512–46517.

74. Hamilton, R.W., et al. 1980. Platelet function, ultrastructure, and survival in the May-Hegglin anomaly. *Am. J. Clin. Pathol.* **74**:663–668.

75. Canobbio, I., Noris, P., Pecci, A., Balduini, C.L., and Torti, M. 2005. Altered cytoskeleton organization in platelets from patients with MYH9-related disease. *J. Thromb. Haemost.* **3**:1026–1035.

76. DiPumpo, M., et al. 2002. Defective expression of GPIb/IX/V complex in platelets from patients with May-Hegglin anomaly and Sebastian syndrome. *Haematologica.* **87**:943–947.

77. Raccuglia, G. 1971. Gray platelet syndrome. *Am. J. Med.* **51**:818–828.

78. Oramer, E.M., Vainchenker, W., Vinci, G., Guichard, J., and Breton-Gorius, J. 1985. Gray platelet syndrome: immunoelectron microscopic localisation of fibrinogen and von Willebrand factor in platelets and megakaryocytes. *Blood.* **66**:1309–1316.

79. Stenberg, P.E., et al. 1998. Prolonged bleeding time with defective platelet filopodia formation in the Wistar Furth rat. *Blood.* **91**:1599–1608.

Platelets in inflammation and atherogenesis

Meinrad Gawaz, Harald Langer, and Andreas E. May

Medizinische Klinik III, Eberhard Karls Universität Tübingen, Tübingen, Germany.

Platelets represent an important linkage between inflammation, thrombosis, and atherogenesis. Inflammation is characterized by interactions among platelets, leukocytes, and ECs. These interactions trigger autocrine and paracrine activation processes that lead to leukocyte recruitment into the vascular wall. Platelet-induced chronic inflammatory processes at the vascular wall result in development of atherosclerotic lesions and atherothrombosis. This Review highlights the molecular machinery and inflammatory pathways used by platelets to initiate and accelerate atherothrombosis.

Platelet interaction with endothelium

At the site of vascular lesions, ECM proteins such as vWF and collagen are exposed to the blood. Platelet adhesion to the exposed matrix is considered to be the initial step in thrombus formation. Platelets adhere to vWF via the membrane adhesion receptor glycoprotein Ib/IX/V (GPIb/IX/V) (1) and to collagen via GPVI (2–4). This results in platelet activation and transformation of the integrin receptors $\alpha_{IIb}\beta_3$ (GPIIb/IIIa, fibrinogen receptor) (5, 6) and $\alpha_2\beta_1$ (collagen receptor) (4, 7), which firmly bind to the respective ECM components. Subsequently, platelets spread and form a surface for the recruitment of additional platelets via fibrinogen bridges between 2 $\alpha_{IIb}\beta_3$ receptors.

In vitro studies with human ECs. In recent years, however, it has become increasingly evident that endothelial denudation is not an absolute prerequisite to allow platelet attachment to the arterial wall. The intact, nonactivated endothelium normally prevents platelet adhesion to the ECM. The adhesion receptors involved in platelet attachment to the subendothelial matrix, e.g., following rupture of an atherosclerotic plaque, have been well defined during the past decade; however, the molecular determinants that promote the interaction between platelets and endothelium are incompletely understood. Whereas endothelium normally controls platelet reactivity through inhibitory and modulating mechanisms involving COX-2, PGI_2, or prostanoid synthetic systems, inflamed ECs develop properties that render them adhesive for platelets. In vitro studies showed that platelets adhere to the intact but activated human EC monolayer (8–10). Platelet adhesion to human umbilical vein ECs (HUVECs) is mediated by a GPIIb/IIIa–dependent bridging mechanism involving platelet-bound fibrinogen, fibronectin, and vWF (10). In HUVECs infected with herpes virus or stimulated with IL-1, platelet adhesion was effectively inhibited by antibodies to vWF or $\alpha_{IIb}\beta_3$ integrin, respectively (9, 11, 12). Furthermore, the involvement of the EC receptors ICAM-1, $\alpha_v\beta_3$ integrin, and GPIb in the binding of activated platelets to HUVECs has been described in vitro (10).

In vivo studies of mouse models. Most in vitro studies have evaluated platelet-endothelium adhesion to human ECs under static conditions with limited attention to the dynamic situation in vivo. Studies using intravital microscopy, however, confirmed that platelet-endothelium adhesion occurs even under high shear stress in vivo (13–19). The results of these studies provide evidence that, similar to interaction with extracellular matrix proteins at the site of vascular lesions, adhesion of platelets to the intact endothelium is coordinated in a multistep process that involves platelet tethering, followed by rolling and subsequent firm adhesion to the vascular wall (Figure 1). These processes involve interactions involving at least 2 types of receptors, selectins and integrins, which induce receptor-specific activation signals in both platelets and the respective adhesive cell type.

The initial loose contact between circulating platelets and vascular endothelium ("platelet rolling") is mediated by selectins, present on both ECs and platelets (20). P-selectin (CD62P) is rapidly expressed on the endothelial surface in response to inflammatory stimuli by translocating from membranes of storage granules (Weibel-Palade bodies) to the plasma membrane within seconds. Endothelial P-selectin has been demonstrated to mediate platelet rolling in both arterioles and venules in acute inflammatory processes (14, 15). E-selectin, which is also expressed on inflamed ECs, allows a loose contact between platelets and endothelium in vivo (15). In line with the concept of endothelial inflammation as a trigger for platelet accumulation, the process of platelet rolling does not require previous platelet activation, since platelets from mice lacking P- and/or E-selectin roll as efficiently as wild-type platelets (17).

So far, few studies have addressed the exact nature of the ligands expressed on platelets that bind to endothelial P-selectin. One candidate that has been identified as a potential counterreceptor for platelet P-selectin is the leucine-rich GPIb/IX/V, also known as the vWF receptor complex. Romo et al. (21) have recently demonstrated that cells expressing P-selectin roll on immobilized GPIbα. Platelets rolling on activated endothelium can be inhibited by antibodies against both P-selectin and GPIbα, indicating that the vWF receptor mediates platelet adhesion to both the subendothelial matrix and the intact endothelium (21). Further, PSGL-1, a glycoprotein that avidly associates with P-selectin, is present on platelets and mediates platelet rolling to the endothelial monolayer under high shear rates (22, 23).

The interaction between P-selectin and PSGL-1 or GPIb/IX/V, however, is rapidly reversible and insufficient for stable adhesion. Rapid conversion to stable adhesion requires additional contacts between the platelet and the endothelium. Integrins are recognized as the major class of surface receptor mediating stable adhesion at high shear in hematopoietic cells (1). Although the role of integrins in mediating firm platelet adhesion to ECM proteins at

Nonstandard abbreviations used: CD40L, CD40 ligand; GP, glycoprotein; HUVEC, human umbilical vein EC; MCP-1, monocyte chemoattractant protein-1; PF4, platelet factor 4; uPAR, urokinase-type plasminogen activator receptor.

Conflict of interest: The authors have declared that no conflict of interest exists.

Citation for this article: *J. Clin. Invest.* **115**:3378–3384 (2005).
doi:10.1172/JCI27196.

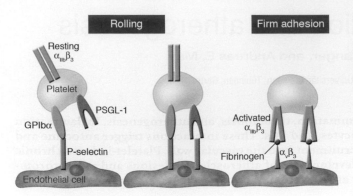

Figure 1
Platelet-endothelium adhesion. Activated endothelium surface expresses P-selectin. Platelet surface receptors GPIbα and PSGL-1 interact with endothelial P-selectin and mediate platelet rolling. Subsequent firm adhesion is mediated through β3 integrins.

the lesion site is well established, their role in platelet-endothelium adhesion in vivo is incompletely understood so far. In vitro, both β3 integrins (αIIbβ3 and αvβ3) have been shown to mediate firm platelet-endothelium adhesion under static conditions (9, 10). In vivo, firm platelet adhesion to the endothelium can be inhibited by anti-αIIbβ3 mAb, and platelets defective in αIIbβ3 do not firmly adhere to activated ECs (19). Thus, although we are just starting to understand the molecular requirements of platelet-endothelium adhesion under dynamic conditions, it seems to be a well-controlled multistep process involving interaction of platelet PSGL-1 or GPIbα with endothelial P-selectin ("rolling") followed by subsequent β3 integrin–mediated firm platelet adhesion. These receptor-dependent platelet-EC interactions allow transcellular communication via soluble mediators and might therefore play an important role in the initiation and progression of vascular inflammation (Figure 2).

Platelet-derived mediators stimulate inflammation
During the adhesion process, platelets become activated and release an arsenal of potent inflammatory and mitogenic substances into the local microenvironment, thereby altering chemotactic, adhesive, and proteolytic properties of ECs (24). These platelet-induced alterations of the endothelial phenotype support chemotaxis, adhesion, and transmigration of monocytes to the site of inflammation (Figure 2).

Released from dense granules, α-granules, lysosomes, the canalicular system, or the cytosol, platelets secrete or expose adhesion proteins (e.g., fibrinogen, fibronectin, vWF, thrombospondin, vitronectin, P-selectin, GPIIb/IIIa), growth factors (e.g., PDGF, TGF-β, EGF, bFGF), chemokines (e.g., RANTES, platelet factor 4 [CXC chemokine ligand 4], epithelial neutrophil-activating protein 78 [CXC chemokine ligand 5]), cytokine-like factors (e.g., IL-1β, CD40 ligand, β-thromboglobulin), and coagulation factors (e.g., factor V, factor XI, PAI-1, plasminogen, protein S). These proteins act in a concerted and finely regulated manner to influence widely differing biological functions such as cell adhesion, cell aggregation, chemotaxis, cell survival and proliferation, coagulation, and proteolysis, all of which accelerate inflammatory processes and cell recruitment. For example, IL-1β has been identified as a major mediator of platelet-induced activation of

ECs (25, 26). The IL-1β activity expressed by platelets appears to be associated with the platelet surface, and coincubation of ECs with thrombin-activated platelets induces IL-1β–dependent secretion of IL-6 and IL-8 from ECs (26). Furthermore, incubation of cultured ECs with thrombin-stimulated platelets significantly enhances the secretion of endothelial monocyte chemoattractant protein-1 (MCP-1) in an IL-1β–dependent manner (12). MCP-1 belongs to the CC family of chemokines and is thought to play a key role in the regulation of monocyte recruitment to inflamed tissue and in atherosclerosis (27, 28).

However, platelet IL-1β does not only modify endothelial release of chemotactic proteins. IL-1β additionally can increase endothelial expression of adhesion molecules. Surface expression of ICAM-1 and αvβ3 on ECs is significantly enhanced by activated platelets via IL-1β (12). Both enhanced chemokine release and upregulation of endothelial adhesion molecules through platelet-derived IL-1β act in concert and promote neutrophil and monocyte adhesion to the endothelium. IL-1β–dependent expression of early inflammatory genes, such as MCP-1 or ICAM-1, involves the activation of the transcription factor NF-κB. Transient adhesion of platelets to the endothelium initiates degradation of IκB and supports activation of NF-κB in ECs, thereby inducing NF-κB–dependent chemokine gene transcription (29, 30). Likewise,

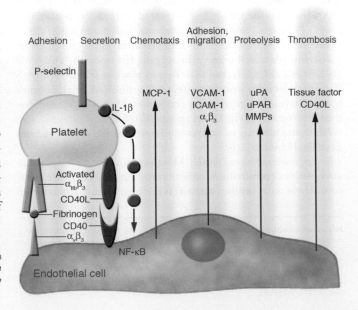

Figure 2
Adherent platelets inflame ECs. Firm platelet adhesion involving αIIbβ3 induces platelet surface exposure of P-selectin (CD62P) and release of CD40L and IL-1β, which stimulate ECs to provide an inflammatory milieu that supports proatherogenic alterations of endothelium.

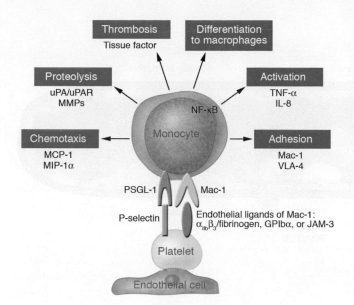

Figure 3
Adherent platelets recruit and inflame monocytes. Adherent and/ or activated platelets mainly interact with monocytic PSGL-1 via P-selectin and with monocytic Mac-1 ($\alpha_M\beta_2$) via $\alpha_{IIb}\beta_3$ (and fibrinogen bridging) or GPIbα. Thereby, platelets initiate monocyte secretion of chemokines, cytokines, and procoagulatory tissue factor, upregulate and activate adhesion receptors and proteases, and induce monocyte differentiation into macrophages. Thus, platelet-monocyte interaction provides an atherogenic milieu at the vascular wall that supports plaque formation.

platelet-induced NF-κB activation was largely reduced by IL-1β antagonists, which supports the notion that platelet IL-1β is the molecular determinant of platelet-dependent activation of the transcription factor. Activation of NF-κB involves a cascade of phosphorylation processes. One family of kinases that is involved in NF-κB–dependent gene expression is the MAPKs, such as p38 MAPK. In a manner similar to that of recombinant human IL-1β, activated platelets have the potential to induce phosphorylation of p38 MAPK. Correspondingly, transfection of a dominant-negative p38 mutant significantly reduced platelet-induced MCP-1 secretion in ECs (31).

Once recruited to the vascular wall, platelets may promote inflammation by chemoattraction of leukocytes through mediators such as platelet-activating factor and macrophage inflammatory protein-1α, may stimulate smooth muscle cell proliferation (TGF-β, PDGF, serotonin) (32), and may contribute to matrix degradation by secretion of MMP-2 (33).

A finely regulated functional interaction of platelets with chemokines has also been implicated in atherogenesis (34). Activated platelets can release chemokines and can induce the secretion of chemokines in various cells of the vascular wall; in turn, certain chemokines can enhance platelet aggregation and adhesion in combination with primary agonists and can trigger monocyte recruitment (35). One such candidate for monocyte recruitment is RANTES, which has been shown to trigger monocyte arrest on inflamed and atherosclerotic endothelium (35). Deposition of platelet-derived RANTES induces monocyte recruitment mediated by P-selectin (36, 37). Another platelet-derived chemokine is platelet factor 4 (PF4), the most abundant protein secreted by activated platelets. First, PF4 acts as a chemoattractant for monocytes promoting their differentiation into macrophages (38). Second, PF4 may directly aggravate the atherogenic actions of hypercholesterolemia by promoting the retention of lipoproteins. Sachais and colleagues have recently shown that PF4 can facilitate the retention of LDL on cell surfaces by inhibition of its degradation by the LDL receptor (39). In addition, PF4 markedly enhances the esterification and uptake of oxidized LDL by macrophages (40). The fact that PF4 has been found in human atherosclerotic lesions and was found associated with macrophages in

early lesions and with foam cells in more advanced lesions (41) supports the concept that PF4 released from locally activated platelets enters the vessel wall and promotes vascular inflammation and atherogenesis.

Furthermore, release of platelet-derived CD40 ligand (CD40L, CD154) induces inflammatory responses in endothelium. Henn et al. (42) showed that platelets store CD40L in high amounts and release it within seconds after activation in vitro. Ligation of CD40 on ECs by CD40L expressed on the surface of activated platelets increased the release of IL-8 and MCP-1, the principal chemoattractants for neutrophils and monocytes (42). In addition, platelet CD40L enhanced the expression of endothelial adhesion receptors including E-selectin, VCAM-1, and ICAM-1, all molecules that mediate the attachment of neutrophils, monocytes, and lymphocytes to the inflamed vessel wall (42). Moreover, CD40L induces endothelial tissue factor expression (43). Hence, like IL-1β, CD40L expressed on platelets induces ECs to release chemokines and to express adhesion molecules, thereby generating signals for the recruitment of leukocytes in the process of inflammation. CD40 ligation on ECs, smooth muscle cells, and macrophages initiates the expression and release of matrix-degrading enzymes, the MMPs. These enzymes, which degrade ECM proteins, significantly contribute to destruction and remodeling of inflamed tissue. Activated platelets release MMP-2 during aggregation (33, 44). Furthermore, adhesion of activated platelets to ECs results in generation and secretion of MMP-9 and of the protease receptor urokinase-type plasminogen activator receptor (uPAR) on cultured endothelium (45). The endothelial release of MMP-9 is dependent on both the fibrinogen receptor GPIIb/IIIa and CD40L, since inhibition of either mechanism resulted in reduction of platelet-induced matrix degradation activity of ECs. Moreover, GPIIb/IIIa ligation results in substantial release of CD40L in the absence of any further platelet agonist (45, 46) (Figure 2). These results suggest that the release of platelet-derived proinflammatory mediators like CD40L is dependent on GPIIb/IIIa–mediated adhesion. This mechanism may be pathophysiologically important to localize platelet-induced inflammation of the endothelium at a site of firm platelet-endothelium adhesion.

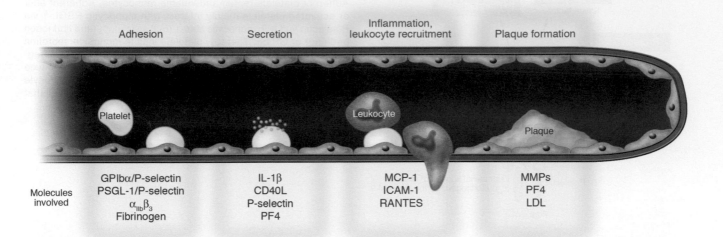

Figure 4
Hypothetical model of atherogenesis triggered by platelets. Activated platelets roll along the endothelial monolayer via GPIbα/P-selectin or PSGL-1/P-selectin. Thereafter, platelets firmly adhere to vascular endothelium via β3 integrins, release proinflammatory compounds (IL-1β, CD40L), and induce a proatherogenic phenotype of ECs (chemotaxis, MCP-1; adhesion, ICAM-1). Subsequently, adherent platelets recruit circulating leukocytes, bind them, and inflame them by receptor interactions and paracrine pathways, thereby initiating leukocyte transmigration and foam cell formation. Thus, platelets provide the inflammatory basis for plaque formation before physically occluding the vessel by thrombosis upon plaque rupture.

Platelets synthesize biologically active proteins

The platelet "secretome" is derived from intracellular storage granules, eicosanoid and phospholipid synthesis (47), and, as recently recognized, synthesis of proteins from constitutive mRNAs (48–50). It is now clear that platelets synthesize biologically relevant proteins in response to physiological stimuli that are regulated via gene expression programs at the translational level (51). A small amount of constitutive protein synthesis (e.g., GPIIb/IIIa, vWF) occurs in nonstimulated platelets (52). However, synthesis of specific proteins is remarkably enhanced in response to activation. It seems that the pattern of protein synthesis is dependent on distinct agonists and requires ligation and engagement of β3 integrins for maximal production of the protein product. For example, platelet activation through platelet-activating factor or thrombin induces rapid and sustained synthesis of pro–IL-1β and processing into active IL-1β (49). The recent integration of proteomics into biochemical and biological platelet research has proved to be a powerful tool in understanding platelet function. Although platelet proteomics is a young field, remarkable advances have already been accomplished. To date, more than 300 proteins released by human platelets after thrombin activation have been identified (53, 54). Coppinger et al. showed that, while absent in normal vasculature, a variety of newly described platelet-derived proteins that may promote atherogenesis (e.g., secretoganin III, calumenin, cyclophilin A) are present in human atherosclerotic tissue (54). Thus, proteomics and analysis of activation-dependent protein synthesis open a new and promising direction of platelet research and may disclose novel molecular mechanisms of platelet-mediated inflammation and atherothrombosis.

Platelet interactions with leukocytes

When activated, platelets coaggregate with circulating leukocytes (55). Once adherent to the vascular wall, platelets also

provide a sticky surface to recruit leukocytes to the vessel wall. Recruitment of circulating leukocytes to the vascular wall requires multistep adhesive and signaling events that result in the infiltration of inflammatory cells into the blood vessel wall, including selectin-mediated attachment and rolling, leukocyte activation, integrin-mediated firm adhesion, and diapedesis. Leukocytes tether to adherent platelets via PSGL-1–P-selectin interactions (56, 57) and, subsequently, firmly adhere via binding of Mac-1 (CD11b/CD18, αMβ2) to GPIbα (58) and/or other receptors of the platelet membrane, including JAM-3 (59) and ICAM-2 (60), or bridging proteins such as fibrinogen (bound to GPIIb/IIIa) (61, 62) or high–molecular weight kininogen (bound to GPIbα) (63) (Figure 3). However, the exact contribution of each system remains to be elucidated in vivo. During this adhesive process, receptor engagement of PSGL-1 and Mac-1 together with platelet-derived inflammatory compounds induces inflammatory cascades in monocytes (64, 65). In addition, engagement of PSGL-1 by P-selectin also drives translationally regulated expression of proteins such as uPAR, a critical surface protease receptor and regulator of integrin-mediated leukocyte adhesion in vivo (66–68) (Figure 3).

Platelets in animal models of atherosclerosis

Abundant recent data support the concept of atherosclerosis as a chronic inflammatory disease (69). However, the contribution of platelets to the process of atherosclerosis was unclear for decades. With the help of intravital microscopy and the availability of appropriate atherosclerotic animal models, it has become evident that platelets adhere to the arterial wall in vivo even in the absence of EC denudation (70, 71). Theilmeier and coworkers found in hypercholesteremic rabbits that platelets adhere to predilection sites of atherosclerosis before lesions are detectable (70). Recently, we showed that platelets adhere to the vascular endothelium of the carotid artery in apoE-deficient mice before the development of manifest atherosclerotic lesions (71). Substantial platelet adhesion to the carotid artery early in atherogenesis involved both platelet GPIbα and α_{IIb}β_3. Platelet adhesion to the carotid wall coincided with inflammatory gene expression and preceded the invasion of leukocytes (71). Prolonged antibody blockade of platelet GPIbα profoundly reduced leukocyte accumulation in the arterial intima and attenuated atherosclerotic lesion formation. Moreover, apoE-deficient mice lacking GPIIb exhibit substantially reduced formation of atherosclerotic lesions (72). Further, circulating activated platelets and platelet–leukocyte/monocyte aggregates promote formation of atherosclerotic lesions (73). The importance of P-selectin for atherosclerotic lesion development has also been described (74–76). The importance of platelets in development of atherosclerosis is also documented by Belton et al., who showed that inhibition of COX-1, an enzyme that is exclusively present in platelets, prevented gross lesion formation in apoE^{-/-} mice (77). Interestingly, effective inhibition of downstream activation cascades can also effectively inhibit atherosclerosis in various models. The body of evidence implicating the CD40-CD40L system in atherogenesis is compelling: Disruption of CD40-CD40L in mouse models of atherosclerosis both downregulated early disease events such as initial plaque formation (78, 79) and could halt the progression of established lesions to more advanced unstable lesions (80). Further, IL-1–dependent mechanisms have been shown to promote atherogenesis in mice in vivo (81, 82).

Together, these in vitro and in vivo data overwhelmingly support the hypothesis that platelets trigger early events of atherogenesis and are critical for atherosclerotic lesion formation in mice (Figure 4). This complex process appears to require adhesive mechanisms mediated, basically, by PSGL-1/P-selectin and β_3 integrins. Adhesion-induced secretion of proatherogenic proteins (e.g., CD40L, IL-1β) and incorporation of platelet-derived products (e.g., PF4) result in inflammation of the vessel wall and subsequent vascular remodeling.

However, the results from mouse models cannot be uncritically transferred to the human situation, since mouse platelets differ substantially from human platelets in several respects (e.g., higher platelet count, different expression profile of surface receptors). What is the evidence that platelets are also linked to atheroprogression in humans?

Platelets and atherosclerosis in humans

The availability of conclusive data obtained in humans is very limited. Nevertheless, there is some evidence that platelets are involved in atheroprogression in humans. An increase in systemic platelet activation has been described for a variety of atherosclerotic diseases, including coronary artery disease (83), transplant vasculopathy (84), and carotid artery disease (85). Recently, it was found that activation of circulating platelets is associated with enhanced wall thickness of the carotid artery in humans (85, 86). Enhanced systemic platelet activation correlates with progression of intima media thickness of the carotid artery in type 2 diabetes (85). Moreover, PF4 (41) and other platelet-derived chemokines and growth factors (54) are found in human atherosclerotic plaques. Previously, we reported that systemic platelet activation is associated with an accelerated progression of transplant vasculopathy (84). Current antiplatelet drugs (aspirin, clopidogrel) do not seem to have a major impact on atheroprogression in humans. However, most antiplatelet strategies in high-risk patients have been applied for secondary prevention at a rather advanced atherosclerotic disease state. Clinical studies are required that evaluate the efficacy of a long-term antiplatelet strategy for primary prevention in high-risk patients at an early stage of atherosclerotic disease.

Future considerations

It has become clear that, besides their role in hemostasis and thrombosis, platelets regulate a variety of inflammatory responses and are key players in atherothrombosis. Thrombosis and inflammation are therefore linked rather than separate entities. Because atherothrombotic diseases are a major cause of morbidity and mortality in developed countries, understanding the role of platelets in vascular inflammation and atherosclerosis is an important challenge. Major achievements have been made in elucidating the molecular mechanisms of platelet interaction with ECs of the arterial wall. Defining the specific requirements for platelets adhering to endothelium may lead to the development of novel therapeutic strategies. The era of genomics and proteomics has recently been introduced in platelet research and will continue to offer major tools to help understand platelet pathology in the course of atherosclerosis.

Address correspondence to: Meinrad Gawaz, Medizinische Klinik III, Eberhard Karls Universität Tübingen, Otfried-Müller-Straße 10, D-72076 Tübingen, Germany. Phone: 49-7071-29-83688; Fax: 49-7071-29-5749; E-mail: meinrad.gawaz@med.uni-tuebingen.de.

1. Ruggeri, Z.M. 2002. Platelets in atherothrombosis. *Nat. Med.* **8**:1227–1234.
2. Massberg, S., et al. 2003. A crucial role of glycoprotein VI for platelet recruitment to the injured arterial wall in vivo. *J. Exp. Med.* **197**:41–49.
3. Nieswandt, B., et al. 2001. Glycoprotein VI but not alpha2beta1 integrin is essential for platelet interaction with collagen. *EMBO J.* **20**:2120–2130.
4. Nieswandt, B., and Watson, S.P. 2003. Platelet-collagen interaction: is GPVI the central receptor? *Blood.* **102**:449–461.
5. Arya, M., et al. 2003. Glycoprotein Ib-IX-mediated activation of integrin alpha(IIb)beta(3): effects of receptor clustering and von Willebrand factor adhesion. *J. Thromb. Haemost.* **1**:1150–1157.
6. Kasirer-Friede, A., et al. 2002. Lateral clustering of platelet GP Ib-IX complexes leads to up-regulation of the adhesive function of integrin alpha IIbbeta 3. *J. Biol. Chem.* **277**:11949–11956.
7. Kahn, M.L. 2004. Platelet-collagen responses:

molecular basis and therapeutic promise. *Semin. Thromb. Hemost.* **30**:419–425.
8. Gawaz, M., Neumann, F.J., Ott, I., Schiessler, A., and Schomig, A. 1996. Platelet function in acute myocardial infarction treated with direct angioplasty. *Circulation.* **93**:229–237.
9. Gawaz, M., et al. 1997. Vitronectin receptor (alpha(v)beta3) mediates platelet adhesion to the luminal aspect of endothelial cells: implications for reperfusion in acute myocardial infarction. *Circulation.* **96**:1809–1818.
10. Bombeli, T., Schwartz, B.R., and Harlan, J.M. 1998. Adhesion of activated platelets to endothelial cells: evidence for a GPIIbIIIa-dependent bridging mechanism and novel roles for endothelial intercellular adhesion molecule 1 (ICAM-1), alphavbeta3 integrin, and GPIbalpha. *J. Exp. Med.* **187**:329–339.
11. Etingin, O.R., Silverstein, R.L., and Hajjar, D.P. 1993. von Willebrand factor mediates platelet

adhesion to virally infected endothelial cells. *Proc. Natl. Acad. Sci. U. S. A.* **90**:5153–5156.
12. Gawaz, M., et al. 2000. Platelets induce alterations of chemotactic and adhesive properties of endothelial cells mediated through an interleukin-1-dependent mechanism. Implications for atherogenesis. *Atherosclerosis.* **148**:75–85.
13. Johnson, R.C., et al. 1995. Blood cell dynamics in P-selectin-deficient mice. *Blood.* **86**:1106–1114.
14. Frenette, P.S., Johnson, R.C., Hynes, R.O., and Wagner, D.D. 1995. Platelets roll on stimulated endothelium in vivo: an interaction mediated by endothelial P-selectin. *Proc. Natl. Acad. Sci. U. S. A.* **92**:7450–7454.
15. Frenette, P.S., et al. 1998. Platelet-endothelial interactions in inflamed mesenteric venules. *Blood.* **91**:1318–1324.
16. Massberg, S., et al. 2004. Enhanced in vivo platelet adhesion in vasodilator-stimulated phosphoprotein (VASP)-deficient mice. *Blood.* **103**:136–142.

17. Massberg, S., et al. 1998. Platelet-endothelial cell interactions during ischemia/reperfusion: the role of P-selectin. *Blood.* **92**:507–515.

18. Massberg, S., et al. 1999. Increased adhesion and aggregation of platelets lacking cyclic guanosine 3′,5′-monophosphate kinase I. *J. Exp. Med.* **189**:1255–1264.

19. Massberg, S., et al. 1999. Fibrinogen deposition at the postischemic vessel wall promotes platelet adhesion during ischemia-reperfusion in vivo. *Blood.* **94**:3829–3838.

20. Subramaniam, M., et al. 1996. Defects in hemostasis in P-selectin-deficient mice. *Blood.* **87**:1238–1242.

21. Romo, G.M., et al. 1999. The glycoprotein Ib-IX-V complex is a platelet counterreceptor for P-selectin. *J. Exp. Med.* **190**:803–814.

22. Frenette, P.S., et al. 2000. P-Selectin glycoprotein ligand 1 (PSGL-1) is expressed on platelets and can mediate platelet-endothelial interactions in vivo. *J. Exp. Med.* **191**:1413–1422.

23. Laszik, Z., et al. 1996. P-selectin glycoprotein ligand-1 is broadly expressed in cells of myeloid, lymphoid, and dendritic lineage and in some nonhematopoietic cells. *Blood.* **88**:3010–3021.

24. Gawaz, M. 2004. Role of platelets in coronary thrombosis and reperfusion of ischemic myocardium. *Cardiovasc. Res.* **61**:498–511.

25. Hawrylowicz, C.M., Howells, G.L., and Feldmann, M. 1991. Platelet-derived interleukin 1 induces human endothelial adhesion molecule expression and cytokine production. *J. Exp. Med.* **174**:785–790.

26. Kaplanski, G., et al. 1994. Interleukin-1 induces interleukin-8 secretion from endothelial cells by a juxtacrine mechanism. *Blood.* **84**:4242–4248.

27. Boring, L., Gosling, J., Cleary, M., and Charo, I.F. 1998. Decreased lesion formation in CCR2–/– mice reveals a role for chemokines in the initiation of atherosclerosis. *Nature.* **394**:894–897.

28. Lu, B., et al. 1998. Abnormalities in monocyte recruitment and cytokine expression in monocyte chemoattractant protein 1-deficient mice. *J. Exp. Med.* **187**:601–608.

29. Gawaz, M., et al. 1998. Activated platelets induce monocyte chemotactic protein 1 secretion and surface expression of intercellular adhesion molecule-1 on endothelial cells. *Circulation.* **98**:1164–1171.

30. Gawaz, M., et al. 2002. Transient platelet interaction induces MCP-1 production by endothelial cells via I kappa B kinase complex activation. *Thromb. Haemost.* **88**:307–314.

31. Dickfeld, T., et al. 2001. Transient interaction of activated platelets with endothelial cells induces expression of monocyte-chemoattractant protein-1 via a p38 mitogen-activated protein kinase mediated pathway. Implications for atherogenesis. *Cardiovasc. Res.* **49**:189–199.

32. Ross, R., Bowen-Pope, D.F., and Raines, E.W. 1985. Platelets, macrophages, endothelium, and growth factors. Their effects upon cells and their possible roles in atherogenesis. *Ann. N. Y. Acad. Sci.* **454**:254–260.

33. Sawicki, G., Salas, E., Murat, J., Miszta-Lane, H., and Radomski, M.W. 1997. Release of gelatinase A during platelet activation mediates aggregation. *Nature.* **386**:616–619.

34. Weber, C. 2005. Platelets and chemokines in atherosclerosis: partners in crime. *Circ. Res.* **96**:612–616.

35. von Hundelshausen, P., et al. 2001. RANTES deposition by platelets triggers monocyte arrest on inflamed and atherosclerotic endothelium. *Circulation.* **103**:1772–1777.

36. Schober, A., et al. 2002. Deposition of platelet RANTES triggering monocyte recruitment requires P-selectin and is involved in neointima formation after arterial injury. *Circulation.* **106**:1523–1529.

37. von Hundelshausen, P., et al. 2005. Heterophilic interactions of platelet factor 4 and RANTES promote monocyte arrest on endothelium. *Blood.* **105**:924–930.

38. Scheuerer, B., et al. 2000. The CXC-chemokine platelet factor 4 promotes monocyte survival and induces monocyte differentiation into macrophages. *Blood.* **95**:1158–1166.

39. Sachais, B.S., et al. 2002. Platelet factor 4 binds to low-density lipoprotein receptors and disrupts the endocytic machinery, resulting in retention of low-density lipoprotein on the cell surface. *Blood.* **99**:3613–3622.

40. Nassar, T., et al. 2003. Platelet factor 4 enhances the binding of oxidized low-density lipoprotein to vascular wall cells. *J. Biol. Chem.* **278**:6187–6193.

41. Pitsilos, S., et al. 2003. Platelet factor 4 localization in carotid atherosclerotic plaques: correlation with clinical parameters. *Thromb. Haemost.* **90**:1112–1120.

42. Henn, V., et al. 1998. CD40 ligand on activated platelets triggers an inflammatory reaction of endothelial cells. *Nature.* **391**:591–594.

43. Slupsky, J.R., et al. 1998. Activated platelets induce tissue factor expression on human umbilical vein endothelial cells by ligation of CD40. *Thromb. Haemost.* **80**:1008–1014.

44. Fernandez-Patron, C., et al. 1999. Differential regulation of platelet aggregation by matrix metalloproteinases-9 and -2. *Thromb. Haemost.* **82**:1730–1735.

45. May, A.E., et al. 2002. Engagement of glycoprotein IIb/IIIa (alpha(IIb)beta3) on platelets upregulates CD40L and triggers CD40L-dependent matrix degradation by endothelial cells. *Circulation.* **106**:2111–2117.

46. Nannizzi-Alaimo, L., Alves, V.L., and Phillips, D.R. 2003. Inhibitory effects of glycoprotein IIb/IIIa antagonists and aspirin on the release of soluble CD40 ligand during platelet stimulation. *Circulation.* **107**:1123–1128.

47. Reed, G.L. 2004. Platelet secretory mechanisms. *Semin. Thromb. Hemost.* **30**:441–450.

48. Lindemann, S., et al. 2001. Integrins regulate the intracellular distribution of eukaryotic initiation factor 4E in platelets. A checkpoint for translational control. *J. Biol. Chem.* **276**:33947–33951.

49. Lindemann, S., et al. 2001. Activated platelets mediate inflammatory signaling by regulated interleukin 1beta synthesis. *J. Cell Biol.* **154**:485–490.

50. Lindemann, S.W., Weyrich, A.S., and Zimmerman, G.A. 2005. Signaling to translational control pathways: diversity in gene regulation in inflammatory and vascular cells. *Trends Cardiovasc. Med.* **15**:9–17.

51. Weyrich, A.S., et al. 2004. Change in protein phenotype without a nucleus: translational control in platelets. *Semin. Thromb. Hemost.* **30**:491–498.

52. Kieffer, N., Guichard, J., Farcet, J.P., Vainchenker, W., and Breton-Gorius, J. 1987. Biosynthesis of major platelet proteins in human blood platelets. *Eur. J. Biochem.* **164**:189–195.

53. McRedmond, J.P., et al. 2004. Integration of proteomics and genomics in platelets: a profile of platelet proteins and platelet-specific genes. *Mol. Cell. Proteomics.* **3**:133–144.

54. Coppinger, J.A., et al. 2004. Characterization of the proteins released from activated platelets leads to localization of novel platelet proteins in human atherosclerotic lesions. *Blood.* **103**:2096–2104.

55. McEver, R.P. 2001. Adhesive interactions of leukocytes, platelets, and the vessel wall during hemostasis and inflammation. *Thromb. Haemost.* **86**:746–756.

56. Evangelista, V., et al. 1999. Platelet/polymorphonuclear leukocyte interaction: P-selectin triggers protein-tyrosine phosphorylation-dependent CD11b/CD18 adhesion: role of PSGL-1 as a signaling molecule. *Blood.* **93**:876–885.

57. Yang, J., Furie, B.C., and Furie, B. 1999. The biol-

ogy of P-selectin glycoprotein ligand-1: its role as a selectin counterreceptor in leukocyte-endothelial and leukocyte-platelet interaction. *Thromb. Haemost.* **81**:1–7.

58. Simon, D.I., et al. 2000. Platelet glycoprotein ib-alpha is a counterreceptor for the leukocyte integrin Mac-1 (CD11b/CD18). *J. Exp. Med.* **192**:193–204.

59. Santoso, S., et al. 2002. The junctional adhesion molecule 3 (JAM-3) on human platelets is a counterreceptor for the leukocyte integrin Mac-1. *J. Exp. Med.* **196**:679–691.

60. Diacovo, T.G., deFougerolles, A.R., Bainton, D.F., and Springer, T.A. 1994. A functional integrin ligand on the surface of platelets: intercellular adhesion molecule-2. *J. Clin. Invest.* **94**:1243–1251.

61. Wright, S.D., et al. 1988. Complement receptor type three (CD11b/CD18) of human polymorphonuclear leukocytes recognizes fibrinogen. *Proc. Natl. Acad. Sci. U. S. A.* **85**:7734–7738.

62. Altieri, D.C., Bader, R., Mannucci, P.M., and Edgington, T.S. 1988. Oligospecificity of the cellular adhesion receptor Mac-1 encompasses an inducible recognition specificity for fibrinogen. *J. Cell Biol.* **107**:1893–1900.

63. Chavakis, T., et al. 2003. High molecular weight kininogen regulates platelet-leukocyte interactions by bridging Mac-1 and glycoprotein Ib. *J. Biol. Chem.* **278**:45375–45381.

64. Weyrich, A.S., et al. 1996. Activated platelets signal chemokine synthesis by human monocytes. *J. Clin. Invest.* **97**:1525–1534.

65. Neumann, F.J., et al. 1997. Induction of cytokine expression in leukocytes by binding of thrombin-stimulated platelets. *Circulation.* **95**:2387–2394.

66. May, A.E., et al. 1998. Urokinase receptor (CD87) regulates leukocyte recruitment via beta 2 integrins in vivo. *J. Exp. Med.* **188**:1029–1037.

67. Preissner, K.T., Kanse, S.M., and May, A.E. 2000. Urokinase receptor: a molecular organizer in cellular communication. *Curr. Opin. Cell Biol.* **12**:621–628.

68. May, A.E., et al. 2002. Urokinase receptor surface expression regulates monocyte adhesion in acute myocardial infarction. *Blood.* **100**:3611–3617.

69. Lusis, A.J. 2000. Atherosclerosis. *Nature.* **407**:233–241.

70. Theilmeier, G., et al. 2002. Endothelial von Willebrand factor recruits platelets to atherosclerosis-prone sites in response to hypercholesterolemia. *Blood.* **99**:4486–4493.

71. Massberg, S., et al. 2002. A critical role of platelet adhesion in the initiation of atherosclerotic lesion formation. *J. Exp. Med.* **196**:887–896.

72. Massberg, S., et al. 2005. Platelet adhesion via glycoprotein IIb integrin is critical for atheroprogression and focal cerebral ischemia: an in vivo study in mice lacking glycoprotein IIb. *Circulation.* **112**:1180–1188.

73. Huo, Y., et al. 2003. Circulating activated platelets exacerbate atherosclerosis in mice deficient in apolipoprotein E. *Nat. Med.* **9**:61–67.

74. Dong, Z.M., et al. 1998. The combined role of P- and E-selectins in atherosclerosis. *J. Clin. Invest.* **102**:145–152.

75. Dong, Z.M., Brown, A.A., and Wagner, D.D. 2000. Prominent role of P-selectin in the development of advanced atherosclerosis in ApoE-deficient mice. *Circulation.* **101**:2290–2295.

76. Burger, P.C., and Wagner, D.D. 2003. Platelet P-selectin facilitates atherosclerotic lesion development. *Blood.* **101**:2661–2666.

77. Belton, O.A., Duffy, A., Toomey, S., and Fitzgerald, D.J. 2003. Cyclooxygenase isoforms and platelet vessel wall interactions in the apolipoprotein E knockout mouse model of atherosclerosis. *Circulation.* **108**:3017–3023.

78. Lutgens, E., et al. 1999. Requirement for CD154 in the progression of atherosclerosis. *Nat. Med.* **5**:1313–1316.

79. Schonbeck, U., Sukhova, G.K., Shimizu, K., Mach, F., and Libby, P. 2000. Inhibition of CD40 signaling limits evolution of established atherosclerosis in mice. *Proc. Natl. Acad. Sci. U. S. A.* **97**:7458–7463.

80. Mach, F., Schonbeck, U., Sukhova, G.K., Atkinson, E., and Libby, P. 1998. Reduction of atherosclerosis in mice by inhibition of CD40 signalling. *Nature.* **394**:200–203.

81. Chi, H., Messas, E., Levine, R.A., Graves, D.T., and Amar, S. 2004. Interleukin-1 receptor signaling mediates atherosclerosis associated with bacterial exposure and/or a high-fat diet in a murine apolipoprotein E heterozygote model: pharmacotherapeutic implications. *Circulation.* **110**:1678–1685.

82. Kirii, H., et al. 2003. Lack of interleukin-1beta decreases the severity of atherosclerosis in ApoE-deficient mice. *Arterioscler. Thromb. Vasc. Biol.* **23**:656–660.

83. Willoughby, S., Holmes, A., and Loscalzo, J. 2002. Platelets and cardiovascular disease. *Eur. J. Cardiovasc. Nurs.* **1**:273–288.

84. Fateh-Moghadam, S., et al. 2000. Changes in surface expression of platelet membrane glycoproteins and progression of heart transplant vasculopathy. *Circulation.* **102**:890–897.

85. Fateh-Moghadam, S., et al. 2005. Platelet degranulation is associated with progression of intima-media thickness of the common carotid artery in patients with diabetes mellitus type 2. *Arterioscler. Thromb. Vasc. Biol.* **25**:1299–1303.

86. Koyama, H., et al. 2003. Platelet P-selectin expression is associated with atherosclerotic wall thickness in carotid artery in humans. *Circulation.* **108**:524–529.

Structure and function of the platelet integrin $\alpha_{IIb}\beta_3$

Joel S. Bennett

Hematology-Oncology Division, Department of Medicine, University of Pennsylvania School of Medicine, Philadelphia, Pennsylvania, USA.

The platelet integrin $\alpha_{IIb}\beta_3$ is required for platelet aggregation. Like other integrins, $\alpha_{IIb}\beta_3$ resides on cell surfaces in an equilibrium between inactive and active conformations. Recent experiments suggest that the shift between these conformations involves a global reorganization of the $\alpha_{IIb}\beta_3$ molecule and disruption of constraints imposed by the heteromeric association of the α_{IIb} and β_3 transmembrane and cytoplasmic domains. The biochemical, biophysical, and ultrastructural results that support this conclusion are discussed in this Review.

Integrins are ubiquitous transmembrane α/β heterodimers that mediate diverse processes requiring cell-matrix and cell-cell interactions such as tissue migration during embryogenesis, cellular adhesion, cancer metastases, and lymphocyte helper and killer cell functions (1). Eighteen integrin α subunits and 8 integrin β subunits have been identified in mammals that combine to form 24 different heterodimers. The resulting heterodimers can then be grouped into subfamilies according to the identity of their β subunit (1). Platelets express 3 members of the β_1 subfamily ($\alpha_{IIb}\beta_1$, $\alpha_v\beta_1$, and $\alpha_{vI}\beta_1$) that support platelet adhesion to the ECM proteins collagen, fibronectin, and laminin, respectively (2–5), and both members of the β_3 subfamily ($\alpha_v\beta_3$ and $\alpha_{IIb}\beta_3$). Although $\alpha_v\beta_3$ mediates platelet adhesion to osteopontin and vitronectin in vitro (6, 7), it is uncertain whether it plays a role in platelet function in vivo. By contrast, $\alpha_{IIb}\beta_3$, a receptor for fibrinogen, vWF, fibronectin, and vitronectin, is absolutely required for platelet aggregation. Consequently, inherited abnormalities in $\alpha_{IIb}\beta_3$ number or function preclude platelet aggregation, resulting in the bleeding disorder Glanzmann thrombasthenia (8). Conversely, thrombi that arise in the arterial circulation result from the $\alpha_{IIb}\beta_3$-mediated formation of platelet aggregates (9). Because $\alpha_{IIb}\beta_3$ plays an indispensable role in hemostasis and thrombosis, it is among the most intensively studied integrins. Thus, there is a wealth of new information relating $\alpha_{IIb}\beta_3$ structure and function, the subject of this Review.

Expression of $\alpha_{IIb}\beta_3$ is restricted to cells of the megakaryocyte lineage. In megakaryocytes, $\alpha_{IIb}\beta_3$ is assembled from α_{IIb} and β_3 precursors in the endoplasmic reticulum (10) and undergoes posttranslational processing in the Golgi complex, where α_{IIb} is cleaved into heavy and light chains (11). There are approximately 80,000 copies of $\alpha_{IIb}\beta_3$ on the surface of unstimulated platelets (12), and additional heterodimers in the membranes of platelet granules are translocated to the platelet surface during platelet secretion (13). A critical feature of $\alpha_{IIb}\beta_3$ function is that it is modulated by platelet agonists. Thus, while $\alpha_{IIb}\beta_3$ can support the adhesion of unstimulated platelets to many of its ligands when they are immobilized in vitro, platelet stimulation is required to enable $\alpha_{IIb}\beta_3$ to mediate platelet aggregation by binding soluble fibrinogen and vWF (14). EM images of rotary-shadowed $\alpha_{IIb}\beta_3$ reveal that the heterodimer consists of an 8-by-12-nm nodular head containing its

ligand-binding site and two 18-nm flexible stalks containing its transmembrane (TM) and cytoplasmic domains (15).

Crystal structures for the extracellular portions of $\alpha_v\beta_3$ and $\alpha_{IIb}\beta_3$

A major advance in understanding the structure and function of $\alpha_{IIb}\beta_3$ resulted from the reports of crystal structures for the extracellular portions of $\alpha_{IIb}\beta_3$ (16) and the closely related integrin $\alpha_v\beta_3$ (17). Xiong and coworkers prepared crystals of a presumably activated conformation of the $\alpha_v\beta_3$ extracellular region grown in the presence of Ca^{2+} (17). Surprisingly, the crystals revealed that the head region was severely bent over 2 nearly parallel tails (Figure 1). When the structure was extended, its appearance and dimensions were consistent with rotary-shadowed EM images of $\alpha_{IIb}\beta_3$. The structure itself revealed that the amino terminus of α_v was folded into a β-propeller configuration, followed by a "thigh" and 2 "calf" domains, constituting the extracellular portion of the α_v stalk. The α_v "knee" or "genu," the site at which the head region bends, was located between the thigh and first calf domain. The β_3 head consists of a βA domain whose fold resembles that of integrin α subunit "I-domains" and contains a metal ion–dependent adhesion site (MIDAS) motif, as well as a hybrid domain whose fold is similar to that of I-set Ig domains. The interface between the α_v β-propeller and the β_3 βA domain, the site at which the α_v head interacts with the β_3 head, resembles the interface between the $G\alpha$ and $G\beta$ subunits of G proteins. The β_3 stalk consists of a PSI (plexin, semaphorin, integrin) domain, 4 tandem EGF repeats, and a unique carboxyterminal βTD domain. A cyclic Arg-Gly-Asp–containing (RGD-containing) pentapeptide, soaked into the crystal in the presence of Mn^{2+} (18), inserted into a crevice between the β-propeller and βA domains with the Arg side chain located in a groove on the upper surface of the propeller and the Asp carboxylate protruding into a cleft between loops on the βA surface, implying that the crevice constitutes at least a portion of the binding site for RGD-containing $\alpha_v\beta_3$ ligands.

Subsequently, Xiao et al. reported 2 crystal structures of a complex consisting of the α_{IIb} β-propeller and the β_3 βA, hybrid, and PSI domains (16). The structures revealed an open, presumably high-affinity conformation, similar to EM images of the $\alpha_v\beta_3$ extracellular domain–containing ligand, with a 62° angle of separation between the α and β subunits due in part to a 10-Å downward movement of the α_7 helix of the βA domain relative to the hybrid and the PSI domain. Reorganization of hydrogen bonds in the interface between the α_7 helix and βC strand of the hybrid domain allowed the hybrid domain and the rigidly connected PSI domain to swing out, causing a 70-Å separation of the α_{IIb} and β_3 stalks at their "knees," a feature noted in EM images of active forms of $\alpha_{IIb}\beta_3$ in the presence or absence of ligand (19).

Nonstandard abbreviations used: GpA, glycophorin A; MIDAS, metal ion–dependent adhesion site; PSI, plexin, semaphorin, integrin; RGD, Arg-Gly-Asp; TM, transmembrane.

Conflict of interest: The author has declared that no conflict of interest exists.

Citation for this article: *J. Clin. Invest.* **115**:3363–3369 (2005). doi:10.1172/JCI26989.

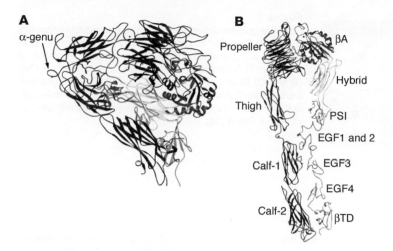

Figure 1
Ribbon diagram of the structure of the extracellular portion of
$\alpha_v\beta_3$. (**A**) Bent conformation of $\alpha_v\beta_3$ as it was present in the
crystal. (**B**) Extension of the structure to reveal its domains.
Adapted with permission from *Annual Review of Cell and
Developmental Biology* (97).

Ligand binding to $\alpha_{IIb}\beta_3$

Fibrinogen, the major $\alpha_{IIb}\beta_3$ ligand, is composed of pairs of Aα, Bβ, and γ chains folded into 3 nodular domains. Although peptides corresponding to either the carboxyterminal 10–15 amino acids of the γ chain (20) or the 2 α chain RGD motifs inhibit fibrinogen binding to $\alpha_{IIb}\beta_3$ (21), only the γ chain sequence is required for fibrinogen binding to $\alpha_{IIb}\beta_3$ (22). Nonetheless, RGD-based peptides and peptidomimetics inhibit $\alpha_{IIb}\beta_3$ function in vitro and are clinically effective antagonists of $\alpha_{IIb}\beta_3$ function in vivo (23). The structural basis for these observations is not entirely clear, but competitive binding measurements indicate that γ chain and RGD peptides cannot bind to $\alpha_{IIb}\beta_3$ at the same time (24), implying that RGD peptides inhibit fibrinogen binding by preventing the interaction of the γ chain with $\alpha_{IIb}\beta_3$.

Ligand binding to $\alpha_{IIb}\beta_3$ involves specific regions of the aminoterminal portions of both α_{IIb} and β_3. In the crystal structure of the $\alpha_{IIb}\beta_3$ head domain, ligand binds to a "specificity-determining" loop in the β_3 βA domain and to a "cap" composed of 4 loops on the upper surface of the α_{IIb} β-propeller domain (16). The α_{IIb} β-propeller results from the folding of 7 contiguous aminoterminal repeats (17, 25). Each blade of the propeller is formed from 4 antiparallel β strands located in each repeat; loops connecting the strands are located on either the upper or the lower surface of the propeller. A number of naturally occurring and laboratory-induced mutations distributed between α_{IIb} residues 145 and 224 and located in loops on the upper surface of the propeller impair $\alpha_{IIb}\beta_3$ function, implying that these residues interact with ligand (26–28). Further, Kamata et al. replaced each of the 27 loops in the α_{IIb} propeller with the corresponding loops from α_4 or α_5 (29). They found that 8 replacements, all located on the upper surface of the second, third, and fourth repeats, abrogated fibrinogen binding to $\alpha_{IIb}\beta_3$, suggesting that fibrinogen binds to the upper surface of the propeller in a region centered around the third repeat. Previous chemical cross-linking experiments suggested that the fibrinogen γ chain binds to α_{IIb} in the vicinity of its second calmodulin-like motif near amino acids 294–314 (30), but these residues are located on the lower surface of the propeller and are unlikely to interact with ligands such as fibrinogen (16). It is noteworthy that ligand binding itself induces conformational changes in $\alpha_{IIb}\beta_3$, most often detected by the appearance of neoepitopes for mAbs. In fact, such ligand-induced changes or LIBSs (ligand-induced binding sites) may be responsible for the

immune-mediated thrombocytopenia associated with the clinical use of $\alpha_{IIb}\beta_3$ antagonists (31).

Ligand binding to $\alpha_{IIb}\beta_3$ requires divalent cations (32). Eight divalent cation-binding sites were identified in the $\alpha_v\beta_3$ crystal structure (17, 18). Four were located in the α_v β-propeller domain, 1 at the α_v genu, and 3 in the β_3 βA domain, but only those located in the βA domain appeared to participate in ligand binding. In the absence of ligand, only the βA ADMIDAS (adjacent to the metal ion–dependent adhesion site) motif was occupied, but when Mn^{2+} and a cyclic RGD ligand were present, each of the βA sites contained a cation. One site was the βA MIDAS; Mn^{2+} present at this site was in direct contact with ligand. A second Mn^{2+}, located 6 Å from the MIDAS, was bound to a site designated ligand-induced metal-binding site (LIMBS), but the cation at this site did not interact with ligand. It had been postulated that Mn^{2+} induces integrin activation by antagonizing inhibitory effects of Ca^{2+} (33), but the $\alpha_v\beta_3$ crystal structure suggests that cations bound to the MIDAS and LIMBS motifs act by stabilizing the ligand-occupied conformation of the βA domain (18).

Regulation of $\alpha_{IIb}\beta_3$ ligand-binding activity

Integrins reside on cell surfaces in an equilibrium between inactive and active conformations (34). In experiments where the cytoplasmic domains of $\alpha_L\beta_2$ and $\alpha_5\beta_1$ were replaced by acidic and basic peptides (35, 36), purified integrins were inactive when their stalks were in proximity and active when the stalks were farther apart. This was corroborated by measurements of fluorescence resonance energy transfer (FRET) efficiency between cyan and yellow fluorescent proteins fused to the cytoplasmic domains of α_L and β_2 expressed in K562 cells (37). FRET efficiency decreased when $\alpha_L\beta_2$ interacted with immobilized or soluble ligand, implying that bidirectional signaling resulted from the coupling of conformational changes in the $\alpha_L\beta_2$ extracellular domain to the spatial separation of the α_L and β_2 cytoplasmic domains, a result consistent with EM images of $\alpha_{IIb}\beta_3$ in which scissor-like movements of the α_{IIb} and β_3 stalks differentiate active and inactive molecules (19).

Nonetheless, the relationship of these observations to the $\alpha_{IIb}\beta_3$ and $\alpha_v\beta_3$ crystal structures is controversial. Takagi et al., supported by negatively stained EM images of active and inactive integrins, suggested that the bent conformation of $\alpha_v\beta_3$ in crystals corresponds to low-affinity $\alpha_v\beta_3$ and the shift to a high-affinity conformation occurs when the integrin undergoes a global reorganization characterized by a "switchblade-like" opening to an extended structure and scissor-like separation of the α and β subunit stalks (34). Xiong et al., however, suggested that the bent conformation resulted from flexibility at the α_v and β_3 genua and from crystal contacts not likely to occur in nature (17). This possibility was supported by cryo-EM reconstructions of intact inactive $\alpha_{IIb}\beta_3$ molecules, which revealed a collapsed but unbent structure consisting of a large globular head and an L-shaped stalk whose axis was rotated approximately 60° with respect to the head and was

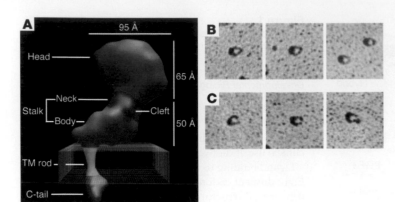

Figure 2
Cryo-EM reconstruction and rotary-shadowed EM images of $\alpha_{IIb}\beta_3$. (**A**) Cryo-EM reconstruction. The resolution is 20 Å. Adapted with permission from *Proceedings of the National Academy of Sciences of the United States of America* (38). (**B** and **C**) Rotary-shadowed EM images. The images in **B** were obtained in the presence of 1 mM Ca^{2+} and the images in **C** in the presence of 1 mM Mn^{2+}. Reproduced with permission from *Blood* (19).

connected at an angle of approximately 90° to a rod containing the TM domains of the integrin (Figure 2A) (38). They also suggested that extension at the "knees" may be a post-ligand-binding "outside-in" signaling event and that the transition of $\alpha_v\beta_3$ from its inactive to its active conformation results when the CD loop of the β_3 βTD domain moves away from the βA domain, allowing the latter to assume its active conformation (39). How to reconcile each of these models with the rotary-shadowed EM images of demonstrably inactive and active $\alpha_{IIb}\beta_3$ shown in Figure 2, B and C, is not obvious.

The α_{IIb} and β_3 cytoplasmic domains constrain $\alpha_{IIb}\beta_3$ function

Cytoplasmic domain sequences, most convincingly demonstrated for conserved membrane-proximal sequences, constrain integrins in their low-affinity (inactive) conformations. Thus, truncation of the α_{IIb} cytoplasmic domain at Gly991 or the β_3 cytoplasmic domain at Leu717 or deletion of the conserved membrane-proximal α_{IIb} GFFKR or β_3 LLITIHD motifs (Table 1) shifts $\alpha_{IIb}\beta_3$ to its active state (40). Similarly, constitutive $\alpha_{IIb}\beta_3$ function can be induced by replacement of α_{IIb} residue F992, F993, or R995 or β_3 residue D723 with alanine, whereas heterodimers containing simultaneous R995→D and D723→R substitutions are inactive (41). These observations led to the suggestion that the membrane-proximal sequences form an activation-constraining "clasp," an essential feature of which is a salt bridge between α_{IIb} R995 and β_3 D723. Paradoxically, replacing the α_{IIb} cytoplasmic domain with the cytoplasmic domain of α_2, α_5, α_{6A}, or α_{6B}, each of which contains a GFFKR motif, activates $\alpha_{IIb}\beta_3$ (40). This implies that additional cytoplasmic domain sequences modulate $\alpha_{IIb}\beta_3$ function, consistent with the inhibitory effects observed for the β_3 mutation Ser752Pro (42), β_3 truncation at Arg724 (43), and mutations involving the β_3 sequences EFAKFEEE, NPLY, and NITY (44–46) and the α_{IIb} sequence Pro998/Pro999 (47, 48).

Interaction between the α_{IIb} and β_3 cytoplasmic domains has been studied experimentally using peptides dissolved in aqueous buffer or anchored to phospholipid micelles via aminoterminal myristoylation. Using terbium luminescence and electrospray ionization mass spectroscopy, Haas and Plow observed the formation of a cation-containing complex involving α_{IIb} residues 999–1,008 and β_3 residues 721–740 (49). Similarly, Vallar et al. used surface plasmon resonance to detect a weak (K_d ~50 μM) KVGFFKR-dependent, calcium-stabilized

complex between soluble α_{IIb} cytoplasmic domain and immobilized β_3 cytoplasmic domain peptides (50). Further, Weljie et al. determined an NMR structure for a heterodimer that formed at low ionic strength between an 11-residue GFFKR-containing α_{IIb} peptide and a 25-residue LLITIHD-containing β_3 peptide (51). They identified 2 conformers differing in the conformation of the β_3 backbone: one had an elongated β_3 structure; the other was bent back at D723–A728, causing the peptide to adopt a closed L shape. Nonetheless, both conformers were predominantly helical with significant hydrophobic interactions between V990 and F993 of α_{IIb} and L717–I721 of β_3. Although there was no NMR evidence of an R995–D723 salt bridge, modeling suggested that a salt bridge was possible if the β_3 backbone was elongated. Vinogradova et al. also used NMR to characterize complexes between full-length $\alpha_{IIb}\beta_3$ cytoplasmic domain peptides (48, 52, 53). Despite low affinity, they identified interfaces for the complexes that included hydrophobic and electrostatic interactions between membrane-proximal helices (Figure 3A) (52). When the experiments were repeated in the presence of diphosphocholine micelles, α_{IIb} residues 989–993 and β_3 residues 716–721 were embedded in lipid and there was interaction between β_3 residues 741 and 747 and micelle lipid (53). Talin binding to β_3 disrupted the complex of α_{IIb} with β_3 as well as β_3 interaction with lipid (Figure 3B). On the other hand, Li et al. were unable to detect heteromeric interaction between proteins corresponding to the α_{IIb} and β_3 TM and cytoplasmic domains in diphosphocholine micelles at physiologic salt concentrations using a number of biophysical techniques, perhaps because heteromeric interaction is substantially weaker than homomeric interaction (54). Similarly, Ulmer et al. did not detect heteromeric

Table 1

Amino acid sequences of the TM and cytoplasmic domains of α_{IIb} and β_3[A]

TM domains

α_{IIb}	W_{968}VLV<u>GVLGG</u>LLLLTILVLAMW_{988}
β_3	I_{693}LVVLLSVMGAILL<u>IG</u>LAALLIW_{715}

Cytoplasmic domains

α_{IIb}	<u>K_{989}</u>VGFFKRNRPPLEEDDEEGE_{1008}
β_3	<u>K_{716}</u>LLITIHDRKEFAKFEEERARAKWDTAN<u>NPLY</u>KEATSTFT<u>NITY</u>RGT_{762}

[A]The amino acids in the α_{IIb} and β_3 TM and cytoplasmic domains are designated in the single-letter code; the subscript numbers correspond to the position of the preceding amino acid in the sequence for mature α_{IIb} and β_3 (98, 99). The α_{IIb} GxxxG motif, β_3 G708, the membrane-proximal regions of both cytoplasmic domains, and both β_3 NxxY motifs are underlined.

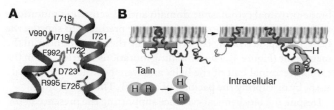

Figure 3
Interaction of the α_{IIb} and β_3 cytoplasmic domains. (**A**) Backbone ribbon diagram of the $\alpha_{IIb}\beta_3$ membrane-proximal cytoplasmic domain clasp showing hydrophobic and electrostatic interactions. Reproduced with permission from *Cell* (52). (**B**) Model of the changes that may occur in the clasp following talin binding to the β_3 cytoplasmic domain. Adapted with permission from *Proceedings of the National Academy of Sciences of the United States of America* (53).

interactions of α_{IIb} with β_3 in an NMR analysis of a coiled-coil construct containing the α_{IIb} and β_3 cytoplasmic domains (55).

Proteins that interact with the α_{IIb} and β_3 cytoplasmic domains

Proteins have been identified, most often using yeast 2-hybrid screens, that bind to the cytoplasmic domains of integrin α and β subunits. These proteins include CIB (calcium- and integrin-binding protein) (56), Aup1 (ancient ubiquitous protein 1) (57), ICln (a chloride channel regulatory protein) (58), and PP1c (the catalytic subunit of protein phosphatase 1) (59), each of which binds to the membrane-proximal α_{IIb} sequence KVGFFKR. However, because a substantial portion of this sequence is likely embedded in the plasma membrane (60, 61), the physiologic importance of these interactions is uncertain. Proteins that interact with the β_3 cytoplasmic domain include the cytoskeletal proteins talin, α-actinin, filamin, myosin, and skelemin; various members of the Src family of kinases; the kinases integrin-linked kinase (ILK), Syk, and Shc; the adaptor Grb2; the scaffold RACK1; CD98 (62); and β_3-endonexin (63). Binding of myosin, Shc, and Grb2 requires platelet aggregation and spreading, as well as the Fyn-mediated phosphorylation of β_3 tyrosines 747 and 759, and has been implicated in post-receptor-binding cytoskeleton-mediated events such as clot retraction (64).

Binding of β_3-endonexin or talin to the β_3 cytoplasmic domain is noteworthy because it can activate $\alpha_{IIb}\beta_3$. β_3-Endonexin, a 14-kDa protein of unknown function, induces $\alpha_{IIb}\beta_3$ activation when coexpressed with $\alpha_{IIb}\beta_3$ in tissue culture cells by interacting with residues located in both the aminoterminal and the carboxyterminal regions of the β_3 cytoplasmic domain, in particular the carboxyterminal NITY motif (65–67). Nonetheless, there is no evidence as yet that β_3-endonexin regulates $\alpha_{IIb}\beta_3$ function in platelets. One explanation for the presence of 2 discontinuous β_3-endonexin–binding sites in the β_3 cytoplasmic domain has been provided by an NMR analysis of a protein encompassing the β_3 TM and cytoplasmic domains (Figure 4) (68). This analysis revealed that the β_3 TM helix extended into the membrane-proximal region of the cytoplasmic domain, ending at an apparent hinge at residues H722–D723 (Table 1). Two additional helical stretches, extending from residues 725 to 735 and 748 to 755, were also present (Figure 4). Because the latter helices can interact with each other, they can place the proximal and distal regions of the β_3 cytoplasmic domain in proximity.

Talin, an abundant 250-kDa cytoskeletal protein, forms antiparallel homodimers that bind to the cytoplasmic domain of integrin

β subunits as well as to other cytoskeletal proteins such as actin and vinculin (69). Talin is composed of a 50-kDa head domain containing its principal integrin-binding site and a 220-kDa rod domain that binds to integrins with lesser affinity (69). The talin head itself contains an approximately 300-residue FERM (four-point-one, ezrin, radixin, moesin) domain that folds into F1, F2, and F3 subdomains (69). F2 and F3 bind to the β_3 cytoplasmic domain, although the affinity of F3 binding is substantially greater (70). A crystal structure for a fusion protein composed of the F2 and F3 subdomains and a contiguous aminoterminal peptide corresponding to the midportion of the β_3 cytoplasmic domain, including its NPLY motif, revealed that the interaction of the β_3 peptide with F3 was mainly hydrophobic and that NPLY interacted with F3 in a manner that resembled that of canonical PTB domain ligands (71). However, studies using NMR also revealed that F3 and F2-F3 interact with the membrane-proximal region of the β_3 cytoplasmic domain (71, 72), consistent with previous observations that talin binds to peptides corresponding to this portion of β_3 (73).

Overexpressing the talin head domain in $\alpha_{IIb}\beta_3$-expressing CHO cells induces $\alpha_{IIb}\beta_3$ activation (74), either directly because talin disrupts the clasp between α_{IIb} and β_3 (Figure 3B) or indirectly via conformational changes induced by F3 binding to the β_3 NPLY motif (70). Conversely, reducing talin expression using short hairpin RNAs decreases ligand binding to $\alpha_{IIb}\beta_3$ in CHO cells and in ES cell–derived agonist-stimulated megakaryocytes (75). Taken together, these results imply that talin binding to the β_3 cytoplasmic domain may be a final step in $\alpha_{IIb}\beta_3$ activation. Nonetheless, how talin binding to the β_3 cytoplasmic domain is regulated remains to be determined. The integrin-binding domain in intact talin appears to be masked (76). Although the enzyme calpain can cleave talin, releasing its head domain (77), calpain activation in platelets is a relatively late step after platelet stimulation (78) and would be unlikely to contribute to integrin-activating inside-out signaling. On the other hand, talin binds to membrane-associated phosphoinositol 4,5-bisphosphate, inducing a conformational change that enables it to bind to the β_1 cytoplasmic domain (79). By analogy, talin binding to phosphoinositol 4,5-bisphosphate may enable it to bind to β_3.

Regulation of $\alpha_{IIb}\beta_3$ function by TM domain interaction
TM domain–mediated protein oligomerization is a common mechanism for the assembly of membrane proteins and regula-

Extracellular domain of β_3

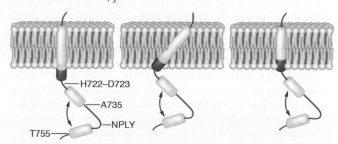

Figure 4
Model of the structure of the β_3 TM and cytoplasmic domains. Helices are shown as cylinders. Three different orientations of the β_3 TM domain in the plasma membrane are shown. The membrane-proximal region of the cytoplasmic domain is shaded. Arrows indicate possible interactions between helices. Adapted with permission from *Biochemistry* (68).

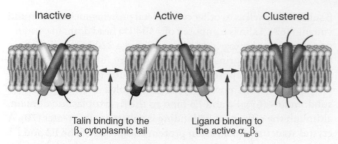

Inactive　　　Active　　　Clustered

Talin binding to the　　Ligand binding to
β₃ cytoplasmic tail　　the active α_IIbβ₃

Figure 5

Diagram illustrating the "push-pull" hypothesis for regulation of the α_IIbβ₃ activation state. The white and blue cylinders represent the α_IIb TM and membrane-proximal cytoplasmic domain helices, respectively. The red and green cylinders represent the β₃ TM and membrane-proximal cytoplasmic domain helices, respectively.

tion of protein function (80). Specificity is achieved via specific sequence motifs superimposed on more general oligomerization frameworks (81–83). For example, the sequence motif GxxxG, first recognized as a framework for the homomeric association of the glycophorin A (GpA) TM helix (81), has been identified as the most overrepresented sequence motif in TM domain databases (82).

With regard to integrin TM domains, Li and coworkers reported that peptides corresponding to the α_IIb and β₃ TM domains readily undergo homodimeric and homotrimeric association, respectively, in phospholipid micelles (54), and Schneider and Engelman found that fusion proteins containing the α₂β₁, α₄β₇, and α_IIbβ₃ TM domains undergo integrin-specific TM domain–mediated homomeric and heteromeric association in bacterial membranes (84). Subsequently, Li et al. reported that facilitating the homomeric association of the β₃ TM helix by replacing either G708 or M701 with a polar asparagine induced α_IIbβ₃ activation and clustering when the mutants were expressed in CHO cells (85). They also found that mutation of the α_IIb GxxxG motif located at residues 972–975 disrupted the homomeric association of α_IIb TM helix (86) and paradoxically induced α_IIbβ₃ activation and clustering (87). These observations suggested the "push-pull" mechanism for α_IIbβ₃ activation shown in Figure 5. Processes that destabilize the association of the α_IIb and β₃ TM helices, such as talin binding to the β₃ cytoplasmic domain, would be expected to promote dissociation of the helices with concomitant α_IIbβ₃ activation. Conversely, intermolecular interactions that either require separation of the α_IIb and β₃ TM helices, such as homo-oligomerization, or are more favorable when they separate, such as ligand-induced α_IIbβ₃ clustering (88), would be expected to pull the equilibrium toward the activated state.

The ability of homomeric TM helix interactions to induce α_IIbβ₃ activation and clustering remains controversial (89, 90), but there is compelling evidence that heterodimeric interactions constrain α_IIbβ₃ in a low-affinity state. By simultaneously scanning the α_IIb and β₃ TM helices with cysteine residues, Luo et al. detected the formation of disulfide bonds with a helical periodicity in a region corresponding to α_IIb residues 966–974 and β₃ residues 693–702, consistent with the presence of a unique α_IIbβ₃ TM heterodimer (91). They also scanned the α_IIb and β₃ helices with leucines, confirming that mutation of the α_IIb GxxxG motif induces α_IIbβ₃ activation (90). Partridge et al. used random mutagenesis of the β₃ TM and cytoplasmic domains to search for interactions constraining α_IIbβ₃ activation (92). They detected 12 activating mutations in the mem-

brane-proximal cytoplasmic domain and 13 activating mutations in the β₃ TM helix. Nine of the latter were predicted to shorten the helix, perhaps activating α_IIbβ₃ by altering the tilt of the helix in the membrane (Figure 4). The remaining mutations were located in the carboxyterminal half of the helix and were postulated to activate α_IIbβ₃ by disrupting the packing of an α_IIbβ₃ TM heterodimer.

Despite the biochemical evidence supporting the presence of α_IIb and β₃ TM domain oligomers, their existence has not been confirmed by NMR spectroscopy or x-ray crystallography because of difficulty in obtaining high-resolution structures for TM proteins using these techniques. However, computational methods have been used to construct TM domain models incorporating the constraints imposed by mutational data. Based on cryo-EM images (Figure 2A), Adair and Yeager proposed that the TM domains of inactive α_IIbβ₃ associate in a parallel α-helical coiled coil (38). Using the R995–D723 salt bridge as the primary constraint, they found that a right-handed coiled coil based on the GpA TM dimer (93) placed more conserved residues in the helix-helix interface than a coiled coil based on the canonical left-handed leucine zipper. Gottschalk and coworkers proposed that the α_IIb and β₃ TM helices remain in close contact in the activated state and that the helix-helix interface is a GpA-like structure containing the α_IIb G972xxxG975 and β₃ S699xxxA703 motifs (94). Moreover, simulated annealing and molecular dynamics supported a model in which the α_IIb and β₃ TM domains interact weakly in a right-handed coiled coil when the integrin is in its low-affinity conformation (95). Subsequently, in order to account for both aminoterminal and carboxyterminal restraints, Gottschalk proposed that the α_IIbβ₃ TM and membrane-proximal cytoplasmic domains form a right-handed coiled coil in which the helices interact over their entire length, placing the α_IIb GxxxG motif, but not β₃ S699xxxA703, in the helix-helix interface (96). By contrast, Luo et al. used their disulfide cross-linking data to construct a model based on the GpA TM dimer; however, in this model, the α_IIb GxxxG-like motif corresponded to residues 968–972, rather than 972–975 (91). DeGrado and coworkers used a Monte Carlo–simulated annealing algorithm to obtain atomic models for an α_IIb TM homodimer (86) and an α_IIbβ₃ heterodimer (87). In each case, a family of structures was found that satisfied mutational constraints. For the α_IIb homodimer, all structures had right-handed crossing angles ranging from 40° to 60°, but with an interface rotated by 50° relative to the GpA homodimer. In the case of the α_IIbβ₃ heterodimer, initial docking identified local minima with both right- and left-handed crossing angles. However, the right-handed structures had lower energies and more extensive interactions, and the α_IIb GxxxG motif was in intimate contact with the β₃ TM domain. Lastly, Partridge et al., using a Monte Carlo simulation, obtained 2 structures for an α_IIbβ₃ TM heterodimer with helix packing near either the amino or the carboxyl termini of the helices, respectively; of the 2 models, carboxyterminal helix packing was more consistent with their mutational data (92). It is obvious that there is wide disparity among these models, making it clear that obtaining actual structures for α_IIb and β₃ TM domain hetero- and homo-oligomers will be the next major advance in our understanding of the structural basis for the regulation of platelet integrin function.

Address correspondence to: Joel S. Bennett, Hematology-Oncology Division, Department of Medicine, University of Pennsylvania School of Medicine, 914 BRB II/III, 421 Curie Boulevard, Philadelphia, Pennsylvania 19104-6058, USA. Phone: (215) 573-3280; Fax: (215) 573-7039; E-mail: bennetts@mail.med.upenn.edu.

1. Hynes, R.O. 2002. Integrins: bidirectional, allosteric signaling machines. *Cell.* **110**:673–687.

2. Piotrowicz, R.S., Orchekowski, R.P., Nugent, D.J., Yamada, K.Y., and Kunicki, T.J. 1988. Glycoprotein Ic-IIa functions as an activation-independent fibronectin receptor on human platelets. *J. Cell Biol.* **106**:1359–1364.

3. Ill, C.R., Engvall, E., and Ruoslahti, E. 1984. Adhesion of platelets to laminin in the absence of activation. *J. Cell Biol.* **99**:2140–2145.

4. Sonnenberg, A., Modderman, P., and Hogervorst, F. 1988. Laminin receptor on platelets is the integrin VLA-6. *Nature.* **336**:487–488.

5. Staatz, W.D., Rajpara, S.M., Wayner, E.A., Carter, W.G., and Santoro, S.A. 1989. The membrane glycoprotein Ia-IIa (VLA-2) complex mediates the Mg^{++}-dependent adhesion of platelets to collagen. *J. Cell Biol.* **108**:1917–1924.

6. Bennett, J.S., Chan, C., Vilaire, G., Mousa, S.A., and DeGrado, W.F. 1997. Agonist-activated αvβ3 on platelets and lymphocytes binds to the matrix protein osteopontin. *J. Biol. Chem.* **272**:8137–8140.

7. Paul, B.Z.S., Vilaire, G., Kunapuli, S.P., and Bennett, J.S. 2003. Concurrent signaling from Gα$_q$- and Gα$_i$-coupled pathways is essential for agonist-induced αvβ3 activation on human platelets. *J. Thromb. Haemost.* **1**:814–820.

8. George, J.N., Caen, J.P., and Nurden, A.T. 1990. Glanzmann's thrombasthenia: the spectrum of clinical disease. *Blood.* **75**:1383–1395.

9. Lefkovits, J., Plow, E., and Topol, E. 1995. Platelet glycoprotein IIb/IIIa receptors in cardiovascular medicine. *N. Engl. J. Med.* **332**:1553–1559.

10. Duperray, A., et al. 1987. Biosynthesis and processing of platelet GPIIb-IIIa in human megakaryocytes. *J. Cell Biol.* **104**:1665–1673.

11. Kolodziej, M.A., Vilaire, G., Gonder, D., Poncz, M., and Bennett, J.S. 1991. Study of the endoproteolytic cleavage of platelet glycoprotein IIb using oligonucleotide-mediated mutagenesis. *J. Biol. Chem.* **266**:23499–23504.

12. Wagner, C.L., et al. 1996. Analysis of GPIIb/IIIa receptor number by quantitation of 7E3 binding to human platelets. *Blood.* **88**:907–914.

13. Niiya, K., et al. 1987. Increased surface expression of the membrane glycoprotein IIb/IIIa complex induced by platelet activation. Relationship to the binding of fibrinogen and platelet aggregation. *Blood.* **70**:475–483.

14. Bennett, J.S. 1996. Structural biology of glycoprotein IIb-IIIa. *Trends Cardiovasc. Med.* **6**:31–37.

15. Weisel, J.W., Nagaswami, C., Vilaire, G., and Bennett, J.S. 1992. Examination of the platelet membrane glycoprotein IIb/IIIa complex and its interaction with fibrinogen and other ligands by electron microscopy. *J. Biol. Chem.* **267**:16637–16643.

16. Xiao, T., Takagi, J., Coller, B.S., Wang, J.H., and Springer, T.A. 2004. Structural basis for allostery in integrins and binding to fibrinogen-mimetic therapeutics. *Nature.* **432**:59–67.

17. Xiong, J.P., et al. 2001. Crystal structure of the extracellular segment of integrin alpha Vbeta3. *Science.* **294**:339–345.

18. Xiong, J.P., et al. 2002. Crystal structure of the extracellular segment of integrin alpha Vbeta3 in complex with an Arg-Gly-Asp ligand. *Science.* **296**:151–155.

19. Litvinov, R.I., et al. 2004. Functional and structural correlations of individual alphaIIbbeta3 molecules. *Blood.* **104**:3979–3985.

20. Kloczewiak, M., Timmons, S., Lukas, T.J., and Hawiger, J. 1984. Platelet receptor recognition site on human fibrinogen. Synthesis and structure-function relationships of peptides corresponding to the carboxy-terminal segment of the γ chain. *Biochemistry.* **23**:1767–1774.

21. Gartner, T.K., and Bennett, J.S. 1985. The tetrapeptide analogue of the cell attachment site of fibro-

nectin inhibits platelet aggregation and fibrinogen binding to activated platelets. *J. Biol. Chem.* **260**:11891–11894.

22. Farrell, D.H., Thiagarajan, P., Chung, D.W., and Davie, E.W. 1992. Role of fibrinogen α and γ chain sites in platelet aggregation. *Proc. Natl. Acad. Sci. U. S. A.* **89**:10729–10732.

23. Bennett, J.S. 2001. Novel platelet inhibitors. *Annu. Rev. Med.* **52**:161–184.

24. Bennett, J.S., Shattil, S.J., Power, J.W., and Gartner, T.K. 1988. Interaction of fibrinogen with its platelet receptor. Differential effects of α and γ chain fibrinogen peptides on the glycoprotein IIb-IIIa complex. *J. Biol. Chem.* **263**:12948–12953.

25. Springer, T.A. 1997. Folding of the N-terminal, ligand-binding region of integrin α-subunits into a β-propeller domain. *Proc. Natl. Acad. Sci. U. S. A.* **94**:65–72.

26. Kamata, T., Irie, A., Tokuhira, M., and Takada, Y. 1996. Critical residues of integrin alphaIIb subunit for binding of alphaIIbbeta3 (glycoprotein IIb-IIIa) to fibrinogen and ligand-mimetic antibodies (PAC-1, OP-G2, and LJ-CP3). *J. Biol. Chem.* **271**:18610–18615.

27. Tozer, E.C., Baker, E.K., Ginsberg, M.H., and Loftus, J.C. 1999. A mutation in the α subunit of the platelet integrin α$_{IIb}$β$_3$ identifies a novel region important for ligand binding. *Blood.* **93**:918–924.

28. Basani, R.B., et al. 2000. A naturally occurring mutation near the amino terminus of αIIb defines a new region involved in ligand binding to αIIbβ3. *Blood.* **95**:180–188.

29. Kamata, T., Tieu, K.K., Irie, A., Springer, T.A., and Takada, Y. 2001. Amino acid residues in the alpha IIb subunit that are critical for ligand binding to integrin alpha IIbbeta 3 are clustered in the beta-propeller model. *J. Biol. Chem.* **276**:44275–44283.

30. D'Souza, S.E., Ginsberg, M.H., Burke, T.A., and Plow, E.F. 1990. The ligand binding site of the platelet integrin receptor GPIIb-IIIa is proximal to the second calcium binding domain of its α subunit. *J. Biol. Chem.* **265**:3440–3446.

31. Bougie, D.W., et al. 2002. Acute thrombocytopenia after treatment with tirofiban or eptifibatide is associated with antibodies specific for ligand-occupied GPIIb/IIIa. *Blood.* **100**:2071–2076.

32. Bennett, J.S., and Vilaire, G. 1979. Exposure of platelet fibrinogen receptors by ADP and epinephrine. *J. Clin. Invest.* **64**:1393–1401.

33. Smith, J., Piotrowicz, R., and Mathis, D. 1994. A mechanism for divalent cation regulation of beta 3-integrins. *J. Biol. Chem.* **269**:960–967.

34. Takagi, J., Petre, B., Walz, T., and Springer, T. 2002. Global conformational rearrangements in integrin extracellular domains in outside-in and inside-out signaling. *Cell.* **110**:599–611.

35. Lu, C., Takagi, J., and Springer, T.A. 2001. Association of the membrane proximal regions of the alpha and beta subunit cytoplasmic domains constrains an integrin in the inactive state. *J. Biol. Chem.* **276**:14642–14648.

36. Takagi, J., Erickson, H.P., and Springer, T.A. 2001. C-terminal opening mimics 'inside-out' activation of integrin α5β1. *Nat. Struct. Biol.* **8**:412–416.

37. Kim, M., Carman, C.V., and Springer, T.A. 2003. Bidirectional transmembrane signaling by cytoplasmic domain separation in integrins. *Science.* **301**:1720–1725.

38. Adair, B.D., and Yeager, M. 2002. Three-dimensional model of the human platelet integrin alpha IIbbeta 3 based on electron cryomicroscopy and x-ray crystallography. *Proc. Natl. Acad. Sci. U. S. A.* **99**:14059–14064.

39. Xiong, J.P., Stehle, T., Goodman, S.L., and Arnaout, M.A. 2003. New insights into the structural basis of integrin activation. *Blood.* **102**:1155–1159.

40. O'Toole, T.E., et al. 1994. Integrin cytoplasmic domains mediate inside-out signal transduction.

J. Cell Biol. **124**:1047–1059.

41. Hughes, P.E., et al. 1996. Breaking the integrin hinge. A defined structural constraint regulates integrin signaling. *J. Biol. Chem.* **271**:6571–6574.

42. Chen, Y., et al. 1992. Ser752→Pro mutation in the cytoplasmic domain of integrin β$_3$ subunit and defective activation of platelet integrin α$_{IIb}$β$_3$ (glycoprotein IIb-IIIa) in a variant of Glanzmann thrombasthenia. *Proc. Natl. Acad. Sci. U. S. A.* **89**:10169–10173.

43. Wang, R., Shattil, S.J., Ambruso, D.R., and Newman, P.J. 1997. Truncation of the cytoplasmic domain of β$_3$ in a variant form of Glanzmann thrombasthenia abrogates signaling through the integrin α$_{IIb}$β$_3$ complex. *J. Clin. Invest.* **100**:2393–2403.

44. O'Toole, T.E., Ylanne, J., and Culley, B.M. 1995. Regulation of integrin affinity states through an NPXY motif in the beta subunit cytoplasmic domain. *J. Biol. Chem.* **270**:8553–8558.

45. Ylanne, J., et al. 1995. Mutation of the cytoplasmic domain of the integrin beta 3 subunit. Differential effects on cell spreading, recruitment to adhesion plaques, endocytosis, and phagocytosis. *J. Biol. Chem.* **270**:9550–9557.

46. Xi, X., Bodnar, R.J., Li, Z., Lam, S.C., and Du, X. 2003. Critical roles for the COOH-terminal NITY and RGT sequences of the integrin beta3 cytoplasmic domain in inside-out and outside-in signaling. *J. Cell Biol.* **162**:329–339.

47. Leisner, T.M., Wencel-Drake, J.D., Wang, W., and Lam, S.C.-T. 1999. Bidirectional transmembrane modulation of integrin α$_{IIb}$β$_3$ conformations. *J. Biol. Chem.* **274**:12945–12949.

48. Vinogradova, O., Haas, T., Plow, E.F., and Qin, J. 2000. A structural basis for integrin activation by the cytoplasmic tail of the alpha IIb-subunit. *Proc. Natl. Acad. Sci. U. S. A.* **97**:1450–1455.

49. Haas, T.A., and Plow, E.F. 1996. The cytoplasmic domain of alphaIIb beta3. A ternary complex of the integrin alpha and beta subunits and a divalent cation. *J. Biol. Chem.* **271**:6017–6026.

50. Vallar, L., et al. 1999. Divalent cations differentially regulate integrin α$_{IIb}$ cytoplasmic tail binding to β$_3$ and to calcium- and integrin-binding protein. *J. Biol. Chem.* **274**:17257–17266.

51. Weljie, A.M., Hwang, P.M., and Vogel, H.J. 2002. Solution structures of the cytoplasmic tail complex from platelet integrin alpha IIb- and beta 3-subunits. *Proc. Natl. Acad. Sci. U. S. A.* **99**:5878–5883.

52. Vinogradova, O., et al. 2002. A structural mechanism of integrin alpha(IIb)beta(3) "inside-out" activation as regulated by its cytoplasmic face. *Cell.* **110**:587–597.

53. Vinogradova, O., et al. 2004. Membrane-mediated structural transitions at the cytoplasmic face during integrin activation. *Proc. Natl. Acad. Sci. U. S. A.* **101**:4094–4099.

54. Li, R., et al. 2001. Oligomerization of the integrin alphaIIbbeta3: roles of the transmembrane and cytoplasmic domains. *Proc. Natl. Acad. Sci. U. S. A.* **98**:12462–12467.

55. Ulmer, T.S., Yaspan, B., Ginsberg, M.H., and Campbell, I.D. 2001. NMR analysis of structure and dynamics of the cytosolic tails of integrin alpha IIb beta 3 in aqueous solution. *Biochemistry.* **40**:7498–7508.

56. Naik, U.P., Patel, P.M., and Parise, L.V. 1997. Identification of a novel calcium-binding protein that interacts with the integrin α$_{IIb}$ cytoplasmic domain. *J. Biol. Chem.* **272**:4651–4654.

57. Kato, A., et al. 2002. Ancient ubiquitous protein 1 binds to the conserved membrane-proximal sequence of the cytoplasmic tail of the integrin alpha subunits that plays a crucial role in the inside-out signaling of alpha IIbbeta 3. *J. Biol. Chem.* **277**:28934–28941.

58. Larkin, D., et al. 2004. ICln, a novel integrin alphaIIb-beta3-associated protein, functionally regulates

platelet activation. *J. Biol. Chem.* **279**:27286–27293.

59. Vijayan, K.V., Liu, Y., Li, T.T., and Bray, P.F. 2004. Protein phosphatase 1 associates with the integrin alphaIIb subunit and regulates signaling. *J. Biol. Chem.* **279**:33039–33042.

60. Armulik, A., Nilsson, I., von Heijne, G., and Johansson, S. 1999. Determination of the border between the transmembrane and cytoplasmic domains of human integrin subunits. *J. Biol. Chem.* **274**:37030–37034.

61. Stefansson, A., Armulik, A., Nilsson, I., von Heijne, G., and Johansson, S. 2004. Determination of N- and C-terminal borders of the transmembrane domain of integrin subunits. *J. Biol. Chem.* **279**:21200–21205.

62. Feral, C.C., et al. 2005. CD98hc (SLC3A2) mediates integrin signaling. *Proc. Natl. Acad. Sci. U. S. A.* **102**:355–360.

63. Buensuceso, C.S., Arias-Salgado, E.G., and Shattil, S.J. 2004. Protein-protein interactions in platelet alphaIIbbeta3 signaling. *Semin. Thromb. Hemost.* **30**:427–439.

64. Phillips, D.R., Prasad, K.S., Manganello, J., Bao, M., and Nannizzi-Alaimo, L. 2001. Integrin tyrosine phosphorylation in platelet signaling. *Curr. Opin. Cell Biol.* **13**:546–554.

65. Shattil, S.J., et al. 1995. β3-Endonexin, a novel polypeptide that interacts specifically with the cytoplasmic tail of the integrin β3 subunit. *J. Cell Biol.* **131**:807–816.

66. Kashiwagi, H., et al. 1997. Affinity modulation of platelet integrin $\alpha_{IIb}\beta_3$ by β3-endonexin, a selective binding partner of the β3 integrin cytoplasmic tail. *J. Cell Biol.* **137**:1433–1443.

67. Eigenthaler, M., Hofferer, L., Shattil, S.J., and Ginsberg, M.H. 1997. A conserved sequence motif in the integrin β3 cytoplasmic domain is required for its specific interaction with β3-endonexin. *J. Biol. Chem.* **272**:7693–7698.

68. Li, R., et al. 2002. Characterization of the monomeric form of the transmembrane and cytoplasmic domains of the integrin beta 3 subunit by NMR spectroscopy. *Biochemistry.* **41**:15618–15624.

69. Calderwood, D.A. 2004. Talin controls integrin activation. *Biochem. Soc. Trans.* **32**:434–437.

70. Calderwood, D.A., et al. 2002. The phosphotyrosine binding (PTB)-like domain of talin activates integrins. *J. Biol. Chem.* **277**:21749–21758.

71. Garcia-Alvarez, B., et al. 2003. Structural determinants of integrin recognition by talin. *Mol. Cell.* **11**:49–58.

72. Ulmer, T.S., Calderwood, D.A., Ginsberg, M.H.,

and Campbell, I.D. 2003. Domain-specific interactions of talin with the membrane-proximal region of the integrin beta3 subunit. *Biochemistry.* **42**:8307–8312.

73. Patil, S., et al. 1999. Identification of a talin-binding site in the integrin beta(3) subunit distinct from the NPLY regulatory motif of post-ligand binding functions. *J. Biol. Chem.* **274**:28575–28583.

74. Calderwood, D.A., et al. 1999. The talin head domain binds to integrin beta subunit cytoplasmic tails and regulates integrin activation. *J. Biol. Chem.* **274**:28071–28074.

75. Tadokoro, S., et al. 2003. Talin binding to integrin beta tails: a final common step in integrin activation. *Science.* **302**:103–106.

76. Pearson, M.A., Reczek, D., Bretscher, A., and Karplus, P.A. 2000. Structure of the ERM protein moesin reveals the FERM domain fold masked by an extended actin binding tail domain. *Cell.* **101**:259–270.

77. Yan, B., Calderwood, D.A., Yaspan, B., and Ginsberg, M.H. 2001. Calpain cleavage promotes talin binding to the beta 3 integrin cytoplasmic domain. *J. Biol. Chem.* **276**:28164–28170.

78. Fox, J.E., Taylor, R.G., Taffarel, M., Boyles, J.K., and Goll, D.E. 1993. Evidence that activation of platelet calpain is induced as a consequence of binding of adhesive ligand to the integrin, glycoprotein IIb-IIIa. *J. Cell Biol.* **120**:1501–1507.

79. Martel, V., et al. 2001. Conformation, localization, and integrin binding of talin depend on its interaction with phosphoinositides. *J. Biol. Chem.* **276**:21217–21227.

80. Popot, J.L., and Engelman, D.M. 2000. Helical membrane protein folding, stability, and evolution. *Annu. Rev. Biochem.* **69**:881–922.

81. Russ, W.P., and Engelman, D.M. 2000. The GxxxG motif: a framework for transmembrane helix-helix association. *J. Mol. Biol.* **296**:911–919.

82. Senes, A., Gerstein, M., and Engelman, D.M. 2000. Statistical analysis of amino acid patterns in transmembrane helices: the GxxxG motif occurs frequently and in association with beta-branched residues at neighboring positions. *J. Mol. Biol.* **296**:921–936.

83. Dawson, J.P., Melnyk, R.A., Deber, C.M., and Engelman, D.M. 2003. Sequence context strongly modulates association of polar residues in transmembrane helices. *J. Mol. Biol.* **331**:255–262.

84. Schneider, D., and Engelman, D.M. 2004. Involvement of transmembrane domain interactions in signal transduction by alpha/beta integrins. *J. Biol. Chem.* **279**:9840–9846.

85. Li, R., et al. 2003. Activation of integrin alphaIIbbeta3 by modulation of transmembrane helix associations. *Science.* **300**:795–798.

86. Li, R., Bennett, J.S., and Degrado, W.F. 2004. Structural basis for integrin alphaIIbbeta3 clustering. *Biochem. Soc. Trans.* **32**:412–415.

87. Li, W., et al. 2005. A push-pull mechanism for regulating integrin function. *Proc. Natl. Acad. Sci. U. S. A.* **102**:1424–1429.

88. Fox, J.E., et al. 1996. The platelet cytoskeleton stabilizes the interaction between alphaIIbbeta3 and its ligand and induces selective movements of ligand-occupied integrin. *J. Biol. Chem.* **271**:7004–7011.

89. Kim, M., Carman, C.V., Yang, W., Salas, A., and Springer, T.A. 2004. The primacy of affinity over clustering in regulation of adhesiveness of the integrin $\alpha_L\beta_2$. *J. Cell Biol.* **167**:1241–1253.

90. Luo, B.H., Carman, C.V., Takagi, J., and Springer, T.A. 2005. Disrupting integrin transmembrane domain heterodimerization increases ligand binding affinity, not valency or clustering. *Proc. Natl. Acad. Sci. U. S. A.* **102**:3679–3684.

91. Luo, B.H., Springer, T.A., and Takagi, J. 2004. A specific interface between integrin transmembrane helices and affinity for ligand. *PLoS Biol.* **2**:e153.

92. Partridge, A.W., Liu, s., Kim, S., Bowie, J.U., and Ginsberg, M.H. 2005. Transmembrane domain helix packing stabilizes integrin alphaIIbbeta3 in the low affinity state. *J. Biol. Chem.* **280**:7294–7300.

93. MacKenzie, K.R., Prestegard, J.H., and Engelman, D.M. 1997. A transmembrane helix dimer: structure and implications. *Science.* **276**:131–133.

94. Gottschalk, K.E., Adams, P.D., Brunger, A.T., and Kessler, H. 2002. Transmembrane signal transduction of the alpha(IIb)beta(3) integrin. *Protein Sci.* **11**:1800–1812.

95. Gottschalk, K.E., and Kessler, H. 2004. Evidence for hetero-association of transmembrane helices of integrins. *FEBS Lett.* **557**:253–258.

96. Gottschalk, K.E. 2005. A coiled-coil structure of the alphaIIbbeta3 integrin transmembrane and cytoplasmic domains in its resting state. *Structure (Camb.).* **13**:703–712.

97. Arnaout, M.A., Mahalingam, B., and Xiong, J.P. 2005. Integrin structure, allostery, and bidirectional signaling. *Annu. Rev. Cell Dev. Biol.* doi:10.1146/annurev.cellbio.21.090704.151217.

98. Poncz, M., et al. 1987. Structure of the platelet membrane glycoprotein IIb. *J. Biol. Chem.* **262**:8476–8482.

99. Zimrin, A.B., et al. 1988. Structure of platelet glycoprotein IIIa. *J. Clin. Invest.* **81**:1470–1475.

Minding the gaps to promote thrombus growth and stability

Lawrence F. Brass, Li Zhu, and Timothy J. Stalker

Departments of Medicine and Pharmacology, University of Pennsylvania, Philadelphia, Pennsylvania, USA.

Efforts to understand the role of platelets in hemostasis and thrombosis have largely focused on the earliest events of platelet activation, those that lead to aggregation. Although much remains to be learned about those early events, this Review examines a later series of events: the interactions between platelets that can only occur once aggregation has begun, bringing platelets into close contact with each other, creating a protected environment in the gaps between aggregated platelets, and fostering the continued growth and stability of the hemostatic plug.

Introduction

The mechanisms of platelet activation have grown increasingly familiar. Vascular injury exposes tissue factor within the vessel wall (or allows the local accumulation of circulating tissue factor–containing microparticles), which leads to local generation of thrombin and fibrin. Simultaneously, newly exposed collagen fibrils within the vessel wall become decorated with plasma-derived vWF, circulating platelets adhere long enough to be activated by collagen and thrombin, and a growing mound of platelets is formed as additional circulating platelets are recruited by soluble agonists, including thrombin, ADP, and thromboxane A_2 (TxA_2). Activated platelets stick to each other via bridges formed by the $\alpha_{IIb}\beta_3$ integrin and multivalent adhesive proteins such as fibrinogen, fibrin, and vWF. They also accelerate thrombin formation. The net result is a hemostatic plug or thrombus composed of activated platelets embedded within a fibrin mesh — a structure that can withstand the shear forces generated by flowing blood in the arterial circulation.

In general terms, this description has withstood the test of time, evolving as new molecules are shown to be participants in or regulators of hemostatic plug formation. Equally important, a number of successful drugs have been developed that target key events in platelet activation with the goal of preventing pathological thrombus formation. These include drugs that block ADP receptors, impair TxA_2 synthesis, or prevent the binding of adhesive proteins to activated $\alpha_{IIb}\beta_3$. However, even as it accounts for the initiating events of platelet activation, this model leaves some intriguing questions unanswered. For example, what happens *after* activated platelets stick to each other? Are interactions between platelets required beyond those mediated by $\alpha_{IIb}\beta_3$? What prevents destabilization of the hemostatic plug during the time needed for wound healing to occur? What keeps the integrin engaged with its ligands? What is happening in the gap between activated platelets? How do events on the surface of opposing platelets compare with events on the unopposed surfaces of activated platelets? This Review will address some of these issues, focusing on events within and across the gap between activated platelets and the role that those events play in thrombus growth and stability.

Nonstandard abbreviations used: CD40L, CD40 ligand; ESAM, endothelial cell–specific adhesion molecule; Gas6, growth arrest–specific gene 6; JAM, junctional adhesion molecule; PEAR1, platelet endothelial aggregation receptor 1; sCD40L, soluble CD40L; SLAM, signaling lymphocytic activation molecule; TxA_2, thromboxane A_2.

Conflict of interest: The authors have declared that no conflict of interest exists.

Citation for this article: *J. Clin. Invest.* **115**:3385–3392 (2005).
doi:10.1172/JCI26869.

Gaps, contacts, and the role of clot retraction

In contrast to most cells, platelets are not normally in stable contact with each other but develop such contacts once aggregation has begun. Electron micrographs show the close proximity of the plasma membranes of adjacent platelets but do not show adherens or tight junctions such as those formed by epithelial and endothelial cells (1–3). Estimates for the width of the gap between adjacent platelets range from as little as 0 to as much as 50 nm (4). This short distance between platelets should make it possible for molecules on the surface of one platelet to bind to molecules on an adjacent platelet. This could be a direct interaction, as when one cell adhesion molecule binds to another in trans, or an indirect interaction, such as occurs when multivalent adhesive proteins link activated $\alpha_{IIb}\beta_3$ on adjacent platelets. In either case, these interactions can theoretically provide both an adhesive force and a secondary source of intracellular signaling. Junctional molecules have been detected in platelets, as have some of the molecules associated with junction formation in other types of cells (5). Close contacts between platelets not only allow platelet/platelet interactions to occur but can also limit the diffusion of plasma molecules into the gap between platelets and prevent the escape of platelet activators from within the gap. This might, for example, limit the access of plasmin to embedded fibrin, thereby helping to prevent premature dissolution of the hemostatic plug. It might also foster the accumulation of platelet activators within a protected environment in which higher concentrations can be reached and maintained. Seen in this context, clot retraction, which is dependent on the interaction between actin/myosin complexes and the cytoplasmic domain of $\alpha_{IIb}\beta_3$ as well as the binding of fibrinogen or vWF to the extracellular domain of the integrin, can be viewed as a mechanism for narrowing the gaps between platelets and increasing the local concentration of soluble ligands for platelet receptors (6).

Any list of molecules that participate in contact-dependent and contact-facilitated interactions between platelets is likely to be incomplete, but a number of participants and potential participants have been identified. Those that will be discussed here include integrins and other cell adhesion molecules, direct interactions between receptors and surface-attached ligands, and bioactive molecules that are either secreted or proteolytically shed from the surface of activated platelets (Figure 1 and Table 1). Some of these molecules also play a role in the interaction of platelets with other types of cells, but the focus here will be on platelet/platelet interactions.

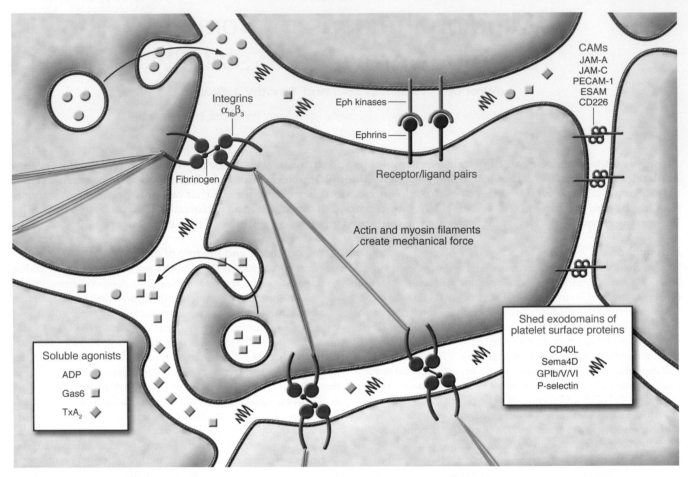

Figure 1
Events in the gap. The gap between activated platelets in a growing thrombus is small enough for integrins and other cell adhesion molecules (CAMs) to interact and for interactions to occur between receptors such as Eph kinases and their cell surface ligands, known as ephrins. The space between platelets also provides a protected environment in which soluble agonists (ADP and TxA$_2$) and the proteolytically shed exodomains of platelet surface proteins can accumulate. Signaling by ephrins and Eph kinases promotes integrin engagement and outside-in signaling. The mechanical forces generated by the contraction of actin/myosin filaments may compress the space between platelets, improving contacts and increasing the concentration of soluble agonists. The binding of cell adhesion molecules to their partners may limit diffusion into and out of the space between platelets in addition to stabilizing platelet/platelet interactions.

Integrins, adhesion, and outside-in signaling

Outside-in signaling refers to the intracellular signaling events that occur downstream of activated integrins once ligand binding has occurred. Since this topic has been reviewed by others (7) and is covered in another Review in this series (8), only a few points will be made here. Integrin signaling depends in large part on the formation of protein complexes that link to the integrin cytoplasmic domain. Some of the protein/protein interactions that involve the cytoplasmic domains of $\alpha_{IIb}\beta_3$ help regulate integrin activation; others participate in outside-in signaling and clot retraction. Proteins capable of binding directly to the cytoplasmic domains of $\alpha_{IIb}\beta_3$ include β_3-endonexin (9), CIB (10), talin (11), myosin (12), Shc (13), and the tyrosine kinases Src (14) and Syk (15, 16). While some of these interactions require the phosphorylation of tyrosine residues Y747 and Y759 in the β_3 cytoplasmic domain, others do not. Shc, for example, requires Y759 phosphorylation (13). Myosin binding requires phosphorylation of both Y747 and Y759 (12). Fibrinogen binding to the extracellular domain of activated $\alpha_{IIb}\beta_3$ stimulates a rapid increase in the activity of Src family members and Syk. Studies of platelets from

mice lacking these kinases suggest that these events are required for the initiation of outside-in signaling and for full platelet spreading, irreversible aggregation, and clot retraction (17–20).

The phosphorylation of the β_3 cytoplasmic domain is an event of particular relevance to this Review. Phosphorylation is mediated by 1 or more Src family members and can require both activation of the integrin and its engagement with an adhesive protein (18, 19). Mutation of Y747 and Y759 in β_3 to phenylalanine produces mice whose platelets disaggregate and which show reduced clot retraction and a tendency to rebleed from sites on the tail where bleeding times were previously measured (21). Loss of clot retraction is also a hallmark of $\alpha_{IIb}\beta_3$-deficient platelets from patients with Glanzmann thrombasthenia — reflecting the dependence of clot retraction on the interaction of $\alpha_{IIb}\beta_3$ with extracellular fibrin and with intracellular actin/myosin filaments.

Junctional adhesion molecules in the Ig superfamily

JAM-A and JAM-C. Junctional adhesion molecules (JAMs) are Ig domain–containing, Ca^{++}-independent cell adhesion molecules.

Three have been identified thus far: JAM-A (also known as JAM-1 and F11R), JAM-B (JAM-2, VE-JAM), and JAM-C (JAM-3) (22, 23). Platelets express JAM-A and JAM-C. JAMs have an extracellular domain with 2 Ig domains, a single transmembrane region, and a short cytoplasmic tail that terminates in a binding site for cytosolic proteins with an appropriate PDZ domain. JAM-A localizes to tight junctions of endothelial and epithelial cells and is also found on monocytes, neutrophils, and lymphocytes. JAM-C has been found on endothelial cells, lymphatic vessels, DCs, and NK cells. JAM-A contributes to cell-cell adhesion by forming *trans* interactions involving the N-terminal Ig domain. However, JAMs also support heterotypic interactions, in particular integrin binding via their membrane-proximal Ig domain. For example, JAM-A binds $\alpha_L\beta_2$ integrin on leukocytes and contributes to leukocyte extravasation across the endothelium (24). JAM-C forms adhesive interactions with JAM-B and binds to leukocyte $\alpha_M\beta_2$ (25).

Platelet JAM-A was originally described as the antigen for a platelet-activating antibody (26). Subsequent studies showed that the activating effects of the antibody are dependent on activation of platelet FcγRII receptors, a property shared with mAbs directed against other platelet surface proteins as well (27). However, JAM-A has also been shown to induce platelet adhesion and spreading via interactions with immobilized JAM-A in vitro, suggesting that it might play a role in homotypic platelet/platelet and platelet/endothelial interactions as well as mediating heterophilic interactions between platelets and leukocytes by binding to $\alpha_L\beta_2$ (24, 28). Similarly, JAM-C was found to bind $\alpha_M\beta_2$ integrin on leukocytes, suggesting a role for JAM-C in platelet/leukocyte interactions (25). The contribution of JAMs to platelet-platelet adhesion has yet to be determined. Interestingly, the cytoplasmic tail of JAM-A becomes phosphorylated via a PKC-dependent mechanism following platelet aggregation (29), suggesting that JAM-A may have a role in contact-dependent signaling.

ESAM. Endothelial cell–specific adhesion molecule (ESAM) is structurally similar to the JAMs (30) with an extracellular domain consisting of 2 Ig domains and a single transmembrane domain. The cytoplasmic tail of ESAM is longer than any of the JAMs and contains several proline-rich regions, as well as a C-terminal PDZ target domain (30). ESAM was identified as a cell adhesion molecule in endothelial cells (30), but it is also expressed by platelets. In endothelial cells ESAM binds via homotypic interactions (30, 31), colocalizes with tight junction proteins (31), and binds to the PDZ domain–containing adaptor protein MAGI-1 (32). ESAM is readily detected on the surface of activated platelets, but not on resting platelets, suggesting that it may be located on an intracellular structure, such as the membranes of α-granules (31). Its function on platelets is unknown.

CD226. CD226, or DNAM-1, is a fourth Ig superfamily adhesion molecule on the surface of platelets (33). CD226 consists of an extracellular domain with 2 Ig domains, a single transmembrane domain, and a cytoplasmic tail (34). CD226-mediated adhesion of NK cells is dependent on phosphorylation of Ser329 within the CD226 cytoplasmic tail by PKC (35). Cross-linking of CD226 on NK cells results in tyrosine phosphorylation of the cytoplasmic domain and increased cytolytic activity, indicating a possible role in signal transduction. In addition to an association with the integrin $\alpha_L\beta_2$, a recent study identified the polio virus receptor (PVR; CD155) and the adherens junction protein nectin-2 (36) as potential ligands for CD226. CD226 has been shown to participate in the binding of activated platelets to endothelial cells in vitro (34).

Endothelial cells express nectin-2, so it is possible that it serves as a ligand for platelet CD226 as well. It is unknown whether platelets express nectins, or whether CD226 also mediates platelet-platelet adhesion. However, CD226-mediated adhesion and CD226 tyrosine phosphorylation are dependent on platelet activation (34), again suggesting the possibility that CD226 participates in platelet signal transduction pathways in the late stages of platelet activation and thrombus formation.

PECAM-1. Platelet–endothelial cell adhesion molecule (PECAM-1; CD31) is known for its high level of expression on endothelial cells, where it accumulates at junctions between cells. However, as its name suggests, PECAM-1 is also expressed on the surface of resting and activated platelets. PECAM-1 is a type 1 transmembrane protein with 6 extracellular Ig domains and an extended cytoplasmic domain of approximately 118 residues (reviewed in ref. 37). The most membrane-distal Ig domain can support homotypic interactions in *trans*. The C-terminus contains phosphorylatable tyrosine residues that represent tandem ITIM domains capable of binding the tyrosine phosphatases SHP-2 and, possibly, SHP-1 (38). Loss of PECAM-1 expression in mice increases responsiveness to collagen via the GPVI collagen receptor, consistent with a model in which platelet activation leads to tyrosine phosphorylation of PECAM-1, allowing SHP-1/2 to bind and bringing the phosphatase near its substrates, including the Fcγ chain partner of GPVI (39, 40). This may normally provide a braking effect on collagen-induced signaling and prevent unwarranted platelet activation, but how this fits within the context of contact-induced signaling once platelets begin to aggregate remains to be clarified.

SLAM and CD84. Signaling lymphocytic activation molecule (SLAM; CD150) and CD84 are members of the CD2 family of homophilic adhesion molecules that have been studied extensively in lymphocytes, but have now been shown to be expressed in platelets as well (41–43). The members of the family are type 1 membrane glycoproteins in the Ig superfamily. Notable differences among them are in the cytoplasmic domain, which supports binding interactions with a variety of adaptor/partner proteins. SLAM and CD84 are expressed on the surface of resting as well as activated platelets and become tyrosine phosphorylated during platelet activation, but only if aggregation is allowed to occur (41). Immobilized CD84 causes platelet activation in vitro. Mice that lack SLAM have a defect in platelet aggregation in response to collagen or a PAR4-activating peptide, but a normal response to ADP and a normal bleeding time. In a mesenteric vascular injury model, female *SLAM*$^{-/-}$ mice showed a marked decrease in platelet accumulation. Male mice were normal (41). The presence of SLAM, CD84, and 2 of their known adaptor proteins (SAP and EAT-2) in platelets provides a novel mechanism by which close contacts between platelets can support thrombus stability.

Contact-dependent signaling

Eph kinases and ephrins. Direct contacts between platelets promote signaling by more than 1 mechanism. In addition to signaling events that occur downstream from cell adhesion molecules, there are receptors that interact in *trans* with cell surface ligands. This type of interaction is illustrated by Eph kinases and their membrane-bound ligands, known as ephrins. Eph kinases are a large family of receptor tyrosine kinases with an extracellular ligand-binding domain and an intracellular kinase domain and PDZ target domain. Like signaling by other receptor tyrosine kinases, signaling by the Eph kinases is promoted by the clus-

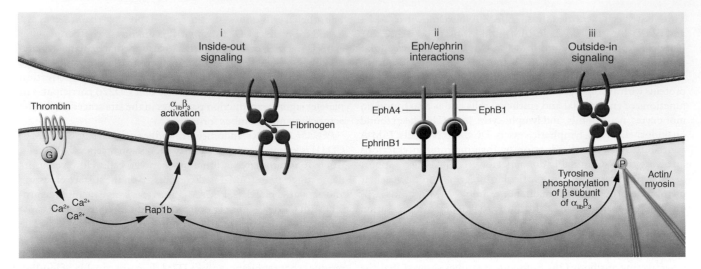

Figure 2
Eph/ephrin interactions between platelets. Agonists such as thrombin, ADP, and TxA$_2$ are ligands for G protein–coupled receptors, causing a rapid increase in the cytosolic Ca^{2+} concentration and producing the inside-out signaling events that lead to integrin $\alpha_{IIb}\beta_3$ activation (i). The increasingly stable contacts between platelets permit ephrinB1 to bind to EphA4 and EphB1. As discussed in the text, Eph/ephrin interactions support inside-out signaling by activating Rap1b (ii) and also promote outside-in signaling by promoting the tyrosine phosphorylation of the β subunit of $\alpha_{IIb}\beta_3$ (iii).

tering and transphosphorylation induced by ligand binding. The ligands for Eph kinases are cell surface proteins known as ephrins that fall into 2 groups depending on whether they have a glycophosphatidylinositol anchor (the ephrin A family) or a transmembrane domain (the ephrin B family). In turn, Eph kinases are divided into 2 groups by ligand preference. For the most part, A ligands bind to A kinases and B ligands bind to B kinases, although exceptions do exist (44, 45). The cytoplasmic domains of the ephrin B family members are highly related. Each is composed of 90–100 residues with 5 conserved tyrosines, at least 3 of which can become phosphorylated (46), and a C-terminus that can serve as a binding site for cytoplasmic proteins with appropriate PDZ domains (47–50).

Eph kinases and ephrins are best known for their role in neuronal organization and brain development (51–53), and as markers distinguishing arteries from veins during vasculogenesis (54). Ephrins can also serve as viral entry receptors (55). Eph/ephrin interactions are particularly relevant to contact-dependent signaling because the binding of an ephrin to an Eph kinase triggers signaling in both the receptor-expressing cell and the ligand-expressing cell (47, 48, 56–60). These events can be dependent on the tyrosine phosphorylation of Eph and ephrin, but phosphorylation-independent interactions have also been described, including those mediated by the PDZ target domains (47–49, 57, 61–65) and the sterile α motif domains (66–69). Taken together, these binding domains make possible the regulated formation of large signaling complexes around clustered Eph kinases and ephrins (45, 51, 70, 71).

Human platelets express 2 Eph kinases, EphA4 and EphB1, and at least 1 ligand, ephrinB1, that can bind to both (72). Forced clustering of either EphA4 or ephrinB1 causes platelets to adhere to immobilized fibrinogen. Clustering of ephrinB1 also causes the activation of Rap1B, a known intermediate in integrin activation in platelets (73, 74), and promotes platelet aggregation initiated by suboptimal concentrations of agonists (72, 75). During platelet activation the signaling complexes that form with EphA4 include

the Src family members Lyn and Fyn (72). Blockade of Eph/ephrin interactions leads to reversible platelet aggregation at low agonist concentrations and limits the growth of platelet thrombi on collagen-coated surfaces under arterial flow conditions (72, 76). It also impairs phosphorylation of the $\beta3$ cytoplasmic domain, inhibiting the association of myosin with $\alpha_{IIb}\beta_3$ and, as a result, impairing clot retraction (76). EphA4 is constitutively associated with $\alpha_{IIb}\beta_3$ in both resting and activated platelets and colocalizes with the integrin at sites of contact between aggregated platelets (76). Collectively, these observations suggest a model in which the onset of aggregation brings platelets into close proximity and allows ephrinB1 to bind to EphA4 and EphB1. Signals downstream of both the receptors and the kinases then promote further integrin activation (in part by activating Rap1b) and integrin signaling (in part by promoting $\beta3$ phosphorylation). In turn, these events promote thrombus growth and stability (Figure 2).

PEAR1. Platelet endothelial aggregation receptor 1 (PEAR1) is a newly described molecule expressed on platelets and endothelial cells (77). It is a 140- to 150-kDa type 1 transmembrane protein with 15 EGF repeats that is equally expressed on the surface of resting and activated platelets and has, as yet, no known ligand. Platelet activation leads to the phosphorylation of tyrosine and serine residues within the PEAR1 cytoplasmic domain and supports association with Shc. Phosphorylation in activated platelets can be inhibited by prevention of aggregation, suggesting that PEAR1-dependent events are dependent on contact and not just activation.

Shedding and secreting into the gap

More than just an empty space, the gap between platelets provides a safe harbor in which platelet-derived molecules can accumulate. In addition to the numerous plasma- and megakaryocyte-derived proteins that are secreted from platelet α-granules (78), activated platelets release agonists such as ADP and TxA$_2$ and presumably continue to do so even after thrombus formation has begun. Platelets also shed surface molecules, including GPIbα (79), GPV (80), GPVI

Table 1

Molecules that participate in contact-facilitated events in platelets

Integrins and outside-in signaling

$\alpha_{IIb}\beta_3$

Junctional adhesion (and signaling) molecules

JAM-A

JAM-C

ESAM

CD226

PECAM-1

SLAM and CD84

Contact-dependent signaling molecules

EphrinB1 and EphA4/B1

PEAR1

Ligands that are shed or secreted

ADP

TxA$_2$

CD40L

Sema4D

Gas6

(81, 82), and P-selectin (83). Cleavage of each of these proteins can be prevented with inhibitors of metalloproteases, and, in at least 2 cases (GPIbα and GPV), a role for a particular metalloprotease, ADAM17, has been established through studies on platelets from mice that lack it (79, 80). The advantage that the platelet derives from shedding surface proteins can be surmised but is not entirely clear. Downregulation of responsiveness to collagen has been proposed as the benefit derived from shedding GPVI (82). Loss of the exodomain of GPIbα, which serves as the primary vWF receptor on platelets, would also impair platelet interactions with collagen, but since cleavage occurs after platelet activation, the impact on thrombus prevention is not readily apparent. It is possible that the reduction in net surface charge that would accompany loss of the negatively charged GPIbα exodomain allows platelets to pack more closely together in a thrombus. It may also expose interaction sites on the remaining extracellular portion of GPIbα that are occluded in the intact molecule, but this remains speculative. In contrast to GPIb and GPVI, where cleavage of the exodomain may predominantly impair function, cleavage of 2 other molecules on the surface of activated platelets, CD40 ligand and sema4D, gives rise to bioactive fragments that can stimulate platelets as well as other nearby cells.

CD40L. CD40 ligand (CD40L; CD154) is a 33-kDa transmembrane protein that is present on the surface of activated platelets, but not resting platelets (84–86). Its appearance on the platelet surface is followed by the gradual release of an 18-kDa exodomain fragment of CD40L (85). Both the surface-bound form and the soluble form of CD40L (sCD40L) are trimers (87). Platelets also express CD40, the 48-kDa transmembrane protein that acts as a receptor for CD40L (85, 88, 89). Unlike CD40L, CD40 is detectable on the surface of resting as well as activated platelets (85, 88). CD40L is a member of the TNF family, and platelet-derived sCD40L or activated, CD40L-expressing platelets can elicit responses from endothelial cells and monocytes that appear to be proatherogenic (84, 85, 90–92). The extracellular portion of CD40L includes a binding domain for CD40 and a KGD (RGD in mice) integrin-recognition sequence. The bind-

ing of sCD40L to activated platelets is blocked by mutations in the KGD sequence and by antibodies to $\alpha_{IIb}\beta_3$. Binding occurs to platelets that lack CD40, but not to platelets that lack $\beta3$, all of which suggests that sCD40L can bind to activated $\alpha_{IIb}\beta_3$ on platelets and that the KGD sequence is involved (88). Consistent with a role in thrombus formation, *CD40L$^{-/-}$* mice show delayed occlusion following vascular injury and decreased thrombus stability (88). Loss of the gene encoding CD40 has no apparent effect (88). Infusion of sCD40L into *CD40L$^{-/-}$* mice reverses the phenotype observed after FeCl$_3$-induced vascular injury, but sCD40L with an altered KGD sequence does not. Notably, platelet aggregation is normal in *CD40L$^{-/-}$* platelets, although growth of platelet plugs on collagen-coated surfaces under shear is affected (88, 93). All of this suggests that the effects of CD40L on platelets are mediated by $\alpha_{IIb}\beta_3$, not CD40, and that platelet behavior in the relatively low-shear conditions present in an aggregometer cuvette may not fully reflect the events that are relevant to thrombus stability.

Sema4D and its receptors. Semaphorins are a large family of structurally related proteins (94, 95). Like Eph kinases and ephrins, semaphorins are best known for their role in the CNS, but individual family members have been found elsewhere, including hematopoietic cells. Semaphorins have in common a 500–amino acid residue extracellular "sema" domain that forms a 7-bladed propeller structure (96). Sema4D, or CD100, is a 150-kDa type 1 glycoprotein that forms a disulfide-linked homodimer. Each subunit has an N-terminal sema domain, an Ig domain, a lysine-rich stretch, a transmembrane domain, and a cytoplasmic tail with consensus tyrosine and serine phosphorylation sites whose role is unclear. Sema4D is best known for its role in lymphocyte biology. Soluble sema4D released from activated lymphocytes inhibits immune cell migration, acts as a costimulator for CD40-induced B cell proliferation and Ig production, and induces release of proinflammatory cytokines by monocytes (97–100). *Sema4D$^{-/-}$* mice are viable but show defective B cell development, impaired T cell activation, and blunted immune responses (101). Two receptors have been identified for sema4D: plexin-B1, a high-affinity receptor expressed on endothelial cells (102, 103), and CD72, which is expressed on lymphocytes (104–106). Plexin-B1 was recently shown to elicit proangiogenic responses and signaling in endothelial cells (102, 107, 108).

We have recently identified sema4D on the surface of human and mouse platelets (109). Platelet activation causes progressive loss of intact sema4D and the shedding of its disulfide-linked exodomain. Shedding is sensitive to metalloprotease inhibitors (110). These results suggest that at least some of the sema4D expressed in platelets is initially located on the cell surface and that platelet activation leads to cleavage of the protein by a platelet-associated metalloprotease, releasing a larger fragment and leaving a smaller fragment behind in the membrane. Western blots show that platelets also express CD72 and, therefore, may be capable of responding to as well as releasing sema4D (110). The rate at which sema4D is shed from activated platelets, like the rate of CD40L shedding, lags behind the rate of platelet aggregation. This suggests that at least some of the shed sema4D exodomain should be protected in the gaps between platelets, while the rest is washed downstream where it can bind to endothelial cells.

Gas6 and its receptors. Growth arrest–specific gene 6 (Gas6) is a 75-kDa secreted protein related to protein S. Like protein S and other proteins whose synthesis requires vitamin K as a cofactor,

Gas6 contains γ-carboxylated glutamic acid residues that allow it to bind to negatively charged phospholipids in a Ca^{++}-dependent manner (111). Gas6 is expressed in a number of tissues, including vascular smooth muscle, and levels of expression are upregulated following vascular injury. In mouse platelets, Gas6 is found in the α-granules (112, 113). Upon activation by thrombin, ADP, or collagen, Gas6 is secreted from rat platelets (112). Gas6 is a ligand for the related receptor tyrosine kinases Axl, Sky, and Mer (114, 115), all of which are expressed on platelets (113). Like other growth factor receptors, Axl family members have been shown to stimulate PI3K and phospholipase Cγ (111). A reasonable hypothesis is that secreted Gas6 could bind to receptors on the platelet surface and cause signaling that would promote platelet plug formation and stability, perhaps while binding to negatively charged phospholipids on adjacent platelets. Consistent with this hypothesis, platelets from $Gas6^{-/-}$ mice were found to have an aberrant response to agonists in which aggregation begins but ends prematurely (113). Furthermore, although the tail bleeding time of the $Gas6^{-/-}$ mice was normal, the mice were resistant to thrombosis (113), as are mice lacking any 1 of the 3 Gas6 receptors. Platelets from the receptor-deleted mice also failed to aggregate normally in response to agonists (116–118). Biochemical studies showed that Gas6 signaling promotes β3 phosphorylation and, therefore, clot retraction (117).

Collectively, these results are consistent with a role for secreted Gas6 in promoting integrin outside-in signaling and helping to perpetuate and stabilize platelet plug formation. Most granule secretion occurs after the onset of aggregation, and Gas6 secreted into the confined spaces between platelets might achieve high local concentrations and cause receptor activation. However, a number of issues remain to be settled before this hypothesis can be fully accepted. In particular, the biology of the Gas6 system needs to be explored further in humans, as most current evidence is from experiments with mice. At least 1 recent study suggests that human platelets do not store and secrete Gas6 (119).

Conclusion

There is now ample evidence that the signaling events that support platelet activation continue after integrin activation, granule secretion, and platelet aggregation have begun. Some of these events can reasonably be expected to promote the growth and stability of the hemostatic plug, support clot retraction, and help to maintain the plug in place until wound healing is complete or at least well under way. In contrast to the initiating events of platelet activation, these "late" events can take advantage of the close proximity between platelets once aggregation begins and may even extend to the contacts that develop between platelets, endothelial cells, and leukocytes as the hemostatic plug evolves. The models of contact-dependent signaling discussed in this Review are just examples of what will probably turn out to be a more complex process involving additional cell surface molecules whose function remains to be identified. The picture that emerges is that the gaps and contact points between activated platelets in a thrombus are sites of highly relevant activity.

Acknowledgments

This work was supported by grants from the National Heart, Lung, and Blood Institute (to L.F. Brass) and the American Heart Association (to T.J. Stalker).

Address correspondence to: Lawrence Brass, University of Pennsylvania, Room 915 Biomedical Research Building II, 421 Curie Boulevard, Philadelphia, Pennsylvania 19104, USA. Phone: (215) 573-3540; Fax: (215) 573-2189; E-mail: Brass@mail.med.upenn.edu.

1. Humbert, M., et al. 1996. Ultrastructural studies of platelet aggregates from human subjects receiving clopidogrel and from a patient with an inherited defect of an ADP-dependent pathway of platelet activation. *Arterioscler. Thromb. Vasc. Biol.* **16**:1532–1543.
2. White, J.G. 1972. Interaction of membrane systems in blood platelets. *Am. J. Pathol.* **66**:295–312.
3. White, J.G. 1988. Platelet membrane ultrastructure and its changes during platelet activation. *Prog. Clin. Biol. Res.* **283**:1–32.
4. Skaer, R.J., Emmines, J.P., and Skaer, H.B. 1979. The fine structure of cell contacts in platelet aggregation. *J. Ultrastruct. Res.* **69**:28–42.
5. Elrod, J.W., et al. 2003. Expression of junctional proteins in human platelets. *Platelets.* **14**:247–251.
6. Tschumperlin, D.J., et al. 2004. Mechanotransduction through growth-factor shedding into the extracellular space. *Nature.* **429**:83–86.
7. Shattil, S.J., and Newman, P.J. 2004. Integrins: dynamic scaffolds for adhesion and signaling in platelets. *Blood.* **104**:1606–1615.
8. Bennett, J.S. 2005. Structure and function of the platelet integrin $\alpha_{IIb}\beta_3$. *J. Clin. Invest.* **115**:3363–3369. doi:10.1172/JCI26989.
9. Shattil, S.J., et al. 1995. β3-Endonexin, a novel polypeptide that interacts specifically with the cytoplasmic tail of the integrin β3 subunit. *J. Cell Biol.* **131**:807–816.
10. Naik, U.P., Patel, P.M., and Parise, L.V. 1997. Identification of a novel calcium-binding protein that interacts with the integrin alphaIIb cytoplasmic domain. *J. Biol. Chem.* **272**:4651–4654.
11. Calderwood, D.A., et al. 1999. The talin head domain binds to integrin beta subunit cytoplasmic tails and regulates integrin activation. *J. Biol. Chem.*

274:28071–28074.
12. Jenkins, A.L., et al. 1998. Tyrosine phosphorylation of the beta3 cytoplasmic domain mediates integrin-cytoskeletal interactions. *J. Biol. Chem.* **273**:13878–13885.
13. Cowan, K.J., Law, D.A., and Phillips, D.R. 2000. Identification of Shc as the primary protein binding to the tyrosine-phosphorylated beta3 subunit of alphaIIbbeta3 during outside-in integrin platelet signaling. *J. Biol. Chem.* **275**:36423–36429.
14. Arias-Salgado, E.G., et al. 2003. Src kinase activation by direct interaction with the integrin beta cytoplasmic domain. *Proc. Natl. Acad. Sci. U. S. A.* **100**:13298–13302.
15. Gao, J., Zoller, K.E., Ginsberg, M.H., Brugge, J.S., and Shattil, S.J. 1997. Regulation of the pp72syk protein tyrosine kinase by platelet integrin alphaIIbbeta3. *EMBO J.* **16**:6414–6425.
16. Woodside, D.G., et al. 2001. Activation of Syk protein tyrosine kinase through interaction with integrin beta cytoplasmic domains. *Curr. Biol.* **11**:1799–1804.
17. Payrastre, B., et al. 2000. The integrin alphaIIb/beta3 in human platelet signal transduction. *Biochem. Pharmacol.* **60**:1069–1074.
18. Philips, D.R., Prasad, K.S.S., Manganello, J., Bao, M., and Nannizzi-Alaimo, L. 2001. Integrin tyrosine phosphorylation in platelet signaling. *Curr. Opin. Cell Biol.* **13**:546–554.
19. Phillips, D.R., Nannizzi-Alamio, L., and Prasad, K.S.S. 2001. β3 Tyrosine phosphorylation in αIIbβ3 (platelet membrane GP IIb-IIIa) outside-in integrin signaling. *Thromb. Haemost.* **86**:246–258.
20. Obergfell, A., et al. 2002. Coordinate interactions of Csk, Src and Syk kinases with alphaIIbbeta3 ini-

tiate integrin signaling to the cytoskeleton. *J. Cell Biol.* **157**:265–275.
21. Law, D.A., et al. 1999. Integrin cytoplasmic tyrosine motif is required for outside-in alphaIIbbeta3 signalling and platelet function. *Nature.* **401**:808–811.
22. Muller, W.A. 2003. Leukocyte-endothelial-cell interactions in leukocyte transmigration and the inflammatory response. *Trends Immunol.* **24**:327–334.
23. Bazzoni, G. 2003. The JAM family of junctional adhesion molecules. *Curr. Opin. Cell Biol.* **15**:525–530.
24. Ostermann, G., Weber, K.S., Zernecke, A., Schroder, A., and Weber, C. 2002. JAM-1 is a ligand of the beta(2) integrin LFA-1 involved in transendothelial migration of leukocytes. *Nat. Immunol.* **3**:151–158.
25. Santoso, S., et al. 2002. The junctional adhesion molecule 3 (JAM-3) on human platelets is a counterreceptor for the leukocyte integrin Mac-1. *J. Exp. Med.* **196**:679–691.
26. Kornecki, E., Walkowiak, B., Naik, U.P., and Ehrlich, Y.H. 1990. Activation of human platelets by a stimulatory monoclonal antibody. *J. Biol. Chem.* **265**:10042–10048.
27. Naik, U.P., Ehrlich, Y.H., and Kornecki, E. 1995. Mechanisms of platelet activation by a stimulatory antibody: cross-linking of a novel platelet receptor for monoclonal antibody F11 with the Fc gamma RII receptor. *Biochem. J.* **310**:155–162.
28. Babinska, A., et al. 2002. F11-receptor (F11R/JAM) mediates platelet adhesion to endothelial cells: role in inflammatory thrombosis. *Thromb. Haemost.* **88**:843–850.
29. Ozaki, H., et al. 2000. Junctional adhesion molecule (JAM) is phosphorylated by protein kinase C upon platelet activation. *Biochem. Biophys. Res. Commun.* **276**:873–878.

30. Hirata, K., et al. 2001. Cloning of an immunoglobulin family adhesion molecule selectively expressed by endothelial cells. *J. Biol. Chem.* **276**:16223–16231.

31. Nasdala, I., et al. 2002. A transmembrane tight junction protein selectively expressed on endothelial cells and platelets. *J. Biol. Chem.* **277**:16294–16303.

32. Wegmann, F., Ebnet, K., Du Pasquier, L., Vestweber, D., and Butz, S. 2004. Endothelial adhesion molecule ESAM binds directly to the multidomain adaptor MAGI-1 and recruits it to cell contacts. *Exp. Cell Res.* **300**:121–133.

33. Scott, J.L., et al. 1989. Characterization of a novel membrane glycoprotein involved in platelet activation. *J. Biol. Chem.* **264**:13475–13482.

34. Kojima, H., et al. 2003. CD226 mediates platelet and megakaryocytic cell adhesion to vascular endothelial cells. *J. Biol. Chem.* **278**:36748–36753.

35. Shibuya, A., Lanier, L.L., and Phillips, J.H. 1998. Protein kinase C is involved in the regulation of both signaling and adhesion mediated by DNAX accessory molecule-1 receptor. *J. Immunol.* **161**:1671–1676.

36. Bottino, C., et al. 2003. Identification of PVR (CD155) and Nectin-2 (CD112) as cell surface ligands for the human DNAM-1 (CD226) activating molecule. *J. Exp. Med.* **198**:557–567.

37. Newman, P.J., and Newman, D.K. 2003. Signal transduction pathways mediated by PECAM-1: new roles for an old molecule in platelet and vascular cell biology. *Arterioscler. Thromb. Vasc. Biol.* **23**:953–964.

38. Newman, P.J. 1999. Switched at birth: a new family for PECAM-1. *J. Clin. Invest.* **103**:5–9.

39. Patil, S., Newman, D.K., and Newman, P.J. 2001. Platelet endothelial cell adhesion molecule-1 serves as an inhibitory receptor that modulates platelet responses to collagen. *Blood.* **97**:1727–1732.

40. Jones, K.L., et al. 2001. Platelet endothelial cell adhesion molecule-1 is a negative regulator of platelet-collagen interactions. *Blood.* **98**:1456–1463.

41. Nanda, N., et al. 2005. Platelet aggregation induces platelet aggregate stability via SLAM family receptor signaling. *Blood.* doi:10.1182/blood-2005-01-0333.

42. Krause, S.W., Rehli, M., Heinz, S., Ebner, R., and Andreesen, R. 2000. Characterization of MAX.3 antigen, a glycoprotein expressed on mature macrophages, dendritic cells and blood platelets: identity with CD84. *Biochem. J.* **346**:729–736.

43. Martin, M., et al. 2001. CD84 functions as a homophilic adhesion molecule and enhances IFN-gamma secretion: adhesion is mediated by Ig-like domain 1. *J. Immunol.* **167**:3668–3676.

44. Gale, N.W., et al. 1996. Eph receptors and ligands comprise two major specificity subclasses and are reciprocally compartmentalized during embryogenesis. *Neuron.* **17**:9–19.

45. Klein, R. 2001. Excitatory Eph receptors and adhesive ephrin ligands. *Curr. Opin. Cell Biol.* **13**:196–203.

46. Kalo, M.S., Yu, H.H., and Pasquale, E.B. 2001. In vivo tyrosine phosphorylation sites of activated ephrin-B1 and EphB2 from neural tissues. *J. Biol. Chem.* **276**:38940–38948.

47. Torres, R., et al. 1998. PDZ proteins bind, cluster, and synaptically colocalize with Eph receptors and their ephrin ligands. *Neuron.* **21**:1453–1463.

48. Lin, D., Gish, G.D., Songyang, Z., and Pawson, T. 1999. The carboxyl terminus of B class ephrins constitutes a PDZ binding motif. *J. Biol. Chem.* **274**:3726–3733.

49. Bruckner, K., et al. 1999. EphrinB ligands recruit GRIP family PDZ adaptor proteins into raft membrane microdomains. *Neuron.* **22**:511–524.

50. Della Rocca, G.J., et al. 1997. Ras-dependent mitogen-activated protein kinase activation by G protein-coupled receptors: convergence of Gi- and Gq-mediated pathways on calcium/calmodulin, Pyk2, and Src kinase. *J. Biol. Chem.* **272**:19125–19132.

51. Kullander, K., and Klein, R. 2002. Mechanisms and function of Eph and ephrin signaling. *Nat. Rev. Mol. Cell Biol.* **3**:475–486.

52. Holmberg, J., and Frisen, J. 2002. Ephrins are not only unattractive. *Trends Neurosci.* **25**:239–243.

53. Depaepe, V., et al. 2005. Ephrin signalling controls brain size by regulating apoptosis of neural progenitors. *Nature.* **435**:1244–1250.

54. Adams, R.H., and Klein, R. 2000. Eph receptors and ephrin ligands: essential mediators of vascular development. *Trends Cardiovasc. Med.* **10**:183–188.

55. Negrete, O.A., et al. 2005. EphrinB2 is the entry receptor for Nipah virus, an emergent deadly paramyxovirus. *Nature.* **436**:401–405.

56. Holland, S.J., et al. 1997. Juxtamembrane tyrosine residues couple the Eph family receptor EphB2/Nuk to specific SH2 domain proteins in neuronal cells. *EMBO J.* **16**:3877–3888.

57. Hock, B., et al. 1998. PDZ-domain-mediated interaction of the Eph-related receptor tyrosine kinase EphB3 and the ras-binding protein AF6 depends on the kinase activity of the receptor. *Proc. Natl. Acad. Sci. U. S. A.* **95**:9779–9784.

58. Dodelet, V.C., Pazzagli, C., Zisch, A.H., Hauser, C.A., and Pasquale, E.B. 1999. A novel signaling intermediate, SHEP1, directly couples Eph receptors to R-Ras and Rap1A. *J. Biol. Chem.* **274**:31941–31946.

59. Pandey, A., Duan, H., and Dixit, V.M. 1995. Characterization of a novel src-like adapter protein that associates with the Eck receptor tyrosine kinase. *J. Biol. Chem.* **270**:19201–19204.

60. Cowan, C.A., and Henkemeyer, M. 2001. The SH2/SH3 adaptor Grb4 transduces B-ephrin reverse signals. *Nature.* **413**:174–179.

61. Birgbauer, E., Cowan, C.A., Sretavan, D.W., and Henkemeyer, M. 2000. Kinase independent function of EphB receptors in retinal axon parhfinding to the optic disc from dorsal but not ventral retina. *Development.* **127**:1231–1241.

62. Buchert, M., et al. 1999. The junction-associated protein AF-6 interacts and clusters with specific Eph receptor tyrosine kinases at specialized sites of cell-cell contact in the brain. *J. Cell Biol.* **144**:361–371.

63. Holland, S.J., et al. 1996. Bidirectional signaling through the EPH-family receptor Nuk and its transmembrane ligands. *Nature.* **383**:722–725.

64. Palmer, A., et al. 2002. Ephrin B phosphorylation and reverse signaling: regulation by Src kinases and PTP-BL phosphatase. *Mol. Cell.* **9**:725–737.

65. Xu, Z., Kwok-On, L., Zhou, H.-M., Lin, S.-C., and Ip, N.Y. 2003. Ephrin-B1 reverse signaling activates JNK through a novel mechanism that is independent of tyrosine phosphorylation. *J. Biol. Chem.* **278**:24767–24775.

66. Schultz, J., Ponting, C.P., Hofmann, K., and Bork, P. 1997. SAM as a protein interaction domain involved in developmental regulation. *Protein Sci.* **6**:249–253.

67. Thanos, C.D., Goodwill, K.E., and Bowie, J.U. 1999. Oligomeric structure of the human Ephb2 receptor SAM domain. *Science.* **283**:833–836.

68. Stapleton, D., Balan, I., Pawson, T., and Sicheri, F. 1999. The crystal structure of an Eph receptor SAM domain reveals a mechanism for modular dimerization. *Nat. Struct. Biol.* **6**:44–49.

69. Smalla, M., et al. 1999. Solution structure of the receptor tyrosine kinase EphB2 SAM domain and identification of two distinct homotypic interaction sites. *Protein Sci.* **8**:1954–1961.

70. Wilkinson, D.G. 2000. Eph receptors and ephrins: regulators of guidance and assembly. *Int. Rev. Cytol.* **196**:177–244.

71. Gale, N.W., and Yancopoulos, G.D. 1999. Growth factors acting via endothelial cell-specific receptor tyrosine kinases: VEGFs, angiopoietins, and ephrins in vascular development. *Genes Dev.* **13**:1055–1066.

72. Prevost, N., Woulfe, D., Tanaka, T., and Brass, L.F. 2002. Interactions between Eph kinases and ephrins provide a mechanism to support platelet aggregation once cell-to-cell contact has occurred. *Proc. Natl. Acad. Sci. U. S. A.* **99**:9219–9224.

73. Bertoni, A., et al. 2002. Relationships between Rap1b, affinity modulation of integrin $\alpha_{IIb}\beta_3$, and the actin cytoskeleton. *J. Biol. Chem.* **277**:25715–25721.

74. Chrzanowska-Wodnicka, M., Smyth, S.S., Schoenwaelder, S.M., Fischer, T.H., and White, G.C., 2nd. 2005. Rap1b is required for normal platelet function and hemostasis in mice [corrigendum 2005, **115**:2296]. *J. Clin. Invest.* **115**:680–687. doi:10.1172/JCI200522793.

75. Prevost, N., et al. 2004. Signaling by ephrinB1 and Eph kinases in platelets promotes Rap1 activation, platelet adhesion, and aggregation via effector pathways that do not require phosphorylation of ephrinB1. *Blood.* **103**:1348–1355.

76. Prevost, N., et al. 2005. Eph kinases and ephrins support thrombus growth and stability by regulating integrin outside-in signaling in platelets. *Proc. Natl. Acad. Sci. U. S. A.* **102**:9820–9825.

77. Nanda, N., et al. 2005. Platelet endothelial aggregation receptor 1 (PEAR1), a novel epidermal growth factor repeat-containing transmembrane receptor, participates in platelet contact-induced activation. *J. Biol. Chem.* **280**:24680–24689.

78. Coppinger, J.A., et al. 2004. Characterization of the proteins released from activated platelets leads to localization of novel platelet proteins in human atherosclerotic lesions. *Blood.* **103**:2096–2104.

79. Bergmeier, W., et al. 2004. Tumor necrosis factor-alpha-converting enzyme (ADAM17) mediates GPI-balpha shedding from platelets in vitro and in vivo. *Circ. Res.* **95**:677–683.

80. Rabie, T., Strehl, A., Ludwig, A., and Nieswandt, B. 2005. Evidence for a role of ADAM17 (TACE) in the regulation of platelet glycoprotein V. *J. Biol. Chem.* **280**:14462–14468.

81. Bergmeier, W., et al. 2004. GPVI down-regulation in murine platelets through metalloproteinase-dependent shedding. *Thromb. Haemost.* **91**:951–958.

82. Stephens, G., et al. 2005. Platelet activation induces metalloproteinase-dependent GP VI cleavage to down-regulate platelet reactivity to collagen. *Blood.* **105**:186–191.

83. Berger, G., Hartwell, D.W., and Wagner, D.D. 1998. P-selectin and platelet clearance. *Blood.* **92**:4446–4452.

84. Henn, V., et al. 1998. CD40 ligand on activated platelets triggers an inflammatory reaction of endothelial cells. *Nature.* **391**:591–594.

85. Henn, V., Steinbach, S., Buchner, K., Presek, P., and Kroczek, R.A. 2001. The inflammatory action of CD40 ligand (CD154) expressed on activated human platelets is temporally limited by coexpressed CD40. *Blood.* **98**:1047–1054.

86. Hermann, A., Rauch, B.H., Braun, M., Schror, K., and Weber, A.A. 2001. Platelet CD40 ligand (CD40L): subcellular localization, regulation of expression, and inhibition by clopidogrel. *Platelets.* **12**:74–82.

87. Locksley, R.M., Killeen, N., and Lenardo, M.J. 2001. The TNF and TNF receptor superfamilies: integrating mammalian biology. *Cell.* **104**:487–501.

88. Andre, P., et al. 2002. CD40L stabilizes arterial thrombi by a β3 integrin-dependent mechanism. *Nat. Med.* **8**:247–252.

89. Inwald, D.P., McDowall, A., Peters, M.J., Callard, R.E., and Klein, N.J. 2003. CD40 is constitutively expressed on platelets and provides a novel mechanism for platelet activation. *Circ. Res.* **92**:1041–1048.

90. Mach, F., Schonbeck, U., Sukhova, G.K., Akinson, E., and Libby, P. 1998. Reduction of atherosclerosis in mice by inhibition of CD40 signaling. *Nature.* **394**:200–203.

91. May, A.E., et al. 2002. Engagement of glycoprotein IIb/IIIa on platelets upregulates CD40L and triggers CD40L-dependent matrix degradation by endothelial cells. *Circulation.* **106**:2111–2117.

92. Danese, S., and Fiocchi, C. 2005. Platelet activation

and the CD40/CD40 ligand pathway: mechanisms and implications for human disease. *Crit. Rev. Immunol.* **25**:103–122.

93. Crow, A.R., Leytin, V., Starkey, A.F., Rand, M.L., and Lazarus, A.H. 2003. CD154 (CD40 ligand)-deficient mice exhibit prolonged bleeding time and decreased shear-induced platelet aggregates. *J. Thromb. Haemost.* **1**:850–852.

94. Tamagnone, L., and Comoglio, P.M. 2000. Signalling by semaphorin receptors: cell guidance and beyond. *Trends Cell Biol.* **10**:377–383.

95. Yu, H.H., and Kolodkin, A.L. 1999. Semaphorin signaling: a little less per-plexin. *Neuron.* **22**:11–14.

96. Love, C.A., et al. 2003. The ligand-binding face of the semaphorins revealed by the high-resolution crystal structure of SEMA4D. *Nat. Struct. Biol.* **10**:843–848.

97. Kumanogoh, A., et al. 2000. Identification of CD72 as a lymphocyte receptor for the class IV semaphorin CD100: a novel mechanism for regulating B cell signaling. *Immunity.* **13**:621–631.

98. Ishida, I., et al. 2003. Involvement of CD100, a lymphocyte semaphorin, in the activation of the human immune system via CD72: implications for the regulation of immune and inflammatory responses. *Int. Immunol.* **15**:1027–1034.

99. Delaire, S., et al. 2001. Biological activity of soluble CD100. II. Soluble CD100, similarly to H-SemaIII, inhibits immune cell migration. *J. Immunol.* **166**:4348–4354.

100. Wang, X., et al. 2001. Functional soluble CD100/Sema4D released from activated lymphocytes: possible role in normal and pathologic immune responses. *Blood.* **97**:3498–3504.

101. Shi, W., et al. 2000. The class IV semaphorin CD100 plays nonredundant roles in the immune system: defective B and T cell activation in CD100-deficient mice. *Immunity.* **13**:633–642.

102. Basile, J.R., Barac, A., Zhu, T., Guan, K.L., and Gutkind, J.S. 2004. Class IV semaphorins promote angiogenesis by stimulating Rho-initiated pathways through plexin-B. *Cancer Res.* **64**:5212–5224.

103. Conrotto, P., et al. 2005. Sema4D induces angiogenesis through Met recruitment by Plexin B1. *Blood.* **105**:4321–4329.

104. Kumanogoh, A., and Kikutani, H. 2001. The CD100-CD72 interaction: a novel mechanism of immune regulation. *Trends Immunol.* **22**:670–676.

105. Adachi, T., Flaswinkel, H., Yakura, H., Reth, M., and Tsubata, T. 1998. The B cell surface protein CD72 recruits the tyrosine phosphatase SHP-1 upon tyrosine phosphorylation. *J. Immunol.* **160**:4662–4665.

106. Baba, T., et al. 2005. Dual regulation of BCR-mediated growth inhibition signaling by CD72. *Eur. J. Immunol.* **35**:1634–1642.

107. Tamagnone, L., et al. 1999. Plexins are a large family of receptors for transmembrane, secreted, and GPI-anchored semaphorins in vertebrates. *Cell.* **99**:71–80.

108. Conrotto, P., Corso, S., Gamberini, S., Comoglio, P.M., and Giordano, S. 2004. Interplay between scatter factor receptors and B plexins controls invasive growth. *Oncogene.* **23**:5131–5137.

109. Zhu, L., et al. 2003. Semaphorin 4D (CD100) is expressed on the surface of human platelets and proteolytically shed during platelet activation [abstract]. *Blood.* **102**:292a.

110. Zhu, L., et al. 2005. Regulated shedding of sema4D from the platelet surface produces a bioactive second messenger in thrombotic disorders [abstract]. *Blood.* In press.

111. Melaragno, M.G., Fridell, Y.-W., and Berk, B.C. 1999. The Gas6/Axl system: a novel regulator of vascular cell function. *Trends Cardiovasc. Med.* **9**:250–253.

112. Ishimoto, Y., and Nakano, T. 2000. Release of a product of growth arrest-specific gene 6 from rat platelets. *FEBS Lett.* **466**:197–199.

113. Angelillo-Scherrer, A., et al. 2001. Deficiency or inhibition of Gas6 causes platelet dysfunction and protects mice against thrombosis. *Nat. Med.* **7**:215–221.

114. Stitt, T.N., et al. 1995. The anticoagulation factor protein and its relative, Gas6, are ligands for the Tyro3/Axl family of receptor tyrosine kinases. *Cell.* **80**:661–670.

115. Varnum, B.C., et al. 1995. Axl receptor tyrosine kinase stimulated by the vitamin K-dependent protein encoded by the growth-arrest-specific gene 6. *Nature.* **373**:623–626.

116. Chen, C., et al. 2004. Mer receptor tyrosine kinase signaling participates in platelet function. *Arterioscler. Thromb. Vasc. Biol.* **24**:1118–1123.

117. Angelillo-Scherrer, A., et al. 2005. Role of Gas6 receptors in platelet signaling during thrombus stabilization and implications for antithrombotic therapy. *J. Clin. Invest.* **115**:237–246. doi:10.1172/JCI200522079.

118. Gould, W.R., et al. 2005. Gas6 receptors Axl, Sky and Mer enhance platelet activation and regulate thrombotic responses. *J. Thromb. Haemost.* **3**:733–741.

119. Balogh, I., Hafizi, S., Stenhoff, J., Hansson, K., and Dahlback, B. 2005. Analysis of Gas6 in human platelets and plasma. *Arterioscler. Thromb. Vasc. Biol.* **25**:1280–1286.

Untying the Gordian knot: policies, practices, and ethical issues related to banking of umbilical cord blood

Joanne Kurtzberg,[1] Anne Drapkin Lyerly,[2] and Jeremy Sugarman[3]

[1]Pediatric Blood and Marrow Transplant Program, Department of Obstetrics and Gynecology, Duke University Medical Center, Durham, North Carolina, USA. [2]Center for the Study of Medical Ethics and Humanities and Department of Obstetrics and Gynecology, Duke University Medical Center, Durham, North Carolina, USA. [3]Phoebe R. Berman Bioethics Institute and Department of Medicine, Johns Hopkins University, Baltimore, Maryland, USA.

Since the first successful transplantation of umbilical cord blood in 1988, cord blood has become an important source of hematopoietic stem and progenitor cells for the treatment of blood and genetic disorders. Significant progress has been accompanied by challenges for scientists, ethicists, and health policy makers. With the recent recognition of the need for a national system for the collection, banking, distribution, and use of cord blood and the increasing focus on cord blood as an alternative to embryos as a source of tissue for regenerative medicine, cord blood has garnered significant attention. We review the development of cord blood banking and transplantation and then discuss the scientific and ethical issues influencing both established and investigational practices surrounding cord blood collection, banking, and use.

In 1988, a 6-year-old boy from North Carolina with Fanconi anemia was transplanted in Paris with HLA-matched umbilical cord blood from his baby sister (1). Most scientists and physicians at the time were highly skeptical, doubting that a few ounces of cord blood contained sufficient stem and progenitor cells to rescue bone marrow after myeloablative therapy. However, this child engrafted without incident, fully reconstituting his blood, bone marrow, and immune system with donor cells. He remains well and durably engrafted with donor cells 17 years following the original transplant (J. Kurtzberg, personal communication).

From experimentation to practice: development of cord blood transplantation

Over the 5–6 years following the first cord blood transplant, approximately 60 additional transplants between HLA-matched siblings were performed worldwide. Reports of results to a volunteer registry (2) demonstrated that cord blood contained sufficient numbers of stem and progenitor cells to reconstitute the entire hematopoietic system of a child after myeloablative therapy and that the incidence of graft-versus-host disease (GVHD) was 10-fold lower than that seen after transplantation with HLA-matched bone marrow obtained from a sibling.

At this time it was becoming apparent that the diversity of HLA alleles and antigens was vast and that it was never going to be possible to find fully matched related and unrelated adult donors for all patients in need of allogeneic transplantation therapy from then-available sources. The National Marrow Donor Foundation (NMDP) and other international registries successfully recruited, typed, and listed millions of volunteer unrelated adult donors, but only 25–50% of patients in need could locate sufficiently matched donors in a timely fashion.

Donors for patients of minority ethnic backgrounds were even scarcer and more difficult to locate. To provide donors for all patients in need, transplant physicians needed to find a way to transplant partially mismatched grafts. Transplants using partially HLA-mismatched adult hematopoietic stem cells from mobilized blood or bone marrow, with or without T cell depletion, were failing because of high rates of graft failure, severe GVHD, and failure of the immune system to properly reconstitute for several years after transplantation, leading to death from opportunistic infections (3–5).

The observation that transplantation of HLA-matched umbilical cord blood from donors that were related family members caused less GVHD led to the hypothesis that this graft source might be transplantable in the unrelated-donor setting. To this end, in 1991, the first public cord blood bank in the world was created at the New York Blood Center (6). Cord blood, the residual blood from the baby remaining in the placenta or "afterbirth" delivered in the third stage of labor, was collected ex utero, tested for blood-borne pathogens, cryopreserved, and stored under liquid nitrogen until selected for a transplant patient in need.

In 1993, the first unrelated-donor umbilical cord blood transplant in the world, using a cord blood unit from the bank at the New York Blood Center, was performed in a 3-year-old child with recurrent T cell acute lymphoblastic leukemia. In 1996, the outcomes of this transplant and the next consecutive 24 unrelated-donor cord blood transplants performed at Duke University Medical Center using cord blood units banked at the New York Blood Center were reported (7). Important observations in these patients and subsequent reports from other centers and registries including the New York Blood Center and the European Cord Blood Registry, Eurocord (8–13) demonstrated that unrelated-donor cord blood could engraft in the bone marrow of children undergoing myeloablative therapy for leukemias and genetic diseases (14–18), that reasonable outcomes could be achieved using partially HLA-mismatched grafts, that the incidence and severity of acute and chronic GVHD were lower and milder than those seen in recipients of bone marrow transplants from unrelated donors

Nonstandard abbreviations used: GVHD, graft-versus-host disease; HRSA, Health Resources and Services Administration; IOM, Institute of Medicine.

Conflict of interest: The authors have declared that no conflict of interest exists.

Citation for this article: *J. Clin. Invest.* **115**:2592–2597 (2005). doi:10.1172/JCI26690.

Table 1

Reported engraftment and survival rates following unrelated-donor cord blood transplant

Condition	Engraftment (%)	Survival (%)
Infants with leukemia	80	55
Children with leukemia	75	49
Adults with leukemia	28–78	75
Patients with nonmalignant diseases	70–80	80

(8, 11), but that graft-versus-leukemia effects were retained (13). It also became apparent that cell dose strongly correlated with clinical outcomes, including time to engraftment and probability of overall engraftment and survival (7, 10). Engraftment times were observed to be slower than those of bone marrow or mobilized peripheral blood (8, 12).

Over the 12 years since the first unrelated-donor cord blood transplant was performed at Duke University Medical Center, there have been more than 6,000 unrelated-donor transplants performed in more than 150 locations around the world. In the vast majority of these transplants, HLA mismatching between donor and recipient was present at 1 or 2 HLA antigens. Efficacy has been demonstrated in both children and adults with leukemias (7–13, 19–23) and children with hemoglobinopathies (14, 15), immunodeficiency syndromes (16), bone marrow failure syndromes (24), and inborn errors of metabolism (17, 18). Reported survival rates (Table 1) are similar to those seen in patients transplanted with matched bone marrow from unrelated donors despite the fact that the cord blood was generally mismatched at 1 or 2 HLA loci. The strong correlation of cell dose with engraftment and survival following cord blood transplant has been confirmed. Several retrospective registry analyses of more than 3,000 patients published over the past few years have shown that cord blood engrafts more slowly than bone marrow and that the cumulative incidence of engraftment is slightly lower than that of bone marrow; that the incidence and severity of acute GVHD is lower after cord blood transplantation as compared with bone marrow transplantation (11); and that overall survival rates are comparable. In one study, results obtained following transplant of cord blood from a 5/6 mismatched unrelated donor were found to be equivalent to those obtained following transplant of bone marrow from a 6/6 matched unrelated donor (12). No prospective trials comparing transplant outcomes after bone marrow or cord blood transplantation have been conducted to date.

How much is enough? Initially, because of early results correlating cell dose with engraftment and survival, cord blood transplantation was restricted to use in children and small adults, generally weighing less than 40 kg. More recently, however, the use of cord blood has been extended to include adults, allowing for better definition of cell dose limitations and thresholds (19–24). The results of these transplants have helped define a requirement for a minimum cell dose of 3×10^7 to 3.5×10^7 nucleated cells/kg in order to obtain acceptable clinical outcomes (25). Since only 12% of the current inventory in established public cord blood banks contains sufficient cells to deliver this dose to patients weighing more than 60 kg, alternative strategies to increase cell dose for adults and larger pediatric patients have been explored. Ex vivo expansion from

bulk cord blood or CD34+ cord blood cells selected using various cytokines or supporting cells promotes 10- to 200-fold increases in nucleated cells, clonal progenitor cells, and CD34+ cells in vitro. However, augmentation of conventional, unmanipulated cord blood for transplantation has not yielded significant improvements in the observed time to engraftment or overall probability of engraftment (26, 27), and it remains unclear why. No current testing procedures are capable of adequately measuring the stem cell population, and in vitro and animal studies currently use surrogates for such assays, which may not correlate well with clinical outcomes — a clear technological deficiency in the field. Alternatively, we may simply be unable to expand true cord blood stem cells with these procedures.

Pilot trials combining 2 cord blood units for a single patient look promising (28). The use of reduced-intensity preparative regimens in adult patients who have had prior cycles of standard chemotherapy to treat a malignant condition with engraftment of lower-dosed single umbilical cord blood grafts has also been reported (29, 30). However, this approach has not been successful in children because of graft rejection in these younger, immunocompetent hosts. Recently, CD34+-selected, haploidentical cells from mobilized peripheral blood of adult donors have been used to augment cord blood transplantation in adults, leading to early neutrophil recovery and the cord blood graft providing durable marrow and immune reconstitution (31).

The only prospective multicenter trial of unrelated-donor cord blood transplantation to date — the Cord Blood Transplantation Study — was sponsored by the National Heart, Lung, and Blood Institute and performed from 1997 to 2004. The study funded the establishment of 3 unrelated cord blood banks that followed common quality standards and operating procedures with respect to donor recruiting; screening (donors are screened for events in their medical history that would exclude them as donors, e.g., multiple pregnancy, prematurity, placental deformity, self or sibling diagnosed with cancer, prior receipt of a transplant, or demonstrating risk behaviors likely to increase the chance of infection with blood-borne infectious diseases); collection of informed donor consent; obtaining medical histories; obtaining blood samples from maternal donors; and cord blood collection, processing, testing, cryopreservation, and storage, as well as searching for and releasing cord blood units for transplantation (32). Twenty-six transplant centers participated in a prospective clinical trial designed to examine the safety and efficacy of unrelated cord blood transplantation in infants, children, and adults with malignancies; children with congenital immunodeficiency disorders; and children with inborn errors of metabolism. The study participants employed common preparative regimens, prophylaxis against GVHD, and supportive care measures (32–35). Results in children with malignant and nonmalignant conditions were favorable, with 55% survival in children with malignancies and 78% survival in children with nonmalignant conditions. Results in a very high-risk group of adults were inferior to those seen in children and in individuals receiving bone marrow from an unrelated donor. Subsequent studies in adults, reported by single centers or registries, revealed more encouraging results (19–23, 28). The cumulative incidence of engraftment by day 42 after transplantation was approximately 80% in all study strata including adults and children as well as children with malignant diseases, inborn errors of metabolism, and immunodeficiency syndromes. Factors adversely affecting engraftment or survival included lower cell doses, pretransplant CMV seropositivity in the recipient, non-white descent, and greater HLA mismatching.

The major obstacles to the success of unrelated cord blood transplantation today include slower engraftment times resulting in longer hospitalizations and increased resource utilization (e.g., packed red blood cells and platelet transfusions), lack of sufficient numbers of larger cord blood units containing enough cells for transplantation in an adult, and an increasing need for ethnic diversity among donors to achieve closer HLA matching.

Cord blood collection

The cord blood, which typically would be discarded at birth with the placenta, can be collected without physical risk to the mother or baby donor. Cord blood can be collected from the delivered placenta (ex utero) (Figure 1) or during the third stage of labor (in utero). Many public cord blood banks employ dedicated staff to perform ex utero collections away from the delivery room so that the privacy of the family is preserved and so clinicians are not distracted from their usual practices. Alternatively, obstetricians or midwives perform in utero collections while waiting for the placenta to deliver. In either case, after sterile preparation, the umbilical vein is punctured with a 17-gauge needle attached to a sterile, closed system collection bag containing citrate phosphate dextrose anticoagulant, which is positioned lower than the placenta. Blood flows from the placenta through the cord into the bag over approximately 9–10 minutes. Experienced collectors harvest an average of 110 ml from a single placenta. The cord blood unit is labeled and subsequently shipped to the bank for processing, testing, cryopreservation, and storage.

The issue of obtaining consent for collection of cord blood has been controversial in the field of cord blood transplantation (36, 37). Historically, the cord blood was considered to be the property of the hospital in which the baby was born, to be used, if desired, without patients' express consent. This practice, however, neglected the fact that for some women the placenta would not necessarily be considered a medical waste product, perhaps for some very important cultural reasons (30). Furthermore, once it was discovered that cord blood was rich in hematopoietic stem and progenitor cells that could be used in human transplantation, express consent for the collection of cord blood has been required for public banking. The necessity of consent was further recognized with the appreciation that collection for banking and transplantation required additional medical and personal information about newborns and their mothers. After being given relevant information and the opportunity to have any questions answered, if the cord blood is to be collected and saved, mothers currently must sign a consent form that indicates: (a) the donation of her baby's cord blood is voluntary; (b) she gives permission for her blood and the cord blood to be tested for blood-borne pathogens, e.g., HIV, hepatitis B and C, syphilis, human T cell lymphotropic virus, and West Nile Virus, and agrees to provide a detailed family medical history to the bank; (c) the cord blood is not being stored for personal use by the baby or other relatives and instead will be listed on a registry of unrelated donors and made available to patients in need of donors for unrelated-donor transplantation; (d) she may be contacted in the future by the bank to obtain follow-up information on the baby's health; and (e) she understands what measures will be used to protect her confidentiality and that of the baby. Unlike adult donor registries, the identity of a cord blood donor is not revealed to the recipient, and the recipient may not contact the donor in the future.

The timing of the provision of maternal consent also remains controversial. The vast majority of banks believe that consent should be obtained from the mother before collection of the cord blood. A minority of banks collect the cord blood without the mother's knowledge or consent and only subsequently ask the mother for permission to keep the cord blood if the collection was successful (36). The majority of practitioners and blood bank personnel also agree that consent should not be obtained from a woman in active labor or in other circumstances where her decision-making capacity is compromised due to narcotic or other mind-altering analgesics. However, consent from a woman in early

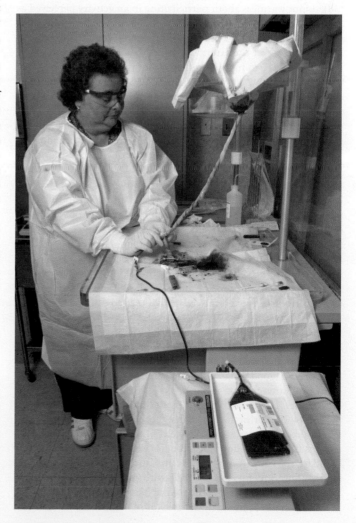

Figure 1

Cord blood harvest following delivery. The placenta is suspended, fetal side down, in a stand lined with a clean chuck pad. The cord is brought down and gently tethered onto a work surface below. The cord is prepped with betadine and alcohol. The umbilical vein is punctured with a 17-gauge needle attached to a sterile, pre-barcoded collection bag containing 25 ml citrate phosphate dextrose anticoagulant, which is placed on a rocking scale. The cord blood flows into the bag by gravity over approximately 9 minutes. Cessation of flow is indicated by stabilization of the bag weight on the scale. After completion of the collection, the tubing is stripped and heat sealed. The cord blood collection bag is transported to the processing laboratory.

labor has been allowed and advocated by some (37). Despite efforts to recruit women during the third trimester of pregnancy and obtain their informed consent well before the onset of labor, many pregnant women in labor come to the hospital interested in cord blood donation but without having given prior consent. Some centers have addressed this situation by having pregnant women in early labor sign a short or "mini" consent form allowing collection of the cord blood and maternal samples and then meet with her after she recovers from the delivery to educate her and obtain full informed consent for subsequent cord blood banking.

Although there is a general consensus that informed consent for the collection, storage, and use of cord blood should be obtained prior to labor and delivery, variations on the practice have raised further questions about justice and the allocation of resources. As has been previously noted, justice requires the establishment of mechanisms to ensure that all participants have a fair opportunity to participate in umbilical cord blood banking and use (39). The few banks that obtain consent after admission to the hospital or following collection of bankable units have in fact had considerable success in banking units from otherwise underrepresented populations (37). The data highlight the fact that policy making regarding informed consent involves a balance of protecting the interests of donor families and the challenges of equitable recruitment. Thus, recruitment and consent should proceed with sensitivity to the wide range of cultural beliefs about the placenta and umbilical cord (38), donor wariness about participating in an activity perceived as investigational, and suspicion that members of the donor's community will not reap the benefits of this technology.

Concerns about the extent of commercialization of cord blood banking have been highlighted by recent patent litigation. A private biotechnology company, PharmaStem Therapeutics Inc., has claimed that its patents covering collection, cryopreservation, storage, and use of cord blood entitle it to licensing fees (40). To this end, the company brought suit against private banks that had not signed their licensing agreement and sent letters to approximately 25,000 obstetricians asking them not to collect cord blood for 5 private cord blood banks that had not obtained a license with PharmaStem Therapeutics Inc., claiming potential liability for patent infringement. Following a US federal court order in July 2004 prohibiting PharmaStem from further contacting obstetricians due to the fact that it had made "false and misleading" statements to obstetricians, a September 2004 ruling found that PharmaStem failed to prove infringement of its patents because the banks did not sell or offer to sell cryopreserved cord blood (41). Most recently, the US patent office reexamined 1 of the 2 PharmaStem patents in question and rejected the company's related claims (42).

Cord blood banking

Cord blood transplantation has been made possible in large part by the creation of worldwide umbilical cord blood blanks. There are 2 types of cord blood banks, public and private. Unrelated-donor transplant programs employ public banks as their source of donor cord blood units. These cord blood units are donated on a volunteer basis by women delivering healthy babies at term. Private banks, which are for-profit entities, store "directed donations" collected by obstetricians from babies born into families who intend to use the cord blood for the baby from whom it came (autologous donation) or for another family member in need of future transplantation therapy.

As commercial cord blood banking has proliferated, its ethical justification has been widely debated (39, 43–46). In addition to the

issues such as patent infringement that can accompany commercialization, private banking has raised a number of other concerns.

Generally, an initial storage fee of $1,000–1,500 is charged followed by a yearly storage fee of approximately $100. While there are a few clear indications for this practice (e.g., a sibling with cancer, a hemoglobinopathy, marrow failure, congenital immunodeficiency syndrome, or inborn error of metabolism), the vast majority of families who store with private banks pay to have access to stem cells in the future for use in treating degenerative diseases or problems related to injuries or aging. Currently there is no evidence that this use will be feasible or efficacious in such circumstances. Many private banks aggressively advertise their services, promising, for example, "peace of mind and a powerful medical resource used to treat many severe illnesses for your child and loved ones" (47). References to rare and yet-to-be-tested applications for cord blood transplantation play on the fears of new parents wanting to provide every possible advantage for their newborn child (48). In addition to concerns about exploitation, marketing may be inaccurate or misleading: one common reason offered by these private banks for storing autologous cord blood is to have a source of stem cells for transplantation if the child were to develop leukemia. However, most children with leukemias can be cured with conventional chemotherapy alone, and in those for whom this approach fails, allogeneic transplantation is the treatment of choice. Furthermore, leukemic cells have been found in autologous cord blood of children presenting with leukemia from 1–11 years of age (49).

With increasing use of cord blood transplantation, public banks face other challenges. Procuring and providing units for public use in unrelated allogeneic transplantation has involved funding the creation of their inventory from third-party sources. Currently there are approximately 14 public cord blood banks in the United States and approximately 30 more worldwide. All of these banks struggle financially because the revenues gained from the sale of cord blood units for transplantation are not sufficient to support basic operations of a bank that is in the process of building inventory. There is also no requirement for public cord blood banks to list their inventories on a single registry available to all transplant centers. Thus, transplant center staff must be knowledgeable about and search multiple banks and registries to find the best donor for their patients.

In 2004, after appropriation of $20,000,000 by the US Congress to increase the inventory of cord blood units in US public banks, the Health Resources and Services Administration (HRSA) asked the Institute of Medicine (IOM) to perform a study to determine the best way to organize public cord blood banking and distribution to patients undergoing unrelated transplantation, the results of which were published in April 2005 (39, 50). In short, the IOM recommended that HRSA contract with eligible banks to procure approximately 150,000 new, ethnically diverse, unrelated donor cord blood units over the next 5 years. The units will need to meet quality standards as defined by an advisory board, the FDA, and other accrediting agencies. They will also have to be listed on a computerized, web-based, system created to allow searching of all unrelated cord blood and blood obtained from adult donors from a single-point of access. Since the publication of the IOM study, new legislation has been passed by the US House of Representatives (51) and is pending in the Senate (bill number S. 1317) to appropriate funds to establish a National Cord Blood Program, funded and administered through HRSA, and to build a high-quality, ethnically diverse inventory of 150,000 new cord blood units to be listed on a combined registry for adult and cord blood donors. Under this new program, patient advocacy must be provided and transplant outcomes collected and evaluated. This funding should enable selected banks to increase operations to build inventory, thereby increasing the number and quality of cord blood units available for unrelated transplantation.

Cord blood use

Currently, cord blood is used as a source of unrelated hematopoietic stem and progenitor cells for allogeneic transplantation used in the treatment of patients with leukemia, lymphoma, hemoglobinopathies, bone marrow failure syndromes, congenital immunodeficiency syndromes, and inborn errors of metabolism. The fact that cord blood can be transplanted without an identical HLA match increases access to transplantation therapy for patients unable to find a perfectly matched adult donor.

Directed donor cord blood banking is indicated in the small population of women delivering healthy babies where the parents know they carry mutations associated with genetic diseases of the blood or immune system or inborn errors of metabolism or where there is an older affected sibling with one of these diseases who would be a candidate for transplantation therapy. This would include families with children diagnosed with a pediatric malignancy, congenital immunodeficiency syndrome, hemoglobinopathy, or lysosomal storage disease. In these cases, a fully matched or haploidentical sibling cord blood sample of sufficient size could serve as the source for allogeneic transplantation of the known or future affected children in that family.

If, in the future, gene therapy is shown to be safe and effective, autologous cells from an affected child may also be of use. Unfortunately, the initial experience utilizing gene therapy for the treatment of X-linked SCID (resulting from γ chain gene mutations), although successful in obtaining gene transfection, expression, and clinical correction of the immune defect, resulted in the unanticipated, aberrant insertion of the vector near an *LMO2* oncogene, inducing T cell acute lymphoblastic leukemia in 4 of 11 patients (32).

Some families carrying mutations for lethal genetic diseases elect to perform preimplantation diagnosis and embryo selection to conceive a healthy child and/or a healthy and HLA-matched donor for their affected child (53). While most of these cases are successful, a few have resulted in misdiagnosis and birth of a baby who is either affected or not a match. Families undergoing this procedure are counseled extensively so that they are aware of the potential risks of the procedure. Some elect to have chorionic villus sampling or amniocentesis to confirm the disease state and HLA typing of the fetus during pregnancy to allow for elective termination if the baby is affected with the genetic disease. Others, for whom termination is not an option, carry the pregnancy to term and obtain confirmatory testing at birth.

In addition to the potential for misdiagnosis, the use of preimplantation genetic diagnosis (PGD) to facilitate creation of a healthy child who could also become an HLA-compatible source of hematopoietic stem cells has raised a number of concerns (54, 55). New reproductive technologies are often met with religious and philosophical objections associated with respect for the sanctity of human life and objections to the discard of human embryos, particularly healthy embryos that are not HLA compatible. Others have questioned parental motivations — whether having a child to save its sibling is the right reason to create a human being. But it has been argued that parents that create a child for such purposes stand on higher moral ground than a host of other parents, who may procreate to carry on their family legacy, "balance" gender in their family, or provide siblings with playmates. Nevertheless, some have argued that PGD to create a stem cell donor be treated as research, in order to ensure institutional oversight, collection of data about adverse effects, and federal standards for minimizing risks to children (56).

Finally, umbilical cord blood cells may have potential applications in the field of regenerative medicine. Cord blood has been shown in some studies to transdifferentiate to a limited extent into nonhematopoietic cells, including those of the brain, heart, liver, pancreas, bone, and cartilage, in experimental culture and animal systems (57, 58). Recently it has been demonstrated that both cardiac and glial cell differentiation of cord blood donor cells occurred in recipients of unrelated donor cord blood transplantation as part of a treatment regime for Krabbe disease and Sanfilippo syndrome (59, 60). These observations raise the possibility that cord blood may serve as a source of cells to facilitate tissue repair and regeneration in the distant future. While this is purely speculative at this time, developments over the next decade are expected to clarify the potential role of both allogeneic and autologous cord blood in this emerging field.

Concluding comments

Technology and ethics surrounding the collection, banking, and use of umbilical cord blood has been marked by significant progress over the last 15 years. However, despite rapid progress in many areas, the current US inventory is poorly organized, fragmented, and of varying quality. Recent federal legislation should provide the impetus and resources for the cord blood banking community to organize and adopt uniform high-quality standards, to list on a common national registry, and also to list adult unrelated donors (e.g., similar to the approach currently practiced by the NMDP) while increasing the pool of cord blood donors available for unrelated transplantation.

Acknowledgments

There are many individuals responsible for the development of cord blood banking and transplantation. These include, but are not limited to, Eliane Gluckman, who performed the first related cord blood transplant in 1988; Gordon Douglas of New York Hospital, who lived in Salisbury, North Carolina, for 1 month, in order to collect the cord blood from the sibling of this child; Ted Boyce, Hal Broxmeyer, and Judy Bard, who hypothesized that cord blood was enriched for hematopoietic stem and progenitor cells; Nancy Kernan and John Wagner, who formed the first voluntary cord blood transplant registry to report outcomes of related cord blood transplants; Pablo Rubinstein, who established the first public cord blood bank in the world; the National Heart, Lung, and Blood Institute, which has supported research in cord blood banking and transplantation; and the National Marrow Donor Program, which has supported public cord blood banks and listed cord blood donors in the unrelated donor registry; the physicians, nurses, and allied healthcare professionals caring for these patients; and the patients and their families, whose courage, dignity, and determination motivate us all.

Address correspondence to: Joanne Kurtzberg, Box 3350, Duke University Medical Center, Durham, North Carolina 27710, USA. Phone: (919) 668-1100; Fax: (919) 681-8942; E-mail: kurtz001@mc.duke.edu.

1. Gluckman, E., et al. 1989. Hematopoietic reconstitution in a patient with Fanconi's anemia by means of umbilical-cord blood from an HLA-identical sibling. *N. Engl. J. Med.* **321**:1174–1178.

2. Wagner, J.E., Kernan, N.A., Steinbuch, M., Broxmeyer, H.E., and Gluckman, E. 1995. Allogeneic sibling umbilical-cord-blood transplantation in children with malignant and non-malignant disease. *Lancet.* **346**:214–219.

3. Kernan, N.A., et al. 1993. Analysis of 462 transplantations from unrelated donors facilitated by the National Marrow Donor Program. *N. Engl. J. Med.* **328**:593–602.

4. To, L.B., et al. 1992. Comparison of haematological recovery times and supportive care requirements of autologous recovery phase peripheral blood stem cell transplants, autologous bone marrow transplants, and allogeneic bone marrow transplants. *Bone Marrow Transplant.* **9**:277–284.

5. Drobyski, W.R., et al. 1994. Effect of T-cell depletion as graft-versus-host disease prophylaxis on engraftment, relapse, and disease-free survival in unrelated marrow transplantation for chronic myelogenous leukemia. *Blood.* **83**:1980–1987.

6. Rubinstein, P., Rosenfield, R.E., Adamson, J.W., and Stevens, C.E. 1993. Stored placental blood for unrelated bone marrow reconstitution. *Blood.* **81**:1679–1690.

7. Kurtzberg, J., et al. 1996. Placental blood as a source of hematopoietic stem cells for transplantation into unrelated recipients. *N. Engl. J. Med.* **335**:157–166.

8. Wagner, J., Kernan, N., Broxmeyer, H., and Gluckman, E. 1996. Successful transplantation of HLA-matched and HLA-mismatched umbilical cord blood from unrelated donors: analysis of engraftment and acute graft-versus-host disease. *Blood.* **88**:795–802.

9. Gluckman, E., et al. 1997. Outcome of cord-blood transplantation from related and unrelated donors. *N. Engl. J. Med.* **337**:373–381.

10. Rubinstein, P., et al. 1998. Outcomes among 562 recipients of placental-blood transplants from unrelated donors. *N. Engl. J. Med.* **339**:1565–1577.

11. Rocha, V., et al. 2000. Graft-versus-host disease in children who have received a cord-blood or bone marrow transplant from an HLA-identical sibling. *N. Engl. J. Med.* **342**:1846–1854.

12. Barker, J.N., et al. 2001. Survival after transplantation of unrelated donor umbilical cord blood is comparable to that of human leukocyte antigen–matched unrelated donor bone marrow: results of a matched-pair analysis. *Blood.* **97**:2957–2961.

13. Rocha, V., et al. 2001. Comparison of outcomes of unrelated bone marrow and umbilical cord blood transplants in children with acute leukemia. *Blood.* **97**:2962–2971.

14. Locatelli, F., et al. 2003. Related umbilical cord blood transplantation in patients with thalassemia and sickle cell disease. *Blood.* **101**:2137–2143.

15. Hall, J.G., Martin, P.L., Wood, S., and Kurtzberg, J. 2004. Unrelated umbilical cord blood transplantation for an infant with B-thalassemia major. *J. Pediatr. Hematol. Oncol.* **26**:382–385.

16. Staba, S., et al. 2004. Cord-blood transplants from unrelated donor in patients with Hurler's syndrome. *N. Engl. J. Med.* **350**:1960–1969.

17. Myers, L.A., Hershfield, M.S., Neale, W.T., Escolar, M., and Kurtzberg, J. 2005. Purine nucleoside phosphorylase deficiency (PNP-Def) presenting with lymphopenia and developmental delay: successful correction with umbilical cord blood transplantation. *J. Pediatr.* **145**:710–712.

18. Escolar, M.L., et al. 2005. Transplantation of umbilical-cord blood in babies with infantile Krabbe's disease. *N. Engl. J. Med.* **352**:2069–2081.

19. Laughlin, M.J., et al. 2001. Hematopoietic engraftment and survival in adult recipients of umbilical-cord blood from unrelated donors. *N. Engl. J. Med.* **344**:1815–1822.

20. Long, G.D., et al. 2003. Unrelated umbilical cord blood transplantation in adult patients. *Biol. Blood Marrow Transplant.* **9**:772–780.

21. Laughlin, M.J., et al. 2004. Outcomes after transplantation of cord blood or bone marrow from unrelated donors in adults with leukemia. *N. Engl. J. Med.* **351**:2265–2275.

22. Rocha, V., et al. 2004. Transplants of umbilical-cord blood or bone marrow from unrelated donors in adults with acute leukemia. *N. Engl. J. Med.* **351**:2276–2285.

23. Takahashi, S., et al. 2004. Single-institute comparative analysis of unrelated bone marrow transplantation and cord blood transplantation for adult patients with hematologic malignancies. *Blood.* **104**:3813–3820.

24. Fruchtman, S.M., et al. 2004. The successful treatment of severe aplastic anemia with autologous cord blood transplantation. *Biol. Blood Marrow Transplant.* **10**:741–742.

25. Gluckman, E., et al. 2004. Factors associated with outcomes of unrelated cord blood transplant: guidelines for donor choice. *Exp. Hematol.* **32**:397–407.

26. Jaroscak, J., et al. 2003. Augmentation of umbilical cord blood (UCB) transplantation with ex vivo-expanded UCB cells: results of a phase 1 trial using the AastromReplicell system. *Blood.* **101**:5061–5067.

27. McNiece, I., Harrington, J., Turney, J., Kellner, J., and Shpall, E.J. 2004. Ex vivo expansion of cord blood mononuclear cells on mesencymal stem cells. *Cytotherapy.* **6**:311–317.

28. Barker, J.N., et al. 2005. Transplantation of 2 partially HLA-matched umbilical cord blood units to enhance engraftment in adults with hematologic malignancy. *Blood.* **105**:1343–1347.

29. Barker, J.N., et al. 2003. Rapid and complete donor chimerism in adult recipients of unrelated donor umbilical cord blood transplantation after reduced-intensity conditioning. *Blood.* **102**:1915–1919.

30. Chao, N.J., et al. 2004. Adult recipients of umbilical cord blood transplants after nonmyeloablative preparative regimens. *Biol. Blood Marrow Transplant.* **10**:569–575.

31. Fernandez, M.N., et al. 2001. Cord blood transplants: early recovery of neutrophils from co-transplanted sibling haploidenitcal progenitor cells and lack of engraftment of cultured cord blood cells, as ascertained by analysis of DNA polymorphisms. *Bone Marrow Transplant.* **28**:355–364.

32. Fraser, J.K., et al. 1998. Cord blood transplantation study (COBLT): cord blood bank standard operating procedures. *J. Hematother.* **7**:521–561.

33. The Cord Blood Transplantation Study. The EMMES Corp. http://www.emmes.com.

34. Kurtzberg, J., et al. 2005. Results of the cord blood transplantation (COBLT) study unrelated donor banking program. *Transfusion.* **45**:842–855.

35. Cairo, M.S., et al. 2005. Characterization of banked umbilical cord blood hematopoietic progenitor cells and lymphocyte subsets and correlation with ethnicity, birth weight, sex, and type of delivery: a cord blood transplantation (COBLT) study report. *Transfusion.* **45**:856–866.

36. Sugarman, J., Kurtzberg, J., Box, T.L., and Horner, R.D. 2002. Optimization of informed consent for umbilical cord blood banking. *Am. J. Obstet. Gynecol.* **187**:1642–1646.

37. Vawter, E.R., et al. 2002. A phased consent policy for cord blood donation. *Transfusion.* **42**:1268–1274.

38. Jenkins, G.L., and Sugarman, J. 2004. The importance of cultural considerations in the promotion of ethical research with human biological material [review]. *J. Lab. Clin. Med.* **145**:118–124.

39. Burgio, B.R., Gluckman, E., and Locatelli, F. 2003. Ethical reappraisal of 15 years of cord blood transplantation. *Lancet.* **361**:250–252.

40. Steinbrook, R. 2004. The cord-blood-bank controversies [comment]. *N. Engl. J. Med.* **351**:2255–2257.

41. Anonymous. 2004. Pharmastem Therapeutics, Inc. v. ViaCell, Inc. WL 2127192. C.A. No. 02-148 GMS (D. Del.).

42. Anonymous. 2005. US Patent Office rejects PharmaStem cord blood stem cell patents. *Today's stem cell research: stem cell research medical and health news.* http://www.stemnews.com/archives/000430.html.

43. Chapman, A.R., Frankel, M.S., and Garfinkel, M.S. 1999. Stem cell research and applications: monitoring the frontiers of biomedical research. American Association for the Advancement of Science. Washington, DC, USA. http://www.aaas.org/spp/sfrl/projects/stem/report.pdf.

44. Fisk, N.M., Roberts, I.A.G., Markwald, R., and Mironov, V. 2005. Can routine commercial cord blood banking be scientifically and ethically justified? *PLoS Med.* **2**:87–90.

45. Ecker, J.L., and Greene, M.F. 2005. The case against private umbilical cord blood banking. *Obstet. Gynecol.* **105**:1282–1284.

46. Sugarman, J., et al. 1997. Ethical issues in umbilical cord blood banking. Working Group on Ethical Issues in Umbilical Cord Blood Banking [review]. *JAMA.* **278**:938–943.

47. Cord Blood Registry. http://www.cordblood.com/index.asp.

48. Assas, G.I. 1999. Waste and longing – the legal status of placental-blood banking. *N. Engl. J. Med.* **340**:1521–1524.

49. Rowley, J. 1998. Backtracking leukemia to birth. *Nat. Med.* **4**:150–151.

50. Committee on Establishing a National Cord Blood Stem Cell Bank Program. 2005. *Cord blood: establishing a national hematopoietic stem cell bank program.* National Academies Press. Washington, DC, USA. 320 pp.

51. Anonymous. 2005. Stem Cell Research Enhancement Act of 2005. HR 810. 109th Congress, 1st sess. *Congr. Rec. (Dly. Ed.).* **151**:H3795–H3809.

52. Hacein-Bey-Abina, S., et al. 2003. A serious adverse event after successful gene therapy for X-linked severe combined immunodeficiency. *N. Engl. J. Med.* **348**:255–256.

53. Verlinsky, Y., et al. 2004. Preimplantation HLA testing. *JAMA.* **291**:2079–2085.

54. Fost, N.C. 2004. Conception for donation. *JAMA.* **291**:2125–2126.

55. Burgio, G.R., and Locatelli, F. 2000. Ethics of creating programmed stem-cell donors. *Lancet.* **356**:1868–1869.

56. Wolf, S.M., Kahn, J.P., and Wagner, J.E. 2003. Using preimplantation genetic diagnosis to create a stem cell donor: issues, guidelines and limits. *J. Law Med. Ethics.* **31**:327–339.

57. Porada, G.A., Porada, C., and Zanjani, E.D. 2004. The fetal sheep: a unique model system for assessing the full differentiative potential of human stem cells. *Yonsei Med. J.* **45**(Suppl.):7–14.

58. Kogler, G., et al. 2004. A new human somatic stem cell from placental cord blood with intrinsic pluripotent differentiation potential. *J. Exp. Med.* **200**:123–135.

59. Hall, J., Crapnell, K.B., Staba, S., and Kurtzberg, J. 2004. Isolation of oligodendrocyte precursors from umbilical cord blood [abstract]. *Biol. Blood Marrow Transplant.* **10**:67.

60. Crapnell, K.B., Turner, K., Hall, J.G., Staba, S.L., and Kurtzberg, J. 2003. Umbilical cord blood cells engraft and differentiate in cardiac tissues after human transplantation [abstract]. *Blood.* **102**:153b.

Part 9

Gastrointestinal System

Intestinal ion transport and the pathophysiology of diarrhea

Michael Field

Department of Medicine, and Department of Physiology and Cellular Biophysics,
College of Physicians and Surgeons, Columbia University, New York, New York, USA.

Worldwide, diarrhea claims several million lives annually, mostly those of infants. Poverty, crowding, and contaminated water supplies all contribute. Almost all of these deaths could have been prevented with adequate fluid replacement. Although its incidence is much lower in the more affluent nations, diarrhea remains one of the two most common reasons for visits to pediatric emergency departments and is also common among the institutionalized elderly. Chronic diarrheas, while less common, often present diagnostic dilemmas and can be difficult to manage. This article will begin by reviewing the relevant physiology; then, with that as context, it will consider pathophysiology, both general aspects and specific diarrheal syndromes. Finally, rationales for management will be briefly considered. The review is amply referenced with some of the references listed at the end of the article and some at the *JCI* website (http://www.jci.org/cgi/content/full/111/7/931/DC1).

Physiology

All segments of intestine from duodenum to distal colon have mechanisms for both absorbing and secreting water and electrolytes. This review begins with a historical approach to the development of knowledge in the field, considering absorption and secretion separately, although, inevitably, the processes involved are interconnected.

Absorption

Over the last half-century, two key sets of observations have energized and organized research in this area. The first was the discovery in the 1960s that sugars (glucose and galactose) and amino acids are absorbed across the small-intestinal brush border membrane via carriers that couple their movements to that of Na^+ (1). Na^+ coupling permits the organic solute to be transported uphill, i.e., from low luminal to higher cell concentration, a gradient opposite to that for Na^+. The organic solutes then move downhill from enterocyte to blood via basolateral membrane carriers that operate independently of ion movements. As discovered subsequently, some oligopeptides, instead of being first hydrolyzed into amino acids, are absorbed intact across the intestinal brush border by a proton-coupled mechanism. This absorptive process is indirectly coupled to Na^+ transport, since the needed protons are provided by Na/H exchange, which acidifies the unstirred

layer abutting the brush border membrane (2, S1). The Na^+ gradient, therefore, is the driving force for amino acid, oligopeptide, and sugar absorption. As these organic solutes are absorbed, salt is absorbed with them, and water follows osmotically — transport from enterocyte to lateral intercellular space creates a local osmotic gradient that initiates water flow. As described below, the coupled transport of Na^+ and organic solute is the theoretical basis for oral rehydration therapy in severe diarrhea.

The second key set of observations arose from studies, performed in the late 1960s on human subjects, of intestinal salt absorption in the absence of nutrients (3, 4). Segments of jejunum and ileum were perfused with solutions of varied electrolyte composition, and the appearance or disappearance of ions and water was recorded. The data suggested different transport mechanisms in jejunum and ileum. In the jejunum, $NaHCO_3$ is absorbed via Na/H exchange (the secreted H^+ neutralizes an equivalent amount of luminal HCO_3^-) and Cl^- movement is purely passive. In the ileum (and, as shown later, also in the proximal colon) (S2), NaCl is absorbed via equal rates of Na/H and Cl/HCO_3 exchanges.

Three Na/H exchangers (NHEs) have since been localized to intestinal brush border membranes and cloned; NHE2 and NHE3 are found in both small intestine and colon (5, S3). NHE3 appears to be quantitatively more important, since the NHE2 knockout mouse suffers gastric dysfunction but no intestinal disability (S4) whereas the NHE3 knockout mouse suffers from chronic diarrhea (6). A third, Cl-dependent NHE is found in crypt cells of rat distal colon (S5). The NHE first identified in intestine, NHE1, is present only in the basolateral membrane of enterocytes and is involved in HCO_3^- secretion (see below).

Two anion exchangers have also been localized to small-intestinal and colonic brush border membranes and cloned (7, S6). They have been given curious names: downregulated in adenoma (DRA) and putative anion transporter 1 (PAT1). DRA was first cloned from colonic mucosa; it was found to be downregulated in villus adenomas and carcinomas (S7) and subsequently was found to incur mutations in the rare diarrheal disorder familial chloride diarrhea (8) (see "Congenital diarrheas" below). Both DRA and PAT1 are abundant in the duodenum and present at higher density there than NHE2 and NHE3, suggesting a role in duodenal alkalinization (S8). In the colon, DRA appears to predominate over PAT1 (S6).

More than two brush border ion exchangers are required, of course, for the enterocyte to engage in transcellular salt absorption. Increased turnover of the Na/K pump and the opening of Cl^- and K^+ channels are also necessary, the latter to counteract associated cell swelling, to permit serosal exit of Cl^- taken up from the lumen, and to dissipate the added uptake of K^+ through the pump. The cellular mechanisms involved are summarized in Figure 1.

Nonstandard abbreviations used: Na/H exchanger (NHE); downregulated in adenoma (DRA); putative anion transporter 1 (PAT1); cholera toxin (CT); protein kinase A (PKA); cystic fibrosis (CF); cystic fibrosis transmembrane conductance regulator (CFTR); NHE3 kinase A regulatory protein (E3KARP); vasoactive intestinal peptide (VIP); enteric nervous system (ENS); heat-stable toxin (STa); cGMP-dependent protein kinase (PKG); oral rehydration solution (ORS).

Conflict of interest: The author has declared that no conflict of interest exists.

Citation for this article: *J. Clin. Invest.* **111**:931–943 (2003). doi:10.1172/JCI200318326.

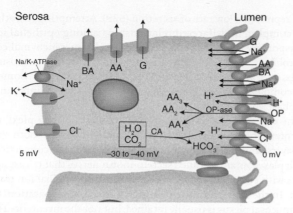

Figure 1

Ileal absorptive cell. Multiple brush border transporters couple ion influxes (Na^+ and, in one instance, H^+) to organic solute influxes or exchange one ion for another. Basolateral-membrane carriers facilitate diffusion of organic solutes and are not coupled to ion movements. Na/K-ATPase in the basolateral membrane uses energy from ATP hydrolysis to drive Na^+ extrusion and K^+ uptake (3:2 stoichiometry), both moving against their electrochemical gradients. K^+ and Cl^- channels in the basolateral membrane open in response to cell swelling and elevations of intracellular Ca^{2+}. Some of the cellular H^+ and HCO_3^- extruded in exchange for Na^+ and Cl^- is provided through the action of carbonic anhydrase (CA). Additional HCO_3^- may enter the cell through the basolateral membrane NHE1 or the $Na(HCO_3)_3$ cotransporter (not shown). The cell electric potential is 30–40 mV negative relative to the lumen, providing an electric, as well as a chemical, driving force for Na^+ entry via Na–organic solute cotransport. G, glucose or galactose; AA, amino acid (there are actually several amino acid carriers); BA, bile acid anion; OP, oligopeptide; OPase, oligopeptidase (almost all oligopeptides entering the cell intact are quickly hydrolyzed).

Secretion

Knowledge about secretion arose from two sources: university research laboratories preoccupied with intestinal physiology, or with epithelial physiology more generally, and field stations in Asia dedicated to cholera research. As investigators moved back and forth between field stations and universities, these initially separate groups came together. Until the late 1960s, despite emerging research into cholera (see below), the gut was generally thought of as an absorptive epithelium. Rare reports of active secretion by intestinal epithelium were largely ignored (S9). In 1965, an *Annual Review of Physiology* article entitled "Secretory mechanisms of the digestive tract" did not even mention intestinal secretion (S10), whereas, in 1970, the same annual published a review devoted to intestinal secretion (S11).

Critical to cholera research were two essentially simultaneous reports in 1959, one from Calcutta (9), the other from Bombay (10), that a cell-free filtrate of a *Vibrio cholerae* culture will elicit fluid secretion when instilled into the intestines of rabbits, proving that a substance released by the bacteria (cholera toxin) is responsible. A year later, in a report on small-bowel biopsies taken from both acutely ill and convalescent cholera patients, the intestinal epithelium was demonstrated to remain intact, seemingly uninjured by the cholera organism and its toxins (11). These two reports set the stage for the investigation of cholera as an active-transport disease, or, more specifically, for exploration of the possibility that cholera toxin (CT) behaves like a luminally active hormone, stimulating the intestinal epithelium to actively secrete electrolytes and water.

However, because cholera researchers, like other gastrointestinal tract physiologists, were not yet prepared to think of the intestine as a secretory epithelium, alternative explanations were initially proposed that ultimately proved wrong.

In the late 1960s and into the 1970s, the ion transport properties of rabbit ileum were systematically examined in vitro using the isolated mucosa (stripped of muscle) mounted in so-called Ussing chambers. cAMP and agents that increase its intracellular concentration were shown to have two striking and mutually reinforcing effects on ion transport: (a) stimulation of active Cl^- secretion ("active" refers here to uphill transport, i.e., transport against an electrochemical gradient), and (b) inhibition of active, electroneutral NaCl absorption (12, S12). Coupled absorption of Na^+ and glucose was not affected. The implication of these observations for the pathophysiology of diarrhea was obvious, and these investigators quickly began a collaboration with cholera researchers to explore the possible role of cAMP in cholera in particular and diarrheal disease more generally. The collaboration soon established that CT, when added to rabbit ileal mucosa in Ussing chambers, produces the same effects as cAMP, albeit with a slower time course, and does so by stimulating adenylyl cyclase activity in enterocytes (13, 14).

Over the next decade, the individual cell membrane transporters contributing to active Cl^- secretion were functionally identified, giving rise to a model of general applicability to several electrolyte-secreting organs (Figure 2). cAMP, via protein kinase A (PKA), stimulates secretion by activating or enhancing the transport activities of three membrane proteins:

1. The apical anion channel. cAMP opens the apical anion channel (to be precise, it increases the probability of its opening), thereby initiating secretion (15). In the intestine, at least, cAMP also promotes insertion of additional channels (recruited from the ER) into the apical membrane (S13, S14).

2. A basolateral membrane K^+ channel. The opening of a basolateral membrane K^+ channel repolarizes the cell, counteracting the depolarizing effect of opening the Cl^- channel and thus sustaining the electrical driving force for Cl^- extrusion into the lumen. A cAMP-activatable K^+ channel has recently been identified and cloned (16, S15, S16).

3. A basolateral membrane NaK2Cl cotransporter. The turnover of a basolateral membrane NaK2Cl cotransporter (17, S17) determines the maximal rate of serosal Cl^- entry (once the apical anion channel and the basolateral membrane K^+ channel have been activated). cAMP appears to enhance cotransporter activity not via PKA-induced phosphorylation of the transport protein, but indirectly by opening the apical anion channel and thereby decreasing cell $[Cl^-]$, and also by enhancing cotransporter insertion into the basolateral membrane from an intracellular storage site (S17, S18).

The inherited disease cystic fibrosis (CF) develops when dysfunctional mutations occur in the gene for the cAMP-responsive apical anion channel. Accordingly, this channel, which is present in a number of tissues, has been dubbed the cystic fibrosis transmembrane conductance regulator or CFTR for short. CFTR is a member of the superfamily of ABC proteins and has a single-channel conductance of 7–10 pS and a linear current-voltage relationship (18). Studies of intestine from both humans and transgenic mice with CF show a near total absence of electrolyte secretion (19, 20), which perhaps explains why patients with CF occasionally develop intestinal obstruction (known, in neonates, as meconium ileus).

Not only is cAMP's secretory effect lost in CF, but its inhibition of electroneutral NaCl absorption is also lost, indicating a critical role for normal CFTR in the latter. The mechanism for cAMP's (and also Ca^{2+}'s) inhibition of electroneutral NaCl absorption is not fully elucidated but is known to involve several proteins, including NHE3 kinase A regulatory protein (E3KARP), NHE regulatory factor (NHERF), ezrin, and α-actinin-4 (S19–S21). NHE3, DRA, and CFTR all possess affinity for a discrete binding domain on E3KARP. It has recently been proposed that CFTR, NHE3, and either DRA or PAT1 bind to one domain of E3KARP (which dimerizes in the process) (S19, S20) and that signal transduction proteins such as PKA (via ezrin) (S21) bind to a second domain. Whether PKA inhibits electroneutral NaCl absorption by activating the bound CFTR or through a more direct effect on E3KARP remains to be determined.

Regulation of absorption and secretion

Once a role for cAMP in gut electrolyte secretion was established, the hunt was on for regulatory endocrine/paracrine/neural compounds. Two classes of compounds emerged: those that stimulate active secretion and inhibit active absorption, and those with the opposite effects (Table 1). The former group includes four kinds of agents: (a) neurotransmitters, including vasoactive intestinal peptide (VIP), acetylcholine, substance P, and the nucleotides ATP and UTP; (b) the paracrine agents serotonin and neurotensin, which are released by endocrine (enterochromaffin) cells in the intestinal epithelium; (c) agents released by inflammatory cells, including mainly prostaglandins and leukotrienes but also platelet-activating factor (which causes other inflammatory cells to release prostaglandins) (S22), histamine, and serotonin; and (d) guanylin, a luminally active peptide released into the gut lumen by goblet cells (see "Toxigenic E. coli" below).

Among these compounds, cAMP mediates the action of VIP, UTP/UDP, and prostaglandins (at high concentration). The action of guanylin is mediated by cGMP (see below). For all of the others (including lower concentrations of prostaglandins), activation of their receptors increases intracellular Ca^{2+} and activates PKC. The Ca^{2+} increase opens a basolateral K^+ channel different from that opened by cAMP (21). Based on studies with other tissues, PKC activation may either directly open the CFTR anion channel or heighten its sensitivity to PKA (S23–S25).

A striking difference between the effects of cyclic nucleotide-dependent secretagogues and Ca-dependent ones is evident in vitro. The former cause persistent secretion, whereas the latter cause transient secretion and the swift development of tachyphylaxis. This difference may be less pronounced in vivo, since the relevant effectors are quickly removed by hydrolysis and blood flow, perhaps thereby preventing or reversing tachyphylaxis.

The group of compounds that both inhibit active secretion (HCO_3^- as well as Cl^-) and enhance active absorption includes norepinephrine (via α_2-receptors), neuropeptide Y, enkephalins, all neurotransmitters, and somatostatin, both a neurotransmitter and a paracrine agent (S26). Among leukotriene precursors, 12-hydroxyeicosatetraenoic acid is antisecretory, blocking the basolateral membrane Ca^{2+}-activated K^+ channel (S27). Other leukotriene precursors are secretory stimuli (S28).

With a multitude of locally produced compounds providing bidirectional regulation of intestinal ion transport, it is not surprising that the system can be fine-tuned for a variety of circumstances. In the small intestine, so-called basal conditions may in fact represent a low rate of secretion (S29). Attempting to outline the complex, multifaceted relationships among epithelial ion-transporting cells, epithelial endocrine cells, mesenchymal cells (fibroblasts, mast cells, neutrophils, T and B lymphocytes, eosinophils, etc.), and nerves is beyond the scope of this review. Some of the complexity can be gleaned from Table 1. A few points can be emphasized as follows.

The enteric nervous system (ENS) consists of two plexi, the myenteric (between the circular and the longitudinal smooth muscle layers) and the submucosal, along with their interconnections. Both plexi contain so-called secreto-motor nerves that release agonists with either direct or indirect effects on epithelial ion transport. In the isolated (muscle-stripped) mucosal preparation, the submucosal plexus is usually retained, but not the myenteric. That the former alone profoundly influences ion transport is evident from the effect of adding tetrodotoxin to this preparation (S30).

VIP and acetylcholine are the predominant transmitters with direct epithelial secretory and/or antiabsorptive effects released by secreto-motor nerves; acetylcholine also acts on other neurons and on mesenchymal cells.

The mast cell is a key mediator for epithelial-neural interactions in allergic reactions and some parasitic infections (S31). Activation of mucosal mast cells by IgE or a variety of agonists including substance P, acetylcholine, and neurotensin causes release of histamine, serotonin, rat mast cell protease II (which appears to enhance absorption of macromolecules), prostaglandins, leukotrienes, and platelet-activating factor.

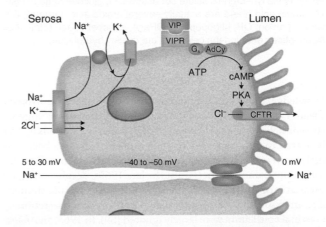

Figure 2

Intestinal Cl^- secretion. CFTR, the apical anion channel, mutations in which cause cystic fibrosis, is opened by PKA-catalyzed phosphorylations (its activity can also be enhanced through protein kinase G– and PKC-induced phosphorylations). The VIPR receptor is a G protein–linked activator of adenylyl cyclase (AdCy). The electroneutral NaK2Cl cotransporter facilitates the coupled movement of these ions across the basolateral membrane. The drop in cell [Cl^-] that follows opening of the apical anion channel enables an inward flux of the three ions. Cotransporter activity also increases in response to cell shrinkage. When CFTR is opened, the resulting anion efflux depolarizes the cell's electric potential. The repolarization required to maintain secretion is accomplished through the opening of a cAMP-sensitive, Ca^{2+}-insensitive K^+ channel in the basolateral membrane. During secretion, the cell electric potential is 40–50 mV negative and the serosal potential is 5–30 mV positive, both relative to the lumen (the serosal potential tends to be closer to 5 mV in the small intestine and 30 mV in the distal colon). The transmural potential difference drives Na^+ movement into the lumen through the tight junctions via the lateral intercellular spaces.

Table 1

Local control of intestinal ion transport

Agonist	Source	Target cell(s)	Intracellular mediator(s)	Secondary mediator(s) of transporting epithelial cell effects
Secretory/Antiabsorptive stimuli				
VIP (S91)	Neural	Epith	cAMP	
ACh (S92, S93)	Neural	Epith, neural, mes	Ca^{2+}, PKC	Eicos, ? VIP
SP (S94–S96)	Neural	Neural, mast, epith	Ca^{2+}, PKC	Histamine, eicos
ATP/ADP (S97)	Neural	Mes, epith	Ca^{2+}, PKC	Eicos
UTP/UDP (S97)	Neural	Mes, epith	cAMP	Eicos
5-HT (S98, S99)	EC, mast cell	Neural, epith	Ca^{2+}, PKC	VIP and ACh
NT (S94, S100)	EC	Neural	Ca^{2+}, PKC	Eicos, adeno, VIP, and ACh
Guanylin (39)	Goblet cells	Epith	cGMP	
PGs (13, S101)	Mes	Epith, neural	Ca^{2+}, PKC; cAMP	
LTs, pre-LTs (S28, S102)	Mes	Epith, neural	Ca^{2+}, PKC	
PAF (S22, S103)	Mes	Mes	Ca^{2+}, PKC; cAMP	Eicos
Histamine (S104, S105)	Mast cell	Mes, neural	Ca^{2+}, PKC; cAMP	Eicos, VIP, ACh
Bradykinin (S106)	Vascular	Mes	Ca^{2+}, PKC; cAMP	Eicos
Adenosine (S107–S111)	Epith, lumen	Epith, mes	A1 R: Ca^{2+}, PKC; A2 R: cAMP	PGs (not LTs)
Endothelin-1 (S112)	Vascular	Mes	Ca^{2+}, PKC; cAMP	PGs
Antisecretory/Proabsorptive stimuli				
Norepi (22, S32, S113)	Neural	Epith, neural	Activate G$_i$	
NPY (S32, S114–S117)	Neural, EC	Epith, neural	?	? Norepi
SST (S26, S118, S119)	EC, neural	Epith, neural	Activate G$_i$ (SST R2)	
Enkeph (S120, S121)	Neural	Epith, neural	?	
12-HETE (S27)	Mes	Epith	Block BL K$^+$ channel	

Epith, epithelial; ACh, acetylcholine; mes, mesenchymal cell (fibroblast, neutrophil, eosinophil, lymphocyte, mast cell, etc.); SP, substance P; eicos, eicosanoids (prostaglandins, leukotrienes, and their precursors); 5-HT, 5-hydroxy-tryptamine (serotonin); EC, enterochromaffin or epithelial endocrine cell; NT, neurotensin; adeno, adenosine; PG, prostaglandin; LT, leukotriene; PAF, platelet-activating factor; R, receptor; norepi, norepinephrine; G$_i$, inhibitory G protein; NPY, neuropeptide Y; SST, somatostatin; enkeph, enkephalins (opioids); 12-HETE, 12-hydroxyeicosatetraenoic acid; BL, basolateral. Numbers in parentheses indicate references.

Norepinephrine, the predominant antisecretory/proabsorptive agonist in the intestines, activates α_2-receptors on both enterocytes (22) and nerves (S32, S33). The neural effect is inhibitory, blocking release of secretion-inducing neurotransmitters. The direct epithelial effect (and perhaps also the neural effect) is exerted, in part at least, through inhibition of adenylyl cyclase. Adrenergic fibers are present in the submucosal plexus.

Other transport and permeability properties

Barrier function. In addition to transporting ions, nutrients, and water, the intestinal epithelium, consisting of both enterocytes and their encircling tight junctions (zona occludens), functions as a barrier that restricts the flow of luminal contents into the blood and lymphatics and vice versa. In the small intestine, tight junctions are, on the average, of the low-resistance type, meaning that most of the passive permeability of the epithelium to small monovalent ions and water resides in these junctional complexes (tight junctions in villi have higher resistance than do those in crypts) (S34). Colonic intercellular junctions are tighter, their resistance increasing steadily from proximal to distal portions. Although these junctions, which comprise a number of discrete proteins, are extracellular, their permeability properties are regulated by intracellular structures, especially actin filaments (23).

Osmolarity of intestinal contents. Unlike renal tubules, the intestines neither dilute nor concentrate their contents, the osmolarity of which, except in duodenum and proximal jejunum shortly after eating, is the same as plasma osmolarity (24). Although it may seem counterintuitive, the osmolarity of ions and molecules in stool water is also identical to plasma osmolarity (S35).

Other ion transport mechanisms and their geography. Mechanisms vary both longitudinally, along the length of the intestines, and horizontally, along the crypt-to-villus or crypt-to-surface cell axis.

Longitudinally, differences exist in Na$^+$ entry mechanisms, sites of HCO$_3^-$ secretion, and sites of active K$^+$ transport. Na$^+$ crosses small-intestinal and colonic brush borders via Na/H exchange and, in the small bowel, also by Na–organic solute cotransport. In the distal colon, luminal Na$^+$ is also absorbed via an aldosterone-sensitive Na$^+$ channel that belongs to the same molecular family as does the channel in the distal renal tubule (S36). In hyperaldosteronemic states (dehydration, liver disease, and heart failure), colonic Na$^+$ absorption (and therefore also Cl$^-$ and water absorption) is enhanced.

HCO$_3^-$ is absorbed in the jejunum (via Na/H exchange) and secreted in the duodenum, ileum, and colon. Its secretion can be explained in part by Cl/HCO$_3$ exchange (when it is not offset by Na/H exchange), but most intestinal HCO$_3^-$ secretion appears to occur via CFTR, which is permeable to both HCO$_3^-$ and Cl$^-$; the relative rates of secretion of HCO$_3^-$ and Cl$^-$ appear to be determined by cell-to-lumen concentration differences and by their relative conductivity through CFTR (25). At the enterocyte's basolateral membrane, HCO$_3^-$ entry involves both Na/H exchange and Na(HCO$_3$)$_3$ cotransport (S37–S39).

Active absorption and secretion of K+ both occur in the colon, but neither occurs in the small intestine, where K+ movement is strictly diffusional. K+ is actively absorbed in distal colon via an H/K-ATPase homologous to the gastric H/K-ATPase responsible for acid secretion and is actively secreted throughout the colon via apical K+ channels (26, S40, S41). K+ absorption is enhanced by Na+ and K+ depletion and inhibited by cAMP. K+ secretion is enhanced by chronic dietary K+ loading and by cAMP.

Horizontally, the anatomic sites for secretion and absorption have traditionally been thought to differ, secretion occurring in crypts and absorption in small-intestinal villi and colonic surface cells. This undoubtedly overreaching generalization is consistent with some studies but not others. In support, (a) hypertonic disruption of small-intestinal villus cells does not reduce CT-induced secretion (27); (b) in distal colon, intracellular microelectrode recordings have shown appropriate electrical responses to secretory stimuli only in crypt cells, not in surface cells (S42, S43); and (c) almost all CFTR molecules have been localized to small-intestinal and colonic crypts by immunofluorescence and in situ hybridization (28, 29). Intriguingly, however, occasional villus cells, representing 3% of the total, immunostain positive for CFTR (30). These cells, which superficially resemble neighboring villus cells but lack brush border disaccharidases, may be secretory. Alternatively, they may represent a high-conductance route for passive Cl- movement.

On the other hand, microperfusion of individual crypts dissected from distal colon has demonstrated base-line absorption and agonist-inducible secretion (31). Does this dual functionality also apply to small-intestinal crypts? Probably not. Colonic crypts are much larger, their cells constituting most of the epithelium; in contrast, small-intestinal crypt cells normally constitute only about 20% of the epithelium. Furthermore, an apical NHE has not been found in small-intestinal crypt cells (32, S44); an apical anion exchanger is present (32), but it alone cannot enable fluid absorption.

Not all investigators will agree (S45, S46), but the following conclusions seem best supported by the evidence: (a) in small intestine, most, but conceivably not all, secretion arises from crypts, and all, or nearly all, absorption arises from villi; and (b) in colon, crypts display both absorption and secretion, but surface cells only absorb. In distal colon, Na+ channel–mediated absorption is confined to surface cells (S47).

Due to this at-least partial separation of absorptive and secretory functions, the latter predominate in diseases that selectively damage villar or surface epithelium (e.g., enteric infection, inflammatory bowel disease, celiac sprue). This predominance is further accentuated in celiac disease, because crypt hypertrophy also develops.

Table 2, which lists the various cell membrane transport proteins that collectively account for the intestinal tract's absorptive and secretory processes for electrolytes, provides a somewhat cryptic summary of much of the above. One entry in the table is not referred to above: ClC-4, a second apical Cl- channel in mouse and human intestine that belongs to the family of ClC anion channels (S48) and that colocalizes with CFTR (S49). It is activated by depolarization but the effects, if any, of intracellular mediators or extracellular regulators on its activity are unknown. Whether its variable expression can explain the far more severe intestinal disease that occurs in CF knockout mice than in humans with CF is unclear. ClC-4 has been detected in both murine and human intestine. Only minimal secretion, which is not increased by traditional effectors, has been observed in CF intestinal preparations. Since

CF gene mutations generally interfere with CFTR traffic to the apical membrane (S50), and since normal CFTR is a regulator of other membrane transporters and ion channels, apical placement of ClC-4 may also be affected in CF. It will be interesting to look at ClC-4 expression in CF intestine.

Pathophysiology: general aspects
Patients may use the word diarrhea for increases in stool mass (in adults, up to 250 g/d is considered normal), stool liquidity, or stool frequency. The first criterion truly defines diarrhea, but patients rarely provide quantitative information on their daily stool output. The other two criteria suggest the first, but occasionally stool frequency or liquidity increases without an appreciable increase in stool mass. Most such instances constitute a purely functional disorder known as irritable bowel syndrome, which is responsible for at least 25% of all visits to gastroenterologists. Although moderate increases in stool mass (500–1,000 g/d) need medical attention if they persist, this review is mainly concerned with more severe diarrheas — those that can lead to dehydration.

Two caveats about diarrhea are worth stressing. First, significant malabsorption of nutrients does not always cause diarrhea, since the normal adult colon's reabsorptive capacity is about 5 l/d. Second, patients may develop profuse diarrhea in the absence of nutrient malabsorption or even histologically evident mucosal damage or inflammation, as, for example, when the small intestine's immense secretory capacity is triggered by a bacterial enterotoxin.

Diarrheal driving forces
Osmosis, active secretion, exudation, and altered motility can all drive diarrhea. Specific diarrheal illnesses often involve more than one of these forces.

Osmotic diarrhea. When poorly absorbable, low–molecular weight aqueous solutes are ingested, their osmotic force quickly pulls water and, secondarily, ions into the intestinal lumen. Individuals with normal gut function will develop osmotic diarrhea when they ingest large amounts of poorly absorbable solutes, such as lactulose (if they are being treated for hepatic encephalopathy), sorbitol (if they continually chew sugar-free gum), or Mg^{2+} (if they take certain antacids or bowel purgatives).

Osmotic diarrhea can also develop when an ordinarily absorbable nutrient is ingested by an individual with an absorptive defect, for example, lactose by someone with congenital lactase deficiency, or carbohydrate by someone with gluten-sensitive enteropathy (celiac disease). Maldigestion, as seen in pancreatic insufficiency, may cause osmotic diarrhea of colonic origin: unabsorbed carbohydrates and lipids, on reaching the colon, are hydrolyzed by bacteria into short-chain organic acids, the quantity of which may overwhelm the colon's capacity for their absorption.

Secretory diarrhea. Diarrhea resulting from overstimulation of the intestinal tract's secretory capacity can develop in "pure" form (e.g., cholera) or as a component of a more complex disease process (e.g., celiac disease, Crohn disease). "Pure" secretory diarrhea is characterized by (a) large stool volumes (which can exceed 1 liter per hour in well hydrated adults), (b) absence of red or white blood cells in the stool, (c) absence of fever or other systemic symptoms (except those due to dehydration), (d) persistence of diarrhea with fasting (volume may diminish, however), and (e) lack of excess osmotic gap in stool electrolytes. Osmotic gap (OG) is defined as follows: $OG = 290 - 2\{[Na^+] + [K^+]\}$, where 290 is the assumed osmolarity of blood plasma. A gap greater than 50 mM is considered abnormal

Table 2

Intestinal transport proteins

Transport process	Ions transported	Species	Location A/BL	Horizontal/vertical distribution	Associated gene defect (m or h)
NaCl abs	Na/H	NHE2 (5, S3, S4)	A	SI, DC (V and SC > Cr)	Gastric atrophy (m)
		NHE3 (5, 6, S3, S4)	A	SI, PC (V, SC >> Cr)	Chronic diarrhea (m)
		Cl-dep. NHE (S5)	A	DC (Cr)	
	Cl/HCO₃	dra (7, 8, S8)	A	Duod, ileum, C (V and Cr)	Familial Cl⁻ diarrhea (h)
		PAT1 (S6)	A	Duod, ileum > C	
	K⁺	High G; Ca²⁺, stretch-activated (21, S122–S125)	RI	Throughout	
	Cl⁻	ClC-2; PKC-activated (S126, S127)	BL	Throughout	
Na⁺ abs	Na⁺	ENa (~4 pS) (S36, S47)	A	DC (SC)	
	K⁺	High G; Ca²⁺, stretch-activated	BL		
Cl⁻ secr	Cl⁻	CFTR (~7 pS) (15, S13, S14)	A	Throughout (cr >> V, SC)	Cystic fibrosis (m, h)
		ClC-4 (S49)	A	Same as for CFTR	
	K⁺	High G (200–300 pS); Ca²⁺, stretch-activated	BL	Throughout	
		Low G, cAMP-activated; (16, S15, S16, S121, S125, S126)	BL	Throughout (cr >> V)	
	Na⁺, K⁺, Cl⁻	NKCC1 (17, S17, S18)	BL	Throughout	
HCO₃⁻ abs	Na⁺, H⁺	NHE2, NHE3	A	Jej (V)	
	K⁺	High G (200–300 pS); Ca²⁺, stretch-activated	BL		
HCO₃⁻ secr	Cl/HCO₃	Dra, PAT1	A	Duod, ileum, C (V, SC, cr)	
	HCO₃⁻	CFTR (25)	A	Duod, ileum, C (cr >> V)	
	Na/H	NHE1	BL	Throughout (cr, V, SC)	
	Na(HCO₃)₃	NBC1 (S37, S38)	BL	Duod, PC	
K⁺ abs	K/H	H⁺/K⁺-ATPase (26)	A	DC	
	K⁺	High G; Ca²⁺, stretch-activated	BL		
K⁺ secr	K⁺	High G, aldo-activated (S40, S41)	A	C (cr)	
	Na⁺, K⁺, 2Cl⁻	NKCC1	BL	Throughout	
	Na/K	Na/K-ATPase	BL	Throughout	

Since the action of Na/K-ATPase on the basolateral membrane is central to all of these ion transport processes, its involvement is assumed without mention, except in the instance of K⁺ secretion, where it serves as a K⁺-loading mechanism. A, apical membrane; BL, basolateral membrane; m, transgenic mouse; h, human; abs, absorption; SI, small intestine; PC, proximal colon; V, villus; SC, colonic surface cell; Cr, crypt; dep., dependent; DC, descending colon; dra, downregulated in adenoma; duod, duodenum; C, colon; G, conductance; ENa, NA⁺ channel; secr, secretion; aldo, aldosterone.

(S51); the normal gap is made up of Mg^{2+}, Ca^{2+}, NH_4^+, and perhaps organic cations. The pattern of stool electrolytes in patients with acute cholera shows Na^+, K^+, and Cl^- concentrations not very different from those in plasma and an HCO_3^- concentration somewhat higher then in plasma. In contrast, normal stool shows low $[Na^+]$ and high $[K^+]$ concentrations, due mainly to the colon's reabsorption of Na^+ and secretion (both active and passive) of K^+; and a low $[Cl^-]$ concentration, due to the replacement of $[Cl^-]$ by short-chain organic acid anions generated by colonic bacteria. Normally, $[HCO_3^-]$ concentration is similar to that in plasma.

A number of secretory stimuli can cause diarrhea. These include bacterial enterotoxins, hormones generated by endocrine neoplasms, dihydroxy bile acids, hydroxylated fatty acids, and inflammatory mediators. In the absence of vigorous fluid replacement, marked dehydration leading to vascular collapse can occur, as can severe losses of K^+ and HCO_3^-.

Exudative diarrhea. If the intestinal epithelium's barrier function is compromised by loss of epithelial cells or disruption of tight junctions, hydrostatic pressure in blood vessels and lymphatics will cause water and electrolytes, mucus, protein, and sometimes even red and white cells to accumulate luminally (e.g., ulcerative colitis, shigellosis, intestinal lymphangiectasia).

If the condition is chronic, the continuing protein loss can lead to hypoalbuminemia and hypoglobulinemia.

Diarrhea resulting from motility disturbances. Both increases and decreases in gut motility can lead to diarrhea. Examples of the former are thyrotoxicosis and opiate withdrawal. Decreases in effective motility in the small intestine due to large diverticula, smooth muscle damage (scleroderma, dermatomyositis, amyloidosis, muscular dystrophy, or radiation injury), or autonomic neuropathy (diabetic, idiopathic) can result in bacterial overgrowth. As specified below, bacterial overgrowth can lead to diarrhea.

Pathophysiology of acute diarrheas

Enteric infection with enterotoxin-producing bacteria

Summarized here is information on three bacterial enterotoxins produced by two enteric organisms, *V. cholerae* and *Escherichia coli*. In fact, similar toxic peptides are produced by *Yersinia* (S52) and *Salmonella* (S53) and perhaps other enteropathogenic bacteria.

V. cholerae. We are in the midst of a worldwide pandemic of cholera, due to biotype El Tor, that apparently began in the 1960s (S54). The disease is now or has been recently prevalent in India, Bangladesh, Pakistan, Indochina, Indonesia, Afghanistan, Africa,

South America, and Mexico. Although it is rare in the US, 61 proven cases were reported to the Centers for Disease Control from 1995–2000 (S55). More than half were acquired during travel abroad, and most of the others were acquired from undercooked imported seafood. The organism appears to inhabit estuaries (in a form that is difficult to culture) and to spread to human populations under conditions of contaminated water supplies, inadequate sewage disposal, crowding, and malnutrition. Wars and earthquakes are inciting factors.

Every medical student now learns the mechanism of action of CT, the *V. cholerae* enterotoxin. CT is an 84-kDa protein consisting of a dimeric A subunit and five identical B subunits (Figure 3). The A subunit consists of two peptides linked by an S-S bond, the larger, A1, containing the toxic activity. Each of the B subunits binds tightly to the GM1 ganglioside abundant in the intestinal brush border membrane (33); binding is so tight that only the normal 5- to 7-day turnover of the epithelium eliminates toxicity. The toxin's A1 subunit, which is endocytosed following toxin binding to GM1 (S56), covalently modifies the α subunit of G_s, the adenylyl cyclase–stimulating G protein. More specifically, the endocytosed A1 subunit catalyzes covalent bonding of adenosine diphosphoribose from NAD to the α subunit of G_s ($G_s\alpha$), inhibiting the intrinsic GTPase activity of $G_s\alpha$ thereby preventing self-inactivation of its adenylyl cyclase-stimulating activity (34, 35).

V. cholerae produces an additional physiologically active peptide, zona occludens toxin, that increases the permeability of tight junctions (S57); however, its relevance to diarrhea is unclear.

Toxigenic E. coli. Toxigenic E. coli, which proliferate in institutional settings and commonly infect American travelers abroad (traveler's diarrhea), elaborate at least two secretion-stimulating enterotoxins. The first, heat-labile toxin, is a high–molecular weight peptide that is immunologically related to CT and has the same cellular mechanism of action (S58). The second, heat-stable toxin (STa), is a cysteine-rich, low–molecular weight, heat-stable peptide (S59) that activates guanylate cyclase in the intestinal epithelium (36), thereby activating cGMP-dependent protein kinase (PKG), which, like PKA, can open the CFTR anion channel (37) (Figure 4). Twelve years ago, the brush border receptor for STa was cloned and identified as a guanylyl cyclase, which is activated upon STa binding (38). The existence in mammalian intestine of an enzymatically active receptor for a bacterial enterotoxin raised the question of whether an endogenous ligand for this receptor, one with structural homology to STa, also exists. This question was answered about a year later when a low–molecular weight peptide that mimics STa action and has structural homology to STa was extracted from rat intestinal contents and dubbed "guanylin" (39). This luminally active hormone is, like STa, rich in cysteine (although its four cysteines are two less than in STa). Since it is synthesized in and secreted from goblet cells, and its synthesis is enhanced by oral salt loading, it may play roles in both fluid homeostasis and the moistening of goblet cell mucoid secretions (S60, S61).

As an aside, a close homolog of guanylin, known as uroguanylin, also originates in the gut epithelium and is secreted into the blood by enterochromaffin cells (S62). It is a diuretic (at least in the kidneys of rodents) and may explain why an oral salt load induces a greater diuresis than the same salt load given intravenously.

Role of the ENS in enterotoxin-induced secretion. Although the direct effects of CT, heat-labile toxin, and STa on the ion-transporting cells of the intestinal epithelium would seem sufficient to explain the resulting secretion (and in the in vitro mucosal preparation

they are sufficient), there is ample evidence from in vivo experiments that the ENS also plays a role — in the mind of some, a dominant role. When segments of mammalian small bowel are perfused with toxin-containing solutions, most of the resulting fluid transport changes can be reversed with neural poisons such as tetrodotoxin and hexamethonium (40). Recent experiments with a specific VIP antagonist demonstrate that much (but not all) of the secretion induced in these enterotoxin perfusions is mediated by VIP (41).

The sequence of events, summarized in Figure 5, seems to be as follows. The enterotoxins attach to endocrine (enterochromaffin) cells on the villus surface of the intestinal epithelium, causing an increase of cAMP or cGMP. cAMP or cGMP triggers these cells to release serotonin, neurotensin, and possibly other peptides into the subepithelial space. These paracrine mediators set in motion a secretory neural reflex that involves (a) afferent enteric neurons that begin subepithelially and end in the myenteric plexus, (b) interneurons that connect the myenteric to the submucosal plexus, and (c) efferent neurons in the submucosal plexus that release VIP (and probably additional secretory stimuli) near the basal surface of the epithelium. The interneuronal effect appears to be mediated by acetylcholine and/or substance P. Finally, the VIP (and perhaps another agonist) attaches to receptors on both villus and crypt enterocytes, activating adenylyl cyclase and inducing cAMP-mediated alterations of ion transport (inhibition of absorption in villus cells and stimulation of secretion in crypt cells).

One caveat is that the role of the ENS in enterotoxin-induced secretion may be more striking in the animal studies than in the disease itself; in segmental perfusions, enterotoxin does not penetrate into the crypts, whereas, in cholera patients, a direct effect of CT on crypt cells is likely since *Vibrio* organisms proliferate in the crypts as well as elsewhere in the gut lumen.

Other enteric infections

Common bacterial causes of enteritis and/or colitis are *Shigella*, *Salmonella*, *Yersinia*, enteroinvasive *E. coli*, *Aeromonas*, and *Campylobacter*. These organisms invade the epithelium and multiply intracellularly, damaging the surface epithelium and causing inflammation. Diarrhea is due both to the epithelial damage (exudation and decreased absorptive capacity) and to the action of inflammatory mediators.

Some viruses also invade the intestinal epithelium, causing enterocyte destruction, inflammation, and a temporary sprue-like syndrome (partial villus atrophy and crypt hypertrophy). These include rotovirus (worldwide, probably the most common cause of diarrhea in infants), certain adenoviruses, and the so-called Norwalk-like agents (S63). These last agents have recently caused a flurry of brief episodes of vomiting and diarrhea among passengers on cruise ships (S64).

Some enteric bacteria produce cytolytic toxins that destroy epithelial cells, interfering with absorption and causing inflammation. These include *Shigella dysenteriae* (Shiga toxin), a protein synthesis inhibitor; *Clostridium difficile*, whose A toxin alters cytoskeletal structure, and whose proliferation is induced by broadspectrum antibiotics; *Vibrio parahaemolyticus*; *Clostridium perfringens*; and enteroadherent and enterohemorrhagic *E. coli* (S65–S68). Most of these bacterial toxins cause breakdown of the cytoskeleton through ADP ribosylation or glucosylation of proteins from the Rho family, which regulate actin polymerization.

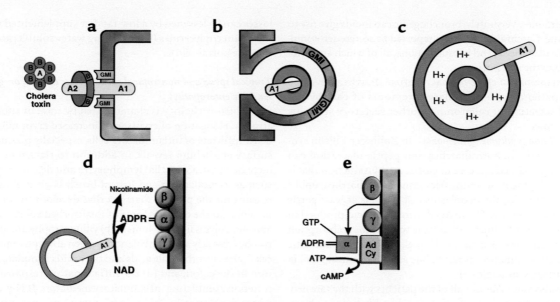

Figure 3

Cellular mechanism of action of cholera toxin. (**a**) The toxin, through its B subunits, which encircle the enzymically active A subunit, binds to the ubiquitous glycolipid membrane receptor, the monosialoganglioside GM1. (**b**) The entire complex is endocytosed by both clathrin-dependent and clathrin-independent means. (**c**) Inwardly directed proton pumps acidify the CT-containing endocytic vesicle, causing the toxin subunits to dissociate. Then the enzymically active A1 peptide is inserted into the vesicle membrane, with its catalytic site exposed to cytoplasm. (**d**) The A1 peptide is an ADP-ribosyltransferase that cleaves NAD into adenosine diphosphoribose (ADPR) and nicotinamide and covalently bonds the former to the α subunit of the G_s adenylyl cyclase–stimulatory G protein. (**e**) The intrinsic GTPase activity of the ADPR-modified α subunit of G_s is markedly inhibited, allowing GTP to remain bound to it; the $G_s\alpha$-GTP complex separates from its membrane mooring to $G_s\beta$ and $G_s\gamma$, attaching to and activating adenylyl cyclase (AdCy).

Cryptosporidia, a unicellular parasite, caused an epidemic of diarrhea in Milwaukee ten years ago through contamination of the city water supply (S69). Persistent cryptosporidial enteritis can occur in immunocompromised individuals. *Cryptosporidia* gain entry to the apical cytoplasm of enterocytes by first fusing with the enterocyte microvillus membrane (S70, S71). Heavy infestation with *Giardia lamblia* leads to disruption of the enterocyte microvillus membrane, which, in turn, leads to malabsorption and diarrhea (S72).

Acute T cell activation

Parenteral administration of cytolytic antibodies to T cell surface antigens causes massive acute, but short-lived, diarrhea, as seen in humans treated for rejection of transplanted organs and, experimentally, in mice (S73). The critical cytokine released in these circumstances appears to be TNF-α, presumably because it causes the massive release of prostaglandins and other secretory stimuli.

Pathophysiology of chronic diarrheas

Congenital diarrheas

Two rare congenital autosomal-recessive defects in intestinal fluid absorption are associated with diarrhea: congenital chloride diarrhea, which results from a defect in brush border Cl/HCO_3 exchange (see above), and congenital sodium-secretory diarrhea, which results from an as-yet undefined defect in brush border Na/H exchange (8, S74). In both cases, nutrient absorption is normal. Also in both, abnormal fluid absorption begins in utero, manifesting itself as maternal polyhydramnios. Congenital chloride diarrhea has been more extensively reported, with clusters of cases found in Finland, Saudi Arabia, Kuwait, and Poland and

scattered cases elsewhere (8). In this syndrome, stool pH is acidic, and metabolic alkalosis develops. Similar acid-base changes are seen in carbohydrate malabsorption and Zollinger-Ellison syndrome. In contrast, in most severe diarrheas, stool pH is alkaline, and metabolic acidosis develops.

Several other rare congenital defects in nutrient absorption or in intestinal epithelial structure can also cause neonatal diarrhea (S75): congenital sucrase-isomaltase deficiency; congenital glucose-galactose malabsorption; lysinuric protein intolerance, a defect in dibasic amino acid transport across the enterocyte basolateral membrane; acrodermatitis enteropathica, a defect in zinc absorption; and microvillus inclusion disease, which includes severe villus atrophy and crypt hypoplasia with the characteristic finding of microvilli contained within vesicles inside the enterocytes.

Hormone-secreting neoplasms

In several uncommon tumors, hormones are produced and released that directly stimulate intestinal secretion, causing profuse diarrhea or, in one instance (gastrinoma), interfering with nutrient absorption. For a comprehensive review, see ref. 42.

Pancreatic cholera. In patients with pancreatic cholera, certain endocrine neoplasms that occur most commonly in pancreatic islets but occasionally in the proximal intestinal mucosa secrete large quantities of VIP, the enteric secretory neurotransmitter. Some ganglioneuromas (neural crest tumors seen in early childhood) and pheochromocytomas do so also.

Carcinoid syndrome. In carcinoid syndrome patients, cutaneous flushing and profuse diarrhea develop when carcinoid tumors in intestine metastasize to the liver, thereby not only amplifying the tumor mass but also bypassing hepatic detoxification of tumor-released hormones. Very rarely, bronchogenic carcinoids give rise to the syndrome. Carcinoids have been reported to secrete serotonin, bradykinin, substance P, and prostaglandins, all of which are secretory stimuli in the intestine.

Medullary carcinoma of the thyroid. Profuse diarrhea develops in 30% of patients with medullary carcinoma of the thyroid because of secretion of calcitonin, another secretory stimulus in the intestine.

Zollinger-Ellison syndrome (gastrinoma). In Zollinger-Ellison syndrome (gastrinoma), both diarrhea and peptic ulceration can result from the marked increase in gastric acid production that is associated with gastrin-secreting neoplasms. The diarrhea, unlike those associated with the neoplasms just cited, is at least partly malabsorptive in nature. The marked increase in acid production both presents the intestines with a large volume for reabsorption and lowers pH in the proximal small bowel, inactivating pancreatic digestive enzymes and precipitating bile acids. Diarrheal stool in these patients has an acidic pH.

Systemic mastocytosis. About half of the patients with the rare neoplasm systemic mastocytosis develop diarrhea, likely due to histamine-induced gastric hypersecretion, a cause similar to that of the diarrhea in Zollinger-Ellison syndrome.

Ileal disease or resection

Nearly all conjugated bile acids are normally reabsorbed in the distal ileum via Na–bile acid cotransport, with very little reaching the colon (S76–S77). Small-bowel Crohn disease may lead to bile acid malabsorption, since distal ileum is frequently affected; also, obstruction there may develop, necessitating ileal resection. Unabsorbed dihydroxy bile acids such as deoxycholate are amphophiles that, when not incorporated into lipid-containing micelles in the gut lumen, tend to insert into the lipid matrix of the enterocyte or colonocyte brush border membrane. Deconjugation of these bile acids by colonic bacteria enhances their amphophilicity. They activate secretion primarily in the colon but, in the absence of lipids, can also do so in the small bowel. The mechanism for their secretory effect is complex, involving both intracellular Ca^{2+} (probably secondary to membrane phospholipase activation) and cAMP (S78, S79). Bile acid–induced increase in cAMP appears to involve the ENS or mast cells (S80). Diarrhea that results from moderate malabsorption of bile acids can be somewhat reduced by cholestyramine, a resin that binds bile acid anions. But if more than 100 cm of distal ileum is resected, so much of the enteric bile acid load fails to be reabsorbed that the body's bile acid pool shrinks (43). Then, trapping of additional bile acids with cholestyramine shrinks the bile acid pool further, causing fat malabsorption and thereby worsening the diarrhea. This is because fatty acids that reach the colon undergo bacterial hydroxylation, which renders them amphophilic. Like amphophilic bile acids, and probably by a similar mechanism, they induce secretion (44).

Intestinal lymphectasia

Obstruction of intestinal lymphatics can be congenital or neoplastic. The resulting increase in interstitial hydrostatic pressure disrupts tight junctions, thereby greatly increasing epithelial permeability not only to water and electrolytes but also to serum proteins. Since chylomicrons are absorbed through lymphatics, fat ingestion by patients with intestinal lymphectasia will increase pressure in their partially obstructed intestinal lymphatics, thereby increasing enteric protein loss and diarrhea. These losses can be lessened by a low-fat diet supplemented with medium-chain triglycerides (which, being water soluble, are absorbed into the portal vein).

Tropical sprue and nontropical sprue (celiac disease, or gluten-sensitive enteropathy)

Sprue histopathology is characterized by total or subtotal villus atrophy, elongation of crypts with increased crypt mitoses, and a dense infiltrate of inflammatory cells, especially plasma cells. The surface epithelium reveals, in addition to flattening, a marked increase in intraepithelial lymphocytes and degenerative changes such as vacuolization and loss of brush border. There are four reasons for the profuse diarrhea that develops in some of these patients: (a) the osmotic force of unabsorbed solutes pulls water and electrolytes into the lumen; (b) villus atrophy and crypt hypertrophy adversely alter the balance between absorptive and secretory cells; (c) unabsorbed bile acids and fatty acids stimulate fluid secretion in the colon; and (d) the inflammatory response generates secretion-stimulating inflammatory mediators. IFN-γ and TNF-α, whose concentrations in the lamina propria increase in celiac disease, also downregulate nutrient absorption (S73, S81–S83).

Diabetes mellitus

Diarrhea accompanied by rectal incontinence is an occasional complication of long-standing, insulin-dependent diabetes. It typically occurs in patients with poor diabetic control and peripheral neuropathy. Intestinal biopsies in such patients are usually normal, and nutrient malabsorption or bacterial overgrowth is present only in a minority of cases. In most instances, the diarrhea is secondary to degeneration of adrenergic nerves that, as mentioned above, are antisecretory and/or proabsorptive in intestinal fluid homeostasis (45. S84). Other evidence of

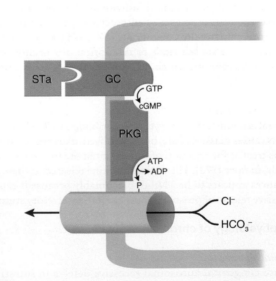

Figure 4

Cellular mechanism of action of heat-stable *E. coli* enterotoxin (STa). Luminal toxin binds to and activates guanylyl cyclase C (GC), a brush border enzyme with a high rate of expression in both small intestine and proximal colon. The resulting increase in [cGMP] activates membrane-bound protein kinase G (PKG), which, in turn, opens neighboring anion channels or inhibits neighboring Na/H and Cl/HCO$_3$ exchangers (not shown). Guanylin, a mammalian homolog of STa, is the physiologic agonist for the STa-sensitive GC. It is secreted into the lumen by goblet cells. In the duodenum, its action results in an HCO$_3^-$-rich secretion.

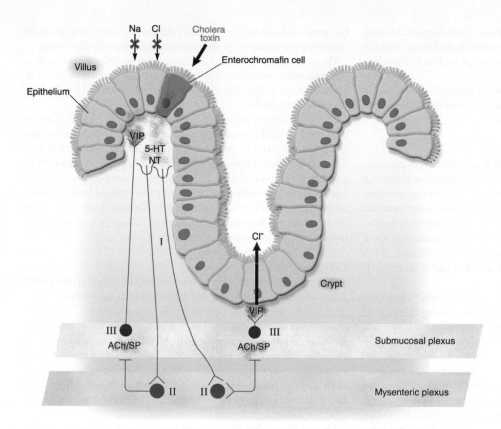

Figure 5
CT induces secretion, in part, through activation of an intramural neural reflex. Toxin attaches to enterochromaffin cells in the epithelium, causing an increase in [cAMP] there. In response to the latter, serotonin (5-HT), neurotensin (NT), and possibly additional peptides are released. These activate afferent neurons (I), whose axons course from close to the epithelium to the myenteric neural plexus, where they connect with interneurons (II) that, in turn, through release of acetylcholine (ACh) and/or substance P (SP), activate secreto-motor neurons (III) in the submucosal neural plexus. Axons from these secreto-motor neurons reach the epithelial surface, in both the villus and the crypt regions, releasing VIP and thereby stimulating secretion in crypts and inhibiting nutrient-independent salt absorption in villi. For a comprehensive review, see ref. 42.

autonomic neuropathy can usually be elicited. There is anecdotal evidence that diabetic diarrhea can sometimes be controlled with an α_2 agonist such as clonidine or with a somatostatin analog such as octreotide (S85, S86).

Bacterial overgrowth
Normally, only a relatively small number of bacteria proliferate in the small bowel. In the presence of local stasis or recirculation of luminal contents, however, bacteria will proliferate in the affected segment of small bowel (e.g., afferent loop dysfunction after gastro-jejunostomy; jejunal diverticula; strictures or enterocolic fistulas in Crohn disease; and ineffective peristalsis in scleroderma or diabetes). If the stasis occurs proximally, the bacteria will deconjugate and partially dehydroxylate bile acids, leading to their rapid diffusional reabsorption. The resulting decrease in concentration of effective bile salts can cause fat malabsorption. As mentioned above, lipids that are hydroxylated by bacteria are thereby converted into secretion-stimulating amphophiles. In addition, bacterial proteases can inactivate brush border oligosaccharidases, leading to carbohydrate malabsorption.

Inflammatory bowel disease
In regional enteritis (Crohn disease), damage to or resection of the distal ileum can lead to bile acid malabsorption, which, as outlined above, causes diarrhea. Also in Crohn disease, enterocolic fistulas may develop, resulting in bacterial overgrowth of the small bowel, and this, as outlined just above, causes diarrhea. In both ulcerative colitis and Crohn colitis, inflammatory mediators such as prostaglandins stimulate colonic secretion, contributing to diarrhea. Cytokines generated in the inflamed mucosa may also downregulate fluid-absorptive mechanisms (S73, S81–S83).

Principles of management

Oral rehydration therapy
Cholera and *E. coli* enterotoxins stimulate fluid secretion without injuring the intestinal epithelium; in particular, they do not impair the cellular processes for nutrient absorption. Even in viral enteritis, which does cause epithelial injury, considerable absorptive capacity remains. Long ago it was noted that, in cholera, diarrheal losses could not be replaced by oral administration of a simple salt solution — it only added to diarrheal volume. We now know why: the cAMP response to CT inhibits nutrient-independent salt absorption (i.e., Na/H and Cl/HCO₃ exchanges). In the 1960s, investigators in Dacca and Calcutta, mindful of the newly emerging concept of Na-coupled absorption of sugars and amino acids, found that, when they added glucose to an orally administered salt solution (an isosmotic solution with approximately equimolar amounts of Na⁺ and glucose), cholera patients were able to absorb the solution (46, 47). In areas where the incidence of severe diarrhea is high and health workers and sterile intravenous solutions are scarce, the use of an oral rehydration solution administered by family members has saved many lives.

Although such solutions can replace diarrheal losses, they do not facilitate reabsorption of secreted fluid and therefore do not lessen diarrhea. Some reabsorption of secreted fluid can be achieved by substitution of glucose polymers for free glucose in the oral rehydration solution. Normally, the rate of hydrolysis of starch to maltose is slow compared with the rates of hydrolysis of maltose and of absorption of free glucose. Thus, as fast as glucose becomes available, it is absorbed, and hyperosmolarity of the luminal contents never develops. A glucose polymer can be thought of as a "glucose battery," providing a continuous supply of glucose for absorption under isosmotic conditions. Solutions containing extracts from rice or other grains are increasingly being used for oral rehydration and appear not only to replace fluid losses but also to lessen diarrhea (S87).

The same principle may apply to proteins and to the amino acids and oligopeptides generated from them, but there is relatively little experimental evidence of their efficacy.

Pharmacologic intervention

Antibiotics. To minimize development of drug-resistant organisms, antibiotics should not be used for relatively mild, self-limited infectious diarrheas. When the diarrhea is hemorrhagic, seriously dehydrating, associated with serious systemic signs and symptoms, or of longer than 5 days' duration without improvement, antibiotic use is appropriate. Fluoroquinolones are commonly used. *C. difficile*-induced diarrhea should be treated by withdrawal of the offending antibiotic and oral administration of either metronidazole or vancomycin.

Antidiarrheals. In mild diarrheas, antimotility agents (e.g., loperamide) will lessen stool frequency and, by increasing the time of contact with the gut epithelium, will also lessen stool volume. Such agents are contraindicated in severe diarrhea, because they may cause pooling of large fluid volumes in paralyzed bowel loops.

Since intestinal ion transport mechanisms operate within a complex regulatory framework, drugs that either mimic or potentiate the effects of antisecretory/proabsorptive hormones or neurotransmitters are proving useful. Recently, an enkephalinase inhibitor, racecedotril (also known as acetorphan), has been shown to lessen the volume of acute infectious diarrhea in children, presumably by preventing breakdown of enkephalins in the mucosa, which are anti-secretory (S88, S89). Its action is confined to intestine and, unlike loperamide, it doesn't reduce transit time.

Somatostatin analogues (e.g., octreotide) will sometimes control diarrhea. In hormone-secreting neoplasms, they block hormone production by the tumor (42). Since they also appear to have a direct antisecretory effect on the gut epithelium (S90), they have been employed for treating cancer chemotherapy–induced diarrhea. Anecdotal experience with an α_2 agonist, clonidine, suggests its utility in some diarrheas, especially in diabetic diarrhea, but severe side effects limit its usefulness; one side effect, postural hypotension, would be problematic in dehydrated individuals.

Thus far, drugs that block the actions of secretagogues, such as prostaglandin synthase inhibitors, have not proven effective as antidiarrheals perhaps because of the multiplicity of locally produced secretagogues.

Two cellular sites provide targets for the development of transport inhibitors: the apical anion channel (CFTR) and the basolateral cAMP-sensitive K^+ channel. In the laboratory, specific blockers for each have been identified. Of particular interest, a class of high potency inhibitors of the CFTR anion channel (thiazolidinones) has recently been identified through screening a large number of compounds in transfected cells (48). One of these, on a single intraperitoneal injection into mice (250 µg/kg), reduced CT-induced secretion by more than 90%.

Overview of treatment

Understandably, therapies for epidemic infectious diarrheas must differ somewhat from those for chronic diarrheas. Epidemic cholera is most effectively dealt with at the contamination level. Cholera vaccines were not covered in this review; they are only partially effective and then only temporarily. An effective oral vaccine is not available. Efforts to improve sanitation are of more practical value. In a recent study, simple filtration of drinking water through cloth, such as a flat fragment of an old sari, reduced cholera incidence in Bengladeshi villages by 50% (49). Once infection has occurred, ORS remains the mainstay of treatment.

Several drugs that either have proven or may prove useful in treating chronic and some acute diarrheas were cited above. In addition, diarrheas with an inflammatory component can be treated with anti-inflammatory agents, most notably glucocorticoids.

Note: Due to space constraints, a number of important references could not be included in this reference list. Interested readers can find a supplementary reference list at (http://www.jci.org/cgi/content/full/111/7/931/DC1).

Address correspondence to: Michael Field, 299 Riverside Drive, Apartment 9D, New York, New York 10025, USA. E-mail: mf9@columbia.edu.

1. Schultz, S.G., Fuisz, R.E., and Curran, P.F. 1966. Amino acid and sugar transport in rabbit ileum. *J. Gen. Physiol.* **49**:849–866.
2. Ganapathy, V., and Leibach, F.H. 1985. Is intestinal peptide transport energized by a proton gradient? *Am. J. Physiol.* **249**:G153–G160.
3. Turnberg, L.A., Fordtran, J.S., Carter, N.W., and Rector, F.C., Jr. 1970. Mechanism of bicarbonate absorption and its relationship to sodium transport in the human jejunum. *J. Clin. Invest.* **49**:548–556.
4. Turnberg, L.A., Bieberdorf, F.A., Morawski, S.G., and Fordtran, J.S. 1970. Interrelationships of chloride, bicarbonate, sodium, and hydrogen transport in the human ileum. *J. Clin. Invest.* **49**:557–567.
5. Hoogerwerf, W.A., et al. 1996. NHE2 and NHE3 are human and rabbit intestinal brush-border proteins. *Am. J. Physiol.* **270**:G29–G41.
6. Schultheis, P.J., et al. 1998. Renal and intestinal absorptive defects in mice lacking the NHE3 Na+/H+ exchanger. *Nat. Genet.* **19**:282–285.
7. Melvin, J.E., Park, K., Richardson, L., Schultheis, P.J., and Shull, G.E. 1999. Mouse down-regulated in adenoma (DRA) is an intestinal Cl-/HCO3- exchanger and is up-regulated in colon of mice lacking the NHE3 Na+/H+ exchanger. *J. Biol. Chem.* **274**:22855–22861.

8. Kere, J., Lohi, H., and Hoglund, P. 1999. Genetic disorders of membrane transport. III. Congenital chloride diarrhea. *Am. J. Physiol.* **276**:G7–G13.
9. De, S.N. 1959. Enterotoxicity of bacteria-free culture-filtrate of *Vibrio cholerae*. *Nature.* **183**:1533–1534.
10. Dutta, N.K., Panse, M.W., and Kulkarni, D.R. 1959. Role of cholera toxin in experimental cholera. *J. Bacteriol.* **78**:594–595.
11. Gangarosa, E.J., Beisel, W.R., Benyajatic, C., Sprinz, H., and Piyaratn, P. 1960. The nature of the gastrointestinal lesion in Asiatic cholera and its relation to pathogenesis: a biopsy study. *American Journal of Tropical Medicine and Hygiene.* **9**:125–135.
12. Field, M. 1971. Ion transport in rabbit ileal mucosa. II. Effects of cyclic 3′, 5′-AMP. *Am. J. Physiol.* **221**:992–997.
13. Kimberg, D.V., Field, M., Johnson, J., Henderson, A., and Gershon, E. 1971. Stimulation of intestinal mucosal adenyl cyclase by cholera enterotoxin and prostaglandins. *J. Clin. Invest.* **50**:1218–1230.
14. Field, M., Fromm, D., Al-Awqati, Q., and Greenough, W.B., 3rd. 1972. Effect of cholera enterotoxin on ion transport across isolated ileal mucosa. *J. Clin. Invest.* **51**:796–804.
15. Tabcharani, J.A., Chang, X.B., Riordan, J.R., and Hanrahan, J.W. 1991. Phosphorylation-regulated

Cl- channel in CHO cells stably expressing the cystic fibrosis gene. *Nature.* **352**:628–631.
16. Kunzelmann, K., et al. 2001. Cloning and function of the rat colonic epithelial K+ channel KVLQT1. *J. Membr. Biol.* **179**:155–164.
17. Payne, J.A., et al. 1995. Primary structure, functional expression, and chromosomal localization of the bumetanide-sensitive Na-K-Cl cotransporter in human colon. *J. Biol. Chem.* **270**:17977–17985.
18. Sheppard, D.N., and Welsh, M.J. 1999. Structure and function of the CFTR chloride channel. *Physiol. Rev.* **79**(Suppl. 1):S23–S45.
19. O'Loughlin, E.V., et al. 1991. Abnormal epithelial transport in cystic fibrosis jejunum. *Am. J. Physiol.* **260**:G758–G756.
20. Grubb, B.R. 1997. Ion transport across the murine intestine in the absence and presence of CFTR. *Comp. Biochem. Physiol. A Physiol.* **118**:277–282.
21. Neilson, M.S., Warth, R., Bleich, M., Weyland, B., and Greger, R. 1998. The basolateral Ca2+-dependent K+ channel in rat colonic crypt cells. *Pflugers Arch.* **435**:267–272.
22. Chang, E.B., Field, M., and Miller, R.J. 1982. Alpha-2-adrenergic receptor regulation of ion transport in rabbit ileum. *Am. J. Physiol.* **242**:G237–G242.
23. Madara, J.L. 1987. Intestinal absorptive cell tight

junctions are linked to cytoskeleton. *Am. J. Physiol.* **253**:C171–C175.

24. Fordtran, J.S., Rector, F.C., Jr., Ewton, M.F., Soter, N., and Kinney, J. 1965. Permeability characteristics of the human small intestine. *J. Clin. Invest.* **44**:1935–1944.

25. Poulsen, J.H., Fischer, H., Illek, B., and Machen, T.E. 1994. Bicarbonate conductance and pH regulatory capability of cystic fibrosis transmembrane conductance regulator. *Proc. Natl. Acad. Sci. U. S. A.* **91**:5340–5344.

26. Binder, H.J., Sangan, P., and Rajendran, V.M. 1999. Physiological and molecular studies of colonic H+,K+-ATPase. *Semin. Nephrol.* **19**:405–414.

27. Roggin, G.M., Banwell, J.G., Yardley, J.H., and Hendrix, T.R. 1972. Unimpaired response of rabbit jejunum to cholera toxin after selective damage to villus epithelium. *Gastroenterology.* **63**:981–988.

28. Crawford, I., et al. 1991. Immunocytochemical localization of the cystic fibrosis gene product CFTR. *Proc. Natl. Acad. Sci. U. S. A.* **88**:9262–8266.

29. Trezise, A.E.O., and Buchwald, M. 1991. In vivo cell-specific expression of the cystic fibrosis transmembrane conductance regulator. *Nature.* **353**:434–437.

30. Ameen, N.A., Ardito, T., Kashgarian, M., and Marino, C.R. 1995. A unique subset of rat and human intestinal villus cells express the cyctic fibrosis transmembrane conductance regulator. *Gastroenterology.* **108**:1016–1023.

31. Singh, S.K., Binder, H.J., Boron, W.F., and Geibel, J.P. 1995. Fluid absorption in isolated perfused colonic crypts. *J. Clin. Invest.* **96**:2373–2379.

32. Knickelbein, R.G., Aronson, P.S., and Dobbins, J.W.

1988. Membrane distribution of sodium-hydrogen and chloride-bicarbonate exchangers in crypt and villus cell membranes from rabbit ileum. *J. Clin. Invest.* **82**:2158–2163.

33. King, C.A., and van Heyningen, W.E. 1973. Deactivation of cholera toxin by a sialidase-resistant monosialosyl ganglioside. *J. Infect. Dis.* **127**:639–647.

34. Gill, D.M., and King, C.A. 1975. The mechanism of action of cholera toxin in pigeon erythrocyte lysates. *J. Biol. Chem.* **250**:6424–6432.

35. Cassel, D., and Selinger, Z. 1977. Mechanism of cholera toxin activation of adenylate cyclase: inhibition of GTP hydrolysis at the regulatory site. *Proc. Natl. Acad. Sci. U. S. A.* **74**:3307–3311.

36. Field, M., Graf, L.H., Jr., Laird, W.J., and Smith, P.L. 1978. Heat-stable enterotoxin of *Escherichia coli*: in vitro effects on guanylate cyclase activity, cyclic GMP concentration and ion transport in small intestine. *Proc. Natl. Acad. Sci. U. S. A.* **75**:2800–2804.

37. Vaandrager, A.B., et al. 1998. Membrane targeting of cGMP-dependent protein kinase is required for cystic fibrosis transmembrane conductance regulator Cl- channel activation. *Proc. Natl. Acad. Sci. U. S. A.* **95**:1466–1471.

38. Schulz, S., Green, C.K., Yuen, P.S., and Garbers, D.L. 1990. Guanylyl cyclase is a heat-stable enterotoxin receptor. *Cell.* **63**:941–948.

39. Currie, M.G., et al. 1992. Guanylin: an endogenous activator of intestinal guanylate cyclase. *Proc. Natl. Acad. Sci. U. S. A.* **89**:947–951.

40. Lundgren, O. 2002. Enteric nerves and diarrhea. *Pharmacol. Toxicol.* **90**:109–120.

41. Mourad, F.H., and Nassar, C.F. 2000. Effect of vaso-

active intestinal polypeptide (VIP) antagonism on rat jejunal fluid and electrolyte secretion induced by cholera and *Escherichia coli* enterotoxins. *Gut.* **47**:382–386.

42. Jensen, R.T. 1999. Overview of chronic diarrhea caused by functional neuroendocrine neoplasms. *Semin. Gastrointest. Dis.* **10**:156–172.

43. Hofmann, A.F., and Poley, J.R. 1972. Role of bile acid malabsorption in pathogenesis of diarrhea and steatorrhea in patients with ileal resection. I. Response to cholestyramine or replacement of dietary long chain triglyceride by medium chain triglyceride. *Gastroenterology.* **62**:918–934.

44. Racusen, L.C., and Binder, H.J. 1979. Ricinoleic acid stimulation of active anion secretion in colonic mucosa of the rat. *J. Clin. Invest.* **63**:743–749.

45. Chang, E.B., Bergenstal, R.M., and Field, M. 1985. Diarrhea in streptozocin-treated rats. Loss of adrenergic regulation of intestinal fluid and electrolyte transport. *J. Clin. Invest.* **75**:1666–1670.

46. Pierce, N.F., et al. 1968. Effect of intra-gastric glucose-electrolyte infusion upon water and electrolyte balance in Asiatic cholera. *Gastroenterology.* **55**:333–343.

47. Hirschhorn, N. 1968. Decrease in net stool output in cholera during intestinal perfusion with glucose-containing solution. *N. Engl. J. Med.* **279**:176–188.

48. Ma, T., et al. 2002. Thiazolidinone CFTR inhibitor identified by high-throughput screening blocks cholera toxin-induced intestinal fluid secretion. *J. Clin. Invest.* **110**:1651–1658.

49. Colwell, R.R., et al. 2003. Reduction of cholera in Bengladeshi villages by simple filtration. *Proc. Natl. Acad. Sci. U. S. A.* **100**:1051–1055.

Molecular mediators of hepatic steatosis and liver injury

Jeffrey D. Browning[1] and Jay D. Horton[1,2]

[1]Departments of Internal Medicine and [2]Molecular Genetics, University of Texas Southwestern Medical Center at Dallas, Dallas, Texas, USA.

Obesity and its associated comorbidities are among the most prevalent and challenging conditions confronting the medical profession in the 21st century. A major metabolic consequence of obesity is insulin resistance, which is strongly associated with the deposition of triglycerides in the liver. Hepatic steatosis can either be a benign, noninflammatory condition that appears to have no adverse sequelae or can be associated with steatohepatitis: a condition that can result in end-stage liver disease, accounting for up to 14% of liver transplants in the US. Here we highlight recent advances in our understanding of the molecular events contributing to hepatic steatosis and nonalcoholic steatohepatitis.

Nonalcoholic fatty liver disease (NAFLD) is a clinicopathological term that encompasses a disease spectrum ranging from simple triglyceride accumulation in hepatocytes (hepatic steatosis) to hepatic steatosis with inflammation (steatohepatitis), fibrosis, and cirrhosis (1). NAFLD is the most frequent cause of abnormal liver function tests (LFTs) in the US (2, 3), affecting approximately 30 million Americans. Excess hepatic triglyceride accumulation is associated with various drugs, nutritional factors, and multiple genetic defects in energy metabolism. However, the most common disorder associated with hepatic steatosis is insulin resistance (3). As such, it has been proposed that NAFLD be included as a component of the metabolic syndrome (4).

Day et al. (5) initially proposed a "two-hit" model to explain the progression of NAFLD. The "first hit" constitutes the deposition of triglycerides in the cytoplasm of the hepatocyte. The disease does not progress unless additional cellular events occur (the "second hit") that promote inflammation, cell death, and fibrosis, which are the histologic hallmarks of nonalcoholic steatohepatitis (NASH). Recent studies in animal models of NAFLD have provided new insights into the molecular and physiologic alterations that constitute the first and second hits in the progression of NAFLD to end-stage liver disease and will be the focus of this review.

Epidemiology

The prevalence of NAFLD in the general population is estimated to be between 14% and 24% (6–8). NAFLD used to be almost exclusively a disease of adults. However, the estimated prevalence of the disorder has increased markedly in all segments of the population and now extends to children. The rising prevalence of obesity and type 2 diabetes in the population is likely responsible for the burgeoning number of individuals with hepatic steatosis (9, 10).

The progression of NAFLD to cirrhosis may differ significantly among ethnic groups. Hispanics with NAFLD appear to progress to NASH and cirrhosis more frequently than either blacks or whites. In contrast to Hispanics, blacks may be at reduced risk for the development of NASH and end-stage liver disease (11, 12).

Development of hepatic steatosis

As summarized in Figure 1, a series of molecular and physiologic alterations occur in the setting of insulin resistance that results in the accumulation of triglycerides in liver. The conventional explanation for hepatic triglyceride accumulation is that obesity and insulin resistance result in increased release of FFAs from adipocytes. Increased adipocyte mass and increased hydrolysis of triglycerides through increased hormone-sensitive lipase activity contribute to elevated plasma levels of FFAs (reviewed in ref. 13). The rate of hepatic FFA uptake is unregulated and therefore directly proportional to plasma FFA concentrations (14).

FFAs taken up by the liver are metabolized by one of two pathways: oxidation to generate ATP or esterification to produce triglycerides, which are either incorporated into VLDL particles for export or stored within the hepatocyte. As discussed below, defects in one or both of these pathways can lead to hepatic steatosis.

Molecular mediators of lipogenesis and their role in hepatic steatosis

A central metabolic function of the liver is to maintain plasma glucose levels regardless of the nutritional state of the animal. In the setting of energy excess, glucose is converted to fatty acids via the conversion of glucose to pyruvate, which enters the Krebs cycle in the mitochondria (Figure 1). Citrate formed in the Krebs cycle is shuttled to the cytosol where it is converted to acetyl-CoA by ATP citrate lyase. Acetyl-CoA carboxylase 1 (ACC1) then converts acetyl-CoA to malonyl-CoA, which is used by fatty acid synthase to form palmitic acid (C16:0). Palmitic acid is then either desaturated by stearoyl-CoA desaturase (SCD) to palmitoleic acid, or further elongated by the long chain fatty acyl elongase to form stearic acid (C18:0), which also can be desaturated to form oleic acid (C18:1) (15). These fatty acids are used to synthesize triglycerides — the primary source of energy storage and transport. Humans (16) and mice (17) with hepatic steatosis accumulate excess oleic acid, the end-product of de novo fatty acid synthesis. This suggests that fatty acid synthetic rates are increased in the insulin-resistant liver.

De novo synthesis of fatty acids in liver is regulated independently by insulin and glucose (18, 19). Insulin's ability to activate lipogenesis is transcriptionally mediated by the membrane-bound transcription

---

Nonstandard abbreviations used: acetyl-CoA carboxylase (ACC); AMP-activated protein kinase (AMPK); basic helix-loop-helix-leucine zipper (bHLH-Zip); carbohydrate response element binding protein (ChREBP); carnitine palmitoyl transferase-1 (CPT-1); diethylaminoethoxyhexestrol (DEAEH); liver function test (LFT); liver-type pyruvate kinase (L-PK); malondialdehyde (MDA); mitochondrial respiratory chain (MRC); nonalcoholic fatty-liver disease (NAFLD); nonalcoholic steatohepatitis (NASH); polyunsaturated fatty acid (PUFA); reactive oxygen species (ROS); stearoyl-CoA desaturase (SCD); *trans*-4-hydroxy-2-nonenal (HNE).

Conflict of interest: The authors have declared that no conflict of interest exists.

Citation for this article: *J. Clin. Invest.* **114**:147–152 (2004). doi:10.1172/JCI200422422.

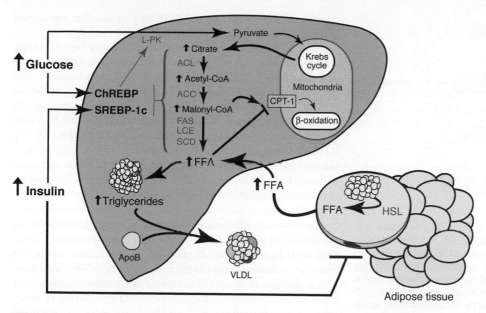

Figure 1

Metabolic alterations resulting in hepatic triglyceride accumulation in insulin-resistant states. Insulin resistance is manifested by hyperinsulinemia, increased hepatic glucose production, and decreased glucose disposal. In adipocytes, insulin resistance increases hormone-sensitive lipase (HSL) activity, resulting in elevated rates of triglyceride lipolysis and enhanced FFA flux to the liver. FFAs can either be oxidized in the mitochondria to form ATP or esterified to produce triglycerides for storage or incorporation into VLDL particles. In liver, hyperinsulinemia induces SREBP-1c expression, leading to the transcriptional activation of all lipogenic genes. Simultaneously, hyperglycemia activates ChREBP, which transcriptionally activates L-PK and all lipogenic genes. The synergistic actions of SREBP-1c and ChREBP coordinately activate the enzymatic machinery necessary for the conversion of excess glucose to fatty acids. A consequence of increased fatty acid synthesis is increased production of malonyl-CoA, which inhibits CPT-1, the protein responsible for fatty acid transport into the mitochondria. Thus, in the setting of insulin resistance, FFAs entering the liver from the periphery, as well as those derived from de novo lipogenesis, will be preferentially esterified to triglycerides. ACL, ATP citrate lyase; CPT-1, carnitine palmitoyl transferase-1; FAS, fatty acid synthase; LCE, long-chain fatty acyl elongase.

factor, sterol regulatory element–binding protein-1c (SREBP-1c) (20, 21). SREBP-1c is one of three SREBP isoforms that belong to the basic helix-loop-helix-leucine zipper (bHLH-Zip) family of transcription factors (22). In the nucleus, SREBP-1c transcriptionally activates all genes required for lipogenesis (15, 23). Importantly, the overexpression of SREBP-1c in transgenic mouse livers leads to the development of a classic fatty liver due to increased lipogenesis (24). We (25), and others (26, 27) have demonstrated that increased rates of hepatic fatty acid synthesis contribute to the development of fatty livers in rodent models of insulin-resistant diabetes and obesity.

Hyperinsulinemia and elevated hepatic glucose production are hallmarks of insulin resistance (28). It might be anticipated that SREBP-1c would not be activated in states of insulin resistance. Surprisingly, even in the presence of profound insulin resistance, insulin stimulates hepatic SREBP-1c transcription, resulting in increased rates of de novo fatty acid biosynthesis (25). The contribution SREBP-1c makes to triglyceride accumulation in insulin-resistant livers has been explored in *ob/ob* mice. *Ob/ob* mice are severely obese and insulin resistant due to a mutation in the leptin gene and, as a consequence, these mice have hepatic steatosis (29). Inactivation of the *Srebp-1* gene in the livers of *ob/ob* mice results in an approximately 50% reduction in hepatic triglycerides (30). Thus, SREBP-1 plays a significant role in the development of hepatic steatosis in this animal model of insulin resistance.

SREBP-1c also activates ACC2 (23), an isoform of ACC that produces malonyl-CoA at the mitochondrial membrane (31). Increases in malonyl-CoA result in decreased oxidation of fatty acids due to inhibition of carnitine palmitoyl transferase-1 (CPT-1), which shuttles fatty acids into mitochondria (32). The critical role of ACC2 in hepatic fatty acid metabolism was revealed in mice that harbored the genetic deletion of the *Acc2* gene. The *Acc2* knockout mice were resistant to obesity, owing to increased activity of CPT-1, resulting in an increased rate of fatty acid oxidation (33, 34). Adenoviral-mediated expression of malonyl-CoA decarboxylase, an enzyme that degrades malonyl-CoA, also results in increased fatty acid β oxidation and reduced hepatic triglyceride stores (35).

Carbohydrate (glucose)-mediated stimulation of lipogenesis is transcriptionally mediated by a second bHLH-Zip transcription factor, designated carbohydrate response element binding protein (ChREBP) (36). Glucose activates ChREBP by regulating the entry of ChREBP from the cytosol into the nucleus and by activating the binding of the transcription factor to DNA (37). Glucose stimulates ChREBP to bind to an E-box motif in the promoter of liver-type pyruvate kinase (L-PK), a key regulatory enzyme in glycolysis. L-PK catalyzes the conversion of phosphoenolpyruvate to pyruvate, which enters the Krebs cycle to generate citrate, the principal source of acetyl-CoA used for fatty acid synthesis. Recently, ChREBP knockout mice have been developed and characterized (38). As predicted from in vitro studies, the expression of L-PK was reduced by approximately 90% in livers of ChREBP knockout mice. An unexpected finding was that the mRNA levels of all fatty acid synthesis enzymes also were reduced by approximately 50% (38). This suggests that ChREBP can independently stimulate the transcription of all lipogenic genes. Thus, activation of L-PK stimulates both glycolysis and lipogenesis, thereby facilitating the conversion of glucose to fatty acids under conditions of energy excess. Whether inactivation of ChREBP will attenuate the development of fatty livers in insulin-resistant states is currently under investigation; however, it seems likely that excessive stimulation of lipogenesis by ChREBP stimulation would be important only after the development of hyperglycemia.

A third transcription factor that participates in the development of hepatic steatosis in rodents is PPAR-γ. PPAR-γ is a member of the nuclear hormone receptor superfamily that is required for normal adipocyte differentiation (39). Normally, PPAR-γ is expressed at very low levels in the liver; however, in animal models with insulin resistance and fatty livers, the expression of PPAR-γ is markedly increased (40, 41). Previous studies have demonstrated that SREBP-1c can transcriptionally activate PPAR-γ, and it has been

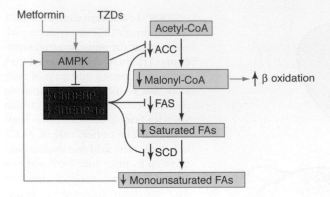

Figure 2
Consequences of hepatic AMPK activation. The pharmacologic agents, metformin and thiazolidinediones (TZDs), activate AMPK in the liver. In addition, the deletion of SCD results in AMPK activation through an undetermined mechanism. The activation of AMPK reduces lipogenesis through three independent mechanisms. Activated AMPK phosphorylates and inhibits the activity of ACC, which reduces malonyl-CoA formation. ChREBP is phosphorylated by activated AMPK, which inhibits its entry into the nucleus, thus suppressing L-PK and lipogenic gene expression. SREBP-1c expression is reduced by activated AMPK through undefined mechanisms. The cumulative result of AMPK activation, whether by drugs or through the deletion of SCD, is a reduction in fatty acid synthesis, decreased malonyl-CoA concentrations, and increased CPT-1 activity, resulting in increased fatty acid oxidation.

suggested that SREBP-1c may activate PPAR-γ by stimulating production of an activating ligand for the nuclear receptor (42, 43).

The importance of PPAR-γ expression in the development of fatty livers has been demonstrated by the development of liver-specific gene deletions of *Ppar-γ* in two different insulin-resistant mouse models, the *ob/ob* mouse and the lipodystrophic transgenic mouse, named AZIP/F-1. AZIP-F-1 mice are insulin resistant due to a near absence of white adipose tissue and leptin deficiency (41). The genetic deletion of hepatic PPAR-γ in livers of either *ob/ob* (44) or AZIP-F-1 (45) mice markedly attenuates the development of hepatic steatosis, independent of the presence of hyperinsulinemia or hyperglycemia.

The precise molecular events mediated by PPAR-γ that promote triglyceride deposition in the liver have not been fully defined. It is also not known whether PPAR-γ expression is increased in human livers with steatosis.

AMP-activated protein kinase and hepatic steatosis

AMP-activated protein kinase (AMPK) is a heterotrimeric protein that serves as a sensor of cellular energy levels (46). AMPK is activated by increased cellular AMP levels, a marker of decreased cellular energy stores. Activated AMPK stimulates ATP-producing catabolic pathways, such as fatty acid β oxidation, and inhibits ATP-consuming processes, such as lipogenesis, directly by phosphorylating regulatory proteins and indirectly by affecting expression levels of genes in these pathways (46).

The fatty acid composition of liver can also influence the amount of triglyceride that accumulates by altering AMPK activity. The genetic deletion of SCD-1, an enzyme responsible for the synthesis of monounsaturated fatty acids, protects against the development of fatty livers and insulin resistance in mice (47, 48). In the absence of SCD-1, AMPK is activated (49), resulting in phosphorylation and

inhibition of both ACC (50) and ChREBP (51) as well as a reduction in the expression levels of SREBP-1c (52).

The antidiabetic drug metformin also activates hepatic AMPK (52). Treatment of *ob/ob* mice with metformin markedly reduced hepatic steatosis (53), and its administration to humans with NASH improved LFT numbers and decreased liver size (54). A second class of antidiabetic drugs, the thiazolidinediones, are principally recognized as drugs that activate PPAR-γ; however, recent data suggest that they also can activate AMPK (55, 56). Pilot studies in humans using pioglitazone (57, 58) and rosiglitazone (59) have demonstrated the efficacy of these agents in reducing hepatic fat, presumably as a consequence of the molecular events summarized in Figure 2.

Taken together, these studies suggest that increased hepatic lipogenesis is an important metabolic abnormality underlying the pathogenesis of hepatic steatosis in insulin-resistant livers. Increased lipogenesis may actually cause dual metabolic alterations that lead to increased hepatic triglyceride content. The first alteration is direct — through the increased synthesis of triglycerides. The second is indirect — through increased production of malonyl-CoA, which inhibits CPT-1 and fatty acid entry into the mitochondria, thus reducing β oxidation and enhancing fatty acid and triglyceride accumulation (Figure 1).

It is important to note that the concept that endogenous fatty acid synthesis contributes significantly to hepatic steatosis is based on data from the studies of mice. Stable-isotope studies in humans suggest that de novo hepatic fatty acid synthesis contributes only modestly to the amount of triglycerides synthesized in liver (60). Although there is evidence that de novo hepatic fatty acid synthesis is increased in humans with insulin resistance (61), the overall importance of this pathway in the development of hepatic steatosis remains to be determined.

Disease progression: steatosis to NASH

According to the two-hit hypothesis, hepatic steatosis is a prerequisite for subsequent events that lead to liver injury (5). Despite the high prevalence of NAFLD and its potential for serious sequelae, the underlying etiological factors that determine disease progression to cirrhosis remain poorly understood. Studies to clearly define the molecular and physiologic changes that mediate the presumed transition from hepatic steatosis to NASH have been limited by several factors. First, no animal models incorporate all features of human steatohepatitis. Second, the available noninvasive techniques to study hepatic metabolism in humans are limited. Third, liver biopsies are required to identify individuals with NASH, precluding large population-based studies. Therefore, our current understanding of the mechanisms by which hepatic steatosis progresses to NASH is based almost exclusively on correlative data from animal models. How well these animal models reflect the human pathophysiology of NASH is not known.

NASH is histologically similar to alcohol-induced steatohepatitis, a disease that can progress to cirrhosis and liver failure. Many of the factors implicated in the development of alcoholic steatohepatitis are also associated with NASH. These factors can be grouped into two broad categories: factors causing an increase in oxidative stress and factors promoting expression of proinflammatory cytokines. Although there is considerable evidence implicating cytokines in the development of NASH, the focus here will be on the potential role of lipid-induced cellular injury in the development of NASH.

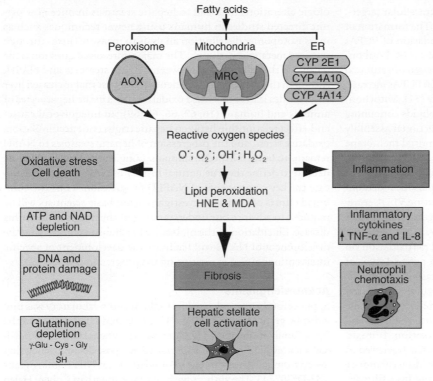

Figure 3
Mechanisms of lipid-induced cellular injury in NAFLD. ROS are formed through oxidative processes within the cell. In the mitochondria, impaired MRC activity leads to the formation of superoxide anions and hydrogen peroxide. The accumulation of fatty acids in the cytosol increases fatty acid oxidation in peroxisomes and the ER. The initial reaction in peroxisomal β oxidation is catalyzed by acyl-CoA oxidase (AOX) that forms hydrogen peroxide through the donation of electrons to molecular oxygen. Microsomal ω oxidation is catalyzed by cytochrome P450 (CYP) enzymes 2E1, 4A10, and 4A14, which form ROS through flavoprotein-mediated donation of electrons to molecular oxygen. PUFAs are extremely susceptible to lipid peroxidation by ROS. By-products of PUFA peroxidation are aldehydes, such as HNE and MDA. These aldehydes are themselves cytotoxic and can freely diffuse into the extracellular space to affect distant cells. ROS and aldehydes induce oxidative stress and cell death via ATP and NAD depletion, DNA and protein damage, and glutathione depletion. Additionally, they induce inflammation through the production of proinflammatory cytokines, leading to neutrophil chemotaxis. Within the extracellular space, HNE and MDA are themselves potent chemoattractants for neutrophils. Finally, ROS and products of lipid peroxidation can lead to fibrosis by activating hepatic stellate cells, which synthesize collagen and perpetuate the inflammatory response.

Oxidative stress

Oxidative stress results from an imbalance between pro-oxidant and antioxidant chemical species that leads to oxidative damage of cellular macromolecules (62). The predominant pro-oxidant chemicals in fatty livers are singlet oxygen molecules, superoxide anions, hydrogen peroxide, and hydroxyl radicals: molecules collectively referred to as reactive oxygen species (ROS). As depicted in Figure 3, the oxidation of fatty acids is an important source of ROS in fatty livers (63–66). Some of the consequences of increased ROS include depletion of ATP and nicotinamide dinucleotide, DNA damage, alterations in protein stability, the destruction of membranes via lipid peroxidation, and the release of proinflammatory cytokines (62, 67). Increased production of ROS in the presence of excess FFAs has been validated in animal models of NASH (66, 68). Human livers with NASH have increased levels of by-products of lipid peroxidation, providing further evidence of an increase in oxidative stress in this condition (69).

Mitochondrial dysfunction

Mitochondrial β oxidation is the dominant oxidative pathway for the disposition of fatty acids under normal physiologic conditions but can also be a major source of ROS (70). Several lines of evidence suggest that mitochondrial function is impaired in patients with NASH. Ultrastructural mitochondrial abnormalities have been documented in patients with NASH (71). Similar mitochondrial lesions are found in liver biopsy specimens from patients treated with 4,4′-diethylaminoethoxyhexestrol, a drug that inhibits mitochondrial respiratory chain (MRC) activity and mitochondrial β oxidation (72). Prolonged treatment with this agent is associated with hepatic steatosis and steatohepatitis that is histologically indistinguishable from NAFLD in humans (72). The ultrastructural mitochondrial defects in patients with NAFLD may be indicative of defective oxidative-phosphorylation, inasmuch as these patients also have reduced MRC activity (73) and impaired ATP synthesis after a fructose challenge (74). MRC dysfunction can directly lead to the production of ROS. If electron flow is interrupted at any point in the respiratory chain, the preceding respiratory intermediates can transfer electrons to molecular oxygen to produce superoxide anions and hydrogen peroxide (65, 66).

As the oxidative capacity of the mitochondria becomes impaired, cytosolic fatty acids accumulate. Alternative pathways in the peroxisomes (β oxidation) and in microsomes (ω oxidation) are activated, resulting in the formation of additional ROS (72, 75, 76). In the initial step of peroxisomal β oxidation, hydrogen peroxide is formed by the action of acyl-CoA oxidase, which donates electrons directly to molecular oxygen (64). Microsomal ω oxidation of fatty acids, catalyzed primarily by cytochrome P450 enzymes 2E1, 4A10, and 4A14, forms ROS through flavoprotein-mediated donation of electrons to molecular oxygen (63). Additionally, dicarboxylic acids, another product of microsomal fatty acid ω oxidation, can impair mitochondrial function by uncoupling oxidative-phosphorylation (77). Protonated dicarboxylic acids cycle from the inner to the outer mitochondrial membrane, resulting in dissipation of the mitochondrial proton gradient without concomitant ATP production (78). The cumulative effect of extramitochondrial fatty acid oxidation is a further increase in oxidative stress and mitochondrial impairment.

Lipid peroxidation

ROS are relatively short-lived molecules that exert local effects (79). However, they can attack polyunsaturated fatty acids (PUFAs) and initiate lipid peroxidation within the cell (79), which results in the formation of aldehyde by-products such as *trans*-4-hydroxy-2-nonenal (HNE) and malondialdehyde (MDA). These molecules have longer half-lives than ROS and have the potential to diffuse from their site of origin to reach distant intracellular and extracellular targets, thereby amplifying the effects of oxidative stress. The formation of HNE and MDA occurs only through the peroxidation of PUFAs (79), which are preferentially oxidized owing to decreased carbon-hydrogen bond strength in methylene groups between unsaturated carbon pairs (80). As the number of double bonds in PUFAs increase, their rate of peroxidation increases exponentially (81). Mitochondria have a substantial concentration of phospholipids containing docosahexaenoic, which may be essential for functional assembly of the MRC (82). Peroxidation of these mitochondrial membrane components could lead to further diminution of MRC activity and increased cellular oxidative stress. Additionally, the peroxidation of PUFAs has also been shown to enhance postendoplasmic reticulum presecretory proteolysis of ApoB, thereby attenuating VLDL secretion in rodents (83). The reduction in VLDL secretion may further contribute to triglyceride accumulation in the liver.

In addition to the deleterious effects of lipid peroxidation on organelle function, aldehydes formed through peroxidation of PUFAs also are detrimental to cellular homeostasis. They impair nucleotide and protein synthesis, deplete the natural antioxidant glutathione, increase production of the proinflammatory cytokine TNF-α, promote influx of inflammatory cells into the liver, and activate stellate cells, leading to collagen deposition, fibrosis, and the perpetuation of the inflammatory response (reviewed in refs. 79, 84). These effects have the potential to directly induce hepatocyte death and necrosis, inflammation, and liver fibrosis: all of the histologic hallmarks of NASH.

Conclusions

Over the past 5 years, substantial progress has been made in identifying the molecular and physiologic changes that cause hepatic steatosis. The transcription factors that control hepatic lipid metabolism have been identified. The elucidation of these physi-ologic alterations that lead to hepatic steatosis in mice now permit directed studies in humans using newer techniques such as mass isotopomer distribution analysis to measure fluxes through involved metabolic pathways. The major unresolved question is the nature of the relationship between hepatic steatosis and NASH. Although there is abundant evidence to suggest that increased liver triglycerides lead to increased oxidative stress in the hepatocytes of animals and humans (16, 63, 66, 68, 69), an unequivocal cause-and-effect relationship between hepatic triglyceride accumulation, oxidative stress, and the progression of hepatic steatosis to NASH remains to be established in humans. Longitudinal studies also are needed to define the true natural history of NAFLD and to delineate the key components of NAFLD progression. Thus the combined efforts of clinician-investigators and basic scientists will be required to advance our understanding of the progression of this disease. Elucidation of the molecular mechanisms underlying the development of NASH will facilitate the development of specific interventions aimed at preventing the progression of NAFLD.

Acknowledgments

Support for the research cited from the authors' laboratory was provided by grants from the NIH (HL-20948 and HL-38049) and the Perot Family Foundation. J.D. Horton is a Pew Scholar in the biomedical sciences. J.D. Browning is supported by a post-doctoral fellowship award from the American Liver Foundation and NIH training grant T32-DK-07745. The authors wish to thank Jonathan Cohen, Helen Hobbs, and Joseph Goldstein for critical reading of the manuscript.

Address correspondence to: Jay D. Horton, Departments of Internal Medicine and Molecular Genetics, University of Texas Southwestern Medical Center at Dallas, 5323 Harry Hines Boulevard, Room L5-238, Dallas, Texas 75390-9046, USA. Phone: (214) 648-9677; Fax: (214) 648-8804; E-mail: jay.horton@utsouthwestern.edu.

1. Neuschwander-Tetri, B., and Caldwell, S. 2003. Nonalcoholic steatohepatitis:summary of an AASLD single topic conference. *Hepatology.* **37**:1202–1219.

2. Clark, J.M., Brancati, F.L., and Diehl, A.M. 2003. The prevalence and etiology of elevated aminotransferase levels in the United States. *Am. J. Gastroenterol.* **98**:960–967.

3. Angulo, P. 2002. Nonalcoholic fatty liver disease. *N. Engl. J. Med.* **346**:1221–1231.

4. Marchesini, G., et al. 2001. Nonalcoholic fatty liver disease a feature of the metabolic syndrome. *Diabetes.* **50**:1844–1850.

5. Day, C.P., and James, O.F.W. 1998. Steatohepatitis: a tale of two "hits"? *Gastroenterology.* **114**:842–845.

6. Nomura, H., et al. 1988. Prevalence of fatty liver in a general population in Okinawa, Japan. *Jpn. J. Med.* **27**:142–149.

7. Hilden, M., Christoffersen, P., Juhl, E., and Dalgaard, J.B. 1977. Liver histology in a "normal" population — examination of 503 consecutive fatal traffic casualties. *Scand. J. Gastroenterol.* **12**:593–597.

8. Bellentani, S., et al. 2000. Prevalence of and risk factors for hepatic steatosis in northern Italy. *Ann. Intern. Med.* **132**:112–117.

9. Flegal, K.M., Ogden, C.L., Wei, R., Kuczmarski, R.L., and Johnson, C.L. 2001. Prevalence of overweight in US children: comparison of US growth charts from the Centers for Disease Control and Prevention with other reference values for body mass index. *Am. J. Clin. Nutr.* **73**:1086–1093.

10. Rashid, M., and Roberts, E.A. 2000. Nonalcoholic steatohepatitis in children. *J. Pediatr. Gastroenterol. Nutr.* **30**:48–52.

11. Browning, J.D., Kumar, K.S., Saboorian, M.H., and Thiele, D.L. 2004. Ethnic differences in the prevalence of cryptogenic cirrhosis. *Am. J. Gastroenterol.* **99**:292–298.

12. Caldwell, S.H., Harris, D.M., Patrie, J.T., and Hespenheide, E.E. 2002. Is NASH underdiagnosed among African Americans? *Am. J. Gastroenterol.* **97**:1496–1500.

13. Lewis, G.F., Carpentier, A., Adeli, K., and Giacca, A. 2002. Disordered fat storage and mobilization in the pathogenesis of insulin resistance and type 2 diabetes. *Endocr. Rev.* **23**:201–229.

14. Wahren, J., Sato, Y., Ostman, J., Hagenfeldt, L., and Felig, P. 1984. Turnover and splanchnic metabolism of free fatty acids and ketones in insulin-dependent diabetics at rest and in response to exercise. *J. Clin. Invest.* **73**:1367–1376.

15. Horton, J.D., Goldstein, J.L., and Brown, M.S. 2002. SREBPs: activators of the complete program of cholesterol and fatty acid synthesis in the liver. *J. Clin. Invest.* **109**:1125–1131. doi:10.1172/JCI200215593.

16. Araya, J., et al. 2004. Increase in long-chain polyunsaturated fatty acid n-6/n-3 ratio in relation to hepatic steatosis in patients with non-alcoholic fatty liver disease. *Clin. Sci. (Lond).* **106**:635–643.

17. Shimomura, I., Shimano, H., Korn, B.S., Bashmakov, Y., and Horton, J.D. 1998. Nuclear sterol regulatory element-binding proteins activate genes responsible for the entire program of unsaturated fatty acid biosynthesis in transgenic mouse liver. *J. Biol. Chem.* **273**:35299–35306.

18. Koo, S.-H., Dutcher, A.K., and Towle, H.C. 2001. Glucose and insulin function through two distinct transcription factors to stimulate expression of lipogenic enzyme genes in liver. *J. Biol. Chem.* **276**:9437–9445.

19. Stoeckman, A.K., and Towle, H.C. 2002. The role of SREBP-1c in nutritional regulation of lipogenic enzyme gene expression. *J. Biol. Chem.* **277**:27029–27035.

20. Shimomura, I., et al. 1999. Insulin selectively increases SREBP-1c mRNA in the livers of rats with streptozotocin-induced diabetes. *Proc. Natl. Acad. Sci. U. S. A.* **96**:13656–13661.

21. Foretz, M., Guichard, C., Ferre, P., and Foufelle, F. 1999. Sterol regulatory element binding protein-1c is a major mediator of insulin action on the hepatic expression of glucokinase and lipogenesis-related genes. *Proc. Natl. Acad. Sci. U. S. A.* **96**:12737–12742.

22. Brown, M.S., and Goldstein, J.L. 1997. The SREBP pathway: regulation of cholesterol metabolism by proteolysis of a membrane-bound transcription factor. *Cell.* **89**:331–340.

23. Horton, J.D., et al. 2003. Combined analysis of oligonucleotide microarray data from transgenic and knockout mice identifies direct SREBP target genes. *Proc. Natl. Acad. Sci. U. S. A.* **100**:12027–12032.

24. Shimano, H., et al. 1997. Isoform 1c of sterol regulatory element binding protein is less active than isoform 1a in livers of transgenic mice and in cultured cells. *J. Clin. Invest.* **99**:846–854.

25. Shimomura, I., Bashmakov, Y., and Horton, J.D. 1999. Increased levels of nuclear SREBP-1c associated with fatty livers in two mouse models of diabetes mellitus. *J. Biol. Chem.* **274**:30028–30032.

26. Martin, R.J. 1974. In vivo lipogenesis, and enzyme levels in adipose tissue and liver tissues from

pair-fed genetically obese and lean rats. *Life Sci.* **14**:1447–1453.

27. Memon, R.A., Grunfeld, C., Moser, A.H., and Feingold, K.R. 1994. Fatty acid synthesis in obese insulin resistant diabetic mice. *Horm. Metab. Res.* **26**:85–87.

28. Lam, T.K.T., et al. 2003. Mechanisms of the free fatty acid-induced increase in hepatic glucose production. *Am. J. Physiol. Endocrinol. Metab.* **284**:E863–E873.

29. Halaas, J.L., et al. 1995. Weight-reducing effects of the plasma protein encoded by the obese gene. *Science.* **269**:543–546.

30. Yahagi, N., et al. 2002. Absence of sterol regulatory element-binding protein-1 (SREBP-1) ameliorates fatty livers but not obesity or insulin resistance in Lepob/Lepob mice. *J. Biol. Chem.* **277**:19353–19357.

31. Abu-Elheiga, L., et al. 2000. The subcellular localization of acetyl-CoA carboxylase 2. *Proc. Natl. Acad. Sci. U. S. A.* **97**:1444–1449.

32. McGarry, J.D., Mannaerts, G.P., and Foster, D.W. 1977. A possible role for malonyl-CoA in the regulation of hepatic fatty acid oxidation and ketogenesis. *J. Clin. Invest.* **60**:265–270.

33. Abu-Elheiga, L., Matzuk, M.M., Abo-Hashema, K.A.H., and Wakil, S.J. 2001. Continuous fatty acid oxidation and reduced fat storage in mice lacking acetyl-CoA carboxylase 2. *Science.* **291**:2613–2616.

34. Abu-Elheiga, L., Oh, W., Kordari, P., and Wakil, S.J. 2003. Acetyl-CoA carboxylase 2 mutant mice are protected against obesity and diabetes induced by high-fat/high-carbohydrate diets. *Proc. Natl. Acad. Sci. U. S. A.* **100**:10207–10212.

35. An, J., et al. 2004. Hepatic expression of malonyl-CoA decarboxylase reverses muscle, liver and whole-animal insulin resistance. *Nat. Med.* **10**:268–274.

36. Yamashita, H., et al. 2001. A glucose-responsive transcription factor that regulates carbohydrate metabolism in the liver. *Proc. Natl. Acad. Sci. U. S. A.* **98**:9116–9121.

37. Kawaguchi, T., Osatomi, K., Yamashita, H., Kabashima, T., and Uyeda, K. 2002. Mechanism for fatty acid "sparing" effect on glucose-induced transcription. Regulation of carbohydrate-responsive element-binding protein by Amp-activated protein kinase. *J. Biol. Chem.* **277**:3829–3835.

38. Iizuka, K., Bruick, R.K., Liang, G., Horton, J.D., and Uyeda, K. 2004. Deficiency of carbohydrate response element binding protein (ChREBP) reduces lipogenesis as well as glycolysis. *Proc. Natl. Acad. Sci. U. S. A.* In press.

39. Tontonoz, P., Hu, E., and Spiegelman, B.M. 1994. Stimulation of adipogenesis in fibroblasts by PPAR gamma 2, a lipid-activated transcription factor. *Cell.* **79**:1147–1156.

40. Edvardsson, U., et al. 1999. Rosiglitazone (BRL49653), a PPARgamma-selective agonist, causes peroxisome proliferator-like liver effects in obese mice. *J. Lipid. Res.* **40**:1177–1184.

41. Chao, L., et al. 2000. Adipose tissue is required for the antidiabetic, but not for the hypolipidemic, effect of thiazolidinediones. *J. Clin. Invest.* **106**:1221–1228.

42. Kim, J.B., Wright, H.M., Wright, M., and Spiegelman, B.M. 1998. ADD1/SREBP-1 activates PPAR gamma through the production of endogenous ligand. *Proc. Natl. Acad. Sci. U. S. A.* **95**:4333–4337.

43. Fajas, L., et al. 1999. Regulation of peroxisome proliferator-activated receptor gamma expression by adipocyte differentiation and determination factor 1/sterol regulatory element binding protein 1: implications for adipocyte differentiation and metabolism. *Mol. Cell. Biol.* **19**:5495–5503.

44. Matsusue, K., et al. 2003. Liver-specific disruption of PPARgamma in leptin-deficient mice improves fatty liver but aggravates diabetic phenotypes. *J. Clin. Invest.* **111**:737–747. doi:10.1172/JCI200317223.

45. Gavrilova, O., et al. 2003. Liver peroxisome proliferator-activated receptor gamma contributes to hepatic steatosis, triglyceride clearance, and regulation of body fat mass. *J. Biol. Chem.* **278**:34268–34276.

46. Hardie, D.G. 2003. Minireview: The AMP-activated protein kinase cascade: the key sensor of cellular energy status. *Endocrinology.* **144**:5179–5183.

47. Cohen, P., et al. 2002. Role for stearoyl-CoA desaturase-1 in leptin-mediated weight loss. *Science.* **297**:240–243.

48. Ntambi, J.M., et al. 2002. Loss of stearoyl-CoA desaturase-1 function protects mice against adiposity. *Proc. Natl. Acad. Sci. U. S. A.* **99**:11482–11486.

49. Dobrzyn, P., et al. 2004. Stearoyl-CoA desaturase 1 deficiency increases fatty acid oxidation by activating AMP-activated protein kinase in liver. *Proc. Natl. Acad. Sci. U. S. A.* **101**:6409–6414.

50. Hardie, D.G., Scott, J.W., Pan, D.A., and Hudson, E.R. 2003. Management of cellular energy by the AMP-activated protein kinase system. *FEBS Lett.* **546**:113–120.

51. Kawaguchi, T., Takenoshita, M., Kabashima, T., and Uyeda, K. 2001. Glucose and cAMP regulate the L-type pyruvate kinase gene by phosphorylation/dephosphorylation of the carbohydrate response element binding protein. *Proc. Natl. Acad. Sci. U. S. A.* **98**:13710–13715.

52. Zhou, G., et al. 2001. Role of AMP-activated protein kinase in mechanism of metformin action. *J. Clin. Invest.* **108**:1167–1174. doi:10.1172/JCI200113505.

53. Lin, H.Z., et al. 2000. Metformin reverses fatty liver disease in obese, leptin-deficient mice. *Nat. Med.* **6**:998–1003.

54. Marchesini, G., et al. 2001. Metformin in non alcoholic steatosis. *Lancet.* **358**:893–894.

55. Fryer, L.G., Parbu-Patel, A., and Carling, D. 2002. The anti-diabetic drugs rosiglitazone and metformin stimulate AMP-activated protein kinase through distinct signaling pathways. *J. Biol. Chem.* **277**:25226–25232.

56. Saha, A.K., et al. 2004. Pioglitazone treatment activates AMP-activated protein kinase in rat liver and adipose tissue in vivo. *Biochem. Biophys. Res. Commun.* **314**:580–585.

57. Bajaj, M., et al. 2003. Pioglitazone reduces hepatic fat content and augments splanchnic glucose uptake in patients with type 2 diabetes. *Diabetes.* **52**:1364–1370.

58. Promrat, K., et al. 2004. A pilot study of pioglitazone treatment for nonalcoholic steatohepatitis. *Hepatology.* **39**:188–196.

59. Neuschwander-Tetri, B.A., et al. 2003. Improved nonalcoholic steatohepatitis after 48 weeks of treatment with the PPAR-gamma ligand rosiglitazone. *Hepatology.* **38**:1008–1017.

60. Hellerstein, M.K., Schwarz, J.M., and Neese, R.A. 1996. Regulation of hepatic de novo lipogenesis in humans. *Annu. Rev. Nutr.* **16**:523–557.

61. Diraison, F., Dusserre, E., Vidal, H., Sothier, M., and Beylot, M. 2002. Increased hepatic lipogenesis but decreased expression of lipogenic gene in adipose tissue in human obesity. *Am. J. Physiol. Endocrinol. Metab.* **282**:E46–E51.

62. Robertson, G., Leclercq, I., and Farrell, G.C. 2001. Nonalcoholic steatosis and steatohepatitis: II. Cytochrome P-450 enzymes and oxidative stress. *Am. J. Physiol. Gastrointest. Liver Physiol.* **281**:G1135–G1139.

63. Lieber, C.S. 2004. CYP2E1: from ASH to NASH. *Hepatol. Res.* **28**:1–11.

64. Mannaerts, G.P., Van Veldhoven, P.P., and Casteels, M. 2000. Peroxisomal lipid degradation via beta- and alpha-oxidation in mammals. *Cell Biochem. Biophys.* **32**:73–87.

65. Garcia-Ruiz, C., Colell, A., Morales, A., Kaplowitz, N., and Fernandez-Checa, J.C. 1995. Role of oxidative stress generated from the mitochondrial electron transport chain and mitochondrial glutathione status in loss of mitochondrial function and activation of transcription factor nuclear factor-kappa B: studies with isolated mitochondria and rat hepatocytes. *Mol. Pharmacol.* **48**:825–834.

66. Hensley, K., et al. 2000. Dietary choline restriction causes complex I dysfunction and increased H(2)O(2) generation in liver mitochondria. *Carcinogenesis.* **21**:983–989.

67. Bergamini, C.M., Gambetti, S., Dondi, A., and Cervellati, C. 2004. Oxygen, reactive oxygen species and tissue damage. *Curr. Pharm. Des.* **10**:1611–1626.

68. Yang, S., et al. 2000. Mitochondrial adaptations to obesity-related oxidant stress. *Arch. Biochem. Biophys.* **378**:259–268.

69. Seki, S., et al. 2002. In situ detection of lipid peroxidation and oxidative DNA damage in non-alcoholic fatty liver diseases. *J. Hepatol.* **37**:56–62.

70. Reddy, J.K., and Mannaerts, G.P. 1994. Peroxisomal lipid metabolism. *Annu. Rev. Nutr.* **114**:343–370.

71. Caldwell, S.H., et al. 1999. Mitochondrial abnormalities in non-alcoholic steatohepatitis. *J. Hepatol.* **31**:430–434.

72. Berson, A., et al. 1998. Steatohepatitis-inducing drugs cause mitochondrial dysfunction and lipid peroxidation in rat hepatocytes. *Gastroenterology.* **114**:764–774.

73. Perez-Carreras, M., et al. 2003. Defective hepatic mitochondrial respiratory chain in patients with non-alcoholic steatohepatitis. *Hepatology.* **38**:999–1007.

74. Cortez-Pinto, H., Zhi Lin, H., Qi Yang, S., Odwin Da Costa, S., and Diehl, A.M. 1999. Lipids up-regulate uncoupling protein 2 expression in rat hepatocytes. *Gastroenterology.* **116**:1184–1193.

75. Johnson, E.F., Palmer, C.N., Griffin, K.J., and Hsu, M.H. 1996. Role of the peroxisome proliferator-activated receptor in cytochrome P450 4A gene regulation. *FASEB J.* **10**:1241–1248.

76. Kersten, S., et al. 1999. Peroxisome proliferator-activated receptor alpha mediates the adaptive response to fasting. *J. Clin. Invest.* **103**:1489–1498.

77. Tonsgard, J.H., and Getz, G.S. 1985. Effect of Reye's syndrome serum on isolated chinchilla liver mitochondria. *J. Clin. Invest.* **76**:816–825.

78. Hermesh, O., Kalderon, B., and Bar Tana, J. 1990. Mitochondria uncoupling by a long chain fatty acyl analogue. *J. Biol. Chem.* **273**:3937–3942.

79. Esterbauer, H., Schaur, R.J., and Zollner, H. 1991. Chemistry and biochemistry of 4-hydroxynonenal, malonaldehyde and related aldehydes. *Free Radic. Biol. Med.* **11**:81–128.

80. Gardner, H.W. 1989. Oxygen radical chemistry of polyunsaturated fatty acids. *Free Radic. Biol. Med.* **7**:65–86.

81. Wagner, B.A., Buettner, G.R., and Burns, C.P. 1994. Free radical-mediated lipid peroxidation in cells: oxidizability is a function of cell lipid bis-allylic hydrogen content. *Biochemistry.* **33**:4449–4453.

82. Infante, J.P., and Huszagh, V.A. 2000. Secondary carnitine deficiency and impaired docosahexaenoic (22:6n-3) acid synthesis: a common denominator in the pathophysiology of diseases of oxidative phosphorylation and beta-oxidation. *FEBS Lett.* **468**:1–5.

83. Pan, M., et al. 2004. Lipid peroxidation and oxidant stress regulate hepatic apolipoprotein B degradation and VLDL production. *J. Clin. Invest.* **113**:1277–1287. doi:10.1172/JCI200419197.

84. Yamauchi, T., et al. 2003. Globular adiponectin protected ob/ob mice from diabetes and apoE-deficient mice from atherosclerosis. *J. Biol. Chem.* **278**:2461–2468.

Liver fibrosis

Ramón Bataller[1] and David A. Brenner[2]

[1]Liver Unit, Institut de Malalties Digestives i Metabòliques, Hospital Clinic, Institut d'Investigació Biomèdiques August Pi i Sunyer (IDIBAPS), Barcelona, Catalonia, Spain. [2]Department of Medicine, Columbia University, New York, New York, USA.

Liver fibrosis is the excessive accumulation of extracellular matrix proteins including collagen that occurs in most types of chronic liver diseases. Advanced liver fibrosis results in cirrhosis, liver failure, and portal hypertension and often requires liver transplantation. Our knowledge of the cellular and molecular mechanisms of liver fibrosis has greatly advanced. Activated hepatic stellate cells, portal fibroblasts, and myofibroblasts of bone marrow origin have been identified as major collagen-producing cells in the injured liver. These cells are activated by fibrogenic cytokines such as TGF-β1, angiotensin II, and leptin. Reversibility of advanced liver fibrosis in patients has been recently documented, which has stimulated researchers to develop antifibrotic drugs. Emerging antifibrotic therapies are aimed at inhibiting the accumulation of fibrogenic cells and/or preventing the deposition of extracellular matrix proteins. Although many therapeutic interventions are effective in experimental models of liver fibrosis, their efficacy and safety in humans is unknown. This review summarizes recent progress in the study of the pathogenesis and diagnosis of liver fibrosis and discusses current antifibrotic strategies.

Historical perspective

Liver fibrosis results from chronic damage to the liver in conjunction with the accumulation of ECM proteins, which is a characteristic of most types of chronic liver diseases (1). The main causes of liver fibrosis in industrialized countries include chronic HCV infection, alcohol abuse, and nonalcoholic steatohepatitis (NASH). The accumulation of ECM proteins distorts the hepatic architecture by forming a fibrous scar, and the subsequent development of nodules of regenerating hepatocytes defines cirrhosis. Cirrhosis produces hepatocellular dysfunction and increased intrahepatic resistance to blood flow, which result in hepatic insufficiency and portal hypertension, respectively (2).

Hepatic fibrosis was historically thought to be a passive and irreversible process due to the collapse of the hepatic parenchyma and its substitution with a collagen-rich tissue (3, 4). Currently, it is considered a model of the wound-healing response to chronic liver injury (5). Early clinical reports in the 1970s suggested that advanced liver fibrosis is potentially reversible (6). However, liver fibrosis received little attention until the 1980s, when hepatic stellate cells (HSCs), formerly known as lipocytes, Ito cells, or perisinusoidal cells, were identified as the main collagen-producing cells in the liver (7). This cell type, first described by von Kupffer in 1876, undergoes a dramatic phenotypic activation in chronic liver diseases with the acquisition of fibrogenic properties (8). Methods to obtain HSCs from both rodent and human livers were rapidly standardized in the 1980s (9, 10), and prolonged culture of HSCs on plastic was widely accepted as a model for the study of activated HSCs (11). Key signals that modulate HSCs' fibrogenic actions were delineated (12). Experimental models for studying liver fibrogenesis in rats and in transgenic mice were developed, which corroborated the cell culture studies and led to the identification of key fibrogenic mediators (13). Besides HSCs, portal myofibroblasts and cells of bone marrow

origin have been recently shown to exhibit fibrogenic potential (14, 15). At the clinical level, the natural history of liver fibrosis, from early changes to liver cirrhosis, was delineated in patients with chronic HCV infection (16, 17). Rapid and slower fibrosers were identified, and genetic and environmental factors influencing fibrosis progression were partially uncovered (18). Since the demonstration, in the 1990s, that even advanced liver fibrosis is reversible, researchers have been stimulated to identify antifibrotic therapies (19). Biotechnology and pharmaceutical companies are increasingly interested in developing antifibrotic programs, and clinical trials are currently underway. However, the most effective therapy for treating hepatic fibrosis to date is still to remove the causative agent (20). A number of drugs are able to reduce the accumulation of scar tissue in experimental models of chronic liver injury. Renin-angiotensin system blockers and antioxidants are the most promising drugs, although their efficacy has not been tested in humans. Lack of clinical trials is due to the requirement of long follow-up studies and to the fact that liver biopsy, an invasive procedure, is still the gold-standard method for detecting changes in liver fibrosis. The current effort to develop noninvasive markers to assess liver fibrosis is expected to facilitate the design of clinical trials.

Recently, NASH has been recognized as a major cause of liver fibrosis (21). First described by Ludwig et al., it is considered part of the spectrum of nonalcoholic fatty liver diseases (22). These range from steatosis to cirrhosis and can eventually lead to hepatocellular carcinoma. NASH is a component of the metabolic syndrome, which is characterized by obesity, type 2 diabetes mellitus, and dyslipidemia, with insulin resistance as a common feature. As the prevalence of obesity is rapidly increasing, a rise in the prevalence of NASH is anticipated.

This review outlines recent progress in the pathogenesis, diagnosis, and treatment of liver fibrosis, summarizes recent data on the mechanisms leading to fibrosis resolution, and discusses future prospects aimed at developing effective antifibrotic therapies.

Natural history and diagnosis

The onset of liver fibrosis is usually insidious, and most of the related morbidity and mortality occur after the development of cirrhosis (16). In the majority of patients, progression to cirrhosis

Nonstandard abbreviations used: CTLA, cytotoxic T lymphocyte antigen; HSC, hepatic stellate cell; NASH, nonalcoholic steatohepatitis; PBC, primary biliary cirrhosis; TIMP-1, tissue inhibitor of metalloproteinase type 1.

Conflict of interest: The authors have declared that no conflict of interest exists.

Citation for this article: *J. Clin. Invest.* **115**:209–218 (2005). doi:10.1172/JCI200524282.

Table 1

Genetic and nongenetic factors associated with fibrosis progression in different types of chronic liver diseases

Type of liver disease	Candidate genes	Candidate genes (full name)	Nongenetic factors
Chronic HCV infection	HFE	Hereditary hemochromatosis gene	Alcohol intake
	Angiotensinogen	Angiotensinogen	Coinfection HIV and/or hepatitis B virus
	TGF-β1	Transforming growth factor β1	Age at time of acute infection
	TNF-α	Tumor necrosis factor α	Liver transplantation
	ApoE	Apolipoprotein E	Diabetes mellitus
	MEH	Microsomal epoxide hydroxylase	No response to therapy
	MCP-1	Monocyte chemotactic protein type 1	
	MCP-2	Monocyte chemotactic protein type 2	
	Factor V	Factor V (Leiden)	
Alcohol-induced	IL-10	Interleukin 10	Alcohol intake
	IL-1β	Interleukin 1β	Episodes of alcoholic hepatitis
	ADH	Alcohol dehydrogenase	
	ALDH	Aldehyde dehydrogenase	
	CYP2E1	cytochrome P450, family 2, subfamily e, polypeptide 1	
	TNF-α	Tumor necrosis factor α	
	CTLA-4	Cytotoxic T lymphocyte antigen type 4	
	TAP2	Transporter-associated antigen-processing type 2	
	MnSOD	Manganese superoxide dismutase	
NASH	HFE	Hereditary hemochromatosis gene	Age
	Angiotensinogen	Angiotensinogen	Severity of obesity
	TGF-β1	Transforming growth factor β1	Diabetes mellitus
			Hypertriglyceridemia
PBC	IL-1β	Interleukin 1β	
	TNF-α	Tumor necrosis factor α	
	ApoE	Apolipoprotein E	
Autoimmune hepatitis	HLA-II	Human leukocyte antigen type II haplotypes	Type II autoimmune hepatitis
			No response to therapy

occurs after an interval of 15–20 years. Major clinical complications of cirrhosis include ascites, renal failure, hepatic encephalopathy, and variceal bleeding. Patients with cirrhosis can remain free of major complications for several years (compensated cirrhosis). Decompensated cirrhosis is associated with short survival, and liver transplantation is often indicated as the only effective therapy (23). Cirrhosis is also a risk factor for developing hepatocellular carcinoma. Liver fibrosis progresses rapidly to cirrhosis in several clinical settings, including repeated episodes of severe acute alcoholic hepatitis, subfulminant hepatitis, and fibrosing cholestasis in patients with HCV reinfection after liver transplantation (24). The natural history of liver fibrosis is influenced by both genetic and environmental factors (Table 1). Epidemiological studies have identified polymorphisms in a number of candidate genes that may influence the progression of liver fibrosis in humans (18). These genetic factors may explain the broad spectrum of responses to the same etiological agent found in patients with chronic liver diseases. However, some studies have yielded contradictory results due to poor study design, and further research is required to clarify the actual role of genetic variants in liver fibrosis.

Liver biopsy is considered the gold-standard method for the assessment of liver fibrosis (25). Histologic examination is useful in identifying the underlying cause of liver disease and assessing the necroinflammatory grade and the stage of fibrosis. Fibrosis stage is assessed by using scales such as Metavir (stages I–IV) and Ishak score (stages I–V). Specific staining of ECM proteins (e.g., with Sirius red) can be used to quantify the degree of fibrosis, using computer-guided morphometric analysis. Liver biopsy is an invasive procedure, with pain and major complications occurring in 40% and 0.5% of patients, respectively (26). Sampling error can occur, especially when small biopsies are analyzed. Histologic examination is prone to intra- and interobserver variation and does not predict disease progression (27). Therefore, there is a need for reliable, simple, and noninvasive methods for assessing liver fibrosis. Scores that include routine laboratory tests, such as platelet count, aminotransferase serum levels, prothrombin time, and serum levels of acute phase proteins have been proposed (28, 29). Serum levels of proteins directly related to the hepatic fibrogenic process are also used as surrogate markers of liver fibrosis (30), including N-terminal propeptide of type III collagen, hyaluronic acid, tissue inhibitor of metalloproteinase type 1 (TIMP-1), and YKL-40. Although these scores are useful in detecting advanced fibrosis (cirrhosis) in patients, as well as minimal or no fibrosis, they are not effective for differentiating intermediate grades of fibrosis. Also, fibrosis-specific markers may reflect fibrogenesis in other organs (i.e., pancreatic fibrosis in alcoholic patients). Finally, hepatic fibrosis can be estimated by imaging techniques. Ultrasonography, computed tomography, and MRI can detect changes in the hepatic parenchyma due to moderate to severe fibrosis (31). Due to its low cost, ultrasonography is an appealing technique. It is able to detect liver cirrhosis based on changes in liver echogenicity and nodularity as well as signs of portal hypertension. However, ultrasound is highly operator-dependent, and the presence of increased liver echogenicity does not reliably differentiate hepatic steatosis from fibrosis. Noninvasive methods currently in development include blood protein

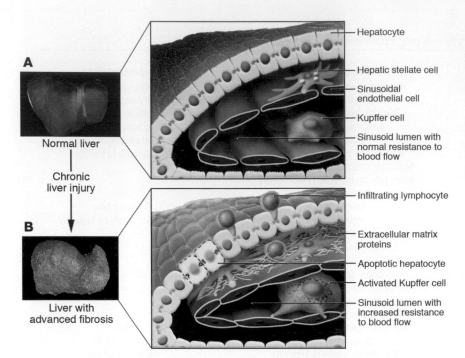

Figure 1
Changes in the hepatic architecture (**A**) associated with advanced hepatic fibrosis (**B**). Following chronic liver injury, inflammatory lymphocytes infiltrate the hepatic parenchyma. Some hepatocytes undergo apoptosis, and Kupffer cells activate, releasing fibrogenic mediators. HSCs proliferate and undergo a dramatic phenotypical activation, secreting large amounts of extracellular matrix proteins. Sinusoidal endothelial cells lose their fenestrations, and the tonic contraction of HSCs causes increased resistance to blood flow in the hepatic sinusoid. Figure modified with permission from *Science & Medicine* (S28).

profiling using proteomic technology and new clinical glycomics technology, which is based on DNA sequencer/fragment analyzers able to generate profiles of serum protein *N*-glycans (32). As the technology becomes validated, the noninvasive diagnosis of liver disease may become routine clinical practice.

Pathogenesis of liver fibrosis
Hepatic fibrosis is the result of the wound-healing response of the liver to repeated injury (1) (Figure 1). After an acute liver injury (e.g., viral hepatitis), parenchymal cells regenerate and replace the necrotic or apoptotic cells. This process is associated with an inflammatory response and a limited deposition of ECM. If the hepatic injury persists, then eventually the liver regeneration fails, and hepatocytes are substituted with abundant ECM, including fibrillar collagen. The distribution of this fibrous material depends on the origin of the liver injury. In chronic viral hepatitis and chronic cholestatic disorders, the fibrotic tissue is initially located around portal tracts, while in alcohol-induced liver disease, it locates in pericentral and perisinusoidal areas (33). As fibrotic liver diseases advance, disease progression from collagen bands to bridging fibrosis to frank cirrhosis occurs.

Liver fibrosis is associated with major alterations in both the quantity and composition of ECM (34). In advanced stages, the liver contains approximately 6 times more ECM than normal, including collagens (I, III, and IV), fibronectin, undulin, elastin, laminin, hyaluronan, and proteoglycans. Accumulation of ECM results from both increased synthesis and decreased degradation (35). Decreased activity of ECM-removing MMPs is mainly due to an overexpression of their specific inhibitors (TIMPs).

HSCs are the main ECM-producing cells in the injured liver (36). In the normal liver, HSCs reside in the space of Disse and are the major storage sites of vitamin A. Following chronic injury, HSCs activate or transdifferentiate into myofibroblast-like cells, acquiring contractile, proinflammatory, and fibrogenic properties (37, 38) (Figure 2A). Activated HSCs migrate and

accumulate at the sites of tissue repair, secreting large amounts of ECM and regulating ECM degradation. PDGF, mainly produced by Kupffer cells, is the predominant mitogen for activated HSCs. Collagen synthesis in HSCs is regulated at the transcriptional and posttranscriptional levels (39). Increased collagen mRNA stability mediates the increased collagen synthesis in activated HSCs. In these cells, posttranscriptional regulation of collagen is governed by sequences in the 3′ untranslated region via the RNA-binding protein αCP2 as well as a stem-loop structure in the 5′ end of collagen mRNA (40). Interestingly, HSCs express a number of neuroendocrine markers (e.g., reelin, nestin, neurotrophins, synaptophysin, and glial-fibrillary acidic protein) and bear receptors for neurotransmitters (8, 41, 42).

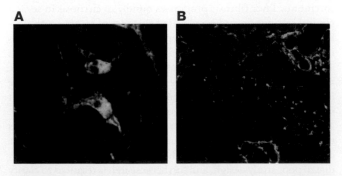

Figure 2
Expression of collagen α1(I) in a model of cholestasis-induced liver fibrosis. Transgenic mice with green fluorescence protein reporter gene under the direction of the collagen α1(I) promoter/enhancers were subjected to bile duct ligation for 2 weeks. (**A**) Collagen α1(I) was markedly expressed by activated HSCs, but not hepatocytes, in the hepatic parenchyma. Magnification, ×200. (**B**) Collagen α1(I) is markedly expressed by myofibroblasts around proliferating bile ducts. HSCs proliferate to initiate collagen deposition in the hepatic parenchyma. Magnification, ×40.

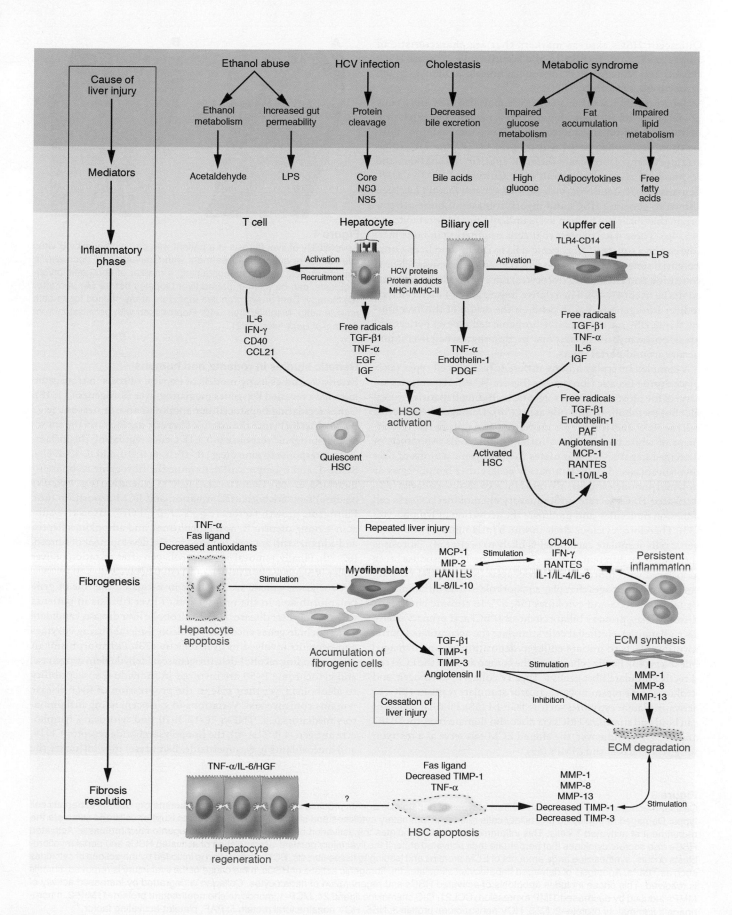

Quiescent HSCs express markers that are characteristic of adipocytes (PPARγ, SREBP-1c, and leptin), while activated HSCs express myogenic markers (α smooth muscle actin, c-myb, and myocyte enhancer factor–2).

Hepatic cell types other than HSCs may also have fibrogenic potential. Myofibroblasts derived from small portal vessels proliferate around biliary tracts in cholestasis-induced liver fibrosis to initiate collagen deposition (43, 44) (Figure 2B). HSCs and portal myofibroblasts differ in specific cell markers and response to apoptotic stimuli (45). Culture of CD34⁺CD38⁻ hematopoietic stem cells with various growth factors has been shown to generate HSCs and myofibroblasts of bone marrow origin that infiltrate human livers undergoing tissue remodeling (15, 46). These data suggest that cells originating in bone marrow can be a source of fibrogenic cells in the injured liver. Other potential sources of fibrogenic cells (i.e., epithelial-mesenchymal transition and circulating fibrocytes) have not been demonstrated in the liver (47, 48). The relative importance of each cell type in liver fibrogenesis may depend on the origin of the liver injury. While HSCs are the main fibrogenic cell type in pericentral areas, portal myofibroblasts may predominate when liver injury occurs around portal tracts.

A complex interplay among different hepatic cell types takes place during hepatic fibrogenesis (Figure 3) (49). Hepatocytes are targets for most hepatotoxic agents, including hepatitis viruses, alcohol metabolites, and bile acids (50). Damaged hepatocytes release ROS and fibrogenic mediators and induce the recruitment of white blood cells by inflammatory cells. Apoptosis of damaged hepatocytes stimulates the fibrogenic actions of liver myofibroblasts (51). Inflammatory cells, either lymphocytes or polymorphonuclear cells, activate HSCs to secrete collagen (52). Activated HSCs secrete inflammatory chemokines, express cell adhesion molecules, and modulate the activation of lymphocytes (53). Therefore, a vicious circle in which inflammatory and fibrogenic cells stimulate each other is likely to occur (54). Fibrosis is influenced by different T helper subsets, the Th2 response being associated with more active fibrogenesis (55). Kupffer cells are resident macrophages that play a major role in liver inflammation by releasing ROS and cytokines (56, 57). In chronic cholestatic disorders (i.e., primary biliary cirrhosis [PBC] and primary sclerosis cholangitis), epithelial cells stimulate the accumulated portal myofibroblasts to initiate collagen deposition around damaged bile ducts (43). Finally, changes in the composition of the ECM can directly stimulate fibrogenesis. Type IV collagen, fibrinogen, and urokinase type plasminogen activator stimulate resident HSCs by activating latent cytokines such as TGF-β1 (58). Fibrillar collagens can bind and stimulate HSCs via discoidin domain receptor DDR2 and integrins. Moreover, the altered ECM can serve as a reservoir for growth factors and MMPs (59).

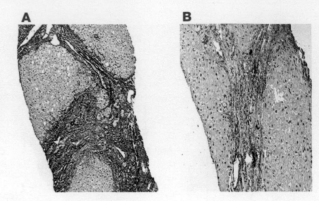

Figure 4
Reversibility of liver fibrosis in a patient with chronic hepatitis B virus infection after successful treatment with lamivudine. A decrease in smooth muscle actin immunostaining, a marker of fibrogenic myofibroblasts, can be seen in paired liver biopsies before (**A**) and after (**B**) therapy. Dark brown granules represent areas stained for smooth muscle actin. Magnification, ×40. Reproduced with permission from *Journal of Hepatology* (S2).

Genetic studies in rodents and humans

Extensive studies using models of hepatic fibrosis in transgenic mice have revealed key genes mediating liver fibrogenesis (1, 18). Genes regulating hepatocellular apoptosis and/or necrosis (e.g., Bcl-xL, Fas) influence the extent of hepatic damage and the subsequent fibrogenic response (60, 61). Genes regulating the inflammatory response to injury (e.g., IL-1β, IL-6, IL-10, and IL-13, IFN-γ, SOCS-1, and osteopontin) determine the fibrogenic response to injury (55, 62–65). Genes mediating ROS generation (e.g., NADPH oxidase) regulate both inflammation and ECM deposition (66). Fibrogenic growth factors (e.g., TGF-β1, FGF), vasoactive substances (angiotensin II, norepinephrine), and adipokines (leptin and adiponectin) are each required for the development of fibrosis (67–70). Finally, removal of excess collagen after cessation of liver injury is regulated by TIMP-1 and TGF-β1 (71, 72).

Association genetic studies have investigated the role of gene polymorphisms in the progression of liver fibrosis in patients with chronic liver diseases (18). In alcoholic liver disease, candidate genes include genes encoding for alcohol-metabolizing enzymes and proteins involved in liver toxicity (73). Polymorphisms in genes encoding alcohol-dehydrogenase, aldehyde-dehydrogenase, and cytochrome P450 are involved in individual susceptibility to alcoholism, yet their role in the progression of liver disease remains controversial. Variations in genes encoding inflammatory mediators (e.g., TNF-α, IL-1β, Il-10, and cytotoxic T lymphocyte antigen–4 [CTLA-4]), the lipopolysaccharide receptor CD14, and antioxidants (e.g., superoxide dismutase) may influence the

Figure 3
Cellular mechanisms of liver fibrosis. Different types of hepatotoxic agents produce mediators that induce inflammatory actions in hepatic cell types. Damaged hepatocytes and biliary cells release inflammatory cytokines and soluble factors that activate Kupffer cells and stimulate the recruitment of activated T cells. This inflammatory milieu stimulates the activation of resident HSCs into fibrogenic myofibroblasts. Activated HSCs also secrete cytokines that perpetuate their activated state. If the liver injury persists, accumulation of activated HSCs and portal myofibroblasts occurs, synthesizing large amounts of ECM proteins and leading to tissue fibrosis. ECM degradation is inhibited by the actions of cytokines such as TIMPs. Apoptosis of damaged hepatocytes stimulates the fibrogenic actions of HSCs. If the cause of the liver injury is removed, fibrosis is resolved. This phase includes apoptosis of activated HSCs and regeneration of hepatocytes. Collagen is degraded by increased activity of MMPs induced by decreased TIMP expression. CCL21, C-C chemokine ligand 21; MCP-1, monocyte chemoattractant protein–1; MIP-2, macrophage inflammatory protein–2; NS3, HCV nonstructural protein 3; NS5, HCV nonstructural protein 5; PAF, platelet-activating factor.

progression of alcohol-induced liver disease (74, 75). In chronic cholestatic disorders such as PBC, polymorphisms in IL-1β, IL-1 receptor antagonists, and TNF-α genes are associated with faster disease progression (76). Some alleles of the apolipoprotein E gene influence the response to therapy of PBC with ursodeoxycholic acid, which suggests that genetic polymorphisms may predict therapeutic response (77). In HCV liver disease, genetic variations are involved in susceptibility to persistent HCV infection, response to antiviral therapy, and progression of liver disease (78). Polymorphisms in genes involved in the immune response to HCV infection (e.g., transporter associated with antigen processing 2, mannose-binding lectin, and specific HLA-II alleles) and fibrogenic agonists (angiotensinogen and TGF-β1) influence fibrosis progression (79–81). The fibrogenic effect of heterozygosity in the C282Y mutation of the hemochromatosis gene in patients with chronic hepatitis C is controversial (82, 83). Finally, little is known about genetic factors and NASH (84), and polymorphisms in fibrogenic mediators such as angiotensinogen and TGF-β1 may be associated with more severe liver disease.

Key cytokines involved in liver fibrosis

Cytokines regulating the inflammatory response to injury modulate hepatic fibrogenesis in vivo and in vitro (85). Monocyte chemotactic protein type 1 and RANTES stimulate fibrogenesis while IL-10 and IFN-γ exert the opposite effect (55, 86). Among growth factors, TGF-β1 appears to be a key mediator in human fibrogenesis (58). In HSCs, TGF-β favors the transition to myofibroblast-like cells, stimulates the synthesis of ECM proteins, and inhibits their degradation. Strategies aimed at disrupting TGF-β1 synthesis and/or signaling pathways markedly decreased fibrosis in experimental models (87). PDGF is the most potent mitogen for HSCs and is upregulated in the fibrotic liver (12); its inhibition attenuates experimental liver fibrogenesis (88).

Cytokines with vasoactive properties also regulate liver fibrogenesis. Vasodilator substances (e.g., nitric oxide, relaxin) exert antifibrotic effects while vasoconstrictors (e.g., norepinephrine, angiotensin II) have opposite effects (67, 89). Endothelin-1, a powerful vasoconstrictor, stimulates fibrogenesis through its type A receptor (90). Among vasoactive cytokines, angiotensin II seems to play a major role in liver fibrogenesis. Angiotensin II is the effector peptide of the renin-angiotensin system, which is a major regulator of arterial pressure homeostasis in humans. Key components of this system are locally expressed in chronically injured livers, and activated HSCs de novo generate angiotensin II (91, 92). Importantly, pharmacological and/or genetic ablation of the renin-angiotensin system markedly attenuates experimental liver fibrosis (70, 93–98). Angiotensin II induces hepatic inflammation and stimulates an array of fibrogenic actions in activated HSCs, including cell proliferation, cell migration, secretion of proinflammatory cytokines, and collagen synthesis (66, 99, 100). These actions are largely mediated by ROS generated by a nonphagocytic form of NADPH oxidase. Unlike the phagocytic type, NADPH oxidases present in fibrogenic cell types are constitutively active, producing relatively low levels of ROS under basal conditions and generating higher levels of oxidants in response to cytokines, stimulating redox-sensitive intracellular pathways. NADPH oxidase also plays a key role in the inflammatory actions of Kupffer cells (101). Disruption of an active NADPH oxidase protects mice from developing severe liver injury following prolonged alcohol intake and/or bile duct ligation (66, 102).

Adipokines, which are cytokines mainly derived from the adipose tissue, regulate liver fibrogenesis. Leptin is required for HSC activation and fibrosis development (103, 104). In contrast, adiponectin markedly inhibits liver fibrogenesis in vitro and in vivo (69). The actions of these cytokines may explain why obesity influences fibrosis development in patients with chronic hepatitis C (105).

Intracellular signaling pathways mediating liver fibrogenesis

Data on intracellular pathways regulating liver fibrogenesis are mainly derived from studies using cultured HSCs, while understanding of their role in vivo is progressing through experimental fibrogenesis studies using knockout mice (106). Several mitogen-

Table 2
Main antifibrotic drugs in development for the treatment of liver fibrosis

Agent	Main mechanism	Antifibrotic effects in HSCs	Antifibrotic effects in experimental fibrosis	Antifibrotic effect in humans
Angiotensin inhibitors	Inhibits HSC activation	Consistent positive data	Consistent positive data	Retrospective study
Colchicine	Inhibits inflammatory response	Limited data	Limited data	Discrepant results
Corticosteroids	Inhibits inflammatory response	Limited data	Limited data	Effective in autoimmune hepatitis
Endothelin inhibitors	Inhibits HSC function	Limited data	Limited data	Not tested
Interferon-α	Inhibits HSC activation	Consistent positive data	Consistent positive data	Effective in chronic hepatitis C
Interleukin 10	Inhibits inflammatory response	Limited data	Consistent positive data	Isolated reports in chronic hepatitis C
Pentoxifylline	Inhibits HSC activation	Consistent positive data	Consistent positive data	Not tested
Phosphatidylcholine	Decreases oxidative stress	Limited data	Consistent positive data	Not proven in alcohol-induced fibrosis
PPAR agonists	Inhibits HSC activation	Consistent data	Consistent positive data	Isolated reports in NASH
S-adenosyl-methionine	Antioxidant	Limited data	Not tested	Effective in alcohol-induced fibrosis
Sho-saiko-to	Antioxidant	Consistent positive data	Consistent	Isolated reports in chronic hepatitis C
TGF-β1 inhibitors	Inhibits HSC activation and function	Consistent positive data	Consistent positive data	Not tested
Tocopherol	Antioxidant	Consistent positive data	Limited data	Isolated reports in NASH

activated protein kinases modulate major fibrogenic actions of HSCs. Extracellular-regulated kinase, which is stimulated in experimentally induced liver injury, mediates proliferation and migration of HSCs (107). In contrast, c-Jun N-terminal kinase regulates apoptosis of hepatocytes as well as the secretion of inflammatory cytokines by cultured HSCs (66, 108, 109). The focal adhesion kinase PI3K-Akt–signaling pathway mediates agonist-induced fibrogenic actions in HSCs (107). The TGF-β1–activated Smad-signaling pathway stimulates experimental hepatic fibrosis and is a potential target for therapy (110, 111). The PPAR pathway regulates HSC activation and experimental liver fibrosis. PPAR-γ ligands inhibit the fibrogenic actions in HSCs and attenuate liver fibrosis in vivo (112, 113). NF-κB may have an inhibitory action on liver fibrosis (114, 115). Other transcription factors are involved in HSC activation and may participate in liver fibrogenesis (116). Recent studies suggest a role for intracellular pathways signaled by Toll-like receptors and β-cathepsin (117, 118).

Pathogenesis of fibrosis in different liver diseases

The pathogenesis of liver fibrosis depends on the underlying etiology. In alcohol-induced liver disease, alcohol alters the population of gut bacteria and inhibits intestinal motility, resulting in an overgrowth of Gram-negative flora. Lipopolysaccharide is elevated in portal blood and activates Kupffer cells through the CD14/Toll-like receptor–4 complex to produce ROS via NADPH oxidase (101). Oxidants activate Kupffer cell NF-κB, causing an increase in TNF-α production. TNF-α induces neutrophil infiltration and stimulates mitochondrial oxidant production in hepatocytes, which are sensitized to undergo apoptosis. Acetaldehyde, the major alcohol metabolism product, and ROS activate HSCs and stimulate inflammatory and fibrogenic signals (119). The pathogenesis of HCV-induced liver fibrosis is poorly understood due to the lack of a rodent model of persistent HCV infection (78). HCV escapes surveillance of the HLA-II–directed immune response and infects hepatocytes, causing oxidative stress and inducing the recruitment of inflammatory cells. Both factors lead to HSC activation and collagen deposition. Moreover, several HCV proteins directly stimulate the inflammatory and fibrogenic actions of HSCs (120). In chronic cholestatic disorders such as PBC, T lymphocytes and cytokines mediate persistent bile duct damage (14). Biliary cells secrete fibrogenic mediators activating neighboring portal myofibroblasts to secrete ECM. Eventually, perisinusoidal HSCs become activated, and fibrotic bands develop. The pathogenesis of liver fibrosis due to NASH is poorly understood. Obesity, type 2 diabetes mellitus, and dyslipidemia are the most common associated conditions (121). A 2-hit model has been proposed: hyperglycemia and insulin resistance lead to elevated serum levels of free fatty acids, resulting in hepatic steatosis. In the second hit, oxidative stress and proinflammatory cytokines promote hepatocyte apoptosis and the recruitment of inflammatory cells, leading to progressive fibrosis.

Is liver fibrosis reversible?

In contrast with the traditional view that cirrhosis is an irreversible disease, recent evidence indicates that even advanced fibrosis is reversible (122). In experimentally induced fibrosis, cessation of liver injury results in fibrosis regression (123). In humans, spontaneous resolution of liver fibrosis can occur after successful treatment of the underlying disease. This observation has been described in patients with iron and copper overload,

alcohol-induced liver injury, chronic hepatitis C, B, and D, hemochromatosis, secondary biliary cirrhosis, NASH, and autoimmune hepatitis (19, 122, 124, 125, S1, S2) (Figure 4). It may take years for significant regression to be achieved; the time varies depending on the underlying cause of the liver disease and its severity. Chronic HCV infection is the most extensively studied condition, and therapy (IFN-α plus ribavirin) with viral clearance results in fibrosis improvement. Importantly, nearly half of patients with cirrhosis exhibit reversal to a significant degree (90). Whether this beneficial effect is associated with improvements in long-term clinical outcome, including decreased portal hypertension, is unknown.

Increased collagenolytic activity is a major mechanism of fibrosis resolution (122). Fibrillar collagens (I and III) are degraded by interstitial MMPs (MMP-1, -8, and -13 in humans and MMP-13 in rodents). During fibrosis resolution, MMP activity increases due to a rapid decrease in the expression of TIMP-1. Partial degradation of fibrillar collagen occurs, and the altered interaction between activated HSCs and ECM favors apoptosis (123). Removal of activated HSCs by apoptosis precedes fibrosis resolution. Stimulation of death receptors in activated HSCs and a decrease in survival factors, including TIMP-1, can precipitate HSC apoptosis (S3).

Several questions remain unanswered: Can we pharmacologically accelerate fibrosis resolution in humans? Can a fibrotic liver completely regress to a normal liver? Does fibrosis reverse similarly in all types of liver diseases? Although isolated cases of complete fibrosis resolution have been reported, it is conceivable that some degree of fibrosis cannot be removed (S4). Resolution may be limited by ECM cross-linking and a failure of activated HSCs to undergo apoptosis.

Therapeutic approaches to the treatment of liver fibrosis

There is no standard treatment for liver fibrosis. Although experimental studies have revealed targets to prevent fibrosis progression in rodents (20) (Table 2), the efficacy of most treatments has not been proven in humans. This is due to the need to perform serial liver biopsies to accurately assess changes in liver fibrosis, the necessity of long-term follow-up studies, and the fact that humans are probably less sensitive to hepatic antifibrotic therapies than rodents. The development of reliable noninvasive markers of liver fibrosis should have a positive impact on the design of clinical trials. The ideal antifibrotic therapy would be one that is liver-specific, well tolerated when administered for prolonged periods of time, and effective in attenuating excessive collagen deposition without affecting normal ECM synthesis.

The removal of the causative agent is the most effective intervention in the treatment of liver fibrosis. This strategy has been shown effective in most etiologies of chronic liver diseases (19, 122, 124, 125, S1, S2). For patients with cirrhosis and clinical complications, liver transplantation is currently the only curative approach (S5). Transplantation improves both survival and quality of life. However, in patients with HCV-induced cirrhosis, viral infection recurs after transplantation (S6), aggressive chronic hepatitis develops, and progression to cirrhosis is common.

Because inflammation precedes and promotes the progression of liver fibrosis, the use of antiinflammatory drugs has been proposed. Corticosteroids are only indicated for the treatment of hepatic fibrosis in patients with autoimmune hepatitis and acute alcoholic hepatitis (S1). Inhibition of the accumulation of activated HSCs by modulating either their activation and/or proliferation or promoting their apoptosis is another strategy.

Antioxidants such as vitamin E, silymarin, phosphatidylcholine, and S-adenosyl-L-methionine inhibit HSC activation, protect hepatocytes from undergoing apoptosis, and attenuate experimental liver fibrosis (S7). Antioxidants exert beneficial effects in patients with alcohol-induced liver disease and NASH (S8, S9). Disrupting TGF-β synthesis and/or signaling pathways prevents scar formation in experimental liver fibrosis (58). Moreover, administration of growth factors (e.g., IGF, hepatocyte growth factor, and cardiotrophin) or their delivery by gene therapy attenuates experimental liver fibrosis (S10, S11). However, these latter approaches have not been tested in humans and may favor cancer development. Substances that inhibit key signal transduction pathways involved in liver fibrogenesis also have the potential to treat liver fibrosis (20). They include pentoxifylline (phosphodiesterase inhibitor), amiloride (Na^+/H^+ pump inhibitor), and S-farnesylthiosalicylic acid (Ras antagonist). Ligands of PPARα and/or PPARγ such as thiazolindiones exert beneficial effects in experimental liver fibrosis and in patients with NASH (S12, S13). The inhibition of the renin-angiotensin system is probably the most promising strategy in treating liver fibrosis. Renin-angiotensin inhibitors are widely used as antifibrotic agents in patients with chronic renal and cardiac diseases and appear to be safe when administered for prolonged periods of time (S14). Little information is available on the use of this approach in patients with chronic liver diseases. Preliminary pilot studies in patients with chronic hepatitis C and NASH suggest that renin-angiotensin blocking agents may have beneficial effects on fibrosis progression (S15). Transplanted patients receiving renin-angiotensin system inhibitors as antihypertensive therapy show less fibrosis progression than patients receiving other types of drugs (S16). However, this approach cannot be recommended in clinical practice until the results of ongoing clinical trials become available. The blockade of endothelin-1 type A receptors and the administration of vasodilators (prostaglandin E2 and nitric oxide donors) exert antifibrotic activity in rodents, yet the effects in humans are unknown (90). Different herbal compounds, many of them traditionally used in Asian countries to treat liver diseases, have been demonstrated to have antifibrotic effects (S17). They include *Sho-saiko-to*, glycyrrhizin, and *savia miltiorhiza*. An alternative approach is the inhibition of collagen production and/or the promotion of its degradation (20). Inhibitors of prolyl-4 hydroxylase and halofuginone prevent the development of experimental liver cirrhosis by inhibiting collagen synthesis. MMP-8 and urokinase-type plasminogen activator stimulate collagen degradation in vivo. The efficacy of these drugs in humans is unknown, and they may result in undesirable side effects. Finally, infusion of mesenchymal stem cells ameliorates experimentally induced fibrosis, which suggests a potential for this approach in the treatment of chronic liver diseases (S18, S19).

A limitation of the current antifibrotic approaches is that antifibrotic drugs are not efficiently taken up by activated HSCs and may produce unwanted side effects. Cell-specific delivery to HSCs could provide a solution to these problems. Promising preliminary results have been recently obtained using different carriers (e.g., cyclic peptides coupled to albumin recognizing collagen type VI receptor and/or PDGFR) (S20). Antifibrotic therapy may differ depending on the type of liver disease. In patients with chronic HCV infection, current antiviral treatments (pegylated IFN plus ribavirin) clear viral infection in more than half of the patients (S21). Sustained virological response is associated with an improvement in liver fibrosis (122). Patients with no sustained response may also experience improvement of liver fibrosis, which suggests that IFN-α has an intrinsic antifibrotic effect (S22). For nonresponder patients, the use of renin-angiotensin system inhibitors is a promising approach. Treatment of the metabolic syndrome in patients with chronic hepatitis C may also decrease fibrosis progression (S23). In patients with alcohol-induced liver disease, the most effective approach is alcohol abstinence (124). Antioxidants (e.g., S-adenosyl-L-methionine and phosphatidylcholine) and hepatocyte protectors (e.g., silymarin) slow down the progression of liver fibrosis and can improve survival (S24). For patients with autoimmune hepatitis, immunosuppressant therapy not only decreases inflammation but also exerts antifibrotic effects (S25). No antifibrotic therapy is available for patients with chronic cholestatic disorders (i.e., primary sclerosing cholangitis and PBC). Ursodeoxycholic acid improves biochemical tests in these patients, but its impact on fibrosis is not consistently proven (S26). In patients with NASH, weight loss and specific treatments of the metabolic syndrome can reduce fibrosis development (125). Recent reports have revealed than antioxidants and insulin sensitizers (e.g., thiazolindiones) may exert antifibrogenic effects in these patients (S27). Large clinical trials are needed to confirm these results.

Future directions

The translation of basic research into improved therapeutics for the management of patients with chronic liver diseases is still poor. The role of pluripotential stem cells in hepatic wound healing is one of the most promising fields. Perfusion of these cells may be a potential approach to promoting fibrosis resolution and liver regeneration. Approaches to removing fibrogenic cells are being evaluated, including development of drug delivery systems that target activated HSCs. Translational research should investigate the molecular mechanisms that cause fibrosis in different types of human liver diseases in order to identify new targets for therapy. In the clinical setting, the identity of the genetic determinants that influence fibrosis progression should be uncovered. Well-designed large-scale epidemiological genetic studies are clearly required. Patients at a high risk of progression to cirrhosis should be identified. Developing simple and reliable noninvasive markers of hepatic fibrosis is an important goal in clinical hepatology and will facilitate the design of clinical trials. Most importantly, the efficacy of antifibrotic drugs known to attenuate experimental liver fibrosis should be tested in humans.

Acknowledgments

The authors' work is supported by grants from the NIH, the Ministerio de Ciencia y Tecnología de España, and the Instituto de Investigación Carlos III (SAF2002-03696 and BFI2002-01202).

Due to space constraints, a number of important references could not be included in this article. References S1–S27 are available online with this article; doi:10.1172/JCI200524282DS1.

Address correspondence to: David A. Brenner, Department of Medicine, Columbia University Medical Center, College of Physicians and Surgeons, 622 West 168th Street, PH 8E-105J, New York, New York 10032, USA. Phone: (212) 305-5838; Fax: (212) 305-8466; E-mail: dab2106@columbia.edu.

1. Friedman, S.L. 2003. Liver fibrosis - from bench to bedside. *J. Hepatol.* **38**(Suppl. 1):S38–S53.

2. Gines, P., Cardenas, A., Arroyo, V., and Rodes, J. 2004. Management of cirrhosis and ascites. *N. Engl. J. Med.* **350**:1646–1654.

3. Popper, H., and Uenfriend, S. 1970. Hepatic fibrosis. Correlation of biochemical and morphologic investigations. *Am. J. Med.* **49**:707–721.

4. Schaffner, F., and Klion, F.M. 1968. Chronic hepatitis. *Annu. Rev. Med.* **19**:25–38.

5. Albanis, E., and Friedman, S.L. 2001. Hepatic fibrosis. Pathogenesis and principles of therapy. *Clin. Liver Dis.* **5**:315–334, v–vi.

6. Soyer, M.T., Ceballos, R., and Aldrete, J.S. 1976. Reversibility of severe hepatic damage caused by jejunoileal bypass after re-establishment of normal intestinal continuity. *Surgery.* **79**:601–604.

7. Friedman, S.L., Roll, F.J., Boyles, J., and Bissell, D.M. 1985. Hepatic lipocytes: the principal collagen-producing cells of normal rat liver. *Proc. Natl. Acad. Sci. U. S. A.* **82**:8681–8685.

8. Geerts, A. 2001. History, heterogeneity, developmental biology, and functions of quiescent hepatic stellate cells. *Semin. Liver Dis.* **21**:311–335.

9. Friedman, S.L., et al. 1992. Isolated hepatic lipocytes and Kupffer cells from normal human liver: morphological and functional characteristics in primary culture. *Hepatology.* **15**:234–243.

10. Otto, D.A., and Veech, R.L. 1980. Isolation of a lipocyte-rich fraction from rat liver nonparenchymal cells. *Adv. Exp. Med. Biol.* **132**:509–517.

11. Rockey, D.C., Boyles, J.K., Gabbiani, G., and Friedman, S.L. 1992. Rat hepatic lipocytes express smooth muscle actin upon activation in vivo and in culture. *J. Submicrosc. Cytol. Pathol.* **24**:193–203.

12. Pinzani, M., Gesualdo, L., Sabbah, G.M., and Abboud, H.E. 1989. Effects of platelet-derived growth factor and other polypeptide mitogens on DNA synthesis and growth of cultured rat liver fat-storing cells. *J. Clin. Invest.* **84**:1786–1793.

13. Wasser, S., and Tan, C.E. 1999. Experimental models of hepatic fibrosis in the rat. *Ann. Acad. Med. Singapore.* **28**:109–111.

14. Ramadori, G., and Saile, B. 2004. Portal tract fibrogenesis in the liver. *Lab. Invest.* **84**:153–159.

15. Forbes, S.J., et al. 2004. A significant proportion of myofibroblasts are of bone marrow origin in human liver fibrosis. *Gastroenterology.* **126**:955–963.

16. Poynard, T., et al. 2000. Natural history of HCV infection. *Baillieres Best Pract. Res. Clin. Gastroenterol.* **14**:211–228.

17. Poynard, T., Bedossa, P., and Opolon, P. 1997. Natural history of liver fibrosis progression in patients with chronic hepatitis C. The OBSVIRC, META-VIR, CLINIVIR, and DOSVIRC groups. *Lancet.* **349**:825–832.

18. Bataller, R., North, K.E., and Brenner, D.A. 2003. Genetic polymorphisms and the progression of liver fibrosis: a critical appraisal. *Hepatology.* **37**:493–503.

19. Hammel, P., et al. 2001. Regression of liver fibrosis after biliary drainage in patients with chronic pancreatitis and stenosis of the common bile duct. *N. Engl. J. Med.* **344**:418–423.

20. Bataller, R., and Brenner, D.A. 2001. Hepatic stellate cells as a target for the treatment of liver fibrosis. *Semin. Liver Dis.* **21**:437–451.

21. Brunt, E.M. 2004. Nonalcoholic steatohepatitis. *Semin. Liver Dis.* **24**:3–20.

22. Ludwig, J., Viggiano, T.R., McGill, D.B., and Oh, B.J. 1980. Nonalcoholic steatohepatitis: Mayo Clinic experiences with a hitherto unnamed disease. *Mayo Clin. Proc.* **55**:434–438.

23. Davis, G.L., Albright, J.E., Cook, S.F., and Rosenberg, D.M. 2003. Projecting future complications of chronic hepatitis C in the United States. *Liver Transpl.* **9**:331–338.

24. Berenguer, M., et al. 2003. Severe recurrent hepatitis C after liver retransplantation for hepatitis C virus-related graft cirrhosis. *Liver Transpl.* **9**:228–235.

25. Afdhal, N.H., and Nunes, D. 2004. Evaluation of liver fibrosis: a concise review. *Am. J. Gastroenterol.* **99**:1160–1174.

26. Thampanitchawong, P., and Piratvisuth, T. 1999. Liver biopsy: complications and risk factors. *World J. Gastroenterol.* **5**:301–304.

27. Regev, A., et al. 2002. Sampling error and intra-observer variation in liver biopsy in patients with chronic HCV infection. *Am. J. Gastroenterol.* **97**:2614–2618.

28. Imbert-Bismut, F., et al. 2001. Biochemical markers of liver fibrosis in patients with hepatitis C virus infection: a prospective study. *Lancet.* **357**:1069–1075.

29. Forns, X., et al. 2002. Identification of chronic hepatitis C patients without hepatic fibrosis by a simple predictive model. *Hepatology.* **36**:986–992.

30. Fontana, R.J., and Lok, A.S. 2002. Noninvasive monitoring of patients with chronic hepatitis C. *Hepatology.* **36**:S57–S64.

31. Hirata, M., Akbar, S.M., Horiike, N., and Onji, M. 2001. Noninvasive diagnosis of the degree of hepatic fibrosis using ultrasonography in patients with chronic liver disease due to hepatitis C virus. *Eur. J. Clin. Invest.* **31**:528–535.

32. Callewaert, N., et al. 2004. Noninvasive diagnosis of liver cirrhosis using DNA sequencer-based total serum protein glycomics. *Nat. Med.* **10**:429–434.

33. Pinzani, M. 1999. Liver fibrosis. *Springer Semin. Immunopathol.* **21**:475–490.

34. Benyon, R.C., and Iredale, J.P. 2000. Is liver fibrosis reversible? *Gut.* **46**:443–446.

35. Arthur, M.J. 2000. Fibrogenesis II. Metalloproteinases and their inhibitors in liver fibrosis. *Am. J. Physiol. Gastrointest. Liver Physiol.* **279**:G245–G249.

36. Gabele, E., Brenner, D.A., and Rippe, R.A. 2003. Liver fibrosis: signals leading to the amplification of the fibrogenic hepatic stellate cell. *Front. Biosci.* **8**:D69–D77.

37. Milani, S., et al. 1990. Procollagen expression by nonparenchymal rat liver cells in experimental biliary fibrosis. *Gastroenterology.* **98**:175–184.

38. Marra, F. 1999. Hepatic stellate cells and the regulation of liver inflammation. *J. Hepatol.* **31**:1120–1130.

39. Lindquist, J.N., Marzluff, W.F., and Stefanovic, B. 2000. Fibrogenesis. III. Posttranscriptional regulation of type I collagen. *Am. J. Physiol. Gastrointest. Liver Physiol.* **279**:G471–G476.

40. Lindquist, J.N., Parsons, C.J., Stefanovic, B., and Brenner, D.A. 2004. Regulation of alpha1(I) collagen messenger RNA decay by interactions with alphaCP at the 3'-untranslated region. *J. Biol. Chem.* **279**:23822–23829.

41. Sato, M., Suzuki, S., and Senoo, H. 2003. Hepatic stellate cells: unique characteristics in cell biology and phenotype. *Cell Struct. Funct.* **28**:105–112.

42. Oben, J.A., Yang, S., Lin, H., Ono, M., and Diehl, A.M. 2003. Norepinephrine and neuropeptide Y promote proliferation and collagen gene expression of hepatic myofibroblastic stellate cells. *Biochem. Biophys. Res. Commun.* **302**:685–690.

43. Kinnman, N., and Housset, C. 2002. Peribiliary myofibroblasts in biliary type liver fibrosis. *Front. Biosci.* **7**:d496–d503.

44. Magness, S.T., Bataller, R., Yang, L., and Brenner, D.A. 2004. A dual reporter gene transgenic mouse demonstrates heterognity in hepatic fibrogenic cell populations. *Hepatology.* **40**:1151–1159.

45. Knittel, T., et al. 1999. Rat liver myofibroblasts and hepatic stellate cells: different cell populations of the fibroblast lineage with fibrogenic potential. *Gastroenterology.* **117**:1205–1221.

46. Suskind, D.L., and Muench, M.O. 2004. Searching for common stem cells of the hepatic and hematopoietic systems in the human fetal liver: CD34+ cytokeratin 7/8+ cells express markers for stellate cells. *J. Hepatol.* **40**:261–268.

47. Phillips, R.J., et al. 2004. Circulating fibrocytes traffic to the lungs in response to CXCL12 and mediate fibrosis. *J. Clin. Invest.* **114**:438–446. doi:10.1172/JCI200420997.

48. Kalluri, R., and Neilson, E.G. 2003. Epithelial-mesenchymal transition and its implications for fibrosis. *J. Clin. Invest.* **112**:1776–1784. doi:10.1172/JCI200320530.

49. Kmiec, Z. 2001. Cooperation of liver cells in health and disease. *Adv. Anat. Embryol. Cell Biol.* **161**:III–XIII, 1–151.

50. Higuchi, H., and Gores, G.J. 2003. Mechanisms of liver injury: an overview. *Curr. Mol. Med.* **3**:483–490.

51. Canbay, A., Friedman, S., and Gores, G.J. 2004. Apoptosis: the nexus of liver injury and fibrosis. *Hepatology.* **39**:273–278.

52. Casini, A., et al. 1997. Neutrophil-derived superoxide anion induces lipid peroxidation and stimulates collagen synthesis in human hepatic stellate cells: role of nitric oxide. *Hepatology.* **25**:361–367.

53. Vinas, O., et al. 2003. Human hepatic stellate cells show features of antigen-presenting cells and stimulate lymphocyte proliferation. *Hepatology.* **38**:919–929.

54. Maher, J.J. 2001. Interactions between hepatic stellate cells and the immune system. *Semin. Liver Dis.* **21**:417–426.

55. Shi, Z., Wakil, A.E., and Rockey, D.C. 1997. Strain-specific differences in mouse hepatic wound healing are mediated by divergent T helper cytokine responses. *Proc. Natl. Acad. Sci. U. S. A.* **94**:10663–10668.

56. Naito, M., Hasegawa, G., Ebe, Y., and Yamamoto, T. 2004. Differentiation and function of Kupffer cells. *Med. Electron Microsc.* **37**:16–28.

57. Thurman, R.G. 1998. Alcoholic liver injury involves activation of Kupffer cells by endotoxin. *Am. J. Physiol.* **275**:G605–G611.

58. Gressner, A.M., Weiskirchen, R., Breitkopf, K., and Dooley, S. 2002. Roles of TGF-beta in hepatic fibrosis. *Front. Biosci.* **7**:d793–d807.

59. Olaso, E., et al. 2001. DDR2 receptor promotes MMP-2-mediated proliferation and invasion by hepatic stellate cells. *J. Clin. Invest.* **108**:1369–1378. doi:10.1172/JCI200112373.

60. Takehara, T., et al. 2004. Hepatocyte-specific disruption of Bcl-xL leads to continuous hepatocyte apoptosis and liver fibrotic responses. *Gastroenterology.* **127**:1189–1197.

61. Canbay, A., et al. 2002. Fas enhances fibrogenesis in the bile duct ligated mouse: a link between apoptosis and fibrosis. *Gastroenterology.* **123**:1323–1330.

62. Safadi, R., et al. 2004. Immune stimulation of hepatic fibrogenesis by CD8 cells and attenuation by transgenic interleukin-10 from hepatocytes. *Gastroenterology.* **127**:870–882.

63. Sahai, A., Malladi, P., Melin-Aldana, H., Green, R.M., and Whitington, P.F. 2004. Upregulation of osteopontin expression is involved in the development of nonalcoholic steatohepatitis in a dietary murine model. *Am. J. Physiol. Gastrointest. Liver Physiol.* **287**:G264–G273.

64. Yoshida, T., et al. 2004. SOCS1 is a suppressor of liver fibrosis and hepatitis-induced carcinogenesis. *J. Exp. Med.* **199**:1701–1707.

65. Streetz, K.L., et al. 2003. Interleukin 6/gp130-dependent pathways are protective during chronic liver diseases. *Hepatology.* **38**:218–229.

66. Bataller, R., et al. 2003. NADPH oxidase signal transduces angiotensin II in hepatic stellate cells and is critical in hepatic fibrosis. *J. Clin. Invest.* **112**:1383–1394. doi:10.1172/JCI200318212.

67. Oben, J.A., et al. 2004. Hepatic fibrogenesis requires sympathetic neurotransmitters. *Gut.* **53**:438–445.

68. Yu, C., et al. 2003. Role of fibroblast growth factor type 1 and 2 in carbon tetrachloride-induced hepatic injury and fibrogenesis. *Am. J. Pathol.*

163:1653–1662.

69. Kamada, Y., et al. 2003. Enhanced carbon tetrachloride-induced liver fibrosis in mice lacking adiponectin. *Gastroenterology.* **125**:1796–1807.

70. Kanno, K., Tazuma, S., and Chayama, K. 2003. AT1A-deficient mice show less severe progression of liver fibrosis induced by CCl(4). *Biochem. Biophys. Res. Commun.* **308**:177–183.

71. Ueberham, E., et al. 2003. Conditional tetracycline-regulated expression of TGF-beta1 in liver of transgenic mice leads to reversible intermediary fibrosis. *Hepatology.* **37**:1067–1078.

72. Yoshiji, H., et al. 2002. Tissue inhibitor of metalloproteinases-1 attenuates spontaneous liver fibrosis resolution in the transgenic mouse. *Hepatology.* **36**:850–860.

73. Agarwal, D.P. 2001. Genetic polymorphisms of alcohol metabolizing enzymes. *Pathol. Biol.* **49**:703–709.

74. Degoul, F., et al. 2001. Homozygosity for alanine in the mitochondrial targeting sequence of superoxide dismutase and risk for severe alcoholic liver disease. *Gastroenterology.* **120**:1468–1474.

75. Jarvelainen, H.A., et al. 2001. Promoter polymorphism of the CD14 endotoxin receptor gene as a risk factor for alcoholic liver disease. *Hepatology.* **33**:1148–1153.

76. Donaldson, P., et al. 2001. HLA and interleukin 1 gene polymorphisms in primary biliary cirrhosis: associations with disease progression and disease susceptibility. *Gut.* **48**:397–402.

77. Corpechot, C., et al. 2001. Apolipoprotein E polymorphism, a marker of disease severity in primary biliary cirrhosis? *J. Hepatol.* **35**:324–328.

78. Schuppan, D., Krebs, A., Bauer, M., and Hahn, E.G. 2003. Hepatitis C and liver fibrosis. *Cell Death Differ.* **10**(Suppl. 1):S59–S67.

79. Powell, E.E., et al. 2000. Host genetic factors influence disease progression in chronic hepatitis C. *Hepatology.* **31**:828–833.

80. Sasaki, K., et al. 2000. Mannose-binding lectin polymorphisms in patients with hepatitis C virus infection. *Scand. J. Gastroenterol.* **35**:960–965.

81. Akuta, N., et al. 2001. Risk factors of hepatitis C virus-related liver cirrhosis in young adults: positive family history of liver disease and transporter associated with antigen processing 2(TAP2)*0201 allele. *J. Med. Virol.* **64**:109–116.

82. Martinelli, A.L., et al. 2000. Are haemochromatosis mutations related to the severity of liver disease in hepatitis C virus infection? *Acta Haematol.* **102**:152–156.

83. Thorburn, D., et al. 2002. The role of iron and haemochromatosis gene mutations in the progression of liver disease in chronic hepatitis C. *Gut.* **50**:248–252.

84. Day, C.P. 2004. The potential role of genes in nonalcoholic fatty liver disease. *Clin. Liver Dis.* **8**:673–691, xi.

85. Marra, F. 2002. Chemokines in liver inflammation and fibrosis. *Front. Biosci.* **7**:d1899–d1914.

86. Schwabe, R.F., Bataller, R., and Brenner, D.A. 2003. Human hepatic stellate cells express CCR5 and RANTES to induce proliferation and migration. *Am. J. Physiol. Gastrointest. Liver Physiol.* **285**:G949–G958.

87. Shek, F.W., and Benyon, R.C. 2004. How can transforming growth factor beta be targeted usefully to combat liver fibrosis? *Eur. J. Gastroenterol. Hepatol.* **16**:123–126.

88. Borkham-Kamphorst, E., Stoll, D., Gressner, A.M., and Weiskirchen, R. 2004. Antisense strategy against PDGF B-chain proves effective in preventing experimental liver fibrogenesis. *Biochem. Biophys. Res. Commun.* **321**:413–423.

89. Williams, E.J., et al. 2001. Relaxin inhibits effective collagen deposition by cultured hepatic stellate cells and decreases rat liver fibrosis in vivo. *Gut.* **49**:577–583.

90. Cho, J.J., et al. 2000. An oral endothelin-A receptor antagonist blocks collagen synthesis and deposition in advanced rat liver fibrosis. *Gastroenterology.* **118**:1169–1178.

91. Paizis, G., et al. 2002. Up-regulation of components of the renin-angiotensin system in the bile duct–ligated rat liver. *Gastroenterology.* **123**:1667–1676.

92. Bataller, R., et al. 2003. Activated human hepatic stellate cells express the renin-angiotensin system and synthesize angiotensin II. *Gastroenterology.* **125**:117–125.

93. Jonsson, J.R., et al. 2001. Angiotensin-converting enzyme inhibition attenuates the progression of rat hepatic fibrosis. *Gastroenterology.* **121**:148–155.

94. Paizis, G., et al. 2001. Effect of angiotensin II type 1 receptor blockade on experimental hepatic fibrogenesis. *J. Hepatol.* **35**:376–385.

95. Ramalho, L.N., et al. 2002. Effect of losartan, an angiotensin II antagonist, on secondary biliary cirrhosis. *Hepatogastroenterology.* **49**:1499–1502.

96. Wei, H.S., et al. 2000. The regulatory role of AT 1 receptor on activated HSCs in hepatic fibrogenesis: effects of RAS inhibitors on hepatic fibrosis induced by CCl(4). *World J. Gastroenterol.* **6**:824–828.

97. Wei, H.S., et al. 2000. Effects of AT1 receptor antagonist, losartan, on rat hepatic fibrosis induced by CCl(4). *World J. Gastroenterol.* **6**:540–545.

98. Tuncer, I., Ozbek, H., Ugras, S., and Bayram, I. 2003. Anti-fibrogenic effects of captopril and candesartan cilexetil on the hepatic fibrosis development in rat. The effect of AT1-R blocker on the hepatic fibrosis. *Exp. Toxicol. Pathol.* **55**:159–166.

99. Bataller, R., et al. 2000. Angiotensin II induces contraction and proliferation of human hepatic stellate cells. *Gastroenterology.* **118**:1149–1156.

100. Bataller, R., et al. 2003. Prolonged infusion of angiotensin II into normal rats induces stellate cell activation and proinflammatory events in liver. *Am. J. Physiol. Gastrointest. Liver Physiol.* **285**:G642–G651.

101. Wheeler, M.D., et al. 2001. The role of Kupffer cell oxidant production in early ethanol-induced liver disease. *Free Radic. Biol. Med.* **31**:1544–1549.

102. Kono, H., et al. 2000. NADPH oxidase–derived free radicals are key oxidants in alcohol-induced liver disease. *J. Clin. Invest.* **106**:867–872.

103. Marra, F. 2002. Leptin and liver fibrosis: a matter of fat. *Gastroenterology.* **122**:1529–1532.

104. Ikejima, K., et al. 2002. Leptin receptor–mediated signaling regulates hepatic fibrogenesis and remodeling of extracellular matrix in the rat. *Gastroenterology.* **122**:1399–1410.

105. Ortiz, V., Berenguer, M., Rayon, J.M., Carrasco, D., and Berenguer, J. 2002. Contribution of obesity to hepatitis C–related fibrosis progression. *Am. J. Gastroenterol.* **97**:2408–2414.

106. Pinzani, M. 2002. PDGF and signal transduction in hepatic stellate cells. *Front. Biosci.* **7**:d1720–d1726.

107. Marra, F., et al. 1999. Extracellular signal-regulated kinase activation differentially regulates platelet-derived growth factor's actions in hepatic stellate cells, and is induced by in vivo liver injury in the rat. *Hepatology.* **30**:951–958.

108. Schwabe, R.F., et al. 2004. Differential requirement for c-Jun NH2-terminal kinase in TNFalpha- and Fas-mediated apoptosis in hepatocytes. *FASEB J.* **18**:720–722.

109. Schwabe, R.F., Schnabl, B., Kweon, Y.O., and Brenner, D.A. 2001. CD40 activates NF-kappa B and c-Jun N-terminal kinase and enhances chemokine secretion on activated human hepatic stellate cells. *J. Immunol.* **166**:6812–6819.

110. Schnabl, B., et al. 2001. The role of Smad3 in mediating mouse hepatic stellate cell activation. *Hepatology.* **34**:89–100.

111. Dooley, S., et al. 2003. Smad7 prevents activation of hepatic stellate cells and liver fibrosis in rats. *Gastroenterology.* **125**:178–191.

112. Marra, F., et al. 2000. Ligands of peroxisome proliferator-activated receptor gamma modulate profibrogenic and proinflammatory actions in hepatic stellate cells. *Gastroenterology.* **119**:466–478.

113. Galli, A., et al. 2002. Antidiabetic thiazolidinediones inhibit collagen synthesis and hepatic stellate cell activation in vivo and in vitro. *Gastroenterology.* **122**:1924–1940.

114. Boya, P., et al. 2001. Nuclear factor-kappa B in the liver of patients with chronic hepatitis C: decreased RelA expression is associated with enhanced fibrosis progression. *Hepatology.* **34**:1041–1048.

115. Rippe, R.A., Schrum, L.W., Stefanovic, B., Solis-Herruzo, J.A., and Brenner, D.A. 1999. NF-kappaB inhibits expression of the alpha1(I) collagen gene. *DNA Cell Biol.* **18**:751–761.

116. Mann, D.A., and Smart, D.E. 2002. Transcriptional regulation of hepatic stellate cell activation. *Gut.* **50**:891–896.

117. Paik, Y.H., et al. 2003. Toll-like receptor 4 mediates inflammatory signaling by bacterial lipopolysaccharide in human hepatic stellate cells. *Hepatology.* **37**:1043–1055.

118. Canbay, A., et al. 2003. Cathepsin B inactivation attenuates hepatic injury and fibrosis during cholestasis. *J. Clin. Invest.* **112**:152–159. doi:10.1172/JCI200317740.

119. Maher, J.J., Zia, S., and Tzagarakis, C. 1994. Acetaldehyde-induced stimulation of collagen synthesis and gene expression is dependent on conditions of cell culture: studies with rat lipocytes and fibroblasts. *Alcohol Clin. Exp. Res.* **18**:403–409.

120. Bataller, R., Paik, Y.H., Lindquist, J.N., Lemasters, J.J., and Brenner, D.A. 2004. Hepatitis C virus core and nonstructural proteins induce fibrogenic effects in hepatic stellate cells. *Gastroenterology.* **126**:529–540.

121. Wanless, I.R., and Shiota, K. 2004. The pathogenesis of nonalcoholic steatohepatitis and other fatty liver diseases: a four-step model including the role of lipid release and hepatic venular obstruction in the progression to cirrhosis. *Semin. Liver Dis.* **24**:99–106.

122. Arthur, M.J. 2002. Reversibility of liver fibrosis and cirrhosis following treatment for hepatitis C. *Gastroenterology.* **122**:1525–1528.

123. Issa, R., et al. 2004. Spontaneous recovery from micronodular cirrhosis: evidence for incomplete resolution associated with matrix cross-linking. *Gastroenterology.* **126**:1795–1808.

124. Pares, A., Caballeria, J., Bruguera, M., Torres, M., and Rodes, J. 1986. Histological course of alcoholic hepatitis. Influence of abstinence, sex and extent of hepatic damage. *J. Hepatol.* **2**:33–42.

125. Dixon, J.B., Bhathal, P.S., Hughes, N.R., and O'Brien, P.E. 2004. Nonalcoholic fatty liver disease: Improvement in liver histological analysis with weight loss. *Hepatology.* **39**:1647–1654.

SREBPs: activators of the complete program of cholesterol and fatty acid synthesis in the liver

Jay D. Horton,[1,2] Joseph L. Goldstein,[1] and Michael S. Brown[1]

[1]Department of Molecular Genetics, and [2]Department of Internal Medicine, University of Texas Southwestern Medical Center, Dallas, Texas, USA.

Lipid homeostasis in vertebrate cells is regulated by a family of membrane-bound transcription factors designated sterol regulatory element–binding proteins (SREBPs). SREBPs directly activate the expression of more than 30 genes dedicated to the synthesis and uptake of cholesterol, fatty acids, triglycerides, and phospholipids, as well as the NADPH cofactor required to synthesize these molecules (1–4). In the liver, three SREBPs regulate the production of lipids for export into the plasma as lipoproteins and into the bile as micelles. The complex, interdigitated roles of these three SREBPs have been dissected through the study of ten different lines of gene-manipulated mice. These studies form the subject of this review.

SREBPs: activation through proteolytic processing

SREBPs belong to the basic helix-loop-helix–leucine zipper (bHLH-Zip) family of transcription factors, but they differ from other bHLH-Zip proteins in that they are synthesized as inactive precursors bound to the endoplasmic reticulum (ER) (1, 5). Each SREBP precursor of about 1150 amino acids is organized into three domains: (a) an NH2-terminal domain of about 480 amino acids that contains the bHLH-Zip region for binding DNA; (b) two hydrophobic transmembrane–spanning segments interrupted by a short loop of about 30 amino acids that projects into the lumen of the ER; and (c) a COOH-terminal domain of about 590 amino acids that performs the essential regulatory function described below.

In order to reach the nucleus and act as a transcription factor, the NH2-terminal domain of each SREBP must be released from the membrane proteolytically (Figure 1). Three proteins required for SREBP processing have been delineated in cultured cells, using the tools of somatic cell genetics (see ref. 5 for review). One is an escort protein designated SREBP cleavage–activating protein (SCAP). The other two are proteases, designated Site-1 protease (S1P) and Site-2 protease (S2P). Newly synthesized SREBP is inserted into the membranes of the ER, where its COOH-terminal regulatory domain binds to the COOH-terminal domain of SCAP (Figure 1). SCAP is both an escort for SREBPs and a sensor of sterols. When cells become depleted in cholesterol, SCAP escorts the SREBP from the ER to the Golgi apparatus, where the two proteases reside. In the Golgi apparatus, S1P, a membrane-bound serine protease, cleaves the SREBP in the luminal loop between its two membrane-spanning segments, dividing the SREBP molecule in half (Figure 1). The NH2-terminal bHLH-Zip domain is then released from the membrane via a second cleavage mediated by S2P, a membrane-bound zinc metalloproteinase. The NH2-terminal domain, designated nuclear SREBP (nSREBP), translocates to the nucleus, where it activates transcription by binding to non-palindromic sterol response elements (SREs) in the promoter/enhancer regions of multiple target genes.

When the cholesterol content of cells rises, SCAP senses the excess cholesterol through its membranous sterol-sensing domain, changing its conformation in such a way that the SCAP/SREBP complex is no longer incorporated into ER transport vesicles. The net result is that SREBPs lose their access to S1P and S2P in the Golgi apparatus, so their bHLH-Zip domains cannot be released from the ER membrane, and the transcription of target genes ceases (1, 5). The biophysical mechanism by which SCAP senses sterol levels in the ER membrane and regulates its movement to the Golgi apparatus is not yet understood. Elucidating this mechanism will be fundamental to understanding the molecular basis of cholesterol feedback inhibition of gene expression.

SREBPs: two genes, three proteins

The mammalian genome encodes three SREBP isoforms, designated SREBP-1a, SREBP-1c, and SREBP-2. SREBP-2 is encoded by a gene on human chromosome 22q13. Both SREBP-1a and -1c are derived from a single gene on human chromosome 17p11.2 through the use of alternative transcription start sites that produce alternate forms of exon 1, designated 1a and 1c (1). SREBP-1a is a potent activator of all SREBP-responsive genes, including those that mediate the synthesis of cholesterol, fatty acids, and triglycerides. High-level transcriptional activation is dependent on exon 1a, which encodes a longer acidic transactivation segment than does the first exon of SREBP-1c. The roles of SREBP-1c and SREBP-2 are more restricted than that of SREBP-1a. SREBP-1c preferentially enhances transcription of genes required for fatty acid synthesis but not cholesterol synthesis. Like SREBP-1a, SREBP-2 has a long transcriptional activation domain, but it preferentially activates cholesterol synthesis (1). SREBP-1a and SREBP-2 are the predominant isoforms of SREBP in most cultured cell lines, whereas SREBP-1c and SREBP-2 predominate in the liver and most other intact tissues (6).

When expressed at higher than physiologic levels, each of the three SREBP isoforms can activate all enzymes indicated in Figure 2, which shows the biosynthetic pathways used to generate cholesterol and fatty acids. However, at normal levels of expression, SREBP-1c favors the fatty acid biosynthetic pathway and SREBP-2 favors cholesterologenesis. SREBP-2–responsive genes in the cholesterol biosynthetic pathway include those for the enzymes HMG-CoA synthase, HMG-CoA reductase, farnesyl diphosphate synthase, and squalene synthase. SREBP-1c–responsive genes include those for ATP citrate lyase (which produces acetyl-CoA) and acetyl-CoA carboxylase and fatty acid synthase (which together produce palmitate [C16:0]). Other SREBP-1c target genes encode a rate-limiting enzyme of the fatty acid elongase complex, which converts palmitate to stearate (C18:0) (ref. 7); stearoyl-CoA desaturase, which converts stearate to oleate (C18:1); and glycerol-3-phosphate acyltransferase, the first committed enzyme in triglyceride and phospholipid synthesis (3).

Citation for this article: *J. Clin. Invest.* **109**:1125–1131 (2002). DOI:10.1172/JCI200215593.

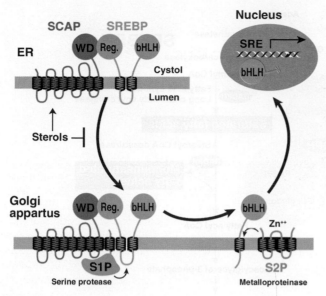

Figure 1

Model for the sterol-mediated proteolytic release of SREBPs from membranes. SCAP is a sensor of sterols and an escort of SREBPs. When cells are depleted of sterols, SCAP transports SREBPs from the ER to the Golgi apparatus, where two proteases, Site-1 protease (S1P) and Site-2 protease (S2P), act sequentially to release the NH₂-terminal bHLH-Zip domain from the membrane. The bHLH-Zip domain enters the nucleus and binds to a sterol response element (SRE) in the enhancer/promoter region of target genes, activating their transcription. When cellular cholesterol rises, the SCAP/SREBP complex is no longer incorporated into ER transport vesicles, SREBPs no longer reach the Golgi apparatus, and the bHLH-Zip domain cannot be released from the membrane. As a result, transcription of all target genes declines. Reprinted from ref. 5 with permission.

Finally, SREBP-1c and SREBP-2 activate three genes required to generate NADPH, which is consumed at multiple stages in these lipid biosynthetic pathways (8) (Figure 2).

Knockout and transgenic mice

Ten different genetically manipulated mouse models that either lack or overexpress a single component of the SREBP pathway have been generated in the last 6 years (9–16). The key molecular and metabolic alterations observed in these mice are summarized in Table 1.

Knockout mice that lack all nSREBPs die early in embryonic development. For instance, a germline deletion of *S1p*, which prevents the processing of all SREBP isoforms, results in death before day 4 of development (15, 17). Germline deletion of *Srebp2* leads to 100% lethality at a later stage of embryonic development than does deletion of *S1p* (embryonic day 7–8). In contrast, germline deletion of *Srebp1*, which eliminates both the 1a and the 1c transcripts, leads to partial lethality, in that about 15–45% of *Srebp1⁻ᐟ⁻* mice survive (13). The surviving homozygotes manifest elevated levels of SREBP-2 mRNA and protein (Table 1), which presumably compensates for the loss of SREBP-1a and -1c. When the SREBP-1c transcript is selectively eliminated, no embryonic lethality is observed, suggesting that the partial embryonic lethality in the *Srebp1⁻ᐟ⁻* mice is due to the loss of the SREBP-1a transcript (16).

To bypass embryonic lethality, we have produced mice in which all SREBP function can be disrupted in adulthood through induction of Cre recombinase. For this purpose, *loxP* recombination

sites were inserted into genomic regions that flank crucial exons in the *Scap* or *S1p* genes (so-called floxed alleles) (14, 15). Mice homozygous for the floxed gene and heterozygous for a Cre recombinase transgene, which is under control of an IFN-inducible promoter (*MX1*-Cre), can be induced to delete *Scap* or *S1p* by stimulating IFN expression. Thus, following injection with polyinosinic acid–polycytidylic acid, a double-stranded RNA that provokes antiviral responses, the Cre recombinase is produced in liver and disrupts the floxed gene by recombination between the *loxP* sites.

Cre-mediated disruption of *Scap* or *S1p* dramatically reduces nSREBP-1 and nSREBP-2 levels in liver and diminishes expression of all SREBP target genes in both the cholesterol and the fatty acid synthetic pathways (Table 1). As a result, the rates of synthesis of cholesterol and fatty acids fall by 70–80% in Scap- and S1p-deficient livers.

In cultured cells, the processing of SREBP is inhibited by sterols, and the sensor for this inhibition is SCAP (5). To learn whether SCAP performs the same function in liver, we have produced transgenic mice that express a mutant SCAP with a single amino acid substitution in the sterol-sensing domain (D443N) (12). Studies in tissue culture show that SCAP(D443N) is resistant to inhibition by sterols. Cells that express a single copy of this mutant gene overproduce cholesterol (18). Transgenic mice that express this mutant version of SCAP in the liver exhibit a similar phenotype (12). These livers manifest elevated levels of nSREBP-1 and nSREBP-2, owing to constitutive SREBP processing, which is not suppressed when the animals are fed a cholesterol-rich diet. nSREBP-1 and -2 increase the expression of all SREBP target genes shown in Figure 2, thus stimulating cholesterol and fatty acid synthesis and causing a marked accumulation of hepatic cholesterol and triglycerides (Table 1). This transgenic model provides strong in vivo evidence that SCAP activity is normally under partial inhibition by endogenous sterols, which keeps the synthesis of cholesterol and fatty acids in a partially repressed state in the liver.

Function of individual SREBP isoforms in vivo

To study the functions of individual SREBPs in the liver, we have produced transgenic mice that overexpress truncated versions of SREBPs (nSREBPs) that terminate prior to the membrane attachment domain. These nSREBPs enter the nucleus directly, bypassing the sterol-regulated cleavage step. By studying each nSREBP isoform separately, we could determine their distinct activating properties, albeit when overexpressed at nonphysiologic levels.

Overexpression of nSREBP-1c in the liver of transgenic mice produces a triglyceride-enriched fatty liver with no increase in cholesterol (10). mRNAs for fatty acid synthetic enzymes and rates of fatty acid synthesis are elevated fourfold in this tissue, whereas the mRNAs for cholesterol synthetic enzymes and the rate of cholesterol synthesis are not increased (8). Conversely, overexpression of nSREBP-2 in the liver increases the mRNAs encoding all cholesterol biosynthetic enzymes; the most dramatic is a 75-fold increase in HMG-CoA reductase mRNA (11). mRNAs for fatty acid synthesis enzymes are increased to a lesser extent, consistent with the in vivo observation that the rate of cholesterol synthesis increases 28-fold in these transgenic nSREBP-2 livers, while fatty acid synthesis increases only fourfold. This increase in cholesterol synthesis is even more remarkable when one considers the extent of cholesterol overload in this tissue, which would ordinarily reduce SREBP processing and essentially abolish cholesterol synthesis (Table 1).

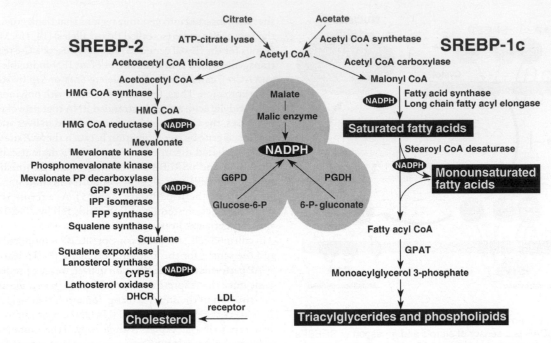

Figure 2
Genes regulated by SREBPs. The diagram shows the major metabolic intermediates in the pathways for synthesis of cholesterol, fatty acids, and triglycerides. In vivo, SREBP-2 preferentially activates genes of cholesterol metabolism, whereas SREBP-1c preferentially activates genes of fatty acid and triglyceride metabolism. DHCR, 7-dehydrocholesterol reductase; FPP, farnesyl diphosphate; GPP, geranylgeranyl pyrophosphate synthase; CYP51, lanosterol 14α-demethylase; G6PD, glucose-6-phosphate dehydrogenase; PGDH, 6-phosphogluconate dehydrogenase; GPAT, glycerol-3-phosphate acyltransferase.

We have also studied the consequences of overexpressing SREBP-1a, which is expressed only at low levels in the livers of adult mice, rats, hamsters, and humans (6). nSREBP-1a transgenic mice develop a massive fatty liver engorged with both cholesterol and triglycerides (9), with heightened expression of genes controlling cholesterol biosynthesis and, still more dramatically, fatty acid synthesis (Table 1). The preferential activation of fatty acid synthesis (26-fold increase) relative to cholesterol synthesis (fivefold increase) explains the greater accumulation of triglycerides in their livers. The relative representation of the various fatty acids accumulating in this tissue is also unusual. Transgenic nSREBP-1a livers contain about 65% oleate (C18:1), markedly higher levels than the 15–20% found in typical wild-type livers (8) — a result of the induction of fatty acid elongase and stearoyl-CoA desaturase-1 (7). Considered together, the overexpression studies indicate that both SREBP-1 isoforms show a relative preference for activating fatty acid synthesis, whereas SREBP-2 favors cholesterol.

The phenotype of animals lacking the *Srebp1* gene, which encodes both the SREBP-1a and -1c transcripts, also supports the notion of distinct hepatic functions for SREBP-1 and SREBP-2 (13). Most homozygous SREBP-1 knockout mice die in utero. The surviving *Srebp1⁻/⁻* mice show reduced synthesis of fatty acids, owing to reduced expression of mRNAs for fatty acid synthetic enzymes (Table 1). Hepatic nSREBP-2 levels increase in these mice, presumably in compensation for the loss of nSREBP-1. As a result, transcription of cholesterol biosynthetic genes increases, producing a threefold increase in hepatic cholesterol synthesis (Table 1).

The studies in genetically manipulated mice clearly show that, as in cultured cells, SCAP and S1P are required for normal SREBP processing in the liver. SCAP, acting through its sterol-sensing

domain, mediates feedback regulation of cholesterol synthesis. The SREBPs play related but distinct roles: SREBP-1c, the predominant SREBP-1 isoform in adult liver, preferentially activates genes required for fatty acid synthesis, while SREBP-2 preferentially activates the LDL receptor gene and various genes required for cholesterol synthesis. SREBP-1a and SREBP-2, but not SREBP-1c, are required for normal embryogenesis.

Transcriptional regulation of SREBP genes
Regulation of SREBPs occurs at two levels — transcriptional and posttranscriptional. The posttranscriptional regulation discussed above involves the sterol-mediated suppression of SREBP cleavage, which results from sterol-mediated suppression of the movement of the SCAP/SREBP complex from the ER to the Golgi apparatus (Figure 1). This form of regulation is manifest not only in cultured cells (1), but also in the livers of rodents fed cholesterol-enriched diets (19).

The transcriptional regulation of the SREBPs is more complex. SREBP-1c and SREBP-2 are subject to distinct forms of transcriptional regulation, whereas SREBP-1a appears to be constitutively expressed at low levels in liver and most other tissues of adult animals (6). One mechanism of regulation shared by SREBP-1c and SREBP-2 involves a feed-forward regulation mediated by SREs present in the enhancer/promoters of each gene (20, 21). Through this feed-forward loop, nSREBPs activate the transcription of their own genes. In contrast, when nSREBPs decline, as in *Scap* or *S1p* knockout mice, there is a secondary decline in the mRNAs encoding SREBP-1c and SREBP-2 (14, 15).

Three factors selectively regulate the transcription of SREBP-1c: liver X-activated receptors (LXRs), insulin, and glucagon.

Table 1
Alterations in hepatic lipid metabolism in gene-manipulated mice overexpressing or lacking SREBPs

Genetic manipulation	Amount of nSREBPs	Expression of target genes			Lipid synthesis		Liver content		Plasma levels	
		HMGR	FAS	LDLR	Chol	FA	Chol	TG	Chol	TG
		Fold difference relative to values in wild-type mice								
Transgenic mice										
SREBP-1a (9, 11)	↑1a	↑37	↑20	↑6	↑5	↑26	↑6	↑22	↓0.7	↓0.4
SREBP-1c (10, 11)	↑1c	n.c.	↑4	n.c.	n.c.	↑4	n.c.	↑4	n.c.	↓0.6
SREBP-2 (11)	↑2	↑75	↑15	↑6	↑28	↑4	↑3	↑4	n.c.	↓0.5
SCAP (D443N) (12)	↑1a, ↑1c, ↑2	↑18	↑11	↑2	↑5	↑7	↑6	↑9	↓0.5	↓0.5
Liver-specific knockout mice										
SCAP (14)	↓1a, ↓1c, ↓2	↓0.1	↓0.1	↓0.3	↓0.3	↓0.2	↓0.8	↓0.4	↓0.8	↓0.4
S1P (15)	↓1a, ↓1c, ↓2	↓0.6	↓0.3	↓0.5	↓0.3	↓0.3	n.c.	↓0.5	↓0.6	↓0.6
Germline knockout mice										
SREBP-1a & 1c (13)	↓1a, ↓1c, ↑2	↑2.2	↓0.7	↑1.3	↑3	↓0.6	↑1.5	n.c	↓0.7	↓0.7
SREBP-1c (16)	↓1a, ↓1c, ↑2	↑1.3	↓0.3	n.c	↑3	↓0.5	↑1.2	n.c	↓0.8	↓0.4
SREBP-2 (13)	Embryonic lethal	—	—	—	—	—	—	—	—	—
S1P (15, 17)	Embryonic lethal	—	—	—	—	—	—	—	—	—

HMGR, HMG-CoA reductase; FAS, fatty acid synthase; LDLR, LDL receptor; Chol, cholesterol; TG, triglycerides; n.c., no change.

LXRα and LXRβ, nuclear receptors that form heterodimers with retinoid X receptors, are activated by a variety of sterols, including oxysterol intermediates that form during cholesterol biosynthesis (22–24). An LXR-binding site in the *SREBP-1c* promoter activates SREBP-1c transcription in the presence of LXR agonists (23). The functional significance of LXR-mediated SREBP-1c regulation has been confirmed in two animal models. Mice that lack both LXRα and LXRβ express reduced levels of SREBP-1c and its lipogenic target enzymes in liver and respond relatively weakly to treatment with a synthetic LXR agonist (23). Because a similar blunted response is found in mice that lack SREBP-1c, it appears that LXR increases fatty acid synthesis largely by inducing SREBP-1c (16). LXR-mediated activation of SREBP-1c transcription provides a mechanism for the cell to induce the synthesis of oleate when sterols are in excess (23). Oleate is the preferred fatty acid for the synthesis of cholesteryl esters, which are necessary for both the transport and the storage of cholesterol.

LXR-mediated regulation of SREBP-1c appears also to be one mechanism by which unsaturated fatty acids suppress SREBP-1c transcription and thus fatty acid synthesis. Rodents fed diets enriched in polyunsaturated fatty acids manifest reduced SREBP-1c mRNA expression and low rates of lipogenesis in liver (25). In vitro, unsaturated fatty acids competitively block LXR activation of SREBP-1c expression by antagonizing the activation of LXR by its endogenous ligands (26). In addition to LXR-mediated transcriptional inhibition, polyunsaturated fatty acids lower SREBP-1c levels by accelerating degradation of its mRNA (27). These combined effects may contribute to the long-recognized ability of polyunsaturated fatty acids to lower plasma triglyceride levels.

SREBP-1c and the insulin/glucagon ratio
The liver is the organ responsible for the conversion of excess carbohydrates to fatty acids to be stored as triglycerides or burned in muscle. A classic action of insulin is to stimulate fatty acid synthesis in liver during times of carbohydrate excess. The action

of insulin is opposed by glucagon, which acts by raising cAMP. Multiple lines of evidence suggest that insulin's stimulatory effect on fatty acid synthesis is mediated by an increase in SREBP-1c. In isolated rat hepatocytes, insulin treatment increases the amount of mRNA for SREBP-1c in parallel with the mRNAs of its target genes (28, 29). The induction of the target genes can be blocked if a dominant negative form of SREBP-1c is expressed (30). Conversely, incubating primary hepatocytes with glucagon or dibutyryl cAMP decreases the mRNAs for SREBP-1c and its associated lipogenic target genes (30, 31).

In vivo, the total amount of SREBP-1c in liver and adipose tissue is reduced by fasting, which suppresses insulin and increases glucagon levels, and is elevated by refeeding (32, 33). The levels of mRNA for SREBP-1c target genes parallel the changes in SREBP-1c expression. Similarly, SREBP-1c mRNA levels fall when rats are treated with streptozotocin, which abolishes insulin secretion, and rise after insulin injection (29). Overexpression of nSREBP-1c in livers of transgenic mice prevents the reduction in lipogenic mRNAs that normally follows a fall in plasma insulin levels (32). Conversely, in livers of *Scap* knockout mice that lack all nSREBPs in the liver (14) or knockout mice lacking either nSREBP-1c (16) or both SREBP-1 isoforms (34), there is a marked decrease in the insulin-induced stimulation of lipogenic gene expression that normally occurs after fasting/refeeding. It should be noted that insulin and glucagon also exert a posttranslational control of fatty acid synthesis though changes in the phosphorylation and activation of acetyl-CoA carboxylase. The posttranslational regulation of fatty acid synthesis persists in transgenic mice that overexpress nSREBP-1c (10). In these mice, the rates of fatty acid synthesis, as measured by [³H]water incorporation, decline after fasting even though the levels of the lipogenic mRNAs remain high (our unpublished observations).

Taken together, the above evidence suggests that SREBP-1c mediates insulin's lipogenic actions in liver. Recent in vitro and in vivo studies involving adenoviral gene transfer suggest that SREBP-1c may also contribute to the regulation of glucose uptake and glucose synthesis. When overexpressed in hepatocytes, nSREBP-1c induces expression of glucokinase, a key enzyme in glucose utilization. It also suppresses phosphoenolpyruvate carboxykinase, a key gluconeogenic enzyme (35, 36).

SREBPs in disease

Many individuals with obesity and insulin resistance also have fatty livers, one of the most commonly encountered liver abnormalities in the US (37). A subset of individuals with fatty liver go on to develop fibrosis, cirrhosis, and liver failure. Evidence indicates that the fatty liver of insulin resistance is caused by SREBP-1c, which is elevated in response to the high insulin levels. Thus, SREBP-1c levels are elevated in the fatty livers of obese (*ob/ob*) mice with insulin resistance and hyperinsulinemia caused by leptin deficiency (38, 39). Despite the presence of insulin resistance in peripheral tissues, insulin continues to activate SREBP-1c transcription and cleavage in the livers of these insulin-resistant mice. The elevated nSREBP-1c increases lipogenic gene expression, enhances fatty acid synthesis, and accelerates triglyceride accumulation (31, 39). These metabolic abnormalities are reversed with the administration of leptin, which corrects the insulin resistance and lowers the insulin levels (38).

Metformin, a biguanide drug used to treat insulin-resistant diabetes, reduces hepatic nSREBP-1 levels and dramatically lowers the lipid accumulation in livers of insulin-resistant *ob/ob* mice (40). Metformin stimulates AMP-activated protein kinase (AMPK), an enzyme that inhibits lipid synthesis through phosphorylation and inactivation of key lipogenic enzymes (41). In rat hepatocytes, metformin-induced activation of AMPK also leads to decreased mRNA expression of SREBP-1c and its lipogenic target genes (41), but the basis of this effect is not understood.

The incidence of coronary artery disease increases with increasing plasma LDL-cholesterol levels, which in turn are inversely proportional to the levels of hepatic LDL receptors. SREBPs stimulate LDL receptor expression, but they also enhance lipid synthesis (1), so their net effect on plasma lipoprotein levels depends on a balance between opposing effects. In mice, the plasma levels of lipoproteins tend to fall when SREBPs are either overexpressed or underexpressed. In transgenic mice that overexpress nSREBPs in liver, plasma cholesterol and triglycerides are generally lower than in control mice (Table 1), even though these mice massively overproduce fatty acids, cholesterol, or both. Hepatocytes of nSREBP-1a transgenic mice overproduce VLDL, but these particles are rapidly removed through the action of LDL receptors, and they do not accumulate in the plasma. Indeed, some nascent VLDL particles are degraded even before secretion by a process that is mediated by LDL receptors (42). The high levels of nSREBP-1a in these animals support continued expression of the LDL receptor, even in cells whose cholesterol concentration is elevated. In LDL receptor–deficient mice carrying the nSREBP-1a transgene, plasma cholesterol and triglyceride levels rise tenfold (43).

Mice that lack all SREBPs in liver as a result of disruption of *Scap* or *S1p* also manifest lower plasma cholesterol and triglyceride levels (Table 1). In these mice, hepatic cholesterol and triglyceride synthesis is markedly reduced, and this likely causes a decrease in VLDL production and secretion. LDL receptor mRNA and LDL clearance from plasma is also significantly reduced in these mice, but the reduction in LDL clearance is less than the overall reduction in VLDL secretion, the net result being a decrease in plasma

lipid levels (15). However, because humans and mice differ substantially with regard to LDL receptor expression, LDL levels, and other aspects of lipoprotein metabolism, it is difficult to predict whether human plasma lipids will rise or fall when the SREBP pathway is blocked or activated.

SREBPs in liver: unanswered questions

The studies of SREBPs in liver have exposed a complex regulatory system whose individual parts are coming into focus. Major unanswered questions relate to the ways in which the transcriptional and posttranscriptional controls on SREBP activity are integrated so as to permit independent regulation of cholesterol and fatty acid synthesis in specific nutritional states. A few clues regarding these integration mechanisms are discussed below.

Whereas cholesterol synthesis depends almost entirely on SREBPs, fatty acid synthesis is only partially dependent on these proteins. This has been shown most clearly in cultured nonhepatic cells such as Chinese hamster ovary cells. In the absence of SREBP processing, as when the Site-2 protease is defective, the levels of mRNAs encoding cholesterol biosynthetic enzymes and the rates of cholesterol synthesis decline nearly to undetectable levels, whereas the rate of fatty acid synthesis is reduced by only 30% (44). Under these conditions, transcription of the fatty acid biosynthetic genes must be maintained by factors other than SREBPs. In liver, the gene encoding fatty acid synthase (*FASN*) can be activated transcriptionally by upstream stimulatory factor, which acts in concert with SREBPs (45). The *FASN* promoter also contains an LXR element that permits a low-level response to LXR ligands even when SREBPs are suppressed (46). These two transcription factors may help to maintain fatty acid synthesis in liver when nSREBP-1c is low.

Another mechanism of differential regulation is seen in the ability of cholesterol to block the processing of SREBP-2, but not SREBP-1, under certain metabolic conditions. This differential regulation has been studied most thoroughly in cultured cells such as human embryonic kidney (HEK-293) cells. When these cells are incubated in the absence of fatty acids and cholesterol, the addition of sterols blocks processing of SREBP-2, but not SREBP-1, which is largely produced as SREBP-1a in these cells (47). Inhibition of SREBP-1 processing requires an unsaturated fatty acid, such as oleate or arachidonate, in addition to sterols (47). In the absence of fatty acids and in the presence of sterols, SCAP may be able to carry SREBP-1 proteins, but not SREBP-2, to the Golgi apparatus. Further studies are necessary to document this apparent independent regulation of SREBP-1 and SREBP-2 processing and to determine its mechanism.

Acknowledgments

Support for the research cited from the authors' laboratories was provided by grants from the NIH (HL-20948), the Moss Heart Foundation, the Keck Foundation, and the Perot Family Foundation. J.D. Horton is a Pew Scholar in the Biomedical Sciences and is the recipient of an Established Investigator Grant from the American Heart Association and a Research Scholar Award from the American Digestive Health Industry.

Address correspondence to: Jay D. Horton, Department of Molecular Genetics, University of Texas Southwestern Medical Center, 5323 Harry Hines Boulevard, Room L5.238, Dallas, Texas 73590-9046, USA. Phone: (214) 648-9677; Fax: (214) 648-8804; E-mail: jay.horton@utsouthwestern.edu.

1. Brown, M.S., and Goldstein, J.L. 1997. The SREBP pathway: regulation of cholesterol metabolism by proteolysis of a membrane-bound transcription factor. *Cell.* **89**:331–340.

2. Horton, J.D., and Shimomura, I. 1999. Sterol regulatory element-binding proteins: activators of cholesterol and fatty acid biosynthesis. *Curr. Opin. Lipidol.* **10**:143–150.

3. Edwards, P.A., Tabor, D., Kast, H.R., and Venkateswaran, A. 2000. Regulation of gene expression by SREBP and SCAP. *Biochim. Biophys. Acta.* **1529**:103–113.

4. Sakakura, Y., et al. 2001. Sterol regulatory element-binding proteins induce an entire pathway of cholesterol synthesis. *Biochem. Biophys. Res. Comm.* **286**:176–183.

5. Goldstein, J.L., Rawson, R.B., and Brown, M.S. 2002. Mutant mammalian cells as tools to delineate the sterol regulatory element-binding protein pathway for feedback regulation of lipid synthesis. *Arch. Biochem. Biophys.* **397**:139–148.

6. Shimomura, I., Shimano, H., Horton, J.D., Goldstein, J.L., and Brown, M.S. 1997. Differential expression of exons 1a and 1c in mRNAs for sterol regulatory element binding protein-1 in human and mouse organs and cultured cells. *J. Clin. Invest.* **99**:838–845.

7. Moon, Y.-A., Shah, N.A., Mohapatra, S., Warrington, J.A., and Horton, J.D. 2001. Identification of a mammalian long chain fatty acyl elongase regulated by sterol regulatory element-binding proteins. *J. Biol. Chem.* **276**:45358–45366.

8. Shimomura, I., Shimano, H., Korn, B.S., Bashmakov, Y., and Horton, J.D. 1998. Nuclear sterol regulatory element binding proteins activate genes responsible for entire program of unsaturated fatty acid biosynthesis in transgenic mouse liver. *J. Biol. Chem.* **273**:35299–35306.

9. Shimano, H., et al. 1996. Overproduction of cholesterol and fatty acids causes massive liver enlargement in transgenic mice expressing truncated SREBP-1a. *J. Clin. Invest.* **98**:1575–1584.

10. Shimano, H., et al. 1997. Isoform 1c of sterol regulatory element binding protein is less active than isoform 1a in livers of transgenic mice and in cultured cells. *J. Clin. Invest.* **99**:846–854.

11. Horton, J.D., et al. 1998. Activation of cholesterol synthesis in preference to fatty acid synthesis in liver and adipose tissue of transgenic mice overproducing sterol regulatory element-binding protein-2. *J. Clin. Invest.* **101**:2331–2339.

12. Korn, B.S., et al. 1998. Blunted feedback suppression of SREBP processing by dietary cholesterol in transgenic mice expressing sterol-resistant SCAP(D443N). *J. Clin. Invest.* **102**:2050–2060.

13. Shimano, H., et al. 1997. Elevated levels of SREBP-2 and cholesterol synthesis in livers of mice homozygous for a targeted disruption of the SREBP-1 gene. *J. Clin. Invest.* **100**:2115–2124.

14. Matsuda, M., et al. 2001. SREBP cleavage-activating protein (SCAP) is required for increased lipid synthesis in liver induced by cholesterol deprivation and insulin elevation. *Genes Dev.* **15**:1206–1216.

15. Yang, J., et al. 2001. Decreased lipid synthesis in livers of mice with disrupted Site-1 protease gene. *Proc. Natl. Acad. Sci. USA.* **98**:13607–13612.

16. Liang, G., et al. 2002. Diminished hepatic response to fasting/refeeding and liver X receptor agonists in mice with selective deficiency of sterol regulatory element-binding protein-1c. *J. Biol. Chem.* **277**:9520–9528.

17. Mitchell, K.J., et al. 2001. Functional analysis of secreted and transmembrane proteins critical to mouse development. *Nat. Genet.* **28**:241–249.

18. Hua, X., Nohturfft, A., Goldstein, J.L., and Brown, M.S. 1996. Sterol resistance in CHO cells traced to point mutation in SREBP cleavage activating protein (SCAP). *Cell.* **87**:415–426.

19. Shimomura, I., et al. 1997. Cholesterol feeding reduces nuclear forms of sterol regulatory element binding proteins in hamster liver. *Proc. Natl. Acad. Sci. USA.* **94**:12354–12359.

20. Sato, R., et al. 1996. Sterol-dependent transcriptional regulation of sterol regulatory element-binding protein-2. *J. Biol. Chem.* **271**:26461–26464.

21. Amemiya-Kudo, M., et al. 2000. Promoter analysis of the mouse sterol regulatory element-binding protein-1c gene. *J. Biol. Chem.* **275**:31078–31085.

22. Janowski, B.A., et al. 1999. Structural requirements of ligands for the oxysterol liver X receptors LXRα and LXRβ. *Proc. Natl. Acad. Sci. USA.* **96**:266–271.

23. Repa, J.J., et al. 2000. Regulation of mouse sterol regulatory element-binding protein-1c gene (SREBP-1c) by oxysterol receptors, LXRα and LXRβ. *Genes Dev.* **14**:2819–2830.

24. DeBose-Boyd, R.A., Ou, J., Goldstein, J.L., and Brown, M.S. 2001. Expression of sterol regulatory element-binding protein 1c (SREBP-1c) mRNA in rat hepatoma cells requires endogenous LXR ligands. *Proc. Natl. Acad. Sci. USA.* **98**:1477–1482.

25. Xu, J., Nakamura, M.T., Cho, H.P., and Clarke, S.D. 1999. Sterol regulatory element binding protein-1 expression is suppressed by dietary polyunsaturated fatty acids. *J. Biol. Chem.* **274**:23577–23583.

26. Ou, J., et al. 2001. Unsaturated fatty acids inhibit transcription of the sterol regulatory element-binding protein-1c (SREBP-1c) gene by antagonizing ligand-dependent activation of the LXR. *Proc. Natl. Acad. Sci. USA.* **98**:6027–6032.

27. Xu, J., Teran-Garcia, M., Park, J.H.Y., Nakamura, M.T., and Clarke, S.D. 2001. Polyunsaturated fatty acids suppress hepatic sterol regulatory element-binding protein-1 expression by accelerating transcript decay. *J. Biol. Chem.* **276**:9800–9807.

28. Foretz, M., Guichard, C., Ferre, P., and Foufelle, F. 1999. Sterol regulatory element binding protein-1c is a major mediator of insulin action on the hepatic expression of glucokinase and lipogenesis-related genes. *Proc. Natl. Acad. Sci. USA.* **96**:12737–12742.

29. Shimomura, I., et al. 1999. Insulin selectively increases SREBP-1c mRNA in livers of rats with streptozotocin-induced diabetes. *Proc. Natl. Acad. Sci. USA.* **96**:13656–13661.

30. Foretz, M., et al. 1999. ADD1/SREBP-1c is required in the activation of hepatic lipogenic gene expression by glucose. *Mol. Cell. Biol.* **19**:3760–3768.

31. Shimomura, I., et al. 2000. Decreased IRS-2 and increased SREBP-1c lead to mixed insulin resistance and sensitivity in livers of lipodystrophic and *ob/ob* mice. *Mol. Cell.* **6**:77–86.

32. Horton, J.D., Bashmakov, Y., Shimomura, I., and Shimano, H. 1998. Regulation of sterol regulatory element binding proteins in livers of fasted and refed mice. *Proc. Natl. Acad. Sci. USA.* **95**:5987–5992.

33. Kim, J.B., et al. 1998. Nutritional and insulin regulation of fatty acid synthetase and leptin gene expression through ADD1/SREBP1. *J. Clin. Invest.* **101**:1–9.

34. Shimano, H., et al. 1999. Sterol regulatory element-binding protein-1 as a key transcription factor for nutritional induction of lipogenic enzyme genes. *J. Biol. Chem.* **274**:35832–35839.

35. Becard, D., et al. 2001. Adenovirus-mediated overexpression of sterol regulatory element binding protein-1c mimics insulin effects on hepatic gene expression and glucose homeostasis in diabetic mice. *Diabetes.* **50**:2425–2430.

36. Chakravarty, K., et al. 2001. Sterol regulatory element-binding protein-1c mimics the negative effect of insulin on phosphoenolpyruvate carboxykinase (GTP) gene transcription. *J. Biol. Chem.* **276**:34816–34823.

37. Marchesini, G., et al. 2001. Nonalcoholic fatty liver disease. *Diabetes.* **50**:1844–1850.

38. Shimomura, I., Hammer, R.E., Ikemoto, S., Brown, M.S., and Goldstein, J.L. 1999. Leptin reverses insulin resistance and diabetes mellitus in mice with congenital lipodystrophy. *Nature.* **401**:73–76.

39. Shimomura, I., Bashmakov, Y., and Horton, J.D. 1999. Increased levels of nuclear SREBP-1c associated with fatty livers in two mouse models of diabetes mellitus. *J. Biol. Chem.* **274**:30028–30032.

40. Lin, H.Z., et al. 2000. Metformin reverses fatty liver disease in obese, leptin-deficient mice. *Nat. Med.* **6**:998–1003.

41. Zhou, G., et al. 2001. Role of AMP-activated protein kinase in mechanism of metformin action. *J. Clin. Invest.* **108**:1167–1174. DOI:10.1172/JCI200113505.

42. Gillian-Daniel, D.L., Bates, P.W., Tebon, A., and Attie, A.D. 2002. Endoplasmic reticulum localization of the low density lipoprotein receptor mediates pre-secretory degradation of apolipoprotein B. *Proc. Natl. Acad. Sci. USA.* **99**:4337–4342.

43. Horton, J.D., Shimano, H., Hamilton, R.L., Brown, M.S., and Goldstein, J.L. 1999. Disruption of LDL receptor gene in transgenic SREBP-1a mice unmasks hyperlipidemia resulting from production of lipid-rich VLDL. *J. Clin. Invest.* **103**:1067–1076.

44. Pai, J., Guryev, O., Brown, M.S., and Goldstein, J.L. 1998. Differential stimulation of cholesterol and unsaturated fatty acid biosynthesis in cells expressing individual nuclear sterol regulatory element binding proteins. *J. Biol. Chem.* **273**:26138–26148.

45. Latasa, M.-J., Moon, Y.S., Kim, K.-H., and Sul, H.S. 2000. Nutritional regulation of the fatty acid synthase promoter *in vivo*: sterol regulatory element binding protein functions through an upstream region containing a sterol regulatory element. *Proc. Natl. Acad. Sci. USA.* **97**:10619–10624.

46. Joseph, S.B., et al. 2002. Direct and indirect mechanisms for regulation of fatty acid synthase gene expression by LXRs. *J. Biol. Chem.* **277**:11019–11025.

47. Hannah, V.C., Ou, J., Luong, A., Goldstein, J.L., and Brown, M.S. 2001. Unsaturated fatty acids down-regulate SREBP isoforms 1a and 1c by two mechanisms in HEK-293 cells. *J. Biol. Chem.* **276**:4365–4372.

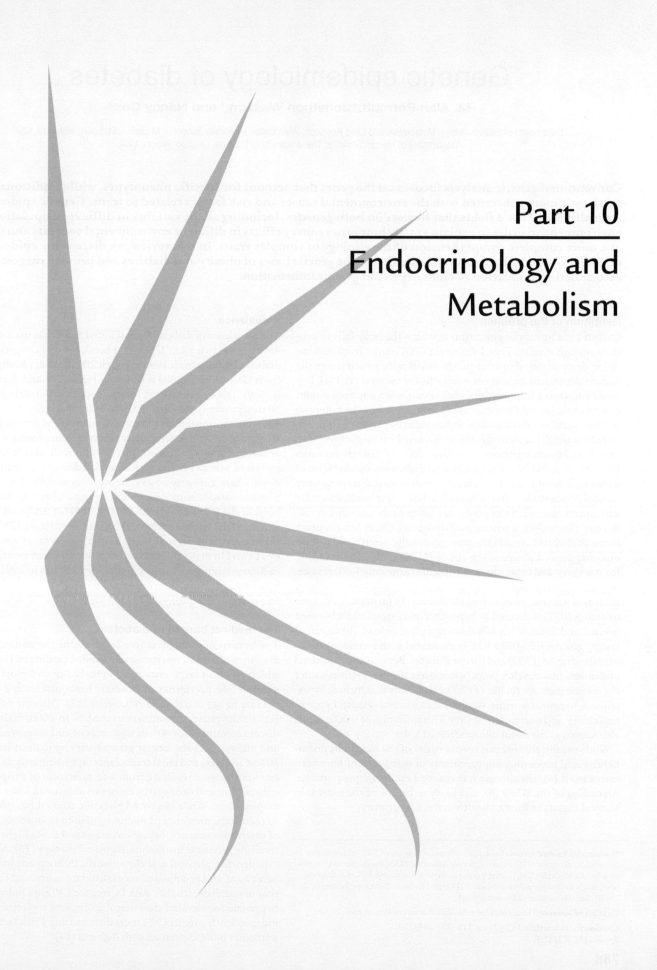

Part 10
Endocrinology and Metabolism

Genetic epidemiology of diabetes

M. Alan Permutt,[1] Jonathon Wasson,[1] and Nancy Cox[2]

[1]Department of Endocrinology, Metabolism and Lipid Research, Washington University School of Medicine, St. Louis, Missouri, USA.
[2]Department of Human Genetics, The University of Chicago, Chicago, Illinois, USA.

Conventional genetic analysis focuses on the genes that account for specific phenotypes, while traditional epidemiology is more concerned with the environmental causes and risk factors related to traits. Genetic epidemiology is an alliance of the 2 fields that focuses on both genetics, including allelic variants in different populations, and environment, in order to explain exactly how genes convey effects in different environmental contexts and to arrive at a more complete comprehension of the etiology of complex traits. In this review, we discuss the epidemiology of diabetes and the current understanding of the genetic bases of obesity and diabetes and provide suggestions for accelerated accumulation of clinically useful genetic information.

Definition of the problem

Diabetes is a metabolic condition in which the body fails to produce enough insulin. Type 1 diabetes (T1D) results from autoimmune destruction of insulin-producing β cells, which leaves the patient dependent on insulin injections for survival (1) T2D, formerly known as adult-onset diabetes, occurs when impaired insulin effectiveness (insulin resistance) is accompanied by the failure to produce sufficient β cell insulin. Patients can be placed on regimens to reduce weight or manage diet or treated with medication and, less often, insulin injections. This latter form of diabetes accounts for as much as 95% of cases. Gestational diabetes is another form of diabetes, defined as a state of glucose intolerance during pregnancy that usually subsides after delivery but has major implications for subsequent risk of T2D, as pregnancy serves as an "environmental" stressor that reveals a genetic predisposition. Other less common forms of diabetes include the rare, genetically determined disease maturity onset diabetes of the young (MODY), diabetes resulting from surgery, and other illnesses that constitute only 1–5% of cases. Based on plasma glucose measurements, 2 conditions have been identified with increased risk of the disease (2): (a) impaired glucose tolerance (IGT) is defined as hyperglycemia intermediate between normal and diabetic levels following a glucose load; (b) impaired fasting glucose (IFG), like IGT, is associated with increased cardiovascular disease (CVD) and future diabetes. Because complications of diabetes may develop years before overt disease, many consider the disease part of a cluster of CVD risk factors that include hypertension, hyperinsulinemia, dyslipidemia, visceral obesity, hypercoagulability, and microalbuminuria. This collection of risk factors is also known as the metabolic syndrome (3, 4).

While insulin therapy can reverse many of the metabolic disturbances, and numerous improvements in management have been introduced (5), the disease has reached epidemic proportions. According to the WHO (6), it is likely to be one of the most substantial threats to human health in the 21st century.

Prevalence

The prevalence of diabetes in the United States has risen 40%, from 4.9% in 1990 to 6.9% in 1999 (7). A breakdown of the prevalence of diabetes by state from 1990 through 2001 and of obesity by state from 1991 through 2003 is shown in Figure 1, A and B, respectively (8, 9). The disease affects various groups differently, occurring 10 times more commonly in those older than 65 years compared with those younger than 45 years. Minority racial groups including Hispanics, African Americans, and Native Americans are generally affected at a rate 2–4 times that for white individuals. The recent increased prevalence has also been noted in children and adolescents, where T2D may now occur more commonly than T1D (10). The estimated lifetime risk of developing diabetes for individuals born in the United States in 2000 is 33% for males and 39% for females (7). It is highest among Hispanic females, at 53%. Diabetes is associated with large reductions in life expectancy, on the order of 11 years in males diagnosed at age 40. While an estimated 18.2 million persons had diabetes in the United States in 2002 (11), diabetes worldwide has been estimated to affect 151 million persons, and that number projected to increase to 324 million by 2025 (2).

The medical burden of diabetes

The burden of diabetes is to a large extent the consequence of macrovascular and microvascular complications of the disease, which result in large increases in morbidity and mortality. For example, the prevalence of ischemic heart disease is 2–14 times the rate in age-matched nondiabetics (12). Diabetic retinopathy is the chief cause of blindness in the US. In 2000, diabetic renal disease accounted for 40% of new cases of end-stage renal disease, and diabetics are the largest group receiving dialysis (more than 50% of all cases) and renal transplants (approximately 25%). Lower extremity disease resulting from a combination of peripheral vascular disease and neuropathy causes an increase in lower extremity amputations. While improved glycemic control has been shown to reduce the incidence of microvascular complications, episodes of severe symptomatic hypoglycemia were 3 times higher in those receiving intensive insulin management therapy (13). Along with experiencing physical and cognitive disabilities, adults with diabetes have an age-adjusted mortality rate estimated to be twice that of nondiabetics (12). Risk factors for CVD, including systolic hypertension, elevated cholesterol levels, and cigarette smoking, independently predict CVD mortality, and any 1 risk factor affects outcomes more in persons with diabetes (14).

Nonstandard abbreviations used: ASP, affected sibling pair; CVD, cardiovascular disease; DZ, dizygotic; IGT, impaired glucose tolerance; IRAS, Insulin Resistance Atherosclerosis Study; MODY, maturity onset diabetes of the young; MZ, monozygotic; SNP, single nucleotide polymorphism; T1D, type 1 diabetes; T2D, type 2 diabetes; VNTR, variable number of tandem repeats.

Conflict of interest: The authors have declared that no conflict of interest exists.

Citation for this article: *J. Clin. Invest.* **115**:1431–1439 (2005).
doi:10.1172/JCI24758.

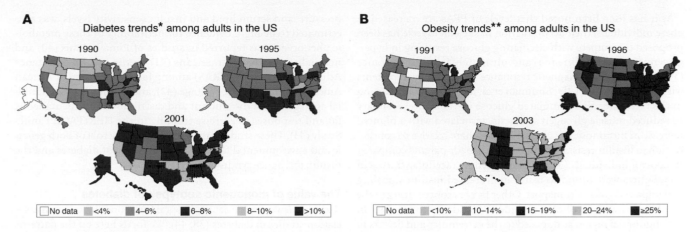

Figure 1
(**A**) Diabetes trends among adults in the US. *Includes gestational diabetes. Adapted from ref. 8. (**B**) Obesity trends among US adults. **BMI ≥ 30 (about 30 pounds overweight for a 5-ft 4-in. individual). Adapted from ref. 9.

Although increase in diabetes prevalence occurs mostly in middle-aged and older adults, there is strong evidence of an increase in the prevalence of T2D in children (10). For example, in Japan the incidence in school children (6–15 years old) has doubled over a 20-year period, such that T2D is now more common than T1D (15). In the US, up to 45% of the newly diagnosed diabetics in the pediatric age group have T2D (10). This rise in diabetes rates in children reflects, at least in part, the growing prevalence of obesity in this age group (16).

Direct medical expenditures and lost productivity due to diabetes were estimated to cost the US $132 billion in 2002 (17). The per capita expenditures were twice those for individuals without the disease. While the prevalence of diagnosed diabetes is less than 5% of the population, almost $1 of every $5 spent on health care in the US is for patients with diabetes. As the prevalence of diabetes increases with age, and because of the increasing diabetes-prone populations, it has been estimated that the number of diagnosed cases will increase. Thus the projected total cost in 2002 dollars could be as high as $192 billion by 2020.

Etiology of the diabetes epidemic

The sudden increase in diabetes in the last few years is due not to genetic factors but rather to the increase in obesity. This phenomenon is currently being documented in Africa, where the incidence of diabetes is rising with urbanization. The incidence is also rising among Africans who have immigrated to the US (18, 19). Epidemiological studies have regularly shown the relationship between diabetes and obesity, mediated in part by nutritional and lifestyle factors (20, 21). The most common measure of obesity, body mass index (BMI), combines measurements of height and weight. People with a BMI greater than 25 are said to be overweight, while those with a BMI greater than 30 are defined as obese (22). The Nurses Health Study showed that the risk for developing diabetes increased sharply for individuals observed as having a BMI greater than 23 for 16 years and was increased 20-fold for those with a BMI greater than 30 (18). In a recent study of measures of obesity and CVD risk factors in Australian adults, the prevalence of T2D rose from 5% in normal-weight to 16% in obese males; of hypertension, from 20% to 49%; and of dyslipidemia, from 18% to 61%, with even higher prevalence in females (23).

The molecular and physiological relationships between obesity and diabetes are not fully understood, and this subject is an area of intense investigation (see ref. 22 for review). The "thrifty genotype" hypothesis was proposed to account for a genetic advantage of accelerated fat deposition during times of restricted availability of calories, which leaves individuals faced with harmful consequences given the abundant food supply and reduced levels of physical activity in developed countries today (24). Noting an association between low birth weight and increased incidence of diabetes in later life, Hales and Barker have hypothesized that intrauterine malnutrition result in reduced birth weight and to subsequent changes leading to disease in adults (25, 26). This phenomenon, also known as the "thrifty phenotype" hypothesis, proposes that fetal malnutrition results in impaired pancreatic β cell development and insulin resistance. Offspring are subsequently more prone to diabetes and the metabolic syndrome when exposed to abundant nutrition later in life. In this regard, the increased prevalence of T2D in offspring of diabetic mothers may be a consequence of environmental factors operating on a genetic background, i.e., an altered intrauterine environment superimposed on a genetic predisposition in the fetus. While epidemiological studies have confirmed these observations, virtually nothing is known of their mechanisms, and this is an active area of investigation (27). If the relationship between obesity and diabetes could be understood, or obesity effectively prevented with treatment, then therapies directed at these mechanisms might curtail the increasing incidence of the disease.

The relationship between obesity and diabetes has been extensively studied in inbred strains of mice (28). Mice from a single inbred strain fed a high-fat diet all became insulin resistant, yet only about half became both obese and diabetic. Interestingly, 10% became diabetic but resisted obesity, and 10% became obese but not diabetic. The mechanisms responsible are unlikely to be purely genetic, and the results are consistent with the hypothesis that epigenetic changes and stochastic factors contribute to the phenotypic diversity. More recently, ER stress was shown to be the etiology of obesity-induced insulin resistance and diabetes in experimental mouse models, and this mechanism promises to be a rewarding area of investigation in the near future (29). If the degree of ER stress varies among mice, then perhaps this could explain the phenotypic differences in mice that are genetically identical.

As it has long been noted that levels of FFAs are increased in obese individuals, their accumulation in skeletal muscle has been proposed to compete with circulating glucose resulting in hyperglycemia, hyperinsulinemia, and ultimately insulin resistance (30). Recently, using magnetic resonance spectroscopy in patients with obesity and/or T2D, Shulman et al. have shown a reduction in the rate of insulin-stimulated glucose metabolism secondary to reduced muscle glycogen synthesis, associated with a blunted increase in intramuscular glucose 6 phosphate relative to concentration in insulin-resistant offspring of diabetic parents compared to control individuals (31, 32). Increased intracellular fatty acid metabolites were found to result in decreased insulin signaling and impaired glucose transport. Other factors contributing to the insulin resistance of obesity include the tendency to store fat in the abdominal region as opposed to the extremities and defects in adipocyte fatty acid metabolism and mitochondrial fatty acid oxidation. While impaired mitochondrial activity in insulin-resistant offspring of patients with T2D has been observed (33), the genetic basis for reduced mitochondrial biogenesis has not been elucidated. Reduced mitochondrial activity results in reduced energy expenditure, obesity, increased intramuscular fatty acid accumulation, and insulin resistance and has therefore been incorporated into the thrifty gene hypothesis (32).

Heritability of diabetes

The recent increased prevalence of obesity and diabetes must be largely attributable to changes in nongenetic risk factors. Yet environmental aspects must certainly accelerate the disease in those with genetic predisposition. There is a clear need to understand the genetic basis for the regulation of food intake, energy expenditure, and variations in energy balance in various individuals. In the long run, it may be more beneficial to develop treatments based on these genetic mechanisms than to rely on the use of will power to modify lifestyle. Moreover, different aspects of environment may be more critical in different subsets of individuals. We know from the natural history of diabetes complications that when patients are first diagnosed, there may already be marked progression of microvascular and macrovascular complications (34). The overwhelming majority of obese individuals have insulin resistance, yet only 5–10% develop pancreatic β cell failure and diabetes. Discovering the genetic risk factors for the disease will likely have many positive consequences.

The familial occurrences of both T1D and T2D have been long noted. A sibling's risk of developing T1D (5–10%) is perhaps 12- to 100-fold greater than the risk in the general population (0.1–0.4%) (35). Concordance in monozygotic (MZ) twins has been consistently shown to be greater than that in dizygotic (DZ) twins (36). For T2D, the concordance among MZ twins has been observed to be 50–92%, higher than the 37% concordance in DZ twins (36). Thus while the relative risk to a sibling, a measure of the genetic contribution, is considerably greater for T1D than for T2D, the concordance and absolute risk are substantially greater for T2D, which perhaps underscores the importance of the environmental contribution to the latter (37).

Quantitative phenotypes related to glucose homeostasis are also known to be heritable (38). In families with an increased genetic susceptibility to T2D, heritability estimates for β cell function and features of the insulin resistance syndrome of 72% and 78%, respectively, were calculated (39). The heritability of other features of the insulin resistance syndrome, including BMI, blood pressure, and serum lipid and insulin sensitivity levels, was also estimated to be high. Evidence for heritability of these metabolic phenotypes was reported in studies of Pima Indians (40) and nondiabetic Japanese Americans (41); in the Insulin Resistance Atherosclerosis Study (IRAS) among family members of African American and Hispanic heritage (42); and in a study of the familial aggregation of the amount and distribution of subcutaneous fat and responses to exercise training in the HERITAGE Family Study (43). These studies strongly support the role of both genetic and environmental factors in the etiology of diabetes and the insulin resistance syndrome.

The value of monogenic subtypes of diabetes

Recent reviews provide detailed appraisals of both linkage and association studies in diabetes (38, 44). We focus here on the patterns that have emerged in the search for genetic risk factors for diabetes. Linkage mapping, positional cloning, and candidate gene studies have been most successful in the forms of diabetes with the simplest genetic models. Early studies characterizing diabetes resulting from mutations in *insulin* (*INS*) and *insulin receptor* (45) genes as well as the mitochondrial genome (46) provided important insights into glucose homeostasis, but it is the studies on MODY that provide the classic example of the successful application of genetics to diabetes (47, 48). Linkage mapping studies on MODY were quite successful, and within a few years, a combination of positional cloning and studies of positional candidate genes led to the identification of *glucokinase* (*GCK*) and *hepatocyte nuclear factors 4α and 1α* (*HNF4A* and *TCF1*) as genes in which a single mutation could lead to the complex metabolic phenotype of diabetes (see ref. 49 for a recent review). *GCK* is the rate-limiting enzyme in glucose metabolism; *HNF4A* and *TCF1* are transcription factors expressed in a variety of tissues, including the liver and pancreas where they work in a transcription regulatory network to regulate the expression of genes involved in glucose transport and metabolism (49).

Once these genes had been implicated in MODY, a number of other transcription factors within the same regulatory network were also found to be MODY genes. As noted above, mutations in MODY genes, at the *INS* and *INS receptor* genes, or in the mitochondrial genome collectively account for only a small proportion of diabetes — about 1–5% (50). However, the monogenic forms of diabetes provide important insights into how we should be thinking about the genetic components of a phenotype as broad and metabolically complex as that of diabetes. The genes in which a single mutation is sufficient to generate this phenotype play an absolutely central role in glucose homeostasis (e.g., *INS*, the *INS receptor*, *GCK*) or are capable of affecting the regulation of many genes that act within this overall pathway (e.g., transcription factors). From these studies we learned that glucose homeostasis is a balance between insulin production, determined by β cell mass and/or function, and insulin action. Some of the genes that have been shown to affect these processes either in humans or experimental animals are illustrated in Figure 2.

The general observations on the genes implicated in monogenic forms of diabetes fit well within the emerging theories for how biological networks might be expected to perform and suggest that a systematic application of network theory to the metabolic and regulatory pathways underlying glucose homeostasis may provide a fruitful avenue for prioritizing genes for future studies. Recent research suggests that a scale-free topology is a nearly universal feature of networks, whether we are considering air traffic patterns,

Normal glucose homeostasis

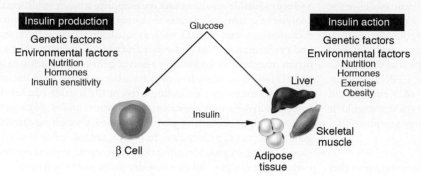

T1D

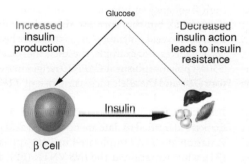

Early T2D

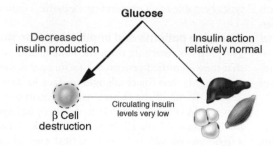

Late T2D

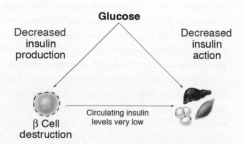

Figure 2
Diabetes results from an imbalance between the insulin-producing capacity of the islet β cell and the requirement for insulin action in insulin target tissues such as liver, adipose tissue. and skeletal muscle. Some of the many genes that have been shown or could possibly contribute to the imbalance are illustrated.

network theory with even the information we already have about the pathways implicated in glucose homeostasis, we could predict that genes/proteins located at hubs that are the most highly interconnected are those most vulnerable to degrading the overall system. Thus, genetic variation at hub genes may be more likely to lead to detectable perturbations in glucose homeostasis. While we might use network theory to prioritize genes for study simply based on the relative connectedness of the genes at the hubs in the networks, it might also be useful to overlay such an analysis on existing information we have about potential genetic risk factors for diabetes via linkage mapping or linkage disequilibrium mapping studies. A similar strategy was recently applied in studies of Alzheimer disease, with promising results (54).

Genetic studies on T1D

Genetic linkage studies of T1D and T2D have been quite variable, spanning the spectrum of results generally observed for disorders with complex inheritance. T1D is unique among complex disorders in the magnitude of the familial risk attributable to a single locus, HLA. While HLA was originally implicated through association studies as a candidate gene (55), the magnitude of the evidence for linkage at HLA in T1D is larger than has been observed for a linkage in any other complex disorder, although some other autoimmune disorders also have strong evidence for linkage in the HLA region (56). Because of the major contribution of HLA to the familial risk of T1D, identification of the other genetic risk factors may be more akin to identifying modifier loci for monogenic disorders than identifying primary susceptibility loci for complex disorders. Indeed, the non-HLA genes that have been reproducibly characterized as T1D susceptibility loci have been identified largely through candidate gene studies. Evidence for linkage at the non-HLA loci implicated in T1D has been uneven at best. For example, the very common class I alleles at *INS* variable number of tandem repeats (VNTR) are significantly overtransmitted from parents heterozygous for this allele to offspring affected with T1D (57). There was no evidence for linkage of the INS region to T1D in 100 affected sibling pair (ASP) families (58) and only modest evidence in more than 200 families (59). In a

the Internet, or biological pathways in complex organisms (51–53). A network with scale-free topology is characterized by that there is a relatively small number of hubs that have substantially more connections than average, along with a much larger number of nodes that have a very limited number of other connections. Such networks can be very stable and robust but are most vulnerable at those hubs that have the most connections to other nodes. Using large sample (767 families) including ASP families from both the US and the United Kingdom, there was significant evidence for linkage near INS (60). Similarly, initial evidence for linkage near CTLA4, another locus with support for affecting susceptibility to T1D through association studies, was detected in some individual samples, but little evidence for linkage in the region of CTLA4 was found in larger, combined samples (60). Finally, PTPN22, a locus recently implicated as a candidate gene in T1D (61), rheumatoid arthritis (62), and systemic lupus erythematosus (63), shows little evidence for linkage to T1D even in relatively large samples (60).

Genetic studies on T2D

The success of linkage mapping for T2D has been similar to that observed for other complex disorders, which has been, regrettably, quite limited. More than 25 genome-wide screens have been conducted on samples from all over the world (for review, see ref. 44). Despite the number of studies, there are only a few regions with replicated evidence for linkage: 1q, 3q, 8p, 10q, 12q, and 20q. Even in these regions, however, evidence for linkage is far from universal (3–7 of the more than 25 studies show nominally significant evidence for linkage), peaks are broad, and it is unlikely that all studies with linkage signal in a given region reflect the contribution of the same susceptibility genes.

Factors contributing to the complexity of analysis

Why has linkage mapping been relatively unsuccessful in localizing susceptibility genes for T1D and T2D in even relatively large, combined data sets? The complexity of the underlying genetic model is clearly a contributing factor. We have almost certainly underestimated the number of different genetic risk factors for both disorders and overestimated the magnitude of effect that might be expected for any one of these loci, excepting, perhaps, HLA in T1D. It might be argued that the initial success with HLA in T1D, the first susceptibility gene successfully linked to a complex disorder, encouraged us to establish a series of unrealistic expectations for how genetic risk factors contribute to diabetes and other complex disorders. Among the complexities of the genetic models for complex disorders that are likely to contribute to the difficulties in linkage mapping are gene-gene and gene-environment interactions. Such interactions are difficult to accommodate in primary linkage mapping studies and yet are a requisite part of the definition of a complex trait. It is not clear whether these problems could be solved by increasing either the sample sizes for linkage studies or the number of phenotypes examined. In particular, the failure to adequately measure and account for nongenetic factors affecting risk of diabetes almost certainly has reduced our ability to successfully map genetic risk factors.

The diagnosis of diabetes has long been standardized and is both reliably and inexpensively achieved with a simple blood test. These factors contributed substantially to making diabetes the first complex disorder to be widely studied using genetic tools. But the diagnosis of diabetes is designed to focus on the clinical consequences of elevated blood glucose levels rather than the underlying genetic liability to this very complex metabolic disease. The simple dichotomous diagnosis masks a tremendous amount of clinical heterogeneity, and it is likely that the genetic heterogeneity of diabetes is at least as great as the clinical heterogeneity. Thus, efforts to specify more genetically homogeneous samples according to clinical characteristics might also be fruitful. For example, stratifying T1D families for linkage analyses according to antibody positiv-

ity, or patients with T2D by BMI, might lead to more consistent and reproducible results in linkage mapping studies. Additionally, analysis of quantitative traits that may be related to the primary dichotomous trait of T2D, such as insulin resistance, β cell mass and performance, BMI and other features of the metabolic syndrome, may lead to the identification of genes contributing to risk of T2D (64). Several such studies have been already been initiated (see ref. 65 for review), including a recent IRAS study of quantitative traits in African American and Hispanic families (66), and the results point to promising genomic regions, though no causative genes have yet been identified. In this regard, use of animal models could help in gene identification, as syntenic regions are being evaluated in congenic strains in order to narrow regions conveying genetic risk for T2D, as, for example, in obese mice (67). We must also recognize, however, that the quantitative phenotypes that we now know how to measure easily are not necessarily the phenotypes best able to characterize the genetic liability to T2D.

Emerging patterns and implications for study design

The patterns emerging from the linkage and association studies that have identified genetic risk factors for diabetes offer intriguing insights into the challenges we face in improving our study designs. Some of the factors are common but so low risk that they would be quite difficult to detect in linkage mapping studies. For example, the allele increasing risk of T2D at PPARG has a frequency of 0.85–0.95 in most of the world's populations (68) and is associated with very modest increase in risk. Similarly, the class I alleles (or polymorphisms in linkage disequilibrium with them) increasing risk for T1D at the INS VNTR are found at very high frequency in populations of European and Asian descent (0.70–0.85), but they increase risk only modestly (69). There are rarer amino acid polymorphisms that have been reliably associated with diabetes. For example, the allele increasing risk of T2D at the T504A polymorphism at CAPN10 ranges in frequency from 0.04 to 0.16 (70), and the allele increasing risk of T1D at the R620W polymorphism at PTPN22 ranges in frequency from 0.08 to 0.14 (69). But many of the polymorphisms associated with increased risk of diabetes identified to date are not amino acid polymorphisms. The variation at CTLA4 implicated in T1D appears to affect splicing (71), while variation at the INS VNTR (69), at CAPN10 (72), and at HNF4A (73, 74) (as discussed in more detail below) may affect gene expression. Linkage mapping, even with very large samples, will miss many of these risk factors. Similarly, genome-wide association mapping focused exclusively on common haplotypes will miss many of the rarer risk alleles. Strategies targeting known amino acid polymorphisms will miss rare, unknown susceptibility variants and may not detect the effects of the more common noncoding sequence polymorphisms either. Until it becomes clearer whether there will be a predominant frequency spectrum or polymorphism type in the genetic variation affecting susceptibility to diabetes (whether type 1 or 2), it seems prudent to adopt strategies that enable detection of susceptibility alleles with a wide range of frequencies and effects. Some known genes associated or linked with diabetes are listed in Table 1 (61, 70, 73, 74, 90–117).

Sample size for low-risk genes associations

Identification of susceptibility alleles for T1D, outside the HLA locus, and T2D, whether through positional cloning or in the context of studies on functional candidates, has been challenging. Initial positive results are usually only inconsistently replicated. For

Table 1

Some known genes associated or linked with diabetes by replication in at least 2 studies[A]

Type	Gene	Gene name	Function	SNP or allele or locus or marker	Refs.
MODY 1	HNF4A	Hepatocyte nuclear factor 4 α	Transcription factor	Mutations in 13 families	90, 91
MODY 2	GCK	Glucokinase	Glucose metabolism	130 different mutations described	92, 93
MODY 3	TCF1	Hepatocyte nuclear factor 1α	Transcription factor	120 different mutations described in all racial ethnic backgrounds	93, 94
MODY 4	IPF1	Insulin promoter factor 1	Transcription factor	Rare mutations; 1 family described	95
MODY 5	TCF2	Hepatocyte nuclear factor 1β	Transcription factor	Rare mutations	96, 97
MODY 6	NEUROD1	Neurogenic differentiation 1	Transcription factor	Mutations described in 2 families with autosomal dominant form	98
T1D	HLA	Human leukocyte antigen	Immune system regulation	Variants in multiple genes	99, 100
T1D	INS	Insulin	Involved in numerous aspects of metabolism	VNTR	99, 101
T1D	CTLA4	Cytoxic T-lymphocyte–associated protein 4	Immune system regulation	T17A	102, 103
T1D	PTPN22	Protein tyrosine phosphate, non-receptor type 22	Immune system regulation	SNP C1858T	61, 104
T2D	ABCC8	ATP-binding cassette, subfamily C, sulfonylurea receptor	Regulator of potassium channels and insulin release	SNPs in various exons	105, 106
T2D	CAPN10	Calpain 10	Protease	Various intronic SNP haplotypes	70, 107
T2D	GCGR	Glucagon receptor	Controls hepatic glucose production and insulin secretion	G40S	108, 109
T2D	GCK	Glucokinase	Glucose metabolism	Microsatellite in 3′ end of gene	110, 111
T2D	KCNJ11	Potassium inwardly-rectifying channel, subfamily J, member 11	Regulation of insulin secretion	E23K	112, 113
T2D	PPARG	Peroxisome proliferator-activated receptor γ	Transcription factor	P12A	114, 115
T2D	HNF4A	Hepatocyte nuclear factor 4 α	Transcription factor	P2 promoter SNPs	73, 74
T2D	SLC2A1	Glut 1	Glucose transporter	XbaI(−) restriction site	116, 117

[A]with the exception of MODY 4 and MODY 6.

example, conflicting results of different studies on the Pro12Ala polymorphism in the PPARG2 gene were resolved by analysis of large family and case control samples; and a meta-analysis of all published studies further demonstrated that this polymorphism does affect risk of T2D, but only to a small degree (60). Similarly the E23K polymorphism in the Kir6.2 subunit of the ATP-regulated potassium channel has been shown by meta-analysis to contribute a small but significant risk to the disease in the populations studied (75, 76). Except for HLA in T1D, the susceptibility alleles for T1D and T2D quite modestly affect risk of disease, which mandates the study of large sample sizes. This argues for large collaborative studies, wherein sample sizes will be on the order of thousands and replication will be conducted during the primary investigation rather than through the time-consuming process of publication of multiple individual studies.

Investigating potential regulatory regions of candidate genes through haplotype-tagged SNPs

While every nonsynonymous coding single nucleotide polymorphism (SNP) in candidate genes should be tested for possible contribution to disease susceptibility, 2 recent studies highlight the importance of conducting association studies with markers in potential regulatory regions. Earlier studies with SNPs in or near the coding region of HNF4A, a gene previously shown to be mutated in rare cases of MODY (77), had yielded no association with T2D (78). More recently, it was discovered that a second promoter exists 40 kb upstream of the gene (79, 80) and that SNPs in the region of this second promoter and in other parts of the

noncoding sequence of HNF4A were associated with T2D in Ashkenazi Jews (74) and in a sample in Finland (73). Results of follow-up studies in other populations may provide some confirmation of the association between T2D and noncoding SNPs at HNF4A and its regulatory regions (81), but we should not be surprised if some studies in even large replication samples fail to observe associations and should be equally prepared for the possibility that not all studies will identify the very same polymorphisms as showing association. The nature of regulatory variation virtually insures that effects attributable to one polymorphism might be attenuated by effects of a second polymorphism — thus, the cumulative effects of regulatory variants may be poorly predicted by the marginal effects measured for any individual variant.

DNA diagnostics and pharmacogenetics in clinical trials

The use of genomic tools provided by the Human Genome Project offers the opportunity to identify individuals at risk, classify subtypes of the disease, choose therapy based on more accurate diagnosis (82), more precisely delineate the environmental factors that contribute to the onset and progression of the disease and its complications, and monitor responses to therapy (83, 84). Recently, genetic information was applied to clinical diabetes management in a randomized crossover trial of gliclazide, an agent affecting insulin secretion, and metformin, an agent that enhances insulin action. Compared with patients with typical T2D, patients with diabetes caused by a particular MODY mutation in TCF1 (85) had a 4- to 5-fold greater response to gliclazide than to metformin. Another example was the recent finding of heterozygous mutations in the ATP-sensitive K$^+$ channel subunit of the *Kir6.2* gene in 7 of 11 patients with neonatal diabetes (86). This gene plays a critical role in glucose-stimulated insulin secretion. Remarkably, several patients who previously required insulin injections were taken off insulin and treated with oral medication, which again illustrates the efficacy of pharmacogenetics for treatment of some diabetics.

A large number of clinical trials for both T1D and T2D are currently being conducted. The Type 1 Diabetes Genetics Consortium (T1DGC; http://www.t1dgc.org) will organize international efforts to identify genes that determine an individual's risk of T1D through the identification of 2,500 new families with 2 or more affected siblings. To explore approaches to treatment of T2D in youth, the TODAY (Treatment Options for Type 2 Diabetes in Adolescents and Youth; http://www.TODAYstudy.org) study will enroll 750 children and teenagers that have recently been diagnosed with T2D. Participants will be assigned to groups for treatment aimed at weight reduction and increasing physical activity. The Look AHEAD (Action for Health in Diabetes) trial is a multicenter, randomized clinical trial that will examine the consequences of a lifestyle intervention designed to achieve and maintain weight loss over the long term through decreased caloric intake and increased exercise in 5,000 obese patients with T2D. The National Heart, Lung, and Blood Institute–led Action to Control Cardiovascular Risk in Diabetes (ACCORD) trial is designed to test the effects of glycemia and blood pressure control on major CVD events and the use of fibrate treatment to increase HDL cholesterol and lower triglycerides (http://www.accordtrial.org/public/index.cfm). The Bypass Angioplasty Revascularization Investigation 2 Diabetes (BARI 2D) trial (http://www.bari2d.org) addresses questions about therapy in adults with T2D and stable CAD who might be candidates for revascularization. These studies represent ideal opportunities to incorporate DNA diagnostic testing and assessment of variable responses to therapeutic interventions.

Large prospective cohort study for the effects of genes and environment on diabetes

While case-control studies have much to offer for the assessment of the interactions between genes and environmental factors, Francis Collins noted that clinically diagnosed cases may represent only the more severely affected individuals with the disease and highlighted the difficulties of selecting an unbiased control group (87). To more accurately quantify genetic contribution and population-wide risk, he proposed prospective, population-based cohort studies (88). In a case-control study, there would be no samples available from cases prior to the onset of disease to search for predictive markers. A prospective study of 200,000 people would likely yield more than 5,000 cases of diabetes, based on current prevalence, in addition to numerous other features of the metabolic syndrome such as obesity, hyperinsulinism, dyslipidemia, hypertension, and CVD. This population, if also studied with high-throughput, low-cost genotyping, could represent a major resource for physician-scientists to accelerate the incorporation of genetics into clinical medicine.

Unanswered questions and future opportunities

Many of the questions raised here cannot yet be answered. Will genome-wide association studies work in a way that genome-wide linkage mapping studies did not? Will the identification of more homogeneous subsets of patients be the key to making any real headway in either association or linkage mapping studies? Are nongenetic risk factors largely uniform, or are they perhaps as variable as genetic risk factors? Will we really get improvement in understanding with a "better" version of more of the same, or do we need to move to something qualitatively different? Genomic technology is advancing rapidly, and larger, higher-powered studies will soon be possible — these studies should allow us to address these questions. The challenge now is for clinical scientists to provide well-characterized populations with carefully recorded phenotypic and environmental data. This challenge will extend to the acquisition of new organizational skills to collate these data from many centers and provide integration with the large volume of genetic data soon to be generated (89). The opportunities are great for future diabetes genetic epidemiology research to provide clinically useful information, the most vital goal of the Human Genome Project.

Acknowledgments

The authors wish to acknowledge NIH grants DK049583 ("Metabolic Basis of NIDDM: A Sib Pair Analysis") and DK58026 ("International Type 2 Linkage Consortium"); and Ping An, Richard Bergman, Ernesto Bernal-Mizrachi, Rudy Leibel, and Mike Province for helpful discussions and suggestions regarding the manuscript. Corentin Cras-Meneure helped with illustrations as well as discussions.

Address correspondence to: M. Alan Permutt, Washington University School of Medicine, 660 South Euclid Avenue, Campus Box 8127, St. Louis, Missouri 63110-1010, USA. Phone: (314) 362-8680; Fax: (314) 747-2692; E-mail: apermutt@im.wustl.edu.

1. Porte, D., Sherwin, R.S., and Baron, A. 2003. *Ellenberg & Rifkin's diabetes mellitus*. McGraw-Hill. New York, New York, USA. 1047 pp.
2. Zimmet, P., Shaw, J., and Alberti, K.G. 2003. Preventing Type 2 diabetes and the dysmetabolic syndrome in the real world: a realistic view. *Diabet. Med.* **20**:693–702.
3. Misra, A., and Vikram, N.K. 2004. Insulin resistance syndrome (metabolic syndrome) and obesity in Asian Indians: evidence and implications. *Nutrition.* **20**:482–491.
4. Meigs, J.B., et al. 2003. Prevalence and characteristics of the metabolic syndrome in the San Antonio Heart and Framingham Offspring Studies. *Diabetes.* **52**:2160–2167.
5. Hirsch, I.B. 2004. Blood glucose monitoring technology: translating data into practice. *Endocr. Pract.* **10**:67–76.
6. WHO. Diabetes program. http://www.who.int/diabetes/en/.
7. Narayan, K.M., Boyle, J.P., Thompson, T.J., Sorensen, S.W., and Williamson, D.F. 2003. Lifetime risk for diabetes mellitus in the United States. *JAMA.* **290**:1884–1890.
8. Centers for Disease Control and Prevention. 2005. Diabetes maps. Maps — diabetes and gestational diabetes trends among adults in the united states, behavioral risk factor surveillance system: 1990, 1995 and 2001. http://www.cdc.gov/diabetes/statistics/maps/index.htm.
9. Centers for Disease Control and Prevention. 2005. Overweight and obesity: obesity trends: U.S. obesity trends 1985–2003. http://www.cdc.gov/nccdphp/dnpa/obesity/trend/maps/index.htm.
10. Alberti, G., et al. 2004. Type 2 diabetes in the young: the evolving epidemic: the international diabetes federation consensus workshop. *Diabetes Care.* **27**:1798–1811.
11. Centers for Disease Control and Prevention. 2005. National diabetes fact sheet: national estimates on diabetes. http://www.cdc.gov/diabetes/pubs/estimates.htm#prev.
12. Engelgau, M.M., et al. 2004. The evolving diabetes burden in the United States. *Ann. Intern. Med.* **140**:945–950.
13. The Diabetes Control and Complications Trial Research Group.1993. The effect of intensive treatment of diabetes on the development and progression of long-term complications in insulin-dependent diabetes mellitus. *N. Engl. J. Med.* **329**:977–986.
14. Stamler, J., Vaccaro, O., Neaton, J.D., and Wentworth, D. 1993. Diabetes, other risk factors, and 12-yr cardiovascular mortality for men screened in the Multiple Risk Factor Intervention Trial. *Diabetes Care.* **16**:434–444.
15. Kitagawa, T., Owada, M., Urakami, T., and Yamauchi, K. 1998. Increased incidence of non-insulin dependent diabetes mellitus among Japanese schoolchildren correlates with an increased intake of animal protein and fat. *Clin. Pediatr. (Phila.)* **37**:111–115.
16. Bhargava, S.K., et al. 2004. Relation of serial chang-

es in childhood body-mass index to impaired glucose tolerance in young adulthood. *N. Engl. J. Med.* **350**:865–875.

17. Hogan, P., Dall, T., and Nikolov, P. 2003. Economic costs of diabetes in the US in 2002. *Diabetes Care.* **26**:917–932.

18. Motala, A.A., Omar, M.A., and Pirie, F.J. 2003. Diabetes in Africa. Epidemiology of type 1 and type 2 diabetes in Africa. *J. Cardiovasc. Risk.* **10**:77–83.

19. Rotimi, C.N., et al. 2001. In search of susceptibility genes for type 2 diabetes in West Africa: the design and results of the first phase of the AADM study. *Ann. Epidemiol.* **11**:51–58.

20. Zimmet, P.Z. 1999. Diabetes epidemiology as a tool to trigger diabetes research and care. *Diabetologia.* **42**:499–518.

21. Hu, F.B., et al. 2001. Diet, lifestyle, and the risk of type 2 diabetes mellitus in women. *N. Engl. J. Med.* **345**:790–797.

22. Speakman, J.R. 2004. Obesity: the integrated roles of environment and genetics [review]. *J. Nutr.* **134**(8 Suppl.):2090S–2105S.

23. Dalton, M., et al. 2003. Waist circumference, waist-hip ratio and body mass index and their correlation with cardiovascular disease risk factors in Australian adults. *J. Intern. Med.* **254**:555–563.

24. Neel, J.V. 1999. The "thrifty genotype" in 1998 [review]. *Nutr. Rev.* **57**:S2–S9.

25. Hales, C.N., et al. 1991. Fetal and infant growth and impaired glucose tolerance at age 64. *BMJ.* **303**:1019–1022.

26. Hales, C.N., and Barker, D.J. 2001. The thrifty phenotype hypothesis. *Br. Med. Bull.* **60**:5–20.

27 NIH, Office of Extramural Research. 2003. The fetal basis of adult disease: role of the environment [program announcement PAR-03-121]. http://grants2.nih.gov/grants/guide/pa-files/PAR-03-121.html.

28. Burcelin, R., Crivelli, V., Dacosta, A., Roy-Tirelli, A., and Thorens, B. 2002. Heterogeneous metabolic adaptation of C57Bl/6J mice to high-fat diet. *Am. J. Physiol. Endocrinol. Metab.* **282**:E834–E842.

29. Ozcan, U., et al. 2004. Endoplasmic reticulum stress links obesity, insulin action, and type 2 diabetes. *Science.* **306**:457–461.

30. Randle, P.J., Garland, P.B., Hales, C.N., and Newsholme, E.A. 1963. The glucose fatty-acid cycle. Its role in insulin sensitivity and the metabolic disturbances of diabetes mellitus. *Lancet.* **1**:785–789.

31. Rothman, D.L., Shulman, R.G., and Shulman, G.I. 1992. 31P nuclear magnetic resonance measurements of muscle glucose-6-phosphate. Evidence for reduced insulin-dependent muscle glucose transport or phosphorylation activity in non-insulin-dependent diabetes mellitus. *J. Clin. Invest.* **89**:1069–1075.

32. Shulman, G.I. 2004. Unraveling the cellular mechanism of insulin resistance in humans: new insights from magnetic resonance spectroscopy. *Physiology (Bethesda).* **19**:183–190.

33. Petersen, K.F., Dufour, S., Befroy, D., Garcia, R., and Shulman, G.I. 2004. Impaired mitochondrial activity in the insulin-resistant offspring of patients with type 2 diabetes. *N. Engl. J. Med.* **350**:664–671.

34. Matthews, D.R. 1999. The natural history of diabetes-related complications: the UKPDS experience. United Kingdom Prospective Diabetes Study. *Diabetes Obes. Metab.* **1**(Suppl. 2):S7–S13.

35. Redondo, M.J., Fain, P.R., and Eisenbarth, G.S. 2001. Genetics of type 1A diabetes. *Recent Prog. Horm. Res.* **56**:69–89.

36. Beck-Nielsen, H., Vaag, A., Poulsen, P., and Gaster, M. 2003. Metabolic and genetic influence on glucose metabolism in type 2 diabetic subjects — experiences from relatives and twin studies. *Best Pract. Res. Clin. Endocrinol. Metab.* **17**:445–467.

37. Florez, J.C., Hirschhorn, J., and Altshuler, D. 2003. The inherited basis of diabetes mellitus: implications for the genetic analysis of complex traits.

Annu. Rev. Genomics Hum. Genet. **4**:257–291.

38. Poulsen, P., Kyvik, K.O., Vaag, A., and Beck-Nielsen, H. 1999. Heritability of type II (non-insulin-dependent) diabetes mellitus and abnormal glucose tolerance — a population-based twin study. *Diabetologia.* **42**:139–145.

39. Mills, G.W., et al. 2004. Heritability estimates for beta cell function and features of the insulin resistance syndrome in UK families with an increased susceptibility to type 2 diabetes. *Diabetologia.* **47**:732–738.

40. Hanson, R.L., et al. 2001. Family and genetic studies of indices of insulin sensitivity and insulin secretion in Pima Indians. *Diabetes Metab. Res. Rev.* **17**:296–303.

41. Austin, M.A., et al. 2004. Heritability of multivariate factors of the metabolic syndrome in nondiabetic Japanese Americans. *Diabetes.* **53**:1166–1169.

42. Henkin, L., et al. 2003. Genetic epidemiology of insulin resistance and visceral adiposity. The IRAS Family Study design and methods. *Ann. Epidemiol.* **13**:211–217.

43. Perusse, L., et al. 2000. Familial aggregation of amount and distribution of subcutaneous fat and their responses to exercise training in the HERITAGE family study. *Obes. Res.* **8**:140–150.

44. McCarthy, M.I. 2003. Growing evidence for diabetes susceptibility genes from genome scan data. *Curr. Diab. Rep.* **3**:159–167.

45. Musso, C., et al. 2004. Clinical course of genetic diseases of the insulin receptor (type A and Rabson-Mendenhall syndromes): a 30-year prospective. *Medicine (Baltimore).* **83**:209–222.

46. Maassen, J.A., et al. 2004. Mitochondrial diabetes: molecular mechanisms and clinical presentation. *Diabetes.* **53**(Suppl. 1):S103–S109.

47. Permutt, M.A., and Hattersley, A.T. 2000. Searching for type 2 diabetes genes in the post-genome era. *Trends Endocrinol. Metab.* **11**:383–393.

48. Shih, D.Q., and Stoffel, M. 2002. Molecular etiologies of MODY and other early-onset forms of diabetes. *Curr. Diab. Rep.* **2**:125–134.

49. Fajans, S.S., Bell, G.I., and Polonsky, K.S. 2001. Molecular mechanisms and clinical pathophysiology of maturity-onset diabetes of the young. *N. Engl. J. Med.* **345**:971–980.

50. Ledermann, H.M. 1995. Maturity-onset diabetes of the young (MODY) at least ten times more common in Europe than previously assumed? [letter.] *Diabetologia.* **38**:1482.

51. Jeong, H., Tombor, B., Albert, R., Oltvai, Z.N., and Barabasi, A.L. 2000. The large-scale organization of metabolic networks. *Nature.* **407**:651–654.

52. Barabasi, A.L., and Oltvai, Z.N. 2004. Network biology: understanding the cell's functional organization. *Nat. Rev. Genet.* **5**:101–113.

53. Almaas, E., Kovacs, B., Vicsek, T., Oltvai, Z.N., and Barabasi, A.L. 2004. Global organization of metabolic fluxes in the bacterium Escherichia coli. *Nature.* **427**:839–843.

54. Krauthammer, M., Kaufmann, C.A., Gilliam, T.C., and Rzhetsky, A. 2004. Molecular triangulation: bridging linkage and molecular-network information for identifying candidate genes in Alzheimer's disease. *Proc. Natl. Acad. Sci. U. S. A.* **101**:15148–15153.

55. Cudworth, A.G., and Woodrow, J.C. 1975. Evidence for HL-A-linked genes in "juvenile" diabetes mellitus. *Br. Med. J.* **3**:133–135.

56. John, S., et al. 2004. Whole-genome scan, in a complex disease, using 11,245 single-nucleotide polymorphisms: comparison with microsatellites. *Am. J. Hum. Genet.* **75**:54–64.

57. Spielman, R.S., McGinnis, R.E., and Ewens, W.J. 1993. Transmission test for linkage disequilibrium: the insulin gene region and insulin-dependent diabetes mellitus (IDDM). *Am. J. Hum. Genet.* **52**:506–516.

58. Cox, N.J., and Spielman, R.S. 1989. The insulin gene and susceptibility to IDDM. *Genet. Epidemiol.* **6**:65–69.

59. Concannon, P., et al. 1998. A second-generation screen of the human genome for susceptibility to insulin-dependent diabetes mellitus. *Nat. Genet.* **19**:292–296.

60. Cox, N.J., et al. 2001. Seven regions of the genome show evidence of linkage to type 1 diabetes in a consensus analysis of 767 multiplex families. *Am. J. Hum. Genet.* **69**:820–830.

61. Bottini, N., et al. 2004. A functional variant of lymphoid tyrosine phosphatase is associated with type I diabetes. *Nat. Genet.* **36**:337–338.

62. Begovich, A.B., et al. 2004. A missense single-nucleotide polymorphism in a gene encoding a protein tyrosine phosphatase (PTPN22) is associated with rheumatoid arthritis. *Am. J. Hum. Genet.* **75**:330–337.

63. Kyogoku, C., et al. 2004. Genetic association of the R620W polymorphism of protein tyrosine phosphatase PTPN22 with human SLE. *Am. J. Hum. Genet.* **75**:504–507.

64. Watanabe, R.M., et al. 1999. Genome wide linkage analysis of type 2 diabetes-related quantitative traits in the FUSION study. *Diabetes Abstract Book.* **48**(Suppl. 1):A46.

65. Hanson, R.L., and Knowler, W.C. 2003. Quantitative trait linkage studies of diabetes-related traits. *Curr. Diab. Rep.* **3**:176–183.

66. Rich, S.S., et al. 2004. Identification of quantitative trait loci for glucose homeostasis: the Insulin Resistance Atherosclerosis Study (IRAS) Family Study. *Diabetes.* **53**:1866–1875.

67. Valdar, W.S., Flint, J., and Mott, R. 2003. QTL fine-mapping with recombinant-inbred and in vitro heterogeneous stocks. *Mamm. Genome.* **14**:830–838.

68. Altshuler, D., et al. 2000. The common PPARg pro12ala polymorphism is associated with decreased risk of type 2 diabetes. *Nat. Genet.* **26**:76–80.

69. Barratt, B.J., et al. 2004. Remapping the insulin gene/IDDM2 locus in type 1 diabetes. *Diabetes.* **53**:1884–1889.

70. Weedon, M.N., et al. 2003. Meta-analysis and a large association study confirm a role for calpain-10 variation in type 2 diabetes susceptibility. *Am. J. Hum. Genet.* **73**:1208–1212.

71. Ueda, H., et al. 2003. Association of the T-cell regulatory gene CTLA4 with susceptibility to autoimmune disease. *Nature.* **423**:506–511.

72. Horikawa, Y., et al. 2000. Genetic variation in the calpain 10 gene (CAPN10) is associated with type 2 diabetes mellitus [erratum 2000, **26**:502]. *Nat. Genet.* **26**:163–175.

73. Silander, K., et al. 2004. Genetic variation near the hepatocyte nuclear factor-4α gene predicts susceptibility to type 2 diabetes. *Diabetes.* **53**:1141–1149.

74. Love-Gregory, L.D., et al. 2004. A common polymorphism in the upstream promoter region of the hepatocyte nuclear factor-4α gene on chromosome 20q is associated with type 2 diabetes and appears to contribute to the evidence for linkage in an Ashkenazi Jewish population. *Diabetes.* **53**:1134–1140.

75. Love-Gregory, L., et al. 2003. E23K single nucleotide polymorphism in the islet ATP-sensitive potassium channel gene (Kir6.2) contributes as much to the risk of type II diabetes in Caucasians as the PPARgamma Pro12Ala variant. *Diabetologia.* **46**:136–137.

76. Gloyn, A.L., et al. 2003. Large-scale association studies of variants in genes encoding the pancreatic beta-cell KATP channel subunits Kir6.2 (KCNJ11) and SUR1 (ABCC8) confirm that the KCNJ11 E23K variant is associated with type 2 diabetes. *Diabetes.* **52**:568–572.

77. Lausen, J., et al. 2000. Naturally occurring mutations in the human HNF4α gene impair the func-

tion of the transcription factor to a varying degree. *Nucleic Acids Res.* **28**:430–437.

78. Ghosh, S., et al. 1999. Type 2 diabetes: evidence for linkage on chromosome 20 in 716 Finnish affected sib pairs. *Proc. Natl. Acad. Sci. U. S. A.* **96**:2198–2203.

79. Thomas, H., et al. 2001. A distant upstream promoter of the HNF-4alpha gene connects the transcription factors involved in maturity-onset diabetes of the young. *Hum. Mol. Genet.* **10**:2089–2097.

80. Hansen, S.K., et al. 2002. Genetic evidence that HNF-1α-dependent transcriptional control of HNF-4α is essential for human pancreatic β cell function. *J. Clin. Invest.* **110**:827–833. doi:10.1172/JCI200215085.

81. Weedon, M.N., et al. 2004. Common variants of the hepatocyte nuclear factor-4 P2 promoter are associated with type 2 diabetes in the U.K. population. *Diabetes.* **53**:3002–3006.

82. Shepherd, M., and Hattersley, A.T. 2004. 'I don't feel like a diabetic any more': the impact of stopping insulin in patients with maturity onset diabetes of the young following genetic testing. *Clin. Med.* **4**:144–147.

83. Bentley, D.R. 2004. Genomes for medicine. *Nature.* **429**:440–445.

84. Bell, J. 2004. Predicting disease using genomics. *Nature.* **429**:453–456.

85. Pearson, E.R., et al. 2003. Genetic cause of hyperglycaemia and response to treatment in diabetes. *Lancet.* **362**:1275–1281.

86. Sagen, J.V., et al. 2004. Permanent neonatal diabetes due to mutations in KCNJ11 encoding Kir6.2: patient characteristics and initial response to sulfonylurea therapy. *Diabetes.* **53**:2713–2718.

87. Collins, F.S. 2004. The case for a US prospective cohort study of genes and environment. *Nature.* **429**:475–477.

88. Whelton, P.K., and Gordis, L. 2000. Epidemiology of clinical medicine. *Epidemiol. Rev.* **22**:140–144.

89. Altshuler, J.S., and Altshuler, D. 2004. Organizational challenges in clinical genomic research. *Nature.* **429**:478–481.

90. Yamagata, K., et al. 1996. Mutations in the hepatocyte nuclear factor-4α gene in maturity-onset diabetes of the young (MODY1). *Nature.* **384**:458–460.

91. Lindner, T., et al. 1997. Hepatic function in a family with a nonsense mutation (R154X) in the hepatocyte nuclear factor-4α/MODY1 gene. *J. Clin. Invest.* **100**:1400–1405.

92. Gidh-Jain, M., et al. 1993. Glucokinase mutations

associated with non-insulin-dependent (type 2) diabetes mellitus have decreased enzymatic activity: implications for structure/function relationships. *Proc. Natl. Acad. Sci. U. S. A.* **90**:1932–1936.

93. Barrio, R., et al. 2002. Nine novel mutations in maturity-onset diabetes of the young (MODY) candidate genes in 22 Spanish families. *J. Clin. Endocr. Metab.* **87**:2532–2539.

94. Yamagata, K., et al. 1996. Mutations in the hepatocyte nuclear factor-1α gene in maturity-onset diabetes of the young (MODY3). *Nature.* **384**:455–457.

95. Stoffers, D.A., Ferrer, J., Clarke, W.L., and Habener, J.F. 1997. Early-onset type-II diabetes mellitus (MODY4) linked to IPF1 [letter]. *Nat. Genet.* **17**:138–141.

96. Lindner, T.H., et al. 1999. A novel syndrome of diabetes mellitus, renal dysfunction and genital malformation associated with a partial deletion of the pseudo-POU domain of hepatocyte nuclear factor-1-beta. *Hum. Mol. Genet.* **8**:2001–2008.

97. Horikawa, Y., et al. 1997. Mutation in hepatocyte nuclear factor-1-beta gene (TCF2) associated with MODY [letter]. *Nat. Genet.* **17**:384–385.

98. Malecki, M.T., et al. 1999. Mutations in NEUROD1 are associated with the development of type 2 diabetes mellitus. *Nat. Genet.* **23**:323–328.

99. Davies, J.L., et al. 1994. A genome-wide search for human type 1 diabetes susceptibility genes. *Nature.* **371**:130–136.

100. Hashimoto, L., et al. 1994. Genetic mapping of a susceptibility locus for insulin-dependent diabetes mellitus on chromosome 11q. *Nature.* **371**:161–164.

101. Mein, C.A., et al. 1998. A search for type 1 diabetes susceptibility genes in families from the United Kingdom. *Nat. Genet.* **19**:297–300.

102. Donner, H., et al. 1997. CTLA4 Alanine-17 confers genetic susceptibility to Graves' disease and to type 1 diabetes mellitus. *J. Clin. Endocrinol. Metab.* **82**:143–146.

103. Marron, M.P., et al. 1997. Insulin-dependent diabetes mellitus (IDDM) is associated with CTLA4 polymorphisms in multiple ethnic groups. *Hum. Mol. Genet.* **6**:1275–1282.

104. Smyth, D., et al. 2004. Replication of an association between the lymphoid tyrosine phosphatase locus (LYP/PTPN22) with type 1 diabetes, and evidence for its role as a general autoimmunity locus. *Diabetes.* **53**:3020–3023.

105. Inoue, H., et al. 1996. Sequence variants in the sulfonylurea receptor (SUR) gene are associated with

NIDDM in Caucasians. *Diabetes.* **45**:825–831.

106. Hani, E.H., et al. 1997. Genetic studies of the sulfonylurea receptor gene locus in NIDDM and in morbid obesity among French Caucasians. *Diabetes.* **46**:688–694.

107. del Bosque-Plata, L., et al. 2004 Association of the calpain-10 gene with type 2 diabetes mellitus in a Mexican population. *Mol. Genet. Metab.* **81**:122–126.

108. Hager, J., et al. 1995. A missense mutation in the glucagon receptor gene is associated with non-insulin-dependent diabetes mellitus. *Nat. Genet.* **9**:299–304.

109. Gough, S.C., et al. 1995. Mutation of the glucagon receptor gene and diabetes mellitus in the UK: association or founder effect? *Hum. Mol. Genet.* **4**:1609–1612.

110. Chiu, K.C., et al. 1992. A genetic marker at the glucokinase gene locus for type 2 (non-insulin-dependent) diabetes mellitus in Mauritian Creoles. *Diabetologia.* **35**:632–638.

111. McCarthy, M.I., et al. 1994. Glucokinase gene polymorphisms: a genetic marker for glucose intolerance in a cohort of elderly Finnish men. *Diabetes Med.* **11**:198–204

112. Hani, E.H., et al. 1998. Missense mutations in the pancreatic islet beta cell inwardly rectifying K+ channel gene (KIR6.2/BIR): a meta-analysis suggests a role in the polygenic basis of type II diabetes mellitus in Caucasians. *Diabetologia.* **41**:1511–1115.

113. Gloyn, A.L., et al. 2001. Association studies of variants in promoter and coding regions of beta-cell ATP-sensitive K-channel genes SUR1 and Kir6.2 with Type 2 diabetes mellitus (UKPDS 53). *Diabetes Med.* **18**:206–212.

114. Deeb, S.S., et al. 1998. A Pro12Ala substitution in PPARgamma2 associated with decreased receptor activity, lower body mass index and improved insulin sensitivity. *Nat. Genet.* **20**:284–287.

115. Hara, K., et al. 2000. The Pro12Ala polymorphism in PPAR gamma2 may confer resistance to type 2 diabetes. *Biochem. Biophys. Res. Commun.* **271**:212–216.

116. Li, S.R., Baroni, M.G., Oelbaum, R.S., Stock, J., and Galton, D.J. 1988. Association of genetic variant of the glucose transporter with non-insulin-dependent diabetes mellitus. *Lancet.* **2**:368–370.

117. Tao, T., et al. 1995. HepG2/erythrocyte glucose transporter (GLUT1) gene in NIDDM: a population association study and molecular scanning in Japanese subjects. *Diabetologia.* **38**:942–947.

ATP-sensitive potassium channelopathies: focus on insulin secretion

Frances M. Ashcroft

University Laboratory of Physiology, Oxford University, Oxford, United Kingdom.

ATP-sensitive potassium (K$_{ATP}$) channels, so named because they are inhibited by intracellular ATP, play key physiological roles in many tissues. In pancreatic β cells, these channels regulate glucose-dependent insulin secretion and serve as the target for sulfonylurea drugs used to treat type 2 diabetes. This review focuses on insulin secretory disorders, such as congenital hyperinsulinemia and neonatal diabetes, that result from mutations in K$_{ATP}$ channel genes. It also considers the extent to which defective regulation of K$_{ATP}$ channel activity contributes to the etiology of type 2 diabetes.

General properties of ATP-sensitive potassium channels

Physiological roles

ATP-sensitive potassium (K$_{ATP}$) channels couple cell metabolism to electrical activity of the plasma membrane by regulating membrane K$^+$ fluxes (1). A reduction in metabolism opens K$_{ATP}$ channels, producing K$^+$ efflux, membrane hyperpolarization, and suppression of electrical activity. Conversely, increased metabolism closes K$_{ATP}$ channels. The consequent membrane depolarization stimulates electrical activity and may thereby trigger cellular responses such as the release of hormones and neurotransmitters, or muscle contraction.

Studies on isolated cells and tissues, and more recently on genetically modified mice and patients with mutations in K$_{ATP}$ channel genes, have demonstrated that K$_{ATP}$ channels play a multitude of physiological roles (1). They contribute to glucose homeostasis by regulating insulin secretion from pancreatic β cells (2–7), glucagon secretion from pancreatic α cells (8), somatostatin secretion from D cells (9), and GLP-1 secretion from L cells (10). In ventromedial hypothalamic neurons they mediate the counter-regulatory response to glucose (11), and in arcuate nucleus neurons they may be involved in appetite regulation (12). In these glucose-sensing cells, K$_{ATP}$ channels respond to fluctuating changes in blood glucose concentration. In many other tissues, however, they are largely closed under resting conditions and open only in response to ischemia, hormones, or neurotransmitters. In cardiac muscle and central neurons the resulting reduction in electrical activity helps protect against cardiac stress and brain seizures (13–17). K$_{ATP}$ channels are involved in ischemic preconditioning in heart (18) and the regulation of vascular smooth muscle tone (opening of K$_{ATP}$ channels leads to relaxation) (19–21). They also modulate electrical activity and neurotransmitter release at synapses

in many brain regions, including the hippocampus, substantia nigra, and hypothalamus (12, 22–27).

Given their critical role in regulating electrical excitability in many cells, it is perhaps not surprising that disruption of K$_{ATP}$ channel function can lead to disease. To date, mutations in K$_{ATP}$ channel genes have been shown to cause neonatal diabetes (7, 28–33), hyperinsulinemia (6, 34–40), and dilated cardiomyopathy (41) in humans. Studies on genetically modified mice have also implicated impaired K$_{ATP}$ channel function in Prinzmetal angina (20–21). This review will focus on recent advances in our understanding of the role of the K$_{ATP}$ channel in insulin secretory disorders.

The central importance of K$_{ATP}$ channels in insulin secretion was first established over 20 years ago (Figure 1) (2, 42). At substimulatory glucose concentrations, K$^+$ efflux through open K$_{ATP}$ channels keeps the β cell membrane at a negative potential at which voltage-gated Ca^{2+} channels are closed. An increase in plasma glucose stimulates glucose uptake and metabolism by the β cell, producing changes in cytosolic nucleotide concentrations that result in closure of K$_{ATP}$ channels. This leads to a membrane depolarization that opens voltage-gated Ca^{2+} channels, initiating β cell electrical activity, Ca^{2+} influx, a rise in intracellular calcium concentration ([Ca^{2+}]$_i$), and thereby exocytosis of insulin granules. Glucose metabolism has additional effects on insulin release (5, 43), but under physiological conditions K$_{ATP}$ channel closure is an essential step in stimulus-secretion coupling in β cells. This explains why sulfonylurea drugs, such as tolbutamide and glibenclamide, which close K$_{ATP}$ channels (44), are such effective insulin secretagogues and are widely used to treat type 2 diabetes. They bypass β cell metabolism by binding directly to the K$_{ATP}$ channel. Conversely, the K channel opener diazoxide opens K$_{ATP}$ channels and inhibits insulin secretion, independently of blood glucose levels (45).

Metabolic regulation

K$_{ATP}$ channels are subject to complex regulation by numerous cytosolic factors, the most important being the adenine nucleotides ATP and ADP, which interact with 2 sites on the channel, one inhibitory and the other stimulatory (46–49). These can be distinguished experimentally because nucleotide binding to the stimulatory, but not the inhibitory, site requires Mg^{2+} (47, 49). Thus, in the absence of Mg^{2+} only the inhibitory effect is observed. Within the cell, however, where Mg^{2+} is always present, channel activity is determined by the balance between the inhibitory and stimulatory effects of

Nonstandard abbreviations used: DEND syndrome, developmental delay, muscle weakness, epilepsy, dysmorphic features, and neonatal diabetes; GCK, glucokinase; GDH, glutamate dehydrogenase; HI, congenital hyperinsulinism of infancy; K$_{ATP}$, ATP-sensitive potassium (channel); MIDD, maternally inherited diabetes with deafness; MODY, maturity-onset diabetes of the young; PNDM, permanent neonatal diabetes mellitus; Po, open probability; SCHAD, L-3-hydroxyacyl-CoA dehydrogenase; SUR, sulfonylurea receptor; TNDM, transient neonatal diabetes mellitus.

Conflict of interest: The author has declared that no conflict of interest exists.

Citation for this article: *J. Clin. Invest.* **115**:2047–2058 (2005).
doi:10.1172/JCI25495.

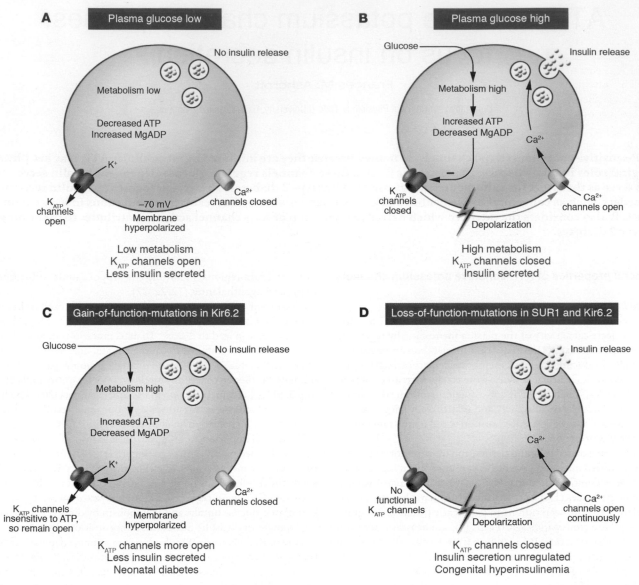

Figure 1

The K_{ATP} channel couples glucose metabolism to insulin secretion. Glucose enters the cell via the GLUT2 transporter, and via glycolytic and mitochondrial metabolism leads to an increase in ATP and a fall in MgADP in the immediate vicinity of the K_{ATP} channel. This results in K_{ATP} channel closure, membrane depolarization, opening of voltage-gated Ca^{2+} channels, Ca^{2+} influx, and exocytosis of insulin granules.

nucleotides. At nucleotide concentrations found in cells, ATP is predicted to block the channel and MgADP to reverse channel inhibition by ATP (42, 49). Consequently, reciprocal changes in the intracellular concentrations of ATP and MgADP are suggested to mediate metabolic regulation of the K_{ATP} channel.

Molecular structure

The K_{ATP} channel is an octameric complex of 4 Kir6.x and 4 SURx subunits (Figure 2) (50, 51). The pore-forming Kir6.x subunit belongs to the inwardly rectifying family of potassium channels (52–54). There are 2 isoforms: Kir6.1, which is found in vascular smooth muscle (54), and Kir6.2, which has a widespread tissue distribution (53). Binding of ATP to Kir6.x causes K_{ATP} channel closure (48, 55).

The sulfonylurea receptor (SUR) belongs to the ABC transporter family (56). It functions as a regulatory subunit, endowing the

channel with sensitivity to (a) stimulation by Mg nucleotides, via its 2 cytosolic nucleotide-binding domains (57, 58); (b) activation by K channel opener drugs (e.g., diazoxide); and (c) inhibition by sulfonylureas (e.g., glibenclamide, tolbutamide) (48, 56). Variations in SUR subunit composition produce K_{ATP} channels with different sensitivities to metabolism and drug regulation (22, 44). SUR1 is found in pancreas and brain (56), SUR2A in cardiac and skeletal muscle (59, 60), and SUR2B in a variety of tissues including brain and smooth muscle (61).

A digression on β cell electrical activity

Electrical activity of the β cell membrane, in the form of Ca^{2+}-dependent action potentials, is essential for insulin secretion: no insulin is secreted in its absence, and the extent of insulin secretion and that of electrical activity are directly correlated (62). The

K_{ATP} channel thus both initiates electrical activity and regulates its extent at suprathreshold glucose concentrations (2, 63). This is because the resting potential in pancreatic β cells is primarily controlled by the K_{ATP} channel.

The K_{ATP} channel also determines the electrical resistance of the β cell membrane, which is low when K_{ATP} channels are open and high when they are closed. The membrane potential is given by the product of the electrical resistance of the membrane and the current flowing across it. This means that when K_{ATP} channels are open and membrane resistance is low, small currents will affect the membrane potential (and thus insulin secretion) only minimally, whereas when K_{ATP} channels are (largely) closed and membrane resistance is high, the same current will elicit β cell depolarization, electrical activity, and insulin secretion. This explains why potentiators of insulin secretion such as acetylcholine and arginine, which produce *small* inward currents (64, 65), are effective secretagogues

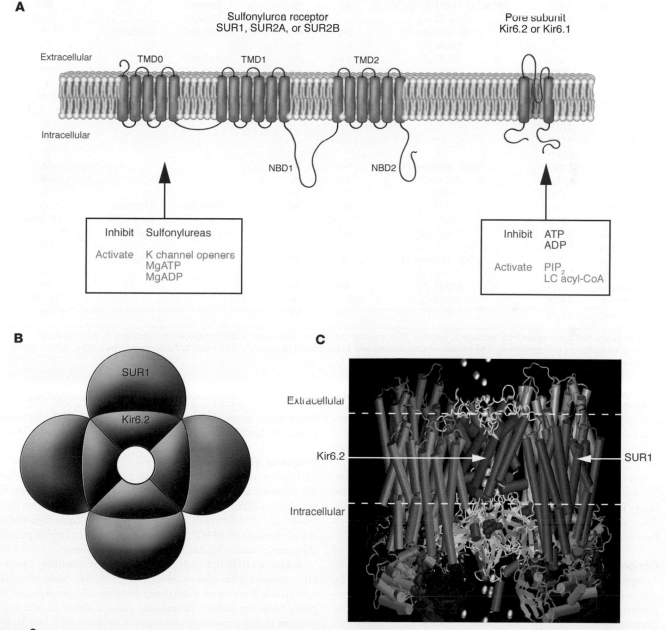

Figure 2
Molecular structure of the K_{ATP} channel. (**A**) Schematic representation of the transmembrane topology of a single SURx (left) or Kir6.x (right) subunit. Mg-nucleotide binding/hydrolysis at the nucleotide-binding domains (NBD1, NBD2) of SUR stimulates channel activity. Sulfonylureas (stimulatory) and K channel openers (inhibitory) also bind to SUR1. Binding of ATP or ADP to Kir6.2 closes the pore, an effect that does not require Mg^{2+}. Conversely, binding of phospholipids such as PIP_2, or long-chain (LC) acyl-coAs, stimulates K_{ATP} channel activity and decreases its ATP sensitivity. (**B**) Schematic representation of the octameric K_{ATP} channel complex viewed in cross section. Four Kir6.2 subunits come together to form the pore through which K^+ ions move, and each is associated with a regulatory SURx subunit. (**C**) Model of how SUR1 and Kir6.2 might assemble to form the K_{ATP} channel. The SUR model is described in ref. 115 and the Kir6.2 model in ref. 116. The model illustrates that the channel complex contains 4 ATP-binding sites (on Kir6.2) and 8 Mg nucleotide–binding sites (on SUR1).

Table 1

Kir6.2 mutations associated with insulin secretory disorders

Position	Mutation	Clinical phenotype			Functional effect		Reference
		Insulin secretion	Neurol. features	Epilepsy	ATP sensitivity	Po	
N terminus	Y12STOP	HI					36
N terminus	F35L	PNDM (1)	No	No			29
N terminus	F35V	PNDM (1)	No	No			28
N terminus	C42R	PNDM (1), TNDM (1), MODY (2)	No	No	Reduced		86
N terminus	R50P	PNDM (1)			Reduced		30, 89
N terminus	Q52R	DEND syndrome (1)	Yes	Yes	Reduced	Increased	7, 83
N terminus	G53N	PNDM (1)	No	No			29
N terminus	G53R	TNDM (2)	No	No	Reduced		33
N terminus	G53S	TNDM (3)	No	No	Reduced		33
Slide helix	V59G	DEND syndrome (1)	Yes	Yes	Reduced	Increased	7
Slide helix	V59M	Intermediate DEND (10)	Yes (7/10)	No	Reduced	Increased	7, 28, 30, 83
TMs	K67N	HI					117, 118
TMs	A101D	HI					38
TMs	R136L	HI					38
TMs	L147P	HI					37
TMs	C166F	DEND syndrome (1)	Yes	Yes			85
TMs	K170R	PNDM (1)	No	No			30
TMs	K170N	PNDM (1)	Yes	No			30
C terminus	I182V	TNDM (1)	No	No	Reduced	Increased	33
C terminus	R201C	PNDM (6)	Yes (2/6)	No	Reduced	Unaffected	7, 29, 30–32
C terminus	R201H	PNDM (12)	No	No	Reduced	Unaffected	7, 28, 29, 31, 83
C terminus	P254L	HI					118
C terminus	I296L	DEND syndrome (1)	Yes	Yes	Reduced	H increased	7, 84
C terminus	R301H	HI					38
C terminus	E322K	PNDM (1)	No	No			29
C terminus	Y330C	PNDM (3)	Yes (2/3)	No			28, 29
C terminus	F333I	PNDM (1)	No	No			28

Mutations in Kir6.2 that cause HI or monogenic diabetes. TMs, transmembrane domains; neurol. features, neurological features (i.e., developmental delay, epilepsy, or other symptoms). The numbers in parentheses refer to the number of cases. Position refers to the position of the residue in the primary sequence of the protein.

only at glucose concentrations that shut most K_{ATP} channels. It also implies that these agents will be *more* effective in disease states that lead to decreased K_{ATP} channel activity (congenital hyperinsulinism of infancy) and *less* effective under conditions that enhance K_{ATP} channel activity (diabetes).

The K_{ATP} channel plays a lesser role in regulating electrical activity of muscle and nerve cells because additional channels contribute to the resting membrane potential and membrane resistance. Thus, equivalent changes in K_{ATP} channel activity are expected to have less dramatic effects in these cell types.

Congenital hyperinsulinism of infancy

Congenital hyperinsulinism of infancy (HI) is characterized by continuous, unregulated insulin secretion despite severe hypoglycemia (6), and, without therapy, the hypoglycemia may cause irreversible brain damage. The disease usually presents at birth or within the first year of life. In the general population, the incidence is estimated as 1 in 50,000 live births, but this can be higher in isolated communities (e.g., 1 in 2,500 in the Arabian peninsula) (6, 34). Most cases of HI are sporadic, but familial forms have been also described, and the disease may result from homozygous, compound heterozygous, or heterozygous mutations (35, 36, 39, 40). Mild cases can be managed with diazoxide or even diet, but more severe forms require subtotal pancreatectomy (commonly about

90–95%). This results in pancreatic insufficiency, and a high incidence of iatrogenic diabetes. The main features of HI are summarized here; further details can be found in other reviews (6, 34, 66).

Functional effects of HI mutations

Mutations in 5 different genes can produce HI: the K_{ATP} channel subunits *SUR1* (6, 35) and Kir6.2 (*KCNJ11*) (36, 37), and the metabolic enzymes glucokinase (*GCK*) (66, 67), glutamate dehydrogenase (*GLUD1*) (68), and short-chain L-3-hydroxyacyl-CoA dehydrogenase (*SCHAD*) (69). However, in about half of patients the genetic basis of HI has not yet been determined.

Mutations in SUR1 (*ABCC8*) are the most common cause of HI, accounting for almost 50% of cases (6, 34). More than 100 mutations have been described, distributed throughout the gene. They comprise 2 functional classes: those in which the protein is not present in the surface membrane (class I), and those in which the channel is present in the plasma membrane but is always closed, independent of the metabolic state of the cell (class II). Class I mutations are characterized by loss of K_{ATP} channels in the plasma membrane, which may result from impaired SUR1 synthesis, abnormal SUR1 maturation, defective channel assembly, or faulty surface membrane trafficking (6, 70–72) (since SUR1 is required for surface expression of Kir6.2 [ref. 73], Kir6.2 is also absent). Class II mutations impair the ability of MgADP to stimulate channel activity (6, 39, 57), so that ATP inhibition becomes dominant and the K$_{ATP}$ channel is permanently closed even at low glucose concentrations. Many, but not all, class II mutations reside in the nucleotide-binding domains of SUR1. In general, class I mutations produce more severe disease, whereas some class II mutations cause a milder phenotype because a residual response to MgADP remains (39, 40). However, there is no precise genotype-phenotype correlation, and the same mutation can produce HI with differing degrees of severity in different people.

Mutations in Kir6.2 (*KCNJ11*) that cause HI are much rarer than those in SUR1 (Table 1) (36–38). They also act by reducing, or abolishing, K$_{ATP}$ channel activity in the surface membrane. Because the K$_{ATP}$ channel sets the β cell membrane potential, decreased channel activity produces a persistent membrane depolarization and continuous insulin secretion, irrespective of the blood glucose level (Figure 1).

The glycolytic enzyme glucokinase (GCK) and the mitochondrial enzyme glutamate dehydrogenase (GDH) play key roles in β cell metabolism. It is therefore reasonable to postulate that the gain-of-function mutations in these genes produce HI by increasing ATP synthesis and shifting the K$_{ATP}$ channel inhibition curve to lower glucose concentrations. As a result, K$_{ATP}$ channel activity would decrease, stimulating insulin secretion. Mutations in GDH cause an additional hyperammonemia and an enhanced insulin response to a protein meal or the amino acid leucine (68). The latter may be explained by allosteric stimulation of GDH by leucine, leading to enhanced ATP generation. With one exception (74), the mutations in *GCK* and *GLUD1* (encoding GDH) reported to date produce a mild form of HI that does not require pancreatectomy. It is still unclear how mutations in *SCHAD*, which is involved in mitochondrial fatty acid oxidation, cause HI.

Certain mutations in SUR1, for example R1353H, produce familial leucine-sensitive HI, in which hypoglycemia is provoked by leucine (40). The phenotype is less severe, consistent with the fact that mutant K$_{ATP}$ channels show a partial response to MgADP. As leucine stimulates β cell ATP production, it seems possible that some SUR1 (and GDH) mutations cause a partial β cell depolarization, which is insufficient to elicit electrical activity and insulin release in the resting state. However, further K$_{ATP}$ channel closure as a result of leucine-stimulated ATP production enhances membrane depolarization, thus eliciting insulin secretion. This hypothesis is supported by the fact that leucine can trigger insulin secretion in normal subjects, although not enough to provoke hypoglycemia, and that enhanced leucine sensitivity is produced by prior treatment with tolbutamide, which closes K$_{ATP}$ channels (75).

Although HI is often associated with neurological problems, it is difficult to assess whether these result from hypoglycemia prior to diagnosis and treatment, or are a direct consequence of the mutation itself. In many extrapancreatic tissues, K$_{ATP}$ channels are closed at rest, so loss-of-function mutations might be expected to have little effect. Furthermore, sulfonylurea therapy in type 2 diabetic patients has few side effects (though it is not clear whether these drugs access brain K$_{ATP}$ channels).

Implications for therapy

In general, mutations in SUR1 and Kir6.2 cause a severe form of HI that is refractory to diazoxide (6, 38) and requires subtotal pancreatectomy. This arises because K$_{ATP}$ channels are either absent or drug-resistant. In contrast, HI caused by mutations in *GCK*, *GLUD1*, or *SCHAD* responds well to diazoxide (6), as K$_{ATP}$ channel properties are normal. In these patients, opening of K$_{ATP}$ channels by diazoxide hyperpolarizes the β cell and so reduces electrical activity and insulin secretion despite high plasma glucose levels. Patients with *GCK* mutations may even be able to control their symptoms by eating regularly (76). This is because *GCK* mutations merely reset the threshold for glucose-stimulated insulin secretion, so that insulin release is still regulated, although the amount is abnormally high for the blood glucose level. Thus, genotyping HI patients may help determine correct therapy.

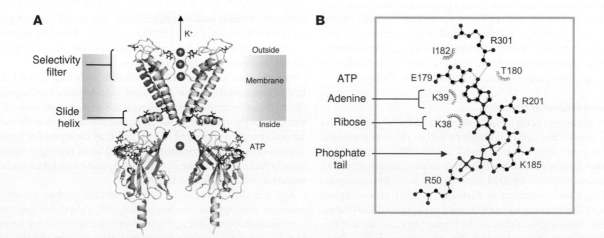

Figure 3
Location of disease-causing mutations in Kir6.2. (**A**) Structural model of Kir6.2 (116) viewed from the side. For clarity, only 2 transmembrane domains, and 2 separate cytosolic domains, are shown. Residues mutated in neonatal diabetes are shown in red, and those that cause hyperinsulinism of infancy in blue. ATP (green) is docked into its binding site. Of the residues implicated in neonatal diabetes, R50, R201, Y330C, and F333I lie close to the ATP-binding site; F35, C42, and E332K at the interface between Kir6.2 subunits; Q52 and G53 in a region postulated to interface with SUR1; and V59, C166, and I296L within regions of the channel involved in gating. (**B**) Close-up of the putative ATP-binding site with residues lying within 3.5 Å of ATP indicated. Residues mutated in neonatal diabetes are shown in red. Part **B** is adapted with permission from *The EMBO Journal* (116).

Table 2
Disease severity correlates with extent of MgATP block

Mutation	Phenotype	Fraction unblocked I-K_{ATP} at 1 mM MgATP	Reference
WT	None	2.4%	33
G53S	TNDM	6%	33
G53R	TNDM	11%	33
I182V	TNDM	12%	33
R201H	PNDM	22%	33
I296L	DEND syndrome	36%	84

Fraction of unblocked K_{ATP} current measured in the heterozygous state in the presence of 1 mM ATP for WT (Kir6.2/SUR1) and Kir6.2 mutant channels.

Both sulfonylureas and K channel openers can act as chemical chaperones and correct surface trafficking defects associated with some SUR1 mutations (71, 72). As the resulting K_{ATP} channels have normal nucleotide sensitivity, drugs with similar chaperone activity that do not block the channel could potentially be useful for treating HI. However, not all mutations can be rescued in this way, and the pharmacology varies with the mutation. Thus, the K channel opener diazoxide corrected trafficking of SUR-R1349H (an effect reversed by glibenclamide) (71), whereas the sulfonylureas glibenclamide and tolbutamide, but not diazoxide, restored surface expression of SUR1-A116P and SUR1-V187D (72). It may not be simple, therefore, to develop drugs that correct surface trafficking generically.

HI and diabetes

There is accumulating evidence that HI caused by some SUR1 mutations can progress to type 2 diabetes in later life (39, 40). This is hypothesized to occur because continuous β cell membrane depolarization, due to reduced K_{ATP} channel activity, causes a maintained influx of Ca^{2+} that activates apoptosis, thereby reducing β cell mass and insulin secretion. Studies on Kir6.2 knockout mice support this idea (4, 77). Except where no surgery has been carried out (39), it is difficult to distinguish whether SUR1 mutations predispose human patients to type 2 diabetes, or whether the diabetes results from early pancreatectomy. It is also uncertain whether HI mutations in other genes enhance susceptibility to diabetes.

Permanent neonatal diabetes

Neonatal diabetes mellitus is characterized by hyperglycemia requiring insulin therapy that develops within the first 6 months of life. It is a rare disorder affecting 1 in 400,000 live births (78), and it may be either transient (TNDM) or permanent (PNDM). Approximately 50% of PNDM cases result from heterozygous gain-of-function mutations in Kir6.2 (7, 28–33). Homozygous loss-of-function mutations in GCK are a rare cause of the disease (66, 79, 80) and are thought to act by impairing K_{ATP} channel closure indirectly, via reduced metabolic generation of ATP. Transient neonatal diabetes (TNDM), which resolves with a median of 3 months, is usually associated with an abnormality of imprinting in chromosome 6q24 (81). However, heterozygous mutations in Kir6.2 can produce a remitting relapsing form of neonatal diabetes that resembles TNDM (33).

To date, 20 mutations in 14 different residues have been reported to cause neonatal diabetes (Table 1). Their location in a structural model of Kir6.2 is shown in Figure 3. They form striking clusters in the putative ATP-binding site, at the inner end of the transmembrane domains (where the second transmembrane helices come together to form the inner mouth of the channel), and in cytosolic loops that link these 2 regions. This suggests that the PNDM mutations may provide clues to how ATP binding is translated into closure of the Kir6.2 pore. Two residues, V59 and R201, are mutated more commonly than others.

A range of phenotypes

Gain-of-function mutations in Kir6.2 cause a range of phenotypes, distinguished by increasing severity (Table 1). The most common class of mutation produces PNDM alone. These patients show minimal insulin secretion in response to i.v. glucose but may secrete insulin in response to sulfonylureas (7, 28, 32, 82, 83). Other classes of mutation cause more severe phenotypes. Some patients have delayed speech and walking and may show muscle weakness in addition to neonatal diabetes (7, 29, 30). A third class of mutations cause DEND syndrome, which is characterized by marked developmental delay, muscle weakness, epilepsy, dysmorphic features, and neonatal diabetes (7, 83–85). Mutations that cause less severe phenotypes than PNDM are also found. Thus, some mutations cause TNDM (33) and 1 mutation has been identified that causes diabetes of variable severity, ranging from TNDM to diabetes of adult onset (age 22–26) that resembles maturity-onset diabetes of the young (86).

Functional effects of the mutations

All PNDM mutations studied to date are gain-of-function mutations that act by reducing the ability of ATP to block the K_{ATP} channel and thereby increasing the K_{ATP} current amplitude under resting conditions (7, 33, 83, 84). There is a reasonable correlation between the clinical phenotype and the extent to which MgATP inhibition is reduced, when the heterozygous state is simulated in heterologous expression systems. Table 2 shows that, at physiologically relevant concentrations of MgATP (1–5 mM), mutations that cause small increases in K_{ATP} current result in TNDM, while larger increases in current cause PNDM alone, and an even greater increase is associated with DEND syndrome. It will be important to determine whether this correlation between K_{ATP} current and disease severity is a consistent feature.

In β cells, an increase in K_{ATP} current leads to a smaller depolarization in response to increased metabolism. Consequently, electrical activity and insulin secretion will be diminished, and the greater the increase in K_{ATP} current, the more severely insulin secretion will be impaired. The β cell may be especially sensitive to gain-of-function mutations in Kir6.2 as its resting potential is largely determined by the K_{ATP} channel, and its metabolism is very sensitive to blood glucose levels.

Kir6.2 is also expressed in skeletal muscle, cardiac muscle, and neurons throughout the brain (52, 53), a distribution consistent with the neurological symptoms found in severe forms of the disease. Based on the functional data, it seems likely that in these tissues a greater reduction in ATP sensitivity is required to increase the K_{ATP} current sufficiently to influence electrical activity. This might arise because K_{ATP} channels contribute less to the electrical activity of these cells, perhaps because of a low density of K_{ATP} channels; or because of contributions to membrane current from other ion channels; or because Kir6.2 associates with SUR2 (which may reduce the response to metabolism [ref. 22]);

or because of differences in cell metabolism or other channel regulators. It is worth noting that K_{ATP} channels are normally closed in many of these tissues and open only under conditions of metabolic stress (25, 27, 87, 88).

Precisely how severe mutations in Kir6.2 lead to epilepsy, developmental delay, muscle weakness, and dysmorphic features remains to be established, and sorting it out is likely to require animal models. At first sight, it appears paradoxical that epileptic symptoms can be caused by a *gain*-of-function mutation in a potassium channel. However, this is easily explained by overactivity of K_{ATP} channels in *inhibitory* neurons, which would decrease inhibitory tone and enhance excitability of target neurons. Consistent with this idea, K_{ATP} channels are expressed in most GABAergic inhibitory interneurons in the hippocampus, but only in a minority of excitatory pyramidal neurons (24); and diazoxide inhibits firing of interneurons but not pyramidal cells (25). Furthermore, K_{ATP} channel openers prevent GABA release in slices of substantia nigra (26, 27). The muscle weakness could be of neural or muscle origin, as both skeletal muscle (53) and nerve terminals (89) express Kir6.2. Interestingly, the electrocardiogram of patients with PNDM mutations is superficially normal (7, 28), as observed for transgenic mice overexpressing a cardiac-targeted gain-of-function mutation in Kir6.2 (90).

Although gain-of-function Kir6.2 mutations cause epilepsy in humans, transgenic mice overexpressing WT Kir6.2 or SUR1 in forebrain are *protected* against seizure and ischemic injury (15, 16), and knockout of Kir6.2 enhances susceptibility to generalized seizure (14). This suggests that K_{ATP} current magnitude may be critical for brain function, and that too little current as well as too much can predispose to seizure: this might reflect the relative amounts of K_{ATP} current in inhibitory and excitatory neurons. Alternatively, the transgenic mice may not provide a perfect model of the human disease.

Molecular basis of reduced ATP sensitivity
As discussed above, ATP has 2 effects on the K_{ATP} channel: ATP binding to Kir6.2, in a process that does not require Mg^{2+}, blocks the channel, whereas binding of MgATP to SUR1 stimulates channel activity (49). It is therefore essential to analyze the molecular mechanism of ATP block both in Mg^{2+}-free solution (to avoid the complicating effects of MgATP stimulation), and in the presence of Mg^{2+} to determine whether coupling of SUR1 to Kir6.2 is affected.

Studies in Mg-free solution have shown that although all PNDM mutations reduce the ability of ATP to block the K_{ATP} channel, they do so by a variety of molecular mechanisms (83, 84). To date, mutations that cause neonatal diabetes alone (e.g., R201H/C, I182V) impair ATP binding directly (7, 33, 83, 91), consistent with their location within the predicted ATP-binding site (Figures 3 and 4). In contrast, mutations that cause DEND syndrome (e.g., V59G, Q52R, I296L) affect ATP inhibition indirectly. They bias the channel toward the open state and impair its ability to close (83, 84) (Figure 5). Consequently, the fraction of time the channel spends open in the absence of ATP (the intrinsic open probability, Po) is increased. Because the affinity of the open state for ATP is less than that of the closed state, mutant channels are less inhibited by ATP (92). These mutations lie in regions predicted to be involved in channel gating (Figure 3). Finally, some mutations (V59G, I296L) appear to influence both intrinsic gating and ATP binding (or the mechanism by which binding is translated into channel closure) (83, 84).

In the presence of Mg^{2+}, the ATP sensitivity of the channel is reduced (49). This effect appears to be greater for several PNDM mutant channels (33, 84), suggesting that MgATP activation via SUR1 may be enhanced. However, studies to date are limited, and this area requires more work.

The importance of heterozygosity
All PNDM patients identified to date are heterozygotes. Because Kir6.2 is a tetramer (50, 51), their cells will express a mixed pop-

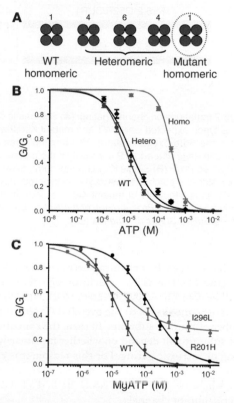

Figure 4
PNDM mutations reduce channel inhibition by ATP. (**A**) Schematic of the mixture of channels with different subunit compositions expected when WT and mutant Kir6.2 are coexpressed (as in the heterozygous state). The relative numbers of the channel types expected if WT and mutant subunits segregate independently (i.e., follow a binomial distribution) are indicated above the figure. The circle indicates the only channel type predicted to show a substantial change in ATP sensitivity if the mutation affects ATP binding (see text). (**B**) Mean relationship between [ATP] and K_{ATP} current (G), expressed relative to the conductance in the absence of nucleotide (G_C) for Kir6.2/SUR1 (red, $n = 6$) and heterozygous (black, $n = 6$) and homomeric (blue, $n = 6$) Kir6.2-R201H/SUR1 channels. The smooth curves are the best fit to the Hill equation. The IC_{50} was 7 μM, 12 μM, and 300 μM for WT, heterozygous R201H, and homomeric R201H channels, respectively. Data were obtained in the absence of Mg^{2+}. Part **B** reproduced with permission from *Proceedings of the National Academy of Sciences of the United States of America* (83). (**C**) Mean relationship between [MgATP] and K_{ATP} current, expressed relative to the conductance in the absence of nucleotide for Kir6.2/SUR1 (red, $n = 6$) and heterozygous Kir6.2-R201H (black, $n = 5$) and heterozygous Kir6.2-I296L/SUR1 (blue, $n = 5$) channels. The smooth curves are the best fit to the Hill equation. The IC_{50} was 13 μM, 140 μM, and 50 μM for WT, heterozygous R201H, and heterozygous I296L channels, respectively. Data in **C** are from refs. 33 and 84.

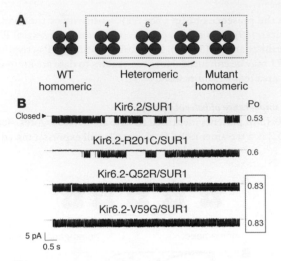

Figure 5
Effects of Kir6.2 mutations on channel gating. (**A**) Schematic of the different channel types expected when WT and mutant Kir6.2 are coexpressed (as in the heterozygous state). The box indicates channel types expected to have altered ATP sensitivity if the mutation affects channel gating (see text). (**B**) Single K_{ATP} channel currents recorded at –60 mV from inside-out patches from oocytes coinjected with mRNAs encoding SUR1 plus either WT or mutant Kir6.2 as indicated. Reproduced with permission from *Proceedings of the National Academy of Sciences of the United States of America* (83).

ulation of channels, each containing between 0 and 4 mutant subunits (Figure 4A). The ATP sensitivity of each of these 5 channel types will be determined by the extent to which each subunit (WT or mutant) contributes to the overall ATP sensitivity, and by the number of mutant subunits. In turn, the contribution of the mutant subunit will depend on whether it primarily affects ATP binding or intrinsic gating. For this reason, it is essential to study the effect of a mutation on the heterozygous channel population, whose behavior is not easily predicted from studies of homomeric mutant channels.

Because binding of a single ATP molecule closes the K_{ATP} channel (93), a mutation that reduces ATP binding will substantially impair ATP block only when all 4 subunits are mutant. If WT and mutant Kir6.2 subunits distribute randomly (51, 93), only one-sixteenth of channels in the heterozygous population will contain 4 mutant subunits (Figure 4). Thus the shift in the ATP sensitivity of the mean channel population will be small, precisely as seen for heterozygous R201C and R201H channels (7, 33, 83).

Mutations that affect intrinsic gating cause larger shifts in the ATP sensitivity of the heterozygous channel population. This can be explained if the energy of the open state scales with the number of mutant subunits (83); in this case, the ATP sensitivity of more than 90% of channels in the heterozygous population will be affected (Figure 5A). This explains why mutations that affect the intrinsic Po (such as Q52R, V59G, and I296L) cause a greater shift in the ATP sensitivity of heterozygous channels. The ATP sensitivity of heterozygous Q52R channels is well fit by assuming that the entire shift in ATP sensitivity results from the observed change in channel gating (83). However, other mutations (V5G, I296L) cause even greater shifts in the ATP sensitivity of both homomeric and heterozygous channels and appear to have additional effects on ATP binding and/or transduction (83, 84).

The importance of heterozygosity in determining the severity of a mutation appears to be a novel feature of gain-of-function K_{ATP} channelopathies. Accordingly, we may expect to find similar phenotypic variability in other tetrameric ion channels where channel function can be influenced differentially by the presence of 1, or more, mutant subunits. It is also worth noting that if mutant and WT subunits were to express at different levels, the composition of the heterozygous population might deviate from a binomial distribution and thus influence the channel ATP sensitivity in a less quantitatively predictable fashion.

Implications for therapy

Prior to the discovery that PNDM could be caused by mutations in Kir6.2, it was supposed that patients with this disorder suffered from early-onset, type 1 diabetes, and they were treated with insulin. Recent studies, however, demonstrate that patients with mutations that cause PNDM alone can be managed on sulfonylureas (7, 28, 82). It remains to be seen whether sulfonylurea therapy is effective in the long term. However, it is important to remember that while insulin therapy may control the diabetes, it does not mitigate the effects of enhanced K_{ATP} channel activity in nonpancreatic tissue: this requires drugs that close K_{ATP} channels. Whether sulfonylurea therapy will be as suitable for patients with mutations that increase the intrinsic Po of the channel remains unclear, because these channels are less sensitive to the drugs (83). This is expected, as mutations that stabilize the channel in the open state impair sulfonylurea inhibition (92). However, it may be possible to manage such patients, on a combination of insulin, to control their diabetes, and sulfonylureas, to control the extrapancreatic effects. A key question is whether sulfonylureas will be effective at treating CNS symptoms, as the extent to which they cross the blood/brain barrier is unclear. Drugs that enter the brain and selectively block K_{ATP} channels might be beneficial in this respect.

The efficacy of sulfonylurea therapy in PNDM will also depend on the particular gene that is mutated. Loss-of-function mutations in metabolic genes, such as *GCK*, reduce ATP production and enhance K_{ATP} channel activity indirectly. Blocking the K_{ATP} channel with sulfonylureas will restore Ca^{2+} influx but, as ATP is also required for the secretory process, may not be sufficient to reinstate insulin secretion fully.

Impaired metabolic regulation and early-onset diabetes

Impaired metabolic regulation of K_{ATP} channels, resulting from mutations in genes that influence β cell metabolism, can cause both monogenic HI and diabetes. As described above, mutations in *GCK* and *GLUD1* cause HI, probably by enhancing metabolic ATP generation and decreasing K_{ATP} channel activity. Conversely, mutations that reduce ATP synthesis, and thus the extent of K_{ATP} channel inhibition in response to glucose metabolism, may be expected to give rise to diabetes.

Maturity-onset diabetes of the young (MODY) is characterized by early onset, autosomal dominant inheritance, and pancreatic β cell dysfunction (66, 94). It is caused by mutations in at least 7 different genes. Heterozygous loss-of-function mutations in GCK cause MODY2, which is associated with moderate hyperglycemia and is often asymptomatic and detected only during routine screening: as described above, homozygous inactivating mutations in GCK impair β cell metabolism more severely and cause permanent neonatal diabetes. Five other MODY genes encode the transcription factors HNF4α, HNF1α, IPF1, HNF1β, and NEUROD, which are not only important for β cell development but can also regulate expression of genes critical for glucose metabolism. For example, knockout of HNF1α, which causes MODY3, impairs β cell glucose metabolism and thereby K_{ATP} channel closure in mice (95). Whether other transcription factors influence K_{ATP} channel function has yet to be established. Finally, as described above, mutations in Kir6.2 itself can also give rise to MODY (86).

Defective mitochondrial metabolism can also produce diabetes, as in maternally inherited diabetes with deafness (MIDD), which results from a mutation in the mitochondrial DNA that encodes a leucine transfer RNA (96, 97). It seems plausible to speculate that enhanced K_{ATP} channel activity, due to reduced ATP production, contributes to the impaired insulin secretion.

Implications for type 2 diabetes

Given that mutations in Kir6.2 cause neonatal diabetes and MODY, and that mutations in genes that regulate K_{ATP} channel activity lead to MODY and MIDD, which present in childhood or early adult life, it is natural to conjecture that common genetic variations in the same genes, which produce smaller functional effects, might predispose to type 2 diabetes in later life. Accumulating evidence supports this view.

Large-scale association studies indicate that a common variant (E23K) in Kir6.2 is strongly associated with an enhanced susceptibility to type 2 diabetes (98–100). Although the effect is statistically small (the odds ratio associated with the K allele is only ~1.2), the high prevalence of the K allele (34%) makes this a significant population risk. Precisely how the E23K polymorphism enhances diabetes susceptibility remains unclear. Data obtained by expressing recombinant K_{ATP} channels in heterologous systems are conflicting. Some studies demonstrate an increase in intrinsic Po, with a consequent reduction in ATP sensitivity, and enhanced activation by Mg nucleotides (101, 102), whereas others report no effect on ATP sensitivity but enhanced sensitivity to the stimulatory action of long-chain acyl-CoAs (103). It is also possible that the E23K variant may cause a greater reduction in ATP block of the K_{ATP} channel in pancreatic β cells than in heterologous systems. Interestingly, in mice, a 4-fold reduction in K_{ATP} channel ATP sensitivity is enough to cause neonatal diabetes (3). Furthermore, large-scale studies in humans have shown that the E23K variant is associated with reduced insulin secretion in glucose-tolerant subjects (104, 105). It is also worth remembering that because Kir6.2 is expressed in multiple tissues, diabetes susceptibility may involve tissues other

than the β cell: for example, glucagon secretion is increased in individuals carrying 2 K alleles (106).

Polymorphisms in HNF1α, HNF4α, and GCK (107–109), and in genes involved in mitochondrial metabolism (110, 111), have also been associated with an increased risk of type 2 diabetes. They probably influence disease susceptibility by impairing metabolic regulation of K_{ATP} channel activity.

This leads to the conclusion that good candidate genes for polygenic disease (in this case, type 2 diabetes) are those that cause monogenic disease (PNDM, MODY). As they manifest only in later life, the functional effects of individual gene variants associated with polygenic disease are likely to be small. It also suggests that the increased risk of diabetes with age may reflect a deteriorating β cell metabolism, leading to enhanced K_{ATP} channel activity and reduced insulin secretion (62). This idea is supported by the fact that insulin secretion is critically dependent on mitochondrial metabolism (112), which is known to decline with age (113) and has been implicated in age-related disease (112–114).

Conclusions

K_{ATP} channel activity in the β cell is finely balanced, reflecting its physiological role as a key regulator of insulin release. Increased activity leads to reduced insulin secretion, and, conversely, reduced K_{ATP} channel activity decreases insulin release. Consequently, mutations that either directly or indirectly alter β cell K_{ATP} channel activity produce a spectrum of diseases that ranges from severe HI to DEND syndrome (Figure 6).

In general, there is a good correlation between the magnitude of the K_{ATP} current and disease severity. Thus mutations that cause a total loss of K_{ATP} channel in the surface membrane produce severe HI, whereas those that impair channel function only partially produce a milder phenotype that can be treated with diazoxide or results in leucine-sensitive HI. Likewise, the extent to which the ATP sensitivity of Kir6.2 is decreased determines the severity of the diabetes phenotype. Mutations that cause the greatest reduction in ATP inhibition, and the largest increase in K_{ATP} channel activity, manifest effects in multiple tissues and lead to developmental delay, epilepsy, muscle weakness, and neonatal diabetes. Lesser shifts in ATP sensitivity give rise to neonatal diabetes with developmental delay, while even smaller changes manifest primarily in the β cell and cause neonatal diabetes alone.

The range of phenotypes associated with monogenic K_{ATP} channelopathies, which present at birth, suggests a similar spectrum

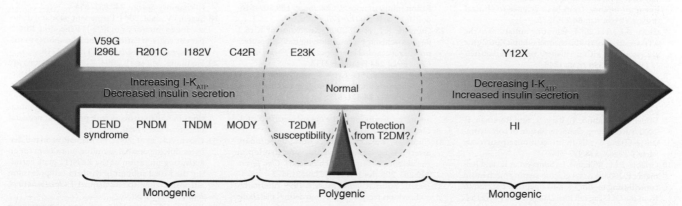

Figure 6
Schematic illustrating the relationship between K_{ATP} channel activity and insulin secretion. T2DM, type 2 diabetes mellitus.

of K_{ATP} channel polymorphisms that have less severe effects on channel function and predispose to late-onset, polygenic, disease. The association of the E23K polymorphism in Kir6.2 with susceptibility to type 2 diabetes supports this idea. Because of a gradual decline in β cell function with age, polymorphisms that cause a small decrease in K_{ATP} channel activity are unlikely to predispose to late development of HI but rather may help to protect against type 2 diabetes.

It is noteworthy that mutations in Kir6.2 cause both diabetes and HI, whereas mutations in SUR1 are, to date, entirely associated with hyperinsulinism. This is not unexpected, however. A total loss of protein (Kir6.2 or SUR1) will necessarily result in loss of K_{ATP} channel activity and thus HI. Loss of metabolic regulation, however, will have different effects in Kir6.2 and SUR1. For SUR1, impaired binding/hydrolysis and/or transduction of Mg nucleotides will decrease K_{ATP} channel activity, producing HI. Mutations that enhance Mg-nucleotide activation seem inherently far less likely. Thus, diabetes due to SUR1 mutations will be rare. Mutations that impair ATP binding to Kir6.2, on the other hand, will enhance K_{ATP} channel activity and cause diabetes. It is not understood why mutations that affect Kir6.2 gating invariably increase channel opening (thereby lowering ATP sensitivity), but this may suggest that the channel is most stable in the open state.

Finally, a similar spectrum of insulin secretory disorders can be expected from mutations in genes that regulate K_{ATP} channel activity, with mutations that indirectly enhance K_{ATP} currents mediating diabetes and those that reduce channel activity producing HI. As expected, loss-of-function mutations in metabolic genes cause PNDM or MODY depending on the extent to which they impair β cell metabolism, while gain-of-function mutations cause HI of varying severity. Naturally occurring polymorphisms in genes that cause minor variations in metabolic regulation, or density, of K_{ATP} channels can be predicted to produce small differences in K_{ATP} channel activity and insulin secretion, so influencing diabetes susceptibility. Some such polymorphisms have been identified, and more are likely to follow.

In conclusion, individuals may be predisposed to either HI or diabetes according to the magnitude of their β cell K_{ATP} current under resting conditions, with their position on this delicate K_{ATP} channel see-saw being determined by their unique combination of polymorphisms in multiple genes.

Acknowledgments

The author wishes to thank those working in the field who have contributed greatly to the ideas expressed in this review, especially Jennifer Antcliff, Christophe Girard, Peter Proks, and Paolo Tammaro in the author's group, and colleagues Roger Cox, Anna Gloyn, Fiona Gribble, Andrew Hattersley, and Patrik Rorsman. The author also thanks the Wellcome Trust, Diabetes UK, GrowBeta, and the Royal Society for support. F.M. Ashcroft is the Royal Society GlaxoSmithKline Research Professor.

Address correspondence to: Frances Ashcroft, University Laboratory of Physiology, Parks Road, Oxford OX1 3PT, United Kingdom. Phone: 01865-285810; Fax: 01865-285813; E-mail: frances.ashcroft@physiol.ox.ac.uk.

1. Seino, S., and Miki, T. 2004. Gene targeting approach to clarification of ion channel function: studies of Kir6.x null mice. *J. Physiol.* **554**:295–300.

2. Ashcroft, F.M., Harrison, D.E., and Ashcroft, S.J.H. 1984. Glucose induces closure of single potassium channels in isolated rat pancreatic β-cells. *Nature.* **312**:446–448.

3. Koster, J.C., Marshall, B.A., Ensor, N., Corbett, J.A., and Nichols, C.G. 2000. Targeted overactivity of beta cell K(ATP) channels induces profound neonatal diabetes. *Cell.* **100**:645–654.

4. Miki, T., et al. 1998. Defective insulin secretion and enhanced insulin action in K_{ATP} channel-deficient mice. *Proc. Natl. Acad. Sci. U. S. A.* **95**:10402–10406.

5. Seghers, V., Nakazaki, M., DeMayo, F., Aguilar-Bryan, L., and Bryan, J. 2000. Sur1 knockout mice. A model for K(ATP) channel-independent regulation of insulin secretion. *J. Biol. Chem.* **275**:9270–9277.

6. Dunne, M.J., Cosgrove, K.E., Shepherd, R.M., Aynsley-Green, A., and Lindley, K.J. 2004. Hyperinsulinism in infancy: from basic science to clinical disease. *Physiol. Rev.* **84**:239–275.

7. Gloyn, A.L., et al. 2004. Activating mutations in the ATP-sensitive potassium channel subunit Kir6.2 gene are associated with permanent neonatal diabetes. *N. Engl. J. Med.* **350**:1838–1849.

8. Gopel, S.O., et al. 2000. Regulation of glucagon release in mouse alpha-cells by K_{ATP} channels and inactivation of TTX-sensitive Na+ channels. *J. Physiol.* **528**:509–520.

9. Gopel, S.O., Kanno, T., Barg, S., and Rorsman, P. 2000. Patch-clamp characterisation of somatostatin-secreting δ-cells in intact mouse pancreatic islets. *J. Physiol.* **528**:497–507.

10. Gribble, F.M., Williams, L., Simpson, A.K., and Reimann, F. 2003. A novel glucose-sensing mechanism contributing to glucagon-like peptide-1 secretion from the GLUTag cell line. *Diabetes.* **52**:1147–1154.

11. Miki, T., et al. 2001. ATP-sensitive K+ channels in the hypothalamus are essential for the maintenance of glucose homeostasis. *Nat. Neurosci.* **4**:507–512.

12. Wang, R., et al. 2004. The regulation of glucose-excited neurons in the hypothalamic arcuate nucleus by glucose and feeding-relevant peptides. *Diabetes.* **53**:1959–1965.

13. Zingman, L.V., et al. 2002. Kir6.2 is required for adaptation to stress. *Proc. Natl Acad. Sci. U. S. A.* **99**:13278 13283.

14. Yamada, K., et al. 2001. Protective role of ATP-sensitive potassium channels in hypoxia-induced generalized seizure. *Science.* **292**:1543–1546.

15. Hernandez-Sanchez, C., et al. 2001. Mice transgenically overexpressing sulfonylurea receptor 1 in forebrain resist seizure induction and excitotoxic neuron death. *Proc. Natl Acad. Sci. U. S. A.* **98**:3549–3554.

16. Heron-Milhavet, L., et al. 2004. Protection against hypoxic-ischemic injury in transgenic mice overexpressing Kir6.2 channel pore in forebrain. *Mol. Cell. Neurosci.* **25**:585–593.

17. Suzuki, M., et al. 2002. Role of sarcolemmal K_{ATP} channels in cardioprotection against ischemia/reperfusion injury in mice. *J. Clin. Invest.* **109**:509–516. doi:10.1172/JCI200214270.

18. Gumina, R.J., et al. 2003. Knockout of Kir6.2 negates ischemic preconditioning-induced protection of myocardial energetics. *Am. J. Physiol. Heart Circ. Physiol.* **284**:H2106–H2113.

19. Daut, J., Klieber, H.G., Cyrys, S., and Noack, T. 1994. K_{ATP} channels and basal coronary vascular tone. *Cardiovasc. Res.* **28**:811–817.

20. Miki, T., et al. 2002. Mouse model of Prinzmetal angina by disruption of the inward rectifier Kir6.1. *Nat. Med.* **8**:466–472.

21. Chutkow, W.A., et al. 2002. Episodic coronary artery vasospasm and hypertension develop in the absence of Sur2 K_{ATP} channels. *J. Clin. Invest.* **110**:203–208. doi:10.1172/JCI200215672.

22. Liss, B., Bruns, R., and Roeper, J. 1999. Alternative sulfonylurea receptor expression defines metabolic sensitivity of K_{ATP} channels in dopaminergic midbrain neurons. *EMBO J.* **18**:833–846.

23. Avshalumov, M.V., and Rice, M.E. 2003. Activation of ATP-sensitive K+ (K_{ATP}) channels by H_2O_2 underlies glutamate-dependent inhibition of striatal dopamine release. *Proc. Natl. Acad. Sci. U. S. A.* **100**:11729–11734.

24. Zawar, C., Plant, T.D., Schirra, C., Konnerth, A., and Neumcke, B. 1999. Cell-type specific expression of ATP-sensitive potassium channels in the rat hippocampus. *J. Physiol.* **514**:327–341.

25. Griesemer, D., Zawar, C., and Neumcke, B. 2002. Cell-type specific depression of neuronal excitability in rat hippocampus by activation of ATP-sensitive potassium channels. *Eur. Biophys. J.* **31**:467–477.

26. Schmid-Antomarchi, H., Amoroso, S., Fosset, M., and Lazdunski, M. 1990. K+ channel openers activate brain sulfonylurea-sensitive K+ channels and block neurosecretion. *Proc. Natl. Acad. Sci. U. S. A.* **87**:3489–3492.

27. Amoroso, S., Schmid-Antomarchi, H., Fosset, M., and Lazdunski, M. 1990. Glucose, sulfonylureas, and neurotransmitter release: role of ATP-sensitive K+ channels. *Science.* **247**:852–854.

28. Sagen, J.V., et al. 2004. Permanent neonatal diabetes due to mutations in KCNJ11 encoding Kir6.2: patient characteristics and initial response to sulfonylurea therapy. *Diabetes.* **53**:2713–2718.

29. Vaxillaire, M., et al. 2004. Kir6.2 mutations are a common cause of permanent neonatal diabetes mellitus in a large cohort of French patients. *Diabetes.* **53**:2719–2722.

30. Massa, O., et al. 2005. KCNJ11 activating mutations in Italian patients with permanent neonatal diabetes. *Hum. Mutat.* **25**:22–27.

31. Gloyn, A.L., et al. 2004. Permanent neonatal diabetes due to paternal germline mosaicism for an activating mutation of the KCNJ11 gene encoding the Kir6.2 subunit of the beta-cell potassium adenosine triphosphate channel. *J. Clin. Endocrinol. Metab.* **89**:3932–3935.

32. Edghill, E.L., et al. 2004. Activating mutations in the KCNJ11 gene encoding the ATP-sensitive K+ channel subunit Kir6.2 are rare in clinically defined

type 1 diabetes diagnosed before 2 years. *Diabetes.* **53**:2998–3001.

33. Gloyn, A.L., et al. 2005. Relapsing diabetes can result from moderately activating mutations in *KCNJ11. Hum. Mol. Genet.* **14**:925–934.

34. Glaser, B., Thornton, P., Otonkoski, T., and Junien, C. 2000. Genetics of neonatal hyperinsulinism. *Arch. Dis. Child. Fetal Neonatal Ed.* **82**:F79–F86.

35. Thomas, P.M., et al. 1995. Mutations in the sulfonylurea receptor gene in familial persistent hyperinsulinemic hypoglycemia of infancy. *Science.* **268**:426–429.

36. Nestorowicz, A., et al. 1997. A nonsense mutation in the inward rectifier potassium channel gene, Kir6.2, is associated with familial hyperinsulinism. *Diabetes.* **46**:1743–1748.

37. Thomas, P., Ye, Y., and Lightner, E. 1996. Mutation of the pancreatic islet inward rectifier Kir6.2 also leads to familial persistent hyperinsulinemic hypoglycemia of infancy. *Hum. Mol. Genet.* **5**:1809–1812.

38. Henwood, M.J., et al. 2005. Genotype-phenotype correlations in children with congenital hyperinsulinism due to recessive mutations of the adenosine triphosphate-sensitive potassium channel genes. *J. Clin. Endocrinol. Metab.* **90**:789–794.

39. Huopio, H., et al. 2000. Dominantly inherited hyperinsulinism caused by a mutation in the sulfonylurea receptor type 1. *J. Clin. Invest.* **106**:897–906.

40. Magge, S.N., et al. 2004. Familial leucine-sensitive hypoglycemia of infancy due to a dominant mutation of the beta-cell sulfonylurea receptor. *J. Clin. Endocrinol. Metab.* **89**:4450–4456.

41. Bienengraeber, M., et al. 2004. ABCC9 mutations identified in human dilated cardiomyopathy disrupt catalytic KATP channel gating. *Nat. Genet.* **36**:382–387.

42. Ashcroft, F.M., and Rorsman, P. 1989. Electrophysiology of the pancreatic β-cell. *Prog. Biophys. Mol. Biol.* **54**:87–143.

43. Henquin, J.C. 2000. Triggering and amplifying pathways of regulation of insulin secretion by glucose. *Diabetes.* **49**:1751–1760.

44. Gribble, F.M., and Reimann, F. 2003. Sulphonylurea action revisited: the post-cloning era. *Diabetologia.* **46**:875–891.

45. Gribble, F., and Ashcroft, F.M. 2000. New windows on the mechanism of action of potassium channel openers [review]. *Trends Pharmacol. Sci.* **21**:439–445.

46. Bokvist, K., et al. 1991. Separate processes mediate nucleotide-induced inhibition and stimulation of the ATP-regulated K(+)-channels in mouse pancreatic beta-cells. *Proc. R. Soc. Lond., B, Biol. Sci.* **243**:139–144.

47. Hopkins, W.F., Fatherazi, S., Peter-Riesch, B., Corkey, B.E., and Cook, D.L. 1992. Two sites for adenine-nucleotide regulation of ATP-sensitive potassium channels in mouse pancreatic beta-cells and HIT cells. *J. Membr. Biol.* **129**:287–295.

48. Tucker, S.J., Gribble, F.M., Zhao, C., Trapp, S., and Ashcroft, F.M. 1997. Truncation of Kir6.2 produces ATP-sensitive K⁺ channels in the absence of the sulphonylurea receptor. *Nature.* **387**:179–183.

49. Gribble, F.M., Tucker, S.J., Haug, T., and Ashcroft, F.M. 1998. MgATP activates the β-cell KATP channel by interaction with its SUR1 subunit. *Proc. Natl. Acad. Sci. U. S. A.* **95**:7185–7190.

50. Clement, J.P., 4th, et al. 1997. Association and stoichiometry of KATP channel subunits. *Neuron.* **18**:827–838.

51. Shyng, S.L., and Nichols, C.G. 1997. Octameric stoichiometry of the KATP channel complex. *J. Gen. Physiol.* **110**:655–664.

52. Inagaki, N., et al. 1995. Reconstitution of IKATP: an inward rectifier subunit plus the sulfonylurea receptor. *Science.* **270**:1166–1169.

53. Sakura, H., Ammala, C., Smith, P.A., Gribble, F.M., and Ashcroft, F.M. 1995. Cloning and functional expression of the cDNA encoding a novel ATP-sensitive potassium channel subunit expressed in pancreatic β-cells, brain, heart and skeletal muscle. *FEBS Lett.* **377**:338–344.

54. Inagaki, N., et al. 1995. Cloning and functional characterization of a novel ATP-sensitive potassium channel ubiquitously expressed in rat tissues, including pancreatic islets, pituitary, skeletal muscle, and heart. *J. Biol. Chem.* **270**:5691–5694.

55. Tanabe, K., et al. 1999. Direct photoaffinity labeling of the Kir6.2 subunit of the ATP-sensitive K⁺ channel by 8-azido-ATP. *J. Biol. Chem.* **274**:3931–3933.

56. Aguilar-Bryan, L., et al. 1995. Cloning of the β-cell high-affinity sulfonylurea receptor: a regulator of insulin secretion. *Science.* **268**:423–426.

57. Nichols, C.G., et al. 1996. Adenosine diphosphate as an intracellular regulator of insulin secretion. *Science.* **272**:1785–1787.

58. Gribble, F.M., Tucker, S.J., and Ashcroft, F.M. 1997. The essential role of the Walker A motifs of SUR1 in KATP channel activation by MgADP and diazoxide. *EMBO J.* **16**:1145–1152.

59. Chutkow, W.A., Simon, M.C., Beau, M.M.L., and Burant, C.F. 1996. Coning, tissue expression, and chromosomal localization of SUR2, the putative drug-binding subunit of cardiac, skeletal muscle, and vascular KATP channels. *Diabetes.* **45**:1439–1445.

60. Inagaki, N., et al. 1996. A family of sulfonylurea receptors determines the pharmacological properties of ATP-sensitive K⁺ channels. *Neuron.* **16**:1011–1017.

61. Isomoto, S., et al. 1996. A novel sulfonylurea receptor forms with BIR (Kir6.2) a smooth muscle type ATP-sensitive K⁺ channel. *J. Biol. Chem.* **271**:24321–24324.

62. Ashcroft, F.M., and Rorsman, P. 2004. Type-2 diabetes mellitus: not quite exciting enough? *Hum. Mol. Genet.* **13**:R21–R31.

63. Kanno, T., Rorsman, P., and Gopel, S.O. 2002. Glucose-dependent regulation of rhythmic action potential firing in pancreatic beta-cells by K(ATP)-channel modulation. *J. Physiol.* **545**:501–507.

64. Smith, P.A., et al. 1997. Electrogenic arginine transport mediates stimulus-secretion coupling in mouse pancreatic β-cells. *J. Physiol.* **499**:625–635.

65. Rolland, J.F., Henquin, J.C., and Gilon, P. 2002. G protein-independent activation of an inward Na(+) current by muscarinic receptors in mouse pancreatic beta-cells. *J. Biol. Chem.* **277**:38373–38380.

66. Gloyn, A.L. 2003. Glucokinase (GCK) mutations in hyper- and hypoglycemia: maturity-onset diabetes of the young, permanent neonatal diabetes, and hyperinsulinemia of infancy. *Hum. Mutat.* **22**:353–362.

67. Glaser, B., et al. 1998. Familial hyperinsulinism caused by an activating glucokinase mutation. *N. Engl. J. Med.* **338**:226–230.

68. Stanley, C.A., et al. 1998. Hyperinsulinism and hyperammonemia in infants with regulatory mutations of the glutamate dehydrogenase gene. *N. Engl. J. Med.* **338**:1352–1357.

69. Molven, A., et al. 2004. Familial hyperinsulinemic hypoglycemia caused by a defect in the SCHAD enzyme of mitochondrial fatty acid oxidation. *Diabetes.* **53**:221–227.

70. Taschenberger, G., et al. 2002. Identification of a familial hyperinsulinism-causing mutation in the sulfonylurea receptor 1 that prevents normal trafficking and function of KATP channels. *J. Biol. Chem.* **277**:17139–17146.

71. Partridge, C.J., Beech, D.J., and Sivaprasadarao, A. 2001. Identification and pharmacological correction of a membrane trafficking defect associated with a mutation in the sulfonylurea receptor causing familial hyperinsulinism. *J. Biol. Chem.* **276**:35947–35952.

72. Yan, F., et al. 2004. Sulfonylureas correct trafficking defects of ATP-sensitive potassium channels caused by mutations in the sulfonylurea receptor. *J. Biol. Chem.* **279**:11096–11105.

73. Zerangue, N., Schwappach, B., Jan, Y.N., and Jan, L.Y. 1999. A new ER trafficking signal regulates the subunit stoichiometry of plasma membrane K(ATP) channels. *Neuron.* **22**:537–548.

74. Cuesta-Munoz, A.L., et al. 2004. Severe persistent hyperinsulinemic hypoglycemia due to a de novo glucokinase mutation. *Diabetes.* **53**:2164–2168.

75. Fajans, S.S., Floyd, J.C., Knopf, R.F., and Conn, F.W. 1967. Effect of amino acids and proteins on insulin secretion in man. *Recent Prog. Horm. Res.* **23**:617–662.

76. Gloyn, A.L., et al. 2003. Insights into the biochemical and genetic basis of glucokinase activation from naturally occurring hypoglycemia mutations. *Diabetes.* **52**:2433–2440.

77. Miki, T., Iwanaga, T., Nagashima, K., Ihara, Y., and Seino, S. 2001. Roles of ATP-sensitive K⁺ channels in cell survival and differentiation in the endocrine pancreas. *Diabetes.* **50**(Suppl. 1):S48–S51.

78. Shield, J.P.H. 1996. Neonatal diabetes. In *Childhood diabetes.* Volume 4. J.H. Shield and J.D. Baum, editors. Baillière Tindall. London, United Kingdom. 681–740.

79. Gloyn, A.L., et al. 2002. Complete glucokinase deficiency is not a common cause of permanent neonatal diabetes [letter]. *Diabetologia.* **45**:290.

80. Njølstad, P.R., et al. 2001. Neonatal diabetes mellitus due to complete glucokinase deficiency. *N. Engl. J. Med.* **344**:1588–1592.

81. Temple, I.K., et al. 1995. An imprinted gene(s) for diabetes? *Nat. Genet.* **9**:110–112.

82. Zung, A., Glaser, B., Nimri, R., and Zadik, Z. 2004. Glibenclamide treatment in permanent neonatal diabetes mellitus due to an activating mutation in Kir6.2. *J. Clin. Endocrinol. Metab.* **89**:5504–5507.

83. Proks, P., et al. 2004. Molecular basis of Kir6.2 mutations associated with neonatal diabetes or neonatal diabetes plus neurological features. *Proc. Natl. Acad. Sci. U. S. A.* **101**:17539–17544.

84. Peter Proks, P., et al. 2005. A gating mutation at the internal mouth of the Kir6.2 pore is associated with DEND syndrome. *EMBO Rep.* **6**:470–475.

85. Nabi-Buisson, N., et al. 2004. Early epileptic encephalopathy and neonatal diabetes, associated with mutation in ATP-sensitive potassium channel, Kir6.2. *Am. J. Hum. Genet.* **75**(Suppl. 1):323.

86. Yorifuji, T., et al. 2005. The C42R mutation in the Kir6.2 (KCNJ11) gene as a cause of transient neonatal diabetes, childhood diabetes, or later-onset, apparently type 2 diabetes mellitus. *J. Clin. Endocrinol. Metab.* doi:10.1210/jc.2005-0096.

87. Nichols, C.G., and Lederer, W.J. 1990. The regulation of ATP-sensitive K⁺ channel activity in intact and permeabilized rat ventricular myocytes. *J. Physiol.* **423**:91–110.

88. Watts, A.E., Hicks, G.A., and Henderson, G. 1995. Putative pre- and postsynaptic ATP-sensitive potassium channels in the rat substantia nigra *in vitro. J. Neurosci.* **15**:3065–3074.

89. Deist, M., Repp, H., and Dreyer, F. 1992. Sulfonylurea-sensitive K⁺ channels and their probable role for the membrane potential of mouse motor nerve endings. *Pflugers Arch.* **421**:292–294.

90. Koster, J.C., et al. 2001. Tolerance for ATP-insensitive K(ATP) channels in transgenic mice. *Circ. Res.* **89**:1022–1029.

91. John, S.A., Weiss, J.N., Xie, L.H., and Ribalet, B. 2003. Molecular mechanism for ATP-dependent closure of the K+ channel Kir6.2. *J. Physiol.* **552**:23–34.

92. Trapp, S., Proks, P., Tucker, S.J., and Ashcroft, F.M. 1998. Molecular analysis of KATP channel gating and implications for channel inhibition by ATP. *J. Gen. Physiol.* **112**:333–349.

93. Markworth, E., Schwanstecher, C., and Schwanstecher, M. 2000. ATP⁴⁻ mediates closure of pancreatic beta-cell ATP-sensitive potassium channels by interaction with 1 of 4 identical sites. *Diabetes.* **49**:1413–1418.

94. Fajans, S.S., Bell, G.I., and Polonsky, K.S. 2001. Molecular mechanisms and clinical pathophysiology of maturity-onset diabetes of the young. *N. Engl. J. Med.* **345**:971–980.

95. Dukes, I.D., et al. 1998. Defective pancreatic (β-cell glycolytic signaling in hepatocyte nuclear factor-1α-deficient mice. *J. Biol. Chem.* **273**:24457–24464.

96. van den Ouweland, J.M., et al. 1992. Mutation in mitochondrial tRNA(Leu)(UUR) gene in a large pedigree with maternally transmitted type II diabetes mellitus and deafness. *Nat. Genet.* **1**:368–371.

97. van den Ouweland, J.M., Maechler, P., Wollheim, C.B., Attardi, G., and Maassen J.A. 1999. Functional and morphological abnormalities of mitochondria harbouring the tRNA(Leu)(UUR) mutation in mitochondrial DNA derived from patients with maternally inherited diabetes and deafness (MIDD) and progressive kidney disease. *Diabetologia.* **42**:485–492.

98. Hani, E.H., et al. 1998. Missense mutations in the pancreatic islet beta cell inwardly rectifying K⁺ channel gene (Kir6.2/BIR): a meta-analysis suggests a role in the polygenic basis of type II diabetes mellitus in Caucasians. *Diabetologia.* **41**:1511–1515.

99. Gloyn, A.L., et al. 2003. Large-scale association studies of variants in genes encoding the pancreatic β-cell K$_{ATP}$ channel subunits Kir6.2 (KCNJ11) and SUR1 (ABCC8) confirm that the KCNJ11 E23K variant is associated with type 2 diabetes. *Diabetes.* **52**:568–572.

100. Barroso, I., et al. 2003. Candidate gene association study in type 2 diabetes indicates a role for genes involved in β-cell function as well as insulin action. *PLoS Biol.* **1**:41–55.

101. Schwanstecher, C., Meyer, U., and Schwanstecher, M. 2003. K(IR)6.2 polymorphism predisposes to type 2 diabetes by inducing overactivity of pancreatic β-cell ATP-sensitive K⁺ channels. *Diabetes.* **51**:875–879.

102. Schwanstecher, C., Neugebauer, B., Schulz, M., and Schwanstecher, M. 2002. The common single nucleotide polymorphism E23K in K(IR)6.2 sensitizes pancreatic β-cell ATP-sensitive potassium channels toward activation through nucleoside diphosphates. *Diabetes.* **51**(Suppl. 3):S363–S367.

103. Reidel, M.J., Boora, P., Steckley, D., de Vries, G., and Light, P.E. 2003. Kir6.2 polymorphisms sensitize beta-cell ATP-sensitive potassium channels to activation by acyl CoAs: a possible cellular mechanism for increased susceptibility to type 2 diabetes? *Diabetes.* **52**:2630–2635.

104. Nielsen, E.M., et al. 2003. The E23K variant of Kir6.2 associates with impaired post-OGTT serum insulin response and increased risk of type 2 diabetes. *Diabetes.* **52**:573–577.

105. Florez, J.C., et al. 2004. Haplotype structure and genotype-phenotype correlations of the sulfonylurea receptor and the islet ATP-sensitive potassium channel gene region. *Diabetes.* **53**:1360–1368.

106. Tschritter, O., et al. 2002. The prevalent Glu23Lys polymorphism in the potassium inward rectifier 6.2 (KIR6.2) gene is associated with impaired glucagon suppression in response to hyperglycemia. *Diabetes.* **51**:2854–2860.

107. Triggs-Raine, B.L., et al. 2002. HNF-1α G319S, a transactivation-deficient mutant, is associated with altered dynamics of diabetes onset in an Oji-Cree community. *Proc. Natl. Acad. Sci. U. S. A.* **99**:4614–4619.

108. Stone, L.M., Kahn, S.E., Fujimoto, W.Y., Deeb, S.S., and Porte, D., Jr. 1996. A variation at position -30 of the β-cell glucokinase gene promoter is associated with reduced β-cell function in middle-aged Japanese-American men. *Diabetes.* **45**:422–428.

109. Weedon, M.N., et al. 2004. Common variants of the hepatocyte nuclear factor-4alpha P2 promoter are associated with type 2 diabetes in the U.K. population. *Diabetes.* **53**:3002–3006.

110. Sesti, G., et al. 2003. A common polymorphism in the promoter of UCP2 contributes to the variation in insulin secretion in glucose-tolerant subjects. *Diabetes.* **52**:1280–1283.

111. Poulton, J., et al. 2002. Type 2 diabetes is associated with a common mitochondrial variant: evidence from a population-based case-control study. *Hum. Mol. Genet.* **11**:1581–1583.

112. Maechler, P., and Wollheim, C.B. 2001. Mitochondrial function in normal and diabetic β-cells. *Nature.* **414**:807–812.

113. Petersen, K.F., et al. 2003. Mitochondrial dysfunction in the elderly: possible role in insulin resistance. *Science.* **300**:1140–1142.

114. Michikawa, Y., Mazzucchelli, F., Bresolin, N., Scarlato, G., and Attardi, G. 1999. Aging-dependent large accumulation of point mutations in the human mtDNA control region for replication. *Science.* **286**:774–779.

115. Campbell, J.D., Proks, P., Lippiat, J.D., Sansom, M.S.P., and Ashcroft, F.M. 2004. Identification of a functionally important negatively charged residue within the second catalytic site of the SUR1 nucleotide binding domains. *Diabetes.* **53**(Suppl. 3): S123–S127.

116. Antcliff, J., Haider, S., Proks, P., Sansom, M., and Ashcroft, F.M. 2005. Functional analysis of a structural model of the ATP-binding site of the K$_{ATP}$ channel Kir6.2 subunit. *EMBO J.* **24**:229–239.

117. Reimann, F., et al. 2003. Characterisation of new K$_{ATP}$-channel mutations associated with congenital hyperinsulinism in the Finnish population. *Diabetologia.* **46**:241–249.

118. Huopio, H., et al. 2002. Acute insulin response tests for the differential diagnosis of congenital hyperinsulinism. *J. Clin. Endocrinol. Metab.* **87**:4502–4507.

Gestational diabetes mellitus

Thomas A. Buchanan[1] and Anny H. Xiang[2]

[1]Departments of Medicine, Obstetrics and Gynecology, and Physiology and Biophysics, and [2]Department of Preventive Medicine, University of Southern California Keck School of Medicine, Los Angeles, California, USA.

Gestational diabetes mellitus (GDM) is defined as glucose intolerance of various degrees that is first detected during pregnancy. GDM is detected through the screening of pregnant women for clinical risk factors and, among at-risk women, testing for abnormal glucose tolerance that is usually, but not invariably, mild and asymptomatic. GDM appears to result from the same broad spectrum of physiological and genetic abnormalities that characterize diabetes outside of pregnancy. Indeed, women with GDM are at high risk for having or developing diabetes when they are not pregnant. Thus, GDM provides a unique opportunity to study the early pathogenesis of diabetes and to develop interventions to prevent the disease.

Historical perspective

For more than a century, it has been known that diabetes antedating pregnancy can have severe adverse effects on fetal and neonatal outcomes (1). As early as in the 1940s, it was recognized that women who developed diabetes years after pregnancy had experienced abnormally high fetal and neonatal mortality (2). By the 1950s the term "gestational diabetes" was applied to what was thought to be a transient condition that affected fetal outcomes adversely, then abated after delivery (3). In the 1960s, O'Sullivan found that the degree of glucose intolerance during pregnancy was related to the risk of developing diabetes after pregnancy. He proposed criteria for the interpretation of oral glucose tolerance tests (OGTTs) during pregnancy that were fundamentally statistical, establishing cut-off values — approximately 2 standard deviations — for diagnosing glucose intolerance during pregnancy (4). In the 1980s those cut-off points were adapted to modern methods for measuring glucose and applied to the modern definition of gestational diabetes — glucose intolerance with onset or first recognition during pregnancy (5). While based on O'Sullivan's values for predicting diabetes after pregnancy, the diagnosis of gestational diabetes mellitus (GDM) also identifies pregnancies at increased risk for perinatal morbidity (6–8) and long-term obesity and glucose intolerance in offspring (9–11).

Population perspective

Clinical detection of GDM is carried out to identify pregnancies at increased risk for perinatal morbidity and mortality. Available data do not identify a threshold of maternal glycemia at which such risk begins or increases rapidly. A multinational study, the Hyperglycemia and Adverse Pregnancy Outcome study, is underway to explore this issue in a large multiethnic cohort. In the absence of a defined glucose threshold for perinatal risk, many different sets of glycemic criteria have been proposed and are employed worldwide for the diagnosis of GDM. The criteria

Nonstandard abbreviations used: DPP, Diabetes Prevention Program; GAD, glutamic acid decarboxylase; GDM, gestational diabetes mellitus; MODY, maturity-onset diabetes of the young; OGTT, oral glucose tolerance test; TRIPOD, Troglitazone in Prevention of Diabetes.

Conflict of interest: T.A. Buchanan and A.H. Xiang receive grant support from Takeda Pharmaceuticals North America Inc. T.A. Buchanan is also a consultant to Takeda Pharmaceuticals North America Inc. and is on the company's Actos Speakers' Bureau.

Citation for this article: *J. Clin. Invest.* **115**:485–491 (2005). doi:10.1172/JCI200524531.

currently recommended by the American Diabetes Association (12) are based on O'Sullivan's criteria (see above). The detection of GDM, a condition that is generally asymptomatic, involves screening in 2 sequential steps (Tables 1 and 2), followed by administration of a 2- or 3-hour OGTT (Table 3) to women determined to be at risk by screening. Women with very high clinical risk characteristics may be diagnosed with probable pregestational (preexisting) diabetes based on the criteria provided in Table 4. When the diagnostic criteria for a 3-hour OGTT presented in Table 3 was applied to a group of Caucasian women in Toronto, approximately 7% had GDM (6). The frequency of GDM may vary among ethnic groups (higher in groups with increased prevalence of hyperglycemia) (13–16) and with the use of different diagnostic criteria (higher when lower glucose thresholds are applied and vice versa) (4). Nonetheless, all approaches to GDM detection pinpoint — and thereby allow diagnosis of — women with glucose tolerance in the upper end of the population distribution during pregnancy. A small minority of those women have glucose levels that would be diagnostic of diabetes outside of pregnancy (Table 4). The great majority have lower glucose levels. Both groups impart to their offspring an increased risk of perinatal morbidity and long-term obesity and diabetes that appear to be related at least in part to fetal overnutrition in utero. They also incur for themselves a risk of diabetes after pregnancy that is the main focus of this paper.

Etiology and pathogenesis

Normal pregnancy

A detailed discussion of glucose regulation in pregnancy is beyond the scope of this paper. However, 2 points are important for the discussion that follows. First, pregnancy is normally attended by progressive insulin resistance that begins near midpregnancy and progresses through the third trimester to levels that approximate the insulin resistance seen in individuals with type 2 diabetes. The insulin resistance appears to result from a combination of increased maternal adiposity and the insulin-desensitizing effects of hormonal products of the placenta. The fact that insulin resistance rapidly abates following delivery suggests that the major contributors to this state of resistance are placental hormones. The second point is that pancreatic β cells normally increase their insulin secretion to compensate for the insulin resistance of pregnancy (see Figure 1, for example). As a result, changes in circulating glucose levels over the course of

Table 1

Screening for GDM, step 1: clinical risk assessment[A,B]

Risk category	Clinical characteristics	Recommended screening
High risk (presence of any is sufficient)	Marked obesity Diabetes in first-degree relative(s) Personal history of glucose intolerance Prior delivery of macrosomic infant Current glycosuria	Blood glucose screening at initial antepartum visit or as soon as possible thereafter; repeat at 24–28 wk if not already diagnosed with GDM by that time
Average risk	Fits neither low- nor high-risk profile	Blood glucose screening between 24 and 28 wk gestation
Low risk (all required)	Age <25 Low-risk ethnicity[C] No diabetes in first-degree relatives Normal prepregnancy weight and pregnancy weight gain No personal history of abnormal glucose levels No prior poor obstetrical outcomes	Blood glucose screening not required

[A]Based on recommendations of the American Diabetes Association (5). [B]Performed at initial antepartum visit. [C]Ethnicities other than Hispanic, African, Native American, South or East Asian, Pacific Islander, or indigenous Australian, all of which have relatively high rates of GDM. Adapted with permission from *N. Engl. J. Med.* (72). Copyright 1999, Massachussets Medical Society. All rights reserved.

pregnancy are quite small compared with the large changes in insulin sensitivity. Robust plasticity of β cell function in the face of progressive insulin resistance is the hallmark of normal glucose regulation during pregnancy.

Gestational diabetes

GDM is a form of hyperglycemia. In general, hyperglycemia results from an insulin supply that is inadequate to meet tissue demands for normal blood glucose regulation. Studies conducted during late pregnancy, when, as discussed below, insulin requirements are high and differ only slightly between normal and gestational diabetic women, consistently reveal reduced insulin responses to nutrients in women with GDM (17–23). Studies conducted before or after pregnancy, when women with prior GDM are usually more insulin resistant than normal women (also discussed below), often reveal insulin responses that are similar in the 2 groups or reduced only slightly in women with prior GDM (18, 22–26). However, when insulin levels and responses are expressed relative to each individual's degree of insulin resistance, a large defect in pancreatic β cell function is a consistent finding in women with prior GDM (23, 25, 27).

Potential causes of inadequate β cell function are myriad and not fully described. Outside of pregnancy, there are 3 general settings that are recognized — through classification as distinct forms of diabetes mellitus (12) — as separate categories of β cell dysfunction: (a) autoimmune; (b) monogenic; and (c) occurring on a background of insulin resistance. There is evidence that β cell dysfunction in GDM can occur in all 3 major settings, a fact that is not surprising given that GDM is detected by what is, in essence, population screening for elevated glucose levels among pregnant women.

Autoimmune diabetes and GDM

Type 1 diabetes results from autoimmune destruction of pancreatic β cells. It accounts for approximately 5–10% of diabetes in the general population (12). Prevalence rates vary by ethnicity, with the highest rates in Scandinavians and the lowest rates (i.e., 0%) in full-blooded Native Americans. Type 1 diabetes is characterized by circulating immune markers directed against pancreatic islets (anti–islet cell antibodies) or β cell antigens (such as glutamic acid

decarboxylase [GAD]). A small minority (less than 10% in most studies) of women with GDM have the same markers present in their circulation (17, 28–31). Although detailed physiological studies of these women are lacking, they most likely have inadequate insulin secretion resulting from autoimmune damage to and destruction of pancreatic β cells. They appear to have evolving type 1 diabetes, which comes to clinical attention through routine glucose screening during pregnancy. The frequency of anti–islet cell and anti-GAD antibodies detected in GDM tends to parallel ethnic trends in the prevalence of type 1 diabetes outside of pregnancy. Patients with anti–islet cell or anti-GAD antibodies often, but not invariably, are lean, and they can rapidly develop overt diabetes after pregnancy (30).

Monogenic diabetes and GDM

Monogenic diabetes mellitus has been identified outside of pregnancy in 2 general forms. Some patients have mutations in autosomes (autosomal dominant inheritance pattern, commonly referred to as maturity-onset diabetes of the young [MODY], with genetic subtypes denoted as MODY 1, MODY 2, etc.). Others have mutations in mitochondrial DNA, often with distinct clinical syndromes such as deafness. In both instances, onset tends to occur

Table 2

Screening for GDM, step 2: blood glucose screening[A]

Glucose cut point[B]	Percentage of women with positive test[C]	Sensitivity for GDM[C]
≥140 mg/dl (7.8 mmol/l)	14–18%	~80%
≥130 mg/dl (7.2 mmol/l)	20–25%	~90%

[A]The test is a 50-g oral glucose challenge performed in patients with high or average clinical risk characteristics. No preparation is needed; the test can be done any time of day. Women with very high clinical risk characteristics may proceed directly to measurement of fasting glucose or to diagnostic OGTT (see Tables 3 and 4). [B]Venous serum or plasma glucose measured by certified clinical laboratory. Cut point identifies women recommended for 2- or 3-hour OGTTs as described in Table 3. [C]May vary with ethnicity and with diagnostic OGTT employed. Adapted with permission from *N. Engl. J. Med.* (72). Copyright 1999, Massachussets Medical Society. All rights reserved.

Table 3

Diagnosis of GDM during pregnancy[A]

Procedure	Glucose cut points[B]		
	Time (h)	mg/dl	mmol/l
100-g, 3-h OGTT[C]	Fasting	95	5.3
	1	180	10.0
	2	155	8.6
	3	140	7.8
75-g, 2-h OGTT[C]	Fasting	95	5.3
	1	180	10.0
	2	155	8.6

[A]Based on recommendations of the American Diabetes Association (5). [B]Venous serum or plasma glucose measured by a certified clinical laboratory. [C]Two or more values meeting or exceeding the cut points are required for diagnosis. Adapted with permission from *N. Engl. J. Med.* (72). Copyright 1999, Massachussets Medical Society. All rights reserved.

at an early age relative to other forms of nonimmune diabetes (e.g., type 2 diabetes, described below), and patients tend not to be obese or insulin resistant. Both features point to abnormalities in the regulation of β cell mass and/or function. Indeed, detailed metabolic studies have revealed abnormalities in glucose-mediated insulin secretion in some forms of MODY (32). Mutations that cause several subtypes of MODY have been found in women with GDM. These include mutations in genes coding for: (a) glucokinase (MODY 2) (29, 33–35); (b) hepatocyte nuclear factor 1α (MODY 3) (29); (c) and insulin promoter factor 1 (MODY 4) (29). Together, these monogenic forms of GDM account for less than 10% of GDM cases (29, 33–36). They likely represent cases of preexisting diabetes that are first detected by routine glucose screening during pregnancy.

Insulin resistance, β cell dysfunction, and GDM

The majority of women with GDM appear to have β cell dysfunction that occurs on a background of chronic insulin resistance. As noted above, pregnancy normally induces quite marked insulin resistance. This physiological insulin resistance also occurs in women with GDM. However, it occurs on a background of chronic insulin resistance to which the insulin resistance of pregnancy is partially additive. As a result, pregnant women with GDM tend to have even greater insulin resistance than normal pregnant women. Differences in whole-body insulin sensitivity tend to be small in the third trimester, owing to the marked effects of pregnancy itself on insulin resistance. Nonetheless, precise and direct measures of insulin sensitivity applied during the third trimester have identified, in women with GDM, exaggerated resistance to insulin's ability to stimulate glucose utilization (17, 18) and to suppress both glucose production (17, 18) and fatty acid levels (17). After delivery, when the acquired insulin resistance of pregnancy abates, women who had GDM end up, on average, with considerably greater insulin resistance than normal women. This finding, which has been consistent across studies in which whole-body insulin sensitivity has been measured directly (22, 23, 25, 26, 37–40), indicates that most women who develop GDM have chronic insulin resistance. Sequential measurements of insulin sensitivity performed in the same women before pregnancy, early in the second trimester, and in the third trimester have documented insulin resistance in both lean and obese women who go on to develop GDM (18, 24).

Only a small number of potential biochemical mediators of the chronic insulin resistance that frequently accompanies GDM and that likely contributes to the high risk of type 2 diabetes have been examined. It is likely that there is not a single underlying biochemical etiology. Women with GDM tend to be obese, so mechanisms promoting obesity and/or linking obesity to insulin resistance are likely to play a role. Small studies have revealed increased circulating levels of leptin (41) and the inflammatory markers TNF-α (42) and C-reactive protein (43) and decreased levels of adiponectin (44, 45) in women with GDM. Increased content of fat in liver (46) and muscle (47) has also been reported in women with previous gestational diabetes. All of these findings are consistent with the current understanding of some potential causes of obesity-related insulin resistance.

Defects in the binding of insulin to its receptor in skeletal muscle do not appear to be involved in the exaggerated insulin resistance of GDM (48). Alterations in the insulin signaling pathway (49–52), abnormal subcellular localization of GLUT4 transporters (53), reduced expression of PPARγ (49), increased expression of the membrane glycoprotein PC-1 (51), and reduced insulin-mediated glucose transport (52, 53) have been found in skeletal muscle or fat cells of women with GDM or a history thereof compared with normal women. Whether any of these defects is primary or the result of more fundamental defects in insulin action is currently unknown. Given that GDM represents a cross-section of young women with glucose intolerance, mechanisms that lead to chronic insulin resistance in GDM are likely to be as varied as they are in the general population.

It has long been thought (and taught) that GDM develops in women who cannot increase their insulin secretion when faced with the increased insulin needs imposed by late pregnancy. Serial studies of women who develop GDM do not support that concept. As illustrated in Figure 1, insulin secretion in obese women who develop GDM can increase considerably over weeks or months in association with the acquired insulin resistance of pregnancy. However, the increase occurs along an insulin sensitivity-secretion curve that is approximately 50% lower (i.e., 50% less insulin for any degree of insulin resistance) than that of normal women. These short-term responses appear to occur on a background of long-term deterioration of β cell function that, over years, leads to progressive hyperglycemia and diabetes (see "Link to diabetes after pregnancy," below). Longitudinal studies of lean and obese women before pregnancy, at the beginning of the second trimester, and in the third trimester also reveal an increase in insulin secretion in association with the acquired insulin resistance of pregnancy (18, 24). However, the increase is less than that which occurs in normal pregnant women despite somewhat greater insulin resistance in individuals with GDM. These small

Table 4

Probable pregestational diabetes[A]

Timing of sample	Serum or plasma glucose[B]
After overnight fast	≥126 mg/dl (7.0 mmol/l)
Random	≥200 mg/dl (11.1 mmol/l)

[A]Based on recommendations of the American Diabetes Association (5). [B]Venous serum or plasma glucose measured by a certified clinical laboratory. Adapted with permission from *N. Engl. J. Med.* (72). Copyright 1999, Massachussets Medical Society. All rights reserved.

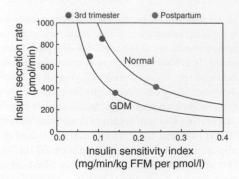

Figure 1
Insulin sensitivity-secretion relationships in women with GDM and normal women during the third trimester and remote from pregnancy. Values were measured at the end of 3-hour hyperglycemic clamps (plasma glucose, about 180 mg/dl) (22). Prehepatic insulin secretion rates were calculated from steady-state plasma insulin and C-peptide levels. Insulin sensitivity index was calculated as steady-state glucose infusion rate divided by steady-state plasma insulin concentration. FFM, fat-free mass. Figure reproduced with permission from *J. Clin. Endocrinol. Metab.* (27). Copyright 2001, The Endocrine Society.

but elegant physiological studies reveal that the limitation in insulin secretion in women with GDM is not necessarily fixed. Rather, in at least some of them, insulin secretion is low relative to their insulin sensitivity but responsive to changing sensitivity. One approach to the prevention of diabetes after GDM has taken advantage of this responsiveness (discussed below in "Link to diabetes after pregnancy").

Very little is known about the genetics of GDM in women with chronic insulin resistance. The few studies that have been done have compared allele frequencies of candidate genes in women with and without GDM, with no selection for specific phenotypic subtypes of GDM. Variants that differed in frequency between control and GDM subjects were found in genes coding for: (a) the islet-specific promoter of glucokinase (54), known to be important for glucose sensing by β cells; (b) calpain-10 (55), a gene associated with type 2 diabetes in Hispanic Americans and some other ethnic groups; (c) the sulfonylurea receptor 1 (56), which is involved in glucose-stimulated insulin secretion; and (d) the β3 adrenoreceptor, which may regulate body composition. Whether these findings will be confirmed in larger studies with broader representation among women with GDM remains to be determined.

Link to diabetes after pregnancy
The hyperglycemia of GDM is detected at one point in a women's life. If glucose levels are not already in the diabetic range, GDM could represent glucose intolerance that is limited to pregnancy, is chronic but stable, or is at a stage in the progression to diabetes. Long-term follow-up studies, recently reviewed by Kim et al. (57), reveal that most, but not all, women with GDM do progress to diabetes after pregnancy. Only approximately 10% of patients have diabetes soon after delivery (58). Incident cases appear to occur at a relatively constant rate during the first 10 years thereafter (57), and the few studies that have been conducted over a period of more than 10 years reveal a stable long-term risk of approximately 70% (57).

Most studies of risk factors for the development of diabetes after GDM fail to distinguish among the possible subtypes of GDM and diabetes discussed above. They generally reveal risk factors, such

as obesity, weight gain, and increased age, that are known to be associated with type 2 diabetes. Relatively high glucose levels during and soon after pregnancy also correlate with increased risk of diabetes, perhaps because they identify women who are relatively close to developing diabetes when the diagnosis of GDM is made.

Longitudinal studies of the pathophysiology of diabetes that develops after GDM are limited to Hispanic women with clinical characteristics suggesting a risk for type 2 diabetes. Those studies have revealed much about the β cell defect that leads to type 2 diabetes after GDM in 1 ethnic group. First, weight gain and additional pregnancies, factors associated with chronic and acute insulin resistance, respectively, independently increase the risk of developing diabetes (59). Second, decreasing β cell function is associated with increasing hyperglycemia (Figure 2) (60). The impact of reduced β cell function on glucose levels is relatively small until the disposition index, which reflects acute insulin responses to glucose in relation to insulin resistance, is very low (approximately 10–15% of normal). Thereafter, relatively small differences in β cell function are associated with relatively large increases in glucose levels (60). Third, treatment of insulin resistance at the stage of impaired glucose tolerance results in a reciprocal downregulation of insulin secretion (61), which in turn is associated with a reduction in the risk of diabetes and with preservation of β cell function (62). Taken together, these 3 findings reveal a progressive loss of insulin secretion that appears to be caused by high insulin secretory demands imposed by chronic insulin resistance. Glycemia in the diabetic range is a relatively late consequence of the loss of insulin secretion. That loss can be slowed or stopped through the treatment of insulin resistance in order to reduce of high insulin secretory demands (62). Whether the same or similar mechanisms of progressive β cell dysfunction and opportunities for β cell preservation occur in other ethnic groups remains to be determined.

Implications for clinical care
To date, insights into the mechanisms underlying impaired glucose regulation in GDM have not had an important impact on clinical management during pregnancy. The focus for antepartum care is on the use of standard antidiabetic treatments, mostly appropriate nutrition and exogenous insulin delivery but more recently administration of selected oral antidiabetic agents (63–65), to normalize maternal pre- and postprandial glucose levels and minimize fetal overnutrition. Fetal ultrasound measurements have also been used to refine the identification of pregnancies in which the fetus demonstrates signs of excessive adiposity — pregnancies in which the need to aggressively lower maternal glucose is the greatest (66, 67).

After pregnancy, the main focus of clinical care should be on reducing the risk of diabetes and detecting and treating diabetes that does develop. Measurement of fasting glucose in the immediate postpartum period will identify women with persistent fasting hyperglycemia in the diabetic range. Other women should have an OGTT sometime during the first 2–6 months postpartum and, if not diabetic, annual testing for diabetes. Family planning is important to reduce the occurrence of unplanned pregnancies in the presence of poorly controlled diabetes (68), a scenario that leads to serious birth defects in offspring (69).

Classification of patients into 1 of the 3 major subtypes of GDM discussed in this review can aid in clinical management. Lean patients are less likely to be insulin resistant than overweight or obese patients, so autoimmune and monogenic forms of diabetes should be considered in such patients. Screening for evolving autoimmune diabetes

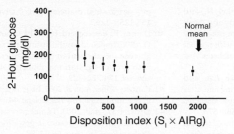

Figure 2
Relationship between pancreatic β cell function and post-challenge glucose levels in women with prior GDM. Data are from 71 nonpregnant Hispanic women who had at least 2 (86% had at least 3) sets of oral and frequently sampled i.v. glucose tolerance tests that were scheduled at 15-month intervals between 15 and 75 months after the index pregnancy (totaling 280 sets of tests). Participants had fasting plasma glucose of less than 140 mg/dl at entry into the study and were followed until that value was exceeded. Disposition index (*x* axis) is the product of minimal model insulin sensitivity (S_I) and the acute insulin response to i.v. glucose (AIRg), a measure of pancreatic β cell compensation for insulin resistance. The *y* axis shows glucose values at hour 2 of 75-g OGTTs. Symbols represent mean values for disposition index and corresponding 2-hour glucose values (± 1 SD) in each octile of disposition index. The mean disposition index was 2018 in Hispanic women without a history of GDM (arrow). Figure based on data from ref. 60.

by measuring antibodies to GAD may be warranted, particularly in women with no strong family history of diabetes who are from ethnic groups in which type 1 diabetes is relatively common. Although there are no established treatments to modify the progression to type 1 diabetes, careful monitoring of glucose levels is advised because patients can rapidly develop diabetes after pregnancy (30). Genotyping for monogenic diabetes is still primarily a research tool, but clinical tests are being developed. Early-onset diabetes with a relevant family history (autosomal dominant inheritance for MODY; maternal inheritance for mitochondrial mutations) may provide a clue to the diagnosis. In addition, Ellard et al. (34) have provided 4 clinical criteria that have relatively high specificity for identifying women with the glucokinase mutations that cause 1 form of MODY, MODY2: (a) persisting fasting hyperglycemia (105–145 mg/dl) after pregnancy; (b) a small (less than 82 mg/dl) increment in glucose above the fasting level during a 75-g, 2-hour OGTT; (c) insulin treatment during at least 1 pregnancy but subsequently controlled on diet; and (d) a first-degree relative with type 2 diabetes, GDM, or fasting serum or plasma glucose greater than 100 mg/dl. The constellation was infrequent in patients in the United Kingdom, but 80% of women who met all 4 criteria had glucokinase mutations. Identification of monogenic forms of diabetes is important for genetic counseling.

There is currently no clinical role for genetic testing for variants that have been associated with polygenic forms of type 2 diabetes (see "Insulin resistance, β cell dysfunction, and GDM," above). The variants cannot be used reliably to discriminate between normal individuals and individuals affected with diabetes and, just as importantly, testing is not available to clinicians. On the other hand, recent advances in the understanding of mechanistic links between GDM and type 2 diabetes have been translated into clinical care aimed at reducing the risk of diabetes. At least 2 studies of diabetes prevention in high-risk individuals have included women with a history of GDM. In the US Diabetes Prevention Program (DPP) (70), intensive lifestyle modification to promote weight loss and increase physical activity resulted in a 58% reduction in the risk of type 2 diabetes in adults with

impaired glucose tolerance. GDM was one of the risk factors that led to inclusion in the study. Protection against diabetes was observed in all ethnic groups. Treatment with metformin in the same study also reduced the risk of diabetes, but to a lesser degree and primarily in the youngest and most overweight participants. To date, specific results from women with a history of GDM have not been published.

The Troglitazone in Prevention of Diabetes (TRIPOD) study was conducted exclusively on Hispanic women with recent GDM. Assignment to treatment with the insulin-sensitizing drug troglitazone was associated with a 55% reduction in the incidence of diabetes. Protection from diabetes was closely linked to initial reductions in endogenous insulin requirements and ultimately associated with stabilization of pancreatic β cell function (62). Stabilization of β cell function was also observed when troglitazone treatment was started at the time of initial detection of diabetes by annual glucose tolerance testing (71). The DPP and TRIPOD studies support clinical management that focuses on aggressive treatment of insulin resistance to reduce the risk of type 2 diabetes and, at least in Hispanic women, to preserve pancreatic β cell function.

Taken together, these 2 studies suggest that postpartum management of women with clinical characteristics suggesting a risk for type 2 diabetes should focus on treatment of insulin resistance and monitoring of glycemia both to assess success (as reflected by stabilization of glucose levels) and to detect diabetes if it develops.

Future directions
Considerable work is needed to dissect the various mechanisms underlying maternal GDM and its evolution to diabetes after pregnancy. Large studies of screening for evolving autoimmune diabetes are necessary to more accurately define the clinical characteristics of women who need such screening as part of routine GDM management. Genetic studies may help identify women whose β cells will tolerate insulin resistance poorly, as well as women who develop poor insulin secretion for reasons unrelated to insulin resistance. Studies of gene-environment interactions and additional studies of insulin action in muscle and fat may identify causes of insulin resistance, especially as they relate to obesity. Better understanding of mechanisms that can lead to GDM should allow more rational development and administration of therapy during pregnancy, as well as more rational approaches to prevention of both GDM and diabetes after pregnancy. GDM is an especially attractive target for such studies because the disease is detected in the course of routine clinical care and it provides an opportunity to study relatively early stages of glucose dysregulation that may be fundamental to the long-term pathobiology of diabetes.

Acknowledgments
We thank our long-term collaborators Siri Kjos, Ruth Peters, and Richard Bergman for their contributions to studies on the pathogenesis of type 2 diabetes after GDM in Hispanic women. Our work cited in this paper was supported by research grants from the NIH (R01-DK46374, R01-DK61628, and M01-RR00043), the American Diabetes Association (Clinical Research Award and Distinguished Clinical Scientist Award), and Parke-Davis Pharmaceutical Research (the TRIPOD study).

Address correspondence to: Thomas A. Buchanan, Room 6602 GNH, 1200 North State Street, Los Angeles, California 90089-9317, USA. Phone: (323) 226-4632; Fax: (323) 226-2796; E-mail: buchanan@usc.edu.

1. Duncan, M. 1882. On puerperal diabetes. *Trans. Obstet. Soc. Lond.* **24**:256–285.

2. Miller, H.C. 1946. The effect of diabetic and prediabetic pregnancies on the fetus and newborn infant. *J. Pediatr.* **26**:455–461.

3. Carrington, E.R., Shuman, C.R., and Reardon, H.S. 1957. Evaluation of the prediabetic state during pregnancy. *Obstet. Gynecol.* **9**:664–669.

4. O'Sullivan, J.B., and Mahan, C.M. 1964. Criteria for the oral glucose tolerance test in pregnancy. *Diabetes.* **13**:278–285.

5. American Diabetes Association. 2003. Report of the expert committee on the diagnosis and classification of diabetes mellitus. *Diabetes Care.* **26**(Suppl. 1):S5–S20.

6. Naylor, C.D., Sermer, M., Chen, E., and Sykora, K. 1996. Cesarean delivery in relation to birth weight and gestational glucose tolerance: pathophysiology or practice style? *JAMA.* **275**:1165–1170.

7. Magee, M.S., Walden, C.E., Benedetti, T.J., and Knopp, R.H. 1993. Influence of diagnostic criteria on the incidence of gestational diabetes and perinatal morbidity. *JAMA.* **269**:609–615.

8. Schmidt, M.I., et al. 2001. Gestational diabetes mellitus diagnosed with a 2-h 75-g oral glucose tolerance test and adverse pregnancy outcomes. *Diabetes Care.* **24**:1151–1155.

9. Pettitt, D.J., and Knowler, W.C. 1998. Long-term effects of the intrauterine environment, birth weight, and breast-feeding on Pima Indians. *Diabetes Care.* **21**(Suppl. 2):B138–B141.

10. Silverman, B.L., Rizzo, T.A., Cho, N.H., and Metzger, B.E. 1998. Long-term effects of the intrauterine environment. The Northwestern University Diabetes in Pregnancy Center. *Diabetes Care.* **21**(Suppl. 2):B142–B149.

11. Vohr, B.R., McGarvey, S.T., and Tucker, R. 1999. Effects of maternal gestational diabetes on offspring adiposity at 4-7 years of age. *Diabetes Care.* **22**:1284–1291.

12. American Diabetes Association. 2004. Diagnosis and classification of diabetes mellitus. *Diabetes Care.* **27**(Suppl. 1):S5–S10.

13. King, H. 1998. Epidemiology of glucose intolerance and gestational diabetes in women of childbearing age. *Diabetes Care.* **21**(Suppl. 2):B9–B13.

14. Ben-Haroush, A., Yogev, Y., and Hod, M. 2004. Epidemiology of gestational diabetes mellitus and its association with Type 2 diabetes. *Diabet. Med.* **21**:103–113.

15. Weijers, R.N., Bekedam, D.J., and Smulders, Y.M. 2002. Determinants of mild gestational hyperglycemia and gestational diabetes mellitus in a large Dutch multiethnic cohort. *Diabetes Care.* **25**:72–77.

16. Gunton, J.E., Hitchman, R., and McElduff, A. 2001. Effects of ethnicity on glucose tolerance, insulin resistance and beta cell function in 223 women with an abnormal glucose challenge test during pregnancy. *Aust. N. Z. J. Obstet. Gynecol.* **41**:182–186.

17. Xiang, A.H., et al. 1999. Multiple metabolic defects during late pregnancy in women at high risk for type 2 diabetes mellitus. *Diabetes.* **48**:848–854.

18. Catalano, P.M., Huston, L., Amini, S.B., and Kalhan, S.C. 1999. Longitudinal changes in glucose metabolism during pregnancy in obese women with normal glucose tolerance and gestational diabetes. *Am. J. Obstet. Gynecol.* **180**:903–916.

19. Catalano, P.M., Tyzbir, E.D., Roman, N.M., Amini, S.B., and Sims, E.A. 1991. Longitudinal changes in insulin release and insulin resistance in nonobese pregnant women. *Am. J. Obstet. Gynecol.* **165**:1667–1672.

20. Yen, S.C.C., Tsai, C.C., and Vela, P. 1971. Gestational diabetogenesis: quantitative analysis of glucose-insulin interrelationship between normal pregnancy and pregnancy with gestational diabetes. *Am. J. Obstet. Gynecol.* **111**:792–800.

21. Buchanan, T.A., Metzger, B.E., Freinkel, N., and Bergman, R.N. 1990. Insulin sensitivity and B-cell responsiveness to glucose during late pregnancy in lean and moderately obese women with normal glucose tolerance or mild gestational diabetes. *Am. J. Obstet. Gynecol.* **162**:1008–1014.

22. Homko, C., Sivan, E., Chen, X., Reece, E.A., and Boden, G. 2001. Insulin secretion during and after pregnancy in patients with gestational diabetes mellitus. *J. Clin. Endocrinol. Metab.* **86**:568–573.

23. Kautzky-Willer, A., et al. 1997. Pronounced insulin resistance and inadequate betacell secretion characterize lean gestational diabetes during and after pregnancy. *Diabetes Care.* **20**:1717–1723.

24. Catalano, P.M., et al. 1993. Carbohydrate metabolism during pregnancy in control subjects and women with gestational diabetes. *Am. J. Physiol.* **264**:E60–E67.

25. Ryan, E.A., et al. 1995. Defects in insulin secretion and action in women with a history of gestational diabetes. *Diabetes.* **44**:506–512.

26. Osei, K., Gaillard, T.R., and Schuster, D.P. 1998. History of gestational diabetes leads to distinct metabolic alterations in nondiabetic African-American women with a parental history of type 2 diabetes. *Diabetes Care.* **21**:1250–1257.

27. Buchanan, T.A. 2001. Pancreatic B-cell defects in gestational diabetes: implications for the pathogenesis and prevention of type 2 diabetes. *J. Clin. Endocrinol. Metab.* **86**:989–993.

28. Petersen, J.S., et al. 1996. GAD65 autoantibodies in women with gestational or insulin dependent diabetes mellitus diagnosed during pregnancy. *Diabetologia.* **39**:1329–1333.

29. Weng, J., et al. 2002. Screening for MODY mutations, GAD antibodies, and type 1 diabetes–associated HLA genotypes in women with gestational diabetes mellitus. *Diabetes Care.* **25**:68–71.

30. Mauricio, D., et al. 1992. Islet cell antibodies identify a subset of gestational diabetic women with higher risk of developing diabetes shortly after pregnancy. *Diabetes Nutr. Metab.* **5**:237–241.

31. Catalano, P.M., Tyzbir, E.D., and Sims, E.A.H. 1990. Incidence and significance of islet cell antibodies in women with previous gestational diabetes. *Diabetes Care.* **13**:478–482.

32. Fajans, S.S., Bell, G.I., and Polonsky, K.S. 2001. Molecular mechanisms and clinical pathophysiology of maturity-onset diabetes of the young. *N. Engl. J. Med.* **345**:971–980.

33. Kousta, E., et al. 2001. Glucokinase mutations in a phenotypically selected multiethnic group of women with a history of gestational diabetes. *Diabet. Med.* **18**:683–684.

34. Ellard, S., et al. 2000. A high prevalence of glucokinase mutations in gestational diabetic subjects selected by clinical criteria. *Diabetologia.* **43**:250–253.

35. Saker, P.J., et al. 1996. High prevalence of a missense mutation of the glucokinase gene in gestational diabetic patients due to a founder-effect in a local population. *Diabetologia.* **39**:1325–1328.

36. Chen, Y., Liao, W.X., Roy, A.C., Loganath, A., and Ng, S.C. 2000. Mitochondrial gene mutations in gestational diabetes mellitus. *Diabetes Res. Clin. Pract.* **48**:29–35.

37. Ward, W.K., et al. 1985. Insulin resistance and impaired insulin secretion in subjects with a history of gestational diabetes mellitus. *Diabetes.* **34**:861–869.

38. Ward, W.K., Johnston, C.L.W., Beard, J.C., Benedetti, T.J., and Porte, D., Jr. 1985. Abnormalities of islet B cell function, insulin action and fat distribution in women with a history of gestational diabetes: relation to obesity. *J. Clin. Endocrinol. Metab.* **61**:1039–1045.

39. Catalano, P.M., et al. 1986. Subclinical abnormalities of glucose metabolism in subjects with previous gestational diabetes. *Am. J. Obstet. Gynecol.* **155**:1255–1263.

40. Damm, P., Vestergaard, H., Kuhl, C., and Pedersen, O. 1996. Impaired insulin-stimulated nonoxidative glucose metabolism in glucose-tolerant women with previous gestational diabetes. *Am. J. Obstet. Gynecol.* **174**:722–729.

41. Kautzky-Willer, A., et al. 2001. Increased plasma leptin in gestational diabetes. *Diabetologia.* **44**:164–172.

42. Winkler, G., et al. 2002. Tumor necrosis factor system and insulin resistance in gestational diabetes. *Diabetes Res. Clin. Pract.* **56**:93–99.

43. Retnakaran, R., et al. 2003. C-reactive protein and gestational diabetes: the central role of maternal obesity. *J. Clin. Endocrinol. Metab.* **88**:3507–3512.

44. Retnakaran, R., et al. 2004. Reduced adiponectin concentration in women with gestational diabetes: a potential factor in progression to type 2 diabetes. *Diabetes Care.* **27**:799–800.

45. Williams, M.A., et al. 2004. Plasma adiponectin concentrations in early pregnancy and subsequent risk of gestational diabetes mellitus. *J. Clin. Endocrinol. Metab.* **89**:2306–2311.

46. Tiikkainen, M., et al. 2002. Liver-fat accumulation and insulin resistance in obese women with previous gestational diabetes. *Obes. Res.* **10**:859–867.

47. Kautzky-Willer, A., et al. 2003. Increased intramyocellular lipid concentration identifies impaired glucose metabolism in women with previous gestational diabetes *Diabetes.* **52**:244–251.

48. Damm, P., et al. 1993. Insulin receptor binding and tyrosine kinase activity in skeletal muscle from normal pregnant women and women with gestational diabetes. *Obstet. Gynecol.* **82**:251–259.

49. Catalano, P.M., et al. 2002. Downregulated IRS-1 and PPARgamma in obese women with gestational diabetes: relationship to FFA during pregnancy. *Am. J. Physiol.* **282**:E522–E533.

50. Shao, J., Yamashita, H., Qiao, L., Draznin, B., and Friedman, J.E. 2002. Phosphatidylinositol 3-kinase redistribution is associated with skeletal muscle insulin resistance in gestational diabetes mellitus. *Diabetes.* **51**:19–29.

51. Shao, J., et al. 2000. Decreased insulin receptor tyrosine kinase activity and plasma cell membrane glycoprotein-1 over expression in skeletal muscle from obese women with gestational diabetes (GDM): evidence for increased serine/threonine phosphorylation in pregnancy and GDM. *Diabetes.* **49**:603–610.

52. Friedman, J.E., et al. 1999. Impaired glucose transport and insulin receptor tyrosine phosphorylation in skeletal muscle from obese women with gestational diabetes. *Diabetes.* **48**:1807–1814.

53. Garvey, W.T., et al. 1993. Multiple defects in the adipocyte glucose transport system cause cellular insulin resistance in gestational diabetes. *Diabetes.* **42**:1773–1785.

54. Zaidi, F.K., et al. 1997. Homozygosity for a common polymorphism in the islet-specific promoter of the glucokinase gene is associated with a reduced early insulin response to oral glucose in pregnant women. *Diabet. Med.* **14**:228–234.

55. Leipold, H., et al. 2004. Calpain-10 haplotype combination and association with gestational diabetes mellitus. *Obstet. Gynecol.* **103**:1235–1240.

56. Rissanen, J., et al. 2000. Sulfonylurea receptor 1 gene variants are associated with gestational diabetes and type 2 diabetes but not with altered secretion of insulin. *Diabetes Care.* **23**:70–73.

57. Kim, C., Newton, K.M., and Knopp, R.H. 2002. Gestational diabetes and the incidence of type 2 diabetes. *Diabetes Care.* **25**:1862–1868.

58. Kjos, S.L., et al. 1990. Gestational diabetes mellitus: the prevalence of glucose intolerance and diabetes mellitus in the first two months postpartum. *Am. J. Obstet. Gynecol.* **163**:93–98.

59. Peters, R.K., Kjos, S.L., Xiang, A., and Buchanan,

T.A. 1996. Long-term diabetogenic effect of a single pregnancy in women with prior gestational diabetes mellitus. *Lancet.* **347**:227–230.

60. Buchanan, T.A., et al. 2003. Changes in insulin secretion and sensitivity during the development of type 2 diabetes after gestational diabetes in Hispanic women. *Diabetes.* **52**(Suppl. 1):A34.

61. Buchanan, T.A., et al. 2000. Response of pancreatic B-cells to improved insulin sensitivity in women at high risk for type 2 diabetes. *Diabetes.* **49**:782–788.

62. Buchanan, T.A., et al. 2002. Preservation of pancreatic B-cell function and prevention of type 2 diabetes by pharmacological treatment of insulin resistance in high-risk Hispanic women. *Diabetes.* **51**:2769–2803.

63. Langer, O., Conway, D.L., Berkus, M.D., Xenakis, E.M., and Gonzales, O. 2000. A comparison of glyburide and insulin in women with gestational diabetes mellitus. *N. Engl. J. Med.* **343**:1134–1138.

64. Hellmuth, E., Damm, P., and Molsted-Pedersen, L. 2000. Oral hypoglycaemic agents in 118 diabetic pregnancies. *Acta. Obstet. Gynecol. Scand.* **79**:958–962.

65. Glueck, C.J., Goldenberg, N., Streicher, P., and Wang, P. 2003. Metformin and gestational diabetes. *Curr. Diab. Rep.* **3**:303–312.

66. Kjos, S.L., et al. 2001. A randomized controlled trial utilizing glycemic plus fetal ultrasound parameters vs glycemic parameters to determine insulin therapy in gestational diabetes with fasting hyperglycemia. *Diabetes Care.* **24**:1904–1910.

67. Buchanan, T.A., et al. 1994. Use of fetal ultrasound to select metabolic therapy for pregnancies complicated by mild gestational diabetes. *Diabetes Care.* **17**:275–283.

68. Kjos, S.L., Peters, R.K., Xiang, A., Schaefer, U., and Buchanan, T.A. 1998. Hormonal choices after gestational diabetes: subsequent pregnancy, contra-

ception and hormone replacement. *Diabetes Care.* **21**(Suppl. 2):B50–B57.

69. Towner, D., et al. 1995. Congenital malformations in pregnancies complicated by non-insulin-dependent diabetes mellitus: increased risk from poor maternal metabolic control but not from exposure to sulfonylurea drugs. *Diabetes Care.* **18**:1446–1451.

70. Diabetes Prevention Program Research Group. 2002. Reduction in the incidence of type 2 diabetes with lifestyle intervention or metformin. *N. Engl. J. Med.* **346**:393–403.

71. Xiang, A.H., et al. 2004. Pharmacological treatment of insulin resistance at two different stages in the evolution of type 2 diabetes: impact on glucose tolerance and β-cell function. *J. Clin. Endocrinol. Metab.* **89**:2846–2851.

72. Kjos, S.L., and Buchanan, T.A. 1999. Gestational diabetes mellitus. *N. Engl. J. Med.* **341**:1749–1756.

Regeneration of the pancreatic β cell

Massimo Trucco

Division of Immunogenetics, Department of Pediatrics, University of Pittsburgh School of Medicine,
Children's Hospital of Pittsburgh, Pittsburgh, Pennsylvania, USA.

Type 1 diabetes is the result of an autoimmune attack against the insulin-producing β cells of the endocrine pancreas. Current treatment for patients with type 1 diabetes typically involves a rigorous and invasive regimen of testing blood glucose levels many times a day along with subcutaneous injections of recombinant DNA–derived insulin. Islet transplantation, even with its substantially improved outcome in recent years, is still not indicated for pediatric patients. However, in light of the fact that some regenerative capabilities of the endocrine pancreas have been documented and recent research has shown that human ES cell lines can be derived in vitro, this review discusses whether it is practical or even possible to combine these lines of research to more effectively treat young diabetic patients.

In vertebrates, the process of gastrulation takes place very early during the development of the embryo. This process reorganizes the embryo's cells into 3 layers: ectoderm, endoderm, and mesoderm. The ectoderm forms the skin and the central nervous system; the mesoderm gives rise to the cells from which blood, bone, and muscle are derived; and the endoderm forms the respiratory and digestive tracts (1, 2). The embryonic endoderm, taking the shape of the primitive gut tube, serves as a template for the gastrointestinal tract from which the embryonic pancreas eventually buds. It has been shown that the branching morphogenesis of the pancreatic bud gives rise to the ducts and the acinar components of the gland. Endocrine progenitors, proliferating from the budding ducts, then form aggregates of differentiated cells known as the islets of Langerhans (Figure 1). While the pancreatic acini, composed of cells dedicated to the secretion into the intestine of enzymes that will participate in the digestive process of the ingested food, constitute the exocrine component of the gland, the islets are made up of 4 cell populations, organized in a stereotypical topological order, which constitute instead the endocrine component of the gland. In the islet, the α cells produce glucagon; the β cells, insulin; the γ cells, pancreatic polypeptide; and the δ cells, somatostatin (1, 2) (Figure 1).

It is in large part due to this knowledge of pancreatic embryogenesis that residual pancreatic endocrine progenitors, able to guarantee islet homeostasis, were originally postulated to be still present in the pancreatic ducts of the adult gland. The identification and characterization of these putative progenitors is of paramount importance, not only for a better understanding of endocrine pancreatic physiology and pathology, but also for the development of potential therapeutic approaches that their correct exploitation may offer.

Type 1 diabetes is the clinical consequence of the destruction of the insulin-producing β cells of the pancreas, mediated by autoreactive T cells specifically directed against β cell determinants (3). The loss of the majority of the β cell population, evident at the onset of the disease, requires daily subcutaneous injections of quantities of insulin that should be proportional to the quantity of glucose present in the blood at each moment in time. The physical replacement of the β cell mass constitutes the rationale for which islet transplantation was originally proposed by Paul Lacy (4). Although it was recently demonstrated that islet cell transplants can be performed with greater chances of success than just a few years ago (5), the constrains under which this is clinically possible are still too numerous to allow the broad application of this procedure to permanently cure the disease. The immunosuppressive drug regimen necessary to protect islets from a recurrent autoimmune response and allorejection may, with time, irreversibly damage kidney function, while the process of islet isolation itself, even if drastically improved during the last few years, damages transplantable islets and, consequently, two to three donors are necessary in order to obtain the minimal cell mass sufficient for transplantation into a single recipient (6).

In light of recent discoveries showing the regenerative capabilities of the endocrine pancreas and the in vitro derivation of human ES cell lines, we have to consider possible, non–mutually exclusive alternatives to allogeneic islet transplantation.

Does the elusive pancreatic stem cell exist?

There exists, in humans, a stem cell committed to a specific lineage that is capable of giving rise to all types of differentiated cells and tissues, including extraembryonic tissues: the totipotent cell. The mammalian zygote perhaps should be considered the preeminent totipotent stem cell by antonomasia. However, in utero, this stem cell continues to divide and becomes an amalgam of similar, but not identical, daughter cells. We do not yet know how to distinguish among these daughter cells the few that continue to have the capacity to regenerate the whole, multivariate, final product — if these totipotent cells still exist at all. Consequently, we do not yet have specific markers capable of characterizing the totipotent cell. Once the various tissues and organs begin to form, we do not know whether any totipotent cells are preserved within them and, if they are, how long they could continue to be functional. Intuitively, we can argue that precursors of some kind are present and active within our body throughout life, because even elderly people are able to repair damaged tissues, albeit with reduced efficiency. However, we still do not know where these hypothetical precursor cells may be hiding and which final differentiated cells they can in fact generate.

Teleologically, our regenerative system should have developed according to the same rationale that underlies the stationing of firehouses throughout an entire city to allow each unit to be able to more rapidly reach the fire location and efficiently intervene. That is, it should have deployed into each organ not necessarily totipotent cells, but at least precursors with self-maintenance capabilities, as well as those necessary to replace worn-out

Nonstandard abbreviations used: ALS, anti-lymphocyte serum; BMP4, bone morphogenetic protein 4; GIP, glucose-dependent insulinotropic polypeptide; MDC, muscle-derived cell; PMP, pancreas-derived multipotent precursor.

Conflict of interest: The author has declared that no conflict of interest exists.

Citation for this article: *J. Clin. Invest.* **115**:5–12 (2005).
doi:10.1172/JCI200523935.

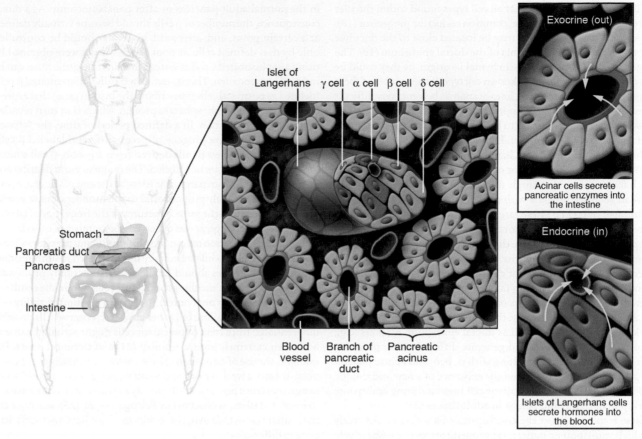

Figure 1

Cross section of the pancreas. The pancreas houses 2 distinctly different tissues. Its bulk comprises exocrine tissue, which is made up of acinar cells that secrete pancreatic enzymes delivered to the intestine to facilitate the digestion of food. Scattered throughout the exocrine tissue are many thousands of clusters of endocrine cells known as islets of Langerhans. Within the islet, α cells produce glucagon; β cells, insulin; δ cells, somatostatin; and γ cells, pancreatic polypeptide — all of which are delivered to the blood.

cells. These precursors should be able to support a process that, while progressing relatively slowly in the maintenance of tissue homeostasis, could be converted, in case of crisis, into a rapidly operating system. Consequently, this system would be highly effective at responding promptly to a number of stimulations, promoted by nervous signaling, microarchitectural remnants of destroyed tissues, metabolites of energy generating processes, peptide growth factors, etc. (7–9). Physiologically, in the endocrine pancreas, hormone-producing cells that are at the end of their life span should be continuously, albeit quite slowly, replaced by newly generated cells. We need to learn more about this regenerative process if we wish to take advantage of it for therapeutic purposes.

In considering the utility of stem cells for the regeneration of the β pancreatic cell, we currently face some major questions: in the adult endocrine pancreas, do multipotent progenitors still exist? If so, where exactly are they located? What are the best markers for recognizing and isolating them? What are the stimuli able to activate their differentiation pathway(s)? How large is the time window in which they can still respond to the needs of an aging tissue? In the absence of pancreatic progenitors, are there adult pluripotent stem cells — cells not committed to a specific lineage that may differentiate into all types of cells and tissues, with the exception of extraembryonic tissues — located elsewhere in the body that may

have the capability to regenerate β cell mass? If these cells are all lineage-committed precursor cells, can we instead use ES cell lines derived in vitro to substitute the lost β cells of the pancreas?

The road to regeneration

Increases in β cell mass may occur through increased β cell replication, increased β cell size, decreased β cell death, and differentiation of possibly existing β cell progenitors (10). It has been shown that occasional endocrine cells can be found embedded in normal pancreatic ducts. However, these cells are few and far between (11). The number of these duct-associated endocrine cells physiologically increases as the consequence of severe insulin resistance in obese individuals or during pregnancy (12, 13). Similar histological changes are observed under conditions of tissue injury and repair after partial pancreatectomy, duct ligation, cellophane wrapping of the gland, or IFN-γ overexpression driven by the insulin promoter (14–17). Even then, within the ducts, only a small number of cells become insulin positive. This suggests that even if some hypothetical precursors exist, the process of formation of endocrine cells out of the islet (neogenesis) would not be a frequently observed property of the duct epithelium. On the other hand, the fact that α and β cells develop from a possibly common, non–hormone-expressing, yet *Pdx1*-positive precursor (Pdx1 being a transcription factor required for pancreatic development) suggests that all cell types found within the islet may originate from a bona fide, common endocrine progenitor (18). These endocrine progenitors may be located close to the duct but may not actually be components of the ductal epithelium (19). The progenitor cells could be mesenchymal in origin, or they could be cells differentiated from an unknown cell type. If the number of these progenitors is extremely small, lineage analysis becomes very difficult because of the lack of known appropriate markers. Moreover, if these cells are as rare as they appear to be, it becomes difficult to quantify their contribution to normal endocrine cell turnover.

These are some of the conclusions discussed by Weir and Bonner-Weir (20) in commenting on the study by Seaberg et al., in which it was shown that single murine adult pancreatic precursor cells can generate progeny with characteristics of pancreatic cells, including β cells (21). These rare (1 in 3,000–9,000 cells) pancreas-derived multipotent precursors (PMPs) do not seem to be bona fide pluripotent ES cells, since they lack, for example, the Oct4 and Nanog markers that direct the propagation of undifferentiated ES cells; nor are these cells of clear ectodermal, mesodermal, or endodermal origin, since they failed to express other markers considered specific for precursors of each of the embryonic cell types (20, 21). Because, surprisingly, these PMPs also lacked some β cell markers (e.g., HNF3β) as well as ductal epithelium markers (e.g., cytokeratin), but were able to generate differentiation products with neural characteristics along with α, β, δ, and acinar pancreatic cells, the authors proposed the existence of a new and unique ectodermal/endodermal precursor cell present during embryonic development that could persist in adult tissues (21).

These results support the conclusions of another recent study in which multipotent pancreatic progenitors were prospectively isolated using flow-cytometric cell sorting (22). The marker used in this case was c-Met, the HGF receptor. The rationale for this choice was the known signal exchange between epithelial and mesenchymal cells, promoting the interaction between c-Met and HGF, which plays an important role in the development of the pancreas. The authors suggest that c-Met–HGF interaction is critically responsible for growth and differentiation of pancreatic stem and progenitor cells not only during development but also in the adult, where they maintain homeostasis and promote regeneration. Colonies derived from single c-Met–positive cells, sorted from neonatal and adult mouse pancreatic tissues, contained cells expressing several markers for endocrine, acinar, and ductal lineage cells. While neuroectodermal markers were not evaluated, the isolated pancreatic stem cells in the study by Suzuki et al. (22) were also able to generate offspring cells expressing hepatocyte and gastrointestinal cell markers, possibly due to the selection marker used. Seaberg's PMP-derived cells were grown instead in the serum-free medium conditions normally used for neural stem cell culture (21).

All of these studies, even with their somewhat divergent outcomes, seem to support the conclusion that endocrine precursor cells of some kind exist in the pancreas. They are present not only in the duct, but also within the islets themselves, since both subpopulations were independently used as the source of the isolated single cell precursors (21). On the one hand, this conclusion supports the working hypothesis of those who propose that pancreatic ductal cells can transdifferentiate into β cells and that this is a physiologic process generally more efficiently activated by increased metabolic demand and tissue injuries (23); on the other hand, it may also accommodate the most recent results of Dor and colleagues (24), who propose instead that no β cell can arise from non–β cell progenitors, whether in the normal adult pancreas or after pancreatectomy. As a direct consequence, the number of β cells should become virtually defined at a certain point, and, afterward, glycemia should be controlled only by that defined cellular pool. Dor's results were obtained by using a sophisticated Cre/lox system that, in transgenic mice, can be induced by tamoxifen. This system labels fully differentiated β cells (defined as postnatal cells transcribing the insulin gene) that express the human alkaline phosphatase protein, which is in turn revealed by a histochemical stain. In a defined period of time, the "chase," only the cells that are progeny of preexisting and labeled β cells are newly labeled. New β cells derived from any non–β cell source, including stem cells, are not labeled. The frequency and distribution of labeled β cells within pancreatic islets, at the end of the chase period, should be inversely proportional to the number of new, nonlabeled cells present in the same structures. If the frequency of labeled β cells does not change, as was observed, the number of cells derived from the differentiation of non–insulin-producing precursors must be minimal or null, while terminally differentiated insulin-producing β cells themselves should be the cells that actually proliferate and give rise to other insulin-producing β cells. While the results of Seaberg et al. (21) do not contest the proven yet limited ability of a β cell to divide, the failure of Dor et al. (24) to observe cells possibly differentiated from stem or precursor cells might actually be due to both their extremely limited number (21) and technical issues. For example, the use of tamoxifen injected intraperitoneally or subcutaneously twice a week for two and a half weeks (24) may have blocked neogenesis from precursor cells mainly in pancreatic ducts adjacent to involuted islets, as observed by Pelengaris et al. (25), and once the tamoxifen was withdrawn, these cells may not have had sufficient time to differentiate.

Controlling the autoimmune response

On this basis, we can tentatively conclude that precursors of a perhaps unconventional type can be located both in close proximity to and inside the endocrine tissue and that they can be activated by increased metabolic demand or by still-unknown secreted factors, normally able to accelerate the process that guarantees homeostasis of islets of Langerhans under normal conditions. The physiologic equilibrium between lost and newly generated cells can be altered by the action of β cell–specific, autoreactive T cells in instances in which autoimmunity develops (3). Once T cell killing activity overcomes the regenerative compensatory activity of the organ, the number of functional β cells progressively decreases until they become too few to maintain the glucohomeostasis of the entire body. The time of transition over this metabolic threshold becomes immediately evident with the presentation of the characteristic signs of the clinical onset of type 1 diabetes. During the course of disease, even if the regenerative properties of the pancreas remain functional, the continued presence of diabetogenic, autoreactive T cells consistently nullifies the reparative effort. The fact that these autoreactive T cells remain present in the body of the diabetic patient for a long time is proven by experiments in which healthy islet cells transplanted into syngeneic, long-term diabetic mice or humans were quickly killed by these same autoreactive T cells (26).

The autoimmune response is successfully averted in the NOD mouse either by directly eliminating the majority of the autoreactive T cells with anti–T cell antibodies or by substituting all or part of the immunocompetent cell repertoire with bone marrow cells obtained from diabetic-resistant donors.

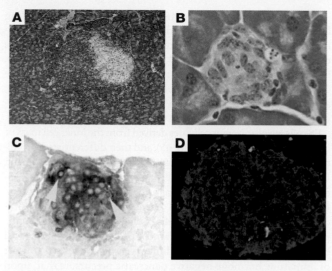

Figure 2
Regeneration of the β cell in diabetic NOD mice. (**A**) In NOD mice, the infiltration of autoreactive T cells into the islets of Langerhans (resulting in insulitis) begins at around 4 weeks of age. At 20 to 23 weeks, approximately 85% of female mice are diabetic, i.e., their glycemia is greater than 300 mg/dl. Magnification, ×200. (**B**) When it is successfully transplanted with bone marrow from a non–diabetes-prone donor and hematopoietic chimerism is established, the NOD mouse no longer show signs of autoimmune activity. However, while there is no more evidence of insulitis in the endogenous pancreas, there is also no sign of insulin production (no red staining). Magnification, ×400. (**C**) Insulin-positive cells in the islets can be seen to be dividing (yellow arrows); i.e., they are insulin (blue) and BrdU (red) positive. Magnification, ×400. (**D**) Three to 4 months after bone marrow transplantation, new insulin-positive cells (shown in red) are present throughout the endogenous pancreas. Magnification, ×200. Thus, when the islets transplanted under the kidney capsule in order to maintain euglycemia while regeneration takes place are removed by nephrectomy, the mice remain nondiabetic. Figure reproduced with permission from *Stem Cells* (33).

In a study by Ogawa et al., the treatment of overtly diabetic NOD mice with anti-lymphocyte serum (ALS) abrogated autoimmunity but achieved only partial clinical remission (27). Transient treatment of overtly diabetic NOD mice with ALS and exendin-4, a potent insulinotropic hormone that promotes replication and differentiation of β cells in vitro and in vivo, achieved instead complete remission of 88% of the treated animals within 75 days, accompanied by progressive normalization of glucose tolerance, improved islet histology, increased insulin content in the pancreas, and almost normal insulin release in response to a glucose challenge. These results show that exendin-4 synergistically augments the remission-inducing effect of ALS, possibly by promoting differentiation of β cell precursors (27).

Figure 3
Using a GFP-transgenic mouse as donor, it is possible to observe how the majority of the transplanted bone marrow cells do not directly participate in the regeneration of the endogenous pancreas. As shown here, there are no double-positive (orange) cells in the newly formed islets. The donor cells (green) appear to be located close to possibly existing juxta-ductal precursor cells, which may be activated by bone marrow cell–secreted factors. Insulin-positive cells are red. Magnification, ×400. See also refs. 32 and 33.

Also, the successful induction of a mixed allogeneic chimerism obtained after transplanting bone marrow from a diabetes-resistant donor into a diabetic animal following a sublethal dose of irradiation is sufficient to block and eventually also revert the systematic invasion and inflammation of the islets by the autoreactive T cells that result in insulitis (Figure 2) (28–30). Within the endocrine pancreas, once the insult of autoimmunity is abrogated, the physiologic process of regeneration can continue efficiently, eventually replenishing the population of insulin-producing cells to a number sufficient to maintain euglycemia, thus curing the diabetic recipient (Figure 2D) (31–33). While this process takes place — and it is still debatable whether this occurs over weeks (32) or months (33) — the recipient's glycemia must be controlled by additional, independent measures. The most commonly used technique is to transplant into the recipient islets from the same marrow donor. However, the successful engraftment of the transplanted bone marrow, or the establishment of a steady hematopoietic chimerism, would have to be maintained without the use of immunosuppressive agents. These potent drugs may kill not only the still-present autoreactive T cells of the recipient, but also the β cells themselves, thereby defeating the purpose of the transplant (34–36). The use of immunosuppressive agents may also interfere with the observed rise of regulatory T cells, a possible explanation for the long-lasting immunoregulatory cell–dominant condition observed in cured animals. Adoptive transfer experiments in which both diabetogenic lymphocytes and splenocytes from ALS-treated, long-term diabetes–free NOD mice were transplanted in NOD/SCID mice with no signs of diabetes induction support this hypothesis (27).

A subject of ongoing debate is whether either or both the transplanted bone marrow and the cotransplanted β cells are necessary for promoting an efficient regenerative process, independent of their ability to block autoimmunity or preserve euglycemia, respectively. They may, for example, secrete factors such as glucagon-like peptides, which are useful in order to sustain an efficient regenerative process (27, 37, 38). Strong evidence suggests that the hematopoietic precursors present in the bone marrow cell population do not directly participate in the reparative process of the insulin-producing cell population (Figure 3) (32, 33). In the cured recipient, insulin-producing cells that are genetically marked (by GFP or the Y chromosome of the male donor in the female recipient) to indicate they are of donor origin are extremely rare, occurring in no more than 2 of more than 100,000 β cells. These cells may actually be the result of sporadic cell fusion processes (39). A different source of donor cells, for example, the spleen, might be able to block autoimmunity and also provide mesenchymal β cell precursors (40). However, the hypothesized presence in the mouse spleen of embryonic mesenchymal cells that

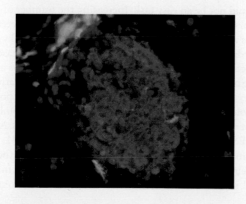

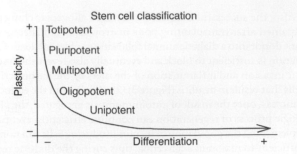

Figure 4
Classification of stem cells based on their developmental potential according to Wagers and Weissman (44). Totipotent, able to give rise to all embryonic and extraembryonic cell types; pluripotent, able to give rise to all cell types of the embryo proper; multipotent, able to give rise to a subset of cell lineages; oligopotent, able to give rise to a restricted subset of cell lineages; unipotent, able to contribute only one mature cell type.

lack surface expression of CD45 and are able to differentiate into endothelial and endodermal cells remains to be confirmed.

Further questions

Even when regeneration of the β cells from precursor cells is definitively proven, issues still to be resolved will include the time frame and physiological conditions necessary for regeneration to occur and reach completion, as well as the circumstances that facilitate or limit the regeneration process and the ability of clinicians to promote or avoid these, to more efficiently achieve the desired therapeutic results. Also, assuming the existence of β cell precursors, we still do not know whether these cells are immortal or subject to senescence, a situation that would leave a narrow window of time for intervention. This matter may be especially relevant in diabetic individuals in whom the reparative process has been repressed by autoimmune surveillance for a long period of time. If successful immunoregulatory intervention cannot be initiated immediately after the clinical onset of the disease, a full recovery of the endocrine function of the gland via the physiologic regeneration route may become impossible. However, if the regenerative process is proven to be irreversibly compromised at a certain point, it may still be possible to transplant into diabetic patients functional precursor cells from nondiabetic donors or to artificially convert the patient's own cells from other tissues or lineages into insulin-producing β cells.

The potential for ES cells

Even if specific markers necessary to recognize β cell precursors were available, the physical isolation of these cells from a patient's pancreas would not be an easy task. Increasing the number of possible precursors ex vivo while avoiding the activation of differentiation pathways would also be problematic, as would be eventually facilitating their differentiation toward the final product, in this case, a functional β cell. Culturing and expanding β cells in vitro has also been shown to be difficult and perhaps limited to a few proliferative cycles, since they are terminally differentiated cells. It has been postulated that it would be easier to derive precursor cells from the embryo and use them to regenerate the damaged endocrine pancreas. Although human ES cell lines have been successfully derived (41, 42) and recently made available to the scientific community (43), the need to direct their differentiation toward a specific final product, in this case the β cell, still remains a major hurdle that must be overcome.

A stem cell is, by definition, the one cell capable of duplicating itself and resuming its undifferentiated status, while also originating progeny that can differentiate into one or more final products that are physiologically defined by their specific functions (44). Proceeding through the differentiation pathway, stem cells can be categorized as totipotent, pluripotent, multipotent, oligopotent, and unipotent, depending upon all their possibly reversible, progressively acquired characteristics (Figure 4) (44).

ES cells are pluripotent cell lines derived from the inner cell mass of blastocyst-stage embryos (41–43, 45), and their differentiation in culture may reproduce characteristics of early embryonic development. Based on similarities between mechanisms that control the development of both the adult pancreas and the central nervous system, Lumelsky et al. (46) hypothesized that strategies able to induce production of neural cells from ES cells could be adapted for endocrine pancreatic cell induction. The filament protein nestin is expressed by neural ES cells during neural differentiation and also in a subset of immature, hormone-negative pancreatic precursors that, upon differentiation in vitro, give rise to insulin- and glucagon-expressing cells. Lumelsky et al. implemented a working protocol that began with the enrichment of the nestin-positive cell population, the population of cells derived from embryoid bodies. Embryoid bodies are structures comprising an inner layer of columnar ectoderm surrounding a proamniotic-like cavity and an outer layer of primitive endoderm. The nestin-enriched population was then expanded in serum-free medium in the presence of FGF. After withdrawal of this mitogen from the culture medium to reduce the stimulus for cell division, other factors, such as nicotinamide, previously shown to be useful in directing the differentiation and/or proliferation of precursors found in fetal pancreatic cells (47), were added. The result was the production of aggregates of cells expressing insulin.

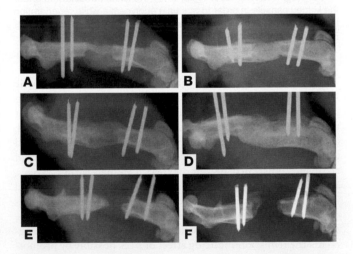

Figure 5
Radiographic evaluation at 4 and 12 weeks after surgery. The critical-sized (i.e., non–spontaneously reparable) defect in the femora treated with bone marrow–derived stromal cells transfected with retro-BMP4 exhibited a notable bridging callus (i.e., the white mass between the 2 extremes of the fracture interval) at both 4 (**A**) and 12 (**B**) weeks after surgery. The defect in the femora treated with MDCs transfected with BMP4 had also developed a bridging callus at 4 (**C**) and 12 (**D**) weeks. No bone formation was radiographically evident in the control — i.e., femora treated with MDCs transfected with the *LacZ* gene — at both 4 (**E**) and 12 (**F**) weeks after surgery. Figure reproduced with permission from *Langenbecks Archives of Surgery* (55).

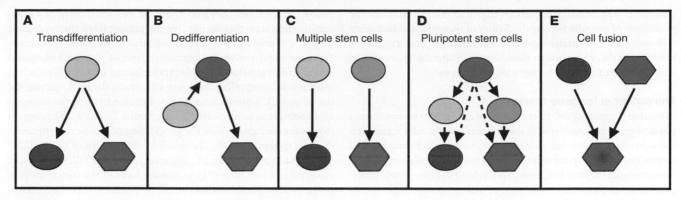

Figure 6
Schematic diagram depicting potential and known mechanisms of adult stem cell plasticity (**A–E**). Orange or green ovals, tissue-specific stem cells; blue ovals, pluripotent stem cells; red ovals and green hexagons, differentiated cells of tissue-specific (orange or green) lineages. Figure reproduced with permission from *Cell* (44).

However, in other hands, similar manipulations have failed to produce a pancreatic endocrine phenotype in nestin-positive–selected cells. Nestin-positive cells seem to contribute to the extra- and intra-microvasculature of the islet rather than to the hormone-producing cells of the pancreas themselves (48, 49). Further criticism of these results originated from the evidence that cells in culture may actually concentrate from the medium the insulin they seem to have produced (50). Definite answers may be obtained by repeating the same experiments using ES cells carrying transgenic fluorescent reporter genes controlled by the insulin promoter (51, 52).

The potential of adult stem cells

While the approach discussed above intends to recreate, using specific feeder cell layers and known growth factors, an environment similar to that existing in vivo and also to facilitate and guide a desired type of cellular differentiation, the approach in another study was to isolate in vitro adult pluripotent stem cells from any manageable source (e.g., the bone marrow) and physically introduce these cells into the already existing, appropriate environment of living recipients. By receiving spatially and temporally restricted signals from the environment, the precursors may differentiate into the cells constitutive of that specific target tissue. Lagasse et al. (53) astutely took advantage of this approach when they successfully generated hepatocytes in vivo from purified HSCs. Using a cocktail of specific antibodies able to recognize a combination of distinct cell surface markers (c-kit^highThy^lowLin^-Sca-1^+) on a single cell, the authors physically isolated HSCs from bone marrow and transplanted them into the fumarylacetoacetate hydrolase–deficient mouse, an animal model of fatal hereditary tyrosinemia type I. Liver function in 4 out of 9 mutant mice was restored to near normal (transaminase and tyrosine levels were slightly increased), when, 3 weeks after transplantation, the standard treatment with 2-(2-nitro-4-trifluoro-methylbenzoyl)-1,3-cyclohexane-dione was discontinued, and these mice survived for an additional 6 months without signs of progressive liver failure or renal tubular damage. When the experiment was interrupted at 7 months after transplantation, 30–50% of the liver mass showed cells expressing donor-derived markers (53). Similarly, by transferring precursors into the brain of a recipient mouse, Weimann et al. (54) were able to see purified HSCs differentiate into Purkinje neurons. Like HSCs, once transduced with a retrovirus to express bone morphogenetic protein 4 (BMP4), muscle-derived cell

(MDC) precursors were able to dramatically improve the healing of a spontaneously irreparable bone fracture (Figure 5) (55).

These data suggest that the signals sent via host-secreted factors or by cell-to-cell contacts are powerful enough to guide the transplanted precursors to differentiate into the same type of cells surrounding them, even across different lineages. This ability to transdifferentiate (i.e., generate a progeny belonging to a tissue lineage different from the one of origin) certainly constituted, at the time of publication, an astonishing discovery (56, 57). No one had previously even speculated that a mammalian stem cell, present in a mature individual, could possess such an impressive plasticity (Figure 6) (44). The possibility that new β cells could be generated from adult stem cells isolated from tissues belonging to other lineages is particularly appealing, because it would avoid the potential ethical problems associated with the use of ES cells. However, the same authors who originally proposed transdifferentiation processes to explain their results soon realized that much of the sensational data they interpreted as results of an efficient transdifferentiation process were most likely the result of cell-to-cell fusion (Figure 6) (58–61). Others have argued that not all the results used to support the transdifferentiation capabilities of adult stem cells should be dismissed or could be solely explained by the cell fusion theory. Ianus et al. (62) deliberately entitled their article "In vivo derivation of glucose-competent pancreatic endocrine cells from bone marrow *without* evidence of cell fusion" (ital-

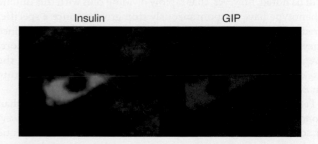

Figure 7
Duodenal sections from transgenic mice harboring the *GIP/Ins* transgene. The K cells of the gut examined by immunofluorescence microscopy showed both human insulin production (green) and expression of the glucose-dependent insulinotropic polypeptide (red). Figure reproduced with permission from *Science* (63).

ics added). Regardless, it is perhaps useful to instead consider the possibility of actually forcing the fusion of transplanted precursor cells with the recipient damaged cells as a means of producing newly functional cells, even though the endocrine pancreas appears to be an environment difficult to physically reach in vivo.

The potential for gene therapy

Instead of promoting cell-to-cell fusion in an effort to restore certain physiologic functions lost with the death of specific cells, it has been considered perhaps more efficient to transfect cells belonging to a certain lineage with genes able to convert them into cells carrying the characteristics of those lost, even if they belong to a completely different lineage. In particular, in the field of diabetes research, gut K cells of the mouse were induced to produce human insulin by transfecting the human insulin gene linked to the 5'-regulatory region of the gene encoding glucose-dependent insulinotropic polypeptide (GIP) (Figure 7) (63). Also, research demonstrating that hepatocytes transfected with the *Pdx1* gene under the control of the rat insulin 1 promoter were able to produce insulin (64, 65) attracted significant attention and has inspired new hope. In these studies, sufficient levels of insulin were secreted to satisfy the needs of a diabetic mouse, which, when treated, became and remained steadily euglycemic. These studies, however, have not yet been successfully repeated by other groups.

The future problems we face

The ability to clone human embryos and to derive from them human ES cell lines is already a reality (41, 43, 45). Even if tomorrow's scientists could derive, by cloning, cell lines able to be differentiated toward cells with the phenotype of a β cell and in sufficient numbers to satisfy the needs of the diabetic transplant recipient, in order that they can be used for clinical purposes, these cell lines should carry the diploid genome of somatic cell nuclei derived from individuals unrelated to the oocyte donor, i.e., those of the recipient. Further, we would still need to isolate from each patient his or her own stem cell line or develop a bank of cell lines representing the broad diversity of human MHC polymorphisms in order to avoid allorejection. Under these circumstances, however, the presence of autoimmunity will remain a concern. If we opt to avoid this demanding preparative effort to match donor and recipient, it would be necessary to reduce the likelihood of allorejection by other means. It may be easier to achieve this goal by using ES cells rather than any other adult cell for transplant, since the already-studied human ES cell lines appear able to downregulate the expression of MHC antigens at their surface (66, 67). It should still be noted, however, that rapidly dividing cells with this unique phenotype have to spontaneously stop growing once a specific, predetermined, total population cell mass has been reached, since they appear not to be controlled by NK cells (66). Furthermore, even assuming that we could overcome all of these challenges, the use of human ES cell lines tailored *ad personam* would constitute an extremely demanding and expensive proposition (68).

Perhaps equally demanding would be the use of means other than bone marrow transplant to evade an autoimmune response. However, some promising strategies, possibly more suitable for therapeutic applications than bone marrow transplant, have already been pro-

posed. Once refined, they may allow the reestablishment of a tolerant immune state that would free the patient from the protracted attack of T cells against their β cell structures and consequently from the requirement for administration of powerful immunosuppressive regimens even when transplanted with MHC-matched insulin-producing cells. While very efficient in the NOD mouse, the use of anti-CD3 monoclonal antibodies able to skew the response of diabetogenic, autoreactive T cells from a Th1 to a Th2 program (69) seems only able to slow the progression of disease to permanent diabetes in humans (70). The possible involvement of CD4+CD25+ T cells in the preservation of tolerance, postulated following results obtained in cured mice (71), is also the basis of the treatment proposed by Zheng and collaborators (72), which involves the administration of rapamycin, agonistic IL-2–Ig fusion protein, and a mutant, antagonist-type IL-15–related cytolytic Ig fusion protein to obtain long-term engraftment and tolerance of allogeneic islets transplanted into overtly diabetic NOD mice. The CD4+CD25+ T cell population is also increased once CD40, CD80, and CD86 cell surface molecules are specifically downregulated by ex vivo treatment of NOD mice DCs with a mixture of antisense oligonucleotides targeting CD40, CD80, and CD86 primary transcripts (73). The incidence of diabetes is significantly delayed by a single injection of the engineered NOD mouse–derived DCs into syngeneic recipients. Assuming then that autoimmunity could be overcome, the potential regeneration-based treatment would still require that we determine the length of time the β cell regenerative potential could be exploited and the characterization of possible triggering factors that would allow quantitative and chronological limits to be bypassed.

Conclusion

Our young diabetic patients must check their blood glucose levels and be injected with insulin at least 4 times a day. Concurrently, they live with the constant threat of incidents secondary to unpredictable acute hypoglycemic episodes and the ever-present worry of chronic complications. Although human ES cell research carries with it enormous scientific potential in the treatment and possible cure of many diseases, in the near future, the advances in the realm of immunoregulation may precede those in the stem cell arena. To me, finding safe ways to block autoimmunity seems to be the first goal we should achieve in order to give our patients a reliable solution to their heavy, lifelong burden, since it is a prerequisite for both the efficient use of an ES cell–based therapy and the reestablishment of euglycemia capitalizing on the pancreatic regenerative pathway.

Acknowledgments

I want to thank Len Harrison and the confidential invited peer reviewers for critically reading the paper; and Timothy Kieffer, Johnny Huard, YiGang Chang, and Tatiana Zorina for allowing me to use their photographs.

Address correspondence to: Massimo Trucco, Division of Immunogenetics, Children's Hospital of Pittsburgh, Rangos Research Center, 3460 5th Avenue, Pittsburgh, Pennsylvania 15213-3205, USA. Phone: (412) 692-6570; Fax: (412) 692-5809; E-mail: mnt@pitt.edu.

1. Wells, J.M. 2003. Genes expressed in the developing endocrine pancreas and their importance for stem cell and diabetes research. *Diabetes Metab. Res. Rev.* **19**:191–201.
2. Jensen, J. 2004. Gene regulatory factors in pancreatic development. *Dev. Dyn.* **229**:176–200.
3. Bach, J.F. 1994. Insulin-dependent diabetes mellitus as an autoimmune disease. *Endocr. Rev.* **15**:516–542.
4. Lacy, P.E. 1982. Pancreatic transplantation as a means of insulin delivery. *Diabetes Care.* **5**(Suppl. 1):93–97.
5. Shapiro, A.M., et al. 2000. Islet transplantation in seven patients with type 1 diabetes mellitus using a glucocorticoid-free immunosuppressive regimen. *N. Eng. J. Med.* **343**:230–238.
6. Ryan, E.A., et al. 2001. Clinical outcomes and insulin

secretion after islet transplantation with the Edmonton protocol. *Diabetes.* **50**:710–719.

7. Steffensen, I., Dulin, M.F., Walters, E.T., and Morris, C.E. 1995. Peripheral regeneration and central sprouting of sensory neurone axons in *Aplysia californica* following nerve injury. *J. Exp. Biol.* **198**:2067–2078.

8. Ramachandra, A.V., Swamy, M.S., and Kurup, A.K. 1996. Local and systemic alterations in cyclic 3',5' AMP phosphodiesterase activity in relation to tail regeneration under hypothyroidism and T4 replacement in the lizard, Mabuya carinata. *Mol. Reprod. Dev.* **45**:48–51.

9. Stocum, D.L. 2002. A tail of transdifferentiation. *Science.* **298**:1901–1903.

10. Lipsett, M., and Finegood, D.T. 2002. β-cell neogenesis during prolonged hyperglycemia in rats. *Diabetes.* **51**:1834–1841.

11. Gu, D., Lee, M.S., Krahl, T., and Sarvetnick, N. 1994. Transitional cells in the regenerating pancreas. *Development.* **120**:1873–1881.

12. Bernard-Kargar, C., and Ktorza, A. 2001. Endocrine pancreas plasticity under physiological and pathological conditions. *Diabetes.* **50** (Suppl. 1):S30–S35.

13. Brelje, T.C., et al. 1993. Effect of homologous placental lactogens, prolactins, and growth hormones on islet β-cell division and insulin secretion in rat, mouse, and human islets: implication for placental lactogen regulation of islet function during pregnancy. *Endocrinology.* **132**:879–887.

14. Rosenberg, L. 1995. In vivo cell transformation: neogenesis of beta cells from pancreatic ductal cells. *Cell Transplant.* **4**:371–383.

15. Bouwens, L. 1998. Transdifferentiation versus stem cell hypothesis for the regeneration of islet beta-cells in the pancreas. *Microsc. Res. Tech.* **15**:332–336.

16. Bonner-Weir, S., Deery, D., Leahy, J.L., and Weir, G.C. 1989. Compensatory growth of pancreatic beta-cells in adult rats after short-term glucose infusion. *Diabetes.* **38**:49–53.

17. Arnush, M., et al. 1996. Growth factors in the regenerating pancreas of gamma-interferon transgenic mice. *Lab. Invest.* **74**:985–990.

18. Herrera, P.L. 2000. Adult insulin- and glucagon-producing cells differentiate from two independent cell lineages. *Development.* **127**:2317–2322.

19. Gu, G., Dubauskaite, J., and Melton, D.A. 2002. Direct evidence for the pancreatic lineage: NGN3+ cells are islet progenitors and are distinct from duct progenitors. *Development.* **129**:2447–2457.

20. Weir, G.C., and Bonner-Weir, S. 2004. Beta-cell precursors — a work in progress. *Nat. Biotechnol.* **22**:1–2.

21. Seaberg, R.M., et al. 2004. Clonal identification of multipotent precursors from adult mouse pancreas that generate neural and pancreatic lineages. *Nat. Biotechnol.* **22**:1115–1124.

22. Suzuki, A., Nakauchi, H., and Taniguchi, H. 2004. Prospective isolation of multipotent pancreatic progenitors using flow-cytometric cell sorting. *Diabetes.* **53**:2143–2152.

23. Bonner-Wier, S., and Sharma, A. 2002. Pancreatic stem cells. *J. Pathol.* **197**:519–526.

24. Dor, Y., Brown, J., Matinez, O.I., and Melton, D.A. 2004. Adult pancreatic beta-cells are formed by self-duplication rather than stem-cell differentiation. *Nature.* **429**:41–46.

25. Pelengaris, S., Khan, M., and Evan, G.I. 2002. Suppression of Myc-induced apoptosis in β cells exposes multiple oncogenic properties of Myc and triggers carcinogenic progression. *Cell.* **109**:321–334.

26. Sutherland, D.E., et al. 1984. Twin-to-twin pancreas transplantation: reversal and reenactment of the pathogenesis of type I diabetes. *Trans. Assoc. Am. Physicians.* **97**:80–87.

27. Ogawa, N., List, J.F., Habener, J.F., and Maki, T. 2004. Cure of overt diabetes in NOD mice by transient treatment with anti-lymphocyte serum and exendin-4. *Diabetes.* **53**:1700–1705.

28. Ikehara, S., et al. 1985. Prevention of type 1 diabetes in nonobese diabetic mice by allogeneic bone marrow transplantation. *Proc. Natl. Acad. Sci. U. S. A.* **82**:7743–7747.

29. Li, H., et al. 1996. Mixed allogeneic chimerism induced by a sublethal approach prevents autoimmune diabetes and reverses insulitis in nonobese diabetic (NOD) mice. *J. Immunol.* **156**:380–388.

30. Zorina, T.D., et al. 2002. Distinct characteristics and features of allogeneic chimerism in the NOD mouse model of autoimmune diabetes. *Cell Transplant.* **11**:113–123.

31. Ryu, S., Kodama, S., Ryu, K., Schoenfeld, D.A., and Faustman, D.L. 2001. Reversal of established autoimmune diabetes by restoration of endogenous β cell function. *J. Clin. Invest.* **108**:63–72. doi:10.1172/JCI200112335.

32. Hess, D., et al. 2003. Bone marrow-derived stem cells initiate pancreatic regeneration. *Nat. Biotechnol.* **21**:763–770.

33. Zorina, T.D., et al. 2003. Recovery of the endogenous beta cell function in autoimmune diabetes. *Stem Cells.* **21**:377–388.

34. Ricordi, C., et al. 1991. In vivo effect of FK506 on human pancreatic islets. *Transplantation.* **52**:519–526.

35. Tamura, K., et al. 1995. Transcriptional inhibition of insulin by FK506 and possible involvement of FK506 binding protein-12 in pancreatic β-cell. *Transplantation.* **59**:1606–1615.

36. Bell, E., et al. 2003. Rapamycin has a deleterious effect on MIN-6 cells and rat and human islets. *Diabetes.* **52**:2731–2739.

37. Paris, M., Bernard-Kargar, C., Berthault, M.F., Bouwens, L., and Ktorza, A. 2003. Specific and combined effects of insulin and glucose on functional pancreatic β-cell mass in vivo and in adult rats. *Endocrinology.* **144**:2717–2727.

38. Farilla, L., et al. 2002. Glucagon-like peptide-1 promotes islet cell growth and inhibits apoptosis in Zucker diabetic rats. *Endocrinology.* **143**:4397–4408.

39. Lechner, A., et al. 2004. No evidence for significant transdifferentiation of bonemarrow into pancreatic β-cells in vivo. *Diabetes.* **53**:616–623.

40. Kodama, S., Kühtreiber, W., Fujimura, S., Dale, E.A., and Faustman, D.L. 2003. Islet regeneration during the reversal of autoimmune diabetes in NOD mice. *Science.* **302**:1223–1227.

41. Thomson, J.A., et al. 1998. Embryonic stem cell lines derived from human blastocysts. *Science.* **282**:1145–1147.

42. Shamblott, M.J., et al. 1998. Derivation of pluripotent stem cells from cultured human primordial germ cells. *Proc. Natl. Acad. Sci. U. S. A.* **95**:13726–13731.

43. Cowan, C.A., et al. 2004. Derivation of embryonic stem-cell lines from human blastocysts. *N. Engl. J. Med.* **350**:1353–1356.

44. Wagers, A.J., and Weissman, I.L. 2004. Plasticity of adult stem cells. *Cell.* **116**:639–648.

45. Hwang, W.S., et al. 2004. Evidence of a pluripotent human embryonic stem cell line derived from a cloned blastocyst. *Science.* **303**:1669–1674.

46. Lumelsky, N., et al. 2001. Differentiation of embryonic stem cells to insulin-secreting structures similar to pancreatic islets. *Science.* **292**:1389–1394.

47. Beattie, G.M., Rubin, J.S., Mally, M.I., Otonkoski, T., and Hayek, A. 1996. Regulation of proliferation and differentiation of human fetal pancreatic islet cells by extracellular matrix, hepatocyte growth factor, and cell-cell contact. *Diabetes.* **45**:1223–1228.

48. Selander, L., and Edlund, H. 2002. Nestin is expressed in mesenchymal and not epithelial cells of the developing mouse pancreas. *Mech. Dev.* **113**:189–192.

49. Treutelaar, M.K., et al. 2003. Nestin-lineage cells contribute to the microvasculature but not endocrine cells of the islet. *Diabetes.* **52**:2503–2512.

50. Rajagopal, J., Anderson, W.J., Kume, S., Martinez, O.I., and Melton, D.A. 2003. Insulin staining of ES cell progeny from insulin uptake. *Science.* **299**:363.

51. Hara, M., et al. 2003. Transgenic mice with green fluorescent protein-labeled pancreatic beta -cells. *Am. J. Physiol. Endocrinol. Metab.* **284**:E177–E183.

52. Bertera, S., et al. 2003. Body window-enabled in vivo multicolor imaging of transplanted mouse islets expressing an insulin-timer fusion protein. *BioTechniques.* **35**:718–722.

53. Lagasse, E., et al. 2000. Purified hematopoietic stem cells can differentiate into hepatocytes in vivo. *Nat. Med.* **6**:1229–1234.

54. Weimann, J.M., Charlton, C.A., Brazelton, T.R., Hackman, R.C., and Blau, H.M. 2003. Contribution of transplanted bone marrow cells to Purkinje neurons in human adult brains. *Proc. Natl. Acad. Sci. U. S. A.* **100**:2088–2093.

55. Rose, T., et al. 2003. The role of cell type in bone healing mediated by ex vivo gene therapy. *Langenbecks Arch. Surg.* **388**:347–355.

56. Kahn, A. 2000. Converting hepatocytes to β-cells — a new approach for diabetes? (News & Views). *Nat. Med.* **6**:505–506.

57. Perry, D. 2000. Patients' voices: the powerful sound in stem cell debate. *Science.* **287**:1423.

58. Wagers, A.J., Sherwood, R.I., Christensen, J.L., and Weissman, I.L. 2002. Little evidence for developmental plasticity of adult hematopoietic stem cells. *Science.* **297**:2256–2259.

59. Wang, X., et al. 2003. Cell fusion is the principal source of bone-marrow-derived hepatocytes. *Nature.* **422**:897–901.

60. Choi, J.B., et al. 2003. Little evidence of transdifferentiation of bone marrow-derived cells into pancreatic beta cells. *Diabetologia.* **46**:1366–1374.

61. Alvarez-Dolado, M., et al. 2003. Fusion of bone marrow-derived cells with Purkinje neurons, cardiomyocytes and hepatocytes. *Nature.* **425**:968–973.

62. Ianus, A., Holz, G.G., Theise, N.D., and Hussain, M.A. 2003. In vivo derivation of glucose-competent pancreatic endocrine cells from bone marrow without evidence of cell fusion. *J. Clin. Invest.* **111**:843–850. doi:10.1172/JCI200316502.

63. Cheung, A.T., et al. 2000. Glucose dependent insulin release from genetically engineered K cells. *Science.* **290**:1959–1962.

64. Ferber, S., et al. 2000. Pancreatic and duodenal homeobox gene 1 induces expression of insulin genes in liver and ameliorates streptozotocin-induced hyperglycemia. *Nat. Med.* **6**:568–572.

65. Ber, I., et al. 2003. Functional, persistent, and extended liver to pancreas transdifferentiation. *J. Biol. Chem.* **278**:31950–31957.

66. Drukker, M., et al. 2002. Characterization of the expression of MHC proteins in human embryonic stem cells. *Proc. Natl. Acad. Sci. U. S. A.* **99**:9864–9869.

67. Li, L., et al. 2004. Human embryonic stem cells possess immune-privileged properties. *Stem Cells.* **22**:448–456.

68. Kennedy, D. 2004. Stem cells, redux. *Science.* **303**:1581.

69. Chatenoud, L., Primo, J., and Bach, J.-F. 1997. CD3 antibody-induced dominant self tolerance in overt autoimmunity in non-obese diabetic mice. *J. Immunol.* **158**:2947–2954.

70. Herold, K.C., et al. 2002. Anti-CD3 monoclonal antibody in new-onset type 1 diabetes mellitus. *N. Engl. J. Med.* **346**:1692–1698.

71. Chatenoud, L. 2003. CD3-specific antibody-induced active tolerance. *Nat. Rev. Immunol.* **3**:123–132.

72. Zheng, X.X., et al. 2003. Favorably tipping the balance between cytopathic and regulatory T cells to create transplantation tolerance. *Immunity.* **19**:503–514.

73. Machen, J., et al. 2004. Antisense oligonucleotide downregulating co-stimulation confer diabetes-preventive properties to NOD dendritic cells. *J. Immunol.* **173**:4331–4341.

Challenges facing islet transplantation for the treatment of type 1 diabetes mellitus

Kristina I. Rother and David M. Harlan

Islet and Autoimmunity Branch, National Institutes of Diabetes and Digestive and Kidney Diseases, NIH, Bethesda, Maryland, USA.

Islet transplantation represents a most impressive recent advance in the search for a type 1 diabetes mellitus cure. While several hundred patients have achieved at least temporary insulin independence after receiving the islet "mini-organs" (containing insulin-producing β cells), very few patients remain insulin independent beyond 4 years after transplantation. In this review, we describe historic as well as technical details about the procedure and provide insight into clinical and basic research efforts to overcome existing hurdles for this promising therapy.

Worldwide, more than 750 individuals with type 1 diabetes mellitus (T1DM) have received allogeneic islet transplants since 1974, in an effort to cure their chronic condition. Though this is still a small number (especially when compared with the estimated 1 million afflicted with T1DM and an additional 17 million with type 2 diabetes in the US, not to mention the estimated 140 million with diabetes worldwide), much has been learned, especially since the promising results of the Edmonton group were published in 2000 (1, 2). This report described 7 consecutive patients with T1DM who became insulin independent after receiving islet allografts, which reflects a success rate never previously achieved. The initial enthusiasm over the observation that islet transplantation can restore insulin-independent euglycemia to patients with long-standing T1DM has been dampened by complications associated with the procedure itself and the immunosuppression necessary to prevent rejection of the transplanted islets, as well as by the gradual loss of islet function and other problems arising from the placement of allogeneic islets in the liver (3, 4).

At the same time, our understanding about the natural history of T1DM has changed. Epidemiological data indicate that the prognosis for survival among patients with T1DM is good and improving (5–7). Additional evidence strongly suggests that a significant minority with even long-standing T1DM continue to display islet function (8). For instance, we found that at least 40% of individuals with chronic T1DM (mean duration of 23 years) screened for our islet transplantation protocol had measurable circulating C-peptide levels, an equimolar by-product of endogenous insulin production (B.J. Digon et al., unpublished results). Similar percentages of patients with persistent C-peptide secretion have been documented in the literature (9–11), contrasted by a smaller number of individuals (11%) reported in the Diabetes Control and Complications Trial (12). Further, old (13) as well as more recent sporadic case reports (14) and other small studies (15) have suggested that diabetes can sometimes resolve and/or that pancreatic insulin production can at least be promoted in patients with long-standing T1DM. These studies have raised heretofore underexplored avenues for clinical investigation, which we will return to.

Brief history

In 1924, after approximately 40 years of unsuccessful attempts by various investigators to control diabetes using partial pancreas transplantation, the English surgeon Charles Pybus (1882–1975) made a statement that resonates even today: "Not much can be said about the principles of grafting, but it seems that until we are able to understand them (and I feel we do not understand them at present, especially the chemical factors), then we must continue to fail in such operations, although they may appear the most rational treatment for the diseases for which they are attempted" (16, 17). Almost 50 years later, Ballinger and Lacy reported their results isolating and transplanting islets into rats (18). In the late 1970s, various groups including Najarian, Sutherland, et al. (19), and Largiader et al. (20) described their experience with intraportal and intrasplenic human islet allotransplants in patients with nonautoimmune diabetes, one of which was successful for at least a 10-month follow-up period. In 1990 Scharp et al. reported similar success in a patient with T1DM (21); their results were made possible in part by improved islet isolation techniques developed by Ricordi and colleagues (22). It was, however, only after the Edmonton group's report (2) that islet transplantation appeared as a true alternative to conservative medical management or whole-organ pancreas transplantation.

The Edmonton report stimulated several medical centers worldwide to rejuvenate or establish islet transplant facilities. However, islet transplantation efforts have confronted logistical limitations including the costs associated with cadaveric pancreas procurement, the supplies and equipment required to perform islet isolation, and an experienced team with the relevant expertise in cell processing and invasive radiology, as well as general medicine and nursing. However real these mundane fiscal constraints may be, several other and rather significant biological hurdles currently being addressed on several fronts also limit the field. The remainder of this review will focus upon those hurdles and ongoing research efforts to overcome them.

Supply of cells capable of physiologically regulated insulin secretion

As of today, the only cells known to be capable of sensing a human being's ambient blood glucose and converting that information into appropriately regulated insulin secretion are the β cells found in intact islets of Langerhans. As a source of physiologically regulated insulin, β cells can be transplanted as the small minority of cells present within a whole pancreas (as is currently performed by

Nonstandard abbreviations used: ES cell, embryonic stem cell; IBMIR, instant blood-mediated inflammatory reaction; T1DM, type 1 diabetes mellitus.

Conflict of interest: The authors have declared that no conflict of interest exists.

Citation for this article: *J. Clin. Invest.* **114**:877–883 (2004). doi:10.1172/JCI200423235.

whole-organ pancreas transplantation) or as the enriched fraction of cell clusters we call "isolated islets." Two points are worth emphasizing with regard to isolated islets. First, all such preparations contain significant if variable amounts of contaminating vascular, ductal, and/or acinar tissue despite the purification procedure. The effect of such impurities on patient outcome is not known; however, no adverse reactions to non-islet transplanted tissue have been described to date. Second, isolated islets are best described as mini-organs in that they are highly organized cell clusters with intricate paracrine crosstalk among insulin-producing β cells, glucagon-producing α cells, and other hormone-producing cells (23). Islets also have a rich capillary network (24) and contain resident cells with presumed immunological function. Regardless how islets are "packaged" for transplant, i.e., within a pancreas for a vascularized whole-organ graft or as an isolated islet preparation, for all practical purposes, the source must be a brain-dead donor, and the number of such potential donors is quite limited. After unsuitable organ donors are eliminated (e.g., those suffering from malignant or infectious diseases or others with sudden cardiovascular collapse and unacceptable warm ischemia time, which is associated with a rapid decline of organ function), only about 12,000 individuals in the United States remain as potential organ donors each year. However, despite considerable time, effort, and money directed at improving the donation rate, family consent for donation is given only about half the time. Thus, only approximately 6,000 human pancreata are available in the United States each year for transplantation. Further exacerbating the problem, human islet isolation efforts successfully yield preparations with sufficient islet numbers of suitable quality only about half the time; thus current practice could yield approximately 3,000 transplantation-quality preparations per year. Last, nearly all published islet transplantation experience suggests that for most recipients, islets from 2–4 donors are required in order to promote the engraftment of sufficient insulin-producing cells to achieve insulin independence. Most islet recipients have been selected in part due to their small physical size and normal insulin sensitivity: they require fewer donor islets, which increases the likelihood that the limited islet mass transplanted will confer insulin independence (25, 26). If patients with greater body mass indices and/ or with insulin resistance were also considered for an islet transplant, the 3,000 transplantable islet preparations presently achievable would likely be sufficient to restore euglycemia to fewer than 1,000 patients per year, or less than 0.1% of patients with T1DM, or approximately 0.005% of those with either form of diabetes.

Alternative sources of cells with physiologically regulated insulin secretion

Other than cadaveric pancreata, many potential cellular sources for physiologically regulated insulin secretion have been widely discussed in the scientific literature, but the chasm between the promise each holds and practical reality remains quite broad. Three potential sources in particular have been widely discussed: "growing" islets in vitro; using islets from species other than humans; and promoting β cell differentiation from precursor stem cells.

Expanding islet cellular mass in vitro. While a variety of culture conditions have been published for propagating islet cells from adult donors, those techniques invariably result in an inexorable decline in insulin production by the cultured cells (see refs. 27–29). Stated another way, while many groups have reported a favorable stoichiometry with regard to islet cell numbers, no group has developed a strategy for reliably expanding islet populations with dependable and robust insulin-producing capacity. Further, as islets are essentially mini-organs, in order for propagation to occur in vitro, an islet (or the appropriate precursor cells within the islet) must presumably — in no particular order — divide, leave the original islet, then reaggregate with other appropriate cells to create a new islet, while leaving the original mini-organ structure sufficiently intact to remain functional. Last, it is not yet possible to identify the cell(s) within the islet preparation that are responsible for new islet growth. Several reports have suggested that islet progenitor cells exist within the pancreatic ducts (30–32), and yet a recent publication cogently argues that the only cell capable of differentiating into a β cell in the adult mouse is an existing β cell (33).

Islets from species other than humans. Others have studied a xenogeneic source for the desired physiologically regulated insulin-secreting cell. Pig islets, in particular, have been widely studied for a variety of reasons, including: (a) the fact that humans had been treated with pig insulin for more than 60 years; (b) favorable husbandry — in that the species has large litters with offspring that attain adult size rapidly and with relatively robust islet numbers; (c) the fact that pig islets respond to glucose in the same physiological glucose range as human islets; and (d) the existence of a suitable societal-cultural relationship between the species. That is, since pigs are currently widely bred, then slaughtered for our food supply, the use of their islets to restore health may be an option that satisfies ethical concerns in a more widely accepted way than would, for example, islets isolated from dogs. Two main concerns continue to limit the use of pig tissues as the xenogeneic source for transplant, however. First, humans express high titers of antibodies against a galactose α(1, 3) galactose residue that is present on most pig cells, resulting in a hyper-acute rejection response whenever pig tissues are transplanted into humans. Second, pig cells, like all mammalian cells, contain endogenous retrovirus. There is also the possibility that the pig endogenous retrovirus might infect the human host, especially under the circumstance of administration of a large inoculum (as might occur when transplanting tissue) into a host with a weakened immune system (as would be expected when administering immunosuppressive drugs to prevent graft rejection). In order to address these concerns, some investigators have created genetically modified pigs that do not express the galactose α(1, 3) galactose residue (34) and plan to breed these animals in special pathogen-free conditions in an effort to minimize the risk for zoonotic disease (35–39).

Promotion of β cell differentiation from stem cells. A third, widely studied potential source for insulin-producing cells is a strategy aimed at promoting the differentiation of a precursor cell into a "β-like cell," with the embryonic stem cell (ES cell) being perhaps the most promising precursor candidate. Clearly, as all mammals generate all cell types from the originally fertilized egg, ES cells are known to have pluripotent capacity. Further, groups have published culture techniques promoting differentiation of ES cells into what appear to be insulin-producing cell clusters (40, 41). More recent studies have suggested that at least some of these anti-insulin antibody–staining cell clusters may be artifactual (a result of insulin uptake from the growth media) and that in any case such cultured cell clusters fall short of β cells; e.g., they lack insulin granules and the appropriate capacity for regulated insulin secretion (42). Still other research has suggested that additional adult progenitor cell populations can, under some circumstances, differentiate into functional β cells in rodents (43, 44). Thus, studies by internationally recognized investigators have yielded data that is difficult to synthesize; i.e., some data suggest that β cells

Table 1

Systemic side effects commonly associated with the immunosuppressive agents typically administered following islet transplant

Immunosuppressant (brand name)	Drug classification	Common and important side effects (Phase of drug administration)
Rapamycin, also known as Sirolimus (Rapamune)	Macrocyclic lactone	Hyperlipidemia, antiproliferation (e.g., anemia, diarrhea) (Maintenance)
FK506, also known as Tacrolimus (Prograf)	Calcineurin inhibitor	Hypertension, nephrotoxicity, CNS effects (e.g., tremor), diabetogenicity (Maintenance)
Daclizumab (Zenapax)	mAb-binding IL-2 receptor α subunit	May increase risk of infections; hypersensitivity (Induction)

in mice can arise only from existing β cells, while others suggest a capacity for stem cell–to–β cell transdifferentiation (see refs. 33, 43, 44). The outcome of this scientific debate will become evident through additional study.

Several additional points are relevant to the islet supply question; for instance:

(a) If stem cells capable of differentiating into functional β cells exist in vivo (in the pancreas, the bone marrow, the spleen, or elsewhere), then what limits the natural occurrence of that process? Why, for instance, do patients with T1DM so rarely, if ever, recover from the loss of β cell function that caused their disease?

(b) Why do functional islets form only in islets of Langerhans and the latter only in the pancreas? If our goal is to recapitulate in vitro the process that promoted islet development in vivo, then what hormonal, cell-, and/or matrix-mediated structural signals are generated within the pancreas that promote islet development?

(c) As in all situations where demand outstrips supply, one way to more efficiently utilize the precious resource would be to limit islet loss during isolation or following transplantation. Nilsson, Korsgren, Bennet, et al., for instance, have reported several studies indicating that purified islets, when mixed with whole blood, result in activation of platelets and of both the coagulation and complement cascades in a process they named the instant blood-mediated inflammatory reaction (IBMIR) (45–48). Their studies suggest that IBMIR leads not only to islet destruction (albeit incomplete), but also to activation of an anti-islet adaptive immune response by promoting an innate immune response. Further, IBMIR may contribute to the bleeding and/or thrombotic complications associated with the islet transplant procedure (see below); and

(d) Any insulin-producing cell generated from progenitors in vitro will have to overcome rather stringent regulatory hurdles with regard to human safety. That is, regulatory agencies exist to help navigate the difficult waters that lie between a potential new therapy's initial early successes and the dissemination of that therapy for more widespread application. For instance, when short stature was treated with growth hormone isolated from human pituitary glands, the therapy was shown to be effective but also later to carry with it some risk for Creutzfeldt-Jakob disease. Similarly, we must recognize that cells engineered for physiologically regulated insulin secretion might yield not only the desired consequence once transplanted (insulin independence), but also undesired consequences. For example, rodent studies have suggested that islets transplanted into the liver display an increased propensity to develop into insulin-producing tumors (48), and serious concerns about the role of stem cells in malignancy have been raised (49).

Immunosuppression

The publication of the Edmonton islet transplantation study (2) addressed an important scientific question: is there a reasonable likelihood that, through transplantation of a sufficient number of islets (see above) and with an immunosuppressive regimen that avoids the use of diabetogenic glucocorticoids, patients will be restored to insulin-independent euglycemia, at least temporarily? For a therapy as expensive and invasive as islet transplantation to be considered for more widespread clinical application, however, an endpoint more rigorous than insulin independence at 1 year after transplant needs to be met. Most important, what do current studies suggest regarding the impact of islet transplantation on patient survival and quality of life?

As shown in Table 1, several systemic side effects are commonly associated with the typically administered immunosuppressive agents (rapamycin and FK506) (4, 50), including pneumonitis (51), suppression of bone marrow function, mouth ulcers, deteriorating renal function, peripheral edema, tremor, hyperlipidemia, hypertension, weight loss, diarrhea, and fatigue. While several of these side effects can be countered medically (e.g., hyperlipidemia, hypertension, or infection), the small but significantly increased malignancy risk and the deteriorating renal function require special comment.

T1DM-associated mortality has been associated with several clinical variables; most notably, the traditional cardiovascular risk factors (smoking, blood pressure, lipid levels), autonomic neuropathy, and the degree of renal dysfunction. For the traditional cardiovascular risk factors, while glycemia control decreases microvascular complication rates, no large prospective study has shown a statistically significant effect on macrovascular complications or survival (52, 53). On the other hand, studies have demonstrated that both blood pressure control (54, 55) and lipid control (56) decrease diabetes-associated mortality with both high statistical and clinical relevance. However, while autonomic dysfunction has been widely discussed as predictive for diabetes prognosis, more recent and prospective studies have concluded that the apparent association is explained by other factors (57, 58). Last, and most important, kidney dysfunction has consistently been found to independently predict mortality among patients with diabetes (59–61). Thus, the increasingly appreciated risk of chronic renal failure after an other-than-kidney transplant (estimated incidence as high as 21%) (62–64), presumably accounted for at least in part by the calcineurin phosphatase inhibitors — currently the mainstay of most immunosuppressive regimens — has raised concerns regarding the overall impact on survival of a transplant-based approach to the treatment of a diabetic patient. That is, the net effect of improved glycemia control produced by the transplant, when balanced against the immunosuppressive agent–associated hypertension, hyperlipidemia, and decreased renal function, may actually increase mortality. Indeed, our recent evaluation of the US pancreas transplant population has suggested statistically greater mortality for at least the first 4 years after transplant for those individuals with preserved kidney function compared with patients with chronic diabetes that is similarly difficult to control but who have not received a transplant (65). Clearly, a critical need exists for strategies to prevent the rejection of cell-based insulin delivery vehicles (isolated islets, whole pancreas, or other strategies) such that kidney function is not impaired. Perhaps antibody-based approaches that are designed to promote anti-graft immune tolerance and therefore minimize the need for chronic administration of immunosuppressive agents (e.g., as in recent studies reported by Herold et al.; refs. 66, 67) will fill this void.

Immunosuppressed organ transplant recipients are also known to be at greater risk for both cancer (relative risk: 3–4) and infection (for instance, see ref. 68) as representative of many studies indicating that well over half of immunosuppressed kidney allograft recipients suffer infectious complications — fortunately almost always easily treated. Among cancers, some relatively rare types predominate in organ transplant recipients as compared with the general population, including lymphomas and lymphoproliferative disorders, Kaposi sarcoma, renal carcinomas, cervical carcinomas in situ, hepatobiliary carcinomas, anogenital carcinomas, and various other sarcomas. In contrast, most of the common malignancies, except skin and lip cancers (69), occur with only marginally increased incidence in immunosuppressed patients compared with nonimmunosuppressed individuals. The risk of developing cancer or infection as a result of the newer antibody-based regimens remains unknown (70).

It is somewhat ironic that current immunosuppressive agents are themselves associated with insulin resistance and/or decreased β cell function and as such promote diabetes onset. The incidence of new-onset diabetes in organ transplant recipients ranges between 2%–53% (71). The particular immunosuppressive agent employed appears to influence rates of islet engraftment and survival and/or insulin production and action. The diabetogenic effects of glucocorticoids have been the most widely appreciated, in fact since the early descriptions of Cushing syndrome (72, 73). Calcineurin phosphatase inhibitors also clearly interfere with normal β cell function (74–76). While FK506 appears to confer greater diabetogenic risk than cyclosporin, that observation must be weighed against its preferred use by many transplant teams to prevent allograft loss. Last, studies in Sprague-Dawley rats have demonstrated that rapamycin, with its known antiproliferative properties, is also associated with insulin resistance (77) and when combined with FK506 (which serves to decrease β cell function) induces diabetes. In our experience and according to similar data reported from the Edmonton group (50), patients achieving insulin independence following an islet transplant nevertheless displayed deficient first-phase insulin secretory responses (10%–30% of normal) in intravenous glucose tolerance tests. Insulin secretion in such patients is also delayed and deficient after an oral glucose load — reminiscent of responses seen in type 2 diabetes. Clearly, one cannot easily dissect the contribution of each possibly responsible factor such as immunosuppression, abnormal islet innervation and vascularization (24), hepatic localization (78–80), engraftment of a subnormal islet number, and others. We have, however, observed that islet recipients in whom islet function was measurable (from circulating C-peptide) yet insufficient to render them insulin independent had declining glucose control with elevated immunosuppressive drug levels (Rother et al., unpublished results). These data

suggest that higher concentrations of immunosuppressive agents can further impair islet function when that function is already challenged. However, with the introduction of novel immunosuppressive regimens, we may be able to avoid such diabetogenic effects.

Safety concerns with the islet transplant procedure

While almost certainly safer than whole-organ pancreas transplantation, the islet transplant procedure as it currently stands is not risk free. We must remember that while the procedure is not yet deemed a suitable replacement for the current standard treatment for the majority of patients with T1DM, there exists much hope for and possibility of improved efficacy and safety. Some complications must be addressed further. A recent review from the Edmonton group, for instance, reported on complications associated with 68 consecutive islet infusion procedures (81), some of which are potentially serious. These include partial portal vein thrombosis with subsequent anticoagulation (the latter resulting in an expanding hepatic hematoma in 1 patient that required surgery) and/or significant intra-abdominal hemorrhage. Their complication rate can be reported using procedure number as the denominator, in which case the prevalence was 9% per procedure; or, in recognition of the fact that most patients require more than 1 islet infusion (in the Edmonton report, the mean number of procedures was 2 for the 34 patients reported), the rate of potentially serious complications can be reported as 18% per patient.

As with any transplantation procedure, although the risk is small, infectious agents can be transmitted via an allograft. An illustration of this point is that despite the extensive screening applied to all blood donors, occasional recipients develop a blood-borne illness from transfusion. Organ and/or tissue donors are extensively screened using known serologic tests, but very recent infections can certainly escape detection.

Long-term safety concerns. Additional safety concerns require consideration, since long-term follow-up of islet recipients remains quite limited. For instance, we and others have observed hepatic structural changes following islet transplantation in both non-human primates (79) and in patients (78, 80). Both glycogen accumulation and localized steatosis have been demonstrated immunohistochemically and/or by chemical shift MRI. No long-term adverse effects of the transplant procedure on liver function have been reported, but the association between fatty liver disease, obesity, and type 2 diabetes leading to progressive fibrosis and eventual cirrhosis is sobering (82). In fact, it has been speculated that secondary hepatic effects can be used as indirect determinants of islet function, as a recent report described resolution of hepatic imaging abnormalities when the islet allografts failed (78).

Gradually increasing portal venous pressures following islet infusions have also been reported, with each subsequent infusion leading to a greater pressure increase than the last (83). Again, the clinical impact of these pressure increases is unknown.

Finally, while current immunosuppressive approaches are superior to past therapies, both immediate and chronic rejection of transplanted organs and tissues persist as major problems. Since iatrogenic immunosuppression does not completely eliminate the body's immune responses, the remaining mitigated immune activity not only threatens the transplanted organ or tissue but can sensitize the recipient to donor tissues and thus complicate the search for a suitable donor match for any subsequently required transplant. Such sensitization following islet allograft rejection has recently been reported (84). We therefore strongly recommend carefully screening

Table 2

The most important criteria upon which islet transplant recipients have been chosen for existing clinical protocols

Indication	Contraindication
T1DM > 5 years	Cardiac disease
Age > 18 and < 65 years	Active infection
BMI < 28 kg/m²	Liver abnormalities (e.g., ultrasound evidence of portal hypertension)
Suboptimal glycemia control specifically	Any history of malignancy except squamous or basal skin cancer
due to lack of awareness of severe hypoglycemia	
Insulin requirement < 0.7 units/kg/d and HgbA1C[A] < 10%	Creatinine clearance < 60 ml/min/m², or macroalbuminuria
No endogenous insulin production	Untreated proliferative retinopathy
(C-peptide values below detection limits, even after provocative testing	Untreated hyperlipidemia
with agents that stimulate β cell insulin secretion)	Previous transplant or serologic evidence of
	anti–donor tissue specific Ab's

[A]HgbA1C is the measure of glycosylated hemoglobin that is used to provide an estimate of the average blood glucose control for the previous 3 months.

of islet-transplant candidates and exclusion of individuals with early kidney damage from this as-yet-experimental procedure.

Selecting the ideal candidate: victims of our own success

In a certain sense, improvements in diabetes care have made more difficult the design of clinical trials to overcome the biological hurdles still limiting islet transplantation. The prognosis for patients with even long-standing and brittle T1DM is better than for any other condition treated by transplantation (65), and that prognosis appears to be steadily improving (5–7). Patients with brittle diabetes and end-stage kidney disease may be the best candidates, since the benefits of kidney transplantation in this population have been clearly demonstrated (65); they will thus require immunosuppressive agents to preserve their kidney allograft — adding an islet allograft to the therapeutic mix would seem appropriate. Other patients with brittle diabetes — even those with preserved kidney function — who are plagued by severe hypoglycemia despite optimized care, may also be good candidates (85), but such patients are quite rare, since optimized conservative care frequently eliminates the worst of these symptoms (86). Further, the perception of the term "optimized care" can vary widely, and we urge referral of individuals with diabetes that is difficult to control to specialists well versed in multidisciplinary and the most up-to-date insulin-based regimens before concluding that any patient's diabetes is uncontrollable. We do not support considering individuals with progressive microvascular complications (especially nephropathy) as good candidates due to the known nephrotoxicity of current immunosuppressive regimens. For the same reason, we counsel against islet transplantation in children until more is known about the long-term safety of this approach.

Defining clinical success

As our intent with this review is to identify those barriers still limiting the more widespread application of islet transplantation for the treatment of diabetes, we wish to emphasize our support for this therapeutic approach. That said, one of the largest hurdles limiting development of this therapy is the difficult interplay of the factors described above. In any experimental technique, one attempts to control all variables except the one under investigation. For the patient given an islet transplant, however, many factors can affect the outcome, with the relative contribution from those disparate

variables impossible to tease out. For instance, present technology does not allow investigators to assess with any degree of accuracy the factors that might affect islet function, including: islet quality prior to transplant, the autoimmune response that initially caused the T1DM, the alloimmune response, immunosuppressive effects on islet function, islet dysfunction that might result from metabolic demands placed upon a limited islet mass, etc. Furthermore, when islets are infused into the portal vein, their widespread and diffuse distribution renders impractical any attempt to follow an anti-islet immune response using current biopsy techniques. Thus, if an investigator wishes to test, for example, a novel immunomodulatory approach to prevent islet allograft rejection, and islet function is lost after transplant, it is not yet possible to understand the mechanisms underlying that failure. The many unknowns, considered within the context of the great costs associated with islet transplantation and the relatively excellent prognosis now associated with even long-term diabetes, conspire to create a difficult path for future clinical investigation that falls within well-established medical, ethical, and economical guidelines.

The future

How and why then do we as a community continue in our pursuit of islet transplantation as a cure for T1DM? Quoting John F. Kennedy's speech given at Rice University in 1962 when he announced the government's plan to land a man on the moon, we choose to pursue islet transplantation research "not because [it is] easy, but because [it is] hard, because that goal will serve to organize and measure the best of our energies and skills, because that challenge is one that we are willing to accept, one we are unwilling to postpone, and one which we intend to win . . . " These inspiring words must be considered in the context of what we learn every day from those with T1DM: despite improvements in care, the disease still takes a terrible toll on quality of life.

To be more specific, we suggest the following as potential avenues of research that should be further pursued in the continued development of this most promising therapy. First, we believe an appropriately safe and effective means of preventing immune system–mediated islet destruction after transplant needs to be found. Toward this end, we suggest the following:

(a) While criteria for patient enrollment employed in several protocols evaluating the safety and efficacy of islet transplantation are shown in Table 2, we advocate efforts to define a cohort with high-risk diabetes, as such patients (currently difficult to identify) would better fulfill the requirements of appropriate risk/benefit analyses in the evaluation of newer islet transplantation approaches.

(b) Another approach might be to test novel immunomodulatory strategies in, for example, individuals receiving a kidney transplant due to renal failure. Such a population not only has a worse prognosis for survival than do patients with diabetes (and so a greater tolerance for the new immune intervention's potential toxicity), but following the outcome of such patients is much easier and more objective. That is, patients without an autoimmune etiology underlying their renal failure can be selected, a kidney from a single donor is sufficient to treat the patient (whereas islet recipients typically require islets from 2 or more donors), kidney biopsies to follow graft function are now routine, techniques for overcoming an acute kidney rejection episode exist, and donor kidney function can reasonably be assessed. If a novel immune therapy proves sufficiently safe and effective in the kidney transplant recipient population, then such a strategy could be rapidly tested in a selection of islet allograft recipients. As a first step, patients with kidney failure from T1DM and thus requiring a kidney transplant could also receive allogeneic islets.

(c) Still another approach might be to test a novel immunomodulatory regimen in patients with T1DM of some duration yet who still have some measurable level of β cell function. As discussed herein, a surprising number of patients with T1DM display some evidence of either an anti-β cell immune response (at least anti-islet antibody titers) and/or β cell function (detectable levels of circulating C peptide) even years after diagnosis. These observations suggest the as-yet-untested hypothesis that through control of the autoimmune response, and perhaps other factors, some β cell functional recovery could ensue (87). An advantage of such an approach is that it would be more scientifically rigorous with less uncontrollable variables (like allogeneic islet quality and alloimmune responses that are difficult to monitor) and also achieves a more supportable risk/benefit balance than current immunosuppression-based protocols that enroll patients with new-onset disease. That is, the testing of an immune intervention of unknown long-term safety in an adult with long-standing T1DM to see whether the β cell function might improve seems to fulfill the high

standards of clinical investigation in a more balanced fashion than would testing that agent in a patient with new-onset disease (typically a child). We are actively engaged in testing this hypothesis.

(d) We recommend research to develop practical tools for the immunological monitoring of the islet recipient so that both allo- and autoimmune responses can be followed.

Second, before we can hope to "cure" patients with diabetes by transplanting cells capable of physiologically regulated insulin secretion, we need a reliable and preferably renewable source of such cells. In general, this means we will need: control of the material and methods required to generate such cells; and reliable assays for determining the cells' quality (viability, insulin producing capacity or "potency," genetic stability, etc.).

At present, allogeneic islets isolated from cadaveric donors fall short of this appropriately high standard, but by clearly recognizing and stating the problems, investigators in the field can focus on overcoming them. The use of islets from cadaveric pancreata will always be hampered by lack of control over the starting material and also remains constrained by their limited supply, the imperfect means of isolating those islets, and the great expense associated with that exercise. The ability to propagate, in vitro or in a surrogate animal species, cells capable of physiologically regulated insulin secretion would be a major step forward.

Last, we advise careful follow-up of those patients already given allogeneic islets by transplant such that many variables are monitored, including the function of those allogeneic islets but also effects the islets may have on liver structure or function, effects of the immunosuppressive agents of diabetes complications, survival, and quality of life. The first annual report of the Collaborative Islet Transplant Registry (see http://spitfire.emmes.com/study/isl/reports/reports.htm) is an important step forward. We also encourage longer-duration studies using large-animal models to address some of these questions.

Address correspondence to: David M. Harlan, Islet and Autoimmunity Branch, National Institutes of Diabetes and Digestive and Kidney Diseases, NIH, Building 10, Room 8N307, Bethesda, Maryland 20892, USA. Phone: (301) 594-3407; Fax: (301) 480-4518; E-mail: davidmh@intra.niddk.nih.gov.

1. Hirshberg, B., Rother, K.I., Digon, B.J., III, Venstrom, J., and Harlan, D.M. 2003. State of the art: islet transplantation for the cure of type 1 diabetes mellitus. *Rev. Endocr. Metab. Disord.* **4**:381–389.

2. Shapiro, A.M., et al. 2000. Islet transplantation in seven patients with type 1 diabetes mellitus using a glucocorticoid-free immunosuppressive regimen. *N. Engl. J. Med.* **343**:230–238.

3. Hirshberg, B., Rother, K.I., and Harlan, D.M. 2003. Islet transplantation: where do we stand now? *Diabetes Metab. Res. Rev.* **19**:175–178.

4. Hirshberg, B., et al. 2003. Solitary islet transplantation for type 1 diabetes mellitus using steroid sparing immunosuppression: the NIH experience. *Diabetes Care.* **26**:3288–3295.

5. Hovind, P., et al. 2003. Decreasing incidence of severe diabetic microangiopathy in type 1 diabetes. *Diabetes Care.* **26**:1258–1264.

6. Nordwall, M., Bojestig, M., Arnqvist, H.J., and Ludvigsson, J. 2004. Declining incidence of severe retinopathy and persisting decrease of nephropathy in an unselected population of Type 1 diabetes – the Linkoping Diabetes Complications Study. *Diabetologia.* **47**:1266–1272.

7. Nishimura, R., et al. 2001. Mortality trends in type 1 diabetes. The Allegheny County (Pennsylvania) Registry 1965-1999. *Diabetes Care.* **24**:823–827.

8. Palmer, J.P., et al. 2004. C-peptide is the appropriate outcome measure for type 1 diabetes clinical trials to preserve beta-cell function: report of an ADA workshop, 21–22 October 2001. *Diabetes.* **53**:250–264.

9. Lohr, M., and Kloppel, G. 1987. Residual insulin positivity and pancreatic atrophy in relation to duration of chronic type 1 (insulin-dependent) diabetes mellitus and microangiopathy. *Diabetologia.* **30**:757–762.

10. Madsbad, S. 1983. Factors of importance for residual beta-cell function in type 1 diabetes mellitus. A review. *Acta Med. Scand. Suppl.* **671**:61–67.

11. Eff, C., Faber, O., and Deckert, T. 1978. Persistent insulin secretion, assessed by plasma C-peptide estimation in long-term juvenile diabetics with a low insulin requirement. *Diabetologia.* **15**:169–172.

12. Steffes, M.W., Sibley, S., Jackson, M., and Thomas, W. 2003. Beta-cell function and the development of diabetes-related complications in the diabetes control and complications trial. *Diabetes Care.* **26**:832–836.

13. Williams, P.W. 1894. Notes on diabetes treated with extract and by grafts of sheep's pancreas. *BMJ.* **2**:1303–1304.

14. Karges, B., et al. 2004. Complete long-term recovery of beta-cell function in autoimmune type 1 diabetes after insulin treatment. *Diabetes Care.* **27**:1207–1208.

15. Ilic, S., Jovanovic, L., and Wollitzer, A.O. 2000. Is the paradoxical first trimester drop in insulin requirement due to an increase in C-peptide concentration in pregnant Type I diabetic women? *Diabetologia.* **43**:1329–1330.

16. Benedum, J. 1999. The early history of endocrine cell transplantation. *J. Mol. Med.* **77**:30–35.

17. Pybus, F.C. 1924. Notes on suprarenal and pancreatic grafting. *Lancet.* 550–551.

18. Ballinger, W.F., and Lacy, P.E. 1972. Transplantation of intact pancreatic islets in rats. *Surgery.* **72**:175–186.

19. Najarian, J.S., et al. 1977. Human islet transplantation: a preliminary report. *Transplant. Proc.* **9**:233–236.

20. Largiader, F., Kolb, E., and Binswanger, U. 1980. A long-term functioning human pancreatic islet allotransplant. *Transplantation.* **29**:76–77.

21. Scharp, D.W., et al. 1990. Insulin independence after islet transplantation into type I diabetic patient. *Diabetes.* **39**:515–518.

22. Ricordi, C., Lacy, P.E., Finke, E.H., Olack, B.J., and Scharp, D.W. 1988. Automated method for isolation of human pancreatic islets. *Diabetes.* **37**:413–420.

23. Flattem, N., et al. 2001. Alpha- and beta-cell responses to small changes in plasma glucose in the conscious dog. *Diabetes.* **50**:367–375.

24. Brissova, M., et al. 2004. Intraislet endothelial cells contribute to revascularization of transplanted

pancreatic islets. *Diabetes.* **53**:1318–1325.

25. Hering, B.J., et al. 2004. Transplantation of cultured islets from two-layer preserved pancreases in type 1 diabetes with anti-CD3 antibody. *Am. J. Transplant.* **4**:390–401.

26. Shapiro, A.M., and Ricordi, C. 2004. Unraveling the secrets of single donor success in islet transplantation. *Am. J. Transplant.* **4**:295–298.

27. Beattie, G.M., et al. 2002. A novel approach to increase human islet cell mass while preserving beta-cell function. *Diabetes.* **51**:3435–3439.

28. Beattie, G.M., Hayek, A., and Levine, F. 2000. Growth and genetic modification of human beta-cells and beta-cell precursors. *Genet. Eng.* **22**:99–120.

29. Hayek, A., and Beattie, G.M. 2002. Alternatives to unmodified human islets for transplantation. *Curr. Diab. Rep.* **2**:371–376.

30. Bonner-Weir, S., et al. 2000. In vitro cultivation of human islets from expanded ductal tissue. *Proc. Natl. Acad. Sci. U. S. A.* **97**:7999–8004.

31. Dutta, S., et al. 2001. PDX:PBX complexes are required for normal proliferation of pancreatic cells during development. *Proc. Natl. Acad. Sci. U. S. A.* **98**:1065–1070.

32. Sharma, A., et al. 1999. The homeodomain protein IDX-1 increases after an early burst of proliferation during pancreatic regeneration. *Diabetes.* **48**:507–513.

33. Dor, Y., Brown, J., Martinez, O.I., and Melton, D.A. 2004. Adult pancreatic beta-cells are formed by self-duplication rather than stem-cell differentiation. *Nature.* **429**:41–46.

34. Kolber-Simonds, D., et al. 2004. Production of alpha-1,3-galactosyltransferase null pigs by means of nuclear transfer with fibroblasts bearing loss of heterozygosity mutations. *Proc. Natl. Acad. Sci. U. S. A.* **101**:7335–7340.

35. Blusch, J.H., Patience, C., and Martin, U. 2002. Pig endogenous retroviruses and xenotransplantation. *Xenotransplantation.* **9**:242–251.

36. Magre, S., Takeuchi, Y., and Bartosch, B. 2003. Xenotransplantation and pig endogenous retroviruses. *Rev. Med. Virol.* **13**:311–329.

37. Quinn, G., et al. 2004. Genotyping of porcine endogenous retroviruses from a family of miniature swine. *J. Virol.* **78**:314–319.

38. Martin, U., et al. 2000. Productive infection of primary human endothelial cells by pig endogenous retrovirus (PERV). *Xenotransplantation.* **7**:138–142.

39. Tucker, A., et al. 2002. The production of transgenic pigs for potential use in clinical xenotransplantation: microbiological evaluation. *Xenotransplantation.* **9**:191–202.

40. Hori, Y., et al. 2002. Growth inhibitors promote differentiation of insulin-producing tissue from embryonic stem cells. *Proc. Natl. Acad. Sci. U. S. A.* **99**:16105–16110.

41. Lumelsky, N., et al. 2001. Differentiation of embryonic stem cells to insulin-secreting structures similar to pancreatic islets. *Science.* **292**:1389–1394.

42. Rajagopal, J., Anderson, W.J., Kume, S., Martinez, O.I., and Melton, D.A. 2003. Insulin staining of ES cell progeny from insulin uptake. *Science.* **299**:363.

43. Ianus, A., Holz, G.G., Theise, N.D., and Hussain, M.A. 2003. In vivo derivation of glucose-competent pancreatic endocrine cells from bone marrow without evidence of cell fusion. *J. Clin. Invest.* **111**:843–850. doi:10.1172/JCI200316502.

44. Kodama, S., Kuhtreiber, W., Fujimura, S., Dale, E.A., and Faustman, D.L. 2003. Islet regeneration during the reversal of autoimmune diabetes in NOD mice. *Science.* **302**:1223–1227.

45. Bennet, W., Groth, C.G., Larsson, R., Nilsson, B., and Korsgren, O. 2000. Isolated human islets trigger an instant blood mediated inflammatory reaction: implications for intraportal islet transplantation as a treatment for patients with type 1 diabetes. *Ups. J. Med. Sci.* **105**:125–133.

46. Johansson, U., Olsson, A., Gabrielsson, S., Nilsson, B., and Korsgren, O. 2003. Inflammatory mediators expressed in human islets of Langerhans: implications for islet transplantation. *Biochem. Biophys. Res. Commun.* **308**:474–479.

47. Moberg, L., et al. 2002. Production of tissue factor by pancreatic islet cells as a trigger of detrimental thrombotic reactions in clinical islet transplantation. *Lancet.* **360**:2039–2045.

48. Dombrowski, F., Klingmuller, D., and Pfeifer, U. 1998. Insulinomas derived from hyperplastic intra-hepatic islet transplants. *Am. J. Pathol.* **152**:1025–1038.

49. Marx, J. 2003. Cancer research. Mutant stem cells may seed cancer. *Science.* **301**:1308–1310.

50. Ryan, E.A., et al. 2002. Successful islet transplantation: continued insulin reserve provides long-term glycemic control. *Diabetes.* **51**:2148–2157.

51. Digon, B.J., III, Rother, K.I., Hirshberg, B., and Harlan, D.M. 2003. Sirolimus-induced interstitial pneumonitis in an islet transplant recipient. *Diabetes Care.* **26**:3191.

52. The Diabetes Control and Complications Trial Research Group. 1993. The effect of intensive treatment of diabetes on the development and progression of long-term complications in insulin-dependent diabetes mellitus. *N. Engl. J. Med.* **329**:977–986.

53. UK Prospective Diabetes Study Group. 1998. Intensive blood-glucose control with sulphonylureas or insulin compared with conventional treatment and risk of complications in patients with type 2 diabetes (UKPDS 33). *Lancet.* **352**:837–853.

54. UK Prospective Diabetes Study Group. 1998. Tight blood pressure control and risk of macrovascular and microvascular complications in type 2 diabetes: UKPDS 38. *BMJ.* **317**:703–713.

55. Adler, A.I., et al. 2000. Association of systolic blood pressure with macrovascular and microvascular complications of type 2 diabetes (UKPDS 36): prospective observational study. *BMJ.* **321**:412–419.

56. Collins, R., Armitage, J., Parish, S., Sleigh, P., and Peto, R. 2003. MRC/BHF Heart Protection Study of cholesterol-lowering with simvastatin in 5963 people with diabetes: a randomised placebo-controlled trial. *Lancet.* **361**:2005–2016.

57. Orchard, T.J., Lloyd, C.E., Maser, R.E., and Kuller, L.H. 1996. Why does diabetic autonomic neuropathy predict IDDM mortality? An analysis from the Pittsburgh Epidemiology of Diabetes Complications Study. *Diabetes Res. Clin. Pract.* **34**(Suppl.):S165–S171.

58. Soedamah-Muthu, S.S., et al. 2004. Risk factors for coronary heart disease in type 1 diabetic patients in Europe: the EURODIAB Prospective Complications Study. *Diabetes Care.* **27**:530–537.

59. Allen, K.V., and Walker, J.D. 2003. Microalbuminuria and mortality in long-duration type 1 diabetes. *Diabetes Care.* **26**:2389–2391.

60. Sarnak, M.J., et al. 2003. Kidney disease as a risk factor for development of cardiovascular disease: a statement from the American Heart Association Councils on Kidney in Cardiovascular Disease, High Blood Pressure Research, Clinical Cardiology, and Epidemiology and Prevention. *Circulation.* **108**:2154–2169.

61. Strippoli, G.F., et al. 2003. Clinical and therapeutic aspects of diabetic nephropathy. *J. Nephrol.* **16**:487–499.

62. Mazur, M.J., et al. 2004. Decline in native renal function early after bladder-drained pancreas transplantation alone. *Transplantation.* **77**:844–849.

63. Ojo, A.O., et al. 2003. Chronic renal failure after transplantation of a nonrenal organ. *N. Engl. J. Med.* **349**:931–940.

64. Rubel, J.R., Milford, E.L., McKay, D.B., and Jarcho, J.A. 2004. Renal insufficiency and end-stage renal disease in the heart transplant population. *J. Heart Lung Transplant.* **23**:289–300.

65. Venstrom, J.M., et al. 2003. Pancreas transplantation decreases survival for patients with diabetes and preserved kidney function. *JAMA.* **290**:2817–2823.

66. Herold, K.C., et al. 2002. Anti-CD3 monoclonal antibody in new-onset type 1 diabetes mellitus. *N. Engl. J. Med.* **346**:1692–1698.

67. Steele, C., et al. 2004. Insulin Secretion in Type 1 Diabetes. *Diabetes.* **53**:426–433.

68. Boggi, U., et al. 2004. A benefit-risk assessment of basiliximab in renal transplantation. *Drug Saf.* **27**:91–106.

69. Penn, I. 2000. Post-transplant malignancy: the role of immunosuppression. *Drug Saf.* **23**:101–113.

70. Couzin, J. 2003. Clinical trials. Diabetes' brave new world. *Science.* **300**:1862–1865.

71. Montori, V.M., et al. 2002. Posttransplantation diabetes: a systematic review of the literature. *Diabetes Care.* **25**:583–592.

72. Davani, B., et al. 2004. Aged transgenic mice with increased glucocorticoid sensitivity in pancreatic beta-cells develop diabetes. *Diabetes.* **53**(Suppl. 1):S51–S59.

73. Delaunay, F., et al. 1997. Pancreatic beta cells are important targets for the diabetogenic effects of glucocorticoids. *J. Clin. Invest.* **100**:2094–2098.

74. Bloom, R.D., et al. 2002. Association of hepatitis C with posttransplant diabetes in renal transplant patients on tacrolimus. *J. Am. Soc. Nephrol.* **13**:1374–1380.

75. First, M.R., Gerber, D.A., Hariharan, S., Kaufman, D.B., and Shapiro, R. 2002. Posttransplant diabetes mellitus in kidney allograft recipients: incidence, risk factors, and management. *Transplantation.* **73**:379–386.

76. Weir, M.R., and Fink, J.C. 1999. Risk for posttransplant Diabetes mellitus with current immunosuppressive medications. *Am. J. Kidney Dis.* **34**:1–13.

77. Lopez-Talavera, J.C., et al. 2004. Hepatocyte growth factor gene therapy for pancreatic islets in diabetes: reducing the minimal islet transplant mass required in a glucocorticoid-free rat model of allogeneic portal vein islet transplantation. *Endocrinology.* **145**:467–474.

78. Bhargava, R., et al. 2004. Prevalence of hepatic steatosis after islet transplantation and its relation to graft function. *Diabetes.* **53**:1311–1317.

79. Hirshberg, B., Mog, S., Patterson, N., Leconte, J., and Harlan, D.M. 2002. Histopathological study of intrahepatic islets transplanted in the nonhuman primate model using Edmonton protocol immunosuppression. *J. Clin. Endocrinol. Metab.* **87**:5424–5429.

80. Markmann, J.F., et al. 2003. Magnetic resonance-defined periportal steatosis following intraportal islet transplantation: a functional footprint of islet graft survival? *Diabetes.* **52**:1591–1594.

81. Owen, R.J., et al. 2003. Percutaneous transhepatic pancreatic islet cell transplantation in type 1 diabetes mellitus: radiologic aspects. *Radiology.* **229**:165–170.

82. Clouston, A.D., and Powell, E.E. 2004. Nonalcoholic fatty liver disease: is all the fat bad? *Intern. Med. J.* **34**:187–191.

83. Casey, J.J., et al. 2002. Portal venous pressure changes after sequential clinical islet transplantation. *Transplantation.* **74**:913–915.

84. Han, D., et al. 2004. Assessment of cytotoxic lymphocyte gene expression in the peripheral blood of human islet allograft recipients: elevation precedes clinical evidence of rejection. *Diabetes.* **53**:2281–2290.

85. Ryan, E.A., et al. 2004. Assessment of the severity of hypoglycemia and glycemic lability in type 1 diabetic subjects undergoing islet transplantation. *Diabetes.* **53**:955–962.

86. Cryer, P.E. 2004. Diverse causes of hypoglycemia-associated autonomic failure in diabetes. *N. Engl. J. Med.* **350**:2272–2279.

87. Ruggles, J.A., Kelemen, D., and Baron, A. 2004. Emerging therapies: controlling glucose homeostasis, immunotherapy, islet transplantation, gene therapy, and islet cell neogenesis and regeneration [review]. *Endocrinol. Metab. Clin. North Am.* **33**:239–252, xii.

How does blood glucose control with insulin save lives in intensive care?

Greet Van den Berghe

Department of Intensive Care Medicine, Catholic University of Leuven, Leuven, Belgium.

Patients requiring prolonged intensive care are at high risk for multiple organ failure and death. Insulin resistance and hyperglycemia accompany critical illness, and the severity of this "diabetes of stress" reflects the risk of death. Recently it was shown that preventing hyperglycemia with insulin substantially improves outcome of critical illness. This article examines some potential mechanisms underlying prevention of glucose toxicity as well as the effects of insulin independent of glucose control. Unraveling the molecular mechanisms will provide new insights into the pathogenesis of multiple organ failure and open avenues for novel therapeutic strategies.

Historical introduction

The 1952 Scandinavian epidemic of poliomyelitis necessitated mechanical ventilation of a large number of patients with respiratory failure, an intervention that reduced mortality from 80% to 40% (1). Since then, development of sophisticated mechanical devices to support all vital organ functions, a wide array of powerful drugs, and high-tech monitoring systems have revolutionized modern intensive-care medicine. This evolution improved short-term survival of previously lethal conditions such as multiple trauma, extensive burns, major surgery, and severe sepsis. Many patients nowadays indeed survive the initial shock phase of such conditions but often subsequently enter a chronic phase of critical illness. Mortality among such patients requiring intensive care for more than a few days has remained around 20% worldwide, to a large extent irrespective of the initial disease or trauma for which they were admitted to the intensive care unit (ICU). Most deaths in the ICU occurring beyond the first few days of critical illness are attributable to nonresolving failure of multiple organ systems, either due to or coinciding with sepsis. An increased susceptibility to infectious complications and the functional and structural sequelae of the systemic inflammatory response to infection and cellular injury play a role (2). Several lines of evidence support the concept that disturbed cellular energy metabolism contributes to organ failure (3, 4). This was originally ascribed exclusively to inadequate tissue perfusion and cellular hypoxia. Recent studies, however, also point to a disturbance in oxygen utilization rather than delivery, which has been termed cytopathic hypoxia (3, 5, 6). Although extensive research efforts during the last decade focused on strategies to prevent or reverse the potentially lethal multiple organ failure, only few of them revealed positive results (7–10). One of these strategies is tight blood glucose control with insulin (10).

Nonstandard abbreviations used: CRP, C-reactive protein; GABA, γ-aminobutyric acid; GH, growth hormone; GHBP, GH-binding protein; GIK, glucose, insulin, and potassium; HXK-II, hexokinase II; ICU, intensive care unit; IGFBP-1, IGF-binding protein–1; IRS-1, insulin receptor substrate–1; MnSOD, manganese superoxide dismutase; PEPCK, phosphoenolpyruvate carboxykinase; VDAC, voltage-dependent anion channel.

Conflict of interest: The author holds an unrestrictive Catholic University of Leuven – Novo Nordisk Chair of Research.

Citation for this article: *J. Clin. Invest.* **114**:1187–1195 (2004).
doi:10.1172/JCI200423506.

Insulin resistance and hyperglycemia in the critically ill

One hundred fifty years ago, Reyboso observed glucosuria, a condition induced by ether anesthesia, in which glucose is discharged in the urine, and in 1877 Claude Bernard described hyperglycemia during hemorrhagic shock (11). Today, it is well known that any type of acute illness or injury results in insulin resistance, glucose intolerance, and hyperglycemia, a constellation termed "diabetes of injury" (12, 13). Illness or trauma increases hepatic glucose production with ongoing gluconeogenesis despite hyperglycemia and abundantly released insulin. Hepatic insulin resistance is further characterized by elevated circulating levels of IGF–binding protein–1 (IGFBP-1) (14, 15). Also, in skeletal muscle and heart, insulin-stimulated glucose uptake is impaired (16, 17). Overall, glucose uptake in critically ill patients, however, is increased but takes place mainly in the tissues that are not dependent on insulin for glucose uptake, such as, among others, the nervous system and the blood cells (13, 18). The most severe cases of stress-induced hyperglycemia (13) and highest levels of circulating IGFBP-1 (14, 15) are observed in patients with the highest risk of death. Orchestrated "counterregulatory" hormonal responses, cytokine release, and signals from the nervous system, all affecting glucose metabolic pathways, bring about the diabetes of injury. The hormones involved include catecholamines, cortisol, glucagon, and growth hormone (GH). Proinflammatory cytokines affect glucose homeostasis indirectly, by stimulating counterregulatory hormone secretion, and directly, by altering insulin receptor signaling (Figure 1) (13, 19, 20). Although insulin receptor signaling is still incompletely understood, generation of SOCS-1 and SOCS-3 may be involved. Indeed, IL-6–stimulated SOCS-3 generation has been shown to inhibit insulin receptor tyrosine phosphorylation and downstream signal transduction (21), and both SOCS-1 and SOCS-3 have been shown to degrade insulin receptor substrate–1 (IRS-1) and IRS-2 (22). Furthermore, both endogenous and exogenous catecholamines promptly inhibit insulin secretion from β cells, and catecholamines as well as angiotensin II exert anti-insulin effects. Abnormalities in insulin signaling have been described in a rat model of critical illness resulting from prolonged administration of a nonlethal dose of endotoxin and concomitant starvation partly mimicking the condition of human sepsis (23). In the liver, abundance and tyrosine phosphorylation of the insulin receptor, IRS-1, and IRS-2 were reduced. Furthermore, there was reduced association of PI3K with IRS-1. In skeletal muscle, similar abnormalities were observed, except that the number of

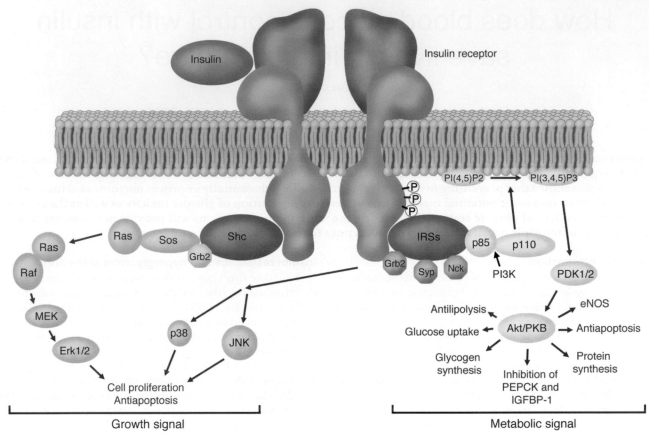

Figure 1

Simplified model of insulin signaling. Insulin binding to the extracellular domain of the insulin receptor elicits a conformational change, which in turn leads to receptor autophosphorylation (P) and tyrosine phosphorylation of intracellular protein substrates. Two main branching pathways are activated by insulin: (a) One is the MAPK signaling cascade, in which the Grb2/Sos pathway leads to activation of Ras signaling, affecting cell proliferation and apoptosis. In view of their mitogenic nature, these can be characterized as "growth signal" effects. (b) The other is the IRS pathway, which leads to activation of kinases dependent upon the heterodimeric (p85/p110) PI3K, such as Akt, also referred to as protein kinase B (PKB); Akt modulates enzyme activities that, besides affecting NO generation and apoptosis, control glucose, lipid, and protein metabolism. This PI3K-branching pathway is termed the "metabolic signal." PI(4,5)P2, phosphoinositide 4,5 di-phosphate; PI(3,4,5)P3, phosphoinositide 3,4,5 tri-phosphate; PDK1 phosphoinositide–dependent kinase–1; MEK, MAPK kinase.

insulin receptors and the abundance of the IRSs were normal. Furthermore, the insulin signaling defects in the muscle were not seen when rats were studied shortly after injection of endotoxin or under conditions of adequate nutrition (23). Apart from these insights generated from in vitro or animal models, little is known about the exact molecular basis of insulin resistance in critically ill patients. Type 2 diabetes mellitus, and to a lesser extent obesity, are also characterized by hyperglycemia, reduced glucose uptake and oxidation, unsuppressed gluconeogenesis, and impaired glycogen and NO synthesis. Here, the metabolic consequences of insulin resistance are mediated predominantly by abnormalities along the IRS-1–PI3K pathway of insulin signaling (Figure 1). However, a disrupted PI3K pathway does not necessarily mean that the other insulin signaling pathways are equally unresponsive. Indeed, signaling through the Ras-MAPK cascade, for example, via Erk 1 and Erk 2, may retain normal sensitivity. Compensatory hyperinsulinemia may thus still exert mitogenic actions in certain cell types, while the PI3K-dependent metabolic actions of insulin are suppressed (24–28). This discrepancy may occur in vascular smooth muscle cells and in specific capillary endothelial cells of patients with type 2 diabetes and obesity. Proliferation of retinal

capillary endothelial cells results in microaneurysms and neovascularization. Excessive proliferation of arterial smooth muscle cells and increased extracellular matrix lead to atherosclerosis. Hence, compensatory hyperinsulinemia, due to metabolic insulin resistance, may contribute to some of the vascular complications of type 2 diabetes and obesity through overstimulation of the mitogenic, insulin-sensitive MAPK signaling pathway. Whether or not insulin resistance is similarly "selective" during critical illness, and whether hyperinsulinemia exerts deleterious effects through this pathway in the acute setting of severe illness, is at present unknown. The diabetes of injury used to be interpreted as an adaptive stress response and as such important for survival. Particularly, the overall increase in glucose turnover and the fact that hyperglycemia persists despite abundantly released insulin were considered arguments in favor of tolerating moderately elevated blood glucose levels during critical illness. Indeed, if one considers hyperglycemia of injury as beneficial in promoting cellular glucose uptake in non–insulin-dependent tissues, tolerating modest degrees of hyperglycemia is beneficial. Consequently, blood glucose concentrations of 160–200 mg/dl were recommended to maximize cellular glucose uptake while avoiding hyperosmolarity (18).

In addition, moderate hyperglycemia was often viewed as a buffer against hypoglycemia-induced brain damage. In 2001, however, the critical care community was forced to reconsider this dogma (29), as a large, randomized, controlled, clinical study — hereafter referred to as "the Leuven study" — showed that preventing even moderate hyperglycemia during critical illness substantially improved outcome (10).

Intensive insulin therapy in critical illness: clinical benefits

The Leuven study of critically ill patients, the majority of whom did not previously have diabetes, showed that titrating insulin infusion during intensive care to strict normoglycemia (below 110 mg/dl) strikingly reduced mortality when compared with the conventional insulin treatment (Figure 2). The latter comprised insulin infusion only when blood glucose exceeded 200 mg/dl, leading to average blood glucose levels of 150–160 mg/dl (10). Although the Diabetes Mellitus, Insulin Glucose Infusion in Acute Myocardial Infarction study previously showed that avoiding excessive hyperglycemia (blood glucose >200 mg/dl) after acute myocardial infarction in patients with diabetes mellitus improved long-term outcome (30), the Leuven study (10) on nondiabetic ICU patients aimed for a much lower level of blood glucose. The benefit of intensive insulin therapy in the ICU was particularly apparent among patients with prolonged critical illness, requiring intensive care for more than 5 days, with mortality reduced from 20.2% to 10.6%. Besides saving lives, intensive insulin therapy also prevented complications such as severe nosocomial infections, acute renal failure, liver dysfunction, critical illness polyneuropathy, muscle weakness, and anemia and thus reduced the time that patients were dependent on intensive care. Although intensive insulin therapy induced a slightly higher incidence of hypoglycemia as compared with the conventional approach, these episodes were never associated with clinically relevant sequelae. Indeed, the use of an insulin titration algorithm guaranteed that when hypoglycemia occurred, it was always quickly resolved. Although the Leuven study included a large number of patients recovering from cardiac surgery, the beneficial effects of strict glucose control were equally present in most other diagnostic subgroups. Furthermore, Krinsley recently confirmed the survival benefit of implementing tight blood glucose control with insulin in a mixed medical-surgical intensive care population (31). The substantial improvement of outcome with such a simple measure was considered major progress in the modern era of intensive care.

Mechanisms explaining the acute life-saving effects of intensive insulin therapy in the ICU

One immediately wonders how such a simple intervention — preventing a moderate degree of hyperglycemia with insulin — during the relatively short time a patient is in intensive care was able to prevent the most feared complications such as sepsis, multiple organ failure, and death. Normal cells are relatively protected from deleterious effects of brief exposure to moderate hyperglycemia by a downregulation of glucose transporters (32). In diabetes mellitus, prolonged untreated hyperglycemia contributes to the development of chronic debilitating complications. However, except in embryonic development, where hyperglycemia causes acute toxicity, the time required for hyperglycemia to cause disorders in patients with diabetes is several orders of magnitude longer than the time it took to prevent life-threatening complications with

insulin therapy in the ICU. In order to understand how metabolic control with insulin proved to be so acutely protective in the critically ill, the following key questions must be answered. Is glycemic control crucial in bringing about the clinical benefits of intensive insulin therapy, or is blood glucose control epiphenomenal to the other metabolic and nonmetabolic effects of insulin? Are there factors predisposing the critically ill to hyperglycemia-induced toxicity? What are the common pathways mediating the plethora of clinical benefits of intensive insulin therapy?

Preventing direct glucose toxicity plays a crucial role

Clinical evidence. A post hoc analysis of the Leuven study (33) revealed a linear correlation between the degree of hyperglycemia and the risk of death, which persisted after correction for insulin dose and severity of illness scores. Patients in the conventional insulin treatment group who showed only moderate hyperglycemia (110–150 mg/dl or 6.1–8.3 mmol/l) had a lower risk of death than those with frank hyperglycemia (150–200 mg/dl or >8.3 mmol/l) but a higher risk of death than those who were intensively treated with insulin to restore blood glucose levels to below 110 mg/dl (6.1 mmol/l) (33). Similarly, for the prevention of morbidity effects such as acute renal failure, bacteremia, and anemia, it appeared crucial to reduce blood glucose to below 110 mg/dl. The risk of developing critical illness polyneuropathy in particular correlated linearly with blood glucose levels (34). Multivariate logistic regression analysis confirmed the independent role of blood glucose control in achieving most of the clinical benefits of intensive insulin therapy and underlined the importance of lowering the blood glucose level to strict normoglycemia.

Target tissue responsiveness to insulin in the critically ill. Since critically ill patients suffer from hepatic and skeletal muscle insulin resistance, the mechanism by which insulin lowered blood glucose in these patients is not obvious. Analysis of liver and skeletal muscle

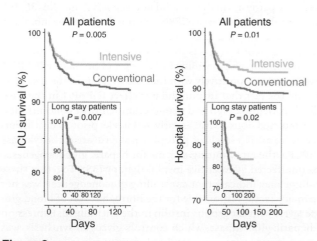

Figure 2
Intensive insulin therapy saves lives in the intensive care unit. Kaplan-Meier curves show cumulative survival of 1,548 patients from the Leuven study who received intensive insulin treatment (blood glucose maintained below 110 mg/dl; yellow) or conventional insulin treatment (insulin only given when blood glucose exceeded 200 mg/dl, resulting in mean blood glucose levels of 150–160 mg/dl; green) during their ICU or hospital stay. The upper panels display results from all patients; the lower panels display results for long-stay (>5 days) ICU patients only. *P* values were determined with the use of the Mantel-Cox log-rank test. Adapted with permission from the *New England Journal of Medicine* (10).

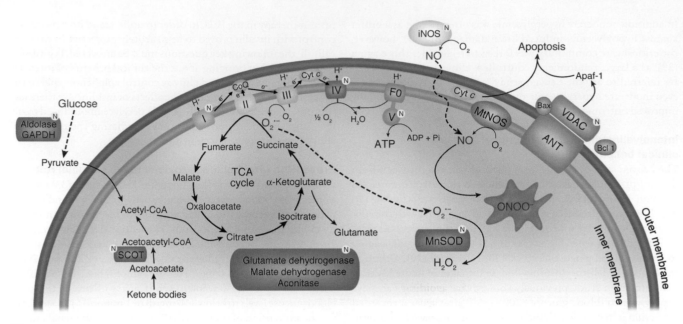

Figure 3

A diagrammatic representation of energy production in mitochondria and the mechanism of peroxynitrite generation. Excessive glycolysis and oxidative phosphorylation may result in more peroxynitrite generation in the critically ill. The ensuing nitration of mitochondrial complexes I and IV, MnSOD, GAPDH, and VDAC may suppress the activity of the mitochondrial electron transfer chain, impair detoxification of superoxide, shuttle glucose into toxic pathways, and increase apoptosis, respectively. These toxic effects may explain organ and cellular system failure related to adverse outcome in the critically ill. Proteins that are nitrated are indicated by the letter N in a yellow circle. Figure adapted with permission from *American journal of physiology. Heart and circulatory physiology* (44). TCA, tricarboxylic acid cycle; CoQ, coenzyme Q; Cyt *c*, cytochrome *c*; mtNOS, mitochondrial NO synthase; ANT, adenine nucleotide translocase; SCOT, succinyl-CoA:3-oxoacid CoA-transferase, ONOO⁻, peroxynitrite; F0, the portion of the mitochondrial ATP synthase that channels protons through the membrane.

biopsies, obtained immediately after death from nonsurvivors in the Leuven study, indicated that the classical insulin-regulated metabolic pathways in the liver did not respond to insulin (35, 36). For example, expression of *IGFBP-1*, normally under the inhibitory control of insulin, was unaltered by insulin in the critically ill (35). Circulating IGFBP-1 levels in both survivors and nonsurvivors were also unaltered by insulin at all times, and circulating IGFBP-1 levels correlated positively with the mRNA levels in the liver (35). Despite the fact that insulin did not effect IGFBP-1 in the critically ill patients, a high and rising serum IGFBP-1 concentration still predicted nonsurvival as early as 3 weeks prior to death. Furthermore, mRNA levels of phosphoenolpyruvate carboxykinase (PEPCK), the rate-limiting enzyme for hepatic gluconeogenesis, was unaffected by intensive insulin therapy (35). Together these findings may indicate that controlling gluconeogenesis was not the major factor responsible for the normalization of blood glucose levels with exogenous insulin in the critically ill. Expression of hepatic glucokinase, which controls glycogen synthesis, was also unaltered by intensive insulin therapy in critically ill patients, which, along with previous findings, confirms severe hepatic insulin resistance. In contrast, analysis of snap-frozen skeletal muscle biopsies of nonsurvivors in the Leuven study showed that skeletal muscle steady-state mRNA levels of the glucose transporter GLUT-4 and hexokinase II (HXK-II) were upregulated by insulin (36). Since GLUT-4 controls insulin-stimulated glucose uptake in muscle and since HXK-II is the rate-limiting step in intracellular insulin-stimulated glucose metabolism, the data suggest that in the critically ill patient, insulin lowers blood glucose predominantly through increased skeletal muscle glucose uptake (36).

However, true glucose kinetics can only be estimated from glucose turnover studies. Such a study, using a well-designed canine model of critical illness, recently confirmed a more severe insulin resistance in the liver as compared with other peripheral tissues (37). The transcription factor FOXO-1 is a forkhead protein expressed in liver and β cells but not in adipose tissue or skeletal muscle (38). Dephosphorylated FOXO-1 translocates to the nucleus, where it stimulates transcription of PEPCK and glucose-6-phosphatase. Insulin, via the PI3K/Akt branch of its signaling cascade, phosphorylates FOXO-1, whereby it is retained in the cytosol, blocking transcription of PEPCK or glucose-6-phosphatase. In the critically ill patient, adipose tissue and skeletal muscle remain relatively responsive to insulin, whereas the liver is much more resistant, and the β cells appear unable to compensate fully for hyperglycemia. Although impaired phosphorylation of FOXO-1 could theoretically explain the difference among tissues, this possibility remains to be investigated.

Mechanisms of accelerated glucose toxicity in the critically ill

There are two possible explanations, not mutually exclusive, for hyperglycemia to be more acutely toxic in ICU patients than in healthy individuals or patients with diabetes mellitus. The first is accentuated cellular glucose overload, and the second is more pronounced toxic side effects of glycolysis and oxidative phosphorylation.

Cellular glucose overload in the critically ill. The central and peripheral nervous system, hepatocytes, and endothelial, epithelial, and immune cells are cellular compartments that take up glucose independently of insulin. Three glucose transporters, GLUT-1,

GLUT-2, and GLUT-3, facilitate insulin-independent glucose transport in these tissues. In virtually all cell types, the GLUT-1 transporter ensures basal glucose uptake due to its low K_m (≈ 2 mM). In hepatocytes, renal tubular cells, pancreatic β cells, and the gastrointestinal mucosa, the high K_m (≈ 66 mM) and high V_{max} of the GLUT-2 transporter allow glucose to enter the cell directly in equilibrium with the level of extracellular glucose. In neurons, GLUT-3 transporters ($K_m \approx 9$ mM) are predominant. In normal cells, hyperglycemia downregulates GLUT-1 transporters, thereby protecting cells against glucose overload (32).

Cytokines, angiotensin II, endothelin-1, VEGF, TGF-β, and hypoxia have been shown to upregulate expression and membrane localization of GLUT-1 and GLUT-3 in different cell types, such as the endothelium, neurons, astrocytes, alveolar epithelial cells, and vascular smooth muscle cells (39–43). This "stress response" may overrule the normal protection of the cells against hyperglycemia, thus allowing cellular glucose overload. Hence, particularly in critical illness, characterized by high circulating levels of all these regulators, all organ systems that take up glucose passively may theoretically be at high risk for direct glucose toxicity. In contrast, skeletal muscle and the myocardium, which normally take up glucose predominantly via the insulin-dependent GLUT-4 transporter, may be relatively protected against toxic effects of circulating glucose.

More pronounced toxic side effects of oxidative phosphorylation in the critically ill. Besides cellular glucose overload, vulnerability to glucose toxicity may be due to increased generation of and/or deficient scavenging systems for ROS produced by activated glycolysis and oxidative phosphorylation (Figure 3). Glucose in the cytosol undergoes glycolysis, and its metabolite pyruvate is further transformed into acetyl-CoA, after which, in the presence of O$_2$, oxidative phosphorylation generates ATP. Along with the generation of ATP by the mitochondrial respiratory chain complexes I–V, a small amount of superoxide is concomitantly produced. Normally, 2–5% of O$_2$ used in the mitochondria is metabolized into superoxide, which is subsequently detoxified by manganese superoxide dismutase (MnSOD). When more glucose enters the cell and more pyruvate is being used for oxidative phosphorylation, more superoxide will be generated. Superoxide interacts with NO to form peroxynitrite, which nitrates proteins, such as mitochondrial complexes I and IV, MnSOD, GAPDH, and the voltage-dependent anion channel (VDAC). During critical illness as compared with the non–critically ill condition, more peroxynitrite may be generated due to cytokine-induced iNOS activation and hypoxia/reperfusion-associated superoxide production (44). Hence, when cells of critically ill patients are overloaded with glucose, more superoxide and peroxynitrite production is to be expected. The ensuing nitration of mitochondrial complexes I and IV, MnSOD, GAPDH, and VDAC may theoretically suppress the activity of the mitochondrial electron transfer chain; impair detoxification of superoxide; shuttle glucose into toxic pathways — such as the polyol and hexosamine pathways — or induce advanced glycation end-product formation; and increase apoptosis, respectively.

Singer et al. showed that tyrosine nitration relates to suppressed activity of complex I in a cellular model of cytokine overload (45). Brownlee reported GAPDH inhibition linked to vascular damage of organs and tissues of patients with diabetes (46). We recently showed that preventing hyperglycemia with intensive insulin therapy beneficially affects the hepatocytic mitochondrial compartment of critically ill patients (47). Liver biopsies revealed profound ultrastructural abnormalities in hepatocytic mitochondria

of patients to whom conventional insulin therapy was randomly allocated and thus with moderate hyperglycemia, whereas these abnormalities were virtually absent when normoglycemia was maintained during intensive care. The prevention or reversal of these morphological abnormalities had a functional correlate, as reflected by a higher activity of respiratory chain complex I and complex IV. In contrast to what was observed in the liver, electron microscopy revealed no major abnormalities in the mitochondria of skeletal muscle, and morphology as well as respiratory chain activity were not detectably affected by intensive insulin therapy in this tissue (47). The lack of effect on skeletal muscle mitochondria suggests a direct effect of glucose control, rather than of insulin, as the likely explanation. Whether these mitochondrial changes shown in critically ill patients are associated with alterations in protein nitration is currently being investigated.

Prevention of hyperglycemia-induced mitochondrial dysfunction in other cellular systems that allow glucose to enter passively, such as immune and endothelial cells, and in the central and peripheral nervous system would theoretically explain some of the protective effects of insulin therapy in critically ill patients. Increased blood glucose levels indeed have previously been shown to be associated with an increased risk of postoperative infection in patients with diabetes (48, 49). The Leuven study of nondiabetic ICU patients showed that maintaining normoglycemia with insulin prevents severe nosocomial infections and lethal sepsis (10). Polymorphonuclear neutrophil dysfunction (50), decreased intracellular bactericidal (51, 52) and opsonic activity (50, 53) following exposure to high concentrations of glucose, as well as nonenzymatic glycosylation of immunoglobulins (54) may play a role in the increased incidence of infections in patients with hyperglycemia. In an animal model of prolonged critical illness induced by trauma (55), it was recently shown that maintaining normoglycemia with insulin indeed rapidly affected innate immunity by preserving phagocytosis and oxidative burst function of monocytes (56). Ongoing studies will differentiate between the direct impact of preventing hyperglycemia and that of hyperinsulinemia on innate immunity.

Direct glucose toxicity may also explain why preventing moderate hyperglycemia with insulin clearly protects the central and the peripheral nervous system against secondary insults, as shown in the Leuven study (10, 34). Protection of the peripheral nervous system was evidenced by prevention of the development of critical illness polyneuropathy, which had immediate functional consequences such as shorter duration of mechanical ventilation and intensive care stay. Protection of the central nervous system of patients with isolated brain injury was revealed by the fact that intracranial pressure was lowered. Clinically, this finding coincided with fewer seizures, decreased incidence of diabetes insipidus, and improved long-term rehabilitation (34). Alternatively, a direct effect of insulin could have mediated some of these benefits, as insulin receptors are expressed at high levels in many areas of the brain and in different cell types (57). Indeed, local administration of insulin has shown to decrease uptake of γ-aminobutyric acid (GABA) by astrocytes and to increase the number and sensitivity of GABA receptors postsynaptically (58). This increased availability of GABA results in higher amplitude of miniature inhibitory postsynaptic currents. By inducing these GABA-mediated, neuroinhibitory effects, insulin may be directly neuroprotective. The reduced incidence of seizures observed in the Leuven study (34) corroborates such an explanation.

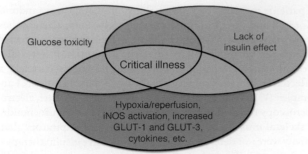

Figure 4
Venn diagram modeling the effect of the interaction between glucose toxicity and lack of insulin on the vulnerable state of critical illness. Complications of type 1 and type 2 diabetes are explained by hyperglycemia and/or lack of insulin effect. Critical illness is also characterized by hyperglycemia and lack of insulin effect, but additional risk factors render both of these effects more acutely toxic, as indicated by the blue shading. These risk factors include the post-hypoxia reperfused state, iNOS-activated NO generation, increased expression of GLUT-1 and GLUT-3 transporters, and cytokine-, neurological-, and hormone-induced alterations in cellular processes. Hence, improved outcome of critical illness with insulin-titrated maintenance of normoglycemia is likely to be explained by the prevention of both direct glucose toxicity and insulin-induced effects that are independent of glucose control.

Other metabolic effects of insulin in the critically ill

Improvement of dyslipidemia. Insulin exerts other metabolic effects besides the control of blood glucose. As in patients with diabetes mellitus (59), abnormal serum lipid profiles are observed in critically ill patients (60–62). Most characteristically, circulating triglyceride levels are elevated, whereas the levels of HDL and LDL cholesterol are very low (63). On the other hand, the numbers of circulating small, dense LDL particles, which presumably are more proatherogenic than the medium and large LDL particles (64), are increased (65). Interestingly, this dyslipidemia could be partially restored by intensive insulin therapy, with almost complete reversal of the hypertriglyceridemia and a substantial increase in, but not normalization of, the serum levels of HDL and LDL (36). The roles of triglycerides in energy provision and of lipoproteins in coordinating transportation of lipid components (cholesterol, triglycerides, phospholipids, lipid-soluble vitamins) are well established (66). In addition, it has recently been shown that lipoproteins can scavenge endotoxins and by doing so are able to prevent death in animal models (67, 68). Multivariate logistic regression analysis demonstrated that the improvement of dyslipidemia with insulin in the Leuven study explained a significant part of the beneficial effect on mortality and organ failure and, surprisingly, its effects surpassed those of glycemic control and insulin dose (36). The latter was an important observation, because when controlling for all metabolic effects of insulin, including the lipid effect, the risk associated with a high dose of exogenously administered insulin (32, 69) disappeared (36). This finding negates the notion that high doses of insulin would be deleterious in the acute setting of critical illness (69). Rather, the data provide a strong argument in favor of titrating insulin to the doses that are required in order to achieve its metabolic effects. Evidently, blood glucose level is the easiest value to measure, and when insulin is titrated to normoglycemia, the other metabolic effects occur concomitantly. A molecular explanation for the dominant effect of serum lipid correction, however, still needs to be delineated.

Anabolic effects. In view of the catabolic state of critically ill patients with loss of lean body mass, despite adequate enteral or parenteral nutrition, the beneficial effects of insulin administration to the critically ill may have been mediated in part by its anabolic actions. Indeed, the binding of insulin to its receptor normally suppresses proteolysis and activates protein synthesis, both through the PI3K signaling pathway, and evokes cell proliferation through the MAPK pathway. Although in the Leuven study, no clinically obvious anabolic effects were observed, analysis of the skeletal muscle biopsies revealed a higher protein content with intensive insulin therapy (47). Also, in a rabbit model of prolonged critical illness, intensive insulin therapy appeared to prevent weight loss (56). Poor blood glucose control in patients with type 1 diabetes mellitus has been associated with low serum IGF-I levels, which can be increased by insulin therapy (70). Thus, the anabolic effects of insulin could be partly mediated by a rise in serum IGF-I. However, contrary to expectations, in the critically ill patients, serum IGF-I, acid-labile subunit, IGFBP-3, and GH-binding protein (GHBP) levels were further suppressed instead of increased by intensive insulin therapy, and circulating growth hormone (GH) levels were elevated instead of lowered (71). A molecular explanation for this transformation to a more "GH-resistant" state with insulin therapy in the critically ill is still lacking. Growth-promoting and anabolic effects of insulin are presumably mediated largely by its suppressive effect on IGFBP-1 (72, 73), whereby bioavailable IGF-I increases. Hence, the fact that intensive insulin therapy did not affect IGFBP-1 generation and serum levels in the critically ill patient may explain why such anabolic effects of insulin did not appear to play a major role in its beneficial effect on outcome.

Other nonmetabolic effects of insulin in the critically ill

Anti-inflammatory effects. Critical illness also resembles diabetes mellitus in the activation of the inflammatory cascade, although the inflammation, as reflected by a high circulating level of C-reactive protein (CRP), in the critically ill is several times more pronounced than in diabetic patients. Intensive insulin therapy prevented excessive inflammation in critically ill patients, as indicated by lowered serum CRP and mannose-binding lectin levels (10, 74). The anti-inflammatory effects of intensive insulin therapy were present independently of its preventive effect on infections. This finding was recently confirmed in an experimental rabbit model of prolonged critical illness (56). The exact mechanisms explaining the anti-inflammatory effects of intensive insulin therapy have not yet been unraveled. Insulin may exert direct anti-inflammatory effects through its suppression of NF-κB–regulated pathways, including the production of inflammatory cytokines such as TNF-α, macrophage migration–inhibitory factor, and the generation of superoxide (75, 76). Alternatively, prevention of hyperglycemia may also play a role. The effect of insulin-titrated maintenance of normoglycemia on inflammation in the critically ill (74) was no longer independently related to the outcome benefit when the changes in lipid metabolism were taken into account (36). This observation is suggestive of a link between the anti-inflammatory effect of intensive insulin therapy and its amelioration of the lipid profile.

Preventing endothelial dysfunction and hypercoagulation. Diabetes mellitus and critical illness are both hypercoagulable states (77, 78). In the critically ill, this may contribute to the risk of organ failure. Putative causes in diabetes include vascular endothelium dysfunction (79), increased circulating levels of several clotting factors (80, 81), elevated platelet activation (82, 83), and inhibition of the fibrinolytic system (84). Levels of the anticoagulant protein C are also decreased (85). The similarities between critical illness and diabetes (86, 87); the powerful preventive effect of intensive insulin therapy on septicemia, multiple organ failure, and mortality (10); the influence of intensive insulin therapy on endothelial activation; and the balance between coagulation and fibrinolysis in the critically ill should be investigated in detail. On the one hand, insulin has been shown to activate Ca^{2+}-independent eNOS generation in endothelial cells (88). On the other hand, the recent concept of "selective insulin resistance," involving unresponsiveness of the PI3K post-receptor signaling pathway but responsiveness of the Ras-MAP kinase pathways in endothelial cells, raises the possibility of insulin-induced aggravation of endothelial activation, the latter evoked by hyperglycemia, angiotensin II, and vascular growth factors such as VEGF. Since VEGF is known to be highly expressed in the inflammatory phase of critical illness, this hypothesis was recently tested in samples obtained in the Leuven study. In line with the clinical benefits observed, circulating levels of the adhesion molecules soluble ICAM-I and E-selectin were downregulated by intensive insulin therapy (G. Van den Berghe, unpublished data). This suggests that in the critically ill, the PI3K pathway in the endothelium, as in skeletal muscle, remains at least partially responsive, in contrast to the metabolic pathways in the liver, which are highly insulin resistant.

Anti-apoptotic effects. More than 4 decades ago the concept was introduced that glucose, insulin, and potassium (GIK), administered concomitantly, could protect the ischemic myocardium (89). Opie suggested that the mechanism behind this cardioprotective effect of GIK is ATP generation when O_2 supply is limited by promoting glycolysis instead of FFA oxidation (90). Currently, experimental data support the hypothesis that insulin itself has direct cardioprotective effects during reperfusion, mainly via anti-apoptotic properties that are independent of glucose uptake (91–93). The insulin signaling pathways involved include PI3K, Akt, and eNOS phosphorylation (92). Whether such direct insulin-induced cell survival plays a role in mediating the observed protection of organ function in the critically ill remains unclear.

Future directions
It is clear that several mechanisms are involved and interrelated in explaining the clinical benefits of intensive insulin therapy in the critically ill (Figure 4). Accelerated toxicity of hyperglycemia and lack of insulin effect during critical illness explain why the consequences are so rapidly deleterious in this condition. The direct effects of preventing hyperglycemia as well as the distinct metabolic and nonmetabolic effects of insulin that occur concomitantly with the glycemic control are likely to play a role. The relative contribution of those different mechanisms, however, is presently unknown. Furthermore, several questions await an answer. The exact molecular basis for the increased susceptibility for these toxic

effects in the critically ill patient remains to be explored. It also remains unclear which of the insulin signal transduction pathways in different target tissues respond to insulin treatment and how these effects mediate the protection on organ function and overall outcome. Various hypotheses have been advanced, and these can be tested by analysis of tissue samples obtained from patients and from experimental animals in carefully designed studies. Further investigation of the mitochondrial abnormalities in tissues other than the liver and skeletal muscle, such as neurons, kidneys, and endothelial and epithelial cells, as well as the selective impact of insulin and glucose control on this cellular compartment should be investigated in great detail. Furthermore, exploring the molecular link between improved glycemic and lipid control and insulin, endothelial function, inflammation, and clinical outcome will be required. Finally, manipulation of those pathways that are linked with outcome but appear unresponsive to intensive insulin therapy may point to the potential for other therapeutic strategies to further improve survival in the ICU.

Conclusion
The results of the clinical study performed in Leuven demonstrated that a simple metabolic intervention, maintaining normoglycemia with insulin, improved survival and reduced morbidity of critically ill patients. This reflected major clinical progress in the modern era of intensive care. Furthermore, this clinical study has opened a whole new area of basic research besides the ongoing search for additional clinical applications. Indeed, only some of the underlying mechanisms have been studied, and many more pathways need to be investigated in great detail. Results from fundamental research carefully designed to elucidate how the clinical benefits were brought about are likely to set off development of new strategies for further improving outcome of critically ill patients and perhaps also of other target populations within and outside the hospital setting.

Acknowledgments
I thank Roger Bouillon, Ilse Vanhorebeek, and Lies Langouche for critically reviewing the manuscript. This work was supported by the Fund for Scientific Research, Flanders, Belgium (G.0278.03), the Research Council of the Catholic University of Leuven (OT 03/56), and the Belgian Foundation for Research in Congenital Heart Diseases. The author is a Fundamental Clinical Research Investigator (G.3C05.95N) for the Fund for Scientific Research, Flanders, Belgium, and holds an unrestrictive Catholic University of Leuven – Novo Nordisk Chair of Research.

Address correspondence to: Greet Van den Berghe, Department of Intensive Care Medicine, Catholic University of Leuven, B-3000 Leuven, Belgium. Phone: 32-16-34-40-21; Fax: 32-16-34-40-15; E-mail: greta.vandenberghe@med.kuleuven.ac.be.

1. Lassen, H.C.A. 1953. A preliminary report on the 1952 epidemic of poliomyelitis in Copenhagen with special reference to the treatment of respiratory insufficiency. *Lancet.* **i**:37–40.
2. Fink, M.P., and Evans, T.W. 2002. Mechanisms of organ dysfunction in critical illness: report from a round table conference held in Brussels. *Intensive Care Med.* **28**:369–375.
3. Fink, M.P. 2001. Cytopathic hypoxia. Mitochondrial dysfunction as mechanism contributing to organ dysfunction in sepsis. *Crit. Care Clin.* **17**:219–237.
4. Brealey, D., et al. 2002. Association between

mitochondrial dysfunction and severity and outcome of septic shock. *Lancet.* **360**:219–223.
5. Singer, M., and Brealey, D. 1999. Mitochondrial dysfuntion in sepsis. *Biochem. Soc. Symp.* **66**:149–166.
6. Crouser, E.D., Julian, M.W., Blaho, D.V., and Pfeiffer, D.R. 2002. Endotoxin-induced mitochondrial damage correlates with impaired respiratory activity. *Crit. Care Med.* **30**:276–284.
7. Amato, M.B., et al. 1998. Effect of a protective-ventilation strategy on mortality in the acute respiratory distress syndrome. *N. Engl. J. Med.* **338**:347–354.
8. Rivers, E., et al. 2001. Early goal-directed therapy

in the treatment of severe sepsis and septic shock. *N. Engl. J. Med.* **345**:1368–1377.
9. Bernard, G.R., et al. 2001. Efficacy and safety of recombinant human activated protein C for severe sepsis. *N. Engl. J. Med.* **344**:699–709.
10. Van den Berghe, G., et al. 2001. Intensive insulin therapy in critically ill patients. *N. Engl. J. Med.* **345**:1359–1367.
11. Bernard, C. 1878. *Leçons sur les phénomènes de la vie communs aux animaux et aux végétaux.* Volume 1. J.B. Baillière et Fils. Paris, France. 564 pp.
12. Thorell, A., Nygren, J., and Ljungqvist, O. 1999.

Insulin resistance: a marker of surgical stress. *Curr. Opin. Clin. Nutr. Metab. Care.* **21**:69–78.

13. McCowen, K.C., Malhotra, A., and Bistrian, B.R. 2001. Stress-induced hyperglycaemia. *Crit. Care Clin.* **17**:107–124.

14. Van den Berghe, G., et al. 1999. Reactivation of pituitary hormone release and metabolic improvement by infusion of growth hormone-releasing peptide and thyrotropin-releasing hormone in patients with protracted critical illness. *J. Clin. Endocrinol. Metab.* **84**:1311–1323.

15. Van den Berghe, G., et al. 2000. A paradoxical gender dissociation within the growth hormone/insulin-like growth factor I axis during protracted critical illness. *J. Clin. Endocrinol. Metab.* **85**:183–192.

16. Wolfe, R.R., Durkot, M.J., Allsop, J.R., and Burke, J.F. 1979. Glucose metabolism in severely burned patients. *Metabolism.* **28**:210–220.

17. Wolfe, R.R., Herndon, D.N., Jahoor, F., Miyoshi, H., and Wolfe, M. 1987. Effects of severe burn injury on substrate cycling by glucose and fatty acids. *N. Engl. J. Med.* **317**:403–408.

18. Mizock, B.A. 1995. Alterations in carbohydrate metabolism during stress: a review of the literature. *Am. J. Med.* **98**:75–84.

19. Grimble, R.F. 2002. Inflammatory status and insulin resistance. *Curr. Opin. Clin. Nutr. Metab. Care.* **5**:551–559.

20. Marette, A. 2002. Mediators of cytokine-induced insulin resistance in obesity and other inflammatory settings. *Curr. Opin. Clin. Nutr. Metab. Care.* **5**:377–383.

21. Senn, J.J., et al. 2003. Suppressor of cytokine signaling-3 (SOCS-3), a potential mediator of interleukin-6-dependent insulin resistance in hepatocytes. *J. Biol. Chem.* **278**:13740–13746.

22. Rui, L., Yuan, M., Frantz, D., Shoelson, S., and White, M.F. 2002. SOCS-1 and SOCS-3 block insulin signaling by ubiquitin-mediated degradation of IRS1 and IRS2. *J. Biol. Chem.* **277**:42394–42398.

23. McCowen, K.C., et al. 2001. Sustained endotoxemia leads to marked down-regulation of early steps in the insulin-signaling cascade. *Crit. Care Med.* **29**:839–846.

24. Jiang, Z.Y., et al. 1999. Characterization of selective resistance to insulin signaling in the vasculature of obese Zucker (fa/fa) rats. *J. Clin. Invest.* **104**:447–457.

25. Cusi, K., et al. 2000. Insulin resistance differentially affects the PI 3-kinase-and MAP kinase-mediated signaling in human muscle. *J. Clin. Invest.* **105**:311–320.

26. Draznin, B., et al. 2000. Effects of insulin on prenylation as a mechanism of potential detrimental influence of hyperinsulinemia. *Endocrinology.* **141**:1310–1316.

27. Golovchenko, I., Goalstone, M.L., Watson, P., Brownlee, M., and Draznin, B. 2000. Hyperinsulinemia enhances transcriptional activity of Nuclear Factor-kB induced by angiotensin II, hyperglycemia, and advanced glycosylation end products in vascular smooth muscle cells. *Circ. Res.* **87**:746–752.

28. Montagnani, M., et al. 2002. Inhibition of phosphatidylinositol 3-kinase enhances mitogenic actions of insulin in endothelial cells. *J. Biol. Chem.* **277**:1794–1799.

29. Mizock, B.A. 2001. Alterations in fuel metabolism in critical illness: hyperglycaemia. *Best Pract. Res. Clin. Endocrinol. Metab.* **15**:533–551.

30. Malmberg, K., et al. 1995. Randomized trial of insulin-glucose infusion followed by subcutaneous insulin treatment in diabetic patients with acute myocardial infarction (DIGAMI study): effects on mortality at 1 year. *J. Am. Coll. Cardiol.* **26**:57–65.

31. Krinsley, J.S. 2004. Effect of an intensive glucose management protocol on the mortality of critically ill adult patients. *Mayo Clin. Proc.* **79**:992–1000.

32. Klip, A., Tsakiridis, T., Marette, A., and Ortiz, P.A. 1994. Regulation of expression of glucose trans-

porters by glucose: a review of studies in vivo and in cell cultures. *FASEB J.* **8**:43–53.

33. Van den Berghe, G., et al. 2003. Outcome benefit of intensive insulin therapy in the critically ill: insulin dose versus glycemic control. *Crit. Care Med.* **31**:359–366.

34. Van den Berghe, G., Schoonheydt, K., Becx, P., Bruyninckx, F., and Wouters, P.J. 2004. Intensive insulin therapy protects the central and peripheral nervous system of intensive care patients. In: Program and abstracts of the 86th Annual Meeting of the Endocrine Society. New Orleans, Louisiana, USA. June 16–19, 2004. Abstract OR8-2, p. 79.

35. Mesotten, D., et al. 2002. Regulation of insulin-like growth factor binding protein-1 during protracted critical illness. *J. Clin. Endocrinol. Metab.* **87**:5516–5523.

36. Mesotten, D., Swinnen, J., Vanderhoydonc, F., Wouters, P.J., and Van den Berghe, G. 2004. Contribution of circulating lipids to the improved outcome of critical illness by glycemic control with intensive insulin therapy. *J. Clin. Endocrinol. Metab.* **89**:219–226.

37. Donmoyer, C.M., et al. 2003. Infection impairs insulin-dependent hepatic glucose uptake during total parenteral nutrition. *Am. J. Physiol. Endocrinol. Metab.* **284**:E574–E582.

38. Puigserver, P., et al. 2003. Insulin-regulated hepatic gluconeogenesis through FOXO1-PGC-1alpha interaction. *Nature.* **423**:550–555.

39. Pekala, P., Marlow, M., Heuvelman, D., and Connolly, D. 1990. Regulation of hexose transport in aortic endothelial cells by vascular permeability factor and tumor necrosis factor-alpha, but not by insulin. *J. Biol. Chem.* **265**:18051–18054.

40. Shikhman, A.R., Brinson, D.C., Valbracht, J., and Lotz, M.K. 2001. Cytokine regulation of facilitated glucose transport in human articular chondrocytes. *J. Immunol.* **167**:7001–7008.

41. Quinn, L.A., and McCumbee, W.D. 1998. Regulation of glucose transport by angiotensin II and glucose in cultured vascular smooth muscle cells. *J. Cell. Physiol.* **177**:94–102.

42. Clerici, C., and Matthay, M.A. 2000. Hypoxia regulates gene expression of alveolar epithelial transport proteins. *J. Appl. Physiol.* **88**:1890–1896.

43. Sanchez-Alvarez, R., Tabernero, A., and Medina, J.M. 2004. Endothelin-1 stimulates the translocation and upregulation of both glucose transporter and hexokinase in astrocytes: relationship with gap junctional communication. *J. Neurochem.* **89**:703–714.

44. Aulak, K.S., Koeck, T., Crabb, J.W., and Stuehr, D.J. 2004. Dynamics of protein nitration in cells and mitochondria. *Am. J. Physiol. Heart Circ. Physiol.* **286**:H30–H38.

45. Frost, M., Wang, Q., Moncada, S., and Singer, M. 2004. Bi-phasic, oxygen- and NO-dependent modulation of complex I activity in activated macrophages by S-nitrosylation and nitration. *Am. J. Physiol.* In press.

46. Brownlee, M. 2001. Biochemistry and molecular cell biology of diabetic complications. *Nature.* **414**:813–820.

47. Vanhorebeek, I., et al. 2004. Strict blood glucose control with insulin in critically ill patients protects hepatocytic mitochondrial ultrastructure and function. *Lancet.* In press.

48. Pozzilli, P., and Leslie, R.D. 1994. Infections and diabetes: mechanisms and prospects for prevention. *Diabet. Med.* **11**:935–941.

49. Funari, A.P., Zerr, K.J., Grunkemeier, G.L., and Starr, A. 1999. Continuous intravenous insulin infusion reduces the incidence of deep sternal wound infection in diabetic patients after cardiac surgical procedures. *Ann. Thorac. Surg.* **67**:352–360.

50. Rassias, A.J., et al. 1999. Insulin infusion improves neutrophil function in diabetic cardiac surgery patients. *Anesth. Analg.* **88**:1011–1016.

51. Nielson, C.P., and Hindson, D.A. 1989. Inhibition of polymorphonuclear leukocyte respiratory burst by elevated glucose concentrations in vitro. *Diabetes.* **38**:1031–1035.

52. Perner, A., Nielsen, S.E., and Rask-Madsen, J. 2003. High glucose impairs superoxide production from isolated blood neutrophils. *Intensive Care Med.* **29**:642–645.

53. Rayfield, E.J., et al. 1982. Infection and diabetes: the case for glucose control. *Am. J. Med.* **72**:439–450.

54. Black, C.T., Hennessey, P.J., and Andrassy, R.J. 1990. Short-term hyperglycemia depresses immunity through nonenzymatic glycosylation of circulating immunoglobulin. *J. Trauma.* **30**:830–832.

55. Weekers, F., et al. 2002. A novel *in vivo* rabbit model of hypercatabolic critical illness reveals a bi-phasic neuroendocrine stress response. *Endocrinology.* **143**:764–774.

56. Weekers, F., et al. 2003. Metabolic, endocrine and immune effects of stress hyperglycemia in a rabbit model of prolonged critical illness. *Endocrinology.* **144**:5329–5338.

57. Bruning, J.C., et al. 2000. Role of brain insulin receptor in control of body weight and reproduction. *Science.* **289**:2122–2155.

58. Vincent, A.M., Brownlee, M., and Russell, J.W. 2002. Oxidative stress and programmed cell death in diabetic neuropathy. *Ann. N. Y. Acad. Sci.* **959**:368–383.

59. Taskinen, M.R. 2001. Pathogenesis of dyslipidemia in type 2 diabetes. *Exp. Clin. Endocrinol. Diabetes.* **109**:S180–S188.

60. Lanza-Jacoby, S., Wong, S.H., Tabares, A., Baer, D., and Schneider, T. 1992. Disturbances in the composition of plasma lipoproteins during gram-negative sepsis in the rat. *Biochim. Biophys. Acta.* **1124**:233–240.

61. Khovidhunkit, W., Memon, R.A., Feingold, K.R., and Grunfeld, C. 2000. Infection and inflammation-induced proatherogenic changes of lipoproteins [review]. *J. Infect. Dis.* **181**:S462–S472.

62. Carpentier, Y.A., and Scruel, O. 2002. Changes in the concentration and composition of plasma lipoproteins during the acute phase response. *Curr. Opin. Clin. Nutr. Metab. Care.* **5**:153–158.

63. Gordon, B.R., et al. 1996. Low lipid concentrations in critical illness: implications for preventing and treating endotoxemia. *Crit. Care Med.* **24**:584–589.

64. Kwiterovich, P.O. 2002. Lipoprotein heterogeneity: diagnostic and therapeutic implications. *Am. J. Cardiol.* **90**:1i–10i.

65. Feingold, K.R., et al. 1993. The hypertriglyceridemia of acquired immunodeficiency syndrome is associated with an increased prevalence of low density lipoprotein subclass pattern B. *J. Clin. Endocrinol. Metab.* **76**:1423–1427.

66. Tulenko, T.N., and Sumner, A.E. 2002. The physiology of lipoproteins. *J. Nucl. Cardiol.* **9**:638–649.

67. Harris, H.W., Grunfeld, C., Feingold, K.R., and Rapp, J.H. 1990. Human very low density lipoproteins and chylomicrons can protect against endotoxin-induced death in mice. *J. Clin. Invest.* **86**:696–702.

68. Harris, H.W., et al. 1993. Chylomicrons alter the fate of endotoxin, decreasing tumor necrosis factor release and preventing death. *J. Clin. Invest.* **91**:1028–1034.

69. Finney, S.J., Zekveld, C., Elia, A., and Evans, T.W. 2003. Glucose control and mortality in critically ill patients. *JAMA.* **290**:2041–2047.

70. Brismar, K., Fernqvist-Forbes, E., Wahren, J., and Hall, K. 1994. Effect of insulin on the hepatic production of insulin-like growth factor-binding protein-1 (IGFBP-1), IGFBP-3, and IGF-I in insulin-dependent diabetes. *J. Clin. Endocrinol. Metab.* **79**:872–878.

71. Mesotten, D., et al. 2004. Regulation of the somatotropic axis by intensive insulin therapy during protracted critical illness. *J. Clin. Endocrinol. Metab.*

89:3105–3113.

72. Suikkari, A.M., et al. 1988. Insulin regulates the serum levels of low molecular weight insulin-like growth factor-binding protein. *J. Clin. Endocrinol. Metab.* **66**:266–272.

73. Suwanichkul, A., Morris, S.L., and Powell, D.R. 1993. Identification of an insulin-responsive element in the promoter of the human gene for insulin-like growth factor binding protein-1. *J. Biol. Chem.* **268**:17063–17068.

74. Hansen, T.K., Thiel, S., Wouters, P.J., Christiansen, J.S., and Van den Berghe, G. 2003. Intensive insulin therapy exerts antiinflammatory effects in critically ill patients and counteracts the adverse effect of low mannose-binding lectin levels. *J. Clin. Endocrinol. Metab.* **88**:1082–1088.

75. Das, U.N. 2001. Is insulin an antiinflammatory molecule? *Nutrition.* **17**:409–413.

76. Dandona, P., et al. 2001. Insulin inhibits intranuclear factor kappaB and stimulates IkappaB in mononuclear cells in obese subjects: evidence for an anti-inflammatory effect? *J. Clin. Endocrinol. Metab.* **86**:3257–3265.

77. Carr, M.E. 2001. Diabetes mellitus: a hypercoagulable state. *J. Diabetes Complicat.* **15**:44–54.

78. Calles-Escandon, J., Garcia-Rubi, E., Mirza, S., and Mortensen, A. 1999. Type 2 diabetes: one disease, multiple cardiovascular risk factors. *Coron. Artery Dis.* **10**:23–30.

79. Williams, E., Timperley, W.R., Ward, J.D., and Duckworth, T. 1980. Electron microscopical studies of vessels in diabetic peripheral neuropathy. *J. Clin. Pathol.* **33**:462–470.

80. Patrassi, G.M., Vettor, R., Padovan, D., and Girolami, A. 1982. Contact phase of blood coagulation in diabetes mellitus. *Eur. J. Clin. Invest.* **12**:307–311.

81. Carmassi, F., et al. 1992. Coagulation and fibrinolytic system impairment in insulin dependent diabetes mellitus. *Thromb. Res.* **67**:643–654.

82. Hughes, A., et al. 1983. Diabetes, a hypercoagulable state? Hemostatic variables in newly diagnosed type 2 diabetic patients. *Acta Haematol.* **69**:254–259.

83. Garcia Frade, L.J., et al. 1987. Diabetes mellitus as a hypercoagulable state: its relationship with fibrin fragments and vascular damage. *Thromb. Res.* **47**:533–540.

84. Carmassi, F., et al. 1992. Coagulation and fibrinolytic system impairment in insulin dependent diabetes mellitus. *Thromb. Res.* **67**:643–654.

85. Vukovich, T.C., and Schernthaner, G. 1986. Decreased protein C levels in patients with insulin-dependent type I diabetes mellitus. *Diabetes.* **35**:617–619.

86. Garcia Frade, L.J., et al. 1987. Changes in fibrinolysis in the intensive care patient. *Thromb. Res.* **47**:593–599.

87. Mavrommatis, A.C., et al. 2001. Activation of the fibrinolytic system and utilization of the coagulation inhibitors in sepsis: comparison with severe sepsis and septic shock. *Intensive Care Med.* **27**:1853–1859.

88. Montagnani, M., Chen, H., Barr, V.A., and Quon, M.J. 2001. Insulin-stimulated activation of e-NOS is independent of Ca2+ but requires phosphorylation by Akt at Ser(1179). *J. Biol. Chem.* **276**:30392–30398.

89. Sodi-Pallares, D., Testelli, M.R., and Fishleder, B. 1962. Effects of an intravenous infusion of a potassium-insulin-glucose solution on the electrocardiographic signs of myocardial infarction. A preliminary clinical report. *Am. J. Cardiol.* **9**:166–181.

90. Opie, L. 1970. The glucose hypothesis: relation to acute myocardial ischemia. *J. Mol. Cell. Cardiol.* **1**:107–114.

91. Jonassen, A., Aasum, E., Riemersma, R., Mjos, O., and Larsen, T. 2000. Glucose-insulin-potassium reduces infarct size when administered during reperfusion. *Cardiovasc. Drugs Ther.* **14**:615–623.

92. Gao, F., et al. 2002. Nitric oxide mediates the antiapoptotic effect of insulin in myocardial ischemia-reperfusion: the role of PI3-kinase, Akt, and eNOS phosphorylation. *Circulation.* **105**:1497–1502.

93. Jonassen, A., Sack, M., Mjos, O., and Yellon, D. 2001. Myocardial protection by insulin at reperfusion requires early administration and is mediated via Akt and p70s6 kinase cell-survival signalling. *Circ. Res.* **89**:1191–1198.

Thyrotropin receptor–associated diseases: from adenomata to Graves disease

Terry F. Davies, Takao Ando, Reigh-Yi Lin, Yaron Tomer, and Rauf Latif

Department of Medicine, Mount Sinai School of Medicine, New York, New York, USA.

The thyroid-stimulating hormone receptor (TSHR) is a G protein–linked, 7–transmembrane domain (7-TMD) receptor that undergoes complex posttranslational processing unique to this glycoprotein receptor family. Due to its complex structure, TSHR appears to have unstable molecular integrity and a propensity toward over- or underactivity on the basis of point genetic mutations or antibody-induced structural changes. Hence, both germline and somatic mutations, commonly located in the transmembrane regions, may induce constitutive activation of the receptor, resulting in congenital hyperthyroidism or the development of actively secreting thyroid nodules. Similarly, mutations leading to structural alterations may induce constitutive inactivation and congenital hypothyroidism. The TSHR is also a primary antigen in autoimmune thyroid disease, and some TSHR antibodies may activate the receptor, while others inhibit its activation or have no influence on signal transduction at all, depending on how they influence the integrity of the structure. Clinical assays for such antibodies have improved significantly and are a useful addition to the investigative armamentarium. Furthermore, the relative instability of the receptor can result in shedding of the TSHR ectodomain, providing a source of antigen and activating the autoimmune response. However, it may also provide decoys for TSHR antibodies, thus influencing their biological action and clinical effects. This review discusses the role of the TSHR in the physiological and pathological stimulation of the thyroid.

The master switch in the regulation of the thyroid gland, including its growth and differentiation, is the thyroid-stimulating hormone (TSH) receptor (TSHR). The TSHR is a 7-transmembrane domain (7-TDM) G protein–coupled receptor anchored to the surface of the plasma membrane of thyrocytes and a variety of other cell types (1). In addition, the TSHR has been implicated in a range of thyroid diseases (Table 1). For example, certain TSHR mutations cause constitutive overactivity of thyroid cells, leading to active nodule formation or rare cases of congenital hyperthyroidism. In contrast, other TSHR mutations have resulted in receptor inactivation or rare cases of congenital hypothyroidism (2). The TSHR is also a major autoantigen in autoimmune thyroid disease (AITD). In particular, the TSHR is the target of the immune response in patients with Graves disease, who exhibit unique TSHR-stimulating antibodies (1, 3). This review, therefore, encompasses those diseases involving TSHR structural variants and in which TSHR is a major antigenic target.

An overview of the TSHR

TSHR structure. Prior to successful cloning, the subunit structure of the TSHR had been deduced by affinity labeling of thyrocyte membranes using radiolabeled and photoactivated TSH (4). The cloning of the canine TSHR in 1989 resulted from cross-hybridization procedures with a luteinizing hormone (LH; also known as lutropin) receptor probe (5) and was soon followed by the

cloning of the human gene (6–8). The deduced protein structure established its membership in the family of G protein–coupled receptors having sequence similarity with the adrenergic-rhodopsin receptors (Figure 1). The *TSHR* gene on chromosome 14q3 (9) codes for a 764-aa protein, which comprises a signal peptide of 21 aa; a large, glycosylated ectodomain of 394 residues encoded by 9 exons; and 349 residues encoded by the tenth and largest exon, which constitute the 7 TMDs and cytoplasmic tail. The sequence also revealed 2 nonhomologous segments within the TSHR ectodomain (residues 38–45 and 316–366) not found in otherwise closely related glycoprotein hormone receptors such as those for LH and follicle-stimulating hormone (FSH; also known as follitropin) (3). The initial TSH cross-linking studies described above indicated that the mature TSHR contained 2 subunits (4), and the subsequent molecular cloning of the TSHR indicated that both subunits were encoded by a single gene, which indicated that intramolecular cleavage must have occurred (4, 10, 11), something not observed with the LH and FSH receptors. One TSHR subunit consists of a large extracellular domain (or ectodomain; mostly the α, or A, subunit), and the second contains the short membrane-anchored and intracellular portion of the receptor (the β, or B, subunit) (Figure 1).

The TSH-binding pocket on the TSHR. Expression on the plasma membrane of the TSHR ectodomain with a short lipid tail is sufficient for high-affinity binding of TSH (12–14). Hence, the TSHR ectodomain, consisting mainly of 9 leucine-rich repeats (LRRs) and an N-terminal tail, encoded by exons 2–8, forms the binding domain for TSH. The 7 TMDs are joined intracellularly by connecting loops that interact with G proteins when the receptor is activated (15), whereas the exoplasmic loops, outside the cell, have ancillary roles in receptor structure and activation (16). Investigation of the 2 nonhomologous segments within the ectodomain showed that deletion of residues 38–45 abrogated TSH binding (17), whereas deletion of residues 316–366 did not (17, 18), and the TSHR-transfected cells were still capable of TSH-mediated sig-

Nonstandard abbreviations used: AITD, autoimmune thyroid disease; FSH, follicle-stimulating hormone; hCG, human chorionic gonadotropin; LH, luteinizing hormone; LRR, leucine-rich repeat; NIS, sodium iodine symporter; PAX8, paired box gene 8; Tg, thyroglobulin; TMD, transmembrane domain; TPO, thyroid peroxidase; TSH, thyroid-stimulating hormone; TSHR, TSH receptor; TTF, thyroid transcription factor.

Conflict of interest: Terry F. Davies is a consultant to Kronus Inc. and RSR Ltd., which manufacture and distribute tests for the measurement of thyroid antibodies including TSHR antibodies, and to the Abbott Corp., which manufactures a thyroid hormone preparation.

Citation for this article: *J. Clin. Invest.* **115**:1972–1983 (2005). doi:10.1172/JCI26031.

Table 1

Diseases of the TSHR

Site	Abnormality	Phenotype
Germline	TSHR mutations	Gain-of-function (congenital hyperthyroidism)
		Loss-of-function (congenital hypothyroidism)
	Non-TSHR polymorphisms	Susceptibility to TSHR autoimmunity (Graves disease; environmental interactions)
Somatic	TSHR mutations	Gain of function (hot nodules)

naling. A detailed mutational analysis of residues 38–45 showed that Cys41 was the critical residue required for TSHR expression (19, 20), and data also indicated that Cys41 interaction with other neighboring Cys residues in the TSHR ectodomain, via disulfide bonds, was essential for high-affinity TSH binding (20). Hence, cysteine bonding helped restrain the structure critical for TSH ligand binding (17). In order to better define the TSH-binding sites on the TSHR, mutagenesis of functional TSHRs was undertaken together with the use of synthetic peptides. Such studies showed that there were multiple TSH-binding sites in the region of the LRRs of the TSHR, which is consistent with the existence of a conformational binding site (21, 22). Another approach employed a panel of epitope-mapped TSHR antibodies to block labeled TSH binding to the native TSHR. This approach defined TSH binding based on TSHR sequences recognized by the antibodies and suggested 3 distinct TSH binding regions in the TSHR: aa 246–260, 277–296, and 381–385 (23, 24). These regions may fold together to form a complex TSH-binding pocket (Figure 2).

TSHR cleavage and shedding. A unique posttranslational proteolytic event clips the TSHR into 2 subunits (3, 4) as indicated in Figure 1. Intramolecular cleavage results in removal of the unique, intervening, approximately 50-aa polypeptide segment in the ectodomain (aa 316–366) (10, 25). This cleavage step may involve an uncharacterized MMP-like enzyme acting at the cell surface (26, 27). Following cleavage, the α and β subunits are disulfide-bonded by cysteine residues flanking the now-absent cleaved 50-aa region, a structure compatible with molecular modeling of the TSHR (28). Subsequently these α-β disulfide links are broken, by protein disulfide isomerase (29) and possibly by progression of β subunit degradation toward the membrane (30), which leads to shedding of the α subunit from the membrane-bound receptor. This explains the high ratio of TSHR β to TSHR α subunits (up to 3:1) found in normal thyroid membrane preparations (10, 11). However, TSHR signal transduction is not dependent on ectodomain cleavage, as has been demonstrated with a noncleavable construct (18).

TSHR multimerization. Recent studies have documented the propensity of G protein–coupled receptors to form homo- and heterodimeric forms, and these forms may have functional roles

in protein trafficking (31), internalization (32), receptor stability (33), and signaling (34). We have found that the TSHR also forms oligomers in both TSHR-transfected cells and native thyrocytes (35). While unstimulated TSHRs were found in multimeric forms (36), this multimeric state was reversed by TSH (37). We still do not understand the physiologic consequences of TSHR multimerization, intramolecular cleavage, and subunit shedding. These fascinating processes most likely have a major effect on thyroid cell function and the interface between the TSHR and the immune repertoire, and they may play a role in the susceptibility of the thyroid to immune attack.

The open and closed TSHR hypothesis. Unlike other glycoprotein receptors, the TSHR is constitutively active and susceptible to enhanced constitutive activation by mutation, deletions, and even mild trypsin digestion (38, 39). Studies using mutational analyses have suggested that the putative electrostatic interactions between the ectodomain and the extracellular loops of the TMDs in the TSHR may be critical for the maintenance of a relatively inactive "closed" state (16). When these constraints are absent or removed, for example by a mutation or ligand binding, an "open" conformation ensues. This 2-state model predicts that

Figure 1

TSHR structure. This computer model of the TSHR shows the 7 TMDs (spirals) embedded within the plasma membrane and a short cytoplasmic tail, which together make up the β/B subunit. The unique 50-aa–long cleaved region (about 316–366 aa) is shown in gray. Forming a long array, the 9 LRRs, each consisting of 20–24 aa, are depicted as spirals (α helices and β pleated sheets) on the ectodomain of the receptor and make up the major portion of the α/A subunit. The LRRs have a characteristic horseshoe shape with a concave inner surface. C, C-terminus; N, N-terminus. Figure adapted with permission from *Thyroid* (28).

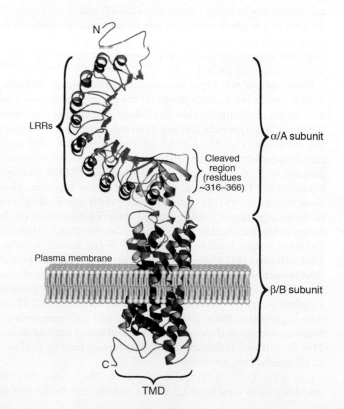

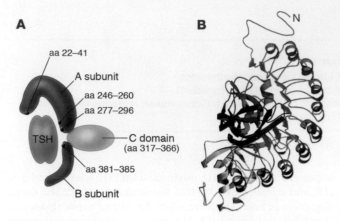

Figure 2
The TSH-binding pocket and TSHR antibody epitopes. (**A**) Schematic representation of the TSHR ectodomain showing the major regions (black dots) of TSH binding. Figure adapted with permission from *Thyroid* (24). (**B**) Model of the TSH-binding pocket, with TSH ligand making contact with the epitopes within and outside of the LRRs. Figure adapted with permission from *Thyroid* (28).

the open format of the receptor, when stabilized, would lead to full activation. Further support for this was provided by the finding that constitutive activation developed when the TSHR ectodomain was truncated, which suggested that the presence of the ectodomain dampened a constitutively active β subunit (40, 41). Additionally, a recent TSHR computer modeling and docking study (28) has shown that several mutations in the TMD that are associated with increased TSHR basal activity are caused by the formation of new interactions that stabilize the open, activated form of the receptor. Therefore, the TSHR must maintain an equilibrium between the inactive (closed) and active (opened) formats at any given time, and any factor (such as a mutation, high-affinity ligand binding, or high-affinity antibody binding) may shift this equilibrium from a quiescent structure (closed) to an active (opened) format.

TSHR and lipid rafts. Lipid rafts are sphingolipid- and cholesterol-rich membrane microdomains on the plasma membrane. The association of sphingolipids with cholesterol condenses the packing of the sphingolipids, leading to an enhanced mobility within the membrane (42, 43). Lipid rafts have been associated with signal transduction within cells because they sequester signaling proteins. Our observation of the dissociation of TSHR multimers following TSH binding, as discussed above, suggested that these dissociated TSHRs may move more easily into lipid rafts to facilitate signaling (1). Using labeled cholera toxin, which binds to lipid rafts enriched in G_{M1} gangliosides, we showed that the TSHRs that were localized to lipid rafts in fact moved out with TSH activation (44). Other researchers, using classical differential lipid extraction methodology, were unable to detect either stimulated TSHRs or their main $G_s\alpha$ (called *gsp*; G–stimulatory protein) signaling partner, within caveolin-restricted lipid rafts (45). These data suggest that the size of monomeric and multimeric receptors may determine their movement in and out of such domains. This, in turn, may influence signal transduction both by TSH and TSHR-stimulating antibodies.

Role of the TSHR in thyroid development. The mouse thyroid gland begins to develop at E8.5 as an endodermal thickening in

the floor of the primitive pharynx (46). Thyroid precursor cells express a combination of transcription factors (Figure 3), and by E14, *TSHR* mRNA appears and the molecular and morphological differentiation of the thyroid gland occurs (47). At E17, the formation of thyroglobulin (Tg; a colloidal protein stored in the lumen of thyroid follicles and the substrate for synthesis of thyroid hormones by iodination) then begins (48). Since the activation of the TSH/TSHR signaling pathway is concurrent with the expression of genes needed for thyroid hormone synthesis and secretion, the TSHR must play an important role in the growth of the thyroid gland (49, 50). This has been confirmed by studies in TSHR-defective mice. In several mutant mouse lines derived to possess a nonfunctional TSHR (e.g., *Tshr^{hyt}Tshr^{hyt}*) (51), the thyroid gland developed to a much smaller size than normal at 2 months of age in the absence of a functional TSHR. Furthermore, the expression levels of thyroid peroxidase (TPO; the enzyme responsible for Tg iodination) and the sodium iodine symporter (NIS; which transports iodine into the thyroid cell) were greatly reduced. Conversely, no significant changes were detected in the amounts of Tg and transcription factors such as paired box gene 8 (PAX8) and thyroid transcription factors 1 and 2 (TTF1 and TTF2). Separately, in a TSHR-KO mouse model, developed by homologous recombination (52), we found that the thyroid glands were smaller than those of control littermates. Histology of these thyroid glands showed that the TSHR-KO thyroid had fewer follicles and more non–follicle-associated interstitial cells within the gland compared with wild-type thyroids. In humans, fetal thyroid gland formation occurs during weeks 7–9 of gestation, and thyroid embryogenesis is largely complete by 12 weeks of gestation (Figure 3). At 12 weeks, the fetal thyroid is capable of concentrating iodide and synthesizing thyroid hormones, which lessens the dependence on maternal thyroid hormone in the second trimester. Fetal TSH is first detectable at 10 weeks of gestation by bioassay and radioimmunoassay (53). The level of serum TSH at this time is low until a structurally matured thyroid gland is developed at approximately week 18 (53).

Extrathyroidal TSHRs. TSHR expression is not unique to thyroid tissue, as originally thought. *TSHR* mRNA and protein have been detected in a variety of extrathyroidal cell types (1) (Table 2). For example, recent studies using TSHR antibodies have demonstrated the presence of TSHR-specific immunoreactivity in fibroblasts and adipose tissue from the retro-orbital and pretibial space of patients with Graves ophthalmopathy and pretibial myxedema (54, 55). TSHR signal transduction and its consequences in nonthyroidal cells have been poorly investigated to date but appear to modulate a variety of cellular processes. For example, TSH ligand has been implicated in the induction of lipolysis (56) and in the modulation of osteoclast activity (57), and clearly such widespread expression of the TSHR must impact self tolerance to such an important antigen and the immune repertoire.

Genetic alterations in the *TSHR*

Germline mutations. The apparent instability of the TSHR means that point mutations could be expected to result in changes in TSHR function either due to a change in its constitutive signaling ability or, as with any such glycoprotein, secondary to a change in receptor expression. With inactivating mutations, there may be only a partial reduction in TSH function causing partial TSH resistance, and inappropriately increased serum TSH levels, or a more marked loss of TSHR function, resulting in congenital hypo-

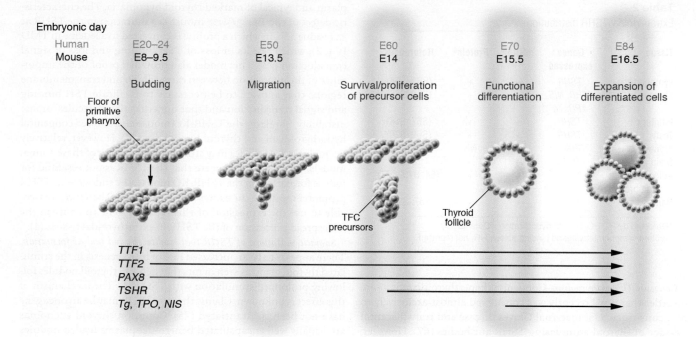

Figure 3
Schematic representation of the stages of development of the thyroid follicular cells and the expression of relevant genes. At mouse E8 (E20 in human), the median thyroid bud appears as a thickening in the floor of the pharynx and expresses a combination of transcription factors such as *PAX8*, the transcription factor essential for the thyrocyte promoter activation of *TPO*; *Tg* and *NIS*; and *TTF1* and *TTF2*, responsible for morphogenesis of the thyroid gland and maintenance of the thyrocyte cell type. At about E13.5 in mouse and E50 in human, the thyroid diverticulum starts its migration from the pharyngeal floor and reaches its definitive pretracheal position. By E14 (E60 in human), the thyroid follicular cell (TFC) precursors express the *TSHR*. By E15.5, the thyroid follicular organization appears with the expression of a series of proteins that are essential for thyroid hormone biosynthesis, including *TPO*, *Tg*, and *NIS*. Figure modified with permission from *Clinical Genetics* (47).

thyroidism. The syndromes of TSH resistance, therefore, are heterogeneous in phenotype, genotype, and mode of inheritance.

Compensated partial TSH resistance. Resistance to TSH is assessed by measuring serum TSH levels (50). Partial TSH resistance is defined as the maintenance of adequate thyroid function resulting in mild (6–15 mU/l) to moderate (20–50 mU/l) elevations in serum TSH associated with normal free thyroid hormone concentrations. In patients suffering from this type of resistance, thyroid volume may be normal, or there may be a modest degree of hypoplasia. While inactivating mutations of the TSHR may be inherited from one or both parents, partial TSH resistance is most commonly inherited from one parent. In cases where there are defects inherited from both parents, the TSH resistance more often becomes more severe. For example, a compound heterozygote (with a P162A mutation on the maternal allele and a C600R mutation on the paternal allele) exhibited a high TSH level, while in cases where only 1 allele was mutated, the TSH levels were lower (59). Figure 4 shows some of the reported mutations resulting in compensated partial TSH resistance. Such mutant TSHR sequences have also been studied using transient expression systems. Their constitutive TSH-independent cAMP activities were markedly reduced compared with those of wild-type TSHRs, and inadequate increases in cAMP generation were seen following TSH stimulation (60). In addition, flow cytometry experiments on the transfected cells showed reduced expression of some mutant TSHRs at the cell membrane. Hence, there is a variety of loss-of-function mutations of the *TSHR* gene resulting in inappropriately elevated TSH levels in patients. Such patients often

have normal thyroid hormone levels, so their condition thus meets the definition of mild thyroid failure (58). However, serum thyroid autoantibodies are usually not detected in these patients.

Congenital hypothyroidism. Mutations in a variety of genes, particularly those involved in thyroid ontogeny, cause thyroid dysgenesis and congenital hypothyroidism at a rate of 1 in 4,000 births. Thyroid dysgenesis can take the form of complete, ectopic, or hypoplastic thyroid development, or the hypothyroidism may be secondary to dyshormonogenesis. Genes known to be involved in this disorder in humans, in addition to *TSHR*, include *TTF1* and *TTF2*, *PAX8*, *NIS*, *TPO*, and *Tg* (Figure 3). The *sonic hedgehog* gene has also been implicated in congenital hypothyroidism in mice (46, 61, 62). Large elevations in TSH levels, along with decreased thyroid hormone levels, are generally found in babies with thyroid hypoplasia or agenesis along with low or absent serum Tg. Such children may have either homozygous or compound heterozygous mutations in their TSHR, which cause uncompensated TSH resistance (63, 64) (Figure 4A). For example, a substitution of threonine in place of a highly conserved alanine at position 553 in the fourth TMD was found to be homozygous in 2 congenitally hypothyroid siblings and heterozygous in both parents and 2 unaffected siblings, resulting in compensated partial TSH resistance in these individuals (60, 65). Functional analysis of cells expressing transfected receptors with this mutation showed extremely low levels of TSHR expression at the surface of these cells compared with cells expressing wild-type receptors. However, not all TSH resistance is secondary to mutations in the TSHR itself (66).

Table 2

Extrathyroidal TSHR distribution

Tissue	Genes expressed	mRNA[A]	Protein[B]	References
Lymphocytes	TSHR	+	+	(S63–S65)
Thymus	TSHR, NIS, TPO, Tg	+	+	(S58, S66)
Pituitary	TSHR	+	+	(S67, S68)
Testis	TSHR	+	+	(S69, S70)
Kidney	TSHR, Tg	+	NR	(S71)
Brain	TSHR	+	+	(S72)
Adipose cells and fibroblasts	TSHR	+	+	(54, S73–S77)
Bone	TSHR	+	+	(57, S78)

[A]Detected by RT-PCR/in situ hybridization. [B]Detected by immunohistochemistry/ligand binding assay. NR, not reported.

Congenital hyperthyroidism. Congenital hyperthyroidism is a rare disorder that until recently was considered almost exclusively to be a consequence of maternal Graves disease and transplacental passage of thyroid-stimulating TSHR antibodies (67). However, sporadic cases and a few familial cases of nonautoimmune congenital hyperthyroidism have been reported in which the affected individuals possess activating mutations of the TSHR — often, but certainly not exclusively, in exon 10, which encodes the transmembrane region and intracellular tail (2, 68–70) (Figure 4B). The treatment of congenital hyperthyroidism in the absence of an autoimmune cause involves controlling thyroid function with antithyroid drugs and performing a total thyroidectomy when the child is older than 5 years of age.

Models of congenital hypothyroidism in the mouse. Two models of complete uncompensated TSH resistance with high TSH levels and low thyroid hormone levels have been developed in the mouse: the *Tshr^hyt^Tshr^hyt* mouse and the TSHR-KO mouse. Defective thyroid ontogenesis and inherited primary hypothyroidism were first described in the *Tshr^hyt^Tshr^hyt* mouse in 1981 (71). These mice have thyroids with small and sparse follicles poorly endowed with cyto-

plasm and typical of marked thyroid hypoplasia. The characteristic defect of the *Tshr^hyt^Tshr^hyt* mouse is a mutation in the TSHR at aa residue 556, in which a proline is converted to a leucine in TMD IV (72), which results in loss of TSH binding and loss of signal transduction (73). This model also provided proof of the importance of the interaction between extracellular and transmembrane regions that appears to be necessary to facilitate TSH binding and signal transduction and that underlies the molecules' innate instability. Similarly, the TSHR-KO mouse also showed congenital hypothyroidism and thyroid hypoplasia (52). However, relatively normal thyroid follicles may still be found in both of these animal models, which would indicate that the TSHR is not essential for follicle formation. In the TSHR-KO mouse, the endogenous TSHR promoter results in the expression of GFP (52). Therefore, we were able to use the expression of this reporter gene to confirm the widespread expression of the TSHR in extrathyroidal tissues (1).

Somatic mutations of TSHR that lead to thyroid nodule formation. There appears to be an increased rate of mutagenesis in the stimulated thyroid gland as seen in rats that develop thyroid nodules following prolonged stimulation with TSH (74). The mechanism of this effect is unknown. Claims that the TSHR may be an oncogene have not been substantiated (75). Common thyroid adenomas are usually well encapsulated benign neoplasms (called nodules or adenomata), which sometimes (approximately 5–10% of the time) may appear to take up excess amounts of radioiodine ("hot" nodules) but more commonly exhibit normal or decreased uptake and retention of radioiodine compared with normal tissue (called "cold" nodules or nontoxic adenomata) (76).

Activating TSHR somatic mutations. Only somatic mutations with a positive effect on both growth and cell function would be expected to induce clonal expansion and thyroid nodule formation. Autonomous benign and malignant thyroid nodules have been shown to have a variety of point mutations leading to constitutive activation of the TSHR — often referred to as gain-of-function mutations and once again most commonly occurring in the TMDs (as summarized in Figure 4C) (77, 78). Additional tumors may have defects in $G_s\alpha$ rather than the TSHR itself (76). However, hyperthyroidism resulting from single or multiple adenomas (toxic multinodular goiters) is less common than hyperthyroidism

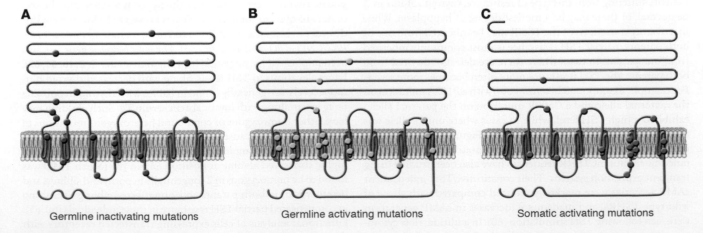

A   Germline inactivating mutations

B   Germline activating mutations

C   Somatic activating mutations

Figure 4

Examples of mutation in human and mouse TSHR. The locations of constitutively germline inactivating (**A**), germline activating (**B**), and somatic activating (**C**) mutations are represented. Most activating mutations (shown in **B** and **C**) have been localized to exon 10, which codes for the transmembrane and cytoplasmic regions of the receptor. Figure adapted with permission from *The New England Journal of Medicine* (S62).

resulting from Graves disease. Indeed, the degree of disparity in the prevalence of these entities may be dependent upon the iodine intake of the population at risk (76). Hence, nodules with activating TSHR mutations are more likely to cause hyperthyroidism in the presence of normal or high iodine intake and may be clinically uncommon in geographic areas of iodine deficiency.

Inactivating TSHR somatic mutations. In contrast to toxic adenomas, inactivating mutations of the TSHR would reduce the uptake of radioiodine into the thyroid follicular cells and would be unlikely to cause nodule formation, since the mutation would fail to induce thyroid cell proliferation. Hence, inactivating somatic TSHR mutations are unusual in these tumors (76). Thyroid cancers frequently exhibit reduced TSHR expression, which is presumably secondary to ongoing dedifferentiation, and this reduces expression of all thyroid-specific genes (76).

Multinodular goiter. Examination of toxic multinodular goiters (Plummer disease) has revealed separate and distinct activating mutations in different hot nodules within the same thyroid gland. Hence, the autonomous nodules are polyclonal in origin. Activating mutations of the TSHR are present before the formation of an actual nodule (78, 79) and can be seen throughout the gland in toxic multinodular goiters, indicative of their role in true nodule formation.

G protein mutations. Although outside the scope of this review, the classic example of activating defects of the thyroid arising as a result of mutant $G_s\alpha$ in sporadic cases of thyroid adenomas (76). In these adenomas, somatic point mutations in, for example, Arg201 or Gln227, inhibited GTP hydrolysis and led to the development of autonomous hyperfunctioning nodules (75).

"Specificity crossover" at the TSHR

Promiscuity of the TSHR. Ligand specificity between TSH, FSH, LH, and human chorionic gonadotropin (hCG) and their cognate receptors is determined by the conformation of the horseshoe structure in the LRRs of their receptor ectodomains, as discussed above (22, 28). The N-terminal portion of the β subunit is additionally important for ligand binding, as indicated, for example, by loss of TSH binding after mutation of Y385 (80) or disruption of sulfation of Tyr385 and Tyr387 (81). However, it has been known for many years that both LH and hCG can interact with the TSHR — a phenomenon termed *specificity crossover* (82, 83) — and can initiate second signal (84). TSH, in contrast, has poor crossover with the gonadotropin receptors, once again suggestive of the innate instability of the TSHR itself and the ease with which it can be stimulated or blocked. Furthermore, we also showed that normal hCG was able to induce thyroid cell growth (85) as well as activate a recombinant TSHR expressed in mammalian cells (86).

Gestational thyrotoxicosis. Some degree of increased thyroid stimulation is normal during early pregnancy and is associated with increasing concentrations of serum placental hCG, which peaks at 10–12 weeks (10–40 IU/ml) (Figure 5) (87). In multiple pregnancies, serum hCG levels tend to be even higher, with greater potential impact on thyroid function. Thyroid enlargement, low TSH levels — a sensitive reflection of enhanced thyroid hormone levels — and sometimes a detectable increase in free thyroid hormone levels may all be seen transiently in apparently normal pregnancy (87). This first-trimester phenomenon is often referred to as gestational thyrotoxicosis and is secondary to specificity crossover by hCG at the TSHR. As hCG levels fall, results of thyroid function tests return to normal. However, this degree of transient, mild thyroid overactivity does not require treatment.

Hyperemesis gravidarum. Excessive vomiting in early pregnancy is usually self-limited, but 30–50% of women with this disorder have increased thyroid function and may develop mild clinical thyrotoxicosis (82, 83). Such cases are usually associated with high hCG levels, and hCG may show increased thyroid-stimulating activity due to posttranslational changes including reduced sialic acid content. However, not all women with severe hyperemesis have such high hCG levels. A missense mutation in the germline TSHR ectodomain has been reported to cause the TSHR to become supersensitive to hCG and accounts for the development of thyrotoxicosis (88). This mutation resulted in a lysine being replaced with an arginine at position 183 in the fifth LRR. In fact, many mutations at position 183 increase the affinity of hCG for the TSHR (89), and these may occur more commonly in hyperemesis than previously thought (90).

Hydatidiform mole, trophoblastic tumors, and choriocarcinoma. Enormous quantities of hCG may be secreted by hydatidiform moles, trophoblastic tumors, and choriocarcinoma tissue. When the level exceeds approximately 200 IU/ml, patients develop hyperthyroidism (82, 83). Effective chemotherapy of the tumor returns the patient to euthyroidism. Molecular variants of hCG with increased thyrotropic potency have been described in patients with choriocarcinoma (91, 92). In such patients, the hCG with increased thyroid-stimulating potency may have reduced sialic acid content or a truncated C-terminal tail, among other possibilities (93).

TSHR autoimmunity

AITD. AITD traditionally includes autoimmune thyroiditis (Hashimoto disease) and autoimmune hyperthyroidism (Graves disease). Graves disease, in particular, is the result of a complex autoimmune response to the TSHR that results in the production of TSHR-specific T cells and TSHR autoantibodies, thought to be secondary to an interaction between susceptibility genes and environmental triggers.

Genetic susceptibility. The paradigm of a genetic contribution to TSHR autoimmunity is supported by abundant epidemiological and experimental data (for a review, see ref. 94). Graves disease has been known for many years to be familial (95–97), and in a recent study, we have shown that the sibling risk ratio (referred to as 8s) in Graves disease was high, at 11.6 (98). Furthermore, twin studies have shown a concordance rate for Graves disease in monozygotic twins of 20–35% compared with 2–7% in dizygotic twins (99, 100). Several genes and genetic regions have been identified as being linked or associated with AITD (see *Potential factors causing susceptibility to TSHR autoimmunity*). However, evidence suggesting the association of polymorphisms in the most obvious candidate, the *TSHR* gene, with Graves disease has been surprisingly weak (101). In contrast, *HLA-DR3* was the first Graves disease susceptibility gene to be identified, with an odds ratio of 2–3 (102). More recently, we have shown that the presence of arginine at position 74 of the HLA-DR β1 chain confers most of this susceptibility (102). CTLA-4 is an important costimulatory molecule that participates in the interaction between T cells and APCs (103) and serves mainly to downregulate T cell activation by APCs. The *CTLA4* gene has been shown to be weakly and functionally associated with several autoimmune diseases including Graves disease (104–109). Another general autoimmunity gene that was recently identified is *protein tyrosine phosphatase non-receptor type 22 (PTPN22)*. The *PTPN22* gene was originally found to be associated with rheumatoid arthritis (110) and was later found to be associated with systemic lupus erythematosus, type 1 diabetes, and Graves disease, with an odds

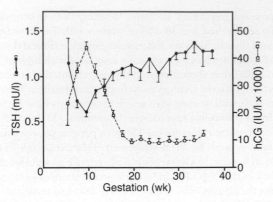

Figure 5
Gestational thyrotoxicosis. Shown here is the inverse relationship between serum hCG and TSH levels in early pregnancy. The level of TSH falls as thyroid function increases. Simultaneously, hCG levels increase. Adapted with permission from *The Journal of Clinical Endocrinology and Metabolism* (87).

ratio of 1.9 (111–113). Another costimulatory molecule that was recently found to be associated with Graves disease is CD40 (114, 115); however, some investigators have been unable to show this association (116). In addition to these immune response–related genes, we (117) and others (118) have recently shown that the gene encoding Tg is a major gene involved in TSHR-related autoimmunity. Amino acid substitutions in this gene predisposed individuals to Graves disease and autoimmune thyroiditis (117). Most likely, these susceptibility genes interact and influence disease phenotype and severity. For example, we have reported a marked synergistic increase in the risk ratio for individuals carrying a specific HLA-DR3/Tg haplotype (117).

Environmental susceptibility. While the significantly higher concordance rate for Graves disease in monozygotic twins points to a strong genetic susceptibility, the fact that the concordance rate in monozygotic twins was in fact only approximately 30% rather than 90% may point to significant environmental influences (99, 119). The most important nongenetic potential risk factors contributing to the etiology of AITD, including TSHR autoimmunity, are thought to be dietary iodine intake (95, 120), smoking (121, 122), stress (123, 124), pregnancy, exposure to radiation, and infection (119, 124, 125). All of these suggest environmental triggers; however, their association with AITD requires further investigation and validation.

The TSHR as an autoantigen

Graves disease, T cells, and the discovery of TSHR autoimmunity. Although a restricted set of T cells reacts to TSHR antigen in patients with TSHR autoimmunity (S1, S2), the fact that a stimulating antibody to the TSHR is the primary cause of the hyperthyroidism has tended to dominate research on this disease. The presence of long-acting thyroid stimulator in the sera of patients with hyperthyroid Graves disease was discovered almost 50 years ago (S3). Patient sera stimulated radioiodine release from prelabeled guinea pig thyroid for a much longer time period than did a pituitary TSH preparation. This prolonged stimulating activity was then found to reside in the IgG fraction of serum from many patients with Graves disease (S4). With the availability of biologically active radiolabeled TSH, it became possible to probe thyroid membranes

for the TSHR; the Graves IgG was shown to compete with TSH for receptor occupancy (S5) and contained TSHR antibodies acting as TSH agonists. Autoantibodies that bind to the TSHR and initiate a second signal are TSHR-*stimulating*, while those that only bind may be TSHR-*blocking* or TSHR-*neutral* (Figure 6). The original, and brave, self-infusion of plasma from patients with Graves disease by Adams and colleagues and the resulting thyroid stimulation (S6) was the absolute confirmation of the role of TSHR antibody in the induction of human hyperthyroidism and was one of the first demonstrations of antibody transfer causing autoimmune disease.

Animal models of TSHR autoimmunity. Initial attempts to generate a Graves disease animal model used recombinant human TSHR ectodomain protein from prokaryotic and eukaryotic cell expression systems (S7, S8). The resulting TSHR antibodies blocked the action of TSH on the TSHR (S9, S10). The Shimojo mouse model of Graves disease (S11, S12) was the first system for generating hyperthyroid mice. Fibroblasts expressing both the TSHR and MHC class II antigens were injected intraperitoneally into syngeneic mice, which resulted in thyroid hypertrophy and the presence of TSHR-stimulating antibodies in the serum. The first 260 aa of the TSHR ectodomain, chimeric with the C-terminal ectodomain of the LH receptor, were essential to induce the production of TSHR antibodies (S13), which indicated that TSHR antibody-binding sites existed in the LRR-containing domain (22). This approach also revealed the necessity of immunizing with the intact conformation of the TSHR ectodomain in order to induce TSHR-stimulating antibodies and was later confirmed utilizing TSHR-expressing plasmid and viral vectors (S14– S16). Similarly, the first approximately 289 aa residues of the TSHR α subunit, expressed by an adenovirus vector, induced Graves hyperthyroidism more frequently than did a noncleaving receptor (S17).

TSHR-stimulating antibodies. TSHR antibodies are usually of the IgG$_1$ subclass, which has been suggested as evidence of their oligoclonality (S18). At least in animal models of immunization-induced Graves disease, the role of the Th1 immune response may be primary in directing the formation of TSHR antibodies (S19). The confor-

---

Potential factors causing susceptibility to TSHR autoimmunity

External factors
    Infection
    Trauma
    Stress
    Iodine intake
    Irradiation
Internal factors
    Thyroid autoantibodies
    Sex steroids
    Pregnancy
    Fetal microchimerism
Increased genetic susceptibility
    HLA class II genes with an arginine at position 74
    CTLA4
    CD40
    Tg
    Unknown genes on a variety of confirmed loci

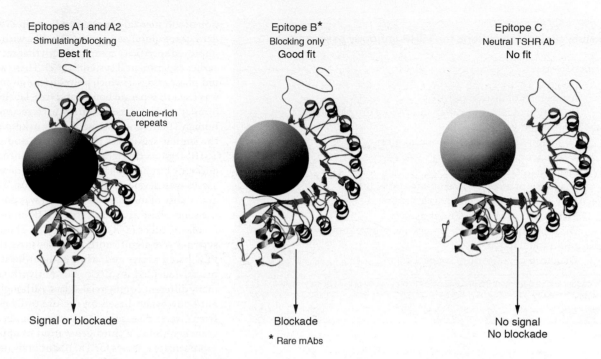

Figure 6
A hypothetical model simplified to explain the effect of structural changes in the TSH-binding site by diverse TSHR antibodies. The TSH-binding pocket, represented by the LRRs, is shown by spirals representing the α helix and the β pleated sheets represented by the wide red arrows. The gray region represents the unique cleaved region (316–366 aa) of the receptor. (**A**) Epitope A1 represents the site where thyroid-stimulating antibodies bind in part to the LRRs, bringing about a structural change in the receptor that leads to signal transduction; Epitope A2 represents a similar competing site, where TSH-blocking antibodies bind (both illustrated as a best fit). (**B**) Epitope B is the least common site, where TSHR-blocking antibodies may bind but do not compete with antibodies binding to Epitope A. They bind in part to the LRR region but do not bring about the required structural change for signal transduction yet are still able to hinder TSH binding to this site (illustrated as a good fit). (**C**) Epitope C is where neutral antibodies bind to the cleaved region and/or the N terminus of the TSHR ectodomain, bringing no appropriate structural alteration to the TSHR and thus leaving the LRR region free for TSH, and other TSHR antibodies, to bind. Thus, neutral antibodies result in no signal transduction and do not block TSH binding (illustrated as no fit).

mational binding site of TSHR-stimulating monoclonal antibodies generated from human (S20), hamster (S21), and mouse (S22) is shared and on the TSHR α subunit (approximately 316 aa) (designated as Epitope A on the α/A subunit) (Figure 6A) (S23). Recently, Costagliola et al. have localized the conformational binding site of 1 mouse monoclonal stimulating antibody specific to the TSHR as the N-terminal region of the LRRs in the ectodomain (S24), and this is likely to be the major stimulatory epitope (Figures 2B and 6A). Of additional importance is an observation that shed α subunits preferentially recognize TSHR antibodies compared with intact membrane-bound receptor (S25), which indicates easier access to the epitope. Hence, shed TSHR α subunits have the potential to act as a systemic antigenic stimulus and may contribute to the breakdown of TSHR tolerance in susceptible individuals (S17, S26). However, such fragments may also have the potential to act as "decoys" to divert stimulating antibodies away from the intact TSHR.

TSHR-blocking antibodies. This class of antibody is able to induce hypothyroidism by blocking the thyrotropic action of TSH. Both TSHR-blocking and -stimulating antibodies are seen in patients with Graves disease and in animal models. In addition, TSHR-blocking antibodies can also be seen in individuals with atrophic Hashimoto disease (S27). In Graves disease, blocking antibodies may play a role in the fluctuation of thyroid function and may contribute to the poor correlation between thyroid function and serum titers of TSHR autoantibodies (S28–S30). During the

course of treatment of patients with Graves disease, TSHR-blocking antibodies may become the more prevalent antibody, thus contributing to the development of hypothyroidism (S31). Cases of transient congenital hypothyroidism have been well documented as secondary to placental transfer of maternal TSHR-blocking antibodies (67, S32). It has also been shown that TSHR antibodies from Graves disease patients and atrophic Hashimoto thyroiditis patients can compete for binding to the N terminus of the TSHR β subunit (S33). Recently, a human TSHR-stimulating mAb has been isolated and also found to compete with TSHR autoantibodies in both Graves disease and atrophic Hashimoto thyroiditis (S20). These studies indicate that the binding site for TSHR-stimulating mAbs (designated as Epitope A on the TSHR) is also similar to that of some TSHR blocking antibodies (Figure 6) (S34) and that routine binding competition assays cannot, therefore, distinguish between stimulating and blocking varieties of TSHR antibodies. However, the majority of TSHR-blocking antibodies do not compete with TSHR-stimulating antibodies and bind to an area we have designated as Epitope B (Figure 6).

TSHR-neutral autoantibodies. The neutral TSHR antibodies, by definition, do not affect TSH binding to the TSHR and bind to a site designated as Epitope C (Figure 6). Their presence was first clearly reported in a patient with Graves disease and monoclonal gammopathy (S35). Results of studies using affinity-purified TSHR have suggested that normal control sera may also contain IgG that binds

Possible clinical indications for TSHR antibody assessment

Graves disease
    Diagnosis
    Assessment of disease activity
    Prediction of remission after antithyroid drug administration
Graves ophthalmopathy
    Detection of underlying autoimmunity
    Assessment of disease activity
Hypothyroidism
    Diagnosis of atrophic Hashimoto thyroiditis
Pregnancy and neonatal thyroid dysfunction[A]
    Prediction of neonatal Graves disease
    Diagnosis of transient hyper- and hypothyroidism in the newborn
Toxic multinodular goiter
    Diagnosis of coexisting Graves disease

[A]The function of persistent TSHR antibodies in pregnancy should be determined using a bioassay.

the TSHR and lacks stimulating or blocking activity and may represent a natural TSHR antibody (S36). In the animal models of Graves disease (S11, S14–S16), the major linear epitopes (small binding regions of aa not influenced by the conformation of the receptor), which were recognized by neutral TSHR antibodies, were mainly the approximately 20 N-terminal aa and/or the TSHR cleaved region (S37) (Figures 1 and 2). Neither of these epitopes appear to be critical for TSH binding to the TSHR (i.e., they are outside the TSH binding pocket; see ref. 24). The clinical or pathophysiological relevance of neutral TSHR antibodies remains unclear. Our recent data, however, indicate that they are capable of inhibiting TSHR cleavage and as a consequence prolong the half-life of the TSHR (S38).

Detecting TSHR antibodies. TSHR antibodies in the serum of patients with AITD are most commonly measured by receptor protein–binding assays in which labeled TSH is competed with monoclonal TSHR antibodies for binding to target TSHRs. TSH competition assays (112, S39) compete TSHR antibodies with radioiodine-labeled bovine TSH binding to solubilized porcine TSHRs and are available in most commercial clinical laboratories. Hence, the levels of TSHR antibodies can be routinely determined in patients with Graves disease as a measure of disease activity. However, the biological activity of stimulating antibodies must still be measured in a thyroid cell or another mammalian cell with transfected TSHRs in order to detect cAMP generated in response to TSHR antibody stimulation via the TSHR. Levels of TSHR-blocking antibodies can be measured in a similar way by detecting a reduction in TSH-mediated cAMP-generated response. When the patient is hyperthyroid, there is no need for a bioassay to determine the stimulating or blocking activity of the TSHR antibodies present, since the patient serves as the "bioassay" (S40). As discussed earlier, there are also autoantibodies specific to the TSHR that do not compete for TSH binding and, therefore, are not detectable in the current clinical assays (the neutral antibodies).

The new generations of clinical assays for TSHR antibodies. The original 1974 protein-binding assay for TSHR antibodies used a crude thyroid membrane preparation with a crude Ig fraction of the patients serum but was considerably nonspecific (S41). The first genera-

tion of commercial assays, introduced in 1982, used detergent-solubilized porcine TSHRs, with greatly improved specificity compared with that seen in the earlier experimental systems (S42). These were liquid-phase assays in which polyethylene glycol (PEG) was used to separate bound and free, labeled TSH. Similar results were obtained with recombinant human TSHRs and porcine TSHRs, in keeping with the similar sequences of the human and porcine TSHR regions thought to be important for TSHR antibody binding (S43). The second generation of assays was developed in the late 1990s. With the availability of mAbs to the TSHR, it was possible to use solid-phase assays with the TSHR immobilized on plastic tubes (S44) or plates (S45). These serum systems were significantly more sensitive than the PEG-based assays and, when standardized appropriately, yielded results consistent with those of many different commercial assays. Although TSHR antibody assays based on the use of labeled IgG from Graves disease patients were described many years ago (S46), a third generation of apparently more sensitive assays for TSHR antibody detection has been developed only recently (S39). These assays followed the isolation of thyroid-stimulating mAbs to the TSHR. Such solid-phase systems use competition for labeled monoclonal TSHR-stimulating mAb binding to immobilized TSHRs and appear to be more stable than assays employing labeled TSH. Using these newer techniques, many investigators have found that more than 95% of untreated patients with Graves disease have detectable TSHR antibodies at the time of diagnosis.

The clinical utility of TSHR antibody measurement

TSHR antibody measurements serve as a marker for TSHR autoimmunity, particularly in Graves disease. The putative clinical indications for the measurement of TSHR antibodies are summarized in *Possible clinical indications for TSHR antibody assessment*.

Changes in TSHR antibody levels with administration of antithyroid drugs. TSHR antibody levels fall after treatment of patients with antithyroid drugs. However, the majority of patients with hyperthyroid Graves disease relapse after stopping a 12-month course of antithyroid drugs, with the actual percentage varying among populations and with their levels of iodine intake (40–70%) (S47). The fall in TSHR antibody levels with antithyroid drug treatment was most clearly demonstrated in a study on patients with Hashimoto thyroiditis (S5). This effect is secondary to a combination of their mild immunosuppressive action and their induction of a decrease in thyroid antigen expression (S48–S50). The measurement of TSHR antibodies in patients with Graves disease has proven to be a useful predictor of relapse and remission after antithyroid drug treatment only if the levels of TSHR antibody remain high (S40, S47, S50, S51). Unfortunately, many studies of the efficacy of using TSHR antibody levels to predict relapse and remission of Graves disease have used TSHR antibody assays of dubious sensitivity, precision, and specificity (S52). Recent improvements in TSHR antibody assays with the introduction of solid-phase systems described above (S39, S44, S45, S53) should improve the accuracy of relapse prediction and treatment evaluation in patients with Graves disease.

Potential factors interfering with the action of TSHR antibodies. There are a number of potential reasons why measurement of TSHR antibodies may be unreliable in the prediction of thyrotoxic recurrence after antithyroid drug administration in addition to problems with the assay (S40). These include: (a) low antibody quantity (S54); (b) variable antibody affinity; and (c) reduced thyroid function, for example, secondary to concurrent thyroiditis or iodine deficiency. Hence, a negative TSHR antibody assay result is of little help to the clinician. The loss of previously detectable TSHR antibodies indicates that the immune system may have become more tolerant, but such patients may still have a high recurrence rate after discontinuing antithyroid drugs because the TSHR antibodies were undetectable.

Changes in TSHR antibody levels after surgery and radioiodine treatment. TSHR antibody levels also fall after removal or ablation of the thyroid gland. Since the TSHR is expressed in a variety of extrathyroidal tissues, this effect cannot be secondary to removal of the antigen. More likely, the release of a large quantity of antigen at the time of surgery may induce widespread apoptosis of thyroid-specific T cells and B cells, while radioiodine may be cytotoxic to intrathyroidal immune cells as well as thyroid cells. However, after radioiodine ablation of the thyroid, there is first a marked, but transient, increase in TSHR antibody levels, possibly due to the radiosensitivity of regulatory T cells, before the long-term decrease in levels is seen (S47, S55). This transient phenomenon has been implicated in the deterioration of Graves ophthalmic eye disease after radioiodine administration (S56, S57).

The role of the TSHR in the extrathyroidal manifestations of Graves disease. As discussed earlier (Table 2), the TSHR is expressed in a wide variety of tissues (1) including the thymus, where it normally allows deletion of TSHR-specific T cells (S58). However, the TSHR is also considered a major antigen leading to the manifestation of extrathyroidal diseases including Graves ophthalmopathy and pretibial myxedema (S59, S60). There is highly suggestive evidence that links the TSHR as a shared thyroidal, retrobulbar, and dermal antigen in patients with Graves disease. For example, TSHR expression in retro-orbital fibroblasts is exacerbated in patients with ophthalmopathy (S59, S61).

Conclusions

Diseases of the thyroid are common, and the TSHR is associated with a great many of these disorders. The TSHR undergoes complex posttranslational processing, which results in a unique structure of 2 covalently linked subunits in its ectodomain. This structure appears to provide inherent instability, since point mutations, mostly in the transmembrane regions, confer constitutive overactivity or underactivity indicative of the conformational changes that ensue. Similarly, the TSHR is highly susceptible to specificity crossover by other glycoprotein hormones such as hCG, which suggests that the receptor can be easily switched from a closed to an open state. The TSHR-stimulating antibodies of Graves disease are similarly dependent on the correct conformation of the TSH and have no interaction with nonglycosylated or reduced receptor, while shed TSHR ectodomains may act as immune stimulants or antibody decoys. Hence, unique posttranslational processing appears to explain the propensity of the TSHRs to be involved in human disease.

Acknowledgments

We thank the NIH, the Marvin Sinkoff Endowment, the David Owen Segal Endowment, and the Joseph and Arita Steckler Fund for continuing support.

Note: References S1–S78 are available online with this article; doi:10.1172/JCI26031DS1.

Address correspondence to: Terry F. Davies, One Gustave L. Levy Place, Box 1055, Mount Sinai School of Medicine, New York, New York 10029, USA. Phone: (212) 241-6627; Fax: (212) 241-4218; E-mail: terry.davies@mssm.edu.

1. Davies, T.F., Marians, R., and Latif, R. 2002. The TSH receptor reveals itself. *J. Clin. Invest.* **110**:161–164. doi:10.1172/JCI200216234.
2. Duprez, L., et al. 1999. Pathology of the TSH receptor. *J. Pediatr. Endocrinol. Metab.* **12**(Suppl. 1):295–302.
3. Rapoport, B., Chazenbalk, G.D., Jaume, J.C., and McLachlan, S.M. 1998. The thyrotropin (TSH) receptor: interaction with TSH and autoantibodies [review]. *Endocr. Rev.* **19**:673–716.
4. Kajita, Y., Rickards, C.R., Buckland, P.R., Howell, R.D., and Rees Smith, B. 1985. Analysis of thyrotropin receptors by photoaffinity labeling. Orientation of receptor subunits in the cell membrane. *Biochem. J.* **227**:413–420.
5. Parmentier, M., et al. 1989. Molecular cloning of the thyrotropin receptor. *Science.* **246**:1620–1622.
6. Libert, F., et al. 1989. Cloning, sequencing and expression of the human thyrotropin (TSH) receptor: evidence for binding of autoantibodies. *Biochem. Biophys. Res. Commun.* **165**:1250–1255.
7. Nagayama, Y., Kaufman, K.D., Seto, P., and Rapoport, B. 1989. Molecular cloning, sequence and functional expression of the cDNA for the human thyrotropin receptor. *Biochem. Biophys. Res. Commun.* **165**:1250–1255.
8. Misrahi, M., et al. 1990. Cloning, sequencing and expression of human TSH receptor. *Biochem. Biophys. Res. Commun.* **166**:394–403.
9. Libert, F., Passage, E., Lefort, A., Vassart, G., and Mattei, M.G. 1990. Localization of human thyrotropin receptor gene to chromosome region 14q3 by in situ hybridization. *Cytogenet. Cell Genet.* **54**:82–83.
10. Loosfelt, H., et al. 1992. Two-subunit structure of the human thyrotropin receptor. *Proc. Natl. Acad. Sci. U. S. A.* **89**:3765–3769.
11. Misrahi, M., et al. 1994. Processing of the precursors of the human thyroid-stimulating hormone receptor in various eukaryotic cells (human thyrocytes, transfected L cells and baculovirus-infected insect cells). *Eur. J. Biochem.* **222**:711–719.
12. Shi, Y., Zou, M., Parhar, R.S., and Farid, N.R. 1993. High-affinity binding of thyrotropin to the extracellular domain of its receptor transfected in Chinese hamster ovary cells. *Thyroid.* **3**:129–133.
13. Da Costa, C.R., and Johnstone, A.P. 1998. Production of the thyrotrophin receptor extracellular domain as a glycosylphosphatidylinositol-anchored membrane protein and its interaction with thyrotrophin and autoantibodies. *J. Biol. Chem.* **273**:11874–11880.
14. Costagliola, S., Khoo, D., and Vassart, G. 1998. Production of bioactive amino-terminus domain of the thyrotropin receptor via insertion in the plasma membrane by a glycosylphosphatidylinositol anchor. *FEBS Lett.* **436**:427–433.
15. Chazenbalk, G.D., Nagayama, Y., Russo, D., Wadsworth, H.L., and Rapoport, B. 1990. Functional analysis of the cytoplasmic domains of the human thyrotropin receptor by site-directed mutagenesis. *J. Biol. Chem.* **265**:20970–20975.
16. Vlaeminck-Guillem, V., Ho, S.C., Rodien, P., Vassart, G., and Costagliola, S. 2002. Activation of the cAMP pathway by the TSH receptor involves switching of the ectodomain from a tethered inverse agonist to an agonist. *Mol. Endocrinol.* **16**:736–746.
17. Wadsworth, H.L., Chazenbalk, G.D., Nagayama, Y., Russo, D., and Rapoport, B. 1990. An insertion in the human thyrotropin receptor critical for high affinity hormone binding. *Science.* **249**:1423–1425.
18. Chazenbalk, G.D., Tanaka, K., McLachlan, S.M., and Rapoport, B. 1999. On the functional importance of thyrotropin receptor intramolecular cleavage. *Endocrinology.* **140**:4516–4520.
19. Wadsworth, H.L., Russo, D., Nagayama, Y., Chazenbalk, G.D., and Rapoport, B. 1992. Studies on the role of amino acids 38–45 in the expression of a functional thyrotropin receptor. *Mol. Endocrinol.* **6**:394–398.
20. Chen, C.R., Tanaka, K., Chazenbalk, G.D., McLachlan, S.M., and Rapoport, B. 2001. A full biological response to autoantibodies in Graves' disease requires a disulfide-bonded loop in the thyrotropin receptor N terminus homologous to a laminin epidermal growth factor-like domain. *J. Biol. Chem.* **276**:14767–14772.
21. Nagayama, Y., Russo, D., Wadsworth, H.L., Chazenbalk, G.D., and Rapoport, B. 1991. Eleven amino acids (Lys-201 to Lys-211) and 9 amino acids (Gly-222 to Leu-230) in the human thyrotropin receptor are involved in ligand binding. *J. Biol. Chem.* **266**:14926–14930.
22. Smits, G., et al. 2003. Glycoprotein hormone receptors: determinants in leucine-rich repeats responsible for ligand specificity. *EMBO J.* **22**:2692–2703.
23. Nagayama, Y., Wadsworth, H.L., Russo, D., Chazenbalk, G.D., and Rapoport, B. 1991. Binding domains of stimulatory and inhibitory thyrotropin (TSH) receptor autoantibodies determined with chimeric TSH-lutropin/chorionic gonadotropin receptors. *J. Clin. Invest.* **88**:336–340.

24. Jeffreys, J., et al. 2002. Characterization of the thyrotropin binding pocket. *Thyroid.* **12**:1051–1061.

25. Chazenbalk, G.D., et al. 1997. Evidence that the thyrotropin receptor ectodomain contains not one, but two, cleavage sites. *Endocrinology.* **138**:2893–2899.

26. Couet, J., et al. 1996. Shedding of human thyrotropin receptor ectodomain. Involvement of a matrix metalloprotease. *J. Biol. Chem.* **271**:4545–4552.

27. de Bernard, S., et al. 1999. Sequential cleavage and excision of a segment of the thyrotropin receptor ectodomain. *J. Biol. Chem.* **274**:101–107.

28. Nunez Miguel, R.N., et al. 2004. Analysis of the TSH receptor-TSH interaction by comparative modeling. *Thyroid.* **14**:991–1011.

29. Couet, J., et al. 1996. Cell surface protein disulfide-isomerase is involved in the shedding of human thyrotropin receptor ectodomain. *Biochemistry.* **35**:14800–14805.

30. Tanaka, K., Chazenbalk, G.D., McLachlan, S.M., and Rapoport, B. 1999. Subunit structure of thyrotropin receptors expressed on the cell surface. *J. Biol. Chem.* **274**:33979–33984.

31. Jordan, B.A., Trapaidze, N., Gomes, I., Nivarthi, R., and Devi, L.A. 2001. Oligomerization of opioid receptors with beta 2-adrenergic receptors: a role in trafficking and mitogen-activated protein kinase activation. *Proc. Natl. Acad. Sci. U. S. A.* **98**:343–348.

32. Devi, L.A., and Brady, L.S. 2000. Dimerization of G-protein coupled receptors [review]. *Neuropsychopharmacology.* **23**(4 Suppl.):S3–S4.

33. Trapaidze, N., Gomes, I., Cvejic, S., Bansinath, M., and Devi, L.A. 2000. Opioid receptor endocytosis and activation of MAP kinase pathway. *Brain Res. Mol. Brain Res.* **76**:220–228.

34. Overton, M.C., and Blumer, K.J. 2000. G-protein-coupled receptors function as oligomers in vivo. *Curr. Biol.* **10**:341–344.

35. Graves, P.N., Vlase, H., Bobovnikova, Y., and Davies, T.F. 1996. Multimeric complex formation by the thyrotropin receptor in solubilized thyroid membranes. *Endocrinology.* **137**:3915–3920.

36. Latif, R., Graves, P., and Davies, T.F. 2001. Oligomerization of the human thyrotropin receptor: fluorescent protein-tagged hTSHR reveals post-translational complexes. *J. Biol. Chem.* **276**:45217–45224.

37. Latif, R., Graves, P., and Davies, T.F. 2002. Ligand-dependent inhibition of oligomerization at the human thyrotropin receptor. *J. Biol. Chem.* **277**:45059–45067.

38. Van Sande, J., et al. 1996. Specific activation of the thyrotropin receptor by trypsin. *Mol. Cell. Endocrinol.* **119**:161–168.

39. Vassart, G., Pardo, L., and Costagliola, S. 2004. A molecular dissection of the glycoprotein hormone receptors. *Trends Biochem. Sci.* **29**:119–126.

40. Szkudlinski, M.W., Fremont, V., Ronin, C., and Weintraub, B.D. 2002. Thyroid-stimulating hormone and thyroid-stimulating hormone receptor structure-function relationships. *Physiol. Rev.* **82**:473–502.

41. Zhang, M., et al. 2000. The extracellular domain suppresses constitutive activity of the transmembrane domain of the human TSH receptor: implications for hormone-receptor interaction and antagonist design. *Endocrinology.* **141**:3514–3517.

42. Brown, D.A., and London, E. 1998. Functions of lipid rafts in biological membranes. *Annu. Rev. Cell Dev. Biol.* **14**:111–136.

43. Simons, K., and Toomre, D. 2000. Lipid rafts and signal transduction [review]. *Nat. Rev. Mol. Cell Biol.* **1**:31–39.

44. Latif, R., Ando, T., Daniel, S., and Davies, T.F. 2003. Localization and regulation of thyrotropin receptors within lipid rafts. *Endocrinology.* **144**:4725–4728.

45. Costa, M.J., et al. 2004. Sphingolipid-cholesterol domains (lipid rafts) in normal human and dog thyroid follicular cells are not involved in thyrotropin receptor signaling. *Endocrinology.* **145**:1464–1472.

46. Di Lauro, R., and De Felice, M. 2001. *Thyroid gland: anatomy and development in endocrinology.* 4th edition. L.J. DeGroot and J.L. Jameson, editors. WB Saunders Co. Philadelphia, Pennsylvania, USA. 1268–1277.

47. Van Vliet, G. 2003. Development of the thyroid gland: lessons from congenitally hypothyroid mice and men [review]. *Clin. Genet.* **63**:445–455.

48. De Felice, M., et al. 1998. A mouse model for hereditary thyroid dysgenesis and cleft palate. *Nat. Genet.* **19**:395–398.

49. Brown, R.S. 2004. Minireview: developmental regulation of thyrotropin receptor gene expression in the fetal and newborn thyroid. *Endocrinology.* **145**:4058–4061.

50. De Felice, M., Postiglione, M.P., and Di Lauro, R. 2004. Minireview: thyrotropin receptor signaling in development and differentiation of the thyroid gland: insights from mouse models and human diseases. *Endocrinology.* **145**:4062–4067.

51. Postiglione, M.P., et al. 2002. Role of the thyroid-stimulating hormone receptor signaling in development and differentiation of the thyroid gland. *Proc. Natl. Acad. Sci. U. S. A.* **99**:15462–15467.

52. Marians, R.C., et al. 2002. Defining thyrotropin-dependent and -independent steps of thyroid hormone synthesis by using thyrotropin receptor-null mice. *Proc. Natl. Acad. Sci. U. S. A.* **99**:15776–15781.

53. Burrow, G.N., Fisher, D.A., and Larsen, P.R. 1994. Maternal and fetal thyroid function. *N. Engl. J. Med.* **331**:1072–1078.

54. Bahn, R.S., et al. 1998. Thyrotropin receptor expression in Graves' orbital adipose/connective tissues: potential autoantigen in Graves' ophthalmopathy. *J. Clin. Endocrinol. Metab.* **83**:998–1002.

55. Daumerie, C., Ludgate, M., Costagliola, S., and Many, M.C. 2002. Evidence for thyrotropin receptor immunoreactivity in pretibial connective tissue from patients with thyroid-associated dermopathy. *Eur. J. Endocrinol.* **146**:35–38.

56. Vizek, K., Razova, M., and Melichar, V. 1979. Lipolytic effect of TSH, glucagon and hydrocortisone on the adipose tissue of newborns and adults in vitro. *Physiol. Bohemoslov.* **28**:325–331.

57. Abe, E., et al. 2003. TSH is a negative regulator of skeletal remodeling. *Cell.* **115**:151–162.

58. Refetoff, S. 2003. The syndrome of resistance to thyroid stimulating hormone. *J. Chin. Med. Assoc.* **66**:441–452.

59. Alberti, L., et al. 2002. Germline mutations of TSH receptor gene as cause of nonautoimmune subclinical hypothyroidism. *J. Clin. Endocrinol. Metab.* **87**:2549–2555.

60. Park, S.M., Clifton-Bligh, R.J., Betts, P., and Chatterjee, V.K. 2004. Congenital hypothyroidism and apparent athyreosis with compound heterozygosity or compensated hypothyroidism with probable hemizygosity for inactivating mutations of the TSH receptor. *Clin. Endocrinol. (Oxf.).* **60**:220–227.

61. Fagman, H., Grande, M., Gritli-Linde, A., and Nilsson, M. 2004. Genetic deletion of sonic hedgehog causes hemiagenesis and ectopic development of the thyroid in mouse. *Am. J. Pathol.* **164**:1865–1872.

62. Damante, G., Tell, G., and Di Lauro, R. 2001. A unique combination of transcription factors controls differentiation of thyroid cells. *Prog. Nucleic Acid Res. Mol. Biol.* **66**:307–356.

63. Sunthornthepvarakui, T., Gottschalk, M.E., Hayashi, Y., and Refetoff, S. 1995. Brief report: resistance to thyrotropin caused by mutations in the thyrotropin-receptor gene. *N. Engl. J. Med.* **332**:155–160.

64. Refetoff, S. 2003. Resistance to thyrotropin. *J. Endocrinol. Invest.* **26**:770–779.

65. Abramowicz, M.J., Duprez, L., Parma, J., Vassart, G., and Heinrichs, C. 1997. Familial congenital hypothyroidism due to inactivating mutation of the thyrotropin receptor causing profound hypoplasia of the thyroid gland. *J. Clin. Invest.* **99**:3018–3024.

66. Xie, J., et al. 1997. Resistance to thyrotropin (TSH) in three families is not associated with mutations in the TSH receptor or TSH. *J. Clin. Endocrinol. Metab.* **82**:3933–3940.

67. McKenzie, J.M., and Zakarija, M. 1992. Fetal and neonatal hyperthyroidism and hypothyroidism due to maternal TSH receptor antibodies. *Thyroid.* **2**:155–159.

68. Duprez, L., et al. 1994. Germline mutations in the thyrotropin receptor gene cause non-autoimmune autosomal dominant hyperthyroidism. *Nat. Genet.* **7**:396–401.

69. Kopp, P., et al. 1995. Brief report: congenital hyperthyroidism caused by a mutation in the thyrotropin-receptor gene. *N. Engl. J. Med.* **332**:150–154.

70. Tonacchera, M., et al. 1996. Functional characteristics of three new germline mutations of the thyrotropin receptor gene causing autosomal dominant toxic thyroid hyperplasia. *J. Clin. Endocrinol. Metab.* **81**:547–554.

71. Beamer, W.J., Eicher, E.M., Maltais, L.J., and Southard, J.L. 1981. Inherited primary hypothyroidism in mice. *Science.* **212**:61–63.

72. Stein, S.A., et al. 1994. Identification of a point mutation in the thyrotropin receptor of the hyt/hyt hypothyroid mouse. *Mol. Endocrinol.* **8**:129–138.

73. Stein, S.A., et al. 1991. The site of the molecular defect in the thyroid gland of the hyt/hyt mouse: abnormalities in the TSH receptor-G protein complex. *Thyroid.* **1**:257–266.

74. Fernandez Rodriguez, A., et al. 1991. Induction of thyroid proliferative changes in rats treated with antithyroid compound. *Anat. Histol. Embryol.* **20**:289–298.

75. Matsuo, K., Friedman, E., Gejman, P.V., and Fagin, J.A. 1993. The thyrotropin receptor (TSH-R) is not an oncogene for thyroid tumors: structural studies of the TSH-R and the alpha-subunit of Gs in human thyroid neoplasms. *J. Clin. Endocrinol. Metab.* **76**:1446–1451.

76. Krohn, K., and Paschke, R. 2002. Somatic mutations in thyroid nodular disease. *Mol. Genet. Metab.* **75**:202–208.

77. Rodien, P., Ho, S.C., Vlaeminck, V., Vassart, G., and Costagliola, S. 2003. Activating mutations of TSH receptor. *Ann. Endocrinol. (Paris).* **64**:12–16.

78. Krohn, K., Wohlgemuth, S., Gerber, H., and Paschke, R. 2000. Hot microscopic areas of iodine-deficient euthyroid goitres contain constitutively activating TSH receptor mutations. *J. Pathol.* **192**:37–42.

79. Tonacchera, M., et al. 2000. Activating thyrotropin receptor mutations are present in nonadenomatous hyperfunctioning nodules of toxic or autonomous multinodular goiter. *J. Clin. Endocrinol. Metab.* **85**:2270–2274.

80. Kosugi, S., Ban, T., Akamizu, T., and Kohn, L.D. 1991. Site-directed mutagenesis of a portion of the extracellular domain of the rat thyrotropin receptor important in autoimmune thyroid disease and nonhomologous with gonadotropin receptors. Relationship of functional and immunogenic domains. *J. Biol. Chem.* **266**:19413–19418.

81. Costagliola, S., et al. 2002. Tyrosine sulfation is required for agonist recognition by glycoprotein hormone receptors. *EMBO J.* **21**:504–513.

82. Hershman, J.M. 1999. Human chorionic gonadotropin and the thyroid: hyperemesis gravidarum and trophoblastic tumors [review]. *Thyroid.* **9**:653–657.

83. Hershman, J.M. 2004. Physiological and pathological aspects of the effect of human chorionic gonadotropin on the thyroid. *Best Pract. Res. Clin. Endocrinol. Metab.* **18**:249–265.

84. Davies, T., Taliadouros, G., Catt, K., and Nisula, B. 1979. Assessment of urinary thyrotropin-competing activity in choriocarcinoma and thyroid disease: further evidence for human chorionic gonadotropin interacting at the thyroid cell membrane. *J. Clin. Endocrinol. Metab.* **49**:353–357.

85. Davies, T.F., and Platzer, M. 1986. hCG-induced TSH receptor activation and growth acceleration in FRTL-5 thyroid cells. *Endocrinology.* **118**:2149–2151.

86. Tomer, Y., Huber, G.K., and Davies, T.F. 1992. Human chorionic gonadotropin (hCG) interacts directly with recombinant human TSH receptors. *J. Clin. Endocrinol. Metab.* **74**:1477–1479.

87. Glinoer, D., et al. 1990. Regulation of maternal thyroid during pregnancy. *J. Clin. Endocrinol. Metab.* **71**:276–287.

88. Rodien, P., et al. 1998. Familial gestational hyperthyroidism caused by a mutant thyrotropin receptor hypersensitive to human chorionic gonadotropin. *N. Engl. J. Med.* **339**:1823–1826.

89. Smits, G., et al. 2002. Lysine 183 and glutamic acid 157 of the TSH receptor: two interacting residues with a key role in determining specificity toward TSH and human CG. *Mol. Endocrinol.* **16**:722–735.

90. Rodien, P., et al. 2004. Abnormal stimulation of the thyrotrophin receptor during gestation. *Hum. Reprod. Update.* **10**:95–105.

91. Tsuruta, E., et al. 1995. Pathogenic role of asialo human chorionic gonadotropin in gestational thyrotoxicosis. *J. Clin. Endocrinol. Metab.* **80**:350–355.

92. Pekary, A.E., et al. 1993. Increased in vitro thyrotropic activity of partially sialated human chorionic gonadotropin extracted from hydatidiform moles of patients with hyperthyroidism. *J. Clin. Endocrinol. Metab.* **76**:70–74.

93. Kraiem, Z., Lahat, N., Sadeh, O., Blithe, D.L., and Nisula, B.C. 1997. Desialylated and deglycosylated human chorionic gonadotropin are superagonists of native human chorionic gonadotropin in human thyroid follicles. *Thyroid.* **7**:783–788.

94. Tomer, Y., and Davies, T.F. 2003. Searching for the autoimmune thyroid disease susceptibility genes: from gene mapping to gene function. *Endocr. Rev.* **24**:694–717.

95. Bartels, E.D. 1941. *Heredity in Graves' disease.* Munksgaard. Copenhagen, Denmark. 384 pp.

96. Martin, L. 1945. The heredity and familial aspects of exophthalmic goitre and nodular goitre. *Q. J. Med.* **14**:207–219.

97. Hall, R., and Stanbury, J.B. 1967. Familial studies of autoimmune thyroiditis. *Exp. Clin. Endocrinol.* **2**:719–725.

98. Villanueva, R., Greenberg, D.A., Davies, T.F., and Tomer, Y. 2003. Sibling recurrence risk in autoimmune thyroid disease. *Thyroid.* **13**:761–764.

99. Brix, T.H., Kyvik, K.O., Christensen, K., and Hegedus, L. 2001. Evidence for a major role of heredity in Graves' disease: a population-based study of two Danish twin cohorts. *J. Clin. Endocrinol. Metab.* **86**:930–934.

100. Ringold, D.A., et al. 2002. Further evidence for a strong genetic influence on the development of autoimmune thyroid disease: the California twin study. *Thyroid.* **12**:647–653.

101. Ban, Y., Greenberg, D.A., Concepcion, E.S., and Tomer, Y. 2002. A germline single nucleotide polymorphism at the intracellular domain of the human thyrotropin receptor does not have a major effect on the development of Graves' disease. *Thyroid.* **12**:1079–1083.

102. Ban, Y., et al. 2004. Arginine at position 74 of the HLA-DR beta1 chain is associated with Graves' disease. *Genes Immun.* **5**:203–208.

103. Chambers, C.A., and Allison, J.P. 1997. Co-stimulation in T cell responses. *Curr. Opin. Immunol.* **9**:396–404.

104. Yanagawa, T., Hidaka, Y., Guimaraes, V., Soliman, M., and DeGroot, L.J. 1995. CTLA-4 gene polymorphism associated with Graves' disease in a Caucasian population. *J. Clin. Endocrinol. Metab.* **80**:41–45.

105. Kouki, T., et al. 2000. CTLA-4 gene polymorphism at position 49 in exon 1 reduces the inhibitory function of CTLA-4 and contributes to the pathogenesis of Graves' disease. *J. Immunol.* **165**:6606–6611.

106. Tomer, Y., Greenberg, D.A., Barbesino, G., Concepcion, E., and Davies, T.F. 2001. CTLA-4 and not CD28 is a susceptibility gene for thyroid autoantibody production. *J. Clin. Endocrinol. Metab.* **86**:1687–1693.

107. Ban, Y., and Tomer, Y. 2003. The contribution of immune regulatory and thyroid specific genes to the etiology of Graves' and Hashimoto's diseases. *Autoimmunity.* **36**:367–379.

108. Ueda, H., et al. 2003. Association of the T-cell regulatory gene CTLA4 with susceptibility to autoimmune disease. *Nature.* **423**:506–511.

109. Ban, Y., Tozaki, T., Taniyama, M., and Tomita, M. 2004. Association of a thyroglobulin gene polymorphism with Hashimoto's thyroiditis in the Japanese population. *Clin. Endocrinol. (Oxf.).* **61**:263–268.

110. Begovich, A.B., et al. 2004. A missense single-nucleotide polymorphism in a gene encoding a protein tyrosine phosphatase (PTPN22) is associated with rheumatoid arthritis. *Am. J. Hum. Genet.* **75**:330–337.

111. Bottini, N., et al. 2004. A functional variant of lymphoid tyrosine phosphatase is associated with type I diabetes. *Nat. Genet.* **36**:337–338.

112. Smyth, D., et al. 2004. Replication of an association between the lymphoid tyrosine phosphatase locus (LYP/PTPN22) with type 1 diabetes, and evidence for its role as a general autoimmunity locus. *Diabetes.* **53**:3020–3023.

113. Velaga, M.R., et al. 2004. The codon 620 tryptophan allele of the lymphoid tyrosine phosphatase (LYP) gene is a major determinant of Graves' disease. *J. Clin. Endocrinol. Metab.* **89**:5862–5865.

114. Tomer, Y., Concepcion, E., and Greenberg, D.A. 2002. A C/T single-nucleotide polymorphism in the region of the CD40 gene is associated with Graves' disease. *Thyroid.* **12**:1129–1135.

115. Kim, T.Y., et al. 2003. A C/T polymorphism in the 5'-untranslated region of the CD40 gene is associated with Graves' disease in Koreans. *Thyroid.* **13**:919–925.

116. Houston, F.A., et al. 2004. Role of the CD40 locus in Graves' disease. *Thyroid.* **14**:506–509.

117. Ban, Y., et al. 2003. Amino acid substitutions in the thyroglobulin gene are associated with susceptibility to human and murine autoimmune thyroid disease. *Proc. Natl. Acad. Sci. U. S. A.* **100**:15119–15124.

118. Collins, J.E., et al. 2003. Association of a rare thyroglobulin gene microsatellite variant with autoimmune thyroid disease. *J. Clin. Endocrinol. Metab.* **88**:5039–5042.

119. Tomer, Y., and Davies, T.F. 1993. Infection, thyroid disease, and autoimmunity. *Endocr. Rev.* **14**:107–120.

120. Bulow Pedersen, I., et al. 2002. Large differences in incidences of overt hyper- and hypothyroidism associated with a small difference in iodine intake: a prospective comparative register-based population survey. *J. Clin. Endocrinol. Metab.* **87**:4462–4469.

121. Prummel, M.F., and Wiersinga, W.M. 1993. Smoking and risk of Graves' disease. *JAMA.* **269**:479–482.

122. Vestergaard, P., et al. 2002. Smoking as a risk factor for Graves' disease, toxic nodular goiter, and autoimmune hypothyroidism. *Thyroid.* **12**:69–75.

123. Santos, A.M., et al. 2002. Grave's disease and stress [abstract]. *Acta Med. Port.* **15**:423–427.

124. Winsa, B., et al. 1991. Stressful life events and Graves' disease. *Lancet.* **338**:1475–1479.

125. Sonino, N., et al. 1993. Life events in the pathogenesis of Graves' disease. A controlled study. *Acta Endocrinol. (Copenh.).* **128**:293–296.

Inflammation, stress, and diabetes

Kathryn E. Wellen and Gökhan S. Hotamisligil

Department of Genetics & Complex Diseases, Harvard School of Public Health, Boston, Massachusetts, USA.

Over the last decade, an abundance of evidence has emerged demonstrating a close link between metabolism and immunity. It is now clear that obesity is associated with a state of chronic low-level inflammation. In this article, we discuss the molecular and cellular underpinnings of obesity-induced inflammation and the signaling pathways at the intersection of metabolism and inflammation that contribute to diabetes. We also consider mechanisms through which the inflammatory response may be initiated and discuss the reasons for the inflammatory response in obesity. We put forth for consideration some hypotheses regarding important unanswered questions in the field and suggest a model for the integration of inflammatory and metabolic pathways in metabolic disease.

Inflammation, stress, and diabetes

Survival of multicellular organisms depends on the ability to fight infection and heal damage and the ability to store energy for times of low nutrient availability or high energy need. Metabolic and immune systems are therefore among the most basic requirements across the animal kingdom, and many nutrient and pathogen-sensing systems have been highly conserved from organisms such as *Caenorhabditis elegans* and *Drosophila* to mammals. Perhaps not surprisingly, metabolic and immune pathways have also evolved to be closely linked and interdependent. Many hormones, cytokines, signaling proteins, transcription factors, and bioactive lipids can function in both metabolic and immune roles. In addition to using some of the same cellular machinery, metabolic and immune systems also regulate each other. The normal inflammatory response relies upon metabolic support, and energy redistribution, particularly the mobilization of stored lipid, plays an important role in fighting infection during the acute-phase response (1). The basic inflammatory response thus favors a catabolic state and suppresses anabolic pathways, such as the highly conserved and powerful insulin signaling pathway.

The integration of metabolism and immunity, which under normal conditions is beneficial for the maintenance of good health, can become deleterious under conditions of metabolic challenge, as exemplified by the immunosuppression characteristic of malnourished or starving individuals (1–3). Famine has been a prominent hazard to human health throughout history, and for thousands of years the link between infection and poor nutrition has been well recognized. Today this threat is as widespread as ever, and there are approximately 1 billion undernourished individuals worldwide (3). In the past century, however, the pendulum has also swung in the opposite direction, and now as many if not more people are overweight or obese (4). With the advent of this chronic metabolic overload, a new set of problems and complications at the intersection of metabolism and immunity has emerged, including the obesity-linked inflammatory diseases diabetes, fatty liver disease, airway inflammation, and atherosclerosis (5).

There is now a wealth of evidence indicating close ties between metabolic and immune systems. Among the many reasons to maintain a healthy weight is the emerging paradigm that metabolic imbalance leads to immune imbalance, with starvation and immunosuppression on one end of the spectrum and obesity and inflammatory diseases on the other end (Figure 1). In this article, we will discuss the molecular and cellular links between metabolism and inflammation, particularly in the context of obesity and diabetes. Common inflammatory mediators, stress responses, and signaling pathways will be highlighted. Finally, we will consider the origin of and the reasons for the inflammatory response in obesity.

Obesity is characterized by inflammation

Factors at the crossroads of inflammation and metabolic disease. A little more than a decade ago, the first molecular link between inflammation and obesity — TNF-α — was identified when it was discovered that this inflammatory cytokine is overexpressed in the adipose tissues of rodent models of obesity (6, 7). As is the case in mice, TNF-α is overproduced in the adipose as well as muscle tissues of obese humans (8–10). Administration of recombinant TNF-α to cultured cells or to whole animals impairs insulin action, and obese mice lacking functional TNF-α or TNF receptors have improved insulin sensitivity compared with wild-type counterparts (6, 11). Thus, particularly in experimental models, it is clear that overproduction of TNF-α in adipose tissue is an important feature of obesity and contributes significantly to insulin resistance.

It rapidly became clear, however, that obesity is characterized by a broad inflammatory response and that many inflammatory mediators exhibit patterns of expression and/or impact insulin action in a manner similar to that of TNF-α during obesity, in animals ranging from mice and cats to humans (12–14). Transcriptional profiling studies have revealed that inflammatory and stress-response genes are among the most abundantly regulated gene sets in adipose tissue of obese animals (15–17). A list of many of these genes, which have been identified through a variety of approaches, is provided in Table 1.

In addition to inflammatory cytokines regulating metabolic homeostasis, molecules that are typical of adipocytes, with well-established metabolic functions, can regulate the immune response. Leptin is one such hormone that plays important roles in both adaptive and innate immunity, and both mice and humans lacking leptin function exhibit impaired immunity (18–20). Indeed, reduced leptin levels may be responsible, at least in part, for immunosuppression associated with starvation, as leptin administration has been shown to reverse the immunosuppression of mice starved for 48 hours (21). Adiponectin, resistin, and visfatin are

Nonstandard abbreviations used: AP-1, activator protein-1; DAG, diacylglycerol; FABP, fatty acid–binding protein; IκB, inhibitor of NF-κB; IKK, inhibitor of NF-κB kinase; IRS, insulin receptor substrate; JIP1, JNK-interacting protein-1; LXR, liver X receptor; TLR, Toll-like receptor; TZD, thiazolidinedione.

Conflict of interest: The authors have declared that no conflict of interest exists.

Citation for this article: *J. Clin. Invest.* **115**:1111–1119 (2005).
doi:10.1172/JCI200525102.

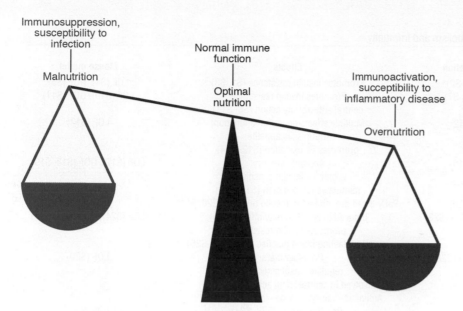

Figure 1
Metabolism and immunity are closely linked. Both overnutrition and undernutrition have implications for immune function. Starvation and malnutrition can suppress immune function and increase susceptibility to infections. Obesity is associated with a state of aberrant immune activity and increasing risk for associated inflammatory diseases, including atherosclerosis, diabetes, airway inflammation, and fatty liver disease. Thus, optimal nutritional and metabolic homeostasis is an important part of appropriate immune function and good health.

also examples of molecules with immunological activity that are produced in adipocytes (22–26).

Finally, lipids themselves also participate in the coordinate regulation of inflammation and metabolism. Elevated plasma lipid levels are characteristic of obesity, infection, and other inflammatory states. Hyperlipidemia in obesity is responsible in part for inducing peripheral tissue insulin resistance and dyslipidemia and contributes to the development of atherosclerosis. It is interesting to note that metabolic changes characteristic of the acute-phase response are also proatherogenic; thus, altered lipid metabolism that is beneficial in the short term in fighting against infection is harmful if maintained chronically (1). The critical importance of bioactive lipids is also evident in their regulation of lipid-targeted signaling pathways through fatty acid–binding proteins (FABPs) and nuclear receptors (see "Regulation of inflammatory pathways," below).

Macrophages and the link between inflammation and metabolism. The high level of coordination of inflammatory and metabolic pathways is highlighted by the overlapping biology and function of macrophages and adipocytes in obesity (Figure 2). Gene expression is highly similar; macrophages express many, if not the majority of "adipocyte" gene products such as the adipocyte/macrophage FABP aP2 (also known as FABP4) and PPARγ, while adipocytes can express many "macrophage" proteins such as TNF-α, IL-6, and MMPs (6, 27–29). Functional capability of these 2 cell types also overlaps. Macrophages can take up and store lipid to become atherosclerotic foam cells. Preadipocytes under some conditions can exhibit phagocytic and antimicrobial properties and appear to even be able to differentiate into macrophages in the right environment, which suggests a potential immune role for preadipocytes (30, 31). Furthermore, macrophages and adipocytes colocalize in adipose tissue in obesity. The recent finding that obesity is characterized by macrophage accumulation in white adipose tissue has added another dimension to our understanding of the development of adipose tissue inflammation in obesity (16, 17). Macrophages in adipose tissue are likely to contribute to the production of inflammatory mediators either alone or in concert with adipocytes, which suggests a potentially important influence of macrophages in promoting insulin resistance. However, no direct evidence has been offered to establish this connection thus far.

In terms of the immune response, integration between macrophages and adipocytes makes sense, given that both cell types participate in the innate immune response: macrophages in their role as immune cells by killing pathogens and secreting inflammatory cytokines and chemokines; and adipocytes by releasing lipids that may modulate the inflammatory state or participate in neutralization of pathogens. While it is not yet known whether macrophages are drawn to adipose tissue in other inflammatory conditions, it is conceivable that macrophage accumulation in adipose tissue is a feature not only of obesity, but of other inflammatory states as well.

Inflammatory pathways to insulin resistance

As discussed above, it is now apparent that obesity is associated with a state of chronic, low-grade inflammation, particularly in white adipose tissue. How do inflammatory cytokines and/or fatty acids mediate insulin resistance? How do the stresses of obesity manifest inside of cells? In recent years, much has been learned about the intracellular signaling pathways activated by inflammatory and stress responses and how these pathways intersect with and inhibit insulin signaling.

Insulin affects cells through binding to its receptor on the surface of insulin-responsive cells. The stimulated insulin receptor phosphorylates itself and several substrates, including members of the insulin receptor substrate (IRS) family, thus initiating downstream signaling events (32, 33). The inhibition of signaling downstream of the insulin receptor is a primary mechanism through which inflammatory signaling leads to insulin resistance. Exposure of cells to TNF-α or elevated levels of free fatty acids stimulates inhibitory phosphorylation of serine residues of IRS-1 (34–36). This phosphorylation reduces both tyrosine phosphorylation of IRS-1 in response to insulin and the ability of IRS-1 to associate with the insulin receptor and thereby inhibits downstream signaling and insulin action (35, 37, 38).

Recently it has become clear that inflammatory signaling pathways can also become activated by metabolic stresses originating

Table 1

Factors that mediate the intersection of metabolism and immunity

Factors	Metabolic regulation	Effects	Mouse model
TNF-α	↑ in obesity (S1, S2)[A]	Promotes insulin resistance (S1, S3)	LOF (S4), GOF (S5)
IL-6	↑ in obesity (S6, S7)	Promotes insulin resistance; central anti-obesity action (S8–S11)	LOF (S10), GOF (S11)
Leptin	↑ in obesity (S12)	Multiple effects on immune function; suppresses appetite; promotes FA oxidation (S12–S14)	LOF (S12)
Adiponectin	↓ in obesity (S15)	Antiinflammatory; promotes insulin sensitivity; stimulates FA oxidation (S15, S16)	LOF (S17), GOF (S18, S19)
Visfatin	↑ in obesity (S20)	Early B cell growth factor; insulin mimetic (S20, S21)	LOF (S20)
Resistin	Variable in obesity (S22, S23)	Induced in endotoxemia/inflammation; promotes insulin resistance; regulates fasting blood glucose level (S23–S26)	LOF (S24), GOF (S25, S27)
IL-1	↑ by hyperglycemia (S28)	Proinflammatory; regulates insulin secretion; involved in central leptin action (S29, S30)	LOF (S30)
IL-1Rα	↑ in obesity (S31)	Antiinflammatory; opposes leptin action (S29)	
IL-8	↑ in obesity (S32)	Proatherogenic (S33, S34)	LOF (S33)
IL-10	↑ in obesity; ↓ in metabolic syndrome (S35)	Antiinflammatory; promotes insulin sensitivity (S36)	
IL-18	↑ in obesity (S37, S38)	Proatherogenic (S39, S40)	LOF (S40) GOF (S39)
MCP-1	↑ in obesity (S41)	Proatherogenic; promotes insulin resistance (S34, S41, S42)	LOF (S42)
MIF	↑ in obesity (S43)	Inhibits macrophage migration	
M-CSF		Monocyte/macrophage differentiation; stimulates adipose growth (S44)	GOF (S44)
TGF-β	↑ in obesity (S45, S46)	Inhibits adipocyte differentiation and adipose tissue development; regulates atherosclerosis (S47–S49)	LOF (S50, S51), GOF (S49)
Soluble TNFR	↑ in obesity (S52–S54)	Proinflammatory	
C-reactive protein	↑ in obesity (S55, S56)	Proinflammatory; atherogenic; risk factor for diabetes (S55, S57–S59)	GOF (S57)
Haptoglobin	↑ in obesity (S60)	Proinflammatory	

↑, increase; ↓, decrease; FA, fatty acid; GOF, gain-of-function; IL-1Rα, IL-1 receptor α; LOF, loss-of-function; MCP-1, monocyte chemotactic protein-1; MIF, macrophage migration inhibitory factor; TNFR, TNF receptor. [A]See Supplemental References; supplemental material available online with this article; doi:10.1172/JCI200525102DS1.

from inside the cell as well as by extracellular signaling molecules. It has been demonstrated that obesity overloads the functional capacity of the ER and that this ER stress leads to the activation of inflammatory signaling pathways and thus contributes to insulin resistance (39–41). Additionally, increased glucose metabolism can lead to a rise in mitochondrial production of ROS. ROS production is elevated in obesity, which causes enhanced activation of inflammatory pathways (42, 43).

Several serine/threonine kinases are activated by inflammatory or stressful stimuli and contribute to inhibition of insulin signaling, including JNK, inhibitor of NF-κB kinase (IKK), and PKC-θ (44). Again, the activation of these kinases in obesity highlights the overlap of metabolic and immune pathways; these are the same kinases, particularly IKK and JNK, that are activated in the innate immune response by Toll-like receptor (TLR) signaling in response to LPS, peptidoglycan, double-stranded RNA, and other microbial products (45). Hence it is likely that components of TLR signaling pathways will also exhibit strong metabolic activities.

JNK. The 3 members of the JNK group of serine/threonine kinases, JNK-1, -2, and -3, belong to the MAPK family and regulate multiple activities in development and cell function, in large part through their ability to control transcription by phosphorylating activator protein-1 (AP-1) proteins, including c-Jun and JunB (46). JNK has recently emerged as a central metabolic regulator, playing an important role in the development of insulin resistance in obesity (47). In response to stimuli such as ER stress, cytokines, and fatty acids, JNK is activated, whereupon it associates with and phosphorylates IRS-1 on Ser307, impairing insulin action (36, 39, 48). In obesity, JNK activity is elevated in liver, muscle, and fat tissues, and loss of JNK1 prevents the development of insulin resistance and diabetes in both genetic and dietary mouse models of obesity (47). Modulation of hepatic JNK1 in adult animals also produces systemic effects on glucose metabolism, which underscores the importance of this pathway in the liver (49). The contribution of the JNK pathway in adipose, muscle, or other tissues to systemic insulin resistance is currently unclear. In addition, a mutation in JNK-interacting protein-1 (JIP1), a protein that binds JNK and regulates its activity, has been identified in diabetic humans (50). The phenotype of the JIP1 loss-of-function model is very

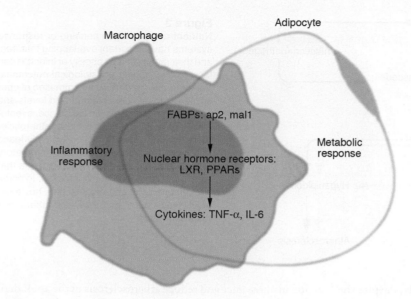

Macrophage   Adipocyte

FABPs: ap2, mal1

Inflammatory
response

Nuclear hormone receptors:
LXR, PPARs

Metabolic
response

Cytokines: TNF-α, IL-6

Figure 2
Lipids and inflammatory mediators: integration of metabolic and immune responses in adipocytes
and macrophages through shared mechanisms. Under normal conditions, adipocytes store
lipids and regulate metabolic homeostasis, and macrophages function in the inflammatory
response, although each cell type has the capacity to perform both functions. In obesity, adipose tissue becomes inflamed, both via infiltration of adipose tissue by macrophages and as a
result of adipocytes themselves becoming producers of inflammatory cytokines. Inflammation of
adipose tissue is a crucial step in the development of peripheral insulin resistance. In addition, in
proatherosclerotic conditions such as obesity and dyslipidemia, macrophages accumulate lipid
to become foam cells. Adipocytes and macrophages share common features such as expression of cytokines, FABPs, nuclear hormone receptors, and many other factors. As evidenced by
genetic loss-of-function models, adipocyte/macrophage FABPs modulate both lipid accumulation in adipocytes and cholesterol accumulation in macrophages, as well as the development
of insulin resistance and atherosclerosis. PPARγ and LXR pathways oppose inflammation and
promote cholesterol efflux from macrophages and lipid storage in adipocytes.

to lipid infusion, high-fat diet, or genetic obesity (59, 60). Moreover, inhibition of IKKβ in human diabetics by high-dose aspirin treatment also improves insulin signaling, although at this dose, it is not clear whether other kinases are also affected (61). Recent studies have also begun to tease out the importance of IKK in individual tissues or cell types to the development of insulin resistance. Activation of IKK in liver and myeloid cells appears to contribute to obesity-induced insulin resistance, though this pathway may not be as important in muscle (62–64).

Other pathways. In addition to serine/threonine kinase cascades, other pathways contribute to inflammation-induced insulin resistance. For example, at least 3 members of the SOCS family, SOCS1, -3, and -6, have been implicated in cytokine-mediated inhibition of insulin signaling (65–67). These molecules appear to inhibit insulin signaling either by interfering with IRS-1 and IRS-2 tyrosine phosphorylation or by targeting IRS-1 and IRS-2 for proteosomal degradation (65, 68). SOCS3 has also been demonstrated to regulate central leptin action, and both whole body reduction in SOCS3 expression ($SOCS3^{+/-}$) and neural SOCS3 disruption result in resistance to high-fat diet–induced obesity and insulin resistance (69, 70).

Inflammatory cytokine stimulation can also lead to induction of iNOS. Overproduction of nitric oxide also appears to contribute to impairment of both muscle cell insulin action and β cell function in obesity (71, 72). Deletion of iNOS prevents impairment of insulin signaling in muscle caused by a high-fat diet (72). Thus, induction of SOCS proteins and iNOS represent 2 additional and potentially important mechanisms that contribute to cytokine-mediated insulin resistance. It is likely that additional mechanisms linking inflammation with insulin resistance remain to be uncovered.

Regulation of inflammatory pathways
Lipids and lipid targets. The role of lipids in metabolic disease is complex. As discussed above, hyperlipidemia leads to increased uptake of fatty acids by muscle cells and production of fatty acid metabolites that stimulate inflammatory cascades and inhibit insulin signaling (54). On the other hand, intracellular lipids can also be antiinflammatory. Ligands of the liver X receptor (LXR) and PPAR families of nuclear hormone receptors are oxysterols and fatty acids, respectively, and activation of these transcription factors inhibits inflammatory gene expression in macrophages and adipocytes, in large part through suppression of NF-κB (73–79).

LXR function is also regulated by innate immune pathways. Signaling from TLRs inhibits LXR activity in macrophages, causing enhanced cholesterol accumulation and accounting, at least in part, for the proatherogenic effects of infection (80). Indeed, lack of MyD88, a critical mediator of TLR signaling, reduces

similar to that of JNK1 deficiency in mice, with reduced JNK activity and increased insulin sensitivity (51). Interestingly, the JNK2 isoform plays a significant nonredundant role in atherosclerosis (52), though apparently not in type 2 diabetes. Recent studies in mice demonstrate that JNK inhibition in established diabetes or atherosclerosis might be a viable therapeutic avenue for these diseases in humans (52, 53).

PKC and IKK. Two other inflammatory kinases that play a large role in counteracting insulin action, particularly in response to lipid metabolites, are IKK and PKC-θ. Lipid infusion has been demonstrated to lead to a rise in levels of intracellular fatty acid metabolites, such as diacylglycerol (DAG) and fatty acyl CoAs. This rise is correlated with activation of PKC-θ and increased Ser307 phosphorylation of IRS-1 (54). PKC-θ may impair insulin action by activation of another serine/threonine kinase, IKKβ, or JNK (55). Other PKC isoforms have also been reported to be activated by lipids and may also participate in inhibition of insulin signaling (56).

IKKβ can impact on insulin signaling through at least 2 pathways. First, it can directly phosphorylate IRS-1 on serine residues (34, 57). Second, it can phosphorylate inhibitor of NF-κB (IκB), thus activating NF-κB, a transcription factor that, among other targets, stimulates production of multiple inflammatory mediators, including TNF-α and IL-6 (58). Mice heterozygous for IKKβ are partially protected against insulin resistance due

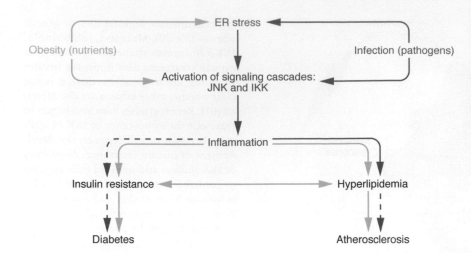

Figure 3
Nutrient and pathogen sensing or response systems have important overlapping features, and their modulation by obesity or infection can lead to overlapping physiological outcomes. For example, the chronic inflammation of obesity leads to elevated plasma lipid levels and the development of insulin resistance, eventually resulting in fatty liver disease, atherosclerosis, and diabetes. Infection typically leads to a more transient and robust inflammatory response and short-term hyperlipidemia that aids in the resolution of the infection. In some circumstances of chronic infection, however, insulin resistance, diabetes, and atherosclerosis can result.

atherosclerosis in *apoE*−/− mice (81). Interestingly, despite the inhibitory effects of TLR signaling on LXR cholesterol metabolism, LXR appears to be necessary for the complete response of macrophages to infection. In the absence of LXR, macrophages undergo accelerated apoptosis and are thus unable to appropriately respond to infection (82). Unliganded PPARδ also seems to have proinflammatory functions, mediated at least in part through its association with the transcriptional repressor B cell lymphoma 6 (BCL-6) (83).

The activity of these lipid ligands is influenced by cytosolic FABPs. Animals lacking the adipocyte/macrophage FABPs ap2 and mal1 are strongly protected against type 2 diabetes and atherosclerosis, a phenotype reminiscent of that of thiazolidinedione-treated (TZD-treated) mice and humans (27, 84, 85). One mechanism for this phenotype is potentially related to the availability of endogenous ligands for these receptors that stimulate storage of lipids in adipocytes and suppress inflammatory pathways in macrophages (86). In general, it appears that location in the body, the composition of the surrounding cellular environment, and coupling to target signaling pathways are critical for determining whether lipids promote or suppress inflammation and insulin resistance. Accumulation of cholesterol in macrophages promotes atherosclerosis and of lipid in muscle and liver promotes insulin resistance, while, as seen in TZD-treated and FABP-deficient mice, if lipids are forced to remain in adipose tissue, insulin resistance in the context of obesity can be reduced (85). Thus, lipids and their targets clearly play both metabolic and inflammatory roles; however, the functions that they assume are dependent on multiple factors.

Pharmacological manipulation of inflammation. In corroboration of genetic evidence in mice that loss of inflammatory mediators or signaling molecules prevents insulin resistance (11, 47, 59), pharmacological targeting of inflammatory pathways also improves insulin action. Effective treatment has been demonstrated both with inhibitors of inflammatory kinases and with agonists of relevant transcription factors. As discussed above, salicylates promote insulin signaling by inhibiting inflammatory kinase cascades within the cell. Through inhibition of IKK and possibly other kinases, salicylates are able to improve glucose metabolism in both obese mice and diabetic humans (53, 55, 74). Targeting of JNK using a synthetic inhibitor and/or an inhibitory peptide has been demonstrated to improve insulin

action in obese mice and reduce atherosclerosis in the apoE-deficient rodent model (52, 53). These results directly demonstrate the therapeutic potential of JNK inhibitors in diabetes.

Synthetic ligands have been produced to all 3 PPAR isoforms as well as LXR-α, though only PPARγ and PPARα ligands have been approved for clinical treatment (87, 88). TZDs, high-affinity ligands of PPARγ, which are given clinically as insulin-sensitizing agents, likely improve insulin action through multiple mechanisms, including both activating lipid metabolism and reducing production of inflammatory mediators such as TNF-α (85, 89–93). Synthetic PPARα ligands, fibrates, are used to treat hyperlipidemia. These drugs appear to work predominantly through stimulation of fatty acid oxidation, though they also have antiinflammatory actions that contribute to their effects (87, 94). LXR ligands have been demonstrated to improve glucose metabolism in experimental animals (88), and it remains to be seen whether suppression of inflammation contributes to this action.

In targeting inflammation to treat insulin resistance and diabetes, it is possible that seeking inhibitors for individual inflammatory mediators may not be a maximally effective strategy, as other redundant components may be sufficient to continue to propagate inflammatory pathways. For example, targeting individual inflammatory cytokines may not be highly effective, whereas targeting the inflammatory kinases JNK and IKK generates a robust antidiabetic action because these factors integrate signals from multiple inflammatory mediators. On the other hand, if a more central process or mediator can be identified, this may provide an even more attractive target. The ER stress pathway could potentially be one such central process, in that this pathway is able to activate both JNK and IKK; thus, inhibiting the ER stress response through addition of chaperones or other mechanisms could potentially disable both of these arms of the inflammatory response and rescue insulin action (39). It has recently been demonstrated that mice in which the chaperone ORP150 is transgenically or adenovirally overexpressed exhibited reduced ER stress and improved insulin tolerance compared with controls, whereas reduction of the expression of this molecule in liver results in increased ER stress and insulin resistance (40, 41).

Origin of inflammation in obesity
While we are now aware of many of the inflammatory factors that mediate insulin resistance and have some understanding of the intracellular pathways involved, there is still much that remains

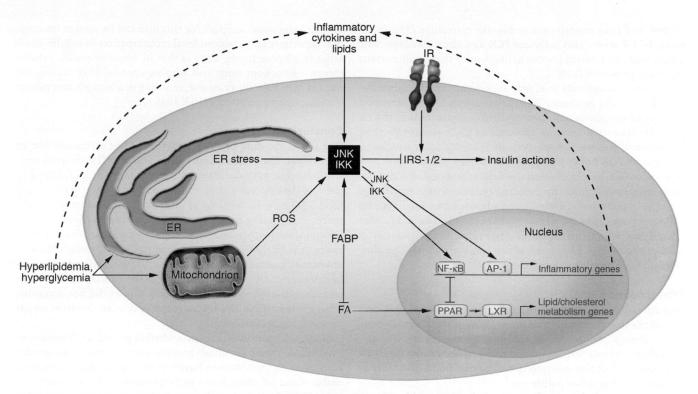

Figure 4
Model of overlapping metabolic and inflammatory signaling and sensing pathways in adipocytes or macrophages. Inflammatory pathways can be initiated by extracellular mediators such as cytokines and lipids or by intracellular stresses such as ER stress or excess ROS production by mitochondria. Signals from all of these mediators converge on inflammatory signaling pathways, including the kinases JNK and IKK. These pathways lead to the production of additional inflammatory mediators through transcriptional regulation as well as to the direct inhibition of insulin signaling. Other pathways such as those mediated through the SOCS proteins and iNOS are also involved in inflammation-mediated inhibition of insulin action. Opposing the inflammatory pathways are transcription factors from the PPAR and LXR families, which promote nutrient transport and metabolism and antagonize inflammatory activity. More proximal regulation is provided by FABPs, which likely sequester ligands of these transcription factors, thus promoting a more inflammatory environment. The absence of FABPs is antiinflammatory. The cell must strike a balance between metabolism and inflammation. In conditions of overnutrition, this becomes a particular challenge, as the very processes required for response to nutrients and nutrient utilization, such as mitochondrial oxidative metabolism and increasing protein synthesis in the ER, can induce the inflammatory response. IR, insulin receptor.

poorly understood. Crucial questions that are currently open regard the initiation of the inflammatory response. Is inflammation the primary event linking obesity with insulin resistance, or does the inflammatory response begin only after the onset of resistance to insulin? How and why does the body initiate an inflammatory response to obesity? Does obesity per se induce an inflammatory response, or is inflammation initiated as a secondary event by hyperlipidemia or hyperglycemia?

In reviewing the facts, it is fairly clear that obesity promotes states of both chronic low-grade inflammation and insulin resistance. However, even in the absence of obesity, infusion of animals with inflammatory cytokines or lipids can cause insulin resistance (54). Additionally, humans with some other chronic inflammatory conditions are at increased risk for diabetes; for example, about one-third of patients with chronic hepatitis C develop type 2 diabetes, and elevated TNF-α levels are implicated in this link (95, 96). Rheumatoid arthritis also predisposes patients to diabetes and particularly cardiovascular disease, and some evidence indicates a link between inflammatory lung diseases and risk of cardiovascular disease and diabetes (97–99). Finally, removal of inflammatory mediators or pathway components, such as TNF-α, JNK, and IKK, protects against insulin resistance in obese mouse models, and

treatment of humans with drugs that target these pathways, such as salicylates, improves insulin sensitivity (6, 11, 47, 59, 61). Thus, the available evidence strongly suggests that type 2 diabetes is an inflammatory disease and that inflammation is a primary cause of obesity-linked insulin resistance, hyperglycemia, and hyperlipidemia rather than merely a consequence (Figure 3).

But how does the inflammatory response begin? Though this question cannot currently be answered, we can suggest some reasonable speculations based on the available data. It seems likely that the inflammatory response is initiated in the adipocytes themselves, as they are the first cells affected by the development of obesity, or potentially in neighboring cells that may be affected by adipose growth. How might expanding adipocytes trigger an inflammatory response?

One mechanism that, based on newly emerging data, appears to be of central importance is the activation of inflammatory pathways by ER stress. Obesity generates conditions that increase the demand on the ER (39–41). This is particularly the case for adipose tissue, which undergoes severe changes in tissue architecture, increases in protein and lipid synthesis, and perturbations in intracellular nutrient and energy fluxes. In both cultured cells and whole animals, ER stress leads to activation of JNK and thus contributes to insulin resistance (39). Interestingly, ER stress also activates IKK and thus may represent a common mechanism for the activation of these 2 important signaling pathways (100).

A second mechanism that may be relevant in the initiation of inflammation in obesity is oxidative stress. Due to increased delivery of glucose to adipose tissue, endothelial cells in the fat pad may take up increasing amounts of glucose through their constitutive glucose transporters. Increased glucose uptake by endothelial cells in hyperglycemic conditions causes excess production of ROS in mitochondria, which inflicts oxidative damage and activates inflammatory signaling cascades inside endothelial cells (101). Endothelial injury in the adipose tissue might attract inflammatory cells such as macrophages to this site and further exacerbate the local inflammation. Hyperglycemia also stimulates ROS production in adipocytes, which leads to increased production of proinflammatory cytokines (42).

Why inflammation?

Perhaps one of the most difficult questions to answer is why obesity elicits an inflammatory response. Why, if the ability to store excess energy has been preserved through the course of evolution, does the body react in a manner that is harmful to itself? We hypothesize that this reaction is tied to the interdependency of metabolic and immune pathways.

Could obesity-induced inflammation simply be a side effect of this interaction that was never selected against since chronic obesity and its associated disorders have been so rare over time for people in their reproductive years? Perhaps the stresses of obesity are similar enough to the stresses of an infection that the body reacts to obesity as it would to an infection. For example, in both infection and obesity, intracellular stress pathways such as the JNK and IKK–NF-κB pathways are activated. Could these pathways be activated by similar mechanisms in both conditions? One mechanism that appears to be critical for initiation of this response in both situations is ER stress. During viral infection, stress pathways are activated by an excess of viral proteins in the ER (102). Similarly, the demands of obesity also result in an overloaded ER and activation of these pathways (39). Another scenario might be related to the capturing of components of the insulin-signaling pathway by microorganisms. Some pathogens activate host intracellular signaling cascades, including the PI3K-Akt pathway, which is also critical for insulin signaling (102). Perhaps in a situation in which this pathway becomes overstimulated by an increased need to take up glucose, the cell begins to interpret the signal as an indication of infection and responds by resisting the anabolic insulin signal and instead activating catabolic and inflammatory pathways.

On the other hand, perhaps the inflammatory response to obesity is not simply an undesirable byproduct, but rather a homeostatic mechanism to prevent the organism from reaching a point at which excess fat accumulation impairs mobility or otherwise diminishes fitness. Lipid storage and accumulation of fat weight require anabolic processes, exemplified by insulin action, whereas inflammation stimulates catabolism, including lipolysis from adipocytes. It is conceivable that mechanisms such as the activation of catabolism via inflammation (and hence resistance to anabolic signals) may be an attempt to keep body weight within acceptable bounds. While there is no available experimental evidence that addresses the role of low-grade inflammation in such homeostasis, some support for this idea can be seen in findings that experimentally induced local inflammation or insulin resistance in adipose tissue, such as that in adipose-specific insulin receptor knockout mice and adipose-specific TNF transgenic mice, is metabolically favorable, resulting in a lean phenotype and systemic insulin sensitivity (103, 104).

Conclusions

Our understanding of the characteristics of inflammation in obesity and the mechanisms by which this inflammation contributes to insulin resistance has been increasing rapidly over the last decade, such that we can now suggest a synthesized model (Figure 4). While it is clear that inhibition of insulin receptor signaling pathways is a central mechanism through which inflammatory and stress responses mediate insulin resistance, it is likely that other relevant pathways, molecules, and alternative mechanisms involved in this interaction have yet to be uncovered. Of particular interest is the role of alterations in mitochondrial function in diabetes. While we did not cover this topic in this article, the reader is referred to an excellent recent review for more information (105).

Another important question is whether genetic differences can predispose some individuals to inflammation-mediated insulin resistance. Several studies have reported associations between diabetes and polymorphisms in the promoters of TNF-α and IL-6 (106–109). The most well-accepted polymorphism associated with type 2 diabetes is found in the gene encoding PPARγ (110). As it is a transcription factor with some antiinflammatory activities, such as suppressing the production of TNF-α, one could imagine how altered activity of PPARγ could affect susceptibility to inflammation in obesity. Similarly, genetic variations in the FABP, JNK, IKK, or ER stress pathways or any other loci that modulate the extent of inflammation and consequently insulin resistance could define the risk of individuals for developing metabolic complication of obesity.

Finally, in addition to diabetes and cardiovascular disease, inflammation is also known to be important for linking obesity to airway inflammation and asthma, fatty liver disease, and possibly cancer and other pathologies. Understanding the mechanisms leading from obesity to inflammation will have important implications for the design of novel therapies to reduce the morbidity and mortality of obesity through the prevention of its associated chronic inflammatory disorders.

Acknowledgments

We are grateful to the members of the Hotamisligil laboratory for their contributions. Research in the Hotamisligil laboratory has been supported by the NIH, the American Diabetes Association, and the Pew and Sandler Foundations. We regret the omission of many important references by our colleagues in the field due to space limitations.

Note: References S1–S60 are available online with this article; doi:10.1172/JCI200525102DS1.

Address correspondence to: Gökhan S. Hotamisligil, Department of Genetics and Complex Diseases, Harvard School of Public Health, 665 Huntington Avenue, Boston, Massachusetts 02115, USA. Phone: (617) 432-1950; Fax: (617) 432-1941; E-mail: ghotamis@hsph.harvard.edu.

1. Khovidhunkit, W., et al. 2004. Effects of infection and inflammation on lipid and lipoprotein metabolism: mechanisms and consequences to the host [review]. *J. Lipid Res.* **45**:1169–1196.
2. Chandra, R.K. 1996. Nutrition, immunity and infection: from basic knowledge of dietary manipulation of immune responses to practical application of ameliorating suffering and improving survival. *Proc. Natl. Acad. Sci. U. S. A.* **93**:14304–14307.
3. Blackburn, G.L. 2001. Pasteur's Quadrant and malnutrition. *Nature.* **409**:397–401.
4. Cummings, D.E., and Schwartz, M.W. 2003. Genetics and pathophysiology of human obesity. *Annu. Rev. Med.* **54**:453–471.
5. Hotamisligil, G.S. 2004. Inflammation, TNFalpha, and insulin resistance. In *Diabetes mellitus: a fundamental and clinical text.* D.T.S. LeRoith and J.M. Olefsky, editors. 3rd edition. Lippincott, Williams and Wilkins. New York, New York, USA. 953–962.
6. Hotamisligil, G.S., Shargill, N.S., and Spiegelman, B.M. 1993. Adipose expression of tumor necrosis factor-alpha: direct role in obesity-linked insulin resistance. *Science.* **259**:87–91.
7. Sethi, J.K., and Hotamisligil, G.S. 1999. The role of TNF alpha in adipocyte metabolism. *Semin. Cell Dev. Biol.* **10**:19–29.
8. Hotamisligil, G.S., et al. 1995. Increased adipose tissue expression of tumor necrosis factor-alpha in human obesity and insulin resistance. *J. Clin. Invest.* **95**:2409–2415.
9. Kern, P.A., et al. 1995. The expression of tumor necrosis factor in human adipose tissue. Regulation by obesity, weight loss, and relationship to lipoprotein lipase. *J. Clin. Invest.* **95**:2111–2119.
10. Saghizadeh, M., et al. 1996. The expression of TNF alpha by human muscle. Relationship to insulin resistance. *J. Clin. Invest.* **97**:1111–1116.
11. Uysal, K.T., et al. 1997. Protection from obesity-induced insulin resistance in mice lacking TNF-alpha function. *Nature.* **389**:610–614.
12. Pickup, J.C. 2004. Inflammation and activated innate immunity in the pathogenesis of type 2 diabetes. *Diabetes Care.* **27**:813–823.
13. Dandona, P., Aljada, A., and Bandyopadhyay, A. 2004. Inflammation: the link between insulin resistance, obesity and diabetes. *Trends Immunol.* **25**:4–7.
14. Miller, C., et al. 1998. Tumor necrosis factor alpha levels in adipose tissue of lean and obese cats. *J. Nutr.* **128**(Suppl. 12):2751S–2752S.
15. Soukas, A., et al. 2000. Leptin-specific patterns of gene expression in white adipose tissue. *Genes Dev.* **14**:963–980.
16. Weisberg, S.P., et al. 2003. Obesity is associated with macrophage accumulation in adipose tissue. *J. Clin. Invest.* **112**:1796–1808. doi:10.1172/JCI200319246.
17. Xu, H., et al. 2003. Chronic inflammation in fat plays a crucial role in the development of obesity-related insulin resistance. *J. Clin. Invest.* **112**:1821–1830. doi:10.1172/JCI200319451.
18. Fernandes, G., et al. 1978. Immune response in the mutant diabetic C57BL/Ks-dt+ mouse. Discrepancies between in vitro and in vivo immunological assays. *J. Clin. Invest.* **61**:243–250.
19. Farooqi, I.S., et al. 2002. Beneficial effects of leptin on obesity, T cell hyporesponsiveness, and neuroendocrine/metabolic dysfunction of human congenital leptin deficiency. *J. Clin. Invest.* **110**:1093–1103. doi:10.1172/JCI200215693.
20. Chandra, R.K. 1980. Cell-mediated immunity in genetically obese C57BL/6J ob/ob) mice. *Am. J. Clin. Nutr.* **33**:13–16.
21. Lord, G.M., et al. 1998. Leptin modulates the T-cell immune response and reverses starvation-induced immunosuppression. *Nature.* **394**:897–901.
22. Berg, A.H., Combs, T.P., and Scherer, P.E. 2002. ACRP30/adiponectin: an adipokine regulating glucose and lipid metabolism. *Trends Endocrinol. Metab.* **13**:84–89.
23. Ouchi, N., et al. 2003. Obesity, adiponectin and vascular inflammatory disease. *Curr. Opin. Lipidol.* **14**:561–566.
24. Steppan, C.M., and Lazar, M.A. 2004. The current biology of resistin. *J. Intern. Med.* **255**:439–447.
25. Lehrke, M., et al. 2004. An inflammatory cascade leading to hyperresistinemia in humans. *PLoS Med.* **1**:e45. doi:10.1371/journal.pmed.0010045.
26. Fukuhara, A., et al. 2005. Visfatin: a protein secreted by visceral fat that mimics the effects of insulin. *Science.* **307**:426–430.
27. Makowski, L., et al. 2001. Lack of macrophage fatty-acid-binding protein aP2 protects mice deficient in apolipoprotein E against atherosclerosis. *Nat. Med.* **7**:699–705.
28. Tontonoz, P., et al. 1998. PPARgamma promotes monocyte/macrophage differentiation and uptake of oxidized LDL. *Cell.* **93**:241–252.
29. Bouloumie, A., et al. 2001. Adipocyte produces matrix metalloproteinases 2 and 9: involvement in adipose differentiation. *Diabetes.* **50**:2080–2086.
30. Cousin, B., et al. 1999. A role for preadipocytes as macrophage-like cells. *FASEB J.* **13**:305–312.
31. Charriere, G., et al. 2003. Preadipocyte conversion to macrophage. Evidence of plasticity. *J. Biol. Chem.* **278**:9850–9855.
32. White, M.F. 1997. The insulin signalling system and the IRS proteins. *Diabetologia.* **40**(Suppl. 2):S2–S17.
33. Saltiel, A.R., and Pessin, J.E. 2002. Insulin signaling pathways in time and space. *Trends Cell Biol.* **12**:65–71.
34. Yin, M.J., Yamamoto, Y., and Gaynor, R.B. 1998. The anti-inflammatory agents aspirin and salicylate inhibit the activity of I(kappa)B kinase-beta. *Nature.* **396**:77–80.
35. Hotamisligil, G.S., et al. 1996. IRS-1-mediated inhibition of insulin receptor tyrosine kinase activity in TNF-alpha- and obesity-induced insulin resistance. *Science.* **271**:665–668.
36. Aguirre, V., et al. 2000. The c-Jun NH(2)-terminal kinase promotes insulin resistance during association with insulin receptor substrate-1 and phosphorylation of Ser(307). *J. Biol. Chem.* **275**:9047–9054.
37. Aguirre, V., et al. 2002. Phosphorylation of Ser307 in insulin receptor substrate 1 blocks interactions with the insulin receptor and inhibits insulin action. *J. Biol. Chem.* **277**:1531–1537.
38. Paz, K., et al. 1997. A molecular basis for insulin resistance. Elevated serine/threonine phosphorylation of IRS-1 and IRS-2 inhibits their binding to the juxtamembrane region of the insulin receptor and impairs their ability to undergo insulin-induced tyrosine phosphorylation. *J. Biol. Chem.* **272**:29911–29918.
39. Ozcan, U., et al. 2004. Endoplasmic reticulum stress links obesity, insulin action, and type 2 diabetes. *Science.* **306**:457–461.
40. Nakatani, Y., et al. 2005. Involvement of endoplasmic reticulum stress in insulin resistance and diabetes. *J. Biol. Chem.* **280**:847–851.
41. Ozawa, K., et al. 2005. The endoplasmic reticulum chaperone improves insulin resistance in type 2 diabetes. *Diabetes.* **54**:657–663.
42. Lin, Y., et al. 2005. The hyperglycemia-induced inflammatory response in adipocytes: the role of reactive oxygen species. *J. Biol. Chem.* **280**:4617–4626.
43. Furukawa, S., et al. 2004. Increased oxidative stress in obesity and its impact on metabolic syndrome. *J. Clin. Invest.* **114**:1752–1761. doi:10.1172/JCI200421625.
44. Zick, Y. 2003. Role of Ser/Thr kinases in the uncoupling of insulin signaling. *Int. J. Obes. Relat. Metab. Disord.* **27**(Suppl. 3):S56–S60.
45. Medzhitov, R. 2001. Toll-like receptors and innate immunity. *Nat. Rev. Immunol.* **1**:135–145.
46. Davis, R.J. 2000. Signal transduction by the JNK group of MAP kinases. *Cell.* **103**:239–252.
47. Hirosumi, J., et al. 2002. A central role for JNK in obesity and insulin resistance. *Nature.* **420**:333–336.
48. Gao, Z., et al. 2004. Inhibition of insulin sensitivity by free fatty acids requires activation of multiple serine kinases in 3T3-L1 adipocytes. *Mol. Endocrinol.* **18**:2024–2034.
49. Nakatani, Y., et al. 2004. Modulation of the JNK pathway in liver affects insulin resistance status. *J. Biol. Chem.* **279**:45803–45809.
50. Waeber, G., et al. 2000. The gene MAPK8IP1, encoding islet-brain-1, is a candidate for type 2 diabetes. *Nat. Genet.* **24**:291–295.
51. Jaeschke, A., Czech, M.P., and Davis, R.J. 2004. An essential role of the JIP1 scaffold protein for JNK activation in adipose tissue. *Genes Dev.* **18**:1976–1980.
52. Ricci, R., et al. 2004. Requirement of JNK2 for scavenger receptor A-mediated foam cell formation in atherogenesis. *Science.* **306**:1558–1561.
53. Kaneto, H., et al. 2004. Possible novel therapy for diabetes with cell-permeable JNK-inhibitory peptide. *Nat. Med.* **10**:1128–1132.
54. Yu, C., et al. 2002. Mechanism by which fatty acids inhibit insulin activation of insulin receptor substrate-1 (IRS-1)-associated phosphatidylinositol 3-kinase activity in muscle. *J. Biol. Chem.* **277**:50230–50236.
55. Perseghin, G., Petersen, K., and Shulman, G.I. 2003. Cellular mechanism of insulin resistance: potential links with inflammation. *Int. J. Obes. Relat. Metab. Disord.* **27**(Suppl. 3):S6–S11.
56. Schmitz-Peiffer, C. 2002. Protein kinase C and lipid-induced insulin resistance in skeletal muscle. *Ann. N. Y. Acad. Sci.* **967**:146–157.
57. Gao, Z., et al. 2002. Serine phosphorylation of insulin receptor substrate 1 by inhibitor kappa B kinase complex. *J. Biol. Chem.* **277**:48115–48121.
58. Shoelson, S.E., Lee, J., and Yuan, M. 2003. Inflammation and the IKK beta/I kappa B/NF-kappa B axis in obesity- and diet-induced insulin resistance. *Int. J. Obes. Relat. Metab. Disord.* **27**(Suppl. 3):S49–S52.
59. Yuan, M., et al. 2001. Reversal of obesity- and diet-induced insulin resistance with salicylates or targeted disruption of Ikkbeta. *Science.* **293**:1673–1677.
60. Kim, J.K., et al. 2001. Prevention of fat-induced insulin resistance by salicylate. *J. Clin. Invest.* **108**:437–446. doi:10.1172/JCI200111559.
61. Hundal, R.S., et al. 2002. Mechanism by which high-dose aspirin improves glucose metabolism in type 2 diabetes. *J. Clin. Invest.* **109**:1321–1326. doi:10.1172/JCI200214955.
62. Cai, D., et al. 2005. Local and systemic insulin resistance resulting from hepatic activation of IKK-beta and NF-kappaB. *Nat. Med.* **11**:183–190.
63. Arkan, M.C., et al. 2005. IKK-beta links inflammation to obesity-induced insulin resistance. *Nat. Med.* **11**:191–198.
64. Rohl, M., et al. 2004. Conditional disruption of IkappaB kinase 2 fails to prevent obesity-induced insulin resistance. *J. Clin. Invest.* **113**:474–481. doi:10.1172/JCI200418712.
65. Rui, L., et al. 2002. SOCS-1 and SOCS-3 block insulin signaling by ubiquitin-mediated degradation of IRS1 and IRS2. *J. Biol. Chem.* **277**:42394–42398.
66. Mooney, R.A., et al. 2001. Suppressors of cytokine signaling-1 and -6 associate with and inhibit the insulin receptor. A potential mechanism for cytokine-mediated insulin resistance. *J. Biol. Chem.* **276**:25889–25893.
67. Emanuelli, B., et al. 2001. SOCS-3 inhibits insulin signaling and is up-regulated in response to tumor necrosis factor-alpha in the adipose tissue of obese mice. *J. Biol. Chem.* **276**:47944–47949.
68. Ueki, K., Kondo, T., and Kahn, C.R. 2004. Suppressor of cytokine signaling 1 (SOCS-1) and SOCS-3 cause insulin resistance through inhibition of tyrosine phosphorylation of insulin receptor substrate

proteins by discrete mechanisms. *Mol. Cell Biol.* **24**:5434–5446.

69. Mori, H., et al. 2004. Socs3 deficiency in the brain elevates leptin sensitivity and confers resistance to diet-induced obesity. *Nat. Med.* **10**:739–743.

70. Howard, J.K., et al. 2004. Enhanced leptin sensitivity and attenuation of diet-induced obesity in mice with haploinsufficiency of Socs3. *Nat. Med.* **10**:734–738.

71. Shimabukuro, M., et al. 1997. Role of nitric oxide in obesity-induced beta cell disease. *J. Clin. Invest.* **100**:290–295.

72. Perreault, M., and Marette, A. 2001. Targeted disruption of inducible nitric oxide synthase protects against obesity-linked insulin resistance in muscle. *Nat. Med.* **7**:1138–1143.

73. Seo, J.B., et al. 2004. Activated liver X receptors stimulate adipocyte differentiation through induction of peroxisome proliferator-activated receptor gamma expression. *Mol. Cell. Biol.* **24**:3430–3444.

74. Moller, D.E., and Berger, J.P. 2003. Role of PPARs in the regulation of obesity-related insulin sensitivity and inflammation. *Int. J. Obes. Relat. Metab. Disord.* **27**(Suppl. 3):S17–S21.

75. Joseph, S.B., et al. 2003. Reciprocal regulation of inflammation and lipid metabolism by liver X receptors. *Nat. Med.* **9**:213–219.

76. Jiang, C., Ting, A.T., and Seed, B. 1998. PPAR-gamma agonists inhibit production of monocyte inflammatory cytokines. *Nature.* **391**:82–86.

77. Chawla, A., et al. 2001. PPAR-gamma dependent and independent effects on macrophage-gene expression in lipid metabolism and inflammation. *Nat. Med.* **7**:48–52.

78. Chawla, A., et al. 2001. A PPAR gamma-LXR-ABCA1 pathway in macrophages is involved in cholesterol efflux and atherogenesis. *Mol. Cell.* **7**:161–171.

79. Daynes, R.A., and Jones, D.C. 2002. Emerging roles of PPARs in inflammation and immunity. *Nat. Rev. Immunol.* **2**:748–759.

80. Castrillo, A., et al. 2003. Crosstalk between LXR and toll-like receptor signaling mediates bacterial and viral antagonism of cholesterol metabolism. *Mol. Cell.* **12**:805–816.

81. Bjorkbacka, H., et al. 2004. Reduced atherosclerosis in MyD88-null mice links elevated serum cholesterol levels to activation of innate immunity signaling pathways. *Nat. Med.* **10**:416–421.

82. Joseph, S.B., et al. 2004. LXR-dependent gene expression is important for macrophage survival and the innate immune response. *Cell.* **119**:299–309.

83. Lee, C.H., et al. 2003. Transcriptional repression of atherogenic inflammation: modulation by PPARdelta. *Science.* **302**:453–457.

84. Maeda, K., et al. 2005. Adipocyte/macrophage fatty acid binding proteins control integrated metabolic responses in obesity and diabetes. *Cell Metabolism.* **1**:107–119.

85. Spiegelman, B.M. 1998. PPAR-gamma: adipogenic regulator and thiazolidinedione receptor. *Diabetes.* **47**:507–514.

86. Makowski, L., and Hotamisligil, G.S. 2004. Fatty acid binding proteins — the evolutionary crossroads of inflammatory and metabolic responses. *J. Nutr.* **134**:2464S–2468S.

87. Lee, C.H., Olson, P., and Evans, R.M. 2003. Minireview: lipid metabolism, metabolic diseases, and peroxisome proliferator-activated receptors. *Endocrinology.* **144**:2201–2207.

88. Laffitte, B.A., et al. 2003. Activation of liver X receptor improves glucose tolerance through coordinate regulation of glucose metabolism in liver and adipose tissue. *Proc. Natl. Acad. Sci. U. S. A.* **100**:5419–5424.

89. Wellen, K.E., et al. 2004. Interaction of tumor necrosis factor-alpha- and thiazolidinedione-regulated pathways in obesity. *Endocrinology.* **145**:2214–2220.

90. Peraldi, P., Xu, M., and Spiegelman, B.M. 1997. Thiazolidinediones block tumor necrosis factor-alpha-induced inhibition of insulin signaling. *J. Clin. Invest.* **100**:1863–1869.

91. Ruan, H., Pownall, H.J., and Lodish, H.F. 2003. Troglitazone antagonizes tumor necrosis factor-alpha-induced reprogramming of adipocyte gene expression by inhibiting the transcriptional regulatory functions of NF-kappaB. *J. Biol. Chem.* **278**:28181–28192.

92. Miles, P.D., et al. 1997. TNF-alpha-induced insulin resistance in vivo and its prevention by troglitazone. *Diabetes.* **46**:1678–1683.

93. Lehmann, J.M., et al. 1995. An antidiabetic thiazolidinedione is a high affinity ligand for peroxisome proliferator-activated receptor gamma (PPAR gamma). *J. Biol. Chem.* **270**:12953–12956.

94. Staels, B., et al. 1998. Activation of human aortic smooth-muscle cells is inhibited by PPARalpha but not by PPARgamma activators. *Nature.* **393**:790–793.

95. Knobler, H., et al. 2003. Tumor necrosis factor-alpha-induced insulin resistance may mediate the hepatitis C virus-diabetes association. *Am. J. Gastroenterol.* **98**:2751–2756.

96. Bahtiyar, G., et al. 2004. Association of diabetes and hepatitis C infection: epidemiologic evidence and pathophysiologic insights. *Curr. Diab. Rep.* **4**:194–198.

97. Iribarren, C., Tolstykh, I.V., and Eisner, M.D. 2004. Are patients with asthma at increased risk of coronary heart disease? *Int. J. Epidemiol.* **33**:743–748.

98. Rana, J.S., et al. 2004. Chronic obstructive pulmonary disease, asthma, and risk of type 2 diabetes in women. *Diabetes Care.* **27**:2478–2484.

99. Sattar, N., et al. 2003. Explaining how "high-grade" systemic inflammation accelerates vascular risk in rheumatoid arthritis. *Circulation.* **108**:2957–2963.

100. Hung, J.H., et al. 2004. Endoplasmic reticulum stress stimulates the expression of cyclooxygenase-2 through activation of NF-kappaB and pp38 mitogen-activated protein kinase. *J. Biol. Chem.* **279**:46384–46392.

101. Brownlee, M. 2001. Biochemistry and molecular cell biology of diabetic complications. *Nature.* **414**:813–820.

102. Hatada, E.N., Krappmann, D., and Scheidereit, C. 2000. NF-kappaB and the innate immune response. *Curr. Opin. Immunol.* **12**:52–58.

103. Bluher, M., et al. 2002. Adipose tissue selective insulin receptor knockout protects against obesity and obesity-related glucose intolerance. *Dev. Cell.* **3**:25–38.

104. Xu, H., et al. 2002. Exclusive action of transmembrane TNF alpha in adipose tissue leads to reduced adipose mass and local but not systemic insulin resistance. *Endocrinology.* **143**:1502–1511.

105. Lowell, B.B., and Shulman, G.I. 2005. Mitochondrial dysfunction and type 2 diabetes. *Science.* **307**:384–387.

106. Vozarova, B., et al. 2003. The interleukin-6 (-174) G/C promoter polymorphism is associated with type-2 diabetes mellitus in Native Americans and Caucasians. *Hum. Genet.* **112**:409–413.

107. Dalziel, B., et al. 2002. Association of the TNF-alpha -308 G/A promoter polymorphism with insulin resistance in obesity. *Obes. Res.* **10**:401–407.

108. Costa, A., et al. 2003. Lower rate of tumor necrosis factor-alpha -863A allele and higher concentration of tumor necrosis factor-alpha receptor 2 in first-degree relatives of subjects with type 2 diabetes. *Metabolism.* **52**:1068–1071.

109. Furuta, M., et al. 2002. Relationship of the tumor necrosis factor-alpha -308 A/G promoter polymorphism with insulin sensitivity and abdominal fat distribution in Japanese patients with type 2 diabetes mellitus. *Diabetes Res. Clin. Pract.* **56**:141–145.

110. Florez, J.C., Hirschhorn, J., and Altshuler, D. 2003. The inherited basis of diabetes mellitus: implications for the genetic analysis of complex traits. *Annu. Rev. Genomics Hum. Genet.* **4**:257–291.

Adiponectin and adiponectin receptors in insulin resistance, diabetes, and the metabolic syndrome

Takashi Kadowaki,[1,2] Toshimasa Yamauchi,[1,3] Naoto Kubota,[1,2,3]
Kazuo Hara,[1,2] Kohjiro Ueki,[1] and Kazuyuki Tobe[1]

[1]Department of Metabolic Diseases, Graduate School of Medicine, University of Tokyo, Tokyo, Japan. [2]Division of Applied Nutrition,
National Institute of Health and Nutrition, Tokyo, Japan. [3]Department of Integrated Molecular Science on Metabolic Diseases,
Graduate School of Medicine, University of Tokyo, Tokyo, Japan.

Adiponectin is an adipokine that is specifically and abundantly expressed in adipose tissue and directly sensitizes the body to insulin. Hypoadiponectinemia, caused by interactions of genetic factors such as SNPs in the *Adiponectin* gene and environmental factors causing obesity, appears to play an important causal role in insulin resistance, type 2 diabetes, and the metabolic syndrome, which are linked to obesity. The adiponectin receptors, AdipoR1 and AdipoR2, which mediate the antidiabetic metabolic actions of adiponectin, have been cloned and are downregulated in obesity-linked insulin resistance. Upregulation of adiponectin is a partial cause of the insulin-sensitizing and antidiabetic actions of thiazolidinediones. Therefore, adiponectin and adiponectin receptors represent potential versatile therapeutic targets to combat obesity-linked diseases characterized by insulin resistance. This Review describes the pathophysiology of adiponectin and adiponectin receptors in insulin resistance, diabetes, and the metabolic syndrome.

The prevalence of obesity has increased dramatically in recent years (1, 2). It is commonly associated with type 2 diabetes, coronary artery disease, and hypertension, and the coexistence of these diseases has been termed the metabolic syndrome (3–7). Insulin resistance is a key feature of these diseases and is defined as a state that requires more insulin to obtain the biological effects achieved by a lower amount of insulin in the normal state. Thus, any defects in the insulin signaling cascade can cause insulin resistance. Insulin stimulates a signaling network composed of a number of molecules, initiating the activation of insulin receptor tyrosine kinase and phosphorylation of the insulin receptor substrate (IRS) proteins (e.g., IRS-1 and IRS-2) (8). Among several components of the network, the signaling axis of IRS proteins and PI3K, which activates downstream serine/threonine kinases including Akt, regulates most of the metabolic actions of insulin, such as suppression of hepatic glucose production and activation of glucose transport in muscle and adipocytes (9). It is known that this pathway is impaired at the multiple steps through alterations in the protein levels and activities of the signaling molecules, enzymes, and transcription factors in insulin resistance caused by obesity, a state of increased adiposity (9).

White adipose tissue (WAT) is a major site of energy storage and is important for energy homeostasis: it stores energy in the form of triglycerides during nutritional abundance and releases it as FFAs during nutritional deprivation (10, 11). While WAT provides a survival advantage in times of starvation, excess WAT is now linked to obesity-related health problems in the current nutritionally rich environment. Regulated by multiple hormonal signals, nuclear hormone receptors (12, 13), and the CNS (14),

WAT has been increasingly recognized as an important endocrine organ that secretes a number of biologically active "adipokines" (15–19). Some of these adipokines have been shown to directly or indirectly affect insulin sensitivity through modulation of insulin signaling and the molecules involved in glucose and lipid metabolism (20). Of these adipokines, adiponectin has recently attracted much attention because of its antidiabetic and antiatherogenic effects and is expected to be a novel therapeutic tool for diabetes and the metabolic syndrome (21). Indeed, a decrease in the circulating levels of adiponectin by genetic and environmental factors has been shown to contribute to the development of diabetes and the metabolic syndrome. The thiazolidinedione (TZD) class of antidiabetic drugs, which also have pleiotropic effects on cardiovascular diseases and lipid metabolism, is known to exert its effects partly through increasing the levels of the active form of adiponectin, as described below.

In this Review, we describe recent progress in research on the pathophysiological role of adiponectin and adiponectin receptors in insulin resistance, type 2 diabetes, and the metabolic syndrome. Since the length of this Review is limited, we recommend that readers also consult other recent reviews on adiponectin research (21–23).

Association of hypoadiponectinemia with insulin resistance, diabetes, and the metabolic syndrome

Adiponectin, also termed Acrp30 (24), AdipoQ (25), apM1 (26), or GBP28 (27), was originally identified independently by 4 groups using different approaches. The *Adiponectin* gene encodes a secreted protein expressed exclusively in both WAT *and* brown adipose tissue. Adiponectin has a carboxyl-terminal globular domain and an amino-terminal collagen domain and is structurally similar to complement 1q (28, 29), which belongs to a family of proteins that form characteristic multimers (30, 31). Adiponectin exists in a wide range of multimer complexes in plasma and combines via its collagen domain to create 3 major oligomeric forms: a low–molecular weight (LMW) trimer, a middle–molecular weight (MMW) hexamer, and high–molecular weight (HMW) 12- to 18-mer adiponectin (32, 33). In contrast to the expression of adipokines such

Nonstandard abbreviations used: AdipoR, adiponectin receptor; AMPK, AMP-activated protein kinase; APPL, adaptor protein containing pleckstrin homology domain, phosphotyrosine-binding domain, and leucine zipper motif; GPCR, G protein-coupled receptor; HMW, high molecular weight; IRS, insulin receptor substrate; LMW, low molecular weight; MMW, middle molecular weight; PR-5, pathogenesis related-5; TZD, thiazolidinedione; WAT, white adipose tissue.

Conflict of interest: The authors have declared that no conflict of interest exists.

Citation for this article: *J. Clin. Invest.* **116**:1784–1792 (2006). doi:10.1172/JCI29126.

Table 1

Physiological and pathophysiological conditions and treatment modalities associated with either decrease or increase in plasma adiponectin levels

Pathophysiological conditions or treatment modality	Ref.
Hypoadiponectinemia is associated with:	
Genetic variation in *Adiponectin* gene	33, 72–77
Obesity	35–37
Insulin resistance	37, 38
Type 2 diabetes	37, 39–44
Metabolic syndrome	45
Dyslipidemia	45
Cardiovascular disease	46, 47
Hypertension	48, 49
Sex hormones (androgen, testosterone)	51, 52
Oxidative stress	57
Carbohydrate-rich diet	56
Increase in adiponectin levels is observed following:	
Administration of:	
TZDs	58, 79, 104–108
Angiotensin II receptor blocker (ARB)	110
Angiotensin-converting enzyme inhibitors (ACEIs)	111
Heart failure	114
Renal failure	115
Weight loss	112, 113
Dietary factors:	
Soy protein	53
Oils	54, 55

as TNF-α and resistin, which cause insulin resistance, adiponectin expression is reduced in obese, insulin-resistant rodent models (25). Plasma adiponectin levels are also decreased in an obese rhesus monkey model that frequently develops type 2 diabetes (34). Importantly, a decrease in plasma adiponectin levels preceded the onset of diabetes in these animals, in parallel with the observation of decreased insulin sensitivity (34). Plasma adiponectin levels have also been reported to be reduced in obese humans, particularly those with visceral obesity, and to correlate inversely with insulin resistance (35–38) (Table 1). Prospective and longitudinal studies (37, 39–44) have shown that lower adiponectin levels are associated with a higher incidence of diabetes. Adiponectin, but not inflammatory markers such as C-reactive protein and IL-6, has been shown to be significantly related to the development of type 2 diabetes in Pima Indians (44). Hypoadiponectinemia has also been demonstrated to be independently associated with the metabolic syndrome — indeed, more strongly than are any other inflammatory markers (45). Reduced plasma adiponectin levels are also commonly observed in a variety of states frequently associated with insulin resistance, such as cardiovascular disease (46, 47) and hypertension (48, 49).

How is the level of plasma adiponectin physiologically regulated? There is a sexual dimorphism in the circulating levels of adiponectin. Indeed, female humans and rodents have higher plasma adiponectin levels than males, suggesting that sexual hormones regulate the production of adiponectin, although it is controversial how these hormones, such as estrogen and testosterone, are involved in the regulation of plasma adiponectin level (50–52). Nevertheless, this may partly account for the fact that females are more sensitive to insulin than males. Some dietary factors, such as soy protein (53), fish oils (54), and linoleic acid (55), are also suggested to increase plasma adiponectin levels, which is consistent with the fact that intake of these factors is thought to have a protective effect on the development of diabetes. On the other hand, a carbohydrate-rich diet appears to decrease plasma adiponectin level (56). Oxidative stress has also been suggested to inhibit the expression of adiponectin (57). Although the mechanism underlying this regulation is unclear, this may contribute to the decrease in plasma adiponectin in obesity, which is associated with increased oxidative stress in adipose tissue. Thus, the plasma adiponectin level is affected by multiple factors, including gender, aging, and lifestyle.

Discovery of the insulin-sensitizing action of adiponectin

The insulin-sensitizing effect of adiponectin was first identified by 3 independent groups in 2001 (58–60). We assessed whether adiponectin was able to improve insulin resistance in KKAy mice (KK mice overexpressing the agouti protein), as a model of the metabolic syndrome and type 2 diabetes linked to obesity. Plasma adiponectin levels were decreased in KKAy mice fed a high-fat diet. Replenishment of adiponectin significantly ameliorated high-fat diet–induced insulin resistance and hypertriglyceridemia, which led us to propose that adiponectin is an insulin-sensitizing adipokine (58). These data also strongly suggested that the high-fat diet–induced, obesity-linked decrease in adiponectin level is causally involved in obesity-linked insulin resistance and the metabolic syndrome. Scherer and colleagues reported that an acute increase in the level of circulating adiponectin triggers a transient decrease in basal glucose level by inhibiting both the expression of hepatic gluconeogenic enzymes and the rate of endogenous glucose production in both wild-type and type 2 diabetic mice, and they proposed that adiponectin sensitizes the body to insulin (59). A truncated form of adiponectin that includes the globular domain cleaved proteolytically from full-length adiponectin has been reported to exist in plasma, although in very small amounts (60). Lodish and colleagues reported that a proteolytic cleavage product of adiponectin, which structurally resembles globular adiponectin, increases fatty-acid oxidation in muscle, decreases plasma glucose, and causes weight loss in mice (60).

Subsequently, the chronic effects of adiponectin on insulin resistance in vivo were investigated by generation of adiponectin transgenic mice (61, 62) or adiponectin-deficient mice (63–66). Globular adiponectin transgenic *ob/ob* mice showed partial amelioration of insulin resistance and diabetes (61). Full-length adiponectin transgenic mice showed suppression of insulin-mediated endogenous glucose production (62). Our adiponectin-deficient mice showed mild insulin resistance with glucose intolerance while on a standard diet (63). The adiponectin-deficient mice examined by Maeda et al. exhibited a near-normal insulin sensitivity when fed a standard laboratory diet but developed severe insulin resistance, especially in skeletal muscle, in as few as 2 weeks on a high-fat, high-sucrose diet (64). Ma et al. reported that adiponectin-deficient mice displayed increased fatty-acid oxidation in skeletal muscle but showed no effect on either insulin sensitivity or glucose tolerance whether on a standard or a high-fat diet (65). Scherer's group reported that adiponectin-deficient mice showed mild insulin resistance in the liver while on a standard diet (66), and their phenotype was very similar to those of our adiponectin-deficient mice. On the other hand, some discrepancies in phenotypes that

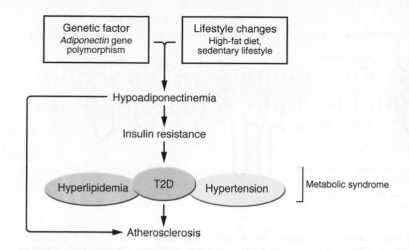

Figure 1
Adiponectin hypothesis for insulin resistance, the metabolic syndrome, and atherosclerosis. Reduced adiponectin levels can be caused by interactions of genetic factors such as SNP 276 in the *Adiponectin* gene itself and environmental factors, i.e., lifestyle changes that cause obesity, such as a high-fat diet and sedentary lifestyle. This reduction in adiponectin levels in turn appears to play an important causal role in the development of insulin resistance, type 2 diabetes (T2D), and metabolic disease, thereby indirectly causing atherosclerosis. Moreover, reduced adiponectin levels also directly play a causal role in the development of atherosclerosis.

have been described among adiponectin-deficient mice are most likely due to differences in genetic background. Adiponectin-deficient mice exhibited other features of the metabolic syndrome, such as hyperlipidemia and hypertension (48, 63).

With respect to the molecular mechanisms underlying the insulin-sensitizing action of adiponectin, we found that full-length adiponectin stimulated AMP-activated protein kinase (AMPK) phosphorylation and activation in the liver, while globular adiponectin did so in both skeletal muscle and the liver (67). Blocking AMPK activation by use of a dominant-negative mutant inhibited these effects of full-length or globular adiponectin, indicating that stimulation of glucose utilization and fatty-acid combustion by adiponectin occurs through activation of AMPK (67). Thus, an adipocyte-derived antidiabetic hormone, adiponectin, activates AMPK, thereby directly regulating glucose metabolism and insulin sensitivity (67). These data also suggested that there may be 2 distinct receptors for adiponectin in the liver and skeletal muscle, with different binding affinities for globular and full-length adiponectin. Lodish, Ruderman, and colleagues also showed that the adiponectin globular domain could enhance muscle fat oxidation and glucose transport via AMPK activation and acetyl-CoA carboxylase inhibition (68). Scherer et al. reported that in adiponectin transgenic mice (62), reduced expression of gluconeogenic enzymes such as phosphoenolpyruvate carboxylase and glucose-6-phosphatase is associated with elevated phosphorylation of hepatic AMPK, which may account for inhibition of endogenous glucose production by adiponectin (59, 67, 69).

Adiponectin also increased fatty-acid combustion and energy consumption, in part via PPARα activation, which led to decreased triglyceride content in the liver and skeletal muscle, and thereby a coordinated increase of in vivo insulin sensitivity (61).

Adiponectin gene SNPs in human insulin resistance and type 2 diabetes

The *Adiponectin* gene is located on chromosome 3q27, which has been reported to be linked to type 2 diabetes and the metabolic syndrome (70–72). Therefore, the *Adiponectin* gene appears to be a promising candidate susceptibility gene for type 2 diabetes. Among the SNPs in the *Adiponectin* gene, 1 SNP located 276 bp downstream of the translational start site (SNP 276) was concomitantly associated with decreased plasma adiponectin level, greater insulin resistance, and an increased risk of type 2 diabetes (73). The subjects, both of whose 2 alleles of SNP 276 are the G (G/G

genotype), had an approximately doubled risk for developing type 2 diabetes as compared with those with the T/T genotype (73). It is noteworthy that more than 40% of Japanese individuals have the "at-risk" G/G genotype, which makes subjects prone to genetically decreased adiponectin levels and thus susceptible to type 2 diabetes (73). Subjects with an I164T missense mutation in the globular domain of adiponectin had significantly lower plasma adiponectin levels than those without, independently of BMI (74).

Similar associations of the *Adiponectin* gene with susceptibility to type 2 diabetes have also been reported in other ethnic groups. In white German and North American subjects, SNP 276, either independently or as a haplotype together with SNP 45 in exon 2, was shown to be associated with obesity and insulin resistance (75, 76). In white French subjects, 2 SNPs in the promoter region of the *Adiponectin* gene, SNP 11377 and SNP 11391, were significantly associated with hypoadiponectinemia and type 2 diabetes (72). Taken together, these data strongly support the hypothesis that adiponectin plays a pivotal role in the pathogenesis of type 2 diabetes. A recent haplotype analysis based on a dense SNP map in a large sample clarified a 2-block linkage disequilibrium structure of the *Adiponectin* gene, the first block including the promoter SNPs and the second spanning the exons and introns (77). It is noteworthy that neither block has more than 1 SNP significantly associated with the plasma adiponectin level. The haplotypes in the first block were associated with increased adiponectin level, whereas the haplotypes in the second block were associated with decreased adiponectin level. This result indicated the existence of at least 2 causal haplotypes or SNPs in the *Adiponectin* gene.

Based on the significant body of evidence discussed above, we have proposed the "adiponectin hypothesis," in which reduced plasma adiponectin levels caused by interactions between genetic factors, such as SNPs in the *Adiponectin* gene itself, and environmental factors causing obesity, such as a sedentary lifestyle, may play a crucial role in the development of insulin resistance, type 2 diabetes, and the metabolic syndrome (21) (Figure 1).

Role of HMW adiponectin in insulin resistance and type 2 diabetes

Several observations support the hypothesis that HMW adiponectin is the more active form of the protein and has a more relevant role in insulin sensitivity and in protecting against diabetes. First, rare mutations — G84R and G90S — in the collagen domain are closely associated with type 2 diabetes (33, 73, 78). Subjects with

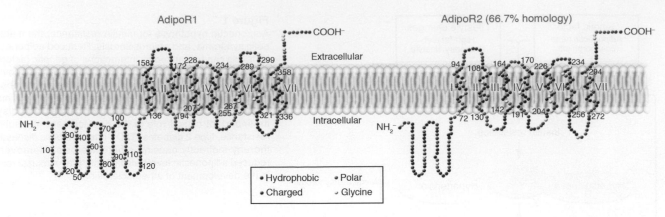

Figure 2
Structure of adiponectin receptors. AdipoR1 and AdipoR2 (66.7% amino acid identity with AdipoR1) are predicted to contain 7 transmembrane domains but are structurally and topologically distinct from GPCRs. Redrawn with permission from *Endocrine Reviews* (21); copyright 2005, The Endocrine Society.

either of these 2 mutations have extremely low levels of HMW adiponectin (Table 1). Moreover, the 2 mutant adiponectins recombinantly expressed in NIH-3T3 fibroblasts were not able to form the HMW form of adiponectin. Second, increases in the ratio of plasma HMW adiponectin levels to total adiponectin levels correlate with improvement in insulin sensitivity during treatment with an insulin-sensitizing drug, TZD, in both mice and human diabetic patients, whereas increases in total serum adiponectin levels do not show good correlations with improvement in insulin sensitivity during treatment with TZD at the individual level (79). Third, the level of plasma HMW adiponectin was reported to be associated with parameters related to glucose homeostasis in a cohort study (80). It is noteworthy that the ratio of plasma HMW adiponectin to total adiponectin correlated more significantly with glucose and insulin levels than did the total adiponectin level (80), suggesting that alterations in plasma HMW adiponectin level may be more relevant to the prediction of insulin resistance than are total plasma adiponectin alterations. Consistent with this, levels of total adiponectin, HMW adiponectin, LMW adiponectin, and the HMW-to-total adiponectin ratio all correlated significantly with key features of central obesity and the insulin-stimulated glucose disposal rate (81). However, HMW adiponectin levels, not total adiponectin levels, are primarily responsible for these relationships, suggesting that measurement of the HMW adiponectin level may be superior to measurement of total adiponectin (81). Using an ELISA system for selective measurement of HMW adiponectin (82), we also found HMW adiponectin and the HMW-to-total adiponectin ratio to have significantly better power for the prediction of insulin resistance and the metabolic syndrome in humans (83). Thus, HMW adiponectin level may be the superior biomarker for insulin resistance, the metabolic syndrome, and type 2 diabetes.

Cloning, function, and regulation of adiponectin receptors

In order to further determine the pathophysiological significance and molecular mechanism of adiponectin action, we isolated cDNA for adiponectin receptors mediating the antidiabetic effects of adiponectin from a human skeletal muscle cDNA library by screening for globular adiponectin binding (84). The cDNA ana-

lyzed encoded a protein designated human adiponectin receptor 1 (AdipoR1) (84). This protein is structurally conserved from yeast to humans (especially in the 7 transmembrane domains). Interestingly, the yeast homologue (YOL002c) plays a key role in metabolic pathways that regulate lipid metabolism, such as fatty-acid oxidation (85). Since at that time there may have been 2 distinct adiponectin receptors, as was described above (67), we searched for a homologous gene in the human and mouse databases. We found only 1 gene that was significantly homologous (67% amino acid identity) with AdipoR1, which was termed AdipoR2 (84). AdipoR1 is ubiquitously expressed, including abundant expression in skeletal muscle, whereas AdipoR2 is most abundantly expressed in the mouse liver. AdipoR1 and AdipoR2 appear to be integral membrane proteins; the N-terminus is internal and the C-terminus is external — opposite to the topology of all other reported G protein–coupled receptors (GPCRs) (84) (Figure 2). Expression of AdipoR1 and AdipoR2 or suppression of AdipoR1 and AdipoR2 expression supports our conclusion that AdipoR1 and AdipoR2 serve as receptors for globular and full-length adiponectin and mediate increased AMPK, PPARα ligand activities, fatty-acid oxidation, and glucose uptake by adiponectin (Figure 3).

Lodish's group reported that T-cadherin was capable of binding adiponectin in C2C12 myoblasts; however, T-cadherin was not expressed in hepatocytes or the liver (86), the most important target organ (66, 69, 87). Moreover, T-cadherin by itself was thought to have no effect on adiponectin cellular signaling or function, since T-cadherin is without an intracellular domain. These data raised the possibility that T-cadherin may be one of the adiponectin-binding proteins.

Most recently, a 2-hybrid study revealed that the C-terminal extracellular domain of AdipoR1 interacted with adiponectin, whereas the N-terminal cytoplasmic domain of AdipoR1 interacted with APPL (adaptor protein containing pleckstrin homology domain, phosphotyrosine-binding domain, and leucine zipper motif) (88). Moreover, interaction of APPL with AdipoR1 in mammalian cells was stimulated by adiponectin binding, and this interaction played important roles in adiponectin signaling and adiponectin-mediated downstream events such as lipid oxidation and glucose uptake. These data clearly indicated that adiponectin receptors directly interacted with adiponectin and mediated adiponectin effects. Furthermore,

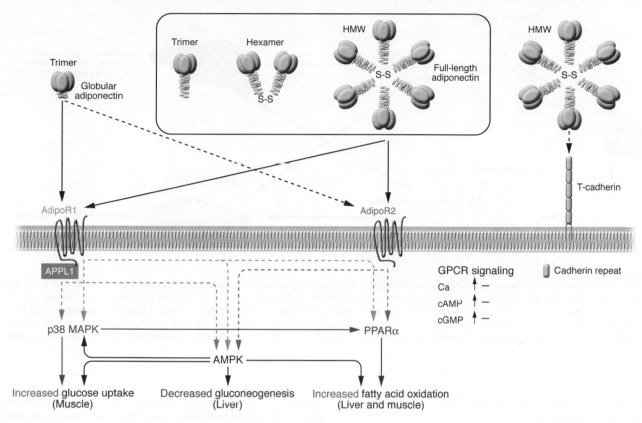

Figure 3

Signal transduction by adiponectin receptors. Globular adiponectin exists as a trimer, whereas full-length adiponectin exists as at least 3 species of multimers: an LMW trimer, an MMW hexamer, and an HMW multimor. Suppression of AdipoR1 by RNA interference markedly reduces globular adiponectin binding, whereas suppression of AdipoR2 by RNA interference largely reduces full-length adiponectin–specific binding (21, 84). The dotted line between AdipoR2 and globular adiponectin reflects that AdipoR2 is a relatively low-affinity receptor for globular adiponectin. AdipoR1 and AdipoR2 do not seem to be coupled with G proteins, since overexpression of AdipoR1/R2 has little effect on cAMP, cGMP, and intracellular calcium levels, but instead these receptors activate unique sets of signaling molecules such as PPARα, AMPK, and p38 MAPK. In C2C12 myocytes overexpressing AdipoR1/R2, adiponectin stimulates PPARα, AMPK, and p38 MAPK activation, glucose uptake, and fatty-acid oxidation (84). Suppression of AMPK or PPARα partially reduces adiponectin-stimulated fatty-acid oxidation, and suppression of AMPK or p38 MAPK partially reduces adiponectin-stimulated glucose uptake. In hepatocytes overexpressing AdipoR1/R2, adiponectin stimulates PPARα or AMPK and fatty-acid oxidation (84). Suppression of AMPK or PPARα in these hepatocytes partially reduces adiponectin-stimulated fatty-acid oxidation. Moreover, treatment with adiponectin reduces plasma glucose levels and molecules involved in gluconeogenesis in the liver, and dominant-negative AMPK partly reduces these effects. These data support the conclusion that AdipoR1 and AdipoR2 serve as receptors for globular and full-length adiponectin and mediate increased AMPK, PPARα ligand activities, p38 MAPK, and adiponectin-induced biological functions. T-cadherin is capable of binding adiponectin but is thought to have no effect on adiponectin cellular signaling, since T-cadherin lacks an intracellular domain (86). Interaction of APPL1 (adaptor protein containing pleckstrin homology domain, phosphotyrosine-binding domain, and leucine zipper motif 1) with AdipoR1 appears to play important roles in adiponectin signaling and adiponectin-mediated downstream events such as lipid oxidation and glucose uptake (88). S-S, disulfide bond. Adapted with permission from *Endocrine Reviews* (21); copyright 2005, The Endocrine Society.

these data strongly supported that the N-terminus of adiponectin receptors is internal and the C-terminus is external (88).

The expression levels of both AdipoR1 and AdipoR2 were significantly decreased in muscle and adipose tissue of insulin-resistant *ob/ob* mice, probably in part because of obesity-linked hyperinsulinemia (89). Moreover, adiponectin-induced activation of AMPK was impaired in the skeletal muscle of *ob/ob* mice. These data suggest that adiponectin resistance is present in *ob/ob* mice, presumably due to decreased expression of AdipoR1 and AdipoR2 (89). Thus, obesity decreases not only plasma adiponectin levels but also AdipoR1/R2 expression, thereby reducing adiponectin sensitivity and leading to insulin resistance, which in turn aggravates hyperinsulinemia, creating a "vicious cycle" (89). Adiponectin receptor expression in the skeletal muscle of type 2 diabetic patients has been reported to be decreased (90). In addition, a correlation has been reported between adiponectin receptor gene expression and insulin sensitivity in nondiabetic Mexican Americans with or without a family history of type 2 diabetes (91). Moreover, AdipoR1 mRNA expression was positively correlated with in vivo insulin and C-peptide concentrations, first-phase insulin secretion, and plasma triglyceride and cholesterol concentrations before and after adjustment for sex, age, waist-to-hip ratio, and body fat. Expression of AdipoR2 mRNA was clearly associated only with plasma triglyceride concentrations. In multivariate linear regression models, mRNA expression of AdipoR1, but not AdipoR2, was a determinant of first-phase insulin secretion independently of insulin sensitivity and body fat (92). Since AdipoR1 and AdipoR2 are expressed in pancreatic β cells, these receptors may play a role in insulin secretion (93).

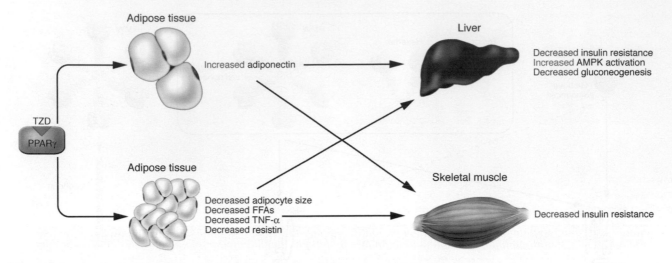

Figure 4
TZDs ameliorate insulin resistance and diabetes by both adiponectin-dependent and -independent pathways. We propose that there are 2 different pathways in the amelioration of insulin resistance induced by the PPARγ agonists TZDs, such as pioglitazone and probably rosiglitazone. One involves an adiponectin-dependent pathway and the other an adiponectin-independent pathway. TZDs increase adiponectin levels, ameliorating insulin resistance, increasing AMPK activation, and decreasing gluconeogenesis in the liver. On the other hand, independently of adiponectin, TZDs decrease adipocyte size, serum FFA levels, and expression of TNF-α and resistin, thus contributing to amelioration of insulin resistance in skeletal muscle.

Adiponectin and adiponectin receptors as therapeutic targets

According to our adiponectin hypothesis (21), a therapeutic strategy for the treatment of insulin resistance, type 2 diabetes, the metabolic syndrome, and cardiovascular disease may include the upregulation of plasma adiponectin levels, the upregulation of adiponectin receptors, or the development of adiponectin receptor agonists.

TZD-mediated upregulation of plasma adiponectin level. TZDs are known to improve systemic insulin sensitivity in animal models of obesity-linked insulin resistance and diabetes, by enhancing glucose disposal in skeletal muscle and suppressing gluconeogenesis in the liver. TZDs have been widely used as therapeutic agents for the treatment of type 2 diabetes (94–99). TZDs have been proposed to ameliorate insulin resistance by binding to and activating PPARγ in adipose tissue, thereby promoting adipocyte differentiation and increasing the number of small adipocytes that are more sensitive to insulin (100–103). Plasma adiponectin levels have been shown to be upregulated by TZDs (58, 79, 104–108) (Table 1), and HMW adiponectin is a predominant form of adiponectin upregulated by TZDs (89). TZDs may upregulate adiponectin by generating small adipocytes that abundantly express and secrete adiponectin (100, 102, 109) and/or directly activating *Adiponectin* gene transcription (106). TZDs may also directly facilitate the generation of HMW adiponectin. Since adiponectin is an insulin-sensitizing adipokine, it is reasonable to speculate that the action whereby TZDs increase insulin sensitivity is mediated, at least in part, by increased plasma adiponectin levels. However, whether the TZD-induced increase in plasma adiponectin level is causally involved in TZD-mediated insulin-sensitizing effects has not been addressed experimentally. Adiponectin-deficient (*Adipo*−/−) *ob/ob* mice with a C57BL/6 background were used to investigate whether the PPARγ agonist pioglitazone is capable of ameliorating insulin resistance in the absence of adiponectin (63, 87). The absence of adiponectin had no effect on either the obesi-

ty or the diabetic phenotype of these mice. The severity of insulin resistance and diabetes observed in *ob/ob* mice was significantly reduced in association with significant upregulation of serum adiponectin levels by low-dose (10 mg/kg) pioglitazone treatment. Amelioration of insulin resistance in *ob/ob* mice was attributed to decreased glucose production and increased AMPK levels in the liver, but not to increased glucose uptake in skeletal muscle. In contrast, the severity of insulin resistance and diabetes was not reduced in *Adipo*−/−*ob/ob* mice (87). With high-dose pioglitazone treatment, the insulin resistance and diabetes of *ob/ob* mice were again significantly ameliorated; this was attributed not only to decreased glucose production in the liver but also to increased glucose uptake in skeletal muscle. Interestingly, *Adipo*−/−*ob/ob* mice also displayed significant amelioration of insulin resistance and diabetes. The serum FFA and triglyceride levels as well as adipocyte sizes in *ob/ob* and *Adipo*−/−*ob/ob* mice were unchanged after low-dose pioglitazone treatment but were significantly reduced to a similar degree after high-dose pioglitazone treatment. Moreover, the expression of TNF-α and resistin in adipose tissues of *ob/ob* and *Adipo*−/−*ob/ob* mice were unchanged after low-dose pioglitazone but were decreased after high-dose pioglitazone. Although both high and low doses of pioglitazone ameliorated insulin resistance and diabetes, the underlying mechanisms may be different (87). We propose that there are 2 different pathways in the amelioration of insulin resistance induced by TZDs such as pioglitazone, and probably rosiglitazone. One involves an adiponectin-dependent pathway and the other an adiponectin-independent pathway (Figure 4). TZDs increase adiponectin levels via activation of *Adiponectin* gene transcription without stimulating adipocyte differentiation (58, 106), thereby increasing AMPK activation, decreasing gluconeogenesis in the liver, and ameliorating insulin resistance and type 2 diabetes. On the other hand, independently of adiponectin, TZDs induce adipocyte differentiation, leading to an increase in the number of small adipocytes, which is associated with decreased serum FFA levels and decreased TNF-α and resistin expression, together contributing to amelioration of insulin resistance in skeletal muscle (87).

Scherer's group demonstrated that *ob/ob* mice showed significantly improved glucose tolerance after rosiglitazone treatment, whereas *Adipo^-/-^ob/ob* mice responded only partially to this treatment and remained severely glucose intolerant (66), suggesting that rosiglitazone ameliorated glucose intolerance via both adiponectin-dependent and -independent pathways. Moreover, rosiglitazone significantly increased AMPK activity in the livers of wild-type mice, whereas it had no effect on *Adipo^-/-^* mice. In skeletal muscle, AMPK activity was also significantly increased in wild-type mice, while no increase was detectable in *Adipo^-/-^* mice. These data are in complete agreement with our data. Other pharmacological agents as well as lifestyle changes have also been reported to be associated with upregulation of plasma adiponectin levels (53, 55, 110–115) (Table 1).

Upregulation of adiponectin receptors and development of adiponectin receptor agonists. Since AdipoR1 and AdipoR2 are downregulated in obesity-linked insulin resistance and diabetes, both upregulation of AdipoR1 and AdipoR2 expression and agonism of AdipoR1 and AdipoR2 may be a logical approach to providing a novel treatment modality for insulin resistance and type 2 diabetes (84, 89). Previously, Staels's group reported that adiponectin receptors are expressed in human macrophages and that *adiponectin receptor* expression levels may be regulated by agonists of the nuclear receptors PPARα, PPARγ, and liver X receptor (116). We have recently shown that, in KKA^y mice, a PPARα agonist reversed decreases in AdipoR1 and AdipoR2 expression, which was lower in white and brown adipose tissue of KKA^y mice than in that of wild-type control KK mice (117). These data suggested that dual activation of PPARγ and PPARα enhanced the action of adiponectin by increasing both total and HMW adiponectin level and adiponectin receptor number, which can ameliorate obesity-linked insulin resistance.

Osmotin is a member of the pathogenesis related-5 (PR-5) family of plant defense proteins (24 members in *Arabidopsis thaliana*) that induce apoptosis in yeast. It is ubiquitous in fruits and vegetables, etc., and the genes encoding the PR-5 protein sequenced from many different species are about 50–95% identical. PR-5 family proteins are also extremely stable and may remain active even when in contact with the human digestive or respiratory system. Bressan's group isolated a yeast clone that exhibited hypersensitivity to osmotin, sequenced the cDNA inserts, and found that PHO36/YOL002c, the yeast homologue of AdipoR1, is a receptor for osmotin (118). X-ray crystallographic studies revealed that both globular adiponectin and osmotin consist of antiparallel β-strands arranged in the shape of a β-barrel. Domain I (lectinlike domain) of osmotin showed similarity to globular adiponectin in 3D structure, suggesting that these 2 proteins share the lectin-like domain (118). Interestingly, osmotin activates AMPK via adiponectin receptors in mammalian C2C12 myocytes (118). These data raise the possibility that further research examining similarities in adiponectin and osmotin may facilitate the development of potential adiponectin receptor agonists (118). Although further studies will be needed to determine the physiological and pathophysiological roles of AdipoR1 and AdipoR2, the enhancement or mimicking of adiponectin action through modulation of expression and/or function of AdipoR1 and AdipoR2 can be a novel therapeutic strategy for the treatment of insulin resistance, the metabolic syndrome, and type 2 diabetes.

In summary, adiponectin is an adipokine that exerts a potent insulin-sensitizing effect by binding to its receptors such as AdipoR1 and AdipoR2, leading to activation of AMPK, PPARα, and presumably some other unknown signaling pathways. Indeed, circulating levels of adiponectin, especially HMW adiponectin, are positively correlated with insulin sensitivity and altered by various genetic and environmental factors, pathological conditions, and medications. Thus, monitoring the levels of HMW adiponectin is a good predictable marker for type 2 diabetes and the metabolic syndrome. Moreover, methods to increase adiponectin levels, such as TZD administration, are expected to be effective for the treatment of these diseases. In the future, enhancing or mimicking adiponectin action through modulation of expression and/or function of the adiponectin receptors may be a novel and promising therapeutic strategy for insulin resistance, type 2 diabetes, and the metabolic syndrome.

Acknowledgments

This work was supported by the Program for Promotion of Fundamental Studies in Health Sciences of the Organization for Pharmaceutical Safety and Research of Japan, a grant from the Human Science Foundation (to T. Kadowaki), a Grant-in-Aid for the Development of Innovative Technology from the Ministry of Education, Culture, Sports, Science, and Technology of Japan (to T. Kadowaki), a Grant-in Aid for Creative Scientific Research (10NP0201) from the Japan Society for the Promotion of Science (to T. Kadowaki), and by Health Science Research Grants (Research on Human Genome and Gene Therapy) from the Ministry of Health, Labour and Welfare of Japan (to T. Kadowaki).

Address correspondence to: Takashi Kadowaki, Department of Metabolic Diseases, Graduate School of Medicine, University of Tokyo, 7-3-1 Hongo, Bunkyo-ku, Tokyo 113-8655, Japan. Phone: 81-3-5800-8818; Fax: 81-3-5800-9797; E-mail: kadowaki-3im@h.u-tokyo.ac.jp.

1. Flier, J.S. 2004. Obesity wars: molecular progress confronts an expanding epidemic. *Cell.* **116**:337–350.
2. Friedman, J.M. 2000. Obesity in the new millennium. *Nature.* **404**:632–634.
3. Reaven, G.M. 1998. Role of insulin resistance in human disease. *Diabetes.* **37**:1595–1607.
4. Kaplan, N.M. 1989. The deadly quartet. Upper-body obesity, glucose intolerance, hypertriglyceridemia, and hypertension. *Arch. Intern. Med.* **149**:1514–1520.
5. Matsuzawa, Y. 1997. Pathophysiology and molecular mechanisms of visceral fat syndrome: the Japanese experience. *Diabetes Metab. Rev.* **13**:3 13.
6. National Cholesterol Education Program. 2001. Executive Summary of the Third Report of the National Cholesterol Education Program (NCEP) Expert Panel on Detection, Evaluation, and Treatment of High Blood Cholesterol in Adults (Adult Treatment Panel III). *JAMA.* **285**:2486–2497.
7. International Diabetes Federation. 2005. A new worldwide definition of the metabolic syndrome. http://www.idf.org/home/index.cfm?unode=32EF2063-B966-468F-928C-A5682A4E3910.
8. Taniguchi, C.M., Emanuelli, B., and Kahn, C.R. 2006. Critical nodes in signaling pathways: insights into insulin action. *Nat. Rev. Mol. Cell Biol.* **7**:85–96.
9. Saltiel, A.R., and Kahn, C.R. 2001. Insulin signaling and the regulation of glucose and lipid metabolism. *Nature.* **414**:799–806.
10. Kahn, C.R. 2000. Triglycerides and toggling the tummy. *Nat. Genet.* **25**:6–7.
11. Spiegelman, B.M., and Flier, J.S. 2001. Obesity and regulation of energy balance. *Cell.* **104**:531–543.
12. Lowell, B.B. 1999. PPARgamma: an essential regulator of adipogenesis and modulator of fat cell function. *Cell.* **99**:239–242.
13. Chawla, A., Saez, E., and Evans, R.M. 2000. Don't know much bile-ology. *Cell.* **103**:1–4.
14. Obici, S., Feng, Z., Arduini, A., Conti, R., and Rossetti, L. 2003. Inhibition of hypothalamic carnitine palmitoyltransferase-1 decreases food intake and glucose production. *Nat. Med.* **9**:756–761.
15. Hotamisligil, G.S., Shargill, N.S., and Spiegelman, B.M.

1993. Adipose expression of tumor necrosis factor-alpha: direct role in obesity-linked insulin resistance. *Science.* **259**:87–91.

16. Zhang, Y., et al. 1994. Positional cloning of the mouse obese gene and its human homologue. *Nature.* **372**:425–432.

17. Steppan, C.M., et al. 2001. The hormone resistin links obesity to diabetes. *Nature.* **409**:307–312.

18. Yang, Q., et al. 2005. Serum retinal binding protein 4 contributes to insulin resistance in obesity and type 2 diabetes. *Nature.* **436**:356–362.

19. Lazar, M.A. 2006. The humoral side of insulin resistance. *Nat. Med.* **12**:43–44.

20. Kershaw, E.E., and Flier, J.S. 2004. Adipose tissue as an endocrine organ. *J. Clin. Endocrinol. Metab.* **89**:2548–2556.

21. Kadowaki, T., and Yamauchi, T. 2005. Adiponectin and adiponectin receptors. *Endocr. Rev.* **26**:439–451.

22. Berg, A.H., and Scherer, P.E. 2005. Adipose tissue, inflammation, and cardiovascular disease. *Circ. Res.* **96**:939–949.

23. Okamoto, Y., Kihara, S., Funahashi, T., Matsuzawa, Y., and Libby, P. 2006. Adiponectin: a key adipocytokine in metabolic syndrome. *Clin. Sci. (Lond.)* **110**:267–278.

24. Scherer, P.E., Williams, S., Fogliano, M., Baldini, G., and Lodish, H.F. 1995. A novel serum protein similar to C1q, produced exclusively in adipocytes. *J. Biol. Chem.* **270**:26746–26749.

25. Hu, E., Liang, P., and Spiegelman, B.M. 1996. AdipoQ is a novel adipose-specific gene dysregulated in obesity. *J. Biol. Chem.* **271**:10697–10703.

26. Maeda, K., et al. 1996. cDNA cloning and expression of a novel adipose specific collagen-like factor, apM1 (AdiPose Most abundant Gene transcript 1). *Biochem. Biophys. Res. Commun.* **221**:286–289.

27. Nakano, Y., Tobe, T., Choi-Miura, N.H., Mazda, T., and Tomita, M. 1996. Isolation and characterization of GBP28, a novel gelatin-binding protein purified from human plasma. *J. Biochem. (Tokyo).* **120**:803–812.

28. Shapiro, L., and Scherer, P.E. 1998. The crystal structure of a complement-1q family protein suggests an evolutionary link to tumor necrosis factor. *Curr. Biol.* **8**:335–338.

29. Yokota, T., et al. 2000. Adiponectin, a new member of the family of soluble defense collagens, negatively regulates the growth of myelomonocytic progenitors and the functions of macrophages. *Blood.* **96**:1723–1732.

30. Crouch, E., Persson, A., Chang, D., and Heuser, J. 1994. Molecular structure of pulmonary surfactant protein D (SP-D). *J. Biol. Chem.* **269**:17311–17319.

31. McCormack, F.X., et al. 1997. The Cys6 intermolecular disulfide bond and the collagen-like region of rat SP-A play critical roles in interactions with alveolar type II cells and surfactant lipids. *J. Biol. Chem.* **272**:27971–27979.

32. Pajvani, U.B., et al. 2003. Structure-function studies of the adipocyte-secreted hormone Acrp30/adiponectin. Implications for metabolic regulation and bioactivity. *J. Biol. Chem.* **278**:9073–9085.

33. Waki, H., et al. 2003. Impaired multimerization of human adiponectin mutants associated with diabetes: molecular structure and multimer formation of adiponectin. *J. Biol. Chem.* **278**:40352–40363.

34. Hotta, K., et al. 2001. Circulating concentrations of the adipocyte protein adiponectin are decreased in parallel with reduced insulin sensitivity during the progression to type 2 diabetes in rhesus monkeys. *Diabetes.* **50**:1126–1133.

35. Arita, Y., et al. 1999. Paradoxical decrease of an adipose-specific protein, adiponectin, in obesity. *Biochem. Biophys. Res. Commun.* **257**:79–83.

36. Ryo, M., et al. 2004. Adiponectin as a biomarker of the metabolic syndrome. *Circ. J.* **68**:975–981.

37. Yatagai, T., et al. 2003. Hypoadiponectinemia is associated with visceral fat accumulation and insu-

lin resistance in Japanese men with type 2 diabetes mellitus. *Metabolism.* **52**:1274–1278.

38. Yamamoto, Y., Hirose, H., Saito, I., Nishikai, K., and Saruta, T. 2004. Adiponectin, an adipocyte-derived protein, predicts future insulin-resistance: two-year follow-up study in Japanese population. *J. Clin. Endocrinol. Metab.* **89**:87–90.

39. Lindsay, R.S., et al. 2002. Adiponectin and development of type 2 diabetes in the Pima Indian population. *Lancet.* **360**:57–58.

40. Daimon, M., et al. 2003. Decreased serum levels of adiponectin are a risk factor for the progression to type 2 diabetes in the Japanese population: the Funagata study. *Diabetes Care.* **26**:2015–2020.

41. Snehalatha, C., et al. 2003. Plasma adiponectin is an independent predictor of type 2 diabetes in Asian Indians. *Diabetes Care.* **26**:3226–3229.

42. Spranger, J., et al. 2003. Adiponectin and protection against type 2 diabetes mellitus. *Lancet.* **361**:226–228.

43. Duncan, B.B., et al. 2004. Adiponectin and the development of type 2 diabetes: the Atherosclerosis Risk in Communities Study. *Diabetes.* **53**:2473–2478.

44. Krakoff, J., et al. 2003. Inflammatory markers, adiponectin, and risk of type 2 diabetes in the Pima Indian. *Diabetes Care.* **26**:1745–1751.

45. Matsushita, K., et al. 2006. Comparison of circulating adiponectin and proinflammatory markers regarding their association with metabolic syndrome in Japanese men. *Arterioscler. Thromb. Vasc. Biol.* **26**:871–876.

46. Kumada, M., et al. 2003. Association of hypoadiponectinemia with coronary artery disease in men. *Arterioscler. Thromb. Vasc. Biol.* **23**:85–89.

47. Pischon, T., et al. 2004. Plasma adiponectin levels and risk of myocardial infarction in men. *JAMA.* **291**:1730–1737.

48. Ouchi, N., et al. 2003. Association of hypoadiponectinemia with impaired vasoreactivity. *Hypertension.* **42**:231–234.

49. Adamczak, M., et al. 2003. Decreased plasma adiponectin concentration in patients with essential hypertension. *Am. J. Hypertens.* **16**:72–75.

50. Combs, T.P., et al. 2003. Sexual differentiation, pregnancy, calorie restriction, and aging affect the adipocyte-specific secretory protein adiponectin. *Diabetes.* **52**:268–276.

51. Nishizawa, H., et al. 2002. Androgens decrease plasma adiponectin, an insulin-sensitizing adipocyte-derived protein. *Diabetes.* **51**:2734–2741.

52. Xu, A., et al. 2005. Testosterone selectively reduces the high molecular weight form of adiponectin by inhibiting its secretion from adipocytes. *J. Biol. Chem.* **280**:18073–18080.

53. Nagasawa, A., et al. 2002. Effects of soy protein diet on the expression of adipose genes and plasma adiponectin. *Horm. Metab. Res.* **34**:635–639.

54. Flachs, P., et al. 2006. Polyunsaturated fatty acids of marine origin induce adiponectin in mice fed a high-fat diet. *Diabetologia.* **49**:394–397.

55. Nagao, K., et al. 2003. Conjugated linoleic acid enhances plasma adiponectin level and alleviates hyperinsulinemia and hypertension in Zucker diabetic fatty (fa/fa) rats. *Biochem. Biophys. Res. Commun.* **310**:562–566.

56. Pischon, T., et al. 2005. Association between dietary factors and plasma adiponectin concentrations in men. *Am. J. Clin. Nutr.* **81**:780–786.

57. Furukawa, S., et al. 2004. Increased oxidative stress in obesity and its impact on metabolic syndrome. *J. Clin. Invest.* **114**:1752–1761. doi:10.1172/JCI200421625.

58. Yamauchi, T., et al. 2001. The fat-derived hormone adiponectin reverses insulin resistance associated with both lipoatrophy and obesity. *Nat. Med.* **7**:941–946.

59. Berg, A.H., et al. 2001. The adipocyte-secreted protein Acrp30 enhances hepatic insulin action. *Nat. Med.* **7**:947–953.

60. Fruebis, J., et al. 2001. Proteolytic cleavage product of 30-kDa adipocyte complement-related protein increases fatty acid oxidation in muscle and causes weight loss in mice. *Proc. Natl. Acad. Sci. U. S. A.* **98**:2005–2010.

61. Yamauchi, T., et al. 2003. Globular adiponectin protected ob/ob mice from diabetes and ApoE-deficient mice from atherosclerosis. *J. Biol. Chem.* **278**:2461–2468.

62. Combs, T.P., et al. 2004. A transgenic mouse with a deletion in the collagenous domain of adiponectin displays elevated circulating adiponectin and improved insulin sensitivity. *Endocrinology.* **145**:367–383.

63. Kubota, N., et al. 2002. Disruption of adiponectin causes insulin resistance and neointimal formation. *J. Biol. Chem.* **277**:25863–25866.

64. Maeda, N., et al. 2002. Diet-induced insulin resistance in mice lacking adiponectin/ACRP30. *Nat. Med.* **8**:731–737.

65. Ma, K., et al. 2002. Increased beta-oxidation but no insulin resistance or glucose intolerance in mice lacking adiponectin. *J. Biol. Chem.* **277**:34658–34661.

66. Nawrocki, A.R., et al. 2006. Mice lacking adiponectin show decreased hepatic insulin sensitivity and reduced responsiveness to peroxisome proliferator-activated receptor gamma agonists. *J. Biol. Chem.* **281**:2654–2660.

67. Yamauchi, T., et al. 2002. Adiponectin stimulates glucose utilization and fatty-acid oxidation by activating AMP-activated protein kinase. *Nat. Med.* **8**:1288–1295.

68. Tomas, E., et al. 2002. Enhanced muscle fat oxidation and glucose transport by ACRP30 globular domain: acetyl-CoA carboxylase inhibition and AMP-activated protein kinase activation. *Proc. Natl. Acad. Sci. U. S. A.* **99**:16309–16313.

69. Combs, T.P., et al. 2001. Endogenous glucose production is inhibited by the adipose-derived protein Acrp30. *J. Clin. Invest.* **108**:1875–1881. doi:10.1172/JCI200114120.

70. Kissebah, A., et al. 2000. Quantitative trait loci on chromosomes 3 and 17 influence phenotypes of the metabolic syndrome. *Proc. Natl. Acad. Sci. U. S. A.* **97**:14478–14483.

71. Mori, Y., et al. 2002. Genome-wide search for type 2 diabetes in Japanese affected sib-pairs confirms susceptibility genes on 3q, 15q, and 20q and identifies two new candidate loci on 7p and 11p. *Diabetes.* **51**:1247–1255.

72. Vasseur, F., et al. 2002. Single-nucleotide polymorphism haplotypes in the both proximal promoter and exon 3 of the APM1 gene modulate adipocyte-secreted adiponectin hormone levels and contribute to the genetic risk for type 2 diabetes in French Caucasians. *Hum. Mol. Genet.* **11**:2607–2614.

73. Hara, K., et al. 2002. Genetic variation in the gene encoding adiponectin is associated with an increased risk of type 2 diabetes in the Japanese population. *Diabetes.* **51**:536–540.

74. Kondo, H., et al. 2002. Association of adiponectin mutation with type 2 diabetes: a candidate gene for the insulin resistance syndrome. *Diabetes.* **51**:2325–2328.

75. Stumvoll, M., et al. 2002. Association of the T-G polymorphism in adiponectin (exon 2) with obesity and insulin sensitivity: interaction with family history of type 2 diabetes. *Diabetes.* **51**:37–41.

76. Menzaghi, C., et al. 2002. A haplotype at the adiponectin locus is associated with obesity and other features of the insulin resistance syndrome. *Diabetes.* **51**:2306–2312.

77. Woo, J.G., et al. 2006. Interactions between non-contiguous haplotypes in the adiponectin gene ACDC are associated with plasma adiponectin. *Diabetes.* **55**:523–529.

78. Tasanen, K., et al. 2000. Collagen XVII is destabilized by a glycine substitution mutation in the cell adhesion

domain Col15. *J. Biol. Chem.* **275**:3093–3099.

79. Pajvani, U.B., et al. 2004. Complex distribution, not absolute amount of adiponectin, correlates with thiazolidinedione-mediated improvement in insulin sensitivity. *J. Biol. Chem.* **279**:12152–12162.

80. Lara-Castro, C., et al. 2006. Adiponectin multimeric complexes and the metabolic syndrome trait cluster. *Diabetes.* **55**:249–259.

81. Fisher, F.F., et al. 2005. Serum high molecular weight complex of adiponectin correlates better with glucose tolerance than total serum adiponectin in Indo-Asian males. *Diabetologia.* **48**:1084–1087.

82. Ebinuma, H., et al. 2006. A novel ELISA system for selective measurement of human adiponectin multimers by using proteases. *Clin. Chim. Acta.* doi:10.1016/j.cca.2006.03.014.

83. Hara, K., et al. 2006. Measurement of the high-molecular weight form of adiponectin in plasma is useful for the prediction of insulin resistance and metabolic syndrome. *Diabetes Care.* **29**:1357–1362.

84. Yamauchi, T., et al. 2003. Cloning of adiponectin receptors that mediate antidiabetic metabolic effects. *Nature.* **423**:762–769.

85. Karpichev, I.V., et al. 2002. Multiple regulatory roles of a novel Saccharomyces cerevisiae protein, encoded by YOL002c, in lipid and phosphate metabolism. *J. Biol. Chem.* **277**:19609–19617.

86. Hug, C., et al. 2004. T-cadherin is a receptor for hexameric and high-molecular-weight forms of Acrp30/adiponectin. *Proc. Natl. Acad. Sci. U. S. A.* **101**:10308–10313.

87. Kubota, N., et al. 2006. Pioglitazone ameliorates insulin resistance and diabetes by both adiponectin-dependent and -independent pathways. *J. Biol. Chem.* **281**:8748–8755.

88. Mao, X., et al. 2006. APPL1 binds to adiponectin receptors and mediates adiponectin signalling and function. *Nat. Cell Biol.* **8**:516–523.

89. Tsuchida, A., et al. 2004. Insulin/Foxo1 pathway regulates expression levels of adiponectin receptors and adiponectin sensitivity. *J. Biol. Chem.* **279**:30817–30822.

90. Civitarese, A.E., et al. 2004. Adiponectin receptors gene expression and insulin sensitivity in non-diabetic Mexican Americans with or without a family history of type 2 diabetes. *Diabetologia.* **47**:816–820.

91. Debard, C., et al. 2004. Expression of key genes of fatty acid oxidation, including adiponectin receptors, in skeletal muscle of type 2 diabetic patients.

Diabetologia. **47**:917–925.

92. Harald, S., et al. 2004. Expression of adiponectin receptor mRNA in human skeletal muscle cells is related to in vivo parameters of glucose and lipid metabolism. *Diabetes.* **53**:2195–2201.

93. Kharroubi, I., et al. 2003. Expression of adiponectin receptors in pancreatic beta cells. *Biochem. Biophys. Res. Commun.* **312**:1118–1122.

94. Bowen, L., et al. 1991. The effect of CP 68,722, a thiazolidinedione derivative, on insulin sensitivity in lean and obese Zucker rats. *Metabolism.* **40**:1025–1030.

95. Nolan, J.J., et al. 1994. Improvement in glucose tolerance and insulin resistance in obese subjects treated with troglitazone. *N. Engl. J. Med.* **331**:1188–1193.

96. Saltiel, A.R. 2001. New perspectives into the molecular pathogenesis and treatment of type 2 diabetes. *Cell.* **104**:517–529.

97. Mauvais-Jarvis, F., et al. 2000. Understanding the pathogenesis and treatment of insulin resistance and type 2 diabetes mellitus: what can we learn from transgenic and knockout mice? *Diabetes Metab.* **26**:433–448.

98. Yki-Jarvinen, H. 2004. Thiazolidinediones. *N. Engl. J. Med.* **351**:1106–1118.

99. Rangwala, S.M., et al. 2004. Peroxisome proliferator-activated receptor gamma in diabetes and metabolism. *Trends Pharmacol. Sci.* **25**:331–336.

100. Okuno, A., et al. 1998. Troglitazone increases the number of small adipocytes without the change of white adipose tissue mass in obese Zucker rats. *J. Clin. Invest.* **101**:1354–1361.

101. Evans, R.M., et al. 2004. PPARs and the complex journey to obesity. *Nat. Med.* **10**:355–361.

102. Kubota, N., et al. 1999. PPAR gamma mediates high-fat diet-induced adipocyte hypertrophy and insulin resistance. *Mol. Cell.* **4**:597–609.

103. Olefsky, J.M., et al. 2000. PPAR gamma and the treatment of insulin resistance. *Trends Endocrinol. Metab.* **11**:362–368.

104. Yu, J.G., et al. 2002. The effect of thiazolidinediones on plasma adiponectin levels in normal, obese, and type 2 diabetic subjects. *Diabetes.* **51**:2968–2974.

105. Chandran, M., et al. 2003. Adiponectin: more than just another fat cell hormone? *Diabetes Care.* **26**:2442–2450.

106. Iwaki, M., et al. 2003. Induction of adiponectin, a fat-derived antidiabetic and antiatherogenic factor, by nuclear receptors. *Diabetes.* **52**:1655–1663.

107. Hirose, H., et al. 2002. Effects of pioglitazone on

metabolic parameters, body fat distribution, and serum adiponectin levels in Japanese male patients with type 2 diabetes. *Metabolism.* **51**:314–317.

108. Maeda, N., et al. 2001. PPARγ ligands increase expression and plasma concentrations of adiponectin, an adipose-derived protein. *Diabetes.* **50**:2094–2099.

109. Yamauchi, T., et al. 2001. The mechanisms by which both heterozygous peroxisome proliferator-activated receptor gamma (PPARgamma) deficiency and PPARgamma agonist improve insulin resistance. *J. Biol. Chem.* **276**:41245–41254.

110. Furuhashi, M., et al. 2003. Blockade of the renin-angiotensin system increases adiponectin concentrations in patients with essential hypertension. *Hypertension.* **42**:76–81.

111. Koh, K.K., et al. 2005. Vascular and metabolic effects of combined therapy with ramipril and simvastatin in patients with type 2 diabetes. *Hypertension.* **45**:1088–1093.

112. Yang, W.S., et al. 2001. Weight reduction increases plasma levels of an adipose-derived anti-inflammatory protein, adiponectin. *J. Clin. Endocrinol. Metab.* **86**:3815–3819.

113. Esposito, K., et al. 2003. Effect of weight loss and lifestyle changes on vascular inflammatory markers in obese women: a randomized trial. *JAMA.* **289**:1799–1804.

114. Kistorp, C., et al. 2005. Plasma adiponectin, body mass index, and mortality in patients with chronic heart failure. *Circulation.* **112**:1756–1762.

115. Tentolouris, N., et al. 2004. Plasma adiponectin concentrations in patients with chronic renal failure: relationship with metabolic risk factors and ischemic heart disease. *Horm. Metab. Res.* **36**:721–727.

116. Chinetti, G., et al. 2004. Expression of adiponectin receptors in human macrophages and regulation by agonists of the nuclear receptors PPARalpha, PPARgamma, and LXR. *Biochem. Biophys. Res. Commun.* **314**:151–158.

117. Tsuchida, A., et al. 2005. Peroxisome proliferator activated receptor (PPAR) alpha activation increases adiponectin receptors and reduces obesity-related inflammation in adipose tissue: comparison of activation of PPARalpha, PPARgamma, and their combination. *Diabetes.* **54**:3358–3370.

118. Narasimhan, M.L., et al. 2005. Osmotin is a homolog of mammalian adiponectin and controls apoptosis in yeast through a homolog of mammalian adiponectin receptor. *Mol. Cell.* **17**:171–180.

The role of cholesterol efflux in regulating the fertilization potential of mammalian spermatozoa

Alexander J. Travis[1] and Gregory S. Kopf[2]

[1]Center for Research on Reproduction and Women's Health, University of Pennsylvania School of Medicine, Philadelphia, Pennsylvania, USA.
[2]Women's Health Research Institute, Wyeth Research, Collegeville, Pennsylvania, USA.

Following spermatogenesis and spermiogenesis, mammalian spermatozoa leaving the testis appear to be morphologically mature but clearly are immature from a functional standpoint; that is, they have acquired neither progressive motility nor the ability to fertilize a metaphase II–arrested egg. Although progressive motility is acquired and signaling pathways mature during sperm transit through the epididymis, complete fertilization capacity in vivo is conferred only during residence in the female reproductive tract. Similar observations have been made using a variety of in vitro assays, suggesting that a series of events, some initiated by environmental cues, confer on sperm the ability to fertilize the egg.

This acquired capacity to fertilize was first observed by Austin (1) and Chang (2), who demonstrated that freshly ejaculated sperm cannot fertilize eggs until they reside in the female reproductive tract for a finite period of time. All of the cellular events that allow the ejaculated sperm to fertilize an egg were subsumed into a single phenomenon that was termed "capacitation." The ability to capacitate sperm in vitro has been of great importance to both scientists and clinicians, who have used it to study the basic biology of fertilization and to develop various assisted reproductive technologies for humans and other species.

Work by many investigators has established that the process of fertilization, not surprisingly, represents a series of elegant intercellular communication and cellular activation events (3–5). Sperm functions such as motility and capacitation in the female reproductive tract are likely modulated by environmental cues in the luminal fluid, as well as by interactions with oviductal epithelium or other female tissues (6). When sperm arrive in the oviduct and encounter the ovulated, metaphase II–arrested egg enclosed in its cumulus cell matrix, a complex series of cell-cell and cell-ECM interactions ensues, initiating cellular signaling events that permit the fusion of the sperm and egg plasma membranes. Several of these cell-matrix and cell-cell interactions involve novel gamete surface proteins and matrices. Signal transduction events leading to gamete activation, in particular sperm acrosomal exocytosis and egg cortical granule secretion, share some features with signaling events described in somatic cells.

Sperm membrane cholesterol efflux contributes to one such novel signaling mechanism that controls sperm capacitation, and the details of this effect are now beginning to be understood at the molecular level. Knowledge of how cholesterol efflux occurs in these cells, as well as how this efflux is integrated with trans-membrane signaling to regulate sperm function, may reveal much about the fertilization process and may also provide insights into the role and dynamics of membrane cholesterol efflux in somatic cell function. Here, we offer a short overview of the role of cholesterol efflux in regulating sperm capacitation, with an aim toward identifying areas of future investigation that may ultimately provide a greater understanding of the role of this sterol in regulating signal transduction. The reader is referred to several extensive reviews of this subject (7, 8).

Molecular basis of capacitation

After attaining morphological maturity in the testis, sperm must undergo two distinct processes of functional maturation to be able to fertilize an egg. The first occurs in the epididymis of the male reproductive tract, as sperm move from the caput to the corpus and then to the caudal regions of this organ, where they are stored prior to ejaculation. During this transit, the signaling pathways that regulate capacitation are enabled. Thus, caput epididymal sperm fail to be capacitated in the presence of molecular stimuli (defined below) that are sufficient to capacitate sperm residing in the cauda epididymis (9).

Several molecular events are likely involved in the acquisition of signaling competence. For example, concomitant with the maturation of these signaling pathways, epididymal sperm undergo dramatic alterations in their membrane sterol content. Such changes are highly species-specific and are also highly specific with regard to the class of sterol that is being changed (10). In addition, intracellular signaling systems that control capacitation mature during epididymal transit (ref. 11; M. Fornes et al. unpublished data). How alterations in membrane sterol composition integrate with the maturation of signaling pathways is still not fully understood.

The majority of alterations of epididymal sperm sterol content probably result from interactions of the sperm with the epididymal epithelium. Epithelial linings of both the epididymis and the vas deferens appear to have a highly developed sterol-producing capacity (12), although the impact of sterol synthetic capacity in the vas deferens on sterol levels in ejaculated sperm is unclear. There are also changes in the content of other sperm lipids during epididymal maturation. In some species, phospholipids are the major source of energy for endogenous oxidative respiration and therefore phospholipid levels decline during epididymal maturation (13). Changes in either sperm sterol or phospholipid levels might serve to alter the membrane cholesterol/phospholipid molar ratio, which has been implicated in the regulation of capacitation, as described below.

Given the species-specific nature of these large-scale alterations in lipid content, it is difficult to speak generally about their function. However, in all species examined thus far, cauda epididymal sperm possess clearly delineated membrane domains that differ

Nonstandard abbreviations used: filipin-sterol complex (FSC); protein kinase A (PKA).

Conflict of interest: No conflict of interest has been declared.

Citation for this article: J. Clin. Invest. 110:731–736 (2002). doi:10.1172/JCI200216392.

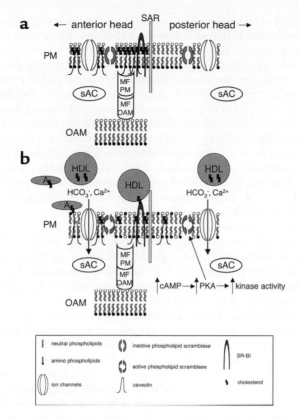

Figure 1
Model for the organization of cholesterol in the membranes of the sperm head, and its role in signaling. Schematic diagrams of a murine sperm incubated under noncapacitating conditions (**a**) and capacitating conditions (**b**). In this model, membrane rafts (regions enriched in sterols) are organized in the plasma membrane of the anterior head. Membrane fusion proteins are inactive due to being tethered by caveolin. The distribution of phospholipids in the membrane leaflets is asymmetrical, as the scramblases are inactive. When exposed to capacitating conditions (bicarbonate, calcium, albumin, and HDLs), cholesterol is removed via a specific pathway (HDL–SR-BI) and a nonspecific pathway (albumin). Increases in sAC activity elevate cAMP levels and stimulate downstream kinases. This signaling results in increased phospholipid scrambling, causing a disordered distribution of amino- and neutral phospholipids. Together with the increase in membrane fluidity caused by the sterol efflux, this change results in lateral movement of some sterols and caveolin from the anterior to the posterior head. The loss of sterols from the membrane causes a disruption of the interaction between caveolin and the membrane fusion proteins, resulting in their activation. The plasma membrane (PM) and outer acrosomal membrane (OAM) are shown immediately adjacent to the subacrosomal ring (SAR). For simplicity, both an anion and a cation are drawn as passing through a single ion channel in **b**. Cholesterol can exist in a free state in the membrane or be associated with cholesterol acceptors such as HDLs or albumin (A). sAC, a novel HCO_3^--activated adenylyl cyclase that associates at least in part with membranes, is shown here in the sperm cytoplasm. Membrane fusion proteins (MF) can associate with either the PM or the OAM.

in their sterol composition. Initially characterized by the presence of filipin-sterol complexes (FSCs) visible by freeze-fracture electron microscopy, these domains impart heterogeneity on the sperm surface within a given region of these cells. Such subdomains suggest the possibility of still more precise compartmentalization of function beyond the obvious polarization of these cells into head and tail domains that contribute to egg interaction and motility regulation, respectively. Indeed, these subdomains have recently been hypothesized to act as scaffolds or foci for signaling pathways regulating sperm capacitation in both the head and flagellum (14).

Signaling and the control of fertilization competence in sperm

Recent studies by several laboratories using in vitro models support the idea that capacitation requires transmembrane signaling and intracellular signal transduction. The development of in vitro capacitation protocols for sperm of several different species has shown the critical importance of three media constituents, namely Ca^{2+}, HCO_3^-, and a protein that can function as a cholesterol acceptor, such as serum albumin.

Work by several laboratories shows that capacitation is likely regulated by a novel signal transduction pathway involving crosstalk between cAMP, protein kinase A (PKA), and tyrosine kinases (Figure 1). This signaling leads to the phosphorylation on tyrosine residues of several proteins, the identities of which are only starting to be elucidated (7). As initially demonstrated by Visconti et al. (11), cauda epididymal mouse sperm, when incubated in vitro in media known to support capacitation, display a time-dependent increase in protein tyrosine phosphorylation that correlates with the onset of functional capacitation, operationally defined by the ability of the sperm population to fertilize eggs. The apparent absence of external stimuli, such as hormones or cytokines, that might initiate the observed tyrosine phosphorylation suggested that signaling might be regulated by a time-dependent mechanism or by specific components in the media. Subsequent work showed that the extracellular Ca^{2+}, HCO_3^-, and serum albumin in the capacitation medium are all absolutely required for these molecular and functional changes (11) and implicated a novel adenylyl cyclase/cAMP/PKA signaling system in sperm capacitation (7).

Just as interesting as this novel mode of signal transduction is the unusual mechanism by which these medium components activate cAMP signaling (Figure 1). Although there is clear evidence that Ca^{2+} can regulate the activity of specific adenylyl cyclase and cyclic nucleotide phosphodiesterase family members, the effects of HCO_3^- on adenylyl cyclase activity have been demonstrated in only a small number of cells or tissues, including ocular ciliary processes, corneal endothelium, choroid plexus, the medullary and cortical regions of the kidney (15), and sperm (4). Presently, most is known about the sperm adenylyl cyclase. This enzyme is not regulated in a manner similar to that seen with the classical 12-transmembrane, G protein–regulated somatic cell adenylyl cyclases (4). A considerable amount of effort has been devoted to characterizing the sperm HCO_3^--activated adenylyl cyclase, which was recently purified and cloned (16). This protein, now termed "sAC," has many novel characteristics and is likely to exist in multiple forms as a consequence of alternative splicing and proteolysis. Its catalytic domains resemble the adenylyl cyclases of Cyanobacteria, enzymes that can also be regulated by HCO_3^- (17, 18).

Role of cholesterol efflux in signal transduction leading to capacitation

The historical requirement for serum albumin in defined media to support capacitation had been hypothesized by several groups to be due to the ability of albumin to serve as a sink for cholesterol removal from the sperm plasma membrane (10, 19). Removal of this sterol likely accounts for the changes in membrane fluidity observed during capacitation and the subsequent decrease in the membrane cholesterol/phospholipid ratios (7). Such changes in membrane dynamics are likely to significantly affect cellular function (see below). Coupled with evidence that sperm membrane sterol levels decline following exposure to albumin, the observation that other cholesterol acceptors like HDL and β-cyclodextrins effectively support capacitation suggests that the primary action of serum albumin is in mediating cholesterol efflux (20, 21). Interestingly, the action of serum albumin, HDL, and β-cyclodextrins as cholesterol acceptors is somehow coupled to the cAMP-dependent pathway described above (Figure 1) (20, 21).

How might changes in the sterol content of the membrane regulate transmembrane signaling leading to capacitation? Biophysical studies demonstrate that cholesterol alters the bulk properties of biological membranes. For example, this sterol can increase the orientation order of the membrane lipid hydrocarbon chains, restricting the ability of membrane proteins to undergo conformational changes by rendering their surrounding membrane less fluid. High concentrations of cholesterol can thereby inhibit capacitation indirectly by diminishing the conformational freedom and hence the biological activity of sperm surface proteins. Alternatively, cholesterol might directly affect specific membrane proteins that function in transmembrane signaling. As shown in Figure 1, either or both of these effects of cholesterol could modulate ion transporters and effector enzymes like sAC. This story, however, is likely to be more complex and to involve additional lipid and sterol species, particularly in light of evidence that phospholipids move between distinct sperm membrane subdomains during capacitation.

Roles of sperm membrane subdomains in signaling

In all species examined thus far by visualization of FSCs the plasma membrane overlying the acrosome has been found to be markedly enriched in sterols, relative to either the post-acrosomal plasma membrane or the acrosomal or nuclear membranes (21–24) (Figure 2). The highly conserved demarcation of these two subdomains in the plasma membrane of the sperm head suggests their importance in the organization or control of signal transduction or cellular metabolism. Membrane subdomains enriched in cholesterol and sphingolipids, as opposed to phospholipids, have been suggested to perform these functions in somatic cells. These domains have been termed "membrane rafts," as they are believed to represent liquid-ordered domains in a "sea" of liquid-disordered membrane (refs. 25, 26; see also Simons and Ehehalt, this Perspective series, ref. 27). Their lipid composition makes the raft fractions of membranes separable from nonraft fractions based on their insolubility in detergents such as Triton X-100 or on their lighter buoyant density when centrifuged through density gradients. Using these two biochemical criteria, we have recently demonstrated that mammalian sperm possess such membrane rafts (14).

What then are the functions of cholesterol and these rafts in sperm membranes? The first potential role is in compartmentalizing pathways to specific regions of the cell. This "prefabricated" ordering of pathway components is critical in sperm because of their extraordinarily polarized design, as well as the fact that they are both transcriptionally and translationally inactive. Sperm must assemble and organize their pathways so that they may function precisely where needed, as they cannot synthesize new proteins to meet changing needs in the female tract. One protein enriched in sperm membrane rafts that might function to compartmentalize pathways is caveolin-1 (14). In somatic cells, this protein has been suggested to anchor a variety of signaling and metabolic proteins to membrane rafts (28–30). By scaffolding such molecules, caveolins have been suggested to tether pathway components in "preassembled complexes" that then can be activated by the dissociation of the interaction between the proteins and caveolin (31). Although specific caveolin-interacting proteins have yet to be identified in sperm, we have localized caveolin to the same regions of the sperm that are enriched in cholesterol (Figure 2) and have determined that proteins believed to be important for acrosomal exocytosis are enriched in membrane raft fractions (ref. 14; A.J. Travis et al., unpublished data). Together, these data suggest a role for membrane rafts and caveolin in mediating the localization and/or organization of specific signaling pathways in sperm.

In addition to physically compartmentalizing specific pathways, membrane rafts may regulate such pathways by facilitating the efflux of cholesterol from the sperm plasma membrane. Cholesterol efflux from rafts might initiate signaling by at least two mechanisms. First, efflux could increase membrane fluidity and thus allow previously partitioned integral membrane proteins or membrane-anchored proteins to interact with one another in order to initiate signaling. In this regard, proteins believed to be important in the fusion of the sperm with the egg plasma membrane have been shown to translocate from the anterior to the posterior sperm head during capacitation, suggesting that the loss of cholesterol and the concomitant increase in membrane fluidity is essential for fertilization (32). Second, cholesterol efflux could activate signaling by disrupting the interaction of caveolin with specific signaling molecules, thereby freeing them to form functional signaling complexes. One argument against this latter possibility being critical to sperm function is the finding that mice carrying a targeted deletion of the *Caveolin1* gene appear to be fertile (33, 34). However, given the essential role of sperm in the propagation of life, a redundancy of systems would not be unexpected. Indeed, knockout models of several genes hypothesized to be critical for fertilization have resulted in only subtle reductions in male fertility (35, 36).

If sperm membrane rafts function in part by mediating cholesterol efflux, then a loss of cholesterol from such regions should be observed when sperm are incubated under capacitating conditions. In fact, cholesterol efflux results in dramatic changes in the pattern and number of FSCs, with loss from the plasma

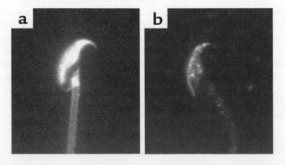

Figure 2
Localization of sterols and caveolin-1 in noncapacitated murine sperm. (**a**) The localization of sterols in murine sperm, as visualized by the inherent fluorescence properties of filipin. (**b**) Immunolocalization of caveolin-1 in murine sperm. As performed in ref. 14.

membrane overlying the acrosome, some diffusion of FSCs into the post-acrosomal plasma membrane, and loss from the plasma membrane of the flagellum (21, 23, 24). A semiquantitative analysis of cholesterol efflux, based on the density of FSCs in different sperm regions before and after capacitation, suggests that efflux occurs from all areas of the sperm that originally contained cholesterol, including both the head and the flagellum (24). Hence, the molecule or molecules that mediate this efflux are likely widespread throughout this cell.

Models of cholesterol efflux from sperm membranes

In somatic cells, several molecular pathways have been proposed to carry out cholesterol efflux (see Tall et al., this Perspective series, ref. 37). These include unmediated aqueous diffusion, interactions with lipid-poor apolipoproteins, membrane micro-solubilization, or efflux that is mediated by specific molecules by either facilitated or active transport mechanisms (38, 39). Molecules proposed to mediate these processes include the scavenger receptors SR-BI and CD36 (40), members of the ATP-binding cassette (ABC) transporter family (41), and caveolin (42), although the role of this latter candidate has been controversial (43). Several of these proteins function most efficiently in concert with a specific class of sterol acceptor. For example, SR-BI mediates efflux to HDL, whereas ABC-A1 mediates efflux to lipid-poor apolipoproteins such as apoA1. Because simple diffusion into an aqueous medium is inefficient, physiological cholesterol efflux from sperm most likely is enhanced by the presence of cholesterol acceptors in the luminal fluid of the female tract. As mentioned above, this phenomenon can be mimicked in vitro by incubating sperm in the presence of a cholesterol acceptor such as serum albumin or β-cyclodextrins.

Several reports suggest that the cholesterol acceptors HDL and albumin, which are both found in oviductal and follicular fluid, can stimulate capacitation (44, 45). Interestingly, human follicular fluid albumin provides a more efficient sink for cellular cholesterol than does the better described serum albumin (44). Moreover, HDL levels in bovine oviductal fluid vary over the estrous cycle, having an inverse log relationship with serum progesterone (45). Thus, the abundance of a known cholesterol acceptor increases at the time of estrus. A better understanding of the nature of these changes and their regulation could shed light on the mechanism of cholesterol efflux from the sperm plasma membrane and might be of great benefit for helping define specific subsets of idiopathic infertility. Such information might also suggest alternative approaches to contraception. However, knockout mice lacking each of the genes encoding the most obvious candidates mentioned above have been generated and are apparently fertile (33, 34, 46, 47). These negative findings may be explained by the effect of oviductal fluid albumin, which is plentiful and can function as a cholesterol acceptor to activate sperm function. Albumin is believed to function nonspecifically by providing a relatively hydrophobic environment in the vicinity of the plasma membrane, facilitating the otherwise inefficient diffusion of cholesterol into an aqueous medium. Efflux to oviductal fluid albumin might provide a redundant, nonspecific mechanism for efflux in vivo, thus safeguarding sperm signaling and fertilization competence when the specific pathways are compromised, as in the knockout models studied.

Recent work on phospholipid scramblases has begun to clarify the relationship between HCO_3^- and Ca^{2+} signaling and the induction of cholesterol efflux during capacitation (48, 49). These enzymes translocate choline phospholipids to the inner leaflet,

and aminophospholipids to the outer leaflet along their concentration gradients, thus reducing the asymmetry of phospholipid distribution across the membrane bilayer (50). Phospholipid scrambling in sperm appears to require both exposure to HCO_3^- and PKA activity (48, 51–53), and it has been proposed that cAMP generated by sAC triggers a downstream increase in phospholipid scramblase activity, which in turn facilitates cholesterol efflux, potentially through a mediator such as SR-BI (53). A variation of this model, reflecting our observation that rafts are dissipated during capacitation, is shown in Figure 1. Either caveolin or the local topography of a membrane subdomain such as a raft might also promote efflux by providing a clustering of cholesterol, raising its local concentration. This would, in turn, promote efflux down a gradient either specifically through SR-BI to HDL, or nonspecifically to albumin.

Other sterols and lipids in the sperm

Despite the focus on cholesterol above, it should be noted that sperm are remarkable for the variety of sterols they possess. The sperm of rodents, primates, and other species contain varying amounts of desmosterol (21, 54), which undergoes efflux from the sperm membrane during capacitation and could function in a manner similar to that of cholesterol (21). Sperm cells of different species also contain variable amounts of sterol sulfates (10, 21). Sterol sulfotransferases, which have been reported to exist within the female tract, are presumed to render sperm membranes more fluid as part of capacitation (55).

Ceramides, another class of membrane lipids that has been implicated in cell signaling, may also contribute to the control of sperm function. For example, increasing sperm ceramide levels has been shown to enhance capacitation by increasing the efflux of cholesterol and desmosterol (56). However, it is unclear whether this effect is through the direct action of the ceramide produced on downstream signaling proteins, or through an increase in lipid disorder, such as that promoted by the phospholipid scramblases. The basic structural component of sphingolipids, ceramide, can be formed by the degradation of sphingomyelin by sphingomyelinase. As recently reviewed by Kolesnick (57), ceramide can exert signaling effects on cells via several independent mechanisms. First, it can increase membrane fluidity by changing lipid packing. In addition, ceramide can directly affect the activity of protein phosphatases and protein kinases. Finally, ceramide can act indirectly by its degradation via ceramidase into sphingosine, which can be phosphorylated by sphingosine kinase into sphingosine-1-phosphate (S1P). This highly reactive compound has been shown to stimulate a G protein–coupled receptor, $S1P_1$ or EDG-1 (58). Given the complex and dynamic sterol and lipid composition of sperm, much work needs to be done to elucidate the pathways regulating sterol/lipid efflux and transducing such efflux into downstream signaling events that ultimately regulate sperm function.

Conclusions and future directions

We are now in a position to make significant advances in our understanding of sperm capacitation at the molecular level, progress that should ultimately lead to new and better techniques for enhancing fertility, identifying and treating certain forms of male infertility, and preventing conception. One remarkable insight from this work has been the importance of membrane cholesterol efflux in initiating transmembrane signaling events that confer fertilization competence. The identity of the physiologically relevant cholesterol acceptors and modulators of cholesterol efflux is therefore of great interest. Still, it is clear that cholesterol efflux represents only a part of this story. The involvement of phospholipid translocation in mediating dynamic changes in the membrane, rendering it conducive to transmembrane signaling, and the modulation of membrane components of signal transduction cascades by cholesterol or phospholipids will yield important insights into the links between environmental sensing and transmembrane signaling in the sperm. How these dynamic events within the membrane are integrated with membrane microdomains and rafts should provide new and exciting avenues of investigation for the foreseeable future.

Acknowledgments

This work was supported by NIH grants 1K01-RR00188 (to A.J. Travis) and P01-HD06274 (to G.S. Kopf) and a CONRAD Mellon Junior Investigator Award (to A.J. Travis).

Address correspondence to: Gregory S. Kopf, Women's Health Research Institute, Wyeth Research, PO Box 8299, Philadelphia, Pennsylvania 19101-8299, USA. Phone: (484) 865-7846; Fax: (484) 865-9367; E-mail: kopfg@wyeth.com.

Alexander J. Travis's present address is: The James A. Baker Institute for Animal Health, College of Veterinary Medicine, Cornell University, Ithaca, New York, USA.

1. Austin, C.R. 1952. The "capacitation" of the mammalian sperm. *Nature.* **170**:326.
2. Chang, M.C. 1951. Fertilizing capacity of spermatozoa deposited into the fallopian tubes. *Nature.* **168**:697–698.
3. Wassarman, P.M. 1999. Mammalian fertilization: molecular aspects of gamete adhesion, exocytosis, and fusion. *Cell.* **96**:175–183.
4. Kopf, G.S. 2002. Signal transduction mechanisms regulating sperm acrosomal exocytosis. In *Fertilization.* D.M. Hardy, editor. Academic Press Inc. San Diego, California, USA. 181–223.
5. Quill, T.A., and Garbers, D.L. 2002. Sperm motility activation and chemoattraction. In *Fertilization.* D.M. Hardy, editor. Academic Press Inc. San Diego, California, USA. 29–55.
6. Suarez, S.S. 1998. The oviductal sperm reservoir in mammals: mechanisms of formation. *Biol. Reprod.* **58**:1105–1107.
7. Kopf, G.S., Visconti, P.E., and Galantino-Homer, H. 1999. Capacitation of the mammalian spermatozoon. *Advances in Developmental Biochemistry.* **5**:81–105.
8. Jaiswal, B.S., and Eisenbach, M. 2002. Capacitation. In *Fertilization.* D.M. Hardy, editor. Academic Press Inc. San Diego, California, USA. 57–117.
9. Yanagimachi, R. 1994. Mammalian fertilization. In *The physiology of reproduction.* E. Knobil and J.D. Neill, editors. Raven Press. New York, New York, USA. 189–317.
10. Cross, N.L. 1998. Role of cholesterol in sperm capacitation. *Biol. Reprod.* **59**:7–11.
11. Visconti, P.E., et al. 1995. Capacitation of mouse spermatozoa. I. Correlation between the capacitation state and protein tyrosine phosphorylation. *Development.* **121**:1129–1137.
12. Hamilton, D.W., Jones, A.L., and Fawcett, D.W. 1969. Cholesterol biosynthesis in the mouse epididymis and ductus deferens: a biochemical and morphological study. *Biol. Reprod.* **1**:167–184.
13. Scott, T.W. 1973. Lipid metabolism of spermatozoa. *J. Reprod. Fertil. Suppl.* **18**:65–76.
14. Travis, A.J., et al. 2001. Expression and localization of caveolin-1, and the presence of membrane rafts, in mouse and Guinea pig spermatozoa. *Dev. Biol.* **240**:599–610.
15. Mittag, T.W., Guo, W.B., and Kobayashi, K. 1993. Bicarbonate-activated adenylyl cyclase in fluid-transporting tissues. *Am. J. Physiol.* **264**:F1060–F1064.
16. Buck, J., Sinclair, M.L., Schapal, L., Cann, M.J., and Levin, L.R. 1999. Cytosolic adenylyl cyclase defines a unique signaling molecule in mammals. *Proc. Natl. Acad. Sci. USA.* **96**:79–84.
17. Chen, Y., et al. 2000. Soluble adenylyl cyclase as an evolutionarily conserved bicarbonate sensor. *Science.* **289**:625–628.
18. Jaiswal, B.S., and Conti, M. 2001. Identification and functional analysis of splice variants of the germ cell soluble adenylyl cyclase. *J. Biol. Chem.* **276**:31698–31708.

19. Langlais, J., and Roberts, K.D. 1985. A molecular membrane model of sperm capacitation and the acrosome reaction of mammalian spermatozoa. *Gamete Res.* **12**:183–224.
20. Visconti, P.E., et al. 1999. Cholesterol efflux-mediated signal transduction in mammalian sperm. β-Cyclodextrins initiate transmembrane signaling leading to an increase in protein tyrosine phosphorylation and capacitation. *J. Biol. Chem.* **274**:3235–3242.
21. Visconti, P.E., et al. 1999. Cholesterol efflux-mediated signal transduction in mammalian sperm: cholesterol release signals an increase in protein tyrosine phosphorylation during mouse sperm capacitation. *Dev. Biol.* **214**:429–443.
22. Friend, D.S. 1982. Plasma-membrane diversity in a highly polarized cell. *J. Cell Biol.* **93**:243–249.
23. Suzuki, F. 1988. Changes in the distribution of intramembranous particles and filipin-sterol complexes during epididymal maturation of golden hamster spermatozoa. *J. Ultrastruct. Mol. Struct. Res.* **100**:39–54.
24. Lin, Y., and Kan, F.W. 1996. Regionalization and redistribution of membrane phospholipids and cholesterol in mouse spermatozoa during in vitro capacitation. *Biol. Reprod.* **55**:1133–1146.
25. Smart, E.J., et al. 1999. Caveolins, liquid-ordered domains, and signal transduction. *Mol. Cell. Biol.* **19**:7289–7304.
26. Simons, K., and Toomre, D. 2000. Lipid rafts and signal transduction. *Nat. Rev. Mol. Cell Biol.* **1**:31–39.
27. Simons, K., and Ehehalt, R. 2002. Cholesterol, lipid rafts, and disease. *J. Clin. Invest.* **110**:597–603. doi:10.1172/JCI200216390.
28. Razani, B., Rubin, C.S., and Lisanti, M.P. 1999. Regulation of cAMP-mediated signal transduction via interaction of caveolins with the catalytic subunit of protein kinase A. *J. Biol. Chem.* **274**:26353–26360.
29. Li, S., et al. 1995. Evidence for a regulated interaction between heterotrimeric G proteins and caveolin. *J. Biol. Chem.* **270**:15693–15701.
30. Scherer, P.E., and Lisanti, M.P. 1997. Association of phosphofructokinase-M with caveolin-3 in differentiated skeletal myotubes. Dynamic regulation by extracellular glucose and intracellular metabolites. *J. Biol. Chem.* **272**:20698–20705.
31. Okamoto, T., Schlegel, A., Scherer, P.E., and Lisanti, M.P. 1998. Caveolins, a family of scaffolding proteins for organizing "preassembled signaling complexes" at the plasma membrane. *J. Biol. Chem.* **273**:5419–5422.
32. Cowan, A.E., Koppel, D.E., Vargas, L.A., and Hunnicutt, G.R. 2001. Guinea pig fertilin exhibits restricted lateral mobility in epididymal sperm and becomes freely diffusing during capacitation. *Dev. Biol.* **236**:502–509.
33. Drab, M., et al. 2001. Loss of caveolae, vascular dysfunction, and pulmonary defects in caveolin-1 gene-disrupted mice. *Science.* **293**:2449–2452.
34. Razani, B., et al. 2001. Caveolin-1 null mice are viable

but show evidence of hyperproliferative and vascular abnormalities. *J. Biol. Chem.* **276**:38121–38138.
35. Baba, T., Azuma, S., Kashiwabara, S., and Toyoda, Y. 1994. Sperm from mice carrying a targeted mutation of the acrosin gene can penetrate the oocyte zona pellucida and effect fertilization. *J. Biol. Chem.* **269**:31845–31849.
36. Lu, Q., and Shur, B.D. 1997. Sperm from beta 1,4-galactosyltransferase-null mice are refractory to ZP3-induced acrosome reactions and penetrate the zona pellucida poorly. *Development.* **124**:4121–4131.
37. Tall, A.R., Costet, P., and Wang, N. 2002. Regulation and mechanisms of macrophage cholesterol efflux. *J. Clin. Invest.* In press. doi:10.1172/JCI200216391.
38. Rothblat, G.H., et al. 1999. Cell cholesterol efflux: integration of old and new observations provides new insights. *J. Lipid Res.* **40**:781–796.
39. Fielding, C.J., and Fielding, P.E. 2001. Cellular cholesterol efflux. *Biochim. Biophys. Acta.* **1533**:175–189.
40. Krieger, M. 1998. The "best" of cholesterols, the "worst" of cholesterols: a tale of two receptors. *Proc. Natl. Acad. Sci. USA.* **95**:4077–4080.
41. Santamarina-Fojo, S., Remaley, A.T., Neufeld, E.B., and Brewer, H.B., Jr. 2001. Regulation and intracellular trafficking of the ABCA1 transporter. *J. Lipid Res.* **42**:1339–1345.
42. Fielding, P.E., and Fielding, C.J. 1995. Plasma membrane caveolae mediate the efflux of cellular free cholesterol. *Biochemistry.* **34**:14288–14292.
43. Matveev, S., Uittenbogaard, A., van Der Westhuyzen, D., and Smart, E.J. 2001. Caveolin-1 negatively regulates SR-BI mediated selective uptake of high-density lipoprotein-derived cholesteryl ester. *Eur. J. Biochem.* **268**:5609–5616.
44. Langlais, J., et al. 1988. Identification of sterol acceptors that stimulate cholesterol efflux from human spermatozoa during in vitro capacitation. *Gamete Res.* **20**:185–201.
45. Ehrenwald, E., Foote, R.H., and Parks, J.E. 1990. Bovine oviductal fluid components and their potential role in sperm cholesterol efflux. *Mol. Reprod. Dev.* **25**:195–204.
46. Febbraio, M., et al. 1999. A null mutation in murine CD36 reveals an important role in fatty acid and lipoprotein metabolism. *J. Biol. Chem.* **274**:19055–19062.
47. McNeish, J., et al. 2000. High density lipoprotein deficiency and foam cell accumulation in mice with targeted disruption of ATP-binding cassette transporter-1. *Proc. Natl. Acad. Sci. USA.* **97**:4245–4250.
48. Harrison, R.A.P., Ashworth, P.J.C., and Miller, N.G.A. 1996. Bicarbonate/CO2, an effector of capacitation, induces a rapid and reversible change in the lipid architecture of boar sperm plasma membranes. *Mol. Reprod. Dev.* **45**:378–391.
49. Gadella, B.M., Miller, N.G., Colenbrander, B., van Golde, L.M., and Harrison, R.A. 1999. Flow cytometric detection of transbilayer movement of fluorescent phospholipid analogues across the boar sperm plasma membrane: elimination of labeling

artifacts. *Mol. Reprod. Dev.* **53**:108–125.

50. Basse, F., Stout, J.G., Sims, P.J., and Wiedmer, T. 1996. Isolation of an erythrocyte membrane protein that mediates Ca2+-dependent transbilayer movement of phospholipid. *J. Biol. Chem.* **271**:17205–17210.

51. Purohit, S.B., Laloraya, M., and Kumar, P.G. 1998. Bicarbonate-dependent lipid ordering and protein aggregation are part of the nongenomic action of progesterone on capacitated spermatozoa. *J. Androl.* **19**:608–618.

52. Gadella, B.M., and Harrison, R.A. 2000. The capacitating agent bicarbonate induces protein kinase A-dependent changes in phospholipid transbilayer behavior in the sperm plasma membrane. *Development.* **127**:2407–2420.

53. Flesch, F.M., et al. 2001. Bicarbonate stimulated phospholipid scrambling induces cholesterol redistribution and enables cholesterol depletion in the sperm plasma membrane. *J. Cell Sci.* **114**:3543–3555.

54. Lin, D.S., Connor, W.E., Wolf, D.P., Neuringer, M., and Hachey, D.L. 1993. Unique lipids of primate spermatozoa: desmosterol and docosahexaenoic acid. *J. Lipid Res.* **34**:491–499.

55. Langlais, J., et al. 1981. Localization of cholesteryl sulfate in human spermatozoa in support of a hypothesis for the mechanism of capacitation. *Proc. Natl. Acad. Sci. USA.* **78**:7266–7270.

56. Cross, N.L. 2000. Sphingomyelin modulates capacitation of human sperm *in vitro. Biol. Reprod.* **63**:1129–1134.

57. Kolesnick, R. 2002. The therapeutic potential of modulating the ceramide/sphingomyelin pathway. *J. Clin. Invest.* **110**:3–8. doi:10.1172/JCI200216127.

58. Hla, T., Lee, M.J., Ancellin, N., Paik, J.H., and Kluk, M.J. 2001. Lysophospholipids: receptor revelations. *Science.* **294**:1875–1878.

Trophoblast differentiation during embryo implantation and formation of the maternal-fetal interface

Kristy Red-Horse,[1] Yan Zhou,[1] Olga Genbacev,[1] Akraporn Prakobphol,[1] Russell Foulk,[2] Michael McMaster,[1] and Susan J. Fisher[1,3,4,5]

[1]Department of Stomatology, University of California San Francisco, San Francisco, California, USA. [2]The Nevada Center for Reproductive Medicine, Reno, Nevada, USA. [3]Department of Obstetrics, Gynecology and Reproductive Sciences, [4]Department of Pharmaceutical Chemistry, and [5]Department of Anatomy, University of California San Francisco, San Francisco, California, USA.

Trophoblasts, the specialized cells of the placenta, play a major role in implantation and formation of the maternal-fetal interface. Through an unusual differentiation process examined in this review, these fetal cells acquire properties of leukocytes and endothelial cells that enable many of their specialized functions. In recent years a great deal has been learned about the regulatory mechanisms, from transcriptional networks to oxygen tension, which control trophoblast differentiation. The challenge is to turn this information into clinically useful tests for monitoring placental function and, hence, pregnancy outcome.

In some societies the critical importance of the placenta in determining pregnancy outcome is acknowledged by its special treatment after birth. For example, the Malay people bury placentas in prominent locations near their homes, a symbolic act in recognition of the fact that the placenta is an essential in utero companion of the baby. We now know that this prescient ritual reflects an important reality. During the last decade and a half, numerous studies in transgenic mice have shown that placentation is a critical regulator of embryonic and fetal development. For example, inactivation of the *retinoblastoma* (*Rb*) tumor suppressor gene in mice resulted in unscheduled cell proliferation, apoptosis, and widespread developmental defects that eventually led to embryonic death. Careful analysis of the placenta showed numerous abnormalities, including disruption of the transport region where nutrient, waste, and gas exchange occurs and a decrease in vascularization. When conditional knockout and tetraploid rescue strategies were used to supply Rb-null embryos with wild-type placentas, the animals survived to term; the neurological and erythroid abnormalities that were thought to lead to the intrauterine demise of the pups were abrogated (1). Thus, Rb function in the murine extraembryonic lineages is required for normal differentiation of cells within the embryo. The principle that embryonic and placental development are tightly linked has been strongly reinforced by studies in the burgeoning field of life-course epidemiology and has led to the proposal of the "developmental origins hypothesis": adult medical conditions such as cardiovascular disease and type 2 diabetes originate in response to undernutrition either in utero or during infancy and early childhood (2). Given that the placenta is an important regulator of fetal growth before birth, it is likely that a subset of the initiating events that eventually lead to the

aforementioned diseases will be traced to faulty placentation at structural and/or functional levels (S1).

In this context it becomes critically important to understand, at a molecular level, placental development, which determines the organ's functional capacity. This review focuses on one part of this process: trophoblast differentiation, a component that is integral to implantation and trophoblast invasion of the uterus.

As discussed below, the trophectoderm layer of the blastocyst is the first embryonic cell type to exhibit highly differentiated functions. Approximately 1 week after fertilization, trophoblasts participate in a complex dialogue with maternal cells that enables implantation, a process that quickly sequesters the human embryo within the uterine wall. Further embryonic development requires the rapid assembly of the basic building blocks of the placenta: floating and anchoring chorionic villi. The unique structure of the human maternal-fetal interface is established by differentiation of cytotrophoblasts in anchoring villi. The latter process entails many unusual elements. For example, these fetal cells, which are derived from the placenta, form elaborate connections with maternal vessels, thereby diverting uterine blood flow to the placenta. Moreover, the hemiallogeneic placental cells that reside within the uterine wall coexist with an unusual population of decidual leukocytes, predominantly CD56^{bright}, CD16$^-$ NK cells. Due to the sheer size and complexity of the published literature, an in-depth analysis of trophoblast interactions with the maternal immune system is beyond the scope of this review.

In humans, defects in formation of the maternal-fetal interface that are associated with a variety of pregnancy complications are helping to pinpoint critical aspects of this process that are particularly vulnerable to failure. Genomic and proteomic approaches will yield a great deal of information about the mechanisms involved. The challenge is to translate this knowledge into clinical tests of placental function for the purpose of predicting, diagnosing, and/or treating the major diseases associated with pregnancy, such as preeclampsia, with an incidence of 7–8% (3), and preterm labor, with an incidence of 10% (4). The application of these types of tests will revolutionize the practice

---

Nonstandard abbreviations used: bHLH, basic helix-loop-helix; E, embryonic day; HIF, hypoxia-inducible factor; LIF, leukemia inhibitory factor; PlGF, placental growth factor; Rb, retinoblastoma; VHL, von Hippel–Lindau.

Conflict of interest: The authors have declared that no conflict of interest exists.

Citation for this article: *J. Clin. Invest.* **114**:744–754 (2004). doi:10.1172/JCI200422991.

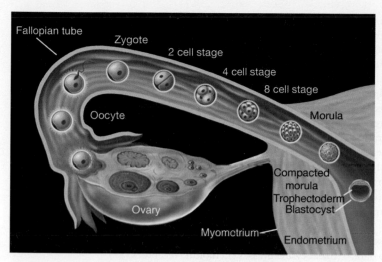

Figure 1
The early stages of human development from fertilization to blastocyst formation. Fertilization occurs in the fallopian tube within 24 to 48 hours of ovulation. The initial stages of development, from fertilized ovum (zygote) to a solid mass of cells (morula), occur as the embryo passes through the fallopian tube encased within a nonadhesive protective shell (the zona pellucida). The morula enters the uterine cavity approximately two to three days after fertilization. The appearance of a fluid-filled inner cavity marks the transition from morula to blastocyst and is accompanied by cellular differentiation: the surface cells become the trophoblast (and give rise to extraembryonic structures, including the placenta) and the inner cell mass gives rise to the embryo. Within 72 hours of entering the uterine cavity, the embryo hatches from the zona, thereby exposing its outer covering of trophectoderm. Figure kindly provided by S.S. Gambhir and J. Strommer, Stanford University (Stanford, California, USA).

of maternal-fetal medicine, with lifelong benefits to the health of both the mother and her offspring, and also have a huge impact on the staggering economic costs associated with these pregnancy complications.

Fertilization and the initial stages of development
Fertilization occurs in the fallopian tube within 24 to 48 hours of ovulation (Figure 1). The initial stages of development, from fertilized ovum (zygote) to a mass of cells (morula), occur as the embryo passes through the fallopian tube encased within a nonadhesive protective coating known as the zona pellucida. The morula enters the uterine cavity approximately 2 to 3 days after fertilization.

Implantation
Trophectoderm differentiation. Implantation is dependent on differentiation of the trophoblast lineage (reviewed in ref. 5). In the mouse, the first signs of this process are evident during the morula stage, when cell division creates two distinct cellular populations. The outer cells of the blastocyst, which are polarized, give rise to the trophectoderm, whereas cells in the interior, which are not polarized, give rise to the inner cell mass (Figure 1) (6). Interestingly, these initial events do not irreversibly establish the extraembryonic and embryonic murine lineages: the polarized outer cells can contribute to the inner cell mass (7), and the inner cells retain the ability to differentiate into trophectoderm (S2). Whether human blastomeres exhibit a similar degree of plasticity is unknown. At a molecular level, this cellular polarization

requires epithelial-cadherin–mediated homotypic cell adhesion (8) as well as protein kinase C signaling (9). The transcriptional regulators that govern segregation of the extraembryonic and embryonic lineages are also beginning to be identified. The homeobox gene *nanog* is more strongly expressed in the inner apolar cells of the morula than in the outer cells that are destined to form the trophectoderm. In the blastocyst, *nanog* expression is confined to the inner cell mass and, in the later embryo, to primordial germ cells. This factor, which is required for embryonic stem cell self-renewal, may play an important role in generation of the intraembryonic lineage (10, 11). Conversely, by the late morula stage, the caudal-related homeobox gene *cdx2* is more strongly expressed in the outer layer of polarized blastomeres (5). Overexpression of this transcription factor in embryonic stem cells induces trophoblast formation (S3). Additionally, the domain transcription factor Oct4 is required to generate the inner cell mass (12) by a mechanism that includes suppression of trophoblast-specific genes (13, 14). The high-mobility group transcription factor Sox2 may play a role similar to that of Oct4 (15). However, neither Oct4 nor Sox2 expression is confined to particular subsets of blastomeres, making it unlikely that these transcription factors are involved in initial decisions that govern the fate of these cells. The above findings led Rossant and colleagues (5) to propose that polarization provides molecular cues (e.g., *cdx2* expression) that specify the trophoblast lineage, with apolarization delivering different signals (e.g., Nanog, Oct4, and Sox2) to cells that are fated to become components of the inner cell mass. In this scenario it is interesting to note the important contribution of both genetic and positional cues. Recent data that were generated using an RNA interference approach to modulate gene expression in human embryonic stem cells suggest that these transcription factors are playing similar roles in this system (16).

The implantation process. The initial stages of embryonic and uterine development, which are spatially separated, must be temporally coordinated for pregnancy to occur. Once the blastocyst emerges from the zona pellucida, approximately 6 days after fertilization, the embryonic and maternal cells enter into a complex dialogue that enables implantation, a process that rapidly sequesters the human embryo within the uterine wall (reviewed in refs. 17 and 18). Hormones play very important roles in implantation. For example, studies using transgenic approaches show that uterine expression of progesterone receptor A and estrogen receptor α, but not progesterone receptor B or estrogen receptor β, is required for implantation (19, 20). Nevertheless, the situation is complex, as decidualization, the specialized response of endometrial stromal cells to pregnancy, can be experimentally induced in estrogen receptor α null females (21, 22). In mice, a very narrow range of estrogen concentrations determines the duration of the window of receptivity, i.e., the period of time during which the uterus is able to support implantation (23). Interestingly, estrogen also has effects on the blastocyst, as its 4-hydroxy catechol metabolite mediates blastocyst activation, a requisite step in implantation (24). As for the uterus, a number of cytokines, growth factors, and transcription factors play important roles in this cascade. For example, maternal expression of leukemia inhibitory factor (LIF) is required to initiate implantation (25). IL-11 (26), insulin-like growth factors and binding proteins (27, S4), Hoxa-10 (28, 29), and a forkhead transcription factor (30) are involved in decidualization. The dialogue between the blastocyst and the uterus involves molecular families that might logically be predicted to function during the peri-implantation period (e.g., EGF family members; ref. 31) as well as unexpected participants, such as endocannabinoids and their G-protein–coupled receptors (32). Embryonic signals such as those generated by chorionic gonadotropin also play important roles by acting on the endometrium (33). In keeping with the results of global gene profiling experiments in other systems, the application of this approach to the study of murine implantation (34, 35) and human uterine receptivity (36, 37) graphically demonstrates the myriad of mechanisms that are involved in these processes.

The current challenge is to use the data summarized above to gain entrée into the pathways that control implantation. To date, evidence from many different sources, including studies in mice and humans, suggests that LIF is at the top of the pyramid of molecules whose maternal expression is required for effective implantation. What are the downstream effectors? The long-standing observation that vascular permeability increases at the site of murine implantation prompted investigators to look for the expression of molecules with functions that are relevant to this process. For example, phospholipase A_2 generates arachidonic acid, a substrate for the COX-2 enzyme, which produces prostaglandins. In some genetic strains of mice, deleting *cox-2* interferes with implantation (38), while in others a compensatory upregulation of COX-1 expression rescues this phenotype, resulting in the birth of live pups, although litter size is reduced (39) because decidual growth is retarded (40). Interestingly, the converse is also true, e.g., COX-2 compensation is observed in the absence of COX-1 (41). With regard to the actions of *cox-2* products, evidence to date suggests that the COX-2–derived prostaglandin, prostacyclin, plays a particularly important role in implantation by activating the nuclear hormone receptor PPARδ (42), which is required for normal (murine) placental development (43). Other PPAR family members, such as PPARγ, play important roles in human trophoblast differentiation and function (S5, S6). Finally, an emerging area of research is the investigation of evolutionarily conserved genes, including those encoding FGFs, IGFs, bone morphogenetic proteins, Wnts, Noggin, and Indian hedgehog proteins and their receptors, which appear to have potential roles in implantation and embryo spacing in the uterus (44, 45). It is important to note that other pathways lie downstream from LIF, e.g., Msx-1 and Wnt4, whose expression is also aberrant in Hoxa10–null mice (46).

In humans, the implantation story has an unexpected twist. Recent evidence suggests that the L-selectin system, which mediates rolling and tethering of leukocytes on blood vessels, plays an important role in implantation (47). This mechanism is particularly well suited for initiating implantation because of its many special attributes (48). For example, the rapid kinetics that characterize interactions between these carbohydrate-binding receptors (on leukocytes) and their specialized oligosaccharide ligands (on blood vessel walls) allows for rolling and tethering of leukocytes when they encounter shear stress. Subsequently, integrin-mediated firm adhesion is triggered by exposure to the chemokine- and cytokine-rich milieu at the vessel wall. At a morphological level, analogies can be drawn between key steps in leukocyte emigration from blood and trophoblast attachment to the uterine wall.

Implantation begins with apposition; the trophectoderm of the originally free-floating blastocyst lies adjacent to the uterine epithelium, but the blastocyst is easily dislodged (Figure 2) (49). Soon thereafter, blastocyst adhesion to the uterine wall is stabilized, and trophoblasts transmigrate across the uterine epithelium, a process that in humans buries the entire embryo beneath the uterine surface. Subsequent development depends on the ability of trophoblasts to adhere to the uterine epithelium under conditions of shear stress that are created when these fetal cells breach uterine vessels, a process that diverts maternal blood flow to the placenta (Figure 3). At a molecular level, trophoblast adhesion from the stage of implantation onwards is an integrin-dependent process (50, 51) that takes place in a chemokine- and cytokine-rich microenvironment analogous to the blood-vascular interface (52, 53). In this context it is interesting to note that in humans uterine expression of chemokines is hormonally regulated and the blastocyst expresses chemokine receptors (S7).

Together, these findings raised the possibility that implantation and placentation utilize other components of the leukocyte emigration system, such as selectins and their ligands (Figure 2). Immunolocalization experiments showed that the trophectoderm, which covers the surface of implantation-competent human embryos, stains brightly for L-selectin but does not express the other members of the selectin family. As the luminal epithelium becomes receptive during the luteal phase of the menstrual cycle, these uterine cells display a dramatic upregulation of the expression of the specialized sulfated oligosaccharides that function as high-affinity receptors for L-selectin. A variety of assays show that the L-selectin system functions by tethering human trophoblasts under conditions of shear stress. It will be very interesting to identify points of intersection between this pathway and the other molecular cascades, summarized above, that are known to regulate implantation.

Finally, there are interesting data from both human and murine systems that highlight the importance of achieving implantation during the period of optimal uterine receptivity. For example, female mice with a null mutation for the gene encoding cytosolic phospholipase $A_2\alpha$ have small litters and often exhibit pregnancy failure (54). Interestingly, the cause is a shifting forward in time of the window of receptivity. The consequences of delaying implantation include retarded fetal-placental growth and abnormal uterine spacing of embryos. Similarly, mice that lack expression of the LDL receptor–related protein, which was originally thought to be required for implantation (55), exhibit delayed development. It is likely that this same effect or a related variant also occurs in humans, as the risk of early pregnancy loss increases as a function of delaying implantation (56).

Trophoblast invasion and maturation of the maternal-fetal interface

Trophoblast stem cell self-renewal. Once the embryo is anchored within the uterine wall, the next major hurdle is rapid formation of the extraembryonic lineages, a necessary prelude to assembly of the maternal-fetal interface. In recent years a great deal of information about trophoblast differentiation has come from a variety of experimental approaches. The generation of mouse trophoblast stem cell lines from early-stage mouse embryos (i.e., prior to embryonic day [E] 7.5) has proved to be an important experimental tool for studying this process as well as the mechanisms that promote self-renewal of the trophoblast stem

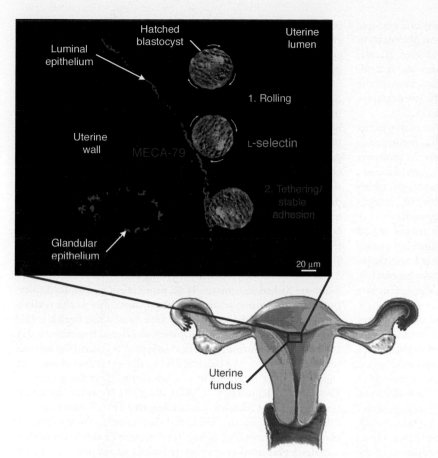

Figure 2

Implantation in humans involves a number of the molecular mechanisms that mediate leukocyte emigration from the blood to sites of inflammation or injury. The diagram was made from a combination of images: MECA-79 antibody staining of uterine tissue sections and L-selectin antibody staining of cultured embryos. Recently acquired evidence suggests that an implantation-competent human blastocyst expresses L-selectin on its surface (green). This receptor interacts with specialized carbohydrate ligands, including sulfated species, recognized by the MECA-79 antibody, which stains the uterine luminal and glandular epithelium. The specialized nature of these interactions translates into an unusual form of cell adhesion: rolling and tethering. In the uterus, MECA-79 immunoreactivity peaks during the window of receptivity. This finding suggests that apposition, the first step in implantation, includes L-selectin–mediated tethering of the blastocyst to the upper portion of the posterior wall of the uterine fundus.

cell population (57, 58). The actions of FGF family members are crucial, as trophoblast stem cells are derived by plating disaggregated extraembryonic cells on mouse embryonic fibroblasts in the presence of FGF4, which binds Fgfr2 IIIc, ultimately leading to MAPK signaling (5). To date, attempts to use an FGF-based strategy to derive human trophoblast stem cells have failed, which suggests that different factors are required for self-renewal of this population. This is not unexpected, as despite the many similarities at a molecular level between murine and human placentation, there are also dramatic morphological differences. Although treatment of human embryonic stem cells with bone morphogenic protein–4 results in the expression of syncytial trophoblast markers, the cells do not continue to divide (59). Thus, the majority of what we know about human trophoblast differentiation has come from studies of early-gestation placental cells using in situ and in vitro approaches (17).

Trophoblast differentiation. Additionally, gene deletion studies in mice have either advertently or inadvertently yielded interesting insights into the molecular requirements of normal placentation (60). Development of the tetraploid rescue technique for providing mutant embryos with wild-type placentas has allowed a careful separation of the embryonic and extraembryonic phenotypes (61, 62). As a result, many aspects of murine trophoblast differentiation can now be explained in terms of specific transcriptional regulators (5, 58, 63). As for those regulators that control early events, deletion of the T-box gene *Eomes* results in embryonic death soon after implantation due to an arrest in trophectoderm development (64). Embryos that lack expression of activating protein–2γ, a member of a family of transcription factors that controls cell pro-

liferation, die at E7.5 due to malformation of the maternal-fetal interface. The defect is likely attributable to negative effects on the stem cell population, which has reduced expression of *Eomes* and *cdx2*, two genes that are upregulated in response to FGF treatment of trophoblast stem cells (65). Deletion of an orphan receptor, estrogen-receptor–related receptor protein β, results in death at E9.5. Demise is attributable to depletion of the trophoblast stem/progenitor population, which is shunted to the giant cell lineage (66), the murine equivalent of the human cytotrophoblast population that carries out interstitial and endovascular invasion (58, 67). Interestingly, deletion of the TGF-β family member Nodal also leads to expansion of the trophoblast giant cell population at the expense of the other subtypes (68).

Basic helix-loop-helix (bHLH) transcription factors play important roles during the later stages of trophoblast differentiation. *Mash2*, a paternally imprinted gene, is required for the production of the precursor population that gives rise to the spongiotrophoblast layer (69). Interestingly, there is evidence that the polycomb group protein, extraembryonic ectoderm development (known as Eed), may be required for lineage-specific Mash2 repression during differentiation (70). Since bHLH transcription factors function as dimers, their binding partners also need to be expressed at the proper time and location. Finally, their activity can be inhibited by interactions with dominant-negative HLH factors that lack DNA-binding sequences, e.g., Id1 and Id2, which also localize to the extraembryonic ectoderm (S8). Human trophoblast progenitors also express Mash2 and Id2, which are downregulated as the cells differentiate along the invasive pathway (71). In keeping with the known functions of these factors, forced expression of Id2 inhibits cytotrophoblast differentiation in vitro (72). This is one example of the many similarities between murine and human placentation at the molecular level. In contrast, another bHLH factor, Hand1, is required for the differentiation of murine primary and secondary giant cells (73, 74). That human trophoblasts beyond the blastocyst stage do not express Hand1 is one of the few examples of divergent placental evolution in the two species (75, S9).

A number of transcription factors also regulate formation of the labyrinth zone, the area of the murine placenta that corresponds to the floating villi of the human placenta, i.e., where the transport of nutrients, wastes, and gasses takes place (58). A screen of human tissues revealed the surprising result that *glial cells missing-1,* which controls the neuronal to glial transition in *Drosophila* (76), is solely and constitutively expressed in placental cytotrophoblasts (71). Mice that lack expression of this transcription factor die at E10 because of a block in the branching of the chorioallantoic interface and an absence of the placental labyrinth (77). Other known regulators of labyrinth development include retinoic acid receptors (78, 79) and PPARγ (80). Wnt2 (81) and growth factor, e.g., hepatocyte growth factor (82), signals are also required.

Placental function regulates many aspects of embryonic and fetal development. When tetraploid (wild-type) and diploid (mutant) cells are aggregated, the hyperdiploid cells are allocated to the placenta. This method is a powerful technique for supplying mutant embryos with normal placentas, thereby separating a molecule's embryonic and extraembryonic effects (61, 62). The widespread application of this technology has revealed the critical importance of normal placental function to embryonic and fetal development. A startling array of effects is propagated downstream from abnormal placentation. For example, targeted mutation of the DNA-binding domain of the Ets2 transcription factor produces numerous defects in the extraembryonic compartment, including a substantial decrease in MMP-9 production, an attendant failure in extracellular matrix remodeling, and a proliferation defect that involves the ectoplacental cone. Ets2-null embryos, which are growth restricted, die before E8.5. Subsequent tetraploid rescue experiments resulted in the birth of viable, fertile mice with hair defects — a dramatically different embryonic phenotype from that observed when the entire conceptus lacks Ets2 expression (83). The general phenomenon of abnormal placental function affecting embryonic and fetal development has been documented in association with the deletion of many other genes from very diverse molecular families. For example, deletion of *JunB*, an immediate early gene product and member of the AP-1 transcription factor family, also causes fetal growth restriction. This ultimately leads to embryonic death between E8.5 and E10 due to defects in both placental and decidual vessels, the latter receiving improper signals from invading trophoblasts. Again, tetraploid rescue resulted in fetuses that were normally grown but in this case osteopenic (84). Likewise, fetal growth restriction secondary to defects in the placental labyrinth of the extracellular signal regulated kinase-2–null mice is rescued by tetraploid aggregation (85). Similarly, tetraploid rescue results in the birth of mice lacking the HGF receptor c-Met (86) or SOCS-3 (87).

In other cases, supplying mutant embryos with normal extraembryonic derivatives allows development to progress further than was previously possible. Mice that lack expression of thrombomodulin (88) and desmoplakin (89) fall into this category, as do homozygous null mutants that lack expression of *Alk2*, which encodes a type I TGF-β family receptor for activins and bone morphogenic protein-7 (90). The latter case provides direct evidence that signals transmitted through this receptor, which is expressed in the extraembryonic region, are required at the time of gastrulation in order to achieve normal mesoderm formation.

Interestingly, the impact of aneuploidy on the conceptus in terms of embryonic development can also be alleviated to varying degrees by tetraploid rescue. For example, the development of mice with a *t*-complex variant of chromosome 17 is sustained for several additional days by wild-type placental cells (91), and tetraploid aggregation allows the birth of mice that carry an additional maternal X chromosome (92). It is important to note that, while in some cases the effects on the embryo of placental malfunction are likely to be directly attributable to defects in the organs' transport functions, in many other instances the connection is likely to involve a higher order of complexity. For example, tetraploid aggregation shows that keratin 8 is a necessary component of the barrier that prevents TNF-mediated apoptosis of trophoblast giant cells (93).

The extent to which extraembryonic and embryonic development are linked in humans is an interesting, unresolved question that has gained additional importance due to the realization that the foundation of many aspects of adult health is laid down in utero (i.e., the developmental origins hypothesis [2]). In this context, normal placental function is critical for normal fetal development. At a biochemical level, a great deal of evidence suggests that alterations in placental transport functions are associated with growth restriction (94, S10). However, the actual cause-and-effect relationship is likely to be much more complicated. For example, uterine blood flow is a critical regulator of placental function and, hence, fetal growth (S11). Additionally, impaired placental transport is linked to reduced umbilical blood flow and attendant changes in the fetal circulation (S12). Thus, in humans, it is difficult to sort out the primary defect from the ripple effects. This problem is further complicated by the fact that many commonly used drugs (e.g., nicotine) negatively affect trophoblast differentiation and formation of the maternal-fetal interface (95, 96) as well as placental transport of amino acids and fetal growth (S13). Additionally, subclinical viral (e.g., cytomegalovirus) and bacterial infections, which are surprisingly common, can inhibit cytotrophoblast differentiation and/or invasion (97, 98). Despite the obvious complexity of these interrelationships it is possible to envision a deterioration in placental function that translates into alterations, at the molecular level, in the fetal circulation that are maintained throughout life, one possible explanation for why a restriction in intrauterine growth is linked to adult cardiovascular disease.

At a genetic level, fetal aneuploidies allow an evaluation of the impact of specific changes in chromosome number, deletion, and/or translocation on placental development. For example, in the case of trisomy 21, the ability of cytotrophoblasts to fuse into syncytiotrophoblasts is impaired (Figure 3) (S14). Additionally, cytotrophoblast differentiation along the pathway that leads to uterine invasion is dysregulated, as shown by abnormal expression of stage-specific antigens that are modulated during this process as well as a high rate of apoptosis among this subpopulation of cells (99). We speculate that the latter observations explain the high rate of fetal loss in these pregnancies (see Figure 3). In other cases, such as confined placental mosaicism, which occurs in approximately 2% of viable pregnancies, the consequences often include unexplained fetal growth restriction (S15). In this context, exploring the effects of specific

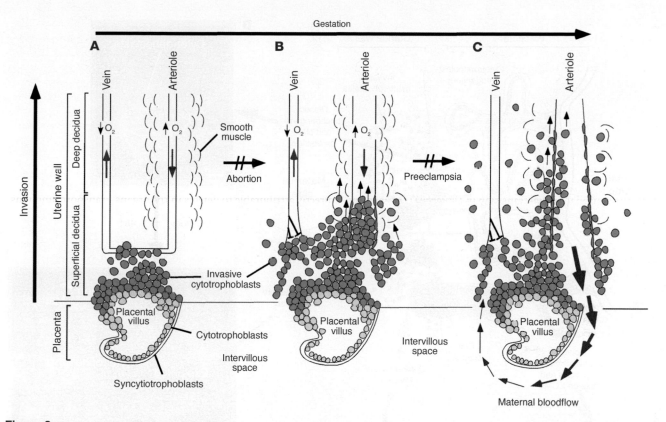

Figure 3

Oxygen tension plays an important role in guiding the differentiation process that leads to cytotrophoblast invasion of the uterus. **(A)** The early stages of placental development take place in a relatively hypoxic environment that favors cytotrophoblast proliferation rather than differentiation along the invasive pathway. Accordingly, this cell population (light green cells) rapidly increases in number as compared with the embryonic lineages. **(B)** As development continues, cytotrophoblasts (dark green cells) invade the uterine wall and plug the maternal vessels, a process that helps maintain a state of physiological hypoxia. As indicated by the blunt arrows, cytotrophoblasts migrate farther up arteries than veins. **(C)** By 10 to 12 weeks of human pregnancy, blood flow to the intervillous space begins. As the endovascular component of cytotrophoblast invasion progresses, the cells migrate along the lumina of spiral arterioles, replacing the maternal endothelial lining. Cytotrophoblasts are also found in the smooth muscle walls of these vessels. In normal pregnancy the process whereby placental cells remodel uterine arterioles involves the decidual and inner third of the myometrial portions of these vessels. As a result, the diameter of the arterioles expands to accommodate the dramatic increase in blood flow that is needed to support rapid fetal growth later in pregnancy. It is likely that failed endovascular invasion leads, in some cases, to abortion, whereas an inability to invade to the appropriate depth is associated with preeclampsia and a subset of pregnancies in which the growth of the fetus is restricted.

aneuploidies on placental development could yield new insights into mechanisms of both normal and abnormal placentation.

Oxygen regulates trophoblast differentiation and proliferation. Physiological factors also play an important role in formation of the maternal-fetal interface and, consequently, fetal growth. Oxygen tension, a function of uterine blood flow, is a prime example (S11). In recent years a great deal has been learned about the fundamental mechanisms that couple the trophoblasts' ability to sense oxygen levels with their differentiative and metabolic status. Important clues about oxygen's effects on the placenta have come from several lines of evidence that suggest that the early stages of placental (and embryonic) development take place in an environment that is hypoxic relative to the uterus. Specifically, direct measurement of uterine oxygen tension demonstrates physiological hypoxia (2–5% O_2; reviewed in ref. 100). Furthermore, blood flow to the human intervillous space does not begin until 10 to 12 weeks of pregnancy (101). Studies in both nonhuman primates (S16) and humans (102) suggest that trophoblasts actively limit their access to uterine blood by plugging the lumina of the decidual vessels. Why?

Our work shows that cytotrophoblasts proliferate in vitro under hypoxic conditions that are comparable to those found during early pregnancy in the uterine cavity and the superficial decidua. As trophoblast invasion of the uterus proceeds, the placental cells encounter increasingly higher oxygen levels, which trigger their exit from the cell cycle and subsequent differentiation (103, 104). Hypoxia also regulates cell fate in the murine placenta (105). We speculate that the paradoxical effects of oxygen in controlling the balance between cytotrophoblast proliferation and differentiation explain in part why the mass of the placenta increases much more rapidly than that of the embryo. Histological sections of early-stage pregnant human uteri show bilaminar embryos surrounded by thousands of trophoblast cells (S17). The fact that hypoxia stimulates cytotrophoblasts, but not most other cells, to undergo mitosis (106) could help account for the difference in size between the embryo and the placenta, a discrepancy that continues well into the second trimester of pregnancy (S18). Thus, the structure of the mature placenta is established in advance of the period of rapid fetal growth that occurs during the latter half of pregnancy.

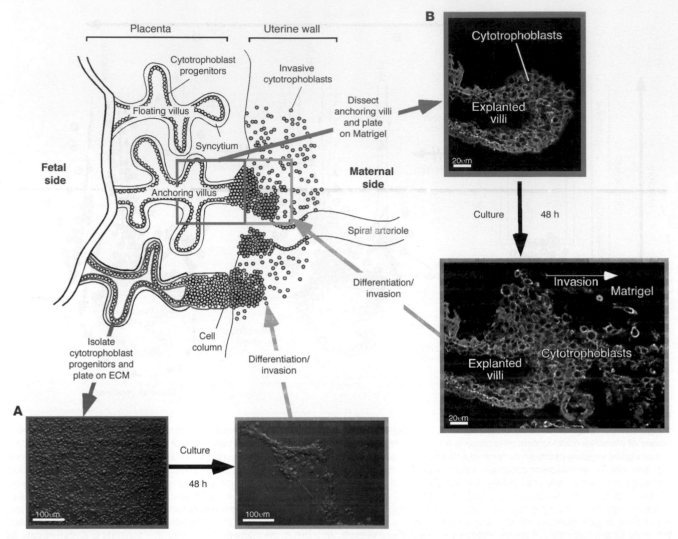

Figure 4

Illustration of two in vitro models for studying human cytotrophoblast invasion. (**A**) When human cytotrophoblasts (light green cells encircled in red) are isolated from early-gestation placentas and plated on an extracellular matrix (ECM) substrate (Matrigel), they differentiate along the pathway that leads to uterine invasion. By 12 hours in culture these cells form aggregates that resemble cell columns of anchoring villi, and by 48 hours they switch on expression of a repertoire of stage-specific antigens that are expressed in cytotrophoblasts within the uterine wall in situ (dark green cells). These molecules facilitate uterine invasion, vascular mimicry, and evasion of the maternal immune response. (**B**) When anchoring villi are dissected from the surfaces of early-gestation human placentas (blue box) and plated on an ECM substrate, cytotrophoblasts in cell columns continue to differentiate. By 48 hours many cytotrophoblasts have left the columns and invaded the substrate (green box). During this process they execute the same phenotypic switch that isolated cells carry out.

What are the molecular underpinnings of this unusual relationship? Many different lines of evidence suggest that the hypoxia-inducible factor (HIF) system that controls cellular responses to oxygen deprivation is involved (reviewed in refs. 107–109). The three HIF-α family members are bHLH transcription factors that also contain a Per/Arnt/Sim domain that facilitates their dimerization with HIF1-β (the aryl hydrocarbon receptor nuclear translocator). The heterodimers activate the transcription of numerous downstream targets by binding to a hypoxia-responsive promoter element (5′-TACGTG-3′) that is present in a variety of relevant genes, including VEGF, glucose transporter–1, and Stra13, the latter a regulator of murine placental development (63, 110). Because these responses need to be extremely rapid, an important element of control occurs at the protein level. Specifi-

cally, enzymatic hydroxylation of certain HIF-α proline residues is required for interactions with the von Hippel–Lindau (VHL) tumor suppressor protein, which under normoxic conditions targets these proteins for polyubiquitination and degradation in the proteosome. Additionally, hydroxylation of specific HIF-α asparagine residues prevents the recruitment of transcriptional coactivators. Interestingly, these enzymatic reactions, which depend on molecular oxygen, do not occur in a low-oxygen environment, providing a direct link between hypoxia, HIF stabilization, and the transcription of downstream target genes. In keeping with the concept that oxygen plays an important role in placental development, deletion of many of the individual components of the cell's machinery for sensing and responding to changes in oxygen tension leads to prenatal lethality secondary to placental

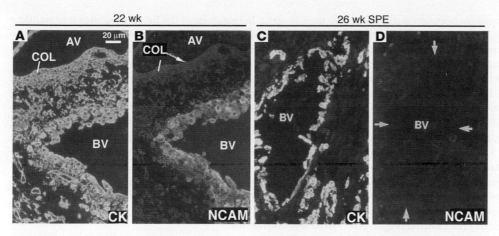

Figure 5
In severe preeclampsia as compared to normal pregnancy, cytotrophoblast invasion is shallow and the cells fail to switch on the expression of stage-specific antigens that are normally upregulated as they penetrate the uterine wall and blood vessels (BVs). (**A** and **C**) Cytokeratin (CK) staining of tissue sections of the maternal-fetal interface allows visualization of cytotrophoblasts that invade the uterine wall. (**B**) In normal pregnancy, the subpopulation of human cytotrophoblasts that carries out endovascular invasion upregulates expression of the neural cell adhesion molecule (NCAM). (**D**) In cases of severe preeclampsia (SPE), NCAM immunoreactivity is either absent or very weak (indicated by the arrows). AV, anchoring villus; COL, cell column of anchoring villus.

defects. For example, mice that lack expression of either VHL (111) or HIF1-β (105) die in utero as a result of faulty placentation. The lack of heat shock protein 90, which also controls HIF availability (112), is associated with placental defects that lead to embryonic death at E9.0/9.5 (113). VHL and HIFs also play important roles in human cytotrophoblast differentiation and/or invasion in vitro (114). The latter data are in accord with studies of normal and abnormal placentation in vivo. For example, trophoblast remodeling of spiral arterioles is restricted at high altitude, suggesting that hypoxia affects placentation in utero (S19). Interestingly, the incidence of preeclampsia (the new and rapid onset of maternal hypertension, proteinuria, and edema), which is associated with reduced blood flow to the placenta and deficient arterial invasion (S20, S21), increases several-fold at high elevations (S22).

Angiogenic and/or vasculogenic factors regulate trophoblast differentiation and proliferation. Angiogenesis is one well-known consequence of hypoxia (115). In the special case of placentation, the combined actions of the aforementioned transcriptional and physiological regulators result in the differentiation of an unusual subpopulation of cytotrophoblasts with many vascular-type attributes. The progenitors, which are components of anchoring chorionic villi, leave the placenta and attach to the uterine wall, the first step in both the interstitial and endovascular components of uterine invasion (Figure 3). A great deal about these processes has been learned by constructing in vitro models that allow functional perturbation of molecules that are regulated during human uterine invasion (Figure 4). For example, as cytotrophoblasts transition from the fetal to the maternal compartment, they modulate expression of a broad repertoire of growth factors and receptors with diverse roles in vasculogenesis, angiogenesis, and lymphangiogenesis: VEGFR1–3, soluble VEGFR-1 (also known as sFlt1), VEGF-A, VEGF-C, and placental growth factor (PlGF). Functional perturbation experiments show that these ligand-receptor interactions promote cytotrophoblast differentiation and/or invasion and survival (116). We speculate that the expression of regulators of lym-

phangiogenesis, e.g., VEGFR-3 and VEGF-C (117), could contribute to the specialized nature of trophoblast-lined uterine vessels, which are able to expand greatly as fetal requirements for maternal blood increase in the latter half of pregnancy. Additionally, these cytotrophoblasts express Ang2, a ligand that is also involved in lymphangiogenesis (118, 119). Since cytotrophoblasts lack expression of Tie receptors (which bind Ang2), maternal cells are the likely targets (118).

Defects in cytotrophoblast differentiation are associated with preeclampsia. As cytotrophoblasts invade the uterine wall they also acquire a vascular-like repertoire of adhesion molecules. The onset of cytotrophoblast differentiation and/or invasion is characterized by reduced staining for receptors characteristic of polarized cytotrophoblast epithelial progenitors —integrin α6β4 and epithelial cadherin— and the onset of expression of adhesion receptors characteristic of endothelium — vascular-endothelial cadherin, IgG family members VCAM-1 and PECAM-1, and integrins αVβ3 and α1β1 (reviewed in ref. 120). Thus, as cytotrophoblasts from anchoring villi invade and remodel the wall of the uterus, these epithelial cells of ectodermal origin acquire an adhesion receptor repertoire characteristic of endothelial cells. We theorize that this switch permits the heterotypic adhesive interactions that allow fetal and maternal cells to cohabit in the uterine vasculature during normal pregnancy.

The preeclampsia syndrome reveals the significance of the differentiation program in which invasive cytotrophoblasts acquire vascular-like properties. Preeclampsia, a serious complication, is the leading cause of maternal mortality in developed countries (reviewed in ref. 121). The mother shows signs of widespread alterations in endothelial function, such as high blood pressure, proteinuria, and edema. In some cases the fetus stops growing, resulting in fetal growth restriction. Compounding the dangers of this condition is the fact that the maternal and fetal signs can appear suddenly at any time from mid–second trimester until term, hence the name preeclampsia (derived from the Greek *eklampsis*, meaning sudden flash or development).

Specific placental defects are associated with preeclampsia, especially the most severe cases that occur during the second and early third trimesters of pregnancy. The anchoring villi that give rise to invasive cytotrophoblasts are most severely affected. The extent of interstitial invasion of the uterine parenchyma is variable but frequently shallow (Figure 3). Endovascular invasion of the blood vessels is consistently rudimentary, making it extremely difficult to find any maternal vessels that contain cytotrophoblasts (S21, S23, 122). These anatomical defects suggested to us that during preeclampsia, cytotrophoblast differentiation along the invasive pathway is abnormal. Biopsies of the uterine wall of women with this syndrome showed that invasive cytotrophoblasts retain expression of adhesion receptors characteristic of the progenitor population and fail to turn on receptors that promote invasion and/or assumption of an endothelial phenotype (122). It is interesting to note that these deficits do not occur in isolation. In the most severely affected patients, immunolocalization on tissue sections of the placenta showed that cytotrophoblast VEGF-A and VEGFR-1 staining decreased; however, staining for PlGF was unaffected. Cytotrophoblast secretion of the soluble form of VEGFR-1 (sFlt-1) in vitro also increased (116), an observation that gains additional importance in light of the recent discovery that excess sFlt-1 produces a preeclampsia-like syndrome in rats (123). However, it is important to note that preeclampsia has a very complex etiology and an equally complex constellation of placental effects. For example, in macaques (S16) and humans, endovascular cytotrophoblasts express the neural cell adhesion molecule, which is significantly downregulated in preeclampsia (Figure 5). Data such as these reinforce the concept that this pregnancy complication is associated with global deficits in cytotrophoblast differentiation and/or invasion.

Conclusions and future directions

In recent years a great deal of progress has been made toward identifying the factors that govern trophoblast differentiation and, consequently, implantation and formation of the maternal-fetal interface. One important source of information has been the surprising number of transgenic mice, produced for other purposes, that have primary placental defects, a trend that was noted a decade ago (60). These analyses have also revealed the critical importance of normal placental function to embryonic development, as there are many examples in which tetraploid aggregation, which supplies a mutant embryo with a normal placenta, either rescues or lessens the severity of the embryonic defects. This principle likely applies to humans, as some aspects of adult health appear to be programmed in utero. Another important source of information has been studies of normal human placental development, which have led in turn to a better understanding of the defects that are associated with common pregnancy complications such as preeclampsia. In this context, the future challenge is to translate our basic knowledge of factors that govern trophoblast differentiation into clinically useful tests of placental function, a process that will inevitably revolutionize the practice of maternal-fetal medicine. Recently, interesting examples of this type of translational research have been published. For example, increased circulating levels of soluble VEGFR-1 and reduced levels of PlGF predict the subsequent development of preeclampsia (124, 125). By building on these types of studies, prenatal tests of placental function will eventually become as common as those tests already in place to assess the health of other important organs. Along the way we will also learn a great deal about possible therapeutic strategies for preventing and/or treating pregnancy complications.

Acknowledgments

We thank the past and present members of our laboratory who have made important contributions to this work and S.K. Dey for his thoughtful comments. We also thank Harbindar Singh for information about birth practices in Malaysia and Mary McKenney for excellent editorial support. Funding sources include The University of California Tobacco-Related Disease Program (8DT-0176), The National Institute of General Medical Sciences Minority Biomedical Research Support Research Initiative for Scientific Enhancement (R25 GM59298 and R25 GM56847), and NIH Grants U01 HD 42283 (part of the Cooperative Program on Trophoblast-Maternal Tissue Interactions), HL 64597, and HD 30367.

Due to space constraints, a number of important references could not be included in this article. Interested readers can find a supplementary reading list at http://www.jci.org/cgi/content/full/114/6/744/DC1.

Address correspondence to: Susan Fisher, HSW 604, University of California, San Francisco, San Francisco, California 94143-0512, USA. Phone: (415) 476-5297; Fax: (415) 476-4204; E-mail: sfisher@cgl.ucsf.edu.

1. Wu, L., et al. 2003. Extra-embryonic function of Rb is essential for embryonic development and viability. *Nature.* **421**:942–947.
2. Barker, D.J. 2004. Developmental origins of adult health and disease. *J. Epidemiol. Community Health.* **58**:114–115.
3. Sibai, B.M., et al. 1997. Risk factors associated with preeclampsia in healthy nulliparous women. The Calcium for Preeclampsia Prevention (CPEP) Study Group. *Am. J. Obstet. Gynecol.* **177**:1003–1010.
4. Goldenberg, R.L., Hauth, J.C., and Andrews, W.W. 2000. Intrauterine infection and preterm delivery. *N. Engl. J. Med.* **342**:1500–1507.
5. Kunath, T., Strumpf, D., and Rossant, J. 2004. Early trophoblast determination and stem cell maintenance in the mouse-a review. *Placenta.* **25**(Suppl.):S32–S38.
6. Johnson, M.H., and Ziomek, C.A. 1981. The foundation of two distinct cell lineages within the mouse morula. *Cell.* **24**:71–80.
7. Rossant, J., and Vijh, K.M. 1980. Ability of outside cells from preimplantation mouse embryos to form inner cell mass derivatives. *Dev. Biol.* **76**:475–482.
8. Larue, L., Ohsugi, M., Hirchenhain, J., and Kemler, R. 1994. E-cadherin null mutant embryos fail to form a trophectoderm epithelium. *Proc. Natl. Acad. Sci. U. S. A.* **91**:8263–8267.
9. Bloom, T.L. 1989. The effects of phorbol ester on mouse blastomeres: a role for protein kinase C in compaction? *Development.* **106**:159–171.

10. Chambers, I., et al. 2003. Functional expression cloning of Nanog, a pluripotency sustaining factor in embryonic stem cells. *Cell.* **113**:643–655.
11. Mitsui, K., et al. 2003. The homeoprotein Nanog is required for maintenance of pluripotency in mouse epiblast and ES cells. *Cell.* **113**:631–642.
12. Nichols, J., et al. 1998. Formation of pluripotent stem cells in the mammalian embryo depends on the POU transcription factor Oct4. *Cell.* **95**:379–391.
13. Liu, L., and Roberts, R.M. 1996. Silencing of the gene for the beta subunit of human chorionic gonadotropin by the embryonic transcription factor Oct-3/4. *J. Biol. Chem.* **271**:16683–16689.
14. Ezashi, T., Ghosh, D., and Roberts, R.M. 2001. Repression of Ets-2-induced transactivation of the tau interferon promoter by Oct-4. *Mol. Cell. Biol.* **21**:7883–7891.
15. Avilion, A.A., et al. 2003. Multipotent cell lineages in early mouse development depend on SOX2 function. *Genes Dev.* **17**:126–140.
16. Hay, D.C., Sutherland, L., Clark, J., and Burdon, T. 2004. Oct-4 knockdown induces similar patterns of endoderm and trophoblast differentiation markers in human and mouse embryonic stem cells. *Stem Cells.* **22**:225–235.
17. Norwitz, E.R., Schust, D.J., and Fisher, S.J. 2001. Implantation and the survival of early pregnancy. *N. Engl. J. Med.* **345**:1400–1408.
18. Paria, B.C., Reese, J., Das, S.K., and Dey, S.K. 2002. Deciphering the cross-talk of implantation:

advances and challenges. *Science.* **296**:2185–2188.
19. Conneely, O.M., Mulac-Jericevic, B., DeMayo, F., Lydon, J.P., and O'Malley, B.W. 2002. Reproductive functions of progesterone receptors. *Recent Prog. Horm. Res.* **57**:339–355.
20. Hewitt, S.C., and Korach, K.S. 2002. Estrogen receptors: structure, mechanisms and function. *Rev. Endocr. Metab. Disord.* **3**:193–200.
21. Paria, B.C., Tan, J., Lubahn, D.B., Dey, S.K., and Das, S.K. 1999. Uterine decidual response occurs in estrogen receptor-alpha-deficient mice. *Endocrinology.* **140**:2704–2710.
22. Curtis Hewitt, S., Goulding, E.H., Eddy, E.M., and Korach, K.S. 2002. Studies using the estrogen receptor alpha knockout uterus demonstrate that implantation but not decidualization-associated signaling is estrogen dependent. *Biol. Reprod.* **67**:1268–1277.
23. Ma, W.G., Song, H., Das, S.K., Paria, B.C., and Dey, S.K. 2003. Estrogen is a critical determinant that specifies the duration of the window of uterine receptivity for implantation. *Proc. Natl. Acad. Sci. U. S. A.* **100**:2963–2968.
24. Paria, B.C., et al. 1998. Coordination of differential effects of primary estrogen and catecholestrogen on two distinct targets mediates embryo implantation in the mouse. *Endocrinology.* **139**:5235–5246.
25. Stewart, C.L., et al. 1992. Blastocyst implantation depends on maternal expression of leukaemia inhibitory factor. *Nature.* **359**:76–79.

26. Robb, L., et al. 1998. Infertility in female mice lacking the receptor for interleukin 11 is due to a defective uterine response to implantation. *Nat. Med.* **4**:303–308.

27. Crossey, P.A., Pillai, C.C., and Miell, J.P. 2002. Altered placental development and intrauterine growth restriction in IGF binding protein-1 transgenic mice. *J. Clin. Invest.* **110**:411–418. doi:10.1172/JCI200210077.

28. Benson, G.V., et al. 1996. Mechanisms of reduced fertility in Hoxa-10 mutant mice: uterine homeosis and loss of maternal Hoxa-10 expression. *Development.* **122**:2687–2696.

29. Taylor, H.S., Arici, A., Olive, D., and Igarashi, P. 1998. HOXA10 is expressed in response to sex steroids at the time of implantation in the human endometrium. *J. Clin. Invest.* **101**:1379–1384.

30. Christian, M., et al. 2002. Cyclic AMP-induced forkhead transcription factor, FKHR, cooperates with CCAAT/enhancer-binding protein beta in differentiating human endometrial stromal cells. *J. Biol. Chem.* **277**:20825–20832.

31. Paria, B.C., Elenius, K., Klagsbrun, M., and Dey, S.K. 1999. Heparin-binding EGF-like growth factor interacts with mouse blastocysts independently of ErbB1: a possible role for heparin sulfate proteoglycans and ErbB4 in blastocyst implantation. *Development.* **126**:1997–2005.

32. Wang, H., et al. 2003. Differential G protein-coupled cannabinoid receptor signaling by anandamide directs blastocyst activation for implantation. *Proc. Natl. Acad. Sci. U. S. A.* **100**:14914–14919.

33. Fazleabas, A.T., Donnelly, K.M., Srinivasan, S., Fortman, J.D., and Miller, J.B. 1999. Modulation of the baboon (Papio anubis) uterine endometrium by chorionic gonadotrophin during the period of uterine receptivity. *Proc. Natl. Acad. Sci. U. S. A.* **96**:2543–2548.

34. Hemberger, M., et al. 2001. UniGene cDNA array-based monitoring of transcriptome changes during mouse placental development. *Proc. Natl. Acad. Sci. U. S. A.* **98**:13126–13131.

35. Reese, J., et al. 2001. Global gene expression analysis to identify molecular markers of uterine receptivity and embryo implantation. *J. Biol. Chem.* **276**:44137–44145.

36. Kao, L.C., et al. 2002. Global gene profiling in human endometrium during the window of implantation. *Endocrinology.* **143**:2119–2138.

37. Carson, D.D., et al. 2002. Changes in gene expression during the early to mid-luteal (receptive phase) transition in human endometrium detected by high-density microarray screening. *Mol. Hum. Reprod.* **8**:871–879.

38. Lim, H., et al. 1997. Multiple female reproductive failures in cyclooxygenase 2-deficient mice. *Cell.* **91**:197–208.

39. Wang, H., et al. 2004. Rescue of female infertility from the loss of cyclooxygenase-2 by compensatory up-regulation of cyclooxygenase-1 is a function of genetic makeup. *J. Biol. Chem.* **279**:10649–10658.

40. Cheng, J.G., and Stewart, C.L. 2003. Loss of cyclooxygenase-2 retards decidual growth but does not inhibit embryo implantation or development to term. *Biol. Reprod.* **68**:401–404.

41. Reese, J., Brown, N., Paria, B.C., Morrow, J., and Dey, S.K. 1999. COX-2 compensation in the uterus of COX-1 deficient mice during the pre-implantation period. *Mol. Cell. Endocrinol.* **150**:23–31.

42. Lim, H., et al. 1999. Cyclo-oxygenase-2-derived prostacyclin mediates embryo implantation in the mouse via PPARdelta. *Genes Dev.* **13**:1561–1574.

43. Barak, Y., et al. 2002. Effects of peroxisome proliferator-activated receptor delta on placentation, adiposity, and colorectal cancer. *Proc. Natl. Acad. Sci. U. S. A.* **99**:303–308.

44. Paria, B.C., et al. 2001. Cellular and molecular responses of the uterus to embryo implantation can be elicited by locally applied growth factors. *Proc. Natl. Acad. Sci. U. S. A.* **98**:1047–1052.

45. Matsumoto, H., Zhao, X., Das, S.K., Hogan, B.L., and Dey, S.K. 2002. Indian hedgehog as a progesterone-responsive factor mediating epithelial-mesenchymal interactions in the mouse uterus. *Dev. Biol.* **245**:280–290.

46. Daikoku, T., et al. 2004. Uterine Msx-1 and Wnt4 signaling becomes aberrant in mice with the loss of leukemia inhibitory factor or Hoxa-10: Evidence for a novel cytokine-homeobox-Wnt signaling in implantation. *Mol. Endocrinol.* **18**:1238–1250.

47. Genbacev, O.D., et al. 2003. Trophoblast L-selectin-mediated adhesion at the maternal-fetal interface. *Science.* **299**:405–408.

48. Rosen, S.D. 2004. Ligands for L-selectin: homing, inflammation, and beyond. *Annu. Rev. Immunol.* **22**:129–156.

49. Carson, D.D., et al. 2000. Embryo implantation. *Dev. Biol.* **223**:217–237.

50. Damsky, C.H., Fitzgerald, M.L., and Fisher, S.J. 1992. Distribution patterns of extracellular matrix components and adhesion receptors are intricately modulated during first trimester cytotrophoblast differentiation along the invasive pathway, in vivo. *J. Clin. Invest.* **89**:210–222.

51. Zhou, Y., et al. 1997. Human cytotrophoblasts adopt a vascular phenotype as they differentiate. A strategy for successful endovascular invasion? *J. Clin. Invest.* **99**:2139–2151.

52. Guleria, I., and Pollard, J.W. 2000. The trophoblast is a component of the innate immune system during pregnancy. *Nat. Med.* **6**:589–593.

53. Red-Horse, K., Drake, P.M., Gunn, M.D., and Fisher, S.J. 2001. Chemokine ligand and receptor expression in the pregnant uterus: reciprocal patterns in complementary cell subsets suggest functional roles. *Am. J. Pathol.* **159**:2199–2213.

54. Song, H., et al. 2002. Cytosolic phospholipase A2alpha is crucial for 'on-time' embryo implantation that directs subsequent development. *Development.* **129**:2879–2889.

55. Herz, J., Couthier, D.E., and Hammer, R.E. 1993. Correction: LDL receptor-related protein internalizes and degrades uPA-PAI-1 complexes and is essential for embryo implantation. *Cell.* **73**:428.

56. Wilcox, A.J., Baird, D.D., and Weinberg, C.R. 1999. Time of implantation of the conceptus and loss of pregnancy. *N. Engl. J. Med.* **340**:1796–1799.

57. Tanaka, S., Kunath, T., Hadjantonakis, A.K., Nagy, A., and Rossant, J. 1998. Promotion of trophoblast stem cell proliferation by FGF4. *Science.* **282**:2072–2075.

58. Rossant, J., and Cross, J.C. 2001. Placental development: lessons from mouse mutants. *Nat. Rev. Genet.* **2**:538–548.

59. Xu, R.H., et al. 2002. BMP4 initiates human embryonic stem cell differentiation to trophoblast. *Nat. Biotechnol.* **20**:1261–1264.

60. Cross, J.C., Werb, Z., and Fisher, S.J. 1994. Implantation and the placenta: key pieces of the development puzzle. *Science.* **266**:1508–1518.

61. Nagy, A., et al. 1990. Embryonic stem cells alone are able to support fetal development in the mouse. *Development.* **110**:815–821.

62. Nagy, A., Rossant, J., Nagy, R., Abramow-Newerly, W., and Roder, J.C. 1993. Derivation of completely cell culture-derived mice from early-passage embryonic stem cells. *Proc. Natl. Acad. Sci. U. S. A.* **90**:8424–8428.

63. Cross, J.C., et al. 2003. Genes, development and evolution of the placenta. *Placenta.* **24**:123–130.

64. Russ, A.P., et al. 2000. Eomesodermin is required for mouse trophoblast development and mesoderm formation. *Nature.* **404**:95–99.

65. Auman, H.J., et al. 2002. Transcription factor AP-2gamma is essential in the extra-embryonic lineages for early postimplantation development. *Development.* **129**:2733–2747.

66. Luo, J., et al. 1997. Placental abnormalities in mouse embryos lacking the orphan nuclear receptor ERR-beta. *Nature.* **388**:778–782.

67. Adamson, S.L., et al. 2002. Interactions between trophoblast cells and the maternal and fetal circulation in the mouse placenta. *Dev. Biol.* **250**:358–373.

68. Ma, G.T., et al. 2001. Nodal regulates trophoblast differentiation and placental development. *Dev. Biol.* **236**:124–135.

69. Guillemot, F., Nagy, A., Auerbach, A., Rossant, J., and Joyner, A.L. 1994. Essential role of Mash-2 in extra-embryonic development. *Nature.* **371**:333–336.

70. Wang, J., Mager, J., Schnedier, E., and Magnuson, T. 2002. The mouse PcG gene eed is required for Hox gene repression and extraembryonic development. *Mamm. Genome.* **13**:493–503.

71. Janatpour, M.J., et al. 1999. A repertoire of differentially expressed transcription factors that offers insight into mechanisms of human cytotrophoblast differentiation. *Dev. Genet.* **25**:146–157.

72. Janatpour, M.J., et al. 2000. Id-2 regulates critical aspects of human cytotrophoblast differentiation, invasion and migration. *Development.* **127**:549–558.

73. Riley, P., Anson-Cartwright, L., and Cross, J.C. 1998. The Hand1 bHLH transcription factor is essential for placentation and cardiac morphogenesis. *Nat. Genet.* **18**:271–275.

74. Firulli, A.B., McFadden, D.G., Lin, Q., Srivastava, D., and Olson, E.N. 1998. Heart and extra-embryonic mesodermal defects in mouse embryos lacking the bHLH transcription factor Hand1. *Nat. Genet.* **18**:266–270.

75. Knofler, M., Meinhardt, G., Vasicek, R., Husslein, P., and Egarter, C. 1998. Molecular cloning of the human Hand1 gene/cDNA and its tissue-restricted expression in cytotrophoblastic cells and heart. *Gene.* **224**:77–86.

76. Jones, B.W., Fetter, R.D., Tear, G., and Goodman, C.S. 1995. glial cells missing: a genetic switch that controls glial versus neuronal fate. *Cell.* **82**:1013–1023.

77. Anson-Cartwright, L., et al. 2000. The glial cells missing-1 protein is essential for branching morphogenesis in the chorioallantoic placenta. *Nat. Genet.* **25**:311–314.

78. Sapin, V., Dolle, P., Hindelang, C., Kastner, P., and Chambon, P. 1997. Defects of the chorioallantoic placenta in mouse RXRalpha null fetuses. *Dev. Biol.* **191**:29–41.

79. Wendling, O., Chambon, P., and Mark, M. 1999. Retinoid X receptors are essential for early mouse development and placentogenesis. *Proc. Natl. Acad. Sci. U. S. A.* **96**:547–551.

80. Barak, Y., et al. 1999. PPAR gamma is required for placental, cardiac, and adipose tissue development. *Mol. Cell.* **4**:585–595.

81. Monkley, S.J., Delaney, S.J., Pennisi, D.J., Christiansen, J.H., and Wainwright, B.J. 1996. Targeted disruption of the Wnt2 gene results in placentation defects. *Development.* **122**:3343–3353.

82. Schmidt, C., et al. 1995. Scatter factor/hepatocyte growth factor is essential for liver development. *Nature.* **373**:699–702.

83. Yamamoto, H., et al. 1998. Defective trophoblast function in mice with a targeted mutation of Ets2. *Genes Dev.* **12**:1315–1326.

84. Kenner, L., et al. 2004. Mice lacking JunB are osteopenic due to cell-autonomous osteoblast and osteoclast defects. *J. Cell Biol.* **164**:613–623.

85. Hatano, N., et al. 2003. Essential role for ERK2 mitogen-activated protein kinase in placental development. *Genes Cells.* **8**:847–856.

86. Dietrich, S., et al. 1999. The role of SF/HGF and c-Met in the development of skeletal muscle. *Development.* **126**:1621–1629.

87. Takahashi, Y., et al. 2003. SOCS3: an essential regulator of LIF receptor signaling in trophoblast giant cell differentiation. *EMBO J.* **22**:372–384.

88. Isermann, B., Hendrickson, S.B., Hutley, K., Wing, M., and Weiler, H. 2001. Tissue-restricted expression of thrombomodulin in the placenta rescues thrombomodulin-deficient mice from early lethality and reveals a secondary developmental block. *Development.* **128**:827–838.

89. Gallicano, G.I., Bauer, C., and Fuchs, E. 2001. Rescuing desmoplakin function in extra-embryonic ectoderm reveals the importance of this protein in embryonic heart, neuroepithelium, skin and vasculature. *Development.* **128**:929–941.

90. Mishina, Y., Crombie, R., Bradley, A., and Behringer, R.R. 1999. Multiple roles for activin-like kinase-2 signaling during mouse embryogenesis. *Dev. Biol.* **213**:314–326.

91. Sugimoto, M., Karashima, Y., Abe, K., Tan, S.S., and Takagi, N. 2003. Tetraploid embryos rescue the early defects of tw5/tw5 mouse embryos. *Genesis.* **37**:162–171.

92. Goto, Y., and Takagi, N. 1998. Tetraploid embryos rescue embryonic lethality caused by an additional maternally inherited X chromosome in the mouse. *Development.* **125**:3353–3363.

93. Jaquemar, D., et al. 2003. Keratin 8 protection of placental barrier function. *J. Cell Biol.* **161**:749–756.

94. Bajoria, R., Sooranna, S.R., Ward, S., D'Souza, S., and Hancock, M. 2001. Placental transport rather than maternal concentration of amino acids regulates fetal growth in monochorionic twins: implications for fetal origin hypothesis. *Am. J. Obstet. Gynecol.* **185**:1239–1246.

95. Genbacev, O., Bass, K.E., Joslin, R.J., and Fisher, S.J. 1995. Maternal smoking inhibits early human cytotrophoblast differentiation. *Reprod. Toxicol.* **9**:245–255.

96. Genbacev, O., et al. 2000. Concordant in situ and in vitro data show that maternal cigarette smoking negatively regulates placental cytotrophoblast passage through the cell cycle. *Reprod. Toxicol.* **14**:495–506.

97. Pereira, L., Maidji, E., McDonagh, S., Genbacev, O., and Fisher, S. 2003. Human cytomegalovirus transmission from the uterus to the placenta correlates with the presence of pathogenic bacteria and maternal immunity. *J. Virol.* **77**:13301–13314.

98. Yamamoto-Tabata, T., McDonagh, S., Chang, H.T., Fisher, S., and Pereira, L. 2004. Human cytomegalovirus interleukin-10 downregulates metalloproteinase activity and impairs endothelial cell migration and placental cytotrophoblast invasiveness in vitro. *J. Virol.* **78**:2831–2840.

99. Wright, A., et al. 2004. Trisomy 21 is associated with variable defects in cytotrophoblast differentiation along the invasive pathway. *Am. J. Med.*

Genet. In press.

100. Maltepe, E., and Simon, M.C. 1998. Oxygen, genes, and development: an analysis of the role of hypoxic gene regulation during murine vascular development. *J. Mol. Med.* **76**:391–401.

101. Jauniaux, E., et al. 2000. Onset of maternal arterial blood flow and placental oxidative stress. A possible factor in human early pregnancy failure. *Am. J. Pathol.* **157**:2111–2122.

102. Jauniaux, E., Gulbis, B., and Burton, G.J. 2003. The human first trimester gestational sac limits rather than facilitates oxygen transfer to the foetus--a review. *Placenta.* **24**(Suppl. A):S86–S93.

103. Genbacev, O., Joslin, R., Damsky, C.H., Polliotti, B.M., and Fisher, S.J. 1996. Hypoxia alters early gestation human cytotrophoblast differentiation/invasion in vitro and models the placental defects that occur in preeclampsia. *J. Clin. Invest.* **97**:540–550.

104. Genbacev, O., Zhou, Y., Ludlow, J.W., and Fisher, S.J. 1997. Regulation of human placental development by oxygen tension. *Science.* **277**:1669–1672.

105. Adelman, D.M., Gertsenstein, M., Nagy, A., Simon, M.C., and Maltepe, E. 2000. Placental cell fates are regulated in vivo by HIF-mediated hypoxia responses. *Genes Dev.* **14**:3191–3203.

106. Douglas, R.M., and Haddad, G.G. 2003. Genetic models in applied physiology: invited review: effect of oxygen deprivation on cell cycle activity: a profile of delay and arrest. *J. Appl. Physiol.* **94**:2068–2083; discussion 2084.

107. Semenza, G.L. 2003. Targeting HIF-1 for cancer therapy. *Nat. Rev. Cancer.* **3**:721–732.

108. Safran, M., and Kaelin, W.G., Jr. 2003. HIF hydroxylation and the mammalian oxygen-sensing pathway. *J. Clin. Invest.* **111**:779–783. doi:10.1172/JCI200318181.

109. Masson, N., and Ratcliffe, P.J. 2003. HIF prolyl and asparaginyl hydroxylases in the biological response to intracellular O(2) levels. *J. Cell. Sci.* **116**:3041–3049.

110. Wykoff, C.C., Pugh, C.W., Maxwell, P.H., Harris, A.L., and Ratcliffe, P.J. 2000. Identification of novel hypoxia dependent and independent target genes of the von Hippel-Lindau (VHL) tumour suppressor by mRNA differential expression profiling. *Oncogene.* **19**:6297–6305.

111. Gnarra, J.R., et al. 1997. Defective placental vasculogenesis causes embryonic lethality in VHL-deficient mice. *Proc. Natl. Acad. Sci. U. S. A.* **94**:9102–9107.

112. Katschinski, D.M., et al. 2002. Heat induction of the unphosphorylated form of hypoxia-inducible factor-1alpha is dependent on heat shock protein-

90 activity. *J. Biol. Chem.* **277**:9262–9267.

113. Voss, A.K., Thomas, T., and Gruss, P. 2000. Mice lacking HSP90beta fail to develop a placental labyrinth. *Development.* **127**:1–11.

114. Genbacev, O., Krtolica, A., Kaelin, W., and Fisher, S.J. 2001. Human cytotrophoblast expression of the von Hippel-Lindau protein is downregulated during uterine invasion in situ and upregulated by hypoxia in vitro. *Dev. Biol.* **233**:526–536.

115. Pugh, C.W., and Ratcliffe, P.J. 2003. Regulation of angiogenesis by hypoxia: role of the HIF system. *Nat. Med.* **9**:677–684.

116. Zhou, Y., et al. 2002. Vascular endothelial growth factor ligands and receptors that regulate human cytotrophoblast survival are dysregulated in severe preeclampsia and hemolysis, elevated liver enzymes, and low platelets syndrome. *Am. J. Pathol.* **160**:1405–1423.

117. Makinen, T., and Alitalo, K. 2002. Molecular mechanisms of lymphangiogenesis. *Cold Spring Harb. Symp. Quant. Biol.* **67**:189–196.

118. Zhou, Y., Bellingard, V., Feng, K.T., McMaster, M., and Fisher, S.J. 2003. Human cytotrophoblasts promote endothelial survival and vascular remodeling through secretion of Ang2, PlGF, and VEGF-C. *Dev. Biol.* **263**:114–125.

119. Gale, N.W., et al. 2002. Angiopoietin-2 is required for postnatal angiogenesis and lymphatic patterning, and only the latter role is rescued by Angiopoietin-1. *Dev. Cell.* **3**:411–423.

120. Damsky, C.H., and Fisher, S.J. 1998. Trophoblast pseudo-vasculogenesis: faking it with endothelial adhesion receptors. *Curr. Opin. Cell Biol.* **10**:660–666.

121. Lain, K.Y., and Roberts, J.M. 2002. Contemporary concepts of the pathogenesis and management of preeclampsia. *JAMA.* **287**:3183–3186.

122. Zhou, Y., Damsky, C.H., and Fisher, S.J. 1997. Preeclampsia is associated with failure of human cytotrophoblasts to mimic a vascular adhesion phenotype. One cause of defective endovascular invasion in this syndrome? *J. Clin. Invest.* **99**:2152–2164.

123. Maynard, S.E., et al. 2003. Excess placental soluble fms-like tyrosine kinase 1 (sFlt1) may contribute to endothelial dysfunction, hypertension, and proteinuria in preeclampsia. *J. Clin. Invest.* **111**:649–658. doi:10.1172/JCI200317189.

124. Levine, R.J., et al. 2004. Circulating angiogenic factors and the risk of preeclampsia. *N. Engl. J. Med.* **350**:672–683.

125. Thadhani, R., et al. 2004. First trimester placental growth factor and soluble fms-like tyrosine kinase 1 and risk for preeclampsia. *J. Clin. Endocrinol. Metab.* **89**:770–775.

Acute renal failure: definitions, diagnosis, pathogenesis, and therapy

Robert W. Schrier, Wei Wang, Brian Poole, and Amit Mitra

Department of Medicine, University of Colorado Health Sciences Center, Denver, Colorado, USA.

Acute renal failure (ARF), characterized by sudden loss of the ability of the kidneys to excrete wastes, concentrate urine, conserve electrolytes, and maintain fluid balance, is a frequent clinical problem, particularly in the intensive care unit, where it is associated with a mortality of between 50% and 80%. In this review, the epidemiology and pathophysiology of ARF are discussed, including the vascular, tubular, and inflammatory perturbations. The clinical evaluation of ARF and implications for potential future therapies to decrease the high mortality are described.

During the bombing of London in World War II, Bywaters and Beall described an acute loss of kidney function that occurred in severely injured crush victims (1). Acute tubular necrosis (ATN) was the term coined to describe this clinical entity, because of histological evidence for patchy necrosis of renal tubules at autopsy. In the clinical setting, the terms ATN and acute renal failure (ARF) are frequently used interchangeably. However, for the purposes of this review, the term ARF, rather than ATN, will be used. ARF will not include increases in blood urea due to reversible renal vasoconstriction (prerenal azotemia) or urinary tract obstruction (postrenal azotemia).

The mortality of ARF approached 100% in World War II, since the development of acute hemodialysis for clinical use had not yet occurred. Acute hemodialysis was first used clinically during the Korean War in 1950 to treat military casualties, and this led to a decrease in mortality of the ARF clinical syndrome from about 90% to about 50% (2, 3). In the half century that has since passed, much has been learned about the pathogenesis of ischemic and nephrotoxic ARF in experimental models, but there has been very little improvement in mortality. This may be explained by changing demographics: the age of patients with ARF continues to rise, and comorbid diseases are increasingly common in this population. Both factors may obscure any increased survival related to improved critical care.

Examining the incidence of ARF in several military conflicts does, however, provide some optimism (4). The incidence of ARF in seriously injured casualties decreased between World War II and the Korean War, and again between that war and the Vietnam War, despite the lack of any obvious difference in the severity of the injuries. What was different was the rapidity of the fluid resuscitation of the patients. Fluid resuscitation on the battlefield with the rapid evacuation of the casualties to hospitals by helicopter began during the Korean War and was optimized further during the Vietnam War. For seriously injured casualties the incidence of ischemic ARF was one in 200 in the Korean War and one in 600 in the Vietnam War (5). This historical sequence of events sug-

gests that early intervention could prevent the occurrence of ARF, at least in military casualties. In experimental studies it has been shown that progression from an azotemic state associated with renal vasoconstriction and intact tubular function (prerenal azotemia) to established ARF with tubular dysfunction occurs if the renal ischemia is prolonged (6). Moreover, early intervention with fluid resuscitation was shown to prevent the progression from prerenal azotemia to established ARF.

Diagnostic evaluation of ARF

One important question, therefore, is how to assure that an early diagnosis of acute renal vasoconstriction can be made prior to the occurrence of tubular dysfunction, thus providing the potential to prevent progression to established ARF. In this regard, past diagnostics relied on observation of the patient response to a fluid challenge: decreasing levels of blood urea nitrogen (BUN) indicated the presence of reversible vasoconstriction, while uncontrolled accumulation of nitrogenous waste products, i.e., BUN and serum creatinine, indicated established ARF. This approach, however, frequently led to massive fluid overload in the ARF patient with resultant pulmonary congestion, hypoxia, and premature need for mechanical ventilatory support and/or hemodialysis. On this background the focus turned to an evaluation of urine sediment and urine chemistries to differentiate between renal vasoconstriction with intact tubular function and established ARF (7). It was well established that if tubular function was intact, renal vasoconstriction was associated with enhanced tubular sodium reabsorption. Specifically, the fraction of filtered sodium that is rapidly reabsorbed by normal tubules of the vasoconstricted kidney is greater than 99%. Thus, when nitrogenous wastes, such as creatinine and urea, accumulate in the blood due to a fall in glomerular filtration rate (GFR) secondary to renal vasoconstriction with intact tubular function, the fractional excretion of filtered sodium (FE_{Na} = [(urine sodium × plasma creatinine) / (plasma sodium × urine creatinine)]) is less than 1%. An exception to this physiological response of the normal kidney to vasoconstriction is when the patient is receiving a diuretic, including mannitol, or has glucosuria, which decreases tubular sodium reabsorption and increases FE_{Na}. It has recently been shown in the presence of diuretics that a rate of fractional excretion of urea (FE_{urea}) of less than 35 indicates intact tubular function, thus favoring renal vasoconstriction rather than established ARF as a cause of the azotemia (8). Also, renal vasoconstriction in a patient with advanced chronic renal failure may not be expected to be associated with an FE_{Na} of less than

Nonstandard abbreviations used: acute renal failure (ARF); acute tubular necrosis (ATN); blood urea nitrogen (BUN); continuous renal replacement therapy (CRRT); fractional excretion of filtered sodium (FE_{Na}); glomerular filtration rate (GFR); intermittent hemodialysis (IHD); lactic dehydrogenase (LDH); α-melanocyte–stimulating hormone (αMSH); NO synthase (NOS); Tamm-Horsfall protein (THP).

Conflict of interest: The authors have declared that no conflict of interest exists.

Citation for this article: *J. Clin. Invest.* **114**:5–14 (2004).
doi:10.1172/JCI200422353.

1 because of chronic adaptation to an increased single-nephron GFR. Specifically, the adaptive decrease in tubular reabsorption to maintain sodium balance in chronic renal disease may make the interpretation of FE_{Na} difficult in this setting.

The approximately 80% diagnostic specificity of FE_{Na} in distinguishing azotemia associated with renal vasoconstriction and intact tubular function from established ARF with tubular dysfunction may result from limited sensitivity of this parameter, or, perhaps more likely, the patient may actually be progressing from a prerenal azotemic state to established ARF. With established ARF the urine-concentrating capacity is abolished; thus measurement of urinary osmolality may complement the use of FE_{Na} in the diagnostic separation of renal vasoconstriction from established ARF in the patient with a rising BUN measurement and serum creatinine level. Since advanced age and low protein intake may diminish maximal urinary osmolality, this diagnostic parameter may be less sensitive than FE_{Na} in the azotemic patient. Increased excretion of tubular epithelial cells, indicated by examination of the urinary sediment, is characteristic of established ARF but also has limitations in diagnostic value particularly in nonoliguric ARF. Table 1 shows guidelines for urinary indices whereby established ARF can be distinguished from renal vasoconstriction with intact tubular function, i.e., prerenal azotemia. It also should be pointed out that some causes of ARF, including radiocontrast media and myoglobinuria, may be associated with an FE_{Na} of less than 1 (9). This may be related to the early presence of severe renal vasoconstriction and intact distal tubule function, which can occur in the presence of proximal tubule injury.

Recent observational results suggest that for patients with a rising BUN level and/or serum creatinine concentration, early consultation with a nephrologist can decrease the occurrence and mortality of ARF (10). The reasons for this observation are not clear but may involve early evaluation by the nephrologist of not only blood but also urinary chemistries, diminished incidence of fluid overload and need for mechanical ventilation, and perhaps earlier initiation of dialysis. On this background a search for sensitive parameters for the early diagnosis of ARF and a uniformly accepted definition of established ARF has emerged. A clinical definition of ARF has been sought that focuses on the degree of nitrogenous-waste accumulation in the blood, whether serum creatinine, BUN, or cystatin C (11), in order to make an early diagnosis. This approach, however, has important limitations. With established ARF the GFR is generally less than 10 ml/min. The amount of nitrogenous accumulation in the blood depends on the duration of the time that the patient has had a GFR of less than 10 ml/min. Thus, a catabolic patient with established ARF may have a serum creatinine concentration of 1.8 mg/dl when seen within the first 24 hours of established ARF, but the same individual may have a concentration of 10 mg/dl when seen after 5 days of established ARF, even though the GFR has been less than 10 ml/min in both circumstances. Nevertheless, acceptance of a 50% acute rise in serum creatinine to a concentration above 2.0 mg/dl as the clinical definition of ARF could increase the likelihood of early intervention and bring more consistency to the patient populations recruited for future intervention trials. However, such a definition does not differentiate among rises in serum creatinine that result (a) from a renal vasoconstriction–mediated prerenal azotemic state with intact tubular function, such as that which occurs with volume depletion, and advanced cardiac or liver failure; (b) from a postrenal azotemic state due to urinary tract obstruction; or (c)

from established ARF due to an acute ischemic and/or nephrotoxic insult. A sensitive approach to ruling out postrenal azotemia is to identify post-void residual bladder urine as less than 50 ml and exclude pyelocalyceal dilatation using renal ultrasonography.

The differentiation of prerenal azotemia and established ARF, however, has proven more difficult. Several biomarkers have been proposed for the early diagnosis of ARF and are currently under study. These include increased urinary excretion of kidney injury molecule-1 (12), IL-18 (13), and tubular enzymes (14). In order for any of these or other potential future biomarkers to become practical in the early clinical diagnosis of established ARF, specificity, sensitivity, rapid availability, cost effectiveness, and advantages as compared with the inexpensive and readily available measure of FE_{Na} must be considered.

From a clinical-diagnostic viewpoint, a potential advancement may be an agreement to use the term ARF, rather than ATN, independent of whether the acute insult was ischemic or nephrotoxic. Such an agreement, however, would also necessitate using the terms prerenal azotemia and postrenal azotemia for those potentially reversible conditions that result from renal vasoconstriction and urinary tract obstruction, respectively. Alternatively, it must be recognized that diagnosis of ARF based solely on a given rise in serum creatinine, BUN, or cystatin C levels would encompass many patients who do not have the clinical syndrome of established ARF, an entity that is not readily reversible by fluid resuscitation, cardiac or hepatic functional improvement, or relief of a urinary tract obstruction.

Mechanisms of ARF

Based on the foregoing comments, this discussion of mechanisms of ARF will not include nitrogenous-waste accumulation due to renal vasoconstriction with intact tubular function (prerenal azotemia) or urinary tract obstruction (postrenal azotemia). The mechanisms of ARF involve both vascular and tubular factors (15). An ischemic insult to the kidney will in general be the cause of the ARF discussed herein. While a decrease in renal blood flow with diminished oxygen and substrate delivery to the tubule cells is an important ischemic factor, it must be remembered that a relative increase in oxygen demand by the tubule is also a factor in renal ischemia.

The term ATN, although frequently used to describe the syndrome of clinical ARF in the absence of prerenal and postrenal

Table 1

Guidelines for urinary indices whereby established ARF can be distinguished from renal vasoconstriction with intact tubular function (prerenal azotemia)

Laboratory test	Prerenal azotemia	ARF
Urine osmolality (mOsm/kg)	>500	<400
Urine sodium level (mEq/l)	<20	>40
Urine/plasma creatinine ratio	>40	<20
Fractional excretion of sodium (%)	<1	>2
Fractional excretion of urea (%)	<35	>35
Urinary sediment	Normal; occasional hyaline or fine granular casts	Renal tubular epithelial cells; granular and muddy brown casts

Osm, osmole; Eq, equivalent.

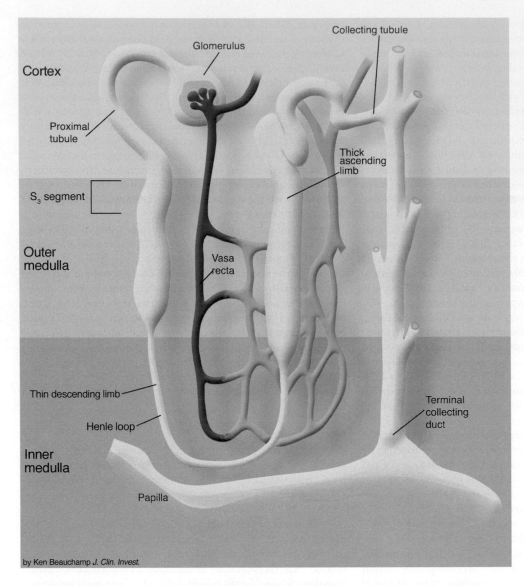

Figure 1
Relative hypoxia in the outer medulla predisposes to ischemic injury in the S_3 segment of the proximal tubule. The thick ascending limb is also located in this hypoxic region of the kidney and, depending on tubular reabsorptive demand, may also undergo ischemic injury. The thick ascending limb may, however, be more protected against ischemic injury, because this nephron segment possesses more glycolytic machinery for ATP synthesis than the S_3 segment.

Cortex
Glomerulus
Collecting tubule
Proximal tubule
Thick ascending limb
S_3 segment
Outer medulla
Vasa recta
Thin descending limb
Henle loop
Terminal collecting duct
Inner medulla
Papilla
by Ken Beauchamp *J. Clin. Invest.*

azotemia, is somewhat of a misnomer. Specifically, in established ARF the presence of tubular necrosis upon histological examination of the kidney is seen in only occasional tubule cells at best and, in some cases, may not even be detectable (16). What is clear with established ARF, however, is that the glomeruli are morphologically normal. This perhaps justifies the investigative focus on renal tubules. The following discussion will describe some of the vascular, tubular, and inflammatory perturbations that occur with an acute renal insult. An attempt will then be made to integrate the important underlying pathogenetic factors.

Renal vascular abnormalities

Loss of autoregulation and increased renal vasoconstriction: the role of increased cytosolic and mitochondrial calcium. Acute ischemic injury has been shown in experimental animals to be associated with a loss of renal autoregulation (17). Moreover, rather than the normal autoregulatory renal vasodilation that occurs during a decrease in renal perfusion pressure, there is evidence that renal vasoconstriction actually occurs in the ischemic kidney. An increase in the response to renal nerve stimulation has also been observed in association with an acute ischemic insult (17). More-

over, the vasoconstrictor response to exogenous norepinephrine and endothelin has been observed to be increased in the acutely ischemic kidney (18). These vascular abnormalities observed in the ischemic kidney may be related to the resultant increase in cytosolic calcium observed in the afferent arterioles of the glomerulus. The observation that intrarenal calcium channel blockers can reverse the loss of autoregulation and the increase in sensitivity to renal nerve stimulation (17) supports a pathogenetic role of increased cytosolic calcium concentration in the afferent arteriole of the ischemic kidney. The mitochondrial calcium accumulation in the ischemic kidney has also been reversed by administration of calcium channel blockers (19–21). Moreover, calcium channel blockers have been shown to attenuate renal dysfunction and toxicity associated with the immunosuppressive drug cyclosporine following cadaveric renal transplantation, when administered prior to the drug and ischemic insults (22).

Outer medullary congestion. Outer medullary congestion of the kidney is another vascular hallmark of acute renal ischemia. This congestion has been proposed to worsen the relative hypoxia in the outer medulla and thus the hypoxic injury in the S_3 segment of the proximal tubule and the thick ascending limb of the Henle

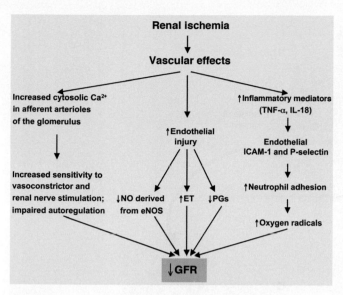

Figure 2
Vascular factors contributing to the pathogenesis of ischemic ARF. ET, endothelin; PG, prostaglandin. Figure modified with permission from the *Journal of Nephrology* (15).

loop (Figure 1) (23). Upregulation of adhesion molecules has been implicated in this outer medullary congestion, and antibodies to ICAMs and P-selectin have been shown to afford protection against acute ischemic injury (24–27). Acute renal ischemia has also been shown to be associated with endothelial damage, perhaps at least in part because of increased oxidant injury. Some evidence in support of this possibility is the observation that activated leukocytes enhance the renal ischemic injury, and this observation could not be duplicated using leukocytes from patients with chronic granulomatous disease that do not produce reactive oxygen species (28). Oxidant injury may also lead to a decrease in eNOS and vasodilatory prostaglandins as well as an increase in endothelin, all of which may enhance the renal vasoconstrictor effect of circulating pressor agents present in ARF (29–31).

With this spectrum of vascular consequences observed in association with an acute ischemic insult (Figure 2), an earlier suggestion was made that the term ATN should be exchanged for the term vasomotor nephropathy. However, renal vasodilators, which return renal blood flow to normal in experimental ARF animal models and humans with established ARF, have not been shown to increase GFR. Thus, in recent years the search for the mechanisms mediating ARF has focused primarily on the renal tubule.

Tubular abnormalities
There is no question that renal tubule dysfunction occurs in established ARF, since tubular sodium reabsorption is decreased, i.e., $FE_{Na} > 2.0$, at a time when normal renal tubules avidly increase tubular sodium reabsorption in response to renal vasoconstriction. However, investigators of tubular abnormalities, which occur following acute renal ischemic insults, must ultimately demonstrate how the observed tubule perturbation mediates the fall in GFR to less than 10% of normal — the hallmark of established ARF.

Structural changes during ischemic ARF. ARF is characterized by tubular dysfunction with impaired sodium and water reabsorption and is associated with the shedding and excretion of proximal

tubule brush border membranes and epithelial tubule cells into the urine (32) (Figure 3). Approximately 30–70% of these shed epithelial tubule cells in the urine are viable and can be grown in culture (33). Recent studies using cellular and molecular techniques have provided information relating to the structural abnormalities of injured renal tubules that occur both in vitro and in vivo. In vitro studies using chemical anoxia have revealed abnormalities in the proximal tubule cytoskeleton that are associated with translocation of Na^+/K^+-ATPase from the basolateral to the apical membrane (34) (Figure 3). A comparison of cadaveric transplanted kidneys with delayed versus prompt graft function has also provided important results regarding the role of Na^+/K^+-ATPase in ischemic renal injury (35). This study demonstrated that, compared with kidneys with prompt graft function, those with delayed graft function had a significantly greater cytoplasmic concentration of Na^+/K^+-ATPase and actin-binding proteins — spectrin (also known as fodrin) and ankyrin — that had translocated from the basolateral membrane to the cytoplasm (Figure 4). Such a translocation of Na^+/K^+-ATPase from the basolateral membrane to the cytoplasm could explain the decrease in tubular sodium reabsorption that occurs with ARF. The mechanisms whereby the critical residence of Na^+/K^+-ATPase in the basolateral membrane, which facilitates vectorial sodium transport, is uncoupled by hypoxia or ischemia have been an important focus of research. The actin-binding proteins, spectrin and ankyrin, serve as substrates for the calcium-activated cysteine protease calpain (36) (Figure 5). In this regard, in vitro studies in proximal tubules have shown a rapid rise in cytosolic calcium concentration during acute hypoxia, which antedates the evidence of tubular injury as assessed by lactic dehydrogenase (LDH) release (37) (Figure 6).

There is further evidence to support the importance of the translocation of Na^+/K^+-ATPase from the basolateral membrane to the cytoplasm during renal ischemia/reperfusion. Specifically, calpain-mediated breakdown products of the actin-binding protein spectrin have been shown to occur with renal ischemia. Calpain activity was also demonstrated to be increased during hypoxia in isolated proximal tubules (38). Measurement of LDH release following calpain inhibition has demonstrated attenuation of hypoxic damage to proximal tubules (39). There was no evidence in proximal tubules during hypoxia of an increase in cathepsin, another cysteine protease. Further studies demonstrated a calcium-independent pathway for calpain activation during hypoxia. Calpastatin, an endogenous cellular inhibitor of calpain activation, was shown to be diminished during hypoxia in association with the rise in another cysteine protease, caspase (40). This effect of diminished calpastatin activity could be reversed by caspase inhibition. In Figure 7, proteolytic pathways that may be involved in calpain-mediated proximal tubule cell injury during hypoxia are illustrated. Calcium activation of phospholipase A has also been shown to contribute to renal tubular injury during ischemia (41).

Tubular obstruction in ischemic ARF. The existence of proteolytic pathways involving cysteine proteases, namely calpain and caspases, may therefore explain the decrease in proximal tubule sodium reabsorption and increased FE_{Na} secondary to proteolytic uncoupling of Na^+/K^+-ATPase from its basolateral membrane anchoring proteins. This tubular perturbation alone, however, does not explain the fall in GFR that leads to nitrogenous-waste retention and thus the rise in BUN and serum creatinine. There are, however, potential pathways whereby loss of brush border membranes, loss of viable and nonviable proximal tubule cells, and decreased proxi-

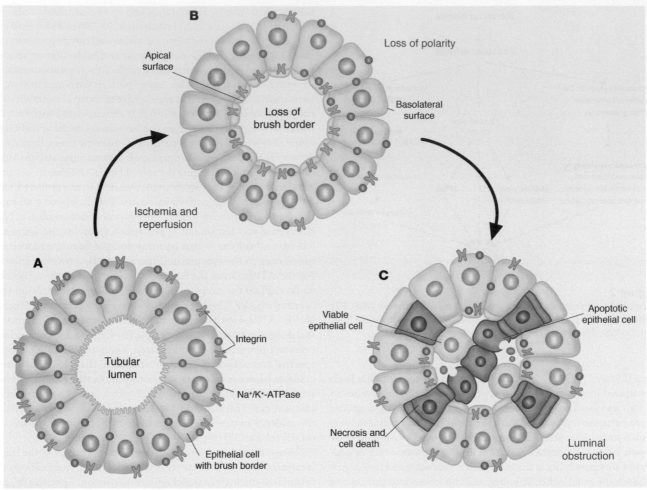

Figure 3
Following ischemia and reperfusion, morphological changes occur in the proximal tubules, including loss of polarity, loss of the brush border, and redistribution of integrins and Na+/K+-ATPase to the apical surface. Calcium and reactive oxygen species may also have a role in these morphological changes, in addition to subsequent cell death resulting from necrosis and apoptosis. Both viable and nonviable cells are shed into the tubular lumen, resulting in the formation of casts and luminal obstruction and contributing to the reduction in the GFR. Figure modified with permission from the *New England Journal of Medicine* (94).

mal tubule sodium reabsorption may lead to a decreased GFR during ARF. First of all, brush border membranes and cellular debris could provide the substrate for intraluminal obstruction in the highly resistant portion of distal nephrons (Figure 3). In fact, microdissection of individual nephrons of kidneys from patients with ARF demonstrated obstructing casts in distal tubules and collecting ducts (42). This observation could explain the dilated proximal tubules that are observed upon renal biopsy of ARF kidneys, even though GFR is less than 10% of normal. The intraluminal casts in ARF kidneys stain prominently for Tamm-Horsfall protein (THP), which is produced in the thick ascending limb (Figure 8). Importantly, THP is secreted into tubular fluid as a monomer but subsequently may become a polymer that forms a gel-like material in the presence of increased luminal Na+ concentration, as occurs in the distal nephron during clinical ARF with the decrease in tubular sodium reabsorption (43). Thus, the THP polymeric gel in the distal nephron provides an intraluminal environment for distal cast formation involving viable, apoptotic, and necrotic tubule epithelial cells, brush border membranes, and ECM (extracellular

matrix) (e.g., fibronectin) (Figure 3) (44). Whether tubular obstruction by casts is alone sufficient to account for the decreased GFR associated with clinical ARF is unknown. Certainly, net transglomerular capillary pressure can be decreased secondary to increased tubule pressure, as has been demonstrated using micropuncture techniques in experimental animals with acute ureteral obstruction (45). However, micropuncture studies in kidneys with acute ischemic injury have demonstrated that normalization of proximal tubular flow rate in a single nephron can dislodge some previously obstructing luminal casts and improve GFR in the same nephron (46). Thus, it can be proposed that at least some of the luminal casts would not cause obstruction to tubular flow if glomerular and thus tubular pressures were normal.

There is some experimental evidence that viable tubular epithelial cells released into the lumen during ischemia may adhere to other tubular cells and ECM and thereby cause intraluminal obstruction. This cellular adhesion during acute renal ischemia has been shown to involve integrin-mediated adhesion molecules via binding to Arg-Gly-Asp (RGD) sequences (47). In support of this

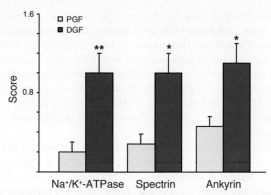

Figure 4
Immunofluorescent staining revealed the cellular location of the actin-binding proteins ankyrin, spectrin, and Na+/K+-ATPase in cadaveric transplanted kidneys with prompt graft function (PGF) and delayed graft function (DGF) (35). The stained sections were examined at ×60–600 magnification using a scanning laser confocal microscope. Scoring was done as follows: 0, continuous staining confined to the basolateral membrane; 0.5, 1.0, and 1.5, interrupted linear staining of basolateral membrane with less than 50%, approximately 50%, and greater than 50% of the staining, respectively, appearing in the cytoplasm. In kidneys with delayed graft function, approximately 50% of the ankyrin, spectrin, and Na+/K+-ATPase was translocated from the basolateral membrane to the cytoplasm, whereas those kidneys with prompt graft function had only minimal translocation of these proteins from the basolateral membrane. *$P < 0.01$ vs. PGF, **$P < 0.05$ vs. PGF.

possibility, synthetic cyclical RGD compounds administered during the reperfusion period have been shown to attenuate tubular obstruction and to reverse the related increase in proximal tubular pressure (48). Relief of tubular obstruction in experimental ARF, as assessed by nephron micropuncture, has also been shown with a solute diuresis induced by mannitol (49).

Tubuloglomerular balance and tubular fluid backleak in ischemic ARF. The decrease in proximal tubular sodium reabsorption that is associated with acute ischemic injury would increase sodium chloride delivery to the macula densa and thereby activate the tubuloglomerular feedback mechanism and decrease GFR (50). Micropuncture perfusion studies delivering increased sodium chloride to the macula densa have demonstrated a decrease in single-nephron GFR by as much as 50% (50). This degree of decline in GFR, however, could not explain the much greater decrease in GFR that is characteristic of clinical ARF. However, since the tone of the afferent arteriole modulates the tubulo-glomerular feedback mechanism, the increased sensitivity of this glomerular arteriole to vasoconstriction, as discussed earlier (18), could enhance the sensitivity of the tubuloglomerular feedback response in patients with clinical ARF. Moreover, the combination of tubular cast formation and activation of the tubuloglomerular feedback mechanism during acute renal ischemia, both of which can be linked to the ARF-related decrease in proximal tubular sodium reabsorption, can provide an adequate explanation for the drastic fall in GFR observed in clinical ARF. With respect to the role of tubuloglomerular feedback in ischemic ARF, a potential beneficial effect should also be considered. Activation of the tubuloglomerular feedback mechanism and the resultant decrease in GFR during acute renal ischemia would decrease sodium chloride delivery to damaged tubules, thereby lessening the demand for ATP-dependent tubular reabsorption.

The loss of the tubular epithelial cell barrier and/or the tight junctions between viable cells (51) during acute renal ischemia could lead to a leak of glomerular filtrate back into the circulation. If this occurs and normally non-reabsorbable substances, such as inulin, leak back into the circulation, then a falsely low GFR will be measured as inulin clearance. It should be noted, however, that the degree of extensive tubular damage observed in experimental studies demonstrating tubular fluid backleak is rarely observed with clinical ARF in humans (52). Moreover, dextran sieving studies in patients with ARF demonstrated that, at best, only a 10% decrease in GFR could be explained by backleak of filtrate (53). Cadaveric transplanted kidneys with delayed graft function, however, may have severe tubular necrosis, and thus backleak of glomerular filtration may be more important in this setting. The various potential mechanisms whereby alterations in tubular factors can decrease GFR in ischemic renal injury are shown in Figure 9.

Inflammation

There is now substantial evidence for the involvement of inflammation in the pathogenesis of the decreased GFR associated with acute renal ischemic injury. In this regard, there is experimental evidence that iNOS may contribute to tubular injury during ARF. Hypoxia in isolated proximal tubules has been shown to increase NO release (54), and Western blot analysis in ischemic kidney homogenates has demonstrated increased iNOS protein expression (55). An antisense oligonucleotide was shown to block the upregulation of iNOS and afford functional protection against acute renal ischemia (55). Moreover, when isolated proximal tubules from iNOS, eNOS, and neuronal NO synthase (nNOS) knockout mice were exposed to hypoxia, only the tubules from the iNOS knockout mice were protected against hypoxia, as assessed by LDH release (56). The iNOS knockout mice were also shown to have lower mortality during ischemia/reperfusion than wild-type mice (57).

There is also evidence that the scavenging of NO by oxygen radicals produces peroxynitrite that causes tubule damage during ischemia (58–60). The administration of α-melanocyte–stimulating hormone (αMSH) affords protection against ischemic/reperfusion renal injury by blocking both the induction of iNOS and leukocyte infiltration

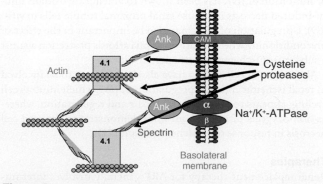

Figure 5
Potential cytoskeletal targets of cysteine proteases during hypoxia/ischemia in proximal tubules. The basolateral membrane of the proximal tubules contains the subunits of Na+/K+-ATPase, which are linked to the actin cytoskeleton by ankyrin and spectrin. This forms a metabolically stable complex. Potential cytoskeletal targets of cysteine proteases during hypoxia/ischemia are shown. Ank, ankyrin; CAM, cell adhesion molecule; 4.1, protein 4.1. Figure reproduced with permission from Taylor & Francis (36).

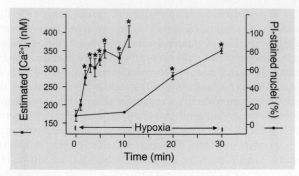

Figure 6
Increased free, intracellular Ca^{2+} concentration in isolated proximal tubules during hypoxia (measured with the fluorescent Ca^{2+} indicator Fura-2) precedes cell membrane damage, as assessed by propidium iodide (PI) staining (37). *Significant vs. time 0.

into the kidney during ischemia (61). Oxygen radical scavengers, such as superoxide dismutase, have also been shown to protect against acute renal injury associated with endotoxemia (62). Caspase inhibitors, IL-18 antibodies, and caspase-1 knockout mice have also been shown to be protective against ischemia/reperfusion injury (63, 64).

While iNOS may contribute to ischemic injury of renal tubules, there is evidence that the vascular effect of eNOS in the glomerular afferent arteriole is protective against ischemic injury. In this regard, eNOS knockout mice have been shown to be more sensitive to endotoxin-related injury than normal mice (65). Moreover, the protective role of vascular eNOS may be more important than the deleterious effect of iNOS at the tubule level during renal ischemia. The basis for this tentative conclusion is the observation that treatment of mice with the nonspecific NO synthase (NOS) inhibitor L-NAME, which blocks both iNOS and eNOS, worsens renal ischemic injury as compared with vehicle treatment (66). It has also been demonstrated that NO may downregulate eNOS (67) and is a potent inducer of heme oxygenase-1, which has been shown to be cytoprotective against renal injury (68).

The MAPK pathway also appears to be involved in renal oxidant injury. Activation of extracellular signal–regulated kinase (ERK) or inhibition of JNK has been shown to ameliorate oxidant injury–induced necrosis in mouse renal proximal tubule cells in vitro (69). Upregulation of ERK may also be important in the effect of preconditioning whereby early ischemia affords protection against a subsequent ischemia/reperfusion insult (70).

Alterations in cell cycling have also been shown to be involved in renal ischemic injury. Upregulation of p21, which inhibits cell cycling, appears to allow cellular repair and regeneration, whereas homozygous p21 knockout mice demonstrate enhanced cell necrosis in response to an ischemic insult (71).

Therapies
Renal replacement therapy for ARF generally involves intermittent hemodialysis (IHD) or continuous renal replacement therapy (CRRT), e.g., continuous veno-veno-hemofiltration (CVVH). Since the hemodynamic stress is less with CVVH than with IHD, it is possible that any additional hemodynamic or nephrotoxic insult, which might prolong the course of ARF and thereby increase mortality, might be less with CRRT. The most recent metaanalysis, however, of randomized results comparing IHD with CRRT in ARF has not shown any difference in survival (72).

As with most disease conditions, the earlier an intervention can be instituted in acute renal ischemia, the more favorable the outcome. Thus, biomarkers more sensitive than the rise in serum creatinine concentration associated with ARF will be necessary to achieve early intervention. As previously discussed, there are several diagnostic markers under study. Presently, however, the determination of FE$_{Na}$ using spot urine and blood sodium and creatinine measurements is the primary and most readily available early marker of established ARF.

Prolonged duration of the ARF clinical course and the need for dialysis are major factors projecting a poor prognosis. Patients with ARF who require dialysis have a 50–70% mortality rate. Infection and cardiopulmonary complications are the major causes of death in patients with ARF. Excessive fluid administration in patients with established ARF may lead to pulmonary congestion, hypoxia, the need for ventilatory support, pneumonia, and multiorgan dysfunction syndrome, which has an 80–90% mortality rate (9, 73, 74). Until means to reverse the diminished host defense mechanisms in azotemic patients with clinical ARF are available, every effort should be made to avoid invasive procedures such as the placement of bladder catheters, intravenous lines, and mechanical ventilation.

Over and above such supportive care, it may be that combination therapy will be necessary to prevent or attenuate the course of ARF. Such combination therapy must involve agents with potential beneficial effects on vascular tone, tubular obstruction, and inflammation. However, vasodilator agents such as calcium channel blockers and natriuretic peptides may induce unwanted side effects such as systemic vasodilation and hypotension (75, 76), which increase sympathetic tone and activity of the renin-angiotensin system (77). These compensatory neurohormonal responses support blood pressure but cause renal vasoconstriction (78), which may obscure beneficial effects of calcium channel blockers and natriuretic peptides on the kidney.

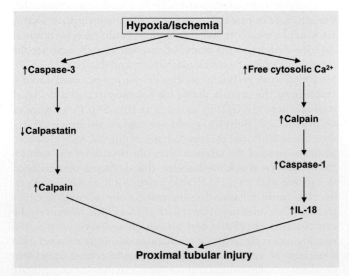

Figure 7
Hypoxic/ischemic proximal tubular necrosis results in activation of cysteine protease pathways involving calpains and both caspase-1 (an inflammatory caspase) and caspase-3 (an executioner caspase involved in apoptosis). Calpain is activated both by increased free cytosolic Ca^{2+} and decreased calpastatin. Calpain then activates caspase-1, which stimulates the inflammatory cytokine IL-18. The executioner caspase-3 cleaves calpastatin (40).

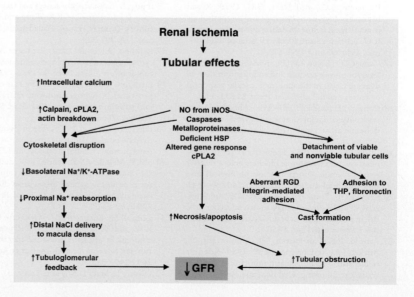

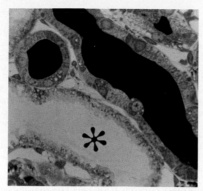

Figure 8
Staining of Tamm-Horsfall protein of casts in renal tubules and dilated tubules in ischemic ARF. The asterisk indicates the tubular lumen. ×100.

Since an improvement of GFR of only 10 ml/min from each kidney of a patient with ARF may circumvent the need for dialysis and potentially improve survival, the bilateral intrarenal infusion of a short-acting vasodilator and/or an impermeant solute such as mannitol is a potential approach that is less invasive than hemodialysis (79). Another approach is to combine a systemic vasodilator such as a natriuretic peptide, with dopamine (80) or mannitol (81), which not only will attenuate the systemic hypotensive effects of the vasodilator but can also increase tubular solute flow and thereby decrease tubular obstruction. This combined therapeutic approach has been shown to be effective in acute renal injury in experimental animals (81).

The effect of antiinflammatory agents, including reactive oxygen species scavengers, in the treatment of ARF should also be investigated. As discussed earlier, the use of inhibitors of NOS must be specific, since ARF is made more severe by nonspecific NOS inhibition (66). For example, a specific iNOS inhibitor, L-NIL, has been shown to afford protection in sepsis-related ARF in the rat (67). In this regard, cytokine-induced tubular damage may occur in sepsis in the absence of a decrease in renal blood flow. The antiinflammatory effect of and inhibition of iNOS by αMSH, which has been shown to be effective up to 8 hours after the insult, also needs to be studied as an approach to alter the course of clinical ARF (61). Inhibition of TNF-α has afforded renal protection in experimental endotoxemic ARF (82). However, the administration of antibodies to TNF-α in septic patients has not improved survival (83). In acute renal ischemia, the cellular breakdown of ATP occurs with leakage of the nucleotide products out of tubular cells. On this background, the administration of exogenous ATP has been shown to afford protection in experimental ischemic ATP. While ATP salvage has been shown to be protective against experimental tubular necrosis (84), recent experimental evidence has suggested that GTP salvage may be preferred in preventing apop-

tosis associated with acute renal injury (85). The effect of the administration of synthetic RGD peptides to attenuate tubular obstruction has yet to be examined in clinical ARF. However, since patients are generally seen after clinical ARF is established with a GFR less than 10% of normal, delivery of these synthetic RGD peptides to the tubular lumen would be difficult in the absence of accompanying intrarenal vasodilation. Experiments examining upregulation of heat shock proteins (57, 86) or the protective effects of preconditioning against a later insult (70) suggest intriguing renal-protective approaches in need of further study.

Since patients with clinical ARF are generally seen after the insult, with the exception of kidney transplant subjects (22) and recipients of radiocontrast media (87), ways to enhance recovery and thereby lessen the duration of clinical ARF have been sought. Unfortunately, the beneficial effects of insulin-derived growth factor observed in experimental animals were not duplicated in a randomized study of patients with clinical ARF (88). A recent exciting approach in this area has been the administration of stem cells or endothelial cells to enhance recovery from acute ischemic renal injury in experimental animals (89–92). There is ongoing research into the mechanisms whereby injured tubules are relined with new cells actively engaged in DNA synthesis. The pathways by which surviving cells reenter the cell cycle and replicate may involve the early immediate response genes (93).

In summary, while much has been learned about acute renal ischemic injury, which is frequent in hospitalized patients, the mortality of clinical ARF remains high. The future, however, holds substantial promise for earlier diagnosis and effective interventions that are able to prevent or shorten the course of acute renal injury and thus improve survival. An ARF network for clinical trials with adequate statistical power would certainly facilitate this process.

Acknowledgments
This work was supported by National Institute of Diabetes and Digestive and Kidney Diseases grant DK52599.

Address correspondence to: Robert W. Schrier, Department of Medicine, University of Colorado Health Sciences Center, 4200 East Ninth Avenue, Box C-281, Denver, Colorado 80262, USA. Phone: (303) 315-8059; Fax: (303) 315-2685; E-mail: robert.schrier@uchsc.edu.

Figure 9
Effects of ischemia on renal tubules in the pathogenesis of ischemic ARF. cPLA2, cytosolic phospholipase A2; HSP, heat shock protein. Figure modified with permission from the *Journal of Nephrology* (15).

1. Bywaters, E.G., and Beall, D. 1941. Crush injuries with impairment of renal function. *Br. Med. J.* 1:427–432.

2. Teschan, P.E., et al. 1955. Post-traumatic renal insufficiency in military casualties. I. Clinical characteristics. *Am. J. Med.* 18:172–186.

3. Smith, L.H., Jr., et al. 1955. Post-traumatic renal insufficiency in military casualties. II. Management, use of an artificial kidney, prognosis. *Am. J. Med.* 18:187–198.

4. Butkus, D.E. 1984. Post-traumatic acute renal failure in combat casualties: a historical review. *Mil. Med.* 149:117–124.

5. Whelton, A., and Donadio, J.V., Jr. 1969. Post-traumatic acute renal failure in Vietnam. A comparison with the Korean war experience. *Johns Hopkins Med. J.* 124:95–105.

6. Reineck, H.J., O'Connor, G.J., Lifschitz, M.D., and Stein, J.H. 1980. Sequential studies on the pathophysiology of glycerol-induced acute renal failure. *J. Lab. Clin. Med.* 96:356–362.

7. Miller, T.R., et al. 1978. Urinary diagnostic indices in acute renal failure: a prospective study. *Ann. Intern. Med.* 89:47–50.

8. Carvounis, C.P., Nisar, S., and Guro-Razuman, S. 2002. Significance of the fractional excretion of urea in the differential diagnosis of acute renal failure. *Kidney Int.* 62:2223–2229.

9. Esson, M.L., and Schrier, R.W. 2002. Diagnosis and treatment of acute tubular necrosis. *Ann. Intern. Med.* 137:744–752.

10. Mehta, R.L., et al. 2002. Nephrology consultation in acute renal failure: does timing matter? *Am. J. Med.* 113:456–461.

11. Coll, E., et al. 2000. Serum cystatin C as a new marker for noninvasive estimation of glomerular filtration rate and as a marker for early renal impairment. *Am. J. Kidney Dis.* 36:29–34.

12. Han, W.K., Bailly, V., Abichandani, R., Thadhani, R., and Bonventre, J.V. 2002. Kidney Injury Molecule-1 (KIM-1): a novel biomarker for human renal proximal tubule injury. *Kidney Int.* 62:237–244.

13. Parikh, C.R., Jani, A., Melnikov, V.Y., Faubel, S., and Edelstein, C.L. 2004. Urinary interleukin-18 is a marker of human acute tubular necrosis. *Am. J. Kidney Dis.* 43:405–414.

14. Westhuyzen, J., et al. 2003. Measurement of tubular enzymuria facilitates early detection of acute renal impairment in the intensive care unit. *Nephrol. Dial. Transplant.* 18:543–551.

15. Kribben, A., Edelstein, C.L., and Schrier, R.W. 1999. Pathophysiology of acute renal failure. *J. Nephrol.* 12(Suppl. 2):S142–S151.

16. Racusen, L.C., and Nast, C.C. 1999. Renal histopathology, urine cytology, and cytopathology in acute renal failure. In *Atlas of diseases of the kidney.* R.W. Schrier, editor. Blackwell Science. Philadelphia, Pennsylvania, USA. 1–9, 12.

17. Conger, J.D., Robinette, J.B., and Schrier, R.W. 1988. Smooth muscle calcium and endothelium-derived relaxing factor in the abnormal vascular responses of acute renal failure. *J. Clin. Invest.* 82:532–537.

18. Conger, J.D., and Falk, S.A. 1993. Abnormal vasoreactivity of isolated arterioles from rats with ischemic acute renal failure (ARF) [abstract]. *J. Am. Soc. Nephrol.* 4:733A.

19. Burke, T.J., et al. 1984. Protective effect of intrarenal calcium membrane blockers before or after renal ischemia. Functional, morphological, and mitochondrial studies. *J. Clin. Invest.* 74:1830–1841.

20. Arnold, P.E., Lumlertgul, D., Burke, T.J., and Schrier, R.W. 1985. In vitro versus in vivo mitochondrial calcium loading in ischemic acute renal failure. *Am. J. Physiol.* 248:F845–F850.

21. Arnold, P.E., Van Putten, V.J., Lumlertgul, D., Burke, T.J., and Schrier, R.W. 1986. Adenine nucleotide metabolism and mitochondrial Ca2+ transport following renal ischemia. *Am. J. Physiol.* 250:F357–F363.

22. Neumayer, H.H., and Wagner, K. 1987. Prevention of delayed graft function in cadaver kidney transplants by diltiazem: outcome of two prospective, randomized clinical trials. *J. Cardiovasc. Pharmacol.* 10(Suppl. 10):S170–S177.

23. Mason, J., Torhorst, J., and Welsch, J. 1984. Role of the medullary perfusion defect in the pathogenesis of ischemic renal failure. *Kidney Int.* 26:283–293.

24. Kelly, K.J., Williams, W.W., Jr., Colvin, R.B., and Bonventre, J.V. 1994. Antibody to intercellular adhesion molecule 1 protects the kidney against ischemic injury. *Proc. Natl. Acad. Sci. U. S. A.* 91:812–816.

25. Molitoris, B.A., and Marrs, J. 1999. The role of cell adhesion molecules in ischemic acute renal failure. *Am. J. Med.* 106:583–592.

26. Zizzi, H.C., et al. 1997. Quantification of P-selectin expression after renal ischemia and reperfusion. *J. Pediatr. Surg.* 32:1010–1013.

27. Singbartl, K., Green, S.A., and Ley, K. 2000. Blocking P-selectin protects from ischemia/reperfusion-induced acute renal failure. *FASEB J.* 14:48–54.

28. Linas, S.L., Shanley, P.F., Whittenburg, D., Berger, E., and Repine, J.E. 1988. Neutrophils accentuate ischemia-reperfusion injury in isolated perfused rat kidneys. *Am. J. Physiol.* 255:F728–F735.

29. Molitoris, B.A., Sandoval, R., and Sutton, T.A. 2002. Endothelial injury and dysfunction in ischemic acute renal failure. *Crit. Care Med.* 30(Suppl. 5):S235–S240.

30. Sutton, T.A., Fisher, C.J., and Molitoris, B.A. 2002. Microvascular endothelial injury and dysfunction during ischemic acute renal failure. *Kidney Int.* 62:1539–1549.

31. Badr, K.F., et al. 1989. Mesangial cell, glomerular and renal vascular responses to endothelin in the rat kidney. Elucidation of signal transduction pathways. *J. Clin. Invest.* 83:336–342.

32. Thadhani, R., Pascual, M., and Bonventre, J.V. 1996. Acute renal failure. *N. Engl. J. Med.* 334:1448–1460.

33. Racusen, L.C. 1998. Epithelial cell shedding in acute renal injury. *Clin. Exp. Pharmacol. Physiol.* 25:273–275.

34. Molitoris, B.A., Chan, L.K., Shapiro, J.I., Conger, J.D., and Falk, S.A. 1989. Loss of epithelial polarity: a novel hypothesis for reduced proximal tubule Na+ transport following ischemic injury. *J. Membr. Biol.* 107:119–127.

35. Alejandro, V.S., et al. 1995. Postischemic injury, delayed function and Na+/K(+)-ATPase distribution in the transplanted kidney. *Kidney Int.* 48:1308–1315.

36. Edelstein, C., and Schrier, R. 1999. The role of calpain in renal proximal tubular and hepatocyte injury. In *Calpain: pharmacology and toxicology of calcium-dependent protease.* K.K. Wang and P.W. Yuen, editors. Taylor & Francis. Philadelphia, Pennsylvania, USA. 307–329.

37. Kribben, A., et al. 1994. Evidence for role of cytosolic free calcium in hypoxia-induced proximal tubule injury. *J. Clin. Invest.* 93:1922–1929.

38. Edelstein, C.L., et al. 1995. The role of cysteine proteases in hypoxia-induced rat renal proximal tubular injury. *Proc. Natl. Acad. Sci. U. S. A.* 92:7662–7666.

39. Edelstein, C.L., et al. 1997. Effect of glycine on prelethal and postlethal increases in calpain activity in rat renal proximal tubules. *Kidney Int.* 52:1271–1278.

40. Shi, Y., Melnikov, V.Y., Schrier, R.W., and Edelstein, C.L. 2000. Downregulation of the calpain inhibitor protein calpastatin by caspases during renal ischemia-reperfusion. *Am. J. Physiol. Renal Physiol.* 279:F509–F517.

41. Choi, K.H., Edelstein, C.L., Gengaro, P., Schrier, R.W., and Nemenoff, R.A. 1995. Hypoxia induces changes in phospholipase A2 in rat proximal tubules: evidence for multiple forms. *Am. J. Physiol.* 269:F846–F853.

42. Oliver, J., Mac, D.M., and Tracy, A. 1951. The pathogenesis of acute renal failure associated with traumatic and toxic injury: renal ischemia, nephrotoxic damage and the ischemic episode. *J. Clin. Invest.* 30:1307–1439.

43. Wangsiripaisan, A., Gengaro, P.E., Edelstein, C.L., and Schrier, R.W. 2001. Role of polymeric Tamm-Horsfall protein in cast formation: oligosaccharide and tubular fluid ions. *Kidney Int.* 59:932–940.

44. Zuk, A., Bonventre, J.V., and Matlin, K.S. 2001. Expression of fibronectin splice variants in the postischemic rat kidney. *Am. J. Physiol. Renal Physiol.* 280:F1037–F1053.

45. Tanner, G.A. 1982. Nephron obstruction and tubuloglomerular feedback. *Kidney Int. Suppl.* 12:S213–S218.

46. Conger, J.D., Robinette, J.B., and Kelleher, S.P. 1984. Nephron heterogeneity in ischemic acute renal failure. *Kidney Int.* 26:422–429.

47. Noiri, E., et al. 1994. Cyclic RGD peptides ameliorate ischemic acute renal failure in rats. *Kidney Int.* 46:1050–1058.

48. Goligorsky, M.S., and DiBona, G.F. 1993. Pathogenetic role of Arg-Gly-Asp-recognizing integrins in acute renal failure. *Proc. Natl. Acad. Sci. U. S. A.* 90:5700–5704.

49. Burke, T.J., Cronin, R.E., Duchin, K.L., Peterson, L.N., and Schrier, R.W. 1980. Ischemia and tubule obstruction during acute renal failure in dogs: mannitol in protection. *Am. J. Physiol.* 238:F305–F314.

50. Schnermann, J. 2003. Homer W. Smith Award lecture. The juxtaglomerular apparatus: from anatomical peculiarity to physiological relevance. *J. Am. Soc. Nephrol.* 14:1681–1694.

51. Molitoris, B.A., Falk, S.A., and Dahl, R.H. 1989. Ischemia-induced loss of epithelial polarity. Role of the tight junction. *J. Clin. Invest.* 84:1334–1339.

52. Edelstein, C., and Schrier, R. 2001. Pathophysiology of ischemic acute renal failure. In *Diseases of the kidney and urinary tract.* 7th edition. R.W. Schrier, editor. Lippincott Williams & Wilkins. Philadelphia, Pennsylvania, USA. 1041–1070.

53. Myers, B.D., Chui, F., Hilberman, M., and Michaels, A.S. 1979. Transtubular leakage of glomerular filtrate in human acute renal failure. *Am. J. Physiol.* 237:F319–F325.

54. Yu, L., Gengaro, P.E., Niederberger, M., Burke, T.J., and Schrier, R.W. 1994. Nitric oxide: a mediator in rat tubular hypoxia/reoxygenation injury. *Proc. Natl. Acad. Sci. U. S. A.* 91:1691–1695.

55. Noiri, E., Peresleni, T., Miller, F., and Goligorsky, M.S. 1996. In vivo targeting of inducible NO synthase with oligodeoxynucleotides protects rat kidney against ischemia. *J. Clin. Invest.* 97:2377–2383.

56. Ling, H., et al. 1998. Effect of hypoxia on proximal tubules isolated from nitric oxide synthase knockout mice. *Kidney Int.* 53:1642–1646.

57. Ling, H., et al. 1999. Attenuation of renal ischemia-reperfusion injury in inducible nitric oxide synthase knockout mice. *Am. J. Physiol.* 277:F383–F390.

58. Xia, Y., Dawson, V.L., Dawson, T.M., Snyder, S.H., and Zweier, J.L. 1996. Nitric oxide synthase generates superoxide and nitric oxide in arginine-depleted cells leading to peroxynitrite-mediated cellular injury. *Proc. Natl. Acad. Sci. U. S. A.* 93:6770–6774.

59. Noiri, E., et al. 2001. Oxidative and nitrosative stress in acute renal ischemia. *Am. J. Physiol. Renal Physiol.* 281:F948–F957.

60. Wangsiripaisan, A., et al. 1999. Effect of nitric oxide donors on renal tubular epithelial cell-matrix adhesion. *Kidney Int.* 55:2281–2288.

61. Chiao, H., et al. 1997. Alpha-melanocyte-stimulating hormone protects against renal injury after ischemia in mice and rats. *J. Clin. Invest.* 99:1165–1172.

62. Wang, W., et al. 2003. Interaction among nitric oxide, reactive oxygen species, and antioxidants during endotoxemia-related acute renal failure. *Am. J. Physiol. Renal Physiol.* 284:F532–F537.

63. Melnikov, V.Y., et al. 2001. Impaired IL-18 process-

ing protects caspase-1-deficient mice from ischemic acute renal failure. *J. Clin. Invest.* **107**:1145–1152.

64. Melnikov, V.Y., et al. 2002. Neutrophil-independent mechanisms of caspase-1– and IL-18–mediated ischemic acute tubular necrosis in mice. *J. Clin. Invest.* **110**:1083–1091. doi:10.1172/JCI200215623.

65. Wang, W., et al. 2004. Endothelial nitric oxide synthase (eNOS) deficient mice exhibit increased susceptibility to endotoxin-induced acute renal failure. *Am. J. Physiol.* In press.

66. Atanasova, I., Burke, T.J., McMurtry, I.F., and Schrier, R.W. 1995. Nitric oxide synthase inhibition and acute renal ischemia: effect on systemic hemodynamics and mortality. *Ren. Fail.* **17**:389–403.

67. Schwartz, D., et al. 1997. Inhibition of constitutive nitric oxide synthase (NOS) by nitric oxide generated by inducible NOS after lipopolysaccharide administration provokes renal dysfunction in rats. *J. Clin. Invest.* **100**:439–448.

68. Sikorski, E.M., Hock, T., Hill-Kapturczak, N., and Agarwal, A. 2004. The story so far: molecular regulation of the heme oxygenase-1 gene in renal injury. *Am. J. Physiol. Renal Physiol.* **286**:F425–F441.

69. Arany, I., Megyesi, J.K., Kaneto, H., Tanaka, S., and Safirstein, R.L. 2004. Activation of ERK or inhibition of JNK ameliorates H_2O_2 cytotoxicity in mouse renal proximal tubule cells. *Kidney Int.* **65**:1231–1239.

70. Park, K.M., Chen, A., and Bonventre, J.V. 2001. Prevention of kidney ischemia/reperfusion-induced functional injury and JNK, p38, and MAPK kinase activation by remote ischemic pretreatment. *J. Biol. Chem.* **276**:11870–11876.

71. Price, P.M., Megyesi, J., and Safirstein, R.L. 2003. Cell cycle regulation: repair and regeneration in acute renal failure. *Semin. Nephrol.* **23**:449–459.

72. Teehan, G.S., et al. 2003. Dialysis membrane and modality in acute renal failure: understanding discordant meta-analyses. *Semin. Dial.* **16**:356–360.

73. Brivet, F.G., Kleinknecht, D.J., Loirat, P., and Landais, P.J. 1996. Acute renal failure in intensive care

units: causes, outcome, and prognostic factors of hospital mortality; a prospective, multicenter study. French Study Group on Acute Renal Failure. *Crit. Care Med.* **24**:192–198.

74. Poole, B., and Schrier, R. 2005. Acute renal failure in the intensive care unit. In *Textbook of critical care.* 5th edition. F. Mitchell, E. Abraham, J.Vincent, and P. Kochanek, editors. Elsevier. Philadelphia, Pennsylvania, USA. In press.

75. Nakamoto, M., Shapiro, J.I., Shanley, P.F., Chan, L., and Schrier, R.W. 1987. In vitro and in vivo protective effect of atriopeptin III on ischemic acute renal failure. *J. Clin. Invest.* **80**:698–705.

76. Allgren, R.L., et al. 1997. Anaritide in acute tubular necrosis. Auriculin Anaritide Acute Renal Failure Study Group. *N. Engl. J. Med.* **336**:828–834.

77. Schrier, R.W., and Wang, W. 2004. Acute renal failure and sepsis. *N. Engl. J. Med.* **351**:159–169.

78. Wang, W., et al. 2002. Protective effect of renal denervation on normotensive endotoxemia-induced acute renal failure in mice. *Am. J. Physiol. Renal Physiol.* **283**:F583–F587.

79. Burke, T.J., Arnold, P.E., and Schrier, R.W. 1983. Prevention of ischemic acute renal failure with impermeant solutes. *Am. J. Physiol.* **244**:F646–F649.

80. Conger, J.D., Falk, S.A., Yuan, B.H., and Schrier, R.W. 1989. Atrial natriuretic peptide and dopamine in a rat model of ischemic acute renal failure. *Kidney Int.* **35**:1126–1132.

81. Lieberthal, W., Sheridan, A.M., and Valeri, C.R. 1990. Protective effect of atrial natriuretic factor and mannitol following renal ischemia. *Am. J. Physiol.* **258**:F1266–F1272.

82. Knotek, M., et al. 2001. Endotoxemic renal failure in mice: role of tumor necrosis factor independent of inducible nitric oxide synthase. *Kidney Int.* **59**:2243–2249.

83. Abraham, E., et al. 1998. Double-blind randomised controlled trial of monoclonal antibody to human tumour necrosis factor in treatment of septic shock. NORASEPT II Study Group. *Lancet.* **351**:929–933.

84. Siegel, N.J., et al. 1980. Enhanced recovery from acute renal failure by the postischemic infusion of adenine nucleotides and magnesium chloride in rats. *Kidney Int.* **17**:338–349.

85. Kelly, K.J., Plotkin, Z., and Dagher, P.C. 2001. Guanosine supplementation reduces apoptosis and protects renal function in the setting of ischemic injury. *J. Clin. Invest.* **108**:1291–1298. doi:10.1172/JCI200113018.

86. Meldrum, K.K., Meldrum, D.R., Sezen, S.F., Crone, J.K., and Burnett, A.L. 2001. Heat shock prevents simulated ischemia-induced apoptosis in renal tubular cells via a PKC-dependent mechanism. *Am. J. Physiol. Regul. Integr. Comp. Physiol.* **281**:R359–R364.

87. Tepel, M., et al. 2000. Prevention of radiographic-contrast-agent-induced reductions in renal function by acetylcysteine. *N. Engl. J. Med.* **343**:180–184.

88. Hirschberg, R., et al. 1999. Multicenter clinical trial of recombinant human insulin-like growth factor I in patients with acute renal failure. *Kidney Int.* **55**:2423–2432.

89. Brodsky, S.V., et al. 2002. Endothelial dysfunction in ischemic acute renal failure: rescue by transplanted endothelial cells. *Am. J. Physiol. Renal Physiol.* **282**:F1140–F1149.

90. Kale, S., et al. 2003. Bone marrow stem cells contribute to repair of the ischemically injured renal tubule. *J. Clin. Invest.* **112**:42–49. doi:10.1172/JCI200317856.

91. Lin, F., et al. 2003. Hematopoietic stem cells contribute to the regeneration of renal tubules after renal ischemia-reperfusion injury in mice. *J. Am. Soc. Nephrol.* **14**:1188–1199.

92. Gupta, S., Verfaillie, C., Chmielewski, D., Kim, Y., and Rosenberg, M.E. 2002. A role for extrarenal cells in the regeneration following acute renal failure. *Kidney Int.* **62**:1285–1290.

93. Safirstein, R., DiMari, J., Megyesi, J., and Price, P. 1998. Mechanisms of renal repair and survival following acute injury. *Semin. Nephrol.* **18**:519–522.

94. Thadhani, R., Pascual, M., and Bonventre, J. 1996. Acute renal failure. *N. Engl. J. Med.* **334**:1448–1460.

Kidney stone disease

Fredric L. Coe,[1] Andrew Evan,[2] and Elaine Worcester[1]

[1]Renal Section, University of Chicago, Chicago, Illinois, USA. [2]Department of Anatomy, Indiana University, Indianapolis, Indiana, USA.

About 5% of American women and 12% of men will develop a kidney stone at some time in their life, and prevalence has been rising in both sexes. Approximately 80% of stones are composed of calcium oxalate (CaOx) and calcium phosphate (CaP); 10% of struvite (magnesium ammonium phosphate produced during infection with bacteria that possess the enzyme urease), 9% of uric acid (UA); and the remaining 1% are composed of cystine or ammonium acid urate or are diagnosed as drug-related stones. Stones ultimately arise because of an unwanted phase change of these substances from liquid to solid state. Here we focus on the mechanisms of pathogenesis involved in CaOx, CaP, UA, and cystine stone formation, including recent developments in our understanding of related changes in human kidney tissue and of underlying genetic causes, in addition to current therapeutics.

Clinical aspects of stone disease

Stone passage

Nonobstructing stones produce no symptoms or signs apart from hematuria. Stone passage produces renal colic that usually begins as a mild discomfort and progresses to a plateau of extreme severity over 30–60 minutes. If the stone obstructs the uretero-pelvic junction, pain localizes to the flank; as the stone moves down the ureter, pain moves downward and anterior. Stones at the uretero-vesicular junction often cause dysuria and urinary frequency mistaken for infection. Colic is independent of body position or motion and is described as a boring or burning sensation associated with nausea and vomiting. Stones less than 5 mm in diameter have a high chance of passage; those of 5–7 mm have a modest chance (50%) of passage, and those greater than 7 mm almost always require urological intervention. Ideally, stone analysis is performed by infrared spectroscopy or x-ray diffraction. Renal stone burden is best gauged using CT radiographs taken with 5-mm cuts, without infusion of contrast agents. The radiographic appearance and density of stones as measured by CT is a guide to their composition (1).

Urological management of stones

Extracorporeal shock wave lithotripsy (ESWL), in which sound waves are used to break the stone into small pieces that can more easily pass into the bladder, is widely used and valuable for small stones (2). Modern instruments facilitate passage of endoscopes up the ureter into the kidney pelvis and permit local stone disruption with high-powered lasers (3). Percutaneous stone removal via instruments introduced into the kidney through a small flank incision permits disruption and removal of even very large stones (4).

Nonstandard abbreviations used: $b^{0,+}AT$, $b^{0,+}$ amino acid transporter; BM, basement membrane; CaOx, calcium oxalate; CaP, calcium phosphate; CaSR, calcium-sensing receptor; CD, collecting duct; CLCN5, chloride channel 5; DCT, distal convoluted tubule; dRTA, distal renal tubular acidosis; ESWL, extracorporeal shock wave lithotripsy; IH, idiopathic hypercalciuria; PH, primary hyperoxaluria; PT, proximal tubule; PTH, parathyroid hormone; SF, stone former; SS, supersaturation; rBAT, related to $b^{0,+}AT$; TAL, thick ascending limb; UA, uric acid; ULM, upper limit of metastability.

Conflict of interest: F.L. Coe has a financial interest in Litholink Corp., a provider of kidney stone testing services. The remaining authors have declared that no conflict of interest exists.

Citation for this article: *J. Clin. Invest.* **115**:2598–2608 (2005). doi:10.1172/JCI26662.

Renal function of stone forming people is reduced

Within the National Health and Nutrition Examination Survey III data set, subjects with a BMI greater than or equal to 27 who had kidney stones had lower estimated glomerular filtration rates than non–stone formers (non-SFs) matched for age, sex, race, and BMI (5). SFs also have higher blood pressures than non-SFs (6). Obstruction of the urinary tract, sequelae of urological interventions, and the processes that cause stone formation may all injure renal tissue, reduce renal function, and raise blood pressure.

Determinants of phase change

Supersaturation

Stones result from a phase change in which dissolved salts condense into solids, and all phase changes are driven by supersaturation (SS), which is usually approximated for such salts by the ratio of their concentration in the urine to their solubilities (7) and calculated by computer algorithms. At SS values less than 1, crystals of a substance will dissolve; at SS values greater than 1, crystals can form and grow. As expected, the composition of stones that patients form correlates with SS values from the urine they produce (8). Although increasing urine volume is an obvious way to lower SS, patients examined in a variety of practice settings have been found to be able to increase their urine volume by an average of only 0.3 l/d (9). Moreover, for unclear reasons, sodium intake and urinary calcium excretion has been found to increase with increased urine volume, partly offsetting the fall in SS. Along with urine volume, urine calcium and oxalate concentrations are the main determinants of calcium oxalate (CaOx) SS; urine calcium concentration and pH are the main determinants of calcium phosphate (CaP) SS; and urinary pH is the main determinant of uric acid (UA) SS.

The upper limit of metastability

Urine with SS greater than 1 is referred to as metastable because the excess dissolved material, being present at a concentration above its solubility, must eventually precipitate. One can add oxalate or calcium to urine and note the SS needed to produce a solid phase of CaOx or CaP. That value, called the upper limit of metastability (ULM), varies with urine SS (10) and is lower among patients with stones than in matched control subjects (11). This suggests that mechanisms that normally protect against solid-phase development are less effective in patients with stones than healthy individuals. Urine contains molecules that retard the for-

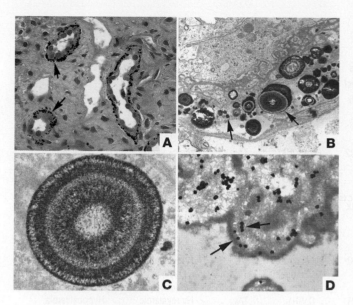

Figure 1
Initial sites of crystal deposition and localization of osteopontin. The initial sites of calcium deposits in the deep papillary tissue of an idiopathic CaOx SF are shown in light (**A**) and transmission electron microscopic (TEM) images (**B–D**). In **A**, Yasue-stained (the Yasue stain detects calcium) biopsy tissue reveals sites of crystal deposits (arrows) within the BM of thin Henle loops and not in nearby inner medullary CDs. By TEM (**B** and **C**), the crystal deposits appear as single spheres with a multi-laminated (6–7 layers) internal morphology consisting of a central light region of crystalline material surrounded by a dark layer of matrix material (arrows). Note that the cells lining this Henle loop appear morphologically normal. Osteopontin (**D**) localizes on either side of the apatite layers, sometimes forming with a clear "tram track"–like appearance (arrows). Magnification: ×1,800 (**A**); ×20,000 (**B**); ×35,000 (**C**); ×37,000 (**D**). **A** and **B** reprinted from ref. 19. **C** and **D** reprinted with permission from *Kidney International* (21).

mation of solid CaOx and CaP phases (12), and it is precisely this retardation that can permit the transient existence of SS. The measurement of ULM is not currently used in clinical practice.

Modulators of the ULM, crystal growth, and aggregation
Urine citrate reduces SS by binding calcium and inhibits nucleation and growth of calcium crystals (13); it is measured clinically, and low levels are treated as a cause of stones. Osteopontin (14), prothrombin F1 fragment (15), the inter-α-trypsin inhibitor molecule (16), calgranulin (17), Tamm Horsfall glycoprotein (18), as well as albumin, RNA and DNA fragments, and glycosaminoglycans have all been identified as urine inhibitors of CaOx and CaP crystallization (12). They have in common long stretches of polyanion chains that can bond with surface calcium atoms and prevent crystal growth. Such blockade prevents measurable phase changes and raises the ULM because newly formed crystals cannot grow beyond extremely minute dimensions. These molecules also prevent clumping of small particles into larger ones (aggregation). Because we do not know which, if any, of these molecules contribute to stone pathogenesis, they are presently of great research interest.

CaOx stones
The vast majority of CaOx SFs suffer from no systemic disease and as such are described as idiopathic CaOx SFs. Some have primary hyperparathyroidism or other disorders of calcium

Figure 2
Accumulation of interstitial crystal deposits as seen in light and TEM images in a papillary biopsy from an idiopathic CaOx SF. (**A**) Light microscopy reveals extensive accumulation of crystalline deposits (green arrow) shown around the Henle loops and nearby vascular bundles and inner medullary CDs. This progressive accumulation of crystalline material in the interstitium results in the formation of incomplete to complete cuffs of plaque. (**B**) TEM reveals a normal thin Henle loop surrounded by a complete cuff of interstitial plaque. (**C**) TEM shows a site of plaque located in the interstitial space, away from a tubular wall. Note that single crystal deposits appear embedded in a sea of matrix. Magnification: ×1,500 (**A**); ×13,000 (**B**); ×13,000 (**C**). **A** reprinted from ref. 19. **B** reprinted with permission from *Urological Research* (123). **C** reprinted with permission from *Kidney International* (21).

metabolism, and others present with hyperoxaluria because of bowel disease (enteric hyperoxaluria) and genetic disorders of oxalate metabolism (primary hyperoxaluria [PH]). We discuss these groups separately below.

Pathogenesis of idiopathic CaOx stones
Stones form on interstitial apatite plaque. CaOx stones form on the surfaces of the renal papillae over collections of interstitial suburothelial CaP particles (19) named Randall plaque. The number of CaOx stones formed, adjusted for duration of stone formation, varies directly with plaque surface coverage (20), as would be expected if plaque were a surface that promotes CaOx overgrowth.

Plaque begins in the basement membranes of the thin Henle loops. Basement membrane (BM) plaque comprises a myriad of particles in which crystal and organic layers alternate (Figure 1, A and B); the outer surface of all particles is the organic layer (Figure 1, B and C). The crystal in plaque is always biological apatite, the mineral phase found in bone. Outside the BM, in the interstitium,

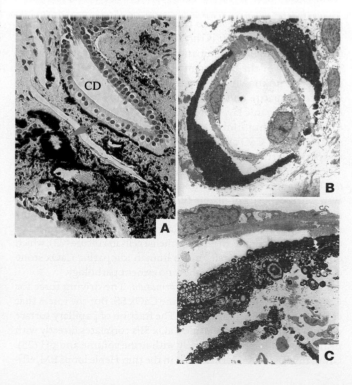

Table 1

Monogenic hypercalciuric stone-forming diseases

Disease	OMIM	Inheritance	Locus/gene	Gene product	Functions	Location	Phenotype Renal/urine	Other
Dent disease complex (37–42)	300009, 310468, 300008	X-linked	Xp11.22/ *CLCN5*	ClC-5	Endosomal chloride channel	PT, TAL, CD (type A intercalated cells)	Hypercalciuria, phosphaturia, LMW proteinuria, CRF	Rickets (variable)
Bartter syndrome type I (49)	601678, 600839	Autosomal recessive	15q15–21/ *SLC12A1*	NKCC2	Na-K-2Cl cotransporter (Na absorption)	TAL	Hypokalemic alkalosis, Na wasting, hypercalciuria	Elevated PGE level
Bartter syndrome type II (49)	241200, 600359	Autosomal recessive	11q24–25/ *KCNJ1*	ROMK	K channel (K supply for NKCC2)	TAL, CD	Hypokalemic alkalosis, Na wasting, hypercalciuria	Elevated PGE level
Bartter syndrome type III (49)	607364, 602023	Autosomal recessive	1p36/ *CLCNKB*	ClC-Kb	Basolateral Cl⁻ channel (Cl⁻ absorption	TAL, DCT, CD	Hypokalemic alkalosis, hypercalciuria	
Bartter syndrome type V (59–60)	601199	Autosomal dominant	3q13.3–q21/ *CASR* (severe gain of function)	CaSR	CaSR)	TAL, DCT, PT, CD	Hypokalemic alkalosis, hypomagnesemia, hypercalciuria, CRF	Hypocalcemia, tetany, seizures
Autosomal dominant hypocalcemic hypercalciuria (55–58)	146200, 601199	Autosomal dominant	3q13.3–q21/ *CASR* (gain of function)	CaSR	CaSR	TAL, DCT, PT, CD	Hypercalciuria, CRF	Hypocalcemia, hyperphosphatemia, low PTH level, possible tetany
Familial hypomagnesemia with hypercalciuria and nephrocalcinosis (61–63)	248250, 603959	Autosomal recessive	3q27/ *PCLN1* (*CLDN16*)	Paracellin-1 (claudin 16)	Tight junction protein	TAL, DCT	Hypercalciuria, hypermagnesuria, polyuria, dRTA, CRF	Low serum magnesium, tetany, seizures

All of the diseases listed cause nephrocalcinosis; stones occur in Dent disease, Bartter syndrome types III and V, and autosomal dominant hypocalcemic hypercalciuria and familial hypomagnesemia. CRF, chronic renal failure; LMW, low molecular weight; NKCC2, sodium-potassium–2 chloride cotransporter; PGE, prostaglandin E; ROMK, renal outer medullary potassium channel. Corresponding references are shown in parentheses.

plaque particles coalesce so that islands of crystal float in an organic sea (Figure 2). It is this coalescent material that extends to the suburothelial region and over which CaOx stones grow. The identity of the organic molecules surrounding apatite in plaque are unknown except for osteopontin, which coats the surface of apatite and positions itself (Figure 1D) precisely at the apatite organic layer interface (21). We envision that plaque forms in the thin limb BM and extends to the suburothelial space and that CaOx stones then form over the organic coating of apatite particles.

Crystals are not found in tubule lumens of the deep or outer medulla or cortex. The lining cells of the tubules are normal in appearance (Figure 2) even when deposits of plaque fill the BM; no lumen crystals have to date been found in tissue from idiopathic CaOx SFs. For this reason, we do not review here the large literature on attachment of CaOx and CaP crystals to renal epithelial cells in culture (22), which does not appear to be relevant to human idiopathic CaOx stone disease. Interstitial cells also show no evident pathology.

Mechanisms of plaque and stone formation. The driving force for CaOx overgrowth on plaque is urine CaOx SS. But the forces that create the plaque are not so clear. The fraction of papillary surface covered by plaque in idiopathic CaOx SFs correlates directly with urine calcium level and inversely with urine volume and pH (23). Given that the initial formation is in the thin Henle loops BM, effi-

cient water extraction in the collecting duct (CD) combined with high deliveries of calcium as a result of idiopathic hypercalciuria (IH) may increase tubule and interstitial calcium concentrations and by as-yet-undetermined mechanisms stimulate apatite deposition in the thin-limb BM.

Clinical implications. High fluid intake may be beneficial not only to prevent CaOx overgrowth, but also to reduce plaque formation itself. Thiazide diuretics, which lower urine calcium, may reduce plaque as well as urine CaOx SS, whereas measures that reduce urine oxalate concentration are important in prevention of overgrowth but have no apparent role with regard to plaque formation. As new technologies permit visualization and quantification of plaque in the kidneys of patients, effects of treatments on plaque may become a matter for prospective trials.

Idiopathic hypercalciuria

Urinary calcium excretion as measured on random diets varies in humans (24), and excretion rates among idiopathic CaOx SFs cluster at the high end of this distribution (25). Lowering urinary calcium levels prevents stone recurrence (12). Conventional upper limits of 250 mg/d (for women) and 300 mg/d (for men) are near the 95th percentiles for our published data sets (26) and are commonly used as diagnostic cutpoints. We believe SFs in the 70th percentiles (170 mg/d for women and 210 mg/d for men)

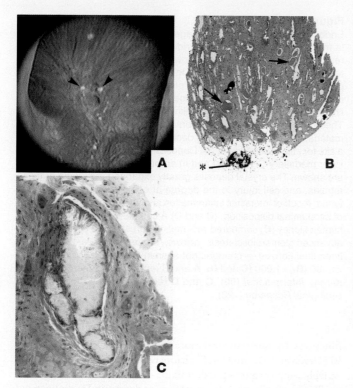

Figure 3
Endoscopic and histological images from a SF following small bowel bypass. The papillary surface (**A**) shows small, round nodular structures (arrowheads) near the openings of the Bellini ducts; distinct sites of Randall plaque material are not found. (**B**) Biopsy through a region that contained nodules reveals crystal deposition in the lumens of a few CDs as far down as the Bellini ducts (indicate by the asterisk). Note dilated CDs (arrows) with cast material in regions of fibrosis around crystal deposit–filled CDs. (**C**) A single CD is shown to be completely filled with crystals with injured lining cells. Magnification: ×100 (**B**); ×550 (**C**). Reprinted with permission from ref. 19.

increased by glucose or sucrose loads (29), high sodium intake (30), and a high-protein diet (31). Although dietary calcium is the main contributor to high urinary calcium levels in IH, low-calcium diets are not widely used clinically because they are not of proven effectiveness in stone prevention (32) and may predispose to reduced bone mineral content (33). Thiazide diuretics lower urinary calcium levels and promote positive bone mineral balances (34), and multiple prospective trials have documented their effectiveness for preventing calcium stones (see "Treatment trials: stone recurrence outcomes").

Monogenic disorders that cause hypercalciuria and stones

These rare disorders illustrate links among genetic variation, urinary calcium excretion, and calcium stones. Two recent reviews (35, 36) have detailed all known mutations that can cause hypercalciuria in humans or animals; here we discuss only those that cause stones in humans. We have excluded mere gene associations with the presence of stones, with the single exception of a cytosolic soluble adenylyl cyclase.

Dent disease. This X-linked hypercalciuric stone-forming disorder (Table 1) is due to mutations in the gene that codes for the voltage-gated endosomal chloride channel 5 (CLCN5) (37–39). Resultant low-molecular-weight proteinuria includes parathyroid hormone (PTH) loss (40, 41). As in IH, hypercalciuria results from increased serum calcitriol levels (42), and thiazide treatment lowers urine calcium concentrations (43). Because some patients with Dent disease lack CLCN5 (44) mutations, other genetic defects may also contribute to this disease. CLCN5 mutations are rarely found in IH patients (45). Gene KO studies in mice suggest that defective proximal tubule (PT) endocytosis increases PT lumen PTH levels, which activates 1α-hydroxylase and raises serum calcitriol levels, causing hypercalciuria (46, 47). In one study of Clcn5-KO mice (48), urinary 25OHD$_3$ levels were high, serum levels of 25OHD$_3$ and calcitriol were

may well benefit from lower calcium excretion rates. Data linking calcium excretion to stone risk (27) are supportive of the idea that it is a graded risk factor.

Diagnosis of IH requires exclusion of hypercalcemia, vitamin D excess, hyperthyroidism, malignant neoplasm, and sarcoidosis. Pathogenesis and systemic manifestations of IH are well described in standard textbooks (12). About 50% of first-degree relatives of people whose urinary calcium excretion level exceeds the 95th percentile also surpass these limits, establishing the familial and therefore inherited nature of IH. We emphasize here the most salient features and new research concerning this condition.

On average, patients with IH have higher than normal serum levels of 1,25(OH)$_2$D$_3$ (calcitriol) and increased intestinal absorption of calcium (28), and vitamin D receptor expression on circulating blood monocytes may also be increased. High intestinal calcium absorption can raise the load of filtered calcium presented to the renal tubules. The reabsorption of filtered calcium by the renal tubules may also be reduced. Either mechanism can increase urinary calcium level, which can be further

Table 2
Primary hyperoxaluria

Disease	OMIM	Locus/gene	Gene product	Functions	Location	Phenotype Renal/Urine
PH type I (77)	259900, 604285	2q36–q37/ *AGXT*	Alanine glyoxylate aminotransferase	Conversion of glyoxylate to glycine; excess glyoxylate oxidized to oxalate	Liver (enzyme may be absent, mistargeted, or nonfunctional)	Hyperoxaluria, CRF; increased urine glycolate excretion
PH type II (77)	260000, 604296	9cen/ *GRHPR*	Glyoxylate reductase/ hydroxypyruvate reductase	Conversion of glyoxylate to glycolate	Liver	Hyperoxaluria, CRF; increased urine L-glyceric acid excretion

Both diseases are autosomal recessive and cause stones and systemic oxalosis. Corresponding references are shown in parentheses.

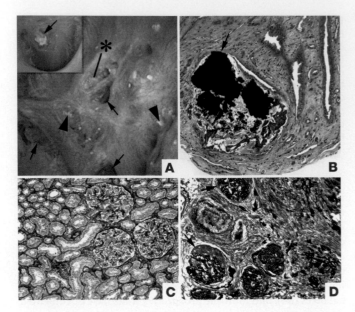

Figure 4
Endoscopic and histological images from a brushite SF. (**A**) Papilla from a brushite SF that was video recorded at the time of stone removal shows depressions (arrows) near the papillary tip and flattening, a phenomenon not seen in CaOx SFs. Like CaOx SFs, the papilla possessed sites of Randall plaque (arrowheads), though lesser in number. In addition, papillae possess sites of a yellowish crystalline deposit at the openings of Bellini ducts (indicated by the asterisk). These ducts were occasionally enlarged and filled with a crystalline material that protruded from the duct (inset, arrow) that might serve as a site for stone attachment. (**B**) Deposits in the lumens of an individual inner medullary CD (arrow) and in an occasional nearby Henle loop are shown. The crystal deposits greatly expanded the lumen of these tubules, and cell injury to the degree of complete cell necrosis was found. A cuff of interstitial inflammation and fibrosis accompanied sites of intraluminal disposition. (**C** and **D**) A cortical sample from a normal human kidney (**C**) compared with that of a brushite SF (**D**) that reveals advanced glomerulosclerosis (arrows), moderate tubular atrophy, and interstitial fibrosis — changes not seen in CaOx SFs. Magnification: ×1,400 (**B**); ×1,000 (**C** and **D**). **A** and **B** reprinted with permission from *Kidney International* (92). **C** and **D** reprinted with permission from *Urological Research* (123).

low, and urine calcium concentration was normal; urinary loss of 25OHD₃ was thought to limit the increase in calcitriol from high lumen PTH.

Bartter syndromes. Of the 5 major genetic variants associated with Bartter syndrome (49), cortical and medullary renal calcifications (50) are found in patients with Bartter syndrome types I and II (Table 1), which arise from transport defects that reduce thick ascending limb (TAL) NaCl and calcium reabsorption, creating hypercalciuria. Defects in a TAL basolateral membrane chloride channel, *CLCNKB*, produce only variable hypercalciuria and uncommon crystal deposits (type III). Defects in a chloride channel subunit known as Barttin, encoded by *BSND*, produce Bartter syndrome type IV, in which deafness, but neither stones nor nephrocalcinosis, are observed. Loop diuretic treatment of premature babies, a pharmacological analog of the reduced TAL reabsorption seen in Bartter syndromes, can cause renal calcifications, 50% of which regress over time (51).

Abnormalities of the calcium-sensing receptor. The calcium-sensing receptor (CaSR) regulates PTH secretion in response to blood calcium levels (52) and is also present on the apical membranes of PT and medullary CD principal cells and the basal membranes of TAL and distal convoluted tubule (DCT) cells, where it regulates transport

functions. Gain-of-function mutations of the CaSR reduce serum PTH levels and TAL and DCT calcium reabsorption (53, 54), causing variably symptomatic hypocalcemia (55–57) with low serum PTH levels. Stones usually result from calcium and vitamin D administered in order to raise serum calcium levels. One particular mutation causes multigland primary hyperparathyroidism (58). A severe gain-of-function defect of the CaSR causes Bartter syndrome type V (Table 1), and its clinical symptoms are ascribed to CaSR activation–mediated downregulation of a renal outer medullary potassium channel (ROMK) that is mutated in type II Bartter syndrome (59, 60).

Familial hypomagnesemia with hypercalciuria and nephrocalcinosis. Paracellin-1, a member of the claudin family (61, 62) found only in the TAL and DCT, is essential for normal, voltage-driven paracellular magnesium reabsorption in the TAL. Patients with paracellin-1 defects develop hypomagnesemia, hypercalciuria, polyuria, nephrocalcinosis, stones, distal renal tubular acidosis (dRTA), and renal failure (Table 1). A striking incidence of hypercalciuria and stones has been found in heterozygous family members (63).

Soluble adenylyl cyclase. Among selected IH patients with reduced vertebral bone mineral density, Reed et al. (64) identified linkage to a locus on chromosome 1 (1q23.3–q24). In a specific gene (*human*

Table 3
Distal renal tubular acidosis

Disease	OMIM	Inheritance	Locus/gene	Gene product	Functions	Location	Phenotype
Autosomal dominant dRTA (96–98)	179800, 109270	Autosomal dominant	17q21–q22/ *SLC4A1*	AE1	Chloride-bicarbonate exchanger	CD (type A intercalated cells)	Osteomalacia, low K level
Autosomal recessive dRTA with hearing loss (99)	267300, 192132	Autosomal recessive	2p13/ *ATP6V1B1*	B1 subunit of vacuolar H⁺-ATPase	Proton secretion	CD (type A intercalated cells); cochlea	Sensorineural hearing loss, growth failure, rickets, low K level
Autosomal recessive dRTA (100)	602722, 605239	Autosomal recessive	7q33–34/ *ATP6V0A4*	a4 subunit of vacuolar H⁺-ATPase	Proton secretion	CD (type A intercalated cells), PT, DCT	Growth failure, rickets, low K level

All 3 diseases cause metabolic acidosis and hypercalciuria, stones, and nephrocalcinosis. Corresponding references are shown in parentheses.

Table 4

Known gene defects in cysteinuria

Disease	OMIM	Inheritance	Locus/gene	Gene product	Functions	Location	Phenotype
Cystinuria type A (type I) (110, 111)	104614	Autosomal recessive	2p16.3/ *SLC3A1*	rBAT	Heavy unit of heteromeric aa transporter	PT S3	Heterozygotes: normal urine cystine excretion
Cystinuria type B (usually type non-I) (110, 111)	604144	Incompletely autosomal recessive	19q13.1/ *SLC7A9*	b⁰,⁺AT	Light subunit of heteromeric aa transporter	PT S1 and S2	Heterozygotes: elevated urine cystine excretion

Both types cause stones and increased urine excretion of cystine and dibasic amino acids. S1–S3, segments 1–3 of the PT. Corresponding references are shown in parentheses.

soluble adenylyl cyclase [hsAC]) (65), 4 base changes segregated with IH, and bone mineral density decreased as the number of base changes increased. hsAC is a cytosolic enzyme sensitive to bicarbonate concentration in the presence of divalent cations and could explain bicarbonate sensitivity of renal and bone cells (66).

Primary hyperparathyroidism

Clinicians readily distinguish this stone-forming condition from IH because blood calcium levels are high in primary hyperparathyroidism and normal in IH. The two disorders share hypercalciuria and generally elevated serum levels of calcitriol. Diagnosis, pathogenesis, treatment approaches, and the genes related to gland enlargement have been well covered elsewhere (67).

Hyperoxaluria

Urine oxalate concentration affects CaOx SS exactly as does urine calcium concentration (68); therefore, any conditions that increase oxalate absorption from food or lead to increased oxalate production can cause CaOx stone formation.

Dietary hyperoxaluria in idiopathic CaOx SFs. Whereas the 95th percentiles for urinary oxalate excretion for females and males are 45 mg/d and 55 mg/d, respectively, and the corresponding 70th percentiles are 31 mg/d and 41 mg/d (12), values up to approximately 80 mg/d are common among idiopathic CaOx SFs and are often unexplained by a systemic cause (69). A low-calcium diet that reduces CaOx crystallization in the gut lumen, thereby facilitating oxalate ion absorption, is easily identified and corrected (70). High protein or oxalate intake are other causes of this condition (71, 72). Oxalate absorption from food may depend upon gastrointestinal transporters (73, 74), and genetic variations in oxalate transporting proteins in red blood cells have been linked to CaOx stone formation (73).

Enteric hyperoxaluria. If colon is present and receiving small bowel effluent, fat malabsorption of any cause (e.g., small bowel disease, resection or bypass, or exocrine pancreatic insufficiency) results in increased colon oxalate absorption and urine oxalate excretion in the range of 80–140 mg/d (69). Undigested fatty acids promote colon oxalate absorption (75). Diarrheal fluid losses and low urine pH and citrate levels resulting from stool bicarbonate losses increase urine CaOx SS and UA SS; stones usually contain 1 or both of these phases (69).

Among patients with enteric hyperoxaluria resulting from small bowel bypass for the treatment of obesity (19), some terminal CDs are plugged with apatite crystals (Figure 3) and show epithelial cell death and surrounding interstitial inflammation. How the apatite plugs form remains unknown; possibly high oxalate concentra-

tions in tubule fluid injure epithelial cells and disturb pH regulation so that lumen pH increases. Plugging and consequent nephron damage may account for renal function loss in such patients and may benefit from urine dilution.

Treatment involves high fluid intake and reducing dietary oxalate and fat intake to the extent that nutritional requirement guidelines permit. Calcium supplements (500–1,000 mg) taken with each meal can bind food oxalate and hinder absorption. Cholestyramine (2–4 g with each meal) binds oxalate and fatty acids and reduces urine oxalate excretion. If stones contain UA, potassium alkali is needed to increase urine pH (76). No formal trials have validated these treatment approaches.

Primary hyperoxaluria. Mutations in 1 of 2 genes — *AGXT* or *GRHPR* (77) — leads to oxalate overproduction and urine oxalate excretion rates above 100 mg/d or even 200 mg/d, resulting in PH (Table 2). When gene alterations divert active enzyme into mitochondria, where it cannot function effectively, pyridoxine may reduce urine oxalate concentration by increasing enzyme activity (78). When loss-of-function mutations inactivate enzymes, actively increasing urine volume and perhaps treatment with orthophosphate supplements may be advised (79). In type I PH, renal failure is frequent and presently treated by combined segmental liver and renal transplantation (80). Type II PH, being milder, seldom causes renal failure.

Hypocitraturia

The 30th percentiles for urine citrate excretion in normal women and men, 424 mg/d and 384 mg/d (26), respectively, correspond well with usual criteria for hypocitraturia (less than 500 mg/d for women and less than 350 mg/d for men). We have already alluded to the role of citrate in binding calcium and inhibiting calcium crystallization (see "Modulators of the ULM, crystal growth, and aggregation"), so it is reasonable to presume that low urine citrate levels could permit CaOx stones.

Regulation of urine citrate. Urine citrate concentration is determined mainly by tubule reabsorption (81), which is increased by acid loads and reduced by alkali loads. Therefore, states of alkali loss such as intestinal malabsorption with diarrhea or any cause of metabolic acidosis reduce urine citrate level (82, 83). Potassium depletion, a common consequence of thiazide intake, also lowers urine citrate level, which can be raised by potassium administration (84). However, in most patients who excrete subnormal amounts of citrate in their urine, no apparent cause can be found, and the underlying mechanisms are presently unknown.

Clinical management. Idiopathic CaOx SFs with low urine citrate levels can benefit from potassium citrate salt administration, which increases urine citrate level and reduces stone recurrence (see "Treatment trials: stone recurrence outcomes"). Suggested dosages of potassium citrate salts range from 20–60 mEq/d, achieved in 2 or 3 doses. Alkali treatment does not reliably reduce urine calcium levels (85, 86) in IH. If hypercalciuria is present and does not abate with alkali, addition of thiazide is prudent to prevent an increase in CaP SS and development of CaP stones.

Hyperuricosuria

High urine UA excretion in men and women (above 800 mg/d and 750 mg/d, respectively) is associated with the formation of idiopathic CaOx stones (12). Dissolved UA salts appear to reduce the solubility of CaOx (87–89), fostering stones. Although hyperuricosuria results from high dietary intake of beef, poultry, and fish and can be abolished by dietary changes, no clinical trials document the effectiveness of this approach. A single trial has validated the use of allopurinol, an inhibitor of xanthine oxidase, which reduces UA production, as effective in reducing stone recurrence (see "Treatment trials: stone recurrence outcomes").

CaP stones

In a majority of kidney stones, CaOx is the main constituent and CaP is present in amounts ranging from 1% to 10% (90). When CaP becomes the main constituent (>50%) of stones, the stones are called CaP stones, and patients who form CaP stones are referred to as CaP SFs. CaP is present in urinary stones as either apatite (the principal constituent of bones and teeth) or brushite (calcium monohydrogen phosphate). Brushite, but not apatite, stones are physically resistant to ESWL, so repeated treatments may be needed (91). Brushite stones are associated with a distinctive renal disease (92). What drives the development of brushite versus apatite stones is not known.

CD plugging and brushite stones

In patients whose stones contain brushite (92), plugs of apatite fill the lumens of the terminal CD (Figure 4A). Epithelial cells are damaged or destroyed (Figure 4B). Around affected tubules, the interstitium is inflamed and scarred. Idiopathic CaOx SFs do not have intratubular crystals, whereas all brushite SFs we have biopsied to date have exhibited CD plugging. Stone type is therefore a clue to renal pathology. In the cortex of the 10 patients studied thus far (92), glomerular obsolescence and interstitial fibrosis exceeds that in cortical tissue from non-SFs and idiopathic CaOx SFs (Figure 4, C and D). Serum creatinine level was generally higher and 24-hour urine creatinine clearance was lower in the brushite SFs than in healthy controls. In our large clinical series (93), we could not find reduced renal function in brushite versus CaOx SFs, so these biopsied patients may have unusually severe renal disease. Compared with idiopathic CaOx SFs, and adjusted for sex, numbers of stones, and duration of stone disease, brushite SFs require a greater number of ESWL treatments, and this may contribute to their renal injury. Although potassium citrate salts are effective as a treatment for CaOx stones with hypocitraturia, they, along with ESWL, may promote the formation of CaP stones, whose prevalence has risen for the past 3 decades. Use of citrates requires attention in order to avoid increasing CaP SS excessively via high urine pH.

High urine pH as a main risk factor for CaP stones

Idiopathic CaOx SFs and CaP SFs both are hypercalciuric because of IH, but CaP SFs produce urine of higher pH, which favors CaP crystallization (93) by increasing the abundance of urine monohydrogen phosphate (pKa ~6.7), the ion that combines with calcium. Urine pH rises progressively with increasing CaP percentage in stones (93). The mechanism of increased urinary pH in CaP SFs is unknown. Blood bicarbonate levels are normal, and net acid excretion is not low. High ammonia production and excretion could raise urinary pH (94) but was not obvious in our patients (93).

Management of CaP SFs

CaP SFs are usually treated with fluids and thiazide diuretics to lower urine calcium excretion. Urine citrate excretion can be reduced, as in idiopathic CaOx SFs, but because potassium citrate salts can increase urine pH and CaP SS, careful follow-up is needed. No clinical trials have documented treatment outcomes for CaP SFs.

Hereditary dRTA

dRTA is a syndrome of metabolic acidosis and hypokalemia with alkaline urine pH, low urine citrate excretion, hypercalciuria, and often bone disease (95). Stones are common and usually ascribed to the combination of hypercalciuria, low urine citrate, and high urine pH. If given an acid load, patients cannot reduce urine pH below 5.5, the conventional benchmark (95). Trials have yet to document stone treatment outcomes.

Autosomal dominant dRTA (Table 3) is caused by mutations in a chloride-bicarbonate exchanger encoded by *SLC4A1* on the basolateral surface of type A intercalated cells (96). The abnormal transporter is not targeted to the basolateral cell surface (97) and may form hetero-oligomers with the normal protein, reducing its function (98). Although dRTA is mild, stones and nephrocalcinosis are common and lead to the diagnosis, often in adult life. Autosomal recessive dRTA with sensory hearing loss (Table 3) causes dehydration and growth failure, beginning in infancy or early childhood (99). Hypercalciuria and high urine pH cause nephrocalcinosis. Autosomal recessive dRTA without hearing loss (Table 3) also causes growth failure and dehydration; some patients develop late-onset hearing loss (100). Autosomal recessive dRTA with osteopetrosis and cerebral calcification, caused by carbonic anhydrase II deficiency, does not cause nephrocalcinosis or stones (101).

Acquired and "incomplete" dRTA

Acquired forms of dRTA with stone formation are usually associated with the autoimmune disease Sjögren syndrome and/or use of the carbonic anhydrase inhibitor acetazolamide (102, 103). "Incomplete" dRTA is used as a diagnostic term when urine citrate excretion rate is low despite high urinary pH (pH >6.5) and blood chemistries are normal; given an oral acid load, such patients may fail to lower urinary pH below 5.5 (104, 105). Treatment of the low urine citrate is as already described for CaOx and CaP SFs. Stone treatment outcomes are not known for this condition.

Management of dRTA

Supplemental alkali in the form of potassium citrate salts is required to correct the metabolic acidosis and hypokalemia. Dosage for adults should range between 1 and 3 mEq of alkali per kilogram body weight (95). No trials document stone outcomes with alkali treatment.

UA stones

UA stones form because urine pH is abnormally low. Only 90 mg/l of undissociated UA dissolves in human urine at 37°C; therefore, at a pH of 5.35 (the pKa for the dissociation of the N9 proton of UA), only 180 mg/l of total urate species can be dissolved, whereas the

Table 5
Treatment trials for nephrolithiasis

Treatment	Type	Enrolled			Completed		New stones	
		%F	No. C	No. T	No. C	No. T	No. C	No. T
K citrate, 30-60 mEq/d (115)	RCD	56	28	27	20	18	14 (70%)	5 (28%)
K Mg citrate, 60 mEq/d (116)	RCD	21	33	31	25	16	16 (64%)	2 (13%)
K citrate (variable dose)[A] (117)	RC	38	25	25	22	16	6 (27%)	5 (31%)
Chlorthalidone, 25 mg/d (118)	RCD	12	31	42	26	28	12 (46%)	4 (14%)
HCTZ, 25 mg 2/d (119)	RCD	24	25	25	25	23	10 (40%)	4 (17%)
Indapamide, 2 5 mg/d (120)	RCD	24	25	50	21	43	9 (43%)	6 (15%)
LCD vs. reduced Na and P (diet) (32)	RC	0	60	60	51	52	23 (45%)	12 (23%)
Allopurinol, 300 mg/d (121)	RCD	n/a	31	31	31	29	11 (35%)	5 (17%)

RCD, randomized, controlled, double-blind; RC randomized, controlled; Enrolled, number of controls (No. C) and treated (No. T) subjects entering the study and percent female (%F); Completed, number of subjects who completed the study; New stones, number (%) of subjects who formed stones. All studies were for a period of 3 years except for the diet study, which was 5 years. [A]Dose of sodium potassium citrate varied to maintain urine pH between 7.0 and 7.2. HCTZ, hydrochlorothiazide; P, protein; L, low-calcium diet; n/a, not available. Corresponding references are shown in parentheses.

total urate concentration of healthy individuals and UA SFs approximates 500 mg/l of urine (12). Low urine pH is due in part to low ammonia excretion (106, 107). Low urinary pH and UA stones are common in patients with gout, diabetes mellitus, and the metabolic syndrome, perhaps as a result of insulin resistance that may reduce renal ammonia excretion (108). Urinary pH falls with increasing body weight (109), probably because of insulin resistance. Chronic diarrhea also lowers urinary pH (69) and causes UA stones.

UA stones are not easily seen on kidney, ureter, or bladder x-rays. On CT images, they resemble calcium stones, from which they can be distinguished by their lower density (1). On an i.v. pyelogram radiograph, UA stones appear as filling defects. UA gravel can occlude ureters and produce acute anuric renal failure; UA stones can fill the entire renal collecting system. UA gravel and stones are often orange or red, having adsorbed uricine, a bilirubin breakdown pigment. Prevention and even dissolution of UA stones depends upon the ability to increase urine pH above 6.0, which is accomplished by administration of 20–30 mEq potassium alkali, 2 or 3 times daily. Allopurinol is not usually required or indicated but can be used if alkali does not suffice and urine UA level is increased.

Cystinuria

Genetic abnormalities
Mutations in renal epithelial cell transporters result in reduced reabsorption and increased urine excretion of the dibasic amino acids, including cystine. High cystine SS because of overexcretion causes cystine stones (Table 4). The main apical resorption system for cystine in the kidney is a heterodimer transporter composed of a light chain catalytic subunit, $b^{0,+}$ amino acid transporter ($b^{0,+}$AT), and rBAT (related to $b^{0,+}$AT), a heavy chain subunit (110). rBAT is responsible for targeting $b^{0,+}$AT to the brush border membrane, and it is $b^{0,+}$AT that acts as the transporting assembly. Although cystinuria can be subtyped genetically (Table 4), patients of either type A or B are clinically indistinguishable (111). In all forms of cystinuria, urine cystine excretion ranges from 350–500 mg/d.

Clinical management of cystinuria
Treatment of cystinuria aims to reduce SS below 1. Cystine solubility approximates 1 mM (243 mg/l) and rises with urine pH but is variable and higher in urine of cystinuric patients as compared with

CaOx SFs (112) and therefore is best measured directly (113) rather than estimated via a pH nomogram. Management (114) begins with achieving a daily urine volume of 3.5–5 liters, which can dissolve the cystine excreted by some patients. Others also require administration of 40–80 mEq/d of potassium alkali to raise pH. Reduction of sodium and protein intake reduces urine cystine excretion. Those measures failing, D-penicillamine, α-mercaptopropionylglycine, or captopril, all of which form soluble heterodimers with cysteine, can be used. Blood pressure reduction can limit captopril dosing so the drug has little practical utility. D-Penicillamine and α-mercaptopropionylglycine have side effects including loss of taste, fever, proteinuria, serum sickness–type reactions, and even frank nephrotic syndrome. Increasing fluid and alkali intake are therefore preferred therapeutic modalities in as many cases as possible.

Treatment trials: stone recurrence outcomes
To date, 7 of 8 prospective trials, each lasting at least 3 years, have shown that selective treatments have a distinct benefit for idiopathic CaOx SFs (Table 5). Table 5 makes clear that dropout rates were high, and we are not aware that any of the investigators followed up with those individuals that did not complete the trial in order to ascertain their stone status. Therefore, additional trials may not be altogether without merit. Even so, and despite these imperfections, it is difficult to conclude that treatments are without important benefits.

Thiazide and citrate: recurrent calcium SFs. Three prospective trials of potassium citrate salts have been performed to date, and 2 (115, 116) indicate therapeutic benefit (Table 5). The third (117) lacks a double-blind design, and its control subjects formed fewer stones than those in the other studies. All subjects were CaOx SFs with hypocitraturia. Thiazide diuretic agents have been studied in 3 randomized, controlled, double-blind trials. Chlorthalidone (118) at doses of 25 mg/d or 50 mg/d gave equivalent results that are pooled (Table 5). Hydrochlorothiazide gave therapeutic effects equivalent to chlorthalidone (119), as did indapamide (120).

Dietary changes for recurrent calcium SFs. Borghi et al. (32) compared a low-calcium diet to a normal calcium diet reduced in protein and sodium in a 5-year randomized trial (Table 5). The 2 diets gave comparable results during the first 3 years, but by 5 years, the normal calcium, reduced sodium, and protein diet proved superior in that only 23% of patients had formed a new stone.

Allopurinol: recurrent calcium SFs. Allopurinol (121) has been shown to reduce new CaOx stone formation in patients with hyperuricosuria (Table 5).

Patients who have formed only 1 calcium stone. Over a 5-year period, 20% of patients who maintained a urine volume above 2.5 l/d had stone recurrence (122) compared with approximately 50% of patients who were counseled to avoid dehydration and to observe moderate salt and protein intakes. Counseled controls did not actually increase urine volume. We do not combine this trial with the others, as it concerns only SFs with a single stone.

A final word about stones

It is difficult to accept recurrent stone formation as incidental in any patient and allow it to continue without efforts to understand its causes and offer such treatments as seem appropriate. Available trials offer physicians excellent treatment strategies for prevention of calcium stones, and since UA stones are a consequence of low urine pH, physicians can treat them confidently despite the lack of prospective trials for additional therapeutics. Even cystine stones can be prevented, albeit with imperfect remedies. But treatments may pose their own problems. Although potassium citrate salts are effective, they, along with ESWL, may promote the formation of CaP stones, the prevalence of which continues to rise with time. Possibly this means that the use of citrates requires special attention to avoid increasing CaP SS excessively via high urine pH. Although we treat urine SS, the tissue processes of stone formation are complex, not as yet obviously related to solution chemistry within specific nephron segments, and not well understood. This is a significant area of interest that requires new research. Abnormal urine pH and calcium excretion rate are predominant findings in SFs that play a major role in the pathogenesis of stone formation. Their biologies are therefore also of particular research importance. Perhaps most important in the long run will be uncovering the links between genetic variability and urine calcium excretion and pH, for these seem at the very center of the problem of kidney stone disease.

Address correspondence to: Fredric L. Coe, Renal Section MC 5100, University of Chicago School of Medicine, 5841 South Maryland Avenue, Chicago, Illinois 60637, USA. Phone: (773) 702-1475; Fax: (773) 702-5818; E-mail: f-coe@uchicago.edu.

1. Zarse, C.A., et al. 2004. Helical computed tomography accurately reports urinary stone composition using attenuation values: in vitro verification using high-resolution micro-computed tomography calibrated to fourier transform infrared microspectroscopy. *Urology.* **63**:828–833.

2. Lingeman, J.E., Kim, S.C., Kuo, R.L., McAteer, J.A., and Evan, A.P. 2003. Shockwave lithotripsy: anecdotes and insights. *J. Endourol.* **17**:687–693.

3. Bagley, D.H. 2002. Expanding role of ureteroscopy and laser lithotripsy for treatment of proximal ureteral and intrarenal calculi. *Curr. Opin. Urol.* **12**:277–280.

4. Clayman, R.V. 2005. Percutaneous nephrolithotomy: an update [comment]. *J. Urol.* **173**:1199.

5. Gillen, D.L., Worcester, E.M., and Coe, F.L. 2005. Decreased renal function among adults with a history of nephrolithiasis: a study of NHANES III. *Kidney Int.* **67**:685–690.

6. Madore, F., Stampfer, M.J., Rimm, E.B., and Curhan, G.C. 1998. Nephrolithiasis and risk of hypertension. *Am. J. Hypertens.* **11**:46–53.

7. Brown, C.M., and Purich, D.L. 1992. Physical-chemical processes in kidney stone formation. In *Disorders of bone and mineral metabolism.* F.L. Coe and M.J. Favus, editors. Raven Press. New York, New York, USA. 613–624.

8. Parks, J.H., Coward, M., and Coe, F.L. 1997. Correspondence between stone composition and urine supersaturation in nephrolithiasis. *Kidney Int.* **51**:894–900.

9. Parks, J.H., Goldfischer, E.R., and Coe, F.L. 2003. Changes in urine volume accomplished by physicians treating nephrolithiasis. *J. Urol.* **169**:863–866.

10. Asplin, J.R., Parks, J.H., and Coe, F.L. 1997. Dependence of upper limit of metastability on supersaturation in nephrolithiasis. *Kidney Int.* **52**:1602–1608.

11. Asplin, J.R., Parks, J.H., Nakagawa, Y., and Coe, F.L. 2002. Reduced crystallization inhibition by urine from women with nephrolithiasis. *Kidney Int.* **61**:1821–1829.

12. Coe, F.L., and Parks, J.H. 2000. Pathogenesis and treatment of nephrolithiasis. In *The kidney.* D. Seldin and G. Giebisch, editors. Lippincott Williams & Wilkins. Philadelphia, Pennsylvania, USA. 1841–1867.

13. Coe, F.L., Parks, J.H., and Nakagawa, Y.N. 2002. Inhibitors and promoters of calcium oxalate crystallization: their relationship to the pathogenesis of nephrolithiasis. In *Disorders of bone and mineral metabolism.* F.L. Coe and M.J. Favus, editors. Lippincott

Williams & Wilkins. Philadelphia, Pennsylvania. 741–775.

14. Asplin, J.R., Arsenault, D., Parks, J.H., Coe, F.L., and Hoyer, J.R. 1998. Contribution of human uropontin to inhibition of calcium oxalate crystallization. *Kidney Int.* **53**:194–199.

15. Stapleton, A.M., and Ryall, R.L. 1995. Blood coagulation proteins and urolithiasis are linked: crystal matrix protein is the F1 activation peptide of human prothrombin. *Br. J. Urol.* **75**:712–719.

16. Marengo, S.R., Resnick, M.I., Yang, L., and Churchill, P.C. 1998. Differential expression of urinary inter-alpha-trypsin inhibitor trimers and dimers in normal compared to active calcium oxalate stone forming men. *J. Urol.* **159**:1444–1450.

17. Pillay, S.N., Asplin, J.R., and Coe, F.L. 1998. Evidence that calgranulin is produced by kidney cells and is an inhibitor of calcium oxalate crystallization. *Am. J. Physiol.* **275**:F255–F261.

18. Hess, B., Nakagawa, Y., Parks, J.H., and Coe, F.L. 1991. Molecular abnormality of Tamm Horsfall glycoprotein in calcium oxalate nephrolithiasis. *Am. J. Physiol.* **29**:F569–F578.

19. Evan, A.P., et al. 2003. Randall's plaque of patients with nephrolithiasis begins in basement membranes of thin loops of Henle. *J. Clin. Invest.* **111**:607–616. doi:10.1172/JCI200317038.

20. Kim, S.C., et al. 2005. Stone formation is proportional to papillary surface coverage by Randall's plaque. *J. Urol.* **173**:117–119.

21. Evan, A.P., et al. 2005. Apatite plaque particles in inner medulla of kidneys of calcium oxalate stone formers: osteopontin localization. *Kidney Int.* **68**:145–154.

22. Sheng, X., Jung, T., Wesson, J.A., and Ward, M.D. 2005. Adhesion at calcium oxalate crystal surfaces and the effect of urinary constituents. *Proc. Natl. Acad. Sci. U. S. A.* **102**:267–272.

23. Kuo, R.L., et al. 2003. Urine calcium and volume predict coverage of renal papilla by Randall's plaque. *Kidney Int.* **64**:2150–2154.

24. Marshall, R.W., Cochran, M., and Hodgkinson, A. 1972. Relationships between calcium and oxalic acid intake in the diet and their excretion in the urine of normal and renal-stone-forming subjects. *Clin. Sci.* **43**:91–99.

25. Lemann, J., Jr., Worcester, E.M., and Gray, R.W. 1991. Hypercalciuria and stones [review]. *Am. J. Kidney Dis.* **17**:386–391.

26. Parks, J.H., and Coe, F.L. 1986. A urinary calcium-citrate index for the evaluation of nephrolithiasis.

Kidney Int. **30**:85–90.

27. Curhan, G.C., Willett, W.C., Speizer, F.E., and Stampfer, M.J. 2001. Twenty-four-hour urine chemistries and the risk of kidney stones among women and men. *Kidney Int.* **59**:2290–2298.

28. Favus, M.J., Karnauskas, A.J., Parks, J.H., and Coe, F.L. 2004. Peripheral blood monocyte vitamin D receptor levels are elevated in patients with idiopathic hypercalciuria. *J. Clin. Endocrinol. Metab.* **89**:4937–4943.

29. Lemann, J., Jr., Lennon, E.J., Piering, W.F., Prien, E.L., Jr., and Ricanati, E.S. 1970. Evidence that glucose ingestion inhibits net renal tubular reabsorption of calcium and magnesium in man. *J. Lab. Clin. Med.* **75**:578–585.

30. Breslau, N.A., Sakhaee, K., and Pak, C.Y. 1985. Impaired adaptation to salt-induced urinary calcium losses in postmenopausal osteoporosis. *Trans. Assoc. Am. Physicians.* **98**:107–115.

31. Hess, B., Ackermann, D., Essig, M., Takkinen, R., and Jaeger, P. 1995. Renal mass and serum calcitriol in male idiopathic calcium renal stone formers: role of protein intake. *J. Clin. Endocrinol. Metab.* **80**:1916–1921.

32. Borghi, L., et al. 2002. Comparison of two diets for the prevention of recurrent stones in idiopathic hypercalciuria. *N. Engl. J. Med.* **346**:77–84.

33. Asplin, J.R., et al. 2003. Bone mineral density and urine calcium excretion among subjects with and without nephrolithiasis. *Kidney Int.* **63**:662–669.

34. Coe, F.L., Parks, J.H., Bushinsky, D.A., Langman, C.B., and Favus, M.J. 1988. Chlorthalidone promotes mineral retention in patients with idiopathic hypercalciuria. *Kidney Int.* **33**:1140–1146.

35. Gambaro, G., et al. 2004. Genetics of hypercalciuria and calcium nephrolithiasis: from the rare monogenic to the common polygenic forms. *Am. J. Kidney Dis.* **44**:963–986.

36. Moe, O.W., and Bonny, O. 2005. Genetic hypercalciuria. *J. Am. Soc. Nephrol.* **16**:729–745.

37. Thakker, R.V. 2000. Pathogenesis of Dent's disease and related syndromes of X-linked nephrolithiasis. *Kidney Int.* **57**:787–793.

38. Gunther, W., Luchow, A., Cluzeaud, F., Vandewalle, A., and Jentsch, T.J. 1998. ClC-5, the chloride channel mutated in Dent's disease, colocalizes with the proton pump in endocytotically active kidney cells. *Proc. Natl. Acad. Sci. U. S. A.* **95**:8075–8080.

39. Luyckx, V.A., et al. 1998. Intrarenal and subcellular localization of rat CLC5. *Am. J. Physiol.* **275**:F761–F769.

40. Reinhart, S.C., et al. 1994. Characterization of carrier females and affected males with x-linked recessive nephrolithiasis. *J. Am. Soc. Nephrol.* **5**:S54–S58.

41. Nakazato, H., et al. 1997. Mutations in the CLCN5 gene in Japanese patients with familial idiopathic low-molecular-weight proteinuria. *Kidney Int.* **52**:895–900.

42. Scheinman, S.J. 1998. X-linked hypercalciuric nephrolithiasis: clinical syndromes and chloride channel mutations. *Kidney Int.* **53**:3–17.

43. Raja, K.A., et al. 2002. Responsiveness of hypercalciuria to thiazide in Dent's disease. *J. Am. Soc. Nephrol.* **13**:2938–2944.

44. Hoopes, R.R., Jr., et al. 2001. Evidence for genetic heterogeneity in Dent's disease. *Kidney Int.* **65**:1615–1620.

45. Scheinman, S.J., et al. 2000. Isolated hypercalciuria with mutation in CLCN5: relevance to idiopathic hypercalciuria. *Kidney Int.* **57**:232–239.

46. Wang, S.S., et al. 2000. Mice lacking renal chloride channel, CLC-5, are a model for Dent's disease, a nephrolithiasis disorder associated with defective receptor-mediated endocytosis. *Hum. Mol. Genet.* **9**:2937–2945.

47. Silva, I.V., et al. 2003. The ClC-5 knockout mouse model of Dent's disease has renal hypercalciuria and increased bone turnover. *J. Bone Miner. Res.* **18**:615–623.

48. Gunther, W., Piwon, N., and Jentsch, T.J. 2003. The ClC-5 chloride channel knock-out mouse — an animal model for Dent's disease. *Pflugers Arch.* **445**:456–462.

49. Naesens, M., Steels, P., Verberckmoes, R., Vanrenterghem, Y., and Kuypers, D. 2004. Bartter's and Gitelman's syndromes: from gene to clinic. *Nephron Physiol.* **96**:65–78.

50. Taugner, R., et al. 1988. The juxtaglomerular apparatus in Bartter's syndrome and related tubulopathies. An immunocytochemical and electron microscopic study. *Virchows Arch. A. Pathol. Anat. Histol.* **412**:459–470.

51. Pope, J.C., et al. 1996. The natural history of nephrocalcinosis in premature infants treated with loop diuretics. *J. Urol.* **156**:709–712.

52. Brown, E.M., Pollak, M., and Hebert, S.C. 1998. The extracellular calcium-sensing receptor: its role in health and disease. *Annu. Rev. Med.* **49**:15–29.

53. Bai, J., and Friedman, P.A. 2001. Calcium sensing receptor regulation of renal mineral ion transport. *Cell Calcium.* **35**:229–237.

54. Hebert, S.C., Brown, E.M., and Harris, H.W. 1997. Role of the Ca(2+)-sensing receptor in divalent mineral ion homeostasis. *J. Exp. Biol.* **200**:295–302.

55. Pearce, S.H., et al. 1996. A familial syndrome of hypocalcemia with hypercalciuria due to mutations in the calcium-sensing receptor. *N. Engl. J. Med.* **335**:1115–1122.

56. Okazaki, R., et al. 1999. A novel activating mutation in calcium-sensing receptor gene associated with a family of autosomal dominant hypocalcemia. *J. Clin. Endocrinol. Metab.* **84**:363–366.

57. Yamamoto, M., Akatsu, T., Nagase, T., and Ogata, E. 2000. Comparison of hypocalcemic hypercalciuria between patients with idiopathic hypoparathyroidism and those with gain-of-function mutations in the calcium-sensing receptor: is it possible to differentiate the two disorders? *J. Clin. Endocrinol. Metab.* **85**:4583–4591.

58. Carling, T., et al. 2000. Familial hypercalcemia and hypercalciuria caused by a novel mutation in the cytoplasmic tail of the calcium receptor. *J. Clin. Endocrinol. Metab.* **85**:2042–2047.

59. Watanabe, S., et al. 2002. Association between activating mutations of calcium-sensing receptor and Bartter's syndrome. *Lancet.* **360**:692–694.

60. Vargas-Poussou, R., et al. 2002. Functional characterization of a calcium-sensing receptor mutation in severe autosomal dominant hypocalcemia with a Bartter-like syndrome. *J. Am. Soc. Nephrol.* **13**:2259–2266.

61. Muller, D., et al. 2003. A novel claudin 16 mutation associated with childhood hypercalciuria abolishes binding to ZO-1 and results in lysosomal mistargeting. *Am. J. Hum. Genet.* **73**:1293–1301.

62. Weber, S., et al. 2000. Familial hypomagnesaemia with hypercalciuria and nephrocalcinosis maps to chromosome 3q27 and is associated with mutations in the PCLN-1 gene. *Eur. J. Hum. Genet.* **8**:414–422.

63. Weber, S., et al. 2001. Novel paracellin-1 mutations in 25 families with familial hypomagnesemia with hypercalciuria and nephrocalcinosis. *J. Am. Soc. Nephrol.* **12**:1872–1881.

64. Reed, B.Y., Heller, H.J., Gitomer, W.L., and Pak, C.Y. 1999. Mapping a gene defect in absorptive hypercalciuria to chromosome 1q23.3-q24. *J. Clin. Endocrinol. Metab.* **84**:3907–3913.

65. Reed, B.Y., et al. 2002. Identification and characterization of a gene with base substitutions associated with the absorptive hypercalciuria phenotype and low spinal bone density. *J. Clin. Endocrinol. Metab.* **87**:1476–1485.

66. Geng, W., et al. 2005. Cloning and characterization of the human soluble adenylyl cyclase. *Am. J. Physiol. Cell Physiol.* **288**:C1305 C1316.

67. Heller, H.J., and Pak, C.Y.C. 2002. Primary hyperparathyroidism. In *Disorders of bone and mineral metabolism.* F.L. Coe and M.J. Favus, editors. Lippincott Williams & Wilkins. Philadelphia, Pennsylvania, USA. 516–534.

68. Pak, C.Y., et al. 2004. Rapid communication: relative effect of urinary calcium and oxalate on saturation of calcium oxalate. *Kidney Int.* **66**:2032–2037.

69. Parks, J.H., Worcester, E.M., O'Connor, R.C., and Coe, F.L. 2003. Urine stone risk factors in nephrolithiasis patients with and without bowel disease. *Kidney Int.* **63**:255–265.

70. von Unruh, G.E., Voss, S., Sauerbruch, T., and Hesse, A. 2004. Dependence of oxalate absorption on the daily calcium intake. *J. Am. Soc. Nephrol.* **15**:1567–1573.

71. Holmes, R.P., Goodman, H.O., and Assimos, D.G. 2001. Contribution of dietary oxalate to urinary oxalate excretion. *Kidney Int.* **59**:270–276.

72. Bataille, P., Presne, C., and Fournier, A. 2002. Prevention of recurrent stones in idiopathic hypercalciuria. *N. Engl. J. Med.* **346**:1007–1009.

73. Baggio, B., et al. 1993. Evidence of a link between erythrocyte band 3 phosphorylation and anion transport in patients with 'idiopathic' calcium oxalate nephrolithiasis. *Miner. Electrolyte Metab.* **19**:17–20.

74. Baggio, B., et al. 1986. An inheritable anomaly of red-cell oxalate transport in "primary" calcium nephrolithiasis correctable with diuretics. *N. Engl. J. Med.* **314**:599–604.

75. Dobbins, J.W., and Binder, H.J. 1997. Importance of the colon in enteric hyperoxaluria. *N. Engl. J. Med.* **296**:298–301.

76. Worcester, E.M. 2002. Stones from bowel disease. *Endocrinol. Metab. Clin. North Am.* **31**:979–999.

77. Danpure, C.J., and Rumsby, G. 2004. Molecular aetiology of primary hyperoxaluria and its implications for clinical management. *Expert Rev. Mol. Med.* **2004**:1–16.

78. Asplin, J.R. 2002. Hyperoxaluric calcium nephrolithiasis. *Endocrinol. Metab. Clin. North Am.* **31**:927–949.

79. Milliner, D.S., et al. 1994. Results of long-term treatment with orthophosphate and pyridoxine in patients with primary hyperoxaluria. *N. Engl. J. Med.* **331**:1553–1558.

80. Jamieson, N.V. 1998. The results of combined liver/kidney transplantation for primary hyperoxaluria (PH1) 1984-1997. The European PH1 transplant registry report. European PH1 Transplantation Study Group. *J. Nephrol.* **11**(Suppl. 1):36–41.

81. Brennan, S., Hering-Smith, K., and Hamm, L.L. 1988. Effect of pH on citrate reabsorption in the proximal convoluted tubule. *Am. J. Physiol.* **255**:F301–F306.

82. Sakhaee, K., et al. 1993. Alkali absorption and citrate excretion in calcium nephrolithiasis. *J. Bone Miner. Res.* **8**:789–794.

83. Donnelly, S., Kamel, K.S., Vasuvattakul, S., Narins, R.G., and Halperin, M.L. 1992. Might distal renal tubular acidosis be a proximal tubular cell disorder? *Am. J. Kidney Dis.* **19**:272–281.

84. Pak, C.Y., et al. 1985. Correction of hypocitraturia and prevention of stone formation by combined thiazide and potassium citrate therapy in thiazide-unresponsive hypercalciuric nephrolithiasis. *Am. J. Med.* **79**:284–288.

85. Lemann, J., Jr., Pleuss, J.A., Gray, R.W., and Hoffmann, R.G. 1991. Potassium administration reduces and potassium deprivation increases urinary calcium excretion in healthy adults. *Kidney Int.* **39**:973–983.

86. Lemann, J., Jr., Gray, R.W., and Pleuss, J.A. 1989. Potassium bicarbonate, but not sodium bicarbonate, reduces urinary calcium excretion and improves calcium balance in healthy men. *Kidney Int.* **35**:688–695.

87. Grover, P.K., Ryall, R.L., and Marshall, V.R. 1990. Effect of urate on calcium oxalate crystallization in human urine: evidence for a promotory role of hyperuricosuria in urolithiasis. *Clin. Sci. (Lond.).* **79**:9–15.

88. Grover, P.K., Ryall, R.L., Potezny, N., and Marshall, V.R. 1990. The effect of decreasing the concentration of urinary urate on the crystallization of calcium oxalate in undiluted human urine. *J. Urol.* **143**:1057–1061.

89. Grover, P.K., Ryall, R.G., and Marshall, V.R. 1992. Calcium oxalate crystallization in urine: role of urate and glycosaminoglycans. *Kidney Int.* **41**:149–154.

90. Mandel, N.S., and Mandel, G.S. 1989. Urinary tract stone disease in the United States veteran population. II. Geographical analysis of variations in composition. *J. Urol.* **142**:1516–1521.

91. Klee, L.W., Brito, C.G., and Lingeman, J.E. 1991. The clinical implications of brushite calculi. *J. Urol.* **145**:715–718.

92. Evan, A.P., et al. 2005. Crystal-associated nephropathy in patients with brushite nephrolithiasis. *Kidney Int.* **67**:576–591.

93. Parks, J.H., Worcester, E.M., Coe, F.L., Evan, A.P., and Lingeman, J.E. 2004. Clinical implications of abundant calcium phosphate in routinely analyzed kidney stones. *Kidney Int.* **66**:777–785.

94. Carlisle, E.J., Donnelly, S.M., and Halperin, M.L. 1991. Renal tubular acidosis (RTA): recognize the ammonium defect and pHorget the urine pH. *Pediatr. Nephrol.* **5**:242–248.

95. DuBose, T.D., Jr. 2004. Acid-base disorders. In *The kidney.* B.M. Brenner, editor. Saunders. Philadelphia, Pennsylvania, USA. 922–996.

96. Karet, F.E., et al. 1998. Mutations in the chloride-bicarbonate exchanger gene AE1 cause autosomal dominant but not autosomal recessive distal renal tubular acidosis. *Proc. Natl. Acad. Sci. U. S. A.* **95**:6337–6342.

97. Devonald, M.A., Smith, A.N., Poon, J.P., Ihrke, G., and Karet, F.E. 2003. Non-polarized targeting of AE1 causes autosomal dominant distal renal tubular acidosis. *Nat. Genet.* **33**:125–127.

98. Quilty, J.A., Cordat, E., and Reithmeier, R.A. 2002. Impaired trafficking of human kidney anion exchanger (kAE1) caused by hetero-oligomer formation with a truncated mutant associated with distal renal tubular acidosis. *Biochem. J.* **368**:895–903.

99. Karet, F.E., et al.1999. Mutations in the gene encoding B1 subunit of H+-ATPase cause renal tubular acidosis with sensorineural deafness. *Nat. Genet.* **21**:84–90.

100. Stehberger, P.A., et al. 2003. Localization and regulation of the ATP6V0A4 (a4) vacuolar H+-ATPase subunit defective in an inherited form of distal renal tubular acidosis. *J. Am. Soc. Nephrol.* **14**:3027–3038.

101. Shah, G.N., Bonapace, G., Hu, P.Y., Strisciuglio, P., and Sly, W.S. 2004. Carbonic anhydrase II deficiency syndrome (osteopetrosis with renal tubular acidosis and brain calcification): novel mutations in CA2 identified by direct sequencing expand the opportunity for genotype-phenotype correlation. *Hum. Mutat.* **24**:272.

102. Cohen, E.P., et al. 1992. Absence of H+ ATPase in cortical collecting tubules of a patient with Sjogren's syndrome and distal renal tubular acidosis. *J. Am. Soc. Nephrol.* **3**:264–271.

103. Ahlstrand, C., and Tiselius, H.G. 1987. Urine composition and stone formation during treatment with acetazolamide. *Scand. J. Urol. Nephrol.* **21**:225–228.

104. Gault, M.H., et al. 1991. Comparison of patients with idiopathic calcium phosphate and calcium oxalate stones. *Medicine.* **70**:345–358.

105. Osther, P.J., Bollerslev, J., Hansen, A.B., Engel, K., and Kildeberg, P. 1993. Pathophysiology of incomplete renal tubular acidosis in recurrent renal stone formers: evidence of disturbed calcium, bone and citrate metabolism. *Urol. Res.* **21**:169–173.

106. Sakhaee, K., Adams-Huet, B., Moe, O.W., and Pak, C.Y. 2002. Pathophysiologic basis for normouricosuric uric acid nephrolithiasis. *Kidney Int.* **62**:971–979.

107. Pak, C.Y., Sakhaee, K., Peterson, R.D., Poindexter, J.R., and Frawley, W.H. 2001. Biochemical profile of idiopathic uric acid nephrolithiasis. *Kidney Int.* **60**:757–761.

108. Abate, N., Chandalia, M., Cabo-Chan, A.V., Jr., Moe, O.W., and Sakhaee, K. 2004. The metabolic syndrome and uric acid nephrolithiasis: novel features of renal manifestation of insulin resistance. *Kidney Int.* **65**:386–392.

109. Maalouf, N.M., et al. 2004. Association of urinary pH with body weight in nephrolithiasis. *Kidney Int.* **65**:1422–1425.

110. Palacin, M., et al. 2005. The genetics of heteromeric amino acid transporters. *Physiology (Bethesda).* **20**:112–124.

111. Font-Llitjos, M., et al. 2005. New insights into cystinuria: 40 new mutations, genotype-phenotype correlation, and digenic inheritance causing partial phenotype. *J. Med. Genet.* **42**:58–68.

112. Nakagawa, Y., Asplin, J.R., Goldfarb, D.S., Parks, J.H., and Coe, F.L. 2000. Clinical use of cystine supersaturation measurements. *J. Urol.* **164**:1481–1485.

113. Coe, F.L., Clark, C., Parks, J.H., and Asplin, J.R. 2001. Solid phase assay of urine cystine supersaturation in the presence of cystine binding drugs. *J. Urol.* **166**:688–693.

114. Shekarriz, B., and Stoller, M.L. 2002. Cystinuria and other noncalcareous calculi. *Endocrinol. Metab. Clin. North Am.* **31**:951–977.

115. Barcelo, P., Wuhl, O., Servitge, E., Rousaud, A., and Pak, C.Y. 1993. Randomized double-blind study of potassium citrate in idiopathic hypocitraturic calcium nephrolithiasis. *J. Urol.* **150**:1761–1764.

116. Ettinger, B., et al. 1997. Potassium-magnesium citrate is an effective prophylaxis against recurrent calcium oxalate nephrolithiasis. *J. Urol.* **158**:2069–2073.

117. Hofbauer, J., Hobarth, K., Szabo, N., and Marberger, M. 1994. Alkali citrate prophylaxis in idiopathic recurrent calcium oxalate urolithiasis — a prospective randomized study. *Br. J. Urol.* **73**:362–365.

118. Ettinger, B., Citron, J.T., Livermore, B., and Dolman, L.I. 1988. Chlorthalidone reduces calcium oxalate calculous recurrence but magnesium hydroxide does not. *J. Urol.* **139**:679–684.

119. Laerum, E., and Larsen, S. 1984. Thiazide prophylaxis of urolithiasis: a double-blind study in general practice. *Acta Med. Scand.* **215**:383–389.

120. Borghi, L., Meschi, T., Guerra, A., and Novarini, A. 1993. Randomized prospective study of a nonthiazide diuretic, indapamide, in preventing calcium stone recurrences. *J. Cardiovasc. Pharmacol.* **22**(Suppl. 6):S78–S86.

121. Ettinger, B., Tang, A., Citron, J.T., Livermore, B., and Williams, T. 1986. Randomized trial of allopurinol in the prevention of calcium oxalate calculi. *N. Engl. J. Med.* **315**:1386–1389.

122. Borghi, L., et al. 1996. Urinary volume, water and recurrences of idiopathic calcium nephrolithiasis: a 5-year randomized prospective study. *J. Urol.* **155**:839–843.

123. Evan, A.P., Coe, F.L., Lingeman, J.E., and Worcester, E. 2005. Insights on the pathology of kidney stone formation. *Urol. Res.* doi:10.1007/s00240-005-0488-0.

Salt handling and hypertension

Kevin M. O'Shaughnessy[1] and Fiona E. Karet[2,3]

[1]Clinical Pharmacology Unit, [2]Division of Renal Medicine, and [3]Department of Medical Genetics, University of Cambridge, Cambridge, United Kingdom.

The kidney plays a central role in our ability to maintain appropriate sodium balance, which is critical to determination of blood pressure. In this review we outline current knowledge of renal salt handling at the molecular level, and, given that Westernized societies consume more salt than is required for normal physiology, we examine evidence that the lowering of salt intake can combat hypertension.

Introduction

Salt was once ascribed magical properties, and its spillage carried ill omen. It has been used as a monetary device throughout human history, and its economic influence has started wars. It should thus come as no surprise that salt consumption is still a controversial topic, especially among medical epidemiologists, health policy makers, and lobbyists for the salt industry (1, 2). An omnivorous diet that includes commercially prepared meals provides several times the amount of sodium needed for normal physiologic function, and debate as to the importance of diminishing this consumption in the general population remains vociferous.

In evolutionary terms, our exposure to a high-salt intake (>6 g/d) is recent. This probably explains features in modern humans such as the very low sodium content of human breast milk (about 10 mM) compared with that of nonprimate mammals. Our hominid ancestors genetically adapted over hundreds of millennia to a very low-salt environment in equatorial savanna, probably consuming less than 0.1 g/d. Preliterate humans show no age-related blood pressure changes until they move into an urbanized high-salt environment.

The observation that salt intake is associated with hypertension is not new: the "hard pulse" resulting from a high salt intake is referred to in the *Nei Ching*, a classic Chinese text probably dating from the first millennium BCE (3). Over the past half-century or so, much energy has been devoted to dissecting this relationship. Work at Duke University in the late 1940s first showed the effectiveness of a very low-salt diet (based on rice and giving an intake less than 0.5 g/d) in reducing blood pressure (BP) in patients with malignant hypertension, a sometimes irreversible condition where uncontrolled severe elevation of BP results in end-organ damage to kidneys, heart, brain, and eyes.

The kidney's contribution to sodium homeostasis is crucial (Figure 1). Prior to the elucidation of the molecular contributors to both renal sodium reabsorption and the linked functions of potassium and chloride handling, it was evident that the kidney was centrally involved in BP determination. For example, it was demonstrated in the 1970s that transplantation of kidneys from genetically normotensive rats into genetically hypertensive recipient strains could prevent or correct hypertension (4). Furthermore, Guyton's pressure natriuresis theory, which remains a cornerstone of our thinking concerning sodium homeostasis, argues that hypertension cannot be sustained without active renal involvement (5, 6).

Human genetics

Attempts to identify the genes whose function or, indeed, dysfunction affects human BP have shed light on the molecular pathways involved (7). It is generally accepted that in any individual, some five or six genes contribute to the final arterial pressure level, which reflects a complex network of gene-gene and gene-environment interactions. In some individuals, however, defects in a single gene cause marked abnormalities of BP. The genetic causes of almost all of these rare Mendelian forms of hypertension and hypotension have now been elucidated, and, remarkably, they converge upon a final common pathway: the regulation of sodium reabsorption in the kidney (Figure 1). Mutations that increase renal sodium reabsorption increase BP, whereas those that decrease renal sodium reabsorption serve to decrease it. This analysis has had a major impact on our understanding of renal homeostatic mechanisms and underscores the importance of salt handling in determination of BP.

Liddle syndrome

Grant Liddle described a three-generation kindred with autosomal dominant inheritance of early-onset hypertension and hypokalemic alkalosis associated with suppressed levels of aldosterone (8). The index patient received a cadaveric renal transplant in 1989, after which her hypertension and biochemical derangements resolved (9). The abnormalities in Liddle syndrome can be ameliorated by a low-salt diet plus amiloride or triamterene, which are antagonists of the epithelial sodium channel (ENaC) of the collecting duct.

Liddle syndrome is associated with mutations in either the β or the γ subunit of the non–voltage-gated ENaC gene (10, 11). Mostly, these mutations result in truncations of the cytoplasmic C-terminal tail of the relevant subunit. Consistent with the dominant inheritance pattern, they are associated with a gain of function in ENaC.

In the kidney, functional ENaC channels are expressed at the apical surface of collecting duct principal cells and are composed of at least three subunits: α, β, and γ. ENaC activity is chiefly regulated by variation in the number of channels present at the cell surface. Normally, removal of ENaC occurs following an interaction between a conserved PY motif in the C-terminal tail of one of its subunits and the E3 ligase Nedd4-2 (13, 14), resulting in ubiquitination, internalization, and proteasome-mediated degradation. Nedd4-2 is phosphorylated by the action of serum and glucocorticoid kinase 1, the function of which is upregulated by aldosterone (15, 16). Once Nedd4-2 is phosphorylated, it loses the ability to interact with ENaC, leading to the observed increase in activity in response to this hormone. Liddle syndrome mutations

Nonstandard abbreviations used: chloride channel, kidney B (CLC-KB); Dietary Approaches to Stop Hypertension (DASH); epithelial sodium channel (ENaC); mineralocorticoid receptor (MR); pseudohypoaldosteronism type 1 (PHA1); renin-angiotensin-aldosterone axis (RAAA); sodium chloride cotransporter (NCCT); sodium-potassium-chloride cotransporter (NKCC2); with no lysine (WNK).

Conflict of interest: The authors have declared that no conflict of interest exists.

Citation for this article: *J. Clin. Invest.* **113**:1075–1081 (2004). doi:10.1172/JCI200421560.

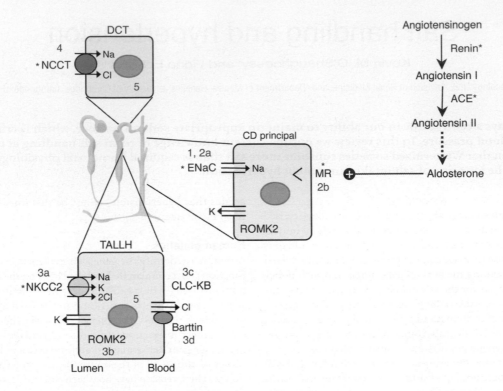

Figure 1
The renin-angiotensin-aldosterone axis and molecular pathways of sodium reabsorption in the nephron. A cartoon of a nephron is shown, with sodium-reabsorbing cells of the thick ascending limb of the loop of Henle (TALLH), distal convoluted tubule (DCT), and collecting duct (CD). The latter is responsible for fine regulation via the aldosterone-activated mineralocorticoid receptor (MR). In all cells, sodium exits the basolateral compartment via the Na/K-ATPase (not shown). Aldosterone synthesis is controlled by angiotensin II. Numbers reflect cellular components affected in disorders of sodium homeostasis referred to in the text: 1, Liddle syndrome; 2a/b, recessive/dominant PHA1; 3a/b/c/d, types I/II/III/IV Bartter syndrome; 4, Gitelman syndrome; 5, Gordon syndrome. *Sites of action of antihypertensive drugs. ROMK2, renal outer medullary potassium channel.

all affect the PY motif, leading to constitutive ENaC expression and, therefore, increased sodium reabsorption. Thus, the kidneys of patients with Liddle syndrome behave as if they were consuming and retaining excessive amounts of salt, and a low-salt diet is an important feature of therapy.

Pseudohypoaldosteronism type 1

Pseudohypoaldosteronism type 1 (PHA1) represents the clinical inverse of Liddle syndrome. It is a rare inherited disorder characterized by renal salt wasting and hyperkalemic metabolic acidosis, despite markedly elevated renin and aldosterone levels, in the setting of otherwise normal renal and adrenal function (17). The clinical picture is thus one of renal resistance to mineralocorticoids. Clinically distinct autosomal recessive and autosomal dominant forms of the disease have been described. Both generally present in the first weeks of life, with dehydration, sodium wasting, hyponatremia, and hyperkalemic metabolic acidosis. In recessive PHA1, patients have severe sodium wasting from the colon, the sweat and salivary glands, and the kidney. These children have recurrent life-threatening episodes of salt wasting and hyperkalemia, requiring lifelong sodium supplementation and treatment with potassium-binding resins.

Recessive PHA1 is caused by homozygous loss-of-function mutations in any one of the ENaC subunits (18, 19), causing a marked reduction of sodium reabsorption in the cortical collecting duct. The linked secretions of potassium and hydrogen ions in this

segment are blocked as well. The ensuing hyperkalemic volume-depleted state stimulates the renin-angiotensin system, resulting in elevated aldosterone levels and maximal activation of the mineralocorticoid receptor (MR). Because of the absence of ENaC, the MR is unable to stimulate sodium reabsorption, and so sodium wasting and hyperkalemic acidosis persist.

The severity of the clinical course of recessive PHA1 highlights the crucial role of ENaC in sodium homeostasis, even in individuals ingesting a high-salt diet. There is some variability in the prognosis of patients with recessive PHA1. For patients with homozygous null mutations, the outcome is often very poor. Even minor illness can bring on rapid deterioration with hypotension and hyperkalemia; nausea and vomiting often herald and then accelerate the clinical decline.

In the dominant form of PHA1, sodium wasting is limited to the kidney. While these patients may be quite ill at birth, they typically have a much milder course, generally responding well to salt supplementation. They may even be able to discontinue treatment after the first few years of life. Heterozygous loss-of-function mutations in the MR gene cause this form of the disease (20).

Inherited metabolic alkaloses:
Bartter and Gitelman syndromes

Bartter and Gitelman syndromes were originally described as variations of a single disease process (21, 22) resulting in hypokalemic metabolic alkalosis. More recent biochemical and latterly genetic studies have permitted their separation into distinct disorders, with separable phenotypic characteristics. The genetic defects involve either the salt transporters that are targets of diuretics, or other transporters that are their essential cellular partners. In both diseases, the mode of inheritance is autosomal recessive. To date, four genes have been implicated in the pathogenesis of Bartter syndrome in different kindreds, whereas all cases of Gitelman syndrome studied, now numbering several hundred, are accounted for by mutations in a single gene. In all these variants, the net effect is renal salt wasting, leading to low BP, reduced serum potassium, and an activated renin-angiotensin system.

Features that differentiate Bartter and Gitelman syndromes involve renal calcium handling and deposition, serum magnesium, and clinical presentation (23). In Bartter syndrome, affected individuals may present in infancy or early childhood with severe volume depletion and failure to thrive. Prematurity and maternal polyhydramnios are common. The metabolic dysfunction is usually accompanied by hypercalciuria with normocalcemia and normomagnesemia. Renal tract calcification is very common and may be present even in neonates. Nephrocalcinosis in infancy suggests type I or II Bartter syndrome. Hyperprostaglandinuria and a therapeutic response to indomethacin are features of type II disease. By contrast, those with type III Bartter syndrome may well be normocalciuric and mildly hypomagnesemic, and devoid of renal calcium deposition. Type IV patients are deaf.

The biochemical picture in untreated Bartter syndrome is reminiscent of that occasionally seen in otherwise normal people on long-term loop diuretic therapy. The target for loop diuretics is the electroneutral sodium-potassium-chloride cotransporter (NKCC2), which is expressed apically in the thick ascending limb of Henle's loop (Figure 1). Defects in the gene encoding NKCC2 cause type I Bartter syndrome (24). Subsequently, two further defective genes that result in loss of NKCC2 function have been identified in different kindreds. These are *ROMK2*, which encodes the inward-rectifier renal outer medullary potassium channel that regulates NKCC2's activity by recycling K^+ ions back into the tubular fluid (type II Bartter syndrome) (25), and *CLCNKB*, which encodes the basolateral chloride channel in the same loop of Henle cells (type III) (26). Most recently, type IV Bartter syndrome has been attributed to loss of function in a novel protein, Barttin, an essential cofactor for chloride channel, kidney B (CLC-KB) function in both the kidney and the inner ear, which explains the concomitant deafness (27, 28).

By far the majority of patients suspected of having Bartter syndrome in fact have the much more common Gitelman syndrome. The phenotype here is often much milder, usually being identified in late childhood or even in adulthood. Some affected individuals are asymptomatic, but others may be more severely affected, with growth problems and, not uncommonly, joint problems, tetany, and/or other neuromuscular abnormalities. A survey of presenting symptoms (29) highlights the differences in perception between patients and their physicians: while the latter often consider Gitelman syndrome to be asymptomatic, most patients would disagree. Anecdotally, it is reported that affected individuals may note a long-standing preference for salty over sweet foods and snacks. Biochemically, Gitelman syndrome is characterized by hypocalciuria and hypomagnesemia with renal magnesium wasting.

Gitelman syndrome patients display many of the biochemical changes seen in individuals on thiazide diuretics, which are commonly used for the treatment of hypertension; indeed, the surreptitious abuse of diuretics (or laxatives) remains the commonest differential diagnosis. All cases of Gitelman syndrome are due to loss of function of the sodium chloride cotransporter (NCCT), the target of thiazides that is present on the apical epithelial surface of distal renal convoluted tubule cells (30) (Figure 1).

The treatment of Bartter and Gitelman syndromes can be difficult, as the degree of salt wasting may be severe. Aggressive replacement of salt, and K^+ in particular, is essential. Some patients respond well to the administration of indomethacin, especially in type II Bartter syndrome. Mg^{2+} supplementation is also a useful adjunct in Gitelman syndrome.

Gordon syndrome

The clinical inverse of Gitelman syndrome is that bearing Richard Gordon's name (31). Here, the hypertension is associated with chloride-dependent sodium retention, accompanied by elevated serum potassium and acidosis. Despite the clinical indication of overactivity of NCCT, linkage to this gene was formally excluded in favor of at least three other loci (32, 33). Two responsible genes have now been identified, encoding the with no lysine (K) kinases WNK1 and WNK4 (34). WNK1 is expressed ubiquitously and is particularly associated with chloride-transporting epithelia at all sites (35), whereas WNK4 expression is limited to the distal nephron. Of particular interest here is recent evidence that WNK4 acts as a negative regulator of NCCT function (36, 37). Disease-causing mutations relieve this inhibition, leading to transporter overactivity. WNK4 also affects the renal outer medullary potassium (ROMK) channel (38) and might represent an intriguing novel antihypertensive target.

From molecular genetics to the population

Clearly, the majority of hypertensive individuals do not have such draconian genetic defects as are outlined above. The single-gene disorders may be rare, but they do highlight an important relationship between salt (of which, in general, we consume vast amounts in excess of our physiologic needs), the renin-angiotensin-aldosterone axis (RAAA), and BP. With the aim of extending the important physiologic insights provided by these disorders, as many as several hundred population- or cohort-based genetic studies have been designed to identify genes implicated in faulty BP regulation (see refs. 39, 40). These studies are of two main types: firstly, association studies, usually comparing polymorphisms (or occasionally haplotypes) in or near particular candidate genes; and secondly, genome-wide linkage analyses seeking disease-causing genetic loci. However, despite enormous efforts to find genes in the wider hypertensive population, the results have often been rather mixed. Two genes do stand out for which the majority of data are positive: *AGT*, encoding angiotensinogen, and the gene encoding the β subunit of ENaC (reviewed in ref. 41). For both of these, there are proven functional consequences to the genetic findings that suggest a subtle alteration in sodium handling in affected individuals. These results from human genetic and physiologic investigations are lent further credence by supporting information from animal models (see below).

Many reasons for the lack of major progress from many other association-based and genome-wide approaches can be and have been offered, often based either on criticisms of statistical methodology or power, or on the subset of individuals selected for study. For example, a recent meta-analysis of genome-wide scans

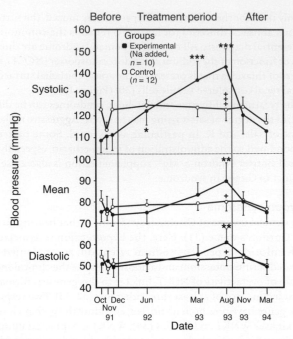

Figure 2
BP changes in chimpanzees fed a salt-supplemented diet. Twenty-two chimpanzees were fed infant formula either alone (*n* = 12) or with added salt (*n* = 10) as follows: 5 g/d for 19 weeks, 10 g/d for 3 weeks, and 15 g/d for 67 weeks. A 20-week period without added dietary salt concluded the experiment. The BP changes were significantly increased over base line (*$P < 0.05$, **$P < 0.01$, ***$P < 0.001$) and significantly different between groups (+$P < 0.05$; +++$P < 0.001$). Reproduced with permission from *Nature Medicine* (48).

included wildly different populations whose difference could easily confound rather than enhance results (42).

Animal data

Given the enormous background genetic variability of the human species, laboratory animals have been an obvious choice to test hypotheses concerning the relationship between salt and BP, at both the genetic and the dietary level. Recently, targeted manipulation of candidate genes in the RAAA in rodents has added to the armamentarium of evidence. For example, mice over- or underexpressing *Agt* (43, 44) or bearing a Liddle syndrome mutation (45, 46) have proven to be useful and tractable models.

At the whole-animal level, studies of salt consumption are also of interest; indeed, Dahl himself developed the now classical Dahl strain of salt-sensitive rat for the purpose (47). Critics of this particular model, however, point to the levels of salt intake required to increase BP as being vastly in excess of the human dietetic range; the Dahl rat is typically given 8% saline to drink. Nevertheless, a careful study in chimpanzees showed that their BP could be raised incrementally by elevation of salt intake from a base line of just 0.5 g/d to 5, 10, and 15 g/d (Figure 2) (48). As we share 98.4% of our genome with the chimpanzee, this can be viewed as a robust and relevant model.

Human data

The modern literature concerning the relationship between salt and BP in humans really begins in the 1960s with the work of Lewis Dahl. He correlated the prevalence of hypertension in five geograph-

ically distinct populations with their average daily salt intakes and proposed that BP rises linearly with salt consumption (49).

Many population studies have followed Dahl's, but the results have often been conflicting. For example, the 1988 National Heart, Lung, and Blood Institute–sponsored Intersalt study, the largest of its kind, was drawn from 52 centers in 32 countries and included more than 10,000 individuals (50). Despite its apparent breadth of recruitment, it failed to find any correlation unless four apparent outliers with very low salt intakes were included. Without their inclusion, the range of salt intakes across the centers telescoped from 0.12–14 g/d to just 6–14 g/d. With hindsight, the goal of demonstrating a significant dose-response relationship within this narrow range was probably unrealistic. However, this study did report that in populations with a high salt intake, the relationship between BP and age was stronger than in populations with a low salt intake: between the ages of 25 and 55 the slope was approximately 0.9 mmHg higher for each 10 mmol (0.6 g) difference in salt intake.

A problem with inter-population studies is their openness to ecological confounders: the salt content of the diet is not the only difference between so-called low- and high-salt populations (see the discussion, below, of the impact of diet in the Dietary Approaches to Stop Hypertension trial). Like spot BP measurements, single measurements of 24-hour sodium excretion are notoriously variable. This renders studies like Intersalt susceptible to the statistical effect of dilutional regression bias that may reduce (or inflate) any real association of salt intake and hypertension. Indeed, the final credibility of the Intersalt study was heavily undermined by attempts to "correct" for this effect to produce much larger estimates of the pressor effect of salt (51).

An alternative to population-based trials is intervention trials. Dozens of these have been published, though they are often small and have been combined by meta-analysis. The difficulty with this approach is that many trials have flawed design: they often are unblinded or include treated hypertensives. Nevertheless, the meta-analysis by Law et al., focusing on studies that lasted more than 5 weeks, reported an average fall in systolic BP of 5 mmHg resulting from a 50-mmol (~3 g/d) reduction in salt intake (52). This is very close to the figures published in a more recent meta-analysis that again focused on the longer-term intervention trials (53).

The single most important intervention trial to date has been the Dietary Approaches to Stop Hypertension (DASH) study, or, more correctly, its sub-study, DASH-Sodium (54). In this study, 412 individuals were assigned to either a normal "American" diet or the DASH diet (high in fruit and vegetables and low-fat dairy products) and one of three different levels of salt intake: high, intermediate, and low, approximating 9, 6, and 3 g/d of salt, respectively. Participants were kept on these diets for 30 days, and the BP reductions were as shown in Figure 3. BP reduction was again in keeping with the values from the meta-analyses above, but the effect of the DASH diet was at least as great. This effect of diet can be explained partly by the well-established interplay of dietary cations. For example, intake of calcium dictated by water "hardness" is inversely related to cardiovascular mortality, and calciuresis is a feature of salt loading (55, 56). Potassium supplementation similarly lowers BP (57) and may even protect against stroke independently of its effect on BP (58). Different levels of dietary folic acid and antioxidant vitamins (C and E) are also relevant, since they may affect BP through effects on endothelial function and bioavailable nitric oxide levels within the vasculature (59, 60). It is

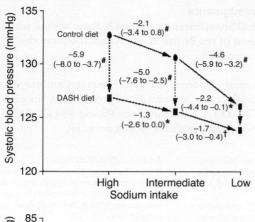

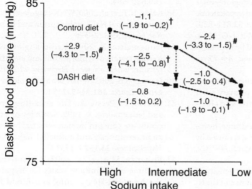

Figure 3
BP changes during the DASH-Sodium trial. Although arrows are uni-directional, the order in which individuals were assigned a given salt level was random with a crossover design. The numbers next to the lines connecting the data points are the mean changes in BP. The 95% confidence intervals are given in parentheses. *$P < 0.05$, †$P < 0.01$, and #$P < 0.001$; symbols indicate significant differences in BP between groups or between dietary sodium categories. Reproduced with permission from the *New England Journal of Medicine* (54).

thus becoming clear that the dietary context in which salt restriction occurs can be as crucial as the restriction itself.

It is therefore interesting that in recent weeks, the US Institute of Medicine has issued a strong across-the-board recommendation that salt consumption be approximately halved — which may be difficult in these fast-food days — and also advised a significant increase in potassium intake "to lower blood pressure [and] blunt the effects of salt," among other benefits (61).

Salt sensitivity

The differences between the population and intervention studies on salt intake can be partly explained by interindividual variation in susceptibility to the pressor effect of salt loading. This heterogeneity in response to salt was obvious in the original human intervention studies (62), and in the recent salt loading study in chimpanzees, discussed above (48). Salt sensitivity is affected by age, race, and disease state and hence is more common in the elderly, in those of Afro-Caribbean races, and in type 2 diabetics. There is a definable genetic influence within families (63). However, there is no universal definition of salt sensitivity, and therefore studies resort to arbitrary cutoffs to define salt "responders" and "nonresponders." This has not perturbed the intense trawls for susceptibility genotypes

even within the well-characterized DASH cohort itself (64). Until simple molecular markers like these become available, it will be impossible to identify easily those individuals in the general population who would benefit most from dietary salt restriction.

That said, salt restriction is worthwhile in subjects taking specific antihypertensive agents. For example, dietary manipulation in patients receiving amiloride or triamterene should be beneficial, because these act competitively with sodium for transport via ENaC. In addition, there is evidence that reducing salt intake increases the efficacy of angiotensin-converting enzyme inhibitors, because of the RAAA activation that salt depletion induces (65).

Is the anion important?

Berghoff and Geraci first reported in 1929 that loading hypertensive individuals with sodium bicarbonate did not have the same pressor effect as loading with sodium chloride (66). This observation has been confirmed in more recent studies using citrate or phosphate as the counter-anion (67, 68). Loading with equimolar amounts of sodium salts causes similar degrees of Na^+ retention, weight gain, and suppression of renin and aldosterone, but, curiously, only sodium chloride causes an expansion of the plasma volume and a rise in BP. Why chloride should have this unique property against other sodium salts is not clear, but from a dietary viewpoint it makes little difference, since the bulk of the sodium we consume is in the form of sodium chloride.

Are there critical periods for salt exposure?

The Barker hypothesis that birth weight and adult BP are reciprocally related has had a considerable impact on how we now view exposure to factors in utero or in early postnatal life, especially the long-term effects on gene expression that may result (69). Rat models of salt-sensitive hypertension clearly show discrete periods of sensitivity to salt loading that program long-term changes in BP and cardiovascular morphology (70). Evidence that a similar phenomenon can occur in humans is lacking thus far (71) apart from the notable feature of PHA1 whereby neonatal salt repletion usually becomes unnecessary as the child grows. It is conceivable that our scrutiny of salt intake in adults is misplaced, and that we should instead be addressing salt intake in infancy and/or early childhood for maximal benefit.

Conclusions

Overall, it seems clear that the body's handling of salt has an important effect on cardiovascular status, and recent evidence strongly supports the use of saluretics, in the shape of thiazide diuretics, as first-line therapy for hypertension and coronary heart disease prevention (72). The real contention that exists is not whether salt affects BP, but whether the current body of evidence supports a reduction in salt intake in the general population. Impressive as the DASH data are, for example, this was a short-term study, and dietary manipulations have poor compliance rates in the long term (73). In other areas of therapeutic intervention, the gold standard of proof is a randomized controlled trial showing an impact on disease outcome. Given the prevalence of hypertension and the strong relationship between stroke risk and BP, an endpoint of stroke prevention represents an important goal. Antihypertensive-drug trials have in fact demonstrated stroke protection with relatively modest BP reductions. The reduction in average BP following a reduction in salt intake to 3 or even just to 6 g/d may be modest at under 5 mmHg, but this is the same level of reduction seen in meta-analyses of antihypertensive-intervention trials (74), which suggests that similar levels of stroke prevention might be achievable. The public health and budgetary implications seem irresistible. Nevertheless, if the highly polarized divide separating those who favor reduction in salt intake from those who oppose it is to be bridged, two complementary approaches are probably necessary: first, much more careful clinical, biochemical, and pharmacologic phenotyping of hypertensive participants in both genetic and in vivo studies; and second, an eventual controlled intervention trial to confirm the effect on outcome.

Acknowledgments

Kevin M. O'Shaughnessy and Fiona E. Karet wish to acknowledge the support of the British Heart Foundation and the Wellcome Trust, respectively.

Address correspondence to: Fiona E. Karet, Cambridge Institute for Medical Research, Box 139 Addenbrooke's Hospital, Cambridge CB2 2XY, United Kingdom. Phone: 44-1223-762617; Fax: 44-1223-331206; E-mail: fek1000@cam.ac.uk.

1. Taubes, G. 1998. The (political) science of salt. *Science.* **281**:898–907.
2. Taubes, G. 2000. A DASH of data in the salt debate. *Science.* **288**:1319.
3. Veith, I. 1949. *Huang Ti Nei Ching Su Wen. The Yellow Emperor's Classic Of Internal Medicine.* Williams & Wilkins. Baltimore, Maryland, USA. 253 pp.
4. Dahl, L.K., and Heine, M. 1975. Primary role of renal homografts in setting chronic blood pressure levels in rats. *Circ. Res.* **36**:692–696.
5. Guyton, A.C. 1991. Abnormal renal function and autoregulation in essential hypertension. *Hypertension.* **18**:49–53.
6. Guyton, A.C. 1991. Blood pressure control: special role of the kidneys and body fluids. *Science.* **252**:1813–1816.
7. Lifton, R.P., Gharavi, A.G., and Geller, D.S. 2001. Molecular mechanisms of human hypertension. *Cell.* **104**:545–556.
8. Liddle, G., Bledsoe, T., and Coppage, W. 1963. A familial renal disorder simulating primary aldosteronism but with negligible aldosterone secretion. *Trans. Assoc. Am. Physicians.* **76**:199–213.
9. Botero-Velez, M., Curtis, J.J., and Warnock, D.G. 1994. Brief report. Liddle's syndrome revisited: a disorder of sodium reabsorption in the distal tubule. *N. Engl. J. Med.* **330**:178–181.
10. Shimkets, R.A., et al. 1994. Liddle's syndrome: heritable human hypertension caused by mutations in the beta subunit of the epithelial sodium channel. *Cell.* **79**:407–414.
11. Hansson, J.H., et al. 1995. Hypertension caused by a truncated epithelial sodium channel gamma subunit: genetic heterogeneity of Liddle syndrome. *Nat. Genet.* **11**:76–82.
12. Canessa, C.M., et al. 1994. Amiloride-sensitive epithelial Na$^+$ channel is made of three homologous subunits. *Nature.* **367**:463–467.
13. Staub, O., et al. 1996. WW domains of Nedd4 bind to the proline-rich PY motifs in the epithelial Na$^+$ channel deleted in Liddle's syndrome. *EMBO J.* **15**:2371–2380.
14. Snyder, P.M., Steines, J.C., and Olson, D.R. 2004. Relative contribution of Nedd4 and Nedd4-2 to ENaC regulation in epithelia determined by RNA interference. *J. Biol. Chem.* **279**:5042–5046.
15. Naray-Fejes-Toth, A., Canessa, C., Cleaveland, E.S., Aldrich, G., and Fejes-Toth, G. 1999. sgk is an aldosterone-induced kinase in the renal collecting duct. Effects on epithelial Na$^+$ channels. *J. Biol. Chem.* **274**:16973–16978.
16. Chen, S.Y., et al. 1999. Epithelial sodium channel regulated by aldosterone-induced protein sgk. *Proc. Natl. Acad. Sci. U. S. A.* **96**:2514–2519.
17. Cheek, D.B., and Perry, J.W. 1958. A salt wasting syndrome in infancy. *Arch. Dis. Child.* **33**:252–256.
18. Chang, S.S., et al. 1996. Mutations in subunits of the epithelial sodium channel cause salt wasting with hyperkalaemic acidosis, pseudohypoaldosteronism type 1. *Nat. Genet.* **12**:248–253.
19. Strautnieks, S.S., Thompson, R.J., Gardiner, R.M., and Chung, E. 1996. A novel splice-site mutation in the gamma subunit of the epithelial sodium channel gene in three pseudohypoaldosteronism type 1 families. *Nat. Genet.* **13**:248–250.

20. Kerem, E., et al. 1999. Pulmonary epithelial sodium-channel dysfunction and excess airway liquid in pseudohypoaldosteronism. *N. Engl. J. Med.* **341**:156–162.
21. Bartter, F.C., Pronove, P., Gill, J.R., Jr., and MacCardle, R.C. 1962. Hyperplasia of the juxtaglomerular complex with hyperaldosteronism and hypokalemic alkalosis. A new syndrome. *Am. J. Med.* **33**:811–828.
22. Gitelman, H.J., Graham, J.B., and Welt, L.G. 1966. A new familial disorder characterized by hypokalemia and hypomagnesemia. *Trans. Assoc. Am. Physicians.* **79**:221–235.
23. Bettinelli, A., et al. 1992. Use of calcium excretion values to distinguish two forms of primary renal tubular hypokalemic alkalosis: Bartter and Gitelman syndromes. *J. Pediatr.* **120**:38–43.
24. Simon, D.B., et al. 1996. Bartter's syndrome, hypokalaemic alkalosis with hypercalciuria, is caused by mutations in the Na-K-2Cl cotransporter NKCC2. *Nat. Genet.* **13**:183–188.
25. Simon, D.B., et al. 1996. Genetic heterogeneity of Bartter's syndrome revealed by mutations in the K$^+$ channel, ROMK. *Nat. Genet.* **14**:152–156.
26. Simon, D.B., et al. 1997. Mutations in the chloride channel gene, *CLCNKB*, cause Bartter's syndrome type III. *Nat. Genet.* **17**:171–178.
27. Estevez, R., et al. 2001. Barttin is a Cl$^-$ channel beta-subunit crucial for renal Cl$^-$ reabsorption and inner ear K$^+$ secretion. *Nature.* **414**:558–561.
28. Birkenhager, R., et al. 2001. Mutation of *BSND* causes Bartter syndrome with sensorineural deafness and kidney failure. *Nat. Genet.* **29**:310–314.
29. Cruz, D.N., Shaer, A.J., Bia, M.J., Lifton, R.P., and Simon, D.B. 2001. Gitelman's syndrome revisited: an evaluation of symptoms and health-related quality of life. *Kidney Int.* **59**:710–717.
30. Simon, D.B., et al. 1996. Gitelman's variant of Bartter's syndrome, inherited hypokalaemic alkalosis, is caused by mutations in the thiazide-sensitive Na-Cl cotransporter. *Nat. Genet.* **12**:24–30.
31. Gordon, R.D., Geddes, R.A., Pawsey, C.G., and O'Halloran, M.W. 1970. Hypertension and severe hyperkalaemia associated with suppression of renin and aldosterone and completely reversed by dietary sodium restriction. *Australas. Ann. Med.* **19**:287–294.
32. Mansfield, T.A., et al. 1997. Multilocus linkage of familial hyperkalaemia and hypertension, pseudohypoaldosteronism type II, to chromosomes 1q31–42 and 17p11-q21. *Nat. Genet.* **16**:202–205.
33. Disse-Nicodeme, S., et al. 2000. A new locus on chromosome 12p13.3 for pseudohypoaldosteronism type II, an autosomal dominant form of hypertension. *Am. J. Hum. Genet.* **67**:302–310.
34. Wilson, F.H., et al. 2001. Human hypertension caused by mutations in WNK kinases. *Science.* **293**:1107–1112.
35. Choate, K.A., Kahle, K.T., Wilson, F.H., Nelson-Williams, C., and Lifton, R.P. 2003. WNK1, a kinase mutated in inherited hypertension with hyperkalemia, localizes to diverse Cl–transporting epithelia. *Proc. Natl. Acad. Sci. U. S. A.* **100**:663–668.
36. Yang, C.-L., Angell, J., Mitchell, R., and Ellison, D.H. 2003. WNK kinases regulate thiazide-sensitive Na-Cl cotransport. *J. Clin. Invest.* **111**:1039–1045.

doi:10.1172/JCI200317443.
37. Wilson, F.H., et al. 2003. Molecular pathogenesis of inherited hypertension with hyperkalemia: the Na-Cl cotransporter is inhibited by wild-type but not mutant WNK4. *Proc. Natl. Acad. Sci. U. S. A.* **100**:680–684.
38. Kahle, K.T., et al. 2003. WNK4 regulates the balance between renal NaCl reabsorption and K$^+$ secretion. *Nat. Genet.* **35**:372–376.
39. Jeunemaitre, X. 2003. Renin-angiotensin-aldosterone system polymorphisms and essential hypertension: where are we? *J. Hypertens.* **21**:2219–2222.
40. Harrap, S.B. 2003. Where are all the blood-pressure genes? *Lancet.* **361**:2149–2151.
41. Corvol, P., Persu, A., Gimenez-Roqueplo, A.P., and Jeunemaitre, X. 1999. Seven lessons from two candidate genes in human essential hypertension: angiotensinogen and epithelial sodium channel. *Hypertension.* **33**:1324–1331.
42. Province, M.A., et al. 2003. A meta-analysis of genome-wide linkage scans for hypertension: the National Heart, Lung and Blood Institute Family Blood Pressure Program. *Am. J. Hypertens.* **16**:144–147.
43. Kim, H.S., et al. 1995. Genetic control of blood pressure and the angiotensinogen locus. *Proc. Natl. Acad. Sci. U. S. A.* **92**:2735–2739.
44. Kim, H.S., et al. 1999. Homeostasis in mice with genetically decreased angiotensinogen is primarily by an increased number of renin-producing cells. *J. Biol. Chem.* **274**:14210–14217.
45. Pradervand, S., et al. 2003. Dysfunction of the epithelial sodium channel expressed in the kidney of a mouse model for Liddle syndrome. *J. Am. Soc. Nephrol.* **14**:2219–2228.
46. Dahlmann, A., et al. 2003. Mineralocorticoid regulation of epithelial Na$^+$ channels is maintained in a mouse model of Liddle's syndrome. *Am. J. Physiol. Renal Physiol.* **285**:F310–F318.
47. Dahl, L.K., Heine, M., and Tassinari, L. 1962. Effects of chronic excess salt ingestion: evidence that genetic factors play an important role in susceptibility to experimental hypertension. *J. Exp. Med.* **115**:1173–1190.
48. Denton, D., et al. 1995. The effect of increased salt intake on blood pressure of chimpanzees. *Nat. Med.* **1**:1009–1016.
49. Dahl, L.K. 1960. Possible role of salt intake in the development of essential hypertension. In *Essential hypertension: an international symposium.* P. Cottier and K.D. Bock, editors. Springer-Verlag. Berlin, Germany. 61–75.
50. 1988. Intersalt: an international study of electrolyte excretion and blood pressure. Results for 24 hour urinary sodium and potassium excretion. Intersalt Cooperative Research Group. *Br. Med. J.* **297**:319–328.
51. Elliott, P., et al. 1996. Intersalt revisited: further analyses of 24 hour sodium excretion and blood pressure within and across populations. Intersalt Cooperative Research Group. *Br. Med. J.* **312**:1249–1253.
52. Law, M.R., Frost, C.D., and Wald, N.J. 1991. By how much does dietary salt reduction lower blood pressure? III. Analysis of data from trials of salt reduction. *Br. Med. J.* **302**:819–824.

53. He, F.J., and MacGregor, G.A. 2003. How far should salt intake be reduced? *Hypertension.* **42**:1093–1099.

54. Sacks, F.M., et al. 2001. Effects on blood pressure of reduced dietary sodium and the dietary approaches to stop hypertension (DASH) diet. *N. Engl. J. Med.* **344**:3–10.

55. Neri, L.C., and Johansen, H.L. 1978. Water hardness and cardiovascular mortality. *Ann. N. Y. Acad. Sci.* **304**:203–221.

56. Lind, L., Lithell, H., Gustafsson, I.B., Pollare, T., and Ljunghall, S. 1993. Calcium metabolism and sodium sensitivity in hypertensive subjects. *J. Hum. Hypertens.* **7**:53–57.

57. Cappuccio, F.P., and MacGregor, G.A. 1991. Does potassium supplementation lower blood pressure? A meta-analysis of published trials. *J. Hypertens.* **9**:465–473.

58. Ascherio, A., et al. 1998. Intake of potassium, magnesium, calcium, and fiber and risk of stroke among US men. *Circulation.* **98**:1198–1204.

59. Verhaar, M.C., Stroes, E., and Rabelink, T.J. 2002. Folates and cardiovascular disease. *Arterioscler. Thromb. Vasc. Biol.* **22**:6–13.

60. Rathaus, M., and Bernheim, J. 2002. Oxygen species in the microvascular environment: regulation of vascular tone and the development of hypertension. *Nephrol. Dial. Transplant.* **17**:216–221.

61. 2004. Dietary reference intakes: water, potassium, sodium, chloride, and sulfate. Report of the Institute of Medicine of the National Academies. Washington, DC, USA. http://www.iom.edu/report.asp?id=18495.

62. Luft, F.C., and Weinberger, M.H. 1997. Heterogeneous responses to changes in dietary salt intake: the salt-sensitivity paradigm. *Am. J. Clin. Nutr.* **65**(Suppl. 2):612S–617S.

63. Luft, F.C., et al. 1987. Influence of genetic variance on sodium sensitivity of blood pressure. *Klin. Wochenschr.* **65**:101–109.

64. Svetkey, L.P., et al. 2001. Angiotensinogen genotype and blood pressure response in the Dietary Approaches to Stop Hypertension (DASH) study. *J. Hypertens.* **19**:1949–1956.

65. Weir, M.R., et al. 1998. Influence of race and dietary salt on the antihypertensive efficacy of an angiotensin-converting enzyme inhibitor or a calcium channel antagonist in salt-sensitive hypertensives. *Hypertension.* **31**:1088–1096.

66. Berghoff, R.S., and Geraci, A.S. 1929. The influence of sodium chloride on blood pressure. *Br. Med. J.* **56**:395–397.

67. Shore, A.C., Markandu, N.D., and MacGregor, G.A. 1988. A randomized crossover study to compare the blood pressure response to sodium loading with and without chloride in patients with essential hypertension. *J. Hypertens.* **6**:613–617.

68. Kurtz, T.W., Al Bander, H.A., and Morris, R.C., Jr. 1987. "Salt-sensitive" essential hypertension in men. Is the sodium ion alone important? *N. Engl. J. Med.* **317**:1043–1048.

69. Dodic, M., Moritz, K., Koukoulas, I., and Wintour, E.M. 2002. Programmed hypertension: kidney, brain or both? *Trends Endocrinol. Metab.* **13**:403–408.

70. Zicha, J., and Kunes, J. 1999. Ontogenetic aspects of hypertension development: analysis in the rat. *Physiol. Rev.* **79**:1227–1282.

71. Holliday, M.A. 1995. Is blood pressure in later life affected by events in infancy? *Pediatr. Nephrol.* **9**:663–666.

72. Antihypertensive and Lipid-Lowering Treatment to Prevent Heart Attack Trial Collaborative Research Group. 2003. Diuretic versus alpha-blocker as first-step antihypertensive therapy: final results from the Antihypertensive and Lipid-Lowering Treatment to Prevent Heart Attack Trial (ALLHAT). *Hypertension.* **42**:239–246.

73. Luft, F.C., Morris, C.D., and Weinberger, M.H. 1997. Compliance to a low-salt diet. *Am. J. Clin. Nutr.* **65**(Suppl. 2):698S–703S.

74. Collins, R., et al. 1990. Blood pressure, stroke, and coronary heart disease. Part 2. Short-term reductions in blood pressure: overview of randomised drug trials in their epidemiological context. *Lancet.* **335**:827–838.

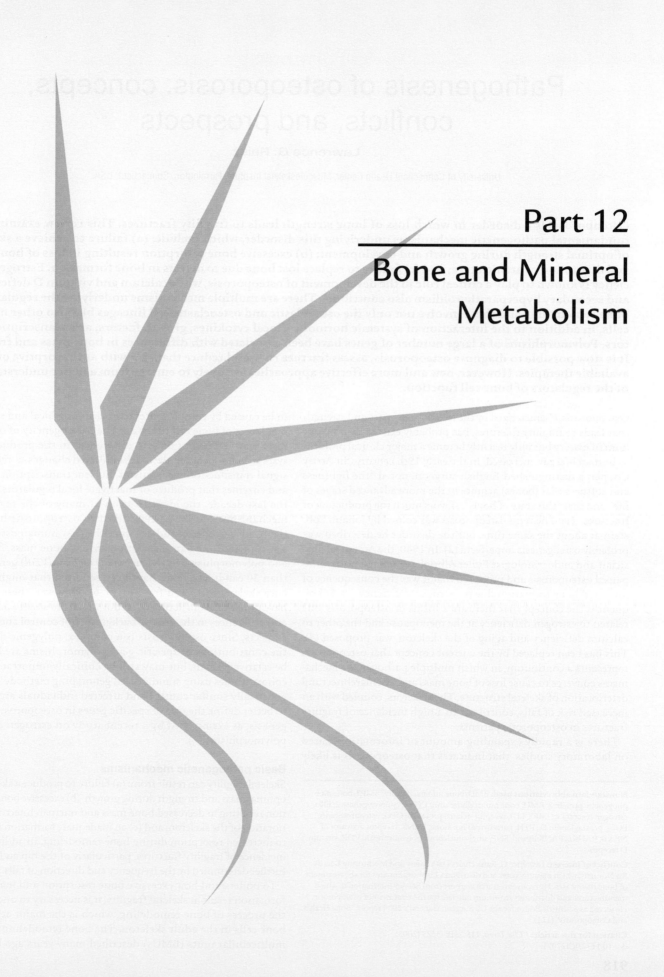

Part 12
Bone and Mineral Metabolism

Pathogenesis of osteoporosis: concepts, conflicts, and prospects

Lawrence G. Raisz

University of Connecticut Health Center, Musculoskeletal Institute, Farmington, Connecticut, USA.

Osteoporosis is a disorder in which loss of bone strength leads to fragility fractures. This review examines the fundamental pathogenetic mechanisms underlying this disorder, which include: (a) failure to achieve a skeleton of optimal strength during growth and development; (b) excessive bone resorption resulting in loss of bone mass and disruption of architecture; and (c) failure to replace lost bone due to defects in bone formation. Estrogen deficiency is known to play a critical role in the development of osteoporosis, while calcium and vitamin D deficiencies and secondary hyperparathyroidism also contribute. There are multiple mechanisms underlying the regulation of bone remodeling, and these involve not only the osteoblastic and osteoclastic cell lineages but also other marrow cells, in addition to the interaction of systemic hormones, local cytokines, growth factors, and transcription factors. Polymorphisms of a large number of genes have been associated with differences in bone mass and fragility. It is now possible to diagnose osteoporosis, assess fracture risk, and reduce that risk with antiresorptive or other available therapies. However, new and more effective approaches are likely to emerge from a better understanding of the regulators of bone cell function.

Osteoporosis, characterized by the loss of bone mass and strength that leads to fragility fractures, has probably existed throughout human history but only recently became a major clinical problem as human lifespan increased. In the early 19th century, Sir Astley Cooper, a distinguished English surgeon, noted "the lightness and softness that (bones) acquire in the more advanced stages of life" and that "this state of bone . . . favors much the production of fractures" (1). The term osteoporosis was coined by Johann Lobstein at about the same time, but the disorder he described was probably osteogenesis imperfecta (2). In 1940, the American physician and endocrinologist Fuller Albright described postmenopausal osteoporosis and proposed that it was the consequence of impaired bone formation due to estrogen deficiency (3). Subsequently, the concept that there are 2 forms of osteoporosis, one related to estrogen deficiency at the menopause and the other to calcium deficiency and aging of the skeleton, was proposed (4). This has been replaced by the current concept that osteoporosis represents a continuum, in which multiple pathogenetic mechanisms converge to cause loss of bone mass and microarchitectural deterioration of skeletal structure. These factors, coupled with an increased risk of falls, contribute to a high incidence of fragility fractures in osteoporotic patients.

There is a rapidly expanding amount of information, based on laboratory studies, that indicates that osteoporosis is likely to be caused by complex interactions among local and systemic regulators of bone cell function. The heterogeneity of osteoporosis may be due not only to differences in the production of systemic and local regulators, but also to changes in receptors, signal transduction mechanisms, nuclear transcription factors, and enzymes that produce or inactivate local regulators. Within the last decade, the identification of many of the regulatory mechanisms that have been linked to osteoporosis has been the result of genetic studies. Since the first human osteoporosis study indicated an association among bone mass, fragility, and polymorphisms in the *vitamin D receptor* (*VDR*) gene, more than 30 candidate genes have been reported that might influence skeletal mass and fragility (5, 6). However, these studies have presented conflicting data, due in part to small sample size and differences in the genetic background of control and disease subjects. Since osteoporosis is a complex, polygenic disorder, the contributions of specific gene polymorphisms are likely to be relatively small, but may still be clinically important. Large cohort studies using standardized genotyping methodology and genetically similar control and affected individuals are needed to better define the role of specific genes in osteoporosis pathogenesis, as exemplified by a recent study on estrogen receptor polymorphisms (7).

Basic pathogenetic mechanisms

Skeletal fragility can result from: (a) failure to produce a skeleton of optimal mass and strength during growth; (b) excessive bone resorption resulting in decreased bone mass and microarchitectural deterioration of the skeleton; and (c) an inadequate formation response to increased resorption during bone remodeling. In addition, the incidence of fragility fractures, particularly of the hip and wrist, is further determined by the frequency and direction of falls.

To understand how excessive bone resorption and inadequate formation result in skeletal fragility, it is necessary to understand the process of bone remodeling, which is the major activity of bone cells in the adult skeleton. The bone remodeling or bone multicellular units (BMUs) described many years ago by Frost

Nonstandard abbreviations used: BMD, bone mineral density; BMP2, bone morphogenetic protein 2; BMU, bone multicellular unit; COX2, cyclooxygenase 2; ERα, estrogen receptor α; LRP5, LDL receptor–related protein 5; OPG, osteoprotegerin; PGE$_2$, prostaglandin E$_2$; PTH, parathyroid hormone; RANK, receptor activator of NF-κB; RANKL, RANK ligand; SNP, single nucleotide polymorphism; VDR, vitamin D receptor.

Conflict of interest: Lawrence G. Raisz chairs Data Safety and Monitoring Boards for Novartis, which manufactures and distributes bisphosphonates for the treatment of bone metastases. He receives research support from Servier International, which manufactures and distributes strontium ranelate for the treatment of osteoporosis. He served as Scientific Editor for the US Surgeon General's 2004 report "Bone Health and Osteoporosis" (111).

Citation for this article: *J. Clin. Invest.* **115**:3318–3325 (2005). doi:10.1172/JCI27071.

and others (8) can occur either on the surface of trabecular bone as irregular Howship lacunae or in cortical bone as relatively uniform cylindrical haversian systems. As illustrated in Figure 1, the process begins with the activation of hematopoietic precursors to become osteoclasts, which normally requires an interaction with cells of the osteoblastic lineage. Because the resorption and reversal phases of bone remodeling are short and the period required for osteoblastic replacement of the bone is long, any increase in the rate of bone remodeling will result in a loss of bone mass. Moreover, the larger number of unfilled Howship lacunae and haversian canals will further weaken the bone. Excessive resorption can also result in complete loss of trabecular structures, so that there is no template for bone formation. Thus, there are multiple ways in which an increase in osteoclastic resorption can result in skeletal fragility. However, high rates of resorption are not always associated with bone loss; for example, during the pubertal growth spurt. Hence an inadequate formation response during remodeling is an essential component of the pathogenesis of osteoporosis.

Central role of estrogen

The concept that estrogen deficiency is critical to the pathogenesis of osteoporosis was based initially on the fact that postmenopausal women, whose estrogen levels naturally decline, are at the highest risk for developing the disease. Morphologic studies and measurements of certain biochemical markers have indicated that bone remodeling is accelerated at the menopause, as both markers of resorption and formation are increased (8, 9). Hence, contrary to Albright's original hypothesis, an increase in bone resorption, and not impaired bone formation, appears to be the driving force for bone loss in the setting of estrogen deficiency. But the rapid and continuous bone loss that occurs for several years after the menopause must indicate an impaired bone formation response, since in younger individuals going through the pubertal growth spurt, even faster rates of bone resorption can be associated with an increase in bone mass. However, the increased bone formation that normally occurs in response to mechanical loading is diminished in estrogen deficiency, suggesting estrogen is both anti-catabolic and anabolic (10).

Estrogen deficiency continues to play a role in bone loss in women in their 70s and 80s, as evidenced by the fact that estrogen treatment rapidly reduces bone breakdown in these older women (11). Moreover, recent studies in humans have shown that the level of estrogen required to maintain relatively normal bone remodeling in older postmenopausal women is lower than that required to stimulate classic target tissues such as the breast and uterus (12). Fracture risk is inversely related to estrogen levels in post-

menopausal women, and as little as one-quarter of the dose of estrogen that stimulates the breast and uterus is sufficient to decrease bone resorption and increase bone mass in older women (13). This greater sensitivity of the skeleton may be age related. In 3-month-old mice, the uterus appeared to be more responsive to estrogen than bone, whereas in 6-month-old mice, the reverse was found (14).

Estrogen is critical for epiphyseal closure in puberty in both sexes and regulates bone turnover in men as well as women. In fact, estrogen has a greater effect than androgen in inhibiting bone resorption in men, although androgen may still play a role (15). Estrogen may also be important in the acquisition of peak bone mass in men (16). Moreover, osteoporosis in older men is more closely associated with low estrogen than with low androgen levels (17).

Estrogen deficiency increases and estrogen treatment decreases the rate of bone remodeling, as well as the amount of bone lost with each remodeling cycle. Studies in animal models and in cell culture have suggested that this involves multiple sites of estrogen action, not only on the cells of the BMU, but on other marrow cells (Figure 1). Estrogen acts through 2 receptors: estrogen receptor α (ERα) and ERβ. ERα appears to be the primary mediator of estrogen's actions on the skeleton (10). Osteoblasts do express

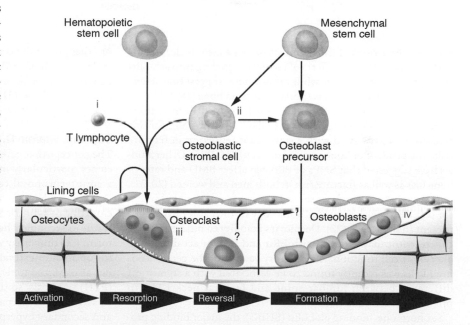

Figure 1

The BMU with possible sites of estrogen action. Bone remodeling on the surface of trabecular bone is illustrated here. The process is similar in the haversian systems in cortical bone. Osteoclast *activation* is ordinarily initiated by an interaction of hematopoietic precursors with cells of the osteoblast lineage but may also be initiated by inflammatory cells, particularly T cells (see Figure 2 for further details). Once osteoclasts are formed, there is a *resorption* phase of limited duration and a brief *reversal* phase, during which the bone surface is covered by mononuclear cells but bone formation has not yet begun. The *formation* phase is then initiated, possibly by factors produced by the osteoclast or reversal cells or released from the bone matrix. The formation phase, which is substantially longer than the first 3 phases, involves the production of matrix by progressive waves of osteoblasts. These then become flat lining cells, are embedded in the bone as osteocytes, or undergo apoptosis. Potential sites of action of estrogen include effects on T cell cytokine production (i); effects on stromal or osteoblastic cells to alter their production of RANKL or OPG (ii); direct inhibition of differentiated osteoclasts (iii); and effects on bone formation mediated by osteoblasts or osteocytes to enhance the response to mechanical forces initiated by these cells (iv). Note that the BMU is shown as being compartmentalized by an overlying layer of cells. It has been proposed that these are separated lining cells, but they may be of vascular origin.

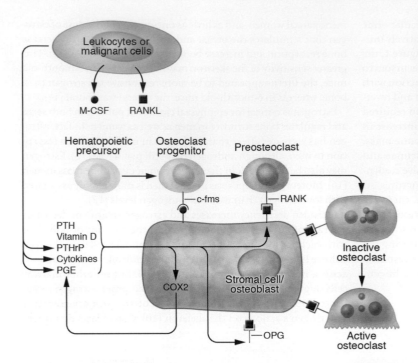

Figure 2
Regulation of osteoclast formation and activity. In physiologic remodeling, activation of bone resorption requires contact between cells of the osteoblast and osteoclast lineages. M-CSF, which may be either membrane bound or secreted, interacts with its receptor, c-fms, to stimulate differentiation and proliferation of hematopoietic progenitors, which then express RANK as preosteoclasts. Osteoclast differentiation and activity are stimulated by RANK/RANKL interaction, and this interaction can be blocked by soluble OPG. Bone-resorbing factors can also stimulate COX2 activity, which may amplify responses to RANKL and OPG by producing prostaglandins. In pathologic conditions, inflammatory and malignant cells can increase osteoclastogenesis by producing soluble or membrane-bound M-CSF and RANKL as well as PTH-related protein (PTHrP), cytokines, and prostaglandins.

ERβ, but the actions of ERβ agonists on bone are less clear. Some studies suggest that the effects of estrogen signaling through ERα and ERβ are in opposition, while other studies suggest that activation of these 2 receptors has similar effects on bone (18, 19).

Single nucleotide polymorphisms (SNPs) of ERα may affect bone fragility. In the largest study to date, 1 of the SNPs for this receptor was associated with a significant reduction in fracture risk, independent of bone mineral density (BMD) (7). Other studies have suggested that SNPs of ERα can affect BMD and rates of bone loss as well as fracture risk in both men and women (20, 21).

An orphan nuclear receptor, estrogen receptor–related receptor α (ERRα), with sequence homology to ERα and ERβ, is also present in bone cells (22). Despite its inability to bind estrogens, this receptor may interact with ERα and ERβ or act directly to alter bone cell function. A regulatory variant of the gene encoding ERRα was recently found to be associated with a significant difference between lumbar spine and femoral neck BMD in premenopausal women (23).

Sex hormone–binding globulin (SHBG), the major binding protein for sex steroids in plasma, may not only alter the bioavailability of estrogen to hormone-responsive tissues but also affect its entry into cells. Epidemiologic studies suggest that SHBG may have an effect on bone loss and fracture risk independent of the effect as a binding protein (24). Local formation of estradiol from aromatase could play an additional role (25).

While estrogen can act on cells of the osteoblastic lineage, its effects on bone may also be dependent on actions on cells of the hematopoietic lineage, including osteoclast precursors, mature osteoclasts, and lymphocytes. Local cytokines and growth factors may mediate these effects. Bone loss after ovariectomy in rodent models can be prevented by inhibiting IL-1 or TNF-α and does not occur in mice deficient in the IL-1 receptor or TNF-α (26, 27). The effects of estrogen on cytokine production may be mediated by T cells (28). A direct effect of estrogen in accelerating osteoclast apoptosis has been attributed to increased TGF-β production (29).

Another possibility is that estrogen exerts its beneficial effects by suppressing ROS (30). In estrogen deficiency, thiol antioxidant defenses may be diminished, and the resultant increase in ROS may induce TNF-α (31). The relevance of these findings for human osteoporosis has yet to be determined.

Calcium, vitamin D, and parathyroid hormone

The concept that osteoporosis is due primarily to calcium deficiency, particularly in the elderly, was initially put forward as a counterproposal to Albright's estrogen deficiency theory. Decreased calcium intake, impaired intestinal absorption of calcium due to aging or disease, as well as vitamin D deficiency can result in secondary hyperparathyroidism. The active hormonal form, 1,25 dihydroxy vitamin D (calcitriol), is not only necessary for optimal intestinal absorption of calcium and phosphorus, but also exerts a tonic inhibitory effect on parathyroid hormone (PTH) synthesis, so that there are dual pathways that can lead to secondary hyperparathyroidism (32). Vitamin D deficiency and secondary hyperparathyroidism can contribute not only to accelerated bone loss and increasing fragility, but also to neuromuscular impairment that can increase the risk of falls (33, 34). Clinical trials involving older individuals at high risk for calcium and vitamin D deficiency indicate that supplementation of both can reverse secondary hyperparathyroidism, decrease bone resorption, increase bone mass, decrease fracture rates, and even decrease the frequency of falling (32). However, in a large recent study, calcium and vitamin D supplementation did not reduce fracture incidence significantly, perhaps because this population was less deficient in vitamin D (35).

Polymorphisms of the VDR have been studied extensively, but the results have been variable. This may be in part because the effect of a given polymorphism in this receptor is dependent on an interaction with the environment, particularly with calcium (36). VDR polymorphisms are also associated with differences in the response to therapy with calcitriol (37). There is also evidence for an effect on fracture risk independent of bone density and bone turnover, which might be due to an alteration in the frequency of falls (38).

Secondary hyperparathyroidism presents when there is relative insufficiency of vitamin D, that is, where the levels of the circulating form — 25-hydroxy vitamin D — fall below 30 ng/ml, suggesting that the target for vitamin D supplementation should be at this level or higher (39). The seasonal decrease in vitamin D level and increase in PTH level during the winter months is associated with an increase in fractures, independent of the increase in rate of falls (40). In addition, increased PTH levels are associated with increased mortality in the frail elderly, independent of bone mass and vitamin D status. The precise mechanisms underlying this relationship have not yet been determined, but the risk of cardiovascular death was increased (41). Polymorphisms of the calcium-sensing receptor, which regulates calcium secretion by suppression of *PTH* translation and PTH secretion, have not yet been associated with any alteration in bone phenotype (42, 43).

Receptor activator of NF-κB, its ligand, and osteoprotegerin

The concept that stimulation of bone resorption requires an interaction between cells of the osteoblastic and osteoclastic lineages was put forward many years ago, but its molecular mechanism was only identified recently (43, 44). Three members of the TNF and TNF receptor superfamily are involved (Figure 2); osteoblasts produce RANKL, a ligand for the receptor activator of NF-κB (RANK) on hematopoietic cells, which activates the differentiation of osteoclasts and maintains their function. Osteoblasts also produce and secrete osteoprotegerin (OPG), a decoy receptor that can block RANKL/RANK interactions. Stimulators of bone resorption have been found to increase RANKL expression in osteoblasts, and some also decrease OPG expression (43). Bone cells appear to express the membrane-bound form of RANKL, and thus, osteoblasts must physically interact with osteoclasts precursors in order to activate RANK. Soluble RANKL can be produced by activated T lymphocytes and is as active as membrane-bound RANKL in binding to RANK (45). Studies in transgenic mice showed that overexpression of OPG produced osteopetrosis, while OPG-knockout mice had a phenotype of severe osteoporosis with a high incidence of fractures (46). Recently a monoclonal antibody against RANKL was shown to produce prolonged inhibition of bone resorption in postmenopausal women (47). It was also shown that RANKL levels were increased on the surface of bone marrow cells from early postmenopausal women who are estrogen deficient (48). However, it has been difficult to demonstrate a role for OPG deficiency in the pathogenesis of osteoporosis, since OPG levels are not consistently altered. OPG levels increase with age, and it is possible that OPG production rises as a homeostatic response to limit the bone loss that occurs with an increase in other bone-resorbing factors (21, 49). Polymorphisms in the *OPG* gene have been associated with osteoporotic fractures and differences in BMD (50). OPG polymorphisms have also been linked to coronary artery disease (51). This is consistent with recent information linking the RANKL/RANK/OPG system to vascular calcification (52).

The RANKL/RANK interaction is critical for both differentiation and maintenance of osteoclast activity and hence represents a final common pathway for any pathogenetic factor in osteoporosis that acts by increasing bone resorption. While it is assumed that cells of the stromal/osteoblastic lineage are the major source of RANKL in physiologic bone remodeling, other cells may act as a source of RANKL in pathologic states; for example, T cell production may play a role in osteoporosis as well as inflammatory bone loss (53).

Recently a second system that might affect the interaction between osteoblasts and osteoclasts has been identified (54). This involves the membrane adapter DNAX-activating protein 12 and the Fc receptor common γ chain. Deletion of these molecules results in severe osteopetrosis in mice. The molecules are involved in signaling through the immunoreceptor tyrosine-based activation motif (ITAM). Cooperation between RANKL and ITAM signaling may be essential for osteoclastogenesis, for which nuclear factor of activated T cells (NFAT) is the master transcription factor.

Genes determining osteoblast differentiation and function

The recent discoveries of signal transduction pathways and transcription factors critical for osteoblast differentiation and function have opened up new approaches to understanding the pathogenesis of osteoporosis. Gene deletion studies have shown that absence of runt-related transcription factor 2 (Runx2) or a downstream factor, osterix, are critical for osteoblast differentiation (55, 56). Interestingly, overexpression of Runx2 leads to a decrease in bone mass (57). A role for polymorphisms of these transcription factors in osteoporosis has not yet been identified.

The recent identification of the critical role for the Wnt signaling pathway in regulating osteoblast function is of particular interest, since it has been shown to play an important role in determining bone mass and strength (58–61) (Figure 3). LDL receptor–related protein 5 (LRP5) interacts with the frizzled receptor to transduce signaling by Wnt ligands. A mutation of LRP5 that leads to constitutive activation can result in an increase in bone density (58, 59). The phenotype of families with LRP5 activating mutations varies considerably, although all show a striking absence of fractures. Some have normal skeletal architecture, while others show abnormalities due to skeletal overgrowth (60). Deletion of *LRP5* results in a severe osteoporotic syndrome

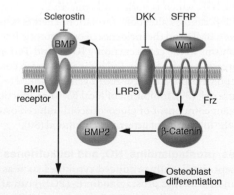

Figure 3

Interaction of the Wnt, BMP, and sclerostin pathways. Differentiation of osteoblasts during both development and remodeling is dependent on the activity of both the Wnt and BMP pathways. Wnt signaling requires the interaction of the LRP5 and frizzled receptors (Frz) and can be inhibited by Dickkopf (DKK; an inhibitor of LRP5) and secreted frizzled-related protein (SFRP). Antagonists such as sclerostin can block both BMP and Wnt signaling. The mediator of the canonical Wnt pathway, β-catenin, can synergize with BMP2 to enhance osteoblast differentiation and bone formation. Consistent with these interactions are the findings that high bone mass can result both from activating mutations of the Wnt pathway and deletion of *SOST*, the gene encoding sclerostin.

associated with abnormal eye development (61). Polymorphisms of LRP5 have been associated with differences in bone mass and fractures (62, 63). Mutations in LRP5 have been identified in a few patients with idiopathic juvenile osteoporosis (64).

Animal and in vitro studies indicate that the Wnt signaling pathway is critical for osteoblast differentiation and function. Studies in mice suggest that the increase in bone mass in animals with activating mutations of LRP5 is due to an increased response to mechanical loading (65). The fact that fluid shear stress activates β-catenin signaling further supports the concept that Wnt signaling is critical in the response to mechanical loading (66). However, Wnt signaling is also critical in bone development and can affect peak bone mass (67). The inhibition of skeletal growth by glucocorticoids may be mediated by effects on Wnt signaling (68).

The precise mechanisms whereby Wnt signaling alters osteoblast function are not fully understood, but there is evidence that the canonical β-catenin pathway is involved and that there is an interaction with bone morphogenetic protein 2 (BMP2) (69) (Figure 3). There are a number of inhibitors that have been shown to interact with BMP2 and with the Wnt signaling pathway. One of these, sclerostin, the product of the *SOST* gene, has been shown to inhibit both BMP2 and Wnt signaling (70, 71). Inactivating mutations of this gene can produce the high-bone-mass disorder Van Buchem disease or sclerosteosis (72, 73). Another potential inhibitory factor is the production of secreted frizzled-related protein (SFRP) by osteoblasts (74).

Local and systemic growth factors

Remodeling imbalance, characterized by an impaired bone formation response to increased activation of bone remodeling, is an essential component of the pathogenesis of osteoporosis (8, 75). This may be due, in part, to an age-related decrease in the capacity of osteoblasts to replicate and differentiate. However, it seems likely that specific defects in the production or activity of local and systemic growth factors will also contribute to impaired bone formation. BMPs as well as other members of the TNF family have been implicated. IGF, which is both a systemic and local regulator, as well as TGF-β, can also alter bone formation. There is some association between BMD and the incidence of osteoporotic fractures and polymorphisms in the genes encoding IGF-1 and TGF-β (76–78), but the largest study to date, in Icelandic and Danish cohorts, suggests that polymorphisms of the *BMP2* gene are linked to low BMD and fracture risk (79). Inhibition of local IGF-1 production may be an important component of glucocorticoid-induced osteoporosis as well as the inhibition of growth in childhood (80).

Cytokines, prostaglandins, NO, and leukotrienes

The concept that locally produced cytokines such as IL-1 and prostaglandins such as prostaglandin E_2 (PGE_2) can affect bone is more than 30 years old (81, 82). Subsequently, many cytokines were found to either stimulate or inhibit bone resorption and formation (83). Prostaglandins have both stimulatory and inhibitory actions; however, the predominant effect of PGE_2, which is the major prostaglandin produced by bone cells, is to stimulate both resorption and formation (84). The possibility that these factors might also be involved in the pathogenesis of osteoporosis is based largely on animal studies of bone loss after ovariectomy (26, 27, 53, 85, 86); however, there is evidence that polymorphisms of IL-1, IL-6, TNF-α, and their receptors can influence bone mass in humans (87–89).

Prostaglandins, particularly PGE_2, are produced by bone cells largely through the action of inducible cyclooxygenase 2 (COX2). COX2 is induced by most of the factors that stimulate bone resorption and thus may enhance the response to these agents (84). Treatment with COX inhibitors blunts the response to impact loading and fluid shear stress, indicating that prostaglandins play an important role in the response mechanical forces, and this may be enhanced by estrogen (90, 91). In epidemiologic studies, small increases in BMD and decreases in fracture risk have been reported in individuals using NSAIDS (92, 93).

NO is produced by bone cells and is a cofactor for the anabolic response to mechanical loading (91, 94). However, unlike prostaglandins, NO may inhibit bone resorption, perhaps by increasing OPG production (95). This effect may be responsible for the increase in BMD that has been demonstrated in patients treated with isosorbide mononitrate and other activators of the NO pathway (96).

Leukotrienes, the products of lipoxygenase, can affect bone by stimulating resorption and inhibiting formation (97). Recently, arachidonate 15-lipoxygenase (encoded by *Alox15*), was identified as a negative regulator of bone density in mice (98), and polymorphisms in the human gene, *ALOX15*, were found to be associated with differences in peak BMD in postmenopausal women (99).

Collagen abnormalities

A polymorphism of the first intron of the gene coding for the type I collagen α1 chain and increased levels of homocysteine can influence fracture risk independent of BMD (100–102). This may be due to differences in helix formation or cross-linking of collagen, challenging the concept that mineral and matrix composition are normal in osteoporosis and that only structural abnormalities account for skeletal fragility.

Leptin and neural pathways

Leptin deficiency or resistance is associated with high BMD in mice despite the fact that gonadal function is diminished (103). This has been attributed to a central effect on adrenergic signaling. Increased β-adrenergic activity can decrease bone mass, but other neural pathways may be involved (104, 105). Some, but not all, epidemiologic studies suggest that β-adrenergic blockers can decrease fracture risk and increase BMD (106, 107). Another neural pathway has recently been implicated by the finding that mice in which the cannabinoid type 1 receptor is inactivated, as well as mice treated with antagonist of this receptor, are protected from ovariectomy-induced bone loss (108).

Implication for the diagnosis and treatment of osteoporosis

Prior to the introduction of BMD measurements, the diagnosis of osteoporosis was only made when fragility fractures occurred in an appropriate clinical setting, largely in postmenopausal women and older men. Today we use BMD to diagnose osteoporosis and osteopenia before fractures occur, as well as to confirm the diagnosis in patients with fragility fractures (109). Our goal is to identify those individuals most likely to have fractures in the future and to treat them appropriately so that their fracture risk will be reduced.

While low BMD is an important risk factor, many other factors have been identified in epidemiologic studies, including age, low body weight, and smoking. Many drugs and diseases that can aggravate bone loss and fragility are also important in predicting future fracture risk, such as glucocorticoid or thyroid hormone excess and chronic inflammatory disorders (109). Another key predictor is the rate of bone remodeling. Increased rates of osteoclastic bone resorption, measured by the level of collagen breakdown products, as well as increased bone formation, measured by bone-specific alkaline phosphatase, osteocalcin, or procollagen peptide levels, are associated with an increase in risk for bone loss and fragility fractures. Techniques that can analyze the microarchitecture of bone are being developed that could further improve diagnosis (110). In the future, it may be possible to identify genetic profiles that predict greater or lesser risk of fractures.

At present the therapy of osteoporosis is directed at the major pathogenetic mechanisms outlined here, with a strong emphasis on prevention of bone loss and fractures (111). While the first major pathogenetic mechanism, failure to achieve optimal peak bone mass, is determined largely by our genes, nutrition and lifestyle can have an effect during growth and development. Moreover, adequate calcium and vitamin D intake and appropriate physical activity may not only increase peak bone mass but also slow bone loss and reduce fracture risk throughout life.

Pharmacotherapy for osteoporosis has been focused mainly on interventions that could reverse the second pathogenetic mechanism, excessive bone resorption (112). In the past, estrogen replacement was the most widely used therapy. Ironically, the first study that truly established the antifracture efficacy of estrogen has also resulted in its being largely abandoned as the primary therapy for osteoporosis (113). Data from the Women's Health Initiative indicated that the estrogen, particularly when combined with progestins, increased the risk of breast cancer and cardiovascular disease, which outweighed the benefits to the skeleton. The fact that much lower doses of estrogen can prevent bone loss in postmenopausal women (13) suggests that the cost/benefit ratio might be reversed with such therapy, but appropriate long-term studies have not yet been carried out. One alternative is the use of selective estrogen receptor modulators. One of these, raloxifene, has been shown to decrease the risk of vertebral fracture and does not appear to increase the risk of breast cancer (114).

The discovery that calcitonin is a selective inhibitor of osteoclastic activity more than 40 years ago led to the concept that calcitonin deficiency might be the cause of osteoporosis and that its administration might be the cure. However, calcitonin deficiency was not found in osteoporotic patients, and calcitonin therapy has been less effective than other antiresorptive agents, possibly because osteoclasts can escape calcitonin inhibition (112, 114).

The clinical trials showing that bisphosphonates could provide effective and relatively safe antiresorptive therapy have had an enormous impact on the management of osteoporosis. The currently FDA-approved bisphosphonates — alendronate, risedronate, and ibandronate — have been shown to reduce the incidence of fractures, including both vertebral and nonvertebral fractures, particularly hip fractures (112, 114). These agents bind to the bone surface and then are taken up by osteoclasts, leading to their inactivation and programmed cell death. Potent bisphosphonates have now been in use for more than 10 years, and there is some concern that prolonged use may cause excessive inhibition of bone remodeling and slow the repair of fractures and of microdamage (115).

Effective treatment for the third pathogenetic mechanism, impaired bone formation, has only become available recently. Although it was shown more than 60 years ago that intermittent administration of low doses of PTH could increase bone mass in animals, an effective clinical approach was only developed and approved by the FDA in 2002. Daily subcutaneous injections of the synthetic N-terminal portion of PTH, teriparatide, can increase bone mass and strength by markedly increasing bone formation, while only modestly increasing bone resorption (116). This therapy is as effective as bisphosphonates in fracture reduction. Extensive studies on the mechanism of action of PTH are underway and should help us in identifying the critical pathways that regulate bone formation and are abnormal in osteoporotic patients. Another putative anabolic agent, strontium ranelate, has recently been approved for treatment of osteoporosis in several countries outside the US (117). Its mechanism of action is not well understood.

New antiresorptive approaches are being investigated, including selective inhibitors of osteoclastic hydrogen ion transport and cathepsin K as well as antagonists to the $\alpha_v\beta_3$ integrins that are necessary for osteoclast adherence and motility (118). New anabolic approaches are likely to emerge from further study of the transcriptional regulation of osteoblasts.

While improvements in diagnosis and therapy are important, it is equally true that we have sufficient tools today to assess fracture risk and prevent or treat osteoporosis to reduce that risk. Thus, there are 2 challenges: (a) to develop better diagnostic tools and treatment; and (b) to apply our current knowledge more broadly in the community. The latter is far from being met. Moreover, it is critical that we meet both these challenges in the coming decades if we are to deal effectively with the increase in osteoporosis and fractures that is projected as our population ages (111).

Address correspondence to: Lawrence G. Raisz, University of Connecticut Health Center, Musculoskeletal Institute, 263 Farmington Avenue, Farmington, Connecticut 06032, USA. Phone: (860) 679-2129; Fax: (860) 679-1258; E-mail: raisz@nso.uchc.edu.

1. Cooper, A., and Cooper, B.B. 1822. A treatise on dislocations, and on fractures of the joints. Churchill. London, United Kingdom. 425 pp.
2. Schapira, D., and Schapira, C. 1992. Osteoporosis: the evolution of a scientific term. *Osteoporos. Int.* **2**:164–167.
3. Albright, F., Bloomberg, E., and Smith, P.H. 1940. Postmenopausal osteoporosis. *Trans. Assoc. Am. Physicians.* **55**:298–305.
4. Riggs, B.L., et al. 1982. Changes in bone mineral density of the proximal femur and spine with aging. Differences between the postmenopausal and senile osteoporosis syndromes. *J. Clin. Invest.* **70**:716–723.
5. Liu, Y.Z., Liu, Y.J., Recker, R.R., and Deng, H.W. 2003. Molecular studies of identification of genes for osteoporosis: the 2002 update. *J. Endocrinol.* **177**:147–196.
6. Baldock, P.A., and Eisman, J.A. 2004. Genetic determinants of bone mass. *Curr. Opin. Rheumatol.* **16**:450–456.
7. Ioannidis, J.P., et al. 2004. Differential genetic effects of ESR1 gene polymorphisms on osteoporosis outcomes. *JAMA.* **292**:2105–2114.
8. Parfitt, A.M., Villanueva, A.R., Foldes, J., and Rao, D.S. 1995. Relations between histologic indices of bone formation: implications for the pathogenesis of spinal osteoporosis. *J. Bone Miner. Res.* **10**:466–473.
9. Ebeling, P.R., et al. 1996. Bone turnover markers and bone density across the menopausal transition. *J. Clin. Endocrinol. Metab.* **81**:3366–3371.
10. Lee, K., Jessop, H., Suswillo, R., Zaman, G., and Lanyon, L. 2003. Endocrinology: bone adaptation requires oestrogen receptor-alpha. *Nature.* **424**:389.
11. Prestwood, K.M., et al. 1994. The short-term effects of conjugated estrogen on bone turnover in older women. *J. Clin. Endocrinol. Metab.* **79**:366–371.
12. Prestwood, K.M., Kenny, A.M., Unson, C., and Kulldorff, M. 2000. The effect of low dose micronized 17ss-estradiol on bone turnover, sex hormone levels, and side effects in older women: a randomized, double blind, placebo-controlled study. *J. Clin. Endocrinol. Metab.* **85**:4462–4469.
13. Prestwood, K.M., Kenny, A.M., Kleppinger, A., and Kulldorff, M. 2003. Ultralow-dose micronized 17beta-estradiol and bone density and bone metabolism in older women: a randomized controlled trial. *JAMA.* **290**:1042–1048.
14. Modder, U.I., et al. 2004. Dose-response of estrogen

on bone versus the uterus in ovariectomized mice. *Eur. J. Endocrinol.* **151**:503–510.

15. Falahati-Nini, A., et al. 2000. Relative contributions of testosterone and estrogen in regulating bone resorption and formation in normal elderly men. *J. Clin. Invest.* **106**:1553–1560.

16. Khosla, S., Melton, L.J., 3rd, Atkinson, E.J., and O'Fallon, W.M. 2001. Relationship of serum sex steroid levels to longitudinal changes in bone density in young versus elderly men. *J. Clin. Endocrinol. Metab.* **86**:3555–3561.

17. Van Pottelbergh, I., Goemaere, S., Zmierczak, H., and Kaufman, J.M. 2004. Perturbed sex steroid status in men with idiopathic osteoporosis and their sons. *J. Clin. Endocrinol. Metab.* **89**:4949–4953.

18. Windahl, S.H., et al. 2001. Female estrogen receptor beta-/- mice are partially protected against age-related trabecular bone loss. *J. Bone Miner. Res.* **16**:1388–1398.

19. Sims, N.A., et al. 2002. Deletion of estrogen receptors reveals a regulatory role for estrogen receptors-beta in bone remodeling in females but not in males. *Bone.* **30**:18–25.

20. Albagha, O.M., et al. 2005. Association of oestrogen receptor alpha gene polymorphisms with postmenopausal bone loss, bone mass, and quantitative ultrasound properties of bone. *J. Med. Genet.* **42**:240–246.

21. Khosla, S., et al. 2004. Relationship of estrogen receptor genotypes to bone mineral density and to rates of bone loss in men. *J. Clin. Endocrinol. Metab.* **89**:1808–1816.

22. Bonnelye, E., and Aubin, J.E. 2005. Estrogen receptor-related receptor alpha: a mediator of estrogen response in bone. *J. Clin. Endocrinol. Metab.* **90**:3115–3121.

23. Laflamme, N., et al. 2005. A frequent regulatory variant of the estrogen-related receptor alpha gene associated with BMD in French-Canadian premenopausal women. *J. Bone Miner. Res.* **20**:938–944.

24. Goderie-Plomp, H.W., et al. 2004. Endogenous sex hormones, sex hormone-binding globulin, and the risk of incident vertebral fractures in elderly men and women: the Rotterdam Study. *J. Clin. Endocrinol. Metab.* **89**:3261–3269.

25. Van Pottelbergh, I., Goemaere, S., and Kaufman, J.M. 2003. Bioavailable estradiol and an aromatase gene polymorphism are determinants of bone mineral density changes in men over 70 years of age. *J. Clin. Endocrinol. Metab.* **88**:3075–3081.

26. Kimble, R.B., et al. 1995. Simultaneous block of interleukin-1 and tumor necrosis factor is required to completely prevent bone loss in the early postovariectomy period. *Endocrinology.* **136**:3054–3061.

27. Lorenzo, J.A., et al. 1998. Mice lacking the type I interleukin-1 receptor do not lose bone mass after ovariectomy. *Endocrinology.* **139**:3022–3025.

28. Gao, Y., et al. 2004. Estrogen prevents bone loss through transforming growth factor beta signaling in T cells. *Proc. Natl. Acad. Sci. U. S. A.* **101**:16618–16623.

29. Hughes, D.E., et al. 1996. Estrogen promotes apoptosis of murine osteoclasts mediated by TGF-beta. *Nat. Med.* **2**:1132–1136.

30. Lean, J.M., et al. 2003. A crucial role for thiol antioxidants in estrogen-deficiency bone loss. *J. Clin. Invest.* **112**:915–923. doi:10.1172/JCI200318859.

31. Lean, J.M., Jagger, C.J., Kirstein, B., Fuller, K., and Chambers, T.J. 2005. Hydrogen peroxide is essential for estrogen-deficiency bone loss and osteoclast formation. *Endocrinology.* **146**:728–735.

32. Lips, P. 2001. Vitamin D deficiency and secondary hyperparathyroidism in the elderly: consequences for bone loss and fractures and therapeutic implications. *Endocr. Rev.* **22**:477–501.

33. Bischoff-Ferrari, H.A., et al. 2004. Effect of vitamin D on falls: a meta-analysis. *JAMA.* **291**:1999–2006.

34. Sambrook, P.N., et al. 2004. Serum parathyroid hormone predicts time to fall independent of vita-

min D status in a frail elderly population. *J. Clin. Endocrinol. Metab.* **89**:1572–1576.

35. Grant, A.M., et al. 2005. Oral vitamin D3 and calcium for secondary prevention of low-trauma fractures in elderly people (Randomised Evaluation of Calcium Or vitamin D, RECORD): a randomised placebo-controlled trial. *Lancet.* **365**:1621–1628.

36. Ferrari, S.L., Rizzoli, R., Slosman, D.O., and Bonjour, J.P. 1998. Do dietary calcium and age explain the controversy surrounding the relationship between bone mineral density and vitamin D receptor gene polymorphisms? *J. Bone Miner. Res.* **13**:363–370.

37. Morrison, N.A., et al. 2005. Vitamin D receptor genotypes influence the success of calcitriol therapy for recurrent vertebral fracture in osteoporosis. *Pharmacogenet. Genomics.* **15**:127–135.

38. Garnero, P., Munoz, F., Borel, O., Sornay-Rendu, E., and Delmas, P.D. 2005. Vitamin D receptor gene polymorphisms are associated with the risk of fractures in postmenopausal women, independently of bone mineral density. The OFELY study. *J. Clin. Endocrinol. Metab.* **90**:4829–4835.

39. Lips, P. 2004. Which circulating level of 25-hydroxyvitamin D is appropriate? *J. Steroid Biochem. Mol. Biol.* **89–90**:611–614.

40. Pasco, J.A., et al. 2004. Seasonal periodicity of serum vitamin D and parathyroid hormone, bone resorption, and fractures: the Geelong Osteoporosis Study. *J. Bone Miner. Res.* **19**:752–758.

41. Sambrook, P.N., et al. 2004. Serum parathyroid hormone is associated with increased mortality independent of 25-hydroxy vitamin d status, bone mass, and renal function in the frail and very old: a cohort study. *J. Clin. Endocrinol. Metab.* **89**:5477–5481.

42. Bollerslev, J., et al. 2004. Calcium-sensing receptor gene polymorphism A986S does not predict serum calcium level, bone mineral density, calcaneal ultrasound indices, or fracture rate in a large cohort of elderly women. *Calcif. Tissue Int.* **74**:12–17.

43. Suda, T., et al. 1999. Modulation of osteoclast differentiation and function by the new members of the tumor necrosis factor receptor and ligand families. *Endocr. Rev.* **20**:345–357.

44. Rodan, G.A., and Martin, T.J. 1982. Role of osteoblasts in hormonal control of bone resorption — a hypothesis [letter]. *Calcif. Tissue Int.* **34**:311.

45. Kanamaru, F., et al. 2004. Expression of membrane-bound and soluble receptor activator of NF-kappaB ligand (RANKL) in human T cells. *Immunol. Lett.* **94**:239–246.

46. Bucay, N., et al. 1998. Osteoprotegerin-deficient mice develop early onset osteoporosis and arterial calcification. *Genes Dev.* **12**:1260–1268.

47. Bekker, P.J., et al. 2004. A single-dose placebo-controlled study of AMG 162, a fully human monoclonal antibody to RANKL, in postmenopausal women. *J. Bone Miner. Res.* **19**:1059–1066.

48. Eghbali-Fatourechi, G., et al. 2003. Role of RANK ligand in mediating increased bone resorption in early postmenopausal women. *J. Clin. Invest.* **111**:1221–1230. doi:10.1172/JCI200317215.

49. Khosla, S., et al. 2002. Correlates of osteoprotegerin levels in women and men. *Osteoporos. Int.* **13**:394–399.

50. Langdahl, B.L., Carstens, M., Stenkjaer, L., and Eriksen, E.F. 2002. Polymorphisms in the osteoprotegerin gene are associated with osteoporotic fractures. *J. Bone Miner. Res.* **17**:1245–1255.

51. Soufi, M., et al. 2004. Osteoprotegerin gene polymorphisms in men with coronary artery disease. *J. Clin. Endocrinol. Metab.* **89**:3764–3768.

52. Collin-Osdoby, P. 2004. Regulation of vascular calcification by osteoclast regulatory factors RANKL and osteoprotegerin. *Circ. Res.* **95**:1046–1057.

53. Weitzmann, M.N., et al. 2001. T cell activation induces human osteoclast formation via receptor activator of nuclear factor kappaB ligand-dependent and -independent mechanisms. *J. Bone Miner.*

Res. **16**:328–337.

54. Takayanagi, H. 2005. Mechanistic insight into osteoclast differentiation in osteoimmunology. *J. Mol. Med.* **83**:170–179.

55. Ducy, P., Zhang, R., Geoffroy, V., Ridall, A.L., and Karsenty, G. 1997. Osf2/Cbfa1: a transcriptional activator of osteoblast differentiation. *Cell.* **89**:747–754.

56. Nakashima, K., et al. 2002. The novel zinc finger-containing transcription factor osterix is required for osteoblast differentiation and bone formation. *Cell.* **108**:17–29.

57. Geoffroy, V., Kneissel, M., Fournier, B., Boyde, A., and Matthias, P. 2002. High bone resorption in adult aging transgenic mice overexpressing cbfa1/runx2 in cells of the osteoblastic lineage. *Mol. Cell. Biol.* **22**:6222–6233.

58. Little, R.D., et al. 2002. A mutation in the LDL receptor-related protein 5 gene results in the autosomal dominant high-bone-mass trait. *Am. J. Hum. Genet.* **70**:11–19.

59. Boyden, L.M., et al. 2002. High bone density due to a mutation in LDL-receptor-related protein 5. *N. Engl. J. Med.* **346**:1513–1521.

60. Van Wesenbeeck, L., et al. 2003. Six novel missense mutations in the LDL receptor-related protein 5 (LRP5) gene in different conditions with an increased bone density. *Am. J. Hum. Genet.* **72**:763–771.

61. Gong, Y., et al. 2001. LDL receptor-related protein 5 (LRP5) affects bone accrual and eye development. *Cell.* **107**:513–523.

62. Koay, M.A., et al. 2004. Influence of LRP5 polymorphisms on normal variation in BMD. *J. Bone Miner. Res.* **19**:1619–1627.

63. Bollerslev, J., et al. 2005. LRP5 gene polymorphisms predict bone mass and incident fractures in elderly Australian women. *Bone.* **36**:599–606.

64. Hartikka, H., et al. 2005. Heterozygous mutations in the LDL receptor-related protein 5 (LRP5) gene are associated with primary osteoporosis in children. *J. Bone Miner. Res.* **20**:783–789.

65. Akhter, M.P., et al. 2004. Bone biomechanical properties in LRP5 mutant mice. *Bone.* **35**:162–169.

66. Norvell, S.M., Alvarez, M., Bidwell, J.P., and Pavalko, F.M. 2004. Fluid shear stress induces beta-catenin signaling in osteoblasts. *Calcif. Tissue Int.* **75**:396–404.

67. Kronenberg, H., and Kobayashi, T. 2004. Transcriptional regulation in development of bone. *Endocrinology.* **146**:1012–1017.

68. Ohnaka, K., Tanabe, M., Kawate, H., Nawata, H., and Takayanagi, R. 2005. Glucocorticoid suppresses the canonical Wnt signal in cultured human osteoblasts. *Biochem. Biophys. Res. Commun.* **329**:177–181.

69. Mbalaviele, G., et al. 2005. beta-Catenin and BMP-2 synergize to promote osteoblast differentiation and new bone formation. *J. Cell. Biochem.* **94**:403–418.

70. Li, X., et al. 2005. Sclerostin binds to LRP5/6 and antagonizes canonical Wnt signaling. *J. Biol. Chem.* **280**:19883–19887.

71. Winkler, D.G., et al. 2004. Sclerostin inhibition of Wnt-3a-induced C3H10T1/2 cell differentiation is indirect and mediated by BMP proteins. *J. Biol. Chem.* **280**:2498–2502.

72. Loots, G.G., et al. 2005. Genomic deletion of a long-range bone enhancer misregulates sclerostin in Van Buchem disease. *Genome Res.* **15**:928–935.

73. Balemans, W., et al. 2001. Increased bone density in sclerosteosis is due to the deficiency of a novel secreted protein (SOST). *Hum. Mol. Genet.* **10**:537–543.

74. Bodine, P.V., et al. 2005. The Wnt antagonist secreted frizzled-related protein-1 controls osteoblast and osteocyte apoptosis. *J. Cell. Biochem.* doi:10.1002/jcb.20599.

75. Eriksen, E.F., et al. 1990. Cancellous bone remodeling in type I (postmenopausal) osteoporosis: quan-

titative assessment of rates of formation, resorption, and bone loss at tissue and cellular levels. *J. Bone Miner. Res.* **5**:311–319.

76. Lau, E.M., et al. 2004. Osteoporosis and transforming growth factor-beta-1 gene polymorphism in Chinese men and women. *J. Bone Miner. Metab.* **22**:148–152.

77. Langdahl, B.L., Carstens, M., Stenkjaer, L., and Eriksen, E.F. 2003. Polymorphisms in the transforming growth factor beta 1 gene and osteoporosis. *Bone.* **32**:297–310.

78. Kim, J.G., Roh, K.R., and Lee, J.Y. 2002. The relationship among serum insulin-like growth factor-I, insulin-like growth factor-I gene polymorphism, and bone mineral density in postmenopausal women in Korea. *Am. J. Obstet. Gynecol.* **186**:345–350.

79. Styrkarsdottir, U., et al. 2003. Linkage of osteoporosis to chromosome 20p12 and association to BMP2. *PLoS Biol.* **1**:E69.

80. Mehls, O., Himmele, R., Homme, M., Kiepe, D., and Klaus, G. 2001. The interaction of glucocorticoids with the growth hormone-insulin-like growth factor axis and its effects on growth plate chondrocytes and bone cells. *J. Pediatr. Endocrinol. Metab.* **14**(Suppl. 6):1475–1482.

81. Klein, D.C., and Raisz, L.G. 1970. Prostaglandins: stimulation of bone resorption in tissue culture. *Endocrinology.* **86**:1436–1440.

82. Raisz, L.G., et al. 1975. Effect of osteoclast activating factor from human leukocytes on bone metabolism. *J. Clin. Invest.* **56**:408–413.

83. Horwitz, M.C., and Lorenzo, J.A. 2002. Local regulators of bone: IL-1, TNF, lymphotoxin, interferon-γ, IL-8, IL-10, IL-4, the LIF/IL-6 family, and additional cytokines. In *Principles of bone biology.* J.P. Bilezikian, L.G. Raisz, and G.A. Rodan, editors. Academic Press. San Diego, California, USA. 961–977.

84. Pilbeam, C.C., Harrison, J.R., and Raisz, L.G. 2002. Prostaglandins and bone metabolism. In *Principles of bone biology.* J.P. Bilezikian, L.G. Raisz, and G.A. Rodan, editors. Academic Press. San Diego, California, USA. 979–994.

85. Kawaguchi, H., et al. 1995. Ovariectomy enhances and estrogen replacement inhibits the activity of bone marrow factors that stimulate prostaglandin production in cultured mouse calvariae. *J. Clin. Invest.* **96**:539–548.

86. Ammann, P., et al. 1997. Transgenic mice expressing soluble tumor necrosis factor-receptor are protected against bone loss caused by estrogen deficiency. *J. Clin. Invest.* **99**:1699–1703.

87. Tasker, P.N., Albagha, O.M., Masson, C.B., Reid, D.M., and Ralston, S.H. 2004. Association between TNFRSF1B polymorphisms and bone mineral density, bone loss and fracture. *Osteoporos. Int.*

15:903–908.

88. Chung, H.W., et al. 2003. Association of interleukin-6 promoter variant with bone mineral density in premenopausal women. *J. Hum. Genet.* **48**:243–248.

89. Ferrari, S.L., et al. 2004. Interactions of interleukin-6 promoter polymorphisms with dietary and lifestyle factors and their association with bone mass in men and women from the Framingham Osteoporosis Study. *J. Bone Miner. Res.* **19**:552–559.

90. Forwood, M.R. 1996. Inducible cyclo-oxygenase (COX-2) mediates the induction of bone formation by mechanical loading in vivo. *J. Bone Miner. Res.* **11**:1688–1693.

91. Bakker, A.D., et al. 2004. Additive effects of estrogen and mechanical stress on nitric oxide and prostaglandin E(2) production by bone cells from osteoporotic donors. *Osteoporos. Int.* **16**:983–989.

92. Raisz, L.G. 2001. Potential impact of selective cyclooxygenase-2 inhibitors on bone metabolism in health and disease. *Am. J. Med.* **110**(Suppl. 3A):43S–45S.

93. Carbone, L.D., et al. 2003. Association between bone mineral density and the use of nonsteroidal anti-inflammatory drugs and aspirin: impact of cyclooxygenase selectivity. *J. Bone Miner. Res.* **18**:1795–1802.

94. Chow, J.W., Fox, S.W., Lean, J.M., and Chambers, T.J. 1998. Role of nitric oxide and prostaglandins in mechanically induced bone formation. *J. Bone Miner. Res.* **13**:1039–1044.

95. Wang, F.S., et al. 2004. Nitric oxide donor increases osteoprotegerin production and osteoclastogenesis inhibitory activity in bone marrow stromal cells from ovariectomized rats. *Endocrinology.* **145**:2148–2156.

96. Jamal, S.A., Cummings, S.R., and Hawker, G.A. 2004. Isosorbide mononitrate increases bone formation and decreases bone resorption in postmenopausal women: a randomized trial. *J. Bone Miner. Res.* **19**:1512–1517.

97. Traianedes, K., Dallas, M.R., Garrett, I.R., Mundy, G.R., and Bonewald, L.F. 1998. 5-Lipoxygenase metabolites inhibit bone formation in vitro. *Endocrinology.* **139**:3178–3184.

98. Klein, R.F., et al. 2004. Regulation of bone mass in mice by the lipoxygenase gene Alox15. *Science.* **303**:229–232.

99. Urano, T., et al. 2005. Association of a single nucleotide polymorphism in the lipoxygenase ALOX15 5′-flanking region (-5229G/A) with bone mineral density. *J. Bone Miner. Metab.* **23**:226–230.

100. Mann, V., and Ralston, S.H. 2003. Meta-analysis of COL1A1 Sp1 polymorphism in relation to bone mineral density and osteoporotic fracture. *Bone.* **32**:711–717.

101. McLean, R.R., et al. 2004. Homocysteine as a predictive factor for hip fracture in older persons. *N. Engl. J. Med.* **350**:2042–2049.

102. van Meurs, J.B., et al. 2004. Homocysteine levels and the risk of osteoporotic fracture. *N. Engl. J. Med.* **350**:2033–2041.

103. Ducy, P., et al. 2000. Leptin inhibits bone formation through a hypothalamic relay: a central control of bone mass. *Cell.* **100**:197–207.

104. Elefteriou, F., et al. 2005. Leptin regulation of bone resorption by the sympathetic nervous system and CART. *Nature.* **434**:514–520.

105. Takeda, S. 2005. Central control of bone remodeling. *Biochem. Biophys. Res. Commun.* **328**:697–699.

106. Pasco, J.A., et al. 2004. Beta-adrenergic blockers reduce the risk of fracture partly by increasing bone mineral density: Geelong Osteoporosis Study. *J. Bone Miner. Res.* **19**:19–24.

107. Reid, I.R., et al. 2005. beta-Blocker use, BMD, and fractures in the study of osteoporotic fractures. *J. Bone Miner. Res.* **20**:613–618.

108. Idris, A.I., et al. 2005. Regulation of bone mass, bone loss and osteoclast activity by cannabinoid receptors. *Nat. Med.* **11**:774–799.

109. Raisz, L.G. 2005. Clinical practice. Screening for osteoporosis. *N. Engl. J. Med.* **353**:164–171.

110. Benito, M., et al. 2003. Deterioration of trabecular architecture in hypogonadal men. *J. Clin. Endocrinol. Metab.* **88**:1497–1502.

111. Office of the Surgeon General. 2004. Bone health and osteoporosis: a report of the Surgeon General. US Department of Health and Human Services. Rockville, Maryland, USA. 404 pp.

112. Rosen, C.J. 2005. Clinical practice. Postmenopausal osteoporosis. *N. Engl. J. Med.* **353**:595–603.

113. Anderson, G.L., et al. 2004. Effects of conjugated equine estrogen in postmenopausal women with hysterectomy: the Women's Health Initiative randomized controlled trial. *JAMA.* **291**:1701–1712.

114. Cranney, A., et al. 2002. Meta-analyses of therapies for postmenopausal osteoporosis. IX: summary of meta-analyses of therapies for postmenopausal osteoporosis. *Endocr. Rev.* **23**:570–578.

115. Ott, S.M. 2005. Long-term safety of bisphosphonates. *J. Clin. Endocrinol. Metab.* **90**:1897–1899.

116. Neer, R.M., et al. 2001. Effect of parathyroid hormone (1-34) on fractures and bone mineral density in postmenopausal women with osteoporosis. *N. Engl. J. Med.* **344**:1434–1441.

117. Meunier, P.J., et al. 2004. The effects of strontium ranelate on the risk of vertebral fracture in women with postmenopausal osteoporosis. *N. Engl. J. Med.* **350**:459–468.

118. Rodan, G.A., and Martin, T.J. 2000. Therapeutic approaches to bone diseases. *Science.* **289**:1508–1514.

Paget disease of bone

G. David Roodman[1,2] and Jolene J. Windle[3]

[1]Department of Medicine, Division of Hematology-Oncology, University of Pittsburgh Medical Center, Pittsburgh, Pennsylvania, USA.
[2]VA Pittsburgh Healthcare System, Medicine/Hematology-Oncology, Pittsburgh, Pennsylvania, USA. [3]Department of Human Genetics,
Virginia Commonwealth University, Richmond, Virginia, USA.

Paget disease of bone (PD) is characterized by excessive bone resorption in focal areas followed by abundant new bone formation, with eventual replacement of the normal bone marrow by vascular and fibrous tissue. The etiology of PD is not well understood, but one PD-linked gene and several other susceptibility loci have been identified, and paramyxoviral gene products have been detected in pagetic osteoclasts. In this review, the pathophysiology of PD and evidence for both a genetic and a viral etiology for PD will be discussed.

Normal bone remodeling

The normal adult skeleton undergoes constant remodeling, with osteoclasts removing bone and osteoblasts forming new bone at sites of previous bone resorption in a closely coupled fashion. Bone remodeling occurs in discrete areas, termed basic multicellular units, and it is estimated that the entire adult skeleton is remodeled every 2–4 years (1, 2).

Osteoclasts are derived from mononuclear precursor cells in the monocyte-macrophage lineage, which fuse to form multinucleated osteoclasts that are then activated to resorb bone. Both systemic and local factors in the bone microenvironment play critical roles in the regulation of osteoclast formation and activity. In particular, receptor activator of NF-κB ligand (RANKL; also referred to as TRANCE, osteoclast differentiation-inducing factor, or osteoprotegerin [OPG] ligand), a member of the TNF superfamily, is a critical regulator of osteoclast formation (3–5). Most osteotropic factors, including $1,25-(OH)_2D_3$, IL-1, IL-11, and parathyroid hormone (PTH), promote osteoclast formation indirectly by binding to marrow stromal cells and inducing expression of RANKL on their surface (5, 6). RANKL then binds the receptor activator of NF-κB (RANK) receptor on osteoclast precursors, leading to activation of a number of downstream signaling pathways, including the NF-κB, AKT, JNK, p38 MAPK, and ERK pathways. Each of these pathways has been implicated in osteoclast differentiation, function, or survival (6–9) (Figure 1). The importance of the RANKL–NF-κB signaling pathway in osteoclastogenesis has been highlighted by the finding that targeted disruption of multiple genes encoding components of this pathway (RANKL, RANK, TNF receptor–associated factor 6 [TRAF6], NF-κB, and $NFATc_1$, an NF-κB–activated transcription factor) in mice causes profound osteopetrosis (5, 9–17). However, additional RANK-activated signaling pathways are also clearly important regulators of osteoclasts, since both c-src– and c-fos–deficient mice display profound osteopetrosis, resulting from impaired osteoclast function in c-src–knockout mice (18) and from the failure of osteoclasts to form in the c-fos

knockouts (19). In addition, the transcription factor PU.1, which is involved in the process of commitment of hematopoietic stem cells, is critical for normal osteoclast formation, since fetal mice lacking this gene do not form osteoclasts (20).

TNF and IL-1 activate many of the same downstream signaling pathways as does RANKL (Figure 1), and both of these cytokines have been shown to play a role in the regulation of osteoclast differentiation and function (21–27). However, neither cytokine appears to be as central to the process of osteoclastogenesis as RANKL, since disruption of either the TNF or IL-1 receptors in mice results in a minimal bone phenotype, as compared with that of the RANK- or RANKL-knockout mice (28–32). Further, to date, neither IL-1 nor TNF has been implicated in the pathogenesis of PD.

Once bone resorption within a basic multicellular unit is complete, osteoblast precursors are then recruited to the site of previous bone resorption and differentiate to become bone-forming cells. Osteoblasts are derived from mesenchymal stem cells, which can form osteoblasts, osteocytes, or muscle cells (1). The transcription factor RUNX-2, also known as core binding factor–α-1 (CBFA-1), is critical for the differentiation of osteoblasts, since bone does not develop in mice lacking CBFA-1 (33, 34). Osteoblasts ultimately terminally differentiate into osteocytes, which are trapped in the calcified bone matrix. Regulators of osteoblast differentiation include the bone morphogenetic proteins, insulin-like growth factors, TGF-β, fibroblast growth factors, and platelet-derived growth factors (35–38).

Abnormal bone remodeling in Paget disease

Paget disease of bone (PD) is the second most common bone disease after osteoporosis (39). The disease is characterized by focal regions of highly exaggerated bone remodeling, with abnormalities in all phases of the remodeling process. The majority of patients with PD are elderly, with the age at diagnosis usually more than 50 years. It affects both males and females, with a slight predominance in males. Although PD is often asymptomatic, 10–30% of patients experience pain, skeletal deformity, neurologic symptoms, pathologic fractures, or deafness (39). Table 1 lists the common symptoms and findings in Paget patients. Patients may have only one affected bone or have pagetic lesions in multiple bones. However, PD remains highly localized, and patients rarely develop new lesions in previously unaffected bones after diagnosis (40). The most serious complication of PD is development of osteosarcoma in the pagetic bone, although this is relatively rare, occurring in less than 1% of patients (41, 42).

Nonstandard abbreviations used: CBFA-1, core binding factor–α-1; CDV, canine distemper virus; MV, measles virus; MVNP, MV nucleocapsid protein; OPG, osteoprotegerin; PD, Paget disease of bone; PTH, parathyroid hormone; RANK, receptor activator of NF-κB; RANKL, RANK ligand; SH2, Src homology 2; SSPE, subacute sclerosing panencephalitis; TRAF6, TNF receptor–associated factor 6; UBA, ubiquitin-associating.

Conflict of interest: The authors have declared that no conflict of interest exists.

Citation for this article: *J. Clin. Invest.* **115**:200–208 (2005). doi:10.1172/JCI200524281.

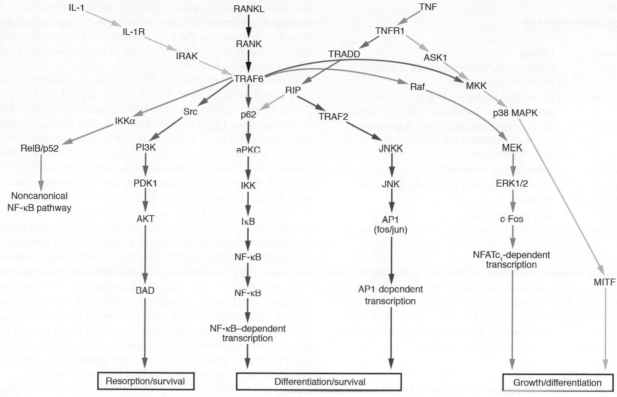

Figure 1

Signaling pathways involved in osteoclast formation and activity. When RANKL binds RANK, multiple signaling pathways can be activated, including NF-κB, AKT, JNK, p38 MAPK, and ERK, resulting in subsequent activation of genes that regulate osteoclast formation, bone resorption, and survival. TRAF6 appears to play a central role in the activation of most of these pathways. AP1, activator protein 1; aPKC, atypical PKC; IκB, inhibitor of κB; ASK1, apoptosis signal–regulating kinase 1; BAD, Bcl-2–associated death promoter; IL-1R, IL-1 receptor; IKKα, IκB kinase α; IRAK, IL-1 receptor–associated kinase; JNKK, JNK kinase; MEK, MAPK/ERK kinase; MITF, microphthalmia transcription factor; MKK, MAPK kinase; NFATc$_1$, nuclear factor of activated T cells cytoplasmic 1; PDK1, phosphoinositide-dependent protein kinase 1; RIP, receptor interacting protein; TNFR1, TNF receptor 1; TRADD, TNF receptor 1–associated death domain.

The initial phase of PD is characterized by excessive bone resorption in a focal region, and radiological examination in the early stages of the disease frequently shows an osteolytic lesion. Subsequently, bone formation is also markedly increased, with increased numbers of osteoblasts that appear hyperactive but normal morphologically. The increased population of osteoblasts rapidly deposits new bone in a chaotic fashion so that the bone formed in pagetic lesions is of poor quality and is disorganized rather than lamellar in character (Figure 2). The poor quality of pagetic bone accounts for the bowing or even fracture of bones affected by PD. As rapid bone formation predominates in the more advanced stages of PD, the lesions become sclerotic, with observed replacement of the bone marrow with vascular and fibrous tissue and thickening of the bone (43).

Blood chemistries of Paget patients usually reflect the increased bone remodeling, with elevated levels of both bone resorption and formation markers (see *Biochemical markers of bone remodeling that are increased in PD*). For example, the level of alkaline phosphatase in the serum, which reflects osteoblast activity, and N-telopeptide of type I collagen in the urine, which is released during bone resorption and reflects osteoclast activity, can both be markedly elevated (up to 10- to 20-fold) in patients with PD (44). Because bone resorption and formation remain coupled in PD, there is a high correlation between the levels of bone resorption and bone formation markers in Paget patients.

Osteoclasts in PD

The osteoclast is the primary cell affected by PD. Osteoclasts in pagetic lesions are increased in both number and size (43), and in cross-section are seen to contain up to 100 nuclei, in contrast to normal osteoclasts, which contain 3–20 nuclei (Figure 3, A and B). A striking feature of pagetic osteoclasts is the characteristic nuclear inclusions, which consist of paracrystalline arrays that are similar to nucleocapsids of paramyxoviruses (Figure 3C) (45). These nuclear inclusions are not present in other bone marrow cells in the pagetic lesion or in nonpagetic bone in patients with PD. Similar nuclear inclusions have been reported in osteoclasts from patients with oxalosis, osteopetrosis, and giant cell tumors of bone (46–48), but this is not a consistent finding in these conditions, as it is in PD.

In addition to the morphologic abnormalities in pagetic osteoclasts, osteoclast precursors are physiologically abnormal. In vitro studies of bone marrow samples obtained from affected bones of Paget patients have identified several unique differences between pagetic and normal osteoclast precursors. This "pagetic phenotype" is characterized by hypersensitivity of osteoclast precursors to several osteoclastogenic factors, including 1,25-(OH)$_2$D$_3$ (49, 50) and RANKL (50, 51). Osteoclasts precursors in bone marrow cultures obtained from pagetic lesions form osteoclasts at concentrations of these factors that are 10- to 100-fold lower than levels required for normal osteoclast formation. In addition, the level

Table 1

Clinical presentation in PD

Symptom	Etiology
Bone pain	Usually results from osteoarthritis in joints adjacent to pagetic bones
Bone deformities	Can result in bowing of a limb or increased skull size due to rapid formation of poor-quality bone
Fracture	Bone in pagetic lesions is weaker than normal bone and can develop characteristic "chalk stick–like" fractures
Hearing loss	Temporal bone involvement
Nerve root compression	Impingement of nerve root by increased bone formation
Headache	Skull affected by PD

of TAF$_{II}$-17, a component of the TAF$_{II}$D transcription factor complex that binds the vitamin D receptor, is increased in osteoclast precursors from affected bones of Paget patients as compared with normal osteoclast precursors (52). The increased level of TAF$_{II}$-17 appears to be responsible, in part, for the hypersensitivity of pagetic osteoclast precursors to 1,25-(OH)$_2$D$_3$ (52).

Treatment of PD

Since the osteoclast is the primary cell affected by PD, treatment has been directed at inhibiting osteoclast formation or osteoclastic bone resorption or inducing osteoclast apoptosis. Table 2 lists agents that are currently used for the treatment of PD. Calcitonin was initially used to treat patients with PD because it inhibits osteoclastic bone resorption and osteoclast formation (53). Calcitonin can induce remission in patients with PD, but more than 50% of patients treated with salmon calcitonin for more than 6 months develop calcitonin antibodies, and 10–20% become resistant to calcitonin. Bisphosphonates, which block osteoclast formation and induce osteoclast apoptosis (54), have supplanted calcitonin as the treatment of choice for PD. First-generation bisphosphonates, such as etidronate, can induce partial or complete remissions in PD patients, and with the development of more potent bisphosphonates, patients can experience prolonged remissions, lasting months to years. In patients with more extensive PD involving many bones, intravenous bisphosphonates such as pamidronate or zoledronate may be used. However, neither calcitonin nor bisphosphonates cure PD; they only control the disease process. Treatment of PD is indicated to control bone pain, prevent fractures, minimize bleeding prior to surgery on a pagetic bone, and decrease local progression in weight-bearing bones or the skull.

Etiology of PD

Genetics of PD. The cause of PD is currently an area of intensive investigation, and both genetic and nongenetic factors have been implicated in the pathogenesis of this disease. Genetic factors are clearly an important component of the etiology of PD, since 15–40% of affected patients have a first-degree relative with PD (55), and numerous studies have described extended families with PD exhibiting an autosomal dominant mode of inheritance (56–58). Although familial PD was initially thought to have a very high penetrance, recent studies have suggested that the penetrance is highly variable (59). Ethnic differences in the incidence of PD have

been noted, and these persist with emigration to other locales. For example, PD is common in persons of Anglo-Saxon origin, but the prevalence is low in the Far East and does not change when populations from this region move to areas of higher prevalence, such as the United Kingdom (60).

Several susceptibility loci for PD have been recently identified, including 2q36, 5q31, 5q35, 10p13, 18q21–22, and 18q23 (61–66) (Table 3). Mutations in the *TNFRSF11A* gene (encoding RANK) on chromosome 18q21–22 have been linked to familial expansile osteolysis, a rare bone disease that shares many clinical features with but is distinct from PD (67). In addition, a *TNFRSF11A* mutation was identified in an Asian family with early-onset PD (68). However, RANK mutations have not been observed in patients with the more typical form of PD, which occurs predominately in elderly patents and rarely occurs in Asians. In 2002 Laurin et al. reported a point mutation (P392L) in *SQSTM1*, which maps to chromosome 5q35, in two French Canadian Paget families and several unrelated patients (69). *SQSTM1* encodes sequestosome 1, also known as p62, which is a ubiquitin-binding protein that is involved in the IL-1, TNF, and RANKL signaling pathways (Figure 1). Subsequently, other groups have identified additional mutations in p62 in both familial and nonfamilial PD, including both amino acid substitutions and mutations that result in total deletion of the ubiquitin-binding domain (70–72). *SQSTM1* is the gene most frequently linked to PD, and mutations of this gene have been detected in up to 30% of familial Paget cases studied.

p62 was first identified as a protein that binds to the Src homology 2 (SH2) domain of p56^{lck} in a phosphotyrosine-independent manner (73). It was named sequestosome 1 because it forms a cytoplasmic complex with ubiquitinated proteins (74, 75). The *SQSTM1* gene is highly conserved, especially in the COOH-terminal region of the protein. This region of the protein interacts noncovalently with polyubiquitin chains and shares structural homology with other proteins containing ubiquitin-associating (UBA) domains. Interestingly, all the PD-associated mutations in p62 identified to date are located in this region (Figure 4). Patients with truncation mutations in p62 exhibit a more severe Paget phenotype than patients with any of the point mutations (72, 76). However, it is not yet clear whether alterations in ubiquitin binding are directly related to the

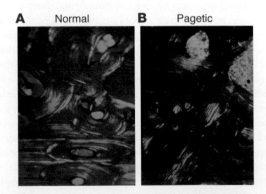

Figure 2

Normal and pagetic bone. Normal (**A**) and pagetic (**B**) bone are shown under polarized light. Pagetic bone is poorly organized and very chaotic in structure and forms "a mosaic pattern." In contrast, normal bone is highly organized with a lamellar structure. Figure reproduced with permission from The Paget's Foundation for Paget's Disease of Bone Related Disorders.

**Biochemical markers of bone remodeling
that are increased in PD**

Markers of bone resorption
    Urinary hydroxyproline
    Serum N-telopeptide of type I collagen
    Serum C-telopeptide of type I collagen
    Serum deoxypyridinoline cross-links of type I collagen
Markers of bone formation
    Serum alkaline phosphatase
    Serum bone-specific alkaline phosphatase
    Osteocalcin
    Serum N-terminal propeptide of type I collagen

mechanism of pathogenesis. In addition to the UBA domain, the p62 protein is characterized by several other motifs that mediate protein-protein interactions that are relevant to signaling pathways involved in osteoclastogenesis, including an atypical PKC–interacting domain, a Zn-finger domain that mediates binding to regulated intramembrane proteolysis, and a TRAF6-interacting domain (77).

As shown in Figure 1, p62 plays a critical role in multiple signaling pathways that regulate osteoclastogenesis. Duran and coworkers reported that the P392L mutation in p62 that is linked to PD results in enhanced NF-κB signaling (78), although the mechanism responsible for this remains to be determined. The PD-associated p62 mutations could potentially affect a number of cell processes, including signaling, ubiquitin-dependent proteolysis, and others.

We have recently shown in preliminary studies that transfection of normal human osteoclast precursors with a P392L mutant p62 construct enhanced the sensitivity of normal human osteoclast precursors to RANKL and increased osteoclast formation (79). Interestingly, we observed neither hypersensitivity of the precursors to 1,25-(OH)$_2$D$_3$ nor an increased number of nuclei per osteoclast in the osteoclasts that formed in vivo, both of which are characteristics of pagetic osteoclasts. We have also targeted the P392L mutant p62 gene to cells in the osteoclast lineage of transgenic mice using the tartrate resistant acid phosphatase (TRAP) promoter, which directs expression to both osteoclasts and osteoclast precursors. Initial characterization of these mice showed that they have increased osteoclast numbers and are osteopenic but do not develop the increased osteoblast activity that is characteristic of pagetic lesions. These preliminary in vitro and in vivo studies suggest that the P392L mutation in p62 enhances osteoclast formation, possibly through increased RANK signaling. The increased osteoclast activity caused by mutations in the p62 gene may explain the increased bone turnover that has been observed in some Paget patients in bones not affected by PD (80).

Studies of families with PD linked to mutations in the p62 gene also suggest that these mutations may not completely account for the pathogenesis of PD. The severity of disease in family members carrying the same mutation can vary widely, and up to 20% of individuals who harbor p62 mutations and are older than 55 years do not have PD (81). These data suggest that additional factors may be affected by the pathogenesis of PD.

A number of additional candidate genes have been evaluated to determine whether they might be linked to PD, including genes known to be involved in osteoclast biology or genes whose deletion results in mice that display an osteoclast phenotype. For example, the gene encoding OPG, a decoy receptor for RANKL, is on chromosome 18q24.2. OPG decreases osteoclast formation by binding to RANKL and interfering with its binding to the RANK receptor (3). Mutations in OPG have been reported in patients with idiopathic hyperphosphatasia, a rare congenital disorder that occurs in childhood and is characterized by deafness and bone lesions that affect the entire skeleton (82), but OPG mutations have not been reported in adults with PD. However, it is likely that additional genes linked to PD will be identified, and it is reasonable to hypothesize that they may be involved in the regulation of osteoclast formation, function, or lifespan, since increased osteoclast activity is central to PD. These genes may be components of the RANK or other signaling pathways that control osteoclast formation or may result in increased expression of transcription factors such as c-fos or NFATc$_1$ that are critical for osteoclastogenesis.

Potential viral etiology of PD

Several observations suggest that environmental factors may also contribute to the pathogenesis of PD. The variable penetrance of PD within families with a genetic predisposition to PD, the observation that the disease remains highly localized to a particular bone or bones rather than affecting the entire skeleton, and the fact that the incidence and severity of the disease has been changing over the last 25 years (83, 84) all support the hypothesis that additional, nongenetic factors are involved in the development of PD. PD affected approximately 2–8% of the population in the United Kingdom and New Zealand 20 years ago, but recent studies of the prevalence of PD in subjects of European origin in 2 New Zealand cities found that the prevalence of PD was about half of what had been estimated 25 years ago (84). Similarly, Van Staa and coworkers recently conducted a radiologic survey in 10 British cities and found a decrease in the incidence of PD compared with the findings of the original studies performed some 20 years earlier (83). These reports suggest that an additional, nongenetic factor(s)

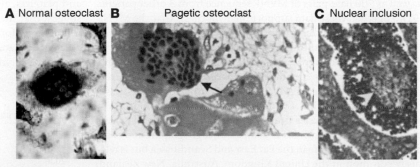

Figure 3
Osteoclasts in normal bone and in Paget's disease. (**A**) Normal osteoclasts are large multinucleated cells that contain between 3 and 20 nuclei per cell. (**B**) In contrast, pagetic osteoclasts are markedly increased in number and size and can contain up to 100 nuclei (arrow). (**C**) On ultrastructural examination, pagetic osteoclasts have characteristic nuclear (arrowhead) and occasional cytoplasmic inclusions containing paracrystalline arrays that are similar to paramyxoviral nucleocapsids.

Table 2

PD therapies

Agent	Dosage
Calcitonin	50–100 units subcutaneously daily for 6–18 months
Etidronate	20–400 mg orally daily for 6 months
Pamidronate	30–90 mg intravenously daily for 1–2 days
Alendronate	40 mg orally daily for 6 months
Tiludronate	400 mg orally daily for 3 months
Risedronate	30 mg orally daily for 2 months

may be involved in the initiation of the disease process in patients with a genetic predisposition to PD.

Ultrastructural, immunohistochemical, in situ hybridization, and biological studies have all suggested a possible viral etiologic factor in PD, although an infectious virus has not been isolated. Abe et al. reported that budding viruses could be detected in osteoclasts from PD patients (85), but this has not been confirmed by other investigators. Mills and coworkers demonstrated that the nuclear inclusions observed in osteoclasts from PD patients cross-reacted with antibodies against respiratory syncytial virus and measles virus nucleocapsid protein (MVNP) (86). Basle and coworkers reported that MVNP mRNA was present in osteoclasts from 5 patients with PD but not in 3 control patients (87). Similarly, Mills and coworkers (86) found that MVNP protein was present in osteoclasts and/or cultured bone cells from patients with PD. In bone biopsy specimens, both MVNP and respiratory syncytial virus nucleocapsid proteins were detected in the same osteoclasts on serial sections. Basle and coworkers (88) also detected MVNP in 6 of 6 specimens isolated from patients with PD but found other paramyxoviral nucleocapsid proteins as well. Mills and coworkers (89) studied long-term marrow cultures of samples from 12 patients with PD and found that in all 12 cultures, MVNP and/or syncytial virus nucleocapsid proteins were present in the mononuclear cells or the osteoclast-like multinucleated cells that formed. In contrast, these viral proteins were detected in less than 5% of osteoclast-like cells from control subjects.

Reddy et al. detected MVNP transcripts in bone marrow samples obtained from affected bones from 9 of 10 patients with PD (90). More recently, Friedrichs et al. reported the full-length sequence of an MVNP gene isolated from marrow cells of a Paget patient, as well as 700 base pairs of MVNP sequence from 3 other patients (91). Together, these data support the hypothesis that the MVNP gene is present in osteoclasts from patients with PD. However, this is not a universal finding, since others have been unable to detect the presence of paramyxoviral transcripts in either freshly isolated bone marrow specimens, osteoclasts, or cultured marrow cells from Paget patients (92, 93). Further, none of these studies have demonstrated that a virus is the cause of PD. Importantly, prior to the era of measles virus (MV) immunization, measles was a ubiquitous infection, while PD has a distinct geographic and racial distribution. PD is rare in the Far East and Scandinavia but is relatively common in the United Kingdom, Australia, New Zealand, and the United States. These results suggest that if involved, a viral infection by itself does not cause PD.

Our laboratory has undertaken a series of studies to determine whether MV could induce pagetic-like osteoclasts and bone lesions. MV consists of 6 genes: the nucleocapsid, matrix, fusion, hemagglutinin, and the P and L genes, which constitute the viral

polymerase. Kurihara et al. showed that transfection of normal human osteoclast precursors with the MVNP gene, but not the matrix or fusion gene, resulted in the formation of osteoclasts that exhibited many of the characteristics of pagetic osteoclasts (94). These characteristics included increased rate of osteoclast formation, increased numbers and size of osteoclasts formed in vitro, increased bone resorbing capacity of the osteoclasts, hypersensitivity of transfected osteoclast precursors to $1,25\text{-}(OH)_2D_3$, and increased expression levels of $TAF_{II}\text{-}17$. These characteristics are also observed in osteoclasts formed in vitro from freshly isolated marrow samples from Paget patients (94). Further, MV infection of marrow cells from transgenic mice in which expression of the human MV receptor CD46 was targeted to cells in the osteoclast lineage resulted in formation of osteoclasts that had many of the characteristics of Paget osteoclasts (95) (normal mice do not express MV receptors and are resistant to MV infection).

Additional support for a potential role for MVNP in the pathogenesis of PD is provided by our preliminary studies, in which expression of the MVNP gene was targeted to the cells in the osteoclast lineage in transgenic mice (MVNP mice) (96). Histomorphometric analysis of bones from 17 MVNP and 16 wild-type mice examined between 3 and 14 months of age showed there was a significant increase in osteoclast numbers and osteoblast activity in MVNP mice. This was accompanied by a marked increase in the amount of woven bone and in the cancellous bone volume. In contrast, bone volume decreased between 3 and 14 months of age in wild-type mice. Ex vivo studies showed that the osteoclasts formed in marrow cultures from MVNP mice were increased in number, were hypersensitive to $1,25\text{-}(OH)_2D_3$, and had an increased bone resorbing capacity compared with wild-type osteoclasts in culture. These results suggest that expression of MVNP in osteoclasts in vivo can induce bone changes that share many of the features of PD.

However, several questions remain concerning the involvement of MV in the pathogenesis of PD. MV infections predominantly occur in children rather than in adults, while PD is usually diagnosed in elderly patients. The osteoclast is not a self-renewing cell but is formed by fusion of postmitotic precursors. Thus, cell types other than osteoclasts must serve as a reservoir for MV to persist for long periods of time in patients with PD. Reddy et al. have reported that cells of other hematopoietic lineages from Paget disease patients, including immature multipotent precursors that give rise to granulocytes, erythrocytes, macrophages, and platelets, also express MVNP transcripts (97). These results suggest that the pluripotent hematopoietic stem cell may be the initial target for MV infection in PD.

Persistent paramyxoviral infections do occur. Chronic MV infection of the nervous system has been reported in patients with sub-

Table 3

Genetic loci linked to PD

Locus	Gene	Protein affected
2q36	?	?
5q31	?	?
5q35	SQSTM1	p62
6p	?	?
10p13	?	?
18q21–22	TNFRSF11A	RANK
18q23	?	?

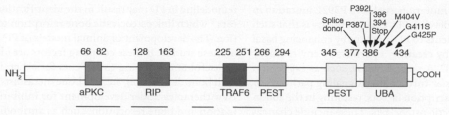

Figure 4

Structure of the p62 protein. The blocks indicate domains that mediate association with other proteins or are hypothesized to mediate these associations based upon homology with other proteins. The solid lines below the protein indicate stretches of sequence identity (of 20 amino acids or more) among the mouse, rat, and human p62 proteins. The arrows above the protein indicate the Paget disease-associated mutations identified to date. The splice donor and stop mutations result in a truncated protein lacking the UBA domain. PEST denotes hydrophobic regions that target proteins for rapid degradation (P, proline; E, glutaric acid; S, serine; T, threonine).

acute sclerosing panencephalitis (SSPE), which develops usually 5 years after the onset of a classic MV infection (98). However, as with PD patients, it has been difficult to rescue infectious virus from SSPE patients. Other RNA viruses can also persist in vivo, including influenza virus (99) and swine vesicular disease virus (100), and result in a carrier state in which the virus cannot be detected or is asymptomatic for long periods of time.

Several groups have also investigated the possible association of another paramyxovirus, canine distemper virus (CDV), with PD. An epidemiologic study in England suggested that patients with PD were more likely to have a pet dog than non-PD controls (101). Gordon and colleagues found that bone specimens from 11 of 25 Paget patients in England expressed CDV mRNA according to in situ hybridization analysis (102), and they also amplified an RT-PCR product for the CDV nucleocapsid gene from pagetic bone cells. Using in situ PCR techniques, Mee and coworkers also detected CDV nucleocapsid transcripts in osteoclasts from bone biopsies from 12 of 12 Paget patients in England (103). Taken together, these studies demonstrate that paramyxoviruses can induce changes in osteoclasts and bone that are similar to those found in PD. However, the role of paramyxoviral transcripts or proteins in the etiology of PD remains controversial.

Other factors that may be involved in the development of PD

Several studies have reported increased levels of IL-6 and/or M-CSF in patients with PD (104–106). Osteoclasts formed in bone

marrow cultures from patients with PD secrete large quantities of IL-6 into the conditioned media, with IL-6 levels reaching up to 2,000 pg/ml (104). IL-6 levels are also increased in the bone marrow plasma of affected bones from Paget patients, as well as in their peripheral blood (104). Since IL-6 has been shown to induce osteoclast formation (107), it is possible that IL-6 plays a role in the enhanced osteoclast formation in PD. Alternatively, the increased levels of IL-6 seen in patients with PD may simply be a marker for the increased osteoclast formation.

Athanasou and coworkers reported that serum levels of M-CSF are also increased in Paget disease patients at diagnosis and fall when the patients are treated effectively with bisphosphonates (106). M-CSF in combination with RANKL is a critical factor for osteoclast formation (108), and rodents deficient in M-CSF develop osteopetrosis (108, 109). The increased levels of M-CSF in PD may reflect the increased numbers of osteoblasts present in the pagetic lesion, since osteoblasts produce M-CSF (110). When PD patients are in remission, osteoblast activity decreases, and M-CSF levels would be expected to fall accordingly. It is possible that the increased levels of IL-6 and M-CSF together could further increase osteoclast activity in the pagetic lesion, thereby amplifying the pagetic process.

A proposed model for the development of PD

Any model for the development of PD must take into account both genetic and nongenetic factors, the highly localized nature of the disease, and its late onset. The involvement of a nongenetic factor in the etiology of PD would explain why some individuals

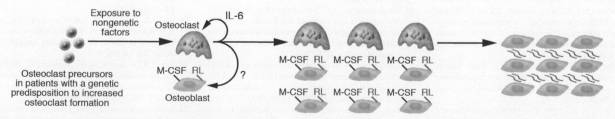

Figure 5

A proposed model for the pathogenesis of PD. Mutations that enhance basal osteoclastogenesis predispose patients to PD by creating a permissive environment for its development. A second factor, such as expression of certain viral proteins, may further alter signaling pathways or expression of specific transcription factors, resulting in the abnormal characteristics of pagetic osteoclasts. For example, the increased sensitivity of osteoclast precursors to low levels of 1,25-$(OH)_2D_3$ and RANKL (RL) enhances osteoclast formation. Further, the increased numbers of osteoclasts would secrete high levels of IL-6, which would further enhance osteoclast formation. Since osteoclast and osteoblast activity remain coupled in PD, the increased osteoclast activity would result in increased osteoblast numbers and rapid formation of new bone. The increased numbers of immature osteoblasts expressing high levels of RANKL and M-CSF would further increase osteoclast formation. As more and more bone is formed, the lesion would eventually become sclerotic.

who have a PD-associated mutation, such as a P392L mutation in the *SQSTM1* gene, do not develop PD. One possibility is that such mutations predispose patients to PD, perhaps by enhancing basal osteoclastogenesis, thereby creating a permissive environment for the development of PD. A second factor, such as expression of certain viral proteins, may further alter signaling pathways or expression of specific transcription factors, resulting in the abnormal characteristics of pagetic osteoclasts. These include changes in the vitamin D receptor transcription complex and changes in NF-κB signaling and other signaling pathways that increase osteoclast formation. For example, the increased sensitivity of osteoclast precursors to low levels of 1,25-$(OH)_2D_3$ and RANKL would enhance osteoclast formation. Further, the increased numbers of osteoclasts would secrete high levels of IL-6, which would further enhance osteoclast formation. Since osteoclast and osteoblast activity remain coupled in PD, the increased osteoclast activity would result in increased osteoblast numbers and rapid formation of new bone. The increased numbers of immature osteoblasts expressing high levels of RANKL and M-CSF would further increase osteoclast formation. As more and more bone is formed, the lesion would eventually become sclerotic. This model for the pathogenesis of PD is depicted in Figure 5.

Summary

There has been a tremendous output of new information on the pathogenesis of PD in recent years. Identification of genes involved in osteoclastogenesis that are mutated in PD and the characterization of other nongenetic factors that may be involved have provided important insights into the control of bone remodeling in PD as well as in normal bone. Future studies of the abnormal bone remodeling in PD may result in the identification of "coupling factors," which link osteoclastic bone resorption to new bone formation. The development of animal models of PD should greatly aid in these studies. If these coupling factors are identified, they may be useful in developing treatments for other diseases associated with bone destruction, such as bone metastasis and osteoporosis. New therapies under development for inhibiting osteoclast formation and bone resorption, such as antibodies to RANKL and inhibitors of cathepsin K, the enzyme secreted by osteoclasts that degrades bone matrix and is required for bone resorption (111), may be useful treatments for PD. Vitamin D receptor antagonists, which could decrease the hypersensitivity of pagetic osteoclast precursors in PD patients to physiologic levels of vitamin D, may also be therapeutic avenues to pursue. Thus, understanding of the pathophysiology of PD should provide important insights into the mechanisms that control normal osteoclast differentiation and osteoblast formation and may lead to development of new anabolic therapies for treating patients with severe bone loss.

Acknowledgments

Studies by the author's laboratory are supported by research funds from the NIH (grant PO1-AR049363). We also thank Donna Gaspich for preparation of the manuscript and Frederick Singer (John Wayne Cancer Institute, Santa Monica, California, USA) for helpful discussions.

Address correspondence to: G. David Roodman, VA Pittsburgh Healthcare System, Medicine/Hematology-Oncology (111-H), University Drive, Pittsburgh, Pennsylvania 15240, USA. Phone: (412) 688-6571; Fax: (412) 688-6960; E-mail: roodmangd@upmc.edu.

1. Jilka, R.L. 2003. Biology of the basic multicellular unit and the pathophysiology of osteoporosis. *Med. Pediatr. Oncol.* **41**:182–185.

2. Parfitt, A.M. 2002. Targeted and nontargeted bone remodeling: relationship to basic multicellular unit origination and progression. *Bone.* **30**:5–7.

3. Yasuda, H., et al. 1998. Osteoclast differentiation factor is a ligand for osteoprotegerin / osteoclastogenesis-inhibitory factor and is identical to TRANCE/RANKL. *Proc. Natl. Acad. Sci. U. S. A.* **95**:3597–3602.

4. Lacey, D.L., et al. 1998. Osteoprotegerin ligand is a cytokine that regulates osteoclast differentiation and activation. *Cell.* **93**:165–176.

5. Kong, Y.Y., et al. 1999. OPGL is a key regulator of osteoclastogenesis, lymphocyte development and lymph-node organogenesis. *Nature.* **397**:315–323.

6. Horwood, N.J., Elliott, J., Martin, T.J., and Gillespie, M.T. 1998. Osteotropic agents regulate the expression of osteoclast differentiation factor and osteoprotegerin in osteoblastic stromal cells. *Endocrinology.* **139**:4743–4746.

7. Hofbauer, L.C., and Heufelder, A.E. 2001. Role of receptor activator of nuclear factor-kappaB ligand and osteoprotegerin in bone cell biology. *J. Mol. Med.* **79**:243–253.

8. Lee, S.E., et al. 2002. The phosphatidylinositol 3-kinase, p38, and extracellular signal-regulated kinase pathways are involved in osteoclast differentiation. *Bone.* **30**:71–77.

9. Kim, N., Odgren, P.R., Kim, D.K., Marks, S.C.J., and Choi, Y. 2002. Diverse roles of the tumor necrosis factor family member TRANCE in skeletal physiology revealed by TRANCE deficiency and partial rescue by a lymphocyte-expressed TRANCE transgene. *Proc. Natl. Acad. Sci. U. S. A.* **97**:10905–10910.

10. Dougall, W.C., et al. 1999. RANK is essential for osteoclast and lymph node development. *Genes Dev.* **13**:2412–2424.

11. Li, J., et al. 2000. RANK is the intrinsic hematopoietic cell surface receptor that controls osteoclastogenesis and regulation of bone mass and calcium metabolism. *Proc. Natl. Acad. Sci. U. S. A.* **97**:1566–1571.

12. Naito, A., et al. 1999. Severe osteopetrosis, defective interleukin-1 signalling and lymph node organogenesis in TRAF6-deficient mice. *Genes Cells.* **4**:353–362.

13. Lomaga, M.A., et al. 1999. TRAF6 deficiency results in osteopetrosis and defective interleukin-1, CD40, and LPS signaling. *Genes Dev.* **13**:1015–1024.

14. Iotsova, V., et al. 1997. Osteopetrosis in mice lacking NF-kappaB1 and NF-kappaB2. *Nat. Med.* **3**:1285–1289.

15. Franzoso, G., et al. 1997. Requirement for NF-kappaB in osteoclast and B-cell development. *Genes Dev.* **11**:3482–3496.

16. Xing, L., et al. 2002. NF-kappaB p50 and p52 expression is not required for RANK-expressing osteoclast progenitor formation but is essential for RANK- and cytokine-mediated osteoclastogenesis. *J. Bone Miner. Res.* **17**:1200–1210.

17. Takayanagi, H., et al. 2002. Induction and activation of the transcription factor NFATc1 (NFAT2) integrate RANKL signaling in terminal differentiation of osteoclasts. *Dev. Cell.* **3**:889–901.

18. Boyce, B.F., Yoneda, T., Lowe, C. Soriano, P., and Mundy, G.R. 1992. Requirement of pp60c-src expression for osteoclasts to form ruffled borders and resorb bone in mice. *J. Clin. Invest.* **90**:1622–1627.

19. Grigoriadis, A.E., et al. 1994. c-Fos: a key regulator of osteoclast-macrophage lineage determination and bone remodeling. *Science.* **266**:443–448.

20. Tondravi, M.M., et al. 1997. Osteopetrosis in mice lacking haematopoietic transcription factor PU.1. *Nature.* **386**:81–84.

21. Pfeilschifter, J., Chenu, C., Bird, A., Mundy, G.R., and Roodman, G.D. 1989. Interleukin-1 and tumor necrosis factor stimulate the formation of human osteoclast-like cells in vitro. *J. Bone Miner. Res.* **4**:113–118.

22. Goldring, S.R. 2003. Pathogenesis of bone and cartilage destruction in rheumatoid arthritis. *Rheumatology.* **42**(Suppl. 2):ii1–ii6.

23. Kobayashi, K., et al. 2000. Tumor necrosis factor alpha stimulates osteoclast differentiation by a mechanism independent of the ODF/RANKL-RANK interaction. *J. Exp. Med.* **191**:275–286.

24. Lam, J., et al. 2000. TNF-α induces osteoclastogenesis by direct stimulation of macrophages exposed to permissive levels of RANK ligand. *J. Clin. Invest.* **106**:1481–1488.

25. Boyce, B.F., Aufdemorte, T.B., Garrett, I.R., Yates, A.J., and Mundy, G.R. 1989. Effects of interleukin-1 on bone turnover in normal mice. *Endocrinology.* **125**:1142–1150.

26. Uy, H.L., et al. 1995. Use of an in vivo model to determine the effects of interleukin-1 on cells at different stages in the osteoclast lineage. *J. Bone Miner. Res.* **10**:295–301.

27. Fox, S.W., Fuller, K., and Chambers, T.J. 2000. Activation of osteoclasts by interleukin-1: divergent responsiveness in osteoclasts formed in vivo and in vitro. *J. Cell Physiol.* **184**:334–340.

28. Erickson, S.L., et al. 1994. Decreased sensitivity to tumour-necrosis factor but normal T-cell development in TNF receptor 2-deficient mice. *Nature.* **372**:560–563.

29. Peschon, J.J., et al. 1998. TNF receptor-deficient mice reveal divergent roles for p55 and p75 in several models of inflammation. *J. Immunol.* **160**:943–952.

30. Abu-Amer, Y., et al. 2000. Tumor necrosis factor receptors types 1 and 2 differentially regulate osteoclastogenesis. *J. Biol. Chem.* **275**:27307–27310.

31. Glaccum, M.B., et al. 1997. Phenotypic and functional characterization of mice that lack the type I

receptor for IL-1. *J. Immunol.* **159**:3364–3371.

32. Lorenzo, J.A., et al. 1998. Mice lacking the type I interleukin-1 receptor do not lose bone mass after ovariectomy. *Endocrinology.* **139**:3022–3025.

33. Ducy, P., Zhang, R., Geoffroy, V., Ridall, A.L., and Karsenty, G. 1995. Osf2/Cbfal: a transcriptional activator of osteoblast differentiation. *Cell.* **89**:747–754.

34. Ducy, P., and Karsenty, G. 1998. Genetic control of cell differentiation in the skeleton. *Curr. Opin. Cell Biol.* **10**:614–619.

35. Marie, P.J. 2003. Fibroblast growth factor signaling controlling osteoblast differentiation. *Gene.* **316**:23–32.

36. Roelen, B.A., and Dijke, P. 2003. Controlling mesenchymal stem cell differentiation by TGFBeta family members. *J. Orthop. Sci.* **8**:740–748.

37. Blair, H.C., Zaidi, M., and Schlesinger, P.H. 2002. Mechanisms balancing skeletal matrix synthesis and degradation. *Biochem. J.* **364**:329–341.

38. Olney, R.C. 2003. Regulation of bone mass by growth hormone. *Med. Pediatr. Oncol.* **41**:228–234.

39. Kanis, J.A. 1998. *Pathophysiology and treatment of Paget's disease of bone.* 2nd edition. Martin Dunitz. London, United Kingdom. 310 pp.

40. Maldague, B., and Malghem, J. 1987. Dynamic radiologic patterns of Paget's disease of bone. *Clin. Orthop.* **217**:126–251.

41. Huvos, A.G. 1986. Osteogenic sarcoma of bones and soft tissues in older persons. A clinicopathologic analysis of 117 patients older than 60 years. *Cancer.* **57**:1442–1449.

42. Hansen, M.F., Nellissery, M.J., and Bhatia, P. 1999. Common mechanisms of osteosarcoma and Paget's disease. *J. Bone Miner. Res.* **14**(Suppl. 2):39–44.

43. Hosking, D.J. 1981. Paget's disease of bone. *Br. Med. J. (Clin. Res. Ed.).* **283**:686–688.

44. Alvarez, L., Peris, P., and Pons, F. 1997. Relationship between biochemical markers of bone turnover and bone scintigraphic indices in assessment of Paget's disease activity. *Arthritis Rheum.* **40**:461–468.

45. Rebel, A., et al. 1981. Towards a viral etiology for Paget's disease of bone. *Metab. Bone Dis. Relat. Res.* **3**:235–238.

46. Mills, B.G., Yabe, H., and Singer, F.R. 1988. Osteoclasts in human osteopetrosis contain viral-nucleocapsid-like nuclear inclusions. *J. Bone Miner. Res.* **3**:101–106.

47. Bianco, P., Silvestrini, G., Ballanti, P., and Bonucci, E. 1992. Paramyxovirus-like nuclear inclusions identical to those of Paget's disease of bone detected in giant cells of primary oxalosis. *Virchows Arch. A, Pathol. Anat. Histopathol.* **421**:427–433.

48. Mills, B.G. 1981. Comparison of the ultrastructure of a malignant tumor of the mandible containing giant cells with Paget's Disease of bone. *J. Oral Pathol.* **10**:203–215.

49. Menaa, C., et al. 2000. 1,25-dihydroxyvitamin D3 sensitivity of osteoclast precursors from patients with Paget's disease. *J. Bone Miner. Res.* **15**:228–236.

50. Neale, S.D., Smith, R., Wass, J.A., and Athanasou, N.A. 2000. Osteoclast differentiation from circulating mononuclear precursors in Paget's disease is hypersensitive to 1,25-dihydroxyvitamin D(3) and RANKL. *Bone.* **27**:409–416.

51. Menaa, C., et al. 2000. Enhanced RANK ligand expression and responsivity of bone marrow cells in Paget's disease of bone. *J. Clin. Invest.* **105**:1833–1838.

52. Kurihara, N., et al. 2004. Role of TAFII-17, a VDR binding protein, in the increased osteoclast formation in Paget's disease. *J. Bone Miner. Res.* **19**:1154–1164.

53. Sexton, P.M., Findlay, D.M., and Martin, T.J. 1999. Calcitonin. *Curr. Med. Chem.* **6**:1067–1093.

54. Rogers, M.J. 2003. New insights into the molecular mechanisms of action of bisphosphonates. *Curr. Pharm. Des.* **9**:2643–2658.

55. Miron-Canelo, J.A., Del Pino-Montes, J., Vicente-Arroyo, M., and Saenz-Gonzalez, M.C. 1997. Epidemiological study of Paget's Disease of bone in a zone of the province of Salamanca (Spain). The Paget's Disease of the Bone Study Group of Salamanca. *Eur. J. Epidemiol.* **13**:801–805.

56. Morales-Piga, A.A., Rey-Rey, J.S., Corres-Gonzalez, J., Garcia-Sagredo, J.M., and Lopez-Abente, G. 1995. Frequency and characteristics of familial aggregation of Paget's disease of bone. *J. Bone Miner. Res.* **10**:663–670.

57. Montagu, M.F.A. 1949. Paget's disease (osteitis deformans) and hereditary. *Am. J. Hum. Genet.* **1**:94–95.

58. Sofaer, J.A., Holloway, S.M., and Emery, A.E. 1983. A family study of Paget's disease of bone. *J. Epidemiol. Community Health.* **37**:226–231.

59. Leach, R.J., Singer, F.R., Cody, J.D., and Roodman, G.D. 1999. Variable disease severity associated with a Paget's disease predisposition gene. *J. Bone Miner. Res.* **14**:17–20.

60. Barker, D.J. 1981. The epidemiology of Paget's disease. *Metab. Bone Dis. Relat. Res.* **3**:231–233.

61. Leach, R.J., Singer, F.R., and Roodman, G.D. 2001. The genetics of Paget's disease of the bone. *J. Clin. Endocrinol. Metab.* **86**:24–28.

62. Cody, J.D., et al. 1997. Genetic linkage of Paget disease of the bone to chromosome 18q. *Am. J. Hum. Genet.* **61**:1117–1122.

63. Haslam, S.I., et al. 1998. Paget's disease of bone: evidence for a susceptibility locus on chromosome 18q and for genetic heterogeneity. *J. Bone Miner. Res.* **13**:911–917.

64. Laurin, N., et al. 2001. Paget disease of bone: mapping of two loci at 5q35-qter and 5q31. *Am. J. Hum. Genet.* **69**:528–543.

65. Hocking, L.J., et al. 2001. Genomewide search in familial Paget disease of bone shows evidence of genetic heterogeneity with candidate loci on chromosomes 2q36, 10p13, and 5q35. *Am. J. Hum. Genet.* **69**:1055–1061.

66. Good, D.A., et al. 2002. Linkage of Paget disease of bone to a novel region on human chromosome 18q23. *Am. J. Hum. Genet.* **70**:517–525.

67. Hughes, A.E., et al. 2000. Mutations in TNFRS-F11A, affecting the signal peptide of RANK, cause familial expansile osteolysis. *Nat. Genet.* **24**:45–48.

68. Nakatsuka, K., Nishizawa, Y., and Ralston, S.H. 2003. Phenotypic characterization of early onset Paget's disease of bone caused by a 27-bp duplication in the TNFRSF11A gene. *J. Bone Miner. Res.* **18**:1381–1385.

69. Laurin, N., Brown, J.P., Morissette, J., and Raymond, V. 2002. Recurrent mutation of the gene encoding sequestosome 1 (SQSTM1/p62) in Paget disease of bone. *Am. J. Hum. Genet.* **70**:1582–1588.

70. Johnson-Pais, T.L., et al. 2003. Three novel mutations in SQSTM1 identified in familial Paget's disease of bone. *J. Bone Miner. Res.* **18**:1748–1753.

71. Hocking, L.J., et al. 2002. Domain-specific mutations in sequestosome 1 (SQSTM1) cause familial and sporadic Paget's disease. *Hum. Mol. Genet.* **11**:2735–2739.

72. Hocking, L.J., et al. 2004. Novel UBA domain mutations of SQSTM1 in Paget's disease of bone: genotype phenotype correlation, functional analysis, and structural consequences. *J. Bone Miner. Res.* **19**:1122–1127.

73. Park, I., et al. 1995. Phosphotyrosine-independent binding of a 62-kDa protein to the src homology 2 (SH2) domain of p56lck and its regulation by phosphorylation of Ser-59 in the lck unique N-terminal region. *Proc. Natl. Acad. Sci. U. S. A.* **92**:12338–12342.

74. Vadlamudi, R.K., Joung, I., Strominger, J.L., and Shin, J. 1996. p62, a phosphotyrosine-independent ligand of the SH2 domain of p56lck, belongs to a new class of ubiquitin-binding proteins. *J. Biol. Chem.* **271**:20235–20237.

75. Shin, J. 1998. P62 and the sequestosome, a novel mechanism for protein metabolism. *Arch. Pharm. Res.* **21**:629–633.

76. Ciani, B., Layfield, R., Cavey, J.R., Sheppard, P.W., and Searle, M.S. 2003. Structure of the ubiquitin-associated domain of p62 (SQSTM1) and implications for mutations that cause Paget's disease of bone. *J. Biol. Chem.* **278**:37409–37412.

77. Geetha, T., and Wooten, M.W. 2002. Structure and functional properties of the ubiquitin binding protein p62. *FEBS Lett.* **512**:19–24.

78. Duran, A., et al. 2004. The atypical PKC-interacting protein p62 is an important mediator of RANK-activated osteoclastogenesis. *Dev. Cell.* **6**:303–309.

79. Kurihara, N., et al. 2004. The p392L mutation in the sequestasome-1 gene (p62) that is linked to Paget's disease (PD) is not sufficient to induce a pagetic phenotype in osteoclast (OCL) precursors [abstract]. *J. Bone Miner. Res.* **19**:553.

80. Meunier, P.J., Coindre, J.M., Edouard, C.M., and Arlot, M.E. 1980. Bone histomorphometry in Paget's disease. Quantitative and dynamic analysis of pagetic and nonpagetic bone tissue. *Arthritis Rheum.* **23**:1095–1103.

81. Laurin, N., Morissette, J., Raymond, V., and Brown, J.P. 2002. Large phenotypic variability of Paget disease of bone caused by the P392L sequestasome 1/p62 mutation [abstract]. *J. Bone Miner. Res.* **17**:S380.

82. Whyte, M.P., et al. 2002. Osteoprotegerin deficiency and juvenile Paget's disease. *N. Engl. J. Med.* **18**:175–184.

83. van Staa, T.P., et al. 2002. Incidence and natural history of Paget's disease of bone in England and Wales. *J. Bone Miner. Res.* **17**:465–471.

84. Doyle, T. Gunn, J., Anderson, G., Gill, M., and Gundy, T. 2002. Paget's disease in New Zealand: evidence for declining prevalence. *Bone.* **31**:616–619.

85. Abe, S., et al. 1995. Viral behavior of paracrystalline inclusions in osteoclasts of Paget's disease of bone. *Ultrastruct. Pathol.* **19**:455–461.

86. Mills, B.G., et al. 1984. Evidence for both respiratory syncytial virus and measles virus antigens in the osteoclasts of patients with Paget's disease of bone. *Clin. Orthop.* **183**:303–311.

87. Basle, M.F., Fournier, J.G., Rozenblatt, S., Rebel, A., and Bouteille, M. 1986. Measles virus RNA detected in Paget's disease bone tissue by in situ hybridization. *J. Gen. Virol.* **67**:907–913.

88. Basle, M.F., et al. 1985. Paramyxovirus antigens in osteoclasts from Paget's bone tissue detected by monoclonal antibodies. *J. Gen. Virol.* **66**:2103–2110.

89. Mills, B.G., et al. 1994. Multinucleated cells formed in vitro from Paget's bone marrow express viral antigens. *Bone.* **15**:443–448.

90. Reddy, S.V., Singer, F.R., and Roodman, G.D. 1995. Bone marrow mononuclear cells from patients with Paget's disease contain measles virus nucleocapsid messenger ribonucleic acid that has mutations in a specific region of the sequence. *J. Clin. Endocrinol. Metab.* **80**:2108–2111.

91. Friedrichs, W., et al. 2002. Sequence analysis of measles virus nucleocapsid transcripts in patients with Paget's disease. *J. Bone Miner. Res.* **17**:145–157.

92. Birch, M.A., et al. 1994. Absence of paramyxovirus RNA in cultures of pagetic bone cells and in pagetic bone. *J. Bone Miner. Res.* **9**:11–16.

93. Helfrich, M.H., et al. 2000. A negative search for a paramyxoviral etiology of Paget's disease of bone: molecular, immunological, and ultrastructural studies in UK patients. *J. Bone Miner. Res.* **15**:2315–2329.

94. Kurihara, N., Reddy, S.V., Menaa, C., Anderson, D., and Roodman, G.D. 2000. Osteoclasts expressing the measles virus nucleocapsid gene display a pagetic phenotype. *J. Clin. Invest.* **105**:607–614.

95. Reddy, S.V., et al. 2001. Osteoclasts formed by measles virus-infected osteoclast precursors from hCD46 transgenic mice express characteristics of

pagetic osteoclasts. *Endocrinology.* **142**:2898–2905.

96. Roodman, G.D., et al. 2004. The measles virus nucleocapsid (MVNP) gene is sufficient to induce a pagetic bone phenotype in vivo [abstract]. *J. Bone Miner. Res.* **19**:54.

97. Reddy, S.V., et al. 1999. Measles virus nucleocapsid transcript expression is not restricted to the osteoclast lineage in patients with Paget's disease of bone. *Exp. Hematol.* **27**:1528–1532.

98. Gascon, G.G. 1996. Subacute sclerosing panencephalitis. *Semin. Pediatr. Neurol.* **3**:260–269.

99. Marschall, M., Schuler, A., and Meier-Ewert, H. 1996. Influenza C virus RNA is uniquely stabilized in a steady state during primary and secondary persistent infections. *J. Gen. Virol.* **77**:681–686.

100. Lin, F., Mackay, D.K., and Knowles, N.J. 1998. The persistence of swine vesicular disease virus infection in pigs. *Epidemiol. Infect.* **121**:459–472.

101. Anderson, D.C., and O'Driscoll, J.B. 1986. Dogs and Paget's disease [letter]. *Lancet.* **1**:41.

102. Gordon, M.T., Anderson, D.C., and Sharpe, P.T. 1991. Canine distemper virus localized in bone cells of patients with Paget's disease. *Bone.* **12**:195–201.

103. Mee, A.P., et al. 1998. Detection of canine distemper virus in 100% of Paget's disease samples by in situ-reverse transcriptase-polymerase chain reaction. *Bone.* **23**:171–175.

104. Roodman, G.D., et al. 1992. Interleukin 6. A potential autocrine/paracrine factor in Paget's disease of bone. *J. Clin. Invest.* **89**:46–52.

105. Hoyland, J.A., Freemont, A.J., and Sharpe, P.T. 1994. Interleukin-6, IL-6 receptor, and IL-6 nuclear factor gene expression in Paget's disease. *J. Bone Miner. Res.* **9**:75–80.

106. Neale, S.D., Schulze, E., Smith, R., and Athanasou, N.A. 2002. The influence of serum cytokines and growth factors on osteoclast formation in Paget's disease. *QJM.* **95**:233–240.

107. Kurihara, N., Bertolini, D., Suda, T., Akiyama, Y., and Roodman, G.D. 1990. IL-6 stimulates osteoclast-like multinucleated cell formation in long-term human marrow cultures by inducing IL-1 release. *J. Immunol.* **144**:4226–4230.

108. Tsurukai, T., Udagawa, N., Matsuzaki, K., Takahashi, N., and Suda, T. 2000. Roles of macrophage-colony stimulating factor and osteoclast differentiation factor in osteoclastogenesis. *J. Bone Miner. Metab.* **18**:177–184.

109. Van Wesenbeeck, L., et al. 2002. The osteopetrotic mutation toothless (tl) is a loss-of-function frameshift mutation in the rat Csf1 gene: evidence of a crucial role for CSF-1 in osteoclastogenesis and endochondral ossification. *Proc. Natl. Acad. Sci. U. S. A.* **99**:14303–14308.

110. Rubin, J., Fan, X., Thornton, D., Bryant, R., and Biskobing, D. 1996. Regulation of murine osteoblast macrophage colony-stimulation factor production by 1,25(OH)2D3. *Calcif. Tissue Int.* **59**:291–296.

111. Troen, B.R. 2004. The role of cathepsin K in normal bone resorption. *Drug News Perspect.* **17**:19–28.

Estrogen deficiency and bone loss: an inflammatory tale

M. Neale Weitzmann[1] and Roberto Pacifici[1,2]

[1]Division of Endocrinology, Metabolism, and Lipids and [2]Molecular Pathogenesis Program, Emory University, Atlanta, Georgia, USA.

Estrogen plays a fundamental role in skeletal growth and bone homeostasis in both men and women. Although remarkable progress has been made in our understanding of how estrogen deficiency causes bone loss, the mechanisms involved have proven to be complex and multifaceted. Although estrogen is established to have direct effects on bone cells, recent animal studies have identified additional unexpected regulatory effects of estrogen centered at the level of the adaptive immune response. Furthermore, a potential role for reactive oxygen species has now been identified in both humans and animals. One major challenge is the integration of a multitude of redundant pathways and cytokines, each apparently capable of playing a relevant role, into a comprehensive model of postmenopausal osteoporosis. This Review presents our current understanding of the process of estrogen deficiency–mediated bone destruction and explores some recent findings and hypotheses to explain estrogen action in bone. Due to the inherent difficulties associated with human investigation, many of the lessons learned have been in animal models. Consequently, many of these principles await further validation in humans.

The term *estrogen* refers to numerous steroidal and nonsteroidal molecules capable of inducing estrus. This hormone family plays a fundamental role in skeletal growth and homeostasis. In addition, estrogens are used as pharmacological agents to prevent postmenopausal bone loss. Research during the last decade has revealed that estrogen regulates bone homeostasis through unexpected regulatory effects on the immune system and on oxidative stress and direct effects on bone cells. Many of these observations derive from studies with inbred mice selected for their rapid response to ovariectomy (ovx), which represent an optimal model to investigate the acute effects of estrogen deficiency. However, the conclusions of these studies await confirmation in additional strains of rodents as well as in humans. Since the response to estrogen deprivation is strain specific (1) and estrogen has a more potent anabolic effect in mice than in humans, it is likely that differences will emerge between the mechanisms of estrogen action in humans and rodents.

Prior to 1987, bone cells were not generally considered direct targets of estrogen. However, it is now firmly established that osteoblasts (OBs) (2), osteocytes (3), and osteoclasts (OCs) (4) express functional estrogen receptors (ERs). These receptors are also expressed in bone marrow stromal cells (SCs), the precursors of OBs, which provide physical support for nascent OCs, T cells, B cells, and most other cells in human and mouse bone marrow (5). Estrogen signals through 2 receptors, ERα and ERβ (6). Bone cells contain both receptors, but their distributions within bone are not homogeneous. In humans, ERα is the predominant isoform in cortical bone, while ERβ is the predominant species in trabecular bone. In general, ERα mediates most actions of estrogen on bone cells (7, 8). In vitro studies suggest that estrogen's bone-sparing effects are mediated by both estrogen and androgen receptors (9), although subsequent in vivo studies showed that estrogen does not prevent bone loss in mice that possess a functional androgen receptor but lack ERα and ERβ (10).

The mechanism through which information is transduced from ligand-bound receptors has been the subject of intense research since 1960. It is now clear that ligand binding to ERs produces a conformational change that promotes receptor dimerization and binding to specific DNA sequences called estrogen response elements (EREs) (11). At the promoter the ligand-bound receptor forms a complex with coactivator proteins, which activates the general transcriptional machinery and increases expression of target genes through chromatin remodeling. ERs can also recruit corepressors, which negatively regulate ER-dependent gene expression. In addition to this classical modality of gene activation, alternative mechanisms have been described that account for estrogen's ability to both stimulate and repress the expression of genes encoding critical osteoclastogenic factors such as IL-6, TNF-α, and M-CSF. For example, activated ERs can bind to transcription factors such as NF-κB and prevent binding to DNA, which explains how estrogen represses IL-6 production (12). Equally relevant for the bone-sparing activity of estrogen are its effects on many families of kinases. Estrogen decreases casein kinase 2 (CK2) activity, leading to reduced phosphorylation of the nuclear protein Egr-1. Dephosphorylated Egr-1 has increased affinity for the transcriptional activator Sp-1, a factor critical for expression of the *MCSF* gene (13). Formation of an Egr-1/Sp-1 complex during estrogen deficiency decreases the nuclear level of free Sp-1, thus blunting *MCSF* transcription. Estrogen is also capable of blunting JNK activity. The resulting decrease in production of activator protein 1 (AP1) factors explains the repressive effects of estrogen on *TNF* gene expression (14) as well as why estrogen decreases the sensitivity of maturing OCs to the osteoclastogenic factor receptor activator of NF-κB (RANK) ligand (RANKL) (15).

Although many estrogenic effects are mediated by nuclear ERs, some responses originate in the plasma membrane. In fact, estrogen produces rapid effects (within seconds or minutes) in various cell types, including bone cells. These nongenomic (or nongenotropic) actions are due to signaling by a membrane receptor. Estrogen's ability to induce OC apoptosis and inhibit OB apoptosis is

---

Nonstandard abbreviations used: CIITA, class II transactivator; ER, estrogen receptor; OB, osteoblast; OC, osteoclast; RANK, receptor activator of NF-κB; RANKL, RANK ligand; ovx, ovariectomy, ovariectomized; SC, stromal cell.

Conflict of interest: The authors have declared that no conflict of interest exists.

Citation for this article: *J. Clin. Invest.* **116**:1186–1194 (2006). doi:10.1172/JCI28550.

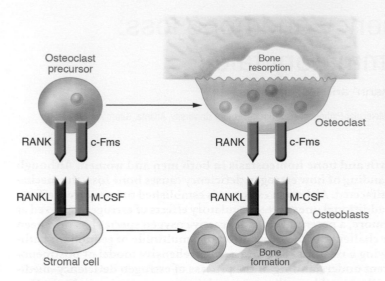

Figure 1
Cells and cytokines responsible for physiological OC renewal. OC precursors may differentiate from the population of monocytes/macrophages, among which they circulate by virtue of their expression of the receptor RANK. When RANKL binds to this receptor in the presence of the trophic factor M-CSF, which in turn binds to its receptor, colony-stimulating factor receptor 1 (c-Fms), OC precursors differentiate and fuse together to form mature, multinucleated bone-resorbing OCs. Under physiological conditions the dominant source of RANKL and M-CSF in the bone marrow microenvironment is from the bone-forming cells, the OBs, and their SC precursors.

linked to its ability to increase ERK1 and ERK2 phosphorylation and repress JNK activity (9, 16). The phosphorylation of these cytoplasmic kinases and their transport to the nucleus modulates the activity of transcription factors required for antiapoptotic actions of estrogen (16).

Effects of estrogen deficiency on bone turnover and architecture

Aging bone is gradually replaced by new tissue through a process called bone remodeling or turnover. Bone remodeling occurs through the coordinated action of OBs and OCs. The activities of OCs and OBs are combined into defined anatomical spaces called basic multicellular units (BMUs) (17). A remodeling cycle begins with the activation of a new BMU on a previously inactive surface of bone. This process involves the disappearance of bone-lining cells and their replacement by OCs that generate resorption lacunae on the endosteal surface of bone over a 2-week interval. The resorption phase is then terminated, probably by OC apoptosis, and after a brief reversal phase, a team of OBs is recruited that fills in the resorption cavity with new bone (18). The net result is the replacement of a packet of old bone with new bone. At menopause there is a transient, accelerated phase of bone loss that is followed by slower, sustained bone loss (18). Although in men there is no abrupt cessation of gonadal function in the sixth decade of life, they do experience an age-related decrease in unbound sex steroids resulting from progressive increases in circulating sex hormone–binding globulin (18). In both men and women there is a steady decline in unbound (bioavailable) estrogen levels with aging, exacerbated in women at menopause by a marked decrease in estrogen levels.

Estrogen deficiency leads to dramatic elevations in the number of BMUs through increased activation frequency, which is the number of new remodeling units activated in each unit of time (19). Enhanced activation frequency expands the remodeling space, increases cortical porosity, and enlarges the resorption area on trabecular surfaces. This phenomenon is caused primarily by increased OC formation, a complex event involving various hematopoietic and immune cells (5), as well as increased OC recruitment to bone surfaces to be remodeled. Estrogen deficiency also augments erosion depth by prolonging the resorption phase of the remodeling cycle through increased OC lifespan due to reduced apoptosis (20).

The net bone loss caused by the combined effects of increased activation frequency and erosion depth is limited in part by a compensatory augmentation of bone formation within each remodeling unit. This event is a consequence of stimulated osteoblastogenesis fueled by an expansion of the pool of early mesenchymal progenitors and by increased commitment of such pluripotent precursors toward the osteoblastic lineage (21). In spite of stimulated osteoblastogenesis, the net increase in bone formation is inadequate to compensate for enhanced bone resorption because of an augmentation in OB apoptosis, a phenomenon also induced by estrogen deficiency (9).

An additional event triggered by estrogen withdrawal, which limits the magnitude of the compensatory elevation in bone formation, is the increased production of inflammatory cytokines such as IL-7 and TNF, which limit the activity of mature OBs (22, 23). Increased bone resorption, trabecular thinning and perforation, and a loss of connection between the remaining trabeculae are the dominant features of the initial phase of rapid bone loss that follows the onset of estrogen deficiency (24). This acute phase is followed by a long-lasting period of slower bone loss where the dominant microarchitectural change is trabecular thinning. This phase is due in part to impaired osteoblastic activity secondary to increased OB apoptosis (25).

Initiation of estrogen replacement therapy (ERT) in experimental animals and humans decreases erosion depth and OC activation frequency by stimulating apoptosis and blocking osteoclastogenesis (19). Long-term ERT at high doses not only blunts bone resorption but also stimulates bone formation, leading to a net anabolic effect (26). A decrease in OB apoptosis resulting from a nongenotropic effect of estrogen is likely a major mechanism driving this effect (27, 28). However, the increase in OB lifespan is offset in part by a repressive effect of estrogen on osteoblastogenesis, a phenomenon that explains why the anabolic effect of estrogen is observed only at high doses and during long-term treatment.

Effects of estrogen deficiency on osteoclast formation
The dominant acute effect of estrogen is the blockade of new OC formation. OCs arise by cytokine-driven proliferation and differentiation of monocyte precursors that circulate within the hematopoietic cell pool (29). This process is facilitated by bone marrow

SCs, which provide physical support for nascent OCs and produce soluble and membrane-associated factors essential for the proliferation and differentiation of OC precursors (Figure 1).

The minimal essential cytokines required for OC formation under basal conditions are RANKL and M-CSF. These factors are produced primarily by bone marrow SCs, OBs, and activated T cells (30). RANKL is a TNF superfamily member that exists in membrane-bound and soluble forms. RANKL binds to the transmembrane receptor RANK expressed on the surface of OCs and OC precursors. RANKL also binds to osteoprotegerin (OPG), a soluble decoy receptor produced by numerous hematopoietic cells. Thus OPG, by sequestering RANKL and preventing its binding to RANK, functions as a potent antiosteoclastogenic cytokine (30). RANKL promotes the differentiation of OC precursors from an early stage of maturation into fully mature, multinucleated OCs. RANKL is also capable of activating mature OCs, thus stimulating these cells to resorb bone. M-CSF induces proliferation of early OC precursors, differentiation of more mature OCs, and fusion of mononucleated pre-OCs and increases the survival of mature OCs.

While basal levels of RANKL and M-CSF are essential for physiological OC renewal, additional cytokines either produced or regulated by T cells are responsible for the upregulation of OC formation observed during estrogen deficiency (5). One such factor is TNF, a cytokine that enhances OC formation directly (31) and by upregulating the SC production of RANKL and the responsiveness of OC precursors to this factor (32, 33). The ability of TNF to increase the osteoclastogenic activity of RANKL is due to synergistic interactions at the level of NF-κB and AP1 signaling (34). In addition, TNF and RANKL synergistically upregulate RANK expression in OC precursors (35). Furthermore, TNF stimulates OC activity (36) and inhibits osteoblastogenesis (37), thus further driving an imbalance between bone formation and resorption.

Like TNF, IL-1 promotes RANKL expression by bone marrow SCs and OBs and stimulates OC lifespan and activity. IL-1 directly targets OC precursors and promotes OC differentiation in the presence of permissive levels of RANKL. Furthermore, IL-1 mediates, in part, the osteoclastogenic effect of TNF by enhancing SC expression of RANKL and directly stimulating differentiation of OC precursors (38). TNF and IL-1 have potent antiapoptotic effects in OCs, prolonging OC lifespan and accelerating bone resorption.

T cells and ovariectomy-induced bone loss

A major focus of osteoporosis research is to understand the reasons for accelerated bone loss following menopause. Thus an important question is how estrogen deficiency leads to increased OC formation. It is recognized that osteoclastogenesis in response to estrogen deficiency is cytokine driven (39). One cytokine responsible for augmented osteoclastogenesis during estrogen deficiency is TNF, and its relevance has been demonstrated in multiple animal models. For example, ovx fails to induce bone loss in TNF KO mice and in mice lacking the p55 TNF receptor (40). Likewise, transgenic mice insensitive to TNF due to the overexpression of a soluble TNF receptor (41) and mice treated with the TNF inhibitor TNF-binding protein (42) are protected from ovx-induced bone loss.

The presence of increased levels of TNF in the bone marrow of ovx animals and in the conditioned media of peripheral blood cells of postmenopausal women is well documented (43–45), although the cells responsible have not been identified conclusively. Recent studies on highly purified bone marrow cells have revealed that ovx increases production of TNF by T cells but not monocytes (33),

and that earlier identifications of TNF production by monocytes were likely due to T cell contamination of monocytes purified by adherence. Thus the ovx-induced increase in TNF levels is likely to be due to T cell TNF production. These findings in the mouse are concordant with those in humans, in which adherent mononuclear blood cells contain CD3⁺CD56⁺ lymphocytes, a TNF-producing subset of adherent T cells (46). In this study the number of CD3⁺CD56⁺ T cells was decreased by estrogen treatment and inversely correlated with bone density. These results are not surprising as T cells can secrete a wide repertoire of cytokines, some pro-osteoclastogenic and some antiosteoclastogenic.

In the absence of strong activation signals, T cells appear to repress OC formation (47), but the relevance of this phenomenon in vivo has not been established. In contrast, activated T cells play a key role in the regulation of OC formation through increased production of RANKL and TNF (48–50). Activated T cells also produce IFN-α and IFN-γ, which in part limit RANKL-induced bone resorption by repressing NF-κB and JNK signaling pathways (51). The net effect of T cells on OC formation may consequently represent the prevailing balance of pro- and antiosteoclastogenic T cell cytokine secretion. However, it appears that during stimulated conditions such as inflammation (48) and estrogen deficiency (33), pro-osteoclastogenic cytokines prevail.

Attesting to the relevance of T cells in estrogen deficiency–induced bone loss in vivo, measurements of trabecular bone by peripheral quantitative CT and μCT revealed that athymic T cell–deficient nude mice are completely protected from the trabecular bone loss induced by ovx (33, 40, 52). T cell–deficient mice also fail to respond to ovx with the expected increase in bone turnover (33, 40, 52). T cells are key inducers of bone wasting because ovx increases T cell TNF production to a level sufficient to augment RANKL-induced osteoclastogenesis (33). The specific relevance of T cell TNF production in vivo was demonstrated by the finding that while reconstitution of nude recipient mice with T cells from wild-type mice restores the capacity of ovx to induce bone loss, reconstitution with T cells from TNF-deficient mice does not (40). T cell–derived TNF may further augment bone loss by stimulating T cell RANKL production.

Mechanisms of estrogen regulation of T cell TNF production

It has been shown that ovx upregulates T cell TNF production primarily by increasing the number of TNF-producing T cells (40). This is the result of a complex pathway that involves the thymus and bone marrow. The upstream mechanisms by which estrogen deficiency expands the pool of TNF-producing T cells are summarized in Figure 2. In the bone marrow, ovx promotes T cell activation, resulting in increased T cell proliferation and life span through antigen presentation by macrophages and DCs (53, 54). This process is due to the ability of estrogen deficiency to upregulate the expression of MHC class II in macrophages and DCs (53–55). The question thus arises as to the nature of the antigens. Estrogen deficiency is likely to increase T cell reactivity to a pool of self and foreign antigens physiologically present in healthy animals and humans. This is consistent with the fact that T cell clones expressing TCRs directed against self antigens not expressed in the thymus survive negative selection during T cell maturation (56). Such clones (autoreactive or self-reactive T cells) reside in peripheral lymphatic organs of adult individuals. In addition, foreign antigens of bacterial origin are physiologically absorbed in

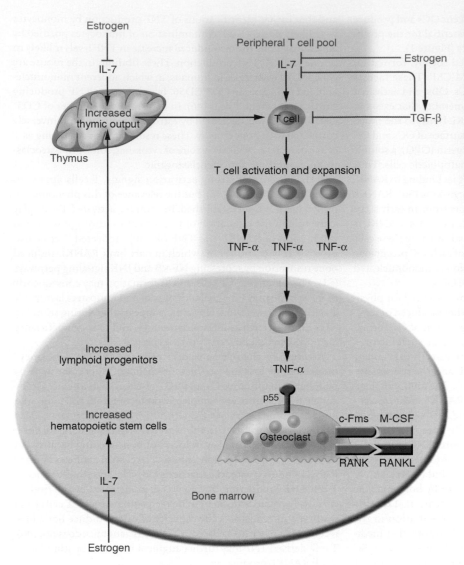

Figure 2
Estrogen suppresses T cell TNF production by regulating T cell differentiation and activity in the bone marrow, thymus, and peripheral lymphoid organs. In the bone marrow, estrogen downregulates the proliferation of hematopoietic stem cells through an IL-7–dependent mechanism, resulting in a smaller pool of lymphoid progenitors. T cell precursors leave the bone marrow and migrate to the thymus, where T cell differentiation, selection, and expansion take place, in large measure under control of IL-7. Following release from the thymus (thymic output), these new T cells home to peripheral lymphoid organs, including the bone marrow itself. Estrogen prevents T cell activation in part by directly blunting antigen presentation and in part via repression of IL-7 and IFN-γ production. This effect is amplified by the upregulation of the IL-7 suppressor TGF-β. The net result of these actions is a decrease in the number of TNF-producing T cells. The blunted levels of TNF diminish RANKL-induced OC formation, ultimately preventing bone loss.

the gut. As these peptides come into contact with immune cells locally and systemically, they induce low-grade T cell activation (57). Thus a moderate immune response is constantly in place in healthy humans and rodents due to presentation by MHC class II and MHC class I molecules of both self and foreign peptides to CD4+ and CD8+ T cells (58). This autoreactive response is thought to be essential for immune cell survival and renewal (59).

The effects of ovx on antigen presentation and the resulting changes in T cell activation, proliferation, and lifespan are explained by a stimulatory effect of ovx on the expression of the gene encoding class II transactivator (CIITA). The product of *CIITA* is a non–DNA-binding factor induced by IFN-γ that functions as a transcriptional coactivator at the MHC class II promoter (60). Increased *CIITA* expression in macrophages derived from ovx mice results from ovx-mediated increases in both T cell IFN-γ production and the responsiveness of *CIITA* to IFN-γ (53). The relevance of IFN-γ to ovx-induced bone loss is suggested by the failure of IFN-γ receptor–null mice to undergo T cell activation and sustain bone loss in response to ovx (53).

The actions of IFN-γ in bone turnover are controversial. Observations in humans and experimental disease models indicate that IFN-γ promotes bone resorption and causes bone loss under various conditions. For example, IFN-γ knockout mice are protected from infection-induced alveolar bone loss (61), while in erosive tuberculoid leprosy and psoriatic arthritis, the number of IFN-γ–producing cells and the levels of IFN-γ in synovial fluid correlates positively with bone destruction (62, 63). Furthermore, IFN-γ is reported to be efficacious in the treatment of osteopetrosis through restoration of bone resorption in humans (64) and rodents (65). In contrast, others have reported that IFN-γ exerts potent antiosteoclastogenic effects in vitro (51), that silencing of IFN-γ receptor signaling leads to more rapid onset of collagen-induced arthritis and bone resorption (66), and that IFN-γ decreases serum calcium concentration and osteoclastic bone resorption in nude mice (67, 68).

These conflicting effects of IFN-γ may be explained by the fact that IFN-γ influences OC formation via both direct and indirect effects. IFN-γ directly blocks OC formation through targeting of maturing OCs, as observed in vitro. However, IFN-γ is also a potent inducer of antigen presentation and thus of T cell activation. Therefore, when IFN-γ levels are increased in vivo, activated T cells secrete pro-osteoclastogenic factors, and this activity offsets the antiosteoclastogenic effect of IFN-γ. For example, IFN-γ–producing human T cells have been reported to directly induce osteoclastogenesis from human monocytes via RANKL expression (69). While under certain conditions the net effect of IFN-γ is inhibition of OC formation, other conditions favor stimulation of OC differentiation. Generally, when T cell activation occurs in response to an innate immune response such as LPS exposure, IFN-γ functions as an antiresorptive agent. Conversely, when T cell activation occurs through an adaptive immune response, as in estrogen deficiency, IFN-γ stimulates bone resorption. The inability of IFN-γ to blunt differentiation of maturing OCs in bone, where RANKL is abundant (70), contributes to the explanation of why in some conditions the in vivo proresorptive effects of IFN-γ are more potent than the suppression of osteoclastogenesis induced by IFN-γ in vitro.

Estrogen deficiency upregulates IFN-γ production through TGF-β downregulation. Estrogen has a direct stimulatory effect on the production of this factor, which is mediated through direct binding of the estrogen/ER complex to an ERE in the TGF-β promoter (71). TGF-β is recognized as a powerful repressor of T cell activation. Indeed, TGF-β exerts strong immunosuppressive effects by inhibiting the activation and proliferation of T cells and their production of proinflammatory cytokines including IFN-γ. Studies in a transgenic mouse model that expresses a dominant-negative form of the TGF-β receptor specifically in T cells have contributed to the understanding of the relevance of the repressive effects of this cytokine on T cell function in bone loss associated with estrogen deficiency (52). This animal model, known as CD4dnTGFβRII, is severely osteopenic due to increased bone resorption. More importantly, mice with T cell–specific blockade of TGF-β signaling are completely resistant to the bone-sparing effects of estrogen (52). This phenotype results from a failure of estrogen to repress IFN-γ production, which in turn leads to increased T cell activation and TNF production. Gain-of-function experiments confirmed that elevation of the systemic levels of TGF-β prevents ovx-induced bone loss and bone turnover (52).

Another mechanism by which estrogen regulates IFN-γ and TNF production is by repressing the production of IL-7, a potent lymphopoietic cytokine and inducer of bone destruction in vivo (72). IL-7 receptor knockout mice display increased bone volume and bone mineral density (72). In contrast, IL-7 transgenic mice have expanded bone marrow cavities with focal osteolysis of cortical bone and eroded bone surfaces (73). IL-7 has been reported to induce production of RANKL by human T cells (74), and injection of IL-7 into mice in vivo induces bone destruction by inducing T cell production of RANKL and TNF (75). Importantly, levels of IL-7 are significantly elevated following ovx (22, 76, 77), and in vivo IL-7 blockade using neutralizing antibodies is effective in preventing ovx-induced bone destruction (22) by suppressing T cell expansion and TNF and IFN-γ production (76). Furthermore, a recent study shows that liver-derived IGF-1 is permissive for ovx-induced trabecular bone loss by modulation of the number of T cells and the expression of IL-7 (77). The relevance of IL-7 in the mechanism of ovx-induced bone loss has been confirmed in part by another recent investigation showing that ovx does not induce cortical bone loss in IL-7 knockout mice (78).

Indeed, the elevated bone marrow levels of IL-7 contribute to the expansion of the T cell population in peripheral lymphoid organs through several mechanisms. First, IL-7 directly stimulates T cell proliferation by lowering tolerance to weak self antigens. Second, IL-7 increases antigen presentation by upregulating the produc-

tion of IFN-γ. Third, IL-7 and TGF-β inversely regulate each other's production (79, 80). The reduction in TGF-β signaling, characteristic of estrogen deficiency, may serve to further stimulate IL-7 production, thus driving the cycle of osteoclastogenic cytokine production and bone wasting. New studies further implicate IL-7 as a downstream effector of IGF-1 action in ovx-induced trabecular bone loss (77).

In estrogen deficiency, IL-7 compounds bone loss by suppressing bone formation, thus uncoupling bone formation from resorption. Recent studies have also identified elevated levels of IL-7 in patients suffering from multiple myeloma and in multiple myeloma–derived cell lines (81) and have suggested a role for IL-7 in the enhanced bone resorption and suppressed bone formation associated with multiple myeloma. Increased IL-7 expression has also been implicated in the bone loss sustained by patients with rheumatoid arthritis (82, 83).

IL-7 is a stimulator of both B and T cell lineages, and it has been suggested that IL-7 also induces bone loss by a mechanism involving the expansion of cells of the B lineage, in particular B220+IgM− B cell precursors (72), a population that is greatly expanded during estrogen deficiency (72, 84). How B lineage cells may lead to bone destruction is not presently understood but may involve overexpression of RANKL, a property of activated B cells (85). Alternatively, early B220+IgM− precursor cells have been found to be capable of differentiating into OCs in response to M-CSF and/or RANKL in vitro (75, 86) and may thus contribute to increasing the pool of early OC precursors. These cells have likewise been suggested to play a potential role in arthritic bone destruction.

T cell thymic output and bone loss

The thymus undergoes progressive structural and functional decline with age, coinciding with increased circulating sex steroid levels at puberty (87). By middle age most parenchymal tissue is replaced by fat, and in both mice and humans fewer T cells are produced and exported to secondary lymphoid organs. However, the thymus continues to generate new T cells even into old age (88, 89). In fact, active lymphocytic thymic tissue has been documented in adults up to 107 years of age (90). Under severe T cell depletion secondary to HIV infection, chemotherapy, or bone marrow transplant, an increase in thymic output (known as thymic rebound) becomes critical for long-term restoration of T cell homeostasis. For example, middle-aged women treated with autologous bone marrow transplants develop thymic hypertrophy and a resurgence of thymic T cell output, which contributes to the restoration of a wide T cell repertoire (91), although the intensity of thymic rebound declines with age. The mechanism driving thymic rebound is not completely understood, but one factor involved is IL-7 (92). Importantly, IL-7 alone is not sufficient to enhance thymopoiesis in young mice (93) but plays a more relevant role in aged mice (94).

Both androgens and estrogen have a profound suppressive effect on thymic function. Accordingly, castration reverses thymic atrophy and increases export of recent thymic emigrants to the periphery (95), while sex steroid inhibits thymus regeneration by promoting thymocyte apoptosis and an arrest of differentiation (96). Restoration of thymic function after castration occurs in young (97) as well as in very old rodents (98).

In accordance with the notion that estrogen deficiency induces a rebound in thymic function, ovx expands the population of thymic T cells and leads to the thymic export of naive T cells (76). Indeed, stimulated thymic T cell output accounts for approximately 50%

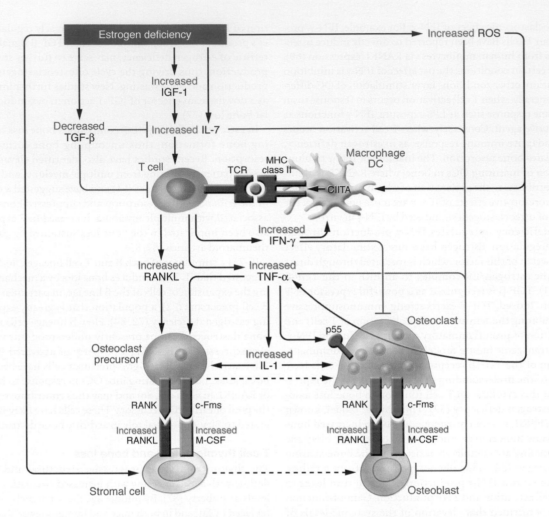

Figure 3
Schematic representation of the main mechanisms and feedback interactions by which estrogen deficiency leads to bone loss. The bone loss induced by estrogen deficiency is due to a complex interplay of hormones and cytokines that converge to disrupt the process of bone remodeling. Estrogen deficiency leads to a global increase in IL-7 production in target organs such as bone, thymus, and spleen, in part through decreases in TGF-β and increased IGF-1 production. This leads to an initial wave of T cell activation. Activated T cells release IFN-γ, which increases antigen presentation by DCs and macrophages (Mφ) by upregulating MHC class II expression through the transcription factor CIITA. Estrogen deficiency also amplifies T cell activation and osteoclastogenesis by downregulating antioxidant pathways, leading to an upswing in ROS. The resulting increase in ROS stimulates antigen presentation and the production of TNF by mature OCs. The combined effect of IFN-γ and ROS markedly enhances antigen presentation, amplifying T cell activation and promoting release of the osteoclastogenic factors RANKL and TNF. TNF further stimulates SC and OB RANKL and M-CSF production, in part via IL-1 upregulation, driving OC formation. TNF and IL-7 further exacerbate bone loss by blunting bone formation through direct repressive effects on OBs.

of the increase in the number of T cells in the periphery, while the remainder is due to enhanced peripheral expansion. Similarly, thymectomy reduces by approximately 50% the bone loss induced by ovx, thus demonstrating that the thymus plays a previously unrecognized causal effect in ovx-induced bone loss in mice. The remaining bone loss is a consequence of the peripheral expansion of naive and memory T cells (76). This finding, which awaits confirmation in humans, suggests that estrogen deficiency–induced thymic rebound may be responsible for the exaggerated bone loss in young women undergoing surgical menopause (99) or for the rapid bone loss characteristic of women in their first 5–7 years after natural menopause (18). Indeed, an age-related decrease in estrogen deficiency–induced thymic rebound could mitigate the stimulatory effects of sex steroid deprivation and explain why the rate of bone loss in postmenopausal women diminishes as aging progresses (18).

Estrogen, oxidative stress, and T cell–dependent bone loss

Recently, it has been suggested that ROS may play a role in postmenopausal bone loss by generating a more oxidized bone microenvironment (100, 101). In vivo support of this hypothesis is found from experiments in which ovx induces oxidative stress and impairs antioxidant expression in adult rats (102). Furthermore, administration of agents that increase the intracellular concentration of the antioxidant glutathione in bone prevents bone loss during estrogen deficiency in mice, while depletion of glutathione by buthionine sulfoximine (BSO), which inhibits glutathione synthesis, enhances bone loss (103). The NO donor nitroglycerin is also reported to prevent bone loss in ovx rats (104, 105), while in the presence of N-nitro-l-arginine methyl ester (L-NAME), an NO synthase inhibitor, estrogen was ineffective in reversing bone loss. This suggests that the protective effect of estrogen may be mediated in part through NO (104). In human studies nitroglycerin significantly prevented osteoporotic fractures in postmenopausal women (106).

The mechanisms of action of ROS and the cellular targets that regulate bone mass are poorly understood. OCs have been shown to both generate and be activated by ROS (107, 108). Glutathione peroxidase, responsible for intracellular degradation of hydrogen peroxide, is the predominant antioxidant enzyme expressed by OCs (109) and is upregulated by estrogen. Overexpression of glutathione peroxidase in the preosteoclastic cell line RAW 264.7 abolishes OC formation (109). This suppression of osteoclastic differentiation by antioxidants is likely to occur through protection of phosphatases from reversible inhibition by the ROS hydrogen peroxide (110) and by suppression of thioredoxin expression. Thioredoxin is induced by oxidative stress and enhances OC formation (111). Consistent with a role for hydrogen peroxide in this pathway, catalase was found to prevent ovx-induced bone loss in mice (109).

Estrogen enhances the levels of antioxidants in many cell lineages including the OC (103, 112). The expression of OC TNF is augmented by ROS. Bone loss caused by BSO has significant similarities to bone loss induced by estrogen deficiency, as both processes are TNF dependent (113). Moreover, soluble TNF receptors prevent both bone loss and the rise in thiol-based antioxidants characteristic of estrogen deficiency (113).

Although the mechanisms of ROS action on bone during estrogen deficiency are poorly understood, it is known that immune cells are biological targets of ROS. ROS are important stimulators of antigen presentation by DCs as well as DC-induced T cell activation. Antioxidants potently inhibit DC differentiation and activation of T cells (114, 115) in part by suppressing expression of MHC class II and costimulatory molecules in response to antigen (116). N-acetyl-cysteine (NAC), which acts as an intracellular scavenger by restoring intracellular concentration of glutathione, can block DC maturation (117) and DC-mediated T cell activation (118). ROS are also generated upon DC interaction with T cells (119) and can reduce T cell lifespan by stimulating T cell apoptosis (120). Interestingly, NAC treatment has been shown to protect against ovx-induced bone loss (103). These data are consistent with studies demonstrating that NAC treatment blunts ovx-induced DC activation in the bone marrow, decreases antigen presentation and expression of costimulatory molecules, and prevents T cell activation and TNF production (121). Taken together, these data suggest a model for ovx-induced bone loss in which estrogen deficiency lowers antioxidant levels, thereby increasing ROS. Additionally, estrogen deficiency augments TNF expression by enhancing OC-mediated TNF production and by stimulating APC-induced expansion of the TNF-producing T cells that are central to bone destruction.

Conclusions

Remarkable progress has been made in the last 2 decades in our understanding of the mechanisms of bone destruction during estrogen deficiency. The directions that future research into postmenopausal osteoporosis will take are hard to predict, and new surprises are likely in store. For example, an intriguing link has recently been made between the sympathetic nervous system and bone loss during gonadal failure (122). Our view of postmenopausal osteoporosis will no doubt continue to evolve over the next decade, and radical new therapies will ultimately follow as we gain new knowledge and understanding of this multifaceted malady.

Most new data are derived from studies in mice and remain to be validated in humans. These validation studies will be essential for defining the role of inflammatory cytokines in postmenopausal bone loss, as selective inhibitors might be developed as new therapeutic agents.

The ovx mouse is an excellent model to investigate the acute effects of estrogen withdrawal, although it is not suitable to study the long-term skeletal effects of menopause, as bone loss subsides within a few weeks after ovx in this model. Thus additional animal models and long-term human studies are needed. Since critical effects of estrogen on bone involve regulation of precursor cell differentiation and signaling pathways, which are few and short-lived, many pivotal effects of estrogen in vivo are difficult to reproduce in vitro. Similarly, regulatory events observed in vitro are often not relevant in vivo. It is therefore essential that in vitro studies are validated using in vivo model systems. For example, while estrogen stimulates IFN-γ production in cell cultures (122), estrogen represses it in vivo (53). Similarly, while IFN-γ blocks OC formation through direct targeting of maturing OCs, IFN-γ stimulates osteoclastogenesis and bone resorption in estrogen-deficient mice.

In summary, the multifaceted activities of estrogen are fully reflected in bone. Of the many surprises encountered investigating estrogen action in bone is the relationship among estrogen, the immune system, and the skeleton (Figure 3). Clearly, if this relationship is equally relevant in humans as in rodents, postmenopausal osteoporosis should be regarded as the product of an inflammatory disease bearing many characteristics of an organ-limited autoimmune disorder, triggered by estrogen deficiency, and brought about by chronic mild decreases in T cell tolerance. Why such a pathway should have emerged is intriguing. One explanation is suggested by the need to stimulate bone resorption in the immediate postpartum period in order to meet the markedly increased maternal demand for calcium brought about by milk production. The signal for this event is the drop in estrogen levels early postpartum. Henry Kronenberg (Harvard University, Boston, Massachusetts, USA) has suggested that postmenopausal bone loss should be regarded as an unintended recapitulation of this phenomenon (personal communication).

Another response to delivery is the restoration of normal immune reactivity and the loss of tolerance to the fetus. It is tempting to speculate that cessation of ovarian function induces bone loss through an adaptive immune response because natural selection has centralized these 2 key adaptations to postpartum within the immune system.

Acknowledgments

M.N. Weitzmann is supported in part by grants from the National Osteoporosis Foundation and the National Institutes of Diabetes and Digestive and Kidney Diseases (DK067389). R. Pacifici is supported in part by grants from the National Institutes of Health (AR 49659). We are grateful to Francesco Grassi (Emory University, Atlanta, Georgia, USA) and Timothy Chambers (St. George's Hospital Medical School, London, United Kingdom) for their helpful suggestions.

Address correspondence to: Roberto Pacifici, Division of Endocrinology, Metabolism, and Lipids, Emory University School of Medicine, 101 Woodruff Circle, Room 1307, Atlanta, Georgia 30322, USA. Phone: (404) 712-8420; Fax: (404) 727-1300; E-mail: roberto.pacifici@emory.edu.

1. Bouxsein, M.L., et al. 2005. Ovariectomy-induced bone loss varies among inbred strains of mice. *J. Bone Miner. Res.* **20**:1085–1092.

2. Komm, B.S., et al. 1988. Estrogen binding, receptor mRNA, and biologic response in osteoblast-like osteosarcoma cells. *Science.* **241**:81–84.

3. Tomkinson, A., Gevers, E.F., Wit, J.M., Reeve, J., and Noble, B.S. 1998. The role of estrogen in the control of rat osteocyte apoptosis. *J. Bone Miner. Res.* **13**:1243–1250.

4. Oursler, M.J., Osdoby, P., Pyfferoen, J., Riggs, B.L., and Spelsberg, T.C. 1991. Avian osteoclasts as estrogen target cells. *Proc. Natl. Acad. Sci. U. S. A.* **88**:6613–6617.

5. Weitzmann, M.N., and Pacifici, R. 2005. The role of T lymphocytes in bone metabolism. *Immunol. Rev.* **208**:154–168.

6. Kuiper, G.G., Enmark, E., Pelto-Huikko, M., Nilsson, S., and Gustafsson, J.A. 1996. Cloning of a novel receptor expressed in rat prostate and ovary. *Proc. Natl. Acad. Sci. U. S. A.* **93**:5925–5930.

7. Barkhem, T., et al. 1998. Differential response of estrogen receptor alpha and estrogen receptor beta to partial estrogen agonists/antagonists. *Mol. Pharmacol.* **54**:105–112.

8. Hall, J.M., and McDonnell, D.P. 1999. The estrogen receptor beta-isoform (ERbeta) of the human estrogen receptor modulates ERalpha transcriptional activity and is a key regulator of the cellular response to estrogens and antiestrogens. *Endocrinology.* **140**:5566–5578.

9. Kousteni, S., et al. 2001. Nongenotropic, sex-nonspecific signaling through the estrogen or androgen receptors: dissociation from transcriptional activity. *Cell.* **104**:719–730.

10. Sims, N.A., et al. 2003. A functional androgen receptor is not sufficient to allow estradiol to protect bone after gonadectomy in estradiol receptor–deficient mice. *J. Clin. Invest.* **111**:1319–1327. doi:10.1172/JCI200317246.

11. Smith, C.L., and O'Malley, B.W. 2004. Coregulator function: a key to understanding tissue specificity of selective receptor modulators. *Endocr. Rev.* **25**:45–71.

12. Stein, B., and Yang, M.X. 1995. Repression of the interleukin-6 promoter by estrogen receptor is mediated by NF-kappa B and C/EBP beta. *Mol. Cell. Biol.* **15**:4971–4979.

13. Srivastava, S., et al. 1998. Estrogen blocks M-CSF gene expression and osteoclast formation by regulating phosphorylation of Egr-1 and its interaction with Sp-1. *J. Clin. Invest.* **102**:1850–1859.

14. Srivastava, S., et al. 1999. Estrogen decreases TNF gene expression by blocking JNK activity and the resulting production of c-jun and junD. *J. Clin. Invest.* **104**:503–513.

15. Srivastava, S., et al. 2001. Estrogen decreases osteoclast formation by down-regulating receptor activator of NF-kappa B ligand (RANKL)-induced JNK activation. *J. Biol. Chem.* **276**:8836–8840.

16. Kousteni, S., et al. 2003. Kinase-mediated regulation of common transcription factors accounts for the bone-protective effects of sex steroids. *J. Clin. Invest.* **111**:1651–1664. doi:10.1172/JCI200317261.

17. Frost, H.M. 1983. Bone histomorphometry: analysis of trabecular bone dynamics. In *Bone histomorphometry: techniques and interpretation.* R.R. Recker, editor. CRC Press. Boca Raton, Florida, USA. 109–131.

18. Riggs, B.L., Khosla, S., and Melton, L.J., 3rd. 2002. Sex steroids and the construction and conservation of the adult skeleton. *Endocr. Rev.* **23**:279–302.

19. Eriksen, E.F., Langdahl, B., Vesterby, A., Rungby, J., and Kassem, M. 1999. Hormone replacement therapy prevents osteoclastic hyperactivity: a histomorphometric study in early postmenopausal women. *J. Bone Miner. Res.* **14**:1217–1221.

20. Hughes, D.E., et al. 1996. Estrogen promotes apoptosis of murine osteoclasts mediated by TGF-beta. *Nat. Med.* **2**:1132–1136.

21. Jilka, R.L., et al. 1998. Loss of estrogen upregulates osteoblastogenesis in the murine bone marrow. Evidence for autonomy from factors released during bone resorption. *J. Clin. Invest.* **101**:1942–1950.

22. Weitzmann, M.N., Roggia, C., Toraldo, G., Weitzmann, L., and Pacifici, R. 2002. Increased production of IL-7 uncouples bone formation from bone resorption during estrogen deficiency. *J. Clin. Invest.* **110**:1643–1650. doi:10.1172/JCI200317261.

23. Gilbert, L., et al. 2000. Inhibition of osteoblast differentiation by tumor necrosis factor-alpha. *Endocrinology.* **141**:3956–3964.

24. Eriksen, E.F., et al. 1990. Cancellous bone remodeling in type I (postmenopausal) osteoporosis: quantitative assessment of rates of formation, resorption, and bone loss at tissue and cellular levels. *J. Bone Miner. Res.* **5**:311–319.

25. Riggs, B.L., and Parfitt, A.M. 2005. Drugs used to treat osteoporosis: the critical need for a uniform nomenclature based on their action on bone remodeling. *J. Bone Miner. Res.* **20**:177–184.

26. Vedi, S., et al. 1999. Bone remodeling and structure in postmenopausal women treated with long-term, high-dose estrogen therapy. *Osteoporos. Int.* **10**:52–58.

27. Manolagas, S.C. 2000. Birth and death of bone cells: basic regulatory mechanisms and implications for the pathogenesis and treatment of osteoporosis. *Endocr. Rev.* **21**:115–137.

28. Manolagas, S.C., Kousteni, S., and Jilka, R.L. 2002. Sex steroids and bone. *Recent Prog. Horm. Res.* **57**:385–409.

29. Teitelbaum, S.L. 2000. Bone resorption by osteoclasts. *Science.* **289**:1504–1508.

30. Khosla, S. 2001. Minireview: the OPG/RANKL/RANK system. *Endocrinology.* **142**:5050–5055.

31. Kim, N., et al. 2005. Osteoclast differentiation independent of the TRANCE-RANK-TRAF6 axis. *J. Exp. Med.* **202**:589–595.

32. Hofbauer, L.C., et al. 1999. Interleukin-1beta and tumor necrosis factor-alpha, but not interleukin-6, stimulate osteoprotegerin ligand gene expression in human osteoblastic cells. *Bone.* **25**:255–259.

33. Cenci, S., et al. 2000. Estrogen deficiency induces bone loss by enhancing T-cell production of TNF-α. *J. Clin. Invest.* **106**:1229–1237.

34. Lam, J., et al. 2000. TNF-α induces osteoclastogenesis by direct stimulation of macrophages exposed to permissive levels of RANK ligand. *J. Clin. Invest.* **106**:1481–1488.

35. Zhang, Y.H., Heulsmann, A., Tondravi, M.M., Mukherjee, A., and Abu-Amer, Y. 2001. Tumor necrosis factor-alpha (TNF) stimulates RANKL-induced osteoclastogenesis via coupling of TNF type 1 receptor and RANK signaling pathways. *J. Biol. Chem.* **276**:563–568.

36. Fuller, K., Murphy, C., Kirstein, B., Fox, S.W., and Chambers, T.J. 2002. TNFalpha potently activates osteoclasts, through a direct action independent of and strongly synergistic with RANKL. *Endocrinology.* **143**:1108–1118.

37. Nanes, M.S. 2003. Tumor necrosis factor-alpha: molecular and cellular mechanisms in skeletal pathology. *Gene.* **321**:1–15.

38. Wei, S., Kitaura, H., Zhou, P., Ross, F.P., and Teitelbaum, S.L. 2005. IL-1 mediates TNF-induced osteoclastogenesis. *J. Clin. Invest.* **115**:282–290. doi:10.1172/JCI200317261.

39. Pfeilschifter, J., Koditz, R., Pfohl, M., and Schatz, H. 2002. Changes in proinflammatory cytokine activity after menopause. *Endocr. Rev.* **23**:90–119.

40. Roggia, C., et al. 2001. Up-regulation of TNF-producing T cells in the bone marrow: a key mechanism by which estrogen deficiency induces bone loss in vivo. *Proc. Natl. Acad. Sci. U. S. A.* **98**:13960–13965.

41. Ammann, P., et al. 1997. Transgenic mice expressing soluble tumor necrosis factor-receptor are protected against bone loss caused by estrogen deficiency. *J. Clin. Invest.* **99**:1699–1703.

42. Kimble, R., Bain, S., and Pacifici, R. 1997. The functional block of TNF but not of IL-6 prevents bone loss in ovariectomized mice. *J. Bone Min. Res.* **12**:935–941.

43. Pacifici, R., et al. 1991. Effect of surgical menopause and estrogen replacement on cytokine release from human blood mononuclear cells. *Proc. Natl. Acad. Sci. U. S. A.* **88**:5134–5138.

44. Ralston, S.H., Russell, R.G.G., and Gowen, M. 1990. Estrogen inhibits release of tumor necrosis factor from peripheral blood mononuclear cells in postmenopausal women. *J. Bone Miner. Res.* **5**:983–988.

45. Shanker, G., Sorci-Thomas, M., and Adams, M.R. 1994. Estrogen modulates the expression of tumor necrosis factor alpha mRNA in phorbol ester-stimulated human monocytic THP-1 cells. *Lymphokine Cytokine Res.* **13**:377–382.

46. Abrahamsen, B., Bendtzen, K., and Beck-Nielsen, H. 1997. Cytokines and T-lymphocyte subsets in healthy post-menopausal women: estrogen retards bone loss without affecting the release of IL-1 or IL-1ra. *Bone.* **20**:251–258.

47. Grcevic, D., Lee, S.K., Marusic, A., and Lorenzo, J.A. 2000. Depletion of CD4 and CD8 T lymphocytes in mice in vivo enhances 1, 25- dihydroxyvitamin D(3)-stimulated osteoclast-like cell formation in vitro by a mechanism that is dependent on prostaglandin synthesis. *J. Immunol.* **165**:4231–4238.

48. Kong, Y.Y., et al. 1999. Activated T cells regulate bone loss and joint destruction in adjuvant arthritis through osteoprotegerin ligand. *Nature.* **402**:304–309.

49. Horwood, N.J., et al. 1999. Activated T lymphocytes support osteoclast formation in vitro. *Biochem. Biophys. Res. Commun.* **265**:144–150.

50. Weitzmann, M.N., et al. 2001. T cell activation induces human osteoclast formation via receptor activator of nuclear factor kappaB ligand-dependent and -independent mechanisms. *J. Bone Miner. Res.* **16**:328–337.

51. Takayanagi, H., et al. 2000. T-cell-mediated regulation of osteoclastogenesis by signalling crosstalk between RANKL and IFN-gamma. *Nature.* **408**:600–605.

52. Gao, Y., et al. 2004. Estrogen prevents bone loss through transforming growth factor beta signaling in T cells. *Proc. Natl. Acad. Sci. U. S. A.* **101**:16618–16623.

53. Cenci, S., et al. 2003. Estrogen deficiency induces bone loss by increasing T cell proliferation and lifespan through IFN-gamma-induced class II transactivator. *Proc. Natl. Acad. Sci. U. S. A.* **100**:10405–10410.

54. Grassi, F., and Pacifici, R. 2005. Ovariectomy increases the formation of T cell niches at the resorption surfaces. *J. Bone Miner. Res.* **20**:Abs F395.

55. Adamski, J., Ma, Z., Nozell, S., and Benveniste, E.N. 2004. 17beta-Estradiol inhibits class II major histocompatibility complex (MHC) expression: influence on histone modifications and cbp recruitment to the class II MHC promoter. *Mol. Endocrinol.* **18**:1963–1974.

56. Robey, E.A., et al. 1992. A self-reactive T cell population that is not subject to negative selection. *Int. Immunol.* **4**:969–974.

57. Rammensee, H.G., Falk, K., and Rotzschke, O. 1993. Peptides naturally presented by MHC class I molecules. *Annu. Rev. Immunol.* **11**:213–244.

58. Grossman, Z., and Paul, W.E. 2000. Self-tolerance: context dependent tuning of T cell antigen recognition. *Semin. Immunol.* **12**:197–203; discussion 257–344.

59. Tanchot, C., Lemonnier, F.A., Perarnau, B., Freitas, A.A., and Rocha, B. 1997. Differential requirements for survival and proliferation of CD8 naive or memory T cells. *Science.* **276**:2057–2062.

60. Boss, J.M., and Jensen, P.E. 2003. Transcriptional regulation of the MHC class II antigen presentation pathway. *Curr. Opin. Immunol.* **15**:105–111.

61. Baker, P.J., et al. 1999. CD4(+) T cells and the proinflammatory cytokines gamma interferon and interleukin-6 contribute to alveolar bone loss in mice. *Infect. Immun.* **67**:2804–2809.

62. Arnoldi, J., Gerdes, J., and Flad, H.D. 1990. Immunohistologic assessment of cytokine production of infiltrating cells in various forms of leprosy. *Am. J. Pathol.* **137**:749–753.

63. Firestein, G.S., Alvaro-Gracia, J.M., and Maki, R. 1990. Quantitative analysis of cytokine gene expression in rheumatoid arthritis. *J. Immunol.* **144**:3347–3353.

64. Key, L.L., Jr., et al. 1995. Long-term treatment of osteopetrosis with recombinant human interferon gamma. *N. Engl. J. Med.* **332**:1594–1599.

65. Rodriguiz, R.M., Key, L.L., Jr., and Ries, W.L. 1993. Combination macrophage-colony stimulating factor and interferon-gamma administration ameliorates the osteopetrotic condition in microphthalmic (mi/mi) mice. *Pediatr. Res.* **33**:384–389.

66. Vermeire, K., et al. 1997. Accelerated collagen-induced arthritis in IFN-gamma receptor-deficient mice. *J. Immunol.* **158**:5507–5513.

67. Sato, K., et al. 1992. Prolonged decrease of serum calcium concentration by murine gamma-interferon in hypercalcemic, human tumor (EC-GI)-bearing nude mice. *Cancer Res.* **52**:444–449.

68. Tohkin, M., Kakudo, S., Kasai, H., and Arita, H. 1994. Comparative study of inhibitory effects by murine interferon gamma and a new bisphosphonate (alendronate) in hypercalcemic, nude mice bearing human tumor (LJC-1-JCK). *Cancer Immunol. Immunother.* **39**:155–160.

69. Kotake, S., et al. 2005. IFN-gamma-producing human T cells directly induce osteoclastogenesis from human monocytes via the expression of RANKL. *Eur. J. Immunol.* **35**:3353–3363.

70. Huang, W., O'Keefe, R.J., and Schwarz, E.M. 2003. Exposure to receptor-activator of NFkappaB ligand renders pre-osteoclasts resistant to IFN-gamma by inducing terminal differentiation. *Arthritis. Res. Ther.* **5**:R49–R59.

71. Yang, N.N., Venugopalan, M., Hardikar, S., and Glasebrook, A. 1996. Identification of an estrogen response element activated by metabolites of 17b-estradiol and raloxifene. *Science*. **273**:1222–1225.

72. Miyaura, C., et al. 1997. Increased B-lymphopoiesis by interleukin 7 induces bone loss in mice with intact ovarian function: similarity to estrogen deficiency. *Proc. Natl. Acad. Sci. U. S. A.* **19**:9360–9365.

73. Valenzona, H.O., Pointer, R., Ceredig, R., and Osmond, D.G. 1996. Prelymphomatous B cell hyperplasia in the bone marrow of interleukin 7 transgenic mice: precursor B cell dynamics, microenvironmental organization and osteolysis. *Exp. Hematol.* **24**:1521–1529.

74. Weitzmann, M.N., Cenci, S., Rifas, L., Brown, C., and Pacifici, R. 2000. Interleukin-7 stimulates osteoclast formation by up-regulating the T- cell production of soluble osteoclastogenic cytokines. *Blood*. **96**:1873–1878.

75. Toraldo, G., Roggia, C., Qian, W.P., Pacifici, R., and Weitzmann, M.N. 2003. IL-7 induces bone loss in vivo by induction of receptor activator of nuclear factor kappa B ligand and tumor necrosis factor alpha from T cells. *Proc. Natl. Acad. Sci. U. S. A.* **100**:125–130.

76. Ryan, M.R., et al. 2005. An IL-7-dependent rebound in thymic T cell output contributes to the bone loss induced by estrogen deficiency. *Proc. Natl. Acad. Sci. U. S. A.* **102**:16735–16740.

77. Lindberg, M.K., et al. 2006. Liver-derived IGF-I is permissive for ovariectomy-induced trabecular bone loss. *Bone*. **38**:85–92.

78. Lee, S.K., et al. 2006. Interleukin-7 influences osteoclast function in vivo but is not a critical factor in ovariectomy-induced bone loss. *J. Bone Miner. Res.* doi:10.1359/jbmr.060117.

79. Huang, M., et al. 2002. IL-7 inhibits fibroblast TGF-β production and signaling in pulmonary fibrosis. *J. Clin. Invest.* **109**:931–937. doi:10.1172/JCI200214685.

80. Dubinett, S.M., et al. 1995. Down-regulation of murine fibrosarcoma transforming growth factor-beta 1 expression by interleukin 7. *J. Natl. Cancer Inst.* **87**:593–597.

81. Giuliani, N., et al. 2002. Human myeloma cells stimulate the receptor activator of nuclear factor-kappa B ligand (RANKL) in T lymphocytes: a potential role in multiple myeloma bone disease. *Blood*. **100**:4615–4621.

82. van Roon, J.A., Glaudemans, K.A., Bijlsma, J.W., and Lafeber, F.P. 2003. Interleukin 7 stimulates tumour necrosis factor alpha and Th1 cytokine production in joints of patients with rheumatoid arthritis. *Ann. Rheum. Dis.* **62**:113–119.

83. De Benedetti, F., et al. 1995. Elevated circulating interleukin-7 levels in patients with systemic juvenile rheumatoid arthritis. *J. Rheumatol.* **22**:1581–1585.

84. Masuzawa, T., et al. 1994. Estrogen deficiency stimulates B lymphopoiesis in mouse bone marrow. *J. Clin. Invest.* **94**:1090–1097.

85. Manabe, N., et al. 2001. Connection between B lymphocyte and osteoclast differentiation pathways. *J. Immunol.* **167**:2625–2631.

86. Sato, T., Shibata, T., Ikeda, K., and Watanabe, K. 2001. Generation of bone-resorbing osteoclasts from B220+ cells: its role in accelerated osteoclastogenesis due to estrogen deficiency. *J. Bone Miner. Res.* **16**:2215–2221.

87. Haynes, B.F., Sempowski, G.D., Wells, A.F., and Hale, L.P. 2000. The human thymus during aging. *Immunol. Res.* **22**:253–261.

88. Douek, D.C., and Koup, R.A. 2000. Evidence for thymic function in the elderly. *Vaccine*. **18**:1638–1641.

89. Jamieson, B.D., et al. 1999. Generation of functional thymocytes in the human adult. *Immunity*. **10**:569–575.

90. Steinmann, G.G., Klaus, B., and Muller-Hermelink, H.K. 1985. The involution of the ageing human thymic epithelium is independent of puberty. A morphometric study. *Scand. J. Immunol.* **22**:563–575.

91. Hakim, F.T., et al. 2005. Age-dependent incidence, time course, and consequences of thymic renewal in adults. *J. Clin. Invest.* **115**:930–939. doi:10.1172/JCI200317261.

92. Mackall, C.L., et al. 2001. IL-7 increases both thymic-dependent and thymic-independent T-cell regeneration after bone marrow transplantation. *Blood*. **97**:1491–1497.

93. Chu, Y.W., et al. 2004. Exogenous IL-7 increases recent thymic emigrants in peripheral lymphoid tissue without enhanced thymic function. *Blood*. **101**:1110–1119.

94. Alpdogan, O., et al. 2001. Administration of interleukin-7 after allogeneic bone marrow transplantation improves immune reconstitution without aggravating graft-versus-host disease. *Blood*. **98**:2256–2265.

95. Utsuyama, M., and Hirokawa, K. 1989. Hypertrophy of the thymus and restoration of immune functions in mice and rats by gonadectomy. *Mech. Ageing Dev.* **47**:175–185.

96. Okasha, S.A., et al. 2001. Evidence for estradiol-induced apoptosis and dysregulated T cell maturation in the thymus. *Toxicology*. **163**:49–62.

97. Roden, A.C., et al. 2004. Augmentation of T cell levels and responses induced by androgen deprivation. *J. Immunol.* **173**:6098–6108.

98. Sutherland, J.S., et al. 2005. Activation of thymic regeneration in mice and humans following androgen blockade. *J. Immunol.* **175**:2741–2753.

99. Hreshchyshyn, M.M., Hopkins, A., Zylstra, S., and Anbar, M. 1988. Effects of natural menopause, hysterectomy, and oophorectomy on lumbar spine and femoral neck bone densities. *Obstet. Gynecol.* **72**:631–638.

100. Basu, S., Michaelsson, K., Olofsson, H., Johansson, S., and Melhus, H. 2001. Association between oxidative stress and bone mineral density. *Biochem. Biophys. Res. Commun.* **288**:275–279.

101. Maggio, D., et al. 2003. Marked decrease in plasma antioxidants in aged osteoporotic women: results of a cross-sectional study. *J. Clin. Endocrinol. Metab.* **88**:1523–1527.

102. Muthusami, S., et al. 2005. Ovariectomy induces oxidative stress and impairs bone antioxidant system in adult rats. *Clin. Chim. Acta.* **360**:81–86.

103. Lean, J.M., et al. 2003. A crucial role for thiol antioxidants in estrogen-deficiency bone loss. *J. Clin. Invest.* **112**:915–923. doi:10.1172/JCI200317261.

104. Wimalawansa, S.J., De Marco, G., Gangula, P., and Yallampalli, C. 1996. Nitric oxide donor alleviates ovariectomy-induced bone loss. *Bone*. **18**:301–304.

105. Hao, Y.J., Tang, Y., Chen, F.B., and Pei, F.X. 2005. Different doses of nitric oxide donor prevent osteoporosis in ovariectomized rats. *Clin. Orthop. Relat. Res.* **435**:226–231.

106. Jamal, S.A., Cummings, S.R., and Hawker, G.A. 2004. Isosorbide mononitrate increases bone formation and decreases bone resorption in postmenopausal women: a randomized trial. *J. Bone Miner. Res.* **19**:1512–1517.

107. Steinbeck, M.J., Appel, W.H., Jr., Verhoeven, A.J., and Karnovsky, M.J. 1994. NADPH-oxidase expression and in situ production of superoxide by osteoclasts actively resorbing bone. *J. Cell Biol.* **126**:765–772.

108. Ha, H., et al. 2004. Reactive oxygen species mediate RANK signaling in osteoclasts. *Exp. Cell Res.* **301**:119–127.

109. Lean, J.M., Jagger, C.J., Kirstein, B., Fuller, K., and Chambers, T.J. 2005. Hydrogen peroxide is essential for estrogen-deficiency bone loss and osteoclast formation. *Endocrinology*. **146**:728–735.

110. Reth, M. 2002. Hydrogen peroxide as second messenger in lymphocyte activation. *Nat. Immunol.* **3**:1129–1134.

111. Lean, J., Kirstein, B., Urry, Z., Chambers, T., and Fuller, K. 2004. Thioredoxin-1 mediates osteoclast stimulation by reactive oxygen species. *Biochem. Biophys. Res. Commun.* **321**:845–850.

112. Chen, J.R., et al. 2005. Transient versus sustained phosphorylation and nuclear accumulation of ERKs underlie anti-versus pro-apoptotic effects of estrogens. *J. Biol. Chem.* **280**:4632–4638.

113. Jagger, C.J., Lean, J.M., Davies, J.T., and Chambers, T.J. 2005. Tumor necrosis factor-alpha mediates osteopenia caused by depletion of antioxidants. *Endocrinology*. **146**:113–118.

114. Mizuashi, M., Ohtani, T., Nakagawa, S., and Aiba, S. 2005. Redox imbalance induced by contact sensitizers triggers the maturation of dendritic cells. *J. Invest. Dermatol.* **124**:579–586.

115. Rutault, K., Alderman, C., Chain, B.M., and Katz, D.R. 1999. Reactive oxygen species activate human peripheral blood dendritic cells. *Free Radic. Biol. Med.* **26**:232–238.

116. Maemura, K., et al. 2005. Reactive oxygen species are essential mediators in antigen presentation by Kupffer cells. *Immunol. Cell Biol.* **83**:336–343.

117. Vosters, O., et al. 2003. Dendritic cells exposed to nacystelyn are refractory to maturation and promote the emergence of alloreactive regulatory T cells. *Transplantation*. **75**:383–389.

118. Verhasselt, V., et al. 1999. N-acetyl-L-cysteine inhibits primary human T cell responses at the dendritic cell level: association with NF-kappaB inhibition. *J. Immunol.* **162**:2569–2574.

119. Matsue, H., et al. 2003. Generation and function of reactive oxygen species in dendritic cells during antigen presentation. *J. Immunol.* **171**:3010–3018.

120. Hildeman, D.A., et al. 1999. Reactive oxygen species regulate activation-induced T cell apoptosis. *Immunity*. **10**:735–744.

121. Grassi, F., and Pacifici, R. 2005. Oxidative stress induced dendritic cell-dependent T cell activation. A novel mechanism by which estrogen deficiency causes bone loss. *J. Bone Min. Res.* **20**:a1144.

122. Elefteriou, F., et al. 2005. Leptin regulation of bone resorption by the sympathetic nervous system and CART. *Nature*. **434**:514–520.

123. Fox, H.S., Bond, B.L., and Parslow, T.G. 1991. Estrogen regulates the IFN-gamma promoter. *J. Immunol.* **146**:4362–4367.

Regulation of bone mass by Wnt signaling

Venkatesh Krishnan,[1] Henry U. Bryant,[1] and Ormond A. MacDougald[2]

[1]Musculoskeletal Research, Lilly Research Laboratories, Indianapolis, Indiana, USA.
[2]Departments of Molecular and Integrative Physiology and Internal Medicine, University of Michigan, Ann Arbor, Michigan, USA.

Wnt proteins are a family of secreted proteins that regulate many aspects of cell growth, differentiation, function, and death. Considerable progress has been made in our understanding of the molecular links between Wnt signaling and bone development and remodeling since initial reports that mutations in the Wnt coreceptor low-density lipoprotein receptor–related protein 5 (LRP5) are causally linked to alterations in human bone mass. Of the pathways activated by Wnts, it is signaling through the canonical (i.e., Wnt/β-catenin) pathway that increases bone mass through a number of mechanisms including renewal of stem cells, stimulation of preosteoblast replication, induction of osteoblastogenesis, and inhibition of osteoblast and osteocyte apoptosis. This pathway is an enticing target for developing drugs to battle skeletal diseases as Wnt/β-catenin signaling is composed of a series of molecular interactions that offer potential places for pharmacological intervention. In considering opportunities for anabolic drug discovery in this area, one must consider multiple factors, including (a) the roles of Wnt signaling for development, remodeling, and pathology of bone; (b) how pharmacological interventions that target this pathway may specifically treat osteoporosis and other aspects of skeletal health; and (c) whether the targets within this pathway are amenable to drug intervention. In this Review we discuss the current understanding of this pathway in terms of bone biology and assess whether targeting this pathway might yield novel therapeutics to treat typical bone disorders.

Wnt/β-catenin signaling

Wnt signaling plays an important role in development and maintenance of many organs and tissues, including bone (1). Although Wnt proteins signal through several pathways to regulate cell growth, differentiation, function, and death, the Wnt/β-catenin or canonical pathway appears to be particularly important for bone biology (reviewed in refs. 2, 3). The complexities of the Wnt/β-catenin signaling pathway in multiple cell types have been reviewed elsewhere (4, 5), and an outline of the pathway is shown in Figure 1. If Wnts are not expressed or if their binding to receptors is inhibited, degradation of β-catenin is facilitated via interactions with a protein complex consisting of adenomatous polyposis coli (APC), axin, and glycogen synthase kinase 3 (GSK3). APC and axin act as scaffold proteins allowing GSK3 to bind and phosphorylate β-catenin, identifying it for degradation by the β-TrCP–mediated ubiquitin/proteasome pathway.

Activation of Wnt/β-catenin signaling occurs upon binding of Wnt to the 7-transmembrane domain–spanning frizzled receptor and low-density lipoprotein receptor–related protein 5 and 6 (LRP5/6) coreceptors (Figure 1). Signals are generated through the proteins Disheveled, Axin, and Frat-1, which disrupt the protein complex and inhibit the activity of GSK3, thus causing hypophosphorylation of its substrate, β-catenin (6). Stabilized β-catenin then accumulates in the cytosol and translocates to the nucleus, where this transcriptional coactivator interacts with T cell factor/lymphoid enhancer binding factor (TCF/LEF) transcription fac-

tors to mediate many of the effects of Wnts on gene transcription. Binding of β-catenin displaces transcriptional corepressors (e.g., silencing mediator of retinoid and thyroid receptors and nuclear receptor corepressor [SMRT/NCoR]) bound to TCF/LEF and recruits transcriptional coactivators (e.g., p300 and cAMP response element–binding protein [p300/CBP]) (7).

Wnt signaling is tightly regulated by members of several families of secreted antagonists. Interactions between Wnts and frizzled receptors are inhibited by members of the secreted frizzled-related protein (sFRP) family and Wnt inhibitory factor 1 (WIF-1; Figure 1). LRP5/6 coreceptor activity is inhibited by members of the sclerostin (SOST gene product) and Dickkopf (Dkk) families, all of which bind LRP5/6. Dkk1, -2, and -4 bind with various affinities to LRP5 and LRP6. Interaction of the Dkk/LRP complex with kremen internalizes the complex for degradation, thus diminishing the number of Wnt coreceptors available for signaling (8).

Wnt signaling regulates bone mass

Bone mass is influenced by the balance achieved between bone-forming cells (osteoblasts) and bone-resorbing cells (osteoclasts). Loss-of-function mutations in human LRP5 are associated with osteoporosis-pseudoglioma syndrome, which is characterized by low bone mineral density and skeletal fragility (9). In contrast, mutations in the N terminus of human LRP5 (e.g., G171V) that reduce affinity of LRP5 for Dkk1 are associated with high bone mass (10–12). These human bone phenotypes are largely supported by animal models with altered expression of LRP5. For example, Lrp5−/− mice have a low bone mass phenotype due to reduced proliferation of precursor cells (13). Furthermore, mice that overexpress the G171V LRP5 mutant in osteoblasts have enhanced osteoblast activity, reduced osteoblast apoptosis (Figure 2), and a high bone mass phenotype reminiscent of that observed in humans with this mutation (14). Interestingly, overexpression of wild-type Lrp5 leads to a more subtle bone phenotype, suggesting that the G171V mutant has a gain-in-function phenotype suggestive of a dominant-positive mechanism. Loss of bone mineral density in

Nonstandard abbreviations used: APC, adenomatous polyposis coli; BMP, bone morphogenic protein; C/EBP, CCAAT/enhancer binding protein; Dkk, Dickkopf; GSK, glycogen synthase kinase; LEF, lymphoid enhancer binding factor; LiCl, lithium chloride; LRP, low-density lipoprotein receptor–related protein; RANKL, receptor activator of NF-κB ligand; Runx2, runt-related transcription factor 2; sFRP, secreted frizzled-related protein; TCF, T cell factor; WIF-1, Wnt inhibitory factor 1; WISP, Wnt-induced secreted protein.

Conflict of interest: V. Krishnan and H.U. Bryant own stock in Eli Lilly & Co. O.A. MacDougald has declared that no conflict of interest exists.

Citation for this article: *J. Clin. Invest.* **116**:1202–1209 (2006). doi:10.1172/JCI28551.

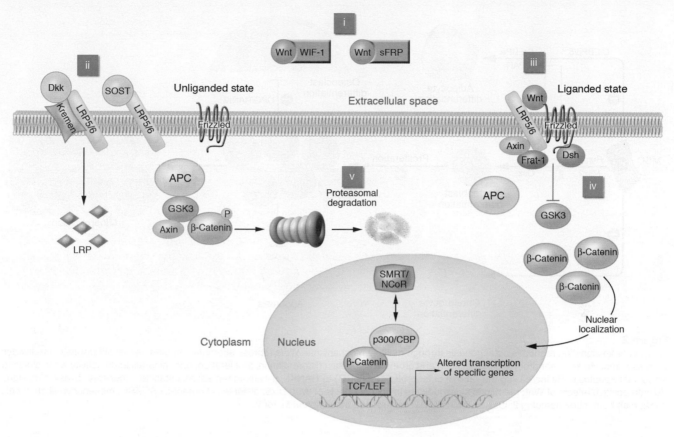

Figure 1

Elements of Wnt/β-catenin signaling. In the liganded state, binding of Wnt to the frizzled receptor inhibits GSK3 activity through mechanisms involving Axin, Frat-1, and Disheveled (Dsh). β-Catenin accumulates and is translocated to the nucleus, where it binds to TCF/LEF, causing displacement of transcriptional corepressors (e.g., silencing mediator of retinoid and thyroid receptors and nuclear receptor corepressor; SMRT/NCoR) with transcriptional coactivators (e.g., p300 and cAMP response element–binding protein; p300/CBP). Wnt signaling can be blocked by interactions of Wnt with inhibitory factors including WIF-1 and sFRP or the interaction of LRP5/6 with the Dkk/Kremen complex or sclerostin (SOST gene product). Phosphorylation of β-catenin by GSK3 stimulates β-catenin degradation. Potential intervention points for drug therapy (i–v) are indicated.

Lrp5⁻/⁻ mice is further exacerbated by loss of an *Lrp6* allele, suggesting that Wnts signal through both the LRP5 and LRP6 coreceptors to influence bone mass (15). Finally, disruption of endogenous LRP inhibitors such as Dkk1 (11) or sclerostin (16) increases the ability of Wnts to stabilize β-catenin and stimulate osteogenesis, further cementing the evidence that signaling from LRP coreceptors is important for bone development.

Direct roles for Wnt signaling in the regulation of trabecular bone formation and bone mass were further supported by studies of mice lacking the soluble Wnt inhibitor sFRP1 (17). These mice show reduced osteoblast and osteocyte apoptosis in vivo, and results from work with marrow-derived cells from *Sfrp1⁻/⁻* mice suggest that in addition to preventing apoptosis, Wnt signaling may also increase bone by stimulating differentiation and replication of osteoblasts (Figure 2). While bone phenotypes observed in mice with altered expression or activity of Wnt coreceptors or inhibitors support a simple and direct relationship between Wnt signaling and bone mass, the relationship between members of the Wnt signaling pathway and bone biology will undoubtedly be more complex. For example, a recent report indicates that, as expected, activation of Wnt/β-catenin signaling induces osteoblastogenesis and that these effects are blocked by Dkk1 (18). However, during osteoblastogenesis Wnt/β-catenin signaling — presumably initiat-

ed by Wnt7b — induces expression of Dkk2, which is then surprisingly required for subsequent mineralization. Thus *Dkk2⁻/⁻* mice have increased secreted matrix (osteoid) but impaired mineralization, culminating in an osteopenic phenotype (18). While mineralization is partially rescued in vitro by Dkk2, another inhibitor of Wnt signaling, sFRP3, fails to stimulate mineralization, suggesting that Dkk2 may act through a mechanism distinct from Wnt antagonizing activity. This concept is not unprecedented as sFRP1 inhibits osteoclastogenesis through binding of receptor activator of NF-κB ligand (RANKL; ref. 19) and reorients axonal growth through interactions with frizzled 2 (20). Although the functional significance is unknown, other Wnt inhibitors such as WIF-1 and sFRP2 may also be induced in osteoblasts (21).

Taken together, these studies suggest that endogenous Wnt signaling plays an important role in osteoblastogenesis and bone formation (10–18); however, which of the 19 Wnts are involved has yet to be delineated. It is likely that Wnt activity in bone marrow varies throughout stages of development and has important contributions from several Wnts. One of these may be Wnt7b, which is induced during osteoblastogenesis (18, 22). Another is Wnt10b, which is expressed in bone marrow (23, 24), and deficiency of which leads to reduced trabecular bone mass, bone mineral density, and serum osteocalcin level (24). In addition, Wnt1, Wnt4,

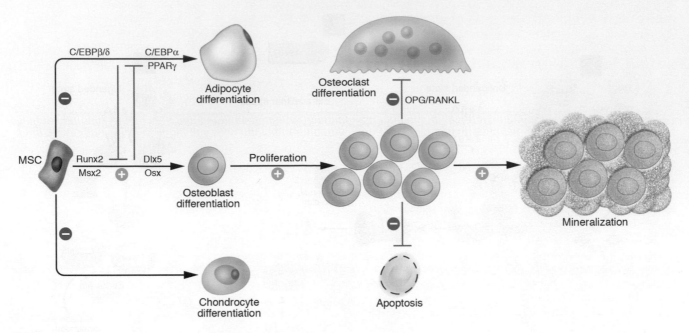

Figure 2
Wnt/β-catenin signaling regulates osteogenesis through multiple mechanisms. Wnts repress alternative mesenchymal differentiation pathways such as adipocyte and chondrocyte differentiation and promote osteoblast differentiation, proliferation, and mineralization activity while blocking osteoblast apoptosis. By increasing the ratio of osteoprotegerin (OPG) to RANKL, β-catenin represses osteoclastogenesis. Green plus signs indicate positive effects of Wnt; red minus signs indicate inhibitory effects of Wnt. Dlx5, distal-less homeobox 5; MSC, mesenchymal stem cell; Msx2, msh homeobox homolog 2; Osx, osterix; Runx2, runt-related transcription factor 2.

and Wnt14 are expressed in calvarial tissue and osteoblast cultures (13), and Wnt1 and Wnt3a are induced by bone morphogenic protein 2 (BMP-2) in a mesenchymal precursor cell line (25). However, because Wnts function through autocrine and paracrine mechanisms, analysis of those Wnts that specifically contribute to bone formation, as well as the frizzled receptors mediating their effects, will require in situ analysis of gene expression within bone and marrow and confirmation by genetic approaches.

Wnt regulates osteoblastogenesis through the canonical pathway

One of the mechanisms whereby Wnt signaling increases bone formation is via stimulation of the development of osteoblasts, and there is considerable in vitro evidence supporting a role for Wnt/β-catenin (i.e., canonical) signaling in this process (Figure 2). For example, inhibition of GSK3 enzymatic activity with lithium chloride (LiCl; ref. 26) or small molecules (e.g., Chir99021 and LY603281-31-8) stimulates mesenchymal precursors to differentiate into osteoblasts (24, 27, 28). This concept is supported by observations with Wnt3a, Wnt1, Wnt10b, and constitutively active β-catenin, all of which activate signaling through β-catenin and stimulate osteoblastogenesis, while Dkk1, which inhibits this pathway, reduces osteoblastogenesis (24, 28, 29). Importantly, activation of Wnt/β-catenin signaling also inhibits adipogenesis of mesenchymal precursors (30, 31), which may have clinical importance given the positive correlation reported between marrow adipose content and bone fractures (32).

Further evidence that Wnt signaling increases bone mass through the Wnt/β-catenin pathway comes from the results of in vivo studies using pharmacological inhibitors of GSK3β. For example, administration of LiCl for 4 weeks dramatically increased bone formation rate and number of osteoblasts in C57BL/6 mice (33). Similar results were obtained in osteopenic $Lrp5^{-/-}$ mice, indicating that LiCl acts downstream of LRP5. Consistent with the in vitro results described above, inhibition of GSK3 reduced the number of marrow adipocytes over this period. LiCl influences other signaling pathways besides Wnt, and GSK3 regulates many proteins besides β-catenin. However, the fact that LiCl stabilizes β-catenin and increases TCF-based reporter gene activity and expression of Wnt-responsive genes strongly supports a mechanism mediated by the Wnt/β-catenin signaling pathway (33).

Role of β-catenin at various stages of osteoblast development

During embryonic development, the level of β-catenin is increased in differentiating osteoblasts (34), and pharmacological and genetic approaches have indicated that Wnt signaling increases bone mass through a number of mechanisms including renewal of stem cells (35), stimulation of preosteoblast replication (13), induction of osteoblastogenesis (13), and inhibition of osteoblast and osteocyte apoptosis (Figure 2) (17). These variable results likely arise because Wnt/β-catenin signaling regulates bone development and accrual through different mechanisms at different stages of life. This concept is supported by the results of studies using mouse models in which targeted deletion of β-catenin occurs early or late in osteoblastogenesis. For example, *dermo*-Cre mice have a targeted deletion of β-catenin in mesenchymal precursors of chondrogenesis and osteogenesis (36, 37). These mice show a reduction in all relevant markers of osteogenesis and an absence of both endochondral and intramembranous bone at E18.5 in the developing embryo. Thus β-catenin is required for the early stages of osteoblastogenesis, and indeed its absence steers the fate of mesenchymal precursors toward chondrogenesis (34, 37).

To examine the importance of β-catenin later in osteoblast development, constitutively active β-catenin was overexpressed in osteoblasts expressing collagen α_I-CRE (38). These mice manifest an osteopetrotic phenotype; however, no change in osteoblast activity or histomorphometric evidence of bone formation was observed. Instead, bone resorption and osteoclastogenesis were defective due to increased expression of osteoprotegerin, a decoy receptor for RANKL (38). On the other hand, targeted deletion of β-catenin in mature osteoblasts with collagen α_I-Cre caused increased bone resorption and a marked increase in the number of tartrate-resistant acid phosphatase–positive (TRAP-positive) multinucleated osteoclasts due to reduced expression of osteoprotegerin (38). Consistent with these observations in mice, autosomal-dominant osteopetrosis type I patients with a gain-of-function T253I mutation in LRP5 have decreased numbers of small osteoclasts, although osteoclastogenesis in response to RANKL was normal in vitro (39). Finally, mice in which β-catenin has been deleted using osteocalcin-CRE and mice in which β-catenin has been activated with conditional *Apc* mutants provide further support for the finding that β-catenin regulates osteoblast differentiation. In addition, these mice also demonstrate that β-catenin regulates osteoclastogenesis through effects on expression of osteoprotegerin and RANKL (40).

A role for β-catenin in regulation of osteoclastogenesis has been clearly delineated through multiple genetic approaches; however, there is also considerable evidence that altering Wnt signaling upstream of β-catenin does not increase bone formation through altered resorption. For example, alterations in osteoclast variables were not observed in *Lrp5*$^{-/-}$, *Sfrp1*$^{-/-}$, or *Wnt10b*$^{-/-}$ mice or with LiCl treatment (13, 17, 24). One can speculate that complete loss or overexpression of β-catenin are more extreme perturbations of this signaling system than are normally observed through alterations of Wnt activity earlier in the pathway. In addition, constitutively active β-catenin may lack autoregulatory pathways triggered by the Wnt pathway.

Mechanisms whereby Wnt signaling regulates bone mass

As described above, Wnt signaling increases bone mass through diverse mechanisms. While effects on osteoblastogenesis and apoptosis have been studied in some mechanistic detail and will be elaborated upon here, this is not to diminish the potential importance of other mechanisms mentioned earlier that are less well studied, including renewal of stem cells (35), stimulation of preosteoblast replication (13), and enhancement of osteoblast activity (Figure 2) (13, 17).

Osteoblastogenesis versus adipogenesis. There is considerable evidence for the existence of a mesenchymal stem cell that gives rise to both osteogenic and adipogenic cells, and in vitro and in vivo experimental models have provided compelling evidence for a reciprocal relationship between these cell lineages (41–43). For example, cultures of bone marrow stromal cells as well as immortalized clonal lines (e.g., ST2) are capable of both osteogenic and adipogenic differentiation, depending upon culture conditions. Furthermore, single cell clones from bone marrow can differentiate in vitro into either adipocytes or osteoblasts (44). In addition to signaling by Wnt/β-catenin, a number of factors influence the fate of these marrow-derived mesenchymal stem cells, including retinoic acid, BMPs, vitamin D3, glucocorticoids, notch, sonic hedgehog, parathyroid hormone, parathyroid hormone–related peptide, and PPARγ ligands (24, 43, 45–47). Indeed, Wnt signaling may be required for or even mediate a subset of effects of BMP, parathyroid hormone, and hedgehog on cell fate decisions toward osteoblastogenesis (25, 48).

Pharmacological and genetic treatments that activate Wnt/β-catenin signaling in mesenchymal precursors repress adipogenesis and stimulate osteoblastogenesis (Figure 2). In preadipocyte models expression of Wnt does not influence induction of the transcription factors CCAAT/enhancer binding protein β (C/EBPβ) and C/EBPδ, but Wnt signaling blocks induction of master adipogenic transcription factors C/EBPα and PPARγ (30). Suppression of Wnt/β-catenin signaling with dominant-negative TCFs or sFRPs stimulates spontaneous adipogenesis, indicating that endogenous Wnts inhibit preadipocyte differentiation (30, 31). Wnt signaling is initiated in part by Wnt10b. Its expression is high in dividing and confluent preadipocytes, and Wnt10b is rapidly suppressed upon induction of differentiation (30, 31). In addition, ectopic expression of Wnt10b stabilizes free cytosolic β-catenin and is a potent inhibitor of adipogenesis. Most conclusively, Wnt10b antisera promotes adipogenesis when added to media of 3T3-L1 preadipocytes. Interestingly, expression of Wnt5b is transiently induced during differentiation of 3T3-L1 cells, and adenoviral expression of Wnt5b causes a slight increase in adipogenesis, presumably due to destabilization of β-catenin (49, 50). Wnt5b may activate noncanonical Wnt signaling, which has been reported to antagonize Wnt/β-catenin signaling (51), or Wnt5b may compete with other Wnts for binding to frizzled receptors. Further work is required to assess whether Wnt5b inhibits osteoblastogenesis.

Mesenchymal precursors such as ST2 cells express low but biologically relevant levels of adipogenic transcription factors C/EBPα and PPARγ and osteoblast transcription factors such as runt-related transcription factor 2 (Runx2), msh homeobox homolog 2 (Msx2), distal-less homeobox 5 (Dlx5), and osterix (24). Expression of these 2 classes of transcription factors is maintained at low levels due to negative feedback, and imbalance leads to differentiation. For example, Msx2 binds to C/EBPα and inhibits its ability to transactivate the PPARγ promoter, and Msx2 represses adipogenesis (52, 53). Similarly, PPARγ binds to Runx2 and inhibits transactivation of the osteocalcin promoter, and activation of PPARγ represses osteoblastogenesis (54). Constitutive Wnt/β-catenin signaling favors expression of osteoblast genes at the expense of adipocyte genes (24). Wnt signaling could regulate the fate of mesenchymal precursors by repressing adipocyte transcription factors, stimulating osteoblast transcription factors, or both (Figure 2). Increased bone mass in *Pparg*$^{+/-}$ mice, increased osteogenesis in precursor cells from PPARγ-null mice, and decreased bone density following treatment of mice with a PPARγ agonist make this factor an attractive target (55–57). Indeed, suppression of PPARγ is required for Wnt10b to stimulate osteoblastogenesis (24). A recent report indicated that a transcriptional regulator, transcription coactivator with PDZ domain (TAZ), mediates the effects of BMP-2 on mesenchymal cell fate by inhibiting PPARγ activity while stimulating that of Runx2; however, the potential role of TAZ-mediated effects of Wnt/β-catenin signaling has not been reported (58).

Apoptosis. Induction of bone accrual in mouse models with increased Wnt signaling is due in part to reduced apoptosis of osteoblasts and osteocytes (14, 17, 59). Wnt signaling inhibits apoptosis in response to a wide variety of cellular insults, including chemotherapeutic agents and serum deprivation (60–62). Prevention of apoptosis occurs in a wide variety of cell models, including mesenchymal precursors, preosteoblasts, and osteoblasts. While signaling by canonical Wnts appears to universally protect against apoptosis through mechanisms involving β-catenin and activation of PI3K/Akt, other mechanistic aspects are dependent on cell type. For example, in rat intestinal epithelial cells, induction of cyclooxygenase-2 and Wnt-induced secreted protein 1 (WISP-1), but not Bcl-2, are critical for repression of apoptosis caused by c-myc (60). In preadipocytes, increased production of insulin-like growth factors feeds back through an autocrine/paracrine mechanism to block apoptosis due to serum deprivation (61). Finally, in preosteoblasts, activation of Src, ERK, and Akt by Wnt3a is required for prevention of apoptosis. In this cell model, Wnt signaling induces expression of Bcl-2 through a process requiring active ERK (62). The mechanism or mechanisms by which Wnt/β-catenin signaling brings about an increase in the number of osteoblasts and osteocytes in vivo have yet to be determined.

Wnt signaling as cause and treatment for bone diseases

Historically, diseases of bone loss have been treated with agents that block bone resorption; however, this type of therapy stimulates only a modest increase in bone mineral density, and osteoporotic patients retain an elevated risk for fracture. With the recent introduction of teriparatide (human parathyroid hormone 1–34) into clinical practice, the potential to treat patients with an anabolic therapy was introduced (63). This drug is proven to decrease risk of vertebral and nonvertebral fractures in patients with postmenopausal osteoporosis (64). Pharmaceuticals that specifically activate the Wnt/β-catenin pathway in bone also hold tremendous promise as anabolic agents that may add to or complement treatment with teriparatide (3). Potential patient populations may include those with osteopenia or osteoporosis due to (a) causes unrelated to Wnt signaling and causes that do not impair effects of Wnt signaling on bone formation and (b) defects in Wnt signaling, as long as the drug acts downstream of the defect.

In any drug discovery program, issues of safety are paramount, especially for treatment of chronic disease of bone that will likely involve long-term therapy. This is particularly true for activators of Wnt/β-catenin signaling, since Wnts were first identified as insertion sites for mouse mammary tumor virus (64) and since mutations in APC and β-catenin that increase Wnt signaling are associated with colon and other cancers (65). When considering how best to target drug discovery in the Wnt/β-catenin pathway, identification and screening upstream in the pathway is more promising than targeting β-catenin and downstream events. For example, humans and mice with altered expression of LRP5, sFRP1, and Wnt10b all have alterations in bone mass with relatively few effects elsewhere (9, 17, 24). Side effects of drug therapy targeting the Wnt/β-catenin pathway are unknown. Functional haploinsufficiency for LRP5 may cause familial exudative vitreoretinopathy (66), and activation of Wnt10b signaling in fat decreases adiposity and increases skin thickness (67). In contrast, altering expression of β-catenin causes profound developmental effects (68, 69) and in bone regulates osteoblastogenesis, osteoclastogenesis, and probability of benign osteomatas (34, 38, 40). This underscores also that drugs should be selected for moderate effects on the pathway, as strong activators will have a much higher probability of effects in nontarget tissues. Despite the risks, the paucity of anabolic drugs

for regulating bone mass and the compelling evidence demonstrating that Wnt/β-catenin signaling stimulates bone formation justify the considerable effort being put forth by the pharmaceutical industry to target this pathway.

Osteoporosis. Osteoporosis is a prevalent skeletal disorder characterized by compromised bone strength and consequent increased risk of fractures. Postmenopausal women are at higher risk for developing osteoporosis and osteoporosis-related fractures. There are multiple etiologies for this complex metabolic bone disease, and, with the exception of osteoporosis-pseudoglioma syndrome due to mutations in LRP5 (9), it is unknown whether Wnt signaling plays a role. Interestingly, dexamethasone increases expression of Dkk1 and sFRP1 and represses Wnt/β-catenin signaling in human osteoblasts, suggesting a role for this pathway in glucocorticoid-induced osteoporosis (70–72). Further mechanistic work in human osteoporosis will be important to fully understand the relevance of Wnt signaling pathways in this disease.

To address the effects of increasing Wnt signaling on bone mass under normal and osteoporosis conditions, expression of Wnt10b was directed to bone marrow using the fatty acid–binding protein 4 (FABP4) promoter (24, 67). Wnt10b increased bone mineral density throughout the weight-bearing skeleton. Increased trabecular bone was observed throughout the endocortical compartment, with a 4-fold increase in bone volume fraction in the femoral distal metaphysis and improved material properties including strength. Although there was a trend toward decreased bone volume fraction and mineral density in ovariectomized FABP4-Wnt10b mice, these mice were protected due to their higher initial bone mass. Thus the potential health benefit from increasing Wnt/β-catenin signaling by Wnt10b is underscored by resistance to bone loss associated with estrogen depletion as well as aging (24).

Transgenic models such as FABP4-Wnt10b mice provide supporting evidence that Wnt signaling can impair development of osteoporosis; however, expression of Wnt10b in marrow of these transgenic mice is not inducible and may have altered bone development (24). Thus it is more desirable to evaluate approaches in skeletally mature animals with pharmacological activators of Wnt/β-catenin signaling. Recent work indicates that inhibition of GSK3 increases bone formation, density, and strength in an ovariectomized rat model (27). Ovariectomy of rats at 6 months of age leads to significant trabecular bone loss within 4 weeks, with a high turnover signature that resembles bone loss observed in postmenopausal women (73). Oral administration of LY603281-31-8, a GSK3α and -3β dual inhibitor, to ovariectomized rats for 2 months resulted in an increase in trabecular area, thickness, and number that was accompanied by improved trabecular connectivity as evidenced by decreased trabecular separation (27). Accordingly, biomechanical analysis found that LY603218-31-8 significantly improved vertebral strength, stiffness, and work to failure relative to ovariectomized controls. In addition, bone mineral density at both cancellous and cortical sites was significantly improved (27). The magnitude of responses to GSK3 inhibition was comparable to that observed with once-daily administration of teriparatide. In addition, genes reflecting enhanced osteoblast activity such as Runx2, collagen α_I, collagen α_V, bone sialoprotein, and biglycan were induced in trabecular bone obtained from distal femur. Increased bone mass was also observed with LiCl treatment of SAMP6 mice, which have premature osteoporosis due to impaired osteoblastogenesis (33). Taken together, these observations offer strong evidence for an increase in bone formation in response to inhibitors of GSK3 and suggest that activators of Wnt/β-catenin signaling show promise as therapeutic agents for osteoporosis.

Bone-related cancers. Given the important role of Wnt signaling for bone development, it is possible that agents modifying this pathway could be of value to skeletal disorders other than osteoporosis. For example, Tian and coworkers recently analyzed the bone marrow of patients with newly diagnosed multiple myeloma and identified an increase in Dkk1 in the serum of these patients (74). Notably, the severity of the bone lesion was correlated with increased Dkk1 levels in these patients. The authors indicate that not all newly diagnosed patients show elevated levels of Dkk1 in their serum and that this finding may be restricted to a subset of end-stage severe multiple myeloma patients. The authors propose that Dkk1 produced by myeloma cells blocks differentiation of osteoblasts and promotes the early proliferation leading to reduced viability of pluripotent stem cells, later shifting the balance between osteoblasts and osteoclasts in favor of osteoclasts (74). This then facilitates the lytic lesions in bone that are a hallmark of this painful disease. Although expression of Dkk1 is limited to a subset of severe multiple myeloma patients, it is conceivable that early intervention with activators of Wnt/β-catenin signaling could slow development of bone lesions in these patients. Again, while activation of Wnt signaling may decrease some of the painful symptoms caused by excessive secretion of Dkk1, great care will need to be taken in targeting this pathway in patients because of the potential to increase progression of cancer.

Potential for Wnt signaling as a pharmacological target: "druggable" interventions?

Wnt/β-catenin signaling offers multiple steps that may be considered for pharmacological intervention, and some of these are highlighted in Figure 1 (i–v). Important features to consider in selecting drug discovery targets include the type of target (i.e., G protein–coupled receptors, enzymes, protein-protein interactions, or transcriptional factors), cellular location, role of the target in the pathway (central regulator versus fine tuning), and selectivity of the target for the pathway of interest. Historically, the best targets for small molecules are receptors or enzymes, particularly those at extracellular sites. Protein or antibody strategies can be useful to target protein-protein interactions extracellularly. Obviously, selectivity of the target for bone in this case is an important consideration to limit off-target tissue toxicities.

A review of the canonical Wnt signaling pathway suggests several interesting potential intervention points (Figure 1). (i) Availability of Wnt for binding to frizzled receptors is regulated by binding to Wnt inhibitory proteins such as sFRPs and WIF-1, and it is conceivable that small molecules or peptides could inhibit these interactions. Support for this approach comes from results of studies of *Sfrp1*−/− mice, which have increased bone formation without other obvious phenotypes (17). (ii) Availability of the LRP5 complex for Wnt/β-catenin signaling is also regulated by proteins from the Dkk and sclerostin families. Dkk1 interferes with canonical Wnt signaling in vertebrates by binding directly to LRP5. Simultaneously, Dkk interacts with a transmembrane protein, kremen, which causes internalization of the Dkk/LRP complex and a loss of Wnt signaling. Thus if interactions with Dkk1 were inhibited, more LRP5 would be available for activation of the Wnt pathway. Mutations in LRP5 that decrease affinity for Dkk and increase bone mass in humans suggest that this approach might be suc-

cessful (10). In addition to Dkk, recent evidence suggests that sclerostin may also bind and inhibit signaling by LRP5/6 (75). Thus disruption of these interactions may also yield an increase in bone formation as evidenced by individuals with van Buchem disease (76). Protein therapeutic strategies offer the greatest chance of success at disrupting interactions between LRP and binding proteins, as there has been limited success at building small molecule inhibitors of protein-protein interactions. (iii) Since the frizzled receptor is a member of the G protein–coupled receptor family, which has been a highly successful family for generation of small molecule pharmacologic agents, it may be possible to foresee small molecule screening strategies having some degree of success. However, identification of small molecule mimics for type II G protein–coupled receptors has only been marginally successful (77). In addition, frizzled receptors are atypical members of the 7-transmembrane–spanning domain family of G protein–coupled receptors, and little is known about how to identify molecular agonists for this type of receptor. In addition, identities of those frizzled receptors that influence bone mass are unknown. (iv) Wnt/β-catenin signaling stabilizes β-catenin by inhibiting GSK3, and a variety of small molecule inhibitors increase osteoblastogenesis in vitro and bone formation in vivo. Although characterization of small molecule inhibitors of GSK3 is still underway, safety issues have not been reported for LiCl (78), which is widely used by adult patients to treat bipolar disorder. (v) In looking at targets further downstream of GSK3 (Figure 1), the degradation of β-catenin is mediated by the ubiquitin/proteasome pathway, and inhibiting these enzymes increases bone formation (79). However, specificity of these protease inhibitors remains a challenge in the area of pharmaceutical intervention. Although speculative, interaction of β-catenin/TCF with transcriptional coactivators is increased by acetylation of β-catenin. Thus histone deacetylase inhibitors could conceivably be used to increase expression of specific genes pertinent to bone cells, although specificity is likely to be an issue (7).

Safety considerations in targeting the Wnt pathway. Treatment of chronic disorders such as osteoporosis require heightened awareness of safety considerations, and given the important role of the Wnt pathways in development, the toxicologic potential of molecules modulating the Wnt pathway should be given thorough consideration. One area of speculative concern with regard to targeting of drugs to the Wnt pathway has been induction of cancer. While to date no reports connecting human tumors to mutation or dysregulation of genes encoding Wnt ligands or receptors have been made, certain components within the Wnt pathway have been implicated. For example, nuclear β-catenin functions to maintain the proliferative potential of keratinocytes in culture (80). A more direct relationship with human tumors is suggested by elevation of β-catenin levels in various cancers (81), including some types of skin cancer, as a moderate increase of β-catenin nuclear staining was observed with basal cell carcinomas (82). Mutations in APC or AXIN2 leading to accumulation of β-catenin have also been associated with colorectal cancer, as have activating mutations in β-catenin (83–87).

In addition to β-catenin, other players in the canonical Wnt signaling cascade have been linked to tumorigenesis. Inhibition of GSK3 results in increased cyclin D1, cyclin E, and c-Myc, and overexpression of these cell cycle regulators has been linked with tumor cell formation (88, 89), leading to concern that long-term inhibition of GSK3 may increase the risk of carcinogenesis. This of course will be an important safety consideration for development of GSK3 inhibitors. However, it should be possible to generate inhibitors of this enzyme without a significant cancer risk, as long-term use of the nonspecific GSK3 inhibitor lithium is not known to be associated with increased risk of cancer in bipolar patients (78). Furthermore, activation of GSK3 by histone deacetylase inhibitors has been associated with targeting tumor cells for elimination by natural killer cells (90). As molecules emerge from ongoing drug discovery efforts that target aspects of the Wnt signaling pathway, attention to tumor potential and other toxicities will be of paramount importance. Hopefully the worldwide efforts currently underway to target Wnt/β-catenin signaling will be successful and generate therapeutics that positively impact human skeletal health.

Address correspondence to: Venkatesh Krishnan, Eli Lilly and Company, Lilly Corporate Center, Building 98C/3/3338, Indianapolis, Indiana 46285, USA. Phone: (317) 276-0603; Fax: (317) 651-6333; E-mail: krishnan_gary@lilly.com. Or to: Ormond A. MacDougald, Department of Molecular and Integrative Physiology, 7620 Medical Science II, 1301 East Catherine Drive, Ann Arbor, Michigan 48109-0622, USA. Phone: (734) 647-4880; Fax: (734) 936-8813; E-mail: macdouga@umich.edu.

1. Cadigan, K.M., and Nusse, R. 1997. Wnt signaling: a common theme in animal development. *Genes Dev.* **11**:3286–3305.
2. Westendorf, J.J., Kahler, R.A., and Schroeder, T.M. 2004. Wnt signaling in osteoblasts and bone diseases. *Gene.* **341**:19–39.
3. Rawadi, G., and Roman-Roman, S. 2005. Wnt signalling pathway: a new target for the treatment of osteoporosis. *Expert Opin. Ther. Targets.* **9**:1063–1077.
4. Logan, C.Y., and Nusse, R. 2004. The Wnt signaling pathway in development and disease. *Annu. Rev. Cell Dev. Biol.* **20**:781–810.
5. Moon, R.T., Kohn, A.D., De Ferrari, G.V., and Kaykas, A. 2004. WNT and beta-catenin signalling: diseases and therapies. *Nat. Rev. Genet.* **5**:691–701.
6. Hay, E., et al. 2005. Interaction between LRP5 and Frat1 mediates the activation of the Wnt canonical pathway. *J. Biol. Chem.* **280**:13616–13623.
7. Levy, L., et al. 2004. Acetylation of beta-catenin by p300 regulates beta-catenin-Tcf4 interaction. *Mol. Cell. Biol.* **24**:3404–3414.
8. Mao, B., et al. 2002. Kremen proteins are Dickkopf receptors that regulate Wnt/beta-catenin signalling. *Nature.* **417**:664–667.
9. Gong, Y., et al. 2001. LDL receptor-related protein 5 (LRP5) affects bone accrual and eye development. *Cell.* **107**:513–523.
10. Ai, M., Holmen, S.L., Van Hul, W., Williams, B.O., and Warman, M.L. 2005. Reduced affinity to and inhibition by DKK1 form a common mechanism by which high bone mass-associated missense mutations in LRP5 affect canonical Wnt signaling. *Mol. Cell. Biol.* **25**:4946–4955.
11. Boyden, L.M., et al. 2002. High bone density due to a mutation in LDL-receptor-related protein 5. *N. Engl. J. Med.* **346**:1513–1521.
12. Van Wesenbeeck, L., et al. 2003. Six novel missense mutations in the LDL receptor-related protein 5 (LRP5) gene in different conditions with an increased bone density. *Am. J. Hum. Genet.* **72**:763–771.
13. Kato, M., et al. 2002. Cbfa1-independent decrease in osteoblast proliferation, osteopenia, and persistent embryonic eye vascularization in mice deficient in Lrp5, a Wnt coreceptor. *J. Cell Biol.* **157**:303–314.
14. Babij, P., et al. 2003. High bone mass in mice expressing a mutant LRP5 gene. *J. Bone Miner. Res.* **18**:960–974.
15. Holmen, S.L., et al. 2004. Decreased BMD and limb deformities in mice carrying mutations in both Lrp5 and Lrp6. *J. Bone Miner. Res.* **19**:2033–2040.
16. Li, X., et al. 2005. Sclerostin binds to LRP5/6 and antagonizes canonical Wnt signaling. *J. Biol. Chem.* **280**:19883–19887.
17. Bodine, P.V., et al. 2004. The Wnt antagonist secreted frizzled-related protein-1 is a negative regulator of trabecular bone formation in adult mice. *Mol. Endocrinol.* **18**:1222–1237.
18. Li, X., et al. 2005. Dkk2 has a role in terminal osteoblast differentiation and mineralized matrix formation. *Nat. Genet.* **37**:945–952.
19. Hausler, K.D., et al. 2004. Secreted frizzled-related protein-1 inhibits RANKL-dependent osteoclast

20. Rodriguez, J., et al. 2005. SFRP1 regulates the growth of retinal ganglion cell axons through the Fz2 receptor. *Nat. Neurosci.* **8**:1301–1309.
21. Vaes, B.L., et al. 2002. Comprehensive microarray analysis of bone morphogenetic protein 2-induced osteoblast differentiation resulting in the identification of novel markers for bone development. *J. Bone Miner. Res.* **17**:2106–2118.
22. Zhang, Y., et al. 2004. The LRP5 high-bone-mass G171V mutation disrupts LRP5 interaction with Mesd. *Mol. Cell. Biol.* **24**:4677–4684.
23. Reya, T., et al. 2000. Wnt signaling regulates B lymphocyte proliferation through a LEF-1 dependent mechanism. *Immunity.* **13**:15–24.
24. Bennett, C.N., et al. 2005. Regulation of osteoblastogenesis and bone mass by Wnt10b. *Proc. Natl. Acad. Sci. U. S. A.* **102**:3324–3329.
25. Rawadi, G., Vayssiere, B., Dunn, F., Baron, R., and Roman-Roman, S. 2003. BMP-2 controls alkaline phosphatase expression and osteoblast mineralization by a Wnt autocrine loop. *J. Bone Miner. Res.* **18**:1842–1853.
26. Stambolic, V., Ruel, L., and Woodgett, J.R. 1996. Lithium inhibits glycogen synthase kinase-3 activity and mimics wingless signalling in intact cells. *Curr. Biol.* **6**:1664–1668.
27. Kulkarni, N.E., et al. 2006. Orally bioavailable GSK-3alpha/beta dual inhibitor increases markers of cellular differentiation in vitro and bone mass in vivo. *J. Bone Miner. Res.* In press.
28. Jackson, A., et al. 2005. Gene array analysis of Wnt-regulated genes in C3H10T1/2 cells. *Bone.* **36**:585–598.
29. Gregory, C.A., et al. 2005. How wnt signaling affects bone repair by mesenchymal stem cells from the bone marrow. *Ann. N. Y. Acad. Sci.* **1049**:97–106.
30. Ross, S.E., et al. 2000. Inhibition of adipogenesis by Wnt signaling. *Science.* **289**:950–953.
31. Bennett, C.N., et al. 2002. Regulation of Wnt signaling during adipogenesis. *J. Biol. Chem.* **277**:30998–31004.
32. Nuttall, M.E., and Gimble, J.M. 2000. Is there a therapeutic opportunity to either prevent or treat osteopenic disorders by inhibiting marrow adipogenesis? *Bone.* **27**:177–184.
33. Clement-Lacroix, P., et al. 2005. Lrp5-independent activation of Wnt signaling by lithium chloride increases bone formation and bone mass in mice. *Proc. Natl. Acad. Sci. U. S. A.* **102**:17406–17411.
34. Day, T.F., Guo, X., Garrett-Beal, L., and Yang, Y. 2005. Wnt/beta-catenin signaling in mesenchymal progenitors controls osteoblast and chondrocyte differentiation during vertebrate skeletogenesis. *Dev. Cell.* **8**:739–750.
35. Reya, T., and Clevers, H. 2005. Wnt signalling in stem cells and cancer. *Nature.* **434**:843–850.
36. Hu, H., et al. 2005. Sequential roles of Hedgehog and Wnt signaling in osteoblast development. *Development.* **132**:49–60.
37. Hill, T.P., Spater, D., Taketo, M.M., Birchmeier, W., and Hartmann, C. 2005. Canonical Wnt/beta-catenin signaling prevents osteoblasts from differ-

entiating into chondrocytes. *Dev. Cell.* **8**:727–738.
38. Glass, D.A., 2nd, et al. 2005. Canonical Wnt signaling in differentiated osteoblasts controls osteoclast differentiation. *Dev. Cell.* **8**:751–764.
39. Henriksen, K., et al. 2005. Osteoclasts from patients with autosomal dominant osteopetrosis type I caused by a T253I mutation in low-density lipoprotein receptor-related protein 5 are normal in vitro, but have decreased resorption capacity in vivo. *Am. J. Pathol.* **167**:1341–1348.
40. Holmen, S.L., et al. 2005. Essential role of beta-catenin in postnatal bone acquisition. *J. Biol. Chem.* **280**:21162–21168.
41. Sottile, V., Halleux, C., Bassilana, F., Keller, H., and Seuwen, K. 2002. Stem cell characteristics of human trabecular bone-derived cells. *Bone.* **30**:699–704.
42. Pereira, R.C., Delany, A.M., and Canalis, E. 2002. Effects of cortisol and bone morphogenetic protein-2 on stromal cell differentiation: correlation with CCAAT-enhancer binding protein expression. *Bone.* **30**:685–691.
43. Lecka-Czernik, B., et al. 2002. Divergent effects of selective peroxisome proliferator-activated receptor-gamma 2 ligands on adipocyte versus osteoblast differentiation. *Endocrinology.* **143**:2376–2384.
44. Park, S.R., Oreffo, R.O., and Triffitt, J.T. 1999. Interconversion potential of cloned human marrow adipocytes in vitro. *Bone.* **24**:549–554.
45. Canalis, E., and Delany, A.M. 2002. Mechanisms of glucocorticoid action in bone. *Ann. N. Y. Acad. Sci.* **966**:73–81.
46. Sciaudone, M., Gazzerro, E., Priest, L., Delany, A.M., and Canalis, E. 2003. Notch 1 impairs osteoblastic cell differentiation. *Endocrinology.* **144**:5631–5639.
47. Chan, G.K., et al. 2003. Parathyroid hormone-related peptide interacts with bone morphogenetic protein 2 to increase osteoblastogenesis and decrease adipogenesis in pluripotent C3H10T 1/2 mesenchymal cells. *Endocrinology.* **144**:5511–5520.
48. Kulkarni, N.H., et al. 2005. Effects of parathyroid hormone on Wnt signaling pathway in bone. *J. Cell. Biochem.* **95**:1178–1190.
49. Kanazawa, A., et al. 2005. Wnt5b partially inhibits canonical Wnt/beta-catenin signaling pathway and promotes adipogenesis in 3T3-L1 preadipocytes. *Biochem. Biophys. Res. Commun.* **330**:505–510.
50. Kanazawa, A., et al. 2004. Association of the gene encoding wingless-type mammary tumor virus integration-site family member 5B (WNT5B) with type 2 diabetes. *Am. J. Hum. Genet.* **75**:832–843.
51. Topol, L., et al. 2003. Wnt-5a inhibits the canonical Wnt pathway by promoting GSK-3-independent beta-catenin degradation. *J. Cell Biol.* **162**:899–908.
52. Cheng, S.L., Shao, J.S., Charlton-Kachigian, N., Loewy, A.P., and Towler, D.A. 2003. MSX2 promotes osteogenesis and suppresses adipogenic differentiation of multipotent mesenchymal progenitors. *J. Biol. Chem.* **278**:45969–45977.
53. Ichida, F., et al. 2004. Reciprocal roles of MSX2 in regulation of osteoblast and adipocyte differentiation. *J. Biol. Chem.* **279**:34015–34022.
54. Jeon, M.J., et al. 2003. Activation of peroxisome proliferator-activated receptor-gamma inhibits

the Runx2-mediated transcription of osteocalcin in osteoblasts. *J. Biol. Chem.* **278**:23270–23277.

55. Akune, T., et al. 2004. PPARγ insufficiency enhances osteogenesis through osteoblast formation from bone marrow progenitors. *J. Clin. Invest.* **113**:846–855. doi:10.1172/JCI200419900.

56. Ali, A.A., et al. 2005. Rosiglitazone causes bone loss in mice by suppressing osteoblast differentiation and bone formation. *Endocrinology.* **146**:1226–1235.

57. Kawaguchi, H., et al. 2005. Distinct effects of PPAR-gamma insufficiency on bone marrow cells, osteoblasts, and osteoclastic cells. *J. Bone Miner. Metab.* **23**:275–279.

58. Hong, J.H., et al. 2005. TAZ, a transcriptional modulator of mesenchymal stem cell differentiation. *Science.* **309**:1074–1078.

59. Bodine, P.V., et al. 2005. The Wnt antagonist secreted frizzled-related protein-1 controls osteoblast and osteocyte apoptosis. *J. Cell. Biochem.* **96**:1212–1230.

60. You, Z., et al. 2002. Wnt signaling promotes oncogenic transformation by inhibiting c-Myc-induced apoptosis. *J. Cell Biol.* **157**:429–440.

61. Longo, K.A., et al. 2002. Wnt signaling protects 3T3-L1 preadipocytes from apoptosis through induction of insulin-like growth factors. *J. Biol. Chem.* **277**:38239–38244.

62. Almeida, M., Han, L., Bellido, T., Manolagas, S.C., and Kousteni, S. 2005. Wnt proteins prevent apoptosis of both uncommitted osteoblast progenitors and differentiated osteoblasts by beta-catenin-dependent and -independent signaling cascades involving Src/ERK and phosphatidylinositol 3-kinase/AKT. *J. Biol. Chem.* **280**:41342–41351.

63. Neer, R.M., et al. 2001. Effect of parathyroid hormone (1-34) on fractures and bone mineral density in postmenopausal women with osteoporosis. *N. Engl. J. Med.* **344**:1434–1441.

64. Nusse, R., and Varmus, H.E. 1982. Many tumors induced by the mouse mammary tumor virus contain a provirus integrated in the same region of the host genome. *Cell.* **31**:99–109.

65. Bienz, M., and Clevers, H. 2000. Linking colorectal cancer to Wnt signaling. *Cell.* **103**:311–320.

66. Toomes, C., et al. 2004. Mutations in LRP5 or FZD4 underlie the common familial exudative vit-

reoretinopathy locus on chromosome 11q. *Am. J. Hum. Genet.* **74**:721–730.

67. Longo, K.A., et al. 2004. Wnt10b inhibits development of white and brown adipose tissues. *J. Biol. Chem.* **279**:35503–35509.

68. Haegel, H., et al. 1995. Lack of beta-catenin affects mouse development at gastrulation. *Development.* **121**:3529–3537.

69. Huelsken, J., et al. 2000. Requirement for beta-catenin in anterior-posterior axis formation in mice. *J. Cell Biol.* **148**:567–578.

70. Ohnaka, K., Taniguchi, H., Kawate, H., Nawata, H., and Takayanagi, R. 2004. Glucocorticoid enhances the expression of dickkopf-1 in human osteoblasts: novel mechanism of glucocorticoid-induced osteoporosis. *Biochem. Biophys. Res. Commun.* **318**:259–264.

71. Ohnaka, K., Tanabe, M., Kawate, H., Nawata, H., and Takayanagi, R. 2005. Glucocorticoid suppresses the canonical Wnt signal in cultured human osteoblasts. *Biochem. Biophys. Res. Commun.* **329**:177–181.

72. Wang, F.S., et al. 2005. Secreted frizzled-related protein 1 modulates glucocorticoid attenuation of osteogenic activities and bone mass. *Endocrinology.* **146**:2415–2423.

73. Kalu, D.N. 1991. The ovariectomized rat model of postmenopausal bone loss. *Bone Miner.* **15**:175–191.

74. Tian, E., et al. 2003. The role of the Wnt-signaling antagonist DKK1 in the development of osteolytic lesions in multiple myeloma. *N. Engl. J. Med.* **349**:2483–2494.

75. Semenov, M., Tamai, K., and He, X. 2005. SOST is a ligand for LRP5/LRP6 and a Wnt signaling inhibitor. *J. Biol. Chem.* **280**:26770–26775.

76. Brunkow, M.E., et al. 2001. Bone dysplasia sclerosteosis results from loss of the SOST gene product, a novel cystine knot-containing protein. *Am. J. Hum. Genet.* **68**:577–589.

77. Bagger, Y.Z., et al. 2005. Oral salmon calcitonin induced suppression of urinary collagen type II degradation in postmenopausal women: a new potential treatment of osteoarthritis. *Bone.* **37**:425–430.

78. Cohen, Y., Chetrit, A., Cohen, Y., Sirota, P., and Modan, B. 1998. Cancer morbidity in psychiatric patients: influence of lithium carbonate treatment.

Med. Oncol. **15**:32–36.

79. Garrett, I.R., et al. 2003. Selective inhibitors of the osteoblast proteasome stimulate bone formation in vivo and in vitro. *J. Clin. Invest.* **111**:1771–1782. doi:10.1172/JCI200316198.

80. Zhu, A.J., and Watt, F.M. 1999. Beta-catenin signalling modulates proliferative potential of human epidermal keratinocytes independently of intercellular adhesion. *Development.* **126**:2285–2298.

81. Karim, R., Tse, G., Putti, T., Scolyer, R., and Lee, S. 2004. The significance of the Wnt pathway in the pathology of human cancers. *Pathology.* **36**:120–128.

82. Doglioni, C., et al. 2003. Alterations of beta-catenin pathway in non-melanoma skin tumors: loss of alpha-ABC nuclear reactivity correlates with the presence of beta-catenin gene mutation. *Am. J. Pathol.* **163**:2277–2287.

83. Lustig, B., and Behrens, J. 2003. The Wnt signaling pathway and its role in tumor development. *J. Cancer Res. Clin. Oncol.* **129**:199–221.

84. Rubinfeld, B., et al. 1996. Binding of GSK3beta to the APC-beta-catenin complex and regulation of complex assembly. *Science.* **272**:1023–1026.

85. Korinek, V., et al. 1997. Constitutive transcriptional activation by a β-catenin-TCF complex in APC-/- colon carcinoma. *Science.* **275**:1784–1787.

86. Liu, W., et al. 2000. Mutations in AXIN2 cause colorectal cancer with defective mismatch repair by activating beta-catenin/TCF signalling. *Nat. Genet.* **26**:146–147.

87. Morin, P.J., et al. 1997. Activation of beta-catenin-Tcf signaling in colon cancer by mutations in beta-catenin or APC. *Science.* **275**:1787–1790.

88. Dong, J., et al. 2005. Role of glycogen synthase kinase 3beta in rapamycin-mediated cell cycle regulation and chemosensitivity. *Cancer Res.* **65**:1961–1972.

89. van Noort, M., Meeldijk, J., van der Zee, R., Destree, O., and Clevers, H. 2002. Wnt signaling controls the phosphorylation status of beta-catenin. *J. Biol. Chem.* **277**:17901–17905.

90. Skov, S., et al. 2005. Cancer cells become susceptible to natural killer cell killing after exposure to histone deacetylase inhibitors due to glycogen synthase kinase-3-dependent expression of MHC class I-related chain A and B. *Cancer Res.* **65**:11136–11145.

Part 13
Neurological Disorders

VEGF: a critical player in neurodegeneration

Erik Storkebaum and Peter Carmeliet

Center for Transgene Technology and Gene Therapy, Flanders Interuniversity Institute for Biotechnology, University of Leuven, Leuven, Belgium.

VEGF is a prototype angiogenic factor, but recent evidence indicates that this growth factor also has direct effects on neural cells. Abnormal regulation of VEGF expression has now been implicated in several neurodegenerative disorders, including motoneuron degeneration. This has stimulated an increasing interest in assessing the therapeutic potential of VEGF as a neuroprotective agent for such neurodegenerative disorders.

No factor is better known for its angiogenic effects than VEGF — this molecule has been implicated in virtually every type of angiogenic disorder, including those associated with cancer, ischemia, and inflammation (1). Recent studies have revealed, however, that VEGF is also involved in neurodegeneration. How can we explain this unexpected finding? It turns out that the role of VEGF in the nervous system is not restricted only to regulating vessel growth: VEGF also has direct effects on different types of neural cells — including even neural stem cells (NSCs). This link between angiogenesis and neurogenesis offers novel opportunities to better decipher the insufficiently understood molecular pathogenesis of many neurodegenerative disorders, and promises to open future avenues for improved treatment.

Vascular effect of VEGF in neurodegeneration

How can VEGF, and other vascular factors in general, affect neurodegeneration? Recent genetic studies have revealed that reduced VEGF levels cause neurodegeneneration in part by impairing neural tissue perfusion. Indeed, mice with reduced VEGF levels due to a subtle targeted deletion of the hypoxia response element in the promoter of the *VEGF* gene (*VEGF$^{\delta/\delta}$* mice) develop adult-onset motoneuron degeneration (2), reminiscent of amyotrophic lateral sclerosis (ALS). This condition is also known as Lou Gehrig's disease, after the famous baseball player who died of this incurable disorder (3). A follow-up human genetic study documented that "low-VEGF" haplotypes in the *VEGF* promoter and leader sequence are associated with lower VEGF plasma levels (due to impaired transcription and translation of several VEGF isoforms, including the novel large L-VEGF isoform) and an increased risk of ALS, at least in three European populations (4). As with any other genetic association study, independent high-power replication studies and, even more importantly, additional functional evidence would support the involvement of VEGF in motoneuron degeneration, although the genetic demonstration of a modifier role of VEGF for such a complex disease might be more challenging than is currently anticipated. Reduction of VEGF levels by 50% is known to significantly impair angiogenesis (5). Somewhat unexpectedly, *VEGF$^{\delta/\delta}$* mice, with VEGF levels that are suppressed by only 25% in the spinal cord, had no major defects in angiogenesis, as measured by capillary density in the spinal cord (2). Yet, neural perfusion was impaired, causing chronic ischemia of motoneurons in the spinal cord (Figure 1) (2). Intriguingly, decreased regional cerebral blood flow has also been reported in

patients with ALS (6, 7). These findings, together with the facts that large motoneurons are particularly vulnerable to free radicals generated during ischemia and that release of reactive oxygen species increases with age, suggest that chronic and/or repetitive neural perfusion deficits may set the stage for the development of neuronal damage and, ultimately, neurodegeneration at adult onset. The mechanisms by which low VEGF levels reduce neural perfusion remain to be determined, but they might involve impaired vasoregulation. One possibility is that VEGF affects vascular tone by controlling the release of the vasorelaxant nitric oxide by endothelial cells (ECs). Alternatively, VEGF may be required for the normal functioning of perivascular autonomic nerves, which critically regulate vascular tone and, hence, tissue perfusion. Ongoing studies in *VEGF$^{\delta/\delta}$* mice suggest indeed that low VEGF levels cause adult-onset degeneration of these perivascular autonomic nerves (8).

Perfusion deficits have also been documented in other neurodegenerative disorders, including Alzheimer disease and Huntington disease. Evidence of the presence of perfusion deficits often precedes onset of clinical symptoms, suggesting that they causally contribute to the pathogenesis of these disorders (9, 10). Various types of vascular defects have been documented in neurodegenerative disorders (Table 1), including fibrosis and deposition of amorphous substances within the vessel wall and thickening of the basement membrane and interstitial matrix. These abnormalities lead to vessel narrowing, loss of vasoregulation, hypoperfusion, and impaired oxygen diffusion to the neural tissue. Notably, a progressive loss of the microvasculature and reduction of blood supply to the nervous system with increasing age may explain the late onset of many neurodegenerative disorders (11). To what extent VEGF — or other molecules — are involved in the deregulation of neural perfusion in these disorders remains to be elucidated.

In addition to the above-described etiologic role of VEGF in the initiation phase of neurodegeneration, VEGF may also have additional effects which can occur secondarily in response to the neurodegeneration process itself. These secondary effects likely relate to the fact that VEGF expression is upregulated by ischemic and inflammatory stimuli, which often accompany neurodegeneration. The elevated neural VEGF levels in humans suffering acute focal neural ischemia, diabetes, Alzheimer disease, and vascular dementia may thus be secondary to the ischemia and inflammation in these disorders (12–14), although a primary role cannot be formally excluded to date. Such elevated VEGF levels could initiate a vicious cycle. Indeed, excess VEGF may cause hemangioma formation, microvascular leakage, and fragility. These abnormalities may lead to edema formation and bleeding, and further impair perfusion and tissue oxygenation. The resultant hypoxia may then, in turn, further upregulate VEGF levels again.

Nonstandard abbreviations used: neural stem cell (NSC); amyotrophic lateral sclerosis (ALS); endothelial cell (EC).

Conflict of interest: The authors have declared that no conflict of interest exists.

Citation for this article: *J. Clin. Invest.* **113**:14–18 (2004). doi:10.1172/JCI200420682.

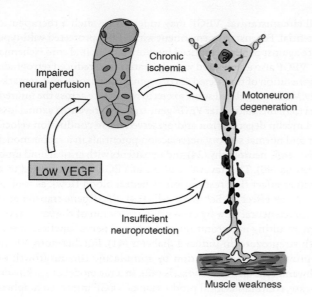

Impaired
neural perfusion

Chronic
ischemia

Motoneuron
degeneration

Low VEGF

Insufficient
neuroprotection

Muscle weakness

Figure 1
Low VEGF levels cause amyotrophic lateral sclerosis. Low VEGF levels impair spinal cord perfusion and cause chronic ischemia of motoneurons, but also deprive these cells of vital VEGF-dependent survival and neuroprotective signals. Both mechanisms result in adult-onset progressive degeneration of motoneurons, with associated muscle weakness, paralysis, and death — as is typical in amyotrophic lateral sclerosis.

Neurotrophic effects of VEGF in neurodegeneration

There is growing evidence that insufficient neuroprotection by VEGF may be a novel etiologic mechanism of motoneuron degeneration. First, VEGF is expressed in the spinal cord in neurons and glial cells, and is rapidly elevated by hypoxia, whereas VEGFR-2 and neuropilin-1 are expressed in motoneurons (2), indicating that all molecular players required to coordinate a neuroprotective effect by VEGF are present in vivo. Second, in $VEGF^{\delta/\delta}$ mice with ALS, insufficient VEGF may deprive motoneurons of critical survival and neuroprotective signals, especially in conditions of hypoxic or excitotoxic stress, and thereby cause motoneuron degeneration

(Figure 1) (2). This may explain why $VEGF^{\delta/\delta}$ mice are unusually sensitive to transient spinal cord ischemia and remain paralyzed after a minor ischemic insult, while wild-type mice show only a transient clinical deficit (4). Third, neuronal overexpression of VEGFR-2 delays onset of motoneuron degeneration in SOD-1[G93A] mice — an established mouse model of ALS (8). Fourth, in a mouse model of spinal and bulbar muscular atrophy (also known as SBMA, or Kennedy's disease — a lower motor neuron disease caused by CAG repeat expansion in the androgen receptor gene), reduced spinal cord VEGF protein levels were observed, even in presymptomatic mice (15). Finally, several recent in vitro studies have documented that VEGF protects cultured motoneurons against death in conditions of hypoxia, oxidative stress, and serum deprivation (2), glutamate-induced excitotoxicity (16, 17), or mutant SOD-1–induced toxicity (18), thus indicating that VEGF has direct neuroprotective effects on motoneurons.

The neuroprotective activity of VEGF might also play a more important (etiologic) role in other neurodegenerative disorders than originally anticipated, although the evidence thus far is largely based on in vitro studies. Various neural cells express one or more of the VEGF receptors (e.g., VEGFR-1, -2 and neuropilin-1) and can thus directly respond to VEGF released by neighboring neural cells (2, 14, 19). For instance, VEGF stimulates axonal outgrowth in explant cultures of retinal or superior cervical

Table 1
Vascular defects in neurodegenerative disorders

Neurodegenerative disease	Vascular defect
Amyotrophic lateral sclerosis	Normal vascular density in the spinal cord but reduced neural perfusion causes chronic ischemia due to low VEGF levels (2, 4).
Alzheimer disease	β-amyloid causes microvascular degeneration (focal swelling, atrophy, and death of vascular cells, disruption of the blood-brain barrier), cerebral amyloid angiopathy (CAA), abnormal vasoregulation, vessel wall rupture, cerebral perfusion deficits, hemorrhages, and infarcts (10). Tortuous thickening of the basement membrane, collagen deposition, CAA, and the presence of plaques/tangles impede diffusion of oxygen. Reduced vascular density correlates with the degree of vascular dementia, but reactive angiogenesis may be induced by cerebral ischemia and upregulation of VEGF (50).
Parkinson disease	Microvascular basement membrane deposits and capillary dysfunction (51).
Huntington disease	Perfusion deficits due to vascular hyporeactivity (9).
Diabetic neuropathy	Epineural vessels: arteriolar attenuation, venous distension, arteriovenous shunting, new vessel formation. Endoneural vessels: hyalin thickening and increased deposition of amorphous substances in the vessel wall, in association with endothelial cell growth and basement membrane thickening, pericyte loss and — in some cases — capillary pruning and obstruction (52).
CADASIL	Cerebral autosomal dominant arteriopathy with subcortical infarcts and leukoencephalopathy (CADASIL): vascular malformations and degeneration due to smooth muscle cell dysfunction, resulting from Notch3 mutations (53).
Stroke	Angiogenic cerebral scores correlate with the recovery from stroke (54).
Prion disease	Mobile cells in vessel walls such as dendritic cells and monocytes are implicated in the spread of disease-associated prion protein (55).

and dorsal root ganglia (14). Furthermore, under conditions of hypoxic, excitotoxic, or oxidative stress, VEGF increases the survival of hippocampal, cortical, cerebellar granule, dopaminergic, autonomic, and sensory neurons (2, 14, 16–18, 20–23). VEGF also stimulates the growth (14) and survival of Schwann cells in hypoxic conditions (24), and increases proliferation and migration of astrocytes (23, 25) and microglial cells (26). Future genetic and functional studies in vivo will have to elucidate how important the neuroprotective effect of VEGF is in the pathogenesis of other neurodegenerative disorders.

Neurogenic effects of VEGF

An impaired regenerative response of neural tissue by adult NSCs has been proposed as a contributory factor in the pathogenesis of neurodegenerative disorders. There is increasing evidence that blood vessels and angiogenic factors such as VEGF play an important role in the control of neurogenesis via crosstalk pathways. For example, ECs and NSCs appear at similar locations and developmental stages in the neural germinal zones (27). In the adult mouse, NSCs are neighbors to ECs and proliferate in small clusters around dividing capillaries ("vascular niches") in the subependymal zone of the lateral ventricle (28). Moreover, when neural progenitors are cocultured with ECs, the maturation, neurite outgrowth, and migration of neurons are enhanced (29). Conditions which increase neural activity and stimulate neurogenesis also trigger angiogenesis (30, 31), whereas cranial irradiation causes a decrease in both hippocampal neurogenesis and angiogenesis (32). VEGF is produced by ependymal cells at neurogenic sites, stimulates the proliferation of neuronal precursors in cerebral cortical cultures in vitro, and, upon intracerebroventricular infusion, increases neural progenitor growth in the subventricular and subgranular zone of the hippocampal dentate gyrus (33–36), effects that may be mediated via VEGFR-2 (36). VEGF may also promote neurogenesis by stimulating ECs to release neurogenic signals, such as brain-derived neurotrophic factor (37). Notably, intracerebroventricular administration of VEGF reduces infarct size, in part by stimulating neurogenesis and angiogenesis (38). Despite this suggestive evidence, there is currently no formal proof that impaired neurogenesis due to low VEGF levels contributes to neurodegeneration. Another outstanding but challenging question is also whether delivery of VEGF would beneficially affect the outcome of neurodegeneration by stimulating neurogenesis.

VEGF for the treatment of neurodegeneration?

The emerging evidence for an etiologic role of VEGF in (at least some types of) neurodegeneration provides a rationale for considering the therapeutic potential of VEGF for neurodegenerative disorders, which are mostly incurable. Although the evidence is still circumstantial, VEGF may indeed have such a therapeutic potential. For instance, treatment with VEGF-protected wild-type mice against paralysis induced by short-term spinal cord ischemia (4). VEGF also augmented vascularization and reduced retrograde degeneration of transected corticospinal tract axons, and, remarkably, VEGF stimulated some axons to regenerate across the injured area (39). Intramuscular *VEGF* gene transfer prevented axonal loss and myelin degeneration and preserved nerve conduction velocities and normal sensory nerve action potentials in a rabbit model of ischemic neuropathy (24) and in patients with critical hind limb ischemia (40). This favorable effect of VEGF was attributed to a vascular effect (i.e., restoration of neural blood flow), as well as to a direct effect on Schwann cells (24). *VEGF* gene transfer also preserved neural flow by preventing regression of the vasa nervorum, resulting in restoration of peripheral nerve function in rats with streptozotocin-induced diabetes (41). Furthermore, VEGF improved nerve regeneration by stimulating the outgrowth of Schwann cells and blood vessels (42). In a rat model of Parkinson disease, continuous local production of VEGF improved amphetamine-induced rotational behavior (43). VEGF can also have favorable effects on the recovery of an ischemic brain insult. Administration of VEGF via intravenous or intracerebrovascular infusion, or via topical application on the cortex, enhanced angiogenesis in the penumbra and improved neural recovery (44), reduced edema formation and infarct volume (45, 46), and amplified neurogenesis (38), while knockdown of VEGF after ischemic stroke enlarged the infarct volume (47). Some caution, however, is warranted, as other studies suggest that delivery of a VEGF trap reduces edema formation and spares ischemic brain tissue (48), while infusion of VEGF induces vascular leakage, with resultant hemorrhagic transformation of the ischemic lesions (44). VEGF has also been documented to have beneficial effects in epilepsy, as infusion of VEGF into rat brains protected CA1 and CA3 hippocampal regions against neuronal cell damage induced by epilepsy (49). Overall, although additional work is required, these studies highlight the important biological role of VEGF in neurodegenerative disorders and provide sufficient rationale for evaluating the potential therapeutic utility of VEGF in their management.

Acknowledgments

The authors wish to thank Ann Vandenhoeck for the artwork, and all previous and current collaborators for assistance in this study.

Address correspondence to: Peter Carmeliet, Center for Transgene Technology and Gene Therapy, Flanders Interuniversity Institute for Biotechnology, University of Leuven, Campus Gasthuisberg, Herestraat 49, B-3000 Leuven, Belgium. Phone: 32-16-34-57-74; Fax: 32-16-34-59-90; E-mail: peter.carmeliet@med.kuleuven.ac.be.

1. Carmeliet, P. 2003. Angiogenesis in health and disease. *Nat. Med.* **9**:653–660.
2. Oosthuyse, B., et al. 2001. Deletion of the hypoxia-response element in the vascular endothelial growth factor promoter causes motor neuron degeneration. *Nat. Genet.* **28**:131–138.
3. Cleveland, D.W., and Rothstein, J.D. 2001. From Charcot to Lou Gehrig: deciphering selective motor neuron death in ALS. *Nat. Rev. Neurosci.* **2**:806–819.
4. Lambrechts, D., et al. 2003. VEGF is a modifier of amyotrophic lateral sclerosis in mice and humans and protects motoneurons against ischemic death. *Nat. Genet.* **34**:383–394.
5. Carmeliet, P., et al. 1996. Abnormal blood vessel development and lethality in embryos lacking a single VEGF allele. *Nature.* **380**:435–439.
6. Kobari, M., Obara, K., Watanabe, S., Dembo, T., and Fukuuchi, Y. 1996. Local cerebral blood flow in motor neuron disease: correlation with clinical findings. *J. Neurol. Sci.* **144**:64–69.
7. Waldemar, G., Vorstrup, S., Jensen, T.S., Johnsen, A., and Boysen, G. 1992. Focal reductions of cerebral blood flow in amyotrophic lateral sclerosis: a [99mTc]-d,l-HMPAO SPECT study. *J. Neurol. Sci.* **107**:19–28.
8. Storkebaum, E., et al. 2003. VEGF, a modifier of motor neuron degeneration in SOD1 G93A mice, protects against motor neuron loss after spinal cord ischemia: evidence for a vascular hypothesis. Program No. 602.11. 2003. Abstract Viewer/Itinerary Planner. Society for Neuroscience. Washington, DC, USA. http://sfn.scholarone.com.
9. Deckel, A.W., and Duffy, J.D. 2000. Vasomotor hyporeactivity in the anterior cerebral artery during motor activation in Huntington's disease patients. *Brain Res.* **872**:258–261.
10. Kalaria, R.N. 2002. Small vessel disease and Alzheimer's dementia: pathological considerations. *Cerebrovasc. Dis.* **13**:48–52.
11. Buijs, P.C., et al. 1998. Effect of age on cerebral blood flow: measurement with ungated two-dimensional phase-contrast MR angiography in 250 adults. *Radi-*

ology. **209**:667–674.

12. Tarkowski, E., et al. 2002. Increased intrathecal levels of the angiogenic factors VEGF and TGF-beta in Alzheimer's disease and vascular dementia. *Neurobiol. Aging.* **23**:237–243.

13. Samii, A., Unger, J., and Lange, W. 1999. Vascular endothelial growth factor expression in peripheral nerves and dorsal root ganglia in diabetic neuropathy in rats. *Neurosci. Lett.* **262**:159–162.

14. Sondell, M., Lundborg, G., and Kanje, M. 1999. Vascular endothelial growth factor has neurotrophic activity and stimulates axonal outgrowth, enhancing cell survival and Schwann cell proliferation in the peripheral nervous system. *J. Neurosci.* **19**:5731–5740.

15. Thomas, P.S., et al. 2003. SBMA motor neuronopathy in AR YAC CAG100 transgenic mice may involve altered expression of VEGF. Program No. 413.2. 2003. Abstract Viewer/Itinerary Planner. Society for Neuroscience. Washington, DC, USA. http://sfn.scholarone.com.

16. Matsuzaki, H., et al. 2001. Vascular endothelial growth factor rescues hippocampal neurons from glutamate-induced toxicity: signal transduction cascades. *FASEB J.* **15**:1218–1220.

17. Svensson, B., et al. 2002. Vascular endothelial growth factor protects cultured rat hippocampal neurons against hypoxic injury via an antiexcitotoxic, caspase-independent mechanism. *J. Cereb. Blood Flow Metab.* **22**:1170–1175.

18. Li, B., Xu, W., Luo, C., Gozal, D., and Liu, R. 2003. VEGF-induced activation of the PI3-K/Akt pathway reduces mutant SOD1-mediated motor neuron cell death. *Brain Res. Mol. Brain Res.* **111**:155–164.

19. Sondell, M., and Kanje, M. 2001. Postnatal expression of VEGF and its receptor flk-1 in peripheral ganglia. *Neuroreport.* **12**:105–108.

20. Jin, K.L., Mao, X.O., and Greenberg, D.A. 2000. Vascular endothelial growth factor: direct neuroprotective effect in in vitro ischemia. *Proc. Natl. Acad. Sci. U. S. A.* **97**:10242–10247.

21. Jin, K.L., Mao, X.O., and Greenberg, D.A. 2000. Vascular endothelial growth factor rescues HN33 neural cells from death induced by serum withdrawal. *J. Mol. Neurosci.* **14**:197–203.

22. Ogunshola, O.O., et al. 2002. Paracrine and autocrine functions of neuronal vascular endothelial growth factor (VEGF) in the central nervous system. *J. Biol. Chem.* **277**:11410–11415.

23. Silverman, W.F., Krum, J.M., Mani, N., and Rosenstein, J.M. 1999. Vascular, glial and neuronal effects of vascular endothelial growth factor in mesencephalic explant cultures. *Neuroscience.* **90**:1529–1541.

24. Schratzberger, P., et al. 2000. Favorable effect of VEGF gene transfer on ischemic peripheral neuropathy. *Nat. Med.* **6**:405–413.

25. Krum, J.M., Mani, N., and Rosenstein, J.M. 2002. Angiogenic and astroglial responses to vascular endothelial growth factor administration in adult rat brain. *Neuroscience.* **110**:589–604.

26. Forstreuter, F., Lucius, R., and Mentlein, R. 2002. Vascular endothelial growth factor induces chemotaxis and proliferation of microglial cells. *J. Neuroimmunol.* **132**:93–98.

27. Zerlin, M., and Goldman, J.E. 1997. Interactions between glial progenitors and blood vessels during early postnatal corticogenesis: blood vessel contact represents an early stage of astrocyte differentiation. *J. Comp. Neurol.* **387**:537–546.

28. Palmer, T.D., Willhoite, A.R., and Gage, F.H. 2000. Vascular niche for adult hippocampal neurogenesis. *J. Comp. Neurol.* **425**:479–494.

29. Leventhal, C., Rafii, S., Rafii, D., Shahar, A., and Goldman, S.A. 1999. Endothelial trophic support of neuronal production and recruitment from the adult mammalian subependyma. *Mol. Cell. Neurosci.* **13**:450–464.

30. Black, J.E., Isaacs, K.R., Anderson, B.J., Alcantara, A.A., and Greenough, W.T. 1990. Learning causes synaptogenesis, whereas motor activity causes angiogenesis, in cerebellar cortex of adult rats. *Proc. Natl. Acad. Sci. U. S. A.* **87**:5568–5572.

31. Kokaia, Z., and Lindvall, O. 2003. Neurogenesis after ischaemic brain insults. *Curr. Opin. Neurobiol.* **13**:127–132.

32. Monje, M.L., Mizumatsu, S., Fike, J.R., and Palmer, T.D. 2002. Irradiation induces neural precursor-cell dysfunction. *Nat. Med.* **8**:955–962.

33. Yang, K., and Cepko, C.L. 1996. Flk-1, a receptor for vascular endothelial growth factor (VEGF), is expressed by retinal progenitor cells. *J. Neurosci.* **16**:6089–6099.

34. Yourey, P.A., Gohari, S., Su, J.L., and Alderson, R.F. 2000. Vascular endothelial cell growth factors promote the in vitro development of rat photoreceptor cells. *J. Neurosci.* **20**:6781–6788.

35. Jin, K., et al. 2002. Vascular endothelial growth factor (VEGF) stimulates neurogenesis in vitro and in vivo. *Proc. Natl. Acad. Sci. U. S. A.* **99**:11946–11950.

36. Zhu, Y., Jin, K., Mao, X.O., and Greenberg, D.A. 2003. Vascular endothelial growth factor promotes proliferation of cortical neuron precursors by regulating E2F expression. *FASEB J.* **17**:186–193.

37. Louissaint, A., Jr., Rao, S., Leventhal, C., and Goldman, S.A. 2002. Coordinated interaction of neurogenesis and angiogenesis in the adult songbird brain. *Neuron.* **34**:945–960.

38. Sun, Y., et al. 2003. VEGF-induced neuroprotection, neurogenesis, and angiogenesis after focal cerebral ischemia. *J. Clin. Invest.* **111**:1843–1851. doi:10.1172/JCI200317977.

39. Facchiano, F., et al. 2002. Promotion of regeneration of corticospinal tract axons in rats with recombinant vascular endothelial growth factor alone and combined with adenovirus coding for this factor. *J. Neurosurg.* **97**:161–168.

40. Simovic, D., Isner, J.M., Ropper, A.H., Pieczek, A., and Weinberg, D.H. 2001. Improvement in chronic ischemic neuropathy after intramuscular phVEGF165 gene transfer in patients with critical limb ischemia. *Arch. Neurol.* **58**:761–768.

41. Schratzberger, P., et al. 2001. Reversal of experimental diabetic neuropathy by VEGF gene transfer. *J. Clin. Invest.* **107**:1083–1092.

42. Sondell, M., Lundborg, G., and Kanje, M. 1999. Vascular endothelial growth factor stimulates Schwann cell invasion and neovascularization of acellular nerve grafts. *Brain Res.* **846**:219–228.

43. Yasuhara, T., et al. 2003. Vascular endothelial growth factor has a neuroprotective effect on Parkinson's disease. Program No. 233.6. 2003. Abstract Viewer/Itinerary Planner. Society for Neuroscience. Washington, DC, USA. http://sfn.scholarone.com.

44. Zhang, Z.G., et al. 2000. VEGF enhances angiogenesis and promotes blood-brain barrier leakage in the ischemic brain. *J. Clin. Invest.* **106**:829–838.

45. Harrigan, M.R., Ennis, S.R., Sullivan, S.E., and Keep, R.F. 2003. Effects of intraventricular infusion of vascular endothelial growth factor on cerebral blood flow, edema, and infarct volume. *Acta Neurochir. (Wien).* **145**:49–53.

46. Hayashi, T., Abe, K., and Itoyama, Y. 1998. Reduction of ischemic damage by application of vascular endothelial growth factor in rat brain after transient ischemia. *J. Cereb. Blood Flow Metab.* **18**:887–895.

47. Yang, Z.J., et al. 2002. Role of vascular endothelial growth factor in neuronal DNA damage and repair in rat brain following a transient cerebral ischemia. *J. Neurosci. Res.* **70**:140–149.

48. van Bruggen, N., et al. 1999. VEGF antagonism reduces edema formation and tissue damage after ischemia/reperfusion injury in the mouse brain. *J. Clin. Invest.* **104**:1613–1620.

49. Croll, S.D., Goodman, J.H., and Scharfman, H.E. 2003. Vascular endothelial growth factor (VEGF) in seizures: a double-edged sword. In *Recent advances in epilepsy research.* D.K. Binder and H.E. Scharfman, editors. Kluwer Academic/Plenum Publishers. New York, New York, USA. In press.

50. Vagnucci, A.H., Jr., and Li, W.W. 2003. Alzheimer's disease and angiogenesis. *Lancet.* **361**:605–608.

51. Farkas, E., De Jong, G.I., de Vos, R.A., Jansen Steur, E.N., and Luiten, P.G. 2000. Pathological features of cerebral cortical capillaries are doubled in Alzheimer's disease and Parkinson's disease. *Acta Neuropathol. (Berl).* **100**:395–402.

52. Boulton, A.J., and Malik, R.A. 1998. Diabetic neuropathy. *Med. Clin. North Am.* **82**:909–929.

53. Kalimo, H., Ruchoux, M.M., Viitanen, M., and Kalaria, R.N. 2002. CADASIL: a common form of hereditary arteriopathy causing brain infarcts and dementia. *Brain Pathol.* **12**:371–384.

54. Krupinski, J., Kaluza, J., Kumar, P., Kumar, S., and Wang, J.M. 1994. Role of angiogenesis in patients with cerebral ischemic stroke. *Stroke.* **25**:1794–1798.

55. Koperek, O., et al. 2002. Disease-associated prion protein in vessel walls. *Am. J. Pathol.* **161**:1979–1984.

Neurogenesis and brain injury: managing a renewable resource for repair

Anna F. Hallbergson, Carmen Gnatenco, and Daniel A. Peterson

Neural Repair and Neurogenesis Laboratory, Department of Neuroscience, The Chicago Medical School, North Chicago, Illinois, USA.

The brain shows limited ability to repair itself, but neurogenesis in certain areas of the adult brain suggests that neural stem cells may be used for structural brain repair. It will be necessary to understand how neurogenesis in the adult brain is regulated to develop strategies that harness neural stem cells for therapeutic use.

In adult centres the nerve paths are something fixed, ended, immutable. Everything may die, nothing may be regenerated. It is for the science of the future to change, if possible, this harsh decree.
—S. Ramon y Cajal, *Degeneration and regeneration of the nervous system*, 1928 (1)

Updating Cajal's decree

As observed by the pioneering neuroscientist Santiago Ramon y Cajal, the mature CNS was distinguished from the developing nervous system by the lack of growth and cellular regeneration. The fixed neuronal population of the adult brain was understood to be necessary to maintain the functional stability of adult brain circuitry. This explanation has also been offered to account for the lack of endogenous CNS repair following injury or disease.

Ever perceptive, Cajal left open the possibility for future advances to alter this "harsh decree," and in the last several decades, mounting evidence has led to the view of the CNS as a dynamic, plastic organ, endowed with some potential for self-repair and regeneration. Recent progress in understanding continued neurogenesis in the adult brain has raised hopes that self-renewal leading to structural repair by new neurons may even be possible in the mature CNS. Nevertheless, under normal conditions, neurogenesis in the adult brain appears to be restricted to the discrete germinal centers: the subventricular zone and the hippocampal dentate gyrus (2) (Figure 1). While some reports indicate that neurogenesis in the adult CNS may be more widespread and include the cerebral cortex (3, 4), other reports cast doubt on these observations (5–7). More study is needed to establish the origin, extent, survival, and function of new neurons in these other regions. This Spotlight summarizes our current understanding of the regulation of adult neurogenesis and its relevance to structural brain repair. We propose strategies for harnessing the potential of neural stem cells for brain repair and consider how to apply these strategies to the aging brain.

Plasticity in adult neurogenesis

Neurogenesis in the adult brain can be divided into three phases in accordance with the sequence of neurogenesis during CNS development: (a) proliferation, when new cells are generated; (b) migration toward target areas; and (c) terminal differentiation into distinct phenotypes (Figure 1). The use of the term "neurogenesis" implies progression through differentiation and should not be used in cases where only proliferation is studied. It is not yet fully known whether these phases in adult neurogenesis are regulated by the same mechanisms that regulate development, or even whether the same mechanisms regulate neurogenesis in the two adult germinal centers. Nevertheless, new, functional neurons are generated in these areas, expanding the definition of plasticity in the adult brain to now include cellular addition to circuitry (8). New hippocampal neurons may participate in the processing of memory in the hippocampus (9), while new olfactory bulb neurons appear to participate in processing olfactory input (10). Newly generated hippocampal neurons may also contribute to the response of the nervous system to antidepressant administration (11).

Interestingly, adult neurogenesis is not static, but its rate may fluctuate in response to environmental change, even subtle macroenvironmental alterations (2). Examples of positive regulators of neurogenesis include physical activity, environmental enrichment, caloric restriction, and modulation of neural activity (12). The responsiveness of adult neurogenesis to environmental influences suggests that its regulation may be under the control of expressed factors whose level of availability dictates the rate of neurogenesis. In normal development, a vast repertoire of proneural genes guide stem cells to a neural fate, and these may continue to be expressed or repressed in neurogenic regions of the adult brain in response to stimuli. Normal CNS development is also guided by the spatial and temporal expression of various trophic or growth factors that guide cell-fate choices and determine the size of the neuronal population. Indeed, the elevated expression of these factors observed in the mature brain following injury has been thought to represent the brain's attempt to protect injured neurons by activating developmental programs (13).

In addition to protecting neurons, trophic factors have been shown to stimulate proliferation of adult-derived neural stem cells and to instruct their differentiation (14–18). For example, FGF-2 is a potent mitogen for a variety of cells and modulates embryonic development and differentiation, adult angiogenesis, wound healing, and tissue repair. In the adult brain, FGF-2 is a survival factor and is neuroprotective against a variety of insults (13). While the basal rate of neurogenesis is the same in wild-type and FGF-2–null mutant mice, Yoshimura et al. (19) found that gene delivery of FGF-2 to the null mutant mice produced an elevation in neurogenesis. This observation demonstrates that neurogenesis is not simply a cell-autonomous property of resident stem cells but is determined by the environmental milieu. Their present study, appearing in this issue of the *JCI* (20), extends this work by showing that enhanced expression of exogenous FGF-2 by gene delivery after injury attenuated hippocampal cell loss following lesion while simultaneously upregulating neurogenesis in this region.

Conflict of interest: The authors have declared that no conflict of interest exists.

Citation for this article: *J. Clin. Invest.* **112**:1128–1133 (2003). doi:10.1172/JCI200320098.

958

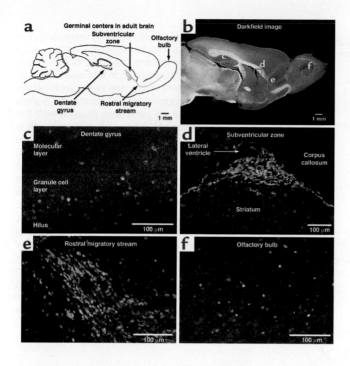

Figure 1
Germinal centers in the adult brain. Neurogenesis in the adult brain is largely confined to two germinal centers: the dentate gyrus and the subventricular zone, shown schematically (**a**) and in a corresponding sagittal section of the rodent brain (**b**). Insets in **b** show the position of high-resolution micrographs in **c–f**. In the dentate gyrus (**c**), newly generated cells are detected through incorporation of the thymidine analog BrdU and labeled with a green fluorophore (Cy2). These cells differentiate into mature neurons, as seen by their coexpression of the marker NeuN (red) but not S100β (blue), a marker for mature astrocytes. In contrast, cells generated in the subventricular zone (**d**) do not differentiate into mature neurons (red) but migrate away through the rostral migratory stream (RMS). Within the RMS (**e**), newly generated cells are surrounded by astrocytes (glial fibrillary acidic protein [GFAP], blue) and begin to express immature neuronal markers (polysialylated neural cell adhesion molecule [PSA-NCAM], red) as they migrate to the olfactory bulb. Upon arrival in the olfactory bulb (**f**), newly generated cells differentiate into mature neurons (NeuN, red), but not astrocytes (S100β, blue).

Limitations of self-repair by neurogenesis

Various injury models produce increased neurogenesis in germinal centers (Figure 1), but these regions represent a small portion of the total CNS. What are the prospects for self-repair following injury in other brain structures? As discussed above, some cortical regions may contain newly generated neurons, but these appear to be limited, transient populations, and their confirmation awaits further study (5, 6). Most studies using widespread brain injury models, such as ischemic or traumatic brain injury, have reported neurogenesis only within the germinal structures and not in cortical or striatal structures. Even when neurogenesis in the striatum (but not cortex) has been reported following ischemic injury, the resulting neurons survive only a short time (21). Thus, while a growing number of studies report injury-induced enhancement of neurogenesis, it is clear that under normal circumstances such responses do not lead to complete structural or functional recovery in non-neurogenic brain regions.

As in development, the specific temporal and spatial expression of appropriate trophic factors may be necessary not only to achieve the initial generation of new neural progenitor cells, but also to direct their migration, differentiation, and survival. Functional integration of new cells into existing circuitry may require additional signals. In the adult brain, these signals may normally exist only in germinal centers. By performing a targeted cortical lesion, Magavi et al. (22) have shown that progenitor cells can be recruited into the adult cerebral cortex, suggesting that this environment is not intrinsically inhibitory to neuronal differentiation. The identification and expression of appropriate signals may lead to structural repair in brain structures other than the germinal centers (23).

Therapeutic prospects for structural brain repair

Although much has been learned from observing endogenous responses to injuries, it will be necessary to develop targeted therapeutic intervention for brain injury or disease. At the present time, there are two possible therapeutic strategies to achieve structural brain repair: recruitment and replacement (Figure 2).

Recruitment strategies

Recruitment of endogenous progenitors presents the most elegant strategy for replacing neurons lost to injury or disease. Proliferation of neural progenitors in germinal centers can be induced by delivery of mitogenic trophic factors, such as FGF-2 (2, 20, 24). Recruitment of neural progenitors to adjacent brain structures has also been achieved through delivery of other trophic factors, such as brain-derived neurotrophic factor (BDNF), that may instruct cell differentiation (16, 17). By delivering another factor, TGF-α, into the striatum, Fallon et al. (25) were able to induce proliferation in the adjacent subventricular zone with subsequent migration of these newly generated cells into the striatum. However, these cells only differentiated following striatal injury, suggesting that additional, unknown signals were required. Recruitment of peripherally derived progenitor cells, such as blood-borne progenitor cells, may provide another cell population for brain repair.

While studies such as those discussed above have generated considerable enthusiasm for the eventual recruitment of endogenous progenitor cells as a therapeutic strategy, the successful development of stem cell recruitment therapy will depend on our ability to manage the proliferation, migration, differentiation, and functional integration of recruited cells. The first step toward this goal is to investigate the regulation of the events that specify cell fate both in cell culture and in vivo. For our understanding to advance, it is critical that a wide range of cell types be studied, from embryonic stem cells to adult-derived cells, to test the generality of the regulatory mechanisms. However, characterizing stem cells will not be sufficient for the development of therapeutic strategies. It will also be necessary to understand how the brain microenvironment changes as a result of injury, disease, and aging, in order to develop therapies that can recruit endogenous progenitors for structural brain repair.

Stem cell recruitment therapy will require targeted and regulated delivery of the key environmental signals or factors to the brain region where repair is needed (Figure 2). One possible route of delivery is the systemic administration of an identified factor or its synthetic precursor. This delivery approach requires that the factor cross the blood-brain barrier and reach the target region at the necessary dose. As factors delivered this way may act upon a wide variety of stem cells, this approach will likely be limited by the nonspecific activation of stem cells in other organ systems or

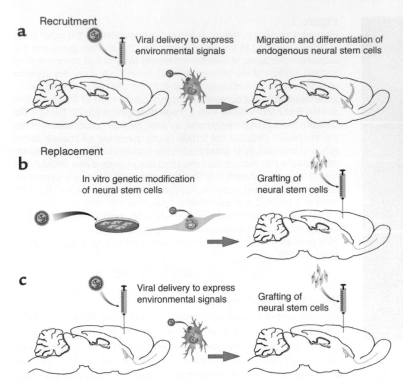

Figure 2

Therapeutic strategies for brain repair by stem cells. For brain repair in regions outside of the germinal centers, such as the cerebral cortex, stem cells may contribute to repair through recruitment or replacement. In the case of recruitment (**a**), the environment in the non-neurogenic region must be enhanced with appropriate environmental signals to attract endogenous neural stem cells, possibly from the germinal centers, to expand this population of cells, and to instruct their differentiation into appropriate neurons. This environmental enhancement could most likely be achieved using in vivo gene therapy leading to transgene expression by endogenous cells and may require a precise temporal and spatial delivery of appropriate transgenes to achieve the desired outcome. In the case of replacement (**b**), neural stem cells derived from embryonic, fetal, or adult sources could be expanded in vitro, directed down specific neuronal lineages, and genetically modified to express the necessary environmental signals. Thus committed to the correct phenotype and expressing necessary environmental signals to ensure their survival, these cells could then be grafted to the target region. Alternatively (**c**), the environment could be enhanced by gene delivery before (or possibly after) the grafting of the replacement neural stem cells.

even brain regions other than the intended target. An alternative strategy is the direct, targeted delivery of the factor by intracranial infusion. While this approach is suitable for experimental studies, chronic intracranial cannulation of patients will likely present unacceptable health risks. Perhaps the most promising delivery strategy for managing stem cell recruitment is gene therapy (26). While still requiring intracranial injection into the target region, development of viral vectors that can infect postmitotic cells (such as neurons) and that maintain stable, long-term gene expression offer the possibility of requiring only a single delivery episode. With continued advances in regulation of gene expression, it may be possible to turn transgene expression on during the appropriate time frame, or to shut it down entirely when recruitment is not desired. By expressing the transgene directly within the targeted cell, it is possible to extend management of therapeutic factors to directly modify the cell through, for example, the expression of intracellular signals or surface receptors. Finally, it may be possible to construct viral-delivery systems that allow for sequential or simultaneous expression of different environmental signals to accommodate the need to guide recruited stem cells through several stages of cell-fate specification.

Figure 3

Endogenous cortical proliferation is enhanced following FGF-2 gene delivery. (**a**) Endogenous proliferation occurs infrequently in the naive entorhinal cortex. Newly generated cells labeled by BrdU administration over 48 hours are indicated by the arrow and shown at higher magnification in the inset. (**b**) Intracerebral saline injection prior to BrdU treatment produced no visible increase in proliferation in the entorhinal cortex. (**c**) Injection of adenovirus expressing the reporter gene LacZ produced little effect on proliferation in the entorhinal cortex. (**d**) Adenovirus-mediated gene delivery of FGF-2 produced a substantially increased proliferation in the entorhinal cortex.

Replacement strategies

Even if all of the key environmental signals necessary to recruit endogenous progenitor cells were determined, the migratory distance from germinal centers would still present an obstacle for repopulation of some brain regions by recruitment. Another challenge to a recruitment strategy is that the different signals needed to achieve migration followed by differentiation into the desired neuronal phenotype may require distinct spatial and temporal delivery of factors that may be difficult to achieve. In such cases, an alternative strategy (Figure 2) is to generate a population of defined progenitor cells in culture that could be tailored to a specific neuronal lineage. These cells would then be grafted into the injured brain region (27). Such cell-replacement strategies have been used for some time, with success primarily in experimental models of Parkinson and Huntington diseases (28). Replacement

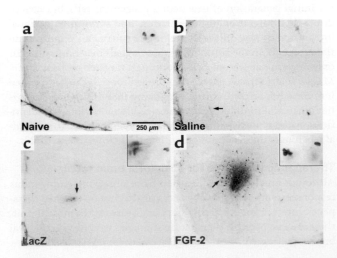

of lost or injured neurons may be successful where the environment can specify the appropriate fate of the grafted progenitor cells. This is true for progenitor cells grafted into germinal centers but not necessarily for other brain regions (29). While neonatal brains can readily incorporate grafted progenitor cells, relatively few graft-derived neurons are typically found following intrastriatal or intracortical grafting to the mature brain. These results suggest that, in the adult brain, non-neurogenic environments may support only limited differentiation (28). The state of cell differentiation may also be an important factor, as improved survival with intrastriatal transplantation has been obtained using undifferentiated embryonic stem cells (30). Thus success with cell-replacement strategies for structural brain repair may require an appropriate combination of choosing the correct cell-lineage state and creating a supportive brain environment.

The goal of creating a supportive environment for the grafting of replacement cells will likely use the same strategies for trophic factor delivery that are discussed above for recruitment of endogenous cells. In developing therapies, it will be important to understand the brain's response to trophic factor delivery, irrespective of whether this delivery is intended for recruitment or replacement strategies. For example, in studies aimed at preparing the cortical environment for subsequent stem cell grafts, we evaluated the response of the entorhinal cortex to gene delivery of a trophic factor, FGF-2. Lying adjacent to the hippocampus, the entorhinal cortex is important for processing learning and memory, and it suffers neuronal loss early in Alzheimer disease. Normally, there is limited endogenous proliferation in the entorhinal cortex, but delivery of a trophic factor gene, FGF-2, caused an increase in local cortical proliferation (Figure 3) (A.F. Hallbergson and D.A. Peterson, unpublished observations). Similar proliferative responses are found in other cortical regions following trophic factor gene delivery (31). It remains to be determined whether the newly generated endogenous cells are transitory, or whether they survive and differentiate. These observations illustrate that, in the grafting of stem cells to the brain, it may be difficult to discriminate the relative contributions of the specific environmental manipulation (i.e., trophic factor delivery) and the local cellular response (proliferation of endogenous cells) to the survival and incorporation of the grafted stem cells.

Structural repair in the aging brain

Most animal studies investigating the regulation of neurogenesis and the grafting of neural stem cells have been conducted on neonatal or young adult brains, and much of this work has used injury models that approximate neurodegenerative diseases, such as Alzheimer or Parkinson disease or stroke. However, the patient populations targeted by these experiments are older, typically at or beyond the sixth decade of life. Does the aging brain provide the same environment for neurogenesis as does the young adult brain? In aged rodents, there are significantly fewer new neurons generated in the germinal centers (32, 33). However, this age-related decline can be reversed, at least in the hippocampus, by environ-

mental enrichment (34), N-methyl-D-aspartate receptor antagonist treatment (35), or reduction of corticosteroid levels (36). Reduction of corticosteroid levels may not be a useful therapeutic approach, as lowering levels also leads to neuronal cell death.

The reversal of age-related decline in neurogenesis suggests that there is no intrinsic limitation with aging, and that, instead, the environment may change to provide less support for neurogenesis with aging. One possibility is that the age-related decline in neurogenesis results from reduced trophic factor levels. This possibility is supported by the reported enhancement of neurogenesis in aged rodent brains following exogenous delivery of IGF-I (37), or of FGF-2 or heparin-binding EGF-like growth factor (38). Furthermore, pretreatment with trophic factors improved the survival of fetal neurons grafted to the aged hippocampus (39, 40), suggesting that the availability of trophic factors in the aged hippocampus may be limited.

An age-related limitation in the brain environment is also suggested by the clinical experience with fetal-tissue grafts to the brains of Parkinson patients. Freed et al. (41) found that the prospects for clinical improvement decrease as a function of patient age and that fetal transplants may contribute to the development of persistent dyskinesia. Another recent report found no improvement in patient outcome following fetal-neuron transplantation to Parkinson patients (42). Thus, stem cell therapies for the aged brain may hold great clinical potential, but the capacity of the aged brain to support the recruitment or survival of transplanted stem cells remains to be determined. As discussed above, it is likely that the environment of the aged brain will need to be augmented, and investigation into regulation of the capacity of the aged brain to support structural brain repair should receive high priority.

Conclusions

It is possible in certain circumstances to recruit neural stem cells in the adult brain to contribute new neurons into areas depleted by experimental injury. Much work is required to elucidate the signals involved in recruiting stem cells for this purpose. Not only must environmental changes resulting from disease pathology or injury response be better understood, but the capacity of undifferentiated stem cells to respond to the spatial and temporal presentation of these signals must be determined. The prospects for developing therapeutic applications from advancing our understanding of how stem cell fate is managed relies on continued and unfettered basic research into fundamental properties of embryonic and adult-derived stem cells. Finally, age-related changes in the brain's capacity to support both endogenous and grafted neural stem cells need to be better understood to advance the possible harnessing of neurogenesis for structural brain repair.

Address correspondence to: Daniel A. Peterson, Neural Repair and Neurogenesis Laboratory, Department of Neuroscience, The Chicago Medical School, 3333 Green Bay Road, North Chicago, Illinois 60064, USA. Phone: (847) 578-3411; Fax: (847) 578-8545; E-mail: daniel.peterson@finchcms.edu.

1. Ramon y Cajal, S. 1928. *Degeneration and regeneration of the nervous system.* Volume 2. Haffner Publishing Co. New York, New York, USA. p. 750.
2. Peterson, D.A. 2002. Stem cells in brain plasticity and repair. *Curr. Opin. Pharmacol.* 2:34–42.
3. Gould, E., Vail, N., Wagers, M., and Gross, C.G. 2001. Adult-generated hippocampal and neocortical neurons in macaques have a transient existence.

Proc. Natl. Acad. Sci. U. S. A. 98:10910–10917.
4. Lie, D.C., et al. 2002. The adult substantia nigra contains progenitor cells with neurogenic potential. *J. Neurosci.* 22:6639–6649.
5. Kornack, D.R., and Rakic, P. 2001. Cell proliferation without neurogenesis in adult primate neocortex. *Science.* 294:2127–2130.
6. Koketsu, D., Mikami, A., Miyamoto, Y., and Hisat-

sune, T. 2003. Nonrenewal of neurons in the cerebral neocortex of adult macaque monkeys. *J. Neurosci.* 23:937–942.
7. Ehninger, D., and Kempermann, G. 2003. Regional effects of wheel running and environmental enrichment on cell genesis and microglia proliferation in the adult murine neocortex. *Cereb. Cortex.* 13:845–851.

8. van Praag, H., et al. 2002. Functional neurogenesis in the adult hippocampus. *Nature.* **415**:1030–1034.

9. Shors, T.J., et al. 2001. Neurogenesis in the adult is involved in the formation of trace memories. *Nature.* **410**:372–376.

10. Carleton, A., Petreanu, L.T., Lansford, R., Alvarez-Buylla, A., and Lledo, P.M. 2003. Becoming a new neuron in the adult olfactory bulb. *Nat. Neurosci.* **6**:507–518.

11. Santarelli, L., et al. 2003. Requirement of hippocampal neurogenesis for the behavioral effects of antidepressants. *Science.* **301**:805–809.

12. Ray, J., and Peterson, D.A. 2002. Neural stem cells in the adult hippocampus. In *Neural stem cells for brain and spinal cord repair.* T. Zigova, E.Y. Snyder, and P.R. Sanberg, editors. Humana Press. Totawa, New Jersey, USA. 269–286.

13. Peterson, D.A., and Gage, F.H. 1999. Trophic factor therapy for neuronal death. In *Alzheimer disease.* R.D. Terry, R. Katzman, K.L. Bick, and S.S. Sisodia, editors. Lippincott-Raven Publishers. Philadelphia, Pennsylvania, USA. 373–388.

14. Kuhn, H.G., Winkler, J., Kempermann, G., Thal, L.J., and Gage, F.H. 1997. Epidermal growth factor and fibroblast growth factor-2 have different effects on neural progenitors in the adult rat brain. *J. Neurosci.* **17**:5820–5829.

15. Zigova, T., Pencea, V., Wiegand, S.J., and Luskin, M.B. 1998. Intraventricular administration of BDNF increases the number of newly generated neurons in the adult olfactory bulb. *Mol. Cell. Neurosci.* **11**:234–245.

16. Pencea, V., Bingaman, K.D., Wiegand, S.J., and Luskin, M.B. 2001. Infusion of brain-derived neurotrophic factor into the lateral ventricle of the adult rat leads to new neurons in the parenchyma of the striatum, septum, thalamus, and hypothalamus. *J. Neurosci.* **21**:6706–6717.

17. Benraiss, A., Chmielnicki, E., Lerner, K., Roh, D., and Goldman, S.A. 2001. Adenoviral brain-derived neurotrophic factor induces both neostriatal and olfactory neuronal recruitment from endogenous progenitor cells in the adult forebrain. *J. Neurosci.* **21**:6718–6731.

18. Aberg, M.A., Aberg, N.D., Hedbacker, H., Oscarsson, J., and Eriksson, P.S. 2000. Peripheral infusion of IGF-I selectively induces neurogenesis in the adult rat hippocampus. *J. Neurosci.* **20**:2896–2903.

19. Yoshimura, S., et al. 2001. FGF-2 regulation of neurogenesis in adult hippocampus after brain injury. *Proc. Natl. Acad. Sci. U. S. A.* **98**:5874–5879.

20. Yoshimura, S., et al. 2003. FGF-2 regulates neurogenesis and degeneration in the dentate gyrus after traumatic brain injury in mice. *J. Clin. Invest.* **112**:1218–1226. doi:10.1172/JCI200316618.

21. Arvidsson, A., Collin, T., Kirik, D., Kokaia, Z., and Lindvall, O. 2002. Neuronal replacement from endogenous precursors in the adult brain after stroke. *Nat. Med.* **8**:963–970.

22. Magavi, S.S., Leavitt, B.R., and Macklis, J.D. 2000. Induction of neurogenesis in the neocortex of adult mice. *Nature.* **405**:951–955.

23. Arlotta, P., Magavi, S.S., and Macklis, J.D. 2003. Molecular manipulation of neural precursors in situ: induction of adult cortical neurogenesis. *Exp. Gerontol.* **38**:173–182.

24. Nakatomi, H., et al. 2002. Regeneration of hippocampal pyramidal neurons after ischemic brain injury by recruitment of endogenous neural progenitors. *Cell.* **110**:429–441.

25. Fallon, J., et al. 2000. In vivo induction of massive proliferation, directed migration, and differentiation of neural cells in the adult mammalian brain. *Proc. Natl. Acad. Sci. U. S. A.* **97**:14686–14691.

26. Peterson, D.A., Ray, J., and Gage, F.H. 2000. Future prospects of gene therapy for treating CNS diseases. In *Innovative animal models of CNS diseases: from molecule to therapy.* D.F. Emrich, R.L. Dean, and P.R. Sanberg, editors. Humana Press. Totawa, New Jersey, USA. 378–388.

27. Wu, P., et al. 2002. Region-specific generation of cholinergic neurons from fetal human neural stem cells grafted in adult rat. *Nat. Neurosci.* **5**:1271–1278.

28. Bjorklund, A., and Lindvall, O. 2000. Cell replacement therapies for central nervous system disorders. *Nat. Neurosci.* **3**:537–544.

29. Suhonen, J.O., Peterson, D.A., Ray, J., and Gage, F.H. 1996. Differentiation of adult hippocampus-derived progenitors into olfactory neurons in vivo. *Nature.* **383**:624–627.

30. Bjorklund, L.M., et al. 2002. Embryonic stem cells develop into functional dopaminergic neurons after transplantation in a Parkinson rat model. *Proc. Natl. Acad. Sci. U. S. A.* **99**:2344–2349.

31. Hallbergson, A.F., Vega, C.J., Peterson, L.D., and Peterson, D.A. 2003. In vivo gene delivery of FGF-2 stimulates proliferation in the adult cerebral cortex: implications for therapeutic recruitment of neural progenitor cells. *Proceedings of the Keystone Symposium: from stem cells to therapy.* Steamboat Springs, Colorado, USA. E3:3025. (Abstr.)

32. Kuhn, H.G., Dickinson-Anson, H., and Gage, F.H. 1996. Neurogenesis in the dentate gyrus of the adult rat: age-related decrease of neuronal progenitor proliferation. *J. Neurosci.* **16**:2027–2033.

33. Peterson, L.D., and Peterson, D.A. 2002. Age-related environmental impairment of olfactory bulb neurogenesis. *Proceedings of the Federation of European Neuroscience Societies.* **3**:69.10. (Abstr.)

34. Kempermann, G., Kuhn, H.G., and Gage, F.H. 1998. Experience-induced neurogenesis in the senescent dentate gyrus. *J. Neurosci.* **18**:3206–3212.

35. Nacher, J., Alonso-Llosa, G., Rosell, D.R., and McEwen, B.S. 2003. NMDA receptor antagonist treatment increases the production of new neurons in the aged rat hippocampus. *Neurobiol. Aging.* **24**:273–284.

36. Cameron, H.A., and McKay, R.D. 1999. Restoring production of hippocampal neurons in old age. *Nat. Neurosci.* **2**:894–897.

37. Lichtenwalner, R.J., et al. 2001. Intracerebroventricular infusion of insulin-like growth factor-I ameliorates the age-related decline in hippocampal neurogenesis. *Neuroscience.* **107**:603–613.

38. Jin, K., et al. 2003. Neurogenesis and aging: FGF-2 and HB-EGF restore neurogenesis in hippocampus and subventricular zone of aged mice. *Aging Cell.* **2**:175–183.

39. Zaman, V., and Shetty, A.K. 2002. Combined neurotrophic supplementation and caspase inhibition enhances survival of fetal hippocampal CA3 cell grafts in lesioned CA3 region of the aging hippocampus. *Neuroscience.* **109**:537–553.

40. Zaman, V., and Shetty, A.K. 2003. Fetal hippocampal CA3 cell grafts enriched with fibroblast growth factor-2 exhibit enhanced neuronal integration into the lesioned aging rat hippocampus in a kainate model of temporal lobe epilepsy. *Hippocampus.* **13**:618–632.

41. Freed, C.R., et al. 2001. Transplantation of embryonic dopamine neurons for severe Parkinson's disease. *N. Engl. J. Med.* **344**:710–719.

42. Olanow, C.W., et al. 2003. A double-blind controlled trial of bilateral fetal nigral transplantation in Parkinson's disease. *Ann. Neurol.* **54**:403–414.

Oxidative stress, cell cycle, and neurodegeneration

Jeffrey A. Klein and Susan L. Ackerman

The Jackson Laboratory, Bar Harbor, Maine, USA.

While numerous studies have examined the existence of increased reactive oxygen species (ROS) in later-onset neurodegenerative disorders, the mechanism by which neurons die under conditions of oxidative stress remains largely unknown. Fairly recent evidence has suggested that one mechanism linked to the death of terminally differentiated neurons is aberrant reentry into the cell cycle. This phenomenon has been reported in Alzheimer disease (AD) patients (1), Down syndrome patients (2), and several mouse neurodegenerative models (3–5). We will discuss recent findings regarding the influence of oxidative stress on neurodegeneration and possible connections between oxidative stress and unscheduled cell cycle reentry, the understanding of which could lead to new strategies in the development of therapeutic agents for neurodegenerative disorders.

Oxidative stress and neuron death

Under normal physiological conditions, it is estimated that up to 1% of the mitochondrial electron flow leads to the formation of superoxide ($O_2^{\cdot-}$), the primary oxygen free radical produced by mitochondria; and interference with electron transport can dramatically increase $O_2^{\cdot-}$ production. While these partially reduced oxygen species can attack iron sulfur centers in a variety of enzymes, $O_2^{\cdot-}$ is rapidly converted within the cell to hydrogen peroxide (H_2O_2) by the superoxide dismutases (SOD1, SOD2, and SOD3). However, H_2O_2 can react with reduced transition metals, via the Fenton reaction, to produce the highly reactive hydroxyl radical ($\cdot$OH), a far more damaging molecule to the cell. In addition to forming H_2O_2, $O_2^{\cdot-}$ radicals can rapidly react with nitric oxide (NO) to generate cytotoxic peroxynitrite anions ($ONOO^-$). Peroxynitrite can react with carbon dioxide, leading to protein damage via the formation of nitrotyrosine and lipid oxidation.

The generation of ROS in normal cells, including neurons, is under tight homeostatic control. To help detoxify ROS, biological antioxidants, including glutathione, α-tocopherol (vitamin E), carotenoids, and ascorbic acid, will react with most oxidants. In addition, the antioxidant enzymes catalase and glutathione peroxidase detoxify H_2O_2 by converting it to O_2 and H_2O. However, when ROS levels exceed the antioxidant capacity of a cell, a deleterious condition known as oxidative stress occurs. Unchecked, excessive ROS can lead to the destruction of cellular components including lipids, protein, and DNA, and ultimately cell death via apoptosis or necrosis (6).

Nonstandard abbreviations used: reactive oxygen species (ROS); Alzheimer disease (AD); superoxide ($O_2^{\cdot-}$); hydrogen peroxide (H_2O_2); superoxide dismutase (SOD); β-amyloid (Aβ); substantia nigra (SNc); Parkinson disease (PD); cyclin-dependent kinase (CDK); proliferating cell nuclear antigen (PCNA); 8-hydroxydeoxyguanosine (8-OHdG); extracellular signal–regulated kinase (ERK).

Conflict of interest: The authors have declared that no conflict of interest exists.

Citation for this article: *J. Clin. Invest.* **111**:785–793 (2003). doi:10.1172/JCI200318182.

Although numerous in vitro studies have implicated ROS in neuronal death (7), the relative lack of in vivo evidence has contributed to some controversy surrounding the role of ROS in the pathophysiology of later-onset neurodegenerative disorders. Mice homozygous for a targeted mutation of the α-tocopherol transfer protein gene (α-*Ttp*) develop retinal degeneration and gait abnormalities at 1 year of age, correlated with degeneration of the posterior column of the spinal cord (8). These behavioral and pathological changes were largely prevented with α-tocopherol supplementation. In addition, *Sod2*−/− mice develop gait abnormalities at 12 days of age, associated with vacuolization in brainstem and cortical regions (9). These results indicate that the loss of antioxidant genes can lead to neuron loss.

Markers of oxidative stress are found in postmortem examination of brains from patients with many neurodegenerative disorders (10). However, whether oxidative stress is involved in the development and/or progression of these disorders, or is merely associated with end-stage disease, is in dispute. DNA oxidation, protein oxidation, and lipid peroxidation have been reported in regions containing neurofibrillary tangles and senile plaques of brains from AD patients (10). As an apparent compensatory response, increased levels of catalase, glutathione peroxidase, and glutathione reductase were observed by RT-PCR from the hippocampus and inferior parietal lobe of brains of patients with AD (11). Biochemical assays for salicylate hydroxylation using β-amyloid (Aβ) fragments, the primary constituent of senile plaques, suggest that these peptides can generate free radicals that may underlie some of the molecular alterations observed in AD brains (12, 13). In agreement, increases in Aβ deposition resulted in the induction of oxidative stress in transgenic mice overexpressing the mutant amyloid precursor protein and presenilin 1, two proteins implicated in the progression of AD (14). These results suggest that Aβ may be involved in free radical generation. However, other studies suggest that Aβ may be a cellular response to oxidative stress or an antioxidant and implicate other processes as primary in generating free radical damage (15).

Dopaminergic neurons in the substantia nigra (SNc) of brains of Parkinson disease (PD) patients also exhibit hallmarks of oxidative stress, including lipid peroxidation, nucleic acid and protein oxidation, and changes in some antioxidants (16). Furthermore, α-synuclein, the main deposition product in inclusions in the SNc of PD patients, is a specific target of nitration, suggestive of the role of oxidative damage in the formation of these inclusions (17). In agreement, in vitro studies have shown that these aggregates are stabilized by oxidation. These findings, and the intrinsic potential for the oxidative metabolism of dopamine to generate ROS, have suggested that oxidative stress may be involved in the death of neurons in the SNc of these patients. In addition, 1-methyl-4-phenyl-1,2,3,6-tetrahydropyridine, rotenone, and paraquat, compounds that can generate ROS and/or disrupt the electron transport chain, have been found to induce symptoms of PD in animal models, including, in some cases, the deposition of α-synuclein aggregates (18–20). These substances also represent environmental risk factors for PD (21).

Oxidative damage has also been reported in several other age-related neurodegenerative diseases, including Huntington disease, progressive supranuclear palsy, amyotrophic lateral sclerosis, and prion disorders, all of which have abnormal protein aggregation as a major component of their pathology (22–24). Although recent research had questioned the relationship between protein fibril formation and disease, it is also clear that some sort of protein aggregation plays an important role in the pathologies of these various diseases and that oxidative stress and protein inclusions may cooperate in neuronal death in several neurodegenerative disorders.

Intrinsic oxidative stress, and presumably cell damage, increase with age due to either diminished antioxidant defenses or the increase in mitochondrial dysfunction (25). These age-related effects, compounded with genetic and environmental risk factors, and the high energy dependence but relatively low level of antioxidants in the brain, may provide a unifying mechanism for the high incidence of neurodegenerative disorders in the aged population.

Cell cycle and neuron death

The misregulation of the cell cycle in many cell types can lead to unchecked proliferation and neoplastic disease. As in other cell types, the cell cycle in the CNS is tightly regulated. Neuronal precursors proceed through the cell cycle to produce larger numbers of neurons than are needed, and excess neurons are eliminated by selective apoptosis. However, once neurons are terminally differentiated, they are maintained in a quiescent G0 state and no longer cycle.

A decade ago it was hypothesized that cell cycle abnormalities may be intimately connected to the death of terminally differentiated neurons (26, 27). The bases for this hypothesis included the observation that tumors arising from terminally differentiated neurons are extremely rare, indicating that these cells are resistant to neoplastic transformation. Experimental evidence for this hypothesis included the demonstration that forced expression of oncogenes in terminally differentiated cells, including neurons, can cause cell death rather than cell proliferation (27).

In vitro studies have examined the link between aberrant cell cycle reentry and neuronal cell death in an attempt to better understand the mechanisms by which neurons can be forced back into the cell cycle. Postmitotic chick retinal neurons cultured in the presence of nerve growth factor, insulin, and neurotrophin-3 underwent apoptosis via cell cycle reentry, characterized by an increase in cyclin B2 expression (28). Treatment of these cells with roscovitine, a cyclin-dependent kinase (CDK) inhibitor, prevented cell cycle reentry and apoptosis. These results demonstrate that although at least some postmitotic neurons retain the capacity to respond to growth factors by reentering into the cell cycle, such stimulation causes apoptosis rather than proliferation.

Unscheduled cell cycle reentry can also be induced by neurotoxic insults. For example, kainic acid–treated cerebellar granule cells upregulate the cell cycle proteins CDK2, cyclin E, and E2F-1 and replicate their DNA prior to apoptosis (29). Similarly, aberrant cell cycle reentry and subsequent apoptosis were observed when rat embryonic cortical neurons were cultured with toxic concentrations of Aβ peptides (30). These neurons showed increased expression of cyclin D1, cyclin E, and cyclin A, as well as increased phosphorylation of the retinoblastoma (Rb) protein. In addition, Aβ treatment increased the amount of the ganglioside GD3, a component of the sphingolipid signaling system known to affect neuronal proliferation and differentiation, and reduction of GD3 prevented cell cycle reentry and apoptosis (31).

Several animal models directly or indirectly support the hypothesis that aberrant cell cycle reentry leads to neuronal death. Expression of SV40 T antigen in cerebellar Purkinje cells or retinal photoreceptors of transgenic mice causes unscheduled cell cycle reentry and cell death (27, 32, 33). Dying cells expressed high levels of proliferating cell nuclear antigen (PCNA) and had undergone DNA synthesis. Activation of G1- and S-phase proteins occurs prior to cell death in dorsal root ganglion cells from mice lacking neurotrophin-3 (34). Apoptosis could be almost entirely eliminated by injections of olomucine, a G1-phase inhibitor. In the ischemia/reperfusion stroke model in both mice and rats, cerebral neurons were observed to reenter the cell cycle prior to death (35, 36). Cyclin D1 activation and a loss of cyclin-dependent kinase inhibitors accompanied the cell death. The treatment of primary rat neuron cultures from stroke-induced animals with cyclin-dependent kinase inhibitors significantly reduced in vitro cell death. In the spontaneous early-onset neurodegenerative mouse mutants staggerer (Rora^{sg}), Lurcher (Grid2^{Lc}), and weaver (Kcnj6^{wv}), cell cycle reentry has been observed to occur in postmigratory granule cells prior to apoptosis (3, 4). This has been correlated with the expression of cell cycle proteins PCNA, CDK4, cyclin D, and cyclin A, as well as with incorporation of the DNA analog BrdU. These models support the idea that cell cycle reentry underlies neuronal apoptosis in vivo as well as in vitro.

Recent evidence suggests that cell cycle reentry precedes neuronal apoptosis in human neurodegenerative diseases as well. Recently, increases in the cell cycle regulators cyclin D1, CDK4, hyperphosphorylated pRb, and E2F-1 were observed in both spinal motor neuron and the motor cortex of postmortem tissue of ALS patients (37). Neuronal loss in the cerebral cortex of Down syndrome patients has also been correlated with overexpression of both mitogenic and differentiation signals (2). Hippocampal pyramidal neurons from the brains of these patients exhibited increased levels of CDK4, an important regulator of the cell cycle. These same neurons with increased CDK4 expression also contained neurofibrillary tangles and granulovacuolar degeneration, suggesting a correlation between cell cycle reentry and pathology.

An increase in cell cycle proteins, including PCNA, cyclin D, CDK4, and cyclin B1, has been observed in the hippocampus, as well as other diseased brain regions, from postmortem tissue of AD patients (38, 39, 40). These proteins were not observed in nondiseased brain regions or in control-patient brains. In addition, FISH analysis of the hippocampus and basal forebrain of AD patient brains revealed many tetraploid neurons, indicating that these neurons had undergone approximately a full round of DNA synthesis (41). Furthermore, the cell cycle–associated protein Ki-67, detected in numerous types of tumors, has been observed in AD patient brains, as well as in patients with Pick disease and intractable temporal lobe epilepsy (42).

Data from these studies and others support the idea that mitogenic signals may lead to an increase in cell cycle proteins, thus driving neurons into aberrant cell cycle reentry. Furthermore, it is clear that when terminally differentiated neurons are forced into the cell cycle, they do not proliferate — they die. Although the mechanisms by which cell cycle reentry causes cell death are unknown, a direct link between cell cycle and neuronal death was recently made with the observation that CDC2, a cell cycle regulator, induces the phosphorylation and activation of BAD, a trigger of apoptosis (43).

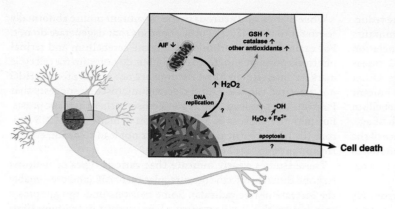

Figure 1
Oxidative stress and cell cycle reentry lead to cell death in the harlequin mouse. An 80% reduction of AIF protein expression in the Hq mutant mouse is associated with increased activity of total glutathione (GSH) and catalase, presumably through increases in hydrogen peroxide (H_2O_2). Surviving neurons may express additional antioxidant pathways. Increases in H_2O_2 could lead to cellular damage via formation of hydroxyl radicals ($\cdot$OH) by the Fenton reaction. In addition, oxidative stress may trigger cell cycle reentry in some terminally differentiated neurons. These neurons undergo apoptosis after DNA replication. Possible mechanisms for cell cycle reentry are shown in Figure 2.

Connections between oxidative stress and cell cycle

The multiple demonstrations of biomarkers of oxidative stress in many age-related neurodegenerative disorders, combined with the more recent reports of cell cycle abnormalities in neurons from these patients, suggest that these processes may be intertwined at the molecular level. At first, the idea of oxidative stress and cell cycle reentry seems counterintuitive. In fact, examination of the current literature on the effect of oxidative stress on the cell cycle reveals that increases in ROS-induced DNA damage are correlated with cell cycle arrest (44). However, whether ROS-exposed cells undergo growth arrest or apoptosis may depend in part on where the cell resides in the cell cycle when insulted. For example, human fibroblasts treated with H_2O_2 underwent either cell cycle arrest or apoptosis. The majority of the apoptotic fibroblasts were in the S phase of the cell cycle, whereas growth-arrested cells were predominantly in the G1 or the G2/M phase (45).

This apoptotic death of fibroblasts in the S phase is consistent with the death of neurons that have aberrantly reentered the cell cycle. Dorsal root ganglion neurons from neurotrophin-3 mutant mice override the G1 phase cell cycle restriction point but die by apoptosis in the S phase (34). In addition, hippocampal pyramidal and basal forebrain neurons from AD brains exhibit chromosomal duplication but do not undergo mitosis prior to cell death, consistent with cell death in the S or G2 phase of the cell cycle (41). These results suggest that neurons are influenced by the same cell cycle checkpoints that govern apoptosis in other cell types.

Recently, we described a novel genetic mouse model that may begin to unravel the mechanisms by which oxidative stress stimulates quiescent neurons to leave the G0 phase and reenter the cell cycle prior to S-phase removal. The X-linked harlequin ($Pdcd8^{Hq}$; Hq) mutation was first identified by the almost complete absence of hair in hemizygous males and homozygous females carrying this spontaneous mutation (46). In addition to alopecia, these mice develop progressive ataxia beginning at 4–5 months of age (5, 47). The onset of ataxia is correlated with apoptosis of cerebellar granule cells, the major interneurons in the cerebellum. In addition to cerebellar neuron loss, aged Hq mutant mice have progressive retinal degeneration, as demonstrated clinically by electroretinography (ERG) and also by histological examination. Cells in the ganglion cell layer (both retinal ganglion cells and displaced amacrine cells) are lost beginning at 3 months of age. The loss of ganglion cells leads to hypoplasia of the optic nerve. Subsequently, there is loss of both horizontal and amacrine cells in the inner nuclear layer, and of photoreceptors in the outer nuclear layer. Clinically, both rod and cone ERG responses are greatly attenuated in mutant mice by 4 months of age.

A positional cloning strategy identified the Hq mutation as an ecotropic leukemia proviral insertion into the first intron of the apoptosis-inducing factor gene, Aif, also known as programmed cell death 8 ($Pdcd8$). This insertion causes aberrant splicing from exon 1 into the provirus and a corresponding decrease of 80–90% in Aif mRNA and protein. AIF is a ubiquitously expressed 57-kDa flavoprotein, containing a flavin adenine dinucleotide prosthetic group (48). The large C-terminal domain of AIF is similar to several plant and bacterial oxidoreductases, particularly semidehydroascorbate and ascorbate reductases and NADH-dependent ferredoxin reductases. Recent determination of the crystal structure of AIF demonstrates a glutathione reductase–like fold and high similarity to biphenyl dioxygenase (49).

Under normal physiological conditions, AIF is located in the intermembrane space of the mitochondria. However, during apoptosis, AIF translocates from the mitochondria to the cytoplasm and nucleus (48). This redistribution of AIF can be induced by several apoptotic paradigms in many different cell types, including primary neurons (50). Furthermore, cultured cells undergo apoptosis when transfected with AIF lacking the N-terminal mitochondrial targeting sequence or when recombinant AIF is injected into the nucleus or cytoplasm (48). The apoptogenic activity of AIF is independent of its redox activity and is not suppressed by caspase inhibitors in some mammalian cells (51–53), although genetic evidence in *Caenorhabditis elegans*, isolated mammalian mitochondria, and transiently transfected cells suggests that AIF release is caspase-dependent.

In further support of a proapoptotic role for AIF, exposed targeted ES cells hemizygous for a null mutation in the Aif gene were resistant to serum starvation but not to other apoptosis inducers (54). Examination of embryoid bodies produced by the in vitro aggregation of these cells demonstrated that these structures failed to undergo cavitation, suggesting a proapoptotic role for AIF in early development. However, chimeric mice could not be produced from $Aif^{-/Y}$ ES cells, so the normal physiological function of mitochondrially localized AIF remains unknown.

As mentioned above, AIF has strong structural homology to glutathione reductase, well established as an essential component of the H_2O_2-scavenging glutathione system. This, combined with previous observations of upregulation of oxidative stress markers in many types of adult-onset neurodegeneration, suggested that the reduction in AIF in Hq animals may cause oxidative stress, via direct or, more likely, indirect regulation of H_2O_2. Consistent with an increase in intracellular H_2O_2, we observed increased activity of the hydrogen peroxide scavenger catalase, and total glutathione, an essential electron donor for the reduction of hydroperoxides, in cerebellar extracts from *Hq* mutants at 4 weeks of age, months before the onset of neurodegeneration (Figure 1). However, increases in levels of SOD1 and SOD2, superoxide scavengers, were not detected in mice at either age. Other hallmarks of oxidative stress were also observed in *Hq* mutant mice. Increased lipid peroxidation was found in the cerebellum and the rest of the brain in *Hq* mutant mice at 1 month of age. Furthermore, DNA oxidation, as evidenced by the presence of the modified base 8-hydroxydeoxyguanosine (8-OHdG), was noted in many neuron types, including cerebellar granule cells and retinal ganglion, amacrine, and horizontal cells.

In vitro studies of cerebellar granule cell cultures from *Hq* mutant mice demonstrated that while these cells were less sensitive to serum starvation–induced apoptosis, as predicted by studies on AIF-null ES cells, they were more sensitive to H_2O_2. This sensitivity could be rescued by retroviral transfection with *Aif* cDNA containing the N-terminal mitochondrial localization signal. Interestingly, wild-type granule cells transfected with the same retrovirus were more resistant to peroxide-induced cell death, suggesting that increased amounts of AIF may actually be protective against oxidative stress–induced cell death in some instances.

The mechanism by which AIF can act to reduce oxidative stress in vivo is still puzzling at the biochemical level. In vitro, AIF has NADPH oxidase activity that, like other NADH oxidases, generates superoxide radicals (51). The loss of this activity would not be expected to generate an increase in free radicals but could in fact reduce free radical production. However, whether NADPH is the in vivo substrate of AIF is unknown (55). Although not yet tested, it has been hypothesized that AIF may play a role in the mitochondrial respiratory chain (56). If so, a dramatic reduction of AIF could increase free radical production via defects in mitochondrial respiration. Alternatively, AIF may participate in an unknown redox cycling or coupling pathway (56). Such a role would be consistent with the antioxidant activity of glutathione reductase, a structural relative of AIF, which can mediate H_2O_2 activity both through the recycling of glutathione or by coupling with the antioxidant functions of the glutaredoxin system (57).

Because of the association of oxidative stress and abnormal cell cycle checkpoint function and reports of aberrant neuronal cell cycle reentry in progressive human neurodegenerative diseases, we investigated whether neurons in the *Hq* mutant cerebellum and retina had reentered into the cell cycle. Indeed, in older *Hq* animals many cerebellar granule cells and retinal ganglion, amacrine, and horizontal cells had newly synthesized nuclear DNA, as demonstrated by BrdU incorporation (Figure 1). These cells also expressed PCNA and CDC47, both markers of S phase. Consistent with studies of postmortem tissue from patients with neurodegenerative disorders, no evidence of mitotic figures in any of these cell types was observed. The nuclei of many of these cells were clearly pyknotic, and most of the cells were also positive for activated caspase-3, indicating that cell cycle reentry in these neurons was associated with apoptosis. Furthermore, all cycling neurons were positive for 8-OHdG, demonstrating oxidative stress–induced DNA damage. However, not all 8-OHdG–positive cells were in S phase, a result that may indicate a temporal relationship between oxidative stress and cell cycle. These observations and the detection of oxidative stress markers long before neurodegeneration demonstrate that induction of unscheduled cell cycle reentry is highly correlated with, and may be induced by, oxidative stress.

While many dying neurons in the *Hq* mutant mouse abnormally reenter into the cell cycle, other neurons that degenerate do not. For example, both Purkinje cells in the cerebellum and retinal photoreceptors are progressively lost but do not stain for cell cycle markers. Interestingly, these neurons are not positive for 8-OHdG nor activated caspase-3, and electron microscopy suggests that Purkinje cells, at least, undergo necrosis rather than apoptosis. Furthermore, the loss of AIF expression in many other CNS neurons (including cortical neurons) does not lead to death, even in *Hq* mutant mice aged to 2 years.

These studies clearly indicate that various types of neurons respond differently to the downregulation of AIF (and presumably the increase in free radicals). Some neurons undergo apoptosis correlated with cell cycle reentry, whereas others die without signs of either process. Furthermore, those neurons that survive may be more resistant to oxidative stress. We (5) and others (58) have observed that sensitivity to oxidative stress differs between neuron types. Cerebellar granule cells exposed to a 15-minute pulse of 100 µM H_2O_2 displayed only a 25% survival rate after 15 hours in culture. In contrast, a similar number of cortical neurons survived when exposed to 100 µM H_2O_2 for 24 hours (58), suggesting that granule cells are more sensitive to oxidative stress than are cortical neurons. Therefore, *Hq* mutant mice provide a model to examine both the influence of oxidative stress on cell cycle reentry and the mechanisms underlying the differential response of different types of neurons to oxidative stress.

Mechanisms of oxidative stress–induced cell cycle reentry

Although both oxidative stress and cell cycle reentry have been implicated in the onset of later-onset neurodegenerative diseases and clearly occur together at the cellular level in *Hq* mutant mice, the mechanism by which oxidative stress may lead to cell cycle abnormalities remains unknown.

Cumulative DNA damage caused by endogenous free radicals has been suggested to underlie cancer and other age-related disorders, including neurodegeneration (44, 59). Can this accumulation of mutations over time play a major role in cell cycle–induced neuronal apoptosis? The progressive accumulation of oxidative-damaged DNA and the temporal increase in cell cycle reentry in the retina and cerebellum of aging *Hq* mutant mice are consistent with such a theory. The amount of 8-OHdG, a major DNA lesion resulting from free radical attack that has been shown to alter the base-pairing properties of guanine in in vitro assays, is elevated in nuclear and mitochondrial DNA in neurons in diseased brain regions of patients with neurodegenerative disorders. Furthermore, studies have shown that increases in this modified base are correlated with increased incidence of cancer (and therefore cell cycle abnormalities) (60). In addition to base modifications, oxidative stress can cause other potentially mutational events including strand breaks, discontinuous loss of heterozygosity, and large deletions (61). Thus, if the oxidation of DNA surpasses the DNA-repair capacity of the cell, mutations could accumulate, leading to the loss of genome stability. Like malignant transformation, cell cycle reentry in neurons could require somatic mutation of a complex group of genes. Furthermore, like cancer, the genetic alterations required to cause cell cycle reentry may be specific to particular cell types.

On the other hand, there is little direct evidence for mutation accumulations in aging neurons. Most direct measurements of

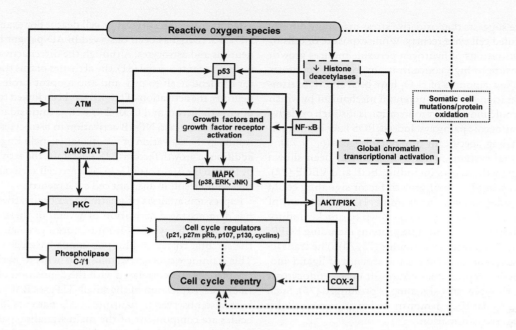

Figure 2

Possible mechanisms for the induction of unscheduled cell cycle reentry in terminally differentiated neurons by oxidative stress. We suggest that oxidative stress, mediated through ROS, could lead to aberrant cell cycle reentry through several different pathways. First, the presence of ROS can result in either cumulative DNA or protein damage that could result in cell cycle reentry. Second, a decrease in histone deacetylase activity can result in chromatin remodeling, leading to transcriptional activation of genes, the products of some of which may lead to cell cycle reentry. Third, many different ROS-activated pathways that have been previously shown to influence the cell cycle in non-neuronal cells could directly or indirectly manipulate a quiescent G0 neuron back into the cell cycle. Several of these pathways also appear to be activated in neurodegenerative disorders. Considerable cross-talk exists between these pathways, as indicated by arrows. Solid arrows represent published pathways, while broken arrows represent hypothesized mechanisms of action.

mutation frequencies have quantified rates in replicating cell types using specific locus tests, such as the mutation rate of the selectable hypoxanthine phosphoribosyl transferase gene, *Hprt* (59). Recently, however, a method for the measurement of mutation frequency during aging in postmitotic tissues has been performed in transgenic mice bearing copies of lacZ reporter plasmids at defined positions in the genome (62). As measured by this system, the frequency of spontaneous mutations was increased in liver between 4 and 23 months of age, and large deletions of adjacent mouse genomic material exponentially increased in animals over 23 months of age. However, such age-related increases in mutation frequency were not seen in plasmids rescued from the brains of these mice. This study did not find evidence for an increased mutation rate in the aging mouse brain; however, only alterations at, or adjacent to, the integrated transgenes were surveyed. Furthermore, large deletions in neurons may be more detrimental to the survival of neurons, and thus, cells harboring such events could be eliminated by apoptosis or other cell death mechanisms. Lastly, these studies were performed in wild-type mice. Perhaps cumulative damage in neurons is contingent on other susceptibility factors, such as exposure to environmental genotoxins or polymorphisms in genes involved in either the metabolism of these toxins or other cellular functions. In any case, whether oxidative stress induces mutations that would result in cell cycle reentry of aging neurons is unclear.

Another potential indirect mechanism for oxidative stress in the induction of aberrant cell cycle reentry is through histone deacetylation. Under normal circumstances, an increase in histone deacetylase activity has been associated with transcrip-

tional repression (63). In agreement, treatment of granule cell cultures with histone deacetylase inhibitors led to an increase in E2F-1 transcription (64). This increase was associated with a rise in the levels of cyclin E and the cell death molecules APAF-1 and caspase-3, recently shown to be a direct transcriptional target of E2F-1 (64, 65). Oxidative stress has been shown to decrease the activity of histone deacetylases 1–10 (66). This change in deacetylase activity could lead to a global inactivation of transcriptional repressors, leading to activation of numerous genes (including cell cycle–inducing genes) and subsequent cell death. An NAD$^+$-dependent histone deacetylase, SIR2, may make an important contribution to longevity in yeast by regulating metabolism and chromatin silencing (67). The homolog of the *Sir2* gene, named *SIR2α*, has been found in numerous mammals, including mice. Because of the dependence of SIR2α activity on its cofactor NAD$^+$, decreases in NAD$^+$ levels, occurring because of disruption of metabolic processes, would limit SIR2α activity. This could be one mechanism by which cell cycle reentry is mediated in the *Hq* mutant mouse. Decreased levels of *Aif*, an NADH oxidase, would decrease the amounts of NAD$^+$ available for metabolic processes. In addition, mitochondrial damage would further decrease the available amounts of NAD$^+$. This in turn, could decrease the activity of SIR2α and result in an increase of transcriptional activation, thereby leading to aberrant cell cycle reentry. Furthermore, SIR2 has been observed to negatively regulate p53 to promote cell survival under conditions of stress — including oxidative stress (68). Therefore, decreased SIR2α activity could result in an upregulation of p53 activity, which can in turn signal to downstream effectors of the cell cycle (see below).

Other evidence suggests that oxidative stress may more directly trigger unscheduled cell cycle reentry. While exposure of cells to moderate concentrations of hydrogen peroxide induces growth arrest, and exposure to high concentrations induces apoptosis and/or necrosis, low amounts of H_2O_2 have been shown to stimulate cell proliferation (69). One potential mechanism by which oxidative stress could induce cell proliferation is through oxidative stimulation of mitogenic pathways. Indeed, ROS have been implicated in cell signaling, specifically through mitogens (6, 70, 71).

In non-neuronal systems, oxidative stress has been shown to upregulate growth factors including EGF and VEGF (72). Oxidative stress stimulation of growth factor signaling could also be ligand independent. For example, H_2O_2 treatment of cell lines or cultured T cells has been shown to quickly induce tyrosine phosphorylation of multiple proteins including EGFR, PDGFR, and the T cell receptor complex (73, 74). The transactivation of these receptors occurs in the absence of ligand and, in the case of EGF receptor, does not result in the endocytosis of the receptor; thus receptor signaling is prolonged (75). Such enhanced signaling has been demonstrated to augment cellular proliferation and transformation.

The transactivation of growth factor receptors appears to signal through several pathways, including the extracellular signal–regulated kinases (ERKs; a subset of the MAPKs), PI3K/AKT, and phospholipase C-γ1 (74). The activation of each of these signaling pathways has been shown to induce either apoptosis or cell survival depending on the cell type and oxidative insult. Several other enzymatic pathways are modulated either directly or indirectly by oxidative stress. These include the stress-activated protein kinases JNK and p38, JAK/STAT, PKC, and ataxia telangiectasia mutated, which also participate in signals leading to either cell survival or cell death (74). There is significant cross-talk between these signals, which complicates the task of establishing pathways of oxidative stress–induced apoptosis, and presumably also oxidative stress–induced cell cycle reentry (74) (Figure 2).

ROS can also activate the transcription factor and tumor suppressor protein p53 (76). Although this protein plays an important role as a sensor of genotoxic stress and regulator of genes necessary for growth arrest and cell death, recent evidence suggests that p53 activation can in turn activate genes, including antioxidants and heparin-binding EGF-like factor, that function in compensatory survival pathways (77). p53 can also activate ERK and AKT, leading to an induction of COX-2, an important mediator of various proliferative diseases, including cancer (77).

In neuronal cultures, some evidence supports the rapid upregulation of cell cycle–related genes by oxidative stress. Dopamine, the oxidative metabolism of which generates free radicals, or hydrogen peroxide treatment of postmitotic chick sympathetic neurons, resulted in the expression of cell cycle–related genes, including cyclin B and CDK5, prior to apoptosis (78). Treatment of these cells with antioxidants disrupted both the rise in cell cycle proteins and the subsequent apoptosis. Low potassium levels in cerebellar granule cell cultures induce ROS generation and apoptosis, concomitant with the activation of MAPKK7, one of the activators of the JNK pathway (79). In addition, upregulation of MAPKs, including ERK1/2 and JNK, as well as the serine-threonine kinase Akt, has been observed in H_2O_2-treated cultured cortical neurons (80).

ROS and/or oxidative damage can activate gene transcription. However, as discussed previously, transcribed genes may be implicated in either cell survival or cell death. For example, increased NF-κB activity has been observed in AD patient brains, in both neurons and astrocytes. Although the direct activation of NF-κB by ROS is controversial (81), the activities of this transcription factor can lend itself to pro- and antiapoptotic roles in the cell. In neurons, the activation of NF-κB has been linked to the activation of growth factors and COX-2 as well as antioxidants, specifically SOD2. In contrast, NF-κB activation in astrocytes and microglia results in the activation of pro-oxidants, including nitric oxide, in addition to growth factors (82). Therefore, the activation of NF-κB by ROS in the brain could lead either to cell survival or to signaling that could result in aberrant cell cycle reentry.

Expression analysis of postmortem brains from AD patients has demonstrated activation of several of these ROS-activated signaling pathways. EGFR-immunoreactive neuritic plaques in the cerebral cortex and hippocampus have been observed (83). This staining was observed mainly in neurons in the periphery of the plaques and correlated with the deposition of paired helical filaments. Expression of the small GTPases RAC and CDC42 has been also observed in neurons in AD patients (84). These molecules are components of the main signaling pathway between external stressors (such as oxidative stress) and the MAPK pathways. In addition, both RAC and CDC42 have been implicated in progression of the cell cycle, providing a potential link between oxidative stress and cell cycle reentry (82). MAPK p38 expression has been localized to tau-positive neurofibrillary tangles (85, 86). Since p38 has previously been implicated in cell proliferation, apoptosis, and differentiation, this finding could also provide a link between oxidative stress and cell cycle reentry in AD brains.

Lastly, oxidative stress is well known for causing protein damage via nitration or oxidation (23). Such protein modifications have been repeatedly shown in postmortem brain tissue from patients with neurodegenerative disorders, and these reactions can occur generally or have protein specificity. Oxidation could gradually disable proteins necessary for cell cycle repression.

In summary, although unscheduled cell cycle reentry may be mediated by many different insults, evidence is mounting that both cell cycle and oxidative stress may be team players in neurodegenerative disorders. We have discussed some of the potential pathways by which the repression of the cell cycle in neurons may be modulated by ROS (Figure 2). Given the distinctive pathology of different neurodegenerative diseases, it is likely that specific types of neurons may respond to different signaling pathways, or need activation of multiple signaling pathways. The testing of the roles of these and other potential pathways in neurodegeneration will require the development and analysis of additional animal models, in addition to further correlative human studies. However, the careful dissection of the interplay between oxidative stress and the cell cycle will greatly advance our understanding of several debilitating neurodegenerative disorders.

Acknowledgments

We thank Edward Leiter for comments on the manuscript and Jennifer Smith for graphical assistance with the manuscript figures. This work was supported by NIH grants AG-19358 and NS-42613 to S.L. Ackerman.

Address correspondence to: Susan L. Ackerman, The Jackson Laboratory, 600 Main Street, Bar Harbor, Maine 04609, USA. Phone: (207) 288-6494; Fax: (207) 288-6077; E-mail: sla@jax.org.

1. Nagy, Z., Esiri, M.M., and Smith, A.D. 1998. The cell division cycle and the pathophysiology of Alzheimer's disease. *Neuroscience.* **87**:731–739.

2. Nagy, Z. 1999. Mechanisms of neuronal death in Down's syndrome. *J. Neural Transm. Suppl.* **57**:233–245.

3. Herrup, K., and Busser, J.C. 1995. The induction of multiple cell cycle events precedes target-related neuronal death. *Development.* **121**:2385–2395.

4. Migheli, A., et al. 1999. A cell cycle alteration precedes apoptosis of granule cell precursors in the weaver mouse cerebellum. *Am. J. Pathol.* **155**:365–373.

5. Klein, J.A., et al. 2002. The harlequin mouse mutation downregulates apoptosis-inducing factor. *Nature.* **419**:367–374.

6. Kannan, K., and Jain, S.K. 2000. Oxidative stress and apoptosis. *Pathophysiology.* **7**:153–163.

7. Leonardi, E.T., and Mytilineou, C. 1998. Cell culture models of neuronal degeneration and neuroprotection. Implications for Parkinson's disease. *Adv. Exp. Med. Biol.* **446**:203–222.

8. Yokota, T., et al. 2001. Delayed-onset ataxia in mice lacking alpha-tocopherol transfer protein: model for neuronal degeneration caused by chronic oxidative stress. *Proc. Natl. Acad. Sci. U. S. A.* **98**:15185–15190.

9. Melov, S., et al. 1998. A novel neurological phenotype in mice lacking mitochondrial manganese superoxide dismutase. *Nat. Genet.* **18**:159–163.

10. Sayre, L.M., Smith, M.A., and Perry, G. 2001. Chemistry and biochemistry of oxidative stress in neurodegenerative disease. *Curr. Med. Chem.* **8**:721–738.

11. Aksenov, M.Y., et al. 1998. The expression of key oxidative stress-handling genes in different brain regions in Alzheimer's disease. *J. Mol. Neurosci.* **11**:151–164.

12. Hensley, K., et al. 1994. A model for beta-amyloid aggregation and neurotoxicity based on free radical generation by the peptide: relevance to Alzheimer disease. *Proc. Natl. Acad. Sci. U. S. A.* **91**:3270–3274.

13. Jang, J.H., and Surh, Y.J. 2002. Beta-amyloid induces oxidative DNA damage and cell death through activation of c-Jun N terminal kinase. *Ann. N. Y. Acad. Sci.* **973**:228–236.

14. Matsuoka, Y., Picciano, M., La Francois, J., and Duff, K. 2001. Fibrillar beta-amyloid evokes oxidative damage in a transgenic mouse model of Alzheimer's disease. *Neuroscience.* **104**:609–613.

15. Smith, M.A., Rottkamp, C.A., Nunomura, A., Raina, A.K., and Perry, G. 2000. Oxidative stress in Alzheimer's disease. *Biochim. Biophys. Acta.* **1502**:139–144.

16. Jenner, P., and Olanow, C.W. 1996. Oxidative stress and the pathogenesis of Parkinson's disease. *Neurology.* **47**(Suppl. 3):S161–S170.

17. Giasson, B.I., Ischiropoulos, H., Lee, V.M., and Trojanowski, J.Q. 2002. The relationship between oxidative/nitrative stress and pathological inclusions in Alzheimer's and Parkinson's diseases. *Free Radic. Biol. Med.* **32**:1264–1275.

18. Fukushima, T., et al. 1995. Radical formation site of cerebral complex I and Parkinson's disease. *J. Neurosci. Res.* **42**:385–390.

19. Cassarino, D.S., et al. 1997. Elevated reactive oxygen species and antioxidant enzyme activities in animal and cellular models of Parkinson's disease. *Biochim. Biophys. Acta.* **1362**:77–86.

20. Sherer, T.B., et al. 2002. An in vitro model of Parkinson's disease: linking mitochondrial impairment to altered alpha-synuclein metabolism and oxidative damage. *J. Neurosci.* **22**:7006–7015.

21. Chun, H.S., et al. 2001. Dopaminergic cell death induced by MPP(+), oxidant and specific neurotoxicants shares the common molecular mechanism. *J. Neurochem.* **76**:1010–1021.

22. Albers, D.S., and Augood, S.J. 2001. New insights into progressive supranuclear palsy. *Trends Neurosci.* **24**:347–353.

23. Butterfield, D.A., and Kanski, J. 2001. Brain protein oxidation in age-related neurodegenerative disorders that are associated with aggregated proteins. *Mech. Ageing Dev.* **122**:945–962.

24. Kim, J.I., et al. 2001. Oxidative stress and neurodegeneration in prion diseases. *Ann. N. Y. Acad. Sci.* **928**:182–186.

25. Cassarino, D.S., and Bennett, J.P., Jr. 1999. An evaluation of the role of mitochondria in neurodegenerative diseases: mitochondrial mutations and oxidative pathology, protective nuclear responses, and cell death in neurodegeneration. *Brain Res. Brain Res. Rev.* **29**:1–25.

26. Heintz, N. 1993. Cell death and the cell cycle: a relationship between transformation and neurodegeneration? *Trends Biochem. Sci.* **18**:157–159.

27. Feddersen, R.M., Ehlenfeldt, R., Yunis, W.S., Clark, H.B., and Orr, H.T. 1992. Disrupted cerebellar cortical development and progressive degeneration of Purkinje cells in SV40 T antigen transgenic mice. *Neuron.* **9**:955–966.

28. Frade, J.M. 2000. Unscheduled re-entry into the cell cycle induced by NGF precedes cell death in nascent retinal neurones. *J. Cell Sci.* **113**:1139–1148.

29. Verdaguer, E., et al. 2002. Kainic acid-induced apoptosis in cerebellar granule neurons: an attempt at cell cycle re-entry. *Neuroreport.* **13**:413–416.

30. Copani, A., et al. 1999. Mitotic signaling by beta-amyloid causes neuronal death. *FASEB J.* **13**:2225–2234.

31. Copani, A., et al. 2002. Beta-amyloid-induced synthesis of the ganglioside GD3 is a requisite for cell cycle reactivation and apoptosis in neurons. *J. Neurosci.* **22**:3963–3968.

32. Feddersen, R.M., Clark, H.B., Yunis, W.S., and Orr, H.T. 1995. In vivo viability of postmitotic Purkinje neurons requires pRb family member function. *Mol. Cell. Neurosci.* **6**:153–167.

33. Al-Ubaidi, M.R., et al. 1997. Unscheduled DNA replication precedes apoptosis of photoreceptors expressing SV40 T antigen. *Exp. Eye Res.* **64**:573–585.

34. ElShamy, W.M., Fridvall, L.K., and Ernfors, P. 1998. Growth arrest failure, G1 restriction point override, and S phase death of sensory precursor cells in the absence of neurotrophin-3. *Neuron.* **21**:1003–1015.

35. Katchanov, J., et al. 2001. Mild cerebral ischemia induces loss of cyclin-dependent kinase inhibitors and activation of cell cycle machinery before delayed neuronal cell death. *J. Neurosci.* **21**:5045–5053.

36. Osuga, H., et al. 2000. Cyclin-dependent kinases as a therapeutic target for stroke. *Proc. Natl. Acad. Sci. U. S. A.* **97**:10254–10259.

37. Rangathan, S., and Bowser, R. 2003. Alterations in G(1) to S phase cell-cycle regulators during amyotrophic lateral sclerosis. *Am. J. Pathol.* **162**:823–835.

38. McShea, A., Harris, P.L., Webster, K.R., Wahl, A.F., and Smith, M.A. 1997. Abnormal expression of the cell cycle regulators P16 and CDK4 in Alzheimer's disease. *Am. J. Pathol.* **150**:1933–1939.

39. Busser, J., Geldmacher, D.S., and Herrup, K. 1998. Ectopic cell cycle proteins predict the sites of neuronal cell death in Alzheimer's disease brain. *J. Neurosci.* **18**:2801–2807.

40. Herrup, K., and Arendt, T. 2002. Re-expression of cell cycle proteins induces neuronal cell death during Alzheimer's disease. *J. Alzheimers Dis.* **4**:243–247.

41. Yang, Y., Geldmacher, D.S., and Herrup, K. 2001. DNA replication precedes neuronal cell death in Alzheimer's disease. *J. Neurosci.* **21**:2661–2668.

42. Nagy, Z., Esiri, M.M., and Smith, A.D. 1997. Expression of cell division markers in the hippocampus in Alzheimer's disease and other neurodegenerative conditions. *Acta Neuropathol. (Berl.)* **93**:294–300.

43. Konishi, Y., Lehtinen, M., Donovan, N., and Bonni, A. 2002. Cdc2 phosphorylation of BAD links the cell cycle to the cell death machinery. *Mol. Cell.* **9**:1005–1016.

44. Migliore, L., and Coppede, F. 2002. Genetic and environmental factors in cancer and neurodegenerative diseases. *Mutat. Res.* **512**:135–153.

45. Chen, Q.M., Liu, J., and Merrett, J.B. 2000. Apoptosis or senescence-like growth arrest: influence of cell-cycle position, p53, p21 and bax in H_2O_2 response of normal human fibroblasts. *Biochem. J.* **347**:543–551.

46. Barber, B.R. 1971. Research news. *Mouse News Letter.* **45**:34–35.

47. Bronson, R.T., Lane, P.W., Harris, B.S., and Davisson, M.T. 1990. Harlequin (*Hq*) produces progressive cerebellar atrophy. *Mouse Genome.* **87**:110.

48. Susin, S.A., et al. 1999. Molecular characterization of mitochondrial apoptosis-inducing factor. *Nature.* **397**:441–446.

49. Mate, M.J., et al. 2002. The crystal structure of the mouse apoptosis-inducing factor AIF. *Nat. Struct. Biol.* **9**:442–446.

50. Cande, C., Cecconi, F., Dessen, P., and Kroemer, G. 2002. Apoptosis-inducing factor (AIF): key to the conserved caspase-independent pathways of cell death? *J. Cell Sci.* **115**:4727–4734.

51. Miramar, M.D., et al. 2001. NADH oxidase activity of mitochondrial apoptosis-inducing factor. *J. Biol. Chem.* **276**:16391–16398.

52. Wang, X., Yang, C., Chai, J., Shi, Y., and Xue, D. 2002. Mechanisms of AIF-mediated apoptotic DNA degradation in *Caenorhabditis elegans*. *Science.* **298**:1587–1592.

53. Arnoult, D., et al. 2002. Mitochondrial release of apoptosis-inducing factor occurs downstream of cytochrome c release in response to several pro-apoptotic stimuli. *J. Cell Biol.* **159**:923–929.

54. Joza, N., et al. 2001. Essential role of the mitochondrial apoptosis-inducing factor in programmed cell death. *Nature.* **410**:549–554.

55. Bonni, A. 2003. Neurodegeneration: a non-apoptotic role for AIF in the brain. *Curr. Biol.* **13**:R19–R21.

56. Lipton, S.A., and Bossy-Wetzel, E. 2002. Dueling activities of AIF in cell death versus survival: DNA binding and redox activity. *Cell.* **111**:147–150.

57. Holmgren, A. 2000. Antioxidant function of thioredoxin and glutaredoxin systems. *Antioxid. Redox Signal.* **2**:811–820.

58. White, A.R., et al. 1998. Survival of cultured neurons from amyloid precursor protein knock-out mice against Alzheimer's amyloid-beta toxicity and oxidative stress. *J. Neurosci.* **18**:6207–6217.

59. Turker, M.S. 2000. Somatic cell mutations: can they provide a link between aging and cancer? *Mech. Ageing Dev.* **117**:1–19.

60. Halliwell, B. 1998. Can oxidative DNA damage be used as a biomarker of cancer risk in humans? Problems, resolutions and preliminary results from nutritional supplementation studies. *Free Radic. Res.* **29**:469–486.

61. Sekiguchi, M., and Tsuzuki, T. 2002. Oxidative nucleotide damage: consequences and prevention. *Oncogene.* **21**:8895–8904.

62. Dolle, M.E., et al. 1997. Rapid accumulation of genome rearrangements in liver but not in brain of old mice. *Nat. Genet.* **17**:431–434.

63. Pazin, M.J., and Kadonaga, J.T. 1997. What's up and down with histone deacetylation and transcription? *Cell.* **89**:325–328.

64. Boutillier, A.L., Trinh, E., and Loeffler, J.P. 2003. Selective E2F-dependent gene transcription is controlled by histone deacetylase activity during neuronal apoptosis. *J. Neurochem.* **84**:814–828.

65. Nahle, Z., et al. 2002. Direct coupling of the cell cycle and cell death machinery by E2F. *Nat. Cell Biol.* **4**:859–864.

66. Rahman, I. 2002. Oxidative stress, transcription factors and chromatin remodeling in lung inflammation. *Biochem. Pharmacol.* **64**:935–942.

67. Guarente, L. 2000. Sir2 links chromatin silencing, metabolism, and aging. *Genes Dev.* **14**:1021–1026.

68. Luo, J., et al. 2001. Negative control of p53 by

Sir2alpha promotes cell survival under stress. *Cell.* **107**:137–148.

69. Clement, M.V., and Pervaiz, S. 1999. Reactive oxygen intermediates regulate cellular response to apoptotic stimuli: an hypothesis. *Free Radic. Res.* **30**:247–252.

70. Chan, P.H. 2001. Reactive oxygen radicals in signaling and damage in the ischemic brain. *J. Cereb. Blood Flow Metab.* **21**:2–14.

71. Droge, W. 2002. Free radicals in the physiological control of cell function. *Physiol. Rev.* **82**:47–95.

72. Maulik, N., and Das, D.K. 2002. Redox signaling in vascular angiogenesis. *Free Radic. Biol. Med.* **33**:1047–1060.

73. Huang, R.P., Wu, J.X., Fan, Y., and Adamson, E.D. 1996. UV activates growth factor receptors via reactive oxygen intermediates. *J. Cell Biol.* **133**:211–220.

74. Martindale, J.L., and Holbrook, N.J. 2002. Cellular response to oxidative stress: signaling for suicide and survival. *J. Cell. Physiol.* **192**:1–15.

75. Ravid, T., Sweeney, C., Gee, P., Carraway, K.L., 3rd, and Goldkorn, T. 2002. Epidermal growth factor receptor activation under oxidative stress fails to promote c-Cbl mediated down-regulation. *J. Biol. Chem.* **277**:31214–31219.

76. Nakamura, T., and Sakamoto, K. 2001. Reactive oxygen species up-regulates cyclooxygenase-2, p53, and Bax mRNA expression in bovine luteal cells. *Biochem. Biophys. Res. Commun.* **284**:203–210.

77. Han, J.A., et al. 2002. P53-mediated induction of Cox-2 counteracts p53-or genotoxic stress-induced apoptosis. *EMBO J.* **21**:5635–5644.

78. Shirvan, A., et al. 1998. Expression of cell cycle-related genes during neuronal apoptosis: is there a distinct pattern? *Neurochem. Res.* **23**:767–777.

79. Trotter, L., et al. 2002. Mitogen-activated protein kinase kinase 7 is activated during low potassium-induced apoptosis in rat cerebellar granule neurons. *Neurosci. Lett.* **320**:29–32.

80. Crossthwaite, A.J., Hasan, S., and Williams, R.J. 2002. Hydrogen peroxide-mediated phosphorylation of ERK1/2, Akt/PKB and JNK in cortical neurones: dependence on Ca(2+) and PI3-kinase. *J. Neurochem.* **80**:24–35.

81. Li, N., and Karin, M. 1999. Is NF-kappaB the sensor of oxidative stress? *FASEB J.* **13**:1137–1143.

82. Mattson, M.P., and Camandola, S. 2001. NF-κB in neuronal plasticity and neurodegenerative disorders. *J. Clin. Invest.* **107**:247–254.

83. Birecree, E., Whetsell, W.O., Jr., Stoscheck, C., King, L.E., Jr., and Nanney, L.B. 1988. Immunoreactive epidermal growth factor receptors in neuritic plaques from patients with Alzheimer's disease. *J. Neuropathol. Exp. Neurol.* **47**:549–560.

84. Zhu, X., et al. 2000. Activation of oncogenic pathways in degenerating neurons in Alzheimer disease. *Int. J. Dev. Neurosci.* **18**:433–437.

85. Zhu, X., et al. 2000. Activation of p38 kinase links tau phosphorylation, oxidative stress, and cell cycle-related events in Alzheimer disease. *J. Neuropathol. Exp. Neurol.* **59**:880–888.

86. Atzori, C., et al. 2001. Activation of the JNK/p38 pathway occurs in diseases characterized by tau protein pathology and is related to tau phosphorylation but not to apoptosis. *J. Neuropathol. Exp. Neurol.* **60**:1190–1197.

Neuronal degeneration and mitochondrial dysfunction

Eric A. Schon[1,2] and Giovanni Manfredi[3]

[1]Department of Neurology, and [2]Department of Genetics and Development, Columbia University, New York, New York, USA.
[3]Department of Neurology and Neuroscience, Weill Medical College of Cornell University, New York, New York, USA.

Over the last decade, the underlying genetic bases of several neurodegenerative disorders, including Huntington disease (HD), Friedreich ataxia, hereditary spastic paraplegia, and rare familial forms of Parkinson disease (PD), Alzheimer disease (AD), and amyotrophic lateral sclerosis (ALS), have been identified. However, the etiologies of sporadic AD, PD, and ALS, which are among the most common neurodegenerative diseases, are still unclear, as are the pathogenic mechanisms giving rise to the various, and often highly stereotypical, clinical features of these diseases. Despite the differential clinical features of the various neurodegenerative disorders, the fact that neurons are highly dependent on oxidative energy metabolism has suggested a unified pathogenetic mechanism of neurodegeneration, based on an underlying dysfunction in mitochondrial energy metabolism.

Mitochondria are the seat of a number of important cellular functions, including essential pathways of intermediate metabolism, amino acid biosynthesis, fatty acid oxidation, steroid metabolism, and apoptosis. Of key importance to our discussion here is the role of mitochondria in oxidative energy metabolism. Oxidative phosphorylation (OXPHOS) generates most of the cell's ATP, and any impairment of the organelle's ability to produce energy can have catastrophic consequences, not only due to the primary loss of ATP, but also due to indirect impairment of "downstream" functions, such as the maintenance of organellar and cellular calcium homeostasis. Moreover, deficient mitochondrial metabolism may generate reactive oxygen species (ROS) that can wreak havoc in the cell. It is for these reasons that mitochondrial dysfunction is such an attractive candidate for an "executioner's" role in neuronal degeneration.

The mitochondrion is the only organelle in the cell, aside from the nucleus, that contains its own genome and genetic machinery. The human mitochondrial genome (1) is a tiny 16.6-kb circle of double-stranded mitochondrial DNA (mtDNA) (Figure 1). It encodes 13 polypeptides, all of which are components of the respiratory chain/OXPHOS system, plus 24 genes, specifying two ribosomal RNAs (rRNAs) and 22 transfer RNAs (tRNAs), that are required to synthesize the 13 polypeptides. Obviously, an organelle as complex as a mitochondrion requires more than 37 gene products; in fact, about 850 polypeptides, all encoded by nuclear DNA (nDNA), are required to build and maintain a functioning organelle. These proteins are synthesized in the cytoplasm and are imported into the organelle, where they are partitioned into the mitochondrion's four main compartments — the outer mitochondrial membrane (OMM), the inner mitochondrial membrane (IMM), the intermembrane space (IMS), and the matrix, located in the interior (the organelle's "cytoplasm"). Of the 850 proteins, approximately 75 are structural components of the respiratory complexes (Figure 2) and at least another 20 are required to assemble and maintain them in working order. The five complexes of the respiratory chain/OXPHOS system — complexes I (NADH ubiquinone oxidoreductase), II (succinate ubiquinone oxidoreductase), III (ubiquinone–cytochrome c reductase), IV (cytochrome c oxidase), and V (ATP synthase) — are all located in the IMM. There are also two electron carriers, ubiquinone (also called coenzyme Q), located in the IMM, and cytochrome c, located in the IMS.

Besides the fact that it operates under dual genetic control, four other features unique to the behavior of this organelle are important in understanding mitochondrial function in neurodegeneration. First, as opposed to the nucleus, in which there are two sets of chromosomes, there are thousands of mtDNAs in each cell, with approximately five mtDNAs per organelle. Second, organellar division and mtDNA replication operate independently of the cell cycle, both in dividing cells (such as glia) and in postmitotic nondividing cells (such as neurons). Third, upon cell division, the mitochondria (and their mtDNAs) are partitioned randomly between the daughter cells (mitotic segregation). Finally, the number of organelles varies among cells, depending in large part on the metabolic requirements of that cell. Thus, skin fibroblasts contain a few hundred mitochondria, whereas neurons may contain thousands and cardiomyocytes tens of thousands of organelles. Taken together, these features highlight the fact that mitochondria obey the laws of population genetics, not mendelian genetics.

Mitochondrial diseases

Mitochondria and their DNA are inherited exclusively from the mother. Thus, pathogenic mutations in mtDNA can cause maternally inherited syndromes. Since these mutations arise initially in only one mtDNA, the population of mtDNAs that was originally homoplasmic (i.e., only one mtDNA genotype) becomes heteroplasmic (two or more coexisting mtDNA genotypes). Obviously, the ratio of normal to mutated mtDNAs in a heteroplasmic population, and their spatial and temporal distributions, will be critical in determining whether and when a subpopulation

Nonstandard abbreviations used: Huntington disease (HD); Parkinson disease (PD); familial Parkinson disease (FPD); Alzheimer disease (AD); familial Alzheimer disease (FAD); amyotrophic lateral sclerosis (ALS); sporadic amyotrophic lateral sclerosis (SALS); oxidative phosphorylation (OXPHOS); reactive oxygen species (ROS); mitochondrial DNA (mtDNA); ribosomal RNA (rRNA); transfer RNA (tRNA); nuclear DNA (nDNA); inner mitochondrial membrane (IMM); intermembrane space (IMS); mitochondrial encephalomyopathy with lactic acidosis and strokelike episodes (MELAS); myoclonus epilepsy with ragged-red fibers (MERRF); ragged-red fiber (RRF); Kearns-Sayre syndrome (KSS); progressive external ophthalmoplegia (PEO); cytochrome c oxidase (CCO); Leber hereditary optic neuropathy (LHON); Leigh syndrome (LS); Friedreich ataxia (FRDA); mitochondrial superoxide dismutase (Mn-SOD); Cu,Zn-superoxide dismutase (Cu,Zn-SOD); β-amyloid peptide (Aβ); progressive supranuclear palsy (PSP); terminal deoxynucleotidyl transferase–mediated dUTP nick end-labeling (TUNEL).

Conflict of interest: The authors have declared that no conflict of interest exists.

Citation for this article: *J. Clin. Invest.* **111**:303–312 (2003). doi:10.1172/JCI200317741.

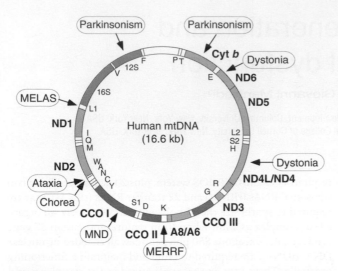

Figure 1
Map of the human mitochondrial genome (1). Polypeptide-coding genes (boldface) are outside the circle and specify seven subunits of NADH dehydrogenase–coenzyme Q oxidoreductase (ND), one subunit of coenzyme Q–cytochrome *c* oxidoreductase (Cyt *b*), three subunits of cytochrome *c* oxidase (CCO), and two subunits of ATP synthase (A) (see also Figure 2). Protein synthesis genes (12S and 16S rRNAs, and 22 tRNAs [one-letter code]) are inside the circle. Mutations in mtDNA associated with MELAS and MERRF, and mutations with features of neurodegenerative disorders, such as ataxia, chorea, dystonia, motor neuron disease (MND), and parkinsonism, are boxed. A complete list of pathogenic mtDNA mutations may be found in ref. 64.

and cytochrome *c* oxidase (CCO) deficiency. In muscle biopsies, the mutated mtDNAs accumulate preferentially in the RRFs, and RRFs are typically negative for CCO activity.

On the other hand, there are some mtDNA mutations that tend to cause rather selective neuronal degeneration (Figure 1 and Table 1). This situation applies, for example, to mutations in the mtDNA genes coding for subunits of complexes I and V. Mutations in complex I subunits have been associated specifically with Leber hereditary optic neuropathy (LHON), a subacute degeneration of the optic nerve causing bilateral visual failure and blindness. Remarkably, LHON affects males far more than females. Equally remarkably, many patients are homoplasmic for the mutation (i.e., the mutation affects all mtDNAs), which is present in all tissues, and yet the disease is essentially confined to the optic nerve. Complex I activity in cells from these patients ranges from extremely mild to severe. Mutations in complex I genes are also associated with an "LHON-plus" syndrome, characterized pathologically by basal ganglia degeneration and dystonia, and with a maternally inherited form of levodopa-responsive parkinsonism.

Mutations in the mtDNA-encoded subunit 6 of complex V (ATPase 6) are associated with a syndrome characterized clinically by neuropathy, ataxia, and retinitis pigmentosa (NARP) at mutation loads of 70–90% in brain, and with a form of subacute necrotizing encephalomyelopathy known as Leigh syndrome (maternally inherited Leigh syndrome [MILS]) at mutation loads above 90%. These mutations cause varying degrees of impairment of mitochondrial ATP synthesis. Other mutations in mtDNA that cause Leigh syndrome (LS) have been found in ND5, tRNA^Trp, tRNA^Val, and CCO III; patients with the latter two were atypical, in that the disease was of relatively late onset.

of mutant mtDNAs will have overt phenotypic consequences. In this regard, it is important to note that most mitochondrial diseases due to maternally inherited mutations in mtDNA are recessive, that is, a very high amount of mutated mtDNA must be present (typically more than 70% of the total population of mtDNAs) in order to cause overt dysfunction. Since mtDNA encodes subunits of the respiratory complexes, pathogenic mutations in mtDNA cause diseases arising from problems in OXPHOS. On the other hand, mutations in nucleus-encoded proteins that are targeted to mitochondria, or that affect organellar function indirectly, can affect almost any aspect of cellular function, not just oxidative energy metabolism.

Neurodegenerative diseases associated with mutations in mitochondrial genes
Diseases associated with mtDNA mutations are typically heterogeneous and are often multisystemic. Since heart, skeletal muscle, and brain are among the most energy-dependent tissues of the body, it is not surprising that many mitochondrial disorders are encephalo-cardiomyopathies. For example, mutations in the tRNA^Leu(UUR) gene cause mitochondrial encephalomyopathy with lactic acidosis and strokelike episodes (MELAS), mutations in tRNA^Lys cause myoclonus epilepsy with ragged-red fibers (MERRF), and large-scale spontaneous deletions cause Kearns-Sayre syndrome (KSS) and progressive external ophthalmoplegia (PEO) (2). While these diseases are clinically distinct, they share a number of features found in many (but by no means all) mitochondrial diseases, including lactic acidosis, massive mitochondrial proliferation in muscle (resulting in ragged-red fibers [RRFs]),

Neurodegenerative diseases associated with mutations in nuclear gene products that are targeted to mitochondria
Mutations in a small but growing subset of the 850-odd nDNA-encoded gene products that are present in mitochondria have been implicated in neurodegenerative disease (Table 1). While some, such as LS, are essentially the mendelian counterpart of diseases previously associated with mutations in mtDNA, the others are *sui generis* and have highly diverse genetic origins.

Leigh syndrome
LS is a fatal neurodegenerative condition pathologically characterized by subacute symmetrical necrotic lesions in the subcortical regions of the CNS, including thalamus, basal ganglia, brainstem, and spinal cord, accompanied by demyelination, gliosis, and vascular proliferation in affected areas. Onset is most frequently in infancy or early childhood but may sometimes be in adult life. Symptoms include motor and mental regression, dystonia, ataxia, and abnormal breathing. Death generally occurs within 2 years after onset. It is now clear that LS results from impaired mitochondrial energy metabolism, which can arise from a variety of causes. As noted above, LS can be caused by maternally inherited mutations, but inheritance of LS can also be autosomal-recessive or X-linked. In the latter group are mutations in subunits of the pyruvate dehydrogenase complex, mutations in nuclear-encoded respiratory chain subunits (mainly of complex I but sometimes also of complex II), and mutations in proteins involved in respiratory chain assembly.

CCO deficiency is one of the most common causes of autosomal-recessive LS, but interestingly, all the mutations described to date have been in genes that are required for CCO assembly. Only one sporadic mutation causing LS was found in a CCO structural subunit, in mtDNA-encoded CCO III. CCO-deficient LS is most commonly due to mutations in SURF1, a mitochondrially targeted CCO assembly protein of unknown function, and in SCO2, a mitochondrially targeted protein required for insertion of copper into the CCO holoprotein.

Friedreich ataxia

Friedreich ataxia (FRDA) is the most common of the hereditary ataxias. It is defined clinically by progressive limb and gait ataxia, axonal sensory neuropathy, absent tendon reflexes, and pyramidal signs. FRDA, which is autosomal-recessive, results from mutations in FRDA, which encodes frataxin, a mitochondrially targeted iron-storage protein (3, 4). Trinucleotide repeat (GAA) expansions in intron 1, sometimes in combination with point mutations on the other allele, reduce frataxin expression, resulting in decreased activities of holoproteins containing iron-sulfur groups, including complexes I, II, and III, and aconitase, a Krebs cycle enzyme. Notably, frataxin is particularly abundant in the cerebellum and spinal cord. Mitochondrial enzyme deficiencies and late-stage iron accumulation were also found in a frataxin knockout mouse (5). While the pathogenesis of FRDA is still unclear, one possibility is that the presence of unbound (free) reactive iron, via the Fenton reaction, generates free radicals within the mitochondria, leading to oxidative damage and inactivation of mitochondrial enzymes. For unknown reasons, mitochondrial superoxide dismutase (Mn-SOD), a key antioxidant protein, is deficient in cultured cells from FRDA patients, making them even more prone to oxidative damage. The role of oxidative damage in the pathogenesis of FRDA is also supported by the finding that idebenone, an antioxidant similar to ubiquinone, can reduce myocardial hypertrophy and also decrease markers of oxidative stress in FRDA patients (6, 7).

Hereditary spastic paraplegia

Hereditary spastic paraplegia is a progressive disorder resulting in paraparesis, with onset in childhood or early adulthood. Upper motor neurons are involved selectively, and ataxia and retinitis pigmentosa are also common. Autosomal-dominant, X-linked, and autosomal-recessive forms have been described. Patients with a recessive form of hereditary spastic paraplegia harbor mutations in SPG7, which encodes paraplegin (8), a mitochondrially targeted protein with strong similarity to yeast metalloproteases. Although the underlying pathogenic mechanism is still unclear, mutations in paraplegin cause OXPHOS dysfunction with CCO deficiency and mitochondrial proliferation (i.e., RRFs) in muscle (8). Recently, an autosomal-dominant form of hereditary spastic paraplegia (SPG13) has been attributed to mutations in the mitochondrial import chaperonin HSP60 (9).

Deafness-dystonia syndrome

Deafness-dystonia syndrome, also called Mohr-Tranebjaerg syndrome, is an X-linked recessive disorder characterized by progressive sensorineural deafness, dystonia, cortical blindness, and psychiatric illness. It results from mutations in TIMM8A, which encodes deafness/dystonia protein-1 (DDP1) (10), a component of the mitochondrial protein import machinery located in the IMS. In fibroblasts from patients, mitochondrial protein import is impaired (although not in a global fashion), probably because the DDP1 mutation impairs assembly of TIMM8A into a functional subcomplex of the import machinery (11).

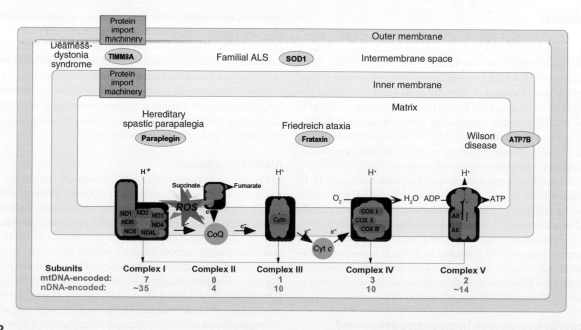

Figure 2
Schematic representation of the mitochondrion with its electron transport chain (ETC). The ETC is the principal source of ROS in the cell. In addition to mutations in mtDNA- and nDNA-encoded components of the ETC, a number of mutant mitochondrial proteins that do not belong to the ETC have been associated with neurodegenerative disorders. The intramitochondrial localization and the clinical phenotypes associated with mutations of some of these proteins (in ovals) are indicated.

Table 1

Neurodegenerative disorders with mitochondrial involvement

Neurodegenerative diseases due to:	Mutated gene product
Primary mutations in mtDNA	
Leigh syndrome	CCO III, ND5, tRNA^{Trp}, tRNA^{Val}
LHON/Parkinsonism/Dystonia	Complex I mtDNA-encoded subunits
Motor neuron disease	CCO I
NARP/MILS	ATPase 6
Parkinsonism	12S rRNA
Nuclear gene mutations in mitochondrion-targeted proteins affecting OXPHOS	
Leigh syndrome with complex I deficiency	Complex I nDNA-encoded subunits
Leigh syndrome with complex II deficiency	SDH flavoprotein
Leigh syndrome with complex IV deficiency	SURF1, SCO2
Leigh syndrome with PDH deficiency	PDH E1-α subunit
Nuclear gene mutations in other mitochondrion-targeted proteins	
ALS	Cu,Zn-SOD
Friedreich ataxia	Frataxin
Hereditary spastic paraplegia	Paraplegin, HSP60
Mohr-Tranebjaerg syndrome	Deafness/dystonia protein-1 (TIMM8A)
Wilson disease	Cu-transporting ATPase (ATP7B)
Nuclear gene mutations in non–mitochondrion-targeted proteins	
AD	ABPP, presenilin-1, presenilin-2
HD	Huntingtin
PD	Parkin, α-synuclein
PSP	Tau protein
Putative secondary mitochondrial involvement	
Sporadic AD	Unknown
Sporadic ALS	Unknown
Sporadic PD	Unknown

LHON, Leber hereditary optic neuropathy; NARP, neuropathy, ataxia, and retinitis pigmentosa; MILS, maternally inherited Leigh syndrome.

Wilson disease

Wilson disease is an early-onset autosomal-recessive disease characterized by movement disorders (e.g., parkinsonism and dystonia), psychiatric symptoms, and liver failure. The defect results from mutations in *ATP7B*, which encodes a copper-transporting P-type ATPase and results in copper accumulation in kidney, liver, and the basal ganglia of the brain. ATP7B exists in two isoforms, a 159-kDa form that localizes to the *trans*-Golgi network, and a 140-kDa form that localizes to mitochondria (12). The mitochondrial localization of the 140-kDa isoform (presumably it is in the inner membrane, but this has not yet been demonstrated) suggests that it might play a role in the functions of copper-dependent enzymes, such as CCO. In fact, mitochondria in affected tissues have morphological abnormalities, as well as deficiency of liver mitochondrial enzymes, especially complex I and aconitase. Thus, oxidative damage mediated by mitochondrial copper accumulation may play a role in the pathogenesis of Wilson disease. Since ATP7B is present in multiple subcellular compartments, the role of mitochondria in the pathogenesis of the disease may be secondary rather than primary, and the issue of mitochondrial involvement in Wilson disease remains controversial.

Mitochondrial dysfunction in disorders associated with mutations in nonmitochondrial proteins, and in neurodegenerative disorders with unknown causes

The most problematic and contentious connections between mitochondrial dysfunction and neurodegeneration involve those disorders that result either from primary mutations in gene products not targeted to mitochondria or from still unknown causes (Table 1). Because, *in toto*, these disorders account for the vast majority of neurodegenerative disease, the possibility of a unifying principle based on mitochondrial dysfunction as a primary cause of neuronal cell death (13) has become not only attractive, but even seductive. The key question, of course, is whether the data supporting this hypothesis are persuasive. We summarize here the evidence for and against such connections.

Alzheimer disease

AD is the most common form of dementia in the elderly. Approximately 5% of AD cases are familial (FAD) with autosomal-dominant transmission; these result mainly from mutations in the amyloid β precursor protein (ABPP) or in the ABPP processing proteins presenilin-1 (PS1) and presenilin-2 (PS2). Most patients with AD have sporadic cases (SAD) without a known genetic defect. The neuropathology in both FAD and SAD is characterized by neuronal loss and by the presence of amyloid plaques and neurofibrillary tangles. Aggregated β-amyloid peptide (Aβ), a 40- to 42–amino acid subfragment of the much larger ABPP, is considered by many to be the principal culprit in the pathogenesis of AD, but the mechanism by which Aβ40-42 causes neurodegeneration is highly controversial. It is not clear whether plaques and tangles are a cause or an effect, and it has even been proposed that the accumulation of Aβ40-42 is a *protective* response to oxidative damage caused by mitochondrial dysfunction (14).

Although the findings of mutated forms of APP or presenilins in FAD point toward a pathogenetic role for Aβ40-42, the lack of a clear-cut mechanism for pathogenesis has caused hypotheses involving mitochondrial dysfunction to flourish. In fact, there are several disparate lines of evidence pointing to mitochondrial involvement in the disease. For example, it has long been known that patients with trisomy 21 (Down syndrome) develop the neuropathology of AD, presumably because of an increased dosage of the *ABPP* gene, which resides on chromosome 21. Not only did cultured astrocytes from Down syndrome patients show increased intracellular levels of Aβ42 associated with impaired mitochondrial membrane potential, but the converse was also true, namely, that impairment of mitochondrial OXPHOS in normal astrocytes resulted in alterations of ABPP processing and elevated levels of Aβ42 (15). These data are consistent with the finding that Aβ can inhibit respiration when added to isolated mitochondria (16).

Several groups have found either reduced CCO activity (17) or reduced levels of immunohistochemically detectable CCO polypeptides (18, 19) in AD brains. At this point, there is no conclusive evidence that mtDNA mutations are higher in AD patients than in normal elderly individuals, or that any particular mtDNA mutation is pathogenic in AD. Therefore, it is more likely that the respiratory chain defects observed in AD brains derive from nuclear gene defects or from acquired damage to respiratory chain components.

Amyotrophic lateral sclerosis

ALS is a neurodegenerative disorder affecting cortical motor neurons and the anterior horn cells of the spinal cord. The disease generally starts in the fourth or fifth decade of life and progresses rapidly, leading to paralysis and premature death. While most cases are sporadic and of unknown etiology, about 15–20% are familial (FALS); of these, about 10% are caused by dominant mutations in *SOD1*, encoding Cu,Zn-superoxide dismutase (Cu,Zn-SOD). Consistent with the dominant nature of the disorder, mutations in *SOD1* do not decrease Cu,Zn-SOD activity significantly, implying that the mutations confer some as-yet unknown toxic "gain of function" on the protein.

Mutant mice carrying the common G93A mutation on a transgene develop severe motor neuron disease reminiscent of ALS, and in addition, the motor neurons develop massive mitochondrial degeneration. In *SOD1^G93A^* mice, the motor symptoms were preceded by a transient explosive increase in vacuoles derived from degenerating mitochondria in the motor neurons (20). These mice also had reduced respiratory chain function (21, 22), and, when fed creatine, a key component of the mitochondrial energy buffering and transfer system, their motor performance improved, as did survival (23).

Interestingly, some patients with sporadic ALS (SALS) accumulate abnormal mitochondria in anterior horn cells (24). Reductions in respiratory chain activity have been observed in spinal cords from SALS patients (25, 26), and this reduction appeared to correlate with a concomitant decline in total mtDNA in the tissue, implying that total numbers of organelles were reduced in the spinal cords of these patients (26).

It had long been thought that Cu,Zn-SOD was confined to the cytosol, and that only Mn-SOD (SOD2) was present in mitochondria. Recently, however, it has been determined that SOD1 is also located in the IMS of mitochondria. Thus, SOD activity in mitochondria is compartmentalized within the organelle itself, with SOD1 in the IMS and SOD2 in the matrix. This observation now provides a stronger rationale to explain some of the mitochondrion-related pathology in ALS, as mutant SOD1 could damage mitochondria directly. Transgenic mice expressing the G93A mutation in SOD1 show early activation of the mitochondrion-dependent apoptotic pathway (see below), with release of cytochrome *c* to the cytosol and translocation of the proapoptotic protein Bax to mitochondria, and similar effects have been observed in the spinal cords of SALS patients. Furthermore, mouse neuroblastoma cells expressing mutant SOD1 had impaired mitochondrial calcium buffering capacity, leading to elevated cytoplasmic calcium, a potential stimulus for apoptosis. Another proposed mitochondrion-related mechanism is based on the finding that cytosolic chaperones, such as HSP70, block the import of mutant SOD1, but not wild-type SOD1, into mitochondria (27). In this scenario, the mutant SOD1 forms aggregates with these HSPs and at the same time titrates them out of the cytosol, thereby killing the cell. Finally, mutant SOD1, but not wild-type SOD1, has been found to interact with at least four proteins, one of which — lysyl-tRNA synthetase — has an isoform that is targeted to mitochondria (28).

Taken together, these findings suggest that mitochondrial dysfunction might play a role in the pathogenesis of both sporadic and familial ALS (29). The mechanisms, however, appear to be complex and may very well involve not only OXPHOS deficiency, but also apoptosis and oxidative damage, as culprits. Notably, treatment of SOD1-mutant mice with the antibiotic minocycline, which inhibits mitochondrial permeability transition-mediated cytochrome *c* release, extended survival (30). Recent data, however, have failed to support the hypothesis that the gain of function of mutant Cu,Zn-SOD is an increase in copper-mediated free radical damage (31). Specifically, mice that lacked dismutase activity still developed the features of ALS, implying that the gain of function is not mediated by an increase in ROS that results from elevated or altered SOD1 activity.

Huntington disease

HD is a chronic autosomal-dominant disease characterized by choreoathetotic movements and progressive cognitive and emotional disturbances. Pathologically, there is selective degeneration of striatal neurons, with marked atrophy of the putamen and caudate nucleus. HD is a trinucleotide repeat disorder, with expansions of a CAG repeat near the N-terminus of huntingtin, the product of the *IT15* gene on chromosome 4. The mutated protein accumulates as nuclear inclusions in striatum and cortex.

Nevertheless, a defect in energy metabolism has been proposed as a potential pathogenic mechanism underlying HD, based on evidence obtained both in vivo and at autopsy. MRI spectroscopy in occipital cortex and basal ganglia of HD patients showed elevated lactate (32), as well as a reduced ratio of phosphocreatine to inorganic phosphate (PCr/Pi) in muscle (33). Animals treated with malonate, which inhibits respiratory chain complex II, develop pathological lesions in the striatum that closely resemble those of HD (34). Defects in complexes II and III, and in aconitase, have been described in postmortem HD brains, especially in the basal ganglia. In transgenic mouse models of HD, similar defects have been observed by some groups, but not by others. Lymphoblasts from HD patients showed increased cyanide-induced depolarization of mitochondria, as well as a greater vulnerability to staurosporin, an inducer of mitochondrion-mediated apoptosis. Recently, it was demonstrated by electron microscopy that mutant huntingtin localizes on neuronal mitochondrial membranes. Furthermore, it was shown that lymphoblast mitochondria from patients with HD have a lower membrane potential and a tendency to become depolarized at lower calcium levels than control mitochondria. Brain mitochondria from transgenic mice expressing mutant huntingtin showed similar mitochondrial defects (35).

Despite evidence in support of mitochondrial involvement in the pathogenesis of HD, the mechanism by which mutated huntingtin could cause bioenergetic dysfunction is still unknown, but some hypotheses have been formulated. Mutant huntingtin is known to bind to, and presumably downregulate, the activity of several polyglutamine-containing transcription factors that modulate nuclear gene transcription, such as Sp1, p53, TATA-binding protein, and cAMP-responsive element–binding (CREB) protein. These factors are thought to be involved in regulating the transcription of proteins involved in mitochondrial energy metabolism and in mitochondrion-initiated apoptosis. For example, mitochondrial dysfunction was shown to be associated with increased activated CREB in cultured cells (36). Therefore, by blocking the activity of critical transcription factors, mutant huntingtin might cause mitochondrial dysfunction. This hypothesis is partially weakened by the recent observation that, in a transgenic mouse model of HD, polyglutamine-containing transcription factors did not seem to be critically depleted in the nuclei of brain cells (37).

The issue of cause and effect has been highlighted by the recent finding that the expanded polyglutamine stretch in mutated huntingtin prevents it from binding to Hip-1 (huntingtin-interacting protein-1), as normal huntingtin does, thereby allowing free Hip-1 to bind to Hippi (Hip-1 protein interactor). Since Hip-1/Hippi heterodimers can recruit procaspase-8 (cysteine aspartic acid–specific protease 8) (see below), the weakened interaction between mutated huntingtin and Hip-1 is a proapoptotic event (38). As apoptosis, including that associated with caspase-8, can occur via both a mitochondrial (i.e., cytochrome c–induced) and a nonmitochondrial ("extrinsic") pathway, it is unclear at present whether the caspase-Hippi pathway is upstream or downstream of the known mitochondrial alterations in HD.

Parkinson disease

Nowhere is the role of mitochondrial respiratory chain dysfunction more persuasive, but also more contentious, than in the field of PD. PD is a neurodegenerative disorder characterized clinically by rigidity, tremor, and bradykinesia. The pathological hallmark of the disease is the loss of dopaminergic neurons in the substantia nigra pars compacta. As with ALS and AD, most PD cases are sporadic, but there are a few rare cases in which PD is inherited as a mendelian trait (familial PD [FPD]).

The mitochondrial connection to PD began almost 20 years ago, when it was discovered that 1-methyl-4-phenyl-1,2,3,6-tetrahydropyridine (MPTP) causes parkinsonism in humans and in laboratory animals. MPTP inhibits complex I of the respiratory chain, but only indirectly, as it is the MPTP metabolite N-methyl-4-phenylpyridinium ion (MPP$^+$), produced by the mitochondrial outer membrane protein monoamine oxidase B (MAO-B), that causes the damage. Soon thereafter, investigators found complex I deficiency and oxidative damage in the substantia nigra of PD patients, as well as reduced immunoreactivity for complex I subunits in affected neurons. Because complexes I and III are

the principal sources of free radicals in the cell, altered complex I function in the substantia nigra pars compacta could be responsible for the increased DNA damage and lipid peroxidation found in PD brains (39).

Mutations in FPD have been found in the genes encoding α-synuclein and parkin, but the role of these polypeptides in neurodegeneration is unknown. It is noteworthy, however, that both α-synuclein and parkin, which are components of Lewy bodies (abnormal aggregates found in PD neurons), can bind to each other in vitro (40), that α-synuclein can bind to CCO (41), and that mitochondrial respiratory chain inhibitors can stimulate aggregation of α-synuclein into Lewy bodies in vitro (42, 43).

As with AD, the role of mtDNA mutations in the respiratory chain defects of PD is quite controversial: while some groups showed that cybrids containing mtDNA derived from PD platelets had reduced complex I activity, others could not replicate this finding. Only three mutations in mtDNA were associated specifically with parkinsonism, and they did not present as "classical" PD. Patients with a heteroplasmic mutation in the 12S rRNA gene had parkinsonism as part of a syndrome characterized by deafness and neuropathy, while a boy with a heteroplasmic microdeletion in the cytochrome b gene had an overlap syndrome of parkinsonism and MELAS. A parkinsonian syndrome was also associated with one of the LHON mutations in complex I. On the other hand, it was recently reported that about 10% of patients diagnosed with clinical, biochemical, and/or morphological features of mitochondrial disease had features of parkinsonism, and that these patients responded positively to anti-Parkinson medications (44). Unfortunately, however, none of the 76 patients studied were diagnosed at the genetic level.

Progressive supranuclear palsy

Progressive supranuclear palsy (PSP) is a late-onset (age 50–70 years) neurological disorder with rapid progression, characterized by extrapyramidal symptoms, palsy of vertical gaze of supranuclear origin, and cognitive impairment. The pathological hallmark of PSP is the presence of neurofibrillary filaments containing hyperphosphorylated tau protein in the subcortical regions of the brain, with diffuse neuronal degeneration and gliosis. Genetic studies have linked an extended *tau* haplotype (H1) with the disease.

OXPHOS defects have been reported in muscle from PSP patients. Postmortem PSP brains had an increased content of malondialdehyde, a marker of lipid peroxidation, and reduced activity of α-ketoglutarate dehydrogenase, the rate-limiting enzyme of the Krebs cycle, but they also had normal respiratory chain activities (45, 46). While it is conceivable that a combination of mitochondrial dysfunction and oxidative stress could generate a vicious cycle leading to further oxidative damage and neuronal degeneration, at this point it is difficult to hypothesize a model consistent with both OXPHOS defects in muscle and normal complex I and IV activities in brain, or to explain why the Krebs cycle should be affected in particular.

Mechanisms of mitochondrion-mediated cell death

In neurodegenerative diseases, specific populations of neurons die, but by what mechanism? Apoptotic cell death has been invoked as the underlying basis for neuronal loss, but the data are both conflicting and confusing. Apoptosis in mammalian cells is a highly regulated process that operates to sculpt cell populations during the normal course of embryonic and fetal development, but it can also operate during adult life to eliminate cells that are under various stresses. In broad view, apoptosis can operate via two pathways, one mitochondrion-mediated and the other receptor-mediated but mitochondrion-independent (Figure 3). In the mitochondrion-dependent pathway, an external insult — elevated cytosolic calcium, to cite but one example — acts to cause release of cytochrome c, which is located in the IMS. Cytosolic cytochrome c can then bind APAF-1 (apoptotic protease-activating factor 1), which then binds to the inactive form of caspase-9. This "initiator" caspase complex, or "apoptosome," can now activate a cascade of events, beginning with activation of downstream "effector" caspases, such as caspase-3, followed by activation of caspases further downstream, ultimately resulting in the hallmarks of apoptosis (condensation of nuclear and cytoplasmic contents, nDNA fragmentation, cell blebbing, and autophagy of membrane-bound bodies). Mitochondrion-mediated activation of caspase-9 can also occur via external (extracellular) receptor-mediated signals (e.g., TNF, growth factor deprivation, etc.) to target various ligands — e.g., Bad, Bax, Bik, Noxa — to the mitochondrion, thereby causing cytochrome c release and binding of APAF-1. Under other circumstances, a separate mitochondrion-independent pathway also operates (Figure 3).

Cell death in mitochondrial and in neurodegenerative disease

Among the numerous stresses known to participate in mitochondrion-mediated apoptosis, at least in vitro, bioenergetic failure and elevated ROS figure prominently (47). Thus, two key hypotheses immediately present themselves. First, one would expect a causal relationship between mitochondrial dysfunction and neurodegeneration via apoptosis, perhaps due, at least in part, to changes in ROS levels. Second, if neurodegeneration is related in any way to defects in respiratory chain function or in OXPHOS, then one would expect authentic mitochondrial diseases to show many, if not all, of the features of cell death that are postulated to occur in neurodegenerative diseases. So what do we know about apoptosis and ROS in mitochondrial disease and in neurodegeneration? Not unexpectedly, the literature is conflicting.

It is well known that muscle biopsies from patients with "classic" mitochondrial disorders, including those with RRFs, show little or no evidence of necrosis, fiber loss, elevated circulating creatine kinase, or inflammation. Nevertheless, markers of increased ROS have been found in muscle biopsies from patients with mitochondrial disease (48–50). It is clear that elevated ROS are injurious to mitochondria. For example, ablation of the mitochondrial *SOD2* gene in mice caused rapid death via a fatal cardiomyopathy (51), and even loss of only one *SOD2* allele was harmful (52). Encouragingly, pathology could be ameliorated, and lifespan extended, in *SOD2*−/− mice upon pharmacological intervention to reduce free radicals (53).

An analysis of muscle biopsies from patients with CCO deficiency due to various known pathogenic mutations in mtDNA (e.g., those causing MERRF, MELAS, KSS, and PEO) and in nucleus-encoded mitochondrially targeted proteins (e.g., those causing LS) showed features of apoptosis in CCO-negative RRFs — DNA fragmentation by terminal deoxynucleotidyl transferase–mediated dUTP nick end-labeling (TUNEL), condensed chroma-

tin, and elevated immunostaining for caspase-1, caspase-3, and Fas ligand (54, 55). However, another group found no evidence for apoptosis (56). In a third study, some apoptotic markers (Bax, p53, Fas, and caspase-3) were *not* increased in CCO-negative RRFs, in spite of the fact that other apoptosis-related proteins (Bcl-2, Bcl-X$_L$, DNase I) were increased, as was the amount of cytosolic cytochrome c (50). What little evidence there was for apoptosis in muscle seemed to be confined to a few RRFs (57).

If markers of elevated ROS and apoptosis can be found in these muscle biopsies, why do these muscle fibers not degenerate? This apparent paradox points out that a marker of apoptosis is merely that — a marker — and does not necessarily imply that apoptosis is actually occurring. In skeletal muscle, there are at least two reasons why apoptosis is not occurring in muscle. First, APAF-1 is either present at very low levels (58) or missing entirely (59) in skeletal muscle. In the absence of APAF-1, cytosolic cytochrome c has no partner with which to bind in order to activate caspase-9 and induce apoptosis (60) (although in some circumstances the absence of APAF-1 may trigger a "default pathway" of necrosis [ref. 61]). In other words, cytosolic cytochrome c in this situation is a marker of, but cannot induce, apoptosis. In support of this view, Nonaka's group could detect APAF-1 and released cytochrome c only in RRFs from patients with MERRF, MELAS, or PEO, but not in the far more abundant relatively normal muscle fibers (57). Second, an inhibitor of apoptosis called ARC (apoptotic repressor with caspase recruitment domain), which is expressed almost exclusively in skeletal and heart muscle (62), not only inhibits apoptosis in these tissues but can also protect mitochondria from free radical damage (63).

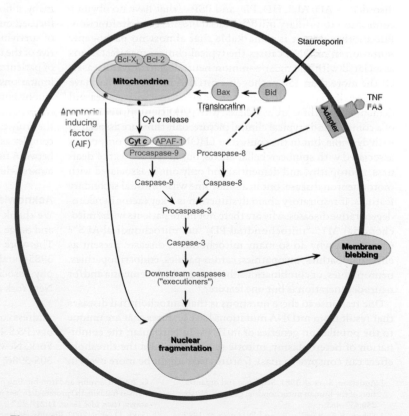

Figure 3
A schematic view of the main pathways of apoptosis: mitochondrion-mediated and mitochondrion-independent. See text for details.

What about neuronal death in mitochondrial disease? There is neuronal loss in KSS, MERRF, and MELAS, and the regions that are affected differ in each disease, but no studies have been published regarding apoptosis in these cells. Intriguingly, TUNEL assays in postmortem brains from these patients showed little if any positive staining in neurons, but staining was observed in non-neuronal cells, such as microglia (E. Bonilla, personal communication). Moreover, there are regions of the brain — the choroid plexus in KSS, for example — that are nearly homoplasmic for mutated mtDNA and are devoid of CCO immunoreactivity, and yet are both morphologically intact and TUNEL-negative (E. Bonilla, personal communication). Therefore, it is still unclear how neurons die in mitochondrial disorders.

Perspective

It is clear that impairment of mitochondrial energy metabolism is the key pathogenic factor in a number of neurodegenerative disorders (Table 1). This is most clearly seen in maternally inherited diseases that result from pathogenic mutations in mtDNA that interdict the normal functioning of the respiratory chain and OXPHOS. Examples of this group of diseases include LHON and maternally inherited LS. Similarly, mutations in nuclear genes that encode mitochondrially targeted proteins required for respiratory chain assembly and homeostasis can cause neurodegenerative disease. Examples in this class are autosomal-recessive LS, FRDA, hereditary spastic paraplegia, and Mohr-Tranebjaerg syndrome.

What is striking about both of these groups of neurodegenerative disorders is how diverse they are in their clinical manifestations, and how little they resemble the "classic" neurodegenerative disorders — AD, ALS, HD, PD, and PSP — that have no obvious connection to primary mitochondrial bioenergetic dysfunction. Put another way, it is remarkable that almost no pathogenic mutation in mtDNA causes the typical clinical manifestations associated with the most common neurodegenerative diseases. Of the more than 150 pathogenic mtDNA mutations that have been described to date (64), only one is associated with chorea and dementia; only three are associated with parkinsonism, but as part of a complex and atypical clinical picture; only three are associated with dystonia, but in the context of LHON or MELAS; only one is associated with spinocerebellar ataxia, but in the context of deafness, neuropathy, and dementia; and only one is associated with motor neuron disease, but in a syndrome with unusual secondary features. If respiratory chain dysfunction is a key factor in neurodegenerative diseases, why are there not more patients with "mitochondrial AD," "mitochondrial PD," or "mitochondrial ALS"? Conversely, why do so many mitochondrial diseases present as encephalopathies, myopathies, cardio-pathies, endocrinopathies, neuropathies, or combinations thereof, in which dementia and/or neurodegeneration is but one feature?

One response to these questions is that mitochondrial diseases that result from mtDNA mutations have features that are unique to the population genetics of mtDNA. In particular, the combination of heteroplasmy, mitotic segregation, and the threshold effect can conspire to mask features that might be more overt in disorders that obey the rules of mendelian genetics. In addition, it could be argued that many mitochondrial disorders strike early in life, thus precluding the features of dementia, chorea, and the like to evince themselves as they do in AD, ALS, or HD, which have later onset. While formally these are valid objections, they are not particularly convincing. There are many mitochondrial patients with extremely high levels of mutations in multiple tissues (and especially in the brain), and there are enough patients who live into their fourth and even fifth decades of life in whom the symptoms of AD, ALS, and PD should eventually be recapitulated.

This does not mean, however, that OXPHOS defects play no role in the late-onset neurodegenerative diseases. Rather, mitochondrial dysfunction probably plays an important, but secondary, role in these disorders. For example, it is clear that there is respiratory chain deficiency in hippocampal neurons in AD patients, as measured by biochemistry, histochemistry, and immunohistochemistry. It is likely, however, that these deficits are secondary to some underlying dysfunction, of which Aβ deposition is but one of many possible primary causes.

The failure to detect true cell death in muscle from patients with mitochondrial disease tells us little about neuronal cell death in these patients. It does, however, send a cautionary message regarding a too-eager willingness to equate cell death markers with the event itself. Conversely, the massive amounts of cell death in the neurodegenerative disorders mean that the very cells that are postulated to be dying due to mitochondrial dysfunction — e.g., motor neurons in ALS, striatal cells in HD, and, most pointedly, nigral cells in PD — are, in the final analysis, unavailable for examination, especially in the terminal stages of the disease and in autopsy tissue. Thus, many, if not all, studies that rely on patient material have, a fortiori, focused on the function, or lack of same, in a skewed population of surviving cells, such as glia, that are likely to be unrepresentative of the true pathogenetic course of events. Furthermore, studies of patient neuronal tissue or of model cells in vitro have their own limitations, not the least of which is the absence of the interactions among entire populations and subpopulations of cells that define the behavior of a vascularized tissue in vivo. These caveats point out the truly difficult nature of deducing pathogenesis in diseases as complex as neurodegenerative disorders. A mechanistic relationship between neurodegeneration and mitochondrial dysfunction still awaits a direct, clear, and unequivocal demonstration.

Acknowledgments

We thank Salvatore DiMauro, Michio Hirano, Michael T. Lin, and Serge Przedborski for insight and advice on the manuscript. This work was supported by grants from the NIH (NS-28828, NS-39854, and HD-32062, to E.A. Schon), from the Muscular Dystrophy Association (to E.A. Schon and G. Manfredi), and from the New York Academy of Medicine Speaker's Fund (to G. Manfredi).

Address correspondence to: Eric A. Schon, Department of Neurology, P&S 4-431, Columbia University, 630 West 168th Street, New York, New York 10032, USA. Phone: (212) 305-1665; Fax: (212) 305-3986; E-mail: eas3@columbia.edu.

1. Anderson, S., et al. 1981. Sequence and organization of the human mitochondrial genome. *Nature.* **290**:457–465.

2. Schon, E.A., Bonilla, E., and DiMauro, S. 1997. Mitochondrial DNA mutations and pathogenesis. *J. Bioenerg. Biomembr.* **29**:131–149.

3. Cavadini, P., O'Neill, H.A., Benada, O., and Isaya, G. 2002. Assembly and iron-binding properties of human frataxin, the protein deficient in Friedreich ataxia. *Hum. Mol. Genet.* **11**:217–227.

4. Gakh, O., et al. 2002. Physical evidence that yeast frataxin is an iron storage protein. *Biochemistry.* **41**:6798–6804.

5. Puccio, H. 2001. Mouse models for Friedreich atax-ia exhibit cardiomyopathy, sensory nerve defect and Fe-S enzyme deficiency followed by intramitochondrial iron deposits. *Nat. Genet.* **27**:181–186.

6. Rustin, P., et al. 1999. Effect of idebenone on cardiomyopathy in Friedreich's ataxia: a preliminary study. *Lancet.* **354**:477–479.

7. Schulz, J.B., et al. 2000. Oxidative stress in patients

with Friedreich ataxia. *Neurology.* **55**:1719–1721.

8. Casari, G., et al. 1998. Spastic paraplegia and OXPHOS impairment caused by mutations in paraplegin, a nuclear-encoded mitochondrial metalloprotease. *Cell.* **93**:973–983.

9. Hansen, J.J., et al. 2002. Hereditary spastic paraplegia SPG13 is associated with a mutation in the gene encoding the mitochondrial chaperonin Hsp60. *Am. J. Hum. Genet.* **70**:1328–1332.

10. Jin, H., et al. 1996. A novel X-linked gene, DDP, shows mutations in families with deafness (DFN-1), dystonia, mental deficiency and blindness. *Nat. Genet.* **14**:177–180.

11. Roesch, K., Curran, S.P., Tranebjaerg, L., and Koehler, C.M. 2002. Human deafness dystonia syndrome is caused by a defect in assembly of the DDP1/TIMM8a-TIMM13 complex. *Hum. Mol. Genet.* **11**:477–486.

12. Lutsenko, S., and Cooper, M.J. 1998. Localization of the Wilson's disease protein product to mitochondria. *Proc. Natl. Acad. Sci. USA.* **95**:6004–6009.

13. Orth, M., and Schapira, A.H. 2001. Mitochondria and degenerative disorders. *Am. J. Med. Genet.* **106**:27–36.

14. Smith, M.A., et al. 2002. Amyloid-beta, tau alterations and mitochondrial dysfunction in Alzheimer disease: the chickens or the eggs? *Neurochem. Int.* **40**:527–531.

15. Busciglio, J., et al. 2002. Altered metabolism of the amyloid β precursor protein is associated with mitochondrial dysfunction in Down's syndrome. *Neuron.* **33**:677–688.

16. Casley, C.S., et al. 2002. Beta-amyloid inhibits integrated mitochondrial respiration and key enzyme activities. *J. Neurochem.* **80**:91–100.

17. Mutisya, E.M., Bowling, A.C., and Beal, M.F. 1994. Cortical cytochrome oxidase activity is reduced in Alzheimer's disease. *J. Neurochem.* **63**:2179–2184.

18. Bonilla, E., et al. 1999. Mitochondrial involvement in Alzheimer's disease. *Biochim. Biophys. Acta.* **1410**:171–182.

19. Ojaimi, J., et al. 1999. Irregular distribution of cytochrome *c* oxidase protein subunits in aging and Alzheimer's disease. *Ann. Neurol.* **46**:656–660.

20. Kong, J., and Xu, Z. 1998. Massive mitochondrial degeneration in motor neurons triggers the onset of amyotrophic lateral sclerosis in mice expressing a mutant SOD1. *J. Neurosci.* **18**:3241–3250.

21. Browne, S.E., et al. 1998. Metabolic dysfunction in familial, but not sporadic, amyotrophic lateral sclerosis. *J. Neurochem.* **71**:281–287.

22. Mattiazzi, M., et al. 2002. Mutated human SOD1 causes dysfunction of oxidative phosphorylation in mitochondria of transgenic mice. *J. Biol. Chem.* **277**:29626–29633.

23. Klivenyi, P., et al. 1999. Neuroprotective effects of creatine in a transgenic animal model of amyotrophic lateral sclerosis. *Nat. Med.* **5**:347–350.

24. Sasaki, S., and Iwata, M. 1996. Impairment of fast axonal transport in the proximal axons of anterior horn neurons in amyotrophic lateral sclerosis. *Neurology.* **47**:535–540.

25. Borthwick, G.M., et al. 1999. Mitochondrial enzyme activity in amyotrophic lateral sclerosis: implications for the role of mitochondria in neuronal cell death. *Ann. Neurol.* **46**:787–790.

26. Wiedemann, F.R., et al. 2002. Mitochondrial DNA and respiratory chain function in spinal cords of ALS patients. *J. Neurochem.* **80**:616–625.

27. Okado-Matsumoto, A., and Fridovich, I. 2002. Amyotrophic lateral sclerosis: a proposed mechanism. *Proc. Natl. Acad. Sci. USA.* **99**:9010–9014.

28. Kunst, C.B., Mezey, E., Brownstein, M.J., and Patterson, D. 1997. Mutations in SOD1 associated with amyotrophic lateral sclerosis cause novel protein interactions. *Nat. Genet.* **15**:91–94.

29. Menzies, F.M., Ince, P.G., and Shaw, P.J. 2002. Mitochondrial involvement in amyotrophic lateral sclerosis. *Neurochem. Int.* **40**:543–551.

30. Zhu, S., et al. 2002. Minocycline inhibits cytochrome *c* release and delays progression of amyotrophic lateral sclerosis in mice. *Nature.* **417**:74–78.

31. Subramaniam, J.R., et al. 2002. Mutant SOD1 causes motor neuron disease independent of copper chaperone-mediated copper loading. *Nat. Neurosci.* **5**:301–307.

32. Jenkins, B.G., Koroshetz, W.J., Beal, M.F., and Rosen, B.R. 1993. Evidence for impairment of energy metabolism *in vivo* in Huntington's disease using localized ¹H NMR spectroscopy. *Neurology.* **43**:2689–2695.

33. Koroshetz, W.J., Jenkins, B.G., Rosen, B.R., and Beal, M.F. 1997. Energy metabolism defects in Huntington's disease and effects of coenzyme Q10. *Ann. Neurol.* **41**:160–165.

34. Beal, M.F., et al. 1993. Age-dependent striatal excitotoxic lesions produced by the endogenous mitochondrial inhibitor malonate. *J. Neurochem.* **61**:1147–1150.

35. Panov, A.V., et al. 2002. Early mitochondrial calcium defects in Huntington's disease are a direct effect of polyglutamines. *Nat. Neurosci.* **5**:731–736.

36. Arnould, T., et al. 2002. CREB activation induced by mitochondrial dysfunction is a new signaling pathway that impairs cell proliferation. *EMBO J.* **21**:53–63.

37. Yu, Z.X., Li, S.H., Nguyen, H.P., and Li, X.J. 2002. Huntingtin inclusions do not deplete polyglutamine-containing transcription factors in HD mice. *Hum. Mol. Genet.* **11**:905–914.

38. Gervais, F.G., et al. 2002. Recruitment and activation of caspase-8 by the Huntingtin-interacting protein Hip-1 and a novel partner Hippi. *Nat. Cell Biol.* **4**:95–105.

39. Dexter, D.T., et al. 1994. Indices of oxidative stress and mitochondrial function in individuals with incidental Lewy body disease. *Ann. Neurol.* **35**:38–44.

40. Choi, P., et al. 2001. Co-association of parkin and alpha-synuclein. *Neuroreport.* **12**:2839–2843.

41. Elkon, H., et al. 2002. Mutant and wild-type alpha-synuclein interact with mitochondrial cytochrome *c* oxidase. *J. Mol. Neurosci.* **18**:229–238.

42. Lee, H.J., Choi, C., and Lee, S.J. 2002. Membrane-bound alpha-synuclein has a high aggregation propensity and the ability to seed the aggregation of the cytosolic form. *J. Biol. Chem.* **277**:671–678.

43. Lee, H.J., et al. 2002. Formation and removal of alpha-synuclein aggregates in cells exposed to mitochondrial inhibitors. *J. Biol. Chem.* **277**:5411–5417.

44. Finsterer, J. 2002. Parkinson syndrome as a manifestation of mitochondriopathy. *Acta Neurol. Scand.* **105**:384–389.

45. Albers, D.S., et al. 2000. Frontal lobe dysfunction in progressive supranuclear palsy: evidence for oxidative stress and mitochondrial impairment. *J. Neurochem.* **74**:878–881.

46. Park, L.C., et al. 2001. Mitochondrial impairment in the cerebellum of the patients with progressive supranuclear palsy. *J. Neurosci. Res.* **66**:1028–1034.

47. Kroemer, G., and Reed, J.C. 2000. Mitochondrial control of cell death. *Nat. Med.* **6**:513–519.

48. Mitsui, T., et al. 1996. Oxidative damage to skeletal muscle DNA from patients with mitochondrial encephalomyopathies. *J. Neurol. Sci.* **139**:111–116.

49. Filosto, M., et al. 2002. Antioxidant agents have a different expression pattern in muscle fibers of patients with mitochondrial diseases. *Acta Neuropathol. (Berl.)* **103**:215–220.

50. Umaki, Y., et al. 2002. Apoptosis-related changes in skeletal muscles of patients with mitochondrial diseases. *Acta Neuropathol. (Berl.)* **103**:163–170.

51. Li, Y., et al. 1995. Dilated cardiomyopathy and neonatal lethality in mutant mice lacking manganese superoxide dismutase. *Nat. Genet.* **11**:376–381.

52. Murakami, K., et al. 1998. Mitochondrial susceptibility to oxidative stress exacerbates cerebral infarction that follows permanent focal cerebral ischemia in mutant mice with manganese superoxide dismutase deficiency. *J. Neurosci.* **18**:205–213.

53. Melov, S., et al. 2001. Lifespan extension and rescue of spongiform encephalopathy in superoxide dismutase 2 nullizygous mice treated with superoxide dismutase-catalase mimetics. *J. Neurosci.* **21**:8348–8353.

54. Mirabella, M., et al. 2000. Apoptosis in mitochondrial encephalomyopathies with mitochondrial DNA mutations: a potential pathogenic mechanism. *Brain.* **123**:93–104.

55. Di Giovanni, S., et al. 2001. Apoptosis and ROS detoxification enzymes correlate with cytochrome c oxidase deficiency in mitochondrial encephalomyopathies. *Mol. Cell. Neurosci.* **17**:696–705.

56. Sciacco, M., et al. 2001. Lack of apoptosis in mitochondrial encephalomyopathies. *Neurology.* **56**:1070–1074.

57. Ikezoe, K., et al. 2002. Apoptosis is suspended in muscle of mitochondrial encephalomyopathies. *Acta Neuropathol. (Berl.)* **103**:531–540.

58. Zou, H., et al. 1997. Apaf-1, a human protein homologous to C. elegans CED-4, participates in cytochrome c-dependent activation of caspase-3. *Cell.* **90**:405–413.

59. Burgess, D.H., et al. 1999. Human skeletal muscle cytosols are refractory to cytochrome c-dependent activation of type-II caspases and lack APAF-1. *Cell Death Differ.* **6**:256–261.

60. Yoshida, H., et al. 1998. Apaf1 is required for mitochondrial pathways of apoptosis and brain development. *Cell.* **94**:739–750.

61. Miyazaki, K., et al. 2001. Caspase-independent cell death and mitochondrial disruptions observed in the Apaf1-deficient cells. *J. Biochem. (Tokyo).* **129**:963–969.

62. Koseki, T., Inohara, N., Chen, S., and Nunez, G. 1998. ARC, an inhibitor of apoptosis expressed in skeletal muscle and heart that interacts selectively with caspases. *Proc. Natl. Acad. Sci. USA.* **95**:5156–5160.

63. Neuss, M., et al. 2001. The apoptotic regulatory protein ARC (apoptosis repressor with caspase recruitment domain) prevents oxidant stress-mediated cell death by preserving mitochondrial function. *J. Biol. Chem.* **276**:33915–33922.

64. Servidei, S. 2002. Mitochondrial encephalomyopathies: gene mutation. *Neuromuscul. Disord.* **12**:334–339.

The role of cerebral amyloid β accumulation in common forms of Alzheimer disease

Sam Gandy

Farber Institute for Neurosciences, Thomas Jefferson University, Philadelphia, Pennsylvania, USA.

For approximately 80 years following Alzheimer's description of the disease that bears his name, a gulf divided researchers who believed that extracellular deposits of the amyloid β (Aβ) peptide were pathogenic from those who believed that the deposits were secondary detritus. Since 1990, the discoveries of missense mutations in the Aβ peptide precursor (APP) and the APP-cleaving enzyme presenilin 1 (PS1) have enabled much progress in understanding the molecular, cellular, and tissue pathology of the aggregates that accumulate in the interstices of the brains of patients with autosomal dominant familial Alzheimer disease (AD). Clarification of the molecular basis of common forms of AD has been more elusive. The central questions in common AD focus on whether cerebral and cerebrovascular Aβ accumulation is (a) a final neurotoxic pathway, common to all forms of AD; (b) a toxic by-product of an independent primary metabolic lesion that, by itself, is also neurotoxic; or (c) an inert by-product of an independent primary neurotoxic reaction. Antiamyloid medications are entering clinical trials so that researchers can evaluate whether abolition of cerebral amyloidosis can mitigate, treat, or prevent the dementia associated with common forms of AD. Successful development of antiamyloid medications is critical for elucidating the role of Aβ in common AD.

Alzheimer disease (AD) is the most common cause of dementia, a clinicopathological state whose name literally means "loss of the ability to think." There is much disagreement even among AD specialists about the basic nature of the disease, and that controversy is no better illustrated than in contemplation of how the disease is initiated at the molecular level. Histologically, the neuronal cytoskeleton twists, literally, into structures called neurofibrillary tangles (NFTs). Outside the cell, the amyloid β (Aβ) peptide aggregates into clumps called oligomers, which accumulate and form deposits called amyloid plaques. Based on studies of a syndrome known as mild cognitive impairment (MCI) (a possible prodrome to dementia), the development of detectable entorhinal NFTs is considered to be the histological correlate of MCI and, many believe, the harbinger of incipient AD (1). Still, levels of cortical synaptic markers correlate with cognitive status at time of death better than do either plaque load or tangle load (2), which is consistent with the concept that neurotransmission failure is the proximate cause of cognitive decline.

Amyloid is a highly ordered precipitate of extracellular protein, misnamed "starch-like" by Rudolf Virchow because of its reactivity to the PAS stain (3). Systemic amyloid deposits can occur in any organ and are often large and amorphous; cerebral amyloid deposits take the form of delimited, miliary spheres called plaques.

Plaques contain a trace amount of glycosaminoglycans, which explains the PAS positivity. To the neuropathologist, the diagnosis of "amyloid" is applied to any proteinaceous tissue precipitate that binds the dye Congo red. Congophilia is a property of all amyloids and is related to the defining ability of these precipitates to form β-pleated sheets that subsequently assemble into fibrils (4).

In AD, brain amyloid is composed almost entirely of a 4 kDa amyloid β (Aβ) peptide (5) that exhibits microheterogeneity in amino acid sequence and in a variety of biophysical states (Figure 1). Most Aβ is comprised of a peptide designated Aβ40, $Aβ_{40}$, $Aβ_{1-40}$, or, in some cases, $Aβ_{x-40}$. Peptides with various amino termini, all bearing an identical carboxyl terminus, form a major proportion (greater than 95%) of the total Aβ produced by cells (6). A minor fraction (less than 5%) of the newly generated Aβ ends at residue 42 (6). This "long Aβ" (also abbreviated as $Aβ_{1-42}$, Aβ42, $Aβ_{42}$, or $Aβ_{x-42}$, the latter 3 representing species with heterogeneous amino termini) is much more aggregatable than $Aβ_{40}$; hence, "long Aβ" is believed to initiate the formation of oligomers, fibrils, and plaques (7).

The generation of Aβ from its precursor, the Aβ peptide precursor (APP), is illustrated in Figure 2. APP is first cleaved at the amino terminus of Aβ by a membrane-bound aspartyl protease (β-secretase). This cleavage generates a large secreted derivative (soluble $APP_β$ [$sAPP_β$]) and a membrane-bound β-cleaved carboxyterminal fragment of APP (CTFβ; also known as C99). Cleavage of CTFβ by γ-secretase results in the production of the Aβ40 and Aβ42 species described above. The term "soluble Aβ" generally is applied either to newly generated, cell-secreted Aβ or to that fraction of tissue or synthetic Aβ that is taken into the aqueous phase of a non–detergent-containing extraction buffer. "Misfolded" and "aggregated" Aβ are terms used to describe very early, nonspecific changes in Aβ folding states or solubility states, respectively (e.g., aggregated Aβ solutions usually scatter light to a greater extent than do solutions of soluble Aβ). "Oligomeric" Aβ refers to peptide assemblies with limited stoichiometry (e.g., dimers, trimers, etc.), while protofibrils (PFs) are structures of intermediate order between aggregates and fibrils. The term "Aβ-

Nonstandard abbreviations used: Aβ, amyloid β; ABri, British amyloid; AD, Alzheimer disease; ADAM, a disintegrin and metalloproteinase; ADan, Danish amyloid; ADDL, Aβ-derived diffusible ligand; AICD, APP intracellular domain; APP, Aβ peptide precursor; BACE, β-APP site–cleaving enzyme; C89, β'-cleaved carboxyterminal fragment of APP; CTFβ, β-cleaved carboxyterminal fragment of APP (also known as C99); FBD, familial British dementia; FDD, familial Danish dementia; FTD, frontotemporal dementia; HRT, hormone replacement therapy; MCI, mild cognitive impairment; NFT, neurofibrillary tangle; PF, protofibril; PM, plasma membrane; PP, protein phosphatase; PS, presenilin; ROCK, rho-associated protein kinase; $sAPP_α$, soluble $APP_α$.

Conflict of interest: The author currently receives grant support from the Robert C. Atkins Foundation and Neurochem Inc. and consultant fees from representatives of the Women's Health Research Institute of Wyeth Inc. and Elan Pharmaceuticals Inc.

Citation for this article: *J. Clin. Invest.* **115**:1121–1129 (2005).
doi:10.1172/JCI200525100.

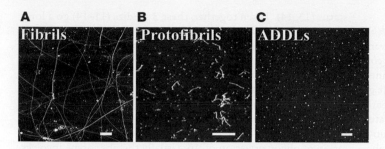

Figure 1
Different assembly (biophysical) states of Aβ. The assembled forms obtained from incubation of synthetic Aβ are highly sensitive to preparation and incubation. Widely differing proportions of insoluble fibrils (**A**), soluble PFs (**B**), and oligomers (also known as ADDLs) are revealed by atomic force microscopy. Typical PF and fibril preparations contain varying levels of small globular molecules, putatively Aβ oligomers (ADDLs). ADDL preparations (**C**) initiated from monomeric dimethyl sulfoxide stock solutions are fibril- and PF-free and uniquely comprise oligomers. Scale bars: 200 nm. Figure reproduced with permission from *Trends in Neurosciences* (8).

derived diffusible ligands" (ADDLs) is also applied to pre-protofibrillar intermediates (Figure 1), based less on a structural definition than on the neurotoxic activity of these oligomers. Indeed, oligomers, PFs, and ADDLs are believed to be the assembly states of Aβ with the most potent toxicity and are believed by many in the field to be the proximate mediators of Aβ-induced neurotoxicity, especially in primary neuronal culture models (8, 9). The final assemblies, called fibrils, are the basic building blocks of the amyloid plaque (Figure 3) and are so named because of their characteristic ultrastructural appearance.

Amyloids exhibit a typical exponential growth property known as seeding, which means that once a few fibrils are formed, they instruct the misfolding of other amyloidogenic peptides. The transmissibility of another amyloidosis, prion disease, can be viewed as an extreme example of seeding. In that case, the activation energy and favorability of alternative, neurotoxic forms of the prion protein are believed to permit aggressive propagation of fibrillogenesis. Of note, prion diseases, such as Creutzfeldt-Jakob disease, provide the clearest heuristic evidence that amyloids can be neurotoxic and that amyloid plaques per se are not required for neurotoxicity and clinical disease (10).

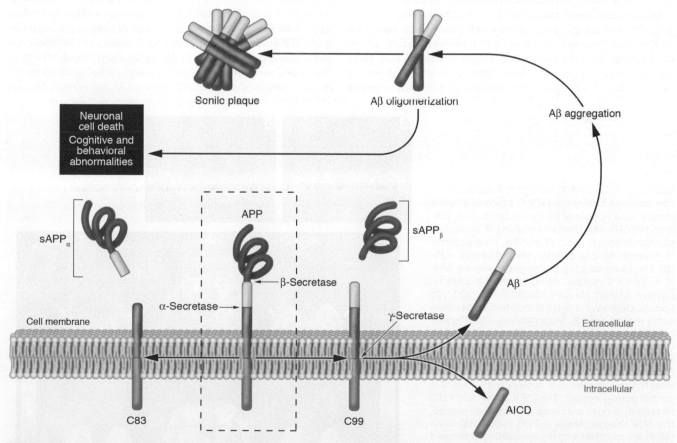

Figure 2
APP processing and Aβ accumulation. Mature APP (center, inside dashed box) is metabolized by 2 competing pathways, the α-secretase pathway that generates sAPPα and C83 (also known as CTFα; left) and the β-secretase pathway that generates sAPPβ and C99 (right). Some β-secretase cleavage is displaced by 10 amino acid residues and generates sAPPβ' and C89 (see Figure 4). All carboxyterminal fragments (C83, C99, and C89) are substrates for γ-secretase, generating the APP intracellular domain (AICD) and, respectively, the secreted peptides p3 (not shown), Aβ (right), and Glu11 Aβ (see Figure 4). Aβ aggregates into small multimers (dimers, trimers, etc.) known as oligomers. Oligomers appear to be the most potent neurotoxins, while the end stage senile plaque is relatively inert.

What is the relationship between amyloidosis and cognitive decline?

Neuropathologists have long recognized that — among the various structural markers of AD — plaque burden is the poorest correlate of cognitive status at the time of death. As more refined techniques for measuring Aβ levels have become available (e.g., ELISA), this correlation has been revisited again and again. Early ELISA correlations between brain Aβ levels and cognitive status were not much better than the histological correlation of cognitive status and plaque load, due, at least in part, to the seeding phenomenon: once fibrils and plaques begin to form, their concentrations rise rapidly, over several orders of magnitude. The massive levels of plaque Aβ achieved by exponential fibrillogenesis are not readily quantitatively solubilizable, and the interindividual differences in solubilizable Aβ are believed to be a source of much variability and noise in studies aimed at determining clinicopathological correlation between amyloid load and cognitive status, since small amounts of differences in amyloid solubilizability can cause dramatic variability in Aβ levels measured by ELISA. More recently, Naslund et al. revisited the correlation between Aβ and cognitive status using brains from well-characterized subjects at the threshold between MCI and dementia, where contamination by plaque Aβ is least problematic (11). In this study, for the first time, a strong correlation between Aβ concentration and cognitive status was documented. These results dovetail well with data obtained using transgenic mice that possess a mutant form of APP (Tg2576 mice), which show that memory deficits correlate with Aβ elevation and precede plaque formation (12). More recently, brain and cerebrospinal fluid levels of Aβ oligomers (ADDLs) have been reported to correlate closely with cognitive

status (13, 14). Taken together, these 3 studies (11–14) go a long way toward resolving the heretofore apparently poor correlation between cognition and either Aβ levels or amyloid load.

What is the strongest evidence that AD can ever begin with amyloidosis?

Genetic evidence links altered Aβ metabolism to the rare subset of AD that is autosomal dominant and completely penetrant. These forms of AD share a single common feature, and that is the facilitation of oligomerization of Aβ. Usually the phenotype involves an obvious change in the proteolytic processing of APP. Missense mutations within APP are clustered around the N and C termini of the Aβ domain, and the usual phenotype involves increased generation of the hyperaggregatable Aβ species that end(s) at residue 42 (Figure 4; reviewed in ref. 15). Familial AD pedigrees have been identified worldwide, and the geography of the initial discovery is often used to name the mutant molecule (e.g., Sweden, Holland, Britain, Indiana, Flanders, etc). A few mutations occur within the Aβ domain that result in increased intrinsic aggregatability of all Aβ species. A recently discovered pathogenic mutation, found in AD patients from northern Sweden, near the Arctic Circle, actually *decreases* the levels of total Aβ generation, yet the molecules that are generated have an increased propensity to oligomerize (16, 17). Pathogenic APP mutations are the rarest known genetic cause of familial AD, being responsible for disease in only about 50 families worldwide. Still, the existence of familial AD due to proamyloidogenic APP mutations around the Aβ domain and the final common phenotype (promotion of Aβ aggregation) provides the most compelling evidence that the entire clinicopathological picture of AD *can* — at least sometimes — begin with Aβ aggregation. None of

Figure 3

Amyloid plaque–forming transgenic mice and positron emission tomography (PET) scans of amyloid plaque load in normal human subjects and subjects with AD. (**A**) Thioflavin staining of subiculum of control mouse, aged 14 months. Fluorescence is nonspecific and cellular. Magnification, ×20. (**B**) Thioflavin staining of littermate, mutant APP X mutant PS mouse, demonstrating thioflavin-positive amyloid plaques. Magnification, ×20. (**C**) Immunostaining of amyloid plaque from cortex from same mouse as in **B**. Magnification, ×40. Figures courtesy of Michelle Ehrlich (Thomas Jefferson University). (**D**) [18F]FDDNP PET scan (to examine amyloid plaque and NFT load), MRI, and fluorodeoxy-glucose (FDG) PET (to examine glucose metabolism) images of a subject with AD and a control normal subject. The [18F]FDDNP and FDG (summed) images are coregistered to their respective MRI images. Areas of FDG hypometabolism (blue) are matched with the localization of amyloid plaques and NFTs as visualized by [18F]FDDNP binding. The color bar represents the scaling of the [18F]FDDNP and FDG images. FDDNP, 2-(1-[6-[(2-[18F]fluoroethyl)(methyl)amino]-2-naphthyl]ethylidene)malononitrile; max, maximum; min, minimum. Figure reproduced with permission from the *American Journal of Geriatric Psychiatry* (62).

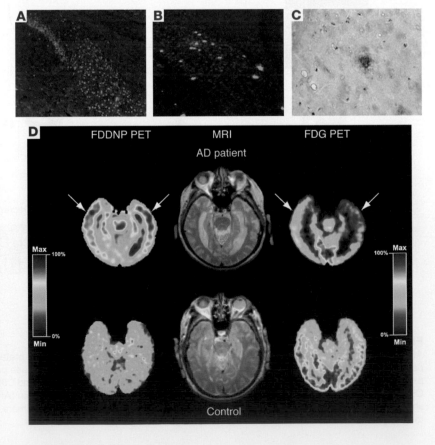

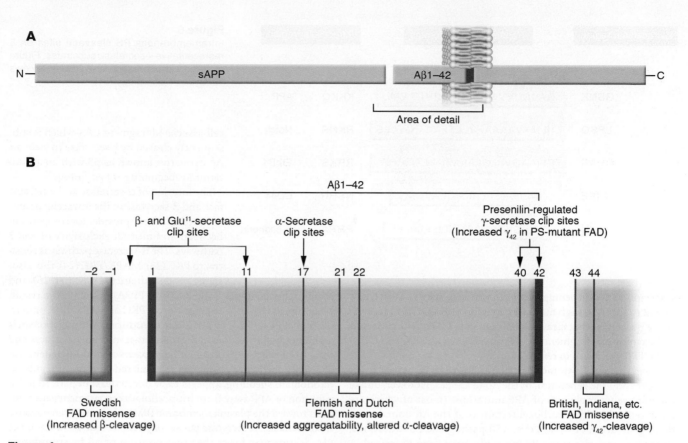

Figure 4

(**A**) Structure and topology of APP. (**B**) The fine structure around the Aβ domain, secretase cleavage sites, and locations of some selected familial AD missense mutations. Figure modified with permission from *Trends in Endocrinology and Metabolism: TEM* (15).

these pathogenic mutations has ever been identified in any individual who failed to develop dementia, with the notable exception of a handful of members of a mutant *APP* family who possessed not only a mutant form of *APP* but also the ε2 allele of the *APOE* gene (18). *APOE* polymorphisms are discussed in more detail below, but the rare cases in which both mutant APP and the *APOE ε2* polymorphism coexist in the same individuals provide the best evidence that *APOE ε2* can protect against the development of AD.

Additional support for the role of amyloid in the process of neurodegeneration comes from 2 newly described hereditary conditions, familial British and Danish dementias (FBD and FDD, respectively), in which patients suffering from either condition present with extensive amyloid deposition in the CNS that is colocalized with NFTs and associated with neurodegeneration. Although the biochemical properties of NFTs in both FBD and FDD are indistinguishable from those found in AD, the amyloid deposits in the former 2 disorders are composed of 2 new molecules, British amyloid (ABri) and Danish amyloid (ADan), which are structurally unrelated to the Aβ associated with AD (19). ABri and ADan are generated by missense mutations in the stop codons in the *BRI* and *DAN* genes and therefore only exist in nature in humans with these genetic disorders. These illnesses caused by generation of otherwise nonexistent proteins also argue against disease models that describe the origin of cerebral amyloidosis as being due to perturbation of the normal function of the respective amyloid precursors, since the gene for the amyloidogenic peptide is only present in affected individuals.

What is the usual cause of autosomal dominant, familial AD?

The preponderance of the familial AD for which the defect is known is attributable to completely penetrant missense mutations in the catalytic subunit of γ-secretase. The mutation is in a complex, 8-transmembrane–domain protein called presenilin (PS) that

Figure 5

Topology of the 4 components that comprise the high molecular weight γ-secretase complex. Black bar represents the cleavage site for processing of the zymogen form of PS1 into the amino and carboxyterminal fragments that self associate and form the active enzyme. Figure modified from an image courtesy of Jan Naslund (Karolinska Institute, Stockholm, Sweden).

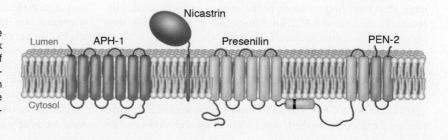

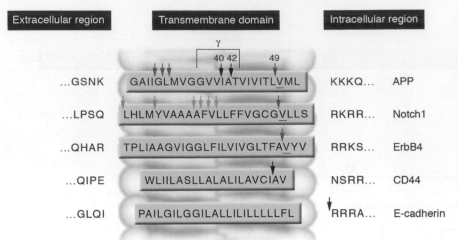

Figure 6
Intramembranous PS cleavage sites on 5 representative γ-secretase substrates. Figure modified with permission from *Nature Reviews Molecular Cell Biology* (61).

cell-associated fragment C83, which is subsequently cleaved by γ-secretase to yield an Aβ derivative known as p3 with an amino terminus beginning at Leu[17] of Aβ.

The location of α-secretase at the cell surface and β-secretase in the intracellular and endocytic pathway provides for competition between and mutual exclusivity of the 2 pathways. The α-secretase pathway is sensitive to PKC (23), MAPK/ERK (24), rho-associated protein kinase (ROCK) (ref. 25), and protein phosphatases 1 and 2A (PP1, PP2A) (ref. 23). As a result, when signals that involve activation of PKC or MAPK/ERK or inactivation of PP1, PP2A, or ROCK are transduced, APP metabolism is dramatically shifted toward the α-secretase–mediated pathway and away from the β-secretase–mediated pathway and Aβ generation (26). This phenomenon, known as regulated cleavage or regulated ectodomain shedding, appears to be due, at least in part, to redistribution of APP away from intracellular compartments and outward toward the plasma membrane (PM) where APP can encounter α-secretase. α-secretase per se may also be redistributed to the PM. Statins may lower the concentration of Aβ by stimulating α-secretase to yield a processing pattern that is indistinguishable from regulated cleavage modulated by PKC and MAPK/ERK (27). Recent evidence from Postina and colleagues using transgenic mice validates α-secretase activation as a viable therapeutic opportunity that can antagonize brain Aβ accumulation (28). These investigators showed that upregulation and downregulation of α-secretase activity can modulate amyloid burden in plaque-forming mice in the predicted, bidirectional manner.

associates with other components known as nicastrin, APH-1, and PEN-2 to form the high molecular weight (>10⁶ kDa; see Figure 5) active γ-secretase that cleaves APP fragments CTFα (the α-cleaved carboxyterminal fragment of APP, also known as C83) and CTFβ and CTFβ′ (reviewed in ref. 20).

As with most APP mutations, the mutations in PS appear to act by increasing the generation of the Aβ42 species. However, unlike the relatively simple case of APP mutations (most of which are clustered around the carboxyl terminus of the Aβ domain and favor generation of Aβ42), over 150 pathogenic PS mutations, scattered throughout the PS molecule, have been described. A dozen more substrates, in addition to APP, exist for γ-secretase (Figure 6), and even within APP, γ-secretase can cleave at several sites, including a site distinct from the canonical γ-secretase site. Cleavage at this second, so-called "ε-site" liberates the cytoplasmic APP intracellular domain (AICD), which, in analogy with other PS substrates, may traffic to the nucleus and act as a transcription factor (reviewed in ref. 20).

In addition to γ-secretase, which contains mutations that can cause AD, 2 other secretases control the cleavages that initiate APP processing. An integral aspartyl proteinase, known as β-secretase or β-APP–site cleaving enzyme (BACE; ref. 21), generates APP carboxyl fragments known as C99 (or CTFβ) and C89 (β′-cleaved carboxyterminal fragment of APP; also known as CTFβ′), which bear either Aβ [Asp¹] or Aβ [Glu¹¹] at their amino termini (see Figure 4B). These cleavage sites are sometimes known as β and β′ sites, respectively, and may be differentially generated in endosomes and the *trans*-Golgi network, respectively. The large ectodomain fragment generated by BACE-mediated cleavage is sAPPβ. APP is one of only 2 known BACE substrates, which makes BACE inhibition a popular anti-Aβ therapeutic strategy, and the development of safe, orally active BACE inhibitors is being aggressively pursued. To date, success has been limited; the BACE catalytic pocket is especially large, thus requiring relatively bulky molecules for efficient inhibition. To date, these BACE inhibitors have been limited by their toxicity and/or their exclusion from the CNS.

Another APP secretase proteinase, α-secretase, is comprised of cell-surface metalloproteinases, a disintegrin and metalloproteinase–10 (ADAM-10) and –17 (22). These proteases cleave several known important substrates in addition to APP, including pro–TNF-α and pro–TGF-α. The APP derivatives generated by α-secretase are the released (or shed) ectodomain fragment sAPPα and the

What is the role of Aβ in common forms of AD?

Until 2004, one could justify the formulation that all autosomal dominant forms of AD could be linked to the accumulation of Aβ oligomers. A serious challenge to that model has arisen with the discovery of a family of individuals with a PS mutation manifesting clinically as frontotemporal dementia (FTD) and pathologically as "pure" NFT disease, i.e., there is no parenchymal or cerebrovascular amyloidosis in FTD (29). FTD is usually due to mutations involving the cytoskeletal protein tau, and these diseases are therefore called tauopathies (30). Since FTDs are NFT diseases, their existence has provided the strongest evidence that tangles cannot cause amyloidosis. No pathogenic PS1 mutation has heretofore been described as associating with pure NFT pathology. Furthermore, no generation of excess levels of Aβ42 was observed when the FTD mutant PS was transfected into cultured cells (29). The possibility remains that the PS1 mutation is acting by elevating the levels of toxic Aβ oligomers that induce tangle formation but not plaques, but no precedent exists for such a pathomechanism. Interest is now focused on the overexpression of this FTD mutant PS1 in transgenic mice, in order to assess further the nature of its pathogenicity.

The vast majority of all AD lacks a predictable, autosomal dominant mode of inheritance. While APP and the PSs constitute the

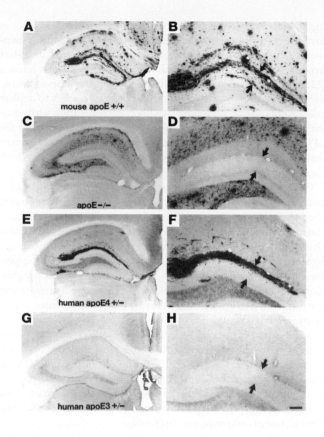

Figure 7

APOE isoform–specific regulation of Aβ plaque burden. Aβ plaque load is highest in the hippocampi of mice expressing murine apoE (**A**, magnified in **B**) and lowest in *APOE* knockout animals (**C**, magnified in **D**). Plaque load is moderate in the hippocampi of human *APOE* ε4–expressing mice (**E**, magnified in **F**) as compared to murine apoE–expressing mice. Plaque load in the hippocampi of human *APOE* ε4–expressing mice is greater than plaque load in the hippocampi from human *APOE* ε3–expressing mice (**G**, magnified in **H**). Amyloid deposits in the dentate gyrus, indicated by arrows, never develop in the absence of apoE. Scale bar: 150 μm (**A**, **C**, **E**, and **G**) and 60 μm (**B**, **D**, **F**, and **H**). Figure modified with permission from *Proceedings of the National Academy of Sciences of the United States of America* (40).

More recent studies, however, have raised questions about the relevance of these SDS-resistant apoE:Aβ complexes under physiological conditions since immunoprecipitation assays indicate that apoE ε3 and apoE ε4 bind (36) Aβ and stimulate fibrillogenesis (38) equivalently, at least under the conditions studied. This is bolstered by data obtained from purification and characterization of apoE:Aβ complexes from the brains of subjects with various *APOE* genotypes, in which we demonstrated no effect of *APOE* genotype on the quality or composition of the apoE:Aβ complex (39).

The best evidence linking *APOE* genotype to Aβ accumulation comes from studies of human *APOE* isoform–specific transgenic mice created on an *APOE* null background and crossed with mice bearing a pathogenic mutant form of APP (40). Plaque load was clearly enhanced in ε4 mice as compared with ε3 mice (ref. 40; Figure 7). We recently demonstrated that, unlike human ε3, human ε4 is associated with elevated levels of endogenous murine brain Aβ as a function of aging and gender, even in mice that will therefore never develop cerebral amyloidosis. These data support the conclusion that there exists an effect of ε4 on Aβ upstream of fibrillogenesis (41).

Are there other plausible hypotheses for the pathomechanisms of *APOE* polymorphisms?

Another model that could explain, at least in part, the effects of ε4 involves the relative deficiency of ε4 as an antioxidant, since ε4 lacks the cysteine residues that are present in ε2 and ε3 and help buffer oxidant stress, when ε4 is compared with ε3 or ε2 (42). This dovetails with evidence that antioxidants can slow (43) the progression of AD and perhaps even prevent familial AD due to APP mutations (see above discussion about *APOE* ε2 and APP mutations; ref. 18).

We included an assessment of this possibility in our evaluation of the effect of aging and *APOE* genotype on levels of nonamyloidogenic endogenous mouse Aβ (41), extending the growing body of evidence indicating that oxidized prostaglandins known as isoprostanes might be useful markers for AD (reviewed in ref. 44). $F_{2\alpha}$-isoprostanes have been reported to rise in the tissues and/or body fluids of patients with AD and of transgenic plaque-forming mice at around the age of incipient amyloidosis (44–46). $F_{2\alpha}$-isoprostanes have also been shown to stimulate both generation (47) and aggregation (48) of Aβ, possibly placing isoprostane accumulation upstream of Aβ aggregation. This is an important point because human (but not murine) Aβ is reported to be a pro-oxidant (49). We were able to dissociate the isoprostane elevation from the formation of histological amyloid by using mice bearing only endogenous murine APP and Aβ (41). Thus, plaque formation is not a prerequisite for elevation of brain isoprostane levels. Indeed, one model that we propose (41) suggests that oxidized lipids might sometimes initiate Aβ aggregation.

only known AD genes, at least 1 important genetic risk factor is known for about 25% of the population with AD, and that is the *APOE* ε4 genotype (31). apoE (encoded by *APOE*) is the body's major cholesterol transport protein and binds primarily to the LDL receptor (LDLR) and the LDLR-related protein (LRP). The most common form of *APOE* is the ε3 type, bearing a Cys at residue 112 and an Arg at residue 158. However, about 15% of the *APOE* alleles in the general population are of the ε4 type, in which the Cys at 112 is changed to an Arg. In the population of AD patients, the allele frequency is roughly tripled, with 45% of *APOE* alleles being of the ε4 type (31). This translates into a tripling of the risk of developing AD for every ε4 allele present, such that ε4 homozygotes have a 9- to 10-fold increased risk of developing dementia (31). However, the "ε4 effect" is modified by gender and is age-specific, with its peak effect observed at around 70 years of age (32). There are octogenarian ε4 homozygotes who appear to have escaped the "ε4 effect", and it is this fact that causes ε4 to be described as a risk factor variant rather than a disease gene. Somewhere between 25 and 35% of the population with AD carry at least one ε4 allele (ref. 33 and references therein). Conversely, the *APOE* ε2 allele appears to protect against the development of AD (18). This protection is consistent with the oxidation model of AD (see "Are there other plausible hypotheses for the pathomechanisms of *APOE* polymorphisms?", below).

How does *APOE* ε4 increase the risk for AD?

Efforts to link *APOE* ε4 with Aβ accumulation have met with mixed results. apoE ε3, but not ε4, forms complexes with Aβ that are resistant to denaturation, as discovered by LaDu et al. (34), and this has been confirmed when the source of apoE is cell-culture medium, human cerebrospinal fluid, or human plasma (34–37).

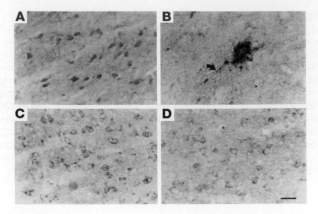

Figure 8

Comparison of the immunocytochemical distributions of APP (**A** and **B**) and PS1 (**C** and **D**) in control (**A** and **C**) and thiamine-deficient (**B** and **D**) mouse brain. Note the accumulation of APP immunoreactivity in a dystrophic neuritic cluster during thiamine deficiency (**B**). Arrow shows APP accumulation in an abnormal neurite arising from the neuritic cluster. PS1-immunostained sections adjacent to **A** and **B** reveal no accumulation of PS1 immunoreactivity in abnormal structures during thiamine deficiency except in areas of severe cell loss, in which the immunostaining is pale (**D**). Scale bar: 25 μm. Figure reproduced with permission from *American Journal of Pathology* (50).

Abnormal oxidation is one of the most consistent themes in all aging-related diseases of all organ systems, and it is interesting to note that the brain lesions resulting from oxidative toxicity induced by thiamine deficiency include abnormal, amyloid precursor-containing neuritic lesions (Figure 8; ref. 50). In ongoing studies, these lesions are now being induced in plaque-competent transgenic mice in order to determine whether the neuritic brain lesions induced by the oxidative reactions of thiamine deficiency accelerate amyloid deposition.

What is the role of cholesterol in AD?

The involvement of *APOE* polymorphisms in AD immediately raised the question of whether cholesterol might play a role in the risk for the disease. In evaluating this possibility, Sparks and colleagues determined that Watanabe rabbits on a 2% cholesterol diet developed increased levels of Aβ immunoreactivity in vesicles inside the neurons of their brains (51). Elucidation of the mechanism underly-

ing this observation began to be vigorously pursued in plaque-forming transgenic mouse models, and soon several groups demonstrated that fat feeding increases brain plaque load and that hypocholesterolemic agents such as statins lower plaque burden (52, 53). In cultured neurons, both simvastatin and atorvastatin as well as cyclodextrin, a compound used in the laboratory to deplete cellular cholesterol, were shown to lower Aβ generation while activating the alternate, nonamyloidogenic α-secretase pathway for APP metabolism (27, 54). At least some of the actions of statins on α-secretase appear to occur via the rho/ROCK pathway, which can modulate sAPPα generation (25).

The epidemiological evidence linking cholesterol and AD is confusing. Some studies have concluded that statin use lowers the risk for dementia (55), while others conclude that this association is artifactual (56). Several epidemiological studies also show that lipid levels are *lower* in persons who develop AD, not *higher* (57). Perhaps the most encouraging evidence, albeit preliminary, is the apparent ability of atorvastatin to slow progression of dementia in AD (58). The publication of this report and the results of a large National Institute on Aging–sponsored prevention trial of statins are eagerly anticipated. At the moment, caution is recommended, especially when one reflects upon the ongoing efforts to reconcile epidemiological evidence that hormone replacement therapy (HRT) could delay or prevent AD with the disappointment of clinical trials of HRT. A primary prevention trial of HRT is now underway, focusing on institution of replacement during the perimenopausal period rather than the typical, late postmenopausal protocol.

What antiamyloid medicines are in current clinical trials?

Antiamyloid strategies fall into 3 basic categories: immunotherapy agents, antiaggregants, and secretase modulators. Active immunization (vaccination) has been limited by the side effect of

Figure 9

Possible Aβ-dependent and Aβ-independent mechanisms of neurotoxicity. In this model, 1 route toward neurotoxicity begins with either APP mutations or PS1 mutations (top center of figure). The pathway toward toxicity flows downward via elevated Aβ42 levels and elevated oligomer levels. An alternative pathway is shown on the right side of the figure, whereby superoxide ($O_2^{\cdot-}$) or hydrogen peroxide (H_2O_2) oxidize lipids such as prostaglandins, forming $F_{2\alpha}$-isoprostanes. Both H_2O_2 and $F_{2\alpha}$-isoprostanes are known to accelerate Aβ aggregation and presumably its oligomerization. This author proposes that toxicity in this pathway occurs *both* directly from reactive oxygen species (H_2O_2 and $O_2^{\cdot-}$) *and* via acceleration of Aβ oligomerization by these reactive oxygen species.

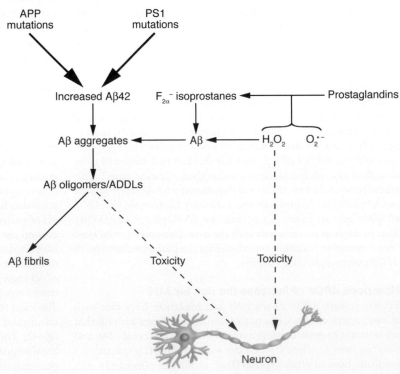

acute allergic encephalitis in 5% of 300 subjects (59), but phase I trials of passive immunization are now underway. The antiaggregant approach is represented by commercial efforts at blocking the promotion of amyloidogenesis by glycosaminoglycans of the extracellular matrix. Glycosaminoglycan mimetics are currently in phases II and III of clinical trials aiming to evaluate their effectiveness at either slowing the progression of AD or preventing rebleeding due to cerebral amyloid angiopathy. Statins, as mentioned above, represent the best examples of α-secretase activators that are currently in treatment and prevention trials. Flurbiprofen, a derivative of ibuprofen, is a modulator of Aβ42 generation that is believed to allosterically modify γ-secretase so that generation of Aβ42 is inhibited while an otherwise trace species, Aβ38, is generated (60). A large AD prevention trial of conventional NSAIDs was aborted in 2005 when excess cardiac mortality became associated with their use. Other γ-secretase modulators are in the pipeline at several pharmaceutical companies, but progress is especially slow with such compounds because of the known risk factor of mechanism-based toxicity due to inhibition of Notch processing (61). The website www.clinicaltrials.gov remains the best source of up-to-date information on which trials are enrolling and whom to contact locally for further information.

What is the role of Aβ accumulation in common forms of AD?

Genetic evidence indicates that accumulation of Aβ in some biophysical form is injurious to the brain. However, this formulation is most defensible in the approximately 50 families worldwide who have early onset AD due to APP mutations. Also, data from studies

of other cerebral amyloidoses (e.g., FBD and FDD) support a key role for amyloidosis in causing neurodegeneration. Metabolism and trafficking of APP are tightly controlled events, and it is entirely possible that excess Aβ42 generation and/or accumulation is an initial symptom of some other primary problem (such as excess oxidation) that is itself independently neurotoxic. In such a scenario, however, it remains possible (and likely) that Aβ accumulation is a second, parallel pathway that is also toxic (Figure 9). This suggests that antiamyloid therapeutic strategies are likely to be beneficial in the treatment of AD even if they are not curative. Dissection of the genesis of the Aβ accumulation phenotype is important in order to point the way to any upstream, primary lesions. The successful development of effective antiamyloid therapies — immunomodulators, antiaggregants, secretase modulators, or some combination of these — remains a key goal, the attainment of which is required in order to elucidate the role played by cerebral accumulation of toxic Aβ oligomers in the clinical picture of AD.

Acknowledgments

The author acknowledges support from United States Public Health Service grants P01 AG10491 and R01 NS41017. In the past, the author has also received consultant fees and/or grant support from Parke-Davis, Wyeth, F. Hoffman-LaRoche Ltd., Pfizer Inc., Eisai Inc., and Andrx Corp.

Address correspondence to: Sam Gandy, Farber Institute for Neurosciences, Thomas Jefferson University, 900 Walnut Street, Suite 467, Philadelphia, Pennsylvania 19107, USA. Phone: (215) 503-4200; Fax: (215) 503-4358; E-mail: samgandy@earthlink.net.

1. Mitchell, T.W., et al. 2002. Parahippocampal tau pathology in healthy aging, mild cognitive impairment, and early Alzheimer's disease. *Ann. Neurol.* **51**:182–189.
2. Alford, M.F., Masliah, E., Hansen, L.A., and Terry, R.D. 1994. A simple dot-immunobinding assay for quantification of synaptophysin-like immunoreactivity in human brain. *J. Histochem. Cytochem.* **42**:283–287.
3. Virchow, R. 1854. Zur cellulosefrage. *Virchows Arch. Pathol. Anat. Physiol.* **6**:416–426.
4. Glenner, G.G., and Terry, W.D. 1974. Characterization of amyloid. *Annu. Rev. Med.* **25**:131–135.
5. Glenner, G.G., and Wong, C.W. 1984. Alzheimer's disease and Down's syndrome: sharing of a unique cerebrovascular amyloid fibril protein. *Biochem. Biophys. Res. Commun.* **122**:1131–1135.
6. Naslund, J., et al. 1994. Relative abundance of Alzheimer A beta amyloid peptide variants in Alzheimer disease and normal aging. *Proc. Natl. Acad. Sci. U. S. A.* **91**:8378–8382.
7. Younkin, S.G. 1995. Evidence that A beta 42 is the real culprit in Alzheimer's disease. *Ann. Neurol.* **37**:287–288.
8. Klein, W.L., Krafft, G.A., and Finch, C.E. 2001. Targeting small Abeta oligomers: the solution to an Alzheimer's disease conundrum? *Trends Neurosci.* **24**:219–224.
9. Kayed, R., et al. 2004. Permeabilization of lipid bilayers is a common conformation-dependent activity of soluble amyloid oligomers in protein misfolding diseases. *J. Biol. Chem.* **279**:46363–46366.
10. Harper, J.D., and Lansbury, P.T., Jr. 1997. Models of amyloid seeding in Alzheimer's disease and scrapie: mechanistic truths and physiological consequences of the time-dependent solubility of amyloid proteins. *Annu. Rev. Biochem.* **66**:385–407.
11. Naslund, J., et al. 2000. Correlation between elevated levels of amyloid beta-peptide in the brain and cognitive decline. *JAMA.* **83**:1571–1577.
12. Hsiao, K., et al. 1996. Correlative memory deficits, Abeta elevation, and amyloid plaques in transgenic mice. *Science.* **274**:99–102.
13. Gong, Y., et al. 2003. Alzheimer's disease-affected brain: presence of oligomeric A beta ligands (ADDLs) suggests a molecular basis for reversible memory loss. *Proc. Natl. Acad. Sci. U. S. A.* **100**:10417–10422.
14. Georganopoulou, D.G., et al. 2005. Nanoparticle-based detection in cerebral spinal fluid of a soluble pathogenic biomarker for Alzheimer's disease. *Proc. Natl. Acad. Sci. U. S. A.* **102**:2273–2276.
15. Gandy, S. 1999. Neurohormonal signaling pathways and the regulation of Alzheimer beta-amyloid precursor metabolism. *Trends Endocrinol. Metab.* **10**:273–279.
16. Nilsberth, C., et al. 2001. The 'Arctic' APP mutation (E693G) causes Alzheimer's disease by enhanced Abeta protofibril formation. *Nat. Neurosci.* **4**:887–893.
17. Paivio, A., Jarvet, J., Graslund, A., Lannfelt, L., and Westlind-Danielsson, A. 2004. Unique physicochemical profile of beta-amyloid peptide variant Abeta1-40E22G protofibrils: conceivable neuropathogen in Arctic mutant carriers. *J. Mol. Biol.* **339**:145–159.
18. St. George-Hyslop, P., et al. 1994. Alzheimer's disease and possible gene interaction [letter]. *Science.* **263**:537.
19. Ghiso, J., and Frangione, B. 2002. Amyloidosis and Alzheimer's disease. *Adv. Drug Deliv. Rev.* **54**:1539–1551.
20. Carter, T.L., et al. 2004. Alzheimer amyloid precursor aspartyl proteinase activity in CHAPSO homogenates of *Spodoptera frugiperda* cells. *Alzheimer Dis. Assoc. Disord.* **18**:261–263.
21. Vassar, R., et al. 1999. Beta-secretase cleavage of Alzheimer's amyloid precursor protein by the transmembrane aspartic protease BACE. *Science.* **286**:735–741.
22. Buxbaum, J.D., et al. 1998. Evidence that tumor necrosis factor alpha converting enzyme is involved in regulated alpha-secretase cleavage of the Alzheimer amyloid protein precursor. *J. Biol. Chem.* **273**:27765–27767.
23. Caporaso, G.L., Gandy, S.E., Buxbaum, J.D., Ramabhadran, T.V., and Greengard, P. 1992. Protein phosphorylation regulates secretion of Alzheimer beta/A4 amyloid precursor protein. *Proc. Natl. Acad. Sci. U. S. A.* **89**:3055–3059.
24. Mills, J., et al. 1997. Regulation of amyloid precursor protein catabolism involves the mitogen-activated protein kinase signal transduction pathway. *J. Neurosci.* **17**:9415–9422.
25. Pedrini, S., et al. 2005. Modulation of statin-activated shedding of Alzheimer APP ectodomain by ROCK. *PLoS Med.* **2**:e18. doi:10.1371/journal.pmed.0020018.
26. Buxbaum, J.D., Koo, E.H., and Greengard, P. 1993. Protein phosphorylation inhibits production of Alzheimer amyloid beta/A4 peptide. *Proc. Natl. Acad. Sci. U. S. A.* **90**:9195–9198.
27. Kojro, E., Gimpl, G., Lammich, S., Marz, W., and Fahrenholz, F. 2001. Low cholesterol stimulates the nonamyloidogenic pathway by its effect on the alpha-secretase ADAM 10. *Proc. Natl. Acad. Sci. U. S. A.* **98**:5815–5820.
28. Postina, R., et al. 2004. A disintegrin-metalloproteinase prevents amyloid plaque formation and hippocampal defects in an Alzheimer disease mouse model. *J. Clin. Invest.* **113**:1456–1464. doi:10.1172/JCI200420864.
29. Dermaut, B., et al. 2004. A novel presenilin 1 mutation associated with Pick's disease but not β-amyloid plaques. *Ann. Neurol.* **55**:617–626.
30. Hutton, M. 2004. Presenilin mutations associated with fronto-temporal dementia [review]. *Ann. Neurol.* **55**:604–606.

31. Mayeux, R., et al. 1993. The apolipoprotein epsilon 4 allele in patients with Alzheimer's disease. *Ann. Neurol.* **34**:752–754.

32. Farrer, L.A., et al. 1997. Effects of age, sex, and ethnicity on the association between apolipoprotein E genotype and Alzheimer disease. A meta-analysis. APOE and Alzheimer Disease Meta Analysis Consortium. *JAMA.* **278**:1349–1356.

33. Martins, R.N., et al. 1995. ApoE genotypes in Australia: roles in early and late onset Alzheimer's disease and Down's syndrome. *Neuroreport.* **6**:1513–1516.

34. LaDu, M.J., et al. 1994. Isoform-specific binding of apolipoprotein E to beta-amyloid. *J. Biol. Chem.* **269**:23403–23406.

35. Zhou, Z., Smith, J.D., Greengard, P., and Gandy, S. 1996. Alzheimer amyloid-beta peptide forms denaturant-resistant complex with type epsilon 3 but not type epsilon 4 isoform of native apolipoprotein E. *Mol. Med.* **2**:175–180.

36. Zhou, Z., Relkin, N., Ghiso, J., Smith, J.D., and Gandy, S. 2002. Human cerebrospinal fluid apolipoprotein E isoforms are apparently inefficient at complexing with synthetic Alzheimer's amyloid-[beta] peptide (A[beta] 1–40) in vitro. *Mol. Med.* **8**:376–381.

37. Yang, D.S., Smith, J.D., Zhou, Z., Gandy, S.E., and Martins, R.N. 1997. Characterization of the binding of amyloid-beta peptide to cell culture-derived native apolipoprotein E2, E3, and E4 isoforms and to isoforms from human plasma. *J. Neurochem.* **68**:721–725.

38. Sweeney, D., Martins, R., LeVine, H., 3rd, Smith, J.D., and Gandy, S. 2004. Similar promotion of Abeta1-42 fibrillogenesis by native apolipoprotein E epsilon 3 and epsilon 4 isoforms. *J. Neuroinflammation.* **1**:15–20.

39. Naslund, J., et al. 1995. Characterization of stable complexes involving apolipoprotein E and the amyloid beta peptide in Alzheimer's disease brain. *Neuron.* **15**:219–228.

40. Holtzman, D.M., et al. 2000. Apolipoprotein E isoform-dependent amyloid deposition and neuritic degeneration in a mouse model of Alzheimer's disease. *Proc. Natl. Acad. Sci. U. S. A.* **97**:2892–2897.

41. Yao, J., et al. 2004. Aging, gender and APOE isotype modulate Alzheimer's Aβ peptides and F_2-isoprostanes in the absence of detectable amyloid deposits. *J. Neurochem.* **90**:1011–1018.

42. Miyata, M., and Smith, J.D. 1996. Apolipoprotein E allele-specific antioxidant activity and effects on cytotoxicity by oxidative insults and beta-amyloid peptides. *Nat. Genet.* **14**:55–61.

43. Mayeux, R., and Sano, M. 1999. Treatment of Alzheimer's disease. *N. Engl. J. Med.* **341**:1670–1679.

44. Pratico, D., and Sung, S. 2004. Lipid peroxidation and oxidative imbalance: early functional events in Alzheimer's disease. *J. Alzheimers. Dis.* **6**:171–175.

45. Pratico, D., Uryu, K., Leight, S., Trojanoswki, J.Q., and Lee, V.M. 2001. Increased lipid peroxidation precedes amyloid plaque formation in an animal model of Alzheimer amyloidosis. *J. Neurosci.* **21**:4183–4187.

46. Yao, Y., et al. 2004. Brain inflammation and oxidative stress in a transgenic mouse model of Alzheimer-like brain amyloidosis [serial online]. *J. Neuroinflammation.* **1**:21. http://www.jneuroinflammation.com/content/1/1/21.

47. Qin, W., et al. 2003. Cyclooxygenase (COX)-1 and COX-2 potentiate beta-amyloid peptide generation through mechanisms that involve gamma-secretase activity. *J. Biol. Chem.* **278**:50970–50977.

48. Boutoud, O., et al. 2002. Prostaglandin H2 (PGH2) accelerates formation of amyloid beta 1-42 oligomers. *J. Neurochem.* **82**:1003–1006.

49. Huang, X., et al. 1999. The A beta peptide of Alzheimer's disease directly produces hydrogen peroxide through metal ion reduction. *Biochemistry.* **38**:7609–7616.

50. Calingasan, N.Y., et al. 1996. Novel neuritic clusters with accumulations of amyloid precursor protein and amyloid precursor-like protein 2 immunoreactivity in brain regions damaged by thiamine deficiency. *Am. J. Pathol.* **149**:1063–1071.

51. Sparks, D.L., et al. 1994. Induction of Alzheimer-like beta-amyloid immunoreactivity in the brains of rabbits with dietary cholesterol. *Exp. Neurol.* **126**:88–94.

52. Refolo, L.M., et al. 2000. Hypercholesterolemia accelerates the Alzheimer's amyloid pathology in a transgenic mouse model. *Neurobiol. Dis.* **7**:321–331.

53. Refolo, L.M., et al. 2001. A cholesterol-lowering drug reduces beta-amyloid pathology in a transgenic mouse model of Alzheimer's disease. *Neurobiol Dis.* **8**:890–899.

54. Simons, M., et al. 1998. Cholesterol depletion inhibits the generation of beta-amyloid in hippocampal neurons. *Proc. Natl. Acad. Sci. U. S. A.* **95**:6460–6464.

55. Wolozin, B., Kellman, W., Ruosseau, P., Celesia, G.G., and Siegel, G. 2000. Decreased prevalence of Alzheimer disease associated with 3-hydroxy-3-methyglutaryl coenzyme A reductase inhibitors. *Arch. Neurol.* **57**:1439–1443.

56. Li, G., et al. 2004. Statin therapy and risk of dementia in the elderly: a community-based prospective cohort study. *Neurology.* **63**:1624–1628.

57. Romas, S.N., Tang, M.X., Berglund, L., and Mayeux, R. 1999. APOE genotype, plasma lipids, lipoproteins, and AD in community elderly. *Neurology.* **53**:517–521.

58. Sparks, D.L., et al. 2004. Benefit of atorvastatin in the treatment of Alzheimer disease. *Neurobiol. Aging.* **25**(Suppl. 1):S24.

59. Nicoll, J.A., et al. 2003. Neuropathology of human Alzheimer disease after immunization with amyloid-beta peptide: a case report. *Nat. Med.* **9**:448–452.

60. Weggen, S., et al. 2001. A subset of NSAIDs lower amyloidogenic Abeta42 independently of cyclooxygenase activity. *Nature.* **414**:212–216.

61. Fortini, M.E. 2002. Gamma-secretase-mediated proteolysis in cell-surface-receptor signalling. *Nat. Rev. Mol. Cell Biol.* **3**:673–684.

62. Shoghi-Jadid, K., et al. 2002. Localization of neurofibrillary tangles and beta-amyloid plaques in the brains of living patients with Alzheimer disease. *Am. J. Geriatr. Psychiatry.* **10**:24–35.

Alzheimer disease therapy:
Can the amyloid cascade be halted?

Todd E. Golde

Department of Neuroscience, Mayo Clinic Jacksonville, Jacksonville, Florida, USA.

In the not too distant future, clinical management of Alzheimer disease (AD) is likely to resemble the present management of atherosclerotic disease. Sometime before an individual reaches age 50, an internist will initiate a screening program to determine that person's risk for developing AD. This assessment will include a comprehensive genetic screen for AD-risk loci, determination of plasma amyloid β peptide (Aβ) levels, family history of AD, and, perhaps, potential environmental risks. Depending on the risk prediction, a follow-up visit with an Alzheimer specialist may be scheduled. During this visit, an amyloid-binding agent will be injected and used to evaluate the extent of amyloid deposition in the brain. Based on the amount of deposition present and the initial risk assessment, the specialist will then develop a personalized therapeutic regimen. This regimen might consist of an Aβ vaccination, an amyloid-lowering drug, an anti-inflammatory agent, a neuronal growth factor, an antioxidant, or a combination of these approaches. The efficacy of therapy will be monitored by measurement of plasma Aβ levels, imaging of amyloid in the brain, and volumetric scanning of the brain. Primary screening, along with monitoring of the presymptomatic indicators of disease, and appropriate intervention will significantly reduce one's risk for developing AD.

Although proposal of such a scenario might seem highly speculative, recent and expected advances in (a) understanding the pathogenesis of AD, (b) identifying the genetic factors that confer risk for AD, (c) validating potential biomarkers for AD, and (d) developing therapeutic agents that target both Aβ and downstream pathological changes greatly increase the likelihood that AD will be managed successfully in the future.

AD is the leading cause of dementia in the elderly

Estimates of prevalence vary, but 1–5% of the population over age 65, and 20–40% of the population over age 85, may be affected by AD (1). Given the increasing number of elderly individuals in industrialized societies, AD represents a burgeoning epidemic that exacts a tremendous toll on the individuals it affects, along with their families and caregivers. Moreover, AD has a tremendous negative economic impact amounting to over $100 billion a year. Treatment of AD in the US reportedly costs more per patient than management of other major age-associated diseases (2, 3).

Beginning with short-term memory loss, and continuing with more widespread cognitive and emotional dysfunction, typical late-onset AD (LOAD) occurs after age 65 and follows an insidious 5- to 15-year course. Although AD usually presents without motor or sensory alterations, rare variants (such as spastic paraparesis) with atypical clinical presentations are occasionally recognized (4, 5). Even today, definitive diagnosis of AD is only possible through postmortem analysis of the brain (1). This histopathological analysis of the brain demonstrates the classic triad of AD pathology: (a) senile plaques containing Aβ, (b) neurofibrillary tangles (NFTs) containing tau, and (c) widespread neuronal loss in the hippocampus and select cortical and subcortical areas.

Toward a complete understanding of AD pathogenesis: is Aβ the cholesterol of AD?

Aβ accumulation as the initiating factor in AD pathogenesis. Much of the Aβ that accumulates in the AD brain is deposited as amyloid within senile plaques and cerebral vessels. Although numerous proteins are associated with the amyloid deposits in AD, the principal proteinaceous component of AD amyloid is the approximately 4-kDa Aβ. Aβ is produced from the amyloid β protein precursor (APP) through two sequential proteolytic cleavages made by enzymes referred to as secretases (Figure 1) (6). APP is first cleaved at the amino-terminus of Aβ by a membrane-bound aspartyl protease (referred to as β-secretase). This cleavage generates a large secreted derivative (sAPPβ) and a membrane-bound APP carboxy-terminal fragment (CTFβ). Cleavage of CTFβ by γ-secretase results in the production of Aβ peptides of varying length. The two species of most interest are a 40–amino acid Aβ peptide (Aβ40) and a 42–amino acid Aβ peptide (Aβ42). At the same time, a cognate CTFγ is produced. Two homologous polytopic membrane proteases, referred to as presenilins 1 and 2 (PS1 and PS2), are likely γ-secretases. If they are not γ-secretases, PSs are at least essential cofactors for this cleavage (7).

A great deal of evidence that Aβ is not a disease marker, but that it plays a causal role in the development of AD pathology, emerges from a variety of genetic, pathological, and biochemical studies (8–10). These studies demonstrate, first, that mutations in the APP, PS1, and PS2 genes that are linked to the early-onset forms of familial AD increase total Aβ, specifically increase the relative amount of long Aβ peptides ending at Aβ42, or alter the primary sequence of Aβ so that it is more fibrillogenic (11, 12). Second, apoE4, a major risk factor in typical LOAD, increases Aβ deposition (13). Third, Aβ42 deposits as amyloid more rapidly than do shorter species (14). Fourth, overexpression of familial AD–linked mutant APPs, or coexpression of mutant APP and familial AD–linked PSs in the brains of transgenic mice, leads to Aβ deposition and other AD-like pathology (15). Fifth, Aβ can self-aggregate in vitro, and these aggregates can directly and indirectly mediate neurotoxicity (16).

Recent evidence from the study of British familial dementia (BFD) and familial Danish dementia (FDD) provides further support for the hypothesis that Aβ accumulation in the brain is the

Nonstandard abbreviations used: Alzheimer disease (AD); amyloid β peptide (Aβ); late-onset AD (LOAD); amyloid β protein precursor (APP); secreted derivative of amyloid β protein precursor (sAPP); carboxy-terminal fragment (CTF); presenilin (PS); British familial dementia (BFD); familial Danish dementia (FDD); neurofibrillary tangle (NFT); frontotemporal dementia with parkinsonism (FTDP); β-secretase (BACE1, for β-site APP-cleaving enzyme); Food and Drug Administration (FDA); cerebrospinal fluid (CSF).

Conflict of interest: The author has declared that no conflict of interest exists.

Citation for this article: *J. Clin. Invest.* **111**:11–18 (2003). doi:10.1172/JCI200317527.

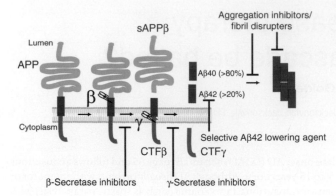

Figure 1

Aβ generation, aggregation, and sites for therapeutic intervention. APP is a type I transmembrane protein that is processed in several different pathways. The Aβ generation pathway is shown. Generation of Aβ in the β-secretase pathway requires two proteolytic events, a proteolytic cleavage at the amino-terminus of the Aβ sequence, referred to as β-secretase cleavage, and a cleavage at the carboxy-terminus, known as γ-secretase cleavage, which results in another carboxy-terminal fragment (CTFγ). Although many Aβ peptides of various lengths can be produced in this fashion, the two of most interest are Aβ40, which is the predominant Aβ peptide, and Aβ42, which is typically produced at much lower levels than Aβ40. Although both peptides can aggregate, Aβ42 is thought to aggregate much more rapidly and to seed the aggregation of Aβ40. Sites for anti-Aβ intervention are indicated. Scissors indicate proteolytic cleavages. "sAPPβ" refers to the large secreted derivative generated by β-secretase cleavage of APP.

cause of AD. BFD and FDD are late-onset dementing disorders characterized by non-Aβ plaques, tangles, and neuronal loss. As in AD, the primary defect in BFD appears to be the abnormal generation of a peptide with amyloidogenic properties (17, 18). Thus, in two additional disorders, accumulation of an amyloidogenic peptide is linked to dementia.

Collectively, these data support a modified version of the amyloid cascade hypothesis (19). This modified cascade is schematically illustrated in Figure 2. Initially, investigators proposed that Aβ accumulation as amyloid triggered a pathological cascade that ultimately produced the complete pathological and clinical symptoms of AD. Today, this hypothesis remains valid, except that it is less clear whether Aβ deposited as amyloid, or some other less well characterized Aβ aggregate, initiates the cascade leading to neuronal death and dysfunction. Indeed, small Aβ oligomers (20), also referred to as Aβ-derived diffusible ligands (ADDLs), and protofibrils (21, 22) have emerged as alternative aggregated forms of Aβ that may mediate toxicity. Moreover, there is some evidence that intracellular accumulations of Aβ may also be neurotoxic (23, 24). In any case, the evidence that Aβ accumulation initiates AD pathology provides a framework in which to develop rational approaches for AD therapy based on altering Aβ accumulation (Figure 1).

Downstream effects of Aβ accumulation. At the present time, the Aβ-induced pathological cascades leading to neuronal dystrophy and death are not nearly as well defined as the pathways leading to Aβ generation and accumulation. This might be attributable to the failure of animal models to fully recapitulate the pathological features of AD, especially with respect to the development of NFTs and neuronal cell death (25). Multiple pathways are also likely to mediate Aβ toxicity. There is evidence that Aβ can be directly neurotoxic, induce oxidative stress, incite an inflammatory response, and alter calcium

homeostasis. These events might be mediated by direct interaction of Aβ aggregates with cellular membranes, or by the binding of Aβ to microglial and neuronal cellular receptors (26–28).

One important downstream event in the Aβ-induced cascade appears to be the development of neurofibrillary pathology. NFTs are intracellular inclusions composed of approximately 10-nm paired helical filaments (PHFs). PHFs themselves are composed of hyperphosphorylated bundles of the microtubule-binding protein tau (29). Abnormal accumulations of hyperphosphorylated tau are also seen in the swollen, tortuous, neuritic processes often, but not exclusively, found in association with senile plaques. Genetic evidence from the study of frontotemporal dementia with parkinsonism linked to chromosome 17 (FTDP-17) indicates that mutations in tau cause this disease, which is characterized by cell death and neurofibrillary pathology (30, 31). Moreover, overexpression of FTDP mutant tau is sufficient to induce NFT pathology and neurofibrillary degeneration in transgenic mice (32, 33). In such mice, Aβ injection or crossing of the tau mutant mice with APP transgenic mice results in enhanced NFT formation (33, 34). Significantly, FTDP-linked tau mutations do not cause alterations in Aβ. Thus, it appears that tau pathology is an important aspect of the pathological cascade induced by Aβ. Therapies aimed at modulating the Aβ-induced changes that lead to NFT formation are likely to be of some benefit in AD. Unfortunately, these pathways are not well defined, and no inhibitors of this process have been identified.

The rationale for anti-Aβ therapy. Based on this evidence, it is not unreasonable to propose that the role Aβ plays in AD is akin to the role cholesterol plays in atherosclerotic disease. Age-associated accumulation of either triggers a complex pathological lesion that, after a long period of time, results in clinical symptoms. Moreover, just as numerous factors contribute to cholesterol deposition, numerous factors can also influence Aβ accumulation. Finally, just as lowering cholesterol levels has proven beneficial in the management of atherosclerotic disease, therapies aimed at reducing Aβ accumulation are likely to be effective in preventing AD. It is likely that the reduction in total Aβ levels need not be complete in order to have some benefit if preventive therapy can be initiated. In AD caused by most mutations in APP, PS1, or PS2, Aβ42 levels are increased by as little as 30% (12). Such an elevation can result in the onset of AD 30–40 years earlier than typical LOAD. By inference, it is likely that reducing total Aβ levels by 30%, or effecting similar selective reductions in the highly pathogenic Aβ42, may delay the development of AD to such an extent that it is no longer a major health care problem.

Anti-Aβ therapies under development

Secretase inhibitors. Research clarifying the metabolic pathways that regulate Aβ production has revealed that the secretases that produce the Aβ may be good therapeutic targets since inhibition of either β- or γ-secretase decreases Aβ production. More progress has been made in developing γ-secretase inhibitors, because high-throughput screens carried out in the pharmaceutical industry have identified numerous γ-secretase inhibitors. Multiple classes of potent γ-secretase inhibitors have now been described, and several of these have been shown to target both PS1 and PS2 (refs. 35–38; reviewed in ref. 7). At least one γ-secretase inhibitor is in clinical trials. Moreover, treatment of mice with a γ-secretase inhibitor reduces Aβ levels in the brain and attenuates Aβ deposition (39). However, despite these advances,

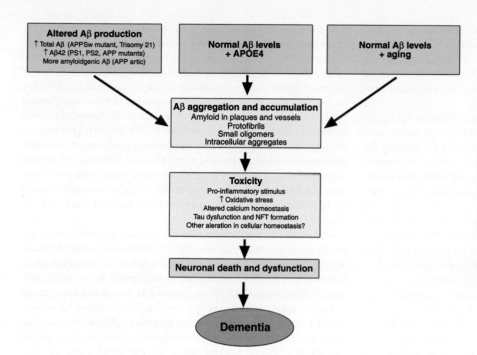

Figure 2

Aβ aggregation as the cause of AD. A modified version of the amyloid cascade hypothesis is shown. This version takes into account the possibility that Aβ aggregates other than those found in classic amyloid deposits initiate the pathological cascade. It is possible that Aβ-induced toxicity in turn results in alterations in the brain, such as increased APP and apoE expression, that enhance Aβ deposition, although this is not shown in the figure. Besides known genetic pathways, a pathway in which normal Aβ levels in the context of normal aging may lead to Aβ accumulation is shown. "APPSw" refers to the APP Swedish mutant linked to familial AD; this mutation alters the lysine-methionine sequence immediately preceding Aβ to asparagine-leucine. Trisomy 21 is also known as Down syndrome.

numerous concerns over the use of γ-secretase inhibitors as AD therapeutics remain. These concerns center on target-mediated toxicity caused by interference with γ-secretase–mediated Notch signaling (40, 41); inhibition of signaling mediated by newly recognized γ-secretase substrates (such as the epidermal growth factor receptor ErbB4) or unrecognized substrates (42); or accumulation of potentially neurotoxic APP CTFβ, which invariably occurs when γ-secretase is inhibited (43, 44).

Although the development of β-secretase inhibitors has lagged behind the development of γ-secretase inhibitors, many believe that β-secretase is likely to be a better therapeutic target. β-Secretase (BACE1, for β-site APP-cleaving enzyme) knockout mice produce no Aβ, yet they have no obvious pathological phenotype (45, 46). Significantly, the crystal structure of BACE1 has been solved (47, 48). Such structural information will surely speed the drug discovery efforts, currently underway, to develop potent nonpeptidic BACE1 inhibitors. Although the knockout studies partially allayed fears that BACE1 inhibition might be problematic due to inhibition of cleavage of non-APP substrates, concerns remain regarding target-mediated toxicity. Moreover, the crystal structure of BACE1 reveals a wide, active-site gorge that may be difficult to target with small-molecule inhibitors (47, 48).

Very recently, several Food and Drug Administration–approved (FDA-approved) NSAIDs, including ibuprofen, sulindac, and indomethacin, have been shown to be selective Aβ42-lowering agents (49). Moreover, long-term treatment of APP transgenic mice with ibuprofen attenuates Aβ deposition (50). Although the mechanisms by which these NSAIDs lower Aβ42 have not been established, the effect is independent of cyclooxygenase inhibition, which is the primary anti-inflammatory target of these compounds (49). These substances do not change the total level of Aβ produced but, rather, shift cleavage from Aβ42 to a shorter 38–amino acid Aβ peptide (Aβ38). This finding suggests that they are interacting with γ-secretase. Although the contribution of Aβ38 to AD pathology is not known, it is generally accepted that Aβ42 is the pathogenic Aβ species (9). Therefore, investigators believe that lowering Aβ42 levels is

a therapeutic strategy worthy of further investigation. The implications of these findings, with respect to the therapeutic potential of anti-inflammatory agents, will be discussed shortly.

Cholesterol-altering drugs. Epidemiologic data and data from model systems indicate that cholesterol-altering drugs may have an impact on the development of AD, and that this effect could be attributed to effects on Aβ accumulation. Retrospective studies on β-hydroxy-β-methylglutaryl-coenzyme A (HMG-CoA) reductase inhibitors (statins) show a large reduction in the risk for developing AD in individuals taking these drugs (51, 52), whereas individuals with elevated cholesterol are at higher risk for the development of AD (53–56). In culture and animal model systems, statins and other cholesterol-lowering agents decrease Aβ levels and Aβ deposition (57–59), whereas high-cholesterol diets in APP transgenic mice increase Aβ deposition (60). In addition, inhibitors of acetyl coenzyme A:cholesterol acyltransferase (ACAT), the enzyme that converts free cholesterol to cholesterol esters, also appear to decrease Aβ production (61).

Cholesterol's role in Aβ metabolism appears to be quite complex and is the subject of recent reviews (62, 63). Cholesterol-modulating drugs could influence Aβ deposition by (a) directly influencing Aβ production through alterations in secretase activity, (b) directly altering Aβ deposition, or (c) indirectly influencing Aβ deposition by altering levels of factors such as apoE. Alternatively, it is possible that the beneficial effect of cholesterol-lowering drugs on AD is related not to effects on Aβ, but rather to the fact that a CNS ischemic event can convert preclinical AD to clinically diagnosable dementia (64). It is worth noting that in a prospective population-based study, high systolic blood pressure was associated with a higher relative risk for AD than elevated serum cholesterol levels were (56). Nevertheless, regardless of the mechanism, treatment with statins or other cholesterol-altering agents may have a significant clinical benefit in the prevention of AD.

The complex interaction of cholesterol with Aβ indicates that there are many potential ways to alter Aβ metabolism. Other examples of the complex effects of drugs on Aβ metabolism include the action of the PI3K inhibitor wortmannin (65). Wortmannin inhibits Aβ production, both in cells and in vivo, apparently by altering APP trafficking. Although such drugs do not selectively target Aβ, if these compounds are relatively nontoxic (which is not the case for wortmannin), they are reasonable candidates for anti-Aβ therapy.

Therapies targeting Aβ aggregation or removal. Because Aβ aggregation appears essential for the initiation of the AD pathogenic cascade, it may also be possible to prevent AD by altering Aβ aggregation or removing aggregates that are already formed. A number of research groups are currently exploring the development of direct Aβ aggregation inhibitors (66). While some encouraging results have been reported in animal models (67, 68), these compounds are peptide-like and unlikely to make good drugs. An alternative strategy for altering Aβ aggregation was reported recently. In APP transgenic mice treated with clioquinol (an antibiotic and bioavailable Cu/Zn chelator), marked reduction in Aβ deposition occurred after several months of treatment (69). Zinc and other divalent cations appear necessary for Aβ aggregation (70). Thus, metal chelation may have some therapeutic benefit in the treatment of AD, either by preventing Aβ aggregation or by disrupting preformed aggregates. Clioquinol is a reasonably well tolerated drug in humans and is currently in a phase II clinical trial for AD.

One of the most surprising developments in anti-Aβ therapy is Aβ immunization. Direct immunization with aggregated Aβ42 was originally shown to attenuate Aβ deposition significantly in APP transgenic mice (71). Aβ immunization now appears to be effective in reducing amyloid deposition in multiple mouse models when mice are immunized, either actively with Aβ, or passively with intact anti-Aβ antibodies (72–76). However, it appears that there are some limits to the ability of immunization to clear existing plaques. Immunization of mice with large initial amyloid loads does not have a significant impact on amyloid deposition (76). Whether this lack of clearance can be attributed to an inherent limitation of the immunization approach or to the lack of production of sufficient amounts of anti-Aβ to clear large amounts of Aβ is unknown. In the latter case, one would postulate that simply increasing the amount of anti-Aβ would cause more Aβ to be cleared. Significantly, several groups have shown that, even in the apparent absence of any effect on Aβ load in the brain, Aβ immunization can ameliorate a cognitive deficit in reference memory and working spatial memory in APP transgenic mice (74, 77). This suggests that, even in the absence of Aβ reduction, immunization may have some therapeutic effect. However, given that the relationship between memory deficits observed in these mice and those in humans with AD is unknown, the significance of this behavioral correction in mice is unclear.

Of interest are experiments showing that the local application of anti-Aβ to the brain can result in rapid clearance and resolution of the plaques, along with a robust microglial infiltration and activation (78). Based on these observations, it may be possible to rapidly clear existing Aβ deposits, at least in mouse models, given sufficient local concentrations of anti-Aβ in the brain. While questions regarding mechanisms abound, it is thought that antibodies to Aβ do one or more of the following: (a) enhance clearance of Aβ, (b) disrupt Aβ fibrils, (c) prevent Aβ fibril formation, and/or (d) block the toxic effects of Aβ aggregates.

Although the initial phase I trial of Aβ42 immunization in humans was well tolerated, the discontinuation of the phase II trial due to meningio-encephalitic presentation in about 5% of the study group represents a severe setback for direct immunization strategies. Unfortunately, due to the paucity of information on the nature of the side effects, all hypotheses regarding the nature of the postvaccination syndrome remain highly speculative (79). Very recent data now show that one patient with the postvaccination syndrome did have modest anti-Aβ titers and high levels of anti-Aβ antibodies in the cerebrospinal fluid (CSF) (80, 81). This patient did respond to steroid-induced immunosuppression, but the anti-Aβ titers remained unchanged after recovery. Although it has been suggested that the high CSF anti-Aβ titers caused disease in this individual, it is equally likely that the high CSF titers were the result of the meningio-encephalitic presentation, perhaps due to a T cell response against Aβ or APP.

Studies of Aβ metabolism reveal a number of potential therapeutic strategies that may alter Aβ accumulation in the AD brain. Agents targeting Aβ-induced cascades are also being evaluated; however, it is much more difficult to determine the potential efficacy of these, since the APP mouse models do not demonstrate all of the pathological features apparent in the AD brain. Moreover, because of the lack of clarity regarding how Aβ leads to neuronal dysfunction and death, most therapeutic modalities targeting downstream effects of Aβ are not necessarily specific to AD. Thus, agents such as antioxidants, neurotrophic factors, apoptosis inhibitors, and other neuroprotective agents may all be of benefit in the treatment of AD. They are also likely to be of general utility in other neurodegenerative conditions.

One intriguing modality currently being considered for AD treatment is the use of NSAIDs. Multiple epidemiologic studies support a role for the use of NSAIDs in preventing the development of AD. Based on the known mechanism of action of NSAIDs and the evidence for an Aβ-induced inflammatory cascade in the AD brain, it is proposed that the anti-inflammatory property of these drugs is responsible for their apparent benefit to patients with AD. The recent data demonstrating that some NSAIDs can selectively lower Aβ42 raise the possibility that this mechanism, rather than the anti-inflammatory property of these compounds, confers protection (49). Alternatively, it may be the anti-inflammatory property, or a combination of the anti-inflammatory and Aβ42-lowering properties, that confers protection. Significantly, several NSAIDs are currently being tested for efficacy in either treating or preventing AD. However, only one of the current trials is using an NSAID, ibuprofen, that potentially lowers Aβ42.

Primary prevention or therapeutic intervention?

The epidemiologic data supporting a protective role for both statins and NSAIDs in AD indicate that primary prevention of AD may be feasible. Whether therapeutic intervention is likely to have a disease-modifying effect is much more controversial. Definitive insight into the temporal progression of AD is lacking. However, evidence (gathered from the study of AD-like pathological features in the brains of patients with Down syndrome) suggests that Aβ accumulation precedes clinical cognitive impairment by many years, or even decades (82). Moreover, recent data using registered volumetric MRI show that brain atrophy begins prior to the onset of symptoms in carriers of AD-linked PS mutations (83, 84). If Aβ is deposited for years before symptoms appear, then what is the trigger for the cognitive decline? Since the trigger is unknown, it is quite likely that anti-Aβ therapy will be more effective when implemented prior to the onset of disease symptoms.

Predicting the risk for development of AD

Although it is possible that anti-Aβ agents will demonstrate disease modification when they are given to patients with early stages of AD or mild cognitive impairment, failure to show efficacy in these patients should not prevent further study of these drugs' usefulness in the prevention of AD. However, several obstacles will impede the rapid testing of agents in preventive trials if they fail in initial therapeutic trials. The first is that any drug used for primary prevention must be safe. Because of this, drugs such as statins and NSAIDs, which are well tolerated, will likely be tested in a preventative paradigm before novel secretase inhibitors or immunization approaches are tested. The second obstacle centers on the lack of ability to predict who is likely to develop AD. Thousands of subjects are needed for primary prevention studies, which are therefore both difficult to conduct and very expensive. Improvement in the ability to predict who is likely to develop the disease will allow trials on only those individuals at high risk for AD, clearly helping primary prevention studies. Although predictive tests for AD are lagging behind therapeutic advances, future developments in AD genetics, discovery or validation of predictive biomarkers, and advances in imaging are likely to identify individuals at high risk for development of AD, thus enabling primary prevention trials to be conducted in a more cost-effective and efficient manner.

Identifying genes that confer risk for LOAD. Concordance rates in LOAD between identical twins are quite high, suggesting that AD has a very significant genetic component (85, 86). Moreover, family history of AD is a very important risk factor. ApoE4, located on chromosome 19, is currently the only established genetic risk factor for LOAD. However, more than 50% of individuals who develop LOAD lack an apoE4 allele, and some apoE4 homozygotes do not develop AD (87). Additional potential AD-risk genes have been identified on chromosomes 6, 9, 10, and 12; association studies also implicate other candidate genes as risk factors for AD (88). A locus on chromosome 10 is currently under intense scrutiny, because three independent analyses identify this region as one likely to contain a gene or genes that contributes significantly to AD risk (89–91). Of note, one of these studies found evidence for linkage using plasma Aβ42 levels as an intermediate phenotype for AD (90). New data on the human genome will speed discovery of genes involved in LOAD. In the future, analyses of multiple genes will play a significant role in identifying individuals at high risk for developing AD.

Predictive biomarkers. CSF Aβ, tau, and isoprostane measurements are receiving attention as potential diagnostic biomarkers for AD (92–95). Given the accuracy of current clinical diagnoses of AD, such diagnostic procedures are not likely to significantly impact the management of AD. What is really needed is a predictive biomarker that can be used to predict risk for developing AD, and also to monitor treatment. Although measurement of plasma Aβ levels is still a long way from finding a place in the clinical management of AD, there is emerging evidence that such measurements may be useful, both in predicting risk for developing AD and for monitoring anti-Aβ therapy (90, 96). Plasma Aβ42 levels are elevated in patients with AD-causing mutations in APP, PS1, or PS2 (12). In at least one study, patients with higher plasma Aβ levels developed AD more rapidly than those with lower Aβ levels (97, 98). Moreover, it appears that plasma Aβ levels are at least partially genetically determined (96). Ongoing prospective studies should enable a more complete evaluation of the potential role of plasma Aβ as a predictive biomarker for the development of AD.

Imaging. If Aβ deposition precedes clinical AD by many years, it should be possible to monitor presymptomatic amyloid deposition. Recent studies show that Aβ can be imaged in transgenic mice, following peripheral administration of amyloid-binding compounds, which suggests that imaging of amyloid deposits in living patients will be feasible in the near future (78, 99, 100). Moreover, advances in MRI resolution may one day make imaging amyloid as routine as MRI scans are today (101, 102). Such advances will be extremely useful, because they may identify individuals at risk for imminent development of AD.

Monitoring disease progression using biomarkers and imaging

Any disease-modifying therapy must show clinical efficacy in order to be approved for the treatment or prevention of AD. This means that the rate of cognitive decline must be decreased or halted. Until that hurdle is overcome, ancillary studies examining effects of any therapy on Aβ or other biomarker levels in the plasma or CSF, on amyloid load in the brain, or on brain atrophy are largely meaningless. Once a convincing link is established between changes in any of these biomarkers and a clinically appreciable disease-modifying effect, approval of future drugs, which work through a similar mechanism, may require only that they modify linked biomarkers. For this reason, it is extremely important to monitor these parameters in current and future clinical trials. Studies with novel agents that assess both biomarkers and cognitive outcomes may be much more informative than studies that only assess clinical outcomes.

Summary

Cognitive enhancers that target acetylcholinesterase remain the only FDA-approved therapies for the treatment of the cognitive decline in AD. Such therapy is unlikely to modify the course of the disease to any significant extent. In contrast, therapies currently being developed that are based on an increased understanding of the pathogeneses of AD are likely to have disease-modifying effects. Given the plethora of potential targets, it is likely that successful anti-Aβ therapies will emerge. The major challenge that remains is to show that such therapies actually alter cognitive decline in humans. The medical community should be cautious in evaluating the efficacy of anti-Aβ drugs, as they may not show such disease-modifying effects when given in therapeutic trials. To restate the analogy to atherosclerotic disease, by the time a patient is experiencing angina, the patient needs a bypass or angioplasty, not a cholesterol-lowering agent (although after intervention such an agent would be appropriate). Similarly, in AD, by the time a patient is symptomatic, Aβ-lowering therapies may not be effective. We must hope that advances in diagnostic prediction and monitoring of disease progression proceed with a pace that equals the advances currently being made in developing AD therapeutics that target Aβ. If they do, then it is likely that AD will become manageable through a combination of presymptomatic screening, early therapeutic intervention, and vigilant monitoring of the effectiveness of treatment.

Address correspondence to: Todd E. Golde, Department of Neuroscience, Mayo Clinic Jacksonville, 4500 San Pablo Road, Jacksonville, Florida 32224, USA. Phone: (904) 953-2538; Fax: (904) 953-7370; E-mail: tgolde@mayo.edu.

1. Small, G.W., et al. 1997. Diagnosis and treatment of Alzheimer disease and related disorders. Consensus statement of the American Association for Geriatric Psychiatry, the Alzheimer's Association, and the American Geriatrics Society. *JAMA.* **278**:1363–1371.

2. Rice, D.P., et al. 2001. Prevalence, costs, and treatment of Alzheimer's disease and related dementia: a managed care perspective. *Am. J. Manag. Care.* **7**:809–818.

3. Fillit, H., Hill, J.W., and Fufferman, R. 2002. Health care utilization and costs of Alzheimer's disease: the role of co-morbid conditions, disease stage, and pharmacotherapy. *Fam. Med.* **34**:528–535.

4. Crook, R., et al. 1998. A variant of Alzheimer's disease with spastic paraparesis and unusual plaques due to deletion of exon 9 of presenilin 1. *Nat. Med.* **4**:452–455.

5. Kwok, J.B., et al. 1997. Two novel (M233T and R278T) presenilin-1 mutations in early-onset Alzheimer's disease pedigrees and preliminary evidence for association of presenilin-1 mutations with a novel phenotype. *Neuroreport.* **8**:1537–1542.

6. Golde, T.E., Eckman, C.B., and Younkin, S.G. 2000. Biochemical detection of Aβ isoforms: implications for pathogenesis, diagnosis, and treatment of Alzheimer's disease. *Biochim. Biophys. Acta.* **1502**:172–187.

7. Golde, T.E., and Younkin, S.G. 2001. Presenilins as therapeutic targets for the treatment of Alzheimer's disease. *Trends Mol. Med.* **7**:264–269.

8. Hardy, J. 1997. Amyloid, the presenilins and Alzheimer's disease. *Trends Neurosci.* **20**:154–159.

9. Younkin, S.G. 1998. The role of A beta 42 in Alzheimer's disease. *J. Physiol. Paris.* **92**:289–292.

10. Selkoe, D.J. 2001. Alzheimer's disease: genes, proteins, and therapy. *Physiol. Rev.* **81**:741–766.

11. Nilsberth, C., et al. 2001. The 'Arctic' APP mutation (E693G) causes Alzheimer's disease by enhanced Abeta protofibril formation. *Nat. Neurosci.* **4**:887–893.

12. Scheuner, D., et al. 1996. Secreted amyloid beta-protein similar to that in the senile plaques of Alzheimer's disease is increased in vivo by the presenilin 1 and 2 and APP mutations linked to familial Alzheimer's disease. *Nat. Med.* **2**:864–870.

13. Holtzman, D.M., et al. 2000. Apolipoprotein E facilitates neuritic and cerebrovascular plaque formation in an Alzheimer's disease model. *Ann. Neurol.* **47**:739–747.

14. Jarrett, J.T., Berger, E.P., and Lansbury, P.T., Jr. 1993. The carboxy terminus of β amyloid protein is critical for the seeding of amyloid formation: implications for pathogenesis of Alzheimer's disease. *Biochemistry.* **32**:4693–4697.

15. Price, D.L., and Sisodia, S.S. 1998. Mutant genes in familial Alzheimer's disease and transgenic models. *Annu. Rev. Neurosci.* **21**:479–505.

16. Yankner, B.A. 1996. Mechanisms of neuronal degeneration in Alzheimer's disease. *Neuron.* **16**:921–932.

17. Vidal, R., et al. 1999. A stop-codon mutation in the BRI gene associated with familial British dementia. *Nature.* **399**:776–781.

18. Vidal, R., et al. 2000. A decamer duplication in the 3′ region of the BRI gene originates an amyloid peptide that is associated with dementia in a Danish kindred. *Proc. Natl. Acad. Sci. USA.* **97**:4920–4925.

19. Hardy, J.A., and Higgins, G.A. 1992. Alzheimer's disease: the amyloid cascade hypothesis. *Science.* **256**:184–185.

20. Lambert, M.P., et al. 1998. Diffusible, nonfibrillar ligands derived from Abeta1-42 are potent central nervous system neurotoxins. *Proc. Natl. Acad. Sci. USA.* **95**:6448–6453.

21. Walsh, D.M., Lomakin, A., Benedek, G.B., Condron, M.M., and Teplow, D.B. 1997. Amyloid beta-protein fibrillogenesis. Detection of a protofibrillar

intermediate. *J. Biol. Chem.* **272**:22364–22372.

22. Lashuel, H.A., Hartley, D., Petre, B.M., Walz, T., and Lansbury, P.T., Jr. 2002. Neurodegenerative disease: amyloid pores from pathogenic mutations. *Nature.* **418**:291.

23. LaFerla, F.M., Tinkle, B.T., Bieberich, C.J., Haudenschild, C.C., and Jay, G. 1995. The Alzheimer's A beta peptide induces neurodegeneration and apoptotic cell death in transgenic mice. *Nat. Genet.* **9**:21–30.

24. Gouras, G.K., et al. 2000. Intraneuronal Abeta42 accumulation in human brain. *Am. J. Pathol.* **156**:15–20.

25. Price, D.L., Tanzi, R.E., Borchelt, D.R., and Sisodia, S.S. 1998. Alzheimer's disease: genetic studies and transgenic models. *Annu. Rev. Genet.* **32**:461–493.

26. Small, D.H., Mok, S.S., and Bornstein, J.C. 2001. Alzheimer's disease and Abeta toxicity: from top to bottom. *Nat. Rev. Neurosci.* **2**:595–598.

27. Akiyama, H., et al. 2000. Inflammation and Alzheimer's disease. *Neurobiol. Aging.* **21**:383–421.

28. Varadarajan, S., Yatin, S., Aksenova, M., and Butterfield, D.A. 2000. Review: Alzheimer's amyloid beta-peptide-associated free radical oxidative stress and neurotoxicity. *J. Struct. Biol.* **130**:184–208.

29. Lee, V.M.-Y., and Trojanowski, J.Q. 1992. The disordered neuronal cytoskeleton in Alzheimer's disease. *Curr. Opin. Neurobiol.* **2**:653–656.

30. Hutton, M., et al. 1998. Association of missense and 5′-splice-site mutations in tau with the inherited dementia FTDP-17. *Nature.* **393**:702–705.

31. Spillantini, M.G., et al. 1998. Mutation in the tau gene in familial multiple system tauopathy with presenile dementia. *Proc. Natl. Acad. Sci. USA.* **95**:7737–7741.

32. Lewis, J., et al. 2000. Neurofibrillary tangles, amyotrophy and progressive motor disturbance in mice expressing mutant (P301L) tau protein. *Nat. Genet.* **25**:402–405.

33. Gotz, J., Chen, F., van Dorpe, J., and Nitsch, R.M. 2001. Formation of neurofibrillary tangles in P301l tau transgenic mice induced by Abeta 42 fibrils. *Science.* **293**:1491–1495.

34. Lewis, J., et al. 2001. Enhanced neurofibrillary degeneration in transgenic mice expressing mutant tau and APP. *Science.* **293**:1487–1491.

35. Evin, G., et al. 2001. Aspartyl protease inhibitor pepstatin binds to the presenilins of Alzheimer's disease. *Biochemistry.* **40**:8359–8368.

36. Li, Y.M., et al. 2000. Photoactivated gamma-secretase inhibitors directed to the active site covalently label presenilin 1. *Nature.* **405**:689–694.

37. Esler, W.P., et al. 2000. Transition-state analogue inhibitors of gamma-secretase bind directly to presenilin-1. *Nat. Cell Biol.* **2**:428–434.

38. Seiffert, D., et al. 2000. Presenilin-1 and -2 are molecular targets for gamma-secretase inhibitors. *J. Biol. Chem.* **275**:34086–34091.

39. Dovey, H., Varghese, J., and Anderson, J.P. 2000. Functional gamma-secretase inhibitors reduce beta-amyloid peptide levels in the brain. *J. Neurochem.* **76**:1–10.

40. De Strooper, B., et al. 1999. A presenilin-1-dependent gamma-secretase-like protease mediates release of Notch intracellular domain. *Nature.* **398**:518–522.

41. Hadland, B.K., et al. 2001. Gamma-secretase inhibitors repress thymocyte development. *Proc. Natl. Acad. Sci. USA.* **98**:7487–7491.

42. Ni, C.Y., Murphy, M.P., Golde, T.E., and Carpenter, G. 2001. γ-Secretase cleavage and nuclear localization of ErbB-4 receptor tyrosine kinase. *Science.* **294**:2179–2181.

43. Kammesheidt, A., et al. 1992. Deposition of β/A4 immunoreactivity and neuronal pathology in transgenic mice expressing the carboxyl-terminal fragment of the Alzheimer amyloid precursor in the brain. *Proc. Natl. Acad. Sci. USA.* **89**:10857–10861.

44. Yankner, B.A. 1989. Neurotoxicity of a fragment of the amyloid precursor associated with Alzheimer's disease. *Science.* **245**:417–420.

45. Cai, H., et al. 2001. BACE1 is the major beta-secretase for generation of Abeta peptides by neurons. *Nat. Neurosci.* **4**:233–234.

46. Luo, Y., et al. 2001. Mice deficient in BACE1, the Alzheimer's beta-secretase, have normal phenotype and abolished beta-amyloid generation. *Nat. Neurosci.* **4**:231–232.

47. Hong, L., et al. 2002. Crystal structure of memapsin 2 (beta-secretase) in complex with an inhibitor OM00-3. *Biochemistry.* **41**:10963–10967.

48. Hong, L., et al. 2000. Structure of the protease domain of memapsin 2 (beta-secretase) complexed with inhibitor. *Science.* **290**:150–153.

49. Weggen, S., et al. 2001. A subset of NSAIDs lower amyloidogenic Abeta42 independently of cyclooxygenase activity. *Nature.* **414**:212–216.

50. Lim, G.P., et al. 2000. Ibuprofen suppresses plaque pathology and inflammation in a mouse model for Alzheimer's disease. *J. Neurosci.* **20**:5709–5714.

51. Wolozin, B., Kellman, W., Ruosseau, P., Celesia, G.G., and Siegel, G. 2000. Decreased prevalence of Alzheimer disease associated with 3-hydroxy-3- methyglutaryl coenzyme A reductase inhibitors. *Arch. Neurol.* **57**:1439–1443.

52. Jick, H., Zornberg, G.L., Jick, S.S., Seshadri, S., and Drachman, D.A. 2000. Statins and the risk of dementia. *Lancet.* **356**:1627–1631.

53. Jarvik, G.P., et al. 1995. Interactions of apolipoprotein E genotype, total cholesterol level, age, and sex in prediction of Alzheimer's disease: a case-control study. *Neurology.* **45**:1092–1096.

54. Notkola, I.L., et al. 1998. Serum total cholesterol, apolipoprotein E epsilon 4 allele, and Alzheimer's disease. *Neuroepidemiology.* **17**:14–20.

55. Hofman, A., et al. 1997. Atherosclerosis, apolipoprotein E, and prevalence of dementia and Alzheimer's disease in the Rotterdam Study. *Lancet.* **349**:151–154.

56. Kivipelto, M., et al. 2001. Midlife vascular risk factors and Alzheimer's disease in later life: longitudinal, population based study. *BMJ.* **322**:1447–1451.

57. Simons, M., et al. 1998. Cholesterol depletion inhibits the generation of beta-amyloid in hippocampal neurons. *Proc. Natl. Acad. Sci. USA.* **95**:6460–6464.

58. Refolo, L.M., et al. 2001. A cholesterol-lowering drug reduces beta-amyloid pathology in a transgenic mouse model of Alzheimer's disease. *Neurobiol. Dis.* **8**:890–899.

59. Fassbender, K., et al. 2001. Simvastatin strongly reduces levels of Alzheimer's disease beta-amyloid peptides Abeta 42 and Abeta 40 in vitro and in vivo. *Proc. Natl. Acad. Sci. USA.* **98**:5856–5861.

60. Refolo, L.M., et al. 2000. Hypercholesterolemia accelerates the Alzheimer's amyloid pathology in a transgenic mouse model. *Neurobiol. Dis.* **7**:321–331.

61. Puglielli, L., et al. 2001. Acyl-coenzyme A: cholesterol acyltransferase modulates the generation of the amyloid beta-peptide. *Nat. Cell Biol.* **3**:905–912.

62. Golde, T.E., and Eckman, C.B. 2001. Cholesterol modulation as an emerging strategy for the treatment of Alzheimer's disease. *Drug Discov. Today.* **6**:1049–1055.

63. Wolozin, B. 2001. A fluid connection: cholesterol and Abeta. *Proc. Natl. Acad. Sci. USA.* **98**:5371–5373.

64. Snowdon, D.A., et al. 1997. Brain infarction and the clinical expression of Alzheimer disease. The Nun Study. *JAMA.* **277**:813–817.

65. Haugabook, S.J., et al. 2001. Reduction of Abeta accumulation in the Tg2576 animal model of Alzheimer's disease after oral administration of the phosphatidyl-inositol kinase inhibitor wortmannin. *FASEB J.* **15**:16–18.

66. Soto, C. 1999. Plaque busters: strategies to inhibit amyloid formation in Alzheimer's disease. *Mol. Med. Today.* **5**:343–350.

67. Permanne, B., et al. 2002. Reduction of amyloid load and cerebral damage in a transgenic mouse model of Alzheimer's disease by treatment with a beta-sheet breaker peptide. *FASEB J.* **16**:860–862.
68. Sigurdsson, E.M., Permanne, B., Soto, C., Wisniewski, T., and Frangione, B. 2000. In vivo reversal of amyloid-beta lesions in rat brain. *J. Neuropathol. Exp. Neurol.* **59**:11–17.
69. Cherny, R.A., et al. 2001. Treatment with a copper-zinc chelator markedly and rapidly inhibits beta-amyloid accumulation in Alzheimer's disease transgenic mice. *Neuron.* **30**:665–676.
70. Curtain, C.C., et al. 2001. Alzheimer's disease amyloid-beta binds copper and zinc to generate an allosterically ordered membrane-penetrating structure containing superoxide dismutase-like subunits. *J. Biol. Chem.* **276**:20466–20473.
71. Schenk, D., et al. 1999. Immunization with amyloid-beta attenuates Alzheimer-disease-like pathology in the PDAPP mouse. *Nature.* **400**:173–177.
72. Lemere, C.A., et al. 2000. Nasal A beta treatment induces anti-A beta antibody production and decreases cerebral amyloid burden in PD-APP mice. *Ann. NY Acad. Sci.* **920**:328–331.
73. DeMattos, R.B., et al. 2001. Peripheral anti-A beta antibody alters CNS and plasma A beta clearance and decreases brain A beta burden in a mouse model of Alzheimer's disease. *Proc. Natl. Acad. Sci. USA.* **98**:8850–8855.
74. Janus, C., et al. 2000. A beta peptide immunization reduces behavioural impairment and plaques in a model of Alzheimer's disease. *Nature.* **408**:979–982.
75. Bard, F., et al. 2000. Peripherally administered antibodies against amyloid beta-peptide enter the central nervous system and reduce pathology in a mouse model of Alzheimer disease. *Nat. Med.* **6**:916–919.
76. Das, P., Murphy, M.P., Younkin, L.H., Younkin, S.G., and Golde, T.E. 2001. Reduced effectiveness of Abeta1-42 immunization in APP transgenic mice with significant amyloid deposition. *Neurobiol. Aging.* **22**:721–727.
77. Morgan, D., et al. 2000. A beta peptide vaccination prevents memory loss in an animal model of Alzheimer's disease. *Nature.* **408**:982–985.
78. Bacskai, B.J., et al. 2001. Imaging of amyloid-beta deposits in brains of living mice permits direct observation of clearance of plaques with immunotherapy. *Nat. Med.* **7**:369–372.
79. Das, P., and Golde, T.E. 2002. Open peer commentary regarding Abeta immunization and CNS inflammation by Pasinetti et al. *Neurobiol. Aging.* **23**:671–674.
80. Hock, C., et al. 2002. Generation of antibodies specific for beta-amyloid by vaccination of patients with Alzheimer disease. *Nat. Med.* **8**:1270–1275.
81. Schenk, D. 2002. Opinion. Amyloid-beta immunotherapy for Alzheimer's disease: the end of the beginning. *Nat. Rev. Neurosci.* **3**:824–828.
82. Lemere, C.A., et al. 1996. Sequence of deposition of heterogeneous amyloid beta-peptides and APO E in Down syndrome: implications for initial events in amyloid plaque formation. *Neurobiol. Dis.* **3**:16–32.
83. Chan, D., et al. 2001. Rates of global and regional cerebral atrophy in AD and frontotemporal dementia. *Neurology.* **57**:1756–1763.
84. Fox, N.C., et al. 2001. Imaging of onset and progression of Alzheimer's disease with voxel-compression mapping of serial magnetic resonance images. *Lancet.* **358**:201–205.
85. Breitner, J.C., et al. 1994. Alzheimer's disease in the NAS-NRC registry of aging twin veterans. II. Longitudinal findings in a pilot series. National Academy of Sciences. National Research Council Registry. *Dementia.* **5**:99–105.
86. Rubinsztein, D.C. 1997. The genetics of Alzheimer's disease. *Prog. Neurobiol.* **52**:447–454.
87. Roses, A.D. 1997. Apolipoprotein E, a gene with complex biological interactions in the aging brain. *Neurobiol. Dis.* **4**:170–185.
88. Kehoe, P., et al. 1999. A full genome scan for late onset Alzheimer's disease. *Hum. Mol. Genet.* **8**:237–245.
89. Myers, A., et al. 2000. Susceptibility locus for Alzheimer's disease on chromosome 10. *Science.* **290**:2304–2305.
90. Ertekin-Taner, N., et al. 2000. Linkage of plasma Abeta42 to a quantitative locus on chromosome 10 in late-onset Alzheimer's disease pedigrees. *Science.* **290**:2303–2304.
91. Bertram, L., et al. 2000. Evidence for genetic linkage of Alzheimer's disease to chromosome 10q. *Science.* **290**:2302–2303.
92. Galasko, D. 2001. Biological markers and the treatment of Alzheimer's disease. *J. Mol. Neurosci.* **17**:119–125.
93. Boss, M.A. 2000. Diagnostic approaches to Alzheimer's disease. *Biochim. Biophys. Acta.* **1502**:188–200.
94. Pratico, D., et al. 2000. Increased 8,12-iso-iPF2alpha-VI in Alzheimer's disease: correlation of a noninvasive index of lipid peroxidation with disease severity. *Ann. Neurol.* **48**:809–812.
95. Montine, T.J., et al. 2001. Cerebrospinal fluid abeta42, tau, and f2-isoprostane concentrations in patients with Alzheimer disease, other dementias, and in age-matched controls. *Arch. Pathol. Lab. Med.* **125**:510–512.
96. Ertekin-Taner, N., et al. 2001. Heritability of plasma amyloid beta in typical late-onset Alzheimer's disease pedigrees. *Genet. Epidemiol.* **21**:19–30.
97. Mayeux, R., et al. 1999. Plasma amyloid beta-peptide 1-42 and incipient Alzheimer's disease. *Ann. Neurol.* **46**:412–416.
98. Schupf, N., et al. 2001. Elevated plasma amyloid beta-peptide 1-42 and onset of dementia in adults with Down syndrome. *Neurosci. Lett.* **301**:199–203.
99. Skovronsky, D.M., et al. 2000. In vivo detection of amyloid plaques in a mouse model of Alzheimer's disease. *Proc. Natl. Acad. Sci. USA.* **97**:7609–7614.
100. Klunk, W.E., et al. 2002. Imaging Abeta plaques in living transgenic mice with multiphoton microscopy and methoxy-X04, a systemically administered Congo red derivative. *J. Neuropathol. Exp. Neurol.* **61**:797–805.
101. Jack, C.R., Jr., et al. 2002. Antemortem MRI findings correlate with hippocampal neuropathology in typical aging and dementia. *Neurology.* **58**:750–757.
102. Bacskai, B.J., Klunk, W.E., Mathis, C.A., and Hyman, B.T. 2002. Imaging amyloid-beta deposits in vivo. *J. Cereb. Blood Flow Metab.* **22**:1035–1041.

Diagnosis and treatment of Parkinson disease: molecules to medicine

Joseph M. Savitt,[1,2] Valina L. Dawson,[1,2,3,4] and Ted M. Dawson[1,2,3]

[1]Institute for Cell Engineering, [2]Department of Neurology, [3]Department of Neuroscience, and
[4]Department of Physiology, Johns Hopkins University School of Medicine, Baltimore, Maryland, USA.

Parkinson disease (PD) is a relatively common disorder of the nervous system that afflicts patients later in life with tremor, slowness of movement, gait instability, and rigidity. Treatment of these cardinal features of the disease is a success story of modern science and medicine, as a great deal of disability can be alleviated through the pharmacological correction of brain dopamine deficiency. Unfortunately these therapies only provide temporary, though significant, relief from early symptoms and do not halt disease progression. In addition, pathological changes outside of the motor system leading to cognitive, autonomic, and psychiatric symptoms are not sufficiently treated by current therapies. Much as the discovery of dopamine deficiency led to powerful treatments for motor symptoms, recent discoveries concerning the role of specific genes in PD pathology will lead to the next revolution in disease therapy. Understanding why and how susceptible cells in motor and nonmotor regions of the brain die in PD is the first step toward preventing this cell death and curing or slowing the disease. In this review we discuss recent discoveries in the fields of diagnosis and treatment of PD and focus on how a better understanding of disease mechanisms gained through the study of monogenetic forms of PD has provided novel therapeutic targets.

History

Parkinson disease (PD) is a chronic, progressive neurodegenerative disorder that affects at least 1% of people by age 70 (1–3). James Parkinson provided the first detailed description of the disease in his 1817 monograph "An Essay on the Shaking Palsy." In the latter part of the nineteenth century, Charcot further refined the description of this disorder and identified the cardinal clinical features of PD including rest tremor, rigidity, balance impairment, and slowness of movement (reviewed in ref. 4). An early clue to the pathology of the disease came from Brissaud, who speculated that damage in the substantia nigra (SN) might lead to PD (5, 6). Eosinophilic inclusions (Lewy bodies) later were identified in the brains of PD patients (7) and, along with abnormalities in the SN, became a recognized pathologic marker of the disease (8).

A major advance in the understanding of PD came when dopamine deficiency was discovered in the corpus striatum and SN of brains taken from patients (9). Later studies demonstrated the connection between the SN and the striatum, thus suggesting that dopaminergic cell loss in the SN directly leads to dopaminergic deficiency in the striatum (10). The determination that PD is a disease of dopamine loss led to the development of rational therapies aimed at correcting this deficiency (11). After some

Nonstandard abbreviations used: COMT, catechol-O-methyl transferase; CoQ10, coenzyme Q10; DBS, deep brain stimulation; GDNF, glial cell line–derived neurotrophic factor; LRRK2, leucine-rich repeat kinase 2; MAO-B, monoamine oxidase type B; MPTP, 1-methyl-4-phenyl-1,2,3,6-tetrahydropyridine; PD, Parkinson disease; PINK-1, PTEN-induced putative kinase 1; rTMS, repetitive TMS; SN, substantia nigra; SPECT, single photon emission computed tomography; TMS, transcranial magnetic stimulation; UCH-L1, ubiquitin carboxyterminal hydrolase L1.

Conflict of interest: J.M. Savitt receives support from Kyowa and Cephalon Inc. for participation in clinical research trials. Under a licensing agreement between MCI Pharmaceuticals and Johns Hopkins University, T.M. Dawson and V.L. Dawson are entitled to a share of the royalty received by the university from MCI Pharmaceuticals. The terms of this arrangement are being managed by Johns Hopkins University in accordance with its conflict-of-interest policies. T.M. Dawson is a consultant for AnGes Inc., Mylan Laboratories Inc., and Boehringer Ingelheim Pharmaceuticals, Inc.

Citation for this article: *J. Clin. Invest.* **116**:1744–1754 (2006). doi:10.1172/JCI29178.

initial uncertainty, the dopamine precursor levodopa proved to be a powerful PD treatment (12). Subsequent advances in therapy included combining levodopa with a peripheral decarboxylase inhibitor, such as carbidopa or benserazide (13, 14). This combination significantly reduced the nausea and vomiting associated with levodopa therapy and allowed a greater proportion of levodopa to enter the brain. Similarly, catechol-O-methyltransferase (COMT) inhibitors, which prolong the half-life of levodopa and dopamine were found to enhance the effect of a given levodopa dose (15–17). In addition to increasing the level of dopamine precursors, the focus of therapeutic design was also on limiting the breakdown of endogenous dopamine. The monoamine oxidase type B (MAO-B) inhibitor selegiline works in this fashion and provides symptomatic benefit (18). Finally, the development of dopamine agonists that directly stimulate postsynaptic dopamine receptors, thus bypassing dopamine synthesis completely, further illustrates how novel therapies can be borne from knowledge of pathology (19, 20).

Surgical therapies that reduce tremor and rigidity in PD patients were used prior to the advent of levodopa treatment. Meyers pioneered surgical lesioning procedures that targeted symptoms and spared patients from the hemiparesis that resulted from earlier surgical approaches. These procedures largely were abandoned once levodopa therapy became more common (21). Recent advances in the understanding of basal ganglia physiology and the development of new technologies has led to a reemergence of surgical PD therapies in the form of deep brain stimulation (DBS) (22, 23). DBS has become increasingly common in patients whose disease is difficult to manage with medical therapy alone.

Despite these landmark advances in symptomatic PD therapy, the ability of these treatments to facilitate an acceptable quality of life for the patient wanes with time. This is due to the development of motor complications including wearing-off (the return of PD symptoms too soon after a given levodopa dose), the presence of involuntary abnormal movements (dyskinesias and dystonia), and the emergence of treatment-resistant symp-

Table 1

Diagnosing PD

Cardinal features of PD

Bradykinesia (slow movements, decrement of frequency and amplitude of repetitive movements)
Rest tremor (most commonly beginning in the hand)
Cogwheel rigidity
Postural instability (seen in later stage disease)

Associated features in support of the diagnosis

Sustained and significant levodopa effect
Reduced armswing
Difficulty rising from a low chair
Impaired olfaction
Hypophonia
Stooped posture
Depression/anxiety
Shuffling/festinating/freezing gait
Unilateral symptom onset
Difficulty turning in bed
Drooling
Constipation
Micrographia
Sleep disturbance/REM sleep behavior disorder
Seborrhea

Atypical features suggesting alternative diagnoses

Limited, nonsustained, or no response to levodopa and:

Symptom/sign	Consider
Early dementia with cognitive fluctuations, hallucinations, and delusions	Dementia with Lewy bodies
Early autonomic impairment and/or cerebellar ataxia, nocturnal stridor, rapid progression, dysarthria	Multiple system atrophy
Stepwise progression, prominent lower extremity involvement, known vascular disease, possible levodopa responsiveness	Vascular Parkinsonism
Cortical sensory impairment, automatic movements (alien hand phenomenon), apraxia, prominent akinetic limb dystonia	Cortical-basal ganglionic degeneration
Supranuclear gaze palsy, early falls, axial rigidity, pseudobulbar palsy, frontal lobe deficits	Progressive supranuclear palsy
Kinetic or postural tremor especially involving the head or voice, alcohol responsiveness, autosomal dominant pattern of inheritance	Essential tremor
Urinary incontinence, gait disorder, cognitive impairment, and brain imaging suggesting communicating hydrocephalus	Normal pressure hydrocephalus
Symmetric onset with previous exposure to neuroleptics or other PD-mimicking drugs	Drug-induced Parkinsonism

---

toms such as gait impairment, cognitive decline, autonomic dysfunction, and medication-induced psychosis. Clearly, the current symptomatic therapies cannot completely ameliorate later-stage symptoms, nor can they address the ongoing degeneration in the dopaminergic and nondopaminergic systems. For this reason, a good deal of current research has focused on finding the cause of dopaminergic cell loss and on exploring protective, restorative, and replacement therapies. Much in the same manner that understanding the cellular pathology of PD led to a revolution in symptomatic therapy, a better understanding of the molecular pathology of PD will lead to prevention and cure.

Diagnosis

Even with recent advances in our understanding of disease mechanisms, the diagnosis of PD is usually made based on patient history and physical examination alone. PD should be considered if a person exhibits one or more of the cardinal features of the disease, including tremor at rest, bradykinesia, rigidity, and, in more advanced cases, postural instability.

Perhaps more than any other single feature, the presence of a typical rest tremor increases the likelihood of pathologically supported PD, although approximately 20% of patients fail to develop a typical rest tremor (24). Supporting evidence also may come from a history that includes associated symptoms and the absence of findings that would suggest an alternative diagnosis (reviewed in ref. 25; see Table 1). Many of these associated symptoms include nonmotor complaints including disrupted sleep, depression, fatigue, constipation, and anxiety. Additionally, the early PD patient may complain of stiffness, slowness, tremor, and imbalance even when the neurological exam is normal (26). Finally, a significant and lasting clinical response to dopaminergic therapy is characteristic of PD, and the lack of such a response should prompt a search for alternative diagnoses. Despite careful examination, the rate of PD misdiagnosis is approximately 10-25% (27-29). Complicating diagnosis is the clinical heterogeneity of PD. Patients may present with a tremor-predominant clinical picture or lack tremor completely. Indeed patients presenting with early postural instability/gait difficulty

Table 2

Potential biomarkers of PD

Imaging

Reduced striatal metabolism on ^{18}F-deoxyglucose imaging
Reduced dopa uptake and decarboxylation on ^{18}F-dopa PET
Altered dopamine receptor binding on ^{11}C-raclopride PET
Reduced level of dopamine transporter ligand uptake
    on [123]I β-CIT SPECT
Mineral deposition in the SN on transcranial ultrasound
Cardiac denervation on cardiac scintigraphy

Clinical

Personality questionnaires
Neuropsychiatric testing
Sleep evaluation
TMS/evoked potentials
Olfactory screening

Biochemical

Genetic screening
Mitochondrial complex I measurement
α-Synuclein levels and isoforms in blood

[123]I β-CIT, iodine 123-2β-carbomethoxy-3β-(4-iodophenyl)tropane.

(PIGD) or rigidity/bradykinesia follow a more rapid course of disease than do those presenting with early tremor (30).

Inaccuracy in PD diagnosis and the desire to identify presymptomatic patients have prompted the search for disease biomarkers that include imaging techniques and laboratory-based or clinical assays. Brain imaging studies using both PET and single photon emission computed tomography (SPECT) are able to distinguish those subjects with PD from normal controls with greater than 95% sensitivity (31). The results are not as good, however, when imaging is used to distinguish PD from similar disorders such as progressive supranuclear palsy or multiple system atrophy. In these cases multiple techniques and careful data analysis may be required to clearly identify whether idiopathic PD or another disorder is the cause of a Parkinsonian syndrome (32). Studies are being conducted to search for other possible diagnostic aids in PD, of which some of the most promising use transcranial ultrasound (33), examine deficits in olfaction (34), and determine oligomeric α-synuclein in blood from PD patients (35). A list of potential biomarkers under study is shown in Table 2 (for a more detailed review see ref. 36).

The screening of affected and presymptomatic individuals for known genetic mutations may aid in PD diagnosis. With the discovery that mutations in the *leucine-rich repeat kinase 2* (*LRRK2*) gene are more common than expected in certain populations (37, 38), there is little doubt that screening patients at risk will become increasingly common. The research implications are great, as investigators will be able to follow presymptomatic patients over time to assess biomarkers, risk factors, and potential protective therapies. Until disease- and risk-modifying therapies are available, and until more is known about penetrance rates, mutation-phenotype correlations, and gene frequency, genetic screening for known mutations likely will have little clinical impact on the average PD patient.

Pathogenesis

The pathologic examination of brains from PD patients demonstrates neuronal cell loss, especially of the dopamine-rich, pig-

mented neurons in the SN, and the presence of Lewy bodies and Lewy neurites in multiple brain regions (39, 40). Lewy bodies and neurites stain with antibodies to α-synuclein, ubiquitin, and a variety of other biochemical markers and are found in many areas of the PD brain: not only the SN, but also the dorsal motor nucleus of the vagus, locus ceruleus, raphe and reticular formation nuclei, thalamus, amygdala, olfactory nuclei, pediculopontine nucleus, and cerebral cortex, among others (39, 41). Indeed pathology is present outside of the brain as well, involving autonomic and submucosal ganglia (41, 42). Despite this widespread pathology, much of the research into the PD pathogenesis has focused on the cell loss and Lewy bodies seen in the dopaminergic SN. The past focus on dopaminergic deficits is related to the prominent motor manifestations of PD for which patients seek treatment. However, it is clear that PD is more than just a syndrome of dopaminergic deficiency and that future research and therapy will need to address the multiple neuronal systems affected in PD.

Some studies suggest that environmental factors lead to PD. The occurrence of postencephalitic Parkinsonism supports this view, as a particular viral infection in the early twentieth century placed patients at a higher risk for developing nigral cell loss and some Parkinsonian clinical features (43). Some pathologic features were atypical of PD, however, including a lack of Lewy bodies and a prominence of neurofibrillary tangles (44). In addition, the discovery of toxins that induce a Parkinsonian condition both in animal models and in humans further supports the possibility of an environmental trigger. The most widely studied toxin is a meperidine analog, 1-methyl-4-phenyl-1,2,3,6-tetrahydropyridine (MPTP), which, when mistakenly injected, leads to the clinical features of PD (45, 46). In addition, exposure to other toxins such as rotenone, paraquat, maneb, and epoxomicin can induce a Parkinsonian syndrome in experimental animals (47–49).

Genetic causes or predispositions also play an important role in PD. The disease is inherited in well-characterized kindreds (50, 51), and patients by chance have an affected family member more often than expected (52). Population-based association studies have identified genetic loci that may contribute to the development of "sporadic PD" (53). Thus far, genetic variability in tau, semaphorin 5A, α-synuclein, fibroblast growth factor 20, and nuclear receptor-related 1 genes are associated with increased disease risk (54–56). Perhaps more intriguing is the discovery of single gene mutations responsible for causing disease phenotypes that can be indistinguishable from sporadic PD. Understanding and comparing how mutations in specific genes can lead to Parkinsonism will provide new model systems and a better knowledge of the more common (i.e., sporadic) form of the disease (Table 3). These specific genes, their syndromes, and how their discovery impacts the search for the cause and treatment of PD are described below.

α-Synuclein and the role of protein aggregation

The first genetic mutation causing PD was found in the gene encoding α-synuclein and consisted of an alanine-to-threonine substitution (A53T) (57). This and 2 other point mutations (A30P and E46K) in the coding region of this gene, as well as gene duplications and triplications, cause a very rare, autosomal-dominant form of the disease (58–61). The clinical syndrome resulting from an A30P mutation can be indistinguishable from sporadic PD, including typical motor findings and reported late-stage dementia and psychosis (62). On the other hand, the A53T mutation can lead to unusual features such as earlier onset (less than 45 years

Table 3
Monogenetic causes of PD

Gene/protein	Pattern	Prevalence	Pathology	Common features	Notes
α-Synuclein	AD	Very rare	Lewy bodies	Early-onset dementia; presentation variable with mutation type	Aggregation of protein in Lewy bodies from genetic and sporadic forms of PD
Parkin	AR (mostly)	18% EOPD (50% with family history)	Rare Lewy bodies, if any	Early onset, slow progression	Protein is involved in ubiquitination
DJ-1	AR	<1% EOPD	Unknown	Early onset, slow progression	Protein is involved in the cellular stress response
PINK-1	AR (carriers may be at increased risk)	2–3% EOPD	Unknown	Early onset, slow progression	Protein is a mitochondrial kinase
LRRK2	AD	Highly variable	Lewy bodies	Typical PD (mostly)	Protein is a kinase with multiple putative domains

AH, autosomal recessive; AD, autosomal dominant; EOPD, early-onset PD (usually before 50 years of age).

of age), myoclonus, and more severe autonomic dysfunction (63). The E46K mutation causes Parkinsonism with the added features of dementia and hallucinations. Patients harboring gene triplications have an early age of onset with prominent autonomic and cognitive dysfunction, a rapidly progressive course, and broader brain pathology compared with sporadic cases (64, 65). Those patients with gene duplications have a less severe phenotype and a later age of onset than those with triplications, suggesting that the level of expression of α-synuclein correlates with disease severity (61, 66). In addition, polymorphisms within the α-synuclein promoter are associated with an increased PD risk, further implicating altered α-synuclein expression as a disease mechanism (67–69). This gene dosage effect, coupled with the important finding of α-synuclein immunoreactivity in Lewy bodies (70), provides strong evidence that this protein plays an important role in the pathogenesis of both sporadic and certain inherited forms of the disease. Furthermore, mice engineered to lack α-synuclein show resistance to the dopaminergic toxin MPTP (71), implicating α-synuclein in the pathogenic mechanism that leads to MPTP-induced Parkinsonism.

α-Synuclein is a highly conserved, abundant presynaptic phosphoprotein that adopts an elongated, unstructured shape in solution (72). In Lewy bodies, however, α-synuclein — the prominent structural component — is present in aggregated and insoluble filaments that are hyperphosphorylated and ubiquitinated (73, 74). It is likely that the abnormal aggregation of α-synuclein into a toxic, misfolded form contributes to neuronal cell death in both overexpressed wild-type and missense mutated proteins (75, 76). Factors such as the presence of a pathologic α-synuclein mutation, oxidative and nitrosative stress, phosphorylation, mitochondrial and proteasomal dysfunction, as well as dopamine can influence aggregation and folding of α-synuclein into a variety of forms including protofibrils, fibrils, and filaments. α-Synuclein is normally processed and cleaved at its C terminus by unidentified synucleinases (77). Truncation of α-synuclein appears to correlate with disease severity and with its propensity to oligomerize. It appears that the protofibrils and fibrils are the most toxic forms, and the creation and stabilization of these forms by mutation or cellular milieu may be a central pathologic mechanism (Figure 1).

α-Synuclein likely is involved in synaptic vesicle function (78), and its intracellular distribution and metabolism involve axonal transport (79) and degradation via the autophagic and proteasomal

systems (80). Recent studies suggest that it may act as a co-chaperone with cysteine-string protein α in the maintenance of nerve terminals (81). The association of α-synuclein with vesicles has led to speculation that the oligomerized protein may rupture cellular membranes through a pore-forming mechanism, leading to neurotransmitter leakage and toxicity (82). Other theories suggest that the abnormally aggregated protein may inhibit a range of normal cellular functions, such as axonal transport and protein turnover, via the ubiquitin-proteasomal or chaperone-mediated autophagic systems (83–85). Methods to interrupt synuclein aggregation by reducing its expression, increasing its degradation, impairing the formation of toxic aggregates, or inhibiting its truncation are logical therapeutic targets and are actively being explored (84).

Parkin and the role of the ubiquitin-proteasomal system

Mutations in the gene encoding parkin cause a form of autosomal-recessive, early-onset PD (86). Affected patients, though Parkinsonian and highly responsive to levodopa, do show several less typical features including early age of onset, prominent dystonia, severe motor fluctuation, and a more symmetric onset of symptoms (87, 88). Though initially described as a recessive disorder, there is controversial evidence that possessing a single parkin mutation does make one more likely to develop PD and show evidence of nigrostriatal dysfunction on imaging studies (89–92). Moreover, there are certain missense mutations that seem to be inherited in an autosomal-dominant manner (89).

Mutations in the gene encoding parkin are the most common genetic cause for early-onset PD, with prevalence rates approaching 50% for those with an autosomal-recessive family history and perhaps 18% of all those developing early-onset disease (less than 45 years of age) without a clear family history (93). Patients with parkin-related PD show loss of nigrostriatal and locus ceruleus neurons but, with rare exception, do not develop classic Lewy bodies (94–96).

Parkin is an E3 ligase that participates in the addition of ubiquitin molecules to target proteins, thereby marking them for proteasomal degradation (97). Loss of normal parkin function is postulated to lead to the abnormal accumulation of toxic substrates and resultant cell death. Numerous putative parkin substrates have been identified (75). The candidate substrates CDCrel-1, CDCrel-2, Pael-R, cyclin E, p38/JTV-1 (also known as AIMP2), and far upstream element–binding protein-1 (FBP-1) appear to accu-

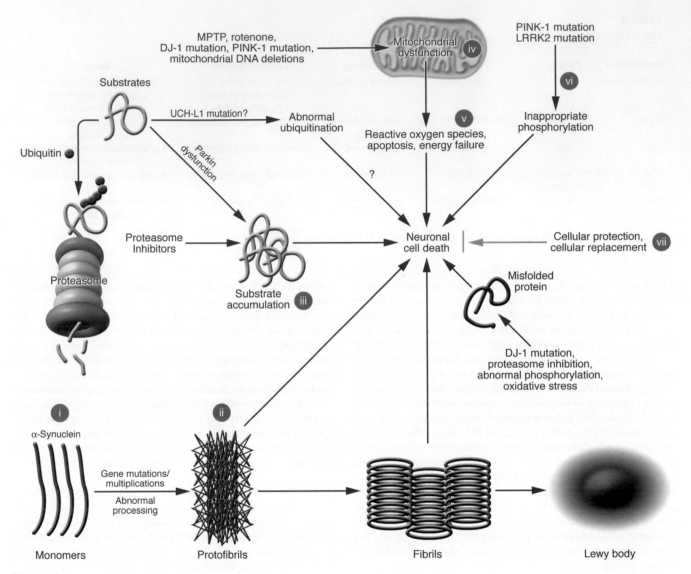

Figure 1

Model of dopaminergic cell death and possible sites for therapeutic intervention in PD. Studies on inherited forms of PD have led to the identification of genes that, when mutated, lead to dopaminergic cell loss. These genes are involved in a variety of cellular processes that include protein ubiquitination and degradation via the proteasome, response to oxidative stress, protein phosphorylation, mitochondrial function, and protein folding. Potential points of therapeutic intervention are highlighted: gene silencing therapies to reduce synuclein levels (i); inhibitors of synuclein aggregation and/or processing (ii); interventions to downregulate toxic substrates or upregulate parkin or proteasomal function (iii); interventions to enhance mitochondrial function with factors such as CoQ10, DJ-1, or PINK-1 (iv); free radical scavengers and antioxidants (v); kinase inhibitors to block LRRK2 activity or interventions to increase PINK-1 function (vi); and other therapies using trophic factors such as GDNF, survival genes, or fetal/stem cell replacement that would protect or replace susceptible cells (vii).

mulate in patients with parkin-associated PD, but only p38/JTV-1 and its interacting partner, FBP-1, accumulate in sporadic PD as well as in parkin knockout mice, suggesting that these proteins are true parkin substrates (98, 99).

Additional evidence for the importance of the ubiquitin-proteasomal system in PD pathogenesis comes from genetic association studies implicating the gene encoding ubiquitin-specific protease 24 (100). Moreover, mutations in ubiquitin carboxy-terminal hydrolase L1 (UCH-L1) in 2 individuals from a single affected German family (101) provide further support for the role of the ubiquitin-proteasome system in PD. The ability of mutations in this gene to cause or increase the risk of PD is high-

ly controversial, and no additional families have been reported (75, 102, 103). Indeed, recent studies suggest that UCH-L1 is not a PD susceptibility gene (75, 102, 103). The exact function of UCH-L1 is unclear, but it does involve ubiquitin hydrolase and ligase activities, and, interestingly, the protein is localized to Lewy bodies (see ref. 75). In addition, UCH-L1 can promote an alternative form of synuclein ubiquitination that uses the lysine 63 (K63) residue of ubiquitin to form polyubiquitin chains (104). This linkage is not directly related to proteasomal degradation, but may be involved in α-synuclein inclusion formation as well as in the function of parkin (105). The role of K63 ubiquitination in the development of inclusions and ultimately the pathogenesis of PD and other ubiquitin-positive neurodegenerative diseases is a promising field of study.

The link between sporadic PD and proteasomal dysfunction is supported by studies showing that parkin is nitrosylated in sporadic PD patients and that this modification impairs parkin's E3 ligase activity (106). Moreover, dopamine also may directly inhibit this activity through covalent modification (107). Additional evidence for the importance of the proteasome comes from pathologic studies that have found decreased proteasomal activity and expression in the SN of PD patients and the intriguing, but controversial, finding that exposure to proteasomal toxins can induce PD-like pathology in animal models (47, 108, 109). Interventions that increase proteasomal function or perhaps induce parkin expression may be a means of neuroprotection in PD.

DJ-1 and the role of oxidative stress
Mutations in the gene encoding DJ-1 similar to those in parkin lead to early-onset, autosomal-recessive PD (110). Clinically, patients are levodopa responsive with asymmetric onset of symptoms, slow progression, and variable severity. Behavioral, psychiatric, and dystonic features occur in patients with DJ-1 mutations (111). The prevalence of DJ-1 mutations is likely less than 1% in the early-onset patient subgroup (112).

The exact function of the DJ-1 protein is unknown, though evidence suggests it may act as an antioxidant, oxidation/reduction sensor, chaperone, and/or protease (75). Pathologically DJ-1 is not present in Lewy bodies, though it colocalizes with α-synuclein in other neurodegenerative diseases such as multiple system atrophy and Alzheimer disease (113). Some fraction of the protein is found in mitochondria (114). Oxidized and insoluble forms of the protein accumulate in the brains of patients with sporadic PD (113, 115). Interestingly, DJ-1 can interact with parkin in vitro under conditions of oxidative stress (115). In addition, when in a particular oxidized state, DJ-1 interacts with α-synuclein to prevent formation of fibrils (116). This latter activity suggests that DJ-1 acts as a redox-sensitive chaperone that protects cells from α-synuclein misfolding and toxicity under conditions of oxidative stress. In vivo studies have documented increased susceptibility of mice deficient in DJ-1 to the toxic effects of MPTP, and in culture, neurons from these mice were more sensitive to oxidative stress than those expressing DJ-1 (117). In addition, similar to parkin, DJ-1 is susceptible to protein S-nitrosylation that may be involved in the control of protein activity (118). There is speculation as well that DJ-1 functions as a regulator of apoptosis by modification of PTEN function and through interactions with several apoptosis-regulating proteins (119). The biology of DJ-1 links synuclein and parkin function with the phenomenon of oxidative stress, apoptosis, and the mitochondrion — all of which play a role in the pathogenesis of PD.

PINK-1 and the role of the mitochondrion
Mutations in the gene encoding PTEN-induced putative kinase 1 (PINK-1) are a cause of early-onset, autosomal-recessive PD (120). The clinical phenotype of point mutations is difficult to distinguish from other forms of early-onset disease, whereas a deletion mutation can induce a broader phenotype including dystonia and cognitive impairment (121). In general, patients with PINK-1 mutations show onset at an average of 35 years, slow progression, and levodopa responsiveness. The phenotype of PINK-1–, parkin–, and DJ-1–associated PD are indistinguishable, and atypical features may be a result of early onset rather than the etiology of disease (122). Heterozygous PINK-1 mutations were found in a cohort of early-onset patients in excess of the number seen in controls (123). This suggests that possessing a single PINK-1 mutation may predispose an individual to PD. In addition, imaging studies have demonstrated altered dopaminergic function in single PINK-1 mutation carriers (124). The prevalence of PINK-1 mutation is between that of parkin and DJ-1 and is present in the homozygous state in 2–3% of early-onset patients (122, 125).

PINK-1 is a mitochondrial protein kinase whose substrates are unknown (S1). PINK-1 is induced by PTEN, the same protein whose activity is suppressed by DJ-1. PINK-1 mutation may lead to mitochondrial dysfunction and increased sensitivity to cellular stress through a defect in the apoptosis pathway (S2). PINK-1 appears to play an important role in mitochondrial function, as *Drosophila* lacking PINK-1 have substantial mitochondrial defects resulting in apoptotic muscle degeneration and male sterility. Interestingly, parkin rescues the PINK-1 loss-of-function phenotype, suggesting that parkin and PINK-1 function in a common biochemical pathway (S3–S5).

The presence of PINK-1 and DJ-1 in the mitochondrion underscores the role that this cellular organelle plays in PD pathogenesis. A common feature of sporadic PD is evidence of complex I mitochondrial dysfunction (119). This component of the electron transport chain also is affected by rotenone and MPTP, 2 toxins whose effects can model PD (S6). Rotenone, when given chronically to animals by infusion, produces SN cell and dopaminergic fiber loss with α-synuclein accumulation and the formation of Lewy body–like inclusions (S7). Similarly, MPTP exposure in both patients and animal models leads to nigral cell loss and Parkinsonian symptoms (S6) and under conditions of chronic administration leads to the formation of α-synuclein–containing inclusions (S8). Furthermore, animals deficient in α-synuclein are resistant to MPTP, implying the participation of this protein in the pathogenic mechanism of MPTP (71). Additionally, the toxins 6-hydroxydopamine and paraquat produce dopaminergic cell damage by inducing oxidative stress that may mimic the mitochondrial toxicity seen with rotenone and MPTP (S9).

Two recent studies provide a clue as to how oxidative stress may lead to cellular dysfunction in PD. These studies showed that in PD, SN neurons accumulate mitochondrial DNA deletion mutations at an abnormally high rate (S10, S11). The deletion in each individual cell appears clonal and likely results from a single mutation event and subsequent expansion of the affected mitochondrion. The authors suggest that this mutation load is sufficient to cause impaired cellular respiration, as determined by the loss of cytochrome *c* oxidase staining. These findings suggest a possible mechanism that begins with increased oxidative stress in the SN and leads to mitochondrial DNA mutation and subsequent failure of mitochondrial function. The accumulation of impaired mitochondria within a cell in turn leads to respiratory chain deficiency and SN cellular pathology. Interestingly, a mutation in the polymerase responsible for mitochondrial DNA replication has been associated with the accumulation of deletions in mitochondrial DNA, SN cell loss, and early-onset Parkinsonism (S12). The role that mitochondrial dysfunction and oxidative stress play in PD pathogenesis has led to trials of antioxidant and promitochondrial compounds, including coenzyme Q10 (CoQ10), vitamin E, and creatine, as possible neuroprotectants in the disease.

LRRK2 and phosphorylation

Mutation in the gene for LRRK2 (also called dardarin) was identified in 2004 as a cause of autosomal-dominant PD (S13, S14). Most cases of LRRK2-related disease show typical late and asymmetric onset of symptoms, levodopa responsiveness, and other features that are indistinguishable from sporadic disease, although rarely, early onset, amyotrophy, gaze palsy, dementia, and psychiatric symptoms are observed (S13–S15). The prevalence of LRRK2 mutations in familial and sporadic PD is highly dependent on the particular study population and the nature of the mutation. In a Spanish cohort, 5.3% of all PD patients (9.6% of those with a family history) tested positive for the mutation, with G2019S being most common (4.3% of all patients; 6.4% of PD patients with a family history) (S15). The R1441G mutation was found in 8% of PD patients from the Basque region of Spain (S14), but only 0.7% of a more diverse Spanish cohort (S15). Remarkably, the G2019S mutation was found in 18.3% of all PD cases from an Ashkenazi Jewish cohort and in 30% of those with a family history (37). The lack of a family history in many of these patients suggested an estimated penetrance rate of 32%. Even more striking is the finding that in a North African Arab PD cohort, the prevalence of the G2019S mutation was approximately 40% in both sporadic and familial cases, again suggesting possible reduced penetrance (38). Mirroring sporadic PD, the prevalence rate of the G2019S mutation increases with age, reaching 25% by age 55 and 85% by age 70 (S16). Autopsy material from the majority of affected patients demonstrates typical PD pathology, including SN and locus ceruleus cell loss with Lewy bodies and Lewy neurites. Atypical pathology also occurs and likely is more common with particular mutants such as R1441C (S13, S17). This pathology has included the absence of Lewy bodies and the presence of synuclein-negative, ubiquitin-positive inclusions, LRRK2-filled neurites, and abnormal tau pathology, the latter of which is suggestive more of progressive supranuclear palsy than of PD (S13). Additionally the Y1699C mutation may lead to abnormal pathology including features similar to motor neuron disease, though a subsequent report found typical PD pathology in a patient carrying the same mutation from another family (S13, S18).

LRRK2 is a large protein that includes Roc (Ras in complex protein), COR (C-terminal of Roc), leucine-rich repeat, mixed lineage kinase, WD40 (a putative protein-protein interaction domain terminating in a tryptophan-aspartic acid dipeptide), and ankryin domains (S13, S14, S19). Function and localization data for this protein is limited, but there is evidence that it is cytoplasmic, associated with mitochondria, and capable of autophosphorylation (S20). Recent data suggests that an increase in LRRK2 kinase activity plays a role in the pathogenesis of at least 3 of the known mutations (S20, S21). In addition, LRRK2 binds to parkin and leads to increased protein aggregation and ubiquitination in a cell culture model, and disease-associated LRRK2 mutants are toxic to SH-SY5Y cells and cortical neurons (S22). Furthermore, LRRK2 may, as do other WD40 proteins, participate in mitochondrial fission or interact with the cytoskeleton, suggesting a role in membrane, vesicular, or protein trafficking (S23).

The biology of LRRK2 and PINK-1 adds to other data suggesting that protein phosphorylation plays a vital role in PD pathogenesis. Clearly if abnormal kinase activity is responsible for LRRK2-mediated disease, kinase inhibitors, particularly LRRK2 kinase inhibitors, may have therapeutic relevance. Protein phosphorylation involving the mixed lineage kinase pathway may play a role in PD

and remains an important therapeutic target despite the recent failure of the kinase inhibitor CEP-1347 in clinical trials (S24). In addition, abnormal phosphorylation of proteins such as α-synuclein is seen in PD and in models of the disease (S25), suggesting that inhibitors of α-synuclein phosphorylation may be therapeutic. Finally, the neurotrophic effects of glial cell line–derived neurotrophic factor (GDNF) on dopaminergic neurons involve the activation of the protein kinase Ret and a myriad of downstream targets including Akt that are important in a cell's response to oxidative stress and to DJ-1 function (S26, S27). These data suggest that other inducers of Akt signaling may be therapeutic.

Symptomatic therapy

Current medical and surgical therapies for PD are symptomatic and lack significant disease-modifying effect. Indeed the most effective medical therapy continues to be levodopa mixed with a peripheral decarboxylase inhibitor (S28–S30). The most recent advances in available symptomatic medical therapies in the United States have revolved around prolonging the effect of levodopa through the use of COMT inhibition and by changing the availability and formulation of older medications such as selegiline and apomorphine. The latter, given subcutaneously with an oral antiemetic, can temporarily rescue patients during disabling motor "off" time (time spent with reduced levodopa effect) (S31). Looking beyond dopaminergic therapy, other symptomatic medical therapies showing promise in published clinical trials include istradefylline, an adenosine A2 antagonist that reduces "off" time (S32), and sarizotan, a serotonin agonist that reduces levodopa-induced dyskinesias (S33, S34).

There are data suggesting that deferring levodopa therapy in favor of dopamine agonists such as ropinerole or pramipexole may delay motor complications such as dyskinesias (S29, S30). The dopamine agonists are less effective than levodopa in symptomatic relief, and nearly all patients with advancing PD will require levodopa at some point. The reduced risk of dyskinesias with the dopamine agonists may be related to their longer half-life, and thus more stable receptor stimulation, relative to levodopa. Supporting these data are studies showing that continuous infusion of levodopa or an agonist has benefit over oral interval dosing (S35–S37). These findings have led to the development of experimental therapies including the transdermal patch (for delivery of rotigotine) and infusion delivery systems (for Duodopa) that reduce pulsatile drug delivery. In addition, this concept has given rise to the theory that COMT inhibitors given with levodopa may reduce the risk of future dyskinesias by prolonging brain exposure to a given levodopa dose (S38).

Surgical DBS is perhaps the most influential development in symptomatic PD therapy since levodopa and is reviewed in detail elsewhere (S39). In this procedure, an electrode is inserted through the skull to reach and stimulate the globus pallidus, subthalamic nucleus (STN), or ventral intermediate thalamus. A pacemaker-like device is implanted and connected to the electrode through wires buried beneath the skin. The results of this therapy can be quite marked and include the reduction of "off" time, increased "on" time without dyskinesias, reduction of levodopa dose, and improved tremor. The STN has recently emerged as the preferential target due to improved symptom control and a reduced energy requirement (S40). Currently this procedure is largely reserved for advanced cases of PD in which motor complications and/or medication intolerance have led to an unacceptable decline in quality of life. In addition, disabling tremor that is not responsive to medical therapy may respond well to DBS. Very encouraging are the recent, albeit small, studies showing the benefits of electrode implantation into the pedunculopontine nucleus with resultant improvement in bradykinesia, gait freezing, and postural stability (S41).

A newer, experimental field of study is the use of transcranial magnetic stimulation (TMS). In this process a current is passed through a coil to generate a magnetic field. The coil is placed near the head to induce stimulation in the nearby brain structures. TMS studies show that PD patients have a measurable abnormality in the inhibitory control of the cortex that results in a shortened "silent period." Also seen are inconsistently measurable changes in motor threshold and abnormal activation during voluntary input (S42). Some of these abnormalities normalize with medical and surgical PD therapies. The effect of repetitive TMS (rTMS) is now being explored as a possible therapeutic intervention. A meta-analysis of available data has shown a small but significant effect of rTMS (approximately 20% improvement on the motor Unified Parkinson's Disease Rating Scale [UPDRS]). Other large studies have failed, prompting concern regarding the standardization of technique. In addition, a measurable placebo effect, as detected by PET scan, can be seen with rTMS that may be confounding study results (S43). A well-defined protocol including rTMS applied while in the "on" state, inclusion of a placebo control, and specific parameters for intensity and frequency of stimulation was used in a recent study, showing improvement of limb bradykinesia and gait for at least one month after a course of rTMS therapy (S44). The magnitude of improvement was thought to be similar to the effect of a single levodopa dose. These results require duplication, and the cumulative effect and clinical significance of this therapy must be subjected to further study before rTMS can be recommended for routine treatment. The paucity of adverse effects of rTMS does make the optimization of this treatment an attractive field of study.

A review of symptomatic PD therapies would not be complete without consideration of progress made in the treatment of nonmotor sequelae. Given the good response of motor symptoms to medical and surgical therapies, it is often poor balance, sleep interruption, cognitive impairments, anxiety, depression, and drooling that become most disabling (S45). Indeed the recognition of widespread pathology in PD suggests that these affected areas lying outside of the dopaminergic motor pathway are contributing to patients' symptoms. Specifically, abnormalities in the noradrenergic and serotonin nuclei may lead to anxiety and depression as well as the autonomic, sleep, and visual disturbances seen in PD, while changes in the neocortex, limbic system, and cholinergic nucleus basalis may be involved in cognitive decline later in the disease. Treatment of these symptoms can be rewarding and involves interventions including agents such as midodrine and pyridostigmine for blood pressure support (S46), atropine drops for symptomatic control of salivation (S47), cholinesterase antagonists for cognitive decline (S48), antidepressants (serotonin selective reuptake inhibitors and possibly others) for treatment of depression (S49), and atypical antipsychotics (most commonly clozapine and quetiapine) for treatment of psychosis (S50). A recent study by Ondo et al. has cast some doubt on the effectiveness of quetiapine for the treatment of psychosis, underscoring the need for more rigorous trials of therapies aimed at the treatment of nonmotor PD symptoms (S51).

Neuroprotection and future therapies

The search for compounds that can slow or halt the progression of PD is an active area of clinical research. The neuroprotective properties of selegiline, an MAO-B inhibitor, continue to be debated, with studies showing that even after years of treatment, individuals receiving selegiline appear to demonstrate slower PD progression (S52, S53). Rasagiline is an irreversible MAO-B inhibitor that likely works via several pathways to provide symptomatic relief, but may also slow disease progression (S54, S55). Like selegiline, rasagiline's symptomatic effect has made it difficult to convincingly extract neuroprotective data, and the magnitude of this effect will likely rely on further observation. The protective effect of both selegiline and rasagiline may occur through prevention of the GAPDH cell death cascade by blocking the S-nitrosylation of GAPDH, the binding of GAPDH to Siah (a protein E3 ligase that aids in the translocation of GAPDH to the nucleus), and the subsequent Siah-mediated degradation of nuclear proteins that leads to cell death (S56).

Another neuroprotective candidate is CoQ10. It was hypothesized that this antioxidant and electron transport chain component might correct the mitochondrial complex I dysfunction and CoQ10 deficiency seen in PD patients. In a pilot study CoQ10 appeared to slow disease progression at the highest dose of 1,200 mg/d, although the subject numbers were small and the results await confirmation (S57). Based on a futility study design, the antiinflammatory, anticaspase drug minocycline and the promitochondrial compound creatine were both recently deemed worthy of future consideration as neuroprotectants in PD (S58).

Mixed results have been obtained in looking at the disease-modifying effects of GDNF. This potent dopaminergic neuron survival factor was infused into PD patients in 3 recent studies. A double-blind, placebo-controlled study failed to demonstrate clinical improvement and was discontinued due to lack of efficacy, the development of antibodies in patients, and safety concerns raised over the development of cerebellar toxicity in a group of exposed nonhuman primates (S59). Two prior open-label studies showed a beneficial effect of infusion, and the reason behind the disparity needs further investigation. In addition, GDNF and related compounds are being examined for efficacy using alternative means of delivery including implantation of capsules, engineered cells, and viral vectors.

One step beyond neuroprotection is cell replacement therapy, wherein cells lost in PD are replaced. There have been mixed results in clinical trials attempting this strategy: some open-label studies showed benefit, while 2 double-blind, placebo-controlled studies failed to meet their primary endpoints and noted the worrisome development of dyskinesias (S60–S63). In these studies, ventral midbrain tissue was isolated from fetal human tissue and ectopically transplanted into the striatum of PD patients. Autopsy and imaging studies do verify that the transplanted tissue can survive and functionally integrate. Studies continue to explore the reasons for variable clinical response with possible explanations involving surgical technique, level of immunosuppression, cell preparation and survivability, and patient selection (S64). In addition, these cells lack normal synaptic input since they are not properly localized to the SN. A fully integrated graft may need to be placed in the SN and recapitulate functional connectivity to the striatum. Until these problems are addressed and better understood, fetal cell transplantation cannot be recommended as a routine therapeutic option.

The isolation of human embryonic stem cells has provided a potential source for transplantation material. Over the past few years a good deal of effort has gone into developing protocols to induce the proper dopaminergic characteristics in these undifferentiated cells to make them suitable candidates for transplant. Several protocols exist that can induce a dopaminergic phenotype, but poor cell survival after transplant into animal models has led to disappointing results (S65). Several other roadblocks stand in the way of stem cell therapy, including concerns regarding exposing cells to xenogenic factors during the expansion and differentiation phases, the possibility of tumor formation if cells are not properly differentiated, the possibility of tissue rejection, and purification of the transplanted cells. In addition, there is the concern that even if ample numbers of the appropriate cells are produced, these cells may suffer from the same difficulties seen with fetal cell transplants. Finally, there is little reason to believe that transplantation of dopaminergic cells will alleviate the symptoms related to the degeneration seen in non-nigrostriatal brain areas and with other neurotransmitter types. Despite these caveats, if a way can be found to reliably reproduce the symptomatic benefit seen in some fetal transplant studies, the availability of large numbers of differentiated stem cells may one day make such transplantation a very attractive therapy.

Conclusions

Just as landmark discoveries that identified PD as a disease of dopamine deficiency led to the development of rational symptomatic therapies such as levodopa and dopamine agonists, so will the understanding of disease mechanisms spurred on by the study of genetics lead to novel neuroprotective and restorative therapies. The identification of monogenetic forms of PD has uncovered a role for proteasomal and mitochondrial dysfunction, oxidative stress, protein misfolding, and aberrant phosphorylation in the pathophysiology of PD. While the interplay and temporal relationship among these pathologic processes are unclear at present, further research to determine which factors are most proximate to the cell death process and which are most amenable to pharmaceutical intervention are the challenges for the future.

Acknowledgments

T.M. Dawson and V.L. Dawson are supported by grants from the NIH (National Institute of Neurological Disorders and Stroke grants NS38377, NS051468, NS048206, and NS054207), the Lee Martin Trust, the Sylvia Nachlas Trust, the National Parkinson Foundation, and the Michael J. Fox Foundation. T.M. Dawson is the Leonard and Madlyn Abramson Professor in Neurodegenerative Diseases. J.M. Savitt is supported by the American Parkinson Disease Foundation and NIH National Institute of Neurological Disorders and Stroke grant NS052624.

Note: References S1–S65 are available online with this article; doi:10.1172/JCI29178DS1.

Address correspondence to: Ted M. Dawson, Institute for Cell Engineering, Department of Neurology and Neuroscience, Johns Hopkins University School of Medicine, 733 North Broadway, BRB Suite 731, Baltimore, Maryland 21205, USA. Phone: (410) 614-3359; Fax: (410) 614-9568; E-mail: tdawson@jhmi.edu.

1. Nutt, J.G., and Wooten, G.F. 2005. Clinical practice. Diagnosis and initial management of Parkinson's disease. *N. Engl. J. Med.* **353**:1021–1027.
2. Elbaz, A., et al. 2002. Risk tables for parkinsonism and Parkinson's disease. *J. Clin. Epidemiol.* **55**:25–31.
3. de Rijk, M.C., et al. 2000. Prevalence of Parkinson's disease in Europe: a collaborative study of population-based cohorts. Neurologic Diseases in the Elderly Research Group. *Neurology.* **54**(Suppl. 5):S21–S23.
4. Goetz, C.G. 2002. Charcot and Parkinson's disease. In *Parkinson's disease diagnosis and clinical management.* S. Factor and W. Weiner, editors. Demos Medical Publishing. New York, New York, USA. 19–26.
5. Brissaud, E. 1895. Nature et pathogenie de la maladie de Parkinson. In *Leçons sur les malades nerveuses.* Masson. Paris, France. 488–501.
6. Kapp, W. 1992. The history of drugs for the treatment of Parkinson's disease. *J. Neural Transm. Suppl.* **38**:1–6.
7. Lewy, F.H. 1912. Paralysis agitans. I. Pathologische anatomie. In *Handbuch der neurologie.* M. Lewandowsky, editor. Springer. Berlin, Germany. 920–933.
8. Greenfield, J.G., and Bosanquet, F.D. 1953. The brain-stem lesions in Parkinsonism. *J. Neurol. Neurosurg. Psychiatry.* **16**:213–226.
9. Ehringer, H., and Hornykiewicz, O. 1960. Verteilung von noradrenalin und dopamin (3-hydroxytyramin) ingerhirn des menschen und ihr verhalten bei erkrankugen des extrapyramidalen systems. *Klin. wochenschr.* **38**:1236–1239.
10. Poirier, L.J., and Sourkes, T.L. 1965. Influence of the substantia nigra on the catecholamine content of the striatum. *Brain.* **88**:181–192.
11. Carlsson, A., Lindqvist, M., and Magnusson, T. 1957. 3,4-Dihydroxyphenylalanine and 5-hydroxytryptophan as reserpine antagonists. *Nature.* **180**:1200.
12. Cotzias, G.C., Papavasiliou, P.S., and Gellene, R. 1969. Modification of Parkinsonism--chronic treatment with L-dopa. *N. Engl. J. Med.* **280**:337–345.
13. Rinne, U.K., and Sonninen, V. 1973. Brain catecholamines and their metabolites in Parkinsonian patients. Treatment with levodopa alone or combined with a decarboxylase inhibitor. *Arch. Neurol.* **28**:107–110.
14. Rinne, U.K., Sonninen, V., and Sirtola, T. 1972. Treatment of Parkinson's disease with L-DOPA and decarboxylase inhibitor. *Z. Neurol.* **202**:1–20.
15. Ericsson, A.D. 1971. Potentiation of the L-Dopa effect in man by the use of catechol-O-methyltransferase inhibitors. *J. Neurol. Sci.* **14**:193–197.
16. Myllyla, V.V., Sotaniemi, K.A., Illi, A., Suominen, K., and Keranen, T. 1993. Effect of entacapone, a COMT inhibitor, on the pharmacokinetics of levodopa and on cardiovascular responses in patients with Parkinson's disease. *Eur. J. Clin. Pharmacol.* **45**:419–423.
17. Roberts, J.W., et al. 1993. Catechol-O-methyltransferase inhibitor tolcapone prolongs levodopa/carbidopa action in parkinsonian patients. *Neurology.* **43**:2685–2688.
18. Chrisp, P., Mammen, G.J., and Sorkin, E.M. 1991. Selegiline. A review of its pharmacology, symptomatic benefits and protective potential in Parkinson's disease. *Drugs Aging.* **1**:228–248.
19. Gopinathan, G., et al. 1981. Lisuride in parkinsonism. *Neurology.* **31**:371–376.
20. Calne, D.B., Teychenne, P.F., Leigh, P.N., Bamji, A.N., and Greenacre, J.K. 1974. Treatment of parkinsonism with bromocriptine. *Lancet.* **2**:1355–1356.
21. Brophy, B.P. 1998. Surgical palliation of dyskinesiae in Parkinson's disease. *Stereotact. Funct. Neurosurg.* **70**:107–113.
22. Bergman, H., Wichmann, T., and DeLong, M.R. 1990. Reversal of experimental parkinsonism by lesions of the subthalamic nucleus. *Science.* **249**:1436–1438.
23. Benabid, A.L., et al. 2000. Future prospects of brain stimulation. *Neurol. Res.* **22**:237–246.
24. Hughes, A.J., Daniel, S.E., Blankson, S., and Lees, A.J. 1993. A clinicopathologic study of 100 cases of Parkinson's disease. *Arch. Neurol.* **50**:140–148.
25. Pal, P., Samii, A., and Calne, D. 2002. Cardinal features of early Parkinson's disease. In *Parkinson's disease: diagnosis and clinical management.* S. Factor and W. Weiner, editors. Demos Medical Publishing. New York, New York, USA. 41–56.
26. de Lau, L.M., Koudstaal, P.J., Hofman, A., and Breteler, M.M. 2006. Subjective complaints precede Parkinson disease: the rotterdam study. *Arch. Neurol.* **63**:362–365.
27. Hughes, A.J., Daniel, S.E., Kilford, L., and Lees, A.J. 1992. Accuracy of clinical diagnosis of idiopathic Parkinson's disease: a clinico-pathological study of 100 cases. *J. Neurol. Neurosurg. Psychiatry.* **55**:181–184.
28. Rajput, A.H., Rozdilsky, B., and Rajput, A. 1991. Accuracy of clinical diagnosis in parkinsonism--a prospective study. *Can. J. Neurol. Sci.* **18**:275–278.
29. Hughes, A.J., Daniel, S.E., and Lees, A.J. 2001. Improved accuracy of clinical diagnosis of Lewy body Parkinson's disease. *Neurology.* **57**:1497–1499.
30. Suchowersky, O., et al. 2006. Practice parameter: diagnosis and prognosis of new onset Parkinson disease (an evidence-based review). *Neurology.* **66**:968–975.
31. Marek, K., Jennings, D., and Seibyl, J. 2003. Neuroimaging in Parkinson's disease. In *Handbook of Parkinson's disease.* 3rd edition. R. Pahwa, K. Lyons, and W. Koller, editors. Marcel Dekker, Inc. New York, New York, USA. 179–202.
32. Van Laere, K., et al. 2006. Dual-tracer dopamine transporter and perfusion SPECT in differential diagnosis of parkinsonism using template-based discriminant analysis. *J. Nucl. Med.* **47**:384–392.
33. Berg, D., Hochstrasser, H., Schweitzer, K.J., and Riess, O. 2006. Disturbance of iron metabolism in Parkinson's disease -- ultrasonography as a biomarker. *Neurotox. Res.* **9**:1–13.

34. Ponsen, M.M., et al. 2004. Idiopathic hyposmia as a preclinical sign of Parkinson's disease. *Ann. Neurol.* **56**:173–181.

35. El-Agnaf, O.M., et al. 2006. Detection of oligomeric forms of alpha-synuclein protein in human plasma as a potential biomarker for Parkinson's disease. *FASEB J.* **20**:419–425.

36. Michell, A.W., Lewis, S.J., Foltynie, T., and Barker, R.A. 2004. Biomarkers and Parkinson's disease. *Brain.* **127**:1693–1705.

37. Ozelius, L.J., et al. 2006. LRRK2 G2019S as a cause of Parkinson's disease in Ashkenazi Jews. *N. Engl. J. Med.* **354**:424–425.

38. Lesage, S., et al. 2006. LRRK2 G2019S as a cause of Parkinson's disease in North African Arabs. *N. Engl. J. Med.* **354**:422–423.

39. Braak, H., et al. 2003. Staging of brain pathology related to sporadic Parkinson's disease. *Neurobiol. Aging.* **24**:197–211.

40. Marsden, C. 1983. Neuromelanin and Parkinson's disease. *J. Neural Transm. Suppl.* **19**:121–141.

41. Jellinger, K. 2005. The pathology of parkinson's disease-recent advances. In *Scientific basis for the treatment of Parkinson's disease.* 2nd edition. N. Galvez-Jimenez, editor. Taylor & Francis. New York, New York, USA/London, United Kingdom. 53–85.

42. Braak, H., de Vos, R.A., Bohl, J., and Del Tredici, K. 2006. Gastric alpha-synuclein immunoreactive inclusions in Meissner's and Auerbach's plexuses in cases staged for Parkinson's disease-related brain pathology. *Neurosci. Lett.* **396**:67–72.

43. Poskanzer, D.C., and Schwab, R.S. 1963. Cohort analysis of Parkinson's syndrome: evidence for a single etiology related to subclinical infection about 1920. *J. Chronic Dis.* **16**:961–973.

44. Reid, A.H., McCall, S., Henry, J.M., and Taubenberger, J.K. 2001. Experimenting on the past: the enigma of von Economo's encephalitis lethargica. *J. Neuropathol. Exp. Neurol.* **60**:663–670.

45. Burns, R.S., et al. 1983. A primate model of parkinsonism: selective destruction of dopaminergic neurons in the pars compacta of the substantia nigra by N-methyl-4-phenyl-1,2,3,6-tetrahydropyridine. *Proc. Natl. Acad. Sci. U. S. A.* **80**:4546–4550.

46. Langston, J.W., Ballard, P., Tetrud, J.W., and Irwin, I. 1983. Chronic Parkinsonism in humans due to a product of meperidine-analog synthesis. *Science.* **219**:979–980.

47. McNaught, K.S., Perl, D.P., Brownell, A.L., and Olanow, C.W. 2004. Systemic exposure to proteasome inhibitors causes a progressive model of Parkinson's disease. *Ann. Neurol.* **56**:149–162.

48. Betarbet, R., et al. 2000. Chronic systemic pesticide exposure reproduces features of Parkinson's disease. *Nat. Neurosci.* **3**:1301–1306.

49. Landrigan, P.J., et al. 2005. Early environmental origins of neurodegenerative disease in later life. *Environ. Health Perspect.* **113**:1230–1233.

50. Bell, J., and Clark, A. 1926. A pedigee of paralysis agitans. *Ann. Eugen.* **1**:455–462.

51. Allen, W. 1937. Inheritence of the shaking palsy. *Arch. Intern. Med.* **60**:424–436.

52. Lazzarini, A.M., et al. 1994. A clinical genetic study of Parkinson's disease: evidence for dominant transmission. *Neurology.* **44**:499–506.

53. Maraganore, D.M., et al. 2005. High-resolution whole-genome association study of Parkinson disease. *Am. J. Hum. Genet.* **77**:685–693.

54. Mizuta, I., et al. 2006. Multiple candidate gene analysis identifies alpha-synuclein as a susceptibility gene for sporadic Parkinson's disease. *Hum. Mol. Genet.* **15**:1151–1158.

55. van der Walt, J.M., et al. 2004. Fibroblast growth factor 20 polymorphisms and haplotypes strongly influence risk of Parkinson disease. *Am. J. Hum. Genet.* **74**:1121–1127.

56. Le, W.D., et al. 2003. Mutations in NR4A2 associated with familial Parkinson disease. *Nat. Genet.*

33:85–89.

57. Polymeropoulos, M.H., et al. 1997. Mutation in the alpha-synuclein gene identified in families with Parkinson's disease. *Science.* **276**:2045–2047.

58. Kruger, R., et al. 1998. Ala30Pro mutation in the gene encoding alpha-synuclein in Parkinson's disease. *Nat. Genet.* **18**:106–108.

59. Zarranz, J.J., et al. 2004. The new mutation, E46K, of alpha-synuclein causes Parkinson and Lewy body dementia. *Ann. Neurol.* **55**:164–173.

60. Singleton, A.B., et al. 2003. alpha-Synuclein locus triplication causes Parkinson's disease. *Science.* **302**:841.

61. Nishioka, K., et al. 2006. Clinical heterogeneity of alpha-synuclein gene duplication in Parkinson's disease. *Ann. Neurol.* **59**:298–309.

62. Langston, J.W., et al. 1998. Novel alpha-synuclein-immunoreactive proteins in brain samples from the Contursi kindred, Parkinson's, and Alzheimer's disease. *Exp. Neurol.* **154**:684–690.

63. Spira, P.J., Sharpe, D.M., Halliday, G., Cavanagh, J., and Nicholson, G.A. 2001. Clinical and pathological features of a Parkinsonian syndrome in a family with an Ala53Thr alpha-synuclein mutation. *Ann. Neurol.* **49**:313–319.

64. Farrer, M., et al. 2004. Comparison of kindreds with parkinsonism and alpha-synuclein genomic multiplications. *Ann. Neurol.* **55**:174–179.

65. Muenter, M.D., et al. 1998. Hereditary form of parkinsonism--dementia. *Ann. Neurol.* **43**:768–781.

66. Eriksen, J.L., Przedborski, S., and Petrucelli, L. 2005. Gene dosage and pathogenesis of Parkinson's disease. *Trends Mol. Med.* **11**:91–96.

67. Pals, P., et al. 2004. alpha-Synuclein promoter confers susceptibility to Parkinson's disease. *Ann. Neurol.* **56**:591–595.

68. Hadjigeorgiou, G.M., et al. 2005. Association of alpha-synuclein Rep1 polymorphism and Parkinson's disease: influence of Rep1 on age at onset. *Mov. Disord.* **21**:534–539.

69. Tan, E.K., et al. 2004. Alpha-synuclein haplotypes implicated in risk of Parkinson's disease. *Neurology.* **62**:128–131.

70. Spillantini, M.G., et al. 1997. Alpha-synuclein in Lewy bodies. *Nature.* **388**:839–840.

71. Dauer, W., et al. 2002. Resistance of alpha-synuclein null mice to the parkinsonian neurotoxin MPTP. *Proc. Natl. Acad. Sci. U. S. A.* **99**:14524–14529.

72. Weinreb, P.H., Zhen, W., Poon, A.W., Conway, K.A., and Lansbury, P.T., Jr. 1996. NACP, a protein implicated in Alzheimer's disease and learning, is natively unfolded. *Biochemistry.* **35**:13709–13715.

73. Hasegawa, M., et al. 2002. Phosphorylated alpha-synuclein is ubiquitinated in alpha-synucleinopathy lesions. *J. Biol. Chem.* **277**:49071–49076.

74. Spillantini, M.G., Crowther, R.A., Jakes, R., Hasegawa, M., and Goedert, M. 1998. alpha-Synuclein in filamentous inclusions of Lewy bodies from Parkinson's disease and dementia with lewy bodies. *Proc. Natl. Acad. Sci. U. S. A.* **95**:6469–6473.

75. Moore, D.J., West, A.B., Dawson, V.L., and Dawson, T.M. 2005. Molecular pathophysiology of Parkinson's disease. *Annu. Rev. Neurosci.* **28**:57–87.

76. Mizuno, Y., Mochizuki, H., and Hattori, N. 2005. Alpha-synuclein, nigral degeneration and parkinsonism. In *Scientific basis for the treatment of Parkinson's disease.* 2nd edition. N. Galvez-Jimenez, editor. Taylor & Francis. New York, New York, USA/London, United Kingdom. 87–104.

77. Li, W., et al. 2005. Aggregation promoting C-terminal truncation of alpha-synuclein is a normal cellular process and is enhanced by the familial Parkinson's disease-linked mutations. *Proc. Natl. Acad. Sci. U. S. A.* **102**:2162–2167.

78. Abeliovich, A., et al. 2000. Mice lacking alpha-synuclein display functional deficits in the nigrostriatal dopamine system. *Neuron.* **25**:239–252.

79. Jensen, P.H., Li, J.Y., Dahlstrom, A., and Dotti, C.G.

1999. Axonal transport of synucleins is mediated by all rate components. *Eur. J. Neurosci.* **11**:3369–3376.

80. Webb, J.L., Ravikumar, B., Atkins, J., Skepper, J.N., and Rubinsztein, D.C. 2003. Alpha-synuclein is degraded by both autophagy and the proteasome. *J. Biol. Chem.* **278**:25009–25013.

81. Chandra, S., Gallardo, G., Fernandez-Chacon, R., Schluter, O.M., and Sudhof, T.C. 2005. Alpha-synuclein cooperates with CSPalpha in preventing neurodegeneration. *Cell.* **123**:383–396.

82. Rochet, J.C., et al. 2004. Interactions among alpha-synuclein, dopamine, and biomembranes: some clues for understanding neurodegeneration in Parkinson's disease. *J. Mol. Neurosci.* **23**:23–34.

83. Cuervo, A.M., Stefanis, L., Fredenburg, R., Lansbury, P.T., and Sulzer, D. 2004. Impaired degradation of mutant alpha-synuclein by chaperone-mediated autophagy. *Science.* **305**:1292–1295.

84. Mukaetova-Ladinska, E.B., and McKeith, I.G. 2006. Pathophysiology of synuclein aggregation in Lewy body disease. *Mech. Ageing Dev.* **127**:188–202.

85. Snyder, H., et al. 2003. Aggregated and monomeric alpha-synuclein bind to the S6' proteasomal protein and inhibit proteasomal function. *J. Biol. Chem.* **278**:11753–11759.

86. Kitada, T., et al. 1998. Mutations in the parkin gene cause autosomal recessive juvenile parkinsonism. *Nature.* **392**:605–608.

87. Lohmann, E., et al. 2003. How much phenotypic variation can be attributed to parkin genotype? *Ann. Neurol.* **54**:176–185.

88. Bonifati, V., et al. 2001. The parkin gene and its phenotype. Italian PD Genetics Study Group, French PD Genetics Study Group and the European Consortium on Genetic Susceptibility in Parkinson's Disease. *Neurol. Sci.* **22**:51–52.

89. Oliveira, S.A., et al. 2003. Parkin mutations and susceptibility alleles in late-onset Parkinson's disease. *Ann. Neurol.* **53**:624–629.

90. Khan, N.L., et al. 2005. Dopaminergic dysfunction in unrelated, asymptomatic carriers of a single parkin mutation. *Neurology.* **64**:134–136.

91. Lincoln, S.J., et al. 2003. Parkin variants in North American Parkinson's disease: cases and controls. *Mov. Disord.* **18**:1306–1311.

92. Oliveira, S.A., et al. 2003. Association study of Parkin gene polymorphisms with idiopathic Parkinson disease. *Arch. Neurol.* **60**:975–980.

93. Lucking, C.B., et al. 2000. Association between early-onset Parkinson's disease and mutations in the parkin gene. French Parkinson's Disease Genetics Study Group. *N. Engl. J. Med.* **342**:1560–1567.

94. Mori, H., et al. 1998. Pathologic and biochemical studies of juvenile parkinsonism linked to chromosome 6q. *Neurology.* **51**:890–892.

95. Farrer, M., et al. 2001. Lewy bodies and parkinsonism in families with parkin mutations. *Ann. Neurol.* **50**:293–300.

96. Sasaki, S., Shirata, A., Yamane, K., and Iwata, M. 2004. Parkin-positive autosomal recessive juvenile Parkinsonism with alpha-synuclein-positive inclusions. *Neurology.* **63**:678–682.

97. Zhang, Y., et al. 2000. Parkin functions as an E2-dependent ubiquitin- protein ligase and promotes the degradation of the synaptic vesicle-associated protein, CDCrel-1. *Proc. Natl. Acad. Sci. U. S. A.* **97**:13354–13359.

98. Ko, H.S., et al. 2005. Accumulation of the authentic parkin substrate aminoacyl-tRNA synthetase cofactor, p38/JTV-1, leads to catecholaminergic cell death. *J. Neurosci.* **25**:7968–7978.

99. Ko, H.S., Kim, S.W., Sriram, S.R., Dawson, V.L., and Dawson, T.M. 2006. Identification of far up stream element binding protein-1 as an authentic parkin substrate. *J. Biol. Chem.* doi:10.1074/jbc.C600041200.

100. Oliveira, S.A., et al. 2005. Identification of risk and age-at-onset genes on chromosome 1p in Parkin-

son disease. *Am. J. Hum. Genet.* **77**:252–264.

101. Leroy, E., et al. 1998. The ubiquitin pathway in Parkinson's disease. *Nature.* **395**:451–452.

102. Healy, D.G., et al. 2006. UCHL-1 is not a Parkinson's disease susceptibility gene. *Ann. Neurol.* **59**:627–633.

103. Maraganore, D.M., et al. 2004. UCHL1 is a Parkinson's disease susceptibility gene. *Ann. Neurol.* **55**:512–521.

104. Liu, Y., Fallon, L., Lashuel, H.A., Liu, Z., and Lansbury, P.T., Jr. 2002. The UCH-L1 gene encodes two opposing enzymatic activities that affect alpha-synuclein degradation and Parkinson's disease susceptibility. *Cell.* **111**:209–218.

105. Lim, K.L., Dawson, V.L., and Dawson, T.M. 2006. Parkin-mediated lysine 63-linked polyubiquitination: a link to protein inclusions formation in Parkinson's and other conformational diseases? *Neurobiol. Aging.* **27**:524–529.

106. Chung, K.K., et al. 2004. S-nitrosylation of parkin regulates ubiquitination and compromises parkin's protective function. *Science.* **304**:1328–1331.

107. LaVoie, M.J., Ostaszewski, B.L., Weihofen, A., Schlossmacher, M.G., and Selkoe, D.J. 2005. Dopamine covalently modifies and functionally inactivates parkin. *Nat. Med.* **11**:1214–1221.

108. McNaught, K.S., and Jenner, P. 2001. Proteasomal function is impaired in substantia nigra in Parkinson's disease. *Neurosci. Lett.* **297**:191–194.

109. McNaught, K.S., Belizaire, R., Isacson, O., Jenner, P., and Olanow, C.W. 2003. Altered proteasomal function in sporadic Parkinson's disease. *Exp. Neurol.* **179**:38–46.

110. Bonifati, V., et al. 2003. Mutations in the DJ-1 gene associated with autosomal recessive early-onset parkinsonism. *Science.* **299**:256–259.

111. Dekker, M., et al. 2003. Clinical features and neuroimaging of PARK7-linked parkinsonism. *Mov. Disord.* **18**:751–757.

112. Lockhart, P.J., et al. 2004. DJ-1 mutations are a rare cause of recessively inherited early onset parkinsonism mediated by loss of protein function [letter]. *J. Med. Genet.* **41**:e22.

113. Choi, J., et al. 2006. Oxidative damage of DJ-1 is linked to sporadic Parkinson's and Alzheimer's diseases. *J. Biol. Chem.* **281**:10816–10824.

114. Zhang, L., et al. 2005. Mitochondrial localization of the Parkinson's disease related protein DJ-1: implications for pathogenesis. *Hum. Mol. Genet.* **14**:2063–2073.

115. Moore, D.J., et al. 2005. Association of DJ-1 and parkin mediated by pathogenic DJ-1 mutations and oxidative stress. *Hum. Mol. Genet.* **14**:71–84.

116. Zhou, W., Zhu, M., Wilson, M.A., Petsko, G.A., and Fink, A.L. 2006. The oxidation state of DJ-1 regulates its chaperone activity toward alpha-synuclein. *J. Mol. Biol.* **356**:1036–1048.

117. Kim, R.H., et al. 2005. Hypersensitivity of DJ-1-deficient mice to 1-methyl-4-phenyl-1,2,3,6-tetrahydropyrindine (MPTP) and oxidative stress. *Proc. Natl. Acad. Sci. U. S. A.* **102**:5215–5220.

118. Ito, G., Ariga, H., Nakagawa, Y., and Iwatsubo, T. 2006. Roles of distinct cysteine residues in S-nitrosylation and dimerization of DJ-1. *Biochem. Biophys. Res. Commun.* **339**:667–672.

119. Abou-Sleiman, P.M., Muqit, M.M., and Wood, N.W. 2006. Expanding insights of mitochondrial dysfunction in Parkinson's disease. *Nat. Rev. Neurosci.* **7**:207–219.

120. Valente, E.M., et al. 2004. Hereditary early-onset Parkinson's disease caused by mutations in PINK1. *Science.* **304**:1158–1160.

121. Li, Y., et al. 2005. Clinicogenetic study of PINK1 mutations in autosomal recessive early-onset parkinsonism. *Neurology.* **64**:1955–1957.

122. Klein, C., et al. 2005. PINK1, Parkin, and DJ-1 mutations in Italian patients with early-onset parkinsonism. *Eur. J. Hum. Genet.* **13**:1086–1093.

123. Valente, E.M., et al. 2004. PINK1 mutations are associated with sporadic early-onset parkinsonism. *Ann. Neurol.* **56**:336–341.

124. Ibanez, P., et al. 2006. Mutational analysis of the PINK1 gene in early-onset parkinsonism in Europe and North Africa. *Brain.* **129**:686–694.

125. Tan, E.K., et al. 2006. *PINK1* mutations in sporadic early-onset Parkinson's disease. *Mov. Disord.* doi:10.1002/mds.20810.

Huntingtin in health and disease

Anne B. Young

Neurology Service, Massachusetts General Hospital, Boston, Massachusetts, USA.

After linkage of the Huntington disease (HD) gene was found in 1983, it took ten years of work by an international group to identify the mutation in the gene *interesting transcript 15* (*IT15*) that causes the disease (1, 2). HD is an autosomal dominant inherited neurodegenerative disease that becomes manifest in midlife and causes progressive motor, psychiatric, and cognitive dysfunction. It is invariably terminal. HD symptoms can begin as early as 2 years or as late as 90 years, although the average age of onset is in the late 30s and early 40s. If a child inherits the gene from his or her father, a phenomenon called anticipation frequently occurs, whereby the child's age of onset is lower than the father's (3).

The *IT15* gene is composed of 67 exons and encodes a protein of 3,144 amino acids, called huntingtin (2). Exon 1 contains a CAG trinucleotide repeat that encodes the amino acid glutamine, followed by another repeat that encodes proline. In unaffected individuals, there are 10–34 CAG repeats. In those affected by HD, there are more than 40 repeats. In those with 35–39 repeats, the disease is variably penetrant (3–5). The age of onset of the disease varies inversely with the number of CAG repeats. Individuals with juvenile onset usually have over 55 repeats, and they usually inherit the gene from their father. Men occasionally have expanded repeats in their sperm (6). The expansion is thought to occur via slippage during the DNA replication process.

Wild-type huntingtin

Huntingtin contains a few domains that suggest particular functions, including WW domains and caspase cleavage sites (7, 8), but the function of the protein remains unknown. Huntingtin is expressed in the cytoplasm of most cells in the body. In the brain, expression is found predominantly in neurons (2, 9–11). Within the cell, the protein is associated with the endoplasmic reticulum, the microtubules, and organelles such as the mitochondria and synaptic vesicles (12). The association of huntingtin with these cellular components is loose, since during the process of differential centrifugation huntingtin can be separated from them (10, 12). The protein contains several caspase cleavage sites (8). Yeast two-hybrid screening studies have shown that huntingtin binds to a number of proteins, including Hap-1, α-adaptin, and several others (7, 13, 14). Additional studies found that huntingtin binds to GAPDH (15). Huntingtin was also recently shown to associate with PSD95, a protein found at the postsynaptic membrane and involved in anchoring of receptor proteins, particularly the *N*-methyl-D-aspartate (NMDA) receptor (16).

The huntingtin protein is necessary for normal development in mice, as knockouts do not survive beyond day 7–8 of embryogenesis (17–19). Conditional knockouts, in which the gene is turned off during adulthood, develop a neurodegenerative disease (20). Recently, cell models of huntingtin expression revealed that the wild-type protein partially protects cells from noxious stimuli or from mutant huntingtin (21). Huntingtin may also be important for cell survival through mechanisms such as growth factor stimulation, as BDNF rescued cells expressing mutant huntingtin (22). Removal of the caspase sites in the protein is beneficial for cells, and wild-type huntingtin protects against mutant-induced cell death (23).

Mutant huntingtin

There are no antibodies that can distinguish mutant from wild-type protein, although there are antibodies that bind selectively to the polyglutamine repeat (24, 25). Presumed mutant huntingtin is found not only in the cytoplasm but also in the nucleus, where it forms aggregates (or neuronal intranuclear inclusions [NIIs]) (26–28). Aggregates also develop in neurites. The aggregates are ubiquitinated, although antibodies against huntingtin appear to stain more aggregates than do antibodies against ubiquitin. Western blot analysis of HD brain tissue shows full-length huntingtin protein in the nuclear fraction as well as abundant immunopositive bands at lower molecular weight, suggesting proteolytic products in the nucleus. In contrast, in control brains there was full-length protein in the total homogenate but no nuclear protein and few huntingtin fragments in any fraction (26).

Studies of HD brains show that there are more inclusions in the cortex than in the striatum, and that cortical and striatal neurites contain numerous aggregates (11, 27). Postmortem studies of HD brains also show differential loss of projection neurons containing enkephalin, adenosine A2a, and dopamine D2 receptors compared with cells containing substance P, dynorphin, and dopamine D1 receptors (29–31). In juvenile HD, both types of striatal projection neurons are equally affected (32). In the cortex, neurons in the deeper layers (layers V and VI), which use the neurotransmitter glutamate, develop nuclear and neurite aggregates (27).

Aggregates in HD were first observed in an electromicroscopy study of in vivo biopsies of HD brains (33). This observation was not pursued at the time, but it was remembered in 1996 when studies of the first HD transgenic mouse (28) (expressing the first exon of human huntingtin driven by the huntingtin promoter) were reported. These mice develop normally until around 5 weeks of age, when they begin to lose weight and to perform less well on the Rotorod test. Both brain weight and body weight diminish subsequently. The animals develop diabetes and tremors and become less active. They are finally moribund and die at around 13 weeks (34). Extensive early studies of the brains of these animals showed no clear neurochemical abnormalities like those seen in postmortem HD brains. Electromicroscopy studies, however, showed intranuclear inclusions. In the transgenic mice, it then became obvious from immunocytochemical studies

Nonstandard abbreviations used: Huntington disease (HD); *N*-methyl-D-aspartate (NMDA); neuronal intranuclear inclusion (NII).

Conflict of interest: The author has declared that no conflict of interest exists.

Citation for this article: *J. Clin. Invest.* **111**:299–302 (2003). doi:10.1172/JCI200317742.

that these NIIs were positive for the HD protein and for ubiquitin. Furthermore, virtually all neurons in the brains of these so-called R6/2 mice contained NIIs (28). These studies led scientists to revisit the examination of human brains in which NIIs were also found (11, 26). The frequency of NIIs in human HD brains was lower than in the transgenic mice, and the aggregates appeared as described above.

The formation of aggregates was subsequently thought to be the sine qua non of HD pathogenesis. Two papers then appeared that suggested that the aggregates were an epiphenomenon, since cell death did not necessarily result from neuronal huntingtin aggregation, yet cell death did arise after the expression of mutant huntingtin (35, 36). It appeared that, to be toxic, the mutant protein had to get into the nucleus, since constructs with nuclear-export signals attenuated death resulting from exon 1 overexpression. Furthermore, inhibition of caspases rescued cells from death.

Advocates of the hypothesis that aggregates cause cell death have been studying the phenomenon in various in vitro and cell culture assays (37, 38). Polyglutamine peptides are not soluble and need to be tied to other proteins to be studied in solution. Tight aggregates apparently form into polar zippers that are held together by hydrogen bonding (39). Some have also hypothesized that transglutaminases link glutamines to lysines covalently (40). No one has yet purified enough of the N-terminal fragment expressing the polyglutamine stretch to perform crystallization. The purified full-length protein has not been isolated. Aggregation, however, can be studied by fusing a polyglutamine peptide with a GST protein with a sequence that can be broken with trypsin (41). Once the protein is in solution, the GST protein is cleaved off with trypsin, and then the aggregation process can be followed by filter assays. Aggregation is dependent on the length of the polyglutamine repeat. A transition seems to occur in the range of 39–40 repeats. Peptides with fewer than 39–40 glutamines aggregate less robustly than peptides with more than 40 repeats. Cell culture models also show length-dependent polyglutamine aggregation. High-throughput screens using in vitro and cell culture assays are now being employed to identify compounds that interfere with the aggregation process.

Transgenic mice

The first transgenic mouse model of HD is described above (34). Numerous additional models have since been created, and each appears to lend new information about the disease (42–45). The ones that have been studied in most detail are the R6/2 and R6/1 mice (34), the N171 mice (42), the YAC72 mice (44), and the conditional exon 1 transgenics and knockouts (20, 46). Both strains of R6 mice exhibit little cell death and neuritic pathology but widespread NIIs. N171 mice show striatal cell death and widespread NIIs. N171 mice develop normally but have the onset of progressive neurological decline at about 4 months. In N171 mice, the transgene is driven by the prion promoter. YAC72 mice, which contain the entire human gene and the human promoter, show variable amounts of hyperactivity at about 1 year and then become less active, displaying striatal pathology and evidence of apoptotic cell death.

Examination of signaling proteins in the brains of R6 animals showed changes consistent with an alteration in gene transcription (47). Genes encoding proteins, such as receptors, that had previously been found to change in HD were expressed at lower levels, in a pattern consistent with the human studies. Microarray analysis of the brains of R6 mice, interrogating over 6,500 mouse genes, revealed changes in expression levels mainly of genes involved in signaling pathways (48). R6 mice have been treated with a variety of agents, including free radical scavengers, glutamate antagonists, creatine, and caspase inhibitors, and these are seen to prolong life by about 10–20% (49, 50). Indeed, even environmental enrichment prolongs life by 20–50% (51). R6/2 mice have been crossed with dominant negative IL-1–converting enzyme (ICE) knockdowns, and offspring live longer than controls (50). Intraventricular infusions of caspase inhibitors also prolong life, as does minocycline (an ICE inhibitor) (52). Electrophysiological studies of cells in brain slices revealed that, early in the disease, there are enhanced NMDA receptor responses, but eventually cells become resistant to NMDA agonists (53, 54).

N171 mice exhibited abnormalities that are similar to those found in R6 mice, but they have not been subjected to the same level of examination (48). YAC mice have been examined pathologically but not neurochemically (44).

Conditional exon 1 transgenic mice show behavioral abnormalities, brain atrophy, NIIs, striatal gliosis, and reduction in dopamine D1 receptors when the transgene is expressed, and a reversal of these changes when the transgene is turned off in adulthood (46). Conditional knockouts show a progressive neurodegenerative disorder when the wild-type mouse gene is turned off in adulthood (20).

Other whole-organism models

Worm, fruit fly, and yeast models of HD have been created primarily by transgene approaches using glutamine-encoding trinucleotide expansions (55–57). These models have allowed the identification of suppressors and enhancers of expanded polyglutamine-induced pathology. Heat shock proteins, for example, appear to repress polyQ-induced pathology (58). In PC12 cells (see below), histone deacetylase inhibitors were found to attenuate polyQ-induced toxicity, suggesting that drugs that attenuate histone deacetylation might have therapeutic potential (57). Studies of such compounds are now underway in several transgenic animals.

Cell culture models

Several cell models of HD are available. Mouse neurons have been fused with mouse teratoma cells (35), and the resulting hybrids have been stably transfected with various polyQ-containing peptides. Another model is based on stably transfected, temperature-sensitive, immortalized mouse striatal neurons (21). PC12 cells have been stably transfected with various genes encoding polyglutamines (59, 60). One study of transfected PC12 cells found that CBP protein was sequestered in aggregates in a polyQ-dependent manner, suggesting that genes controlled by CBP may be predominantly affected in the disease (61). Since CBP has histone acetylase activity, these findings suggested that histone deacetylase inhibitors might be used as treatment (57). Indeed, histone deacetylase inhibitors improve phenotype in transgenic flies (57).

Human studies

In 1996, the Huntington Study Group was formed (62). It is a multi-institutional organization of sites involved in collaborative studies of individuals with HD. Clinical trials of a variety of putative neuroprotective agents have been carried out. These include trials of the glutamate antagonists lamotrigine, remacemide, and riluzole (63). Riluzole and remacemide improved motor function, but lamotragine had no effect. Vitamin E, a free radical scavenger, was not an effective treatment; neither was idebenone, a booster of energy metabolism. Recently, a two-by-two factorial study of coenzyme Q10 and remacemide showed a 13% slowing of progression with coenzyme Q10 and an improvement in motor function with remacemide, although the results were not statistically significant (64, 65). Studies of riluzole, creatine, and minocycline in HD patients are now underway. As we look toward treating individuals in the presymptomatic or early symptomatic phases of the disease, biomarkers or other characteristics of disease onset and progression need to be defined.

Summary

HD has received at great deal of attention in the field of neuroscience as a model of neurodegeneration. Because it is relatively common and presymptomatic individuals can be identified, investigators have focused on the disease in the hope of finding therapies that can be given to gene carriers presymptomatically to prevent disease development. The results to date suggest that huntingtin is necessary for developing and sustaining normal brain function. In HD, the protein with the expanded polyQ may not function as effectively as the wild-type protein, and this may put neurons (especially in the striatum) under stress. Striatal neurons then become vulnerable to the abundant glutamatergic input from the cerebral cortex. Excitotoxicity, mitochondrial stress, and free radicals increase, and caspases within the cell are activated. Mutant huntingtin is then cleaved, resulting in polyQ-containing fragments that are susceptible to aggregate formation. The mutant fragments and aggregates recruit transcriptional factors vital to the normal function of neurons. Cells survive in a dysfunctional state for some time, and it appears that turning off the mutant gene can result in reversal of neurodegeneration. As many steps in the process of functional decline and cell death represent potential drug targets, we should eventually find a cure.

Acknowledgments

This work was supported by US Public Health Service grants NS38106 and AG13617, and the Hereditary Disease Foundation.

Address correspondence to: Anne B. Young, Neurology Service, VBK-915, Massachusetts General Hospital, 32 Fruit Street, Boston, Massachusetts 02114, USA. Phone: (617) 726-2385; Fax: (617) 726-2353; E-mail: young@helix.mgh.harvard.edu.

1. Gusella, J.F., et al. 1983. A polymorphic DNA marker genetically linked to Huntington's disease. *Nature.* **306**:234–238.
2. 1993. A novel gene containing a trinucleotide repeat that is unstable in Huntington's disease chromosomes. The Huntington's Disease Collaborative Research Group. *Cell.* **72**:971–983.
3. Young, A.B. 1998. Huntington's disease and other trinucleotide repeat disorders. In *Molecular neurology.* J.B. Martin, editor. Scientific American Medicine. New York, New York, USA. 35–54.
4. Rubinsztein, D.C., et al. 1996. Phenotypic characterization of individuals with 30-40 CAG repeats in the Huntington disease (HD) gene reveals HD cases with 36 repeats and apparently normal elderly individuals with 36-39 repeats. *Am. J. Hum. Genet.* **59**:16–22.
5. Snell, R.G., et al. 1993. Relationship between trinucleotide repeat expansion and phenotypic variation in Huntington's disease. *Nat. Genet.* **4**:393–397.
6. MacDonald, M.E., et al. 1993. Gametic but not somatic instability of CAG repeat length in Huntington's disease. *J. Med. Genet.* **30**:982–986.
7. Faber, P.W., et al. 1998. Huntingtin interacts with a family of WW domain proteins. *Hum. Mol. Genet.* **7**:1463–1474.
8. Wellington, C.L., et al. 1998. Caspase cleavage of gene products associated with triplet expansion disorders generates truncated fragments containing the polyglutamine tract. *J. Biol. Chem.* **273**:9158–9167.
9. Landwehrmeyer, G.B., et al. 1995. Huntington's disease gene: regional and cellular expression in brain of normal and affected individuals. *Ann. Neurol.* **37**:218–230.
10. DiFiglia, M., et al. 1995. Huntingtin is a cytoplasmic protein associated with vesicles in human and rat brain neurons. *Neuron.* **14**:1075–1081.
11. Ferrante, R.J., et al. 1997. Heterogeneous topographic and cellular distribution of Huntington expression in the normal human neostriatum. *J. Neurosci.* **17**:3052–3063.
12. Velier, J., et al. 1998. Wild-type and mutant huntingtins function in vesicle trafficking in the secretory and endocytic pathways. *Exp. Neurol.* **152**:34–40.
13. Li, X.J., et al. 1995. A huntingtin-associated protein enriched in brain with implications for pathology. *Nature.* **378**:398–402.
14. Kalchman, M.A., et al. 1997. HIP1, a human homologue of S. cerevisiae Sla2p, interacts with membrane-associated huntingtin in the brain. *Nat. Genet.* **16**:44–53.
15. Burke, J.R., et al. 1996. Huntingtin and DRPLA proteins selectively interact with the enzyme GAPDH. *Nat. Med.* **2**:347–350.
16. Sun, Y., Savanenin, A., Reddy, P.H., and Liu, Y.F. 2001. Polyglutamine-expanded huntingtin promotes sensitization of N-Methyl-D-aspartate receptors via post-synaptic density 95. *J. Biol. Chem.* **276**:24713–24718.
17. Duyao, M.P., et al. 1995. Inactivation of the mouse Huntington's disease gene homolog Hdh. *Science.* **269**:407–410.
18. O'Kusky, J.R., Nasir, J., Cicchetti, F., Parent, A., and Hayden, M.R. 1999. Neuronal degeneration in the basal ganglia and loss of pallido-subthalamic synapses in mice with targeted disruption of the Huntington's disease gene. *Brain Res.* **818**:468–479.
19. Dragatsis, I., Efstratiadis, A., and Zeitlin, A. 1998. Mouse mutant embryos lacking huntingtin are rescued from lethality by wild type extraembryonic tissues. *Development.* **125**:1529–1539.
20. Dragatsis, I., Levine, M.S., and Zeitlin, S. 2000. Inactivation of Hdh in the brain and testis results in progressive neurodegeneration and sterility in mice. *Nat. Genet.* **26**:300–306.
21. Rigamonti, D., et al. 2000. Wild-type huntingtin protects from apoptosis upstream of caspase-3. *J. Neurosci.* **20**:3705–3713.
22. Zuccato, C., et al. 2001. Loss of huntingtin-mediated BDNF gene transcription in Huntington's disease. *Science.* **293**:493–498.
23. Leavitt, B.R., et al. 2001. Wild-type huntingtin reduces the cellular toxicity of mutant huntingtin in vivo. *Am. J. Hum. Genet.* **68**:313–324.
24. Persichetti, F., et al. 1996. Differential expression of normal and mutant Huntington's disease gene alleles. *Neurobiol. Dis.* **3**:183–190.
25. Trottier, Y., et al. 1995. Cellular localization of the Huntington's disease protein and discrimination of the normal and mutated form. *Nat. Genet.* **10**:104–110.
26. DiFiglia, M., et al. 1997. Aggregation of huntingtin in neuronal intranuclear inclusions and dystrophic neurites in brain. *Science.* **277**:1990-1993.
27. Sapp, E., et al. 1997. Huntingtin localization in brains of normal and Huntington's disease patients. *Ann. Neurol.* **42**:604–612.
28. Davies, S.W., et al. 1997. Formation of neuronal intranuclear inclusions underlies the neurological dysfunction in mice transgenic for the HD mutation. *Cell.* **90**:537–548.
29. Reiner, A., et al. 1988. Differential loss of striatal projection neurons in Huntington disease. *Proc. Natl. Acad. Sci. USA.* **85**:5733–5737.
30. Albin, R.L., et al. 1990. Abnormalities of striatal projection neurons and N-methyl-D-aspartate receptors in presymptomatic Huntington's disease. *N. Engl. J. Med.* **322**:1293–1298.
31. Glass, M., Dragunow, M., and Faull, R.L. 2000. The pattern of neurodegeneration in Huntington's disease: a comparative study of cannabinoid, dopamine, adenosine and GABA(A) receptor alterations in the human basal ganglia in Huntington's disease. *Neuroscience.* **97**:505–519.
32. Albin, R.L., Reiner, A., Anderson, K.D., Penney, J.B., and Young, A.B. 1990. Striatal and nigral neuron subpopulations in rigid Huntington's disease: implications for the functional anatomy of chorea and rigidity-akinesia. *Ann. Neurol.* **27**:357–365.
33. Roizin, L., Stellar, S., and Liu, J.C. 1979. Neuronal nuclear-cytoplasmic changes in Huntington's chorea: electron microscope investigations. In *Advances in neurology.* Volume 23. T.N. Chase, N.S. Wexler, and A. Barbeau, editors. Raven Press. New York, New York, USA. 95–122.
34. Mangiarini, L., et al. 1996. Exon 1 of the *HD* gene with an expanded CAG repeat is sufficient to cause a progressive neurological phenotype in transgenic mice. *Cell.* **87**:493–506.
35. Kim, M., et al. 1999. Mutant huntingtin expression in clonal striatal cells: dissociation of inclusion formation and neuronal survival by caspase inhibition. *J. Neurosci.* **19**:964–973.
36. Saudou, F., Finkbeiner, S., Devys, D., and Greenberg, M.E. 1998. Huntingtin acts in the nucleus to induce apoptosis but death does not correlate with the formation of intranuclear inclusions. *Cell.* **95**:55–66.
37. Scherzinger, E., et al. 1997. Huntingtin-encoded polyglutamine expansions form amyloid-like protein aggregates in vitro and in vivo. *Cell.* **90**:549–558.
38. Huang, C.C., et al. 1998. Amyloid formation by mutant huntingtin: threshold, progressivity and recruitment of normal polyglutamine proteins. *Somat. Cell Mol. Genet.* **24**:217–233.
39. Perutz, M.F., Johnson, T., Suzuki, M., and Finch, J.T. 1994. Glutamine repeats as polar zippers: their possible role in inherited neurodegenerative diseases. *Proc. Natl. Acad. Sci. USA.* **91**:5355–5358.
40. Green, D.R. 1999. Harm's way: polyglutamine

repeats and the activation of an apoptotic pathway. *Neuron.* **22**:416–417.

41. Hollenbach, B., et al. 1999. Aggregation of truncated GST-HD exon 1 fusion proteins containing normal range and expanded glutamine repeats. *Philos. Trans. R. Soc. Lond. B Biol. Sci.* **354**:991–994.

42. Schilling, G., et al. 1999. Intranuclear inclusions and neuritic aggregates in transgenic mice expressing a mutant N-terminal fragment of huntingtin. *Hum. Mol. Genet.* **8**:397–407.

43. Reddy, P.H., et al. 1999. Transgenic mice expressing mutated full-length HD cDNA: a paradigm for locomotor changes and selective neuronal loss in Huntington's disease. *Philos. Trans. R. Soc. Lond. B Biol. Sci.* **354**:1035–1045.

44. Hodgson, J.G., et al. 1999. A YAC mouse model for Huntington's disease with full-length mutant huntingtin, cytoplasmic toxicity, and selective striatal neurodegeneration. *Neuron.* **23**:181–192.

45. Ordway, J.M., et al. 1997. Ectopically expressed CAG repeats cause intranuclear inclusions and a progressive late onset neurological phenotype in the mouse. *Cell.* **91**:753–763.

46. Yamamoto, A., Lucas, J.J., and Hen, R. 2000. Reversal of neuropathology and motor dysfunction in a conditional model of Huntington's disease. *Cell.* **101**:57–66.

47. Cha, J.-H.J., et al. 1998. Altered brain neurotransmitter receptors in transgenic mice expressing a portion of an abnormal human Huntington disease gene. *Proc. Natl. Acad. Sci. USA.* **95**:6480–6485.

48. Luthi-Carter, R., et al. 2000. Decreased expression of striatal signaling genes in a mouse model of Huntington's disease. *Hum. Mol. Genet.* **9**:1259–1271.

49. Ferrante, R.J., et al. 2000. Neuroprotective effects of creatine in a transgenic mouse model of Huntington's disease. *J. Neurosci.* **20**:4389–4397.

50. Ona, V.O., et al. 1999. Inhibition of caspase-1 slows disease progression in a mouse model of Huntington's disease. *Nature.* **399**:263–267.

51. Carter, R.J., Hunt, M.J., and Morton, A.J. 2000. Environmental stimulation increases survival in mice transgenic for exon 1 of the Huntington's disease gene. *Mov. Disord.* **15**:925–937.

52. Chen, M., et al. 2000. Minocycline inhibits caspase-1 and caspase-3 expression and delays mortality in a transgenic mouse model of Huntington disease. *Nat. Med.* **6**:797–801.

53. Levine, M.S., et al. 1999. Enhanced sensitivity to N-methyl-D-aspartate receptor activation in transgenic and knockin mouse models of Huntington's disease. *J. Neurosci. Res.* **58**:515–532.

54. Hansson, O., et al. 1999. Transgenic mice expressing a Huntington's disease mutation are resistant to quinolinic acid–induced striatal excitotoxicity. *Proc. Natl. Acad. Sci. USA.* **96**:8727–8732.

55. Sipione, S., and Cattaneo, E. 2001. Modeling Huntington's disease in cells, flies, and mice. *Mol. Neurobiol.* **23**:21–51.

56. Faber, P.W., Alter, J.R., MacDonald, M.E., and Hart, A.C. 1999. Polyglutamine-mediated dysfunction and apoptotic death of a Caenorhabditis elegans sensory neuron. *Proc. Natl. Acad. Sci. USA.* **96**:179–184.

57. Steffan, J.S., et al. 2001. Histone deacetylase inhibitors arrest polyglutamine-dependent neurodegeneration in Drosophila. *Nature.* **413**:739–743.

58. Satyal, S.H., et al. 2000. Polyglutamine aggregates alter protein folding homeostasis in Caenorhabditis elegans. *Proc. Natl. Acad. Sci. USA.* **97**:5750–5755.

59. Li, S.H., Cheng, A.L., Li, H., and Li, X.J. 1999. Cellular defects and altered gene expression in PC12 cells stably expressing mutant huntingtin. *J. Neurosci.* **19**:5159–5172.

60. Kazantsev, A., Preisinger, E., Dranovsky, A., Goldgaber, D., and Housman, D. 1999. Insoluble detergent-resistant aggregates form between pathological and nonpathological lengths of polyglutamine in mammalian cells. *Proc. Natl. Acad. Sci. USA.* **96**:11404–11409.

61. Steffan, J.S., et al. 2000. The Huntington's disease protein interacts with p53 and CBP and represses transcription. *Proc. Natl. Acad. Sci. USA.* **97**:6763–6768.

62. Kieburtz, K., et al. 1996. Unified Huntington's disease rating scale: reliability and consistency. *Mov. Disord.* **11**:136–142.

63. Kieburtz, K. 1999. Antiglutamate therapies in Huntington's disease. *J. Neural Transm. Suppl.* **55**:97–102.

64. Schilling, G., Coonfield, M.L., Ross, C.A., and Borchelt, D.R. 2001. Coenzyme Q10 and remacemide hydrochloride ameliorate motor deficits in a Huntington's disease transgenic mouse model. *Neurosci. Lett.* **315**:149–153.

65. 2001. A randomized, placebo-controlled trial of coenzyme Q10 and remacemide in Huntington's disease. *Neurology.* **57**:397–404.

Programmed cell death in amyotrophic lateral sclerosis

Christelle Guégan[1,2] and Serge Przedborski[1,3,4]

[1]Department of Neurology, Columbia University, New York, New York, USA. [2]Institut National de la Santé et de la Recherche Médicale, Unit 421, Institut Mondor de Médecine Moléculaire, Créteil, France. [3]Department of Pathology, and [4]Center of Neurobiology and Behavior, Columbia University, New York, New York, USA

Amyotrophic lateral sclerosis (ALS) is a relentless fatal paralytic disorder confined to the voluntary motor system (1). Its prevalence is about three to five in 100,000 individuals, making it the most frequent paralytic disease in adults. Although ALS can strike anyone at any age, generally the onset of the disease is in the fourth or fifth decade of life. Common clinical features of ALS include muscle weakness, fasciculations, brisk (or depressed) reflexes, and extensor plantar responses. Even though motor deficit usually predominates in the limbs, bulbar enervation can also be severely affected, leading to atrophy of the tongue, dysphagia, and dysarthria. Other cranial nerves (e.g., oculomotor nerves) are usually spared. The progressive decline of muscular function results in paralysis, speech and swallowing disabilities, emotional disturbance, and, ultimately, respiratory failure causing death among the vast majority of ALS patients within 2–5 years after the onset of the disease. Pathologically, ALS is characterized by a loss of upper motor neurons in the cerebral cortex and of the lower motor neurons in the spinal cord. Often, there is also a profound degeneration of the corticospinal tracts, which is most evident at the level of the spinal cord. The few remaining motor neurons are generally atrophic, and many demonstrate abnormal accumulation of neurofilament, in both their cell bodies and axons. To date, only a few approved treatments (e.g., mechanical ventilation and riluzole) prolong survival in ALS patients to some extent. However, the development of more effective neuroprotective therapies remains impeded by our limited knowledge of the actual mechanisms by which neurons die in ALS, and of how the disease progresses and propagates.

ALS, like other common neurodegenerative disorders, is sporadic in the vast majority of patients, and familial in only a few (1). The clinical and pathological expressions of ALS are almost indistinguishable between the familial and sporadic forms, although often in the former the age at onset is younger, the course of the disease more rapid, and the survival after diagnosis shorter (1). The cause of sporadic ALS remains unknown, while that of at least some familial forms has been identified (see below). Although the identified gene defects responsible for ALS account for a minute fraction of cases, most experts believe that unraveling the molecular basis by which those mutant gene products cause neurodegeneration may shed light on the etiopathogenesis of the common sporadic form of ALS.

Nonstandard abbreviations used: amyotrophic lateral sclerosis (ALS); copper/zinc superoxide dismutase (SOD1); pheochromocytoma-12 (PC-12); programmed cell death (PCD); extracellular signal–regulated kinase (ERK); neurotrophin receptor (NTR); apoptotic protease-activating factor-1 (Apaf-1).

Conflict of interest: The authors have declared that no conflict of interest exists.

Citation for this article: *J. Clin. Invest.* **111**:153–161 (2003). doi:10.1172/JCI200317610.

Genetic forms of ALS

Although familial ALS is often referred to as a single entity, genetic evidence actually reveals at least four different types that have been assigned to distinct loci of the human genome (2). This review will focus on a form that is responsible for the disease in approximately 20% of all familial cases and that is linked to mutations in the gene for the cytosolic free radical–scavenging enzyme superoxide dismutase 1 (SOD1) (3, 4). To date, approximately 100 different point mutations in SOD1 throughout the entire gene have been identified in ALS families, and all but one is dominant. Many of these mutations lead to the substitution of an amino acid within regions of the enzyme with very distinct structural and functional roles. It is thus fascinating to note that so many discrete SOD1 alterations share a similar clinical phenotype, even though disease duration and, to a lesser extent, age at onset vary among patients with different SOD1 mutations (5). Also astonishing is the fact that SOD1 mutations, which are present at birth and, by virtue of SOD1's ubiquitous expression, in all tissues, produce a rapidly progressive adult-onset degenerative condition in which motor neurons are almost exclusively affected.

Most of these mutations have apparently reduced enzymatic activity (3, 6), a finding that has prompted investigators to test whether a loss of SOD1 activity can kill neurons. It was unequivocally shown that reducing SOD1 activity to about 50% using antisense oligonucleotides kills pheochromocytoma-12 (PC-12) cells and motor neurons in spinal cord organotypic cultures (7, 8). However, mutant mice deficient in SOD1 do not develop any motor neuron disease (9), and the transgenic expression of different SOD1 mutants in both mice (10–12) and rats (13) causes an ALS-like syndrome in these animals, whether SOD1 free radical–scavenging catalytic activity is increased, normal, or almost absent (10–14). These observations provide compelling evidence that the cytotoxicity of mutant SOD1 is mediated not by a loss-of-function but rather by a gain-of-function effect (15).

Transgenic mutant SOD1 mouse model of ALS

As indicated above, the transgenic expression of different SOD1 mutants in both mice (10–12) and rats (13) produces a paralytic syndrome in these animals that replicates the clinical and pathological hallmarks of ALS. The age at onset of symptoms and the lifespan of these transgenic rodents vary among the different lines, depending on the mutation expressed and its level of expression, but when they become symptomatic they invariably show motor abnormalities that progress with the same pattern (11, 16).

The first motor abnormality, at least in mice, is the development of a fine tremor in at least one limb when the animal is held in the air by the tail (16). Thereafter, weakness and atrophy of proximal muscles, predominantly in the hind limbs, develop progressively. At the end stage, transgenic mutant SOD1 mice are severely paralyzed and can no longer feed or drink on their own (16). Neither their nontransgenic littermates nor age-matched transgenic mice expressing wild-type SOD1 enzyme develop any of these motor abnormalities.

The first neuropathological changes seen in transgenic mutant SOD1 mice are perikarya, axonal and dendritic vacuoles in motor neurons with little involvement in the surrounding neuropil, and undetectable neuronal loss or gliosis (11, 17). In the transgenic mutant SOD1 mice that express a glycine-to-alanine substitution at position 93 (G93A) (10), these changes are observed in 4- to 6-week-old asymptomatic animals. By the time the first symptom, fine limb tremor, arises (about 90 days), vacuolization is prominent, and some neuronal loss, especially of large motor neurons (>25 μm), is observed in the spinal cord. At the end stage, dramatic paralysis (about 140 days), there is still some degree of vacuolization, but the prominent features are the dramatic loss of motor neurons (~50%), an abundance of dystrophic neurites, a marked gliosis (18), some globular Lewy body–like intracellular inclusions, and a dearth of motor neurons filled with phosphorylated neurofilaments (10, 16, 17, 19). Despite the close similarities between the phenotype of transgenic mutant SOD1 rodents and ALS, this experimental model departs from the human disease in a few important ways. First, vacuolar degeneration has not been a well-recognized component of motor neuron pathology in ALS. Second, neurofilamentous accumulation in cell bodies and proximal axons is infrequent in the lines of transgenic animals that express mutant SOD1, while it is conspicuous in ALS. Third, none of the transgenic lines show degeneration in the rodent equivalent of the human corticospinal tract. Notably, these transgenic animals replicate in rodents the effect of mutant SOD1, but how relevant this is to the sporadic form of ALS — which is not linked to SOD1 mutations — is unknown. Despite these imperfections and limitations, transgenic mutant SOD1 rodents unquestionably represent an excellent experimental model of ALS, one which has already generated valuable insights into the pathogenesis of ALS and opened new therapeutic avenues for this dreadful disease.

Hypothesis for mutant SOD1 cytotoxicity

Despite the explosion of ALS research engendered by the discovery of the SOD1 mutations, the actual nature of the gained function by which mutant SOD1 kills motor neurons in ALS remains elusive. Multiple mechanisms have been implicated in the demise of motor neurons in ALS (20), but only a few may be directly relevant to the form linked to mutant SOD1. For instance, the known free radical–scavenging function of SOD1 led researchers to believe that mutant SOD1–induced neurodegeneration was due to an oxidative stress. This idea was, at least initially, received with enthusiasm due to the fact that a variety of markers of oxidative damage are indeed increased in ALS spinal cords (20). Currently, it is thought that, if SOD1 mutants were to generate oxidative stress, they could do so by two distinct and not mutually exclusive mechanisms. In the first mechanism, the point mutations would relax SOD1 conformation, hence allowing abnormal kinds or amounts of substrates to reach and react with the transitional metal — copper — contained in the catalytic site of the enzyme. Among the aberrant substrates to be proposed are peroxynitrite (21) and hydrogen peroxide (22), both of which can directly or indirectly mediate serious tissue damage. In the second mechanism, it is speculated that SOD1 mutations are associated with a

labile binding of zinc to the protein (23), and that, by having lost zinc, mutant SOD1, in the presence of nitric oxide, will catalyze the production of peroxynitrite (24), which can inflict serious oxidative damage to virtually all cellular elements.

Alternatively, mutant SOD1 cytotoxicity may result from the propensity of this mutant protein to form intracellular proteinaceous aggregates (25), which are a prominent pathological feature of several of the transgenic lines (12, 13, 19), and of various cultured cell types expressing mutant SOD1, including motor neurons (26). As in other neurodegenerative disorders with intracellular inclusions, whether or not these proteinaceous aggregates are actually noxious remains uncertain. Nevertheless, it may be speculated that their presence in the cytosol of motor neurons may be deleterious, by, for example, impairing the microtubule-dependent axonal transport of vital nutriments, or by perturbing the normal turnover of intracellular proteins (27).

Early on in the effort to determine the nature of mutant SOD1's gained function, it was discovered that transfected neuronal cells expressing mutant SOD1 cDNA were dying by apoptosis (28), a form of programmed cell death (PCD). Similar observations were subsequently made in transfected PC-12 cells (29) and in primary neurons grown from transgenic mice expressing mutant SOD1 (30). Collectively, these in vitro data have led many investigators to consider that mutant SOD1 may kill motor neurons by activating PCD, a term that we here use in the sense of cell death mediated by specific signaling pathways. The possible implication of PCD in ALS has been rather appealing to the field of motor neuron diseases ever since the neuronal apoptosis inhibitory protein (NAIP) was identified as a candidate gene for an inherited ALS-related disorder, spinal muscular atrophy (31). The remainder of this review will focus on PCD in ALS. Because most of the published mechanistic investigations of that topic have been performed in transgenic mutant SOD1 mice, this appraisal will emphasize this mouse model of ALS, but human data will be cited whenever possible to support the relevance of the animal findings to the human condition.

Morphology of dying motor neurons

In light of the presumed proapoptotic properties of mutant SOD1 observed in vitro, it may be wondered whether, in transgenic mutant SOD1 mice, dying spinal cord motor neurons would also exhibit features of apoptosis, whose morphological hallmarks include cytoplasmic and nuclear condensation, compaction of nuclear chromatin into sharply circumscribed masses along the inside of the nuclear membrane, and structural preservation of organelles (at least until the cell is broken into membrane-bound fragments called apoptotic bodies that are phagocytized). This question has been examined in several careful morphological studies performed in transgenic mutant SOD1 mice (17, 32–34). In these animals, most of the sick neurons are atrophic, and their cytoplasm is occupied with vacuoles corresponding to dilated rough ER, Golgi apparatus, and mitochondria (17). From our own ultrastructural studies in these mice (S. Przedborski, unpublished observations), we can add that many sick neurons have diffusely condensed cytoplasm and nuclei and irregular shapes. Although the actual type of this cell death remains to be determined, these dying neurons exhibit a rather nonapoptotic morphology with some features reminiscent of autophagic or cytoplasmic neuronal death (35). Yet, in our experience, definitely apoptotic cells are seen but are rare in the spinal cord of affected transgenic mutant SOD1

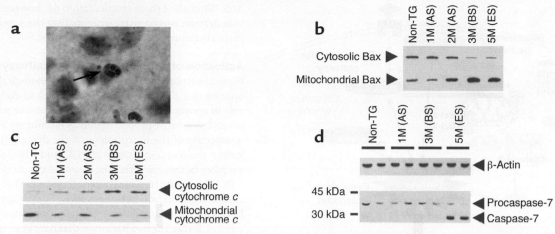

Figure 1
Illustrations of PCD alterations in spinal cord of transgenic mutant SOD1 mice. (**a**) Photomicrograph of definitely apoptotic cells found in the anterior horn of an end-stage transgenic SOD1^{G93A} mouse. The arrow shows several typical round chromatin clumps. (**b**) Western blot analysis of spinal cord extracts, demonstrating the relocation of Bax from the cytosol (top) to the mitochondria (bottom) over the course of the disease. (**c**) Coincidental changes of cytochrome *c* in the opposite direction. (**d**) Later in the disease, effector caspases such as caspase-7 are activated. 1M (AS): 1 month, asymptomatic stage; 2M (AS): 2 months, asymptomatic stage; 3M (BS): 3 months, beginning of symptoms; 5M (ES): 5 months, end stage; Non-TG: nontransgenic littermates. Modified from ref. 63.

mice (Figure 1a). For instance, in end-stage transgenic SOD1^{G93A} mice, which have lost about 50% of their anterior-horn motor neurons, it can be estimated that about two apoptotic cells will be seen per 40-μm-thick section of the lumbar spinal cord. We have also observed that the vast majority of these apoptotic cells no longer exhibit definite morphological characteristics or express phenotypic markers that allow their identification as neurons or glia. However, some (less than 15%) of the spinal cord apoptotic cells are still immunoreactive for specific proteins such as neurofilament or glial fibrillary acid protein (33), suggesting that both neuronal and glial cells are dying by apoptosis in the mutant SOD1 model (34). In our opinion, the paucity of apoptotic dying motor neurons in this mouse model of ALS reflects the difficulty in detecting these cells by morphological means due to the presumed low daily rate of motor neuron loss (16) and the notoriously rapid disappearance of apoptotic cells. Forms of PCD with morphological features distinct from apoptosis also exist (35–37), making it difficult to exclude the possibility that a nonapoptotic form of PCD underlies mutant SOD1–related cellular degeneration.

In human ALS cases, using morphological criteria including size, shape, and aggregates of Nissl substance, Martin (38) has arranged residual spinal cord motor neurons in ALS postmortem samples in three categories that he believes reflect different stages of degeneration. In the chromatolysis stage, motor neurons still resemble their normal counterparts except that the cell body appears swollen and round, the Nissl substance dispersed, and the nucleus eccentrically placed. Some chromatolytic neurons have prominent cytoplasmic hyaline body inclusions. In the attritional stage, the cytoplasm and the nucleus appear homogenous and condensed, and the cell body appears shrunken and with hazy multipolar shape. In the so-called apoptotic stage, the affected motor neuron is approximately one-fifth of its normal diameter, the cytoplasm and nucleus are extremely condensed, and the cell body adopts a fusiform or round shape devoid of any process. Notably, in none of the three stages do residual motor neurons show appreciable cytoplasmic vacuoles or nuclear condensation accompanied by round chromatin clumps.

Taken together, these findings suggest that, while degenerating neurons in both human ALS and its experimental models do exhibit some features reminiscent of apoptosis, the vast majority of dying cells cannot confidently be labeled as typical apoptotic.

Expression of apoptotic markers
Besides exhibiting singular morphological features, apoptotic cells may also show a variety of cellular alterations. The detection of internucleosomal DNA cleavage by either gel electrophoresis or in situ methods has emerged as a popular means of supporting the occurrence of apoptosis in all sorts of pathological situations, including ALS. However, like many of these apoptotic markers, DNA fragmentation detected by in situ methods (e.g., terminal deoxynucleotidyl transferase-mediated nick end labeling) is now well recognized as also occurring in nonapoptotic cell death, including necrosis (35). So the value of DNA cleavage evidenced by in situ techniques as a specific marker of apoptosis may be limited. In addition to this caveat, the search for DNA fragmentation in ALS postmortem samples has generated conflicting results. In one autopsy study, DNA fragmentation was detected by an in situ method in spinal cord motor neurons in ALS but not in control specimens (39). In two other similar studies, DNA fragmentation was detected not only in the motor cortex and spinal cord of ALS specimens, but also, though to a lesser degree, in control specimens (40, 41). In a subsequent study, internucleosomal DNA fragmentation was detected in affected (e.g., motor cortex and spinal cord) but not in spared brain regions (e.g., somatosensory cortex) of ALS cases (38), and, in diseased motor neurons, only at the somatodendritic attrition and apoptotic stages and not at the chromatolytic stage (38). The author of that study has also documented DNA fragmentation in anterior-horn gray matter of the spinal cord and motor cortex of ALS cases by gel electrophoresis (38), a technique not frequently used in the nervous system to identify apoptosis since, in many neurological situations, it is difficult to obtain samples with a sufficiently high proportion of dying cells. In contrast to all these positive findings, other

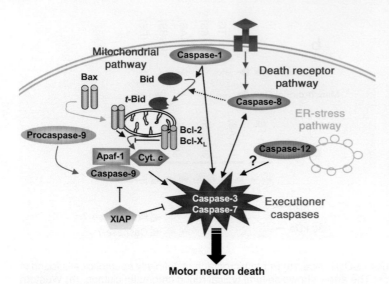

Figure 2

Molecular pathways of PCD. To date, at least three different PCD molecular pathways have been recognized: the mitochondrial pathway (also called intrinsic), the death receptor pathway (also called extrinsic), and the ER pathway. In the mitochondrial PCD pathway, translocation of the proapoptotic protein Bax and the BH3-domain-only protein Bid from the cytosol to the mitochondria promotes cell death by inducing the release of cytochrome c (Cyt. c) from the mitochondria to the cytosol; although both full-length and truncated Bid (t-Bid) translocate to the mitochondria, t-Bid is the most biologically active form. Once in the cytosol, cytochrome c activates caspase-9 in the presence of Apaf-1, which, in turn, activates downstream executioner caspases. This pathway can be inhibited by the antiapoptotic protein Bcl-2 and by the protein caspase inhibitor X chromosome–linked inhibitor of apoptosis (XIAP). In the death receptor pathway, caspase-8 is activated by death receptors (members of the TNFR family) in the plasma membrane via the intermediary of adapter proteins. Death receptors include Fas (CD95) and the low-affinity neurotrophin receptor (p75[NTR]). Activated caspase-8 then activates executioner caspases, directly or indirectly, through the activation of Bid. Stress in the ER, including disruption of ER-calcium homeostasis and accumulation of excess proteins in the ER, can also result in apoptosis through activation of caspase-12. This caspase is not activated by membrane- or mitochondria-targeted apoptotic signals. Activation of the upstream caspase-1, the key enzyme responsible for the activation of IL-1, also results in activation of executioner caspases and enhances, at least in part through the cleavage of Bid, the activation of the mitochondria-dependent apoptotic pathway.

groups, using similar techniques and tissue samples, have failed to provide any evidence of internucleosomal cleavage of DNA in postmortem tissue from human ALS cases or from animal models of the disease (32, 41, 42). Although the actual reason for these divergent results is unclear, they cast doubt on the reliability and even the specificity of such findings.

Two other apoptotic markers, the Le[Y] antigen (43) and fractin (44), were also studied in ALS, and here the picture seems less ambiguous. Neither marker was detected in spinal cords of controls, but were highly expressed in spinal cords of, respectively, ALS cases (39) and transgenic SOD1[G93A] mice (34). Likewise, the levels of the apoptosis-related protein prostate apoptosis response-4 (45) were increased in spinal cord samples from both ALS patients and transgenic mutant SOD1 mice compared with their respective controls (46). Together with the morphological data summarized above, these findings support the view that apoptosis occurs in

ALS. What all of these studies fail to do, however, is to provide definite mechanistic insights into the significance of these alterations in the pathogenesis of ALS.

Activation of apoptotic molecular pathways

Given the ambiguous results of the morphological studies, it appears that a more convincing approach to evaluating the role of apoptosis in ALS may be to determine whether the neurodegenerative process in transgenic mutant SOD1 mice, irrespective of the morphology of the dying cells, involves known molecular mediators of PCD, and whether targeting such key factors can affect the course of the disease.

PCD is a multistep machinery (Figure 2) that involves a complex interaction between survival pathways, activated by trophic factors, and death pathways, activated by various stresses. So far, the two pathways that have been most implicated in neuronal survival are the PI3K pathway, which activates Akt (also known as protein kinase B) to suppress the activation of proapoptotic proteins, and the extracellular signal–regulated kinase/MAPK (ERK/MAPK) pathway, which activates antiapoptotic proteins (47).

The best-known PCD-mediating pathways are those involved in the activation of caspase-3. The caspases are a family of cysteine-aspartate proteases (Figure 2; see below for details), many of which are involved in PCD either at the level of upstream signaling (notably caspase-8 and caspase-9) or more downstream at the effector level (notably caspase-3). Caspase-8 and caspase-9 both cleave procaspase-3 to activate it. Caspase-9 is activated by a signal derived from mitochondria under the control of the Bcl-2 family of proteins (Figure 2; see below for details). Caspase-8 is activated by death receptors (members of the TNF receptor family) in the plasma membrane via the intermediary of adapter proteins (48). Death receptors include the low-affinity neurotrophin receptor (p75[NTR]) and Fas (CD95); the latter seems to participate in the death of embryonic motor neurons in primary cultures (49), but whether Fas contributes to the death of mature motor neurons and to the neurodegenerative process in transgenic mice expressing mutant SOD1 remains to be demonstrated. Other key molecules in PCD signaling include ceramide, MAPKs (JNK and p38), and the transcription factors activator protein-1 and NF-κB (47, 48).

In light of the presumed proapoptotic properties of mutant SOD1 (28), it is tempting to suggest that the mutant protein may be a death-signaling molecule in itself, either directly, by setting in motion the PCD cascade, or indirectly, by interacting with a variety of intracellular targets such as trophic factors, Bcl-2 family members, or even mitochondria. Mitochondria are a particularly appealing target, because they not only contain mutant SOD1 (50) but are structurally and functionally altered in transgenic mutant SOD1 mice (51, 52), and because they play a pivotal role in PCD (53). Also relevant to the issue of death and survival signals in the mutant SOD1 model are the Western blot and immunohistochemical demonstrations of the weakening surviving signal mediated by PI3K/Akt in spinal cords of transgenic mutant SOD1 mice even before overt neuropathological features arise (54). Once the mutant SOD1–mediated neurodegenerative process has been initiated, several secondary alterations develop in spinal cords of transgenic mutant SOD1 mice, including microglial cell activation (18) and T cell infiltration (55), both of which may release a plethora of cytokines and other pro-PCD mediators. Accordingly, while the nature of the initial death signal in transgenic mutant SOD1 mice remains elusive, in a more advanced stage of the disease the increased expression of several extracellular inflammation-related factors such as IL-1β, IL-6, and TNF-α (56) may amplify the death signals that are already reaching motor neurons in this mouse model of ALS, by activating death receptors such as Fas (49). IL-1β content is also elevated in human ALS spinal cords (57).

The role of the Bcl-2 family
in motor-neuronal cell death in ALS

The Bcl-2 family, implicated in the regulation of PCD (Figure 2), is composed of both cell-death suppressors such as Bcl-2 and Bcl-X_L and promoters such as Bax, Bad, Bak, and Bcl-xS (58). Many of these molecules are present and active within the nervous system and appear to be potent modulators of neuronal death. In human ALS cases and affected transgenic SOD1^{G93A} mice, Bcl-2 mRNA content appears significantly decreased and Bax mRNA content significantly increased in the lumbar cord compared with those of controls (59, 60). This is consistent with the finding in both human ALS cases and symptomatic transgenic SOD1^{G93A} mice that the spinal cord expressions of the anti-apoptotic proteins Bcl-2 and Bcl-X_L are either unchanged (40, 61) or decreased (38, 60), whereas that of the proapoptotic Bax and Bad proteins is increased (38, 40, 60). Different SOD1 mutations do not cause exactly the same neuropathology. It is important to note that a very similar pattern of changes of selected pro- and anti-cell death Bcl-2 family members was found in spinal cords of affected transgenic SOD1^{G86R} mice compared with their wild-type counterparts (62). None of these alterations, however, is seen in young asymptomatic transgenic SOD1^{G93A} mice; but they clearly become progressively more conspicuous as the neurodegenerative process progresses (60). With regard to function, in the spinal cords of both ALS patients and affected transgenic mice expressing mutant SOD1, Bax is not only upregulated but is also expressed mainly in its active homodimeric conformation (38, 60). As illustrated in Figure 1b, it is markedly relocated from the cytosol to the mitochondria (38, 63); this relocation is, in many cellular settings, a prerequisite to the recruitment of the mitochondria-dependent apoptosis pathway. So it seems that, in ALS, during the neurodegenerative process, the fine-tuned balance between cell-death antagonist and agonist of the Bcl-2 family is upset in favor of pro-cell death forces. In support of this view is the finding that overexpression of Bcl-2, presumably by buffering some of the pro-cell death drive (60), mitigates neurodegeneration and prolongs survival in transgenic SOD1^{G93A} mice (64); a similar beneficial effect of Bcl-2 was reported in mutant SOD1-transfected PC-12 cells (29).

Other meaningful Bcl-2 family members that appear to be in play in ALS include Bid and Harakiri, two potent pro-PCD peptides, which can participate in the cell death process, either directly or indirectly, by potentiating the effect of Bax. Bid appears to be highly expressed in the spinal cord of transgenic SOD1^{G93A} mice and is cleaved into its most active form during the progression of the disease (65). Harakiri's expression has been detected in motor neurons of ALS, but not of control, spinal cord specimens, specifically in spared neurons, of which some exhibited an abnormal morphology reminiscent of that labeled by Troost et al. (61) as apoptotic (66).

The quest to elucidate how Bcl-2 family members are deregulated in ALS is fascinating. As in other pathological situations, it is unlikely that mutant SOD1 directly produces the observed changes in Bax. It is more likely that mutant SOD1 ignites intracellular signaling pathways, which, in turn, cause Bax upregulation and translocation. This scenario would be consistent with what we currently know about the regulation of Bax and how Bax is usually brought into action in PCD. The tumor suppressor protein p53 counts among the rare molecules known to regulate Bax expression (67). In normal situations, p53 basal levels in the cell are very low, but upon activation, as seen in pathological situations, there is a rapid rise in p53 mRNA and protein levels, as well as posttranslational modifications that stabilize the protein (68). Activation of the p53 pathway in ALS is evidenced by the demonstration that p53 is increased in the nuclear fraction of affected brain regions in ALS patients (69), as is p53 immunostaining in neuron nuclei of transgenic SOD1^{G86R} mice (62). Despite the compelling evidence that p53 is activated in ALS, two independent studies have failed to provide any supportive data for an actual role for this transcriptional factor in mutant SOD1-mediated neurodegeneration (70, 71).

Caspases in the ALS neurodegenerative process

Caspases are members of a distinct family of cysteine proteases that share the ability to cleave their substrates after specific aspartic acid residues and that are present in cells as inactive zymogens, called procaspases. So far, 14 different mammalian caspases have been identified that differ in primary sequence and substrate specificity. An instrumental role for caspases in ALS neurodegen-eration is supported by the demonstration that the irreversible broad-caspase inhibitor benzyloxycarbonyl-Val-Ala-Asp(O-methyl)-fluoromethyl-ketone attenuates mutant SOD1-mediated cell death in transfected PC-12 cells (29) and in transgenic SOD1^{G93A} mice (57).

All of the identified caspases are grouped based on their function. One group includes caspases -1, -4, -5, -11, -12, and -14, which are now believed to play a primary role in cytokine maturation. Among these, in ALS, the lion's share of attention until now has been given to caspase-1, the key enzyme responsible for the activation of IL-1. Procaspase-1 is highly expressed in spinal cord motor neurons, and its activation in the spinal cord of transgenic mutant SOD1 mice coincides with the development of the glial response and with the very beginning of the loss of motor neurons (33, 34, 57, 63, 72). Despite caspase-1's likely indirect role in PCD, chronic inhibition of caspase-1 by a dominant negative mutant of the enzyme has been proven effective in prolonging the life of transgenic SOD1^{G93A} mice (73). So far, the status of the other members of the caspase-1 subfamily in ALS is unknown. Some preliminary investigations show that caspase-12, which is known to be activated following ER stress (74), is expressed in motor neurons of nontransgenic mice, and even more so in those of symptomatic transgenic SOD1^{G93A} mice (C. Guégan et al., unpublished observations). In symptomatic transgenic SOD1^{G93A} mice, most of the motor neurons immunopositive for caspase-12 appear condensed, shrunken, and vacuolized. Although more work on caspase-12 remains to be done in this model of ALS, our preliminary data argue that sick cells are the site of an ER stress whose occurrence could well contribute to the overall cascade of deleterious events that ultimately underlies the demise of spinal cord motor neurons in the mutant SOD1 model.

By contrast, caspases -2, -3, -6, -7, -8, -9, and -10 have been implicated in apoptosis per se, although their roles can be further divided into "initiator" and "effector."

Initiator caspases include procaspases -2, -8, -9, and -10, all of which have long prodomains and protein-protein interaction motifs, such as the death-effector domain and the caspase-activation and -recruitment domain, that contribute to the transduction of various signals into proteolytic activity. Procaspase-8 is activated after ligation of certain cell surface receptors, such as the TNF receptors. Interestingly, while significant glial response and production of IL-1β occur early in transgenic mutant SOD1 mice (see above), activation of procaspase-8, like induction of TNF-α (56), is only detected in spinal cords near the end stage (65). This suggests that, in this ALS model, the TNF/caspase-8 machinery may be a late contributor to the degenerative process. Caspase-2 is another initiator of PCD whose activation occurs in the spinal cord of affected transgenic mutant SOD1 mice (S. Vukosavic et al., unpublished observations). Yet ablation of caspase-2 in transgenic SOD1^{G93A} mice has been reported to be of no consequence to the expression of the disease (75), indicating that whatever the role of caspase-2 is in ALS, it is dispensable. A third caspase initiator is caspase-9, whose role is pivotal in the so-called mitochondria-dependent PCD pathway (53). Here, after a death stimulus, released mitochondrial cytochrome c interacts in the cytosol with apoptotic protease-activating factor-1 (Apaf-1) in the presence of dATP, which stimulates the processing of procaspase-9 into its active form, which in turn can activate the downstream executioner caspases (see below). Evidence of prominent recruitment of this mitochondrial pathway has been documented in spinal cord specimens of both ALS patients and transgenic SOD1^{G93A} mice (63). In that study, it is shown that, while cytochrome c is confined to the mitochondria in cells in the control samples, it is diffusely distributed in the cytosol in several of the spared cells, especially neurons, in the pathological samples. It is also demonstrated, at least in transgenic mutant SOD1^{G93A} mice, that the mitochondrial cytochrome c translocation to the cytosol occurs at the same time as the cytosolic Bax translocation to the mitochondria and activation of procaspase-9, and before activation of downstream caspase executioners such as procaspase-3 and procaspase-7 (Figure 1, b–d). Because caspase-9 is thought to be so critical in many cell-death settings, it is very likely that the observed translocation of cytochrome c and activation of procaspase-9 in ALS represent significant pathological events. Consistent with this view is the finding that prevention of mitochondrial cytochrome c release lengthens the lifespan of transgenic SOD1^{G93A} mice (76).

Effector caspases include procaspases -3, -6, and -7, all of which have short prodomains and lack intrinsic enzymatic activity. However, upon their cleavage, which is triggered by, for example, initiator caspases, effector caspases acquire the capacity to cleave a large number of intracellular substrates, which probably results in the eventual death of the cell. Consistent with this scenario, it has been reported that key effector caspases such as caspase-3 and caspase-7 (see Figure 1d) are indeed activated in spinal cords of transgenic mutant SOD1 mice in a time-dependent manner that parallels the time course of the neurodegenerative process (33, 34); activation of procaspase-3 has also been observed in spinal cord samples from ALS patients (38). Yet current data on the sequence of events in the PCD cascade indicate that, once effector caspases have been activated, the cell death process, at least in certain pathological settings, has reached a point of no return. This would suggest that, in these specific conditions, the death commitment point is situated upstream of these caspases, and,

consequently, interventions aimed at inhibiting these downstream caspases may fail to provide any real neuroprotective benefit (77). Whether this applies to the demise of motor neurons in ALS remains to be determined.

Conclusion

In this review we have described evidence that numerous key molecular components of PCD are recruited in ALS. We have also shown that, while precious data on PCD in ALS have been obtained thanks to the study of postmortem human samples, information regarding the temporal relationships of these changes and their significance in the pathological cascade emanates essentially from the use of transgenic mutant SOD1 mouse models. In light of the above-described PCD-related changes, it would appear that this active form of cell death is not the sole pathological mediator of cell demise in ALS but rather one key component within a coalition of deleterious factors ultimately responsible for the degenerative process. As discussed above, however, the actual relationships between mutant SOD1 and the various other presumed culprits represented by protein aggregates, oxidant production, and PCD activation are still unknown, and a better understanding of the pathogenic cascade in ALS will require their elucidation.

In our opinion, one of the most important take-home messages from the body of work summarized above is that an apoptotic morphology should not be used as the sole criterion of whether molecular pathways of PCD have been recruited. Indeed, we can not stress enough that the PCD molecular pathways may be activated in a neurodegenerative process such as that seen in ALS, even when the prevalent morphology of the dying cells is nonapoptotic. Relatedly, caspase-9 is instrumental in paraptosis (37), a specific morphological form of nonapoptotic cell death.

Apart from the question of whether the morphology of dying neurons in ALS is apoptotic, but still relevant to our discussion, is the contrast between the paucity of morphologically identified dying cells and the rather robust spinal cord molecular PCD alterations. How can this striking discrepancy be reconciled? First, it is possible that the morphological expression of PCD is much more ephemeral than its molecular translation. Therefore, since in ALS the degenerative process is asynchronous, small lasting differences in the expression of these markers may have significant impact on the total number of cells that exhibit a given marker at a given time point. Second, it is also possible that, since apoptotic morphological features are confined to the cell body while PCD molecular alterations may be found not only in cell but also in cell processes, axons, and nerve terminals. Thus, the detection of PCD morphology may be much more challenging than the detection of PCD molecular events. Third, the molecular tools used in all of the cited studies see not only the rare cells that are truly dying but also the numerous sick cells that may or may not ultimately die and that thus may or may not show the typical apoptotic morphology.

Another important point that derives from the work in transgenic mutant SOD1 mice is that not only neurons but also glial cells appear to be the site of PCD-cascade activation. This observation does not undermine the potential pathogenic role of PCD in the ALS death process, but it raises the possibility that PCD may not only kill neurons in this disease. However, since SOD1 is expressed in all cells, not only in motor neurons, it is possible that the activation of PCD in both neurons and glia reflects the ubiquitous nature of the mutant protein expression. Whether

PCD is also activated in neuron and glial cells in the forms of ALS that are not linked to mutant SOD1 is unknown at this point.

Clearly, the overall mechanism of neurodegeneration in ALS is still incompletely known. Nevertheless, the available evidence indicates that PCD is in play in ALS and thus warrants further investigation of the role of the PCD cascade in ALS pathogenesis and treatment. The most effective therapeutic strategies tested so far in transgenic mutant SOD1 mice target very distinct molecular pathways. We can therefore imagine that, ultimately, the best therapy for ALS will come from a combination of several interventions and not from a single treatment. In keeping with this view, unraveling the sequence of key PCD factors recruited during ALS neurodegeneration should enable us to identify the most significant molecules to be targeted by this therapeutic cocktail to produce optimal neuroprotection.

Acknowledgments

The authors wish to thank Robert E. Burke and Miquel Vila for their insightful comments on the manuscript, and Pat White and Brian Jones for their help in its preparation. The authors also acknowledge the support of National Institute of Neurological Disorders and Stroke grants R29 NS37345, RO1 NS38586, NS42269, and P50 NS38370, US Department of Defense grant DAMD 17-99-1-9471, the Lowenstein Foundation, the Lillian Goldman Charitable Trust, the Parkinson's Disease Foundation, the Muscular Dystrophy Association, the ALS Association, and Project ALS.

Address correspondence to: Serge Przedborski, BB-307, Columbia University, 650 West 168th Street, New York, New York 10032, USA. Phone: (212) 305-1540; Fax: (212) 305-5450; E-mail: SP30@columbia.edu.

1. Rowland, L.P. 1995. Hereditary and acquired motor neuron diseases. In *Merritt's textbook of neurology*. L.P. Rowland, editor. Williams & Wilkins. Philadelphia, Pennsylvania, USA. 742-749.

2. Brown, R.H., Jr. 1995. Amyotrophic lateral sclerosis: recent insights from genetics and transgenic mice. *Cell.* **80**:687-692.

3. Deng, H.-X., et al. 1993. Amyotrophic lateral sclerosis and structural defects in Cu,Zn superoxide dismutase. *Science.* **261**:1047-1051.

4. Rosen, D.R., et al. 1993. Mutations in Cu/Zn superoxide dismutase gene are associated with familial amyotrophic lateral sclerosis. *Nature.* **362**:59-62.

5. Cudkowicz, M.E., et al. 1997. Epidemiology of mutations in superoxide dismutase in amyotrophic lateral sclerosis. *Ann. Neurol.* **41**:210-221.

6. Przedborski, S., et al. 1996. Blood superoxide dismutase, catalase and glutathione peroxidase activities in familial and sporadic amyotrophic lateral sclerosis. *Neurodegeneration.* **5**:57-64.

7. Troy, C.M., and Shelanski, M.L. 1994. Down-regulation of copper/zinc superoxide dismutase causes apoptotic death in PC12 neuronal cells. *Proc. Natl. Acad. Sci. USA.* **91**:6384-6387.

8. Rothstein, J.D., Bristol, L.A., Hosler, B., Brown, R.H., Jr., and Kuncl, R.W. 1994. Chronic inhibition of superoxide dismutase produces apoptotic death of spinal neurons. *Proc. Natl. Acad. Sci. USA.* **91**:4155-4159.

9. Reaume, A.G., et al. 1996. Motor neurons in Cu/Zn superoxide dismutase-deficient mice develop normally but exhibit enhanced cell death after axonal injury. *Nat. Genet.* **13**:43-47.

10. Gurney, M.E., et al. 1994. Motor neuron degeneration in mice that express a human Cu, Zn superoxide dismutase mutation. *Science.* **264**:1772-1775.

11. Wong, P.C., et al. 1995. An adverse property of a familial ALS-linked SOD1 mutation causes motor neuron disease characterized by vacuolar degeneration of mitochondria. *Neuron.* **14**:1105-1116.

12. Bruijn, L.I., et al. 1997. ALS-linked SOD1 mutant G85R mediated damage to astrocytes and promotes rapidly progressive disease with SOD1-containing inclusions. *Neuron.* **18**:327-338.

13. Nagai, M., et al. 2001. Rats expressing human cytosolic copper-zinc superoxide dismutase transgenes with amyotrophic lateral sclerosis: associated mutations develop motor neuron disease. *J. Neurosci.* **21**:9246-9254.

14. Subramaniam, J.R., et al. 2002. Mutant SOD1 causes motor neuron disease independent of copper chaperone-mediated copper loading. *Nat. Neurosci.* **5**:301-307.

15. Brown, R.H., Jr. 1995. Superoxide dismutase in familial amyotrophic lateral sclerosis: models for gain of function. *Curr. Opin. Neurobiol.* **5**:841-846.

16. Chiu, A.Y., et al. 1995. Age-dependent penetrance of disease in a transgenic mouse model of familial amyotrophic lateral sclerosis. *Mol. Cell. Neurosci.* **6**:349-362.

17. Dal Canto, M.C., and Gurney, M.E. 1995. Neuropathological changes in two lines of mice carrying a transgene for mutant human Cu,Zn SOD, and in mice overexpressing wild type human SOD: a model of familial amyotrophic lateral sclerosis (FALS). *Brain Res.* **676**:25-40.

18. Almer, G., Vukosavic, S., Romero, N., and Przedborski, S. 1999. Inducible nitric oxide synthase upregulation in a transgenic mouse model of familial amyotrophic lateral sclerosis. *J. Neurochem.* **72**:2415-2425.

19. Tu, P.H., et al. 1996. Transgenic mice carrying a human mutant superoxide dismutase transgene develop neuronal cytoskeletal pathology resembling human amyotrophic lateral sclerosis lesions. *Proc. Natl. Acad. Sci. USA.* **93**:3155-3160.

20. Cleveland, D.W., and Rothstein, J.D. 2001. From Charcot to Lou Gehrig: deciphering selective motor neuron death in ALS. *Nat. Rev. Neurosci.* **2**:806-819.

21. Beckman, J.S., Carson, M., Smith, C.D., and Koppenol, W.H. 1993. ALS, SOD and peroxynitrite. *Nature.* **364**:584.

22. Wiedau-Pazos, M., et al. 1996. Altered reactivity of superoxide dismutase in familial amyotrophic lateral sclerosis. *Science.* **271**:515-518.

23. Crow, J.P., Sampson, J.B., Zhuang, Y.X., Thompson, J.A., and Beckman, J.S. 1997. Decreased zinc affinity of amyotrophic lateral sclerosis-associated superoxide dismutase mutants leads to enhanced catalysis of tyrosine nitration by peroxynitrite. *J. Neurochem.* **69**:1936-1944.

24. Estevez, A.G., et al. 1999. Induction of nitric oxide-dependent apoptosis in motor neurons by zinc-deficient superoxide dismutase. *Science.* **286**:2498-2500.

25. Bruijn, L.I., et al. 1998. Aggregation and motor neuron toxicity of an ALS-linked SOD1 mutant independent from wild-type SOD1. *Science.* **281**:1851-1854.

26. Durham, H.D., Roy, J., Dong, L., and Figlewicz, D.A. 1997. Aggregation of mutant Cu/Zn superoxide dismutase proteins in a culture model of ALS. *J. Neuropathol. Exp. Neurol.* **56**:523-530.

27. Johnson, W.G. 2000. Late-onset neurodegenerative diseases: the role of protein insolubility. *J. Anat.* **196**:609-616.

28. Rabizadeh, S., et al. 1995. Mutations associated with amyotrophic lateral sclerosis convert superoxide dismutase from an antiapoptotic gene to a proapoptotic gene: studies in yeast and neural cells. *Proc. Natl. Acad. Sci. USA.* **92**:3024-3028.

29. Ghadge, G.D., et al. 1997. Mutant superoxide dismutase-1-linked familial amyotrophic lateral sclerosis: molecular mechanisms of neuronal death and protection. *J. Neurosci.* **17**:8756-8766.

30. Mena, M.A., et al. 1997. Effects of wild-type and mutated copper/zinc superoxide dismutase on neuronal survival and L-DOPA-induced toxicity in postnatal midbrain culture. *J. Neurochem.* **69**:21-33.

31. Roy, N., et al. 1995. The gene for neuronal apoptosis inhibitory protein is partially deleted in individuals with spinal muscular atrophy. *Cell.* **80**:167-178.

32. Migheli, A., et al. 1999. Lack of apoptosis in mice with ALS. *Nat. Med.* **5**:966-967.

33. Pasinelli, P., Houseweart, M.K., Brown, R.H., Jr., and Cleveland, D.W. 2000. Caspase-1 and -3 are sequentially activated in motor neuron death in Cu,Zn superoxide dismutase-mediated familial amyotrophic lateral sclerosis. *Proc. Natl. Acad. Sci. USA.* **97**:13901-13906.

34. Vukosavic, S., et al. 2000. Delaying caspase activation by Bcl-2: a clue to disease retardation in a transgenic mouse model of amyotrophic lateral sclerosis. *J. Neurosci.* **20**:9119-9125.

35. Clarke, P.G.H. 1999. Apoptosis versus necrosis. In *Cell death and diseases of the nervous system*. V.E. Koliatsos and R.R. Ratan, editors. Humana Press. Totowa, New Jersey, USA. 3-28.

36. Yaginuma, H., et al. 1996. A novel type of programmed neuronal death in the cervical spinal cord of the chick embryo. *J. Neurosci.* **16**:3685-3703.

37. Sperandio, S., de Belle, I., and Bredesen, D.E. 2000. An alternative, nonapoptotic form of programmed cell death. *Proc. Natl. Acad. Sci. USA.* **97**:14376-14381.

38. Martin, L.J. 1999. Neuronal death in amyotrophic lateral sclerosis is apoptosis: possible contribution of a programmed cell death mechanism. *J. Neuropathol. Exp. Neurol.* **58**:459-471.

39. Yoshiyama, Y., Yamada, T., Asanuma, K., and Asahi, T. 1994. Apoptosis related antigen, Le(Y) and nick-end labeling are positive in spinal motor neurons in amyotrophic lateral sclerosis. *Acta Neuropathol. (Berl.)* **88**:207-211.

40. Ekegren, T., Grundstrom, E., Lindholm, D., and Aquilonius, S.M. 1999. Upregulation of Bax protein and increased DNA degradation in ALS spinal cord motor neurons. *Acta Neurol. Scand.* **100**:317-321.

41. Migheli, A., Cavalla, P., Marino, S., and Schiffer, D. 1994. A study of apoptosis in normal and pathologic nervous tissue after in situ end-labeling of DNA strand breaks. *J. Neuropathol. Exp. Neurol.* **53**:606-616.

42. He, B.P., and Strong, M.J. 2000. Motor neuronal death in sporadic amyotrophic lateral sclerosis (ALS) is not apoptotic. A comparative study of ALS and chronic aluminium chloride neurotoxicity in New Zealand white rabbits. *Neuropathol. Appl. Neurobiol.* **26**:150-160.

43. Hiraishi, K., Suzuki, K., Hakomori, S., and Adachi, M. 1993. Le(y) antigen expression is correlated with apoptosis (programmed cell death). *Glycobiology.* **3**:381-390.

44. Suurmeijer, A.J., van der Wijk, J., van Veldhuisen, D.J., Yang, F., and Cole, G.M. 1999. Fractin immunostaining for the detection of apoptotic cells and apoptotic bodies in formalin-fixed and paraffin-embedded tissue. *Lab. Invest.* **79**:619–620.

45. Rangnekar, V.M. 1998. Apoptosis mediated by a novel leucine zipper protein Par-4. *Apoptosis.* **3**:61–66.

46. Pedersen, W.A., Luo, H., Kruman, I., Kasarskis, E., and Mattson, M.P. 2000. The prostate apoptosis response-4 protein participates in motor neuron degeneration in amyotrophic lateral sclerosis. *FASEB J.* **14**:913–924.

47. Harper, S.J., and LoGrasso, P. 2001. Signaling for survival and death in neurones. The role of stress-activated kinases, JNK and p38. *Cell. Signal.* **13**:299–310.

48. Gupta, S. 2001. Molecular steps of death receptor and mitochondrial pathways of apoptosis. *Life Sci.* **69**:2957–2964.

49. Raoul, C., et al. 2002. Motoneuron death triggered by a specific pathway downstream of Fas. Potentiation by ALS-Linked SOD1 mutations. *Neuron.* **35**:1067–1083.

50. Higgins, C.M., Jung, C., Ding, H., and Xu, Z. 2002. Mutant Cu, Zn superoxide dismutase that causes motoneuron degeneration is present in mitochondria in the CNS. *J. Neurosci.* **22**:RC215.

51. Kong, J.M., and Xu, Z.S. 1998. Massive mitochondrial degeneration in motor neurons triggers the onset of amyotrophic lateral sclerosis in mice expressing a mutant SOD1. *J. Neurosci.* **18**:3241–3250.

52. Browne, S.E., et al. 1998. Metabolic dysfunction in familial, but not sporadic, amyotrophic lateral sclerosis. *J. Neurochem.* **71**:281–287.

53. Kroemer, G., and Reed, J.C. 2000. Mitochondrial control of cell death. *Nat. Med.* **6**:513–519.

54. Warita, H., et al. 2001. Early decrease of survival signal-related proteins in spinal motor neurons of presymptomatic transgenic mice with a mutant SOD1 gene. *Apoptosis.* **6**:345–352.

55. Alexianu, M.E., Kozovska, M., and Appel, S.H. 2001. Immune reactivity in a mouse model of familial ALS correlates with disease progression. *Neurology.* **57**:1282–1289.

56. Nguyen, M.D., Julien, J.P., and Rivest, S. 2001. Induction of proinflammatory molecules in mice with amyotrophic lateral sclerosis: no requirement for proapoptotic interleukin-1beta in neurodegeneration. *Ann. Neurol.* **50**:630–639.

57. Li, M., et al. 2000. Functional role of caspase-1 and caspase-3 in an ALS transgenic mouse model. *Science.* **288**:335–339.

58. Chao, D.T., and Korsmeyer, S.J. 1998. BCL-2 family: regulators of cell death. *Annu. Rev. Immunol.* **16**:395–419.

59. Mu, X., He, J., Anderson, D.W., Trojanowski, J.Q., and Springer, J.E. 1996. Altered expression of bcl-2 and bax mRNA in amyotrophic lateral sclerosis spinal cord motor neurons. *Ann. Neurol.* **40**:379–386.

60. Vukosavic, S., Dubois-Dauphin, M., Romero, N., and Przedborski, S. 1999. Bax and Bcl-2 interaction in a transgenic mouse model of familial amyotrophic lateral sclerosis. *J. Neurochem.* **73**:2460–2468.

61. Troost, D., Aten, J., Morsink, F., and De Jong, J.M.B.V. 1995. Apoptosis in amyotrophic lateral sclerosis is not restricted to motor neurons. Bcl-2 expression is increased in unaffected post-central gyrus. *Neuropathol. Appl. Neurobiol.* **21**:498–504.

62. Gonzalez de Aguilar, J.L., et al. 2000. Alteration of the Bcl-x/Bax ratio in a transgenic mouse model of amyotrophic lateral sclerosis: evidence for the implication of the p53 signaling pathway. *Neurobiol. Dis.* **7**:406–415.

63. Guégan, C., Vila, M., Rosoklija, G., Hays, A.P., and Przedborski, S. 2001. Recruitment of the mitochondrial-dependent apoptotic pathway in amyotrophic lateral sclerosis. *J. Neurosci.* **21**:6569–6576.

64. Kostic, V., Jackson-Lewis, V., De Bilbao, F., Dubois-Dauphin, M., and Przedborski, S. 1997. Bcl-2: prolonging life in a transgenic mouse model of familial amyotrophic lateral sclerosis. *Science.* **277**:559–562.

65. Guégan, C., et al. 2002. Instrumental activation of Bid by caspase-1 in a transgenic mouse model of ALS. *Mol. Cell. Neurosci.* **20**:553–562.

66. Shinoe, T., et al. 2001. Upregulation of the proapoptotic BH3-only peptide harakiri in spinal neurons of amyotrophic lateral sclerosis patients. *Neurosci. Lett.* **313**:153–157.

67. Miyashita, T., et al. 1994. Tumor suppressor p53 is a regulator of bcl-2 and bax gene expression in vitro and in vivo. *Oncogene.* **9**:1799–1805.

68. Appella, E., and Anderson, C.W. 2001. Post-translational modifications and activation of p53 by genotoxic stresses. *Eur. J. Biochem.* **268**:2764–2772.

69. Martin, L.J. 2000. p53 is abnormally elevated and active in the CNS of patients with amyotrophic lateral sclerosis. *Neurobiol. Dis.* **7**:613–622.

70. Prudlo, J., et al. 2000. Motor neuron cell death in a mouse model of FALS is not mediated by the p53 cell survival regulator. *Brain Res.* **879**:183–187.

71. Kuntz, C., Kinoshita, Y., Beal, M.F., Donehower, L.A., and Morrison, R.S. 2000. Absence of p53: no effect in a transgenic mouse model of familial amyotrophic lateral sclerosis. *Exp. Neurol.* **165**:184–190.

72. Pasinelli, P., Borchelt, D.R., Houseweart, M.K., Cleveland, D.W., and Brown, R.H.J. 1998. Caspase-1 is activated in neural cells and tissue with amyotrophic lateral sclerosis-associated mutations in copper-zinc superoxide dismutase. *Proc. Natl. Acad. Sci. USA.* **95**:15763–15768.

73. Friedlander, R.M., Brown, R.H., Gagliardini, V., Wang, J., and Yuan, J. 1997. Inhibition of ICE slows ALS in mice. *Nature.* **388**:31.

74. Nakagawa, T., et al. 2000. Caspase-12 mediates endoplasmic-reticulum-specific apoptosis and cytotoxicity by amyloid-beta. *Nature.* **403**:98–103.

75. Bergeron, L., et al. 1998. Defects in regulation of apoptosis in caspase-2-deficient mice. *Genes Dev.* **12**:1304–1314.

76. Zhu, S., et al. 2002. Minocycline inhibits cytochrome c release and delays progression of amyotrophic lateral sclerosis in mice. *Nature.* **417**:74–78.

77. Zheng, T.S., et al. 2000. Deficiency in caspase-9 or caspase-3 induces compensatory caspase activation. *Nat. Med.* **6**:1241–1247.

Multiple sclerosis

David A. Hafler

Laboratory of Molecular Immunology, Center for Neurologic Diseases, Brigham and Women's Hospital and Harvard Medical School,
Boston, Massachusetts, USA. The Broad Institute, Massachusetts Institute of Technology and Harvard University, Cambridge, Massachusetts, USA.

Multiple sclerosis is a complex genetic disease associated with inflammation in the CNS white matter thought to be mediated by autoreactive T cells. Clonal expansion of B cells, their antibody products, and T cells, hallmarks of inflammation in the CNS, are found in MS. This review discusses new methods to define the molecular pathology of human disease with high-throughput examination of germline DNA haplotypes, RNA expression, and protein structures that will allow the generation of a new series of hypotheses that can be tested to develop better understanding of and therapies for this disease.

Historical perspective

A French neurologist at the Salpetrière in Paris, Jean Martin Charcot, first described multiple sclerosis (MS) in 1868, noting the accumulation of inflammatory cells in a perivascular distribution within the brain and spinal cord white matter of patients with intermittent episodes of neurologic dysfunction (1–3). This led to the term *sclérose en plaques disseminées*, or multiple sclerosis. The more recent observation in 1948 by Elvin Kabat of increases in oligoclonal immunoglobulin in the cerebrospinal fluid of patients with MS provided further evidence of an inflammatory nature to the disease (4, 5). In the past half-century, several large population–based MS twin studies demonstrated a strong genetic basis to this clinical-pathologic entity (6–13). Lastly, the demonstration of an autoimmune, at times demyelinating, disease in mammals with immunization of CNS myelin (experimental autoimmune encephalomyelitis, or EAE), first made by Thomas Rivers at the Rockefeller Institute in 1933 with the repeated injection of rabbit brain and spinal cord into primates (14), has led to the generally accepted hypothesis that MS is secondary to an autoimmune response to self-antigens in a genetically susceptible host (see box, *What we know about MS*). It should be pointed out that although the inflammation found in the CNS of patients with MS is thought to represent an autoimmune response, this is based on negative experiments where investigators have not been able to consistently isolate a microbial agent from the tissue of diseased patients. Nevertheless, primary viral infections in the CNS may induce an autoimmune response (15), and the recurring lesson from the EAE model is that the minimal requirement for inducing inflammatory, autoimmune CNS demyelinating disease is the activation of myelin-reactive T cells in the peripheral immune system (16, 17).

Advances in immunology have provided clinicians with powerful tools to better understand the underlying causes of MS, leading to new therapeutic advances. The future calls for extending the original observations of Charcot and Kabat by defining the molecular pathology of MS at the level of DNA haplotype structure, CNS and peripheral mRNA and protein expression, leading to the generation of a new series of disease-related hypotheses.

Nonstandard abbreviations used: acute disseminated encephalomyelitis (ADEM); altered peptide ligand (APL); experimental autoimmune encephalomyelitis (EAE); glatiramer acetate (GA); myelin basic protein peptide p85–99 (MBP p85–99); primary progressive multiple sclerosis (PPMS); relapsing-remitting multiple sclerosis (RRMS); single nucleotide polymorphism (SNP); secondary progressive multiple sclerosis (SPMS); T cell receptor (TCR).

Conflict of interest: The author has declared that no conflict of interest exists.

Citation for this article: *J. Clin. Invest.* **113**:788–794 (2004). doi:10.1172/JCI200421357.

Pathology

Gross examination of brain tissue of individuals with MS reveals multiple sharply demarcated plaques in the CNS white matter with a predilection to the optic nerves and white matter tracts of the periventricular regions, brain stem, and spinal cord. As was recognized early on and so elegantly investigated in more recent studies, substantial axonal injury with axonal transections is abundant throughout active MS lesions (18).

The inflammatory cell profile of active lesions is characterized by perivascular infiltration of oligoclonal T cells (19) consisting of CD4[+]/CD8 α/β (20, 21) and γ/δ (22) T cells and monocytes with occasional B cells and infrequent plasma cells (23). Lymphocytes may be found in normal-appearing white matter beyond the margin of active demyelination (24). Macrophages are most prominent in the center of the plaques and are seen to contain myelin debris, while oligodendrocyte counts are reduced. In chronic-active lesions, the inflammatory cell infiltrate is less prominent and may be largely restricted to the rim of the plaque, suggesting the presence of ongoing inflammatory activity along the lesion edge. Recently four pathologic categories of the disease were defined on the basis of myelin protein loss, the geography and extension of plaques, the patterns of oligodendrocyte destruction, and the immunopathological evidence of complement activation. Two patterns (I and II) showed close similarities to T cell–mediated or T cell plus antibody–mediated autoimmune encephalomyelitis, respectively. The other patterns (III and IV) were highly suggestive of a vasculopathy or primary oligodendrocyte dystrophy, reminiscent of virus- or toxin-induced demyelination rather than autoimmunity (25). It was of interest that the pattern of pathology tended to be the same in multiple lesions from any single individual with MS.

Natural history

MS, like other presumed autoimmune diseases, is more common in females and often first manifests clinical symptoms during young adulthood. At its onset, MS can be clinically categorized as either relapsing-remitting MS (RRMS, observed in 85–90% of patients) or primary progressive MS (PPMS). Relapses or "attacks" typically present subacutely, with symptoms developing over hours to several days, persisting for several days or weeks, and then gradually dissipating. The attacks are likely caused by the traffic of activated, myelin-reactive T cells into the CNS, causing acute inflammation with associated edema. The ability of high dose steroids to so quickly abrogate MS symptoms suggests that the acute edema and its subsequent resolution underlie the clinical relapse and remission, respectively. Studies in acute disseminated encephalomyelitis (ADEM) in humans (26) and EAE in rodents (27) suggest that immunologically, acute attacks are self-limited by regulatory T cells.

The outcome in patients with RRMS is variable; untreated, approximately 50% of all MS patients require the use of a walking aid within ten years after clinical onset (28), although the consequences on prognosis of newer treatment regiments are not as yet clear. Increased attack frequency and poor recovery from attacks in the first years of clinical disease predict a more rapid deterioration. Multiple MRI lesions, particularly those that gadolinium enhance on the first MRI scan, also predict a more severe subsequent course. Early on in the disease, there are frequent gadolinium-enhanced MRI lesions, consistent with acute fluxes of activated, autoreactive T cells into the CNS causing a breakdown of the blood-brain barrier which may be associated with clinical events. However, with time, the extent of recovery from attacks is often decreased, and baseline neurological disability accrues. Ultimately, approximately 40% of relapsing-remitting patients stop having attacks and develop what may be a progressive neurodegenerative secondary disorder related to the chronic CNS inflammation, known as secondary progressive multiple sclerosis (SPMS) (29). The evolution to this secondary progressive form of the disease is associated with significantly fewer gadolinium-enhanced lesions and a decrease in brain parenchymal volume (30, 31). Similarly, while earlier RRMS is sensitive to immunosuppression (32), as times goes on, responsiveness to immunotherapy decreases and may in fact disappear in later forms of SPMS. Thus, rather than conceiving MS as first a relapsing-remitting and then a secondary progressive disease, it could be hypothesized that MS is a continuum where there are acute inflammatory events early on with secondary induction of a neurodegenerative process refractory to immunologic intervention. This hypothesis awaits experimental verification where early immunotherapy prevents the onset of secondary progressive disease. Such critical investigations require new models of investigation using natural history studies that can be performed over decades.

The primary progressive form of MS is characterized from the onset by the absence of acute attacks and instead involves a gradual clinical decline. Clinically, this form of the disease is associated with a lack of response to any form of immunotherapy (32). This leads to the notion that PPMS may in fact be a very different disease as compared to RRMS. A recent commentary points out the similarities between PPMS and human T lymphotropic virus type I–associated myelopathy (15), where there is a progressive decline in neurologic function from the disease onset.

Diagnosis

In the absence of a specific immune-based assay, the diagnosis of MS continues to be predicated on the clinical history and neurological exam; that is, finding multiple lesions in time and space in the CNS. The use of MRI has had a major impact on allowing the early and more precise diagnosis of the disease (33). In a recent prospective study, patients experiencing their first episode suggestive of CNS demyelination and having MRI evidence of at least three typical lesions were followed for an average of 42 months. Within that period, a significant proportion of patients developed an additional relapse, thus qualifying for the diagnosis of clinically definite MS. If there were no MRI lesions, the probability of developing MS was substantially less. More than half of those developing MS experienced the additional relapse within one year of their first episode (34). Thus, it seems reasonable to label the first attack of what appears to be MS as "singular sclerosis" and to explain to patients that there is a high likelihood of developing MS. This indicates to the patient that you have an understanding of the underlying problem but that the prognosis is not as yet clear, allowing patients who never have another attack to be saved from carrying a diagnosis of MS.

ADEM is a monophasic demyelinating illness that can present with clinical, imaging, and laboratory manifestations indistinguishable from an acute MS attack (35). However, typical ADEM is seen in pediatric populations and has more of an explosive course associated with alterations in mental status, and a post-viral or post-vaccination history is often elicited. This disease is associated with significant responses to myelin proteins, indicated both by T cell and antibody measurements (26, 36). It is usually not difficult to distinguish this disease entity from "singular sclerosis," and the laboratory measurement of circulating high affinity antimyelin basic protein antibodies in ADEM but not MS may aid in the diagnosis. It should be noted that there are many reports demonstrating low affinity antimyelin autoantibodies by ELISA in sera and CSF of patients with MS, though their role either in the disease's pathogenesis or in predicting outcome is still not well defined. However, high affinity antimyelin basic protein or myelin oligodendrocyte antibodies appear to be more difficult to detect in the serum or CSF of patients with MS while antimyelin oligodendrocyte antibodies can be found in MS CNS plaque tissue (37).

MRI is the optimal imaging modality for MS. From a diagnostic standpoint, it is important to keep in mind that the typical appearance of multiple lesions on MRIs is not specific for MS. In the appropriate clinical setting, however, this appearance provides an important ancillary diagnostic tool that may establish the

What we know about MS	
Inflammation in CNS white matter	Different pathological subtypes. Early γδ T cell infiltration. Later CD4 and CD8 T cells with loss of myelin and axons.
Increase in IgG in CSF	CSF Ig is oligoclonal. B and T cells in CNS and CSF are also oligoclonal.
Complex genetic disease	λs of 20–40. High degree of inheritability. So far, only MHC region on chromosome 6 clearly associated with MS.
Antibody autoreactivity	Variable results among different studies. Agreement on presence of autoantibodies in MS plaque tissue. No high affinity autoantibodies in serum.
T cell autoreactivity	Autoreactivity to self-myelin antigens in patients with MS and normal subjects. In patients with MS, modest increase in frequency of myelin-reactive, activated T cells.

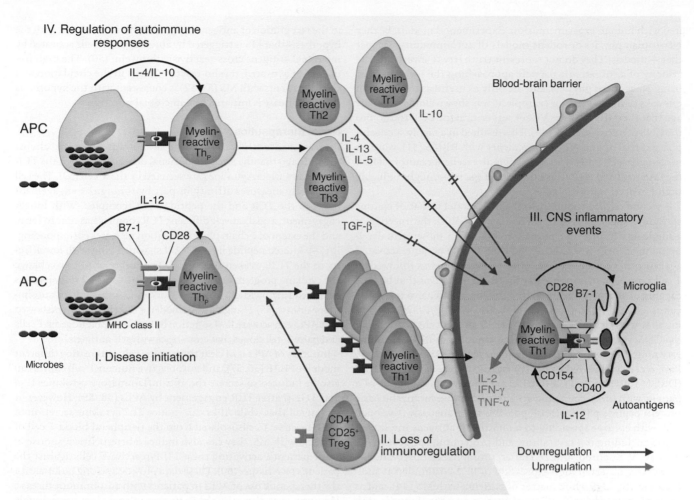

Figure 1
Working hypothesis as to the cause of MS. (I) In a genetically susceptible host, common microbes both activate the APCs through toll receptors and contain protein sequences cross-reactive with self myelin antigens. This leads to what can be defined as the minimal requirement for inducing an autoimmune, inflammatory CNS disease in mammals. (II) Underlying immunoregulatory defects, such as decreases of regulatory T cells in the circulation of patients with MS, allow the further pathologic activation of autoreactive T cells (96). (III) Activated myelin-reactive T cells migrate into the CNS and recognize antigen presented by microglia, local APCs. Th1 cytokines are secreted and an inflammatory cascade is initiated. (IV) Regulation of autoimmune responses. Naturally occurring mechanisms may exist to regulate autoimmune responses including the induction of autoreactive Th2 (IL-4, IL-5, IL-13), Th3 (TGF-β), or Tr1 (IL-10) cytokine–secreting T cells that migrate to the CNS and downregulate (red arrow) inflammatory Th1 autoreactive T cells (green arrow). Therapies may attempt to induce Th2 (Copaxone, altered peptide ligands), Th3 (mucosal antigen), or Tr1 (β-IFNs, steroids). Thp, precursor T cell.

multifocality of CNS involvement. MRI has also been used to assess MS disease activity, disease burden, and the dynamic evolution in these parameters over time (38). MRI is 4–10 times more sensitive than the clinical evaluation in capturing CNS lesions (39), and serial studies have unequivocally demonstrated that clinically apparent changes reflect only a minor component of disease activity. Lesions in the cerebrum are much more likely to be clinically silent, as compared to lesions in the brainstem or spinal cord.

Therapy
Therapies for MS have emerged over the last two decades with the demonstration of efficacy of three classes of immunomodulating therapies that impact the course of early MS: immunosuppressive drugs such as mitoxantrone and cyclophosphamide; β-IFNs; and an MHC-binding protein that engages the T cell receptor (TCR), glatiramer acetate (GA). The underlying pathology of MS as an

inflammatory CNS disease was instrumental in leading to the drug treatments presently used. While these drug therapies were not prospectively designed based on a detailed understanding of the disease's pathophysiology, examination of these drug's mechanisms of action has provided insight into the etiology of MS. Newer therapies in clinical trials are based on a more rational understanding of the disease, and these will be discussed in more detail.

Disease mechanisms
Immunopathophysiology of MS. It is often asked whether EAE, briefly discussed above, is really MS. Our laboratory investigates the pathophysiology of MS by directly studying patients with the disease and does not directly investigate mouse models. Some of our colleagues will even note in good humor that MS is a superb model of EAE! (V. Kuchroo and S. Miller, personal communication.) The truth is, we are and we will always be appropriately lim-

ited with human experimentation. Experimental models, be they Newtonian physics or rodent models of autoimmunity, are just that — models. They do not represent truth (try to investigate the velocity of a sub-atomic particle approaching the speed of light using Newtonian physics), and are only as useful as the question one asks of the model. For example, it was shown almost a decade ago that the α4β7 integrin, VLA-4, was critical for T cell traffic into the CNS of mice with EAE (40). This resulted in a highly successful phase II trial of anti–VLA-4 in patients with RRMS (41), which is now in phase III investigations. This is an excellent example of how the EAE model, if used to ask the correct question, might be highly useful in developing therapies for MS.

A second critical lesson from the EAE model is that of epitope spreading, first observed by Eli Sercarz (42). With the injection of a single myelin protein epitope into mice with subsequent development of EAE, it was observed that T cells became activated against other epitopes of the same protein; this was followed by T cell activation in response to other myelin proteins that become capable of adoptively transferring the disease to naive mice. The epitope spreading requires costimulation with B7/CD28, suggesting that with tissue damage in the CNS an adjuvant is created in the CNS with the expression of high amounts of B7.1 costimulatory molecules associated with antigen release (43). Moreover, we have recently observed that a transgenic mouse expressing DR2 (DRB1*1501) and a TCR (Ob1A12) cloned from the blood of a patient with MS recognizing an immunodominant myelin basic protein peptide p85–99 (MBP p85–99) spontaneously developed EAE with epitope spreading to a number of antigens implicated in MS including α-β crystalline, and proteolipid apoprotein (D. Altman, V. Kuchroo, and D. Hafler, unpublished observations). As we have observed high expression of B7.1 costimulatory molecules in the CNS white matter of patients with MS (44), and as most patients exhibit T cell reactivity to a number of myelin antigens (45), it is likely that by the time a patient develops clinical MS there has been epitope spreading with reactivity to multiple myelin epitopes. However, the presence of clonally expanded T cells in the CSF and brain tissue of patients with the disease raises the issue that there may be clonal reactivity to just a few myelin antigens. Single cell cloning of T cells from the inflamed CNS tissue screened against combinatorial and protein libraries may allow new insights into the pathophysiology of MS.

Data from a number of laboratories combined with experimental data from the EAE model where myelin antigen is injected with adjuvant into mammals indicate that there are autoreactive T cells recognizing myelin antigens in the circulation of mammals. It appears that the activation of these T cells is the critical event in inducing autoimmune disease (Figure 1). We and others first demonstrated over a decade ago that T cell clones isolated from the blood of patients with MS frequently exhibit exquisite specificity for the immunodominant p85–99 epitope of myelin basic protein (45–47). However, while the TCR appears to be highly specific in recognizing this peptide, altering the peptide ligand can change the TCR conformation to yield a higher degree of T cell cross-reactivity (48). Using combinatorial chemistry, even a greater degree of cross-reactivity could be demonstrated, and a number of viral epitopes were identified that could trigger autoreactive T cell clones in a manner that would not be predicted by simple algorithms (49). Indeed, one MBP-reactive T cell clone recognized an epitope of an entirely different self-protein, the myelin oligodendrocyte glycoprotein. Hence, a significant degree of functional degeneracy exists in the recognition of antigens by T cells. This is consistent with the hypothesis that MS is triggered by autoreactive T cells activated by microbial antigens cross-reactive with myelin (50). The high frequency of activated, myelin-reactive T cells in the circulation and CSF of patients with MS (51, 52) is consistent with the hypothesis that the disease is initiated by a microbial infection.

Novel therapeutics

Peptides bound to MHC as therapeutic options. It was recognized almost a decade ago that the strength of signal delivered through the TCR determines which cytokines are secreted by the T cell (53). The cell apparently measures affinity in part by timing the engagement between the TCR and the peptide/MHC complex. With longer engagement, a qualitatively different TCR complex has time to form, and the extent of ζ chain phosphorylation increases correspondingly (54). Altered peptide ligands (APLs), which bind with low affinity to the TCR, weaken this signal. The ability of APLs to change the cytokine program of a T cell from a Th1 to a Th2 response was exploited first by Kuchroo and coworkers as a therapy for autoimmune disease (55). Using the EAE model of MS, these authors showed that APLs can activate IL-4 secretion by both encephalitogenic T cells and naive T cell clones that cross-react with self-antigens.

Injection of APLs is of clear therapeutic value in treating different models of EAE (56, 57), and autoreactive human T cell clones can also be induced to secrete the anti-inflammatory cytokines IL-4 and TGF-β after TCR engagement by APLs (58, 59). However, it was noted that while APLs can induce Th2 cytokine secretion of MBP-reactive T cells isolated from the peripheral blood T cell of patients with MS, they can also induce a heteroclitic response in some patients, activating these MBP-reactive T cells against the patient's own tissues (60). These data provide a strong rationale for the therapeutic use of APLs in patients with autoimmune disease. However, they also raise the issue that in some instances, highly degenerate TCRs can recognize APLs as self-antigens.

A recently published phase II clinical trial testing an altered MBP p85–99 peptide confirms both of these conclusions. At the higher peptide dosage tested, two of seven MS patients developed remarkably high frequencies of myelin basic protein–reactive T cells, and these responses were likely associated with significant increases in MRI-detectable lesions (61) and perhaps even disease exacerbations. In contrast, patients treated with lower doses of the APL showed no such disease flare-ups and may have indeed exhibited some degree of immune deviation towards increases in IL-4 secretion of MBP-reactive T cells (61, 62). Thus, APLs represent a classic double-edged sword. In our outbred population, given the high degree of degeneracy in the immune system, it is unclear whether it is possible to find APLs of self-peptides that pose no risk of cross-reactivity with self.

An alternative approach to the use of a single APL is the administration of peptide mixtures that contain many different antigen specificities. Random copolymers that contain amino acids commonly used as MHC anchors and TCR contact residues have been proposed as possible "universal APLs." GA (Copaxone) is a random sequence polypeptide consisting of four amino acids (alanine (A), lysine (K), glutamate (E), and tyrosine (Y) at a final molar ratio of A:K:E:Y of 4.5:3.6:1.5:1) with an average length of 40–100 amino acids (63). Directly labeled GA binds efficiently to different murine H-2 I-A molecules, as well as to their human counterparts, the MHC class II DR molecules, but does not bind MHC class II DQ or MHC class I molecules in vitro (64). In phase III clinical trials, GA subcutaneously administered to patients with RRMS decreases the rate of exacerbations and prevents the appearance of new lesions detectable by MRI (65, 66). This represents perhaps the first successful use of an agent that ameliorates autoimmune disease by altering signals through the TCR.

A "universal antigen" containing multiple epitopes would be expected to induce proliferation of naive T cells isolated from the circulation, due to its expected high degree of cross-reactivity with other peptide antigens. Indeed, GA induces strong MHC class II DR-restricted proliferative responses in T cells isolated from MS patients or from healthy controls (64). In most patients, daily injection with GA causes a striking loss of responsiveness to this random polypeptide antigen, accompanied by greater secretion of IL-5 and IL-13 by $CD4^+$ T cells, indicating a shift toward a Th2 response (67–70). In addition, the surviving GA-reactive T cells exhibit a high degree of degeneracy, as measured by their ability to cross-react with a large variety of peptides represented in a combinatorial library (68).

Thus, in vivo administration of GA induces highly cross-reactive $CD4^+$ T cells that are immune-deviated to secrete Th2 cytokines. We have proposed that GA-induced migration of highly cross-reactive Th2 (and perhaps Th3) cells to sites of inflammation allows their highly degenerate TCRs to contact self-antigens, which they recognize as weak agonists, much like APLs. These T cells then apparently secrete suppressive, Th2/Th3 cytokines, thus restricting local inflammation. Thus, knowledge of the strong genetic association for MHC in patients with MS has indirectly led to a number of therapeutic trials and new insights into the disease.

Cytokines and costimulatory signals. β-IFN has similarly had a major impact on the treatment of RRMS, though whether it can prevent the transition to SPMS is still not as yet known. The mechanism of action of β-IFN is also not as yet clear, and likely involves alterations of a number of different pathways including induction of IL-10 and inhibition of T cell traffic by blocking metalloproteinases (71). Clinical trials that block the common IL-12 and IL-23 p40 chain are about to begin, as are efforts to block costimulatory signals provided by B7-CD28 interactions with CTLA-4 Ig.

What remains unknown: the genetic basis of MS

In summary of over a century of research on MS, the scientific community has demonstrated that MS is a complex genetic inflammatory disease of the CNS white matter accompanied by T cell, B cell, and macrophage infiltration; the antigenic target of these immune cells is not certain, but are likely to be common myelin antigens shown to be encephalitogenic in the EAE model. To date, the MHC gene region is the only area of the human genome clearly associated with the disease, though the precise genes in that region responsible for MS are not as yet known. In the same way that Charcot and then others defined the key features of MS by simply examining brain pathology and observing inflammation, it is critical to redefine the *molecular pathology* of inflammatory human disease in terms of germline DNA sequence based on the haplotype map, transcription products by RNA microarrays (72), and translation products by tandem mass spectrometry. The combination of such approaches will likely generate a new series of hypotheses that can be examined by both animal and in vitro models of human disease. As MS is a complex genetic disease, understanding which combinations of genes provide the multitude of perhaps relatively minor risk factors which in the population as a whole provide protection from microbial disease but together, in unfortunate random combinations, result in human autoimmune disease is a central goal of present research efforts.

New approaches to understanding the genetic basis of MS

Approximately 15–20% of MS patients have a family history of MS, but large extended pedigrees are uncommon, with most MS families having no more than two or three affected individuals. Studies in twins (6–10, 12, 13) and conjugal pairs (73) indicate that much of this familial clustering is the result of shared genetic risk factors, while studies of migrants (74) and apparent epidemics (75) indicate a clear role for environmental factors. Detailed population-based studies of familial recurrence risk (76–78) have provided estimates for familial clustering with λs, the ratio of the risk of disease in the siblings of an affected individual compared with the general population equal to approximately 20–40 (79, 80). It has become clear that this represents a complex genetic disease with no clear mode of inheritance.

Genetic diseases may fundamentally be divided into two types. First are the "gene disruptions," where there is a gene mutation or deletion, which exhibits high penetrance, and where there is the emergence of a clear clinical phenotype. Sickle cell anemia and muscular dystrophy are two such examples with mutations of the hemoglobin and dystrophin genes, respectively. In these diseases, linkage studies, i.e., linking rather large segments of the human genome identified by so-called microsatellite markers among family members with the disease, followed by positional cloning of the disease gene, have been a powerful tool in human genetics. Such studies in families with multiple sib pairs with MS have been less successful. Specifically, to date, the only confirmed genetic feature to emerge from these efforts is the association and linkage of the disease with alleles and haplotypes from the MHC on chromosome 6p21 (81–86). In the mid 1990s, whole genome screens for linkage (87–89) were published. While these investigations have continued to accumulate whole genome linkage data and almost all of these screens have found more regions of potential linkage than would be expected by chance alone, no other clearly statistically significant region has emerged by linkage investigations.

The other types of genetic diseases are more complex; an alternative hypothesis emerging from the linkage studies is that MS, as a common disease, is caused by common allelic variants each with only subtle but important variations in function. Put another way, crude theoretical modeling of human population history suggested that variants which have a high population frequency as a whole, and are likely to be responsible for complex traits (the common disease–common variant hypothesis), will generally be very old and therefore accompanied by rather little linkage disequilibrium (90). Quantitatively, this may translate to dozens of gene regions each with risk factors of less than ×1.1–×1.4 but which in concert lead to major risk for disease development. It may be postulated that as populations emerged out of Africa 30,000 to 50,000 years ago, exposure to new microbes resulted in what are thought to be major population bottlenecks, with survival of individuals with allelic variants allowing for resistance to the novel infectious event. These combinations of different genes providing resistance to the population, when randomly coming together, result in a hyper-responsive immune system, with subsequent autoimmune diseases the price an individual may pay for protection of the general population. Organ specificity may have emerged because each infectious agent evolved with a population bottleneck would select for a single "MHC restricting" element and subsequent antigen specificity.

Identifying the common allelic variants that may underlie such common diseases requires a different approach from linkage studies. One method might be to actually sequence the whole genome among a group of 5,000 patients with MS as compared to an equal number of healthy controls. While this would be the most sensitive approach, as all variants would be identified, at this stage of technology it would be impossible to even consider. It could be argued that as there appear to be only about 10 million variant, single nucleotide polymorphisms (SNPs) in the population, we could just examine those in the patients with disease compared to control subjects. This would also be far beyond present technologies. The possible emerging solution is both elegant and simple, and is based on a recent observation that was in fact suggested by studies of the MHC region over a decade ago. The discovery is that genetic variants tend to occur together in what are called "haplotype blocks." That is, recent investigations (91) have shown that recombination is not uniformly distributed along chromosomes, as previously assumed, but rather is concentrated in hot spots that are on average some 20 to 40kb apart (haplotype blocks). It has also been shown that in Europeans and Americans of European descent there is very little haplotype diversity within these genomic haplotype blocks (92). Again, this extensive linkage disequilibrium is most probably the consequence of a severe population bottleneck affecting Europeans some 30,000 to 50,000 years ago (93). The European population is thus ideal for screening for association of allelic variants with disease, since very few SNP markers from each of these linkage disequilibrium blocks will be required to screen the entire genome (94). It is expected that there will be approximately 100,000 such haplotype blocks. Assuming that three SNPs are required to interrogate fully the haplotype diversity associated with each block, the whole genome could be screened using approximately 300,000 SNPs (~10% of all SNPs). This approach has been used by Rioux, Daly, Lander, and coworkers to identify the *IBD5* locus in a previously identified linkage peak in patients with inflammatory bowel disease (95).

A whole genome association scan, while attractive, is only beginning to be feasible as the cost of genotyping continues to decrease. It is also possible that such an approach may fail because MS may be the result of more than the one genetic syndrome that it is generally believed to be or that hundreds or even thousands of genes, each representing only a fractional risk factor, are associated with the occurrence of MS. Epistatic effects of genes will also complicate the analysis. Nevertheless, large, properly powered experiments will definitively answer the question as to issues of disease heterogeneity and relative risk factors, and will prevent the wasting of resources on underpowered investigations that may provide no definitive answers.

The formation of international consortiums, which allow significant collections of patients, combined with high-throughput genotyping will be critical in performing whole genome scans based on the haplotype map. These collaborative efforts, although using many resources, will be necessary in providing a true road map for rational drug discovery. In this regard, the International MS Genetic Consortium was created two years ago by institutions around the globe including the University of Cambridge, the University of California at San Francisco, Duke University, Vanderbilt University, Harvard Medical School, the Massachusetts Institute of Technology, and the Brigham and Women's Hospital. These new partnerships in medical science requiring collaborations across scientific disciplines and medical institutions will challenge the fabric of funding, authorships, and scientific credit that have traditionally defined academic success. Finally, unlike "gene knockout diseases" which require gene therapy that has been difficult to achieve clinically, elucidation of specific pathways will likely require only minor modification of allelic gene functions. Studies in the EAE model have indicated that modification of only a few gene loci are required to eliminate disease risk. Thus, pharmacologic targeting of relatively few pathways (with proper safeguards for privacy) in populations screened for disease risk may be the ultimate treatment for both the inflammatory and degenerative components of MS.

Acknowledgments

The author wishes to thank all of the present and past laboratory members and colleagues for so many of the ideas and concepts used in this article. In particular, I would like to acknowledge Amit Bar-Or, Kevin O'Connor, John Rioux, and Phil De Jager who provided specific assistance, and my long-term partnerships with Howard Weiner, Samia Khoury, and Vijay Kuchroo.

Address correspondence to: David A. Hafler, Jack, Sadie and David Breakstone Professor of Neurology (Neuroscience), NRB, 77 Avenue Louis Pasteur, Harvard Medical School, Boston, Massachusetts 02115, USA. Phone: (617) 525-5330; Fax: (617) 525-5333; E-mail: dhafler@rics.bwh.harvard.edu.

1. Charcot, J. 1868. *Comptes rendus des séances et mémoires lus à la société de Biologie.*
2. Charcot, J. 1868. Histologie de la sclérose en plaque. *Gazette des Hôpitaux.* **41**:554–566.
3. Charcot, J. 1877. *Lectures on the diseases of the nervous system.* The New Sydenham Society. London, United Kingdom. 157–222.
4. Kabat, E.A., Glusman, M., and Knaub, V. 1948. Quantitative estimation of the albumin and gamma globulin in normal and pathologic cerebrospinal fluid by immunochemical methods. *Am. J. Med.* **4**:653-662.
5. Kabat, E.A., Freedman, D.A., Murray, J.P., and Knaub, V. 1950. A study of the crystalline albumin, gamma globulin and the total protein in the cerebrospinal fluid of one hundred cases of multiple sclerosis and other diseases. *Am. J. Med. Sci.* **219**:55-64.
6. Mackay, R.P., and Myrianthopoulos, N.C. 1966. Multiple sclerosis in twins and their relatives. *Arch. Neurol.* **15**:449-462.
7. Williams, A., et al. 1980. Multiple sclerosis in twins. *Neurology.* **30**:1139-1147.
8. Ebers, G.C., et al. 1986. A population-based study of multiple sclerosis in twins. *N. Engl. J. Med.* **315**:1638–1642.
9. Heltberg, A., and Holm, N. 1982. Concordance in twins and recurrence in sibships in MS. *Lancet.* **1**:1068.
10. Kinnunen, E., Koskenvuo, M., Kaprio, J., and Aho, K. 1987. Multiple sclerosis in a nationwide series of twins. *Neurology.* **37**:1627-1629.
11. Utz, U., et al. 1993. Skewed T-cell receptor repertoire in genetically identical twins correlates with multiple sclerosis. *Nature.* **364**:243-247.
12. Mumford, C., et al. 1992. The UK study of MS in twins. *J. Neurol.* **239**:62.
13. French Research Group on Multiple Sclerosis. 1992. MS in 54 twinships: concordance rate is independent of zygosity. *Ann. Neurol.* **32**:724-727.
14. Rivers, T.M., Sprunt, D.H., and Berry, G.P. 1933. Observations on attempts to produce acute disseminated encephalomyelitis in monkeys. *J. Exp. Med.* **58**:39-53.
15. Hafler, D.A. 1999. The distinction blurs between an autoimmune versus microbial hypothesis in multiple sclerosis. *J. Clin. Invest.* **104**:527-529.
16. Ben-Nun, A., Wekerle, H., and Cohen, I.R. 1981. The rapid isolation of clonable antigen-specific T lymphocyte lines capable of mediating autoimmune encephalomyelitis. *Eur. J. Immunol.* **11**:195-199.
17. Goverman, J., et al. 1993. Transgenic mice that express a myelin basic protein-specific T cell receptor develop spontaneous autoimmunity. *Cell.* **72**:551-560.
18. Trapp, B.D., et al. 1998. Axonal transection in the lesions of multiple sclerosis. *N. Engl. J. Med.* **338**:278-285.
19. Wucherpfennig, K.W., et al. 1992. T cell receptor V alpha-V beta repertoire and cytokine gene expression in active multiple sclerosis lesions. *J. Exp. Med.* **175**:993-1002.
20. Traugott, U., Reinherz, E.L., and Raine, C.S. 1983. Multiple sclerosis: distribution of T cell subsets within active chronic lesions. *Science.* **219**:308-310.
21. Hauser, S.L., et al. 1986. Immunohistochemical analysis of the cellular infiltrate in multiple sclerosis lesions. *Ann. Neurol.* **19**:578-587.
22. Wucherpfennig, K.W., et al. 1992. Gamma delta T-cell receptor repertoire in acute multiple sclerosis

lesions. *Proc. Natl. Acad. Sci. U. S. A.* **89**:4588–4592.

23. Prineas, J.W., and Wright, R.G. 1978. Macrophages, lymphocytes, and plasma cells in the perivascular compartment in chronic multiple sclerosis. *Lab. Invest.* **38**:409–421.

24. Prineas, J. 1975. Pathology of the early lesion in multiple sclerosis. *Hum. Pathol.* **6**:531–554.

25. Lucchinetti, C.F., Bruck, W., Rodriguez, M., and Lassmann, H. 1996. Distinct patterns of multiple sclerosis pathology indicates heterogeneity on pathogenesis. *Brain Pathol.* **6**:259–274.

26. Pohl-Koppe, A., Burchett, S.K., Thiele, E.A., and Hafler, D.A. 1998. Myelin basic protein reactive Th2 T cells are found in acute disseminated encephalomyelitis. *J. Neuroimmunol.* **91**:19–27.

27. Khoury, S.J., Hancock, W.W., and Weiner, H.L. 1992. Oral tolerance to myelin basic protein and natural recovery from experimental autoimmune encephalomyelitis are associated with downregulation of inflammatory cytokines and differential upregulation of transforming growth factor beta, interleukin 4, and prostaglandin E expression in the brain. *J. Exp. Med.* **176**:1355–1364.

28. Weinshenker, B.G. 1994. Natural history of multiple sclerosis. *Ann. Neurol.* **36**(Suppl):S6–S11.

29. Confavreux, C., Vukusic, S., Moreau, T., and Adeleine, P. 2000. Relapses and progression of disability in multiple sclerosis. *N. Engl. J. Med.* **343**:1430–1438.

30. Khoury, S.J., et al. 1994. Longitudinal MRI imaging in multiple sclerosis: correlation between disability and lesion burden. *Neurology.* **44**:2120–2124.

31. Filippi, M., et al. 1995. Correlations between changes in disability and T2-weighted brain MRI activity in multiple sclerosis: a follow-up study. *Neurology.* **45**:255–260.

32. Hohol, M.J., et al. 1999. Treatment of progressive multiple sclerosis with pulse cyclophosphamide/ methylprednisolone: response to therapy is linked to the duration of progressive disease. *Mult. Scler.* **5**:403–409.

33. McDonald, W.I., et al. 2001. Recommended diagnostic criteria for multiple sclerosis: guidelines from the International Panel on the Diagnosis of Multiple Sclerosis. *Ann. Neurol.* **50**:121–127.

34. Achiron, A., and Barak, Y. 2000. Multiple sclerosis — from probable to definite diagnosis: a 7-year prospective study. *Arch. Neurol.* **57**:974–979.

35. Griffin, D.E. 1990. Monophasic autoimmune inflammatory diseases of the CNS and PNS. *Res. Publ. Assoc. Res. Nerv. Ment. Dis.* **68**:91–104.

36. O'Connor, K.C., et al. 2003. Myelin basic protein-reactive autoantibodies in the serum and cerebrospinal fluid of multiple sclerosis patients are characterized by low-affinity interactions. *J. Neuroimmunol.* **136**:140–148.

37. Genain, C.P., Cannella, B., Hauser, S.L., and Raine, C.S. 1999. Identification of autoantibodies associated with myelin damage in multiple sclerosis. *Nat. Med.* **5**:170–175.

38. Bourdette, D., Antel, J., McFarland, H., and Montgomery, E. 1999. Monitoring relapsing remitting MS patients. *J. Neuroimmunol.* **98**:16–21.

39. Barkhof, F., et al. 1992. Relapsing-remitting multiple sclerosis: sequential enhanced MR imaging vs clinical findings in determining disease activity. *AJR Am. J. Roentgenol.* **159**:1041–1047.

40. Yednock, T.A., et al. 1992. Prevention of experimental autoimmune encephalomyelitis by antibodies against alpha 4 beta 1 integrin. *Nature.* **356**:63–66.

41. Miller, D.H., et al. 2003. A controlled trial of natalizumab for relapsing multiple sclerosis. *N. Engl. J. Med.* **348**:15–23.

42. Lehmann, P.V., Forsthuber, T., Miller, A., and Sercarz, E.E. 1992. Spreading of T-cell autoimmunity to cryptic determinants of an autoantigen. *Nature.* **358**:155–157.

43. Miller, S.D., et al. 1995. Blockade of CD28/B7-1

interaction prevents epitope spreading and clinical relapses of murine EAE. *Immunity.* **3**:739–745.

44. Windhagen, A., et al. 1995. Expression of costimulatory molecules B7-1 (CD80), B7-2 (CD86), and interleukin 12 cytokine in multiple sclerosis lesions. *J. Exp. Med.* **182**:1985–1996.

45. Ota, K., et al. 1990. T-cell recognition of an immunodominant myelin basic protein epitope in multiple sclerosis. *Nature.* **346**:183–187.

46. Pette, M., et al. 1990. Myelin basic protein-specific T lymphocyte lines from MS patients and healthy individuals. *Neurology.* **40**:1770–1776.

47. Martin, R., et al. 1990. Fine specificity and HLA restriction of myelin basic protein-specific cytotoxic T cell lines from multiple sclerosis patients and healthy individuals. *J. Immunol.* **145**:540–548.

48. Ausubel, L.J., Kwan, C.K., Sette, A., Kuchroo, V., and Hafler, D.A. 1996. Complementary mutations in an antigenic peptide allow for crossreactivity of autoreactive T-cell clones. *Proc. Natl. Acad. Sci. U. S. A.* **93**:15317–15322.

49. Hemmer, B., et al. 1997. Identification of high potency microbial and self ligands for a human autoreactive class II-restricted T cell clone. *J. Exp. Med.* **185**:1651–1659.

50. Wucherpfennig, K.W., and Strominger, J.L. 1995. Molecular mimicry in T cell-mediated autoimmunity: viral peptides activate human T cell clones specific for myelin basic protein. *Cell.* **80**:695–705.

51. Scholz, C., Patton, K.T., Anderson, D.E., Freeman, G.J., and Hafler, D.A. 1998. Expansion of autoreactive T cells in multiple sclerosis is independent of exogenous B7 costimulation. *J. Immunol.* **160**:1532–1538.

52. Lovett-Racke, A.E., et al. 1998. Decreased dependence of myelin basic protein-reactive T cells on CD28-mediated costimulation in multiple sclerosis patients. A marker of activated/memory T cells. *J. Clin. Invest.* **101**:725–730.

53. Evavold, B.D., and Allen, P.M. 1991. Separation of IL-4 production from Th cell proliferation by an altered T cell receptor ligand. *Science.* **252**:1308–1310.

54. Sloan-Lancaster, J., Shaw, A.S., Rothbard, J.B., and Allen, P.M. 1994. Partial T cell signaling: altered phospho-zeta and lack of zap70 recruitment in APL-induced T cell anergy. *Cell.* **79**:913–922.

55. Nicholson, L.B., Murtaza, A., Hafler, B.P., Sette, A., and Kuchroo, V.K. 1997. A T cell receptor antagonist peptide induces T cells that mediate bystander suppression and prevent autoimmune encephalomyelitis induced with multiple myelin antigens. *Proc. Natl. Acad. Sci. U. S. A.* **94**:9279–9284.

56. Nicholson, L.B., Greer, J.M., Sobel, R.A., Lees, M.B., and Kuchroo, V.K. 1995. An altered peptide ligand mediates immune deviation and prevents autoimmune encephalomyelitis. *Immunity.* **3**:397–405.

57. Brocke, S., et al. 1996. Treatment of experimental encephalomyelitis with a peptide analogue of myelin basic protein. *Nature.* **379**:343–346.

58. Windhagen, A., et al. 1995. Modulation of cytokine patterns of human autoreactive T cell clones by a single amino acid substitution of their peptide ligand. *Immunity.* **2**:373–380.

59. Ausubel, L.J., Krieger, J.I., and Hafler, D.A. 1997. Changes in cytokine secretion induced by altered peptide ligands of myelin basic protein peptide 85-99. *J. Immunol.* **159**:2502–2512.

60. Ausubel, L.J., Bieganowska, K.D., and Hafler, D.A. 1999. Cross-reactivity of T-cell clones specific for altered peptide ligands of myelin basic protein. *Cell. Immunol.* **193**:99–107.

61. Bielekova, B., et al. 2000. Encephalitogenic potential of the myelin basic protein peptide (amino acids 83-99) in multiple sclerosis: results of a phase II clinical trial with an altered peptide ligand. *Nat. Med.* **6**:1167–1175.

62. Kappos, L., et al. 2000. Induction of a non-encephalitogenic type 2 T helper-cell autoimmune

response in multiple sclerosis after administration of an altered peptide ligand in a placebo-controlled, randomized phase II trial. The Altered Peptide Ligand in Relapsing MS Study Group. *Nat. Med.* **6**:1176–1182.

63. Bornstein, M.B., et al. 1987. A pilot trial of Cop 1 in exacerbating-remitting multiple sclerosis. *N. Engl. J. Med.* **41**:533–539.

64. Fridkis-Hareli, M., and Strominger, J.L. 1998. Promiscuous binding of synthetic copolymer 1 to purified HLA-DR molecules. *J. Immunol.* **160**:4386–4397.

65. Johnson, K.P., et al. 1998. Extended use of glatiramer acetate (Copaxone) is well tolerated and maintains its clinical effect on multiple sclerosis relapse rate and degree of disability. Copolymer 1 Multiple Sclerosis Study Group. *Neurology.* **50**:701–708.

66. Filippi, M., et al. 2001. Glatiramer acetate reduces the proportion of new MS lesions evolving into "black holes." *Neurology.* **57**:731–733.

67. Duda, P.W., Krieger, J.I., Schmied, M.C., Balentine, C., and Hafler, D.A. 2000. Human and murine CD4 T cell reactivity to a complex antigen: recognition of the synthetic random polypeptide glatiramer acetate. *J. Immunol.* **165**:7300–7307.

68. Duda, P.W., Schmied, M.C., Cook, S.L., Krieger, J.I., and Hafler, D.A. 2000. Glatiramer acetate (Copaxone) induces degenerate, Th2-polarized immune responses in patients with multiple sclerosis. *J. Clin. Invest.* **105**:967–976.

69. Qin, Y., Zhang, D.Q., Prat, A., Pouly, S., and Antel, J. 2000. Characterization of T cell lines derived from glatiramer-acetate–treated multiple sclerosis patients. *J. Neuroimmunol.* **108**:201–206.

70. Dabbert, D., et al. 2000. Glatiramer acetate (copolymer-1)-specific, human T cell lines: cytokine profile and suppression of T cell lines reactive against myelin basic protein. *Neurosci. Lett.* **289**:205–208.

71. Stuve, O., et al. 1996. Interferon beta-1b decreases the migration of T lymphocytes in vitro: effects on matrix metalloproteinase-9. *Ann. Neurol.* **40**:853–863.

72. Dyment, D.A., and Ebers, G.C. 2002. An array of sunshine in multiple sclerosis. *N. Engl. J. Med.* **347**:1445–1447.

73. Robertson, N.P., et al. 1997. Offspring recurrence rates and clinical characteristics of conjugal multiple sclerosis. *Lancet.* **349**:1587–1590.

74. Dean, G., McLoughlin, H., Brady, R., Adelstein, A.M., and Tallett-Williams, J. 1976. Multiple sclerosis among immigrants in greater London. *Br. Med. J.* **1**:861–864.

75. Kurtzke, J.F., Gudmundsson, K.R., and Bergmann, S. 1982. MS in Iceland. 1. Evidence of a post-war epidemic. *Neurology.* **32**:143–150.

76. Carton, H., et al. 1997. Risks of multiple sclerosis in relatives of patients in Flanders, Belgium. *J. Neurol. Neurosurg. Psychiatr.* **62**:329–333.

77. Robertson, N.P., Clayton, D., Fraser, M., Deans, J., and Compston, D.A. 1996. Clinical concordance in sibling pairs with multiple sclerosis. *Neurology.* **47**:347–352.

78. Sadovnick, A., Baird, P., and Ward, R. 1988. Multiple sclerosis: updated risks for relatives. *Am. J. Med. Genet.* **39**:533–541.

79. Risch, N. 1990. Linkage strategies for genetically complex traits. II. The power of affected relative pairs. *Am. J. Hum. Genet.* **46**:229–241.

80. Sawcer, S., and Goodfellow, P.N. 1998. Inheritance of susceptibility to multiple sclerosis. *Curr. Opin. Immunol.* **10**:697–703.

81. Allen, M., et al. 1994. Association of susceptibility to multiple sclerosis in Sweden with HLA class II DRB1 and DQB1 alleles. *Hum. Immunol.* **39**:41–48.

82. Coraddu, F., et al. 1998. HLA associations with multiple sclerosis in the Canary Islands. *J. Neuroimmunol.* **87**:130–135.

83. Haegert, D.G., et al. 1993. HLA-DQA1 and -DQB1 associations with multiple sclerosis in Sardinia and French Canada: evidence for immunogenetically

distinct patient groups. *Neurology.* **43**:548–552.

84. Hauser, S.L., et al. 1989. Extended major histocompatibility complex haplotypes in patients with multiple sclerosis. *Neurology.* **39**:275–277.

85. Kellar-Wood, H.F., et al. 1995. Multiple sclerosis and the HLA-D region: linkage and association studies. *J. Neuroimmunol.* **58**:183–190.

86. Olerup, O., Hillert, J., and Fredrikson, S. 1990. The HLA-D region-associated MS-susceptibility genes may be located telomeric to the HLA-DP subregion. *Tissue Antigens.* **36**:37–39.

87. Haines, J.L., et al. 1996. A complete genomic screen for multiple sclerosis underscores a role for the major histocompatability complex. The Multiple Sclerosis Genetics Group. *Nat. Genet.* **13**:469–471.

88. Sawcer, S., et al. 1996. A genome screen in multiple sclerosis reveals susceptibility loci on chromosome 6p21 and 17q22. *Nat. Genet.* **13**:464–468.

89. Ebers, G.C., et al. 1996. A full genome search in multiple sclerosis. *Nat. Genet.* **13**:472–476.

90. Kruglyak, L. 1999. Prospects for whole-genome linkage disequilibrium mapping of common disease genes. *Nat. Genet.* **22**:139–144.

91. Daly, M.J., Rioux, J.D., Schaffner, S.F., Hudson, T.J., and Lander, E.S. 2001. High-resolution haplotype structure in the human genome. *Nat. Genet.* **29**:229–232.

92. Gabriel, S.B., et al. 2002. The structure of haplotype blocks in the human genome. *Science.* **296**:2225–2229.

93. Reich, D.E., et al. 2001. Linkage disequilibrium in the human genome. *Nature.* **411**:199–204.

94. Johnson, G.C., et al. 2001. Haplotype tagging for the identification of common disease genes. *Nat. Genet.* **29**:233–237.

95. Rioux, J.D., et al. 2001. Genetic variation in the 5q31 cytokine gene cluster confers susceptibility to Crohn disease. *Nat. Genet.* **29**:223–228.

96. Viglietta, V., Baecher-Allan, C., and Hafler, D. 2004. Loss of functional suppression by CD4+CD25+ regulatory T cells in patients with multiple sclerosis. *J. Exp. Med.* In press.

Progress and problems in the biology, diagnostics, and therapeutics of prion diseases

Adriano Aguzzi, Mathias Heikenwalder, and Gino Miele

Institute of Neuropathology, University Hospital Zurich, Zurich, Switzerland.

The term "prion" was introduced by Stanley Prusiner in 1982 to describe the atypical infectious agent that causes transmissible spongiform encephalopathies, a group of infectious neurodegenerative diseases that include scrapie in sheep, Creutzfeldt-Jakob disease in humans, chronic wasting disease in cervids, and bovine spongiform encephalopathy in cattle. Over the past twenty years, the word "prion" has been taken to signify various subtly different concepts. In this article, we refer to the prion as the transmissible principle underlying prion diseases, without necessarily implying any specific biochemical or structural identity. When Prusiner started his seminal work, the study of transmissible spongiform encephalopathies was undertaken by only a handful of scientists. Since that time, the "mad cow" crisis has put prion diseases on the agenda of both politicians and the media. Significant progress has been made in prion disease research, and many aspects of prion pathogenesis are now understood. And yet the diagnostic procedures available for prion diseases are not nearly as sensitive as they ought to be, and no therapeutic intervention has been shown to reliably affect the course of the diseases. This article reviews recent progress in the areas of pathogenesis of, diagnostics of, and therapy for prion diseases and highlights some conspicuous problems that remain to be addressed in each of these fields.

Prion pathogenesis, diagnostics, and therapy: where do we stand?

Prion diseases, also known as transmissible spongiform encephalopathies (TSEs), are invariably fatal neurodegenerative disorders affecting a broad spectrum of host species and arise via genetic, infectious, or sporadic mechanisms (Table 1). In humans, prion diseases result from infectious modes of transmission (variant Creutzfeldt-Jakob disease [vCJD], iatrogenic CJD, Kuru); inherited modes of transmission in which there is nonconservative germ line mutation of the *PRNP* gene open reading frame (familial CJD, Gerstmann-Sträussler-Scheinker Syndrome, Fatal Familial Insomnia) (1, 2); and modes of transmission that have as yet been neither determined nor understood (sporadic CJD [sCJD]). The clinical symptoms associated with each of the human prion disease forms vary dramatically (2).

Nomenclature applied to prion biology continues to be complex and confusing to nonspecialists. Here we utilize the term "prion" to denote the causative agent of prion diseases, without implying associated structural properties. We refer to the disease-associated prion protein (PrP^{Sc}), a disease-specific isoform of the host-encoded cellular prion protein (PrP^{C}), which accumulates in individuals affected with most forms of TSE (Figure 1) (3). While PrP^{Sc} is classically defined as partially protease-resistant, aggregated PrP, it has recently been shown that PrP^{C} may undergo disease-associated

structural modifications that do not impart properties of inherent protease resistance (4). In light of this, it is advisable that PrP^{Sc} be defined on the basis of disease-associated structural modifications rather than properties of protease resistance.

Prion diseases are conceptually recent; the first cases of Creutzfeldt-Jakob disease were described eight decades ago (5, 6), yet the protein-only theory of prion infection was originally formulated in 1967 (7) and later refined and the term "prion" coined in 1982 (8). The precise physical nature of the prion agent is still the subject of intense scientific controversy. PrP^{Sc} may or may not be congruent with the infectious agent. It remains to be formally proven whether the infectious unit consists primarily or exclusively of: (a) a subspecies of PrP^{Sc}; (b) an intermediate form of PrP (PrP*) (9); (c) other host-derived proteins (10); or (d) nonprotein compounds (which may include glycosaminoglycans and maybe even nucleic acids) (11). We still do not know, therefore, whether the prion hypothesis is correct in its entirety.

As with any other disease, a thorough mechanistic understanding of pathogenesis is the best foundation for devising sensitive predictive diagnostics and efficacious therapeutic regimens.

The purpose of the present article is to discuss some aspects of the state of the art in prion science and their impact on prion diagnostics, primarily with respect to peripherally acquired prion disease. As of now, no causal therapies can be offered to prion disease victims. Yet we are witnessing the emergence of an impressive wealth of therapeutic approaches, some of which certainly deserve to be tested for their validity.

Progress in understanding prion pathogenesis

Prion pathogenesis is a dynamic process that can be broken down into spatially and temporally distinct phases: (a) infection and peripheral replication; (b) transmigration from the periphery to the CNS (also termed "neuroinvasion"); and (c) neurodegeneration. But what are the mechanisms underlying neuroinvasion, and which cellular compartments are involved in replication and neuroinvasion of prions?

Nonstandard abbreviations used: bovine spongiform encephalopathy (BSE); cellular prion protein (PrP^{C}); Creutzfeldt-Jakob disease (CJD); C-terminal transmembrane PrP (^{Ctm}PrP); CXC chemokine receptor 5 (CXCR5); cytidyl-guanyl oligodeoxynucleotide (CpG-ODN); days post-inoculation (dpi); disease-associated prion protein (PrP^{Sc}); erythroid differentiation–related factor (EDRF); follicular dendritic cell (FDC); LTβ receptor immunoglobulin fusion protein (LTβR-Ig); lymphoreticular system (LRS); lymphotoxin (LT); *Prnp* knockout (*Prnp$^{o/o}$*); Rocky Mountain Laboratory (RML); sporadic Creutzfeldt-Jakob disease (sCJD); toll-like receptor 9 (TLR9); transmissible spongiform encephalopathy (TSE); tyrosine-tyrosine-arginine (YYR); variant Creutzfeldt-Jakob disease (vCJD).

Conflict of interest: The authors have declared that no conflict of interest exists.

Citation for this article: *J. Clin. Invest.* **114**:153–160 (2004). doi:10.1172/JCI200422438.

Table 1

Spectrum of prion diseases of humans and animals

Prion disease	Natural host species	Etiology
sCJD	Humans	Unknown (somatic *PRNP* mutation?)
Familial Creutzfeldt-Jakob disease (fCJD)	Humans	Familial (germ line *PRNP* mutation)
Iatrogenic Creutzfeldt-Jakob disease (iCJD)	Humans	Surgical procedures (infection)
vCJD	Humans	Ingestion of BSE-contaminated food; transfusion medicine (infection)
Kuru	Humans	Ingestion, ritualistic cannibalism (infection)
Fatal Familial Insomnia (FFI)	Humans	Familial (germ line *PRNP* mutation)
Gerstmann-Sträussler-Scheinker Syndrome	Humans	Familial (germ line *PRNP* mutation)
Scrapie	Sheep, goats	Infection, natural; mode of transmission unclear
Chronic Wasting Disease (CWD)	Deer, Elk	Infection; mode of transmission unclear
BSE	Cattle	Ingestion of BSE-contaminated feed (infection)
Transmissible mink encephalopathy	Mink	Ingestion (infection); Origin unclear
Feline spongiform encephalopathy	Cats	Ingestion of BSE-contaminated feed (infection)
Spongiform encephalopathy of zoo animals	Zoologic bovids, primates	Ingestion of BSE-contaminated feed (infection)

Peripheral replication

Cell tropism of prions varies dramatically among animal species and is also in part dependent on the particular strain of prion agent. For example, prions are lymphotropic in sheep scrapie and vCJD (12) but less so in sCJD (13) and bovine spongiform encephalopathy (BSE). Different prion strains can lead to different routes of peripheral replication in experimental models of scrapie (14, 15), and, therefore, strain-encoded properties might also determine the route of peripheral replication. With respect to peripheral pathogenesis of prion diseases, it is well established that replication of the prion agent occurs in high titers in lymphoid tissues such as spleen and lymph nodes well before neuroinvasion and subsequent detection in the CNS (16).

Upon oral challenge, an early rise in prion infectivity is observed in the distal ileum of infected organisms. This applies to several species but has been most extensively investigated in sheep, and Western blot analyses and bioassays have shown that Peyer's patches accumulate PrPSc and contain high titers of prion infectivity. This is true also in the mouse model of scrapie, where administration of mouse-adapted scrapie prions (Rocky Mountain Laboratory [RML] strain) induces a surge in intestinal prion infectivity as early as a few days after inoculation (17, 18). Indeed, immune cells are crucially involved in the process of neuroinvasion after oral application: mature follicular dendritic cells (FDCs), located in Peyer's patches, may be critical for the transmission of scrapie from the gastrointestinal tract (16, 18).

Neuroinvasion

The resistance to prions of mice that lack expression of PrPC, encoded by *Prnp* (a single-copy gene located on chromosome 2 in mice and 20 in humans), is well documented (19, 20). While the precise physiological function of PrPC is unclear, expression of it is absolutely required for transportation of the infectious agent both from the peripheral sites to the CNS (21) and within the CNS (22). However, reconstitution of *Prnp* knockout (*Prnp*o/o) mice with WT bone marrow is insufficient to restore neuroinvasion in *Prnp*o/o mice (21). Hence one could argue that the elemental compartment required for prion neuroinvasion is stromal and must express PrPC. Nevertheless, in adoptive

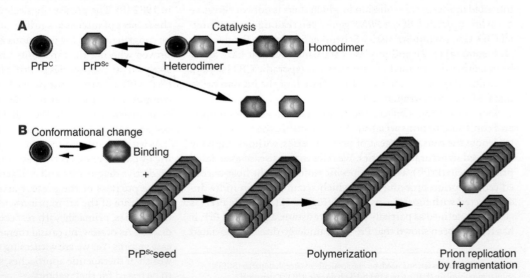

Figure 1

Models of PrPC to PrPSc conversion. (**A**) The heterodimer model proposes that upon infection of an appropriate host cell, the incoming PrPSc (orange) starts a catalytic cascade using PrPC (blue) or a partially unfolded intermediate arising from stochastic fluctuations in PrPC conformations as a substrate, converting it by a conformational change into a new β-sheet–rich protein. The newly formed PrPSc (green-orange) will in turn convert new PrPC molecules. (**B**) The noncatalytic nucleated polymerization model proposes that the conformational change of PrPC into PrPSc is thermodynamically controlled: the conversion of PrPC to PrPSc is a reversible process but at equilibrium strongly favors the conformation of PrPC. Converted PrPSc is established only when it adds onto a fibril-like seed or aggregate of PrPSc. Once a seed is present, further monomer addition is accelerated.

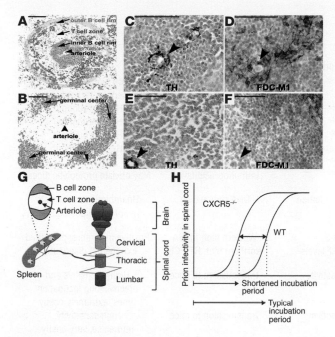

Figure 2
Positioning of FDCs in spleens of WT and CXCR5⁻/⁻ mice. (**A** and **B**) Diagrammatic representation of white pulp follicles in prion-infected CXCR5⁻/⁻ and WT mice. Anti-CD21 immunostaining was performed to visualize the lymphoid white pulp follicle microarchitecture. (**C**) Atypically localized perivascular FDCs in lymph follicles in CXCR5⁻/⁻ mice. Sympathetic nerves, visualized with antibodies to tyrosine hydroxylase (TH), are in close vicinity to FDCs (visualized by FDC-M1 immunostaining) (**D**). Scale bar: 50 mm. In contrast, sympathetic nerves in WT FDCs are localized in B cell areas at the periphery of the follicles (**E** and **F**). Arrowheads indicate TH and FDC-M1 positive areas. (**G**) Sympathetic nerves lining the thoracic spinal cord connect lymphoid organs and the CNS. (**H**) Shortened prion disease incubation period in CXCR5⁻/⁻ mice inoculated intraperitoneally, relative to WT controls.

transfer experiments on *Prnp*°/° mice with WT bone marrow, the capability of the spleen to accumulate prions of the RML strain is restored (21, 23). This suggests that hematopoietic cells transport prions from the entry site to the lymphoreticular system (LRS), which accumulates and replicates prions. B lymphocytes (not necessarily expressing PrPᶜ) are crucial for peripheral prion spread and neuroinvasion (24–26).

This dependence of FDCs on lymphotoxin (LT) signaling by B cells likely may explain — at least in part — the apparent requirement for B cells in peripheral pathogenesis: FDCs have been reported to accumulate PrPˢᶜ following scrapie infection (27). Indeed, blockade of LT signaling via administration of a soluble, dimeric LTβ receptor immunoglobulin fusion protein (LTβR-Ig) ablates mature FDCs and significantly impairs peripheral prion pathogenesis (28, 29).

FDCs are bifunctional cells: they support the formation and maintenance of the lymphoid microarchitecture but also play a role in antigen trapping — capturing immune complexes by Fcγ receptors and binding opsonized antigens to the CD21/CD35 complement receptors. Two studies have demonstrated that the complement system is relevant to prion pathogenesis. Mice genetically engineered to lack complement factors (30) or mice depleted of the C3 complement component (31) exhibited enhanced resistance to peripheral prion inoculation. However, FDCs are most

likely immobile cells and therefore unlikely to be responsible for prion transport into the CNS.

But just which cell types are involved in neuroinvasion? The innervation pattern of lymphoid organs is primarily sympathetic (32). Sympathectomy delays the onset of scrapie, while sympathetic hyperinnervation enhances splenic prion replication and neuroinvasion, which suggests that innervation of secondary lymphoid organs is the rate-limiting step to neuroinvasion (33). However, there is no physical synapse between FDCs and sympathetic nerve endings (34). So how can prions transmigrate from FDCs to sympathetic nerve fibers? A series of recent experiments (discussed below) may go some way toward providing answers.

FDC positioning is a primary determinant of velocity of neuroinvasion

We investigated how the distance between FDCs and splenic nerves affects the velocity of neuroinvasion, utilizing mice deficient in the CXC chemokine receptor 5 (CXCR5), which directs lymphocytes toward specific microcompartments (35). While density, distribution, and branching patterns of sympathetic nerve processes in CXCR5⁻/⁻ spleens are normal, the distance between FDCs and nerve endings is greatly reduced (36).

After peripheral administration of high doses of prions, velocity of pathogenesis was similar in CXCR5⁻/⁻ and WT mice; however, delivery of smaller inocula resulted in a dose-dependent increase in incubation periods in WT mice that was not evident in CXCR5⁻/⁻ mice. Peripheral prion pathogenesis in CXCR5⁻/⁻ mice is therefore more efficient upon incremental reduction of the inoculum.

What is the basis of this reduced incubation period? Kinetics measurements of prion infectivity titers in the thoracic spinal cord provided the answer: following peripheral administration, only traces of infectivity were found in WT spinal cords at 80 days postinoculation (dpi), whereas infectivity rose to measurable levels in the spinal cords of CXCR5⁻/⁻ mice already at 60 dpi. This suggests that increased velocity of prion entry into the CNS in CXCR5⁻/⁻ mice is due to the repositioning of FDCs near highly innervated, splenic arterioles (Figure 2). This was validated by the finding that incubation periods were prolonged in CXCR5⁻/⁻ mice treated with soluble LTβR-Ig to deplete mature FDCs.

Hence topographical relationships within lymphoid organs contribute to prion neuroinvasion. However, it remains to be determined whether this results from passive diffusion of prions or whether mobile cells (e.g., germinal center B cells) are involved in an active transport process.

This study also raises the possibility that spread of infection to peripheral nerves occurs more rapidly when FDCs are in close proximity to nerves in lymphoid tissue other than spleen, such as Peyer's patches. Indeed, FDCs are crucial to disease progression but only during a short window of time following oral scrapie challenge (17). This implicates the efficiency of neuroimmune transfer of prions as a primary determinant of neuroinvasion. The detection of PrPˢᶜ in spleens of sCJD patients (12) suggests that the interface between cells of the immune system and peripheral nerves (the neuroimmune connection) might also be of relevance in sporadic prion disease.

The neurodegeneration issue

There has certainly been progress in understanding the events underlying peripheral prion pathogenesis and neuroinvasion (37). However, prions exert their destructive effects exclusively within

Table 2

Molecular diagnosis of prion disease and prion infectivity

	Method	Diagnostic principle	Platform	Properties
Disease-associated PrP	Prionics–Check, Western blot	PrP^{Sc}	Western blot	Very high specificity, medium throughput
	Prionics–Check LIA	PrP^{Sc}	ELISA	High throughput
	Bio-Rad ELISA	PrP^{Sc}	ELISA	High throughput
	Enfer Scientific TSE kit	PrP^{Sc}	ELISA	High throughput
	InPro Biotechnology CDI	PrP^{Sc} and protease-sensitive PrP^{Sc}	ELISA	High throughput
	Anti-YYR motif mAb's	PrP^{Sc}	Immunoprecipitation	May obviate proteolytic digest
Genetic tests		*PRNP* ORF mutation	DNA sequencing	Unambiguous diagnosis of predisposition
Surrogate markers		14-3-3	Western blot, ELISA	Positive in cerebrospinal fluid
		Blood EDRF levels	Northern blot, RT-PCR	Extraneural diagnostics
Prion infectivity assays	Endpoint bioassay	Prion infectivity	Transmission to mice	Highly sensitive and precise, long incubation times, extremely costly
	Bioassay by incubation time	Prion infectivity	Transmission to mice	High sensitivity, imprecise, very costly
	Scrapie cell assay	Prion infectivity	Transmission to cells	High sensitivity, currently medium throughput, only available for mouse prions
	Scrapie cell assay in endpoint format (SCEPA)	Prion infectivity	Transmission to cells	High sensitivity and precision, low throughput

All currently reported methods of assessing the presence of prion disease are listed, including validated methods and nonvalidated experimental approaches. Most commercial assays rely on the direct detection of disease-associated PrP^{Sc}. LIA, luminescence immunoassay; CDI, conformation-dependent immunoassay. ORF, open reading frame.

the CNS. The precise cause of neurodegeneration remains poorly understood and is a point of contention among prionologists. It seems unlikely that PrP^{Sc} is directly toxic, since tissue devoid of PrP^{C} that subsequently accumulates PrP^{Sc} remains healthy and free of pathology (20, 38). During the conversion process, PrP^{C} levels may be depleted, yet this is also an unlikely cause of pathology, since ablation of PrP^{C} does not result in scrapie-like symptoms (39), even when ablated postnatally (40).

Lindquist and colleagues have suggested a mechanism that may account for prion-associated toxicity: (a) expression of a PrP variant resident in the cytosol was strongly neurotoxic in cultured cells and transgenic mice, which suggests a common framework for diverse PrP neurodegenerative disorders (41); and (b) PrP, retrogradely transported out of the endoplasmic reticulum, produced amorphous aggregates of PrP possessing partial proteinase K resistance in the cytosol. Once conversion occurred, it was self-sustaining (42). It will be interesting to determine whether the disease generated in these mice is, in some way, transmissible. However, while the results obtained here are certainly intriguing, it should be noted that reports elsewhere, although not refuting these observations, argue against the contribution of such potential neurotoxic PrP species (43, 44). Similarly, it has been reported that PrP^{C} in some forms of prion disease assumes a transmembrane topology, C-terminal transmembrane PrP (^{Ctm}PrP), and that the extent of neurotoxicity is a result of concentration of ^{Ctm}PrP, thereby arguing that ^{Ctm}PrP may represent a major toxic moiety (45, 46). However, while we still do not understand the biochemical events involved in cytosolic or ^{Ctm}PrP-induced neurotoxicity, elucidation of this

may aid in the much-needed identification of therapeutic targets. Additionally, in-depth characterization of transgenic mice expressing amino-terminally truncated PrP^{C} (47), in which cerebellar neurodegeneration occurs, may not only aid in the elucidation of the molecular events responsible for potentially common neurodegenerative processes but perhaps also provide clues to the physiological function of PrP^{C} itself.

Prion diagnostics

The ability to secure early diagnosis is vital for therapeutic interventions to be of real value. With respect to animals destined for the human food chain, there is the additional demand to determine presence of the prion agent in tissues in asymptomatic organisms, well before the appearance of any clinical symptoms. This applies equally to the detection of prions in humans, who may participate in tissue donation programs.

Prions were transmitted via blood transfusion in sheep using blood obtained from infected animals prior to the onset of clinical symptoms (48, 49). If the same route applies to humans, this could represent a nightmare scenario for the blood transfusion services (50). A transfusion recipient received blood from an individual harboring the vCJD agent 3.5 years prior to the development of any clinical signs of prion disease in the donor. The unfortunate recipient developed disease 6.5 years after the transfusion.

Detection of PrP^{Sc}

To be truly useful, prion diagnostics should identify "suspect" cases as early during pathogenesis as possible. However, the currently available methods are quite insensitive when compared with those available for other infectious diseases. PrPSc represents the only disease-specific macromolecule identified to date, and all approved commercial testing procedures are based on the immunological detection of PrPSc. While around 50 companies are reported to be developing prion diagnostic assays, all commercial test kits validated for use by the European Union rely on proteolytic removal of endogenous PrPC prior to detection of PrPSc (Table 2). In addition, the conformation-dependent immunoassay (4) utilizes the differential binding of antibodies to native or denatured PrPSc.

Circumvention of the protease digestion step might theoretically yield increased sensitivity of PrPSc-based detection methods and make these methods more amenable to high-throughput technologies. However, it has proved difficult to discriminate between PrPC and PrPSc with antibodies, despite some early reports (51). Interestingly, tyrosine-tyrosine-arginine (YYR) motifs (52) were reasoned to be more solvent-accessible in the pathological isoform of PrP, and a monoclonal antibody directed against these motifs was reported to be capable of selectively detecting PrPSc across a variety of platforms. However, YYR motifs are certainly not unique to pathological prion proteins, and it remains to be determined whether this reagent can really improve the sensitivity of detection of prion infections.

Deposition of PrPSc in lymphoid tissues of human prion disease patients has long been believed to be restricted to vCJD. However, recent results (12) imply that PrPSc is present in spleens and muscle tissue from as much as one third of patients with sCJD. It is presently unclear whether the patients with extraneural PrP represent a specific subset of CJD patients or whether the extraneural-negative patients may simply deposit PrPSc in muscle and spleen at levels that are below the detectability threshold of the assay. If the latter scenario proves true, and if the assay sensitivity can be raised, minimally invasive muscle biopsies may replace brain biopsy in clinical CJD diagnostics.

Surrogate markers and prion infectivity

While presence of PrPSc secures diagnostic association with the presence of prion disease, PrPSc is not always easily detectable in several forms of prion disease (53–55). In order to enhance the safety of the blood supply and of products of bovine origin, absolute specificity in securing diagnosis of asymptomatic prion disease may not be required. Instead, it may be prudent to accommodate less than 100% specificity with a panel of surrogate markers capable of identifying donated blood units from "suspect" individuals rather than requiring definitive diagnosis. It could be envisaged that wide-scale primary screens accommodate a certain degree of loss of specificity to identify samples to be re-tested in a secondary screen utilizing more specific (and likely labor-intensive) criteria.

Several research efforts have been directed at identifying transcripts and proteins differentially expressed in tissues of prion-infected animals relative to disease-free controls (56–58). However, these have mostly focused primarily either on prion-infected neural cell lines or on CNS tissue, frequently with emphasis on late-stage disease. Ideally, these surrogate markers should be detectable (and differentially expressed) in easily accessible body fluids, such as blood or urine. At present, only one extraneural gene was reported to be differentially expressed during prion infection (59): erythroid differentiation–related factor (EDRF; also known as erythroid-associated factor) levels were progressively reduced in spleens of prion-infected mice throughout pathogenesis and also in blood of experimentally infected mice, cattle with BSE, and sheep with clinically manifest scrapie.

Assessment of the levels of surrogate markers in healthy individuals is crucial in order to define the normal range of expression (according to age, sex, etc.) in order to determine what represents abnormal levels. In this respect, it is worth noting that determination of normal expression range must utilize appropriate controls. For example, EDRF transcript levels have recently been reported to show a broad range of variation in healthy human subjects (60). However, since EDRF is an erythroid-specific transcript, it would be imperative to utilize other erythroid transcripts as internal controls to normalize for variations in numbers of circulating cells in which the transcript under study is expressed relative to total cells. More searches for surrogate markers would certainly be useful, and it is likely that surrogate markers of prion disease, particularly if they are detectable in body fluids, will expand the panel of tools available for screening for prion infections.

It is also worth noting here that recent advances in neuroimaging techniques, particularly with respect to MRI, may lead to clinically useful methods of assessment of prion disease in humans, perhaps even the ability to distinguish between sCJD and vCJD (61). For example, in vCJD the pulvinar sign (a high T2 MRI signal in the posterior thalamus) has been suggested to be relatively specific for vCJD, being present in approximately 75% of vCJD patients tested (62). In sCJD, fluid-attenuated inversion recovery and diffusion-weighted MRI sequences appear to be associated with high sensitivity and specificity. MRI imaging techniques such as these may therefore represent a relatively noninvasive method to corroborate suspicion of clinical presentation of human prion disease.

While surrogate markers such as S-100, neuron-specific enolase, and 14-3-3 protein have been suggested as potential biomarkers of prion disease using body fluids such as cerebrospinal fluid (CSF) (63, 64), it is worth remembering that these are clearly surrogate markers of general neurodegenerative disease and are not therefore predictive for human prion disease. For example, one study reported false positives of 14-3-3 detection in CSF samples of patients with herpes simplex encephalitis, hypoxic brain damage, atypical encephalitis, intracerebral metastases of a bronchial carcinoma, and metabolic encephalopathy (65).

It should not be forgotten that there is no ultimate consensus on the nature of the prion: PrPSc itself might represent a surrogate marker of prion disease (66). The real gold standard of prion diagnostics is the detection of prion infectivity (whether or not PrPSc is present). Until recently, the only method available to assay for prion infectivity was the use of the mouse bioassay, in which serial dilutions of test material are inoculated into mice and onset of disease noted. However, this procedure suffers from inaccuracy and is limited by the requirements for scores of mice and significant lengths of time. Recently, the use of highly susceptible cloned neural cell lines has provided what appears to be an assay that delivers a substantial reduction in both cost and time required to perform prion bioassays and may lend itself to high-throughput automation (67). Such assays may advance methodologies aimed at diagnostic assessment of the presence of the prion agent. However, it should be noted that these cell lines are currently reported only to be permissive to murine prions. It is to be expected that the spectrum of prion strains that can be assayed using this technology will expand.

Table 3

Approaches to prion therapy

Therapeutic approach	Target	Properties	References
Polyanions	Possibly membrane-resident PrP^{Sc}	Efficient in cultured cells; relatively toxic in vivo	(70, 88)
Curcumin	Unknown	Efficacy in vivo unproven	(74)
Soluble lymphotoxin receptor	FDCs	Effective in vivo, but only on peripheral pathogenesis; moderate untoward effects	(28, 29)
CpG oligodeoxynucleotides	FDCs, DCs; B cells and macrophages	Severely immunoclastic at doses effective in vivo	(76, 79)
Anti-PrP antibodies	PrP^C	Effective in vivo only if administered in massive doses	(80, 83, 84)
Amyloidotropic intercalators (e.g., Congo red)	PrP^{Sc}	Toxic; questionable efficacy in vivo	(89, 90)
Chemical or immunological sympathectomy	Peripheral nerves involved in neuroinvasion	Very efficacious, but unacceptable toxicity in vivo	(33)
Polyene antibiotics	Possibly membrane-resident PrP^{Sc}	Low efficacy in vivo	(91, 92)
Chlorpromazine and quinacrine	Unknown	Questionable efficacy in vivo; hepatotoxicity	(93–95)
Soluble dimeric PrP^C immunoadhesin	PrP^{Sc}	Effective as transgene, but efficacy upon injection unproven	(87)

While most substances investigated so far may possess prion-curing potential in vitro, no effective therapeutic substance has been identified so far for actual in-vivo therapy in humans. However, fusion proteins either antagonizing soluble prion protein or depleting mature FDCs have been shown to efficiently prolong the incubation time of infected animals.

Prion therapy

For all the promising approaches that are being explored (Table 3), no therapy for prion diseases is available as of yet. Many substances appear to possess prion-curing properties in vitro, including Congo red (68), amphotericin B, anthracyclines (69), sulfated polyanions (70), porphyrins (71), branched polyamines (72), β-sheet breakers (73), the spice curcumin (74), and recently even small interfering RNAs (75). The majority of these molecules exert their biological effects by directly or indirectly interfering with conversion of PrP^C to PrP^{Sc}, thereby also aiding clearance of PrP^{Sc}. Yet none of these compounds have proved very effective for actual therapy.

In a recent report, results obtained in mice have led to the theory that administration of cytidyl-guanyl oligodeoxynucleotides (CpG-ODNs), which stimulate the innate immune system via toll-like receptor 9 (TLR9) signaling receptors on a variety of immune cells, may represent an applicable treatment regimen to delay prion disease in humans (76). Here it was shown that the incubation period of prion disease was extended in mice multidose treated with CpG-ODNs for twenty days. It was concluded that stimulation of innate immunity accounts for this apparent anti-prion effect, possibly through induction of anti-PrP antibodies. However, this is difficult to reconcile with several studies indicating that immune deficiencies of various sorts inhibit prion pathogenesis (24, 25, 30, 77). In addition, prion pathogenesis is unhampered in MyD88$^{-/-}$ mice, in which there is impaired TLR9 signaling (78). In fact, repeated CpG-ODN treatment has severe side effects, ranging from lymphoid follicle destruction and impaired antibody class switch to the development of ascites, hepatotoxicity, and thrombocytopenia (79). In addition, anti-PrP antibodies are not detectable in CpG-ODN–treated mice (79). It is likely therefore that the anti-prion effects of repeated CpG-ODN treatment arise via indiscriminate and undesirable follicular destruction.

Is vaccination against prion disease possible?

Anti-PrP antibodies (30) and F(ab) fragments to PrP (80, 81) can suppress prion replication in cultured cells. However, the mammalian immune system is essentially tolerant to PrP^C (82). Ablation

of *Prnp* (39) renders mice highly susceptible to immunization with prions (22). Tolerance can be circumvented by transgenic expression of an immunoglobulin μ chain containing the epitope-interacting region of a high-affinity anti-PrP monoclonal antibody. This sufficed to block prion pathogenesis upon intraperitoneal prion inoculation (83). Passive immunization may be a useful strategy for prophylaxis of prion diseases, since it has been shown that passive transfer of anti-PrP monoclonal antibodies prior to the onset of clinical symptoms is able to delay the onset of prion disease in mice inoculated intraperitoneally (84). Unfortunately, several efforts aimed at active immunization strategies have met with little success due to the robust immune tolerance to PrP^C. In this respect, it is certainly worth noting that extensive neuronal apoptosis in hippocampus and cerebellum has been shown following intracranial delivery of monoclonal antibodies reactive against a subset of PrP epitopes (85). The implications here are obvious; clearly, exhaustive in-vivo safety trials must be performed prior to the utilization of such strategies in humans.

Soluble prionostatic candidates

Do any serious candidates for prion therapeutic strategies exist? It is well established that expression of two PrP^C moieties that differ subtly from each other are able to inhibit prion replication (10). For example, humans heterozygous for a common *PRNP* polymorphism at codon 129 are largely protected from CJD (86). Although the precise molecular basis for this effect is unclear, it is possible that heterologous PrP^C may exert inhibitory action on prion replication by sequestration. This has been addressed directly by fusion of PrP^C to an immunoglobulin Fcγ domain (87), allowing for ligand dimerization, expression of the molecule as a soluble moiety, and also, therefore, increased overall stability in body fluids. Transgenetic expression of this PrP-Fc$_2$ fusion protein resulted in significant prolongation of incubation period upon prion inoculation via competition with PrP^{Sc}. It remains to be established whether PrP-Fc$_2$ is acting as an anti-prion compound when delivered exogenously. If so, soluble prion protein mutants may well represent anti-prion compounds.

An outlook for prion therapy

Prion diseases continue to present a diagnostic and therapeutic challenge to clinicians and researchers worldwide. There are many aspects of prion biology that remain unclear; we still do not know the precise physical nature of the infectious agent, the molecular and biochemical mechanisms underlying associated neurodegeneration, or the physiological function of PrP^C. The diagnostic tools currently available for prion diseases are significantly less sensitive and satisfactory than those available for other infectious diseases. Additionally, there is a dearth of therapeutic intervention strategies available for these diseases.

However, that said, the last decade or so of prion research has witnessed astounding advances in our knowledge and understanding of basic prion biology, and the field has attracted increasing numbers of researchers from diverse disciplines. Undoubtedly, this trend will trigger further important advances in prion science.

Address correspondence to: Adriano Aguzzi, Institut für Neuropathologie, UniversitätsSpital Zürich, Schmelzbergstrasse 12, CH-8091 Zurich, Switzerland. Phone: 41-1-255-2107; Fax: 41-1-255-4402; E-mail: Adriano.Aguzzi@usz.ch.

1. Glatzel, M., et al. 2003. Human prion diseases: epidemiology and integrated risk assessment. *Lancet Neurol.* **2**:757–763.
2. Collins, P.S., Lawson, V.A., and Masters, P.C. 2004. Transmissible spongiform encephalopathies. *Lancet.* **363**:51–61.
3. Bolton, D.C., McKinley, M.P., and Prusiner, S.B. 1982. Identification of a protein that purifies with the scrapie prion. *Science.* **218**:1309–1311.
4. Safar, J.G., et al. 2002. Measuring prions causing bovine spongiform encephalopathy or chronic wasting disease by immunoassays and transgenic mice. *Nat. Biotechnol.* **20**:1147–1150.
5. Creutzfeldt, H.G. 1920. Über eine eigenartige herdförmige Erkrankung des Zentralnervensystems. *Zeitschrift für die gesamte Neurologie und Psychiatrie.* **57**:1–19.
6. Jakob, A. 1921. Über eigenartige Erkrankungen des Zentralnervensystems mit bemerkenswertem anatomischem Befunde. (Spastische Pseudosklerose-Encephalomyelopathie mit disseminierten Degenerationsherden). *Zeitschrift für die gesamte Neurologie und Psychiatrie.* **64**:147–228.
7. Griffith, J.S. 1967. Self-replication and scrapie. *Nature.* **215**:1043–1044.
8. Prusiner, S.B. 1982. Novel proteinaceous infectious particles cause scrapie. *Science.* **216**:136–144.
9. Weissmann, C. 1991. A 'unified theory' of prion propagation. *Nature.* **352**:679–683.
10. Telling, G.C., et al. 1995. Prion propagation in mice expressing human and chimeric PrP transgenes implicates the interaction of cellular PrP with another protein. *Cell.* **83**:79–90.
11. Priola, S.A., Chesebro, B., and Caughey, B. 2003. Biomedicine. A view from the top--prion diseases from 10,000 feet. *Science.* **300**:917–919.
12. Glatzel, M., Abela, E., Maissen, M., and Aguzzi, A. 2003. Extraneural pathologic prion protein in sporadic Creutzfeldt-Jakob disease. *N. Engl. J. Med.* **349**:1812–1820.
13. Hill, A.F., et al. 2003. Molecular classification of sporadic Creutzfeldt-Jakob disease. *Brain.* **126**:1333–1346.
14. Aguzzi, A., Montrasio, F., and Kaeser, P.S. 2001. Prions: health scare and biological challenge. *Nat. Rev. Mol. Cell Biol.* **2**:118–126.
15. Brown, K.L., et al. 1999. Scrapie replication in lymphoid tissues depends on prion protein- expressing follicular dendritic cells. *Nat. Med.* **5**:1308–1312.
16. Aguzzi, A. 2003. Prions and the immune system: a journey through gut, spleen, and nerves. *Adv. Immunol.* **81**:123–171.
17. Mabbott, N.A., Young, J., McConnell, I., and Bruce, M.E. 2003. Follicular dendritic cell dedifferentiation by treatment with an inhibitor of the lymphotoxin pathway dramatically reduces scrapie susceptibility. *J. Virol.* **77**:6845–6854.
18. Prinz, M., et al. 2003. Oral prion infection requires normal numbers of Peyer's patches but not of enteric lymphocytes. *Am. J. Pathol.* **162**:1103–1111.
19. Büeler, H.R., et al. 1993. Mice devoid of PrP are resistant to scrapie. *Cell.* **73**:1339–1347.
20. Brandner, S., et al. 1996. Normal host prion protein necessary for scrapie-induced neurotoxicity. *Nature.* **379**:339–343.
21. Blättler, T., et al. 1997. PrP-expressing tissue required for transfer of scrapie infectivity from spleen to brain. *Nature.* **389**:69–73.
22. Brandner, S., et al. 1996. Normal host prion protein (PrPC) is required for scrapie spread within the central nervous system. *Proc. Natl. Acad. Sci. U. S. A.* **93**:13148–13151.
23. Kaeser, P.S., Klein, M.A., Schwarz, P., and Aguzzi, A. 2001. Efficient lymphoreticular prion propagation requires prp(c) in stromal and hematopoietic cells. *J. Virol.* **75**:7097–7106.
24. Klein, M.A., et al. 1997. A crucial role for B cells in neuroinvasive scrapie. *Nature.* **390**:687–690.
25. Klein, M.A., et al. 1998. PrP expression in B lymphocytes is not required for prion neuroinvasion. *Nat. Med.* **4**:1429–1433.
26. Montrasio, F., et al. 2001. B lymphocyte-restricted expression of prion protein does not enable prion replication in prion protein knockout mice. *Proc. Natl. Acad. Sci. U. S. A.* **98**:4034–4037.
27. Kitamoto, T., Muramoto, T., Mohri, S., Doh-Ura, K., and Tateishi, J. 1991. Abnormal isoform of prion protein accumulates in follicular dendritic cells in mice with Creutzfeldt-Jakob disease. *J. Virol.* **65**:6292–6295.
28. Montrasio, F., et al. 2000. Impaired prion replication in spleens of mice lacking functional follicular dendritic cells. *Science.* **288**:1257–1259.
29. Mabbott, N.A., Mackay, F., Minns, F., and Bruce, M.E. 2000. Temporary inactivation of follicular dendritic cells delays neuroinvasion of scrapie [letter]. *Nat. Med.* **6**:719–720.
30. Klein, M.A., et al. 2001. Complement facilitates early prion pathogenesis. *Nat. Med.* **7**:488–492.
31. Mabbott, N.A., Bruce, M.E., Botto, M., Walport, M.J., and Pepys, M.B. 2001. Temporary depletion of complement component C3 or genetic deficiency of C1q significantly delays onset of scrapie. *Nat. Med.* **7**:485–487.
32. Felten, D.L., and Felten, S.Y. 1988. Sympathetic noradrenergic innervation of immune organs. *Brain Behav. Immun.* **2**:293–300.
33. Glatzel, M., Heppner, F.L., Albers, K.M., and Aguzzi, A. 2001. Sympathetic innervation of lymphoreticular organs is rate limiting for prion neuroinvasion. *Neuron.* **31**:25–34.
34. Heinen, E., Bosseloir, A., and Bouzahzah, F. 1995. Follicular dendritic cells: origin and function. *Curr. Top. Microbiol. Immunol.* **201**:15–47.
35. Forster, R., et al. 1996. A putative chemokine receptor, BLR1, directs B cell migration to defined lymphoid organs and specific anatomic compartments of the spleen. *Cell.* **87**:1037–1047.
36. Prinz, M., et al. 2003. Positioning of follicular dendritic cells within the spleen controls prion neuroinvasion. *Nature.* **425**:957–962.
37. Aguzzi, A., and Polymenidou, M. 2004. Mammalian prion biology. One century of evolving concepts. *Cell.* **116**:313–327.
38. Mallucci, G., et al. 2003. Depleting neuronal PrP in prion infection prevents disease and reverses spongiosis. *Science.* **302**:871–874.
39. Büeler, H.R., et al. 1992. Normal development and behaviour of mice lacking the neuronal cell-surface PrP protein. *Nature.* **356**:577–582.
40. Mallucci, G.R., et al. 2002. Post-natal knockout of prion protein alters hippocampal CA1 properties, but does not result in neurodegeneration. *EMBO J.* **21**:202–210.
41. Ma, J., Wollmann, R., and Lindquist, S. 2002. Neurotoxicity and neurodegeneration when PrP accumulates in the cytosol. *Science.* **298**:1781–1785.
42. Ma, J., and Lindquist, S. 2002. Conversion of PrP to a self-perpetuating PrPSc-like conformation in the cytosol. *Science.* **298**:1785–1788.
43. Drisaldi, B., et al. 2003. Mutant PrP is delayed in its exit from the endoplasmic reticulum, but neither wild-type nor mutant PrP undergoes retrotranslocation prior to proteasomal degradation. *J. Biol. Chem.* **278**:21732–21743.
44. Roucou, X., Guo, Q., Zhang, Y., Goodyer, C.G., and LeBlanc, A.C. 2003. Cytosolic prion protein is not toxic and protects against Bax-mediated cell death in human primary neurons. *J. Biol. Chem.* **278**:40877–40881.
45. Hegde, R.S., et al. 1998. A transmembrane form of the prion protein in neurodegenerative disease. *Science.* **279**:827–834.
46. Hegde, R.S., et al. 1999. Transmissible and genetic prion diseases share a common pathway of neurodegeneration. *Nature.* **402**:822–826.
47. Shmerling, D., et al. 1998. Expression of amino-terminally truncated PrP in the mouse leading to ataxia and specific cerebellar lesions. *Cell.* **93**:203–214.
48. Houston, F., Foster, J.D., Chong, A., Hunter, N., and Bostock, C.J. 2000. Transmission of BSE by blood transfusion in sheep. *Lancet.* **356**:999–1000.
49. Hunter, N., et al. 2002. Transmission of prion diseases by blood transfusion. *J. Gen. Virol.* **83**:2897–2905.
50. Aguzzi, A. 2000. Prion diseases, blood and the immune system: concerns and reality. *Haematologica.* **85**:3–10.
51. Korth, C., et al. 1997. Prion (PrPSc)-specific epitope defined by a monoclonal antibody. *Nature.* **390**:74–77.
52. Paramithiotis, E., et al. 2003. A prion protein epitope selective for the pathologically misfolded conformation. *Nat. Med.* **9**:893–899.
53. Hsiao, K.K., et al. 1994. Serial transmission in rodents of neurodegeneration from transgenic mice expressing mutant prion protein. *Proc. Natl. Acad. Sci. U. S. A.* **91**:9126–9130.
54. Lasmezas, C.I., et al. 1997. Transmission of the BSE agent to mice in the absence of detectable abnormal prion protein. *Science.* **275**:402–405.
55. Tagliavini, F., et al. 1994. Amyloid fibrils in Gerstmann-Straussler-Scheinker disease (Indiana and Swedish kindreds) express only PrP peptides encoded by the mutant allele. *Cell.* **79**:695–703.
56. Duguid, J.R., and Dinauer, M.C. 1990. Library subtraction of in vitro cDNA libraries to identify differentially expressed genes in scrapie infection. *Nucleic Acids Res.* **18**:2789–2792.
57. Duguid, J., and Trzepacz, C. 1993. Major histocom-

patibility complex genes have an increased brain expression after scrapie infection. *Proc. Natl. Acad. Sci. U. S. A.* **90**:114–117.

58. Dandoy-Dron, F., et al. 1998. Gene expression in scrapie. Cloning of a new scrapie-responsive gene and the identification of increased levels of seven other mRNA transcripts. *J. Biol. Chem.* **273**:7691–7697.

59. Miele, G., Manson, J., and Clinton, M. 2001. A novel erythroid-specific marker of transmissible spongiform encephalopathies. *Nat. Med.* **7**:361–364.

60. Glock, B., et al. 2003. Transcript level of erythroid differentiation-related factor, a candidate surrogate marker for transmissible spongiform encephalopathy diseases in blood, shows a broad range of variation in healthy individuals. *Transfusion.* **43**:1706–1710.

61. Tribl, G.G., et al. 2002. Sequential MRI in a case of Creutzfeldt-Jakob disease. *Neuroradiology.* **44**:223–226.

62. Zeidler, M., Collie, D.A., Macleod, M.A., Sellar, R.J., and Knight, R. 2001. FLAIR MRI in sporadic Creutzfeldt-Jakob disease. *Neurology.* **56**:282.

63. Hsich, G., Kinney, K., Gibbs, C.J., Lee, K.H., and Harrington, M.G. 1996. The 14-3-3 brain protein in cerebrospinal fluid as a marker for transmissible spongiform encephalopathies. *N. Engl. J. Med.* **335**:924–930.

64. Beaudry, P., et al. 1999. 14-3-3 protein, neuron-specific enolase, and S-100 protein in cerebrospinal fluid of patients with Creutzfeldt-Jakob disease. *Dement. Geriatr. Cogn. Disord.* **10**:40–46.

65. Zerr, I., et al. 1998. Detection of 14-3-3 protein in the cerebrospinal fluid supports the diagnosis of Creutzfeldt-Jakob disease. *Ann. Neurol.* **43**:32–40.

66. Aguzzi, A., and Weissmann, C. 1997. Prion research: the next frontiers. *Nature.* **389**:795–798.

67. Klohn, P.C., Stoltze, L., Flechsig, E., Enari, M., and Weissmann, C. 2003. A quantitative, highly sensitive cell-based infectivity assay for mouse scrapie prions. *Proc. Natl. Acad. Sci. U. S. A.* **100**:11666–11671.

68. Caughey, B., and Race, R.E. 1992. Potent inhibition of scrapie-associated PrP accumulation by congo red. *J. Neurochem.* **59**:768–771.

69. Tagliavini, F., et al. 1997. Effectiveness of anthracycline against experimental prion disease in Syrian hamsters. *Science.* **276**:1119–1122.

70. Caughey, B., and Raymond, G.J. 1993. Sulfated polyanion inhibition of scrapie-associated PrP accumulation in cultured cells. *J. Virol.* **67**:643–650.

71. Priola, S.A., Raines, A., and Caughey, W.S. 2000. Porphyrin and phthalocyanine antiscrapie compounds. *Science.* **287**:1503–1506.

72. Supattapone, S., et al. 2001. Branched polyamines cure prion-infected neuroblastoma cells. *J. Virol.* **75**:3453–3461.

73. Soto, C., et al. 2000. Reversion of prion protein conformational changes by synthetic beta-sheet breaker peptides. *Lancet.* **355**:192–197.

74. Caughey, B., et al. 2003. Inhibition of protease-resistant prion protein accumulation in vitro by curcumin. *J. Virol.* **77**:5499–5502.

75. Daude, N., Marella, M., and Chabry, J. 2003. Specific inhibition of pathological prion protein accumulation by small interfering RNAs. *J. Cell. Sci.* **116**:2775–2779.

76. Sethi, S., Lipford, G., Wagner, H., and Kretzschmar, H. 2002. Postexposure prophylaxis against prion disease with a stimulator of innate immunity. *Lancet.* **360**:229–230.

77. Frigg, R., Klein, M.A., Hegyi, I., Zinkernagel, R.M., and Aguzzi, A. 1999. Scrapie pathogenesis in subclinically infected B-cell-deficient mice. *J. Virol.* **73**:9584–9588.

78. Prinz, M., et al. 2003. Prion pathogenesis in the absence of Toll-like receptor signalling. *EMBO Rep.* **4**:195–199.

79. Heikenwalder, M., et al. 2004. Lymphoid follicle destruction and immunosuppression after repeated CpG oligodeoxynucleotide administration. *Nat. Med.* **10**:187–192.

80. Peretz, D., et al. 2001. Antibodies inhibit prion propagation and clear cell cultures of prion infectivity. *Nature.* **412**:739–743.

81. Enari, M., Flechsig, E., and Weissmann, C. 2001. Scrapie prion protein accumulation by scrapie-infected neuroblastoma cells abrogated by exposure to a prion protein antibody. *Proc. Natl. Acad. Sci. U. S. A.* **98**:9295–9299.

82. Souan, L., et al. 2001. Modulation of proteinase-K resistant prion protein by prion peptide immunization. *Eur. J. Immunol.* **31**:2338–2346.

83. Heppner, F.L., et al. 2001. Prevention of scrapie pathogenesis by transgenic expression of anti-prion protein antibodies. *Science.* **294**:178–182.

84. White, A.R., et al. 2003. Monoclonal antibodies inhibit prion replication and delay the development of prion disease. *Nature.* **422**:80–83.

85. Solforosi, L., et al. 2004. Cross-linking cellular prion protein triggers neuronal apoptosis in vivo. *Science.* **303**:1514–1516.

86. Mead, S., et al. 2003. Balancing selection at the prion protein gene consistent with prehistoric kurulike epidemics. *Science.* **300**:640–643.

87. Meier, P., et al. 2003. Soluble dimeric prion protein binds PrP(Sc) in vivo and antagonizes prion disease. *Cell.* **113**:49–60.

88. Ladogana, A., et al. 1992. Sulphate polyanions prolong the incubation period of scrapie-infected hamsters. *J. Gen. Virol.* **73**:661–665.

89. Prusiner, S.B., et al. 1983. Scrapie prions aggregate to form amyloid-like birefringent rods. *Cell.* **35**:349–358.

90. Caughey, B., Ernst, D., and Race, R.E. 1993. Congo red inhibition of scrapie agent replication. *J. Virol.* **67**:6270–6272.

91. Pocchiari, M., Schmittinger, S., and Masullo, C. 1987. Amphotericin B delays the incubation period of scrapie in intracerebrally inoculated hamsters. *J. Gen. Virol.* **68**:219–223.

92. Mange, A., et al. 2000. Amphotericin B inhibits the generation of the scrapie isoform of the prion protein in infected cultures. *J. Virol.* **74**:3135–3140.

93. Korth, C., May, B.C., Cohen, F.E., and Prusiner, S.B. 2001. Acridine and phenothiazine derivatives as pharmacotherapeutics for prion disease. *Proc. Natl. Acad. Sci. U. S. A.* **98**:9836–9841.

94. Barret, A., et al. 2003. Evaluation of quinacrine treatment for prion diseases. *J. Virol.* **77**:8462–8469.

95. Bach, S., et al. 2003. Isolation of drugs active against mammalian prions using a yeast-based screening assay. *Nat. Biotechnol.* **21**:1075–1081.

The genetic epidemiology
of neurodegenerative disease

Lars Bertram and Rudolph E. Tanzi

Genetics and Aging Research Unit, MassGeneral Institute for Neurodegenerative Diseases, Department of Neurology,
Massachusetts General Hospital, Harvard Medical School, Charlestown, Massachusetts, USA.

Gene defects play a major role in the pathogenesis of degenerative disorders of the nervous system. In fact, it has been the very knowledge gained from genetic studies that has allowed the elucidation of the molecular mechanisms underlying the etiology and pathogenesis of many neurodegenerative disorders. In this review, we discuss the current status of genetic epidemiology of the most common neurodegenerative diseases: Alzheimer disease, Parkinson disease, Lewy body dementia, frontotemporal dementia, amyotrophic lateral sclerosis, Huntington disease, and prion diseases, with a particular focus on similarities and differences among these syndromes.

The complexities of common diseases

Familial aggregation had been recognized as a prominent characteristic of many neurodegenerative disorders decades before the underlying molecular genetic or biochemical properties were known. It was often the identification of specific, disease-segregating mutations in previously unknown genes that directed the attention to certain proteins and pathways that are now considered crucial in the pathogenesis of these diseases. These include mutations in the β-amyloid (Aβ) precursor protein, causing Alzheimer disease (AD); in α-synuclein, causing Parkinson disease (PD); or in microtubule-associated protein tau, causing frontotemporal dementia (FTD) with parkinsonism. Another feature observed in most common neurodegenerative diseases — as well as in other common disorders — is a dichotomy between familial (rare) and seemingly nonfamilial (common) forms. The latter are also frequently described as "sporadic" or "idiopathic," although there is a growing body of evidence suggesting that a large proportion of these cases are also significantly influenced by genetic factors. These risk genes are likely to be numerous, displaying intricate patterns of interaction with each other as well as with nongenetic variable, and — unlike classical Mendelian ("simplex") disorders — exhibit no simple or single mode of inheritance. Hence, the genetics of these diseases has been labeled "complex."

A popular conception regarding the genetic makeup of complex diseases is the "common disease/common variant" (CD/CV) hypothesis (1). According to this theory common disorders are also governed by common DNA variants (such as single nucleotide polymorphisms). These variants significantly increase disease risk but are insufficient to actually *cause* a specific disorder. Current empirical and theoretical data support this hypothesis, although there remains great uncertainty as to the number of the underlying risk factors and their specific effect sizes. In this context, it is noteworthy that even recent genetic advances in the study of complex diseases such as AD or diabetes likely only represent the most obvi-

ous, most extreme cases of the underlying risk spectrum (Figure 1; ref. 2). In AD, for instance, rare, fully penetrant autosomal dominant mutations in 3 genes (i.e. *APP*, *PSEN1*, and *PSEN2*) have been shown to cause the disease, while a common, incompletely penetrant susceptibility variant (i.e., ε4 in *APOE*; see below) significantly increases the risk for AD. The identification of these genes early in the study of AD genetics was possible due to the combination of several favorable circumstances, such as the presence of multiple independent mutations in the same locus and the availability of extended, multigenerational pedigrees for DNA genotyping and sequencing (in the case of *PSEN1*) or a large attributable fraction to the overall genetic variance, resulting from relatively high allele frequency and pronounced effect size (in the case of *APOE*). However, identification of disease genes that make smaller overall contributions to the genetic spectrum (because of only few mutational events; e.g., *PSEN2*), or risk factors with smaller effect sizes (i.e. odds ratios [ORs] ranging between 2 and 3), will require much larger samples and possibly more sensitive and efficient analytic tools to enable consistent detection across study populations (2).

Additional, and commonly cited, problems in finding complex disease genes beyond the most obvious are multiple testing, publication bias, and questionable replication (3–5). Multiple testing can be placed under the larger category of "avoidable false positive" findings, which are also caused by testing insufficiently sized samples, using inappropriate matching of cases versus controls, stratifying populations, and choosing inadequate analysis strategies, etc. Publication bias, on the other hand, which indicates the higher a priori likelihood of a positive finding being published as opposed to a negative one, may have been a possible source of serious bias in the early days. However, there is only relatively little empirical evidence that publication bias actually represents a common or significant source of error in current publications investigating the genetics of a number of disorders (e.g., refs. 6–12). The sheer number of publications in the AD genetics literature, for example, reveals that nearly two-thirds represent "negative" articles, with the rest being "positive" or "suggestive," so that one can hardly speak of a preponderance of the positive (13). Finally, independent replication of a positive genetic finding is one of the essential requirements to distinguish a genuine from a false-positive gene effect (Figure 2). However, replication — just like the primary detection of disease association — is affected by a number of factors, which include locus heterogeneity, small effect size, high risk allele fre-

---

Nonstandard abbreviations used: AD, Alzheimer disease; ALS, amyotrophic lateral sclerosis; CJD, Creutzfeld-Jakob disease; EOFAD, early-onset familial AD; FALS, familial ALS; FFI, fatal familial insomnia; FTD, frontotemporal dementia; FTDP-17, FTD with parkinsonism linked to chromosome 17; HD, Huntington disease; LBD, Lewy body dementia; OR, odds ratio; PD, Parkinson disease; PrP, prion protein.

Conflict of interest: The authors have declared that no conflict of interest exists.

Citation for this article: *J. Clin. Invest.* **115**:1449–1457 (2005).
doi:10.1172/JCI24761.

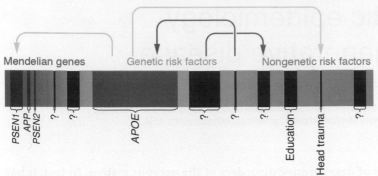

Figure 1

This scheme depicts the risk spectrum predisposing to common diseases as one continuum, using AD as an example. The continuum extends from the most extreme genetic form ("Mendelian genes"; green) to cases influenced by genetic susceptibility factors ("Genetic risk factors"; orange), until reaching into a less well-defined area of cases that may be caused by genes of lesser penetrance/lower effect size and/or altogether nongenetic factors ("Nongenetic risk factors"; gray). Established Mendelian genes (*APP*, *PSEN1*, and *PSEN2*) or genetic risk factors (*APOE*-ε4) are represented by shaded boxes and represent the most obvious candidates of AD genetics; the width of these boxes approximately represents the relative contribution to the overall risk for disease. Black boxes indicate still-elusive disease genes/risk factors ("?"). Colored arrows indicate possible gene-gene and gene-environment interaction patterns: yellow arrows represent previously suggested interactions (e.g., between *PSEN1* and *APOE*-ε4). Note that some interactions (red arrows) as well as the number of elusive genes are entirely hypothetical and are depicted for didactic purposes only.

quency, population stratification, and poor case-control matching. Thus, the failure to provide independent replication may be meaningless if the association study has not been carefully designed. In the case of multiple conflicting reports, metaanalysis across all published studies and/or the evidence for a biochemical/functional consequence of the putative risk allele can help distinguish real disease genes from their harmless counterparts (Figure 2).

Notwithstanding these difficulties, genetic analyses have laid the foundation for understanding a variety of disease mechanisms leading to neurodegeneration and associated symptoms. Likewise, a detailed understanding of their genetic basis will be essential for the development of effective strategies aimed at the early prediction and early prevention/treatment of these devastating diseases. In the following sections, we briefly outline the status of genetic research across a number of common neurodegenerative conditions, with a particular focus on the similarities and differences among disorders.

Alzheimer disease

AD is one of the most serious health problems in the industrialized world. It is an insidious and progressive neurodegenerative disorder that accounts for the vast majority of age-related dementia and is characterized by global cognitive decline and the accumulation of Aβ deposits and neurofibrillary tangles in the brain (Figure 3). Family history is the second-greatest risk factor for the disease after age, and the growing understanding of AD genetics has been central to the knowledge of the pathogenic mechanisms leading to the disease. Genetically, AD is complex and heterogenous and appears to follow an age-related dichotomy: rare and highly penetrant early-onset familial AD (EOFAD) mutations in different genes are transmitted in an autosomal dominant fashion, while late-onset AD (LOAD) without obvious

familial segregation is thought to be explained by the CD/CV hypothesis (14).

EOFAD represents only a small fraction of all AD cases (≤5%) and typically presents with onset ages younger than 65 years, showing autosomal dominant transmission within affected families. To date, more than 160 mutations in 3 genes have been reported to cause EOFAD. These include the *Aβ precursor protein* (*APP*) on chromosome 21 (15), *presenilin 1* (*PSEN1*) on chromosome 14 (16), and *presenilin 2* (*PSEN2*) on chromosome 1 (17, 18). The most frequently mutated gene, *PSEN1*, accounts for the majority of AD cases with onset prior to age 50. While these AD-causing mutations occur in 3 different genes located on 3 different chromosomes, they all share a common biochemical pathway, i.e., the altered production of Aβ leading to a relative overabundance of the $A\beta_{42}$ species, which eventually results in neuronal cell death and dementia. An up-to-date overview of disease-causing mutations in these genes can be found at the Alzheimer Disease & Frontotemporal Dementia Mutation Database (19).

LOAD, on the other hand, is classically defined as AD with onset at age 65 years or older and represents the vast majority of all AD cases. While segregation and twin studies conclusively suggest a major role of genetic factors in this form of AD (20), to date, only 1 such factor has been established, the ε4 allele of the *apolipoprotein E* gene on chromosome 19q13 (*APOE*; Table 1; refs. 21, 22). In contrast to all other association-based findings in AD, the risk effect of *APOE*-ε4 has been consistently replicated in a large number of studies across many ethnic groups, yielding ORs between approximately 3 and approximately 15 for heterozygous and homozygous carriers, respectively, of the ε4 allele in white individuals (for metaanalysis, see ref. 23). In addition to the increased risk exerted by the ε4-allele, a weak, albeit significant, protective effect for the minor allele, ε2, has also been reported in several studies. Unlike the mutations in the known EOFAD genes, *APOE*-ε4 is neither necessary nor sufficient to cause AD but instead operates as a genetic risk modifier by decreasing the age of onset in a dose-dependent manner. Despite its long-known and well-established genetic association, the biochemical consequences of *APOE*-ε4 in AD pathogenesis are not yet fully understood but likely encompass Aβ-aggregation/clearance and/or cholesterol homeostasis (Table 1).

Several lines of evidence suggest that numerous additional LOAD loci (24) — and probably also EOFAD loci (25, 26) — remain to be identified, since the 4 known genes together account for probably less than 50% of the genetic variance of AD. As outlined above, it is currently unclear how many of these loci will prove to be risk factors as opposed to causative variants. As candidates for the former, more than 3 dozen genes have been significantly associated with AD in the past (27, 28). Despite the more than 500 independent association studies, however, no single gene has been shown to be a risk factor with even nearly the same degree of replication or consistency as has *APOE*-ε4. An up-to-date overview of the status of these and other potential AD candidate genes, including metaanalyses across published association studies, can be found at the Alzheimer Research Forum genetic database (13). One of the conclusions to be drawn from currently available data, as well as from the few independently performed metaanalyses on putative AD risk factors, is that even if some of the published associations

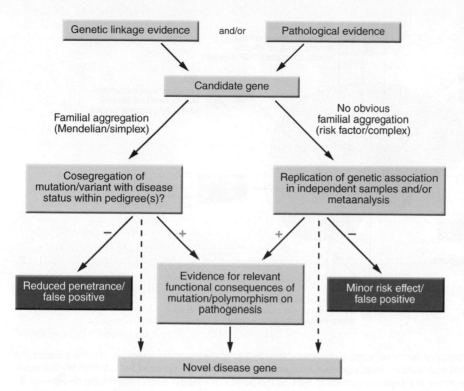

Figure 2
Flow chart of current strategies used to identify novel disease genes. This scheme outlines strategies for identifying mutations and/or polymorphisms causing or predisposing to disease. Candidate genes are chosen based on genetic linkage data and/or known or hypothesized pathobiological relevance to disease mechanisms. This procedure is referred to as the "candidate gene approach." An alternative and inherently similar strategy is based on the detection of formerly unknown genes/proteins according to genetic linkage data and is referred to as "positional cloning." Dashed lines indicate "shortcuts" allowing the definition of a novel disease gene based on the genetic evidence alone, e.g., *APOE-ε4* in AD, of which the precise functional consequences remain unknown despite an established genetic role. Note that there are examples of genes/mutations with reduced penetrance or minor risk effects (red boxes) within bona fide disease genes (e.g., certain mutations in *PSEN1* in AD).

were genuine, their overall effect size is likely to be only minor, i.e. with ORs not exceeding 2.

Parkinson disease

PD is the second most common neurodegenerative disease of adult onset. Histopathologically, it is characterized by a severe loss of dopaminergic neurons in the substantia nigra and cytoplasmic inclusions consisting of insoluble protein aggregates (Lewy bodies; Figure 3), which lead to a progressive movement disorder including the classic triad of tremor, bradykinesia, and rigidity, with an average onset age between 50 and 60 years. Although the heritability – and thus the contribution of genetic factors to the overall prevalence – of PD is likely smaller than that of AD, genetics has played a major role in elucidating the causes of nigrostriatal neuronal loss across a wide spectrum of clinically and histopathologically heterogenous PD cases. As in AD, there appears to be an age-dependent dichotomy: the majority of individuals with an early or even juvenile onset show typical Mendelian inheritance. However, unlike in AD, these cases show a predominantly autosomal-recessive mode of inheritance, and there is an ongoing debate as to whether genetic factors play any substantial role in contributing to disease risk in cases with onset beyond approximately 50 years (29–31).

Notwithstanding these uncertainties, there is a plethora of genetic studies on both forms of the disease, and mutations in at least 5 genes have now been shown to cause familial early-onset parkinsonism (α-synuclein [*SNCA* or *PARK1*; ref. 32]; parkin [*PRKN* or *PARK2*; ref. 33]; DJ-1 [*DJ1* or *PARK7*; ref. 34]; PTEN-induced putative kinase I [*PINK1* or *PARK6*; ref. 35]; and leucine-rich repeat kinase 2 or dardarin [*LRRK2* or *PARK8*; refs. 36, 37]), with several other linkage regions pending characterization and/or replication. As was the case in the study of AD, the first locus to be characterized – *PARK1*, on chromosome 4q21 – involves the protein that is the major constituent of one of the classic neuropathological hall-

marks of the disease, i.e., α-synuclein (32), which can be found at the core of Lewy bodies. While the exact mechanisms underlying α-synuclein toxicity currently remain only incompletely understood, recent evidence suggests that some *SNCA* mutations may change normal protein function *quantitatively* rather than qualitatively, via duplication or triplication of the α-synuclein gene (38, 39). Very recently, mutations in a second gene with dominant inheritance have been identified by several different laboratories (*LRRK2*; refs. 36, 37). While the functional consequences of *LRRK2* mutations are still unknown, it was suggested that at least some mutations could interfere with the protein's kinase activity (40).

While changes in *SNCA* and *LRRK2* are the leading causes of autosomal-dominant forms of PD, the majority of affected pedigrees actually show a recessive mode of inheritance (Table 1). The most frequently involved gene in recessive parkinsonism is parkin (*PRKN*) on chromosome 6q25 (33, 41), which causes nearly half of all early-onset PD cases. Parkin is a ubiquitin ligase that is involved in the ubiquitination of proteins targeted for degradation by the proteasomal system. The spectrum of parkin mutations ranges from amino acid–changing single base mutations to complex genomic rearrangements and exon deletions, which probably result in a loss of protein function. It has been speculated that this may trigger cell death by rendering neurons more vulnerable to cytotoxic insults, e.g., the accumulation of glycosylated α-synuclein (42). In addition to parkin mutations, genetic analyses of 2 non-parkin early-onset, autosomal-recessive PD pedigrees revealed 2 independent, homozygous mutations in *DJ1* (34) on chromosome 1p36 (43). Both mutations result in a loss of function of DJ-1, a protein that is suggested to be involved in oxidative stress response. While several studies have independently confirmed the presence of DJ-1 mutations in other PD cases, the frequency of disease-causing variants in this gene is estimated to be low (~1%; ref. 44). Less than 13 Mb toward the long arm of the same chro-

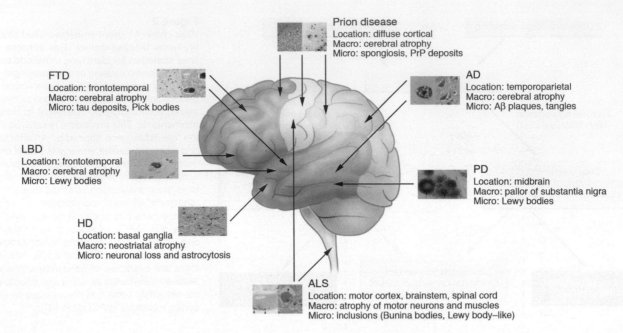

Figure 3
Overview of the anatomical location of and macroscopic and microscopic changes characteristic of the neurodegenerative disorders discussed in this review. Note that the full neuropathological spectrum of these disorders is much more complex than depicted here. When there is more than one characteristic histopathological feature, these are depicted from left to right, as indicated in the labels listing microscopic changes (e.g., the 2 panels for AD depict an Aβ plaque [left] and neurofibrillary tangles [right]). All histopathological images are reprinted with permission from ISN Neuropath Press (ref. 99).

mosome, additional PD-causing mutations were subsequently discovered in *PINK1* (35) following positive linkage evidence to this region (45). This enzyme is expressed with particularly high levels in brain, and the first 2 identified mutations (G309D and W437ter) were predicted to lead to a loss of function that may render neurons more vulnerable to cellular stress, similar to the effects of parkin mutations. While Lewy bodies are typically not found in brains of patients bearing parkin mutations, it is currently unclear whether these are present in PD cases with mutations in *DJ1* and *PINK1*.

At least 6 additional candidate PD loci have been described, including putative disease-causing mutations in the *ubiquitin carboxy-terminal hydrolase L1* (*UCHL1*) on chromosome 4p14 (46), and in a *nuclear receptor of subfamily 4* (*NR4A2*, or *NURR1*; ref. 47) located on 2q22. However, and unlike the previously outlined PD genes, neither of these maps to known PD linkage regions, nor were they independently confirmed beyond the initial reports. However, polymorphisms in both genes have been – albeit inconsistently – associated with PD in some case-control studies. A recent metaanalysis of the S18Y polymorphism in *UCHL1* showed a modest but significant protective effect of the Y allele (11), which suggests that this gene may actually be a susceptibility factor rather than a causal PD gene.

Unlike early-onset PD, the heritability of late-onset PD is probably low (29). Despite this caveat, while a number of whole-genome screens across several late-onset PD family samples have been performed, only a few overlapping genomic intervals have been identified. One of the more extensively studied regions is 17q21, near the gene encoding the microtubule-associated protein tau (*MAPT*; ref. 48). Previously, it had been shown that rare missense mutations in *MAPT* lead to a syndrome of frontotemporal dementia

with parkinsonism (FTD with parkinsonism linked to chromosome 17 [FTDP-17]; see below), but to date no mutations have been identified as causing parkinsonism without frontotemporal degeneration. However, haplotype analyses of the tau gene have revealed some evidence of genetic association of the H1 haplotype with both PD (ref. 49; for metaanalysis see ref. 50) and a related syndrome, progressive supranuclear palsy (PSP; ref. 51). Despite the lack of evidence for genetic linkage to chromosome 19q13, variants in *APOE* have also been tested for a role in PD and related syndromes. Across the nearly 3 dozen different studies available to date, some authors report a significant risk effect of *APOE*-ε4 for PD, while others only see association with certain PD phenotypes or even a risk effect of the ε2 allele, which is protective in AD (see above). A recent metaanalysis on the effects of *APOE* in PD concluded that only the ε2-related increase in PD risk remains significant when all published studies are considered jointly (12). Finally, and in addition to the findings in autosomal-dominant familial PD, there is also some support for a potential role of *SNCA* variants on the risk for late-onset PD (52).

Lewy body dementia

According to some investigators, Lewy body dementia (LBD) is the second most common type of degenerative dementia in the elderly, possibly accounting for up to 15% of all dementia cases in autopsy samples (53). Clinically, LBD is characterized by progressive cognitive impairment with fluctuating course, recurrent visual hallucinations, and parkinsonism. Although formal clinical criteria have been proposed (53), there is a pronounced clinical and neuropathological overlap with AD as well as PD with dementia (PDD). The predominant histological feature of LBD is the presence of cortical and subcortical Lewy bodies (Figure 3), but many patients with LBD also have AD pathology, i.e., cortical amyloid plaques and neurofibrillary tangles. Conversely, Lewy bodies are also frequently observed in cases of classic AD, including in patients with mutations in *APP*, *PSEN1*, and *PSEN2* (54).

While a familial aggregation of LBD has been described (55), the identification of specific LBD genetic factors is complicated by its still-uncertain phenotypic classification, in particular its distinction from AD and PDD. The little genetic evidence that has accrued to date shows — not unexpectedly — substantial overlap with that for AD and PD. For instance, follow-up analyses to the original AD full-genome screen that led to the description of linkage to chromosome 12 (56) found evidence for considerable genetic heterogeneity in the original study population (57). In particular, the authors found that the better part of the AD linkage signal on chromosome 12q13 near 50 Mb was actually caused by a subset of families fulfilling neuropathological criteria for LBD (i.e., 8 of 54 families). However, these families were linked to a more proximal region, i.e., 12p11 with a maximum linkage signal at approximately 27 Mb. Interestingly, this same region was also implicated in a large Japanese pedigree with autosomal dominant PD (*PARK8* [ref. 58], which has now been identified as being caused by mutations in *LRRK2* [refs. 36, 37]) and lies only slightly distal to an AD linkage region on 12p12, which was found by 2 different research groups in a sample different from the original chromosome 12 linkage report (59, 60). While these observations could indicate the presence of common genetic risk factors across these 3 syndromes, they could also be purely accidental or even artificial. As for several of the neurodegenerative disorders, a potential association has also been observed with *APOE-ε4*, albeit inconsistently. However, a moderate risk effect of this allele was recently supported by metaanalysis on LBD case-control studies from 2000 to 2004 (61). Finally and not surprisingly, recent reports also indicate a potential role of α-synuclein in LBD pathogenesis, based on the observation that the occurrence of cortical Lewy bodies and dementia in PD may be dependent on α-synuclein gene dose (see above; 38, 39).

Frontotemporal dementia

FTD is a heterogenous group of syndromes defined clinically by a gradual and progressive change in behavior and personal conduct and/or by a gradual and progressive language dysfunction (62). The initial symptoms typically occur without affecting other cognitive domains, such as memory, and rarely present with an onset age beyond 75 years. In some instances, deficits in behavior and language are also accompanied by parkinsonism or progressive motor neuron disease. Neuropathologically, FTD is caused by neurodegeneration in the frontal and/or temporal lobes (Figure 3). Affected neurons frequently display intracellular, tau-positive inclusions that are distinct from the neurofibrillary tangles observed in AD (63). While 25–40% of all FTD cases are believed to be familial (64), the clinical and neuropathological variability of the syndrome suggests the existence of several distinct genetic factors underlying or modifying pathogenesis. On the other hand, recent advances in the genetics of FTDP-17 (see below) have shown that different mutations in the same gene (or even exon) can lead to a diverse spectrum of FTD-type syndromes, which provides genetic support for merging the apparently diverse clinical entities into 1 overarching category.

The first FTD mutations were identified in cases accompanied by parkinsonism and showing genetic linkage to chromosome 17q21, near the tau gene (FTDP-17). Subsequently, disease-causing mutations were identified in tau (gene: *MAPT*; Table 1) (ref. 65), currently more than 30 in over 100 families worldwide (for an up to date overview, see the Alzheimer Disease & Frontotemporal Dementia Mutation Database; ref. 19). The phenotype observed with mutations in *MAPT* is variable and ranges from classic FTDP-17 to corticobasal degeneration (CBD), PSP, and frontotemporal lobar degeneration (DLDH; ref. 64). Although, to date, no mutations in *MAPT* have been reported in pathologically proven AD cases, there have been sporadic observations of early progressive memory loss reminiscent of AD, e.g., segregating with the R406W mutation in exon 13 (66, 67). However, the

---

Table 1
Overview of established neurodegenerative disease genes

Disease	Gene (first ref.)	Protein	Location	Inheritance	Relevance to pathogenesis
AD	*APP* (15)	Aβ precursor protein	21q21	Dominant	Altered Aβ production (Aβ$_{42}$/Aβ$_{40}$ ratio ↑) and aggregation
AD	*APOE* (21, 22)	Apolipoprotein E	19q13	Risk factor	Unknown (Aβ aggregation? lipid metabolism?)
AD	*PSEN1* (16)	Presenilin 1	14q24	Dominant	Altered Aβ production (Aβ$_{42}$/Aβ$_{40}$ ratio ↑)
AD	*PSEN2* (17, 18)	Presenilin 2	1q31	Dominant	Altered Aβ production (Aβ$_{42}$/Aβ$_{40}$ ratio ↑)
PD	*SNCA* (32)	α-Synuclein	4q21	Dominant	Neurotoxicity by aggregation of α-synuclein (?)
PD	*PRKN* (33)	Parkin	6q25	Recessive	Impaired protein degradation via proteasome
PD	*DJ1* (34)	DJ-1	1p36	Recessive	Impaired oxidative stress response (?)
PD	*PINK1* (35)	PTEN induced putative kinase 1	1p36	Recessive	Mitochondrial dysfunction (?)
PD	*LRRK2* (36, 37)	Leucine-rich repeat kinase 2; dardarin	12q12	Dominant	Unknown
FTD	*MAPT* (53)	Microtubule-associated protein tau	17q21	Dominant	Altered tau-production (4R/3R ratio ↑), and/or altered binding to microtubules
ALS	*SOD1* (63)	Superoxide dismutase 1	21q22	Dominant and recessive	Protein misfolding/aggregation and/or impaired oxidative stress response (?)
ALS	*ALS2* (68, 69)	Alsin	2q33	Recessive	Impaired neuroprotection (?)
HD	*HD* (76)	Huntingtin	4p16	Dominant	Unknown
Prion	*PRNP* (84)	Prion protein	20p13	Dominant and risk	Transformation of PrP^c into PrP^{sc}

Diseases are listed in the order in which they are discussed in the text. Prion, familial prion disease (see text for more details); first ref., publication showing first evidence of disease involvement; ↑, increase; 4R, 4-repeat; PrPc, normal form of PrP; PrPsc, disease-associated (scrapie) form of PrP. For an up-to-date overview on AD association results, see ref. 13; for mutation findings in AD and FTD, see ref. 19; for mutations in ALS, see ref. 82.

same mutation was also found in families segregating more typical FTD (65). Molecular genetic studies have shown that the biochemical consequences of the various mutations at the protein level are quite diverse, including reduction or increase in the binding of tau to microtubules, enhancement of tau aggregation, and alterations in the ratio of specific tau isoforms (i.e., an increased ratio of 4-repeat to 3-repeat tau owing to changes in alternative splicing; reviewed in ref. 68; Table 1). Interestingly, it appears that *MAPT* mutations almost exclusively lead to FTD, with immunohistochemical evidence of both 3- and 4-repeat tau, while classic Pick disease (PiD), in which the 4-repeat isoform is lacking, has not yet been conclusively linked to *MAPT* or any other genetic defect (69). The correlation between 4-repeat tau and genetic variants in *MAPT* is further supported by genetic association studies showing almost unanimous support for an *MAPT* risk haplotype (H1) in samples from patients with PSP or CBD, both characterized by the abundance of 4-repeat tau. This is the same haplotype that has also frequently been associated with PD (for a recent metaanalysis, see ref. 50), which possibly suggests common and as-yet-uncharacterized tau-related pathogenic mechanisms shared by FTD and late-onset PD.

Similar to the examples in the neurodegenerative diseases, tau mutations likely represent the first and most obvious candidates in the puzzle of FTD genetics. They probably account for less than half of the genetic variance in familial FTD (64). In addition to linkage to chromosome 9q21 in a syndrome of FTD coupled with familial amyotrophic lateral sclerosis (ALS; see below), association has been observed between FTD and *APOE*, albeit with highly variable results. Interestingly, and similar to equivalent studies done in PD, a recent metaanalysis on all data published for *APOE* in FTD detected a significant risk effect associated with the ε2-allele but no significant results with ε4 (70). While this observation may be purely incidental, it is similar to findings on the H1-tau haplotype, which has also been associated in some FTD syndromes as well as PD. Collectively, these findings are still too preliminary to allow speculation on any functional consequences of the underlying genetic variants in the pathogenesis of FTD. Finally, recent reports have suggested that some cases of FTD may also be caused by mutations in *PSEN1* (71). However, a more rigorous proof of familial segregation and pathogenetic mechanism of these variants is needed before they can be considered established.

Amyotrophic lateral sclerosis

ALS (also known as motor neuron disease or Lou Gehrig's disease) is characterized by a rapidly progressive degeneration of motor neurons in the brain and spinal chord, which ultimately leads to paralysis and premature death. Overall, the prevalence of ALS is low (approximately 5 in 100,000 individuals), but incidence increases with age, showing a peak between 55 and 75 years. Neuropathological features of ALS include intracellular accumulations and perikaryal inclusions of neurofilament (NF) and intracellular inclusions such as Bunina bodies and Lewy body–like cytoplasmic inclusions (Figure 3). Cognitive impairment and dementia coexist with ALS in at least 5% of the cases, and it was actually in a family displaying FTD with parkinsonism and amyotrophy where evidence of linkage to chromosome 17 was first described (see above).

Familial ALS (FALS) is observed in approximately 10% of all cases, but substantially more ALS cases are suspected to be influenced by genetic factors (72). In addition to the variants in *MAPT* (see above), mutations in 2 genes (*SOD1* and *ALS2*; Table 1) have been shown

to cause FALS. Two years after genetic linkage to chromosome 21q was described in 1991 (73), ALS-causing mutations were identified in the gene encoding superoxide dismutase 1 (*SOD1*; ref. 74), which catalyzes the conversion of superoxide radicals into hydrogen peroxide. Meanwhile, more than 100 mutations in *SOD1* have been described in over 200 pedigrees with FALS worldwide, and all but 1 of the known *SOD1* mutations are inherited in an autosomal dominant fashion (75). Collectively, these mutations account for approximately 20% of FALS cases and for up to 10% of the sporadic cases of ALS, i.e., those not showing an obvious familial segregation (76). Mutations in *SOD1* have been hypothesized as leading to neurodegeneration through protein misfolding, impaired oxidative stress response, cytoskeletal dysfunctions, and glutamatergic excitotoxicity (Table 1; for review, see refs. 77, 78). Recently, mutations in a second gene (*ALS2*; encoding alsin) were identified in independent families with a rare, juvenile-onset autosomal recessive form of ALS and primary lateral sclerosis, a syndrome restricted to upper motor neuron degeneration (79, 80). Additional mutations in *ALS2* have also been described in families suffering from infantile-onset ascending hereditary spastic paralysis, which suggests considerable phenotypic variability of the *ALS2* mutations. Functionally, there is evidence that physiologic expression of alsin is neuroprotective in the presence of SOD1 mutations (81); thus, it is conceivable that *ALS2* mutations abrogate the protective role of this protein. An up-to-date overview of the status of these and other potential ALS genes can be found at the ALS Online Database (82).

Several other putative FALS loci have been detected by means of linkage analysis in individual families or larger FALS samples, but none of the underlying gene defects have been conclusively proven to be causal. A recent full-genome screen in FALS has pinpointed significant linkage to chromosome 9q21 in families with ALS and FTD (83). This overlaps with a location shown to be linked to AD (84), which possibly indicates a common pathophysiological basis for neurodegeneration or dementia in these 2 disorders. Furthermore, various genetic associations with mostly nonfamilial ALS have been claimed but have been met with only inconsistent replication to date. Specifically — unlike for most of the neurodegenerative disorders discussed in this review— there is virtually no evidence for an association between *APOE*-ε4 and risk or disease progression of ALS. One potential ALS-specific candidate is the gene encoding the NF heavy chain gene (*NFH*) on chromosome 22q12, a component of the neuronal inclusions observed histopathologically. However, the genetic data supporting the postulated association between ALS and variants in *NFH* remain scarce, despite the fact that the association initially was described more than a decade ago (85).

Huntington disease

Huntington disease (HD) is caused by degeneration of neurons in the basal ganglia and then in cortical regions (Figure 3), leading clinically to involuntary movements (chorea), psychiatric symptoms, and dementia. Its prevalence is similar to that of ALS but much less than that of most of the other dementing illnesses discussed above. Approximately 90% of HD cases are hereditary and transmitted in an autosomal dominant fashion. As a matter of fact, the HD gene was the first autosomal disease locus to be mapped by genetic linkage analysis (to chromosome 4q16), in 1983 (86). It took 10 more years to identify the underlying gene defect, which proved to be a poly-CAG (encoding glutamine [Q]) repeat in exon 1 of a 350-kDa protein (huntingtin; gene: *HD*; Table 1; ref. 87). The mean repeat length in HD patients is 40–45, although variability is quite wide, ranging from 35 to 120 repeats (88), displaying an inverse correlation with onset age. Interestingly, approximately 10% of all HD cases are considered "de novo," i.e., these cases originate from asymptomatic parents with normal repeat lengths that have expanded to symptomatic range (see below). The precise function of huntingtin remains elusive, but cloning experiments show that it is highly conserved throughout evolution, which suggests an essential functional role of this protein in neuronal development and homoeostasis.

In contrast to all other diseases reviewed here, HD is virtually always attributable to a defect in a single gene, i.e., poly-Q expansion in *huntingtin*, although such defects only account for 50% of the interindividual onset age variation. Thus, recent genetic analyses of HD have mainly focused on the search for factors affecting the onset of the disease. A recent full-genome screen aimed at identifying these genes has revealed several suggestive linkage regions (89). The most promising of these is located on chromosome 6q25, close to the *glutamate receptor, ionotropic, kainate 2 (GRIK2)*, which has been associated with a younger HD onset age in some studies; this potentially supports the notion of glutamate-induced excitotoxicity in the pathogenesis of HD (90, 91). However, this finding awaits further replication and functional characterization. Only a small number of studies have investigated a potential onset-age effect of the *APOE* polymorphisms in HD, and as is the case with ALS, the results have been largely negative.

Creutzfeld-Jacob disease and other prion diseases

Prion diseases include a rare and heterogenous spectrum of clinical and histopathological phenotypes, which are unique in the group of neurodegenerative diseases, as they can be familial (e.g., familial Creutzfeld-Jakob disease [fCJD], fatal familial insomnia [FFI], Gerstmann-Sträussler-Scheinker syndrome [GSS]), sporadic (e.g., Creutzfeld-Jakob disease [CJD], sporadic fatal insomnia [sFI]), or acquired (e.g., kuru, iatrogenic CJD, variant CJD). Most forms are characterized by a rapidly progressing neurodegeneration with spongiosis and amyloid plaques consisting of prion protein (PrP) aggregates, probably created via self-propagation of aberrant or misfolded PrP (92, 93; Figure 3). While only a relatively small subset of cases with prion disease exhibits familial aggregation, genetics has played a crucial role in elucidating the molecular mechanisms underlying these disorders and has facilitated the clinical classification of their various subtypes (94).

As in AD, both causative mutations and risk-conferring gene variants have been identified for the different prion diseases. However, both mutations and risk variants are located within the same locus, i.e., the gene encoding PrP (*PRNP*), on chromosome 20p13 at approximately 5 Mb (Table 1; ref. 95). First, more than 2 dozen different amino acid–changing mutations in the coding region of *PRNP* have been identified as causing familial prion diseases, transmitted in an autosomal dominant fashion with nearly 100% penetrance. There is remarkable heterogeneity in the sense that different mutations throughout the gene can give rise to a variety of different phenotypes associated with all 3 familial forms of prion diseases, i.e. fCJD, FFI, and GSS (reviewed in ref. 94). In addition to these point mutations, there are also rare cases of fCJD and GSS caused by variable numbers of octapeptide (i.e., 24 base-pair) repeats within the coding sequence of *PRNP*. Second, both clinical presentation and disease progression of these familial forms are further modified by a common polymorphism at codon 129, which leads to a nonsynonymous amino acid substitution (from methionine [M] to valine [V]). Most mutated *PRNP* missense alleles are on the same haplotype as the 129M allele, which occurs in virtually all forms of fCJD. In the rare cases where they co-occur with the 129V allele, they lead to a distinct clinical phenotype, and at D178N, even to a different disease entity within the complex of prion diseases: while the D178N-129V haplotype leads to typical fCJD, the 178N-129M haplotype represents the only currently known genetic cause of FFI, which presents with a quite distinct clinical picture.

In addition to its effects on familial forms of prion diseases, the M129V polymorphism also increases the risk for sporadic forms of CJD (sCJD; refs. 96, 97). Specifically, it was found that the homozygous state for either allele (i.e. M/M or V/V) is disoportionally more frequent in sCJD than the M/V genotype (96). Furthermore, homozygosity at this polymorphism leads to a faster disease progression than heterozygosity in nearly all genetic as well as sporadic (including iatrogenic) forms of prion diseases, and in most instances, the M/M genotype is associated with the most aggressive course of disease (98). Interestingly, almost all individuals thus far known to be affected by the newly described "variant" form of CJD (vCJD), which is characterized by a prion protein isotype resembling that found in BSE, also carry only the M/M genotype. Furthermore, there is some evidence suggesting an overrepresentation of the M allele in AD cases versus controls as well (13). Only a few other genetic risk factors for the nonfamilial forms of CJD have been investigated, and none of them has shown any noteworthy results to date. This includes *APOE-ε4*, which was found to increase risk and/or accelerate disease progression in some studies, although the majority of samples failed to replicate either of these effects.

Conclusions

While displaying a diverse array of clinical and histopathological characteristics, the neurodegenerative disorders discussed in this review share a variety of epidemiologic and genetic aspects. First, with the exception of HD, they all feature an etiologic dichotomy, with relatively rare familial forms on the one hand and more frequent multifactorial — and usually later-onset — forms on the other. It is possible (and likely) that a substantial number of cases that were hitherto considered nonfamilial and sporadic will eventually prove to originate from specific disease-causing mutations or genetic risk factors (like *APOE-ε4* in AD). Second, in some cases, the same mutations and polymorphisms have been linked and associated across clinically and neuropathologically diverse disease entities. For instance, according to recent metaanalyses, the *APOE* polymorphism may contribute to risk not only for AD, but also for PD and FTD (albeit with different alleles). If confirmed, these observations could point to 1 or several common genetic and mechanistic denominators for neuronal cell death. Finally, genetics has been essential for elucidating the molecular and biochemical pathways leading to neurodegeneration for almost all of the discussed syndromes and disease entities. Likewise, a detailed understanding of the genetic basis of neurodegeneration will be essential for the design and development of effective early prediction and early prevention/treatment strategies, with the prospect of largely decreasing the incidence of these devastating disorders.

Acknowledgments

This work was sponsored by grants from the National Institute of Mental Health, the National Institute on Aging (Alzheimer's

Disease Research Center), and the Alzheimer's Association. L. Bertram was funded by Deutsche Forschungsgemeinschaft (DFG), Harvard Center for Neurodegeneration and Repair (HCNR), and the National Alliance for Research on Schizophrenia and Depression (NARSAD).

Address correspondence to: Lars Bertram, Genetics and Aging Research Unit, Department of Neurology, MassGeneral Institute for Neurodegenerative Disease, Massachusetts General Hospital, 114 16th Street, Charlestown, Massachusetts 02129, USA. Phone: (617) 724-5567; Fax: (617) 724-1823; E-mail: bertram@helix.mgh.harvard.edu.

1. Lander, E.S. 1996. The new genomics: global views of biology. *Science.* **274**:536–539.
2. Risch, N.J. 2000. Searching for genetic determinants in the new millennium. *Nature.* **405**:847–856.
3. Hardy, J., Myers, A., and Wavrant-De Vriese, F. 2004. Problems and solutions in the genetic analysis of late-onset Alzheimer's disease. *Neurodegenerative Diseases.* **1**:213–217.
4. Munafo, M.R., Clark, T.G., and Flint, J. 2004. Assessing publication bias in genetic association studies: evidence from a recent meta-analysis. *Psychiatry Res.* **129**:39–44.
5. Colhoun, H.M., McKeigue, P.M., and Davey Smith, G. 2003. Problems of reporting genetic associations with complex outcomes. *Lancet.* **361**:865–872.
6. Kuznetsova, T., et al. 2000. Antihypertensive treatment modulates the association between the D/I ACE gene polymorphism and left ventricular hypertrophy: a meta-analysis. *J. Hum. Hypertens.* **14**:447–454.
7. Faraone, S.V., Doyle, A.E., Mick, E., and Biederman, J. 2001. Meta-analysis of the association between the 7-repeat allele of the dopamine D(4) receptor gene and attention deficit hyperactivity disorder. *Am. J. Psychiatry.* **158**:1052–1057.
8. Lohmueller, K.E., Pearce, C.L., Pike, M., Lander, E.S., and Hirschhorn, J.N. 2003. Meta-analysis of genetic association studies supports a contribution of common variants to susceptibility to common disease. *Nat. Genet.* **33**:177–182.
9. Lasky-Su, J.A., Faraone, S.V., Glatt, S.J., and Tsuang, M.T. 2005. Meta-analysis of the association between two polymorphisms in the serotonin transporter gene and affective disorders. *Am. J. Med. Genet. B Neuropsychiatr. Genet.* **133**:110–115.
10. Ioannidis, J.P., Ntzani, E.E., Trikalinos, T.A., and Contopoulos-Ioannidis, D.G. 2001. Replication validity of genetic association studies. *Nat. Genet.* **29**:306–309.
11. Maraganore, D.M., et al. 2004. UCHL1 is a Parkinson's disease susceptibility gene. *Ann. Neurol.* **55**:512–521.
12. Huang, X., Chen, P.C., and Poole, C. 2004. APOE-epsilon2 allele associated with higher prevalence of sporadic Parkinson disease. *Neurology.* **62**:2198–2202.
13. Alzheimer Research Forum. 2005. AlzGene Database for published Alzheimer disease candidate genes. http://www.alzgene.org.
14. Tanzi, R.E. 1999. A genetic dichotomy model for the inheritance of Alzheimer's disease and common age-related disorders. *J. Clin. Invest.* **104**:1175–1179.
15. Goate, A., et al. 1991. Segregation of a missense mutation in the amyloid precursor protein gene with familial Alzheimer's disease. *Nature.* **349**:704–706.
16. Sherrington, R., et al. 1995. Cloning of a gene bearing missense mutations in early-onset familial Alzheimer's disease. *Nature.* **375**:754–760.
17. Rogaev, E.I., et al. 1995. Familial Alzheimer's disease in kindreds with missense mutations in a gene on chromosome 1 related to the Alzheimer's disease type 3 gene. *Nature.* **376**:775–778.
18. Levy-Lahad, E., et al. 1995. Candidate gene for the chromosome 1 familial Alzheimer's disease locus. *Science.* **269**:973–977.
19. Alzheimer Disease & Frontotemporal Dementia Mutation Database. http://www.molgen.ua.ac.be/admutations/.
20. Mayeux, R., Sano, M., Chen, J., Tatemichi, T., and Stern, Y. 1991. Risk of dementia in first-degree relatives of patients with Alzheimer's disease and related disorders. *Arch. Neurol.* **48**:269–273.
21. Strittmatter, W.J., et al. 1993. Apolipoprotein E: high-avidity binding to beta-amyloid and increased frequency of type 4 allele in late-onset familial Alzheimer disease. *Proc. Natl. Acad. Sci. U. S. A.* **90**:1977–1981.
22. Schmechel, D.E., et al. 1993. Increased amyloid beta-peptide deposition in cerebral cortex as a consequence of apolipoprotein E genotype in late-onset Alzheimer disease. *Proc. Natl. Acad. Sci. U. S. A.* **90**:9649–9653.
23. Farrer, L.A., et al. 1997. Effects of age, sex, and ethnicity on the association between apolipoprotein E genotype and Alzheimer disease. A meta-analysis. APOE and Alzheimer Disease Meta Analysis Consortium. *JAMA.* **278**:1349–1356.
24. Daw, E.W., et al. 2000. The number of trait loci in late-onset Alzheimer disease. *Am. J. Hum. Genet.* **66**:196–204.
25. Cruts, M., et al. 1998. Estimation of the genetic contribution of presenilin-1 and -2 mutations in a population-based study of presenile Alzheimer disease. *Hum. Mol. Genet.* **7**:43–51.
26. Campion, D., et al. 1999. Early-onset autosomal dominant Alzheimer disease: prevalence, genetic heterogeneity, and mutation spectrum. *Am. J. Hum. Genet.* **65**:664–670.
27. Finckh, U., et al. 2003. Association of late-onset Alzheimer disease with a genotype of PLAU, the gene encoding urokinase-type plasminogen activator on chromosome 10q22.2. *Neurogenetics.* **4**:213–217.
28. Tanzi, R.E., and Bertram, L. 2001. New frontiers in Alzheimer's disease genetics. *Neuron.* **32**:181–184.
29. Tanner, C.M., et al. 1999. Parkinson disease in twins: an etiologic study. *JAMA.* **281**:341–346.
30. Maher, N.E., et al. 2002. Epidemiologic study of 203 sibling pairs with Parkinson's disease: the GenePD study. *Neurology.* **58**:79–84.
31. de la Fuente-Fernandez, R. 2003. A note of caution on correlation between sibling pairs [author reply]. *Neurology.* **60**:1561.
32. Polymeropoulos, M.H., et al. 1997. Mutation in the alpha-synuclein gene identified in families with Parkinson's disease. *Science.* **276**:2045–2047.
33. Kitada, T., et al. 1998. Mutations in the parkin gene cause autosomal recessive juvenile parkinsonism. *Nature.* **392**:605–608.
34. Bonifati, V., et al. 2003. Mutations in the DJ-1 gene associated with autosomal recessive early-onset parkinsonism. *Science.* **299**:256–259.
35. Valente, E.M., et al. 2004. Hereditary early-onset Parkinson's disease caused by mutations in PINK1. *Science.* **304**:1158–1160.
36. Zimprich, A., et al. 2004. Mutations in LRRK2 cause autosomal-dominant parkinsonism with pleomorphic pathology. *Neuron.* **44**:601–607.
37. Paisan-Ruiz, C., et al. 2004. Cloning of the gene containing mutations that cause PARK8-linked Parkinson's disease. *Neuron.* **44**:595–600.
38. Singleton, A.B., et al. 2003. alpha-Synuclein locus triplication causes Parkinson's disease. *Science.* **302**:841.
39. Chartier-Harlin, M.C., et al. 2004. Alpha-synuclein locus duplication as a cause of familial Parkinson's disease. *Lancet.* **364**:1167–1169.
40. Albrecht, M. 2005. LRRK2 mutations and parkinsonism. *Lancet.* **365**:1230.
41. Matsumine, H., et al. 1997. Localization of a gene for an autosomal recessive form of juvenile Parkinsonism to chromosome 6q25.2-27. *Am. J. Hum. Genet.* **60**:588–596.
42. Petrucelli, L., et al. 2002. Parkin protects against the toxicity associated with mutant alpha-synuclein: proteasome dysfunction selectively affects catecholaminergic neurons. *Neuron.* **36**:1007–1019.
43. van Duijn, C.M., et al. 2001. Park7, a novel locus for autosomal recessive early-onset parkinsonism, on chromosome 1p36. *Am. J. Hum. Genet.* **69**:629–634.
44. Abou-Sleiman, P.M., Healy, D.G., Quinn, N., Lees, A.J., and Wood, N.W. 2003. The role of pathogenic DJ-1 mutations in Parkinson's disease. *Ann. Neurol.* **54**:283–286.
45. Valente, E.M., et al. 2001. Localization of a novel locus for autosomal recessive early-onset parkinsonism, PARK6, on human chromosome 1p35-p36. *Am. J. Hum. Genet.* **68**:895–900.
46. Leroy, E., et al. 1998. The ubiquitin pathway in Parkinson's disease. *Nature.* **395**:451–452.
47. Le, W.D., et al. 2003. Mutations in NR4A2 associated with familial Parkinson disease. *Nat. Genet.* **33**:85–89.
48. Scott, W.K., et al. 2001. Complete genomic screen in Parkinson disease: evidence for multiple genes. *JAMA.* **286**:2239–2244.
49. Martin, E.R., et al. 2001. Association of single-nucleotide polymorphisms of the tau gene with late-onset Parkinson disease. *JAMA.* **286**:2245–2250.
50. Healy, D.G., et al. 2004. Tau gene and Parkinson's disease: a case-control study and meta-analysis. *J. Neurol. Neurosurg. Psychiatr.* **75**:962–965.
51. Conrad, C., et al. 1997. Genetic evidence for the involvement of tau in progressive supranuclear palsy. *Ann. Neurol.* **41**:277–281.
52. Farrer, M., et al. 2001. alpha-Synuclein gene haplotypes are associated with Parkinson's disease. *Hum. Mol. Genet.* **10**:1847–1851.
53. McKeith, I.G., et al. 1996. Consensus guidelines for the clinical and pathologic diagnosis of dementia with Lewy bodies (DLB): report of the consortium on DLB international workshop. *Neurology.* **47**:1113–1124.
54. Hamilton, R.L. 2000. Lewy bodies in Alzheimer's disease: a neuropathological review of 145 cases using alpha-synuclein immunohistochemistry. *Brain Pathol.* **10**:378–384.
55. Tsuang, D.W., DiGiacomo, L., and Bird, T.D. 2004. Familial occurrence of dementia with Lewy bodies. *Am. J. Geriatr. Psychiatry.* **12**:179–188.
56. Pericak-Vance, M.A., et al. 1997. Complete genomic screen in late-onset familial Alzheimer disease. Evidence for a new locus on chromosome 12. *JAMA.* **278**:1237–1241.
57. Scott, W.K., et al. 2000. Fine mapping of the chromosome 12 late-onset Alzheimer disease locus: potential genetic and phenotypic heterogeneity. *Am. J. Hum. Genet.* **66**:922–932.
58. Funayama, M., et al. 2002. A new locus for Parkinson's disease (PARK8) maps to chromosome 12p11.2-q13.1. *Ann. Neurol.* **51**:296–301.
59. Kehoe, P., et al. 1999. A full genome scan for late onset Alzheimer's disease. *Hum. Mol. Genet.* **8**:237–245.
60. Curtis, D., North, B.V., and Sham, P.C. 2001. A novel method of two-locus linkage analysis applied to a genome scan for late onset Alzheimer's disease. *Ann. Hum. Genet.* **65**:473–481.
61. Bang, O.Y., Kwak, Y.T., Joo, I.S., and Huh, K. 2003. Important link between dementia subtype and apolipoprotein E: a meta-analysis. *Yonsei. Med. J.* **44**:401–413.

62. McKhann, G.M., Albert, M.S., Grossman, M., Miller, B., Dickson, D., and Trojanowski, J.Q. 2001. Clinical and pathological diagnosis of frontotemporal dementia: report of the Work Group on Frontotemporal Dementia and Pick's Disease. *Arch. Neurol.* **58**:1803–1809.

63. Brun, A. 1987. Frontal lobe degeneration of non-Alzheimer type. I. Neuropathology. *Arch. Gerontol. Geriatr.* **6**:193–208.

64. Bird, T., et al. 2003. Epidemiology and genetics of frontotemporal dementia/Pick's disease. *Ann. Neurol.* **54**(Suppl 5):S29–S31.

65. Hutton, M., et al. 1998. Association of missense and 5′-splice site mutations in tau with the inherited dementia FTDP-17. *Nature.* **393**:702–705.

66. Reed, L.A., Wszolek, Z.K., and Hutton, M. 2001. Phenotypic correlations in FTDP-17. *Neurobiol. Aging.* **22**:89–107.

67. Rademakers, R., et al. 2003. Tau (MAPT) mutation Arg406Trp presenting clinically with Alzheimer disease does not share a common founder in Western Europe. *Hum. Mutat.* **22**:409–411.

68. Ingelsson, M., and Hyman, B.T. 2002. Disordered proteins in dementia. *Ann. Med.* **34**:259–271.

69. Morris, H.R., et al. 2002. Analysis of tau haplotypes in Pick's disease. *Neurology.* **59**:443–445.

70. Verpillat, P., et al. 2002. Apolipoprotein E gene in frontotemporal dementia: an association study and meta-analysis. *Eur. J. Hum. Genet.* **10**:399–405.

71. Dermaut, B., et al. 2004. A novel presenilin 1 mutation associated with Pick's disease but not beta-amyloid plaques. *Ann. Neurol.* **55**:617–626.

72. Majoor-Krakauer, D., Willems, P.J., and Hofman, A. 2003. Genetic epidemiology of amyotrophic lateral sclerosis. *Clin. Genet.* **63**:83–101.

73. Siddique, T., et al. 1991. Linkage of a gene causing familial amyotrophic lateral sclerosis to chromosome 21 and evidence of genetic-locus heterogeneity. *N. Engl. J. Med.* **324**:1381–1384.

74. Rosen, D.R., et al. 1993. Mutations in Cu/Zn superoxide dismutase gene are associated with familial amyotrophic lateral sclerosis. *Nature.* **362**:59–62.

75. Andersen, P.M., et al. 1995. Amyotrophic lateral sclerosis associated with homozygosity for an Asp90Ala mutation in CuZn-superoxide dismutase. *Nat. Genet.* **10**:61–66.

76. Kato, S., Shaw, P., Wood-Allum, C., Leigh, P.N., and Shaw, C. 2003. Amyotrophic Lateral Sclerosis. In *Neurodegeneration — the molecular pathology of dementia and movement disorders.* D. Dickson, editor. ISN Neuropath Press. Basel, Switzerland. 350–368.

77. Stathopulos, P.B., et al. 2003. Cu/Zn superoxide dismutase mutants associated with amyotrophic lateral sclerosis show enhanced formation of aggregates in vitro. *Proc. Natl. Acad. Sci. U. S. A.* **100**:7021–7026.

78. Wood, J.D., Beaujeux, T.P., and Shaw, P.J. 2003. Protein aggregation in motor neurone disorders. *Neuropathol. Appl. Neurobiol.* **29**:529–545.

79. Yang, Y., et al. 2001. The gene encoding alsin, a protein with three guanine-nucleotide exchange factor domains, is mutated in a form of recessive amyotrophic lateral sclerosis. *Nat. Genet.* **29**:160–165.

80. Hadano, S., et al. 2001. A gene encoding a putative GTPase regulator is mutated in familial amyotrophic lateral sclerosis 2. *Nat. Genet.* **29**:166–173.

81. Kanekura, K., et al. 2004. Alsin, the product of ALS2 gene, suppresses SOD1 mutant neurotoxicity through RhoGEF domain by interacting with SOD1 mutants. *J. Biol. Chem.* **279**:19247–19256.

82. Institute of Psychiatry. 2004. The ALS online database for mutations in ALS related genes. http://www.alsod.org.

83. Hosler, B.A., et al. 2000. Linkage of familial amyotrophic lateral sclerosis with frontotemporal dementia to chromosome 9q21-q22. *JAMA.* **284**:1664–1669.

84. Blacker, D., et al. 2003. Results of a high-resolution genome screen of 437 Alzheimer's disease families. *Hum. Mol. Genet.* **12**:23–32.

85. Figlewicz, D.A., et al. 1994. Variants of the heavy neurofilament subunit are associated with the development of amyotrophic lateral sclerosis. *Hum. Mol. Genet.* **3**:1757–1761.

86. Gusella, J.F., et al. 1983. A polymorphic DNA marker genetically linked to Huntington's disease. *Nature.* **306**:234–238.

87. The Huntington's Disease Collaborative Research Group. 1993. A novel gene containing a trinucleotide repeat that is expanded and unstable on Huntington's disease chromosomes. *Cell.* **72**:971–983.

88. Gusella, J.F., and MacDonald, M.E. 1995. Huntington's disease: CAG genetics expands neurobiology. *Curr. Opin. Neurobiol.* **5**:656–662.

89. Li, J.L., et al. 2003. A genome scan for modifiers of age at onset in Huntington disease: the HD MAPS study. *Am. J. Hum. Genet.* **73**:682–687.

90. Rubinsztein, D.C., et al. 1997. Genotypes at the GluR6 kainate receptor locus are associated with variation in the age of onset of Huntington disease. *Proc. Natl. Acad. Sci. U. S. A.* **94**:3872–3876.

91. MacDonald, M.E., et al. 1999. Evidence for the GluR6 gene associated with younger onset age of Huntington's disease. *Neurology.* **53**:1330–1332.

92. Prusiner, S.B. 1998. Prions. *Proc. Natl. Acad. Sci. U. S. A.* **95**:13363–13383.

93. Bieschke, J., et al. 2004. Autocatalytic self-propagation of misfolded prion protein. *Proc. Natl. Acad. Sci. U. S. A.* **101**:12207–12211.

94. Gambetti, P., Kong, Q., Zou, W., Parchi, P., and Chen, S.G. 2003. Sporadic and familial CJD: classification and characterisation. *Br. Med. Bull.* **66**:213–239.

95. Hsiao, K., et al. 1989. Linkage of a prion protein missense variant to Gerstmann-Straussler syndrome. *Nature.* **338**:342–345.

96. Palmer, M.S., Dryden, A.J., Hughes, J.T., and Collinge, J. 1991. Homozygous prion protein genotype predisposes to sporadic Creutzfeldt-Jakob disease. *Nature.* **352**:340–342.

97. Collinge, J., Palmer, M.S., and Dryden, A.J. 1991. Genetic predisposition to iatrogenic Creutzfeldt-Jakob disease. *Lancet.* **337**:1441–1442.

98. Pocchiari, M., et al. 2004. Predictors of survival in sporadic Creutzfeldt-Jakob disease and other human transmissible spongiform encephalopathies. *Brain.* **127**:2348–2359.

99. Dickson, D., editor. 2003. *Neurodegeneration: the molecular pathology of dementia and movement disorders.* ISN Neuropath Press. Basel, Switzerland. 414 pp.

Sodium channel mutations in epilepsy and other neurological disorders

Miriam H. Meisler and Jennifer A. Kearney

Department of Human Genetics, University of Michigan, Ann Arbor, Michigan, USA.

Since the first mutations of the neuronal sodium channel *SCN1A* were identified 5 years ago, more than 150 mutations have been described in patients with epilepsy. Many are sporadic mutations and cause loss of function, which demonstrates haploinsufficiency of *SCN1A*. Mutations resulting in persistent sodium current are also common. Coding variants of *SCN2A*, *SCN8A*, and *SCN9A* have also been identified in patients with seizures, ataxia, and sensitivity to pain, respectively. The rapid pace of discoveries suggests that sodium channel mutations are significant factors in the etiology of neurological disease and may contribute to psychiatric disorders as well.

Introduction

Voltage-gated sodium channels are essential for the initiation and propagation of action potentials in neurons. The sodium channel α subunits are large, transmembrane proteins with approximately 2,000 amino acid residues, composed of 4 homologous domains containing well-characterized voltage sensor and pore regions (Figure 1). The transmembrane segments are highly conserved through evolution. The 4 domains associate within the membrane to form a sodium-permeable pore, through which sodium ions flow down a concentration gradient during propagation of an action potential. The transmembrane sodium gradient is subsequently restored by the activity of the ATP-dependent sodium/potassium pump. The 3-dimensional structures of related bacterial potassium channels have recently been elucidated (1, 2).

Each sodium channel α subunit is associated with 1 or more β subunits, β1 to β4, that are transmembrane proteins with a single extracellular IgG loop and a short intracellular C terminus (Figure 1). Association with β subunits influences the level of cell surface expression, voltage dependence, and kinetics of the α subunit, as well as association with other signaling and cytoskeletal molecules (3, 4).

Duplication of the α subunit genes during evolution generated 9 mammalian genes encoding active channels that differ in tissue specificity and biophysical properties (Table 1) (5, 6). Many disease mutations have been characterized in the skeletal muscle and cardiac channels, but exploration of the role of the 7 neuronal sodium channels in disease is in an early stage.

β1 Subunit mutations and GEFS+

Generalized epilepsy with febrile seizures plus (GEFS+) (OMIM 604233) is a mild, dominantly inherited epilepsy characterized by febrile seizures in childhood progressing to generalized epilepsy in adults (7, 8). The first connection between sodium channels and epilepsy was the discovery of a β1 subunit mutation in a large Australian family with GEFS+ (9). Affected family members are heterozygous for the missense mutation C121W in the extracellular Ig domain of the β1 subunit. The mutant channel promotes cell surface expression of the α subunit but exhibits

impaired modulation of sodium channel function and cell adhesion (10). A 5–amino acid deletion in the extracellular domain of β1 was subsequently found in a family with febrile seizures and early-onset absence epilepsy (11). Impaired inactivation of sodium channel α subunits is the likely mechanism relating β1 mutations to neuronal hyperexcitability in epilepsy.

Inherited and de novo mutations of *SCN1A* in GEFS+ severe myoclonic epilepsy of infancy

In 1999, linkage analysis in 2 large families localized a second GEFS+ locus to an interval of chromosome 2q24 that includes a sodium channel gene cluster (12, 13). Sequencing of *SCN1A* demonstrated that affected individuals are heterozygous for missense mutations in highly evolutionarily conserved amino acid residues, T875M in 1 family and R1648H in the other (14). Since the initial report, 11 additional *SCN1A* missense mutations have been reported in GEFS+ families (Figure 2A), approximately 10% of cases tested (14–25).

In 2001, Peter De Jonghe and colleagues discovered mutations of *SCN1A* in 7 patients with severe myoclonic epilepsy of infancy (SMEI) (26). This disorder is characterized by early onset, usually within the first 6 months of life, followed by progressive worsening of seizures often accompanied by mental deterioration (OMIM 182389). More than 150 mutations have been identified in children with this disorder (Table 2), approximately 50% of SMEI patients tested. As in GEFS+, the SMEI patients are heterozygous for the mutant alleles. Among 75 cases in which both parents have been tested, in 69 cases, or 90%, the mutations arose de novo in the affected children.

The mutation spectrum in SMEI differs from that in GEFS+. Approximately half of the SMEI mutations are nonsense or frameshift mutations that result in protein truncation (Figure 2B). The truncation mutations are randomly distributed across the SCN1A protein, including the N-terminal domain, transmembrane segments, cytoplasmic loops, and C terminus. The remaining SMEI mutations are missense mutations that, as in GEFS+, are concentrated within the transmembrane segments of the protein (Figure 2C). Within the large cytoplasmic loops of the α subunits, the only missense mutations are those located very close to the adjacent transmembrane segments.

The large number of de novo *SCN1A* mutations in children with SMEI demonstrate the importance of considering mutation in the etiology of neurological disease, even in the absence of a positive family history.

Nonstandard abbreviations used: GEFS+, generalized epilepsy with febrile seizures plus; SMEI, severe myoclonic epilepsy of infancy.

Conflict of interest: The authors have declared that no conflict of interest exists.

Citation for this article: *J. Clin. Invest.* **115**:2010–2017 (2005).
doi:10.1172/JCI25466.

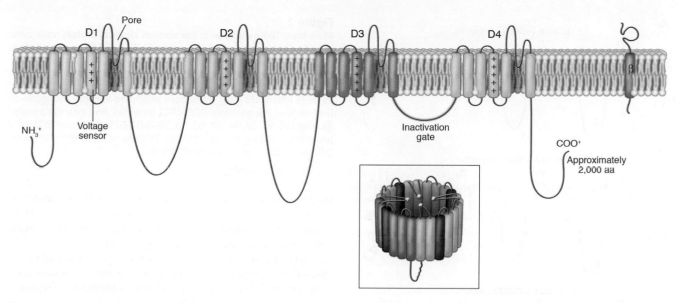

Figure 1
The sodium channel α and β subunits are transmembrane proteins. The 4 homologous domains of the α subunit are represented in different colors. The transmembrane segments associate in the membrane to form an Na⁺-permeable pore lined by the re-entrant S5–S6 pore-loop segments (inset).

Recurrent mutations in *SCN1A*

Approximately 25% of reported mutations in *SCN1A* have occurred independently more than once (27, 28). Examination of the altered nucleotides at the sites of recurrent mutation revealed 2 frequent mechanisms. Fifteen mutations resulted from deamination of ^{m}CpG dinucleotides, including 5 CpGs located in an arginine codon of the CGA class. Deamination converts this codon to UGA, a nonsense codon, resulting in an R-to-X mutation at the amino acid level. Two tetranucleotide direct repeats in *SCN1A* have been the sites of frameshift mutations due to deletion of 1 copy of the repeat. The likely mechanism is slipped-strand mispairing. The frameshift mutations are at amino acid residues K1846 and L1670.

Rare mutations of *SCN2A*

Although *SCN1A* and *SCN2A* are closely related genes, similar in size and exon organization, only a few epilepsy mutations have been detected in *SCN2A* (Figure 2D) (29–32). The difference may be partly accounted for by the likelihood that many more patients have been screened for *SCN1A* mutations. Most missense mutations of *SCN2A* were found in patients with benign familial neonatal-infantile seizures (OMIM 607745), mild syndromes that present within the first year of life but do not progress to adult epilepsy (Table 2). One truncation mutation of *SCN2A* has been identified in a patient with intractable epilepsy resembling SMEI (30).

Haploinsufficiency of *SCN1A*

A small proportion of human genes exhibit haploinsufficiency, meaning that abnormal function results from quantitative reduction of gene expression to 50% of normal levels. Among the epilepsy mutations, the most severe phenotypes are found in individuals who are heterozygous for loss-of-function mutations of *SCN1A*. This is indicated by the high proportion of truncation mutations among SMEI patients, and the lack of truncation mutations in patients with GEFS+, which is milder (Figure 2, A and B). In SMEI, the clinical features of early onset, seizure severity, and

progressive mental deterioration do not differ between patients with truncations located near the N terminus or the C terminus, which indicates that loss of function is the common cause. Rhodes et al. examined the functional effect of 5 missense mutations from patients with severe SMEI and found that 2 of those also result in complete loss of channel activity (33).

Unlike *SCN1A*, *SCN2A* appears not to exhibit haploinsufficiency. In contrast to the many null mutations of *SCN1A*, only a single truncation mutation has been identified in *SCN2A*, in a patient with a severe form of epilepsy resembling SMEI (30) (Figure 2D). In this case the truncated protein was reported to have a dominant-negative effect, rather than loss of function (30).

Analysis of mice with knockout alleles supports the conclusion that *SCN1A* exhibits haploinsufficiency while *SCN2A* does not. Heterozygotes for the *SCN1A* knockout allele exhibit spontaneous seizures and reduced lifespan (34), while heterozygotes for the *SCN2A*-null allele are viable without visible abnormalities (35). *SCN8A*-heterozygous-null mice lack visible abnormalities, but

Table 1
Mammalian voltage-gated sodium channel genes

Gene	Protein	Human chromosome	Major expression	Minor expression
SCN1A	Na$_v$1.1	2q24	CNS, PNS	Cardiac muscle
SCN2A	Na$_v$1.2	2q24	CNS, PNS	
SCN3A	Na$_v$1.3	2q24	CNS, PNS	
SCN4A	Na$_v$1.4	17q23	Skeletal muscle	
SCN5A	Na$_v$1.5	3p21	Cardiac muscle	Skeletal muscle, brain
SCN8A	Na$_v$1.6	12q13	CNS, PNS	Cardiac muscle
SCN9A	Na$_v$1.7	2q24	PNS	
SCN10A	Na$_v$1.8	3p21	PNS	
SCN11A	Na$_v$1.9	3p21	PNS	

PNS, peripheral nervous system.

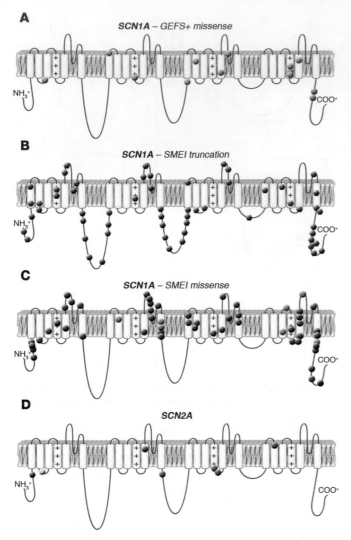

A SCN1A – GEFS+ missense

B SCN1A – SMEI truncation

C SCN1A – SMEI missense

D SCN2A

Figure 2
More than 150 mutations in the sodium channel protein have been identified in patients with GEFS+ and SMEI. (**A**) Missense mutations of SCN1A identified in families with GEFS+ (14, 16, 17, 19, 21–25). (**B**) Truncation mutations of SCN1A identified in SMEI patients (18, 26, 27, 44, 47, 50, 53, 54, 79–84). (**C**) Missense mutations of SCN1A in patients with SMEI (red), intractable childhood epilepsy with generalized tonic-clonic seizures (ICEGTC) (orange), and infantile spasms (purple) (18, 26, 27, 44, 47, 50, 53, 79–84). (**D**) Mutations of SCN2A in patients with benign familial neonatal-infantile seizures (BFNIS) (blue), GEFS+ (yellow), and SMEI (red) (29–32).

ron, this persistent current is thought to reduce the depolarization threshold required for firing, resulting directly in hyperexcitability. Another common mechanism is demonstrated by the GEFS+ mutation D188V, which spends less time in the inactivated state than the wild-type channel (Figure 3B). The result is greater availability of channels for opening in response to depolarization, another route to hyperexcitability. The altered biophysical properties of representative mutant channels are summarized in Table 3.

A unique biochemical mechanism was observed for the missense mutation D1866Y, located in the C-terminal domain of SCN1A, in a family with GEFS+ (23). In the *Xenopus* oocyte system, the mutant channel exhibited a depolarized shift in voltage dependence of fast inactivation; the effect was tenfold greater in the presence of the β1 subunit (Figure 3C). Modeling with the program NEURON (http://www.neuron.yale.edu/neuron/) indicated that this shift is sufficient to produce neuronal hyperexcitability. Because the difference between the wild-type and mutant channels was increased by the presence of the β1 subunit, the effect of the mutation on subunit interaction was tested. Yeast 2-hybrid screen and co-immunoprecipitation demonstrated direct interaction between the C-terminal cytoplasmic domains of the α and β subunits, which was impaired by the D1866Y mutation (23). The D1866Y mutation thus defines an intracellular interaction domain that appears to be required, in combination with the extracellular interaction domain (23), to form the stable α/β complex. Since mutations in either SCN1A or β1 can result in GEFS+, it is not surprising that impaired interaction between the 2 subunits could also cause the disease.

The GEFS+ mutation R1648H has been examined in 3 expression systems with different outcomes. When the mutation was introduced into the rat SCN1A cDNA and examined in the *Xenopus*

subtle deficiencies have been detected in tests of learning and anxiety (36). The β1 and β2 genes do not exhibit haploinsufficiency in mice heterozygous for targeted knockout alleles (37, 38).

Functional effects of SCN1A missense mutations

Twenty SCN1A missense mutations have been evaluated in functional assays (23, 24, 33, 39–46). Functional analysis is complicated by the difficulty of cloning neuronal sodium channel cDNAs, which are uniquely unstable during propagation in bacterial cultures (unlike the muscle sodium channel cDNAs or the calcium and potassium channel cDNAs). Functional studies in the *Xenopus* oocyte system and in transfected mammalian cells do not always agree, and there is little experimental basis for extrapolation to in vivo effects. Nonetheless, some interesting patterns have emerged relating altered channel function to the neuronal hyperexcitability that is thought to underlie seizure disorders.

One common functional abnormality is impaired channel inactivation leading to increased persistent current. Normally, the voltage-gated sodium channels open rapidly in response to altered membrane potential and inactivate rapidly, declining to 1% of maximal current within a few milliseconds (Figure 3A). In 3 different GEFS+ mutations (14, 19), persistent current was increased to 2–5% of peak current (39) (Figure 3A). In the context of the neu-

Table 2
Sodium channel mutations in patients with various types of epilepsy

	SCN1A Na$_v$1.1	SCN2A Na$_v$1.2	SCN1B β1
SMEI (at least 90% sporadic)	150	1	-
GEFS+	13	1	2
ICEGTC	7	-	-
Infantile spasms	1	-	-
Benign familial neonatal-infantile	-	6	-

ICEGTC, intractable childhood epilepsy with generalized tonic-clonic seizures.

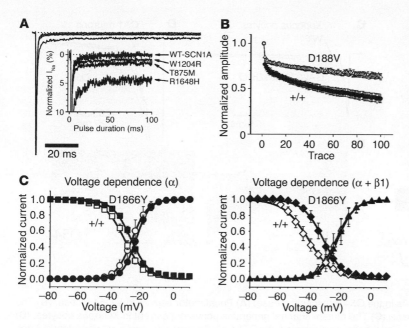

Figure 3

The effects of GEFS+ mutations on *SCN1A* channel properties have been studied in the *Xenopus* oocyte system and in transfected mammalian cells. (**A**) Whole-cell recordings from HEK tsA201 cells transfected with the indicated mutant SCN1A cDNAs demonstrate increased persistent current from the mutant channels (39). (**B**) Mean normalized amplitudes of sodium currents elicited by 80-Hz pulse trains in HEK cells expressing SCN2A cDNA containing the GEFS+ mutation D188V demonstrate reduced cumulative inactivation of the mutant channel during high-frequency trains of channel activation (40). (**C**) Voltage-dependent gating of SCN1A cDNA constructs in *Xenopus* oocytes expressed in the absence (left) or presence (right) of the β1 cDNA (85). See text for discussion.

oocyte expression system, accelerated recovery from inactivation was observed (45). In the human cDNA in transfected mammalian cells, persistent current was the major effect (Figure 3A) (39). Alekov et al. introduced R1648H into the SCN4A cDNA and expressed the clone in mammalian HEK tsA201 cells. In this context, they observed slowed inactivation and accelerated recovery from inactivation, leading to increased channel availability, but no persistent current (42). A second substitution at the same residue, R1648C, was identified in a patient with SMEI (47). In transfected cells, persistent current was generated by R1648C at a level indistinguishable from that of the mutation R1648H that causes GEFS+ (33). Thus, the heterologous expression systems are not able to distinguish between missense mutations that lead to mild disease in vivo and those that lead to severe disease.

The data indicate that seizures can result from increased *SCN1A* channel activity, as in the missense mutations described above, and from reduced activity, as in the truncation mutations. Neuronal firing patterns appear to be extremely sensitive to subtle changes in sodium channel function. In the future, the most physiologically relevant data are likely to be obtained from measurements of neuronal currents in knock-in mouse models carrying human mutations.

We investigated the in vivo effect of an *SCN2A* mutation with impaired inactivation in the Q54 transgenic mouse (48). Analysis of the mutation $SCN2A^{GAL879-881QQQ}$ in the *Xenopus* oocyte system revealed an increase in persistent current and in the percentage of current that inactivated with a slow time constant (Figure 4, B and C) (49). When the mutant cDNA was expressed in transgenic mice under the control of the neuron-specific enolase promoter, the mice exhibited progressive seizures of hippocampal origin accompanied by loss of neurons called hippocampal sclerosis (Figure 4, A and E). Persistent sodium current was detected in recordings from CA1 neurons of the transgenic mice (Figure 4D), demonstrating agreement between the *Xenopus* assay and the in vivo effect. The phenotype of the Q54 mice most closely resembles human mesial temporal lobe epilepsy.

Toward mutation-specific therapy

Selecting the appropriate antiepileptic drug for newly diagnosed epilepsy patients is a difficult process, and many drugs have adverse side effects. An important goal of mutation characterization is the development of individualized treatments tailored to each patient mutation. It is already clear that sodium channel blockers are con-

Table 3

A variety of functional abnormalities in mutant alleles of *SCN1A* encoding the sodium channel Na$_v$1.1

Mutation	Major biophysical abnormalities observed in indicated experimental system
D188V	Decreased entry into and increased rate of recovery from slow inactivated state (*m*) (40)
T875M	Enhanced entry into slow inactivated state (*X*) (45); enhanced entry and hyperpolarizing shift into steady-state slow inactivation (*m*) (39, 42); increased persistent current (*m*) (39, 42)
W1204R	Hyperpolarized shift in voltage dependence of activation and inactivation (*X*) (46); increased persistent current (*m*) (39)
R1648H	Increased rate of recovery from fast inactivation (*X*) (45); increased persistent current (*m*) (39); slowed inactivation and increased recovery from inactivation (*m*) (41)
R1648C	Increased persistent current (*m*) (33)
I1656M	Depolarized shift in voltage dependence of activation (*m*) (43)
G1674R	No measurable sodium current (*m*) (33)
D1866Y	Depolarized shift in voltage dependence of fast inactivation (*X*) due to reduced association with β1 subunit (*m*) (23)

The most prominent effects observed in each study are indicated; most mutations altered several biophysical properties of the channel. *m*, transfected mammalian cells; *X*, *Xenopus* oocyte system.

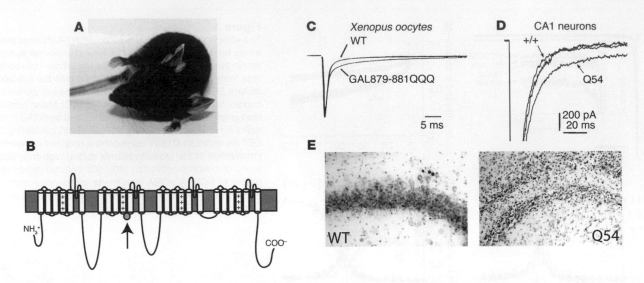

Figure 4
An *SCN2A* mutation with persistent current causes seizures in the Q54 transgenic mouse. (**A**) Focal motor seizure in a Q54 mouse. (**B**) The GAL879–881QQQ mutation is located in the D2S4–S5 linker. (**C**) The mutant channel generates persistent current in *Xenopus* oocytes. (**D**) Whole-cell sodium currents recorded from CA1 hippocampal neurons from presymptomatic Q54 mice demonstrate increased persistent current. (**E**) Nissl-stained sections of hippocampus area CA3 reveal extensive neuronal cell loss in a Q54 mouse compared with a wild-type littermate. Adapted with permission from *Neuroscience* (48).

traindicated for patients with SMEI who are heterozygous for null alleles of *SCN1A*, since further reduction of the level of functional channel worsens their condition. In the case of mutations causing persistent current, open-state blockers may be appropriate. Pharmacological analysis of mutant *SCN5A* channels in LQT-3 provides a model for development of mutation-targeted therapy for a sodium channel defect (50, 51). Genetic variation in drug-metabolizing enzymes may also be predictive of pharmacological effectiveness of antiepileptic drugs (52).

Variable expressivity and incomplete penetrance
Variable expressivity, differences in clinical severity among individuals carrying the same mutation, is a striking feature of both GEFS+ and SMEI. For example, in the GEFS+ family carrying the mutation T875M, approximately half of the heterozygotes experienced febrile seizures in childhood only, while the other half had a variety of seizure types as adults (14). As another example, although most SMEI mutations arise de novo in affected individuals, a few patients are known to have inherited the *SCN1A* mutation from a mildly affected parent (53, 54). Several factors may be considered as contributing to differences between individuals carrying the same sodium channel mutation. There is *intrinsic stochastic variability* in developmental processes even between inbred animals homozygous throughout the genome, as demonstrated by the variable phenotypes of inbred mice with certain mutant genotypes. These processes may be especially important during development of neuronal connectivity in the CNS. Second, it has been suggested that the *accumulation of somatic mutations* during the lifetime of an individual may contribute to the development of neurological disease, and especially focal disorders such as epilepsy (55). Third, *environmental insults* such as trauma may exacerbate the phenotype of individuals carrying a mild mutation. Fourth, multiple differences in genetic background, or *modifier genes*, are segregating in families,

and the combined effect of multiple susceptibility factors may result in greater clinical severity in some individuals.

The influence of modifier genes is amenable to analysis using genetic differences between inbred strains of mice as models. Modifier genes that influence susceptibility to spontaneous seizures (56), kainate-induced seizures (57), electroconvulsive shock (58), and pentylenetetrazol-induced seizures (59) have been mapped to specific mouse chromosomes. Recently, we mapped 2 modifier loci, *Moe1* (*modifier of epilepsy*) and *Moe2*, that influence clinical severity of epilepsy due to a sodium channel mutation in *Scn2a^{Q54}* mice (60). With current genomic technology, it is feasible to clone modifier genes and evaluate their role in human disease. For example, a gene that modifies the severity of an *SCN8A*-induced movement disorder was recently cloned and identified as a putative RNA splicing factor (61).

SCN8A and movement disorders
Mutations of *Scn8a* in the mouse result in movement disorders such as ataxia, dystonia, and tremor (62). Conditional inactivation of *Scn8a* in cerebellar neurons results in a milder ataxia (63, 64). Screening of 150 patients with sporadic or inherited ataxia identified 1 protein truncation mutation that is likely to cause loss of function of *SCN8A* (36). Further screening will be worthwhile to determine the prevalence of *SCN8A* mutations in patients with inherited and sporadic movement disorders.

SCN9A and inherited pain
SCN9A encodes the sodium channel Na$_v$1.7, which is expressed in peripheral sensory and sympathetic neurons and is localized in the nerve terminals of sensory neurons (65). The missense mutation L858H, in a transmembrane segment, was identified in a Chinese kindred with autosomal dominantly inherited primary erythermalgia (OMIM 113020), a disorder characterized by intermittent pain, redness, heat, and swelling in the extremities (66). A second mutation

Table 4
Mouse models of neurological disorders with sodium channel mutations

Gene	Channel	Mutation	Disorder	Reference
SCN1A	Na$_v$1.1	Targeted null	Seizures in heterozygote; homozygous lethal at P16	(34)
SCN2A	Na$_v$1.2	Targeted null	No abnormalities in heterozygote; homozygous lethal at P1	(35)
		Q54 transgene (GAL → QQQ)	Dominantly inherited temporal lobe epilepsy	(48)
SCN8A	Na$_v$1.6	Spontaneous null	Paralysis, lethal	(62)
		Null heterozygote	Learning deficits	(36)
		A1071T	Ataxia, tremor	(86)
		Hypomorph (medJ)	Dystonia	(87)
		Conditional null	CRE-dependent lethal	(64)
SCN9A	Na$_v$1.7	Targeted null	Lethal	(70)
		Conditional null	Pain resistant	(70)
SCN10A	Na$_v$1.8	Targeted null	Pain resistant	(88)
SCN1B	β1	Targeted null	Seizures in heterozygote; lethal seizures in homozygote	(38)
SCN2B	β2	Targeted null	Heterozygote normal; homozygote susceptible to seizures	(37)

in a nearby residue, I848T, was found in a sporadic case of erythermalgia. Functional analysis detected a hyperpolarizing shift in activation and slowed deactivation in both mutants, as well as increased current amplitude in response to slow, small depolarizations, consistent with predicted hyperactivity at the cellular level (67). A third mutation, F1449V, close to D3S6, lowers the current threshold required for action potential generation and repetitive firing (68).

The syndrome familial rectal pain (OMIM 167400) is characterized by brief episodes of intense pain of the submandibular, ocular, and rectal areas with flushing of the surrounding skin. The pain is responsive to the sodium channel blocker carbamazepine. Missense mutations in SCN9A have recently been identified in several families with this disorder (69).

Conditional inactivation of Scn9a in sensory neurons of the mouse resulted in increased threshold for mechanical, thermal, and inflammatory pain (70). The association of SCN9A mutations with inherited pain syndromes, and the pain resistance of the knockout mouse, indicate that specific inhibitors of this sodium channel would be valuable for the treatment of chronic pain. Thus far, the evolutionary conservation of the mammalian α subunit genes has made it difficult to develop inhibitors with specificity for a single channel.

Sodium channels and psychiatric disease?
There is a biological continuum between neurological diseases such as epilepsy and ataxia, and disorders that have been traditionally classified as psychiatric. Because of the essential role of sodium channels in propagating neuronal signals throughout the CNS, and the sensitivity of neuronal firing patterns to subtle mutations in sodium channels, it seems that sodium channel mutations could directly affect emotional and cognitive function. To test this possibility, we screened for mutations in SCN1A, SCN2A, and SCN3A in multiplex families with autism, a psychiatric disorder with high heritability (71). One splice branchpoint and 5 missense mutations were observed among 117 families; these may be con-

sidered candidates for a causal role in autism. The missense mutation R1902C, located within a calmodulin-binding domain of SCN2A, was subsequently found to confer a calcium-dependent conformational change on the complex of calmodulin with the C terminus of Na$_v$1.2 (72). In this context it is interesting that mutations of the voltage-gated calcium channel Ca$_v$1.2 were recently identified in families with Timothy syndrome, a multiorgan disorder with autism as a feature (73).

Preliminary indication that sodium channel variants may influence cognitive function comes from heterozygotes for a null allele of SCN8A, who exhibit a variety of cognitive defects (36). Impaired learning was also observed in heterozygous-null mice (36). An influence of SCN8A on attempted suicide was indicated by preferential transmission of 1 allele of a single-nucleotide polymorphism in intron 21 (74). Autistic features have also been reported in patients with SMEI. Clearly, much work will be required to follow up on these intriguing observations regarding the potential influence of sodium channel and other neuronal ion channel variants on cognitive and psychiatric traits.

Mice with neuronal sodium channel mutations
The development of mouse models is keeping pace with the identification of human mutations, and several lines are available for analysis of disease mechanisms. The features and availability are summarized in Table 4. Four lines with various types of seizures are already available, the Q54 transgenic line with a missense mutation of SCN2A and the knockout lines for SCN1A, SCN1B (β1), and SCN2B (β2). Generation of tissue-specific and inducible knock-in mice carrying specific epilepsy-related sodium channel mutations is in progress in several laboratories. Electrophysiological recordings of neurons from knock-in mice are likely to provide the most reliable information for understanding the relationship between specific mutations and the severity of disease. Because of the effort involved in generating these models, only a small number of selected mutations can be analyzed in vivo. Recordings from mice with SCN8A mutations have revealed different effects on sodium currents in different types of neurons (75–77). In Purkinje cells, for example, SCN8A deficiency reduces bursting activity and resurgent current (77), while in cortical pyramidal neurons the major effect is reduced persistent sodium current (76). Analysis of SCN1A-null homozygotes revealed reduced current density in interneurons but not in excitatory neurons of the hippocampus (34). In addition to elucidating the pathophysiology of epileptogenesis, mouse models will permit evaluation of therapies directed toward specific classes of sodium channel mutations.

Persistent issues: sodium channels and disease
The identification of nearly 200 sodium channel mutations in patients with epilepsy raises several issues for future research. The relatively subtle functional effects of some of the epilepsy mutations

suggest that the nervous system is intolerant of even minor variation in the properties of these channel proteins, which have a direct role in neuronal firing. The large number of de novo mutations in the sporadic disorder SMEI suggests that mutation should be considered as a possible cause of neurological disease even in the absence of family history. A de novo mutation of *SCN9A* was identified in a sporadic case of erythermalgia. Development of genotyping technology has made it easier to screen for mutations, and a commercial sequencing test for *SCN1A* mutations was recently introduced. However, the cost of several thousand dollars per test, as well as ethical and practical concerns about the implications of mutation detection, may slow the application of genetic testing for neurological disease.

Evaluation of coding variants detected in genetic screens poses significant challenges. Which missense mutations cause disease and which are neutral? Large numbers of unaffected controls (approximately 1,000) must be tested to provide statistically significant evidence that a variant discovered in a patient is not also found in unaffected individuals. Functional assays can be helpful in recognizing disease mutations, but the difficulty of generating expression constructs and interpreting assays in exogenous expression systems makes these tests problematic for clinical applications.

Does quantitative variation in sodium channel abundance due to mutation in noncoding regulatory sequences contribute to neurological disease? The haploinsufficiency of *SCN1A* revealed by the loss-of-function coding mutations suggests that reduced expression itself may be sufficient to cause disease. Screening for variation in expression requires access to neuronal RNA samples. Recent studies using cDNA microarray hybridization or allele-specific primer extension reactions suggest that as many as 20% of human genes may vary in expression levels among individuals (78). This interesting area remains essentially unexplored.

If we could identify variants in modifier genes underlying the variable expressivity of sodium channel mutations, perhaps using animal models carrying human mutations, new targets for treatment intervention might result. The growing role of sodium channel mutations in neurological disease provides increased incentive for developing new and more specific sodium channel inhibitors. We anticipate that large-scale sequencing of ion channel genes from patients with epilepsy and other neurological disorders will be carried out within the next few years, with major impact on our understanding of the issues discussed here.

Acknowledgments

This work was supported by NIH research grants R01 NS34509 (to M.H. Meisler) and R21 NS046315 (to J.A. Kearney). We thank Jorgen de Haan, Valerie Drews, Lori Isom, and Christoph Lossin for helpful discussions.

Address correspondence to: Miriam H. Meisler, Department of Human Genetics, University of Michigan Medical School, 4909 Buhl Box 0618, Ann Arbor, Michigan 48109-0618, USA. Phone: (734) 763-5546; Fax: (734) 763-9691; E-mail: meislerm@umich.edu.

1. Jiang, Y., et al. 2003. X-ray structure of a voltage-dependent K+ channel. *Nature.* **423**:33–41.
2. Jiang, Y., Ruta, V., Chen, J., Lee, A., and MacKinnon, R. 2003. The principle of gating charge movement in a voltage-dependent K+ channel. *Nature.* **423**:42–48.
3. Isom, L.L., De Jongh, K.S., and Catterall, W.A. 1994. Auxiliary subunits of voltage-gated ion channels. *Neuron.* **12**:1183–1194.
4. Yu, F.H., et al. 2003. Sodium channel β4, a new disulfide-linked auxiliary subunit with similarity to β2. *J. Neurosci.* **23**:7577–7585.
5. Plummer, N.W., and Meisler, M.H. 1999. Evolution and diversity of mammalian sodium channel genes. *Genomics.* **57**:323–331.
6. Goldin, A.L. 2001. Resurgence of sodium channel research. *Annu. Rev. Physiol.* **63**:871–894.
7. Scheffer, I., and Berkovic, S.F. 1997. Generalized epilepsy with febrile seizures plus. A genetic disorder with heterogeneous clinical phenotypes. *Brain.* **120**:479–490.
8. Singh, R., Scheffer, I.E., Crossland, K., and Berkovic, S.F. 1999. Generalized epilepsy with febrile seizures plus: a common childhood-onset genetic epilepsy syndrome. *Ann. Neurol.* **45**:75–81.
9. Wallace, R., et al. 1998. Febrile seizures and generalized epilepsy associated with a mutation in the Na+-channel beta1 subunit gene SCN1B. *Nat. Genet.* **19**:366–370.
10. Meadows, L.S., et al. 2002. Functional and biochemical analysis of a sodium channel beta 1 subunit mutation responsible for generalized epilepsy with febrile seizures plus type 1. *J. Neurosci.* **22**:10699–10709.
11. Audenaert, D., et al. 2003. A deletion in SCN1B is associated with febrile seizures and early-onset absence epilepsy. *Neurology.* **61**:854–856.
12. Baulac, S., et al. 1999. A second locus for familial generalized epilepsy with febrile seizures plus maps to chromosome 2q21-q33. *Am. J. Hum. Genet.* **65**:1078–1085.
13. Moulard, B., et al. 1999. Identification of a new locus for generalized epilepsy with febrile seizures plus (GEFS+) on chromosome 2q24-q33. *Am. J. Hum. Genet.* **65**:1396–1400.
14. Escayg, A., et al. 2000. Mutations of SCN1A, encoding a neuronal sodium channel, in two families with GEFS+2. *Nat. Genet.* **24**:343–345.
15. Bonanni, P., et al. 2004. Generalized epilepsy with febrile seizures plus (GEFS+): clinical spectrum in seven Italian families unrelated to SCN1A, SCN1B, and GABRG2 gene mutations. *Epilepsia.* **45**:149–158.
16. Abou-Khalil, B., et al. 2001. Partial and generalized epilepsy with febrile seizures plus and a novel SCN1A mutation. *Neurology.* **57**:2265–2272.
17. Annesi, G., et al. 2003. Two novel SCN1A missense mutations in generalized epilepsy with febrile seizures plus. *Epilepsia.* **44**:1257–1258.
18. Ceulemans, B.P., Claes, L.R., and Lagae, L.G. 2004. Clinical correlations of mutations in the SCN1A gene: from febrile seizures to severe myoclonic epilepsy in infancy. *Pediatr. Neurol.* **30**:236–243.
19. Escayg, A., et al. 2001. A novel SCN1A mutation associated with generalized epilepsy with febrile seizures plus — and prevalence of variants in patients with epilepsy. *Am. J. Hum. Genet.* **68**:866–873.
20. Gerard, F., Pereira, S., Robaglia-Schlupp, A., Genton, P., and Szepetowski, P. 2002. Clinical and genetic analysis of a new multigenerational pedigree with GEFS+ (generalized epilepsy with febrile seizures plus). *Epilepsia.* **43**:581–586.
21. Ito, M., et al. 2002. Autosomal dominant epilepsy with febrile seizures plus with missense mutations of the (Na+)-channel alpha 1 subunit gene, SCN1A. *Epilepsy Res.* **48**:15–23.
22. Lerche, H., et al. 2001. Generalized epilepsy with febrile seizures plus: further heterogeneity in a large family. *Neurology.* **57**:1191–1198.
23. Spampanato, J., et al. 2004. A novel epilepsy mutation in the sodium channel SCN1A identifies a cytoplasmic domain for beta subunit interaction. *J. Neurosci.* **24**:10022–10034.
24. Sugawara, T., et al. 2001. Nav1.1 mutations cause febrile seizures associated with afebrile partial seizures. *Neurology.* **57**:703–705.
25. Wallace, R.H., et al. 2001. Neuronal sodium-channel alpha1-subunit mutations in generalized epilepsy with febrile seizures plus. *Am. J. Hum. Genet.* **68**:859–865.
26. Claes, L., et al. 2001. De novo mutations in the sodium-channel gene SCN1A cause severe myoclonic epilepsy of infancy. *Am. J. Hum. Genet.* **68**:1327–1332.
27. Wallace, R.H., et al. 2003. Sodium channel alpha1-subunit mutations in severe myoclonic epilepsy of infancy and infantile spasms. *Neurology.* **61**:765–769.
28. Fukuma, G., et al. 2004. Mutations of neuronal voltage-gated Na+ channel alpha1 subunit gene SCN1A in core severe myoclonic epilepsy in infancy (SMEI) and in borderline SMEI (SMEB). *Epilepsia.* **45**:140–148.
29. Sugawara, T., et al. 2001. A missense mutation of the Na+ channel alpha II subunit gene Na(v)1.2 in a patient with febrile and afebrile seizures causes channel dysfunction. *Proc. Natl. Acad. Sci. U. S. A.* **98**:6381–6389.
30. Kamiya, K., et al. 2004. A nonsense mutation of the sodium channel gene SCN2A in a patient with intractable epilepsy and mental decline. *J. Neurosci.* **24**:2690–2698.
31. Heron, S.E., et al. 2002. Sodium-channel defects in benign familial neonatal-infantile seizures. *Lancet.* **360**:851–852.
32. Berkovic, S.F., et al. 2004. Benign familial neonatal-infantile seizures: characterization of a new sodium channelopathy. *Ann. Neurol.* **55**:550–557.
33. Rhodes, T.H., Lossin, C., Vanoye, C.G., Wang, D.W., and George, A.L., Jr. 2004. Noninactivating voltage-gated sodium channels in severe myoclonic epilepsy of infancy. *Proc. Natl. Acad. Sci. U. S. A.* **101**:11147–11152.
34. Yu, F.H., et al. 2004. Deletion of the Nav1.1 channel: a mouse model for severe myoclonic epilepsy of infancy. Program no. 479.5 presented at the 34th

Annual Society for Neuroscience Meeting. October 25. San Diego, California, USA. http://sfn.scholar-one.com/itin2004/index.html.

35. Planells-Cases, R., et al. 2000. Neuronal death and perinatal lethality in voltage-gated sodium channel alpha(II)-deficient mice. *Biophys. J.* **78**:2878–2891.

36. Trudeau, M.M., Dalton, J.D., Day, J.W., Ranum, L.P.W., and Meisler, M.H. 2004. Heterozygosity for a truncation allele of sodium channel SCN8A in a family with ataxia and cognitive impairment. Program no. 205 presented at the 54th Annual Meeting of the American Society of Human Genetics. October 30. Bethesda, Maryland, USA. http://www.ashg.org/genetics/ashg04s/index.shtml.

37. Chen, C., et al. 2002. Reduced sodium channel density, altered voltage dependence of inactivation, and increased susceptibility to seizures in mice lacking sodium channel beta 2-subunits. *Proc. Natl. Acad. Sci. U. S. A.* **99**:17072–17077.

38. Chen, C., et al. 2004. Mice lacking sodium channel beta1 subunits display defects in neuronal excitability, sodium channel expression, and nodal architecture. *J. Neurosci.* **24**:4030–4042.

39. Lossin, C., Wang, D.W., Rhodes, T.H., Vanoye, C.G., and George, A.L., Jr. 2002. Molecular basis of an inherited epilepsy. *Neuron.* **34**:877–884.

40. Cossette, P., et al. 2003. Functional characterization of the D188V mutation in neuronal voltage-gated sodium channel causing generalized epilepsy with febrile seizures plus (GEFS). *Epilepsy Res.* **53**:107–117.

41. Alekov, A., Rahman, M.M., Mitrovic, N., Lehmann-Horn, F., and Lerche, H. 2000. A sodium channel mutation causing epilepsy in man exhibits subtle defects in fast inactivation and activation in vitro. *J. Physiol.* **529**:533–539.

42. Alekov, A.K., Rahman, M.M., Mitrovic, N., Lehmann-Horn, F., and Lerche, H. 2001. Enhanced inactivation and acceleration of activation of the sodium channel associated with epilepsy in man. *Eur. J. Neurosci.* **13**:2171–2176.

43. Lossin, C., et al. 2003. Epilepsy-associated dysfunction in the voltage-gated neuronal sodium channel SCN1A. *J. Neurosci.* **23**:11289–11295.

44. Sugawara, T., et al. 2003. Nav1.1 channels with mutations of severe myoclonic epilepsy in infancy display attenuated currents. *Epilepsy Res.* **54**:201–207.

45. Spampanato, J., Escayg, A., Meisler, M.H., and Goldin, A.L. 2001. Functional effects of two voltage-gated sodium channel mutations that cause generalized epilepsy with febrile seizures plus type 2. *J. Neurosci.* **21**:7481–7490.

46. Spampanato, J., Escayg, A., Meisler, M.H., and Goldin, A.L. 2003. Generalized epilepsy with febrile seizures plus type 2 mutation W1204R alters voltage-dependent gating of Nav1.1 sodium channels. *Neuroscience.* **116**:37–48.

47. Ohmori, I., Ouchida, M., Ohtsuka, Y., Oka, E., and Shimizu, K. 2002. Significant correlation of the SCN1A mutations and severe myoclonic epilepsy in infancy. *Biochem. Biophys. Res. Commun.* **295**:17–23.

48. Kearney, J.A., et al. 2001. A gain-of-function mutation in the sodium channel gene Scn2a results in seizures and behavioral abnormalities. *Neuroscience.* **102**:307–317.

49. Smith, M.R., and Goldin, A.L. 1997. Interaction between the sodium channel inactivation linker and domain III S4-S5. *Biophys. J.* **73**:1885–1895.

50. Abriel, H., Wehrens, X.H.T., Benhorin, J., Kerem, B., and Kass, R.S. 2000. Molecular pharmacology of the sodium channel mutation D1790G linked to the long-QT syndrome. *Circulation.* **102**:921–925.

51. Benhorin, J., et al. 2000. Effects of flecainide in patients with new SCN5A mutation: mutation-specific therapy for long-QT syndrome? *Circulation.* **101**:1698–1706.

52. Tate, S.K., et al. 2005. Genetic predictors of the maximum doses patients receive during clinical use of the anti-epileptic drugs carbamazepine and phenytoin. *Proc. Natl. Acad. Sci. U. S. A.* **102**:5507–5512.

53. Gennaro, E., et al. 2003. Familial severe myoclonic epilepsy of infancy: truncation of Nav1.1 and genetic heterogeneity. *Epileptic Disord.* **5**:21–25.

54. Fujiwara, T., et al. 2003. Mutations of sodium channel α subunit type 1 (SCN1A) in intractable childhood epilepsies with frequent generalized tonic-clonic seizures. *Brain.* **126**:531–546.

55. Weiss, K.M. 2005. Cryptic causation of human disease: reading between the (germ) lines. *Trends Genet.* **21**:82–88.

56. Legare, M.E., Bartlett, F.S., 2nd, and Frankel, W.N. 2000. A major effect QTL determined by multiple genes in epileptic EL mice. *Genome Res.* **10**:42–48.

57. Schauwecker, P.E., Williams, R.W., and Santos, J.B. 2004. Genetic control of sensitivity to hippocampal cell death induced by kainic acid: a quantitative trait loci analysis. *J. Comp. Neurol.* **477**:96–107.

58. Ferraro, T.N., et al. 2004. Fine mapping of a seizure susceptibility locus on mouse chromosome 1: nomination of Kcnj10 as a causative gene. *Mamm. Genome.* **15**:239–251.

59. Ferraro, T.N., et al. 1999. Mapping loci for pentyl-enetetrazol-induced seizure susceptibility in mice. *J. Neurosci.* **19**:6733–6739.

60. Bergren, S.K., Chen, S., Galecki, A., and Kearney, J.A. 2005. Genetic modifiers affecting severity of epilepsy caused by mutation of sodium channel Scn2a. *Mamm. Genome.* In press.

61. Buchner, D.A., Trudeau, M., and Meisler, M.H. 2003. SCNM1, a putative RNA splicing factor that modifies disease severity in mice. *Science.* **301**:967–969.

62. Meisler, M.H., Kearney, J., Escayg, A., MacDonald, B.T., and Sprunger, L.K. 2001. Sodium channels and neurological disease: insights from Scn8a mutations in the mouse. *Neuroscientist.* **7**:136–145.

63. Levin, S.I., and Meisler, M.H. 2004. Conditional inactivation of the voltage-gated sodium channel Scn8a (Nav1.6) in cerebellar Purkinje and granule cells. Program no. 397.1 presented at the Annual Society for Neuroscience Meeting. October 23–27. San Diego, California, USA. http://sfn.scholarone.com/itin2004/index.html.

64. Levin, S.I., and Meisler, M.H. 2004. Floxed allele for conditional inactivation of the voltage-gated sodium channel Scn8a (Nav1.6). *Genesis.* **39**:234–239.

65. Wood, J.N., Boorman, J.P., Okuse, K., and Baker, M.D. 2004. Voltage-gated sodium channels and pain pathways. *J. Neurobiol.* **61**:55–71.

66. Yang, Y., et al. 2004. Mutations in SCN9A, encoding a sodium channel alpha subunit, in patients with primary erythermalgia. *J. Med. Genet.* **41**:171–174.

67. Cummins, T.R., Dib-Hajj, S.D., and Waxman, S.G. 2004. Electrophysiological properties of mutant Nav1.7 sodium channels in a painful inherited neuropathy. *J. Neurosci.* **24**:8232–8236.

68. Dib-Hajj, S.D., et al. 2005. Gain-of-function mutation in Nav1.7 in familial erythromelalgia induces bursting of sensory neurons. *Brain.* doi:10.1093/brain/awh514.

69. Fertleman, C.R., Rees, M., Parker, K.A., Barlow, E., and Gardiner, R.M. 2004. Identification of the gene underlying an inherited disorder of pain sensation. Program no. 197 presented at the 54th Annual Meeting of the American Society of Human Genetics.

October 26–30. Bethesda, Maryland, USA. http://www.ashg.org/genetics/ashg04s/index.shtml.

70. Nassar, M.A., et al. 2004. Nociceptor-specific gene deletion reveals a major role for Nav1.7 (PN1) in acute and inflammatory pain. *Proc. Natl. Acad. Sci. U. S. A.* **101**:12706–12711.

71. Weiss, L.A., et al. 2003. Sodium channels SCN1A, SCN2A and SCN3A in familial autism. *Mol. Psychiatry.* **8**:186–194.

72. Kim, J., et al. 2004. Calmodulin mediates Ca2+ sensitivity of sodium channels. *J. Biol. Chem.* **279**:45004–45012.

73. Splawski, I., et al. 2004. Ca(V)1.2 calcium channel dysfunction causes a multisystem disorder including arrhythmia and autism. *Cell.* **119**:19–31.

74. Wasserman, D., Geijer, T., Rozanov, V., and Wasserman, J. 2005. Suicide attempt and basic mechanisms in neural conduction: relationships to the SCN8A and VAMP4 genes. *Am. J. Med. Genet. B Neuropsychiatr. Genet.* **133**:116–119.

75. Do, M.T., and Bean, B.P. 2004. Sodium currents in subthalamic nucleus neurons from Nav1.6-null mice. *J. Neurophysiol.* **92**:726–733.

76. Maurice, N., Tkatch, T., Meisler, M., Sprunger, L.K., and Surmeier, D.J. 2001. D1/D5 dopamine receptor activation differentially modulates rapidly inactivating and persistent sodium currents in prefrontal cortex pyramidal neurons. *J. Neurosci.* **21**:2268–2277.

77. Raman, I.M., Sprunger, L.K., Meisler, M.H., and Bean, B.P. 1997. Altered subthreshold sodium currents and disrupted firing patterns in Purkinje neurons of Scn8a mutant mice. *Neuron.* **19**:881–891.

78. Yan, H., Yuan, W., Velculescu, V.E., Vogelstein, B., and Kinzler, K.W. 2002. Allelic variation in human gene expression. *Science.* **297**:1143.

79. Claes, L., et al. 2003. De novo SCN1A mutations are a major cause of severe myoclonic epilepsy of infancy. *Hum. Mutat.* **21**:615–621.

80. Guerrini, R., and Aicardi, J. 2003. Epileptic encephalopathies with myoclonic seizures in infants and children (severe myoclonic epilepsy and myoclonic-astatic epilepsy). *J. Clin. Neurophysiol.* **20**:449–461.

81. Nabbout, R., et al. 2003. Spectrum of SCN1A mutations in severe myoclonic epilepsy of infancy. *Neurology.* **60**:1961–1967.

82. Oguni, H., et al. 2005. Severe myoclonic epilepsy in infancy: clinical analysis and relation to SCN1A mutations in a Japanese cohort. *Adv. Neurol.* **95**:103–117.

83. Ohmori, I., et al. 2003. Is phenotype difference in severe myoclonic epilepsy in infancy related to SCN1A mutations? *Brain Dev.* **25**:488–493.

84. Sugawara, T., et al. 2002. Frequent mutations of SCN1A in severe myoclonic epilepsy in infancy. *Neurology.* **58**:1122–1124.

85. Spampanato, J., Aradi, I., Soltesz, I., and Goldin, A.L. 2004. Increased neuronal firing in computer simulations of sodium channel mutations that cause generalized epilepsy with febrile seizures plus. *J. Neurophysiol.* **91**:2040–2050.

86. Kohrman, D.C., Smith, M.R., Goldin, A.L., Harris, J., and Meisler, M.H. 1996. A missense mutation in the sodium channel Scn8a is responsible for cerebellar ataxia in the mouse mutant jolting. *J. Neurosci.* **16**:5993–5999.

87. Kearney, J.A., et al. 2002. Molecular and pathological effects of a modifier gene on deficiency of the sodium channel Scn8a (Na(v)1.6). *Hum. Mol. Genet.* **11**:2765–2775.

88. Akopian, A.N., et al. 1999. The tetrodotoxin-resistant sodium channel SNS has a specialized function in pain pathways. *Nat. Neurosci.* **2**:541–548.

Finding schizophrenia genes

George Kirov, Michael C. O'Donovan, and Michael J. Owen

Department of Psychological Medicine, Wales College of Medicine, Cardiff University, Cardiff, United Kingdom.

Genetic epidemiological studies suggest that individual variation in susceptibility to schizophrenia is largely genetic, reflecting alleles of moderate to small effect in multiple genes. Molecular genetic studies have identified a number of potential regions of linkage and 2 associated chromosomal abnormalities, and accumulating evidence favors several positional candidate genes. These findings are grounds for optimism that insight into genetic factors associated with schizophrenia will help further our understanding of this disease and contribute to the development of new ways to treat it.

Schizophrenia is a disorder characterized by delusions, hallucinations, reduced interest and drive, altered emotional reactivity, and disorganized behavior. Often, cognitive and behavioral signs are present from early childhood, but the characteristic features generally start in the late teens and early twenties. Although outcomes are variable, even with treatment, the typical course is one of relapses followed by only partial remission as well as a marked reduction in social and occupational function such that sufferers are often the most vulnerable, isolated, and disadvantaged individuals in society. In a recent metaanalysis the median incidence rate was 15.2 per 100,000, with the 10% to 90% quantiles between 7.7 and 43.0 per 100,000 (1).

Genetic epidemiology

The results of numerous family, twin, and adoption studies show conclusively that risk of schizophrenia is increased among the relatives of affected individuals and that it is the result largely of genes rather than shared environment (2–4). In the children and siblings of individuals with schizophrenia, the increase in risk is approximately 10-fold; it is somewhat less than this in parents (Figure 1).

The latter finding is probably explained by a reduction in the reproductive opportunities, drive, and possibly the fertility of affected individuals. Five recent systematically ascertained studies report monozygotic concordance estimated at 41–65% compared with dizygotic concordance of 0–28% and an estimate of broad heritability of 85% (5). Among complex genetic disorders, schizophrenia has one of the highest heritabilities. To place this in perspective, the heritability of schizophrenia is similar to that of type I diabetes (72%–88%; refs. 6, 7), but greater than that of breast cancer (30%; ref. 8), coronary heart disease in males (57%; ref. 9), and type II diabetes (26%; ref. 10).

While the twin and adoption literature leave little doubt that genes are important, they also point to the importance of environmental factors, since the concordance for schizophrenia in monozygotic twins is typically around 50% and heritability estimates are less than 100%. Moreover, we should also note that

risks resulting from some gene-environment interactions tend to be attributed to genes in most genetic epidemiological studies.

It is clear from genetic epidemiology that the mode of transmission is complex (11, 12). However, the number of susceptibility loci, the disease risk conferred by each locus, the extent of genetic heterogeneity, and the degree of interaction among loci all remain unknown. Risch (13) has calculated that the data are incompatible with the existence of a single locus conferring a sibling relative risk (λ_S) of more than 3 and, unless extreme epistasis exists, models with 2 or 3 loci of λ_S less than or equal to 2 are more plausible. It should be emphasized that these calculations are based upon the assumption of homogeneity and refer to population-wide λ_S. It is quite possible that alleles of larger effect are operating in some groups of patients, for example, families with a high density of illness. However, high density families are expected to occur by chance even under polygenic inheritance, and their existence does not prove the existence of disease alleles of large effect (12).

The phenotype of schizophrenia

Different combinations of symptoms exist in different individuals, and there is considerable heterogeneity of disease course and outcome. However, it is not known whether schizophrenia is a single disorder with variable clinical manifestations or a group of syndromes, each with a unique or overlapping pathophysiology. In spite of this, structured and semistructured interviews together with explicit operational diagnostic criteria permit high diagnostic reliability. Given that clinically diagnosed schizophrenia has a high heritability, it should, in principle, be possible to subject it to molecular genetic analysis. However, schizophrenia, as defined by current diagnostic criteria, may well include a number of heterogeneous disease processes, in which case attempts at identifying genes and other etiological factors would be greatly facilitated if it were possible to distinguish reliably among them (14).

The situation is further complicated by the fact that we are unable to define the limits of the clinical phenotype to which genetic liability can lead. It clearly extends beyond the core diagnosis of schizophrenia to encompass a spectrum of disorders, including schizoaffective disorder and schizotypal personality disorder (15, 16). However, the limits of this spectrum and its relationship to other psychoses, affective and nonaffective, and nonpsychotic affective disorders remain uncertain (17, 18).

One of the earliest benefits of the identification of susceptibility genes is that the validity of current nosological categories can be further explored. Knowing the genes involved should illuminate heterogeneity within the current concept of schizophrenia and help us to understand its relationship to other diagnostic groups.

Nonstandard abbreviations used: *ARVCF*, armadillo repeat deleted in velocardiofacial syndrome gene; *COMT*, gene encoding catecholamine O-methyl transferase; *DAO*, gene encoding D–amino acid oxidase; *DAOA*, gene encoding D–amino acid oxidase activator; DTNBP1, dystrobrevin-binding protein 1; ErbB, v-erb-b2 erythroblastic leukemia viral oncogene homologue; *GRM3*, gene encoding metabotropic glutamate receptor type 3; LD, linkage disequilibrium; NMDA, N-methly-D-aspartate; *NRG1*, gene encoding neuregulin 1; *PRODH*, gene encoding proline dehydrogenase; *RGS4*, gene encoding regulator of G protein signaling 4; λ_S, sibling relative risk; SNP, single nucleotide polymorphism; *TRAR4*, gene encoding trace amine receptor 4; *ZDHHC8*, gene encoding zinc finger– and DHHC domain–containing protein 8.

Conflict of interest: The authors have declared that no conflict of interest exists.

Citation for this article: *J. Clin. Invest.* **115**:1440–1448 (2005). doi:10.1172/JCI24759.

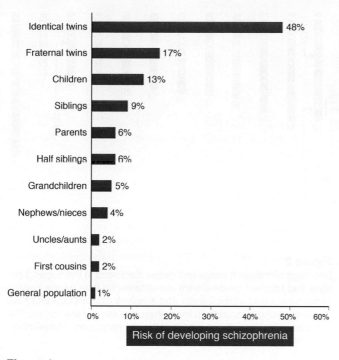

Figure 1
Risk of developing schizophrenia in relatives of schizophrenic probands. The data are based on the review of studies compiled by Gottesman (3).

Epidemiology, pathophysiology, and neurobiology

Epidemiological studies have identified a number of environmental factors associated with schizophrenia: pregnancy and delivery complications, delayed development, winter and spring birth, low IQ, urban birth and domicile, immigration, and the use of illegal drugs (especially cannabis; ref. 19). However, the effect sizes are small, and in most cases the direction of causation is uncertain.

For many years, ideas about the pathophysiology of schizophrenia were based on pharmacological studies. The classic "dopamine hypothesis," in which schizophrenia was attributed to a hyperdopaminergic state, derived from the realization that the potency of the first generation of antipsychotic drugs was proportional to their ability to block dopamine receptors (20). Modern refinements suggest a more complex picture, with increased dopaminergic transmission in the basal ganglia associated with acute psychosis (21) and a prefrontal cortical dopamine deficit responsible for chronic cognitive impairments (22). Psychopharmacological studies have also implicated altered serotonergic, and, more compellingly, glutamatergic transmission (23, 24). However, as these systems interact, the hypotheses are not mutually exclusive, and at least some of the dopaminergic changes may be secondary to altered cortical glutamatergic transmission (25).

The fundamental causes, whatever they are, appear to act during neurodevelopment rather than simply at the onset of the disorder (26, 27). This is supported by many of the epidemiological findings summarized above, by observations that children who subsequently develop schizophrenia have higher rates of neuropsychological and motor deficits, and by neuroimaging studies showing ventricular enlargement and reductions in cortical volume at illness onset (28) and in those at high risk of the disorder by virtue of increased genetic loading (29). The reduction in size of some brain structures shown by neuroimaging could in princi-

ple be due to neurodevelopmental or neurodegenerative processes or a combination of the 2. The existence of a neurodevelopmental mechanism is supported by the failure of neuropathological studies to detect markers of neurodegeneration (30). Neuropathological studies have also failed to establish a clear diagnostic neuropathology, and findings have often been conflicting. However, fairly consistent evidence has emerged for widespread nonuniform reductions in the neuropil and neuronal size, with the temporal lobe, prefrontal cortex, and dorsal thalamus particularly affected (30). These changes, together with reductions seen in synaptic and dendritic markers and abnormalities in white matter (31), suggest deficits in synaptic structure, function, and connectivity between neurons (30).

Epidemiological, pharmacological, and neurobiological studies have certainly brought about progress in our general understanding of schizophrenia. However, it is impossible confidently to implicate specific pathophysiological processes from the rather vague concepts of altered neurodevelopment, synaptic dysfunction, or neuronal connectivity. The more specific neurotransmitter-based hypotheses are very likely to be relevant to some of the overt clinical manifestations of schizophrenia, but with few exceptions, molecular genetic studies predicated on these hypotheses have met with disappointing results. This, together with the evidence for high heritability, has encouraged a number of groups to apply positional genetic approaches as these do not depend upon knowledge of disease pathophysiology.

Three main approaches have been used to seek genes for schizophrenia: positional genetics based on genome-wide linkage analysis, identification of chromosomal abnormalities associated with schizophrenia, and studies of functional candidate genes. In the following sections, we review the main findings that have emerged from each.

Molecular genetic studies

Linkage analysis. In contrast with several other complex disorders, such as breast cancer, Alzheimer disease, Parkinson disease, and epilepsy, no forms of schizophrenia following Mendelian inheritance patterns have yet been discovered. Indeed the results of linkage studies in schizophrenia have been disappointing, with positive studies often falling short of stringent genome-wide levels of significance and abundant failures to replicate. This is probably attributable to a combination of small genetic effects and inadequate sample sizes (32). However, as more than 20 genome-wide studies have been reported and sample sizes increased, some consistent patterns have emerged. Linkages that reached genome-wide significance on their own according to the criteria set forth by Lander and Kruglyak (33) and those that have received strong support from more than one sample are illustrated in Figure 2 (34–54).

Recently, 2 metaanalyses of schizophrenia linkage have been reported. Each used different methods and obtained overlapping but somewhat different results. The study of Badner and Gershon (55) supported the existence of susceptibility genes on chromosomes 8p, 13q, and 22q, while that of Lewis et al. (56) most strongly favored 2q but also found that the number of loci meeting the aggregate criteria for significance was much greater than the number of loci expected by chance ($P < 0.001$). Evidence was also obtained for susceptibility gene regions on chromosomes 5q, 3p, 11q, 6p, 1q, 22q, 8p, 20q, and 14p. The 8p and 22q regions were supported by both metaanalyses, but 8 other regions were supported by only one. Furthermore the region most strongly sup-

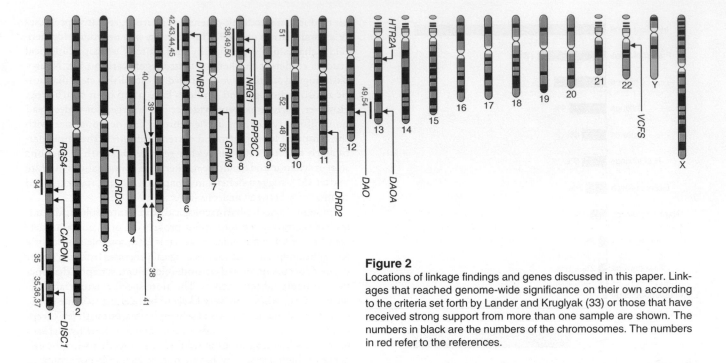

Figure 2
Locations of linkage findings and genes discussed in this paper. Linkages that reached genome-wide significance on their own according to the criteria set forth by Lander and Kruglyak (33) or those that have received strong support from more than one sample are shown. The numbers in black are the numbers of the chromosomes. The numbers in red refer to the references.

ported by the Lewis et al. (56) analysis, that on chromosome 2q, is not one that had received strong support previously; now it is clearly worthy of further investigation.

The linkage data therefore support the predictions made by Risch (13) on the basis of genetic epidemiological findings: it is highly unlikely that there exists a locus of effect size λ_S greater than 3. However, there is evidence that implicates a number of regions, which is consistent with the existence of susceptibility alleles of moderate effect (λ_S 1.2–3) and possibly uncommon loci of larger effect that can be identified in specific samples of large multiply affected families. Of course, the proof that a positive linkage is correct comes when the disease gene, or genes, is identified, and a number of the linked regions are currently being subjected to detailed analysis with, as we shall see, encouraging results.

Positional candidate genetics. Previously, the economics and practicalities of hunting a susceptibility gene within a linked chromosomal region dictated that one should have virtually definitive evidence that the linkage was a true positive before beginning. With recent improvements in our knowledge of genome anatomy and in genome analysis technology, the task of positional cloning has been transformed and now favors bolder endeavors. These considerations, together with the convergence of some positive linkage findings, have led to a number of detailed mapping studies of linked regions which have in turn implicated specific genes. However, the quality of the data has been variable, and a number of putative susceptibility genes have yet to be clearly replicated. Here we focus on the genes for which, at the time of this writing, there are published follow-up studies or for which we are aware of data that allow judgment about whether the gene is likely to be a true positive. In making these judgments, we have primarily been influenced by the strength of the genetic evidence for association to the clinical phenotype. We have given no weight to claims that a single nucleotide polymorphism (SNP) is functional or that the gene has plausibility by virtue of patterns of expression or participation in relevant

pathophysiological pathways. These are key properties of true susceptibility variants, but the number of genes that meet the above criteria is so vast that such evidence cannot be used significantly to increase the prior probability that a gene is implicated in schizophrenia and to prop up weak genetic findings. We have also given little weight to associations with endophenotypes, be they based on electrophysiology, pharmacology, psychology, or neuroimaging or performed in humans or modeled in animals. Such studies are likely to be essential for understanding how variation in a proven susceptibility gene leads to the clinical phenotype. However, the use of endophenotypes for the identification of disease genes requires that the measures in question are trait, rather than state, variables as well as compelling evidence from genetic epidemiology that they reflect genetic vulnerability to that disease rather than environmental factors. Promising data are accumulating for some potential endophenotypes for schizophrenia (57), but, in most cases, uncertainties remain.

Dystrobrevin-binding protein 1. Evidence that the gene encoding dystrobrevin-binding protein 1 (*DTNBP1*) or dysbindin has a role in schizophrenia was first reported by Straub et al. (58), who undertook association mapping across the linkage region on chromosome 6p22.3. Significant associations were found between schizophrenia and several SNPs and multimarker haplotypes spanning *DTNBP1*. A second study based on 203 families, most of which came from Germany (59), provided further evidence for association between *DTNBP1* and schizophrenia as did 2 large studies from our own group, which included approximately 900 schizophrenic patients from the United Kingdom and Ireland (60) and 488 parent-proband trios from Bulgaria (61). Although there have also been some studies in which no association was found (62, 63), significant associations in a total of 10 samples have now been published (64, 65). Moreover, significant association was found in the initially negative sample of Morris and colleagues (63), when 1 additional marker that defined the risk haplotype in our own UK sample was genotyped (60).

Thus, the evidence in favor of *DTNBP1* as a susceptibility gene for schizophrenia is strong. However, there are inconsistencies between the specific risk alleles and haplotypes among studies, which suggests the presence of multiple susceptibility and protective alleles (60) or the possibility that a single susceptibility allele is carried on a remarkable diversity of haplotypes even in closely related populations.

As yet, no causative variant has been identified, and the absence of associated nonsynonymous alleles after systematic mutation detection of the exons (60) suggests that disease susceptibility depends upon variation affecting mRNA expression or processing. The latter possibility is indirectly supported by evidence for as yet unknown *cis*-acting polymorphisms affecting *DTNBP1* expression in the human brain (66) and more directly supported by 2 recent studies showing reduced levels of expression of the mRNA (67) and protein (68) in postmortem brain samples from patients with schizophrenia.

Dysbindin 1 binds both α- and β-dystrobrevin, which are components of the dystrophin glycoprotein complex. The dystrophin complex is found in the sarcolemma of muscle but is also located in postsynaptic densities in a number of brain areas, particularly mossy fiber synaptic terminals in the cerebellum and hippocampus. Although its functions are largely unknown, its location initially suggested that variation in *DTNBP1* might confer risk of schizophrenia by mediating effects on postsynaptic structure and function (58). However, Talbot and colleagues (68) have recently shown that the presynaptic dystrobrevin-independent fraction of dysbindin is reduced in the schizophrenic brain within certain intrinsic glutamatergic neurones of the hippocampus and that this is associated with increased expression of vesicular glutamate transporter type 1. Moreover, a reduction in glutamate release has been demonstrated in cultured neurons with reduced *DTNBP1* expression (64). These data suggest that variation in *DTNBP1* might confer risk by altering presynaptic glutamate function.

Neuregulin 1. Neuregulin 1 (*NRG1*) was implicated in schizophrenia by a study in an Icelandic sample (50). Association analysis across 8p21-22 revealed highly significant evidence for association between schizophrenia and a multimarker haplotype at the 5' end of *NRG1*. Strong evidence for association with the same haplotype was subsequently found in a large sample from Scotland (69) and weakly replicated in a United Kingdom sample (70). Further positive findings have emerged from Irish (71), Chinese (72–75), and South African (76) samples. However, some negative findings have also been reported (76–79). Only the 3 studies of Icelandic, Scottish, and UK samples (50, 69, 70) have implicated the specific Icelandic haplotype, perhaps reflecting differences in the linkage disequilibrium (LD) structure across *NRG1* in European and Asian samples (74).

Despite detailed resequencing (50), it has not yet proven possible to identify specific susceptibility variants, but the Icelandic haplotype points to the 5' end of the gene, once again suggesting that altered expression or perhaps mRNA splicing might be involved. It is even possible at this stage that *NRG1* is not itself the susceptibility gene, as intron 1 contains another expressed sequence (71) whose function is unknown. However, insofar as it is possible to model schizophrenia in animals, behavioral analyses of *NRG1* hypomorphic mice support the view that the association is related to altered *NRG1* function or expression (80). More direct evidence suggesting that altered expression might be involved is also beginning to accumulate (81). Just as

for *DTNBP1*, the mechanisms by which altered *NRG1* function might lead to schizophrenia are unclear. *NRG1* hypomorphic mice have reduced expression of N-methyl-D-aspartate (NMDA) receptors, suggesting that the effects might be mediated through glutamatergic mechanisms (50). However, *NRG1* encodes many mRNA species and proteins; it is thought to encode approximately 15 proteins with a diverse range of functions in the brain, including cell signaling, v-erb-b2 erythroblastic leukemia viral oncogene homologue (ErbB) receptor interactions, axon guidance, synaptogenesis, glial differentiation, myelination, and neurotransmission (82). Any of these could potentially influence susceptibility to schizophrenia.

D–amino acid oxidase and D–amino acid oxidase activator. Chumakov and colleagues (83) undertook association mapping in the linkage region on chromosome 13q22-34. They found associations in French Canadian and Russian populations in markers around 2 novel genes, *G72* and *G30*, which are overlapping but transcribed in opposite directions. *G72* is a primate-specific gene expressed in the caudate and amygdala. With the use of yeast 2-hybrid analysis, evidence for physical interaction was found between *G72* and *D–amino acid oxidase* (*DAO*). DAO is expressed in the human brain, where it oxidizes D-serine, a potent activator of NMDA glutamate receptor. Coincubation of G72 and DAO in vitro revealed a functional interaction between the 2, with G72 enhancing the activity of DAO. Consequently, *G72* has now been named D–amino acid oxidase activator (DAOA). In the same study, *DAO* polymorphisms were shown to be associated with schizophrenia in one of the samples, and analysis of *DAOA* and *DAO* variants revealed modest evidence for a statistical interaction between the loci and disease risk. Given the 3 levels of interaction, the authors concluded that both genes influence risk of schizophrenia through a similar pathway and that this effect is likely to be mediated through altered NMDA-receptor function.

Associations between *DAOA* and schizophrenia have subsequently been reported in samples from Germany (84), China (85), and both the US and South Africa (76) as well as in a small sample of very early onset psychosis subjects from the US (86). As before, and conceivably for similar reasons, there is no consensus concerning the specific risk alleles or haplotypes across studies. The German group also reported association between *DAO* and schizophrenia, although with no evidence for a statistical interaction between the loci and risk of schizophrenia. At present, the published genetic evidence in support of this gene is weaker than that for *DTNBP1* and *NRG1*.

Regulator of G protein signaling 4. Another gene associated with schizophrenia is *Regulator of G protein signaling 4* (*RGS4*), which maps to the putative linkage region on chromosome 1q22. It was targeted for genetic analysis (87) following a microarray-based gene-expression study in which decreased *RGS4* expression was found in schizophrenic postmortem brain. Independent evidence for association between schizophrenia and a haplotype at the 5' end of the gene was found in 2 samples from the US, and while it did not provide significant evidence alone, a sample from India that was included added to the overall level of support. Positive findings have subsequently been reported by several other groups (88–90), but the level of support for each has been modest and the pattern of association different among samples. RGS4 is a negative regulator of G protein–coupled receptors. However, the evidence that RGS4 modulates activity at certain serotonergic (91) and metabotropic glutamatergic receptors (92) while its own expression is modulated

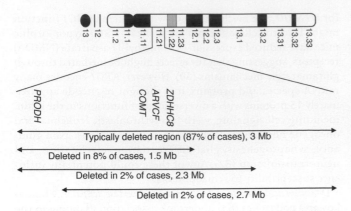

Figure 3
Chromosome 22 and the location of the VCFS deletions. The positions of candidate genes within the typically deleted region that are discussed in this paper are also indicated. The frequencies of the deleted regions are taken from Shaikh et al. (103).

by dopaminergic neurotransmission is of relevance to its possible role in schizophrenia (93). Moreover, RGS4 interacts with ErbB3 (94), which may be of relevance as ErbB3 is a NRG1 receptor whose expression is downregulated in schizophrenic brains (82).

Other positional candidate genes. Association analysis in single studies of *CAPON* (C terminal PDZ domain ligand of neuronal nitric oxide synthase) (95), *PPP3CC* (protein phosphatase 3, catalytic subunit) (96), and trace amine receptor 4 (*TRAR4*) (97), which map to the 1q22, 8p21.3, and 6q23.2 linkage regions, respectively, has suggested these as possible susceptibility genes. As discussed in the original articles, each of these genes can be plausibly related to candidate pathophysiologies of schizophrenia, but at this writing, we are not aware of any replication data to support these hypotheses.

Chromosomal abnormalities

There have been numerous reports of associations between schizophrenia and chromosomal abnormalities (98), but with 2 exceptions, none provides convincing evidence for the location of a susceptibility gene. Several studies have shown that adults with 22q11 deletions have a high risk of schizophrenia (99–101), with the largest study of adult patients to date (*n* = 50) estimating this at 24% (101). The deletion cannot account for a high proportion of schizophrenia cases (102), but reports of linkage to 22q11 (55, 56) suggest that variants in genes mapping to this region might contribute to cases of schizophrenia that do not have 22q11 deletions. Current candidates include catecholamine O-methyl transferase (*COMT*), proline dehydrogenase (*PRODH*), and zinc finger– and DHHC domain–containing protein 8 (*ZDHHC8*). The positions of the deletions on 22q11 and their frequencies (103) are shown in Figure 3.

COMT has been intensively studied because of its key role in dopamine catabolism. Most studies have focused upon a valine-to-methionine change at codon 158 of the brain-predominant membrane-bound form of COMT (MB-COMT) and codon 108 of the soluble form (S-COMT). The valine allele confers higher activity and thermal stability to COMT (104) and has been fairly consistently associated with reduced performance in tests of frontal lobe function (105, 106). The results for schizophrenia have been mixed, with a recent metaanalysis (107) reporting no overall evidence for association with the valine allele.

Since the preparation of the metaanalysis, an Israeli study of over 700 cases reported strong evidence for association between haplotypes, including the val158 allele and 2 flanking, noncoding SNPs (108). As in an earlier study (109), the evidence from haplotypes was stronger than for the valine allele alone, which suggests that *COMT* may well be a susceptibility gene for schizophrenia, but that the effect is not attributable to the valine/methionine variant. We have been unable to replicate association with any of the SNPs or halotypes, including the valine/methionine polymorphism, in a study of more than 2800 individuals including almost 1200 schizophrenics (Williams, et al. in press; ref. 110), but 2 other groups (104, 111) have recently reported rather different haplotype associations at *COMT* in Irish and US samples. As for the Israeli study, haplotypes carrying the val158 allele exhibited stronger evidence for association than did that allele alone while in the second study, the strongest findings included markers spanning the 3′ end of the armadillo repeat deleted in velocardiofacial syndrome gene (*ARVCF*). The latter has also been implicated in an earlier study (109), and its transcribed genomic sequence overlaps with that of *COMT* (112). While the picture is confused, we believe that the evidence does not support a role for valine/methionine 158 in susceptibility to schizophrenia, although a small effect cannot be excluded nor can a role in phenotype modification. However, it remains a strong possibility that variation elsewhere in *COMT* or in *ARVCF* confers susceptibility.

PRODH is another functional candidate gene, given that a loss-of-function mutant mouse exhibits behavioral abnormalities in sensorimotor gating that are analogous to those observed in patients with schizophrenia (113) and that PRODH influences the availability of glutamate. In addition, a heterozygous deletion of the entire *PRODH* gene was found in a family that included 2 subjects with schizophrenia, and 2 heterozygous *PRODH* missense variants were detected in 3 of 63 patients with schizophrenia studied by Jacquet et al. (114). Evidence in favor of association between *PRODH* and schizophrenia has been reported by Liu and colleagues (115) and by Li and colleagues (116). We have been unable to replicate either of these findings in large case-control and family-based association samples (117). Moreover, we have also observed a range of missense mutations that are equally common in patients with schizophrenia and controls (117); and in a very large sample of Japanese subjects, *PRODH* deletions were not associated with schizophrenia (118).

Finally, there is evidence that a SNP in *ZDHHC8*, a gene which encodes a putative transmembrane palmitoyl transferase, directly confers susceptibility to schizophrenia in females (119). The genetic evidence is not strong but has gained circumstantial support from data suggesting that this SNP may affect the splicing of an mRNA expressed in brain regions relevant to schizophrenia (119). Unfortunately, the only published attempt at replication so far found the opposite allele to be associated and no evidence for a gender effect (120).

The other major finding based upon a chromosomal abnormality comes from an extended pedigree in which a balanced chromosomal translocation (1, 11) (q42;q14.3) showed strong evidence for linkage to a fairly broad phenotype consisting of schizophrenia, bipolar disorder, and recurrent depression (36). The translocation was found to disrupt 2 genes on chromosome 1: *DISC1* and *DISC2* (36, 121). *DISC2* contains no open reading frame and may regulate *DISC1* expression via antisense RNA (121). Interestingly, *DISC1* and *DISC2* are located close to the chromosome 1 markers implicated in 2 Finnish linkage studies (35, 37) (Figure 1). It has been suggested that truncation of DISC1 in the translocation family might contribute to schizophrenia by affecting neuronal functions dependent upon intact cytoskeletal regulation, such as neuronal migration, neurite architecture, and intracellular transport (122, 123). While these are interesting hypotheses, it is important to remember that translocations can exert effects on genes other than those directly disrupted. For example, there are several mechanisms by which a translocation can influence the expression of neighboring genes. Thus, in order to unequivocally implicate *DISC1* and/or *DISC2* in the pathogenesis of schizophrenia, it is necessary to identify in another population mutations or polymorphisms that are associated with schizophrenia that are not in linkage disequilibrium with neighboring genes. Four published studies have attempted to do this; negative findings were reported by the Edinburgh group that originally identified *DISC1* and *DISC2* (124) and by a group that focused on the 5′ end of the gene in a large Japanese sample (125). However, positive findings have been reported in a large Finnish sample (see supplemental ref. S1; supplemental references available online with this article; doi:10.1172/JCI24759DS1) and in US samples with schizophrenia, schizoaffective disorder, or bipolar disorder (S2). At present, therefore, the genetic evidence in favor of *DISC1* as a susceptibility gene for schizophrenia is gaining weight but, in our view, is not yet compelling. In addition, there are suggestions that *DISC1* variants might confer susceptibility to a range of phenotypes, including schizoaffective disorder, bipolar disorder, and recurrent major depressive disorder as well as schizophrenia.

Functional candidate genes

There is a huge schizophrenia candidate-gene literature consisting of negative findings or positive findings that have either not been replicated or have seemed so unconvincing that there have been no attempts to replicate them. While most of the reported positives are unlikely to stand the test of time, neither can we conclude that any gene has been effectively excluded, given that most have not been studied exhaustively in large samples through a combination of detailed sequencing of exons, introns, and large regions of 5′ and 3′ flanking sequences combined with exhaustive genotyping. However, metaanalyses do at least suggest that the dopamine receptors *DRD3* (S3) and *DRD2* (S4) and the serotonergic receptor *HTR2A* (S5) might confer risk. The effect sizes, if any, are extremely small with odds ratios (ORs) less than 1.2 and difficult to confirm, but if the associations exist, the findings provide some support for the view that altered dopaminergic and serotonergic function might be a primary event in susceptibility to schizophrenia. If the putative associations are the result of linkage disequilibrium between the assayed markers and the true susceptibility variant, it is possible that the latter might contribute somewhat more, depending upon the degree of LD between the 2.

There is accumulating evidence for abnormalities of glutamatergic neurotransmission in schizophrenia. By and large, analysis of glutamatergic genes as functional candidates for schizophrenia has failed to produce consistent positive findings. However, recently claims have been made that variation in *GRM3* (the gene encoding metabotropic glutamate receptor–type 3) might confer risk of schizophrenia. *GRM3* was first implicated in a case-control study of German patients in which one SNP was significantly associated with schizophrenia (P = 0.0022)

in 1 out of the 3 groups of patients studied (S6). Fujii and colleagues (S7) genotyped 6 SNPs in a small case-control sample from Japan. They found significant association with one SNP (P = 0.011, uncorrected) and a 3-marker haplotype containing this SNP (P = 8.30 × 10^{-4}). Egan et al. (S8) genotyped 7 SNPs, including the 2 implicated in the previous reports, in a family-based association study. Weak evidence for overtransmission to cases was observed for one SNP (P = 0.02, uncorrected). However this was not one of the SNPs implicated previously, and the result failed to replicate in a second sample. A haplotype including this SNP was overtransmitted to cases (global P = 0.01, allele specific P = 0.0001), but again, this did not replicate in a second sample. The authors also presented data relating the single putatively associated GRM3 SNP to cognitive, fMRI, magnetic resonance spectroscopy, and neurochemical variables. They argued that these data all support the conclusion that *GRM3* affects prefrontal and hippocampal physiology, cognition, and risk for schizophrenia by altering glutamate neurotransmission. However, the genetic data implicating *GRM3* as a susceptibility gene for schizophrenia remain weak, although the findings of Egan and colleagues (S8) clearly suggest that further study is warranted.

Functional implications of susceptibility genes. In our view, the evidence most strongly implicates *DTNBP1* and *NRG1* as susceptibility genes for schizophrenia, while the data for *DAO, DAOA, DISC1,* and *RGS4* are promising but not yet compelling. Even in the most convincing cases, the risk haplotypes appear to be associated with small effect sizes (OR < 2.5) and, although this is difficult to determine, do not appear to fully explain the linkage findings that prompted each study. This could suggest that the associated polymorphisms/haplotypes are only in weak LD with the true pathogenic variants, that the linkages reflect variation at more than one susceptibility site in the same gene (or in multiple genes in the area of linkage), or that in some cases, despite the statistical evidence, the associations are spurious. Detailed follow-up studies, including de novo mutation detection and detailed genotyping in large samples drawn from different populations, with the aim of answering these questions are now required.

For most geneticists, the purpose of gene identification is the enhancement of our understanding of the pathogenesis of the disorder in question, and in this respect, the genes identified by positional genetics offer the most unbiased insights. Given the strength of the evidence for some genes, it is now important to identify the specific mechanisms by which the implicated genes alter schizophrenia risk and to identify the molecular and cognitive processes that link these primary events to psychopathology. It has been noted that several of the genes encode proteins that potentially have an impact on the function of glutamatergic synapses (S9–S11), leading to the hypothesis that the primary abnormality in schizophrenia is the synapse, particularly the glutamatergic synapse. The possible importance of synaptic abnormalities in schizophrenia has already been recognized (S12), and we now appear to be at a point of convergence between the genetics of schizophrenia and its neurobiology. However, there are strong reasons for remaining cautious: (a) The genetic evidence is not yet definitive, and we have not identified the specific pathogenic variants, much less the pathogenic mechanisms; (b) If we consider the 2 best-supported genes, *NRG1* encodes proteins with multiple functions of potential relevance to alternative hypotheses of schizophrenia, for example, aberrant myelination (82), while the function(s) of dysbindin are still obscure; (c) The widely held assumption that schizophrenia is a heterogeneous disorder with more that 1 core pathophysiology may well be correct. The desire, then, to fit the data into a unified theory is attractive and parsimonious, in which case it will not be possible to fit the data into a unified theory. It is therefore vital that there be no letup in the hunt for novel schizophrenia genes, the finding of which will allow us to test existing and to generate novel hypotheses of schizophrenia pathogenesis.

Conclusions

Molecular genetic studies of schizophrenia are built upon the firm foundations of reliable diagnostic methodology and a wealth of genetic epidemiological data. The fact that most, if not all, disease genes apparently have only moderate or small effect sizes has proven challenging, as with other common diseases. However, the resources and reagents generated by the human genome project have begun to make adequately powered studies feasible. A number of potential regions of linkage and 2 associated chromosomal abnormalities have been identified, and accumulating evidence favors several positional candidate genes, although in no case has the causative variant(s) been unequivocally identified. These findings are grounds for optimism and suggest that, while we still have far to travel, we are at least moving in the right direction.

Address correspondence to: Michael Owen, Department of Psychological Medicine, Wales College of Medicine, Cardiff University, Henry Wellcome Building, Heath Park, Cardiff CF14 4XN, United Kingdom. Phone: 44-2920-743248; Fax: 44-2920-746554; E-mail: owenmj@cardiff.ac.uk.

1. McGrath, J., et al. 2004. A systematic review of the incidence of schizophrenia: the distribution of rates and the influence of sex, urbanicity, migrant status and methodology. *BMC Med.* **2**:13.
2. McGuffin, P., Owen, M.J., O'Donovan, M.C., Thapar, A., and Gottesman, I.I. 1994. Schizophrenia. In *Seminars in psychiatric genetics, Royal College of Psychiatrists.* London, United Kingdom. 87–109.
3. Gottesman, I.I. 1991. *Schizophrenia genesis: the origins of madness.* Henry Holt & Company Inc. New York, New York, USA. 296 pp.
4. Riley, B.C., and Kendler, K.S. 2004. Schizophrenia: genetics. In *Kaplan and Sadock's comprehensive textbook psychiatry.* 8th edition. B.J. Sadock and V.A. Sadock, editors. Lippincott Williams and Wilkins. Philadelphia, Pennsylvania, USA. 1354–1371.
5. Cardno, A.G., and Gottesman, I.I. 2000. Twin studies of schizophrenia: from bow-and-arrow concordances to star war Mx and functional genomics. *Am. J. Med. Genet.* **97**:12–17.
6. Kyvik, K.O., Green, A., and Beck-Nielsen, H. 1995. Concordance rates of insulin dependent diabetes mellitus: a population based study of young Danish twins. *BMJ.* **311**:913–917.
7. Hyttinen, V., Kaprio, J., Kinnunen, L., Koskenvuo, M., and Tuomilehto, J. 2003. Genetic liability of type 1 diabetes and the onset age among 22,650 young Finnish twin pairs: a nationwide follow-up study. *Diabetes.* **52**:1052–1055.
8. Locatelli, I., Lichtenstein, P., and Yashin, A.I. 2004. The heritability of breast cancer: a bayesian correlated frailty model applied to Swedish twins data. *Twin Res.* **7**:182–191.
9. Zdravkovic, S., et al. 2002. Heritability of death from coronary heart disease: a 36-year follow-up of 20 966 Swedish twins. *J. Intern. Med.* **252**:247–254.
10. Poulsen, P., Kyvik, K.O., Vaag, A., and Beck-Nielsen, H. 1999. Heritability of type II (non-insulin-dependent) diabetes mellitus and abnormal glucose tolerance–a population-based twin study. *Diabetologia.* **42**:139–145.
11. Gottesman, I.I., and Shields, J. 1967. A polygenic theory of schizophrenia. *Proc. Natl. Acad. Sci. U. S. A.* **58**:199–205.
12. McGue, M., and Gottesman, I.I. 1989. A single dominant gene still cannot account for the transmission of schizophrenia. *Arch. Gen. Psychiatry.* **46**:478–480.
13. Risch, N. 1990. Genetic linkage and complex diseases, with special reference to psychiatric disorders [review]. *Genet. Epidemiol.* **7**:3–16; discussion 17–45.
14. Gottesman, I.I., and Shields, J. 1982. *Schizophrenia: the epigenetic puzzle.* Cambridge University Press. Cambridge, United Kingdom. 258 pp.
15. Farmer, A.E., McGuffin, P., and Gottesman I.I. 1987. Twin concordance for DSM-III schizophrenia. Scrutinizing the validity of the definition. *Arch. Gen. Psychiatry.* **44**:634–641.

16. Kendler, K.S., Neale, M.C., and Walsh, D. 1995. Evaluating the spectrum concept of schizophrenia in the Roscommon Family Study. *Am. J. Psychiatry.* **152**:749–754.
17. Kendler, K.S., Karkowski, L.M., and Walsh, D. 1998. The structure of psychosis: latent class analysis of probands from the Roscommon Family Study. *Arch. Gen. Psychiatry.* **55**:492–499.
18. Tienari, P. et al. 2000. Finnish adoptive family study: sample selection and adoptee DSM-III-R diagnoses. *Acta Psychiatr. Scand.* **101**:433–443.
19. Murray, R.M., Jones, P.B., Susser, E., van Os, J., and Cannon, M. 2002. *The epidemiology of schizophrenia.* Cambridge University Press. Cambridge, United Kingdom. 454 pp.
20. Seeman, P., and Lee, T. 1975. Antipsychotic drugs: direct correlation between clinical potency and presynaptic action on dopamine neurons. *Science.* **188**:1217–1219.
21. Abi-Dargham, A., et al. 2000. Increased baseline occupancy of D2 receptors by dopamine in schizophrenia. *Proc. Natl. Acad. Sci. U. S. A.* **97**:8104–8109.
22. Weinberger, D.R., et al. 2001. Prefrontal neurons and the genetics of schizophrenia. *Biol. Psychiatry.* **50**:825–844.
23. Meltzer, H.Y., Matsubara, S., and Lee, J.C. 1989. Classification of typical and atypical antipsychotic drugs on the basis of dopamine D-1, D-2 and serotonin2 pKi values. *J. Pharmacol. Exp. Ther.* **251**:238–246.
24. Javitt, D.C., and Zukin, S.R. 1991. Recent advances in the phencyclidine model of schizophrenia. *Am. J. Psychiatry.* **148**:1301–1308.
25. Laruelle, M., Kegeles, L.S., and Abi-Dargham, A. 2003. Glutamate, dopamine, and schizophrenia: from pathophysiology to treatment. *Ann. N. Y. Acad. Sci.* **1003**:138–158.
26. Weinberger, D.R. 1995. From neuropathology to neurodevelopment. *Lancet.* **346**:552–557.
27. Murray, R.M., and Lewis, S.W. 1987. Is schizophrenia a neurodevelopmental disorder? *Br. Med. J.* **295**:681–682.
28. Pantelis, C., et al. 2003. Neuroanatomical abnormalities before and after onset of psychosis: a cross-sectional and longitudinal MRI comparison. *Lancet.* **361**:281–288.
29. Lawrie, S.M., et al. 1999. Magnetic resonance imaging of brain in people at high risk of developing schizophrenia. *Lancet.* **353**:30–33.
30. Harrison, P.J. 1999. The neuropathology of schizophrenia. A critical review of the data and their interpretation. *Brain.* **122**:593–624.
31. Davis, K.L., et al. 2003. White matter changes in schizophrenia: evidence for myelin-related dysfunction. *Arch. Gen. Psychiatry.* **60**:443–456.
32. Suarez, B.K., Hampe, C.L., and Van Eerdewegh, P. 1994. Problems of replicating linkage claims in psychiatry. In *Genetic approaches to mental disorders.* E.S. Gershon and C.R. Cloninger, editors. American

Psychiatric Press. Washington, DC, USA. 23–46.
33. Lander, E., and Kruglyak, L. 1995. Genetic dissection of complex traits: guidelines for interpreting and reporting linkage results. *Nat. Genet.* **11**:241–247.
34. Brzustowicz, L.M., Hodgkinson, K.A., Chow, E.W., Honer, W.G., and Bassett, A.S. 2000. Location of a major susceptibility locus for familial schizophrenia on chromosome 1q21-q22. *Science.* **288**:678–682.
35. Ekelund, J., et al. 2001. Chromosome 1 loci in Finnish schizophrenia families. *Hum. Mol. Genet.* **10**:1611–1617.
36. Blackwood, D.H., et al. 2001. Schizophrenia and affective disorders-cosegregation with a translocation at chromosome 1q42 that directly disrupts brain-expressed genes: clinical and P300 findings in a family. *Am. J. Hum. Genet.* **69**:428–433.
37. Ekelund, J., et al. 2004. Replication of 1q42 linkage in Finnish schizophrenia pedigrees. *Mol. Psychiatry.* **9**:1037–1041.
38. Gurling, H.M., et al. 2001. Genomewide genetic linkage analysis confirms the presence of susceptibility loci for schizophrenia, on chromosomes 1q32.2, 5q33.2, and 8p21-22 and provides support for linkage to schizophrenia, on chromosomes 11q23.3-24 and 20q12.1-11.23. *Am. J. Hum. Genet.* **68**:661–673.
39. Straub, R.E., MacLean, C.J., O'Neil, F.A., Walsh, D., and Kendler, K.S. 1997. Support for a possible schizophrenia vulnerability locus in region 5q22-31 in Irish families. *Mol. Psychiatry.* **2**:148–155.
40. Devlin, B., et al. 2002. Genome-wide multipoint linkage analyses of multiplex schizophrenia pedigrees from the oceanic nation of Palau. *Mol. Psychiatry.* **7**:689–694.
41. Paunio, T., et al. 2001. Genome-wide scan in a nationwide study sample of schizophrenia families in Finland reveals susceptibility loci on chromosomes 2q and 5q. *Hum. Mol. Genet.* **10**:3037–3048.
42. Straub, R.E., et al. 1995. A potential vulnerability locus for schizophrenia on chromosome 6p24-22: evidence for genetic heterogeneity. *Nat. Genet.* **11**:287–293.
43. Schizophrenia Linkage Collaborative Group for Chromosomes 3, 6, and 8. 1996. Additional support for schizophrenia linkage on chromosomes 6 and 8: a multicenter study. *Am. J. Med. Genet.* **67**:580–594.
44. Maziade, M., Bissonnette, L., Rouillard, E., Roy, M.A., and Merette, C. 1997. Further evidence of a susceptibility gene for schizophrenia in 6p22-p24: a contribution from the eastern Quebec population. *Am. J. Med. Genet.* **74**:666–667.
45. Schwab, S.G., et al. 2000. A genome-wide autosomal screen for schizophrenia susceptibility loci in 71 families with affected siblings: support for loci on chromosome 10p and 6. *Mol. Psychiatry.* **5**:638–649.

46. Martinez, M., et al. 1999. Follow-up study on a susceptibility locus for schizophrenia on chromosome 6q. *Am. J. Med. Genet.* **88**:337–343.

47. Lindholm, E. et al. 2001. A schizophrenia-susceptibility locus at 6q25, in one of the world's largest reported pedigrees. *Am. J. Hum. Genet.* **69**:96–105.

48. Lerer, B., et al. 2003. Genome scan of Arab Israeli families maps a schizophrenia susceptibility gene to chromosome 6q23 and supports a locus at chromosome 10q24. *Mol. Psychiatry.* **8**:488–498.

49. Blouin, J.L., et al. 1998. Schizophrenia susceptibility loci on chromosomes 13q32 and 8p21. *Nat. Genet.* **20**:70–73.

50. Stefansson, H., et al. 2002. Neuregulin 1 and susceptibility to schizophrenia. *Am. J. Hum. Genet.* **71**:877–892.

51. DeLisi, L.E., et al. 2002. A genome-wide scan for linkage to chromosomal regions in 382 sibling pairs with schizophrenia or schizoaffective disorder. *Am. J. Psychiatry.* **159**:803–812.

52. Fallin, M.D., et al. 2003. Genomewide linkage scan for schizophrenia susceptibility loci among Ashkenazi Jewish families shows evidence of linkage on chromosome 10q22. *Am. J. Hum. Genet.* **73**:601–611.

53. Williams, N., et al. 2003. A systematic genomewide linkage study in 353 sib pairs with schizophrenia. *Am. J. Hum. Genet.* **73**:1355–1367.

54. Brzustowicz, L.M., et al. 1999. Linkage of familial schizophrenia to chromosome 13q32. *Am. J. Hum. Genet.* **65**:1096–1103.

55. Badner, J.A., and Gershon, E.S. 2002. Meta-analysis of whole-genome linkage scans of bipolar disorder and schizophrenia. *Mol. Psychiatry.* **7**:405–411.

56. Lewis, C.M., et al. 2003. Genome scan meta-analysis of schizophrenia and bipolar disorder, part II: Schizophrenia. *Am. J. Hum. Genet.* **73**:34–48.

57. Gottesman, I.I., and Gould, T.D. 2003. The endophenotype concept in psychiatry: etymology and strategic intentions. *Am. J. Psychiatry.* **160**:636–645.

58. Straub, R.E., et al. 2002. Genetic variation in the 6p22.3 gene DTNBP1, the human ortholog of the mouse dysbindin gene, is associated with schizophrenia. *Am. J. Hum. Genet.* **71**:337–348.

59. Schwab, S.G., et al. 2003. Support for association of schizophrenia with genetic variation in the 6p22.3 gene, dysbindin, in sib-pair families with linkage and in an additional sample of triad families. *Am. J. Hum. Genet.* **72**:185–190.

60. Williams, N.M., et al. 2004. Identification in 2 independent samples of a novel schizophrenia risk haplotype of the dystrobrevin binding protein gene (DTNBP1). *Arch. Gen. Psychiatry.* **61**:336–344.

61. Kirov, G., et al. 2004. Strong evidence for association between the dystrobrevin binding protein 1 (DTNBP1) gene and schizophrenia in 488 parent-offspring trios from Bulgaria. *Biol. Psychiatry.* **55**:971–975.

62. Van Den Bogaert, A., et al. 2003. The DTNBP1 (dysbindin) gene contributes to schizophrenia, depending on family history of the disease. *Am. J. Hum. Genet.* **73**:1438–1443.

63. Morris, D.W., et al. 2003. No evidence for association of the dysbindin gene [DTNBP1] with schizophrenia in an Irish population-based study. *Schizophr. Res.* **60**:167–172.

64. Numakawa, T., et al. 2004. Evidence of novel neuronal functions of dysbindin, a susceptibility gene for schizophrenia. *Hum. Mol. Genet.* **13**:2699–2708.

65. Funke, B., et al. 2004. Association of the DTNBP1 locus with schizophrenia in a U.S. Population. *Am. J. Hum. Genet.* **75**:891–898.

66. Bray, N.J., Buckland, P.R., Owen, M.J., and O'Donovan, M.C. 2003. Cis-acting variation in the expression of a high proportion of genes in human brain. *Hum. Genet.* **113**:149–153.

67. Weickert, C.S., et al. 2004. Human dysbindin (DTNBP1) gene expression in normal brain and in schizophrenic prefrontal cortex and midbrain. *Arch. Gen. Psychiatry.* **61**:544–555.

68. Talbot, K., et al. 2004. Dysbindin-1 is reduced in intrinsic, glutamatergic terminals of the hippocampal formation in schizophrenia. *J. Clin. Invest.* **113**:1353–1363. doi:10.1172/JCI200420425.

69. Stefansson, H., et al. 2003. Association of neuregulin 1 with schizophrenia confirmed in a Scottish population. *Am. J. Hum. Genet.* **72**:83–87.

70. Williams, N.M., et al. 2003. Support for genetic variation in neuregulin 1 and susceptibility to schizophrenia. *Mol. Psychiatry.* **8**:485–487.

71. Corvin, A.P., et al. 2004. Confirmation and refinement of an 'at-risk' haplotype for schizophrenia suggests the EST cluster, Hs.97362, as a potential susceptibility gene at the neuregulin-1 locus. *Mol. Psychiatry.* **9**:208–213.

72. Yang, J.Z., et al. 2003. Association study of neuregulin 1 gene with schizophrenia. *Mol. Psychiatry.* **8**:706–709.

73. Tang, J.X., et al. 2004. Polymorphisms within 5′ end of the neuregulin 1 gene are genetically associated with schizophrenia in the Chinese population. *Mol. Psychiatry.* **9**:11–12.

74. Zhao, X., et al. 2004. A case control and family based association study of the neuregulin1 gene and schizophrenia. *J. Med. Genet.* **41**:31–34.

75. Li, T., et al. 2004. Identification of a novel neuregulin 1 at-risk haplotype in Han schizophrenia Chinese patients, but no association with the Icelandic/Scottish risk haplotype. *Mol. Psychiatry.* **9**:698–704.

76. Hall, D., Gogos, J.A., and Karayiorgou, M. 2004. The contribution of three strong candidate schizophrenia susceptibility genes in demographically distinct populations. *Genes Brain Behav.* **3**:240–248.

77. Iwata, N., et al. 2004. No association with the neuregulin 1 haplotype to Japanese schizophrenia. *Mol. Psychiatry.* **9**:126–127.

78. Thiselton, D.L., et al. 2004. No evidence for linkage or association of neuregulin-1 (NRG1) with disease in the Irish study of high-density schizophrenia families (ISHDSF). *Mol. Psychiatry.* **9**:777–783.

79. Hong, C.J., et al. 2004. Case-control and family-based association studies between the neuregulin 1 (Arg38Gln) polymorphism and schizophrenia. *Neurosci. Lett.* **366**:158–161.

80. Stefansson, H., Steinthorsdottir, V., Thorgeirsson, T.E., Gulcher, J.R., and Stefansson, K. 2004. Neuregulin 1 and schizophrenia. *Ann. Med.* **36**:62–71.

81. Hashimoto, R., et al. 2004. Expression analysis of neuregulin-1 in the dorsolateral prefrontal cortex in schizophrenia. *Mol. Psychiatry.* **9**:299–307.

82. Corfas, G., Roy, K., and Buxbaum, J.D. 2004. Neuregulin 1-erbB signaling and the molecular/cellular basis of schizophrenia. *Nat. Neurosci.* **7**:575–580.

83. Chumakov, I., et al. 2002. Genetic and physiological data implicating the new human gene G72 and the gene for D-amino acid oxidase in schizophrenia. *Proc. Natl. Acad. Sci. U. S. A.* **99**:13675–13680.

84. Schumacher, J., et al. 2004. Examination of G72 and D-amino-acid oxidase as genetic risk factors for schizophrenia and bipolar affective disorder. *Mol. Psychiatry.* **9**:203–207.

85. Wang, X., et al. 2004. Association of G72/G30 with schizophrenia in the Chinese population. *Biochem. Biophys. Res. Commun.* **319**:1281–1286.

86. Addington, A.M., et al. 2004. Polymorphisms in the 13q33.2 gene G72/G30 are associated with childhood-onset schizophrenia and psychosis not otherwise specified. *Biol. Psychiatry.* **55**:976–980.

87. Chowdari, K.V., et al. 2002. Association and linkage analyses of RGS4 polymorphisms in schizophrenia. *Hum. Mol. Genet.* **11**:1373–1380.

88. Williams, N.M., et al. 2004. Support for RGS4 as a susceptibility gene for schizophrenia. *Biol. Psychiatry.* **55**:192–195.

89. Chen, X., et al. 2004. Regulator of G-protein signaling 4 (RGS4) gene is associated with schizophrenia in Irish high density families. *Am. J. Med. Genet. B Neuropsychiatr. Genet.* **129B**:23–26.

90. Morris, D.W., et al. 2004. Confirming RGS4 as a susceptibility gene for schizophrenia. *Am. J. Med. Genet. B Neuropsychiatr. Genet.* **125B**:50–53.

91. Beyer, C.E., et al. 2004. Regulators of G-protein signaling 4: modulation of 5-HT(1A)-mediated neurotransmitter release in vivo. *Brain Res.* **1022**:214–220.

92. De Blasi, A., Conn, P.J., Pin, J., and Nicoletti, F. 2001. Molecular determinants of metabotropic glutamate receptor signaling. *Trends Pharmacol. Sci.* **22**:114–120.

93. Taymans, J.M., et al. 2004. Dopamine receptor-mediated regulation of RGS2 and RGS4 mRNA differentially depends on ascending dopamine projections and time. *Eur. J. Neurosci.* **19**:2249–2260.

94. Thaminy, S., Auerbach, D., Arnoldo, A., and Stagljar, I. 2003. Identification of novel ErbB3-interacting factors using the split-ubiquitin membrane yeast two-hybrid system. *Genome Res.* **13**:1744–1753.

95. Brzustowicz, L.M., et al. 2004. Linkage disequilibrium mapping of schizophrenia susceptibility to the CAPON region of chromosome 1q22. *Am. J. Hum. Genet.* **74**:1057–1063.

96. Gerber, D.J., et al. 2003. Evidence for association of schizophrenia with genetic variation in the 8p21.3 gene, PPP3CC, encoding the calcineurin gamma subunit. *Proc. Natl. Acad. Sci. U. S. A.* **100**:8993–8998.

97. Duan, J., et al. 2004. Polymorphisms in the trace amine receptor 4 (TRAR4) gene on chromosome 6q23.2 are associated with susceptibility to schizophrenia. *Am. J. Hum. Genet.* **75**:624–638.

98. MacIntyre, D.J., Blackwood, D.H., Porteous, D.J., Pickard, B.S., and Muir, W.J. 2003. Chromosomal abnormalities and mental illness. *Mol. Psychiatry.* **8**:275–287.

99. Pulver, A.E., et al. 1994. Psychotic illness in patients diagnosed with velo-cardio-facial syndrome and their relatives. *J. Nerv. Ment. Dis.* **182**:476–478.

100. Bassett, A.S., et al. 1998. 22q11 deletion syndrome in adults with schizophrenia. *Am. J. Med. Genet.* **81**:328–337.

101. Murphy, K.C., Jones, L.A., and Owen, M.J. 1999. High rates of schizophrenia in adults with velo-cardio-facial syndrome. *Arch. Gen. Psychiatry.* **56**:940–945.

102. Ivanov, D., et al. 2003. A molecular genetic study of the vcfs region in patients with early onset psychosis. *Br. J. Psychiatry.* **183**:409–413.

103. Shaikh, T.H., et al. 2000. Chromosome 22-specific low copy repeats and the 22q11.2 deletion syndrome: genomic organisation and deletion endpoint analysis. *Hum. Mol. Genet.* **9**:489–501.

104. Chen, X., Wang, X., O'Neill, A.F., Walsh, D., and Kendler, K.S. 2004. Variants in the catechol-o-methyltransferase (COMT) gene are associated with schizophrenia in Irish high-density families. *Mol. Psychiatry.* **9**:962–967.

105. Egan, M.F., et al. 2001. Effect of COMT Val108/158 Met genotype on frontal lobe function and risk for schizophrenia. *Proc. Natl. Acad. Sci. U. S. A.* **98**:6917–6922.

106. Malhotra, A.K., et al. 2002. A functional polymorphism in the COMT gene and performance on a test of prefrontal cognition. *Am. J. Psychiatry.* **159**:652–654.

107. Glatt, S.J., Faraone, S.V., and Tsuang, M.T. 2003. Association between a functional catechol O-methyltransferase gene polymorphism and schizophrenia: meta-analysis of case-control and family-based studies. *Am. J. Psychiatry.* **160**:469–476.

108. Shifman, S., et al. 2002. A highly significant association between a comt haplotype and schizophrenia.

Am. J. Hum. Genet. **71**:1296–1302.

109. Li, T., et al. 2000. Family-based linkage disequilibrium mapping using SNP marker haplotypes: application to a potential locus for schizophrenia at chromosome 22q11. *Mol. Psychiatry.* **5**:77–84.

110. Williams, H., et al. 2005. No association between polymorphisms in COMT and schizophrenia in two large samples. *Am. J. Psychiatry.* In press.

111. Sanders, A.R., et al. 2005. Haplotypic association spanning the 22q11.21 genes COMT and ARVCF with schizophrenia. *Mol. Psychiatry.* **10**:353–365.

112. Bray, N.J., et al. 2003. A haplotype implicated in schizophrenia susceptibility is associated with reduced COMT expression in human brain. *Am. J. Hum. Genet.* **73**:152–161.

113. Gogos, J.A., et al. 1999. The gene encoding proline dehydrogenase modulates sensorimotor gating in mice. *Nat. Genet.* **21**:434–439.

114. Jacquet, H., et al. 2002. PRODH mutations and hyperprolinemia in a subset of schizophrenic patients. *Hum. Mol. Genet.* **11**:2243–2249.

115. Liu, H., et al. 2002. Genetic variation at the 22q11 PRODH/DGCR6 locus presents an unusual pattern and increases susceptibility to schizophrenia. *Proc. Natl. Acad. Sci. U. S. A.* **99**:3717–3722.

116. Li, T., et al. 2004. Evidence for association between novel polymorphisms in the PRODH gene and schizophrenia in a Chinese population. *Am. J. Med. Genet. B Neuropsychiatr. Genet.* **129B**:13–15.

117. Williams, H.J., et al. 2003. Association between PRODH and schizophrenia is not confirmed. *Mol. Psychiatry.* **8**:644–645.

118. Ohtsuki, T., et al. 2004. Failure to find association between PRODH deletion and schizophrenia. *Schizophr. Res.* **67**:111–113.

119. Mukai, J., et al. 2004. Evidence that the gene encoding ZDHHC8 contributes to the risk of schizophrenia. *Nat. Genet.* **36**:725–731.

120. Chen, W.Y., et al. 2004. Case-control study and transmission disequilibrium test provide consistent evidence for association between schizophrenia and genetic variation in the 22q11 gene ZDHHC8. *Hum. Mol. Genet.* **13**:2991–2995.

121. Millar, J.K., et al. 2000. Disruption of two novel genes by a translocation co-segregating with schizophrenia. *Hum. Mol. Genet.* **9**:1415–1423.

122. Miyoshi, K., et al. 2003. Disrupted-In-Schizophrenia 1, a candidate gene for schizophrenia, participates in neurite outgrowth. *Mol. Psychiatry.* **8**:685–694.

123. Ozeki, Y, et al. 2003. Disrupted-in-schizophrenia-1 (DISC-1): mutant truncation prevents binding to NudE-like (NUDEL) and inhibits neurite outgrowth. *Proc. Natl. Acad. Sci. U. S. A.* **100**:289–294.

124. Devon, R.S., et al. 2001. Identification of polymorphisms within Disrupted in Schizophrenia 1 and Disrupted in Schizophrenia 2, and an investigation of their association with schizophrenia and bipolar disorder. *Psychiatr. Genet.* **11**:71–78.

125. Kockelkorn, T.T., et al. 2004. Association study of polymorphisms in the 5′ upstream region of human DISC1 gene with schizophrenia. *Neurosci. Lett.* **368**:41–45.

The addicted human brain: insights from imaging studies

Nora D. Volkow,[1,2] Joanna S. Fowler,[3] and Gene-Jack Wang[1]

[1]Department of Medicine, Brookhaven National Laboratory, Upton, New York, USA. [2]Department of Psychiatry, State University of New York at Stony Brook, Stony Brook, New York, USA. [3]Department of Chemistry, Brookhaven National Laboratory, Upton, New York, USA.

Imaging studies have revealed neurochemical and functional changes in the brains of drug-addicted subjects that provide new insights into the mechanisms underlying addiction. Neurochemical studies have shown that large and fast increases in dopamine are associated with the reinforcing effects of drugs of abuse, but also that after chronic drug abuse and during withdrawal, brain dopamine function is markedly decreased and these decreases are associated with dysfunction of prefrontal regions (including orbitofrontal cortex and cingulate gyrus). The changes in brain dopamine function are likely to result in decreased sensitivity to natural reinforcers since dopamine also mediates the reinforcing effects of natural reinforcers and on disruption of frontal cortical functions, such as inhibitory control and salience attribution. Functional imaging studies have shown that during drug intoxication, or during craving, these frontal regions become activated as part of a complex pattern that includes brain circuits involved with reward (nucleus accumbens), motivation (orbitofrontal cortex), memory (amygdala and hippocampus), and cognitive control (prefrontal cortex and cingulate gyrus). Here, we integrate these findings and propose a model that attempts to explain the loss of control and compulsive drug intake that characterize addiction. Specifically, we propose that in drug addiction the value of the drug and drug-related stimuli is enhanced at the expense of other reinforcers. This is a consequence of conditioned learning and of the resetting of reward thresholds as an adaptation to the high levels of stimulation induced by drugs of abuse. In this model, during exposure to the drug or drug-related cues, the memory of the expected reward results in overactivation of the reward and motivation circuits while decreasing the activity in the cognitive control circuit. This contributes to an inability to inhibit the drive to seek and consume the drug and results in compulsive drug intake. This model has implications for therapy, for it suggests a multi-prong approach that targets strategies to decrease the rewarding properties of drugs, to enhance the rewarding properties of alternative reinforcers, to interfere with conditioned-learned associations, and to strengthen cognitive control in the treatment of drug addiction.

Introduction

Addiction is a disorder that involves complex interactions between biological and environmental variables (1). This has made treatment particularly elusive, since attempts to categorize addiction have usually concentrated on one level of analysis.

Nonstandard abbreviations used: positron emission tomography (PET); functional magnetic resonance imaging (fMRI); dopamine (DA); nucleus accumbens (NAc); orbitofrontal cortex (OFC); cingulate gyrus (CG).

Conflict of interest: The authors have declared that no conflict of interest exists.

Citation for this article: *J. Clin. Invest.* **111**:1444–1451 (2003). doi:10.1172/JCI200318533.

Attempts to understand and treat addiction as a purely biological or a purely environmental problem have not been very successful. Recently, important discoveries have increased our knowledge about how drugs of abuse affect biological factors such as genes, protein expression, and neuronal circuits (2, 3); however, much less is known about how these biological factors affect human behavior. Nor do we know much about how environmental factors affect these biological factors and how these in turn alter behavior. Relatively new imaging technologies such as positron emission tomography (PET) and functional magnetic resonance imaging (fMRI) have provided new ways to investigate how the biological factors integrate with one another, how they relate to behavior, and how biological and environmental variables interact in drug addiction (Figure 1).

PET imaging is based on the use of radiotracers labeled with short-lived positron-emitting isotopes (carbon-11, oxygen-15, nitrogen-13, and fluorine-18), which it can measure at very low concentrations (nanomolar to picomolar range) (4). Therefore, PET can be used to measure labeled compounds that selectively bind to specific receptors, transporters, or enzyme types at concentrations that do not perturb function (Figure 2). fMRI is based on the measurement of the changes in magnetic properties in neuronal tissue (4). It is generally believed that the activation signal generated from fMRI results from differences in the magnetic properties of oxygenated versus deoxygenated hemoglobin (blood oxygen level dependant contrast). During activation of a brain region there is an excess of arterial blood delivered into the area, with concomitant changes in the ratio of deoxyhemoglobin to oxyhemoglobin.

Most PET studies of drug addiction have concentrated on the brain dopamine (DA) system, since this is considered to be the neurotransmitter system through which most drugs of abuse exert their reinforcing effects (5). A reinforcer is operationally defined as an event that increases the probability of a subsequent response, and drugs of abuse are considered to be much stronger reinforcers than natural reinforcers (e.g., sex and food) (6). The brain DA system also regulates motivation and drive for everyday activities (7). These imaging studies have revealed that acute and chronic drug consumption have different effects on proteins involved in DA synaptic transmission (Figure 2). Whereas acute drug administration increases DA neurotransmission, chronic drug consumption results in a marked decrease in DA activity, which persists months after detoxification and which is associated with deregulation of frontal brain regions (8). PET and MRI studies have characterized the brain areas and circuits involved in various states of the drug addiction process (intoxication, withdrawal, and craving) and have linked the activity in these neural circuits to behavior (Figure 3). Acute drug intoxication results in a complex and dynamic pattern of activation and deactivation that includes regions neuroanatomically connected with the DA

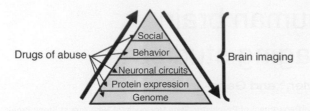

Figure 1
Drugs of abuse have effects at multiple biological and environmental levels. The environmental level is identified as "social," since this is the most relevant of the environmental factors that influence drug abuse in humans. Imaging techniques allow one to assess the effects of drugs of abuse at the protein and the brain circuit levels and to assess how these effects relate to behavior. Imaging also offers a way to start to assess the impact of environmental factors on these biological levels, as well as the impact of gene polymorphisms on protein expression and brain function.

system and known to be involved in reward, memory, motivation/drive, and control (9, 10). The same imaging methods have been used to demonstrate how environmental factors can influence these neuronal circuits, which in turn affect behavior related to drug addiction (e.g., drug consumption). For example, a recent study in nonhuman primates showed that social status affects DA D2 receptor expression in the brain, which in turn affects the propensity for cocaine self-administration (11) (Figure 4).

Here we analyze the results from our imaging program in drug addiction, and from the rich literature, and integrate this body of knowledge with preclinical findings to develop a model that could explain the loss of control and compulsive drug intake observed in the addicted individual.

Drug addiction involves multiple brain circuits

The aforementioned model proposes a network of four circuits involved in drug abuse and addiction: (a) reward, located in the nucleus accumbens (NAc) and the ventral pallidum; (b) motivation/drive, located in the orbitofrontal cortex (OFC) and the subcallosal cortex; (c) memory and learning, located in the amygdala and the hippocampus; and (d) control, located in the prefrontal cortex and the anterior cingulate gyrus (CG) (Figure 5). These four circuits receive direct innervations from DA neurons but are also connected with one another through direct or indirect projections (mostly glutamatergic). Though we have identified specific brain regions associated with each circuit, we have realized that other brain regions are involved in these circuits (e.g., the thalamus and insula), that one region may participate in more than one circuit (e.g., the CG in both control and motivation/drive circuits), and that other brain regions (e.g., the cerebellum) and circuits (e.g., attention and emotion circuits) are likely to be affected in drug addiction. Though our model focuses on DA, it is evident from preclinical studies that modifications in glutamatergic projections mediate many of the adaptations observed with addiction (12). Unfortunately, the lack of radiotracers available to image glutamate function in the human brain has precluded its investigation in drug-addicted subjects.

We propose that the pattern of activity in the four-circuit network outlined in Figure 5 influences how an individual makes choices among behavioral alternatives. These choices are influenced systematically by the reward, memory, motivation, and con-

trol circuits. The response to a stimulus is affected by its momentary saliency — i.e., expected reward, which is processed in part by DA neurons projecting into the NAc (13) — in a hierarchical matrix where the saliency value of stimuli changes as a function of the context and the previous experience of the individual. If the individual has been previously exposed to the stimulus, its saliency value is affected by memory, processed in part by the amygdala and hippocampus. Memories are stored as associations between the stimulus and the positive (pleasant) or negative (aversive) experience it elicited and are facilitated by DA activation (14). The value of the stimulus is weighted against that of other alternative stimuli and changes as a function of the internal needs of the individual, which are processed in part by the OFC (15, 16). For example, the saliency value of food is increased by hunger but decreased by satiety. The stronger the saliency value of the stimulus, which is in part conveyed by the prediction of reward from previously memorized experiences, the greater the activation of the motivational circuit and the stronger the drive to procure it. The cognitive decision to act (or not) to procure the stimulus is processed in part by the prefrontal cortex and the CG (17).

The model proposes that, in the addicted subject, the saliency value of the drug of abuse and its associated cues is enhanced in the reward and motivation/drive circuits but that of other reinforcers is markedly decreased. The enhanced saliency value of the drug of abuse is initiated partly by the much higher intrinsic reward properties of drugs of abuse: increases in DA induced by drugs in the NAc are three- to fivefold higher than those of natural reinforcers (7). Another cause of the enhanced saliency is the lack of habituation of drugs of abuse as compared with that of natural reinforcers (18). It is postulated that the high reward value of drugs leads to a resetting of reward thresholds, which

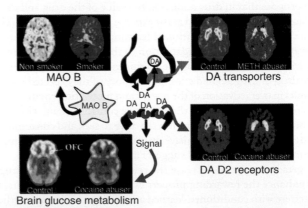

Figure 2
Images obtained with PET (axial sections) that show the effects of chronic drug exposure on various proteins involved in dopamine (DA) neurotransmission and on brain function (as assessed by brain glucose metabolism). While some effects are common to many drugs of abuse, such as decreases in DA D2 receptors in striatal neurons and decreased metabolic activity in the orbitofrontal cortex (OFC), others are more specific. These include the decrease in DA transporters in striatum in methamphetamine (METH) abusers (possibly the result of neurotoxicity to DA terminals) and the decrease in brain monoamine oxidase B (MAO B; the enzyme involved in DA metabolism) in cigarette smokers. The rainbow scale was used to code the PET images; radiotracer concentration is displayed from higher to lower as red > yellow > green > blue. Images from methamphetamine use are adapted from ref. 61. Images from smokers are adapted with permission from ref. 62.

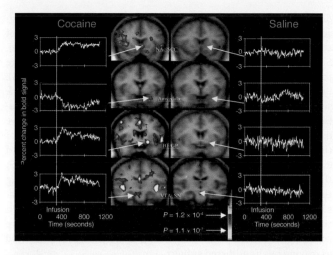

Figure 3
Images of coronal sections obtained with fMRI, showing areas of brain activation and deactivation during cocaine intoxication compared with those after saline administration. During intoxication there is a complex pattern of activation and/or deactivation that includes the ventral tegmental area (VTA) and the substantia nigra (SN), where DA cells are located, as well as regions involved with reward (nucleus accumbens, NAc; basal forebrain, BF; globus pallidus, GP), with memory (amygdala), and with motivation (subcallosal cortex, SCC). The color scale indicates the level of significance (*P* value) of the change in activation of the bold signal. Reproduced with permission from *Neuron* (9).

then results in decreased sensitivity to the reinforcing properties of naturally occurring stimuli (19). Through conditioned learning and a lack of competition by other reinforcers, acquisition of the drug becomes the main motivational drive for the individual. We hypothesize that, during intoxication, the qualitative difference in activity in the DA-regulated reward circuit (greater and longer-lasting activation compared with the activation by nondrug stimuli) (18) produces a corresponding overactivation of the motivational/drive and memory circuits, which deactivate and remove the control exerted by the frontal cortex. Without the inhibitory control, a positive-feedback loop is set forth that results in compulsive drug intake (Figure 5). Because the interactions between the circuits are bidirectional, the activation of the network during intoxication serves to further strengthen the saliency value of the drug.

Reward circuit in drug addiction

The reinforcing effects of drugs during intoxication create an environment that, if perpetuated, triggers the neuronal adaptations that result in addiction. Imaging studies in drug abusers as well as non–drug abusers have shown that drugs of abuse increase the extracellular concentration of DA in the striatum (where the NAc is located) and that these increases were associated with their reinforcing effects. The subjects who had the greatest increases in DA were the ones who experienced drug effects such as "high," "rush," or "euphoria" most intensely (20–22). These studies also showed that the reinforcing effects appeared to be associated not only with the magnitude but also with the abruptness of the DA increase. Thus, for an equivalent increase in DA, the drug was experienced as reinforcing when it was injected intravenously (21), which leads to fast drug uptake in the brain and presumably very fast changes in DA concentration, but not when it was given orally (23), which

leads to a slow rate of brain uptake and presumably slow increases in DA concentration. The dependency of the reinforcing effects of drugs on fast and large increases in DA concentration is reminiscent of the changes in DA concentration induced by phasic DA cell firing (fast-burst firing > 30 Hz) (6), which also leads to fast changes in DA concentration and whose function is to highlight the saliency of stimuli (24). This contrasts with tonic DA cell firing (slow firing at frequencies around 5 Hz) (6), which maintains baseline steady-state DA levels and whose function is to set the overall responsiveness of the DA system. This led us to speculate that the ability of drugs of abuse to induce changes in DA concentration that mimic but exceed those produced by phasic DA cell firing results in overactivation of the neuronal processes that highlight saliency, and that this is one of the relevant variables underlying their high reinforcing value.

However, studies show that increases in DA concentration during intoxication occur in both addicted and nonaddicted subjects, so this by itself cannot explain the process of addiction. Since drug addiction requires chronic drug administration, we suggest that addiction results from the repeated perturbation of reward circuits (marked DA increases followed by DA decreases) and the consequent disruption of the circuits that it regulates (motivation/drive, memory/learning, and control). Indeed, imaging studies in drug-addicted subjects have consistently shown long-lasting decreases in the numbers of DA D2 receptors in drug abusers compared with controls (Figure 2) (reviewed in ref. 8). In addition, studies have shown that cocaine abusers also have decreased DA cell activity, as evidenced by reduced DA release in response to a pharmacological challenge with a stimulant drug (25). We postulate that the decrease in the number of DA D2 receptors, coupled with the decrease in DA cell activity, in the drug abusers would result in a decreased sensitivity of reward circuits to stimulation by natural reinforcers. This decreased sensitivity would lead to decreased

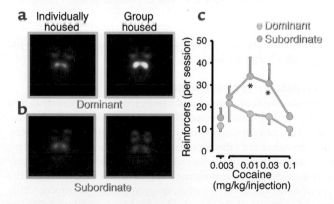

Figure 4
Images of axial sections obtained with PET, showing DA D2 receptors in nonhuman primates that were initially tested while housed in separate cages and then retested after being housed in a group. Animals that became dominant when placed in a group (**a**) showed increased numbers of DA D2 receptors in striatum, whereas subordinate animals (**b**) did not. (**c**) The levels of cocaine administration in the subordinate and the dominant animals. Note the much lower intake of cocaine by dominant animals which possessed higher numbers of DA D2 receptors. The temperature scale was used to code the PET images; radiotracer concentration is displayed from higher to lower as yellow > red. Asterisks indicate significant differences in drug intake between groups. Adapted with permission from ref. 11.

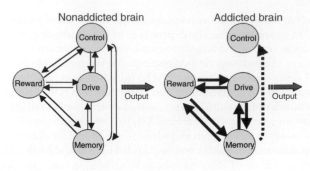

Figure 5

Model proposing a network of four circuits involved with addiction: reward, motivation/drive, memory, and control. These circuits work together and change with experience. Each is linked to an important concept: saliency (reward), internal state (motivation/drive), learned associations (memory), and conflict resolution (control). During addiction, the enhanced value of the drug in the reward, motivation, and memory circuits overcomes the inhibitory control exerted by the prefrontal cortex, thereby favoring a positive-feedback loop initiated by the consumption of the drug and perpetuated by the enhanced activation of the motivation/drive and memory circuits.

interest in ordinary (day-to-day) environmental stimuli, possibly predisposing subjects for seeking drug stimulation as a means to temporarily activate these reward circuits. Imaging studies provide evidence of disrupted sensitivity to natural reinforcers in addiction. For example, in a study by Martin-Solch and colleagues (25), the meso-striatal and meso-corticolimbic circuits of opiate addicts were not activated in response to natural reinforcers, whereas they were in controls subjects. Similarly, in a second study by the same group, DA-regulated reward centers in tobacco smokers failed to activate in response to monetary reward (26). Interestingly, decreased sensitivity of reward circuits to acute alcohol administration has also been documented in cocaine abusers compared with control subjects (27). These findings suggest an overall reduction in the sensitivity of reward circuits in drug-addicted individuals to natural reinforcers, but also possibly to drugs besides the one to which they are addicted.

Motivation/drive circuit in addiction

We postulate that, during addiction, the value of the drug as a reinforcer is so much greater than that of any natural reinforcer that these can no longer compete as viable alternative choices, and the enhanced saliency value of the drug becomes fixed. This contrasts with natural reinforcers, whose saliency is momentary and decreases with exposure to the reinforcer (18) or with the presentation of an alternative, more appealing reinforcer. One area of the brain that is involved in shifting the relative value of reinforcers is the OFC (15, 16).

Imaging studies have provided evidence of disruption of the OFC during addiction (reviewed in ref. 28) (Figure 2). The OFC appears to be hypoactive in drug-addicted subjects tested during withdrawal (29, 30); we postulate that this results from the lack of stimulation by salient stimuli during detoxification. In contrast, in active cocaine abusers, the OFC has been shown to be hypermetabolic in proportion to the intensity of the craving experienced by the subjects (31). We therefore postulated that exposure to the drug or drug-related stimuli in the withdrawal state reactivates the OFC and results in compulsive drug intake. Indeed, activation

of the OFC has been reported during drug intoxication in drug-addicted, but not in non–drug-addicted, subjects, and the level of activation predicted the intensity of drug-induced craving (32, 33). Similarly, activation of the OFC has been reported during exposure to drug-related cues when these elicit craving (reviewed in ref. 28). Since increased OFC activation has been associated with compulsive disorders (reviewed in ref. 34), we postulated that the activation of the OFC in addicted subjects contributes to the compulsive drug intake. Indeed, preclinical studies have shown that damage of the OFC results in a behavioral compulsion to procure the reward even when it is no longer reinforcing (16). This is consistent with the accounts of drug addicts who claim that once they start taking the drug they cannot stop, even when the drug is no longer pleasurable. Since the OFC also processes information associated with the prediction of reward (15), its activation during cue exposure could signal reward prediction, which could then be experienced as craving by the addicted subject.

In detoxified drug abusers, the decreased activity in the OFC is associated with reductions in the numbers of DA D2 receptors in striatum (35, 36). Since DA D2 receptors transmit reward signals into the OFC, this association could be interpreted as a disruption of the OFC, secondary to changes in striatal DA activity (such as lack of stimulation during withdrawal and enhanced stimulation with exposure to drugs or drug-related cues). However, since striatal-frontal connections are bidirectional, this association could also reflect the disruption of the OFC, which then deregulates DA cell activity.

Learning/memory circuit in addiction

The relevance of learning and memory to addiction is made evident by the pernicious effect that a place, a person, or a cue that brings back memories of the drug can have on the addict who is trying to stay clean. These factors trigger an intense desire for the drug (a craving) and, not infrequently, relapse. Multiple memory systems have been proposed in drug addiction, including conditioned-incentive learning (mediated in part by the NAc and the amygdala), habit learning (mediated in part by the caudate and the putamen), and declarative memory (mediated in part by the hippocampus) (reviewed in ref. 37). Through conditioned-incen-

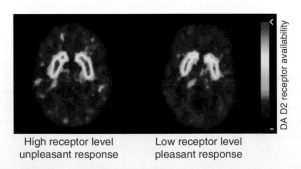

High receptor level   Low receptor level
unpleasant response   pleasant response

Figure 6

Images of axial sections obtained with PET to measure the numbers of DA D2 receptors in subjects who reported the effects of the stimulant drug methylphenidate as pleasant versus those that reported its effects as unpleasant. Subjects with high numbers of DA D2 receptors tended to report the effects of methylphenidate as unpleasant, whereas subjects with low numbers of DA D2 receptors tended to report it as pleasant. The rainbow scale was used to code the PET images; radiotracer concentration is displayed from higher to lower as red > yellow > green > blue. Adapted with permission from ref. 53.

tive learning, the neutral stimuli, coupled with the drug of abuse, acquire reinforcing properties and motivational salience even in the absence of the drug. Through habit learning, well-learned sequences of behavior are elicited automatically by the appropriate stimuli. Finally, declarative memory is related to the learning of affective states in relationship to drug intake.

Memory circuits are likely to influence the effects of the drug during intoxication, since they set the expectations of the drug's effects in the addicted subject (38). Activation of regions linked with memory has been reported during drug intoxication (9, 10) and during craving induced by drug exposure, video, or recall (39–42). Also, studies in drug abusers during withdrawal have shown evidence of decreased D2 receptor expression and decreased DA release in the dorsal striatum (25). In animal studies, the drug-induced changes in the dorsal striatum are observed after longer drug exposures than those observed in the NAc and have been interpreted to reflect further progression into the addicted state (43). This is relevant because involvement of the dorsal striatum, which is a region associated with habit learning, indicates that in drug addiction the routine associated with drug consumption may be triggered automatically by exposure to the drug or drug-related cues (44).

Control circuit in addiction

One of the most consistent findings from imaging studies is that of abnormalities in the prefrontal cortex, including the anterior CG, in drug-addicted subjects (reviewed in ref. 45). The prefrontal cortex is involved in decision making and in inhibitory control (reviewed in ref. 46). Thus its disruption could lead to inadequate decisions that favor immediate rewards over delayed but more favorable responses. It could also account for the impaired control over the intake of the drug even when the addicted subject expresses the desire to refrain from taking the drug (45). Thus, one might expect that the disruptions of self-monitoring and decision-making processes that are observed in drug-addicted subjects (47, 48) are in part related to disrupted prefrontal functions. Moreover, preclinical studies show that chronic administration of cocaine or amphetamine results in a significant increase in dendritic branching and the density of dendritic spines in the prefrontal cortex (49). These changes in synaptic connectivity could be involved in the changes in decision making, judgment, and cognitive control that occur during addiction. Indeed, imaging studies have shown evidence of changes in prefrontal activation during a working-memory task in smokers compared with ex-smokers (50).

We propose that disruption of the prefrontal cortex could lead to loss of self-directed/willed behavior in favor of automatic sensory-driven behavior (45). Moreover, the disruption of self-controlled behavior is likely to be exacerbated during drug intoxication from the loss of inhibitory control that the prefrontal cortex exerts over the amygdala (51). The inhibition of top-down control would release behaviors that are normally kept under close monitoring and would simulate stress-like reactions in which control is inhibited and stimulus-driven behavior is facilitated (45).

Vulnerability to drug addiction

A challenging problem in the neurobiology of drug addiction is to understand why some individuals become addicted to drugs while others do not. The model we propose offers some guidance as to specific disruptions that could make a subject more or less vulnerable to addiction. For example, one could hypothesize that decreased sensitivity of reward circuits to natural reinforcers, decreased

activity of control circuits, or an increased sensitivity of memory/learning or motivation/drive circuits to drug or drug-related stimuli could make an individual more vulnerable to addiction.

In fact, imaging studies have provided evidence that differences in reward circuits may be one of the mechanisms underlying the variability in responsiveness to drugs of abuse, which in turn could influence vulnerability. These studies assessed the extent to which the variability in the number of DA D2 receptors in non–drug-abusing subjects affected their sensitivity to stimulant drugs (52). The data showed that subjects with low numbers of DA D2 receptors tended to describe the effects of the stimulant drug methylphenidate as pleasant, whereas subjects with high numbers of DA D2 receptors tended to describe it as unpleasant (Figure 6). Another study documented that the numbers of DA D2 receptors predicted how much subjects liked the effects of methylphenidate (53). These findings suggest that one of the mechanisms underlying the differences between subjects in their vulnerability to stimulant abuse may be the variability in the expression of DA D2 receptors. Subjects with low numbers of D2 receptors may be at higher risk of abusing stimulant drugs than those with high numbers of D2 receptors, in whom drugs such as methylphenidate may produce unpleasant effects that limit its abuse. A causal association between DA D2 receptor numbers and propensity to self-administer drugs was corroborated by a parallel preclinical study that showed that insertion of the DA D2 receptor gene via a viral vector to increase DA D2 receptor expression in the NAc of rats previously trained to self-administer alcohol resulted in marked reductions in alcohol intake (54). Alcohol intake recovered as the number of DA D2 receptors returned to baseline levels. These results could be taken as indirect evidence of a protective role of high DA D2 receptor numbers against drug abuse. Baseline levels of DA D2 receptors in the brain, which have been shown to be affected by stress (55) and social hierarchy (11), provide a molecular mechanism that could explain the influence of the environment and genetics on predisposition to drug abuse.

Recently, imaging studies showed that offspring of alcoholic families who were considered to be at high risk for alcoholism showed smaller amygdala volumes in comparison with control subjects (56). Moreover, the volume of the amygdala was associated with the amplitude of the P300 in the evoked potential (wave occurring between 300 and 500 ms after a rare target stimulus), which is considered to be a phenotypic marker for vulnerability to alcoholism. Also, a recent imaging study reported structural changes in the OFC of cocaine-addicted subjects (57), and the possibility was discussed that this might have preceded drug use and might have made these subjects more vulnerable to addiction.

Access to transgenic and knockout animals now provides a means to directly evaluate the role that specific genes may play in vulnerability to, or protection against, drug abuse and addiction (58). Thus, information from imaging studies regarding abnormalities in specific proteins in the brains of drug-addicted subjects (e.g., DA D2 receptors and monoamine oxidase B) can now be tested in preclinical models to determine whether these abnormalities reflect changes that preceded drug use and are genetically determined, or whether they are a consequence of chronic drug use.

Conclusion

Here we provide a model that conceptualizes addiction as a state initiated by the qualitatively different and larger reward value of the drug, which triggers a series of adaptations in the reward, motivation/drive, memory, and control circuits of the brain. These changes result in an enhanced and permanent saliency value for the drug, and in the loss of inhibitory control, favoring the emergence of compulsive drug administration. The model has treatment implications, for it suggests strategies to combat drug addiction — specifically (a) interventions to decrease the rewarding value of drugs, such as pharmacological treatments that interfere with the drug's reinforcing effects as well as treatments that make the effects unpleasant; (b) interventions to increase the value of nondrug reinforcers, such as pharmacological and behavioral treatments that increase sensitivity to natural reinforcers and establish alternative reinforcing behaviors; (c) interventions to weaken learned drug responses, such as behavioral treatments to extinguish the learned positive associations with the drug and drug cues but also promote differential reinforcement of other behaviors; and (d) interventions to strengthen frontal control, such as cognitive therapy. The model also highlights the need for therapeutic approaches that include pharmacological as well as behavioral interventions in the treatment of drug addiction (59).

This analysis brings to light the paucity of PET radiotracers currently available for use in imaging of the human brain. Further research on the development of radiotracers that can be used to target other neurotransmitter systems affected by drugs of abuse (e.g., glutamate and γ-aminobutyric acid) will in the future provide a more complete picture of the neurochemical changes that underlie drug addiction. Moreover, access to a wider array of radiotracers will enable researchers to start to investigate the role that gene polymorphisms may play in protein expression, and how this in turn relates to behavioral responses to drugs of abuse (60).

Acknowledgments

The authors are indebted to the Department of Energy (Office of Biological and Environmental Research; DE-ACO2-98CH10886), the National Institute on Drug Abuse (DA-06278, DA-09490, and DA-06891), the National Institute on Alcohol Abuse and Alcoholism (AA/OD-09481), and the Office of National Drug Control Policy for support of our research. We are also indebted to our scientific and technical colleagues and our research volunteers, without whom our efforts on drug addiction would not have been able to proceed.

Address correspondence to: Nora D. Volkow, Department of Medicine, Brookhaven National Laboratory, Upton, New York 11973, USA. Phone: (631) 344-3335; Fax: (631) 344-5260; E-mail: volkow@bnl.gov.

1. Leshner, A.I. 1997. Addiction is a brain disease, and it matters. *Science.* **278**:45–47.
2. Nestler, E.J. 2001. Molecular basis of long-term plasticity underlying addiction. *Nat. Rev. Neurosci.* **2**:119–128.
3. Hyman, S.E., and Malenka, R.C. 2001. Addiction and the brain: the neurobiology of compulsion and its persistence. *Nat. Rev. Neurosci.* **2**:695–703.
4. Volkow, N.D., Rosen, B., and Farde, L. 1997. Imaging the living human brain: magnetic resonance imaging and positron emission tomography. *Proc. Natl. Acad. Sci. U. S. A.* **94**:2787–2788.
5. Koob, G.F., and Bloom, F.E. 1988. Cellular and molecular mechanism of drug dependence. *Science.* **242**:715–723.
6. Wightman, R.M., and Robinson, D.L. 2002. Transient changes in mesolimbic dopamine and their association with 'reward'. *J. Neurochem.* **82**:721–735.
7. Wise, R.A. 2002. Brain reward circuitry: insights from unsensed incentives. *Neuron.* **36**:229–240.
8. Volkow, N.D., Fowler, J.S., and Wang, G.J. 2002. Role of dopamine in drug reinforcement and addiction in humans: results from imaging studies. *Behav. Pharmacol.* **13**:355–366.
9. Breiter, H.C., et al. 1997. Acute effects of cocaine on human brain activity and emotion. *Neuron.* **19**:591–611.
10. Stein, E.A., et al. 1998. Nicotine-induced limbic cortical activation in the human brain: a functional MRI study. *Am. J. Psychiatry.* **155**:1009–1015.
11. Morgan, D., et al. 2002. Social dominance in monkeys: dopamine D2 receptors and cocaine self-administration. *Nat. Neurosci.* **5**:169–174.
12. Cornish, J.L., and Kalivas, P.W. 2001. Cocaine sensitization and craving: differing roles for dopamine and glutamate in the nucleus accumbens. *J. Addict. Dis.* **20**:43–54.
13. Dehaene, S., and Changeux, J.P. 2000. Reward-dependent learning in neuronal networks for planning and decision making. *Prog. Brain Res.* **126**:217–229.
14. Di Chiara, G. 1999. Drug addiction as dopamine-dependent associative learning disorder. *Eur. J. Pharmacol.* **375**:13–30.
15. Schultz, W., Tremblay, L., and Hollerman, J.R. 2000. Reward processing in primate orbitofrontal cortex and basal ganglia. *Cereb. Cortex.* **10**:272–284.

16. Rolls, E.T. 2000. The orbitofrontal cortex and reward. *Cereb. Cortex.* **10**:284–294.
17. Miller, E.K., and Cohen, J.D. 2001. An integrative theory of prefrontal cortex function. *Annu. Rev. Neurosci.* **24**:167–202.
18. Di Chiara, G. 2002. Nucleus accumbens shell and core dopamine: differential role in behavior and addiction. *Behav. Brain Res.* **137**:75–114.
19. Koob, G.F., and Le Moal, M. 2001. Drug addiction, dysregulation of reward, and allostasis. *Neuropsychopharmacology.* **24**:97–129.
20. Laruelle, M., et al. 1995. SPECT imaging of striatal dopamine release after amphetamine challenge. *J. Nucl. Med.* **36**:1182–1190.
21. Volkow, N.D., et al. 1999. Reinforcing effects of psychostimulants in humans are associated with increases in brain dopamine and occupancy of D(2) receptors. *J. Pharmacol. Exp. Ther.* **291**:409–415.
22. Drevets, W.C., et al. 2001. Amphetamine-induced dopamine release in human ventral striatum correlates with euphoria. *Biol. Psychiatry.* **49**:81–96.
23. Volkow, N.D., et al. 2001. Therapeutic doses of oral methylphenidate significantly increase extracellular dopamine in the human brain. *J. Neurosci.* **21**:RC121.
24. Grace, A.A. 2000. The tonic/phasic model of dopamine system regulation and its implications for understanding alcohol and psychostimulant craving. *Addiction.* **95**:S119–S128.
25. Volkow, N.D., et al. 1997. Decreased striatal dopaminergic responsivity in detoxified cocaine abusers. *Nature.* **386**:830–833.
26. Martin-Solch, C., et al. 2001. Changes in brain activation associated with reward processing in smokers and nonsmokers. A positron emission tomography study. *Exp. Brain Res.* **139**:278–286.
27. Volkow, N.D., et al. 2000. Cocaine abusers show a blunted response to alcohol intoxication in limbic brain regions. *Life Sci.* **66**:PL161–PL167.
28. Volkow, N.D., and Fowler, J.S. 2000. Addiction, a disease of compulsion and drive: involvement of the orbitofrontal cortex. *Cereb. Cortex.* **10**:318–325.
29. Volkow, N.D., et al. 1992. Long-term frontal brain metabolic changes in cocaine abusers. *Synapse.* **11**:184–190.
30. Adinoff, B., et al. 2001. Limbic responsiveness to procaine in cocaine-addicted subjects. *Am. J. Psy-

chiatry.* **158**:390–398.
31. Volkow, N.D., et al. 1991. Changes in brain glucose metabolism in cocaine dependence and withdrawal. *Am. J. Psychiatry.* **148**:621–626.
32. Volkow, N.D., et al. 1999. Association of methylphenidate-induced craving with changes in right striato-orbitofrontal metabolism in cocaine abusers: implications in addiction. *Am. J. Psychiatry.* **156**:19–26.
33. Brody, A.L. 2002. Brain metabolic changes during cigarette craving. *Arch. Gen. Psychiatry.* **59**:1162–1172.
34. Insel, T.R. 1992. Towards a neuroanatomy of obsessive-compulsive disorder. *Arch. Gen. Psychiatry.* **49**:739–744.
35. Volkow, N.D., et al. 1993. Decreased dopamine D2 receptor availability is associated with reduced frontal metabolism in cocaine abusers. *Synapse.* **14**:169–177.
36. Volkow, N.D., et al. 2001. Low level of brain dopamine D(2) receptors in methamphetamine abusers: association with metabolism in the orbitofrontal cortex. *Am. J. Psychiatry.* **158**:2015–2021.
37. White, N.M. 1996. Addictive drugs as reinforcers: multiple partial actions on memory systems. *Addiction.* **91**:921–949.
38. Kirk, J.M., Doty, P., and De Wit, H. 1998. Effects of expectancies on subjective responses to oral delta9-tetrahydrocannabinol. *Pharmacol. Biochem. Behav.* **59**:287–293.
39. Grant, S., et al. 1996. Activation of memory circuits during cue-elicited cocaine craving. *Proc. Natl. Acad. Sci. U. S. A.* **93**:12040–12045.
40. Childress, A.R., et al. 1999. Limbic activation during cue-induced cocaine craving. *Am. J. Psychiatry.* **156**:11–18.
41. Kilts, C.D., et al. 2001. Neural activity related to drug craving in cocaine addiction. *Arch. Gen. Psychiatry.* **58**:334–341.
42. Wang, G.-J., et al. 1999. Regional brain metabolic activation during craving elicited by recall of previous drug experiences. *Life Sci.* **64**:775–784.
43. Letchworth, S.R., Nader, M.A., Smith, H.R., Friedman, D.P., and Porrino, L.J. 2001. Progression of changes in dopamine transporter binding site density as a result of cocaine self-administration in rhesus monkeys. *J. Neurosci.* **21**:2799–2807.
44. Ito, R., Dalley, J.W., Robbins, T.W., and Everitt, B.J.

2002. Dopamine release in the dorsal striatum during cocaine-seeking behavior under the control of a drug-associated cue. *J. Neurosci.* **22**:6247–6253.

45. Goldstein, R.Z., and Volkow, N.D. 2002. Drug addiction and its underlying neurobiological basis: neuroimaging evidence for the involvement of the frontal cortex. *Am. J. Psychiatry.* **159**:1642–1652.

46. Royall, D.R., et al. 2002. Executive control function: a review of its promise and challenges for clinical research. A report from the Committee on Research of the American Neuropsychiatric Association. *J. Neuropsychiatry Clin. Neurosci.* **14**:377–405.

47. Bechara, A., and Damasio, H. 2002. Decision-making and addiction (part I): impaired activation of somatic states in substance dependent individuals when pondering decisions with negative future consequences. *Neuropsychologia.* **40**:1675–1689.

48. Ernst, M., et al. 2003. Decision making in adolescents with behavior disorders and adults with substance abuse. *Am. J. Psychiatry.* **160**:33–40.

49. Robinson, T.E., Gorny, G., Mitton, E., and Kolb, B. 2001. Cocaine self-administration alters the morphology of dendrites and dendritic spines in the nucleus accumbens and neocortex. *Synapse.* **39**:257–266.

50. Ernst, M., et al. 2001. Effect of nicotine on brain activation during performance of a working memory task. *Proc. Natl. Acad. Sci. U. S. A.* **98**:4728–4733.

51. Rosenkranz, J.A., and Grace, A.A. 2001. Dopamine attenuates prefrontal cortical suppression of sensory inputs to the basolateral amygdala of rats. *J. Neurosci.* **21**:4090–4103.

52. Volkow, N.D., et al. 1999. Prediction of reinforcing responses to psychostimulants in humans by brain dopamine D2 receptor levels. *Am. J. Psychiatry.* **156**:1440–1443.

53. Volkow, N.D., et al. 2002. Brain DA D2 receptors predict reinforcing effects of stimulants in humans: replication study. *Synapse.* **46**:79–82.

54. Thanos, P.K., et al. 2001. Overexpression of dopamine D2 receptors reduces alcohol self-administration. *J. Neurochem.* **78**:1094–1103.

55. Papp, M., Klimek, V., and Willner, P. 1994. Parallel changes in dopamine D2 receptor binding in limbic forebrain associated with chronic mild stress-induced anhedonia and its reversal by imipramine. *Psychopharmacology.* **115**:441–446.

56. Hill, S.Y., et al. 2001. Right amygdala volume in adolescent and young adult offspring from families at high risk for developing alcoholism. *Biol. Psychiatry.* **49**:894–905.

57. Franklin, T.R., et al. 2002. Decreased gray matter concentration in the insular, orbitofrontal, cingulate, and temporal cortices of cocaine patients. *Biol. Psychiatry.* **51**:134–142.

58. Sora, I., et al. 2001. Molecular mechanisms of cocaine reward: combined dopamine and serotonin transporter knockouts eliminate cocaine place preference. *Proc. Natl. Acad. Sci. U. S. A.* **98**:5300–5305.

59. Kreek, M.J., LaForge, K.S., and Butelman, E. 2002. Pharmacotherapy of addictions. *Nat. Rev. Drug Discov.* **1**:710–726.

60. Miller, G.M., Yatin, S.M., De La Garza, R., II, Goulet, M., and Madras, B.K. 2001. Cloning of dopamine, norepinephrine and serotonin transporters from monkey brain: relevance to cocaine sensitivity. *Brain Res. Mol. Brain Res.* **87**:124–143.

61. Volkow, N.D., et al. 2001. Dopamine transporter losses in methamphetamine abusers are associated with psychomotor impairment. *Am. J. Psychiatry.* **158**:377–382.

62. Fowler, J.S., et al. 1996. Inhibition of monoamine oxidase B in the brains of smokers. *Nature.* **379**:733–738.

Part 14
Genetics and Disease

Mapping the new frontier: complex genetic disorders

Richard Mayeux

Gertrude H. Sergievsky Center and Taub Institute for Research on Alzheimer's Disease and the Aging Brain, Columbia University, New York, New York, USA.

The remarkable achievements in human genetics over the years have been due to technological advances in gene mapping and in statistical methods that relate genetic variants to disease. Nearly every Mendelian genetic disorder has now been mapped to a specific gene or set of genes, but these discoveries have been limited to high-risk, variant alleles that segregate in rare families. With a working draft of the human genome now in hand, the availability of high-throughput genotyping, a plethora of genetic markers, and the development of new analytical methods, scientists are now turning their attention to common complex disorders such as diabetes, obesity, hypertension, and Alzheimer disease. In this issue, the *JCI* provides readers with a series dedicated to complex genetic disorders, offering a view of genetic medicine in the 21st century.

Unraveling complex genetic disorders

The identification of genes underlying Mendelian disorders, named after Gregor Mendel and defined by the occurrence of a disorder in fixed proportions among the offspring of specific matings, has been greatly enhanced over the last few decades by remarkable achievements in gene mapping and the development of rigorous statistical methods. Most of the progress in human genetics during this time has come from the studies of families with rare segregating high-risk alleles. With at least 30,000 genes in the human genome and the identification and characterization of these genes underway, the challenge now is to dissect common complex genetic disorders such as obesity, diabetes, schizophrenia, and cancer. As a group, the majority of these disorders have a tendency to aggregate in families but rarely in the classical Mendelian fashion. While researchers have made some progress in the genetics of complex disorders over the last decade, gaps clearly remain. It is likely that, with characterization of the genetic influences underlying these complex disorders, there will be even greater opportunities for improving the lives of affected individuals.

The ability to genetically map complex disorders has been facilitated by technological improvement in identifying and genotyping polymorphic DNA markers (Table 1). The current trend is to use single-nucleotide polymorphisms (SNPs), the most frequently seen type of genetic polymorphisms, with an estimated 3 million SNPs present in the human genome. Though somewhat less informative than other types of DNA markers, SNPs are technically easier and less expensive to genotype because they have only 2 alleles and require less DNA. For example, a set of 2 arrays can genotype more than 100,000 SNPs with a single primer.

Most researchers believe that complex disorders are oligogenic, the cumulative result of variants in several genes, or polygenic, resulting from a large number of genetic variants, each contributing small effects. Still others have proposed that these disorders result from an interaction between one or more genetic variants and environmental or nongenetic disease risk factors. The motivation for unraveling these complex genetic disorders is clear.

Not only will this shed new light on disease pathogenesis, but it may also provide potential targets for effective treatment, screening, and prevention and increase the understanding of why some patients do not respond to currently available treatments while others do. The difficulty facing researchers who work on these complex genetic disorders is in designing appropriate studies to merge the richness of modern genome science with the vast potential of population-based, epidemiological research.

In this issue, a series of reviews describes the current state of the art in methods for gene mapping of complex disorders, including statistical methods for association studies and linkage disequilibrium mapping. We also include reviews that offer examples of the application of these methods in 3 complex genetic disorders: diabetes (see Permutt et al.; ref. 1), schizophrenia (see Kirov et al.; ref. 2), and neurodegeneration (see Bertram and Tanzi; ref. 3). Clearly, this is an exciting and rapidly evolving area of science in which the elucidation of the human genome can now be applied to common complex genetic problems. However, it is also worth briefly reviewing the traditional application of methods to assess genetic contributions to human disorders.

Estimating the heritability of a disorder

A higher concordance of disease among monozygotic compared with dizygotic twins or a higher risk among relatives (e.g., siblings) of patients with disease than among relatives of controls or those in the general population are usually the observations that lead researchers to believe that a disease is familial or at least possibly under genetic influence. However, there are statistical methods available to determine the degree to which a disease or trait is heritable. These estimates reflect the proportion of genetic variance over the total phenotypic variance from members within the family (4–6). The residual variance is the proportion reflecting environmental or nongenetic risk factors. Studies of heritability provide an estimate of the degree to which the variability in the phenotype is related to genetic variation, but it is difficult to separate shared genetic from shared environmental influences. Siblings, especially twins, share their childhood environment in addition to some portion of their genetic background. Genetic epidemiologists view heritability estimates as approximations of the genetic variance in disease risk because heritability depends on all contributing genetic and environmental or nongenetic compo-

---

Nonstandard abbreviations used: SNP, single-nucleotide polymorphism.

Conflict of interest: The author has declared that no conflict of interest exists.

Citation for this article: *J. Clin. Invest.* **115**:1404–1407 (2005). doi:10.1172/JCI25421.

Table 1

DNA markers for gene mapping

DNA markers	Type of genetic marker	Advantages	Disadvantages	Use
Restriction fragment length polymorphisms	Restriction enzymes at specific sites	Two alleles; site is present or absent	Limited information; expensive; time consuming	Linkage; analysis
Variable number of tandem repeats (VNTR)	Minisatellites	Many alleles; highly informative	Required Southern blotting and radioactive probes; not evenly spread throughout genome	Linkage; analysis
VNTR	Microsatellites	Many alleles; highly informative; widely distributed	Requires polymerase chain reaction (PCR); expensive	Linkage; analysis
Single-nucleotide polymorphisms	SNPs	Two alleles; widely distributed; typing can be automated	Less informative than microsatellites	Linkage and association analyses

nents. As described by Kirov et al. in this series, a high heritability score does not always mean that gene mapping will be easy (2). A change in any one factor can influence the overall estimate. Heritability estimates do not effectively separate shared genetic from shared environmental influences and cannot effectively apportion the degree of gene-environment interaction. This is most certainly true in studies of diabetes (see Permutt et al.; ref. 1).

Segregation analysis is a statistical tool that can model the inheritance pattern. It is useful in the analysis of non-Mendelian or complex genetic disorders that may be polygenic or the result of gene-environment interaction (4). Segregation analysis estimates the appropriate mix of genetic and environmental factors using information from a series of families identified by the researcher. Certain assumptions regarding gene mechanism, the frequency of the variant form of the gene, and its suspected penetrance are provide by the researcher who must also specify the model of inheritance: sporadic, polygenic, dominant, or recessive. A maximum likelihood analysis, the probability of obtaining the observed results given the distribution of data in the population, provides results that reflect the mix of parameters that best fit the observed data compared with a general or mixed model. Segregation analysis estimates genetic contributions by aggregating a set of genes, but it is not specific to a single gene, and the types of families recruited can affect results. For example, very large families will contribute more toward the specified model than smaller ones. Nonetheless, this approach informs the investigator regarding the degree to which the disease is genetic and can also provide some of the parameters of inheritance. For simpler diseases, these genetic parameters can be used in subsequent genetic analyses, such as linkage analysis, to provide greater power in identifying the variant gene or genes.

Identifying disease genes

For the investigation of many inherited Mendelian diseases, researchers have used linkage analysis in families with several affected family members to identify putative involved genes. Linkage analysis attempts to identify a region (locus) of the chromosome or regions (loci) in the genome associated with the disease or trait by identifying which alleles in the loci are segregating with the disease in families. Geneticists use genetic markers that are evenly distributed throughout the genome to reduce the number of chromosomal regions to a handful that may harbor a disease gene. Simply put, this method exploits the biological reality that in meiosis I, genes located close to each other on the same chromosome are inherited together more often than expected by chance. The genes that are far apart will not inherit together because recombination will break up segments of the chromosome. Thus, if a set of marker alleles are segregating with the disease, those markers are assumed to be located near the disease gene (7, 8). Using linkage analysis, scientists determine the likelihood that the loci (genetic marker and disease gene) are linked by calculating the logarithm of the odds or lod score, which is a ratio of 2 likelihoods: the odds that the loci are linked and the odds that the loci are not linked or are independent. To take into account multiple testing and the likelihood of linkage prior to considering the genetic evidence, a lod score of 3 or more is used as an indication of statistically significant linkage with a 5% chance of error, though more stringent criteria have been recommended for genome-wide scans (9). Two-point (ratio of the likelihoods that 2 loci are linked) and multipoint linkage (ratio of likelihoods at each location across the genome) analyses are standard analyses used in gene mapping. Once a location or set of locations suggestive of linkage are identified, researchers turn to finer mapping methods using either a more dense set of additional microsatellites or SNPs in a smaller region underlying the high lod score.

While linkage analysis remains a mainstay of gene mapping, it does have shortcomings. Both genotyping and phenotyping errors have devastating effects on the validity of the lod score. Locus heterogeneity (more than one causal gene) and clinical heterogeneity (multiple forms of the same disease with different etiology) can also pose serious problems. A pattern of inheritance or model must be assumed, and the researcher must estimate the frequency and penetrance of the disease gene. Therefore, the analysis is parametric. In late onset diseases, additional complications can arise when individuals with putative variant allele develop the disease later in life or in a much milder form (incomplete penetrance). Therefore, linkage analysis is best suited for Mendelian disorders, not common complex genetic disorders, unless the correlation between the genotype and the phenotype is known to be very robust (10). Occasionally, a rare, high risk allele is found in patients with a rare, familial form of a common disorder, such as Alzheimer disease. Though the findings often have implications in the disease pathogenesis, the role of the rare mutation is limited for common sporadic forms of disease in the general population. In this series, an example of this is described by Bertram and Tanzi in their discussion of Alzheimer disease, in which mutations in the amyloid pre-

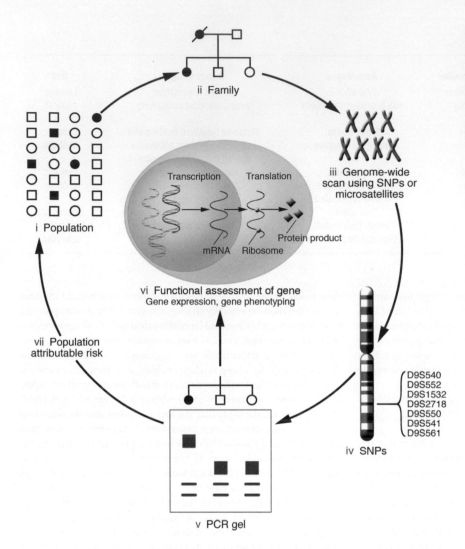

ii Family

iii Genome-wide scan using SNPs or microsatellites

i Population

Transcription   Translation

mRNA   Ribosome

Protein product

vi Functional assessment of gene
Gene expression, gene phenotyping

vii Population attributable risk

D9S540
D9S552
D9S1532
D9S2718
D9S550
D9S541
D9S561

iv SNPs

v PCR gel

Figure 1
Progression of gene mapping in genetic epidemiological studies. (i) Population from which the complex genetic disorder arose. (ii) One of several families included in the genome-wide scan. However, more recently, genome-wide association studies of unrelated patients and controls have been advocated. (iii) Genome-wide scan using microsatellite DNA markers or SNPs. (iv) Fine mapping using a dense collection of SNPs in a region that segregates with disease. (v) Variant allele detection using sequencing. (vi) Functional assessment of the protein product. (vii) Determination of population attributable risk.

cursor protein and presenilin I and II lead to an overproduction of amyloid β protein, which is deposited in the brains of all patients with Alzheimer disease, regardless of the etiology (3).

While linkage analysis is arguably the most powerful method for identifying rare, high-risk alleles in Mendelian disease, many consider genetic association analysis to be the best method for identifying genetic variants related to common complex diseases (11, 12). In contrast to linkage analysis, which involves scanning the entire genome or a very large segment, association analyses are best suited to interrogating smaller regions or segments of the genome. Association analyses are generally model free, or nonparametric, so the researcher does not have to assume a mode of inheritance is unknown. Unlike linkage analysis, where markers are identified, association studies determine whether or not a specific allele within a marker is associated with disease. Association studies can be conducted in a group of randomly selected patients and controls as well as in small families or affected sibling pairs. Thus, this approach is sometimes added to ongoing epidemiological or clinical trials and can be adapted for use with relatively small-sized families. Association analyses of candidate genes underlying quantitative traits such as body mass index as related to obesity or blood pressure in relation to hypertension are also feasible, as will be clear from the discussion by Majumder and Ghosh in this series (ref. 13).

There is at least one important similarity between linkage and association analyses. Linkage analysis involves association within families, while genetic association analysis examines whether affected individuals share the common allele more often than do controls. Patients who share the variant allele may also share a common ancestor from whom the allele originated. In reality, researchers often do both linkage and association analyses. Linkage analysis is used for the genome-wide screen to identify candidate loci. The region is subsequently narrowed using linkage disequilibrium mapping, which is reviewed by Morton in this series (ref. 14). Genome-wide association studies are now feasible and can provide an additional means for identifying genes related to complex disorders. This approach combines the best features of linkage with the strength of association approaches (12). Figure 1 illustrates the progression from the study of a population to the identification of a variant allele and subsequent functional analysis. Genetic epidemiologists often go back to the population in order to determine the population attributable risk, which is defined as the proportion of disease in the population that can be ascribed to the variant allele or risk factor of concern. It is based both on the relative risk (see Gordon and Finch; ref. 15) and the prevalence of the variant in the population.

Association studies also have limitations. Because linkage disequilibrium, cosegregation of a series of genetic markers or alleles, is sustained over only a short chromosomal segment, a large number of loci need to be tested to cover a region (or the genome if a genome-wide association is conducted). This increases the possibility of false-positive findings. Therefore, one cannot rely on the conventional threshold P value of 0.05. With each test, the possibility of a false-positive result increases, requiring the need either for replication in an independent study or computer simulation (11). For complex genetic disease studies, researchers can use computer simulation of 1,000 replicates of the family collection based on observed allele frequencies and recombination fractions to determine the threshold for statistical significance in order to reduce the possibility of false-positive results. For case-control studies, patients with disease and the comparison group of controls can differ in genetic background, introducing variables unrelated to the disease and causing a type of spurious association or confounding termed *population stratification*. Finally, the number of subjects required for these studies can be large, particularly if the heritability or relative risk of the disorder or trait is low. In this series, Gordon and Finch review both the benefits and limitations of using association analysis in family-based and population-based studies to identify genes related to complex disorders (15).

Conclusions

Researchers, clinicians, patients, and their families are likely to reap the benefits of continued application of progress in human genetics to various disciplines in medicine. The relatively new field of genetic epidemiology, a hybrid of genetics and epidemiology, is already capitalizing on this progress by enabling researchers to focus both on genetic variations in different populations (unrelated patients and controls or small families and sibling pairs) and their exposure to environmental or nongenetic risk factors in order to explain how their joint effects lead to disease. With this continued payoff has come the need for a better understanding of the intricacies of genetic exploration and genome science. Designing the appropriate studies, using the correct analytic approach, and appreciating the strengths and weaknesses of genetic methods as applied to common complex disorders is essential. It is our hope that this series, *Complex genetic disorders*, in the *JCI* will facilitate that process for our readers.

Address correspondence to: Richard Mayeux, the Gertrude H. Sergievsky Center, 630 West 168th Street, Columbia University, New York, New York 10032, USA. Phone: (212) 305-2391; Fax: (212) 305-2518; E-mail: rpm2@columbia.edu.

1. Permutt, M.A., Wasson, J., and Cox, N. 2005. Genetic epidemiology of diabetes. *J. Clin. Invest.* **115**:1431–1439. doi:10.1172/JCI24758.

2. Kirov, G., O'Donovan, M.C., and Owen, M.J. 2005. Finding schizophrenia genes. *J. Clin. Invest.* **115**:1440–1448. doi:10.1172/JCI24759.

3. Bertram, L., and Tanzi, R.E. 2005. The genetic epidemiology of neurodegenerative disease. *J. Clin. Invest.* **115**:1449–1457. doi:10.1172/JCI24761.

4. Khoury, M.J., Beaty, T.H., and Cohen, B.H. 1993. *Fundamentals of genetic epidemiology.* Oxford University Press. New York, New York, USA. 383 pp.

5. Grilo, C.M., and Pogue-Geile, M.F. 1991. The nature of environmental influences on weight and obesity: a behavior genetic analysis. *Psychol. Bull.* **110**:520–537.

6. Falconer, D.S. 1989. *Introduction to quantitative genetics.* 3rd edition. John Wiley & Sons. New York, New York, USA. 438 pp.

7. Ott, J. 1999. *Analysis of human genetic linkage.* 3rd edition. Johns Hopkins University Press. Baltimore, Maryland, USA. 382 pp.

8. Terwilliger, J.D., and Ott, J. 1994. *Handbook of human genetic linkage.* Johns Hopkins University Press. Baltimore, Maryland, USA. 320 pp.

9. Lander, E.S., and Schork, N.J. 1994. Genetic dissection of complex traits. *Science.* **265**:2037–2048.

10. Terwilliger, J.D., and Goring, H.H. 2000. Gene mapping in the 20th and 21st centuries: statistical methods, data analysis, and experimental design. *Hum. Biol.* **72**:63–132.

11. Neale, B.M., and Sham, P.C. 2004. The future of association studies: gene-based analysis and replication. *Am. J. Hum. Genet.* **75**:353–362.

12. Hirschhorn, J.N., and Daly, M.J. 2005. Genome-wide association studies for common diseases and complex traits. *Nat. Rev. Genet.* **6**:95–108.

13. Majumder, P.P., and Ghosh, S. 2005. Mapping quantitative trait loci in humans: achievements and limitations. *J. Clin. Invest.* **115**:1419–1424. doi:10.1172/JCI24757.

14. Morton, N.E. 2005. Linkage disequilibrium maps and association mapping. *J. Clin. Invest.* **115**:1425–1430. doi:10.1172/JCI25032.

15. Gordon, D., and Finch, S.J. 2005. Factors affecting statistical power in the detection of genetic association. *J. Clin. Invest.* **115**:1408–1418. doi:10.1172/JCI24756.

Mapping quantitative trait loci in humans: achievements and limitations

Partha P. Majumder and Saurabh Ghosh

Human Genetics Unit, Indian Statistical Institute, Kolkata, India

Recent advances in statistical methods and genomic technologies have ushered in a new era in mapping clinically important quantitative traits. However, many refinements and novel statistical approaches are required to enable greater successes in this mapping. The possible impact of recent findings pertaining to the structure of the human genome on efforts to map quantitative traits is yet unclear.

Mapping quantitative trait loci

Clinical end points are usually binary — affected or unaffected. Such binary end points almost invariably have quantitative precursor states. Myocardial infarction — a binary end point — for example, has many known quantitative precursors, such as blood pressure and cholesterol levels, which determine the end-point risk. Such quantitative traits (QTs) almost always have strong genetic determinants (that is, are highly heritable). It is, therefore, of considerable interest to map the genes underlying a QT. The traditional viewpoint has been that tens or even hundreds of genes determine a QT, each gene contributing a tiny fraction to the overall variation of the QT. If this were true, then efforts to map a QT locus (QTL) would be futile. However, with the availability of precisely mapped high-density DNA markers for many species, early QTL mapping efforts revealed a much simpler genetic architecture for many QTs (1–4). The emerging paradigm was that even if there were many genes determining the value of a QT, some would have major effects, and hence their chromosomal locations could potentially be determined. Figure 1 illustrates the effects of a single locus with 2 alleles on a QT.

Unfortunately, consistent successes in QTL mapping have been achieved only in species in which inbred strains or lines could be developed. In inbred or experimental populations, the parental origin of each allele is known unambiguously, and all offspring have parents with the same genotypes. These 2 features enable pooling of data across families and testing for equality of mean values of the QT in different genotype classes, using standard ANOVA procedures. Even environmental heterogeneity can be largely controlled experimentally. In humans, unambiguous identification of parental origin of alleles or control for environmental heterogeneity are not possible. Even in inbred strains, efforts at fine mapping of QTLs have revealed unforeseen complexities and have resulted in many failures (5–6), and there are many unresolved issues pertaining to study design and statistical analyses (7–8). In humans, and in other outbred species, QTL mapping has only had limited success (see *Taste sensitivity to phenylthiocarbamide* for a success story). In this review, we shall focus only on QTL mapping in humans.

Nonstandard abbreviations used: IBD, identical by descent; QT, quantitative trait; QTL, QT locus; VC, variance components.

Conflict of interest: The authors have declared that no conflict of interest exists.

Citation for this article: *J. Clin. Invest.* **115**:1419–1424 (2005). doi:10.1172/JCI24757.

Approaches to human QTL mapping

The 2 broad approaches — not mutually exclusive, but complementary — are the candidate gene approach and the marker locus approach. In the candidate gene approach, genes that are physiologically or biochemically relevant to the QT (candidate genes) are screened, and the effects of variant alleles on the QT are investigated. This approach cannot lead to the detection of new QTLs. Further, it is often difficult to choose candidate genes. Although this approach seems attractive, there is not yet sufficient evidence to support its general utility.

The availability of polymorphic markers and refinements of statistical methods (9) have made the marker locus approach very popular. The density of markers and the throughput of marker genotyping have increased over the years, and the cost of marker genotyping has decreased, further facilitating QTL mapping by marker locus approach. Thus, availability of dense markers, high-throughput genotyping, and cost are no longer limiting factors for performing genome-wide scans for positional mapping of QTLs (10–13). The major problems at this time seem to be the difficulty of gathering high-quality phenotype data in a sample of adequate size using an appropriate study design and the analysis of these data using a method with high statistical power.

Study designs

The study designs for QTs are, in the main, similar to those for binary complex traits, that is, binary traits with multilocus determination and possibly with environmental influences. Many conceptual and statistical issues are also similar.

Broadly, there are 2 classes of study designs: study designs in which large sets of relatives from extended or nuclear families are sampled and study designs in which pairs of relatives are sampled (e.g., sibling pairs). Often, sampling is not done randomly. For example, when a sibpair design is adopted, often both siblings are chosen from one tail (upper or lower) of the distribution of the QT (concordant siblings) or one sibling is chosen from the upper tail and the other sibling is chosen from the lower tail (discordant siblings). Another sampling design could include a pair of siblings, one chosen from the upper or lower tail of the distribution and the other chosen randomly from among the remaining siblings (single selection; ref. 14). Even when nuclear or extended families are sampled, the ascertainment of a family may be through an individual who belongs to the upper tail or exceeds a predetermined cutoff point of the distribution of the QT. Alternatively, if the study pertains to a QT that is known to be a precursor of a clinical end point (e.g., blood pressure level and myocardial infarction), a family may be ascertained through an individual who

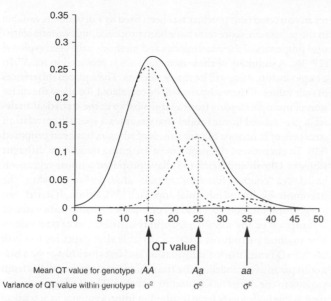

Figure 1
Genetic effects of a single QTL with 2 alleles. The variances of QT values within genotypes can be unequal. The differences in mean values between *AA* and *Aa* genotypes need not be the same as the difference in the mean values between *Aa* and *aa* genotypes. If the mean value of the QT for genotype *Aa* is exactly in the middle of the mean values for genotypes *AA* and *aa*, then the 2 alleles *A* and *a* have additive effects. If the heterozygote mean is shifted toward the mean value of either of the homozygotes, then there is a dominance effect.

has encountered the clinical end point. Any nonrandom sampling scheme obviously entails the screening of a large number of potential sampling units to obtain the requisite number of units that satisfy the inclusion criteria. This is expensive in terms of time, effort, and money. However, the adoption of a nonrandom sampling strategy is statistically more powerful for QTL mapping. Such sampling strategies often require modifications of the standard statistical methods for QTL mapping because the resulting distribution of trait values is no longer the same as in the original source population.

Central issues in QTL mapping

The ability to map a QTL depends on the magnitude of its effect, as measured by the proportion of genetic variance of the QT explained by the putative QTL. Whether or not a QTL can be successfully mapped also depends on the study design, sample size, and statistical method used to analyze the data. In general, even in experimental populations, it has been estimated that under the second filial generation (F_2) design, a QTL with an effect of 5–15% can be detected with a reasonable (80–90%) statistical power if the sample size is between 200 and 300 individuals (8, 15). In natural populations, such as in humans, sample-size estimation is difficult and involves a lot of assumptions, some of which are discussed below.

The second major issue is the nature and distribution of markers on chromosomes. The usefulness of a marker increases with its level of polymorphism (as estimated by the proportion of heterozygous individuals at the locus; see *Locus heterozygosity and marker choice*). As the density of markers is increased, the precision of estimation of the location and effect of the QTL increases. The recent finding that the human genome has a block-like structure with respect to levels of association among loci (16, 17) may potentially reduce the number of markers required in genome-wide scans for QTL mapping to attain the same level of statistical power, although there are many unresolved issues (17, 18).

There are other issues that are also central, such as the extent of gene-gene interaction (epistasis) and genotype-environment interaction in the determination of trait values. However, these issues have received little attention with respect to human QTL mapping because the statistical intricacies of even the 2 central issues listed above are still being worked out.

Statistical methods

Since the positions of the QTLs in the genome are unknown, one can gather genotype data at a large number of marker loci and analyze these data statistically to test whether there is increased allele sharing at the marker loci among individuals who show similar trait values (see *Human QTL mapping: key principles*). If there is increased allele sharing, then the QTL probably lies in the vicinity of this marker locus; that is, the QT and marker loci are linked (2-point mapping). If there is increased allele sharing at several *consecutive* marker loci, as revealed by the joint analysis of the QT data with multiple markers (multipoint mapping), then there is a higher probability that the QTL lies in the interval spanned by these marker loci.

There are then 2 major goals: (a) measuring the expected level of allele sharing at marker loci (based on genotype data) among the sampled sets of relatives and (b) testing whether there is an increased

Taste sensitivity to phenylthiocarbamide
Differences in the ability of humans to taste the sulfur-containing compound phenylthiocarbamide (PTC) were first reported in 1931. The ability to taste PTC is quantitative. Some individuals find even low concentrations of the compound extremely bitter, while others report bitter taste only at high concentrations. A substantial fraction of humans find the compound tasteless even at very high concentrations. The ability to taste PTC is heritable.

Using QTL mapping techniques, Kim et al. (63) have found a gene that accounts for 55–85% of the variance in taste sensitivity. The gene (designated *PTC*) is on the long arm of chromosome 7 and encodes a member of the TAS2R bitter taste receptor family. The *PTC* gene consists of 1002 bp in a single exon, encoding a 7-transmembrane domain, heterotrimeric guanine nucleotide-binding protein–coupled (G protein–coupled) receptor. Within this gene there are 3 common single nucleotide polymorphisms (SNPs), all of which result in amino acid changes. These changes at the amino acid positions 49, 262, and 296 are, respectively, proline (P) to alanine (A), alanine to valine (V), and valine to isoleucine (I). The 2 haplotypes PAV and AVI are predominant. The PAV homozygotes are most taste sensitive (mean PTC score: 10.69–10.00), followed by the PAV/AVI heterozygotes (9.65–8.81), and then the AVI homozygotes (4.31–1.86).

level of sharing among individuals with similar trait values (and, therefore, inferring that the marker and the trait loci are linked).

A pair of relatives can share 0, 1, or 2 alleles that are identical by descent (IBD). (For an allele to be IBD in a pair of relatives, the allele in both the relatives must have been the same allele transmitted by the same ancestor.) A general method for calculating the probabilities of sharing 0, 1, or 2 alleles at a locus was given by Li and Sacks (19); this was then extended by Campbell and Elston (20), and a more general method was developed by Donnelly (21). For pedigrees also, methods for estimating IBD probabilities from genotype data have been developed (22).

Regression

Haseman-Elston regression was the first statistical method that was developed for human QTL mapping (23). This method is applicable to human sibpair data. This *linear* regression model ($Y = a + bX$) includes the squared difference in trait values between members of a sibling pair (Y) as the dependent variable and the number of alleles shared IBD between them (X) at the marker locus (marker IBD score) as the independent variable. If any parent is homozygous at the marker locus or if parental genotypes are missing, then the marker IBD score cannot be determined with certainty. In such a case, the IBD score is estimated by conditioning on either the marker genotypes of parents (if available) and the sibs at the given marker locus (single-point estimate) or an integrated genotype profile based on all available marker loci (multipoint estimate) on a chromosome (24, 25). Under the null hypothesis of no linkage, the regression coefficient is 0, while under linkage it is less than 0. The null hypothesis is easily tested by a standard Student's t test. This method was extended to other relative pairs and pairs drawn from larger pedigrees (26). However, there are statistical limitations in using the Haseman-Elston regression on pedigree data, and it is not considered to be a method of choice.

The choice of the squared trait value difference between members of a sibling pair wastes valuable information by not using the trait values of the individual siblings. Twenty-five years later, it was shown (27) that the inclusion of the sum of trait values of the siblings, along with the squared trait value difference, in the analysis results in gain of statistical power. It was suggested (28) that these 2 variables (squared trait difference and trait sum) be used as dependents in 2 separate linear regression equations with the estimated IBD score as the independent variable and that the estimated slopes be averaged to draw inferences on linkage. This method relies on several assumptions that have been relaxed to develop statistically more sound methods of combining the 2 slope estimates; use of a

mean corrected trait product has been used as a dependent variable in the regression, score tests have been proposed, and various statistical properties of these estimators and methods have been explored (29–36). A summary of these new statistics is provided in ref. 37. In a large sibship, there will be many sibpairs. The squared differences in trait values of these sibpairs will be correlated. To allow the inclusion of multiple sibpairs from a large sibship in the statistical analysis, a generalized linear model that assumes a specific correlation structure of functions of trait values of sibpairs has been proposed (29). To circumvent the problem of assigning weights to different sibpairs, Ghosh and Reich (38) have proposed a linear regression based on a "contrast function" of trait values within a sibship. The maximum-likelihood binomial approach (39), although strictly not a regression method, can also accommodate sibship data without assumption of any specific probability distribution of trait values. The method introduces a latent variable that captures the link between QTs and marker information and tests for linkage via a Bernoulli parameter modeling the transmission of marker alleles from parents to the different sibs within a sibship. These advances in statistical methodologies have resulted in improvements in statistical power to map QTLs, but the regression-based method is applicable only to sibpairs and, under some assumptions, to sibships.

Recently, a novel approach has been proposed (40), in which the IBD scores have been modeled as a function of observed trait values instead of the usual modeling of trait values as a function of IBD scores. This method is applicable to large sibships and also to general pedigrees, but does not necessarily have more statistical power (41) than a competing method called variance components (VC) (discussed below).

In these regression models, the relationship between the dependent and independent variables being *linear* is an *assumption*. This assumption is valid when there is no dominance at the trait locus; but in the presence of dominance, the regression can deviate from linearity. This assumption has been relaxed and nonparametric alternatives have been proposed, as discussed later.

Regression methods continue to be widely used because they are computationally easy and efficient, and the standard deviations of parameters can easily be estimated using resampling techniques (42). However, there is no strong statistical reason for using regression methods for QTL mapping, except when the collected data are from pairs of relatives, such as sibling pairs (discussed below).

Variance components

Another popular statistical approach is called the VC method, which is applicable to large sibships or pedigrees. In the framework underlying this method, the trait value of an individual is assumed to be determined by a major gene, random polygenic, and environmental effects and covariates. The covariance between trait values of a pair of relatives is an increasing function of the extent of allele sharing, IBD, at the marker locus. The general framework and methodology that is currently popular was developed by Amos (43), although Goldgar (44) first proposed this method in the context of human QTL mapping. Amos (43) derived expressions for the covariances in trait values for a number of common pairs of relatives. The trait values of individuals in a pedigree are assumed to follow a multivariate normal distribution, with the variance-covariance matrix determined by the expressions given in Amos (43) or their straightforward generalizations. The likelihood of the observations on a pedigree or any other set of relatives can then be written down by standard statistical methods. The likelihood is maximized to obtain parameter estimates, and a likelihood ratio test is used to test for linkage.

Human QTL mapping: key principles

The key principles that underlie all statistical methods for QTL mapping are as follows:

(1) Persons sharing similar trait values should share alleles at the trait locus at levels higher than those expected by chance or by virtue of the persons' biological relationship. (Conversely, persons who have dissimilar trait values should show a decreased level of sharing of alleles at the trait locus.)

(2) Because chromosomes are passed on more-or-less intact — except for recombinations — from parents to their offspring, persons sharing similar trait values should, in addition to showing increased sharing of alleles at the trait locus, also show increased sharing of alleles at loci around the trait locus.

(3) Because recombinations occur in every generation, the level of increased sharing of alleles at loci around the trait locus decreases in every generation.

Various extensions of the basic model and methodology of Amos (43) have been made. These include extensions to permit likelihood calculations to pedigrees of arbitrary sizes and complexity (25), inclusion of gene-gene (45, 46) and gene-environment (47) interactions in the model, and analysis of multiple correlated traits (48). When the model assumptions hold, especially multivariate normality, the VC method is very powerful, considerably more powerful than the Haseman-Elston regression. Further, it is readily applicable to large and complex pedigrees. Thus, for QTL mapping, the method of choice is VC. For sibling pairs, however, it has been shown — both by theory and by simulation — that the computationally simpler regression methods are as powerful as the VC method (32, 33).

We emphasize here that the statistical power and efficiency of the VC method critically depend on whether the assumption that the trait values are normally distributed in the population is satisfied. However, it is often not feasible to verify distributional and other model assumptions. Further, even when the distribution in the population from which sampling units are drawn is normal, if the sampling design is nonrandom, then the distribution of the QT in the data so obtained may be nonnormal, thus violating the assumptions underlying the VC method. When underlying assumptions are violated, parametric methods (that is, methods — such as VC — that rely on models that assume specific forms of the probability distribution of trait values) can result in a high proportion of either false-positive or false-negative inferences. For example, if the trait distribution has a sharp peak (leptokurtic) and if gene-environment interactions are present, then one can get inflated false-positive error rates (49). Some methods based on permutation tests — which do not rely on normality of the test distribution in drawing inferences — have been proposed to obtain P values (50), but these methods entail enormous increase in the computational load. The problems associated with the possible violation of normality continue to be a limiting factor in practical applications of the VC method. Some novel methods have recently been proposed to deal with these problems (51), but the difficulties are far from resolved. VC methods for mapping QTLs have been implemented in several software packages, including Genehunter (52, 53), Merlin (54), Mx (55), and SOLAR (25).

Nonparametric alternatives

When the assumptions underlying the regression (linearity of relationship between the dependent and the independent variables) and the VC (multivariate normality of QT values of family members) methods hold, these methods are statistically quite powerful for QTL mapping. However, it is difficult to ensure that these assumptions are met, especially for pedigree data. Deviations from these assumptions can adversely affect linkage inferences. Therefore, alternative methods that do not rely critically on these model assumptions (model-free approaches) have begun to be developed. In these model-free methods, there is inevitably a loss of statistical power, but these methods provide safeguards against high rates of both false positives and false negatives.

Since the nature of dependence of estimated marker IBD scores on the squared difference in sibpair trait values is a function of the recombination fraction between the marker and the trait loci and other biological parameters, such as interference and dominance at the trait locus, the assumption of a specific form of functional relationship between them may not be a robust strategy. Rank-based statistics (23, 56) have been proposed to deal with this problem. A proposed nonparametric regression procedure based on kernel smoothing (in which the relationship between the dependent and the independent variables is estimated empirically) has been shown to perform well (57, 58) both in simulations and in practical applications. The available nonparametric methods are useful only for sibpair data. Such methods need to be developed for pedigree data also.

A summary of choices

VC method is statistically the method of choice for QTL mapping, provided that the assumption of multivariate normality of trait values within family members is satisfied. This assumption is hard to test, and more importantly, if it is violated, then it is hard to rectify even by using mathematical transformations of QT values, e.g., logarithmic or power transformations. Further, in families that are selected through a member possessing an extreme QT value, there is an even bigger problem of noncompliance with the normality assumption. Fortunately, there are indications (49) that when this assumption is not met, it is the type I error, rather that the type II error, that is inflated to a greater degree. Thus, with the VC method, if linkage is detected, then chances are good that it is not a false inference.

When the normality assumption is not met, then it may be better to use a nonparametric regression method based on sibpair data, even though there will be loss of statistical power. In this case, the false-positive error rate will be lower. However, no results are yet available on the statistical properties of this method when siblings are selected based on some inclusion criteria, e.g., siblings belonging to opposite extremes of the trait distribution — discordant sibpairs (see, however, Peng and Siegmund; ref. 59).

Unselected samples have low statistical power. Selection of discordant sibpairs yields a high statistical power. This property is also true of families ascertained through a member with an extreme QT value. These selection strategies can be very expensive and difficult to implement in practice. A compromise solution is to select 1 sibling with an extreme value and choose another sibling randomly from among the remaining siblings in the sibship. This selection strategy — less expensive and easier to implement than selecting discordant sibpairs — has comparable, albeit slightly lower, statistical power (14). However, in studies based on sibling pairs in which the focus of interest is on trait-allele relationships at an individual level rather than on allele-sharing in families (association analysis), a crucial criterion for success of QTL mapping is that the frequencies of marker and trait alleles should be in the same ballpark. This means that generating polymorphic markers with high frequencies may not result in greater success of QTL mapping unless the trait alleles have matching frequencies (14). Similar results have also been obtained with respect to association study designs that pertain to unrelated individuals (not siblings) selected from opposite tails of the distribution of QT values (60). This lack of greater success in QTL mapping unless the trait alleles have matching frequencies is not encouraging. While considerable efforts are being made to generate markers that will ease genome-wide association mapping of QTLs, if the allele frequencies at the QTL are very skewed, efforts in mapping the QTL may be unsuccessful. This is in addition to the fact that a QTL that explains less than 10% of the variance of trait values is very hard to map. The recent efforts of the HapMap project (16) to provide markers that may be the most informative for association mapping will not be a panacea for overcoming these limitations of human QTL mapping. As we have discussed, there is also a great need to devise statistical methods for human QTL mapping that do not critically depend on model assumptions. In association mapping, population stratification (61) is a major issue, and therefore, although designs involving unrelated individuals are easier to implement, these are best avoided for human QTL mapping. Further, the statistical power of QTL mapping using association analysis declines very rapidly with the decrease of nonrandom association between the QTL and the marker locus (62). Notwithstanding the caveats listed above, efforts to map human QTLs using a combination of family-based association and linkage analysis methods are continuing and should continue. Successes in practice will crucially depend on refinements of statistical methods and developments of novel approaches to handle interactions among QTLs as well as the effects of environmental factors.

Address correspondence to: Partha P. Majumder, Human Genetics Unit, Indian Statistical Institute, 203 Barrackpore Trunk Road, Kolkata 700108, India. Phone: 91-33-25753209; Fax: 91-33-25773049; E-mail: ppm@isical.ac.in.

1. Paterson, A.H., et al. 1988. Resolution of quantitative traits into Mendelian factors using a complete RFLP linkage map. *Nature.* **335**:721–726.
2. Todd, J.A., et al. 1991. Genetic analysis of autoimmune type 1 diabetes mellitus in mice. *Nature.* **351**:542–547.
3. Hilbert, P., et al. 1991. Chromosomal mapping of two genetic loci associated with blood pressure regulation in hereditary hypertensive rats. *Nature.* **353**:521–529.
4. Jacob, H.J., et al. 1991. Genetic mapping of a gene causing hypertension in the stroke-prone spontaneously hypertensive rat. *Cell.* **67**:213–224.
5. Podolin, P.L., et al. 1998. Localization of two insulin-dependent diabetes (*Idd*) genes to the *Idd10* region on mouse chromosome 3. *Mamm. Genome.* **9**:283–286.
6. Legare, M.E., Bartlett, F.S., and Frankel, W.N. 2000. A major effect QTL determined by multiple genes in epileptic EL mice. *Genome Res.* **10**:42–48.
7. Darvasi, A. 1998. Experimental strategies for the genetic dissection of complex traits in animal models. *Nat. Genet.* **18**:19–24.
8. Doerge, R.W. 2002. Mapping and analysis of quantitative trait loci in experimental populations. *Nat. Rev. Genet.* **3**:43–52.
9. Lander, E.S., and Botstein, D. 1986. Strategies for studying heterogeneous genetic traits in humans by using a linkage map of restriction fragment length polymorphisms. *Proc. Natl. Acad. Sci. U. S. A.* **83**:7353–7357.
10. Perusse, L., et al. 2001. A genome-wide scan for abdominal fat assessed by computed tomography in the Québec Family Study. *Diabetes.* **50**:614–621.
11. Deng, H.W., et al. 2002. A genomewide linkage scan for quantitative-trait loci for obesity phenotypes. *Am. J. Hum. Genet.* **70**:1138–1151.
12. Hirschhorn, J.N., et al. 2001. Genomewide linkage analysis of stature in multiple populations reveals several regions with evidence of linkage to adult height. *Am. J. Hum. Genet.* **69**:106–116.
13. Bosse, Y., et al. 2004. Genome-wide linkage scan reveals multiple susceptibility loci influencing lipid and lipoprotein levels in the Québec Family Study. *J. Lipid Res.* **45**:419–426.
14. Abecasis, G.R., Cookson, W.O., and Cardon, L.R. 2001. The power to detect linkage disequilibrium with quantitative traits in selected samples. *Am. J. Hum. Genet.* **68**:1463–1474.
15. Erickson, D.L., Fenster, C.B., Stenoien, H.K., and

Price, D. 2004. Quantitative trait locus analyses and the study of evolutionary process. *Mol. Ecol.* **13**:2505–2522.
16. International HapMap Consortium. 2003. The International HapMap Project. *Nature.* **426**:789–796.
17. Wall, J.D., and Pritchard, J.K. 2003. Haplotype blocks and linkage disequilibrium in the human genome. *Nat. Rev. Genet.* **4**:587–597.
18. Schulze, T.G., et al. 2004. Defining haplotype blocks and tag single-nucleotide polymorphisms in the human genome. *Hum. Mol. Genet.* **13**:335–342.
19. Li, C.C., and Sacks, J. 1954. The derivation of joint distribution and correlation between relatives by the use of stochastic matrices. *Biometrics.* **10**:347–360.
20. Campbell, M.A., and Elston, R.C. 1971. Relatives of probands: models for preliminary genetic analysis. *Ann. Hum. Genet.* **35**:225–236.
21. Donnelly, K.P. 1983. The probability that related individuals share some section of the genome identical by descent. *Theor. Popul. Biol.* **23**:34–63.
22. Sobel, E., and Lange, K. 1996. Descent graphs in pedigree analysis: applications to haplotyping, location scores, and marker-sharing statistics. *Am. J. Hum. Genet.* **58**:1323–1337.
23. Haseman, J.K., and Elston, R.C. 1972. The investigation of linkage between a quantitative trait and marker loci. *Behav. Genet.* **2**:3–19.
24. Lander, E.S., and Green, P. 1987. Construction of multilocus genetic linkage maps in humans. *Proc. Natl. Acad. Sci. U. S. A.* **84**:2363–2367.
25. Almasy, L., and Blangero, J. 1998. Multipoint quantitative-trait linkage analysis in general pedigrees. *Am. J. Hum. Genet.* **62**:1198–1211.
26. Olson, J.M., and Wijsman, E.M. 1993. Linkage between quantitative trait and marker loci: methods using all relative pairs. *Genet. Epidemiol.* **10**:87–102.
27. Wright, F.A. 1997. The phenotypic difference discards sib-pair QTL information. *Am. J. Hum. Genet.* **60**:740–742.
28. Drigalenko, E. 1998. How sib-pairs reveal linkage. *Am. J. Hum. Genet.* **63**:1242–1245.
29. Elston, R.C., Buxbaum, S., Jacobs, K.B., and Olson, J.M. 2000. Haseman and Elston revisited. *Genet. Epidemiol.* **19**:1–17.
30. Xu, X., Weiss, S., Xu, X., and Wei, I.J. 2000. A unified Haseman-Elston method for testing linkage with quantitative traits. *Am. J. Hum. Genet.* **67**:1025–1028.
31. Forrest, W. 2001. Weighting improves the "new

Haseman Elston" method. *Hum. Hered.* **52**:47–54.
32. Sham, P.C., and Purcell, S. 2001. Equivalence between Haseman-Elston and variance components QTL linkage analysis of selected and nonnormal samples: conditioning on trait values. *Am. J. Hum. Genet.* **68**:1527–1532.
33. Visscher, P.M., and Hopper, J.L. 2001. Power of regression and maximum likelihood methods to map QTL from sib-pair and DZ twin data. *Ann. Hum. Genet.* **65**:583–601.
34. Tang, H.-K., and Siegmund, D. 2001. Mapping quantitative trait loci in oligogenic models. *Biostatistics.* **2**:147–162.
35. Wang, K., and Huang, J. 2002. A score-statistic approach for the mapping of quantitative-trait loci with sibships of arbitrary size. *Am. J. Hum. Genet.* **70**:412–424.
36. Putter, H., Sandkuijl, L.A., and van Houwelingen, J.C. 2002. Score test for detecting linkage to quantitative traits. *Genet. Epidemiol.* **22**:345–355.
37. Cuenco, K.T., Szatkiewicz, J.P., and Feingold, E. 2003. Recent advances in human quantitative-trait-locus mapping: comparison of methods for discordant sibling pairs. *Am. J. Hum. Genet.* **73**:874–885.
38. Ghosh, S., and Reich, T. 2002. Integrating sibship data for mapping quantitative trait loci. *Ann. Hum. Genet.* **66**:169–182.
39. Alcais, A., and Abel, L. 1999. Maximum-likelihood-binomial method for genetic model-free linkage analysis of quantitative traits in sibships. *Genet. Epidemiol.* **17**:102–117.
40. Sham, P.C., Purcell, S., Cherny, S.S. and Abecasis, G.R. 2002. Powerful regression-based quantitative-trait linkage analysis of general pedigrees. *Am. J. Hum. Genet.* **71**:238–253.
41. Yu, X., Knott, S.A., and Visscher, P.M. 2004. Theoretical and empirical power of regression and maximum-likelihood methods to map quantitative trait loci in general pedigrees. *Am. J. Hum. Genet.* **75**:17–26.
42. Visscher, P.M., Thompson, R., and Haley, C.S. 1996. Confidence intervals in QTL mapping by bootstrapping. *Genetics.* **143**:1013–1020.
43. Amos, C.I. 1994. Robust variance-components approach for assessing genetic linkage in pedigrees. *Am. J. Hum. Genet.* **54**:535–543.
44. Goldgar, D.E. 1990. Multipoint analysis of human quantitative genetic variation. *Am. J. Hum. Genet.* **47**:957–967.
45. Stern, M.P., et al. 1996. Evidence for linkage of

regions on chromosomes 6 and 11 to plasma glucose concentrations in Mexican Americans. *Genome Res.* **6**:724–734.

46. Mitchell, B.D., Ghosh, S., Schneider, J.L., Birznicks, G., and Blangero, J. 1997. Power of variance component linkage analysis to detect epistasis. *Genet. Epidemiol.* **14**:1017–1022.

47. Towne, B., Siervogel, R.M., and Blangero, J. 1997. Effects of genotype-by-sex interaction on quantitative trait linkage analysis. *Genet. Epidemiol.* **14**:1053–1058.

48. Almasy, L., Dyer, T.D., and Blangero, J. 1997. Bivariate quantitative trait linkage analysis: pleiotropy versus co-incident linkages. *Genet. Epidemiol.* **14**:953–958.

49. Allison, D.B., Fernandez, J.R., Heo, M., and Beasley, T.M. 2000. Testing the robustness of the new Haseman-Elston quantitative-trait loci-mapping procedure. *Am. J. Hum. Genet.* **67**:249–252.

50. Itturia, S.J., Williams, J.T., Almasy, L., Dyer, T.D., and Blangero, J. 1999. An empirical test of the significance of an observed quantitative trait locus effect that preserves additive genetic variation. *Genet. Epidemiol.* **17**(Suppl.):S169–S173.

51. Shete, S., et al. 2004. Effect of winsorization on power and type 1 error of variance components and related methods of QTL detection. *Behav. Genet.* **34**:153–159.

52. Kruglyak, L., Daly, M.J., Reeve-Daly, M.P., and Lander, E.S. 1996. Parametric and nonparametric linkage analysis: a unified multipoint approach. *Am. J. Hum. Genet.* **58**:1347–1363.

53. Pratt, S.C., Daly, M.J., and Kruglyak, L. 2000. Exact multipoint quantitative-trait linkage analysis in pedigrees by variance components. *Am. J. Hum. Genet.* **66**:1153–1157.

54. Abecasis, G., Cherny, S., Conkson, W., and Cardon, L. 2002. Merlin: rapid analysis of dense genetic maps using sparse gene flow trees. *Nat. Genet.* **30**:97–101.

55. Neale, M.C., Boker, S.M., Xie, G., and Mael, H.H. 2002. *Mx: statistical modeling*. 6th edition. Department of Psychiatry, Virginia Commonwealth University. Richmond, Virginia, USA. 217 pp.

56. Kruglyak, L., and Lander, E.S. 1995. A nonparametric approach for mapping quantitative trait loci. *Genetics*. **139**.1421–1428.

57. Ghosh, S., and Majumder, P.P. 2000. A two-stage variable-stringency semiparametric method for mapping quantitative trait loci with the use of genomewide scan data on sib-pairs. *Am. J. Hum. Genet.* **66**:1046–1061.

58. Ghosh, S., et al. 2003. Linkage mapping of beta 2 EEG waves via non-parametric regression. *Am. J. Med. Genet.* **118**:66–71.

59. Peng, J., and Siegmund, D. 2004. Mapping quantitative traits with random and with ascertained sibships. *Proc. Natl. Acad. Sci. U. S. A.* **101**:7845–7850.

60. Schork, N.J., Nath, S.K., Fallin, D., and Chakravarti, A. 2000. Linkage disequilibrium analysis of biallelic DNA markers, human quantitative trait loci, and threshold-defined case and control subjects. *Am. J. Hum. Genet.* **67**:1208–1218.

61. Cardon, L.R., and Palmer, L.J. 2003. Population stratification and spurious allelic association. *Lancet*. **361**:598–604.

62. Sham, P.C., Cherny, S.S., Purcell, S., and Hewitt, J.K. 2000. Power of linkage versus association analysis of quantitative traits, by use of variance-components models, for sibship data. *Am. J. Hum. Genet.* **66**:1616–1630.

63. Kim, U., et al. 2003. Positional cloning of the human quantitative trait locus underlying taste sensitivity to phenylthiocarbamide. *Science*. **299**:1221–1225.

Linkage disequilibrium maps and association mapping

Newton E. Morton

Human Genetics Division, Southampton General Hospital, Southampton, United Kingdom.

The causal chain between a gene and its effect on disease susceptibility cannot be understood until the effect has been localized in the DNA sequence. Recently, polymorphisms incorporated in the HapMap Project have made linkage disequilibrium (LD) the most powerful tool for localization. The genetics of LD, the maps and databases that it provides, and their use for association mapping, as well as alternative methods for gene localization, are briefly described.

Introduction

Since its origin a generation ago, genetic epidemiology has dealt with etiology, distribution, and control of disease in groups of relatives and with inherited causes of disease in populations, where inheritance may occur through culture, biology, or interactions between the 2 (1). Advances in molecular genetics have altered the emphasis from family resemblance to identification of genes affecting disease susceptibility, the first step in understanding how the actions of these genes can be ameliorated. At the end of the last century, localization of the gene usually occurred by linkage to a chromosome region within which a random fragment was incorporated, amplified, and then sequenced in a vector such as a bacterial plasmid. This process, called positional cloning (2), evolved into association mapping when the Human Genome Project provided a trustworthy DNA sequence (3) that allowed closely linked polymorphisms to be localized by unique flanking sequences without cloning.

Most of these markers for a short nucleotide sequence are single nucleotide polymorphisms (SNPs), nearly all of which are dichotomous (diallelic). Small insertions and deletions play the same role in mapping as SNPs and are loosely included with them. The number of diallelic markers in the human genome has been estimated to be between 10 and 20 million, depending on exclusion or inclusion of uncommon polymorphisms (4). Detectable association of such polymorphisms with susceptibility to a particular disease (or other phenotype) extends over a much shorter distance than linkage, but the high density of SNPs provides much greater resolution within a candidate region. Because of the short history of association mapping, terminology remains imprecise and is a common source of confusion (see Glossary).

Definitions and successes in association mapping

Association between a pair of linked markers is also called linkage disequilibrium (LD) or, less frequently, gametic disequilibrium. However, association has a broader meaning that includes combinations of 3 or more linked markers, at least some of which are in LD. These combinations are called haplotypes if specified for a single chromosome. The sex chromosomes in males and certain chromosomal aberrations are monosomic, with each individual carrying only 1 haplotype. With these exceptions (and some chromosomal aberrations) haplotypes occur in pairs, called diplotypes, consisting of 1 haplotype from each parent. The 2 haplotypes of a diplotype cannot be ascertained with certainty if 2 or more markers are heterozygous except in special cases that include family studies, physical separation of chromosomes, and zero frequency of alternative haplotypes.

This review will focus on methods currently being used for association mapping, but the literature is full of occasional successes by less powerful methods. Association studies have identified many rare, single-gene disorders after localization to a candidate region by linkage or cytogenetic abnormality. Among the first disease gene localizations were cystic fibrosis, Huntington disease, and hemochromatosis. The latter took 16 years because the physical map greatly exceeds the LD map. If the latter had been available, localization could easily have been made in a tenth of the time, with a corresponding reduction in cost. Gene localization in multifactorial inheritance proved more difficult, especially for linkage and cytogenetics, because multiple genes and environmental factors contribute to disease risk, but genes for Alzheimer disease, deep vein thrombosis, inflammatory bowel disease, hypertriglyceridemia, diabetes, schizophrenia, asthma, stroke, myocardial infarction, and a host of other diseases have been identified by association studies. Genome scanning with SNPs has been successful with myocardial infarction, and gene identification is competing with studies of candidate regions. Association studies are expensive, and it is worthwhile to seek the most powerful approaches, even if their language and methods are unfamiliar.

Physical and genetic maps

Association mapping depends on the choice of map taken to represent LD. Physical maps specify distance in the DNA sequence, ideally measured in bp. The closest approach to this ideal is by the DNA sequence nominally finished, although errors in many relatively small areas remain and, of course, polymorphisms affecting the DNA sequence are represented by 1 arbitrary allele. For association mapping, it is convenient to represent location in kb to 3 decimal places, retaining full precision in the finished maps. Two physical maps at lower resolution are derived from chromosome breakage in radiation hybrids, the utility of which is limited to organisms without a finished DNA sequence, and chromosome bands that project cytogenetic assignments to the physical map. At all levels of resolution, physical maps have the additivity that defines a linear map. Thus, if the distance in the ith interval between 2 adjacent markers is d_i and if the inter-

Nonstandard abbreviations used: LD, linkage disequilibrium; LDU, LD unit; SNP, single nucleotide polymorphism.

Conflict of interest: The author has declared that no conflict of interest exists.

Citation for this article: *J. Clin. Invest.* **115**:1425–1430 (2005).
doi:10.1172/JCI25032.

Glossary

Association mapping	Gene localization by linkage disequilibrium, without cloning.
Association probability	Probability that a random haplotype at 2 specified diallelic loci is descended without crossing-over from an ancestral haplotype at maximal disequilibrium.
DNA pooling	Combination of DNA from 2 or more individuals in order to simplify testing at the expense of accuracy and haplotype identification.
Gametic disequilibrium	Linkage disequilibrium.
Genetic epidemiology	Study of etiology, distribution, and control of disease in groups of relatives and of inherited causes of disease in populations.
Genetic maps	Maps specifying distance in crossover counts (linkage maps) or linkage disequilibrium units.
Haplotype	Set of closely linked genetic markers present on 1 chromosome, which tend to be inherited together.
Linkage disequilibrium	Relationship between 2 alleles, which arises more often than can be accounted for by chance, because those alleles are physically close on a chromosome and infrequently separated from one another by recombination.
Malecot parameters	Parameters (M, L, ε) predicting linkage disequilibrium among m markers in a physical map or the larger set of parameters with ε replaced by ε_i for $i = 1, \ldots, n-1$ that predict LD more accurately.
Marker	Short DNA sequence that is polymorphic and useful for mapping by linkage by association.
Oligogene	Gene with small but identifiable effect on disease risk, as contrasted with a large effect for a (usually rare) major gene and an individually unidentifiable effect for a polygene. Different methods are required for studying these 3 classes.
Physical map	Map specifying distance in the DNA sequence, ideally measured in bp. Less reliable physical maps are provided by chromosome bands and breakage in radiation hybrids.
Positional cloning	Gene localization by linkage or cytogenetic assignment to a candidate region, within which a random fragment is incorporated, amplified, and then sequenced in a vector or bacterial plasmid.

vals are mutually exclusive and jointly exhaustive, the distance between any 2 markers is Σd_i, just as in any road map.

Genetic maps also have additivity, but the distance in the ith interval is proportional to $\varepsilon_i d_i$, where ε_i is not a constant but an interval-specific scaling factor such that $\varepsilon_i \geq 0$, and not all ε_i are equal. The distance between any 2 markers if the intervals are mutually exclusive and jointly exhaustive is proportional to $\Sigma \varepsilon_i d_i$. There are 2 types of genetic maps. Linkage maps long antedated physical maps; their development began in 1913, when Sturtevant (5) elaborated the concept of linear arrangement of genes separated by crossing-over. In 1919, Haldane (6) introduced the Morgan (w) as the length of a chromatid that on average has experienced one crossover event per meiosis, thereby taking $w_i = \varepsilon_i d_i/t$ as his measurement of distance, where t is the number of generations observed. Until recently, linkage maps have been estimated directly from recombination, since values of $\varepsilon_i = w_i/d_i$ could not be determined with accuracy until the physical map was finished. In contrast, LD maps determine distance not from recombination, but from LD, and so distance in the ith interval is expected to be $\varepsilon_i d_i = w_i t$. The number of generations is large and can be reliably determined only from a population genetics model that allows ε_i to be estimated directly. Whereas haplotype inference from diplotypes complicates association mapping, the 2 types of data provide virtually identical LD maps (7).

Population genetics of LD

In the middle of the last century, population genetics was revolutionized by Gustave Malecot, a professor of applied mathematics at the University of Lyon. His use of probability theory has elucidated evolution, population structure, forensic application of DNA, and LD, where his retrospective approach led others to coalescent theory, which tries to trace current genotypes to a putative ancestor. These contributions were summarized several years after his death (8) and continue to stimulate population genetics. The basic theory for LD assumes a pair of diallelic loci that underwent a population bottleneck of reduced size because of war, famine, epidemic, migration, or other factors. Let the founder haplotype frequencies be a mixture of extreme LD with probability ρ_0 and a complementary frequency with LD = 0:

Probability 1

$$\rho_0 \begin{bmatrix} Q & 0 \\ R - Q & 1 - R \end{bmatrix} + (1 - \rho_0) \begin{bmatrix} QR & Q(1-R) \\ (1-Q)R & (1-Q)(1-R) \end{bmatrix}$$

The frequency of the rarest allele is Q, and R is the frequency of the associated allele. Then ρ_0 is defined as the association probability in founders, and the expected decay of ρ in t generations gives $\rho_t = (1 - L)Me^{-\theta t} + L$ where $M = (\rho_0 - L)e^{(v + 1/2N)t}/(1 - L)$ and v, N, and L are the mutation rate, effective population size, and asymptotic value of ρ_t as $e^{-\theta t} \to 0$, respectively (9). Random variation including later bottlenecks causes departure from this expectation, but attempts to anticipate its effects are unsuccessful because the ancestral frequencies and the values of v, N, and t are unknown.

The history of this Malecot equation has been reviewed (10). The value of t is so large that excess of ρ_t over L is usually negligible unless recombination θ is so small that 2 or more crossovers in the same small region are almost always separated by generation and therefore independent. Then the Haldane mapping function is appropriate, with $\theta = (1 - e^{-2w})/2 \to w$ and $wt = \Sigma \varepsilon_i d_i$. Therefore the resolution of linkage maps can be enhanced by interpolating dense locations from LD maps, and (more importantly) association mapping can be

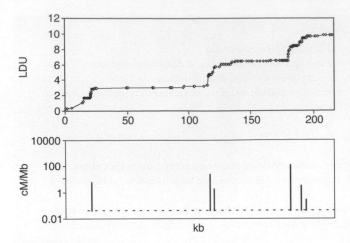

Figure 1
Graph of an LD map for a 216-kb segment of class II region of MHC (7) with corresponding recombination hotspots (12).

based on LD maps that have additive distances measured by $\Sigma\varepsilon$, with $\Sigma\varepsilon_i d_i = 1$ defining 1 LD unit (LDU). The effective bottleneck time in generations is therefore the ratio of LDU to the corresponding distance on the sex-average linkage map in Morgans.

There is an infinity of metrics that describe association for pairs of diallelic markers or affection by marker in a 2 × 2 table. Designating an arbitrary metric by ψ with information K_ψ under the null hypothesis that $\psi = 0$, all of these metrics have the property that in large-sample theory $\chi^2_1 = \psi^2 K_\psi$. Tests on pairs of associated markers are not independent, but n tests give a composite likelihood (lk) with $-2\ln lk = \Sigma_i K_{\psi i} (_1 - \psi_i)^2$ and variance $V = -2\ln lk/(n - m)$ under a hypothesis with m parameters estimated. This does not have the precision of true likelihood, but it gives a good estimate of relative efficiency V/V' of an alternative hypothesis with the same data, but a greater variance V'. This is one of several ways to compare different approaches to LD maps and association mapping.

LD maps
The initial impetus for LD maps came from the discovery that the association probability ρ is predicted by the Malecot equation and provides a higher relative efficiency than other metrics that describe association for pairs of diallelic loci (9, 11). This early research antedated the physical map, the first ("finished") version of which was preceded by evidence from small sequences, that LD is expressed as alternating regions of high and low disequilibrium (blocks and steps) that are fairly stable among populations and are therefore ascribed to low and high recombination, respectively. An ingenious technique for typing recombination in large numbers of sperm has confirmed that ascription in 1 sequence (12), but the method is currently too laborious for use with whole chromosomes. Figure 1 shows how closely the LD map (portrayed here not as coordinates but as a graph) corresponds with recombination. The idea arose that blocks and steps might be so precisely defined that they could be the basis for maps of haplotypes in single blocks delimited by flanking steps. Unfortunately, the frequencies of sequences punctuated in this way varies from 0.02 to 0.40, making haplotype delimitation both arbitrary (13) and acutely sensitive to SNP density (14). These preliminary results led to LD maps delimited in LDU by estimation of the ε_i simultaneously with the Malecot parameters (7). These LD maps were shown to have

a much higher relative efficiency than the kb map, both in describing LD (15) and in association mapping (16). The age of a diallelic polymorphism has been shown to be correlated with its minor allele frequency (17), but time t since the last major bottleneck follows a different pattern with constancy except for alleles that were rare at the last bottleneck or have arisen since (18). Estimates of t are about 1,500 generations in Eurasians and greater in Africans, but this is less than the conventional time assigned to migration out of Africa. Perhaps later bottlenecks tended to reverse progress toward equilibrium, or uncommon alleles tended to be restricted to a few isolates with consequent reduction in effective population size.

One of the attractive features of an LD map is that it can evolve with high relative efficiency from low to high density, from a simple to a more complex model, or from a cosmopolitan map based on several populations to a map specific for a given population, just by beginning iterative estimation with the ε_i of the preliminary map. This convenience may be exploited in a location database to obtain LD maps more specific and more accurate than the contents of the database. In principle, linkage maps could be developed in the same way if the computer programs for linkage mapping accepted trial values for marker locations instead of merely marker order.

Association mapping with an LD map
Association mapping localizes genes affecting disease susceptibility or other phenotypes through association with nearby polymorphisms, usually SNPs. Often association mapping is oriented by previous low-resolution linkage mapping with multiallelic markers to a chromosomal region that is too large for identification of the causal locus, but falling costs of SNP typing have made whole genome scans feasible even for linkage, and the lesser power of single SNPs is compensated for by their much greater density and lower typing costs than multiallelic markers. Successful association mapping at high density leads to causal SNPs that may be confirmed by functional assays with allowance for association. This classical problem of discriminating causality from correlation, the highest level of epidemiology and other nonexperimental sciences, is common to association and functional tests. Zhang et al. (19) described 4 stages from linkage to function during which the search region shrinks and SNP density approaches saturation.

The rigor with which typing and analysis should be carried out has discouraged some scientists (20), but the goal is potentially precious and much has been achieved. Half a century ago, linkage showed that clinically inseparable types of elliptocytosis are determined by different genes (21). This stimulated linkage studies that have shown unanticipated diversity, with at least 8 loci for recessive limb girdle muscular dystrophy instead of the single locus that was once disputed and hundreds of loci for mental retardation and profound deafness and blindness. The immediate impact has been diagnostic for genetic counselling and in some cases for specific treatment. The long-term goal is to develop human genetics to the point where the function and dysfunction of every gene is well understood and to develop pharmacogenetics to the point where every gene for disease susceptibility has a maximally efficient treatment with minimal side effects. Obviously this goal has not been reached only 2 years after substantial completeness of the Human Genome Project made an LD map possible.

A serious obstacle to association mapping is the inverse relation between allele frequency and effect modeled by exponential risk (Figure 2) (22). The major genes of large effect that are the basis of classical genetics are mostly rare. Polygenes of extremely small

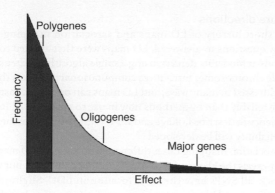

Figure 2
Allele frequency by effect.

effect on fitness account for nearly all alleles subject to selection and are currently beyond cost-effective study. Between them are oligogenes with effects large enough to be detected in a feasible study and perhaps elucidate clinically important metabolic patterns or even be useful targets for pharmacogenetics.

Whereas Figure 1 deals only with alleles subject to selection, the much larger number of neutral polymorphisms has a nearly uniform distribution, with an excess of small frequencies arising from mutation during population expansion since the Pleistocene era. It is currently inconceivable that a genome scan could be based on enough polymorphisms to ensure inclusion of a particular causal SNP, although selection of nonsynonymous substitutions that alter an amino acid would to some extent increase the inclusion probability (4) whereas attempts to reduce the number of SNPs by selection to conserve diversity reduces power (19). Several papers advocate pooling DNA from several individuals for association mapping, but so far the technique has been too delicate to compete with pooling single samples (23).

Extension of association mapping to haplotypes

Most association mapping with an LD map has used composite likelihood for single SNPs, replacing d_i with δ_i (S_i-S_D), where S_i is the location on the LD map of the ith SNP, S is the location of a putative causal SNP, and δ_i is the Kronecker δ that takes the value 1 if $S_i \geq S$ and –1 otherwise, thereby assuring that the derivative of the composite likelihood takes the appropriate sign. Experience has shown that estimates of L and ε are seldom significantly better than for the LD map, and so iteration is often limited to M and S. Good results have been obtained with real and simulated data (16). So far, the association probability ρ has been limited to dichotomous traits but can be extended to quantitative traits.

It is possible to apply an LD map to haplotypes, which may well model the frequency of a causal SNP more accurately than a neighboring associated SNP. Among the problems to be overcome are estimation of haplotype probabilities for each diplotype, perhaps including inference (imputation) of untyped SNPs. Imputation introduces error while failure to impute discards a proportion of participants that increases with the number of SNPs in a haplotype, which also governs the number of haplotypes to be tested. The optimal number of SNPs in a haploset defined by a given set of SNPs appears to be small but must depend on their density and the LD

map. The simplest haploset for association mapping is of size 2 with the causal SNP intermediate on the LD map. The most significant haplotype is identified by the same logic as multiple alleles (11). This design is unique in that overlapping windows do not overlap intervals (e.g., SNP pairs 1,2 and 2,3 do not overlap). Haplotypes with an even number of SNPs take the LD location of the midpoint of their median pair whereas haplosets with an odd number of SNPs take the LD location of their medial SNP. Windows of 3 or more SNPs overlap, generating progressively higher autocorrelations as the number of SNPs increases. These problems have not been solved, but research on association mapping with haplotypes and an LD map is being actively pursued.

Alternative methods for association mapping

Association mapping is possible without an LD map, most simply by selecting the most significant single SNP or haploset. This avoids composite likelihood at the high cost of losing all information about other markers, dispensing with a support interval, and accepting a heavy correction that is prohibitive in a genome scan (24). Neglecting approaches that have not been applied since the physical map was nominally finished, mathematicians have proposed 3 alternatives to LD maps, all using haplotypes. One is non-Bayesian and uses logistic regression based on a similarity graph to select the most significant haploset with an appropriate correction (25). Since only 1 point is specified, the kb and LD maps provide identical results. Comparison with methods that provide a support interval and relative efficiency is not possible.

The other 2 methods are at once coalescent, Bayesian, and haplotypic. Coalescent theory assumes equilibrium, sacrificing an estimate of time (t) that is the principal discriminant between sex-averaged linkage and LD. Interpolation into one of these maps is necessary if the coalescent is to be used for linkage or association mapping, ignoring concern that coalescent theory provides "estimates of the recombination rate from polymorphism data [that] are extremely unreliable" (26). Bayesian methods use prior probabilities based on evidence in the sample and are therefore posterior probabilities, making the number of degrees of freedom unclear and residual variance ambiguous. It is difficult to compare results with non-Bayesian LD maps

Table 1
An abbreviated location database

Point	Locus	Band	kb	m_cM	f_cM	LDU	Q
pter	-	pter	0	0	0	0	-
SNP1	-	p3.1	5.314	-	-	0	0.42
SNP2	AC1D4	p3.1	5.641	-	-	0.44	0.15
AC1D4	AC1D4	p3.1	6.103	6.37	4.15	-	-
M1736	-	p3.0	10.163	8.71	9.03	-	-
•	•	•	•	•	•	•	•
•	•	•	•	•	•	•	•
SNP2419	-	q4.2	75000.451	-	-	983.41	0.13
qter	-	qter	75015.611	60.87	80.02	983.41	-

A prototype of this database, by W.J. Tapper, may be accessed at http://cedar.genetics.soton.ac.uk/public_html/. Filled circles indicate omission of all data except beginning and end. Point, a named sequence of 1 or more nucleotides; locus, an internationally approved name to which points are assigned; band, cytogenetic band to which points and loci are assigned; kb, median location of a point in kb; m_cM, cM location in a male linkage map (underscore indicates data not available on entry, but in general capable of interpolation from other variables); f_cM, cM location in a female linkage map; LDU, genetic location in LDUs; Q, minor allele frequency of diallelic point.

based on a defined set of parameters estimated without preconceptions. Objective criteria for this comparison must be sought, although in other applications the conflict between a few mathematicians who use Bayesian models and most scientists who reject them remains dogmatic. However these issues are resolved, the utility of efforts to improve construction of LD maps or find a more efficient substitute for association mapping should be measured in 4 ways: (a) correspondence with the sex-averaged linkage map; (b) residual variance of alternative LD maps; (c) capability of identifying systematic departures from the scaled linkage map due to selection and other evolutionary events; and (d) power for association mapping. So far, only composite likelihood on LD maps provides all these data and therefore yields a benchmark against which alternatives may be measured. One Bayesian method has a much larger support interval and estimation of error than association mapping with an LD map in the single example for which both were tried (27, 28). The other Bayesian example was scaled to a linkage map but has not yet been applied to an LD map or association mapping (29).

Location databases

During the last century, genetic information was obtained for *Drosophila melanogaster*, *Mus musculus*, and *Homo sapiens*. *Drosophila* led the way by giving each identified locus its position on the linkage map and assignment to a band on a salivary chromosome (30). The human map developed after isozymes and somatic cell hybrids supplied a substantial number of markers (31). As the number of map loci increased, the workshops in which maps were updated became chromosome-specific and then were abandoned in favor of computer databases that initially focused on linkage (32), now extended to physical location and LD. At present, the minimal location database (33) has an entry for each point (SNP, expressed locus, microsatellite, insertions and deletions, etc.), the locus if assigned therein, the chromosome band assignment, kb location, location in cM in the male and female linkage maps, location in one or more LD maps, and minor allele frequency (Table 1). Such a location database allows replacement of kb location as the "finished" map is refined, updating of location in the LD map(s), interpolation from LD to linkage maps, and use of a standard LD map to give starting values for a local map at higher density or for interpolation from such a local map into the standard map. Discrepancies between linkage and LD maps can be detected and effort made to explain them. The utility of a location database is so great that in time it may be adopted by national and international centers of bioinformatics. Perhaps even the forgettably long accession numbers of SNPs will be replaced by a more informative nomenclature.

Future directions

The short history of LD maps and association mapping leaves many questions unanswered. LD maps were first applied to small regions at moderate density, using a single algorithm. Extension to whole chromosomes introduces computational problems that can be addressed in many ways, and LD maps can be created faster and more reliably than by methods now in use to create them. Relative efficiency to describe linkage and LD is critical, but other measures of reliability will be developed.

Association mapping raises different problems. An LD map obviously excels the kb map for SNP assignment to a block, but within a block, all SNPs have the same location in LDU. Slight inclination of a critical block improves association mapping, but the best algorithm has not been established, and the most effective use of haplotypes has not been determined. Association mapping shares with LD maps the problem of recognizing and discarding inferior methods, but the evidence from support intervals and location error is clearly different from reliability measures for LD maps.

The utility of a location database can be enhanced if LD maps are progressively improved rather than beginning each update with the kb map. Operations on the database should allow interpolating high resolution LD into linkage maps without confounding the 2 sources of information. The challenge is 2-fold: first to geneticists, who must identify the most useful contents and operations for local databases; and second, to large centers, such as the European Bioinformatics Institute and the National Centre for Biotechnology Information, which must recognize that the problem is numerical rather than pictorial. Unless it is solved, the Human Genome Project will not realize its potential.

LD maps and association mapping are separated by 90 years from the development of linkage, and their applications are just beginning. Peripheral details and controversial enterprises such as selection of tagging SNPs and definition of blocks should not distract us from the central aim: to fulfill the promise of LD maps and association mapping in medicine, molecular biology, and the understanding of evolution.

Acknowledgments

The contribution of the University of Southampton to the research reported herein was supported in part by grants from the United Kingdom Medical Research Council and Applied Biosystems.

Address correspondence to: Newton Morton, Human Genetics Division, Duthic Building (MP 808), Southampton General Hospital, Tremona Road, Southampton SO16 6YD, United Kingdom. Phone: 44-0-23-8079-6536; Fax: 44-0-23-8079-4264; E-mail: nem@soton.ac.uk.

1. Morton, N.E., and Chung, C.S. 1978. Preface. In *Genetic epidemiology*. N.E. Morton and C.S. Chung, editors. Academic Press. New York, New York, USA. IX–X.
2. Collins, F.S. 1992. Positional cloning: let's not call it reverse anymore. *Nat. Genet.* **1**:3–6.
3. Abramowicz, M. 2003. The Human Genome Project in retrospect. *Adv. Genet.* **50**:231–261, discussion 507–510.
4. Botstein, D., and Risch, N. 2003. Discovering genotypes underlying human phenotypes: past success for Mendelian disease, future approaches for complex disease. *Nat. Genet.* **33**(Suppl.):228–237.
5. Sturtevant, A.H. 1913. The linear arrangement of six sex-linked factors in Drosophila, as shown by their mode of association. *J. Exp. Zool.* **14**:43–59.
6. Haldane, J.B.S. 1919. The combination of linkage values, and the calculation of distances between the loci of linked factors. *J. Genet.* **8**:299–309.
7. Maniatis, N., et al. 2002. The first linkage disequilibrium (LD) maps: delineation of hot and cold blocks by diplotype analysis. *Proc. Natl. Acad. Sci. U. S. A.* **99**:2228–2233.
8. Slatkin, M., and Veuille, M. 2002. *Modern developments in theoretical population genetics. The legacy of Gustave Malecot*. Oxford University Press. Oxford, United Kingdom/New York, New York, USA. 280 pp.
9. Morton, N.E., et al. 2001. The optimal measure of allelic association. *Proc. Natl. Acad. Sci. U. S. A.* **98**:5217–5221.
10. Morton, N.E. 2002. Applications and extensions of Malecot's work in human genetics. In *Modern developments in theoretical population genetics*. M. Slatkin and M. Veuille, editors. Oxford University Press. Oxford, United Kingdom. 20–36.
11. Collins, A., and Morton, N.E. 1998. Mapping a disease locus by allelic association. *Proc. Natl. Acad. Sci. U. S. A.* **95**:1741–1745.
12. Jeffreys, A.J., Kauppi, L., and Neumann, R. 2002. Intensely punctate meiotic recombination in the class II region of the major histocompatibility complex. *Nat. Genet.* **29**:217–222.
13. Daly, M.J., Rioux, J.D., Schaffner, S.F., Hudson, T.J., and Lander, E.S. 2001. High-resolution haplotype structure in the human genome. *Nat. Genet.* **29**:229–232.
14. Ke, X., et al. 2004. The impact of SNP density on fine-scale patterns of linkage disequilibrium. *Hum.*

Mol. Genet. **13**:577–588.

15. Zhang, W., Collins, A., Maniatis, N., Tapper, W., and Morton, N.E. 2002. Properties of linkage disequilibrium (LD) maps. *Proc. Natl. Acad. Sci. U. S. A.* **99**:17004–17007.

16. Maniatis, N., et al. 2004. Positional cloning by linkage disequilibrium. *Am. J. Hum. Genet.* **74**:846–855.

17. Kimura, M., and Ohta, T. 1973. The age of a neutral mutant persisting in a finite population. *Genetics.* **75**:199–212.

18. Zhang, W., et al. 2004. Impact of population structure, effective bottleneck time, and allele frequency on linkage disequilibrium maps. *Proc. Natl. Acad. Sci. U. S. A.* **101**:18075–18080.

19. Zhang, W., Collins, A., and Morton, N.E. 2004. Does haplotype diversity predict power for association mapping of disease susceptibility? *Hum. Genet.* **15**:157–164.

20. Couzin, J. 2002. New mapping project splits the community. *Science.* **296**:1391–1393.

21. Morton, N.E. 1956. The detection and estimation of linkage between the genes for elliptocytosis and the Rh blood type. *Am. J. Hum. Genet.* **8**:80–96.

22. Morton, N.E. 1998. Significance levels in complex inheritance. *Am. J. Hum. Genet.* **62**:690–697.

23. Godde, R., et al. 2004. Refining the results of a whole-genome screen based on 4666 microsatellite markers for defining predisposition factors for multiple sclerosis. *Electrophoresis.* **25**:2212–2218.

24. Risch, N., and Merikangas, K. 1996. The future of genetic studies of complex human diseases. *Science.* **273**:1516–1517.

25. Durrant, C., et al. 2004. Linkage disequilibrium mapping via cladistic analyses of single-nucleotide polymorphism haplotypes. *Am. J. Hum. Genet.* **75**:35–43.

26. Nordborg, M. 2001. Coalescent theory. In *Handbook of statisticalgenetics*. D.J. Balding, M. Bishop, and C. Cannings, editors. John Willey & Sons, Ltd. Chichester, United Kingdom. 179–212.

27. Morris, A.P., Whittaker, J.C., Xu, C.F., Hosking, L.K., and Balding, D.J. 2003. Multipoint linkage-disequilibrium mapping narrows location interval and identifies mutation heterogeneity. *Proc. Natl. Acad. Sci. U. S. A.* **100**:13442–13446.

28. Maniatis, N., et al. 2005. The optimal measure of linkage disequilibrium minimizes error in association mapping of affection status. *Hum. Mol. Genet.* **14**:187–195.

29. McVean, G.A., et al. 2004. The fine-scale structure of recombination rate variance in the human genome. *Science.* **304**:581–584.

30. Bridges, C.B., and Brehme, K.S. 1944. *The mutants of drosophila melangaster*. Carnegie Institution of Washington Publications. Washington, DC, USA. 552 pp.

31. D. Bergsma, editor. 1974. *Human gene mapping.* Vol. X, issue 3 of *Birth defects original article series.* The National Foundation, March of Dimes. Intercontinental Medical Book Corporation. New York, New York, USA. 216 pp.

32. Collins, A., Frezal, J., Teague, J., and Morton, N.E. 1996. A metric map of humans: 23,500 loci in 850 bands. *Proc. Natl. Acad. Sci. U. S. A.* **93**:14771–14775.

33. Ke, X., Tapper, W., and Collins, A. 2001. LDB2000: sequence-based integrated maps of the human genome. *Bioinformatics.* **17**:581–586.

Genetic counselors: translating genomic science into clinical practice

Robin L. Bennett,[1] Heather L. Hampel,[2] Jessica B. Mandell,[3] and Joan H. Marks[4]

[1]Medical Genetics Clinic, University of Washington, Seattle, Washington, USA. [2]The Ohio State University, College of Medicine and Public Health, Columbus, Ohio, USA. [3]King Laboratory, Departments of Medicine and Genome Sciences, University of Washington, Seattle, Washington, USA. [4]Human Genetics Program, Sarah Lawrence College, Bronxville, New York, USA.

In a time of emerging genetic tests and technologies, genetic counselors are faced with the challenge of translating complex genomic data into information that will aid their client's ability to learn about, understand, make, and cope with decisions relating to genetic diagnoses. This article examines the role of the genetic counselor, particularly in counseling individuals at risk for or diagnosed with breast cancer, in an era of high-tech health care and gene patents.

While 2003 marks the 50th anniversary of Watson and Crick's landmark discovery of the DNA helix, it also marks the 34th anniversary of the beginning of a new breed of masters-trained health professionals — genetic counselors. The field of genetic counseling developed from a need to educate, manage, and counsel individuals and families diagnosed with, or at risk for, genetic diseases with respect to how these conditions affect the psychological, medical, financial, and social aspects of one's life. Genetic counselors are at the forefront of introducing and applying the advances from genome science to the lives of individuals and their families.

Genetic counselors' work involves all stages of the life cycle, from preconception counseling to prenatal diagnosis, the diagnosis of newborns or pediatric genetic disorders, and the diagnosis of elderly individuals with inherited predisposition to diseases such as cancer, presenile dementia, psychiatric disorders, and heart disease. Genetic counseling sessions with clients and their families may involve a one-time crisis intervention dealing with a new genetic diagnosis or may develop into a relationship over many years if the client is treated in a specialty clinic for diseases like muscular dystrophy, fragile X syndrome, cystic fibrosis, or Huntington disease.

Genetic counselors are expert educators, skilled in translating the complex language of genomic medicine into terms that are easy to understand. According to the 2002 Professional Status Survey of the National Society of Genetic Counselors (NSGC) (www.nsgc.org), over 75% of genetic counselors are involved in educating physicians, medical students, nurses and other health professionals, community organizations, undergraduates, and grammar and high school students. While the majority of genetic counselors work in university medical centers, hospitals (private and public), or large medical facilities, an increasing number work with diagnostic laboratories and pharmaceutical companies and in positions related to the development of government and public policy. Genetic counselors are also uniquely trained to work as research coordinators for genetic research studies.

The term "genetic counselor" describes a masters-level health professional with extensive training in human genetics and counseling skills. The first group of ten genetic associates graduated from Sarah Lawrence College in 1971 (1). Today there are 28 programs accredited by the American Board of Genetic Counseling (ABGC) in the US and Canada, and similar programs are in place in the United Kingdom, South Africa, and Australia. The ABGC has certified genetic counselors since 1993. Graduates from genetic counseling programs accredited by the ABGC demonstrate competency in 27 areas within four critical domains: communication; critical thinking; interpersonal counseling and psychosocial assessment; and professional ethics and values (2). Coursework in genetic counseling training programs involves human genetics, cytogenetics, developmental biology and embryology, teratology, statistics, qualitative and quantitative research, counseling and communication skills, interviewing, ethics, and public health. Skills are obtained by a combination of role playing and comprehensive fieldwork in a variety of practice settings.

The practice and process of genetic counseling

The definition of genetic counseling adopted by the American Society of Human Genetics in 1975 (3) is under transition as the field of genetic counseling has grown with the demands of genomic medicine (4–6). Traditionally, genetic counseling has centered on prenatal and pediatric genetic diagnoses and decision-making involving reproduction. However, counseling for individuals at risk for common disorders is increasing (Figure 1). Genetic counseling focuses on complex issues related to the value of genetic testing itself, and on medical interventions and health care practices that have varying degrees of efficacy and success (5). The traditional dogma that genetic counseling must be nondirective is being challenged in favor of a psychosocial approach that emphasizes shared deliberation and decision-making between the counselor and the client. The genetic counseling intervention is designed to reduce the client's anxiety, enhance the client's sense of control and mastery over life circumstances, increase the client's understanding of the genetic disease and options for testing and disease management, and provide the individual and family with the tools required to adjust to potential outcomes (7–10). The information provided during genetic counseling helps the individual and family personalize often threatening information in order to clarify their values and strengthen their coping mechanisms (11). A major tenet of all approaches to genetic counseling is that it should be noncoercive.

Establishing and analyzing a pedigree

The process of genetic counseling begins with the merging of the counselor's and the client's expectations, referred to as "contract-

Nonstandard abbreviations used: National Society of Genetic Counselors (NSGC); American Board of Genetic Counseling (ABGC); hereditary breast-ovarian cancer syndrome (HBOC).

Conflict of interest: The authors have declared that no conflict of interest exists.

Citation for this article: *J. Clin. Invest.* **112**:1274–1279 (2003). doi:10.1172/JCI200320113.

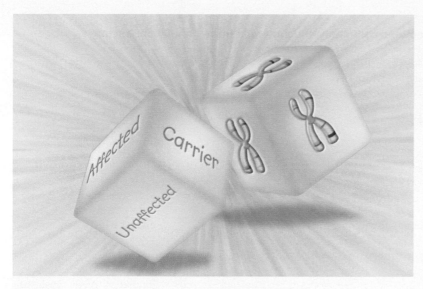

Figure 1
Genetic counseling for individuals at risk for common disorders is increasing.

ing." Early in the session, a medical family history is obtained and graphically recorded as a pedigree. Until the early 1990s, several methods of symbolization were used to record a pedigree (12). In 1995, the Pedigree Standardization Task Force of the NSGC established an international standard system of pedigree nomenclature (13). Confirmation of family medical information through review of medical records and death certificates is often essential to assure that the pedigree, and thus the risk assessment provided to the client, is based on accurate information. Several studies document the potential inaccuracy of much medical family lore, particularly health information about second- and third-degree relatives (14, 15). Pedigrees often need updating, as births, deaths, and illness can alter the assessment of genetic risk.

A pedigree is not only essential for confirming a pattern of inheritance in a family but also serves as a tool for determining genetic-testing strategies, distinguishing genetic from other risk factors, and identifying medical screening needs for healthy individuals. A pedigree is an important method of establishing patient rapport. It also serves as a visual demonstration for providing patient education on variation in disease expression in the family as well as identifying other relatives at risk for disease (14).

Statistical risk assessment based on pedigree analysis and the results of any genetic testing in the patient or other relatives is central to genetic counseling. Counselors explore clients' preconceived notions about patterns of inheritance, chances of testing positive or developing the family disorder, and the burden (e.g., morbidity, mortality, financial and psychological difficulties) of the disease. If genetic counseling information diverges from a client's perceptions, the client may have difficulty incorporating it or implementing disease-management recommendations. Risk information must be provided in multiple forms. For example, the client may perceive a tenfold increase in the relative risk of developing disease as quite high, when the absolute risk may be only 2%. Likewise, a 1 in 20 chance of developing a condition may be perceived as much more significant than a 95% chance that they will be free of the disease.

While pedigree analysis and risk assessment may seem the most difficult aspect of genetic counseling, in reality, addressing the myriad

reactions and emotions of clients responding to a genetic diagnosis can be far more challenging (14–17) (see *Psychological dilemmas, emotions, and reactions commonly encountered during genetic counseling*). This may involve more than one appointment, or referral for psychotherapy. Because a genetic diagnosis may have serious ramifications for a client's extended family, it is paramount that the client be enabled to make an informed decision as to whether to be tested. There is a general trend for health professionals to refer clients to a genetic counselor only *after* the client has received a positive test result. In reality, the most appropriate stage at which to make a referral is early in the consideration of initial genetic testing or diagnosis.

Challenges to the paradigm of health and disease

The era of genomic medicine challenges traditional definitions of "healthy" and "diseased." Traditionally, individuals have sought medical attention out of concern regarding a present illness. Genetic testing facilitates diagnosis of a healthy individual who is expected to develop or has an increased susceptibility to develop a disorder. A class of individuals is created who may be perceived as "genetically unwell" (18). The emotional consequences of diagnosing a healthy person as "diseased" can be profound. There are also instances of survivor guilt, where a healthy individual feels guilty upon testing negative for a family-related genetic condition because he or she has "escaped" the disease whereas other family members have not. Presymptomatic or susceptibility tests have been readily available only since the early 1990s, so the data on their impact are limited. The data suggest that most individuals cope well with the results of these tests, perhaps partly because the tests, to date, have been primarily requested by genetic counselors and medical geneticists specifically trained to help patients interpret the complex information, often with the luxury of 60- to 90-minute visits. With the push to provide genetic testing in primary care settings where appointments are often limited in time, as well as to market tests directly to consumers, it will be interesting to see whether genetic information continues to be well handled.

Testing minors for adult-onset disorders

Genetic testing of healthy children for predisposition for adult-onset disorders is a further area of controversy. Is it the parents' right to know whether a child will develop Huntington disease or familial ovarian cancer? Perhaps the parents wish to make informed choices regarding their child's academic track and insurance coverage and be aware of possible factors in their child's future reproductive choices. Statements from several professional societies suggest that genetic testing for disorders with no medical interventions is unwarranted in healthy children (19–21). This protection of the child's rights is based on the observation that many adults at risk for conditions such as cancer or Huntington disease choose not to be tested, which makes it unfair for parents or health professionals to take this decision from the child. This reasoning is extended to conclude that women have the right to choose prenatal diagnosis for adult-onset conditions only if they are considering terminating the pregnancy as a result of the diagnosis. That is, a woman should not have prenatal diagnosis "just

Psychological dilemmas, emotions, and reactions commonly encountered during genetic counseling

Anger
Anxiety
Anxious preoccupation (defining self by disease or high-risk status)
Grief and mourning
Shattered expectations of normality
Reaction formation (relabeling a loss as a gain)
Intellectualization
Rationalization
Denial
Disbelief
Displacement (finding fault with others)
Magical thinking/superstitious beliefs
Cognitive avoidance (suppressing thoughts about a frightening or overwhelming topic)
Parental guilt
Threat to perceived parental role (fantasy of parenthood, role as protector)
Survivor guilt
Helplessness
Repression
Shame
Hopelessness
Fatalism
Change in perception of self
Change in family belief systems
Challenge to religious beliefs
Change in social functioning
Marital/relationship discord
Search for meaning
Fear of discrimination (insurance, employment, societal)

to know," if she would not terminate the pregnancy, since then a child would be born who would not normally be tested for the condition. Some individuals who do not want to pass a genetic condition to a child, but also do not want to terminate a mutation-positive pregnancy, choose to have genetic diagnosis before implantation. By implanting only embryos that will not have the genetic disease in question, the couple avoids the possibility of having to terminate an affected fetus.

Cancer: a new paradigm for genetic counseling

Genetic counseling for cancer is one of the newest subspecialties in the field. It began at major cancer centers in the early 1990s, mainly through the identification of families suitable for gene-hunting studies, since few cancer susceptibility genes were known. As inherited syndromes involving common cancers such as breast and colon cancer were identified, the genetic counseling field grew rapidly. Counseling positions were created at most cancer hospitals and in large oncology groups — areas previously uninhabited by the profession. In 2002, the NSGC Professional Status Survey indicated that 42% of genetic counselors worked in the field of cancer genetics, a fourfold increase from just 10.4% of counselors in 1994. This expansion of the cancer genetic counseling field is likely due to strong public interest, as cancer is a common disease, and to medical advances leading to the early detection, risk reduction, or prevention of cancer in those at the highest hereditary risk. Cancer genetics will serve as a paradigm for other adult-onset conditions as more predisposition genes are identified.

Counseling for cancer: hereditary breast-ovarian cancer syndrome

Cancer genetic counseling differs significantly from the traditional prenatal and pediatric consultation process (22). At least a three-generation pedigree is required to identify a hereditary pattern of cancers in a family, typically obtained prior to the counseling appointment. Emphasis is also placed on collecting medical records to document all cancer diagnoses, as the accuracy of a risk assessment depends on knowledge of the anatomical location of the disease and the ages of diagnosis of family members (23–29).

During the initial consultation, cancer genetic counselors obtain information on lifestyle risk factors and cancer-screening practices. They provide tailored cancer-surveillance recommendations based on the patient's risk level. Recommendations are provided in a directive manner, a departure from traditional nondirective genetic counseling. Since genetic counseling is a consultation service and is not involved in long-term management, patients return to their referring physicians and other medical specialists after counseling for the coordination of cancer surveillance. It is common for patients to have multiple genetic counseling sessions as part of a cancer genetics consultation, since patients who elect to pursue testing often return for a blood-draw appointment and are required to return for a result-disclosure session.

The leading source of referrals in most cancer genetics programs is the evaluation of patients with a personal or family history of breast cancer. The differential diagnosis for these families includes several known cancer syndromes (Table 1), with hereditary breast-ovarian

Table 1
Differential diagnosis for familial breast cancer

Syndrome	Gene(s)	Breast cancer risk	Other associated cancers
Hereditary breast-ovarian cancer syndrome	BRCA1 and BRCA2	45–85% (14)	Ovarian, prostate
Cowden syndrome	PTEN	25–50% (47, 48)	Thyroid, endometrial
Li-Fraumeni syndrome	TP53	50% (49)	Sarcoma, adrenocortical, brain, leukemia
Peutz-Jeghers syndrome	STK11	RR = 20.3 (breast and gynecologic cancers) (50, 51)	Gastrointestinal, pancreatic, ovarian, uterine
Carriers of ataxia telangiectasia	ATM	RR = 1.5–7.1 (52, 53)	
Familial breast cancer/Hereditary breast and colorectal cancer syndrome	CHK2 1100delC	RR = 2 (54)	Colon (55)

RR, relative risk.

cancer syndrome (HBOC) (30–32) having the highest incidence. Mutations in the *BRCA1* and *BRCA2* genes are most commonly associated with individuals and/or families with HBOC. *BRCA1* and *BRCA2* are tumor-suppressor genes that serve protective functions in breast and ovarian cells. The protective functions of the BRCA1 protein include regulation of estrogen-receptor activity related to the control of estrogen-induced proliferation of breast tissue; involvement in homologous recombination and repair of transcription-coupled oxidation-induced DNA damage; cell-cycle checkpoint control through interactions with *RB*, *ESF1*, and *TP53*; and involvement in chromatin remodeling (31). BRCA2 functions in homologous recombination, interacts with *TP53* in cell-cycle checkpoint control, and is involved in chromatin remodeling (31). Individuals inheriting a germline mutation in *BRCA1* or *BRCA2* need only acquire a mutation in the working copy of their *BRCA* gene in an at-risk cell during their lifetime to lose these protective functions and to eventually develop cancer, also known as the multi-hit model of tumorigenesis as first proposed by Knudson in 1971 (33).

BRCA gene mutations are inherited in an autosomal dominant pattern, so that children of an affected individual have a 50% chance of inheriting HBOC. Individuals with a *BRCA1* gene mutation appear to have a 65–85% lifetime risk of breast cancer and a 39–60% lifetime risk of ovarian cancer (34). Women with a mutation in their *BRCA2* gene have a 45–85% lifetime risk of breast cancer and an 11–25% lifetime risk of ovarian cancer (34). An increased risk for prostate cancer has been noted in men with mutations in either gene (35). Male breast cancer and pancreatic cancer have also been associated with *BRCA2* gene mutations (36, 37). Management for individuals with HBOC includes monthly breast self-examination beginning at age 18, clinical breast examination every 6 months beginning at age 25, and annual mammography beginning at age 25 (38, 39). Ovarian cancer surveillance, the efficacy of which is unproven, consists of transvaginal ultrasound with color-Doppler and cancer antigen-125 (CA-125) tumor marker tests every 6 months beginning between the ages of 25 and 35 (38, 39). Chemoprevention may also hold promise for reducing the risk of cancer among women with HBOC. Many individuals with *BRCA* mutations elect to undergo risk-reducing surgeries such as mastectomy or bilateral salpingo-oophorectomy. While these surgeries significantly reduce the risk of cancer, they do not eliminate it (40, 41).

Genetic testing for the *BRCA1* and *BRCA2* genes has been clinically available in the US since late 1996 from a single laboratory. There are three possible results to a full-sequence test for these genes: (a) a positive result means that a cancer-causing mutation was identified; (b) a negative result means that no mutations were

identified; and (c) an ambiguous result means that a genetic variant of uncertain significance was identified. Variants of uncertain significance (usually missense mutations) create a counseling challenge, since they may or may not lead to an increased risk for cancer in the family. These mutations warrant further analysis within the family and in research laboratories before predictive testing can be offered to other at-risk family members. In addition, a negative sequence result does not completely rule out the presence of mutations in *BRCA1* or *BRCA2*, as sequencing cannot detect large gene rearrangements or rare regulatory mutations.

Gene patents and the future of diagnostic testing

The laboratory that performs *BRCA* gene testing has held US patents on the *BRCA* test since 1999 and European patents on the *BRCA1* test since 2001 (42). The issue of gene patents remains controversial given that not all testing techniques can detect all mutations and that the presence of a patent deters researchers from developing new and perhaps better testing methods and therapeutics. Furthermore, patients must use one laboratory, which has little incentive to offer competitive pricing. In late August 2002, several scientists, genetic testing laboratories, and genetics societies in France, Belgium, The Netherlands, Sweden, the United Kingdom, Spain, and Germany joined together to file opposition with the European Patent Office (42, 43). In Canada, several provincial governments have told their health authorities to continue *BRCA1* and *BRCA2* testing while they fight the patent through legal channels (44). However, proponents of gene patents note that without the ability to recoup costs through patents, laboratories will have no incentive to invest in the research and development necessary to find new genes involved in disease. In 2002, the laboratory currently offering *BRCA1* and *BRCA2* testing made another controversial move by initiating a campaign of marketing directly to consumers, following in the footsteps of pharmaceutical companies. The Atlanta and Denver areas were targeted by print, radio, and television advertisements encouraging women to consider genetic testing for HBOC (45). Critics feel that this could lead to increased health care costs resulting from inappropriate testing, and to a false sense of reassurance in individuals found not to have a gene mutation who are not informed about the limitations of these tests (45, 46). The laboratory also provides training to health care providers whom it does not employ and who do not necessarily have a background in genetics or genetic counseling. Many consider this to entail a conflict of interest, since these providers are trained to give cancer genetic consultations that will likely lead to an increase in requests for this laboratory's tests. These issues, among others, underscore the importance of thorough cancer genetic counseling for all patients who are considering genetic testing so that they can understand what the test can and cannot tell them, determine whether they are appropriate candidates for testing, and give truly informed consent.

Cancer genetic counseling and predictive genetic testing are complex and time consuming, so most physicians prefer to refer patients to specialists in this area. Genetic counseling will likely become more widely used in many fields of medicine as genes for other common adult-onset conditions are discovered — for instance, genes related to cardiovascular disease or psychiatric conditions. The model of cancer genetics can be applied to these new subspecialties as an effective way to integrate counseling services with the services of other medical professionals and researchers.

Address correspondence to: Robin L. Bennett, University of Washington, Medical Genetics, Box 357720, Seattle, Washington 98195-7720, USA. Phone: (206) 616-2135; Fax: (206) 616-2414; E-mail: robinb@u.washington.edu.

1. Marks, J.H., and Richter, M.L. 1976. The genetic associate: a new health professional. *Am. J. Public Health.* **66**:388–390.
2. Walker, A.P. 1998. The practice of genetic counseling. In *A guide to genetic counseling.* D.L. Baker, J.L. Schuette, and W.R. Uhlmann, editors. Wiley-Liss Inc. New York, New York, USA. 1–26.
3. Ad Hoc Committee on Genetic Counseling. 1975. Report to the American Society of Human Genetics. *Am. J. Hum. Genet.* **27**:240–242.
4. Biesecker, B., and Peter, K. 2002. Genetic counseling: ready for a new definition? *J. Genet. Couns.* **11**:536–537. (Abstr.)
5. Weil, J. 2003. Psychosocial genetic counseling in the post-nondirective era: a point of view. *J. Genet. Couns.* **12**:199–211.
6. Biesecker, B.B. 2003. Back to the future of genetic counseling: commentary on "psychosocial genetic counseling in the post-nondirective era." *J. Genet. Couns.* **12**:213–217.
7. Burke, W., Pinsky, L.E., and Press, N.A. 2001. Categorizing genetic tests to identify their ethical, legal and social implications. *Am. J. Med. Genet.* **106**:233–240.
8. Elwyn, G., Gray, J., and Clarke, A. 2000. Shared decision making and non-directiveness in genetic counseling. *J. Med. Genet.* **37**:135–138.
9. Jansen, L.A. 2001. Role of the nurse in clinical genetics. In *Genetics in the clinic: clinical, ethical and social implications for primary care.* M.B. Mahowald, A.S. Scheurle, V.A. McKusick, and T.J. Aspinwall, editors. Mosby Publishing. St. Louis, Missouri, USA. 133–141.
10. McConkie-Rosell, A., and Sullivan, J.A. 1999. Genetic counseling: stress, coping, and the empowerment perspective. *J. Genet. Couns.* **8**:345–357.
11. Shiloh, S., Berkenstadt, M., Meiran, N., Bat-Miriam-Katznelson, M., and Goldman, B. 1997. Mediating effects of perceived personal control in coping with a health threat: the case of genetic counseling. *J. Appl. Soc. Psychol.* **27**:1146–1174.
12. Bennett, R.L., Steinhaus, K.A., Uhrich, S.B., and O'Sullivan, C. 1993. The need for developing standardized pedigree nomenclature. *J. Genet. Couns.* **2**:261–273.
13. Bennett, R.L., et al. 1995. Recommendations for standardized human pedigree nomenclature. *Am. J. Hum. Genet.* **56**:745–752.
14. Bennett, R.L. 1999. *The practical guide to the genetic family history.* Wiley-Liss Inc. New York, New York, USA. 251 pp.
15. Schneider, K. 2002. *Counseling about cancer: strategies for genetic counseling.* Wiley-Liss Inc. New York, New York, USA. 333 pp.
16. 2000. *Psyche and helix: psychological aspects of genetic counseling.* R.G. Resta, editor. Wiley-Liss Inc. New York, New York, USA. 180 pp.
17. Weil, J. 2000. *Psychosocial genetic counseling.* Oxford University Press. New York, New York, USA. 291 pp.
18. Jonsen, A.R., Durfy, S.J., Burke, W., and Motulsky, A.G. 1996. The advent of the "unpatients." *Nat. Med.* **2**:622–624.
19. American Society of Human Genetics/American College of Medical Genetics Reports. 1995. Points to consider: ethical, legal and psychological implications of genetic testing in children and adolescents. *Am. J. Hum. Genet.* **57**:1233–1241.
20. National Society of Genetic Counselors. 1994. Position statement on prenatal and childhood testing for adult-onset disorders. *Perspectives in Genetic Counseling.* **17**:5.
21. American Academy of Pediatrics, Committee on Genetics. 2000. Molecular genetic strategies in pediatric practice: a subject review. *Pediatrics.* **106**:1494–1497.
22. Peters, J.A., and Biesecker, B.B. 1997. Genetic counseling and hereditary cancer. *Cancer Supplement.* **80**:576–586.
23. Douglas, F., O'Dair, L., Robinson, M., Evans, D.G., and Lynch, S.A. 1999. The accuracy of diagnoses as reported in families with cancer: a retrospective study. *J. Med. Genet.* **36**:309–312.
24. Schmidt, S., Becher, H., and Chang-Claude, J. 1998. Breast cancer risk assessment: use of complete pedigree information and the effect of misspecified ages at diagnosis of affected relatives. *Hum. Genet.* **102**:348–356.
25. Bergmann, M., et al. 1998. Validity of self-reported cancers in a prospective cohort study in comparison with data from state cancer registries. *Am. J. Epidemiol.* **147**:556–562.
26. Parent, M., Ghadirian, P., Lacroix, A., and Perret, C. 1997. The reliability of recollections of family history: implications for the medical provider. *J. Cancer Educ.* **12**:114–120.
27. Aitken, J., Bain, C., Ward, M., Siskind, V., and MacLennan, R. 1995. How accurate is self-reported family history of colorectal cancer? *Am. J. Epidemiol.* **141**:863–871.
28. Love, R., Evans, A., and Josten, D. 1985. The accuracy of patient reports of a family history of cancer. *J. Chronic Dis.* **38**:289–293.
29. Novakovic, B., Goldstein, A.M., and Tucker, M.A. 1996. Validation of family history of cancer in deceased family members. *J. Natl. Cancer Inst.* **88**:1492–1493.
30. Lynch, H.T., Snyder, C.L., Lynch, J.F., Riley, B.D., and Rubinstein, W.S. 2003. Hereditary breast-ovarian cancer at the bedside: role of the medical oncologist. *J. Clin. Oncol.* **21**:740–753.
31. Narod, S.A. 2002. Modifiers of risk of hereditary breast and ovarian cancer. *Nat. Rev. Cancer.* **2**:113–123.
32. Robson, M.E. 2002. Clinical considerations in the management of individuals at risk for hereditary breast and ovarian cancer. *Cancer Control.* **9**:457–465.
33. Knudson, A.G. 1971. Mutation and cancer: statistical study of retinoblastoma. *Proc. Natl. Acad. Sci. U. S. A.* **68**:820–823.
34. Antoniou, A., et al. 2003. Average risks of breast and ovarian cancer associated with BRCA1 or BRCA2 mutations detected in case series unselected for family history: a combined analysis of 22 studies. *Am. J. Hum. Genet.* **72**:1117–1130.
35. Struewing, J.P., et al. 1997. The risk of cancer associated with specific mutations of BRCA1 and BRCA2 among Ashkenazi Jews. *N. Engl. J. Med.* **336**:1401–1408.
36. Wooster, R., et al. 1994. Localization of a breast cancer susceptibility gene, BRCA2, to chromosome 13q12-13. *Science.* **265**:2088–2090.
37. 1999. Cancer risks in BRCA2 mutation carriers. The Breast Cancer Linkage Consortium. *J. Natl. Cancer Inst.* **91**:1310–1316.
38. Burke, W., et al. 1997. Recommendations for follow-up care of individuals with an inherited predisposition to cancer. II. BRCA1 and BRCA2. Cancer Genetics Studies Consortium. *JAMA.* **277**:997–1003.
39. Daly, M. 2002. National Comprehensive Cancer Network Practice Guidelines. Genetic/Familial high-risk assessment: breast and ovarian. Version 1.2003. http://www.nccn.org/physician_gls/f_guidelines.html.
40. Rebbeck, T.R., et al. 2002. Prophylactic oophorectomy in carriers of BRCA1 or BRCA2 mutations. *N. Engl. J. Med.* **346**:1616–1622.
41. Hartmann, L., et al. 2001. Efficacy of bilateral prophylactic mastectomy in BRCA1 and BRCA2 gene mutation carriers. *J. Natl. Cancer Inst.* **93**:1633–1637.
42. Benowitz, S. 2003. European groups oppose Myriad's latest patent on BRCA1. *J. Natl. Cancer Inst.* **95**:8–9.
43. Matthijs, G., et al. 2002. Patents and monopolies on diagnostic tests: Europe's opposition against the BRCA1 patents. *Am. J. Hum. Genet.* **71**:182. (Abstr. 93).
44. 2003. Gene patents and the public good. *Nature.* **423**:207.
45. Pearson, H. 2003. Genetic test adverts under scrutiny. *Nature Science Update.* http://www.nature.com/nsu/030317/030317-3.html.
46. Moreno, J. 2002. Selling genetic tests: shades of gray in your DNA. *ABCNEWS.com.* http://abcnews.go.com/sections/living/DailyNews/ONCALL_DTC_brca_tests020923.html.
47. Starink, T.M., et al. 1986. The Cowden syndrome: a clinical and genetic study in 21 patients. *Clin. Genet.* **29**:222–233.
48. Brownstein, M.H., Wolf, M., and Bilowski, J.B. 1978. Cowden's disease: a cutaneous marker of breast cancer. *Cancer.* **41**:2393–2398.
49. Li, F.P., et al. 1988. A cancer family syndrome in twenty-four kindreds. *Cancer Res.* **51**:6094–6097.
50. Boardman, L., et al. 1998. Increased risk for cancer in patients with the Peutz-Jeghers syndrome. *Ann. Intern. Med.* **128**:896–899.
51. Giardiello, F.M., et al. 1987. Increased risk of cancer in the Peutz-Jeghers syndrome. *N. Engl. J. Med.* **316**:1511–1514.
52. Olsen, J.H., et al. 2001. Cancer in patients with ataxia-telangiectasia and in their relatives in the Nordic countries. *J. Natl. Cancer Inst.* **93**:84–85.
53. Swift, M., Morrell, D., Massey, R.B., and Chase, C.L. 1991. Incidence of cancer in 161 families affected by ataxia-telangiectasia. *N. Engl. J. Med.* **325**:1831–1836.
54. Meijers-Heijboer, H., et al. 2002. Low-penetrance susceptibility to breast cancer due to CHEK2(*)1100delC in noncarriers of BRCA1 or BRCA2 mutations. *Nat. Genet.* **31**:55–59.
55. Meijers-Heijboer, H., et al. 2003. The CHEK2 1100delC mutation identifies families with a hereditary breast and colorectal cancer phenotype. *Am. J. Hum. Genet.* **72**:1308–1314.

Genetic counseling throughout the life cycle

Leslie J. Ciarleglio,[1] Robin L. Bennett,[2] Jennifer Williamson,[3]
Jessica B. Mandell,[4] and Joan H. Marks[5]

[1]Yale University School of Medicine, New Haven, Connecticut, USA. [2]Medical Genetics Clinic, University of Washington, Seattle, Washington, USA.
[3]Gertrude H. Sergievsky Center, Columbia University College of Physicians and Surgeons, New York, New York, USA. [4]King Laboratory,
Departments of Medicine and Genome Sciences, University of Washington, Seattle, Washington, USA. [5]Human Genetics Program,
Sarah Lawrence College, Bronxville, New York, USA.

As the definition of genetic counseling continues to evolve, so does the application of genetic counseling services in all areas of medicine and throughout the human life cycle. While governmental policy, economics, ethics, and religion continue to influence society's views regarding the necessity of testing germ cells for mutations to prevent the birth of an affected child or predicting whether healthy adults will develop future life-threatening illness, patient autonomy in the choice of whether to know, or not know, one's genetic make-up remains a core principle of genetic counseling.

Genetic counseling is an expanding field in the era of genomic medicine. This unique medical specialty provides clinical health care, education, and emotional support to individuals and families facing genetic and inherited diseases. Genetic counselors provide services to patients across the lifespan by assessing family and environmental history to determine disease risk; assisting in genetic testing, diagnosis, and disease prevention and management; and offering psychosocial and ethical guidance to help patients make informed, autonomous health care and reproductive decisions (Figure 1). The purpose of this article is to provide an authoritative review of genetic counseling, its role in the front lines of genetic health, and its impact on medical research, education, and patient care.

Prenatal diagnosis: congenital heart disease

There was a time when prenatal genetic counseling was reserved for high-risk patients. For many years, the term "high-risk" was generally defined by a maternal age of 35 years or older at the time of expected delivery. That definition has evolved significantly as an increasing number of tests have become available to pregnant patients and the list of conditions amenable to prenatal diagnosis has grown. Now, all pregnant women and women planning pregnancy are potential candidates for genetic counseling. Yet the array of testing options can be mind-boggling for patients and their health care providers, and the associated anxiety may be magnified logarithmically when test results are abnormal. The genetic counselor plays a crucial role in deciphering this information for families, while simultaneously providing emotional support.

Heart defects are among the most common birth defects and are the leading cause of death in the first year of life (1, 2). Congenital heart disease (CHD) occurs with a frequency of about 8 per 1,000 births. Approximately 25% of all infant deaths are due to congenital malformations, 30% of which are related to CHD (3). There is an increased incidence of CHD in stillborns (4), and autopsy studies suggest that the incidence of fetal CHD may approach 30 per

1,000 (5). Because the majority of infants with CHD are born to mothers with no well-defined risk factors, an increasing number of affected fetuses are identified during routine obstetric scanning (5–8). Consequently, genetic counseling for patients facing the prenatal identification of CHD is common.

Although the majority of CHD cases are sporadic in nature, the current understanding of the major steps of cardiac development allows for categorization of defects by common embryonic origin. The cardiovascular system is the first major system to function during fetal development. The primitive heart, derived from embryonic mesoderm and neural crest cells, begins to form around embryonic day 18 and beats by day 22. The early heart begins as the endocardial tube, which bends, loops, and ultimately partitions itself into the four well-recognized chambers, establishing the basis for separation of pulmonary and systemic blood flow at birth (9). In the case of CHD that is not clearly related to known risk factors (see below), insight has been gained into the underlying genetic mechanisms of abnormal heart formation through an understanding of the embryologic development of the heart (10–13).

Risk factors related to CHD are well described. Maternal risk factors include diabetes, phenylketonuria (14, 15), viral infection such as rubella, and specific exposure to agents such as alcohol, antiseizure medications, Accutane, and lithium. Fetal risk factors include suspected heart disease by routine ultrasound, a known fetal chromosome abnormality, extracardiac structural anomalies, fetal hydrops, and fetal arrhythmia (6, 7, 16). The presence of any risk factor significantly increases the likelihood of developing fetal heart disease, and genetic counseling and fetal echocardiography are clearly indicated in these cases.

Prenatal genetic counseling

For pregnancy-related issues, there are two general types of genetic counseling sessions — those where parents or prospective parents are concerned about potential risks and outcomes of pregnancy, and those where patients are dealing with a specific fetal diagnosis during pregnancy (17). Despite the vastly different nature of these sessions, their courses are similar. All clients come to genetic counseling with some level of anxiety, even in the most routine of circumstances. Fear of the investigation of one's own basic genetic make-up is common, particularly in the prenatal setting. The risk of a problem, or the diagnosis of a specific anomaly during

---

Nonstandard abbreviations used: congenital heart disease (CHD); Risk Evaluation and Education for Alzheimer's Disease (REVEAL).

Conflict of interest: The authors have declared that no conflict of interest exists.

Citation for this article: *J. Clin. Invest.* **112**:1280–1286 (2003). doi:10.1172/JCI200320170.

pregnancy, is often a prospective parent's biggest fear. This anxiety can be overwhelming and may significantly impact a patient's self-image. Such issues need to be explored, acknowledged, and accounted for throughout the counseling session.

Prior to the initial counseling appointment for an already identified fetal anomaly like heart disease, the counselor will ideally have had the opportunity to contact the family to prepare them for the discussion and to begin to develop a rapport. At the initial visit, parents are often asked to recount how the diagnosis was made and what information they have obtained thus far. This allows the family members to "tell their story" in their own words. Such recounting of family history helps the counselor assess their understanding of the diagnosis, their informational resources, and their coping strategies. It also fosters trust between the counselor and the clients.

Immediately after the diagnosis of a fetal cardiac anomaly, most families are not in the frame of mind to construct a detailed three-generation pedigree. The counselor obtains an abbreviated version by asking about family and obstetrical history, use of medications, and illnesses or other exposures in pregnancy (18). The specific fetal cardiac diagnosis, including implications, severity, and natural history, is discussed in detail, as well as options for further testing and for prenatal management. All of these are influenced by gestational age at the time of the diagnosis (Table 1).

After considering the available options, the counselor and the clients determine a management plan. In addition to providing appropriate medical and emotional support, a management plan gives families some sense of control where many feel like victims in a situation beyond their control (13). Further testing, consultations with pediatric cardiologists, nurses, and surgeons, and tours of delivery suites and special-care nurseries may be planned (5, 6). Additionally, lay literature and guidance about appropriate Internet sites and other resources are critical for families and reinforce issues covered in the session (19).

Treating CHD

No specific treatment is currently available to improve the course of structural heart disease in utero. However, prenatal diagnosis of a fetal cardiac defect provides the opportunity for counseling, education, and discussion of perinatal options. Although several studies have shown no significant difference between prenatal and postnatal diagnosis in terms of the subsequent costs of initial hospitalization, length of hospital stay, and neonatal survival (3, 20), early detection is helpful in allowing the family to meet the health care team and to discuss the diagnosis in a non-crisis situation. More importantly, parents may prepare emotionally for the birth of a child with CHD, rendering them better able to make informed decisions about medical and surgical options following delivery. Ideally, CHD-related morbidity will decrease and the clinical outcome will improve for the affected child if treatment is started before the infant develops severe symptoms (5, 21). One study demonstrated fewer seizures in neonates with hypoplastic left heart syndrome diagnosed prenatally, compared with those diagnosed at birth (22), suggesting that prenatal diagnosis might improve long-term neurologic outcome.

Exploring postdiagnosis options

The option of pregnancy termination is necessarily integrated into the counseling of families with fetal anomalies. The hope of surgical correction or palliation for CHD makes this discussion particularly difficult. Parents need to be fully aware of detailed information regarding the risks, benefits, and time constraints of the different methods of termination. Counselors are often asked, "What would you do in this situation?" It is unfair to the patient to answer that question directly, as there is such great variation in individuals' perspectives, needs, and goals. The genetic counselor may instead review issues that would help patients to reach their own decisions. Another difficult issue is cost. Many families have financial concerns about medical bills for a child that may require multiple major surgeries. They also fear the emotional cost to siblings of the affected child. These complex aspects of prenatal genetic counseling may be discussed over several sessions.

As current medical technology affords detailed evaluation of the developing fetus, and as the underlying genetic mechanisms of fetal maldevelopment are better understood, the lines defining high and low risk, and normal and abnormal, will continue to blur. Genetic counselors are uniquely trained to help patients deal effectively with the heavy emotional burdens and subsequent choices imposed by the availability of this information.

Pediatric genetic disease and newborn screening

Pediatricians and other health professionals who work with children with birth defects and genetic disorders with dysmorphic features have long been referring parents for genetic counseling. These families are usually most concerned about natural history, prognosis, consumer networks, and resources related to their child's condition. Eventually the family's thoughts turn to risks for future pregnancies.

Some pediatric genetic conditions have clinical features that, however, are not so apparent. More and more children whose genetic disorders would otherwise go undetected are diagnosed through newborn screening. If diagnosed early enough in life, many diseases can be treated or prevented. Genetic counseling for affected children and families is integral in medical management, family planning, and emotional coping.

The challenges of screening and counseling

The palette of diseases that should be screened in newborns is murky with the advent of new technologies such as tandem mass spectrometry, which allows simultaneous testing for multiple genetic diseases (23). With hundreds of diseases affecting newborns, how can parents and health professionals decide which genetic disorders should be included in a screen? Traditionally, policy regarding newborn screening has been defined by basic principles (see *Principles of newborn screening for the detection of genetic diseases*), although decisions on which disorders to screen for are regulated by individual states. In March 2003, the US General Accounting Office, the body responsible for the audit, evaluation, and investigation of congressional policy and funding decisions, provided a report to congressional requesters, reviewing the variation among state newborn-screening programs and including information on the criteria considered in selecting disorders to include in state programs (see *Principles of newborn screening for the detection of genetic diseases*; and the National Newborn Screening and Genetics Resource Center website at http://genes-r-us.uthscsa.edu) (24).

Determining which disorders should be included in newborn screening entails considerable difficulty, as is apparent from the variety of diseases screened by different states. Currently, the num-

Table 1
Options for prenatal testing

Gestational age (wks)	Test	Description	Screening vs. diagnostic	Information obtained	Risks to pregnancy
Pre-conception	Parental-carrier testing	A variety of tests for genetic diseases more common in specific ethnic groups. Includes cystic fibrosis, fragile X syndrome, Jewish panel, sickle cell anemia, multiple thalassemias	Diagnostic if test is positive. A negative test does not exclude the possibility that the individual is a gene carrier	Identifies all or most gene carriers. Allows planning if further testing indicated or to plan for prenatal diagnostic testing	None
8–14	First-trimester ultrasound	Often accomplished vaginally. Increasingly used in pregnancy evaluation and management. Often involves nuchal translucency measurement	Screening and/or diagnostic	Fetal viability, pregnancy dating, limited evaluation of fetal anatomy. Increased nuchal translucency measurement is associated with increased risk of aneuploidy and CHD	None
10–14	First-trimester fetal echo-cardiography	Emerging technology that attempts to visualize cardiac view routinely obtained in the second-trimester fetal echocardiogram, at earlier gestational ages. Often accomplished by vaginal ultrasound	Screening	Four-chamber view of the fetal heart. Adequate views of outflow tracts are less likely to be obtained at these earlier gestational ages	None
11–14	First-trimester screening	Based on three parameters: (a) maternal age; (b) fetal nuchal translucency, measured by a specially trained and certified sonographer; and (c) biochemical analysis of maternal serum	Screening	Estimated risk of fetal Down syndrome and trisomy 18	None
11–14	Chorionic villus sampling	Biopsy of developing placenta. Available to all pregnant women; routinely offered to women of ages 35 and over at time of delivery, and women determined to be at sufficient risk based on other screening tests	Diagnostic	Fetal karyotype, DNA diagnostics	~1% risk of pregnancy loss
15–22	Maternal serum screening	Blood test offered to all pregnant women to assess risk of certain birth defects	Screening	Estimated risk of fetal Down syndrome, trisomy 18, and open neural tube defects	None
16–24	Targeted fetal ultrasound	Available to all pregnant women. A commonly relied upon tool in pregnancy evaluation and management	Screening and diagnostic	Fetal anatomy, fetal biometry and growth, assessment of placenta and amniotic fluid volume	None
15–22	Genetic amniocentesis	Process of obtaining an amniotic fluid sample. Available to all pregnant women; routinely offered to women of ages 35 and over at delivery, and women determined to be at sufficient risk based on other screening tests	Diagnostic	Fetal karyotyping, DNA diagnostics, biochemical analysis of amniotic fluid	~0.5% risk of pregnancy loss
18–24[A]	Fetal echo-cardiography	Ultrasound of the fetal heart, performed by practitioners experienced with fetal heart anatomy. Recommended for pregnant patients with known maternal or fetal risk factors	Diagnostic	Detailed views of fetal cardiac structure, rhythm, and blood flow	None
≥18	Fetal blood sampling	Typically reserved for specific high-risk situations	Diagnostic	Fetal karyotyping, DNA diagnostics, fetal blood parameters	1–5% risk of pregnancy loss

[A]This test is increasingly being used in the late first and early second trimesters.

ber of diseases screened ranges from four to 36, with the majority being eight or fewer (24). The disorders most commonly screened include phenylketonuria, congenital hypothyroidism, galactosemia, sickle cell disease, and congenital adrenal hyperplasia.

Some policy makers maintain that newborn screening for untreatable genetic disorders is reasonable only for the purpose of preventing the birth of another affected child. When a child has a genetic disorder identified by newborn screening, genetic counseling is paramount for a woman and her partner in order to address the complicated reproductive decisions that may follow. For example, a boy with Duchenne muscular dystrophy (DMD) may go undiagnosed until he is four years old. As this is an X-linked disease transmitted by the mother, the boy's mother may already have at least one other at-risk son. If the diagnosis of DMD had been made with her first newborn, she might have had the opportunity, through genetic counseling, to plan for future pregnancies using prenatal diagnosis, permanent forms of birth control, preimplantation diagnosis, or even ovum donation.

Another issue in newborn screening is that limited data exist to determine when interventions should begin in newborns identified with genetic disease. Interventions may be quite expensive, with enzyme-replacement therapy for lysosomal-storage disorders such as Fabry disease and Gaucher disease costing between $70,000 and $200,000 per year in affected adults (25). Evidence suggests that the effects of storage disorders are manifest in childhood (26, 27). Yet the question remains: is this reason enough to provide newborn screening so that treatment may begin at birth? Is it acceptable to test healthy minors even if no therapy or intervention is available or if best age at intervention, e.g., enzyme replacement theory, is unknown (27–30)? Natural history registries of genetic diseases could provide some of this helpful information, but few exist. Genetic counselors serve critical functions in developing and implementing such registries, to elucidate the implications of genetic diseases throughout the lifespan. Most natural history registries are funded through grants to individual researchers or research consortiums working with patient advocacy groups or with pharmaceutical companies (31–33).

Genetic predisposition to adult-onset disease

The identification of susceptibility genes for common adult-onset genetic diseases is moving the field of genetic counseling in a new and challenging direction. Common diseases such as diabetes, certain cancers, and adult-onset neurodegenerative diseases have an identified genetic component or have been linked to specific chromosomal regions through family linkage and association studies (Table 2). Genetic counseling for these diseases is difficult when the disease is linked to susceptibility genes, which are known to confer an increased risk of disease, rather than deterministic genes, which are predictive of disease onset. With the identification of genes associated with Alzheimer disease, there is considerable interest in the clinical application of this genetic information through genetic counseling and testing.

Alzheimer disease

Alzheimer disease is characterized by gradual onset and progressive cognitive decline, with motor and sensory dysfunction occurring in the later stages (34). Common symptoms include memory loss, disorientation, confusion, language disturbance, and behavioral changes such as agitation, wandering, psychosis (hallucinations, delusions), depression, anxiety, and sleep disturbance (34). Alzheimer disease can be divided into early-onset sporadic and early-onset familial disease, occurring before 65 years of age, and late-onset sporadic and late-onset familial disease, occurring after 65 years of age. The majority of Alzheimer disease cases are associated with late-onset clinical presentation, while early-onset familial Alzheimer disease accounts for less than 5% of all cases (35).

Genetic studies have revealed several genes linked to Alzheimer disease. Three determinative genes, *PSEN1* and *PSEN2*, encoding presenilin-1 and -2, respectively, and *APP*, encoding amyloid-β precursor protein, are associated with autosomal dominant early-onset Alzheimer disease, and one susceptibility gene, *APOE*, encoding apolipoprotein E, is associated with increased risk for late-onset sporadic and familial Alzheimer disease (36). Clinical genetic testing is available for individuals with Alzheimer disease symptoms and at-risk children or siblings of patients with early-onset disease and a known mutation. *APOE* testing is not currently available to asymptomatic individuals with a family history of late-onset disease, unless they participate in research studies. The American College of Medical Genetics and the American Society of Human Genetics do not currently recommend *APOE* genotyping for presymptomatic identification of Alzheimer disease (37).

Principles of newborn screening for the detection of genetic diseases

The disorder should be common.

The disease should manifest in childhood with severe complications, e.g., high burden.

An accurate diagnostic test should exist with a minimal percentage of false positives and/or false negatives. It is considered more acceptable to identify a false positive than to miss an affected child.

The mode of sampling and screening should not be harmful, e.g., a heel stick for obtaining blood.

Timely results should be available. Some states do not offer newborn screening for galactosemia because hospital staff often diagnose symptomatic infants in the first days of life before newborn-screening results are available.

An immediate intervention that significantly changes the natural history of the disease should be available if the diagnosis is made soon enough after birth. For example, implementation of a phenylalanine-restricted diet significantly alters the natural history of phenylketonuria. This service is often provided by specialized metabolic centers with access to geneticists, genetic counselors, nutritionists, and social workers.

Screening should be cost-effective, which may be difficult to determine.

The screening program should have a comprehensive system in place to provide education to parents and health care providers.

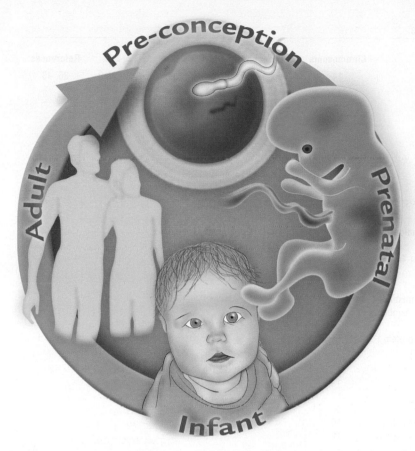

Figure 1
Genetic counseling is a unique medical specialty that provides clinical health care for patients across the lifespan facing genetic and inherited diseases.

Currently, *APOE* genotyping for presymptomatic individuals with a family history of late-onset Alzheimer disease is only available through the Risk Evaluation and Education for Alzheimer's Disease study (REVEAL), an NIH-funded research project underway at Boston, Cornell, Case Western Reserve, and Howard Universities. This study is examining the impact of risk assessment and genetic testing for late-onset Alzheimer disease by providing *APOE* genotyping and results (38). The National Institute on Aging has also established the Alzheimer's Disease Genetics Initiative: The Multiplex Family Study, creating federally funded Alzheimer disease centers throughout the US to facilitate collaboration between researchers conducting family-based linkage studies (www.alzheimers.org). This collaboration should identify more Alzheimer disease risk-factor genes.

To test or not to test

Risk evaluation for Alzheimer disease is challenging. Evaluation of an individual's risk depends on the family history of disease and whether it is early- or late-onset and familial or sporadic. In an early-onset family, a known mutation in an affected patient puts the siblings and children at a 50% risk of inheriting the same mutation. For a person with a family history of late-onset disease, risk is based on information gathered through family-based studies, and epidemiological research has shown that first-degree relatives' risk is two- to threefold greater than

the background population's 10–15% risk of developing the disease (39). Many individuals seeking genetic counseling and testing for late-onset Alzheimer disease assume that genetic-test results are absolute or predictive of disease onset. While the presence of the *APOE4* allele may confer greater risk for disease, this information is not conclusive. Until more useful risk figures using the *APOE* genotype are established, knowing the family history may be more informative than knowing the *APOE* genotype in risk assessment. Ultimately, genetic counseling facilitates an understanding of the importance of family history, other susceptibility factors such as the presence of *APOE*, and the limitations of risk assessment.

For individuals diagnosed with Alzheimer disease, genetic counseling may involve family members in the decision of whether or not to pursue genetic testing or participate in genetic research studies. In such cases, the patient's capacity to consent to genetic testing may be in question. However, the family member serving as health care proxy or legal guardian may be an at-risk daughter, son, or sibling with conflicting motivations for testing. Genetic counselors are instrumental in working with multiple family members to help them formulate a decision that serves the family unit and not just the at-risk individual or the patient with Alzheimer disease.

With any presymptomatic genetic test, genetic counseling also includes a discussion of the risks and benefits of testing. In the case of Alzheimer disease, risks include the psychological impact of finding that one carries an associated genetic factor for the disease, and the possibility of misunderstanding the risk associated with the *APOE4* allele. Insurance costs and employment discrimination are additional concerns for many presymptomatic individuals. While there seems to be some debate as to whether genetic discrimination really occurs, genetic counseling addresses the possibility (40). Current state and federal legislation may not adequately protect people with a genetic susceptibility from high premiums or rejection from long-term care insurance and life insurance. While the Health Insurance Portability and Accountability Act (HIPAA) prohibits group health insurers from excluding presymptomatic individuals from coverage based on genetic-test results, this legislation does not apply to long-term care insurance or life insurance (41). Over 40 states have enacted laws regarding genetic discrimination, which may provide protection equal to or greater than that offered by HIPAA (www.nhgri.nih.gov/Policy_and_public_affairs/Legislation/insure.htm). In the workplace, there is concern that genetic information could be used in hiring, firing, and promotion decisions. The Americans with Disabilities Act prohibits employers from discriminating against individuals with disabilities, but it is unclear how this law applies to those genetically predisposed to illness (42). Existing state laws address genetic discrimination and employment, and several federal genetic nondiscrimination bills addressing both insurance and employment are pending in the current congress (www.

Table 2

Examples of adult-onset diseases and associated genes

Disease	Known genes; proteins	Chromosome	Other possible loci	References
Alzheimer disease	*PSEN1*; Presenilin-1	14	9pter–p32.2, 12p, 12, 17q23, 3q26.1–q26.2, 14q32.1, 10q23–q25, 3q21, 11p15.5, 17q11.1–q11.2, 19q13.1–q13.3, 19q32.2m, 7q35, 20p11.2, 14q24	36, 43
	PSEN2; Presenilin-2	1		
	APP; Amyloid-β precursor protein	21		
	APOE; Apolipoprotein E	19		
Parkinson disease	*SNCA*; α-Synuclein	4	1p35, 2p13, 4p15, 12p11.2–q13.1	44
	DJ1; Oncogene DJ1	1		
	UCHL1; Ubiquitin C-terminal esterase L1	4		
	PARK2; Parkin	6		
Diabetes mellitus, type 2	*NEUROD1*; Neurogenic differentiation 1	2	1q21–q24, 2q, 3q27–qter, 4p, 5q34–q35.2, 12q, 20q12–q13.1	45
	HNF4A; Hepatocyte nuclear factor 4α	20		
	TCF1; Transcription factor 1	12		
	TCF2; Transcription factor 2	17		
	SLC2A4; Solute carrier family 2, member 4	17		
	SLC2A2; Solute carrier family 2, member 2	3		
	GPD2; Mitochondria GAPDH 2	2		
	MAPK8IP1; MAPK 8–interacting protein 1	11		
	IRS1; Insulin receptor substrate 1	2		
Breast cancer	*BRCA1*; Breast cancer 1	17	13q21, 11q, 11p15.5, 17q22, 8q11, 20q	45
	BRCA2; Breast cancer 2	13		
	TP53; Tumor protein p53	17		
	AR; Dihydrotestosterone receptor	X		
	PTEN; Phosphatase and tensin homolog	10		
	STK11; Serine/threonine protein kinase 11	19		
	ATM; Ataxia telangiectasia mutated	11		

nhgri.nih.gov/Policy_and_public_affairs/Legislation/workplace. htm and http://www.genome.gov).

Benefits of predisposition genetic testing

A primary motivation for a person to seek predictive testing is the ability to plan for the future. Individuals at risk for adult-onset diseases often have been shaped by the challenge of serving as caregiver for an affected parent or other family member. In the Alzheimer disease genetic counseling session, individuals discuss the emotional, physical, and financial difficulties of providing care for a loved one who cannot make decisions, complete day-to-day activities such as cooking, dressing, and paying bills, or enjoy activities such as family events or hobbies. Motivation for testing often stems from a desire to prevent passing the burden of this caregiving to the at-risk individual's children. Current information from the Alzheimer's Association (www.alz.org) on long-term care for dementia patients states that the average cost of nursing home care is $42,000 per year and the average lifetime cost per patient is $174,000. Individuals seeking genetic counseling hope that genetic testing can assist in planning future care.

At-risk individuals also seek information that may help them prevent the onset of disease. Prevention studies, however, are in clinical trials, and there are no current medical recommendations. Other topics discussed during genetic counseling include the possibility of participating in research studies like REVEAL and the Alzheimer's Disease Genetics Initiative. Participation in

genetic research may be an alternative for individuals with a family history who do not want to pursue genetic testing or for whom genetic testing is not yet available, but who want to contribute to the understanding of the disease.

While the availability of genetic tests for Alzheimer disease is limited, the future may bring testing that targets specific populations for prevention and treatment. The ultimate goal of genetic research is the identification of at-risk individuals in order to facilitate early and effective treatments in the symptomatic person based on an individual's genotype and strategies to delay the onset of disease in the presymptomatic person. Such advances will enhance the practice of genetic counseling not only for Alzheimer disease but also for other complex adult-onset genetic diseases.

Genetic counselors fill a distinctive position in the complicated and varied arena of genomic medicine and health. Advances in genetic medicine create an even greater demand for expert health care services. Genetic counselors help meet this need, serving in almost every major medical center and across the globe as an increasingly important resource for medical referral and quality patient care. For an international list of genetic counselors and further information about genetic counseling, visit the website of the National Society of Genetic Counselors (www.nsgc.org).

Address correspondence to: Robin L. Bennett, University of Washington, Medical Genetics, Box 357720, Seattle, Washington 98195-7720, USA. Phone: (206) 616-2135; Fax: (206) 616-2414; E-mail: robinb@u.washington.edu.

1. MOD Birth Defects Foundation. 1999. Congenital heart defects. Fact Sheet.

2. Haak, M.C., and van Vugt, J.M.G. 2003. Echocardiography in early pregnancy: review of the literature. *J. Ultrasound Med.* **22**:271–280.

3. Jaeggi, E.T., Sholler, G.F., Jones, O.D.H., and Cooper, S.G. 2001. Comparative analysis of pattern, management and outcome of pre- versus postnatally diagnosed major congenital heart disease: a population-based study. *Ultrasound Obstet. Gynecol.* **17**:380–385.

4. Hoffman, J.I.E. 1995. Incidence of congenital heart disease. II. Prenatal incidence. *Pediatr. Cardiol.* **16**.155–165.

5. Brumund, M.R., and Lutin, W.A. 1998. Advances in antenatal diagnosis and management of the fetus with a heart problem. *Pediatr. Ann.* **27**:486–490.

6. Kleinman, C.S., and Copel, J.A. 1994. Prenatal diagnosis of structural heart disease. In *Maternal fetal medicine. Principles and practice.* R.K. Creasy, R. Resnik, L. Bralow, editors. W.B. Saunders Co. Philadelphia, Pennsylvania, USA. 233–242.

7. Carvalho, J.S., Mavrides, E., Shinebourne, E.A., Campbell, S., and Thilaganathan, B. 2002. Improving the effectiveness of routine prenatal screening for major congenital heart defects. *Heart.* **88**:387–391.

8. Todros, T. 2000. Prenatal diagnosis and management of fetal cardiovascular malformations. *Curr. Opin. Obstet. Gynecol.* **12**:105–109.

9. Moore, K.L., and Persaud, T.V.N. 1998. *Before we are born: essentials of embryology and birth defects.* W.B. Saunders Co. Philadelphia, Pennsylvania, USA. 259–300.

10. Bruneau, B.G. 2003. The developing heart and congenital heart defects: a make or break situation. *Clin. Genet.* **63**:252–267.

11. Goldmuntz, E. 1999. Recent advances in understanding the genetic etiology of congenital heart disease. *Curr. Opin. Pediatr.* **11**:437–448.

12. Srivastava, D. 2001. Genetic assembly of the heart: implications for congenital heart disease. *Annu. Rev. Physiol.* **63**:451–469.

13. Stauffer, N.R., and Murphy, K. 2002. Prenatal diagnosis of congenital heart disease: the beginning. *Crit. Care Nurs. Q.* **25**:1–7.

14. Levy, H.L., et al. 2001. Congenital heart disease in maternal phenylketonuria: report from the Maternal PKU Collaborative Study. *Pediatr. Res.* **49**:636–642.

15. NIH. 2000. *Phenylketonuria: screening and management.* Consensus statement developed at: Consensus Development Conference on Phenylketonuria (PKU): Screening and Management. October 16–18. Bethesda, Maryland, USA. http://consensus.nih.gov/cons/113/113_intro.htm.

16. Freedom, R.M., Leland, N.B., and Smallhorn, J.F. 1992. *Neonatal heart disease.* Springer-Verlag. London, United Kingdom. 7–17.

17. Berman, M.R. 2001. *Parenthood lost: healing the pain after miscarriage, stillbirth, and infant death.* Bergin & Garvey Trade, Greenwood Publishing Group. Westport, Connecticut, USA. 224 pp.

18. Bennett, R.L. 1999. *The practical guide to the genetic family history.* Wiley-Liss Inc. New York, New York, USA. 251 pp.

19. Friedman, A.H., and Kopf, G. *Advances in management of congenital heart disease: helping children lead healthier, happier lives.* Yale-New Haven Children's Hospital booklet. New Haven, Connecticut, USA. 6 pp.

20. Copel, J.A., Tan, A.S.A., and Kleinman, C.S. 1997. Does a prenatal diagnosis of congenital heart disease alter short-term outcome? *Ultrasound Obstet. Gynecol.* **10**:237–241.

21. DeVore, G.R. 1998. Influence of prenatal diagnosis on congenital heart defects. *Ann. N. Y. Acad. Sci.* **847**:46–52.

22. Mahle, W.T., Clancy, R.R., McGaurn, S.P., Goin, J.F., and Clarke, B.J. 2001. Impact of prenatal diagnosis on survival and early neurologic morbidity in neonates with the hypoplastic left heart syndrome. *Pediatrics.* **107**:1277–1282.

23. American College of Medical Genetics/American Society of Human Genetics Test and Technology Transfer Committee Working Group. 2000. Tandem mass spectrometry in newborn screening. *Genet. Med.* **2**:267–269.

24. US General Accounting Office. 2003. *Newborn screening: characteristics of state programs.* Washington, DC, USA. GAO-03-449, 2003. www.gao.gov/new.items/d03449.pdf.

25. Hesselgrave, B.L. 2003. Helping to manage the high cost of rare diseases. *Manag. Care Q.* **11**:1–6.

26. Brady, R.O. 2003. Enzyme replacement therapy: conception, chaos and culmination. *Philos. Trans. R. Soc. Lond. B Biol. Sci.* **358**:915–919.

27. Bennett, R.L., et al. 2002. Fabry disease in genetic counseling practice: recommendations of the National Society of Genetic Counselors. *J. Genet. Couns.* **11**:121–146.

28. American Society of Human Genetics Board of Directors, American College of Medical Genetics Board of Directors. 1995. Points to consider: ethical, legal and psychosocial implications of genetic testing in children and adolescents. *Am. J. Hum. Genet.* **57**:1233–1241.

29. National Society of Genetic Counselors. 1994. Position statement on prenatal and childhood testing for adult-onset disorders. *Perspectives in Genetic Counseling.* **17**:5.

30. American Academy of Pediatrics, Committee on Genetics. 2000. Molecular genetic strategies in pediatric practice: a subject review (RE0023). *Pediatrics.* **106**:1494–1497.

31. Szudek, J., Birch, P., Riccardi, V.M., Evans, D.G., and Friedman, J.M. 2000. Association of clinical features in neurofibromatosis 1 (NF1). *Genet. Epidemiol.* **19**:429–439.

32. 2002. *Cystic Fibrosis Foundation Patient Registry, Annual Data Report, 2001.* 2001. Cystic Fibrosis Foundation. Bethesda, Maryland, USA. www.cff.org/publications.

33. Charrow, J., et al. 2000. The Gaucher registry: demographics and disease characteristics of 1698 patients. *Arch. Intern. Med.* **160**:2835–2843.

34. Small, G.W., et al. 1997. Diagnosis and treatment of Alzheimer disease and related disorders. Consensus statement of the American Association for Geriatric Psychiatry, the Alzheimer's Association, and the American Geriatrics Society. *JAMA.* **278**:1363–1371.

35. Bird, T.D., et al. 1989. Phenotypic heterogeneity in familial Alzheimer's disease: a study of 24 kindreds. *Ann. Neurol.* **25**:12–25.

36. St. George-Hyslop, P.H. 2000. Molecular genetics of Alzheimer's disease. *Biol. Psychiatry.* **47**:183–199.

37. 1995. Statement on use of apolipoprotein E testing for Alzheimer disease. American College of Medical Genetics/American Society of Human Genetics Working Group on ApoE and Alzheimer disease. *JAMA.* **274**:1627–1629.

38. Roberts, J.S., et al. 2003. Reasons for seeking genetic susceptibility testing among first-degree relatives of people with Alzheimer disease. *Alzheimer Dis. Assoc. Disord.* **17**:86–93.

39. Silverman, J.M., et al. 1994. Patterns of risk in first-degree relatives of patients with Alzheimer's disease. *Arch. Gen. Psychiatry.* **51**:577–586.

40. Hall, M.A. 2003. *Genetic discrimination.* North Carolina Genomics and Bioinformatics Consortium. Research Triangle Park, North Carolina, USA. Information sheet. 1–4.

41. *Health Insurance Portability and Accountability Act,* Public Law 104-191, *US Statutes at Large* 110 (1996).

42. *Americans with Disabilities Act, US Code* 42 (1994), 12101–12213.

43. Rocchi, A., Pellegrini, S., Siciliano, G., and Murri, L. 2003. Causative and susceptibility genes for Alzheimer's disease: a review. *Brain Res. Bull.* **61**:1–24.

44. Nussbaum, R., and Ellis, C.E. 2003. Alzheimer's disease and Parkinson's disease. *N. Engl. J. Med.* **348**:1356–1364.

45. OMIM: Online Mendelian Inheritance in Man. http://www.ncbi.nlm.nih.gov/Omim/.

Payment of clinical research subjects

Christine Grady

Department of Clinical Bioethics, Clinical Center, NIH, Bethesda, Maryland, USA.

Offering payment to clinical research subjects, in an effort to enhance recruitment by providing an incentive to take part or enabling subjects to participate without financial sacrifice, is a common yet uneven and contentious practice in the US. Concern exists regarding the potential for payment to unduly influence participation and thus obscure risks, impair judgment, or encourage misrepresentation. Heightening these concerns is the participation not only of adults but also of children in pediatric research trials. Thorough assessment of risks, careful eligibility screening, and attention to a participant's freedom to refuse all serve to reduce the possibility of compensation adversely affecting the individual and/or the study. Institutional review boards currently evaluate payment proposals with minimal guidance from federal regulations. Here, reasons for providing payment, payment models, ethical concerns, and areas for further research are examined.

The payment of human subjects for their participation in scientific research in the US is a common and longstanding practice that has been documented for well over 100 years. As far back as the 1820s, William Beaumont, whom many consider to be the father of gastric physiology, gave patient Alexis St. Martin — a French Canadian voyageur suffering from an incompletely healed gunshot wound to the stomach — food, lodging, clothing, and $150 for the opportunity to study his stomach contents for 1 year (1). In 1900, renowned American military surgeon Walter Reed paid study participants $100 in US gold to allow themselves to be bitten by infected mosquitoes in the famous yellow fever experiments and an additional $100 if they consequently contracted the viral disease (1). According to Susan E. Lederer, author of *Subjected to science: human experimentation in America before the Second World War* (1), in the US, "paying human subjects for their participation in research . . . became routine in the 1920s and 1930s." Other nonmonetary forms of compensation were also common, such as meals, transportation, and burial costs. From the early 1950s, when the world's largest clinical research complex, the NIH Clinical Center, opened, documents show that "normal" healthy volunteers were regularly paid for their participation in biomedical research or money was given to the church or group that organized and recruited these volunteers (2).

Today, newspaper advertisements describing studies that offer "free treatment" for depression or asthma, for example, often state "financial compensation provided." Websites list possible research trials for prospective subjects to review and commonly mention that compensation is provided, often naming a specific dollar amount (3). Although it is unclear exactly how common this practice is, data suggest that a sizable subset of research studies at most organizations or institutions that conduct clinical research pay subjects for participation (4). In addition, studies that offer payment to subjects cover a wide spectrum of types of research, from short-term physiological studies offering no benefit to subjects to longer, phase 3 clinical trials that may offer the prospect of direct therapeutic benefit to subjects (5). Interestingly, there appears to be some variation according to disease or medical subspecialty in the frequency with which payment is offered to research subjects. For example, it is more common in the US to offer payment in asthma, HIV, diabetes, or dermatological research trials than in oncology or cardiovascular trials (6). There are currently no data to explain why this variation exists. Although decisions about offering payment could be influenced by disease severity, sociodemographic characteristics, or the availability of treatment alternatives, some of the variation probably reflects the culture of the subspecialty.

Although the practice of paying subjects for research participation in the US is widespread, it remains a contentious issue (Figure 1). Some commentators believe that the act of paying research subjects is wrong (7), maybe even coercive, while others find it an acceptable and perhaps necessary part of recruitment for clinical investigation (8, 9); others see payment of at least healthy subjects as fair and appropriate (10, 11). In addition, only minimal guidance exists to help investigators determine whether or how much to pay participants in a particular study. The section of the US Code of Federal Regulations governing clinical research does not specifically address the issue of payment of research subjects (12). Institutional review boards (IRBs) — committees designated to review, approve the initiation of, and conduct periodic review of research involving human subjects — are responsible for ensuring that the amount and schedule of proposed payment is ethically acceptable; however, these bodies also operate with minimal and general guidance (4, 12, 13). Consequently, payment practices vary widely in the US (5).

Reasons for offering payment to research subjects and common ethical concerns about offering payment are discussed herein. Models of payment, payment of healthy subjects versus patient-subjects, payment in pediatric research, and practical issues related to the payment of research subjects are also examined.

Payment of research subjects: why or why not?

Biomedical and behavioral research necessary to improve human health and medical care depends on the participation of human subjects. The usual justification offered for paying research subjects is that payment facilitates the timely recruitment of an adequate number and type of subject (9). Payment may be important to research to the extent that it encourages participation. Several studies have shown that response to written surveys is influenced by payment (14–19), as is willingness to participate in hypothetical studies (20, 21), but less is known about the real effect of payment on recruitment in clinical research.

Nonstandard abbreviations used: IRB, institutional review board.

Conflict of interest: The author has declared that no conflict of interest exists.

Citation for this article: *J. Clin. Invest.* **115**:1681–1687 (2005).
doi:10.1172/JCI25694.

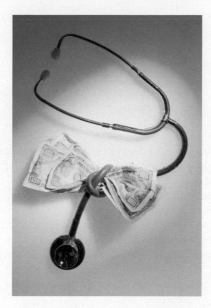

Figure 1
Offering money to clinical research subjects for their participation is a fairly common practice in the US. Yet there is little consensus about why, when, or how much to pay these individuals. Getty Images.

of awareness or distrust. Consequently, money may not only be important to general recruitment but also helpful in achieving the goals of racial, ethnic, gender, and social diversity of subjects participating in biomedical research (22, 23).

However, empirical evidence demonstrating that payment is necessary and/or effective for the recruitment of clinical research subjects is limited. People appear to be attracted to research and motivated to participate for a variety of reasons. Healthy volunteers, who are frequently paid and unlikely to benefit medically from research participation, are often attracted to research and motivated to participate by money. Yet they appear to have a variety of other motives besides those of a financial nature for participation in research, including curiosity, altruism, sensation seeking, and desire for attention provided by physicians (24–26). Patient-subjects — those who suffer the disease or condition under study in a particular research protocol — are often motivated to participate by the hope of personal therapeutic benefit. When patient-subjects enroll in clinical studies and they understand the likelihood of direct medical benefit to be remote or nonexistent, their motivations are likely to be similar to those of healthy volunteers. Although patient-subjects are often offered payment for their participation in clinical studies, little research has been done to evaluate the extent to which money influences recruitment or their willingness to participate (21). Given the diverse motivations of subjects in clinical research and a lack of relevant data, it is difficult to know how much paying subjects helps to accomplish recruitment goals.

Similarly, there are few data to support the idea that paying subjects ensures diversity. In fact, other, less fungible factors, for example, lack of child care for women and long-standing

There are several ways in which payment might enhance recruitment. First, the offer of money as reimbursement for time or expenses incurred could help to make research participation a revenue-neutral activity for participants. In this regard, money may enable individuals to take part who otherwise could not afford to participate or who are not willing to make a financial sacrifice in order to do so. Alternatively, payment could help recruit individuals who believe they should be fairly compensated for their time and effort. Money can also serve as a recruitment incentive, especially if the amount offered is high enough to attract subjects to research and overcome inertia, lack of interest, as well as financial and other barriers (see *Why pay clinical research subjects?*). Some say that financial incentives are also necessary to overcome barriers unique to certain subgroups in the population, such as lack

Why pay clinical research subjects?

Reasons to pay research subjects	Related comments and questions
Incentive necessary to recruit adequate numbers of research subjects	Limited data exist to document the extent to which money works as a recruitment incentive for clinical research or to show the value of money for recruitment compared with other powerful incentives, such as treatment or access to care
Incentive necessary to overcome opportunity costs, inertia, and distrust and recruit hard-to-reach subjects, especially underrepresented groups	Limited data to support the claim that money increases diversity; Money may not overcome all barriers and in fact could contribute to, rather than diminish, suspicion or lack of trust
Reimbursement for expenses (and possibly lost wages) to reduce the barrier of financial sacrifice for participants	Not all research participants are reimbursed; Payment to individuals would vary based on their expenses and the value of their time
Fair compensation or remuneration for the time and inconvenience of research participation	Data suggest that time may not be the main criterion for determining amounts of payment to subjects; Inconvenience is hard to quantify

Ethical concerns about the payment of research participants

Payment may be perceived as:	Related comments and questions
Coercion	Coercion involves a threat of harm. Thus, an offer of money in return for research participation is not coercion.
Undue inducement 　May compromise informed consent by: 　(a) reducing interest in understanding 　risks related to research; (b) reducing the 　voluntary nature of the decision to participate 　Money may unduly influence individuals 　to misrepresent themselves	Undue inducement is not well defined 　Limited data suggest that payment does not affect understanding or 　perception of risk. The adequacy of understanding as part of informed 　consent can be assessed. 　Voluntary decisions can be made when inducements are offered, even in 　the setting of limited or poor financial options 　There are limited data on the influence of money on informed consent 　Other incentives may be as powerful as money 　Careful eligibility screening can minimize risk of misrepresentation
Disproportionate research burden on the poor	Sociodemographics of research participants are not well known 　Inadequate financial reimbursement might 　disproportionately exclude the poor 　Paid participation may be an opportunity, not a burden
Commodification	Services offered as part of research participation are not the same as "selling" body parts or sex and may be of little risk to the health and well-being of the participant

distrust of the research establishment by certain ethnic groups (27–30), may be obstacles to participation that money cannot overcome. Distrust, in fact, could plausibly be exacerbated by an offer of money.

Ethical concerns about offering payment to research subjects

Several ethical concerns have been raised regarding the payment of research participants (see *Ethical concerns about the payment of research participants*). The most commonly expressed concern is that payment could be coercive or serve as undue inducement to research participants. By definition, coercion is understood to involve a threat of physical, psychological, or social harm in order to compel an individual to do something, such as participate in research (31). However, money for research participation is an offer or an opportunity and not a threat and therefore cannot be perceived as coercion. But can money be considered an undue inducement? Existing guidelines warn against undue inducement and its potential to compromise informed consent, although there is disagreement about what exactly constitutes undue inducement and consequently disagreement about the extent to which it is a valid problem in research (32). The US Code of Federal Regulations requires that informed consent be obtained "under circumstances . . . that minimize the possibility of coercion or undue influence" (12, 13). An inducement in clinical research, as defined in *The official IRB guidebook* (33), is deemed undue and therefore troublesome if it is so ". . . attractive that [it can] blind prospective subjects to potential risks or impair their ability to exercise proper judgment . . ."

Do financial incentives blind potential research participants to the risks of research when making decisions regarding participation?

Motivated by cash payments or an attractive financial package, an individual could have less interest in evaluating or understanding study details, reading the consent form, or attempting to understand the goals, purposes, and risks associated with a study. This may be of little concern, however, if a clinical research protocol has almost no risks or has been approved by an IRB that has judged the level of risk to be acceptable (32). If, in addition, there are other mechanisms in place during the informed consent process to assure that participants adequately understand relevant risks (34), then this seems like a misplaced worry and may even represent "unwarranted paternalism" (35). Further, limited evidence suggests that the offer of payment does not obscure the risk perception of potential research participants (20, 21), and there are no data to suggest that it does.

Others worry that money can impair judgment or compromise voluntary decision making. But voluntary decisions are motivated by various factors, sometimes including money, and are not necessarily motivated by altruism alone. When people are choosing a job, making purchases, or making other voluntary decisions, they often consider the monetary aspects of their choice in the form of salary, benefits, or sales price. Decisions are generally complex and multifaceted, however, and are rarely based solely on monetary considerations. Similarly, people participate in clinical research for multiple reasons, and money may be one among those reasons or even the main reason. Limited data suggest that the offer of money is one factor in the decision making of some, but not all, potential participants (21, 24, 36–40).

Even if money is one reason or the main reason to participate in research, does the offer of money impair judgment? In one study, most respondents (75%) thought an offer of $500 for research participation could impair the judgment of others, but many

Table 1

Models of payment for the participation of research subjects

Model	Payment serves as	Amount determined by	Potential advantages	Potential disadvantages
Market	Incentive	Supply and demand; market rates	(a) More rapid recruitment. (b) Completion bonuses encourage subject retention and high completion rate. (c) Possibility of profit for participants. (d) Little or no financial sacrifice by subject.	(a) Undue inducement possibly resulting in: incomplete assessment of risks and benefits by subject; subject concealing information to ensure enrollment/retention. (b) Competition between studies; better-funded studies more likely to meet recruitment goals. (c) Different levels of payment at different locations for multicenter trials.
Wage-payment	Compensation	Standardized "wage" for time and effort, suggested to be commensurate with wages for unskilled, but essential jobs; additional payment for extra burdens such as endurance of uncomfortable procedures	(a) Recognizes contributions of participants. (b) Uniform payment across studies. (c) Equal pay for equal work. (d) Less risk of undue inducement.	(a) May have little impact on recruitment. (b) Might undercompensate some subjects in relation to regular wage and preferentially attract others.
Reimbursement	Reimbursement	Expenses incurred (transport, meals, lodging); with or without reimbursement for lost wages	(a) Makes research participation revenue neutral. (b) Little risk of undue inducement. (c) Little or no financial sacrifice for subject if lost wages are reimbursed.	(a) May have little impact on recruitment. (b) Uneven reimbursement from subject to subject. (c) Reimbursement costs for high-salaried subjects may result in the targeting of low-income populations. (d) Financial sacrifice for subject if lost wages are not reimbursed.
Appreciation	Reward	Token of appreciation given at the conclusion of study	(a) Expresses gratitude for contribution made. (b) Not market dependent. (c) Avoids undue inducement.	(a) Likely to have no impact on recruitment. (b) No basis for consistency.

Table modified from *The New England Journal of Medicine* (46).

fewer (20%) thought it would impair their own judgment (11). Presumably, fewer people would think smaller amounts, say $25 or even $100, would impair judgment, yet most clinical research studies offer considerably less than $500 for participation (6). Some worry that individuals with limited opportunities for earning money may be most susceptible to impaired judgment when faced with an offer of money (23). But, as Wilkinson and Moore argue, even people with few options may still have the ability to make decisions for themselves and thus be capable of autonomous consent (42). In fact, denying the possibility of payment to autonomous research subjects with limited opportunities for earning money further restricts their options rather than protecting them from a situation in which their judgment might be impaired, especially if they would still be invited to participate in research that did not offer compensation. Careful attention, during the process of obtaining informed consent, to subjects' understanding and expectations of clinical research and their sense of freedom to choose to participate or not may be more appropriate, albeit imperfect, than limiting the opportunity to receive payment for participation.

On the other hand, if offering a large amount of money could cause some people to agree to participate in research for which they would otherwise have a profound reluctance, the offer demonstrates disrespect for their deep reservations or preferences.

Limiting the amount of money offered for research participation might minimize the chances that it will unduly influence participants in this way (43, 44). In my view, offering modest amounts of money is unlikely to obscure risks or impair the judgment of most individuals. However, investigators and IRBs should review the offer of money and other inducements outlined in clinical research proposals, especially for research on the margin of reasonable risk or with groups of people that are more likely to be attracted by an offer of money. Additional empirical research would increase our understanding of the extent to which money influences decisions about research participation in relation to other factors; and to what extent, if at all, people actually do agree to participate in research that compromises their deeply held values or interests.

Concern has also been expressed about the potential for money to unduly influence ". . . subjects to lie or conceal information that if known would disqualify them from enrolling or continuing as participants in a research project" (43). Misrepresentation of previous or current medical problems could jeopardize both the safety of the subject and the quality or interpretability of the data. For example, an individual interested in a well-paying MRI study could jeopardize his safety by concealing the history of a shrapnel injury that otherwise would exclude him. A participant in a phase 1 drug study could fabricate side effects in order to stop study participation early without loss of payment, consequently jeopardizing the quality of the science. One study showed a willingness of subjects to conceal information from investigators in lower-risk studies, but this willingness was not associated with payment (20). It is unknown how often such misrepresentation occurs in clinical research and also unclear whether money is uniquely capable of inducing this kind of deception. Perhaps we should worry more about the possibility of desperate patients engaging in deception if they perceive the therapeutic intervention or agent under study to be their best or only remaining therapeutic option. Careful attention to eligibility criteria in the screening history, physical examination, and laboratory tests can minimize, although not eliminate, the possibility of misrepresentation in order to enroll in research trials. In addition, mechanisms such as prorating payments over time might help minimize the possibility of misrepresentation during a study.

Additional concerns about the ethics of offering payment to research subjects have received less attention. Some worry that payment might be more attractive to individuals with low socioeconomic status, and thus the payment of subjects could result in a disproportionate research burden on this population. In addition to worries about distributive justice, a skewed subject pool could confound the generalizability of data. Unfortunately, research subject sociodemographic information is not well documented; when documentation is available, it has been shown that subjects in at least some studies tend to be primarily insured and not economically disadvantaged (45). Interestingly, offering no money or such a small amount of money that participation in research is inaccessible to those who are economically disadvantaged also has the potential to skew the subject pool and contravene principles of distributive justice, especially for research perceived as beneficial to participants.

Models of payment made to research subjects

Assuming that paying research subjects is ethically acceptable, there still remain questions regarding how to pay subjects. Several possible models of payment capture the various ways that payment could be conceptualized and the amount of payment determined: a market model, a wage-payment model, a reimbursement model, and an appreciation model (46) (Table 1). In a market model, payment is designed to be a straightforward incentive. The amount of payment is determined by the market; that is, the value necessary to recruit the number and type of subjects needed in a given time frame. Consequently, studies that need to recruit individuals with rare conditions or characteristics may offer more money, while studies for which there are many willing participants may offer no money. According to the market concept, the amount of money could be increased to overcome aversion to risk or inconvenience. Completion bonuses and escalating incentives would be commonly employed as incentives for subjects to meet data points or complete a study. In contrast, in a wage-payment model, subjects are offered payment as compensation for the time and contribution they make to the research. Money offered to subjects is calculated by a standardized hourly "wage" offered to compensate for their hours of participation, with reasonable additions made for added inconvenience. In a reimbursement model, payment is offered to research participants to reimburse them for actual expenses, such as travel, meals, and parking. One version of a reimbursement model would also offer reimbursement for lost wages. If the latter were adopted, subjects in the same study might be reimbursed at radically different rates, depending on their occupation and normal salary. An appreciation model conceives of money as a reward or token of appreciation for a subject's contribution to research. Appreciation can be shown by awarding a wide range of amounts of money as well as non-monetary gifts. Unlike the other 3 models, appreciation payments may have little impact on study recruitment, as appreciation is often reserved until the study ends. Elsewhere, my colleagues and I have argued that conceptualizing payment as compensation in the form of a wage payment is the most ethically appropriate model because it recognizes the contribution subjects make to research and is relatively standardized across studies. While still possibly able to facilitate recruitment, wage-like payments are unlikely to unduly influence individuals to enroll in research to which they object (46). Finally, offering fair compensation for research participation also demonstrates that society values clinical research and is grateful for subjects' contribution to the common good.

Payment of healthy subjects versus patient-subjects

Commentators sometimes assume or assert that it is legitimate to pay healthy subjects but not patient-subjects for their participation in research (10). Healthy subjects are often motivated by money to participate in research, receive little or no benefit from participation, and may appropriately be considered independent contractors in research (40). Paying money to healthy volunteers is widely accepted, although concerns about undue inducement and distributive justice may still pertain. In contrast, although patient-subjects are often paid to participate in research (5), commentators worry about paying patient-subjects because of their "vulnerability" (10). Certainly illness can make people vulnerable in multiple ways. Presumably, patient-subjects are considered more vulnerable in research studies than healthy subjects because of the nature of the relationship with their physician and because of possible confusion about the difference between participation in clinical research and the receipt of clinical care — the so-called therapeutic misconception (47). Although this is an empirical question, it is at least plausible that offering payment to patient-subjects in research could help them distinguish participation in a research study from the receipt of clinical care and thus actually decrease their vulnerability. Offering money in return for participation might also enable a patient to say no to the physician instead of feeling obligated to do what the physician suggests.

Payment may not be necessary for recruiting patient-subjects into research, especially if they are motivated by an opportunity for therapeutic benefit. However, if the goal of payment is to reduce the financial sacrifice that research subjects have to make, to compensate people for their time, or to show appreciation for their contribution, patient-subjects are equally deserving and should be paid comparably to healthy subjects. When patient-subjects participate in research that offers them desirable therapeutic benefits, money may seem irrelevant and unnecessary, even though not morally objectionable. However, when patient-subjects and healthy subjects are both asked to undergo certain identical study procedures for research purposes, in the interest of fairness, the 2 sets of individuals should be compensated similarly, as both are contributing to the development of generalizable knowledge to benefit others.

Payment of children in research

Offering payment in pediatric research involves special challenges not found in research with consenting adults. First, research with children is vital and promoted by both the FDA (48) and the NIH (49). However, children do not provide their own consent to research but are enrolled by their parents or legal guardians, generally in accord with the child's best interests. Payment to parents for their child's research participation could potentially sway parental decisions in favor of participation since there is no personal risk to themselves. To avoid making children commodities, some argue that parents should not receive money as incentive for their child's research participation (50, 51). However, making it possible for a child to participate in research can be inconvenient and costly for parents, and the amount of risk children can be exposed to in research is strictly limited by federal regulations (52). Consequently, some find carefully calculated payment to compensate parents for time and inconvenience acceptable and unlikely to contribute to significant distortions in parental judgment (53), while others believe that compensation to parents should be limited to reimbursement for expenses (51). The American Academy of Pediatrics recommends the giving of gifts instead of money to children in a post-trial appreciation model (54), although many institutions do not appear to follow these recommendations (55). Giving money or non-cash gifts to children directly instead of to their parents is also difficult because children appreciate money and gifts differently depending on their age. Further empirical and conceptual research is needed to resolve when and how payment should be offered in pediatric research.

Practical suggestions

When deciding whether to offer payment to research participants in a study, investigators should take into account the nature of the study, the nature of participant contributions and vulnerabilities, institutional or organizational guidelines, and local societal and cultural norms. In the research proposal submitted to their IRB, investigators should describe the rationale for payment, how the dollar amount was calculated, and how and when payment will be made. Payment information should also be included in consent forms.

Although payment may be a factor in a subject's decision regarding research participation, IRBs do not consider payment a benefit to offset research risks when deciding whether or not to approve a study. IRBs evaluate whether the risks in a research study are justified by potential benefits; otherwise unacceptable risks cannot be made acceptable by offering money to subjects. Therefore, discussion of payment should only arise after the risk-benefit ratio of a study is found ethically acceptable. IRBs should review the justification for and the amount and schedule of payment and decide whether these variables are appropriate given the particular study and the population to be recruited. In making this determination, an IRB should consider study risks, potential vulnerabilities of the targeted subject population, eligibility criteria and screening plans, proposed methods for assessing subjects' knowledge of risks and ability to make voluntary autonomous decisions, and local norms. An IRB should also review the presentation of information about payment in consent documents as well as related advertisements and information sheets.

Plans for how and when money will be disbursed are also important. Prorating payment for studies involving multiple visits minimizes the possibility of inappropriately influencing someone to remain in a study just to receive a lump sum payment at the end. Payment according to actual time and procedures completed is consistent with offering money as compensation for a subject's time and inconvenience. On the other hand, in longitudinal or long-term studies, where certain data points are critical to the study, it may be appropriate to use escalating incentives or completion bonuses, as long as they are not offered to compensate for increasing risk. These strategies should be approved by the relevant IRB.

Empirical studies and unanswered questions

Although there is a growing volume of empirical research addressing issues related to the ethics of paying research subjects, many more questions need attention. As mentioned above, evaluating whether or under what research circumstances money might impair a subject's judgment would be important, as well as the extent to which payment leads people to participate against deep objections. Payment to clinical research populations who are particularly vulnerable to exploitation, such as children, substance users, or those unable to consent for themselves, warrants further attention. Sociodemographic data on research subjects in both paid and unpaid studies would be useful. More specific guidance, including benchmarks, would greatly assist investigators and IRBs in making decisions about payment for participation. Additional understanding of variation in local or regional norms and participants' values as they relate to money, as well as how to consider the economic conditions in communities in which research will be conducted in formulating an approach to payment, would be useful, especially for multicenter and multinational studies. Additionally, more attention to the influence of nonmonetary incentives and compensation is warranted, although many of the concerns are similar.

Concluding remarks

Payment to research subjects for their participation is a pervasive yet uneven practice in the US. Although there is nothing inherently unethical about paying clinical research subjects, knowing more about its effect on recruitment and its use in different research circumstances is critical. Investigators might offer money to research subjects as an incentive to participate, as fair compensation for their contribution, and/or as a way to reduce any related financial sacrifice. Worries about undue inducement can be reduced by careful assessment of risks as well as attention to eligibility criteria and to the informed and voluntary consent of research subjects. Further dialogue, conceptual analysis, and empirical work about payment made to clinical research subjects may serve to reduce the divide between those who object to such payment and those who promote payment as a sign of respect for the contributions of research subjects and a way to facilitate valuable research.

Acknowledgments

I would like to thank Neal Dickert for important insights, for many fruitful discussions and collaborations on this topic, and for his critical review of this manuscript. The views expressed here are those of the author and do not necessarily reflect the views or policies of the NIH or the US Department of Health and Human Services.

Address correspondence to: Christine Grady, Department of Clinical Bioethics, Clinical Center, NIH, Building 10/Room 1C118, Bethesda, Maryland 20892, USA. Phone: (301) 435-8710; Fax: (301) 496-0760; E-mail: cgrady@nih.gov.

1. Lederer, S. 1995. *Subjected to science: human experimentation in America before the Second World War.* The Johns Hopkins University Press. Baltimore, Maryland, USA. 192 pp.

2. NIH. 1959. *Healthy volunteers help scientists conquer disease.* U.S. Department of Health, Education and Welfare, Public Health Service, NIH. Bethesda, Maryland, USA. 15 pp.

3. Center Watch Clinical Trials Listing Service. http://www.centerwatch.com.

4. Dickert, N., Emanuel, E., and Grady, C. 2002. Paying research subjects: an analysis of current policies. *Ann. Intern. Med.* **136**:368–373.

5. Grady, C., Dickert, N., Jawetz, T., Gensler, G., and Emanuel, E. 2005. An analysis of U.S. practices of paying research participants. *Contemp. Clin. Trials.* **26**:365–375.

6. Center Watch Clinical Trials Listing Service. Clinical trial listings by medical areas. http://www.centerwatch.com/patient/trials.html.

7. McNeill, P. 1997. Paying people to participate in research: why not? *Bioethics.* **11**:390–396.

8. US FDA. 1998. Guidance for institutional review boards and clinical investigators: payment to research subjects. http://www.fda.gov/oc/ohrt/irbs/default.htm.

9. Dunn, L., and Gordon, N. 2005. Improving informed consent and enhancing recruitment for research by understanding economic behavior. *J. Am. Med. Assoc.* **293**:609–612.

10. Lemmens, T., and Elliott, C. 2001. Justice for the professional guinea pig. *Am. J. Bioeth.* **1**:51–53.

11. Wilkinson, M., and Moore, A. 1999. Inducements Revisited. *Bioethics.* **13**:114–130.

12. US Department of Health and Human Services and NIH Office for Protection from Research Risks. 2001. Protection of human subjects. US Code of Federal Regulations, title 45, part 46. http://www.hhs.gov/ohrp/humansubjects/guidance/45cfr46.htm.

13. US FDA. 2004. Protection of human subjects. US Code of Federal Regulations, title 21, part 50. http://www.accessdata.fda.gov/scripts/cdrh/cfdocs/cfcfr/CFRSearch.cfm?CFRPart=50.

14. Asch, D., Christakis, N., and Ubel, P. 1998. Conducting physician mail surveys on a limited budget: a randomized trial comparing $2 bill versus $5 bill incentives. *Med. Care.* **36**:95–99.

15. Church, A. 1993. Estimating the effect of incentives on mail survey response rates: a meta-analysis. *Public Opin. Q.* **5**:62–79.

16. Doody, M., et al. 2003. Randomized trial of financial incentives and delivery methods for improving response to a mailed questionnaire. *Am. J. Epidemiol.* **157**:643–651.

17. Halpern, S., Ubel, P., Berlin, J., and Asch, D. 2002. Randomized trial of $5 versus $10 monetary incentives, envelope size, and candy to increase physician response rates to mailed questionnaires. *Med. Care.* **40**:834–839.

18. VanGeest, J., Wynia, M., Cummins, D., and Wilson, B. 2001. Effects of different monetary incentives on the return rate of a national mail survey of physicians. *Med. Care.* **39**:197–201.

19. Warriner, K., Goyder, J., Gjertsen, H., Hohner, P., and McSpurren, K. 1996. Charities, no; lotteries, no; cash, yes: main effects and interactions in a Canadian incentives experiment. *Public Opin. Q.* **60**:542–562.

20. Bentley, J., and Thacker, P. 2004. The influence of risk and monetary payment on the research participation decision making process. *J. Med. Ethics.* **30**:293–298.

21. Halpern, S., Karlawish, J., Casarett, D., Berlin, J., and Asch, D. 2004. Empirical assessment of whether moderate payments are undue or unjust inducements for participation in clinical trials. *Arch. Intern. Med.* **164**:801–803.

22. Russell, M., Moralejo, D., and Burgess, E. 2000. Paying research subjects: participants' perspectives. *J. Med. Ethics.* **26**:126–130.

23. Grant, R., and Sugarman, J. 2004. Ethics in human subjects research: do incentives matter? *J. Med. Philos.* **29**:717–738.

24. Bigorra, J., and Banos, J. 1990. Weight of financial reward in the decision by medical students and experienced healthy volunteers to participate in clinical trials. *Eur. J. Clin. Pharmacol.* **38**:443–446.

25. Farre, M., Lamas, X., and Cami, J. 1995. Sensation seeking among healthy volunteers participating in phase I clinical trials. *Br. J. Clin. Pharmacol.* **39**:405–409.

26. Cunny, K., and Miller, H. 1994. Participation in clinical drug studies: motivations and barriers. *Clin. Ther.* **16**:273–282.

27. Corbie-Smith, G., Thomas, S.B., and St. George, D. 2002. Distrust, race, and research. *Arch. Intern. Med.* **162**:2458–2463.

28. Corbie-Smith, G., et al. 2003. Influence of race, clinical, and other socio-demographic features on trial participation. *J. Clin. Epidemiol.* **56**:304–309.

29. Harris, Y., et al. 1996. Why African Americans may not be participating in clinical trials. *J. Natl. Med. Assoc.* **88**:630–634.

30. Shavers, V.L., Lynch, C.F., and Burmeister, L.F. 2000. Knowledge of the Tuskegee study and its impact on the willingness to participate in medical research studies. *J. Natl. Med. Assoc.* **92**:563–572.

31. Faden, R., and Beauchamp, T. 1986. *History and theory of informed consent.* Oxford University Press. New York, New York, USA. 392 pp.

32. Emanuel, E.J. 2004. Ending concerns about undue inducement. *J.Law Med. Ethics.* **32**:100–105.

33. Office for Protection from Research Risks. 1993. IRB guidebook. Washington, DC, USA. http://www.hhs.gov/ohrp/irb/irb_guidebook.htm.

34. Grady, C. 2001. Money for research participation: does it jeopardize informed consent? *Am. J. Bioeth.* **1**:40–44.

35. Newton, L. 1982. Inducement, due and otherwise *IRB.* **4**:4–6.

36. Aby, J., Pheley, A., and Steinberg, P. 1996. Motivations for participation in clinical trials of drugs for the treatment of asthma, seasonal allergic rhinitis, and perennial nonallergic rhinitis. *Ann. Allergy Asthma Immunol.* **76**:348–354.

37. Hassar, M., et al. 1977. Free-living volunteer's motivations and attitudes toward pharmacological studies in man. *Clin. Pharmacol. Ther.* **21**:515–519.

38. Korn, J., and Hogan, K. 1992. Effect of incentives and aversiveness of treatment on willingness to participate in research. *Teach. Psychol.* **19**:21–24.

39. van Gelderen, C., Savelkoul, T., van Dokkum, W., and Meulenbelt, J. 1993. Motives and perceptions of healthy volunteers who participate in experiments. *Eur. J. Clin. Pharmacol.* **45**:15–21.

40. Tishler, C., and Bartholomae, S. 2002. The recruitment of normal healthy volunteers: a review of the literature on the use of financial incentives. *J. Clin. Pharmacol.* **42**:363–373.

41. Casarett, D., Karlawish, J., and Asch, D. 2002. Paying hypertension research subjects. *J. Gen. Intern. Med.* **17**:650–652.

42. Wilkinson, M., and Moore, A. 1997. Inducement in research. *Bioethics.* **11**:373–389.

43. Macklin, R. 1981. On paying money to research subjects: 'due' and 'undue' inducements. *IRB.* **3**:1–6.

44. Ackerman, T. 1989. An ethical framework for the practice of paying research subjects. *IRB.* **11**:1–4.

45. Pace, C., Miller, F., and Danis, M. 2003. Enrolling the uninsured in clinical trials: an ethical perspective. *Crit. Care Med.* **31**:S121–S125.

46. Dickert, N., and Grady, C. 1999. What's the price of a research subject? Approaches to payment for research participation. *N. Engl. J. Med.* **341**:198–203.

47. Appelbaum, P., et al. 1987. False hopes and best data: consent to research and the therapeutic misconception. *Hastings Cent. Rep.* **17**:20–24.

48. US FDA. 2003. The Pediatric Research Equity Act of 2003. http://www.fda.gov/opacom/laws/prea.html.

49. NIH. 1998. NIH policy and guidelines on the inclusion of children as participants in research involving human subjects. *NIH Guide Grants Contracts.* http://grants.nih.gov/grants/guide/notice-files/not98-024.html.

50. Medicines and healthcare products regulatory agency. 2001. Clinical trials directive (2001/20/EC). *Official Journal of the European Communities.* **147**:41.

51. Institute of Medicine. 2004. Payments related to children's participation in clinical research. Chapter 6. *Ethical conduct of clinical research involving children.* M. Field and R. Berman, editors. The National Academies Press. Washington, DC, USA. 211–228.

52. US Department of Health and Human Services and NIH. 2001. Additional DHHS protections for children involved as subjects in research. US Code of Federal Regulations, title 45, part 46, subpart D. http://www.hhs.gov/ohrp/humansubjects/guidance/45cfr46.htm#subpartd.

53. Wendler, D., Rackoff, J., Emanuel, E., and Grady, C. 2002. The ethics of paying for children's participation in research. *J. Pediatr.* **141**:166–171.

54. American Academy of Pediatrics, Committee on Drugs. 1995. Guidelines for the ethical conduct of studies to evaluate drugs in pediatric populations. *Pediatrics.* **95**:286–294.

55. Weise, K.L., et al. 2002. National practices regarding payment to research subjects for participating in pediatric research. *Pediatrics.* **110**:577–582.

Embryonic death and the creation of human embryonic stem cells

Donald W. Landry[1] and Howard A. Zucker[2]

[1]Department of Medicine and [2]Department of Pediatrics, Columbia University College of Physicians and Surgeons, New York, New York, USA.

The creation of human embryonic stem cells through the destruction of a human embryo pits the value of a potential therapeutic tool against that of an early human life. This contest of values has resulted in a polarized debate that neglects areas of common interest and perspective. We suggest that a common ground for pursuing research on human embryonic stem cells can be found by reconsidering the death of the human embryo and by applying to this research the ethical norms of essential organ donation.

Introduction

Human embryonic stem cells may or may not possess the innate capacity to provide unique treatments for human disease. But at least for now, live human embryos must be destroyed in the process of creating stem cells. The questionable ethics of such destruction led the US Congress to enact a broad ban on federal funding for human embryonic stem cell research. In 2001, President Bush modified this ban by removing restrictions for work with existing lines, but he continued prohibition of funding for the creation of new lines. However, existing lines may prove inadequate, and the eventual need for new embryonic lines cannot be discounted. Also, research on nonhuman embryonic cells is often not easily translated to humans; thus active research on human embryonic cells is desirable. Nonetheless, the duty to heal the sick cannot override the moral imperative to treat human beings as subjects and not objects. We propose herein a paradigm for research involving embryos that protects nascent human life, is consistent with current federal policy, and yet advances the interests of biomedical science and therapeutic innovation.

Precisely when the life of a human begins remains for some a complicated question, but a general consensus has been achieved on when life ends: life ends when the criteria for brain death are met. However, the criteria for determining the death of the developing human before the onset of neural development have not been formulated. We believe that when the condition of developing human life at the stage of a few-celled embryo is reconsidered, a significant fraction of embryos generated for in vitro fertilization (IVF), heretofore misclassified as nonviable, will be found to be organismically dead. If this is so, the ethical framework currently used for obtaining essential organs for transplantation from deceased adults and children could be extended to cover obtaining stem cells from dead human embryos.

Background on embryonic stem cells

Methods for preparing embryonic stem cells were originally developed in mice in 1981, but human lines were successfully isolated only recently (1); the time lag underscores the problem posed by the interspecific transfer of such technology. In order for human cells to be obtained, donated embryos or embryos generated from

Nonstandard abbreviations used: ICM, inner cell mass; IVF, in vitro fertilization.

Conflict of interest: The authors have declared that no conflict of interest exists.

Citation for this article: *J. Clin. Invest.* **114**:1184–1186 (2004).
doi:10.1172/JCI200423065.

donated eggs and sperm are grown in culture to the blastocyst stage (Figure 1). The blastocyst forms at approximately 4 or 5 days after fertilization and contains from 64 to several hundred cells organized in an outer shell, the trophectoderm, and a collection of polarized inner cells termed the inner cell mass (ICM). The ICM is the locus of pluripotent cells destined to yield all the tissues of the developed organism. In the process of obtaining embryonic stem cells, the trophectoderm is removed by immunosurgery, and the ICM is disaggregated and plated on feeder cells. The resulting cell colonies are mechanically isolated and replated until homogenous colonies are obtained (2). The colonies are then screened for the presence of stem cells. The desired cells are clonogenic and capable of dividing symmetrically for unlimited self-renewal. These cells are also capable of dividing asymmetrically to yield one cell identical to the parent and one able to develop into diverse cell types. Once established, human embryonic stem cell lines can persist stably, apparently for years. At present there is no source of new embryonic human stem cell lines that does not involve the destruction of human embryos.

Definition of death and the ethics of essential organ donation

The evolution of the definition of death for developed human beings can illuminate the consideration of death for developing humans. Prior to the 1960s, the legal definition of death focused on irreversible cessation of cardiopulmonary function (3). Advances in respiratory technology that could maintain a body indefinitely highlighted the inadequacy of this definition and an ad hoc committee at Harvard Medical School in 1968 formulated a new criterion for determining whether death had occurred: the state of irreversible coma (4). US Federal and State courts treat the determination of human death as a specific question of fact to be decided case by case rather than a general question of law (4). Thus implementing the 1968 criterion was not contingent upon statutory enactment, but to avoid confusion, laws were passed. The first legislative attempt in 1970 to frame "[a]n Act relating to and defining death" was criticized on the grounds that it provided for two separate definitions for death and that the definition used in individual cases was contingent only upon the presence or absence of a desire to harvest organs for transplantation (5). Capron and Kass in 1972 proposed that there be a single concept of death but that, depending on the absence or presence of the machinery of life support that made recognizing it uncertain, either irreversible cessation of spontaneous circulatory and respiratory functions or irreversible cessation

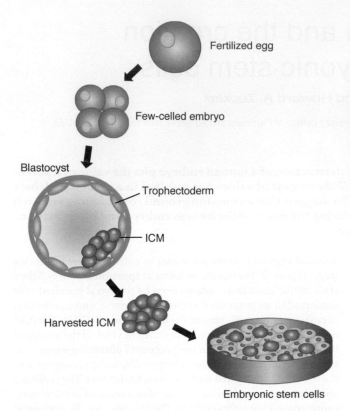

Figure 1
Human embryonic stem cells. The steps from fertilized egg obtained by IVF to embryonic stem cells are depicted. A few-celled embryo gives rise to the blastocyst, a structure comprised of an outer cell layer, the trophectoderm, and the ICM. The ICM is harvested and plated on feeder cells, then replated (not shown) to yield a population of embryonic stem cells.

The concept of organismic death for the human embryo

For a developed human organism, brain death marks the irreversible loss of the capacity for all ongoing and integrated organic functioning. We propose that the defining capacity of a 4- or 8-cell human embryo is continued and integrated cellular division, growth, and differentiation. We further propose that an embryo that has irreversibly lost this capacity, even as its individual cells are alive, is properly considered organismically dead. Even at its earliest stages, the life of the developing organism is more than the sum of the lives of its constituent cells. Each cell is involved in directing the gene expression and differentiation of other cells and, thereby, the growth and development of the whole. Just as the nervous system integrates the activities of the tissues and organs of the developed organism, so a system of chemical communication and surface recognition integrates the cells of the developing organism. The determinants of growth and differentiation, such as Oct4, are now being elucidated (11). If an embryo is deprived of the necessary internal signals, irreversible arrest of cleavage can result. Objective criteria for diagnosing death require straightforward methods for determining irreversible arrest. The determination of irreversibility could be inductive, based on observing the natural history of cleavage arrest in large numbers of embryos, or deductive, based on elucidating biochemical defects that preclude further growth and development — or both. Studies of IVF embryos suggest that irreversibly arrested embryos could contain cells with normal developmental potential.

The prospect for producing normal cells from dead embryos

Approximately 60% of IVF embryos fail to meet criteria for viability and are rejected for uterine transfer. Nonviability, defined as the incapacity to develop to live birth, differs from organismic death; all dead embryos are, of course, nonviable, but most nonviable embryos are not yet dead. Morphological criteria for nonviability include abnormal cleavage, loss of cells, and loss of cytoplasm. But it is the functional criterion — failure to cleave at 24 hours — that, while not by itself proof of irreversibility, likely correlates. Cleavage arrest most often reflects severe genetic abnormalities but — and this is the crucial point — not all cells of arrested embryos need be abnormal for arrest to occur. In a study by Laverge et al., out of 166 frozen embryos thawed for further growth, 78 embryos remained arrested at 24 hours after thawing, and 71 showed no sign of further cleavage at 48 hours (12). Fluorescence in situ hybridization with probes for chromosomes X, Y, and 1 was performed on all blastomeres of 63 arrested embryos, and 80% showed genetic abnormalities. Aneuploidy was common, but defects were both meiotic and mitotic in origin, and mosaic embryos containing at least one normal diploid blastomere were identified. Voullaire et al. performed comparative genomic hybridization on 65 blastomeres from 12 unselected embryos (13). Frozen human embryos, scheduled to be discarded after 5 years of storage, were thawed, observed in culture for 24 hours for cleavage, allowed to expire at room temperature, and then disaggregated for analysis. Marked aneuploidy was noted, but once again mosaic embryos (5 out of 12) with at least one normal blastomere were observed. Thus an unexpectedly high level of mosaicism due to postzygotic nondisjunction, i.e., defective segregation of chromosomes in a single cell after the first cell division, could provide a potential source of normal blastomeres from organismically dead embryos.

of brain functions could be used to determine whether the person was in fact dead (6). Medical, legal, and ethical opinion coalesced around the concept of brain death, and the Capron-Kass proposal combined with the 1975 American Bar Association (ABA) Definition of Death Act ultimately led to the 1981 Uniform Determination of Death Act (UDDA) (7), widely adopted by individual states. It is "the law's determination that brain death is the legal equivalent of death because ... the capacity for life is irretrievably lost when the entire brain, including the brain stem, has ceased functioning" (5). Legal challenges to this definition of death have been unsuccessful (5, 8). More recently, in 1995, the definitions regarding clinical and confirmatory test requirements for brain death were reexamined by the American Academy of Neurology (9), and the criteria set forth by the Harvard committee were reaffirmed. Thus, at the time of brain death, the removal of vital organs for transplant is legally and ethically permitted. As stated by the Omnibus Reconciliation Act of 1986 (10), after brain death criteria have been met, the next of kin may be approached regarding organ donation.

A fully developed human being is considered organismically dead — even as the cells of the various organs remain alive — if the function of the brain is irretrievably lost. The transition from life to organismic death is determined to have occurred if the criteria for brain death are met. But how can one determine organismic death at the earliest stages of a human life, at a time still weeks from the inception of the central nervous system?

Several studies provide insights into the possible developmental potential of blastomeres derived from dead human embryos. In a study in frogs by Byrne et al., tadpole embryos that were derived by nuclear transplantation (i.e., cloned) rarely developed into tadpoles (less than 1%) because a majority underwent irreversible arrest and began to decompose within 24 hours (14). However, if the cells of an arrested blastocyst were disaggregated and injected into normal tadpole blastocysts, those extracted cells frequently resumed division (25%) and were stably incorporated into differentiated tissues. Further, in ethically troubling experiments by Alikani and Willadsen, 107 IVF embryos designated as "nonviable" by the authors were disaggregated, and 247 intact cells were isolated and combined in 36 aggregates (15). Among these aggregates, 33% formed normally organized blastocysts with defined inner cell masses. These experiments indicate that it is likely that blastomeres from organismically dead embryos can maintain their development potential. Detailed studies will be needed to assess the presence of any chromosomal damage in these blastomeres that might affect their long-term function in differentiated tissues.

Damage that must be excluded in the surviving blastomeres includes subtle chromosomal fragmentation, methylation defects, and telomerase anomalies. A natural history study to define human embryonic death could be organized as follows: previously frozen early embryos that have failed to divide within 24 hours of thawing and are no longer wanted for their original reproductive purpose are observed every few hours for several additional 24-hour periods. After observing several hundred embryos, the time beyond which no arrested embryo resumes division can be determined. One can reasonably conclude that embryos that have not divided by this period after thawing will not divide at any later time, i.e., they are organismically dead. Embryos declared dead could then be characterized for secreted or cell surface markers or spectroscopic signatures that correlate with the arrest of cell division. These markers and signatures could then be tested for their predictive value. In this manner the criteria for determining the death of a human embryo could be refined.

Conclusion

A reality of human embryonic life whether in utero or in vitro is a high incidence of death in the first few days after fertilization. We propose that an irreversible arrest of cell division rather than the death of each and every cell is the appropriate measure of organismic death for the embryo. Criteria for determining irreversibility are lacking, but the approach is clear; natural history studies of cleavage arrest will provide an initial definition that can be refined as elucidation of the mechanisms regulating growth and arrest yield biochemical markers for irreversibility. Based on our analysis, we believe that many embryos generated for IVF are dead at the organismic level and yet, due to mosaicism, are likely donors of normal blastomeres. Although the signals required for transforming blastomeres into stem cells — without recourse to the formation and destruction of human blastocysts — remain to be elucidated, relatively unfettered access to human blastomeres would advance this research program.

The legal and administrative process that led to criteria for establishing the death of the developed human provides insights into the comparable process necessary for determining death during embryonic development. If the donation of embryonic cells from organismically dead embryos is considered as analogous to the donation of essential organs from cadavers, the result could be widespread acceptance of the use of such cells for research and development. Species specificity renders research on human cells all the more important, and access to blastomeres from organismically dead embryos would facilitate innovation. As an example, human blastomeres may be suitable for somatic cell cloning with the potential to circumvent the destruction of live human embryos that characterizes the current protocols for "therapeutic cloning." The harvesting of organs from executed prisoners or aborted fetuses is criticized for encouraging executions and abortions. In contrast, the harvesting of cells from organismically dead embryos would not seem to significantly encourage the practice of in vitro fertilization for infertility, a practice in which death is undesired rather than the primary objective. In sum, application of the ethical framework for essential organ donation to the harvesting of human embryonic cells from dead embryos could provide a common ground in which the imperative to safeguard human dignity and the drive for biomedical research are not in conflict.

Acknowledgments

We thank Virginia Papaioannou for helpful discussions.

Address correspondence to: Donald W. Landry, Columbia University, P&S Building 10th floor, Room 445, 630 West 168th Street, New York, New York 10032, USA. Phone: (212) 305-2436; Fax: (212) 305-3475; E-mail: dwl1@columbia.edu.

1. Thomson, J.A., et al. 1998. Embryonic stem cell lines derived from human blastocysts. *Science.* **282**:1145–1147.

2. Rosenthal, N. 2003. Prometheus's vulture and the stem-cell promise. *N. Engl. J. Med.* **349**:267–274.

3. Black, H.C. 1990. *Black's law dictionary: definitions of the terms and phrases of American and English jurisprudence, ancient and modern.* 6th edition. West Publishing Co. St. Paul, Minnesota, USA. 1657 pp.

4. Ad Hoc Committee of the Harvard Medical School. 1968. A definition of irreversible coma. *JAMA.* **205**:337–340.

5. *Kansas Statutes Annotated.* sec. 77–202 (Supp. 1971).

6. Capron, A., and Kass, L. 1972. A statutory definition of the standards for determining human death; an appraisal and a proposal. *University of Pennsylvania Law Review.* **121**:87.

7. President's Commission for the Study of Ethical Problems in Medicine and Biomedical and Behavioral Research. 1981. Defining death: a report on the medical, legal and ethical issues in the determination of death. The Commission: for sale by the Superintendent of Documents, US Government Printing Office. Washington, D.C., USA. 166 pp.

8. In Re Long Island Jewish Medical Center, 641 *New York Supplement* 2d 989 (New York Superior Court 1996).

9. Report of the Quality Standards Subcommittee of the American Academy of Neurology. 1995. Practice parameters for determining brain death in adults. *Neurology.* **45**:1912.

10. *Omnibus Reconciliation Act of 1986.* 42 United States Code Sec. 1320b-8.

11. Nichols, J., et al. 1998. Formation of pluripotent stem cells in the mammalian embryo depends on the POU transcription factor Oct4. *Cell.* **95**:379–391.

12. Laverge, H., et al. 1998. Fluorescent in-situ hybridization on human embryos showing cleavage arrest after freezing and thawing. *Hum. Reprod.* **13**:425–429.

13. Voullaire, L., Slater, H., Williamson, R., and Wilton, L. 2000. Chromosome analysis of blastomeres from human embryos by using comparative genomic hybridization. *Hum. Genet.* **106**:210–217.

14. Byrne, J.A., Simonsson, S., and Gurdon, J.B. 2002. From intestine to muscle: Nuclear reprogramming through defective cloned embryos. *Proc. Natl. Acad. Sci. U. S. A.* **99**:6059–6063.

15. Alikani, M., and Willadsen, S. 2002. Human blastocysts from aggregated mononucleated cells of two or more non-viable zygote-derived embryos. *Reprod. Biomed. Online.* **5**:56–58.

Stem cells: science, policy, and ethics

Gerald D. Fischbach[1] and Ruth L. Fischbach[2]

[1]Faculty of Medicine and [2]Center for Bioethics and Department of Psychiatry, Columbia University College of Physicians and Surgeons, New York, New York, USA.

Human embryonic stem cells offer the promise of a new regenerative medicine in which damaged adult cells can be replaced with new cells. Research is needed to determine the most viable stem cell lines and reliable ways to promote the differentiation of pluripotent stem cells into specific cell types (neurons, muscle cells, etc.). To create new cell lines, it is necessary to destroy preimplantation blastocysts. This has led to an intense debate that threatens to limit embryonic stem cell research. The profound ethical issues raised call for informed, dispassionate debate.

The promise of stem cell research

Few subjects in biomedical science have captured the imagination of both the scientific community and the public as has the use of stem cells for the repair of damaged tissues. Because they may be able to replace cells that have atrophied or have been lost entirely, stem cells offer the hope of restoration of cellular function and relief from suffering associated with many disabling disorders. Beyond tissue repair, cultured stem cells might also find application in the analyses of disease mechanisms and normal development, as assays for screening new drugs, and as vehicles for gene therapy (1).

Each potential use of stem cells promises revolutionary advances. However, the word "promise" must be underscored — to date, no cures have been realized, no disease mechanisms have been uncovered, and no new drugs have been developed. Many in the international scientific community believe that the promise of stem cell–based studies or therapies will be realized only if we can derive new human embryonic stem cell (hESC) lines.

At the present time, the production of new cell lines involves destruction of preimplantation embryos at the 100–200 cell (blastocyst) stage. Debate currently centers on the moral status of these embryos, which are now stored at in vitro fertilization (IVF) clinics or created by somatic cell nuclear transfer (SCNT; discussed in detail below). What is the moral status of the blastocyst? Should blastocysts be protected under the same laws that govern research on human subjects? These and related questions are at the center of a debate that involves the lay public, the scientific community, the press, and the United States Congress.

The outcome of this debate will have an impact on the way we conduct the science of hESCs, a field still very much in its infancy. Indeed, the integrity of the scientific process and its independence from politics and from fundamentalist dogma are at stake. It is important, therefore, to define the relevant terminology and discuss it objectively. Along the way we must reduce the emotional valence of phrases such as "therapeutic cloning" and "destruction of embryos." To engage in this debate, it is important to have an overview of stem cell biology.

Stem cells defined

A stem cell is defined by two properties (see *A stem cell research lexicon*). First, it is a cell that can divide indefinitely, producing a population of identical offspring. Second, stem cells can, on cue, undergo an asymmetric division to produce two dissimilar daughter cells. One is identical to the parent and continues to contribute to the original stem cell line. The other varies in some way. This cell contains a different set of genetic instructions (resulting in an alternative pattern of gene expression) and is characterized by a reduced proliferative capacity and more restricted developmental potential than its parent. Eventually a stem cell becomes known as a "progenitor" or "precursor" cell, committed to producing one or a few terminally differentiated cells such as neurons or muscle cells.

The different types of stem cell populations can be illustrated by considering the earliest stages of embryogenesis (Figure 1). Soon after fertilization, the haploid nuclei of the egg and sperm merge to form a single nucleus with the diploid number of chromosomes. The zygote divides and its progeny also divide several times thereafter to form a compact ball of cells called the morula (likened in appearance to a mulberry). Each of the 32–128 cells in the morula is totipotent in that each one can give rise to all cell types in the embryo plus all of the extraembryonic tissues necessary for implantation in the uterine wall. These cells are also at the center of preimplantation genetic testing (see *Totipotent cells and genetic testing*).

As the morula is swept along the oviduct, the cells continue to proliferate and the morula enlarges to form a hollow sphere called a blastocyst (or blastula). During the final days in the oviduct and the first days in the uterus, a few cells delaminate from the surface layer of the blastula to form an inner cell mass (ICM) within the cavity. This cluster of cells is the source of embryonic stem cells. It is important to emphasize that the ICM forms *prior* to implantation. Blastocysts created in vitro contain an ICM even though the embryo was created and maintained in a test tube. It is possible to isolate cells from the ICM of human blastocysts and grow them in tissue culture (Figure 2), using techniques first developed 20 years ago for the manipulation of mouse embryos. Cells dissociated from the ICM are pluripotent in that they can become any of the hundreds of cell types in the adult body. They are not totipotent because they do not contribute to extraembryonic membranes or the formation of the placenta.

The time from fertilization to implantation in the uterine wall is approximately 14 days in humans. Soon after implantation, the blastocyst invaginates, much like a finger pressing into a round rubber balloon. A critical series of cell movements known as gastrulation results in the formation of the three germ layers of the

This article is adapted from a lecture delivered to the faculty and staff of Columbia University Medical Center on September 23, 2003, by Gerald D. Fischbach at the invitation of Ruth L. Fischbach.

Nonstandard abbreviations used: hESC, human embryonic stem cell; ICM, inner cell mass; IVF, in vitro fertilization; SCNT, somatic cell nuclear transfer.

Conflict of interest: The authors have declared that no conflict of interest exists.

Citation for this article: *J. Clin. Invest.* **114**:1364–1370 (2004). doi:10.1172/JCI200423549.

A stem cell research lexicon

Cell line	Homogeneous population of cells capable of self renewal
Cloned cell line	Population of cells that derives from replication of a single cell
Stem cell	Cell that can divide indefinitely to produce a population of identical offspring
Zygote	Diploid cell resulting from the fusion of male and female gametes at fertilization
Morula	Spheroidal mass of cells resulting from early cleavage divisions of the zygote
Blastocyst (or blastula)	4–5 day–old embryo formed prior to implantation in the uterus; consists of a hollow mass of only a few undifferentiated stem cells
Progenitor (or precursor) cell	Parent cell that is committed to dividing and multiplying in order to produce a specific cell type
Totipotent cell	Cell not committed to a specific lineage that is capable of giving rise to all types of differentiated cells and tissues, including extraembryonic tissues
Pluripotent cell	Cell not committed to a specific lineage that may differentiate into all types of cells and tissues, with the exception of extraembryonic tissues
Multipotent cell	Progenitor cell that can give rise to diverse cell types in response to appropriate environmental cues

developing embryo: the ectoderm, the endoderm, and the mesoderm. The basic plan of the human body is laid out during this remarkable process as the fate of many cells is determined: the endoderm gives rise to the vasculature and blood-forming organs; the mesoderm produces muscle; and the ectoderm gives rise to the skin and the nervous system.

Stem cells are present in each of the three germ layers. The spectrum of offspring from these stem cells is more restricted than that of cells derived from the ICM, so they are described as multipotent rather than pluripotent. Cells in one germ layer breed true; they do not ordinarily transdifferentiate to form derivatives of other germ layers. Indeed, there is strong evidence for a restriction of developmental potential with time throughout embryogenesis (2, 3). However, the plasticity of adult stem cells is an issue of great interest, and it merits further investigation.

Stem cells in adult tissues

Stem cells have been identified in adult tissues including skin, intestine, liver, brain, and bone marrow. Bone marrow stem cells have been studied most extensively because a variety of cell surface and genetic markers have helped delineate various stages of their differentiation during hematopoiesis.

But there are several drawbacks that, a priori, make adult stem cells less attractive than embryonic stem cells as sources for most of the uses described above. It has been difficult to isolate stem cells from adult tissues. The cells are few in number, and it is difficult to keep them proliferating in culture. To date, it appears that cultured adult stem cells give rise to only a limited number of cell types. Finally, they are adult cells and have been exposed to a lifetime of environmental toxins and have also accumulated a lifetime of genetic mutations.

Despite these apparent drawbacks, research on adult stem cells should be pursued vigorously because these problems may be overcome with new techniques and insights. The therapeutic value of partially purified hematopoietic stem cells in repopulating the bone marrow following high-dose chemotherapy is based on the discovery of growth factors that promote the multiplication of blood precursor cells. We need the same type of information about the differentiation of other types of adult stem cells.

Embryonic stem cells

The ability of hESCs to proliferate indefinitely in tissue culture and the wide range of cell types to which they give rise make these cells unique. They become even more valuable as new molecules that trigger their differentiation in vivo are discovered. It has proven easier to mimic the normal sequence of development than to reverse this process in an attempt to have cells dedifferentiate.

In 1998, capitalizing on nearly twenty years of experience with mouse embryonic stem cells, scientists at the University of Wisconsin isolated stem cells from the ICM of human blastocysts

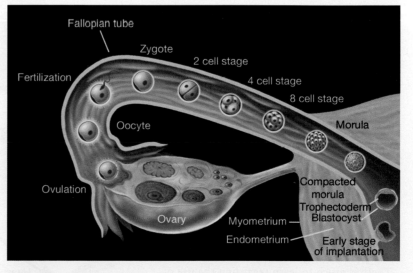

Figure 1
From zygote to blastula: the early stages of human development. Shortly after fertilization, the zygote repeatedly divides to form a solid mass of cells known as the morula. Two to three days after fertilization, the morula enters into the uterine cavity and forms a hollow sphere: the blastocyst. The surface cells form the trophoblast and give rise to extraembryonic tissues, while the inner cell mass is the source of embryonic stem cells and ultimately gives rise to the embryo, following implantation in the uterine wall.

Totipotent cells and genetic testing

The fact that each cell within the morula is totipotent means that one or more cells can be sacrificed without harm. This has led to the remarkable new technique of preimplantation genetic diagnosis, in which one of the morula cells is drawn into a micropipette and subjected to a sensitive PCR assay using cDNA primers now available for an increasing number of genetic disorders in order to screen for potential disease-associated mutations. This remarkable technology allows physicians to inform parents of the health of the embryo before implantation of the still-viable morula in the uterus.

and grew them in tissue culture for prolonged periods of time (4). Under the right conditions, several types of mature cells appeared in the cultures, including nerve cells, muscle cells, bone cells, and pancreatic islet cells (Figure 2). This paper has led to an explosion of research on hESCs.

Results obtained from studies with mouse ESCs raise the possibility that clinical trials with hESCs are not far off. Mouse ESCs have been steered to become spinal cord motor neurons (5), dopaminergic neurons (6, 7), and many other types of cells. One example must suffice here to emphasize their therapeutic promise. In one of the most thorough and elegant studies published to date, mouse ESCs were steered to differentiate into spinal cord motor neurons by successive exposure to retinoic acid and sonic hedgehog, a protein known to trigger the differentiation of motor neurons in developing embryos (5). When treated cells were injected into the spinal cord of a chick embryo, they migrated to their proper location in the ventral horn. Some cells sent axons out of the spinal cord to invade the developing limb (Figure 3) and form synapses on target muscle fibers. This type of research lends hope to individuals suffering from amyotrophic lateral sclerosis, spinal muscular atrophy, spinal cord injury, and related disorders.

Another argument for support of stem cell research follows from the success of transplanting intact human tissues. Pancreatic islets have been implanted into patients with type 1 diabetes to restore them to insulin independence (8). Islet transplantation, according to the Edmonton protocol, works (9, 10). Likewise, implantation of fetal mesencephalic brain tissue into the brains of patients with Parkinson disease resulted in measurable improvement in some indices of motor performance (11). Both implantation studies, however, were limited by tissue availability and, in the Parkinson disease study, there were serious side effects (e.g., dyskinesias). Both studies call for further work with hESCs, with the hope of moving to Phase 1 clinical trials.

There is much to learn regarding the use of stem cells for the treatment of disease. We need additional information about how to keep ESCs dividing until they are called on to differentiate. We must learn more about the growth factors that influence their differentiation into diverse cell types. Most importantly, we must endeavor

to devise stem cell therapy protocols that are safe. This will be greatly facilitated by our understanding of how to turn these cells off in vivo in the event that toxicity develops. In addition, the risk of immune rejection remains a problem. Given the limited genetic diversity of available cell lines, transplantation of stem cell products is subject to the same immune barriers as organ transplantation. At the present time, our only defense against rejection is the administration of long-term immunosuppression therapy, which increases the patients' risk of infection and is associated with nephrotoxicity. In the future, immune rejection might be minimized without the need for toxic drugs, using cells obtained from blastulae that have been created by SCNT.

Somatic cell nuclear transfer

In SCNT, the nucleus from a mature cell is injected into the cytoplasm of an oocyte from which the original (haploid) nucleus has been removed. As in the union of haploid sperm and egg nuclei, one ends up with a diploid number of chromosomes, but in the case of SCNT, all of the chromosomes originate from the donor nucleus. The great advantage of SCNT is that the ESCs derived from blastocysts so created will be genetically similar to the cells of the individual who donated the nucleus. It is less likely, therefore, that the expressed proteins will be recognized as foreign and evoke an immune response in the host (12).

It is difficult to reproduce the course of early embryonic development by reprogramming an adult nucleus. Factors in the egg cytoplasm that regulate gene expression in the hours after a zygote is formed must act in the adult nucleus to produce the same patterns of gene activation that are critical for early embryonic development. The striking success of Korean investigators in deriving new hESC lines from embryos created by SCNT has been widely noted (13–15).

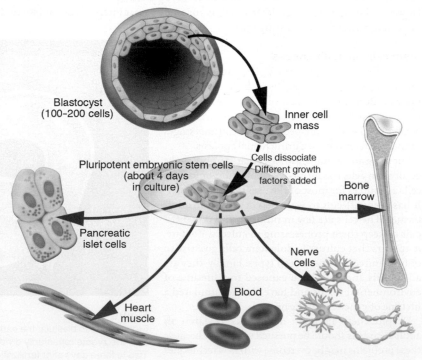

Figure 2
Pluripotent stem cells, isolated from the ICM in the blastocyst, have the ability to give rise to all types of cells in the human body, but not the placenta and other supporting tissues.

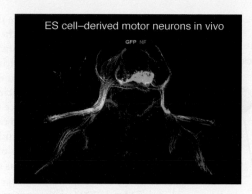

Figure 3
Integration of transplanted mouse embryonic cell–derived motor neurons into the spinal cord in vivo. Transverse section through the lumbar region of the spinal cord reveals that enhanced GFP+ axons exit the spinal cord via the ventral root and project along nerve branches that supply dorsal and ventral limb muscles. The pathway of axons is detected by neurofilament (NF) expression. Reprinted with permission from *Cell Press* (5).

These ESCs may not be entirely "normal," in the sense that subsequent coordinated development is impaired, but they can serve as starting points for the production of specific cell types.

The use of SCNT for the purposes of creating stem cell lines seems to be an innocuous process, but intrauterine implantation of blastocysts created by SCNT might lead to a live birth, a process known as reproductive cloning (Figure 4). Dolly the sheep was generated using this technology. However, successful reproductive cloning is an extremely improbable event. Most embryos created by SCNT are malformed and die in utero. It required 277 attempts to create Dolly, and there is strong evidence that Dolly exhibited many pathologies (e.g., arthritis, obesity) throughout her life (16–18). No one knows how many attempts it would take to create a live human being or what genetic abnormalities such an individual would bear (19–22).

Mice have been cloned from adult nuclei, even from postmitotic, terminally differentiated olfactory neurons (23). Although the full diversity of olfactory neurons was present in the offspring, more work is needed to define the limits of normalcy in such animals.

Years of experience with animals makes it clear that attempts at reproductive cloning of humans is scientifically unjustifiable at this time. Moreover, there are no compelling medical reasons to pursue this research. "Cloning" means to copy, and the word evokes an image of an identical replica. Given all of the epigenetic events that must occur during differentiation, it is inconceivable that an exact replica of an individual animal let alone a human being can be made. Parents of identical twins easily recognize the enormous number of differences between them.

The terms "research cloning" and "therapeutic cloning" have been applied to the creation of blastocysts by SCNT. These terms are unfortunate as they have become confused in the public's mind with reproductive cloning. They share a common word, and they (wrongly) evoke the worst connotations of the oversimplified image of cloning.

Current regulations

At the present time, no research on human embryos, including preimplantation blastocysts, can be supported with federal funds. Researchers must perform such studies with the aid of funding obtained from businesses, private foundations, or other philanthropic sources. This ban includes embryos stored frozen in IVF clinics and embryos created by SCNT. It sets us apart from many countries throughout the world.

The *Code of Federal Regulations* (*CFR*) is a collection of regulations issued by agencies of the federal government. Title 45 of the *CFR* covers the Department of Health and Human Services (DHHS). Part 46 — Protection of Human Subjects — of Title 45 covers research on human subjects and mandates the review of federally funded research involving human subjects by an institutional review board. The regulations originated in 1981 following revelations regarding the Tuskegee Syphilis Trial (1932–1972) in which a group of 400 indigent, black Americans, exploited and prevented from receiving penicillin during the 1940s, 1950s, and 1960s, were allowed to undergo the ravages of tertiary syphilis during the length of the trial. The study prompted the US Congress to establish the National Commission for the Protection of Human Subjects of Biomedical and Behavioral Research. The Commission was asked to develop the basic ethical principles that should govern research using human subjects. The result was the Belmont Report (24), one of the most influential documents in the field of bioethics since it defined the basic ethical principles relevant to research involving human subjects: the principles of respect for persons (autonomy), beneficence, and justice. The protections of 45CFR46 became known as the Common Rule after adoption in 1991 by all federal agencies conducting research with human subjects. Subpart A deals with the basic policies for human subjects protection. Subpart B (now called Additional Protection for Pregnant Women, Human Fetuses, and Neonates Involved in Research) relates to research on viable fetuses, pregnant women, and human IVF. Subpart C pertains to studies involving prisoners, while Subpart D describes special requirements for experiments involving children.

In Subpart B, protections are extended to "the product of conception from implantation until delivery." Recall that the ICM of the blastocyst is formed prior to implantation, which occurs on about day 14 after fertilization. Therefore, dissociation of the ICM is legal according to 45CFR46.

In 1996, Representatives Jay Dickey (R-AR) and Roger Wicker (R-MS) introduced an amendment to the *DHHS Appropriations Bill* (the source of NIH funds) that overrides Subpart B of 45CFR46 by extending protection to "any organism not protected as a human subject under 45CFR46 that is derived by fertilization, parthenogenesis, cloning, or any other means from one or more human gametes or diploid cells." The Dickey Amendment includes "research in which a human embryo is destroyed, discarded, or knowingly subjected to risk of injury or death greater than that allowed for research on fetuses in utero." Thus, preimplantation blastulae are included in the Dickey Amendment. Dissociation of the ICM necessarily destroys the blastocyst and hence, places it above minimal risk. This amendment is attached to the Appropriations Bill each year. Both President Clinton and President George W. Bush have signed bills containing the language of the Dickey Amendment. This amendment blocks investigators in the United States from using federal funds to derive new stem cell lines from early embryos.

In early 2000, Harriet Raab, then General Council of the DHHS, adopted the view that stem cells are not organisms (embryos) and hence are not covered by 45CFR46 or by the Dickey Amendment. Research on hESCs could, therefore, be supported by government funds, provided that the cells were derived from embryos using private funds. This opinion was adopted by Harold Varmus, then-

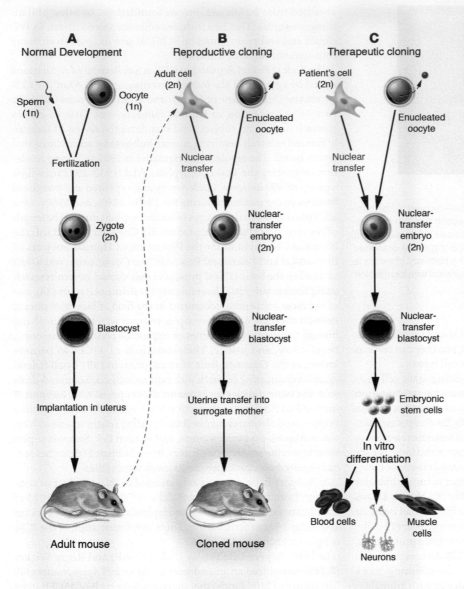

Figure 4
Normal development versus development during reproductive cloning and therapeutic cloning. During normal development (**A**), after fertilization, a diploid zygote is formed, which then undergoes cleavage to form a blastocyst that may be implanted in the uterus and result in a live birth. During reproductive cloning (**B**), the diploid nucleus of an adult donor cell is introduced into the enucleated oocyte. Following artificial activation, division results in a cloned blastocyst. Upon transfer into a surrogate mother, a small number of cloned blastocysts give rise to a clone. Therapeutic cloning (**C**) requires the explantation of cloned blastocysts in culture to yield an ESC line able to differentiate in vitro into any type of cell for therapeutic purposes. Figure modified with permission from the *New England Journal of Medicine* (31).

purposes; (b) the stem cells were in excess of clinical need, meaning that the donors had achieved a successful pregnancy or had simply decided not to proceed with IVF; (c) the stem cells were derived from embryos that were frozen, allowing sufficient time between the emotional experience of creating the embryos and the decision regarding donation; (d) informed consent and institutional review board approval was obtained; and (e) no exchange of money was made, in order to avoid a financial influence.

As of March 2004, there were more than 400,000 frozen embryos stored in IVF clinics nationwide. The options open to donors of these embryos are to destroy them, offer them up for adoption, continue to store them, or donate them for medical research.

These NIH guidelines were accepted by President Clinton but rejected by President George W. Bush soon thereafter. In his speech of August 9, 2001, President Bush recognized the value of research on hESCs and the promise of successful cell replacement therapies. However, he said that he would not condone the destruction of additional embryos to create new hESC lines. At the time, he believed that 62 hESC lines were available in labs around the world, and he made it clear that all subsequent federally supported research would be confined to these existing lines.

Reaction in the scientific community was mixed. Some were relieved that the President recognized the importance of hESC research. Other investigators, skeptical about the existence of 62 cell lines, were disappointed. In the following months it became obvious that there were not 62 usable cell lines; there were fewer than 5. The number of available lines has since grown, and 21 lines are currently listed in the NIH registry (25). However, the tangle of intellectual property requirements and the fact that most of these hESC lines were cultured in contact with mouse cells and bovine serum limits their utility. Moreover, many of them still have not been well characterized in terms of viability and their ability to differentiate.

Elias Zerhouni, Director of the NIH, has made sure that the NIH has done its part to

Director of the NIH, but it caused an uproar in Congress as many members felt that the Raab opinion was a legalism that violated the spirit of the law. Nevertheless, the NIH formed a committee of scientists, lawyers, patient advocates, and clergy to consider guidelines for the use of hESCs. They labored for many months. The final document stated that an NIH grantee could use these hESCs provided several criteria were met, including the following: (a) the stem cells were derived from embryos produced in IVF clinics for reproductive

promote research using the approved cell lines (26). Special training grants for new hESC investigators have been created; supplements to existing research grants have been offered as lures; and many different calls for special grants (requests for applications and requests for proposals) have been issued. The formation of multidisciplinary teams has been encouraged, and research infrastructure grants in support of hESC-based studies have been proposed. hESC-related conferences have been supported, and an excellent hESC web site — the official NIH resource for stem cell research — has been created (26).

Whether it is 5, 21, or 62, the number of available hESC lines is simply not sufficient to provide for the genetic diversity among the recipient population. In developing a new medicine, one would not stop with the first chemical that produced an effect. Efficacy must be optimized and safety must be taken into account.

Creation of new cell lines from human embryos can proceed thanks to support from nongovernment sources. Recently 17 new hESC lines were derived with private funds (27), and more are sure to follow. However, in the long run, the talent represented by the community of scientists supported by the NIH and other federal agencies will be needed for this field to move forward.

Currently in the House of Representatives a bill introduced by D. Weldon (R–FL) and B. Stupak (D–MI) and in the Senate a bill introduced by S. Brownback (R–KS) and M. Landrieu (R–LA) would outlaw the formation of human embryos by SCNT in the private sector as well as by researchers receiving federal funds. This extraordinary legislation would criminalize scientific research, making it punishable by a $1 million fine and 10 years in prison. Effects of this chilling attack on the scientific process extend beyond hESC research. It casts a pall over all science. It indicates a widening gulf between those in public office and the scientific community — a reversal of the coming together of political and scientific minds over the stem cell debate that we are observing in other nations.

Ethical issues

When does life begin? The answer to this question has enormous consequences for the future study of hESCs. Defining life as the moment of conception is certainly a convenient starting point, but this relies on an assumption about the value of a potential life. In this argument, value is placed on function (potential for future development) rather than structure (current state of development). This starting point, conception, is also promoted by many of those who rely on revealed Scripture. For those holding such beliefs, research on stem cells and the destruction of human blastocysts are simply unacceptable.

To many, implantation of the blastocyst in the uterine wall is the best landmark for the definition of life. Indeed, this is the first stage at which the individual is defined (e.g., the blastula is past the stage in which it can split to form twins). This is the point described in Subpart B of 45CFR46 as the first stage covered by human protections regulations. This is also the last developmental stage accepted in the United Kingdom and in many other countries throughout the world. For research on human embryos, gastrulation is another strong candidate, as it is reasonable to consider the phase in which the nervous system is formed and the possibility of sensation first exists as the beginning of human life.

One of the most dangerous trends in this debate is that of offering religious opinions cloaked in the language and veneer of science (e.g., using systems theory to justify the belief that life begins at conception). We have emphasized differences between embryonic and adult stem cells because many in the public and in Congress have claimed, arbitrarily, that the two sources are identical. Richard Dorflinger, Deputy Director of the Secretariat of the pro-life activities of the US Conference of Catholic Bishops, has claimed that adult stem cells hold more promise than embryonic stem cells and that research on embryonic stem cells is therefore unnecessary. The passion behind Dorflinger's statement is laudable, but it must be recognized that it is based on religious conviction, not on scientific induction or verified data.

Several commissions have explored the difficult ethical issues surrounding the definition of the beginning of life, including the National Bioethics Advisory Commission (28), the National Academy of Sciences Advisory Committee, and most recently, the President's Council on Bioethics (29, 30). Although each of these committees condemns research on reproductive cloning, research on SCNT has been upheld even by the President's Council on Bioethics, widely considered to be the most conservative group of the three. While the leader of the President's Council, Leon Kass, is opposed to the creation of embryos by SCNT for research purposes, the committee did not vote to ban SCNT research; rather they called for a four-year moratorium. At the time of the vote, the President's Council on Bioethics contained 17 members; 6 were scientists but only 3 were involved in the fields of cell biology or molecular biology. There were dissenting opinions, but the moratorium carried the day. If one thinks of the time it takes to restart a program once it is dismantled, a four-year moratorium might well turn into a six-year hiatus. Careers would be difficult to maintain, and our best young scientists would probably enter different fields. A similar moratorium initiated at the Asilomar Conference in the 1970s threatened the development of recombinant DNA technology (see *Remembering recombinant DNA*). In retrospect, a prolonged moratorium would have changed the course of science and industry in this country. In like manner, a prolonged moratorium on SCNT research would be a major setback for individuals interested in maintaining our international preeminence in cell biology and biotechnology and might lead to a brain drain from the US to countries more supportive of this line of research.

Those opposed to research on embryos are concerned that we are on a slippery slope, facing a creeping moral degradation fostered by unbridled biotechnology (30). If we agree to destroy an organism that has the potential to develop into a human being, it may be easy to move on to other destructive acts. This zeal poses the danger of depriving millions who suffer from degenerative disorders of the hope and benefits that might derive from stem cell–based research.

There is no absolute right answer to the debate regarding the dissociation of blastocysts to produce more hESC lines. Here we present several considerations that convince us of the ethical validity of using embryos up until the 14th day after fertilization.

Up to embryonic day 14, the blastocyst has no central nervous system and, in our view, cannot be considered sensate. We now remove organs from patients who have been declared brain dead but who are still alive in some sense (e.g., they are warm, breathing, making urine). The use of these organs has saved many lives. We view these two hundred–cell embryos as cell donors certainly at the same moral status or less than these individuals.

The slippery slope argument that the use of blastocysts created by SCNT will lead to reproductive cloning is not compelling. With appropriate federal regulations and oversight, such as the *Hatch-Feinstein Bill*, introduced in the Senate by Orrin Hatch (R–UT) and Dianne Feinstein (D–CA), which seeks to prohibit human reproductive cloning while preserving the use of blastocysts to enhance stem cell research, the scientific community can proceed in an orderly fashion. The UK is now succeeding in this vein under the watchful eye of its Human Fertilisation and Embryology Authority, a nongovernmental body that regulates and inspects all UK clinics providing IVF, donor insemination, and embryo storage while also licensing and monitoring all human embryo research conducted in the UK. The guidepost — implantation into a uterus — is an unambiguous barrier.

The need for hESC research is extraordinary. We are on the doorstep of a new type of restorative therapy that goes beyond treating disease symptoms. Disorders in which the lesions are focal will be the first to undergo stem cell therapy. Replacing β cells in the pancreas, motor neurons in the spinal cord, and dopaminergic cells in the basal ganglia are the most obvious examples. We must weigh the obligations of the moral imperative to help suffering individuals against the inherent value of preimplantation blastocysts.

We have many examples in history where attempts to outlaw fields of study have led to terrible and terrifying consequences (from Galileo to Lysenko). Conversely, many technological breakthroughs now highly valued by both the scientific and lay communities, such as IVF or heart transplants, were once thought to be too dangerous or were seen as "playing God".

Finally, this effort should go forward because we simply will not know the answers unless we do the research. The desire to know is absolutely intrinsic to humans and has a survival value as well as a moral one.

Conclusion

Arguments are often made that hESCs have not cured a single disease. Of course not. Research is hampered by current regulations, and it is difficult to succeed with one hand tied behind one's back. As in all great scientific advances, it takes time and a great deal of money to translate fundamental discoveries into clinically useful treatments.

Scientific advances over the last decade have been extraordinary, but the process of discovery is a fragile one. Each advance raises new questions with new ramifications. It will take all of us — scientists, physicians, health care workers, and patient advocates — a certain amount of effort and courage in the face of contrary views to justify public trust and, thereby, enhance funding for basic research and for applied research.

We can be certain that without research, including federally funded research, we will remain in our current state of ignorance. The public must be kept informed about what the research community is doing. We hope that our great universities and research centers remain at the forefront of this effort.

Acknowledgments
We thank Ross Frommer and Joyce Plaza for careful reading and thoughtful comments on this manuscript.

Address correspondence to: Gerald D. Fischbach, Department of Pharmacology, Dean of the Faculty of Medicine, Columbia University College of Physicians and Surgeons, P&S 2-401, 630 West 168th Street, New York, New York 10032, USA. Phone: (212) 305-2752; Fax: (212) 305-3617; E-mail: gdf@columbia.edu.

1. Evers, B.M., Weissman, I.L., Flake, A.W., Tabar, V., and Weisel, R.D. 2003. Stem cells in clinical practice. *J. Am. Coll. Surg.* **197**:458–478.
2. Raff, M. 2003. Adult stem cell plasticity: fact or artifact? *Annu. Rev. Cell Dev. Biol.* **19**:1–22.
3. Wagers, A.J., and Weissman, I.L. 2004. Plasticity of adult stem cells. *Cell.* **116**:639–648.
4. Thomson, J.A., et al. 1998. Embryonic stem cell lines derived from human blastocysts. *Science.* **282**:1145–1147.
5. Wichterle, H., Lieberam, I., Porter, J.A., and Jessell, T.M. 2002. Directed differentiation of embryonic stem cells into motor neurons. *Cell.* **110**:385–397.
6. Isacson, O. 2004. Problems and solutions for circuits and synapses in Parkinson's disease. *Neuron.* **43**:165–168.
7. Kim, J.H., et al. 2002. Dopamine neurons derived from embryonic stem cells function in an animal model of Parkinson's disease. *Nature.* **418**:50–56.
8. Ricordi, C., and Strom, T.B. 2004. Clinical islet transplantation: advances and immunological challenges. *Nat. Rev. Immunol.* **4**:259–268.
9. Liu, E.H., and Herold, K.C. 2000. Transplantation of the islets of Langerhans: new hope for treatment of type 1 diabetes mellitus. *Trends Endocrinol. Metab.* **11**:379–382.
10. Shapiro, A.M., et al. 2000. Islet transplantation in seven patients with type 1 diabetes mellitus using a glucocorticoid-free immunosuppressive regimen. *N. Engl. J. Med.* **343**:230–238.
11. Freed, C.R., et al. 2001. Transplantation of embryonic dopamine neurons for severe Parkinson's disease. *N. Engl. J. Med.* **344**:710–719.
12. Hochedlinger, K., et al. 2004. Nuclear transplantation, embryonic stem cells and the potential for cell therapy. *Hematol. J.* **5**(Suppl. 3):S114–S117.
13. Hwang, W.S., et al. 2004. Evidence of a pluripotent human embryonic stem cell line derived from a cloned blastocyst. *Science.* **303**:1669–1674.
14. Normile, D. 2004. Research ethics. South Korean cloning team denies improprieties. *Science.* **304**:945.
15. Vogel, G. 2004. Human cloning. Scientists take step toward therapeutic cloning. *Science.* **303**:937–939.
16. Campbell, K.H., McWhir, J., Ritchie, W.A., and Wilmut, I. 1996. Sheep cloned by nuclear transfer from a cultured cell line. *Nature.* **380**:64–66.
17. Wilmut, I. 1999. Dolly's false legacy. *Time.* **74**:76–77.
18. Wilmut, I. 2003. Dolly — her life and legacy. *Cloning Stem Cells.* **5**:99–100.
19. Jaenisch, R. and Wilmut, I. 2001. Developmental biology. Don't clone humans! *Science.* **291**:2552.
20. Rideout, W.M., Eggan, K., and Jaenisch, R. 2001. Nuclear cloning and epigenetic reprogramming of the genome. *Science.* **293**:1093–1097.
21. Wilmut, I., et al. 2002. Somatic cell nuclear transfer. *Nature.* **419**:583–586.
22. Jaenisch, R. and Bird, A. 2003. Epigenetic regulation of gene expression: how the genome integrates intrinsic and environmental signals [review]. *Nat Genet.* **33**(Suppl.):245–254.
23. Eggan, K., et al. 2004. Mice cloned from olfactory sensory neurons. *Nature.* **428**:44–49.
24. The National Commission for the Protection of Human Subjects of Biomedical and Behavioral Research. 1978. *The Belmont report: ethical principles and guidelines for the protection of human subjects of research.* The National Commission. Bethesda, Maryland, USA.
25. National Institutes of Health. 2004. Stem cell information. Research topics. http://stemcells.nih.gov/research.
26. Zerhouni, E. 2003. Stem cell programs. *Science.* **300**:911–912.
27. Cowan, C.A., et al. 2004. Derivation of embryonic stem-cell lines from human blastocysts. *N. Engl. J. Med.* **350**:1353–1356.
28. National Bioethics Advisory Commission. 1999. *Ethical issues in human stem cell research.* National Bioethics Advisory Commission. Rockville, Maryland, USA.
29. President's Council on Bioethics. 2002. *Human cloning and human dignity: an ethical inquiry.* The President's Council on Bioethics. Washington, DC, USA.
30. President's Council on Bioethics. 2003. *Beyond therapy: biotechnology and the pursuit of happiness. A report of the President's Council on Bioethics.* 1st edition. Regan Books. New York, New York, USA. 328 pp.
31. Hochedlinger, K., and Jaenisch, R., 2003. Nuclear transplantation, embryonic stem cells, and the potential for cell therapy. *N. Engl. J. Med.* **349**:275–286.

Index

-3

S

U

V